PhET SIMULATIONS

Available in the Pearson eText and in the Study Area of MasteringPhysics (MP)

Extended Edition includes Chapters 1–44. Standard Edition includes Chapters 1–37.
Three-volume edition: Volume 1 includes Chapters 1–20, Volume 2 includes Chapters 21–37,
and Volume 3 includes Chapters 37–44.

Section		Page
1.6	Estimation	10
1.7	Vector Addition	13
2.4	*Forces in 1 Dimension	47
2.4	*The Moving Man	49
2.5	Lunar Lander	52
3.2	Maze Game	76
3.3	*Projectile Motion	79
3.4	Ladybug Revolution, Motion in 2D	87
5.2	Lunar Lander	146
5.3	Forces in 1 Dimension, Friction, *The Ramp	149
6.2	*The Ramp	181
6.3	Molecular Motors, Stretching DNA	188
7.3	*The Ramp	222
7.5	*Energy Skate Park	229
9.3	Ladybug Revolution	286
10.6	Torque	326
12.3	Balloons & Buoyancy	380
13.2	Lunar Lander	406
13.4	My Solar System	412
14.2	Motion in 2D	443
14.3	*Masses & Springs	446
14.5	*Pendulum Lab	453
15.8	Fourier: Making Waves, Waves on a String	495
16.6	Sound, Wave Interference	529
17.6	States of Matter	566
17.7	The Greenhouse Effect	570
18.3	Balloons & Buoyancy, Friction, Gas Properties	599
18.6	States of Matter	612
21.2	Balloons and Static Electricity, John Travoltage	691
21.6	*Charges and Fields, Electric Field of Dreams, Electric Field Hockey	708
21.7	Microwaves	711
23.2	*Charges & Fields	761
24.5	Molecular Motors, Optical Tweezers and Applications, Stretching DNA	806
25.3	Resistance in a Wire	825
25.4	Battery Voltage, Signal Circuit	829
25.5	Battery-Resistor Circuit, *Circuit Construction Kit (AC+DC), *Circuit Construction Kit (DC Only), Ohm's Law	834

Section		Page
25.6	Conductivity	838
26.4	*Circuit Construction Kit (AC+DC), *Circuit Construction Kit (DC Only)	866
27.3	Magnet and Compass, Magnets and Electromagnets	891
28.5	Faraday's Electromagnetic Lab, Magnets and Electromagnets	933
29.2	Faraday's Electromagnetic Lab, Faraday's Law, Generator	962
31.3	*Circuit Construction Kit (AC+DC), Faraday's Electromagnetic Lab	1031
32.3	Radio Waves & Electromagnetic Fields	1061
32.5	Microwaves	1070
34.4	*Geometric Optics	1131
34.6	Color Vision	1142
35.2	*Wave Interference	1168
36.2	*Wave Interference	1192
38.1	Photoelectric Effect	1262
38.4	Fourier: Making Waves, Quantum Wave Interference	1274
39.2	Davisson-Germer: Electron Diffraction	1287
39.2	Rutherford Scattering	1294
39.3	Models of the Hydrogen Atom	1297
39.3	Neon Lights and Other Discharge Lamps	1304
39.4	Lasers	1307
39.5	Blackbody Spectrum, The Greenhouse Effect	1310
40.1	Fourier: Making Waves	1328
40.1	Quantum Tunneling and Wave Packets	1337
40.3	Double Wells & Covalent Bonds, Quantum Bound States	1343
40.4	Quantum Tunneling and Wave Packets	1347
41.5	Stern-Gerlach Experiment	1383
42.1	Double Wells and Covalent Bonds	1406
42.2	The Greenhouse Effect	1409
42.4	Band Structure, Conductivity	1417
42.6	Semiconductors, Conductivity	1422
43.1	Simplified MRI	1444
43.3	Alpha Decay	1450
43.7	Nuclear Fission	1464

*Indicates an associated tutorial available in the MasteringPhysics Item Library.

ActivPhysics OnLine™
ACTIVITIES

(MP) www.masteringphysics.com

1.1 Analyzing Motion Using Diagrams
1.2 Analyzing Motion Using Graphs
1.3 Predicting Motion from Graphs
1.4 Predicting Motion from Equations
1.5 Problem-Solving Strategies for Kinematics
1.6 Skier Races Downhill
1.7 Balloonist Drops Lemonade
1.8 Seat Belts Save Lives
1.9 Screeching to a Halt
1.10 Pole-Vaulter Lands
1.11 Car Starts, Then Stops
1.12 Solving Two-Vehicle Problems
1.13 Car Catches Truck
1.14 Avoiding a Rear-End Collision
2.1.1 Force Magnitudes
2.1.2 Skydiver
2.1.3 Tension Change
2.1.4 Sliding on an Incline
2.1.5 Car Race
2.2 Lifting a Crate
2.3 Lowering a Crate
2.4 Rocket Blasts Off
2.5 Truck Pulls Crate
2.6 Pushing a Crate Up a Wall
2.7 Skier Goes Down a Slope
2.8 Skier and Rope Tow
2.9 Pole-Vaulter Vaults
2.10 Truck Pulls Two Crates
2.11 Modified Atwood Machine
3.1 Solving Projectile Motion Problems
3.2 Two Balls Falling
3.3 Changing the x-Velocity
3.4 Projectile x- and y-Accelerations
3.5 Initial Velocity Components
3.6 Target Practice I
3.7 Target Practice II
4.1 Magnitude of Centripetal Acceleration
4.2 Circular Motion Problem Solving
4.3 Cart Goes Over Circular Path
4.4 Ball Swings on a String
4.5 Car Circles a Track
4.6 Satellites Orbit
5.1 Work Calculations
5.2 Upward-Moving Elevator Stops
5.3 Stopping a Downward-Moving Elevator
5.4 Inverse Bungee Jumper
5.5 Spring-Launched Bowler
5.6 Skier Speed
5.7 Modified Atwood Machine
6.1 Momentum and Energy Change
6.2 Collisions and Elasticity
6.3 Momentum Conservation and Collisions
6.4 Collision Problems
6.5 Car Collision: Two Dimensions
6.6 Saving an Astronaut
6.7 Explosion Problems
6.8 Skier and Cart
6.9 Pendulum Bashes Box
6.10 Pendulum Person-Projectile Bowling
7.1 Calculating Torques
7.2 A Tilted Beam: Torques and Equilibrium
7.3 Arm Levers
7.4 Two Painters on a Beam
7.5 Lecturing from a Beam

7.6 Rotational Inertia
7.7 Rotational Kinematics
7.8 Rotoride–Dynamics Approach
7.9 Falling Ladder
7.10 Woman and Flywheel Elevator–Dynamics Approach
7.11 Race Between a Block and a Disk
7.12 Woman and Flywheel Elevator–Energy Approach
7.13 Rotoride–Energy Approach
7.14 Ball Hits Bat
8.1 Characteristics of a Gas
8.2 Maxwell-Boltzmann Distribution–Conceptual Analysis
8.3 Maxwell-Boltzmann Distribution–Quantitative Analysis
8.4 State Variables and Ideal Gas Law
8.5 Work Done By a Gas
8.6 Heat, Internal Energy, and First Law of Thermodynamics
8.7 Heat Capacity
8.8 Isochoric Process
8.9 Isobaric Process
8.10 Isothermal Process
8.11 Adiabatic Process
8.12 Cyclic Process–Strategies
8.13 Cyclic Process–Problems
8.14 Carnot Cycle
9.1 Position Graphs and Equations
9.2 Describing Vibrational Motion
9.3 Vibrational Energy
9.4 Two Ways to Weigh Young Tarzan
9.5 Ape Drops Tarzan
9.6 Releasing a Vibrating Skier I
9.7 Releasing a Vibrating Skier II
9.8 One-and Two-Spring Vibrating Systems
9.9 Vibro-Ride
9.10 Pendulum Frequency
9.11 Risky Pendulum Walk
9.12 Physical Pendulum
10.1 Properties of Mechanical Waves
10.2 Speed of Waves on a String
10.3 Speed of Sound in a Gas
10.4 Standing Waves on Strings
10.5 Tuning a Stringed Instrument: Standing Waves
10.6 String Mass and Standing Waves
10.7 Beats and Beat Frequency
10.8 Doppler Effect: Conceptual Introduction
10.9 Doppler Effect: Problems
10.10 Complex Waves: Fourier Analysis
11.1 Electric Force: Coulomb's Law
11.2 Electric Force: Superposition Principle
11.3 Electric Force: Superposition Principle (Quantitative)
11.4 Electric Field: Point Charge
11.5 Electric Field Due to a Dipole
11.6 Electric Field: Problems
11.7 Electric Flux
11.8 Gauss's Law
11.9 Motion of a Charge in an Electric Field: Introduction
11.10 Motion in an Electric Field: Problems
11.11 Electric Potential: Qualitative Introduction

11.12 Electric Potential, Field, and Force
11.13 Electrical Potential Energy and Potential
12.1 DC Series Circuits (Qualitative)
12.2 DC Parallel Circuits
12.3 DC Circuit Puzzles
12.4 Using Ammeters and Voltmeters
12.5 Using Kirchhoff's Laws
12.6 Capacitance
12.7 Series and Parallel Capacitors
12.8 RC Circuit Time Constants
13.1 Magnetic Field of a Wire
13.2 Magnetic Field of a Loop
13.3 Magnetic Field of a Solenoid
13.4 Magnetic Force on a Particle
13.5 Magnetic Force on a Wire
13.6 Magnetic Torque on a Loop
13.7 Mass Spectrometer
13.8 Velocity Selector
13.9 Electromagnetic Induction
13.10 Motional emf
14.1 The RL Circuit
14.2 The RLC Oscillator
14.3 The Driven Oscillator
15.1 Reflection and Refraction
15.2 Total Internal Reflection
15.3 Refraction Applications
15.4 Plane Mirrors
15.5 Spherical Mirrors: Ray Diagrams
15.6 Spherical Mirror: The Mirror Equation
15.7 Spherical Mirror: Linear Magnification
15.8 Spherical Mirror: Problems
15.9 Thin-Lens Ray Diagrams
15.10 Converging Lens Problems
15.11 Diverging Lens Problems
15.12 Two-Lens Optical Systems
16.1 Two-Source Interference: Introduction
16.2 Two-Source Interference: Qualitative Questions
16.3 Two-Source Interference: Problems
16.4 The Grating: Introduction and Qualitative Questions
16.5 The Grating: Problems
16.6 Single-Slit Diffraction
16.7 Circular Hole Diffraction
16.8 Resolving Power
16.9 Polarization
17.1 Relativity of Time
17.2 Relativity of Length
17.3 Photoelectric Effect
17.4 Compton Scattering
17.5 Electron Interference
17.6 Uncertainty Principle
17.7 Wave Packets
18.1 The Bohr Model
18.2 Spectroscopy
18.3 The Laser
19.1 Particle Scattering
19.2 Nuclear Binding Energy
19.3 Fusion
19.4 Radioactivity
19.5 Particle Physics
20.1 Potential Energy Diagrams
20.2 Particle in a Box
20.3 Potential Wells
20.4 Potential Barriers

SEARS AND ZEMANSKY'S

UNIVERSITY PHYSICS

WITH MODERN PHYSICS
13TH EDITION

HUGH D. YOUNG
CARNEGIE MELLON
UNIVERSITY

ROGER A. FREEDMAN
UNIVERSITY
OF CALIFORNIA,
SANTA BARBARA

CONTRIBUTING AUTHOR
A. LEWIS FORD
TEXAS A&M UNIVERSITY

Addison-Wesley

Boston Columbus Indianapolis
New York San Francisco Upper Saddle River
Amsterdam Cape Town Dubai London
Madrid Milan Munich Paris Montréal Toronto
Delhi Mexico City São Paulo Sydney
Hong Kong Seoul Singapore Taipei Tokyo

Publisher: Jim Smith
Executive Editor: Nancy Whilton
Project Editor: Chandrika Madhavan
Director of Development: Michael Gillespie
Editorial Manager: Laura Kenney
Senior Development Editor: Margot Otway
Editorial Assistant: Steven Le
Associate Media Producer: Kelly Reed
Managing Editor: Corinne Benson
Production Project Manager: Beth Collins

Production Management and Composition: Nesbitt Graphics
Copyeditor: Carol Reitz
Interior Designer: Elm Street Publishing Services
Cover Designer: Derek Bacchus
Illustrators: Rolin Graphics
Senior Art Editor: Donna Kalal
Photo Researcher: Eric Shrader
Manufacturing Buyer: Jeff Sargent
Senior Marketing Manager: Kerry Chapman

Cover Photo Credits: Getty Images/Mirko Cassanelli; Mirko Cassanelli

Credits and acknowledgments borrowed from other sources and reproduced, with permission, in this textbook appear on the appropriate page within the text or on p. C-1.

Library of Congress Cataloging-in-Publication Data

Young, Hugh D.
 Sears and Zemansky's university physics : with modern physics. -- 13th ed.
/ Hugh D. Young, Roger A. Freedman ; contributing author, A. Lewis Ford.
    p. cm.
 Includes bibliographical references and index.
 ISBN-13: 978-0-321-69686-1 (student ed. : alk. paper)
 ISBN-10: 0-321-69686-7 (student ed. : alk. paper)
 ISBN-13: 978-0-321-69685-4 (exam copy)
 ISBN-10: 0-321-69685-9 (exam copy)
 1. Physics--Textbooks. I. Freedman, Roger A. II. Ford, A. Lewis (Albert
Lewis) III. Sears, Francis Weston, 1898-1975. University physics. IV. Title.
V. Title: University physics.
 QC21.3.Y68 2012
 530--dc22

                                                    2010044896

ISBN 13: 978-0-321-69686-1; ISBN 10: 0-321-69686-7 (Student edition)
ISBN 13: 978-0-321-69685-4; ISBN 10: 0-321-69685-9 (Exam copy)

Addison-Wesley
is an imprint of

www.pearsonhighered.com   3 4 5 6 7 8 9 10—VHC—14 13 12

BRIEF CONTENTS

MECHANICS

1   Units, Physical Quantities, and Vectors   1
2   Motion Along a Straight Line   35
3   Motion in Two or Three Dimensions   69
4   Newton's Laws of Motion   104
5   Applying Newton's Laws   134
6   Work and Kinetic Energy   176
7   Potential Energy and Energy Conservation   207
8   Momentum, Impulse, and Collisions   241
9   Rotation of Rigid Bodies   278
10   Dynamics of Rotational Motion   308
11   Equilibrium and Elasticity   344
12   Fluid Mechanics   373
13   Gravitation   402
14   Periodic Motion   437

WAVES/ACOUSTICS

15   Mechanical Waves   472
16   Sound and Hearing   509

THERMODYNAMICS

17   Temperature and Heat   551
18   Thermal Properties of Matter   590
19   The First Law of Thermodynamics   624
20   The Second Law of Thermodynamics   652

ELECTROMAGNETISM

21   Electric Charge and Electric Field   687
22   Gauss's Law   725
23   Electric Potential   754
24   Capacitance and Dielectrics   788
25   Current, Resistance, and Electromotive Force   818
26   Direct-Current Circuits   850

27   Magnetic Field and Magnetic Forces   883
28   Sources of Magnetic Field   923
29   Electromagnetic Induction   957
30   Inductance   991
31   Alternating Current   1021
32   Electromagnetic Waves   1051

OPTICS

33   The Nature and Propagation of Light   1080
34   Geometric Optics   1114
35   Interference   1163
36   Diffraction   1190

MODERN PHYSICS

37   Relativity   1223
38   Photons: Light Waves Behaving as Particles   1261
39   Particles Behaving as Waves   1286
40   Quantum Mechanics   1328
41   Atomic Structure   1364
42   Molecules and Condensed Matter   1405
43   Nuclear Physics   1439
44   Particle Physics and Cosmology   1480

APPENDICES

A   The International System of Units   A-1
B   Useful Mathematical Relations   A-3
C   The Greek Alphabet   A-4
D   Periodic Table of Elements   A-5
E   Unit Conversion Factors   A-6
F   Numerical Constants   A-7

Answers to Odd-Numbered Problems   A-9

VOLUME 1: Chapters 1–20 • VOLUME 2: Chapters 21–37 • VOLUME 3: Chapters 37–44

Build **Skills**

Learn basic and advanced skills that help
solve a broad range of physics problems.

Problem-Solving Strategies coach students in
how to approach specific types of problems.

▼

This text's **uniquely extensive set**
of Examples enables students
to explore problem-solving
challenges in exceptional detail.

Consistent

The **Identify / Set Up /
Execute / Evaluate** format, used in
all Examples, encourages students
to tackle problems thoughtfully
rather than skipping to the math.

Focused

**All Examples and Problem-
Solving Strategies** are revised
to be more concise and focused.

Visual

Most Examples employ a diagram—
often a **pencil sketch** that shows
what a student should draw.

Problem-Solving Strategy 5.2 | **Newton's Second Law: Dynamics of Particles**

IDENTIFY *the relevant concepts:* You have to use Newton's second law for *any* problem that involves forces acting on an accelerating

accelerate in different directions, you can use a different set of axes for each body.

Example 5.17 | **Toboggan ride with friction II**

The same toboggan with the same coefficient of friction as in Example 5.16 *accelerates* down a steeper hill. Derive an expression for the acceleration in terms of g, α, μ_k, and w.

SOLUTION

IDENTIFY and SET UP: The toboggan is accelerating, so we must use Newton's second law as given in Eqs. (5.4). Our target variable is the downhill acceleration.

Our sketch and free-body diagram (Fig. 5.23) are almost the same as for Example 5.16. The toboggan's y-component of acceleration a_y is still zero but the x-component a_x is not, so we've drawn the downhill component of weight as a longer vector than the (uphill) friction force.

EXECUTE: It's convenient to express the weight as $w = mg$. Then Newton's second law in component form says

$$\sum F_x = mg \sin \alpha + (-f_k) = ma_x$$
$$\sum F_y = n + (-mg \cos \alpha) = 0$$

5.23 Our sketches for this problem.

(a) The situation   (b) Free-body diagram for toboggan

From the second equation and Eq. (5.5) we get an expression for f_k:

$$n = mg \cos \alpha$$
$$f_k = \mu_k n = \mu_k mg \cos \alpha$$

We substitute this into the x-component equation and solve for a_x:

$$mg \sin \alpha + (-\mu_k mg \cos \alpha) = ma_x$$
$$a_x = g(\sin \alpha - \mu_k \cos \alpha)$$

EVALUATE: As for the frictionless toboggan in Example 5.10, the acceleration doesn't depend on the mass m of the toboggan. That's because all of the forces that act on the toboggan (weight, normal force, and kinetic friction force) are proportional to m.

Let's check some special cases. If the hill is vertical ($\alpha = 90°$) so that $\sin \alpha = 1$ and $\cos \alpha = 0$, we have $a_x = g$ (the toboggan falls freely). For a certain value of α the acceleration is zero; this happens if

$$\sin \alpha = \mu_k \cos \alpha \quad \text{and} \quad \mu_k = \tan \alpha$$

This agrees with our result for the constant-velocity toboggan in Example 5.16. If the angle is even smaller, $\mu_k \cos \alpha$ is greater than $\sin \alpha$ and a_x is *negative;* if we give the toboggan an initial downhill push to start it moving, it will slow down and stop. Finally, if the hill is frictionless so that $\mu_k = 0$, we retrieve the result of Example 5.10: $a_x = g \sin \alpha$.

Notice that we started with a simple problem (Example 5.10) and extended it to more and more general situations. The general result we found in this example includes *all* the previous ones as special cases. Don't memorize this result, but do make sure you understand how we obtained it and what it means.

Suppose instead we give the toboggan an initial push *up* the hill. The direction of the kinetic friction force is now reversed, so the acceleration is different from the downhill value. It turns out that the expression for a_x is the same as for downhill motion except that the minus sign becomes plus. Can you show this?

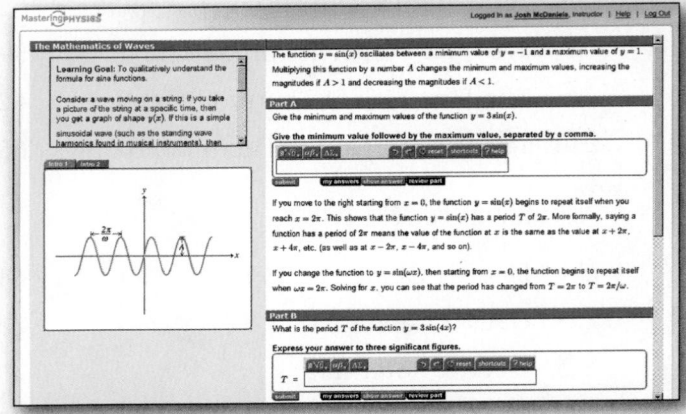

▲

NEW! Video Tutor Solution for Every Example
Each Example is explained and solved by an instructor
in a Video Tutor solution provided in the Study Area
of MasteringPhysics® and in the Pearson eText.

NEW! Mathematics Review Tutorials
MasteringPhysics offers an extensive set of assignable mathematics
review tutorials—covering differential and integral calculus as well
as algebra and trigonometry.

Build **Confidence**

Develop problem-solving confidence through a range of practice options—from guided to unguided.

NEW! Bridging Problems

At the start of each problem set, a Bridging Problem helps students make the leap from routine exercises to challenging problems with confidence and ease.

Each Bridging Problem poses a moderately difficult, multi-concept problem, which often draws on earlier chapters. In place of a full solution, it provides a skeleton solution guide consisting of questions and hints.

A full solution is explained in a Video Tutor, provided in the Study Area of MasteringPhysics® and in the Pearson eText.

BRIDGING PROBLEM **Billiard Physics**

A cue ball (a uniform solid sphere of mass m and radius R) is at rest on a level pool table. Using a pool cue, you give the ball a sharp, horizontal hit of magnitude F at a height h above the center of the ball (Fig. 10.37). The force of the hit is much greater than the friction force f that the table surface exerts on the ball. The hit lasts for a short time Δt. (a) For what value of h will the ball roll without slipping? (b) If you hit the ball dead center ($h = 0$), the ball will slide across the table for a while, but eventually it will roll without slipping. What will the speed of its center of mass be then?

SOLUTION GUIDE

See MasteringPhysics® study area for a Video Tutor solution. (MP)

IDENTIFY and SET UP

1. Draw a free-body diagram for the ball for the situation in part (a), including your choice of coordinate axes. Note that the cue exerts both an impulsive force on the ball and an impulsive torque around the center of mass.
2. The cue force applied for a time Δt gives the ball's center of mass a speed v_{cm}, and the cue torque applied for that same time gives the ball an angular speed ω. What must be the relationship between v_{cm} and ω for the ball to roll without slipping?

3. Draw two free-body diagrams for the ball in part (b): one showing the forces during the hit and the other showing the forces after the hit but before the ball is rolling without slipping.
4. What is the angular speed of the ball in part (b) just after the hit? While the ball is sliding, does v_{cm} increase or decrease? Does ω increase or decrease? What is the relationship between v_{cm} and ω when the ball is finally rolling without slipping?

EXECUTE

5. In part (a), use the impulse–momentum theorem to find the speed of the ball's center of mass immediately after the hit. Then use the rotational version of the impulse–momentum theorem to find the angular speed immediately after the hit. (*Hint:* To write down the rotational version of the impulse–momentum theorem, remember that the relationship between torque and angular momentum is the same as that between force and linear momentum.)
6. Use your results from step 5 to find the value of h that will cause the ball to roll without slipping immediately after the hit.
7. In part (b), again find the ball's center-of-mass speed and angular speed immediately after the hit. Then write Newton's second law for the translational motion and rotational motion of the ball as it is sliding. Use these equations to write expressions for v_{cm} and ω as functions of the elapsed time t since the hit.
8. Using your results from step 7, find the time t when v_{cm} and ω have the correct relationship for rolling without slipping. Then find the value of v_{cm} at this time.

EVALUATE

9. If you have access to a pool table, test out the results of parts (a) and (b) for yourself!
10. Can you show that if you used a hollow cylinder rather than a solid ball, you would have to hit the top of the cylinder to cause rolling without slipping as in part (a)?

10.37

◄ In response to professors, the Problem Sets now include more biomedically oriented problems (BIO), more difficult problems requiring calculus (CALC), and more cumulative problems that draw on earlier chapters (CP).

About 20% of problems are new or revised. These revisions are driven by detailed student-performance data gathered nationally through MasteringPhysics.

Problem difficulty is now indicated by a three-dot ranking system based on data from MasteringPhysics.

14.95 · **CP** In Fig. P14.95 the upper ball is released from rest, collides with the stationary lower ball, and sticks to it. The strings are both 50.0 cm long. The upper ball has mass 2.00 kg, and it is initially 10.0 cm higher than the lower ball, which has mass 3.00 kg. Find the frequency and maximum angular displacement of the motion after the collision.

Figure **P14.95**

10.0 cm

14.96 ·· **CP BIO** *T. rex.* Model the leg of the *T. rex* in Example 14.10 (Section 14.6) as two uniform rods, each 1.55 m long, joined rigidly end to end. Let the lower rod have mass M and the upper rod mass $2M$. The composite object is pivoted about the top of the upper rod. Compute the oscillation period of this object for small-amplitude oscillations. Compare your result to that of Example 14.10.

14.97 ·· **CALC** A slender, uniform, metal rod with mass M is pivoted without friction about an axis through its midpoint and perpendicular to the rod. A horizontal spring with force constant k is attached to the lower end of the rod, with the other end of the spring attached to a rigid support. If the rod is displaced by a small angle Θ from the vertical (Fig. P14.97) and released, show that it moves in angular SHM and calculate the period. (*Hint:* Assume that the angle Θ is small enough for the approximations $\sin \Theta \approx \Theta$ and $\cos \Theta \approx 1$ to be valid. The motion is simple harmonic if $d^2\theta/dt^2 = -\omega^2\theta$, and the period is then $T = 2\pi/\omega$.)

Figure **P14.97**

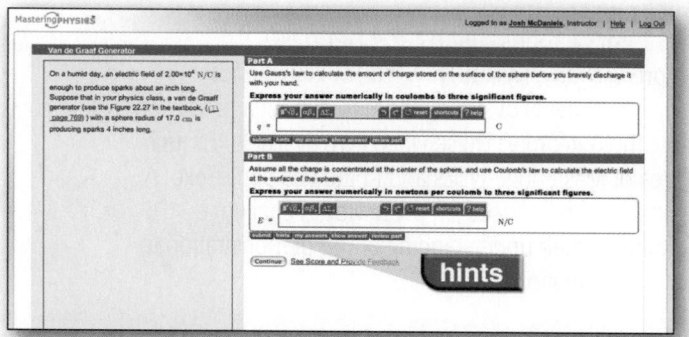

NEW! Enhanced End-of-Chapter Problems in MasteringPhysics
Select end-of-chapter problems will now offer additional support such as problem-solving strategy hints, relevant math review and practice, and links to the eText. These new enhanced problems bridge the gap between guided tutorials and traditional homework problems.

Bring Physics **to Life**

Deepen knowledge of physics by building connections to the real world.

NEW! Applications of Physics
Throughout the text, free-standing captioned photos apply physics to real situations, with particular emphasis on applications of biomedical and general interest. ▶

Application Moment of Inertia of a Bird's Wing
When a bird flaps its wings, it rotates the wings up and down around the shoulder. A hummingbird has small wings with a small moment of inertia, so the bird can make its wings move rapidly (up to 70 beats per second). By contrast, the Andean condor (*Vultur gryphus*) has immense wings that are hard to move due to their large moment of inertia. Condors flap their wings at about one beat per second on takeoff, but at most times prefer to soar while holding their wings steady.

Application Tendons Are Nonideal Springs
Muscles exert forces via the tendons that attach them to bones. A tendon consists of long, stiffly elastic collagen fibers. The graph shows how the tendon from the hind leg of a wallaby (a small kangaroo) stretches in response to an applied force. The tendon does not exhibit the simple, straight-line behavior of an ideal spring, so the work it does has to be found by integration [Eq. (6.7)]. Note that the tendon exerts less force while relaxing than while stretching. As a result, the relaxing tendon does only about 93% of the work that was done to stretch it.

Application Listening for Turbulent Flow
Normal blood flow in the human aorta is laminar, but a small disturbance such as a heart pathology can cause the flow to become turbulent. Turbulence makes noise, which is why listening to blood flow with a stethoscope is a useful diagnostic technique.

◀ **NEW! PhET Simulations and Tutorials**
Sixteen assignable PhET Tutorials enable students to make connections between real-life phenomena and the underlying physics. 76 PhET simulations are provided in the Study Area of MasteringPhysics® and in the Pearson eText.

The comprehensive library of ActivPhysics applets and applet-based tutorials is also available.

NEW! Video Tutor Demonstrations and Tutorials
"Pause and predict" demonstration videos of key physics concepts engage students by asking them to submit a prediction before seeing the outcome. These videos are available through the Study Area of MasteringPhysics and in the Pearson eText. A set of assignable tutorials based on these videos challenge students to transfer their understanding of the demonstration to a related problem situation.

Biomedically Based End-of-Chapter Problems ▶
To serve biosciences students, the text adds a substantial number of problems based on biological and biomedical situations.

Figure **E21.23**

21.24 •• **BIO Base Pairing in DNA, II.** Refer to Exercise 21.23. Figure E21.24 shows the bonding of the cytosine and guanine molecules. The O—H and H—N distances are each 0.110 nm. In this case, assume that the bonding is due only to the forces along the O—H—O, N—H—N, and O—H—N combinations, and assume also that these three combinations are parallel to each other. Calculate the *net* force that cytosine exerts on guanine due to the preceding three combinations. Is this force attractive or repulsive?

Make a Difference with
MasteringPhysics®

www.masteringphysics.com

MasteringPhysics is the most effective and widely used online science tutorial, homework, and assessment system available.

NEW! Pre-Built Assignments ▶

For every chapter in the book, MasteringPhysics now provides pre-built assignments that cover the material with a tested mix of tutorials and end-of-chapter problems of graded difficulty. Professors may use these assignments as-is or take them as a starting point for modification.

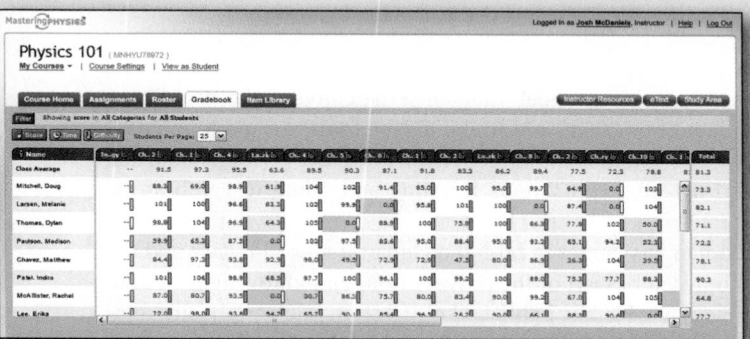

◀ **Gradebook**
- Every assignment is graded automatically.
- Shades of red highlight vulnerable students and challenging assignments.

Class Performance on Assignment ▶

Click on a problem to see which step your students struggled with most, and even their most common wrong answers. Compare results at every stage with the national average or with your previous class.

◀ **Gradebook Diagnostics**

This screen provides your favorite weekly diagnostics. With a single click, charts summarize the most difficult problems, vulnerable students, grade distribution, and even improvement in scores over the course.

ABOUT THE AUTHORS

Hugh D. Young is Emeritus Professor of Physics at Carnegie Mellon University. He earned both his undergraduate and graduate degrees from that university. He earned his Ph.D. in fundamental particle theory under the direction of the late Richard Cutkosky. He joined the faculty of Carnegie Mellon in 1956 and retired in 2004. He also had two visiting professorships at the University of California, Berkeley.

Dr. Young's career has centered entirely on undergraduate education. He has written several undergraduate-level textbooks, and in 1973 he became a coauthor with Francis Sears and Mark Zemansky for their well-known introductory texts. In addition to his role on Sears and Zemansky's *University Physics,* he is also author of Sears and Zemansky's *College Physics*.

Dr. Young earned a bachelor's degree in organ performance from Carnegie Mellon in 1972 and spent several years as Associate Organist at St. Paul's Cathedral in Pittsburgh. He has played numerous organ recitals in the Pittsburgh area. Dr. Young and his wife, Alice, usually travel extensively in the summer, especially overseas and in the desert canyon country of southern Utah.

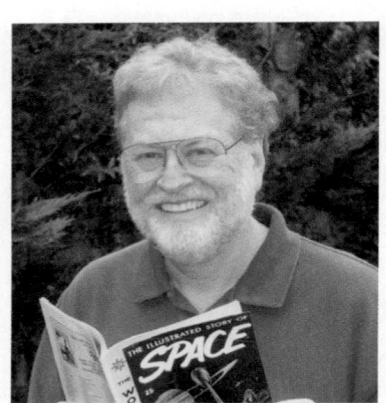

Roger A. Freedman is a Lecturer in Physics at the University of California, Santa Barbara. Dr. Freedman was an undergraduate at the University of California campuses in San Diego and Los Angeles, and did his doctoral research in nuclear theory at Stanford University under the direction of Professor J. Dirk Walecka. He came to UCSB in 1981 after three years teaching and doing research at the University of Washington.

At UCSB, Dr. Freedman has taught in both the Department of Physics and the College of Creative Studies, a branch of the university intended for highly gifted and motivated undergraduates. He has published research in nuclear physics, elementary particle physics, and laser physics. In recent years, he has worked to make physics lectures a more interactive experience through the use of classroom response systems.

In the 1970s Dr. Freedman worked as a comic book letterer and helped organize the San Diego Comic-Con (now the world's largest popular culture convention) during its first few years. Today, when not in the classroom or slaving over a computer, Dr. Freedman can be found either flying (he holds a commercial pilot's license) or driving with his wife, Caroline, in their 1960 Nash Metropolitan convertible.

A. Lewis Ford is Professor of Physics at Texas A&M University. He received a B.A. from Rice University in 1968 and a Ph.D. in chemical physics from the University of Texas at Austin in 1972. After a one-year postdoc at Harvard University, he joined the Texas A&M physics faculty in 1973 and has been there ever since. Professor Ford's research area is theoretical atomic physics, with a specialization in atomic collisions. At Texas A&M he has taught a variety of undergraduate and graduate courses, but primarily introductory physics.

HOW TO SUCCEED IN PHYSICS BY REALLY TRYING

Mark Hollabaugh *Normandale Community College*

Physics encompasses the large and the small, the old and the new. From the atom to galaxies, from electrical circuitry to aerodynamics, physics is very much a part of the world around us. You probably are taking this introductory course in calculus-based physics because it is required for subsequent courses you plan to take in preparation for a career in science or engineering. Your professor wants you to learn physics and to enjoy the experience. He or she is very interested in helping you learn this fascinating subject. That is part of the reason your professor chose this textbook for your course. That is also the reason Drs. Young and Freedman asked me to write this introductory section. We want you to succeed!

The purpose of this section of *University Physics* is to give you some ideas that will assist your learning. Specific suggestions on how to use the textbook will follow a brief discussion of general study habits and strategies.

Preparation for This Course

If you had high school physics, you will probably learn concepts faster than those who have not because you will be familiar with the language of physics. If English is a second language for you, keep a glossary of new terms that you encounter and make sure you understand how they are used in physics. Likewise, if you are farther along in your mathematics courses, you will pick up the mathematical aspects of physics faster. Even if your mathematics is adequate, you may find a book such as Arnold D. Pickar's *Preparing for General Physics: Math Skill Drills and Other Useful Help (Calculus Version)* to be useful. Your professor may actually assign sections of this math review to assist your learning.

Learning to Learn

Each of us has a different learning style and a preferred means of learning. Understanding your own learning style will help you to focus on aspects of physics that may give you difficulty and to use those components of your course that will help you overcome the difficulty. Obviously you will want to spend more time on those aspects that give you the most trouble. If you learn by hearing, lectures will be very important. If you learn by explaining, then working with other students will be useful to you. If solving problems is difficult for you, spend more time learning how to solve problems. Also, it is important to understand and develop good study habits. Perhaps the most important thing you can do for yourself is to set aside adequate, regularly scheduled study time in a distraction-free environment.

Answer the following questions for yourself:
- Am I able to use fundamental mathematical concepts from algebra, geometry and trigonometry? (If not, plan a program of review with help from your professor.)
- In similar courses, what activity has given me the most trouble? (Spend more time on this.) What has been the easiest for me? (Do this first; it will help to build your confidence.)

- Do I understand the material better if I read the book before or after the lecture? (You may learn best by skimming the material, going to lecture, and then undertaking an in-depth reading.)
- Do I spend adequate time in studying physics? (A rule of thumb for a class like this is to devote, on the average, 2.5 hours out of class for each hour in class. For a course meeting 5 hours each week, that means you should spend about 10 to 15 hours per week studying physics.)
- Do I study physics every day? (Spread that 10 to 15 hours out over an entire week!) At what time of the day am I at my best for studying physics? (Pick a specific time of the day and stick to it.)
- Do I work in a quiet place where I can maintain my focus? (Distractions will break your routine and cause you to miss important points.)

Working with Others

Scientists or engineers seldom work in isolation from one another but rather work cooperatively. You will learn more physics and have more fun doing it if you work with other students. Some professors may formalize the use of cooperative learning or facilitate the formation of study groups. You may wish to form your own informal study group with members of your class who live in your neighborhood or dorm. If you have access to e-mail, use it to keep in touch with one another. Your study group is an excellent resource when reviewing for exams.

Lectures and Taking Notes

An important component of any college course is the lecture. In physics this is especially important because your professor will frequently do demonstrations of physical principles, run computer simulations, or show video clips. All of these are learning activities that will help you to understand the basic principles of physics. Don't miss lectures, and if for some reason you do, ask a friend or member of your study group to provide you with notes and let you know what happened.

Take your class notes in outline form, and fill in the details later. It can be very difficult to take word for word notes, so just write down key ideas. Your professor may use a diagram from the textbook. Leave a space in your notes and just add the diagram later. After class, edit your notes, filling in any gaps or omissions and noting things you need to study further. Make references to the textbook by page, equation number, or section number.

Make sure you ask questions in class, or see your professor during office hours. Remember the only "dumb" question is the one that is not asked. Your college may also have teaching assistants or peer tutors who are available to help you with difficulties you may have.

Examinations

Taking an examination is stressful. But if you feel adequately prepared and are well-rested, your stress will be lessened. Preparing for an exam is a continual process; it begins the moment the last exam is over. You should immediately go over the exam and understand any mistakes you made. If you worked a problem and made substantial errors, try this: Take a piece of paper and divide it down the middle with a line from top to bottom. In one column, write the proper solution to the problem. In the other column, write what you did and why, if you know, and why your solution was incorrect. If you are uncertain why you made your mistake, or how to avoid making it again, talk with your professor. Physics continually builds on fundamental ideas and it is important to correct any misunderstandings immediately. *Warning:* While cramming at the last minute may get you through the present exam, you will not adequately retain the concepts for use on the next exam.

PREFACE

This book is the product of more than six decades of leadership and innovation in physics education. When the first edition of *University Physics* by Francis W. Sears and Mark W. Zemansky was published in 1949, it was revolutionary among calculus-based physics textbooks in its emphasis on the fundamental principles of physics and how to apply them. The success of *University Physics* with generations of several million students and educators around the world is a testament to the merits of this approach, and to the many innovations it has introduced subsequently.

In preparing this new Thirteenth Edition, we have further enhanced and developed *University Physics* to assimilate the best ideas from education research with enhanced problem-solving instruction, pioneering visual and conceptual pedagogy, the first systematically enhanced problems, and the most pedagogically proven and widely used online homework and tutorial system in the world.

New to This Edition

- Included in each chapter, **Bridging Problems** provide a transition between the single-concept Examples and the more challenging end-of-chapter problems. Each Bridging Problem poses a difficult, multiconcept problem, which often incorporates physics from earlier chapters. In place of a full solution, it provides a skeleton **Solution Guide** consisting of questions and hints, which helps train students to approach and solve challenging problems with confidence.
- **All Examples, Conceptual Examples, and Problem-Solving Strategies are revised** to enhance conciseness and clarity for today's students.
- The **core modern physics chapters** (Chapters 38–41) are revised extensively to provide a more idea-centered, less historical approach to the material. Chapters 42–44 are also revised significantly.
- **The fluid mechanics chapter now precedes the chapters on gravitation and periodic motion,** so that the latter immediately precedes the chapter on mechanical waves.
- **Additional bioscience applications** appear throughout the text, mostly in the form of marginal photos with explanatory captions, to help students see how physics is connected to many breakthroughs and discoveries in the biosciences.
- The **text has been streamlined** for tighter and more focused language.
- **Using data from MasteringPhysics, changes to the end-of-chapter content** include the following:
  - **15%–20% of problems are new.**
  - The number and level of **calculus-requiring problems** has been increased.
  - Most chapters include **five to seven biosciences-related problems.**
  - The number of **cumulative problems** (those incorporating physics from earlier chapters) has been increased.
- **Over 70 PhET simulations** are linked to the Pearson eText and provided in the Study Area of the MasteringPhysics website (with icons in the print text). These powerful simulations allow students to interact productively with the physics concepts they are learning. PhET clicker questions are also included on the Instructor Resource DVD.
- **Video Tutors bring key content to life throughout the text:**
  - **Dozens of Video Tutors feature "pause-and-predict" demonstrations of key physics concepts** and incorporate assessment as the student progresses to actively engage the student in understanding the key conceptual ideas underlying the physics principles.

Standard, Extended, and Three-Volume Editions

With MasteringPhysics:
- **Standard Edition:** Chapters 1–37 (ISBN 978-0-321-69688-5)
- **Extended Edition:** Chapters 1–44 (ISBN 978-0-321-67546-0)

Without MasteringPhysics:
- **Standard Edition:** Chapters 1–37 (ISBN 978-0-321-69689-2)
- **Extended Edition:** Chapters 1–44 (ISBN 978-0-321-69686-1)
- **Volume 1:** Chapters 1–20 (ISBN 978-0-321-73338-2)
- **Volume 2:** Chapters 21–37 (ISBN 978-0-321-75121-8)
- **Volume 3:** Chapters 37–44 (ISBN 978-0-321-75120-1)

- **Every Worked Example in the book is accompanied by a Video Tutor Solution** that walks students through the problem-solving process, providing a virtual teaching assistant on a round-the-clock basis.
- **All of these Video Tutors play directly through links within the Pearson eText.** Many also appear in the Study Area within MasteringPhysics.

Key Features of *University Physics*

- Deep and extensive **problem sets** cover a wide range of difficulty and exercise both physical understanding and problem-solving expertise. Many problems are based on complex real-life situations.
- This text offers a larger number of **Examples** and **Conceptual Examples** than any other leading calculus-based text, allowing it to explore problem-solving challenges not addressed in other texts.
- A research-based **problem-solving approach (Identify, Set Up, Execute, Evaluate)** is used not just in every Example but also in the Problem-Solving Strategies and throughout the Student and Instructor Solutions Manuals and the Study Guide. This consistent approach teaches students to tackle problems thoughtfully rather than cutting straight to the math.
- **Problem-Solving Strategies** coach students in how to approach specific types of problems.
- The **Figures** use a simplified graphical style to focus on the physics of a situation, and they incorporate **explanatory annotation.** Both techniques have been demonstrated to have a strong positive effect on learning.
- Figures that illustrate Example solutions often take the form of black-and-white **pencil sketches,** which directly represent what a student should draw in solving such a problem.
- The popular **Caution paragraphs** focus on typical misconceptions and student problem areas.
- End-of-section **Test Your Understanding** questions let students check their grasp of the material and use a multiple-choice or ranking-task format to probe for common misconceptions.
- **Visual Summaries** at the end of each chapter present the key ideas in words, equations, and thumbnail pictures, helping students to review more effectively.

Instructor Supplements

Note: For convenience, all of the following instructor supplements (except for the Instructor Resource DVD) can be downloaded from the Instructor Area, accessed via the left-hand navigation bar of MasteringPhysics (www.masteringphysics.com).

Instructor Solutions, prepared by A. Lewis Ford (Texas A&M University) and Wayne Anderson, contain complete and detailed solutions to all end-of-chapter problems. All solutions follow consistently the same Identify/Set Up/Execute/Evaluate problem-solving framework used in the textbook. Download only from the MasteringPhysics Instructor Area or from the Instructor Resource Center (www.pearsonhighered.com/irc).

The cross-platform **Instructor Resource DVD** (ISBN 978-0-321-69661-8) provides a comprehensive library of more than 420 applets from ActivPhysics OnLine as well as all line figures from the textbook in JPEG format. In addition, all the key equations, problem-solving strategies, tables, and chapter summaries are provided in editable Word format. In-class weekly multiple-choice questions for use with various Classroom Response Systems (CRS) are also provided, based on the Test Your Understanding questions in the text. Lecture outlines in PowerPoint are also included along with over 70 PhET simulations.

MasteringPhysics® (www.masteringphysics.com) is the most advanced, educationally effective, and widely used physics homework and tutorial system in the world. Eight years in development, it provides instructors with a library of extensively pre-tested end-of-chapter problems and rich, multipart, multistep tutorials that incorporate a wide variety of answer types, wrong answer feedback, individualized help (comprising hints or simpler sub-problems upon request), all driven by the largest metadatabase of student problem-solving in the world. NSF-sponsored published research (and subsequent studies) show that Mastering-Physics has dramatic educational results. MasteringPhysics allows instructors to build wide-ranging homework assignments of just the right difficulty and length and provides them with efficient tools to analyze both class trends, and the work of any student in unprecedented detail.

MasteringPhysics routinely provides instant and individualized feedback and guidance to more than 100,000 students every day. A wide range of tools and support make MasteringPhysics fast and easy for instructors and students to learn to use. Extensive class tests show that by the end of their course, an unprecedented eight of nine students recommend MasteringPhysics as their preferred way to study physics and do homework.

MasteringPhysics enables instructors to:
- Quickly build homework assignments that combine regular end-of-chapter problems and tutoring (through additional multi-step tutorial problems that offer wrong-answer feedback and simpler problems upon request).
- Expand homework to include the widest range of automatically graded activities available—from numerical problems with randomized values, through algebraic answers, to free-hand drawing.
- Choose from a wide range of nationally pre-tested problems that provide accurate estimates of time to complete and difficulty.
- After an assignment is completed, quickly identify not only the problems that were the trickiest for students but the individual problem types where students had trouble.
- Compare class results against the system's worldwide average for each problem assigned, to identify issues to be addressed with just-in-time teaching.
- Check the work of an individual student in detail, including time spent on each problem, what wrong answers they submitted at each step, how much help they asked for, and how many practice problems they worked.

ActivPhysics OnLine™ (which is accessed through the Study Area within www.masteringphysics.com) provides a comprehensive library of more than 420 tried and tested ActivPhysics applets updated for web delivery using the latest online technologies. In addition, it provides a suite of highly regarded applet-based tutorials developed by education pioneers Alan Van Heuvelen and Paul D'Alessandris. Margin icons throughout the text direct students to specific exercises that complement the textbook discussion.

The online exercises are designed to encourage students to confront misconceptions, reason qualitatively about physical processes, experiment quantitatively, and learn to think critically. The highly acclaimed ActivPhysics OnLine companion workbooks help students work through complex concepts and understand them more clearly. More than 420 applets from the ActivPhysics OnLine library are also available on the Instructor Resource DVD for this text.

The **Test Bank** contains more than 2,000 high-quality problems, with a range of multiple-choice, true/false, short-answer, and regular homework-type questions. Test files are provided both in TestGen (an easy-to-use, fully networkable program for creating and editing quizzes and exams) and Word format. Download only from the MasteringPhysics Instructor Area or from the Instructor Resource Center (www.pearsonhighered.com/irc).

Five Easy Lessons: Strategies for Successful Physics Teaching (ISBN 978-0-805-38702-5) by Randall D. Knight (California Polytechnic State University, San Luis Obispo) is packed with creative ideas on how to enhance any physics course. It is an invaluable companion for both novice and veteran physics instructors.

Student Supplements

The **Study Guide** by Laird Kramer reinforces the text's emphasis on problem-solving strategies and student misconceptions. The *Study Guide for Volume 1* (ISBN 978-0-321-69665-6) covers Chapters 1–20, and the *Study Guide for Volumes 2 and 3* (ISBN 978-0-321-69669-4) covers Chapters 21–44.

The **Student Solutions Manual** by Lewis Ford (Texas A&M University) and Wayne Anderson contains detailed, step-by-step solutions to more than half of the odd-numbered end-of-chapter problems from the textbook. All solutions follow consistently the same Identify/Set Up/Execute/Evaluate problem-solving framework used in the textbook. The *Student Solutions Manual for Volume 1* (ISBN 978-0-321-69668-7) covers Chapters 1–20, and the *Student Solutions Manual for Volumes 2 and 3* (ISBN 978-0-321-69667-0) covers Chapters 21–44.

MasteringPhysics® (www.masteringphysics.com) is a homework, tutorial, and assessment system based on years of research into how students work physics problems and precisely where they need help. Studies show that students who use MasteringPhysics significantly increase their scores compared to hand-written homework. MasteringPhysics achieves this improvement by providing students with instantaneous feedback specific to their wrong answers, simpler sub-problems upon request when they get stuck, and partial credit for their method(s). This individualized, 24/7 Socratic tutoring is recommended by nine out of ten students to their peers as the most effective and time-efficient way to study.

Pearson eText is available through MasteringPhysics, either automatically when MasteringPhysics is packaged with new books, or available as a purchased upgrade online. Allowing students access to the text wherever they have access to the Internet, Pearson eText comprises the full text, including figures that can be enlarged for better viewing. With eText, students are also able to pop up definitions and terms to help with vocabulary and the reading of the material. Students can also take notes in eText using the annotation feature at the top of each page.

Pearson Tutor Services (www.pearsontutorservices.com). Each student's subscription to MasteringPhysics also contains complimentary access to Pearson Tutor Services, powered by Smarthinking, Inc. By logging in with their MasteringPhysics ID and password, students will be connected to highly qualified e-instructors who provide additional interactive online tutoring on the major concepts of physics. Some restrictions apply; offer subject to change.

ActivPhysics OnLine™ (which is accessed through the Study Area within www.masteringphysics.com) provides students with a suite of highly regarded applet-based tutorials (see above). The following workbooks help students work through complex concepts and understand them more clearly.

ActivPhysics OnLine Workbook, Volume 1: Mechanics * Thermal Physics * Oscillations & Waves (978-0-805-39060-5)

ActivPhysics OnLine Workbook, Volume 2: Electricity & Magnetism * Optics * Modern Physics (978-0-805-39061-2)

Acknowledgments

We would like to thank the hundreds of reviewers and colleagues who have offered valuable comments and suggestions over the life of this textbook. The continuing success of *University Physics* is due in large measure to their contributions.

Edward Adelson (Ohio State University), Ralph Alexander (University of Missouri at Rolla), J. G. Anderson, R. S. Anderson, Wayne Anderson (Sacramento City College), Alex Azima (Lansing Community College), Dilip Balamore (Nassau Community College), Harold Bale (University of North Dakota), Arun Bansil (Northeastern University), John Barach (Vanderbilt University), J. D. Barnett, H. H. Barschall, Albert Bartlett (University of Colorado), Marshall Bartlett (Hollins University), Paul Baum (CUNY, Queens College), Frederick Becchetti (University of Michigan), B. Bederson, David Bennum (University of Nevada, Reno), Lev I. Berger (San Diego State University), Robert Boeke (William Rainey Harper College), S. Borowitz, A. C. Braden, James Brooks (Boston University), Nicholas E. Brown (California Polytechnic State University, San Luis Obispo), Tony Buffa (California Polytechnic State University, San Luis Obispo), A. Capecelatro, Michael Cardamone (Pennsylvania State University), Duane Carmony (Purdue University), Troy Carter (UCLA), P. Catranides, John Cerne (SUNY at Buffalo), Tim Chupp (University of Michigan), Shinil Cho (La Roche College), Roger Clapp (University of South Florida), William M. Cloud (Eastern Illinois University), Leonard Cohen (Drexel University), W. R. Coker (University of Texas, Austin), Malcolm D. Cole (University of Missouri at Rolla), H. Conrad, David Cook (Lawrence University), Gayl Cook (University of Colorado), Hans Courant (University of Minnesota), Bruce A. Craver (University of Dayton), Larry Curtis (University of Toledo), Jai Dahiya (Southeast Missouri State University), Steve Detweiler (University of Florida), George Dixon (Oklahoma State University), Donald S. Duncan, Boyd Edwards (West Virginia University), Robert Eisenstein (Carnegie Mellon University), Amy Emerson Missouri (Virginia Institute of Technology), William Faissler (Northeastern University), William Fasnacht (U.S. Naval Academy), Paul Feldker (St. Louis Community College), Carlos Figueroa (Cabrillo College), L. H. Fisher, Neil Fletcher (Florida State University), Robert Folk, Peter Fong (Emory University), A. Lewis Ford (Texas A&M University), D. Frantszog, James R. Gaines (Ohio State University), Solomon Gartenhaus (Purdue University), Ron Gautreau (New Jersey Institute of Technology), J. David Gavenda (University of Texas, Austin), Dennis Gay (University of North Florida), James Gerhart (University of Washington), N. S. Gingrich, J. L. Glathart, S. Goodwin, Rich Gottfried (Frederick Community College), Walter S. Gray (University of Michigan), Paul Gresser (University of Maryland), Benjamin Grinstein (UC San Diego), Howard Grotch (Pennsylvania State University), John Gruber (San Jose State University), Graham D. Gutsche (U.S. Naval Academy), Michael J. Harrison (Michigan State University), Harold Hart (Western Illinois University), Howard Hayden (University of Connecticut), Carl Helrich (Goshen College), Laurent Hodges (Iowa State University), C. D. Hodgman, Michael Hones (Villanova University), Keith Honey (West Virginia Institute of Technology), Gregory Hood (Tidewater Community College), John Hubisz (North Carolina State University), M. Iona, John Jaszczak (Michigan Technical University), Alvin Jenkins (North Carolina State University), Robert P. Johnson (UC Santa Cruz), Lorella Jones (University of Illinois), John Karchek (GMI Engineering & Management Institute), Thomas Keil (Worcester Polytechnic Institute), Robert Kraemer (Carnegie Mellon University), Jean P. Krisch (University of Michigan), Robert A. Kromhout, Andrew Kunz (Marquette University), Charles Lane (Berry College), Thomas N. Lawrence (Texas State University), Robert J. Lee, Alfred Leitner (Rensselaer Polytechnic University), Gerald P. Lietz (De Paul University), Gordon Lind (Utah State University), S. Livingston, Elihu Lubkin (University of Wisconsin, Milwaukee), Robert Luke (Boise State University), David Lynch (Iowa State University), Michael Lysak (San Bernardino Valley College), Jeffrey Mallow (Loyola University), Robert Mania (Kentucky State University), Robert Marchina (University of Memphis), David Markowitz (University of Connecticut), R. J. Maurer, Oren Maxwell (Florida International University), Joseph L. McCauley (University of Houston), T. K. McCubbin, Jr. (Pennsylvania State University), Charles McFarland (University of Missouri at Rolla), James Mcguire (Tulane University), Lawrence McIntyre (University of Arizona), Fredric Messing (Carnegie-Mellon University), Thomas Meyer (Texas A&M University), Andre Mirabelli (St. Peter's College, New Jersey), Herbert Muether (S.U.N.Y., Stony Brook), Jack Munsee (California State University, Long Beach), Lorenzo Narducci (Drexel University), Van E. Neie (Purdue University), David A. Nordling (U. S. Naval Academy), Benedict Oh (Pennsylvania State University), L. O. Olsen, Jim Pannell (DeVry Institute of Technology), W. F. Parks (University of Missouri), Robert Paulson (California State University, Chico), Jerry Peacher (University of Missouri at Rolla), Arnold Perlmutter (University of Miami), Lennart Peterson (University of Florida), R. J. Peterson (University of Colorado, Boulder), R. Pinkston, Ronald Poling (University of Minnesota), J. G. Potter, C. W. Price (Millersville University), Francis Prosser (University of Kansas), Shelden H. Radin, Roberto Ramos (Drexel University), Michael Rapport (Anne Arundel Community College), R. Resnick, James A. Richards, Jr., John S. Risley (North Carolina State University), Francesc Roig (University of California, Santa Barbara), T. L. Rokoske, Richard Roth (Eastern Michigan University), Carl Rotter (University of West Virginia), S. Clark Rowland (Andrews University), Rajarshi Roy (Georgia Institute of Technology), Russell A. Roy (Santa Fe Community College), Dhiraj Sardar (University of Texas, San Antonio), Bruce Schumm (UC Santa Cruz), Melvin Schwartz (St. John's University), F. A. Scott, L. W. Seagondollar, Paul Shand (University of Northern Iowa), Stan Shepherd (Pennsylvania State University), Douglas Sherman (San Jose State), Bruce Sherwood (Carnegie Mellon University), Hugh Siefkin (Greenville College), Tomasz Skwarnicki (Syracuse University), C. P. Slichter, Charles W. Smith (University of Maine, Orono), Malcolm Smith (University of Lowell), Ross Spencer (Brigham Young University), Julien Sprott (University of Wisconsin), Victor Stanionis (Iona College), James Stith (American Institute of Physics), Chuck Stone (North Carolina A&T State

University), Edward Strother (Florida Institute of Technology), Conley Stutz (Bradley University), Albert Stwertka (U.S. Merchant Marine Academy), Kenneth Szpara-DeNisco (Harrisburg Area Community College), Martin Tiersten (CUNY, City College), David Toot (Alfred University), Somdev Tyagi (Drexel University), F. Verbrugge, Helmut Vogel (Carnegie Mellon University), Robert Webb (Texas A & M), Thomas Weber (Iowa State University), M. Russell Wehr, (Pennsylvania State University), Robert Weidman (Michigan Technical University), Dan Whalen (UC San Diego), Lester V. Whitney, Thomas Wiggins (Pennsylvania State University), David Willey (University of Pittsburgh, Johnstown), George Williams (University of Utah), John Williams (Auburn University), Stanley Williams (Iowa State University), Jack Willis, Suzanne Willis (Northern Illinois University), Robert Wilson (San Bernardino Valley College), L. Wolfenstein, James Wood (Palm Beach Junior College), Lowell Wood (University of Houston), R. E. Worley, D. H. Ziebell (Manatee Community College), George O. Zimmerman (Boston University)

In addition, we both have individual acknowledgments we would like to make.

I want to extend my heartfelt thanks to my colleagues at Carnegie Mellon, especially Professors Robert Kraemer, Bruce Sherwood, Ruth Chabay, Helmut Vogel, and Brian Quinn, for many stimulating discussions about physics pedagogy and for their support and encouragement during the writing of several successive editions of this book. I am equally indebted to the many generations of Carnegie Mellon students who have helped me learn what good teaching and good writing are, by showing me what works and what doesn't. It is always a joy and a privilege to express my gratitude to my wife Alice and our children Gretchen and Rebecca for their love, support, and emotional sustenance during the writing of several successive editions of this book. May all men and women be blessed with love such as theirs. — H. D. Y.

I would like to thank my past and present colleagues at UCSB, including Rob Geller, Carl Gwinn, Al Nash, Elisabeth Nicol, and Francesc Roig, for their whole-hearted support and for many helpful discussions. I owe a special debt of gratitude to my early teachers Willa Ramsay, Peter Zimmerman, William Little, Alan Schwettman, and Dirk Walecka for showing me what clear and engaging physics teaching is all about, and to Stuart Johnson for inviting me to become a co-author of *University Physics* beginning with the 9th edition. I want to express special thanks to the editorial staff at Addison-Wesley and their partners: to Nancy Whilton for her editorial vision; to Margot Otway for her superb graphic sense and careful development of this edition; to Peter Murphy for his contributions to the worked examples; to Jason J. B. Harlow for his careful reading of the page proofs; and to Chandrika Madhavan, Steven Le, and Cindy Johnson for keeping the editorial and production pipeline flowing. Most of all, I want to express my gratitude and love to my wife Caroline, to whom I dedicate my contribution to this book. Hey, Caroline, the new edition's done at last — let's go flying! — R. A. F.

Please Tell Us What You Think!

We welcome communications from students and professors, especially concerning errors or deficiencies that you find in this edition. We have devoted a lot of time and effort to writing the best book we know how to write, and we hope it will help you to teach and learn physics. In turn, you can help us by letting us know what still needs to be improved! Please feel free to contact us either electronically or by ordinary mail. Your comments will be greatly appreciated.

December 2010

Hugh D. Young
Department of Physics
Carnegie Mellon University
Pittsburgh, PA 15213
hdy@andrew.cmu.edu

Roger A. Freedman
Department of Physics
University of California, Santa Barbara
Santa Barbara, CA 93106-9530
airboy@physics.ucsb.edu
http://www.physics.ucsb.edu/~airboy/

DETAILED CONTENTS

MECHANICS

1 UNITS, PHYSICAL QUANTITIES, AND VECTORS 1

1.1	The Nature of Physics	2
1.2	Solving Physics Problems	2
1.3	Standards and Units	4
1.4	Unit Consistency and Conversions	6
1.5	Uncertainty and Significant Figures	8
1.6	Estimates and Orders of Magnitude	10
1.7	Vectors and Vector Addition	10
1.8	Components of Vectors	14
1.9	Unit Vectors	19
1.10	Products of Vectors	20
	Summary	26
	Questions/Exercises/Problems	27

2 MOTION ALONG A STRAIGHT LINE 35

2.1	Displacement, Time, and Average Velocity	36
2.2	Instantaneous Velocity	38
2.3	Average and Instantaneous Acceleration	42
2.4	Motion with Constant Acceleration	46
2.5	Freely Falling Bodies	52
2.6	Velocity and Position by Integration	55
	Summary	58
	Questions/Exercises/Problems	59

3 MOTION IN TWO OR THREE DIMENSIONS 69

3.1	Position and Velocity Vectors	70
3.2	The Acceleration Vector	72
3.3	Projectile Motion	77
3.4	Motion in a Circle	85
3.5	Relative Velocity	88
	Summary	94
	Questions/Exercises/Problems	95

4 NEWTON'S LAWS OF MOTION 104

4.1	Force and Interactions	105
4.2	Newton's First Law	108
4.3	Newton's Second Law	112
4.4	Mass and Weight	117
4.5	Newton's Third Law	120
4.6	Free-Body Diagrams	124
	Summary	126
	Questions/Exercises/Problems	127

5 APPLYING NEWTON'S LAWS 134

5.1	Using Newton's First Law: Particles in Equilibrium	134
5.2	Using Newton's Second Law: Dynamics of Particles	140
5.3	Frictional Forces	146
5.4	Dynamics of Circular Motion	154
5.5	The Fundamental Forces of Nature	159
	Summary	161
	Questions/Exercises/Problems	162

6 WORK AND KINETIC ENERGY 176

6.1	Work	177
6.2	Kinetic Energy and the Work–Energy Theorem	181
6.3	Work and Energy with Varying Forces	187
6.4	Power	193
	Summary	196
	Questions/Exercises/Problems	197

7 POTENTIAL ENERGY AND ENERGY CONSERVATION 207

7.1	Gravitational Potential Energy	208
7.2	Elastic Potential Energy	216
7.3	Conservative and Nonconservative Forces	221
7.4	Force and Potential Energy	225
7.5	Energy Diagrams	228
	Summary	230
	Questions/Exercises/Problems	231

8 MOMENTUM, IMPULSE, AND COLLISIONS 241

8.1	Momentum and Impulse	241
8.2	Conservation of Momentum	247
8.3	Momentum Conservation and Collisions	251
8.4	Elastic Collisions	255
8.5	Center of Mass	258
8.6	Rocket Propulsion	262
	Summary	266
	Questions/Exercises/Problems	267

9 ROTATION OF RIGID BODIES 278

9.1	Angular Velocity and Acceleration	278
9.2	Rotation with Constant Angular Acceleration	283
9.3	Relating Linear and Angular Kinematics	285
9.4	Energy in Rotational Motion	288

9.5   Parallel-Axis Theorem   293
9.6   Moment-of-Inertia Calculations   294
       Summary   297
       Questions/Exercises/Problems   298

10   DYNAMICS OF ROTATIONAL MOTION   308

10.1   Torque   308
10.2   Torque and Angular Acceleration
       for a Rigid Body   311
10.3   Rigid-Body Rotation About a Moving Axis   314
10.4   Work and Power in Rotational Motion   320
10.5   Angular Momentum   322
10.6   Conservation of Angular Momentum   325
10.7   Gyroscopes and Precession   328
       Summary   331
       Questions/Exercises/Problems   332

11   EQUILIBRIUM AND ELASTICITY   344

11.1   Conditions for Equilibrium   345
11.2   Center of Gravity   345
11.3   Solving Rigid-Body Equilibrium Problems   348
11.4   Stress, Strain, and Elastic Moduli   352
11.5   Elasticity and Plasticity   357
       Summary   359
       Questions/Exercises/Problems   360

12   FLUID MECHANICS   373

12.1   Density   373
12.2   Pressure in a Fluid   375
12.3   Buoyancy   380
12.4   Fluid Flow   382
12.5   Bernoulli's Equation   385
12.6   Viscosity and Turbulence   389
       Summary   392
       Questions/Exercises/Problems   393

13   GRAVITATION   402

13.1   Newton's Law of Gravitation   402
13.2   Weight   406
13.3   Gravitational Potential Energy   409
13.4   The Motion of Satellites   411
13.5   Kepler's Laws and the Motion of Planets   414
13.6   Spherical Mass Distributions   418
13.7   Apparent Weight and the Earth's Rotation   421
13.8   Black Holes   423
       Summary   427
       Questions/Exercises/Problems   428

14   PERIODIC MOTION   437

14.1   Describing Oscillation   437
14.2   Simple Harmonic Motion   439

14.3   Energy in Simple Harmonic Motion   446
14.4   Applications of Simple Harmonic Motion   450
14.5   The Simple Pendulum   453
14.6   The Physical Pendulum   455
14.7   Damped Oscillations   457
14.8   Forced Oscillations and Resonance   459
       Summary   461
       Questions/Exercises/Problems   462

WAVES/ACOUSTICS

15   MECHANICAL WAVES   472

15.1   Types of Mechanical Waves   473
15.2   Periodic Waves   474
15.3   Mathematical Description of a Wave   477
15.4   Speed of a Transverse Wave   482
15.5   Energy in Wave Motion   486
15.6   Wave Interference, Boundary Conditions,
       and Superposition   489
15.7   Standing Waves on a String   491
15.8   Normal Modes of a String   495
       Summary   499
       Questions/Exercises/Problems   500

16   SOUND AND HEARING   509

16.1   Sound Waves   509
16.2   Speed of Sound Waves   514
16.3   Sound Intensity   518
16.4   Standing Sound Waves
       and Normal Modes   522
16.5   Resonance and Sound   527
16.6   Interference of Waves   529
16.7   Beats   531
16.8   The Doppler Effect   533
16.9   Shock Waves   538
       Summary   541
       Questions/Exercises/Problems   542

THERMODYNAMICS

17   TEMPERATURE AND HEAT   551

17.1   Temperature and Thermal Equilibrium   552
17.2   Thermometers and Temperature Scales   553
17.3   Gas Thermometers and the Kelvin Scale   554
17.4   Thermal Expansion   557
17.5   Quantity of Heat   562
17.6   Calorimetry and Phase Changes   565
17.7   Mechanisms of Heat Transfer   570
       Summary   578
       Questions/Exercises/Problems   579

18 THERMAL PROPERTIES OF MATTER 590

18.1	Equations of State	591
18.2	Molecular Properties of Matter	596
18.3	Kinetic-Molecular Model of an Ideal Gas	599
18.4	Heat Capacities	605
18.5	Molecular Speeds	608
18.6	Phases of Matter	610
	Summary	614
	Questions/Exercises/Problems	615

19 THE FIRST LAW OF THERMODYNAMICS 624

19.1	Thermodynamic Systems	624
19.2	Work Done During Volume Changes	625
19.3	Paths Between Thermodynamic States	628
19.4	Internal Energy and the First Law of Thermodynamics	629
19.5	Kinds of Thermodynamic Processes	634
19.6	Internal Energy of an Ideal Gas	636
19.7	Heat Capacities of an Ideal Gas	636
19.8	Adiabatic Processes for an Ideal Gas	639
	Summary	643
	Questions/Exercises/Problems	644

20 THE SECOND LAW OF THERMODYNAMICS 652

20.1	Directions of Thermodynamic Processes	652
20.2	Heat Engines	654
20.3	Internal-Combustion Engines	657
20.4	Refrigerators	659
20.5	The Second Law of Thermodynamics	661
20.6	The Carnot Cycle	663
20.7	Entropy	669
20.8	Microscopic Interpretation of Entropy	675
	Summary	678
	Questions/Exercises/Problems	679

ELECTROMAGNETISM

21 ELECTRIC CHARGE AND ELECTRIC FIELD 687

21.1	Electric Charge	688
21.2	Conductors, Insulators, and Induced Charges	691
21.3	Coulomb's Law	693
21.4	Electric Field and Electric Forces	698
21.5	Electric-Field Calculations	703
21.6	Electric Field Lines	708
21.7	Electric Dipoles	709
	Summary	714
	Questions/Exercises/Problems	715

22 GAUSS'S LAW 725

22.1	Charge and Electric Flux	725
22.2	Calculating Electric Flux	728
22.3	Gauss's Law	732
22.4	Applications of Gauss's Law	736
22.5	Charges on Conductors	741
	Summary	746
	Questions/Exercises/Problems	747

23 ELECTRIC POTENTIAL 754

23.1	Electric Potential Energy	754
23.2	Electric Potential	761
23.3	Calculating Electric Potential	767
23.4	Equipotential Surfaces	771
23.5	Potential Gradient	774
	Summary	777
	Questions/Exercises/Problems	778

24 CAPACITANCE AND DIELECTRICS 788

24.1	Capacitors and Capacitance	789
24.2	Capacitors in Series and Parallel	793
24.3	Energy Storage in Capacitors and Electric-Field Energy	796
24.4	Dielectrics	800
24.5	Molecular Model of Induced Charge	805
24.6	Gauss's Law in Dielectrics	807
	Summary	809
	Questions/Exercises/Problems	810

25 CURRENT, RESISTANCE, AND ELECTROMOTIVE FORCE 818

25.1	Current	819
25.2	Resistivity	822
25.3	Resistance	825
25.4	Electromotive Force and Circuits	828
25.5	Energy and Power in Electric Circuits	834
25.6	Theory of Metallic Conduction	838
	Summary	841
	Questions/Exercises/Problems	842

26 DIRECT-CURRENT CIRCUITS 850

26.1	Resistors in Series and Parallel	850
26.2	Kirchhoff's Rules	855
26.3	Electrical Measuring Instruments	860
26.4	R-C Circuits	864
26.5	Power Distribution Systems	868
	Summary	873
	Questions/Exercises/Problems	874

27 MAGNETIC FIELD
AND MAGNETIC FORCES 883

27.1 Magnetism 883
27.2 Magnetic Field 885
27.3 Magnetic Field Lines and Magnetic Flux 889
27.4 Motion of Charged Particles
in a Magnetic Field 892
27.5 Applications of Motion
of Charged Particles 896
27.6 Magnetic Force on a
Current-Carrying Conductor 898
27.7 Force and Torque on a Current Loop 901
27.8 The Direct-Current Motor 907
27.9 The Hall Effect 909
Summary 911
Questions/Exercises/Problems 912

28 SOURCES OF MAGNETIC FIELD 923

28.1 Magnetic Field of a Moving Charge 923
28.2 Magnetic Field of a Current Element 926
28.3 Magnetic Field of a Straight
Current-Carrying Conductor 928
28.4 Force Between Parallel Conductors 931
28.5 Magnetic Field of a Circular Current Loop 932
28.6 Ampere's Law 935
28.7 Applications of Ampere's Law 938
28.8 Magnetic Materials 941
Summary 947
Questions/Exercises/Problems 949

29 ELECTROMAGNETIC
INDUCTION 957

29.1 Induction Experiments 958
29.2 Faraday's Law 959
29.3 Lenz's Law 967
29.4 Motional Electromotive Force 969
29.5 Induced Electric Fields 971
29.6 Eddy Currents 974
29.7 Displacement Current and
Maxwell's Equations 975
29.8 Superconductivity 979
Summary 981
Questions/Exercises/Problems 982

30 INDUCTANCE 991

30.1 Mutual Inductance 991
30.2 Self-Inductance and Inductors 994
30.3 Magnetic-Field Energy 998
30.4 The R-L Circuit 1001
30.5 The L-C Circuit 1005
30.6 The L-R-C Series Circuit 1009
Summary 1012
Questions/Exercises/Problems 1013

31 ALTERNATING CURRENT 1021

31.1 Phasors and Alternating
Currents 1021
31.2 Resistance and Reactance 1024
31.3 The L-R-C Series Circuit 1030
31.4 Power in Alternating-Current
Circuits 1034
31.5 Resonance in Alternating-Current
Circuits 1037
31.6 Transformers 1040
Summary 1043
Questions/Exercises/Problems 1044

32 ELECTROMAGNETIC WAVES 1051

32.1 Maxwell's Equations and
Electromagnetic Waves 1052
32.2 Plane Electromagnetic Waves
and the Speed of Light 1055
32.3 Sinusoidal Electromagnetic Waves 1060
32.4 Energy and Momentum
in Electromagnetic Waves 1064
32.5 Standing Electromagnetic Waves 1069
Summary 1073
Questions/Exercises/Problems 1074

OPTICS

33 THE NATURE AND
PROPAGATION OF LIGHT 1080

33.1 The Nature of Light 1080
33.2 Reflection and Refraction 1082
33.3 Total Internal Reflection 1088
33.4 Dispersion 1091
33.5 Polarization 1093
33.6 Scattering of Light 1100
33.7 Huygens's Principle 1102
Summary 1105
Questions/Exercises/Problems 1106

34 GEOMETRIC OPTICS 1114

34.1 Reflection and Refraction at a
Plane Surface 1114
34.2 Reflection at a Spherical Surface 1118
34.3 Refraction at a Spherical Surface 1126
34.4 Thin Lenses 1131
34.5 Cameras 1139
34.6 The Eye 1142
34.7 The Magnifier 1146
34.8 Microscopes and Telescopes 1147
Summary 1152
Questions/Exercises/Problems 1153

35 INTERFERENCE 1163

35.1	Interference and Coherent Sources	1164
35.2	Two-Source Interference of Light	1166
35.3	Intensity in Interference Patterns	1170
35.4	Interference in Thin Films	1173
35.5	The Michelson Interferometer	1179
	Summary	1182
	Questions/Exercises/Problems	1183

36 DIFFRACTION 1190

36.1	Fresnel and Fraunhofer Diffraction	1191
36.2	Diffraction from a Single Slit	1192
36.3	Intensity in the Single-Slit Pattern	1195
36.4	Multiple Slits	1199
36.5	The Diffraction Grating	1201
36.6	X-Ray Diffraction	1205
36.7	Circular Apertures and Resolving Power	1208
36.8	Holography	1211
	Summary	1214
	Questions/Exercises/Problems	1215

MODERN PHYSICS

37 RELATIVITY 1223

37.1	Invariance of Physical Laws	1223
37.2	Relativity of Simultaneity	1227
37.3	Relativity of Time Intervals	1228
37.4	Relativity of Length	1233
37.5	The Lorentz Transformations	1237
37.6	The Doppler Effect for Electromagnetic Waves	1241
37.7	Relativistic Momentum	1243
37.8	Relativistic Work and Energy	1246
37.9	Newtonian Mechanics and Relativity	1249
	Summary	1252
	Questions/Exercises/Problems	1253

38 PHOTONS: LIGHT WAVES BEHAVING AS PARTICLES 1261

38.1	Light Absorbed as Photons: The Photoelectric Effect	1261
38.2	Light Emitted as Photons: X-Ray Production	1266
38.3	Light Scattered as Photons: Compton Scattering and Pair Production	1269
38.4	Wave–Particle Duality, Probability, and Uncertainty	1273
	Summary	1280
	Questions/Exercises/Problems	1281

39 PARTICLES BEHAVING AS WAVES 1286

39.1	Electron Waves	1286
39.2	The Nuclear Atom and Atomic Spectra	1292
39.3	Energy Levels and the Bohr Model of the Atom	1297
39.4	The Laser	1307
39.5	Continuous Spectra	1310
39.6	The Uncertainty Principle Revisited	1314
	Summary	1318
	Questions/Exercises/Problems	1319

40 QUANTUM MECHANICS 1328

40.1	Wave Functions and the One-Dimensional Schrödinger Equation	1328
40.2	Particle in a Box	1338
40.3	Potential Wells	1343
40.4	Potential Barriers and Tunneling	1347
40.5	The Harmonic Oscillator	1350
	Summary	1355
	Questions/Exercises/Problems	1356

41 ATOMIC STRUCTURE 1364

41.1	The Schrödinger Equation in Three Dimensions	1365
41.2	Particle in a Three-Dimensional Box	1366
41.3	The Hydrogen Atom	1372
41.4	The Zeeman Effect	1379
41.5	Electron Spin	1383
41.6	Many-Electron Atoms and the Exclusion Principle	1387
41.7	X-Ray Spectra	1393
	Summary	1397
	Questions/Exercises/Problems	1399

42 MOLECULES AND CONDENSED MATTER 1405

42.1	Types of Molecular Bonds	1405
42.2	Molecular Spectra	1408
42.3	Structure of Solids	1412
42.4	Energy Bands	1416
42.5	Free-Electron Model of Metals	1418
42.6	Semiconductors	1422
42.7	Semiconductor Devices	1425
42.8	Superconductivity	1430
	Summary	1431
	Questions/Exercises/Problems	1432

43 NUCLEAR PHYSICS 1439

43.1	Properties of Nuclei	1439
43.2	Nuclear Binding and Nuclear Structure	1444

43.3   Nuclear Stability and Radioactivity   1449
43.4   Activities and Half-Lives   1456
43.5   Biological Effects of Radiation   1459
43.6   Nuclear Reactions   1462
43.7   Nuclear Fission   1464
43.8   Nuclear Fusion   1469
        Summary   1472
        Questions/Exercises/Problems   1473

44   PARTICLE PHYSICS AND COSMOLOGY   1480

44.1   Fundamental Particles—A History   1480
44.2   Particle Accelerators and Detectors   1485
44.3   Particles and Interactions   1490
44.4   Quarks and the Eightfold Way   1496
44.5   The Standard Model and Beyond   1499

44.6   The Expanding Universe   1501
44.7   The Beginning of Time   1508
        Summary   1517
        Questions/Exercises/Problems   1518

APPENDICES

A   The International System of Units   A-1
B   Useful Mathematical Relations   A-3
C   The Greek Alphabet   A-4
D   Periodic Table of Elements   A-5
E   Unit Conversion Factors   A-6
F   Numerical Constants   A-7

        Answers to Odd-Numbered Problems   A-9
        Photo Credits   C-1
        Index   I-1

UNITS, PHYSICAL QUANTITIES, AND VECTORS

1

? Being able to predict the path of a thunderstorm is essential for minimizing the damage it does to lives and property. If a thunderstorm is moving at 20 km/h in a direction 53° north of east, how far north does the thunderstorm move in 1 h?

LEARNING GOALS

By studying this chapter, you will learn:

- Three fundamental quantities of physics and the units physicists use to measure them.

- How to keep track of significant figures in your calculations.

- The difference between scalars and vectors, and how to add and subtract vectors graphically.

- What the components of a vector are, and how to use them in calculations.

- What unit vectors are, and how to use them with components to describe vectors.

- Two ways of multiplying vectors.

P hysics is one of the most fundamental of the sciences. Scientists of all disciplines use the ideas of physics, including chemists who study the structure of molecules, paleontologists who try to reconstruct how dinosaurs walked, and climatologists who study how human activities affect the atmosphere and oceans. Physics is also the foundation of all engineering and technology. No engineer could design a flat-screen TV, an interplanetary spacecraft, or even a better mousetrap without first understanding the basic laws of physics.

The study of physics is also an adventure. You will find it challenging, sometimes frustrating, occasionally painful, and often richly rewarding. If you've ever wondered why the sky is blue, how radio waves can travel through empty space, or how a satellite stays in orbit, you can find the answers by using fundamental physics. You will come to see physics as a towering achievement of the human intellect in its quest to understand our world and ourselves.

In this opening chapter, we'll go over some important preliminaries that we'll need throughout our study. We'll discuss the nature of physical theory and the use of idealized models to represent physical systems. We'll introduce the systems of units used to describe physical quantities and discuss ways to describe the accuracy of a number. We'll look at examples of problems for which we can't (or don't want to) find a precise answer, but for which rough estimates can be useful and interesting. Finally, we'll study several aspects of vectors and vector algebra. Vectors will be needed throughout our study of physics to describe and analyze physical quantities, such as velocity and force, that have direction as well as magnitude.

1.1 The Nature of Physics

Physics is an *experimental* science. Physicists observe the phenomena of nature and try to find patterns that relate these phenomena. These patterns are called physical theories or, when they are very well established and widely used, physical laws or principles.

CAUTION   **The meaning of the word "theory"** Calling an idea a theory does *not* mean that it's just a random thought or an unproven concept. Rather, a theory is an explanation of natural phenomena based on observation and accepted fundamental principles. An example is the well-established theory of biological evolution, which is the result of extensive research and observation by generations of biologists.

To develop a physical theory, a physicist has to learn to ask appropriate questions, design experiments to try to answer the questions, and draw appropriate conclusions from the results. Figure 1.1 shows two famous facilities used for physics experiments.

Legend has it that Galileo Galilei (1564–1642) dropped light and heavy objects from the top of the Leaning Tower of Pisa (Fig. 1.1a) to find out whether their rates of fall were the same or different. From examining the results of his experiments (which were actually much more sophisticated than in the legend), he made the inductive leap to the principle, or theory, that the acceleration of a falling body is independent of its weight.

The development of physical theories such as Galileo's often takes an indirect path, with blind alleys, wrong guesses, and the discarding of unsuccessful theories in favor of more promising ones. Physics is not simply a collection of facts and principles; it is also the *process* by which we arrive at general principles that describe how the physical universe behaves.

No theory is ever regarded as the final or ultimate truth. The possibility always exists that new observations will require that a theory be revised or discarded. It is in the nature of physical theory that we can disprove a theory by finding behavior that is inconsistent with it, but we can never prove that a theory is always correct.

Getting back to Galileo, suppose we drop a feather and a cannonball. They certainly do *not* fall at the same rate. This does not mean that Galileo was wrong; it means that his theory was incomplete. If we drop the feather and the cannonball *in a vacuum* to eliminate the effects of the air, then they do fall at the same rate. Galileo's theory has a **range of validity:** It applies only to objects for which the force exerted by the air (due to air resistance and buoyancy) is much less than the weight. Objects like feathers or parachutes are clearly outside this range.

Often a new development in physics extends a principle's range of validity. Galileo's analysis of falling bodies was greatly extended half a century later by Newton's laws of motion and law of gravitation.

1.1 Two research laboratories. **(a)** According to legend, Galileo investigated falling bodies by dropping them from the Leaning Tower in Pisa, Italy, and he studied pendulum motion by observing the swinging of the chandelier in the adjacent cathedral. **(b)** The Large Hadron Collider (LHC) in Geneva, Switzerland, the world's largest particle accelerator, is used to explore the smallest and most fundamental constituents of matter. This photo shows a portion of one of the LHC's detectors (note the worker on the yellow platform).

(a)

(b)

1.2 Solving Physics Problems

At some point in their studies, almost all physics students find themselves thinking, "I understand the concepts, but I just can't solve the problems." But in physics, truly understanding a concept *means* being able to apply it to a variety of problems. Learning how to solve problems is absolutely essential; you don't *know* physics unless you can *do* physics.

How do you learn to solve physics problems? In every chapter of this book you will find *Problem-Solving Strategies* that offer techniques for setting up and solving problems efficiently and accurately. Following each *Problem-Solving Strategy* are one or more worked *Examples* that show these techniques in action. (The *Problem-Solving Strategies* will also steer you away from some *incorrect* techniques that you may be tempted to use.) You'll also find additional examples that aren't associated with a particular *Problem-Solving Strategy*. In addition, at the end of each chapter you'll find a *Bridging Problem* that uses more than one of the key ideas from the chapter. Study these strategies and problems carefully, and work through each example for yourself on a piece of paper.

Different techniques are useful for solving different kinds of physics problems, which is why this book offers dozens of *Problem-Solving Strategies.* No matter what kind of problem you're dealing with, however, there are certain key steps that you'll always follow. (These same steps are equally useful for problems in math, engineering, chemistry, and many other fields.) In this book we've organized these steps into four stages of solving a problem.

All of the *Problem-Solving Strategies* and *Examples* in this book will follow these four steps. (In some cases we will combine the first two or three steps.) We encourage you to follow these same steps when you solve problems yourself. You may find it useful to remember the acronym *I SEE*—short for *Identify, Set up, Execute,* and *Evaluate.*

Problem-Solving Strategy 1.1 | Solving Physics Problems

IDENTIFY *the relevant concepts:* Use the physical conditions stated in the problem to help you decide which physics concepts are relevant. Identify the **target variables** of the problem—that is, the quantities whose values you're trying to find, such as the speed at which a projectile hits the ground, the intensity of a sound made by a siren, or the size of an image made by a lens. Identify the known quantities, as stated or implied in the problem. This step is essential whether the problem asks for an algebraic expression or a numerical answer.

SET UP *the problem:* Given the concepts you have identified and the known and target quantities, choose the equations that you'll use to solve the problem and decide how you'll use them. Make sure that the variables you have identified correlate exactly with those in the equations. If appropriate, draw a sketch of the situation described in the problem. (Graph paper, ruler, protractor, and compass will help you make clear, useful sketches.) As best you can,

estimate what your results will be and, as appropriate, predict what the physical behavior of a system will be. The worked examples in this book include tips on how to make these kinds of estimates and predictions. If this seems challenging, don't worry—you'll get better with practice!

EXECUTE *the solution:* This is where you "do the math." Study the worked examples to see what's involved in this step.

EVALUATE *your answer:* Compare your answer with your estimates, and reconsider things if there's a discrepancy. If your answer includes an algebraic expression, assure yourself that it represents what would happen if the variables in it were taken to extremes. For future reference, make note of any answer that represents a quantity of particular significance. Ask yourself how you might answer a more general or more difficult version of the problem you have just solved.

Idealized Models

In everyday conversation we use the word "model" to mean either a small-scale replica, such as a model railroad, or a person who displays articles of clothing (or the absence thereof). In physics a **model** is a simplified version of a physical system that would be too complicated to analyze in full detail.

For example, suppose we want to analyze the motion of a thrown baseball (Fig. 1.2a). How complicated is this problem? The ball is not a perfect sphere (it has raised seams), and it spins as it moves through the air. Wind and air resistance influence its motion, the ball's weight varies a little as its distance from the center of the earth changes, and so on. If we try to include all these things, the analysis gets hopelessly complicated. Instead, we invent a simplified version of the problem. We neglect the size and shape of the ball by representing it as a point object, or **particle.** We neglect air resistance by making the ball move in a vacuum, and we make the weight constant. Now we have a problem that is simple enough to deal with (Fig. 1.2b). We will analyze this model in detail in Chapter 3.

We have to overlook quite a few minor effects to make an idealized model, but we must be careful not to neglect too much. If we ignore the effects of gravity completely, then our model predicts that when we throw the ball up, it will go in a straight line and disappear into space. A useful model is one that simplifies a problem enough to make it manageable, yet keeps its essential features.

1.2 To simplify the analysis of (a) a baseball in flight, we use (b) an idealized model.

(a) A real baseball in flight

Baseball spins and has a complex shape.

Air resistance and wind exert forces on the ball.

Direction of motion

Gravitational force on ball depends on altitude.

(b) An idealized model of the baseball

Baseball is treated as a point object (particle).

No air resistance.

Direction of motion

Gravitational force on ball is constant.

The validity of the predictions we make using a model is limited by the validity of the model. For example, Galileo's prediction about falling bodies (see Section 1.1) corresponds to an idealized model that does not include the effects of air resistance. This model works fairly well for a dropped cannonball, but not so well for a feather.

Idealized models play a crucial role throughout this book. Watch for them in discussions of physical theories and their applications to specific problems.

1.3 Standards and Units

As we learned in Section 1.1, physics is an experimental science. Experiments require measurements, and we generally use numbers to describe the results of measurements. Any number that is used to describe a physical phenomenon quantitatively is called a **physical quantity.** For example, two physical quantities that describe you are your weight and your height. Some physical quantities are so fundamental that we can define them only by describing how to measure them. Such a definition is called an **operational definition.** Two examples are measuring a distance by using a ruler and measuring a time interval by using a stopwatch. In other cases we define a physical quantity by describing how to calculate it from other quantities that we *can* measure. Thus we might define the average speed of a moving object as the distance traveled (measured with a ruler) divided by the time of travel (measured with a stopwatch).

When we measure a quantity, we always compare it with some reference standard. When we say that a Ferrari 458 Italia is 4.53 meters long, we mean that it is 4.53 times as long as a meter stick, which we define to be 1 meter long. Such a standard defines a **unit** of the quantity. The meter is a unit of distance, and the second is a unit of time. When we use a number to describe a physical quantity, we must always specify the unit that we are using; to describe a distance as simply "4.53" wouldn't mean anything.

To make accurate, reliable measurements, we need units of measurement that do not change and that can be duplicated by observers in various locations. The system of units used by scientists and engineers around the world is commonly called "the metric system," but since 1960 it has been known officially as the **International System,** or **SI** (the abbreviation for its French name, *Système International*). Appendix A gives a list of all SI units as well as definitions of the most fundamental units.

1.3 The measurements used to determine (a) the duration of a second and (b) the length of a meter. These measurements are useful for setting standards because they give the same results no matter where they are made.

(a) Measuring the second

Microwave radiation with a frequency of exactly 9,192,631,770 cycles per second ...

Outermost electron

Cesium-133 atom

... causes the outermost electron of a cesium-133 atom to reverse its spin direction.

Cesium-133 atom

An atomic clock uses this phenomenon to tune microwaves to this exact frequency. It then counts 1 second for each 9,192,631,770 cycles.

(b) Measuring the meter

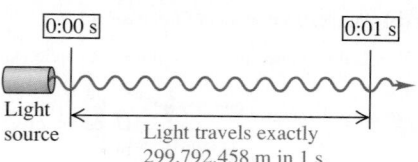

0:00 s   0:01 s

Light source

Light travels exactly 299,792,458 m in 1 s.

Time

From 1889 until 1967, the unit of time was defined as a certain fraction of the mean solar day, the average time between successive arrivals of the sun at its highest point in the sky. The present standard, adopted in 1967, is much more precise. It is based on an atomic clock, which uses the energy difference between the two lowest energy states of the cesium atom. When bombarded by microwaves of precisely the proper frequency, cesium atoms undergo a transition from one of these states to the other. One **second** (abbreviated s) is defined as the time required for 9,192,631,770 cycles of this microwave radiation (Fig. 1.3a).

Length

In 1960 an atomic standard for the meter was also established, using the wavelength of the orange-red light emitted by atoms of krypton (^{86}Kr) in a glow discharge tube. Using this length standard, the speed of light in vacuum was measured to be 299,792,458 m/s. In November 1983, the length standard was changed again so that the speed of light in vacuum was *defined* to be precisely

299,792,458 m/s. Hence the new definition of the **meter** (abbreviated m) is the distance that light travels in vacuum in 1/299,792,458 second (Fig. 1.3b). This provides a much more precise standard of length than the one based on a wavelength of light.

Mass

The standard of mass, the **kilogram** (abbreviated kg), is defined to be the mass of a particular cylinder of platinum–iridium alloy kept at the International Bureau of Weights and Measures at Sèvres, near Paris (Fig. 1.4). An atomic standard of mass would be more fundamental, but at present we cannot measure masses on an atomic scale with as much accuracy as on a macroscopic scale. The *gram* (which is not a fundamental unit) is 0.001 kilogram.

Unit Prefixes

Once we have defined the fundamental units, it is easy to introduce larger and smaller units for the same physical quantities. In the metric system these other units are related to the fundamental units (or, in the case of mass, to the gram) by multiples of 10 or $\frac{1}{10}$. Thus one kilometer (1 km) is 1000 meters, and one centimeter (1 cm) is $\frac{1}{100}$ meter. We usually express multiples of 10 or $\frac{1}{10}$ in exponential notation: $1000 = 10^3$, $\frac{1}{1000} = 10^{-3}$, and so on. With this notation, $1 \text{ km} = 10^3 \text{ m}$ and $1 \text{ cm} = 10^{-2} \text{ m}$.

The names of the additional units are derived by adding a **prefix** to the name of the fundamental unit. For example, the prefix "kilo-," abbreviated k, always means a unit larger by a factor of 1000; thus

$$1 \text{ kilometer} = 1 \text{ km} = 10^3 \text{ meters} = 10^3 \text{ m}$$
$$1 \text{ kilogram} = 1 \text{ kg} = 10^3 \text{ grams} = 10^3 \text{ g}$$
$$1 \text{ kilowatt} = 1 \text{ kW} = 10^3 \text{ watts} = 10^3 \text{ W}$$

A table on the inside back cover of this book lists the standard SI prefixes, with their meanings and abbreviations.

Table 1.1 gives some examples of the use of multiples of 10 and their prefixes with the units of length, mass, and time. Figure 1.5 shows how these prefixes are used to describe both large and small distances.

The British System

Finally, we mention the British system of units. These units are used only in the United States and a few other countries, and in most of these they are being replaced by SI units. British units are now officially defined in terms of SI units, as follows:

Length: 1 inch = 2.54 cm (exactly)

Force: 1 pound = 4.448221615260 newtons (exactly)

1.4 The international standard kilogram is the metal object carefully enclosed within these nested glass containers.

Table 1.1 Some Units of Length, Mass, and Time

Length	Mass	Time
1 nanometer = 1 nm = 10^{-9} m (a few times the size of the largest atom)	1 microgram = 1 μg = 10^{-6} g = 10^{-9} kg (mass of a very small dust particle)	1 nanosecond = 1 ns = 10^{-9} s (time for light to travel 0.3 m)
1 micrometer = 1 μm = 10^{-6} m (size of some bacteria and living cells)	1 milligram = 1 mg = 10^{-3} g = 10^{-6} kg (mass of a grain of salt)	1 microsecond = 1 μs = 10^{-6} s (time for space station to move 8 mm)
1 millimeter = 1 mm = 10^{-3} m (diameter of the point of a ballpoint pen)	1 gram = 1 g = 10^{-3} kg (mass of a paper clip)	1 millisecond = 1 ms = 10^{-3} s (time for sound to travel 0.35 m)
1 centimeter = 1 cm = 10^{-2} m (diameter of your little finger)		
1 kilometer = 1 km = 10^3 m (a 10-minute walk)		

1.5 Some typical lengths in the universe. (f) is a scanning tunneling microscope image of atoms on a crystal surface; (g) is an artist's impression.

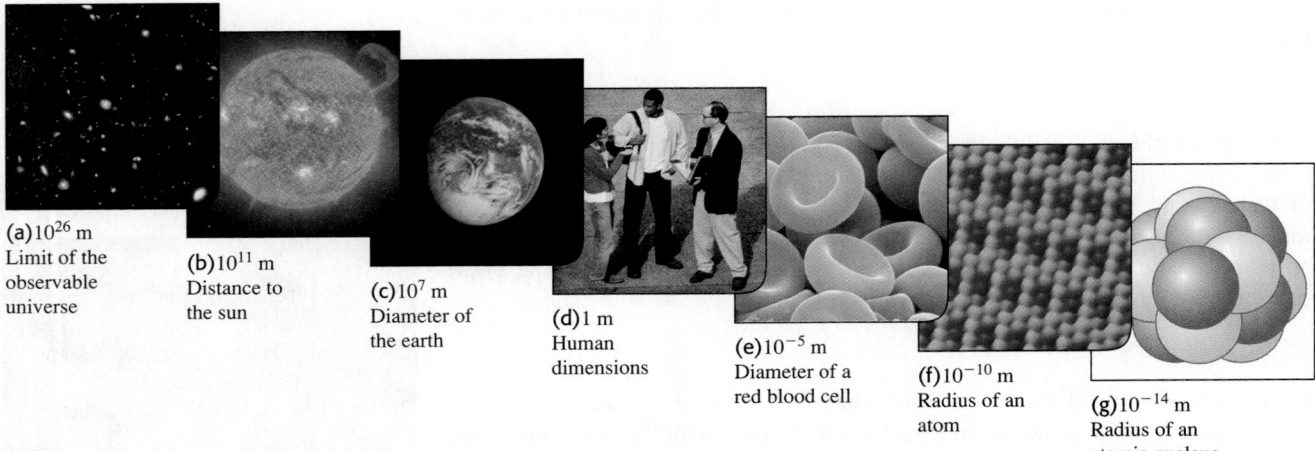

(a) 10^{26} m
Limit of the observable universe

(b) 10^{11} m
Distance to the sun

(c) 10^{7} m
Diameter of the earth

(d) 1 m
Human dimensions

(e) 10^{-5} m
Diameter of a red blood cell

(f) 10^{-10} m
Radius of an atom

(g) 10^{-14} m
Radius of an atomic nucleus

1.6 Many everyday items make use of both SI and British units. An example is this speedometer from a U.S.-built automobile, which shows the speed in both kilometers per hour (inner scale) and miles per hour (outer scale).

The newton, abbreviated N, is the SI unit of force. The British unit of time is the second, defined the same way as in SI. In physics, British units are used only in mechanics and thermodynamics; there is no British system of electrical units.

In this book we use SI units for all examples and problems, but we occasionally give approximate equivalents in British units. As you do problems using SI units, you may also wish to convert to the approximate British equivalents if they are more familiar to you (Fig. 1.6). But you should try to *think* in SI units as much as you can.

1.4 Unit Consistency and Conversions

We use equations to express relationships among physical quantities, represented by algebraic symbols. Each algebraic symbol always denotes both a number and a unit. For example, d might represent a distance of 10 m, t a time of 5 s, and v a speed of 2 m/s.

An equation must always be **dimensionally consistent.** You can't add apples and automobiles; two terms may be added or equated only if they have the same units. For example, if a body moving with constant speed v travels a distance d in a time t, these quantities are related by the equation

$$d = vt$$

If d is measured in meters, then the product vt must also be expressed in meters. Using the above numbers as an example, we may write

$$10 \text{ m} = \left(2 \frac{\text{m}}{\text{s}} \right)(5 \text{ s})$$

Because the unit 1/s on the right side of the equation cancels the unit s, the product has units of meters, as it must. In calculations, units are treated just like algebraic symbols with respect to multiplication and division.

CAUTION **Always use units in calculations** When a problem requires calculations using numbers with units, *always* write the numbers with the correct units and carry the units through the calculation as in the example above. This provides a very useful check. If at some stage in a calculation you find that an equation or an expression has inconsistent units, you know you have made an error somewhere. In this book we will *always* carry units through all calculations, and we strongly urge you to follow this practice when you solve problems. ▌

Problem-Solving Strategy 1.2   Solving Physics Problems

IDENTIFY *the relevant concepts:* In most cases, it's best to use the fundamental SI units (lengths in meters, masses in kilograms, and times in seconds) in every problem. If you need the answer to be in a different set of units (such as kilometers, grams, or hours), wait until the end of the problem to make the conversion.

SET UP *the problem* and **EXECUTE** *the solution:* Units are multiplied and divided just like ordinary algebraic symbols. This gives us an easy way to convert a quantity from one set of units to another: Express the same physical quantity in two different units and form an equality.

For example, when we say that 1 min = 60 s, we don't mean that the number 1 is equal to the number 60; rather, we mean that 1 min represents the same physical time interval as 60 s. For this reason, the ratio $(1 \text{ min})/(60 \text{ s})$ equals 1, as does its reciprocal $(60 \text{ s})/(1 \text{ min})$. We may multiply a quantity by either of these

factors (which we call *unit multipliers*) without changing that quantity's physical meaning. For example, to find the number of seconds in 3 min, we write

$$3 \text{ min} = (3 \text{ min})\left(\frac{60 \text{ s}}{1 \text{ min}}\right) = 180 \text{ s}$$

EVALUATE *your answer:* If you do your unit conversions correctly, unwanted units will cancel, as in the example above. If, instead, you had multiplied 3 min by $(1 \text{ min})/(60 \text{ s})$, your result would have been the nonsensical $\frac{1}{20}$ min²/s. To be sure you convert units properly, you must write down the units at *all* stages of the calculation.

Finally, check whether your answer is reasonable. For example, the result 3 min = 180 s is reasonable because the second is a smaller unit than the minute, so there are more seconds than minutes in the same time interval.

Example 1.1   Converting speed units

The world land speed record is 763.0 mi/h, set on October 15, 1997, by Andy Green in the jet-engine car *Thrust SSC*. Express this speed in meters per second.

SOLUTION

IDENTIFY, SET UP, and EXECUTE: We need to convert the units of a speed from mi/h to m/s. We must therefore find unit multipliers that relate (i) miles to meters and (ii) hours to seconds. In Appendix E (or inside the front cover of this book) we find the equalities 1 mi = 1.609 km, 1 km = 1000 m, and 1 h = 3600 s. We set up the conversion as follows, which ensures that all the desired cancellations by division take place:

$$763.0 \text{ mi/h} = \left(763.0 \frac{\text{mi}}{\text{h}}\right)\left(\frac{1.609 \text{ km}}{1 \text{ mi}}\right)\left(\frac{1000 \text{ m}}{1 \text{ km}}\right)\left(\frac{1 \text{ h}}{3600 \text{ s}}\right)$$
$$= 341.0 \text{ m/s}$$

EVALUATE: Green's was the first supersonic land speed record (the speed of sound in air is about 340 m/s). This example shows a useful rule of thumb: A speed expressed in m/s is a bit less than half the value expressed in mi/h, and a bit less than one-third the value expressed in km/h. For example, a normal freeway speed is about 30 m/s = 67 mi/h = 108 km/h, and a typical walking speed is about 1.4 m/s = 3.1 mi/h = 5.0 km/h.

Example 1.2   Converting volume units

The world's largest cut diamond is the First Star of Africa (mounted in the British Royal Sceptre and kept in the Tower of London). Its volume is 1.84 cubic inches. What is its volume in cubic centimeters? In cubic meters?

SOLUTION

IDENTIFY, SET UP, and EXECUTE: Here we are to convert the units of a volume from cubic inches (in.³) to both cubic centimeters (cm³) and cubic meters (m³). Appendix E gives us the equality 1 in. = 2.540 cm, from which we obtain 1 in.³ = (2.54 cm)³. We then have

$$1.84 \text{ in.}^3 = (1.84 \text{ in.}^3)\left(\frac{2.54 \text{ cm}}{1 \text{ in.}}\right)^3$$

$$= (1.84)(2.54)^3 \frac{\text{in.}^3 \text{ cm}^3}{\text{in.}^3} = 30.2 \text{ cm}^3$$

Appendix F also gives us 1 m = 100 cm, so

$$30.2 \text{ cm}^3 = (30.2 \text{ cm}^3)\left(\frac{1 \text{ m}}{100 \text{ cm}}\right)^3$$

$$= (30.2)\left(\frac{1}{100}\right)^3 \frac{\text{cm}^3 \text{ m}^3}{\text{cm}^3} = 30.2 \times 10^{-6} \text{ m}^3$$

$$= 3.02 \times 10^{-5} \text{ m}^3$$

EVALUATE: Following the pattern of these conversions, you can show that 1 in.³ ≈ 16 cm³ and that 1 m³ ≈ 60,000 in.³. These approximate unit conversions may be useful for future reference.

1.7 This spectacular mishap was the result of a very small percent error—traveling a few meters too far at the end of a journey of hundreds of thousands of meters.

Table 1.2 Using Significant Figures

Multiplication or division:
Result may have no more significant figures than the starting number with the fewest significant figures:

$$\frac{0.745 \times 2.2}{3.885} = 0.42$$

$$1.32578 \times 10^7 \times 4.11 \times 10^{-3} = 5.45 \times 10^4$$

Addition or subtraction:
Number of significant figures is determined by the starting number with the largest uncertainty (i.e., fewest digits to the right of the decimal point):

$$27.153 + 138.2 - 11.74 = 153.6$$

1.8 Determining the value of π from the circumference and diameter of a circle.

The measured values have only three significant figures, so their calculated ratio (π) also has only three significant figures.

1.5 Uncertainty and Significant Figures

Measurements always have uncertainties. If you measure the thickness of the cover of a hardbound version of this book using an ordinary ruler, your measurement is reliable only to the nearest millimeter, and your result will be 3 mm. It would be *wrong* to state this result as 3.00 mm; given the limitations of the measuring device, you can't tell whether the actual thickness is 3.00 mm, 2.85 mm, or 3.11 mm. But if you use a micrometer caliper, a device that measures distances reliably to the nearest 0.01 mm, the result will be 2.91 mm. The distinction between these two measurements is in their **uncertainty.** The measurement using the micrometer caliper has a smaller uncertainty; it's a more accurate measurement. The uncertainty is also called the **error** because it indicates the maximum difference there is likely to be between the measured value and the true value. The uncertainty or error of a measured value depends on the measurement technique used.

We often indicate the **accuracy** of a measured value—that is, how close it is likely to be to the true value—by writing the number, the symbol $\pm$, and a second number indicating the uncertainty of the measurement. If the diameter of a steel rod is given as 56.47 ± 0.02 mm, this means that the true value is unlikely to be less than 56.45 mm or greater than 56.49 mm. In a commonly used shorthand notation, the number 1.6454(21) means 1.6454 ± 0.0021. The numbers in parentheses show the uncertainty in the final digits of the main number.

We can also express accuracy in terms of the maximum likely **fractional error** or **percent error** (also called *fractional uncertainty* and *percent uncertainty*). A resistor labeled "47 ohms $\pm$ 10%" probably has a true resistance that differs from 47 ohms by no more than 10% of 47 ohms—that is, by about 5 ohms. The resistance is probably between 42 and 52 ohms. For the diameter of the steel rod given above, the fractional error is (0.02 mm)/(56.47 mm), or about 0.0004; the percent error is (0.0004)(100%), or about 0.04%. Even small percent errors can sometimes be very significant (Fig. 1.7).

In many cases the uncertainty of a number is not stated explicitly. Instead, the uncertainty is indicated by the number of meaningful digits, or **significant figures,** in the measured value. We gave the thickness of the cover of this book as 2.91 mm, which has three significant figures. By this we mean that the first two digits are known to be correct, while the third digit is uncertain. The last digit is in the hundredths place, so the uncertainty is about 0.01 mm. Two values with the *same* number of significant figures may have *different* uncertainties; a distance given as 137 km also has three significant figures, but the uncertainty is about 1 km.

When you use numbers that have uncertainties to compute other numbers, the computed numbers are also uncertain. When numbers are multiplied or divided, the number of significant figures in the result can be no greater than in the factor with the fewest significant figures. For example, $3.1416 \times 2.34 \times 0.58 = 4.3$. When we add and subtract numbers, it's the location of the decimal point that matters, not the number of significant figures. For example, $123.62 + 8.9 = 132.5$. Although 123.62 has an uncertainty of about 0.01, 8.9 has an uncertainty of about 0.1. So their sum has an uncertainty of about 0.1 and should be written as 132.5, not 132.52. Table 1.2 summarizes these rules for significant figures.

As an application of these ideas, suppose you want to verify the value of π, the ratio of the circumference of a circle to its diameter. The true value of this ratio to ten digits is 3.141592654. To test this, you draw a large circle and measure its circumference and diameter to the nearest millimeter, obtaining the values 424 mm and 135 mm (Fig. 1.8). You punch these into your calculator and obtain the quotient (424 mm)/(135 mm) = 3.140740741. This may seem to disagree with the true value of π, but keep in mind that each of your measurements has three significant figures, so your measured value of π can have only three significant figures. It should be stated simply as 3.14. Within the limit of three significant figures, your value does agree with the true value.

In the examples and problems in this book we usually give numerical values with three significant figures, so your answers should usually have no more than three significant figures. (Many numbers in the real world have even less accuracy. An automobile speedometer, for example, usually gives only two significant figures.) Even if you do the arithmetic with a calculator that displays ten digits, it would be wrong to give a ten-digit answer because it misrepresents the accuracy of the results. Always round your final answer to keep only the correct number of significant figures or, in doubtful cases, one more at most. In Example 1.1 it would have been wrong to state the answer as 341.01861 m/s. Note that when you reduce such an answer to the appropriate number of significant figures, you must *round*, not *truncate*. Your calculator will tell you that the ratio of 525 m to 311 m is 1.688102894; to three significant figures, this is 1.69, not 1.68.

When we calculate with very large or very small numbers, we can show significant figures much more easily by using **scientific notation,** sometimes called **powers-of-10 notation.** The distance from the earth to the moon is about 384,000,000 m, but writing the number in this form doesn't indicate the number of significant figures. Instead, we move the decimal point eight places to the left (corresponding to dividing by 10^8) and multiply by 10^8; that is,

$$384{,}000{,}000 \text{ m} = 3.84 \times 10^8 \text{ m}$$

In this form, it is clear that we have three significant figures. The number 4.00×10^{-7} also has three significant figures, even though two of them are zeros. Note that in scientific notation the usual practice is to express the quantity as a number between 1 and 10 multiplied by the appropriate power of 10.

When an integer or a fraction occurs in a general equation, we treat that number as having no uncertainty at all. For example, in the equation $v_x^2 = v_{0x}^2 + 2a_x(x - x_0)$, which is Eq. (2.13) in Chapter 2, the coefficient 2 is *exactly* 2. We can consider this coefficient as having an infinite number of significant figures $(2.000000\dots)$. The same is true of the exponent 2 in v_x^2 and v_{0x}^2.

Finally, let's note that **precision** is not the same as *accuracy*. A cheap digital watch that gives the time as 10:35:17 A.M. is very *precise* (the time is given to the second), but if the watch runs several minutes slow, then this value isn't very *accurate*. On the other hand, a grandfather clock might be very accurate (that is, display the correct time), but if the clock has no second hand, it isn't very precise. A high-quality measurement is both precise *and* accurate.

Example 1.3 Significant figures in multiplication

The rest energy E of an object with rest mass m is given by Einstein's famous equation $E = mc^2$, where c is the speed of light in vacuum. Find E for an electron for which (to three significant figures) $m = 9.11 \times 10^{-31}$ kg. The SI unit for E is the joule (J); $1 \text{ J} = 1 \text{ kg} \cdot \text{m}^2/\text{s}^2$.

SOLUTION

IDENTIFY and SET UP: Our target variable is the energy E. We are given the value of the mass m; from Section 1.3 (or Appendix F) the speed of light is $c = 2.99792458 \times 10^8$ m/s.

EXECUTE: Substituting the values of m and c into Einstein's equation, we find

$$
\begin{aligned}
E &= (9.11 \times 10^{-31} \text{ kg})(2.99792458 \times 10^8 \text{ m/s})^2 \\
&= (9.11)(2.99792458)^2 (10^{-31})(10^8)^2 \text{ kg} \cdot \text{m}^2/\text{s}^2 \\
&= (81.87659678)(10^{[-31+(2\times8)]}) \text{ kg} \cdot \text{m}^2/\text{s}^2 \\
&= 8.187659678 \times 10^{-14} \text{ kg} \cdot \text{m}^2/\text{s}^2
\end{aligned}
$$

Since the value of m was given to only three significant figures, we must round this to

$$E = 8.19 \times 10^{-14} \text{ kg} \cdot \text{m}^2/\text{s}^2 = 8.19 \times 10^{-14} \text{ J}$$

EVALUATE: While the rest energy contained in an electron may seem ridiculously small, on the atomic scale it is tremendous. Compare our answer to 10^{-19} J, the energy gained or lost by a single atom during a typical chemical reaction. The rest energy of an electron is about 1,000,000 times larger! (We'll discuss the significance of rest energy in Chapter 37.)

Test Your Understanding of Section 1.5 The density of a material is equal to its mass divided by its volume. What is the density (in kg/m³) of a rock of mass 1.80 kg and volume 6.0×10^{-4} m³? (i) 3×10^3 kg/m³; (ii) 3.0×10^3 kg/m³; (iii) 3.00×10^3 kg/m³; (iv) 3.000×10^3 kg/m³; (v) any of these—all of these answers are mathematically equivalent.

1.6 Estimates and Orders of Magnitude

PhET: Estimation

We have stressed the importance of knowing the accuracy of numbers that represent physical quantities. But even a very crude estimate of a quantity often gives us useful information. Sometimes we know how to calculate a certain quantity, but we have to guess at the data we need for the calculation. Or the calculation might be too complicated to carry out exactly, so we make some rough approximations. In either case our result is also a guess, but such a guess can be useful even if it is uncertain by a factor of two, ten, or more. Such calculations are often called **order-of-magnitude estimates.** The great Italian-American nuclear physicist Enrico Fermi (1901–1954) called them "back-of-the-envelope calculations."

Exercises 1.16 through 1.25 at the end of this chapter are of the estimating, or order-of-magnitude, variety. Most require guesswork for the needed input data. Don't try to look up a lot of data; make the best guesses you can. Even when they are off by a factor of ten, the results can be useful and interesting.

Example 1.4 **An order-of-magnitude estimate**

You are writing an adventure novel in which the hero escapes across the border with a billion dollars' worth of gold in his suitcase. Could anyone carry that much gold? Would it fit in a suitcase?

SOLUTION

IDENTIFY, SET UP, and EXECUTE: Gold sells for around $400 an ounce. (The price has varied between $200 and $1000 over the past decade or so.) An ounce is about 30 grams; that's worth remembering. So ten dollars' worth of gold has a mass of $\frac{1}{40}$ ounce, or around one gram. A billion (10^9) dollars' worth of gold

is a hundred million (10^8) grams, or a hundred thousand (10^5) kilograms. This corresponds to a weight in British units of around 200,000 lb, or 100 tons. No human hero could lift that weight!

Roughly what is the *volume* of this gold? The density of gold is much greater than that of water (1 g/cm³), or 1000 kg/m³; if its density is 10 times that of water, this much gold will have a volume of 10 m³, many times the volume of a suitcase.

EVALUATE: Clearly your novel needs rewriting. Try the calculation again with a suitcase full of five-carat (1-gram) diamonds, each worth $100,000. Would this work?

Application **Scalar Temperature, Vector Wind**
This weather station measures temperature, a scalar quantity that can be positive or negative (say, +20°C or −5°C) but has no direction. It also measures wind velocity, which is a vector quantity with both magnitude and direction (for example, 15 km/h from the west).

Test Your Understanding of Section 1.6 Can you estimate the total number of teeth in all the mouths of everyone (students, staff, and faculty) on your campus? (*Hint:* How many teeth are in your mouth? Count them!)

1.7 Vectors and Vector Addition

Some physical quantities, such as time, temperature, mass, and density, can be described completely by a single number with a unit. But many other important quantities in physics have a *direction* associated with them and cannot be described by a single number. A simple example is describing the motion of an airplane: We must say not only how fast the plane is moving but also in what direction. The speed of the airplane combined with its direction of motion together constitute a quantity called *velocity.* Another example is *force,* which in physics means a push or pull exerted on a body. Giving a complete description of a force means describing both how hard the force pushes or pulls on the body and the direction of the push or pull.

When a physical quantity is described by a single number, we call it a **scalar quantity.** In contrast, a **vector quantity** has both a **magnitude** (the "how much" or "how big" part) and a direction in space. Calculations that combine scalar quantities use the operations of ordinary arithmetic. For example, 6 kg + 3 kg = 9 kg, or 4 × 2 s = 8 s. However, combining vectors requires a different set of operations.

To understand more about vectors and how they combine, we start with the simplest vector quantity, **displacement.** Displacement is simply a change in the position of an object. Displacement is a vector quantity because we must state not only how far the object moves but also in what direction. Walking 3 km north from your front door doesn't get you to the same place as walking 3 km southeast; these two displacements have the same magnitude but different directions.

We usually represent a vector quantity such as displacement by a single letter, such as $\vec{A}$ in Fig. 1.9a. In this book we always print vector symbols in ***boldface italic type with an arrow above them.*** We do this to remind you that vector quantities have different properties from scalar quantities; the arrow is a reminder that vectors have direction. When you handwrite a symbol for a vector, *always* write it with an arrow on top. If you don't distinguish between scalar and vector quantities in your notation, you probably won't make the distinction in your thinking either, and hopeless confusion will result.

We always *draw* a vector as a line with an arrowhead at its tip. The length of the line shows the vector's magnitude, and the direction of the line shows the vector's direction. Displacement is always a straight-line segment directed from the starting point to the ending point, even though the object's actual path may be curved (Fig. 1.9b). Note that displacement is not related directly to the total *distance* traveled. If the object were to continue on past P_2 and then return to P_1, the displacement for the entire trip would be *zero* (Fig. 1.9c).

If two vectors have the same direction, they are **parallel.** If they have the same magnitude *and* the same direction, they are *equal,* no matter where they are located in space. The vector $\vec{A}'$ from point P_3 to point P_4 in Fig. 1.10 has the same length and direction as the vector $\vec{A}$ from P_1 to P_2. These two displacements are equal, even though they start at different points. We write this as $\vec{A}' = \vec{A}$ in Fig. 1.10; the boldface equals sign emphasizes that equality of two vector quantities is not the same relationship as equality of two scalar quantities. Two vector quantities are equal only when they have the same magnitude *and* the same direction.

The vector $\vec{B}$ in Fig. 1.10, however, is not equal to $\vec{A}$ because its direction is *opposite* to that of $\vec{A}$. We define the **negative of a vector** as a vector having the same magnitude as the original vector but the *opposite* direction. The negative of vector quantity $\vec{A}$ is denoted as $-\vec{A}$, and we use a boldface minus sign to emphasize the vector nature of the quantities. If $\vec{A}$ is 87 m south, then $-\vec{A}$ is 87 m north. Thus we can write the relationship between $\vec{A}$ and $\vec{B}$ in Fig. 1.10 as $\vec{A} = -\vec{B}$ or $\vec{B} = -\vec{A}$. When two vectors $\vec{A}$ and $\vec{B}$ have opposite directions, whether their magnitudes are the same or not, we say that they are **antiparallel.**

We usually represent the *magnitude* of a vector quantity (in the case of a displacement vector, its length) by the same letter used for the vector, but in *light italic type* with *no* arrow on top. An alternative notation is the vector symbol with vertical bars on both sides:

$$\text{(Magnitude of } \vec{A}) = A = |\vec{A}| \tag{1.1}$$

The magnitude of a vector quantity is a scalar quantity (a number) and is *always positive.* Note that a vector can never be equal to a scalar because they are different kinds of quantities. The expression "$\vec{A} = 6\,\text{m}$" is just as wrong as "2 oranges = 3 apples"!

When drawing diagrams with vectors, it's best to use a scale similar to those used for maps. For example, a displacement of 5 km might be represented in a diagram by a vector 1 cm long, and a displacement of 10 km by a vector 2 cm long. In a diagram for velocity vectors, a vector that is 1 cm long might represent

1.9 Displacement as a vector quantity. A displacement is always a straight-line segment directed from the starting point to the ending point, even if the path is curved.

(a) We represent a displacement by an arrow pointing in the direction of displacement.

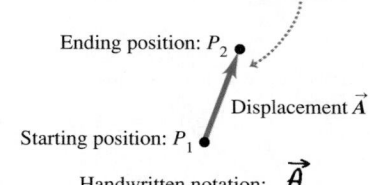

(b) Displacement depends only on the starting and ending positions—not on the path taken.

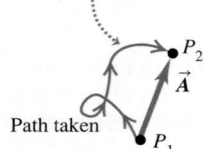

(c) Total displacement for a round trip is 0, regardless of the distance traveled.

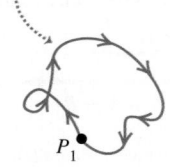

1.10 The meaning of vectors that have the same magnitude and the same or opposite direction.

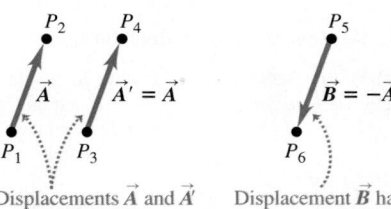

Displacements $\vec{A}$ and $\vec{A}'$ are equal because they have the same length and direction.

Displacement $\vec{B}$ has the same magnitude as $\vec{A}$ but opposite direction; $\vec{B}$ is the negative of $\vec{A}$.

1.11 Three ways to add two vectors. As shown in (b), the order in vector addition doesn't matter; vector addition is commutative.

(a) We can add two vectors by placing them head to tail.

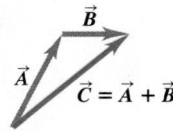

(b) Adding them in reverse order gives the same result.

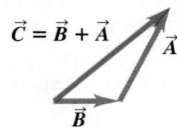

(c) We can also add them by constructing a parallelogram.

1.12 (a) Only when two vectors $\vec{A}$ and $\vec{B}$ are parallel does the magnitude of their sum equal the sum of their magnitudes: $C = A + B$. (b) When $\vec{A}$ and $\vec{B}$ are antiparallel, the magnitude of their sum equals the *difference* of their magnitudes: $C = |A - B|$.

(a) The sum of two parallel vectors

(b) The sum of two antiparallel vectors

a velocity of magnitude 5 m/s. A velocity of 20 m/s would then be represented by a vector 4 cm long.

Vector Addition and Subtraction

Suppose a particle undergoes a displacement $\vec{A}$ followed by a second displacement $\vec{B}$. The final result is the same as if the particle had started at the same initial point and undergone a single displacement $\vec{C}$ (Fig. 1.11a). We call displacement $\vec{C}$ the **vector sum,** or **resultant,** of displacements $\vec{A}$ and $\vec{B}$. We express this relationship symbolically as

$$\vec{C} = \vec{A} + \vec{B} \tag{1.2}$$

The boldface plus sign emphasizes that adding two vector quantities requires a geometrical process and is not the same operation as adding two scalar quantities such as $2 + 3 = 5$. In vector addition we usually place the *tail* of the *second* vector at the *head,* or tip, of the *first* vector (Fig. 1.11a).

If we make the displacements $\vec{A}$ and $\vec{B}$ in reverse order, with $\vec{B}$ first and $\vec{A}$ second, the result is the same (Fig. 1.11b). Thus

$$\vec{C} = \vec{B} + \vec{A} \quad \text{and} \quad \vec{A} + \vec{B} = \vec{B} + \vec{A} \tag{1.3}$$

This shows that the order of terms in a vector sum doesn't matter. In other words, vector addition obeys the commutative law.

Figure 1.11c shows another way to represent the vector sum: If vectors $\vec{A}$ and $\vec{B}$ are both drawn with their tails at the same point, vector $\vec{C}$ is the diagonal of a parallelogram constructed with $\vec{A}$ and $\vec{B}$ as two adjacent sides.

CAUTION **Magnitudes in vector addition** It's a common error to conclude that if $\vec{C} = \vec{A} + \vec{B}$, then the magnitude C should equal the magnitude A plus the magnitude B. In general, this conclusion is *wrong;* for the vectors shown in Fig. 1.11, you can see that $C < A + B$. The magnitude of $\vec{A} + \vec{B}$ depends on the magnitudes of $\vec{A}$ and $\vec{B}$ *and* on the angle between $\vec{A}$ and $\vec{B}$ (see Problem 1.90). Only in the special case in which $\vec{A}$ and $\vec{B}$ are *parallel* is the magnitude of $\vec{C} = \vec{A} + \vec{B}$ equal to the sum of the magnitudes of $\vec{A}$ and $\vec{B}$ (Fig. 1.12a). When the vectors are *antiparallel* (Fig. 1.12b), the magnitude of $\vec{C}$ equals the *difference* of the magnitudes of $\vec{A}$ and $\vec{B}$. Be careful about distinguishing between scalar and vector quantities, and you'll avoid making errors about the magnitude of a vector sum. █

When we need to add more than two vectors, we may first find the vector sum of any two, add this vectorially to the third, and so on. Figure 1.13a shows three vectors $\vec{A}$, $\vec{B}$, and $\vec{C}$. In Fig. 1.13b we first add $\vec{A}$ and $\vec{B}$ to give a vector sum $\vec{D}$; we then add vectors $\vec{C}$ and $\vec{D}$ by the same process to obtain the vector sum $\vec{R}$:

$$\vec{R} = (\vec{A} + \vec{B}) + \vec{C} = \vec{D} + \vec{C}$$

1.13 Several constructions for finding the vector sum $\vec{A} + \vec{B} + \vec{C}$.

(a) To find the sum of these three vectors ...

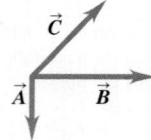

(b) we could add $\vec{A}$ and $\vec{B}$ to get $\vec{D}$ and then add $\vec{C}$ to $\vec{D}$ to get the final sum (resultant) $\vec{R}$, ...

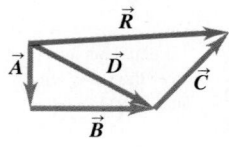

(c) or we could add $\vec{B}$ and $\vec{C}$ to get $\vec{E}$ and then add $\vec{A}$ to $\vec{E}$ to get $\vec{R}$, ...

(d) or we could add $\vec{A}$, $\vec{B}$, and $\vec{C}$ to get $\vec{R}$ directly, ...

(e) or we could add $\vec{A}$, $\vec{B}$, and $\vec{C}$ in any other order and still get $\vec{R}$.

1.14 To construct the vector difference $\vec{A} - \vec{B}$, you can either place the tail of $-\vec{B}$ at the head of $\vec{A}$ or place the two vectors $\vec{A}$ and $\vec{B}$ head to head.

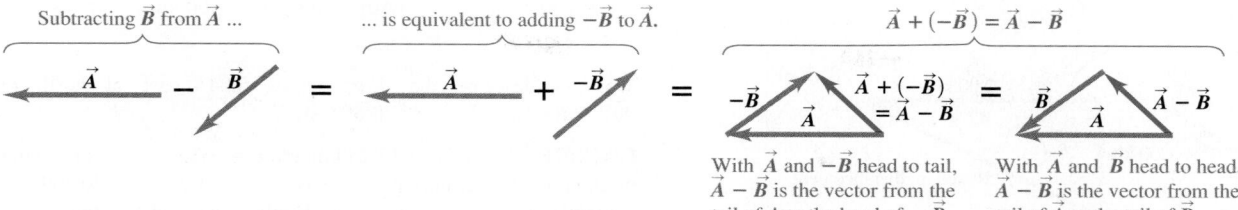

Subtracting $\vec{B}$ from $\vec{A}$ ...   ... is equivalent to adding $-\vec{B}$ to $\vec{A}$. $\vec{A} + (-\vec{B}) = \vec{A} - \vec{B}$

With $\vec{A}$ and $-\vec{B}$ head to tail, $\vec{A} - \vec{B}$ is the vector from the tail of A to the head of $-\vec{B}$.

With $\vec{A}$ and $\vec{B}$ head to head, $\vec{A} - \vec{B}$ is the vector from the tail of $\vec{A}$ to the tail of $\vec{B}$.

Alternatively, we can first add $\vec{B}$ and $\vec{C}$ to obtain vector $\vec{E}$ (Fig. 1.13c), and then add $\vec{A}$ and $\vec{E}$ to obtain $\vec{R}$:

$$\vec{R} = \vec{A} + (\vec{B} + \vec{C}) = \vec{A} + \vec{E}$$

We don't even need to draw vectors $\vec{D}$ and $\vec{E}$; all we need to do is draw $\vec{A}$, $\vec{B}$, and $\vec{C}$ in succession, with the tail of each at the head of the one preceding it. The sum vector $\vec{R}$ extends from the tail of the first vector to the head of the last vector (Fig. 1.13d). The order makes no difference; Fig. 1.13e shows a different order, and we invite you to try others. We see that vector addition obeys the associative law.

We can *subtract* vectors as well as add them. To see how, recall that vector $-\vec{A}$ has the same magnitude as $\vec{A}$ but the opposite direction. We define the difference $\vec{A} - \vec{B}$ of two vectors $\vec{A}$ and $\vec{B}$ to be the vector sum of $\vec{A}$ and $-\vec{B}$:

$$\vec{A} - \vec{B} = \vec{A} + (-\vec{B}) \tag{1.4}$$

Figure 1.14 shows an example of vector subtraction.

A vector quantity such as a displacement can be multiplied by a scalar quantity (an ordinary number). The displacement $2\vec{A}$ is a displacement (vector quantity) in the same direction as the vector $\vec{A}$ but twice as long; this is the same as adding $\vec{A}$ to itself (Fig. 1.15a). In general, when a vector $\vec{A}$ is multiplied by a scalar c, the result $c\vec{A}$ has magnitude $|c|A$ (the absolute value of c multiplied by the magnitude of the vector $\vec{A}$). If c is positive, $c\vec{A}$ is in the same direction as $\vec{A}$; if c is negative, $c\vec{A}$ is in the direction opposite to $\vec{A}$. Thus $3\vec{A}$ is parallel to $\vec{A}$, while $-3\vec{A}$ is antiparallel to $\vec{A}$ (Fig. 1.15b).

A scalar used to multiply a vector may also be a physical quantity. For example, you may be familiar with the relationship $\vec{F} = m\vec{a}$; the net force $\vec{F}$ (a vector quantity) that acts on a body is equal to the product of the body's mass m (a scalar quantity) and its acceleration $\vec{a}$ (a vector quantity). The direction of $\vec{F}$ is the same as that of $\vec{a}$ because m is positive, and the magnitude of $\vec{F}$ is equal to the mass m (which is positive) multiplied by the magnitude of $\vec{a}$. The unit of force is the unit of mass multiplied by the unit of acceleration.

Mastering**PHYSICS**

PhET: Vector Addition

1.15 Multiplying a vector (a) by a positive scalar and (b) by a negative scalar.

(a) Multiplying a vector by a positive scalar changes the magnitude (length) of the vector, but not its direction.

$2\vec{A}$ is twice as long as $\vec{A}$.

(b) Multiplying a vector by a negative scalar changes its magnitude and reverses its direction.

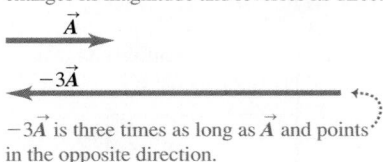

$-3\vec{A}$ is three times as long as $\vec{A}$ and points in the opposite direction.

Example 1.5   **Addition of two vectors at right angles**

A cross-country skier skis 1.00 km north and then 2.00 km east on a horizontal snowfield. How far and in what direction is she from the starting point?

SOLUTION

IDENTIFY and SET UP: The problem involves combining two displacements at right angles to each other. In this case, vector addition amounts to solving a right triangle, which we can do using the Pythagorean theorem and simple trigonometry. The target variables are the skier's straight-line distance and direction from her

starting point. Figure 1.16 is a scale diagram of the two displacements and the resultant net displacement. We denote the direction from the starting point by the angle ϕ (the Greek letter phi). The displacement appears to be about 2.4 km. Measurement with a protractor indicates that ϕ is about $63°$.

EXECUTE: The distance from the starting point to the ending point is equal to the length of the hypotenuse:

$$\sqrt{(1.00 \text{ km})^2 + (2.00 \text{ km})^2} = 2.24 \text{ km}$$

Continued

1.16 The vector diagram, drawn to scale, for a ski trip.

A little trigonometry (from Appendix B) allows us to find angle ϕ:

$$\tan \phi = \frac{\text{Opposite side}}{\text{Adjacent side}} = \frac{2.00 \text{ km}}{1.00 \text{ km}}$$

$$\phi = 63.4°$$

We can describe the direction as 63.4° east of north or $90° - 63.4° = 26.6°$ north of east.

EVALUATE: Our answers (2.24 km and $\phi = 63.4°$) are close to our predictions. In the more general case in which you have to add two vectors *not* at right angles to each other, you can use the law of cosines in place of the Pythagorean theorem and use the law of sines to find an angle corresponding to ϕ in this example. (You'll find these trigonometric rules in Appendix B.) We'll see more techniques for vector addition in Section 1.8.

Test Your Understanding of Section 1.7 Two displacement vectors, $\vec{S}$ and $\vec{T}$, have magnitudes $S = 3$ m and $T = 4$ m. Which of the following could be the magnitude of the difference vector $\vec{S} - \vec{T}$? (There may be more than one correct answer.) (i) 9 m; (ii) 7 m; (iii) 5 m; (iv) 1 m; (v) 0 m; (vi) −1 m.

1.8 Components of Vectors

In Section 1.7 we added vectors by using a scale diagram and by using properties of right triangles. Measuring a diagram offers only very limited accuracy, and calculations with right triangles work only when the two vectors are perpendicular. So we need a simple but general method for adding vectors. This is called the method of *components*.

To define what we mean by the components of a vector $\vec{A}$, we begin with a rectangular (Cartesian) coordinate system of axes (Fig. 1.17a). We then draw the vector with its tail at O, the origin of the coordinate system. We can represent any vector lying in the xy-plane as the sum of a vector parallel to the x-axis and a vector parallel to the y-axis. These two vectors are labeled $\vec{A}_x$ and $\vec{A}_y$ in Fig. 1.17a; they are called the **component vectors** of vector $\vec{A}$, and their vector sum is equal to $\vec{A}$. In symbols,

$$\vec{A} = \vec{A}_x + \vec{A}_y \qquad (1.5)$$

1.17 Representing a vector $\vec{A}$ in terms of (a) component vectors $\vec{A}_x$ and $\vec{A}_y$ and (b) components A_x and A_y (which in this case are both positive).

Since each component vector lies along a coordinate-axis direction, we need only a single number to describe each one. When $\vec{A}_x$ points in the positive x-direction, we define the number A_x to be equal to the magnitude of $\vec{A}_x$. When $\vec{A}_x$ points in the negative x-direction, we define the number A_x to be equal to the negative of that magnitude (the magnitude of a vector quantity is itself never negative). We define the number A_y in the same way. The two numbers A_x and A_y are called the **components** of $\vec{A}$ (Fig. 1.17b).

CAUTION **Components are not vectors** The components A_x and A_y of a vector $\vec{A}$ are just numbers; they are *not* vectors themselves. This is why we print the symbols for components in light italic type with *no* arrow on top instead of in boldface italic with an arrow, which is reserved for vectors.

We can calculate the components of the vector $\vec{A}$ if we know its magnitude A and its direction. We'll describe the direction of a vector by its angle relative to some reference direction. In Fig. 1.17b this reference direction is the positive x-axis, and the angle between vector $\vec{A}$ and the positive x-axis is θ (the Greek letter theta). Imagine that the vector $\vec{A}$ originally lies along the $+x$-axis and that you then rotate it to its correct direction, as indicated by the arrow in Fig. 1.17b on the angle θ. If this rotation is from the $+x$-axis toward the $+y$-axis, as shown in Fig. 1.17b, then θ is *positive;* if the rotation is from the $+x$-axis toward the $-y$-axis, θ is *negative.* Thus the $+y$-axis is at an angle of $90°$, the $-x$-axis at $180°$, and the $-y$-axis at $270°$ (or $-90°$). If θ is measured in this way, then from the definition of the trigonometric functions,

$$\frac{A_x}{A} = \cos\theta \quad \text{and} \quad \frac{A_y}{A} = \sin\theta$$
$$A_x = A\cos\theta \quad \text{and} \quad A_y = A\sin\theta \tag{1.6}$$

(θ measured from the $+x$-axis, rotating toward the $+y$-axis)

In Fig. 1.17b A_x and A_y are positive. This is consistent with Eqs. (1.6); θ is in the first quadrant (between $0°$ and $90°$), and both the cosine and the sine of an angle in this quadrant are positive. But in Fig. 1.18a the component B_x is negative. Again, this agrees with Eqs. (1.6); the cosine of an angle in the second quadrant is negative. The component B_y is positive ($\sin\theta$ is positive in the second quadrant). In Fig. 1.18b both C_x and C_y are negative (both $\cos\theta$ and $\sin\theta$ are negative in the third quadrant).

CAUTION **Relating a vector's magnitude and direction to its components** Equations (1.6) are correct *only* when the angle θ is measured from the positive x-axis as described above. If the angle of the vector is given from a different reference direction or using a different sense of rotation, the relationships are different. Be careful! Example 1.6 illustrates this point. ▮

1.18 The components of a vector may be positive or negative numbers.

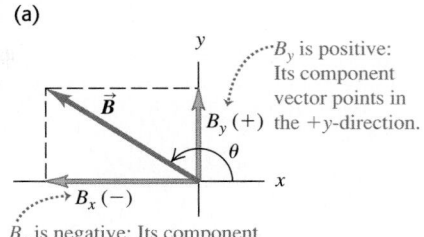

(a)

B_y is positive: Its component vector points in the $+y$-direction.

B_x is negative: Its component vector points in the $-x$-direction.

(b)

Both components of $\vec{C}$ are negative.

Example 1.6 Finding components

(a) What are the x- and y-components of vector $\vec{D}$ in Fig. 1.19a? The magnitude of the vector is $D = 3.00$ m, and the angle $\alpha = 45°$. (b) What are the x- and y-components of vector $\vec{E}$ in Fig. 1.19b? The magnitude of the vector is $E = 4.50$ m, and the angle $\beta = 37.0°$.

SOLUTION

IDENTIFY and SET UP: We can use Eqs. (1.6) to find the components of these vectors, but we have to be careful: Neither of the angles α or β in Fig. 1.19 is measured from the $+x$-axis toward the $+y$-axis. We estimate from the figure that the lengths of the com-

ponents in part (a) are both roughly 2 m, and that those in part (b) are 3m and 4 m. We've indicated the signs of the components in the figure.

EXECUTE: (a) The angle α (the Greek letter alpha) between the positive x-axis and $\vec{D}$ is measured toward the *negative* y-axis. The angle we must use in Eqs. (1.6) is $\theta = -\alpha = -45°$. We then find

$$D_x = D\cos\theta = (3.00\text{ m})(\cos(-45°)) = +2.1\text{ m}$$
$$D_y = D\sin\theta = (3.00\text{ m})(\sin(-45°)) = -2.1\text{ m}$$

Had you been careless and substituted $+45°$ for θ in Eqs. (1.6), your result for D_y would have had the wrong sign.

(b) The x- and y-axes in Fig. 1.19b are at right angles, so it doesn't matter that they aren't horizontal and vertical, respectively. But to use Eqs. (1.6), we must use the angle $\theta = 90.0° - \beta = 90.0° - 37.0° = 53.0°$. Then we find

$$E_x = E\cos 53.0° = (4.50\text{ m})(\cos 53.0°) = +2.71\text{ m}$$
$$E_y = E\sin 53.0° = (4.50\text{ m})(\sin 53.0°) = +3.59\text{ m}$$

EVALUATE: Our answers to both parts are close to our predictions. But ask yourself this: Why do the answers in part (a) correctly have only two significant figures?

1.19 Calculating the x- and y-components of vectors.

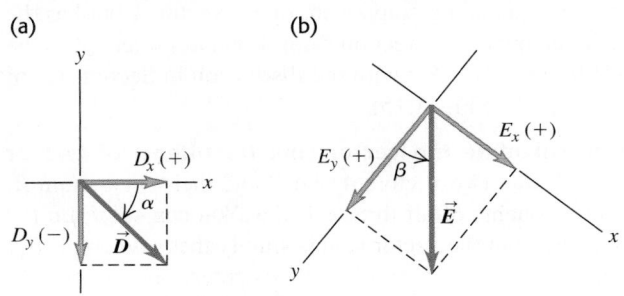

(a)

$D_x(+)$

$D_y(-)$ $\vec{D}$

(b)

$E_y(+)$ $E_x(+)$

β

$\vec{E}$

Doing Vector Calculations Using Components

Using components makes it relatively easy to do various calculations involving vectors. Let's look at three important examples.

1. Finding a vector's magnitude and direction from its components. We can describe a vector completely by giving either its magnitude and direction or its x- and y-components. Equations (1.6) show how to find the components if we know the magnitude and direction. We can also reverse the process: We can find the magnitude and direction if we know the components. By applying the Pythagorean theorem to Fig. 1.17b, we find that the magnitude of vector $\vec{A}$ is

$$A = \sqrt{A_x^2 + A_y^2} \tag{1.7}$$

(We always take the positive root.) Equation (1.7) is valid for any choice of x-axis and y-axis, as long as they are mutually perpendicular. The expression for the vector direction comes from the definition of the tangent of an angle. If θ is measured from the positive x-axis, and a positive angle is measured toward the positive y-axis (as in Fig. 1.17b), then

$$\tan\theta = \frac{A_y}{A_x} \quad \text{and} \quad \theta = \arctan\frac{A_y}{A_x} \tag{1.8}$$

We will always use the notation arctan for the inverse tangent function. The notation $\tan^{-1}$ is also commonly used, and your calculator may have an INV or 2ND button to be used with the TAN button.

CAUTION **Finding the direction of a vector from its components** There's one slight complication in using Eqs. (1.8) to find θ: Any two angles that differ by 180° have the same tangent. Suppose $A_x = 2$ m and $A_y = -2$ m as in Fig. 1.20; then $\tan\theta = -1$. But both 135° and 315° (or −45°) have tangents of −1. To decide which is correct, we have to look at the individual components. Because A_x is positive and A_y is negative, the angle must be in the fourth quadrant; thus $\theta = 315°$ (or −45°) is the correct value. Most pocket calculators give $\arctan(-1) = -45°$. In this case that is correct; but if instead we have $A_x = -2$ m and $A_y = 2$ m, then the correct angle is 135°. Similarly, when A_x and A_y are both negative, the tangent is positive, but the angle is in the third quadrant. You should *always* draw a sketch like Fig. 1.20 to check which of the two possibilities is the correct one. ▌

1.20 Drawing a sketch of a vector reveals the signs of its x- and y-components.

Suppose that $\tan\theta = \dfrac{A_y}{A_x} = -1$. What is θ?

Two angles have tangents of −1: 135° and 315°. Inspection of the diagram shows that θ must be 315°.

2. Multiplying a vector by a scalar. If we multiply a vector $\vec{A}$ by a scalar c, each component of the product $\vec{D} = c\vec{A}$ is the product of c and the corresponding component of $\vec{A}$:

$$D_x = cA_x \qquad D_y = cA_y \qquad \text{(components of } \vec{D} = c\vec{A}) \tag{1.9}$$

For example, Eq. (1.9) says that each component of the vector $2\vec{A}$ is twice as great as the corresponding component of the vector $\vec{A}$, so $2\vec{A}$ is in the same direction as $\vec{A}$ but has twice the magnitude. Each component of the vector $-3\vec{A}$ is three times as great as the corresponding component of the vector $\vec{A}$ but has the opposite sign, so $-3\vec{A}$ is in the opposite direction from $\vec{A}$ and has three times the magnitude. Hence Eqs. (1.9) are consistent with our discussion in Section 1.7 of multiplying a vector by a scalar (see Fig. 1.15).

3. Using components to calculate the vector sum (resultant) of two or more vectors. Figure 1.21 shows two vectors $\vec{A}$ and $\vec{B}$ and their vector sum $\vec{R}$, along with the x- and y-components of all three vectors. You can see from the diagram that the x-component R_x of the vector sum is simply the sum $(A_x + B_x)$

of the x-components of the vectors being added. The same is true for the y-components. In symbols,

$$R_x = A_x + B_x \qquad R_y = A_y + B_y \qquad \text{(components of } \vec{R} = \vec{A} + \vec{B}) \qquad (1.10)$$

Figure 1.21 shows this result for the case in which the components A_x, A_y, B_x, and B_y are all positive. You should draw additional diagrams to verify for yourself that Eqs. (1.10) are valid for *any* signs of the components of $\vec{A}$ and $\vec{B}$.

If we know the components of any two vectors $\vec{A}$ and $\vec{B}$, perhaps by using Eqs. (1.6), we can compute the components of the vector sum $\vec{R}$. Then if we need the magnitude and direction of $\vec{R}$, we can obtain them from Eqs. (1.7) and (1.8) with the A's replaced by R's.

We can extend this procedure to find the sum of any number of vectors. If $\vec{R}$ is the vector sum of $\vec{A}, \vec{B}, \vec{C}, \vec{D}, \vec{E}, \ldots$, the components of $\vec{R}$ are

$$\begin{aligned} R_x &= A_x + B_x + C_x + D_x + E_x + \cdots \\ R_y &= A_y + B_y + C_y + D_y + E_y + \cdots \end{aligned} \qquad (1.11)$$

We have talked only about vectors that lie in the xy-plane, but the component method works just as well for vectors having any direction in space. We can introduce a z-axis perpendicular to the xy-plane; then in general a vector $\vec{A}$ has components A_x, A_y, and A_z in the three coordinate directions. Its magnitude A is

$$A = \sqrt{A_x^2 + A_y^2 + A_z^2} \qquad (1.12)$$

Again, we always take the positive root. Also, Eqs. (1.11) for the components of the vector sum $\vec{R}$ have an additional member:

$$R_z = A_z + B_z + C_z + D_z + E_z + \cdots$$

We've focused on adding *displacement* vectors, but the method is applicable to all vector quantities. When we study the concept of force in Chapter 4, we'll find that forces are vectors that obey the same rules of vector addition that we've used with displacement.

1.21 Finding the vector sum (resultant) of $\vec{A}$ and $\vec{B}$ using components.

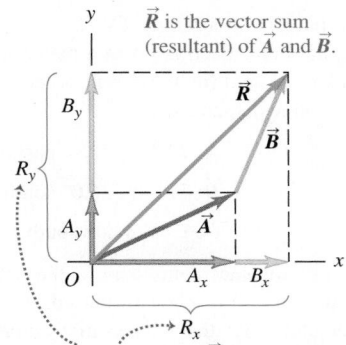

The components of $\vec{R}$ are the sums of the components of $\vec{A}$ and $\vec{B}$:

$$R_y = A_y + B_y \qquad R_x = A_x + B_x$$

Problem-Solving Strategy 1.3   **Vector Addition**

IDENTIFY *the relevant concepts:* Decide what the target variable is. It may be the magnitude of the vector sum, the direction, or both.

SET UP *the problem:* Sketch the vectors being added, along with suitable coordinate axes. Place the tail of the first vector at the origin of the coordinates, place the tail of the second vector at the head of the first vector, and so on. Draw the vector sum $\vec{R}$ from the tail of the first vector (at the origin) to the head of the last vector. Use your sketch to estimate the magnitude and direction of $\vec{R}$. Select the mathematical tools you'll use for the full calculation: Eqs. (1.6) to obtain the components of the vectors given, if necessary, Eqs. (1.11) to obtain the components of the vector sum, Eq. (1.12) to obtain its magnitude, and Eqs. (1.8) to obtain its direction.

EXECUTE *the solution* as follows:
1. Find the x- and y-components of each individual vector and record your results in a table, as in Example 1.7 below. If a vector is described by a magnitude A and an angle θ, measured

from the $+x$-axis toward the $+y$-axis, then its components are given by Eqs. 1.6:

$$A_x = A \cos\theta \qquad A_y = A \sin\theta$$

If the angles of the vectors are given in some other way, perhaps using a different reference direction, convert them to angles measured from the $+x$-axis as in Example 1.6 above.
2. Add the individual x-components algebraically (including signs) to find R_x, the x-component of the vector sum. Do the same for the y-components to find R_y. See Example 1.7 below.
3. Calculate the magnitude R and direction θ of the vector sum using Eqs. (1.7) and (1.8):

$$R = \sqrt{R_x^2 + R_y^2} \qquad \theta = \arctan\frac{R_y}{R_x}$$

EVALUATE *your answer:* Confirm that your results for the magnitude and direction of the vector sum agree with the estimates you made from your sketch. The value of θ that you find with a calculator may be off by 180°; your drawing will indicate the correct value.

Example 1.7 **Adding vectors using their components**

Three players on a reality TV show are brought to the center of a large, flat field. Each is given a meter stick, a compass, a calculator, a shovel, and (in a different order for each contestant) the following three displacements:

$$\vec{A}: 72.4 \text{ m}, 32.0° \text{ east of north}$$

$$\vec{B}: 57.3 \text{ m}, 36.0° \text{ south of west}$$

$$\vec{C}: 17.8 \text{ m due south}$$

The three displacements lead to the point in the field where the keys to a new Porsche are buried. Two players start measuring immediately, but the winner first *calculates* where to go. What does she calculate?

SOLUTION

IDENTIFY and SET UP: The goal is to find the sum (resultant) of the three displacements, so this is a problem in vector addition. Figure 1.22 shows the situation. We have chosen the $+x$-axis as

1.22 Three successive displacements $\vec{A}$, $\vec{B}$, and $\vec{C}$ and the resultant (vector sum) displacement $\vec{R} = \vec{A} + \vec{B} + \vec{C}$.

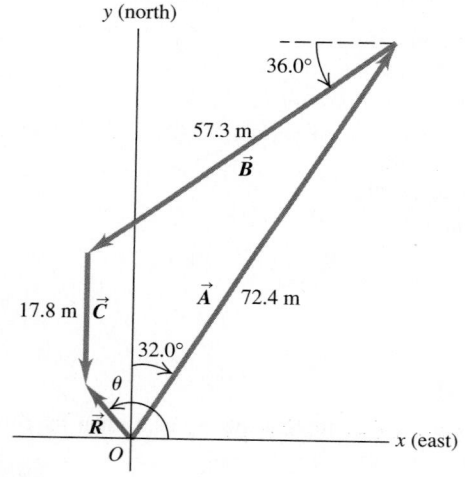

east and the $+y$-axis as north. We estimate from the diagram that the vector sum $\vec{R}$ is about 10 m, 40° west of north (which corresponds to $\theta \approx 130°$).

EXECUTE: The angles of the vectors, measured from the $+x$-axis toward the $+y$-axis, are $(90.0° - 32.0°) = 58.0°$, $(180.0° + 36.0°) = 216.0°$, and $270.0°$, respectively. We may now use Eqs. (1.6) to find the components of $\vec{A}$:

$$A_x = A \cos \theta_A = (72.4 \text{ m})(\cos 58.0°) = 38.37 \text{ m}$$

$$A_y = A \sin \theta_A = (72.4 \text{ m})(\sin 58.0°) = 61.40 \text{ m}$$

We've kept an extra significant figure in the components; we'll round to the correct number of significant figures at the end of our calculation. The table below shows the components of all the displacements, the addition of the components, and the other calculations.

Distance	Angle	x-component	y-component
$A = 72.4$ m	58.0°	38.37 m	61.40 m
$B = 57.3$ m	216.0°	−46.36 m	−33.68 m
$C = 17.8$ m	270.0°	0.00 m	−17.80 m
		$R_x = -7.99$ m	$R_y = 9.92$ m

$$R = \sqrt{(-7.99 \text{ m})^2 + (9.92 \text{ m})^2} = 12.7 \text{ m}$$

$$\theta = \arctan \frac{9.92 \text{ m}}{-7.99 \text{ m}} = -51°$$

Comparing to Fig. 1.22 shows that the calculated angle is clearly off by 180°. The correct value is $\theta = 180° - 51° = 129°$, or 39° west of north.

EVALUATE: Our calculated answers for R and θ agree with our estimates. Notice how drawing the diagram in Fig. 1.22 made it easy to avoid a 180° error in the direction of the vector sum.

Example 1.8 **A simple vector addition in three dimensions**

After an airplane takes off, it travels 10.4 km west, 8.7 km north, and 2.1 km up. How far is it from the takeoff point?

SOLUTION

Let the $+x$-axis be east, the $+y$-axis north, and the $+z$-axis up. Then the components of the airplane's displacement are $A_x = -10.4$ km, $A_y = 8.7$ km, and $A_z = 2.1$ km. From Eq. (1.12), the magnitude of the displacement is

$$A = \sqrt{(-10.4 \text{ km})^2 + (8.7 \text{ km})^2 + (2.1 \text{ km})^2} = 13.7 \text{ km}$$

Test Your Understanding of Section 1.8 Two vectors $\vec{A}$ and $\vec{B}$ both lie in the xy-plane. (a) Is it possible for $\vec{A}$ to have the same magnitude as $\vec{B}$ but different components? (b) Is it possible for $\vec{A}$ to have the same components as $\vec{B}$ but a different magnitude? ❙

1.9 Unit Vectors

A **unit vector** is a vector that has a magnitude of 1, with no units. Its only purpose is to *point*—that is, to describe a direction in space. Unit vectors provide a convenient notation for many expressions involving components of vectors. We will always include a caret or "hat" (^) in the symbol for a unit vector to distinguish it from ordinary vectors whose magnitude may or may not be equal to 1.

In an x-y coordinate system we can define a unit vector $\hat{\imath}$ that points in the direction of the positive x-axis and a unit vector $\hat{\jmath}$ that points in the direction of the positive y-axis (Fig. 1.23a). Then we can express the relationship between component vectors and components, described at the beginning of Section 1.8, as follows:

$$\vec{A}_x = A_x\hat{\imath}$$
$$\vec{A}_y = A_y\hat{\jmath} \qquad (1.13)$$

Similarly, we can write a vector $\vec{A}$ in terms of its components as

$$\vec{A} = A_x\hat{\imath} + A_y\hat{\jmath} \qquad (1.14)$$

Equations (1.13) and (1.14) are vector equations; each term, such as $A_x\hat{\imath}$, is a vector quantity (Fig. 1.23b).

Using unit vectors, we can express the vector sum $\vec{R}$ of two vectors $\vec{A}$ and $\vec{B}$ as follows:

$$\vec{A} = A_x\hat{\imath} + A_y\hat{\jmath}$$
$$\vec{B} = B_x\hat{\imath} + B_y\hat{\jmath}$$
$$\vec{R} = \vec{A} + \vec{B}$$
$$= (A_x\hat{\imath} + A_y\hat{\jmath}) + (B_x\hat{\imath} + B_y\hat{\jmath}) \qquad (1.15)$$
$$= (A_x + B_x)\hat{\imath} + (A_y + B_y)\hat{\jmath}$$
$$= R_x\hat{\imath} + R_y\hat{\jmath}$$

Equation (1.15) restates the content of Eqs. (1.10) in the form of a single vector equation rather than two component equations.

If the vectors do not all lie in the xy-plane, then we need a third component. We introduce a third unit vector $\hat{k}$ that points in the direction of the positive z-axis (Fig. 1.24). Then Eqs. (1.14) and (1.15) become

$$\vec{A} = A_x\hat{\imath} + A_y\hat{\jmath} + A_z\hat{k}$$
$$\vec{B} = B_x\hat{\imath} + B_y\hat{\jmath} + B_z\hat{k} \qquad (1.16)$$

$$\vec{R} = (A_x + B_x)\hat{\imath} + (A_y + B_y)\hat{\jmath} + (A_z + B_z)\hat{k}$$
$$= R_x\hat{\imath} + R_y\hat{\jmath} + R_z\hat{k} \qquad (1.17)$$

1.23 (a) The unit vectors $\hat{\imath}$ and $\hat{\jmath}$. (b) Expressing a vector $\vec{A}$ in terms of its components.

(a)

The unit vectors $\hat{\imath}$ and $\hat{\jmath}$ point in the directions of the x- and y-axes and have a magnitude of 1.

(b)

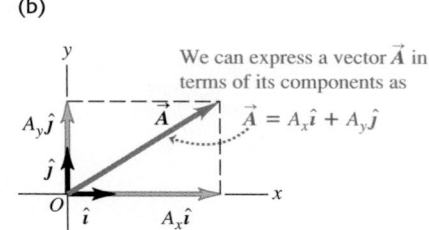

We can express a vector $\vec{A}$ in terms of its components as
$\vec{A} = A_x\hat{\imath} + A_y\hat{\jmath}$

1.24 The unit vectors $\hat{\imath}, \hat{\jmath},$ and $\hat{k}$.

Example 1.9 **Using unit vectors**

Given the two displacements

$$\vec{D} = (6.00\hat{\imath} + 3.00\hat{\jmath} - 1.00\hat{k})\text{ m} \quad \text{and}$$

$$\vec{E} = (4.00\hat{\imath} - 5.00\hat{\jmath} + 8.00\hat{k})\text{ m}$$

find the magnitude of the displacement $2\vec{D} - \vec{E}$.

SOLUTION

IDENTIFY and SET UP: We are to multiply the vector $\vec{D}$ by 2 (a scalar) and subtract the vector $\vec{E}$ from the result, so as to obtain the vector $\vec{F} = 2\vec{D} - \vec{E}$. Equation (1.9) says that to multiply $\vec{D}$ by 2, we multiply each of its components by 2. We can use Eq. (1.17) to do the subtraction; recall from Section 1.7 that subtracting a vector is the same as adding the negative of that vector.

EXECUTE: We have

$$\vec{F} = 2(6.00\hat{\imath} + 3.00\hat{\jmath} - 1.00\hat{k})\text{ m} - (4.00\hat{\imath} - 5.00\hat{\jmath} + 8.00\hat{k})\text{ m}$$

$$= [(12.00 - 4.00)\hat{\imath} + (6.00 + 5.00)\hat{\jmath} + (-2.00 - 8.00)\hat{k}]\text{ m}$$

$$= (8.00\hat{\imath} + 11.00\hat{\jmath} - 10.00\hat{k})\text{ m}$$

From Eq. (1.12) the magnitude of $\vec{F}$ is

$$F = \sqrt{F_x{}^2 + F_y{}^2 + F_z{}^2}$$
$$= \sqrt{(8.00\text{ m})^2 + (11.00\text{ m})^2 + (-10.00\text{ m})^2}$$
$$= 16.9\text{ m}$$

EVALUATE: Our answer is of the same order of magnitude as the larger components that appear in the sum. We wouldn't expect our answer to be much larger than this, but it could be much smaller.

Test Your Understanding of Section 1.9 Arrange the following vectors in order of their magnitude, with the vector of largest magnitude first. (i) $\vec{A} = (3\hat{\imath} + 5\hat{\jmath} - 2\hat{k})$ m; (ii) $\vec{B} = (-3\hat{\imath} + 5\hat{\jmath} - 2\hat{k})$ m; (iii) $\vec{C} = (3\hat{\imath} - 5\hat{\jmath} - 2\hat{k})$ m; (iv) $\vec{D} = (3\hat{\imath} + 5\hat{\jmath} + 2\hat{k})$ m.

1.10 Products of Vectors

Vector addition develops naturally from the problem of combining displacements and will prove useful for calculating many other vector quantities. We can also express many physical relationships by using *products* of vectors. Vectors are not ordinary numbers, so ordinary multiplication is not directly applicable to vectors. We will define two different kinds of products of vectors. The first, called the *scalar product,* yields a result that is a scalar quantity. The second, the *vector product,* yields another vector.

Scalar Product

1.25 Calculating the scalar product of two vectors, $\vec{A} \cdot \vec{B} = AB\cos\phi$.

(a)

Place the vectors tail to tail.

(b) $\vec{A} \cdot \vec{B}$ equals $A(B\cos\phi)$.

(Magnitude of $\vec{A}$) times (Component of $\vec{B}$ in direction of $\vec{A}$)

$B\cos\phi$

(c) $\vec{A} \cdot \vec{B}$ also equals $B(A\cos\phi)$

(Magnitude of $\vec{B}$) times (Component of $\vec{A}$ in direction of $\vec{B}$)

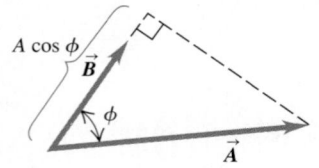

The **scalar product** of two vectors $\vec{A}$ and $\vec{B}$ is denoted by $\vec{A} \cdot \vec{B}$. Because of this notation, the scalar product is also called the **dot product.** Although $\vec{A}$ and $\vec{B}$ are vectors, the quantity $\vec{A} \cdot \vec{B}$ is a scalar.

To define the scalar product $\vec{A} \cdot \vec{B}$ we draw the two vectors $\vec{A}$ and $\vec{B}$ with their tails at the same point (Fig. 1.25a). The angle ϕ (the Greek letter phi) between their directions ranges from 0° to 180°. Figure 1.25b shows the projection of the vector $\vec{B}$ onto the direction of $\vec{A}$; this projection is the component of $\vec{B}$ in the direction of $\vec{A}$ and is equal to $B\cos\phi$. (We can take components along any direction that's convenient, not just the x- and y-axes.) We define $\vec{A} \cdot \vec{B}$ to be the magnitude of $\vec{A}$ multiplied by the component of $\vec{B}$ in the direction of $\vec{A}$. Expressed as an equation,

$$\vec{A} \cdot \vec{B} = AB\cos\phi = |\vec{A}||\vec{B}|\cos\phi \quad \begin{matrix}\text{(definition of the scalar}\\\text{(dot) product)}\end{matrix} \quad (1.18)$$

Alternatively, we can define $\vec{A} \cdot \vec{B}$ to be the magnitude of $\vec{B}$ multiplied by the component of $\vec{A}$ in the direction of $\vec{B}$, as in Fig. 1.25c. Hence $\vec{A} \cdot \vec{B} = B(A\cos\phi) = AB\cos\phi$, which is the same as Eq. (1.18).

The scalar product is a scalar quantity, not a vector, and it may be positive, negative, or zero. When ϕ is between 0° and 90°, $\cos\phi > 0$ and the scalar product is positive (Fig. 1.26a). When ϕ is between 90° and 180° so that $\cos\phi < 0$, the component of $\vec{B}$ in the direction of $\vec{A}$ is negative, and $\vec{A} \cdot \vec{B}$ is negative (Fig. 1.26b). Finally, when $\phi = 90°$, $\vec{A} \cdot \vec{B} = 0$ (Fig. 1.26c). *The scalar product of two perpendicular vectors is always zero.*

For any two vectors $\vec{A}$ and $\vec{B}$, $AB\cos\phi = BA\cos\phi$. This means that $\vec{A} \cdot \vec{B} = \vec{B} \cdot \vec{A}$. The scalar product obeys the commutative law of multiplication; the order of the two vectors does not matter.

We will use the scalar product in Chapter 6 to describe work done by a force. When a constant force $\vec{F}$ is applied to a body that undergoes a displacement $\vec{s}$, the work W (a scalar quantity) done by the force is given by

$$W = \vec{F} \cdot \vec{s}$$

The work done by the force is positive if the angle between $\vec{F}$ and $\vec{s}$ is between 0° and 90°, negative if this angle is between 90° and 180°, and zero if $\vec{F}$ and $\vec{s}$ are perpendicular. (This is another example of a term that has a special meaning in physics; in everyday language, "work" isn't something that can be positive or negative.) In later chapters we'll use the scalar product for a variety of purposes, from calculating electric potential to determining the effects that varying magnetic fields have on electric circuits.

Calculating the Scalar Product Using Components

We can calculate the scalar product $\vec{A} \cdot \vec{B}$ directly if we know the x-, y-, and z-components of $\vec{A}$ and $\vec{B}$. To see how this is done, let's first work out the scalar products of the unit vectors. This is easy, since $\hat{\imath}$, $\hat{\jmath}$, and $\hat{k}$ all have magnitude 1 and are perpendicular to each other. Using Eq. (1.18), we find

$$\hat{\imath} \cdot \hat{\imath} = \hat{\jmath} \cdot \hat{\jmath} = \hat{k} \cdot \hat{k} = (1)(1)\cos 0° = 1$$
$$\hat{\imath} \cdot \hat{\jmath} = \hat{\imath} \cdot \hat{k} = \hat{\jmath} \cdot \hat{k} = (1)(1)\cos 90° = 0$$

(1.19)

Now we express $\vec{A}$ and $\vec{B}$ in terms of their components, expand the product, and use these products of unit vectors:

$$\begin{aligned}
\vec{A} \cdot \vec{B} = {}& (A_x\hat{\imath} + A_y\hat{\jmath} + A_z\hat{k}) \cdot (B_x\hat{\imath} + B_y\hat{\jmath} + B_z\hat{k}) \\
= {}& A_x\hat{\imath} \cdot B_x\hat{\imath} + A_x\hat{\imath} \cdot B_y\hat{\jmath} + A_x\hat{\imath} \cdot B_z\hat{k} \\
& + A_y\hat{\jmath} \cdot B_x\hat{\imath} + A_y\hat{\jmath} \cdot B_y\hat{\jmath} + A_y\hat{\jmath} \cdot B_z\hat{k} \\
& + A_z\hat{k} \cdot B_x\hat{\imath} + A_z\hat{k} \cdot B_y\hat{\jmath} + A_z\hat{k} \cdot B_z\hat{k} \\
= {}& A_xB_x\hat{\imath} \cdot \hat{\imath} + A_xB_y\hat{\imath} \cdot \hat{\jmath} + A_xB_z\hat{\imath} \cdot \hat{k} \\
& + A_yB_x\hat{\jmath} \cdot \hat{\imath} + A_yB_y\hat{\jmath} \cdot \hat{\jmath} + A_yB_z\hat{\jmath} \cdot \hat{k} \\
& + A_zB_x\hat{k} \cdot \hat{\imath} + A_zB_y\hat{k} \cdot \hat{\jmath} + A_zB_z\hat{k} \cdot \hat{k}
\end{aligned}$$

(1.20)

From Eqs. (1.19) we see that six of these nine terms are zero, and the three that survive give simply

$$\vec{A} \cdot \vec{B} = A_xB_x + A_yB_y + A_zB_z \qquad \text{(scalar (dot) product in terms of components)}$$

(1.21)

Thus *the scalar product of two vectors is the sum of the products of their respective components.*

The scalar product gives a straightforward way to find the angle ϕ between any two vectors $\vec{A}$ and $\vec{B}$ whose components are known. In this case we can use Eq. (1.21) to find the scalar product of $\vec{A}$ and $\vec{B}$. Example 1.11 on the next page shows how to do this.

1.26 The scalar product $\vec{A} \cdot \vec{B} = AB\cos\phi$ can be positive, negative, or zero, depending on the angle between $\vec{A}$ and $\vec{B}$.

(a)

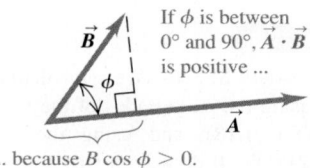

If ϕ is between 0° and 90°, $\vec{A} \cdot \vec{B}$ is positive ...

... because $B\cos\phi > 0$.

(b)

If ϕ is between 90° and 180°, $\vec{A} \cdot \vec{B}$ is negative ...

... because $B\cos\phi < 0$.

(c)

If $\phi = 90°$, $\vec{A} \cdot \vec{B} = 0$ because $\vec{B}$ has zero component in the direction of $\vec{A}$.

$\phi = 90°$

Example 1.10 **Calculating a scalar product**

Find the scalar product $\vec{A} \cdot \vec{B}$ of the two vectors in Fig. 1.27. The magnitudes of the vectors are $A = 4.00$ and $B = 5.00$.

SOLUTION

IDENTIFY and SET UP: We can calculate the scalar product in two ways: using the magnitudes of the vectors and the angle between them (Eq. 1.18), and using the components of the vectors (Eq. 1.21). We'll do it both ways, and the results will check each other.

1.27 Two vectors in two dimensions.

EXECUTE: The angle between the two vectors is $\phi = 130.0° - 53.0° = 77.0°$, so Eq. (1.18) gives us

$$\vec{A} \cdot \vec{B} = AB \cos \phi = (4.00)(5.00) \cos 77.0° = 4.50$$

To use Eq. (1.21), we must first find the components of the vectors. The angles of $\vec{A}$ and $\vec{B}$ are given with respect to the $+x$-axis and are measured in the sense from the $+x$-axis to the $+y$-axis, so we can use Eqs. (1.6):

$$A_x = (4.00) \cos 53.0° = 2.407$$
$$A_y = (4.00) \sin 53.0° = 3.195$$
$$B_x = (5.00) \cos 130.0° = -3.214$$
$$B_y = (5.00) \sin 130.0° = 3.830$$

As in Example 1.7, we keep an extra significant figure in the components and round at the end. Equation (1.21) now gives us

$$\vec{A} \cdot \vec{B} = A_x B_x + A_y B_y + A_z B_z$$
$$= (2.407)(-3.214) + (3.195)(3.830) + (0)(0) = 4.50$$

EVALUATE: Both methods give the same result, as they should.

Example 1.11 **Finding an angle with the scalar product**

Find the angle between the vectors

$$\vec{A} = 2.00\hat{i} + 3.00\hat{j} + 1.00\hat{k} \quad \text{and}$$
$$\vec{B} = -4.00\hat{i} + 2.00\hat{j} - 1.00\hat{k}$$

SOLUTION

IDENTIFY and SET UP: We're given the x-, y-, and z-components of two vectors. Our target variable is the angle ϕ between them (Fig. 1.28). To find this, we'll solve Eq. (1.18), $\vec{A} \cdot \vec{B} = AB \cos \phi$, for ϕ in terms of the scalar product $\vec{A} \cdot \vec{B}$ and the magnitudes A and B. We can evaluate the scalar product using Eq. (1.21),

1.28 Two vectors in three dimensions.

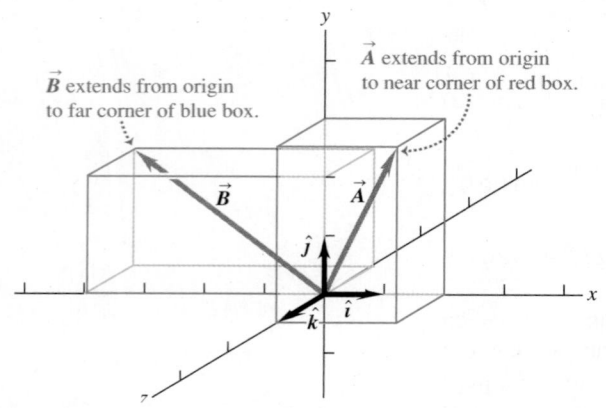

$\vec{B}$ extends from origin to far corner of blue box.

$\vec{A}$ extends from origin to near corner of red box.

$\vec{A} \cdot \vec{B} = A_x B_x + A_y B_y + A_z B_z$, and we can find A and B using Eq. (1.7).

EXECUTE: We solve Eq. (1.18) for $\cos \phi$ and write $\vec{A} \cdot \vec{B}$ using Eq. (1.21). Our result is

$$\cos \phi = \frac{\vec{A} \cdot \vec{B}}{AB} = \frac{A_x B_x + A_y B_y + A_z B_z}{AB}$$

We can use this formula to find the angle between *any* two vectors $\vec{A}$ and $\vec{B}$. Here we have $A_x = 2.00$, $A_y = 3.00$, and $A_z = 1.00$, and $B_x = -4.00$, $B_y = 2.00$, and $B_z = -1.00$. Thus

$$\vec{A} \cdot \vec{B} = A_x B_x + A_y B_y + A_z B_z$$
$$= (2.00)(-4.00) + (3.00)(2.00) + (1.00)(-1.00)$$
$$= -3.00$$
$$A = \sqrt{A_x^2 + A_y^2 + A_z^2} = \sqrt{(2.00)^2 + (3.00)^2 + (1.00)^2}$$
$$= \sqrt{14.00}$$
$$B = \sqrt{B_x^2 + B_y^2 + B_z^2} = \sqrt{(-4.00)^2 + (2.00)^2 + (-1.00)^2}$$
$$= \sqrt{21.00}$$
$$\cos \phi = \frac{A_x B_x + A_y B_y + A_z B_z}{AB} = \frac{-3.00}{\sqrt{14.00}\sqrt{21.00}} = -0.175$$
$$\phi = 100°$$

EVALUATE: As a check on this result, note that the scalar product $\vec{A} \cdot \vec{B}$ is negative. This means that ϕ is between 90° and 180° (see Fig. 1.26), which agrees with our answer.

Vector Product

The **vector product** of two vectors $\vec{A}$ and $\vec{B}$, also called the **cross product,** is denoted by $\vec{A} \times \vec{B}$. As the name suggests, the vector product is itself a vector. We'll use this product in Chapter 10 to describe torque and angular momentum; in Chapters 27 and 28 we'll use it to describe magnetic fields and forces.

To define the vector product $\vec{A} \times \vec{B}$, we again draw the two vectors $\vec{A}$ and $\vec{B}$ with their tails at the same point (Fig. 1.29a). The two vectors then lie in a plane. We define the vector product to be a vector quantity with a direction perpendicular to this plane (that is, perpendicular to both $\vec{A}$ and $\vec{B}$) and a magnitude equal to $AB \sin \phi$. That is, if $\vec{C} = \vec{A} \times \vec{B}$, then

$$C = AB \sin \phi \quad \text{(magnitude of the vector (cross) product of } \vec{A} \text{ and } \vec{B}) \quad (1.22)$$

We measure the angle ϕ from $\vec{A}$ toward $\vec{B}$ and take it to be the smaller of the two possible angles, so ϕ ranges from 0° to 180°. Then $\sin \phi \geq 0$ and C in Eq. (1.22) is never negative, as must be the case for a vector magnitude. Note also that when $\vec{A}$ and $\vec{B}$ are parallel or antiparallel, $\phi = 0$ or 180° and $C = 0$. That is, *the vector product of two parallel or antiparallel vectors is always zero. In particular, the vector product of any vector with itself is zero.*

> **CAUTION** **Vector product vs. scalar product** Be careful not to confuse the expression $AB \sin \phi$ for the magnitude of the vector product $\vec{A} \times \vec{B}$ with the similar expression $AB \cos \phi$ for the scalar product $\vec{A} \cdot \vec{B}$. To see the difference between these two expressions, imagine that we vary the angle between $\vec{A}$ and $\vec{B}$ while keeping their magnitudes constant. When $\vec{A}$ and $\vec{B}$ are parallel, the magnitude of the vector product will be zero and the scalar product will be maximum. When $\vec{A}$ and $\vec{B}$ are perpendicular, the magnitude of the vector product will be maximum and the scalar product will be zero.

There are always *two* directions perpendicular to a given plane, one on each side of the plane. We choose which of these is the direction of $\vec{A} \times \vec{B}$ as follows. Imagine rotating vector $\vec{A}$ about the perpendicular line until it is aligned with $\vec{B}$, choosing the smaller of the two possible angles between $\vec{A}$ and $\vec{B}$. Curl the fingers of your right hand around the perpendicular line so that the fingertips point in the direction of rotation; your thumb will then point in the direction of $\vec{A} \times \vec{B}$. Figure 1.29a shows this **right-hand rule** and describes a second way to think about this rule.

Similarly, we determine the direction of $\vec{B} \times \vec{A}$ by rotating $\vec{B}$ into $\vec{A}$ as in Fig. 1.29b. The result is a vector that is *opposite* to the vector $\vec{A} \times \vec{B}$. The vector product is *not* commutative! In fact, for any two vectors $\vec{A}$ and $\vec{B}$,

$$\vec{A} \times \vec{B} = -\vec{B} \times \vec{A} \quad (1.23)$$

Just as we did for the scalar product, we can give a geometrical interpretation of the magnitude of the vector product. In Fig. 1.30a, $B \sin \phi$ is the component of vector $\vec{B}$ that is *perpendicular* to the direction of vector $\vec{A}$. From Eq. (1.22) the magnitude of $\vec{A} \times \vec{B}$ equals the magnitude of $\vec{A}$ multiplied by the component of $\vec{B}$ perpendicular to $\vec{A}$. Figure 1.30b shows that the magnitude of $\vec{A} \times \vec{B}$ also equals the magnitude of $\vec{B}$ multiplied by the component of $\vec{A}$ perpendicular to $\vec{B}$. Note that Fig. 1.30 shows the case in which ϕ is between 0° and 90°; you should draw a similar diagram for ϕ between 90° and 180° to show that the same geometrical interpretation of the magnitude of $\vec{A} \times \vec{B}$ still applies.

Calculating the Vector Product Using Components

If we know the components of $\vec{A}$ and $\vec{B}$, we can calculate the components of the vector product using a procedure similar to that for the scalar product. First we work out the multiplication table for the unit vectors $\hat{\imath}, \hat{\jmath}$, and $\hat{k}$, all three of which

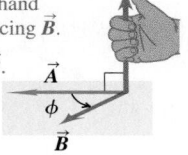

1.29 (a) The vector product $\vec{A} \times \vec{B}$ determined by the right-hand rule. (b) $\vec{B} \times \vec{A} = -\vec{A} \times \vec{B}$; the vector product is anticommutative.

(a) Using the right-hand rule to find the direction of $\vec{A} \times \vec{B}$

① Place $\vec{A}$ and $\vec{B}$ tail to tail.

② Point fingers of right hand along $\vec{A}$, with palm facing $\vec{B}$.

③ Curl fingers toward $\vec{B}$.

④ Thumb points in direction of $\vec{A} \times \vec{B}$.

(b) $\vec{B} \times \vec{A} = -\vec{A} \times \vec{B}$ (the vector product is anticommutative)

Same magnitude but opposite direction

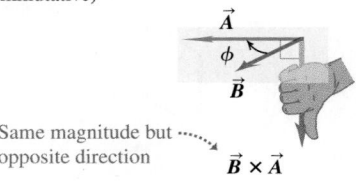

1.30 Calculating the magnitude $AB \sin \phi$ of the vector product of two vectors, $\vec{A} \times \vec{B}$.

(a)

(Magnitude of $\vec{A} \times \vec{B}$) equals $A(B \sin \phi)$.

(Magnitude of $\vec{A}$) times (Component of $\vec{B}$ perpendicular to $\vec{A}$)

$B \sin \phi$

(b)

(Magnitude of $\vec{A} \times \vec{B}$) also equals $B(A \sin \phi)$.

(Magnitude of $\vec{B}$) times (Component of $\vec{A}$ perpendicular to $\vec{B}$)

$A \sin \phi$

1.31 (a) We will always use a right-handed coordinate system, like this one. (b) We will never use a left-handed coordinate system (in which $\hat{\imath} \times \hat{\jmath} = -\hat{k}$, and so on).

(a) A right-handed coordinate system

$$\hat{\imath} \times \hat{\jmath} = \hat{k}$$
$$\hat{\jmath} \times \hat{k} = \hat{\imath}$$
$$\hat{k} \times \hat{\imath} = \hat{\jmath}$$

(b) A left-handed coordinate system; we will not use these.

are perpendicular to each other (Fig. 1.31a). The vector product of any vector with itself is zero, so

$$\hat{\imath} \times \hat{\imath} = \hat{\jmath} \times \hat{\jmath} = \hat{k} \times \hat{k} = \mathbf{0}$$

The boldface zero is a reminder that each product is a zero *vector*—that is, one with all components equal to zero and an undefined direction. Using Eqs. (1.22) and (1.23) and the right-hand rule, we find

$$\hat{\imath} \times \hat{\jmath} = -\hat{\jmath} \times \hat{\imath} = \hat{k}$$
$$\hat{\jmath} \times \hat{k} = -\hat{k} \times \hat{\jmath} = \hat{\imath} \qquad (1.24)$$
$$\hat{k} \times \hat{\imath} = -\hat{\imath} \times \hat{k} = \hat{\jmath}$$

You can verify these equations by referring to Fig. 1.31a.

Next we express $\vec{A}$ and $\vec{B}$ in terms of their components and the corresponding unit vectors, and we expand the expression for the vector product:

$$\vec{A} \times \vec{B} = (A_x\hat{\imath} + A_y\hat{\jmath} + A_z\hat{k}) \times (B_x\hat{\imath} + B_y\hat{\jmath} + B_z\hat{k})$$
$$= A_x\hat{\imath} \times B_x\hat{\imath} + A_x\hat{\imath} \times B_y\hat{\jmath} + A_x\hat{\imath} \times B_z\hat{k}$$
$$+ A_y\hat{\jmath} \times B_x\hat{\imath} + A_y\hat{\jmath} \times B_y\hat{\jmath} + A_y\hat{\jmath} \times B_z\hat{k} \qquad (1.25)$$
$$+ A_z\hat{k} \times B_x\hat{\imath} + A_z\hat{k} \times B_y\hat{\jmath} + A_z\hat{k} \times B_z\hat{k}$$

We can also rewrite the individual terms in Eq. (1.25) as $A_x\hat{\imath} \times B_y\hat{\jmath} = (A_xB_y)\hat{\imath} \times \hat{\jmath}$, and so on. Evaluating these by using the multiplication table for the unit vectors in Eqs. (1.24) and then grouping the terms, we get

$$\vec{A} \times \vec{B} = (A_yB_z - A_zB_y)\hat{\imath} + (A_zB_x - A_xB_z)\hat{\jmath} + (A_xB_y - A_yB_x)\hat{k} \quad (1.26)$$

Thus the components of $\vec{C} = \vec{A} \times \vec{B}$ are given by

$$C_x = A_yB_z - A_zB_y \qquad C_y = A_zB_x - A_xB_z \qquad C_z = A_xB_y - A_yB_x$$
$$(\text{components of } \vec{C} = \vec{A} \times \vec{B}) \qquad (1.27)$$

The vector product can also be expressed in determinant form as

$$\vec{A} \times \vec{B} = \begin{vmatrix} \hat{\imath} & \hat{\jmath} & \hat{k} \\ A_x & A_y & A_z \\ B_x & B_y & B_z \end{vmatrix}$$

If you aren't familiar with determinants, don't worry about this form.

With the axis system of Fig. 1.31a, if we reverse the direction of the z-axis, we get the system shown in Fig. 1.31b. Then, as you may verify, the definition of the vector product gives $\hat{\imath} \times \hat{\jmath} = -\hat{k}$ instead of $\hat{\imath} \times \hat{\jmath} = \hat{k}$. In fact, all vector products of the unit vectors $\hat{\imath}, \hat{\jmath},$ and $\hat{k}$ would have signs opposite to those in Eqs. (1.24). We see that there are two kinds of coordinate systems, differing in the signs of the vector products of unit vectors. An axis system in which $\hat{\imath} \times \hat{\jmath} = \hat{k}$, as in Fig. 1.31a, is called a **right-handed system.** The usual practice is to use *only* right-handed systems, and we will follow that practice throughout this book.

Example 1.12 Calculating a vector product

Vector $\vec{A}$ has magnitude 6 units and is in the direction of the +x-axis. Vector $\vec{B}$ has magnitude 4 units and lies in the xy-plane, making an angle of 30° with the +x-axis (Fig. 1.32). Find the vector product $\vec{C} = \vec{A} \times \vec{B}$.

SOLUTION

IDENTIFY and SET UP: We'll find the vector product in two ways, which will provide a check of our calculations. First we'll use Eq. (1.22) and the right-hand rule; then we'll use Eqs. (1.27) to find the vector product using components.

1.32 Vectors $\vec{A}$ and $\vec{B}$ and their vector product $\vec{C} = \vec{A} \times \vec{B}$. The vector $\vec{B}$ lies in the xy-plane.

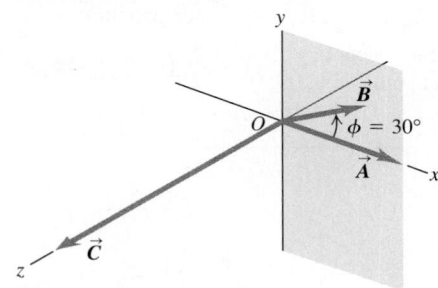

EXECUTE: From Eq. (1.22) the magnitude of the vector product is

$$AB \sin \phi = (6)(4)(\sin 30°) = 12$$

By the right-hand rule, the direction of $\vec{A} \times \vec{B}$ is along the +z-axis (the direction of the unit vector $\hat{k}$), so we have $\vec{C} = \vec{A} \times \vec{B} = 12\hat{k}$.

To use Eqs. (1.27), we first determine the components of $\vec{A}$ and $\vec{B}$:

$A_x = 6$	$A_y = 0$	$A_z = 0$
$B_x = 4 \cos 30° = 2\sqrt{3}$	$B_y = 4 \sin 30° = 2$	$B_z = 0$

Then Eqs. (1.27) yield

$$C_x = (0)(0) - (0)(2) = 0$$
$$C_y = (0)(2\sqrt{3}) - (6)(0) = 0$$
$$C_z = (6)(2) - (0)(2\sqrt{3}) = 12$$

Thus again we have $\vec{C} = 12\hat{k}$.

EVALUATE: Both methods give the same result. Depending on the situation, one or the other of the two approaches may be the more convenient one to use.

Test Your Understanding of Section 1.10 Vector $\vec{A}$ has magnitude 2 and vector $\vec{B}$ has magnitude 3. The angle ϕ between $\vec{A}$ and $\vec{B}$ is known to be 0°, 90°, or 180°. For each of the following situations, state what the value of ϕ must be. (In each situation there may be more than one correct answer.) (a) $\vec{A} \cdot \vec{B} = 0$; (b) $\vec{A} \times \vec{B} = \mathbf{0}$; (c) $\vec{A} \cdot \vec{B} = 6$; (d) $\vec{A} \cdot \vec{B} = -6$; (e) (Magnitude of $\vec{A} \times \vec{B}$) = 6. ❙

Physical quantities and units: Three fundamental physical quantities are mass, length, and time. The corresponding basic SI units are the kilogram, the meter, and the second. Derived units for other physical quantities are products or quotients of the basic units. Equations must be dimensionally consistent; two terms can be added only when they have the same units. (See Examples 1.1 and 1.2.)

Significant figures: The accuracy of a measurement can be indicated by the number of significant figures or by a stated uncertainty. The result of a calculation usually has no more significant figures than the input data. When only crude estimates are available for input data, we can often make useful order-of-magnitude estimates. (See Examples 1.3 and 1.4.)

Significant figures in magenta

$$\pi = \frac{C}{2r} = \frac{0.424 \text{ m}}{2(0.06750 \text{ m})} = 3.14$$

$$123.62 + 8.9 = 132.5$$

Scalars, vectors, and vector addition: Scalar quantities are numbers and combine with the usual rules of arithmetic. Vector quantities have direction as well as magnitude and combine according to the rules of vector addition. The negative of a vector has the same magnitude but points in the opposite direction. (See Example 1.5.)

Vector components and vector addition: Vector addition can be carried out using components of vectors. The x-component of $\vec{R} = \vec{A} + \vec{B}$ is the sum of the x-components of $\vec{A}$ and $\vec{B}$, and likewise for the y- and z-components. (See Examples 1.6–1.8.)

$$R_x = A_x + B_x$$
$$R_y = A_y + B_y \qquad (1.10)$$
$$R_z = A_z + B_z$$

Unit vectors: Unit vectors describe directions in space. A unit vector has a magnitude of 1, with no units. The unit vectors $\hat{\imath}, \hat{\jmath},$ and $\hat{k}$, aligned with the x-, y-, and z-axes of a rectangular coordinate system, are especially useful. (See Example 1.9.)

$$\vec{A} = A_x\hat{\imath} + A_y\hat{\jmath} + A_z\hat{k} \qquad (1.16)$$

Scalar product: The scalar product $C = \vec{A} \cdot \vec{B}$ of two vectors $\vec{A}$ and $\vec{B}$ is a scalar quantity. It can be expressed in terms of the magnitudes of $\vec{A}$ and $\vec{B}$ and the angle ϕ between the two vectors, or in terms of the components of $\vec{A}$ and $\vec{B}$. The scalar product is commutative; $\vec{A} \cdot \vec{B} = \vec{B} \cdot \vec{A}$. The scalar product of two perpendicular vectors is zero. (See Examples 1.10 and 1.11.)

$$\vec{A} \cdot \vec{B} = AB \cos\phi = |\vec{A}||\vec{B}| \cos\phi \quad (1.18)$$
$$\vec{A} \cdot \vec{B} = A_xB_x + A_yB_y + A_zB_z \quad (1.21)$$

Scalar product $\vec{A} \cdot \vec{B} = AB \cos \phi$

Vector product: The vector product $\vec{C} = \vec{A} \times \vec{B}$ of two vectors $\vec{A}$ and $\vec{B}$ is another vector $\vec{C}$. The magnitude of $\vec{A} \times \vec{B}$ depends on the magnitudes of $\vec{A}$ and $\vec{B}$ and the angle ϕ between the two vectors. The direction of $\vec{A} \times \vec{B}$ is perpendicular to the plane of the two vectors being multiplied, as given by the right-hand rule. The components of $\vec{C} = \vec{A} \times \vec{B}$ can be expressed in terms of the components of $\vec{A}$ and $\vec{B}$. The vector product is not commutative; $\vec{A} \times \vec{B} = -\vec{B} \times \vec{A}$. The vector product of two parallel or antiparallel vectors is zero. (See Example 1.12.)

$$C = AB \sin\phi \qquad (1.22)$$
$$C_x = A_yB_z - A_zB_y$$
$$C_y = A_zB_x - A_xB_z \qquad (1.27)$$
$$C_z = A_xB_y - A_yB_x$$

$\vec{A} \times \vec{B}$ is perpendicular to the plane of $\vec{A}$ and $\vec{B}$.

(Magnitude of $\vec{A} \times \vec{B}$) = $AB \sin \phi$

BRIDGING PROBLEM   Vectors on the Roof

An air-conditioning unit is fastened to a roof that slopes at an angle of 35° above the horizontal (Fig. 1.33). Its weight is a force on the air conditioner that is directed vertically downward. In order that the unit not crush the roof tiles, the component of the unit's weight perpendicular to the roof cannot exceed 425 N. (One newton, or 1 N, is the SI unit of force. It is equal to 0.2248 lb.) (a) What is the maximum allowed weight of the unit? (b) If the fasteners fail, the unit slides 1.50 m along the roof before it comes to a halt against a ledge. How much work does the weight force do on the unit during its slide if the unit has the weight calculated in part (a)? As we described in Section 1.10, the work done by a force $\vec{F}$ on an object that undergoes a displacement $\vec{s}$ is $W = \vec{F} \cdot \vec{s}$.

SOLUTION GUIDE

See MasteringPhysics® study area for a Video Tutor solution.

IDENTIFY and SET UP

1. This problem involves vectors and components. What are the known quantities? Which aspect(s) of the weight vector (magnitude, direction, and/or particular components) represent the target variable for part (a)? Which aspect(s) must you know to solve part (b)?
2. Make a sketch based on Fig. 1.33. Add x- and y-axes, choosing the positive direction for each. Your axes don't have to be horizontal and vertical, but they do have to be mutually perpendicular. Make the most convenient choice.
3. Choose the equations you'll use to determine the target variables.

EXECUTE

4. Use the relationship between the magnitude and direction of a vector and its components to solve for the target variable in

1.33 An air-conditioning unit on a slanted roof.

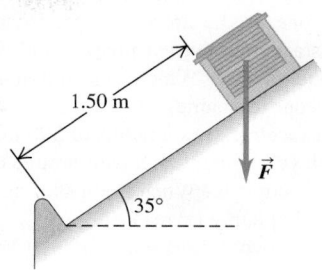

part (a). Be careful: Is 35° the correct angle to use in the equation? (*Hint:* Check your sketch.)
5. Make sure your answer has the correct number of significant figures.
6. Use the definition of the scalar product to solve for the target variable in part (b). Again, make sure to use the correct number of significant figures.

EVALUATE

7. Did your answer to part (a) include a vector component whose absolute value is greater than the magnitude of the vector? Is that possible?
8. There are two ways to find the scalar product of two vectors, one of which you used to solve part (b). Check your answer by repeating the calculation using the other way. Do you get the same answer?

Problems

For instructor-assigned homework, go to www.masteringphysics.com

•, ••, •••: Problems of increasing difficulty. **CP**: Cumulative problems incorporating material from earlier chapters. **CALC**: Problems requiring calculus. **BIO**: Biosciences problems.

DISCUSSION QUESTIONS

Q1.1 How many correct experiments do we need to disprove a theory? How many do we need to prove a theory? Explain.

Q1.2 A guidebook describes the rate of climb of a mountain trail as 120 meters per kilometer. How can you express this as a number with no units?

Q1.3 Suppose you are asked to compute the tangent of 5.00 meters. Is this possible? Why or why not?

Q1.4 A highway contractor stated that in building a bridge deck he poured 250 yards of concrete. What do you think he meant?

Q1.5 What is your height in centimeters? What is your weight in newtons?

Q1.6 The U.S. National Institute of Standards and Technology (NIST) maintains several accurate copies of the international standard kilogram. Even after careful cleaning, these national standard

kilograms are gaining mass at an average rate of about 1 μg/y (y = year) when compared every 10 years or so to the standard international kilogram. Does this apparent change have any importance? Explain.

Q1.7 What physical phenomena (other than a pendulum or cesium clock) could you use to define a time standard?

Q1.8 Describe how you could measure the thickness of a sheet of paper with an ordinary ruler.

Q1.9 The quantity $\pi = 3.14159\ldots$ is a number with no dimensions, since it is a ratio of two lengths. Describe two or three other geometrical or physical quantities that are dimensionless.

Q1.10 What are the units of volume? Suppose another student tells you that a cylinder of radius r and height h has volume given by $\pi r^3 h$. Explain why this cannot be right.

Q1.11 Three archers each fire four arrows at a target. Joe's four arrows hit at points 10 cm above, 10 cm below, 10 cm to the left, and 10 cm to the right of the center of the target. All four of Moe's arrows hit within 1 cm of a point 20 cm from the center, and Flo's four arrows all hit within 1 cm of the center. The contest judge says that one of the archers is precise but not accurate, another archer is accurate but not precise, and the third archer is both accurate and precise. Which description goes with which archer? Explain your reasoning.

Q1.12 A circular racetrack has a radius of 500 m. What is the displacement of a bicyclist when she travels around the track from the north side to the south side? When she makes one complete circle around the track? Explain your reasoning.

Q1.13 Can you find two vectors with different lengths that have a vector sum of zero? What length restrictions are required for three vectors to have a vector sum of zero? Explain your reasoning.

Q1.14 One sometimes speaks of the "direction of time," evolving from past to future. Does this mean that time is a vector quantity? Explain your reasoning.

Q1.15 Air traffic controllers give instructions to airline pilots telling them in which direction they are to fly. These instructions are called "vectors." If these are the only instructions given, is the name "vector" used correctly? Why or why not?

Q1.16 Can you find a vector quantity that has a magnitude of zero but components that are different from zero? Explain. Can the magnitude of a vector be less than the magnitude of any of its components? Explain.

Q1.17 (a) Does it make sense to say that a vector is *negative*? Why? (b) Does it make sense to say that one vector is the negative of another? Why? Does your answer here contradict what you said in part (a)?

Q1.18 If $\vec{C}$ is the vector sum of $\vec{A}$ and $\vec{B}$, $\vec{C} = \vec{A} + \vec{B}$, what must be true about the directions and magnitudes of $\vec{A}$ and $\vec{B}$ if $C = A + B$? What must be true about the directions and magnitudes of $\vec{A}$ and $\vec{B}$ if $C = 0$?

Q1.19 If $\vec{A}$ and $\vec{B}$ are nonzero vectors, is it possible for $\vec{A} \cdot \vec{B}$ and $\vec{A} \times \vec{B}$ *both* to be zero? Explain.

Q1.20 What does $\vec{A} \cdot \vec{A}$, the scalar product of a vector with itself, give? What about $\vec{A} \times \vec{A}$, the vector product of a vector with itself?

Q1.21 Let $\vec{A}$ represent any nonzero vector. Why is $\vec{A}/A$ a unit vector, and what is its direction? If θ is the angle that $\vec{A}$ makes with the $+x$-axis, explain why $(\vec{A}/A) \cdot \hat{\imath}$ is called the *direction cosine* for that axis.

Q1.22 Which of the following are legitimate mathematical operations: (a) $\vec{A} \cdot (\vec{B} - \vec{C})$; (b) $(\vec{A} - \vec{B}) \times \vec{C}$; (c) $\vec{A} \cdot (\vec{B} \times \vec{C})$; (d) $\vec{A} \times (\vec{B} \times \vec{C})$; (e) $\vec{A} \times (\vec{B} \cdot \vec{C})$? In each case, give the reason for your answer.

Q1.23 Consider the two repeated vector products $\vec{A} \times (\vec{B} \times \vec{C})$ and $(\vec{A} \times \vec{B}) \times \vec{C}$. Give an example that illustrates the general rule that these two vector products do not have the same magnitude or direction. Can you choose the vectors $\vec{A}$, $\vec{B}$, and $\vec{C}$ such that these two vector products *are* equal? If so, give an example.

Q1.24 Show that, no matter what $\vec{A}$ and $\vec{B}$ are, $\vec{A} \cdot (\vec{A} \times \vec{B}) = 0$. (*Hint:* Do not look for an elaborate mathematical proof. Rather look at the definition of the direction of the cross product.)

Q1.25 (a) If $\vec{A} \cdot \vec{B} = 0$, does it necessarily follow that $A = 0$ or $B = 0$? Explain. (b) If $\vec{A} \times \vec{B} = 0$, does it necessarily follow that $A = 0$ or $B = 0$? Explain.

Q1.26 If $\vec{A} = 0$ for a vector in the xy-plane, does it follow that $A_x = -A_y$? What *can* you say about A_x and A_y?

EXERCISES

Section 1.3 Standards and Units
Section 1.4 Unit Consistency and Conversions

1.1 • Starting with the definition 1 in. = 2.54 cm, find the number of (a) kilometers in 1.00 mile and (b) feet in 1.00 km.

1.2 •• According to the label on a bottle of salad dressing, the volume of the contents is 0.473 liter (L). Using only the conversions 1 L = 1000 cm^3 and 1 in. = 2.54 cm, express this volume in cubic inches.

1.3 •• How many nanoseconds does it take light to travel 1.00 ft in vacuum? (This result is a useful quantity to remember.)

1.4 •• The density of gold is 19.3 g/cm^3. What is this value in kilograms per cubic meter?

1.5 • The most powerful engine available for the classic 1963 Chevrolet Corvette Sting Ray developed 360 horsepower and had a displacement of 327 cubic inches. Express this displacement in liters (L) by using only the conversions 1 L = 1000 cm^3 and 1 in. = 2.54 cm.

1.6 •• A square field measuring 100.0 m by 100.0 m has an area of 1.00 hectare. An acre has an area of 43,600 ft^2. If a country lot has an area of 12.0 acres, what is the area in hectares?

1.7 • How many years older will you be 1.00 gigasecond from now? (Assume a 365-day year.)

1.8 • While driving in an exotic foreign land you see a speed limit sign on a highway that reads 180,000 furlongs per fortnight. How many miles per hour is this? (One furlong is $\frac{1}{8}$ mile, and a fortnight is 14 days. A furlong originally referred to the length of a plowed furrow.)

1.9 • A certain fuel-efficient hybrid car gets gasoline mileage of 55.0 mpg (miles per gallon). (a) If you are driving this car in Europe and want to compare its mileage with that of other European cars, express this mileage in km/L (L = liter). Use the conversion factors in Appendix E. (b) If this car's gas tank holds 45 L, how many tanks of gas will you use to drive 1500 km?

1.10 • The following conversions occur frequently in physics and are very useful. (a) Use 1 mi = 5280 ft and 1 h = 3600 s to convert 60 mph to units of ft/s. (b) The acceleration of a freely falling object is 32 ft/s^2. Use 1 ft = 30.48 cm to express this acceleration in units of m/s^2. (c) The density of water is 1.0 g/cm^3. Convert this density to units of kg/m^3.

1.11 •• **Neptunium.** In the fall of 2002, a group of scientists at Los Alamos National Laboratory determined that the critical mass of neptunium-237 is about 60 kg. The critical mass of a fissionable material is the minimum amount that must be brought together to start a chain reaction. This element has a density of 19.5 g/cm^3. What would be the radius of a sphere of this material that has a critical mass?

1.12 • **BIO** (a) The recommended daily allowance (RDA) of the trace metal magnesium is 410 mg/day for males. Express this quantity in μg/day. (b) For adults, the RDA of the amino acid lysine is 12 mg per kg of body weight. How many grams per day should a 75-kg adult receive? (c) A typical multivitamin tablet can contain 2.0 mg of vitamin B$_2$ (riboflavin), and the RDA is 0.0030 g/day. How many such tablets should a person take each day to get the proper amount of this vitamin, assuming that he gets none from any other sources? (d) The RDA for the trace element selenium is 0.000070 g/day. Express this dose in mg/day.

Section 1.5 Uncertainty and Significant Figures

1.13 •• Figure 1.7 shows the result of unacceptable error in the stopping position of a train. (a) If a train travels 890 km from Berlin to Paris and then overshoots the end of the track by 10 m, what is the percent error in the total distance covered? (b) Is it correct to write the total distance covered by the train as 890,010 m? Explain.

1.14 • With a wooden ruler you measure the length of a rectangular piece of sheet metal to be 12 mm. You use micrometer calipers to measure the width of the rectangle and obtain the value 5.98 mm. Give your answers to the following questions to the correct number of significant figures. (a) What is the area of the rectangle? (b) What is the ratio of the rectangle's width to its length? (c) What is the perimeter of the rectangle? (d) What is the difference between the length and width? (e) What is the ratio of the length to the width?

1.15 •• A useful and easy-to-remember approximate value for the number of seconds in a year is $\pi \times 10^7$. Determine the percent error in this approximate value. (There are 365.24 days in one year.)

Section 1.6 Estimates and Orders of Magnitude

1.16 • How many gallons of gasoline are used in the United States in one day? Assume that there are two cars for every three people, that each car is driven an average of 10,000 mi per year, and that the average car gets 20 miles per gallon.

1.17 •• **BIO** A rather ordinary middle-aged man is in the hospital for a routine check-up. The nurse writes the quantity 200 on his medical chart but forgets to include the units. Which of the following quantities could the 200 plausibly represent? (a) his mass in kilograms; (b) his height in meters; (c) his height in centimeters; (d) his height in millimeters; (e) his age in months.

1.18 • How many kernels of corn does it take to fill a 2-L soft drink bottle?

1.19 • How many words are there in this book?

1.20 • **BIO** Four astronauts are in a spherical space station. (a) If, as is typical, each of them breathes about 500 cm³ of air with each breath, approximately what volume of air (in cubic meters) do these astronauts breathe in a year? (b) What would the diameter (in meters) of the space station have to be to contain all this air?

1.21 • **BIO** How many times does a typical person blink her eyes in a lifetime?

1.22 • **BIO** How many times does a human heart beat during a lifetime? How many gallons of blood does it pump? (Estimate that the heart pumps 50 cm³ of blood with each beat.)

1.23 • In Wagner's opera *Das Rheingold,* the goddess Freia is ransomed for a pile of gold just tall enough and wide enough to hide her from sight. Estimate the monetary value of this pile. The density of gold is 19.3 g/cm³, and its value is about $10 per gram (although this varies).

1.24 • You are using water to dilute small amounts of chemicals in the laboratory, drop by drop. How many drops of water are in a 1.0-L bottle? (*Hint:* Start by estimating the diameter of a drop of water.)

1.25 • How many pizzas are consumed each academic year by students at your school?

Section 1.7 Vectors and Vector Addition

1.26 •• Hearing rattles from a snake, you make two rapid displacements of magnitude 1.8 m and 2.4 m. In sketches (roughly to scale), show how your two displacements might add up to give a resultant of magnitude (a) 4.2 m; (b) 0.6 m; (c) 3.0 m.

1.27 •• A postal employee drives a delivery truck along the route shown in Fig. E1.27. Determine the magnitude and direction of the resultant displacement by drawing a scale diagram. (See also Exercise 1.34 for a different approach to this same problem.)

Figure **E1.27**

1.28 •• For the vectors $\vec{A}$ and $\vec{B}$ in Fig. E1.28, use a scale drawing to find the magnitude and direction of (a) the vector sum $\vec{A} + \vec{B}$ and (b) the vector difference $\vec{A} - \vec{B}$. Use your answers to find the magnitude and direction of (c) $-\vec{A} - \vec{B}$ and (d) $\vec{B} - \vec{A}$. (See also Exercise 1.35 for a different approach to this problem.)

Figure **E1.28**

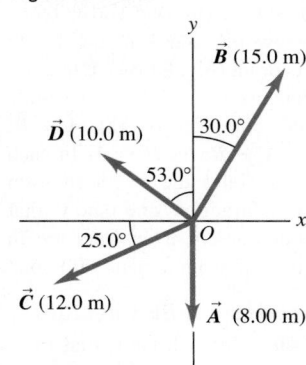

1.29 •• A spelunker is surveying a cave. She follows a passage 180 m straight west, then 210 m in a direction 45° east of south, and then 280 m at 30° east of north. After a fourth unmeasured displacement, she finds herself back where she started. Use a scale drawing to determine the magnitude and direction of the fourth displacement. (See also Problem 1.69 for a different approach to this problem.)

Section 1.8 Components of Vectors

1.30 •• Let the angle θ be the angle that the vector $\vec{A}$ makes with the $+x$-axis, measured counterclockwise from that axis. Find the angle θ for a vector that has the following components: (a) $A_x = 2.00$ m, $A_y = -1.00$ m; (b) $A_x = 2.00$ m, $A_y = 1.00$ m; (c) $A_x = -2.00$ m, $A_y = 1.00$ m; (d) $A_x = -2.00$ m, $A_y = -1.00$ m.

1.31 • Compute the x- and y-components of the vectors $\vec{A}$, $\vec{B}$, $\vec{C}$, and $\vec{D}$ in Fig. E1.28.

1.32 • Vector $\vec{A}$ is in the direction 34.0° clockwise from the $-y$-axis. The x-component of $\vec{A}$ is $A_x = -16.0$ m. (a) What is the y-component of $\vec{A}$? (b) What is the magnitude of $\vec{A}$?

1.33 • Vector $\vec{A}$ has y-component $A_y = +13.0$ m. $\vec{A}$ makes an angle of 32.0° counterclockwise from the $+y$-axis. (a) What is the x-component of $\vec{A}$? (b) What is the magnitude of $\vec{A}$?

1.34 •• A postal employee drives a delivery truck over the route shown in Fig. E1.27. Use the method of components to determine the magnitude and direction of her resultant displacement. In a vector-addition diagram (roughly to scale), show that the resultant displacement found from your diagram is in qualitative agreement with the result you obtained using the method of components.

1.35 • For the vectors $\vec{A}$ and $\vec{B}$ in Fig. E1.28, use the method of components to find the magnitude and direction of (a) the vector sum $\vec{A} + \vec{B}$; (b) the vector sum $\vec{B} + \vec{A}$; (c) the vector difference $\vec{A} - \vec{B}$; (d) the vector difference $\vec{B} - \vec{A}$.

1.36 • Find the magnitude and direction of the vector represented by the following pairs of components: (a) $A_x = -8.60$ cm,

$A_y = 5.20$ cm; (b) $A_x = -9.70$ m, $A_y = -2.45$ m; (c) $A_x = 7.75$ km, $A_y = -2.70$ km.

1.37 •• A disoriented physics professor drives 3.25 km north, then 2.90 km west, and then 1.50 km south. Find the magnitude and direction of the resultant displacement, using the method of components. In a vector-addition diagram (roughly to scale), show that the resultant displacement found from your diagram is in qualitative agreement with the result you obtained using the method of components.

1.38 •• Two ropes in a vertical plane exert equal-magnitude forces on a hanging weight but pull with an angle of 86.0° between them. What pull does each one exert if their resultant pull is 372 N directly upward?

1.39 •• Vector $\vec{A}$ is 2.80 cm long and is 60.0° above the *x*-axis in the first quadrant. Vector $\vec{B}$ is 1.90 cm long and is 60.0° below the *x*-axis in the fourth quadrant (Fig. E1.39). Use components to find the magnitude and direction of (a) $\vec{A} + \vec{B}$; (b) $\vec{A} - \vec{B}$; (c) $\vec{B} - \vec{A}$. In each case, sketch the vector addition or subtraction and show that your numerical answers are in qualitative agreement with your sketch.

Figure **E1.39**

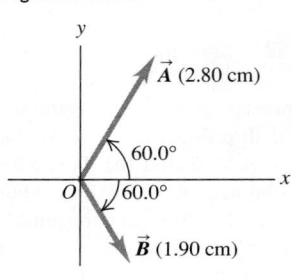

Section 1.9 Unit Vectors

1.40 • In each case, find the *x*- and *y*-components of vector $\vec{A}$: (a) $\vec{A} = 5.0\hat{\imath} - 6.3\hat{\jmath}$; (b) $\vec{A} = 11.2\hat{\jmath} - 9.91\hat{\imath}$; (c) $\vec{A} = -15.0\hat{\imath} + 22.4\hat{\jmath}$; (d) $\vec{A} = 5.0\vec{B}$, where $\vec{B} = 4\hat{\imath} - 6\hat{\jmath}$.

1.41 •• Write each vector in Fig. E1.28 in terms of the unit vectors $\hat{\imath}$ and $\hat{\jmath}$.

1.42 •• Given two vectors $\vec{A} = 4.00\hat{\imath} + 7.00\hat{\jmath}$ and $\vec{B} = 5.00\hat{\imath} - 2.00\hat{\jmath}$, (a) find the magnitude of each vector; (b) write an expression for the vector difference $\vec{A} - \vec{B}$ using unit vectors; (c) find the magnitude and direction of the vector difference $\vec{A} - \vec{B}$. (d) In a vector diagram show $\vec{A}$, $\vec{B}$, and $\vec{A} - \vec{B}$, and also show that your diagram agrees qualitatively with your answer in part (c).

1.43 •• (a) Write each vector in Fig. E1.43 in terms of the unit vectors $\hat{\imath}$ and $\hat{\jmath}$. (b) Use unit vectors to express the vector $\vec{C}$, where $\vec{C} = 3.00\vec{A} - 4.00\vec{B}$. (c) Find the magnitude and direction of $\vec{C}$.

Figure **E1.43**

1.44 •• (a) Is the vector $(\hat{\imath} + \hat{\jmath} + \hat{k})$ a unit vector? Justify your answer. (b) Can a unit vector have any components with magnitude greater than unity? Can it have any negative components? In each case justify your answer. (c) If $\vec{A} = a(3.0\hat{\imath} + 4.0\hat{\jmath})$, where *a* is a constant, determine the value of *a* that makes $\vec{A}$ a unit vector.

Section 1.10 Products of Vectors

1.45 • For the vectors $\vec{A}$, $\vec{B}$, and $\vec{C}$ in Fig. E1.28, find the scalar products (a) $\vec{A} \cdot \vec{B}$; (b) $\vec{B} \cdot \vec{C}$; (c) $\vec{A} \cdot \vec{C}$.

1.46 •• (a) Find the scalar product of the two vectors $\vec{A}$ and $\vec{B}$ given in Exercise 1.42. (b) Find the angle between these two vectors.

1.47 •• Find the angle between each of the following pairs of vectors:

(a) $\vec{A} = -2.00\hat{\imath} + 6.00\hat{\jmath}$ and $\vec{B} = 2.00\hat{\imath} - 3.00\hat{\jmath}$
(b) $\vec{A} = 3.00\hat{\imath} + 5.00\hat{\jmath}$ and $\vec{B} = 10.00\hat{\imath} + 6.00\hat{\jmath}$
(c) $\vec{A} = -4.00\hat{\imath} + 2.00\hat{\jmath}$ and $\vec{B} = 7.00\hat{\imath} + 14.00\hat{\jmath}$

1.48 •• Find the vector product $\vec{A} \times \vec{B}$ (expressed in unit vectors) of the two vectors given in Exercise 1.42. What is the magnitude of the vector product?

1.49 • For the vectors $\vec{A}$ and $\vec{D}$ in Fig. E1.28, (a) find the magnitude and direction of the vector product $\vec{A} \times \vec{D}$; (b) find the magnitude and direction of $\vec{D} \times \vec{A}$.

1.50 • For the two vectors in Fig. E1.39, (a) find the magnitude and direction of the vector product $\vec{A} \times \vec{B}$; (b) find the magnitude and direction of $\vec{B} \times \vec{A}$.

1.51 • For the two vectors $\vec{A}$ and $\vec{B}$ in Fig. E1.43, (a) find the scalar product $\vec{A} \cdot \vec{B}$; (b) find the magnitude and direction of the vector product $\vec{A} \times \vec{B}$.

1.52 • The vector $\vec{A}$ is 3.50 cm long and is directed into this page. Vector $\vec{B}$ points from the lower right corner of this page to the upper left corner of this page. Define an appropriate right-handed coordinate system, and find the three components of the vector product $\vec{A} \times \vec{B}$, measured in cm². In a diagram, show your coordinate system and the vectors $\vec{A}$, $\vec{B}$, and $\vec{A} \times \vec{B}$.

1.53 • Given two vectors $\vec{A} = -2.00\hat{\imath} + 3.00\hat{\jmath} + 4.00\hat{k}$ and $\vec{B} = 3.00\hat{\imath} + 1.00\hat{\jmath} - 3.00\hat{k}$, do the following. (a) Find the magnitude of each vector. (b) Write an expression for the vector difference $\vec{A} - \vec{B}$ using unit vectors. (c) Find the magnitude of the vector difference $\vec{A} - \vec{B}$. Is this the same as the magnitude of $\vec{B} - \vec{A}$? Explain.

PROBLEMS

1.54 • An acre, a unit of land measurement still in wide use, has a length of one furlong ($\frac{1}{8}$ mi) and a width one-tenth of its length. (a) How many acres are in a square mile? (b) How many square feet are in an acre? See Appendix E. (c) An acre-foot is the volume of water that would cover 1 acre of flat land to a depth of 1 foot. How many gallons are in 1 acre-foot?

1.55 •• **An Earthlike Planet.** In January 2006 astronomers reported the discovery of a planet comparable in size to the earth orbiting another star and having a mass about 5.5 times the earth's mass. It is believed to consist of a mixture of rock and ice, similar to Neptune. If this planet has the same density as Neptune (1.76 g/cm³), what is its radius expressed (a) in kilometers and (b) as a multiple of earth's radius? Consult Appendix F for astronomical data.

1.56 •• **The Hydrogen Maser.** You can use the radio waves generated by a hydrogen maser as a standard of frequency. The frequency of these waves is 1,420,405,751.786 hertz. (A hertz is another name for one cycle per second.) A clock controlled by a hydrogen maser is off by only 1 s in 100,000 years. For the following questions, use only three significant figures. (The large number of significant figures given for the frequency simply illustrates the remarkable accuracy to which it has been measured.) (a) What is the time for one cycle of the radio wave? (b) How many cycles occur in 1 h? (c) How many cycles would have occurred during the age of the earth, which is estimated to be 4.6×10^9 years? (d) By how many seconds would a hydrogen maser clock be off after a time interval equal to the age of the earth?

1.57 • **BIO** **Breathing Oxygen.** The density of air under standard laboratory conditions is 1.29 kg/m³, and about 20% of that air consists of oxygen. Typically, people breathe about $\frac{1}{2}$ L of air per breath. (a) How many grams of oxygen does a person breathe in a day? (b) If this air is stored uncompressed in a cubical tank, how long is each side of the tank?

1.58 ••• A rectangular piece of aluminum is 7.60 ± 0.01 cm long and 1.90 ± 0.01 cm wide. (a) Find the area of the rectangle and the uncertainty in the area. (b) Verify that the fractional uncertainty in the area is equal to the sum of the fractional uncertainties in the length and in the width. (This is a general result; see Challenge Problem 1.98.)

1.59 ••• As you eat your way through a bag of chocolate chip cookies, you observe that each cookie is a circular disk with a diameter of 8.50 ± 0.02 cm and a thickness of 0.050 ± 0.005 cm. (a) Find the average volume of a cookie and the uncertainty in the volume. (b) Find the ratio of the diameter to the thickness and the uncertainty in this ratio.

1.60 • **BIO** Biological tissues are typically made up of 98% water. Given that the density of water is 1.0×10^3 kg/m³, estimate the mass of (a) the heart of an adult human; (b) a cell with a diameter of 0.5 μm; (c) a honey bee.

1.61 • **BIO** Estimate the number of atoms in your body. (*Hint:* Based on what you know about biology and chemistry, what are the most common types of atom in your body? What is the mass of each type of atom? Appendix D gives the atomic masses for different elements, measured in atomic mass units; you can find the value of an atomic mass unit, or 1 u, in Appendix E.)

1.62 ••• How many dollar bills would you have to stack to reach the moon? Would that be cheaper than building and launching a spacecraft? (*Hint:* Start by folding a dollar bill to see how many thicknesses make 1.0 mm.)

1.63 ••• How much would it cost to paper the entire United States (including Alaska and Hawaii) with dollar bills? What would be the cost to each person in the United States?

1.64 • **Stars in the Universe.** Astronomers frequently say that there are more stars in the universe than there are grains of sand on all the beaches on the earth. (a) Given that a typical grain of sand is about 0.2 mm in diameter, estimate the number of grains of sand on all the earth's beaches, and hence the approximate number of stars in the universe. It would be helpful to consult an atlas and do some measuring. (b) Given that a typical galaxy contains about 100 billion stars and there are more than 100 billion galaxies in the known universe, estimate the number of stars in the universe and compare this number with your result from part (a).

1.65 ••• Two workers pull horizontally on a heavy box, but one pulls twice as hard as the other. The larger pull is directed at 25.0° west of north, and the resultant of these two pulls is 460.0 N directly northward. Use vector components to find the magnitude of each of these pulls and the direction of the smaller pull.

1.66 •• Three horizontal ropes pull on a large stone stuck in the ground, producing the vector forces $\vec{A}$, $\vec{B}$, and $\vec{C}$ shown in Fig. P1.66. Find the magnitude and direction of a fourth force on the stone that will make the vector sum of the four forces zero.

1.67 •• You are to program a robotic arm on an assembly line to move in the *xy*-plane. Its first displacement is $\vec{A}$; its second displacement is $\vec{B}$, of magnitude 6.40 cm and direction 63.0° measured in the sense from the +*x*-axis toward the −*y*-axis. The resultant $\vec{C} = \vec{A} + \vec{B}$ of the two displacements should also have a magnitude of 6.40 cm, but a direction 22.0° measured in the sense

Figure **P1.66**

from the +*x*-axis toward the +*y*-axis. (a) Draw the vector-addition diagram for these vectors, roughly to scale. (b) Find the components of $\vec{A}$. (c) Find the magnitude and direction of $\vec{A}$.

1.68 ••• **Emergency Landing.** A plane leaves the airport in Galisteo and flies 170 km at 68° east of north and then changes direction to fly 230 km at 48° south of east, after which it makes an immediate emergency landing in a pasture. When the airport sends out a rescue crew, in which direction and how far should this crew fly to go directly to this plane?

1.69 ••• As noted in Exercise 1.29, a spelunker is surveying a cave. She follows a passage 180 m straight west, then 210 m in a direction 45° east of south, and then 280 m at 30° east of north. After a fourth unmeasured displacement she finds herself back where she started. Use the method of components to determine the magnitude and direction of the fourth displacement. Draw the vector-addition diagram and show that it is in qualitative agreement with your numerical solution.

1.70 •• (a) Find the magnitude and direction of the vector $\vec{R}$ that is the sum of the three vectors $\vec{A}$, $\vec{B}$, and $\vec{C}$ in Fig. E1.28. In a diagram, show how $\vec{R}$ is formed from these three vectors. (b) Find the magnitude and direction of the vector $\vec{S} = \vec{C} - \vec{A} - \vec{B}$. In a diagram, show how $\vec{S}$ is formed from these three vectors.

1.71 •• A rocket fires two engines simultaneously. One produces a thrust of 480 N directly forward, while the other gives a 513-N thrust at 32.4° above the forward direction. Find the magnitude and direction (relative to the forward direction) of the resultant force that these engines exert on the rocket.

1.72 •• A sailor in a small sailboat encounters shifting winds. She sails 2.00 km east, then 3.50 km southeast, and then an additional distance in an unknown direction. Her final position is 5.80 km directly east of the starting point (Fig. P1.72). Find the magnitude and direction of the third leg of the journey. Draw the vector-addition diagram and show that it is in qualitative agreement with your numerical solution.

Figure **P1.72**

1.73 ••• **BIO Dislocated Shoulder.** A patient with a dislocated shoulder is put into a traction apparatus as shown in Fig. P1.73. The pulls $\vec{A}$ and $\vec{B}$ have equal magnitudes and must combine to produce an outward traction force of 5.60 N on the patient's arm. How large should these pulls be?

Figure **P1.73**

1.74 ••• On a training flight, a student pilot flies from Lincoln, Nebraska, to Clarinda, Iowa, then to St. Joseph, Missouri, and then to Manhattan, Kansas (Fig. P1.74). The directions are shown relative to north: 0° is north, 90° is east, 180° is south, and 270° is west. Use the method of components to find (a) the distance she has to fly from Manhattan to get back to Lincoln, and (b) the direction (relative to north) she must fly to get there. Illustrate your solutions with a vector diagram.

Figure **P1.74**

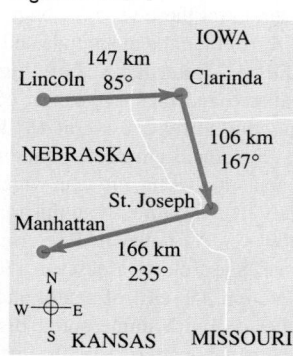

1.75 •• **Equilibrium.** We say an object is in *equilibrium* if all the forces on it balance (add up to zero). Figure P1.75 shows a beam weighing 124 N that is supported in equilibrium by a 100.0-N pull and a force $\vec{F}$ at the floor. The third force on the beam is the 124-N weight that acts vertically downward. (a) Use vector components to find the magnitude and direction of $\vec{F}$. (b) Check the reasonableness of your answer in part (a) by doing a graphical solution approximately to scale.

Figure **P1.75**

100.0-N pull

$\vec{F}$ 30.0°

40.0°

1.76 ••• **Getting Back.** An explorer in the dense jungles of equatorial Africa leaves his hut. He takes 40 steps northeast, then 80 steps 60° north of west, then 50 steps due south. Assume his steps all have equal length. (a) Sketch, roughly to scale, the three vectors and their resultant. (b) Save the explorer from becoming hopelessly lost in the jungle by giving him the displacement, calculated using the method of components, that will return him to his hut.

1.77 ••• A graphic artist is creating a new logo for her company's website. In the graphics program she is using, each pixel in an image file has coordinates (x, y), where the origin $(0, 0)$ is at the upper left corner of the image, the $+x$-axis points to the right, and the $+y$-axis points down. Distances are measured in pixels. (a) The artist draws a line from the pixel location $(10, 20)$ to the location $(210, 200)$. She wishes to draw a second line that starts at $(10, 20)$, is 250 pixels long, and is at an angle of 30° measured clockwise from the first line. At which pixel location should this second line end? Give your answer to the nearest pixel. (b) The artist now draws an arrow that connects the lower right end of the first line to the lower right end of the second line. Find the length and direction of this arrow. Draw a diagram showing all three lines.

1.78 ••• A ship leaves the island of Guam and sails 285 km at 40.0° north of west. In which direction must it now head and how far must it sail so that its resultant displacement will be 115 km directly east of Guam?

1.79 •• **BIO** **Bones and Muscles.** A patient in therapy has a forearm that weighs 20.5 N and that lifts a 112.0-N weight. These two forces have direction vertically downward. The only other significant forces on his forearm come from the biceps muscle (which acts perpendicularly to the forearm) and the force at the elbow. If the biceps produces a pull of 232 N when the forearm is raised 43° above the horizontal, find the magnitude and direction of the force that the elbow exerts on the forearm. (The sum of the elbow force and the biceps force must balance the weight of the arm and the weight it is carrying, so their vector sum must be 132.5 N, upward.)

1.80 ••• You are hungry and decide to go to your favorite neighborhood fast-food restaurant. You leave your apartment and take the elevator 10 flights down (each flight is 3.0 m) and then go 15 m south to the apartment exit. You then proceed 0.2 km east, turn north, and go 0.1 km to the entrance of the restaurant. (a) Determine the displacement from your apartment to the restaurant. Use unit vector notation for your answer, being sure to make clear your choice of coordinates. (b) How far did you travel along the path you took from your apartment to the restaurant, and what is the magnitude of the displacement you calculated in part (a)?

1.81 •• While following a treasure map, you start at an old oak tree. You first walk 825 m directly south, then turn and walk 1.25 km at 30.0° west of north, and finally walk 1.00 km at 40.0° north of east, where you find the treasure: a biography of Isaac Newton! (a) To return to the old oak tree, in what direction should you head and how far will you walk? Use components to solve this problem. (b) To see whether your calculation in part (a) is reasonable, check it with a graphical solution drawn roughly to scale.

1.82 •• A fence post is 52.0 m from where you are standing, in a direction 37.0° north of east. A second fence post is due south from you. What is the distance of the second post from you, if the distance between the two posts is 80.0 m?

1.83 •• A dog in an open field runs 12.0 m east and then 28.0 m in a direction 50.0° west of north. In what direction and how far must the dog then run to end up 10.0 m south of her original starting point?

1.84 ••• Ricardo and Jane are standing under a tree in the middle of a pasture. An argument ensues, and they walk away in different directions. Ricardo walks 26.0 m in a direction 60.0° west of north. Jane walks 16.0 m in a direction 30.0° south of west. They then stop and turn to face each other. (a) What is the distance between them? (b) In what direction should Ricardo walk to go directly toward Jane?

1.85 ••• John, Paul, and George are standing in a strawberry field. Paul is 14.0 m due west of John. George is 36.0 m from Paul, in a direction 37.0° south of east from Paul's location. How far is George from John? What is the direction of George's location from that of John?

1.86 ••• You are camping with two friends, Joe and Karl. Since all three of you like your privacy, you don't pitch your tents close together. Joe's tent is 21.0 m from yours, in the direction 23.0° south of east. Karl's tent is 32.0 m from yours, in the direction 37.0° north of east. What is the distance between Karl's tent and Joe's tent?

1.87 •• Vectors $\vec{A}$ and $\vec{B}$ have scalar product -6.00 and their vector product has magnitude $+9.00$. What is the angle between these two vectors?

1.88 •• **Bond Angle in Methane.** In the methane molecule, CH_4, each hydrogen atom is at a corner of a regular tetrahedron with the carbon atom at the center. In coordinates where one of the C–H bonds is in the direction of $\hat{i} + \hat{j} + \hat{k}$, an adjacent C–H bond is in the $\hat{i} - \hat{j} - \hat{k}$ direction. Calculate the angle between these two bonds.

1.89 •• Vector $\vec{A}$ has magnitude 12.0 m and vector $\vec{B}$ has magnitude 16.0 m. The scalar product $\vec{A} \cdot \vec{B}$ is 90.0 m². What is the magnitude of the vector product between these two vectors?

1.90 •• When two vectors $\vec{A}$ and $\vec{B}$ are drawn from a common point, the angle between them is ϕ. (a) Using vector techniques, show that the magnitude of their vector sum is given by

$$\sqrt{A^2 + B^2 + 2AB \cos \phi}$$

(b) If $\vec{A}$ and $\vec{B}$ have the same magnitude, for which value of ϕ will their vector sum have the same magnitude as $\vec{A}$ or $\vec{B}$?

1.91 •• A cube is placed so that one corner is at the origin and three edges are along the *x*-, *y*-, and *z*-axes of a coordinate system (Fig. P1.91). Use vectors to compute (a) the angle between the edge along the *z*-axis (line *ab*) and the diagonal from the origin to the opposite corner (line *ad*), and (b) the angle between line *ac* (the diagonal of a face) and line *ad*.

Figure **P1.91**

1.92 •• Vector $\vec{A}$ has magnitude 6.00 m and vector $\vec{B}$ has magnitude 3.00 m. The vector product between these two vectors has magnitude 12.0 m². What are the two possible values for the scalar product of these two vectors? For each value of $\vec{A} \cdot \vec{B}$, draw a sketch that shows $\vec{A}$ and $\vec{B}$ and explain why the vector products in the two sketches are the same but the scalar products differ.

1.93 •• The scalar product of vectors $\vec{A}$ and $\vec{B}$ is +48.0 m². Vector $\vec{A}$ has magnitude 9.00 m and direction 28.0° west of south. If vector $\vec{B}$ has direction 39.0° south of east, what is the magnitude of $\vec{B}$?

1.94 ••• Obtain a *unit vector* perpendicular to the two vectors given in Exercise 1.53.

1.95 •• You are given vectors $\vec{A} = 5.0\hat{\imath} - 6.5\hat{\jmath}$ and $\vec{B} = -3.5\hat{\imath} + 7.0\hat{\jmath}$. A third vector $\vec{C}$ lies in the *xy*-plane. Vector $\vec{C}$ is perpendicular to vector $\vec{A}$, and the scalar product of $\vec{C}$ with $\vec{B}$ is 15.0. From this information, find the components of vector $\vec{C}$.

1.96 •• Two vectors $\vec{A}$ and $\vec{B}$ have magnitudes $A = 3.00$ and $B = 3.00$. Their vector product is $\vec{A} \times \vec{B} = -5.00\hat{k} + 2.00\hat{\imath}$. What is the angle between $\vec{A}$ and $\vec{B}$?

1.97 •• Later in our study of physics we will encounter quantities represented by $(\vec{A} \times \vec{B}) \cdot \vec{C}$. (a) Prove that for any three vectors $\vec{A}$, $\vec{B}$, and $\vec{C}$, $\vec{A} \cdot (\vec{B} \times \vec{C}) = (\vec{A} \times \vec{B}) \cdot \vec{C}$. (b) Calculate $(\vec{A} \times \vec{B}) \cdot \vec{C}$ for the three vectors $\vec{A}$ with magnitude $A = 5.00$ and angle $\theta_A = 26.0°$ measured in the sense from the +*x*-axis toward the +*y*-axis, $\vec{B}$ with $B = 4.00$ and $\theta_B = 63.0°$, and $\vec{C}$ with magnitude 6.00 and in the +*z*-direction. Vectors $\vec{A}$ and $\vec{B}$ are in the *xy*-plane.

CHALLENGE PROBLEMS

1.98 ••• The length of a rectangle is given as $L \pm l$ and its width as $W \pm w$. (a) Show that the uncertainty in its area A is $a = Lw + lW$. Assume that the uncertainties l and w are small, so that the product lw is very small and you can ignore it. (b) Show that the fractional uncertainty in the area is equal to the sum of the fractional uncertainty in length and the fractional uncertainty in width. (c) A rectangular solid has dimensions $L \pm l$, $W \pm w$, and $H \pm h$. Find the fractional uncertainty in the volume, and show that it equals the sum of the fractional uncertainties in the length, width, and height.

1.99 ••• **Completed Pass.** At Enormous State University (ESU), the football team records its plays using vector displacements, with the origin taken to be the position of the ball before the play starts. In a certain pass play, the receiver starts at $+1.0\hat{\imath} - 5.0\hat{\jmath}$, where the units are yards, $\hat{\imath}$ is to the right, and

$\hat{\jmath}$ is downfield. Subsequent displacements of the receiver are $+9.0\hat{\imath}$ (in motion before the snap), $+11.0\hat{\jmath}$ (breaks downfield), $-6.0\hat{\imath} + 4.0\hat{\jmath}$ (zigs), and $+12.0\hat{\imath} + 18.0\hat{\jmath}$ (zags). Meanwhile, the quarterback has dropped straight back to a position $-7.0\hat{\jmath}$. How far and in which direction must the quarterback throw the ball? (Like the coach, you will be well advised to diagram the situation before solving it numerically.)

1.100 ••• **Navigating in the Solar System.** The *Mars Polar Lander* spacecraft was launched on January 3, 1999. On December 3, 1999, the day *Mars Polar Lander* touched down on the Martian surface, the positions of the earth and Mars were given by these coordinates:

	x	*y*	*z*
Earth	0.3182 AU	0.9329 AU	0.0000 AU
Mars	1.3087 AU	−0.4423AU	−0.0414 AU

In these coordinates, the sun is at the origin and the plane of the earth's orbit is the *xy*-plane. The earth passes through the +*x*-axis once a year on the autumnal equinox, the first day of autumn in the northern hemisphere (on or about September 22). One AU, or *astronomical unit,* is equal to 1.496×10^8 km, the average distance from the earth to the sun. (a) In a diagram, show the positions of the sun, the earth, and Mars on December 3, 1999. (b) Find the following distances in AU on December 3, 1999: (i) from the sun to the earth; (ii) from the sun to Mars; (iii) from the earth to Mars. (c) As seen from the earth, what was the angle between the direction to the sun and the direction to Mars on December 3, 1999? (d) Explain whether Mars was visible from your location at midnight on December 3, 1999. (When it is midnight at your location, the sun is on the opposite side of the earth from you.)

1.101 ••• **Navigating in the Big Dipper.** All the stars of the Big Dipper (part of the constellation Ursa Major) may appear to be the same distance from the earth, but in fact they are very far from each other. Figure P1.101 shows the distances from the earth to each of these stars. The distances are given in light-years (ly), the distance that light travels in one year. One light-year equals 9.461×10^{15} m. (a) Alkaid and Merak are 25.6° apart in the earth's sky. In a diagram, show the relative positions of Alkaid, Merak, and our sun. Find the distance in light-years from Alkaid to Merak. (b) To an inhabitant of a planet orbiting Merak, how many degrees apart in the sky would Alkaid and our sun be?

Figure **P1.101**

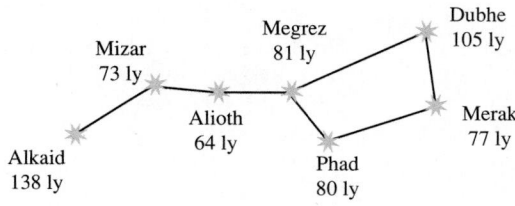

1.102 ••• The vector $\vec{r} = x\hat{\imath} + y\hat{\jmath} + z\hat{k}$, called the *position vector,* points from the origin (0, 0, 0) to an arbitrary point in space with coordinates (x, y, z). Use what you know about vectors to prove the following: All points (x, y, z) that satisfy the equation $Ax + By + Cz = 0$, where A, B, and C are constants, lie in a plane that passes through the origin and that is perpendicular to the vector $A\hat{\imath} + B\hat{\jmath} + C\hat{k}$. Sketch this vector and the plane.

Answers

Chapter Opening Question ?

Take the $+x$-axis to point east and the $+y$-axis to point north. Then what we are trying to find is the y-component of the velocity vector, which has magnitude $v = 20$ km/h and is at an angle $\theta = 53°$ measured from the $+x$-axis toward the $+y$-axis. From Eqs. (1.6) we have $v_y = v \sin \theta = (20 \text{ km/h}) \sin 53° = 16$ km/h. So the thunderstorm moves 16 km north in 1 h.

Test Your Understanding Questions

1.5 Answer: (ii) Density $= (1.80 \text{ kg})/(6.0 \times 10^{-4} \text{ m}^3) = 3.0 \times 10^3$ kg/m^3. When we multiply or divide, the number with the fewest significant figures controls the number of significant figures in the result.

1.6 The answer depends on how many students are enrolled at your campus.

1.7 Answers: (ii), (iii), and (iv) The vector $-\vec{T}$ has the same magnitude as the vector $\vec{T}$, so $\vec{S} - \vec{T} = \vec{S} + (-\vec{T})$ is the *sum* of one vector of magnitude 3 m and one of magnitude 4 m. This sum has magnitude 7 m if $\vec{S}$ and $-\vec{T}$ are parallel and magnitude 1 m if $\vec{S}$ and $-\vec{T}$ are antiparallel. The magnitude of $\vec{S} - \vec{T}$ is 5 m if $\vec{S}$ and $-\vec{T}$ are perpendicular, so that the vectors $\vec{S}$, $\vec{T}$, and $\vec{S} - \vec{T}$ form a 3–4–5 right triangle. Answer (i) is impossible because the magnitude of the sum of two vectors cannot be greater than the sum of the magnitudes; answer (v) is impossible because the sum of two vectors can be zero only if the two vectors are antiparallel and have the same magnitude; and answer (vi) is impossible because the magnitude of a vector cannot be negative.

1.8 Answers: (a) yes, (b) no Vectors $\vec{A}$ and $\vec{B}$ can have the same magnitude but different components if they point in different directions. If they have the same components, however, they are the same vector ($\vec{A} = \vec{B}$) and so must have the same magnitude.

1.9 Answer: all have the same magnitude The four vectors $\vec{A}$, $\vec{B}$, $\vec{C}$, and $\vec{D}$ all point in different directions, but all have the same magnitude:

$$A = B = C = D = \sqrt{(\pm 3 \text{ m})^2 + (\pm 5 \text{ m})^2 + (\pm 2 \text{ m})^2}$$
$$= \sqrt{9 \text{ m}^2 + 25 \text{ m}^2 + 4 \text{ m}^2} = \sqrt{38 \text{ m}^2} = 6.2 \text{ m}$$

1.10 Answers: (a) $\phi = 90°$, **(b)** $\phi = 0°$ or $\phi = 180°$, **(c)** $\phi = 0°$, **(d)** $\phi = 180°$, **(e)** $\phi = 90°$ (a) The scalar product is zero only if $\vec{A}$ and $\vec{B}$ are perpendicular. (b) The vector product is zero only if $\vec{A}$ and $\vec{B}$ are either parallel or antiparallel. (c) The scalar product is equal to the product of the magnitudes ($\vec{A} \cdot \vec{B} = AB$) only if $\vec{A}$ and $\vec{B}$ are parallel. (d) The scalar product is equal to the negative of the product of the magnitudes ($\vec{A} \cdot \vec{B} = -AB$) only if $\vec{A}$ and $\vec{B}$ are antiparallel. (e) The magnitude of the vector product is equal to the product of the magnitudes $[(\text{magnitude of } \vec{A} \times \vec{B}) = AB]$ only if $\vec{A}$ and $\vec{B}$ are perpendicular.

Bridging Problem

Answers: (a) 5.2×10^2 N
(b) 4.5×10^2 N $\cdot$ m

MOTION ALONG A STRAIGHT LINE

? A bungee jumper speeds up during the first part of his fall, then slows to a halt as the bungee cord stretches and becomes taut. Is it accurate to say that the jumper is *accelerating* as he slows during the final part of his fall?

LEARNING GOALS

By studying this chapter, you will learn:

• How to describe straight-line motion in terms of average velocity, instantaneous velocity, average acceleration, and instantaneous acceleration.

• How to interpret graphs of position versus time, velocity versus time, and acceleration versus time for straight-line motion.

• How to solve problems involving straight-line motion with constant acceleration, including free-fall problems.

• How to analyze straight-line motion when the acceleration is not constant.

What distance must an airliner travel down a runway before reaching takeoff speed? When you throw a baseball straight up in the air, how high does it go? When a glass slips from your hand, how much time do you have to catch it before it hits the floor? These are the kinds of questions you will learn to answer in this chapter. We are beginning our study of physics with *mechanics,* the study of the relationships among force, matter, and motion. In this chapter and the next we will study *kinematics,* the part of mechanics that enables us to describe motion. Later we will study *dynamics,* which relates motion to its causes.

In this chapter we concentrate on the simplest kind of motion: a body moving along a straight line. To describe this motion, we introduce the physical quantities *velocity* and *acceleration.* In physics these quantities have definitions that are more precise and slightly different from the ones used in everyday language. Both velocity and acceleration are *vectors:* As you learned in Chapter 1, this means that they have both magnitude and direction. Our concern in this chapter is with motion along a straight line only, so we won't need the full mathematics of vectors just yet. But using vectors will be essential in Chapter 3 when we consider motion in two or three dimensions.

We'll develop simple equations to describe straight-line motion in the important special case when the acceleration is constant. An example is the motion of a freely falling body. We'll also consider situations in which the acceleration varies during the motion; in this case, it's necessary to use integration to describe the motion. (If you haven't studied integration yet, Section 2.6 is optional.)

2.1 Displacement, Time, and Average Velocity

Suppose a drag racer drives her AA-fuel dragster along a straight track (Fig. 2.1). To study the dragster's motion, we need a coordinate system. We choose the *x*-axis to lie along the dragster's straight-line path, with the origin *O* at the starting line. We also choose a point on the dragster, such as its front end, and represent the entire dragster by that point. Hence we treat the dragster as a **particle.**

A useful way to describe the motion of the particle that represents the dragster is in terms of the change in the particle's coordinate *x* over a time interval. Suppose that 1.0 s after the start the front of the dragster is at point P_1, 19 m from the origin, and 4.0 s after the start it is at point P_2, 277 m from the origin. The *displacement* of the particle is a vector that points from P_1 to P_2 (see Section 1.7). Figure 2.1 shows that this vector points along the *x*-axis. The *x*-component of the displacement is the change in the value of *x*, $(277 \text{ m} - 19 \text{ m}) = 258 \text{ m}$, that took place during the time interval of $(4.0 \text{ s} - 1.0 \text{ s}) = 3.0 \text{ s}$. We define the dragster's **average velocity** during this time interval as a *vector quantity whose x-component is the change in x divided by the time interval:* $(258 \text{ m})/(3.0 \text{ s}) = 86 \text{ m/s}$.

In general, the average velocity depends on the particular time interval chosen. For a 3.0-s time interval *before* the start of the race, the average velocity would be zero because the dragster would be at rest at the starting line and would have zero displacement.

Let's generalize the concept of average velocity. At time t_1 the dragster is at point P_1, with coordinate x_1, and at time t_2 it is at point P_2, with coordinate x_2. The displacement of the dragster during the time interval from t_1 to t_2 is the vector from P_1 to P_2. The *x*-component of the displacement, denoted Δx, is the change in the coordinate *x*:

$$\Delta x = x_2 - x_1 \tag{2.1}$$

The dragster moves along the *x*-axis only, so the *y*- and *z*-components of the displacement are equal to zero.

CAUTION **The meaning of Δx** Note that Δx is *not* the product of Δ and *x*; it is a single symbol that means "the change in the quantity *x*." We always use the Greek capital letter Δ (delta) to represent a *change* in a quantity, equal to the *final* value of the quantity minus the *initial* value—never the reverse. Likewise, the time interval from t_1 to t_2 is Δt, the change in the quantity *t*: $\Delta t = t_2 - t_1$ (final time minus initial time). ❚

The *x*-component of average velocity, or **average x-velocity,** is the *x*-component of displacement, Δx, divided by the time interval Δt during which

2.1 Positions of a dragster at two times during its run.

x is positive to the right of the origin (O), negative to the left of it.

When the dragster moves in the +x-direction, the displacement Δx is positive and so is the average x-velocity:

$$v_{\text{av-}x} = \frac{\Delta x}{\Delta t} = \frac{258 \text{ m}}{3.0 \text{ s}} = 86 \text{ m/s}$$

2.2 Positions of an official's truck at two times during its motion. The points P_1 and P_2 now indicate the positions of the truck, and so are the reverse of Fig. 2.1.

the displacement occurs. We use the symbol $v_{av\text{-}x}$ for average x-velocity (the subscript "av" signifies average value and the subscript x indicates that this is the x-component):

$$v_{av\text{-}x} = \frac{x_2 - x_1}{t_2 - t_1} = \frac{\Delta x}{\Delta t} \qquad \text{(average } x\text{-velocity, straight-line motion)} \quad (2.2)$$

As an example, for the dragster $x_1 = 19$ m, $x_2 = 277$ m, $t_1 = 1.0$ s, and $t_2 = 4.0$ s, so Eq. (2.2) gives

$$v_{av\text{-}x} = \frac{277 \text{ m} - 19 \text{ m}}{4.0 \text{ s} - 1.0 \text{ s}} = \frac{258 \text{ m}}{3.0 \text{ s}} = 86 \text{ m/s}$$

The average x-velocity of the dragster is positive. This means that during the time interval, the coordinate x increased and the dragster moved in the positive x-direction (to the right in Fig. 2.1).

If a particle moves in the *negative* x-direction during a time interval, its average velocity for that time interval is negative. For example, suppose an official's truck moves to the left along the track (Fig. 2.2). The truck is at $x_1 = 277$ m at $t_1 = 16.0$ s and is at $x_2 = 19$ m at $t_2 = 25.0$ s. Then $\Delta x = (19 \text{ m} - 277 \text{ m}) = -258$ m and $\Delta t = (25.0 \text{ s} - 16.0 \text{ s}) = 9.0$ s. The x-component of average velocity is $v_{av\text{-}x} = \Delta x / \Delta t = (-258 \text{ m})/(9.0 \text{ s}) = -29$ m/s. Table 2.1 lists some simple rules for deciding whether the x-velocity is positive or negative.

CAUTION **Choice of the positive x-direction** You might be tempted to conclude that positive average x-velocity must mean motion to the right, as in Fig. 2.1, and that negative average x-velocity must mean motion to the left, as in Fig. 2.2. But that's correct *only* if the positive x-direction is to the right, as we chose it to be in Figs. 2.1 and 2.2. Had we chosen the positive x-direction to be to the left, with the origin at the finish line, the dragster would have negative average x-velocity and the official's truck would have positive average x-velocity. In most problems the direction of the coordinate axis will be yours to choose. Once you've made your choice, you *must* take it into account when interpreting the signs of $v_{av\text{-}x}$ and other quantities that describe motion!

With straight-line motion we sometimes call Δx simply the displacement and $v_{av\text{-}x}$ simply the average velocity. But be sure to remember that these are really the x-components of vector quantities that, in this special case, have *only* x-components. In Chapter 3, displacement, velocity, and acceleration vectors will have two or three nonzero components.

Figure 2.3 is a graph of the dragster's position as a function of time—that is, an **x-t graph.** The curve in the figure *does not* represent the dragster's path in space; as Fig. 2.1 shows, the path is a straight line. Rather, the graph is a pictorial way to represent how the dragster's position changes with time. The points p_1 and p_2 on the graph correspond to the points P_1 and P_2 along the dragster's path. Line p_1p_2 is the hypotenuse of a right triangle with vertical side $\Delta x = x_2 - x_1$

Table 2.1 Rules for the Sign of x-Velocity

If the x-coordinate is:	. . . the x-velocity is:
Positive & increasing (getting more positive)	Positive: Particle is moving in $+x$-direction
Positive & decreasing (getting less positive)	Negative: Particle is moving in $-x$-direction
Negative & increasing (getting less negative)	Positive: Particle is moving in $+x$-direction
Negative & decreasing (getting more negative)	Negative: Particle is moving in $-x$-direction

Note: These rules apply to both the average x-velocity $v_{av\text{-}x}$ and the instantaneous x-velocity v_x (to be discussed in Section 2.2).

2.3 The position of a dragster as a function of time.

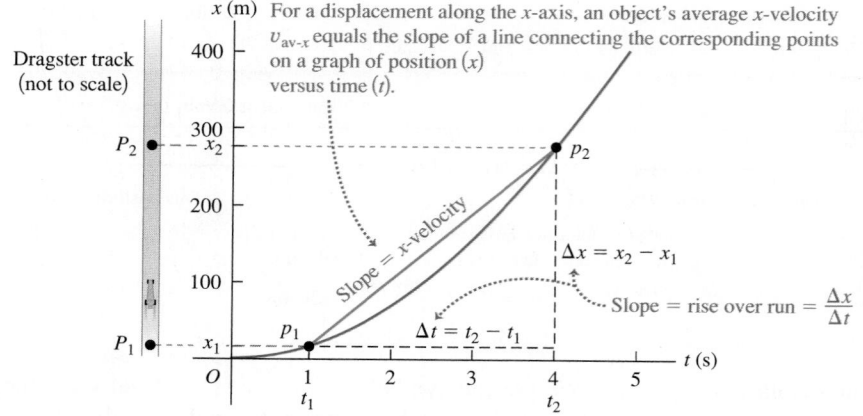

and horizontal side $\Delta t = t_2 - t_1$. The average x-velocity $v_{\text{av-}x} = \Delta x/\Delta t$ of the dragster equals the *slope* of the line p_1p_2—that is, the ratio of the triangle's vertical side Δx to its horizontal side Δt.

The average x-velocity depends only on the total displacement $\Delta x = x_2 - x_1$ that occurs during the time interval $\Delta t = t_2 - t_1$, not on the details of what happens during the time interval. At time t_1 a motorcycle might have raced past the dragster at point P_1 in Fig. 2.1, then blown its engine and slowed down to pass through point P_2 at the same time t_2 as the dragster. Both vehicles have the same displacement during the same time interval and so have the same average x-velocity.

If distance is given in meters and time in seconds, average velocity is measured in meters per second (m/s). Other common units of velocity are kilometers per hour (km/h), feet per second (ft/s), miles per hour (mi/h), and knots (1 knot = 1 nautical mile/h = 6080 ft/h). Table 2.2 lists some typical velocity magnitudes.

Table 2.2 Typical Velocity Magnitudes

A snail's pace	10^{-3} m/s
A brisk walk	2 m/s
Fastest human	11 m/s
Freeway speeds	30 m/s
Fastest car	341 m/s
Random motion of air molecules	500 m/s
Fastest airplane	1000 m/s
Orbiting communications satellite	3000 m/s
Electron orbiting in a hydrogen atom	2×10^6 m/s
Light traveling in a vacuum	3×10^8 m/s

Test Your Understanding of Section 2.1 Each of the following automobile trips takes one hour. The positive x-direction is to the east. (i) Automobile A travels 50 km due east. (ii) Automobile B travels 50 km due west. (iii) Automobile C travels 60 km due east, then turns around and travels 10 km due west. (iv) Automobile D travels 70 km due east. (v) Automobile E travels 20 km due west, then turns around and travels 20 km due east. (a) Rank the five trips in order of average x-velocity from most positive to most negative. (b) Which trips, if any, have the same average x-velocity? (c) For which trip, if any, is the average x-velocity equal to zero?

2.2 Instantaneous Velocity

Sometimes the average velocity is all you need to know about a particle's motion. For example, a race along a straight line is really a competition to see whose average velocity, $v_{\text{av-}x}$, has the greatest magnitude. The prize goes to the competitor who can travel the displacement Δx from the start to the finish line in the shortest time interval, Δt (Fig. 2.4).

But the average velocity of a particle during a time interval can't tell us how fast, or in what direction, the particle was moving at any given time during the interval. To do this we need to know the **instantaneous velocity,** or the velocity at a specific instant of time or specific point along the path.

2.4 The winner of a 50-m swimming race is the swimmer whose average velocity has the greatest magnitude—that is, the swimmer who traverses a displacement Δx of 50 m in the shortest elapsed time Δt.

CAUTION **How long is an instant?** Note that the word "instant" has a somewhat different definition in physics than in everyday language. You might use the phrase "It lasted just an instant" to refer to something that lasted for a very short time interval. But in physics an instant has no duration at all; it refers to a single value of time.

To find the instantaneous velocity of the dragster in Fig. 2.1 at the point P_1, we move the second point P_2 closer and closer to the first point P_1 and compute the average velocity $v_{\text{av-}x} = \Delta x / \Delta t$ over the ever-shorter displacement and time interval. Both Δx and Δt become very small, but their ratio does not necessarily become small. In the language of calculus, the limit of $\Delta x / \Delta t$ as Δt approaches zero is called the **derivative** of x with respect to t and is written dx/dt. *The instantaneous velocity is the limit of the average velocity as the time interval approaches zero; it equals the instantaneous rate of change of position with time.* We use the symbol v_x, with no "av" subscript, for the instantaneous velocity along the x-axis, or the **instantaneous x-velocity:**

$$v_x = \lim_{\Delta t \to 0} \frac{\Delta x}{\Delta t} = \frac{dx}{dt} \quad \text{(instantaneous x-velocity, straight-line motion)} \quad \text{(2.3)}$$

The time interval Δt is always positive, so v_x has the same algebraic sign as Δx. A positive value of v_x means that x is increasing and the motion is in the positive x-direction; a negative value of v_x means that x is decreasing and the motion is in the negative x-direction. A body can have positive x and negative v_x, or the reverse; x tells us where the body is, while v_x tells us how it's moving (Fig. 2.5). The rules that we presented in Table 2.1 (Section 2.1) for the sign of average x-velocity $v_{\text{av-}x}$ also apply to the sign of instantaneous x-velocity v_x.

Instantaneous velocity, like average velocity, is a vector quantity; Eq. (2.3) defines its x-component. In straight-line motion, all other components of instantaneous velocity are zero. In this case we often call v_x simply the instantaneous velocity. (In Chapter 3 we'll deal with the general case in which the instantaneous velocity can have nonzero x-, y-, and z-components.) When we use the term "velocity," we will always mean instantaneous rather than average velocity.

The terms "velocity" and "speed" are used interchangeably in everyday language, but they have distinct definitions in physics. We use the term **speed** to denote distance traveled divided by time, on either an average or an instantaneous basis. Instantaneous *speed*, for which we use the symbol v with *no* subscripts, measures how fast a particle is moving; instantaneous *velocity* measures how fast *and* in what direction it's moving. Instantaneous speed is the magnitude of instantaneous velocity and so can never be negative. For example, a particle with instantaneous velocity $v_x = 25$ m/s and a second particle with $v_x = -25$ m/s are moving in opposite directions at the same instantaneous speed 25 m/s.

2.5 Even when he's moving forward, this cyclist's instantaneous x-velocity can be negative—if he's traveling in the negative x-direction. In any problem, the choice of which direction is positive and which is negative is entirely up to you.

CAUTION **Average speed and average velocity** Average speed is *not* the magnitude of average velocity. When César Cielo set a world record in 2009 by swimming 100.0 m in 46.91 s, his average speed was $(100.0 \text{ m})/(46.91 \text{ s}) = 2.132$ m/s. But because he swam two lengths in a 50-m pool, he started and ended at the same point and so had zero total displacement and zero average *velocity*! Both average speed and instantaneous speed are scalars, not vectors, because these quantities contain no information about direction. ▮

Example 2.1   **Average and instantaneous velocities**

A cheetah is crouched 20 m to the east of an observer (Fig. 2.6a). At time $t = 0$ the cheetah begins to run due east toward an antelope that is 50 m to the east of the observer. During the first 2.0 s of the attack, the cheetah's coordinate x varies with time according to the equation $x = 20 \text{ m} + (5.0 \text{ m/s}^2)t^2$. (a) Find the cheetah's displacement between $t_1 = 1.0$ s and $t_2 = 2.0$ s. (b) Find its average velocity during that interval. (c) Find its instantaneous velocity at $t_1 = 1.0$ s by taking $\Delta t = 0.1$ s, then 0.01 s, then 0.001 s. (d) Derive an expression for the cheetah's instantaneous velocity as a function of time, and use it to find v_x at $t = 1.0$ s and $t = 2.0$ s.

SOLUTION

IDENTIFY and SET UP: Figure 2.6b shows our sketch of the cheetah's motion. We use Eq. (2.1) for displacement, Eq. (2.2) for average velocity, and Eq. (2.3) for instantaneous velocity.

Continued

2.6 A cheetah attacking an antelope from ambush. The animals are not drawn to the same scale as the axis.

(a) The situation

(b) Our sketch

(c) Decisions ① Point axis in direction cheetah runs, so that all values will be positive. ② Place origin at vehicle. ③ Mark initial positions of cheetah and antelope. ④ Mark positions for cheetah at 1 s and 2 s. ⑤ Add the known and unknown quantities.

EXECUTE: (a) At $t_1 = 1.0$ s and $t_2 = 2.0$ s the cheetah's positions x_1 and x_2 are

$$x_1 = 20 \text{ m} + (5.0 \text{ m/s}^2)(1.0 \text{ s})^2 = 25 \text{ m}$$
$$x_2 = 20 \text{ m} + (5.0 \text{ m/s}^2)(2.0 \text{ s})^2 = 40 \text{ m}$$

The displacement during this 1.0-s interval is

$$\Delta x = x_2 - x_1 = 40 \text{ m} - 25 \text{ m} = 15 \text{ m}$$

(b) The average x-velocity during this interval is

$$v_{\text{av-}x} = \frac{x_2 - x_1}{t_2 - t_1} = \frac{40 \text{ m} - 25 \text{ m}}{2.0 \text{ s} - 1.0 \text{ s}} = \frac{15 \text{ m}}{1.0 \text{ s}} = 15 \text{ m/s}$$

(c) With $\Delta t = 0.1$ s the time interval is from $t_1 = 1.0$ s to a new $t_2 = 1.1$ s. At t_2 the position is

$$x_2 = 20 \text{ m} + (5.0 \text{ m/s}^2)(1.1 \text{ s})^2 = 26.05 \text{ m}$$

The average x-velocity during this 0.1-s interval is

$$v_{\text{av-}x} = \frac{26.05 \text{ m} - 25 \text{ m}}{1.1 \text{ s} - 1.0 \text{ s}} = 10.5 \text{ m/s}$$

Following this pattern, you can calculate the average x-velocities for 0.01-s and 0.001-s intervals: The results are 10.05 m/s and 10.005 m/s. As Δt gets smaller, the average x-velocity gets closer to 10.0 m/s, so we conclude that the instantaneous x-velocity at $t = 1.0$ s is 10.0 m/s. (We suspended the rules for significant-figure counting in these calculations.)

(d) To find the instantaneous x-velocity as a function of time, we take the derivative of the expression for x with respect to t. The derivative of a constant is zero, and for any n the derivative of t^n is nt^{n-1}, so the derivative of t^2 is $2t$. We therefore have

$$v_x = \frac{dx}{dt} = (5.0 \text{ m/s}^2)(2t) = (10 \text{ m/s}^2)t$$

At $t = 1.0$ s, this yields $v_x = 10$ m/s, as we found in part (c); at $t = 2.0$ s, $v_x = 20$ m/s.

EVALUATE: Our results show that the cheetah picked up speed from $t = 0$ (when it was at rest) to $t = 1.0$ s ($v_x = 10$ m/s) to $t = 2.0$ s ($v_x = 20$ m/s). This makes sense; the cheetah covered only 5 m during the interval $t = 0$ to $t = 1.0$ s, but it covered 15 m during the interval $t = 1.0$ s to $t = 2.0$ s.

Mastering**PHYSICS**

ActivPhysics 1.1: Analyzing Motion Using Diagrams

Finding Velocity on an x-t Graph

We can also find the x-velocity of a particle from the graph of its position as a function of time. Suppose we want to find the x-velocity of the dragster in Fig. 2.1 at point P_1. As point P_2 in Fig. 2.1 approaches point P_1, point p_2 in the x-t graphs of Figs. 2.7a and 2.7b approaches point p_1 and the average x-velocity is calculated over shorter time intervals Δt. In the limit that $\Delta t \rightarrow 0$, shown in Fig. 2.7c, the slope of the line $p_1 p_2$ equals the slope of the line tangent to the curve at point p_1. Thus, *on a graph of position as a function of time for straight-line motion, the instantaneous x-velocity at any point is equal to the slope of the tangent to the curve at that point.*

If the tangent to the x-t curve slopes upward to the right, as in Fig. 2.7c, then its slope is positive, the x-velocity is positive, and the motion is in the positive x-direction. If the tangent slopes downward to the right, the slope of the x-t graph

2.7 Using an *x-t* graph to go from (a), (b) average *x*-velocity to (c) instantaneous *x*-velocity v_x. In (c) we find the slope of the tangent to the *x-t* curve by dividing any vertical interval (with distance units) along the tangent by the corresponding horizontal interval (with time units).

(a)

As the average *x*-velocity $v_{av\text{-}x}$ is calculated over shorter and shorter time intervals ...

(b)

... its value $v_{av\text{-}x} = \Delta x / \Delta t$ approaches the instantaneous *x*-velocity.

(c)

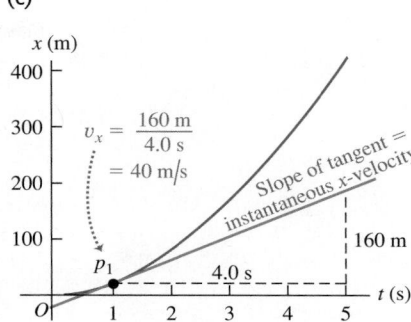

The instantaneous *x*-velocity v_x at any given point equals the slope of the tangent to the *x-t* curve at that point.

2.8 (a) The *x-t* graph of the motion of a particular particle. The slope of the tangent at any point equals the velocity at that point. (b) A motion diagram showing the position and velocity of the particle at each of the times labeled on the *x-t* graph.

(a) *x-t* graph

(b) Particle's motion

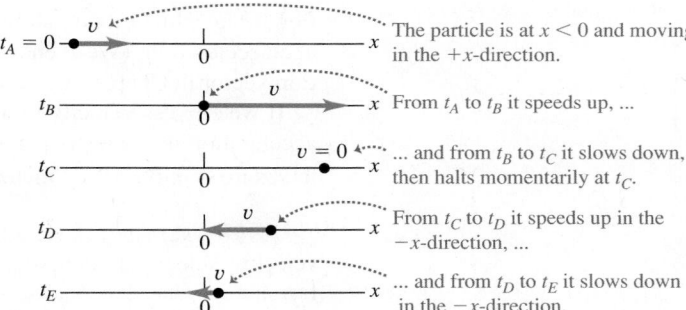

The particle is at $x < 0$ and moving in the +*x*-direction.

From t_A to t_B it speeds up, ...

... and from t_B to t_C it slows down, then halts momentarily at t_C.

From t_C to t_D it speeds up in the −*x*-direction, ...

... and from t_D to t_E it slows down in the −*x*-direction.

The steeper the slope (positive or negative) of an object's *x-t* graph, the greater is the object's speed in the positive or negative *x*-direction.

and the *x*-velocity are negative, and the motion is in the negative *x*-direction. When the tangent is horizontal, the slope and the *x*-velocity are zero. Figure 2.8 illustrates these three possibilities.

Figure 2.8 actually depicts the motion of a particle in two ways: as (a) an *x-t* graph and (b) a **motion diagram** that shows the particle's position at various instants (like frames from a video of the particle's motion) as well as arrows to represent the particle's velocity at each instant. We will use both *x-t* graphs and motion diagrams in this chapter to help you understand motion. You will find it worth your while to draw *both* an *x-t* graph and a motion diagram as part of solving any problem involving motion.

2.9 An *x-t* graph for a particle.

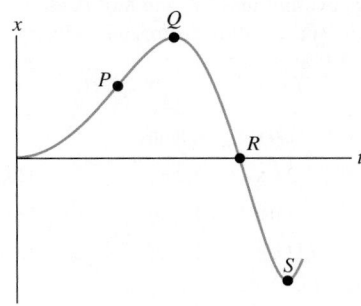

Test Your Understanding of Section 2.2 Figure 2.9 is an *x-t* graph of the motion of a particle. (a) Rank the values of the particle's *x*-velocity v_x at the points *P*, *Q*, *R*, and *S* from most positive to most negative. (b) At which points is v_x positive? (c) At which points is v_x negative? (d) At which points is v_x zero? (e) Rank the values of the particle's *speed* at the points *P*, *Q*, *R*, and *S* from fastest to slowest. ❙

2.3 Average and Instantaneous Acceleration

Just as velocity describes the rate of change of position with time, *acceleration* describes the rate of change of velocity with time. Like velocity, acceleration is a vector quantity. When the motion is along a straight line, its only nonzero component is along that line. As we'll see, acceleration in straight-line motion can refer to either speeding up or slowing down.

Average Acceleration

Let's consider again a particle moving along the *x*-axis. Suppose that at time t_1 the particle is at point P_1 and has *x*-component of (instantaneous) velocity v_{1x}, and at a later time t_2 it is at point P_2 and has *x*-component of velocity v_{2x}. So the *x*-component of velocity changes by an amount $\Delta v_x = v_{2x} - v_{1x}$ during the time interval $\Delta t = t_2 - t_1$.

We define the **average acceleration** of the particle as it moves from P_1 to P_2 to be a vector quantity whose *x*-component $a_{\text{av-}x}$ (called the **average *x*-acceleration**) equals Δv_x, the change in the *x*-component of velocity, divided by the time interval Δt:

$$a_{\text{av-}x} = \frac{v_{2x} - v_{1x}}{t_2 - t_1} = \frac{\Delta v_x}{\Delta t} \qquad \begin{array}{l}\text{(average }x\text{-acceleration,}\\ \text{straight-line motion)}\end{array} \qquad (2.4)$$

For straight-line motion along the *x*-axis we will often call $a_{\text{av-}x}$ simply the average acceleration. (We'll encounter the other components of the average acceleration vector in Chapter 3.)

If we express velocity in meters per second and time in seconds, then average acceleration is in meters per second per second, or $(\text{m/s})/\text{s}$. This is usually written as m/s^2 and is read "meters per second squared."

CAUTION *Acceleration vs. velocity* Be very careful not to confuse acceleration with velocity! Velocity describes how a body's position changes with time; it tells us how fast and in what direction the body moves. Acceleration describes how the velocity changes with time; it tells us how the speed and direction of motion are changing. It may help to remember the phrase "acceleration is to velocity as velocity is to position." It can also help to imagine yourself riding along with the moving body. If the body accelerates forward and gains speed, you feel pushed backward in your seat; if it accelerates backward and loses speed, you feel pushed forward. If the velocity is constant and there's no acceleration, you feel neither sensation. (We'll see the reason for these sensations in Chapter 4.)

Example 2.2 Average acceleration

An astronaut has left an orbiting spacecraft to test a new personal maneuvering unit. As she moves along a straight line, her partner on the spacecraft measures her velocity every 2.0 s, starting at time $t = 1.0$ s:

t	v_x	t	v_x
1.0 s	0.8 m/s	9.0 s	−0.4 m/s
3.0 s	1.2 m/s	11.0 s	−1.0 m/s
5.0 s	1.6 m/s	13.0 s	−1.6 m/s
7.0 s	1.2 m/s	15.0 s	−0.8 m/s

Find the average *x*-acceleration, and state whether the speed of the astronaut increases or decreases over each of these 2.0-s time intervals: (a) $t_1 = 1.0$ s to $t_2 = 3.0$ s; (b) $t_1 = 5.0$ s to $t_2 = 7.0$ s; (c) $t_1 = 9.0$ s to $t_2 = 11.0$ s; (d) $t_1 = 13.0$ s to $t_2 = 15.0$ s.

SOLUTION

IDENTIFY and SET UP: We'll use Eq. (2.4) to determine the average acceleration $a_{\text{av-}x}$ from the change in velocity over each time interval. To find the changes in speed, we'll use the idea that speed v is the magnitude of the instantaneous velocity v_x.

The upper part of Fig. 2.10 is our graph of the x-velocity as a function of time. On this v_x-t graph, the slope of the line connecting the endpoints of each interval is the average x-acceleration $a_{\text{av-}x} = \Delta v_x / \Delta t$ for that interval. The four slopes (and thus the *signs* of the average accelerations) are, respectively, positive, negative, negative, and positive. The third and fourth slopes (and thus the average accelerations themselves) have greater magnitude than the first and second.

2.10 Our graphs of x-velocity versus time (top) and average x-acceleration versus time (bottom) for the astronaut.

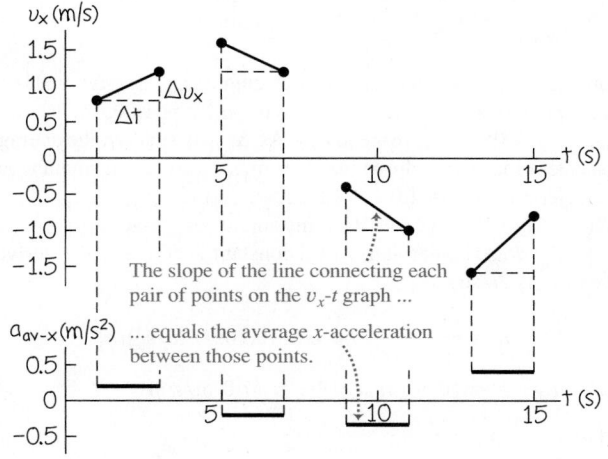

EXECUTE: Using Eq. (2.4), we find:

(a) $a_{\text{av-}x} = (1.2 \text{ m/s} - 0.8 \text{ m/s})/(3.0 \text{ s} - 1.0 \text{ s}) = 0.2 \text{ m/s}^2$. The speed (magnitude of instantaneous x-velocity) increases from 0.8 m/s to 1.2 m/s.

(b) $a_{\text{av-}x} = (1.2 \text{ m/s} - 1.6 \text{ m/s})/(7.0 \text{ s} - 5.0 \text{ s}) = -0.2 \text{ m/s}^2$. The speed decreases from 1.6 m/s to 1.2 m/s.

(c) $a_{\text{av-}x} = [-1.0 \text{ m/s} - (-0.4 \text{ m/s})]/(11.0 \text{ s} - 9.0 \text{ s}) = -0.3 \text{ m/s}^2$. The speed increases from 0.4 m/s to 1.0 m/s.

(d) $a_{\text{av-}x} = [-0.8 \text{ m/s} - (-1.6 \text{ m/s})]/(15.0 \text{ s} - 13.0 \text{ s}) = 0.4 \text{ m/s}^2$. The speed decreases from 1.6 m/s to 0.8 m/s.

In the lower part of Fig. 2.10, we graph the values of $a_{\text{av-}x}$.

EVALUATE: The signs and relative magnitudes of the average accelerations agree with our qualitative predictions. For future reference, note this connection among speed, velocity, and acceleration: Our results show that when the average x-acceleration has the *same* direction (same algebraic sign) as the initial velocity, as in intervals (a) and (c), the astronaut goes faster. When $a_{\text{av-}x}$ has the *opposite* direction (opposite algebraic sign) from the initial velocity, as in intervals (b) and (d), she slows down. Thus positive x-acceleration means speeding up if the x-velocity is positive [interval (a)] but slowing down if the x-velocity is negative [interval (d)]. Similarly, negative x-acceleration means speeding up if the x-velocity is negative [interval (c)] but slowing down if the x-velocity is positive [interval (b)].

Instantaneous Acceleration

We can now define **instantaneous acceleration** following the same procedure that we used to define instantaneous velocity. As an example, suppose a race car driver is driving along a straightaway as shown in Fig. 2.11. To define the instantaneous acceleration at point P_1, we take the second point P_2 in Fig. 2.11 to be closer and closer to P_1 so that the average acceleration is computed over shorter and shorter time intervals. *The instantaneous acceleration is the limit of the average acceleration as the time interval approaches zero.* In the language of calculus, *instantaneous acceleration equals the derivative of velocity with time.* Thus

$$a_x = \lim_{\Delta t \to 0} \frac{\Delta v_x}{\Delta t} = \frac{dv_x}{dt} \qquad \begin{array}{l}\text{(instantaneous } x\text{-acceleration,} \\ \text{straight-line motion)}\end{array} \qquad (2.5)$$

Note that a_x in Eq. (2.5) is really the x-component of the acceleration vector, or the **instantaneous x-acceleration;** in straight-line motion, all other components of this vector are zero. From now on, when we use the term "acceleration," we will always mean instantaneous acceleration, not average acceleration.

2.11 A Grand Prix car at two points on the straightaway.

Example 2.3 **Average and instantaneous accelerations**

Suppose the x-velocity v_x of the car in Fig. 2.11 at any time t is given by the equation

$$v_x = 60 \text{ m/s} + (0.50 \text{ m/s}^3)t^2$$

(a) Find the change in x-velocity of the car in the time interval $t_1 = 1.0$ s to $t_2 = 3.0$ s. (b) Find the average x-acceleration in this time interval. (c) Find the instantaneous x-acceleration at time $t_1 = 1.0$ s by taking Δt to be first 0.1 s, then 0.01 s, then 0.001 s. (d) Derive an expression for the instantaneous x-acceleration as a function of time, and use it to find a_x at $t = 1.0$ s and $t = 3.0$ s.

SOLUTION

IDENTIFY and SET UP: This example is analogous to Example 2.1 in Section 2.2. (Now is a good time to review that example.) In Example 2.1 we found the average x-velocity from the change in position over shorter and shorter time intervals, and we obtained an expression for the instantaneous x-velocity by differentiating the position as a function of time. In this example we have an exact parallel. Using Eq. (2.4), we'll find the average x-acceleration from the change in x-velocity over a time interval. Likewise, using Eq. (2.5), we'll obtain an expression for the instantaneous x-acceleration by differentiating the x-velocity as a function of time.

EXECUTE: (a) Before we can apply Eq. (2.4), we must find the x-velocity at each time from the given equation. At $t_1 = 1.0$ s and $t_2 = 3.0$ s, the velocities are

$$v_{1x} = 60 \text{ m/s} + (0.50 \text{ m/s}^3)(1.0 \text{ s})^2 = 60.5 \text{ m/s}$$

$$v_{2x} = 60 \text{ m/s} + (0.50 \text{ m/s}^3)(3.0 \text{ s})^2 = 64.5 \text{ m/s}$$

The change in x-velocity Δv_x between $t_1 = 1.0$ s and $t_2 = 3.0$ s is

$$\Delta v_x = v_{2x} - v_{1x} = 64.5 \text{ m/s} - 60.5 \text{ m/s} = 4.0 \text{ m/s}$$

(b) The average x-acceleration during this time interval of duration $t_2 - t_1 = 2.0$ s is

$$a_{\text{av-}x} = \frac{v_{2x} - v_{1x}}{t_2 - t_1} = \frac{4.0 \text{ m/s}}{2.0 \text{ s}} = 2.0 \text{ m/s}^2$$

During this time interval the x-velocity and average x-acceleration have the same algebraic sign (in this case, positive), and the car speeds up.

(c) When $\Delta t = 0.1$ s, we have $t_2 = 1.1$ s. Proceeding as before, we find

$$v_{2x} = 60 \text{ m/s} + (0.50 \text{ m/s}^3)(1.1 \text{ s})^2 = 60.605 \text{ m/s}$$

$$\Delta v_x = 0.105 \text{ m/s}$$

$$a_{\text{av-}x} = \frac{\Delta v_x}{\Delta t} = \frac{0.105 \text{ m/s}}{0.1 \text{ s}} = 1.05 \text{ m/s}^2$$

You should follow this pattern to calculate $a_{\text{av-}x}$ for $\Delta t = 0.01$ s and $\Delta t = 0.001$ s; the results are $a_{\text{av-}x} = 1.005 \text{ m/s}^2$ and $a_{\text{av-}x} = 1.0005 \text{ m/s}^2$, respectively. As Δt gets smaller, the average x-acceleration gets closer to 1.0 m/s^2, so the instantaneous x-acceleration at $t = 1.0$ s is 1.0 m/s^2.

(d) By Eq. (2.5) the instantaneous x-acceleration is $a_x = dv_x/dt$. The derivative of a constant is zero and the derivative of t^2 is $2t$, so

$$a_x = \frac{dv_x}{dt} = \frac{d}{dt}[60 \text{ m/s} + (0.50 \text{ m/s}^3)t^2]$$

$$= (0.50 \text{ m/s}^3)(2t) = (1.0 \text{ m/s}^3)t$$

When $t = 1.0$ s,

$$a_x = (1.0 \text{ m/s}^3)(1.0 \text{ s}) = 1.0 \text{ m/s}^2$$

When $t = 3.0$ s,

$$a_x = (1.0 \text{ m/s}^3)(3.0 \text{ s}) = 3.0 \text{ m/s}^2$$

EVALUATE: Neither of the values we found in part (d) is equal to the average x-acceleration found in part (b). That's because the car's instantaneous x-acceleration varies with time. The rate of change of acceleration with time is sometimes called the "jerk."

Finding Acceleration on a v_x-t Graph or an x-t Graph

In Section 2.2 we interpreted average and instantaneous x-velocity in terms of the slope of a graph of position versus time. In the same way, we can interpret average and instantaneous x-acceleration by using a graph with instantaneous velocity v_x on the vertical axis and time t on the horizontal axis—that is, a v_x-t **graph** (Fig. 2.12). The points on the graph labeled p_1 and p_2 correspond to points P_1 and P_2 in Fig. 2.11. The average x-acceleration $a_{\text{av-}x} = \Delta v_x/\Delta t$ during this interval is the slope of the line p_1p_2. As point P_2 in Fig. 2.11 approaches point P_1, point p_2 in the v_x-t graph of Fig. 2.12 approaches point p_1, and the slope of the line p_1p_2 approaches the slope of the line tangent to the curve at point p_1. Thus, on a graph of x-velocity as a function of time, the instantaneous x-acceleration at any point is equal to the slope of the tangent to the curve at that point. Tangents drawn at different points along the curve in Fig. 2.12 have different slopes, so the instantaneous x-acceleration varies with time.

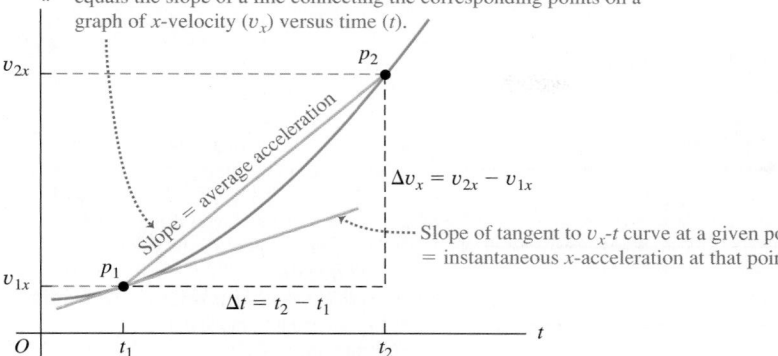

For a displacement along the x-axis, an object's average x-acceleration equals the slope of a line connecting the corresponding points on a graph of x-velocity (v_x) versus time (t).

Slope = average acceleration

$\Delta v_x = v_{2x} - v_{1x}$

Slope of tangent to v_x-t curve at a given point = instantaneous x-acceleration at that point.

$\Delta t = t_2 - t_1$

2.12 A v_x-t graph of the motion in Fig. 2.11.

CAUTION **The signs of x-acceleration and x-velocity** By itself, the algebraic sign of the x-acceleration does *not* tell you whether a body is speeding up or slowing down. You must compare the signs of the x-velocity and the x-acceleration. When v_x and a_x have the *same* sign, the body is speeding up. If both are positive, the body is moving in the positive direction with increasing speed. If both are negative, the body is moving in the negative direction with an x-velocity that is becoming more and more negative, and again the speed is increasing. When v_x and a_x have *opposite* signs, the body is slowing down. If v_x is positive and a_x is negative, the body is moving in the positive direction with decreasing speed; if v_x is negative and a_x is positive, the body is moving in the negative direction with an x-velocity that is becoming less negative, and again the body is slowing down. Table 2.3 summarizes these ideas, and Fig. 2.13 illustrates some of these possibilities.

The term "deceleration" is sometimes used for a decrease in speed. Because it may mean positive or negative a_x, depending on the sign of v_x, we avoid this term.

We can also learn about the acceleration of a body from a graph of its *position* versus time. Because $a_x = dv_x/dt$ and $v_x = dx/dt$, we can write

$$a_x = \frac{dv_x}{dt} = \frac{d}{dt}\left(\frac{dx}{dt}\right) = \frac{d^2x}{dt^2} \qquad (2.6)$$

Table 2.3 Rules for the Sign of x-Acceleration

If x-velocity is:	... x-acceleration is:
Positive & increasing (getting more positive)	Positive: Particle is moving in $+x$-direction & speeding up
Positive & decreasing (getting less positive)	Negative: Particle is moving in $+x$-direction & slowing down
Negative & increasing (getting less negative)	Positive: Particle is moving in $-x$-direction & slowing down
Negative & decreasing (getting more negative)	Negative: Particle is moving in $-x$-direction & speeding up

Note: These rules apply to both the average x-acceleration $a_{\text{av-}x}$ and the instantaneous x-acceleration a_x.

2.13 (a) A v_x-t graph of the motion of a different particle from that shown in Fig. 2.8. The slope of the tangent at any point equals the x-acceleration at that point. (b) A motion diagram showing the position, velocity, and acceleration of the particle at each of the times labeled on the v_x-t graph. The positions are consistent with the v_x-t graph; for instance, from t_A to t_B the velocity is negative, so at t_B the particle is at a more negative value of x than at t_A.

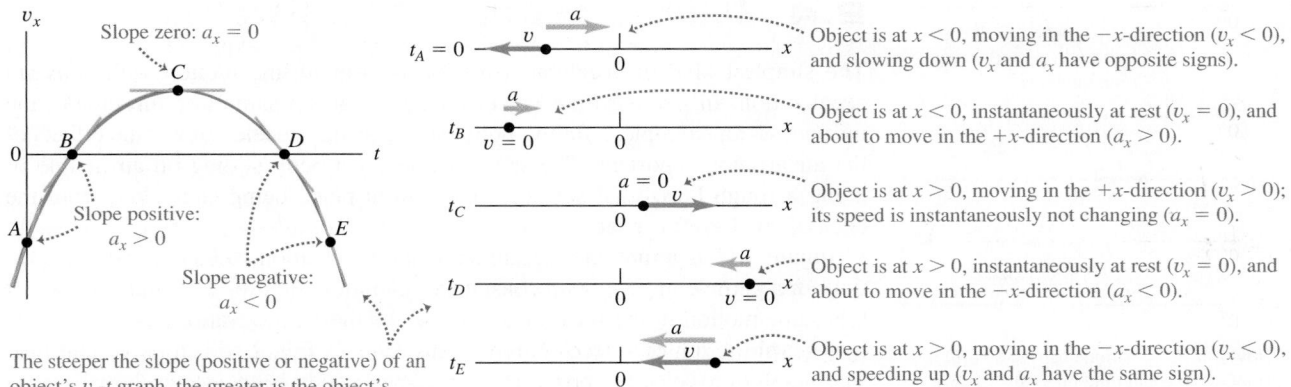

(a) v_x-t graph for an object moving on the x-axis

Slope zero: $a_x = 0$

Slope positive: $a_x > 0$

Slope negative: $a_x < 0$

The steeper the slope (positive or negative) of an object's v_x-t graph, the greater is the object's acceleration in the positive or negative x-direction.

(b) Object's position, velocity, and acceleration on the x-axis

$t_A = 0$ — Object is at $x < 0$, moving in the $-x$-direction ($v_x < 0$), and slowing down (v_x and a_x have opposite signs).

t_B — Object is at $x < 0$, instantaneously at rest ($v_x = 0$), and about to move in the $+x$-direction ($a_x > 0$).

t_C — Object is at $x > 0$, moving in the $+x$-direction ($v_x > 0$); its speed is instantaneously not changing ($a_x = 0$).

t_D — Object is at $x > 0$, instantaneously at rest ($v_x = 0$), and about to move in the $-x$-direction ($a_x < 0$).

t_E — Object is at $x > 0$, moving in the $-x$-direction ($v_x < 0$), and speeding up (v_x and a_x have the same sign).

2.14 (a) The same *x-t* graph as shown in Fig. 2.8a. The *x*-velocity is equal to the *slope* of the graph, and the acceleration is given by the *concavity* or *curvature* of the graph. (b) A motion diagram showing the position, velocity, and acceleration of the particle at each of the times labeled on the *x-t* graph.

(a) *x-t* graph

(b) Object's motion

Slope zero: $v_x = 0$
Curvature downward: $a_x < 0$

Slope negative:
$v_x < 0$
Curvature upward:
$a_x > 0$

Slope negative: $v_x < 0$
Curvature zero: $a_x = 0$

Slope positive: $v_x > 0$
Curvature zero: $a_x = 0$

Slope positive: $v_x > 0$
Curvature upward: $a_x > 0$

$t_A = 0$ — Object is at $x < 0$, moving in the $+x$-direction ($v_x > 0$), and speeding up (v_x and a_x have the same sign).

t_B — Object is at $x = 0$, moving in the $+x$-direction ($v_x > 0$); speed is instantaneously not changing ($a_x = 0$).

t_C — Object is at $x > 0$, instantaneously at rest ($v_x = 0$), and about to move in the $-x$-direction ($a_x < 0$).

t_D — Object is at $x > 0$, moving in the $-x$-direction ($v_x < 0$); speed is instantaneously not changing ($a_x = 0$).

t_E — Object is at $x > 0$, moving in the $-x$-direction ($v_x < 0$), and slowing down (v_x and a_x have opposite signs).

The greater the curvature (upward or downward) of an object's *x-t* graph, the greater is the object's acceleration in the positive or negative *x*-direction.

That is, a_x is the second derivative of x with respect to t. The second derivative of any function is directly related to the *concavity* or *curvature* of the graph of that function (Fig. 2.14). Where the *x-t* graph is concave up (curved upward), the *x*-acceleration is positive and v_x is increasing; at a point where the *x-t* graph is concave down (curved downward), the *x*-acceleration is negative and v_x is decreasing. At a point where the *x-t* graph has no curvature, such as an inflection point, the *x*-acceleration is zero and the velocity is not changing. Figure 2.14 shows all three of these possibilities.

Examining the curvature of an *x-t* graph is an easy way to decide what the *sign* of acceleration is. This technique is less helpful for determining numerical values of acceleration because the curvature of a graph is hard to measure accurately.

Test Your Understanding of Section 2.3 Look again at the *x-t* graph in Fig. 2.9 at the end of Section 2.2. (a) At which of the points *P*, *Q*, *R*, and *S* is the *x*-acceleration a_x positive? (b) At which points is the *x*-acceleration negative? (c) At which points does the *x*-acceleration appear to be zero? (d) At each point state whether the velocity is increasing, decreasing, or not changing.

2.15 A motion diagram for a particle moving in a straight line in the positive *x*-direction with constant positive *x*-acceleration a_x. The position, velocity, and acceleration are shown at five equally spaced times.

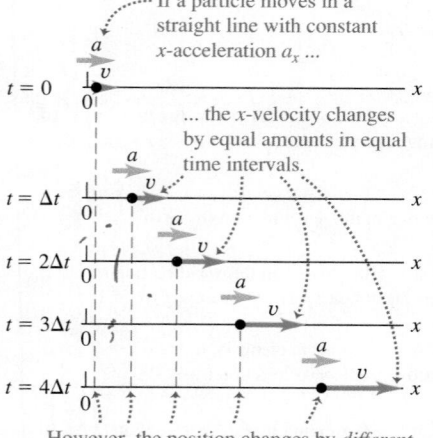

If a particle moves in a straight line with constant *x*-acceleration a_x ...

$t = 0$

... the *x*-velocity changes by equal amounts in equal time intervals.

$t = \Delta t$

$t = 2\Delta t$

$t = 3\Delta t$

$t = 4\Delta t$

However, the position changes by *different* amounts in equal time intervals because the velocity is changing.

2.4 Motion with Constant Acceleration

The simplest kind of accelerated motion is straight-line motion with *constant* acceleration. In this case the velocity changes at the same rate throughout the motion. As an example, a falling body has a constant acceleration if the effects of the air are not important. The same is true for a body sliding on an incline or along a rough horizontal surface, or for an airplane being catapulted from the deck of an aircraft carrier.

Figure 2.15 is a motion diagram showing the position, velocity, and acceleration for a particle moving with constant acceleration. Figures 2.16 and 2.17 depict this same motion in the form of graphs. Since the *x*-acceleration is constant, the a_x-*t* **graph** (graph of *x*-acceleration versus time) in Fig. 2.16 is a horizontal line. The graph of *x*-velocity versus time, or v_x-*t* graph, has a constant *slope* because the acceleration is constant, so this graph is a straight line (Fig. 2.17).

When the x-acceleration a_x is constant, the average x-acceleration $a_{\text{av-}x}$ for any time interval is the same as a_x. This makes it easy to derive equations for the position x and the x-velocity v_x as functions of time. To find an expression for v_x, we first replace $a_{\text{av-}x}$ in Eq. (2.4) by a_x:

$$a_x = \frac{v_{2x} - v_{1x}}{t_2 - t_1} \tag{2.7}$$

Now we let $t_1 = 0$ and let t_2 be any later time t. We use the symbol v_{0x} for the x-velocity at the initial time $t = 0$; the x-velocity at the later time t is v_x. Then Eq. (2.7) becomes

$$a_x = \frac{v_x - v_{0x}}{t - 0} \qquad \text{or}$$

$$v_x = v_{0x} + a_x t \qquad \text{(constant x-acceleration only)} \tag{2.8}$$

In Eq. (2.8) the term $a_x t$ is the product of the constant rate of change of x-velocity, a_x, and the time interval t. Therefore it equals the *total* change in x-velocity from the initial time $t = 0$ to the later time t. The x-velocity v_x at any time t then equals the initial x-velocity v_{0x} (at $t = 0$) plus the change in x-velocity $a_x t$ (Fig. 2.17).

Equation (2.8) also says that the change in x-velocity $v_x - v_{0x}$ of the particle between $t = 0$ and any later time t equals the *area* under the a_x-t graph between those two times. You can verify this from Fig. 2.16: Under this graph is a rectangle of vertical side a_x, horizontal side t, and area $a_x t$. From Eq. (2.8) this is indeed equal to the change in velocity $v_x - v_{0x}$. In Section 2.6 we'll show that even if the x-acceleration is not constant, the change in x-velocity during a time interval is still equal to the area under the a_x-t curve, although in that case Eq. (2.8) does not apply.

Next we'll derive an equation for the position x as a function of time when the x-acceleration is constant. To do this, we use two different expressions for the average x-velocity $v_{\text{av-}x}$ during the interval from $t = 0$ to any later time t. The first expression comes from the definition of $v_{\text{av-}x}$, Eq. (2.2), which is true whether or not the acceleration is constant. We call the position at time $t = 0$ the *initial position*, denoted by x_0. The position at the later time t is simply x. Thus for the time interval $\Delta t = t - 0$ the displacement is $\Delta x = x - x_0$, and Eq. (2.2) gives

$$v_{\text{av-}x} = \frac{x - x_0}{t} \tag{2.9}$$

We can also get a second expression for $v_{\text{av-}x}$ that is valid only when the x-acceleration is constant, so that the x-velocity changes at a constant rate. In this case the average x-velocity for the time interval from 0 to t is simply the average of the x-velocities at the beginning and end of the interval:

$$v_{\text{av-}x} = \frac{v_{0x} + v_x}{2} \qquad \text{(constant x-acceleration only)} \tag{2.10}$$

(This equation is *not* true if the x-acceleration varies during the time interval.) We also know that with constant x-acceleration, the x-velocity v_x at any time t is given by Eq. (2.8). Substituting that expression for v_x into Eq. (2.10), we find

$$v_{\text{av-}x} = \tfrac{1}{2}(v_{0x} + v_{0x} + a_x t)$$

$$= v_{0x} + \tfrac{1}{2}a_x t \qquad \text{(constant x-acceleration only)} \tag{2.11}$$

2.16 An acceleration-time $(a_x\text{-}t)$ graph for straight-line motion with constant positive x-acceleration a_x.

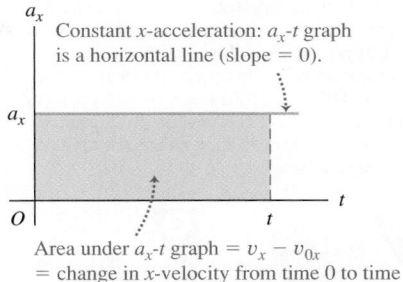

Constant x-acceleration: a_x-t graph is a horizontal line (slope = 0).

Area under a_x-t graph = $v_x - v_{0x}$ = change in x-velocity from time 0 to time t.

2.17 A velocity-time $(v_x\text{-}t)$ graph for straight-line motion with constant positive x-acceleration a_x. The initial x-velocity v_{0x} is also positive in this case.

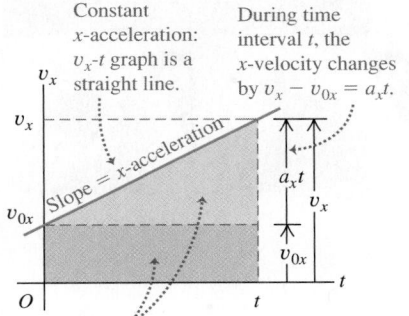

Constant x-acceleration: v_x-t graph is a straight line.

During time interval t, the x-velocity changes by $v_x - v_{0x} = a_x t$.

Slope = x-acceleration

Total area under v_x-t graph = $x - x_0$ = change in x-coordinate from time 0 to time t.

MasteringPHYSICS

PhET: Forces in 1 Dimension
ActivPhysics 1.1: Analyzing Motion Using Diagrams
ActivPhysics 1.2: Analyzing Motion Using Graphs
ActivPhysics 1.3: Predicting Motion from Graphs
ActivPhysics 1.4: Predicting Motion from Equations
ActivPhysics 1.5: Problem-Solving Strategies for Kinematics
ActivPhysics 1.6: Skier Races Downhill

Application Testing Humans at High Accelerations
In experiments carried out by the U.S. Air Force in the 1940s and 1950s, humans riding a rocket sled demonstrated that they could withstand accelerations as great as 440 m/s². The first three photos in this sequence show Air Force physician John Stapp speeding up from rest to 188 m/s (678 km/h = 421 mi/h) in just 5 s. Photos 4–6 show the even greater magnitude of acceleration as the rocket sled braked to a halt.

Finally, we set Eqs. (2.9) and (2.11) equal to each other and simplify:

$$v_{0x} + \tfrac{1}{2}a_x t = \frac{x - x_0}{t} \quad \text{or}$$

$$x = x_0 + v_{0x}t + \tfrac{1}{2}a_x t^2 \quad \text{(constant x-acceleration only)} \quad (2.12)$$

Here's what Eq. (2.12) tells us: If at time $t = 0$ a particle is at position x_0 and has x-velocity v_{0x}, its new position x at any later time t is the sum of three terms—its initial position x_0, plus the distance $v_{0x}t$ that it would move if its x-velocity were constant, plus an additional distance $\tfrac{1}{2}a_x t^2$ caused by the change in x-velocity.

A graph of Eq. (2.12)—that is, an x-t graph for motion with constant x-acceleration (Fig. 2.18a)—is always a *parabola*. Figure 2.18b shows such a graph. The curve intercepts the vertical axis (x-axis) at x_0, the position at $t = 0$. The slope of the tangent at $t = 0$ equals v_{0x}, the initial x-velocity, and the slope of the tangent at any time t equals the x-velocity v_x at that time. The slope and x-velocity are continuously increasing, so the x-acceleration a_x is positive; you can also see this because the graph in Fig. 2.18b is concave up (it curves upward). If a_x is negative, the x-t graph is a parabola that is concave down (has a downward curvature).

If there is zero x-acceleration, the x-t graph is a straight line; if there is a constant x-acceleration, the additional $\tfrac{1}{2}a_x t^2$ term in Eq. (2.12) for x as a function of t curves the graph into a parabola (Fig. 2.19a). We can analyze the v_x-t graph in the same way. If there is zero x-acceleration this graph is a horizontal line (the x-velocity is constant); adding a constant x-acceleration gives a slope to the v_x-t graph (Fig. 2.19b).

2.18 (a) Straight-line motion with constant acceleration. (b) A position-time (x-t) graph for this motion (the same motion as is shown in Figs. 2.15, 2.16, and 2.17). For this motion the initial position x_0, the initial velocity v_{0x}, and the acceleration a_x are all positive.

(a) A race car moves in the x-direction with constant acceleration.

(b) The x-t graph

$$v_x = v_{0x} + a_x t$$

During time interval t, the x-velocity changes by $v_x - v_{0x} = a_x t$.

Slope $= v_x$

Constant x-acceleration: x-t graph is a parabola.

Slope $= v_{0x}$

2.19 (a) How a constant x-acceleration affects a body's (a) x-t graph and (b) v_x-t graph.

(a) An x-t graph for an object moving with positive constant x-acceleration

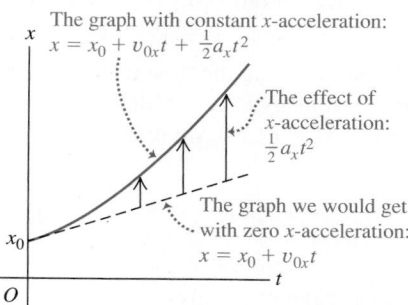

The graph with constant x-acceleration:
$$x = x_0 + v_{0x}t + \tfrac{1}{2}a_x t^2$$

The effect of x-acceleration: $\tfrac{1}{2}a_x t^2$

The graph we would get with zero x-acceleration:
$$x = x_0 + v_{0x}t$$

(b) The v_x-t graph for the same object

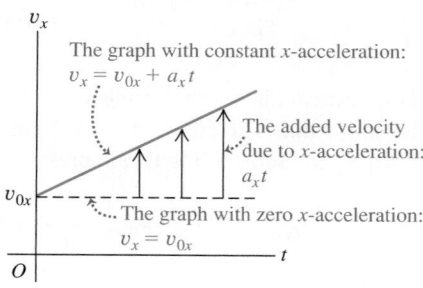

The graph with constant x-acceleration:
$$v_x = v_{0x} + a_x t$$

The added velocity due to x-acceleration: $a_x t$

The graph with zero x-acceleration:
$$v_x = v_{0x}$$

Just as the change in x-velocity of the particle equals the area under the a_x-t graph, the displacement—that is, the change in position—equals the area under the v_x-t graph. To be specific, the displacement $x - x_0$ of the particle between $t = 0$ and any later time t equals the area under the v_x-t graph between those two times. In Fig. 2.17 we divide the area under the graph into a dark-colored rectangle (vertical side v_{0x}, horizontal side t, and area $v_{0x}t$) and a light-colored right triangle (vertical side $a_x t$, horizontal side t, and area $\frac{1}{2}(a_x t)(t) = \frac{1}{2}a_x t^2$). The total area under the v_x-t graph is

Mastering**PHYSICS**

PhET: The Moving Man
ActivPhysics 1.8: Seat Belts Save Lives
ActivPhysics 1.9: Screeching to a Halt
ActivPhysics 1.11: Car Starts, Then Stops
ActivPhysics 1.12: Solving Two-Vehicle
Problems
ActivPhysics 1.13: Car Catches Truck
ActivPhysics 1.14: Avoiding a Rear-End
Collision

$$x - x_0 = v_{0x}t + \tfrac{1}{2}a_x t^2$$

in agreement with Eq. (2.12).

The displacement during a time interval is always equal to the area under the v_x-t curve. This is true even if the acceleration is *not* constant, although in that case Eq. (2.12) does not apply. (We'll show this in Section 2.6.)

It's often useful to have a relationship for position, x-velocity, and (constant) x-acceleration that does not involve the time. To obtain this, we first solve Eq. (2.8) for t and then substitute the resulting expression into Eq. (2.12):

$$t = \frac{v_x - v_{0x}}{a_x}$$

$$x = x_0 + v_{0x}\left(\frac{v_x - v_{0x}}{a_x}\right) + \tfrac{1}{2}a_x\left(\frac{v_x - v_{0x}}{a_x}\right)^2$$

We transfer the term x_0 to the left side and multiply through by $2a_x$:

$$2a_x(x - x_0) = 2v_{0x}v_x - 2v_{0x}^2 + v_x^2 - 2v_{0x}v_x + v_{0x}^2$$

Finally, simplifying gives us

$$v_x^2 = v_{0x}^2 + 2a_x(x - x_0) \qquad \text{(constant } x\text{-acceleration only)} \qquad \text{(2.13)}$$

We can get one more useful relationship by equating the two expressions for $v_{av\text{-}x}$, Eqs. (2.9) and (2.10), and multiplying through by t. Doing this, we obtain

$$x - x_0 = \left(\frac{v_{0x} + v_x}{2}\right)t \qquad \text{(constant } x\text{-acceleration only)} \qquad \text{(2.14)}$$

Note that Eq. (2.14) does not contain the x-acceleration a_x. This equation can be handy when a_x is constant but its value is unknown.

Equations (2.8), (2.12), (2.13), and (2.14) are the *equations of motion with constant acceleration* (Table 2.4). By using these equations, we can solve *any* problem involving straight-line motion of a particle with constant acceleration.

For the particular case of motion with constant x-acceleration depicted in Fig. 2.15 and graphed in Figs. 2.16, 2.17, and 2.18, the values of x_0, v_{0x}, and a_x are all positive. We invite you to redraw these figures for cases in which one, two, or all three of these quantities are negative.

Table 2.4 Equations of Motion with Constant Acceleration

Equation		Includes Quantities		
$v_x = v_{0x} + a_x t$	(2.8)	t		v_x a_x
$x = x_0 + v_{0x}t + \tfrac{1}{2}a_x t^2$	(2.12)	t	x	a_x
$v_x^2 = v_{0x}^2 + 2a_x(x - x_0)$	(2.13)		x	v_x a_x
$x - x_0 = \left(\dfrac{v_{0x} + v_x}{2}\right)t$	(2.14)	t	x	v_x

Problem-Solving Strategy 2.1   Motion with Constant Acceleration

IDENTIFY *the relevant concepts:* In most straight-line motion problems, you can use the constant-acceleration equations (2.8), (2.12), (2.13), and (2.14). If you encounter a situation in which the acceleration *isn't* constant, you'll need a different approach (see Section 2.6).

SET UP *the problem* using the following steps:
1. Read the problem carefully. Make a motion diagram showing the location of the particle at the times of interest. Decide where to place the origin of coordinates and which axis direction is positive. It's often helpful to place the particle at the origin at time $t = 0$; then $x_0 = 0$. Remember that your choice of the positive axis direction automatically determines the positive directions for x-velocity and x-acceleration. If x is positive to the right of the origin, then v_x and a_x are also positive toward the right.
2. Identify the physical quantities (times, positions, velocities, and accelerations) that appear in Eqs. (2.8), (2.12), (2.13), and (2.14) and assign them appropriate symbols — x, x_0, v_x, v_{0x}, and a_x, or symbols related to those. Translate the prose into physics: "*When* does the particle arrive at its highest point" means "What is the value of t when x has its maximum value?" In Example 2.4 below, "Where is the motorcyclist when his velocity is 25 m/s?" means "What is the value of x when $v_x = 25$ m/s?" Be alert for implicit information. For example, "A car sits at a stop light" usually means $v_{0x} = 0$.
3. Make a list of the quantities such as x, x_0, v_x, v_{0x}, a_x, and t. Some of them will be known and some will be unknown.

Write down the values of the known quantities, and decide which of the unknowns are the target variables. Make note of the *absence* of any of the quantities that appear in the four constant-acceleration equations.
4. Use Table 2.4 to identify the applicable equations. (These are often the equations that don't include any of the absent quantities that you identified in step 3.) Usually you'll find a single equation that contains only one of the target variables. Sometimes you must find two equations, each containing the same two unknowns.
5. Sketch graphs corresponding to the applicable equations. The v_x-t graph of Eq. (2.8) is a straight line with slope a_x. The x-t graph of Eq. (2.12) is a parabola that's concave up if a_x is positive and concave down if a_x is negative.
6. On the basis of your accumulated experience with such problems, and taking account of what your sketched graphs tell you, make any qualitative and quantitative predictions you can about the solution.

EXECUTE *the solution:* If a single equation applies, solve it for the target variable, *using symbols only*; then substitute the known values and calculate the value of the target variable. If you have two equations in two unknowns, solve them simultaneously for the target variables.

EVALUATE *your answer:* Take a hard look at your results to see whether they make sense. Are they within the general range of values that you expected?

Example 2.4   Constant-acceleration calculations

A motorcyclist heading east through a small town accelerates at a constant 4.0 m/s² after he leaves the city limits (Fig. 2.20). At time $t = 0$ he is 5.0 m east of the city-limits signpost, moving east at 15 m/s. (a) Find his position and velocity at $t = 2.0$ s. (b) Where is he when his velocity is 25 m/s?

SOLUTION

IDENTIFY and SET UP: The x-acceleration is constant, so we can use the constant-acceleration equations. We take the signpost as the origin of coordinates ($x = 0$) and choose the positive x-axis to point east (see Fig. 2.20, which is also a motion diagram). The known variables are the initial position and velocity, $x_0 = 5.0$ m and $v_{0x} = 15$ m/s, and the acceleration, $a_x = 4.0$ m/s². The unknown target variables in part (a) are the values of the position x and the x-velocity v_x at $t = 2.0$ s; the target variable in part (b) is the value of x when $v_x = 25$ m/s.

EXECUTE: (a) Since we know the values of x_0, v_{0x}, and a_x, Table 2.4 tells us that we can find the position x at $t = 2.0$ s by using

2.20 A motorcyclist traveling with constant acceleration.

Eq. (2.12) and the x-velocity v_x at this time by using Eq. (2.8):

$$x = x_0 + v_{0x}t + \tfrac{1}{2}a_x t^2$$
$$= 5.0 \text{ m} + (15 \text{ m/s})(2.0 \text{ s}) + \tfrac{1}{2}(4.0 \text{ m/s}^2)(2.0 \text{ s})^2$$
$$= 43 \text{ m}$$

$$v_x = v_{0x} + a_x t$$
$$= 15 \text{ m/s} + (4.0 \text{ m/s}^2)(2.0 \text{ s}) = 23 \text{ m/s}$$

(b) We want to find the value of x when $v_x = 25$ m/s, but we don't know the time when the motorcycle has this velocity. Table 2.4 tells us that we should use Eq. (2.13), which involves x, v_x, and a_x but does *not* involve t:

$$v_x^2 = v_{0x}^2 + 2a_x(x - x_0)$$

Solving for x and substituting the known values, we find

$$x = x_0 + \frac{v_x^2 - v_{0x}^2}{2a_x}$$
$$= 5.0 \text{ m} + \frac{(25 \text{ m/s})^2 - (15 \text{ m/s})^2}{2(4.0 \text{ m/s}^2)} = 55 \text{ m}$$

EVALUATE: You can check the result in part (b) by first using Eq. (2.8), $v_x = v_{0x} + a_x t$, to find the time at which $v_x = 25$ m/s, which turns out to be $t = 2.5$ s. You can then use Eq. (2.12), $x = x_0 + v_{0x}t + \tfrac{1}{2}a_x t^2$, to solve for x. You should find $x = 55$ m, the same answer as above. That's the long way to solve the problem, though. The method we used in part (b) is much more efficient.

Example 2.5 Two bodies with different accelerations

A motorist traveling with a constant speed of 15 m/s (about 34 mi/h) passes a school-crossing corner, where the speed limit is 10 m/s (about 22 mi/h). Just as the motorist passes the school-crossing sign, a police officer on a motorcycle stopped there starts in pursuit with a constant acceleration of 3.0 m/s² (Fig. 2.21a). (a) How much time elapses before the officer passes the motorist? (b) What is the officer's speed at that time? (c) At that time, what distance has each vehicle traveled?

SOLUTION

IDENTIFY and SET UP: The officer and the motorist both move with constant acceleration (equal to zero for the motorist), so we can use the constant-acceleration formulas. We take the origin at the sign, so $x_0 = 0$ for both, and we take the positive direction to the right. Let x_P and x_M represent the positions of the officer and the motorist at any time; their initial velocities are $v_{P0x} = 0$ and $v_{M0x} = 15$ m/s, and their accelerations are $a_{Px} = 3.0$ m/s² and $a_{Mx} = 0$. Our target variable in part (a) is the time when the officer passes the motorist—that is, when the two vehicles are at the same position x; Table 2.4 tells us that Eq. (2.12) is useful for this part. In part (b) we're looking for the officer's speed v (the magnitude of his velocity) at the time found in part (a). We'll use Eq. (2.8) for this part. In part (c) we'll use Eq. (2.12) again to find the position of either vehicle at this same time.

Figure 2.21b shows an x-t graph for both vehicles. The straight line represents the motorist's motion, $x_M = x_{M0} + v_{M0x}t = v_{M0x}t$. The graph for the officer's motion is the right half of a concave–up parabola:

$$x_P = x_{P0} + v_{P0x}t + \tfrac{1}{2}a_{Px}t^2 = \tfrac{1}{2}a_{Px}t^2$$

A good sketch will show that the officer and motorist are at the same position ($x_P = x_M$) at about $t = 10$ s, at which time both have traveled about 150 m from the sign.

EXECUTE: (a) To find the value of the time t at which the motorist and police officer are at the same position, we set $x_P = x_M$ by equating the expressions above and solving that equation for t:

$$v_{M0x}t = \tfrac{1}{2}a_{Px}t^2$$

$$t = 0 \quad \text{or} \quad t = \frac{2v_{M0x}}{a_{Px}} = \frac{2(15 \text{ m/s})}{3.0 \text{ m/s}^2} = 10 \text{ s}$$

Both vehicles have the same x-coordinate at *two* times, as Fig. 2.21b indicates. At $t = 0$ the motorist passes the officer; at $t = 10$ s the officer passes the motorist.

(b) We want the magnitude of the officer's x-velocity v_{Px} at the time t found in part (a). Substituting the values of v_{P0x} and a_{Px} into Eq. (2.8) along with $t = 10$ s from part (a), we find

$$v_{Px} = v_{P0x} + a_{Px}t = 0 + (3.0 \text{ m/s}^2)(10 \text{ s}) = 30 \text{ m/s}$$

The officer's speed is the absolute value of this, which is also 30 m/s.

(c) In 10 s the motorist travels a distance

$$x_M = v_{M0x}t = (15 \text{ m/s})(10 \text{ s}) = 150 \text{ m}$$

and the officer travels

$$x_P = \tfrac{1}{2}a_{Px}t^2 = \tfrac{1}{2}(3.0 \text{ m/s}^2)(10 \text{ s})^2 = 150 \text{ m}$$

This verifies that they have gone equal distances when the officer passes the motorist.

EVALUATE: Our results in parts (a) and (c) agree with our estimates from our sketch. Note that at the time when the officer passes the motorist, they do *not* have the same velocity. At this time the motorist is moving at 15 m/s and the officer is moving at 30 m/s. You can also see this from Fig. 2.21b. Where the two x-t curves cross, their slopes (equal to the values of v_x for the two vehicles) are different.

Is it just coincidence that when the two vehicles are at the same position, the officer is going twice the speed of the motorist? Equation (2.14), $x - x_0 = [(v_{0x} + v_x)/2]t$, gives the answer. The motorist has constant velocity, so $v_{M0x} = v_{Mx}$, and the distance $x - x_0$ that the motorist travels in time t is $v_{M0x}t$. The officer has zero initial velocity, so in the same time t the officer travels a distance $\tfrac{1}{2}v_{Px}t$. If the two vehicles cover the same distance in the same amount of time, the two values of $x - x_0$ must be the same. Hence when the officer passes the motorist $v_{M0x}t = \tfrac{1}{2}v_{Px}t$ and $v_{Px} = 2v_{M0x}$—that is, the officer has exactly twice the motorist's velocity. Note that this is true no matter what the value of the officer's acceleration.

2.21 (a) Motion with constant acceleration overtaking motion with constant velocity. (b) A graph of x versus t for each vehicle.

(b)

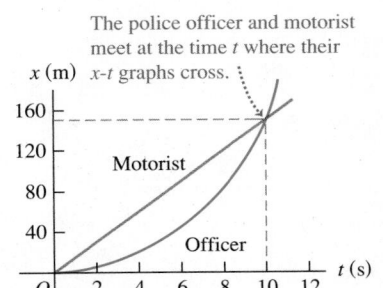

The police officer and motorist meet at the time t where their x-t graphs cross.

(a)

Police officer: initially at rest, constant x-acceleration

$a_{Px} = 3.0$ m/s²

Motorist: constant x-velocity

$v_{M0x} = 15$ m/s

Test Your Understanding of Section 2.4 Four possible v_x-t graphs are shown for the two vehicles in Example 2.5. Which graph is correct?

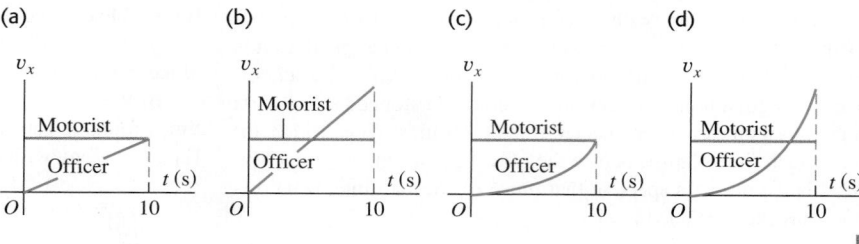

2.5 Freely Falling Bodies

2.22 Multiflash photo of a freely falling ball.

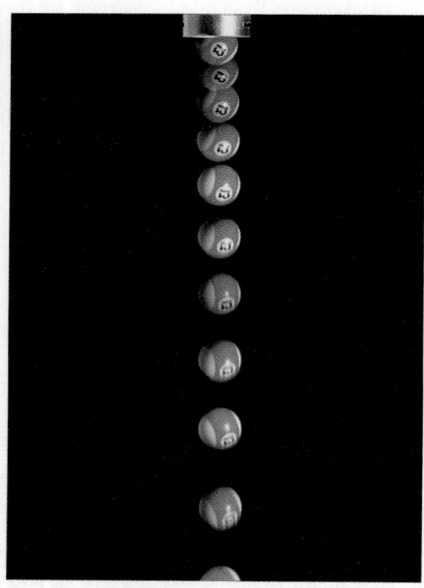

MasteringPHYSICS

PhET: Lunar Lander
ActivPhysics 1.7: Balloonist Drops Lemonade
ActivPhysics 1.10: Pole-Vaulter Lands

The most familiar example of motion with (nearly) constant acceleration is a body falling under the influence of the earth's gravitational attraction. Such motion has held the attention of philosophers and scientists since ancient times. In the fourth century B.C., Aristotle thought (erroneously) that heavy bodies fall faster than light bodies, in proportion to their weight. Nineteen centuries later, Galileo (see Section 1.1) argued that a body should fall with a downward acceleration that is constant and independent of its weight.

Experiment shows that if the effects of the air can be neglected, Galileo is right; all bodies at a particular location fall with the same downward acceleration, regardless of their size or weight. If in addition the distance of the fall is small compared with the radius of the earth, and if we ignore small effects due to the earth's rotation, the acceleration is constant. The idealized motion that results under all of these assumptions is called **free fall,** although it includes rising as well as falling motion. (In Chapter 3 we will extend the discussion of free fall to include the motion of projectiles, which move both vertically and horizontally.)

Figure 2.22 is a photograph of a falling ball made with a stroboscopic light source that produces a series of short, intense flashes. As each flash occurs, an image of the ball at that instant is recorded on the photograph. There are equal time intervals between flashes, so the average velocity of the ball between successive flashes is proportional to the distance between corresponding images. The increasing distances between images show that the velocity is continuously changing; the ball is accelerating downward. Careful measurement shows that the velocity change is the same in each time interval, so the acceleration of the freely falling ball is constant.

The constant acceleration of a freely falling body is called the **acceleration due to gravity,** and we denote its magnitude with the letter g. We will frequently use the approximate value of g at or near the earth's surface:

$$g = 9.8 \text{ m/s}^2 = 980 \text{ cm/s}^2 = 32 \text{ ft/s}^2 \qquad \text{(approximate value near the earth's surface)}$$

The exact value varies with location, so we will often give the value of g at the earth's surface to only two significant figures. On the surface of the moon, the acceleration due to gravity is caused by the attractive force of the moon rather than the earth, and $g = 1.6 \text{ m/s}^2$. Near the surface of the sun, $g = 270 \text{ m/s}^2$.

CAUTION g **is always a positive number** Because g is the *magnitude* of a vector quantity, it is always a *positive* number. If you take the positive direction to be upward, as we do in Example 2.6 and in most situations involving free fall, the acceleration is negative (downward) and equal to $-g$. Be careful with the sign of g, or else you'll have no end of trouble with free-fall problems.

In the following examples we use the constant-acceleration equations developed in Section 2.4. You should review Problem-Solving Strategy 2.1 in that section before you study the next examples.

Example 2.6 A freely falling coin

A one-euro coin is dropped from the Leaning Tower of Pisa and falls freely from rest. What are its position and velocity after 1.0 s, 2.0 s, and 3.0 s?

SOLUTION

IDENTIFY and SET UP: "Falls freely" means "falls with constant acceleration due to gravity," so we can use the constant-acceleration equations. The right side of Fig. 2.23 shows our motion diagram for the coin. The motion is vertical, so we use a vertical

2.23 A coin freely falling from rest.

coordinate axis and call the coordinate y instead of x. We take the origin O at the starting point and the *upward* direction as positive. The initial coordinate y_0 and initial y-velocity v_{0y} are both zero. The y-acceleration is downward (in the negative y-direction), so $a_y = -g = -9.8 \text{ m/s}^2$. (Remember that, by definition, g itself is a positive quantity.) Our target variables are the values of y and v_y at the three given times. To find these, we use Eqs. (2.12) and (2.8) with x replaced by y. Our choice of the upward direction as positive means that all positions and velocities we calculate will be negative.

EXECUTE: At a time t after the coin is dropped, its position and y-velocity are

$$y = y_0 + v_{0y}t + \tfrac{1}{2}a_yt^2 = 0 + 0 + \tfrac{1}{2}(-g)t^2 = (-4.9 \text{ m/s}^2)t^2$$

$$v_y = v_{0y} + a_yt = 0 + (-g)t = (-9.8 \text{ m/s}^2)t$$

When $t = 1.0$ s, $y = (-4.9 \text{ m/s}^2)(1.0 \text{ s})^2 = -4.9$ m and $v_y = (-9.8 \text{ m/s}^2)(1.0 \text{ s}) = -9.8$ m/s; after 1 s, the coin is 4.9 m below the origin (y is negative) and has a downward velocity (v_y is negative) with magnitude 9.8 m/s.

We can find the positions and y-velocities at 2.0 s and 3.0 s in the same way. The results are $y = -20$ m and $v_y = -20$ m/s at $t = 2.0$ s, and $y = -44$ m and $v_y = -29$ m/s at $t = 3.0$ s.

EVALUATE: All our answers are negative, as we expected. If we had chosen the positive y-axis to point downward, the acceleration would have been $a_y = +g$ and all our answers would have been positive.

Example 2.7 Up-and-down motion in free fall

You throw a ball vertically upward from the roof of a tall building. The ball leaves your hand at a point even with the roof railing with an upward speed of 15.0 m/s; the ball is then in free fall. On its way back down, it just misses the railing. Find (a) the ball's position and velocity 1.00 s and 4.00 s after leaving your hand; (b) the ball's velocity when it is 5.00 m above the railing; (c) the maximum height reached; (d) the ball's acceleration when it is at its maximum height.

SOLUTION

IDENTIFY and SET UP: The words "in free fall" mean that the acceleration is due to gravity, which is constant. Our target variables are position [in parts (a) and (c)], velocity [in parts (a) and (b)], and acceleration [in part (d)]. We take the origin at the point where the ball leaves your hand, and take the positive direction to be upward (Fig. 2.24). The initial position y_0 is zero, the initial y-velocity v_{0y} is $+15.0$ m/s, and the y-acceleration is $a_y = -g = -9.80 \text{ m/s}^2$.

In part (a), as in Example 2.6, we'll use Eqs. (2.12) and (2.8) to find the position and velocity as functions of time. In part (b) we must find the velocity at a given *position* (no time is given), so we'll use Eq. (2.13).

Figure 2.25 shows the y-t and v_y-t graphs for the ball. The y-t graph is a concave-down parabola that rises and then falls, and the v_y-t graph is a downward-sloping straight line. Note that the ball's velocity is zero when it is at its highest point.

EXECUTE: (a) The position and y-velocity at time t are given by Eqs. (2.12) and (2.8) with x's replaced by y's:

$$y = y_0 + v_{0y}t + \tfrac{1}{2}a_yt^2 = y_0 + v_{0y}t + \tfrac{1}{2}(-g)t^2$$

$$= (0) + (15.0 \text{ m/s})t + \tfrac{1}{2}(-9.80 \text{ m/s}^2)t^2$$

$$v_y = v_{0y} + a_yt = v_{0y} + (-g)t$$

$$= 15.0 \text{ m/s} + (-9.80 \text{ m/s}^2)t$$

Continued

When $t = 1.00$ s, these equations give $y = +10.1$ m and $v_y = +5.2$ m/s. That is, the ball is 10.1 m above the origin (y is positive) and moving upward (v_y is positive) with a speed of 5.2 m/s. This is less than the initial speed because the ball slows as it ascends. When $t = 4.00$ s, those equations give $y = -18.4$ m and $v_y = -24.2$ m/s. The ball has passed its highest point and is 18.4 m *below* the origin (y is negative). It is moving *downward* (v_y is negative) with a speed of 24.2 m/s. The ball gains speed as it descends; Eq. (2.13) tells us that it is moving at the initial 15.0-m/s speed as it moves downward past the ball's launching point, and continues to gain speed as it descends further.

(b) The y-velocity at any position y is given by Eq. (2.13) with x's replaced by y's:

$$v_y^2 = v_{0y}^2 + 2a_y(y - y_0) = v_{0y}^2 + 2(-g)(y - 0)$$
$$= (15.0 \text{ m/s})^2 + 2(-9.80 \text{ m/s}^2)y$$

When the ball is 5.00 m above the origin we have $y = +5.00$ m, so

$$v_y^2 = (15.0 \text{ m/s})^2 + 2(-9.80 \text{ m/s}^2)(5.00 \text{ m}) = 127 \text{ m}^2/\text{s}^2$$
$$v_y = \pm 11.3 \text{ m/s}$$

We get *two* values of v_y because the ball passes through the point $y = +5.00$ m twice, once on the way up (so v_y is positive) and once on the way down (so v_y is negative) (see Figs. 2.24 and 2.25a).

2.24 Position and velocity of a ball thrown vertically upward.

The ball actually moves straight up and then straight down; we show a U-shaped path for clarity.

$v_y = 0$
$t = ?$
$y = ?$
$y = ?$
$t = 1.00$ s, $v_y = ?$
$t = ?, v_y = ?$
$y = 5.00$ m
$t = ?$
$v_y = ?$
$t = 0, v_{0y} = 15.0$ m/s
$y = 0$

$a_y = -g$
$= -9.80$ m/s^2

$y = ?$
$t = 4.00$ s
$v_y = ?$

(c) At the instant at which the ball reaches its maximum height y_1, its y-velocity is momentarily zero: $v_y = 0$. We use Eq. (2.13) to find y_1. With $v_y = 0$, $y_0 = 0$, and $a_y = -g$, we get

$$0 = v_{0y}^2 + 2(-g)(y_1 - 0)$$
$$y_1 = \frac{v_{0y}^2}{2g} = \frac{(15.0 \text{ m/s})^2}{2(9.80 \text{ m/s}^2)} = +11.5 \text{ m}$$

(d) **CAUTION** **A free-fall misconception** It's a common misconception that at the highest point of free-fall motion, where the velocity is zero, the acceleration is also zero. If this were so, once the ball reached the highest point it would hang there suspended in midair! Remember that acceleration is the rate of change of velocity, and the ball's velocity is continuously changing. At every point, including the highest point, and at any velocity, including zero, the acceleration in free fall is always $a_y = -g = -9.80$ m/s^2.

EVALUATE: A useful way to check any free-fall problem is to draw the y-t and v_y-t graphs as we did in Fig. 2.25. Note that these are graphs of Eqs. (2.12) and (2.8), respectively. Given the numerical values of the initial position, initial velocity, and acceleration, you can easily create these graphs using a graphing calculator or an online mathematics program.

2.25 (a) Position and (b) velocity as functions of time for a ball thrown upward with an initial speed of 15 m/s.

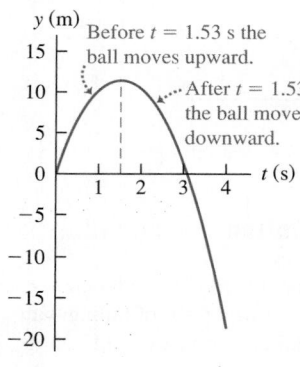

(a) y-t graph (curvature is downward because $a_y = -g$ is negative)

y (m)

Before $t = 1.53$ s the ball moves upward.

After $t = 1.53$ s the ball moves downward.

(b) v_y-t graph (straight line with negative slope because $a_y = -g$ is constant and negative)

v_y (m/s)

Before $t = 1.53$ s the y-velocity is positive.

After $t = 1.53$ s the y-velocity is negative.

Example 2.8 **Two solutions or one?**

At what time after being released has the ball in Example 2.7 fallen 5.00 m below the roof railing?

SOLUTION

IDENTIFY and SET UP: We treat this as in Example 2.7, so y_0, v_{0y}, and $a_y = -g$ have the same values as there. In this example, however, the target variable is the time at which the ball is at $y = -5.00$ m.

The best equation to use is Eq. (2.12), which gives the position y as a function of time t:

$$y = y_0 + v_{0y}t + \tfrac{1}{2}a_yt^2 = y_0 + v_{0y}t + \tfrac{1}{2}(-g)t^2$$

This is a *quadratic* equation for t, which we want to solve for the value of t when $y = -5.00$ m.

EXECUTE: We rearrange the equation so that is has the standard form of a quadratic equation for an unknown x, $Ax^2 + Bx + C = 0$:

$$(\tfrac{1}{2}g)t^2 + (-v_{0y})t + (y - y_0) = At^2 + Bt + C = 0$$

By comparison, we identify $A = \tfrac{1}{2}g$, $B = -v_{0y}$, and $C = y - y_0$. The quadratic formula (see Appendix B) tells us that this equation has *two* solutions:

$$t = \frac{-B \pm \sqrt{B^2 - 4AC}}{2A}$$

$$= \frac{-(-v_{0y}) \pm \sqrt{(-v_{0y})^2 - 4(\tfrac{1}{2}g)(y - y_0)}}{2(\tfrac{1}{2}g)}$$

$$= \frac{v_{0y} \pm \sqrt{v_{0y}^2 - 2g(y - y_0)}}{g}$$

Substituting the values $y_0 = 0$, $v_{0y} = +15.0$ m/s, $g = 9.80$ m/s^2, and $y = -5.00$ m, we find

$$t = \frac{(15.0 \text{ m/s}) \pm \sqrt{(15.0 \text{ m/s})^2 - 2(9.80 \text{ m/s}^2)(-5.00 \text{ m} - 0)}}{9.80 \text{ m/s}^2}$$

You can confirm that the numerical answers are $t = +3.36$ s and $t = -0.30$ s. The answer $t = -0.30$ s doesn't make physical sense, since it refers to a time *before* the ball left your hand at $t = 0$. So the correct answer is $t = +3.36$ s.

EVALUATE: Why did we get a second, fictitious solution? The explanation is that constant-acceleration equations like Eq. (2.12) are based on the assumption that the acceleration is constant for *all* values of time, whether positive, negative, or zero. Hence the solution $t = -0.30$ s refers to an imaginary moment when a freely falling ball was 5.00 m below the roof railing and rising to meet your hand. Since the ball didn't leave your hand and go into free fall until $t = 0$, this result is pure fiction.

You should repeat these calculations to find the times when the ball is 5.00 m *above* the origin ($y = +5.00$ m). The two answers are $t = +0.38$ s and $t = +2.68$ s. These are both positive values of t, and both refer to the real motion of the ball after leaving your hand. At the earlier time the ball passes through $y = +5.00$ m moving upward; at the later time it passes through this point moving downward. [Compare this with part (b) of Example 2.7, and again refer to Fig. 2.25a.]

You should also solve for the times when $y = +15.0$ m. In this case, both solutions involve the square root of a negative number, so there are *no* real solutions. Again Fig. 2.25a shows why; we found in part (c) of Example 2.7 that the ball's maximum height is $y = +11.5$ m, so it *never* reaches $y = +15.0$ m. While a quadratic equation such as Eq. (2.12) always has two solutions, in some situations one or both of the solutions will not be physically reasonable.

Test Your Understanding of Section 2.5 If you toss a ball upward with a certain initial speed, it falls freely and reaches a maximum height h a time t after it leaves your hand. (a) If you throw the ball upward with double the initial speed, what new maximum height does the ball reach? (i) $h\sqrt{2}$; (ii) $2h$; (iii) $4h$; (iv) $8h$; (v) $16h$. (b) If you throw the ball upward with double the initial speed, how long does it take to reach its new maximum height? (i) $t/2$; (ii) $t/\sqrt{2}$; (iii) t; (iv) $t\sqrt{2}$; (v) $2t$.

2.6 Velocity and Position by Integration

This section is intended for students who have already learned a little integral calculus. In Section 2.4 we analyzed the special case of straight-line motion with constant acceleration. When a_x is not constant, as is frequently the case, the equations that we derived in that section are no longer valid (Fig. 2.26). But even when a_x varies with time, we can still use the relationship $v_x = dx/dt$ to find the x-velocity v_x as a function of time if the position x is a known function of time. And we can still use $a_x = dv_x/dt$ to find the x-acceleration a_x as a function of time if the x-velocity v_x is a known function of time.

In many situations, however, position and velocity are not known functions of time, while acceleration is (Fig. 2.27). How can we find the position and velocity in straight-line motion from the acceleration function $a_x(t)$?

We first consider a graphical approach. Figure 2.28 is a graph of x-acceleration versus time for a body whose acceleration is not constant. We can divide the time interval between times t_1 and t_2 into many smaller intervals, calling a typical one Δt. Let the average x-acceleration during Δt be $a_{\text{av-}x}$. From Eq. (2.4) the change in x-velocity Δv_x during Δt is

$$\Delta v_x = a_{\text{av-}x} \, \Delta t$$

Graphically, Δv_x equals the area of the shaded strip with height $a_{\text{av-}x}$ and width Δt—that is, the area under the curve between the left and right sides of Δt. The total change in x-velocity during any interval (say, t_1 to t_2) is the sum of the x-velocity changes Δv_x in the small subintervals. So the total x-velocity change is represented graphically by the *total* area under the a_x-t curve between the vertical

2.26 When you push your car's accelerator pedal to the floorboard, the resulting acceleration is *not* constant: The greater the car's speed, the more slowly it gains additional speed. A typical car takes twice as long to accelerate from 50 km/h to 100 km/h as it does to accelerate from 0 to 50 km/h.

2.27 The inertial navigation system (INS) on board a long-range airliner keeps track of the airliner's acceleration. The pilots input the airliner's initial position and velocity before takeoff, and the INS uses the acceleration data to calculate the airliner's position and velocity throughout the flight.

2.28 An a_x-t graph for a body whose x-acceleration is not constant.

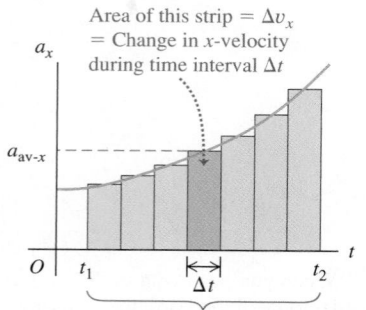

Area of this strip = Δv_x
= Change in x-velocity during time interval Δt

a_x

$a_{\text{av-}x}$

O t_1 Δt t_2 t

Total area under the x-t graph from t_1 to t_2
= Net change in x-velocity from t_1 to t_2

lines t_1 and t_2. (In Section 2.4 we showed this for the special case in which the acceleration is constant.)

In the limit that all the Δt's become very small and their number very large, the value of $a_{\text{av-}x}$ for the interval from any time t to $t + \Delta t$ approaches the instantaneous x-acceleration a_x at time t. In this limit, the area under the a_x-t curve is the *integral* of a_x (which is in general a function of t) from t_1 to t_2. If v_{1x} is the x-velocity of the body at time t_1 and v_{2x} is the velocity at time t_2, then

$$v_{2x} - v_{1x} = \int_{v_{1x}}^{v_{2x}} dv_x = \int_{t_1}^{t_2} a_x \, dt \qquad (2.15)$$

The change in the x-velocity v_x is the time integral of the x-acceleration a_x.

We can carry out exactly the same procedure with the curve of x-velocity versus time. If x_1 is a body's position at time t_1 and x_2 is its position at time t_2, from Eq. (2.2) the displacement Δx during a small time interval Δt is equal to $v_{\text{av-}x} \, \Delta t$, where $v_{\text{av-}x}$ is the average x-velocity during Δt. The total displacement $x_2 - x_1$ during the interval $t_2 - t_1$ is given by

$$x_2 - x_1 = \int_{x_1}^{x_2} dx = \int_{t_1}^{t_2} v_x \, dt \qquad (2.16)$$

The change in position x—that is, the displacement—is the time integral of x-velocity v_x. Graphically, the displacement between times t_1 and t_2 is the area under the v_x-t curve between those two times. [This is the same result that we obtained in Section 2.4 for the special case in which v_x is given by Eq. (2.8).]

If $t_1 = 0$ and t_2 is any later time t, and if x_0 and v_{0x} are the position and velocity, respectively, at time $t = 0$, then we can rewrite Eqs. (2.15) and (2.16) as follows:

$$v_x = v_{0x} + \int_0^t a_x \, dt \qquad (2.17)$$

$$x = x_0 + \int_0^t v_x \, dt \qquad (2.18)$$

Here x and v_x are the position and x-velocity at time t. If we know the x-acceleration a_x as a function of time and we know the initial velocity v_{0x}, we can use Eq. (2.17) to find the x-velocity v_x at any time; in other words, we can find v_x as a function of time. Once we know this function, and given the initial position x_0, we can use Eq. (2.18) to find the position x at any time.

Example 2.9 | **Motion with changing acceleration**

Sally is driving along a straight highway in her 1965 Mustang. At $t = 0$, when she is moving at 10 m/s in the positive x-direction, she passes a signpost at $x = 50$ m. Her x-acceleration as a function of time is

$$a_x = 2.0 \text{ m/s}^2 - (0.10 \text{ m/s}^3)t$$

(a) Find her x-velocity v_x and position x as functions of time. (b) When is her x-velocity greatest? (c) What is that maximum x-velocity? (d) Where is the car when it reaches that maximum x-velocity?

SOLUTION

IDENTIFY and SET UP: The x-acceleration is a function of time, so we *cannot* use the constant-acceleration formulas of Section 2.4. Instead, we use Eq. (2.17) to obtain an expression for v_x as a function of time, and then use that result in Eq. (2.18) to find an expression for x as a function of t. We'll then be able to answer a variety of questions about the motion.

EXECUTE: (a) At $t = 0$, Sally's position is $x_0 = 50$ m and her x-velocity is $v_{0x} = 10$ m/s. To use Eq. (2.17), we note that the integral of t^n (except for $n = -1$) is $\int t^n dt = \frac{1}{n+1}t^{n+1}$. Hence we find

$$v_x = 10 \text{ m/s} + \int_0^t [2.0 \text{ m/s}^2 - (0.10 \text{ m/s}^3)t]\,dt$$

$$= 10 \text{ m/s} + (2.0 \text{ m/s}^2)t - \tfrac{1}{2}(0.10 \text{ m/s}^3)t^2$$

Now we use Eq. (2.18) to find x as a function of t:

$$x = 50 \text{ m} + \int_0^t \left[10 \text{ m/s} + (2.0 \text{ m/s}^2)t - \tfrac{1}{2}(0.10 \text{ m/s}^3)t^2\right] dt$$

$$= 50 \text{ m} + (10 \text{ m/s})t + \tfrac{1}{2}(2.0 \text{ m/s}^2)t^2 - \tfrac{1}{6}(0.10 \text{ m/s}^3)t^3$$

Figure 2.29 shows graphs of a_x, v_x, and x as functions of time as given by the equations above. Note that for any time t, the slope of the v_x-t graph equals the value of a_x and the slope of the x-t graph equals the value of v_x.

(b) The maximum value of v_x occurs when the x-velocity stops increasing and begins to decrease. At that instant, $dv_x/dt = a_x = 0$. So we set the expression for a_x equal to zero and solve for t:

$$0 = 2.0 \text{ m/s}^2 - (0.10 \text{ m/s}^3)t$$

$$t = \frac{2.0 \text{ m/s}^2}{0.10 \text{ m/s}^3} = 20 \text{ s}$$

(c) We find the maximum x-velocity by substituting $t = 20$ s, the time from part (b) when velocity is maximum, into the equation for v_x from part (a):

$$v_{\text{max-}x} = 10 \text{ m/s} + (2.0 \text{ m/s}^2)(20 \text{ s}) - \tfrac{1}{2}(0.10 \text{ m/s}^3)(20 \text{ s})^2$$

$$= 30 \text{ m/s}$$

(d) To find the car's position at the time that we found in part (b), we substitute $t = 20$ s into the expression for x from part (a):

$$x = 50 \text{ m} + (10 \text{ m/s})(20 \text{ s}) + \tfrac{1}{2}(2.0 \text{ m/s}^2)(20 \text{ s})^2$$

$$- \tfrac{1}{6}(0.10 \text{ m/s}^3)(20 \text{ s})^3$$

$$= 517 \text{ m}$$

EVALUATE: Figure 2.29 helps us interpret our results. The top graph shows that a_x is positive between $t = 0$ and $t = 20$ s and negative after that. It is zero at $t = 20$ s, the time at which v_x is maximum (the high point in the middle graph). The car speeds up until $t = 20$ s (because v_x and a_x have the same sign) and slows down after $t = 20$ s (because v_x and a_x have opposite signs).

Since v_x is maximum at $t = 20$ s, the x-t graph (the bottom graph in Fig. 2.29) has its maximum positive slope at this time. Note that the x-t graph is concave up (curved upward) from $t = 0$ to $t = 20$ s, when a_x is positive. The graph is concave down (curved downward) after $t = 20$ s, when a_x is negative.

2.29 The position, velocity, and acceleration of the car in Example 2.9 as functions of time. Can you show that if this motion continues, the car will stop at $t = 44.5$ s?

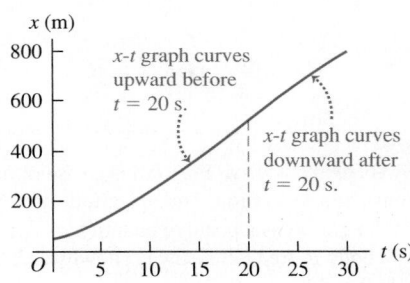

Test Your Understanding of Section 2.6 If the x-acceleration a_x is increasing with time, will the v_x-t graph be (i) a straight line, (ii) concave up (i.e., with an upward curvature), or (iii) concave down (i.e., with a downward curvature)?

CHAPTER 2 SUMMARY

Straight-line motion, average and instantaneous x-velocity: When a particle moves along a straight line, we describe its position with respect to an origin O by means of a coordinate such as x. The particle's average x-velocity $v_{\text{av-}x}$ during a time interval $\Delta t = t_2 - t_1$ is equal to its displacement $\Delta x = x_2 - x_1$ divided by Δt. The instantaneous x-velocity v_x at any time t is equal to the average x-velocity for the time interval from t to $t + \Delta t$ in the limit that Δt goes to zero. Equivalently, v_x is the derivative of the position function with respect to time. (See Example 2.1.)

$$v_{\text{av-}x} = \frac{x_2 - x_1}{t_2 - t_1} = \frac{\Delta x}{\Delta t} \quad (2.2)$$

$$v_x = \lim_{\Delta t \to 0} \frac{\Delta x}{\Delta t} = \frac{dx}{dt} \quad (2.3)$$

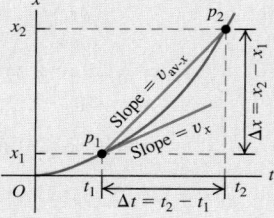

Average and instantaneous x-acceleration: The average x-acceleration $a_{\text{av-}x}$ during a time interval Δt is equal to the change in velocity $\Delta v_x = v_{2x} - v_{1x}$ during that time interval divided by Δt. The instantaneous x-acceleration a_x is the limit of $a_{\text{av-}x}$ as Δt goes to zero, or the derivative of v_x with respect to t. (See Examples 2.2 and 2.3.)

$$a_{\text{av-}x} = \frac{v_{2x} - v_{1x}}{t_2 - t_1} = \frac{\Delta v_x}{\Delta t} \quad (2.4)$$

$$a_x = \lim_{\Delta t \to 0} \frac{\Delta v_x}{\Delta t} = \frac{dv_x}{dt} \quad (2.5)$$

Straight-line motion with constant acceleration: When the x-acceleration is constant, four equations relate the position x and the x-velocity v_x at any time t to the initial position x_0, the initial x-velocity v_{0x} (both measured at time $t = 0$), and the x-acceleration a_x. (See Examples 2.4 and 2.5.)

Constant x-acceleration only:

$$v_x = v_{0x} + a_x t \quad (2.8)$$

$$x = x_0 + v_{0x}t + \tfrac{1}{2}a_x t^2 \quad (2.12)$$

$$v_x^2 = v_{0x}^2 + 2a_x(x - x_0) \quad (2.13)$$

$$x - x_0 = \left(\frac{v_{0x} + v_x}{2}\right)t \quad (2.14)$$

Freely falling bodies: Free fall is a case of motion with constant acceleration. The magnitude of the acceleration due to gravity is a positive quantity, g. The acceleration of a body in free fall is always downward. (See Examples 2.6–2.8.)

Straight-line motion with varying acceleration: When the acceleration is not constant but is a known function of time, we can find the velocity and position as functions of time by integrating the acceleration function. (See Example 2.9.)

$$v_x = v_{0x} + \int_0^t a_x\, dt \quad (2.17)$$

$$x = x_0 + \int_0^t v_x\, dt \quad (2.18)$$

The superhero Green Lantern steps from the top of a tall building. He falls freely from rest to the ground, falling half the total distance to the ground during the last 1.00 s of his fall. What is the height h of the building?

SOLUTION GUIDE

See MasteringPhysics® study area for a Video Tutor solution.

IDENTIFY and SET UP

1. You're told that Green Lantern falls freely from rest. What does this imply about his acceleration? About his initial velocity?
2. Choose the direction of the positive y-axis. It's easiest to make the same choice we used for freely falling objects in Section 2.5.
3. You can divide Green Lantern's fall into two parts: from the top of the building to the halfway point and from the halfway point to the ground. You know that the second part of the fall lasts 1.00 s. Decide what you would need to know about Green

Lantern's motion at the halfway point in order to solve for the target variable h. Then choose two equations, one for the first part of the fall and one for the second part, that you'll use together to find an expression for h. (There are several pairs of equations that you could choose.)

EXECUTE

4. Use your two equations to solve for the height h. Note that heights are always positive numbers, so your answer should be positive.

EVALUATE

5. To check your answer for h, use one of the free-fall equations to find how long it takes Green Lantern to fall (i) from the top of the building to half the height and (ii) from the top of the building to the ground. If your answer for h is correct, time (ii) should be 1.00 s greater than time (i). If it isn't, you'll need to go back and look for errors in how you found h.

Problems   For instructor-assigned homework, go to www.masteringphysics.com

•, ••, •••: Problems of increasing difficulty. **CP**: Cumulative problems incorporating material from earlier chapters. **CALC**: Problems requiring calculus. **BIO**: Biosciences problems.

DISCUSSION QUESTIONS

Q2.1 Does the speedometer of a car measure speed or velocity? Explain.

Q2.2 The top diagram in Fig. Q2.2 represents a series of high-speed photographs of an insect flying in a straight line from left to right (in the positive x-direction). Which of the graphs in Fig. Q2.2 most plausibly depicts this insect's motion?

Figure **Q2.2**

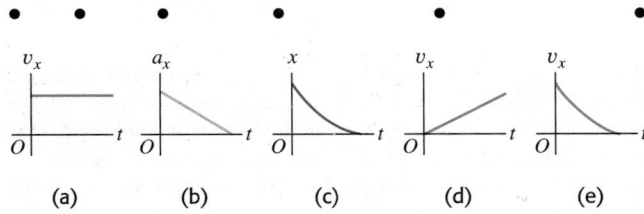

(a)   (b)   (c)   (d)   (e)

Q2.3 Can an object with constant acceleration reverse its direction of travel? Can it reverse its direction *twice*? In each case, explain your reasoning.

Q2.4 Under what conditions is average velocity equal to instantaneous velocity?

Q2.5 Is it possible for an object (a) to be slowing down while its acceleration is increasing in magnitude; (b) to be speeding up while its acceleration is decreasing? In each case, explain your reasoning.

Q2.6 Under what conditions does the magnitude of the average velocity equal the average speed?

Q2.7 When a Dodge Viper is at Elwood's Car Wash, a BMW Z3 is at Elm and Main. Later, when the Dodge reaches Elm and Main,

the BMW reaches Elwood's Car Wash. How are the cars' average velocities between these two times related?

Q2.8 A driver in Massachusetts was sent to traffic court for speeding. The evidence against the driver was that a policewoman observed the driver's car alongside a second car at a certain moment, and the policewoman had already clocked the second car as going faster than the speed limit. The driver argued, "The second car was passing me. I was not speeding." The judge ruled against the driver because, in the judge's words, "If two cars were side by side, you were both speeding." If you were a lawyer representing the accused driver, how would you argue this case?

Q2.9 Can you have a zero displacement and a nonzero average velocity? A nonzero velocity? Illustrate your answers on an x-t graph.

Q2.10 Can you have zero acceleration and nonzero velocity? Explain using a v_x-t graph.

Q2.11 Can you have zero velocity and nonzero average acceleration? Zero velocity and nonzero acceleration? Explain using a v_x-t graph, and give an example of such motion.

Q2.12 An automobile is traveling west. Can it have a velocity toward the west and at the same time have an acceleration toward the east? Under what circumstances?

Q2.13 The official's truck in Fig. 2.2 is at $x_1 = 277$ m at $t_1 = 16.0$ s and is at $x_2 = 19$ m at $t_2 = 25.0$ s. (a) Sketch *two* different possible x-t graphs for the motion of the truck. (b) Does the average velocity v_{av-x} during the time interval from t_1 to t_2 have the same value for both of your graphs? Why or why not?

Q2.14 Under constant acceleration the average velocity of a particle is half the sum of its initial and final velocities. Is this still true if the acceleration is *not* constant? Explain.

Q2.15 You throw a baseball straight up in the air so that it rises to a maximum height much greater than your height. Is the magnitude of the acceleration greater while it is being thrown or after it leaves your hand? Explain.

Q2.16 Prove these statements: (a) As long as you can neglect the effects of the air, if you throw anything vertically upward, it will have the same speed when it returns to the release point as when it was released. (b) The time of flight will be twice the time it takes to get to its highest point.

Q2.17 A dripping water faucet steadily releases drops 1.0 s apart. As these drops fall, will the distance between them increase, decrease, or remain the same? Prove your answer.

Q2.18 If the initial position and initial velocity of a vehicle are known and a record is kept of the acceleration at each instant, can you compute the vehicle's position after a certain time from these data? If so, explain how this might be done.

Q2.19 From the top of a tall building you throw one ball straight up with speed v_0 and one ball straight down with speed v_0. (a) Which ball has the greater speed when it reaches the ground? (b) Which ball gets to the ground first? (c) Which ball has a greater displacement when it reaches the ground? (d) Which ball has traveled the greater distance when it hits the ground?

Q2.20 A ball is dropped from rest from the top of a building of height h. At the same instant, a second ball is projected vertically upward from ground level, such that it has zero speed when it reaches the top of the building. When the two balls pass each other, which ball has the greater speed, or do they have the same speed? Explain. Where will the two balls be when they are alongside each other: at height $h/2$ above the ground, below this height, or above this height? Explain.

Q2.21 An object is thrown straight up into the air and feels no air resistance. How is it possible for the object to have an acceleration when it has stopped moving at its highest point?

Q2.22 When you drop an object from a certain height, it takes time T to reach the ground with no air resistance. If you dropped it from three times that height, how long (in terms of T) would it take to reach the ground?

EXERCISES

Section 2.1 Displacement, Time, and Average Velocity

2.1 • A car travels in the $+x$-direction on a straight and level road. For the first 4.00 s of its motion, the average velocity of the car is $v_{\text{av-}x} = 6.25$ m/s. How far does the car travel in 4.00 s?

2.2 •• In an experiment, a shearwater (a seabird) was taken from its nest, flown 5150 km away, and released. The bird found its way back to its nest 13.5 days after release. If we place the origin in the nest and extend the $+x$-axis to the release point, what was the bird's average velocity in m/s (a) for the return flight, and (b) for the whole episode, from leaving the nest to returning?

2.3 •• **Trip Home.** You normally drive on the freeway between San Diego and Los Angeles at an average speed of 105 km/h (65 mi/h), and the trip takes 2 h and 20 min. On a Friday afternoon, however, heavy traffic slows you down and you drive the same distance at an average speed of only 70 km/h (43 mi/h). How much longer does the trip take?

2.4 •• **From Pillar to Post.** Starting from a pillar, you run 200 m east (the $+x$-direction) at an average speed of 5.0 m/s, and then run 280 m west at an average speed of 4.0 m/s to a post. Calculate (a) your average speed from pillar to post and (b) your average velocity from pillar to post.

2.5 • Starting from the front door of your ranch house, you walk 60.0 m due east to your windmill, and then you turn around and slowly walk 40.0 m west to a bench where you sit and watch the sunrise. It takes you 28.0 s to walk from your house to the windmill and then 36.0 s to walk from the windmill to the bench. For the entire trip from your front door to the bench, what are (a) your average velocity and (b) your average speed?

2.6 •• A Honda Civic travels in a straight line along a road. Its distance x from a stop sign is given as a function of time t by the equation $x(t) = \alpha t^2 - \beta t^3$, where $\alpha = 1.50$ m/s^2 and $\beta = 0.0500$ m/s^3. Calculate the average velocity of the car for each time interval: (a) $t = 0$ to $t = 2.00$ s; (b) $t = 0$ to $t = 4.00$ s; (c) $t = 2.00$ s to $t = 4.00$ s.

Section 2.2 Instantaneous Velocity

2.7 • **CALC** A car is stopped at a traffic light. It then travels along a straight road so that its distance from the light is given by $x(t) = bt^2 - ct^3$, where $b = 2.40$ m/s^2 and $c = 0.120$ m/s^3. (a) Calculate the average velocity of the car for the time interval $t = 0$ to $t = 10.0$ s. (b) Calculate the instantaneous velocity of the car at $t = 0$, $t = 5.0$ s, and $t = 10.0$ s. (c) How long after starting from rest is the car again at rest?

2.8 • **CALC** A bird is flying due east. Its distance from a tall building is given by $x(t) = 28.0$ m $+ (12.4$ m/s$)t - (0.0450$ m/s$^3)t^3$. What is the instantaneous velocity of the bird when $t = 8.00$ s?

2.9 •• A ball moves in a straight line (the x-axis). The graph in Fig. E2.9 shows this ball's velocity as a function of time. (a) What are the ball's average speed and average velocity during the first 3.0 s? (b) Suppose that the ball moved in such a way that the graph segment after 2.0 s was -3.0 m/s instead of $+3.0$ m/s. Find the ball's average speed and average velocity in this case.

Figure **E2.9**

2.10 • A physics professor leaves her house and walks along the sidewalk toward campus. After 5 min it starts to rain and she returns home. Her distance from her house as a function of time is shown in Fig. E2.10. At which of the labeled points is her velocity (a) zero? (b) constant and positive? (c) constant and negative? (d) increasing in magnitude? (e) decreasing in magnitude?

Figure **E2.10**

Figure **E2.11**

2.11 •• A test car travels in a straight line along the x-axis. The graph in Fig. E2.11 shows the car's position x as a function of time. Find its instantaneous velocity at points A through G.

Section 2.3 Average and Instantaneous Acceleration

2.12 • Figure E2.12 shows the velocity of a solar-powered car as a function of time. The driver accelerates from a stop sign, cruises for 20 s at a constant speed of 60 km/h, and then brakes to come to a stop 40 s after leaving the stop sign. (a) Compute the average acceleration during the following time intervals: (i) $t = 0$ to $t = 10$ s; (ii) $t = 30$ s to $t = 40$ s; (iii) $t = 10$ s to $t = 30$ s; (iv) $t = 0$ to $t = 40$ s. (b) What is the instantaneous acceleration at $t = 20$ s and at $t = 35$ s?

Figure **E2.12**

2.13 • **The Fastest (and Most Expensive) Car!** The table shows test data for the Bugatti Veyron, the fastest car made. The car is moving in a straight line (the x-axis).

Time (s)	0	2.1	20.0	53
Speed (mi/h)	0	60	200	253

(a) Make a v_x-t graph of this car's velocity (in mi/h) as a function of time. Is its acceleration constant? (b) Calculate the car's average acceleration (in m/s^2) between (i) 0 and 2.1 s; (ii) 2.1 s and 20.0 s; (iii) 20.0 s and 53 s. Are these results consistent with your graph in part (a)? (Before you decide to buy this car, it might be helpful to know that only 300 will be built, it runs out of gas in 12 minutes at top speed, and it costs $1.25 million!)

2.14 •• **CALC** A race car starts from rest and travels east along a straight and level track. For the first 5.0 s of the car's motion, the eastward component of the car's velocity is given by $v_x(t) = (0.860 \text{ m/s}^3)t^2$. What is the acceleration of the car when $v_x = 16.0$ m/s?

2.15 • **CALC** A turtle crawls along a straight line, which we will call the x-axis with the positive direction to the right. The equation for the turtle's position as a function of time is $x(t) = 50.0$ cm + $(2.00 \text{ cm/s})t - (0.0625 \text{ cm/s}^2)t^2$. (a) Find the turtle's initial velocity, initial position, and initial acceleration. (b) At what time t is the velocity of the turtle zero? (c) How long after starting does it take the turtle to return to its starting point? (d) At what times t is the turtle a distance of 10.0 cm from its starting point? What is the velocity (magnitude and direction) of the turtle at each of these times? (e) Sketch graphs of x versus t, v_x versus t, and a_x versus t, for the time interval $t = 0$ to $t = 40$ s.

2.16 • An astronaut has left the International Space Station to test a new space scooter. Her partner measures the following velocity changes, each taking place in a 10-s interval. What are the magnitude, the algebraic sign, and the direction of the average acceleration in each interval? Assume that the positive direction is to the right. (a) At the beginning of the interval the astronaut is moving toward the right along the x-axis at 15.0 m/s, and at the end of the interval she is moving toward the right at 5.0 m/s. (b) At the beginning she is moving toward the left at 5.0 m/s, and at the end she is moving toward the left at 15.0 m/s. (c) At the beginning she is moving toward the right at 15.0 m/s, and at the end she is moving toward the left at 15.0 m/s.

2.17 • **CALC** A car's velocity as a function of time is given by $v_x(t) = \alpha + \beta t^2$, where $\alpha = 3.00$ m/s and $\beta = 0.100$ m/s^3. (a) Calculate the average acceleration for the time interval $t = 0$ to $t = 5.00$ s. (b) Calculate the instantaneous acceleration for $t = 0$ and $t = 5.00$ s. (c) Draw v_x-t and a_x-t graphs for the car's motion between $t = 0$ and $t = 5.00$ s.

2.18 •• **CALC** The position of the front bumper of a test car under microprocessor control is given by $x(t) = 2.17$ m + $(4.80 \text{ m/s}^2)t^2 - (0.100 \text{ m/s}^6)t^6$. (a) Find its position and acceleration at the instants when the car has zero velocity. (b) Draw x-t, v_x-t, and a_x-t graphs for the motion of the bumper between $t = 0$ and $t = 2.00$ s.

Section 2.4 Motion with Constant Acceleration

2.19 •• An antelope moving with constant acceleration covers the distance between two points 70.0 m apart in 7.00 s. Its speed as it passes the second point is 15.0 m/s. (a) What is its speed at the first point? (b) What is its acceleration?

2.20 •• **BIO Blackout?** A jet fighter pilot wishes to accelerate from rest at a constant acceleration of $5g$ to reach Mach 3 (three times the speed of sound) as quickly as possible. Experimental tests reveal that he will black out if this acceleration lasts for more than 5.0 s. Use 331 m/s for the speed of sound. (a) Will the period of acceleration last long enough to cause him to black out? (b) What is the greatest speed he can reach with an acceleration of $5g$ before blacking out?

2.21 • **A Fast Pitch.** The fastest measured pitched baseball left the pitcher's hand at a speed of 45.0 m/s. If the pitcher was in contact with the ball over a distance of 1.50 m and produced constant acceleration, (a) what acceleration did he give the ball, and (b) how much time did it take him to pitch it?

2.22 •• **A Tennis Serve.** In the fastest measured tennis serve, the ball left the racquet at 73.14 m/s. A served tennis ball is typically in contact with the racquet for 30.0 ms and starts from rest. Assume constant acceleration. (a) What was the ball's acceleration during this serve? (b) How far did the ball travel during the serve?

2.23 •• **BIO Automobile Airbags.** The human body can survive an acceleration trauma incident (sudden stop) if the magnitude of the acceleration is less than 250 m/s^2. If you are in an automobile accident with an initial speed of 105 km/h (65 mi/h) and you are stopped by an airbag that inflates from the dashboard, over what distance must the airbag stop you for you to survive the crash?

2.24 • **BIO** If a pilot accelerates at more than $4g$, he begins to "gray out" but doesn't completely lose consciousness. (a) Assuming constant acceleration, what is the shortest time that a jet pilot starting from rest can take to reach Mach 4 (four times the speed of sound) without graying out? (b) How far would the plane travel during this period of acceleration? (Use 331 m/s for the speed of sound in cold air.)

2.25 • **BIO** **Air-Bag Injuries.** During an auto accident, the vehicle's air bags deploy and slow down the passengers more gently than if they had hit the windshield or steering wheel. According to safety standards, the bags produce a maximum acceleration of $60g$ that lasts for only 36 ms (or less). How far (in meters) does a person travel in coming to a complete stop in 36 ms at a constant acceleration of $60g$?

2.26 • **BIO** **Prevention of Hip Fractures.** Falls resulting in hip fractures are a major cause of injury and even death to the elderly. Typically, the hip's speed at impact is about 2.0 m/s. If this can be reduced to 1.3 m/s or less, the hip will usually not fracture. One way to do this is by wearing elastic hip pads. (a) If a typical pad is 5.0 cm thick and compresses by 2.0 cm during the impact of a fall, what constant acceleration (in m/s² and in g's) does the hip undergo to reduce its speed from 2.0 m/s to 1.3 m/s? (b) The acceleration you found in part (a) may seem rather large, but to fully assess its effects on the hip, calculate how long it lasts.

2.27 • **BIO** **Are We Martians?** It has been suggested, and not facetiously, that life might have originated on Mars and been carried to the earth when a meteor hit Mars and blasted pieces of rock (perhaps containing primitive life) free of the surface. Astronomers know that many Martian rocks have come to the earth this way. (For information on one of these, search the Internet for "ALH 84001.") One objection to this idea is that microbes would have to undergo an enormous lethal acceleration during the impact. Let us investigate how large such an acceleration might be. To escape Mars, rock fragments would have to reach its escape velocity of 5.0 km/s, and this would most likely happen over a distance of about 4.0 m during the meteor impact. (a) What would be the acceleration (in m/s² and g's) of such a rock fragment, if the acceleration is constant? (b) How long would this acceleration last? (c) In tests, scientists have found that over 40% of *Bacillius subtilis* bacteria survived after an acceleration of 450,000g. In light of your answer to part (a), can we rule out the hypothesis that life might have been blasted from Mars to the earth?

2.28 • **Entering the Freeway.** A car sits in an entrance ramp to a freeway, waiting for a break in the traffic. The driver accelerates with constant acceleration along the ramp and onto the freeway. The car starts from rest, moves in a straight line, and has a speed of 20 m/s (45 mi/h) when it reaches the end of the 120-m-long ramp. (a) What is the acceleration of the car? (b) How much time does it take the car to travel the length of the ramp? (c) The traffic on the freeway is moving at a constant speed of 20 m/s. What distance does the traffic travel while the car is moving the length of the ramp?

2.29 •• **Launch of the Space Shuttle.** At launch the space shuttle weighs 4.5 million pounds. When it is launched from rest, it takes 8.00 s to reach 161 km/h, and at the end of the first 1.00 min its speed is 1610 km/h. (a) What is the average acceleration (in m/s²) of the shuttle (i) during the first 8.00 s, and (ii) between 8.00 s and the end of the first 1.00 min? (b) Assuming the acceleration is constant during each time interval (but not necessarily the same in both intervals), what distance does the shuttle travel (i) during the first 8.00 s, and (ii) during the interval from 8.00 s to 1.00 min?

2.30 •• A cat walks in a straight line, which we shall call the x-axis with the positive direction to the right. As an observant physicist, you make measurements of this cat's motion and construct a graph of the feline's velocity as a function of time (Fig. E2.30). (a) Find the cat's velocity at $t = 4.0$ s and at $t = 7.0$ s. (b) What is the cat's acceleration at $t = 3.0$ s? At $t = 6.0$ s? At $t = 7.0$ s? (c) What distance does the cat move during the first 4.5 s? From $t = 0$ to $t = 7.5$ s? (d) Sketch clear graphs of the cat's acceleration and position as functions of time, assuming that the cat started at the origin.

Figure **E2.30**

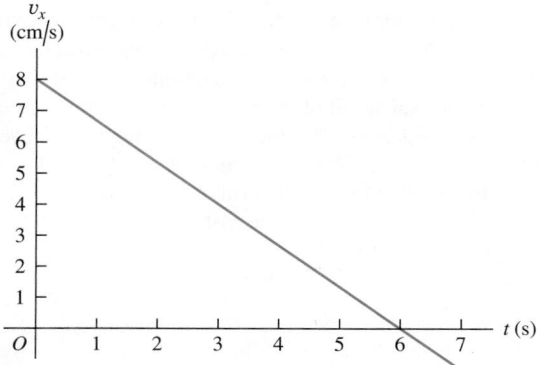

2.31 •• The graph in Fig. E2.31 shows the velocity of a motorcycle police officer plotted as a function of time. (a) Find the instantaneous acceleration at $t = 3$ s, at $t = 7$ s, and at $t = 11$ s. (b) How far does the officer go in the first 5 s? The first 9 s? The first 13 s?

Figure **E2.31**

2.32 • Two cars, A and B, move along the x-axis. Figure E2.32 is a graph of the positions of A and B versus time. (a) In motion diagrams (like Figs. 2.13b and 2.14b), show the position, velocity, and acceleration of each of the two cars at $t = 0$, $t = 1$ s, and $t = 3$ s. (b) At what time(s), if any, do A and B have the same position? (c) Graph velocity versus time for both A and B. (d) At what time(s), if any, do A and B have the same velocity? (e) At what time(s), if any, does car A pass car B? (f) At what time(s), if any, does car B pass car A?

Figure **E2.32**

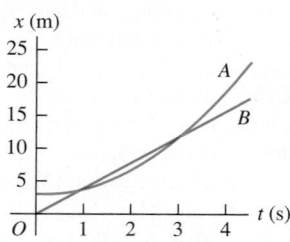

2.33 •• **Mars Landing.** In January 2004, NASA landed exploration vehicles on Mars. Part of the descent consisted of the following stages:

Stage A: Friction with the atmosphere reduced the speed from 19,300 km/h to 1600 km/h in 4.0 min.
Stage B: A parachute then opened to slow it down to 321 km/h in 94 s.
Stage C: Retro rockets then fired to reduce its speed to zero over a distance of 75 m.

Assume that each stage followed immediately after the preceding one and that the acceleration during each stage was constant. (a) Find the rocket's acceleration (in m/s^2) during each stage. (b) What total distance (in km) did the rocket travel during stages A, B, and C?

2.34 • At the instant the traffic light turns green, a car that has been waiting at an intersection starts ahead with a constant acceleration of $3.20 \ m/s^2$. At the same instant a truck, traveling with a constant speed of 20.0 m/s, overtakes and passes the car. (a) How far beyond its starting point does the car overtake the truck? (b) How fast is the car traveling when it overtakes the truck? (c) Sketch an *x-t* graph of the motion of both vehicles. Take $x = 0$ at the intersection. (d) Sketch a v_x-*t* graph of the motion of both vehicles.

Section 2.5 Freely Falling Bodies

2.35 •• (a) If a flea can jump straight up to a height of 0.440 m, what is its initial speed as it leaves the ground? (b) How long is it in the air?

2.36 •• A small rock is thrown vertically upward with a speed of 18.0 m/s from the edge of the roof of a 30.0-m-tall building. The rock doesn't hit the building on its way back down and lands in the street below. Air resistance can be neglected. (a) What is the speed of the rock just before it hits the street? (b) How much time elapses from when the rock is thrown until it hits the street?

2.37 • A juggler throws a bowling pin straight up with an initial speed of 8.20 m/s. How much time elapses until the bowling pin returns to the juggler's hand?

2.38 •• You throw a glob of putty straight up toward the ceiling, which is 3.60 m above the point where the putty leaves your hand. The initial speed of the putty as it leaves your hand is 9.50 m/s. (a) What is the speed of the putty just before it strikes the ceiling? (b) How much time from when it leaves your hand does it take the putty to reach the ceiling?

2.39 •• A tennis ball on Mars, where the acceleration due to gravity is 0.379*g* and air resistance is negligible, is hit directly upward and returns to the same level 8.5 s later. (a) How high above its original point did the ball go? (b) How fast was it moving just after being hit? (c) Sketch graphs of the ball's vertical position, vertical velocity, and vertical acceleration as functions of time while it's in the Martian air.

Figure **E2.40**

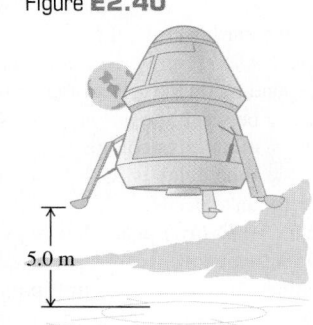

5.0 m

2.40 •• **Touchdown on the Moon.** A lunar lander is making its descent to Moon Base I (Fig. E2.40). The lander descends slowly under the retro-thrust of its descent engine. The engine is cut off when the lander is 5.0 m above the surface and has a downward speed of 0.8 m/s. With the engine off, the lander is in free fall. What is the speed of the lander just before it touches the surface? The acceleration due to gravity on the moon is $1.6 \ m/s^2$.

2.41 •• **A Simple Reaction-Time Test.** A meter stick is held vertically above your hand, with the lower end between your thumb and first finger. On seeing the meter stick released, you grab it with these two fingers. You can calculate your reaction time from the distance the meter stick falls, read directly from the point where your fingers grabbed it. (a) Derive a relationship for your reaction time in terms of this measured distance, *d*. (b) If the measured distance is 17.6 cm, what is the reaction time?

2.42 •• A brick is dropped (zero initial speed) from the roof of a building. The brick strikes the ground in 2.50 s. You may ignore air resistance, so the brick is in free fall. (a) How tall, in meters, is the building? (b) What is the magnitude of the brick's velocity just before it reaches the ground? (c) Sketch a_y-*t*, v_y-*t*, and *y-t* graphs for the motion of the brick.

2.43 •• **Launch Failure.** A 7500-kg rocket blasts off vertically from the launch pad with a constant upward acceleration of $2.25 \ m/s^2$ and feels no appreciable air resistance. When it has reached a height of 525 m, its engines suddenly fail so that the only force acting on it is now gravity. (a) What is the maximum height this rocket will reach above the launch pad? (b) How much time after engine failure will elapse before the rocket comes crashing down to the launch pad, and how fast will it be moving just before it crashes? (c) Sketch a_y-*t*, v_y-*t*, and *y-t* graphs of the rocket's motion from the instant of blast-off to the instant just before it strikes the launch pad.

2.44 •• A hot-air balloonist, rising vertically with a constant velocity of magnitude 5.00 m/s, releases a sandbag at an instant when the balloon is 40.0 m above the ground (Fig. E2.44). After it is released, the sandbag is in free fall. (a) Compute the position and velocity of the sandbag at 0.250 s and 1.00 s after its release. (b) How many seconds after its release will the bag strike the ground? (c) With what magnitude of velocity does it strike the ground? (d) What is the greatest height above the ground that the sandbag reaches? (e) Sketch a_y-*t*, v_y-*t*, and *y-t* graphs for the motion.

Figure **E2.44**

$v = 5.00 \ m/s$

40.0 m to ground

2.45 • **BIO** The rocket-driven sled *Sonic Wind No. 2,* used for investigating the physiological effects of large accelerations, runs on a straight, level track 1070 m (3500 ft) long. Starting from rest, it can reach a speed of 224 m/s (500 mi/h) in 0.900 s. (a) Compute the acceleration in m/s^2, assuming that it is constant. (b) What is the ratio of this acceleration to that of a freely falling body (*g*)? (c) What distance is covered in 0.900 s? (d) A magazine article states that at the end of a certain run, the speed of the sled decreased from 283 m/s (632 mi/h) to zero in 1.40 s and that during this time the magnitude of the acceleration was greater than 40*g*. Are these figures consistent?

2.46 • An egg is thrown nearly vertically upward from a point near the cornice of a tall building. It just misses the cornice on the way down and passes a point 30.0 m below its starting point 5.00 s after it leaves the thrower's hand. Air resistance may be ignored.

(a) What is the initial speed of the egg? (b) How high does it rise above its starting point? (c) What is the magnitude of its velocity at the highest point? (d) What are the magnitude and direction of its acceleration at the highest point? (e) Sketch a_y-t, v_y-t, and y-t graphs for the motion of the egg.

2.47 •• A 15-kg rock is dropped from rest on the earth and reaches the ground in 1.75 s. When it is dropped from the same height on Saturn's satellite Enceladus, it reaches the ground in 18.6 s. What is the acceleration due to gravity on Enceladus?

2.48 • A large boulder is ejected vertically upward from a volcano with an initial speed of 40.0 m/s. Air resistance may be ignored. (a) At what time after being ejected is the boulder moving at 20.0 m/s upward? (b) At what time is it moving at 20.0 m/s downward? (c) When is the displacement of the boulder from its initial position zero? (d) When is the velocity of the boulder zero? (e) What are the magnitude and direction of the acceleration while the boulder is (i) moving upward? (ii) Moving downward? (iii) At the highest point? (f) Sketch a_y-t, v_y-t, and y-t graphs for the motion.

2.49 •• Two stones are thrown vertically upward from the ground, one with three times the initial speed of the other. (a) If the faster stone takes 10 s to return to the ground, how long will it take the slower stone to return? (b) If the slower stone reaches a maximum height of H, how high (in terms of H) will the faster stone go? Assume free fall.

Section 2.6 Velocity and Position by Integration

2.50 • CALC For constant a_x, use Eqs. (2.17) and (2.18) to find v_x and x as functions of time. Compare your results to Eqs. (2.8) and (2.12).

2.51 • CALC A rocket starts from rest and moves upward from the surface of the earth. For the first 10.0 s of its motion, the vertical acceleration of the rocket is given by $a_y = (2.80 \text{ m/s}^3)t$, where the $+y$-direction is upward. (a) What is the height of the rocket above the surface of the earth at $t = 10.0$ s? (b) What is the speed of the rocket when it is 325 m above the surface of the earth?

2.52 •• CALC The acceleration of a bus is given by $a_x(t) = \alpha t$, where $\alpha = 1.2 \text{ m/s}^3$. (a) If the bus's velocity at time $t = 1.0$ s is 5.0 m/s, what is its velocity at time $t = 2.0$ s? (b) If the bus's position at time $t = 1.0$ s is 6.0 m, what is its position at time $t = 2.0$ s? (c) Sketch a_x-t, v_x-t, and x-t graphs for the motion.

2.53 •• CALC The acceleration of a motorcycle is given by $a_x(t) = At - Bt^2$, where $A = 1.50 \text{ m/s}^3$ and $B = 0.120 \text{ m/s}^4$. The motorcycle is at rest at the origin at time $t = 0$. (a) Find its position and velocity as functions of time. (b) Calculate the maximum velocity it attains.

2.54 •• BIO **Flying Leap of the Flea.** High-speed motion pictures (3500 frames/second) of a jumping, 210-μg flea yielded the data used to plot the graph given in Fig. E2.54. (See "The Flying Leap of the Flea" by M. Rothschild, Y. Schlein, K. Parker, C. Neville, and S. Sternberg in the November 1973 *Scientific American.*) This flea was about 2 mm long and jumped at a nearly vertical takeoff angle. Use the graph to answer the questions. (a) Is the acceleration of the flea ever zero? If so, when? Justify your answer. (b) Find the maximum height the flea reached in the first 2.5 ms. (c) Find the flea's acceleration at 0.5 ms, 1.0 ms, and 1.5 ms. (d) Find the flea's height at 0.5 ms, 1.0 ms, and 1.5 ms.

PROBLEMS

2.55 • BIO A typical male sprinter can maintain his maximum acceleration for 2.0 s and his maximum speed is 10 m/s. After reaching this maximum speed, his acceleration becomes zero and then he runs at constant speed. Assume that his acceleration is constant during the first 2.0 s of the race, that he starts from rest, and that he runs in a straight line. (a) How far has the sprinter run when he reaches his maximum speed? (b) What is the magnitude of his average velocity for a race of the following lengths: (i) 50.0 m, (ii) 100.0 m, (iii) 200.0 m?

2.56 •• On a 20-mile bike ride, you ride the first 10 miles at an average speed of 8 mi/h. What must your average speed over the next 10 miles be to have your average speed for the total 20 miles be (a) 4 mi/h? (b) 12 mi/h? (c) Given this average speed for the first 10 miles, can you possibly attain an average speed of 16 mi/h for the total 20-mile ride? Explain.

2.57 •• CALC The position of a particle between $t = 0$ and $t = 2.00$ s is given by $x(t) = (3.00 \text{ m/s}^3)t^3 - (10.0 \text{ m/s}^2)t^2 + (9.00 \text{ m/s})t$. (a) Draw the x-t, v_x-t, and a_x-t graphs of this particle. (b) At what time(s) between $t = 0$ and $t = 2.00$ s is the particle instantaneously at rest? Does your numerical result agree with the v_x-t graph in part (a)? (c) At each time calculated in part (b), is the acceleration of the particle positive or negative? Show that in each case the same answer is deduced from $a_x(t)$ and from the v_x-t graph. (d) At what time(s) between $t = 0$ and $t = 2.00$ s is the velocity of the particle instantaneously not changing? Locate this point on the v_x-t and a_x-t graphs of part (a). (e) What is the particle's greatest distance from the origin ($x = 0$) between $t = 0$ and $t = 2.00$ s? (f) At what time(s) between $t = 0$ and $t = 2.00$ s is the particle *speeding up* at the greatest rate? At what time(s) between $t = 0$ and $t = 2.00$ s is the particle *slowing down* at the greatest rate? Locate these points on the v_x-t and a_x-t graphs of part (a).

2.58 •• CALC A lunar lander is descending toward the moon's surface. Until the lander reaches the surface, its height above the surface of the moon is given by $y(t) = b - ct + dt^2$, where $b = 800$ m is the initial height of the lander above the surface, $c = 60.0$ m/s, and $d = 1.05 \text{ m/s}^2$. (a) What is the initial velocity of the lander, at $t = 0$? (b) What is the velocity of the lander just before it reaches the lunar surface?

2.59 ••• **Earthquake Analysis.** Earthquakes produce several types of shock waves. The most well known are the P-waves (P for *primary* or *pressure*) and the S-waves (S for *secondary* or *shear*). In the earth's crust, the P-waves travel at around 6.5 km/s, while the S-waves move at about 3.5 km/s. The actual speeds vary depending on the type of material they are going through. The time delay between the arrival of these two waves at a seismic recording station tells geologists how far away the earthquake occurred. If the time delay is 33 s, how far from the seismic station did the earthquake occur?

2.60 •• **Relay Race.** In a relay race, each contestant runs 25.0 m while carrying an egg balanced on a spoon, turns around, and comes back to the starting point. Edith runs the first 25.0 m in 20.0 s. On the return trip she is more confident and takes only 15.0 s. What is the magnitude of her average velocity for (a) the

Figure **E2.54**

first 25.0 m? (b) The return trip? (c) What is her average velocity for the entire round trip? (d) What is her average speed for the round trip?

2.61 ••• A rocket carrying a satellite is accelerating straight up from the earth's surface. At 1.15 s after liftoff, the rocket clears the top of its launch platform, 63 m above the ground. After an additional 4.75 s, it is 1.00 km above the ground. Calculate the magnitude of the average velocity of the rocket for (a) the 4.75-s part of its flight and (b) the first 5.90 s of its flight.

2.62 ••• The graph in Fig. P2.62 describes the acceleration as a function of time for a stone rolling down a hill starting from rest. (a) Find the stone's velocity at $t = 2.5$ s and at $t = 7.5$ s. (b) Sketch a graph of the stone's velocity as a function of time.

Figure **P2.62**

2.63 •• Dan gets on Interstate Highway I–80 at Seward, Nebraska, and drives due west in a straight line and at an average velocity of magnitude 88 km/h. After traveling 76 km, he reaches the Aurora exit (Fig. P2.63). Realizing he has gone too far, he turns around and drives due east 34 km back to the York exit at an average velocity of magnitude 72 km/h. For his whole trip from Seward to the York exit, what are (a) his average speed and (b) the magnitude of his average velocity?

Figure **P2.63**

2.64 ••• A subway train starts from rest at a station and accelerates at a rate of 1.60 m/s² for 14.0 s. It runs at constant speed for 70.0 s and slows down at a rate of 3.50 m/s² until it stops at the next station. Find the *total* distance covered.

2.65 •• A world-class sprinter accelerates to his maximum speed in 4.0 s. He then maintains this speed for the remainder of a 100-m race, finishing with a total time of 9.1 s. (a) What is the runner's average acceleration during the first 4.0 s? (b) What is his average

acceleration during the last 5.1 s? (c) What is his average acceleration for the entire race? (d) Explain why your answer to part (c) is not the average of the answers to parts (a) and (b).

2.66 •• A sled starts from rest at the top of a hill and slides down with a constant acceleration. At some later time the sled is 14.4 m from the top, 2.00 s after that it is 25.6 m from the top, 2.00 s later 40.0 m from the top, and 2.00 s later it is 57.6 m from the top. (a) What is the magnitude of the average velocity of the sled during each of the 2.00-s intervals after passing the 14.4-m point? (b) What is the acceleration of the sled? (c) What is the speed of the sled when it passes the 14.4-m point? (d) How much time did it take to go from the top to the 14.4-m point? (e) How far did the sled go during the first second after passing the 14.4-m point?

2.67 • A gazelle is running in a straight line (the *x*-axis). The graph in Fig. P2.67 shows this animal's velocity as a function of time. During the first 12.0 s, find (a) the total distance moved and (b) the displacement of the gazelle. (c) Sketch an a_x-*t* graph showing this gazelle's acceleration as a function of time for the first 12.0 s.

Figure **P2.67**

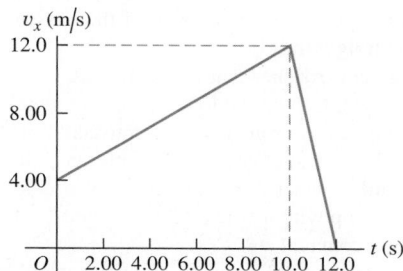

2.68 • A rigid ball traveling in a straight line (the *x*-axis) hits a solid wall and suddenly rebounds during a brief instant. The v_x-*t* graph in Fig. P2.68 shows this ball's velocity as a function of time. During the first 20.0 s of its motion, find (a) the total distance the ball moves and (b) its displacement. (c) Sketch a graph of a_x-*t* for this ball's motion. (d) Is the graph shown really vertical at 5.00 s? Explain.

Figure **P2.68**

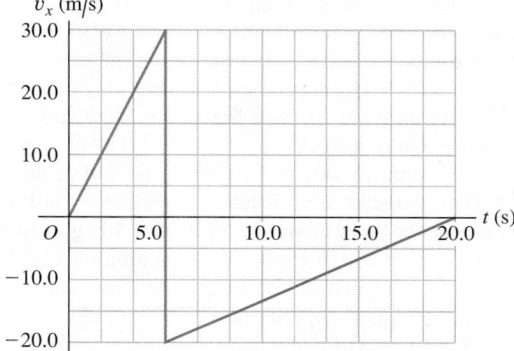

2.69 ••• A ball starts from rest and rolls down a hill with uniform acceleration, traveling 150 m during the second 5.0 s of its motion. How far did it roll during the first 5.0 s of motion?

2.70 •• **Collision.** The engineer of a passenger train traveling at 25.0 m/s sights a freight train whose caboose is 200 m ahead on

Figure **P2.70**

$v_{PT} = 25.0 \text{ m/s}$

$a = -0.100 \text{ m/s}^2$

$v_{FT} = 15.0 \text{ m/s}$

200 m

the same track (Fig. P2.70). The freight train is traveling at 15.0 m/s in the same direction as the passenger train. The engineer of the passenger train immediately applies the brakes, causing a constant acceleration of 0.100 m/s² in a direction opposite to the train's velocity, while the freight train continues with constant speed. Take $x = 0$ at the location of the front of the passenger train when the engineer applies the brakes. (a) Will the cows nearby witness a collision? (b) If so, where will it take place? (c) On a single graph, sketch the positions of the front of the passenger train and the back of the freight train.

2.71 ••• Large cockroaches can run as fast as 1.50 m/s in short bursts. Suppose you turn on the light in a cheap motel and see one scurrying directly away from you at a constant 1.50 m/s. If you start 0.90 m behind the cockroach with an initial speed of 0.80 m/s toward it, what minimum constant acceleration would you need to catch up with it when it has traveled 1.20 m, just short of safety under a counter?

2.72 •• Two cars start 200 m apart and drive toward each other at a steady 10 m/s. On the front of one of them, an energetic grasshopper jumps back and forth between the cars (he has strong legs!) with a constant horizontal velocity of 15 m/s relative to the ground. The insect jumps the instant he lands, so he spends no time resting on either car. What total distance does the grasshopper travel before the cars hit?

2.73 • An automobile and a truck start from rest at the same instant, with the automobile initially at some distance behind the truck. The truck has a constant acceleration of 2.10 m/s², and the automobile an acceleration of 3.40 m/s². The automobile overtakes the truck after the truck has moved 40.0 m. (a) How much time does it take the automobile to overtake the truck? (b) How far was the automobile behind the truck initially? (c) What is the speed of each when they are abreast? (d) On a single graph, sketch the position of each vehicle as a function of time. Take $x = 0$ at the initial location of the truck.

2.74 ••• Two stunt drivers drive directly toward each other. At time $t = 0$ the two cars are a distance D apart, car 1 is at rest, and car 2 is moving to the left with speed v_0. Car 1 begins to move at $t = 0$, speeding up with a constant acceleration a_x. Car 2 continues to move with a constant velocity. (a) At what time do the two cars collide? (b) Find the speed of car 1 just before it collides with car 2. (c) Sketch x-t and v_x-t graphs for car 1 and car 2. For each of the two graphs, draw the curves for both cars on the same set of axes.

2.75 •• A marble is released from one rim of a hemispherical bowl of diameter 50.0 cm and rolls down and up to the opposite rim in 10.0 s. Find (a) the average speed and (b) the average velocity of the marble.

2.76 •• **CALC** An object's velocity is measured to be $v_x(t) = \alpha - \beta t^2$, where $\alpha = 4.00$ m/s and $\beta = 2.00$ m/s³. At $t = 0$ the object is at $x = 0$. (a) Calculate the object's position and acceleration as functions of time. (b) What is the object's maximum *positive* displacement from the origin?

2.77 •• **Passing.** The driver of a car wishes to pass a truck that is traveling at a constant speed of 20.0 m/s (about 45 mi/h). Initially, the car is also traveling at 20.0 m/s and its front bumper is 24.0 m behind the truck's rear bumper. The car accelerates at a constant 0.600 m/s², then pulls back into the truck's lane when the rear of the car is 26.0 m ahead of the front of the truck. The car is 4.5 m long and the truck is 21.0 m long. (a) How much time is required for the car to pass the truck? (b) What distance does the car travel during this time? (c) What is the final speed of the car?

2.78 •• On Planet X, you drop a 25-kg stone from rest and measure its speed at various times. Then you use the data you obtained to construct a graph of its speed v as a function of time t (Fig. P2.78). From the information in the graph, answer the following questions: (a) What is g on Planet X? (b) An astronaut drops a piece of equipment from rest out of the landing module, 3.5 m above the surface of Planet X. How long will it take this equipment to reach the ground, and how fast will it be moving when it gets there? (c) How fast would an astronaut have to project an object straight upward to reach a height of 18.0 m above the release point, and how long would it take to reach that height?

Figure **P2.78**

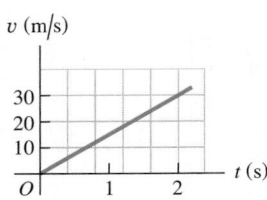

v (m/s)

30
20
10

O 1 2 t (s)

2.79 ••• **CALC** The acceleration of a particle is given by $a_x(t) = -2.00$ m/s² $+ (3.00$ m/s³$)t$. (a) Find the initial velocity v_{0x} such that the particle will have the same x-coordinate at $t = 4.00$ s as it had at $t = 0$. (b) What will be the velocity at $t = 4.00$ s?

2.80 • **Egg Drop.** You are on the roof of the physics building, 46.0 m above the ground (Fig. P2.80). Your physics professor, who is 1.80 m tall, is walking alongside the building at a constant speed of 1.20 m/s. If you wish to drop an egg on your professor's head, where should the professor be when you release the egg? Assume that the egg is in free fall.

Figure **P2.80**

46.0 m

$v = 1.20$ m/s

1.80 m

2.81 • A certain volcano on earth can eject rocks vertically to a maximum height H. (a) How high (in terms of H) would these rocks go if a volcano on Mars ejected them with the same initial velocity? The acceleration due to gravity on Mars is 3.71 m/s², and you can neglect air resistance on both planets. (b) If the rocks are in the air for a time T on earth, for how long (in terms of T) will they be in the air on Mars?

2.82 •• An entertainer juggles balls while doing other activities. In one act, she throws a ball vertically upward, and while it is in the air, she runs to and from a table 5.50 m away at a constant speed of 2.50 m/s, returning just in time to catch the falling ball. (a) With what minimum initial speed must she throw the ball upward to accomplish this feat? (b) How high above its initial position is the ball just as she reaches the table?

2.83 • Visitors at an amusement park watch divers step off a platform 21.3 m (70 ft) above a pool of water. According to the announcer, the divers enter the water at a speed of 56 mi/h (25 m/s). Air resistance may be ignored. (a) Is the announcer correct in this claim? (b) Is it possible for a diver to leap directly upward off the board so that, missing the board on the way down, she enters the water at 25.0 m/s? If so, what initial upward speed is required? Is the required initial speed physically attainable?

2.84 ••• A flowerpot falls off a windowsill and falls past the window below. You may ignore air resistance. It takes the pot 0.420 s to pass from the top to the bottom of this window, which is 1.90 m high. How far is the top of the window below the windowsill from which the flowerpot fell?

2.85 ••• **Look Out Below.** Sam heaves a 16-lb shot straight upward, giving it a constant upward acceleration from rest of 35.0 m/s^2 for 64.0 cm. He releases it 2.20 m above the ground. You may ignore air resistance. (a) What is the speed of the shot when Sam releases it? (b) How high above the ground does it go? (c) How much time does he have to get out of its way before it returns to the height of the top of his head, 1.83 m above the ground?

2.86 ••• **A Multistage Rocket.** In the first stage of a two-stage rocket, the rocket is fired from the launch pad starting from rest but with a constant acceleration of 3.50 m/s^2 upward. At 25.0 s after launch, the second stage fires for 10.0 s, which boosts the rocket's velocity to 132.5 m/s upward at 35.0 s after launch. This firing uses up all the fuel, however, so after the second stage has finished firing, the only force acting on the rocket is gravity. Air resistance can be neglected. (a) Find the maximum height that the stage-two rocket reaches above the launch pad. (b) How much time after the end of the stage-two firing will it take for the rocket to fall back to the launch pad? (c) How fast will the stage-two rocket be moving just as it reaches the launch pad?

2.87 •• **Juggling Act.** A juggler performs in a room whose ceiling is 3.0 m above the level of his hands. He throws a ball upward so that it just reaches the ceiling. (a) What is the initial velocity of the ball? (b) What is the time required for the ball to reach the ceiling? At the instant when the first ball is at the ceiling, the juggler throws a second ball upward with two-thirds the initial velocity of the first. (c) How long after the second ball is thrown do the two balls pass each other? (d) At what distance above the juggler's hand do they pass each other?

2.88 •• A physics teacher performing an outdoor demonstration suddenly falls from rest off a high cliff and simultaneously shouts "Help." When she has fallen for 3.0 s, she hears the echo of her shout from the valley floor below. The speed of sound is 340 m/s. (a) How tall is the cliff? (b) If air resistance is neglected, how fast will she be moving just before she hits the ground? (Her actual speed will be less than this, due to air resistance.)

2.89 ••• A helicopter carrying Dr. Evil takes off with a constant upward acceleration of 5.0 m/s^2. Secret agent Austin Powers jumps on just as the helicopter lifts off the ground. After the two men struggle for 10.0 s, Powers shuts off the engine and steps out of the helicopter. Assume that the helicopter is in free fall after its engine is shut off, and ignore the effects of air resistance. (a) What is the maximum height above ground reached by the helicopter? (b) Powers deploys a jet pack strapped on his back 7.0 s after leaving the helicopter, and then he has a constant downward acceleration with magnitude 2.0 m/s^2. How far is Powers above the ground when the helicopter crashes into the ground?

2.90 •• **Cliff Height.** You are climbing in the High Sierra where you suddenly find yourself at the edge of a fog-shrouded cliff. To find the height of this cliff, you drop a rock from the top and 10.0 s later hear the sound of it hitting the ground at the foot of the cliff. (a) Ignoring air resistance, how high is the cliff if the speed of sound is 330 m/s? (b) Suppose you had ignored the time it takes the sound to reach you. In that case, would you have overestimated or underestimated the height of the cliff? Explain your reasoning.

2.91 ••• **Falling Can.** A painter is standing on scaffolding that is raised at constant speed. As he travels upward, he accidentally nudges a paint can off the scaffolding and it falls 15.0 m to the ground. You are watching, and measure with your stopwatch that it takes 3.25 s for the can to reach the ground. Ignore air resistance. (a) What is the speed of the can just before it hits the ground? (b) Another painter is standing on a ledge, with his hands 4.00 m above the can when it falls off. He has lightning-fast reflexes and if the can passes in front of him, he can catch it. Does he get the chance?

2.92 •• Determined to test the law of gravity for himself, a student walks off a skyscraper 180 m high, stopwatch in hand, and starts his free fall (zero initial velocity). Five seconds later, Superman arrives at the scene and dives off the roof to save the student. Superman leaves the roof with an initial speed v_0 that he produces by pushing himself downward from the edge of the roof with his legs of steel. He then falls with the same acceleration as any freely falling body. (a) What must the value of v_0 be so that Superman catches the student just before they reach the ground? (b) On the same graph, sketch the positions of the student and of Superman as functions of time. Take Superman's initial speed to have the value calculated in part (a). (c) If the height of the skyscraper is less than some minimum value, even Superman can't reach the student before he hits the ground. What is this minimum height?

2.93 ••• During launches, rockets often discard unneeded parts. A certain rocket starts from rest on the launch pad and accelerates upward at a steady 3.30 m/s^2. When it is 235 m above the launch pad, it discards a used fuel canister by simply disconnecting it. Once it is disconnected, the only force acting on the canister is gravity (air resistance can be ignored). (a) How high is the rocket when the canister hits the launch pad, assuming that the rocket does not change its acceleration? (b) What total distance did the canister travel between its release and its crash onto the launch pad?

2.94 •• A ball is thrown straight up from the ground with speed v_0. At the same instant, a second ball is dropped from rest from a height H, directly above the point where the first ball was thrown upward. There is no air resistance. (a) Find the time at which the two balls collide. (b) Find the value of H in terms of v_0 and g so that at the instant when the balls collide, the first ball is at the highest point of its motion.

2.95 • **CALC** Two cars, A and B, travel in a straight line. The distance of A from the starting point is given as a function of time by $x_A(t) = \alpha t + \beta t^2$, with $\alpha = 2.60$ m/s and $\beta = 1.20$ m/s^2. The distance of B from the starting point is $x_B(t) = \gamma t^2 - \delta t^3$, with $\gamma = 2.80$ m/s^2 and $\delta = 0.20$ m/s^3. (a) Which car is ahead just after they leave the starting point? (b) At what time(s) are the cars at the same point? (c) At what time(s) is the distance from A to B neither increasing nor decreasing? (d) At what time(s) do A and B have the same acceleration?

CHALLENGE PROBLEMS

2.96 ••• In the vertical jump, an athlete starts from a crouch and jumps upward to reach as high as possible. Even the best athletes spend little more than 1.00 s in the air (their "hang time"). Treat the athlete as a particle and let y_{max} be his maximum height above the floor. To explain why he seems to hang in the air, calculate the ratio of the time he is above $y_{max}/2$ to the time it takes him to go from the floor to that height. You may ignore air resistance.

2.97 ••• **Catching the Bus.** A student is running at her top speed of 5.0 m/s to catch a bus, which is stopped at the bus stop. When the student is still 40.0 m from the bus, it starts to pull away, moving with a constant acceleration of 0.170 m/s². (a) For how much time and what distance does the student have to run at 5.0 m/s before she overtakes the bus? (b) When she reaches the bus, how fast is the bus traveling? (c) Sketch an *x-t* graph for both the student and the bus. Take $x = 0$ at the initial position of the student. (d) The equations you used in part (a) to find the time have a second solution, corresponding to a later time for which the student and bus are again at the same place if they continue their specified motions. Explain the significance of this second solution. How fast is the bus traveling at this point? (e) If the student's top speed is 3.5 m/s, will she catch the bus? (f) What is the *minimum* speed the student must have to just catch up with the bus? For what time and what distance does she have to run in that case?

2.98 ••• An alert hiker sees a boulder fall from the top of a distant cliff and notes that it takes 1.30 s for the boulder to fall the last third of the way to the ground. You may ignore air resistance.

(a) What is the height of the cliff in meters? (b) If in part (a) you get two solutions of a quadratic equation and you use one for your answer, what does the other solution represent?

2.99 ••• A ball is thrown straight up from the edge of the roof of a building. A second ball is dropped from the roof 1.00 s later. You may ignore air resistance. (a) If the height of the building is 20.0 m, what must the initial speed of the first ball be if both are to hit the ground at the same time? On the same graph, sketch the position of each ball as a function of time, measured from when the first ball is thrown. Consider the same situation, but now let the initial speed v_0 of the first ball be given and treat the height h of the building as an unknown. (b) What must the height of the building be for both balls to reach the ground at the same time (i) if v_0 is 6.0 m/s and (ii) if v_0 is 9.5 m/s? (c) If v_0 is greater than some value v_{max}, a value of h does not exist that allows both balls to hit the ground at the same time. Solve for v_{max}. The value v_{max} has a simple physical interpretation. What is it? (d) If v_0 is less than some value v_{min}, a value of h does not exist that allows both balls to hit the ground at the same time. Solve for v_{min}. The value v_{min} also has a simple physical interpretation. What is it?

Answers

Chapter Opening Question

Yes. Acceleration refers to *any* change in velocity, including both speeding up and slowing down.

Test Your Understanding Questions

2.1 Answer to (a): (iv), (i) and (iii) (tie), (v), (ii); **answer to (b):** (i) and (iii); **answer to (c):** (v) In (a) the average *x*-velocity is $v_{av-x} = \Delta x/\Delta t$. For all five trips, $\Delta t = 1$ h. For the individual trips, we have (i) $\Delta x = +50$ km, $v_{av-x} = +50$ km/h; (ii) $\Delta x = -50$ km, $v_{av-x} = -50$ km/h; (iii) $\Delta x = 60$ km $- 10$ km $= +50$ km, $v_{av-x} = +50$ km/h; (iv) $\Delta x = +70$ km, $v_{av-x} = +70$ km/h; (v) $\Delta x = -20$ km $+ 20$ km $= 0$, $v_{av-x} = 0$. In (b) both have $v_{av-x} = +50$ km/h.

2.2 Answers: (a) *P, Q* and *S* (tie), *R* The *x*-velocity is **(b)** positive when the slope of the *x-t* graph is positive (*P*), **(c)** negative when the slope is negative (*R*), and **(d)** zero when the slope is zero (*Q* and *S*). **(e)** *R, P, Q* and *S* (tie) The speed is greatest when the slope of the *x-t* graph is steepest (either positive or negative) and zero when the slope is zero.

2.3 Answers: (a) *S,* where the *x-t* graph is curved upward (concave up). **(b)** *Q,* where the *x-t* graph is curved downward (concave down). **(c)** *P* and *R,* where the *x-t* graph is not curved either up or down. **(d)** At *P*, $a_x = 0$ (velocity is **not changing**); at *Q*, $a_x < 0$

(velocity is **decreasing,** i.e., changing from positive to zero to negative); at *R*, $a_x = 0$ (velocity is **not changing**); and at *S*, $a_x > 0$ (velocity is **increasing,** i.e., changing from negative to zero to positive).

2.4 Answer: (b) The officer's *x*-acceleration is constant, so her v_x-*t* graph is a straight line, and the officer's motorcycle is moving faster than the motorist's car when the two vehicles meet at $t = 10$ s.

2.5 Answers: (a) (iii) Use Eq. (2.13) with *x* replaced by *y* and $a_y = g$; $v_y^2 = v_{0y}^2 - 2g(y - y_0)$. The starting height is $y_0 = 0$ and the *y*-velocity at the maximum height $y = h$ is $v_y = 0$, so $0 = v_{0y}^2 - 2gh$ and $h = v_{0y}^2/2g$. If the initial *y*-velocity is increased by a factor of 2, the maximum height increases by a factor of $2^2 = 4$ and the ball goes to height $4h$. **(b) (v)** Use Eq. (2.8) with *x* replaced by *y* and $a_y = g$; $v_y = v_{0y} - gt$. The *y*-velocity at the maximum height is $v_y = 0$, so $0 = v_{0y} - gt$ and $t = v_{0y}/g$. If the initial *y*-velocity is increased by a factor of 2, the time to reach the maximum height increases by a factor of 2 and becomes $2t$.

2.6 Answer: (ii) The acceleration a_x is equal to the slope of the v_x-*t* graph. If a_x is increasing, the slope of the v_x-*t* graph is also increasing and the graph is concave up.

Bridging Problem

Answer: $h = 57.1$ m

MOTION IN TWO OR THREE DIMENSIONS

? If a cyclist is going around a curve at constant speed, is he accelerating? If so, in which direction is he accelerating?

What determines where a batted baseball lands? How do you describe the motion of a roller coaster car along a curved track or the flight of a circling hawk? Which hits the ground first: a baseball that you simply drop or one that you throw horizontally?

We can't answer these kinds of questions using the techniques of Chapter 2, in which particles moved only along a straight line. Instead, we need to extend our descriptions of motion to two- and three-dimensional situations. We'll still use the vector quantities displacement, velocity, and acceleration, but now these quantities will no longer lie along a single line. We'll find that several important kinds of motion take place in two dimensions only—that is, in a *plane*. We can describe these motions with two components of position, velocity, and acceleration.

We also need to consider how the motion of a particle is described by different observers who are moving relative to each other. The concept of *relative velocity* will play an important role later in the book when we study collisions, when we explore electromagnetic phenomena, and when we introduce Einstein's special theory of relativity.

This chapter merges the vector mathematics of Chapter 1 with the kinematic language of Chapter 2. As before, we are concerned with describing motion, not with analyzing its causes. But the language you learn here will be an essential tool in later chapters when we study the relationship between force and motion.

3.1 Position and Velocity Vectors

3.1 The position vector $\vec{r}$ from the origin to point P has components x, y, and z. The path that the particle follows through space is in general a curve (Fig. 3.2).

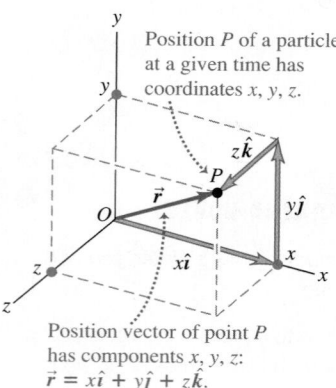

Position P of a particle at a given time has coordinates x, y, z.

Position vector of point P has components x, y, z:
$\vec{r} = x\hat{\imath} + y\hat{\jmath} + z\hat{k}$.

3.2 The average velocity $\vec{v}_{\text{av}}$ between points P_1 and P_2 has the same direction as the displacement $\Delta\vec{r}$.

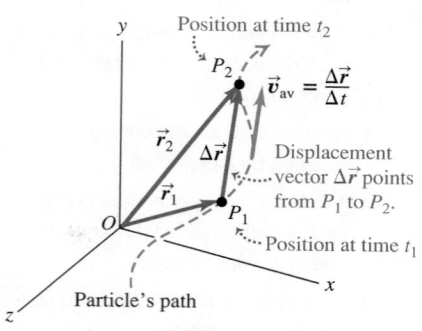

Position at time t_2

$\vec{v}_{\text{av}} = \dfrac{\Delta\vec{r}}{\Delta t}$

Displacement vector $\Delta\vec{r}$ points from P_1 to P_2.

Position at time t_1

Particle's path

3.3 The vectors $\vec{v}_1$ and $\vec{v}_2$ are the instantaneous velocities at the points P_1 and P_2 shown in Fig. 3.2.

The instantaneous velocity vector $\vec{v}$ is tangent to the path at each point.

Particle's path

To describe the *motion* of a particle in space, we must first be able to describe the particle's *position*. Consider a particle that is at a point P at a certain instant. The **position vector** $\vec{r}$ of the particle at this instant is a vector that goes from the origin of the coordinate system to the point P (Fig. 3.1). The Cartesian coordinates x, y, and z of point P are the x-, y-, and z-components of vector $\vec{r}$. Using the unit vectors we introduced in Section 1.9, we can write

$$\vec{r} = x\hat{\imath} + y\hat{\jmath} + z\hat{k} \qquad \text{(position vector)} \qquad (3.1)$$

During a time interval Δt the particle moves from P_1, where its position vector is $\vec{r}_1$, to P_2, where its position vector is $\vec{r}_2$. The change in position (the displacement) during this interval is $\Delta\vec{r} = \vec{r}_2 - \vec{r}_1 = (x_2 - x_1)\hat{\imath} + (y_2 - y_1)\hat{\jmath} + (z_2 - z_1)\hat{k}$. We define the **average velocity** $\vec{v}_{\text{av}}$ during this interval in the same way we did in Chapter 2 for straight-line motion, as the displacement divided by the time interval:

$$\vec{v}_{\text{av}} = \frac{\vec{r}_2 - \vec{r}_1}{t_2 - t_1} = \frac{\Delta\vec{r}}{\Delta t} \qquad \text{(average velocity vector)} \qquad (3.2)$$

Dividing a vector by a scalar is really a special case of *multiplying* a vector by a scalar, described in Section 1.7; the average velocity $\vec{v}_{\text{av}}$ is equal to the displacement vector $\Delta\vec{r}$ multiplied by $1/\Delta t$, the reciprocal of the time interval. Note that the x-component of Eq. (3.2) is $v_{\text{av-}x} = (x_2 - x_1)/(t_2 - t_1) = \Delta x/\Delta t$. This is just Eq. (2.2), the expression for average x-velocity that we found in Section 2.1 for one-dimensional motion.

We now define **instantaneous velocity** just as we did in Chapter 2: It is the limit of the average velocity as the time interval approaches zero, and it equals the instantaneous rate of change of position with time. The key difference is that position $\vec{r}$ and instantaneous velocity $\vec{v}$ are now both vectors:

$$\vec{v} = \lim_{\Delta t \to 0} \frac{\Delta\vec{r}}{\Delta t} = \frac{d\vec{r}}{dt} \qquad \text{(instantaneous velocity vector)} \qquad (3.3)$$

The *magnitude* of the vector $\vec{v}$ at any instant is the *speed* v of the particle at that instant. The *direction* of $\vec{v}$ at any instant is the same as the direction in which the particle is moving at that instant.

Note that as $\Delta t \to 0$, points P_1 and P_2 in Fig. 3.2 move closer and closer together. In this limit, the vector $\Delta\vec{r}$ becomes tangent to the path. The direction of $\Delta\vec{r}$ in this limit is also the direction of the instantaneous velocity $\vec{v}$. This leads to an important conclusion: *At every point along the path, the instantaneous velocity vector is tangent to the path at that point* (Fig. 3.3).

It's often easiest to calculate the instantaneous velocity vector using components. During any displacement $\Delta\vec{r}$, the changes Δx, Δy, and Δz in the three coordinates of the particle are the *components* of $\Delta\vec{r}$. It follows that the components v_x, v_y, and v_z of the instantaneous velocity $\vec{v}$ are simply the time derivatives of the coordinates x, y, and z. That is,

$$v_x = \frac{dx}{dt} \qquad v_y = \frac{dy}{dt} \qquad v_z = \frac{dz}{dt} \qquad \begin{array}{l}\text{(components of}\\ \text{instantaneous velocity)}\end{array} \qquad (3.4)$$

The x-component of $\vec{v}$ is $v_x = dx/dt$, which is the same as Eq. (2.3)—the expression for instantaneous velocity for straight-line motion that we obtained in Section 2.2. Hence Eq. (3.4) is a direct extension of the idea of instantaneous velocity to motion in three dimensions.

We can also get Eq. (3.4) by taking the derivative of Eq. (3.1). The unit vectors $\hat{\imath}$, $\hat{\jmath}$, and $\hat{k}$ are constant in magnitude and direction, so their derivatives are zero, and we find

$$\vec{v} = \frac{d\vec{r}}{dt} = \frac{dx}{dt}\hat{\imath} + \frac{dy}{dt}\hat{\jmath} + \frac{dz}{dt}\hat{k} \qquad (3.5)$$

This shows again that the components of $\vec{v}$ are dx/dt, dy/dt, and dz/dt.

The magnitude of the instantaneous velocity vector $\vec{v}$—that is, the speed—is given in terms of the components v_x, v_y, and v_z by the Pythagorean relation:

$$|\vec{v}| = v = \sqrt{v_x^2 + v_y^2 + v_z^2} \qquad (3.6)$$

Figure 3.4 shows the situation when the particle moves in the xy-plane. In this case, z and v_z are zero. Then the speed (the magnitude of $\vec{v}$) is

$$v = \sqrt{v_x^2 + v_y^2}$$

and the direction of the instantaneous velocity $\vec{v}$ is given by the angle α (the Greek letter alpha) in the figure. We see that

$$\tan\alpha = \frac{v_y}{v_x} \qquad (3.7)$$

(We always use Greek letters for angles. We use α for the direction of the instantaneous velocity vector to avoid confusion with the direction θ of the *position* vector of the particle.)

The instantaneous velocity vector is usually more interesting and useful than the average velocity vector. From now on, when we use the word "velocity," we will always mean the instantaneous velocity vector $\vec{v}$ (rather than the average velocity vector). Usually, we won't even bother to call $\vec{v}$ a vector; it's up to you to remember that velocity is a vector quantity with both magnitude and direction.

3.4 The two velocity components for motion in the xy-plane.

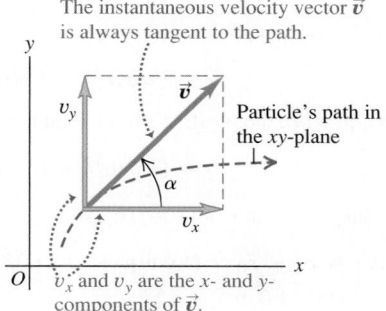

The instantaneous velocity vector $\vec{v}$ is always tangent to the path.

v_x and v_y are the x- and y-components of $\vec{v}$.

Example 3.1   **Calculating average and instantaneous velocity**

A robotic vehicle, or rover, is exploring the surface of Mars. The stationary Mars lander is the origin of coordinates, and the surrounding Martian surface lies in the xy-plane. The rover, which we represent as a point, has x- and y-coordinates that vary with time:

$$x = 2.0\ \text{m} - (0.25\ \text{m/s}^2)t^2$$
$$y = (1.0\ \text{m/s})t + (0.025\ \text{m/s}^3)t^3$$

(a) Find the rover's coordinates and distance from the lander at $t = 2.0$ s. (b) Find the rover's displacement and average velocity vectors for the interval $t = 0.0$ s to $t = 2.0$ s. (c) Find a general expression for the rover's instantaneous velocity vector $\vec{v}$. Express $\vec{v}$ at $t = 2.0$ s in component form and in terms of magnitude and direction.

SOLUTION

IDENTIFY and SET UP: This problem involves motion in two dimensions, so we must use the vector equations obtained in this section. Figure 3.5 shows the rover's path (dashed line). We'll use Eq. (3.1) for position $\vec{r}$, the expression $\Delta\vec{r} = \vec{r}_2 - \vec{r}_1$ for displacement, Eq. (3.2) for average velocity, and Eqs. (3.5), (3.6), and (3.7)

3.5 At $t = 0.0$ s the rover has position vector $\vec{r}_0$ and instantaneous velocity vector $\vec{v}_0$. Likewise, $\vec{r}_1$ and $\vec{v}_1$ are the vectors at $t = 1.0$ s; $\vec{r}_2$ and $\vec{v}_2$ are the vectors at $t = 2.0$ s.

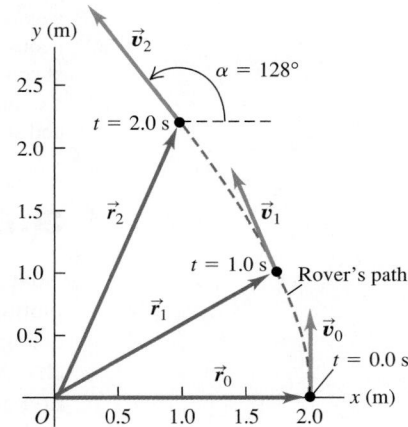

Continued

for instantaneous velocity and its magnitude and direction. The target variables are stated in the problem.

EXECUTE: (a) At $t = 2.0$ s the rover's coordinates are

$$x = 2.0 \text{ m} - (0.25 \text{ m/s}^2)(2.0 \text{ s})^2 = 1.0 \text{ m}$$
$$y = (1.0 \text{ m/s})(2.0 \text{ s}) + (0.025 \text{ m/s}^3)(2.0 \text{ s})^3 = 2.2 \text{ m}$$

The rover's distance from the origin at this time is

$$r = \sqrt{x^2 + y^2} = \sqrt{(1.0 \text{ m})^2 + (2.2 \text{ m})^2} = 2.4 \text{ m}$$

(b) To find the displacement and average velocity over the given time interval, we first express the position vector $\vec{r}$ as a function of time t. From Eq. (3.1) this is

$$\vec{r} = x\hat{\imath} + y\hat{\jmath}$$
$$= [2.0 \text{ m} - (0.25 \text{ m/s}^2)t^2]\hat{\imath}$$
$$+ [(1.0 \text{ m/s})t + (0.025 \text{ m/s}^3)t^3]\hat{\jmath}$$

At $t = 0.0$ s the position vector $\vec{r}_0$ is

$$\vec{r}_0 = (2.0 \text{ m})\hat{\imath} + (0.0 \text{ m})\hat{\jmath}$$

From part (a), the position vector $\vec{r}_2$ at $t = 2.0$ s is

$$\vec{r}_2 = (1.0 \text{ m})\hat{\imath} + (2.2 \text{ m})\hat{\jmath}$$

The displacement from $t = 0.0$ s to $t = 2.0$ s is therefore

$$\Delta\vec{r} = \vec{r}_2 - \vec{r}_0 = (1.0 \text{ m})\hat{\imath} + (2.2 \text{ m})\hat{\jmath} - (2.0 \text{ m})\hat{\imath}$$
$$= (-1.0 \text{ m})\hat{\imath} + (2.2 \text{ m})\hat{\jmath}$$

During this interval the rover moves 1.0 m in the negative x-direction and 2.2 m in the positive y-direction. From Eq. (3.2), the average velocity over this interval is the displacement divided by the elapsed time:

$$\vec{v}_{av} = \frac{\Delta\vec{r}}{\Delta t} = \frac{(-1.0 \text{ m})\hat{\imath} + (2.2 \text{ m})\hat{\jmath}}{2.0 \text{ s} - 0.0 \text{ s}}$$
$$= (-0.50 \text{ m/s})\hat{\imath} + (1.1 \text{ m/s})\hat{\jmath}$$

The components of this average velocity are $v_{av-x} = -0.50$ m/s and $v_{av-y} = 1.1$ m/s.

(c) From Eq. (3.4) the components of *instantaneous* velocity are the time derivatives of the coordinates:

$$v_x = \frac{dx}{dt} = (-0.25 \text{ m/s}^2)(2t)$$
$$v_y = \frac{dy}{dt} = 1.0 \text{ m/s} + (0.025 \text{ m/s}^3)(3t^2)$$

Hence the instantaneous velocity vector is

$$\vec{v} = v_x\hat{\imath} + v_y\hat{\jmath} = (-0.50 \text{ m/s}^2)t\hat{\imath}$$
$$+ [1.0 \text{ m/s} + (0.075 \text{ m/s}^3)t^2]\hat{\jmath}$$

At $t = 2.0$ s the velocity vector $\vec{v}_2$ has components

$$v_{2x} = (-0.50 \text{ m/s}^2)(2.0 \text{ s}) = -1.0 \text{ m/s}$$
$$v_{2y} = 1.0 \text{ m/s} + (0.075 \text{ m/s}^3)(2.0 \text{ s})^2 = 1.3 \text{ m/s}$$

The magnitude of the instantaneous velocity (that is, the speed) at $t = 2.0$ s is

$$v_2 = \sqrt{v_{2x}^2 + v_{2y}^2} = \sqrt{(-1.0 \text{ m/s})^2 + (1.3 \text{ m/s})^2}$$
$$= 1.6 \text{ m/s}$$

Figure 3.5 shows the direction of the velocity vector $\vec{v}_2$, which is at an angle α between 90° and 180° with respect to the positive x-axis. From Eq. (3.7) we have

$$\arctan\frac{v_y}{v_x} = \arctan\frac{1.3 \text{ m/s}}{-1.0 \text{ m/s}} = -52°$$

This is off by 180°; the correct value of the angle is $\alpha = 180° - 52° = 128°$, or 38° west of north.

EVALUATE: Compare the components of *average* velocity that we found in part (b) for the interval from $t = 0.0$ s to $t = 2.0$ s ($v_{av-x} = -0.50$ m/s, $v_{av-y} = 1.1$ m/s) with the components of *instantaneous* velocity at $t = 2.0$ s that we found in part (c) ($v_{2x} = -1.0$ m/s, $v_{2y} = 1.3$ m/s). The comparison shows that, just as in one dimension, the average velocity vector $\vec{v}_{av}$ over an interval is in general *not* equal to the instantaneous velocity $\vec{v}$ at the end of the interval (see Example 2.1).

Figure 3.5 shows the position vectors $\vec{r}$ and instantaneous velocity vectors $\vec{v}$ at $t = 0.0$ s, 1.0 s, and 2.0 s. (You should calculate these quantities for $t = 0.0$ s and $t = 1.0$ s.) Notice that $\vec{v}$ is tangent to the path at every point. The magnitude of $\vec{v}$ increases as the rover moves, which means that its speed is increasing.

Test Your Understanding of Section 3.1 In which of these situations *would* the average velocity vector $\vec{v}_{av}$ over an interval be equal to the instantaneous velocity $\vec{v}$ at the end of the interval? (i) a body moving along a curved path at constant speed; (ii) a body moving along a curved path and speeding up; (iii) a body moving along a straight line at constant speed; (iv) a body moving along a straight line and speeding up.

3.2 The Acceleration Vector

Now let's consider the *acceleration* of a particle moving in space. Just as for motion in a straight line, acceleration describes how the velocity of the particle changes. But since we now treat velocity as a vector, acceleration will describe changes in the velocity magnitude (that is, the speed) *and* changes in the direction of velocity (that is, the direction in which the particle is moving).

In Fig. 3.6a, a car (treated as a particle) is moving along a curved road. The vectors $\vec{v}_1$ and $\vec{v}_2$ represent the car's instantaneous velocities at time t_1, when the car

3.6 (a) A car moving along a curved road from P_1 to P_2. (b) How to obtain the change in velocity $\Delta \vec{v} = \vec{v}_2 - \vec{v}_1$ by vector subtraction. (c) The vector $\vec{a}_{av} = \Delta \vec{v} / \Delta t$ represents the average acceleration between P_1 and P_2.

(a)

This car accelerates by slowing while rounding a curve. (Its instantaneous velocity changes in both magnitude and direction.)

(b)

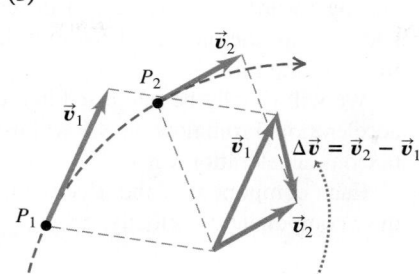

To find the car's average acceleration between P_1 and P_2, we first find the change in velocity $\Delta \vec{v}$ by subtracting $\vec{v}_1$ from $\vec{v}_2$. (Notice that $\vec{v}_1 + \Delta \vec{v} = \vec{v}_2$.)

(c)

The average acceleration has the same direction as the change in velocity, $\Delta \vec{v}$.

is at point P_1, and at time t_2, when the car is at point P_2. The two velocities may differ in both magnitude and direction. During the time interval from t_1 to t_2, the *vector change in velocity* is $\vec{v}_2 - \vec{v}_1 = \Delta \vec{v}$, so $\vec{v}_2 = \vec{v}_1 + \Delta \vec{v}$ (Fig. 3.6b). We define the **average acceleration** $\vec{a}_{av}$ of the car during this time interval as the velocity change divided by the time interval $t_2 - t_1 = \Delta t$:

$$\vec{a}_{av} = \frac{\vec{v}_2 - \vec{v}_1}{t_2 - t_1} = \frac{\Delta \vec{v}}{\Delta t} \qquad \text{(average acceleration vector)} \qquad (3.8)$$

Average acceleration is a *vector* quantity in the same direction as the vector $\Delta \vec{v}$ (Fig. 3.6c). The *x*-component of Eq. (3.8) is $a_{av\text{-}x} = (v_{2x} - v_{1x})/(t_2 - t_1) = \Delta v_x / \Delta t$, which is just Eq. (2.4) for the average acceleration in straight-line motion.

As in Chapter 2, we define the **instantaneous acceleration** $\vec{a}$ (a *vector* quantity) at point P_1 as the limit of the average acceleration vector when point P_2 approaches point P_1, so $\Delta \vec{v}$ and Δt both approach zero (Fig. 3.7). The instantaneous acceleration is also equal to the instantaneous rate of change of velocity with time:

$$\vec{a} = \lim_{\Delta t \to 0} \frac{\Delta \vec{v}}{\Delta t} = \frac{d\vec{v}}{dt} \qquad \text{(instantaneous acceleration vector)} \qquad (3.9)$$

The velocity vector $\vec{v}$, as we have seen, is tangent to the path of the particle. The instantaneous acceleration vector $\vec{a}$, however, does *not* have to be tangent to the path. Figure 3.7a shows that if the path is curved, $\vec{a}$ points toward the concave side of the path—that is, toward the inside of any turn that the particle is making. The acceleration is tangent to the path only if the particle moves in a straight line (Fig. 3.7b).

CAUTION **Any particle following a curved path is accelerating** When a particle is moving in a curved path, it always has nonzero acceleration, even when it moves with constant speed. This conclusion may seem contrary to your intuition, but it's really just contrary to the everyday use of the word "acceleration" to mean that speed is increasing. The more precise definition given in Eq. (3.9) shows that there is a nonzero acceleration whenever the velocity vector changes in any way, whether there is a change of speed, direction, or both. ▮

To convince yourself that a particle has a nonzero acceleration when moving on a curved path with constant speed, think of your sensations when you ride in a car. When the car accelerates, you tend to move inside the car in a

3.7 (a) Instantaneous acceleration $\vec{a}$ at point P_1 in Fig. 3.6. (b) Instantaneous acceleration for motion along a straight line.

(a) Acceleration: curved trajectory

To find the instantaneous acceleration $\vec{a}$ at P_1 ...

... we take the limit of $\vec{a}_{av}$ as P_2 approaches P_1 ...

... meaning that $\Delta \vec{v}$ and Δt approach 0.

$$\vec{a} = \lim_{\Delta t \to 0} \frac{\Delta \vec{v}}{\Delta t}$$

Acceleration points to concave side of path.

(b) Acceleration: straight-line trajectory

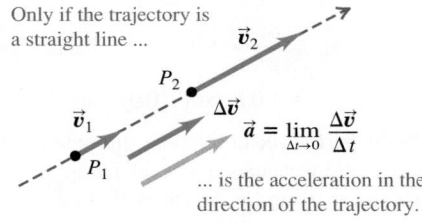

Only if the trajectory is a straight line ...

$$\vec{a} = \lim_{\Delta t \to 0} \frac{\Delta \vec{v}}{\Delta t}$$

... is the acceleration in the direction of the trajectory.

Application Horses on a Curved Path

By leaning to the side and hitting the ground with their hooves at an angle, these horses give themselves the sideways acceleration necessary to make a sharp change in direction.

3.8 When the arrow is released, its acceleration vector has both a horizontal component (a_x) and a vertical component (a_y).

direction *opposite* to the car's acceleration. (We'll discover the reason for this behavior in Chapter 4.) Thus you tend to slide toward the back of the car when it accelerates forward (speeds up) and toward the front of the car when it accelerates backward (slows down). If the car makes a turn on a level road, you tend to slide toward the outside of the turn; hence the car has an acceleration toward the inside of the turn.

We will usually be interested in the instantaneous acceleration, not the average acceleration. From now on, we will use the term "acceleration" to mean the instantaneous acceleration vector $\vec{a}$.

Each component of the acceleration vector is the derivative of the corresponding component of velocity:

$$a_x = \frac{dv_x}{dt} \qquad a_y = \frac{dv_y}{dt} \qquad a_z = \frac{dv_z}{dt} \qquad \text{(components of instantaneous acceleration)} \qquad (3.10)$$

In terms of unit vectors,

$$\vec{a} = \frac{dv_x}{dt}\hat{\imath} + \frac{dv_y}{dt}\hat{\jmath} + \frac{dv_z}{dt}\hat{k} \qquad (3.11)$$

The x-component of Eqs. (3.10) and (3.11), $a_x = dv_x/dt$, is the expression from Section 2.3 for instantaneous acceleration in one dimension, Eq. (2.5). Figure 3.8 shows an example of an acceleration vector that has both x- and y-components.

Since each component of velocity is the derivative of the corresponding coordinate, we can express the components a_x, a_y, and a_z of the acceleration vector $\vec{a}$ as

$$a_x = \frac{d^2x}{dt^2} \qquad a_y = \frac{d^2y}{dt^2} \qquad a_z = \frac{d^2z}{dt^2} \qquad (3.12)$$

The acceleration vector $\vec{a}$ itself is

$$\vec{a} = \frac{d^2x}{dt^2}\hat{\imath} + \frac{d^2y}{dt^2}\hat{\jmath} + \frac{d^2z}{dt^2}\hat{k} \qquad (3.13)$$

Example 3.2 Calculating average and instantaneous acceleration

Let's return to the motions of the Mars rover in Example 3.1. (a) Find the components of the average acceleration for the interval $t = 0.0$ s to $t = 2.0$ s. (b) Find the instantaneous acceleration at $t = 2.0$ s.

SOLUTION

IDENTIFY and SET UP: In Example 3.1 we found the components of the rover's instantaneous velocity at any time t:

$$v_x = \frac{dx}{dt} = (-0.25 \text{ m/s}^2)(2t) = (-0.50 \text{ m/s}^2)t$$

$$v_y = \frac{dy}{dt} = 1.0 \text{ m/s} + (0.025 \text{ m/s}^3)(3t^2)$$

$$= 1.0 \text{ m/s} + (0.075 \text{ m/s}^3)t^2$$

We'll use the vector relationships among velocity, average acceleration, and instantaneous acceleration. In part (a) we determine the values of v_x and v_y at the beginning and end of the interval and

then use Eq. (3.8) to calculate the components of the average acceleration. In part (b) we obtain expressions for the instantaneous acceleration components at any time t by taking the time derivatives of the velocity components as in Eqs. (3.10).

EXECUTE: (a) In Example 3.1 we found that at $t = 0.0$ s the velocity components are

$$v_x = 0.0 \text{ m/s} \qquad v_y = 1.0 \text{ m/s}$$

and that at $t = 2.00$ s the components are

$$v_x = -1.0 \text{ m/s} \qquad v_y = 1.3 \text{ m/s}$$

Thus the components of average acceleration in the interval $t = 0.0$ s to $t = 2.0$ s are

$$a_{\text{av-}x} = \frac{\Delta v_x}{\Delta t} = \frac{-1.0 \text{ m/s} - 0.0 \text{ m/s}}{2.0 \text{ s} - 0.0 \text{ s}} = -0.50 \text{ m/s}^2$$

$$a_{\text{av-}y} = \frac{\Delta v_y}{\Delta t} = \frac{1.3 \text{ m/s} - 1.0 \text{ m/s}}{2.0 \text{ s} - 0.0 \text{ s}} = 0.15 \text{ m/s}^2$$

(b) Using Eqs. (3.10), we find

$$a_x = \frac{dv_x}{dt} = -0.50 \text{ m/s}^2 \qquad a_y = \frac{dv_y}{dt} = (0.075 \text{ m/s}^3)(2t)$$

Hence the instantaneous acceleration vector $\vec{a}$ at time t is

$$\vec{a} = a_x\hat{\imath} + a_y\hat{\jmath} = (-0.50 \text{ m/s}^2)\hat{\imath} + (0.15 \text{ m/s}^3)t\hat{\jmath}$$

At $t = 2.0$ s the components of acceleration and the acceleration vector are

$$a_x = -0.50 \text{ m/s}^2 \qquad a_y = (0.15 \text{ m/s}^3)(2.0 \text{ s}) = 0.30 \text{ m/s}^2$$

$$\vec{a} = (-0.50 \text{ m/s}^2)\hat{\imath} + (0.30 \text{ m/s}^2)\hat{\jmath}$$

The magnitude of acceleration at this time is

$$a = \sqrt{a_x^2 + a_y^2}$$
$$= \sqrt{(-0.50 \text{ m/s}^2)^2 + (0.30 \text{ m/s}^2)^2} = 0.58 \text{ m/s}^2$$

A sketch of this vector (Fig. 3.9) shows that the direction angle β of $\vec{a}$ with respect to the positive x-axis is between 90° and 180°. From Eq. (3.7) we have

$$\arctan \frac{a_y}{a_x} = \arctan \frac{0.30 \text{ m/s}^2}{-0.50 \text{ m/s}^2} = -31°$$

Hence $\beta = 180° + (-31°) = 149°$.

EVALUATE: Figure 3.9 shows the rover's path and the velocity and acceleration vectors at $t = 0.0$ s, 1.0 s, and 2.0 s. (You should use the results of part (b) to calculate the instantaneous acceleration at $t = 0.0$ s and $t = 1.0$ s for yourself.) Note that $\vec{v}$ and $\vec{a}$ are *not* in the same direction at any of these times. The velocity vector $\vec{v}$ is tangent to the path at each point (as is always the case), and the acceleration vector $\vec{a}$ points toward the concave side of the path.

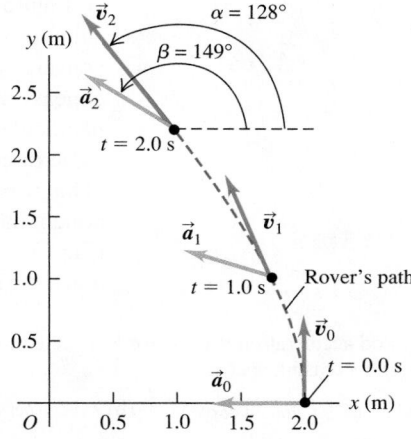

3.9 The path of the robotic rover, showing the velocity and acceleration at $t = 0.0$ s ($\vec{v}_0$ and $\vec{a}_0$), $t = 1.0$ s ($\vec{v}_1$ and $\vec{a}_1$), and $t = 2.0$ s ($\vec{v}_2$ and $\vec{a}_2$).

Parallel and Perpendicular Components of Acceleration

Equations (3.10) tell us about the components of a particle's instantaneous acceleration vector $\vec{a}$ along the x-, y-, and z-axes. Another useful way to think about $\vec{a}$ is in terms of its component *parallel* to the particle's path—that is, parallel to the velocity—and its component *perpendicular* to the path—and hence perpendicular to the velocity (Fig. 3.10). That's because the parallel component $a_\parallel$ tells us about changes in the particle's *speed*, while the perpendicular component $a_\perp$ tells us about changes in the particle's *direction of motion*. To see why the parallel and perpendicular components of $\vec{a}$ have these properties, let's consider two special cases.

In Fig. 3.11a the acceleration vector is in the same direction as the velocity $\vec{v}_1$, so $\vec{a}$ has only a parallel component $a_\parallel$ (that is, $a_\perp = 0$). The velocity change $\Delta\vec{v}$ during a small time interval Δt is in the same direction as $\vec{a}$ and hence in the same direction as $\vec{v}_1$. The velocity $\vec{v}_2$ at the end of Δt is in the same direction as $\vec{v}_1$ but has greater magnitude. Hence during the time interval Δt the particle in Fig. 3.11a moved in a straight line with increasing speed (compare Fig. 3.7b).

In Fig. 3.11b the acceleration is *perpendicular* to the velocity, so $\vec{a}$ has only a perpendicular component $a_\perp$ (that is, $a_\parallel = 0$). In a small time interval Δt, the

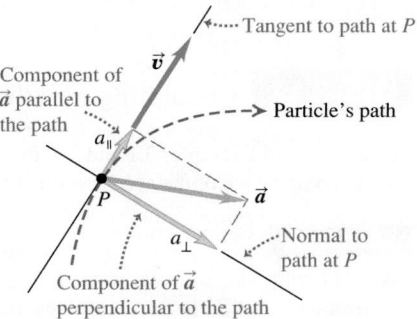

3.10 The acceleration can be resolved into a component $a_\parallel$ parallel to the path (that is, along the tangent to the path) and a component $a_\perp$ perpendicular to the path (that is, along the normal to the path).

3.11 The effect of acceleration directed (a) parallel to and (b) perpendicular to a particle's velocity.

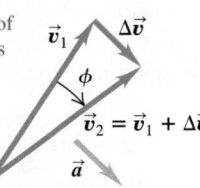

(a) Acceleration parallel to velocity

Changes only *magnitude* of velocity: speed changes; direction doesn't.

$\vec{a}$ $\vec{v}_1$ $\Delta\vec{v}$ $\vec{v}_2 = \vec{v}_1 + \Delta\vec{v}$

(b) Acceleration perpendicular to velocity

Changes only *direction* of velocity: particle follows curved path at constant speed.

$\vec{v}_1$ $\Delta\vec{v}$ ϕ $\vec{v}_2 = \vec{v}_1 + \Delta\vec{v}$ $\vec{a}$

velocity change $\Delta\vec{v}$ is very nearly perpendicular to $\vec{v}_1$, and so $\vec{v}_1$ and $\vec{v}_2$ have different directions. As the time interval Δt approaches zero, the angle ϕ in the figure also approaches zero, $\Delta\vec{v}$ becomes perpendicular to *both* $\vec{v}_1$ and $\vec{v}_2$, and $\vec{v}_1$ and $\vec{v}_2$ have the same magnitude. In other words, the speed of the particle stays the same, but the direction of motion changes and the path of the particle curves.

In the most general case, the acceleration $\vec{a}$ has components *both* parallel and perpendicular to the velocity $\vec{v}$, as in Fig. 3.10. Then the particle's speed will change (described by the parallel component $a_\parallel$) *and* its direction of motion will change (described by the perpendicular component $a_\perp$) so that it follows a curved path.

Figure 3.12 shows a particle moving along a curved path for three different situations: constant speed, increasing speed, and decreasing speed. If the speed is constant, $\vec{a}$ is perpendicular, or *normal,* to the path and to $\vec{v}$ and points toward the concave side of the path (Fig. 3.12a). If the speed is increasing, there is still a perpendicular component of $\vec{a}$, but there is also a parallel component having the same direction as $\vec{v}$ (Fig. 3.12b). Then $\vec{a}$ points ahead of the normal to the path. (This was the case in Example 3.2.) If the speed is decreasing, the parallel component has the direction opposite to $\vec{v}$, and $\vec{a}$ points behind the normal to the path (Fig. 3.12c; compare Fig. 3.7a). We will use these ideas again in Section 3.4 when we study the special case of motion in a circle.

PhET: Maze Game

3.12 Velocity and acceleration vectors for a particle moving through a point P on a curved path with (a) constant speed, (b) increasing speed, and (c) decreasing speed.

(a) When speed is constant along a curved path ...

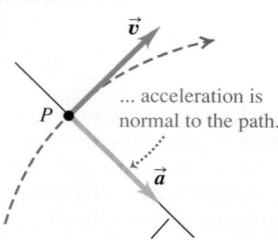

(b) When speed is increasing along a curved path ...

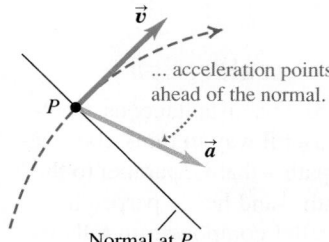

(c) When speed is decreasing along a curved path ...

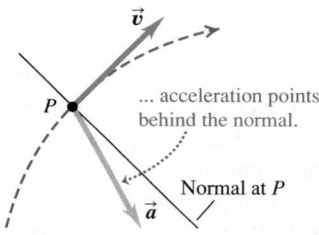

Example 3.3 **Calculating parallel and perpendicular components of acceleration**

For the rover of Examples 3.1 and 3.2, find the parallel and perpendicular components of the acceleration at $t = 2.0$ s.

SOLUTION

IDENTIFY and SET UP: We want to find the components of the acceleration vector $\vec{a}$ that are parallel and perpendicular to the velocity vector $\vec{v}$. We found the directions of $\vec{v}$ and $\vec{a}$ in Examples 3.1 and 3.2, respectively; Fig. 3.9 shows the results. From these directions we can find the angle between the two vectors and the components of $\vec{a}$ with respect to the direction of $\vec{v}$.

EXECUTE: From Example 3.2, at $t = 2.0$ s the particle has an acceleration of magnitude 0.58 m/s² at an angle of $149°$ with respect to the positive x-axis. In Example 3.1 we found that at this time the velocity vector is at an angle of $128°$ with respect to the positive x-axis. The angle between $\vec{a}$ and $\vec{v}$ is therefore $149° - 128° = 21°$ (Fig. 3.13). Hence the components of acceleration parallel and perpendicular to $\vec{v}$ are

$$a_\parallel = a \cos 21° = (0.58 \text{ m/s}^2)\cos 21° = 0.54 \text{ m/s}^2$$
$$a_\perp = a \sin 21° = (0.58 \text{ m/s}^2)\sin 21° = 0.21 \text{ m/s}^2$$

3.13 The parallel and perpendicular components of the acceleration of the rover at $t = 2.0$ s.

EVALUATE: The parallel component $a_\parallel$ is positive (in the same direction as $\vec{v}$), which means that the speed is increasing at this instant. The value $a_\parallel = +0.54$ m/s² tells us that the speed is increasing at this instant at a rate of 0.54 m/s per second. The perpendicular component $a_\perp$ is not zero, which means that at this instant the rover is turning—that is, it is changing direction and following a curved path.

Conceptual Example 3.4   **Acceleration of a skier**

A skier moves along a ski-jump ramp (Fig. 3.14a). The ramp is straight from point A to point C and curved from point C onward. The skier speeds up as she moves downhill from point A to point E, where her speed is maximum. She slows down after passing point E. Draw the direction of the acceleration vector at each of the points B, D, E, and F.

SOLUTION

Figure 3.14b shows our solution. At point B the skier is moving in a straight line with increasing speed, so her acceleration points downhill, in the same direction as her velocity. At points D, E, and F the skier is moving along a curved path, so her acceleration has a component perpendicular to the path (toward the concave side of the path) at each of these points. At point D there is also an acceleration component in the direction of her motion because she is speeding up. So the acceleration vector points *ahead* of the normal to her path at point D, as Fig. 3.14b shows. At point E, the skier's speed is instantaneously not changing; her speed is maximum at this point, so its derivative is zero. There is therefore no parallel component of $\vec{a}$, and the acceleration is perpendicular to her motion. At point F there is an acceleration component *opposite to* the direction of her motion because now she's slowing down. The acceleration vector therefore points *behind* the normal to her path.

In the next section we'll consider the skier's acceleration after she flies off the ramp.

3.14 (a) The skier's path. (b) Our solution.

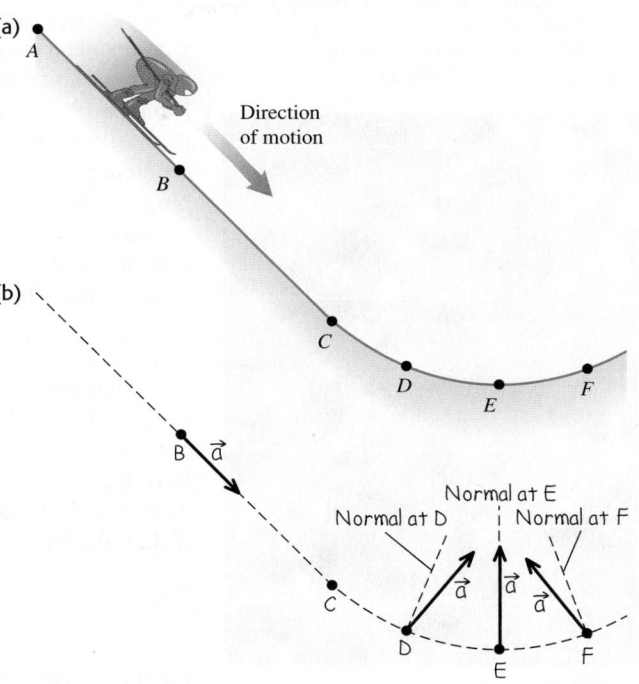

Test Your Understanding of Section 3.2   A sled travels over the crest of a snow-covered hill. The sled slows down as it climbs up one side of the hill and gains speed as it descends on the other side. Which of the vectors (1 through 9) in the figure correctly shows the direction of the sled's acceleration at the crest? (Choice 9 is that the acceleration is zero.)

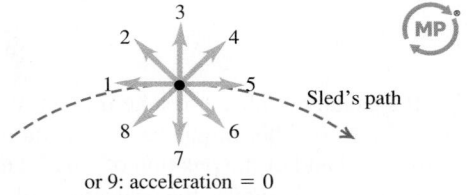

3.3 Projectile Motion

A **projectile** is any body that is given an initial velocity and then follows a path determined entirely by the effects of gravitational acceleration and air resistance. A batted baseball, a thrown football, a package dropped from an airplane, and a bullet shot from a rifle are all projectiles. The path followed by a projectile is called its **trajectory.**

To analyze this common type of motion, we'll start with an idealized model, representing the projectile as a particle with an acceleration (due to gravity) that is constant in both magnitude and direction. We'll neglect the effects of air resistance and the curvature and rotation of the earth. Like all models, this one has limitations. Curvature of the earth has to be considered in the flight of long-range missiles, and air resistance is of crucial importance to a sky diver. Nevertheless, we can learn a lot from analysis of this simple model. For the remainder of this chapter the phrase "projectile motion" will imply that we're ignoring air resistance. In Chapter 5 we will see what happens when air resistance cannot be ignored.

Projectile motion is always confined to a vertical plane determined by the direction of the initial velocity (Fig. 3.15). This is because the acceleration due to

3.15 The trajectory of an idealized projectile.

- A projectile moves in a vertical plane that contains the initial velocity vector $\vec{v}_0$.
- Its trajectory depends only on $\vec{v}_0$ and on the downward acceleration due to gravity.

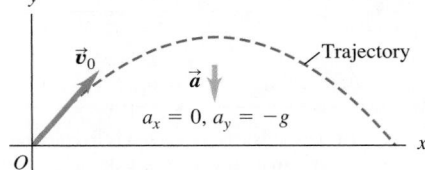

3.16 The red ball is dropped from rest, and the yellow ball is simultaneously projected horizontally; successive images in this stroboscopic photograph are separated by equal time intervals. At any given time, both balls have the same y-position, y-velocity, and y-acceleration, despite having different x-positions and x-velocities.

Mastering**PHYSICS**

ActivPhysics 3.1: Solving Projectile Motion Problems
ActivPhysics 3.2: Two Balls Falling
ActivPhysics 3.3: Changing the x-velocity
ActivPhysics 3.4: Projecting x-y-Accelerations

gravity is purely vertical; gravity can't accelerate the projectile sideways. Thus projectile motion is *two-dimensional*. We will call the plane of motion the xy-coordinate plane, with the x-axis horizontal and the y-axis vertically upward.

The key to analyzing projectile motion is that we can treat the x- and y-coordinates separately. The x-component of acceleration is zero, and the y-component is constant and equal to $-g$. (By definition, g is always positive; with our choice of coordinate directions, a_y is negative.) So *we can analyze projectile motion as a combination of horizontal motion with constant velocity and vertical motion with constant acceleration.* Figure 3.16 shows two projectiles with different x-motion but identical y-motion; one is dropped from rest and the other is projected horizontally, but both projectiles fall the same distance in the same time.

We can then express all the vector relationships for the projectile's position, velocity, and acceleration by separate equations for the horizontal and vertical components. The components of $\vec{a}$ are

$$a_x = 0 \qquad a_y = -g \qquad \text{(projectile motion, no air resistance)} \quad (3.14)$$

Since the x-acceleration and y-acceleration are both constant, we can use Eqs. (2.8), (2.12), (2.13), and (2.14) directly. For example, suppose that at time $t = 0$ our particle is at the point (x_0, y_0) and that at this time its velocity components have the initial values v_{0x} and v_{0y}. The components of acceleration are $a_x = 0$, $a_y = -g$. Considering the x-motion first, we substitute 0 for a_x in Eqs. (2.8) and (2.12). We find

$$v_x = v_{0x} \quad (3.15)$$

$$x = x_0 + v_{0x}t \quad (3.16)$$

For the y-motion we substitute y for x, v_y for v_x, v_{0y} for v_{0x}, and $a_y = -g$ for a_x:

$$v_y = v_{0y} - gt \quad (3.17)$$

$$y = y_0 + v_{0y}t - \tfrac{1}{2}gt^2 \quad (3.18)$$

It's usually simplest to take the initial position (at $t = 0$) as the origin; then $x_0 = y_0 = 0$. This might be the position of a ball at the instant it leaves the thrower's hand or the position of a bullet at the instant it leaves the gun barrel.

Figure 3.17 shows the trajectory of a projectile that starts at (or passes through) the origin at time $t = 0$, along with its position, velocity, and velocity

3.17 If air resistance is negligible, the trajectory of a projectile is a combination of horizontal motion with constant velocity and vertical motion with constant acceleration.

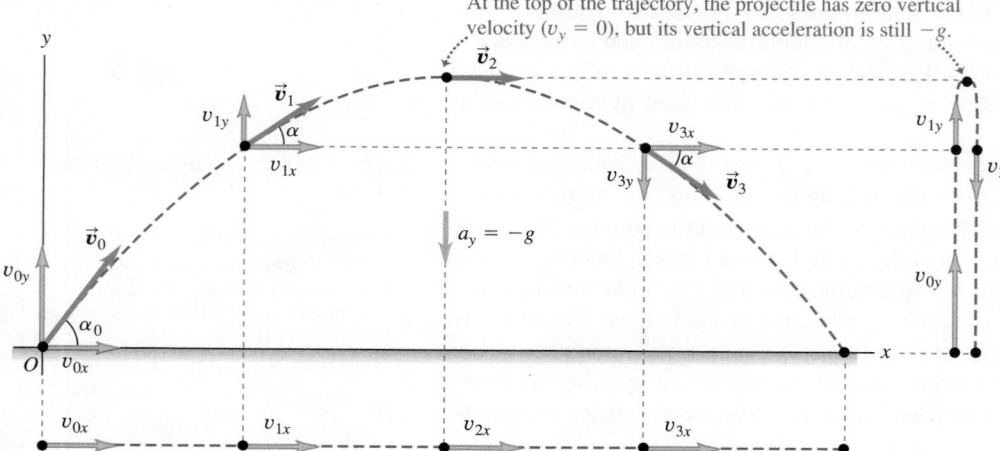

At the top of the trajectory, the projectile has zero vertical velocity ($v_y = 0$), but its vertical acceleration is still $-g$.

Vertically, the projectile is in constant-acceleration motion in response to the earth's gravitational pull. Thus its vertical velocity *changes* by equal amounts during equal time intervals.

Horizontally, the projectile is in constant-velocity motion: Its horizontal acceleration is zero, so it moves equal x-distances in equal time intervals.

components at equal time intervals. The x-component of acceleration is zero, so v_x is constant. The y-component of acceleration is constant and not zero, so v_y changes by equal amounts in equal times, just the same as if the projectile were launched vertically with the same initial y-velocity.

We can also represent the initial velocity $\vec{v}_0$ by its magnitude v_0 (the initial speed) and its angle α_0 with the positive x-axis (Fig. 3.18). In terms of these quantities, the components v_{0x} and v_{0y} of the initial velocity are

$$v_{0x} = v_0\cos\alpha_0 \qquad v_{0y} = v_0\sin\alpha_0 \qquad (3.19)$$

If we substitute these relationships in Eqs. (3.15) through (3.18) and set $x_0 = y_0 = 0$, we find

$$x = (v_0\cos\alpha_0)t \qquad \text{(projectile motion)} \qquad (3.20)$$

$$y = (v_0\sin\alpha_0)t - \tfrac{1}{2}gt^2 \qquad \text{(projectile motion)} \qquad (3.21)$$

$$v_x = v_0\cos\alpha_0 \qquad \text{(projectile motion)} \qquad (3.22)$$

$$v_y = v_0\sin\alpha_0 - gt \qquad \text{(projectile motion)} \qquad (3.23)$$

These equations describe the position and velocity of the projectile in Fig. 3.17 at any time t.

We can get a lot of information from Eqs. (3.20) through (3.23). For example, at any time the distance r of the projectile from the origin (the magnitude of the position vector $\vec{r}$) is given by

$$r = \sqrt{x^2 + y^2} \qquad (3.24)$$

The projectile's speed (the magnitude of its velocity) at any time is

$$v = \sqrt{v_x^2 + v_y^2} \qquad (3.25)$$

The *direction* of the velocity, in terms of the angle α it makes with the positive x-direction (see Fig. 3.17), is given by

$$\tan\alpha = \frac{v_y}{v_x} \qquad (3.26)$$

The velocity vector $\vec{v}$ is tangent to the trajectory at each point.

We can derive an equation for the trajectory's shape in terms of x and y by eliminating t. From Eqs. (3.20) and (3.21), which assume $x_0 = y_0 = 0$, we find $t = x/(v_0\cos\alpha_0)$ and

$$y = (\tan\alpha_0)x - \frac{g}{2v_0^2\cos^2\alpha_0}x^2 \qquad (3.27)$$

Don't worry about the details of this equation; the important point is its general form. Since v_0, $\tan\alpha_0$, $\cos\alpha_0$, and g are constants, Eq. (3.27) has the form

$$y = bx - cx^2$$

where b and c are constants. This is the equation of a *parabola*. In our simple model of projectile motion, the trajectory is always a parabola (Fig. 3.19).

When air resistance *isn't* always negligible and has to be included, calculating the trajectory becomes a lot more complicated; the effects of air resistance depend on velocity, so the acceleration is no longer constant. Figure 3.20 shows a

3.18 The initial velocity components v_{0x} and v_{0y} of a projectile (such as a kicked soccer ball) are related to the initial speed v_0 and initial angle α_0.

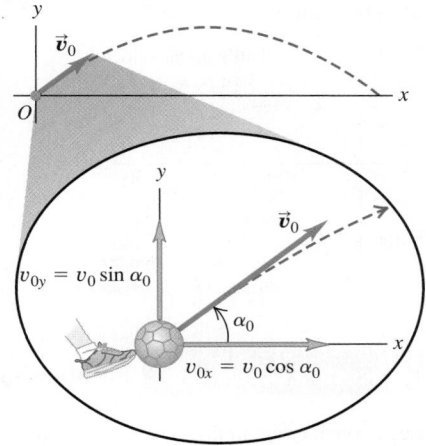

MasteringPHYSICS

PhET: Projectile Motion
ActivPhysics 3.5: Initial Velocity Components
ActivPhysics 3.6: Target Practice I
ActivPhysics 3.7: Target Practice II

3.19 The nearly parabolic trajectories of (a) a bouncing ball and (b) blobs of molten rock ejected from a volcano.

(a) Successive images of ball are separated by equal time intervals.

Successive peaks decrease in height because ball loses energy with each bounce.

(b)

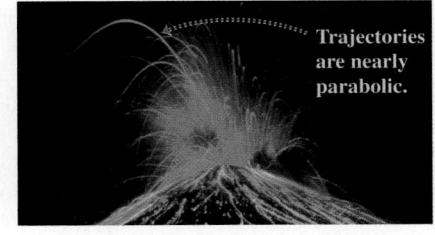

Trajectories are nearly parabolic.

3.20 Air resistance has a large cumulative effect on the motion of a baseball. In this simulation we allow the baseball to fall below the height from which it was thrown (for example, the baseball could have been thrown from a cliff).

computer simulation of the trajectory of a baseball both without air resistance and with air resistance proportional to the square of the baseball's speed. We see that air resistance has a very large effect; the maximum height and range both decrease, and the trajectory is no longer a parabola. (If you look closely at Fig. 3.19b, you'll see that the trajectories of the volcanic blobs deviate in a similar way from a parabolic shape.)

Conceptual Example 3.5 **Acceleration of a skier, continued**

Let's consider again the skier in Conceptual Example 3.4. What is her acceleration at each of the points G, H, and I in Fig. 3.21a *after* she flies off the ramp? Neglect air resistance.

SOLUTION

Figure 3.21b shows our answer. The skier's acceleration changed from point to point while she was on the ramp. But as soon as she leaves the ramp, she becomes a projectile. So at points G, H, and I, and indeed at *all* points after she leaves the ramp, the skier's acceleration points vertically downward and has magnitude g. No matter how complicated the acceleration of a particle before it becomes a projectile, its acceleration as a projectile is given by $a_x = 0$, $a_y = -g$.

3.21 (a) The skier's path during the jump. (b) Our solution.

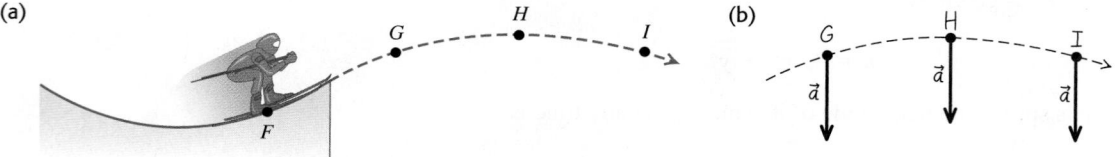

Problem-Solving Strategy 3.1 **Projectile Motion**

NOTE: The strategies we used in Sections 2.4 and 2.5 for straight-line, constant-acceleration problems are also useful here.

IDENTIFY *the relevant concepts:* The key concept to remember is that throughout projectile motion, the acceleration is downward and has a constant magnitude g. Note that the projectile-motion equations don't apply to *throwing* a ball, because during the throw the ball is acted on by both the thrower's hand and gravity. These equations apply only *after* the ball leaves the thrower's hand.

SET UP *the problem* using the following steps:
1. Define your coordinate system and make a sketch showing your axes. Usually it's easiest to make the x-axis horizontal and the y-axis upward, and to place the origin at the initial ($t = 0$) position where the body first becomes a projectile (such as where a ball leaves the thrower's hand). Then the components of the (constant) acceleration are $a_x = 0$, $a_y = -g$, and the initial position is $x_0 = 0$, $y_0 = 0$.
2. List the unknown and known quantities, and decide which unknowns are your target variables. For example, you might be given the initial velocity (either the components or the magnitude and direction) and asked to find the coordinates and velocity components at some later time. In any case, you'll be using

Eqs. (3.20) through (3.23). (Equations (3.24) through (3.27) may be useful as well.) Make sure that you have as many equations as there are target variables to be found.
3. State the problem in words and then translate those words into symbols. For example, *when* does the particle arrive at a certain point? (That is, at what value of t?) *Where* is the particle when its velocity has a certain value? (That is, what are the values of x and y when v_x or v_y has the specified value?) Since $v_y = 0$ at the highest point in a trajectory, the question "When does the projectile reach its highest point?" translates into "What is the value of t when $v_y = 0$?" Similarly, "When does the projectile return to its initial elevation?" translates into "What is the value of t when $y = y_0$?"

EXECUTE *the solution:* Find the target variables using the equations you chose. Resist the temptation to break the trajectory into segments and analyze each segment separately. You don't have to start all over when the projectile reaches its highest point! It's almost always easier to use the same axes and time scale throughout the problem. If you need numerical values, use $g = 9.80 \text{ m/s}^2$.

EVALUATE *your answer:* As always, look at your results to see whether they make sense and whether the numerical values seem reasonable.

Example 3.6 **A body projected horizontally**

A motorcycle stunt rider rides off the edge of a cliff. Just at the edge his velocity is horizontal, with magnitude 9.0 m/s. Find the motorcycle's position, distance from the edge of the cliff, and velocity 0.50 s after it leaves the edge of the cliff.

SOLUTION

IDENTIFY and SET UP: Figure 3.22 shows our sketch of the motorcycle's trajectory. He is in projectile motion as soon as he leaves the edge of the cliff, which we choose to be the origin of coordinates so $x_0 = 0$ and $y_0 = 0$. His initial velocity $\vec{v}_0$ at the edge of the cliff is horizontal (that is, $\alpha_0 = 0$), so its components are $v_{0x} = v_0 \cos \alpha_0 = 9.0$ m/s and $v_{0y} = v_0 \sin \alpha_0 = 0$. To find the motorcycle's position at $t = 0.50$ s, we use Eqs. (3.20) and (3.21); we then find the distance from the origin using Eq. (3.24). Finally, we use Eqs. (3.22) and (3.23) to find the velocity components at $t = 0.50$ s.

EXECUTE: From Eqs. (3.20) and (3.21), the motorcycle's x- and y-coordinates at $t = 0.50$ s are

$$x = v_{0x}t = (9.0 \text{ m/s})(0.50 \text{ s}) = 4.5 \text{ m}$$

$$y = -\tfrac{1}{2}gt^2 = -\tfrac{1}{2}(9.80 \text{ m/s}^2)(0.50 \text{ s})^2 = -1.2 \text{ m}$$

The negative value of y shows that the motorcycle is below its starting point.

From Eq. (3.24), the motorcycle's distance from the origin at $t = 0.50$ s is

$$r = \sqrt{x^2 + y^2} = \sqrt{(4.5 \text{ m})^2 + (-1.2 \text{ m})^2} = 4.7 \text{ m}$$

From Eqs. (3.22) and (3.23), the velocity components at $t = 0.50$ s are

$$v_x = v_{0x} = 9.0 \text{ m/s}$$

$$v_y = -gt = (-9.80 \text{ m/s}^2)(0.50 \text{ s}) = -4.9 \text{ m/s}$$

3.22 Our sketch for this problem.

The motorcycle has the same horizontal velocity v_x as when it left the cliff at $t = 0$, but in addition there is a downward (negative) vertical velocity v_y. The velocity vector at $t = 0.50$ s is

$$\vec{v} = v_x\hat{\imath} + v_y\hat{\jmath} = (9.0 \text{ m/s})\hat{\imath} + (-4.9 \text{ m/s})\hat{\jmath}$$

From Eq. (3.25), the speed (magnitude of the velocity) at $t = 0.50$ s is

$$v = \sqrt{v_x^2 + v_y^2}$$

$$= \sqrt{(9.0 \text{ m/s})^2 + (-4.9 \text{ m/s})^2} = 10.2 \text{ m/s}$$

From Eq. (3.26), the angle α of the velocity vector is

$$\alpha = \arctan \frac{v_y}{v_x} = \arctan\left(\frac{-4.9 \text{ m/s}}{9.0 \text{ m/s}}\right) = -29°$$

The velocity is 29° below the horizontal.

EVALUATE: Just as in Fig. 3.17, the motorcycle's horizontal motion is unchanged by gravity; the motorcycle continues to move horizontally at 9.0 m/s, covering 4.5 m in 0.50 s. The motorcycle initially has zero vertical velocity, so it falls vertically just like a body released from rest and descends a distance $\tfrac{1}{2}gt^2 = 1.2$ m in 0.50 s.

Example 3.7 **Height and range of a projectile I: A batted baseball**

A batter hits a baseball so that it leaves the bat at speed $v_0 = 37.0$ m/s at an angle $\alpha_0 = 53.1°$. (a) Find the position of the ball and its velocity (magnitude and direction) at $t = 2.00$ s. (b) Find the time when the ball reaches the highest point of its flight, and its height h at this time. (c) Find the *horizontal range* R—that is, the horizontal distance from the starting point to where the ball hits the ground.

SOLUTION

IDENTIFY and SET UP: As Fig. 3.20 shows, air resistance strongly affects the motion of a baseball. For simplicity, however, we'll ignore air resistance here and use the projectile-motion equations to describe the motion. The ball leaves the bat at $t = 0$ a meter or so above ground level, but we'll neglect this distance and assume that it starts at ground level ($y_0 = 0$). Figure 3.23 shows our

3.23 Our sketch for this problem.

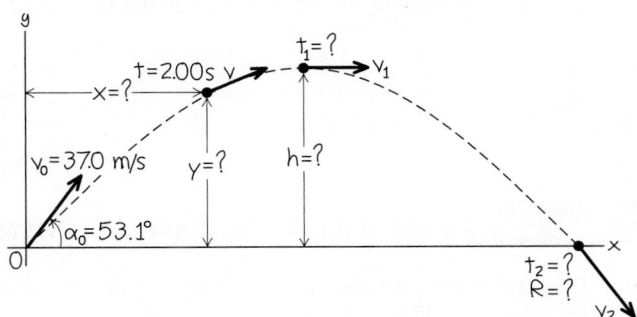

sketch of the ball's trajectory. We'll use the same coordinate system as in Figs. 3.17 and 3.18, so we can use Eqs. (3.20) through

Continued

(3.23). Our target variables are (a) the position and velocity of the ball 2.00 s after it leaves the bat, (b) the time t when the ball is at its maximum height (that is, when $v_y = 0$) and the y-coordinate at this time, and (c) the x-coordinate when the ball returns to ground level ($y = 0$).

EXECUTE: (a) We want to find x, y, v_x, and v_y at $t = 2.00$ s. The initial velocity of the ball has components

$$v_{0x} = v_0 \cos \alpha_0 = (37.0 \text{ m/s}) \cos 53.1° = 22.2 \text{ m/s}$$
$$v_{0y} = v_0 \sin \alpha_0 = (37.0 \text{ m/s}) \sin 53.1° = 29.6 \text{ m/s}$$

From Eqs. (3.20) through (3.23),

$$x = v_{0x}t = (22.2 \text{ m/s})(2.00 \text{ s}) = 44.4 \text{ m}$$

$$y = v_{0y}t - \tfrac{1}{2}gt^2$$
$$= (29.6 \text{ m/s})(2.00 \text{ s}) - \tfrac{1}{2}(9.80 \text{ m/s}^2)(2.00 \text{ s})^2$$
$$= 39.6 \text{ m}$$

$$v_x = v_{0x} = 22.2 \text{ m/s}$$
$$v_y = v_{0y} - gt = 29.6 \text{ m/s} - (9.80 \text{ m/s}^2)(2.00 \text{ s})$$
$$= 10.0 \text{ m/s}$$

The y-component of velocity is positive at $t = 2.00$ s, so the ball is still moving upward (Fig. 3.23). From Eqs. (3.25) and (3.26), the magnitude and direction of the velocity are

$$v = \sqrt{v_x^2 + v_y^2} = \sqrt{(22.2 \text{ m/s})^2 + (10.0 \text{ m/s})^2}$$
$$= 24.4 \text{ m/s}$$

$$\alpha = \arctan\left(\frac{10.0 \text{ m/s}}{22.2 \text{ m/s}}\right) = \arctan 0.450 = 24.2°$$

The direction of the velocity (the direction of the ball's motion) is 24.2° above the horizontal.

(b) At the highest point, the vertical velocity v_y is zero. Call the time when this happens t_1; then

$$v_y = v_{0y} - gt_1 = 0$$
$$t_1 = \frac{v_{0y}}{g} = \frac{29.6 \text{ m/s}}{9.80 \text{ m/s}^2} = 3.02 \text{ s}$$

The height h at the highest point is the value of y at time t_1:

$$h = v_{0y}t_1 - \tfrac{1}{2}gt_1^2$$
$$= (29.6 \text{ m/s})(3.02 \text{ s}) - \tfrac{1}{2}(9.80 \text{ m/s}^2)(3.02 \text{ s})^2$$
$$= 44.7 \text{ m}$$

(c) We'll find the horizontal range in two steps. First, we find the time t_2 when $y = 0$ (the ball is at ground level):

$$y = 0 = v_{0y}t_2 - \tfrac{1}{2}gt_2^2 = t_2\left(v_{0y} - \tfrac{1}{2}gt_2\right)$$

This is a quadratic equation for t_2. It has two roots:

$$t_2 = 0 \quad \text{and} \quad t_2 = \frac{2v_{0y}}{g} = \frac{2(29.6 \text{ m/s})}{9.80 \text{ m/s}^2} = 6.04 \text{ s}$$

The ball is at $y = 0$ at both times. The ball *leaves* the ground at $t_2 = 0$, and it hits the ground at $t_2 = 2v_{0y}/g = 6.04$ s.

The horizontal range R is the value of x when the ball returns to the ground at $t_2 = 6.04$ s:

$$R = v_{0x}t_2 = (22.2 \text{ m/s})(6.04 \text{ s}) = 134 \text{ m}$$

The vertical component of velocity when the ball hits the ground is

$$v_y = v_{0y} - gt_2 = 29.6 \text{ m/s} - (9.80 \text{ m/s}^2)(6.04 \text{ s})$$
$$= -29.6 \text{ m/s}$$

That is, v_y has the same magnitude as the initial vertical velocity v_{0y} but the opposite direction (down). Since v_x is constant, the angle $\alpha = -53.1°$ (below the horizontal) at this point is the negative of the initial angle $\alpha_0 = 53.1°$.

EVALUATE: It's often useful to check results by getting them in a different way. For example, we can also find the maximum height in part (b) by applying the constant-acceleration formula Eq. (2.13) to the y-motion:

$$v_y^2 = v_{0y}^2 + 2a_y(y - y_0) = v_{0y}^2 - 2g(y - y_0)$$

At the highest point, $v_y = 0$ and $y = h$. You should solve this equation for h; you should get the same answer that we obtained in part (b). (Do you?)

Note that the time to hit the ground, $t_2 = 6.04$ s, is exactly twice the time to reach the highest point, $t_1 = 3.02$ s. Hence the time of descent equals the time of ascent. This is *always* true if the starting and end points are at the same elevation and if air resistance can be neglected.

Note also that $h = 44.7$ m in part (b) is comparable to the 52.4-m height above the playing field of the roof of the Hubert H. Humphrey Metrodome in Minneapolis, and the horizontal range $R = 134$ m in part (c) is greater than the 99.7-m distance from home plate to the right-field fence at Safeco Field in Seattle. In reality, due to air resistance (which we have neglected) a batted ball with the initial speed and angle we've used here won't go as high or as far as we've calculated (see Fig. 3.20).

Example 3.8 **Height and range of a projectile II: Maximum height, maximum range**

Find the maximum height h and horizontal range R (see Fig. 3.23) of a projectile launched with speed v_0 at an initial angle α_0 between 0° and 90°. For a given v_0, what value of α_0 gives maximum height? What value gives maximum horizontal range?

SOLUTION

IDENTIFY and SET UP: This is almost the same as parts (b) and (c) of Example 3.7, except that now we want general expressions for h and R. We also want the values of α_0 that give the maximum values of h and R. In part (b) of Example 3.7 we found that the projectile reaches the high point of its trajectory (so that $v_y = 0$) at time $t_1 = v_{0y}/g$, and in part (c) we found that the projectile returns to its starting height (so that $y = y_0$) at time $t_2 = 2v_{0y}/g = 2t_1$. We'll use Eq. (3.21) to find the y-coordinate h at t_1 and Eq. (3.20) to find the x-coordinate R at time t_2. We'll express our answers in terms of the launch speed v_0 and launch angle α_0 using Eqs. (3.19).

EXECUTE: From Eqs. (3.19), $v_{0x} = v_0 \cos \alpha_0$ and $v_{0y} = v_0 \sin \alpha_0$. Hence we can write the time t_1 when $v_y = 0$ as

$$t_1 = \frac{v_{0y}}{g} = \frac{v_0 \sin \alpha_0}{g}$$

Equation (3.21) gives the height $y = h$ at this time:

$$h = (v_0 \sin \alpha_0)\left(\frac{v_0 \sin \alpha_0}{g}\right) - \frac{1}{2}g\left(\frac{v_0 \sin \alpha_0}{g}\right)^2$$

$$= \frac{v_0^2 \sin^2 \alpha_0}{2g}$$

For a given launch speed v_0, the maximum value of h occurs for $\sin \alpha_0 = 1$ and $\alpha_0 = 90°$—that is, when the projectile is launched straight up. (If it is launched horizontally, as in Example 3.6, $\alpha_0 = 0$ and the maximum height is zero!)

The time t_2 when the projectile hits the ground is

$$t_2 = \frac{2v_{0y}}{g} = \frac{2v_0 \sin \alpha_0}{g}$$

The horizontal range R is the value of x at this time. From Eq. (3.20), this is

$$R = (v_0 \cos \alpha_0)t_2 = (v_0 \cos \alpha_0)\frac{2v_0 \sin \alpha_0}{g}$$

$$= \frac{v_0^2 \sin 2\alpha_0}{g}$$

(We used the trigonometric identity $2 \sin \alpha_0 \cos \alpha_0 = \sin 2\alpha_0$, found in Appendix B.) The maximum value of $\sin 2\alpha_0$ is 1; this occurs when $2\alpha_0 = 90°$ or $\alpha_0 = 45°$. This angle gives the maximum range for a given initial speed if air resistance can be neglected.

EVALUATE: Figure 3.24 is based on a composite photograph of three trajectories of a ball projected from a small spring gun at angles of 30°, 45°, and 60°. The initial speed v_0 is approximately the same in all three cases. The horizontal range is greatest for the 45° angle. The ranges are nearly the same for the 30° and 60° angles: Can you prove that for a given value of v_0 the range is the same for both an initial angle α_0 and an initial angle $90° - \alpha_0$? (This is not the case in Fig. 3.24 due to air resistance.)

> **CAUTION** **Height and range of a projectile** We don't recommend memorizing the above expressions for h, R, and $R_{\max}$. They are applicable only in the special circumstances we have described. In particular, the expressions for the range R and maximum range $R_{\max}$ can be used *only* when launch and landing heights are equal. There are many end-of-chapter problems to which these equations do *not* apply.

3.24 A launch angle of 45° gives the maximum horizontal range. The range is shorter with launch angles of 30° and 60°.

A 45° launch angle gives the greatest range; other angles fall shorter.

Launch angle:
$\alpha_0 = 30°$
$\alpha_0 = 45°$
$\alpha_0 = 60°$

Example 3.9   Different initial and final heights

You throw a ball from your window 8.0 m above the ground. When the ball leaves your hand, it is moving at 10.0 m/s at an angle of 20° below the horizontal. How far horizontally from your window will the ball hit the ground? Ignore air resistance.

SOLUTION

IDENTIFY and SET UP: As in Examples 3.7 and 3.8, we want to find the horizontal coordinate of a projectile when it is at a given y-value. The difference here is that this value of y is *not* the same as the initial value. We again choose the x-axis to be horizontal and the y-axis to be upward, and place the origin of coordinates at the point where the ball leaves your hand (Fig. 3.25). We have $v_0 = 10.0$ m/s and $\alpha_0 = -20°$ (the angle is negative because the initial velocity is below the horizontal). Our target variable is the value of x when the ball reaches the ground at $y = -8.0$ m. We'll use Eq. (3.21) to find the time t when this happens, then use Eq. (3.20) to find the value of x at this time.

3.25 Our sketch for this problem.

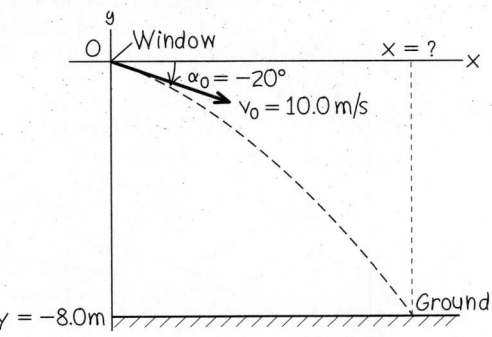

EXECUTE: To determine t, we rewrite Eq. (3.21) in the standard form for a quadratic equation for t:

$$\tfrac{1}{2}gt^2 - (v_0 \sin \alpha_0)t + y = 0$$

Continued

The roots of this equation are

$$t = \frac{v_0 \sin \alpha_0 \pm \sqrt{(-v_0 \sin \alpha_0)^2 - 4\left(\frac{1}{2}g\right)y}}{2\left(\frac{1}{2}g\right)}$$

$$= \frac{v_0 \sin \alpha_0 \pm \sqrt{v_0^2 \sin^2 \alpha_0 - 2gy}}{g}$$

$$= \frac{\left[\begin{array}{c}(10.0 \text{ m/s}) \sin(-20°) \\ \pm \sqrt{(10.0 \text{ m/s})^2 \sin^2(-20°) - 2(9.80 \text{ m/s}^2)(-8.0 \text{ m})}\end{array}\right]}{9.80 \text{ m/s}^2}$$

$$= -1.7 \text{ s} \quad \text{or} \quad 0.98 \text{ s}$$

We discard the negative root, since it refers to a time before the ball left your hand. The positive root tells us that the ball reaches the ground at $t = 0.98$ s. From Eq. (3.20), the ball's x-coordinate at that time is

$$x = (v_0 \cos \alpha_0)t = (10.0 \text{ m/s})[\cos(-20°)](0.98 \text{ s})$$
$$= 9.2 \text{ m}$$

The ball hits the ground a horizontal distance of 9.2 m from your window.

EVALUATE: The root $t = -1.7$ s is an example of a "fictional" solution to a quadratic equation. We discussed these in Example 2.8 in Section 2.5; you should review that discussion.

Example 3.10   The zookeeper and the monkey

A monkey escapes from the zoo and climbs a tree. After failing to entice the monkey down, the zookeeper fires a tranquilizer dart directly at the monkey (Fig. 3.26). The monkey lets go at the instant the dart leaves the gun. Show that the dart will *always* hit the monkey, provided that the dart reaches the monkey before he hits the ground and runs away.

SOLUTION

IDENTIFY and SET UP: We have *two* bodies in projectile motion: the dart and the monkey. They have different initial positions and initial velocities, but they go into projectile motion at the same time $t = 0$. We'll first use Eq. (3.20) to find an expression for the time t when the x-coordinates x_{monkey} and x_{dart} are equal. Then we'll use that expression in Eq. (3.21) to see whether y_{monkey} and y_{dart} are also equal at this time; if they are, the dart hits the monkey. We

make the usual choice for the x- and y-directions, and place the origin of coordinates at the muzzle of the tranquilizer gun (Fig. 3.26).

EXECUTE: The monkey drops straight down, so $x_{\text{monkey}} = d$ at all times. From Eq. (3.20), $x_{\text{dart}} = (v_0 \cos \alpha_0)t$. We solve for the time t when these x-coordinates are equal:

$$d = (v_0 \cos \alpha_0)t \quad \text{so} \quad t = \frac{d}{v_0 \cos \alpha_0}$$

We must now show that $y_{\text{monkey}} = y_{\text{dart}}$ at this time. The monkey is in one-dimensional free fall; its position at any time is given by Eq. (2.12), with appropriate symbol changes. Figure 3.26 shows that the monkey's initial height above the dart-gun's muzzle is $y_{\text{monkey}-0} = d \tan \alpha_0$, so

$$y_{\text{monkey}} = d \tan \alpha_0 - \tfrac{1}{2}gt^2$$

3.26 The tranquilizer dart hits the falling monkey.

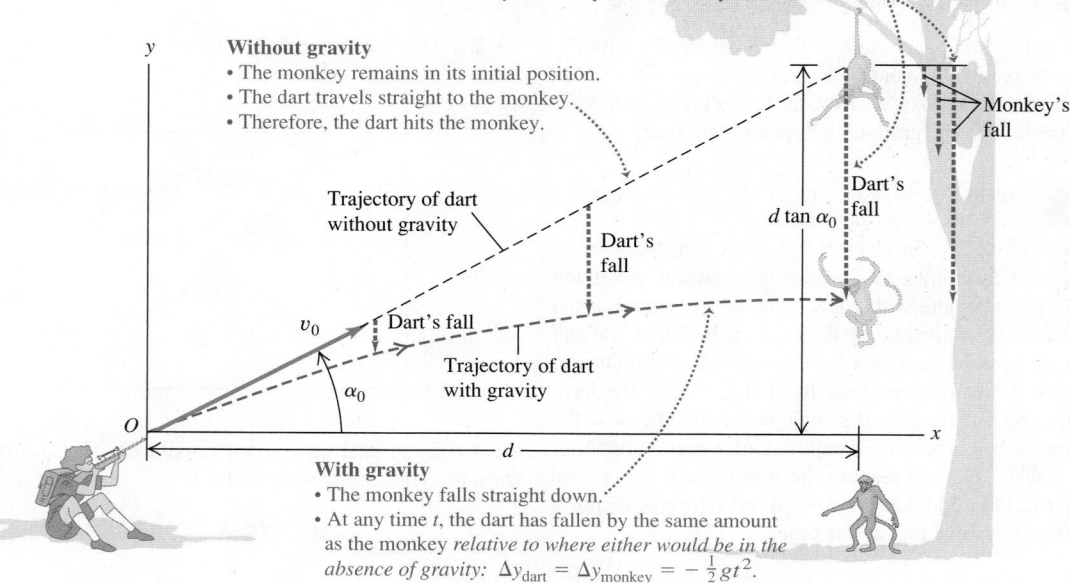

Dashed arrows show how far the dart and monkey have fallen at specific times relative to where they would be without gravity. At any time, they have fallen by the same amount.

Without gravity
• The monkey remains in its initial position.
• The dart travels straight to the monkey.
• Therefore, the dart hits the monkey.

Trajectory of dart without gravity

Dart's fall

Dart's fall

Trajectory of dart with gravity

Monkey's fall

Dart's fall

$d \tan \alpha_0$

v_0

α_0

d

With gravity
• The monkey falls straight down.
• At any time t, the dart has fallen by the same amount as the monkey *relative to where either would be in the absence of gravity*: $\Delta y_{\text{dart}} = \Delta y_{\text{monkey}} = -\tfrac{1}{2}gt^2$.
• Therefore, the dart always hits the monkey.

From Eq. (3.21),

$$y_{\text{dart}} = (v_0 \sin \alpha_0)t - \tfrac{1}{2}gt^2$$

Comparing these two equations, we see that we'll have $y_{\text{monkey}} = y_{\text{dart}}$ (and a hit) if $d \tan \alpha_0 = (v_0 \sin \alpha_0)t$ at the time when the two x-coordinates are equal. To show that this happens, we replace t with $d/(v_0 \cos \alpha_0)$, the time when $x_{\text{monkey}} = x_{\text{dart}}$. Sure enough, we find that

$$(v_0 \sin \alpha_0)t = (v_0 \sin \alpha_0)\,\frac{d}{v_0 \cos \alpha_0} = d \tan \alpha_0$$

EVALUATE: We've proved that the y-coordinates of the dart and the monkey are equal at the same time that their x-coordinates are equal; a dart aimed at the monkey *always* hits it, no matter what v_0 is (provided the monkey doesn't hit the ground first). This result is independent of the value of g, the acceleration due to gravity. With no gravity ($g = 0$), the monkey would remain motionless, and the dart would travel in a straight line to hit him. With gravity, both fall the same distance $gt^2/2$ below their $t = 0$ positions, and the dart still hits the monkey (Fig. 3.26).

Test Your Understanding of Section 3.3
In Example 3.10, suppose the tranquilizer dart has a relatively low muzzle velocity so that the dart reaches a maximum height at a point P before striking the monkey, as shown in the figure. When the dart is at point P, will the monkey be (i) at point A (higher than P), (ii) at point B (at the same height as P), or (iii) at point C (lower than P)? Ignore air resistance.

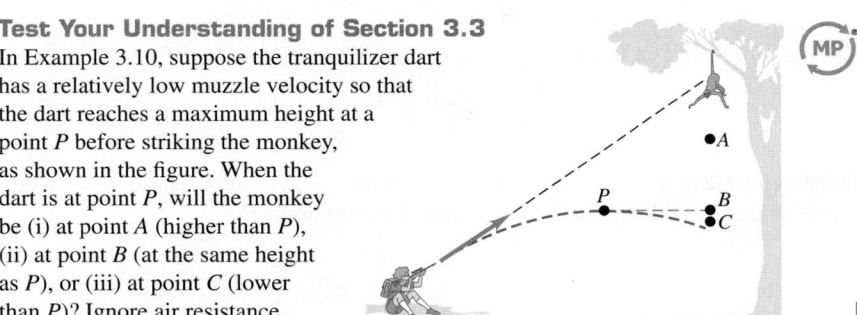

3.4 Motion in a Circle

When a particle moves along a curved path, the direction of its velocity changes. As we saw in Section 3.2, this means that the particle *must* have a component of acceleration perpendicular to the path, even if its speed is constant (see Fig. 3.11b). In this section we'll calculate the acceleration for the important special case of motion in a circle.

Uniform Circular Motion

When a particle moves in a circle with *constant speed,* the motion is called **uniform circular motion.** A car rounding a curve with constant radius at constant speed, a satellite moving in a circular orbit, and an ice skater skating in a circle with constant speed are all examples of uniform circular motion (Fig. 3.27c; compare Fig. 3.12a). There is no component of acceleration parallel (tangent) to the path; otherwise, the speed would change. The acceleration vector is perpendicular (normal) to the path and hence directed inward (never outward!) toward the center of the circular path. This causes the direction of the velocity to change without changing the speed.

3.27 A car moving along a circular path. If the car is in uniform circular motion as in (c), the speed is constant and the acceleration is directed toward the center of the circular path (compare Fig. 3.12).

(a) Car speeding up along a circular path

Component of acceleration parallel to velocity: Changes car's speed

$\vec{v}$

$\vec{a}$

Component of acceleration perpendicular to velocity: Changes car's direction

(b) Car slowing down along a circular path

$\vec{v}$

Component of acceleration perpendicular to velocity: Changes car's direction

$\vec{a}$

Component of acceleration parallel to velocity: Changes car's speed

(c) Uniform circular motion: Constant speed along a circular path

$\vec{v}$

$\vec{a}$

Acceleration is exactly perpendicular to velocity; no parallel component

To center of circle

3.28 Finding the velocity change $\Delta\vec{v}$, average acceleration $\vec{a}_{av}$, and instantaneous acceleration $\vec{a}_{rad}$ for a particle moving in a circle with constant speed.

(a) A particle moves a distance Δs at constant speed along a circular path.

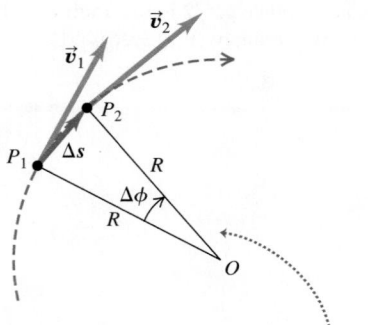

(b) The corresponding change in velocity and average acceleration

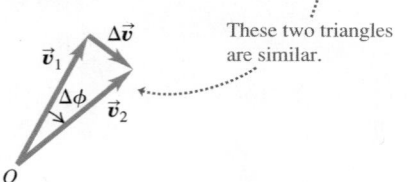

These two triangles are similar.

(c) The instantaneous acceleration

The instantaneous acceleration in uniform circular motion always points toward the center of the circle.

We can find a simple expression for the magnitude of the acceleration in uniform circular motion. We begin with Fig. 3.28a, which shows a particle moving with constant speed in a circular path of radius R with center at O. The particle moves from P_1 to P_2 in a time Δt. The vector change in velocity $\Delta\vec{v}$ during this time is shown in Fig. 3.28b.

The angles labeled $\Delta\phi$ in Figs. 3.28a and 3.28b are the same because $\vec{v}_1$ is perpendicular to the line OP_1 and $\vec{v}_2$ is perpendicular to the line OP_2. Hence the triangles in Figs. 3.28a and 3.28b are *similar*. The ratios of corresponding sides of similar triangles are equal, so

$$\frac{|\Delta\vec{v}|}{v_1} = \frac{\Delta s}{R} \quad \text{or} \quad |\Delta\vec{v}| = \frac{v_1}{R}\Delta s$$

The magnitude a_{av} of the average acceleration during Δt is therefore

$$a_{av} = \frac{|\Delta\vec{v}|}{\Delta t} = \frac{v_1}{R}\frac{\Delta s}{\Delta t}$$

The magnitude a of the *instantaneous* acceleration $\vec{a}$ at point P_1 is the limit of this expression as we take point P_2 closer and closer to point P_1:

$$a = \lim_{\Delta t \to 0}\frac{v_1}{R}\frac{\Delta s}{\Delta t} = \frac{v_1}{R}\lim_{\Delta t \to 0}\frac{\Delta s}{\Delta t}$$

If the time interval Δt is short, Δs is the distance the particle moves along its curved path. So the limit of $\Delta s / \Delta t$ is the speed v_1 at point P_1. Also, P_1 can be any point on the path, so we can drop the subscript and let v represent the speed at any point. Then

$$a_{rad} = \frac{v^2}{R} \quad \text{(uniform circular motion)} \qquad \text{(3.28)}$$

We have added the subscript "rad" as a reminder that the direction of the instantaneous acceleration at each point is always along a radius of the circle (toward the center of the circle; see Figs. 3.27c and 3.28c). So we have found that *in uniform circular motion, the magnitude a_{rad} of the instantaneous acceleration is equal to the square of the speed v divided by the radius R of the circle. Its direction is perpendicular to $\vec{v}$ and inward along the radius.*

Because the acceleration in uniform circular motion is always directed toward the center of the circle, it is sometimes called **centripetal acceleration.** The word "centripetal" is derived from two Greek words meaning "seeking the center." Figure 3.29a shows the directions of the velocity and acceleration vectors at several points for a particle moving with uniform circular motion.

3.29 Acceleration and velocity **(a)** for a particle in uniform circular motion and **(b)** for a projectile with no air resistance.

(a) Uniform circular motion

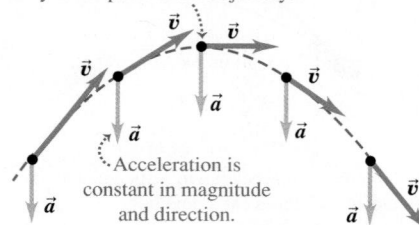

Acceleration has constant magnitude but varying direction.

Velocity and acceleration are always perpendicular.

(b) Projectile motion

Velocity and acceleration are perpendicular only at the peak of the trajectory.

Acceleration is constant in magnitude and direction.

CAUTION **Uniform circular motion vs. projectile motion** The acceleration in uniform circular motion (Fig. 3.29a) has some similarities to the acceleration in projectile motion without air resistance (Fig. 3.29b), but there are also some important differences. In both kinds of motion the *magnitude* of acceleration is the same at all times. However, in uniform circular motion the *direction* of $\vec{a}$ changes continuously so that it always points toward the center of the circle. (At the top of the circle the acceleration points down; at the bottom of the circle the acceleration points up.) In projectile motion, by contrast, the direction of $\vec{a}$ remains the same at all times. ▮

We can also express the magnitude of the acceleration in uniform circular motion in terms of the **period** T of the motion, the time for one revolution (one complete trip around the circle). In a time T the particle travels a distance equal to the circumference $2\pi R$ of the circle, so its speed is

$$v = \frac{2\pi R}{T} \tag{3.29}$$

When we substitute this into Eq. (3.28), we obtain the alternative expression

$$a_{\text{rad}} = \frac{4\pi^2 R}{T^2} \quad \text{(uniform circular motion)} \tag{3.30}$$

MasteringPHYSICS

PhET: Ladybug Revolution
PhET: Motion in 2D

Example 3.11   **Centripetal acceleration on a curved road**

An Aston Martin V8 Vantage sports car has a "lateral acceleration" of $0.96g = (0.96)(9.8 \text{ m/s}^2) = 9.4 \text{ m/s}^2$. This is the maximum centripetal acceleration the car can sustain without skidding out of a curved path. If the car is traveling at a constant 40 m/s (about 89 mi/h, or 144 km/h) on level ground, what is the radius R of the tightest unbanked curve it can negotiate?

SOLUTION

IDENTIFY, SET UP, and EXECUTE: The car is in uniform circular motion because it's moving at a constant speed along a curve that is a segment of a circle. Hence we can use Eq. (3.28) to solve for the target variable R in terms of the given centripetal acceleration

a_{rad} and speed v:

$$R = \frac{v^2}{a_{\text{rad}}} = \frac{(40 \text{ m/s})^2}{9.4 \text{ m/s}^2} = 170 \text{ m (about 560 ft)}$$

This is the *minimum* radius because a_{rad} is the *maximum* centripetal acceleration.

EVALUATE: The minimum turning radius R is proportional to the *square* of the speed, so even a small reduction in speed can make R substantially smaller. For example, reducing v by 20% (from 40 m/s to 32 m/s) would decrease R by 36% (from 170 m to 109 m).

Another way to make the minimum turning radius smaller is to *bank* the curve. We'll investigate this option in Chapter 5.

Example 3.12   **Centripetal acceleration on a carnival ride**

Passengers on a carnival ride move at constant speed in a horizontal circle of radius 5.0 m, making a complete circle in 4.0 s. What is their acceleration?

SOLUTION

IDENTIFY and SET UP: The speed is constant, so this is uniform circular motion. We are given the radius $R = 5.0$ m and the period $T = 4.0$ s, so we can use Eq. (3.30) to calculate the acceleration directly, or we can calculate the speed v using Eq. (3.29) and then find the acceleration using Eq. (3.28).

EXECUTE: From Eq. (3.30),

$$a_{\text{rad}} = \frac{4\pi^2 (5.0 \text{ m})}{(4.0 \text{ s})^2} = 12 \text{ m/s}^2 = 1.3g$$

We can check this answer by using the second, roundabout approach. From Eq. (3.29), the speed is

$$v = \frac{2\pi R}{T} = \frac{2\pi (5.0 \text{ m})}{4.0 \text{ s}} = 7.9 \text{ m/s}$$

The centripetal acceleration is then

$$a_{\text{rad}} = \frac{v^2}{R} = \frac{(7.9 \text{ m/s})^2}{5.0 \text{ m}} = 12 \text{ m/s}^2$$

EVALUATE: As in Example 3.11, the direction of $\vec{a}$ is always toward the center of the circle. The magnitude of $\vec{a}$ is relatively mild as carnival rides go; some roller coasters subject their passengers to accelerations as great as $4g$.

Application Watch Out: Tight Curves Ahead!

These roller coaster cars are in nonuniform circular motion: They slow down and speed up as they move around a vertical loop. The large accelerations involved in traveling at high speed around a tight loop mean extra stress on the passengers' circulatory systems, which is why people with cardiac conditions are cautioned against going on such rides.

3.30 A particle moving in a vertical loop with a varying speed, like a roller coaster car.

Speed slowest, a_{rad} minimum, a_{tan} zero

Speeding up; a_{tan} in same direction as $\vec{v}$

Slowing down; a_{tan} opposite to $\vec{v}$

a_{tan} a_{rad}

$\vec{v}$ $\vec{a}$ $|\vec{a}| = a_{rad}$

Speed fastest, a_{rad} maximum, a_{tan} zero

Nonuniform Circular Motion

We have assumed throughout this section that the particle's speed is constant as it goes around the circle. If the speed varies, we call the motion **nonuniform circular motion.** In nonuniform circular motion, Eq. (3.28) still gives the *radial* component of acceleration $a_{rad} = v^2/R$, which is always *perpendicular* to the instantaneous velocity and directed toward the center of the circle. But since the speed v has different values at different points in the motion, the value of a_{rad} is not constant. The radial (centripetal) acceleration is greatest at the point in the circle where the speed is greatest.

In nonuniform circular motion there is also a component of acceleration that is *parallel* to the instantaneous velocity (see Figs. 3.27a and 3.27b). This is the component $a_\parallel$ that we discussed in Section 3.2; here we call this component a_{tan} to emphasize that it is *tangent* to the circle. The tangential component of acceleration a_{tan} is equal to the rate of change of *speed*. Thus

$$a_{rad} = \frac{v^2}{R} \quad \text{and} \quad a_{tan} = \frac{d|\vec{v}|}{dt} \quad \text{(nonuniform circular motion)} \quad (3.31)$$

The tangential component is in the same direction as the velocity if the particle is speeding up, and in the opposite direction if the particle is slowing down (Fig. 3.30). If the particle's speed is constant, $a_{tan} = 0$.

CAUTION **Uniform vs. nonuniform circular motion** Note that the two quantities

$$\frac{d|\vec{v}|}{dt} \quad \text{and} \quad \left|\frac{d\vec{v}}{dt}\right|$$

are *not* the same. The first, equal to the tangential acceleration, is the rate of change of speed; it is zero whenever a particle moves with constant speed, even when its direction of motion changes (such as in *uniform* circular motion). The second is the magnitude of the vector acceleration; it is zero only when the particle's acceleration *vector* is zero—that is, when the particle moves in a straight line with constant speed. In *uniform* circular motion $|d\vec{v}/dt| = a_{rad} = v^2/r$; in *nonuniform* circular motion there is also a tangential component of acceleration, so $|d\vec{v}/dt| = \sqrt{a_{rad}^2 + a_{tan}^2}$. ∎

Test Your Understanding of Section 3.4 Suppose that the particle in Fig. 3.30 experiences four times the acceleration at the bottom of the loop as it does at the top of the loop. Compared to its speed at the top of the loop, is its speed at the bottom of the loop (i) $\sqrt{2}$ times as great; (ii) 2 times as great; (iii) $2\sqrt{2}$ times as great; (iv) 4 times as great; or (v) 16 times as great?

3.5 Relative Velocity

You've no doubt observed how a car that is moving slowly forward appears to be moving backward when you pass it. In general, when two observers measure the velocity of a moving body, they get different results if one observer is moving relative to the other. The velocity seen by a particular observer is called the velocity *relative* to that observer, or simply **relative velocity.** Figure 3.31 shows a situation in which understanding relative velocity is extremely important.

We'll first consider relative velocity along a straight line, then generalize to relative velocity in a plane.

Relative Velocity in One Dimension

A passenger walks with a velocity of 1.0 m/s along the aisle of a train that is moving with a velocity of 3.0 m/s (Fig. 3.32a). What is the passenger's velocity?

It's a simple enough question, but it has no single answer. As seen by a second passenger sitting in the train, she is moving at 1.0 m/s. A person on a bicycle standing beside the train sees the walking passenger moving at 1.0 m/s + 3.0 m/s = 4.0 m/s. An observer in another train going in the opposite direction would give still another answer. We have to specify which observer we mean, and we speak of the velocity *relative* to a particular observer. The walking passenger's velocity relative to the train is 1.0 m/s, her velocity relative to the cyclist is 4.0 m/s, and so on. Each observer, equipped in principle with a meter stick and a stopwatch, forms what we call a **frame of reference.** Thus a frame of reference is a coordinate system plus a time scale.

Let's use the symbol A for the cyclist's frame of reference (at rest with respect to the ground) and the symbol B for the frame of reference of the moving train. In straight-line motion the position of a point P relative to frame A is given by $x_{P/A}$ (the position of P with respect to A), and the position of P relative to frame B is given by $x_{P/B}$ (Fig. 3.32b). The position of the origin of B with respect to the origin of A is $x_{B/A}$. Figure 3.32b shows that

$$x_{P/A} = x_{P/B} + x_{B/A} \qquad (3.32)$$

In words, the coordinate of P relative to A equals the coordinate of P relative to B plus the coordinate of B relative to A.

The x-velocity of P relative to frame A, denoted by $v_{P/A\text{-}x}$, is the derivative of $x_{P/A}$ with respect to time. The other velocities are similarly obtained. So the time derivative of Eq. (3.32) gives us a relationship among the various velocities:

$$\frac{dx_{P/A}}{dt} = \frac{dx_{P/B}}{dt} + \frac{dx_{B/A}}{dt} \qquad \text{or}$$

$$v_{P/A\text{-}x} = v_{P/B\text{-}x} + v_{B/A\text{-}x} \qquad \text{(relative velocity along a line)} \qquad (3.33)$$

Getting back to the passenger on the train in Fig. 3.32, we see that A is the cyclist's frame of reference, B is the frame of reference of the train, and point P represents the passenger. Using the above notation, we have

$$v_{P/B\text{-}x} = +1.0 \text{ m/s} \qquad v_{B/A\text{-}x} = +3.0 \text{ m/s}$$

From Eq. (3.33) the passenger's velocity $v_{P/A}$ relative to the cyclist is

$$v_{P/A\text{-}x} = +1.0 \text{ m/s} + 3.0 \text{ m/s} = +4.0 \text{ m/s}$$

as we already knew.

In this example, both velocities are toward the right, and we have taken this as the positive x-direction. If the passenger walks toward the *left* relative to the train, then $v_{P/B\text{-}x} = -1.0$ m/s, and her x-velocity relative to the cyclist is $v_{P/A\text{-}x} = -1.0$ m/s + 3.0 m/s = +2.0 m/s. The sum in Eq. (3.33) is always an algebraic sum, and any or all of the x-velocities may be negative.

When the passenger looks out the window, the stationary cyclist on the ground appears to her to be moving backward; we can call the cyclist's velocity relative to her $v_{A/P\text{-}x}$. Clearly, this is just the negative of the *passenger's* velocity relative to the *cyclist*, $v_{P/A\text{-}x}$. In general, if A and B are any two points or frames of reference,

$$v_{A/B\text{-}x} = -v_{B/A\text{-}x} \qquad (3.34)$$

3.31 Airshow pilots face a complicated problem involving relative velocities. They must keep track of their motion relative to the air (to maintain enough airflow over the wings to sustain lift), relative to each other (to keep a tight formation without colliding), and relative to their audience (to remain in sight of the spectators).

3.32 (a) A passenger walking in a train. (b) The position of the passenger relative to the cyclist's frame of reference and the train's frame of reference.

(a)

(b)

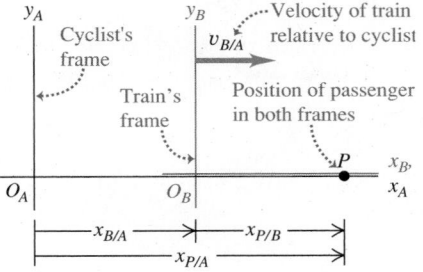

Problem-Solving Strategy 3.2 | Relative Velocity

IDENTIFY *the relevant concepts:* Whenever you see the phrase "velocity relative to" or "velocity with respect to," it's likely that the concepts of relative velocity will be helpful.

SET UP *the problem:* Sketch and label each frame of reference in the problem. Each moving body has its own frame of reference; in addition, you'll almost always have to include the frame of reference of the earth's surface. (Statements such as "The car is traveling north at 90 km/h" implicitly refer to the car's velocity relative to the surface of the earth.) Use the labels to help identify the target variable. For example, if you want to find the *x*-velocity of a car (*C*) with respect to a bus (*B*), your target variable is $v_{C/B\text{-}x}$.

EXECUTE *the solution:* Solve for the target variable using Eq. (3.33). (If the velocities aren't along the same direction, you'll need to use the vector form of this equation, derived later in this section.) It's

important to note the order of the double subscripts in Eq. (3.33): $v_{B/A\text{-}x}$ means "*x*-velocity of *B* relative to *A*." These subscripts obey a kind of algebra, as Eq. (3.33) shows. If we regard each one as a fraction, then the fraction on the left side is the *product* of the fractions on the right side: $P/A = (P/B)(B/A)$. You can apply this rule to any number of frames of reference. For example, if there are three different frames of reference *A*, *B*, and *C*, Eq. (3.33) becomes

$$v_{P/A\text{-}x} = v_{P/C\text{-}x} + v_{C/B\text{-}x} + v_{B/A\text{-}x}$$

EVALUATE *your answer:* Be on the lookout for stray minus signs in your answer. If the target variable is the *x*-velocity of a car relative to a bus ($v_{C/B\text{-}x}$), make sure that you haven't accidentally calculated the *x*-velocity of the *bus* relative to the *car* ($v_{B/C\text{-}x}$). If you've made this mistake, you can recover using Eq. (3.34).

Example 3.13 | Relative velocity on a straight road

You drive north on a straight two-lane road at a constant 88 km/h. A truck in the other lane approaches you at a constant 104 km/h (Fig. 3.33). Find (a) the truck's velocity relative to you and (b) your velocity relative to the truck. (c) How do the relative velocities change after you and the truck pass each other? Treat this as a one-dimensional problem.

SOLUTION

IDENTIFY and SET UP: In this problem about relative velocities along a line, there are three reference frames: you (Y), the truck (T), and the earth's surface (E). Let the positive *x*-direction be north (Fig. 3.33). Then your *x*-velocity relative to the earth is $v_{Y/E\text{-}x} = +88$ km/h. The truck is initially approaching you, so it is moving south and its *x*-velocity with respect to the earth is $v_{T/E\text{-}x} = -104$ km/h. The target variables in parts (a) and (b) are $v_{T/Y\text{-}x}$ and $v_{Y/T\text{-}x}$, respectively. We'll use Eq. (3.33) to find the first target variable and Eq. (3.34) to find the second.

EXECUTE: (a) To find $v_{T/Y\text{-}x}$, we write Eq. (3.33) for the known $v_{T/E\text{-}x}$ and rearrange:

$$v_{T/E\text{-}x} = v_{T/Y\text{-}x} + v_{Y/E\text{-}x}$$
$$v_{T/Y\text{-}x} = v_{T/E\text{-}x} - v_{Y/E\text{-}x}$$
$$= -104 \text{ km/h} - 88 \text{ km/h} = -192 \text{ km/h}$$

The truck is moving at 192 km/h in the negative *x*-direction (south) relative to you.

(b) From Eq. (3.34),

$$v_{Y/T\text{-}x} = -v_{T/Y\text{-}x} = -(-192 \text{ km/h}) = +192 \text{ km/h}$$

3.33 Reference frames for you and the truck.

You are moving at 192 km/h in the positive *x*-direction (north) relative to the truck.

(c) The relative velocities do *not* change after you and the truck pass each other. The relative *positions* of the bodies don't matter. After it passes you the truck is still moving at 192 km/h toward the south relative to you, even though it is now moving away from you instead of toward you.

EVALUATE: To check your answer in part (b), use Eq. (3.33) directly in the form $v_{Y/T\text{-}x} = v_{Y/E\text{-}x} + v_{E/T\text{-}x}$. (The *x*-velocity of the earth with respect to the truck is the opposite of the *x*-velocity of the truck with respect to the earth: $v_{E/T\text{-}x} = -v_{T/E\text{-}x}$.) Do you get the same result?

Relative Velocity in Two or Three Dimensions

We can extend the concept of relative velocity to include motion in a plane or in space by using vector addition to combine velocities. Suppose that the passenger in Fig. 3.32a is walking not down the aisle of the railroad car but from one side of the car to the other, with a speed of 1.0 m/s (Fig. 3.34a). We can again describe the passenger's position *P* in two different frames of reference: *A* for the stationary ground observer and B for the moving train. But instead of coordinates x, we use position vectors $\vec{r}$ because the problem is now two-dimensional. Then, as Fig. 3.34b shows,

$$\vec{r}_{P/A} = \vec{r}_{P/B} + \vec{r}_{B/A} \qquad (3.35)$$

Just as we did before, we take the time derivative of this equation to get a relationship among the various velocities; the velocity of P relative to A is $\vec{v}_{P/A} = d\vec{r}_{P/A}/dt$ and so on for the other velocities. We get

$$\vec{v}_{P/A} = \vec{v}_{P/B} + \vec{v}_{B/A} \qquad \text{(relative velocity in space)} \qquad (3.36)$$

Equation (3.36) is known as the *Galilean velocity transformation*. It relates the velocity of a body P with respect to frame A and its velocity with respect to frame B ($\vec{v}_{P/A}$ and $\vec{v}_{P/B}$, respectively) to the velocity of frame B with respect to frame A ($\vec{v}_{B/A}$). If all three of these velocities lie along the same line, then Eq. (3.36) reduces to Eq. (3.33) for the components of the velocities along that line.

If the train is moving at $v_{B/A} = 3.0$ m/s relative to the ground and the passenger is moving at $v_{P/B} = 1.0$ m/s relative to the train, then the passenger's velocity vector $\vec{v}_{P/A}$ relative to the ground is as shown in Fig. 3.34c. The Pythagorean theorem then gives us

$$v_{P/A} = \sqrt{(3.0 \text{ m/s})^2 + (1.0 \text{ m/s})^2} = \sqrt{10 \text{ m}^2/\text{s}^2} = 3.2 \text{ m/s}$$

Figure 3.34c also shows that the *direction* of the passenger's velocity vector relative to the ground makes an angle ϕ with the train's velocity vector $\vec{v}_{B/A}$, where

$$\tan\phi = \frac{v_{P/B}}{v_{B/A}} = \frac{1.0 \text{ m/s}}{3.0 \text{ m/s}} \qquad \text{and} \qquad \phi = 18°$$

As in the case of motion along a straight line, we have the general rule that if A and B are *any* two points or frames of reference,

$$\vec{v}_{A/B} = -\vec{v}_{B/A} \qquad (3.37)$$

The velocity of the passenger relative to the train is the negative of the velocity of the train relative to the passenger, and so on.

In the early 20th century Albert Einstein showed in his special theory of relativity that the velocity-addition relationship given in Eq. (3.36) has to be modified when speeds approach the speed of light, denoted by c. It turns out that if the passenger in Fig. 3.32a could walk down the aisle at $0.30c$ and the train could move at $0.90c$, then her speed relative to the ground would be not $1.20c$ but $0.94c$; nothing can travel faster than light! We'll return to the special theory of relativity in Chapter 37.

3.34 (a) A passenger walking across a railroad car. (b) Position of the passenger relative to the cyclist's frame and the train's frame. (c) Vector diagram for the velocity of the passenger relative to the ground (the cyclist's frame), $\vec{v}_{P/A}$.

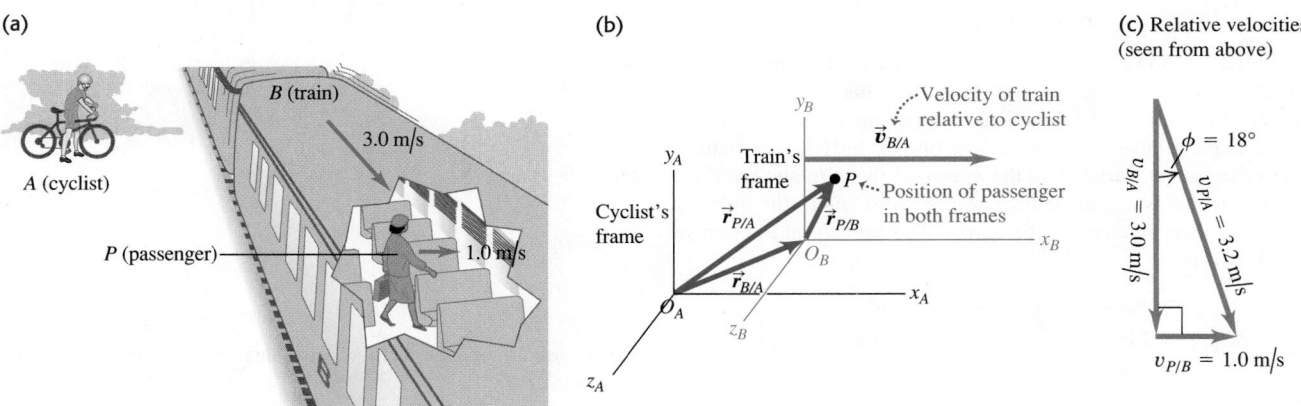

(a)

A (cyclist)

B (train)

3.0 m/s

P (passenger)

1.0 m/s

(b)

y_B

y_A Train's frame

Cyclist's frame

Velocity of train relative to cyclist

$\vec{v}_{B/A}$

P ··· Position of passenger in both frames

$\vec{r}_{P/A}$ $\vec{r}_{P/B}$

O_B x_B

$\vec{r}_{B/A}$ x_A

O_A

z_B

z_A

(c) Relative velocities (seen from above)

$\phi = 18°$

$v_{B/A} = 3.0$ m/s

$v_{P/A} = 3.2$ m/s

$v_{P/B} = 1.0$ m/s

Example 3.14 Flying in a crosswind

An airplane's compass indicates that it is headed due north, and its airspeed indicator shows that it is moving through the air at 240 km/h. If there is a 100-km/h wind from west to east, what is the velocity of the airplane relative to the earth?

SOLUTION

IDENTIFY and SET UP: This problem involves velocities in two dimensions (northward and eastward), so it is a relative velocity problem using vectors. We are given the magnitude and direction of the velocity of the plane (P) relative to the air (A). We are also given the magnitude and direction of the wind velocity, which is the velocity of the air A with respect to the earth (E):

$$\vec{v}_{P/A} = 240 \text{ km/h} \quad \text{due north}$$
$$\vec{v}_{A/E} = 100 \text{ km/h} \quad \text{due east}$$

We'll use Eq. (3.36) to find our target variables: the magnitude and direction of the velocity $\vec{v}_{P/E}$ of the plane relative to the earth.

EXECUTE: From Eq. (3.36) we have

$$\vec{v}_{P/E} = \vec{v}_{P/A} + \vec{v}_{A/E}$$

Figure 3.35 shows that the three relative velocities constitute a right-triangle vector addition; the unknowns are the speed $v_{P/E}$ and the angle α. We find

$$v_{P/E} = \sqrt{(240 \text{ km/h})^2 + (100 \text{ km/h})^2} = 260 \text{ km/h}$$
$$\alpha = \arctan\left(\frac{100 \text{ km/h}}{240 \text{ km/h}}\right) = 23° \text{ E of N}$$

EVALUATE: You can check the results by taking measurements on the scale drawing in Fig. 3.35. The crosswind increases the speed of the airplane relative to the earth, but pushes the airplane off course.

3.35 The plane is pointed north, but the wind blows east, giving the resultant velocity $\vec{v}_{P/E}$ relative to the earth.

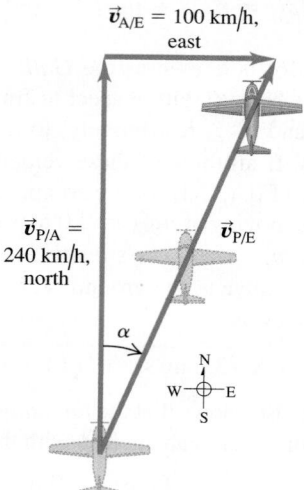

Example 3.15 Correcting for a crosswind

With wind and airspeed as in Example 3.14, in what direction should the pilot head to travel due north? What will be her velocity relative to the earth?

SOLUTION

IDENTIFY and SET UP: Like Example 3.14, this is a relative velocity problem with vectors. Figure 3.36 is a scale drawing of the situation. Again the vectors add in accordance with Eq. (3.36) and form a right triangle:

$$\vec{v}_{P/E} = \vec{v}_{P/A} + \vec{v}_{A/E}$$

As Fig. 3.36 shows, the pilot points the nose of the airplane at an angle β into the wind to compensate for the crosswind. This angle, which tells us the direction of the vector $\vec{v}_{P/A}$ (the velocity of the airplane relative to the air), is one of our target variables. The other target variable is the speed of the airplane over the ground, which is the magnitude of the vector $\vec{v}_{P/E}$ (the velocity of the airplane relative to the earth). The known and unknown quantities are

$$\vec{v}_{P/E} = \text{magnitude unknown} \quad \text{due north}$$
$$\vec{v}_{P/A} = 240 \text{ km/h} \quad \text{direction unknown}$$
$$\vec{v}_{A/E} = 100 \text{ km/h} \quad \text{due east}$$

3.36 The pilot must point the plane in the direction of the vector $\vec{v}_{P/A}$ to travel due north relative to the earth.

We'll solve for the target variables using Fig. 3.36 and trigonometry.

EXECUTE: From Fig. 3.36 the speed $v_{P/E}$ and the angle β are

$$v_{P/E} = \sqrt{(240 \text{ km/h})^2 - (100 \text{ km/h})^2} = 218 \text{ km/h}$$

$$\beta = \arcsin\left(\frac{100 \text{ km/h}}{240 \text{ km/h}}\right) = 25°$$

The pilot should point the airplane 25° west of north, and her ground speed is then 218 km/h.

EVALUATE: There were two target variables—the magnitude of a vector and the direction of a vector—in both this example and Example 3.14. In Example 3.14 the magnitude and direction referred to the *same* vector ($\vec{v}_{P/E}$); here they refer to *different* vectors ($\vec{v}_{P/E}$ and $\vec{v}_{P/A}$).

While we expect a *headwind* to reduce an airplane's speed relative to the ground, this example shows that a *crosswind* does, too. That's an unfortunate fact of aeronautical life.

Test Your Understanding of Section 3.5 Suppose the nose of an airplane is pointed due east and the airplane has an airspeed of 150 km/h. Due to the wind, the airplane is moving due *north* relative to the ground and its speed relative to the ground is 150 km/h. What is the velocity of the air relative to the earth? (i) 150 km/h from east to west; (ii) 150 km/h from south to north; (iii) 150 km/h from southeast to northwest; (iv) 212 km/h from east to west; (v) 212 km/h from south to north; (vi) 212 km/h from southeast to northwest; (vii) there is no possible wind velocity that could cause this.

Position, velocity, and acceleration vectors: The position vector $\vec{r}$ of a point P in space is the vector from the origin to P. Its components are the coordinates x, y, and z.

The average velocity vector $\vec{v}_{av}$ during the time interval Δt is the displacement $\Delta \vec{r}$ (the change in the position vector $\vec{r}$) divided by Δt. The instantaneous velocity vector $\vec{v}$ is the time derivative of $\vec{r}$, and its components are the time derivatives of x, y, and z. The instantaneous speed is the magnitude of $\vec{v}$. The velocity $\vec{v}$ of a particle is always tangent to the particle's path. (See Example 3.1.)

The average acceleration vector $\vec{a}_{av}$ during the time interval Δt equals $\Delta \vec{v}$ (the change in the velocity vector $\vec{v}$) divided by Δt. The instantaneous acceleration vector $\vec{a}$ is the time derivative of $\vec{v}$, and its components are the time derivatives of v_x, v_y, and v_z. (See Example 3.2.)

The component of acceleration parallel to the direction of the instantaneous velocity affects the speed, while the component of $\vec{a}$ perpendicular to $\vec{v}$ affects the direction of motion. (See Examples 3.3 and 3.4.)

$$\vec{r} = x\hat{i} + y\hat{j} + z\hat{k} \tag{3.1}$$

$$\vec{v}_{av} = \frac{\vec{r}_2 - \vec{r}_1}{t_2 - t_1} = \frac{\Delta \vec{r}}{\Delta t} \tag{3.2}$$

$$\vec{v} = \lim_{\Delta t \to 0} \frac{\Delta \vec{r}}{\Delta t} = \frac{d\vec{r}}{dt} \tag{3.3}$$

$$v_x = \frac{dx}{dt} \quad v_y = \frac{dy}{dt} \quad v_z = \frac{dz}{dt} \tag{3.4}$$

$$\vec{a}_{av} = \frac{\vec{v}_2 - \vec{v}_1}{t_2 - t_1} = \frac{\Delta \vec{v}}{\Delta t} \tag{3.8}$$

$$\vec{a} = \lim_{\Delta t \to 0} \frac{\Delta \vec{v}}{\Delta t} = \frac{d\vec{v}}{dt} \tag{3.9}$$

$$a_x = \frac{dv_x}{dt}$$

$$a_y = \frac{dv_y}{dt} \tag{3.10}$$

$$a_z = \frac{dv_z}{dt}$$

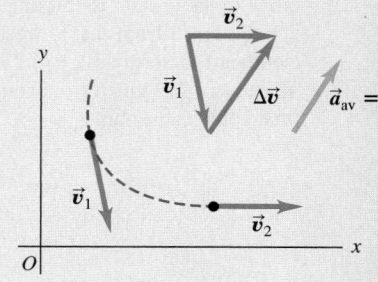

Projectile motion: In projectile motion with no air resistance, $a_x = 0$ and $a_y = -g$. The coordinates and velocity components are simple functions of time, and the shape of the path is always a parabola. We usually choose the origin to be at the initial position of the projectile. (See Examples 3.5–3.10.)

$$x = (v_0 \cos\alpha_0)t \tag{3.20}$$

$$y = (v_0 \sin\alpha_0)t - \tfrac{1}{2}gt^2 \tag{3.21}$$

$$v_x = v_0 \cos\alpha_0 \tag{3.22}$$

$$v_y = v_0 \sin\alpha_0 - gt \tag{3.23}$$

Uniform and nonuniform circular motion: When a particle moves in a circular path of radius R with constant speed v (uniform circular motion), its acceleration $\vec{a}$ is directed toward the center of the circle and perpendicular to $\vec{v}$. The magnitude a_{rad} of the acceleration can be expressed in terms of v and R or in terms of R and the period T (the time for one revolution), where $v = 2\pi R/T$. (See Examples 3.11 and 3.12.)

If the speed is not constant in circular motion (nonuniform circular motion), there is still a radial component of $\vec{a}$ given by Eq. (3.28) or (3.30), but there is also a component of $\vec{a}$ parallel (tangential) to the path. This tangential component is equal to the rate of change of speed, dv/dt.

$$a_{rad} = \frac{v^2}{R} \tag{3.28}$$

$$a_{rad} = \frac{4\pi^2 R}{T^2} \tag{3.30}$$

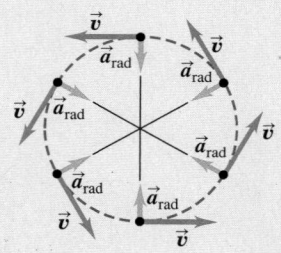

Relative velocity: When a body P moves relative to a body (or reference frame) B, and B moves relative to A, we denote the velocity of P relative to B by $\vec{v}_{P/B}$, the velocity of P relative to A by $\vec{v}_{P/A}$, and the velocity of B relative to A by $\vec{v}_{B/A}$. If these velocities are all along the same line, their components along that line are related by Eq. (3.33). More generally, these velocities are related by Eq. (3.36). (See Examples 3.13–3.15.)

$$v_{P/A\text{-}x} = v_{P/B\text{-}x} + v_{B/A\text{-}x}$$
(relative velocity along a line) $\tag{3.33}$

$$\vec{v}_{P/A} = \vec{v}_{P/B} + \vec{v}_{B/A}$$
(relative velocity in space) $\tag{3.36}$

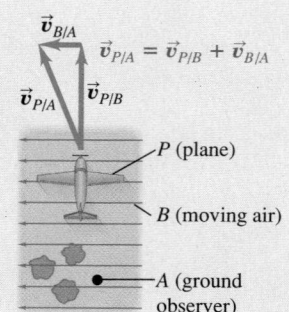

BRIDGING PROBLEM | Launching Up an Incline

You fire a ball with an initial speed v_0 at an angle ϕ above the surface of an incline, which is itself inclined at an angle θ above the horizontal (Fig. 3.37). (a) Find the distance, measured along the incline, from the launch point to the point when the ball strikes the incline. (b) What angle ϕ gives the maximum range, measured along the incline? Ignore air resistance.

SOLUTION GUIDE

See MasteringPhysics® study area for a Video Tutor solution.

IDENTIFY and SET UP

1. Since there's no air resistance, this is a problem in projectile motion. The goal is to find the point where the ball's parabolic trajectory intersects the incline.
2. Choose the x- and y-axes and the position of the origin. When in doubt, use the suggestions given in Problem-Solving Strategy 3.1 in Section 3.3.
3. In the projectile equations from Section 3.3, the launch angle α_0 is measured from the horizontal. What is this angle in terms of θ and ϕ? What are the initial x- and y-components of the ball's initial velocity?
4. You'll need to write an equation that relates x and y for points along the incline. What is this equation? (This takes just geometry and trigonometry, not physics.)

3.37 Launching a ball from an inclined ramp.

EXECUTE

5. Write the equations for the x-coordinate and y-coordinate of the ball as functions of time t.
6. When the ball hits the incline, x and y are related by the equation that you found in step 4. Based on this, at what time t does the ball hit the incline?
7. Based on your answer from step 6, at what coordinates x and y does the ball land on the incline? How far is this point from the launch point?
8. What value of ϕ gives the *maximum* distance from the launch point to the landing point? (Use your knowledge of calculus.)

EVALUATE

9. Check your answers for the case $\theta = 0$, which corresponds to the incline being horizontal rather than tilted. (You already know the answers for this case. Do you know why?)

Problems

For instructor-assigned homework, go to www.masteringphysics.com

•, ••, •••: Problems of increasing difficulty. **CP**: Cumulative problems incorporating material from earlier chapters. **CALC**: Problems requiring calculus. **BIO**: Biosciences problems.

DISCUSSION QUESTIONS

Q3.1 A simple pendulum (a mass swinging at the end of a string) swings back and forth in a circular arc. What is the direction of the acceleration of the mass when it is at the ends of the swing? At the midpoint? In each case, explain how you obtain your answer.

Q3.2 Redraw Fig. 3.11a if $\vec{a}$ is antiparallel to $\vec{v}_1$. Does the particle move in a straight line? What happens to its speed?

Q3.3 A projectile moves in a parabolic path without air resistance. Is there any point at which $\vec{a}$ is parallel to $\vec{v}$? Perpendicular to $\vec{v}$? Explain.

Q3.4 When a rifle is fired at a distant target, the barrel is not lined up exactly on the target. Why not? Does the angle of correction depend on the distance to the target?

Q3.5 At the same instant that you fire a bullet horizontally from a rifle, you drop a bullet from the height of the barrel. If there is no air resistance, which bullet hits the ground first? Explain.

Q3.6 A package falls out of an airplane that is flying in a straight line at a constant altitude and speed. If you could ignore air resistance, what would be the path of the package as observed by the pilot? As observed by a person on the ground?

Q3.7 Sketch the six graphs of the x- and y-components of position, velocity, and acceleration versus time for projectile motion with $x_0 = y_0 = 0$ and $0 < \alpha_0 < 90°$.

Q3.8 If a jumping frog can give itself the same initial speed regardless of the direction in which it jumps (forward or straight up), how is the maximum vertical height to which it can jump related to its maximum horizontal range $R_{max} = v_0^2/g$?

Q3.9 A projectile is fired upward at an angle θ above the horizontal with an initial speed v_0. At its maximum height, what are its velocity vector, its speed, and its acceleration vector?

Q3.10 In uniform circular motion, what are the *average* velocity and *average* acceleration for one revolution? Explain.

Q3.11 In uniform circular motion, how does the acceleration change when the speed is increased by a factor of 3? When the radius is decreased by a factor of 2?

Q3.12 In uniform circular motion, the acceleration is perpendicular to the velocity at every instant. Is this still true when the motion is not uniform—that is, when the speed is not constant?

Q3.13 Raindrops hitting the side windows of a car in motion often leave diagonal streaks even if there is no wind. Why? Is the explanation the same or different for diagonal streaks on the windshield?

Q3.14 In a rainstorm with a strong wind, what determines the best position in which to hold an umbrella?

Q3.15 You are on the west bank of a river that is flowing north with a speed of 1.2 m/s. Your swimming speed relative to the water is 1.5 m/s, and the river is 60 m wide. What is your path relative to the earth that allows you to cross the river in the shortest time? Explain your reasoning.

Q3.16 A stone is thrown into the air at an angle above the horizontal and feels negligible air resistance. Which graph in Fig. Q3.16 best depicts the stone's *speed v* as a function of time *t* while it is in the air?

Figure **Q3.16**

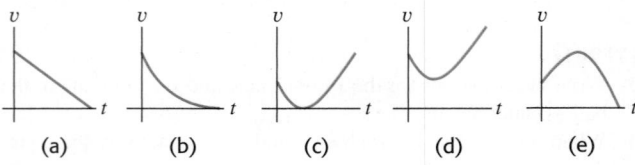

| (a) | (b) | (c) | (d) | (e) |

EXERCISES

Section 3.1 Position and Velocity Vectors

3.1 • A squirrel has *x*- and *y*-coordinates (1.1 m, 3.4 m) at time $t_1 = 0$ and coordinates (5.3 m, −0.5 m) at time $t_2 = 3.0$ s. For this time interval, find (a) the components of the average velocity, and (b) the magnitude and direction of the average velocity.

3.2 • A rhinoceros is at the origin of coordinates at time $t_1 = 0$. For the time interval from $t_1 = 0$ to $t_2 = 12.0$ s, the rhino's average velocity has *x*-component −3.8 m/s and *y*-component 4.9 m/s. At time $t_2 = 12.0$ s, (a) what are the *x*- and *y*-coordinates of the rhino? (b) How far is the rhino from the origin?

3.3 •• **CALC** A web page designer creates an animation in which a dot on a computer screen has a position of $\vec{r} = [4.0 \text{ cm} + (2.5 \text{ cm/s}^2)t^2]\hat{\imath} + (5.0 \text{ cm/s})t\hat{\jmath}$. (a) Find the magnitude and direction of the dot's average velocity between $t = 0$ and $t = 2.0$ s. (b) Find the magnitude and direction of the instantaneous velocity at $t = 0$, $t = 1.0$ s, and $t = 2.0$ s. (c) Sketch the dot's trajectory from $t = 0$ to $t = 2.0$ s, and show the velocities calculated in part (b).

3.4 • **CALC** The position of a squirrel running in a park is given by $\vec{r} = [(0.280 \text{ m/s})t + (0.0360 \text{ m/s}^2)t^2]\hat{\imath} + (0.0190 \text{ m/s}^3)t^3\hat{\jmath}$. (a) What are $v_x(t)$ and $v_y(t)$, the *x*- and *y*-components of the velocity of the squirrel, as functions of time? (b) At $t = 5.00$ s, how far is the squirrel from its initial position? (c) At $t = 5.00$ s, what are the magnitude and direction of the squirrel's velocity?

Section 3.2 The Acceleration Vector

3.5 • A jet plane is flying at a constant altitude. At time $t_1 = 0$ it has components of velocity $v_x = 90$ m/s, $v_y = 110$ m/s. At time $t_2 = 30.0$ s the components are $v_x = -170$ m/s, $v_y = 40$ m/s. (a) Sketch the velocity vectors at t_1 and t_2. How do these two vectors differ? For this time interval calculate (b) the components of the average acceleration, and (c) the magnitude and direction of the average acceleration.

3.6 •• A dog running in an open field has components of velocity $v_x = 2.6$ m/s and $v_y = -1.8$ m/s at $t_1 = 10.0$ s. For the time interval from $t_1 = 10.0$ s to $t_2 = 20.0$ s, the average acceleration of the dog has magnitude 0.45 m/s² and direction 31.0° measured from the +*x*-axis toward the +*y*-axis. At $t_2 = 20.0$ s, (a) what are the *x*- and *y*-components of the dog's velocity? (b) What are the magnitude and direction of the dog's velocity? (c) Sketch the velocity vectors at t_1 and t_2. How do these two vectors differ?

3.7 •• **CALC** The coordinates of a bird flying in the *xy*-plane are given by $x(t) = \alpha t$ and $y(t) = 3.0 \text{ m} - \beta t^2$, where $\alpha = 2.4$ m/s and $\beta = 1.2$ m/s². (a) Sketch the path of the bird between $t = 0$ and $t = 2.0$ s. (b) Calculate the velocity and acceleration vectors of the bird as functions of time. (c) Calculate the magnitude and direction of the bird's velocity and acceleration at $t = 2.0$ s. (d) Sketch the velocity and acceleration vectors at $t = 2.0$ s. At this instant, is the bird speeding up, is it slowing down, or is its speed instantaneously not changing? Is the bird turning? If so, in what direction?

Section 3.3 Projectile Motion

3.8 • **CALC** A remote-controlled car is moving in a vacant parking lot. The velocity of the car as a function of time is given by $\vec{v} = [5.00 \text{ m/s} - (0.0180 \text{ m/s}^3)t^2]\hat{\imath} + [2.00 \text{ m/s} + (0.550 \text{ m/s}^2)t]\hat{\jmath}$. (a) What are $a_x(t)$ and $a_y(t)$, the *x*- and *y*-components of the velocity of the car as functions of time? (b) What are the magnitude and direction of the velocity of the car at $t = 8.00$ s? (b) What are the magnitude and direction of the acceleration of the car at $t = 8.00$ s?

3.9 • A physics book slides off a horizontal tabletop with a speed of 1.10 m/s. It strikes the floor in 0.350 s. Ignore air resistance. Find (a) the height of the tabletop above the floor; (b) the horizontal distance from the edge of the table to the point where the book strikes the floor; (c) the horizontal and vertical components of the book's velocity, and the magnitude and direction of its velocity, just before the book reaches the floor. (d) Draw *x-t*, *y-t*, v_x*-t*, and v_y*-t* graphs for the motion.

3.10 •• A daring 510-N swimmer dives off a cliff with a running horizontal leap, as shown in Fig. E3.10. What must her minimum speed be just as she leaves the top of the cliff so that she will miss the ledge at the bottom, which is 1.75 m wide and 9.00 m below the top of the cliff?

Figure **E3.10**

v_0

9.00 m

1.75 m

Ledge

3.11 • Two crickets, Chirpy and Milada, jump from the top of a vertical cliff. Chirpy just drops and reaches the ground in 3.50 s, while Milada jumps horizontally with an initial speed of 95.0 cm/s. How far from the base of the cliff will Milada hit the ground?

3.12 • A rookie quarterback throws a football with an initial upward velocity component of 12.0 m/s and a horizontal velocity component of 20.0 m/s. Ignore air resistance. (a) How much time is required for the football to reach the highest point of the trajectory? (b) How high is this point? (c) How much time (after it is thrown) is required for the football to return to its original level? How does this compare with the time calculated in part (a)? (d) How far has the football traveled horizontally during this time? (e) Draw *x-t*, *y-t*, v_x*-t*, and v_y*-t* graphs for the motion.

3.13 •• **Leaping the River I.** A car traveling on a level horizontal road comes to a bridge during a storm and finds the bridge washed out. The driver must get to the other side, so he decides to try leaping it with his car. The side of the road the car is on is 21.3 m above the river, while the opposite side is a mere 1.8 m above the river. The river itself is a raging torrent 61.0 m wide. (a) How fast should the car be traveling at the time it leaves the road in order just to clear the river and land safely on the opposite side? (b) What is the speed of the car just before it lands on the other side?

3.14 • **BIO** **The Champion Jumper of the Insect World.** The froghopper, *Philaenus spumarius,* holds the world record for insect jumps. When leaping at an angle of 58.0° above the horizontal, some of the tiny critters have reached a maximum height of 58.7 cm above the level ground. (See *Nature,* Vol. 424, July 31, 2003, p. 509.) (a) What was the takeoff speed for such a leap? (b) What horizontal distance did the froghopper cover for this world-record leap?

3.15 •• Inside a starship at rest on the earth, a ball rolls off the top of a horizontal table and lands a distance D from the foot of the table. This starship now lands on the unexplored Planet X. The commander, Captain Curious, rolls the same ball off the same table with the same initial speed as on earth and finds that it lands a distance $2.76D$ from the foot of the table. What is the acceleration due to gravity on Planet X?

3.16 • On level ground a shell is fired with an initial velocity of 50.0 m/s at 60.0° above the horizontal and feels no appreciable air resistance. (a) Find the horizontal and vertical components of the shell's initial velocity. (b) How long does it take the shell to reach its highest point? (c) Find its maximum height above the ground. (d) How far from its firing point does the shell land? (e) At its highest point, find the horizontal and vertical components of its acceleration and velocity.

3.17 • A major leaguer hits a baseball so that it leaves the bat at a speed of 30.0 m/s and at an angle of 36.9° above the horizontal. You can ignore air resistance. (a) At what *two* times is the baseball at a height of 10.0 m above the point at which it left the bat? (b) Calculate the horizontal and vertical components of the baseball's velocity at each of the two times calculated in part (a). (c) What are the magnitude and direction of the baseball's velocity when it returns to the level at which it left the bat?

3.18 • A shot putter releases the shot some distance above the level ground with a velocity of 12.0 m/s, 51.0° above the horizontal. The shot hits the ground 2.08 s later. You can ignore air resistance. (a) What are the components of the shot's acceleration while in flight? (b) What are the components of the shot's velocity at the beginning and at the end of its trajectory? (c) How far did she throw the shot horizontally? (d) Why does the expression for R in Example 3.8 *not* give the correct answer for part (c)? (e) How high was the shot above the ground when she released it? (f) Draw x-t, y-t, v_x-t, and v_y-t graphs for the motion.

3.19 •• **Win the Prize.** In a carnival booth, you win a stuffed giraffe if you toss a quarter into a small dish. The dish is on a shelf above the point where the quarter leaves your hand and is a horizontal distance of 2.1 m from this point (Fig. E3.19). If you toss the coin with a velocity of 6.4 m/s at an angle of 60° above the horizontal, the coin lands in the dish. You can ignore air resistance. (a) What is the height of the shelf above the point where the

quarter leaves your hand? (b) What is the vertical component of the velocity of the quarter just before it lands in the dish?

3.20 •• Suppose the departure angle α_0 in Fig. 3.26 is 42.0° and the distance d is 3.00 m. Where will the dart and monkey meet if the initial speed of the dart is (a) 12.0 m/s? (b) 8.0 m/s? (c) What will happen if the initial speed of the dart is 4.0 m/s? Sketch the trajectory in each case.

3.21 •• A man stands on the roof of a 15.0-m-tall building and throws a rock with a velocity of magnitude 30.0 m/s at an angle of 33.0° above the horizontal. You can ignore air resistance. Calculate (a) the maximum height above the roof reached by the rock; (b) the magnitude of the velocity of the rock just before it strikes the ground; and (c) the horizontal range from the base of the building to the point where the rock strikes the ground. (d) Draw x-t, y-t, v_x-t, and v_y-t graphs for the motion.

3.22 • Firemen are shooting a stream of water at a burning building using a high-pressure hose that shoots out the water with a speed of 25.0 m/s as it leaves the end of the hose. Once it leaves the hose, the water moves in projectile motion. The firemen adjust the angle of elevation α of the hose until the water takes 3.00 s to reach a building 45.0 m away. You can ignore air resistance; assume that the end of the hose is at ground level. (a) Find the angle of elevation α. (b) Find the speed and acceleration of the water at the highest point in its trajectory. (c) How high above the ground does the water strike the building, and how fast is it moving just before it hits the building?

3.23 •• A 124-kg balloon carrying a 22-kg basket is descending with a constant downward velocity of 20.0 m/s. A 1.0-kg stone is thrown from the basket with an initial velocity of 15.0 m/s perpendicular to the path of the descending balloon, as measured relative to a person at rest in the basket. The person in the basket sees the stone hit the ground 6.00 s after being thrown. Assume that the balloon continues its downward descent with the same constant speed of 20.0 m/s. (a) How high was the balloon when the rock was thrown out? (b) How high is the balloon when the rock hits the ground? (c) At the instant the rock hits the ground, how far is it from the basket? (d) Just before the rock hits the ground, find its horizontal and vertical velocity components as measured by an observer (i) at rest in the basket and (ii) at rest on the ground.

Section 3.4 Motion in a Circle

3.24 •• **BIO** **Dizziness.** Our balance is maintained, at least in part, by the endolymph fluid in the inner ear. Spinning displaces this fluid, causing dizziness. Suppose a dancer (or skater) is spinning at a very fast 3.0 revolutions per second about a vertical axis through the center of his head. Although the distance varies from person to person, the inner ear is approximately 7.0 cm from the axis of spin. What is the radial acceleration (in m/s² and in g's) of the endolymph fluid?

3.25 •• The earth has a radius of 6380 km and turns around once on its axis in 24 h. (a) What is the radial acceleration of an object at the earth's equator? Give your answer in m/s² and as a fraction of g. (b) If a_{rad} at the equator is greater than g, objects will fly off the earth's surface and into space. (We will see the reason for this in Chapter 5.) What would the period of the earth's rotation have to be for this to occur?

3.26 •• A model of a helicopter rotor has four blades, each 3.40 m long from the central shaft to the blade tip. The model is rotated in a wind tunnel at 550 rev/min. (a) What is the linear speed of the blade tip, in m/s? (b) What is the radial acceleration of the blade tip expressed as a multiple of the acceleration of gravity, g?

Figure **E3.19**

3.27 • **BIO Pilot Blackout in a Power Dive.** A jet plane comes in for a downward dive as shown in Fig. E3.27. The bottom part of the path is a quarter circle with a radius of curvature of 350 m. According to medical tests, pilots lose consciousness at an acceleration of 5.5g. At what speed (in m/s and in mph) will the pilot black out for this dive?

Figure **E3.27**

3.28 • The radius of the earth's orbit around the sun (assumed to be circular) is 1.50×10^8 km, and the earth travels around this orbit in 365 days. (a) What is the magnitude of the orbital velocity of the earth, in m/s? (b) What is the radial acceleration of the earth toward the sun, in m/s²? (c) Repeat parts (a) and (b) for the motion of the planet Mercury (orbit radius = 5.79×10^7 km, orbital period = 88.0 days).

3.29 • A Ferris wheel with radius 14.0 m is turning about a horizontal axis through its center (Fig. E3.29). The linear speed of a passenger on the rim is constant and equal to 7.00 m/s. What are the magnitude and direction of the passenger's acceleration as she passes through (a) the lowest point in her circular motion? (b) The highest point in her circular motion? (c) How much time does it take the Ferris wheel to make one revolution?

Figure **E3.29**

3.30 •• **BIO Hypergravity.** At its Ames Research Center, NASA uses its large "20-G" centrifuge to test the effects of very large accelerations ("hypergravity") on test pilots and astronauts. In this device, an arm 8.84 m long rotates about one end in a horizontal plane, and the astronaut is strapped in at the other end. Suppose that he is aligned along the arm with his head at the outermost end. The maximum sustained acceleration to which humans are subjected in this machine is typically 12.5g. (a) How fast must the astronaut's head be moving to experience this maximum acceleration? (b) What is the *difference* between the acceleration of his head and feet if the astronaut is 2.00 m tall? (c) How fast in rpm (rev/min) is the arm turning to produce the maximum sustained acceleration?

Section 3.5 Relative Velocity

3.31 • A "moving sidewalk" in an airport terminal building moves at 1.0 m/s and is 35.0 m long. If a woman steps on at one end and walks at 1.5 m/s relative to the moving sidewalk, how much time does she require to reach the opposite end if she walks (a) in the same direction the sidewalk is moving? (b) In the opposite direction?

3.32 • A railroad flatcar is traveling to the right at a speed of 13.0 m/s relative to an observer standing on the ground. Someone is riding a motor scooter on the flatcar (Fig. E3.32). What is the velocity (magnitude and direction) of the motor scooter relative to the flatcar if its velocity relative to the observer on the ground is (a) 18.0 m/s to the right? (b) 3.0 m/s to the left? (c) zero?

Figure **E3.32**

3.33 •• A canoe has a velocity of 0.40 m/s southeast relative to the earth. The canoe is on a river that is flowing 0.50 m/s east relative to the earth. Find the velocity (magnitude and direction) of the canoe relative to the river.

3.34 • Two piers, A and B, are located on a river: B is 1500 m downstream from A (Fig. E3.34). Two friends must make round trips from pier A to pier B and return. One rows a boat at a constant speed of 4.00 km/h relative to the water; the other walks on the shore at a constant speed of 4.00 km/h. The velocity of the river is 2.80 km/h in the direction from A to B. How much time does it take each person to make the round trip?

Figure **E3.34**

3.35 • **Crossing the River I.** A river flows due south with a speed of 2.0 m/s. A man steers a motorboat across the river; his velocity relative to the water is 4.2 m/s due east. The river is 800 m wide. (a) What is his velocity (magnitude and direction) relative to the earth? (b) How much time is required to cross the river? (c) How far south of his starting point will he reach the opposite bank?

3.36 • **Crossing the River II.** (a) In which direction should the motorboat in Exercise 3.35 head in order to reach a point on the opposite bank directly east from the starting point? (The boat's speed relative to the water remains 4.2 m/s.) (b) What is the velocity of the boat relative to the earth? (c) How much time is required to cross the river?

3.37 •• The nose of an ultralight plane is pointed south, and its airspeed indicator shows 35 m/s. The plane is in a 10-m/s wind blowing toward the southwest relative to the earth. (a) In a vector-addition diagram, show the relationship of $\vec{v}_{P/E}$ (the velocity of the plane relative to the earth) to the two given vectors. (b) Letting x be east and y be north, find the components of $\vec{v}_{P/E}$. (c) Find the magnitude and direction of $\vec{v}_{P/E}$.

3.38 •• An airplane pilot wishes to fly due west. A wind of 80.0 km/h (about 50 mi/h) is blowing toward the south. (a) If the airspeed of the plane (its speed in still air) is 320.0 km/h (about 200 mi/h), in which direction should the pilot head? (b) What is the speed of the plane over the ground? Illustrate with a vector diagram.

3.39 •• **BIO Bird Migration.** Canadian geese migrate essentially along a north–south direction for well over a thousand kilometers in some cases, traveling at speeds up to about 100 km/h. If one such bird is flying at 100 km/h relative to the air, but there is a

40 km/h wind blowing from west to east, (a) at what angle relative to the north–south direction should this bird head so that it will be traveling directly southward relative to the ground? (b) How long will it take the bird to cover a ground distance of 500 km from north to south? (*Note:* Even on cloudy nights, many birds can navigate using the earth's magnetic field to fix the north–south direction.)

PROBLEMS

3.40 •• An athlete starts at point A and runs at a constant speed of 6.0 m/s around a circular track 100 m in diameter, as shown in Fig. P3.40. Find the x- and y-components of this runner's average velocity and average acceleration between points (a) A and B, (b) A and C, (c) C and D, and (d) A and A (a full lap). (e) Calculate the magnitude of the runner's average velocity between A and B. Is his average speed equal to the magnitude of his average velocity? Why or why not? (f) How can his velocity be changing if he is running at constant speed?

Figure **P3.40**

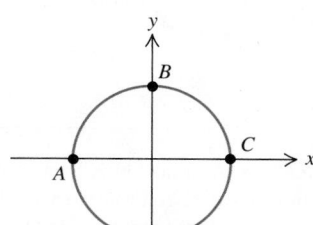

3.41 • **CALC** A rocket is fired at an angle from the top of a tower of height $h_0 = 50.0$ m. Because of the design of the engines, its position coordinates are of the form $x(t) = A + Bt^2$ and $y(t) = C + Dt^3$, where A, B, C, and D are constants. Furthermore, the acceleration of the rocket 1.00 s after firing is $\vec{a} = (4.00\hat{i} + 3.00\hat{j})$ m/s². Take the origin of coordinates to be at the base of the tower. (a) Find the constants A, B, C, and D, including their SI units. (b) At the instant after the rocket is fired, what are its acceleration vector and its velocity? (c) What are the x- and y-components of the rocket's velocity 10.0 s after it is fired, and how fast is it moving? (d) What is the position vector of the rocket 10.0 s after it is fired?

3.42 ••• **CALC** A faulty model rocket moves in the xy-plane (the positive y-direction is vertically upward). The rocket's acceleration has components $a_x(t) = \alpha t^2$ and $a_y(t) = \beta - \gamma t$, where $\alpha = 2.50$ m/s⁴, $\beta = 9.00$ m/s², and $\gamma = 1.40$ m/s³. At $t = 0$ the rocket is at the origin and has velocity $\vec{v}_0 = v_{0x}\hat{i} + v_{0y}\hat{j}$ with $v_{0x} = 1.00$ m/s and $v_{0y} = 7.00$ m/s. (a) Calculate the velocity and position vectors as functions of time. (b) What is the maximum height reached by the rocket? (c) Sketch the path of the rocket. (d) What is the horizontal displacement of the rocket when it returns to $y = 0$?

3.43 •• **CALC** If $\vec{r} = bt^2\hat{i} + ct^3\hat{j}$, where b and c are positive constants, when does the velocity vector make an angle of 45.0° with the x- and y-axes?

3.44 •• **CALC** The position of a dragonfly that is flying parallel to the ground is given as a function of time by $\vec{r} = [2.90 \text{ m} + (0.0900 \text{ m/s}^2)t^2]\hat{i} - (0.0150 \text{ m/s}^3)t^3\hat{j}$. (a) At what value of t does the velocity vector of the insect make an angle of 30.0° clockwise from the +x-axis? (b) At the time calculated in part (a), what are the magnitude and direction of the acceleration vector of the insect?

3.45 •• **CP CALC** A small toy airplane is flying in the xy-plane parallel to the ground. In the time interval $t = 0$ to $t = 1.00$ s, its velocity as a function of time is given by $\vec{v} = (1.20 \text{ m/s}^2)t\hat{i} + [12.0 \text{ m/s} - (2.00 \text{ m/s}^2)t]\hat{j}$. At what

value of t is the velocity of the plane perpendicular to its acceleration?

3.46 •• **CALC** A bird flies in the xy-plane with a velocity vector given by $\vec{v} = (\alpha - \beta t^2)\hat{i} + \gamma t\hat{j}$, with $\alpha = 2.4$ m/s, $\beta = 1.6$ m/s³, and $\gamma = 4.0$ m/s². The positive y-direction is vertically upward. At $t = 0$ the bird is at the origin. (a) Calculate the position and acceleration vectors of the bird as functions of time. (b) What is the bird's altitude (y-coordinate) as it flies over $x = 0$ for the first time after $t = 0$?

3.47 ••• **CP** A test rocket is launched by accelerating it along a 200.0-m incline at 1.25 m/s² starting from rest at point A (Fig. P3.47). The incline rises at 35.0° above the horizontal, and at the instant the rocket leaves it, its engines turn off and it is subject only to gravity (air resistance can be ignored). Find (a) the maximum height above the ground that the rocket reaches, and (b) the greatest horizontal range of the rocket beyond point A.

Figure **P3.47**

3.48 • **Martian Athletics.** In the long jump, an athlete launches herself at an angle above the ground and lands at the same height, trying to travel the greatest horizontal distance. Suppose that on earth she is in the air for time T, reaches a maximum height h, and achieves a horizontal distance D. If she jumped in *exactly* the same way during a competition on Mars, where g_{Mars} is 0.379 of its earth value, find her time in the air, maximum height, and horizontal distance. Express each of these three quantities in terms of its earth value. Air resistance can be neglected on both planets.

3.49 •• **Dynamite!** A demolition crew uses dynamite to blow an old building apart. Debris from the explosion flies off in all directions and is later found at distances as far as 50 m from the explosion. Estimate the maximum speed at which debris was blown outward by the explosion. Describe any assumptions that you make.

3.50 ••• **BIO Spiraling Up.** It is common to see birds of prey rising upward on thermals. The paths they take may be spiral-like. You can model the spiral motion as uniform circular motion combined with a constant upward velocity. Assume a bird completes a circle of radius 6.00 m every 5.00 s and rises vertically at a constant rate of 3.00 m/s. Determine: (a) the speed of the bird relative to the ground; (b) the bird's acceleration (magnitude and direction); and (c) the angle between the bird's velocity vector and the horizontal.

3.51 •• A jungle veterinarian with a blow-gun loaded with a tranquilizer dart and a sly 1.5-kg monkey are each 25 m above the ground in trees 70 m apart. Just as the hunter shoots horizontally at the monkey, the monkey drops from the tree in a vain attempt to escape being hit. What must the minimum muzzle velocity of the dart have been for the hunter to have hit the monkey before it reached the ground?

3.52 ••• A movie stuntwoman drops from a helicopter that is 30.0 m above the ground and moving with a constant velocity whose components are 10.0 m/s upward and 15.0 m/s horizontal and toward the south. You can ignore air resistance. (a) Where on the ground (relative to the position of the helicopter when she drops) should the stuntwoman have placed the foam mats that break her fall? (b) Draw x-t, y-t, v_x-t, and v_y-t graphs of her motion.

3.53 •• In fighting forest fires, airplanes work in support of ground crews by dropping water on the fires. A pilot is practicing by dropping a canister of red dye, hoping to hit a target on the ground below. If the plane is flying in a horizontal path 90.0 m above the ground and with a speed of 64.0 m/s (143 mi/h), at what horizontal distance from the target should the pilot release the canister? Ignore air resistance.

3.54 •• A cannon, located 60.0 m from the base of a vertical 25.0-m-tall cliff, shoots a 15-kg shell at 43.0° above the horizontal toward the cliff. (a) What must the minimum muzzle velocity be for the shell to clear the top of the cliff? (b) The ground at the top of the cliff is level, with a constant elevation of 25.0 m above the cannon. Under the conditions of part (a), how far does the shell land past the edge of the cliff?

3.55 •• An airplane is flying with a velocity of 90.0 m/s at an angle of 23.0° above the horizontal. When the plane is 114 m directly above a dog that is standing on level ground, a suitcase drops out of the luggage compartment. How far from the dog will the suitcase land? You can ignore air resistance.

3.56 ••• As a ship is approaching the dock at 45.0 cm/s, an important piece of landing equipment needs to be thrown to it before it can dock. This equipment is thrown at 15.0 m/s at 60.0° above the horizontal from the top of a tower at the edge of the water, 8.75 m above the ship's deck (Fig. P3.56). For this equipment to land at the front of the ship, at what distance D from the dock should the ship be when the equipment is thrown? Air resistance can be neglected.

Figure **P3.56**

3.57 • CP CALC A toy rocket is launched with an initial velocity of 12.0 m/s in the horizontal direction from the roof of a 30.0-m-tall building. The rocket's engine produces a horizontal acceleration of $(1.60 \text{ m/s}^3)t$, in the same direction as the initial velocity, but in the vertical direction the acceleration is g, downward. Air resistance can be neglected. What horizontal distance does the rocket travel before reaching the ground?

3.58 •• **An Errand of Mercy.** An airplane is dropping bales of hay to cattle stranded in a blizzard on the Great Plains. The pilot releases the bales at 150 m above the level ground when the plane is flying at 75 m/s in a direction 55° above the horizontal. How far in front of the cattle should the pilot release the hay so that the bales land at the point where the cattle are stranded?

3.59 ••• **The Longest Home Run.** According to the *Guinness Book of World Records,* the longest home run ever measured was hit by Roy "Dizzy" Carlyle in a minor league game. The ball traveled 188 m (618 ft) before landing on the ground outside the ballpark. (a) Assuming the ball's initial velocity was in a direction 45° above the horizontal and ignoring air resistance, what did the initial speed of the ball need to be to produce such a home run if the ball was hit at a point 0.9 m (3.0 ft) above ground level? Assume that the ground was perfectly flat. (b) How far

would the ball be above a fence 3.0 m (10 ft) high if the fence was 116 m (380 ft) from home plate?

3.60 ••• A water hose is used to fill a large cylindrical storage tank of diameter D and height $2D$. The hose shoots the water at 45° above the horizontal from the same level as the base of the tank and is a distance $6D$ away (Fig. P3.60). For what *range* of launch speeds (v_0) will the water enter the tank? Ignore air resistance, and express your answer in terms of D and g.

Figure **P3.60**

3.61 •• A projectile is being launched from ground level with no air resistance. You want to avoid having it enter a temperature inversion layer in the atmosphere a height h above the ground. (a) What is the maximum launch speed you could give this projectile if you shot it straight up? Express your answer in terms of h and g. (b) Suppose the launcher available shoots projectiles at twice the maximum launch speed you found in part (a). At what maximum angle above the horizontal should you launch the projectile? (c) How far (in terms of h) from the launcher does the projectile in part (b) land?

3.62 •• **Kicking a Field Goal.** In U.S. football, after a touchdown the team has the opportunity to earn one more point by kicking the ball over the bar between the goal posts. The bar is 10.0 ft above the ground, and the ball is kicked from ground level, 36.0 ft horizontally from the bar (Fig. P3.62). Football regulations are stated in English units, but convert them to SI units for this problem. (a) There is a minimum angle above the ground such that if the ball is launched below this angle, it can never clear the bar, no matter how fast it is kicked. What is this angle? (b) If the ball is kicked at 45.0° above the horizontal, what must its initial speed be if it is to just clear the bar? Express your answer in m/s and in km/h.

Figure **P3.62**

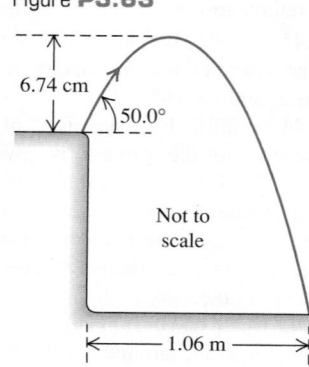

3.63 •• A grasshopper leaps into the air from the edge of a vertical cliff, as shown in Fig. P3.63. Use information from the figure to find (a) the initial speed of the grasshopper and (b) the height of the cliff.

Figure **P3.63**

3.64 •• **A World Record.** In the shot put, a standard track-and-field event, a 7.3-kg object (the shot) is thrown by releasing it at approximately 40° over a straight left leg. The world record for distance, set by Randy Barnes in 1990, is 23.11 m. Assuming that Barnes released the shot put at 40.0° from a height of 2.00 m above the ground, with what speed, in m/s and in mph, did he release it?

3.65 ••• **Look Out!** A snowball rolls off a barn roof that slopes downward at an angle of 40° (Fig. P3.65). The edge of the roof is 14.0 m above the ground, and the snowball has a speed of 7.00 m/s as it rolls off the roof. Ignore air resistance. (a) How far from the edge of the barn does the snowball strike the ground if it doesn't strike anything else while falling? (b) Draw x-t, y-t, v_x-t, and v_y-t graphs for the motion in part (a). (c) A man 1.9 m tall is standing 4.0 m from the edge of the barn. Will he be hit by the snowball?

Figure **P3.65**

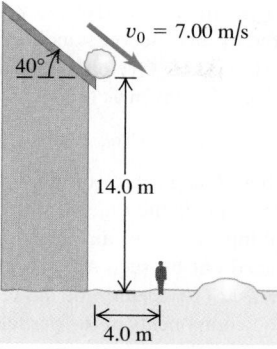

3.66 ••• **On the Flying Trapeze.** A new circus act is called the Texas Tumblers. Lovely Mary Belle swings from a trapeze, projects herself at an angle of 53°, and is supposed to be caught by Joe Bob, whose hands are 6.1 m above and 8.2 m horizontally from her launch point (Fig. P3.66). You can ignore air resistance. (a) What initial speed v_0 must Mary Belle have just to reach Joe Bob? (b) For the initial speed calculated in part (a), what are the magnitude and direction of her velocity when Mary Belle reaches Joe Bob? (c) Assuming that Mary Belle has the initial speed calculated in part (a), draw x-t, y-t, v_x-t, and v_y-t graphs showing the motion of both tumblers. Your graphs should show the motion up until the point where Mary Belle reaches Joe Bob. (d) The night of their debut performance, Joe Bob misses her completely as she flies past. How far horizontally does Mary Belle travel, from her initial launch point, before landing in the safety net 8.6 m below her starting point?

Figure **P3.66**

3.67 •• **Leaping the River II.** A physics professor did daredevil stunts in his spare time. His last stunt was an attempt to jump across a river on a motorcycle (Fig. P3.67). The takeoff ramp was inclined at 53.0°, the river was 40.0 m wide, and the far bank was 15.0 m lower than the top of the ramp. The river itself was 100 m below the ramp. You can ignore air resistance. (a) What should his speed have been at the top of the ramp to have just made it to the edge of the far bank? (b) If his speed was only half the value found in part (a), where did he land?

Figure **P3.67**

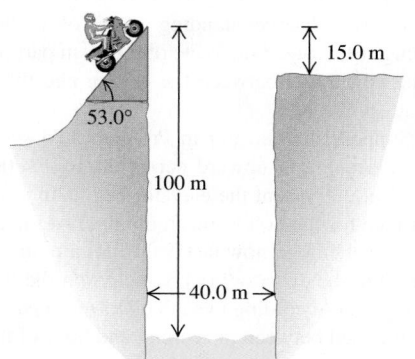

3.68 •• A rock is thrown from the roof of a building with a velocity v_0 at an angle of α_0 from the horizontal. The building has height h. You can ignore air resistance. Calculate the magnitude of the velocity of the rock just before it strikes the ground, and show that this speed is independent of α_0.

3.69 • A 5500-kg cart carrying a vertical rocket launcher moves to the right at a constant speed of 30.0 m/s along a horizontal track. It launches a 45.0-kg rocket vertically upward with an initial speed of 40.0 m/s relative to the cart. (a) How high will the rocket go? (b) Where, relative to the cart, will the rocket land? (c) How far does the cart move while the rocket is in the air? (d) At what angle, relative to the horizontal, is the rocket traveling just as it leaves the cart, as measured by an observer at rest on the ground? (e) Sketch the rocket's trajectory as seen by an observer (i) stationary on the cart and (ii) stationary on the ground.

3.70 • A 2.7-kg ball is thrown upward with an initial speed of 20.0 m/s from the edge of a 45.0-m-high cliff. At the instant the ball is thrown, a woman starts running away from the base of the cliff with a constant speed of 6.00 m/s. The woman runs in a straight line on level ground, and air resistance acting on the ball can be ignored. (a) At what angle above the horizontal should the ball be thrown so that the runner will catch it just before it hits the ground, and how far does the woman run before she catches the ball? (b) Carefully sketch the ball's trajectory as viewed by (i) a person at rest on the ground and (ii) the runner.

3.71 • A 76.0-kg boulder is rolling horizontally at the top of a vertical cliff that is 20 m above the surface of a lake, as shown in Fig. P3.71. The top of the vertical face of a dam is located 100 m from the foot of the cliff, with the top of the dam level with the surface of the water in the lake. A level plain is 25 m below the top of the dam. (a) What must be the minimum speed of the rock just as it leaves the cliff so it will travel to the plain without striking the dam? (b) How far from the foot of the dam does the rock hit the plain?

Figure **P3.71**

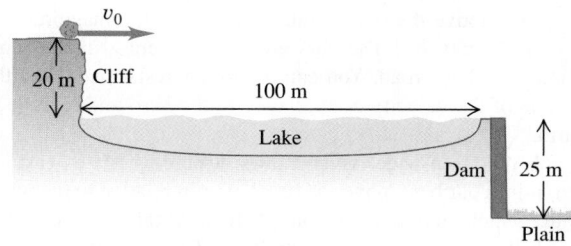

3.72 •• **Tossing Your Lunch.** Henrietta is going off to her physics class, jogging down the sidewalk at 3.05 m/s. Her husband Bruce suddenly realizes that she left in such a hurry that she forgot her lunch of bagels, so he runs to the window of their apartment, which is 38.0 m above the street level and directly above the sidewalk, to throw them to her. Bruce throws them horizontally 9.00 s after Henrietta has passed below the window, and she catches them on the run. You can ignore air resistance. (a) With what initial speed must Bruce throw the bagels so Henrietta can catch them just before they hit the ground? (b) Where is Henrietta when she catches the bagels?

3.73 ••• Two tanks are engaged in a training exercise on level ground. The first tank fires a paint-filled training round with a muzzle speed of 250 m/s at 10.0° above the horizontal while advancing toward the second tank with a speed of 15.0 m/s relative to the ground. The second tank is retreating at 35.0 m/s relative to the ground, but is hit by the shell. You can ignore air resistance and assume the shell hits at the same height above ground from which it was fired. Find the distance between the tanks (a) when the round was first fired and (b) at the time of impact.

3.74 ••• **CP Bang!** A student sits atop a platform a distance h above the ground. He throws a large firecracker horizontally with a speed v. However, a wind blowing parallel to the ground gives the firecracker a constant horizontal acceleration with magnitude a. This results in the firecracker reaching the ground directly under the student. Determine the height h in terms of v, a, and g. You can ignore the effect of air resistance on the vertical motion.

3.75 •• In a Fourth of July celebration, a firework is launched from ground level with an initial velocity of 25.0 m/s at 30.0° from the *vertical*. At its maximum height it explodes in a starburst into many fragments, two of which travel forward initially at 20.0 m/s at ±53.0° with respect to the horizontal, both quantities measured *relative to the original firework just before it exploded*. With what angles with respect to the horizontal do the two fragments initially move right after the explosion, as measured by a spectator standing on the ground?

3.76 • When it is 145 m above the ground, a rocket traveling vertically upward at a constant 8.50 m/s relative to the ground launches a secondary rocket at a speed of 12.0 m/s at an angle of 53.0° above the horizontal, both quantities being measured by an astronaut sitting in the rocket. After it is launched the secondary rocket is in free-fall. (a) Just as the secondary rocket is launched, what are the horizontal and vertical components of its velocity relative to (i) the astronaut sitting in the rocket and (ii) Mission Control on the ground? (b) Find the initial speed and launch angle of the secondary rocket as measured by Mission Control. (c) What maximum height above the ground does the secondary rocket reach?

3.77 ••• In an action-adventure film, the hero is supposed to throw a grenade from his car, which is going 90.0 km/h, to his enemy's car, which is going 110 km/h. The enemy's car is 15.8 m in front of the hero's when he lets go of the grenade. If the hero throws the grenade so its initial velocity relative to him is at an angle of 45° above the horizontal, what should the magnitude of the initial velocity be? The cars are both traveling in the same direction on a level road. You can ignore air resistance. Find the magnitude of the velocity both relative to the hero and relative to the earth.

3.78 • A 400.0-m-wide river flows from west to east at 30.0 m/min. Your boat moves at 100.0 m/min relative to the water no matter which direction you point it. To cross this river, you start from a dock at point A on the south bank. There is a boat landing directly opposite at point B on the north bank, and also one at point C, 75.0 m downstream from B (Fig. P3.78). (a) Where on the north shore will you land if you point your boat perpendicular to the water current, and what distance will you have traveled? (b) If you initially aim your boat directly toward point C and do not change that bearing relative to the shore, where on the north shore will you

Figure **P3.78**

land? (c) To reach point C: (i) at what bearing must you aim your boat, (ii) how long will it take to cross the river, (iii) what distance do you travel, and (iv) and what is the speed of your boat as measured by an observer standing on the river bank?

3.79 • **CALC Cycloid.** A particle moves in the xy-plane. Its coordinates are given as functions of time by

$$x(t) = R(\omega t - \sin \omega t) \qquad y(t) = R(1 - \cos \omega t)$$

where R and ω are constants. (a) Sketch the trajectory of the particle. (This is the trajectory of a point on the rim of a wheel that is rolling at a constant speed on a horizontal surface. The curve traced out by such a point as it moves through space is called a *cycloid*.) (b) Determine the velocity components and the acceleration components of the particle at any time t. (c) At which times is the particle momentarily at rest? What are the coordinates of the particle at these times? What are the magnitude and direction of the acceleration at these times? (d) Does the magnitude of the acceleration depend on time? Compare to uniform circular motion.

3.80 •• A projectile is fired from point A at an angle above the horizontal. At its highest point, after having traveled a horizontal distance D from its launch point, it suddenly explodes into two identical fragments that travel horizontally with equal but opposite velocities as measured *relative to the projectile just before it exploded*. If one fragment lands back at point A, how far from A (in terms of D) does the other fragment land?

3.81 •• An airplane pilot sets a compass course due west and maintains an airspeed of 220 km/h. After flying for 0.500 h, she finds herself over a town 120 km west and 20 km south of her starting point. (a) Find the wind velocity (magnitude and direction). (b) If the wind velocity is 40 km/h due south, in what direction should the pilot set her course to travel due west? Use the same airspeed of 220 km/h.

3.82 •• **Raindrops.** When a train's velocity is 12.0 m/s eastward, raindrops that are falling vertically with respect to the earth make traces that are inclined 30.0° to the vertical on the windows of the train. (a) What is the horizontal component of a drop's velocity with respect to the earth? With respect to the train? (b) What is the magnitude of the velocity of the raindrop with respect to the earth? With respect to the train?

3.83 ••• In a World Cup soccer match, Juan is running due north toward the goal with a speed of 8.00 m/s relative to the ground. A teammate passes the ball to him. The ball has a speed of 12.0 m/s and is moving in a direction 37.0° east of north, relative to the ground. What are the magnitude and direction of the ball's velocity relative to Juan?

3.84 •• An elevator is moving upward at a constant speed of 2.50 m/s. A bolt in the elevator ceiling 3.00 m above the elevator floor works loose and falls. (a) How long does it take for the bolt to fall to the elevator floor? What is the speed of the bolt just as it hits the elevator floor (b) according to an observer in the elevator? (c) According to an observer standing on one of the floor landings of the building? (d) According to the observer in part (c), what distance did the bolt travel between the ceiling and the floor of the elevator?

3.85 • **CP** Suppose the elevator in Problem 3.84 starts from rest and maintains a constant upward acceleration of 4.00 m/s², and the bolt falls out the instant the elevator begins to move. (a) How long does it take for the bolt to reach the floor of the elevator? (b) Just as it reaches the floor, how fast is the bolt moving according to an observer (i) in the elevator? (ii) Standing on the floor landings of the building? (c) According to each observer in part (b), how far has the bolt traveled between ceiling and floor of the elevator?

3.86 •• Two soccer players, Mia and Alice, are running as Alice passes the ball to Mia. Mia is running due north with a speed of 6.00 m/s. The velocity of the ball relative to Mia is 5.00 m/s in a direction 30.0° east of south. What are the magnitude and direction of the velocity of the ball relative to the ground?

3.87 ••• **Projectile Motion on an Incline.** · Refer to the Bridging Problem in Chapter 3. (a) An archer on ground that has a constant upward slope of 30.0° aims at a target 60.0 m farther up the incline. The arrow in the bow and the bull's-eye at the center of the target are each 1.50 m above the ground. The initial velocity of the arrow just after it leaves the bow has magnitude 32.0 m/s. At what angle above the *horizontal* should the archer aim to hit the bull's-eye? If there are two such angles, calculate the smaller of the two. You might have to solve the equation for the angle by iteration— that is, by trial and error. How does the angle compare to that required when the ground is level, with 0 slope? (b) Repeat the problem for ground that has a constant *downward* slope of 30.0°.

CHALLENGE PROBLEMS

3.88 ••• CALC A projectile is thrown from a point *P*. It moves in such a way that its distance from *P* is always increasing. Find the maximum angle above the horizontal with which the projectile could have been thrown. You can ignore air resistance.

3.89 ••• Two students are canoeing on a river. While heading upstream, they accidentally drop an empty bottle overboard. They then continue paddling for 60 minutes, reaching a point 2.0 km farther upstream. At this point they realize that the bottle is missing

and, driven by ecological awareness, they turn around and head downstream. They catch up with and retrieve the bottle (which has been moving along with the current) 5.0 km downstream from the turn-around point. (a) Assuming a constant paddling effort throughout, how fast is the river flowing? (b) What would the canoe speed in a still lake be for the same paddling effort?

3.90 ••• CP A rocket designed to place small payloads into orbit is carried to an altitude of 12.0 km above sea level by a converted airliner. When the airliner is flying in a straight line at a constant speed of 850 km/h, the rocket is dropped. After the drop, the airliner maintains the same altitude and speed and continues to fly in a straight line. The rocket falls for a brief time, after which its rocket motor turns on. Once its rocket motor is on, the combined effects of thrust and gravity give the rocket a constant acceleration of magnitude $3.00g$ directed at an angle of 30.0° above the horizontal. For reasons of safety, the rocket should be at least 1.00 km in front of the airliner when it climbs through the airliner's altitude. Your job is to determine the minimum time that the rocket must fall before its engine starts. You can ignore air resistance. Your answer should include (i) a diagram showing the flight paths of both the rocket and the airliner, labeled at several points with vectors for their velocities and accelerations; (ii) an *x-t* graph showing the motions of both the rocket and the airliner; and (iii) a *y-t* graph showing the motions of both the rocket and the airliner. In the diagram and the graphs, indicate when the rocket is dropped, when the rocket motor turns on, and when the rocket climbs through the altitude of the airliner.

Answers

Chapter Opening Question **?**

A cyclist going around a curve at constant speed has an acceleration directed toward the inside of the curve (see Section 3.2, especially Fig. 3.12a).

Test Your Understanding Questions

3.1 Answer: (iii) If the instantaneous velocity $\vec{v}$ is constant over an interval, its value at any point (including the end of the interval) is the same as the average velocity $\vec{v}_{av}$ over the interval. In (i) and (ii) the direction of $\vec{v}$ at the end of the interval is tangent to the path at that point, while the direction of $\vec{v}_{av}$ points from the beginning of the path to its end (in the direction of the net displacement). In (iv) $\vec{v}$ and $\vec{v}_{av}$ are both directed along the straight line, but $\vec{v}$ has a greater magnitude because the speed has been increasing.

3.2 Answer: vector 7 At the high point of the sled's path, the speed is minimum. At that point the speed is neither increasing nor decreasing, and the parallel component of the acceleration (that is, the horizontal component) is zero. The acceleration has only a perpendicular component toward the inside of the sled's curved path. In other words, the acceleration is downward.

3.3 Answer: (i) If there were no gravity ($g = 0$), the monkey would not fall and the dart would follow a straight-line path (shown as a dashed line). The effect of gravity is to make the

monkey and the dart both fall the same distance $\frac{1}{2}gt^2$ below their $g = 0$ positions. Point *A* is the same distance below the monkey's initial position as point *P* is below the dashed straight line, so point *A* is where we would find the monkey at the time in question.

3.4 Answer: (ii) At both the top and bottom of the loop, the acceleration is purely radial and is given by Eq. (3.28). The radius *R* is the same at both points, so the difference in acceleration is due purely to differences in speed. Since a_{rad} is proportional to the square of *v*, the speed must be twice as great at the bottom of the loop as at the top.

3.5 Answer: (vi) The effect of the wind is to cancel the airplane's eastward motion and give it a northward motion. So the velocity of the air relative to the ground (the wind velocity) must have one 150-km/h component to the west and one 150-km/h component to the north. The combination of these is a vector of magnitude $\sqrt{(150 \text{ km/h})^2 + (150 \text{ km/h})^2} = 212 \text{ km/h}$ that points to the northwest.

Bridging Problem

Answers: **(a)** $R = \dfrac{2v_0^2}{g} \dfrac{\cos(\theta + \phi)\sin\phi}{\cos^2\theta}$ **(b)** $\phi = 45° - \dfrac{\theta}{2}$

4 NEWTON'S LAWS OF MOTION

LEARNING GOALS

By studying this chapter, you will learn:

- What the concept of force means in physics, and why forces are vectors.

- The significance of the net force on an object, and what happens when the net force is zero.

- The relationship among the net force on an object, the object's mass, and its acceleration.

- How the forces that two bodies exert on each other are related.

? This pit crew member is pushing a race car forward. Is the race car pushing back on him? If so, does it push back with the same magnitude of force or a different amount?

We've seen in the last two chapters how to use the language and mathematics of *kinematics* to describe motion in one, two, or three dimensions. But what *causes* bodies to move the way that they do? For example, how can a tugboat push a cruise ship that's much heavier than the tug? Why is it harder to control a car on wet ice than on dry concrete? The answers to these and similar questions take us into the subject of **dynamics,** the relationship of motion to the forces that cause it.

In this chapter we will use two new concepts, *force* and *mass,* to analyze the principles of dynamics. These principles were clearly stated for the first time by Sir Isaac Newton (1642–1727); today we call them **Newton's laws of motion.** The first law states that when the net force on a body is zero, its motion doesn't change. The second law relates force to acceleration when the net force is *not* zero. The third law is a relationship between the forces that two interacting bodies exert on each other.

Newton did not *derive* the three laws of motion, but rather *deduced* them from a multitude of experiments performed by other scientists, especially Galileo Galilei (who died the same year Newton was born). These laws are truly fundamental, for they cannot be deduced or proved from other principles. Newton's laws are the foundation of **classical mechanics** (also called **Newtonian mechanics**); using them, we can understand most familiar kinds of motion. Newton's laws need modification only for situations involving extremely high speeds (near the speed of light) or very small sizes (such as within the atom).

Newton's laws are very simple to state, yet many students find these laws difficult to grasp and to work with. The reason is that before studying physics, you've spent years walking, throwing balls, pushing boxes, and doing dozens of things that involve motion. Along the way, you've developed a set of "common sense" ideas about motion and its causes. But many of these "common sense" ideas don't stand up to logical analysis. A big part of the job of this chapter—and of the rest of our study of physics—is helping you to recognize how "common sense" ideas can sometimes lead you astray, and how to adjust your understanding of the physical world to make it consistent with what experiments tell us.

4.1 Force and Interactions

In everyday language, a **force** is a push or a pull. A better definition is that a force is an *interaction* between two bodies or between a body and its environment (Fig. 4.1). That's why we always refer to the force that one body *exerts* on a second body. When you push on a car that is stuck in the snow, you exert a force on the car; a steel cable exerts a force on the beam it is hoisting at a construction site; and so on. As Fig. 4.1 shows, force is a *vector* quantity; you can push or pull a body in different directions.

When a force involves direct contact between two bodies, such as a push or pull that you exert on an object with your hand, we call it a **contact force.** Figures 4.2a, 4.2b, and 4.2c show three common types of contact forces. The **normal force** (Fig. 4.2a) is exerted on an object by any surface with which it is in contact. The adjective *normal* means that the force always acts perpendicular to the surface of contact, no matter what the angle of that surface. By contrast, the **friction force** (Fig. 4.2b) exerted on an object by a surface acts *parallel* to the surface, in the direction that opposes sliding. The pulling force exerted by a stretched rope or cord on an object to which it's attached is called a **tension force** (Fig. 4.2c). When you tug on your dog's leash, the force that pulls on her collar is a tension force.

In addition to contact forces, there are **long-range forces** that act even when the bodies are separated by empty space. The force between two magnets is an example of a long-range force, as is the force of gravity (Fig. 4.2d); the earth pulls a dropped object toward it even though there is no direct contact between the object and the earth. The gravitational force that the earth exerts on your body is called your **weight.**

To describe a force vector $\vec{F}$, we need to describe the *direction* in which it acts as well as its *magnitude,* the quantity that describes "how much" or "how hard" the force pushes or pulls. The SI unit of the magnitude of force is the *newton,* abbreviated N. (We'll give a precise definition of the newton in Section 4.3.) Table 4.1 lists some typical force magnitudes.

4.1 Some properties of forces.

- A force is a push or a pull.
- A force is an interaction between two objects or between an object and its environment.
- A force is a vector quantity, with magnitude and direction.

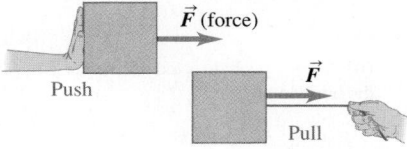

4.2 Four common types of forces.

(a) Normal force $\vec{n}$: When an object rests or pushes on a surface, the surface exerts a push on it that is directed perpendicular to the surface.

(b) Friction force $\vec{f}$: In addition to the normal force, a surface may exert a frictional force on an object, directed parallel to the surface.

(c) Tension force $\vec{T}$: A pulling force exerted on an object by a rope, cord, etc.

(d) Weight $\vec{w}$: The pull of gravity on an object is a long-range force (a force that acts over a distance).

Table 4.1 Typical Force Magnitudes

Sun's gravitational force on the earth	3.5×10^{22} N
Thrust of a space shuttle during launch	3.1×10^{7} N
Weight of a large blue whale	1.9×10^{6} N
Maximum pulling force of a locomotive	8.9×10^{5} N
Weight of a 250-lb linebacker	1.1×10^{3} N
Weight of a medium apple	1 N
Weight of smallest insect eggs	2×10^{-6} N
Electric attraction between the proton and the electron in a hydrogen atom	8.2×10^{-8} N
Weight of a very small bacterium	1×10^{-18} N
Weight of a hydrogen atom	1.6×10^{-26} N
Weight of an electron	8.9×10^{-30} N
Gravitational attraction between the proton and the electron in a hydrogen atom	3.6×10^{-47} N

4.3 Using a vector arrow to denote the force that we exert when **(a)** pulling a block with a string or **(b)** pushing a block with a stick.

(a) A 10-N pull directed 30° above the horizontal

10 N
30°

(b) A 10-N push directed 45° below the horizontal

10 N
45°

4.4 Superposition of forces.

Two forces $\vec{F}_1$ and $\vec{F}_2$ acting on a body at point O have the same effect as a single force $\vec{R}$ equal to their vector sum.

$\vec{F}_2$
$\vec{R}$
O
$\vec{F}_1$

A common instrument for measuring force magnitudes is the *spring balance*. It consists of a coil spring enclosed in a case with a pointer attached to one end. When forces are applied to the ends of the spring, it stretches by an amount that depends on the force. We can make a scale for the pointer by using a number of identical bodies with weights of exactly 1 N each. When one, two, or more of these are suspended simultaneously from the balance, the total force stretching the spring is 1 N, 2 N, and so on, and we can label the corresponding positions of the pointer 1 N, 2 N, and so on. Then we can use this instrument to measure the magnitude of an unknown force. We can also make a similar instrument that measures pushes instead of pulls.

Figure 4.3 shows a spring balance being used to measure a pull or push that we apply to a box. In each case we draw a vector to represent the applied force. The length of the vector shows the magnitude; the longer the vector, the greater the force magnitude.

Superposition of Forces

When you throw a ball, there are at least two forces acting on it: the push of your hand and the downward pull of gravity. Experiment shows that when two forces $\vec{F}_1$ and $\vec{F}_2$ act at the same time at the same point on a body (Fig. 4.4), the effect on the body's motion is the same as if a single force $\vec{R}$ were acting equal to the vector sum of the original forces: $\vec{R} = \vec{F}_1 + \vec{F}_2$. More generally, *any number of forces applied at a point on a body have the same effect as a single force equal to the vector sum of the forces.* This important principle is called **superposition of forces.**

The principle of superposition of forces is of the utmost importance, and we will use it throughout our study of physics. For example, in Fig. 4.5a, force $\vec{F}$ acts on a body at point O. The component vectors of $\vec{F}$ in the directions Ox and Oy are $\vec{F}_x$ and $\vec{F}_y$. When $\vec{F}_x$ and $\vec{F}_y$ are applied simultaneously, as in Fig. 4.5b, the effect is exactly the same as the effect of the original force $\vec{F}$. Hence *any force can be replaced by its component vectors, acting at the same point.*

It's frequently more convenient to describe a force $\vec{F}$ in terms of its x- and y-components F_x and F_y rather than by its component vectors (recall from Section 1.8 that *component vectors* are vectors, but *components* are just numbers). For the case shown in Fig. 4.5, both F_x and F_y are positive; for other orientations of the force $\vec{F}$, either F_x or F_y may be negative or zero.

Our coordinate axes don't have to be vertical and horizontal. Figure 4.6 shows a crate being pulled up a ramp by a force $\vec{F}$, represented by its components F_x and F_y parallel and perpendicular to the sloping surface of the ramp.

4.5 The force $\vec{F}$, which acts at an angle θ from the x-axis, may be replaced by its rectangular component vectors $\vec{F}_x$ and $\vec{F}_y$.

(a) Component vectors: $\vec{F}_x$ and $\vec{F}_y$
Components: $F_x = F\cos\theta$ and $F_y = F\sin\theta$

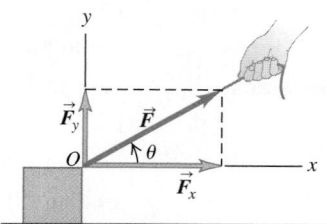

y
$\vec{F}_y$
$\vec{F}$
O
θ
$\vec{F}_x$
x

(b) Component vectors $\vec{F}_x$ and $\vec{F}_y$ together have the same effect as original force $\vec{F}$.

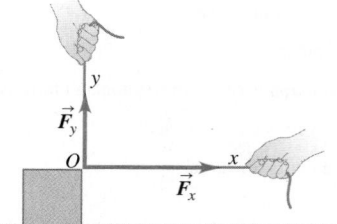

y
$\vec{F}_y$
O
x
$\vec{F}_x$

CAUTION **Using a wiggly line in force diagrams** In Fig. 4.6 we draw a wiggly line through the force vector $\vec{F}$ to show that we have replaced it by its x- and y-components. Otherwise, the diagram would include the same force twice. We will draw such a wiggly line in any force diagram where a force is replaced by its components. Look for this wiggly line in other figures in this and subsequent chapters. ❚

We will often need to find the vector sum (resultant) of *all* the forces acting on a body. We call this the **net force** acting on the body. We will use the Greek letter Σ (capital sigma, equivalent to the Roman S) as a shorthand notation for a sum. If the forces are labeled $\vec{F}_1, \vec{F}_2, \vec{F}_3$, and so on, we abbreviate the sum as

$$\vec{R} = \vec{F}_1 + \vec{F}_2 + \vec{F}_3 + \cdots = \sum \vec{F} \qquad (4.1)$$

We read $\sum \vec{F}$ as "the vector sum of the forces" or "the net force." The component version of Eq. (4.1) is the pair of component equations

$$R_x = \sum F_x \qquad R_y = \sum F_y \qquad (4.2)$$

Here $\sum F_x$ is the sum of the x-components and $\sum F_y$ is the sum of the y-components (Fig. 4.7). Each component may be positive or negative, so be careful with signs when you evaluate these sums. (You may want to review Section 1.8.)

Once we have R_x and R_y we can find the magnitude and direction of the net force $\vec{R} = \sum \vec{F}$ acting on the body. The magnitude is

$$R = \sqrt{R_x^2 + R_y^2}$$

and the angle θ between $\vec{R}$ and the +x-axis can be found from the relationship $\tan \theta = R_y/R_x$. The components R_x and R_y may be positive, negative, or zero, and the angle θ may be in any of the four quadrants.

In three-dimensional problems, forces may also have z-components; then we add the equation $R_z = \sum F_z$ to Eq. (4.2). The magnitude of the net force is then

$$R = \sqrt{R_x^2 + R_y^2 + R_z^2}$$

4.6 F_x and F_y are the components of $\vec{F}$ parallel and perpendicular to the sloping surface of the inclined plane.

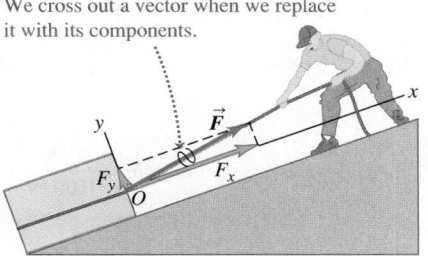

We cross out a vector when we replace it with its components.

4.7 Finding the components of the vector sum (resultant) $\vec{R}$ of two forces $\vec{F}_1$ and $\vec{F}_2$.

$\vec{R}$ is the sum (resultant) of $\vec{F}_1$ and $\vec{F}_2$.

The y-component of $\vec{R}$ equals the sum of the y-components of $\vec{F}_1$ and $\vec{F}_2$.   The same goes for the x-components.

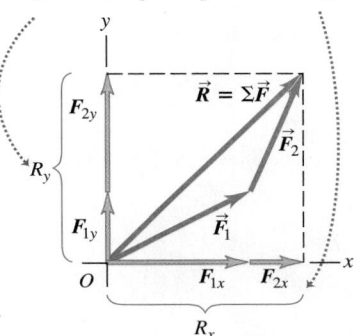

Example 4.1   Superposition of forces

Three professional wrestlers are fighting over a champion's belt. Figure 4.8a shows the horizontal force each wrestler applies to the belt, as viewed from above. The forces have magnitudes $F_1 = 250$ N, $F_2 = 50$ N, and $F_3 = 120$ N. Find the x- and y-components of the net force on the belt, and find its magnitude and direction.

SOLUTION

IDENTIFY and SET UP: This is a problem in vector addition in which the vectors happen to represent forces. We want to find the x- and y-components of the net force $\vec{R}$, so we'll use the component method of vector addition expressed by Eqs. (4.2). Once we know the components of $\vec{R}$, we can find its magnitude and direction.

EXECUTE: From Fig. 4.8a the angles between the three forces $\vec{F}_1$, $\vec{F}_2$, and $\vec{F}_3$ and the +x-axis are $\theta_1 = 180° - 53° = 127°$, $\theta_2 = 0°$, and $\theta_3 = 270°$. The x- and y-components of the three forces are

$$F_{1x} = (250 \text{ N}) \cos 127° = -150 \text{ N}$$
$$F_{1y} = (250 \text{ N}) \sin 127° = 200 \text{ N}$$
$$F_{2x} = (50 \text{ N}) \cos 0° = 50 \text{ N}$$

4.8 (a) Three forces acting on a belt. (b) The net force $\vec{R} = \sum \vec{F}$ and its components.

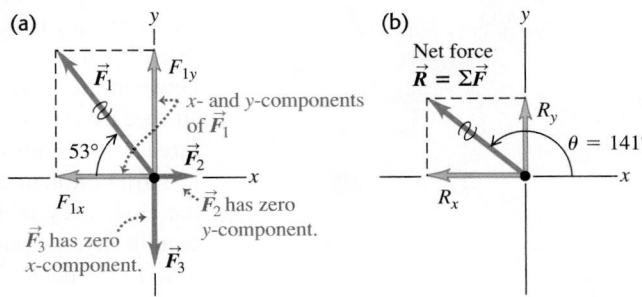

$$F_{2y} = (50 \text{ N}) \sin 0° = 0 \text{ N}$$
$$F_{3x} = (120 \text{ N}) \cos 270° = 0 \text{ N}$$
$$F_{3y} = (120 \text{ N}) \sin 270° = -120 \text{ N}$$

From Eqs. (4.2) the net force $\vec{R} = \sum \vec{F}$ has components

$$R_x = F_{1x} + F_{2x} + F_{3x} = (-150 \text{ N}) + 50 \text{ N} + 0 \text{ N} = -100 \text{ N}$$
$$R_y = F_{1y} + F_{2y} + F_{3y} = 200 \text{ N} + 0 \text{ N} + (-120 \text{ N}) = 80 \text{ N}$$

Continued

The net force has a negative x-component and a positive y-component, as shown in Fig. 4.8b.

The magnitude of $\vec{R}$ is

$$R = \sqrt{R_x^2 + R_y^2} = \sqrt{(-100 \text{ N})^2 + (80 \text{ N})^2} = 128 \text{ N}$$

To find the angle between the net force and the $+x$-axis, we use Eq. (1.8):

$$\theta = \arctan \frac{R_y}{R_x} = \arctan\left(\frac{80 \text{ N}}{-100 \text{ N}}\right) = \arctan(-0.80)$$

The arctangent of -0.80 is $-39°$, but Fig. 4.8b shows that the net force lies in the second quadrant. Hence the correct solution is $\theta = -39° + 180° = 141°$.

EVALUATE: The net force is *not* zero. Your intuition should suggest that wrestler 1 (who exerts the largest force on the belt, $F_1 = 250$ N) will walk away with it when the struggle ends.

You should check the direction of $\vec{R}$ by adding the vectors $\vec{F}_1$, $\vec{F}_2$, and $\vec{F}_3$ graphically. Does your drawing show that $\vec{R} = \vec{F}_1 + \vec{F}_2 + \vec{F}_3$ points in the second quadrant as we found above?

4.9 The slicker the surface, the farther a puck slides after being given an initial velocity. On an air-hockey table (c) the friction force is practically zero, so the puck continues with almost constant velocity.

(a) Table: puck stops short.

(b) Ice: puck slides farther.

(c) Air-hockey table: puck slides even farther.

Test Your Understanding of Section 4.1 Figure 4.6 shows a force $\vec{F}$ acting on a crate. With the x- and y-axes shown in the figure, which statement about the components of the *gravitational* force that the earth exerts on the crate (the crate's weight) is *correct*? (i) The x- and y-components are both positive. (ii) The x-component is zero and the y-component is positive. (iii) The x-component is negative and the y-component is positive. (iv) The x- and y-components are both negative. (v) The x-component is zero and the y-component is negative. (vi) The x-component is positive and the y-component is negative.

4.2 Newton's First Law

How do the forces that act on a body affect its motion? To begin to answer this question, let's first consider what happens when the net force on a body is *zero*. You would almost certainly agree that if a body is at rest, and if no net force acts on it (that is, no net push or pull), that body will remain at rest. But what if there is zero net force acting on a body in *motion*?

To see what happens in this case, suppose you slide a hockey puck along a horizontal tabletop, applying a horizontal force to it with your hand (Fig. 4.9a). After you stop pushing, the puck *does not* continue to move indefinitely; it slows down and stops. To keep it moving, you have to keep pushing (that is, applying a force). You might come to the "common sense" conclusion that bodies in motion naturally come to rest and that a force is required to sustain motion.

But now imagine pushing the puck across a smooth surface of ice (Fig. 4.9b). After you quit pushing, the puck will slide a lot farther before it stops. Put it on an air-hockey table, where it floats on a thin cushion of air, and it moves still farther (Fig. 4.9c). In each case, what slows the puck down is *friction,* an interaction between the lower surface of the puck and the surface on which it slides. Each surface exerts a frictional force on the puck that resists the puck's motion; the difference in the three cases is the magnitude of the frictional force. The ice exerts less friction than the tabletop, so the puck travels farther. The gas molecules of the air-hockey table exert the least friction of all. If we could eliminate friction completely, the puck would never slow down, and we would need no force at all to keep the puck moving once it had been started. Thus the "common sense" idea that a force is required to sustain motion is *incorrect*.

Experiments like the ones we've just described show that when *no* net force acts on a body, the body either remains at rest *or* moves with constant velocity in a straight line. Once a body has been set in motion, no net force is needed to keep it moving. We call this observation *Newton's first law of motion:*

Newton's first law of motion: A body acted on by no net force moves with constant velocity (which may be zero) and zero acceleration.

The tendency of a body to keep moving once it is set in motion results from a property called **inertia.** You use inertia when you try to get ketchup out of a bottle by shaking it. First you start the bottle (and the ketchup inside) moving forward; when you jerk the bottle back, the ketchup tends to keep moving forward and, you hope, ends up on your burger. The tendency of a body at rest to remain at rest is also due to inertia. You may have seen a tablecloth yanked out from under the china without breaking anything. The force on the china isn't great enough to make it move appreciably during the short time it takes to pull the tablecloth away.

It's important to note that the *net* force is what matters in Newton's first law. For example, a physics book at rest on a horizontal tabletop has two forces acting on it: an upward supporting force, or normal force, exerted by the tabletop (see Fig. 4.2a) and the downward force of the earth's gravitational attraction (a long-range force that acts even if the tabletop is elevated above the ground; see Fig. 4.2d). The upward push of the surface is just as great as the downward pull of gravity, so the *net* force acting on the book (that is, the vector sum of the two forces) is zero. In agreement with Newton's first law, if the book is at rest on the tabletop, it remains at rest. The same principle applies to a hockey puck sliding on a horizontal, frictionless surface: The vector sum of the upward push of the surface and the downward pull of gravity is zero. Once the puck is in motion, it continues to move with constant velocity because the *net* force acting on it is zero.

Here's another example. Suppose a hockey puck rests on a horizontal surface with negligible friction, such as an air-hockey table or a slab of wet ice. If the puck is initially at rest and a single horizontal force $\vec{F}_1$ acts on it (Fig. 4.10a), the puck starts to move. If the puck is in motion to begin with, the force changes its speed, its direction, or both, depending on the direction of the force. In this case the net force is equal to $\vec{F}_1$, which is *not* zero. (There are also two vertical forces: the earth's gravitational attraction and the upward normal force exerted by the surface. But as we mentioned earlier, these two forces cancel.)

Now suppose we apply a second force $\vec{F}_2$ (Fig. 4.10b), equal in magnitude to $\vec{F}_1$ but opposite in direction. The two forces are negatives of each other, $\vec{F}_2 = -\vec{F}_1$, and their vector sum is zero:

$$\sum \vec{F} = \vec{F}_1 + \vec{F}_2 = \vec{F}_1 + (-\vec{F}_1) = 0$$

Again, we find that if the body is at rest at the start, it remains at rest; if it is initially moving, it continues to move in the same direction with constant speed. These results show that in Newton's first law, *zero net force is equivalent to no force at all.* This is just the principle of superposition of forces that we saw in Section 4.1.

When a body is either at rest or moving with constant velocity (in a straight line with constant speed), we say that the body is in **equilibrium.** For a body to be in equilibrium, it must be acted on by no forces, or by several forces such that their vector sum—that is, the net force—is zero:

$$\sum \vec{F} = 0 \qquad \text{(body in equilibrium)} \qquad (4.3)$$

For this to be true, each component of the net force must be zero, so

$$\sum F_x = 0 \qquad \sum F_y = 0 \qquad \text{(body in equilibrium)} \qquad (4.4)$$

We are assuming that the body can be represented adequately as a point particle. When the body has finite size, we also have to consider *where* on the body the forces are applied. We will return to this point in Chapter 11.

4.10 (a) A hockey puck accelerates in the direction of a net applied force $\vec{F}_1$. (b) When the net force is zero, the acceleration is zero, and the puck is in equilibrium.

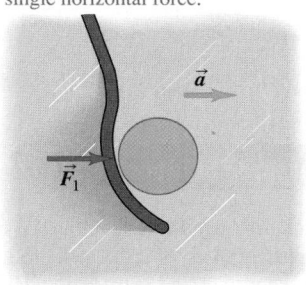

(a) A puck on a frictionless surface accelerates when acted on by a single horizontal force.

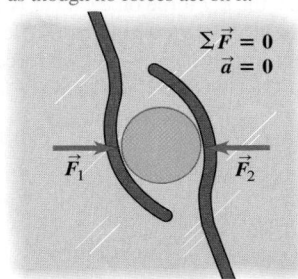

(b) An object acted on by forces whose vector sum is zero behaves as though no forces act on it.

Application Sledding with Newton's First Law

The downward force of gravity acting on the child and sled is balanced by an upward normal force exerted by the ground. The adult's foot exerts a forward force that balances the backward force of friction on the sled. Hence there is no net force on the child and sled, and they slide with a constant velocity.

| **Conceptual Example 4.2** | **Zero net force means constant velocity** |

In the classic 1950 science fiction film *Rocketship X-M*, a spaceship is moving in the vacuum of outer space, far from any star or planet, when its engine dies. As a result, the spaceship slows down and stops. What does Newton's first law say about this scene?

SOLUTION

After the engine dies there are no forces acting on the spaceship, so according to Newton's first law it will *not* stop but will continue to move in a straight line with constant speed. Some science fiction movies are based on accurate science; this is not one of them.

| **Conceptual Example 4.3** | **Constant velocity means zero net force** |

You are driving a Maserati GranTurismo S on a straight testing track at a constant speed of 250 km/h. You pass a 1971 Volkswagen Beetle doing a constant 75 km/h. On which car is the net force greater?

SOLUTION

The key word in this question is "net." Both cars are in equilibrium because their velocities are constant; Newton's first law therefore says that the *net* force on each car is *zero*.

This seems to contradict the "common sense" idea that the faster car must have a greater force pushing it. Thanks to your Maserati's high-power engine, it's true that the track exerts a greater forward force on your Maserati than it does on the Volkswagen. But a *backward* force also acts on each car due to road friction and air resistance. When the car is traveling with constant velocity, the vector sum of the forward and backward forces is zero. There is more air resistance on the fast-moving Maserati than on the slow-moving Volkswagen, which is why the Maserati's engine must be more powerful than that of the Volkswagen.

Inertial Frames of Reference

In discussing relative velocity in Section 3.5, we introduced the concept of *frame of reference*. This concept is central to Newton's laws of motion. Suppose you are in a bus that is traveling on a straight road and speeding up. If you could stand in the aisle on roller skates, you would start moving *backward* relative to the bus as the bus gains speed. If instead the bus was slowing to a stop, you would start moving forward down the aisle. In either case, it looks as though Newton's first law is not obeyed; there is no net force acting on you, yet your velocity changes. What's wrong?

The point is that the bus is accelerating with respect to the earth and is *not* a suitable frame of reference for Newton's first law. This law is valid in some frames of reference and not valid in others. A frame of reference in which Newton's first law *is* valid is called an **inertial frame of reference.** The earth is at least approximately an inertial frame of reference, but the bus is not. (The earth is not a completely inertial frame, owing to the acceleration associated with its rotation and its motion around the sun. These effects are quite small, however; see Exercises 3.25 and 3.28.) Because Newton's first law is used to define what we mean by an inertial frame of reference, it is sometimes called the *law of inertia*.

Figure 4.11 helps us understand what you experience when riding in a vehicle that's accelerating. In Fig. 4.11a, a vehicle is initially at rest and then begins to accelerate to the right. A passenger on roller skates (which nearly eliminate the effects of friction) has virtually no net force acting on her, so she tends to remain at rest relative to the inertial frame of the earth. As the vehicle accelerates around her, she moves backward relative to the vehicle. In the same way, a passenger in a vehicle that is slowing down tends to continue moving with constant velocity relative to the earth, and so moves forward relative to the vehicle (Fig. 4.11b). A vehicle is also accelerating if it moves at a constant speed but is turning (Fig. 4.11c). In this case a passenger tends to continue moving relative to

4.11 Riding in an accelerating vehicle.

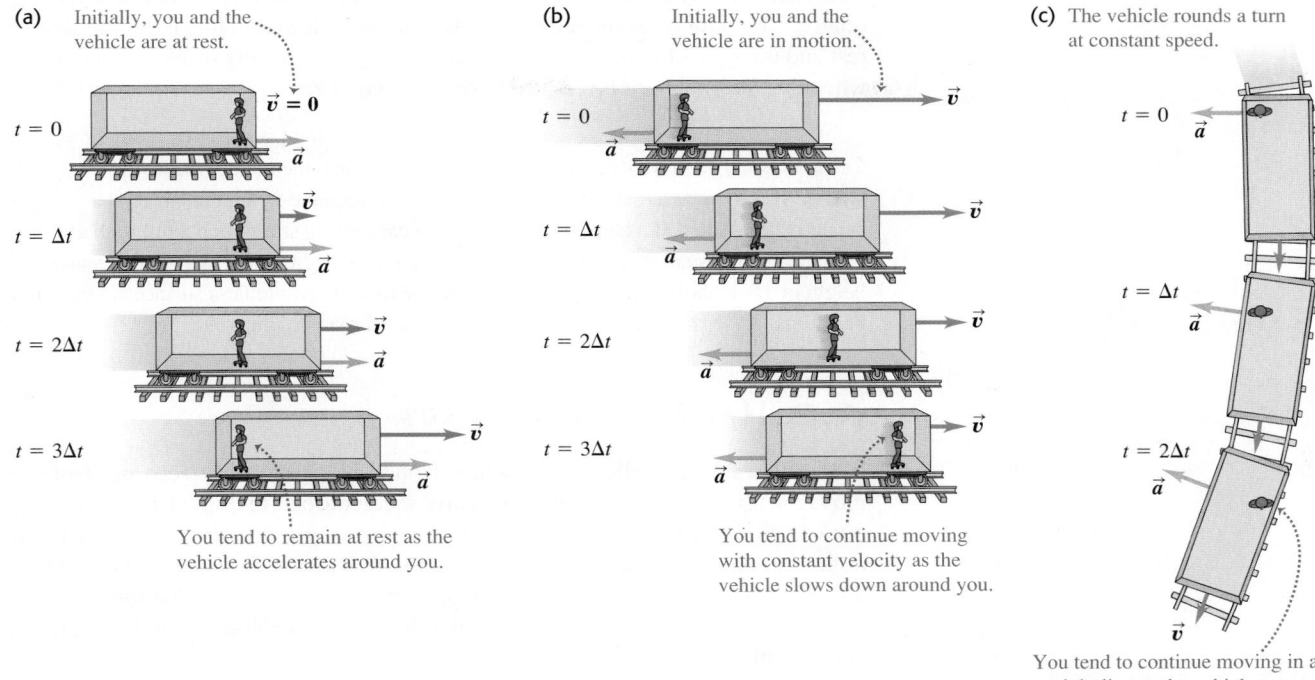

(a) Initially, you and the vehicle are at rest.

$t = 0$ $\vec{v} = 0$ $\vec{a}$

$t = \Delta t$ $\vec{v}$ $\vec{a}$

$t = 2\Delta t$ $\vec{v}$ $\vec{a}$

$t = 3\Delta t$ $\vec{v}$ $\vec{a}$

You tend to remain at rest as the vehicle accelerates around you.

(b) Initially, you and the vehicle are in motion.

$t = 0$ $\vec{v}$ $\vec{a}$

$t = \Delta t$ $\vec{v}$ $\vec{a}$

$t = 2\Delta t$ $\vec{v}$ $\vec{a}$

$t = 3\Delta t$ $\vec{v}$ $\vec{a}$

You tend to continue moving with constant velocity as the vehicle slows down around you.

(c) The vehicle rounds a turn at constant speed.

$t = 0$ $\vec{a}$

$t = \Delta t$ $\vec{a}$

$t = 2\Delta t$ $\vec{a}$

$\vec{v}$

You tend to continue moving in a straight line as the vehicle turns.

the earth at constant speed in a straight line; relative to the vehicle, the passenger moves to the side of the vehicle on the outside of the turn.

In each case shown in Fig. 4.11, an observer in the vehicle's frame of reference might be tempted to conclude that there *is* a net force acting on the passenger, since the passenger's velocity *relative to the vehicle* changes in each case. This conclusion is simply wrong; the net force on the passenger is indeed zero. The vehicle observer's mistake is in trying to apply Newton's first law in the vehicle's frame of reference, which is *not* an inertial frame and in which Newton's first law isn't valid (Fig. 4.12). In this book we will use *only* inertial frames of reference.

We've mentioned only one (approximately) inertial frame of reference: the earth's surface. But there are many inertial frames. If we have an inertial frame of reference A, in which Newton's first law is obeyed, then *any* second frame of reference B will also be inertial if it moves relative to A with constant velocity $\vec{v}_{B/A}$. We can prove this using the relative-velocity relationship Eq. (3.36) from Section 3.5:

$$\vec{v}_{P/A} = \vec{v}_{P/B} + \vec{v}_{B/A}$$

Suppose that P is a body that moves with constant velocity $\vec{v}_{P/A}$ with respect to an inertial frame A. By Newton's first law the net force on this body is zero. The velocity of P relative to another frame B has a different value, $\vec{v}_{P/B} = \vec{v}_{P/A} - \vec{v}_{B/A}$. But if the relative velocity $\vec{v}_{B/A}$ of the two frames is constant, then $\vec{v}_{P/B}$ is constant as well. Thus B is also an inertial frame; the velocity of P in this frame is constant, and the net force on P is zero, so Newton's first law is obeyed in B. Observers in frames A and B will disagree about the velocity of P, but they will agree that P has a constant velocity (zero acceleration) and has zero net force acting on it.

4.12 From the frame of reference of the car, it seems as though a force is pushing the crash test dummies forward as the car comes to a sudden stop. But there is really no such force: As the car stops, the dummies keep moving forward as a consequence of Newton's first law.

There is no single inertial frame of reference that is preferred over all others for formulating Newton's laws. If one frame is inertial, then every other frame moving relative to it with constant velocity is also inertial. Viewed in this light, the state of rest and the state of motion with constant velocity are not very different; both occur when the vector sum of forces acting on the body is zero.

Test Your Understanding of Section 4.2 In which of the following situations is there zero net force on the body? (i) an airplane flying due north at a steady 120 m/s and at a constant altitude; (ii) a car driving straight up a hill with a 3° slope at a constant 90 km/h; (iii) a hawk circling at a constant 20 km/h at a constant height of 15 m above an open field; (iv) a box with slick, frictionless surfaces in the back of a truck as the truck accelerates forward on a level road at 5 m/s². ∎

4.3 Newton's Second Law

Newton's first law tells us that when a body is acted on by zero net force, it moves with constant velocity and zero acceleration. In Fig. 4.13a, a hockey puck is sliding to the right on wet ice. There is negligible friction, so there are no horizontal forces acting on the puck; the downward force of gravity and the upward normal force exerted by the ice surface sum to zero. So the net force $\Sigma \vec{F}$ acting on the puck is zero, the puck has zero acceleration, and its velocity is constant.

But what happens when the net force is *not* zero? In Fig. 4.13b we apply a constant horizontal force to a sliding puck in the same direction that the puck is moving. Then $\Sigma \vec{F}$ is constant and in the same horizontal direction as $\vec{v}$. We find that during the time the force is acting, the velocity of the puck changes at a constant rate;

4.13 Exploring the relationship between the acceleration of a body and the net force acting on the body (in this case, a hockey puck on a frictionless surface).

(a) A puck moving with constant velocity (in equilibrium): $\Sigma \vec{F} = 0$, $\vec{a} = 0$

(b) A constant net force in the direction of motion causes a constant acceleration in the same direction as the net force.

(c) A constant net force opposite the direction of motion causes a constant acceleration in the same direction as the net force.

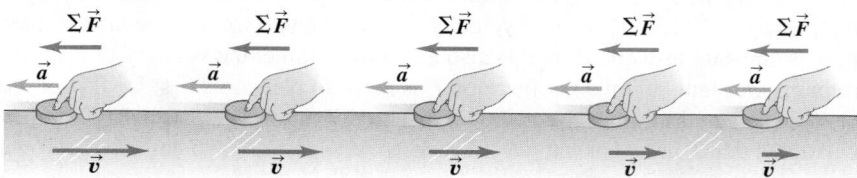

that is, the puck moves with constant acceleration. The speed of the puck increases, so the acceleration $\vec{a}$ is in the same direction as $\vec{v}$ and $\sum\vec{F}$.

In Fig. 4.13c we reverse the direction of the force on the puck so that $\sum\vec{F}$ acts opposite to $\vec{v}$. In this case as well, the puck has an acceleration; the puck moves more and more slowly to the right. The acceleration $\vec{a}$ in this case is to the left, in the same direction as $\sum\vec{F}$. As in the previous case, experiment shows that the acceleration is constant if $\sum\vec{F}$ is constant.

We conclude that *a net force acting on a body causes the body to accelerate in the same direction as the net force.* If the magnitude of the net force is constant, as in Figs. 4.13b and 4.13c, then so is the magnitude of the acceleration.

These conclusions about net force and acceleration also apply to a body moving along a curved path. For example, Fig. 4.14 shows a hockey puck moving in a horizontal circle on an ice surface of negligible friction. A rope is attached to the puck and to a stick in the ice, and this rope exerts an inward tension force of constant magnitude on the puck. The net force and acceleration are both constant in magnitude and directed toward the center of the circle. The speed of the puck is constant, so this is uniform circular motion, as discussed in Section 3.4.

Figure 4.15a shows another experiment to explore the relationship between acceleration and net force. We apply a constant horizontal force to a puck on a frictionless horizontal surface, using the spring balance described in Section 4.1 with the spring stretched a constant amount. As in Figs. 4.13b and 4.13c, this horizontal force equals the net force on the puck. If we change the magnitude of the net force, the acceleration changes in the same proportion. Doubling the net force doubles the acceleration (Fig. 4.15b), halving the net force halves the acceleration (Fig. 4.15c), and so on. Many such experiments show that *for any given body, the magnitude of the acceleration is directly proportional to the magnitude of the net force acting on the body.*

Mass and Force

Our results mean that for a given body, the *ratio* of the magnitude $|\sum\vec{F}|$ of the net force to the magnitude $a = |\vec{a}|$ of the acceleration is constant, regardless of the magnitude of the net force. We call this ratio the *inertial mass,* or simply the **mass,** of the body and denote it by m. That is,

$$m = \frac{|\sum\vec{F}|}{a} \quad \text{or} \quad |\sum\vec{F}| = ma \quad \text{or} \quad a = \frac{|\sum\vec{F}|}{m} \quad (4.5)$$

Mass is a quantitative measure of inertia, which we discussed in Section 4.2. The last of the equations in Eqs. (4.5) says that the greater its mass, the more a body "resists" being accelerated. When you hold a piece of fruit in your hand at the supermarket and move it slightly up and down to estimate its heft, you're applying a force and seeing how much the fruit accelerates up and down in response. If a force causes a large acceleration, the fruit has a small mass; if the same force causes only a small acceleration, the fruit has a large mass. In the same way, if you hit a table-tennis ball and then a basketball with the same force, the basketball has much smaller acceleration because it has much greater mass.

The SI unit of mass is the **kilogram.** We mentioned in Section 1.3 that the kilogram is officially defined to be the mass of a cylinder of platinum–iridium alloy kept in a vault near Paris. We can use this standard kilogram, along with Eqs. (4.5), to define the **newton:**

> **One newton is the amount of net force that gives an acceleration of 1 meter per second squared to a body with a mass of 1 kilogram.**

4.14 A top view of a hockey puck in uniform circular motion on a frictionless horizontal surface.

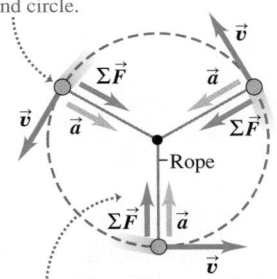

Puck moves at constant speed around circle.

At all points, the acceleration $\vec{a}$ and the net force $\sum\vec{F}$ point in the same direction—always toward the center of the circle.

4.15 For a body of a given mass m, the magnitude of the body's acceleration is directly proportional to the magnitude of the net force acting on the body.

(a) A constant net force $\sum\vec{F}$ causes a constant acceleration $\vec{a}$.

(b) Doubling the net force doubles the acceleration.

(c) Halving the force halves the acceleration.

This definition allows us to calibrate the spring balances and other instruments used to measure forces. Because of the way we have defined the newton, it is related to the units of mass, length, and time. For Eqs. (4.5) to be dimensionally consistent, it must be true that

$$1 \text{ newton} = (1 \text{ kilogram})(1 \text{ meter per second squared})$$

or

$$1 \text{ N} = 1 \text{ kg} \cdot \text{m/s}^2$$

We will use this relationship many times in the next few chapters, so keep it in mind.

We can also use Eqs. (4.5) to compare a mass with the standard mass and thus to *measure* masses. Suppose we apply a constant net force $\Sigma \vec{F}$ to a body having a known mass m_1 and we find an acceleration of magnitude a_1 (Fig. 4.16a). We then apply the same force to another body having an unknown mass m_2, and we find an acceleration of magnitude a_2 (Fig. 4.16b). Then, according to Eqs. (4.5),

$$m_1 a_1 = m_2 a_2$$

$$\frac{m_2}{m_1} = \frac{a_1}{a_2} \quad \text{(same net force)} \tag{4.6}$$

For the same net force, the ratio of the masses of two bodies is the inverse of the ratio of their accelerations. In principle we could use Eq. (4.6) to measure an unknown mass m_2, but it is usually easier to determine mass indirectly by measuring the body's *weight*. We'll return to this point in Section 4.4.

When two bodies with masses m_1 and m_2 are fastened together, we find that the mass of the composite body is always $m_1 + m_2$ (Fig. 4.16c). This additive property of mass may seem obvious, but it has to be verified experimentally. Ultimately, the mass of a body is related to the number of protons, electrons, and neutrons it contains. This wouldn't be a good way to *define* mass because there is no practical way to count these particles. But the concept of mass is the most fundamental way to characterize the quantity of matter in a body.

4.16 For a given net force $\Sigma \vec{F}$ acting on a body, the acceleration is inversely proportional to the mass of the body. Masses add like ordinary scalars.

(a) A known force $\Sigma \vec{F}$ causes an object with mass m_1 to have an acceleration $\vec{a}_1$.

(b) Applying the same force $\Sigma \vec{F}$ to a second object and noting the acceleration allow us to measure the mass.

(c) When the two objects are fastened together, the same method shows that their composite mass is the sum of their individual masses.

Stating Newton's Second Law

We've been careful to state that the *net* force on a body is what causes that body to accelerate. Experiment shows that if a combination of forces $\vec{F}_1$, $\vec{F}_2$, $\vec{F}_3$, and so on is applied to a body, the body will have the same acceleration (magnitude and direction) as when only a single force is applied, if that single force is equal to the vector sum $\vec{F}_1 + \vec{F}_2 + \vec{F}_3 + \cdots$. In other words, the principle of superposition of forces (see Fig. 4.4) also holds true when the net force is not zero and the body is accelerating.

Equations (4.5) relate the magnitude of the net force on a body to the magnitude of the acceleration that it produces. We have also seen that the direction of the net force is the same as the direction of the acceleration, whether the body's path is straight or curved. Newton wrapped up all these relationships and experimental results in a single concise statement that we now call *Newton's second law of motion:*

Newton's second law of motion: If a net external force acts on a body, the body accelerates. The direction of acceleration is the same as the direction of the net force. The mass of the body times the acceleration of the body equals the net force vector.

In symbols,

$$\sum \vec{F} = m\vec{a} \qquad \text{(Newton's second law of motion)} \qquad (4.7)$$

An alternative statement is that the acceleration of a body is in the same direction as the net force acting on the body, and is equal to the net force divided by the body's mass:

$$\vec{a} = \frac{\sum \vec{F}}{m}$$

Newton's second law is a fundamental law of nature, the basic relationship between force and motion. Most of the remainder of this chapter and all of the next are devoted to learning how to apply this principle in various situations.

Equation (4.7) has many practical applications (Fig. 4.17). You've actually been using it all your life to measure your body's acceleration. In your inner ear, microscopic hair cells sense the magnitude and direction of the force that they must exert to cause small membranes to accelerate along with the rest of your body. By Newton's second law, the acceleration of the membranes—and hence that of your body as a whole—is proportional to this force and has the same direction. In this way, you can sense the magnitude and direction of your acceleration even with your eyes closed!

Using Newton's Second Law

There are at least four aspects of Newton's second law that deserve special attention. First, Eq. (4.7) is a *vector* equation. Usually we will use it in component form, with a separate equation for each component of force and the corresponding component of acceleration:

$$\sum F_x = ma_x \qquad \sum F_y = ma_y \qquad \sum F_z = ma_z \qquad \begin{array}{l}\text{(Newton's second}\\ \text{law of motion)}\end{array} \qquad (4.8)$$

This set of component equations is equivalent to the single vector equation (4.7). Each component of the net force equals the mass times the corresponding component of acceleration.

Second, the statement of Newton's second law refers to *external* forces. By this we mean forces exerted on the body by other bodies in its environment. It's impossible for a body to affect its own motion by exerting a force on itself; if it were possible, you could lift yourself to the ceiling by pulling up on your belt! That's why only external forces are included in the sum $\sum \vec{F}$ in Eqs. (4.7) and (4.8).

Third, Eqs. (4.7) and (4.8) are valid only when the mass m is *constant.* It's easy to think of systems whose masses change, such as a leaking tank truck, a rocket ship, or a moving railroad car being loaded with coal. But such systems are better handled by using the concept of momentum; we'll get to that in Chapter 8.

Finally, Newton's second law is valid only in inertial frames of reference, just like the first law. Thus it is not valid in the reference frame of any of the accelerating vehicles in Fig. 4.11; relative to any of these frames, the passenger accelerates even though the net force on the passenger is zero. We will usually assume that the earth is an adequate approximation to an inertial frame, although because of its rotation and orbital motion it is not precisely inertial.

CAUTION *$m\vec{a}$ is not a force* You must keep in mind that even though the vector $m\vec{a}$ is equal to the vector sum $\sum \vec{F}$ of all the forces acting on the body, the vector $m\vec{a}$ is *not* a force. Acceleration is a *result* of a nonzero net force; it is not a force itself. It's "common sense" to think that there is a "force of acceleration" that pushes you back into your seat

4.17 The design of high-performance motorcycles depends fundamentally on Newton's second law. To maximize the forward acceleration, the designer makes the motorcycle as light as possible (that is, minimizes the mass) and uses the most powerful engine possible (thus maximizing the forward force).

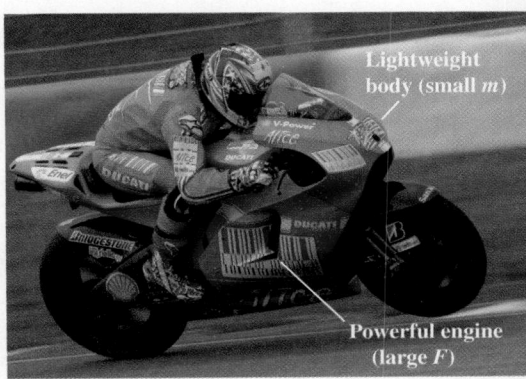

Lightweight body (small m)

Powerful engine (large F)

Application Blame Newton's Second Law
This car stopped because of Newton's second law: The tree exerted an external force on the car, giving the car an acceleration that changed its velocity to zero.

Mastering**PHYSICS**

ActivPhysics 2.1.3: Tension Change
ActivPhysics 2.1.4: Sliding on an Incline

when your car accelerates forward from rest. But *there is no such force;* instead, your inertia causes you to tend to stay at rest relative to the earth, and the car accelerates around you (see Fig. 4.11a). The "common sense" confusion arises from trying to apply Newton's second law where it isn't valid, in the noninertial reference frame of an accelerating car. We will always examine motion relative to *inertial* frames of reference only.

In learning how to use Newton's second law, we will begin in this chapter with examples of straight-line motion. Then in Chapter 5 we will consider more general cases and develop more detailed problem-solving strategies.

Example 4.4 | Determining acceleration from force

A worker applies a constant horizontal force with magnitude 20 N to a box with mass 40 kg resting on a level floor with negligible friction. What is the acceleration of the box?

SOLUTION

IDENTIFY and SET UP: This problem involves force and acceleration, so we'll use Newton's second law. In *any* problem involving forces, the first steps are to choose a coordinate system and to identify all of the forces acting on the body in question. It's usually convenient to take one axis either along or opposite the direction of the body's acceleration, which in this case is horizontal. Hence we take the +x-axis to be in the direction of the applied horizontal force (that is, the direction in which the box accelerates) and the +y-axis to be upward (Fig. 4.18). In most force problems that you'll encounter (including this one), the force vectors all lie in a plane, so the z-axis isn't used.

The forces acting on the box are (i) the horizontal force $\vec{F}$ exerted by the worker, of magnitude 20 N; (ii) the weight $\vec{w}$ of the box—that is, the downward gravitational force exerted by the earth; and (iii) the upward supporting force $\vec{n}$ exerted by the floor. As in Section 4.2, we call $\vec{n}$ a *normal* force because it is normal (perpendicular) to the surface of contact. (We use an italic letter n to avoid confusion with the abbreviation N for newton.) Friction is negligible, so no friction force is present.

The box doesn't move vertically, so the y-acceleration is zero: $a_y = 0$. Our target variable is the x-acceleration, a_x. We'll find it using Newton's second law in component form, Eqs. (4.8).

EXECUTE: From Fig. 4.18 only the 20-N force exerted by the worker has a nonzero x-component. Hence the first of Eqs. (4.8) tells us that

$$\sum F_x = F = 20 \text{ N} = ma_x$$

4.18 Our sketch for this problem. The tiles under the box are freshly waxed, so we assume that friction is negligible.

The box has no vertical acceleration, so the vertical components of the net force sum to zero. Nevertheless, for completeness, we show the vertical forces acting on the box.

The x-component of acceleration is therefore

$$a_x = \frac{\sum F_x}{m} = \frac{20 \text{ N}}{40 \text{ kg}} = \frac{20 \text{ kg} \cdot \text{m/s}^2}{40 \text{ kg}} = 0.50 \text{ m/s}^2$$

EVALUATE: The acceleration is in the +x-direction, the same direction as the net force. The net force is constant, so the acceleration is also constant. If we know the initial position and velocity of the box, we can find its position and velocity at any later time from the constant-acceleration equations of Chapter 2.

To determine a_x, we didn't need the y-component of Newton's second law from Eqs. (4.8), $\sum F_y = ma_y$. Can you use this equation to show that the magnitude n of the normal force in this situation is equal to the weight of the box?

Example 4.5 | Determining force from acceleration

A waitress shoves a ketchup bottle with mass 0.45 kg to her right along a smooth, level lunch counter. The bottle leaves her hand moving at 2.8 m/s, then slows down as it slides because of a constant horizontal friction force exerted on it by the countertop. It slides for 1.0 m before coming to rest. What are the magnitude and direction of the friction force acting on the bottle?

SOLUTION

IDENTIFY and SET UP: This problem involves forces and acceleration (the slowing of the ketchup bottle), so we'll use Newton's second law to solve it. As in Example 4.4, we choose a coordinate system and identify the forces acting on the bottle (Fig. 4.19). We choose the +x-axis to be in the direction that the bottle slides, and

4.19 Our sketch for this problem.

We draw one diagram for the bottle's motion and one showing the forces on the bottle.

take the origin to be where the bottle leaves the waitress's hand. The friction force $\vec{f}$ slows the bottle down, so its direction must be opposite the direction of the bottle's velocity (see Fig. 4.13c).

Our target variable is the magnitude f of the friction force. We'll find it using the x-component of Newton's second law from Eqs. (4.8). We aren't told the x-component of the bottle's acceleration, a_x, but we know that it's constant because the friction force that causes the acceleration is constant. Hence we can calculate a_x using a constant-acceleration formula from Section 2.4. We know the bottle's initial and final x-coordinates ($x_0 = 0$ and $x = 1.0$ m) and its initial and final x-velocity ($v_{0x} = 2.8$ m/s and $v_x = 0$), so the easiest equation to use is Eq. (2.13), $v_x^2 = v_{0x}^2 + 2a_x(x - x_0)$.

EXECUTE: We solve Eq. (2.13) for a_x:

$$a_x = \frac{v_x^2 - v_{0x}^2}{2(x - x_0)} = \frac{(0 \text{ m/s})^2 - (2.8 \text{ m/s})^2}{2(1.0 \text{ m} - 0 \text{ m})} = -3.9 \text{ m/s}^2$$

The negative sign means that the bottle's acceleration is toward the *left* in Fig. 4.19, opposite to its velocity; this is as it must be, because the bottle is slowing down. The net force in the x-direction is the x-component $-f$ of the friction force, so

$$\sum F_x = -f = ma_x = (0.45 \text{ kg})(-3.9 \text{ m/s}^2)$$
$$= -1.8 \text{ kg} \cdot \text{m/s}^2 = -1.8 \text{ N}$$

The negative sign shows that the net force on the bottle is toward the left. The *magnitude* of the friction force is $f = 1.8$ N.

EVALUATE: As a check on the result, try repeating the calculation with the $+x$-axis to the *left* in Fig. 4.19. You'll find that $\sum F_x$ is equal to $+f = +1.8$ N (because the friction force is now in the $+x$-direction), and again you'll find $f = 1.8$ N. The answers for the *magnitudes* of forces don't depend on the choice of coordinate axes!

Some Notes on Units

A few words about units are in order. In the cgs metric system (not used in this book), the unit of mass is the gram, equal to 10^{-3} kg, and the unit of distance is the centimeter, equal to 10^{-2} m. The cgs unit of force is called the *dyne:*

$$1 \text{ dyne} = 1 \text{ g} \cdot \text{cm/s}^2 = 10^{-5} \text{ N}$$

In the British system, the unit of force is the *pound* (or pound-force) and the unit of mass is the *slug* (Fig. 4.20). The unit of acceleration is 1 foot per second squared, so

$$1 \text{ pound} = 1 \text{ slug} \cdot \text{ft/s}^2$$

The official definition of the pound is

$$1 \text{ pound} = 4.448221615260 \text{ newtons}$$

It is handy to remember that a pound is about 4.4 N and a newton is about 0.22 pound. Another useful fact: A body with a mass of 1 kg has a weight of about 2.2 lb at the earth's surface.

Table 4.2 lists the units of force, mass, and acceleration in the three systems.

Test Your Understanding of Section 4.3 Rank the following situations in order of the magnitude of the object's acceleration, from lowest to highest. Are there any cases that have the same magnitude of acceleration? (i) a 2.0-kg object acted on by a 2.0-N net force; (ii) a 2.0-kg object acted on by an 8.0-N net force; (iii) an 8.0-kg object acted on by a 2.0-N net force; (iv) an 8.0-kg object acted on by a 8.0-N net force.

4.20 Despite its name, the English unit of mass has nothing to do with the type of slug shown here. A common garden slug has a mass of about 15 grams, or about 10^{-3} slug.

Table 4.2 Units of Force, Mass, and Acceleration

System of Units	Force	Mass	Acceleration
SI	newton (N)	kilogram (kg)	m/s^2
cgs	dyne (dyn)	gram (g)	cm/s^2
British	pound (lb)	slug	ft/s^2

4.4 Mass and Weight

One of the most familiar forces is the *weight* of a body, which is the gravitational force that the earth exerts on the body. (If you are on another planet, your weight is the gravitational force that planet exerts on you.) Unfortunately, the terms *mass* and *weight* are often misused and interchanged in everyday conversation. It is absolutely essential for you to understand clearly the distinctions between these two physical quantities.

Mass characterizes the *inertial* properties of a body. Mass is what keeps the china on the table when you yank the tablecloth out from under it. The greater the mass, the greater the force needed to cause a given acceleration; this is reflected in Newton's second law, $\Sigma \vec{F} = m\vec{a}$.

Weight, on the other hand, is a *force* exerted on a body by the pull of the earth. Mass and weight are related: Bodies having large mass also have large weight. A large stone is hard to throw because of its large *mass,* and hard to lift off the ground because of its large *weight.*

To understand the relationship between mass and weight, note that a freely falling body has an acceleration of magnitude g. Newton's second law tells us that a force must act to produce this acceleration. If a 1-kg body falls with an acceleration of 9.8 m/s² the required force has magnitude

$$F = ma = (1 \text{ kg})(9.8 \text{ m/s}^2) = 9.8 \text{ kg} \cdot \text{m/s}^2 = 9.8 \text{ N}$$

The force that makes the body accelerate downward is its weight. Any body near the surface of the earth that has a mass of 1 kg *must* have a weight of 9.8 N to give it the acceleration we observe when it is in free fall. More generally, a body with mass m must have weight with magnitude w given by

$$w = mg \qquad \text{(magnitude of the weight of a body of mass } m) \qquad (4.9)$$

4.21 The relationship of mass and weight.

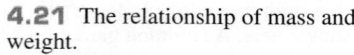
Falling body, mass m Hanging body, mass m

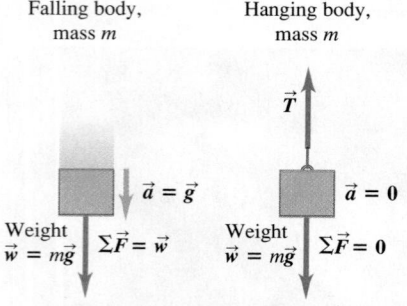

• The relationship of mass to weight: $\vec{w} = m\vec{g}$.
• This relationship is the same whether a body is falling or stationary.

Hence the magnitude w of a body's weight is directly proportional to its mass m. The weight of a body is a force, a vector quantity, and we can write Eq. (4.9) as a vector equation (Fig. 4.21):

$$\vec{w} = m\vec{g} \qquad (4.10)$$

Remember that g is the *magnitude* of $\vec{g}$, the acceleration due to gravity, so g is always a positive number, by definition. Thus w, given by Eq. (4.9), is the *magnitude* of the weight and is also always positive.

CAUTION **A body's weight acts at all times** It is important to understand that the weight of a body acts on the body *all the time,* whether it is in free fall or not. If we suspend an object from a rope, it is in equilibrium, and its acceleration is zero. But its weight, given by Eq. (4.10), is still pulling down on it (Fig. 4.21). In this case the rope pulls up on the object, applying an upward force. The *vector sum* of the forces is zero, but the weight still acts.

Conceptual Example 4.6 | **Net force and acceleration in free fall**

In Example 2.6, a one-euro coin was dropped from rest from the Leaning Tower of Pisa. If the coin falls freely, so that the effects of the air are negligible, how does the net force on the coin vary as it falls?

SOLUTION

In free fall, the acceleration $\vec{a}$ of the coin is constant and equal to $\vec{g}$. Hence by Newton's second law the net force $\Sigma \vec{F} = m\vec{a}$ is also constant and equal to $m\vec{g}$, which is the coin's weight $\vec{w}$ (Fig. 4.22). The coin's velocity changes as it falls, but the net force acting on it is constant. (If this surprises you, reread Conceptual Example 4.3.)

The net force on a freely falling coin is constant even if you initially toss it upward. The force that your hand exerts on the coin to toss it is a contact force, and it disappears the instant the coin

leaves your hand. From then on, the only force acting on the coin is its weight $\vec{w}$.

4.22 The acceleration of a freely falling object is constant, and so is the net force acting on the object.

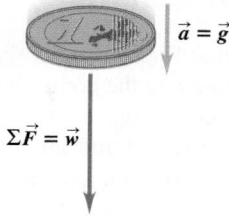
$\vec{a} = \vec{g}$

$\Sigma \vec{F} = \vec{w}$

Variation of g with Location

We will use $g = 9.80 \text{ m/s}^2$ for problems set on the earth (or, if the other data in the problem are given to only two significant figures, $g = 9.8 \text{ m/s}^2$). In fact, the value of g varies somewhat from point to point on the earth's surface—from about 9.78 to 9.82 m/s^2—because the earth is not perfectly spherical and because of effects due to its rotation and orbital motion. At a point where $g = 9.80 \text{ m/s}^2$, the weight of a standard kilogram is $w = 9.80 \text{ N}$. At a different point, where $g = 9.78 \text{ m/s}^2$, the weight is $w = 9.78 \text{ N}$ but the mass is still 1 kg. The weight of a body varies from one location to another; the mass does not.

If we take a standard kilogram to the surface of the moon, where the acceleration of free fall (equal to the value of g at the moon's surface) is 1.62 m/s^2, its weight is 1.62 N, but its mass is still 1 kg (Fig. 4.23). An 80.0-kg astronaut has a weight on earth of $(80.0 \text{ kg})(9.80 \text{ m/s}^2) = 784 \text{ N}$, but on the moon the astronaut's weight would be only $(80.0 \text{ kg})(1.62 \text{ m/s}^2) = 130 \text{ N}$. In Chapter 13 we'll see how to calculate the value of g at the surface of the moon or on other worlds.

Measuring Mass and Weight

In Section 4.3 we described a way to compare masses by comparing their accelerations when they are subjected to the same net force. Usually, however, the easiest way to measure the mass of a body is to measure its weight, often by comparing with a standard. Equation (4.9) says that two bodies that have the same weight at a particular location also have the same mass. We can compare weights very precisely; the familiar equal-arm balance (Fig. 4.24) can determine with great precision (up to 1 part in 10^6) when the weights of two bodies are equal and hence when their masses are equal.

The concept of mass plays two rather different roles in mechanics. The weight of a body (the gravitational force acting on it) is proportional to its mass; we call the property related to gravitational interactions *gravitational mass*. On the other hand, we call the inertial property that appears in Newton's second law the *inertial mass*. If these two quantities were different, the acceleration due to gravity might well be different for different bodies. However, extraordinarily precise experiments have established that in fact the two *are* the same to a precision of better than one part in 10^{12}.

> **CAUTION** **Don't confuse mass and weight** The SI units for mass and weight are often misused in everyday life. Incorrect expressions such as "This box weighs 6 kg" are nearly universal. What is meant is that the *mass* of the box, probably determined indirectly by *weighing,* is 6 kg. Be careful to avoid this sloppy usage in your own work! In SI units, weight (a force) is measured in newtons, while mass is measured in kilograms. ▮

4.23 The weight of a 1-kilogram mass (a) on earth and (b) on the moon.

(a)

$m = 1.00 \text{ kg}$

On earth:
$g = 9.80 \text{ m/s}^2$
$w = mg = 9.80 \text{ N}$

(b)

On the moon:
$g = 1.62 \text{ m/s}^2$ $m = 1.00 \text{ kg}$
$w = mg = 1.62 \text{ N}$

4.24 An equal-arm balance determines the mass of a body (such as an apple) by comparing its weight to a known weight.

w_{unknown} w_{known}

Example 4.7   Mass and weight

A $2.49 \times 10^4 \text{ N}$ Rolls-Royce Phantom traveling in the $+x$-direction makes an emergency stop; the x-component of the net force acting on it is $-1.83 \times 10^4 \text{ N}$. What is its acceleration?

SOLUTION

IDENTIFY and SET UP: Our target variable is the x-component of the car's acceleration, a_x. We use the x-component portion of Newton's second law, Eqs. (4.8), to relate force and acceleration. To do this, we need to know the car's mass. The newton is a unit for

force, however, so $2.49 \times 10^4 \text{ N}$ is the car's *weight*, not its mass. Hence we'll first use Eq. (4.9) to determine the car's mass from its weight. The car has a positive x-velocity and is slowing down, so its x-acceleration will be negative.

EXECUTE: The mass of the car is

$$m = \frac{w}{g} = \frac{2.49 \times 10^4 \text{ N}}{9.80 \text{ m/s}^2} = \frac{2.49 \times 10^4 \text{ kg} \cdot \text{m/s}^2}{9.80 \text{ m/s}^2}$$
$$= 2540 \text{ kg}$$

Continued

Then $\sum F_x = ma_x$ gives

$$a_x = \frac{\sum F_x}{m} = \frac{-1.83 \times 10^4 \text{ N}}{2540 \text{ kg}} = \frac{-1.83 \times 10^4 \text{ kg} \cdot \text{m/s}^2}{2540 \text{ kg}}$$

$$= -7.20 \text{ m/s}^2$$

EVALUATE: The negative sign means that the acceleration vector points in the negative x-direction, as we expected. The magnitude of this acceleration is pretty high; passengers in this car will experience a lot of rearward force from their shoulder belts.

The acceleration is also equal to $-0.735g$. The number -0.735 is also the ratio of -1.83×10^4 N (the x-component of the net force) to 2.49×10^4 N (the weight). In fact, the acceleration of a body, expressed as a multiple of g, is *always* equal to the ratio of the net force on the body to its weight. Can you see why?

Test Your Understanding of Section 4.4 Suppose an astronaut landed on a planet where $g = 19.6$ m/s^2. Compared to earth, would it be easier, harder, or just as easy for her to walk around? Would it be easier, harder, or just as easy for her to catch a ball that is moving horizontally at 12 m/s? (Assume that the astronaut's spacesuit is a lightweight model that doesn't impede her movements in any way.)

4.5 Newton's Third Law

A force acting on a body is always the result of its interaction with another body, so forces always come in pairs. You can't pull on a doorknob without the doorknob pulling back on you. When you kick a football, the forward force that your foot exerts on the ball launches it into its trajectory, but you also feel the force the ball exerts back on your foot. If you kick a boulder, the pain you feel is due to the force that the boulder exerts on your foot.

In each of these cases, the force that you exert on the other body is in the opposite direction to the force that body exerts on you. Experiments show that whenever two bodies interact, the two forces that they exert on each other are always *equal in magnitude* and *opposite in direction*. This fact is called *Newton's third law of motion*:

> **Newton's third law of motion:** If body A exerts a force on body B (an "action"), then body B exerts a force on body A (a "reaction"). These two forces have the same magnitude but are opposite in direction. These two forces act on *different* bodies.

4.25 If body A exerts a force $\vec{F}_{A \text{ on } B}$ on body B, then body B exerts a force $\vec{F}_{B \text{ on } A}$ on body A that is equal in magnitude and opposite in direction: $\vec{F}_{A \text{ on } B} = -\vec{F}_{B \text{ on } A}$.

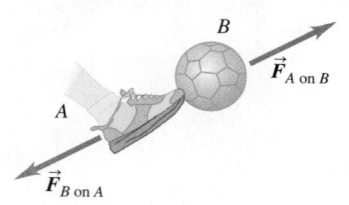

For example, in Fig. 4.25 $\vec{F}_{A \text{ on } B}$ is the force applied *by* body A (first subscript) *on* body B (second subscript), and $\vec{F}_{B \text{ on } A}$ is the force applied *by* body B (first subscript) *on* body A (second subscript). The mathematical statement of Newton's third law is

$$\vec{F}_{A \text{ on } B} = -\vec{F}_{B \text{ on } A} \qquad \text{(Newton's third law of motion)} \qquad [4.11]$$

It doesn't matter whether one body is inanimate (like the soccer ball in Fig. 4.25) and the other is not (like the kicker): They necessarily exert forces on each other that obey Eq. (4.11).

In the statement of Newton's third law, "action" and "reaction" are the two opposite forces (in Fig. 4.25, $\vec{F}_{A \text{ on } B}$ and $\vec{F}_{B \text{ on } A}$); we sometimes refer to them as an **action–reaction pair.** This is *not* meant to imply any cause-and-effect relationship; we can consider either force as the "action" and the other as the "reaction." We often say simply that the forces are "equal and opposite," meaning that they have equal magnitudes and opposite directions.

CAUTION **The two forces in an action–reaction pair act on different bodies** We stress that the two forces described in Newton's third law act on *different* bodies. This is important in problems involving Newton's first or second law, which involve the forces that act on a single body. For instance, the net force on the soccer ball in Fig. 4.25 is the vector sum of the weight of the ball and the force $\vec{F}_{A \text{ on } B}$ exerted by the kicker. You wouldn't include the force $\vec{F}_{B \text{ on } A}$ because this force acts on the kicker, not on the ball.

In Fig. 4.25 the action and reaction forces are *contact* forces that are present only when the two bodies are touching. But Newton's third law also applies to *long-range* forces that do not require physical contact, such as the force of gravitational attraction. A table-tennis ball exerts an upward gravitational force on the earth that's equal in magnitude to the downward gravitational force the earth exerts on the ball. When you drop the ball, both the ball and the earth accelerate toward each other. The net force on each body has the same magnitude, but the earth's acceleration is microscopically small because its mass is so great. Nevertheless, it does move!

Conceptual Example 4.8   **Which force is greater?**

After your sports car breaks down, you start to push it to the nearest repair shop. While the car is starting to move, how does the force you exert on the car compare to the force the car exerts on you? How do these forces compare when you are pushing the car along at a constant speed?

SOLUTION

Newton's third law says that in *both* cases, the force you exert on the car is equal in magnitude and opposite in direction to the force the car exerts on you. It's true that you have to push harder to get the car going than to keep it going. But no matter how hard you push on the car, the car pushes just as hard back on you. Newton's third law gives the same result whether the two bodies are at rest, moving with constant velocity, or accelerating.

You may wonder how the car "knows" to push back on you with the same magnitude of force that you exert on it. It may help to visualize the forces you and the car exert on each other as interactions between the atoms at the surface of your hand and the atoms at the surface of the car. These interactions are analogous to miniature springs between adjacent atoms, and a compressed spring exerts equally strong forces on both of its ends.

Fundamentally, though, the reason we know that objects of different masses exert equally strong forces on each other is that experiment tells us so. Physics isn't merely a collection of rules and equations; rather, it's a systematic description of the natural world based on experiment and observation.

Conceptual Example 4.9   **Applying Newton's third law: Objects at rest**

An apple sits at rest on a table, in equilibrium. What forces act on the apple? What is the reaction force to each of the forces acting on the apple? What are the action–reaction pairs?

SOLUTION

Figure 4.26a shows the forces acting on the apple. $\vec{F}_{\text{earth on apple}}$ is the weight of the apple—that is, the downward gravitational force exerted *by* the earth *on* the apple. Similarly, $\vec{F}_{\text{table on apple}}$ is the upward force exerted *by* the table *on* the apple.

Figure 4.26b shows one of the action–reaction pairs involving the apple. As the earth pulls down on the apple, with force $\vec{F}_{\text{earth on apple}}$, the apple exerts an equally strong upward pull on the earth $\vec{F}_{\text{apple on earth}}$. By Newton's third law (Eq. 4.11) we have

$$\vec{F}_{\text{apple on earth}} = -\vec{F}_{\text{earth on apple}}$$

Also, as the table pushes up on the apple with force $\vec{F}_{\text{table on apple}}$, the corresponding reaction is the downward force $\vec{F}_{\text{apple on table}}$

4.26 The two forces in an action–reaction pair always act on different bodies.

(a) The forces acting on the apple

(b) The action–reaction pair for the interaction between the apple and the earth

$$\vec{F}_{\text{apple on earth}} = -\vec{F}_{\text{earth on apple}}$$

Action–reaction pairs always represent a mutual interaction of two different objects.

(c) The action–reaction pair for the interaction between the apple and the table

$\vec{F}_{\text{table on apple}}$

$\vec{F}_{\text{apple on table}}$

$$\vec{F}_{\text{apple on table}} = -\vec{F}_{\text{table on apple}}$$

(d) We eliminate one of the forces acting on the apple

$\vec{F}_{\text{table on apple}} = \mathbf{0}$

$\vec{F}_{\text{earth on apple}}$

$\vec{F}_{\text{apple on earth}}$

Table removed

The two forces on the apple CANNOT be an action–reaction pair because they act on the same object. We see that if we eliminate one, the other remains.

Continued

exerted by the apple on the table (Fig. 4.26c). For this action–reaction pair we have

$$\vec{F}_{\text{apple on table}} = -\vec{F}_{\text{table on apple}}$$

The two forces acting on the apple, $\vec{F}_{\text{table on apple}}$ and $\vec{F}_{\text{earth on apple}}$, are *not* an action–reaction pair, despite being equal in magnitude and opposite in direction. They do not represent the mutual interaction of two bodies; they are two different forces act-ing on the *same* body. Figure 4.26d shows another way to see this. If we suddenly yank the table out from under the apple, the forces $\vec{F}_{\text{apple on table}}$ and $\vec{F}_{\text{table on apple}}$ suddenly become zero, but $\vec{F}_{\text{apple on earth}}$ and $\vec{F}_{\text{earth on apple}}$ are unchanged (the gravitational interaction is still present). Because $\vec{F}_{\text{table on apple}}$ is now zero, it can't be the negative of the nonzero $\vec{F}_{\text{earth on apple}}$, and these two forces can't be an action–reaction pair. *The two forces in an action–reaction pair **never** act on the same body.*

Conceptual Example 4.10 Applying Newton's third law: Objects in motion

A stonemason drags a marble block across a floor by pulling on a rope attached to the block (Fig. 4.27a). The block is not necessarily in equilibrium. How are the various forces related? What are the action–reaction pairs?

SOLUTION

We'll use the subscripts B for the block, R for the rope, and M for the mason. In Fig. 4.27b the vector $\vec{F}_{\text{M on R}}$ represents the force exerted by the *mason* on the *rope*. The corresponding reaction is the equal and opposite force $\vec{F}_{\text{R on M}}$ exerted by the *rope* on the *mason*. Similarly, $\vec{F}_{\text{R on B}}$ represents the force exerted by the *rope* on the *block*, and the corresponding reaction is the equal and opposite force $\vec{F}_{\text{B on R}}$ exerted by the *block* on the *rope*. For these two action–reaction pairs, we have

$$\vec{F}_{\text{R on M}} = -\vec{F}_{\text{M on R}} \quad \text{and} \quad \vec{F}_{\text{B on R}} = -\vec{F}_{\text{R on B}}$$

Be sure you understand that the forces $\vec{F}_{\text{M on R}}$ and $\vec{F}_{\text{B on R}}$ (Fig. 4.27c) are *not* an action–reaction pair, because both of these forces act on the *same* body (the rope); an action and its reaction *must* always act on *different* bodies. Furthermore, the forces $\vec{F}_{\text{M on R}}$ and $\vec{F}_{\text{B on R}}$ are not necessarily equal in magnitude. Applying Newton's second law to the rope, we get

$$\sum \vec{F} = \vec{F}_{\text{M on R}} + \vec{F}_{\text{B on R}} = m_{\text{rope}}\vec{a}_{\text{rope}}$$

If the block and rope are accelerating (speeding up or slowing down), the rope is not in equilibrium, and $\vec{F}_{\text{M on R}}$ must have a different magnitude than $\vec{F}_{\text{B on R}}$. By contrast, the action–reaction forces $\vec{F}_{\text{M on R}}$ and $\vec{F}_{\text{R on M}}$ are always equal in magnitude, as are $\vec{F}_{\text{R on B}}$ and $\vec{F}_{\text{B on R}}$. Newton's third law holds whether or not the bodies are accelerating.

In the special case in which the rope is in equilibrium, the forces $\vec{F}_{\text{M on R}}$ and $\vec{F}_{\text{B on R}}$ are equal in magnitude, and they are opposite in direction. But this is an example of Newton's *first* law, not his third; these are two forces on the same body, not forces of two bodies on each other. Another way to look at this is that in equilibrium, $\vec{a}_{\text{rope}} = 0$ in the preceding equation. Then $\vec{F}_{\text{B on R}} = -\vec{F}_{\text{M on R}}$ because of Newton's first or second law.

Another special case is if the rope is accelerating but has negligibly small mass compared to that of the block or the mason. In this case, $m_{\text{rope}} = 0$ in the above equation, so again $\vec{F}_{\text{B on R}} = -\vec{F}_{\text{M on R}}$. Since Newton's third law says that $\vec{F}_{\text{B on R}}$ *always* equals $-\vec{F}_{\text{R on B}}$ (they are an action–reaction pair), in this "massless-rope" case $\vec{F}_{\text{R on B}}$ also equals $\vec{F}_{\text{M on R}}$.

For both the "massless-rope" case and the case of the rope in equilibrium, the force of the rope on the block is equal in magnitude and direction to the force of the mason on the rope (Fig. 4.27d). Hence we can think of the rope as "transmitting" to the block the force the mason exerts on the rope. This is a useful point of view, but remember that it is valid *only* when the rope has negligibly small mass or is in equilibrium.

4.27 Identifying the forces that act when a mason pulls on a rope attached to a block.

(a) The block, the rope, and the mason **(b)** The action–reaction pairs **(c)** *Not* an action–reaction pair **(d)** Not necessarily equal

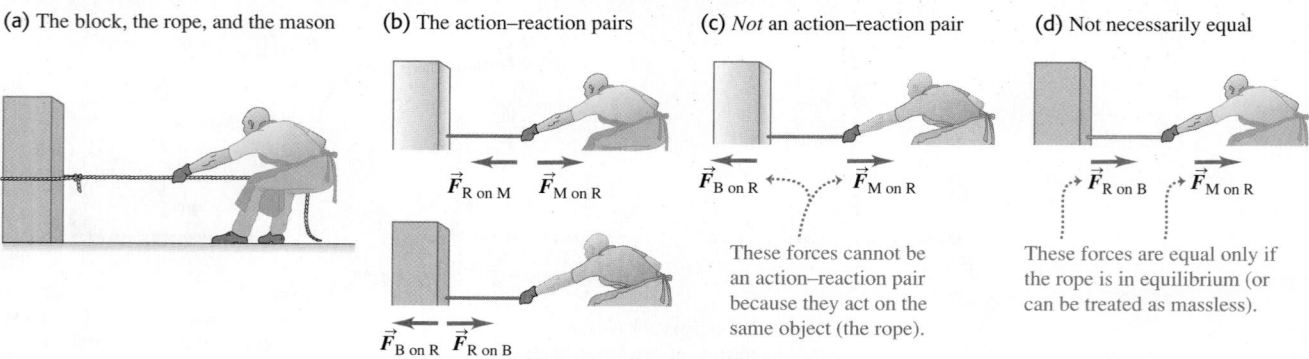

$\vec{F}_{\text{R on M}}$ $\vec{F}_{\text{M on R}}$

$\vec{F}_{\text{B on R}}$ $\vec{F}_{\text{R on B}}$

$\vec{F}_{\text{B on R}}$ $\vec{F}_{\text{M on R}}$

These forces cannot be an action–reaction pair because they act on the same object (the rope).

$\vec{F}_{\text{R on B}}$ $\vec{F}_{\text{M on R}}$

These forces are equal only if the rope is in equilibrium (or can be treated as massless).

Conceptual Example 4.11 **A Newton's third law paradox?**

We saw in Conceptual Example 4.10 that the stonemason pulls as hard on the rope–block combination as that combination pulls back on him. Why, then, does the block move while the stonemason remains stationary?

SOLUTION

To resolve this seeming paradox, keep in mind the difference between Newton's *second* and *third* laws. The only forces involved in Newton's second law are those that act *on* a given body. The vector sum of these forces determines the body's acceleration, if any. By contrast, Newton's third law relates the forces that two *different* bodies exert on *each other*. The third law alone tells you nothing about the motion of either body.

If the rope–block combination is initially at rest, it begins to slide if the stonemason exerts a force $\vec{F}_{\text{M on R}}$ that is *greater* in magnitude than the friction force that the floor exerts on the block (Fig. 4.28). (The block has a smooth underside, which helps to minimize friction.) Then there is a net force to the right on the rope–block combination, and it accelerates to the right. By contrast, the stonemason *doesn't* move because the net force acting on him is *zero*. His shoes have nonskid soles that don't slip on the floor, so the friction force that the floor exerts on him is strong enough to balance the pull of the rope on him, $\vec{F}_{\text{R on M}}$. (Both the block and the stonemason also experience a downward force of gravity and an upward normal force exerted by the floor. These forces balance each other and cancel out, so we haven't included them in Fig. 4.28.)

Once the block is moving at the desired speed, the stonemason doesn't need to pull as hard; he must exert only enough force to balance the friction force on the block. Then the net force on the

4.28 The horizontal forces acting on the block–rope combination (left) and the mason (right). (The vertical forces are not shown.)

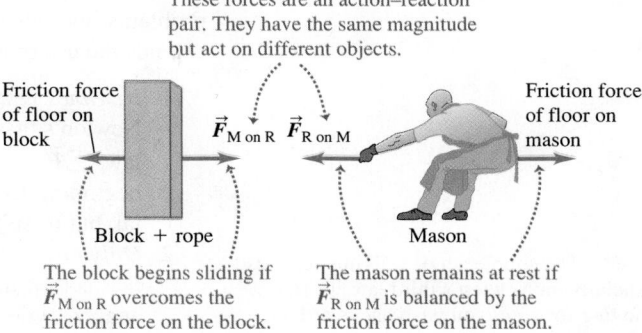

These forces are an action–reaction pair. They have the same magnitude but act on different objects.

Friction force of floor on block

$\vec{F}_{\text{M on R}}$ $\vec{F}_{\text{R on M}}$

Friction force of floor on mason

Block + rope

Mason

The block begins sliding if $\vec{F}_{\text{M on R}}$ overcomes the friction force on the block.

The mason remains at rest if $\vec{F}_{\text{R on M}}$ is balanced by the friction force on the mason.

moving block is zero, and the block continues to move toward the mason at a constant velocity, in accordance with Newton's first law.

So the block accelerates but the stonemason doesn't because different amounts of friction act on them. If the floor were freshly waxed, so that there was little friction between the floor and the stonemason's shoes, pulling on the rope might start the block sliding to the right *and* start him sliding to the left.

The moral of this example is that when analyzing the motion of a body, you must remember that only the forces acting *on* a body determine its motion. From this perspective, Newton's third law is merely a tool that can help you determine what those forces are.

A body that has pulling forces applied at its ends, such as the rope in Fig. 4.27, is said to be in *tension*. The **tension** at any point is the magnitude of force acting at that point (see Fig. 4.2c). In Fig. 4.27b the tension at the right end of the rope is the magnitude of $\vec{F}_{\text{M on R}}$ (or of $\vec{F}_{\text{R on M}}$), and the tension at the left end equals the magnitude of $\vec{F}_{\text{B on R}}$ (or of $\vec{F}_{\text{R on B}}$). If the rope is in equilibrium and if no forces act except at its ends, the tension is the *same* at both ends and throughout the rope. Thus, if the magnitudes of $\vec{F}_{\text{B on R}}$ and $\vec{F}_{\text{M on R}}$ are 50 N each, the tension in the rope is 50 N (*not* 100 N). The *total* force vector $\vec{F}_{\text{B on R}} + \vec{F}_{\text{M on R}}$ acting on the rope in this case is zero!

We emphasize once more a fundamental truth: The two forces in an action–reaction pair *never* act on the same body. Remembering this simple fact can often help you avoid confusion about action–reaction pairs and Newton's third law.

Test Your Understanding of Section 4.5 You are driving your car on a country road when a mosquito splatters on the windshield. Which has the greater magnitude: the force that the car exerted on the mosquito or the force that the mosquito exerted on the car? Or are the magnitudes the same? If they are different, how can you reconcile this fact with Newton's third law? If they are equal, why is the mosquito splattered while the car is undamaged?

MasteringPHYSICS

ActivPhysics 2.1.1: Force Magnitudes

4.6 Free-Body Diagrams

Newton's three laws of motion contain all the basic principles we need to solve a wide variety of problems in mechanics. These laws are very simple in form, but the process of applying them to specific situations can pose real challenges. In this brief section we'll point out three key ideas and techniques to use in any problems involving Newton's laws. You'll learn others in Chapter 5, which also extends the use of Newton's laws to cover more complex situations.

1. *Newton's first and second laws apply to a specific body.* Whenever you use Newton's first law, $\sum \vec{F} = 0$, for an equilibrium situation or Newton's second law, $\sum \vec{F} = m\vec{a}$, for a nonequilibrium situation, you must decide at the beginning to which body you are referring. This decision may sound trivial, but it isn't.

2. *Only forces acting on the body matter.* The sum $\sum \vec{F}$ includes all the forces that act *on* the body in question. Hence, once you've chosen the body to analyze, you have to identify all the forces acting on it. Don't get confused between the forces acting on a body and the forces exerted by that body on some other body. For example, to analyze a person walking, you would include in $\sum \vec{F}$ the force that the ground exerts on the person as he walks, but *not* the force that the person exerts on the ground (Fig. 4.29). These forces form an action–reaction pair and are related by Newton's third law, but only the member of the pair that acts on the body you're working with goes into $\sum \vec{F}$.

3. *Free-body diagrams are essential to help identify the relevant forces.* A **free-body diagram** is a diagram showing the chosen body by itself, "free" of its surroundings, with vectors drawn to show the magnitudes and directions of all the forces applied to the body by the various other bodies that interact with it. We have already shown some free-body diagrams in Figs. 4.18, 4.19, 4.21, and 4.26a. Be careful to include all the forces acting *on* the body, but be equally careful *not* to include any forces that the body exerts on any other body. In particular, the two forces in an action–reaction pair must *never* appear in the same free-body diagram because they never act on the same body. Furthermore, forces that a body exerts on itself are never included, since these can't affect the body's motion.

4.29 The simple act of walking depends crucially on Newton's third law. To start moving forward, you push backward on the ground with your foot. As a reaction, the ground pushes forward on your foot (and hence on your body as a whole) with a force of the same magnitude. This *external* force provided by the ground is what accelerates your body forward.

CAUTION **Forces in free-body diagrams** When you have a complete free-body diagram, you *must* be able to answer this question for each force: What other body is applying this force? If you can't answer that question, you may be dealing with a nonexistent force. Be especially on your guard to avoid nonexistent forces such as "the force of acceleration" or "the $m\vec{a}$ force," discussed in Section 4.3. ▌

When a problem involves more than one body, you have to take the problem apart and draw a separate free-body diagram for each body. For example, Fig. 4.27c shows a separate free-body diagram for the rope in the case in which the rope is considered massless (so that no gravitational force acts on it). Figure 4.28 also shows diagrams for the block and the mason, but these are *not* complete free-body diagrams because they don't show all the forces acting on each body. (We left out the vertical forces—the weight force exerted by the earth and the upward normal force exerted by the floor.)

Figure 4.30 presents three real-life situations and the corresponding complete free-body diagrams. Note that in each situation a person exerts a force on something in his or her surroundings, but the force that shows up in the person's free-body diagram is the surroundings pushing back *on* the person.

Test Your Understanding of Section 4.6   The buoyancy force shown in Fig. 4.30c is one half of an action–reaction pair. What force is the other half of this pair? (i) the weight of the swimmer; (ii) the forward thrust force; (iii) the backward drag force; (iv) the downward force that the swimmer exerts on the water; (v) the backward force that the swimmer exerts on the water by kicking.

4.30 Examples of free-body diagrams. Each free-body diagram shows all of the external forces that act on the object in question.

(a)

The force of the starting block on the runner has a vertical component that counteracts her weight and a large horizontal component that accelerates her.

(b)

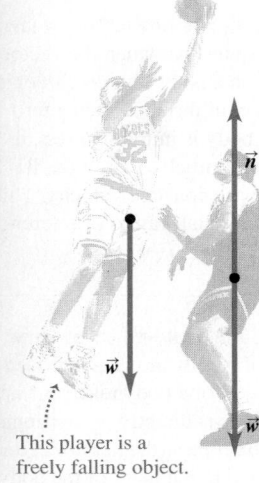

This player is a freely falling object.

To jump up, this player will push down against the floor, increasing the upward reaction force $\vec{n}$ of the floor on him.

(c)

The water exerts a buoyancy force that counters the swimmer's weight.

Kicking causes the water to exert a forward reaction force, or thrust, on the swimmer.

Thrust is countered by drag forces exerted by the water on the moving swimmer.

Force as a vector: Force is a quantitative measure of the interaction between two bodies. It is a vector quantity. When several forces act on a body, the effect on its motion is the same as when a single force, equal to the vector sum (resultant) of the forces, acts on the body. (See Example 4.1.)

$$\vec{R} = \vec{F}_1 + \vec{F}_2 + \vec{F}_3 + \cdots = \sum \vec{F} \quad (4.1)$$

The net force on a body and Newton's first law: Newton's first law states that when the vector sum of all forces acting on a body (the *net force*) is zero, the body is in equilibrium and has zero acceleration. If the body is initially at rest, it remains at rest; if it is initially in motion, it continues to move with constant velocity. This law is valid only in inertial frames of reference. (See Examples 4.2 and 4.3.)

$$\sum \vec{F} = 0 \quad (4.3)$$

Mass, acceleration, and Newton's second law: The inertial properties of a body are characterized by its *mass*. The acceleration of a body under the action of a given set of forces is directly proportional to the vector sum of the forces (the *net force*) and inversely proportional to the mass of the body. This relationship is Newton's second law. Like Newton's first law, this law is valid only in inertial frames of reference. The unit of force is defined in terms of the units of mass and acceleration. In SI units, the unit of force is the newton (N), equal to $1 \text{ kg} \cdot \text{m/s}^2$. (See Examples 4.4 and 4.5.)

$$\sum \vec{F} = m\vec{a} \quad (4.7)$$

$$\sum F_x = ma_x$$

$$\sum F_y = ma_y$$

$$\sum F_z = ma_z \quad (4.8)$$

Weight: The weight $\vec{w}$ of a body is the gravitational force exerted on it by the earth. Weight is a vector quantity. The magnitude of the weight of a body at any specific location is equal to the product of its mass m and the magnitude of the acceleration due to gravity g at that location. While the weight of a body depends on its location, the mass is independent of location. (See Examples 4.6 and 4.7.)

$$w = mg \quad (4.9)$$

Newton's third law and action–reaction pairs: Newton's third law states that when two bodies interact, they exert forces on each other that at each instant are equal in magnitude and opposite in direction. These forces are called action and reaction forces. Each of these two forces acts on only one of the two bodies; they never act on the same body. (See Examples 4.8–4.11.)

$$\vec{F}_{A \text{ on } B} = -\vec{F}_{B \text{ on } A} \quad (4.11)$$

BRIDGING PROBLEM   Links in a Chain

A student suspends a chain consisting of three links, each of mass $m = 0.250$ kg, from a light rope. She pulls upward on the rope, so that the rope applies an upward force of 9.00 N to the chain. (a) Draw a free-body diagram for the entire chain, considered as a body, and one for each of the three links. (b) Use the diagrams of part (a) and Newton's laws to find (i) the acceleration of the chain, (ii) the force exerted by the top link on the middle link, and (iii) the force exerted by the middle link on the bottom link. Treat the rope as massless.

SOLUTION GUIDE

See MasteringPhysics® study area for a Video Tutor solution.

IDENTIFY and SET UP

1. There are four objects of interest in this problem: the chain as a whole and the three individual links. For each of these four objects, make a list of the external forces that act on it. Besides the force of gravity, your list should include only forces exerted by other objects that *touch* the object in question.
2. Some of the forces in your lists form action–reaction pairs (one pair is the force of the top link on the middle link and the force of the middle link on the top link). Identify all such pairs.
3. Use your lists to help you draw a free-body diagram for each of the four objects. Choose the coordinate axes.

4. Use your lists to decide how many unknowns there are in this problem. Which of these are target variables?

EXECUTE

5. Write a Newton's second law equation for each of the four objects, and write a Newton's third law equation for each action–reaction pair. You should have at least as many equations as there are unknowns (see step 4). Do you?
6. Solve the equations for the target variables.

EVALUATE

7. You can check your results by substituting them back into the equations from step 6. This is especially important to do if you ended up with more equations in step 5 than you used in step 6.
8. Rank the force of the rope on the chain, the force of the top link on the middle link, and the force of the middle link on the bottom link in order from smallest to largest magnitude. Does this ranking make sense? Explain.
9. Repeat the problem for the case where the upward force that the rope exerts on the chain is only 7.35 N. Is the ranking in step 8 the same? Does this make sense?

Problems

For instructor-assigned homework, go to www.masteringphysics.com

•, ••, •••: Problems of increasing difficulty. **CP**: Cumulative problems incorporating material from earlier chapters. **CALC**: Problems requiring calculus. **BIO**: Biosciences problems.

DISCUSSION QUESTIONS

Q4.1 Can a body be in equilibrium when only one force acts on it? Explain.

Q4.2 A ball thrown straight up has zero velocity at its highest point. Is the ball in equilibrium at this point? Why or why not?

Q4.3 A helium balloon hovers in midair, neither ascending nor descending. Is it in equilibrium? What forces act on it?

Q4.4 When you fly in an airplane at night in smooth air, there is no sensation of motion, even though the plane may be moving at 800 km/h (500 mi/h). Why is this?

Q4.5 If the two ends of a rope in equilibrium are pulled with forces of equal magnitude and opposite direction, why is the total tension in the rope not zero?

Q4.6 You tie a brick to the end of a rope and whirl the brick around you in a horizontal circle. Describe the path of the brick after you suddenly let go of the rope.

Q4.7 When a car stops suddenly, the passengers tend to move forward relative to their seats. Why? When a car makes a sharp turn, the passengers tend to slide to one side of the car. Why?

Q4.8 Some people say that the "force of inertia" (or "force of momentum") throws the passengers forward when a car brakes sharply. What is wrong with this explanation?

Q4.9 A passenger in a moving bus with no windows notices that a ball that has been at rest in the aisle suddenly starts to move toward the rear of the bus. Think of two different possible explanations, and devise a way to decide which is correct.

Q4.10 Suppose you chose the fundamental SI units to be force, length, and time instead of mass, length, and time. What would be the units of mass in terms of those fundamental units?

Q4.11 Some of the ancient Greeks thought that the "natural state" of an object was to be at rest, so objects would seek their natural state by coming to rest if left alone. Explain why this incorrect view can actually seem quite plausible in the everyday world.

Q4.12 Why is the earth only approximately an inertial reference frame?

Q4.13 Does Newton's second law hold true for an observer in a van as it speeds up, slows down, or rounds a corner? Explain.

Q4.14 Some students refer to the quantity $m\vec{a}$ as "the force of acceleration." Is it correct to refer to this quantity as a force? If so, what exerts this force? If not, what is a better description of this quantity?

Q4.15 The acceleration of a falling body is measured in an elevator traveling upward at a constant speed of 9.8 m/s. What result is obtained?

Q4.16 You can play catch with a softball in a bus moving with constant speed on a straight road, just as though the bus were at rest. Is this still possible when the bus is making a turn at constant speed on a level road? Why or why not?

Q4.17 Students sometimes say that the force of gravity on an object is 9.8 m/s². What is wrong with this view?

Q4.18 The head of a hammer begins to come loose from its wooden handle. How should you strike the handle on a concrete sidewalk to reset the head? Why does this work?

Q4.19 Why can it hurt your foot more to kick a big rock than a small pebble? *Must* the big rock hurt more? Explain.

Q4.20 "It's not the fall that hurts you; it's the sudden stop at the bottom." Translate this saying into the language of Newton's laws of motion.

Q4.21 A person can dive into water from a height of 10 m without injury, but a person who jumps off the roof of a 10-m-tall building and lands on a concrete street is likely to be seriously injured. Why is there a difference?

Q4.22 Why are cars designed to crumple up in front and back for safety? Why not for side collisions and rollovers?

Q4.23 When a bullet is fired from a rifle, what is the origin of the force that accelerates the bullet?

Q4.24 When a string barely strong enough lifts a heavy weight, it can lift the weight by a steady pull; but if you jerk the string, it will break. Explain in terms of Newton's laws of motion.

Q4.25 A large crate is suspended from the end of a vertical rope. Is the tension in the rope greater when the crate is at rest or when it is moving upward at constant speed? If the crate is traveling upward, is the tension in the rope greater when the crate is speeding up or when it is slowing down? In each case explain in terms of Newton's laws of motion.

Q4.26 Which feels a greater pull due to the earth's gravity, a 10-kg stone or a 20-kg stone? If you drop them, why does the 20-kg stone not fall with twice the acceleration of the 10-kg stone? Explain your reasoning.

Q4.27 Why is it incorrect to say that 1.0 kg *equals* 2.2 lb?

Q4.28 A horse is hitched to a wagon. Since the wagon pulls back on the horse just as hard as the horse pulls on the wagon, why doesn't the wagon remain in equilibrium, no matter how hard the horse pulls?

Q4.29 True or false? You exert a push P on an object and it pushes back on you with a force F. If the object is moving at constant velocity, then F is equal to P, but if the object is being accelerated, then P must be greater than F.

Q4.30 A large truck and a small compact car have a head-on collision. During the collision, the truck exerts a force $\vec{F}_{\text{T on C}}$ on the car, and the car exerts a force $\vec{F}_{\text{C on T}}$ on the truck. Which force has the larger magnitude, or are they the same? Does your answer depend on how fast each vehicle was moving before the collision? Why or why not?

Q4.31 When a car comes to a stop on a level highway, what force causes it to slow down? When the car increases its speed on the same highway, what force causes it to speed up? Explain.

Q4.32 A small compact car is pushing a large van that has broken down, and they travel along the road with equal velocities and accelerations. While the car is speeding up, is the force it exerts on the van larger than, smaller than, or the same magnitude as the force the van exerts on it? Which object, the car or the van, has the larger net force on it, or are the net forces the same? Explain.

Q4.33 Consider a tug-of-war between two people who pull in opposite directions on the ends of a rope. By Newton's third law, the force that A exerts on B is just as great as the force that B exerts on A. So what determines who wins? (*Hint:* Draw a free-body diagram showing all the forces that act on each person.)

Q4.34 On the moon, $g = 1.62$ m/s². If a 2-kg brick drops on your foot from a height of 2 m, will this hurt more, or less, or the same if it happens on the moon instead of on the earth? Explain. If a 2-kg brick is thrown and hits you when it is moving horizontally at 6 m/s, will this hurt more, less, or the same if it happens on the moon instead of on the earth? Explain. (On the moon, assume that you are inside a pressurized structure, so you are not wearing a spacesuit.)

Q4.35 A manual for student pilots contains the following passage: "When an airplane flies at a steady altitude, neither climbing nor descending, the upward lift force from the wings equals the airplane's weight. When the airplane is climbing at a steady rate, the upward lift is greater than the weight; when the airplane is descending at a steady rate, the upward lift is less than the weight." Are these statements correct? Explain.

Q4.36 If your hands are wet and no towel is handy, you can remove some of the excess water by shaking them. Why does this get rid of the water?

Q4.37 If you are squatting down (such as when you are examining the books on the bottom shelf in a library or bookstore) and suddenly get up, you can temporarily feel light-headed. What do Newton's laws of motion have to say about why this happens?

Q4.38 When a car is hit from behind, the passengers can receive a whiplash. Use Newton's laws of motion to explain what causes this to occur.

Q4.39 In a head-on auto collision, passengers not wearing seat belts can be thrown through the windshield. Use Newton's laws of motion to explain why this happens.

Q4.40 In a head-on collision between a compact 1000-kg car and a large 2500-kg car, which one experiences the greater force? Explain. Which one experiences the greater acceleration? Explain why. Now explain why passengers in the small car are more likely to be injured than those in the large car, even if the bodies of both cars are equally strong.

Q4.41 Suppose you are in a rocket with no windows, traveling in deep space far from any other objects. Without looking outside the rocket or making any contact with the outside world, explain how you could determine if the rocket is (a) moving forward at a constant 80% of the speed of light and (b) accelerating in the forward direction.

EXERCISES

Section 4.1 Force and Interactions

4.1 • Two forces have the same magnitude F. What is the angle between the two vectors if their sum has a magnitude of (a) $2F$? (b) $\sqrt{2}F$? (c) zero? Sketch the three vectors in each case.

4.2 • Workmen are trying to free an SUV stuck in the mud. To extricate the vehicle, they use three horizontal ropes, producing the force vectors shown in Fig. E4.2. (a) Find the x- and y-components of each of the three pulls. (b) Use the components to find the magnitude and direction of the resultant of the three pulls.

Figure **E4.2**

4.3 • **BIO Jaw Injury.** Due to a jaw injury, a patient must wear a strap (Fig. E4.3) that produces a net upward force of 5.00 N on his chin. The tension is the same throughout the strap. To what tension must the strap be adjusted to provide the necessary upward force?

Figure **E4.3**

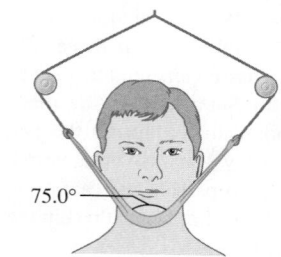

4.4 • A man is dragging a trunk up the loading ramp of a mover's truck. The ramp has a slope angle of 20.0°, and the man pulls upward with a force $\vec{F}$ whose direction makes an angle of 30.0° with the ramp (Fig. E4.4). (a) How large a force $\vec{F}$ is necessary for the component F_x parallel to the ramp to be 60.0 N? (b) How large will the component F_y perpendicular to the ramp then be?

Figure **E4.4**

4.5 •• Two dogs pull horizontally on ropes attached to a post; the angle between the ropes is 60.0°. If dog A exerts a force of 270 N and dog B exerts a force of 300 N, find the magnitude of the resultant force and the angle it makes with dog A's rope.

4.6 • Two forces, $\vec{F}_1$ and $\vec{F}_2$, act at a point. The magnitude of $\vec{F}_1$ is 9.00 N, and its direction is 60.0° above the x-axis in the second quadrant. The magnitude of $\vec{F}_2$ is 6.00 N, and its direction is 53.1° below the x-axis in the third quadrant. (a) What are the x- and y-components of the resultant force? (b) What is the magnitude of the resultant force?

Section 4.3 Newton's Second Law

4.7 •• A 68.5-kg skater moving initially at 2.40 m/s on rough horizontal ice comes to rest uniformly in 3.52 s due to friction from the ice. What force does friction exert on the skater?

4.8 •• You walk into an elevator, step onto a scale, and push the "up" button. You also recall that your normal weight is 625 N. Start answering each of the following questions by drawing a free-body diagram. (a) If the elevator has an acceleration of magnitude 2.50 m/s², what does the scale read? (b) If you start holding a 3.85-kg package by a light vertical string, what will be the tension in this string once the elevator begins accelerating?

4.9 • A box rests on a frozen pond, which serves as a frictionless horizontal surface. If a fisherman applies a horizontal force with magnitude 48.0 N to the box and produces an acceleration of magnitude 3.00 m/s², what is the mass of the box?

4.10 •• A dockworker applies a constant horizontal force of 80.0 N to a block of ice on a smooth horizontal floor. The frictional force is negligible. The block starts from rest and moves 11.0 m in 5.00 s. (a) What is the mass of the block of ice? (b) If the worker stops pushing at the end of 5.00 s, how far does the block move in the next 5.00 s?

4.11 • A hockey puck with mass 0.160 kg is at rest at the origin ($x = 0$) on the horizontal, frictionless surface of the rink. At time $t = 0$ a player applies a force of 0.250 N to the puck, parallel to the x-axis; he continues to apply this force until $t = 2.00$ s. (a) What are the position and speed of the puck at $t = 2.00$ s? (b) If the same force is again applied at $t = 5.00$ s, what are the position and speed of the puck at $t = 7.00$ s?

4.12 • A crate with mass 32.5 kg initially at rest on a warehouse floor is acted on by a net horizontal force of 140 N. (a) What acceleration is produced? (b) How far does the crate travel in 10.0 s? (c) What is its speed at the end of 10.0 s?

4.13 • A 4.50-kg toy cart undergoes an acceleration in a straight line (the x-axis). The graph in Fig. E4.13 shows this acceleration as a function of time. (a) Find the

Figure **E4.13**

maximum net force on this cart. When does this maximum force occur? (b) During what times is the net force on the cart a constant? (c) When is the net force equal to zero?

4.14 • A 2.75-kg cat moves in a straight line (the x-axis). Figure E4.14 shows a graph of the x-component of this cat's velocity as a function of time. (a) Find the maximum net force on this cat. When does this force occur? (b) When is the net force on the cat equal to zero? (c) What is the net force at time 8.5 s?

Figure **E4.14**

4.15 • A small 8.00-kg rocket burns fuel that exerts a time-varying upward force on the rocket as the rocket moves upward from the launch pad. This force obeys the equation $F = A + Bt^2$. Measurements show that at $t = 0$, the force is 100.0 N, and at the end of the first 2.00 s, it is 150.0 N. (a) Find the constants A and B, including their SI units. (b) Find the *net* force on this rocket and its acceleration (i) the instant after the fuel ignites and (ii) 3.00 s after fuel ignition. (c) Suppose you were using this rocket in outer space, far from all gravity. What would its acceleration be 3.00 s after fuel ignition?

4.16 • An electron (mass $= 9.11 \times 10^{-31}$ kg) leaves one end of a TV picture tube with zero initial speed and travels in a straight line to the accelerating grid, which is 1.80 cm away. It reaches the grid with a speed of 3.00×10^6 m/s. If the accelerating force is constant, compute (a) the acceleration; (b) the time to reach the grid; (c) the net force, in newtons. (You can ignore the gravitational force on the electron.)

Section 4.4 Mass and Weight

4.17 • Superman throws a 2400-N boulder at an adversary. What horizontal force must Superman apply to the boulder to give it a horizontal acceleration of 12.0 m/s²?

4.18 • **BIO** (a) An ordinary flea has a mass of 210 μg. How many newtons does it weigh? (b) The mass of a typical froghopper is 12.3 mg. How many newtons does it weigh? (c) A house cat typically weighs 45 N. How many pounds does it weigh, and what is its mass in kilograms?

4.19 • At the surface of Jupiter's moon Io, the acceleration due to gravity is $g = 1.81$ m/s². A watermelon weighs 44.0 N at the surface of the earth. (a) What is the watermelon's mass on the earth's surface? (b) What are its mass and weight on the surface of Io?

4.20 • An astronaut's pack weighs 17.5 N when she is on earth but only 3.24 N when she is at the surface of an asteroid. (a) What is the acceleration due to gravity on this asteroid? (b) What is the mass of the pack on the asteroid?

Section 4.5 Newton's Third Law

4.21 • **BIO** World-class sprinters can accelerate out of the starting blocks with an acceleration that is nearly horizontal and has magnitude 15 m/s². How much horizontal force must a 55-kg sprinter exert on the starting blocks during a start to produce this acceleration? Which body exerts the force that propels the sprinter: the blocks or the sprinter herself?

4.22 A small car (mass 380 kg) is pushing a large truck (mass 900 kg) due east on a level road. The car exerts a horizontal force of 1200 N on the truck. What is the magnitude of the force that the truck exerts on the car?

4.23 Boxes A and B are in contact on a horizontal, frictionless surface, as shown in Fig. E4.23. Box A has mass 20.0 kg and box B has mass 5.0 kg. A horizontal force of 100 N is exerted on box A. What is the magnitude of the force that box A exerts on box B?

Figure **E4.23**

4.24 •• The upward normal force exerted by the floor is 620 N on an elevator passenger who weighs 650 N. What are the reaction forces to these two forces? Is the passenger accelerating? If so, what are the magnitude and direction of the acceleration?

4.25 •• A student with mass 45 kg jumps off a high diving board. Using 6.0×10^{24} kg for the mass of the earth, what is the acceleration of the earth toward her as she accelerates toward the earth with an acceleration of 9.8 m/s^2? Assume that the net force on the earth is the force of gravity she exerts on it.

Section 4.6 Free-Body Diagrams

4.26 • An athlete throws a ball of mass m directly upward, and it feels no appreciable air resistance. Draw a free-body diagram of this ball while it is free of the athlete's hand and (a) moving upward; (b) at its highest point; (c) moving downward. (d) Repeat parts (a), (b), and (c) if the athlete throws the ball at a 60° angle above the horizontal instead of directly upward.

4.27 •• Two crates, A and B, sit at rest side by side on a frictionless horizontal surface. The crates have masses m_A and m_B. A horizontal force $\vec{F}$ is applied to crate A and the two crates move off to the right. (a) Draw clearly labeled free-body diagrams for crate A and for crate B. Indicate which pairs of forces, if any, are third-law action–reaction pairs. (b) If the magnitude of force $\vec{F}$ is less than the total weight of the two crates, will it cause the crates to move? Explain.

4.28 •• A person pulls horizontally on block B in Fig. E4.28, causing both blocks to move together as a unit. While this system is moving, make a carefully labeled free-body diagram of block A if (a) the table is frictionless and (b) there is friction between block B and the table and the pull is equal to the friction force on block B due to the table.

Figure **E4.28**

Horizontal table

4.29 • A ball is hanging from a long string that is tied to the ceiling of a train car traveling eastward on horizontal tracks. An observer inside the train car sees the ball hang motionless. Draw a clearly labeled free-body diagram for the ball if (a) the train has a uniform velocity, and (b) the train is speeding up uniformly. Is the net force on the ball zero in either case? Explain.

4.30 •• CP A .22 rifle bullet, traveling at 350 m/s, strikes a large tree, which it penetrates to a depth of 0.130 m. The mass of the bullet is 1.80 g. Assume a constant retarding force. (a) How much time is required for the bullet to stop? (b) What force, in newtons, does the tree exert on the bullet?

4.31 •• A chair of mass 12.0 kg is sitting on the horizontal floor; the floor is not frictionless. You push on the chair with a force $F = 40.0$ N that is directed at an angle of 37.0° below the horizontal and the chair slides along the floor. (a) Draw a clearly labeled free-body diagram for the chair. (b) Use your diagram and Newton's laws to calculate the normal force that the floor exerts on the chair.

4.32 •• A skier of mass 65.0 kg is pulled up a snow-covered slope at constant speed by a tow rope that is parallel to the ground. The ground slopes upward at a constant angle of 26.0° above the horizontal, and you can ignore friction. (a) Draw a clearly labeled free-body diagram for the skier. (b) Calculate the tension in the tow rope.

PROBLEMS

4.33 CP A 4.80-kg bucket of water is accelerated upward by a cord of negligible mass whose breaking strength is 75.0 N. If the bucket starts from rest, what is the minimum time required to raise the bucket a vertical distance of 12.0 m without breaking the cord?

4.34 ••• A large box containing your new computer sits on the bed of your pickup truck. You are stopped at a red light. The light turns green and you stomp on the gas and the truck accelerates. To your horror, the box starts to slide toward the back of the truck. Draw clearly labeled free-body diagrams for the truck and for the box. Indicate pairs of forces, if any, that are third-law action–reaction pairs. (The bed of the truck is *not* frictionless.)

4.35 • Two horses pull horizontally on ropes attached to a stump. The two forces $\vec{F}_1$ and $\vec{F}_2$ that they apply to the stump are such that the net (resultant) force $\vec{R}$ has a magnitude equal to that of $\vec{F}_1$ and makes an angle of 90° with $\vec{F}_1$. Let $F_1 = 1300$ N and $R = 1300$ N also. Find the magnitude of $\vec{F}_2$ and its direction (relative to $\vec{F}_1$).

4.36 •• CP You have just landed on Planet X. You take out a 100-g ball, release it from rest from a height of 10.0 m, and measure that it takes 2.2 s to reach the ground. You can ignore any force on the ball from the atmosphere of the planet. How much does the 100-g ball weigh on the surface of Planet X?

4.37 •• Two adults and a child want to push a wheeled cart in the direction marked x in Fig. P4.37. The two adults push with horizontal forces $\vec{F}_1$ and $\vec{F}_2$ as shown in the figure. (a) Find the magnitude and direction of the *smallest* force that the child should exert. You can ignore the effects of friction. (b) If the child exerts the minimum force found in part (a), the cart accelerates at 2.0 m/s^2 in the +x-direction. What is the weight of the cart?

Figure **P4.37**

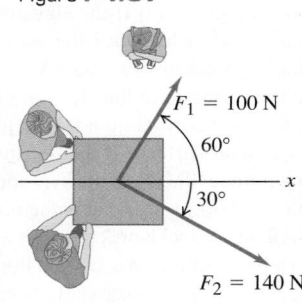

$F_1 = 100$ N
60°
x
30°
$F_2 = 140$ N

4.38 • CP An oil tanker's engines have broken down, and the wind is blowing the tanker straight toward a reef at a constant speed of 1.5 m/s (Fig. P4.38). When the tanker is 500 m from the reef, the wind dies down just as the engineer gets the engines going again. The rudder is stuck, so the only choice is to try to accelerate straight backward away from the reef. The mass of the tanker and cargo is 3.6×10^7 kg, and the engines produce a net horizontal force of 8.0×10^4 N on the tanker. Will the ship hit the reef? If it does, will the oil be safe? The hull can withstand an impact at a speed of 0.2 m/s or less. You can ignore the retarding force of the water on the tanker's hull.

Figure **P4.38**

$F = 8.0 \times 10^4$ N
$v = 1.5$ m/s
3.6×10^7 kg
500 m

4.39 •• **CP BIO** **A Standing Vertical Jump.** Basketball player Darrell Griffith is on record as attaining a standing vertical jump of 1.2 m (4 ft). (This means that he moved upward by 1.2 m after his feet left the floor.) Griffith weighed 890 N (200 lb). (a) What is his speed as he leaves the floor? (b) If the time of the part of the jump before his feet left the floor was 0.300 s, what was his average acceleration (magnitude and direction) while he was pushing against the floor? (c) Draw his free-body diagram (see Section 4.6). In terms of the forces on the diagram, what is the net force on him? Use Newton's laws and the results of part (b) to calculate the average force he applied to the ground.

4.40 ••• **CP** An advertisement claims that a particular automobile can "stop on a dime." What net force would actually be necessary to stop a 850-kg automobile traveling initially at 45.0 km/h in a distance equal to the diameter of a dime, which is 1.8 cm?

4.41 •• **BIO** **Human Biomechanics.** The fastest pitched baseball was measured at 46 m/s. Typically, a baseball has a mass of 145 g. If the pitcher exerted his force (assumed to be horizontal and constant) over a distance of 1.0 m, (a) what force did he produce on the ball during this record-setting pitch? (b) Draw free-body diagrams of the ball during the pitch and just *after* it left the pitcher's hand.

4.42 •• **BIO** **Human Biomechanics.** The fastest served tennis ball, served by "Big Bill" Tilden in 1931, was measured at 73.14 m/s. The mass of a tennis ball is 57 g, and the ball is typically in contact with the tennis racquet for 30.0 ms, with the ball starting from rest. Assuming constant acceleration, (a) what force did Big Bill's tennis racquet exert on the tennis ball if he hit it essentially horizontally? (b) Draw free-body diagrams of the tennis ball during the serve and just after it moved free of the racquet.

4.43 • Two crates, one with mass 4.00 kg and the other with mass 6.00 kg, sit on the frictionless surface of a frozen pond, connected by a light rope (Fig. P4.43). A woman wearing golf shoes (so she can get traction on the ice) pulls horizontally on the 6.00-kg crate with a force F that gives the crate an acceleration of 2.50 m/s². (a) What is the acceleration of the 4.00-kg crate? (b) Draw a free-body diagram for the 4.00-kg crate. Use that diagram and Newton's second law to find the tension T in the rope that connects the two crates. (c) Draw a free-body diagram for the 6.00-kg crate. What is the direction of the net force on the 6.00-kg crate? Which is larger in magnitude, force T or force F? (d) Use part (c) and Newton's second law to calculate the magnitude of the force F.

Figure **P4.43**

4.44 • An astronaut is tethered by a strong cable to a spacecraft. The astronaut and her spacesuit have a total mass of 105 kg, while the mass of the cable is negligible. The mass of the spacecraft is 9.05×10^4 kg. The spacecraft is far from any large astronomical bodies, so we can ignore the gravitational forces on it and the astronaut. We also assume that both the spacecraft and the astronaut are initially at rest in an inertial reference frame. The astronaut then pulls on the cable with a force of 80.0 N. (a) What force does the cable exert on the astronaut? (b) Since $\Sigma \vec{F} = m\vec{a}$, how can a "massless" ($m = 0$) cable exert a force? (c) What is the astronaut's acceleration? (d) What force does the cable exert on the spacecraft? (e) What is the acceleration of the spacecraft?

4.45 • **CALC** To study damage to aircraft that collide with large birds, you design a test gun that will accelerate chicken-sized objects so that their displacement along the gun barrel is given by $x = (9.0 \times 10^3 \text{ m/s}^2)t^2 - (8.0 \times 10^4 \text{ m/s}^3)t^3$. The object leaves the end of the barrel at $t = 0.025$ s. (a) How long must the gun barrel be? (b) What will be the speed of the objects as they leave the end of the barrel? (c) What net force must be exerted on a 1.50-kg object at (i) $t = 0$ and (ii) $t = 0.025$ s?

4.46 •• A spacecraft descends vertically near the surface of Planet X. An upward thrust of 25.0 kN from its engines slows it down at a rate of 1.20 m/s², but it speeds up at a rate of 0.80 m/s² with an upward thrust of 10.0 kN. (a) In each case, what is the direction of the acceleration of the spacecraft? (b) Draw a free-body diagram for the spacecraft. In each case, speeding up or slowing down, what is the direction of the net force on the spacecraft? (c) Apply Newton's second law to each case, slowing down or speeding up, and use this to find the spacecraft's weight near the surface of Planet X.

4.47 •• **CP** A 6.50-kg instrument is hanging by a vertical wire inside a space ship that is blasting off at the surface of the earth. This ship starts from rest and reaches an altitude of 276 m in 15.0 s with constant acceleration. (a) Draw a free-body diagram for the instrument during this time. Indicate which force is greater. (b) Find the force that the wire exerts on the instrument.

4.48 •• Suppose the rocket in Problem 4.47 is coming in for a vertical landing instead of blasting off. The captain adjusts the engine thrust so that the magnitude of the rocket's acceleration is the same as it was during blast-off. Repeat parts (a) and (b).

4.49 •• **BIO** **Insect Dynamics.** The froghopper (*Philaenus spumarius*), the champion leaper of the insect world, has a mass of 12.3 mg and leaves the ground (in the most energetic jumps) at 4.0 m/s from a vertical start. The jump itself lasts a mere 1.0 ms before the insect is clear of the ground. Assuming constant acceleration, (a) draw a free-body diagram of this mighty leaper while the jump is taking place; (b) find the force that the ground exerts on the froghopper during its jump; and (c) express the force in part (b) in terms of the froghopper's weight.

4.50 • A loaded elevator with very worn cables has a total mass of 2200 kg, and the cables can withstand a maximum tension of 28,000 N. (a) Draw the free-body force diagram for the elevator. In terms of the forces on your diagram, what is the net force on the elevator? Apply Newton's second law to the elevator and find the maximum upward acceleration for the elevator if the cables are not to break. (b) What would be the answer to part (a) if the elevator were on the moon, where $g = 1.62$ m/s²?

4.51 •• **CP** **Jumping to the Ground.** A 75.0-kg man steps off a platform 3.10 m above the ground. He keeps his legs straight as he falls, but at the moment his feet touch the ground his knees begin to bend, and, treated as a particle, he moves an additional 0.60 m before coming to rest. (a) What is his speed at the instant his feet touch the ground? (b) Treating him as a particle, what is his acceleration (magnitude and direction) as he slows down, if the acceleration is assumed to be constant? (c) Draw his free-body diagram (see Section 4.6). In terms of the forces on the diagram, what is the net force on him? Use Newton's laws and the results of part (b) to calculate the average force his feet exert on the ground while he slows down. Express this force in newtons and also as a multiple of his weight.

4.52 ••• **CP** A 4.9-N hammer head is stopped from an initial downward velocity of 3.2 m/s in a distance of 0.45 cm by a nail in a pine board. In addition to its weight, there is a 15-N downward force on the hammer head applied by the person using the hammer. Assume that the acceleration of the hammer head is constant while it is in contact with the nail and moving downward. (a) Draw a free-body diagram for the hammer head. Identify the reaction force to each action force in the diagram. (b) Calculate the downward force $\vec{F}$ exerted by the hammer head on the nail while the hammer head is in contact with the nail and moving downward. (c) Suppose the nail is in hardwood and the distance the hammer head travels in coming to rest is only 0.12 cm. The downward forces on the hammer head are the same as in part (b). What then is the force $\vec{F}$ exerted by the hammer head on the nail while the hammer head is in contact with the nail and moving downward?

4.53 •• A uniform cable of weight w hangs vertically downward, supported by an upward force of magnitude w at its top end. What is the tension in the cable (a) at its top end; (b) at its bottom end; (c) at its middle? Your answer to each part must include a free-body diagram. (*Hint:* For each question choose the body to analyze to be a section of the cable or a point along the cable.) (d) Graph the tension in the rope versus the distance from its top end.

4.54 •• The two blocks in Fig. P4.54 are connected by a heavy uniform rope with a mass of 4.00 kg. An upward force of 200 N is applied as shown. (a) Draw three free-body diagrams: one for the 6.00-kg block, one for the 4.00-kg rope, and another one for the 5.00-kg block. For each force, indicate what body exerts that force. (b) What is the acceleration of the system? (c) What is the tension at the top of the heavy rope? (d) What is the tension at the midpoint of the rope?

Figure **P4.54**

F = 200 N

6.00 kg

4.00 kg

5.00 kg

4.55 •• **CP** An athlete whose mass is 90.0 kg is performing weight-lifting exercises. Starting from the rest position, he lifts, with constant acceleration, a barbell that weighs 490 N. He lifts the barbell a distance of 0.60 m in 1.6 s. (a) Draw a clearly labeled free-body force diagram for the barbell and for the athlete. (b) Use the diagrams in part (a) and Newton's laws to find the total force that his feet exert on the ground as he lifts the barbell.

4.56 ••• A hot-air balloon consists of a basket, one passenger, and some cargo. Let the total mass be M. Even though there is an upward lift force on the balloon, the balloon is initially accelerating downward at a rate of $g/3$. (a) Draw a free-body diagram for the descending balloon. (b) Find the upward lift force in terms of the initial total weight Mg. (c) The passenger notices that he is heading straight for a waterfall and decides he needs to go up. What fraction of the total weight must he drop overboard so that the balloon accelerates *upward* at a rate of $g/2$? Assume that the upward lift force remains the same.

Figure **P4.57**

4.57 **CP** Two boxes, A and B, are connected to each end of a light vertical rope, as shown in Fig. P4.57. A constant upward force $F = 80.0$ N is applied to box A. Starting from rest, box B descends 12.0 m in 4.00 s. The tension in the rope connecting the two boxes is 36.0 N. (a) What is the mass of box B? (b) What is the mass of box A?

4.58 ••• **CALC** The position of a 2.75×10^5-N training helicopter under test is given by $\vec{r} = (0.020 \text{ m/s}^3)t^3\hat{\imath} + (2.2 \text{ m/s})t\hat{\jmath} - (0.060 \text{ m/s}^2)t^2\hat{k}$. Find the net force on the helicopter at $t = 5.0$ s.

4.59 • **CALC** An object with mass m moves along the x-axis. Its position as a function of time is given by $x(t) = At - Bt^3$, where A and B are constants. Calculate the net force on the object as a function of time.

4.60 • **CALC** An object with mass m initially at rest is acted on by a force $\vec{F} = k_1\hat{\imath} + k_2t^3\hat{\jmath}$, where k_1 and k_2 are constants. Calculate the velocity $\vec{v}(t)$ of the object as a function of time.

4.61 •• **CP CALC** A mysterious rocket-propelled object of mass 45.0 kg is initially at rest in the middle of the horizontal, frictionless surface of an ice-covered lake. Then a force directed east and with magnitude $F(t) = (16.8 \text{ N/s})t$ is applied. How far does the object travel in the first 5.00 s after the force is applied?

CHALLENGE PROBLEMS

4.62 ••• **CALC** An object of mass m is at rest in equilibrium at the origin. At $t = 0$ a new force $\vec{F}(t)$ is applied that has components

$$F_x(t) = k_1 + k_2y \qquad F_y(t) = k_3t$$

where k_1, k_2, and k_3 are constants. Calculate the position $\vec{r}(t)$ and velocity $\vec{v}(t)$ vectors as functions of time.

Answers

Chapter Opening Question ?

Newton's third law tells us that the car pushes on the crew member just as hard as the crew member pushes on the car, but in the opposite direction. This is true whether the car's engine is on and the car is moving forward partly under its own power, or the engine is off and being propelled by the crew member's push alone. The force magnitudes are different in the two situations, but in either case the push of the car on the crew member is just as strong as the push of the crew member on the car.

Test Your Understanding Questions

4.1 Answer: (iv) The gravitational force on the crate points straight downward. In Fig. 4.6 the x-axis points up and to the right, and the y-axis points up and to the left. Hence the gravitational force has both an x-component and a y-component, and both are negative.

4.2 Answer: (i), (ii), and (iv) In (i), (ii), and (iv) the body is not accelerating, so the net force on the body is zero. [In (iv), the box remains stationary as seen in the inertial reference frame of the ground as the truck accelerates forward, like the skater in Fig. 4.11a.] In (iii), the hawk is moving in a circle; hence it is accelerating and is *not* in equilibrium.

4.3 Answer: (iii), (i) and (iv) (tie), (ii) The acceleration is equal to the net force divided by the mass. Hence the magnitude of the acceleration in each situation is

(i) $a = (2.0 \text{ N})/(2.0 \text{ kg}) = 1.0 \text{ m/s}^2$;

(ii) $a = (8.0 \text{ N})/(2.0 \text{ N}) = 4.0 \text{ m/s}^2$;

(iii) $a = (2.0 \text{ N})/(8.0 \text{ kg}) = 0.25 \text{ m/s}^2$;

(iv) $a = (8.0 \text{ N})/(8.0 \text{ kg}) = 1.0 \text{ m/s}^2$.

4.4 It would take twice the effort for the astronaut to walk around because her weight on the planet would be twice as much as on the earth. But it would be just as easy to catch a ball moving horizontally. The ball's *mass* is the same as on earth, so the horizontal force the astronaut would have to exert to bring it to a stop (i.e., to give it the same acceleration) would also be the same as on earth.

4.5 By Newton's third law, the two forces have equal magnitudes. Because the car has much greater mass than the mosquito, it undergoes only a tiny, imperceptible acceleration in response to the force of the impact. By contrast, the mosquito, with its minuscule mass, undergoes a catastrophically large acceleration.

4.6 Answer: (iv) The buoyancy force is an *upward* force that the *water* exerts on the *swimmer*. By Newton's third law, the other half of the action–reaction pair is a *downward* force that the *swimmer* exerts on the *water* and has the same magnitude as the buoyancy force. It's true that the weight of the swimmer is also downward and has the same magnitude as the buoyancy force; however, the weight acts on the same body (the swimmer) as the buoyancy force, and so these forces aren't an action–reaction pair.

Bridging Problem

Answers: **(a)** *See a Video Tutor solution on MasteringPhysics®*
(b) (i) 2.20 m/s^2; (ii) 6.00 N; (iii) 3.00 N

5 APPLYING NEWTON'S LAWS

? This skydiver is descending under a parachute at a steady rate. In this situation, which has a greater magnitude: the force of gravity or the upward force of the air on the skydiver?

We saw in Chapter 4 that Newton's three laws of motion, the foundation of classical mechanics, can be stated very simply. But *applying* these laws to situations such as an iceboat skating across a frozen lake, a toboggan sliding down a hill, or an airplane making a steep turn requires analytical skills and problem-solving technique. In this chapter we'll help you extend the problem-solving skills you began to develop in Chapter 4.

We'll begin with equilibrium problems, in which we analyze the forces that act on a body at rest or moving with constant velocity. We'll then consider bodies that are not in equilibrium, for which we'll have to deal with the relationship between forces and motion. We'll learn how to describe and analyze the contact force that acts on a body when it rests on or slides over a surface. We'll also analyze the forces that act on a body that moves in a circle with constant speed. We close the chapter with a brief look at the fundamental nature of force and the classes of forces found in our physical universe.

5.1 Using Newton's First Law: Particles in Equilibrium

We learned in Chapter 4 that a body is in *equilibrium* when it is at rest or moving with constant velocity in an inertial frame of reference. A hanging lamp, a kitchen table, an airplane flying straight and level at a constant speed—all are examples of equilibrium situations. In this section we consider only equilibrium of a body that can be modeled as a particle. (In Chapter 11 we'll see how to analyze a body in equilibrium that can't be represented adequately as a particle, such as a bridge that's supported at various points along its span.) The essential physical principle is Newton's first law: When a particle is in equilibrium, the *net* force acting on it—that is, the vector sum of all the forces acting on it—must be zero:

$$\sum \vec{F} = 0 \qquad \text{(particle in equilibrium, vector form)} \qquad (5.1)$$

We most often use this equation in component form:

$$\sum F_x = 0 \qquad \sum F_y = 0 \qquad \text{(particle in equilibrium, component form)} \quad (5.2)$$

This section is about using Newton's first law to solve problems dealing with bodies in equilibrium. Some of these problems may seem complicated, but the important thing to remember is that *all* problems involving particles in equilibrium are done in the same way. Problem-Solving Strategy 5.1 details the steps you need to follow for any and all such problems. Study this strategy carefully, look at how it's applied in the worked-out examples, and try to apply it yourself when you solve assigned problems.

Problem-Solving Strategy 5.1   **Newton's First Law: Equilibrium of a Particle**

IDENTIFY *the relevant concepts:* You must use Newton's *first* law for any problem that involves forces acting on a body in equilibrium—that is, either at rest or moving with constant velocity. For example, a car is in equilibrium when it's parked, but also when it's traveling down a straight road at a steady speed.

If the problem involves more than one body and the bodies interact with each other, you'll also need to use Newton's *third* law. This law allows you to relate the force that one body exerts on a second body to the force that the second body exerts on the first one.

Identify the target variable(s). Common target variables in equilibrium problems include the magnitude and direction (angle) of one of the forces, or the components of a force.

SET UP *the problem* using the following steps:
1. Draw a very simple sketch of the physical situation, showing dimensions and angles. You don't have to be an artist!
2. Draw a free-body diagram for each body that is in equilibrium. For the present, we consider the body as a particle, so you can represent it as a large dot. In your free-body diagram, *do not* include the other bodies that interact with it, such as a surface it may be resting on or a rope pulling on it.
3. Ask yourself what is interacting with the body by touching it or in any other way. On your free-body diagram, draw a force vector for each interaction. Label each force with a symbol for the *magnitude* of the force. If you know the angle at which a force is directed, draw the angle accurately and label it. Include the body's weight, unless the body has negligible mass. If the mass is given, use $w = mg$ to find the weight. A surface in contact with the body exerts a normal force perpendicular to the surface and possibly a friction force parallel to the surface. A rope or chain exerts a pull (never a push) in a direction along its length.
4. *Do not* show in the free-body diagram any forces exerted *by* the body on any other body. The sums in Eqs. (5.1) and (5.2)

include only forces that act *on* the body. For each force on the body, ask yourself "What other body causes that force?" If you can't answer that question, you may be imagining a force that isn't there.
5. Choose a set of coordinate axes and include them in your free-body diagram. (If there is more than one body in the problem, choose axes for each body separately.) Label the positive direction for each axis. If a body rests or slides on a plane surface, it usually simplifies things to choose axes that are parallel and perpendicular to this surface, even when the plane is tilted.

EXECUTE *the solution* as follows:
1. Find the components of each force along each of the body's coordinate axes. Draw a wiggly line through each force vector that has been replaced by its components, so you don't count it twice. The *magnitude* of a force is always positive, but its *components* may be positive or negative.
2. Set the sum of all *x*-components of force equal to zero. In a separate equation, set the sum of all *y*-components equal to zero. (*Never* add *x*- and *y*-components in a single equation.)
3. If there are two or more bodies, repeat all of the above steps for each body. If the bodies interact with each other, use Newton's third law to relate the forces they exert on each other.
4. Make sure that you have as many independent equations as the number of unknown quantities. Then solve these equations to obtain the target variables.

EVALUATE *your answer:* Look at your results and ask whether they make sense. When the result is a symbolic expression or formula, check to see that your formula works for any special cases (particular values or extreme cases for the various quantities) for which you can guess what the results ought to be.

Example 5.1   **One-dimensional equilibrium: Tension in a massless rope**

A gymnast with mass $m_G = 50.0$ kg suspends herself from the lower end of a hanging rope of negligible mass. The upper end of the rope is attached to the gymnasium ceiling. (a) What is the gymnast's weight? (b) What force (magnitude and direction) does the rope exert on her? (c) What is the tension at the top of the rope?

SOLUTION

IDENTIFY and SET UP: The gymnast and the rope are in equilibrium, so we can apply Newton's first law to both bodies. We'll use Newton's third law to relate the forces that they exert on each other. The target variables are the gymnast's weight, w_G; the force that the bottom of the rope exerts on the gymnast (call it $T_{R\,on\,G}$); and the force that the ceiling exerts on the top of the rope (call it $T_{C\,on\,R}$). Figure 5.1 shows our sketch of the situation and free-body diagrams for the gymnast and for the rope. We take the positive y-axis to be upward in each diagram. Each force acts in the vertical direction and so has only a y-component.

The forces $T_{R\,on\,G}$ (the upward force of the rope on the gymnast, Fig. 5.1b) and $T_{G\,on\,R}$ (the downward force of the gymnast on the rope, Fig. 5.1c) form an action–reaction pair. By Newton's third law, they must have the same magnitude.

5.1 Our sketches for this problem.

(a) The situation

Ceiling

Rope

Gymnast

(b) Free-body diagram for gymnast

Y

$T_{R\,on\,G}$

Action–reaction pair

$w_G = m_G g$

(c) Free-body diagram for rope

Y

$T_{C\,on\,R}$

$T_{G\,on\,R}$

Note that Fig. 5.1c includes only the forces that act *on* the rope. In particular, it doesn't include the force that the *rope* exerts on the *ceiling* (compare the discussion of the apple in Conceptual Example 4.9 in Section 4.5). Similarly, the force that the rope exerts on the ceiling doesn't appear in Fig. 5.1c.

EXECUTE: (a) The magnitude of the gymnast's weight is the product of her mass and the acceleration due to gravity, g:

$$w_G = m_G g = (50.0\text{ kg})(9.80\text{ m/s}^2) = 490\text{ N}$$

(b) The gravitational force on the gymnast (her weight) points in the negative y-direction, so its y-component is $-w_G$. The upward force of the rope on the gymnast has unknown magnitude $T_{R\,on\,G}$ and positive y-component $+T_{R\,on\,G}$. We find this using Newton's first law:

Gymnast: $\sum F_y = T_{R\,on\,G} + (-w_G) = 0$ so
$$T_{R\,on\,G} = w_G = 490\text{ N}$$

The rope pulls *up* on the gymnast with a force $T_{R\,on\,G}$ of magnitude 490 N. (By Newton's third law, the gymnast pulls *down* on the rope with a force of the same magnitude, $T_{G\,on\,R} = 490$ N.)

(c) We have assumed that the rope is weightless, so the only forces on it are those exerted by the ceiling (upward force of unknown magnitude $T_{C\,on\,R}$) and by the gymnast (downward force of magnitude $T_{G\,on\,R} = 490$ N). From Newton's first law, the *net* vertical force on the rope in equilibrium must be zero:

Rope: $\sum F_y = T_{C\,on\,R} + (-T_{G\,on\,R}) = 0$ so
$$T_{C\,on\,R} = T_{G\,on\,R} = 490\text{ N}$$

EVALUATE: The *tension* at any point in the rope is the magnitude of the force that acts at that point. For this weightless rope, the tension $T_{G\,on\,R}$ at the lower end has the same value as the tension $T_{C\,on\,R}$ at the upper end. For such an ideal weightless rope, the tension has the same value at any point along the rope's length. (See the discussion in Conceptual Example 4.10 in Section 4.5.)

Example 5.2   **One-dimensional equilibrium: Tension in a rope with mass**

Find the tension at each end of the rope in Example 5.1 if the weight of the rope is 120 N.

SOLUTION

IDENTIFY and SET UP: As in Example 5.1, the target variables are the magnitudes $T_{G\,on\,R}$ and $T_{C\,on\,R}$ of the forces that act at the bottom and top of the rope, respectively. Once again, we'll apply Newton's first law to the gymnast and to the rope, and use Newton's third law to relate the forces that the gymnast and rope exert on each other. Again we draw separate free-body diagrams for the gymnast (Fig. 5.2a) and the rope (Fig. 5.2b). There is now a *third* force acting on the rope, however: the weight of the rope, of magnitude $w_R = 120$ N.

EXECUTE: The gymnast's free-body diagram is the same as in Example 5.1, so her equilibrium condition is also the same. From

Newton's third law, $T_{R\,on\,G} = T_{G\,on\,R}$, and we again have

Gymnast: $\sum F_y = T_{R\,on\,G} + (-w_G) = 0$ so
$$T_{R\,on\,G} = T_{G\,on\,R} = w_G = 490\text{ N}$$

The equilibrium condition $\sum F_y = 0$ for the rope is now

Rope: $\sum F_y = T_{C\,on\,R} + (-T_{G\,on\,R}) + (-w_R) = 0$

Note that the y-component of $T_{C\,on\,R}$ is positive because it points in the +y-direction, but the y-components of both $T_{G\,on\,R}$ and w_R are negative. We solve for $T_{C\,on\,R}$ and substitute the values $T_{G\,on\,R} = T_{R\,on\,G} = 490$ N and $w_R = 120$ N:

$$T_{C\,on\,R} = T_{G\,on\,R} + w_R = 490\text{ N} + 120\text{ N} = 610\text{ N}$$

EVALUATE: When we include the weight of the rope, the tension is *different* at the rope's two ends: 610 N at the top and 490 N at the bottom. The force $T_{\text{C on R}} = 610$ N exerted by the ceiling has to hold up both the 490-N weight of the gymnast and the 120-N weight of the rope.

To see this more clearly, we draw a free-body diagram for a composite body consisting of the gymnast and rope together (Fig. 5.2c). Only two external forces act on this composite body: the force $T_{\text{C on R}}$ exerted by the ceiling and the total weight $w_{\text{G}} + w_{\text{R}} = 490$ N $+ 120$ N $= 610$ N. (The forces $T_{\text{G on R}}$ and $T_{\text{R on G}}$ are *internal* to the composite body. Newton's first law applies only to *external* forces, so these internal forces play no role.) Hence Newton's first law applied to this composite body is

Composite body: $\sum F_y = T_{\text{C on R}} + [-(w_{\text{G}} + w_{\text{R}})] = 0$

and so $T_{\text{C on R}} = w_{\text{G}} + w_{\text{R}} = 610$ N.

Treating the gymnast and rope as a composite body is simpler, but we can't find the tension $T_{\text{G on R}}$ at the bottom of the rope by this method. *Moral: Whenever you have more than one body in a problem involving Newton's laws, the safest approach is to treat each body separately.*

5.2 Our sketches for this problem, including the weight of the rope.

(a) Free-body diagram for gymnast

(b) Free-body diagram for rope

(c) Free-body diagram for gymnast and rope as a composite body

---

Example 5.3   Two-dimensional equilibrium

In Fig. 5.3a, a car engine with weight w hangs from a chain that is linked at ring O to two other chains, one fastened to the ceiling and the other to the wall. Find expressions for the tension in each of the three chains in terms of w. The weights of the ring and chains are negligible compared with the weight of the engine.

SOLUTION

IDENTIFY and SET UP: The target variables are the tension magnitudes T_1, T_2, and T_3 in the three chains (Fig. 5.3a). All the bodies are in equilibrium, so we'll use Newton's first law. We need three independent equations, one for each target variable. However, applying Newton's first law to just one body gives us only *two* equations, as in Eqs. (5.2). So we'll have to consider more than one body in equilibrium. We'll look at the engine (which is acted on by T_1) and the ring (which is acted on by all three chains and so is acted on by all three tensions).

Figures 5.3b and 5.3c show our free-body diagrams and choice of coordinate axes. There are two forces that act on the engine: its weight w and the upward force T_1 exerted by the vertical chain.

Three forces act on the ring: the tensions from the vertical chain (T_1), the horizontal chain (T_2), and the slanted chain (T_3). Because the vertical chain has negligible weight, it exerts forces of the same magnitude T_1 at both of its ends (see Example 5.1). (If the weight of this chain were not negligible, these two forces would have different magnitudes like the rope in Example 5.2.) The weight of the ring is also negligible, which is why it isn't included in Fig. 5.3c.

EXECUTE: The forces acting on the engine are along the y-axis only, so Newton's first law says

Engine: $\sum F_y = T_1 + (-w) = 0$ and $T_1 = w$

The horizontal and slanted chains don't exert forces on the engine itself because they are not attached to it. These forces do appear when we apply Newton's first law to the ring, however. In the free-body diagram for the ring (Fig. 5.3c), remember that T_1, T_2, and T_3 are the *magnitudes* of the forces. We resolve the force with magnitude T_3 into its x- and y-components. The ring is in equilibrium, so using Newton's first law we can write (separate)

5.3 **(a)** The situation. **(b)**, **(c)** Our free-body diagrams.

(a) Engine, chains, and ring

(b) Free-body diagram for engine

(c) Free-body diagram for ring O

Continued

equations stating that the x- and y-components of the net force on the ring are zero:

Ring: $\quad \sum F_x = T_3 \cos 60° + (-T_2) = 0$

Ring: $\quad \sum F_y = T_3 \sin 60° + (-T_1) = 0$

Because $T_1 = w$ (from the engine equation), we can rewrite the second ring equation as

$$T_3 = \frac{T_1}{\sin 60°} = \frac{w}{\sin 60°} = 1.2w$$

We can now use this result in the first ring equation:

$$T_2 = T_3 \cos 60° = w\frac{\cos 60°}{\sin 60°} = 0.58w$$

EVALUATE: The chain attached to the ceiling exerts a force on the ring with a *vertical* component equal to T_1, which in turn is equal to w. But this force also has a horizontal component, so its magnitude T_3 is somewhat larger than w. This chain is under the greatest tension and is the one most susceptible to breaking.

To get enough equations to solve this problem, we had to consider not only the forces on the engine but also the forces acting on a second body (the ring connecting the chains). Situations like this are fairly common in equilibrium problems, so keep this technique in mind.

Example 5.4 An inclined plane

A car of weight w rests on a slanted ramp attached to a trailer (Fig. 5.4a). Only a cable running from the trailer to the car prevents the car from rolling off the ramp. (The car's brakes are off and its transmission is in neutral.) Find the tension in the cable and the force that the ramp exerts on the car's tires.

SOLUTION

IDENTIFY: The car is in equilibrium, so we use Newton's first law. The ramp exerts a separate force on each of the car's tires, but for simplicity we lump these forces into a single force. For a further simplification, we'll neglect any friction force the ramp exerts on the tires (see Fig. 4.2b). Hence the ramp only exerts a force on the car that is *perpendicular* to the ramp. As in Section 4.1, we call this force the *normal* force (see Fig. 4.2a). The two target variables are the magnitude n of the normal force and the magnitude T of the tension in the cable.

SET UP: Figure 5.4 shows the situation and a free-body diagram for the car. The three forces acting on the car are its weight (magnitude w), the tension in the cable (magnitude T), and the normal force (magnitude n). Note that the angle α between the ramp and the horizontal is equal to the angle α between the weight vector $\vec{w}$ and the downward normal to the plane of the ramp. Note also that we choose the x- and y-axes to be parallel and perpendicular to the ramp so that we only need to resolve one force (the weight) into x- and y-components. If we chose axes that were horizontal and vertical, we'd have to resolve both the normal force and the tension into components.

5.4 A cable holds a car at rest on a ramp.

(a) Car on ramp

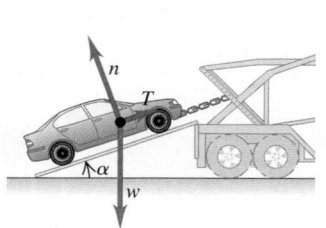

(b) Free-body diagram for car

We replace the weight by its components.

EXECUTE: To write down the x- and y-components of Newton's first law, we must first find the components of the weight. One complication is that the angle α in Fig. 5.4b is *not* measured from the $+x$-axis toward the $+y$-axis. Hence we *cannot* use Eqs. (1.6) directly to find the components. (You may want to review Section 1.8 to make sure that you understand this important point.)

One way to find the components of $\vec{w}$ is to consider the right triangles in Fig. 5.4b. The sine of α is the magnitude of the x-component of $\vec{w}$ (that is, the side of the triangle opposite α) divided by the magnitude w (the hypotenuse of the triangle). Similarly, the cosine of α is the magnitude of the y-component (the side of the triangle adjacent to α) divided by w. Both components are negative, so $w_x = -w \sin \alpha$ and $w_y = -w \cos \alpha$.

Another approach is to recognize that one component of $\vec{w}$ must involve $\sin \alpha$ while the other component involves $\cos \alpha$. To decide which is which, draw the free-body diagram so that the angle α is noticeably smaller or larger than 45°. (You'll have to fight the natural tendency to draw such angles as being close to 45°.) We've drawn Fig. 5.4b so that α is smaller than 45°, so $\sin \alpha$ is less than $\cos \alpha$. The figure shows that the x-component of $\vec{w}$ is smaller than the y-component, so the x-component must involve $\sin \alpha$ and the y-component must involve $\cos \alpha$. We again find $w_x = -w \sin \alpha$ and $w_y = -w \cos \alpha$.

In Fig. 5.4b we draw a wiggly line through the original vector representing the weight to remind us not to count it twice. Newton's first law gives us

$$\sum F_x = T + (-w \sin \alpha) = 0$$
$$\sum F_y = n + (-w \cos \alpha) = 0$$

(Remember that T, w, and n are all *magnitudes* of vectors and are therefore all positive.) Solving these equations for T and n, we find

$$T = w \sin \alpha$$
$$n = w \cos \alpha$$

EVALUATE: Our answers for T and n depend on the value of α. To check this dependence, let's look at some special cases. If the ramp is horizontal ($\alpha = 0$), we get $T = 0$ and $n = w$. As you might expect, no cable tension T is needed to hold the car, and the normal force n is equal in magnitude to the weight. If the ramp is vertical ($\alpha = 90°$), we get $T = w$ and $n = 0$. The cable tension T supports all of the car's weight, and there's nothing pushing the car against the ramp.

> **CAUTION** **Normal force and weight may not be equal** It's a common error to automatically assume that the magnitude n of the normal force is equal to the weight w: Our result shows that this is *not* true in general. It's always best to treat n as a variable and solve for its value, as we have done here.

How would the answers for T and n be affected if the car were being pulled up the ramp at a constant speed? This, too, is an equilibrium situation, since the car's velocity is constant. So the calculation is the same, and T and n have the same values as when the car is at rest. (It's true that T must be greater than $w \sin \alpha$ to *start* the car moving up the ramp, but that's not what we asked.)

Example 5.5   Equilibrium of bodies connected by cable and pulley

Blocks of granite are to be hauled up a 15° slope out of a quarry, and dirt is to be dumped into the quarry to fill up old holes. To simplify the process, you design a system in which a granite block on a cart with steel wheels (weight w_1, including both block and cart) is pulled uphill on steel rails by a dirt-filled bucket (weight w_2, including both dirt and bucket) that descends vertically into the quarry (Fig. 5.5a). How must the weights w_1 and w_2 be related in order for the system to move with constant speed? Ignore friction in the pulley and wheels, and ignore the weight of the cable.

> **SOLUTION**

IDENTIFY and SET UP: The cart and bucket each move with a constant velocity (in a straight line at constant speed). Hence each body is in equilibrium, and we can apply Newton's first law to each. Our target is an expression relating the weights w_1 and w_2.

Figure 5.5b shows our idealized model for the system, and Figs. 5.5c and 5.5d show our free-body diagrams. The two forces on the bucket are its weight w_2 and an upward tension exerted by the cable. As for the car on the ramp in Example 5.4, three forces act on the cart: its weight w_1, a normal force of magnitude n exerted by the rails, and a tension force from the cable. (We're ignoring friction, so we assume that the rails exert no force on the cart parallel to the incline.) Note that we orient the axes differ-

ently for each body; the choices shown are the most convenient ones.

We're assuming that the cable has negligible weight, so the tension forces that the cable exerts on the cart and on the bucket have the same magnitude T. As we did for the car in Example 5.4, we represent the weight of the cart in terms of its x- and y-components.

EXECUTE: Applying $\sum F_y = 0$ to the bucket in Fig. 5.5c, we find

$$\sum F_y = T + (-w_2) = 0 \qquad \text{so} \qquad T = w_2$$

Applying $\sum F_x = 0$ to the cart (and block) in Fig. 5.5d, we get

$$\sum F_x = T + (-w_1 \sin 15°) = 0 \qquad \text{so} \qquad T = w_1 \sin 15°$$

Equating the two expressions for T, we find

$$w_2 = w_1 \sin 15° = 0.26 w_1$$

EVALUATE: Our analysis doesn't depend at all on the direction in which the cart and bucket move. Hence the system can move with constant speed in *either* direction if the weight of the dirt and bucket is 26% of the weight of the granite block and cart. What would happen if w_2 were greater than $0.26w_1$? If it were less than $0.26w_1$?

Notice that we didn't need the equation $\sum F_y = 0$ for the cart and block. Can you use this to show that $n = w_1 \cos 15°$?

5.5 (a) The situation. (b) Our idealized model. (c), (d) Our free-body diagrams.

(a) Dirt-filled bucket pulls cart with granite block

(b) Idealized model of the system

(c) Free-body diagram for bucket

(d) Free-body diagram for cart

Test Your Understanding of Section 5.1 A traffic light of weight w hangs from two lightweight cables, one on each side of the light. Each cable hangs at a 45° angle from the horizontal. What is the tension in each cable? (i) $w/2$; (ii) $w/\sqrt{2}$; (iii) w; (iv) $w\sqrt{2}$; (v) $2w$.

5.6 Correct and incorrect free-body diagrams for a falling body.

(a)

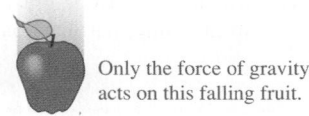

Only the force of gravity acts on this falling fruit.

(b) Correct free-body diagram

◄ RIGHT!
You can safely draw the acceleration vector to one side of the diagram.

(c) Incorrect free-body diagram

◄ WRONG
This vector doesn't belong in a free-body diagram because $m\vec{a}$ is *not* a force.

5.2 Using Newton's Second Law: Dynamics of Particles

We are now ready to discuss *dynamics* problems. In these problems, we apply Newton's second law to bodies on which the net force is *not* zero. These bodies are *not* in equilibrium and hence are accelerating. The net force on the body is equal to the mass of the body times its acceleration:

$$\sum \vec{F} = m\vec{a} \quad \text{(Newton's second law, vector form)} \quad (5.3)$$

We most often use this relationship in component form:

$$\sum F_x = ma_x \qquad \sum F_y = ma_y \qquad \begin{array}{l}\text{(Newton's second law,}\\ \text{component form)}\end{array} \quad (5.4)$$

The following problem-solving strategy is very similar to Problem-Solving Strategy 5.1 for equilibrium problems in Section 5.1. Study it carefully, watch how we apply it in our examples, and use it when you tackle the end-of-chapter problems. You can solve *any* dynamics problem using this strategy.

CAUTION $m\vec{a}$ **doesn't belong in free-body diagrams** Remember that the quantity $m\vec{a}$ is the *result* of forces acting on a body, *not* a force itself; it's not a push or a pull exerted by anything in the body's environment. When you draw the free-body diagram for an accelerating body (like the fruit in Fig. 5.6a), make sure you *never* include the "$m\vec{a}$ force" because *there is no such force* (Fig. 5.6c). You should review Section 4.3 if you're not clear on this point. Sometimes we draw the acceleration vector $\vec{a}$ *alongside* a free-body diagram, as in Fig. 5.6b. But we *never* draw the acceleration vector with its tail touching the body (a position reserved exclusively for the forces that act on the body). █

Problem-Solving Strategy 5.2 **Newton's Second Law: Dynamics of Particles**

IDENTIFY *the relevant concepts:* You have to use Newton's second law for *any* problem that involves forces acting on an accelerating body.

Identify the target variable—usually an acceleration or a force. If the target variable is something else, you'll need to select another concept to use. For example, suppose the target variable is how fast a sled is moving when it reaches the bottom of a hill. Newton's second law will let you find the sled's acceleration; you'll then use the constant-acceleration relationships from Section 2.4 to find velocity from acceleration.

SET UP *the problem* using the following steps:
1. Draw a simple sketch of the situation that shows each moving body. For each body, draw a free-body diagram that shows all the forces acting *on* the body. (The acceleration of a body is determined by the forces that act on it, *not* by the forces that it exerts on anything else.) Make sure you can answer the question "What other body is applying this force?" for each force in your diagram. Never include the quantity $m\vec{a}$ in your free-body diagram; it's not a force!
2. Label each force with an algebraic symbol for the force's *magnitude*. Usually, one of the forces will be the body's weight; it's usually best to label this as $w = mg$.
3. Choose your x- and y-coordinate axes for each body, and show them in its free-body diagram. Be sure to indicate the positive direction for each axis. If you know the direction of the acceleration, it usually simplifies things to take one positive axis along that direction. If your problem involves two or more bodies that

accelerate in different directions, you can use a different set of axes for each body.
4. In addition to Newton's second law, $\sum \vec{F} = m\vec{a}$, identify any other equations you might need. For example, you might need one or more of the equations for motion with constant acceleration. If more than one body is involved, there may be relationships among their motions; for example, they may be connected by a rope. Express any such relationships as equations relating the accelerations of the various bodies.

EXECUTE *the solution* as follows:
1. For each body, determine the components of the forces along each of the body's coordinate axes. When you represent a force in terms of its components, draw a wiggly line through the original force vector to remind you not to include it twice.
2. Make a list of all the known and unknown quantities. In your list, identify the target variable or variables.
3. For each body, write a separate equation for each component of Newton's second law, as in Eqs. (5.4). In addition, write any additional equations that you identified in step 4 of "Set Up." (You need as many equations as there are target variables.)
4. Do the easy part—the math! Solve the equations to find the target variable(s).

EVALUATE *your answer:* Does your answer have the correct units? (When appropriate, use the conversion $1 \text{ N} = 1 \text{ kg} \cdot \text{m/s}^2$.) Does it have the correct algebraic sign? When possible, consider particular values or extreme cases of quantities and compare the results with your intuitive expectations. Ask, "Does this result make sense?"

Example 5.6	Straight-line motion with a constant force

An iceboat is at rest on a frictionless horizontal surface (Fig. 5.7a). A wind is blowing along the direction of the runners so that 4.0 s after the iceboat is released, it is moving at 6.0 m/s (about 22 km/h, or 13 mi/h). What constant horizontal force F_W does the wind exert on the iceboat? The combined mass of iceboat and rider is 200 kg.

SOLUTION

IDENTIFY and SET UP: Our target variable is one of the forces (F_W) acting on the accelerating iceboat, so we need to use Newton's second law. The forces acting on the iceboat and rider (considered as a unit) are the weight w, the normal force n exerted by the surface, and the horizontal force F_W. Figure 5.7b shows the free-body diagram. The net force and hence the acceleration are to the right, so we chose the positive x-axis in this direction. The acceleration isn't given; we'll need to find it. Since the wind is assumed to exert a constant force, the resulting acceleration is constant and we can use one of the constant-acceleration formulas from Section 2.4.

5.7 (a) The situation. (b) Our free-body diagram.

(a) Iceboat and rider on frictionless ice

(b) Free-body diagram for iceboat and rider

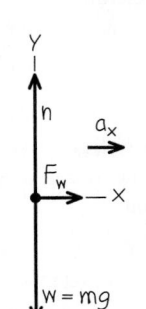

The iceboat starts at rest (its initial x-velocity is $v_{0x} = 0$) and it attains an x-velocity $v_x = 6.0$ m/s after an elapsed time $t = 4.0$ s. To relate the x-acceleration a_x to these quantities we use Eq. (2.8), $v_x = v_{0x} + a_x t$. There is no vertical acceleration, so we expect that the normal force on the iceboat is equal in magnitude to the iceboat's weight.

EXECUTE: The *known* quantities are the mass $m = 200$ kg, the initial and final x-velocities $v_{0x} = 0$ and $v_x = 6.0$ m/s, and the elapsed time $t = 4.0$ s. The three *unknown* quantities are the acceleration a_x, the normal force n, and the horizontal force F_W. Hence we need three equations.

The first two equations are the x- and y-equations for Newton's second law. The force F_W is in the positive x-direction, while the forces n and $w = mg$ are in the positive and negative y-directions, respectively. Hence we have

$$\sum F_x = F_W = ma_x$$
$$\sum F_y = n + (-mg) = 0 \quad \text{so} \quad n = mg$$

The third equation is the constant-acceleration relationship, Eq. (2.8):

$$v_x = v_{0x} + a_x t$$

To find F_W, we first solve this third equation for a_x and then substitute the result into the $\sum F_x$ equation:

$$a_x = \frac{v_x - v_{0x}}{t} = \frac{6.0 \text{ m/s} - 0 \text{ m/s}}{4.0 \text{ s}} = 1.5 \text{ m/s}^2$$
$$F_W = ma_x = (200 \text{ kg})(1.5 \text{ m/s}^2) = 300 \text{ kg} \cdot \text{m/s}^2$$

Since $1 \text{ kg} \cdot \text{m/s}^2 = 1$ N, the final answer is

$$F_W = 300 \text{ N (about 67 lb)}$$

EVALUATE: Our answers for F_W and n have the correct units for a force, and (as expected) the magnitude n of the normal force is equal to mg. Does it seem reasonable that the force F_W is substantially *less* than mg?

Example 5.7	Straight-line motion with friction

Suppose a constant horizontal friction force with magnitude 100 N opposes the motion of the iceboat in Example 5.6. In this case, what constant force F_W must the wind exert on the iceboat to cause the same constant x-acceleration $a_x = 1.5 \text{ m/s}^2$?

SOLUTION

IDENTIFY and SET UP: Again the target variable is F_W. We are given the x-acceleration, so to find F_W all we need is Newton's second law. Figure 5.8 shows our new free-body diagram. The only difference from Fig. 5.7b is the addition of the friction force $\vec{f}$, which points opposite the motion. (Note that the *magnitude* $f = 100$ N is a positive quantity, but the *component* in the x-direction f_x is negative, equal to $-f$ or -100 N.) Because the wind must now overcome the friction force to yield the same acceleration as in Example 5.6, we expect our answer for F_W to be greater than the 300 N we found there.

5.8 Our free-body diagram for the iceboat and rider with a friction force $\vec{f}$ opposing the motion.

Continued

EXECUTE: Two forces now have x-components: the force of the wind and the friction force. The x-component of Newton's second law gives

$$\sum F_x = F_W + (-f) = ma_x$$
$$F_W = ma_x + f = (200 \text{ kg})(1.5 \text{ m/s}^2) + (100 \text{ N}) = 400 \text{ N}$$

EVALUATE: The required value of F_W is 100 N greater than in Example 5.6 because the wind must now push against an additional 100-N friction force.

Example 5.8 Tension in an elevator cable

An elevator and its load have a combined mass of 800 kg (Fig. 5.9a). The elevator is initially moving downward at 10.0 m/s; it slows to a stop with constant acceleration in a distance of 25.0 m. What is the tension T in the supporting cable while the elevator is being brought to rest?

SOLUTION

IDENTIFY and SET UP: The target variable is the tension T, which we'll find using Newton's second law. As in Example 5.6, we'll determine the acceleration using a constant-acceleration formula. Our free-body diagram (Fig. 5.9b) shows two forces acting on the elevator: its weight w and the tension force T of the cable. The elevator is moving downward with decreasing speed, so its acceleration is upward; we chose the positive y-axis to be upward.

The elevator is moving in the negative y-direction, so its initial y-velocity v_{0y} and its y-displacement $y - y_0$ are both negative: $v_{0y} = -10.0 \text{ m/s}$ and $y - y_0 = -25.0 \text{ m}$. The final y-velocity is $v_y = 0$. To find the y-acceleration a_y from this information, we'll use Eq. (2.13) in the form $v_y^2 = v_{0y}^2 + 2a_y(y - y_0)$. Once we have a_y, we'll substitute it into the y-component of Newton's second law from Eqs. (5.4) and solve for T. The net force must be upward to give an upward acceleration, so we expect T to be greater than the weight $w = mg = (800 \text{ kg})(9.80 \text{ m/s}^2) = 7840 \text{ N}$.

EXECUTE: First let's write out Newton's second law. The tension force acts upward and the weight acts downward, so

$$\sum F_y = T + (-w) = ma_y$$

We solve for the target variable T:

$$T = w + ma_y = mg + ma_y = m(g + a_y)$$

5.9 (a) The situation. (b) Our free-body diagram.

(a) Descending elevator

Moving down with decreasing speed

(b) Free-body diagram for elevator

To determine a_y, we rewrite the constant-acceleration equation $v_y^2 = v_{0y}^2 + 2a_y(y - y_0)$:

$$a_y = \frac{v_y^2 - v_{0y}^2}{2(y - y_0)} = \frac{(0)^2 - (-10.0 \text{ m/s})^2}{2(-25.0 \text{ m})} = +2.00 \text{ m/s}^2$$

The acceleration is upward (positive), just as it should be.

Now we can substitute the acceleration into the equation for the tension:

$$T = m(g + a_y) = (800 \text{ kg})(9.80 \text{ m/s}^2 + 2.00 \text{ m/s}^2)$$
$$= 9440 \text{ N}$$

EVALUATE: The tension is greater than the weight, as expected. Can you see that we would get the same answers for a_y and T if the elevator were moving *upward* and *gaining* speed at a rate of 2.00 m/s^2?

Example 5.9 Apparent weight in an accelerating elevator

A 50.0-kg woman stands on a bathroom scale while riding in the elevator in Example 5.8. What is the reading on the scale?

SOLUTION

IDENTIFY and SET UP: The scale (Fig. 5.10a) reads the magnitude of the downward force exerted *by* the woman *on* the scale. By Newton's third law, this equals the magnitude of the upward normal force exerted *by* the scale *on* the woman. Hence our target variable is the magnitude n of the normal force. We'll find n by applying Newton's second law to the woman. We already know her acceleration; it's the same as the acceleration of the elevator, which we calculated in Example 5.8.

Figure 5.10b shows our free-body diagram for the woman. The forces acting on her are the normal force n exerted by the scale and her weight $w = mg = (50.0 \text{ kg})(9.80 \text{ m/s}^2) = 490 \text{ N}$.

5.10 (a) The situation. (b) Our free-body diagram.

(a) Woman in a descending elevator

Moving down with decreasing speed

(b) Free-body diagram for woman

(The tension force, which played a major role in Example 5.8, doesn't appear here because it doesn't act on the woman.) From Example 5.8, the y-acceleration of the elevator and of the woman is $a_y = +2.00 \text{ m/s}^2$. As in Example 5.8, the upward force on the body accelerating upward (in this case, the normal force on the woman) will have to be greater than the body's weight to produce the upward acceleration.

EXECUTE: Newton's second law gives

$$\sum F_y = n + (-mg) = ma_y$$
$$n = mg + ma_y = m(g + a_y)$$
$$= (50.0 \text{ kg})(9.80 \text{ m/s}^2 + 2.00 \text{ m/s}^2) = 590 \text{ N}$$

EVALUATE: Our answer for n means that while the elevator is stopping, the scale pushes up on the woman with a force of 590 N. By Newton's third law, she pushes down on the scale with the same force. So the scale reads 590 N, which is 100 N more than her actual

weight. The scale reading is called the passenger's **apparent weight.** The woman *feels* the floor pushing up harder on her feet than when the elevator is stationary or moving with constant velocity.

What would the woman feel if the elevator were accelerating *downward,* so that $a_y = -2.00 \text{ m/s}^2$? This would be the case if the elevator were moving upward with decreasing speed or moving downward with increasing speed. To find the answer for this situation, we just insert the new value of a_y in our equation for n:

$$n = m(g + a_y) = (50.0 \text{ kg})[9.80 \text{ m/s}^2 + (-2.00 \text{ m/s}^2)]$$
$$= 390 \text{ N}$$

Now the woman feels as though she weighs only 390 N, or 100 N *less* than her actual weight w.

You can feel these effects yourself; try taking a few steps in an elevator that is coming to a stop after descending (when your apparent weight is greater than w) or coming to a stop after ascending (when your apparent weight is less than w).

Apparent Weight and Apparent Weightlessness

Let's generalize the result of Example 5.9. When a passenger with mass m rides in an elevator with y-acceleration a_y, a scale shows the passenger's apparent weight to be

$$n = m(g + a_y)$$

When the elevator is accelerating upward, a_y is positive and n is greater than the passenger's weight $w = mg$. When the elevator is accelerating downward, a_y is negative and n is less than the weight. If the passenger doesn't know the elevator is accelerating, she may feel as though her weight is changing; indeed, this is just what the scale shows.

The extreme case occurs when the elevator has a downward acceleration $a_y = -g$—that is, when it is in free fall. In that case $n = 0$ and the passenger *seems* to be weightless. Similarly, an astronaut orbiting the earth with a spacecraft experiences *apparent weightlessness* (Fig. 5.11). In each case, the person is not truly weightless because a gravitational force still acts. But the person's sensations in this free-fall condition are exactly the same as though the person were in outer space with no gravitational force at all. In both cases the person and the vehicle (elevator or spacecraft) fall together with the same acceleration g, so nothing pushes the person against the floor or walls of the vehicle.

5.11 Astronauts in orbit feel "weightless" because they have the same acceleration as their spacecraft—*not* because they are "outside the pull of the earth's gravity." (If no gravity acted on them, the astronauts and their spacecraft wouldn't remain in orbit, but would fly off into deep space.)

Example 5.10 | **Acceleration down a hill**

A toboggan loaded with students (total weight w) slides down a snow-covered slope. The hill slopes at a constant angle α, and the toboggan is so well waxed that there is virtually no friction. What is its acceleration?

SOLUTION

IDENTIFY and SET UP: Our target variable is the acceleration, which we'll find using Newton's second law. There is no friction, so only two forces act on the toboggan: its weight w and the normal force n exerted by the hill.

Figure 5.12 shows our sketch and free-body diagram. As in Example 5.4, the surface is inclined, so the normal force is not vertical and is not equal in magnitude to the weight. Hence we must use both components of $\sum \vec{F} = m\vec{a}$ in Eqs. (5.4). We take axes parallel

5.12 Our sketches for this problem.

(a) The situation   **(b)** Free-body diagram for toboggan

Continued

and perpendicular to the surface of the hill, so that the acceleration (which is parallel to the hill) is along the positive *x*-direction.

EXECUTE: The normal force has only a *y*-component, but the weight has both *x*- and *y*-components: $w_x = w \sin \alpha$ and $w_y = -w \cos \alpha$. (In Example 5.4 we had $w_x = -w \sin \alpha$. The difference is that the positive *x*-axis was uphill in Example 5.4 but is downhill in Fig. 5.12b.) The wiggly line in Fig. 5.12b reminds us that we have resolved the weight into its components. The acceleration is purely in the +*x*-direction, so $a_y = 0$. Newton's second law in component form then tells us that

$$\sum F_x = w \sin \alpha = ma_x$$
$$\sum F_y = n - w \cos \alpha = ma_y = 0$$

Since $w = mg$, the *x*-component equation tells us that $mg \sin \alpha = ma_x$, or

$$a_x = g \sin \alpha$$

Note that we didn't need the *y*-component equation to find the acceleration. That's part of the beauty of choosing the *x*-axis to lie along the acceleration direction! The *y*-equation tells us the mag-

nitude of the normal force exerted by the hill on the toboggan:

$$n = w \cos \alpha = mg \cos \alpha$$

EVALUATE: Notice that the normal force n is not equal to the toboggan's weight (compare Example 5.4). Notice also that the mass m does not appear in our result for the acceleration. That's because the downhill force on the toboggan (a component of the weight) is proportional to m, so the mass cancels out when we use $\sum F_x = ma_x$ to calculate a_x. Hence *any* toboggan, regardless of its mass, slides down a frictionless hill with acceleration $g \sin \alpha$.

If the plane is horizontal, $\alpha = 0$ and $a_x = 0$ (the toboggan does not accelerate); if the plane is vertical, $\alpha = 90°$ and $a_x = g$ (the toboggan is in free fall).

CAUTION **Common free-body diagram errors** Figure 5.13 shows both the correct way (Fig. 5.13a) and a common *incorrect* way (Fig. 5.13b) to draw the free-body diagram for the toboggan. The diagram in Fig. 5.13b is wrong for two reasons: The normal force must be drawn perpendicular to the surface, and there's no such thing as the "$m\vec{a}$ force." If you remember that "normal" means "perpendicular" and that $m\vec{a}$ is not itself a force, you'll be well on your way to always drawing correct free-body diagrams. ▌

5.13 Correct and incorrect free-body diagrams for a toboggan on a frictionless hill.

(a) Correct free-body diagram for the sled

RIGHT! ▶
Normal force is perpendicular to the surface.

It's OK to draw the acceleration vector adjacent to (but not touching) the body.
◀ **RIGHT!**

(b) Incorrect free-body diagram for the sled

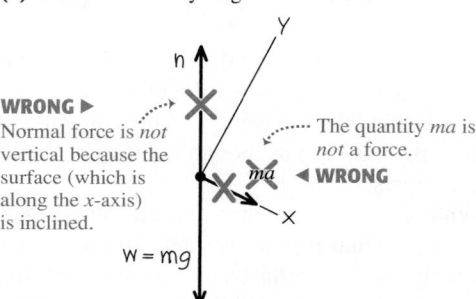

WRONG ▶
Normal force is *not* vertical because the surface (which is along the *x*-axis) is inclined.

The quantity ma is *not* a force.
◀ **WRONG**

Example 5.11 **Two bodies with the same acceleration**

You push a 1.00-kg food tray through the cafeteria line with a constant 9.0-N force. The tray pushes on a 0.50-kg carton of milk (Fig. 5.14a). The tray and carton slide on a horizontal surface so greasy that friction can be neglected. Find the acceleration of the tray and carton and the horizontal force that the tray exerts on the carton.

SOLUTION

IDENTIFY and SET UP: Our *two* target variables are the acceleration of the tray–carton system and the force of the tray on the carton. We'll use Newton's second law to get two equations, one for each target variable. We set up and solve the problem in two ways.

Method 1: We treat the milk carton (mass m_C) and tray (mass m_T) as separate bodies, each with its own free-body diagram (Figs. 5.14b and 5.14c). The force F that you exert on the tray doesn't appear in the free-body diagram for the carton, which is accelerated by the force (of magnitude $F_{T \, on \, C}$) exerted on it by the tray. By Newton's third law, the carton exerts a force of equal magnitude on the tray: $F_{C \, on \, T} = F_{T \, on \, C}$. We take the acceleration to

be in the positive *x*-direction; both the tray and milk carton move with the same *x*-acceleration a_x.

Method 2: We treat the tray and milk carton as a composite body of mass $m = m_T + m_C = 1.50$ kg (Fig. 5.14d). The only horizontal force acting on this body is the force F that you exert. The forces $F_{T \, on \, C}$ and $F_{C \, on \, T}$ don't come into play because they're *internal* to this composite body, and Newton's second law tells us that only *external* forces affect a body's acceleration (see Section 4.3). To find the magnitude $F_{T \, on \, C}$ we'll again apply Newton's second law to the carton, as in Method 1.

EXECUTE: *Method 1:* The *x*-component equations of Newton's second law are

Tray: $\quad \sum F_x = F - F_{C \, on \, T} = F - F_{T \, on \, C} = m_T a_x$

Carton: $\quad \sum F_x = F_{T \, on \, C} = m_C a_x$

These are two simultaneous equations for the two target variables a_x and $F_{T \, on \, C}$. (Two equations are all we need, which means that

5.14 Pushing a food tray and milk carton in the cafeteria line.

(a) A milk carton and a food tray

$m_C = 0.50$ kg

$F = 9.0$ N

$m_T = 1.00$ kg

(b) Free-body diagram for milk carton

(c) Free-body diagram for food tray

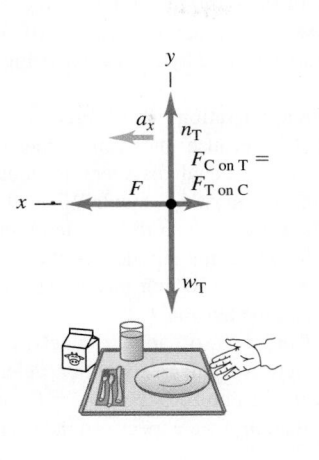

(d) Free-body diagram for carton and tray as a composite body

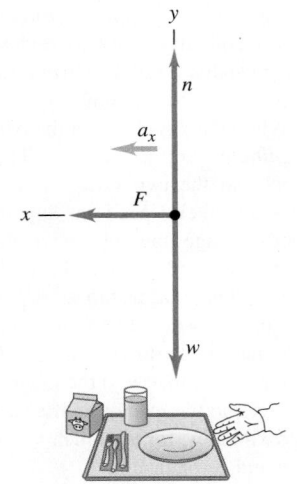

the y-components don't play a role in this example.) An easy way to solve the two equations for a_x is to add them; this eliminates $F_{T \text{ on } C}$, giving

$$F = m_T a_x + m_C a_x = (m_T + m_C)a_x$$

and

$$a_x = \frac{F}{m_T + m_C} = \frac{9.0 \text{ N}}{1.00 \text{ kg} + 0.50 \text{ kg}} = 6.0 \text{ m/s}^2 = 0.61g$$

Substituting this value into the carton equation gives

$$F_{T \text{ on } C} = m_C a_x = (0.50 \text{ kg})(6.0 \text{ m/s}^2) = 3.0 \text{ N}$$

Method 2: The x-component of Newton's second law for the composite body of mass m is

$$\sum F_x = F = ma_x$$

The acceleration of this composite body is

$$a_x = \frac{F}{m} = \frac{9.0 \text{ N}}{1.50 \text{ kg}} = 6.0 \text{ m/s}^2$$

Then, looking at the milk carton by itself, we see that to give it an acceleration of 6.0 m/s² requires that the tray exert a force

$$F_{T \text{ on } C} = m_C a_x = (0.50 \text{ kg})(6.0 \text{ m/s}^2) = 3.0 \text{ N}$$

EVALUATE: The answers are the same with both methods. To check the answers, note that there are different forces on the two sides of the tray: $F = 9.0$ N on the right and $F_{C \text{ on } T} = 3.0$ N on the left. The net horizontal force on the tray is $F - F_{C \text{ on } T} = 6.0$ N, exactly enough to accelerate a 1.00-kg tray at 6.0 m/s².

Treating two bodies as a single, composite body works *only* if the two bodies have the same magnitude *and* direction of acceleration. If the accelerations are different we must treat the two bodies separately, as in the next example.

Example 5.12 **Two bodies with the same magnitude of acceleration**

Figure 5.15a shows an air-track glider with mass m_1 moving on a level, frictionless air track in the physics lab. The glider is connected to a lab weight with mass m_2 by a light, flexible, nonstretching string that passes over a stationary, frictionless pulley. Find the acceleration of each body and the tension in the string.

SOLUTION

IDENTIFY and SET UP: The glider and weight are accelerating, so again we must use Newton's second law. Our three target variables are the tension T in the string and the accelerations of the two bodies.

The two bodies move in different directions—one horizontal, one vertical—so we can't consider them together as we did the bodies in Example 5.11. Figures 5.15b and 5.15c show our free-body diagrams and coordinate systems. It's convenient to have both bodies accelerate in the positive axis directions,

5.15 (a) The situation. (b), (c) Our free-body diagrams.

(a) Apparatus

(b) Free-body diagram for glider

(c) Free-body diagram for weight

Continued

so we chose the positive y-direction for the lab weight to be downward.

We consider the string to be massless and to slide over the pulley without friction, so the tension T in the string is the same throughout and it applies a force of the same magnitude T to each body. (You may want to review Conceptual Example 4.10, in which we discussed the tension force exerted by a massless string.) The weights are m_1g and m_2g.

While the *directions* of the two accelerations are different, their *magnitudes* are the same. (That's because the string doesn't stretch, so the two bodies must move equal distances in equal times and their speeds at any instant must be equal. When the speeds change, they change at the same rate, so the accelerations of the two bodies must have the same magnitude a.) We can express this relationship as $a_{1x} = a_{2y} = a$, which means that we have only *two* target variables: a and the tension T.

What results do we expect? If $m_1 = 0$ (or, approximately, for m_1 much less than m_2) the lab weight will fall freely with acceleration g, and the tension in the string will be zero. For $m_2 = 0$ (or, approximately, for m_2 much less than m_1) we expect zero acceleration and zero tension.

EXECUTE: Newton's second law gives

Glider: $\quad \sum F_x = T = m_1 a_{1x} = m_1 a$

Glider: $\quad \sum F_y = n + (-m_1 g) = m_1 a_{1y} = 0$

Lab weight: $\quad \sum F_y = m_2 g + (-T) = m_2 a_{2y} = m_2 a$

(There are no forces on the lab weight in the x-direction.) In these equations we've used $a_{1y} = 0$ (the glider doesn't accelerate vertically) and $a_{1x} = a_{2y} = a$.

The x-equation for the glider and the equation for the lab weight give us two simultaneous equations for T and a:

Glider: $\qquad T = m_1 a$

Lab weight: $\qquad m_2 g - T = m_2 a$

We add the two equations to eliminate T, giving

$$m_2 g = m_1 a + m_2 a = (m_1 + m_2)a$$

and so the magnitude of each body's acceleration is

$$a = \frac{m_2}{m_1 + m_2} g$$

Substituting this back into the glider equation $T = m_1 a$, we get

$$T = \frac{m_1 m_2}{m_1 + m_2} g$$

EVALUATE: The acceleration is in general less than g, as you might expect; the string tension keeps the lab weight from falling freely. The tension T is *not* equal to the weight $m_2 g$ of the lab weight, but is *less* by a factor of $m_1/(m_1 + m_2)$. If T were equal to $m_2 g$, then the lab weight would be in equilibrium, and it isn't.

As predicted, the acceleration is equal to g for $m_1 = 0$ and equal to zero for $m_2 = 0$, and $T = 0$ for either $m_1 = 0$ or $m_2 = 0$.

CAUTION **Tension and weight may not be equal** It's a common mistake to assume that if an object is attached to a vertical string, the string tension must be equal to the object's weight. That was the case in Example 5.5, where the acceleration was zero, but it's not the case in this example! The only safe approach is *always* to treat the tension as a variable, as we did here. ▌

MasteringPHYSICS

PhET: Lunar Lander
ActivPhysics 2.1.5: Car Race
ActivPhysics 2.2: Lifting a Crate
ActivPhysics 2.3: Lowering a Crate
ActivPhysics 2.4: Rocket Blasts Off
ActivPhysics 2.5: Modified Atwood Machine

Test Your Understanding of Section 5.2 Suppose you hold the glider in Example 5.12 so that it and the weight are initially at rest. You give the glider a push to the left in Fig. 5.15a and then release it. The string remains taut as the glider moves to the left, comes instantaneously to rest, then moves to the right. At the instant the glider has zero velocity, what is the tension in the string? (i) greater than in Example 5.12; (ii) the same as in Example 5.12; (iii) less than in Example 5.12, but greater than zero; (iv) zero. ▌

5.16 The sport of ice hockey depends on having the right amount of friction between a player's skates and the ice. If there were too much friction, the players would move too slowly; if there were too little friction, they would fall over.

5.3 Frictional Forces

We've seen several problems where a body rests or slides on a surface that exerts forces on the body. Whenever two bodies interact by direct contact (touching) of their surfaces, we describe the interaction in terms of *contact forces*. The normal force is one example of a contact force; in this section we'll look in detail at another contact force, the force of friction.

Friction is important in many aspects of everyday life. The oil in a car engine minimizes friction between moving parts, but without friction between the tires and the road we couldn't drive or turn the car. Air drag—the frictional force exerted by the air on a body moving through it—decreases automotive fuel economy but makes parachutes work. Without friction, nails would pull out, light bulbs would unscrew effortlessly, and ice hockey would be hopeless (Fig. 5.16).

Kinetic and Static Friction

When you try to slide a heavy box of books across the floor, the box doesn't move at all unless you push with a certain minimum force. Then the box starts moving, and you can usually keep it moving with less force than you needed to get it started. If you take some of the books out, you need less force than before to get it started or keep it moving. What general statements can we make about this behavior?

First, when a body rests or slides on a surface, we can think of the surface as exerting a single contact force on the body, with force components perpendicular and parallel to the surface (Fig. 5.17). The perpendicular component vector is the normal force, denoted by $\vec{n}$. The component vector parallel to the surface (and perpendicular to $\vec{n}$) is the **friction force,** denoted by $\vec{f}$. If the surface is friction-less, then $\vec{f}$ is zero but there is still a normal force. (Frictionless surfaces are an unattainable idealization, like a massless rope. But we can approximate a surface as frictionless if the effects of friction are negligibly small.) The direction of the friction force is always such as to oppose relative motion of the two surfaces.

The kind of friction that acts when a body slides over a surface is called a **kinetic friction force** $\vec{f}_k$. The adjective "kinetic" and the subscript "k" remind us that the two surfaces are moving relative to each other. The *magnitude* of the kinetic friction force usually increases when the normal force increases. This is why it takes more force to slide a box across the floor when it's full of books than when it's empty. Automotive brakes use the same principle: The harder the brake pads are squeezed against the rotating brake disks, the greater the braking effect. In many cases the magnitude of the kinetic friction force f_k is found experimen-tally to be approximately *proportional* to the magnitude n of the normal force. In such cases we represent the relationship by the equation

$$f_k = \mu_k n \qquad \text{(magnitude of kinetic friction force)} \qquad (5.5)$$

where μ_k (pronounced "mu-sub-k") is a constant called the **coefficient of kinetic friction.** The more slippery the surface, the smaller this coefficient. Because it is a quotient of two force magnitudes, μ_k is a pure number without units.

CAUTION Friction and normal forces are always perpendicular Remember that Eq. (5.5) is *not* a vector equation because $\vec{f}_k$ and $\vec{n}$ are always perpendicular. Rather, it is a scalar relationship between the magnitudes of the two forces.

Equation (5.5) is only an approximate representation of a complex phenome-non. On a microscopic level, friction and normal forces result from the intermol-ecular forces (fundamentally electrical in nature) between two rough surfaces at points where they come into contact (Fig. 5.18). As a box slides over the floor, bonds between the two surfaces form and break, and the total number of such bonds varies; hence the kinetic friction force is not perfectly constant. Smoothing the surfaces can actually increase friction, since more molecules are able to inter-act and bond; bringing two smooth surfaces of the same metal together can cause a "cold weld." Lubricating oils work because an oil film between two surfaces (such as the pistons and cylinder walls in a car engine) prevents them from com-ing into actual contact.

Table 5.1 lists some representative values of μ_k. Although these values are given with two significant figures, they are only approximate, since friction forces can also depend on the speed of the body relative to the surface. For now we'll ignore this effect and assume that μ_k and f_k are independent of speed, in order to concentrate on the simplest cases. Table 5.1 also lists coefficients of static friction; we'll define these shortly.

Friction forces may also act when there is *no* relative motion. If you try to slide a box across the floor, the box may not move at all because the floor exerts an equal and opposite friction force on the box. This is called a **static friction force** $\vec{f}_s$. In Fig. 5.19a, the box is at rest, in equilibrium, under the action of its weight $\vec{w}$ and the upward normal force $\vec{n}$. The normal force is equal in magnitude to the weight ($n = w$) and is exerted on the box by the floor. Now we tie a rope

5.17 When a block is pushed or pulled over a surface, the surface exerts a contact force on it.

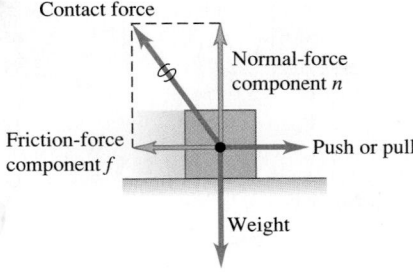

The friction and normal forces are really components of a single contact force.

5.18 The normal and friction forces arise from interactions between molecules at high points on the surfaces of the block and the floor.

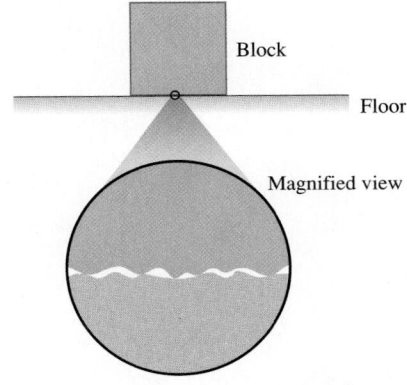

On a microscopic level, even smooth surfaces are rough; they tend to catch and cling.

Table 5.1 Approximate Coefficients of Friction

Materials	Coefficient of Static Friction, μ_s	Coefficient of Kinetic Friction, μ_k
Steel on steel	0.74	0.57
Aluminum on steel	0.61	0.47
Copper on steel	0.53	0.36
Brass on steel	0.51	0.44
Zinc on cast iron	0.85	0.21
Copper on cast iron	1.05	0.29
Glass on glass	0.94	0.40
Copper on glass	0.68	0.53
Teflon on Teflon	0.04	0.04
Teflon on steel	0.04	0.04
Rubber on concrete (dry)	1.0	0.8
Rubber on concrete (wet)	0.30	0.25

5.19 (a), (b), (c) When there is no relative motion, the magnitude of the static friction force f_s is less than or equal to $\mu_s n$. (d) When there is relative motion, the magnitude of the kinetic friction force f_k equals $\mu_k n$. (e) A graph of the friction force magnitude f as a function of the magnitude T of the applied force. The kinetic friction force varies somewhat as intermolecular bonds form and break.

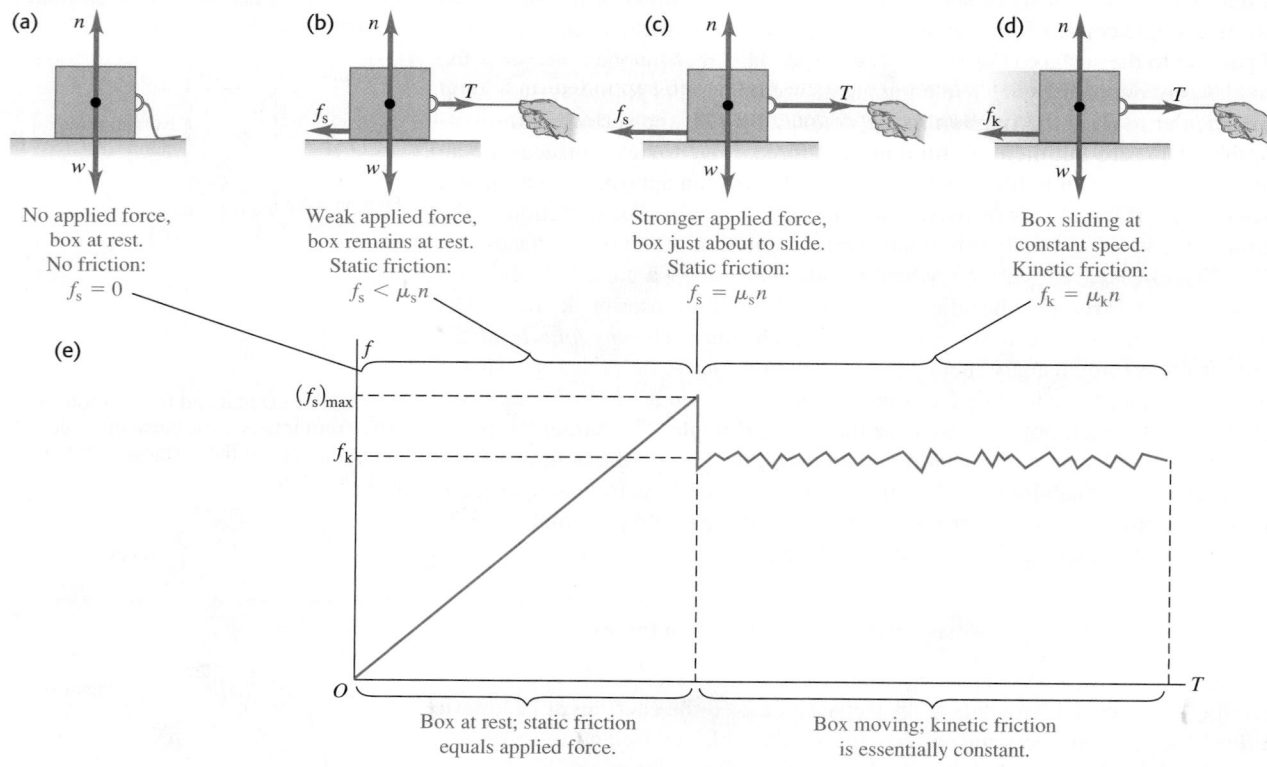

to the box (Fig. 5.19b) and gradually increase the tension T in the rope. At first the box remains at rest because the force of static friction f_s also increases and stays equal in magnitude to T.

At some point T becomes greater than the maximum static friction force f_s the surface can exert. Then the box "breaks loose" (the tension T is able to break the bonds between molecules in the surfaces of the box and floor) and starts to slide. Figure 5.19c shows the forces when T is at this critical value. If T exceeds this value, the box is no longer in equilibrium. For a given pair of surfaces the maximum value of f_s depends on the normal force. Experiment shows that in many cases this maximum value, called $(f_s)_{max}$, is approximately *proportional* to n; we call the proportionality factor μ_s the **coefficient of static friction.** Table 5.1 lists some representative values of μ_s. In a particular situation, the actual force of static friction can have any magnitude between zero (when there is no other force parallel to the surface) and a maximum value given by $\mu_s n$. In symbols,

$$f_s \leq \mu_s n \qquad \text{(magnitude of static friction force)} \qquad (5.6)$$

Like Eq. (5.5), this is a relationship between magnitudes, *not* a vector relationship. The equality sign holds only when the applied force T has reached the critical value at which motion is about to start (Fig. 5.19c). When T is less than this value (Fig. 5.19b), the inequality sign holds. In that case we have to use the equilibrium conditions ($\sum \vec{F} = 0$) to find f_s. If there is no applied force ($T = 0$) as in Fig. 5.19a, then there is no static friction force either ($f_s = 0$).

As soon as the box starts to slide (Fig. 5.19d), the friction force usually *decreases* (Fig. 5.19e); it's easier to keep the box moving than to start it moving. Hence the coefficient of kinetic friction is usually *less* than the coefficient of static friction for any given pair of surfaces, as Table 5.1 shows.

Application **Static Friction and Windshield Wipers**

The squeak of windshield wipers on dry glass is a stick-slip phenomenon. The moving wiper blade sticks to the glass momentarily, then slides when the force applied to the blade by the wiper motor overcomes the maximum force of static friction. When the glass is wet from rain or windshield cleaning solution, friction is reduced and the wiper blade doesn't stick.

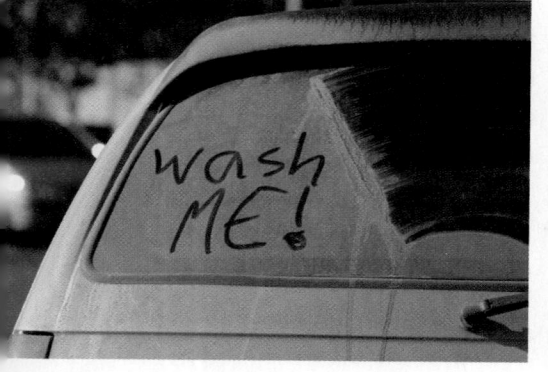

In some situations the surfaces will alternately stick (static friction) and slip (kinetic friction). This is what causes the horrible sound made by chalk held at the wrong angle while writing on the blackboard and the shriek of tires sliding on asphalt pavement. A more positive example is the motion of a violin bow against the string.

When a body slides on a layer of gas, friction can be made very small. In the linear air track used in physics laboratories, the gliders are supported on a layer of air. The frictional force is velocity dependent, but at typical speeds the effective coefficient of friction is of the order of 0.001.

MasteringPHYSICS

PhET: Forces in 1 Dimension
PhET: Friction
PhET: The Ramp
ActivPhysics 2.5: Truck Pulls Crate
ActivPhysics 2.6: Pushing a Crate Up a Wall
ActivPhysics 2.7: Skier Goes Down a Slope
ActivPhysics 2.8: Skier and Rope Tow
ActivPhysics 2.10: Truck Pulls Two Crates

Example 5.13 Friction in horizontal motion

You want to move a 500-N crate across a level floor. To start the crate moving, you have to pull with a 230-N horizontal force. Once the crate "breaks loose" and starts to move, you can keep it moving at constant velocity with only 200 N. What are the coefficients of static and kinetic friction?

SOLUTION

IDENTIFY and SET UP: The crate is in equilibrium both when it is at rest and when it is moving with constant velocity, so we use Newton's first law, as expressed by Eqs. (5.2). We use Eqs. (5.5) and (5.6) to find the target variables μ_s and μ_k.

Figures 5.20a and 5.20b show our sketch and free-body diagram for the instant just before the crate starts to move, when the static friction force has its maximum possible value

$(f_s)_{max} = \mu_s n$. Once the crate is moving, the friction force changes to its kinetic form (Fig. 5.20c). In both situations, four forces act on the crate: the downward weight (magnitude $w = 500$ N), the upward normal force (magnitude n) exerted by the floor, a tension force (magnitude T) to the right exerted by the rope, and a friction force to the left exerted by the ground. Because the rope in Fig. 5.20a is in equilibrium, the tension is the same at both ends. Hence the tension force that the rope exerts on the crate has the same magnitude as the force you exert on the rope. Since it's easier to keep the crate moving than to start it moving, we expect that $\mu_k < \mu_s$.

EXECUTE: Just before the crate starts to move (Fig. 5.20b), we have from Eqs. (5.2)

$$\sum F_x = T + (-(f_s)_{max}) = 0 \quad \text{so} \quad (f_s)_{max} = T = 230 \text{ N}$$
$$\sum F_y = n + (-w) = 0 \quad \text{so} \quad n = w = 500 \text{ N}$$

Now we solve Eq. (5.6), $(f_s)_{max} = \mu_s n$, for the value of μ_s:

$$\mu_s = \frac{(f_s)_{max}}{n} = \frac{230 \text{ N}}{500 \text{ N}} = 0.46$$

After the crate starts to move (Fig. 5.20c) we have

$$\sum F_x = T + (-f_k) = 0 \quad \text{so} \quad f_k = T = 200 \text{ N}$$
$$\sum F_y = n + (-w) = 0 \quad \text{so} \quad n = w = 500 \text{ N}$$

Using $f_k = \mu_k n$ from Eq. (5.5), we find

$$\mu_k = \frac{f_k}{n} = \frac{200 \text{ N}}{500 \text{ N}} = 0.40$$

EVALUATE: As expected, the coefficient of kinetic friction is less than the coefficient of static friction.

5.20 Our sketches for this problem.

(a) Pulling a crate

(b) Free-body diagram for crate just before it starts to move

(c) Free-body diagram for crate moving at constant speed

Example 5.14 Static friction can be less than the maximum

In Example 5.13, what is the friction force if the crate is at rest on the surface and a horizontal force of 50 N is applied to it?

SOLUTION

IDENTIFY and SET UP: The applied force is less than the maximum force of static friction, $(f_s)_{max} = 230$ N. Hence the crate remains at rest and the net force acting on it is zero. The target variable is the magnitude f_s of the friction force. The free-body diagram is the

same as in Fig. 5.20b, but with $(f_s)_{max}$ replaced by f_s and $T = 230$ N replaced by $T = 50$ N.

EXECUTE: From the equilibrium conditions, Eqs. (5.2), we have

$$\sum F_x = T + (-f_s) = 0 \quad \text{so} \quad f_s = T = 50 \text{ N}$$

EVALUATE: The friction force can prevent motion for any horizontal applied force up to $(f_s)_{max} = \mu_s n = 230$ N. Below that value, f_s has the same magnitude as the applied force.

Example 5.15 **Minimizing kinetic friction**

In Example 5.13, suppose you move the crate by pulling upward on the rope at an angle of 30° above the horizontal. How hard must you pull to keep it moving with constant velocity? Assume that $\mu_k = 0.40$.

SOLUTION

IDENTIFY and SET UP: The crate is in equilibrium because its velocity is constant, so we again apply Newton's first law. Since the crate is in motion, the floor exerts a *kinetic* friction force. The target variable is the magnitude T of the tension force.

Figure 5.21 shows our sketch and free-body diagram. The kinetic friction force f_k is still equal to $\mu_k n$, but now the normal

5.21 Our sketches for this problem.

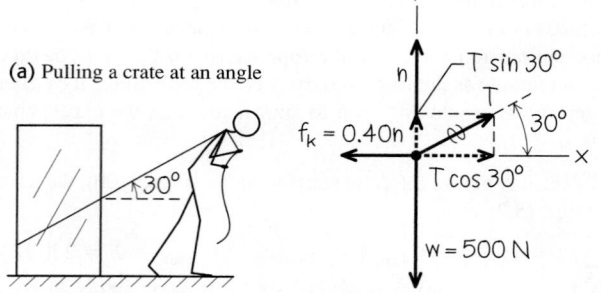

(b) Free-body diagram for moving crate

(a) Pulling a crate at an angle

force n is *not* equal in magnitude to the crate's weight. The force exerted by the rope has a vertical component that tends to lift the crate off the floor; this *reduces* n and so reduces f_k.

EXECUTE: From the equilibrium conditions and the equation $f_k = \mu_k n$, we have

$$\sum F_x = T\cos 30° + (-f_k) = 0 \quad \text{so} \quad T\cos 30° = \mu_k n$$
$$\sum F_y = T\sin 30° + n + (-w) = 0 \quad \text{so} \quad n = w - T\sin 30°$$

These are two equations for the two unknown quantities T and n. One way to find T is to substitute the expression for n in the second equation into the first equation and then solve the resulting equation for T:

$$T\cos 30° = \mu_k(w - T\sin 30°)$$

$$T = \frac{\mu_k w}{\cos 30° + \mu_k \sin 30°} = 188 \text{ N}$$

We can substitute this result into either of the original equations to obtain n. If we use the second equation, we get

$$n = w - T\sin 30° = (500 \text{ N}) - (188 \text{ N})\sin 30° = 406 \text{ N}$$

EVALUATE: As expected, the normal force is less than the 500-N weight of the box. It turns out that the tension required to keep the crate moving at constant speed is a little less than the 200-N force needed when you pulled horizontally in Example 5.13. Can you find an angle where the required pull is *minimum*? (See Challenge Problem 5.121.)

Example 5.16 **Toboggan ride with friction I**

Let's go back to the toboggan we studied in Example 5.10. The wax has worn off, so there is now a nonzero coefficient of kinetic friction μ_k. The slope has just the right angle to make the toboggan slide with constant velocity. Find this angle in terms of w and μ_k.

SOLUTION

IDENTIFY and SET UP: Our target variable is the slope angle α. The toboggan is in equilibrium because its velocity is constant, so we use Newton's first law in the form of Eqs. (5.2).

Three forces act on the toboggan: its weight, the normal force, and the kinetic friction force. The motion is downhill, so the friction force (which opposes the motion) is directed uphill. Figure 5.22 shows our sketch and free-body diagram (compare Fig. 5.12b in Example 5.10). The magnitude of the kinetic friction force is $f_k = \mu_k n$. We expect that the greater the value of μ_k, the steeper will be the required slope.

EXECUTE: The equilibrium conditions are

$$\sum F_x = w\sin\alpha + (-f_k) = w\sin\alpha - \mu_k n = 0$$
$$\sum F_y = n + (-w\cos\alpha) = 0$$

Rearranging these two equations, we get

$$\mu_k n = w\sin\alpha \quad \text{and} \quad n = w\cos\alpha$$

As in Example 5.10, the normal force is *not* equal to the weight. We eliminate n by dividing the first of these equations by the

5.22 Our sketches for this problem.

(a) The situation

(b) Free-body diagram for toboggan

second, with the result

$$\mu_k = \frac{\sin\alpha}{\cos\alpha} = \tan\alpha \quad \text{so} \quad \alpha = \arctan\mu_k$$

EVALUATE: The weight w doesn't appear in this expression. *Any* toboggan, regardless of its weight, slides down an incline with constant speed if the coefficient of kinetic friction equals the tangent of the slope angle of the incline. The arctangent function increases as its argument increases, so it's indeed true that the slope angle α increases as μ_k increases.

Example 5.17   **Toboggan ride with friction II**

The same toboggan with the same coefficient of friction as in Example 5.16 *accelerates* down a steeper hill. Derive an expression for the acceleration in terms of g, α, μ_k, and w.

SOLUTION

IDENTIFY and SET UP: The toboggan is accelerating, so we must use Newton's second law as given in Eqs. (5.4). Our target variable is the downhill acceleration.

Our sketch and free-body diagram (Fig. 5.23) are almost the same as for Example 5.16. The toboggan's y-component of acceleration a_y is still zero but the x-component a_x is not, so we've drawn the downhill component of weight as a longer vector than the (uphill) friction force.

EXECUTE: It's convenient to express the weight as $w = mg$. Then Newton's second law in component form says

$$\sum F_x = mg \sin \alpha + (-f_k) = ma_x$$
$$\sum F_y = n + (-mg \cos \alpha) = 0$$

5.23 Our sketches for this problem.

(a) The situation

(b) Free-body diagram for toboggan

From the second equation and Eq. (5.5) we get an expression for f_k:

$$n = mg \cos \alpha$$
$$f_k = \mu_k n = \mu_k mg \cos \alpha$$

We substitute this into the x-component equation and solve for a_x:

$$mg \sin \alpha + (-\mu_k mg \cos \alpha) = ma_x$$
$$a_x = g(\sin \alpha - \mu_k \cos \alpha)$$

EVALUATE: As for the frictionless toboggan in Example 5.10, the acceleration doesn't depend on the mass m of the toboggan. That's because all of the forces that act on the toboggan (weight, normal force, and kinetic friction force) are proportional to m.

Let's check some special cases. If the hill is vertical ($\alpha = 90°$) so that $\sin \alpha = 1$ and $\cos \alpha = 0$, we have $a_x = g$ (the toboggan falls freely). For a certain value of α the acceleration is zero; this happens if

$$\sin \alpha = \mu_k \cos \alpha \qquad \text{and} \qquad \mu_k = \tan \alpha$$

This agrees with our result for the constant-velocity toboggan in Example 5.16. If the angle is even smaller, $\mu_k \cos \alpha$ is greater than $\sin \alpha$ and a_x is *negative*; if we give the toboggan an initial downhill push to start it moving, it will slow down and stop. Finally, if the hill is frictionless so that $\mu_k = 0$, we retrieve the result of Example 5.10: $a_x = g \sin \alpha$.

Notice that we started with a simple problem (Example 5.10) and extended it to more and more general situations. The general result we found in this example includes *all* the previous ones as special cases. Don't memorize this result, but do make sure you understand how we obtained it and what it means.

Suppose instead we give the toboggan an initial push *up* the hill. The direction of the kinetic friction force is now reversed, so the acceleration is different from the downhill value. It turns out that the expression for a_x is the same as for downhill motion except that the minus sign becomes plus. Can you show this?

Rolling Friction

It's a lot easier to move a loaded filing cabinet across a horizontal floor using a cart with wheels than to slide it. How much easier? We can define a **coefficient of rolling friction** μ_r, which is the horizontal force needed for constant speed on a flat surface divided by the upward normal force exerted by the surface. Transportation engineers call μ_r the *tractive resistance*. Typical values of μ_r are 0.002 to 0.003 for steel wheels on steel rails and 0.01 to 0.02 for rubber tires on concrete. These values show one reason railroad trains are generally much more fuel efficient than highway trucks.

Fluid Resistance and Terminal Speed

Sticking your hand out the window of a fast-moving car will convince you of the existence of **fluid resistance,** the force that a fluid (a gas or liquid) exerts on a body moving through it. The moving body exerts a force on the fluid to push it out of the way. By Newton's third law, the fluid pushes back on the body with an equal and opposite force.

The *direction* of the fluid resistance force acting on a body is always opposite the direction of the body's velocity relative to the fluid. The *magnitude* of the fluid resistance force usually increases with the speed of the body through the fluid.

5.24 A metal ball falling through a fluid (oil).

(a) Metal ball falling through oil

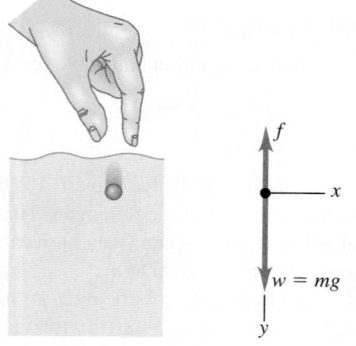

(b) Free-body diagram for ball in oil

$$\uparrow f$$
$$\bullet \longrightarrow x$$
$$\downarrow w = mg$$
$$y$$

Application **Pollen and Fluid Resistance**

These spiky spheres are pollen grains from the ragweed flower (*Ambrosia psilostachya*) and a common cause of hay fever. Because of their small radius (about 10 μm = 0.01 mm), when they are released into the air the fluid resistance force on them is proportional to their speed. The terminal speed given by Eq. (5.9) is only about 1 cm/s. Hence even a moderate wind can keep pollen grains aloft and carry them substantial distances from their source.

This is very different from the kinetic friction force between two surfaces in contact, which we can usually regard as independent of speed. For small objects moving at very low speeds, the magnitude f of the fluid resistance force is approximately proportional to the body's speed v:

$$f = kv \qquad \text{(fluid resistance at low speed)} \qquad (5.7)$$

where k is a proportionality constant that depends on the shape and size of the body and the properties of the fluid. Equation (5.7) is appropriate for dust particles falling in air or a ball bearing falling in oil. For larger objects moving through air at the speed of a tossed tennis ball or faster, the resisting force is approximately proportional to v^2 rather than to v. It is then called **air drag** or simply *drag*. Airplanes, falling raindrops, and bicyclists all experience air drag. In this case we replace Eq. (5.7) by

$$f = Dv^2 \qquad \text{(fluid resistance at high speed)} \qquad (5.8)$$

Because of the v^2 dependence, air drag increases rapidly with increasing speed. The air drag on a typical car is negligible at low speeds but comparable to or greater than rolling resistance at highway speeds. The value of D depends on the shape and size of the body and on the density of the air. You should verify that the units of the constant k in Eq. (5.7) are N · s/m or kg/s, and that the units of the constant D in Eq. (5.8) are N · s²/m² or kg/m.

Because of the effects of fluid resistance, an object falling in a fluid does *not* have a constant acceleration. To describe its motion, we can't use the constant-acceleration relationships from Chapter 2; instead, we have to start over using Newton's second law. As an example, suppose you drop a metal ball at the surface of a bucket of oil and let it fall to the bottom (Fig. 5.24a). The fluid resistance force in this situation is given by Eq. (5.7). What are the acceleration, velocity, and position of the metal ball as functions of time?

Figure 5.24b shows the free-body diagram. We take the positive y-direction to be downward and neglect any force associated with buoyancy in the oil. Since the ball is moving downward, its speed v is equal to its y-velocity v_y and the fluid resistance force is in the $-y$-direction. There are no x-components, so Newton's second law gives

$$\sum F_y = mg + (-kv_y) = ma_y$$

When the ball first starts to move, $v_y = 0$, the resisting force is zero, and the initial acceleration is $a_y = g$. As the speed increases, the resisting force also increases, until finally it is equal in magnitude to the weight. At this time $mg - kv_y = 0$, the acceleration becomes zero, and there is no further increase in speed. The final speed v_t, called the **terminal speed**, is given by $mg - kv_t = 0$, or

$$v_t = \frac{mg}{k} \qquad \text{(terminal speed, fluid resistance } f = kv) \qquad (5.9)$$

Figure 5.25 shows how the acceleration, velocity, and position vary with time. As time goes by, the acceleration approaches zero and the velocity approaches v_t

5.25 Graphs of the motion of a body falling without fluid resistance and with fluid resistance proportional to the speed.

Acceleration versus time

No fluid resistance: constant acceleration.

With fluid resistance: acceleration decreases.

Velocity versus time

No fluid resistance: velocity keeps increasing.

With fluid resistance: velocity has an upper limit.

Position versus time

No fluid resistance: parabolic curve.

With fluid resistance: curve straightens out.

(remember that we chose the positive *y*-direction to be down). The slope of the graph of *y* versus *t* becomes constant as the velocity becomes constant.

To see how the graphs in Fig. 5.25 are derived, we must find the relationship between velocity and time during the interval before the terminal speed is reached. We go back to Newton's second law, which we rewrite using $a_y = dv_y/dt$:

$$m\frac{dv_y}{dt} = mg - kv_y$$

After rearranging terms and replacing mg/k by v_t, we integrate both sides, noting that $v_y = 0$ when $t = 0$:

$$\int_0^v \frac{dv_y}{v_y - v_t} = -\frac{k}{m}\int_0^t dt$$

which integrates to

$$\ln\frac{v_t - v_y}{v_t} = -\frac{k}{m}t \qquad \text{or} \qquad 1 - \frac{v_y}{v_t} = e^{-(k/m)t}$$

and finally

$$v_y = v_t[1 - e^{-(k/m)t}] \tag{5.10}$$

Note that v_y becomes equal to the terminal speed v_t only in the limit that $t \to \infty$; the ball cannot attain terminal speed in any finite length of time.

The derivative of v_y gives a_y as a function of time, and the integral of v_y gives *y* as a function of time. We leave the derivations for you to complete; the results are

$$a_y = ge^{-(k/m)t} \tag{5.11}$$

$$y = v_t\left[t - \frac{m}{k}(1 - e^{-(k/m)t})\right] \tag{5.12}$$

Now look again at Fig. 5.25, which shows graphs of these three relationships.

In deriving the terminal speed in Eq. (5.9), we assumed that the fluid resistance force is proportional to the speed. For an object falling through the air at high speeds, so that the fluid resistance is equal to Dv^2 as in Eq. (5.8), the terminal speed is reached when Dv^2 equals the weight mg (Fig. 5.26a). You can show that the terminal speed v_t is given by

$$v_t = \sqrt{\frac{mg}{D}} \qquad \text{(terminal speed, fluid resistance } f = Dv^2) \tag{5.13}$$

This expression for terminal speed explains why heavy objects in air tend to fall faster than light objects. Two objects with the same physical size but different mass (say, a table-tennis ball and a lead ball with the same radius) have the same value of *D* but different values of *m*. The more massive object has a higher terminal speed and falls faster. The same idea explains why a sheet of paper falls faster if you first crumple it into a ball; the mass *m* is the same, but the smaller size makes *D* smaller (less air drag for a given speed) and v_t larger. Skydivers use the same principle to control their descent (Fig. 5.26b).

Figure 5.27 shows the trajectories of a baseball with and without air drag, assuming a coefficient $D = 1.3 \times 10^{-3}$ kg/m (appropriate for a batted ball at sea level). You can see that both the range of the baseball and the maximum height reached are substantially less than the zero-drag calculation would lead you to believe. Hence the baseball trajectory we calculated in Example 3.8 (Section 3.3) by ignoring air drag is unrealistic. Air drag is an important part of the game of baseball!

5.26 (a) Air drag and terminal speed. (b) By changing the positions of their arms and legs while falling, skydivers can change the value of the constant *D* in Eq. (5.8) and hence adjust the terminal speed of their fall [Eq. (5.13)].

(a) Free-body diagrams for falling with air drag

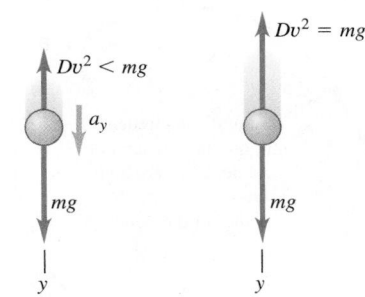

Before terminal speed: Object accelerating, drag force less than weight.

At terminal speed v_t: Object in equilibrium, drag force equals weight.

(b) A skydiver falling at terminal speed

5.27 Computer-generated trajectories of a baseball launched at 50 m/s at 35° above the horizontal. Note that the scales are different on the horizontal and vertical axes.

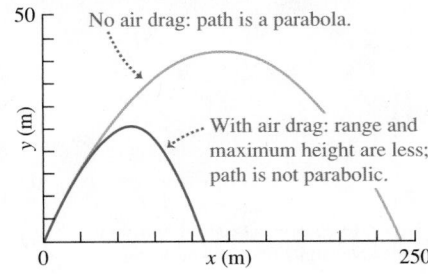

Example 5.18 | Terminal speed of a skydiver

For a human body falling through air in a spread-eagle position (Fig. 5.26b), the numerical value of the constant D in Eq. (5.8) is about 0.25 kg/m. Find the terminal speed for a lightweight 50-kg skydiver.

SOLUTION

IDENTIFY and SET UP: This example uses the relationship among terminal speed, mass, and drag coefficient. We use Eq. (5.13) to find the target variable v_t.

EXECUTE: We find for $m = 50$ kg:

$$v_t = \sqrt{\frac{mg}{D}} = \sqrt{\frac{(50 \text{ kg})(9.8 \text{ m/s}^2)}{0.25 \text{ kg/m}}}$$

$$= 44 \text{ m/s (about 160 km/h, or 99 mi/h)}$$

EVALUATE: The terminal speed is proportional to the square root of the skydiver's mass. A skydiver with the same drag coefficient D but twice the mass would have a terminal speed $\sqrt{2} = 1.41$ times greater, or 63 m/s. (A more massive skydiver would also have more frontal area and hence a larger drag coefficient, so his terminal speed would be a bit less than 63 m/s.) Even the lightweight skydiver's terminal speed is quite high, so skydives don't last very long. A drop from 2800 m (9200 ft) to the surface at the terminal speed takes only $(2800 \text{ m})/(44 \text{ m/s}) = 64$ s.

When the skydiver deploys the parachute, the value of D increases greatly. Hence the terminal speed of the skydiver and parachute decreases dramatically to a much lower value.

5.28 Net force, acceleration, and velocity in uniform circular motion.

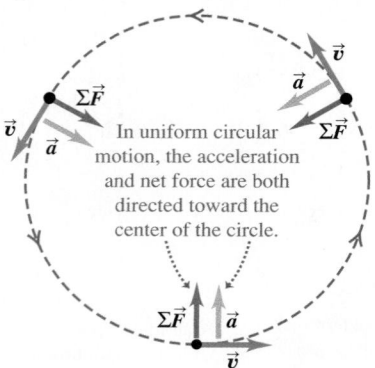

In uniform circular motion, the acceleration and net force are both directed toward the center of the circle.

Test Your Understanding of Section 5.3 Consider a box that is placed on different surfaces. (a) In which situation(s) is there *no* friction force acting on the box? (b) In which situation(s) is there a *static* friction force acting on the box? (c) In which situation(s) is there a *kinetic* friction force on the box? (i) The box is at rest on a rough horizontal surface. (ii) The box is at rest on a rough tilted surface. (iii) The box is on the rough-surfaced flat bed of a truck—the truck is moving at a constant velocity on a straight, level road, and the box remains in the same place in the middle of the truck bed. (iv) The box is on the rough-surfaced flat bed of a truck—the truck is speeding up on a straight, level road, and the box remains in the same place in the middle of the truck bed. (v) The box is on the rough-surfaced flat bed of a truck—the truck is climbing a hill, and the box is sliding toward the back of the truck. ∎

5.4 Dynamics of Circular Motion

We talked about uniform circular motion in Section 3.4. We showed that when a particle moves in a circular path with constant speed, the particle's acceleration is always directed toward the center of the circle (perpendicular to the instantaneous velocity). The magnitude a_{rad} of the acceleration is constant and is given in terms of the speed v and the radius R of the circle by

$$a_{\text{rad}} = \frac{v^2}{R} \qquad \text{(uniform circular motion)} \qquad (5.14)$$

The subscript "rad" is a reminder that at each point the acceleration is radially inward toward the center of the circle, perpendicular to the instantaneous velocity. We explained in Section 3.4 why this acceleration is often called *centripetal acceleration*.

We can also express the centripetal acceleration a_{rad} in terms of the *period T*, the time for one revolution:

$$T = \frac{2\pi R}{v} \qquad (5.15)$$

In terms of the period, a_{rad} is

$$a_{\text{rad}} = \frac{4\pi^2 R}{T^2} \qquad \text{(uniform circular motion)} \qquad (5.16)$$

5.29 What happens if the inward radial force suddenly ceases to act on a body in circular motion?

A ball attached to a string whirls in a circle on a frictionless surface.

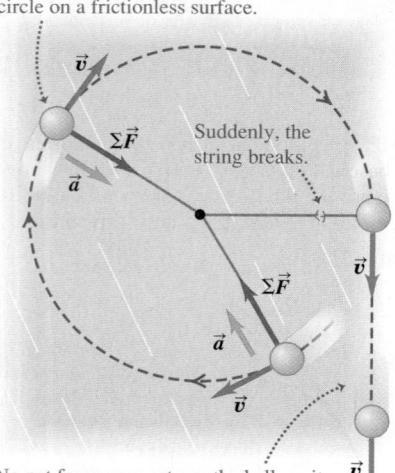

Suddenly, the string breaks.

No net force now acts on the ball, so it obeys Newton's first law—it moves in a straight line at constant velocity.

Uniform circular motion, like all other motion of a particle, is governed by Newton's second law. To make the particle accelerate toward the center of the circle, the net force $\Sigma \vec{F}$ on the particle must always be directed toward the center (Fig. 5.28). The magnitude of the acceleration is constant, so the magnitude F_{net} of the net force must also be constant. If the inward net force stops acting, the particle flies off in a straight line tangent to the circle (Fig. 5.29).

The magnitude of the radial acceleration is given by $a_{rad} = v^2/R$, so the magnitude F_{net} of the net force on a particle with mass m in uniform circular motion must be

$$F_{net} = ma_{rad} = m\frac{v^2}{R} \quad \text{(uniform circular motion)} \quad (5.17)$$

Uniform circular motion can result from *any* combination of forces, just so the net force $\sum\vec{F}$ is always directed toward the center of the circle and has a constant magnitude. Note that the body need not move around a complete circle: Equation (5.17) is valid for *any* path that can be regarded as part of a circular arc.

CAUTION **Avoid using "centrifugal force"** Figure 5.30 shows both a correct free-body diagram for uniform circular motion (Fig. 5.30a) and a common *incorrect* diagram (Fig. 5.30b). Figure 5.30b is incorrect because it includes an extra outward force of magnitude $m(v^2/R)$ to "keep the body out there" or to "keep it in equilibrium." There are three reasons not to include such an outward force, usually called *centrifugal force* ("centrifugal" means "fleeing from the center"). First, the body does *not* "stay out there": It is in constant motion around its circular path. Because its velocity is constantly changing in direction, the body accelerates and is *not* in equilibrium. Second, if there *were* an additional outward force that balanced the inward force, the net force would be zero and the body would move in a straight line, not a circle (Fig. 5.29). And third, the quantity $m(v^2/R)$ is *not* a force; it corresponds to the $m\vec{a}$ side of $\sum\vec{F} = m\vec{a}$ and does not appear in $\sum\vec{F}$ (Fig. 5.30a). It's true that when you ride in a car that goes around a circular path, you tend to slide to the outside of the turn as though there was a "centrifugal force." But we saw in Section 4.2 that what really happens is that you tend to keep moving in a straight line, and the outer side of the car "runs into" you as the car turns (Fig. 4.11c). *In an inertial frame of reference there is no such thing as "centrifugal force."* We won't mention this term again, and we strongly advise you to avoid using it as well.

5.30 (a) Correct and (b) incorrect free-body diagrams for a body in uniform circular motion.

(a) Correct free-body diagram

If you include the acceleration, draw it to one side of the body to show that it's not a force.

(b) Incorrect free-body diagram

The quantity mv^2/R is *not* a force—it doesn't belong in a free-body diagram.

Example 5.19 **Force in uniform circular motion**

A sled with a mass of 25.0 kg rests on a horizontal sheet of essentially frictionless ice. It is attached by a 5.00-m rope to a post set in the ice. Once given a push, the sled revolves uniformly in a circle around the post (Fig. 5.31a). If the sled makes five complete revolutions every minute, find the force F exerted on it by the rope.

SOLUTION

IDENTIFY and SET UP: The sled is in uniform circular motion, so it has a constant radial acceleration. We'll apply Newton's second law to the sled to find the magnitude F of the force exerted by the rope (our target variable).

5.31 (a) The situation. (b) Our free-body diagram.

(a) A sled in uniform circular motion

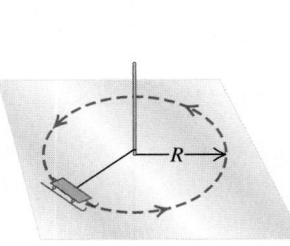

(b) Free-body diagram for sled

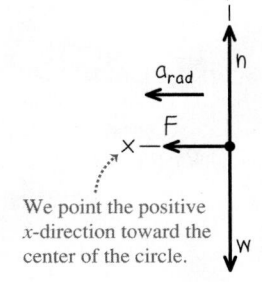

We point the positive x-direction toward the center of the circle.

Figure 5.31b shows our free-body diagram for the sled. The acceleration has only an x-component; this is toward the center of the circle, so we denote it as a_{rad}. The acceleration isn't given, so we'll need to determine its value using either Eq. (5.14) or Eq. (5.16).

EXECUTE: The force F appears in Newton's second law for the x-direction:

$$\sum F_x = F = ma_{rad}$$

We can find the centripetal acceleration a_{rad} using Eq. (5.16). The sled moves in a circle of radius $R = 5.00$ m with a period $T = (60.0\text{ s})/(5\text{ rev}) = 12.0$ s, so

$$a_{rad} = \frac{4\pi^2 R}{T^2} = \frac{4\pi^2(5.00\text{ m})}{(12.0\text{ s})^2} = 1.37\text{ m/s}^2$$

The magnitude F of the force exerted by the rope is then

$$F = ma_{rad} = (25.0\text{ kg})(1.37\text{ m/s}^2)$$
$$= 34.3\text{ kg}\cdot\text{m/s}^2 = 34.3\text{ N}$$

EVALUATE: You can check our value for a_{rad} by first finding the speed using Eq. (5.15), $v = 2\pi R/T$, and then using $a_{rad} = v^2/R$ from Eq. (5.14). Do you get the same result?

A greater force would be needed if the sled moved around the circle at a higher speed v. In fact, if v were doubled while R remained the same, F would be four times greater. Can you show this? How would F change if v remained the same but the radius R were doubled?

Example 5.20 A conical pendulum

An inventor designs a pendulum clock using a bob with mass m at the end of a thin wire of length L. Instead of swinging back and forth, the bob is to move in a horizontal circle with constant speed v, with the wire making a fixed angle β with the vertical direction (Fig. 5.32a). This is called a *conical pendulum* because the suspending wire traces out a cone. Find the tension F in the wire and the period T (the time for one revolution of the bob).

SOLUTION

IDENTIFY and SET UP: To find our target variables, the tension F and period T, we need two equations. These will be the horizontal and vertical components of Newton's second law applied to the bob. We'll find the radial acceleration of the bob using one of the circular motion equations.

Figure 5.32b shows our free-body diagram and coordinate system for the bob at a particular instant. There are just two forces on the bob: the weight mg and the tension F in the wire. Note that the

5.32 (a) The situation. (b) Our free-body diagram.

(a) The situation

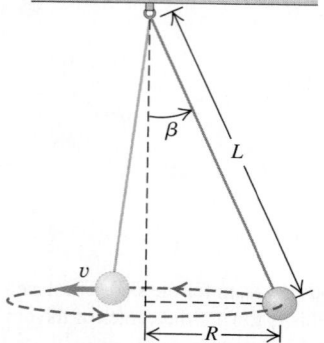

(b) Free-body diagram for pendulum bob

We point the positive x-direction toward the center of the circle.

center of the circular path is in the same horizontal plane as the bob, *not* at the top end of the wire. The horizontal component of tension is the force that produces the radial acceleration a_{rad}.

EXECUTE: The bob has zero vertical acceleration; the horizontal acceleration is toward the center of the circle, which is why we use the symbol a_{rad}. Newton's second law says

$$\sum F_x = F \sin \beta = m a_{\text{rad}}$$
$$\sum F_y = F \cos \beta + (-mg) = 0$$

These are two equations for the two unknowns F and β. The equation for $\sum F_y$ gives $F = mg/\cos \beta$; that's our target expression for F in terms of β. Substituting this result into the equation for $\sum F_x$ and using $\sin \beta / \cos \beta = \tan \beta$, we find

$$a_{\text{rad}} = g \tan \beta$$

To relate β to the period T, we use Eq. (5.16) for a_{rad}, solve for T, and insert $a_{\text{rad}} = g \tan \beta$:

$$a_{\text{rad}} = \frac{4\pi^2 R}{T^2} \quad \text{so} \quad T^2 = \frac{4\pi^2 R}{a_{\text{rad}}}$$

$$T = 2\pi \sqrt{\frac{R}{g \tan \beta}}$$

Figure 5.32a shows that $R = L \sin \beta$. We substitute this and use $\sin \beta / \tan \beta = \cos \beta$:

$$T = 2\pi \sqrt{\frac{L \cos \beta}{g}}$$

EVALUATE: For a given length L, as the angle β increases, $\cos \beta$ decreases, the period T becomes smaller, and the tension $F = mg/\cos \beta$ increases. The angle can never be 90°, however; this would require that $T = 0$, $F = \infty$, and $v = \infty$. A conical pendulum would not make a very good clock because the period depends on the angle β in such a direct way.

Example 5.21 Rounding a flat curve

The sports car in Example 3.11 (Section 3.4) is rounding a flat, unbanked curve with radius R (Fig. 5.33a). If the coefficient of static friction between tires and road is μ_s, what is the maximum speed v_{max} at which the driver can take the curve without sliding?

SOLUTION

IDENTIFY and SET UP: The car's acceleration as it rounds the curve has magnitude $a_{\text{rad}} = v^2/R$. Hence the maximum speed v_{max} (our target variable) corresponds to the maximum acceleration a_{rad} and to the maximum horizontal force on the car toward the center of its circular path. The only horizontal force acting on the car is the friction force exerted by the road. So to solve this problem we'll need Newton's second law, the equations of uniform circular motion, and our knowledge of the friction force from Section 5.3.

The free-body diagram in Fig. 5.33b includes the car's weight $w = mg$ and the two forces exerted by the road: the normal force n and the horizontal friction force f. The friction force must point toward the center of the circular path in order to cause the radial acceleration. The car doesn't slide toward or away from the center

of the circle, so the friction force is *static* friction, with a maximum magnitude $f_{\text{max}} = \mu_s n$ [see Eq. (5.6)].

5.33 (a) The situation. (b) Our free-body diagram.

(a) Car rounding flat curve

(b) Free-body diagram for car

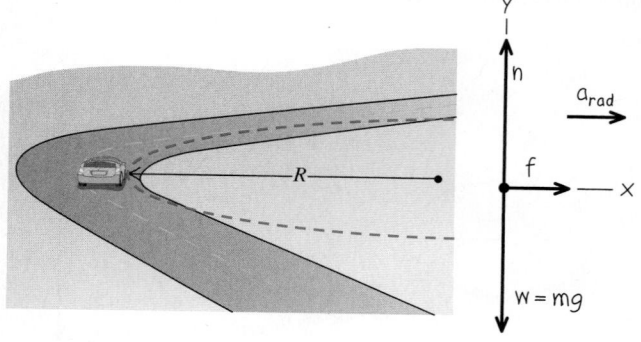

EXECUTE: The acceleration toward the center of the circular path is $a_{rad} = v^2/R$. There is no vertical acceleration. Thus we have

$$\sum F_x = f = ma_{rad} = m\frac{v^2}{R}$$

$$\sum F_y = n + (-mg) = 0$$

The second equation shows that $n = mg$. The first equation shows that the friction force *needed* to keep the car moving in its circular path increases with the car's speed. But the maximum friction force *available* is $f_{max} = \mu_s n = \mu_s mg$, and this determines the car's maximum speed. Substituting $\mu_s mg$ for f and v_{max} for v in the first equation, we find

$$\mu_s mg = m\frac{v^2_{max}}{R} \quad \text{so} \quad v_{max} = \sqrt{\mu_s gR}$$

As an example, if $\mu_s = 0.96$ and $R = 230$ m, we have

$$v_{max} = \sqrt{(0.96)(9.8 \text{ m/s}^2)(230 \text{ m})} = 47 \text{ m/s}$$

or about 170 km/h (100 mi/h). This is the maximum speed for this radius.

EVALUATE: If the car's speed is slower than $v_{max} = \sqrt{\mu_s gR}$, the required friction force is less than the maximum value $f_{max} = \mu_s mg$, and the car can easily make the curve. If we try to take the curve going *faster* than v_{max}, we will skid. We could still go in a circle without skidding at this higher speed, but the radius would have to be larger.

The maximum centripetal acceleration (called the "lateral acceleration" in Example 3.11) is equal to $\mu_s g$. That's why it's best to take curves at less than the posted speed limit if the road is wet or icy, either of which can reduce the value of μ_s and hence $\mu_s g$.

Example 5.22   Rounding a banked curve

For a car traveling at a certain speed, it is possible to bank a curve at just the right angle so that no friction at all is needed to maintain the car's turning radius. Then a car can safely round the curve even on wet ice. (Bobsled racing depends on this same idea.) Your engineering firm plans to rebuild the curve in Example 5.21 so that a car moving at a chosen speed v can safely make the turn even with no friction (Fig. 5.34a). At what angle β should the curve be banked?

SOLUTION

IDENTIFY and SET UP: With no friction, the only forces acting on the car are its weight and the normal force. Because the road is banked, the normal force (which acts perpendicular to the road surface) has a horizontal component. This component causes the car's horizontal acceleration toward the center of the car's circular path. We'll use Newton's second law to find the target variable β.

Our free-body diagram (Fig. 5.34b) is very similar to the diagram for the conical pendulum in Example 5.20 (Fig. 5.32b). The normal force acting on the car plays the role of the tension force exerted by the wire on the pendulum bob.

EXECUTE: The normal force $\vec{n}$ is perpendicular to the roadway and is at an angle β with the vertical (Fig. 5.34b). Thus it has a vertical component $n \cos \beta$ and a horizontal component $n \sin \beta$.

The acceleration in the x-direction is the centripetal acceleration $a_{rad} = v^2/R$; there is no acceleration in the y-direction. Thus the equations of Newton's second law are

$$\sum F_x = n \sin \beta = ma_{rad}$$

$$\sum F_y = n \cos \beta + (-mg) = 0$$

From the $\sum F_y$ equation, $n = mg/\cos \beta$. Substituting this into the $\sum F_x$ equation and using $a_{rad} = v^2/R$, we get an expression for the bank angle:

$$\tan \beta = \frac{a_{rad}}{g} = \frac{v^2}{gR} \quad \text{so} \quad \beta = \arctan \frac{v^2}{gR}$$

EVALUATE: The bank angle depends on both the speed and the radius. For a given radius, no one angle is correct for all speeds. In the design of highways and railroads, curves are often banked for the average speed of the traffic over them. If $R = 230$ m and $v = 25$ m/s (equal to a highway speed of 88 km/h, or 55 mi/h), then

$$\beta = \arctan \frac{(25 \text{ m/s})^2}{(9.8 \text{ m/s}^2)(230 \text{ m})} = 15°$$

This is within the range of banking angles actually used in highways.

5.34 (a) The situation. (b) Our free-body diagram.

(a) Car rounding banked curve

(b) Free-body diagram for car

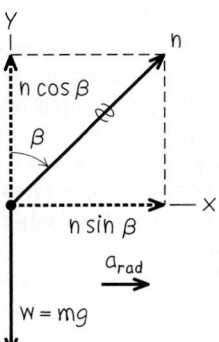

5.35 An airplane banks to one side in order to turn in that direction. The vertical component of the lift force $\vec{L}$ balances the force of gravity; the horizontal component of $\vec{L}$ causes the acceleration v^2/R.

$L \cos \beta$ L

β

$L \sin \beta$

$w = mg$

Mastering**PHYSICS**

ActivPhysics 4.2: Circular Motion Problem Solving
ActivPhysics 4.3: Cart Goes over Circular Path
ActivPhysics 4.4: Ball Swings on a String
ActivPhysics 4.5: Car Circles a Track

Banked Curves and the Flight of Airplanes

The results of Example 5.22 also apply to an airplane when it makes a turn in level flight (Fig. 5.35). When an airplane is flying in a straight line at a constant speed and at a steady altitude, the airplane's weight is exactly balanced by the lift force $\vec{L}$ exerted by the air. (The upward lift force that the air exerts on the wings is a reaction to the downward push the wings exert on the air as they move through it.) To make the airplane turn, the pilot banks the airplane to one side so that the lift force has a horizontal component as Fig. 5.35 shows. (The pilot also changes the angle at which the wings "bite" into the air so that the vertical component of lift continues to balance the weight.) The bank angle is related to the airplane's speed v and the radius R of the turn by the same expression as in Example 5.22: $\tan \beta = v^2/gR$. For an airplane to make a tight turn (small R) at high speed (large v), $\tan \beta$ must be large and the required bank angle β must approach 90°.

We can also apply the results of Example 5.22 to the *pilot* of an airplane. The free-body diagram for the pilot of the airplane is exactly as shown in Fig. 5.34b; the normal force $n = mg/\cos \beta$ is exerted on the pilot by the seat. As in Example 5.9, n is equal to the apparent weight of the pilot, which is greater than the pilot's true weight mg. In a tight turn with a large bank angle β, the pilot's apparent weight can be tremendous: $n = 5.8mg$ at $\beta = 80°$ and $n = 9.6mg$ at $\beta = 84°$. Pilots black out in such tight turns because the apparent weight of their blood increases by the same factor, and the human heart isn't strong enough to pump such apparently "heavy" blood to the brain.

Motion in a Vertical Circle

In Examples 5.19, 5.20, 5.21, and 5.22 the body moved in a horizontal circle. Motion in a *vertical* circle is no different in principle, but the weight of the body has to be treated carefully. The following example shows what we mean.

Example 5.23 Uniform circular motion in a vertical circle

A passenger on a carnival Ferris wheel moves in a vertical circle of radius R with constant speed v. The seat remains upright during the motion. Find expressions for the force the seat exerts on the passenger at the top of the circle and at the bottom.

SOLUTION

IDENTIFY and SET UP: The target variables are n_T, the upward normal force the seat applies to the passenger at the top of the circle, and n_B, the normal force at the bottom. We'll find these using Newton's second law and the uniform circular motion equations.

Figure 5.36a shows the passenger's velocity and acceleration at the two positions. The acceleration always points toward the center of the circle—downward at the top of the circle and upward at the bottom of the circle. At each position the only forces acting are vertical: the upward normal force and the downward force of gravity. Hence we need only the vertical component of Newton's second law. Figures 5.36b and 5.36c show free-body diagrams for the two positions. We take the positive y-direction as upward in both cases (that is, *opposite* the direction of the acceleration at the top of the circle).

EXECUTE: At the top the acceleration has magnitude v^2/R, but its vertical component is negative because its direction is downward.

Hence $a_y = -v^2/R$ and Newton's second law tells us that

Top: $\quad \sum F_y = n_T + (-mg) = -m\dfrac{v^2}{R} \quad$ or

$$n_T = mg\left(1 - \frac{v^2}{gR}\right)$$

5.36 Our sketches for this problem.

(a) Sketch of two positions
(b) Free-body diagram for passenger at top
(c) Free-body diagram for passenger at bottom

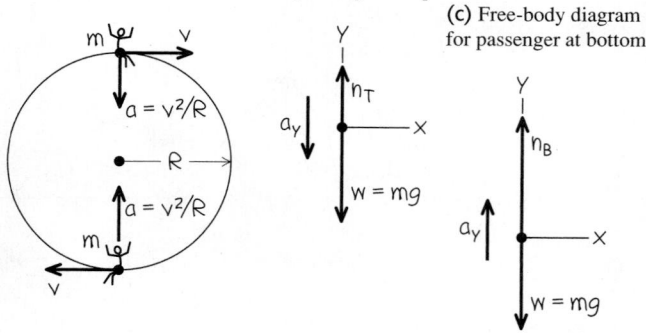

At the bottom the acceleration is upward, so $a_y = +v^2/R$ and Newton's second law says

Bottom: $$\sum F_y = n_B + (-mg) = +m\frac{v^2}{R} \quad \text{or}$$

$$n_B = mg\left(1 + \frac{v^2}{gR}\right)$$

EVALUATE: Our result for n_T tells us that at the top of the Ferris wheel, the upward force the seat applies to the passenger is *smaller* in magnitude than the passenger's weight $w = mg$. If the ride goes fast enough that $g - v^2/R$ becomes zero, the seat applies *no* force, and the passenger is about to become airborne. If v becomes still larger, n_T becomes negative; this means that a *downward* force (such as from a seat belt) is needed to keep the passenger in the seat. By contrast, the normal force n_B at the bottom is always *greater* than the passenger's weight. You feel the seat pushing up on you more firmly than when you are at rest. You can see that n_T and n_B are the values of the passenger's *apparent weight* at the top and bottom of the circle (see Section 5.2).

When we tie a string to an object and whirl it in a vertical circle, the analysis in Example 5.23 isn't directly applicable. The reason is that v is *not* constant in this case; except at the top and bottom of the circle, the net force (and hence the acceleration) does *not* point toward the center of the circle (Fig. 5.37). So both $\sum\vec{F}$ and $\vec{a}$ have a component tangent to the circle, which means that the speed changes. Hence this is a case of *nonuniform* circular motion (see Section 3.4). Even worse, we can't use the constant-acceleration formulas to relate the speeds at various points because *neither* the magnitude nor the direction of the acceleration is constant. The speed relationships we need are best obtained by using the concept of energy. We'll consider such problems in Chapter 7.

Test Your Understanding of Section 5.4 Satellites are held in orbit by the force of our planet's gravitational attraction. A satellite in a small-radius orbit moves at a higher speed than a satellite in an orbit of large radius. Based on this information, what you can conclude about the earth's gravitational attraction for the satellite? (i) It increases with increasing distance from the earth. (ii) It is the same at all distances from the earth. (iii) It decreases with increasing distance from the earth. (iv) This information by itself isn't enough to answer the question. ❚

5.37 A ball moving in a vertical circle.

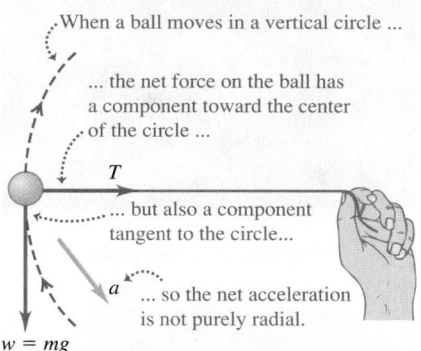

When a ball moves in a vertical circle ...

... the net force on the ball has a component toward the center of the circle ...

T

... but also a component tangent to the circle...

a ... so the net acceleration is not purely radial.

$w = mg$

5.5 The Fundamental Forces of Nature

We have discussed several kinds of forces—including weight, tension, friction, fluid resistance, and the normal force—and we will encounter others as we continue our study of physics. But just how many kinds of forces are there? Our current understanding is that all forces are expressions of just four distinct classes of *fundamental* forces, or interactions between particles (Fig. 5.38). Two are familiar in everyday experience. The other two involve interactions between subatomic particles that we cannot observe with the unaided senses.

Gravitational interactions include the familiar force of your *weight,* which results from the earth's gravitational attraction acting on you. The mutual gravitational attraction of various parts of the earth for each other holds our planet together (Fig. 5.38a). Newton recognized that the sun's gravitational attraction for the earth keeps the earth in its nearly circular orbit around the sun. In Chapter 13 we will study gravitational interactions in greater detail, and we will analyze their vital role in the motions of planets and satellites.

The second familiar class of forces, **electromagnetic interactions,** includes electric and magnetic forces. If you run a comb through your hair, the comb ends up with an electric charge; you can use the electric force exerted by this charge to pick up bits of paper. All atoms contain positive and negative electric charge, so atoms and molecules can exert electric forces on one another (Fig. 5.38b). Contact forces, including the normal force, friction, and fluid resistance, are the combination of all such forces exerted on the atoms of a body by atoms in its surroundings. *Magnetic* forces, such as those between magnets or between a magnet and a piece of iron, are actually the result of electric charges in motion. For example, an electromagnet causes magnetic interactions because electric

5.38 Examples of the fundamental interactions in nature. (a) The moon and the earth are held together and held in orbit by gravitational forces. (b) This molecule of bacterial plasmid DNA is held together by electromagnetic forces between its atoms. (c) The sun shines because in its core, strong forces between nuclear particles cause the release of energy. (d) When a massive star explodes into a supernova, a flood of energy is released by weak interactions between the star's nuclear particles.

(a) Gravitational forces hold planets together.

(b) Electromagnetic forces hold molecules together.

(c) Strong forces release energy to power the sun.

(d) Weak forces play a role in exploding stars.

charges move through its wires. We will study electromagnetic interactions in detail in the second half of this book.

On the atomic or molecular scale, gravitational forces play no role because electric forces are enormously stronger: The electrical repulsion between two protons is stronger than their gravitational attraction by a factor of about 10^{35}. But in bodies of astronomical size, positive and negative charges are usually present in nearly equal amounts, and the resulting electrical interactions nearly cancel out. Gravitational interactions are thus the dominant influence in the motion of planets and in the internal structure of stars.

The other two classes of interactions are less familiar. One, the **strong interaction,** is responsible for holding the nucleus of an atom together. Nuclei contain electrically neutral neutrons and positively charged protons. The electric force between charged protons tries to push them apart; the strong attractive force between nuclear particles counteracts this repulsion and makes the nucleus stable. In this context the strong interaction is also called the *strong nuclear force.* It has much shorter range than electrical interactions, but within its range it is much stronger. The strong interaction plays a key role in thermonuclear reactions that take place at the sun's core and generate the sun's heat and light (Fig. 5.38c).

Finally, there is the **weak interaction.** Its range is so short that it plays a role only on the scale of the nucleus or smaller. The weak interaction is responsible for a common form of radioactivity called beta decay, in which a neutron in a radioactive nucleus is transformed into a proton while ejecting an electron and a nearly massless particle called an antineutrino. The weak interaction between the antineutrino and ordinary matter is so feeble that an antineutrino could easily penetrate a wall of lead a million kilometers thick! Yet when a giant star undergoes a cataclysmic explosion called a supernova, most of the energy is released by way of the weak interaction (Fig. 5.38d).

In the 1960s physicists developed a theory that described the electromagnetic and weak interactions as aspects of a single *electroweak* interaction. This theory has passed every experimental test to which it has been put. Encouraged by this success, physicists have made similar attempts to describe the strong, electromagnetic, and weak interactions in terms of a single *grand unified theory* (GUT), and have taken steps toward a possible unification of all interactions into a *theory of everything* (TOE). Such theories are still speculative, and there are many unanswered questions in this very active field of current research.

Using Newton's first law: When a body is in equilibrium in an inertial frame of reference—that is, either at rest or moving with constant velocity—the vector sum of forces acting on it must be zero (Newton's first law). Free-body diagrams are essential in identifying the forces that act on the body being considered.

Newton's third law (action and reaction) is also frequently needed in equilibrium problems. The two forces in an action–reaction pair *never* act on the same body. (See Examples 5.1–5.5.)

The normal force exerted on a body by a surface is *not* always equal to the body's weight. (See Example 5.3.)

$$\sum \vec{F} = 0 \quad \text{(vector form)} \quad (5.1)$$

$$\sum F_x = 0$$
$$\sum F_y = 0 \quad \text{(component form)} \quad (5.2)$$

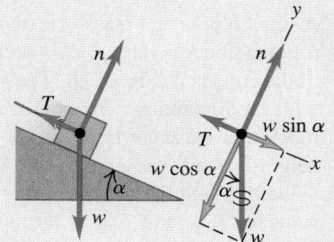

Using Newton's second law: If the vector sum of forces on a body is *not* zero, the body accelerates. The acceleration is related to the net force by Newton's second law.

Just as for equilibrium problems, free-body diagrams are essential for solving problems involving Newton's second law, and the normal force exerted on a body is not always equal to its weight. (See Examples 5.6–5.12.)

Vector form:
$$\sum \vec{F} = m\vec{a} \quad (5.3)$$

Component form:
$$\sum F_x = ma_x \qquad \sum F_y = ma_y \quad (5.4)$$

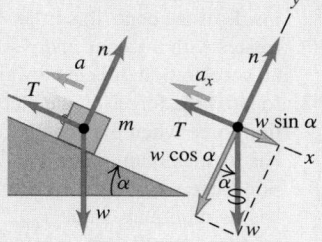

Friction and fluid resistance: The contact force between two bodies can always be represented in terms of a normal force $\vec{n}$ perpendicular to the surface of contact and a friction force $\vec{f}$ parallel to the surface.

When a body is sliding over the surface, the friction force is called *kinetic* friction. Its magnitude f_k is approximately equal to the normal force magnitude n multiplied by the coefficient of kinetic friction μ_k. When a body is *not* moving relative to a surface, the friction force is called *static* friction. The *maximum* possible static friction force is approximately equal to the magnitude n of the normal force multiplied by the coefficient of static friction μ_s. The *actual* static friction force may be anything from zero to this maximum value, depending on the situation. Usually μ_s is greater than μ_k for a given pair of surfaces in contact. (See Examples 5.13–5.17.)

Rolling friction is similar to kinetic friction, but the force of fluid resistance depends on the speed of an object through a fluid. (See Example 5.18.)

Magnitude of kinetic friction force:
$$f_k = \mu_k n \quad (5.5)$$

Magnitude of static friction force:
$$f_s \le \mu_s n \quad (5.6)$$

Forces in circular motion: In uniform circular motion, the acceleration vector is directed toward the center of the circle. The motion is governed by Newton's second law, $\sum \vec{F} = m\vec{a}$. (See Examples 5.19–5.23.)

Acceleration in uniform circular motion:
$$a_{\text{rad}} = \frac{v^2}{R} = \frac{4\pi^2 R}{T^2} \quad (5.14), (5.16)$$

BRIDGING PROBLEM | In a Rotating Cone

A small block with mass m is placed inside an inverted cone that is rotating about a vertical axis such that the time for one revolution of the cone is T (Fig. 5.39). The walls of the cone make an angle β with the horizontal. The coefficient of static friction between the block and the cone is μ_s. If the block is to remain at a constant height h above the apex of the cone, what are (a) the maximum value of T and (b) the minimum value of T? (That is, find expressions for $T_{\max}$ and $T_{\min}$ in terms of β and h.)

SOLUTION GUIDE

See MasteringPhysics® Study Area for a Video Tutor solution.

IDENTIFY and SET UP

1. Although we want the block to not slide up or down on the inside of the cone, this is *not* an equilibrium problem. The block rotates with the cone and is in uniform circular motion, so it has an acceleration directed toward the center of its circular path.
2. Identify the forces on the block. What is the direction of the friction force when the cone is rotating as slowly as possible, so T has its maximum value $T_{\max}$? What is the direction of the friction force when the cone is rotating as rapidly as possible, so T has its minimum value $T_{\min}$? In these situations does the static friction force have its *maximum* magnitude? Why or why not?
3. Draw a free-body diagram for the block when the cone is rotating with $T = T_{\max}$ and a free-body diagram when the cone is rotating with $T = T_{\min}$. Choose coordinate axes, and remember that it's usually easiest to choose one of the axes to be in the direction of the acceleration.
4. What is the radius of the circular path that the block follows? Express this in terms of β and h.
5. Make a list of the unknown quantities, and decide which of these are the target variables.

5.39 A block inside a spinning cone.

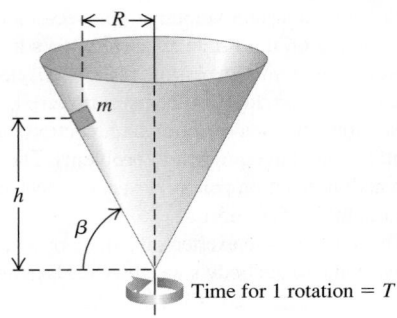

EXECUTE

6. Write Newton's second law in component form for the case in which the cone is rotating with $T = T_{\max}$. Write the acceleration in terms of $T_{\max}$, β, and h, and write the static friction force in terms of the normal force n.
7. Solve these equations for the target variable $T_{\max}$.
8. Repeat steps 6 and 7 for the case in which the cone is rotating with $T = T_{\min}$, and solve for the target variable $T_{\min}$.

EVALUATE

9. You'll end up with some fairly complicated expressions for $T_{\max}$ and $T_{\min}$, so check them over carefully. Do they have the correct units? Is the minimum time $T_{\min}$ less than the maximum time $T_{\max}$, as it must be?
10. What do your expressions for $T_{\max}$ and $T_{\min}$ become if $\mu_s = 0$? Check your results by comparing with Example 5.22 in Section 5.4.

Problems

For instructor-assigned homework, go to www.masteringphysics.com

•, ••, •••: Problems of increasing difficulty. **CP:** Cumulative problems incorporating material from earlier chapters. **CALC:** Problems requiring calculus. **BIO:** Biosciences problems.

DISCUSSION QUESTIONS

Q5.1 A man sits in a seat that is suspended from a rope. The rope passes over a pulley suspended from the ceiling, and the man holds the other end of the rope in his hands. What is the tension in the rope, and what force does the seat exert on the man? Draw a free-body force diagram for the man.

Q5.2 "In general, the normal force is not equal to the weight." Give an example where these two forces are equal in magnitude, and at least two examples where they are not.

Q5.3 A clothesline hangs between two poles. No matter how tightly the line is stretched, it always sags a little at the center. Explain why.

Q5.4 A car is driven up a steep hill at constant speed. Discuss all the forces acting on the car. What pushes it up the hill?

Q5.5 For medical reasons it is important for astronauts in outer space to determine their body mass at regular intervals. Devise a scheme for measuring body mass in an apparently weightless environment.

Q5.6 To push a box up a ramp, is the force required smaller if you push horizontally or if you push parallel to the ramp? Why?

Q5.7 A woman in an elevator lets go of her briefcase but it does not fall to the floor. How is the elevator moving?

Q5.8 You can classify scales for weighing objects as those that use springs and those that use standard masses to balance unknown masses. Which group would be more accurate when used in an accelerating spaceship? When used on the moon?

Q5.9 When you tighten a nut on a bolt, how are you increasing the frictional force? How does a lock washer work?

Q5.10 A block rests on an inclined plane with enough friction to prevent it from sliding down. To start the block moving, is it easier to push it up the plane or down the plane? Why?

Q5.11 A crate of books rests on a level floor. To move it along the floor at a constant velocity, why do you exert a smaller force if you pull it at an angle θ above the horizontal than if you push it at the same angle below the horizontal?

Q5.12 In a world without friction, which of the following activities could you do (or not do)? Explain your reasoning. (a) drive around an unbanked highway curve; (b) jump into the air; (c) start walking on a horizontal sidewalk; (d) climb a vertical ladder; (e) change lanes on the freeway.

Q5.13 Walking on horizontal slippery ice can be much more tiring than walking on ordinary pavement. Why?

Q5.14 When you stand with bare feet in a wet bathtub, the grip feels fairly secure, and yet a catastrophic slip is quite possible. Explain this in terms of the two coefficients of friction.

Q5.15 You are pushing a large crate from the back of a freight elevator to the front as the elevator is moving to the next floor. In which situation is the force you must apply to move the crate the smallest and in which is it the largest: when the elevator is accelerating upward, when it is accelerating downward, or when it is traveling at constant speed? Explain.

Q5.16 The moon is accelerating toward the earth. Why isn't it getting closer to us?

Q5.17 An automotive magazine calls decreasing-radius curves "the bane of the Sunday driver." Explain.

Q5.18 You often hear people say that "friction always opposes motion." Give at least one example where (a) static friction *causes* motion, and (b) kinetic friction *causes* motion.

Q5.19 If there is a net force on a particle in uniform circular motion, why doesn't the particle's speed change?

Q5.20 A curve in a road has the banking angle calculated and posted for 80 km/h. However, the road is covered with ice so you cautiously plan to drive slower than this limit. What may happen to your car? Why?

Q5.21 You swing a ball on the end of a lightweight string in a horizontal circle at constant speed. Can the string ever be truly horizontal? If not, would it slope above the horizontal or below the horizontal? Why?

Q5.22 The centrifugal force is not included in the free-body diagrams of Figs. 5.34b and 5.35. Explain why not.

Q5.23 A professor swings a rubber stopper in a horizontal circle on the end of a string in front of his class. He tells Caroline, in the first row, that he is going to let the string go when the stopper is directly in front of her face. Should Caroline worry?

Q5.24 To keep the forces on the riders within allowable limits, loop-the-loop roller coaster rides are often designed so that the loop, rather than being a perfect circle, has a larger radius of curvature at the bottom than at the top. Explain.

Q5.25 A tennis ball drops from rest at the top of a tall glass cylinder, first with the air pumped out of the cylinder so there is no air resistance, and then a second time after the air has been readmitted to the cylinder. You examine multiflash photographs of the two drops. From these photos how can you tell which one is which, or can you?

Q5.26 If you throw a baseball straight upward with speed v_0, how does its speed, when it returns to the point from where you threw it, compare to v_0 (a) in the absence of air resistance and (b) in the presence of air resistance? Explain.

Q5.27 You throw a baseball straight upward. If air resistance is *not* ignored, how does the time required for the ball to go from the height at which it was thrown up to its maximum height compare to the time required for it to fall from its maximum height back down to the height from which it was thrown? Explain your answer.

Q5.28 You take two identical tennis balls and fill one with water. You release both balls simultaneously from the top of a tall building. If air resistance is negligible, which ball strikes the ground first? Explain. What is the answer if air resistance is *not* negligible?

Q5.29 A ball is dropped from rest and feels air resistance as it falls. Which of the graphs in Fig. Q5.29 best represents its acceleration as a function of time?

Figure **Q5.29**

   (a)   (b)   (c)   (d)   (e)

Q5.30 A ball is dropped from rest and feels air resistance as it falls. Which of the graphs in Fig. Q5.30 best represents its vertical velocity component as a function of time?

Figure **Q5.30**

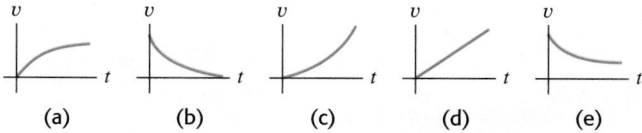

   (a)   (b)   (c)   (d)   (e)

Q5.31 When does a baseball in flight have an acceleration with a positive upward component? Explain in terms of the forces on the ball and also in terms of the velocity components compared to the terminal speed. Do *not* ignore air resistance.

Q5.32 When a batted baseball moves with air drag, does it travel a greater horizontal distance while climbing to its maximum height or while descending from its maximum height back to the ground? Or is the horizontal distance traveled the same for both? Explain in terms of the forces acting on the ball.

Q5.33 "A ball is thrown from the edge of a high cliff. No matter what the angle at which it is thrown, due to air resistance, the ball will eventually end up moving vertically downward." Justify this statement.

EXERCISES

Section 5.1 Using Newton's First Law: Particles in Equilibrium

5.1 • Two 25.0-N weights are suspended at opposite ends of a rope that passes over a light, frictionless pulley. The pulley is attached to a chain that goes to the ceiling. (a) What is the tension in the rope? (b) What is the tension in the chain?

5.2 • In Fig. E5.2 each of the suspended blocks has weight w. The pulleys are frictionless and the ropes have negligible weight. Calculate, in each case, the tension T in the rope in terms of the weight w. In each case, include the free-body diagram or diagrams you used to determine the answer.

Figure **E5.2**

(a)   (b)   (c)

5.3 • A 75.0-kg wrecking ball hangs from a uniform heavy-duty chain having a mass of 26.0 kg. (a) Find the maximum and minimum tension in the chain. (b) What is the tension at a point three-fourths of the way up from the bottom of the chain?

5.4 •• **BIO Injuries to the Spinal Column.** In the treatment of spine injuries, it is often necessary to provide some tension along the spinal column to stretch the backbone. One device for doing this is the Stryker frame, illustrated in Fig. E5.4a. A weight W is attached to the patient (sometimes around a neck collar, as shown in Fig. E5.4b), and friction between the person's body and the bed prevents sliding. (a) If the coefficient of static friction between a 78.5-kg patient's body and the bed is 0.75, what is the maximum traction force along the spinal column that W can provide without causing the patient to slide? (b) Under the conditions of maximum traction, what is the tension in each cable attached to the neck collar?

Figure **E5.4**

5.5 •• A picture frame hung against a wall is suspended by two wires attached to its upper corners. If the two wires make the same angle with the vertical, what must this angle be if the tension in each wire is equal to 0.75 of the weight of the frame? (Ignore any friction between the wall and the picture frame.)

5.6 •• A large wrecking ball is held in place by two light steel cables (Fig. E5.6). If the mass m of the wrecking ball is 4090 kg, what are (a) the tension T_B in the cable that makes an angle of 40° with the vertical and (b) the tension T_A in the horizontal cable?

Figure **E5.6**

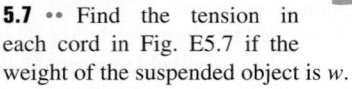

5.7 •• Find the tension in each cord in Fig. E5.7 if the weight of the suspended object is w.

Figure **E5.7**

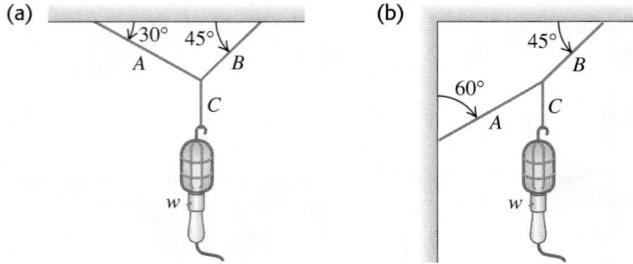

5.8 •• A 1130-kg car is held in place by a light cable on a very smooth (frictionless) ramp, as shown in Fig. E5.8. The cable makes an angle of 31.0° above the surface of the ramp, and the ramp itself rises at 25.0° above the horizontal. (a) Draw a free-body diagram for the car. (b) Find the tension in the cable. (c) How hard does the surface of the ramp push on the car?

Figure **E5.8**

5.9 •• A man pushes on a piano with mass 180 kg so that it slides at constant velocity down a ramp that is inclined at 11.0° above the horizontal floor. Neglect any friction acting on the piano. Calculate the magnitude of the force applied by the man if he pushes (a) parallel to the incline and (b) parallel to the floor.

5.10 •• In Fig. E5.10 the weight w is 60.0 N. (a) What is the tension in the diagonal string? (b) Find the magnitudes of the horizontal forces $\vec{F}_1$ and $\vec{F}_2$ that must be applied to hold the system in the position shown.

Figure **E5.10**

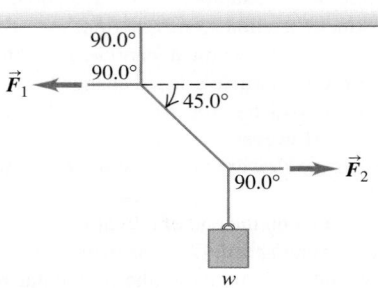

Section 5.2 Using Newton's Second Law: Dynamics of Particles

5.11 •• **BIO Stay Awake!** An astronaut is inside a 2.25×10^6 kg rocket that is blasting off vertically from the launch pad. You want this rocket to reach the speed of sound (331 m/s) as quickly as possible, but you also do not want the astronaut to black out. Medical tests have shown that astronauts are in danger of blacking out at an acceleration greater than $4g$. (a) What is the maximum thrust the engines of the rocket can have to just barely avoid blackout? Start with a free-body diagram of the rocket. (b) What force, in terms of her weight w, does the rocket exert on the astronaut? Start with a free-body diagram of the astronaut. (c) What is the shortest time it can take the rocket to reach the speed of sound?

5.12 •• A 125-kg (including all the contents) rocket has an engine that produces a constant vertical force (the *thrust*) of 1720 N. Inside this rocket, a 15.5-N electrical power supply rests on the floor. (a) Find the acceleration of the rocket. (b) When it has reached an altitude of 120 m, how hard does the floor push on the power supply? (*Hint:* Start with a free-body diagram for the power supply.)

5.13 •• **CP** *Genesis* **Crash.** On September 8, 2004, the *Genesis* spacecraft crashed in the Utah desert because its parachute did not open. The 210-kg capsule hit the ground at 311 km/h and penetrated the soil to a depth of 81.0 cm. (a) Assuming it to be constant, what was its acceleration (in m/s² and in g's) during the crash? (b) What force did the ground exert on the capsule during the crash? Express the force in newtons and as a multiple of the capsule's weight. (c) For how long did this force last?

5.14 • Three sleds are being pulled horizontally on frictionless horizontal ice using horizontal ropes (Fig. E5.14). The pull is of magnitude 125 N. Find (a) the acceleration of the system and (b) the tension in ropes A and B.

Figure **E5.14**

5.15 •• **Atwood's Machine.** A 15.0-kg load of bricks hangs from one end of a rope that passes over a small, frictionless pulley. A 28.0-kg counterweight is suspended from the other end of the rope, as shown in Fig. E5.15. The system is released from rest. (a) Draw two free-body diagrams, one for the load of bricks and one for the counterweight. (b) What is the magnitude of the upward acceleration of the load of bricks? (c) What is the tension in the rope while the load is moving? How does the tension compare to the weight of the load of bricks? To the weight of the counterweight?

Figure **E5.15**

28.0 kg

15.0 kg

5.16 •• **CP** A 8.00-kg block of ice, released from rest at the top of a 1.50-m-long frictionless ramp, slides downhill, reaching a speed of 2.50 m/s at the bottom. (a) What is the angle between the ramp and the horizontal? (b) What would be the speed of the ice at the bottom if the motion were opposed by a constant friction force of 10.0 N parallel to the surface of the ramp?

5.17 •• A light rope is attached to a block with mass 4.00 kg that rests on a frictionless, horizontal surface. The horizontal rope passes over a frictionless, massless pulley, and a block with mass m is suspended from the other end. When the blocks are released, the tension in the rope is 10.0 N. (a) Draw two free-body diagrams, one for the 4.00-kg block and one for the block with mass m. (b) What is the acceleration of either block? (c) Find the mass m of the hanging block? (d) How does the tension compare to the weight of the hanging block?

5.18 •• **CP Runway Design.** A transport plane takes off from a level landing field with two gliders in tow, one behind the other. The mass of each glider is 700 kg, and the total resistance (air drag plus friction with the runway) on each may be assumed constant and equal to 2500 N. The tension in the towrope between the transport plane and the first glider is not to exceed 12,000 N. (a) If a speed of 40 m/s is required for takeoff, what minimum length of runway is needed? (b) What is the tension in the towrope between the two gliders while they are accelerating for the takeoff?

5.19 •• **CP** A 750.0-kg boulder is raised from a quarry 125 m deep by a long uniform chain having a mass of 575 kg. This chain is of uniform strength, but at any point it can support a maximum tension no greater than 2.50 times its weight without breaking. (a) What is the maximum acceleration the boulder can have and still get out of the quarry, and (b) how long does it take to be lifted out at maximum acceleration if it started from rest?

5.20 •• **Apparent Weight.** A 550-N physics student stands on a bathroom scale in an 850-kg (including the student) elevator that is supported by a cable. As the elevator starts moving, the scale reads 450 N. (a) Find the acceleration of the elevator (magnitude and direction). (b) What is the acceleration if the scale reads 670 N? (c) If the scale reads zero, should the student worry? Explain. (d) What is the tension in the cable in parts (a) and (c)?

5.21 •• **CP BIO Force During a Jump.** An average person can reach a maximum height of about 60 cm when jumping straight up from a crouched position. During the jump itself, the person's body from the knees up typically rises a distance of around 50 cm. To keep the calculations simple and yet get a reasonable result, assume that the *entire body* rises this much during the jump. (a) With what initial speed does the person leave the ground to reach a height of 60 cm? (b) Draw a free-body diagram of the person during the jump. (c) In terms of this jumper's weight w, what force does the ground exert on him or her during the jump?

5.22 •• **CP CALC** A 2540-kg test rocket is launched vertically from the launch pad. Its fuel (of negligible mass) provides a thrust force so that its vertical velocity as a function of time is given by $v(t) = At + Bt^2$, where A and B are constants and time is measured from the instant the fuel is ignited. At the instant of ignition, the rocket has an upward acceleration of 1.50 m/s^2 and 1.00 s later an upward velocity of 2.00 m/s. (a) Determine A and B, including their SI units. (b) At 4.00 s after fuel ignition, what is the acceleration of the rocket, and (c) what thrust force does the burning fuel exert on it, assuming no air resistance? Express the thrust in newtons and as a multiple of the rocket's weight. (d) What was the initial thrust due to the fuel?

5.23 •• **CP CALC** A 2.00-kg box is moving to the right with speed 9.00 m/s on a horizontal, frictionless surface. At $t = 0$ a horizontal force is applied to the box. The force is directed to the left and has magnitude $F(t) = (6.00 \text{ N/s}^2)t^2$. (a) What distance does the box move from its position at $t = 0$ before its speed is reduced to zero? (b) If the force continues to be applied, what is the speed of the box at $t = 3.00$ s?

5.24 •• **CP CALC** A 5.00-kg crate is suspended from the end of a short vertical rope of negligible mass. An upward force $F(t)$ is applied to the end of the rope, and the height of the crate above its initial position is given by $y(t) = (2.80 \text{ m/s})t + (0.610 \text{ m/s}^3)t^3$. What is the magnitude of the force F when $t = 4.00$ s?

Section 5.3 Frictional Forces

5.25 • **BIO The Trendelenburg Position.** In emergencies with major blood loss, the doctor will order the patient placed in the Trendelenburg position, in which the foot of the bed is raised to get maximum blood flow to the brain. If the coefficient of static friction between the typical patient and the bedsheets is 1.20, what is the maximum angle at which the bed can be tilted with respect to the floor before the patient begins to slide?

5.26 • In a laboratory experiment on friction, a 135-N block resting on a rough horizontal table is pulled by a horizontal wire. The pull gradually increases until the block begins to move and continues to increase thereafter. Figure E5.26 shows a graph of the friction force on this block as a function of the pull. (a) Identify the

Figure **E5.26**

regions of the graph where static and kinetic friction occur. (b) Find the coefficients of static and kinetic friction between the block and the table. (c) Why does the graph slant upward in the first part but then level out? (d) What would the graph look like if a 135-N brick were placed on the box, and what would the coefficients of friction be in that case?

5.27 •• **CP** A stockroom worker pushes a box with mass 11.2 kg on a horizontal surface with a constant speed of 3.50 m/s. The coefficient of kinetic friction between the box and the surface is 0.20. (a) What horizontal force must the worker apply to maintain the motion? (b) If the force calculated in part (a) is removed, how far does the box slide before coming to rest?

5.28 •• A box of bananas weighing 40.0 N rests on a horizontal surface. The coefficient of static friction between the box and the surface is 0.40, and the coefficient of kinetic friction is 0.20. (a) If no horizontal force is applied to the box and the box is at rest, how large is the friction force exerted on the box? (b) What is the magnitude of the friction force if a monkey applies a horizontal force of 6.0 N to the box and the box is initially at rest? (c) What minimum horizontal force must the monkey apply to start the box in motion? (d) What minimum horizontal force must the monkey apply to keep the box moving at constant velocity once it has been started? (e) If the monkey applies a horizontal force of 18.0 N, what is the magnitude of the friction force and what is the box's acceleration?

5.29 •• A 45.0-kg crate of tools rests on a horizontal floor. You exert a gradually increasing horizontal push on it and observe that the crate just begins to move when your force exceeds 313 N. After that you must reduce your push to 208 N to keep it moving at a steady 25.0 cm/s. (a) What are the coefficients of static and kinetic friction between the crate and the floor? (b) What push must you exert to give it an acceleration of 1.10 m/s^2? (c) Suppose you were performing the same experiment on this crate but were doing it on the moon instead, where the acceleration due to gravity is 1.62 m/s^2. (i) What magnitude push would cause it to move? (ii) What would its acceleration be if you maintained the push in part (b)?

5.30 •• Some sliding rocks approach the base of a hill with a speed of 12 m/s. The hill rises at 36° above the horizontal and has coefficients of kinetic and static friction of 0.45 and 0.65, respectively, with these rocks. (a) Find the acceleration of the rocks as they slide up the hill. (b) Once a rock reaches its highest point, will it stay there or slide down the hill? If it stays there, show why. If it slides down, find its acceleration on the way down.

5.31 •• You are lowering two boxes, one on top of the other, down the ramp shown in Fig. E5.31 by pulling on a rope parallel to the surface of the ramp. Both boxes move together at a constant speed of 15.0 cm/s. The coefficient of kinetic friction between the ramp and the lower box is 0.444, and the coefficient of static friction between the two boxes is 0.800. (a) What force do you need to exert to accomplish this? (b) What are the magnitude and direction of the friction force on the upper box?

Figure **E5.31**

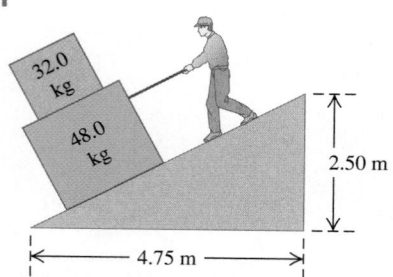

5.32 •• A pickup truck is carrying a toolbox, but the rear gate of the truck is missing, so the box will slide out if it is set moving. The coefficients of kinetic and static friction between the box and the bed of the truck are 0.355 and 0.650, respectively. Starting from rest, what is the shortest time this truck could accelerate uniformly to 30.0 m/s without causing the box to slide? Include a free-body diagram of the toolbox as part of your solution.

5.33 •• **CP Stopping Distance.** (a) If the coefficient of kinetic friction between tires and dry pavement is 0.80, what is the shortest distance in which you can stop an automobile by locking the brakes when traveling at 28.7 m/s (about 65 mi/h)? (b) On wet pavement the coefficient of kinetic friction may be only 0.25. How fast should you drive on wet pavement in order to be able to stop in the same distance as in part (a)? (*Note:* Locking the brakes is *not* the safest way to stop.)

5.34 •• Consider the system shown in Fig. E5.34. Block A weighs 45.0 N and block B weighs 25.0 N. Once block B is set into downward motion, it descends at a constant speed. (a) Calculate the coefficient of kinetic friction between block A and the tabletop. (b) A cat, also of weight 45.0 N, falls asleep on top of block A. If block B is now set into downward motion, what is its acceleration (magnitude and direction)?

Figure **E5.34**

5.35 • Two crates connected by a rope lie on a horizontal surface (Fig. E5.35). Crate A has mass m_A and crate B has mass m_B. The coefficient of kinetic friction between each crate and the surface is μ_k. The crates are pulled to the right at constant velocity by a horizontal force $\vec{F}$. In terms of m_A, m_B, and μ_k, calculate (a) the magnitude of the force $\vec{F}$ and (b) the tension in the rope connecting the blocks. Include the free-body diagram or diagrams you used to determine each answer.

Figure **E5.35**

5.36 •• **CP** A 25.0-kg box of textbooks rests on a loading ramp that makes an angle α with the horizontal. The coefficient of kinetic friction is 0.25, and the coefficient of static friction is 0.35. (a) As the angle α is increased, find the minimum angle at which the box starts to slip. (b) At this angle, find the acceleration once the box has begun to move. (c) At this angle, how fast will the box be moving after it has slid 5.0 m along the loading ramp?

5.37 •• **CP** As shown in Fig. E5.34, block A (mass 2.25 kg) rests on a tabletop. It is connected by a horizontal cord passing over a light, frictionless pulley to a hanging block B (mass 1.30 kg). The coefficient of kinetic friction between block A and the tabletop is 0.450. After the blocks are released from rest, find (a) the speed of each block after moving 3.00 cm and (b) the tension in the cord. Include the free-body diagram or diagrams you used to determine the answers.

5.38 •• A box with mass m is dragged across a level floor having a coefficient of kinetic friction μ_k by a rope that is pulled upward at an angle θ above the horizontal with a force of magnitude F. (a) In terms of m, μ_k, θ, and g, obtain an expression for the magnitude of the force required to move the box with constant speed. (b) Knowing that you are studying physics, a CPR instructor asks you how much force it would take to slide a 90-kg patient across a floor at constant speed by pulling on him at an angle of 25° above the horizontal. By dragging some weights wrapped in an old pair of pants down the hall with a spring balance, you find that $\mu_k = 0.35$. Use the result of part (a) to answer the instructor's question.

5.39 •• A large crate with mass m rests on a horizontal floor. The coefficients of friction between the crate and the floor are μ_s and μ_k. A woman pushes downward at an angle θ below the horizontal on the crate with a force $\vec{F}$. (a) What magnitude of force $\vec{F}$ is required to keep the crate moving at constant velocity? (b) If μ_s is greater than some critical value, the woman cannot start the crate moving no matter how hard she pushes. Calculate this critical value of μ_s.

5.40 •• You throw a baseball straight up. The drag force is proportional to v^2. In terms of g, what is the y-component of the ball's acceleration when its speed is half its terminal speed and (a) it is moving up? (b) It is moving back down?

5.41 • (a) In Example 5.18 (Section 5.3), what value of D is required to make $v_t = 42$ m/s for the skydiver? (b) If the skydiver's daughter, whose mass is 45 kg, is falling through the air and has the same D (0.25 kg/m) as her father, what is the daughter's terminal speed?

Section 5.4 Dynamics of Circular Motion

5.42 •• A small car with mass 0.800 kg travels at constant speed on the inside of a track that is a vertical circle with radius 5.00 m (Fig. E5.42). If the normal force exerted by the track on the car when it is at the top of the track (point B) is 6.00 N, what is the normal force on the car when it is at the bottom of the track (point A)?

Figure **E5.42**

5.43 •• A machine part consists of a thin 40.0-cm-long bar with small 1.15-kg masses fastened by screws to its ends. The screws can support a maximum force of 75.0 N without pulling out. This bar rotates about an axis perpendicular to it at its center. (a) As the bar is turning at a constant rate on a horizontal, frictionless surface, what is the maximum speed the masses can have without pulling out the screws? (b) Suppose the machine is redesigned so that the bar turns at a constant rate in a vertical circle. Will one of the screws be more likely to pull out when the mass is at the top of the circle or at the bottom? Use a free-body diagram to see why. (c) Using the result of part (b), what is the greatest speed the masses can have without pulling a screw?

5.44 • A flat (unbanked) curve on a highway has a radius of 220.0 m. A car rounds the curve at a speed of 25.0 m/s. (a) What is the minimum coefficient of friction that will prevent sliding? (b) Suppose the highway is icy and the coefficient of friction between the tires and pavement is only one-third what you found in part (a). What should be the maximum speed of the car so it can round the curve safely?

5.45 •• A 1125-kg car and a 2250-kg pickup truck approach a curve on the expressway that has a radius of 225 m. (a) At what angle should the highway engineer bank this curve so that vehicles traveling at 65.0 mi/h can safely round it regardless of the condition of their tires? Should the heavy truck go slower than the lighter car? (b) As the car and truck round the curve at find the normal force on each one due to the highway surface.

5.46 •• The "Giant Swing" at a county fair consists of a vertical central shaft with a number of horizontal arms attached at its upper end (Fig. E5.46). Each arm supports a seat suspended from a cable 5.00 m long, the upper end of the cable being fastened to the arm at a point 3.00 m from the central shaft. (a) Find the time of one revolution of the swing if the cable supporting a seat makes an angle of 30.0° with the vertical. (b) Does the angle depend on the weight of the passenger for a given rate of revolution?

Figure **E5.46**

5.47 •• In another version of the "Giant Swing" (see Exercise 5.46), the seat is connected to two cables as shown in Fig. E5.47, one of which is horizontal. The seat swings in a horizontal circle at a rate of 32.0 rpm (rev/min). If the seat weighs 255 N and an 825-N person is sitting in it, find the tension in each cable.

Figure **E5.47**

5.48 •• A small button placed on a horizontal rotating platform with diameter 0.320 m will revolve with the platform when it is brought up to a speed of 40.0 rev/min, provided the button is no more than 0.150 m from the axis. (a) What is the coefficient of static friction between the button and the platform? (b) How far from the axis can the button be placed, without slipping, if the platform rotates at 60.0 rev/min?

5.49 •• **Rotating Space Stations.** One problem for humans living in outer space is that they are apparently weightless. One way around this problem is to design a space station that spins about its center at a constant rate. This creates "artificial gravity" at the outside rim of the station. (a) If the diameter of the space station is 800 m, how many revolutions per minute are needed for the "artificial gravity" acceleration to be 9.80 m/s^2? (b) If the space station is a waiting area for travelers going to Mars, it might be desirable to simulate the acceleration due to gravity on the Martian surface (3.70 m/s^2). How many revolutions per minute are needed in this case?

5.50 • The Cosmoclock 21 Ferris wheel in Yokohama City, Japan, has a diameter of 100 m. Its name comes from its 60 arms, each of which can function as a second hand (so that it makes one revolution every 60.0 s). (a) Find the speed of the passengers when the Ferris wheel is rotating at this rate. (b) A passenger weighs 882 N at the weight-guessing booth on the ground. What is his apparent weight at the highest and at the lowest point on the Ferris wheel? (c) What would be the time for one revolution if the passenger's apparent weight at the highest point were zero? (d) What then would be the passenger's apparent weight at the lowest point?

5.51 •• An airplane flies in a loop (a circular path in a vertical plane) of radius 150 m. The pilot's head always points toward the center of the loop. The speed of the airplane is not constant; the airplane goes slowest at the top of the loop and fastest at the bottom. (a) At the top of the loop, the pilot feels weightless. What is the speed of the airplane at this point? (b) At the bottom of the loop, the speed of the airplane is 280 km/h. What is the apparent weight of the pilot at this point? His true weight is 700 N.

5.52 •• A 50.0-kg stunt pilot who has been diving her airplane vertically pulls out of the dive by changing her course to a circle in a vertical plane. (a) If the plane's speed at the lowest point of the circle is 95.0 m/s, what is the minimum radius of the circle for the acceleration at this point not to exceed $4.00g$? (b) What is the apparent weight of the pilot at the lowest point of the pullout?

5.53 • **Stay Dry!** You tie a cord to a pail of water, and you swing the pail in a vertical circle of radius 0.600 m. What minimum speed must you give the pail at the highest point of the circle if no water is to spill from it?

5.54 •• A bowling ball weighing 71.2 N (16.0 lb) is attached to the ceiling by a 3.80-m rope. The ball is pulled to one side and released; it then swings back and forth as a pendulum. As the rope swings through the vertical, the speed of the bowling ball is 4.20 m/s. (a) What is the acceleration of the bowling ball, in magnitude and direction, at this instant? (b) What is the tension in the rope at this instant?

5.55 •• **BIO Effect on Blood of Walking.** While a person is walking, his arms swing through approximately a 45° angle in $\frac{1}{2}$ s. As a reasonable approximation, we can assume that the arm moves with constant speed during each swing. A typical arm is 70.0 cm long, measured from the shoulder joint. (a) What is the acceleration of a 1.0-g drop of blood in the fingertips at the bottom of the swing? (b) Draw a free-body diagram of the drop of blood in part (a). (c) Find the force that the blood vessel must exert on the drop of blood in part (a). Which way does this force point? (d) What force would the blood vessel exert if the arm were not swinging?

PROBLEMS

5.56 •• An adventurous archaeologist crosses between two rock cliffs by slowly going hand over hand along a rope stretched between the cliffs. He stops to rest at the middle of the rope (Fig. P5.56). The rope will break if the tension in it exceeds 2.50×10^4 N, and our hero's mass is 90.0 kg. (a) If the angle θ is 10.0°, find the tension in the rope. (b) What is the smallest value the angle θ can have if the rope is not to break?

Figure **P5.56**

5.57 ••• Two ropes are connected to a steel cable that supports a hanging weight as shown in Fig. P5.57. (a) Draw a free-body diagram showing all of the forces acting at the knot that connects the two ropes to the steel cable. Based on your force diagram, which of the two ropes will have the greater tension? (b) If the maximum tension either rope can sustain without breaking is 5000 N, determine the maximum value of the hanging weight that these ropes can safely support. You can ignore the weight of the ropes and the steel cable.

Figure **P5.57**

5.58 •• In Fig. P5.58 a worker lifts a weight w by pulling down on a rope with a force $\vec{F}$. The upper pulley is attached to the ceiling by a chain, and the lower pulley is attached to the weight by another chain. In terms of w, find the tension in each chain and the magnitude of the force $\vec{F}$ if the weight is lifted at constant speed. Include the free-body diagram or diagrams you used to determine your answers. Assume that the rope, pulleys, and chains all have negligible weights.

Figure **P5.58**

5.59 ••• A solid uniform 45.0-kg ball of diameter 32.0 cm is supported against a vertical, frictionless wall using a thin 30.0-cm wire of negligible mass, as shown in Fig. P5.59. (a) Draw a free-body diagram for the ball and use it to find the tension in the wire. (b) How hard does the ball push against the wall?

Figure **P5.59**

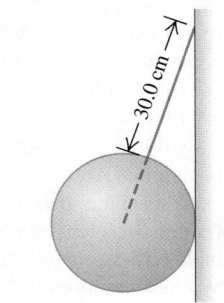

5.60 ••• A horizontal wire holds a solid uniform ball of mass m in place on a tilted ramp that rises 35.0° above the horizontal. The surface of this ramp is perfectly smooth, and the wire is directed away from the center of the ball (Fig. P5.60). (a) Draw a free-body diagram for the ball. (b) How hard does the surface of the ramp push on the ball? (c) What is the tension in the wire?

Figure **P5.60**

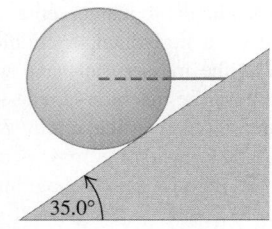

5.61 •• **CP BIO Forces During Chin-ups.** People who do chin-ups raise their chin just over a bar (the chinning bar), supporting themselves with only their arms. Typically, the body below the arms is raised by about 30 cm in a time of 1.0 s, starting from rest. Assume that the entire body of a 680-N person doing chin-ups is raised this distance and that half the 1.0 s is spent accelerating upward and the other half accelerating downward, uniformly in both cases. Draw a free-body diagram of the person's body, and then apply it to find the force his arms must exert on him during the accelerating part of the chin-up.

5.62 •• **CP BIO Prevention of Hip Injuries.** People (especially the elderly) who are prone to falling can wear hip pads to cushion the impact on their hip from a fall. Experiments have shown that if the speed at impact can be reduced to 1.3 m/s or less, the hip will usually not fracture. Let us investigate the worst-case scenario in which a 55-kg person completely loses her footing (such as on icy pavement) and falls a distance of 1.0 m, the distance from her hip to the ground. We shall assume that the person's entire body has the same acceleration, which, in reality, would not quite be true. (a) With what speed does her hip reach the ground? (b) A typical hip pad can reduce the person's speed to 1.3 m/s over a distance of 2.0 cm. Find the acceleration (assumed to be constant) of this person's hip while she is slowing down and the force the pad exerts on it. (c) The force in part (b) is very large. To see whether it is likely to cause injury, calculate how long it lasts.

5.63 ••• **CALC** A 3.00-kg box that is several hundred meters above the surface of the earth is suspended from the end of a short vertical rope of negligible mass. A time-dependent upward force is applied to the upper end of the rope, and this results in a tension in the rope of $T(t) = (36.0 \text{ N/s})t$. The box is at rest at $t = 0$. The only forces on the box are the tension in the rope and gravity. (a) What is the velocity of the box at (i) $t = 1.00$ s and (ii) $t = 3.00$ s? (b) What is the maximum distance that the box descends below its initial position? (c) At what value of t does the box return to its initial position?

5.64 •• **CP** A 5.00-kg box sits at rest at the bottom of a ramp that is 8.00 m long and that is inclined at $30.0°$ above the horizontal. The coefficient of kinetic friction is $\mu_k = 0.40$, and the coefficient of static friction is $\mu_s = 0.50$. What constant force F, applied parallel to the surface of the ramp, is required to push the box to the top of the ramp in a time of 4.00 s?

5.65 •• Two boxes connected by a light horizontal rope are on a horizontal surface, as shown in Fig. P5.35. The coefficient of kinetic friction between each box and the surface is $\mu_k = 0.30$. One box (box B) has mass 5.00 kg, and the other box (box A) has mass m. A force F with magnitude 40.0 N and direction $53.1°$ above the horizontal is applied to the 5.00-kg box, and both boxes move to the right with $a = 1.50$ m/s². (a) What is the tension T in the rope that connects the boxes? (b) What is the mass m of the second box?

5.66 ••• A 6.00-kg box sits on a ramp that is inclined at $37.0°$ above the horizontal. The coefficient of kinetic friction between the box and the ramp is $\mu_k = 0.30$. What *horizontal* force is required to move the box up the incline with a constant acceleration of 4.20 m/s²?

5.67 •• **CP** In Fig. P5.34 block A has mass m and block B has mass 6.00 kg. The coefficient of kinetic friction between block A and the tabletop is $\mu_k = 0.40$. The mass of the rope connecting the blocks can be neglected. The pulley is light and frictionless. When the system is released from rest, the hanging block descends 5.00 m in 3.00 s. What is the mass m of block A?

5.68 •• **CP** In Fig. P5.68 $m_1 = 20.0$ kg and $\alpha = 53.1°$. The coefficient of kinetic friction between the block and the incline is $\mu_k = 0.40$. What must be the mass m_2 of the hanging block if it is to descend 12.0 m in the first 3.00 s after the system is released from rest?

Figure **P5.68**

5.69 ••• **CP** **Rolling Friction.** Two bicycle tires are set rolling with the same initial speed of 3.50 m/s on a long, straight road, and the distance each travels before its speed is reduced by half is measured. One tire is inflated to a pressure of 40 psi and goes 18.1 m; the other is at 105 psi and goes 92.9 m. What is the coefficient of rolling friction μ_r for each? Assume that the net horizontal force is due to rolling friction only.

5.70 •• **A Rope with Mass.** A block with mass M is attached to the lower end of a vertical, uniform rope with mass m and length L. A constant upward force $\vec{F}$ is applied to the top of the rope, causing the rope and block to accelerate upward. Find the tension in the rope at a distance x from the top end of the rope, where x can have any value from 0 to L.

5.71 •• A block with mass m_1 is placed on an inclined plane with slope angle α and is connected to a second hanging block with mass m_2 by a cord passing over a small, frictionless pulley (Fig. P5.68). The coefficient of static friction is μ_s and the coefficient of kinetic friction is μ_k. (a) Find the mass m_2 for which block m_1 moves up the plane at constant speed once it is set in motion. (b) Find the mass m_2 for which block m_1 moves down the plane at constant speed once it is set in motion. (c) For what range of values of m_2 will the blocks remain at rest if they are released from rest?

5.72 •• Block A in Fig. P5.72 weighs 60.0 N. The coefficient of static friction between the block and the surface on which it rests is 0.25. The weight w is 12.0 N and the system is in equilibrium. (a) Find the friction force exerted on block A. (b) Find the maximum weight w for which the system will remain in equilibrium.

Figure **P5.72**

5.73 •• Block A in Fig. P5.73 weighs 2.40 N and block B weighs 3.60 N. The coefficient of kinetic friction between all surfaces is 0.300. Find the magnitude of the horizontal force $\vec{F}$ necessary to drag block B to the left at constant speed (a) if A rests on B and moves with it (Fig. P5.73a). (b) If A is held at rest (Fig. P5.73b).

Figure **P5.73**

(a)

(b)

5.74 ••• A window washer pushes his scrub brush up a vertical window at constant speed by applying a force $\vec{F}$ as shown in Fig. P5.74. The brush weighs 15.0 N and the coefficient of kinetic friction is $\mu_k = 0.150$. Calculate (a) the magnitude of the force $\vec{F}$ and (b) the normal force exerted by the window on the brush.

5.75 •• **BIO** **The Flying Leap of a Flea.** High-speed motion pictures (3500 frames/second) of a jumping 210-μg flea yielded the data to plot the flea's acceleration as a function of time as

Figure **P5.74**

shown in Fig. P5.75. (See "The Flying Leap of the Flea," by M. Rothschild et al. in the November 1973 *Scientific American*.) This flea was about 2 mm long and jumped at a nearly vertical takeoff angle. Use the measurements shown on the graph to answer the questions. (a) Find the *initial* net external force on the flea. How does it compare to the flea's weight? (b) Find the *maximum* net external force on this jumping flea. When does this maximum force occur? (c) Use the graph to find the flea's maximum speed.

Figure **P5.75**

5.76 •• **CP** A 25,000-kg rocket blasts off vertically from the earth's surface with a constant acceleration. During the motion considered in the problem, assume that g remains constant (see Chapter 13). Inside the rocket, a 15.0-N instrument hangs from a wire that can support a maximum tension of 45.0 N. (a) Find the minimum time for this rocket to reach the sound barrier (330 m/s) without breaking the inside wire and the maximum vertical thrust of the rocket engines under these conditions. (b) How far is the rocket above the earth's surface when it breaks the sound barrier?

5.77 ••• **CP CALC** You are standing on a bathroom scale in an elevator in a tall building. Your mass is 64 kg. The elevator starts from rest and travels upward with a speed that varies with time according to $v(t) = (3.0 \text{ m/s}^2)t + (0.20 \text{ m/s}^3)t^2$. When $t = 4.0 \text{ s}$, what is the reading of the bathroom scale?

5.78 ••• **CP Elevator Design.** You are designing an elevator for a hospital. The force exerted on a passenger by the floor of the elevator is not to exceed 1.60 times the passenger's weight. The elevator accelerates upward with constant acceleration for a distance of 3.0 m and then starts to slow down. What is the maximum speed of the elevator?

5.79 •• **CP** You are working for a shipping company. Your job is to stand at the bottom of a 8.0-m-long ramp that is inclined at 37° above the horizontal. You grab packages off a conveyor belt and propel them up the ramp. The coefficient of kinetic friction between the packages and the ramp is $\mu_k = 0.30$. (a) What speed do you need to give a package at the bottom of the ramp so that it has zero speed at the top of the ramp? (b) Your coworker is supposed to grab the packages as they arrive at the top of the ramp, but she misses one and it slides back down. What is its speed when it returns to you?

5.80 •• A hammer is hanging by a light rope from the ceiling of a bus. The ceiling of the bus is parallel to the roadway. The bus is traveling in a straight line on a horizontal street. You observe that the hammer hangs at rest with respect to the bus when the angle between the rope and the ceiling of the bus is 67°. What is the acceleration of the bus?

5.81 ••• A steel washer is suspended inside an empty shipping crate from a light string attached to the top of the crate. The crate slides down a long ramp that is inclined at an angle of 37° above the horizontal. The crate has mass 180 kg. You are sitting inside the crate

(with a flashlight); your mass is 55 kg. As the crate is sliding down the ramp, you find the washer is at rest with respect to the crate when the string makes an angle of 68° with the top of the crate. What is the coefficient of kinetic friction between the ramp and the crate?

5.82 • **CP Lunch Time!** You are riding your motorcycle one day down a wet street that slopes downward at an angle of 20° below the horizontal. As you start to ride down the hill, you notice a construction crew has dug a deep hole in the street at the bottom of the hill. A Siberian tiger, escaped from the City Zoo, has taken up residence in the hole. You apply the brakes and lock your wheels at the top of the hill, where you are moving with a speed of 20 m/s. The inclined street in front of you is 40 m long. (a) Will you plunge into the hole and become the tiger's lunch, or do you skid to a stop before you reach the hole? (The coefficients of friction between your motorcycle tires and the wet pavement are $\mu_s = 0.90$ and $\mu_k = 0.70$.) (b) What must your initial speed be if you are to stop just before reaching the hole?

5.83 ••• In the system shown in Fig. P5.34, block *A* has mass m_A, block *B* has mass m_B, and the rope connecting them has a *nonzero* mass m_{rope}. The rope has a total length *L*, and the pulley has a very small radius. You can ignore any sag in the horizontal part of the rope. (a) If there is no friction between block *A* and the tabletop, find the acceleration of the blocks at an instant when a length *d* of rope hangs vertically between the pulley and block *B*. As block *B* falls, will the magnitude of the acceleration of the system increase, decrease, or remain constant? Explain. (b) Let $m_A = 2.00 \text{ kg}$, $m_B = 0.400 \text{ kg}$, $m_{\text{rope}} = 0.160 \text{ kg}$, and $L = 1.00 \text{ m}$. If there is friction between block *A* and the tabletop, with $\mu_k = 0.200$ and $\mu_s = 0.250$, find the minimum value of the distance *d* such that the blocks will start to move if they are initially at rest. (c) Repeat part (b) for the case $m_{\text{rope}} = 0.040 \text{ kg}$. Will the blocks move in this case?

5.84 ••• If the coefficient of static friction between a table and a uniform massive rope is μ_s, what fraction of the rope can hang over the edge of the table without the rope sliding?

5.85 •• A 40.0-kg packing case is initially at rest on the floor of a 1500-kg pickup truck. The coefficient of static friction between the case and the truck floor is 0.30, and the coefficient of kinetic friction is 0.20. Before each acceleration given below, the truck is traveling due north at constant speed. Find the magnitude and direction of the friction force acting on the case (a) when the truck accelerates at 2.20 m/s² northward and (b) when it accelerates at 3.40 m/s² southward.

5.86 • **CP Traffic Court.** You are called as an expert witness in the trial of a traffic violation. The facts are these: A driver slammed on his brakes and came to a stop with constant acceleration. Measurements of his tires and the skid marks on the pavement indicate that he locked his car's wheels, the car traveled 192 ft before stopping, and the coefficient of kinetic friction between the road and his tires was 0.750. The charge is that he was speeding in a 45-mi/h zone. He pleads innocent. What is your conclusion, guilty or innocent? How fast was he going when he hit his brakes?

5.87 ••• Two identical 15.0-kg balls, each 25.0 cm in diameter, are suspended by two 35.0-cm wires as shown in Fig. P5.87. The entire apparatus is supported by a single 18.0-cm wire, and the surfaces of the balls are perfectly smooth. (a) Find the tension in each of the three wires. (b) How hard does each ball push on the other one?

Figure **P5.87**

18.0 cm

35.0 cm   35.0 cm

5.88 •• **CP** **Losing Cargo.** A 12.0-kg box rests on the flat floor of a truck. The coefficients of friction between the box and floor are $\mu_s = 0.19$ and $\mu_k = 0.15$. The truck stops at a stop sign and then starts to move with an acceleration of 2.20 m/s². If the box is 1.80 m from the rear of the truck when the truck starts, how much time elapses before the box falls off the truck? How far does the truck travel in this time?

5.89 ••• Block A in Fig. P5.89 weighs 1.90 N, and block B weighs 4.20 N. The coefficient of kinetic friction between all surfaces is 0.30. Find the magnitude of the horizontal force $\vec{F}$ necessary to drag block B to the left at constant speed if A and B are connected by a light, flexible cord passing around a fixed, frictionless pulley.

Figure **P5.89**

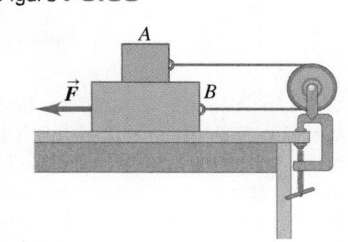

5.90 ••• **CP** You are part of a design team for future exploration of the planet Mars, where $g = 3.7$ m/s². An explorer is to step out of a survey vehicle traveling horizontally at 33 m/s when it is 1200 m above the surface and then fall freely for 20 s. At that time, a portable advanced propulsion system (PAPS) is to exert a constant force that will decrease the explorer's speed to zero at the instant she touches the surface. The total mass (explorer, suit, equipment, and PAPS) is 150 kg. Assume the change in mass of the PAPS to be negligible. Find the horizontal and vertical components of the force the PAPS must exert, and for what interval of time the PAPS must exert it. You can ignore air resistance.

5.91 •• Block A in Fig. P5.91 has a mass of 4.00 kg, and block B has mass 12.0 kg. The coefficient of kinetic friction between block B and the horizontal surface is 0.25. (a) What is the mass of block C if block B is moving to the right and speeding up with an acceleration of 2.00 m/s²? (b) What is the tension in each cord when block B has this acceleration?

Figure **P5.91**

5.92 •• Two blocks connected by a cord passing over a small, frictionless pulley rest on frictionless planes (Fig. P5.92). (a) Which way will the system move when the blocks are released from rest? (b) What is the acceleration of the blocks? (c) What is the tension in the cord?

Figure **P5.92**

5.93 •• In terms of m_1, m_2, and g, find the acceleration of each block in Fig. P5.93. There is no friction anywhere in the system.

Figure **P5.93**

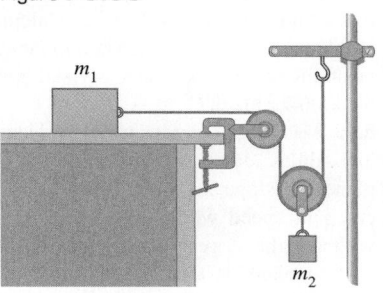

5.94 ••• Block B, with mass 5.00 kg, rests on block A, with mass 8.00 kg, which in turn is on a horizontal tabletop (Fig. P5.94). There is no friction between block A and the tabletop, but the coefficient of static friction between block A and block B is 0.750. A light string attached to block A passes over a frictionless, massless pulley, and block C is suspended from the other end of the string. What is the largest mass that block C can have so that blocks A and B still slide together when the system is released from rest?

Figure **P5.94**

5.95 ••• Two objects with masses 5.00 kg and 2.00 kg hang 0.600 m above the floor from the ends of a cord 6.00 m long passing over a frictionless pulley. Both objects start from rest. Find the maximum height reached by the 2.00-kg object.

5.96 •• **Friction in an Elevator.** You are riding in an elevator on the way to the 18th floor of your dormitory. The elevator is accelerating upward with $a = 1.90$ m/s². Beside you is the box containing your new computer; the box and its contents have a total mass of 28.0 kg. While the elevator is accelerating upward, you push horizontally on the box to slide it at constant speed toward the elevator door. If the coefficient of kinetic friction between the box and the elevator floor is $\mu_k = 0.32$, what magnitude of force must you apply?

5.97 • A block is placed against the vertical front of a cart as shown in Fig. P5.97. What acceleration must the cart have so that block A does not fall? The coefficient of static friction between the block and the cart is μ_s. How would an observer on the cart describe the behavior of the block?

Figure **P5.97**

5.98 ••• Two blocks with masses 4.00 kg and 8.00 kg are connected by a string and slide down a 30.0° inclined plane (Fig. P5.98). The coefficient of kinetic friction between the

Figure **P5.98**

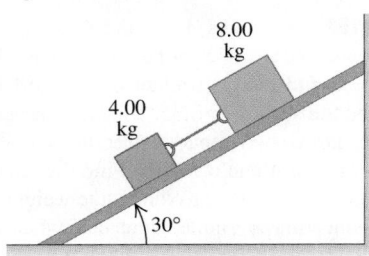

4.00-kg block and the plane is 0.25; that between the 8.00-kg block and the plane is 0.35. (a) Calculate the acceleration of each block. (b) Calculate the tension in the string. (c) What happens if the positions of the blocks are reversed, so the 4.00-kg block is above the 8.00-kg block?

5.99 ••• Block A, with weight 3w, slides down an inclined plane S of slope angle 36.9° at a constant speed while plank B, with weight w, rests on top of A. The plank is attached by a cord to the wall (Fig. P5.99). (a) Draw a diagram of all the forces acting on block A. (b) If the coefficient of kinetic friction is the same between A and B and between S and A, determine its value.

Figure **P5.99**

5.100 •• **Accelerometer.** The system shown in Fig. P5.100 can be used to measure the acceleration of the system. An observer riding on the platform measures the angle θ that the thread supporting the light ball makes with the vertical. There is no friction anywhere. (a) How is θ related to the acceleration of the system? (b) If $m_1 = 250$ kg and $m_2 = 1250$ kg, what is θ? (c) If you can vary m_1 and m_2, what is the largest angle θ you could achieve? Explain how you need to adjust m_1 and m_2 to do this.

Figure **P5.100**

5.101 ••• **Banked Curve I.** A curve with a 120-m radius on a level road is banked at the correct angle for a speed of 20 m/s. If an automobile rounds this curve at 30 m/s, what is the minimum coefficient of static friction needed between tires and road to prevent skidding?

5.102 •• **Banked Curve II.** Consider a wet roadway banked as in Example 5.22 (Section 5.4), where there is a coefficient of static friction of 0.30 and a coefficient of kinetic friction of 0.25 between the tires and the roadway. The radius of the curve is $R = 50$ m. (a) If the banking angle is $\beta = 25°$, what is the *maximum* speed the automobile can have before sliding *up* the banking? (b) What is the *minimum* speed the automobile can have before sliding *down* the banking?

5.103 ••• Blocks A, B, and C are placed as in Fig. P5.103 and connected by ropes of negligible mass. Both A and B weigh 25.0 N each, and the coefficient of kinetic friction between each block and the surface is 0.35. Block C descends with constant velocity. (a) Draw two separate free-body diagrams showing the forces acting on A and on B. (b) Find the tension in the rope connecting blocks A and B. (c) What is the weight of block C? (d) If the rope connecting A and B were cut, what would be the acceleration of C?

Figure **P5.103**

5.104 •• You are riding in a school bus. As the bus rounds a flat curve at constant speed, a lunch box with mass 0.500 kg, suspended from the ceiling of the bus by a string 1.80 m long, is found to hang at rest relative to the bus when the string makes an angle of 30.0° with the vertical. In this position the lunch box is 50.0 m from the center of curvature of the curve. What is the speed v of the bus?

5.105 • **The Monkey and Bananas Problem.** A 20-kg monkey has a firm hold on a light rope that passes over a frictionless pulley and is attached to a 20-kg bunch of bananas (Fig. P5.105). The monkey looks up, sees the bananas, and starts to climb the rope to get them. (a) As the monkey climbs, do the bananas move up, down, or remain at rest? (b) As the monkey climbs, does the distance between the monkey and the bananas decrease, increase, or remain constant? (c) The monkey releases her hold on the rope. What happens to the distance between the monkey and the bananas while she is falling? (d) Before reaching the ground, the monkey grabs the rope to stop her fall. What do the bananas do?

Figure **P5.105**

5.106 •• **CALC** You throw a rock downward into water with a speed of $3mg/k$, where k is the coefficient in Eq. (5.7). Assume that the relationship between fluid resistance and speed is as given in Eq. (5.7), and calculate the speed of the rock as a function of time.

5.107 •• A rock with mass $m = 3.00$ kg falls from rest in a viscous medium. The rock is acted on by a net constant downward force of 18.0 N (a combination of gravity and the buoyant force exerted by the medium) and by a fluid resistance force $f = kv$, where v is the speed in m/s and $k = 2.20$ N·s/m (see Section 5.3). (a) Find the initial acceleration a_0. (b) Find the acceleration when the speed is 3.00 m/s. (c) Find the speed when the acceleration equals $0.1a_0$. (d) Find the terminal speed v_t. (e) Find the coordinate, speed, and acceleration 2.00 s after the start of the motion. (f) Find the time required to reach a speed of $0.9v_t$.

5.108 •• **CALC** A rock with mass m slides with initial velocity v_0 on a horizontal surface. A retarding force F_R that the surface exerts on the rock is proportional to the square root of the instantaneous velocity of the rock $(F_R = -kv^{1/2})$. (a) Find expressions for the velocity and position of the rock as a function of time. (b) In terms of m, k, and v_0, at what time will the rock come to rest? (c) In terms of m, k, and v_0, what is the distance of the rock from its starting point when it comes to rest?

5.109 ••• You observe a 1350-kg sports car rolling along flat pavement in a straight line. The only horizontal forces acting on it are a constant rolling friction and air resistance (proportional to the square of its speed). You take the following data during a time interval of 25 s: When its speed is 32 m/s, the car slows down at a rate of -0.42 m/s^2, and when its speed is decreased to 24 m/s, it slows down at -0.30 m/s^2. (a) Find the coefficient of rolling friction and the air drag constant D. (b) At what constant speed will this car move down an incline that makes a 2.2° angle with the horizontal? (c) How is the constant speed for an incline of angle β related to the terminal speed of this sports car if the car drops off a high cliff? Assume that in both cases the air resistance force is proportional to the square of the speed, and the air drag constant is the same.

5.110 ••• The 4.00-kg block in Fig. P5.110 is attached to a vertical rod by means of two strings. When the system rotates about the axis of the rod, the strings are extended as shown in the diagram and the tension in the upper string is 80.0 N. (a) What is the tension in the lower cord? (b) How many revolutions per minute does the system make? (c) Find the number of revolutions per minute at which the lower cord just goes slack. (d) Explain what happens if the number of revolutions per minute is less than in part (c).

Figure **P5.110**

1.25 m

2.00 m

4.00 kg

1.25 m

5.111 ••• **CALC** Equation (5.10) applies to the case where the initial velocity is zero. (a) Derive the corresponding equation for $v_y(t)$ when the falling object has an initial downward velocity with magnitude v_0. (b) For the case where $v_0 < v_t$, sketch a graph of v_y as a function of t and label v_t on your graph. (c) Repeat part (b) for the case where $v_0 > v_t$. (d) Discuss what your result says about $v_y(t)$ when $v_0 = v_t$.

5.112 ••• **CALC** A small rock moves in water, and the force exerted on it by the water is given by Eq. (5.7). The terminal speed of the rock is measured and found to be 2.0 m/s. The rock is projected *upward* at an initial speed of 6.0 m/s. You can ignore the buoyancy force on the rock. (a) In the absence of fluid resistance, how high will the rock rise and how long will it take to reach this maximum height? (b) When the effects of fluid resistance are included, what are the answers to the questions in part (a)?

5.113 •• **Merry-Go-Round.** One December identical twins Jena and Jackie are playing on a large merry-go-round (a disk mounted parallel to the ground, on a vertical axle through its center) in their school playground in northern Minnesota. Each twin has mass 30.0 kg. The icy coating on the merry-go-round surface makes it frictionless. The merry-go-round revolves at a constant rate as the twins ride on it. Jena, sitting 1.80 m from the center of the merry-go-round, must hold on to one of the metal posts attached to the merry-go-round with a horizontal force of 60.0 N to keep from sliding off. Jackie is sitting at the edge, 3.60 m from the center. (a) With what horizontal force must Jackie hold on to keep from falling off? (b) If Jackie falls off, what will be her horizontal velocity when she becomes airborne?

5.114 •• A 70-kg person rides in a 30-kg cart moving at 12 m/s at the top of a hill that is in the shape of an arc of a circle with a radius of 40 m. (a) What is the apparent weight of the person as the cart passes over the top of the hill? (b) Determine the maximum speed that the cart may travel at the top of the hill without losing contact with the surface. Does your answer depend on the mass of the cart or the mass of the person? Explain.

5.115 •• On the ride "Spindletop" at the amusement park Six Flags Over Texas, people stood against the inner wall of a hollow vertical cylinder with radius 2.5 m. The cylinder started to rotate, and when it reached a constant rotation rate of 0.60 rev/s, the floor on which people were standing dropped about 0.5 m. The people remained pinned against the wall. (a) Draw a force diagram for a person on this ride, after the floor has dropped. (b) What minimum coefficient of static friction is required if the person on the ride is not to slide downward to the new position of the floor? (c) Does your answer in part (b) depend on the mass of the passenger? (*Note:* When the ride is over, the cylinder is slowly brought to rest. As it slows down, people slide down the walls to the floor.)

5.116 •• A passenger with mass 85 kg rides in a Ferris wheel like that in Example 5.23 (Section 5.4). The seats travel in a circle of radius 35 m. The Ferris wheel rotates at constant speed and makes one complete revolution every 25 s. Calculate the magnitude and direction of the net force exerted on the passenger by the seat when she is (a) one-quarter revolution past her lowest point and (b) one-quarter revolution past her highest point.

5.117 • **Ulterior Motives.** You are driving a classic 1954 Nash Ambassador with a friend who is sitting to your right on the passenger side of the front seat. The Ambassador has flat bench seats. You would like to be closer to your friend and decide to use physics to achieve your romantic goal by making a quick turn. (a) Which way (to the left or to the right) should you turn the car to get your friend to slide closer to you? (b) If the coefficient of static friction between your friend and the car seat is 0.35, and you keep driving at a constant speed of 20 m/s, what is the maximum radius you could make your turn and still have your friend slide your way?

5.118 •• A physics major is working to pay his college tuition by performing in a traveling carnival. He rides a motorcycle inside a hollow, transparent plastic sphere. After gaining sufficient speed, he travels in a vertical circle with a radius of 13.0 m. The physics major has mass 70.0 kg, and his motorcycle has mass 40.0 kg. (a) What minimum speed must he have at the top of the circle if the tires of the motorcycle are not to lose contact with the sphere? (b) At the bottom of the circle, his speed is twice the value calculated in part (a). What is the magnitude of the normal force exerted on the motorcycle by the sphere at this point?

5.119 •• A small bead can slide without friction on a circular hoop that is in a vertical plane and has a radius of 0.100 m. The hoop rotates at a constant rate of 4.00 rev/s about a vertical diameter (Fig. P5.119). (a) Find the angle β at which the bead is in vertical equilibrium. (Of course, it has a radial acceleration toward the axis.) (b) Is it possible for the bead to "ride" at the same elevation as the center of the hoop? (c) What will happen if the hoop rotates at 1.00 rev/s?

Figure **P5.119**

←0.100 m→

β

5.120 •• A small remote-controlled car with mass 1.60 kg moves at a constant speed of $v = 12.0$ m/s in a vertical circle inside a hollow metal cylinder that has a radius of 5.00 m (Fig. P5.120). What is the magnitude of the normal force exerted on the car by the walls of the cylinder at

(a) point A (at the bottom of the vertical circle) and (b) point B (at the top of the vertical circle)?

Figure **P5.120**

CHALLENGE PROBLEMS

5.121 ••• **CALC Angle for Minimum Force.** A box with weight w is pulled at constant speed along a level floor by a force $\vec{F}$ that is at an angle θ above the horizontal. The coefficient of kinetic friction between the floor and box is μ_k. (a) In terms of θ, μ_k, and w, calculate F. (b) For $w = 400$ N and $\mu_k = 0.25$, calculate F for θ ranging from $0°$ to $90°$ in increments of $10°$. Graph F versus θ. (c) From the general expression in part (a), calculate the value of θ for which the value of F, required to maintain constant speed, is a minimum. (*Hint:* At a point where a function is minimum, what are the first and second derivatives of the function? Here F is a function of θ.) For the special case of $w = 400$ N and $\mu_k = 0.25$, evaluate this optimal θ and compare your result to the graph you constructed in part (b).

5.122 ••• **Moving Wedge.** A wedge with mass M rests on a frictionless, horizontal tabletop. A block with mass m is placed on the wedge (Fig. P5.122a). There is no friction between the block and the wedge. The system is released from rest. (a) Calculate the acceleration of the wedge and the horizontal and vertical components of the acceleration of the block. (b) Do your answers to part (a) reduce to the correct results when M is very large? (c) As seen by a stationary observer, what is the shape of the trajectory of the block?

Figure **P5.122**

5.123 ••• A wedge with mass M rests on a frictionless horizontal tabletop. A block with mass m is placed on the wedge and a horizontal force $\vec{F}$ is applied to the wedge (Fig. P5.122b). What must the magnitude of $\vec{F}$ be if the block is to remain at a constant height above the tabletop?

5.124 ••• **CALC Falling Baseball.** You drop a baseball from the roof of a tall building. As the ball falls, the air exerts a drag force proportional to the square of the ball's speed ($f = Dv^2$). (a) In a diagram, show the direction of motion and indicate, with the aid of vectors, all the forces acting on the ball. (b) Apply Newton's second law and infer from the resulting equation the general properties of the motion. (c) Show that the ball acquires a terminal speed

that is as given in Eq. (5.13). (d) Derive the equation for the speed at any time. (*Note:*

$$\int \frac{dx}{a^2 - x^2} = \frac{1}{a} \text{arctanh}\left(\frac{x}{a}\right)$$

where

$$\tanh(x) = \frac{e^x - e^{-x}}{e^x + e^{-x}} = \frac{e^{2x} - 1}{e^{2x} + 1}$$

defines the hyperbolic tangent.)

5.125 ••• **Double Atwood's Machine.** In Fig. P5.125 masses m_1 and m_2 are connected by a light string A over a light, frictionless pulley B. The axle of pulley B is connected by a second light string C over a second light, frictionless pulley D to a mass m_3. Pulley D is suspended from the ceiling by an attachment to its axle. The system is released from rest. In terms of m_1, m_2, m_3, and g, what are (a) the acceleration of block m_3; (b) the acceleration of pulley B; (c) the acceleration of block m_1; (d) the acceleration of block m_2; (e) the tension in string A; (f) the tension in string C? (g) What do your expressions give for the special case of $m_1 = m_2$ and $m_3 = m_1 + m_2$? Is this sensible?

Figure **P5.125**

5.126 ••• The masses of blocks A and B in Fig. P5.126 are 20.0 kg and 10.0 kg, respectively. The blocks are initially at rest on the floor and are connected by a massless string passing over a massless and frictionless pulley. An upward force $\vec{F}$ is applied to the pulley. Find the accelerations $\vec{a}_A$ of block A and $\vec{a}_B$ of block B when F is (a) 124 N; (b) 294 N; (c) 424 N.

Figure **P5.126**

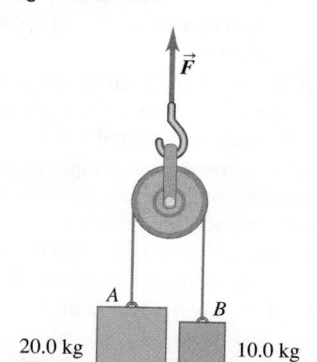

5.127 ••• A ball is held at rest at position A in Fig. P5.127 by two light strings. The horizontal string is cut and the ball starts swinging as a pendulum. Point B is the farthest to the right the ball goes as it swings back and forth. What is the ratio of the tension in the supporting string at position B to its value at A before the horizontal string was cut?

Figure **P5.127**

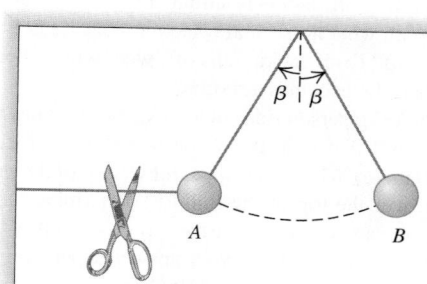

Answers

Chapter Opening Question ?

Neither; the upward force of the air has the *same* magnitude as the force of gravity. Although the skydiver and parachute are descending, their vertical velocity is constant and so their vertical acceleration is zero. Hence the net vertical force on the skydiver and parachute must also be zero, and the individual vertical forces must balance.

Test Your Understanding Questions

5.1 Answer: (ii) The two cables are arranged symmetrically, so the tension in either cable has the same magnitude T. The vertical component of the tension from each cable is $T \sin 45°$ (or, equivalently, $T \cos 45°$), so Newton's first law applied to the vertical forces tells us that $2T \sin 45° - w = 0$. Hence $T = w/(2 \sin 45°) = w/\sqrt{2} = 0.71w$. Each cable supports half of the weight of the traffic light, but the tension is greater than $w/2$ because only the vertical component of the tension counteracts the weight.

5.2 Answer: (ii) No matter what the instantaneous velocity of the glider, its acceleration is constant and has the value found in Example 5.12. In the same way, the acceleration of a body in free fall is the same whether it is ascending, descending, or at the high point of its motion (see Section 2.5).

5.3 Answers to (a): (i), (iii); answers to (b): (ii), (iv); answer to (c): (v) In situations (i) and (iii) the box is not accelerating (so the net force on it must be zero) and there is no other force acting parallel to the horizontal surface; hence no friction force is needed to prevent sliding. In situations (ii) and (iv) the box would start to slide over the surface if no friction were present, so a static friction force must act to prevent this. In situation (v) the box is sliding over a rough surface, so a kinetic friction force acts on it.

5.4 Answer: (iii) A satellite of mass m orbiting the earth at speed v in an orbit of radius r has an acceleration of magnitude v^2/r, so the net force acting on it from the earth's gravity has magnitude $F = mv^2/r$. The farther the satellite is from earth, the greater the value of r, the smaller the value of v, and hence the smaller the values of v^2/r and of F. In other words, the earth's gravitational force decreases with increasing distance.

Bridging Problem

Answers: (a) $T_{\max} = 2\pi \sqrt{\dfrac{h(\cos \beta + \mu_s \sin \beta)}{g \tan \beta (\sin \beta - \mu_s \cos \beta)}}$

(b) $T_{\min} = 2\pi \sqrt{\dfrac{h(\cos \beta - \mu_s \sin \beta)}{g \tan \beta (\sin \beta + \mu_s \cos \beta)}}$

6

WORK AND KINETIC ENERGY

? After finding a piece of breakfast cereal on the floor, this ant picked it up and carried it away. As the ant was lifting the piece of cereal, did the *cereal* do work on the *ant*?

Suppose you try to find the speed of an arrow that has been shot from a bow. You apply Newton's laws and all the problem-solving techniques that we've learned, but you run across a major stumbling block: After the archer releases the arrow, the bow string exerts a *varying* force that depends on the arrow's position. As a result, the simple methods that we've learned aren't enough to calculate the speed. Never fear; we aren't by any means finished with mechanics, and there are other methods for dealing with such problems.

The new method that we're about to introduce uses the ideas of *work* and *energy*. The importance of the energy idea stems from the *principle of conservation of energy*: Energy is a quantity that can be converted from one form to another but cannot be created or destroyed. In an automobile engine, chemical energy stored in the fuel is converted partially to the energy of the automobile's motion and partially to thermal energy. In a microwave oven, electromagnetic energy obtained from your power company is converted to thermal energy of the food being cooked. In these and all other processes, the *total* energy—the sum of all energy present in all different forms—remains the same. No exception has ever been found.

We'll use the energy idea throughout the rest of this book to study a tremendous range of physical phenomena. This idea will help you understand why a sweater keeps you warm, how a camera's flash unit can produce a short burst of light, and the meaning of Einstein's famous equation $E = mc^2$.

In this chapter, though, our concentration will be on mechanics. We'll learn about one important form of energy called *kinetic energy,* or energy of motion, and how it relates to the concept of *work*. We'll also consider *power,* which is the time rate of doing work. In Chapter 7 we'll expand the ideas of work and kinetic energy into a deeper understanding of the concepts of energy and the conservation of energy.

6.1 Work

You'd probably agree that it's hard work to pull a heavy sofa across the room, to lift a stack of encyclopedias from the floor to a high shelf, or to push a stalled car off the road. Indeed, all of these examples agree with the everyday meaning of *work*—any activity that requires muscular or mental effort.

In physics, work has a much more precise definition. By making use of this definition we'll find that in any motion, no matter how complicated, the total work done on a particle by all forces that act on it equals the change in its *kinetic energy*—a quantity that's related to the particle's speed. This relationship holds even when the forces acting on the particle aren't constant, a situation that can be difficult or impossible to handle with the techniques you learned in Chapters 4 and 5. The ideas of work and kinetic energy enable us to solve problems in mechanics that we could not have attempted before.

In this section we'll see how work is defined and how to calculate work in a variety of situations involving *constant* forces. Even though we already know how to solve problems in which the forces are constant, the idea of work is still useful in such problems. Later in this chapter we'll relate work and kinetic energy, and then apply these ideas to problems in which the forces are *not* constant.

The three examples of work described above—pulling a sofa, lifting encyclopedias, and pushing a car—have something in common. In each case you do work by exerting a *force* on a body while that body *moves* from one place to another—that is, undergoes a *displacement* (Fig. 6.1). You do more work if the force is greater (you push harder on the car) or if the displacement is greater (you push the car farther down the road).

The physicist's definition of work is based on these observations. Consider a body that undergoes a displacement of magnitude s along a straight line. (For now, we'll assume that any body we discuss can be treated as a particle so that we can ignore any rotation or changes in shape of the body.) While the body moves, a constant force $\vec{F}$ acts on it in the same direction as the displacement $\vec{s}$ (Fig. 6.2). We define the **work** W done by this constant force under these circumstances as the product of the force magnitude F and the displacement magnitude s:

$$W = Fs \quad \text{(constant force in direction of straight-line displacement)} \quad (6.1)$$

The work done on the body is greater if either the force F or the displacement s is greater, in agreement with our observations above.

CAUTION Work = W, weight = w Don't confuse uppercase W (work) with lowercase w (weight). Though the symbols are similar, work and weight are different quantities. ▌

The SI unit of work is the **joule** (abbreviated J, pronounced "jool," and named in honor of the 19th-century English physicist James Prescott Joule). From Eq. (6.1) we see that in any system of units, the unit of work is the unit of force multiplied by the unit of distance. In SI units the unit of force is the newton and the unit of distance is the meter, so 1 joule is equivalent to 1 *newton-meter* (N·m):

$$1 \text{ joule} = (1 \text{ newton})(1 \text{ meter}) \quad \text{or} \quad 1 \text{ J} = 1 \text{ N} \cdot \text{m}$$

In the British system the unit of force is the pound (lb), the unit of distance is the foot (ft), and the unit of work is the *foot-pound* (ft·lb). The following conversions are useful:

$$1 \text{ J} = 0.7376 \text{ ft} \cdot \text{lb} \qquad 1 \text{ ft} \cdot \text{lb} = 1.356 \text{ J}$$

As an illustration of Eq. (6.1), think of a person pushing a stalled car. If he pushes the car through a displacement $\vec{s}$ with a constant force $\vec{F}$ in the direction

6.1 These people are doing work as they push on the stalled car because they exert a force on the car as it moves.

6.2 The work done by a constant force acting in the same direction as the displacement.

If a body moves through a displacement $\vec{s}$ while a constant force $\vec{F}$ acts on it in the same direction ...

... the work done by the force on the body is $W = Fs$.

Application **Work and Muscle Fibers**
Our ability to do work with our bodies comes from our skeletal muscles. The fiberlike cells of skeletal muscle, shown in this micrograph, have the ability to shorten, causing the muscle as a whole to contract and to exert force on the tendons to which it attaches. Muscle can exert a force of about 0.3 N per square millimeter of cross-sectional area: The greater the cross-sectional area, the more fibers the muscle has and the more force it can exert when it contracts.

6.3 The work done by a constant force acting at an angle to the displacement.

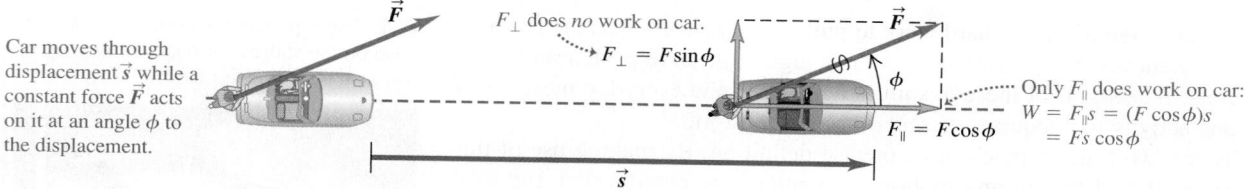

Car moves through displacement $\vec{s}$ while a constant force $\vec{F}$ acts on it at an angle ϕ to the displacement.

$F_\perp$ does *no* work on car.
$F_\perp = F \sin \phi$

Only $F_\parallel$ does work on car:
$W = F_\parallel s = (F \cos \phi)s = Fs \cos \phi$

$F_\parallel = F \cos \phi$

Mastering**PHYSICS**

ActivPhysics 5.1: Work Calculations

of motion, the amount of work he does on the car is given by Eq. (6.1): $W = Fs$. But what if the person pushes at an angle ϕ to the car's displacement (Fig. 6.3)? Then $\vec{F}$ has a component $F_\parallel = F \cos \phi$ in the direction of the displacement and a component $F_\perp = F \sin \phi$ that acts perpendicular to the displacement. (Other forces must act on the car so that it moves along $\vec{s}$, not in the direction of $\vec{F}$. We're interested only in the work that the person does, however, so we'll consider only the force he exerts.) In this case only the parallel component $F_\parallel$ is effective in moving the car, so we define the work as the product of this force component and the magnitude of the displacement. Hence $W = F_\parallel s = (F \cos \phi)s$, or

$$W = Fs \cos \phi \quad \text{(constant force, straight-line displacement)} \quad (6.2)$$

We are assuming that F and ϕ are constant during the displacement. If $\phi = 0$, so that $\vec{F}$ and $\vec{s}$ are in the same direction, then $\cos \phi = 1$ and we are back to Eq. (6.1).

Equation (6.2) has the form of the *scalar product* of two vectors, which we introduced in Section 1.10: $\vec{A} \cdot \vec{B} = AB \cos \phi$. You may want to review that definition. Hence we can write Eq. (6.2) more compactly as

$$W = \vec{F} \cdot \vec{s} \quad \text{(constant force, straight-line displacement)} \quad (6.3)$$

CAUTION **Work is a scalar** Here's an essential point: Work is a *scalar* quantity, even though it's calculated by using two vector quantities (force and displacement). A 5-N force toward the east acting on a body that moves 6 m to the east does exactly the same amount of work as a 5-N force toward the north acting on a body that moves 6 m to the north. ▌

Example 6.1 **Work done by a constant force**

(a) Steve exerts a steady force of magnitude 210 N (about 47 lb) on the stalled car in Fig. 6.3 as he pushes it a distance of 18 m. The car also has a flat tire, so to make the car track straight Steve must push at an angle of 30° to the direction of motion. How much work does Steve do? (b) In a helpful mood, Steve pushes a second stalled car with a steady force $\vec{F} = (160 \text{ N})\hat{\imath} - (40 \text{ N})\hat{\jmath}$. The displacement of the car is $\vec{s} = (14 \text{ m})\hat{\imath} + (11 \text{ m})\hat{\jmath}$. How much work does Steve do in this case?

SOLUTION

IDENTIFY and SET UP: In both parts (a) and (b), the target variable is the work W done by Steve. In each case the force is constant and the displacement is along a straight line, so we can use Eq. (6.2) or (6.3). The angle between $\vec{F}$ and $\vec{s}$ is given in part (a), so we can apply Eq. (6.2) directly. In part (b) both $\vec{F}$ and $\vec{s}$ are given in terms

of components, so it's best to calculate the scalar product using Eq. (1.21): $\vec{A} \cdot \vec{B} = A_x B_x + A_y B_y + A_z B_z$.

EXECUTE: (a) From Eq. (6.2),

$$W = Fs \cos \phi = (210 \text{ N})(18 \text{ m}) \cos 30° = 3.3 \times 10^3 \text{ J}$$

(b) The components of $\vec{F}$ are $F_x = 160$ N and $F_y = -40$ N, and the components of $\vec{s}$ are $x = 14$ m and $y = 11$ m. (There are no z-components for either vector.) Hence, using Eqs. (1.21) and (6.3), we have

$$W = \vec{F} \cdot \vec{s} = F_x x + F_y y$$
$$= (160 \text{ N})(14 \text{ m}) + (-40 \text{ N})(11 \text{ m})$$
$$= 1.8 \times 10^3 \text{ J}$$

EVALUATE: In each case the work that Steve does is more than 1000 J. This shows that 1 joule is a rather small amount of work.

6.4 A constant force $\vec{F}$ can do positive, negative, or zero work depending on the angle between $\vec{F}$ and the displacement $\vec{s}$.

Direction of Force (or Force Component)	Situation	Force Diagram
(a) **Force $\vec{F}$ has a component in direction of displacement:** $W = F_{\parallel}s = (F\cos\phi)s$ Work is *positive*.		
(b) **Force $\vec{F}$ has a component opposite to direction of displacement:** $W = F_{\parallel}s = (F\cos\phi)s$ Work is *negative* (because $F\cos\phi$ is negative for $90° < \phi < 180°$).		
(c) **Force $\vec{F}$ (or force component $F_{\perp}$) is perpendicular to direction of displacement:** The force (or force component) does *no* work on the object.		

Work: Positive, Negative, or Zero

In Example 6.1 the work done in pushing the cars was positive. But it's important to understand that work can also be negative or zero. This is the essential way in which work as defined in physics differs from the "everyday" definition of work. When the force has a component in the *same direction* as the displacement (ϕ between zero and 90°), $\cos\phi$ in Eq. (6.2) is positive and the work W is *positive* (Fig. 6.4a). When the force has a component *opposite* to the displacement (ϕ between 90° and 180°), $\cos\phi$ is negative and the work is *negative* (Fig. 6.4b). When the force is *perpendicular* to the displacement, $\phi = 90°$ and the work done by the force is *zero* (Fig. 6.4c). The cases of zero work and negative work bear closer examination, so let's look at some examples.

There are many situations in which forces act but do zero work. You might think it's "hard work" to hold a barbell motionless in the air for 5 minutes (Fig. 6.5). But in fact, you aren't doing any work at all on the barbell because there is no displacement. You get tired because the components of muscle fibers in your arm do work as they continually contract and relax. This is work done by one part of the arm exerting force on another part, however, *not* on the barbell. (We'll say more in Section 6.2 about work done by one part of a body on another part.) Even when you walk with constant velocity on a level floor while carrying a book, you still do no work on it. The book has a displacement, but the (vertical) supporting force that you exert on the book has no component in the direction of the (horizontal) motion. Then $\phi = 90°$ in Eq. (6.2), and $\cos\phi = 0$. When a body slides along a surface, the work done on the body by the normal force is zero; and when a ball on a string moves in uniform circular motion, the work done on the ball by the tension in the string is also zero. In both cases the work is zero because the force has no component in the direction of motion.

What does it really mean to do *negative* work? The answer comes from Newton's third law of motion. When a weightlifter lowers a barbell as in Fig. 6.6a, his hands and the barbell move together with the same displacement $\vec{s}$. The barbell exerts a force $\vec{F}_{\text{barbell on hands}}$ on his hands in the same direction as the hands' displacement, so the work done by the *barbell* on his *hands* is positive (Fig. 6.6b). But by Newton's third law the weightlifter's hands exert an equal and opposite force $\vec{F}_{\text{hands on barbell}} = -\vec{F}_{\text{barbell on hands}}$ on the barbell (Fig. 6.6c). This force, which keeps the barbell from crashing to the floor, acts opposite to the barbell's displacement. Thus the work done by his *hands* on the *barbell* is negative.

6.5 A weightlifter does no work on a barbell as long as he holds it stationary.

The weightlifter exerts an upward force on the barbell ...

... but because the barbell is stationary (its displacement is zero), he does no work on it.

6.6 This weightlifter's hands do negative work on a barbell as the barbell does positive work on his hands.

(a) A weightlifter lowers a barbell to the floor.

(b) The barbell does *positive* work on the weightlifter's hands.

$\vec{F}_{\text{barbell on hands}}$

The force of the barbell on the weightlifter's hands is in the *same* direction as the hands' displacement.

(c) The weightlifter's hands do *negative* work on the barbell.

$\vec{F}_{\text{hands on barbell}}$

The force of the weightlifter's hands on the barbell is *opposite* to the barbell's displacement.

Because the weightlifter's hands and the barbell have the same displacement, the work that his hands do on the barbell is just the negative of the work that the barbell does on his hands. In general, when one body does negative work on a second body, the second body does an equal amount of *positive* work on the first body.

> **CAUTION** **Keep track of who's doing the work** We always speak of work done *on* a particular body *by* a specific force. Always be sure to specify exactly what force is doing the work you are talking about. When you lift a book, you exert an upward force on the book and the book's displacement is upward, so the work done by the lifting force on the book is positive. But the work done by the *gravitational* force (weight) on a book being lifted is *negative* because the downward gravitational force is opposite to the upward displacement. ▮

Total Work

How do we calculate work when *several* forces act on a body? One way is to use Eq. (6.2) or (6.3) to compute the work done by each separate force. Then, because work is a scalar quantity, the *total* work W_{tot} done on the body by all the forces is the algebraic sum of the quantities of work done by the individual forces. An alternative way to find the total work W_{tot} is to compute the vector sum of the forces (that is, the net force) and then use this vector sum as $\vec{F}$ in Eq. (6.2) or (6.3). The following example illustrates both of these techniques.

Example 6.2 **Work done by several forces**

A farmer hitches her tractor to a sled loaded with firewood and pulls it a distance of 20 m along level ground (Fig. 6.7a). The total weight of sled and load is 14,700 N. The tractor exerts a constant 5000-N force at an angle of 36.9° above the horizontal. A 3500-N friction force opposes the sled's motion. Find the work done by each force acting on the sled and the total work done by all the forces.

SOLUTION

IDENTIFY AND SET UP: Each force is constant and the sled's displacement is along a straight line, so we can calculate the work using the ideas of this section. We'll find the total work in two ways: (1) by adding the work done on the sled by each force and (2) by finding the work done by the net force on the sled. We first draw a free-body diagram showing all of the forces acting on the sled, and we choose a coordinate system (Fig. 6.7b). For each force—weight, normal force, force of the tractor, and friction force—we know the angle between the displacement (in the positive *x*-direction) and the force. Hence we can use Eq. (6.2) to calculate the work each force does.

As in Chapter 5, we'll find the net force by adding the components of the four forces. Newton's second law tells us that because the sled's motion is purely horizontal, the net force can have only a horizontal component.

EXECUTE: (1) The work W_w done by the weight is zero because its direction is perpendicular to the displacement (compare Fig. 6.4c). For the same reason, the work W_n done by the normal force is also zero. (Note that we don't need to calculate the magnitude n to conclude this.) So $W_w = W_n = 0$.

That leaves the work W_T done by the force F_T exerted by the tractor and the work W_f done by the friction force f. From Eq. (6.2),

$$W_T = F_T s \cos\phi = (5000 \text{ N})(20 \text{ m})(0.800) = 80,000 \text{ N} \cdot \text{m}$$
$$= 80 \text{ kJ}$$

The friction force $\vec{f}$ is opposite to the displacement, so for this force $\phi = 180°$ and $\cos\phi = -1$. Again from Eq. (6.2),

$$W_f = fs\cos 180° = (3500 \text{ N})(20 \text{ m})(-1) = -70,000 \text{ N} \cdot \text{m}$$
$$= -70 \text{ kJ}$$

6.7 Calculating the work done on a sled of firewood being pulled by a tractor.

(a)

(b) Free-body diagram for sled

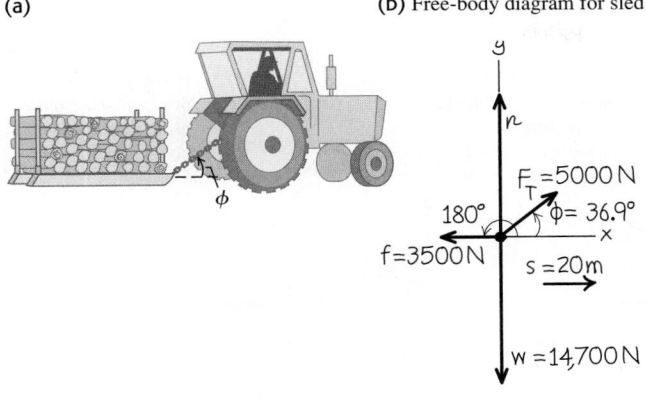

$F_T = 5000$ N

$\phi = 36.9°$

$180°$

$f = 3500$ N

$s = 20$ m

$w = 14,700$ N

The total work W_{tot} done on the sled by all forces is the *algebraic* sum of the work done by the individual forces:

$W_{tot} = W_w + W_n + W_T + W_f = 0 + 0 + 80 \text{ kJ} + (-70 \text{ kJ})$

$= 10 \text{ kJ}$

(2) In the second approach, we first find the *vector* sum of all the forces (the net force) and then use it to compute the total work. The vector sum is best found by using components. From Fig. 6.7b,

$$\sum F_x = F_T \cos\phi + (-f) = (5000 \text{ N}) \cos 36.9° - 3500 \text{ N}$$

$$= 500 \text{ N}$$

$$\sum F_y = F_T \sin\phi + n + (-w)$$

$$= (5000 \text{ N}) \sin 36.9° + n - 14,700 \text{ N}$$

We don't need the second equation; we know that the y-component of force is perpendicular to the displacement, so it does no work. Besides, there is no y-component of acceleration, so $\sum F_y$ must be zero anyway. The total work is therefore the work done by the total x-component:

$$W_{tot} = \left(\sum \vec{F}\right) \cdot \vec{s} = \left(\sum F_x\right)s = (500 \text{ N})(20 \text{ m}) = 10,000 \text{ J}$$

$$= 10 \text{ kJ}$$

EVALUATE: We get the same result for W_{tot} with either method, as we should. Note also that the net force in the x-direction is *not* zero, and so the sled must accelerate as it moves. In Section 6.2 we'll return to this example and see how to use the concept of work to explore the sled's changes of speed.

Test Your Understanding of Section 6.1 An electron moves in a straight line toward the east with a constant speed of 8×10^7 m/s. It has electric, magnetic, and gravitational forces acting on it. During a 1-m displacement, the total work done on the electron is (i) positive; (ii) negative; (iii) zero; (iv) not enough information given to decide.

$\boxed{MP}$

6.2 Kinetic Energy and the Work–Energy Theorem

The total work done on a body by external forces is related to the body's displacement—that is, to changes in its position. But the total work is also related to changes in the *speed* of the body. To see this, consider Fig. 6.8, which shows three examples of a block sliding on a frictionless table. The forces acting on the block are its weight $\vec{w}$, the normal force $\vec{n}$, and the force $\vec{F}$ exerted on it by the hand.

In Fig. 6.8a the net force on the block is in the direction of its motion. From Newton's second law, this means that the block speeds up; from Eq. (6.1), this also means that the total work W_{tot} done on the block is positive. The total work is *negative* in Fig. 6.8b because the net force opposes the displacement; in this case the block slows down. The net force is zero in Fig. 6.8c, so the speed of the block stays the same and the total work done on the block is zero. We can conclude that *when a particle undergoes a displacement, it speeds up if $W_{tot} > 0$, slows down if $W_{tot} < 0$, and maintains the same speed if $W_{tot} = 0$.*

Let's make these observations more quantitative. Consider a particle with mass m moving along the x-axis under the action of a constant net force with magnitude F directed along the positive x-axis (Fig. 6.9). The particle's acceleration is constant and given by Newton's second law, $F = ma_x$. Suppose the speed changes from v_1 to v_2 while the particle undergoes a displacement $s = x_2 - x_1$.

Mastering**PHYSICS**

PhET: The Ramp

6.8 The relationship between the total work done on a body and how the body's speed changes.

(a)

A block slides to the right on a frictionless surface.

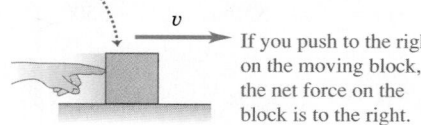

If you push to the right on the moving block, the net force on the block is to the right.

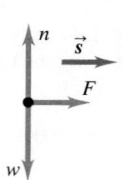

- The total work done on the block during a displacement $\vec{s}$ is positive: $W_{tot} > 0$.
- The block speeds up.

(b)

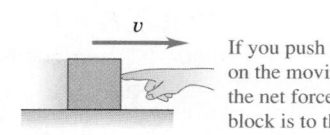

If you push to the left on the moving block, the net force on the block is to the left.

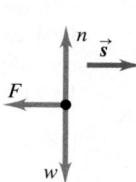

- The total work done on the block during a displacement $\vec{s}$ is negative: $W_{tot} < 0$.
- The block slows down.

(c)

If you push straight down on the moving block, the net force on the block is zero.

- The total work done on the block during a displacement $\vec{s}$ is zero: $W_{tot} = 0$.
- The block's speed stays the same.

6.9 A constant net force $\vec{F}$ does work on a moving body.

from point x_1 to x_2. Using a constant-acceleration equation, Eq. (2.13), and replacing v_{0x} by v_1, v_x by v_2, and $(x - x_0)$ by s, we have

$$v_2^2 = v_1^2 + 2a_x s$$
$$a_x = \frac{v_2^2 - v_1^2}{2s}$$

When we multiply this equation by m and equate ma_x to the net force F, we find

$$F = ma_x = m\frac{v_2^2 - v_1^2}{2s} \qquad \text{and}$$

$$Fs = \tfrac{1}{2}mv_2^2 - \tfrac{1}{2}mv_1^2 \qquad (6.4)$$

The product Fs is the work done by the net force F and thus is equal to the total work W_{tot} done by all the forces acting on the particle. The quantity $\tfrac{1}{2}mv^2$ is called the **kinetic energy** K of the particle:

$$K = \tfrac{1}{2}mv^2 \qquad \text{(definition of kinetic energy)} \qquad (6.5)$$

6.10 Comparing the kinetic energy $K = \tfrac{1}{2}mv^2$ of different bodies.

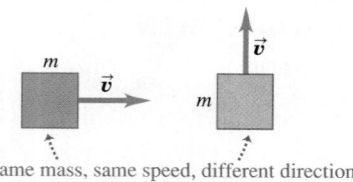

Same mass, same speed, different directions of motion: *same* kinetic energy

Twice the mass, same speed: *twice* the kinetic energy

Same mass, twice the speed: *four times* the kinetic energy

Like work, the kinetic energy of a particle is a scalar quantity; it depends on only the particle's mass and speed, not its direction of motion (Fig. 6.10). A car (viewed as a particle) has the same kinetic energy when going north at 10 m/s as when going east at 10 m/s. Kinetic energy can never be negative, and it is zero only when the particle is at rest.

We can now interpret Eq. (6.4) in terms of work and kinetic energy. The first term on the right side of Eq. (6.4) is $K_2 = \tfrac{1}{2}mv_2^2$, the final kinetic energy of the particle (that is, after the displacement). The second term is the initial kinetic energy, $K_1 = \tfrac{1}{2}mv_1^2$, and the difference between these terms is the *change* in kinetic energy. So Eq. (6.4) says:

The work done by the net force on a particle equals the change in the particle's kinetic energy:

$$W_{tot} = K_2 - K_1 = \Delta K \qquad \text{(work–energy theorem)} \qquad (6.6)$$

This result is the **work–energy theorem.**

The work–energy theorem agrees with our observations about the block in Fig. 6.8. When W_{tot} is *positive,* the kinetic energy *increases* (the final kinetic energy K_2 is greater than the initial kinetic energy K_1) and the particle is going faster at the end of the displacement than at the beginning. When W_{tot} is *negative,* the kinetic energy *decreases* (K_2 is less than K_1) and the speed is less after the displacement. When $W_{tot} = 0$, the kinetic energy stays the same ($K_1 = K_2$) and the speed is unchanged. Note that the work–energy theorem by itself tells us only about changes in *speed,* not velocity, since the kinetic energy doesn't depend on the direction of motion.

From Eq. (6.4) or Eq. (6.6), kinetic energy and work must have the same units. Hence the joule is the SI unit of both work and kinetic energy (and, as we will see later, of all kinds of energy). To verify this, note that in SI units the quantity $K = \frac{1}{2}mv^2$ has units $\mathrm{kg} \cdot (\mathrm{m/s})^2$ or $\mathrm{kg} \cdot \mathrm{m}^2/\mathrm{s}^2$; we recall that $1\ \mathrm{N} = 1\ \mathrm{kg} \cdot \mathrm{m/s}^2$, so

$$1\ \mathrm{J} = 1\ \mathrm{N} \cdot \mathrm{m} = 1\ (\mathrm{kg} \cdot \mathrm{m/s}^2) \cdot \mathrm{m} = 1\ \mathrm{kg} \cdot \mathrm{m}^2/\mathrm{s}^2$$

In the British system the unit of kinetic energy and of work is

$$1\ \mathrm{ft} \cdot \mathrm{lb} = 1\ \mathrm{ft} \cdot \mathrm{slug} \cdot \mathrm{ft/s}^2 = 1\ \mathrm{slug} \cdot \mathrm{ft}^2/\mathrm{s}^2$$

Because we used Newton's laws in deriving the work–energy theorem, we can use this theorem only in an inertial frame of reference. Note also that the work–energy theorem is valid in *any* inertial frame, but the values of W_{tot} and $K_2 - K_1$ may differ from one inertial frame to another (because the displacement and speed of a body may be different in different frames).

We've derived the work–energy theorem for the special case of straight-line motion with constant forces, and in the following examples we'll apply it to this special case only. We'll find in the next section that the theorem is valid in general, even when the forces are not constant and the particle's trajectory is curved.

Problem-Solving Strategy 6.1   **Work and Kinetic Energy**

IDENTIFY *the relevant concepts:* The work–energy theorem, $W_{tot} = K_2 - K_1$, is extremely useful when you want to relate a body's speed v_1 at one point in its motion to its speed v_2 at a different point. (It's less useful for problems that involve the *time* it takes a body to go from point 1 to point 2 because the work–energy theorem doesn't involve time at all. For such problems it's usually best to use the relationships among time, position, velocity, and acceleration described in Chapters 2 and 3.)

SET UP *the problem* using the following steps:
1. Identify the initial and final positions of the body, and draw a free-body diagram showing all the forces that act on the body.
2. Choose a coordinate system. (If the motion is along a straight line, it's usually easiest to have both the initial and final positions lie along one of the axes.)
3. List the unknown and known quantities, and decide which unknowns are your target variables. The target variable may be the body's initial or final speed, the magnitude of one of the forces acting on the body, or the body's displacement.

EXECUTE *the solution:* Calculate the work W done by each force. If the force is constant and the displacement is a straight line, you can use Eq. (6.2) or Eq. (6.3). (Later in this chapter we'll see how to handle varying forces and curved trajectories.) Be sure to check signs; W must be positive if the force has a component in the

direction of the displacement, negative if the force has a component opposite to the displacement, and zero if the force and displacement are perpendicular.

Add the amounts of work done by each force to find the total work W_{tot}. Sometimes it's easier to calculate the vector sum of the forces (the net force) and then find the work done by the net force; this value is also equal to W_{tot}.

Write expressions for the initial and final kinetic energies, K_1 and K_2. Note that kinetic energy involves *mass,* not *weight;* if you are given the body's weight, use $w = mg$ to find the mass.

Finally, use Eq. (6.6), $W_{tot} = K_2 - K_1$, and Eq. (6.5), $K = \frac{1}{2}mv^2$, to solve for the target variable. Remember that the right-hand side of Eq. (6.6) represents the change of the body's kinetic energy between points 1 and 2; that is, it is the *final* kinetic energy minus the *initial* kinetic energy, never the other way around. (If you can predict the sign of W_{tot}, you can predict whether the body speeds up or slows down.)

EVALUATE *your answer:* Check whether your answer makes sense. Remember that kinetic energy $K = \frac{1}{2}mv^2$ can never be negative. If you come up with a negative value of K, perhaps you interchanged the initial and final kinetic energies in $W_{tot} = K_2 - K_1$ or made a sign error in one of the work calculations.

Example 6.3 Using work and energy to calculate speed

Let's look again at the sled in Fig. 6.7 and our results from Example 6.2. Suppose the sled's initial speed v_1 is 2.0 m/s. What is the speed of the sled after it moves 20 m?

SOLUTION

IDENTIFY and SET UP: We'll use the work–energy theorem, Eq. (6.6), $W_{tot} = K_2 - K_1$, since we are given the initial speed $v_1 = 2.0$ m/s and want to find the final speed v_2. Figure 6.11 shows our sketch of the situation. The motion is in the positive x-direction. In Example 6.2 we calculated the total work done by all the forces: $W_{tot} = 10$ kJ. Hence the kinetic energy of the sled and its load must increase by 10 kJ, and the speed of the sled must also increase.

EXECUTE: To write expressions for the initial and final kinetic energies, we need the mass of the sled and load. The combined *weight* is 14,700 N, so the mass is

$$m = \frac{w}{g} = \frac{14{,}700 \text{ N}}{9.8 \text{ m/s}^2} = 1500 \text{ kg}$$

Then the initial kinetic energy K_1 is

$$K_1 = \tfrac{1}{2}mv_1^2 = \tfrac{1}{2}(1500 \text{ kg})(2.0 \text{ m/s})^2 = 3000 \text{ kg} \cdot \text{m}^2/\text{s}^2$$
$$= 3000 \text{ J}$$

6.11 Our sketch for this problem.

The final kinetic energy K_2 is

$$K_2 = \tfrac{1}{2}mv_2^2 = \tfrac{1}{2}(1500 \text{ kg})v_2^2$$

The work–energy theorem, Eq. (6.6), gives

$$K_2 = K_1 + W_{tot} = 3000 \text{ J} + 10{,}000 \text{ J} = 13{,}000 \text{ J}$$

Setting these two expressions for K_2 equal, substituting $1 \text{ J} = 1 \text{ kg} \cdot \text{m}^2/\text{s}^2$, and solving for the final speed v_2, we find

$$v_2 = 4.2 \text{ m/s}$$

EVALUATE: The total work is positive, so the kinetic energy increases $(K_2 > K_1)$ and the speed increases $(v_2 > v_1)$.

This problem can also be solved without the work–energy theorem. We can find the acceleration from $\sum \vec{F} = m\vec{a}$ and then use the equations of motion for constant acceleration to find v_2. Since the acceleration is along the x-axis,

$$a = a_x = \frac{\sum F_x}{m} = \frac{500 \text{ N}}{1500 \text{ kg}} = 0.333 \text{ m/s}^2$$

Then, using Eq. (2.13),

$$v_2^2 = v_1^2 + 2as = (2.0 \text{ m/s})^2 + 2(0.333 \text{ m/s}^2)(20 \text{ m})$$
$$= 17.3 \text{ m}^2/\text{s}^2$$
$$v_2 = 4.2 \text{ m/s}$$

This is the same result we obtained with the work–energy approach, but there we avoided the intermediate step of finding the acceleration. You will find several other examples in this chapter and the next that *can* be done without using energy considerations but that are easier when energy methods are used. When a problem can be done by two methods, doing it by both methods (as we did here) is a good way to check your work.

Example 6.4 Forces on a hammerhead

The 200-kg steel hammerhead of a pile driver is lifted 3.00 m above the top of a vertical I-beam being driven into the ground (Fig. 6.12a). The hammerhead is then dropped, driving the I-beam 7.4 cm deeper into the ground. The vertical guide rails exert a constant 60-N friction force on the hammerhead. Use the work–energy theorem to find (a) the speed of the hammerhead just as it hits the I-beam and (b) the average force the hammerhead exerts on the I-beam. Ignore the effects of the air.

SOLUTION

IDENTIFY: We'll use the work–energy theorem to relate the hammerhead's speed at different locations and the forces acting on it. There are *three* locations of interest: point 1, where the hammerhead starts from rest; point 2, where it first contacts the I-beam; and point 3, where the hammerhead and I-beam come to a halt (Fig. 6.12a). The two target variables are the hammerhead's speed at point 2 and the average force the hammerhead exerts between points 2 and 3. Hence we'll apply the work–energy theorem

twice: once for the motion from 1 to 2, and once for the motion from 2 to 3.

SET UP: Figure 6.12b shows the vertical forces on the hammerhead as it falls from point 1 to point 2. (We can ignore any horizontal forces that may be present because they do no work as the hammerhead moves vertically.) For this part of the motion, our target variable is the hammerhead's final speed v_2.

Figure 6.12c shows the vertical forces on the hammerhead during the motion from point 2 to point 3. In addition to the forces shown in Fig. 6.12b, the I-beam exerts an upward normal force of magnitude n on the hammerhead. This force actually varies as the hammerhead comes to a halt, but for simplicity we'll treat n as a constant. Hence n represents the *average* value of this upward force during the motion. Our target variable for this part of the motion is the force that the *hammerhead* exerts on the I-beam; it is the reaction force to the normal force exerted by the I-beam, so by Newton's third law its magnitude is also n.

EXECUTE: (a) From point 1 to point 2, the vertical forces are the downward weight $w = mg = (200 \text{ kg})(9.8 \text{ m/s}^2) = 1960 \text{ N}$ and the upward friction force $f = 60 \text{ N}$. Thus the net downward force is $w - f = 1900 \text{ N}$. The displacement of the hammerhead from point 1 to point 2 is downward and equal to $s_{12} = 3.00 \text{ m}$. The total work done on the hammerhead between point 1 and point 2 is then

$$W_{\text{tot}} = (w - f)s_{12} = (1900 \text{ N})(3.00 \text{ m}) = 5700 \text{ J}$$

At point 1 the hammerhead is at rest, so its initial kinetic energy K_1 is zero. Hence the kinetic energy K_2 at point 2 equals the total work done on the hammerhead between points 1 and 2:

$$W_{\text{tot}} = K_2 - K_1 = K_2 - 0 = \tfrac{1}{2}mv_2^2 - 0$$

$$v_2 = \sqrt{\frac{2W_{\text{tot}}}{m}} = \sqrt{\frac{2(5700 \text{ J})}{200 \text{ kg}}} = 7.55 \text{ m/s}$$

This is the hammerhead's speed at point 2, just as it hits the I-beam.

(b) As the hammerhead moves downward from point 2 to point 3, its displacement is $s_{23} = 7.4 \text{ cm} = 0.074 \text{ m}$ and the net downward force acting on it is $w - f - n$ (Fig. 6.12c). The total work done on the hammerhead during this displacement is

$$W_{\text{tot}} = (w - f - n)s_{23}$$

The initial kinetic energy for this part of the motion is K_2, which from part (a) equals 5700 J. The final kinetic energy is $K_3 = 0$ (the hammerhead ends at rest). From the work–energy theorem,

$$W_{\text{tot}} = (w - f - n)s_{23} = K_3 - K_2$$

$$n = w - f - \frac{K_3 - K_2}{s_{23}}$$

$$= 1960 \text{ N} - 60 \text{ N} - \frac{0 \text{ J} - 5700 \text{ J}}{0.074 \text{ m}} = 79{,}000 \text{ N}$$

The downward force that the hammerhead exerts on the I-beam has this same magnitude, 79,000 N (about 9 tons)—more than 40 times the weight of the hammerhead.

EVALUATE: The net change in the hammerhead's kinetic energy from point 1 to point 3 is zero; a relatively small net force does positive work over a large distance, and then a much larger net force does negative work over a much smaller distance. The same thing happens if you speed up your car gradually and then drive it into a brick wall. The very large force needed to reduce the kinetic energy to zero over a short distance is what does the damage to your car—and possibly to you.

6.12 (a) A pile driver pounds an I-beam into the ground. (b), (c) Free-body diagrams. Vector lengths are not to scale.

The Meaning of Kinetic Energy

Example 6.4 gives insight into the physical meaning of kinetic energy. The hammerhead is dropped from rest, and its kinetic energy when it hits the I-beam equals the total work done on it up to that point by the net force. This result is true in general: To accelerate a particle of mass m from rest (zero kinetic energy)

6.13 When a billiards player hits a cue ball at rest, the ball's kinetic energy after being hit is equal to the work that was done on it by the cue. The greater the force exerted by the cue and the greater the distance the ball moves while in contact with it, the greater the ball's kinetic energy.

up to a speed , the total work done on it must equal the change in kinetic energy from zero to $K = \frac{1}{2}mv^2$:

$$W_{\text{tot}} = K - 0 = K$$

So *the kinetic energy of a particle is equal to the total work that was done to accelerate it from rest to its present speed* (Fig. 6.13). The definition $K = \frac{1}{2}mv^2$, Eq. (6.5), wasn't chosen at random; it's the *only* definition that agrees with this interpretation of kinetic energy.

In the second part of Example 6.4 the kinetic energy of the hammerhead did work on the I-beam and drove it into the ground. This gives us another interpretation of kinetic energy: *The kinetic energy of a particle is equal to the total work that particle can do in the process of being brought to rest.* This is why you pull your hand and arm backward when you catch a ball. As the ball comes to rest, it does an amount of work (force times distance) on your hand equal to the ball's initial kinetic energy. By pulling your hand back, you maximize the distance over which the force acts and so minimize the force on your hand.

Conceptual Example 6.5 **Comparing kinetic energies**

Two iceboats like the one in Example 5.6 (Section 5.2) hold a race on a frictionless horizontal lake (Fig. 6.14). The two iceboats have masses m and $2m$. The iceboats have identical sails, so the wind exerts the same constant force $\vec{F}$ on each iceboat. They start from rest and cross the finish line a distance s away. Which iceboat crosses the finish line with greater kinetic energy?

SOLUTION

If you use the definition of kinetic energy, $K = \frac{1}{2}mv^2$, Eq. (6.5), the answer to this problem isn't obvious. The iceboat of mass $2m$ has greater mass, so you might guess that it has greater kinetic energy at the finish line. But the lighter iceboat, of mass m, has greater acceleration and crosses the finish line with a greater speed, so you might guess that *this* iceboat has the greater kinetic energy. How can we decide?

The key is to remember that *the kinetic energy of a particle is equal to the total work done to accelerate it from rest.* Both iceboats travel the same distance s from rest, and only the horizontal force F in the direction of motion does work on either iceboat. Hence the total work done between the starting line and the finish line is the *same* for each iceboat, $W_{\text{tot}} = Fs$. At the finish line, each iceboat has a kinetic energy equal to the work W_{tot} done on it, because each iceboat started from rest. So both iceboats have the *same* kinetic energy at the finish line!

6.14 A race between iceboats.

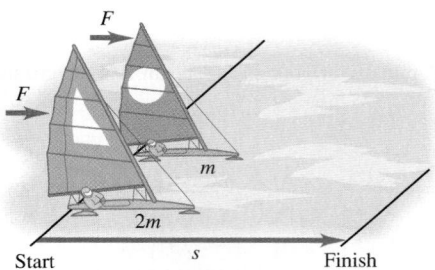

You might think this is a "trick" question, but it isn't. If you really understand the meanings of quantities such as kinetic energy, you can solve problems more easily and with better insight.

Notice that we didn't need to know anything about how much time each iceboat took to reach the finish line. This is because the work–energy theorem makes no direct reference to time, only to displacement. In fact the iceboat of mass m has greater acceleration and so takes less time to reach the finish line than does the iceboat of mass $2m$.

Work and Kinetic Energy in Composite Systems

In this section we've been careful to apply the work–energy theorem only to bodies that we can represent as *particles*—that is, as moving point masses. New subtleties appear for more complex systems that have to be represented as many particles with different motions. We can't go into these subtleties in detail in this chapter, but here's an example.

Suppose a boy stands on frictionless roller skates on a level surface, facing a rigid wall (Fig. 6.15). He pushes against the wall, which makes him move to the right. The forces acting on him are his weight $\vec{w}$, the upward normal forces $\vec{n}_1$ and $\vec{n}_2$ exerted by the ground on his skates, and the horizontal force $\vec{F}$ exerted on him by the wall. There is no vertical displacement, so $\vec{w}$, $\vec{n}_1$, and $\vec{n}_2$ do no work. Force $\vec{F}$ accelerates him to the right, but the parts of his body where that force is applied (the boy's hands) do not move while the force acts. Thus the force $\vec{F}$ also does no work. Where, then, does the boy's kinetic energy come from?

The explanation is that it's not adequate to represent the boy as a single point mass. Different parts of the boy's body have different motions; his hands remain stationary against the wall while his torso is moving away from the wall. The various parts of his body interact with each other, and one part can exert forces and do work on another part. Therefore the *total* kinetic energy of this *composite* system of body parts can change, even though no work is done by forces applied by bodies (such as the wall) that are outside the system. In Chapter 8 we'll consider further the motion of a collection of particles that interact with each other. We'll discover that just as for the boy in this example, the total kinetic energy of such a system can change even when no work is done on any part of the system by anything outside it.

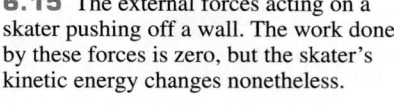

6.15 The external forces acting on a skater pushing off a wall. The work done by these forces is zero, but the skater's kinetic energy changes nonetheless.

Test Your Understanding of Section 6.2 Rank the following bodies in order of their kinetic energy, from least to greatest. (i) a 2.0-kg body moving at 5.0 m/s; (ii) a 1.0-kg body that initially was at rest and then had 30 J of work done on it; (iii) a 1.0-kg body that initially was moving at 4.0 m/s and then had 20 J of work done on it; (iv) a 2.0-kg body that initially was moving at 10 m/s and then did 80 J of work on another body.

6.3 Work and Energy with Varying Forces

So far in this chapter we've considered work done by *constant forces* only. But what happens when you stretch a spring? The more you stretch it, the harder you have to pull, so the force you exert is *not* constant as the spring is stretched. We've also restricted our discussion to *straight-line* motion. There are many situations in which a body moves along a curved path and is acted on by a force that varies in magnitude, direction, or both. We need to be able to compute the work done by the force in these more general cases. Fortunately, we'll find that the work–energy theorem holds true even when varying forces are considered and when the body's path is not straight.

Work Done by a Varying Force, Straight-Line Motion

To add only one complication at a time, let's consider straight-line motion along the x-axis with a force whose x-component F_x may change as the body moves. (A real-life example is driving a car along a straight road with stop signs, so the driver has to alternately step on the gas and apply the brakes.) Suppose a particle moves along the x-axis from point x_1 to x_2 (Fig. 6.16a). Figure 6.16b is a graph of the x-component of force as a function of the particle's coordinate x. To find the work done by this force, we divide the total displacement into small segments Δx_a, Δx_b, and so on (Fig. 6.16c). We approximate the work done by the force during segment Δx_a as the average x-component of force F_{ax} in that segment multiplied by the x-displacement Δx_a. We do this for each segment and then add the results for all the segments. The work done by the force in the total displacement from x_1 to x_2 is approximately

$$W = F_{ax}\Delta x_a + F_{bx}\Delta x_b + \cdots$$

6.16 Calculating the work done by a varying force F_x in the x-direction as a particle moves from x_1 to x_2.

(a) Particle moving from x_1 to x_2 in response to a changing force in the x-direction

(b)

(c)

In the limit that the number of segments becomes very large and the width of each becomes very small, this sum becomes the *integral* of F_x from x_1 to x_2:

$$W = \int_{x_1}^{x_2} F_x\, dx \qquad \text{(varying } x\text{-component of force,} \atop \text{straight-line displacement)}} \tag{6.7}$$

Note that $F_{ax}\Delta x_a$ represents the *area* of the first vertical strip in Fig. 6.16c and that the integral in Eq. (6.7) represents the area under the curve of Fig. 6.16b between x_1 and x_2. *On a graph of force as a function of position, the total work done by the force is represented by the area under the curve between the initial and final positions.* An alternative interpretation of Eq. (6.7) is that the work W equals the average force that acts over the entire displacement, multiplied by the displacement.

In the special case that F_x, the x-component of the force, is constant, it may be taken outside the integral in Eq. (6.7):

$$W = \int_{x_1}^{x_2} F_x\, dx = F_x \int_{x_1}^{x_2} dx = F_x(x_2 - x_1) \qquad \text{(constant force)}$$

But $x_2 - x_1 = s$, the total displacement of the particle. So in the case of a constant force F, Eq. (6.7) says that $W = Fs$, in agreement with Eq. (6.1). The interpretation of work as the area under the curve of F_x as a function of x also holds for a constant force; $W = Fs$ is the area of a rectangle of height F and width s (Fig. 6.17).

Now let's apply these ideas to the stretched spring. To keep a spring stretched beyond its unstretched length by an amount x, we have to apply a force of equal magnitude at each end (Fig. 6.18). If the elongation x is not too great, the force we apply to the right-hand end has an x-component directly proportional to x:

$$F_x = kx \qquad \text{(force required to stretch a spring)} \tag{6.8}$$

6.17 The work done by a constant force F in the x-direction as a particle moves from x_1 to x_2.

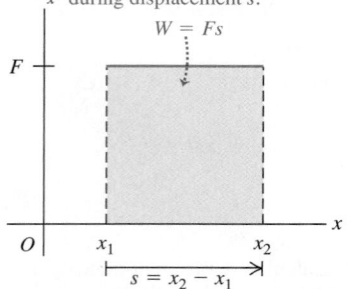

The rectangular area under the graph represents the work done by the constant force of magnitude F during displacement s:

$$W = Fs$$

6.18 The force needed to stretch an ideal spring is proportional to the spring's elongation: $F_x = kx$.

where k is a constant called the **force constant** (or spring constant) of the spring. The units of k are force divided by distance: N/m in SI units and lb/ft in British units. A floppy toy spring such as a Slinky™ has a force constant of about 1 N/m; for the much stiffer springs in an automobile's suspension, k is about 10^5 N/m. The observation that force is directly proportional to elongation for elongations that are not too great was made by Robert Hooke in 1678 and is known as **Hooke's law.** It really shouldn't be called a "law," since it's a statement about a specific device and not a fundamental law of nature. Real springs don't always obey Eq. (6.8) precisely, but it's still a useful idealized model. We'll discuss Hooke's law more fully in Chapter 11.

To stretch a spring, we must do work. We apply equal and opposite forces to the ends of the spring and gradually increase the forces. We hold the left end stationary, so the force we apply at this end does no work. The force at the moving end *does* do work. Figure 6.19 is a graph of F_x as a function of x, the elongation of the spring. The work done by this force when the elongation goes from zero to a maximum value X is

$$W = \int_0^X F_x\, dx = \int_0^X kx\, dx = \tfrac{1}{2}kX^2 \tag{6.9}$$

6.19 Calculating the work done to stretch a spring by a length X.

The area under the graph represents the work done on the spring as the spring is stretched from $x = 0$ to a maximum value X:

$$W = \tfrac{1}{2}kX^2$$

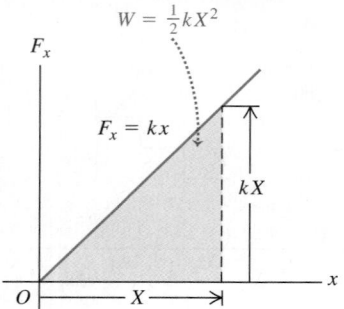

We can also obtain this result graphically. The area of the shaded triangle in Fig. 6.19, representing the total work done by the force, is equal to half the product of the base and altitude, or

$$W = \tfrac{1}{2}(X)(kX) = \tfrac{1}{2}kX^2$$

This equation also says that the work is the *average* force $kX/2$ multiplied by the total displacement X. We see that the total work is proportional to the *square* of the final elongation X. To stretch an ideal spring by 2 cm, you must do four times as much work as is needed to stretch it by 1 cm.

Equation (6.9) assumes that the spring was originally unstretched. If initially the spring is already stretched a distance x_1, the work we must do to stretch it to a greater elongation x_2 (Fig. 6.20a) is

$$W = \int_{x_1}^{x_2} F_x \, dx = \int_{x_1}^{x_2} kx \, dx = \tfrac{1}{2}kx_2^2 - \tfrac{1}{2}kx_1^2 \qquad (6.10)$$

You should use your knowledge of geometry to convince yourself that the trapezoidal area under the graph in Fig. 6.20b is given by the expression in Eq. (6.10).

If the spring has spaces between the coils when it is unstretched, then it can also be compressed, and Hooke's law holds for compression as well as stretching. In this case the force and displacement are in the opposite directions from those shown in Fig. 6.18, and so F_x and x in Eq. (6.8) are both negative. Since both F_x and x are reversed, the force again is in the same direction as the displacement, and the work done by F_x is again positive. So the total work is still given by Eq. (6.9) or (6.10), even when X is negative or either or both of x_1 and x_2 are negative.

> **CAUTION** **Work done *on* a spring vs. work done *by* a spring** Note that Eq. (6.10) gives the work that *you* must do *on* a spring to change its length. For example, if you stretch a spring that's originally relaxed, then $x_1 = 0$, $x_2 > 0$, and $W > 0$: The force you apply to one end of the spring is in the same direction as the displacement, and the work you do is positive. By contrast, the work that the *spring* does on whatever it's attached to is given by the *negative* of Eq. (6.10). Thus, as you pull on the spring, the spring does negative work on you. Paying careful attention to the sign of work will eliminate confusion later on! ▮

6.20 Calculating the work done to stretch a spring from one extension to a greater one.

(a) Stretching a spring from elongation x_1 to elongation x_2

(b) Force-versus-distance graph

The trapezoidal area under the graph represents the work done on the spring to stretch it from $x = x_1$ to $x = x_2$: $W = \tfrac{1}{2}kx_2^2 - \tfrac{1}{2}kx_1^2$

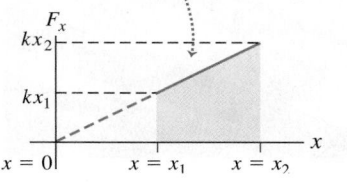

Example 6.6   **Work done on a spring scale**

A woman weighing 600 N steps on a bathroom scale that contains a stiff spring (Fig. 6.21). In equilibrium, the spring is compressed 1.0 cm under her weight. Find the force constant of the spring and the total work done on it during the compression.

SOLUTION

IDENTIFY and SET UP: In equilibrium the upward force exerted by the spring balances the downward force of the woman's weight. We'll use this principle and Eq. (6.8) to determine the force constant k, and we'll use Eq. (6.10) to calculate the work W that the

6.21 Compressing a spring in a bathroom scale.

Because of our choice of axis, both the force component and displacement are negative. The work *on* the spring is positive.

woman does on the spring to compress it. We take positive values of x to correspond to elongation (upward in Fig. 6.21), so that the displacement of the end of the spring (x) and the x-component of the force that the woman exerts on it (F_x) are both negative. The applied force and the displacement are in the same direction, so the work done on the spring will be positive.

EXECUTE: The top of the spring is displaced by $x = -1.0$ cm $= -0.010$ m, and the woman exerts a force $F_x = -600$ N on the spring. From Eq. (6.8) the force constant is then

$$k = \frac{F_x}{x} = \frac{-600 \text{ N}}{-0.010 \text{ m}} = 6.0 \times 10^4 \text{ N/m}$$

Then, using $x_1 = 0$ and $x_2 = -0.010$ m in Eq. (6.10), we have

$$W = \tfrac{1}{2}kx_2^2 - \tfrac{1}{2}kx_1^2$$

$$= \tfrac{1}{2}(6.0 \times 10^4 \text{ N/m})(-0.010 \text{ m})^2 - 0 = 3.0 \text{ J}$$

EVALUATE: The work done is positive, as expected. Our arbitrary choice of the positive direction has no effect on the answer for W. You can test this by taking the positive x-direction to be downward, corresponding to compression. Do you get the same values for k and W as we found here?

Application **Tendons Are Nonideal Springs**
Muscles exert forces via the tendons that attach them to bones. A tendon consists of long, stiff, elastic collagen fibers. The graph shows how the tendon from the hind leg of a wallaby (a small kangaroo) stretches in response to an applied force. The tendon does not exhibit the simple, straight-line behavior of an ideal spring, so the work it does has to be found by integration [Eq. (6.7)]. Note that the tendon exerts less force while relaxing than while stretching. As a result, the relaxing tendon does only about 93% of the work that was done to stretch it.

Force exerted by tendon (N)

Maximum tendon extension

Tendon being stretched

Tendon relaxing

Extension (mm)

Work–Energy Theorem for Straight-Line Motion, Varying Forces

In Section 6.2 we derived the work–energy theorem, $W_{\text{tot}} = K_2 - K_1$, for the special case of straight-line motion with a constant net force. We can now prove that this theorem is true even when the force varies with position. As in Section 6.2, let's consider a particle that undergoes a displacement x while being acted on by a net force with x-component F_x, which we now allow to vary. Just as in Fig. 6.16, we divide the total displacement x into a large number of small segments Δx. We can apply the work–energy theorem, Eq. (6.6), to each segment because the value of F_x in each small segment is approximately constant. The change in kinetic energy in segment Δx_a is equal to the work $F_{ax}\Delta x_a$, and so on. The total change of kinetic energy is the sum of the changes in the individual segments, and thus is equal to the total work done on the particle during the entire displacement. So $W_{\text{tot}} = \Delta K$ holds for varying forces as well as for constant ones.

Here's an alternative derivation of the work–energy theorem for a force that may vary with position. It involves making a change of variable from x to v_x in the work integral. As a preliminary, we note that the acceleration a of the particle can be expressed in various ways, using $a_x = dv_x/dt$, $v_x = dx/dt$, and the chain rule for derivatives:

$$a_x = \frac{dv_x}{dt} = \frac{dv_x}{dx}\frac{dx}{dt} = v_x\frac{dv_x}{dx} \tag{6.11}$$

From this result, Eq. (6.7) tells us that the total work done by the *net* force F_x is

$$W_{\text{tot}} = \int_{x_1}^{x_2} F_x\,dx = \int_{x_1}^{x_2} ma_x\,dx = \int_{x_1}^{x_2} mv_x\frac{dv_x}{dx}\,dx \tag{6.12}$$

Now $(dv_x/dx)dx$ is the change in velocity dv_x during the displacement dx, so in Eq. (6.12) we can substitute dv_x for $(dv_x/dx)\,dx$. This changes the integration variable from x to v_x, so we change the limits from x_1 and x_2 to the corresponding x-velocities v_1 and v_2 at these points. This gives us

$$W_{\text{tot}} = \int_{v_1}^{v_2} mv_x\,dv_x$$

The integral of $v_x\,dv_x$ is just $v_x^2/2$. Substituting the upper and lower limits, we finally find

$$W_{\text{tot}} = \tfrac{1}{2}mv_2^2 - \tfrac{1}{2}mv_1^2 \tag{6.13}$$

This is the same as Eq. (6.6), so the work–energy theorem is valid even without the assumption that the net force is constant.

Example 6.7 **Motion with a varying force**

An air-track glider of mass 0.100 kg is attached to the end of a horizontal air track by a spring with force constant 20.0 N/m (Fig. 6.22a). Initially the spring is unstretched and the glider is moving at 1.50 m/s to the right. Find the maximum distance d that the glider moves to the right (a) if the air track is turned on, so that there is no friction, and (b) if the air is turned off, so that there is kinetic friction with coefficient $\mu_k = 0.47$.

SOLUTION

IDENTIFY and SET UP: The force exerted by the spring is not constant, so we *cannot* use the constant-acceleration formulas of Chapter 2 to solve this problem. Instead, we'll use the work–energy theorem, since the total work done involves the distance moved (our target variable). In Figs. 6.22b and 6.22c we choose the positive x-direction to be to the right (in the direction of the glider's motion). We take $x = 0$ at the glider's initial position (where the spring is unstretched) and $x = d$ (the target variable) at the position where the glider stops. The motion is purely horizontal, so only the horizontal forces do work. Note that Eq. (6.10) gives the work done by the *glider* on the *spring* as it stretches; to use the work–energy theorem we need the work done by the *spring* on the *glider*, which is the negative of Eq. (6.10). We expect the glider to move farther without friction than with friction.

6.22 (a) A glider attached to an air track by a spring. (b), (c) Our free-body diagrams.

(a)

(b) No friction

(c) With friction

EXECUTE: (a) Equation (6.10) says that as the glider moves from $x_1 = 0$ to $x_2 = d$, it does an amount of work $W = \frac{1}{2}kd^2 - \frac{1}{2}k(0)^2 = \frac{1}{2}kd^2$ on the spring. The amount of work that the *spring* does on the *glider* is the negative of this, $-\frac{1}{2}kd^2$. The spring stretches until the glider comes instantaneously to rest, so the final kinetic energy K_2 is zero. The initial kinetic energy is $\frac{1}{2}mv_1^2$, where $v_1 = 1.50$ m/s is the glider's initial speed. From the work–energy theorem,

$$-\tfrac{1}{2}kd^2 = 0 - \tfrac{1}{2}mv_1^2$$

We solve for the distance d the glider moves:

$$d = v_1\sqrt{\frac{m}{k}} = (1.50 \text{ m/s})\sqrt{\frac{0.100 \text{ kg}}{20.0 \text{ N/m}}}$$
$$= 0.106 \text{ m} = 10.6 \text{ cm}$$

The stretched spring subsequently pulls the glider back to the left, so the glider is at rest only instantaneously.

(b) If the air is turned off, we must include the work done by the kinetic friction force. The normal force n is equal in magnitude to the weight of the glider, since the track is horizontal and there are

no other vertical forces. Hence the kinetic friction force has constant magnitude $f_k = \mu_k n = \mu_k mg$. The friction force is directed opposite to the displacement, so the work done by friction is

$$W_{\text{fric}} = f_k d \cos 180° = -f_k d = -\mu_k mgd$$

The total work is the sum of W_{fric} and the work done by the spring, $-\frac{1}{2}kd^2$. The work–energy theorem then says that

$$-\mu_k mgd - \tfrac{1}{2}kd^2 = 0 - \tfrac{1}{2}mv_1^2 \qquad \text{or}$$
$$\tfrac{1}{2}kd^2 + \mu_k mgd - \tfrac{1}{2}mv_1^2 = 0$$

This is a quadratic equation for d. The solutions are

$$d = -\frac{\mu_k mg}{k} \pm \sqrt{\left(\frac{\mu_k mg}{k}\right)^2 + \frac{mv_1^2}{k}}$$

We have

$$\frac{\mu_k mg}{k} = \frac{(0.47)(0.100 \text{ kg})(9.80 \text{ m/s}^2)}{20.0 \text{ N/m}} = 0.02303 \text{ m}$$

$$\frac{mv_1^2}{k} = \frac{(0.100 \text{ kg})(1.50 \text{ m/s})^2}{20.0 \text{ N/m}} = 0.01125 \text{ m}^2$$

so

$$d = -(0.02303 \text{ m}) \pm \sqrt{(0.02303 \text{ m})^2 + 0.01125 \text{ m}^2}$$
$$= 0.086 \text{ m} \quad \text{or} \quad -0.132 \text{ m}$$

The quantity d is a positive displacement, so only the positive value of d makes sense. Thus with friction the glider moves a distance $d = 0.086$ m = 8.6 cm.

EVALUATE: Note that if we set $\mu_k = 0$, our algebraic solution for d in part (b) reduces to $d = v_1\sqrt{m/k}$, the zero-friction result from part (a). With friction, the glider goes a shorter distance. Again the glider stops instantaneously, and again the spring force pulls it toward the left; whether it moves or not depends on how great the *static* friction force is. How large would the coefficient of static friction μ_s have to be to keep the glider from springing back to the left?

Work–Energy Theorem for Motion Along a Curve

We can generalize our definition of work further to include a force that varies in direction as well as magnitude, and a displacement that lies along a curved path. Figure 6.23a shows a particle moving from P_1 to P_2 along a curve. We divide the curve between these points into many infinitesimal vector displacements, and we call a typical one of these $d\vec{l}$. Each $d\vec{l}$ is tangent to the path at its position. Let $\vec{F}$ be the force at a typical point along the path, and let ϕ be the angle between $\vec{F}$ and $d\vec{l}$ at this point. Then the small element of work dW done on the particle during the displacement $d\vec{l}$ may be written as

$$dW = F\cos\phi\, dl = F_\parallel\, dl = \vec{F}\cdot d\vec{l}$$

where $F_\parallel = F\cos\phi$ is the component of $\vec{F}$ in the direction parallel to $d\vec{l}$ (Fig. 6.23b). The total work done by $\vec{F}$ on the particle as it moves from P_1 to P_2 is then

$$W = \int_{P_1}^{P_2} F\cos\phi\, dl = \int_{P_1}^{P_2} F_\parallel\, dl = \int_{P_1}^{P_2} \vec{F}\cdot d\vec{l} \qquad \begin{array}{l}\text{(work done on}\\\text{a curved path)}\end{array} \qquad \text{(6.14)}$$

6.23 A particle moves along a curved path from point P_1 to P_2, acted on by a force $\vec{F}$ that varies in magnitude and direction.

(a)

During an infinitesimal displacement $d\vec{l}$, the force $\vec{F}$ does work dW on the particle:

$$dW = \vec{F} \cdot d\vec{l} = F \cos \phi \, dl$$

(b)

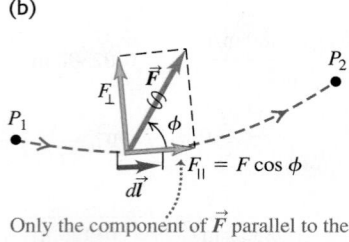

Only the component of $\vec{F}$ parallel to the displacement, $F_{\parallel} = F \cos \phi$, contributes to the work done by $\vec{F}$.

We can now show that the work–energy theorem, Eq. (6.6), holds true even with varying forces and a displacement along a curved path. The force $\vec{F}$ is essentially constant over any given infinitesimal segment $d\vec{l}$ of the path, so we can apply the work–energy theorem for straight-line motion to that segment. Thus the change in the particle's kinetic energy K over that segment equals the work $dW = F_{\parallel} \, dl = \vec{F} \cdot d\vec{l}$ done on the particle. Adding up these infinitesimal quantities of work from all the segments along the whole path gives the total work done, Eq. (6.14), which equals the total change in kinetic energy over the whole path. So $W_{\text{tot}} = \Delta K = K_2 - K_1$ is true *in general,* no matter what the path and no matter what the character of the forces. This can be proved more rigorously by using steps like those in Eqs. (6.11) through (6.13).

Note that only the component of the net force parallel to the path, $F_{\parallel}$, does work on the particle, so only this component can change the speed and kinetic energy of the particle. The component perpendicular to the path, $F_{\perp} = F \sin \phi$, has no effect on the particle's speed; it acts only to change the particle's direction.

The integral in Eq. (6.14) is called a *line integral*. To evaluate this integral in a specific problem, we need some sort of detailed description of the path and of the way in which $\vec{F}$ varies along the path. We usually express the line integral in terms of some scalar variable, as in the following example.

Example 6.8   Motion on a curved path

At a family picnic you are appointed to push your obnoxious cousin Throckmorton in a swing (Fig. 6.24a). His weight is w, the length of the chains is R, and you push Throcky until the chains make an angle θ_0 with the vertical. To do this, you exert a varying horizontal force $\vec{F}$ that starts at zero and gradually increases just enough that Throcky and the swing move very slowly and remain very nearly in equilibrium throughout the process. What is the total work done on Throcky by all forces? What is the work done by the tension T in the chains? What is the work you do by exerting the force $\vec{F}$? (Neglect the weight of the chains and seat.)

SOLUTION

IDENTIFY and SET UP: The motion is along a curve, so we'll use Eq. (6.14) to calculate the work done by the net force, by the tension force, and by the force $\vec{F}$. Figure 6.24b shows our free-body diagram and coordinate system for some arbitrary point in Throcky's motion. We have replaced the sum of the tensions in the two chains with a single tension T.

EXECUTE: There are two ways to find the total work done during the motion: (1) by calculating the work done by each force and then adding those quantities, and (2) by calculating the work done by the net force. The second approach is far easier here because Throcky is in equilibrium at every point. Hence the net force on him is zero, the integral of the net force in Eq. (6.14) is zero, and the total work done on him is zero.

It's also easy to find the work done by the chain tension T because this force is perpendicular to the direction of motion at all points along the path. Hence at all points the angle between the chain tension and the displacement vector $d\vec{l}$ is 90° and the scalar product in Eq. (6.14) is zero. Thus the chain tension does zero work.

6.24 (a) Pushing cousin Throckmorton in a swing. (b) Our free-body diagram.

(a)

(b) Free-body diagram for Throckmorton (neglecting the weight of the chains and seat)

To compute the work done by $\vec{F}$, we need to know how this force varies with the angle θ. The net force on Throcky is zero, so $\Sigma F_x = 0$ and $\Sigma F_y = 0$. From Fig. 6.24b,

$$\sum F_x = F + (-T \sin \theta) = 0$$
$$\sum F_y = T \cos \theta + (-w) = 0$$

By eliminating T from these two equations, we obtain the magnitude $F = w \tan \theta$.

The point where $\vec{F}$ is applied moves through the arc s (Fig. 6.24a). The arc length s equals the radius R of the circular path multiplied by the length θ (in radians), so $s = R\theta$. Therefore the displacement $d\vec{l}$ corresponding to a small change of angle $d\theta$ has a magnitude $dl = ds = R\,d\theta$. The work done by $\vec{F}$ is then

$$W = \int \vec{F} \cdot d\vec{l} = \int F\cos\theta\,ds$$

Now we express F and ds in terms of the angle θ, whose value increases from 0 to θ_0:

$$W = \int_0^{\theta_0} (w\tan\theta)\cos\theta\;(R\,d\theta) = wR\int_0^{\theta_0}\sin\theta\,d\theta$$
$$= wR(1 - \cos\theta_0)$$

EVALUATE: If $\theta_0 = 0$, there is no displacement; then $\cos\theta_0 = 1$ and $W = 0$, as we should expect. If $\theta_0 = 90°$, then $\cos\theta_0 = 0$ and $W = wR$. In that case the work you do is the same as if you had lifted Throcky straight up a distance R with a force equal to his weight w. In fact (as you may wish to confirm), the quantity $R(1 - \cos\theta_0)$ is the increase in his height above the ground during the displacement, so for any value of θ_0 the work done by the force $\vec{F}$ is the change in height multiplied by the weight. This is an example of a more general result that we'll prove in Section 7.1.

We can check our results by writing the forces and the infinitesimal displacement $d\vec{l}$ in terms of their x- and y-components. Figure 6.24a shows that $d\vec{l}$ has a magnitude of ds, an x-component of $ds\cos\theta$, and a y-component of $ds\sin\theta$. Hence $d\vec{l} = $ $\hat{\imath}\,ds\cos\theta + \hat{\jmath}\,ds\sin\theta$. Similarly, we can write the three forces as

$$\vec{T} = \hat{\imath}(-T\sin\theta) + \hat{\jmath}T\cos\theta$$
$$\vec{w} = \hat{\jmath}(-w)$$
$$\vec{F} = \hat{\imath}F$$

We use Eq. (1.21) to calculate the scalar product of each of these forces with $d\vec{l}$:

$$\vec{T} \cdot d\vec{l} = (-T\sin\theta)(ds\cos\theta) + (T\cos\theta)(ds\sin\theta) = 0$$
$$\vec{w} \cdot d\vec{l} = (-w)(ds\sin\theta) = -w\,\sin\theta\,ds$$
$$\vec{F} \cdot d\vec{l} = F(ds\cos\theta) = F\cos\theta\,ds$$

Since $\vec{T} \cdot d\vec{l} = 0$, the integral of this quantity is zero and the work done by the chain tension is zero, just as we found above. Using $ds = R\,d\theta$, we find the work done by the force of gravity is

$$\int \vec{w} \cdot d\vec{l} = \int (-w\sin\theta)R\,d\theta = -wR\int_0^{\theta_0}\sin\theta\,d\theta$$
$$= -wR(1 - \cos\theta_0)$$

Gravity does negative work because this force pulls down while Throcky moves upward. Finally, the work done by the force $\vec{F}$ is the same integral $\int \vec{F} \cdot d\vec{l} = \int F\cos\theta\,ds$ that we calculated above. The method of components is often the most convenient way to calculate scalar products, so use it when it makes your life easier!

Test Your Understanding of Section 6.3   In Example 5.20 (Section 5.4) we examined a conical pendulum. The speed of the pendulum bob remains constant as it travels around the circle shown in Fig. 5.32a. (a) Over one complete circle, how much work does the tension force F do on the bob? (i) a positive amount; (ii) a negative amount; (iii) zero. (b) Over one complete circle, how much work does the weight do on the bob? (i) a positive amount; (ii) a negative amount; (iii) zero.

6.4 Power

The definition of work makes no reference to the passage of time. If you lift a barbell weighing 100 N through a vertical distance of 1.0 m at constant velocity, you do $(100\text{ N})(1.0\text{ m}) = 100$ J of work whether it takes you 1 second, 1 hour, or 1 year to do it. But often we need to know how quickly work is done. We describe this in terms of *power*. In ordinary conversation the word "power" is often synonymous with "energy" or "force." In physics we use a much more precise definition: **Power** is the time *rate* at which work is done. Like work and energy, power is a scalar quantity.

When a quantity of work ΔW is done during a time interval Δt, the average work done per unit time or **average power** P_{av} is defined to be

$$P_{av} = \frac{\Delta W}{\Delta t} \quad \text{(average power)} \tag{6.15}$$

The rate at which work is done might not be constant. We can define **instantaneous power** P as the quotient in Eq. (6.15) as Δt approaches zero:

$$P = \lim_{\Delta t \to 0} \frac{\Delta W}{\Delta t} = \frac{dW}{dt} \quad \text{(instantaneous power)} \tag{6.16}$$

The SI unit of power is the **watt** (W), named for the English inventor James Watt. One watt equals 1 joule per second: $1\text{ W} = 1$ J/s (Fig. 6.25). The kilowatt

6.25 The same amount of work is done in both of these situations, but the power (the rate at which work is done) is different.

$t = 5$ s

Work you do on the box to lift it in 5 s:
$$W = 100\text{ J}$$
Your power output:
$$P = \frac{W}{t} = \frac{100\text{ J}}{5\text{ s}} = 20\text{ W}$$

$t = 0$

$t = 1$ s

Work you do on the same box to lift it the same distance in 1 s:
$$W = 100\text{ J}$$
Your power output:
$$P = \frac{W}{t} = \frac{100\text{ J}}{1\text{ s}} = 100\text{ W}$$

$t = 0$

6.26 The value of the horsepower derives from experiments by James Watt, who measured that a horse could do 33,000 foot-pounds of work per minute in lifting coal from a coal pit.

($1 \text{ kW} = 10^3 \text{ W}$) and the megawatt ($1 \text{ MW} = 10^6 \text{ W}$) are also commonly used. In the British system, work is expressed in foot-pounds, and the unit of power is the foot-pound per second. A larger unit called the *horsepower* (hp) is also used (Fig. 6.26):

$$1 \text{ hp} = 550 \text{ ft} \cdot \text{lb/s} = 33,000 \text{ ft} \cdot \text{lb/min}$$

That is, a 1-hp motor running at full load does 33,000 ft·lb of work every minute. A useful conversion factor is

$$1 \text{ hp} = 746 \text{ W} = 0.746 \text{ kW}$$

The watt is a familiar unit of *electrical* power; a 100-W light bulb converts 100 J of electrical energy into light and heat each second. But there's nothing inherently electrical about a watt. A light bulb could be rated in horsepower, and an engine can be rated in kilowatts.

The *kilowatt-hour* (kW·h) is the usual commercial unit of electrical energy. One kilowatt-hour is the total work done in 1 hour (3600 s) when the power is 1 kilowatt (10^3 J/s), so

$$1 \text{ kW} \cdot \text{h} = (10^3 \text{ J/s})(3600 \text{ s}) = 3.6 \times 10^6 \text{ J} = 3.6 \text{ MJ}$$

The kilowatt-hour is a unit of *work* or *energy,* not power.

In mechanics we can also express power in terms of force and velocity. Suppose that a force $\vec{F}$ acts on a body while it undergoes a vector displacement $\Delta\vec{s}$. If $F_\parallel$ is the component of $\vec{F}$ tangent to the path (parallel to $\Delta\vec{s}$), then the work done by the force is $\Delta W = F_\parallel \Delta s$. The average power is

$$P_{av} = \frac{F_\parallel \Delta s}{\Delta t} = F_\parallel \frac{\Delta s}{\Delta t} = F_\parallel v_{av} \qquad (6.17)$$

Instantaneous power P is the limit of this expression as $\Delta t \rightarrow 0$:

$$P = F_\parallel v \qquad (6.18)$$

where v is the magnitude of the instantaneous velocity. We can also express Eq. (6.18) in terms of the scalar product:

$$P = \vec{F} \cdot \vec{v} \qquad \begin{array}{l}\text{(instantaneous rate at which}\\\text{force } \vec{F} \text{ does work on a particle)}\end{array} \qquad (6.19)$$

Example 6.9 **Force and power**

Each of the four jet engines on an Airbus A380 airliner develops a thrust (a forward force on the airliner) of 322,000 N (72,000 lb). When the airplane is flying at 250 m/s (900 km/h, or roughly 560 mi/h), what horsepower does each engine develop?

SOLUTION

IDENTIFY, SET UP and EXECUTE: Our target variable is the instantaneous power P, which is the rate at which the thrust does work. We use Eq. (6.18). The thrust is in the direction of motion, so $F_\parallel$ is just equal to the thrust. At $v = 250$ m/s, the power developed by each engine is

$$P = F_\parallel v = (3.22 \times 10^5 \text{ N})(250 \text{ m/s}) = 8.05 \times 10^7 \text{ W}$$

$$= (8.05 \times 10^7 \text{ W})\frac{1 \text{ hp}}{746 \text{ W}} = 108,000 \text{ hp}$$

EVALUATE: The speed of modern airliners is directly related to the power of their engines (Fig. 6.27). The largest propeller-driven airliners of the 1950s had engines that developed about 3400 hp

6.27 (a) Propeller-driven and (b) jet airliners.

(a)   (b)

$(2.5 \times 10^6 \text{ W})$, giving them maximum speeds of about 600 km/h (370 mi/h). Each engine on an Airbus A380 develops more than 30 times more power, enabling it to fly at about 900 km/h (560 mi/h) and to carry a much heavier load.

If the engines are at maximum thrust while the airliner is at rest on the ground so that $v = 0$, the engines develop *zero* power. Force and power are not the same thing!

Example 6.10 A "power climb"

A 50.0-kg marathon runner runs up the stairs to the top of Chicago's 443-m-tall Willis Tower, the tallest building in the United States (Fig. 6.28). To lift herself to the top in 15.0 minutes, what must be her average power output? Express your answer in watts, in kilowatts, and in horsepower.

SOLUTION

IDENTIFY and SET UP: We'll treat the runner as a particle of mass m. Her average power output P_{av} must be enough to lift her at constant speed against gravity.

We can find P_{av} in two ways: (1) by determining how much work she must do and dividing that quantity by the elapsed time, as in Eq. (6.15), or (2) by calculating the average upward force she must exert (in the direction of the climb) and multiplying that quantity by her upward velocity, as in Eq. (6.17).

EXECUTE: (1) As in Example 6.8, lifting a mass m against gravity requires an amount of work equal to the weight mg multiplied by the height h it is lifted. Hence the work the runner must do is

$$W = mgh = (50.0 \text{ kg})(9.80 \text{ m/s}^2)(443 \text{ m})$$
$$= 2.17 \times 10^5 \text{ J}$$

She does this work in a time 15.0 min = 900 s, so from Eq. (6.15) the average power is

$$P_{av} = \frac{2.17 \times 10^5 \text{J}}{900 \text{ s}} = 241 \text{ W} = 0.241 \text{ kW} = 0.323 \text{ hp}$$

(2) The force exerted is vertical and the average vertical component of velocity is $(443 \text{ m})/(900 \text{ s}) = 0.492$ m/s, so from Eq. (6.17) the average power is

6.28 How much power is required to run up the stairs of Chicago's Willis Tower in 15 minutes?

$$P_{av} = F_{\parallel}v_{av} = (mg)v_{av}$$
$$= (50.0 \text{ kg})(9.80 \text{ m/s}^2)(0.492 \text{ m/s}) = 241 \text{ W}$$

which is the same result as before.

EVALUATE: The runner's *total* power output will be several times greater than 241 W. The reason is that the runner isn't really a particle but a collection of parts that exert forces on each other and do work, such as the work done to inhale and exhale and to make her arms and legs swing. What we've calculated is only the part of her power output that lifts her to the top of the building.

Test Your Understanding of Section 6.4 The air surrounding an airplane in flight exerts a drag force that acts opposite to the airplane's motion. When the Airbus A380 in Example 6.9 is flying in a straight line at a constant altitude at a constant 250 m/s, what is the rate at which the drag force does work on it? (i) 432,000 hp; (ii) 108,000 hp; (iii) 0; (iv) −108,000 hp; (v) −432,000 hp.

Work done by a force: When a constant force $\vec{F}$ acts on a particle that undergoes a straight-line displacement $\vec{s}$, the work done by the force on the particle is defined to be the scalar product of $\vec{F}$ and $\vec{s}$. The unit of work in SI units is 1 joule = 1 newton-meter (1 J = 1 N · m). Work is a scalar quantity; it can be positive or negative, but it has no direction in space. (See Examples 6.1 and 6.2.)

$$W = \vec{F} \cdot \vec{s} = Fs\cos\phi \quad \text{(6.2), (6.3)}$$
$$\phi = \text{angle between } \vec{F} \text{ and } \vec{s}$$

Kinetic energy: The kinetic energy K of a particle equals the amount of work required to accelerate the particle from rest to speed v. It is also equal to the amount of work the particle can do in the process of being brought to rest. Kinetic energy is a scalar that has no direction in space; it is always positive or zero. Its units are the same as the units of work: 1 J = 1 N · m = 1 kg · m²/s².

$$K = \tfrac{1}{2}mv^2 \quad \text{(6.5)}$$

Doubling m doubles K.

Doubling v quadruples K.

The work–energy theorem: When forces act on a particle while it undergoes a displacement, the particle's kinetic energy changes by an amount equal to the total work done on the particle by all the forces. This relationship, called the work–energy theorem, is valid whether the forces are constant or varying and whether the particle moves along a straight or curved path. It is applicable only to bodies that can be treated as particles. (See Examples 6.3–6.5.)

$$W_{\text{tot}} = K_2 - K_1 = \Delta K \quad \text{(6.6)}$$

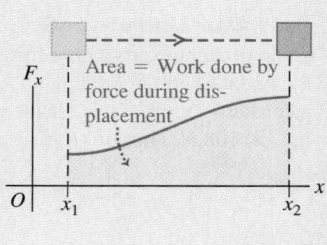

W_{tot} = Total work done on particle along path

$K_1 = \tfrac{1}{2}mv_1^2$

$K_2 = \tfrac{1}{2}mv_2^2 = K_1 + W$

Work done by a varying force or on a curved path: When a force varies during a straight-line displacement, the work done by the force is given by an integral, Eq. (6.7). (See Examples 6.6 and 6.7.) When a particle follows a curved path, the work done on it by a force $\vec{F}$ is given by an integral that involves the angle ϕ between the force and the displacement. This expression is valid even if the force magnitude and the angle ϕ vary during the displacement. (See Example 6.8.)

$$W = \int_{x_1}^{x_2} F_x\, dx \quad \text{(6.7)}$$

$$W = \int_{P_1}^{P_2} F\cos\phi\, dl = \int_{P_1}^{P_2} F_\parallel\, dl$$
$$\quad \text{(6.14)}$$
$$= \int_{P_1}^{P_2} \vec{F} \cdot d\vec{l}$$

Area = Work done by force during displacement

Power: Power is the time rate of doing work. The average power P_{av} is the amount of work ΔW done in time Δt divided by that time. The instantaneous power is the limit of the average power as Δt goes to zero. When a force $\vec{F}$ acts on a particle moving with velocity $\vec{v}$, the instantaneous power (the rate at which the force does work) is the scalar product of $\vec{F}$ and $\vec{v}$. Like work and kinetic energy, power is a scalar quantity. The SI unit of power is 1 watt = 1 joule/second (1 W = 1 J/s). (See Examples 6.9 and 6.10.)

$$P_{\text{av}} = \frac{\Delta W}{\Delta t} \quad \text{(6.15)}$$

$$P = \lim_{\Delta t \to 0} \frac{\Delta W}{\Delta t} = \frac{dW}{dt} \quad \text{(6.16)}$$

$$P = \vec{F} \cdot \vec{v} \quad \text{(6.19)}$$

$t = 5\text{ s}$

$t = 0$

Work you do on the box to lift it in 5 s:
$W = 100$ J
Your power output:
$P = \dfrac{W}{t} = \dfrac{100\text{ J}}{5\text{ s}}$
$= 20$ W

BRIDGING PROBLEM   A Spring That Disobeys Hooke's Law

Consider a hanging spring of negligible mass that does *not* obey Hooke's law. When the spring is extended by a distance x, the force exerted by the spring has magnitude αx^2, where α is a positive constant. The spring is not extended when a block of mass m is attached to it. The block is then released, stretching the spring as it falls (Fig. 6.29). (a) How fast is the block moving when it has fallen a distance x_1? (b) At what rate does the spring do work on the block at this point? (c) Find the maximum distance x_2 that the spring stretches. (d) Will the block *remain* at the point found in part (c)?

6.29 The block is attached to a spring that does not obey Hooke's law.

SOLUTION GUIDE

See MasteringPhysics® study area for a Video Tutor solution.

IDENTIFY and SET UP

1. The spring force in this problem isn't constant, so you have to use the work–energy theorem. You'll also need to use Eq. (6.7) to find the work done by the spring over a given displacement.
2. Draw a free-body diagram for the block, including your choice of coordinate axes. Note that x represents how far the spring is *stretched,* so choose the positive x-axis accordingly. On your coordinate axis, label the points $x = x_1$ and $x = x_2$.
3. Make a list of the unknown quantities, and decide which of these are the target variables.

EXECUTE

4. Calculate the work done on the block by the spring as the block falls an arbitrary distance x. (The integral isn't a difficult one. Use Appendix B if you need a reminder.) Is the work done by the spring positive, negative, or zero?

5. Calculate the work done on the block by any other forces as the block falls an arbitrary distance x. Is this work positive, negative, or zero?
6. Use the work–energy theorem to find the target variables. (You'll also need to use an equation for power.) *Hint:* When the spring is at its maximum stretch, what is the speed of the block?
7. To answer part (d), consider the *net* force that acts on the block when it is at the point found in part (c).

EVALUATE

8. We learned in Chapter 2 that after an object dropped from rest has fallen freely a distance x_1, its speed is $\sqrt{2gx_1}$. Use this to decide whether your answer in part (a) makes sense. In addition, ask yourself whether the algebraic sign of your answer in part (b) makes sense.
9. Find the value of x where the net force on the block would be zero. How does this compare to your result for x_2? Is this consistent with your answer in part (d)?

Problems

For instructor-assigned homework, go to www.masteringphysics.com

•, ••, •••: Problems of increasing difficulty. **CP**: Cumulative problems incorporating material from earlier chapters. **CALC**: Problems requiring calculus. **BIO**: Biosciences problems.

DISCUSSION QUESTIONS

Q6.1 The sign of many physical quantities depends on the choice of coordinates. For example, a_y for free-fall motion can be negative or positive, depending on whether we choose upward or downward as positive. Is the same thing true of work? In other words, can we make positive work negative by a different choice of coordinates? Explain.

Q6.2 An elevator is hoisted by its cables at constant speed. Is the total work done on the elevator positive, negative, or zero? Explain.

Q6.3 A rope tied to a body is pulled, causing the body to accelerate. But according to Newton's third law, the body pulls back on the rope with an equal and opposite force. Is the total work done then zero? If so, how can the body's kinetic energy change? Explain.

Q6.4 If it takes total work W to give an object a speed v and kinetic energy K, starting from rest, what will be the object's speed (in terms of v) and kinetic energy (in terms of K) if we do twice as much work on it, again starting from rest?

Q6.5 If there is a net nonzero force on a moving object, is it possible for the total work done on the object to be zero? Explain, with an example that illustrates your answer.

Q6.6 In Example 5.5 (Section 5.1), how does the work done on the bucket by the tension in the cable compare to the work done on the cart by the tension in the cable?

Q6.7 In the conical pendulum in Example 5.20 (Section 5.4), which of the forces do work on the bob while it is swinging?

Q6.8 For the cases shown in Fig. Q6.8, the object is released from rest at the top and feels no friction or air resistance. In

Figure **Q6.8**

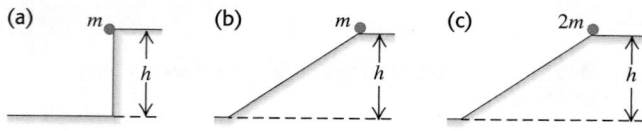

which (if any) cases will the mass have (i) the greatest speed at the bottom and (ii) the most work done on it by the time it reaches the bottom?

Q6.9 A force $\vec{F}$ is in the x-direction and has a magnitude that depends on x. Sketch a possible graph of F versus x such that the force does zero work on an object that moves from x_1 to x_2, even though the force magnitude is not zero at all x in this range.

Q6.10 Does the kinetic energy of a car change more when it speeds up from 10 to 15 m/s or from 15 to 20 m/s? Explain.

Q6.11 A falling brick has a mass of 1.5 kg and is moving straight downward with a speed of 5.0 m/s. A 1.5-kg physics book is sliding across the floor with a speed of 5.0 m/s. A 1.5-kg melon is traveling with a horizontal velocity component 3.0 m/s to the right and a vertical component 4.0 m/s upward. Do these objects all have the same velocity? Do these objects all have the same kinetic energy? For each question, give the reasoning behind your answer.

Q6.12 Can the *total* work done on an object during a displacement be negative? Explain. If the total work is negative, can its magnitude be larger than the initial kinetic energy of the object? Explain.

Q6.13 A net force acts on an object and accelerates it from rest to a speed v_1. In doing so, the force does an amount of work W_1. By what factor must the work done on the object be increased to produce three times the final speed, with the object again starting from rest?

Q6.14 A truck speeding down the highway has a lot of kinetic energy relative to a stopped state trooper, but no kinetic energy relative to the truck driver. In these two frames of reference, is the same amount of work required to stop the truck? Explain.

Q6.15 You are holding a briefcase by the handle, with your arm straight down by your side. Does the force your hand exerts do work on the briefcase when (a) you walk at a constant speed down a horizontal hallway and (b) you ride an escalator from the first to second floor of a building? In each case justify your answer.

Q6.16 When a book slides along a tabletop, the force of friction does negative work on it. Can friction ever do *positive* work? Explain. (*Hint:* Think of a box in the back of an accelerating truck.)

Q6.17 Time yourself while running up a flight of steps, and compute the average rate at which you do work against the force of gravity. Express your answer in watts and in horsepower.

Q6.18 Fractured Physics. Many terms from physics are badly misused in everyday language. In each case, explain the errors involved. (a) A *strong* person is called *powerful*. What is wrong with this use of *power*? (b) When a worker carries a bag of concrete along a level construction site, people say he did a lot of *work*. Did he?

Q6.19 An advertisement for a portable electrical generating unit claims that the unit's diesel engine produces 28,000 hp to drive an electrical generator that produces 30 MW of electrical power. Is this possible? Explain.

Q6.20 A car speeds up while the engine delivers constant power. Is the acceleration greater at the beginning of this process or at the end? Explain.

Q6.21 Consider a graph of instantaneous power versus time, with the vertical P-axis starting at $P = 0$. What is the physical significance of the area under the P-versus-t curve between vertical lines at t_1 and t_2? How could you find the average power from the graph? Draw a P-versus-t curve that consists of two straight-line sections and for which the peak power is equal to twice the average power.

Q6.22 A nonzero net force acts on an object. Is it possible for any of the following quantities to be constant: (a) the particle's speed; (b) the particle's velocity; (c) the particle's kinetic energy?

Q6.23 When a certain force is applied to an ideal spring, the spring stretches a distance x from its unstretched length and does work W. If instead twice the force is applied, what distance (in terms of x) does the spring stretch from its unstretched length, and how much work (in terms of W) is required to stretch it this distance?

Q6.24 If work W is required to stretch a spring a distance x from its unstretched length, what work (in terms of W) is required to stretch the spring an *additional* distance x?

EXERCISES

Section 6.1 Work

6.1 • You push your physics book 1.50 m along a horizontal tabletop with a horizontal push of 2.40 N while the opposing force of friction is 0.600 N. How much work does each of the following forces do on the book: (a) your 2.40-N push, (b) the friction force, (c) the normal force from the tabletop, and (d) gravity? (e) What is the net work done on the book?

6.2 • A tow truck pulls a car 5.00 km along a horizontal roadway using a cable having a tension of 850 N. (a) How much work does the cable do on the car if it pulls horizontally? If it pulls at 35.0° above the horizontal? (b) How much work does the cable do on the tow truck in both cases of part (a)? (c) How much work does gravity do on the car in part (a)?

6.3 • A factory worker pushes a 30.0-kg crate a distance of 4.5 m along a level floor at constant velocity by pushing horizontally on it. The coefficient of kinetic friction between the crate and the floor is 0.25. (a) What magnitude of force must the worker apply? (b) How much work is done on the crate by this force? (c) How much work is done on the crate by friction? (d) How much work is done on the crate by the normal force? By gravity? (e) What is the total work done on the crate?

6.4 •• Suppose the worker in Exercise 6.3 pushes downward at an angle of 30° below the horizontal. (a) What magnitude of force must the worker apply to move the crate at constant velocity? (b) How much work is done on the crate by this force when the crate is pushed a distance of 4.5 m? (c) How much work is done on the crate by friction during this displacement? (d) How much work is done on the crate by the normal force? By gravity? (e) What is the total work done on the crate?

6.5 •• A 75.0-kg painter climbs a ladder that is 2.75 m long leaning against a vertical wall. The ladder makes a 30.0° angle with the wall. (a) How much work does gravity do on the painter? (b) Does the answer to part (a) depend on whether the painter climbs at constant speed or accelerates up the ladder?

6.6 •• Two tugboats pull a disabled supertanker. Each tug exerts a constant force of 1.80×10^6 N, one 14° west of north and the other 14° east of north, as they pull the tanker 0.75 km toward the north. What is the total work they do on the supertanker?

6.7 • Two blocks are connected by a very light string passing over a massless and frictionless pulley (Fig. E6.7). Traveling at constant speed, the 20.0-N block moves 75.0 cm to the right and the 12.0-N block moves 75.0 cm downward. During this process, how much work is done (a) on the 12.0-N block by (i) gravity and (ii) the tension in the string? (b) On the 20.0-N block by (i) gravity,

(ii) the tension in the string, (iii) friction, and (iv) the normal force? (c) Find the total work done on each block.

Figure **E6.7**

6.8 •• A loaded grocery cart is rolling across a parking lot in a strong wind. You apply a constant force $\vec{F} = (30\text{ N})\hat{\imath} - (40\text{ N})\hat{\jmath}$ to the cart as it undergoes a displacement $\vec{s} = (-9.0\text{ m})\hat{\imath} - (3.0\text{ m})\hat{\jmath}$. How much work does the force you apply do on the grocery cart?

6.9 • A 0.800-kg ball is tied to the end of a string 1.60 m long and swung in a vertical circle. (a) During one complete circle, starting anywhere, calculate the total work done on the ball by (i) the tension in the string and (ii) gravity. (b) Repeat part (a) for motion along the semicircle from the lowest to the highest point on the path.

6.10 •• An 8.00-kg package in a mail-sorting room slides 2.00 m down a chute that is inclined at 53.0° below the horizontal. The coefficient of kinetic friction between the package and the chute's surface is 0.40. Calculate the work done on the package by (a) friction, (b) gravity, and (c) the normal force. (d) What is the net work done on the package?

6.11 •• A boxed 10.0-kg computer monitor is dragged by friction 5.50 m up along the moving surface of a conveyor belt inclined at an angle of 36.9° above the horizontal. If the monitor's speed is a constant 2.10 cm/s, how much work is done on the monitor by (a) friction, (b) gravity, and (c) the normal force of the conveyor belt?

6.12 •• You apply a constant force $\vec{F} = (-68.0\text{ N})\hat{\imath} + (36.0\text{ N})\hat{\jmath}$ to a 380-kg car as the car travels 48.0 m in a direction that is 240.0° counterclockwise from the $+x$-axis. How much work does the force you apply do on the car?

Section 6.2 Kinetic Energy and the Work–Energy Theorem

6.13 •• **Animal Energy.** BIO Adult cheetahs, the fastest of the great cats, have a mass of about 70 kg and have been clocked running at up to 72 mph (32 m/s). (a) How many joules of kinetic energy does such a swift cheetah have? (b) By what factor would its kinetic energy change if its speed were doubled?

6.14 •• A 1.50-kg book is sliding along a rough horizontal surface. At point A it is moving at 3.21 m/s, and at point B it has slowed to 1.25 m/s. (a) How much work was done on the book between A and B? (b) If -0.750 J of work is done on the book from B to C, how fast is it moving at point C? (c) How fast would it be moving at C if $+0.750$ J of work were done on it from B to C?

6.15 • **Meteor Crater.** About 50,000 years ago, a meteor crashed into the earth near present-day Flagstaff, Arizona. Measurements from 2005 estimate that this meteor had a mass of about 1.4×10^8 kg (around 150,000 tons) and hit the ground at a speed of 12 km/s. (a) How much kinetic energy did this meteor deliver to the ground? (b) How does this energy compare to the energy released by a 1.0-megaton nuclear bomb? (A megaton bomb releases the same amount of energy as a million tons of TNT, and 1.0 ton of TNT releases 4.184×10^9 J of energy.)

6.16 • **Some Typical Kinetic Energies.** (a) In the Bohr model of the atom, the ground-state electron in hydrogen has an orbital speed of 2190 km/s. What is its kinetic energy? (Consult Appendix F.)

(b) If you drop a 1.0-kg weight (about 2 lb) from a height of 1.0 m, how many joules of kinetic energy will it have when it reaches the ground? (c) Is it reasonable that a 30-kg child could run fast enough to have 100 J of kinetic energy?

6.17 •• In Fig. E6.7 assume that there is no friction force on the 20.0-N block that sits on the tabletop. The pulley is light and frictionless. (a) Calculate the tension T in the light string that connects the blocks. (b) For a displacement in which the 12.0-N block descends 1.20 m, calculate the total work done on (i) the 20.0-N block and (ii) the 12.0-N block. (c) For the displacement in part (b), calculate the total work done on the system of the two blocks. How does your answer compare to the work done on the 12.0-N block by gravity? (d) If the system is released from rest, what is the speed of the 12.0-N block when it has descended 1.20 m?

6.18 • A 4.80-kg watermelon is dropped from rest from the roof of a 25.0-m-tall building and feels no appreciable air resistance. (a) Calculate the work done by gravity on the watermelon during its displacement from the roof to the ground. (b) Just before it strikes the ground, what is the watermelon's (i) kinetic energy and (ii) speed? (c) Which of the answers in parts (a) and (b) would be *different* if there were appreciable air resistance?

6.19 •• Use the work–energy theorem to solve each of these problems. You can use Newton's laws to check your answers. Neglect air resistance in all cases. (a) A branch falls from the top of a 95.0-m-tall redwood tree, starting from rest. How fast is it moving when it reaches the ground? (b) A volcano ejects a boulder directly upward 525 m into the air. How fast was the boulder moving just as it left the volcano? (c) A skier moving at 5.00 m/s encounters a long, rough horizontal patch of snow having coefficient of kinetic friction 0.220 with her skis. How far does she travel on this patch before stopping? (d) Suppose the rough patch in part (c) was only 2.90 m long? How fast would the skier be moving when she reached the end of the patch? (e) At the base of a frictionless icy hill that rises at 25.0° above the horizontal, a toboggan has a speed of 12.0 m/s toward the hill. How high vertically above the base will it go before stopping?

6.20 •• You throw a 20-N rock vertically into the air from ground level. You observe that when it is 15.0 m above the ground, it is traveling at 25.0 m/s upward. Use the work–energy theorem to find (a) the rock's speed just as it left the ground and (b) its maximum height.

6.21 •• You are a member of an Alpine Rescue Team. You must project a box of supplies up an incline of constant slope angle α so that it reaches a stranded skier who is a vertical distance h above the bottom of the incline. The incline is slippery, but there is some friction present, with kinetic friction coefficient μ_k. Use the work–energy theorem to calculate the minimum speed you must give the box at the bottom of the incline so that it will reach the skier. Express your answer in terms of g, h, μ_k, and α.

6.22 •• A mass m slides down a smooth inclined plane from an initial vertical height h, making an angle α with the horizontal. (a) The work done by a force is the sum of the work done by the components of the force. Consider the components of gravity parallel and perpendicular to the surface of the plane. Calculate the work done on the mass by each of the components, and use these results to show that the work done by gravity is exactly the same as if the mass had fallen straight down through the air from a height h. (b) Use the work–energy theorem to prove that the speed of the mass at the bottom of the incline is the same as if it had been dropped from height h, independent of the angle α of the incline. Explain how this speed can be independent of the slope angle. (c) Use the results of part (b) to find the speed of a rock that slides down an icy frictionless hill, starting from rest 15.0 m above the bottom.

6.23 • A sled with mass 8.00 kg moves in a straight line on a frictionless horizontal surface. At one point in its path, its speed is 4.00 m/s; after it has traveled 2.50 m beyond this point, its speed is 6.00 m/s. Use the work–energy theorem to find the force acting on the sled, assuming that this force is constant and that it acts in the direction of the sled's motion.

6.24 •• A soccer ball with mass 0.420 kg is initially moving with speed 2.00 m/s. A soccer player kicks the ball, exerting a constant force of magnitude 40.0 N in the same direction as the ball's motion. Over what distance must the player's foot be in contact with the ball to increase the ball's speed to 6.00 m/s?

6.25 • A 12-pack of Omni-Cola (mass 4.30 kg) is initially at rest on a horizontal floor. It is then pushed in a straight line for 1.20 m by a trained dog that exerts a horizontal force with magnitude 36.0 N. Use the work–energy theorem to find the final speed of the 12-pack if (a) there is no friction between the 12-pack and the floor, and (b) the coefficient of kinetic friction between the 12-pack and the floor is 0.30.

6.26 • A batter hits a baseball with mass 0.145 kg straight upward with an initial speed of 25.0 m/s. (a) How much work has gravity done on the baseball when it reaches a height of 20.0 m above the bat? (b) Use the work–energy theorem to calculate the speed of the baseball at a height of 20.0 m above the bat. You can ignore air resistance. (c) Does the answer to part (b) depend on whether the baseball is moving upward or downward at a height of 20.0 m? Explain.

6.27 • A little red wagon with mass 7.00 kg moves in a straight line on a frictionless horizontal surface. It has an initial speed of 4.00 m/s and then is pushed 3.0 m in the direction of the initial velocity by a force with a magnitude of 10.0 N. (a) Use the work–energy theorem to calculate the wagon's final speed. (b) Calculate the acceleration produced by the force. Use this acceleration in the kinematic relationships of Chapter 2 to calculate the wagon's final speed. Compare this result to that calculated in part (a).

6.28 •• A block of ice with mass 2.00 kg slides 0.750 m down an inclined plane that slopes downward at an angle of 36.9° below the horizontal. If the block of ice starts from rest, what is its final speed? You can ignore friction.

6.29 • **Stopping Distance.** A car is traveling on a level road with speed v_0 at the instant when the brakes lock, so that the tires slide rather than roll. (a) Use the work–energy theorem to calculate the minimum stopping distance of the car in terms of v_0, g, and the coefficient of kinetic friction μ_k between the tires and the road. (b) By what factor would the minimum stopping distance change if (i) the coefficient of kinetic friction were doubled, or (ii) the initial speed were doubled, or (iii) both the coefficient of kinetic friction and the initial speed were doubled?

6.30 •• A 30.0-kg crate is initially moving with a velocity that has magnitude 3.90 m/s in a direction 37.0° west of north. How much work must be done on the crate to change its velocity to 5.62 m/s in a direction 63.0° south of east?

Section 6.3 Work and Energy with Varying Forces

6.31 • **BIO Heart Repair.** A surgeon is using material from a donated heart to repair a patient's damaged aorta and needs to know the elastic characteristics of this aortal material. Tests performed on a 16.0-cm strip of the donated aorta reveal that it stretches 3.75 cm when a 1.50-N pull is exerted on it. (a) What is the force constant of this strip of aortal material? (b) If the maximum distance it will be able to stretch when it replaces the aorta in the damaged heart is 1.14 cm, what is the greatest force it will be able to exert there?

6.32 •• To stretch a spring 3.00 cm from its unstretched length, 12.0 J of work must be done. (a) What is the force constant of this spring? (b) What magnitude force is needed to stretch the spring 3.00 cm from its unstretched length? (c) How much work must be done to compress this spring 4.00 cm from its unstretched length, and what force is needed to compress it this distance?

6.33 • Three identical 6.40-kg masses are hung by three identical springs, as shown in Fig. E6.33. Each spring has a force constant of 7.80 kN/m and was 12.0 cm long before any masses were attached to it. (a) Draw a free-body diagram of each mass. (b) How long is each spring when hanging as shown? (*Hint:* First isolate only the bottom mass. Then treat the bottom two masses as a system. Finally, treat all three masses as a system.)

Figure **E6.33**

6.34 • A child applies a force $\vec{F}$ parallel to the *x*-axis to a 10.0-kg sled moving on the frozen surface of a small pond. As the child controls the speed of the sled, the *x*-component of the force she applies varies with the *x*-coordinate of the sled as shown in Fig. E6.34. Calculate the work done by the force $\vec{F}$ when the sled moves (a) from $x = 0$ to $x = 8.0$ m; (b) from $x = 8.0$ m to $x = 12.0$ m; (c) from $x = 0$ to 12.0 m.

Figure **E6.34**

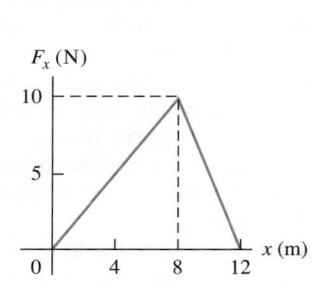

6.35 •• Suppose the sled in Exercise 6.34 is initially at rest at $x = 0$. Use the work–energy theorem to find the speed of the sled at (a) $x = 8.0$ m and (b) $x = 12.0$ m. You can ignore friction between the sled and the surface of the pond.

6.36 • A 2.0-kg box and a 3.0-kg box on a perfectly smooth horizontal floor have a spring of force constant 250 N/m compressed between them. If the initial compression of the spring is 6.0 cm, find the acceleration of each box the instant after they are released. Be sure to include free-body diagrams of each box as part of your solution.

6.37 •• A 6.0-kg box moving at 3.0 m/s on a horizontal, frictionless surface runs into a light spring of force constant 75 N/cm. Use the work–energy theorem to find the maximum compression of the spring.

6.38 •• **Leg Presses.** As part of your daily workout, you lie on your back and push with your feet against a platform attached to two stiff springs arranged side by side so that they are parallel to each other. When you push the platform, you compress the springs. You do 80.0 J of work when you compress the springs 0.200 m from their uncompressed length. (a) What magnitude of force must you apply to hold the platform in this position? (b) How much *additional* work must you do to move the platform 0.200 m *farther,* and what maximum force must you apply?

6.39 •• (a) In Example 6.7 (Section 6.3) it was calculated that with the air track turned off, the glider travels 8.6 cm before it stops instantaneously. How large would the coefficient of static friction μ_s have to be to keep the glider from springing back to the left? (b) If the coefficient of static friction between the glider and the track is $\mu_s = 0.60$, what is the maximum initial speed v_1 that the glider can be given and still remain at rest after it stops instantaneously? With the air track turned off, the coefficient of kinetic friction is $\mu_k = 0.47$.

6.40 • A 4.00-kg block of ice is placed against a horizontal spring that has force constant $k = 200$ N/m and is compressed 0.025 m. The spring is released and accelerates the block along a horizontal surface. You can ignore friction and the mass of the spring. (a) Calculate the work done on the block by the spring during the motion of the block from its initial position to where the spring has returned to its uncompressed length. (b) What is the speed of the block after it leaves the spring?

6.41 • A force $\vec{F}$ is applied to a 2.0-kg radio-controlled model car parallel to the x-axis as it moves along a straight track. The x-component of the force varies with the x-coordinate of the car as shown in Fig. E6.41. Calculate the work done by the force $\vec{F}$ when the car moves from (a) $x = 0$ to $x = 3.0$ m; (b) $x = 3.0$ m to $x = 4.0$ m; (c) $x = 4.0$ m to $x = 7.0$ m; (d) $x = 0$ to $x = 7.0$ m; (e) $x = 7.0$ m to $x = 2.0$ m.

Figure **E6.41**

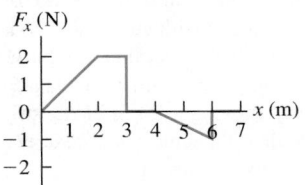

6.42 • Suppose the 2.0-kg model car in Exercise 6.41 is initially at rest at $x = 0$ and $\vec{F}$ is the net force acting on it. Use the work–energy theorem to find the speed of the car at (a) $x = 3.0$ m; (b) $x = 4.0$ m; (c) $x = 7.0$ m.

6.43 •• At a waterpark, sleds with riders are sent along a slippery, horizontal surface by the release of a large compressed spring. The spring with force constant $k = 40.0$ N/cm and negligible mass rests on the frictionless horizontal surface. One end is in contact with a stationary wall. A sled and rider with total mass 70.0 kg are pushed against the other end, compressing the spring 0.375 m. The sled is then released with zero initial velocity. What is the sled's speed when the spring (a) returns to its uncompressed length and (b) is still compressed 0.200 m?

6.44 • **Half of a Spring.** (a) Suppose you cut a massless ideal spring in half. If the full spring had a force constant k, what is the force constant of each half, in terms of k? (*Hint:* Think of the original spring as two equal halves, each producing the same force as the entire spring. Do you see why the forces must be equal?) (b) If you cut the spring into three equal segments instead, what is the force constant of each one, in terms of k?

6.45 •• A small glider is placed against a compressed spring at the bottom of an air track that slopes upward at an angle of 40.0° above the horizontal. The glider has mass 0.0900 kg. The spring has $k = 640$ N/m and negligible mass. When the spring is released, the glider travels a maximum distance of 1.80 m along the air track before sliding back down. Before reaching this maximum distance, the glider loses contact with the spring. (a) What distance was the spring originally compressed? (b) When the glider has traveled along the air track 0.80 m from its initial position against the compressed spring, is it still in contact with the spring? What is the kinetic energy of the glider at this point?

6.46 •• An ingenious bricklayer builds a device for shooting bricks up to the top of the wall where he is working. He places a brick on a vertical compressed spring with force constant $k = 450$ N/m and negligible mass. When the spring is released, the brick is propelled upward. If the brick has mass 1.80 kg and is to reach a maximum height of 3.6 m above its initial position on the compressed spring, what distance must the bricklayer compress the spring initially? (The brick loses contact with the spring when the spring returns to its uncompressed length. Why?)

6.47 •• **CALC** A force in the +x-direction with magnitude $F(x) = 18.0$ N $- (0.530$ N/m$)x$ is applied to a 6.00-kg box that is sitting on the horizontal, frictionless surface of a frozen lake. $F(x)$ is the only horizontal force on the box. If the box is initially at rest at $x = 0$, what is its speed after it has traveled 14.0 m?

Section 6.4 Power

6.48 •• A crate on a motorized cart starts from rest and moves with a constant eastward acceleration of $a = 2.80$ m/s^2. A worker assists the cart by pushing on the crate with a force that is eastward and has magnitude that depends on time according to $F(t) = (5.40$ N/s$)t$. What is the instantaneous power supplied by this force at $t = 5.00$ s?

6.49 • How many joules of energy does a 100-watt light bulb use per hour? How fast would a 70-kg person have to run to have that amount of kinetic energy?

6.50 •• **BIO** **Should You Walk or Run?** It is 5.0 km from your home to the physics lab. As part of your physical fitness program, you could run that distance at 10 km/h (which uses up energy at the rate of 700 W), or you could walk it leisurely at 3.0 km/h (which uses energy at 290 W). Which choice would burn up more energy, and how much energy (in joules) would it burn? Why is it that the more intense exercise actually burns up less energy than the less intense exercise?

6.51 •• **Magnetar.** On December 27, 2004, astronomers observed the greatest flash of light ever recorded from outside the solar system. It came from the highly magnetic neutron star SGR 1806-20 (a *magnetar*). During 0.20 s, this star released as much energy as our sun does in 250,000 years. If P is the average power output of our sun, what was the average power output (in terms of P) of this magnetar?

6.52 •• A 20.0-kg rock is sliding on a rough, horizontal surface at 8.00 m/s and eventually stops due to friction. The coefficient of kinetic friction between the rock and the surface is 0.200. What average power is produced by friction as the rock stops?

6.53 • A tandem (two-person) bicycle team must overcome a force of 165 N to maintain a speed of 9.00 m/s. Find the power required per rider, assuming that each contributes equally. Express your answer in watts and in horsepower.

6.54 •• When its 75-kW (100-hp) engine is generating full power, a small single-engine airplane with mass 700 kg gains altitude at a rate of 2.5 m/s (150 m/min, or 500 ft/min). What fraction of the engine power is being used to make the airplane climb? (The remainder is used to overcome the effects of air resistance and of inefficiencies in the propeller and engine.)

6.55 •• **Working Like a Horse.** Your job is to lift 30-kg crates a vertical distance of 0.90 m from the ground onto the bed of a truck. (a) How many crates would you have to load onto the truck in 1 minute for the average power output you use to lift the crates to equal 0.50 hp? (b) How many crates for an average power output of 100 W?

6.56 •• An elevator has mass 600 kg, not including passengers. The elevator is designed to ascend, at constant speed, a vertical distance of 20.0 m (five floors) in 16.0 s, and it is driven by a motor that can provide up to 40 hp to the elevator. What is the maximum number of passengers that can ride in the elevator? Assume that an average passenger has mass 65.0 kg.

6.57 •• A ski tow operates on a 15.0° slope of length 300 m. The rope moves at 12.0 km/h and provides power for 50 riders at one time, with an average mass per rider of 70.0 kg. Estimate the power required to operate the tow.

6.58 •• The aircraft carrier *John F. Kennedy* has mass 7.4×10^7 kg. When its engines are developing their full power of 280,000 hp, the *John F. Kennedy* travels at its top speed of 35 knots (65 km/h). If 70% of the power output of the engines is applied to pushing the ship through the water, what is the magnitude of the force of water resistance that opposes the carrier's motion at this speed?

6.59 • BIO A typical flying insect applies an average force equal to twice its weight during each downward stroke while hovering. Take the mass of the insect to be 10 g, and assume the wings move an average downward distance of 1.0 cm during each stroke. Assuming 100 downward strokes per second, estimate the average power output of the insect.

PROBLEMS

6.60 ••• CALC A balky cow is leaving the barn as you try harder and harder to push her back in. In coordinates with the origin at the barn door, the cow walks from $x = 0$ to $x = 6.9$ m as you apply a force with x-component $F_x = -[20.0 \text{ N} + (3.0 \text{ N/m})x]$. How much work does the force you apply do on the cow during this displacement?

6.61 •• CALC **Rotating Bar.** A thin, uniform 12.0-kg bar that is 2.00 m long rotates uniformly about a pivot at one end, making 5.00 complete revolutions every 3.00 seconds. What is the kinetic energy of this bar? (*Hint:* Different points in the bar have different speeds. Break the bar up into infinitesimal segments of mass dm and integrate to add up the kinetic energies of all these segments.)

6.62 •• **A Near-Earth Asteroid.** On April 13, 2029 (Friday the 13th!), the asteroid 99942 Apophis will pass within 18,600 mi of the earth—about $\frac{1}{13}$ the distance to the moon! It has a density of 2600 kg/m^3, can be modeled as a sphere 320 m in diameter, and will be traveling at 12.6 km/s. (a) If, due to a small disturbance in its orbit, the asteroid were to hit the earth, how much kinetic energy would it deliver? (b) The largest nuclear bomb ever tested by the United States was the "Castle/Bravo" bomb, having a yield of 15 megatons of TNT. (A megaton of TNT releases 4.184×10^{15} J of energy.) How many Castle/Bravo bombs would be equivalent to the energy of Apophis?

6.63 • A luggage handler pulls a 20.0-kg suitcase up a ramp inclined at 25.0° above the horizontal by a force $\vec{F}$ of magnitude 140 N that acts parallel to the ramp. The coefficient of kinetic friction between the ramp and the incline is $\mu_k = 0.300$. If the suitcase travels 3.80 m along the ramp, calculate (a) the work done on the suitcase by the force $\vec{F}$; (b) the work done on the suitcase by the gravitational force; (c) the work done on the suitcase by the normal force; (d) the work done on the suitcase by the friction force; (e) the total work done on the suitcase. (f) If the speed of the suitcase is zero at the bottom of the ramp, what is its speed after it has traveled 3.80 m along the ramp?

6.64 • BIO **Chin-Ups.** While doing a chin-up, a man lifts his body 0.40 m. (a) How much work must the man do per kilogram of body mass? (b) The muscles involved in doing a chin-up can generate about 70 J of work per kilogram of muscle mass. If the man can just barely do a 0.40-m chin-up, what percentage of his body's mass do these muscles constitute? (For comparison, the *total* percentage of muscle in a typical 70-kg man with 14% body fat is about 43%.) (c) Repeat part (b) for the man's young son, who has arms half as long as his father's but whose muscles can also generate 70 J of work per kilogram of muscle mass. (d) Adults and children have about the same percentage of muscle in their bodies. Explain why children can commonly do chin-ups more easily than their fathers.

6.65 ••• CP A 20.0-kg crate sits at rest at the bottom of a 15.0-m-long ramp that is inclined at 34.0° above the horizontal. A constant horizontal force of 290 N is applied to the crate to push it up the ramp. While the crate is moving, the ramp exerts a constant frictional force on it that has magnitude 65.0 N. (a) What is the total work done on the crate during its motion from the bottom to the top of the ramp? (b) How much time does it take the crate to travel to the top of the ramp?

6.66 ••• Consider the blocks in Exercise 6.7 as they move 75.0 cm. Find the total work done on each one (a) if there is no friction between the table and the 20.0-N block, and (b) if $\mu_s = 0.500$ and $\mu_k = 0.325$ between the table and the 20.0-N block.

6.67 • The space shuttle, with mass 86,400 kg, is in a circular orbit of radius 6.66×10^6 m around the earth. It takes 90.1 min for the shuttle to complete each orbit. On a repair mission, the shuttle is cautiously moving 1.00 m closer to a disabled satellite every 3.00 s. Calculate the shuttle's kinetic energy (a) relative to the earth and (b) relative to the satellite.

6.68 •• A 5.00-kg package slides 1.50 m down a long ramp that is inclined at 24.0° below the horizontal. The coefficient of kinetic friction between the package and the ramp is $\mu_k = 0.310$. Calculate (a) the work done on the package by friction; (b) the work done on the package by gravity; (c) the work done on the package by the normal force; (d) the total work done on the package. (e) If the package has a speed of 2.20 m/s at the top of the ramp, what is its speed after sliding 1.50 m down the ramp?

6.69 •• CP BIO **Whiplash Injuries.** When a car is hit from behind, its passengers undergo sudden forward acceleration, which can cause a severe neck injury known as *whiplash*. During normal acceleration, the neck muscles play a large role in accelerating the head so that the bones are not injured. But during a very sudden acceleration, the muscles do not react immediately because they are flexible, so most of the accelerating force is provided by the neck bones. Experimental tests have shown that these bones will fracture if they absorb more than 8.0 J of energy. (a) If a car waiting at a stoplight is rear-ended in a collision that lasts for 10.0 ms, what is the greatest speed this car and its driver can reach without breaking neck bones if the driver's head has a mass of 5.0 kg (which is about right for a 70-kg person)? Express your answer in m/s and in mph. (b) What is the acceleration of the passengers during the collision in part (a), and how large a force is acting to accelerate their heads? Express the acceleration in m/s^2 and in g's.

6.70 •• CALC A net force along the x-axis that has x-component $F_x = -12.0 \text{ N} + (0.300 \text{ N/m}^2)x^2$ is applied to a 5.00-kg object that is initially at the origin and moving in the $-x$-direction with a speed of 6.00 m/s. What is the speed of the object when it reaches the point $x = 5.00$ m?

6.71 • CALC An object is attracted toward the origin with a force given by $F_x = -k/x^2$. (Gravitational and electrical forces have this distance dependence.) (a) Calculate the work done by the force F_x when the object moves in the x-direction from x_1 to x_2. If $x_2 > x_1$, is the work done by F_x positive or negative? (b) The only other force acting on the object is a force that you exert with your hand to move the object slowly from x_1 to x_2. How much work do you do? If $x_2 > x_1$, is the work you do positive or negative? (c) Explain the similarities and differences between your answers to parts (a) and (b).

6.72 ••• **CALC** The gravitational pull of the earth on an object is inversely proportional to the square of the distance of the object from the center of the earth. At the earth's surface this force is equal to the object's normal weight mg, where $g = 9.8$ m/s^2, and at large distances, the force is zero. If a 20,000-kg asteroid falls to earth from a very great distance away, what will be its minimum speed as it strikes the earth's surface, and how much kinetic energy will it impart to our planet? You can ignore the effects of the earth's atmosphere.

6.73 • **CALC** **Varying Coefficient of Friction.** A box is sliding with a speed of 4.50 m/s on a horizontal surface when, at point P, it encounters a rough section. On the rough section, the coefficient of friction is not constant, but starts at 0.100 at P and increases linearly with distance past P, reaching a value of 0.600 at 12.5 m past point P. (a) Use the work–energy theorem to find how far this box slides before stopping. (b) What is the coefficient of friction at the stopping point? (c) How far would the box have slid if the friction coefficient didn't increase but instead had the constant value of 0.100?

6.74 •• **CALC** Consider a spring that does not obey Hooke's law very faithfully. One end of the spring is fixed. To keep the spring stretched or compressed an amount x, a force along the x-axis with x-component $F_x = kx - bx^2 + cx^3$ must be applied to the free end. Here $k = 100$ N/m, $b = 700$ N/m^2, and $c = 12{,}000$ N/m^3. Note that $x > 0$ when the spring is stretched and $x < 0$ when it is compressed. (a) How much work must be done to stretch this spring by 0.050 m from its unstretched length? (b) How much work must be done to *compress* this spring by 0.050 m from its unstretched length? (c) Is it easier to stretch or compress this spring? Explain why in terms of the dependence of F_x on x. (Many real springs behave qualitatively in the same way.)

6.75 •• **CP** A small block with a mass of 0.0900 kg is attached to a cord passing through a hole in a frictionless, horizontal surface (Fig. P6.75). The block is originally revolving at a distance of 0.40 m from the hole with a speed of 0.70 m/s. The cord is then pulled from below, shortening the radius of the circle in which the block revolves to 0.10 m. At this new distance, the speed of the

Figure **P6.75**

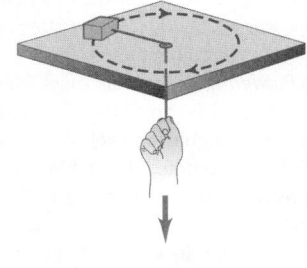

block is observed to be 2.80 m/s. (a) What is the tension in the cord in the original situation when the block has speed $v = 0.70$ m/s? (b) What is the tension in the cord in the final situation when the block has speed $v = 2.80$ m/s? (c) How much work was done by the person who pulled on the cord?

6.76 •• **CALC** **Proton Bombardment.** A proton with mass 1.67×10^{-27} kg is propelled at an initial speed of 3.00×10^5 m/s directly toward a uranium nucleus 5.00 m away. The proton is repelled by the uranium nucleus with a force of magnitude $F = \alpha/x^2$, where x is the separation between the two objects and $\alpha = 2.12 \times 10^{-26}$ N·m^2. Assume that the uranium nucleus remains at rest. (a) What is the speed of the proton when it is 8.00×10^{-10} m from the uranium nucleus? (b) As the proton approaches the uranium nucleus, the repulsive force slows down

the proton until it comes momentarily to rest, after which the proton moves away from the uranium nucleus. How close to the uranium nucleus does the proton get? (c) What is the speed of the proton when it is again 5.00 m away from the uranium nucleus?

6.77 •• **CP CALC** A block of ice with mass 4.00 kg is initially at rest on a frictionless, horizontal surface. A worker then applies a horizontal force $\vec{F}$ to it. As a result, the block moves along the x-axis such that its position as a function of time is given by $x(t) = \alpha t^2 + \beta t^3$, where $\alpha = 0.200$ m/s^2 and $\beta = 0.0200$ m/s^3. (a) Calculate the velocity of the object when $t = 4.00$ s. (b) Calculate the magnitude of $\vec{F}$ when $t = 4.00$ s. (c) Calculate the work done by the force $\vec{F}$ during the first 4.00 s of the motion.

6.78 •• You and your bicycle have combined mass 80.0 kg. When you reach the base of a bridge, you are traveling along the road at 5.00 m/s (Fig. P6.78). At the top of the bridge, you have climbed a vertical distance of 5.20 m and have slowed to 1.50 m/s. You can ignore work done by friction and any inefficiency in the bike or your legs. (a) What is the total work done on you and your bicycle when you go from the base to the top of the bridge? (b) How much work have you done with the force you apply to the pedals?

Figure **P6.78**

$m = 80.0$ kg

5.20 m

6.79 •• You are asked to design spring bumpers for the walls of a parking garage. A freely rolling 1200-kg car moving at 0.65 m/s is to compress the spring no more than 0.090 m before stopping. What should be the force constant of the spring? Assume that the spring has negligible mass.

6.80 •• The spring of a spring gun has force constant $k = 400$ N/m and negligible mass. The spring is compressed 6.00 cm, and a ball with mass 0.0300 kg is placed in the horizontal barrel against the compressed spring. The spring is then released, and the ball is propelled out the barrel of the gun. The barrel is 6.00 cm long, so the ball leaves the barrel at the same point that it loses contact with the spring. The gun is held so the barrel is horizontal. (a) Calculate the speed with which the ball leaves the barrel if you can ignore friction. (b) Calculate the speed of the ball as it leaves the barrel if a constant resisting force of 6.00 N acts on the ball as it moves along the barrel. (c) For the situation in part (b), at what position along the barrel does the ball have the greatest speed, and what is that speed? (In this case, the maximum speed does not occur at the end of the barrel.)

6.81 ••• A 2.50-kg textbook is forced against a horizontal spring of negligible mass and force constant 250 N/m, compressing the spring a distance of 0.250 m. When released, the textbook slides on a horizontal tabletop with coefficient of kinetic friction

$\mu_k = 0.30$. Use the work–energy theorem to find how far the text-book moves from its initial position before coming to rest.

6.82 ••• **Pushing a Cat.** Your cat "Ms." (mass 7.00 kg) is trying to make it to the top of a frictionless ramp 2.00 m long and inclined upward at 30.0° above the horizontal. Since the poor cat can't get any traction on the ramp, you push her up the entire length of the ramp by exerting a constant 100-N force parallel to the ramp. If Ms. takes a running start so that she is moving at 2.40 m/s at the bottom of the ramp, what is her speed when she reaches the top of the incline? Use the work–energy theorem.

6.83 •• **Crash Barrier.** A student proposes a design for an automobile crash barrier in which a 1700-kg sport utility vehicle moving at 20.0 m/s crashes into a spring of negligible mass that slows it to a stop. So that the passengers are not injured, the acceleration of the vehicle as it slows can be no greater than $5.00g$. (a) Find the required spring constant k, and find the distance the spring will compress in slowing the vehicle to a stop. In your calculation, disregard any deformation or crumpling of the vehicle and the friction between the vehicle and the ground. (b) What disadvantages are there to this design?

6.84 ••• A physics professor is pushed up a ramp inclined upward at 30.0° above the horizontal as he sits in his desk chair that slides on frictionless rollers. The combined mass of the professor and chair is 85.0 kg. He is pushed 2.50 m along the incline by a group of students who together exert a constant horizontal force of 600 N. The professor's speed at the bottom of the ramp is 2.00 m/s. Use the work–energy theorem to find his speed at the top of the ramp.

6.85 • A 5.00-kg block is moving at $v_0 = 6.00$ m/s along a frictionless, horizontal surface toward a spring with force constant $k = 500$ N/m that is attached to a wall (Fig. P6.85). The spring has negligible mass.

Figure **P6.85**

(a) Find the maximum distance the spring will be compressed. (b) If the spring is to compress by no more than 0.150 m, what should be the maximum value of v_0?

6.86 •• Consider the system shown in Fig. P6.86. The rope and pulley have negligible mass, and the pulley is frictionless. The coefficient of kinetic friction between the 8.00-kg block and the tabletop is $\mu_k = 0.250$. The blocks are released from rest. Use energy methods to calculate the speed of the 6.00-kg block after it has descended 1.50 m.

Figure **P6.86**

6.87 •• Consider the system shown in Fig. P6.86. The rope and pulley have negligible mass, and the pulley is frictionless. Initially the 6.00-kg block is moving downward and the 8.00-kg block is moving to the right, both with a speed of 0.900 m/s. The blocks come to rest after moving 2.00 m. Use the work–energy theorem to calculate the coefficient of kinetic friction between the 8.00-kg block and the tabletop.

6.88 ••• **CALC Bow and Arrow.** Figure P6.88 shows how the force exerted by the string of a compound bow on an arrow varies as a function of how far back the arrow is pulled (the draw length). Assume that the same force is exerted on the arrow as it moves forward after being released. Full draw for this bow is at a draw length of 75.0 cm. If the bow shoots a 0.0250-kg arrow from full draw, what is the speed of the arrow as it leaves the bow?

Figure **P6.88**

6.89 •• On an essentially frictionless, horizontal ice rink, a skater moving at 3.0 m/s encounters a rough patch that reduces her speed to 1.65 m/s due to a friction force that is 25% of her weight. Use the work–energy theorem to find the length of this rough patch.

6.90 • **Rescue.** Your friend (mass 65.0 kg) is standing on the ice in the middle of a frozen pond. There is very little friction between her feet and the ice, so she is unable to walk. Fortunately, a light rope is tied around her waist and you stand on the bank holding the other end. You pull on the rope for 3.00 s and accelerate your friend from rest to a speed of 6.00 m/s while you remain at rest. What is the average power supplied by the force you applied?

6.91 •• A pump is required to lift 800 kg of water (about 210 gallons) per minute from a well 14.0 m deep and eject it with a speed of 18.0 m/s. (a) How much work is done per minute in lifting the water? (b) How much work is done in giving the water the kinetic energy it has when ejected? (c) What must be the power output of the pump?

6.92 •• **BIO** All birds, independent of their size, must maintain a power output of 10–25 watts per kilogram of body mass in order to fly by flapping their wings. (a) The Andean giant hummingbird (*Patagona gigas*) has mass 70 g and flaps its wings 10 times per second while hovering. Estimate the amount of work done by such a hummingbird in each wingbeat. (b) A 70-kg athlete can maintain a power output of 1.4 kW for no more than a few seconds; the *steady* power output of a typical athlete is only 500 W or so. Is it possible for a human-powered aircraft to fly for extended periods by flapping its wings? Explain.

6.93 ••• A physics student spends part of her day walking between classes or for recreation, during which time she expends energy at an average rate of 280 W. The remainder of the day she is sitting in class, studying, or resting; during these activities, she expends energy at an average rate of 100 W. If she expends a total of 1.1×10^7 J of energy in a 24-hour day, how much of the day did she spend walking?

6.94 ••• The Grand Coulee Dam is 1270 m long and 170 m high. The electrical power output from generators at its base is approximately 2000 MW. How many cubic meters of water must flow from the top of the dam per second to produce this amount of power if 92% of the work done on the water by gravity is converted to electrical energy? (Each cubic meter of water has a mass of 1000 kg.)

6.95 • **BIO** **Power of the Human Heart.** The human heart is a powerful and extremely reliable pump. Each day it takes in and discharges about 7500 L of blood. Assume that the work done by the heart is equal to the work required to lift this amount of blood a height equal to that of the average American woman (1.63 m). The density (mass per unit volume) of blood is 1.05×10^3 kg/m^3. (a) How much work does the heart do in a day? (b) What is the heart's power output in watts?

6.96 ••• Six diesel units in series can provide 13.4 MW of power to the lead car of a freight train. The diesel units have total mass 1.10×10^6 kg. The average car in the train has mass 8.2×10^4 kg and requires a horizontal pull of 2.8 kN to move at a constant 27 m/s on level tracks. (a) How many cars can be in the train under these conditions? (b) This would leave no power for accelerating or climbing hills. Show that the extra force needed to accelerate the train is about the same for a 0.10-m/s^2 acceleration or a 1.0% slope (slope angle $\alpha = \arctan 0.010$). (c) With the 1.0% slope, show that an extra 2.9 MW of power is needed to maintain the 27-m/s speed of the diesel units. (d) With 2.9 MW less power available, how many cars can the six diesel units pull up a 1.0% slope at a constant 27 m/s?

6.97 • It takes a force of 53 kN on the lead car of a 16-car passenger train with mass 9.1×10^5 kg to pull it at a constant 45 m/s (101 mi/h) on level tracks. (a) What power must the locomotive provide to the lead car? (b) How much more power to the lead car than calculated in part (a) would be needed to give the train an acceleration of 1.5 m/s^2, at the instant that the train has a speed of 45 m/s on level tracks? (c) How much more power to the lead car than that calculated in part (a) would be needed to move the train up a 1.5% grade (slope angle $\alpha = \arctan 0.015$) at a constant 45 m/s?

6.98 • **CALC** An object has several forces acting on it. One of these forces is $\vec{F} = axy\hat{i}$, a force in the x-direction whose magnitude depends on the position of the object, with $\alpha = 2.50$ N/m^2. Calculate the work done on the object by this force for the following displacements of the object: (a) The object starts at the point $x = 0$, $y = 3.00$ m and moves parallel to the x-axis to the point $x = 2.00$ m, $y = 3.00$ m. (b) The object starts at the point $x = 2.00$ m, $y = 0$ and moves in the y-direction to the point $x = 2.00$ m, $y = 3.00$ m. (c) The object starts at the origin and moves on the line $y = 1.5x$ to the point $x = 2.00$ m, $y = 3.00$ m.

6.99 •• **Cycling.** For a touring bicyclist the drag coefficient $C(f_{air} = \frac{1}{2}CA\rho v^2)$ is 1.00, the frontal area A is 0.463 m^2, and the coefficient of rolling friction is 0.0045. The rider has mass 50.0 kg, and her bike has mass 12.0 kg. (a) To maintain a speed of 12.0 m/s (about 27 mi/h) on a level road, what must the rider's power output to the rear wheel be? (b) For racing, the same rider uses a different bike with coefficient of rolling friction 0.0030 and mass 9.00 kg. She also crouches down, reducing her drag coefficient to 0.88 and reducing her frontal area to 0.366 m^2. What must her power output to the rear wheel be then to maintain a speed of 12.0 m/s? (c) For the situation in part (b), what power output is required to maintain a speed of 6.0 m/s? Note the great drop in power requirement when the speed is only halved. (For more on aerodynamic speed limitations for a wide variety of human-powered vehicles, see "The Aerodynamics of Human-Powered Land Vehicles," *Scientific American,* December 1983.)

6.100 •• **Automotive Power I.** A truck engine transmits 28.0 kW (37.5 hp) to the driving wheels when the truck is traveling at a constant velocity of magnitude 60.0 km/h (37.3 mi/h) on a level road. (a) What is the resisting force acting on the truck? (b) Assume that 65% of the resisting force is due to rolling friction and the remainder is due to air resistance. If the force of rolling friction is independent of speed, and the force of air resistance is proportional to the square of the speed, what power will drive the truck at 30.0 km/h? At 120.0 km/h? Give your answers in kilowatts and in horsepower.

6.101 •• **Automotive Power II.** (a) If 8.00 hp are required to drive a 1800-kg automobile at 60.0 km/h on a level road, what is the total retarding force due to friction, air resistance, and so on? (b) What power is necessary to drive the car at 60.0 km/h up a 10.0% grade (a hill rising 10.0 m vertically in 100.0 m horizontally)? (c) What power is necessary to drive the car at 60.0 km/h *down* a 1.00% grade? (d) Down what percent grade would the car coast at 60.0 km/h?

CHALLENGE PROBLEMS

6.102 ••• **CALC** On a winter day in Maine, a warehouse worker is shoving boxes up a rough plank inclined at an angle α above the horizontal. The plank is partially covered with ice, with more ice near the bottom of the plank than near the top, so that the coefficient of friction increases with the distance x along the plank: $\mu = Ax$, where A is a positive constant and the bottom of the plank is at $x = 0$. (For this plank the coefficients of kinetic and static friction are equal: $\mu_k = \mu_s = \mu$.) The worker shoves a box up the plank so that it leaves the bottom of the plank moving at speed v_0. Show that when the box first comes to rest, it will remain at rest if

$$v_0^2 \geq \frac{3g \sin^2\alpha}{A \cos\alpha}$$

6.103 ••• **CALC** **A Spring with Mass.** We usually ignore the kinetic energy of the moving coils of a spring, but let's try to get a reasonable approximation to this. Consider a spring of mass M, equilibrium length L_0, and spring constant k. The work done to stretch or compress the spring by a distance L is $\frac{1}{2}kX^2$, where $X = L - L_0$. Consider a spring, as described above, that has one end fixed and the other end moving with speed v. Assume that the speed of points along the length of the spring varies linearly with distance l from the fixed end. Assume also that the mass M of the spring is distributed uniformly along the length of the spring. (a) Calculate the kinetic energy of the spring in terms of M and v. (*Hint:* Divide the spring into pieces of length dl; find the speed of each piece in terms of l, v, and L; find the mass of each piece in terms of dl, M, and L; and integrate from 0 to L. The result is *not* $\frac{1}{2}Mv^2$, since not all of the spring moves with the same speed.) In a spring gun, a spring of mass 0.243 kg and force constant 3200 N/m is compressed 2.50 cm from its unstretched length. When the trigger is pulled, the spring pushes horizontally on a 0.053-kg ball. The work done by friction is negligible. Calculate the ball's speed when the spring reaches its uncompressed length (b) ignoring the mass of the spring and (c) including, using the results of part (a), the mass of the spring. (d) In part (c), what is the final kinetic energy of the ball and of the spring?

6.104 ••• **CALC** An airplane in flight is subject to an air resistance force proportional to the square of its speed v. But there is an additional resistive force because the airplane has wings. Air flowing over the wings is pushed down and slightly forward, so from Newton's third law the air exerts a force on the wings and airplane that is up and slightly backward (Fig. P6.104). The upward force is the lift force that keeps the airplane aloft, and the backward force is called *induced drag*. At flying speeds, induced drag is inversely proportional to v^2, so that the total air resistance force can be expressed by $F_{air} = \alpha v^2 + \beta/v^2$, where α and β are positive constants that depend on the shape and size of the airplane and the density of the air. For a Cessna 150, a small single-engine airplane, $\alpha = 0.30 \text{ N} \cdot \text{s}^2/\text{m}^2$ and $\beta = 3.5 \times 10^5 \text{ N} \cdot \text{m}^2/\text{s}^2$. In steady flight, the engine must provide a forward force that exactly balances the air resistance force. (a) Calculate the speed (in km/h) at which this airplane will have the maximum *range* (that is, travel the greatest distance) for a given quantity of fuel. (b) Calculate the speed (in km/h) for which the airplane will have the maximum *endurance* (that is, remain in the air the longest time).

Figure **P6.104**

Induced drag

Lift Force of air on wings

Answers

Chapter Opening Question

The answer is yes. As the ant was exerting an upward force on the piece of cereal, the cereal was exerting a downward force of the same magnitude on the ant (due to Newton's third law). However, because the ant's body had an upward displacement, the work that the cereal did on the ant was *negative* (see Section 6.1).

Test Your Understanding Questions

6.1 Answer: (iii) The electron has constant velocity, so its acceleration is zero and (by Newton's second law) the net force on the electron is also zero. Therefore the total work done by all the forces (equal to the work done by the net force) must be zero as well. The individual forces may do nonzero work, but that's not what the question asks.

6.2 Answer: (iv), (i), (iii), (ii) Body (i) has kinetic energy $K = \frac{1}{2}mv^2 = \frac{1}{2}(2.0 \text{ kg})(5.0 \text{ m/s})^2 = 25 \text{ J}$. Body (ii) had zero kinetic energy initially and then had 30 J of work done it, so its final kinetic energy is $K_2 = K_1 + W = 0 + 30 \text{ J} = 30 \text{ J}$. Body (iii) had initial kinetic energy $K_1 = \frac{1}{2}mv_1^2 = \frac{1}{2}(1.0 \text{ kg})(4.0 \text{ m/s})^2 = 8.0 \text{ J}$ and then had 20 J of work done on it, so its final kinetic energy is $K_2 = K_1 + W = 8.0 \text{ J} + 20 \text{ J} = 28 \text{ J}$. Body (iv) had initial kinetic energy $K_1 = \frac{1}{2}mv_1^2 = \frac{1}{2}(2.0 \text{ kg})(10 \text{ m/s})^2 = 100 \text{ J}$; when it did 80 J of work on another body, the other body did −80 J of work on body (iv), so the final kinetic energy of body (iv) is $K_2 = K_1 + W = 100 \text{ J} + (-80 \text{ J}) = 20 \text{ J}$.

6.3 Answers: (a) (iii), (b) (iii) At any point during the pendulum bob's motion, the tension force and the weight both act perpendicular to the motion—that is, perpendicular to an infinitesimal displacement $d\vec{l}$ of the bob. (In Fig. 5.32b, the displacement $d\vec{l}$ would be directed outward from the plane of the free-body diagram.) Hence for either force the scalar product inside the integral in Eq. (6.14) is $\vec{F} \cdot d\vec{l} = 0$, and the work done along any part of the circular path (including a complete circle) is $W = \int \vec{F} \cdot d\vec{l} = 0$.

6.4 Answer: (v) The airliner has a constant horizontal velocity, so the net horizontal force on it must be zero. Hence the backward drag force must have the same magnitude as the forward force due to the combined thrust of the four engines. This means that the drag force must do *negative* work on the airplane at the same rate that the combined thrust force does *positive* work. The combined thrust does work at a rate of $4(108,000 \text{ hp}) = 432,000 \text{ hp}$, so the drag force must do work at a rate of −432,000 hp.

Bridging Problem

Answers: (a) $v_1 = \sqrt{\dfrac{2}{m}\left(mgx_1 - \frac{1}{3}\alpha x_1^3\right)} = \sqrt{2gx_1 - \dfrac{2\alpha x_1^3}{3m}}$

(b) $P = -F_{spring-1}v_1 = -\alpha x_1^2 \sqrt{2gx_1 - \dfrac{2\alpha x_1^3}{3m}}$

(c) $x_2 = \sqrt{\dfrac{3mg}{\alpha}}$ (d) No

POTENTIAL ENERGY AND ENERGY CONSERVATION

? As this mallard glides in to a landing, it descends along a straight-line path at a constant speed. Does the mallard's mechanical energy increase, decrease, or stay the same during the glide? If it increases, where does the added energy come from? If it decreases, where does the lost energy go?

LEARNING GOALS

By studying this chapter, you will learn:

- How to use the concept of gravitational potential energy in problems that involve vertical motion.

- How to use the concept of elastic potential energy in problems that involve a moving body attached to a stretched or compressed spring.

- The distinction between conservative and nonconservative forces, and how to solve problems in which both kinds of forces act on a moving body.

- How to calculate the properties of a conservative force if you know the corresponding potential-energy function.

- How to use energy diagrams to understand the motion of an object moving in a straight line under the influence of a conservative force.

When a diver jumps off a high board into a swimming pool, he hits the water moving pretty fast, with a lot of kinetic energy. Where does that energy come from? The answer we learned in Chapter 6 was that the gravitational force (his weight) does work on the diver as he falls. The diver's kinetic energy—energy associated with his *motion*—increases by an amount equal to the work done.

However, there is a very useful alternative way to think about work and kinetic energy. This new approach is based on the concept of *potential energy,* which is energy associated with the *position* of a system rather than its motion. In this approach, there is *gravitational potential energy* even while the diver is standing on the high board. Energy is not added to the earth–diver system as the diver falls, but rather a storehouse of energy is *transformed* from one form (potential energy) to another (kinetic energy) as he falls. In this chapter we'll see how the work–energy theorem explains this transformation.

If the diver bounces on the end of the board before he jumps, the bent board stores a second kind of potential energy called *elastic potential energy.* We'll discuss elastic potential energy of simple systems such as a stretched or compressed spring. (An important third kind of potential energy is associated with the positions of electrically charged particles relative to each other. We'll encounter this potential energy in Chapter 23.)

We will prove that in some cases the sum of a system's kinetic and potential energy, called the *total mechanical energy* of the system, is constant during the motion of the system. This will lead us to the general statement of the *law of conservation of energy,* one of the most fundamental and far-reaching principles in all of science.

7.1 Gravitational Potential Energy

7.1 As a basketball descends, gravitational potential energy is converted to kinetic energy and the basketball's speed increases.

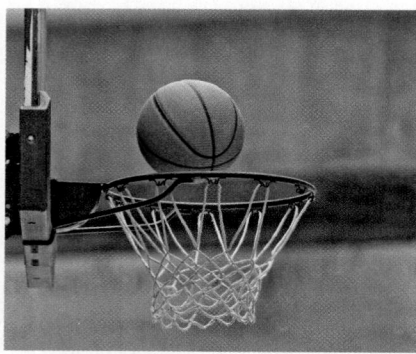

We learned in Chapter 6 that a particle gains or loses kinetic energy because it interacts with other objects that exert forces on it. During any interaction, the change in a particle's kinetic energy is equal to the total work done on the particle by the forces that act on it.

In many situations it seems as though energy has been stored in a system, to be recovered later. For example, you must do work to lift a heavy stone over your head. It seems reasonable that in hoisting the stone into the air you are storing energy in the system, energy that is later converted into kinetic energy when you let the stone fall.

This example points to the idea of an energy associated with the *position* of bodies in a system. This kind of energy is a measure of the *potential* or *possibility* for work to be done; when a stone is raised into the air, there is a potential for work to be done on it by the gravitational force, but only if the stone is allowed to fall to the ground. For this reason, energy associated with position is called **potential energy.** Our discussion suggests that there is potential energy associated with a body's weight and its height above the ground. We call this *gravitational potential energy* (Fig. 7.1).

We now have *two* ways to describe what happens when a body falls without air resistance. One way is to say that gravitational potential energy decreases and the falling body's kinetic energy increases. The other way, which we learned in Chapter 6, is that a falling body's kinetic energy increases because the force of the earth's gravity (the body's weight) does work on the body. Later in this section we'll use the work–energy theorem to show that these two descriptions are equivalent.

7.2 When a body moves vertically from an initial height y_1 to a final height y_2, the gravitational force $\vec{w}$ does work and the gravitational potential energy changes.

(a) A body moves downward

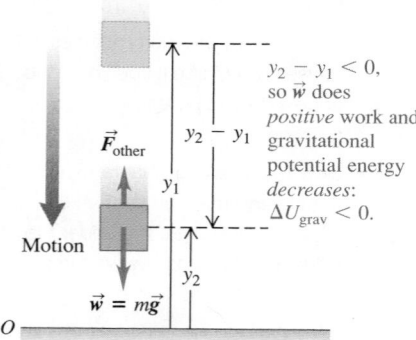

$y_2 - y_1 < 0$, so $\vec{w}$ does *positive* work and gravitational potential energy *decreases*: $\Delta U_{\text{grav}} < 0$.

To begin with, however, let's derive the expression for gravitational potential energy. Suppose a body with mass m moves along the (vertical) y-axis, as in Fig. 7.2. The forces acting on it are its weight, with magnitude $w = mg$, and possibly some other forces; we call the vector sum (resultant) of all the other forces $\vec{F}_{\text{other}}$. We'll assume that the body stays close enough to the earth's surface that the weight is constant. (We'll find in Chapter 13 that weight decreases with altitude.) We want to find the work done by the weight when the body moves downward from a height y_1 above the origin to a lower height y_2 (Fig. 7.2a). The weight and displacement are in the same direction, so the work W_{grav} done on the body by its weight is positive;

$$W_{\text{grav}} = Fs = w(y_1 - y_2) = mgy_1 - mgy_2 \tag{7.1}$$

This expression also gives the correct work when the body moves *upward* and y_2 is greater than y_1 (Fig. 7.2b). In that case the quantity $(y_1 - y_2)$ is negative, and W_{grav} is negative because the weight and displacement are opposite in direction.

Equation (7.1) shows that we can express W_{grav} in terms of the values of the quantity mgy at the beginning and end of the displacement. This quantity, the product of the weight mg and the height y above the origin of coordinates, is called the **gravitational potential energy,** U_{grav}:

$$U_{\text{grav}} = mgy \qquad \text{(gravitational potential energy)} \tag{7.2}$$

(b) A body moves upward

$y_2 - y_1 > 0$, so $\vec{w}$ does *negative* work and gravitational potential energy *increases*: $\Delta U_{\text{grav}} > 0$.

Its initial value is $U_{\text{grav},1} = mgy_1$ and its final value is $U_{\text{grav},2} = mgy_2$. The change in U_{grav} is the final value minus the initial value, or $\Delta U_{\text{grav}} = U_{\text{grav},2} - U_{\text{grav},1}$. We can express the work W_{grav} done by the gravitational force during the displacement from y_1 to y_2 as

$$W_{\text{grav}} = U_{\text{grav},1} - U_{\text{grav},2} = -(U_{\text{grav},2} - U_{\text{grav},1}) = -\Delta U_{\text{grav}} \tag{7.3}$$

The negative sign in front of ΔU_{grav} is *essential*. When the body moves up, y increases, the work done by the gravitational force is negative, and the gravitational potential energy increases ($\Delta U_{\text{grav}} > 0$). When the body moves down, y decreases, the gravitational force does positive work, and the gravitational potential energy decreases ($\Delta U_{\text{grav}} < 0$). It's like drawing money out of the bank (decreasing U_{grav}) and spending it (doing positive work). The unit of potential energy is the joule (J), the same unit as is used for work.

CAUTION **To what body does gravitational potential energy "belong"?** It is *not* correct to call $U_{\text{grav}} = mgy$ the "gravitational potential energy of the body." The reason is that gravitational potential energy U_{grav} is a *shared* property of the body and the earth. The value of U_{grav} increases if the earth stays fixed and the body moves upward, away from the earth; it also increases if the body stays fixed and the earth is moved away from it. Notice that the formula $U_{\text{grav}} = mgy$ involves characteristics of both the body (its mass m) and the earth (the value of g).

Conservation of Mechanical Energy (Gravitational Forces Only)

To see what gravitational potential energy is good for, suppose the body's weight is the *only* force acting on it, so $\vec{F}_{\text{other}} = 0$. The body is then falling freely with no air resistance and can be moving either up or down. Let its speed at point y_1 be v_1 and let its speed at y_2 be v_2. The work–energy theorem, Eq. (6.6), says that the total work done on the body equals the change in the body's kinetic energy: $W_{\text{tot}} = \Delta K = K_2 - K_1$. If gravity is the only force that acts, then from Eq. (7.3), $W_{\text{tot}} = W_{\text{grav}} = -\Delta U_{\text{grav}} = U_{\text{grav},1} - U_{\text{grav},2}$. Putting these together, we get

$$\Delta K = -\Delta U_{\text{grav}} \quad \text{or} \quad K_2 - K_1 = U_{\text{grav},1} - U_{\text{grav},2}$$

which we can rewrite as

$$K_1 + U_{\text{grav},1} = K_2 + U_{\text{grav},2} \quad \text{(if only gravity does work)} \quad (7.4)$$

or

$$\tfrac{1}{2}mv_1^2 + mgy_1 = \tfrac{1}{2}mv_2^2 + mgy_2 \quad \text{(if only gravity does work)} \quad (7.5)$$

The sum $K + U_{\text{grav}}$ of kinetic and potential energy is called E, the **total mechanical energy of the system.** By "system" we mean the body of mass m and the earth considered together, because gravitational potential energy U is a shared property of both bodies. Then $E_1 = K_1 + U_{\text{grav},1}$ is the total mechanical energy at y_1 and $E_2 = K_2 + U_{\text{grav},2}$ is the total mechanical energy at y_2. Equation (7.4) says that when the body's weight is the only force doing work on it, $E_1 = E_2$. That is, E is constant; it has the same value at y_1 and y_2. But since the positions y_1 and y_2 are arbitrary points in the motion of the body, the total mechanical energy E has the same value at *all* points during the motion:

$$E = K + U_{\text{grav}} = \text{constant} \quad \text{(if only gravity does work)}$$

A quantity that always has the same value is called a *conserved* quantity. *When only the force of gravity does work, the total mechanical energy is constant—that is, it is conserved* (Fig. 7.3). This is our first example of the **conservation of mechanical energy.**

When we throw a ball into the air, its speed decreases on the way up as kinetic energy is converted to potential energy; $\Delta K < 0$ and $\Delta U_{\text{grav}} > 0$. On the way back down, potential energy is converted back to kinetic energy and the ball's speed increases; $\Delta K > 0$ and $\Delta U_{\text{grav}} < 0$. But the *total* mechanical energy (kinetic plus potential) is the same at every point in the motion, provided that no force other than gravity does work on the ball (that is, air resistance must be negligible). It's still true that the gravitational force does work on the body as it

Application **Which Egg Has More Mechanical Energy?**
The mechanical energy of each of these identical eggs has the *same* value. The mechanical energy for an egg at rest atop the stone is purely gravitational potential energy. For the falling egg, the gravitational potential energy decreases as the egg descends and the egg's kinetic energy increases. If there is negligible air resistance, the mechanical energy of the falling egg remains constant.

7.3 While this athlete is in midair, only gravity does work on him (if we neglect the minor effects of air resistance). Mechanical energy E—the sum of kinetic and gravitational potential energy—is conserved.

Moving up:
• K decreases.
• U_{grav} increases.
• $E = K + U_{grav}$ stays the same.

Moving down:
• K increases.
• U_{grav} decreases.
• $E = K + U_{grav}$ stays the same.

$\vec{w} = m\vec{g}$

moves up or down, but we no longer have to calculate work directly; keeping track of changes in the value of U_{grav} takes care of this completely.

CAUTION **Choose "zero height" to be wherever you like** When working with gravitational potential energy, we may choose any height to be $y = 0$. If we shift the origin for y, the values of y_1 and y_2 change, as do the values of $U_{grav,1}$ and $U_{grav,2}$. But this shift has no effect on the *difference* in height $y_2 - y_1$ or on the *difference* in gravitational potential energy $U_{grav,2} - U_{grav,1} = mg(y_2 - y_1)$. As the following example shows, the physically significant quantity is not the value of U_{grav} at a particular point, but only the *difference* in U_{grav} between two points. So we can define U_{grav} to be zero at whatever point we choose without affecting the physics.

Example 7.1 **Height of a baseball from energy conservation**

You throw a 0.145-kg baseball straight up, giving it an initial velocity of magnitude 20.0 m/s. Find how high it goes, ignoring air resistance.

SOLUTION

IDENTIFY and SET UP: After the ball leaves your hand, only gravity does work on it. Hence mechanical energy is conserved, and we can use Eqs. (7.4) and (7.5). We take point 1 to be where the ball leaves your hand and point 2 to be where it reaches its maximum height. As in Fig. 7.2, we take the positive y-direction to be upward. The ball's speed at point 1 is $v_1 = 20.0$ m/s; at its maximum height it is instantaneously at rest, so $v_2 = 0$. We take the origin at point 1, so $y_1 = 0$ (Fig. 7.4). Our target variable, the distance the ball moves vertically between the two points, is the displacement $y_2 - y_1 = y_2 - 0 = y_2$.

EXECUTE: We have $y_1 = 0$, $U_{grav,1} = mgy_1 = 0$, and $K_2 = \frac{1}{2}mv_2^2 = 0$. Then Eq. (7.4), $K_1 + U_{grav,1} = K_2 + U_{grav,2}$, becomes

$$K_1 = U_{grav,2}$$

As the energy bar graphs in Fig. 7.4 show, this equation says that the kinetic energy of the ball at point 1 is completely converted to gravitational potential energy at point 2. We substitute $K_1 = \frac{1}{2}mv_1^2$ and $U_{grav,2} = mgy_2$ and solve for y_2:

$$\frac{1}{2}mv_1^2 = mgy_2$$

$$y_2 = \frac{v_1^2}{2g} = \frac{(20.0 \text{ m/s})^2}{2(9.80 \text{ m/s}^2)} = 20.4 \text{ m}$$

7.4 After a baseball leaves your hand, mechanical energy $E = K + U$ is conserved.

Energy at y_2

$E = K + U_{grav}$

... so the mechanical energy $E = K + U$ stays constant.

Energy at y_1

$E = K + U_{grav}$

$v_2 = 0$

After the ball leaves your hand, the only force acting on it is gravity ...

$v_1 = 20.0$ m/s
$m = 0.145$ kg

y_2

$y_1 = 0$

EVALUATE: As a check on our work, use the given value of v_1 and our result for y_2 to calculate the kinetic energy at point 1 and the gravitational potential energy at point 2. You should find that these are equal: $K_1 = \frac{1}{2}mv_1^2 = 29.0$ J and $U_{grav,2} = mgy_2 = 29.0$ J. Note also that we could have found the result $y_2 = v_1^2/2g$ using Eq. (2.13).

What if we put the origin somewhere else? For example, what if we put it 5.0 m below point 1, so that $y_1 = 5.0$ m? Then the total mechanical energy at point 1 is part kinetic and part potential; at point 2 it's still purely potential because $v_2 = 0$. You'll find that this choice of origin yields $y_2 = 25.4$ m, but again $y_2 - y_1 = 20.4$ m. In problems like this, you are free to choose the height at which $U_{grav} = 0$. The physics doesn't depend on your choice, so don't agonize over it.

When Forces Other Than Gravity Do Work

If other forces act on the body in addition to its weight, then $\vec{F}_{\text{other}}$ in Fig. 7.2 is *not* zero. For the pile driver described in Example 6.4 (Section 6.2), the force applied by the hoisting cable and the friction with the vertical guide rails are examples of forces that might be included in $\vec{F}_{\text{other}}$. The gravitational work W_{grav} is still given by Eq. (7.3), but the total work W_{tot} is then the sum of W_{grav} and the work done by $\vec{F}_{\text{other}}$. We will call this additional work W_{other}, so the total work done by all forces is $W_{\text{tot}} = W_{\text{grav}} + W_{\text{other}}$. Equating this to the change in kinetic energy, we have

$$W_{\text{other}} + W_{\text{grav}} = K_2 - K_1 \qquad (7.6)$$

Also, from Eq. (7.3), $W_{\text{grav}} = U_{\text{grav},1} - U_{\text{grav},2}$, so

$$W_{\text{other}} + U_{\text{grav},1} - U_{\text{grav},2} = K_2 - K_1$$

which we can rearrange in the form

$$K_1 + U_{\text{grav},1} + W_{\text{other}} = K_2 + U_{\text{grav},2} \qquad \text{(if forces other than gravity do work)} \qquad (7.7)$$

Finally, using the appropriate expressions for the various energy terms, we obtain

$$\tfrac{1}{2}mv_1^2 + mgy_1 + W_{\text{other}} = \tfrac{1}{2}mv_2^2 + mgy_2 \qquad \text{(if forces other than gravity do work)} \qquad (7.8)$$

The meaning of Eqs. (7.7) and (7.8) is this: *The work done by all forces other than the gravitational force equals the change in the total mechanical energy $E = K + U_{\text{grav}}$ of the system, where U_{grav} is the gravitational potential energy.* When W_{other} is positive, E increases and $K_2 + U_{\text{grav},2}$ is greater than $K_1 + U_{\text{grav},1}$. When W_{other} is negative, E decreases (Fig. 7.5). In the special case in which no forces other than the body's weight do work, $W_{\text{other}} = 0$. The total mechanical energy is then constant, and we are back to Eq. (7.4) or (7.5).

7.5 As this skydiver moves downward, the upward force of air resistance does negative work W_{other} on him. Hence the total mechanical energy $E = K + U$ decreases: The skydiver's speed and kinetic energy K stay the same, while the gravitational potential energy U decreases.

Problem-Solving Strategy 7.1 **Problems Using Mechanical Energy I** (MP)

IDENTIFY *the relevant concepts:* Decide whether the problem should be solved by energy methods, by using $\sum \vec{F} = m\vec{a}$ directly, or by a combination of these. The energy approach is best when the problem involves varying forces or motion along a curved path (discussed later in this section). If the problem involves elapsed time, the energy approach is usually *not* the best choice because it doesn't involve time directly.

SET UP *the problem* using the following steps:
1. When using the energy approach, first identify the initial and final states (the positions and velocities) of the bodies in question. Use the subscript 1 for the initial state and the subscript 2 for the final state. Draw sketches showing these states.
2. Define a coordinate system, and choose the level at which $y = 0$. Choose the positive y-direction to be upward, as is assumed in Eq. (7.1) and in the equations that follow from it.
3. Identify any forces that do work on each body and that *cannot* be described in terms of potential energy. (So far, this means

any forces other than gravity. In Section 7.2 we'll see that the work done by an ideal spring can also be expressed as a change in potential energy.) Sketch a free-body diagram for each body.
4. List the unknown and known quantities, including the coordinates and velocities at each point. Identify the target variables.

EXECUTE *the solution:* Write expressions for the initial and final kinetic and potential energies K_1, K_2, $U_{\text{grav},1}$, and $U_{\text{grav},2}$. If no other forces do work, use Eq. (7.4). If there are other forces that do work, use Eq. (7.7). Draw bar graphs showing the initial and final values of K, $U_{\text{grav},1}$, and $E = K + U_{\text{grav}}$. Then solve to find your target variables.

EVALUATE *your answer:* Check whether your answer makes physical sense. Remember that the gravitational work is included in ΔU_{grav}, so do not include it in W_{other}.

Example 7.2 Work and energy in throwing a baseball

In Example 7.1 suppose your hand moves upward by 0.50 m while you are throwing the ball. The ball leaves your hand with an upward velocity of 20.0 m/s. (a) Find the magnitude of the force (assumed constant) that your hand exerts on the ball. (b) Find the speed of the ball at a point 15.0 m above the point where it leaves your hand. Ignore air resistance.

SOLUTION

IDENTIFY and SET UP: In Example 7.1 only gravity did work. Here we must include the nongravitational, "other" work done by your hand. Figure 7.6 shows a diagram of the situation, including a free-body diagram for the ball while it is being thrown. We let point 1 be where your hand begins to move, point 2 be where the ball leaves your hand, and point 3 be where the ball is 15.0 m above point 2. The nongravitational force $\vec{F}$ of your hand acts only between points 1 and 2. Using the same coordinate system as in Example 7.1, we have $y_1 = -0.50$ m, $y_2 = 0$, and $y_3 = 15.0$ m. The ball starts at rest at point 1, so $v_1 = 0$, and the ball's speed as it leaves your hand is $v_2 = 20.0$ m/s. Our target variables are (a) the magnitude F of the force of your hand and (b) the ball's velocity v_{3y} at point 3.

EXECUTE: (a) To determine F, we'll first use Eq. (7.7) to calculate the work W_{other} done by this force. We have

$$K_1 = 0$$

$$U_{\text{grav},1} = mgy_1 = (0.145 \text{ kg})(9.80 \text{ m/s}^2)(-0.50 \text{ m}) = -0.71 \text{ J}$$

7.6 (a) Applying energy ideas to a ball thrown vertically upward. (b) Free-body diagram for the ball as you throw it.

(a)

(b)

$$K_2 = \tfrac{1}{2}mv_2^2 = \tfrac{1}{2}(0.145 \text{ kg})(20.0 \text{ m/s})^2 = 29.0 \text{ J}$$

$$U_{\text{grav},2} = mgy_2 = (0.145 \text{ kg})(9.80 \text{ m/s}^2)(0) = 0$$

(Don't worry that $U_{\text{grav},1}$ is less than zero; all that matters is the *difference* in potential energy from one point to another.) From Eq. (7.7),

$$K_1 + U_{\text{grav},1} + W_{\text{other}} = K_2 + U_{\text{grav},2}$$

$$W_{\text{other}} = (K_2 - K_1) + (U_{\text{grav},2} - U_{\text{grav},1})$$

$$= (29.0 \text{ J} - 0) + [0 - (-0.71 \text{ J})] = 29.7 \text{ J}$$

But since $\vec{F}$ is constant and upward, the work done by $\vec{F}$ equals the force magnitude times the displacement: $W_{\text{other}} = F(y_2 - y_1)$. So

$$F = \frac{W_{\text{other}}}{y_2 - y_1} = \frac{29.7 \text{ J}}{0.50 \text{ m}} = 59 \text{ N}$$

This is more than 40 times the weight of the ball (1.42 N).

(b) To find v_{3y}, note that between points 2 and 3 only gravity acts on the ball. So between these points mechanical energy is conserved and $W_{\text{other}} = 0$. From Eq. (7.4), we can solve for K_3 and from that solve for v_{3y}:

$$K_2 + U_{\text{grav},2} = K_3 + U_{\text{grav},3}$$

$$U_{\text{grav},3} = mgy_3 = (0.145 \text{ kg})(9.80 \text{ m/s}^2)(15.0 \text{ m}) = 21.3 \text{ J}$$

$$K_3 = (K_2 + U_{\text{grav},2}) - U_{\text{grav},3}$$

$$= (29.0 \text{ J} + 0 \text{ J}) - 21.3 \text{ J} = 7.7 \text{ J}$$

Since $K_3 = \tfrac{1}{2}mv_{3y}^2$, we find

$$v_{3y} = \pm\sqrt{\frac{2K_3}{m}} = \pm\sqrt{\frac{2(7.7 \text{ J})}{0.145 \text{ kg}}} = \pm 10 \text{ m/s}$$

The plus-or-minus sign reminds us that the ball passes point 3 on the way up and again on the way down. The total mechanical energy E is constant and equal to $K_2 + U_{\text{grav},2} = 29.0 \text{ J}$ while the ball is in free fall, and the potential energy at point 3 is $U_{\text{grav},3} = mgy_3 = 21.3 \text{ J}$ whether the ball is moving up or down. So at point 3, the ball's kinetic energy K_3 (and therefore its speed) don't depend on the direction the ball is moving. The velocity v_{3y} is positive $(+10 \text{ m/s})$ when the ball is moving up and negative (-10 m/s) when it is moving down; the speed v_3 is 10 m/s in either case.

EVALUATE: In Example 7.1 we found that the ball reaches a maximum height $y = 20.4$ m. At that point all of the kinetic energy it had when it left your hand at $y = 0$ has been converted to gravitational potential energy. At $y = 15.0$ m, the ball is about three-fourths of the way to its maximum height, so about three-fourths of its mechanical energy should be in the form of potential energy. (The energy bar graphs in Fig. 7.6a show this.) Can you show that this is true from our results for K_3 and $U_{\text{grav},3}$?

Gravitational Potential Energy for Motion Along a Curved Path

In our first two examples the body moved along a straight vertical line. What happens when the path is slanted or curved (Fig. 7.7a)? The body is acted on by the gravitational force $\vec{w} = m\vec{g}$ and possibly by other forces whose resultant we call $\vec{F}_{\text{other}}$. To find the work done by the gravitational force during this displacement, we divide the path into small segments $\Delta\vec{s}$; Fig. 7.7b shows a typical segment. The work done by the gravitational force over this segment is the scalar product of the force and the displacement. In terms of unit vectors, the force is $\vec{w} = m\vec{g} = -mg\hat{j}$ and the displacement is $\Delta\vec{s} = \Delta x\hat{i} + \Delta y\hat{j}$, so the work done by the gravitational force is

$$\vec{w}\cdot\Delta\vec{s} = -mg\hat{j}\cdot(\Delta x\hat{i} + \Delta y\hat{j}) = -mg\Delta y$$

The work done by gravity is the same as though the body had been displaced vertically a distance Δy, with no horizontal displacement. This is true for every segment, so the *total* work done by the gravitational force is $-mg$ multiplied by the *total* vertical displacement $(y_2 - y_1)$:

$$W_{\text{grav}} = -mg(y_2 - y_1) = mgy_1 - mgy_2 = U_{\text{grav},1} - U_{\text{grav},2}$$

This is the same as Eq. (7.1) or (7.3), in which we assumed a purely vertical path. So even if the path a body follows between two points is curved, the total work done by the gravitational force depends only on the difference in height between the two points of the path. This work is unaffected by any horizontal motion that may occur. So *we can use the same expression for gravitational potential energy whether the body's path is curved or straight.*

7.7 Calculating the change in gravitational potential energy for a displacement along a curved path.

(a)

(b)

The work done by the gravitational force depends only on the vertical component of displacement Δy.

In this case Δy is negative.

Conceptual Example 7.3 | **Energy in projectile motion**

A batter hits two identical baseballs with the same initial speed and from the same initial height but at different initial angles. Prove that both balls have the same speed at any height h if air resistance can be neglected.

7.8 For the same initial speed and initial height, the speed of a projectile at a given elevation h is always the same, neglecting air resistance.

SOLUTION

The only force acting on each ball after it is hit is its weight. Hence the total mechanical energy for each ball is constant. Figure 7.8 shows the trajectories of two balls batted at the same height with the same initial speed, and thus the same total mechanical energy, but with different initial angles. At all points at the same height the potential energy is the same. Thus the kinetic energy at this height must be the same for both balls, and the speeds are the same.

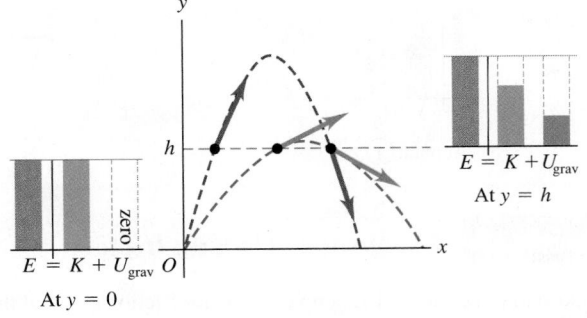

$E = K + U_{\text{grav}}$
At $y = h$

$E = K + U_{\text{grav}}$
At $y = 0$

Example 7.4 | **Speed at the bottom of a vertical circle**

Your cousin Throckmorton skateboards from rest down a curved, frictionless ramp. If we treat Throcky and his skateboard as a particle, he moves through a quarter-circle with radius $R = 3.00$ m (Fig. 7.9). Throcky and his skateboard have a total mass of 25.0 kg. (a) Find his speed at the bottom of the ramp. (b) Find the normal force that acts on him at the bottom of the curve.

SOLUTION

IDENTIFY: We can't use the constant-acceleration equations of Chapter 2 because Throcky's acceleration isn't constant; the slope decreases as he descends. Instead, we'll use the energy approach. Throcky moves along a circular arc, so we'll also use what we learned about circular motion in Section 5.4.

SET UP: The only forces on Throcky are his weight and the normal force $\vec{n}$ exerted by the ramp (Fig. 7.9b). Although $\vec{n}$ acts all along the path, it does zero work because $\vec{n}$ is perpendicular to Throcky's displacement at every point. Hence $W_{\text{other}} = 0$ and mechanical energy is conserved. We take point 1 at the starting point and point 2 at the bottom of the ramp, and we let $y = 0$ be at the bottom of the ramp (Fig. 7.9a). We take the positive y-direction upward; then $y_1 = R$ and $y_2 = 0$. Throcky starts at rest at the top, so $v_1 = 0$. In part (a) our target variable is his speed v_2 at the bottom; in part (b) the target variable is the magnitude n of the normal force at point 2. To find n, we'll use Newton's second law and the relation $a = v^2/R$.

Continued

EXECUTE: (a) The various energy quantities are

$$K_1 = 0 \qquad U_{grav,1} = mgR$$
$$K_2 = \tfrac{1}{2}mv_2^2 \qquad U_{grav,2} = 0$$

From conservation of mechanical energy, Eq. (7.4),

$$K_1 + U_{grav,1} = K_2 + U_{grav,2}$$
$$0 + mgR = \tfrac{1}{2}mv_2^2 + 0$$
$$v_2 = \sqrt{2gR}$$
$$= \sqrt{2(9.80 \text{ m/s}^2)(3.00 \text{ m})} = 7.67 \text{ m/s}$$

This answer doesn't depend on the ramp being circular; Throcky will have the same speed $v_2 = \sqrt{2gR}$ at the bottom of any ramp of height R, no matter what its shape.

(b) To find n at point 2 using Newton's second law, we need the free-body diagram at that point (Fig. 7.9b). At point 2, Throcky is moving at speed $v_2 = \sqrt{2gR}$ in a circle of radius R; his acceleration is toward the center of the circle and has magnitude

$$a_{rad} = \frac{v_2^2}{R} = \frac{2gR}{R} = 2g$$

The y-component of Newton's second law is

$$\sum F_y = n + (-w) = ma_{rad} = 2mg$$
$$n = w + 2mg = 3mg$$
$$= 3(25.0 \text{ kg})(9.80 \text{ m/s}^2) = 735 \text{ N}$$

At point 2 the normal force is three times Throcky's weight. This result doesn't depend on the radius R of the ramp. We saw in Examples 5.9 and 5.23 that the magnitude of n is the *apparent weight*, so at the bottom of the *curved part* of the ramp Throcky feels as though he weighs three times his true weight mg. But when he reaches the *horizontal* part of the ramp, immediately to the right of point 2, the normal force decreases to $w = mg$ and thereafter Throcky feels his true weight again. Can you see why?

EVALUATE: This example shows a general rule about the role of forces in problems in which we use energy techniques: What matters is not simply whether a force *acts*, but whether that force *does work*. If the force does no work, like the normal force $\vec{n}$ here, then it does not appear in Eqs. (7.4) and (7.7).

7.9 (a) Throcky skateboarding down a frictionless circular ramp. The total mechanical energy is constant. (b) Free-body diagrams for Throcky and his skateboard at various points on the ramp.

Example 7.5 **A vertical circle with friction**

Suppose that the ramp of Example 7.4 is not frictionless, and that Throcky's speed at the bottom is only 6.00 m/s, not the 7.67 m/s we found there. What work was done on him by the friction force?

SOLUTION

IDENTIFY and SET UP: Figure 7.10 shows that again the normal force does no work, but now there is a friction force $\vec{f}$ that *does* do work W_f. Hence the nongravitational work W_{other} done on Throcky between points 1 and 2 is equal to W_f and is not zero. We use the same coordinate system and the same initial and final points as in Example 7.4. Our target variable is $W_f = W_{other}$, which we'll find using Eq. (7.7).

EXECUTE: The energy quantities are

$$K_1 = 0$$
$$U_{grav,1} = mgR = (25.0 \text{ kg})(9.80 \text{ m/s}^2)(3.00 \text{ m}) = 735 \text{ J}$$
$$K_2 = \tfrac{1}{2}mv_2^2 = \tfrac{1}{2}(25.0 \text{ kg})(6.00 \text{ m/s})^2 = 450 \text{ J}$$
$$U_{grav,2} = 0$$

7.10 Energy bar graphs and free-body diagrams for Throcky skateboarding down a ramp with friction.

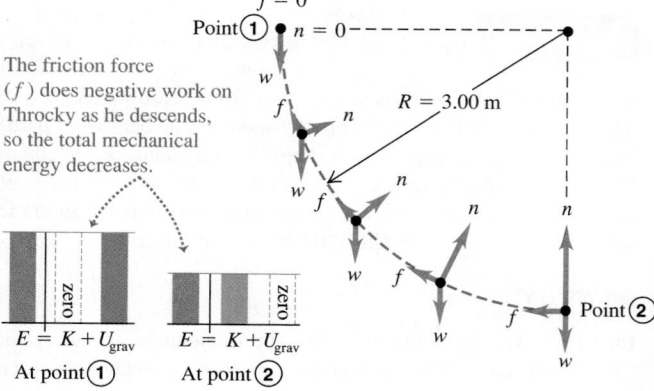

From Eq. (7.7),

$$W_f = W_{\text{other}} = K_2 + U_{\text{grav},2} - K_1 - U_{\text{grav},1}$$
$$= 450\ \text{J} + 0 - 0 - 735\ \text{J} = -285\ \text{J}$$

The work done by the friction force is -285 J, and the total mechanical energy *decreases* by 285 J.

EVALUATE: Our result for W_f is negative. Can you see from the free-body diagrams in Fig. 7.10 why this must be so?

It would be very difficult to apply Newton's second law, $\sum \vec{F} = m\vec{a}$, directly to this problem because the normal and friction forces and the acceleration are continuously changing in both magnitude and direction as Throcky descends. The energy approach, by contrast, relates the motions at the top and bottom of the ramp without involving the details of the motion in between.

Example 7.6 | **An inclined plane with friction**

We want to slide a 12-kg crate up a 2.5-m-long ramp inclined at 30°. A worker, ignoring friction, calculates that he can do this by giving it an initial speed of 5.0 m/s at the bottom and letting it go. But friction is *not* negligible; the crate slides only 1.6 m up the ramp, stops, and slides back down (Fig. 7.11a). (a) Find the magnitude of the friction force acting on the crate, assuming that it is constant. (b) How fast is the crate moving when it reaches the bottom of the ramp?

SOLUTION

IDENTIFY and SET UP: The friction force does work on the crate as it slides. The first part of the motion is from point 1, at the bottom of the ramp, to point 2, where the crate stops instantaneously ($v_2 = 0$). In the second part of the motion, the crate returns to the bottom of the ramp, which we'll also call point 3 (Fig. 7.11a). We take the positive y-direction upward. We take $y = 0$ (and hence $U_{\text{grav}} = 0$) to be at ground level (point 1), so that $y_1 = 0$, $y_2 = (1.6\ \text{m})\sin 30° = 0.80\ \text{m}$, and $y_3 = 0$. We are given $v_1 = 5.0$ m/s. In part (a) our target variable is f, the magnitude of the friction force as the crate slides up; as in Example 7.2, we'll find this using the energy approach. In part (b) our target variable is v_3, the crate's speed at the bottom of the ramp. We'll calculate the work done by friction as the crate slides back down, then use the energy approach to find v_3.

7.11 (a) A crate slides partway up the ramp, stops, and slides back down. (b) Energy bar graphs for points 1, 2, and 3.

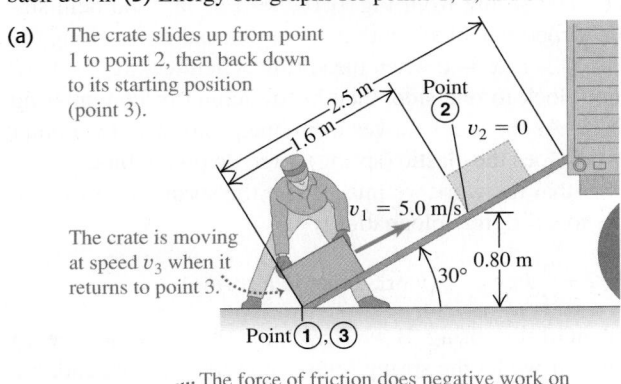

(a) The crate slides up from point 1 to point 2, then back down to its starting position (point 3).

Point ②: $v_2 = 0$
2.5 m
1.6 m
$v_1 = 5.0$ m/s

The crate is moving at speed v_3 when it returns to point 3.

30° 0.80 m

Point ①, ③

(b)

The force of friction does negative work on the crate as it moves, so the total mechanical energy $E = K + U_{\text{grav}}$ decreases.

zero
$E = K + U_{\text{grav}}$
At point ①

zero
$E = K + U_{\text{grav}}$
At point ②

zero
$E = K + U_{\text{grav}}$
At point ③

EXECUTE: (a) The energy quantities are

$$K_1 = \tfrac{1}{2}(12\ \text{kg})(5.0\ \text{m/s})^2 = 150\ \text{J}$$
$$U_{\text{grav},1} = 0$$
$$K_2 = 0$$
$$U_{\text{grav},2} = (12\ \text{kg})(9.8\ \text{m/s}^2)(0.80\ \text{m}) = 94\ \text{J}$$
$$W_{\text{other}} = -fs$$

Here $s = 1.6$ m. Using Eq. (7.7), we find

$$K_1 + U_{\text{grav},1} + W_{\text{other}} = K_2 + U_{\text{grav},2}$$
$$W_{\text{other}} = -fs = (K_2 + U_{\text{grav},2}) - (K_1 + U_{\text{grav},1})$$
$$= (0 + 94\ \text{J}) - (150\ \text{J} + 0) = -56\ \text{J} = -fs$$
$$f = \frac{W_{\text{other}}}{s} = \frac{56\ \text{J}}{1.6\ \text{m}} = 35\ \text{N}$$

The friction force of 35 N, acting over 1.6 m, causes the mechanical energy of the crate to decrease from 150 J to 94 J (Fig. 7.11b).

(b) As the crate moves from point 2 to point 3, the work done by friction has the same negative value as from point 1 to point 2. (The friction force and the displacement both reverse direction but have the same magnitudes.) The total work done by friction between points 1 and 3 is therefore

$$W_{\text{other}} = W_{\text{fric}} = -2fs = -2(56\ \text{J}) = -112\ \text{J}$$

From part (a), $K_1 = 150$ J and $U_{\text{grav},1} = 0$. Equation (7.7) then gives

$$K_1 + U_{\text{grav},1} + W_{\text{other}} = K_3 + U_{\text{grav},3}$$
$$K_3 = K_1 + U_{\text{grav},1} - U_{\text{grav},3} + W_{\text{other}}$$
$$= 150\ \text{J} + 0 - 0 + (-112\ \text{J}) = 38\ \text{J}$$

The crate returns to the bottom of the ramp with only 38 J of the original 150 J of mechanical energy (Fig. 7.11b). Since $K_3 = \tfrac{1}{2}mv_3^2$,

$$v_3 = \sqrt{\frac{2K_3}{m}} = \sqrt{\frac{2(38\ \text{J})}{12\ \text{kg}}} = 2.5\ \text{m/s}$$

EVALUATE: Energy was lost due to friction, so the crate's speed $v_3 = 2.5$ m/s when it returns to the bottom of the ramp is less than the speed $v_1 = 5.0$ m/s at which it left that point. In part (b) we applied Eq. (7.7) to points 1 and 3, considering the round trip as a whole. Alternatively, we could have considered the second part of the motion by itself and applied Eq. (7.7) to points 2 and 3. Try it; do you get the same result for v_3?

MP

Test Your Understanding of Section 7.1 The figure shows two different frictionless ramps. The heights y_1 and y_2 are the same for both ramps. If a block of mass m is released from rest at the left-hand end of each ramp, which block arrives at the right-hand end with the greater speed? (i) block I; (ii) block II; (iii) the speed is the same for both blocks.

7.2 Elastic Potential Energy

There are many situations in which we encounter potential energy that is not gravitational in nature. One example is a rubber-band slingshot. Work is done on the rubber band by the force that stretches it, and that work is stored in the rubber band until you let it go. Then the rubber band gives kinetic energy to the projectile.

This is the same pattern we saw with the pile driver in Section 7.1: Do work on the system to store energy, which can later be converted to kinetic energy. We'll describe the process of storing energy in a deformable body such as a spring or rubber band in terms of *elastic potential energy* (Fig. 7.12). A body is called *elastic* if it returns to its original shape and size after being deformed.

To be specific, we'll consider storing energy in an ideal spring, like the ones we discussed in Section 6.3. To keep such an ideal spring stretched by a distance x, we must exert a force $F = kx$, where k is the force constant of the spring. The ideal spring is a useful idealization because many elastic bodies show this same direct proportionality between force $\vec{F}$ and displacement x, provided that x is sufficiently small.

Let's proceed just as we did for gravitational potential energy. We begin with the work done by the elastic (spring) force and then combine this with the work–energy theorem. The difference is that gravitational potential energy is a shared property of a body and the earth, but elastic potential energy is stored just in the spring (or other deformable body).

Figure 7.13 shows the ideal spring from Fig. 6.18, with its left end held stationary and its right end attached to a block with mass m that can move along the x-axis. In Fig. 7.13a the body is at $x = 0$ when the spring is neither stretched nor compressed. We move the block to one side, thereby stretching or compressing the spring, and then let it go. As the block moves from one position x_1 to another position x_2, how much work does the elastic (spring) force do on the block?

We found in Section 6.3 that the work we must do *on* the spring to move one end from an elongation x_1 to a different elongation x_2 is

$$W = \tfrac{1}{2} kx_2^2 - \tfrac{1}{2} kx_1^2 \qquad \text{(work done on a spring)}$$

where k is the force constant of the spring. If we stretch the spring farther, we do positive work on the spring; if we let the spring relax while holding one end, we do negative work on it. We also saw that this expression for work is still correct if the spring is compressed, not stretched, so that x_1 or x_2 or both are negative. Now we need to find the work done *by* the spring. From Newton's third law the two quantities of work are just negatives of each other. Changing the signs in this equation, we find that in a displacement from x_1 to x_2 the spring does an amount of work W_{el} given by

$$W_{el} = \tfrac{1}{2} kx_1^2 - \tfrac{1}{2} kx_2^2 \qquad \text{(work done by a spring)}$$

7.12 The Achilles tendon, which runs along the back of the ankle to the heel bone, acts like a natural spring. When it stretches and then relaxes, this tendon stores and then releases elastic potential energy. This spring action reduces the amount of work your leg muscles must do as you run.

Mastering**PHYSICS**

ActivPhysics 5.4: Inverse Bungee Jumper
ActivPhysics 5.5: Spring-Launched Bowler

The subscript "el" stands for *elastic*. When x_1 and x_2 are both positive and $x_2 > x_1$ (Fig. 7.13b), the spring does negative work on the block, which moves in the $+x$-direction while the spring pulls on it in the $-x$-direction. The spring stretches farther, and the block slows down. When x_1 and x_2 are both positive and $x_2 < x_1$ (Fig. 7.13c), the spring does positive work as it relaxes and the block speeds up. If the spring can be compressed as well as stretched, x_1 or x_2 or both may be negative, but the expression for W_{el} is still valid. In Fig. 7.13d, both x_1 and x_2 are negative, but x_2 is less negative than x_1; the compressed spring does positive work as it relaxes, speeding the block up.

Just as for gravitational work, we can express the work done by the spring in terms of a given quantity at the beginning and end of the displacement. This quantity is $\frac{1}{2}kx^2$, and we define it to be the **elastic potential energy:**

$$U_{el} = \tfrac{1}{2}kx^2 \qquad \text{(elastic potential energy)} \qquad (7.9)$$

Figure 7.14 is a graph of Eq. (7.9). The unit of U_{el} is the joule (J), the unit used for *all* energy and work quantities; to see this from Eq. (7.9), recall that the units of k are N/m and that $1\ \text{N} \cdot \text{m} = 1\ \text{J}$.

We can use Eq. (7.9) to express the work W_{el} done on the block by the elastic force in terms of the change in elastic potential energy:

$$W_{el} = \tfrac{1}{2}kx_1^2 - \tfrac{1}{2}kx_2^2 = U_{el,1} - U_{el,2} = -\Delta U_{el} \qquad (7.10)$$

When a stretched spring is stretched farther, as in Fig. 7.13b, W_{el} is negative and U_{el} *increases;* a greater amount of elastic potential energy is stored in the spring. When a stretched spring relaxes, as in Fig. 7.13c, x decreases, W_{el} is positive, and U_{el} *decreases;* the spring loses elastic potential energy. Negative values of x refer to a compressed spring. But, as Fig. 7.14 shows, U_{el} is positive for both positive and negative x, and Eqs. (7.9) and (7.10) are valid for both cases. The more a spring is compressed *or* stretched, the greater its elastic potential energy.

CAUTION Gravitational potential energy vs. elastic potential energy An important difference between gravitational potential energy $U_{grav} = mgy$ and elastic potential energy $U_{el} = \frac{1}{2}kx^2$ is that we do *not* have the freedom to choose $x = 0$ to be wherever we wish. To be consistent with Eq. (7.9), $x = 0$ *must* be the position at which the spring is neither stretched nor compressed. At that position, its elastic potential energy and the force that it exerts are both zero. ▌

The work–energy theorem says that $W_{tot} = K_2 - K_1$, no matter what kind of forces are acting on a body. If the elastic force is the *only* force that does work on the body, then

$$W_{tot} = W_{el} = U_{el,1} - U_{el,2}$$

The work–energy theorem, $W_{tot} = K_2 - K_1$, then gives us

$$K_1 + U_{el,1} = K_2 + U_{el,2} \qquad \text{(if only the elastic force does work)} \qquad (7.11)$$

Here U_{el} is given by Eq. (7.9), so

$$\tfrac{1}{2}mv_1^2 + \tfrac{1}{2}kx_1^2 = \tfrac{1}{2}mv_2^2 + \tfrac{1}{2}kx_2^2 \qquad \begin{array}{l}\text{(if only the elastic}\\\text{force does work)}\end{array} \qquad (7.12)$$

In this case the total mechanical energy $E = K + U_{el}$—the sum of kinetic and *elastic* potential energy—is *conserved.* An example of this is the motion of the

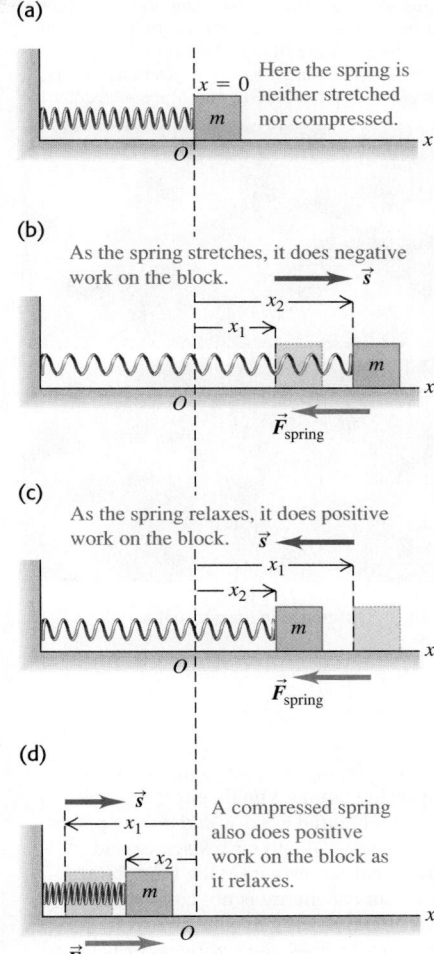

7.13 Calculating the work done by a spring attached to a block on a horizontal surface. The quantity x is the extension or compression of the spring.

(a) Here the spring is neither stretched nor compressed. $x = 0$

(b) As the spring stretches, it does negative work on the block. $\vec{s}$ $\vec{F}_{spring}$

(c) As the spring relaxes, it does positive work on the block. $\vec{s}$ $\vec{F}_{spring}$

(d) A compressed spring also does positive work on the block as it relaxes. $\vec{s}$ $\vec{F}_{spring}$

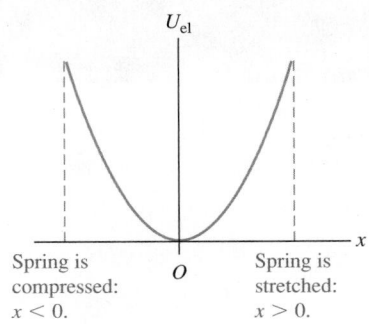

7.14 The graph of elastic potential energy for an ideal spring is a parabola: $U_{el} = \frac{1}{2}kx^2$, where x is the extension or compression of the spring. Elastic potential energy U_{el} is never negative.

U_{el}

Spring is compressed: $x < 0$.

Spring is stretched: $x > 0$.

Application **Elastic Potential Energy of a Cheetah**

When a cheetah gallops, its back flexes and extends by an exceptional amount. Flexion of the back stretches elastic tendons and muscles along the top of the spine and also compresses the spine, storing mechanical energy. When the cheetah launches into its next bound, this energy helps to extend the spine, enabling the cheetah to run more efficiently.

Difference in nose-to-tail length

7.15 Trampoline jumping involves an interplay among kinetic energy, gravitational potential energy, and elastic potential energy. Due to air resistance and frictional forces within the trampoline, mechanical energy is not conserved. That's why the bouncing eventually stops unless the jumper does work with his or her legs to compensate for the lost energy.

block in Fig. 7.13, provided the horizontal surface is frictionless so that no force does work other than that exerted by the spring.

For Eq. (7.12) to be strictly correct, the ideal spring that we've been discussing must also be *massless*. If the spring has a mass, it also has kinetic energy as the coils of the spring move back and forth. We can neglect the kinetic energy of the spring if its mass is much less than the mass m of the body attached to the spring. For instance, a typical automobile has a mass of 1200 kg or more. The springs in its suspension have masses of only a few kilograms, so their mass can be neglected if we want to study how a car bounces on its suspension.

Situations with Both Gravitational and Elastic Potential Energy

Equations (7.11) and (7.12) are valid when the only potential energy in the system is elastic potential energy. What happens when we have *both* gravitational and elastic forces, such as a block attached to the lower end of a vertically hanging spring? And what if work is also done by other forces that *cannot* be described in terms of potential energy, such as the force of air resistance on a moving block? Then the total work is the sum of the work done by the gravitational force (W_{grav}), the work done by the elastic force (W_{el}), and the work done by other forces (W_{other}): $W_{tot} = W_{grav} + W_{el} + W_{other}$. Then the work–energy theorem gives

$$W_{grav} + W_{el} + W_{other} = K_2 - K_1$$

The work done by the gravitational force is $W_{grav} = U_{grav,1} - U_{grav,2}$ and the work done by the spring is $W_{el} = U_{el,1} - U_{el,2}$. Hence we can rewrite the work–energy theorem for this most general case as

$$K_1 + U_{grav,1} + U_{el,1} + W_{other} = K_2 + U_{grav,2} + U_{el,2} \quad \text{(valid in general)} \quad (7.13)$$

or, equivalently,

$$K_1 + U_1 + W_{other} = K_2 + U_2 \quad \text{(valid in general)} \quad (7.14)$$

where $U = U_{grav} + U_{el} = mgy + \frac{1}{2}kx^2$ is the *sum* of gravitational potential energy and elastic potential energy. For short, we call U simply "the potential energy."

Equation (7.14) is *the most general statement* of the relationship among kinetic energy, potential energy, and work done by other forces. It says:

> The work done by all forces other than the gravitational force or elastic force equals the change in the total mechanical energy $E = K + U$ of the system, where $U = U_{grav} + U_{el}$ is the sum of the gravitational potential energy and the elastic potential energy.

The "system" is made up of the body of mass m, the earth with which it interacts through the gravitational force, and the spring of force constant k.

If W_{other} is positive, $E = K + U$ increases; if W_{other} is negative, E decreases. If the gravitational and elastic forces are the *only* forces that do work on the body, then $W_{other} = 0$ and the total mechanical energy (including both gravitational and elastic potential energy) is conserved. (You should compare Eq. (7.14) to Eqs. (7.7) and (7.8), which describe situations in which there is gravitational potential energy but no elastic potential energy.)

Trampoline jumping (Fig. 7.15) involves transformations among kinetic energy, elastic potential energy, and gravitational potential energy. As the jumper descends through the air from the high point of the bounce, gravitational potential energy U_{grav} decreases and kinetic energy K increases. Once the jumper touches the trampoline, some of the mechanical energy goes into elastic potential energy U_{el} stored in the trampoline's springs. Beyond a certain point the jumper's speed and kinetic energy K decrease while U_{grav} continues to decrease and U_{el} continues to increase. At the low point the jumper comes to a momentary halt ($K = 0$) at the lowest point of the trajectory (U_{grav} is minimum) and the springs are maximally stretched (U_{el} is maximum). The springs then convert their energy back into K and U_{grav}, propelling the jumper upward.

Problem-Solving Strategy 7.2   **Problems Using Mechanical Energy II**

Problem-Solving Strategy 7.1 (Section 7.1) is equally useful in solving problems that involve elastic forces as well as gravitational forces. The only new wrinkle is that the potential energy U now includes the elastic potential energy $U_{el} = \frac{1}{2}kx^2$, where x is the dis-

placement of the spring *from its unstretched length*. The work done by the gravitational and elastic forces is accounted for by their potential energies; the work done by other forces, W_{other}, must still be included separately.

Example 7.7   **Motion with elastic potential energy**

A glider with mass $m = 0.200$ kg sits on a frictionless horizontal air track, connected to a spring with force constant $k = 5.00$ N/m. You pull on the glider, stretching the spring 0.100 m, and release it from rest. The glider moves back toward its equilibrium position ($x = 0$). What is its x-velocity when $x = 0.080$ m?

SOLUTION

IDENTIFY and SET UP: As the glider starts to move, elastic potential energy is converted to kinetic energy. The glider remains at the same height throughout the motion, so gravitational potential energy is not a factor and $U = U_{el} = \frac{1}{2}kx^2$. Figure 7.16 shows our sketches. Only the spring force does work on the glider, so $W_{other} = 0$ and we may use Eq. (7.11). We designate the point

7.16 Our sketches and energy bar graphs for this problem.

where the glider is released as point 1 (that is, $x_1 = 0.100$ m) and $x_2 = 0.080$ m as point 2. We are given $v_{1x} = 0$; our target variable is v_{2x}.

EXECUTE: The energy quantities are

$$K_1 = \tfrac{1}{2}mv_{1x}^2 = \tfrac{1}{2}(0.200 \text{ kg})(0)^2 = 0$$
$$U_1 = \tfrac{1}{2}kx_1^2 = \tfrac{1}{2}(5.00 \text{ N/m})(0.100 \text{ m})^2 = 0.0250 \text{ J}$$
$$K_2 = \tfrac{1}{2}mv_{2x}^2$$
$$U_2 = \tfrac{1}{2}kx_2^2 = \tfrac{1}{2}(5.00 \text{ N/m})(0.080 \text{ m})^2 = 0.0160 \text{ J}$$

We use Eq. (7.11) to solve for K_2 and then find v_{2x}:

$$K_2 = K_1 + U_1 - U_2 = 0 + 0.0250 \text{ J} - 0.0160 \text{ J} = 0.0090 \text{ J}$$
$$v_{2x} = \pm\sqrt{\frac{2K_2}{m}} = \pm\sqrt{\frac{2(0.0090 \text{ J})}{0.200 \text{ kg}}} = \pm 0.30 \text{ m/s}$$

We choose the negative root because the glider is moving in the $-x$-direction. Our answer is $v_{2x} = -0.30$ m/s.

EVALUATE: Eventually the spring will reverse the glider's motion, pushing it back in the $+x$-direction (see Fig. 7.13d). The solution $v_{2x} = +0.30$ m/s tells us that when the glider passes through $x = 0.080$ m on this return trip, its speed will be 0.30 m/s, just as when it passed through this point while moving to the left.

Example 7.8   **Motion with elastic potential energy and work done by other forces**

Suppose the glider in Example 7.7 is initially at rest at $x = 0$, with the spring unstretched. You then push on the glider with a constant force $\vec{F}$ (magnitude 0.610 N) in the $+x$-direction. What is the glider's velocity when it has moved to $x = 0.100$ m?

SOLUTION

IDENTIFY and SET UP: Although the force $\vec{F}$ you apply is constant, the spring force isn't, so the acceleration of the glider won't be constant. Total mechanical energy is not conserved because of

Continued

the work done by the force $\vec{F}$, so we must use the generalized energy relationship given by Eq. (7.13). As in Example 7.7, we ignore gravitational potential energy because the glider's height doesn't change. Hence we again have $U = U_{el} = \frac{1}{2}kx^2$. This time, we let point 1 be at $x_1 = 0$, where the velocity is $v_{1x} = 0$, and let point 2 be at $x = 0.100$ m. The glider's displacement is then $\Delta x = x_2 - x_1 = 0.100$ m. Our target variable is v_{2x}, the velocity at point 2.

EXECUTE: The force $\vec{F}$ is constant and in the same direction as the displacement, so the work done by this force is $F\Delta x$. Then the energy quantities are

$$K_1 = 0$$
$$U_1 = \tfrac{1}{2}kx_1^2 = 0$$
$$K_2 = \tfrac{1}{2}mv_{2x}^2$$
$$U_2 = \tfrac{1}{2}kx_2^2 = \tfrac{1}{2}(5.00 \text{ N/m})(0.100 \text{ m})^2 = 0.0250 \text{ J}$$
$$W_{other} = F\Delta x = (0.610 \text{ N})(0.100 \text{ m}) = 0.0610 \text{ J}$$

The initial total mechanical energy is zero; the work done by $\vec{F}$ increases the total mechanical energy to 0.0610 J, of which $U_2 = 0.0250$ J is elastic potential energy. The remainder is kinetic energy. From Eq. (7.13),

$$K_1 + U_1 + W_{other} = K_2 + U_2$$
$$K_2 = K_1 + U_1 + W_{other} - U_2$$

$$= 0 + 0 + 0.0610 \text{ J} - 0.0250 \text{ J} = 0.0360 \text{ J}$$
$$v_{2x} = \sqrt{\frac{2K_2}{m}} = \sqrt{\frac{2(0.0360 \text{ J})}{0.200 \text{ kg}}} = 0.60 \text{ m/s}$$

We choose the positive square root because the glider is moving in the $+x$-direction.

EVALUATE: To test our answer, think what would be different if we disconnected the glider from the spring. Then only $\vec{F}$ would do work, there would be zero elastic potential energy at all times, and Eq. (7.13) would give us

$$K_2 = K_1 + W_{other} = 0 + 0.0610 \text{ J}$$
$$v_{2x} = \sqrt{\frac{2K_2}{m}} = \sqrt{\frac{2(0.0610 \text{ J})}{0.200 \text{ kg}}} = 0.78 \text{ m/s}$$

Our answer $v_{2x} = 0.60$ m/s is less than 0.78 m/s because the spring does negative work on the glider as it stretches (see Fig. 7.13b).

If you stop pushing on the glider when it reaches $x = 0.100$ m, only the spring force does work on it thereafter. Hence for $x > 0.100$ m, the total mechanical energy $E = K + U = 0.0610$ J is constant. As the spring continues to stretch, the glider slows down and the kinetic energy K decreases as the potential energy increases. The glider comes to rest at some point $x = x_3$, at which the kinetic energy is zero and the potential energy $U = U_{el} = \frac{1}{2}kx_3^2$ equals the total mechanical energy 0.0610 J. Can you show that $x_3 = 0.156$ m? (It moves an additional 0.056 m after you stop pushing.) If there is no friction, will the glider remain at rest?

Example 7.9   Motion with gravitational, elastic, and friction forces

A 2000-kg (19,600-N) elevator with broken cables in a test rig is falling at 4.00 m/s when it contacts a cushioning spring at the bottom of the shaft. The spring is intended to stop the elevator, compressing 2.00 m as it does so (Fig. 7.17). During the motion a safety clamp applies a constant 17,000-N frictional force to the elevator. What is the necessary force constant k for the spring?

7.17 The fall of an elevator is stopped by a spring and by a constant friction force.

SOLUTION

IDENTIFY and SET UP: We'll use the energy approach to determine k, which appears in the expression for elastic potential energy. This problem involves *both* gravitational and elastic potential energy. Total mechanical energy is not conserved because the friction force does negative work W_{other} on the elevator. We'll therefore use the most general form of the energy relationship, Eq. (7.13). We take point 1 as the position of the bottom of the elevator when it contacts the spring, and point 2 as its position when it stops. We choose the origin to be at point 1, so $y_1 = 0$ and $y_2 = -2.00$ m. With this choice the coordinate of the upper end of the spring after contact is the same as the coordinate of the elevator, so the elastic potential energy at any point between points 1 and 2 is $U_{el} = \frac{1}{2}ky^2$. The gravitational potential energy is $U_{grav} = mgy$ as usual. We know the initial and final speeds of the elevator and the magnitude of the friction force, so the only unknown is the force constant k (our target variable).

EXECUTE: The elevator's initial speed is $v_1 = 4.00$ m/s, so its initial kinetic energy is

$$K_1 = \tfrac{1}{2}mv_1^2 = \tfrac{1}{2}(2000 \text{ kg})(4.00 \text{ m/s})^2 = 16,000 \text{ J}$$

The elevator stops at point 2, so $K_2 = 0$. At point 1 the potential energy $U_1 = U_{grav} + U_{el}$ is zero; U_{grav} is zero because $y_1 = 0$, and $U_{el} = 0$ because the spring is uncompressed. At point 2 there is both gravitational and elastic potential energy, so

$$U_2 = mgy_2 + \tfrac{1}{2}ky_2^2$$

The gravitational potential energy at point 2 is

$$mgy_2 = (2000 \text{ kg})(9.80 \text{ m/s}^2)(-2.00 \text{ m}) = -39,200 \text{ J}$$

The "other" force is the constant 17,000-N friction force. It acts opposite to the 2.00-m displacement, so

$$W_{\text{other}} = -(17,000 \text{ N})(2.00 \text{ m}) = -34,000 \text{ J}$$

We put these terms into Eq. (7.14), $K_1 + U_1 + W_{\text{other}} = K_2 + U_2$:

$$K_1 + 0 + W_{\text{other}} = 0 + (mgy_2 + \tfrac{1}{2}ky_2^2)$$

$$k = \frac{2(K_1 + W_{\text{other}} - mgy_2)}{y_2^2}$$

$$= \frac{2[16,000 \text{ J} + (-34,000 \text{ J}) - (-39,200 \text{ J})]}{(-2.00 \text{ m})^2}$$

$$= 1.06 \times 10^4 \text{ N/m}$$

This is about one-tenth the force constant of a spring in an automobile suspension.

EVALUATE: There might seem to be a paradox here. The elastic potential energy at point 2 is

$$\tfrac{1}{2}ky_2^2 = \tfrac{1}{2}(1.06 \times 10^4 \text{ N/m})(-2.00 \text{ m})^2 = 21,200 \text{ J}$$

This is *more* than the total mechanical energy at point 1:

$$E_1 = K_1 + U_1 = 16,000 \text{ J} + 0 = 16,000 \text{ J}$$

But the friction force *decreased* the mechanical energy of the system by 34,000 J between points 1 and 2. Did energy appear from nowhere? No. At point 2, which is below the origin, there is also *negative* gravitational potential energy $mgy_2 = -39,200 \text{ J}$. The total mechanical energy at point 2 is therefore not 21,200 J but rather

$$E_2 = K_2 + U_2 = 0 + \tfrac{1}{2}ky_2^2 + mgy_2$$

$$= 0 + 21,200 \text{ J} + (-39,200 \text{ J}) = -18,000 \text{ J}$$

This is just the initial mechanical energy of 16,000 J minus 34,000 J lost to friction.

Will the elevator stay at the bottom of the shaft? At point 2 the compressed spring exerts an upward force of magnitude $F_{\text{spring}} = (1.06 \times 10^4 \text{ N/m})(2.00 \text{ m}) = 21,200 \text{ N}$, while the downward force of gravity is only $w = mg = (2000 \text{ kg})(9.80 \text{ m/s}^2) = 19,600 \text{ N}$. If there were no friction, there would be a net upward force of $21,200 \text{ N} - 19,600 \text{ N} = 1600 \text{ N}$, and the elevator would rebound. But the safety clamp can exert a kinetic friction force of 17,000 N, and it can presumably exert a maximum static friction force greater than that. Hence the clamp will keep the elevator from rebounding.

Test Your Understanding of Section 7.2 Consider the situation in Example 7.9 at the instant when the elevator is still moving downward and the spring is compressed by 1.00 m. Which of the energy bar graphs in the figure most accurately shows the kinetic energy K, gravitational potential energy U_{grav}, and elastic potential energy U_{el} at this instant?

(i)

(ii)

(iii)

(iv)

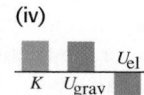

7.3 Conservative and Nonconservative Forces

In our discussions of potential energy we have talked about "storing" kinetic energy by converting it to potential energy. We always have in mind that later we may retrieve it again as kinetic energy. For example, when you throw a ball up in the air, it slows down as kinetic energy is converted to gravitational potential energy. But on the way down, the conversion is reversed, and the ball speeds up as potential energy is converted back to kinetic energy. If there is no air resistance, the ball is moving just as fast when you catch it as when you threw it.

Another example is a glider moving on a frictionless horizontal air track that runs into a spring bumper at the end of the track. The glider stops as it compresses the spring and then bounces back. If there is no friction, the glider ends up with the same speed and kinetic energy it had before the collision. Again, there is a two-way conversion from kinetic to potential energy and back. In both cases we can define a potential-energy function so that the total mechanical energy, kinetic plus potential, is constant or *conserved* during the motion.

Conservative Forces

A force that offers this opportunity of two-way conversion between kinetic and potential energies is called a **conservative force.** We have seen two examples of

7.18 The work done by a conservative force such as gravity depends only on the end points of a path, not on the specific path taken between those points.

Because the gravitational force is conservative, the work it does is the same for all three paths.

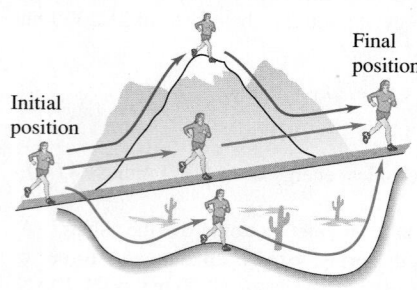

Initial position

Final position

MasteringPHYSICS

PhET: The Ramp

conservative forces: the gravitational force and the spring force. (Later in this book we will study another conservative force, the electric force between charged objects.) An essential feature of conservative forces is that their work is always *reversible*. Anything that we deposit in the energy "bank" can later be withdrawn without loss. Another important aspect of conservative forces is that a body may move from point 1 to point 2 by various paths, but the work done by a conservative force is the same for all of these paths (Fig. 7.18). Thus, if a body stays close to the surface of the earth, the gravitational force $m\vec{g}$ is independent of height, and the work done by this force depends only on the change in height. If the body moves around a closed path, ending at the same point where it started, the *total* work done by the gravitational force is always zero.

The work done by a conservative force *always* has four properties:

1. It can be expressed as the difference between the initial and final values of a *potential-energy* function.
2. It is reversible.
3. It is independent of the path of the body and depends only on the starting and ending points.
4. When the starting and ending points are the same, the total work is zero.

When the *only* forces that do work are conservative forces, the total mechanical energy $E = K + U$ is constant.

Nonconservative Forces

Not all forces are conservative. Consider the friction force acting on the crate sliding on a ramp in Example 7.6 (Section 7.1). When the body slides up and then back down to the starting point, the total work done on it by the friction force is *not* zero. When the direction of motion reverses, so does the friction force, and friction does *negative* work in *both* directions. When a car with its brakes locked skids across the pavement with decreasing speed (and decreasing kinetic energy), the lost kinetic energy cannot be recovered by reversing the motion or in any other way, and mechanical energy is *not* conserved. There is *no* potential-energy function for the friction force.

In the same way, the force of fluid resistance (see Section 5.3) is not conservative. If you throw a ball up in the air, air resistance does negative work on the ball while it's rising *and* while it's descending. The ball returns to your hand with less speed and less kinetic energy than when it left, and there is no way to get back the lost mechanical energy.

A force that is not conservative is called a **nonconservative force.** The work done by a nonconservative force *cannot* be represented by a potential-energy function. Some nonconservative forces, like kinetic friction or fluid resistance, cause mechanical energy to be lost or dissipated; a force of this kind is called a **dissipative force.** There are also nonconservative forces that *increase* mechanical energy. The fragments of an exploding firecracker fly off with very large kinetic energy, thanks to a chemical reaction of gunpowder with oxygen. The forces unleashed by this reaction are nonconservative because the process is not reversible. (The fragments never spontaneously reassemble themselves into a complete firecracker!)

Example 7.10 Frictional work depends on the path

You are rearranging your furniture and wish to move a 40.0-kg futon 2.50 m across the room. A heavy coffee table, which you don't want to move, blocks this straight-line path. Instead, you slide the futon along a dogleg path; the doglegs are 2.00 m and 1.50 m long. How much more work must you do to push the futon along the dogleg path than along the straight-line path? The coefficient of kinetic friction is $\mu_k = 0.200$.

SOLUTION

IDENTIFY and SET UP: Here both you and friction do work on the futon, so we must use the energy relationship that includes "other" forces. We'll use this relationship to find a connection between the work that *you* do and the work that *friction* does. Figure 7.19 shows our sketch. The futon is at rest at both point 1 and point 2, so

7.19 Our sketch for this problem.

$K_1 = K_2 = 0$. There is no elastic potential energy (there are no springs), and the gravitational potential energy does not change because the futon moves only horizontally, so $U_1 = U_2$. From Eq. (7.14) it follows that $W_{other} = 0$. That "other" work done on the futon is the sum of the positive work you do, W_{you}, and the negative work done by friction, W_{fric}. Since the sum of these is zero, we have

$$W_{you} = -W_{fric}$$

Thus we'll calculate the work done by friction to determine W_{you}.

EXECUTE: The floor is horizontal, so the normal force on the futon equals its weight mg and the magnitude of the friction force is $f_k = \mu_k n = \mu_k mg$. The work you do over each path is then

$$\begin{aligned} W_{you} &= -W_{fric} = -(-f_k s) = +\mu_k mgs \\ &= (0.200)(40.0\ \text{kg})(9.80\ \text{m/s}^2)(2.50\ \text{m}) \\ &= 196\ \text{J} \quad \text{(straight-line path)} \end{aligned}$$

$$\begin{aligned} W_{you} &= -W_{fric} = +\mu_k mgs \\ &= (0.200)(40.0\ \text{kg})(9.80\ \text{m/s}^2)(2.00\ \text{m} + 1.50\ \text{m}) \\ &= 274\ \text{J} \quad \text{(dogleg path)} \end{aligned}$$

The extra work you must do is $274\ \text{J} - 196\ \text{J} = 78\ \text{J}$.

EVALUATE: Friction does different amounts of work on the futon, $-196\ \text{J}$ and $-274\ \text{J}$, on these different paths between points 1 and 2. Hence friction is a *nonconservative* force.

Example 7.11 Conservative or nonconservative?

In a region of space the force on an electron is $\vec{F} = Cx\hat{\jmath}$, where C is a positive constant. The electron moves around a square loop in the xy-plane (Fig. 7.20). Calculate the work done on the electron by the force $\vec{F}$ during a counterclockwise trip around the square. Is this force conservative or nonconservative?

SOLUTION

IDENTIFY and SET UP: The force $\vec{F}$ is not constant, and in general it is not in the same direction as the displacement. To calculate the work done by $\vec{F}$, we'll use the general expression for work, Eq. (6.14):

$$W = \int_{P_1}^{P_2} \vec{F} \cdot d\vec{l}$$

where $d\vec{l}$ is an infinitesimal displacement. We'll calculate the work done on each leg of the square separately, and add the results to find the work done on the round trip. If this round-trip work is zero, force $\vec{F}$ is conservative and can be represented by a potential-energy function.

7.20 An electron moving around a square loop while being acted on by the force $\vec{F} = Cx\hat{\jmath}$.

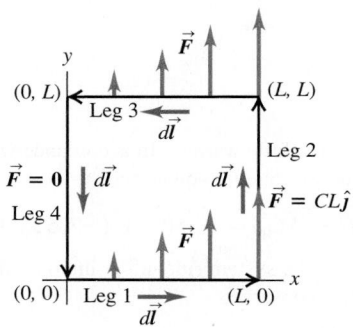

EXECUTE: On the first leg, from $(0, 0)$ to $(L, 0)$, the force is everywhere perpendicular to the displacement. So $\vec{F} \cdot d\vec{l} = 0$, and the work done on the first leg is $W_1 = 0$. The force has the same value $\vec{F} = CL\hat{\jmath}$ everywhere on the second leg, from $(L, 0)$ to (L, L). The displacement on this leg is in the $+y$-direction, so $d\vec{l} = dy\hat{\jmath}$ and

$$\vec{F} \cdot d\vec{l} = CL\hat{\jmath} \cdot dy\hat{\jmath} = CL\ dy$$

The work done on the second leg is then

$$W_2 = \int_{(L, 0)}^{(L, L)} \vec{F} \cdot d\vec{l} = \int_{y=0}^{y=L} CL\ dy = CL \int_0^L dy = CL^2$$

On the third leg, from (L, L) to $(0, L)$, $\vec{F}$ is again perpendicular to the displacement and so $W_3 = 0$. The force is zero on the final leg, from $(0, L)$ to $(0, 0)$, so $W_4 = 0$. The work done by $\vec{F}$ on the round trip is therefore

$$W = W_1 + W_2 + W_3 + W_4 = 0 + CL^2 + 0 + 0 = CL^2$$

The starting and ending points are the same, but the total work done by $\vec{F}$ is not zero. This is a *nonconservative* force; it *cannot* be represented by a potential-energy function.

EVALUATE: Because W is positive, the mechanical energy *increases* as the electron goes around the loop. This is not a mathematical curiosity; it's a much-simplified description of what happens in an electrical generating plant. There, a loop of wire is moved through a magnetic field, which gives rise to a nonconservative force similar to the one here. Electrons in the wire gain energy as they move around the loop, and this energy is carried via transmission lines to the consumer. (We'll discuss how this works in Chapter 29.)

If the electron went *clockwise* around the loop, $\vec{F}$ would be unaffected but the direction of each infinitesimal displacement $d\vec{l}$ would be reversed. Thus the sign of work would also reverse, and the work for a clockwise round trip would be $W = -CL^2$. This is a different behavior than the nonconservative friction force. The work done by friction on a body that slides in any direction over a stationary surface is always negative (see Example 7.6 in Section 7.1).

The Law of Conservation of Energy

Nonconservative forces cannot be represented in terms of potential energy. But we can describe the effects of these forces in terms of kinds of energy other than kinetic and potential energy. When a car with locked brakes skids to a stop, the tires and the road surface both become hotter. The energy associated with this change in the state of the materials is called **internal energy.** Raising the temperature of a body increases its internal energy; lowering the body's temperature decreases its internal energy.

To see the significance of internal energy, let's consider a block sliding on a rough surface. Friction does *negative* work on the block as it slides, and the change in internal energy of the block and surface (both of which get hotter) is *positive*. Careful experiments show that the increase in the internal energy is *exactly* equal to the absolute value of the work done by friction. In other words,

$$\Delta U_{\text{int}} = -W_{\text{other}}$$

where ΔU_{int} is the change in internal energy. If we substitute this into Eq. (7.7) or (7.14), we find

$$K_1 + U_1 - \Delta U_{\text{int}} = K_2 + U_2$$

Writing $\Delta K = K_2 - K_1$ and $\Delta U = U_2 - U_1$, we can finally express this as

$$\Delta K + \Delta U + \Delta U_{\text{int}} = 0 \qquad \text{(law of conservation of energy)} \qquad (7.15)$$

7.21 When 1 liter of gasoline is burned in an automotive engine, it releases 3.3×10^7 J of internal energy. Hence $\Delta U_{\text{int}} = -3.3 \times 10^7$ J, where the minus sign means that the amount of energy stored in the gasoline has decreased. This energy can be converted to kinetic energy (making the car go faster) or to potential energy (enabling the car to climb uphill).

This remarkable statement is the general form of the **law of conservation of energy.** In a given process, the kinetic energy, potential energy, and internal energy of a system may all change. But the *sum* of those changes is always zero. If there is a decrease in one form of energy, it is made up for by an increase in the other forms (Fig. 7.21). When we expand our definition of energy to include internal energy, Eq. (7.15) says: *Energy is never created or destroyed; it only changes form.* No exception to this rule has ever been found.

The concept of work has been banished from Eq. (7.15); instead, it suggests that we think purely in terms of the conversion of energy from one form to another. For example, when you throw a baseball straight up, you convert a portion of the internal energy of your molecules to kinetic energy of the baseball. This is converted to gravitational potential energy as the ball climbs and back to kinetic energy as the ball falls. If there is air resistance, part of the energy is used to heat up the air and the ball and increase their internal energy. Energy is converted back to the kinetic form as the ball falls. If you catch the ball in your hand, whatever energy was not lost to the air once again becomes internal energy; the ball and your hand are now warmer than they were at the beginning.

In Chapters 19 and 20, we will study the relationship of internal energy to temperature changes, heat, and work. This is the heart of the area of physics called *thermodynamics.*

Conceptual Example 7.12   **Work done by friction**

Let's return to Example 7.5 (Section 7.1), in which Throcky skateboards down a curved ramp. He starts with zero kinetic energy and 735 J of potential energy, and at the bottom he has 450 J of kinetic energy and zero potential energy; hence $\Delta K = +450$ J and $\Delta U = -735$ J. The work $W_{\text{other}} = W_{\text{fric}}$ done by the friction forces is -285 J, so the change in internal energy is $\Delta U_{\text{int}} = -W_{\text{other}} = +285$ J. The skateboard wheels and bearings

and the ramp all get a little warmer. In accordance with Eq. (7.15), the sum of the energy changes equals zero:

$$\Delta K + \Delta U + \Delta U_{\text{int}} = +450 \text{ J} + (-735 \text{ J}) + 285 \text{ J} = 0$$

The total energy of the system (including internal, nonmechanical forms of energy) is conserved.

Test Your Understanding of Section 7.3 In a hydroelectric generating station, falling water is used to drive turbines ("water wheels"), which in turn run electric generators. Compared to the amount of gravitational potential energy released by the falling water, how much electrical energy is produced? (i) the same; (ii) more; (iii) less.

7.4 Force and Potential Energy

For the two kinds of conservative forces (gravitational and elastic) we have studied, we started with a description of the behavior of the *force* and derived from that an expression for the *potential energy*. For example, for a body with mass m in a uniform gravitational field, the gravitational force is $F_y = -mg$. We found that the corresponding potential energy is $U(y) = mgy$. To stretch an ideal spring by a distance x, we exert a force equal to $+kx$. By Newton's third law the force that an ideal spring exerts on a body is opposite this, or $F_x = -kx$. The corresponding potential energy function is $U(x) = \frac{1}{2}kx^2$.

In studying physics, however, you'll encounter situations in which you are given an expression for the *potential energy* as a function of position and have to find the corresponding *force*. We'll see several examples of this kind when we study electric forces later in this book: It's often far easier to calculate the electric potential energy first and then determine the corresponding electric force afterward.

Here's how we find the force that corresponds to a given potential-energy expression. First let's consider motion along a straight line, with coordinate x. We denote the x-component of force, a function of x, by $F_x(x)$, and the potential energy as $U(x)$. This notation reminds us that both F_x and U are *functions* of x. Now we recall that in any displacement, the work W done by a conservative force equals the negative of the change ΔU in potential energy:

$$W = -\Delta U$$

Let's apply this to a small displacement Δx. The work done by the force $F_x(x)$ during this displacement is approximately equal to $F_x(x)\,\Delta x$. We have to say "approximately" because $F_x(x)$ may vary a little over the interval Δx. But it is at least approximately true that

$$F_x(x)\,\Delta x = -\Delta U \qquad \text{and} \qquad F_x(x) = -\frac{\Delta U}{\Delta x}$$

You can probably see what's coming. We take the limit as $\Delta x \to 0$; in this limit, the variation of F_x becomes negligible, and we have the exact relationship

$$F_x(x) = -\frac{dU(x)}{dx} \qquad \text{(force from potential energy, one dimension)} \quad \text{(7.16)}$$

This result makes sense; in regions where $U(x)$ changes most rapidly with x (that is, where $dU(x)/dx$ is large), the greatest amount of work is done during a given displacement, and this corresponds to a large force magnitude. Also, when $F_x(x)$ is in the positive x-direction, $U(x)$ *decreases* with increasing x. So $F_x(x)$ and $dU(x)/dx$ should indeed have opposite signs. The physical meaning of Eq. (7.16) is that *a conservative force always acts to push the system toward lower potential energy.*

As a check, let's consider the function for elastic potential energy, $U(x) = \frac{1}{2}kx^2$. Substituting this into Eq. (7.16) yields

$$F_x(x) = -\frac{d}{dx}\left(\tfrac{1}{2}kx^2\right) = -kx$$

which is the correct expression for the force exerted by an ideal spring (Fig. 7.22a). Similarly, for gravitational potential energy we have $U(y) = mgy$; taking care to change x to y for the choice of axis, we get $F_y = -dU/dy = -d(mgy)/dy = -mg$, which is the correct expression for gravitational force (Fig. 7.22b).

7.22 A conservative force is the negative derivative of the corresponding potential energy.

(a) Spring potential energy and force as functions of x

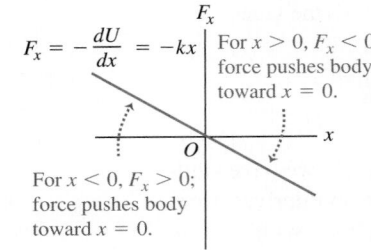

(b) Gravitational potential energy and force as functions of y

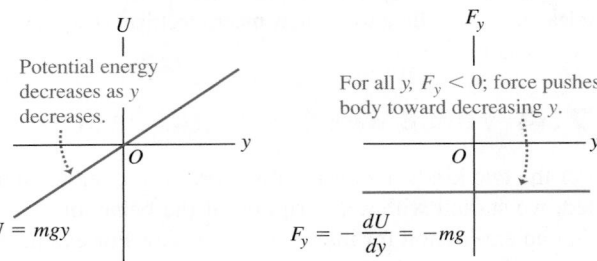

Example 7.13 An electric force and its potential energy

An electrically charged particle is held at rest at the point $x = 0$; a second particle with equal charge is free to move along the positive x-axis. The potential energy of the system is $U(x) = C/x$, where C is a positive constant that depends on the magnitude of the charges. Derive an expression for the x-component of force acting on the movable particle as a function of its position.

SOLUTION

IDENTIFY and SET UP: We are given the potential-energy function $U(x)$. We'll find the corresponding force function using Eq. (7.16), $F_x(x) = -dU(x)/dx$.

EXECUTE: The derivative of $1/x$ with respect to x is $-1/x^2$. So for $x > 0$ the force on the movable charged particle $x > 0$ is

$$F_x(x) = -\frac{dU(x)}{dx} = -C\left(-\frac{1}{x^2}\right) = \frac{C}{x^2}$$

EVALUATE: The x-component of force is positive, corresponding to a repulsion between like electric charges. Both the potential energy and the force are very large when the particles are close together (small x), and both get smaller as the particles move farther apart (large x); the force pushes the movable particle toward large positive values of x, where the potential energy is lower. (We'll study electric forces in detail in Chapter 21.)

Force and Potential Energy in Three Dimensions

We can extend this analysis to three dimensions, where the particle may move in the x-, y-, or z-direction, or all at once, under the action of a conservative force that has components F_x, F_y, and F_z. Each component of force may be a function of the coordinates x, y, and z. The potential-energy function U is also a function of all three space coordinates. We can now use Eq. (7.16) to find each component of force. The potential-energy change ΔU when the particle moves a small distance Δx in the x-direction is again given by $-F_x \Delta x$; it doesn't depend on F_y and F_z, which represent force components that are perpendicular to the displacement and do no work. So we again have the approximate relationship

$$F_x = -\frac{\Delta U}{\Delta x}$$

The y- and z-components of force are determined in exactly the same way:

$$F_y = -\frac{\Delta U}{\Delta y} \qquad F_z = -\frac{\Delta U}{\Delta z}$$

To make these relationships exact, we take the limits $\Delta x \to 0$, $\Delta y \to 0$, and $\Delta z \to 0$ so that these ratios become derivatives. Because U may be a function of all three coordinates, we need to remember that when we calculate each of these derivatives, only one coordinate changes at a time. We compute the derivative of U with respect to x by assuming that y and z are constant and only x varies, and so on. Such a derivative is called a *partial derivative*. The usual notation for a partial derivative is $\partial U/\partial x$ and so on; the symbol ∂ is a modified d. So we write

$$F_x = -\frac{\partial U}{\partial x} \qquad F_y = -\frac{\partial U}{\partial y} \qquad F_z = -\frac{\partial U}{\partial z} \qquad \begin{matrix} \text{(force from} \\ \text{potential energy)} \end{matrix} \qquad (7.17)$$

We can use unit vectors to write a single compact vector expression for the force $\vec{F}$:

$$\vec{F} = -\left(\frac{\partial U}{\partial x}\hat{\imath} + \frac{\partial U}{\partial y}\hat{\jmath} + \frac{\partial U}{\partial z}\hat{k}\right) \qquad \text{(force from potential energy)} \qquad (7.18)$$

The expression inside the parentheses represents a particular operation on the function U, in which we take the partial derivative of U with respect to each coordinate, multiply by the corresponding unit vector, and then take the vector sum. This operation is called the **gradient** of U and is often abbreviated as $\vec{\nabla}U$. Thus the force is the negative of the gradient of the potential-energy function:

$$\vec{F} = -\vec{\nabla}U \qquad (7.19)$$

As a check, let's substitute into Eq. (7.19) the function $U = mgy$ for gravitational potential energy:

$$\vec{F} = -\vec{\nabla}(mgy) = -\left(\frac{\partial(mgy)}{\partial x}\hat{\imath} + \frac{\partial(mgy)}{\partial y}\hat{\jmath} + \frac{\partial(mgy)}{\partial z}\hat{k}\right) = (-mg)\hat{\jmath}$$

This is just the familiar expression for the gravitational force.

Application **Topography and Potential Energy Gradient**
The greater the elevation of a hiker in Canada's Banff National Park, the greater is the gravitational potential energy U_{grav}. Think of an x-axis that runs horizontally from west to east and a y-axis that runs horizontally from south to north. Then the function $U_{grav}(x, y)$ tells us the elevation as a function of position in the park. Where the mountains have steep slopes, $\vec{F} = -\vec{\nabla}U_{grav}$ has a large magnitude and there's a strong force pushing you along the mountain's surface toward a region of lower elevation (and hence lower U_{grav}). There's zero force along the surface of the lake, which is all at the same elevation. Hence U_{grav} is constant at all points on the lake surface, and $\vec{F} = -\vec{\nabla}U_{grav} = \mathbf{0}$.

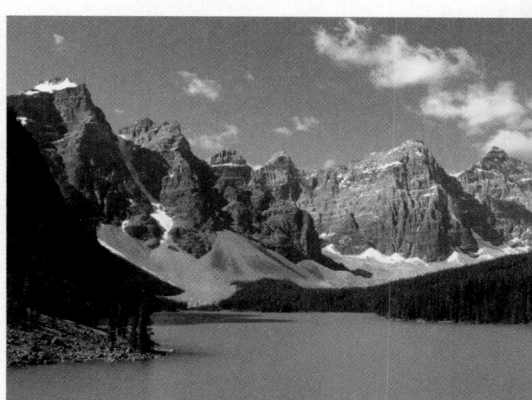

Example 7.14 **Force and potential energy in two dimensions**

A puck with coordinates x and y slides on a level, frictionless air-hockey table. It is acted on by a conservative force described by the potential-energy function

$$U(x, y) = \tfrac{1}{2}k(x^2 + y^2)$$

Find a vector expression for the force acting on the puck, and find an expression for the magnitude of the force.

SOLUTION

IDENTIFY and SET UP: Starting with the function $U(x, y)$, we need to find the vector components and magnitude of the corresponding force $\vec{F}$. We'll find the components using Eq. (7.18). The function U doesn't depend on z, so the partial derivative of U with respect to z is $\partial U/\partial z = 0$ and the force has no z-component. We'll determine the magnitude F of the force using $F = \sqrt{F_x^2 + F_y^2}$.

EXECUTE: The x- and y-components of $\vec{F}$ are

$$F_x = -\frac{\partial U}{\partial x} = -kx \qquad F_y = -\frac{\partial U}{\partial y} = -ky$$

From Eq. (7.18), the vector expression for the force is

$$\vec{F} = (-kx)\hat{\imath} + (-ky)\hat{\jmath} = -k(x\hat{\imath} + y\hat{\jmath})$$

The magnitude of the force is

$$F = \sqrt{(-kx)^2 + (-ky)^2} = k\sqrt{x^2 + y^2} = kr$$

EVALUATE: Because $x\hat{\imath} + y\hat{\jmath}$ is just the position vector $\vec{r}$ of the particle, we can rewrite our result as $\vec{F} = -k\vec{r}$. This represents a force that is opposite in direction to the particle's position vector—that is, a force directed toward the origin, $r = 0$. This is the force that would be exerted on the puck if it were attached to one end of a spring that obeys Hooke's law and has a negligibly small unstretched length compared to the other distances in the problem. (The other end is attached to the air-hockey table at $r = 0$.)

To check our result, note that $U = \tfrac{1}{2}kr^2$, where $r^2 = x^2 + y^2$. We can find the force from this expression using Eq. (7.16) with x replaced by r:

$$F_r = -\frac{dU}{dr} = -\frac{d}{dr}\left(\tfrac{1}{2}kr^2\right) = -kr$$

As we found above, the force has magnitude kr; the minus sign indicates that the force is toward the origin (at $r = 0$).

7.23 (a) A glider on an air track. The spring exerts a force $F_x = -kx$. (b) The potential-energy function.

(a)

The limits of the glider's motion are at $x = A$ and $x = -A$.

(b)

On the graph, the limits of motion are the points where the U curve intersects the horizontal line representing total mechanical energy E.

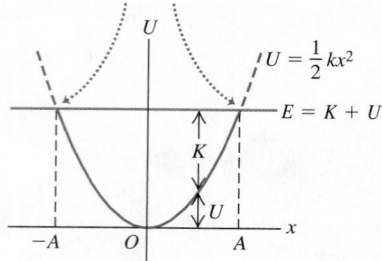

Application Acrobats in Equilibrium

Each of these acrobats is in *unstable* equilibrium. The gravitational potential energy is lower no matter which way an acrobat tips, so if she begins to fall she will keep on falling. Staying balanced requires the acrobats' constant attention.

Test Your Understanding of Section 7.4 A particle moving along the x-axis is acted on by a conservative force F_x. At a certain point, the force is zero. (a) Which of the following statements about the value of the potential-energy function $U(x)$ at that point is correct? (i) $U(x) = 0$; (ii) $U(x) > 0$; (iii) $U(x) < 0$; (iv) not enough information is given to decide. (b) Which of the following statements about the value of the derivative of $U(x)$ at that point is correct? (i) $dU(x)/dx = 0$; (ii) $dU(x)/dx > 0$; (iii) $dU(x)/dx < 0$; (iv) not enough information is given to decide.

7.5 Energy Diagrams

When a particle moves along a straight line under the action of a conservative force, we can get a lot of insight into its possible motions by looking at the graph of the potential-energy function $U(x)$. Figure 7.23a shows a glider with mass m that moves along the x-axis on an air track. The spring exerts on the glider a force with x-component $F_x = -kx$. Figure 7.23b is a graph of the corresponding potential-energy function $U(x) = \frac{1}{2}kx^2$. If the elastic force of the spring is the *only* horizontal force acting on the glider, the total mechanical energy $E = K + U$ is constant, independent of x. A graph of E as a function of x is thus a straight horizontal line. We use the term **energy diagram** for a graph like this, which shows both the potential-energy function $U(x)$ and the energy of the particle subjected to the force that corresponds to $U(x)$.

The vertical distance between the U and E graphs at each point represents the difference $E - U$, equal to the kinetic energy K at that point. We see that K is greatest at $x = 0$. It is zero at the values of x where the two graphs cross, labeled A and $-A$ in the diagram. Thus the speed v is greatest at $x = 0$, and it is zero at $x = \pm A$, the points of *maximum* possible displacement from $x = 0$ for a given value of the total energy E. The potential energy U can never be greater than the total energy E; if it were, K would be negative, and that's impossible. The motion is a back-and-forth oscillation between the points $x = A$ and $x = -A$.

At each point, the force F_x on the glider is equal to the negative of the slope of the $U(x)$ curve: $F_x = -dU/dx$ (see Fig. 7.22a). When the particle is at $x = 0$, the slope and the force are zero, so this is an *equilibrium* position. When x is positive, the slope of the $U(x)$ curve is positive and the force F_x is negative, directed toward the origin. When x is negative, the slope is negative and F_x is positive, again directed toward the origin. Such a force is called a *restoring force;* when the glider is displaced to either side of $x = 0$, the force tends to "restore" it back to $x = 0$. An analogous situation is a marble rolling around in a round-bottomed bowl. We say that $x = 0$ is a point of **stable equilibrium.** More generally, *any minimum in a potential-energy curve is a stable equilibrium position.*

Figure 7.24a shows a hypothetical but more general potential-energy function $U(x)$. Figure 7.24b shows the corresponding force $F_x = -dU/dx$. Points x_1 and x_3 are stable equilibrium points. At each of these points, F_x is zero because the slope of the $U(x)$ curve is zero. When the particle is displaced to either side, the force pushes back toward the equilibrium point. The slope of the $U(x)$ curve is also zero at points x_2 and x_4, and these are also equilibrium points. But when the particle is displaced a little to the right of either point, the slope of the $U(x)$ curve becomes negative, corresponding to a positive F_x that tends to push the particle still farther from the point. When the particle is displaced a little to the left, F_x is negative, again pushing away from equilibrium. This is analogous to a marble rolling on the top of a bowling ball. Points x_2 and x_4 are called **unstable equilibrium** points; *any maximum in a potential-energy curve is an unstable equilibrium position.*

7.24 The maxima and minima of a potential-energy function $U(x)$ correspond to points where $F_x = 0$.

(a) A hypothetical potential-energy function $U(x)$

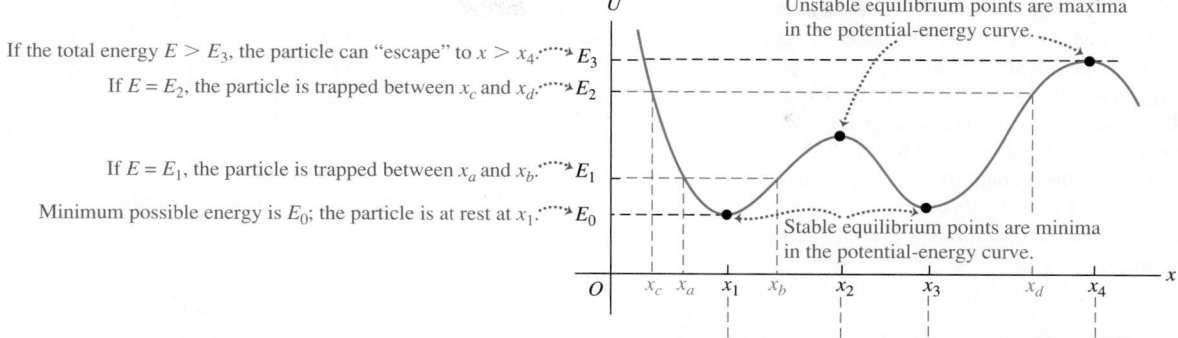

If the total energy $E > E_3$, the particle can "escape" to $x > x_4$. $\longrightarrow E_3$

Unstable equilibrium points are maxima in the potential-energy curve.

If $E = E_2$, the particle is trapped between x_c and x_d. $\longrightarrow E_2$

If $E = E_1$, the particle is trapped between x_a and x_b. $\longrightarrow E_1$

Minimum possible energy is E_0; the particle is at rest at x_1. $\longrightarrow E_0$

Stable equilibrium points are minima in the potential-energy curve.

(b) The corresponding x-component of force $F_x(x) = -dU(x)/dx$

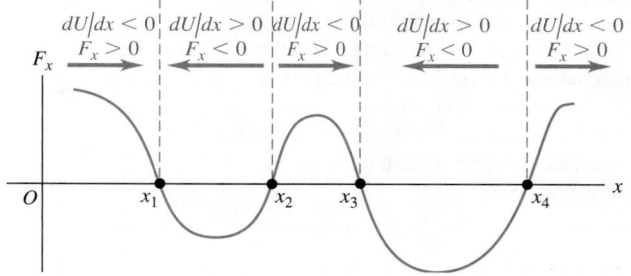

$dU/dx < 0$ $dU/dx > 0$ $dU/dx < 0$ $dU/dx > 0$ $dU/dx < 0$
$F_x > 0$ $F_x < 0$ $F_x > 0$ $F_x < 0$ $F_x > 0$

CAUTION **Potential energy and the direction of a conservative force** The direction of the force on a body is *not* determined by the sign of the potential energy U. Rather, it's the sign of $F_x = -dU/dx$ that matters. As we discussed in Section 7.1, the physically significant quantity is the *difference* in the values of U between two points, which is just what the derivative $F_x = -dU/dx$ measures. This means that you can always add a constant to the potential-energy function without changing the physics of the situation.

Mastering**PHYSICS**

PhET: Energy Skate Park

If the total energy is E_1 and the particle is initially near x_1, it can move only in the region between x_a and x_b determined by the intersection of the E_1 and U graphs (Fig. 7.24a). Again, U cannot be greater than E_1 because K can't be negative. We speak of the particle as moving in a *potential well,* and x_a and x_b are the *turning points* of the particle's motion (since at these points, the particle stops and reverses direction). If we increase the total energy to the level E_2, the particle can move over a wider range, from x_c to x_d. If the total energy is greater than E_3, the particle can "escape" and move to indefinitely large values of x. At the other extreme, E_0 represents the least possible total energy the system can have.

Test Your Understanding of Section 7.5 The curve in Fig. 7.24b has a maximum at a point between x_2 and x_3. Which statement correctly describes what happens to the particle when it is at this point? (i) The particle's acceleration is zero. (ii) The particle accelerates in the positive x-direction; the magnitude of the acceleration is less than at any other point between x_2 and x_3. (iii) The particle accelerates in the positive x-direction; the magnitude of the acceleration is greater than at any other point between x_2 and x_3. (iv) The particle accelerates in the negative x-direction; the magnitude of the acceleration is less than at any other point between x_2 and x_3. (v) The particle accelerates in the negative x-direction; the magnitude of the acceleration is greater than at any other point between x_2 and x_3.

Gravitational potential energy and elastic potential energy: The work done on a particle by a constant gravitational force can be represented as a change in the gravitational potential energy $U_{grav} = mgy$. This energy is a shared property of the particle and the earth. A potential energy is also associated with the elastic force $F_x = -kx$ exerted by an ideal spring, where x is the amount of stretch or compression. The work done by this force can be represented as a change in the elastic potential energy of the spring, $U_{el} = \frac{1}{2}kx^2$.

$$W_{grav} = mgy_1 - mgy_2$$
$$= U_{grav,1} - U_{grav,2}$$
$$= -\Delta U_{grav} \quad (7.1), (7.3)$$

$$W_{el} = \frac{1}{2}kx_1^2 - \frac{1}{2}kx_2^2$$
$$= U_{el,1} - U_{el,2} = -\Delta U_{el} \quad (7.10)$$

When total mechanical energy is conserved: The total potential energy U is the sum of the gravitational and elastic potential energy: $U = U_{grav} + U_{el}$. If no forces other than the gravitational and elastic forces do work on a particle, the sum of kinetic and potential energy is conserved. This sum $E = K + U$ is called the total mechanical energy. (See Examples 7.1, 7.3, 7.4, and 7.7.)

$$K_1 + U_1 = K_2 + U_2 \quad (7.4), (7.11)$$

When total mechanical energy is not conserved: When forces other than the gravitational and elastic forces do work on a particle, the work W_{other} done by these other forces equals the change in total mechanical energy (kinetic energy plus total potential energy). (See Examples 7.2, 7.5, 7.6, 7.8, and 7.9.)

$$K_1 + U_1 + W_{other} = K_2 + U_2 \quad (7.14)$$

Conservative forces, nonconservative forces, and the law of conservation of energy: All forces are either conservative or nonconservative. A conservative force is one for which the work–kinetic energy relationship is completely reversible. The work of a conservative force can always be represented by a potential-energy function, but the work of a nonconservative force cannot. The work done by nonconservative forces manifests itself as changes in the internal energy of bodies. The sum of kinetic, potential, and internal energy is always conserved. (See Examples 7.10–7.12.)

$$\Delta K + \Delta U + \Delta U_{int} = 0 \quad (7.15)$$

As friction slows block, mechanical energy is converted to internal energy of block and ramp.

Determining force from potential energy: For motion along a straight line, a conservative force $F_x(x)$ is the negative derivative of its associated potential-energy function U. In three dimensions, the components of a conservative force are negative partial derivatives of U. (See Examples 7.13 and 7.14.)

$$F_x(x) = -\frac{dU(x)}{dx} \quad (7.16)$$

$$F_x = -\frac{\partial U}{\partial x} \quad F_y = -\frac{\partial U}{\partial y} \quad (7.17)$$

$$F_z = -\frac{\partial U}{\partial z}$$

$$\vec{F} = -\left(\frac{\partial U}{\partial x}\hat{i} + \frac{\partial U}{\partial y}\hat{j} + \frac{\partial U}{\partial z}\hat{k}\right) \quad (7.18)$$

| BRIDGING PROBLEM | A Spring and Friction on an Incline |

A 2.00-kg package is released on a 53.1° incline, 4.00 m from a long spring with force constant $1.20 \times 10^2 \, \text{N/m}$ that is attached at the bottom of the incline (Fig. 7.25). The coefficients of friction between the package and incline are $\mu_s = 0.400$ and $\mu_k = 0.200$. The mass of the spring is negligible.

7.25 The initial situation.

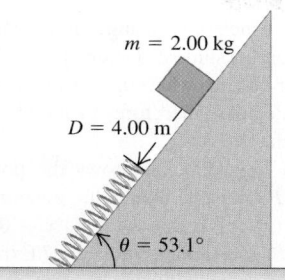

$m = 2.00 \, \text{kg}$

$D = 4.00 \, \text{m}$

$\theta = 53.1°$

(a) What is the maximum compression of the spring? (b) The package rebounds up the incline. How close does it get to its original position? (c) What is the change in the internal energy of the package and incline from when the package is released to when it rebounds to its maximum height?

SOLUTION GUIDE

See MasteringPhysics® study area for a Video Tutor solution.

IDENTIFY and SET UP

1. This problem involves the gravitational force, a spring force, and the friction force, as well as the normal force that acts on the package. Since the spring force isn't constant, you'll have to use energy methods. Is mechanical energy conserved during any part of the motion? Why or why not?

2. Draw free-body diagrams for the package as it is sliding down the incline and sliding back up the incline. Include your choice of coordinate axis. (*Hint:* If you choose $x = 0$ to be at the end of the uncompressed spring, you'll be able to use $U_{\text{el}} = \frac{1}{2}kx^2$ for the elastic potential energy of the spring.)

3. Label the three critical points in the package's motion: its starting position, its position when it comes to rest with the spring maximally compressed, and its position when it's rebounded as far as possible up the incline. (*Hint:* You can assume that the

package is no longer in contact with the spring at the last of these positions. If this turns out to be incorrect, you'll calculate a value of x that tells you the spring is still partially compressed at this point.)

4. Make a list of the unknown quantities and decide which of these are the target variables.

EXECUTE

5. Find the magnitude of the friction force that acts on the package. Does the magnitude of this force depend on whether the package is moving up or down the incline, or on whether or not the package is in contact with the spring? Does the *direction* of the normal force depend on any of these?

6. Write the general energy equation for the motion of the package between the first two points you labeled in step 3. Use this equation to solve for the distance that the spring is compressed when the package is at its lowest point. (*Hint:* You'll have to solve a quadratic equation. To decide which of the two solutions of this equation is the correct one, remember that the distance the spring is compressed is positive.)

7. Write the general energy equation for the motion of the package between the second and third points you labeled in step 3. Use this equation to solve for how far the package rebounds.

8. Calculate the change in internal energy for the package's trip down and back up the incline. Remember that the amount the internal energy *increases* is equal to the amount the total mechanical energy *decreases*.

EVALUATE

9. Was it correct to assume in part (b) that the package is no longer in contact with the spring when it reaches it maximum rebound height?

10. Check your result for part (c) by finding the total work done by the force of friction over the entire trip. Is this in accordance with your result from step 8?

| Problems | For instructor-assigned homework, go to www.masteringphysics.com |

•, ••, •••: Problems of increasing difficulty. **CP**: Cumulative problems incorporating material from earlier chapters. **CALC**: Problems requiring calculus. **BIO**: Biosciences problems.

DISCUSSION QUESTIONS

Q7.1 A baseball is thrown straight up with initial speed v_0. If air resistance cannot be ignored, when the ball returns to its initial height its speed is less than v_0. Explain why, using energy concepts.

Q7.2 A projectile has the same initial kinetic energy no matter what the angle of projection. Why doesn't it rise to the same maximum height in each case?

Q7.3 An object is released from rest at the top of a ramp. If the ramp is frictionless, does the object's speed at the bottom of the ramp depend on the shape of the ramp or just on its height? Explain. What if the ramp is *not* frictionless?

Q7.4 An egg is released from rest from the roof of a building and falls to the ground. Its fall is observed by a student on the roof of the building, who uses coordinates with origin at the roof, and by a student on the ground, who uses coordinates with origin at the ground. Do the two students assign the same or different values to the initial gravitational potential energy, the final gravitational potential energy, the change in gravitational potential energy, and the kinetic energy of the egg just before it strikes the ground? Explain.

Q7.5 A physics teacher had a bowling ball suspended from a very long rope attached to the high ceiling of a large lecture hall. To illustrate his faith in conservation of energy, he would back up to one side of the stage, pull the ball far to one side until the taut rope brought it just to the end of his nose, and then release it. The massive ball would swing in a mighty arc across the stage and then return to stop momentarily just in front of the nose of the stationary, unflinching teacher. However, one day after the demonstration he looked up just in time to see a student at the other side of the stage *push* the ball away from his nose as he tried to duplicate the demonstration. Tell the rest of the story and explain the reason for the potentially tragic outcome.

Q7.6 Lost Energy? The principle of the conservation of energy tells us that energy is never lost, but only changes from one form to another. Yet in many ordinary situations, energy may appear to be lost. In each case, explain what happens to the "lost" energy. (a) A box sliding on the floor comes to a halt due to friction. How did friction take away its kinetic energy, and what happened to that energy? (b) A car stops when you apply the brakes. What happened to its kinetic energy? (c) Air resistance uses up some of the original gravitational potential energy of a falling object. What type of energy did the "lost" potential energy become? (d) When a returning space shuttle touches down on the runway, it has lost almost all its kinetic energy and gravitational potential energy. Where did all that energy go?

Q7.7 Is it possible for a frictional force to *increase* the mechanical energy of a system? If so, give examples.

Q7.8 A woman bounces on a trampoline, going a little higher with each bounce. Explain how she increases the total mechanical energy.

Q7.9 Fractured Physics. People often call their electric bill a *power* bill, yet the quantity on which the bill is based is expressed in *kilowatt-hours*. What are people really being billed for?

Q7.10 A rock of mass m and a rock of mass $2m$ are both released from rest at the same height and feel no air resistance as they fall. Which statements about these rocks are true? (There may be more than one correct choice.) (a) Both have the same initial gravitational potential energy. (b) Both have the same kinetic energy when they reach the ground. (c) Both reach the ground with the same speed. (d) When it reaches the ground, the heavier rock has twice the kinetic energy of the lighter one. (e) When it reaches the ground, the heavier rock has four times the kinetic energy of the lighter one.

Q7.11 On a friction-free ice pond, a hockey puck is pressed against (but not attached to) a fixed ideal spring, compressing the spring by a distance x_0. The maximum energy stored in the spring is U_0, the maximum speed the puck gains after being released is v_0, and its maximum kinetic energy is K_0. Now the puck is pressed so it compresses the spring twice as far as before. In this case, (a) what is the maximum potential energy stored in the spring (in terms of U_0), and (b) what are the puck's maximum kinetic energy and speed (in terms of K_0 and x_0)?

Q7.12 When people are cold, they often rub their hands together to warm them up. How does doing this produce heat? Where did the heat come from?

Q7.13 You often hear it said that most of our energy ultimately comes from the sun. Trace each of the following energies back to the sun: (a) the kinetic energy of a jet plane; (b) the potential energy gained by a mountain climber; (c) the electrical energy used to run a computer; (d) the electrical energy from a hydroelectric plant.

Q7.14 A box slides down a ramp and work is done on the box by the forces of gravity and friction. Can the work of each of these forces be expressed in terms of the change in a potential-energy function? For each force explain why or why not.

Q7.15 In physical terms, explain why friction is a nonconservative force. Does it store energy for future use?

Q7.16 A compressed spring is clamped in its compressed position and then is dissolved in acid. What becomes of its potential energy?

Q7.17 Since only changes in potential energy are important in any problem, a student decides to let the elastic potential energy of a spring be zero when the spring is stretched a distance x_1. The student decides, therefore, to let $U = \frac{1}{2}k(x - x_1)^2$. Is this correct? Explain.

Q7.18 Figure 7.22a shows the potential-energy function for the force $F_x = -kx$. Sketch the potential-energy function for the force $F_x = +kx$. For this force, is $x = 0$ a point of equilibrium? Is this equilibrium stable or unstable? Explain.

Q7.19 Figure 7.22b shows the potential-energy function associated with the gravitational force between an object and the earth. Use this graph to explain why objects always fall toward the earth when they are released.

Q7.20 For a system of two particles we often let the potential energy for the force between the particles approach zero as the separation of the particles approaches infinity. If this choice is made, explain why the potential energy at noninfinite separation is positive if the particles repel one another and negative if they attract.

Q7.21 Explain why the points $x = A$ and $x = -A$ in Fig. 7.23b are called *turning points*. How are the values of E and U related at a turning point?

Q7.22 A particle is in *neutral equilibrium* if the net force on it is zero and remains zero if the particle is displaced slightly in any direction. Sketch the potential-energy function near a point of neutral equilibrium for the case of one-dimensional motion. Give an example of an object in neutral equilibrium.

Q7.23 The net force on a particle of mass m has the potential-energy function graphed in Fig. 7.24a. If the total energy is E_1, graph the speed v of the particle versus its position x. At what value of x is the speed greatest? Sketch v versus x if the total energy is E_2.

Q7.24 The potential-energy function for a force $\vec{F}$ is $U = \alpha x^3$, where α is a positive constant. What is the direction of $\vec{F}$?

EXERCISES

Section 7.1 Gravitational Potential Energy

7.1 • In one day, a 75-kg mountain climber ascends from the 1500-m level on a vertical cliff to the top at 2400 m. The next day, she descends from the top to the base of the cliff, which is at an elevation of 1350 m. What is her change in gravitational potential energy (a) on the first day and (b) on the second day?

7.2 • **BIO How High Can We Jump?** The maximum height a typical human can jump from a crouched start is about 60 cm. By how much does the gravitational potential energy increase for a 72-kg person in such a jump? Where does this energy come from?

7.3 •• **CP** A 120-kg mail bag hangs by a vertical rope 3.5 m long. A postal worker then displaces the bag to a position 2.0 m sideways from its original position, always keeping the rope taut. (a) What horizontal force is necessary to hold the bag in the new position? (b) As the bag is moved to this position, how much work is done (i) by the rope and (ii) by the worker?

7.4 •• **BIO Food Calories.** The *food calorie,* equal to 4186 J, is a measure of how much energy is released when food is metabolized by the body. A certain brand of fruit-and-cereal bar contains

140 food calories per bar. (a) If a 65-kg hiker eats one of these bars, how high a mountain must he climb to "work off" the calories, assuming that all the food energy goes only into increasing gravitational potential energy? (b) If, as is typical, only 20% of the food calories go into mechanical energy, what would be the answer to part (a)? (*Note:* In this and all other problems, we are assuming that 100% of the food calories that are eaten are absorbed and used by the body. This is actually not true. A person's "metabolic efficiency" is the percentage of calories eaten that are actually used; the rest are eliminated by the body. Metabolic efficiency varies considerably from person to person.)

7.5 • A baseball is thrown from the roof of a 22.0-m-tall building with an initial velocity of magnitude 12.0 m/s and directed at an angle of 53.1° above the horizontal. (a) What is the speed of the ball just before it strikes the ground? Use energy methods and ignore air resistance. (b) What is the answer for part (a) if the initial velocity is at an angle of 53.1° *below* the horizontal? (c) If the effects of air resistance are included, will part (a) or (b) give the higher speed?

7.6 •• A crate of mass M starts from rest at the top of a frictionless ramp inclined at an angle α above the horizontal. Find its speed at the bottom of the ramp, a distance d from where it started. Do this in two ways: (a) Take the level at which the potential energy is zero to be at the bottom of the ramp with y positive upward. (b) Take the zero level for potential energy to be at the top of the ramp with y positive upward. (c) Why did the normal force not enter into your solution?

7.7 •• **BIO Human Energy vs. Insect Energy.** For its size, the common flea is one of the most accomplished jumpers in the animal world. A 2.0-mm-long, 0.50-mg critter can reach a height of 20 cm in a single leap. (a) Neglecting air drag, what is the takeoff speed of such a flea? (b) Calculate the kinetic energy of this flea at takeoff and its kinetic energy per kilogram of mass. (c) If a 65-kg, 2.0-m-tall human could jump to the same height compared with his length as the flea jumps compared with its length, how high could the human jump, and what takeoff speed would he need? (d) In fact, most humans can jump no more than 60 cm from a crouched start. What is the kinetic energy per kilogram of mass at takeoff for such a 65-kg person? (e) Where does the flea store the energy that allows it to make such a sudden leap?

7.8 •• An empty crate is given an initial push down a ramp, starting with speed v_0, and reaches the bottom with speed v and kinetic energy K. Some books are now placed in the crate, so that the total mass is quadrupled. The coefficient of kinetic friction is constant and air resistance is negligible. Starting again with v_0 at the top of the ramp, what are the speed and kinetic energy at the bottom? Explain the reasoning behind your answers.

7.9 •• **CP** A small rock with mass 0.20 kg is released from rest at point A, which is at the top edge of a large, hemispherical bowl with radius $R = 0.50$ m (Fig. E7.9). Assume that the size of the rock is small compared to R, so that the rock can be treated as a particle, and assume that the rock slides rather than rolls. The work done by friction on the rock when it moves from point A to point B at the bottom of the bowl has magnitude 0.22 J. (a) Between points A and B, how much work is done on the rock by (i) the normal force and (ii) gravity? (b) What is the speed of the rock as it reaches point B? (c) Of the three forces acting on the rock as it slides down the bowl, which (if any) are constant and which

Figure **E7.9**

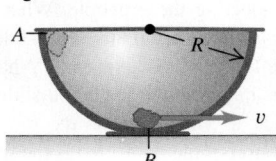

are not? Explain. (d) Just as the rock reaches point B, what is the normal force on it due to the bottom of the bowl?

7.10 •• **BIO Bone Fractures.** The maximum energy that a bone can absorb without breaking depends on its characteristics, such as its cross-sectional area and its elasticity. For healthy human leg bones of approximately 6.0 cm² cross-sectional area, this energy has been experimentally measured to be about 200 J. (a) From approximately what maximum height could a 60-kg person jump and land rigidly upright on both feet without breaking his legs? (b) You are probably surprised at how small the answer to part (a) is. People obviously jump from much greater heights without breaking their legs. How can that be? What else absorbs the energy when they jump from greater heights? (*Hint:* How did the person in part (a) land? How do people normally land when they jump from greater heights?) (c) In light of your answers to parts (a) and (b), what might be some of the reasons that older people are much more prone than younger ones to bone fractures from simple falls (such as a fall in the shower)?

7.11 •• You are testing a new amusement park roller coaster with an empty car of mass 120 kg. One part of the track is a vertical loop with radius 12.0 m. At the bottom of the loop (point A) the car has speed 25.0 m/s, and at the top of the loop (point B) it has speed 8.0 m/s. As the car rolls from point A to point B, how much work is done by friction?

7.12 • **Tarzan and Jane.** Tarzan, in one tree, sights Jane in another tree. He grabs the end of a vine with length 20 m that makes an angle of 45° with the vertical, steps off his tree limb, and swings down and then up to Jane's open arms. When he arrives, his vine makes an angle of 30° with the vertical. Determine whether he gives her a tender embrace or knocks her off her limb by calculating Tarzan's speed just before he reaches Jane. You can ignore air resistance and the mass of the vine.

7.13 •• **CP** A 10.0-kg microwave oven is pushed 8.00 m up the sloping surface of a loading ramp inclined at an angle of 36.9° above the horizontal, by a constant force $\vec{F}$ with a magnitude 110 N and acting parallel to the ramp. The coefficient of kinetic friction between the oven and the ramp is 0.250. (a) What is the work done on the oven by the force $\vec{F}$? (b) What is the work done on the oven by the friction force? (c) Compute the increase in potential energy for the oven. (d) Use your answers to parts (a), (b), and (c) to calculate the increase in the oven's kinetic energy. (e) Use $\Sigma \vec{F} = m\vec{a}$ to calculate the acceleration of the oven. Assuming that the oven is initially at rest, use the acceleration to calculate the oven's speed after traveling 8.00 m. From this, compute the increase in the oven's kinetic energy, and compare it to the answer you got in part (d).

Section 7.2 Elastic Potential Energy

7.14 •• An ideal spring of negligible mass is 12.00 cm long when nothing is attached to it. When you hang a 3.15-kg weight from it, you measure its length to be 13.40 cm. If you wanted to store 10.0 J of potential energy in this spring, what would be its *total* length? Assume that it continues to obey Hooke's law.

7.15 •• A force of 800 N stretches a certain spring a distance of 0.200 m. (a) What is the potential energy of the spring when it is stretched 0.200 m? (b) What is its potential energy when it is compressed 5.00 cm?

7.16 • **BIO Tendons.** Tendons are strong elastic fibers that attach muscles to bones. To a reasonable approximation, they obey Hooke's law. In laboratory tests on a particular tendon, it was found that, when a 250-g object was hung from it, the tendon stretched 1.23 cm. (a) Find the force constant of this tendon in N/m. (b) Because of its thickness, the maximum tension this tendon can support without rupturing is 138 N. By how much can the tendon stretch without rupturing, and how much energy is stored in it at that point?

7.17 • A spring stores potential energy U_0 when it is compressed a distance x_0 from its uncompressed length. (a) In terms of U_0, how much energy does it store when it is compressed (i) twice as much and (ii) half as much? (b) In terms of x_0, how much must it be compressed from its uncompressed length to store (i) twice as much energy and (ii) half as much energy?

7.18 • A slingshot will shoot a 10-g pebble 22.0 m straight up. (a) How much potential energy is stored in the slingshot's rubber band? (b) With the same potential energy stored in the rubber band, how high can the slingshot shoot a 25-g pebble? (c) What physical effects did you ignore in solving this problem?

7.19 •• A spring of negligible mass has force constant $k = 1600$ N/m. (a) How far must the spring be compressed for 3.20 J of potential energy to be stored in it? (b) You place the spring vertically with one end on the floor. You then drop a 1.20-kg book onto it from a height of 0.80 m above the top of the spring. Find the maximum distance the spring will be compressed.

7.20 • A 1.20-kg piece of cheese is placed on a vertical spring of negligible mass and force constant $k = 1800$ N/m that is compressed 15.0 cm. When the spring is released, how high does the cheese rise from this initial position? (The cheese and the spring are *not* attached.)

7.21 •• Consider the glider of Example 7.7 (Section 7.2) and Fig. 7.16. As in the example, the glider is released from rest with the spring stretched 0.100 m. What is the displacement x of the glider from its equilibrium position when its speed is 0.20 m/s? (You should get more than one answer. Explain why.)

7.22 •• Consider the glider of Example 7.7 (Section 7.2) and Fig. 7.16. (a) As in the example, the glider is released from rest with the spring stretched 0.100 m. What is the speed of the glider when it returns to $x = 0$? (b) What must the initial displacement of the glider be if its maximum speed in the subsequent motion is to be 2.50 m/s?

7.23 •• A 2.50-kg mass is pushed against a horizontal spring of force constant 25.0 N/cm on a frictionless air table. The spring is attached to the tabletop, and the mass is not attached to the spring in any way. When the spring has been compressed enough to store 11.5 J of potential energy in it, the mass is suddenly released from rest. (a) Find the greatest speed the mass reaches. When does this occur? (b) What is the greatest acceleration of the mass, and when does it occur?

7.24 •• (a) For the elevator of Example 7.9 (Section 7.2), what is the speed of the elevator after it has moved downward 1.00 m from point 1 in Fig. 7.17? (b) When the elevator is 1.00 m below point 1 in Fig. 7.17, what is its acceleration?

7.25 •• You are asked to design a spring that will give a 1160-kg satellite a speed of 2.50 m/s relative to an orbiting space shuttle. Your spring is to give the satellite a maximum acceleration of $5.00g$. The spring's mass, the recoil kinetic energy of the shuttle, and changes in gravitational potential energy will all be negligible. (a) What must the force constant of the spring be? (b) What distance must the spring be compressed?

7.26 •• A 2.50-kg block on a horizontal floor is attached to a horizontal spring that is initially compressed 0.0300 m. The spring has force constant 840 N/m. The coefficient of kinetic friction between the floor and the block is $\mu_k = 0.40$. The block and spring are released from rest and the block slides along the floor. What is the speed of the block when it has moved a distance of 0.0200 m from its initial position? (At this point the spring is compressed 0.0100 m.)

Section 7.3 Conservative and Nonconservative Forces

7.27 • A 10.0-kg box is pulled by a horizontal wire in a circle on a rough horizontal surface for which the coefficient of kinetic friction is 0.250. Calculate the work done by friction during one complete circular trip if the radius is (a) 2.00 m and (b) 4.00 m. (c) On the basis of the results you just obtained, would you say that friction is a conservative or nonconservative force? Explain.

7.28 • A 75-kg roofer climbs a vertical 7.0-m ladder to the flat roof of a house. He then walks 12 m on the roof, climbs down another vertical 7.0-m ladder, and finally walks on the ground back to his starting point. How much work is done on him by gravity (a) as he climbs up; (b) as he climbs down; (c) as he walks on the roof and on the ground? (d) What is the total work done on him by gravity during this round trip? (e) On the basis of your answer to part (d), would you say that gravity is a conservative or nonconservative force? Explain.

7.29 • A 0.60-kg book slides on a horizontal table. The kinetic friction force on the book has magnitude 1.2 N. (a) How much work is done on the book by friction during a displacement of 3.0 m to the left? (b) The book now slides 3.0 m to the right, returning to its starting point. During this second 3.0-m displacement, how much work is done on the book by friction? (c) What is the total work done on the book by friction during the complete round trip? (d) On the basis of your answer to part (c), would you say that the friction force is conservative or nonconservative? Explain.

7.30 •• **CALC** In an experiment, one of the forces exerted on a proton is $\vec{F} = -\alpha x^2 \hat{\imath}$, where $\alpha = 12$ N/m². (a) How much work does $\vec{F}$ do when the proton moves along the straight-line path from the point (0.10 m, 0) to the point (0.10 m, 0.40 m)? (b) Along the straight-line path from the point (0.10 m, 0) to the point (0.30 m, 0)? (c) Along the straight-line path from the point (0.30 m, 0) to the point (0.10 m, 0)? (d) Is the force $\vec{F}$ conservative? Explain. If $\vec{F}$ is conservative, what is the potential-energy function for it? Let $U = 0$ when $x = 0$.

7.31 • You and three friends stand at the corners of a square whose sides are 8.0 m long in the middle of the gym floor, as shown in Fig. E7.31. You take your physics book and push it from one person to the other. The book has a mass of 1.5 kg, and the coefficient of kinetic friction between the book and the floor is $\mu_k = 0.25$. (a) The book slides from you to Beth and then from Beth to Carlos, along the lines connecting these people. What is the work done by friction during this displacement? (b) You slide the book from you to Carlos along the diagonal of the square. What is the work done by friction during this displacement? (c) You slide the book to Kim, who then slides it back to you. What is the total work done by friction during this motion of the book? (d) Is the friction force on the book conservative or nonconservative? Explain.

Figure **E7.31**

Beth ——— 8.0 m ——— Carlos

8.0 m

You ——— Kim

7.32 • While a roofer is working on a roof that slants at 36° above the horizontal, he accidentally nudges his 85.0-N toolbox, causing it to start sliding downward, starting from rest. If it starts 4.25 m from the lower edge of the roof, how fast will the toolbox be moving just as it reaches the edge of the roof if the kinetic friction force on it is 22.0 N?

7.33 •• A 62.0-kg skier is moving at 6.50 m/s on a frictionless, horizontal, snow-covered plateau when she encounters a rough patch 3.50 m long. The coefficient of kinetic friction between this patch and her skis is 0.300. After crossing the rough patch and returning to friction-free snow, she skis down an icy, frictionless hill 2.50 m high. (a) How fast is the skier moving when she gets to the bottom of the hill? (b) How much internal energy was generated in crossing the rough patch?

Section 7.4 Force and Potential Energy

7.34 •• **CALC** The potential energy of a pair of hydrogen atoms separated by a large distance x is given by $U(x) = -C_6/x^6$, where C_6 is a positive constant. What is the force that one atom exerts on the other? Is this force attractive or repulsive?

7.35 •• **CALC** A force parallel to the x-axis acts on a particle moving along the x-axis. This force produces potential energy $U(x)$ given by $U(x) = \alpha x^4$, where $\alpha = 1.20 \text{ J/m}^4$. What is the force (magnitude and direction) when the particle is at $x = -0.800$ m?

7.36 •• **CALC** An object moving in the xy-plane is acted on by a conservative force described by the potential-energy function $U(x, y) = \alpha(1/x^2 + 1/y^2)$, where α is a positive constant. Derive an expression for the force expressed in terms of the unit vectors $\hat{\imath}$ and $\hat{\jmath}$.

7.37 •• **CALC** A small block with mass 0.0400 kg is moving in the xy-plane. The net force on the block is described by the potential-energy function $U(x, y) = (5.80 \text{ J/m}^2)x^2 - (3.60 \text{ J/m}^3)y^3$. What are the magnitude and direction of the acceleration of the block when it is at the point $x = 0.300$ m, $y = 0.600$ m?

Section 7.5 Energy Diagrams

7.38 • A marble moves along the x-axis. The potential-energy function is shown in Fig. E7.38. (a) At which of the labeled x-coordinates is the force on the marble zero? (b) Which of the labeled x-coordinates is a position of stable equilibrium? (c) Which of the labeled x-coordinates is a position of unstable equilibrium?

Figure **E7.38**

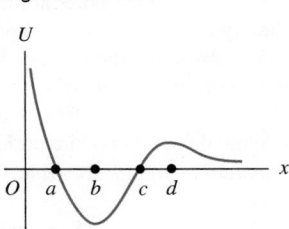

7.39 • **CALC** The potential energy of two atoms in a diatomic molecule is approximated by $U(r) = a/r^{12} - b/r^6$, where r is the spacing between atoms and a and b are positive constants. (a) Find the force $F(r)$ on one atom as a function of r. Draw two graphs: one of $U(r)$ versus r and one of $F(r)$ versus r. (b) Find the equilibrium distance between the two atoms. Is this equilibrium stable? (c) Suppose the distance between the two atoms is equal to the equilibrium distance found in part (b). What minimum energy must be added to the molecule to *dissociate* it—that is, to separate the two atoms to an infinite distance apart? This is called the *dissociation energy* of the molecule. (d) For the molecule CO, the equilibrium distance between the carbon and oxygen atoms is 1.13×10^{-10} m and the dissociation energy is 1.54×10^{-18} J per molecule. Find the values of the constants a and b.

PROBLEMS

7.40 •• Two blocks with different masses are attached to either end of a light rope that passes over a light, frictionless pulley suspended from the ceiling. The masses are released from rest, and the more massive one starts to descend. After this block has descended 1.20 m, its speed is 3.00 m/s. If the total mass of the two blocks is 15.0 kg, what is the mass of each block?

7.41 ••• At a construction site, a 65.0-kg bucket of concrete hangs from a light (but strong) cable that passes over a light, friction-free pulley and is connected to an 80.0-kg box on a horizontal roof (Fig. P7.41). The cable pulls horizontally on the box, and a 50.0-kg bag of gravel rests on top of the box. The coefficients of friction between the box and roof are shown. (a) Find the friction force on the bag of gravel and on the box. (b) Suddenly a worker picks up the bag of gravel. Use energy conservation to find the speed of the bucket after it has descended 2.00 m from rest. (You can check your answer by solving this problem using Newton's laws.)

Figure **P7.41**

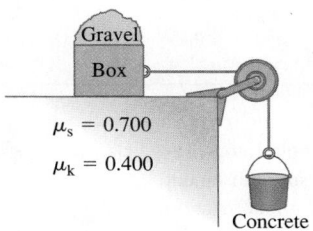

7.42 • A 2.00-kg block is pushed against a spring with negligible mass and force constant $k = 400$ N/m, compressing it 0.220 m. When the block is released, it moves along a frictionless, horizontal surface and then up a frictionless incline with slope 37.0° (Fig. P7.42). (a) What is the speed of the block as it slides along the horizontal surface after having left the spring? (b) How far does the block travel up the incline before starting to slide back down?

Figure **P7.42**

7.43 • A block with mass 0.50 kg is forced against a horizontal spring of negligible mass, compressing the spring a distance of 0.20 m (Fig. P7.43). When released, the block moves on a horizontal tabletop for 1.00 m before coming to rest. The spring constant k is 100 N/m. What is the coefficient of kinetic friction μ_k between the block and the tabletop?

Figure **P7.43**

7.44 • On a horizontal surface, a crate with mass 50.0 kg is placed against a spring that stores 360 J of energy. The spring is released, and the crate slides 5.60 m before coming to rest. What is the speed of the crate when it is 2.00 m from its initial position?

7.45 •• A 350-kg roller coaster starts from rest at point *A* and slides down the frictionless loop-the-loop shown in Fig. P7.45. (a) How fast is this roller coaster moving at point *B*? (b) How hard does it press against the track at point *B*?

Figure **P7.45**

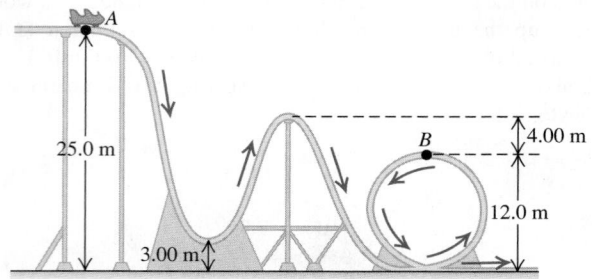

7.46 •• **CP Riding a Loop-the-Loop.** A car in an amusement park ride rolls without friction around the track shown in Fig. P7.46. It starts from rest at point *A* at a height *h* above the bottom of the loop. Treat the car as a particle. (a) What is the minimum value of *h* (in terms of *R*) such that the car moves around the loop without falling off at the top (point *B*)? (b) If *h* = 3.50*R* and *R* = 20.0 m, compute the speed, radial acceleration, and tangential acceleration of the passengers when the car is at point *C*, which is at the end of a horizontal diameter. Show these acceleration components in a diagram, approximately to scale.

Figure **P7.46**

7.47 •• A 2.0-kg piece of wood slides on the surface shown in Fig. P7.47. The curved sides are perfectly smooth, but the rough horizontal bottom is 30 m long and has a kinetic friction coefficient of 0.20 with the wood. The piece of wood starts from rest 4.0 m above the rough bottom. (a) Where will this wood eventually come to rest? (b) For the motion from the initial release until the piece of wood comes to rest, what is the total amount of work done by friction?

Figure **P7.47**

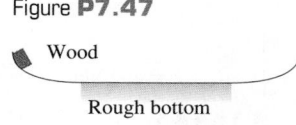

Wood

Rough bottom

7.48 •• **Up and Down the Hill.** A 28-kg rock approaches the foot of a hill with a speed of 15 m/s. This hill slopes upward at a constant angle of 40.0° above the horizontal. The coefficients of static and kinetic friction between the hill and the rock are 0.75 and 0.20, respectively. (a) Use energy conservation to find the maximum height above the foot of the hill reached by the rock. (b) Will the rock remain at rest at its highest point, or will it slide back down the hill? (c) If the rock does slide back down, find its speed when it returns to the bottom of the hill.

7.49 •• A 15.0-kg stone slides down a snow-covered hill (Fig. P7.49), leaving point *A* with a speed of 10.0 m/s. There is no friction on the hill between points *A* and *B*, but there is friction on the level ground at the bottom of the hill, between *B* and the wall. After entering the rough horizontal

Figure **P7.49**

region, the stone travels 100 m and then runs into a very long, light spring with force constant 2.00 N/m. The coefficients of kinetic and static friction between the stone and the horizontal ground are 0.20 and 0.80, respectively. (a) What is the speed of the stone when it reaches point *B*? (b) How far will the stone compress the spring? (c) Will the stone move again after it has been stopped by the spring?

7.50 •• **CP** A 2.8-kg block slides over the smooth, icy hill shown in Fig. P7.50. The top of the hill is horizontal and 70 m higher than its base. What minimum speed must the block have at the base of the hill in order for it to pass over the pit at the far side of the hill?

Figure **P7.50**

7.51 ••• **Bungee Jump.** A bungee cord is 30.0 m long and, when stretched a distance *x*, it exerts a restoring force of magnitude *kx*. Your father-in-law (mass 95.0 kg) stands on a platform 45.0 m above the ground, and one end of the cord is tied securely to his ankle and the other end to the platform. You have promised him that when he steps off the platform he will fall a maximum distance of only 41.0 m before the cord stops him. You had several bungee cords to select from, and you tested them by stretching them out, tying one end to a tree, and pulling on the other end with a force of 380.0 N. When you do this, what distance will the bungee cord that you should select have stretched?

7.52 •• **Ski Jump Ramp.** You are designing a ski jump ramp for the next Winter Olympics. You need to calculate the vertical height *h* from the starting gate to the bottom of the ramp. The skiers push off hard with their ski poles at the start, just above the starting gate, so they typically have a speed of 2.0 m/s as they reach the gate. For safety, the skiers should have a speed no higher than 30.0 m/s when they reach the bottom of the ramp. You determine that for a 85.0-kg skier with good form, friction and air resistance will do total work of magnitude 4000 J on him during his run down the ramp. What is the maximum height *h* for which the maximum safe speed will not be exceeded?

7.53 ••• The Great Sandini is a 60-kg circus performer who is shot from a cannon (actually a spring gun). You don't find many men of his caliber, so you help him design a new gun. This new gun has a very large spring with a very small mass and a force constant of 1100 N/m that he will compress with a force of 4400 N. The inside of the gun barrel is coated with Teflon, so the average friction force will be only 40 N during the 4.0 m he moves in the barrel. At what speed will he emerge from the end of the barrel, 2.5 m above his initial rest position?

7.54 ••• You are designing a delivery ramp for crates containing exercise equipment. The 1470-N crates will move at 1.8 m/s at the top of a ramp that slopes downward at 22.0°. The ramp exerts a 550-N kinetic friction force on each crate, and the maximum static friction force also has this value. Each crate will compress a spring at the bottom of the ramp and will come to rest after traveling a total distance of 8.0 m along the ramp. Once stopped, a crate must not rebound back up the ramp. Calculate the force constant of the spring that will be needed in order to meet the design criteria.

7.55 •• A system of two paint buckets connected by a lightweight rope is released from rest with the 12.0-kg bucket 2.00 m above the floor (Fig. P7.55). Use the principle of conservation of energy to find the speed with which this bucket strikes the floor. You can ignore friction and the mass of the pulley.

Figure **P7.55**

7.56 •• A 1500-kg rocket is to be launched with an initial upward speed of 50.0 m/s. In order to assist its engines, the engineers will start it from rest on a ramp that rises 53° above the horizontal (Fig. P7.56). At the bottom, the ramp turns upward and launches the rocket vertically. The engines provide a constant forward thrust of 2000 N, and friction with

Figure **P7.56**

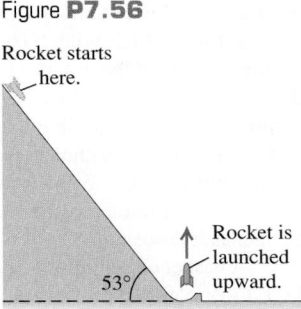

the ramp surface is a constant 500 N. How far from the base of the ramp should the rocket start, as measured along the surface of the ramp?

7.57 • **Legal Physics.** In an auto accident, a car hit a pedestrian and the driver then slammed on the brakes to stop the car. During the subsequent trial, the driver's lawyer claimed that he was obeying the posted 35-mph speed limit, but that the legal speed was too high to allow him to see and react to the pedestrian in time. You have been called in as the state's expert witness. Your investigation of the accident found that the skid marks made while the brakes were applied were 280 ft long, and the tread on the tires produced a coefficient of kinetic friction of 0.30 with the road. (a) In your testimony in court, will you say that the driver was obeying the posted speed? You must be able to back up your conclusion with clear reasoning because one of the lawyers will surely cross-examine you. (b) If the driver's speeding ticket were $10 for each mile per hour he was driving above the posted speed limit, would he have to pay a fine? If so, how much would it be?

7.58 ••• A wooden rod of negligible mass and length 80.0 cm is pivoted about a horizontal axis through its center. A white rat with mass 0.500 kg clings to one end of the stick, and a mouse with mass 0.200 kg clings to the other end. The system is released from rest with the rod horizontal. If the animals can manage to hold on, what are their speeds as the rod swings through a vertical position?

7.59 •• **CP** A 0.300-kg potato is tied to a string with length 2.50 m, and the other end of the string is tied to a rigid support. The potato is held straight out horizontally from the point of support, with the string pulled taut, and is then released. (a) What is the speed of the potato at the lowest point of its motion? (b) What is the tension in the string at this point?

7.60 •• These data are from a computer simulation for a batted baseball with mass 0.145 kg, including air resistance:

t	x	y	v_x	v_y
0	0	0	30.0 m/s	40.0 m/s
3.05 s	70.2 m	53.6 m	18.6 m/s	0
6.59 s	124.4 m	0	11.9 m/s	−28.7 m/s

(a) How much work was done by the air on the baseball as it moved from its initial position to its maximum height? (b) How much work was done by the air on the baseball as it moved from its maximum height back to the starting elevation? (c) Explain why the magnitude of the answer in part (b) is smaller than the magnitude of the answer in part (a).

7.61 •• **Down the Pole.** A fireman of mass m slides a distance d down a pole. He starts from rest. He moves as fast at the bottom as if he had stepped off a platform a distance $h \leq d$ above the ground and descended with negligible air resistance. (a) What average friction force did the fireman exert on the pole? Does your answer make sense in the special cases of $h = d$ and $h = 0$? (b) Find a numerical value for the average friction force a 75-kg fireman exerts, for $d = 2.5$ m and $h = 1.0$ m. (c) In terms of g, h, and d, what is the speed of the fireman when he is a distance y above the bottom of the pole?

7.62 •• A 60.0-kg skier starts from rest at the top of a ski slope 65.0 m high. (a) If frictional forces do -10.5 kJ of work on her as she descends, how fast is she going at the bottom of the slope? (b) Now moving horizontally, the skier crosses a patch of soft snow, where $\mu_k = 0.20$. If the patch is 82.0 m wide and the average force of air resistance on the skier is 160 N, how fast is she going after crossing the patch? (c) The skier hits a snowdrift and penetrates 2.5 m into it before coming to a stop. What is the average force exerted on her by the snowdrift as it stops her?

7.63 • **CP** A skier starts at the top of a very large, frictionless snowball, with a very small initial speed, and skis straight down the side (Fig. P7.63). At what point does she lose contact with the snowball and fly off at a tangent? That is, at the instant she loses contact with the snowball, what angle α does a radial line from the center of the snowball to the skier make with the vertical?

Figure **P7.63**

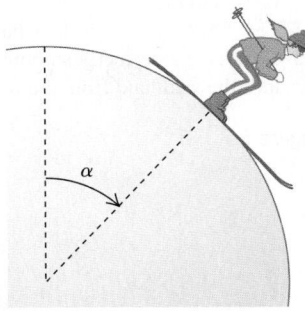

7.64 •• A ball is thrown upward with an initial velocity of 15 m/s at an angle of 60.0° above the horizontal. Use energy conservation to find the ball's greatest height above the ground.

7.65 •• In a truck-loading station at a post office, a small 0.200-kg package is released from rest at point A on a track that is one-quarter of a circle with radius 1.60 m (Fig. P7.65). The size of the package is much less than 1.60 m, so the package can be treated as a particle. It slides down the track and reaches point B with a speed of 4.80 m/s. From point B, it slides on a level surface a distance of

Figure **P7.65**

3.00 m to point C, where it comes to rest. (a) What is the coefficient of kinetic friction on the horizontal surface? (b) How much work is done on the package by friction as it slides down the circular arc from A to B?

7.66 ••• A truck with mass m has a brake failure while going down an icy mountain road of constant downward slope angle α (Fig. P7.66). Initially the truck is moving downhill at speed v_0. After careening downhill a distance L with negligible friction, the truck driver steers the runaway vehicle onto a runaway truck ramp of constant upward slope angle β. The truck ramp has a soft sand surface for which the coefficient of rolling friction is μ_r. What is the distance that the truck moves up the ramp before coming to a halt? Solve using energy methods.

Figure **P7.66**

7.67 •• **CALC** A certain spring is found *not* to obey Hooke's law; it exerts a restoring force $F_x(x) = -\alpha x - \beta x^2$ if it is stretched or compressed, where $\alpha = 60.0$ N/m and $\beta = 18.0$ N/m². The mass of the spring is negligible. (a) Calculate the potential-energy function $U(x)$ for this spring. Let $U = 0$ when $x = 0$. (b) An object with mass 0.900 kg on a frictionless, horizontal surface is attached to this spring, pulled a distance 1.00 m to the right (the $+x$-direction) to stretch the spring, and released. What is the speed of the object when it is 0.50 m to the right of the $x = 0$ equilibrium position?

7.68 •• **CP** A sled with rider having a combined mass of 125 kg travels over the perfectly smooth icy hill shown in Fig. 7.68. How far does the sled land from the foot of the cliff?

Figure **P7.68**

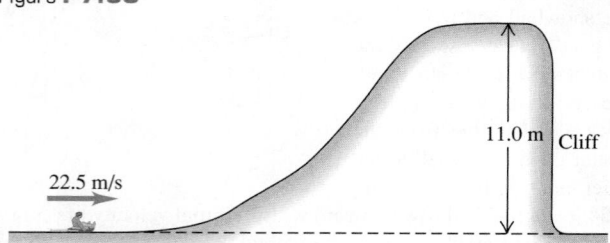

7.69 •• A 0.150-kg block of ice is placed against a horizontal, compressed spring mounted on a horizontal tabletop that is 1.20 m above the floor. The spring has force constant 1900 N/m and is initially compressed 0.045 m. The mass of the spring is negligible. The spring is released, and the block slides along the table, goes off the edge, and travels to the floor. If there is negligible friction between the block of ice and the tabletop, what is the speed of the block of ice when it reaches the floor?

7.70 •• A 3.00-kg block is con- Figure **P7.70**
nected to two ideal horizon-
tal springs having force constants
$k_1 = 25.0$ N/cm and $k_2 =$
20.0 N/cm (Fig. P7.70). The
system is initially in equilibrium on a horizontal, frictionless surface. The block is now pushed 15.0 cm to the right and released

from rest. (a) What is the maximum speed of the block? Where in the motion does the maximum speed occur? (b) What is the maximum compression of spring 1?

7.71 •• An experimental apparatus with mass m is placed on a vertical spring of negligible mass and pushed down until the spring is compressed a distance x. The apparatus is then released and reaches its maximum height at a distance h above the point where it is released. The apparatus is not attached to the spring, and at its maximum height it is no longer in contact with the spring. The maximum magnitude of acceleration the apparatus can have without being damaged is a, where $a > g$. (a) What should the force constant of the spring be? (b) What distance x must the spring be compressed initially?

7.72 •• If a fish is attached to a vertical spring and slowly lowered to its equilibrium position, it is found to stretch the spring by an amount d. If the same fish is attached to the end of the unstretched spring and then allowed to fall from rest, through what maximum distance does it stretch the spring? (*Hint:* Calculate the force constant of the spring in terms of the distance d and the mass m of the fish.)

7.73 ••• **CALC** A 3.00-kg fish is attached to the lower end of a vertical spring that has negligible mass and force constant 900 N/m. The spring initially is neither stretched nor compressed. The fish is released from rest. (a) What is its speed after it has descended 0.0500 m from its initial position? (b) What is the maximum speed of the fish as it descends?

7.74 •• A basket of negligible weight hangs from a vertical spring scale of force constant 1500 N/m. (a) If you suddenly put a 3.0-kg adobe brick in the basket, find the maximum distance that the spring will stretch. (b) If, instead, you release the brick from 1.0 m above the basket, by how much will the spring stretch at its maximum elongation?

7.75 • A 0.500-kg block, attached to a spring with length 0.60 m and force constant 40.0 N/m, is at rest with the back of the block at point A on a frictionless, horizontal air table (Fig. P7.75). The mass of the spring is negligible. You move the block to the right along the surface by pulling with a constant 20.0-N horizontal force. (a) What is the block's speed when the back of the block reaches point B, which is 0.25 m to the right of point A? (b) When the back of the block reaches point B, you let go of the block. In the subsequent motion, how close does the block get to the wall where the left end of the spring is attached?

Figure **P7.75**

7.76 •• **Fraternity Physics.** The brothers of Iota Eta Pi fraternity build a platform, supported at all four corners by vertical springs, in the basement of their frat house. A brave fraternity brother wearing a football helmet stands in the middle of the platform; his weight compresses the springs by 0.18 m. Then four of his fraternity brothers, pushing down at the corners of the platform, compress the springs another 0.53 m until the top of the brave brother's helmet is 0.90 m below the basement ceiling. They then simultaneously release the platform. You can ignore the masses of the springs and platform. (a) When the dust clears, the fraternity asks you to calculate their fraternity brother's speed just before his helmet hit the flimsy ceiling. (b) Without the ceiling, how high would he have gone? (c) In discussing their probation, the dean of students suggests that the next time they try this, they do it outdoors on another planet. Would the answer to part (b) be the same if this stunt were performed on a planet with a different value of g? Assume that the fraternity brothers push the platform down 0.53 m as before. Explain your reasoning.

7.77 ••• **CP** A small block with mass 0.0500 kg slides in a vertical circle of radius $R = 0.800$ m on the inside of a circular track. There is no friction between the track and the block. At the bottom of the block's path, the normal force the track exerts on the block has magnitude 3.40 N. What is the magnitude of the normal force that the track exerts on the block when it is at the top of its path?

7.78 ••• **CP** A small block with mass 0.0400 kg slides in a vertical circle of radius $R = 0.500$ m on the inside of a circular track. During one of the revolutions of the block, when the block is at the bottom of its path, point A, the magnitude of the normal force exerted on the block by the track has magnitude 3.95 N. In this same revolution, when the block reaches the top of its path, point B, the magnitude of the normal force exerted on the block has magnitude 0.680 N. How much work was done on the block by friction during the motion of the block from point A to point B?

7.79 •• A hydroelectric dam holds back a lake of surface area 3.0×10^6 m^2 that has vertical sides below the water level. The water level in the lake is 150 m above the base of the dam. When the water passes through turbines at the base of the dam, its mechanical energy is converted to electrical energy with 90% efficiency. (a) If gravitational potential energy is taken to be zero at the base of the dam, how much energy is stored in the top meter of the water in the lake? The density of water is 1000 kg/m^3. (b) What volume of water must pass through the dam to produce 1000 kilowatt-hours of electrical energy? What distance does the level of water in the lake fall when this much water passes through the dam?

7.80 •• **CALC** How much total energy is stored in the lake in Problem 7.79? As in that problem, take the gravitational potential energy to be zero at the base of the dam. Express your answer in joules and in kilowatt-hours. (*Hint:* Break the lake up into infinitesimal horizontal layers of thickness dy, and integrate to find the total potential energy.)

7.81 ••• A wooden block with mass 1.50 kg is placed against a compressed spring at the bottom of an incline of slope 30.0° (point A). When the spring is released, it projects the block up the incline. At point B, a distance of 6.00 m up the incline from A, the block is moving up the incline at 7.00 m/s and is no longer in contact with the spring. The coefficient of kinetic friction between the block and the incline is $\mu_k = 0.50$. The mass of the spring is negligible. Calculate the amount of potential energy that was initially stored in the spring.

7.82 •• **CP** **Pendulum.** A small rock with mass 0.12 kg is fastened to a massless string with length 0.80 m to form a pendulum. The pendulum is swinging so as to make a maximum angle of 45° with the vertical. Air resistance is negligible. (a) What is the speed of the rock when the string passes through the vertical position? (b) What is the tension in the string when it makes an angle of 45° with the vertical? (c) What is the tension in the string as it passes through the vertical?

7.83 ••• **CALC** A cutting tool under microprocessor control has several forces acting on it. One force is $\vec{F} = -\alpha xy^2 \hat{\jmath}$, a force in the negative y-direction whose magnitude depends on the position of the tool. The constant is $\alpha = 2.50$ N/m^3. Consider the displacement of the tool from the origin to the point $x = 3.00$ m, $y = 3.00$ m. (a) Calculate the work done on the tool by $\vec{F}$ if this displacement is along the straight line $y = x$ that connects these two points. (b) Calculate the work done on the tool by $\vec{F}$ if the tool is first moved out along the x-axis to the point $x = 3.00$ m, $y = 0$ and then moved parallel to the y-axis to the point $x = 3.00$ m, $y = 3.00$ m. (c) Compare the work done by $\vec{F}$ along these two paths. Is $\vec{F}$ conservative or nonconservative? Explain.

7.84 • **CALC** (a) Is the force $\vec{F} = Cy^2 \hat{\jmath}$, where C is a negative constant with units of N/m^2, conservative or nonconservative? Justify your answer. (b) Is the force $\vec{F} = Cy^2 \hat{\imath}$, where C is a negative constant with units of N/m^2, conservative or nonconservative? Justify your answer.

7.85 •• **CALC** An object has several forces acting on it. One force is $\vec{F} = \alpha xy \hat{\imath}$, a force in the x-direction whose magnitude depends on the position of the object. (See Problem 6.98.) The constant is $\alpha = 2.00$ N/m^2. The object moves along the following path: (1) It starts at the origin and moves along the y-axis to the point $x = 0$, $y = 1.50$ m; (2) it moves parallel to the x-axis to the point $x = 1.50$ m, $y = 1.50$ m; (3) it moves parallel to the y-axis to the point $x = 1.50$ m, $y = 0$; (4) it moves parallel to the x-axis back to the origin. (a) Sketch this path in the xy-plane. (b) Calculate the work done on the object by $\vec{F}$ for each leg of the path and for the complete round trip. (c) Is $\vec{F}$ conservative or nonconservative? Explain.

7.86 • A particle moves along the x-axis while acted on by a single conservative force parallel to the x-axis. The force corresponds to the potential-energy function graphed in Fig. P7.86. The particle is released from rest at point A. (a) What is the direction of the force on the particle when it is at point A? (b) At point B? (c) At what value of x is the kinetic energy of the particle a maximum? (d) What is the force on the particle when it is at point C? (e) What is the largest value of x reached by the particle during its motion? (f) What value or values of x correspond to points of stable equilibrium? (g) Of unstable equilibrium?

Figure **P7.86**

CHALLENGE PROBLEM

7.87 ••• **CALC** A proton with mass m moves in one dimension. The potential-energy function is $U(x) = \alpha/x^2 - \beta/x$, where α and β are positive constants. The proton is released from rest at $x_0 = \alpha/\beta$. (a) Show that $U(x)$ can be written as

$$U(x) = \frac{\alpha}{x_0^2} \left[\left(\frac{x_0}{x} \right)^2 - \frac{x_0}{x} \right]$$

Graph $U(x)$. Calculate $U(x_0)$ and thereby locate the point x_0 on the graph. (b) Calculate $v(x)$, the speed of the proton as a function of position. Graph $v(x)$ and give a qualitative description of the motion. (c) For what value of x is the speed of the proton a maximum? What is the value of that maximum speed? (d) What is the force on the proton at the point in part (c)? (e) Let the proton be released instead at $x_1 = 3\alpha/\beta$. Locate the point x_1 on the graph of $U(x)$. Calculate $v(x)$ and give a qualitative description of the motion. (f) For each release point ($x = x_0$ and $x = x_1$), what are the maximum and minimum values of x reached during the motion?

Answers

Chapter Opening Question

The mallard's kinetic energy K remains constant because the speed remains the same, but the gravitational potential energy U_{grav} decreases as the mallard descends. Hence the total mechanical energy $E = K + U_{grav}$ decreases. The lost mechanical energy goes into warming the mallard's skin (that is, an increase in the mallard's internal energy) and stirring up the air through which the mallard passes (an increase in the internal energy of the air). See the discussion in Section 7.3.

Test Your Understanding Questions

7.1 Answer: (iii) The initial kinetic energy $K_1 = 0$, the initial potential energy $U_1 = mgy_1$, and the final potential energy $U_2 = mgy_2$ are the same for both blocks. Mechanical energy is conserved in both cases, so the final kinetic energy $K_2 = \frac{1}{2}mv_2^2$ is also the same for both blocks. Hence the speed at the right-hand end is the *same* in both cases!

7.2 Answer: (iii) The elevator is still moving downward, so the kinetic energy K is positive (remember that K can never be nega-

tive); the elevator is below point 1, so $y < 0$ and $U_{grav} < 0$; and the spring is compressed, so $U_{el} > 0$.

7.3 Answer: (iii) Because of friction in the turbines and between the water and turbines, some of the potential energy goes into raising the temperatures of the water and the mechanism.

7.4 Answers: (a) (iv), (b) (i) If $F_x = 0$ at a point, then the derivative of $U(x)$ must be zero at that point because $F_x = -dU(x)/dx$. However, this tells us absolutely nothing about the *value* of $U(x)$ at that point.

7.5 Answers: (iii) Figure 7.24b shows the x-component of force, F_x. Where this is maximum (most positive), the x-component of force and the x-acceleration have more positive values than at adjacent values of x.

Bridging Problem

Answers: **(a)** 1.06 m
            (b) 1.32 m
            (c) 20.7 J

MOMENTUM, IMPULSE, AND COLLISIONS

? Which could potentially do greater damage to this carrot: a .22-caliber bullet moving at 220 m/s as shown here, or a lightweight bullet of the same length and diameter but half the mass moving at twice the speed?

LEARNING GOALS

By studying this chapter, you will learn:

- The meaning of the momentum of a particle, and how the impulse of the net force acting on a particle causes its momentum to change.

- The conditions under which the total momentum of a system of particles is constant (conserved).

- How to solve problems in which two bodies collide with each other.

- The important distinction among elastic, inelastic, and completely inelastic collisions.

- The definition of the center of mass of a system, and what determines how the center of mass moves.

- How to analyze situations such as rocket propulsion in which the mass of a body changes as it moves.

There are many questions involving forces that cannot be answered by directly applying Newton's second law, $\Sigma \vec{F} = m\vec{a}$. For example, when a moving van collides head-on with a compact car, what determines which way the wreckage moves after the collision? In playing pool, how do you decide how to aim the cue ball in order to knock the eight ball into the pocket? And when a meteorite collides with the earth, how much of the meteorite's kinetic energy is released in the impact?

A common theme of all these questions is that they involve forces about which we know very little: the forces between the car and the moving van, between the two pool balls, or between the meteorite and the earth. Remarkably, we will find in this chapter that we don't have to know *anything* about these forces to answer questions of this kind!

Our approach uses two new concepts, *momentum* and *impulse,* and a new conservation law, *conservation of momentum.* This conservation law is every bit as important as the law of conservation of energy. The law of conservation of momentum is valid even in situations in which Newton's laws are inadequate, such as bodies moving at very high speeds (near the speed of light) or objects on a very small scale (such as the constituents of atoms). Within the domain of Newtonian mechanics, conservation of momentum enables us to analyze many situations that would be very difficult if we tried to use Newton's laws directly. Among these are *collision* problems, in which two bodies collide and can exert very large forces on each other for a short time.

8.1 Momentum and Impulse

In Chapter 6 we re-expressed Newton's second law for a particle, $\Sigma \vec{F} = m\vec{a}$, in terms of the work–energy theorem. This theorem helped us tackle a great number of physics problems and led us to the law of conservation of energy. Let's now return to $\Sigma \vec{F} = m\vec{a}$ and see yet another useful way to restate this fundamental law.

Newton's Second Law in Terms of Momentum

Consider a particle of constant mass m. (Later in this chapter we'll see how to deal with situations in which the mass of a body changes.) Because $\vec{a} = d\vec{v}/dt$, we can write Newton's second law for this particle as

$$\sum \vec{F} = m\frac{d\vec{v}}{dt} = \frac{d}{dt}(m\vec{v}) \qquad (8.1)$$

We can move the mass m inside the derivative because it is constant. Thus Newton's second law says that the net force $\sum\vec{F}$ acting on a particle equals the time rate of change of the combination $m\vec{v}$, the product of the particle's mass and velocity. We'll call this combination the **momentum,** or **linear momentum,** of the particle. Using the symbol $\vec{p}$ for momentum, we have

$$\vec{p} = m\vec{v} \qquad \text{(definition of momentum)} \qquad (8.2)$$

8.1 The velocity and momentum vectors of a particle.

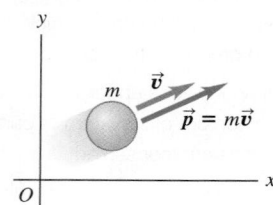

Momentum $\vec{p}$ is a vector quantity; a particle's momentum has the same direction as its velocity $\vec{v}$.

The greater the mass m and speed v of a particle, the greater is its magnitude of momentum mv. Keep in mind, however, that momentum is a *vector* quantity with the same direction as the particle's velocity (Fig. 8.1). Hence a car driving north at 20 m/s and an identical car driving east at 20 m/s have the same *magnitude* of momentum (mv) but different momentum *vectors* ($m\vec{v}$) because their directions are different.

We often express the momentum of a particle in terms of its components. If the particle has velocity components v_x, v_y, and v_z, then its momentum components p_x, p_y, and p_z (which we also call the *x-momentum*, *y-momentum*, and *z-momentum*) are given by

$$p_x = mv_x \qquad p_y = mv_y \qquad p_z = mv_z \qquad (8.3)$$

These three component equations are equivalent to Eq. (8.2).

The units of the magnitude of momentum are units of mass times speed; the SI units of momentum are kg · m/s. The plural of momentum is "momenta."

If we now substitute the definition of momentum, Eq. (8.2), into Eq. (8.1), we get

$$\sum \vec{F} = \frac{d\vec{p}}{dt} \qquad \text{(Newton's second law in terms of momentum)} \qquad (8.4)$$

The net force (vector sum of all forces) acting on a particle equals the time rate of change of momentum of the particle. This, not $\sum\vec{F} = m\vec{a}$, is the form in which Newton originally stated his second law (although he called momentum the "quantity of motion"). This law is valid only in inertial frames of reference.

According to Eq. (8.4), a rapid change in momentum requires a large net force, while a gradual change in momentum requires less net force. This principle is used in the design of automobile safety devices such as air bags (Fig. 8.2).

8.2 If a fast-moving automobile stops suddenly in a collision, the driver's momentum (mass times velocity) changes from a large value to zero in a short time. An air bag causes the driver to lose momentum more gradually than would an abrupt collision with the steering wheel, reducing the force exerted on the driver as well as the possibility of injury.

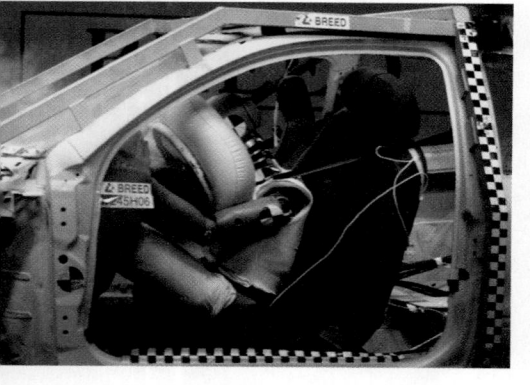

The Impulse–Momentum Theorem

A particle's momentum $\vec{p} = m\vec{v}$ and its kinetic energy $K = \frac{1}{2}mv^2$ both depend on the mass and velocity of the particle. What is the fundamental difference between these two quantities? A purely mathematical answer is that momentum is a vector whose magnitude is proportional to speed, while kinetic energy is a scalar proportional to the speed squared. But to see the *physical* difference between momentum and kinetic energy, we must first define a quantity closely related to momentum called *impulse*.

Let's first consider a particle acted on by a *constant* net force $\sum\vec{F}$ during a time interval Δt from t_1 to t_2. (We'll look at the case of varying forces shortly.) The **impulse** of the net force, denoted by $\vec{J}$, is defined to be the product of the net force and the time interval:

$$\vec{J} = \sum\vec{F}(t_2 - t_1) = \sum\vec{F}\,\Delta t \qquad \text{(assuming constant net force)} \qquad (8.5)$$

Impulse is a vector quantity; its direction is the same as the net force $\sum\vec{F}$. Its magnitude is the product of the magnitude of the net force and the length of time that the net force acts. The SI unit of impulse is the newton-second ($\text{N}\cdot\text{s}$). Because $1\text{ N} = 1\text{ kg}\cdot\text{m/s}^2$, an alternative set of units for impulse is $\text{kg}\cdot\text{m/s}$, the same as the units of momentum.

To see what impulse is good for, let's go back to Newton's second law as restated in terms of momentum, Eq. (8.4). If the net force $\sum\vec{F}$ is constant, then $d\vec{p}/dt$ is also constant. In that case, $d\vec{p}/dt$ is equal to the *total* change in momentum $\vec{p}_2 - \vec{p}_1$ during the time interval $t_2 - t_1$, divided by the interval:

$$\sum\vec{F} = \frac{\vec{p}_2 - \vec{p}_1}{t_2 - t_1}$$

Multiplying this equation by $(t_2 - t_1)$, we have

$$\sum\vec{F}(t_2 - t_1) = \vec{p}_2 - \vec{p}_1$$

Comparing with Eq. (8.5), we end up with a result called the **impulse–momentum theorem:**

$$\vec{J} = \vec{p}_2 - \vec{p}_1 \qquad \text{(impulse–momentum theorem)} \qquad (8.6)$$

> **The change in momentum of a particle during a time interval equals the impulse of the net force that acts on the particle during that interval.**

The impulse–momentum theorem also holds when forces are not constant. To see this, we integrate both sides of Newton's second law $\sum\vec{F} = d\vec{p}/dt$ over time between the limits t_1 and t_2:

$$\int_{t_1}^{t_2} \sum\vec{F}\,dt = \int_{t_1}^{t_2} \frac{d\vec{p}}{dt}\,dt = \int_{\vec{p}_1}^{\vec{p}_2} d\vec{p} = \vec{p}_2 - \vec{p}_1$$

The integral on the left is defined to be the impulse $\vec{J}$ of the net force $\sum\vec{F}$ during this interval:

$$\vec{J} = \int_{t_1}^{t_2} \sum\vec{F}\,dt \qquad \text{(general definition of impulse)} \qquad (8.7)$$

With this definition, the impulse–momentum theorem $\vec{J} = \vec{p}_2 - \vec{p}_1$, Eq. (8.6), is valid even when the net force $\sum\vec{F}$ varies with time.

We can define an *average* net force $\vec{F}_{\text{av}}$ such that even when $\sum\vec{F}$ is not constant, the impulse $\vec{J}$ is given by

$$\vec{J} = \vec{F}_{\text{av}}(t_2 - t_1) \qquad (8.8)$$

When $\sum\vec{F}$ is constant, $\sum\vec{F} = \vec{F}_{\text{av}}$ and Eq. (8.8) reduces to Eq. (8.5).

Figure 8.3a shows the x-component of net force $\sum F_x$ as a function of time during a collision. This might represent the force on a soccer ball that is in contact with a player's foot from time t_1 to t_2. The x-component of impulse during this interval is represented by the red area under the curve between t_1 and t_2. This

Application **Woodpecker Impulse**
The pileated woodpecker (*Dryocopus pileatus*) has been known to strike its beak against a tree up to 20 times a second and up to 12,000 times a day. The impact force can be as much as 1200 times the weight of the bird's head. Because the impact lasts such a short time, the impulse—the product of the net force during the impact multiplied by the duration of the impact—is relatively small. [The woodpecker has a thick skull of spongy bone as well as shock-absorbing cartilage at the base of the lower jaw, and so avoids injury.]

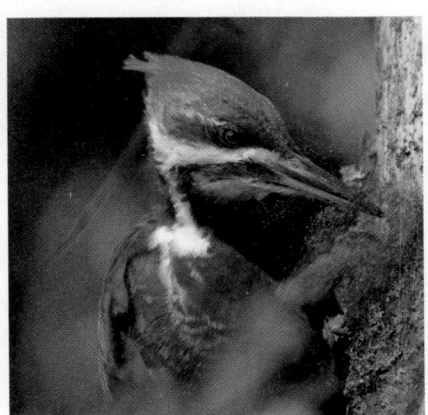

8.3 The meaning of the area under a graph of $\sum F_x$ versus t.

(a)

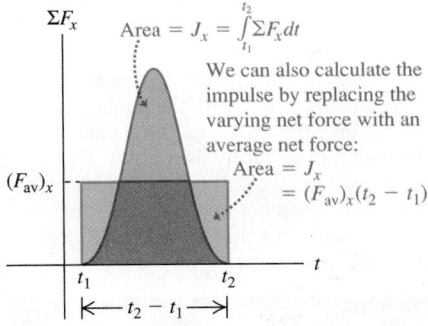

The area under the curve of net force versus time equals the impulse of the net force:

$$\text{Area} = J_x = \int_{t_1}^{t_2}\sum F_x\,dt$$

We can also calculate the impulse by replacing the varying net force with an average net force:

$$\text{Area} = J_x = (F_{\text{av}})_x(t_2 - t_1)$$

(b)

Large force that acts for a short time

The area under both curves is the same, so both forces deliver the same impulse.

Smaller force that acts for a longer time

area is equal to the green rectangular area bounded by t_1, t_2, and $(F_{av})_x$, so $(F_{av})_x(t_2 - t_1)$ is equal to the impulse of the actual time-varying force during the same interval. Note that a large force acting for a short time can have the same impulse as a smaller force acting for a longer time if the areas under the force–time curves are the same (Fig. 8.3b). In this language, an automobile airbag (see Fig. 8.2) provides the same impulse to the driver as would the steering wheel or the dashboard by applying a weaker and less injurious force for a longer time.

Impulse and momentum are both vector quantities, and Eqs. (8.5)–(8.8) are all vector equations. In specific problems, it is often easiest to use them in component form:

$$J_x = \int_{t_1}^{t_2} \Sigma F_x \, dt = (F_{av})_x(t_2 - t_1) = p_{2x} - p_{1x} = mv_{2x} - mv_{1x}$$

$$J_y = \int_{t_1}^{t_2} \Sigma F_y \, dt = (F_{av})_y(t_2 - t_1) = p_{2y} - p_{1y} = mv_{2y} - mv_{1y}$$

(8.9)

and similarly for the z-component.

Momentum and Kinetic Energy Compared

We can now see the fundamental difference between momentum and kinetic energy. The impulse–momentum theorem $\vec{J} = \vec{p}_2 - \vec{p}_1$ says that changes in a particle's momentum are due to impulse, which depends on the *time* over which the net force acts. By contrast, the work–energy theorem $W_{tot} = K_2 - K_1$ tells us that kinetic energy changes when work is done on a particle; the total work depends on the *distance* over which the net force acts. Consider a particle that starts from rest at t_1 so that $\vec{v}_1 = \mathbf{0}$. Its initial momentum is $\vec{p}_1 = m\vec{v}_1 = \mathbf{0}$, and its initial kinetic energy is $K_1 = \frac{1}{2}mv_1^2 = 0$. Now let a constant net force equal to $\vec{F}$ act on that particle from time t_1 until time t_2. During this interval, the particle moves a distance s in the direction of the force. From Eq. (8.6), the particle's momentum at time t_2 is

$$\vec{p}_2 = \vec{p}_1 + \vec{J} = \vec{J}$$

where $\vec{J} = \vec{F}(t_2 - t_1)$ is the impulse that acts on the particle. So *the momentum of a particle equals the impulse that accelerated it from rest to its present speed;* impulse is the product of the net force that accelerated the particle and the *time* required for the acceleration. By comparison, the kinetic energy of the particle at t_2 is $K_2 = W_{tot} = Fs$, the total *work* done on the particle to accelerate it from rest. The total work is the product of the net force and the *distance* required to accelerate the particle (Fig. 8.4).

Here's an application of the distinction between momentum and kinetic energy. Suppose you have a choice between catching a 0.50-kg ball moving at 4.0 m/s or a 0.10-kg ball moving at 20 m/s. Which will be easier to catch? Both balls have the same magnitude of momentum, $p = mv = (0.50 \text{ kg})(4.0 \text{ m/s}) = (0.10 \text{ kg})(20 \text{ m/s}) = 2.0 \text{ kg} \cdot \text{m/s}$. However, the two balls have different values of kinetic energy $K = \frac{1}{2}mv^2$; the large, slow-moving ball has $K = 4.0 \text{ J}$, while the small, fast-moving ball has $K = 20 \text{ J}$. Since the momentum is the same for both balls, both require the same *impulse* to be brought to rest. But stopping the 0.10-kg ball with your hand requires five times more *work* than stopping the 0.50-kg ball because the smaller ball has five times more kinetic energy. For a given force that you exert with your hand, it takes the same amount of time (the duration of the catch) to stop either ball, but your hand and arm will be pushed back five times farther if you choose to catch the small, fast-moving ball. To minimize arm strain, you should choose to catch the 0.50-kg ball with its lower kinetic energy.

Both the impulse–momentum and work–energy theorems are relationships between force and motion, and both rest on the foundation of Newton's laws. They are *integral* principles, relating the motion at two different times separated

MasteringPHYSICS

ActivPhysics 6.1: Momentum and Energy Change

8.4 The *kinetic energy* of a pitched baseball is equal to the work the pitcher does on it (force multiplied by the distance the ball moves during the throw). The *momentum* of the ball is equal to the impulse the pitcher imparts to it (force multiplied by the time it took to bring the ball up to speed).

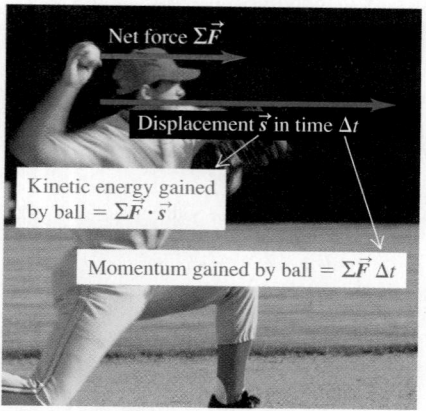

Net force $\Sigma\vec{F}$

Displacement $\vec{s}$ in time Δt

Kinetic energy gained by ball = $\Sigma\vec{F} \cdot \vec{s}$

Momentum gained by ball = $\Sigma\vec{F}\,\Delta t$

by a finite interval. By contrast, Newton's second law itself (in either of the forms $\sum \vec{F} = m\vec{a}$ or $\sum \vec{F} = d\vec{p}/dt$) is a *differential* principle, relating the forces to the rate of change of velocity or momentum at each instant.

Conceptual Example 8.1 Momentum versus kinetic energy

Consider again the race described in Conceptual Example 6.5 (Section 6.2) between two iceboats on a frictionless frozen lake. The boats have masses m and $2m$, and the wind exerts the same constant horizontal force $\vec{F}$ on each boat (see Fig. 6.14). The boats start from rest and cross the finish line a distance s away. Which boat crosses the finish line with greater momentum?

SOLUTION

In Conceptual Example 6.5 we asked how the *kinetic energies* of the boats compare when they cross the finish line. We answered this by remembering that *a body's kinetic energy equals the total work done to accelerate it from rest*. Both boats started from rest, and the total work done was the same for both boats (because the net force and the displacement were the same for both). Hence both boats had the same kinetic energy at the finish line.

Similarly, to compare the *momenta* of the boats we use the idea that *the momentum of each boat equals the impulse that accelerated it from rest*. As in Conceptual Example 6.5, the net force on each boat equals the constant horizontal wind force $\vec{F}$. Let Δt be the time a boat takes to reach the finish line, so that the impulse on the boat during that time is $\vec{J} = \vec{F}\,\Delta t$. Since the boat starts from rest, this equals the boat's momentum $\vec{p}$ at the finish line:

$$\vec{p} = \vec{F}\,\Delta t$$

Both boats are subjected to the same force $\vec{F}$, but they take different times Δt to reach the finish line. The boat of mass $2m$ accelerates more slowly and takes a longer time to travel the distance s; thus there is a greater impulse on this boat between the starting and finish lines. So the boat of mass $2m$ crosses the finish line with a greater magnitude of momentum than the boat of mass m (but with the same kinetic energy). Can you show that the boat of mass $2m$ has $\sqrt{2}$ times as much momentum at the finish line as the boat of mass m?

Example 8.2 A ball hits a wall

You throw a ball with a mass of 0.40 kg against a brick wall. It hits the wall moving horizontally to the left at 30 m/s and rebounds horizontally to the right at 20 m/s. (a) Find the impulse of the net force on the ball during its collision with the wall. (b) If the ball is in contact with the wall for 0.010 s, find the average horizontal force that the wall exerts on the ball during the impact.

SOLUTION

IDENTIFY and SET UP: We're given enough information to determine the initial and final values of the ball's momentum, so we can use the impulse–momentum theorem to find the impulse. We'll then use the definition of impulse to determine the average force. Figure 8.5 shows our sketch. We need only a single axis because the motion is purely horizontal. We'll take the positive x-direction to be to the right. In part (a) our target variable is the x-component of impulse, J_x, which we'll find from the x-components of momentum before and after the impact, using Eqs. (8.9). In part (b), our target variable is the average x-component of force $(F_{av})_x$; once we know J_x, we can also find this force by using Eqs. (8.9).

8.5 Our sketch for this problem.

Before $v_{1x} = -30$ m/s $m = 0.40$ kg

After $v_{2x} = +20$ m/s

EXECUTE: (a) With our choice of x-axis, the initial and final x-components of momentum of the ball are

$$p_{1x} = mv_{1x} = (0.40\ \text{kg})(-30\ \text{m/s}) = -12\ \text{kg}\cdot\text{m/s}$$
$$p_{2x} = mv_{2x} = (0.40\ \text{kg})(+20\ \text{m/s}) = +8.0\ \text{kg}\cdot\text{m/s}$$

From the x-equation in Eqs. (8.9), the x-component of impulse equals the *change* in the x-momentum:

$$J_x = p_{2x} - p_{1x}$$
$$= 8.0\ \text{kg}\cdot\text{m/s} - (-12\ \text{kg}\cdot\text{m/s}) = 20\ \text{kg}\cdot\text{m/s} = 20\ \text{N}\cdot\text{s}$$

(b) The collision time is $t_2 - t_1 = \Delta t = 0.010$ s. From the x-equation in Eqs. (8.9), $J_x = (F_{av})_x(t_2 - t_1) = (F_{av})_x\,\Delta t$, so

$$(F_{av})_x = \frac{J_x}{\Delta t} = \frac{20\ \text{N}\cdot\text{s}}{0.010\ \text{s}} = 2000\ \text{N}$$

EVALUATE: The x-component of impulse J_x is positive—that is, to the right in Fig. 8.5. This is as it should be: The impulse represents the "kick" that the wall imparts to the ball, and this "kick" is certainly to the right.

CAUTION Momentum is a vector Because momentum is a vector, we had to include the negative sign in writing $p_{1x} = -12\ \text{kg}\cdot\text{m/s}$. Had we carelessly omitted it, we would have calculated the impulse to be $8.0\ \text{kg}\cdot\text{m/s} - (12\ \text{kg}\cdot\text{m/s}) = -4\ \text{kg}\cdot\text{m/s}$. This would say that the wall had somehow given the ball a kick to the *left*! Make sure that you account for the *direction* of momentum in your calculations. ▮

The force that the wall exerts on the ball must have such a large magnitude (2000 N, equal to the weight of a 200-kg object) to

Continued

change the ball's momentum in such a short time. Other forces that act on the ball during the collision are comparatively weak; for instance, the gravitational force is only 3.9 N. Thus, during the short time that the collision lasts, we can ignore all other forces on the ball. Figure 8.6 shows the impact of a tennis ball and racket.

Note that the 2000-N value we calculated is the *average* horizontal force that the wall exerts on the ball during the impact. It corresponds to the horizontal line $(F_{av})_x$ in Fig. 8.3a. The horizontal force is zero before impact, rises to a maximum, and then decreases to zero when the ball loses contact with the wall. If the ball is relatively rigid, like a baseball or golf ball, the collision lasts a short time and the maximum force is large, as in the blue curve in Fig. 8.3b. If the ball is softer, like a tennis ball, the collision time is longer and the maximum force is less, as in the orange curve in Fig. 8.3b.

8.6 Typically, a tennis ball is in contact with the racket for approximately 0.01 s. The ball flattens noticeably due to the tremendous force exerted by the racket.

Example 8.3 | Kicking a soccer ball

A soccer ball has a mass of 0.40 kg. Initially it is moving to the left at 20 m/s, but then it is kicked. After the kick it is moving at 45° upward and to the right with speed 30 m/s (Fig. 8.7a). Find the impulse of the net force and the average net force, assuming a collision time $\Delta t = 0.010$ s.

SOLUTION

IDENTIFY and SET UP: The ball moves in two dimensions, so we must treat momentum and impulse as vector quantities. We take the x-axis to be horizontally to the right and the y-axis to be vertically upward. Our target variables are the components of the net

8.7 (a) Kicking a soccer ball. (b) Finding the average force on the ball from its components.

(a) Before-and-after diagram

(b) Average force on the ball

impulse on the ball, J_x and J_y, and the components of the average net force on the ball, $(F_{av})_x$ and $(F_{av})_y$. We'll find them using the impulse–momentum theorem in its component form, Eqs. (8.9).

EXECUTE: Using $\cos 45° = \sin 45° = 0.707$, we find the ball's velocity components before and after the kick:

$$v_{1x} = -20 \text{ m/s} \qquad v_{1y} = 0$$
$$v_{2x} = v_{2y} = (30 \text{ m/s})(0.707) = 21.2 \text{ m/s}$$

From Eqs. (8.9), the impulse components are

$$J_x = p_{2x} - p_{1x} = m(v_{2x} - v_{1x})$$
$$= (0.40 \text{ kg})[21.2 \text{ m/s} - (-20 \text{ m/s})] = 16.5 \text{ kg} \cdot \text{m/s}$$
$$J_y = p_{2y} - p_{1y} = m(v_{2y} - v_{1y})$$
$$= (0.40 \text{ kg})(21.2 \text{ m/s} - 0) = 8.5 \text{ kg} \cdot \text{m/s}$$

From Eq. (8.8), the average net force components are

$$(F_{av})_x = \frac{J_x}{\Delta t} = 1650 \text{ N} \qquad (F_{av})_y = \frac{J_y}{\Delta t} = 850 \text{ N}$$

The magnitude and direction of the average net force $\vec{F}_{av}$ are

$$F_{av} = \sqrt{(1650 \text{ N})^2 + (850 \text{ N})^2} = 1.9 \times 10^3 \text{ N}$$
$$\theta = \arctan \frac{850 \text{ N}}{1650 \text{ N}} = 27°$$

The ball was not initially at rest, so its final velocity does *not* have the same direction as the average force that acted on it.

EVALUATE: $\vec{F}_{av}$ includes the force of gravity, which is very small; the weight of the ball is only 3.9 N. As in Example 8.2, the average force acting during the collision is exerted almost entirely by the object that the ball hit (in this case, the soccer player's foot).

Test Your Understanding of Section 8.1 Rank the following situations according to the magnitude of the impulse of the net force, from largest value to smallest value. In each situation a 1000-kg automobile is moving along a straight east–west road. (i) The automobile is initially moving east at 25 m/s and comes to a stop in 10 s. (ii) The automobile is initially moving east at 25 m/s and comes to a stop in 5 s. (iii) The automobile is initially at rest, and a 2000-N net force toward the east is applied to it for 10 s. (iv) The automobile is initially moving east at 25 m/s, and a 2000-N net force toward the west is applied to it for 10 s. (v) The automobile is initially moving east at 25 m/s. Over a 30-s period, the automobile reverses direction and ends up moving west at 25 m/s.

8.2 Conservation of Momentum

The concept of momentum is particularly important in situations in which we have two or more bodies that *interact*. To see why, let's consider first an idealized system of two bodies that interact with each other but not with anything else—for example, two astronauts who touch each other as they float freely in the zero-gravity environment of outer space (Fig. 8.8). Think of the astronauts as particles. Each particle exerts a force on the other; according to Newton's third law, the two forces are always equal in magnitude and opposite in direction. Hence, the *impulses* that act on the two particles are equal and opposite, and the changes in momentum of the two particles are equal and opposite.

Let's go over that again with some new terminology. For any system, the forces that the particles of the system exert on each other are called **internal forces**. Forces exerted on any part of the system by some object outside it are called **external forces**. For the system shown in Fig. 8.8, the internal forces are $\vec{F}_{B \text{ on } A}$, exerted by particle B on particle A, and $\vec{F}_{A \text{ on } B}$, exerted by particle A on particle B. There are *no* external forces; when this is the case, we have an **isolated system**.

The net force on particle A is $\vec{F}_{B \text{ on } A}$ and the net force on particle B is $\vec{F}_{A \text{ on } B}$, so from Eq. (8.4) the rates of change of the momenta of the two particles are

$$\vec{F}_{B \text{ on } A} = \frac{d\vec{p}_A}{dt} \qquad \vec{F}_{A \text{ on } B} = \frac{d\vec{p}_B}{dt} \tag{8.10}$$

The momentum of each particle changes, but these changes are related to each other by Newton's third law: The two forces $\vec{F}_{B \text{ on } A}$ and $\vec{F}_{A \text{ on } B}$ are always equal in magnitude and opposite in direction. That is, $\vec{F}_{B \text{ on } A} = -\vec{F}_{A \text{ on } B}$, so $\vec{F}_{B \text{ on } A} + \vec{F}_{A \text{ on } B} = 0$. Adding together the two equations in Eq. (8.10), we have

$$\vec{F}_{B \text{ on } A} + \vec{F}_{A \text{ on } B} = \frac{d\vec{p}_A}{dt} + \frac{d\vec{p}_B}{dt} = \frac{d(\vec{p}_A + \vec{p}_B)}{dt} = 0 \tag{8.11}$$

The rates of change of the two momenta are equal and opposite, so the rate of change of the vector sum $\vec{p}_A + \vec{p}_B$ is zero. We now define the **total momentum** $\vec{P}$ of the system of two particles as the vector sum of the momenta of the individual particles; that is,

$$\vec{P} = \vec{p}_A + \vec{p}_B \tag{8.12}$$

Then Eq. (8.11) becomes, finally,

$$\vec{F}_{B \text{ on } A} + \vec{F}_{A \text{ on } B} = \frac{d\vec{P}}{dt} = 0 \tag{8.13}$$

The time rate of change of the *total* momentum $\vec{P}$ is zero. Hence the total momentum of the system is constant, even though the individual momenta of the particles that make up the system can change.

If external forces are also present, they must be included on the left side of Eq. (8.13) along with the internal forces. Then the total momentum is, in general, not constant. But if the vector sum of the external forces is zero, as in Fig. 8.9, these forces have no effect on the left side of Eq. (8.13), and $d\vec{P}/dt$ is again zero. Thus we have the following general result:

> If the vector sum of the external forces on a system is zero, the total momentum of the system is constant.

This is the simplest form of the **principle of conservation of momentum.** This principle is a direct consequence of Newton's third law. What makes this principle useful is that it doesn't depend on the detailed nature of the internal forces that

8.8 Two astronauts push each other as they float freely in the zero-gravity environment of space.

No external forces act on the two-astronaut system, so its total momentum is conserved.

The forces the astronauts exert on each other form an action–reaction pair.

8.9 Two ice skaters push each other as they skate on a frictionless, horizontal surface. (Compare to Fig. 8.8.)

The forces the skaters exert on each other form an action–reaction pair.

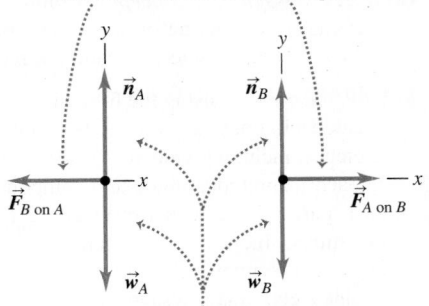

Although the normal and gravitational forces are external, their vector sum is zero, so the total momentum is conserved.

act between members of the system. This means that we can apply conservation of momentum even if (as is often the case) we know very little about the internal forces. We have used Newton's second law to derive this principle, so we have to be careful to use it only in inertial frames of reference.

We can generalize this principle for a system that contains any number of particles A, B, C, . . . interacting only with one another. The total momentum of such a system is

$$\vec{P} = \vec{p}_A + \vec{p}_B + \cdots = m_A \vec{v}_A + m_B \vec{v}_B + \cdots \quad \begin{array}{l}\text{(total momentum of} \\ \text{a system of particles)}\end{array} \quad (8.14)$$

ActivPhysics 6.3: Momentum Conservation and Collisions
ActivPhysics 6.7: Explosion Problems
ActivPhysics 6.10: Pendulum Person-Projectile Bowling

8.10 When applying conservation of momentum, remember that momentum is a vector quantity!

A system of two particles with momenta in different directions

$p_A = 18 \text{ kg} \cdot \text{m/s}$
$p_B = 24 \text{ kg} \cdot \text{m/s}$

You CANNOT find the magnitude of the total momentum by adding the magnitudes of the individual momenta!

$P = p_A + p_B = 42 \text{ kg} \cdot \text{m/s}$ ◀ WRONG

Instead, use vector addition:

$\vec{P} = \vec{p}_A + \vec{p}_B$ ◀ RIGHT!

$P = |\vec{p}_A + \vec{p}_B|$
$= 30 \text{ kg} \cdot \text{m/s at } \theta = 37°$

We make the same argument as before: The total rate of change of momentum of the system due to each action–reaction pair of internal forces is zero. Thus the total rate of change of momentum of the entire system is zero whenever the vector sum of the external forces acting on it is zero. The internal forces can change the momenta of individual particles in the system but not the *total* momentum of the system.

CAUTION **Conservation of momentum means conservation of its components** When you apply the conservation of momentum to a system, remember that momentum is a *vector* quantity. Hence you must use vector addition to compute the total momentum of a system (Fig. 8.10). Using components is usually the simplest method. If p_{Ax}, p_{Ay}, and p_{Az} are the components of momentum of particle A, and similarly for the other particles, then Eq. (8.14) is equivalent to the component equations

$$P_x = p_{Ax} + p_{Bx} + \cdots$$
$$P_y = p_{Ay} + p_{By} + \cdots \quad (8.15)$$
$$P_z = p_{Az} + p_{Bz} + \cdots$$

If the vector sum of the external forces on the system is zero, then P_x, P_y, and P_z are all constant. ▮

In some ways the principle of conservation of momentum is more general than the principle of conservation of mechanical energy. For example, mechanical energy is conserved only when the internal forces are *conservative*—that is, when the forces allow two-way conversion between kinetic and potential energy—but conservation of momentum is valid even when the internal forces are *not* conservative. In this chapter we will analyze situations in which both momentum and mechanical energy are conserved, and others in which only momentum is conserved. These two principles play a fundamental role in all areas of physics, and we will encounter them throughout our study of physics.

Problem-Solving Strategy 8.1 **Conservation of Momentum**

(MP)

IDENTIFY *the relevant concepts:* Confirm that the vector sum of the external forces acting on the system of particles is zero. If it isn't zero, you can't use conservation of momentum.

SET UP *the problem* using the following steps:
1. Treat each body as a particle. Draw "before" and "after" sketches, including velocity vectors. Assign algebraic symbols to each magnitude, angle, and component. Use letters to label each particle and subscripts 1 and 2 for "before" and "after" quantities. Include any given values such as magnitudes, angles, or components.
2. Define a coordinate system and show it in your sketches; define the positive direction for each axis.
3. Identify the target variables.

EXECUTE *the solution:*
1. Write an equation in symbols equating the total initial and final x-components of momentum, using $p_x = mv_x$ for each particle. Write a corresponding equation for the y-components. Velocity components can be positive or negative, so be careful with signs!
2. In some problems, energy considerations (discussed in Section 8.4) give additional equations relating the velocities.
3. Solve your equations to find the target variables.

EVALUATE *your answer:* Does your answer make physical sense? If your target variable is a certain body's momentum, check that the direction of the momentum is reasonable.

Example 8.4 Recoil of a rifle

A marksman holds a rifle of mass $m_R = 3.00$ kg loosely, so it can recoil freely. He fires a bullet of mass $m_B = 5.00$ g horizontally with a velocity relative to the ground of $v_{Bx} = 300$ m/s. What is the recoil velocity v_{Rx} of the rifle? What are the final momentum and kinetic energy of the bullet and rifle?

SOLUTION

IDENTIFY and SET UP: If the marksman exerts negligible horizontal forces on the rifle, then there is no net horizontal force on the system (the bullet and rifle) during the firing, and the total horizontal momentum of the system is conserved. Figure 8.11 shows our sketch. We take the positive x-axis in the direction of aim. The rifle and the bullet are initially at rest, so the initial x-component of total momentum is zero. After the shot is fired, the bullet's x-momentum is $p_{Bx} = m_B v_{Bx}$ and the rifle's x-momentum

8.11 Our sketch for this problem.

is $p_{Rx} = m_R v_{Rx}$. Our target variables are v_{Rx}, p_{Bx}, p_{Rx}, and the final kinetic energies $K_B = \frac{1}{2} m_B v_{Bx}^2$ and $K_R = \frac{1}{2} m_R v_{Rx}^2$.

EXECUTE: Conservation of the x-component of total momentum gives

$$P_x = 0 = m_B v_{Bx} + m_R v_{Rx}$$

$$v_{Rx} = -\frac{m_B}{m_R} v_{Bx} = -\left(\frac{0.00500 \text{ kg}}{3.00 \text{ kg}}\right)(300 \text{ m/s}) = -0.500 \text{ m/s}$$

The negative sign means that the recoil is in the direction opposite to that of the bullet.

The final momenta and kinetic energies are

$$p_{Bx} = m_B v_{Bx} = (0.00500 \text{ kg})(300 \text{ m/s}) = 1.50 \text{ kg} \cdot \text{m/s}$$

$$K_B = \frac{1}{2} m_B v_{Bx}^2 = \frac{1}{2}(0.00500 \text{ kg})(300 \text{ m/s})^2 = 225 \text{ J}$$

$$p_{Rx} = m_R v_{Rx} = (3.00 \text{ kg})(-0.500 \text{ m/s}) = -1.50 \text{ kg} \cdot \text{m/s}$$

$$K_R = \frac{1}{2} m_R v_{Rx}^2 = \frac{1}{2}(3.00 \text{ kg})(-0.500 \text{ m/s})^2 = 0.375 \text{ J}$$

EVALUATE: The bullet and rifle have equal and opposite final *momenta* thanks to Newton's third law: They experience equal and opposite interaction forces that act for the same *time*, so the impulses are equal and opposite. But the bullet travels a much greater *distance* than the rifle during the interaction. Hence the force on the bullet does more work than the force on the rifle, giving the bullet much greater *kinetic energy* than the rifle. The 600:1 ratio of the two kinetic energies is the inverse of the ratio of the masses; in fact, you can show that this always happens in recoil situations (see Exercise 8.26).

Example 8.5 Collision along a straight line

Two gliders with different masses move toward each other on a frictionless air track (Fig. 8.12a). After they collide (Fig. 8.12b), glider B has a final velocity of $+2.0$ m/s (Fig. 8.12c). What is the final velocity of glider A? How do the changes in momentum and in velocity compare?

SOLUTION

IDENTIFY and SET UP: As for the skaters in Fig. 8.9, the total vertical force on each glider is zero, and the net force on each individual glider is the horizontal force exerted on it by the other glider. The net external force on the *system* of two gliders is zero, so their total momentum is conserved. We take the positive x-axis to be to the right. We are given the masses and initial velocities of both gliders and the final velocity of glider B. Our target variables are v_{A2x}, the final x-component of velocity of glider A, and the changes in momentum and in velocity of the two gliders (the value *after* the collision minus the value *before* the collision).

EXECUTE: The x-component of total momentum before the collision is

$$P_x = m_A v_{A1x} + m_B v_{B1x}$$
$$= (0.50 \text{ kg})(2.0 \text{ m/s}) + (0.30 \text{ kg})(-2.0 \text{ m/s})$$
$$= 0.40 \text{ kg} \cdot \text{m/s}$$

8.12 Two gliders colliding on an air track.

(a) Before collision

$v_{A1x} = 2.0$ m/s $v_{B1x} = -2.0$ m/s

$m_A = 0.50$ kg $m_B = 0.30$ kg

(b) Collision

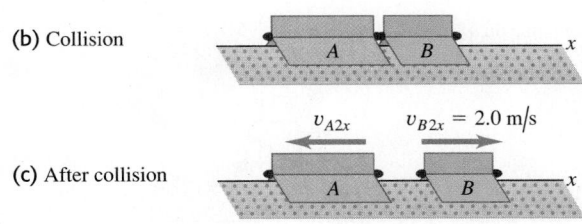

(c) After collision

v_{A2x} $v_{B2x} = 2.0$ m/s

This is positive (to the right in Fig. 8.12) because A has a greater magnitude of momentum than B. The x-component of total momentum has the same value after the collision, so

$$P_x = m_A v_{A2x} + m_B v_{B2x}$$

Continued

We solve for v_{A2x}:

$$v_{A2x} = \frac{P_x - m_B v_{B2x}}{m_A} = \frac{0.40 \text{ kg} \cdot \text{m/s} - (0.30 \text{ kg})(2.0 \text{ m/s})}{0.50 \text{ kg}}$$

$$= -0.40 \text{ m/s}$$

The changes in the x-momenta are

$$m_A v_{A2x} - m_A v_{A1x} = (0.50 \text{ kg})(-0.40 \text{ m/s})$$
$$- (0.50 \text{ kg})(2.0 \text{ m/s}) = -1.2 \text{ kg} \cdot \text{m/s}$$

$$m_B v_{B2x} - m_B v_{B1x} = (0.30 \text{ kg})(2.0 \text{ m/s})$$
$$- (0.30 \text{ kg})(-2.0 \text{ m/s}) = +1.2 \text{ kg} \cdot \text{m/s}$$

The changes in x-velocities are

$$v_{A2x} - v_{A1x} = (-0.40 \text{ m/s}) - 2.0 \text{ m/s} = -2.4 \text{ m/s}$$
$$v_{B2x} - v_{B1x} = 2.0 \text{ m/s} - (-2.0 \text{ m/s}) = +4.0 \text{ m/s}$$

EVALUATE: The gliders were subjected to equal and opposite interaction forces for the same time during their collision. By the impulse–momentum theorem, they experienced equal and opposite impulses and therefore equal and opposite changes in momentum. But by Newton's second law, the less massive glider (B) had a greater magnitude of acceleration and hence a greater velocity change.

Example 8.6 **Collision in a horizontal plane**

Figure 8.13a shows two battling robots on a frictionless surface. Robot A, with mass 20 kg, initially moves at 2.0 m/s parallel to the x-axis. It collides with robot B, which has mass 12 kg and is initially at rest. After the collision, robot A moves at 1.0 m/s in a direction that makes an angle $\alpha = 30°$ with its initial direction (Fig. 8.13b). What is the final velocity of robot B?

SOLUTION

IDENTIFY and SET UP: There are no horizontal external forces, so the x- and y-components of the total momentum of the system are both conserved. Momentum conservation requires that the sum of the x-components of momentum *before* the collision (subscript 1) must equal the sum *after* the collision (subscript 2), and similarly for the sums of the y-components. Our target variable is $\vec{v}_{B2}$, the final velocity of robot B.

8.13 Views from above of the velocities (a) before and (b) after the collision.

(a) Before collision

(b) After collision

EXECUTE: The momentum-conservation equations and their solutions for v_{B2x} and v_{B2y} are

$$m_A v_{A1x} + m_B v_{B1x} = m_A v_{A2x} + m_B v_{B2x}$$

$$v_{B2x} = \frac{m_A v_{A1x} + m_B v_{B1x} - m_A v_{A2x}}{m_B}$$

$$= \frac{\left[\begin{array}{c} (20 \text{ kg})(2.0 \text{ m/s}) + (12 \text{ kg})(0) \\ - (20 \text{ kg})(1.0 \text{ m/s})(\cos 30°) \end{array} \right]}{12 \text{ kg}}$$

$$= 1.89 \text{ m/s}$$

$$m_A v_{A1y} + m_B v_{B1y} = m_A v_{A2y} + m_B v_{B2y}$$

$$v_{B2y} = \frac{m_A v_{A1y} + m_B v_{B1y} - m_A v_{A2y}}{m_B}$$

$$= \frac{\left[\begin{array}{c} (20 \text{ kg})(0) + (12 \text{ kg})(0) \\ - (20 \text{ kg})(1.0 \text{ m/s})(\sin 30°) \end{array} \right]}{12 \text{ kg}}$$

$$= -0.83 \text{ m/s}$$

Figure 8.13b shows the motion of robot B after the collision. The magnitude of $\vec{v}_{B2}$ is

$$v_{B2} = \sqrt{(1.89 \text{ m/s})^2 + (-0.83 \text{ m/s})^2} = 2.1 \text{ m/s}$$

and the angle of its direction from the positive x-axis is

$$\beta = \arctan \frac{-0.83 \text{ m/s}}{1.89 \text{ m/s}} = -24°$$

EVALUATE: We can check our answer by confirming that the components of total momentum before and after the collision are equal. Initially robot A has x-momentum $m_A v_{A1x} = (20 \text{ kg})(2.0 \text{ m/s}) = 40 \text{ kg} \cdot \text{m/s}$ and zero y-momentum; robot B has zero momentum. After the collision, the momentum components are $m_A v_{A2x} = (20 \text{ kg})(1.0 \text{ m/s})(\cos 30°) = 17 \text{ kg} \cdot \text{m/s}$ and $m_B v_{B2x} = (12 \text{ kg})(1.89 \text{ m/s}) = 23 \text{ kg} \cdot \text{m/s}$; the total x-momentum is 40 kg $\cdot$ m/s, the same as before the collision. The final y-components are $m_A v_{A2y} = (20 \text{ kg})(1.0 \text{ m/s})(\sin 30°) = 10 \text{ kg} \cdot \text{m/s}$ and $m_B v_{B2y} = (12 \text{ kg})(-0.83 \text{ m/s}) = -10 \text{ kg} \cdot \text{m/s}$; the total y-component of momentum is zero, the same as before the collision.

Test Your Understanding of Section 8.2 A spring-loaded toy sits at rest on a horizontal, frictionless surface. When the spring releases, the toy breaks into three equal-mass pieces, *A*, *B*, and *C*, which slide along the surface. Piece *A* moves off in the negative *x*-direction, while piece *B* moves off in the negative *y*-direction. (a) What are the signs of the velocity components of piece *C*? (b) Which of the three pieces is moving the fastest?

8.3 Momentum Conservation and Collisions

To most people the term *collision* is likely to mean some sort of automotive disaster. We'll use it in that sense, but we'll also broaden the meaning to include any strong interaction between bodies that lasts a relatively short time. So we include not only car accidents but also balls colliding on a billiard table, neutrons hitting atomic nuclei in a nuclear reactor, the impact of a meteor on the Arizona desert, and a close encounter of a spacecraft with the planet Saturn.

If the forces between the bodies are much larger than any external forces, as is the case in most collisions, we can neglect the external forces entirely and treat the bodies as an *isolated* system. Then momentum is conserved and the total momentum of the system has the same value before and after the collision. Two cars colliding at an icy intersection provide a good example. Even two cars colliding on dry pavement can be treated as an isolated system during the collision if the forces between the cars are much larger than the friction forces of pavement against tires.

Elastic and Inelastic Collisions

If the forces between the bodies are also *conservative,* so that no mechanical energy is lost or gained in the collision, the total *kinetic* energy of the system is the same after the collision as before. Such a collision is called an **elastic collision.** A collision between two marbles or two billiard balls is almost completely elastic. Figure 8.14 shows a model for an elastic collision. When the gliders collide, their springs are momentarily compressed and some of the original kinetic energy is momentarily converted to elastic potential energy. Then the gliders bounce apart, the springs expand, and this potential energy is converted back to kinetic energy.

A collision in which the total kinetic energy after the collision is *less* than before the collision is called an **inelastic collision.** A meatball landing on a plate of spaghetti and a bullet embedding itself in a block of wood are examples of inelastic collisions. An inelastic collision in which the colliding bodies stick together and move as one body after the collision is often called a **completely inelastic collision.** Figure 8.15 shows an example; we have replaced the spring bumpers in Fig. 8.14 with Velcro®, which sticks the two bodies together.

> **CAUTION** **An inelastic collision doesn't have to be *completely* inelastic** It's a common misconception that the *only* inelastic collisions are those in which the colliding bodies stick together. In fact, inelastic collisions include many situations in which the bodies do *not* stick. If two cars bounce off each other in a "fender bender," the work done to deform the fenders cannot be recovered as kinetic energy of the cars, so the collision is inelastic (Fig. 8.16).

Remember this rule: **In any collision in which external forces can be neglected, momentum is conserved and the total momentum before equals the total momentum after; in elastic collisions *only,* the total kinetic energy before equals the total kinetic energy after.**

Completely Inelastic Collisions

Let's look at what happens to momentum and kinetic energy in a *completely* inelastic collision of two bodies (*A* and *B*), as in Fig. 8.15. Because the two bodies stick together after the collision, they have the same final velocity $\vec{v}_2$:

$$\vec{v}_{A2} = \vec{v}_{B2} = \vec{v}_2$$

8.14 Two gliders undergoing an elastic collision on a frictionless surface. Each glider has a steel spring bumper that exerts a conservative force on the other glider.

(a) Before collision

(b) Elastic collision

Kinetic energy is stored as potential energy in compressed springs.

(c) After collision

The system of the two gliders has the same kinetic energy after the collision as before it.

8.15 Two gliders undergoing a completely inelastic collision. The spring bumpers on the gliders are replaced by Velcro®, so the gliders stick together after collision.

(a) Before collision

(b) Completely inelastic collision

The gliders stick together.

(c) After collision

The system of the two gliders has less kinetic energy after the collision than before it.

8.16 Automobile collisions are intended to be inelastic, so that the structure of the car absorbs as much of the energy of the collision as possible. This absorbed energy cannot be recovered, since it goes into a permanent deformation of the car.

Conservation of momentum gives the relationship

$$m_A \vec{v}_{A1} + m_B \vec{v}_{B1} = (m_A + m_B)\vec{v}_2 \quad \text{(completely inelastic collision)} \quad (8.16)$$

If we know the masses and initial velocities, we can compute the common final velocity $\vec{v}_2$.

Suppose, for example, that a body with mass m_A and initial x-component of velocity v_{A1x} collides inelastically with a body with mass m_B that is initially at rest ($v_{B1x} = 0$). From Eq. (8.16) the common x-component of velocity v_{2x} of both bodies after the collision is

$$v_{2x} = \frac{m_A}{m_A + m_B} v_{A1x} \quad \begin{array}{l}\text{(completely inelastic collision,}\\ B \text{ initially at rest)}\end{array} \quad (8.17)$$

Let's verify that the total kinetic energy after this completely inelastic collision is less than before the collision. The motion is purely along the x-axis, so the kinetic energies K_1 and K_2 before and after the collision, respectively, are

$$K_1 = \tfrac{1}{2} m_A v_{A1x}^2$$

$$K_2 = \tfrac{1}{2}(m_A + m_B)v_{2x}^2 = \tfrac{1}{2}(m_A + m_B)\left(\frac{m_A}{m_A + m_B}\right)^2 v_{A1x}^2$$

The ratio of final to initial kinetic energy is

$$\frac{K_2}{K_1} = \frac{m_A}{m_A + m_B} \quad \begin{array}{l}\text{(completely inelastic collision,}\\ B \text{ initially at rest)}\end{array} \quad (8.18)$$

The right side is always less than unity because the denominator is always greater than the numerator. Even when the initial velocity of m_B is not zero, it is not hard to verify that the kinetic energy after a completely inelastic collision is always less than before.

Please note: We don't recommend memorizing Eq. (8.17) or (8.18). We derived them only to prove that kinetic energy is always lost in a completely inelastic collision.

Example 8.7 | A completely inelastic collision

We repeat the collision described in Example 8.5 (Section 8.2), but this time equip the gliders so that they stick together when they collide. Find the common final x-velocity, and compare the initial and final kinetic energies of the system.

SOLUTION

IDENTIFY and SET UP: There are no external forces in the x-direction, so the x-component of momentum is conserved. Figure 8.17 shows our sketch. Our target variables are the final x-velocity v_{2x} and the initial and final kinetic energies K_1 and K_2.

8.17 Our sketch for this problem.

Before $\quad v_{A1x} = 2.0$ m/s $\qquad v_{B1x} = -2.0$ m/s

$\boxed{A} \longrightarrow \qquad \longleftarrow \boxed{B}$
$m_A = 0.50$ kg $\qquad m_B = 0.30$ kg

After $\qquad v_{2x} = ?$
$\boxed{A\ B} \rightarrow$

EXECUTE: From conservation of momentum,

$$m_A v_{A1x} + m_B v_{B1x} = (m_A + m_B)v_{2x}$$

$$v_{2x} = \frac{m_A v_{A1x} + m_B v_{B1x}}{m_A + m_B}$$

$$= \frac{(0.50 \text{ kg})(2.0 \text{ m/s}) + (0.30 \text{ kg})(-2.0 \text{ m/s})}{0.50 \text{ kg} + 0.30 \text{ kg}}$$

$$= 0.50 \text{ m/s}$$

Because v_{2x} is positive, the gliders move together to the right after the collision. Before the collision, the kinetic energies are

$$K_A = \tfrac{1}{2} m_A v_{A1x}^2 = \tfrac{1}{2}(0.50 \text{ kg})(2.0 \text{ m/s})^2 = 1.0 \text{ J}$$

$$K_B = \tfrac{1}{2} m_B v_{B1x}^2 = \tfrac{1}{2}(0.30 \text{ kg})(-2.0 \text{ m/s})^2 = 0.60 \text{ J}$$

The total kinetic energy before the collision is $K_1 = K_A + K_B = 1.6$ J. The kinetic energy after the collision is

$$K_2 = \tfrac{1}{2}(m_A + m_B)v_{2x}^2 = \tfrac{1}{2}(0.50 \text{ kg} + 0.30 \text{ kg})(0.50 \text{ m/s})^2$$

$$= 0.10 \text{ J}$$

EVALUATE: The final kinetic energy is only $\frac{1}{16}$ of the original; $\frac{15}{16}$ is converted from mechanical energy to other forms. If there is a wad of chewing gum between the gliders, it squashes and becomes warmer. If there is a spring between the gliders that is compressed as they lock together, the energy is stored as potential energy of the spring. In both cases the *total* energy of the system is conserved, although *kinetic* energy is not. In an isolated system, however, momentum is *always* conserved whether the collision is elastic or not.

Example 8.8   The ballistic pendulum

Figure 8.18 shows a ballistic pendulum, a simple system for measuring the speed of a bullet. A bullet of mass m_B makes a completely inelastic collision with a block of wood of mass m_W, which is suspended like a pendulum. After the impact, the block swings up to a maximum height y. In terms of y, m_B, and m_W, what is the initial speed v_1 of the bullet?

SOLUTION

IDENTIFY: We'll analyze this event in two stages: (1) the embedding of the bullet in the block and (2) the pendulum swing of the block. During the first stage, the bullet embeds itself in the block so quickly that the block does not move appreciably. The supporting strings remain nearly vertical, so negligible external horizontal force acts on the bullet–block system, and the horizontal component of momentum is conserved. Mechanical energy is *not* conserved during this stage, however, because a nonconservative force does work (the force of friction between bullet and block).

In the second stage, the block and bullet move together. The only forces acting on this system are gravity (a conservative force) and the string tensions (which do no work). Thus, as the block swings, *mechanical energy is conserved*. Momentum is *not* conserved during this stage, however, because there is a net external force (the forces of gravity and string tension don't cancel when the strings are inclined).

SET UP: We take the positive x-axis to the right and the positive y-axis upward. Our target variable is v_1. Another unknown quantity is the speed v_2 of the system just after the collision. We'll use momentum conservation in the first stage to relate v_1 to v_2, and we'll use energy conservation in the second stage to relate v_2 to y.

EXECUTE: In the first stage, all velocities are in the $+x$-direction. Momentum conservation gives

$$m_B v_1 = (m_B + m_W)v_2$$
$$v_1 = \frac{m_B + m_W}{m_B}v_2$$

At the beginning of the second stage, the system has kinetic energy $K = \frac{1}{2}(m_B + m_W)v_2^2$. The system swings up and comes to rest for an instant at a height y, where its kinetic energy is zero and the potential energy is $(m_B + m_W)gy$; it then swings back down. Energy conservation gives

$$\frac{1}{2}(m_B + m_W)v_2^2 = (m_B + m_W)gy$$
$$v_2 = \sqrt{2gy}$$

We substitute this expression for v_2 into the momentum equation:

$$v_1 = \frac{m_B + m_W}{m_B}\sqrt{2gy}$$

EVALUATE: Let's plug in the realistic numbers $m_B = 5.00 \text{ g} = 0.00500 \text{ kg}$, $m_W = 2.00 \text{ kg}$, and $y = 3.00 \text{ cm} = 0.0300 \text{ m}$. We then have

$$v_1 = \frac{0.00500 \text{ kg} + 2.00 \text{ kg}}{0.00500 \text{ kg}}\sqrt{2(9.80 \text{ m/s}^2)(0.0300 \text{ m})}$$

$$= 307 \text{ m/s}$$

The speed v_2 of the block just after impact is

$$v_2 = \sqrt{2gy} = \sqrt{2(9.80 \text{ m/s}^2)(0.0300 \text{ m})}$$

$$= 0.767 \text{ m/s}$$

The speeds v_1 and v_2 seem realistic. The kinetic energy of the bullet before impact is $\frac{1}{2}(0.00500 \text{ kg})(307 \text{ m/s})^2 = 236 \text{ J}$. Just after impact the kinetic energy of the system is $\frac{1}{2}(2.005 \text{ kg})(0.767 \text{ m/s})^2 = 0.590 \text{ J}$. Nearly all the kinetic energy disappears as the wood splinters and the bullet and block become warmer.

8.18 A ballistic pendulum.

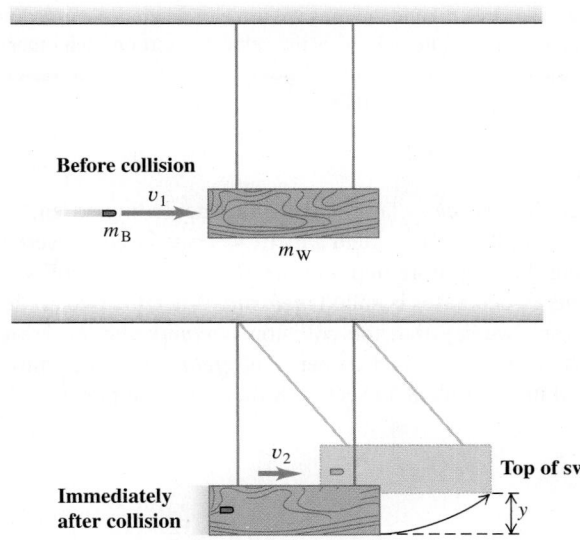

Example 8.9 An automobile collision

A 1000-kg car traveling north at 15 m/s collides with a 2000-kg truck traveling east at 10 m/s. The occupants, wearing seat belts, are uninjured, but the two vehicles move away from the impact point as one. The insurance adjustor asks you to find the velocity of the wreckage just after impact. What is your answer?

SOLUTION

IDENTIFY and SET UP: We'll treat the cars as an isolated system, so that the momentum of the system is conserved. We can do so because (as we show below) the magnitudes of the horizontal forces that the cars exert on each other during the collision are much larger than any external forces such as friction. Figure 8.19 shows our sketch and the coordinate axes. We can find the total momentum $\vec{P}$ before the collision using Eqs. (8.15). The momentum has the same value just after the collision; hence we can find the velocity $\vec{V}$ just after the collision (our target variable) using $\vec{P} = M\vec{V}$, where $M = m_C + m_T = 3000$ kg is the mass of the wreckage.

EXECUTE: From Eqs. (8.15), the components of $\vec{P}$ are

$$P_x = p_{Cx} + p_{Tx} = m_C v_{Cx} + m_T v_{Tx}$$
$$= (1000 \text{ kg})(0) + (2000 \text{ kg})(10 \text{ m/s})$$
$$= 2.0 \times 10^4 \text{ kg} \cdot \text{m/s}$$
$$P_y = p_{Cy} + p_{Ty} = m_C v_{Cy} + m_T v_{Ty}$$
$$= (1000 \text{ kg})(15 \text{ m/s}) + (2000 \text{ kg})(0)$$
$$= 1.5 \times 10^4 \text{ kg} \cdot \text{m/s}$$

The magnitude of $\vec{P}$ is

$$P = \sqrt{(2.0 \times 10^4 \text{ kg} \cdot \text{m/s})^2 + (1.5 \times 10^4 \text{ kg} \cdot \text{m/s})^2}$$
$$= 2.5 \times 10^4 \text{ kg} \cdot \text{m/s}$$

and its direction is given by the angle θ shown in Fig. 8.19:

$$\tan \theta = \frac{P_y}{P_x} = \frac{1.5 \times 10^4 \text{ kg} \cdot \text{m/s}}{2.0 \times 10^4 \text{ kg} \cdot \text{m/s}} = 0.75 \qquad \theta = 37°$$

8.19 Our sketch for this problem.

From $\vec{P} = M\vec{V}$, the direction of the velocity $\vec{V}$ just after the collision is also $\theta = 37°$. The velocity magnitude is

$$V = \frac{P}{M} = \frac{2.5 \times 10^4 \text{ kg} \cdot \text{m/s}}{3000 \text{ kg}} = 8.3 \text{ m/s}$$

EVALUATE: This is an inelastic collision, so we expect the total kinetic energy to be less after the collision than before. As you can show, the initial kinetic energy is 2.1×10^5 J and the final value is 1.0×10^5 J.

We can now justify our neglect of the external forces on the vehicles during the collision. The car's weight is about 10,000 N; if the coefficient of kinetic friction is 0.5, the friction force on the car during the impact is about 5000 N. The car's initial kinetic energy is $\frac{1}{2}(1000 \text{ kg})(15 \text{ m/s})^2 = 1.1 \times 10^5$ J, so -1.1×10^5 J of work must be done to stop it. If the car crumples by 0.20 m in stopping, a force of magnitude $(1.1 \times 10^5 \text{ J})/(0.20 \text{ m}) = 5.5 \times 10^5$ N would be needed; that's 110 times the friction force. So it's reasonable to treat the external force of friction as negligible compared with the internal forces the vehicles exert on each other.

Classifying Collisions

It's important to remember that we can classify collisions according to energy considerations (Fig. 8.20). A collision in which kinetic energy is conserved is called *elastic*. (We'll explore these in more depth in the next section.) A collision in which the total kinetic energy decreases is called *inelastic*. When the two bodies have a common final velocity, we say that the collision is *completely inelastic*. There are also cases in which the final kinetic energy is *greater* than the initial value. Rifle recoil, discussed in Example 8.4 (Section 8.2), is an example.

8.20 Collisions are classified according to energy considerations.

Elastic:
Kinetic energy conserved.

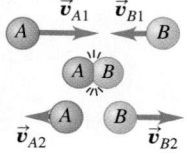

Inelastic:
Some kinetic energy lost.

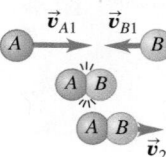

Completely inelastic:
Bodies have same final velocity.

Finally, we emphasize again that we can sometimes use momentum conservation even when there are external forces acting on the system, if the net external force acting on the colliding bodies is small in comparison with the internal forces during the collision (as in Example 8.9)

Test Your Understanding of Section 8.3 For each situation, state whether the collision is elastic or inelastic. If it is inelastic, state whether it is completely inelastic. (a) You drop a ball from your hand. It collides with the floor and bounces back up so that it just reaches your hand. (b) You drop a different ball from your hand and let it collide with the ground. This ball bounces back up to half the height from which it was dropped. (c) You drop a ball of clay from your hand. When it collides with the ground, it stops. ❙

8.4 Elastic Collisions

We saw in Section 8.3 that an *elastic collision* in an isolated system is one in which kinetic energy (as well as momentum) is conserved. Elastic collisions occur when the forces between the colliding bodies are *conservative*. When two billiard balls collide, they squash a little near the surface of contact, but then they spring back. Some of the kinetic energy is stored temporarily as elastic potential energy, but at the end it is reconverted to kinetic energy (Fig. 8.21).

Let's look at an elastic collision between two bodies A and B. We start with a one-dimensional collision, in which all the velocities lie along the same line; we choose this line to be the x-axis. Each momentum and velocity then has only an x-component. We call the x-velocities before the collision v_{A1x} and v_{B1x}, and those after the collision v_{A2x} and v_{B2x}. From conservation of kinetic energy we have

$$\tfrac{1}{2}m_A v_{A1x}^2 + \tfrac{1}{2}m_B v_{B1x}^2 = \tfrac{1}{2}m_A v_{A2x}^2 + \tfrac{1}{2}m_B v_{B2x}^2$$

and conservation of momentum gives

$$m_A v_{A1x} + m_B v_{B1x} = m_A v_{A2x} + m_B v_{B2x}$$

If the masses m_A and m_B and the initial velocities v_{A1x} and v_{B1x} are known, we can solve these two equations to find the two final velocities v_{A2x} and v_{B2x}.

8.21 Billiard balls deform very little when they collide, and they quickly spring back from any deformation they do undergo. Hence the force of interaction between the balls is almost perfectly conservative, and the collision is almost perfectly elastic.

Mastering**PHYSICS**

ActivPhysics 6.2: Collisions and Elasticity
ActivPhysics 6.5: Car Collisions: Two Dimensions
ActivPhysics 6.9: Pendulum Bashes Box

Elastic Collisions, One Body Initially at Rest

The general solution to the above equations is a little complicated, so we will concentrate on the particular case in which body B is at rest before the collision (so $v_{B1x} = 0$). Think of body B as a target for body A to hit. Then the kinetic energy and momentum conservation equations are, respectively,

$$\tfrac{1}{2}m_A v_{A1x}^2 = \tfrac{1}{2}m_A v_{A2x}^2 + \tfrac{1}{2}m_B v_{B2x}^2 \tag{8.19}$$

$$m_A v_{A1x} = m_A v_{A2x} + m_B v_{B2x} \tag{8.20}$$

We can solve for v_{A2x} and v_{B2x} in terms of the masses and the initial velocity v_{A1x}. This involves some fairly strenuous algebra, but it's worth it. No pain, no gain! The simplest approach is somewhat indirect, but along the way it uncovers an additional interesting feature of elastic collisions.

First we rearrange Eqs. (8.19) and (8.20) as follows:

$$m_B v_{B2x}^2 = m_A(v_{A1x}^2 - v_{A2x}^2) = m_A(v_{A1x} - v_{A2x})(v_{A1x} + v_{A2x}) \tag{8.21}$$

$$m_B v_{B2x} = m_A(v_{A1x} - v_{A2x}) \tag{8.22}$$

Now we divide Eq. (8.21) by Eq. (8.22) to obtain

$$v_{B2x} = v_{A1x} + v_{A2x} \tag{8.23}$$

8.22 Collisions between (a) a moving Ping-Pong ball and an initially stationary bowling ball, and (b) a moving bowling ball and an initially stationary Ping-Pong ball.

(a) Ping-Pong ball strikes bowling ball.

BEFORE

AFTER

(b) Bowling ball strikes Ping-Pong ball.

BEFORE

AFTER

8.23 A one-dimensional elastic collision between bodies of equal mass.

When a moving object A has a 1-D elastic collision with an equal-mass, motionless object B …

… all of A's momentum and kinetic energy are transferred to B.

We substitute this expression back into Eq. (8.22) to eliminate v_{B2x} and then solve for v_{A2x}:

$$m_B(v_{A1x} + v_{A2x}) = m_A(v_{A1x} - v_{A2x})$$

$$v_{A2x} = \frac{m_A - m_B}{m_A + m_B} v_{A1x} \qquad (8.24)$$

Finally, we substitute this result back into Eq. (8.23) to obtain

$$v_{B2x} = \frac{2m_A}{m_A + m_B} v_{A1x} \qquad (8.25)$$

Now we can interpret the results. Suppose body A is a Ping-Pong ball and body B is a bowling ball. Then we expect A to bounce off after the collision with a velocity nearly equal to its original value but in the opposite direction (Fig. 8.22a), and we expect B's velocity to be much less. That's just what the equations predict. When m_A is much smaller than m_B, the fraction in Eq. (8.24) is approximately equal to (-1), so v_{A2x} is approximately equal to $-v_{A1x}$. The fraction in Eq. (8.25) is much smaller than unity, so v_{B2x} is much less than v_{A1x}. Figure 8.22b shows the opposite case, in which A is the bowling ball and B the Ping-Pong ball and m_A is much larger than m_B. What do you expect to happen then? Check your predictions against Eqs. (8.24) and (8.25).

Another interesting case occurs when the masses are equal (Fig. 8.23). If $m_A = m_B$, then Eqs. (8.24) and (8.25) give $v_{A2x} = 0$ and $v_{B2x} = v_{A1x}$. That is, the body that was moving stops dead; it gives all its momentum and kinetic energy to the body that was at rest. This behavior is familiar to all pool players.

Elastic Collisions and Relative Velocity

Let's return to the more general case in which A and B have different masses. Equation (8.23) can be rewritten as

$$v_{A1x} = v_{B2x} - v_{A2x} \qquad (8.26)$$

Here $v_{B2x} - v_{A2x}$ is the velocity of B relative to A *after* the collision; from Eq. (8.26), this equals v_{A1x}, which is the *negative* of the velocity of B relative to A *before* the collision. (We discussed relative velocity in Section 3.5.) The relative velocity has the same magnitude, but opposite sign, before and after the collision. The sign changes because A and B are approaching each other before the collision but moving apart after the collision. If we view this collision from a second coordinate system moving with constant velocity relative to the first, the velocities of the bodies are different but the *relative* velocities are the same. Hence our statement about relative velocities holds for *any* straight-line elastic collision, even when neither body is at rest initially. *In a straight-line elastic collision of two bodies, the relative velocities before and after the collision have the same magnitude but opposite sign.* This means that if B is moving before the collision, Eq. (8.26) becomes

$$v_{B2x} - v_{A2x} = -(v_{B1x} - v_{A1x}) \qquad (8.27)$$

It turns out that a *vector* relationship similar to Eq. (8.27) is a general property of *all* elastic collisions, even when both bodies are moving initially and the velocities do not all lie along the same line. This result provides an alternative and equivalent definition of an elastic collision: *In an elastic collision, the relative velocity of the two bodies has the same magnitude before and after the collision.* Whenever this condition is satisfied, the total kinetic energy is also conserved.

When an elastic two-body collision isn't head-on, the velocities don't all lie along a single line. If they all lie in a plane, then each final velocity has two unknown components, and there are four unknowns in all. Conservation of energy and conservation of the x- and y-components of momentum give only three equations. To determine the final velocities uniquely, we need additional information, such as the direction or magnitude of one of the final velocities.

Example 8.10 An elastic straight-line collision

We repeat the air-track collision of Example 8.5 (Section 8.2), but now we add ideal spring bumpers to the gliders so that the collision is elastic. What are the final velocities of the gliders?

SOLUTION

IDENTIFY and SET UP: The net external force on the system is zero, so the momentum of the system is conserved. Figure 8.24 shows our sketch. We'll find our target variables, v_{A2x} and v_{B2x}, using Eq. (8.27), the relative-velocity relationship for an elastic collision, and the momentum-conservation equation.

EXECUTE: From Eq. (8.27),

$$v_{B2x} - v_{A2x} = -(v_{B1x} - v_{A1x})$$
$$= -(-2.0 \text{ m/s} - 2.0 \text{ m/s}) = 4.0 \text{ m/s}$$

From conservation of momentum,

$$m_A v_{A1x} + m_B v_{B1x} = m_A v_{A2x} + m_B v_{B2x}$$
$$(0.50 \text{ kg})(2.0 \text{ m/s}) + (0.30 \text{ kg})(-2.0 \text{ m/s})$$
$$= (0.50 \text{ kg})v_{A2x} + (0.30 \text{ kg})v_{B2x}$$
$$0.50v_{A2x} + 0.30v_{B2x} = 0.40 \text{ m/s}$$

(To get the last equation we divided both sides of the equation just above it by the quantity 1 kg. This makes the units the same as in the first equation.) Solving these equations simultaneously, we find

$$v_{A2x} = -1.0 \text{ m/s} \qquad v_{B2x} = 3.0 \text{ m/s}$$

8.24 Our sketch for this problem.

EVALUATE: Both bodies reverse their directions of motion; A moves to the left at 1.0 m/s and B moves to the right at 3.0 m/s. This is unlike the result of Example 8.5 because that collision was *not* elastic. The more massive glider A slows down in the collision and so loses kinetic energy. The less massive glider B speeds up and gains kinetic energy. The total kinetic energy before the collision (which we calculated in Example 8.7) is 1.6 J. The total kinetic energy after the collision is

$$\tfrac{1}{2}(0.50 \text{ kg})(-1.0 \text{ m/s})^2 + \tfrac{1}{2}(0.30 \text{ kg})(3.0 \text{ m/s})^2 = 1.6 \text{ J}$$

As expected, the kinetic energies before and after this elastic collision are equal. Kinetic energy is transferred from A to B, but none of it is lost.

CAUTION **Be careful with the elastic collision equations** You could *not* have solved this problem using Eqs. (8.24) and (8.25), which apply only if body B is initially *at rest*. Always be sure that you solve the problem at hand using equations that are applicable!

Example 8.11 Moderating fission neutrons in a nuclear reactor

The fission of uranium nuclei in a nuclear reactor produces high-speed neutrons. Before such neutrons can efficiently cause additional fissions, they must be slowed down by collisions with nuclei in the *moderator* of the reactor. The first nuclear reactor (built in 1942 at the University of Chicago) used carbon (graphite) as the moderator. Suppose a neutron (mass 1.0 u) traveling at 2.6×10^7 m/s undergoes a head-on elastic collision with a carbon nucleus (mass 12 u) initially at rest. Neglecting external forces during the collision, find the velocities after the collision. (1 u is the *atomic mass unit,* equal to 1.66×10^{-27} kg.)

SOLUTION

IDENTIFY and SET UP: We neglect external forces, so momentum is conserved in the collision. The collision is elastic, so kinetic

8.25 Our sketch for this problem.

energy is also conserved. Figure 8.25 shows our sketch. We take the x-axis to be in the direction in which the neutron is moving initially. The collision is head-on, so both particles move along this same axis after the collision. The carbon nucleus is initially at rest, so we can use Eqs. (8.24) and (8.25); we replace A by n (for the neutron) and B by C (for the carbon nucleus). We have $m_n = 1.0$ u, $m_C = 12$ u, and $v_{n1x} = 2.6 \times 10^7$ m/s. The target variables are the final velocities v_{n2x} and v_{C2x}.

EXECUTE: You can do the arithmetic. (*Hint:* There's no reason to convert atomic mass units to kilograms.) The results are

$$v_{n2x} = -2.2 \times 10^7 \text{ m/s} \qquad v_{C2x} = 0.4 \times 10^7 \text{ m/s}$$

EVALUATE: The neutron ends up with $|(m_n - m_C)/(m_n + m_C)| = \frac{11}{13}$ of its initial speed, and the speed of the recoiling carbon nucleus is $|2m_n/(m_n + m_C)| = \frac{2}{13}$ of the neutron's initial speed. Kinetic energy is proportional to speed squared, so the neutron's final kinetic energy is $(\frac{11}{13})^2 \approx 0.72$ of its original value. After a second head-on collision, its kinetic energy is $(0.72)^2$, or about half its original value, and so on. After a few dozen collisions (few of which are head-on), the neutron speed will be low enough that it can efficiently cause a fission reaction in a uranium nucleus.

Example 8.12 | A two-dimensional elastic collision

Figure 8.26 shows an elastic collision of two pucks (masses $m_A = 0.500$ kg and $m_B = 0.300$ kg) on a frictionless air-hockey table. Puck A has an initial velocity of 4.00 m/s in the positive x-direction and a final velocity of 2.00 m/s in an unknown direction α. Puck B is initially at rest. Find the final speed v_{B2} of puck B and the angles α and β.

8.26 An elastic collision that isn't head-on.

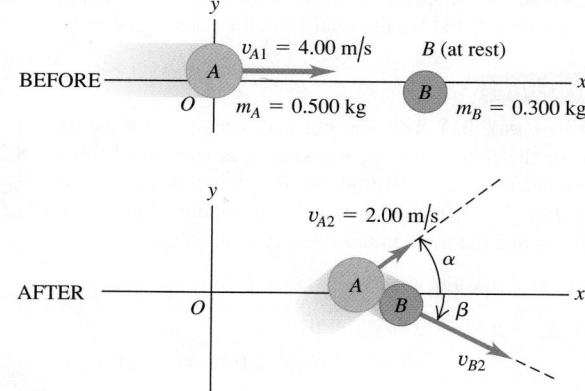

SOLUTION

IDENTIFY and SET UP: We'll use the equations for conservation of energy and conservation of x- and y-momentum. These three equations should be enough to solve for the three target variables given in the problem statement.

EXECUTE: The collision is elastic, so the initial and final kinetic energies of the system are equal:

$$\tfrac{1}{2}m_A v_{A1}^2 = \tfrac{1}{2}m_A v_{A2}^2 + \tfrac{1}{2}m_B v_{B2}^2$$

$$v_{B2}^2 = \frac{m_A v_{A1}^2 - m_A v_{A2}^2}{m_B}$$

$$= \frac{(0.500 \text{ kg})(4.00 \text{ m/s})^2 - (0.500 \text{ kg})(2.00 \text{ m/s})^2}{0.300 \text{ kg}}$$

$$v_{B2} = 4.47 \text{ m/s}$$

Conservation of the x- and y-components of total momentum gives

$$m_A v_{A1x} = m_A v_{A2x} + m_B v_{B2x}$$

$$(0.500 \text{ kg})(4.00 \text{ m/s}) = (0.500 \text{ kg})(2.00 \text{ m/s})(\cos \alpha)$$
$$+ (0.300 \text{ kg})(4.47 \text{ m/s})(\cos \beta)$$

$$0 = m_A v_{A2y} + m_B v_{B2y}$$

$$0 = (0.500 \text{ kg})(2.00 \text{ m/s})(\sin \alpha)$$
$$- (0.300 \text{ kg})(4.47 \text{ m/s})(\sin \beta)$$

These are two simultaneous equations for α and β. We'll leave it to you to supply the details of the solution. (*Hint:* Solve the first equation for $\cos \beta$ and the second for $\sin \beta$; square each equation and add. Since $\sin^2 \beta + \cos^2 \beta = 1$, this eliminates β and leaves an equation that you can solve for $\cos \alpha$ and hence for α. Substitute this value into either of the two equations and solve for β.) The results are

$$\alpha = 36.9° \qquad \beta = 26.6°$$

EVALUATE: To check the answers we confirm that the y-momentum, which was zero before the collision, is in fact zero after the collision. The y-momenta are

$$p_{A2y} = (0.500 \text{ kg})(2.00 \text{ m/s})(\sin 36.9°) = +0.600 \text{ kg} \cdot \text{m/s}$$

$$p_{B2y} = -(0.300 \text{ kg})(4.47 \text{ m/s})(\sin 26.6°) = -0.600 \text{ kg} \cdot \text{m/s}$$

and their sum is indeed zero.

Test Your Understanding of Section 8.4 Most present-day nuclear reactors use water as a moderator (see Example 8.11). Are water molecules (mass $m_w = 18.0$ u) a better or worse moderator than carbon atoms? (One advantage of water is that it also acts as a coolant for the reactor's radioactive core.) ❙

8.5 Center of Mass

We can restate the principle of conservation of momentum in a useful way by using the concept of **center of mass.** Suppose we have several particles with masses m_1, m_2, and so on. Let the coordinates of m_1 be (x_1, y_1), those of m_2 be (x_2, y_2), and so on. We define the center of mass of the system as the point that has coordinates (x_{cm}, y_{cm}) given by

$$x_{cm} = \frac{m_1 x_1 + m_2 x_2 + m_3 x_3 + \cdots}{m_1 + m_2 + m_3 + \cdots} = \frac{\sum_i m_i x_i}{\sum_i m_i}$$

$$y_{cm} = \frac{m_1 y_1 + m_2 y_2 + m_3 y_3 + \cdots}{m_1 + m_2 + m_3 + \cdots} = \frac{\sum_i m_i y_i}{\sum_i m_i}$$

(center of mass) (8.28)

The position vector $\vec{r}_{cm}$ of the center of mass can be expressed in terms of the position vectors $\vec{r}_1$, $\vec{r}_2$, . . . of the particles as

$$\vec{r}_{cm} = \frac{m_1\vec{r}_1 + m_2\vec{r}_2 + m_3\vec{r}_3 + \cdots}{m_1 + m_2 + m_3 + \cdots} = \frac{\sum_i m_i\vec{r}_i}{\sum_i m_i} \quad \text{(center of mass)} \quad (8.29)$$

In statistical language, the center of mass is a *mass-weighted average* position of the particles.

Example 8.13 Center of mass of a water molecule

Figure 8.27 shows a simple model of a water molecule. The oxygen-hydrogen separation is $d = 9.57 \times 10^{-11}$ m. Each hydrogen atom has mass 1.0 u, and the oxygen atom has mass 16.0 u. Find the position of the center of mass.

SOLUTION

IDENTIFY and SET UP: Nearly all the mass of each atom is concentrated in its nucleus, whose radius is only about 10^{-5} times the overall radius of the atom. Hence we can safely represent each atom as a point particle. Figure 8.27 shows our coordinate system, with the x-axis chosen to lie along the molecule's symmetry axis. We'll use Eqs. (8.28) to find x_{cm} and y_{cm}.

EXECUTE: The oxygen atom is at $x = 0$, $y = 0$. The x-coordinate of each hydrogen atom is $d\cos(105°/2)$; the y-coordinates are $\pm d\sin(105°/2)$. From Eqs. (8.28),

$$x_{cm} = \frac{\begin{bmatrix} (1.0\,\text{u})(d\cos 52.5°) + (1.0\,\text{u}) \\ \times (d\cos 52.5°) + (16.0\,\text{u})(0) \end{bmatrix}}{1.0\,\text{u} + 1.0\,\text{u} + 16.0\,\text{u}} = 0.068d$$

$$y_{cm} = \frac{\begin{bmatrix} (1.0\,\text{u})(d\sin 52.5°) + (1.0\,\text{u}) \\ \times (-d\sin 52.5°) + (16.0\,\text{u})(0) \end{bmatrix}}{1.0\,\text{u} + 1.0\,\text{u} + 16.0\,\text{u}} = 0$$

Substituting $d = 9.57 \times 10^{-11}$ m, we find

$$x_{cm} = (0.068)(9.57 \times 10^{-11}\,\text{m}) = 6.5 \times 10^{-12}\,\text{m}$$

8.27 Where is the center of mass of a water molecule?

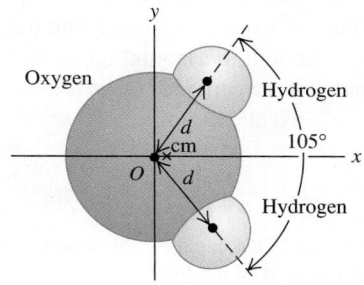

EVALUATE: The center of mass is much closer to the oxygen atom (located at the origin) than to either hydrogen atom because the oxygen atom is much more massive. The center of mass lies along the molecule's *axis of symmetry*. If the molecule is rotated 180° around this axis, it looks exactly the same as before. The position of the center of mass can't be affected by this rotation, so it *must* lie on the axis of symmetry.

For solid bodies, in which we have (at least on a macroscopic level) a continuous distribution of matter, the sums in Eqs. (8.28) have to be replaced by integrals. The calculations can get quite involved, but we can say three general things about such problems (Fig. 8.28). First, whenever a homogeneous body has a geometric center, such as a billiard ball, a sugar cube, or a can of frozen orange juice, the center of mass is at the geometric center. Second, whenever a body has an axis of symmetry, such as a wheel or a pulley, the center of mass always lies on that axis. Third, there is no law that says the center of mass has to be within the body. For example, the center of mass of a donut is right in the middle of the hole.

We'll talk a little more about locating the center of mass in Chapter 11 in connection with the related concept of *center of gravity*.

Motion of the Center of Mass

To see the significance of the center of mass of a collection of particles, we must ask what happens to the center of mass when the particles move. The x- and y-components of velocity of the center of mass, $v_{cm\text{-}x}$ and $v_{cm\text{-}y}$, are the time derivatives of x_{cm} and y_{cm}. Also, dx_1/dt is the x-component of velocity of particle 1,

8.28 Locating the center of mass of a symmetrical object.

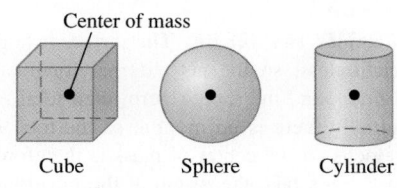

Center of mass

Cube · Sphere · Cylinder

If a homogeneous object has a geometric center, that is where the center of mass is located.

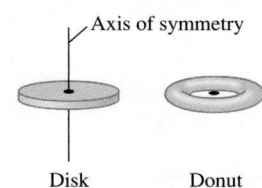

Axis of symmetry

Disk · Donut

If an object has an axis of symmetry, the center of mass lies along it. As in the case of the donut, the center of mass may not be within the object.

8.29 The center of mass of this wrench is marked with a white dot. The net external force acting on the wrench is almost zero. As the wrench spins on a smooth horizontal surface, the center of mass moves in a straight line with nearly constant velocity.

and so on, so $dx_1/dt = v_{1x}$, and so on. Taking time derivatives of Eqs. (8.28), we get

$$
\begin{aligned}
v_{\text{cm-}x} &= \frac{m_1 v_{1x} + m_2 v_{2x} + m_3 v_{3x} + \cdots}{m_1 + m_2 + m_3 + \cdots} \\
v_{\text{cm-}y} &= \frac{m_1 v_{1y} + m_2 v_{2y} + m_3 v_{3y} + \cdots}{m_1 + m_2 + m_3 + \cdots}
\end{aligned}
\tag{8.30}
$$

These equations are equivalent to the single vector equation obtained by taking the time derivative of Eq. (8.29):

$$
\vec{v}_{\text{cm}} = \frac{m_1 \vec{v}_1 + m_2 \vec{v}_2 + m_3 \vec{v}_3 + \cdots}{m_1 + m_2 + m_3 + \cdots}
\tag{8.31}
$$

We denote the *total* mass $m_1 + m_2 + \cdots$ by M. We can then rewrite Eq. (8.31) as

$$
M\vec{v}_{\text{cm}} = m_1 \vec{v}_1 + m_2 \vec{v}_2 + m_3 \vec{v}_3 + \cdots = \vec{P}
\tag{8.32}
$$

The right side is simply the total momentum $\vec{P}$ of the system. Thus we have proved that *the total momentum is equal to the total mass times the velocity of the center of mass.* When you catch a baseball, you are really catching a collection of a very large number of molecules of masses m_1, m_2, m_3, ... The impulse you feel is due to the total momentum of this entire collection. But this impulse is the same as if you were catching a single particle of mass $M = m_1 + m_2 + m_3 + \cdots$ moving with velocity $\vec{v}_{\text{cm}}$, the velocity of the collection's center of mass. So Eq. (8.32) helps to justify representing an extended body as a particle.

For a system of particles on which the net external force is zero, so that the total momentum $\vec{P}$ is constant, the velocity of the center of mass $\vec{v}_{\text{cm}} = \vec{P}/M$ is also constant. Suppose we mark the center of mass of a wrench and then slide the wrench with a spinning motion across a smooth, horizontal tabletop (Fig. 8.29). The overall motion appears complicated, but the center of mass follows a straight line, as though all the mass were concentrated at that point.

Example 8.14   **A tug-of-war on the ice**

James (mass 90.0 kg) and Ramon (mass 60.0 kg) are 20.0 m apart on a frozen pond. Midway between them is a mug of their favorite beverage. They pull on the ends of a light rope stretched between them. When James has moved 6.0 m toward the mug, how far and in what direction has Ramon moved?

SOLUTION

IDENTIFY and SET UP: The surface is horizontal and (we assume) frictionless, so the net external force on the system of James, Ramon, and the rope is zero; their total momentum is conserved. Initially there is no motion, so the total momentum is zero. The velocity of the center of mass is therefore zero, and it remains at rest. Let's take the origin at the position of the mug and let the $+x$-axis extend from the mug toward Ramon. Figure 8.30 shows

our sketch. We use Eq. (8.28) to calculate the position of the center of mass; we neglect the mass of the light rope.

EXECUTE: The initial x-coordinates of James and Ramon are -10.0 m and $+10.0$ m, respectively, so the x-coordinate of the center of mass is

$$
x_{\text{cm}} = \frac{(90.0 \text{ kg})(-10.0 \text{ m}) + (60.0 \text{ kg})(10.0 \text{ m})}{90.0 \text{ kg} + 60.0 \text{ kg}} = -2.0 \text{ m}
$$

When James moves 6.0 m toward the mug, his new x-coordinate is -4.0 m; we'll call Ramon's new x-coordinate x_2. The center of mass doesn't move, so

$$
x_{\text{cm}} = \frac{(90.0 \text{ kg})(-4.0 \text{ m}) + (60.0 \text{ kg})x_2}{90.0 \text{ kg} + 60.0 \text{ kg}} = -2.0 \text{ m}
$$

$$
x_2 = 1.0 \text{ m}
$$

James has moved 6.0 m and is still 4.0 m from the mug, but Ramon has moved 9.0 m and is only 1.0 m from it.

EVALUATE: The ratio of the distances moved, $(6.0 \text{ m})/(9.0 \text{ m}) = \frac{2}{3}$, is the *inverse* ratio of the masses. Can you see why? Because the surface is frictionless, the two men will keep moving and collide at the center of mass; Ramon will reach the mug first. This is independent of how hard either person pulls; pulling harder just makes them move faster.

8.30 Our sketch for this problem.

External Forces and Center-of-Mass Motion

If the net external force on a system of particles is not zero, then total momentum is not conserved and the velocity of the center of mass changes. Let's look at the relationship between the motion of the center of mass and the forces acting on the system.

Equations (8.31) and (8.32) give the *velocity* of the center of mass in terms of the velocities of the individual particles. We take the time derivatives of these equations to show that the *accelerations* are related in the same way. Let $\vec{a}_{cm} = d\vec{v}_{cm}/dt$ be the acceleration of the center of mass; then we find

$$M\vec{a}_{cm} = m_1\vec{a}_1 + m_2\vec{a}_2 + m_3\vec{a}_3 + \cdots \qquad (8.33)$$

Now $m_1\vec{a}_1$ is equal to the vector sum of forces on the first particle, and so on, so the right side of Eq. (8.33) is equal to the vector sum $\Sigma\vec{F}$ of *all* the forces on *all* the particles. Just as we did in Section 8.2, we can classify each force as *external* or *internal*. The sum of all forces on all the particles is then

$$\Sigma\vec{F} = \Sigma\vec{F}_{ext} + \Sigma\vec{F}_{int} = M\vec{a}_{cm}$$

Because of Newton's third law, the internal forces all cancel in pairs, and $\Sigma\vec{F}_{int} = \mathbf{0}$. What survives on the left side is the sum of only the *external* forces:

$$\Sigma\vec{F}_{ext} = M\vec{a}_{cm} \qquad \text{(body or collection of particles)} \qquad (8.34)$$

When a body or a collection of particles is acted on by external forces, the center of mass moves just as though all the mass were concentrated at that point and it were acted on by a net force equal to the sum of the external forces on the system.

This result may not sound very impressive, but in fact it is central to the whole subject of mechanics. In fact, we've been using this result all along; without it, we would not be able to represent an extended body as a point particle when we apply Newton's laws. It explains why only *external* forces can affect the motion of an extended body. If you pull upward on your belt, your belt exerts an equal downward force on your hands; these are *internal* forces that cancel and have no effect on the overall motion of your body.

Suppose a cannon shell traveling in a parabolic trajectory (neglecting air resistance) explodes in flight, splitting into two fragments with equal mass (Fig. 8.31a). The fragments follow new parabolic paths, but the center of mass continues on the original parabolic trajectory, just as though all the mass were still concentrated at that point. A skyrocket exploding in air (Fig. 8.31b) is a spectacular example of this effect.

8.31 (a) A shell explodes into two fragments in flight. If air resistance is ignored, the center of mass continues on the same trajectory as the shell's path before exploding. (b) The same effect occurs with exploding fireworks.

(a)

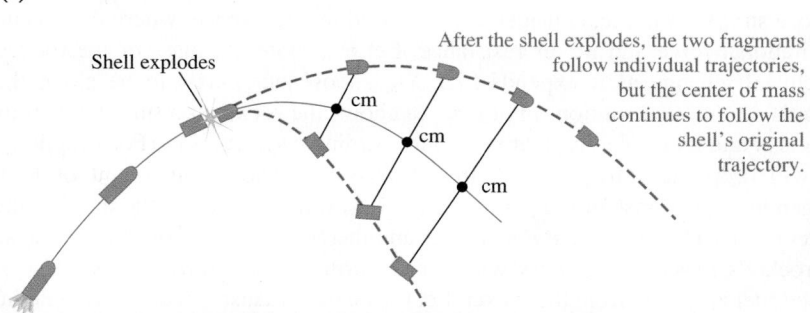

Shell explodes

cm

cm

cm

After the shell explodes, the two fragments follow individual trajectories, but the center of mass continues to follow the shell's original trajectory.

(b)

This property of the center of mass is important when we analyze the motion of rigid bodies. We describe the motion of an extended body as a combination of translational motion of the center of mass and rotational motion about an axis through the center of mass. We will return to this topic in Chapter 10. This property also plays an important role in the motion of astronomical objects. It's not correct to say that the moon orbits the earth; rather, the earth and moon both move in orbits around their center of mass.

There's one more useful way to describe the motion of a system of particles. Using $\vec{a}_{cm} = d\vec{v}_{cm}/dt$, we can rewrite Eq. (8.33) as

$$M\vec{a}_{cm} = M\frac{d\vec{v}_{cm}}{dt} = \frac{d(M\vec{v}_{cm})}{dt} = \frac{d\vec{P}}{dt} \qquad (8.35)$$

The total system mass M is constant, so we're allowed to move it inside the derivative. Substituting Eq. (8.35) into Eq. (8.34), we find

$$\sum\vec{F}_{ext} = \frac{d\vec{P}}{dt} \qquad \text{(extended body or system of particles)} \qquad (8.36)$$

This equation looks like Eq. (8.4). The difference is that Eq. (8.36) describes a *system* of particles, such as an extended body, while Eq. (8.4) describes a single particle. The interactions between the particles that make up the system can change the individual momenta of the particles, but the *total* momentum $\vec{P}$ of the system can be changed only by external forces acting from outside the system.

Finally, we note that if the net external force is zero, Eq. (8.34) shows that the acceleration $\vec{a}_{cm}$ of the center of mass is zero. So the center-of-mass velocity $\vec{v}_{cm}$ is constant, as for the wrench in Fig. 8.29. From Eq. (8.36) the total momentum $\vec{P}$ is also constant. This reaffirms our statement in Section 8.3 of the principle of conservation of momentum.

Test Your Understanding of Section 8.5 Will the center of mass in Fig. 8.31a continue on the same parabolic trajectory even after one of the fragments hits the ground? Why or why not?

❙

Application Jet Propulsion in Squids
Both a jet engine and a squid use variations in their mass to provide propulsion: They increase their mass by taking in fluid at low speed (air for a jet engine, water for a squid), then decrease their mass by ejecting that fluid at high speed. The net result is a propulsive force.

8.6 Rocket Propulsion

Momentum considerations are particularly useful for analyzing a system in which the masses of parts of the system change with time. In such cases we can't use Newton's second law $\sum\vec{F} = m\vec{a}$ directly because m changes. Rocket propulsion offers a typical and interesting example of this kind of analysis. A rocket is propelled forward by rearward ejection of burned fuel that initially was in the rocket (which is why rocket fuel is also called *propellant*). The forward force on the rocket is the reaction to the backward force on the ejected material. The total mass of the system is constant, but the mass of the rocket itself decreases as material is ejected.

As a simple example, consider a rocket fired in outer space, where there is no gravitational force and no air resistance. Let m denote the mass of the rocket, which will change as it expends fuel. We choose our x-axis to be along the rocket's direction of motion. Figure 8.32a shows the rocket at a time t, when its mass is m and its x-velocity relative to our coordinate system is v. (For simplicity, we will drop the subscript x in this discussion.) The x-component of total momentum at this instant is $P_1 = mv$. In a short time interval dt, the mass of the rocket changes by an amount dm. This is an inherently negative quantity because the rocket's mass m *decreases* with time. During dt, a *positive* mass $-dm$ of burned fuel is ejected from the rocket. Let v_{ex} be the exhaust *speed* of this material *relative to the rocket;* the burned fuel is ejected opposite the direction of motion,

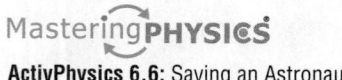

ActivPhysics 6.6: Saving an Astronaut

so its *x*-component of *velocity* relative to the rocket is $-v_{ex}$. The *x*-velocity v_{fuel} of the burned fuel relative to our coordinate system is then

$$v_{fuel} = v + (-v_{ex}) = v - v_{ex}$$

and the *x*-component of momentum of the ejected mass $(-dm)$ is

$$(-dm)v_{fuel} = (-dm)(v - v_{ex})$$

Figure 8.32b shows that at the end of the time interval dt, the *x*-velocity of the rocket and unburned fuel has increased to $v + dv$, and its mass has decreased to $m + dm$ (remember that dm is negative). The rocket's momentum at this time is

$$(m + dm)(v + dv)$$

Thus the *total x*-component of momentum P_2 of the rocket plus ejected fuel at time $t + dt$ is

$$P_2 = (m + dm)(v + dv) + (-dm)(v - v_{ex})$$

According to our initial assumption, the rocket and fuel are an isolated system. Thus momentum is conserved, and the total *x*-component of momentum of the system must be the same at time t and at time $t + dt$: $P_1 = P_2$. Hence

$$mv = (m + dm)(v + dv) + (-dm)(v - v_{ex})$$

This can be simplified to

$$m \, dv = -dm \, v_{ex} - dm \, dv$$

We can neglect the term $(-dm \, dv)$ because it is a product of two small quantities and thus is much smaller than the other terms. Dropping this term, dividing by dt, and rearranging, we find

$$m\frac{dv}{dt} = -v_{ex}\frac{dm}{dt} \qquad (8.37)$$

Now dv/dt is the acceleration of the rocket, so the left side of this equation (mass times acceleration) equals the net force F, or *thrust,* on the rocket:

$$F = -v_{ex}\frac{dm}{dt} \qquad (8.38)$$

The thrust is proportional both to the relative speed v_{ex} of the ejected fuel and to the mass of fuel ejected per unit time, $-dm/dt$. (Remember that dm/dt is negative because it is the rate of change of the rocket's mass, so F is positive.)

The *x*-component of acceleration of the rocket is

$$a = \frac{dv}{dt} = -\frac{v_{ex}}{m}\frac{dm}{dt} \qquad (8.39)$$

8.32 A rocket moving in gravity-free outer space at **(a)** time t and **(b)** time $t + dt$.

(a)   (b)

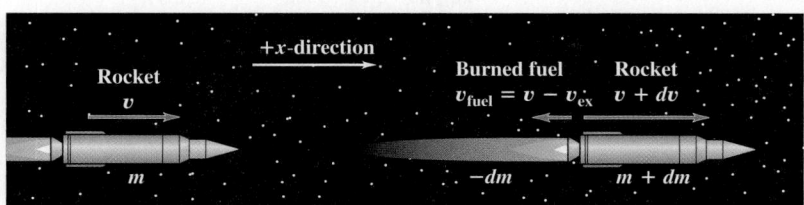

At time *t*, the rocket has mass m and *x*-component of velocity v.

At time $t + dt$, the rocket has mass $m + dm$ (where dm is inherently *negative*) and *x*-component of velocity $v + dv$. The burned fuel has *x*-component of velocity $v_{fuel} = v - v_{ex}$ and mass $-dm$. (The minus sign is needed to make $-dm$ *positive* because dm is negative.)

8.33 To provide enough thrust to lift its payload into space, this Atlas V launch vehicle ejects more than 1000 kg of burned fuel per second at speeds of nearly 4000 m/s.

This is positive because v_{ex} is positive (remember, it's the exhaust *speed*) and dm/dt is negative. The rocket's mass m decreases continuously while the fuel is being consumed. If v_{ex} and dm/dt are constant, the acceleration increases until all the fuel is gone.

Equation (8.38) tells us that an effective rocket burns fuel at a rapid rate (large $-dm/dt$) and ejects the burned fuel at a high relative speed (large v_{ex}), as in Fig. 8.33. In the early days of rocket propulsion, people who didn't understand conservation of momentum thought that a rocket couldn't function in outer space because "it doesn't have anything to push against." On the contrary, rockets work *best* in outer space, where there is no air resistance! The launch vehicle in Fig. 8.33 is *not* "pushing against the ground" to get into the air.

If the exhaust speed v_{ex} is constant, we can integrate Eq. (8.39) to find a relationship between the velocity v at any time and the remaining mass m. At time $t = 0$, let the mass be m_0 and the velocity v_0. Then we rewrite Eq. (8.39) as

$$dv = -v_{ex}\frac{dm}{m}$$

We change the integration variables to v' and m', so we can use v and m as the upper limits (the final speed and mass). Then we integrate both sides, using limits v_0 to v and m_0 to m, and take the constant v_{ex} outside the integral:

$$\int_{v_0}^{v} dv' = -\int_{m_0}^{m} v_{ex}\frac{dm'}{m'} = -v_{ex}\int_{m_0}^{m}\frac{dm'}{m'}$$

$$v - v_0 = -v_{ex}\ln\frac{m}{m_0} = v_{ex}\ln\frac{m_0}{m} \qquad (8.40)$$

The ratio m_0/m is the original mass divided by the mass after the fuel has been exhausted. In practical spacecraft this ratio is made as large as possible to maximize the speed gain, which means that the initial mass of the rocket is almost all fuel. The final velocity of the rocket will be greater in magnitude (and is often *much* greater) than the relative speed v_{ex} if $\ln(m_0/m) > 1$—that is, if $m_0/m > e = 2.71828\ldots$.

We've assumed throughout this analysis that the rocket is in gravity-free outer space. However, gravity must be taken into account when a rocket is launched from the surface of a planet, as in Fig. 8.33 (see Problem 8.112).

Example 8.15 | **Acceleration of a rocket**

The engine of a rocket in outer space, far from any planet, is turned on. The rocket ejects burned fuel at a constant rate; in the first second of firing, it ejects $\frac{1}{120}$ of its initial mass m_0 at a relative speed of 2400 m/s. What is the rocket's initial acceleration?

SOLUTION

IDENTIFY and SET UP: We are given the rocket's exhaust speed v_{ex} and the fraction of the initial mass lost during the first second of firing, from which we can find dm/dt. We'll use Eq. (8.39) to find the acceleration of the rocket.

EXECUTE: The initial rate of change of mass is

$$\frac{dm}{dt} = -\frac{m_0/120}{1\text{ s}} = -\frac{m_0}{120\text{ s}}$$

From Eq. (8.39),

$$a = -\frac{v_{ex}}{m_0}\frac{dm}{dt} = -\frac{2400\text{ m/s}}{m_0}\left(-\frac{m_0}{120\text{ s}}\right) = 20\text{ m/s}^2$$

EVALUATE: The answer doesn't depend on m_0. If v_{ex} is the same, the initial acceleration is the same for a 120,000-kg spacecraft that ejects 1000 kg/s as for a 60-kg astronaut equipped with a small rocket that ejects 0.5 kg/s.

Example 8.16 Speed of a rocket

Suppose that $\frac{3}{4}$ of the initial mass of the rocket in Example 8.15 is fuel, so that the fuel is completely consumed at a constant rate in 90 s. The final mass of the rocket is $m = m_0/4$. If the rocket starts from rest in our coordinate system, find its speed at the end of this time.

SOLUTION

IDENTIFY, SET UP, and EXECUTE: We are given the initial velocity $v_0 = 0$, the exhaust speed $v_{ex} = 2400$ m/s, and the final mass m as a fraction of the initial mass m_0. We'll use Eq. (8.40) to find the final speed v:

$$v = v_0 + v_{ex}\ln\frac{m_0}{m} = 0 + (2400 \text{ m/s})(\ln 4) = 3327 \text{ m/s}$$

EVALUATE: Let's examine what happens as the rocket gains speed. (To illustrate our point, we use more figures than are significant.) At the start of the flight, when the velocity of the rocket is zero, the ejected fuel is moving backward at 2400 m/s relative to our frame of reference. As the rocket moves forward and speeds up, the fuel's speed relative to our system decreases; when the rocket speed reaches 2400 m/s, this relative speed is *zero*. [Knowing the rate of fuel consumption, you can solve Eq. (8.40) to show that this occurs at about $t = 75.6$ s.] After this time the ejected burned fuel moves *forward*, not backward, in our system. Relative to our frame of reference, the last bit of ejected fuel has a forward velocity of 3327 m/s − 2400 m/s = 927 m/s.

Test Your Understanding of Section 8.6 (a) If a rocket in gravity-free outer space has the same thrust at all times, is its acceleration constant, increasing, or decreasing? (b) If the rocket has the same acceleration at all times, is the thrust constant, increasing, or decreasing?

|

Momentum of a particle: The momentum $\vec{p}$ of a particle is a vector quantity equal to the product of the particle's mass m and velocity $\vec{v}$. Newton's second law says that the net force on a particle is equal to the rate of change of the particle's momentum.

$$\vec{p} = m\vec{v} \tag{8.2}$$

$$\sum\vec{F} = \frac{d\vec{p}}{dt} \tag{8.4}$$

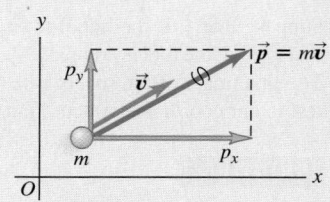

Impulse and momentum: If a constant net force $\sum\vec{F}$ acts on a particle for a time interval Δt from t_1 to t_2, the impulse $\vec{J}$ of the net force is the product of the net force and the time interval. If $\sum\vec{F}$ varies with time, $\vec{J}$ is the integral of the net force over the time interval. In any case, the change in a particle's momentum during a time interval equals the impulse of the net force that acted on the particle during that interval. The momentum of a particle equals the impulse that accelerated it from rest to its present speed. (See Examples 8.1–8.3.)

$$\vec{J} = \sum\vec{F}(t_2 - t_1) = \sum\vec{F}\,\Delta t \tag{8.5}$$

$$\vec{J} = \int_{t_1}^{t_2} \sum\vec{F}\,dt \tag{8.7}$$

$$\vec{J} = \vec{p}_2 - \vec{p}_1 \tag{8.6}$$

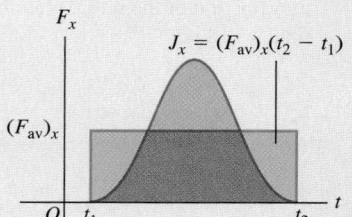

Conservation of momentum: An internal force is a force exerted by one part of a system on another. An external force is a force exerted on any part of a system by something outside the system. If the net external force on a system is zero, the total momentum of the system $\vec{P}$ (the vector sum of the momenta of the individual particles that make up the system) is constant, or conserved. Each component of total momentum is separately conserved. (See Examples 8.4–8.6.)

$$\vec{P} = \vec{p}_A + \vec{p}_B + \cdots$$
$$= m_A\vec{v}_A + m_B\vec{v}_B + \cdots \tag{8.14}$$

If $\sum\vec{F} = 0$, then $\vec{P} = $ constant.

Collisions: In collisions of all kinds, the initial and final total momenta are equal. In an elastic collision between two bodies, the initial and final total kinetic energies are also equal, and the initial and final relative velocities have the same magnitude. In an inelastic two-body collision, the total kinetic energy is less after the collision than before. If the two bodies have the same final velocity, the collision is completely inelastic. (See Examples 8.7–8.12.)

Center of mass: The position vector of the center of mass of a system of particles, $\vec{r}_{cm}$, is a weighted average of the positions $\vec{r}_1, \vec{r}_2, \ldots$ of the individual particles. The total momentum $\vec{P}$ of a system equals its total mass M multiplied by the velocity of its center of mass, $\vec{v}_{cm}$. The center of mass moves as though all the mass M were concentrated at that point. If the net external force on the system is zero, the center-of-mass velocity $\vec{v}_{cm}$ is constant. If the net external force is not zero, the center of mass accelerates as though it were a particle of mass M being acted on by the same net external force. (See Examples 8.13 and 8.14.)

$$\vec{r}_{cm} = \frac{m_1\vec{r}_1 + m_2\vec{r}_2 + m_3\vec{r}_3 + \cdots}{m_1 + m_2 + m_3 + \cdots}$$
$$= \frac{\sum_i m_i\vec{r}_i}{\sum_i m_i} \tag{8.29}$$

$$\vec{P} = m_1\vec{v}_1 + m_2\vec{v}_2 + m_3\vec{v}_3 + \cdots$$
$$= M\vec{v}_{cm} \tag{8.32}$$

$$\sum\vec{F}_{ext} = M\vec{a}_{cm} \tag{8.34}$$

Rocket propulsion: In rocket propulsion, the mass of a rocket changes as the fuel is used up and ejected from the rocket. Analysis of the motion of the rocket must include the momentum carried away by the spent fuel as well as the momentum of the rocket itself. (See Examples 8.15 and 8.16.)

BRIDGING PROBLEM | One Collision After Another

Sphere *A* of mass 0.600 kg is initially moving to the right at 4.00 m/s. Sphere *B*, of mass 1.80 kg, is initially to the right of sphere *A* and moving to the right at 2.00 m/s. After the two spheres collide, sphere *B* is moving at 3.00 m/s in the same direction as before. (a) What is the velocity (magnitude and direction) of sphere *A* after this collision? (b) Is this collision elastic or inelastic? (c) Sphere *B* then has an off-center collision with sphere *C*, which has mass 1.20 kg and is initially at rest. After this collision, sphere *B* is moving at 19.0° to its initial direction at 2.00 m/s. What is the velocity (magnitude and direction) of sphere *C* after this collision? (d) What is the impulse (magnitude and direction) imparted to sphere *B* by sphere *C* when they collide? (e) Is this second collision elastic or inelastic? (f) What is the velocity (magnitude and direction) of the center of mass of the system of three spheres (*A*, *B*, and *C*) after the second collision? No external forces act on any of the spheres in this problem.

SOLUTION GUIDE

See MasteringPhysics® study area for a Video Tutor solution.

IDENTIFY and SET UP

1. Momentum is conserved in these collisions. Can you explain why?
2. Choose the *x*- and *y*-axes, and assign subscripts to values before the first collision, after the first collision but before the second collision, and after the second collision.
3. Make a list of the target variables, and choose the equations that you'll use to solve for these.

EXECUTE

4. Solve for the velocity of sphere *A* after the first collision. Does *A* slow down or speed up in the collision? Does this make sense?
5. Now that you know the velocities of both *A* and *B* after the first collision, decide whether or not this collision is elastic. (How will you do this?)
6. The second collision is two-dimensional, so you'll have to demand that *both* components of momentum are conserved. Use this to find the speed and direction of sphere *C* after the second collision. (*Hint:* After the first collision, sphere *B* maintains the same velocity until it hits sphere *C*.)
7. Use the definition of impulse to find the impulse imparted to sphere *B* by sphere *C*. Remember that impulse is a vector.
8. Use the same technique that you employed in step 5 to decide whether or not the second collision is elastic.
9. Find the velocity of the center of mass after the second collision.

EVALUATE

10. Compare the directions of the vectors you found in steps 6 and 7. Is this a coincidence? Why or why not?
11. Find the velocity of the center of mass before and after the first collision. Compare to your result from step 9. Again, is this a coincidence? Why or why not?

Problems

For instructor-assigned homework, go to www.masteringphysics.com

•, ••, •••: Problems of increasing difficulty. **CP**: Cumulative problems incorporating material from earlier chapters. **CALC**: Problems requiring calculus. **BIO**: Biosciences problems.

DISCUSSION QUESTIONS

Q8.1 In splitting logs with a hammer and wedge, is a heavy hammer more effective than a lighter hammer? Why?

Q8.2 Suppose you catch a baseball and then someone invites you to catch a bowling ball with either the same momentum or the same kinetic energy as the baseball. Which would you choose? Explain.

Q8.3 When rain falls from the sky, what happens to its momentum as it hits the ground? Is your answer also valid for Newton's famous apple?

Q8.4 A car has the same kinetic energy when it is traveling south at 30 m/s as when it is traveling northwest at 30 m/s. Is the momentum of the car the same in both cases? Explain.

Q8.5 A truck is accelerating as it speeds down the highway. One inertial frame of reference is attached to the ground with its origin at a fence post. A second frame of reference is attached to a police car that is traveling down the highway at constant velocity. Is the momentum of the truck the same in these two reference frames? Explain. Is the rate of change of the truck's momentum the same in these two frames? Explain.

Q8.6 (a) When a large car collides with a small car, which one undergoes the greater change in momentum: the large one or the small one? Or is it the same for both? (b) In light of your answer to part (a), why are the occupants of the small car more likely to be hurt than those of the large car, assuming that both cars are equally sturdy?

Q8.7 A woman holding a large rock stands on a frictionless, horizontal sheet of ice. She throws the rock with speed v_0 at an angle α above the horizontal. Consider the system consisting of the woman plus the rock. Is the momentum of the system conserved? Why or why not? Is any component of the momentum of the system conserved? Again, why or why not?

Q8.8 In Example 8.7 (Section 8.3), where the two gliders in Fig. 8.15 stick together after the collision, the collision is inelastic because $K_2 < K_1$. In Example 8.5 (Section 8.2), is the collision inelastic? Explain.

Q8.9 In a completely inelastic collision between two objects, where the objects stick together after the collision, is it possible for the final kinetic energy of the system to be zero? If so, give an example in which this would occur. If the final kinetic energy is zero, what must the initial momentum of the system be? Is the initial kinetic energy of the system zero? Explain.

Q8.10 Since for a particle the kinetic energy is given by $K = \frac{1}{2}mv^2$ and the momentum by $\vec{p} = m\vec{v}$, it is easy to show that $K = p^2/2m$. How, then, is it possible to have an event during which the total momentum of the system is constant but the total kinetic energy changes?

Q8.11 In each of Examples 8.10, 8.11, and 8.12 (Section 8.4), verify that the relative velocity vector of the two bodies has the same magnitude before and after the collision. In each case what happens to the *direction* of the relative velocity vector?

Q8.12 A glass dropped on the floor is more likely to break if the floor is concrete than if it is wood. Why? (Refer to Fig. 8.3b.)

Q8.13 In Fig. 8.22b, the kinetic energy of the Ping-Pong ball is larger after its interaction with the bowling ball than before. From where does the extra energy come? Describe the event in terms of conservation of energy.

Q8.14 A machine gun is fired at a steel plate. Is the average force on the plate from the bullet impact greater if the bullets bounce off or if they are squashed and stick to the plate? Explain.

Q8.15 A net force of 4 N acts on an object initially at rest for 0.25 s and gives it a final speed of 5 m/s. How could a net force of 2 N produce the same final speed?

Q8.16 A net force with x-component $\sum F_x$ acts on an object from time t_1 to time t_2. The x-component of the momentum of the object is the same at t_1 as it is at t_2, but $\sum F_x$ is not zero at all times between t_1 and t_2. What can you say about the graph of $\sum F_x$ versus t?

Q8.17 A tennis player hits a tennis ball with a racket. Consider the system made up of the ball and the racket. Is the total momentum of the system the same just before and just after the hit? Is the total momentum just after the hit the same as 2 s later, when the ball is in midair at the high point of its trajectory? Explain any differences between the two cases.

Q8.18 In Example 8.4 (Section 8.2), consider the system consisting of the rifle plus the bullet. What is the speed of the system's center of mass after the rifle is fired? Explain.

Q8.19 An egg is released from rest from the roof of a building and falls to the ground. As the egg falls, what happens to the momentum of the system of the egg plus the earth?

Q8.20 A woman stands in the middle of a perfectly smooth, frictionless, frozen lake. She can set herself in motion by throwing things, but suppose she has nothing to throw. Can she propel herself to shore *without* throwing anything?

Q8.21 In a zero-gravity environment, can a rocket-propelled spaceship ever attain a speed greater than the relative speed with which the burnt fuel is exhausted?

Q8.22 When an object breaks into two pieces (explosion, radioactive decay, recoil, etc.), the lighter fragment gets more kinetic energy than the heavier one. This is a consequence of momentum conservation, but can you also explain it using Newton's laws of motion?

Q8.23 An apple falls from a tree and feels no air resistance. As it is falling, which of these statements about it are true? (a) Only its momentum is conserved; (b) only its mechanical energy is conserved, (c) both its momentum and its mechanical energy are conserved, (d) its kinetic energy is conserved.

Q8.24 Two pieces of clay collide and stick together. During the collision, which of these statements are true? (a) Only the momentum of the clay is conserved, (b) only the mechanical energy of the clay is conserved, (c) both the momentum and the mechanical energy of the clay are conserved, (d) the kinetic energy of the clay is conserved.

Q8.25 Two marbles are pressed together with a light ideal spring between them, but they are not attached to the spring in any way. They are then released on a frictionless horizontal table and soon move free of the spring. As the marbles are moving away from each other, which of these statements about them are true? (a) Only the momentum of the marbles is conserved, (b) only the mechanical energy of the marbles is conserved, (c) both the momentum and the mechanical energy of the marbles are conserved, (d) the kinetic energy of the marbles is conserved.

Q8.26 A very heavy SUV collides head-on with a very light compact car. Which of these statements about the collision are correct? (a) The amount of kinetic energy lost by the SUV is equal to the amount of kinetic energy gained by the compact, (b) the amount of momentum lost by the SUV is equal to the amount of momentum gained by the compact, (c) the compact feels a considerably greater force during the collision than the SUV does, (d) both cars lose the same amount of kinetic energy.

EXERCISES

Section 8.1 Momentum and Impulse

8.1 • (a) What is the magnitude of the momentum of a 10,000-kg truck whose speed is 12.0 m/s? (b) What speed would a 2000-kg SUV have to attain in order to have (i) the same momentum? (ii) the same kinetic energy?

8.2 • In a certain men's track and field event, the shotput has a mass of 7.30 kg and is released with a speed of 15.0 m/s at 40.0° above the horizontal over a man's straight left leg. What are the initial horizontal and vertical components of the momentum of this shotput?

8.3 •• (a) Show that the kinetic energy K and the momentum magnitude p of a particle with mass m are related by $K = p^2/2m$. (b) A 0.040-kg cardinal (*Richmondena cardinalis*) and a 0.145-kg baseball have the same kinetic energy. Which has the greater magnitude of momentum? What is the ratio of the cardinal's magnitude of momentum to the baseball's? (c) A 700-N man and a 450-N woman have the same momentum. Who has the greater kinetic energy? What is the ratio of the man's kinetic energy to that of the woman?

8.4 • Two vehicles are approaching an intersection. One is a 2500-kg pickup traveling at 14.0 m/s from east to west (the $-x$-direction), and the other is a 1500-kg sedan going from south to north (the $+y$-direction) at 23.0 m/s. (a) Find the x- and y-components of the net momentum of this system. (b) What are the magnitude and direction of the net momentum?

8.5 • One 110-kg football lineman is running to the right at 2.75 m/s while another 125-kg lineman is running directly toward him at 2.60 m/s. What are (a) the magnitude and direction of the net momentum of these two athletes, and (b) their total kinetic energy?

8.6 •• **BIO** **Biomechanics.** The mass of a regulation tennis ball is 57 g (although it can vary slightly), and tests have shown that the ball is in contact with the tennis racket for 30 ms. (This number can also vary, depending on the racket and swing.) We shall assume a 30.0-ms contact time for this exercise. The fastest-known served tennis ball was served by "Big Bill" Tilden in 1931, and its speed was measured to be 73.14 m/s. (a) What impulse and what force did Big Bill exert on the tennis ball in his record serve? (b) If Big Bill's opponent returned his serve with a speed of 55 m/s, what force and what impulse did he exert on the ball, assuming only horizontal motion?

8.7 • **Force of a Golf Swing.** A 0.0450-kg golf ball initially at rest is given a speed of 25.0 m/s when a club strikes. If the club and ball are in contact for 2.00 ms, what average force acts on the ball? Is the effect of the ball's weight during the time of contact significant? Why or why not?

8.8 • **Force of a Baseball Swing.** A baseball has mass 0.145 kg. (a) If the velocity of a pitched ball has a magnitude of 45.0 m/s and the batted ball's velocity is 55.0 m/s in the opposite direction, find the magnitude of the change in momentum of the ball and of the impulse applied to it by the bat. (b) If the ball remains in contact with the bat for 2.00 ms, find the magnitude of the average force applied by the bat.

8.9 • A 0.160-kg hockey puck is moving on an icy, frictionless, horizontal surface. At $t = 0$, the puck is moving to the right at 3.00 m/s. (a) Calculate the velocity of the puck (magnitude and direction) after a force of 25.0 N directed to the right has been applied for 0.050 s. (b) If, instead, a force of 12.0 N directed to the left is applied from $t = 0$ to $t = 0.050$ s, what is the final velocity of the puck?

8.10 • An engine of the orbital maneuvering system (OMS) on a space shuttle exerts a force of $(26,700 \text{ N})\hat{\jmath}$ for 3.90 s, exhausting a negligible mass of fuel relative to the 95,000-kg mass of the shuttle. (a) What is the impulse of the force for this 3.90 s? (b) What is the shuttle's change in momentum from this impulse? (c) What is the shuttle's change in velocity from this impulse? (d) Why can't we find the resulting change in the kinetic energy of the shuttle?

8.11 • **CALC** At time $t = 0$, a 2150-kg rocket in outer space fires an engine that exerts an increasing force on it in the $+x$-direction. This force obeys the equation $F_x = At^2$, where t is time, and has a magnitude of 781.25 N when $t = 1.25$ s. (a) Find the SI value of the constant A, including its units. (b) What impulse does the engine exert on the rocket during the 1.50-s interval starting 2.00 s after the engine is fired? (c) By how much does the rocket's velocity change during this interval?

8.12 •• A bat strikes a 0.145-kg baseball. Just before impact, the ball is traveling horizontally to the right at 50.0 m/s, and it leaves the bat traveling to the left at an angle of 30° above horizontal with a speed of 65.0 m/s. If the ball and bat are in contact for 1.75 ms, find the horizontal and vertical components of the average force on the ball.

8.13 • A 2.00-kg stone is sliding to the right on a frictionless horizontal surface at 5.00 m/s when it is suddenly struck by an object that exerts a large horizontal force on it for a short period of time. The graph in Fig. E8.13 shows the magnitude of this force as a function of time. (a) What impulse does this force exert on the stone? (b) Just after the force stops acting, find the magnitude and direction of the stone's velocity if the force acts (i) to the right or (ii) to the left.

Figure **E8.13**

8.14 •• **BIO** **Bone Fracture.** Experimental tests have shown that bone will rupture if it is subjected to a force density of $1.03 \times 10^8 \text{ N/m}^2$. Suppose a 70.0-kg person carelessly rollerskates into an overhead metal beam that hits his forehead and completely stops his forward motion. If the area of contact with the person's forehead is 1.5 cm², what is the greatest speed with which he can hit the wall without breaking any bone if his head is in contact with the beam for 10.0 ms?

8.15 •• To warm up for a match, a tennis player hits the 57.0-g ball vertically with her racket. If the ball is stationary just

before it is hit and goes 5.50 m high, what impulse did she impart to it?

8.16 •• **CALC** Starting at $t = 0$, a horizontal net force $\vec{F} = (0.280 \text{ N/s})t\hat{\imath} + (-0.450 \text{ N/s}^2)t^2\hat{\jmath}$ is applied to a box that has an initial momentum $\vec{p} = (-3.00 \text{ kg} \cdot \text{m/s})\hat{\imath} + (4.00 \text{ kg} \cdot \text{m/s})\hat{\jmath}$. What is the momentum of the box at $t = 2.00$ s?

Section 8.2 Conservation of Momentum

8.17 •• The expanding gases that leave the muzzle of a rifle also contribute to the recoil. A .30-caliber bullet has mass 0.00720 kg and a speed of 601 m/s relative to the muzzle when fired from a rifle that has mass 2.80 kg. The loosely held rifle recoils at a speed of 1.85 m/s relative to the earth. Find the momentum of the propellant gases in a coordinate system attached to the earth as they leave the muzzle of the rifle.

8.18 • A 68.5-kg astronaut is doing a repair in space on the orbiting space station. She throws a 2.25-kg tool away from her at 3.20 m/s relative to the space station. With what speed and in what direction will she begin to move?

8.19 • **BIO** **Animal Propulsion.** Squids and octopuses propel themselves by expelling water. They do this by keeping water in a cavity and then suddenly contracting the cavity to force out the water through an opening. A 6.50-kg squid (including the water in the cavity) at rest suddenly sees a dangerous predator. (a) If the squid has 1.75 kg of water in its cavity, at what speed must it expel this water to suddenly achieve a speed of 2.50 m/s to escape the predator? Neglect any drag effects of the surrounding water. (b) How much kinetic energy does the squid create by this maneuver?

8.20 •• You are standing on a sheet of ice that covers the football stadium parking lot in Buffalo; there is negligible friction between your feet and the ice. A friend throws you a 0.400-kg ball that is traveling horizontally at 10.0 m/s. Your mass is 70.0 kg. (a) If you catch the ball, with what speed do you and the ball move afterward? (b) If the ball hits you and bounces off your chest, so afterward it is moving horizontally at 8.0 m/s in the opposite direction, what is your speed after the collision?

8.21 •• On a frictionless, horizontal air table, puck A (with mass 0.250 kg) is moving toward puck B (with mass 0.350 kg), which is initially at rest. After the collision, puck A has a velocity of 0.120 m/s to the left, and puck B has a velocity of 0.650 m/s to the right. (a) What was the speed of puck A before the collision? (b) Calculate the change in the total kinetic energy of the system that occurs during the collision.

8.22 •• When cars are equipped with flexible bumpers, they will bounce off each other during low-speed collisions, thus causing less damage. In one such accident, a 1750-kg car traveling to the right at 1.50 m/s collides with a 1450-kg car going to the left at 1.10 m/s. Measurements show that the heavier car's speed just after the collision was 0.250 m/s in its original direction. You can ignore any road friction during the collision. (a) What was the speed of the lighter car just after the collision? (b) Calculate the change in the combined kinetic energy of the two-car system during this collision.

8.23 •• Two identical 1.50-kg masses are pressed against opposite ends of a light spring of force constant 1.75 N/cm, compressing the spring by 20.0 cm from its normal length. Find the speed of each mass when it has moved free of the spring on a frictionless horizontal table.

8.24 • Block A in Fig. E8.24 has mass 1.00 kg, and block B has mass 3.00 kg. The blocks are forced together, compressing a spring

S between them; then the system is released from rest on a level, frictionless surface. The spring, which has negligible mass, is not fastened to either block and drops to the surface after it has expanded. Block B acquires a speed of 1.20 m/s. (a) What is the final speed of block A? (b) How much potential energy was stored in the compressed spring?

Figure **E8.24**

$m_A = 1.00$ kg $m_B = 3.00$ kg

S

8.25 •• A hunter on a frozen, essentially frictionless pond uses a rifle that shoots 4.20-g bullets at 965 m/s. The mass of the hunter (including his gun) is 72.5 kg, and the hunter holds tight to the gun after firing it. Find the recoil velocity of the hunter if he fires the rifle (a) horizontally and (b) at 56.0° above the horizontal.

8.26 • An atomic nucleus suddenly bursts apart (fissions) into two pieces. Piece A, of mass m_A, travels off to the left with speed v_A. Piece B, of mass m_B, travels off to the right with speed v_B. (a) Use conservation of momentum to solve for v_B in terms of m_A, m_B, and v_A. (b) Use the results of part (a) to show that $K_A/K_B = m_B/m_A$, where K_A and K_B are the kinetic energies of the two pieces.

8.27 •• Two ice skaters, Daniel (mass 65.0 kg) and Rebecca (mass 45.0 kg), are practicing. Daniel stops to tie his shoelace and, while at rest, is struck by Rebecca, who is moving at 13.0 m/s before she collides with him. After the collision, Rebecca has a velocity of magnitude 8.00 m/s at an angle of 53.1° from her initial direction. Both skaters move on the frictionless, horizontal surface of the rink. (a) What are the magnitude and direction of Daniel's velocity after the collision? (b) What is the change in total kinetic energy of the two skaters as a result of the collision?

8.28 •• You are standing on a large sheet of frictionless ice and holding a large rock. In order to get off the ice, you throw the rock so it has velocity 12.0 m/s relative to the earth at an angle of 35.0° above the horizontal. If your mass is 70.0 kg and the rock's mass is 15.0 kg, what is your speed after you throw the rock? (See Discussion Question Q8.7.)

8.29 • **Changing Mass.** An open-topped freight car with mass 24,000 kg is coasting without friction along a level track. It is raining very hard, and the rain is falling vertically downward. Originally, the car is empty and moving with a speed of 4.00 m/s. (a) What is the speed of the car after it has collected 3000 kg of rainwater? (b) Since the rain is falling downward, how is it able to affect the horizontal motion of the car?

8.30 • An astronaut in space cannot use a conventional means, such as a scale or balance, to determine the mass of an object. But she does have devices to measure distance and time accurately. She knows her own mass is 78.4 kg, but she is unsure of the mass of a large gas canister in the airless rocket. When this canister is approaching her at 3.50 m/s, she pushes against it, which slows it down to 1.20 m/s (but does not reverse it) and gives her a speed of 2.40 m/s. What is the mass of this canister?

8.31 •• **Asteroid Collision.** Two asteroids of equal mass in the asteroid belt between Mars and Jupiter collide with a glancing blow. Asteroid A, which was initially traveling at 40.0 m/s, is deflected 30.0° from its original direction, while asteroid B,

which was initially at rest, travels at 45.0° to the original direction of A (Fig. E8.31). (a) Find the speed of each asteroid after the collision. (b) What fraction of the original kinetic energy of asteroid A dissipates during this collision?

Figure **E8.31**

A 40.0 m/s A 30.0° 45.0° B

Section 8.3 Momentum Conservation and Collisions

8.32 • Two skaters collide and grab on to each other on frictionless ice. One of them, of mass 70.0 kg, is moving to the right at 2.00 m/s, while the other, of mass 65.0 kg, is moving to the left at 2.50 m/s. What are the magnitude and direction of the velocity of these skaters just after they collide?

8.33 •• A 15.0-kg fish swimming at 1.10 m/s suddenly gobbles up a 4.50-kg fish that is initially stationary. Neglect any drag effects of the water. (a) Find the speed of the large fish just after it eats the small one. (b) How much mechanical energy was dissipated during this meal?

8.34 • Two fun-loving otters are sliding toward each other on a muddy (and hence frictionless) horizontal surface. One of them, of mass 7.50 kg, is sliding to the left at 5.00 m/s, while the other, of mass 5.75 kg, is slipping to the right at 6.00 m/s. They hold fast to each other after they collide. (a) Find the magnitude and direction of the velocity of these free-spirited otters right after they collide. (b) How much mechanical energy dissipates during this play?

8.35 • **Deep Impact Mission.** In July 2005, NASA's "Deep Impact" mission crashed a 372-kg probe directly onto the surface of the comet Tempel 1, hitting the surface at 37,000 km/h. The original speed of the comet at that time was about 40,000 km/h, and its mass was estimated to be in the range $(0.10 - 2.5) \times 10^{14}$ kg. Use the smallest value of the estimated mass. (a) What change in the comet's velocity did this collision produce? Would this change be noticeable? (b) Suppose this comet were to hit the earth and fuse with it. By how much would it change our planet's velocity? Would this change be noticeable? (The mass of the earth is 5.97×10^{24} kg.)

8.36 • A 1050-kg sports car is moving westbound at 15.0 m/s on a level road when it collides with a 6320-kg truck driving east on the same road at 10.0 m/s. The two vehicles remain locked together after the collision. (a) What is the velocity (magnitude and direction) of the two vehicles just after the collision? (b) At what speed should the truck have been moving so that it and the car are both stopped in the collision? (c) Find the change in kinetic energy of the system of two vehicles for the situations of part (a) and part (b). For which situation is the change in kinetic energy greater in magnitude?

8.37 •• On a very muddy football field, a 110-kg linebacker tackles an 85-kg halfback. Immediately before the collision, the linebacker is slipping with a velocity of 8.8 m/s north and the halfback is sliding with a velocity of 7.2 m/s east. What is the velocity (magnitude and direction) at which the two players move together immediately after the collision?

8.38 •• **Accident Analysis.** Two cars collide at an intersection. Car A, with a mass of 2000 kg, is going from west to east, while car B, of mass 1500 kg, is going from north to south at 15 m/s. As a result of this collision, the two cars become enmeshed and move as one afterward. In your role as an expert witness, you inspect the scene and determine that, after the collision, the enmeshed cars moved at an angle of 65° south of east from the point of impact.

(a) How fast were the enmeshed cars moving just after the collision? (b) How fast was car *A* going just before the collision?

8.39 • Two cars, one a compact with mass 1200 kg and the other a large gas-guzzler with mass 3000 kg, collide head-on at typical freeway speeds. (a) Which car has a greater magnitude of momentum change? Which car has a greater velocity change? (b) If the larger car changes its velocity by Δv, calculate the change in the velocity of the small car in terms of Δv. (c) Which car's occupants would you expect to sustain greater injuries? Explain.

8.40 •• **BIO Bird Defense.** To protect their young in the nest, peregrine falcons will fly into birds of prey (such as ravens) at high speed. In one such episode, a 600-g falcon flying at 20.0 m/s hit a 1.50-kg raven flying at 9.0 m/s. The falcon hit the raven at right angles to its original path and bounced back at 5.0 m/s. (These figures were estimated by the author as he watched this attack occur in northern New Mexico.) (a) By what angle did the falcon change the raven's direction of motion? (b) What was the raven's speed right after the collision?

8.41 • At the intersection of Texas Avenue and University Drive, a yellow subcompact car with mass 950 kg traveling east on University collides with a red pickup truck with mass 1900 kg that is traveling north on Texas and has run a red light (Fig. E8.41). The two vehicles stick together as a result of the collision, and the wreckage slides at 16.0 m/s in the direction 24.0° east of north. Calculate the speed of each vehicle before the collision. The collision occurs during a heavy rainstorm; you can ignore friction forces between the vehicles and the wet road.

Figure **E8.41**

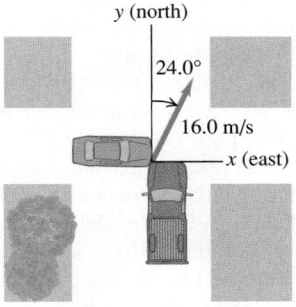

8.42 •• A 5.00-g bullet is fired horizontally into a 1.20-kg wooden block resting on a horizontal surface. The coefficient of kinetic friction between block and surface is 0.20. The bullet remains embedded in the block, which is observed to slide 0.230 m along the surface before stopping. What was the initial speed of the bullet?

8.43 •• **A Ballistic Pendulum.** A 12.0-g rifle bullet is fired with a speed of 380 m/s into a ballistic pendulum with mass 6.00 kg, suspended from a cord 70.0 cm long (see Example 8.8 in Section 8.3). Compute (a) the vertical height through which the pendulum rises, (b) the initial kinetic energy of the bullet, and (c) the kinetic energy of the bullet and pendulum immediately after the bullet becomes embedded in the pendulum.

8.44 •• **Combining Conservation Laws.** A 15.0-kg block is attached to a very light horizontal spring of force constant 500.0 N/m and is resting on a frictionless horizontal table. (Fig. E8.44). Suddenly it is struck by a 3.00-kg stone traveling horizontally at 8.00 m/s to the right, whereupon the stone rebounds at 2.00 m/s horizontally to the left. Find the maximum distance that the block will compress the spring after the collision.

Figure **E8.44**

8.45 •• **CP** A 5.00-kg ornament is hanging by a 1.50-m wire when it is suddenly hit by a 3.00-kg missile traveling horizontally at 12.0 m/s. The missile embeds itself in the ornament during the collision. What is the tension in the wire immediately after the collision?

Section 8.4 Elastic Collisions

8.46 •• A 0.150-kg glider is moving to the right on a frictionless, horizontal air track with a speed of 0.80 m/s. It has a head-on collision with a 0.300-kg glider that is moving to the left with a speed of 2.20 m/s. Find the final velocity (magnitude and direction) of each glider if the collision is elastic.

8.47 •• Blocks *A* (mass 2.00 kg) and *B* (mass 10.00 kg) move on a frictionless, horizontal surface. Initially, block *B* is at rest and block *A* is moving toward it at 2.00 m/s. The blocks are equipped with ideal spring bumpers, as in Example 8.10 (Section 8.4). The collision is head-on, so all motion before and after the collision is along a straight line. (a) Find the maximum energy stored in the spring bumpers and the velocity of each block at that time. (b) Find the velocity of each block after they have moved apart.

8.48 • A 10.0-g marble slides to the left with a velocity of magnitude 0.400 m/s on the frictionless, horizontal surface of an icy New York sidewalk and has a head-on, elastic collision with a larger 30.0-g marble sliding to the right with a velocity of magnitude 0.200 m/s (Fig. E8.48). (a) Find the velocity of each marble (magnitude and direction) after the collision. (Since the collision is head-on, all the motion is along a line.) (b) Calculate the *change in momentum* (that is, the momentum after the collision minus the momentum before the collision) for each marble. Compare the values you get for each marble. (c) Calculate the *change in kinetic energy* (that is, the kinetic energy after the collision minus the kinetic energy before the collision) for each marble. Compare the values you get for each marble.

Figure **E8.48**

8.49 •• **Moderators.** Canadian nuclear reactors use *heavy water* moderators in which elastic collisions occur between the neutrons and deuterons of mass 2.0 u (see Example 8.11 in Section 8.4). (a) What is the speed of a neutron, expressed as a fraction of its original speed, after a head-on, elastic collision with a deuteron that is initially at rest? (b) What is its kinetic energy, expressed as a fraction of its original kinetic energy? (c) How many such successive collisions will reduce the speed of a neutron to 1/59,000 of its original value?

8.50 •• You are at the controls of a particle accelerator, sending a beam of 1.50×10^7 m/s protons (mass *m*) at a gas target of an unknown element. Your detector tells you that some protons bounce straight back after a collision with one of the nuclei of the unknown element. All such protons rebound with a speed of 1.20×10^7 m/s. Assume that the initial speed of the target nucleus is negligible and the collision is elastic. (a) Find the mass of one nucleus of the unknown element. Express your answer in terms of the proton mass *m*. (b) What is the speed of the unknown nucleus immediately after such a collision?

Section 8.5 Center of Mass

8.51 • Three odd-shaped blocks of chocolate have the following masses and center-of-mass coordinates: (1) 0.300 kg, (0.200 m,

0.300 m); (2) 0.400 kg, (0.100 m, −0.400 m); (3) 0.200 kg, (−0.300 m, 0.600 m). Find the coordinates of the center of mass of the system of three chocolate blocks.

8.52 • Find the position of the center of mass of the system of the sun and Jupiter. (Since Jupiter is more massive than the rest of the planets combined, this is essentially the position of the center of mass of the solar system.) Does the center of mass lie inside or outside the sun? Use the data in Appendix F.

8.53 •• **Pluto and Charon.** Pluto's diameter is approximately 2370 km, and the diameter of its satellite Charon is 1250 km. Although the distance varies, they are often about 19,700 km apart, center to center. Assuming that both Pluto and Charon have the same composition and hence the same average density, find the location of the center of mass of this system relative to the center of Pluto.

8.54 • A 1200-kg station wagon is moving along a straight highway at 12.0 m/s. Another car, with mass 1800 kg and speed 20.0 m/s, has its center of mass 40.0 m ahead of the center of mass of the station wagon (Fig. E8.54). (a) Find the position of the center of mass of the system consisting of the two automobiles. (b) Find the magnitude of the total momentum of the system from the given data. (c) Find the speed of the center of mass of the system. (d) Find the total momentum of the system, using the speed of the center of mass. Compare your result with that of part (b).

Figure **E8.54**

8.55 • A machine part consists of a thin, uniform 4.00-kg bar that is 1.50 m long, hinged perpendicular to a similar vertical bar of mass 3.00 kg and length 1.80 m. The longer bar has a small but dense 2.00-kg ball at one end (Fig. E8.55). By what distance will the center of mass of this part move horizontally and vertically if the vertical bar is pivoted counterclockwise through 90° to make the entire part horizontal?

Figure **E8.55**

8.56 • At one instant, the center of mass of a system of two particles is located on the x-axis at $x = 2.0$ m and has a velocity of $(5.0 \text{ m/s})\hat{\imath}$. One of the particles is at the origin. The other particle has a mass of 0.10 kg and is at rest on the x-axis at $x = 8.0$ m. (a) What is the mass of the particle at the origin? (b) Calculate the total momentum of this system. (c) What is the velocity of the particle at the origin?

8.57 •• In Example 8.14 (Section 8.5), Ramon pulls on the rope to give himself a speed of 0.70 m/s. What is James's speed?

8.58 • **CALC** A system consists of two particles. At $t = 0$ one particle is at the origin; the other, which has a mass of 0.50 kg, is on the y-axis at $y = 6.0$ m. At $t = 0$ the center of mass of the system is on the y-axis at $y = 2.4$ m. The velocity of the center of mass is given by $(0.75 \text{ m/s}^3)t^2\hat{\imath}$. (a) Find the total mass of the system. (b) Find the acceleration of the center of mass at any time t. (c) Find the net external force acting on the system at $t = 3.0$ s.

8.59 • **CALC** A radio-controlled model airplane has a momentum given by $[(-0.75 \text{ kg} \cdot \text{m/s}^3)t^2 + (3.0 \text{ kg} \cdot \text{m/s})]\hat{\imath} + (0.25 \text{ kg} \cdot \text{m/s}^2)t\hat{\jmath}$. What are the x-, y-, and z-components of the net force on the airplane?

8.60 •• **BIO** **Changing Your Center of Mass.** To keep the calculations fairly simple, but still reasonable, we shall model a human leg that is 92.0 cm long (measured from the hip joint) by assuming that the upper leg and the lower leg (which includes the foot) have equal lengths and that each of them is uniform. For a 70.0-kg person, the mass of the upper leg would be 8.60 kg, while that of the lower leg (including the foot) would be 5.25 kg. Find the location of the center of mass of this leg, relative to the hip joint, if it is (a) stretched out horizontally and (b) bent at the knee to form a right angle with the upper leg remaining horizontal.

Section 8.6 Rocket Propulsion

8.61 •• A 70-kg astronaut floating in space in a 110-kg MMU (manned maneuvering unit) experiences an acceleration of 0.029 m/s^2 when he fires one of the MMU's thrusters. (a) If the speed of the escaping N_2 gas relative to the astronaut is 490 m/s, how much gas is used by the thruster in 5.0 s? (b) What is the thrust of the thruster?

8.62 • A small rocket burns 0.0500 kg of fuel per second, ejecting it as a gas with a velocity relative to the rocket of magnitude 1600 m/s. (a) What is the thrust of the rocket? (b) Would the rocket operate in outer space where there is no atmosphere? If so, how would you steer it? Could you brake it?

8.63 • A C6-5 model rocket engine has an impulse of $10.0 \text{ N} \cdot \text{s}$ while burning 0.0125 kg of propellant in 1.70 s. It has a maximum thrust of 13.3 N. The initial mass of the engine plus propellant is 0.0258 kg. (a) What fraction of the maximum thrust is the average thrust? (b) Calculate the relative speed of the exhaust gases, assuming it is constant. (c) Assuming that the relative speed of the exhaust gases is constant, find the final speed of the engine if it was attached to a very light frame and fired from rest in gravity-free outer space.

8.64 •• Obviously, we can make rockets to go very fast, but what is a reasonable top speed? Assume that a rocket is fired from rest at a space station in deep space, where gravity is negligible. (a) If the rocket ejects gas at a relative speed of 2000 m/s and you want the rocket's speed eventually to be $1.00 \times 10^{-3}c$, where c is the speed of light, what fraction of the initial mass of the rocket and fuel is *not* fuel? (b) What is this fraction if the final speed is to be 3000 m/s?

8.65 •• A single-stage rocket is fired from rest from a deep-space platform, where gravity is negligible. If the rocket burns its fuel in 50.0 s and the relative speed of the exhaust gas is $v_{ex} = 2100$ m/s, what must the mass ratio m_0/m be for a final speed v of 8.00 km/s (about equal to the orbital speed of an earth satellite)?

PROBLEMS

8.66 •• **CP CALC** A young girl with mass 40.0 kg is sliding on a horizontal, frictionless surface with an initial momentum that is due east and that has magnitude $90.0 \text{ kg} \cdot \text{m/s}$. Starting at $t = 0$, a net force with magnitude $F = (8.20 \text{ N/s})t$ and direction due west is applied to the girl. (a) At what value of t does the girl have a westward momentum of magnitude $60.0 \text{ kg} \cdot \text{m/s}$? (b) How much work has been done on the girl by the force in the time interval from $t = 0$ to the time calculated in part (a)? (c) What is the magnitude of the acceleration of the girl at the time calculated in part (a)?

8.67 •• A steel ball with mass 40.0 g is dropped from a height of 2.00 m onto a horizontal steel slab. The ball rebounds to a height of 1.60 m. (a) Calculate the impulse delivered to the ball during impact. (b) If the ball is in contact with the slab for 2.00 ms, find the average force on the ball during impact.

8.68 • In a volcanic eruption, a 2400-kg boulder is thrown vertically upward into the air. At its highest point, it suddenly explodes (due to trapped gases) into two fragments, one being three times the mass of the other. The lighter fragment starts out with only horizontal velocity and lands 318 m directly north of the point of the explosion. Where will the other fragment land? Neglect any air resistance.

8.69 •• Just before it is struck by a racket, a tennis ball weighing 0.560 N has a velocity of $(20.0 \text{ m/s})\hat{\imath} - (4.0 \text{ m/s})\hat{\jmath}$. During the 3.00 ms that the racket and ball are in contact, the net force on the ball is constant and equal to $-(380 \text{ N})\hat{\imath} + (110 \text{ N})\hat{\jmath}$. (a) What are the x- and y-components of the impulse of the net force applied to the ball? (b) What are the x- and y-components of the final velocity of the ball?

8.70 • Three identical pucks on a horizontal air table have repelling magnets. They are held together and then released simultaneously. Each has the same speed at any instant. One puck moves due west. What is the direction of the velocity of each of the other two pucks?

8.71 •• A 1500-kg blue convertible is traveling south, and a 2000-kg red SUV is traveling west. If the total momentum of the system consisting of the two cars is 7200 kg · m/s directed at 60.0° west of south, what is the speed of each vehicle?

8.72 •• A railroad handcar is moving along straight, frictionless tracks with negligible air resistance. In the following cases, the car initially has a total mass (car and contents) of 200 kg and is traveling east with a velocity of magnitude 5.00 m/s. Find the *final velocity* of the car in each case, assuming that the handcar does not leave the tracks. (a) A 25.0-kg mass is thrown sideways out of the car with a velocity of magnitude 2.00 m/s relative to the car's initial velocity. (b) A 25.0-kg mass is thrown backward out of the car with a velocity of 5.00 m/s relative to the initial motion of the car. (c) A 25.0-kg mass is thrown into the car with a velocity of 6.00 m/s relative to the ground and opposite in direction to the initial velocity of the car.

8.73 • Spheres A (mass 0.020 kg), B (mass 0.030 kg), and C (mass 0.050 kg) are approaching the origin as they slide on a frictionless air table (Fig. P8.73). The initial velocities of A and B are given in the figure. All three spheres arrive at the origin at the same time and stick together. (a) What must the x- and y-components of the initial velocity of C be if all three objects are to end up moving at 0.50 m/s in the $+x$-direction after the collision? (b) If C has the velocity found in part (a), what is the change in the kinetic energy of the system of three spheres as a result of the collision?

8.74 ••• You and your friends are doing physics experiments on a frozen pond that serves as a frictionless, horizontal surface. Sam, with mass 80.0 kg, is given a push and slides eastward. Abigail, with mass 50.0 kg, is sent sliding northward. They collide, and after the collision Sam is moving at 37.0° north of east with a speed of 6.00 m/s and Abigail is moving at 23.0° south of east with a speed of 9.00 m/s. (a) What was the speed of each person before the collision? (b) By how much did the total kinetic energy of the two people decrease during the collision?

8.75 ••• The nucleus of ^{214}Po decays radioactively by emitting an alpha particle (mass 6.65×10^{-27} kg) with kinetic energy 1.23×10^{-12} J, as measured in the laboratory reference frame. Assuming that the Po was initially at rest in this frame, find the recoil velocity of the nucleus that remains after the decay.

8.76 • CP At a classic auto show, a 840-kg 1955 Nash Metropolitan motors by at 9.0 m/s, followed by a 1620-kg 1957 Packard Clipper purring past at 5.0 m/s. (a) Which car has the greater kinetic energy? What is the ratio of the kinetic energy of the Nash to that of the Packard? (b) Which car has the greater magnitude of momentum? What is the ratio of the magnitude of momentum of the Nash to that of the Packard? (c) Let F_N be the net force required to stop the Nash in time t, and let F_P be the net force required to stop the Packard in the same time. Which is larger: F_N or F_P? What is the ratio F_N/F_P of these two forces? (d) Now let F_N be the net force required to stop the Nash in a distance d, and let F_P be the net force required to stop the Packard in the same distance. Which is larger: F_N or F_P? What is the ratio F_N/F_P?

8.77 •• CP An 8.00-kg block of wood sits at the edge of a frictionless table, 2.20 m above the floor. A 0.500-kg blob of clay slides along the length of the table with a speed of 24.0 m/s, strikes the block of wood, and sticks to it. The combined object leaves the edge of the table and travels to the floor. What horizontal distance has the combined object traveled when it reaches the floor?

8.78 ••• CP A small wooden block with mass 0.800 kg is suspended from the lower end of a light cord that is 1.60 m long. The block is initially at rest. A bullet with mass 12.0 g is fired at the block with a horizontal velocity v_0. The bullet strikes the block and becomes embedded in it. After the collision the combined object swings on the end of the cord. When the block has risen a vertical height of 0.800 m, the tension in the cord is 4.80 N. What was the initial speed v_0 of the bullet?

8.79 •• **Combining Conservation Laws.** A 5.00-kg chunk of ice is sliding at 12.0 m/s on the floor of an ice-covered valley when it collides with and sticks to another 5.00-kg chunk of ice that is initially at rest. (Fig. P8.79). Since the valley is icy, there is no friction. After the collision, how high above the valley floor will the combined chunks go?

Figure **P8.73**

Figure **P8.79**

8.80 •• **Automobile Accident Analysis.** You are called as an expert witness to analyze the following auto accident: Car B, of mass 1900 kg, was stopped at a red light when it was hit from behind by car A, of mass 1500 kg. The cars locked bumpers during the collision and slid to a stop with brakes locked on all wheels. Measurements of the skid marks left by the tires showed them to be 7.15 m long. The coefficient of kinetic friction between the tires and the road was 0.65. (a) What was the speed of car *A* just before the collision? (b) If the speed limit was 35 mph, was car *A* speeding, and if so, by how many miles per hour was it *exceeding* the speed limit?

8.81 ·· **Accident Analysis.** A 1500-kg sedan goes through a wide intersection traveling from north to south when it is hit by a 2200-kg SUV traveling from east to west. The two cars become enmeshed due to the impact and slide as one thereafter. On-the-scene measurements show that the coefficient of kinetic friction between the tires of these cars and the pavement is 0.75, and the cars slide to a halt at a point 5.39 m west and 6.43 m south of the impact point. How fast was each car traveling just before the collision?

8.82 ··· **CP** A 0.150-kg frame, when suspended from a coil spring, stretches the spring 0.070 m. A 0.200-kg lump of putty is dropped from rest onto the frame from a height of 30.0 cm (Fig. P8.82). Find the maximum distance the frame moves downward from its initial position.

Figure **P8.82**

8.83 · A rifle bullet with mass 8.00 g strikes and embeds itself in a block with mass 0.992 kg that rests on a frictionless, horizontal surface and is attached to a coil spring (Fig. P8.83).

30.0 cm

The impact compresses the spring 15.0 cm. Calibration of the spring shows that a force of 0.750 N is required to compress the spring 0.250 cm. (a) Find the magnitude of the block's velocity just after impact. (b) What was the initial speed of the bullet?

Figure **P8.83**

|←15.0 cm→|

8.84 ·· **A Ricocheting Bullet.** 0.100-kg stone rests on a frictionless, horizontal surface. A bullet of mass 6.00 g, traveling horizontally at 350 m/s, strikes the stone and rebounds horizontally at right angles to its original direction with a speed of 250 m/s. (a) Compute the magnitude and direction of the velocity of the stone after it is struck. (b) Is the collision perfectly elastic?

8.85 ·· A movie stuntman (mass 80.0 kg) stands on a window ledge 5.0 m above the floor (Fig. P8.85). Grabbing a rope attached to a chandelier, he swings down to grapple with the movie's villain (mass 70.0 kg), who is standing directly under the chandelier. (Assume that the stuntman's center of mass moves downward 5.0 m. He releases the rope just as he

Figure **P8.85**

5.0 m   *m* = 80.0 kg

m = 70.0 kg

reaches the villain.) (a) With what speed do the entwined foes start to slide across the floor? (b) If the coefficient of kinetic friction of their bodies with the floor is $\mu_k = 0.250$, how far do they slide?

8.86 ·· **CP** Two identical masses are released from rest in a smooth hemispherical bowl of radius *R* from the positions shown in Fig. P8.86. You can ignore friction between the masses and the surface of the bowl. If they stick together when they collide, how high above the bottom of the bowl will the masses go after colliding?

Figure **P8.86**

R

8.87 ·· A ball with mass *M*, moving horizontally at 4.00 m/s, collides elastically with a block with mass 3*M* that is initially hanging at rest from the ceiling on the end of a 50.0-cm wire. Find the maximum angle through which the block swings after it is hit.

8.88 ··· **CP** A 20.00-kg lead sphere is hanging from a hook by a thin wire 3.50 m long and is free to swing in a complete circle. Suddenly it is struck horizontally by a 5.00-kg steel dart that embeds itself in the lead sphere. What must be the minimum initial speed of the dart so that the combination makes a complete circular loop after the collision?

8.89 ··· **CP** An 8.00-kg ball, hanging from the ceiling by a light wire 135 cm long, is struck in an elastic collision by a 2.00-kg ball moving horizontally at 5.00 m/s just before the collision. Find the tension in the wire just after the collision.

8.90 ·· A 7.0-kg shell at rest explodes into two fragments, one with a mass of 2.0 kg and the other with a mass of 5.0 kg. If the heavier fragment gains 100 J of kinetic energy from the explosion, how much kinetic energy does the lighter one gain?

8.91 ·· A 4.00-g bullet, traveling horizontally with a velocity of magnitude 400 m/s, is fired into a wooden block with mass 0.800 kg, initially at rest on a level surface. The bullet passes through the block and emerges with its speed reduced to 190 m/s. The block slides a distance of 45.0 cm along the surface from its initial position. (a) What is the coefficient of kinetic friction between block and surface? (b) What is the decrease in kinetic energy of the bullet? (c) What is the kinetic energy of the block at the instant after the bullet passes through it?

8.92 ·· A 5.00-g bullet is shot *through* a 1.00-kg wood block suspended on a string 2.00 m long. The center of mass of the block rises a distance of 0.38 cm. Find the speed of the bullet as it emerges from the block if its initial speed is 450 m/s.

8.93 ·· A neutron with mass *m* makes a head-on, elastic collision with a nucleus of mass *M*, which is initially at rest. (a) Show that if the neutron's initial kinetic energy is K_0, the kinetic energy that it loses during the collision is $4mMK_0/(M + m)^2$. (b) For what value of *M* does the incident neutron lose the most energy? (c) When *M* has the value calculated in part (b), what is the speed of the neutron after the collision?

8.94 ·· **Energy Sharing in Elastic Collisions.** A stationary object with mass m_B is struck head-on by an object with mass m_A that is moving initially at speed v_0. (a) If the collision is elastic, what percentage of the original energy does each object have after the collision? (b) What does your answer in part (a) give for the special cases (i) $m_A = m_B$ and (ii) $m_A = 5m_B$? (c) For what values, if any, of the mass ratio m_A/m_B is the original kinetic energy shared equally by the two objects after the collision?

8.95 ·· **CP** In a shipping company distribution center, an open cart of mass 50.0 kg is rolling to the left at a speed of 5.00 m/s

(Fig. P8.95). You can ignore friction between the cart and the floor. A 15.0-kg package slides down a chute that is inclined at 37° from the horizontal and leaves the end of the chute with a speed of 3.00 m/s. The package lands in the cart and they roll off together. If the lower end of the chute is a vertical distance of 4.00 m above the bottom of

Figure **P8.95**

the cart, what are (a) the speed of the package just before it lands in the cart and (b) the final speed of the cart?

8.96 • A blue puck with mass 0.0400 kg, sliding with a velocity of magnitude 0.200 m/s on a frictionless, horizontal air table, makes a perfectly elastic, head-on collision with a red puck with mass m, initially at rest. After the collision, the velocity of the blue puck is 0.050 m/s in the same direction as its initial velocity. Find (a) the velocity (magnitude and direction) of the red puck after the collision and (b) the mass m of the red puck.

8.97 ••• Jack and Jill are standing on a crate at rest on the frictionless, horizontal surface of a frozen pond. Jack has mass 75.0 kg, Jill has mass 45.0 kg, and the crate has mass 15.0 kg. They remember that they must fetch a pail of water, so each jumps horizontally from the top of the crate. Just after each jumps, that person is moving away from the crate with a speed of 4.00 m/s relative to the crate. (a) What is the final speed of the crate if both Jack and Jill jump simultaneously and in the same direction? (*Hint:* Use an inertial coordinate system attached to the ground.) (b) What is the final speed of the crate if Jack jumps first and then a few seconds later Jill jumps in the same direction? (c) What is the final speed of the crate if Jill jumps first and then Jack, again in the same direction?

8.98 • Suppose you hold a small ball in contact with, and directly over, the center of a large ball. If you then drop the small ball a short time after dropping the large ball, the small ball rebounds with surprising speed. To show the extreme case, ignore air resistance and suppose the large ball makes an elastic collision with the floor and then rebounds to make an elastic collision with the still-descending small ball. Just before the collision between the two balls, the large ball is moving upward with velocity $\vec{v}$ and the small ball has velocity $-\vec{v}$. (Do you see why?) Assume the large ball has a much greater mass than the small ball. (a) What is the velocity of the small ball immediately after its collision with the large ball? (b) From the answer to part (a), what is the ratio of the small ball's rebound distance to the distance it fell before the collision?

8.99 ••• Hockey puck B rests on a smooth ice surface and is struck by a second puck A, which has the same mass. Puck A is initially traveling at 15.0 m/s and is deflected 25.0° from its initial direction. Assume that the collision is perfectly elastic. Find the final speed of each puck and the direction of B's velocity after the collision.

8.100 ••• **Energy Sharing.** An object with mass m, initially at rest, explodes into two fragments, one with mass m_A and the other with mass m_B, where $m_A + m_B = m$. (a) If energy Q is released in the explosion, how much kinetic energy does each fragment have immediately after the explosion? (b) What percentage of the total energy released does each fragment get when one fragment has four times the mass of the other?

8.101 ••• **Neutron Decay.** A neutron at rest decays (breaks up) to a proton and an electron. Energy is released in the decay

and appears as kinetic energy of the proton and electron. The mass of a proton is 1836 times the mass of an electron. What fraction of the total energy released goes into the kinetic energy of the proton?

8.102 •• A ^{232}Th (thorium) nucleus at rest decays to a ^{228}Ra (radium) nucleus with the emission of an alpha particle. The total kinetic energy of the decay fragments is 6.54×10^{-13} J. An alpha particle has 1.76% of the mass of a ^{228}Ra nucleus. Calculate the kinetic energy of (a) the recoiling ^{228}Ra nucleus and (b) the alpha particle.

8.103 • **Antineutrino.** In beta decay, a nucleus emits an electron. A ^{210}Bi (bismuth) nucleus at rest undergoes beta decay to ^{210}Po (polonium). Suppose the emitted electron moves to the right with a momentum of 5.60×10^{-22} kg·m/s. The ^{210}Po nucleus, with mass 3.50×10^{-25} kg, recoils to the left at a speed of 1.14×10^3 m/s. Momentum conservation requires that a second particle, called an antineutrino, must also be emitted. Calculate the magnitude and direction of the momentum of the antineutrino that is emitted in this decay.

8.104 •• Jonathan and Jane are sitting in a sleigh that is at rest on frictionless ice. Jonathan's weight is 800 N, Jane's weight is 600 N, and that of the sleigh is 1000 N. They see a poisonous spider on the floor of the sleigh and immediately jump off. Jonathan jumps to the left with a velocity of 5.00 m/s at 30.0° above the horizontal (relative to the ice), and Jane jumps to the right at 7.00 m/s at 36.9° above the horizontal (relative to the ice). Calculate the sleigh's horizontal velocity (magnitude and direction) after they jump out.

8.105 •• Two friends, Burt and Ernie, are standing at opposite ends of a uniform log that is floating in a lake. The log is 3.0 m long and has mass 20.0 kg. Burt has mass 30.0 kg and Ernie has mass 40.0 kg. Initially the log and the two friends are at rest relative to the shore. Burt then offers Ernie a cookie, and Ernie walks to Burt's end of the log to get it. Relative to the shore, what distance has the log moved by the time Ernie reaches Burt? Neglect any horizontal force that the water exerts on the log and assume that neither Burt nor Ernie falls off the log.

8.106 •• A 45.0-kg woman stands up in a 60.0-kg canoe 5.00 m long. She walks from a point 1.00 m from one end to a point 1.00 m from the other end (Fig. P8.106). If you ignore resistance to motion of the canoe in the water, how far does the canoe move during this process?

Figure **P8.106**

8.107 •• You are standing on a concrete slab that in turn is resting on a frozen lake. Assume there is no friction between the slab and the ice. The slab has a weight five times your weight. If you begin walking forward at 2.00 m/s relative to the ice, with what speed, relative to the ice, does the slab move?

8.108 •• **CP** A 20.0-kg projectile is fired at an angle of 60.0° above the horizontal with a speed of 80.0 m/s. At the highest point of its trajectory, the projectile explodes into two fragments with equal mass, one of which falls vertically with zero initial speed. You can ignore air resistance. (a) How far from the point of firing does the other fragment strike if the terrain is level? (b) How much energy is released during the explosion?

8.109 ••• **CP** A fireworks rocket is fired vertically upward. At its maximum height of 80.0 m, it explodes and breaks into two pieces: one with mass 1.40 kg and the other with mass 0.28 kg. In the explosion, 860 J of chemical energy is converted to kinetic energy of the two fragments. (a) What is the speed of each fragment just after the explosion? (b) It is observed that the two fragments hit the ground at the same time. What is the distance between the points on the ground where they land? Assume that the ground is level and air resistance can be ignored.

8.110 ••• A 12.0-kg shell is launched at an angle of 55.0° above the horizontal with an initial speed of 150 m/s. When it is at its highest point, the shell explodes into two fragments, one three times heavier than the other. The two fragments reach the ground at the same time. Assume that air resistance can be ignored. If the heavier fragment lands back at the same point from which the shell was launched, where will the lighter fragment land, and how much energy was released in the explosion?

8.111 • **CP** A wagon with two boxes of gold, having total mass 300 kg, is cut loose from the horses by an outlaw when the wagon is at rest 50 m up a 6.0° slope (Fig. P8.111). The outlaw plans to have the wagon roll down the slope and across the level ground, and then fall into a canyon where his confederates wait. But in a tree 40 m from the canyon edge wait the Lone Ranger (mass 75.0 kg) and Tonto (mass 60.0 kg). They drop vertically into the wagon as it passes beneath them. (a) If they require 5.0 s to grab the gold and jump out, will they make it before the wagon goes over the edge? The wagon rolls with negligible friction. (b) When the two heroes drop into the wagon, is the kinetic energy of the system of the heroes plus the wagon conserved? If not, does it increase or decrease, and by how much?

Figure **P8.111**

8.112 •• **CALC** In Section 8.6, we considered a rocket fired in outer space where there is no air resistance and where gravity is negligible. Suppose instead that the rocket is accelerating vertically upward from rest on the earth's surface. Continue to ignore air resistance and consider only that part of the motion where the altitude of the rocket is small so that g may be assumed to be constant. (a) How is Eq. (8.37) modified by the presence of the gravity force? (b) Derive an expression for the acceleration a of the rocket, analogous to Eq. (8.39). (c) What is the acceleration of the rocket in Example 8.15 (Section 8.6) if it is near the earth's surface rather than in outer space? You can ignore air resistance. (d) Find the speed of the rocket in Example 8.16 (Section 8.6) after 90 s if the rocket is fired from the earth's surface rather than in outer space.

You can ignore air resistance. How does your answer compare with the rocket speed calculated in Example 8.16?

8.113 •• **A Multistage Rocket.** Suppose the first stage of a two-stage rocket has total mass 12,000 kg, of which 9000 kg is fuel. The total mass of the second stage is 1000 kg, of which 700 kg is fuel. Assume that the relative speed v_{ex} of ejected material is constant, and ignore any effect of gravity. (The effect of gravity is small during the firing period if the rate of fuel consumption is large.) (a) Suppose the entire fuel supply carried by the two-stage rocket is utilized in a single-stage rocket with the same total mass of 13,000 kg. In terms of v_{ex}, what is the speed of the rocket, starting from rest, when its fuel is exhausted? (b) For the two-stage rocket, what is the speed when the fuel of the first stage is exhausted if the first stage carries the second stage with it to this point? This speed then becomes the initial speed of the second stage. At this point, the second stage separates from the first stage. (c) What is the final speed of the second stage? (d) What value of v_{ex} is required to give the second stage of the rocket a speed of 7.00 km/s?

CHALLENGE PROBLEMS

8.114 • **CALC A Variable-Mass Raindrop.** In a rocket-propulsion problem the mass is variable. Another such problem is a raindrop falling through a cloud of small water droplets. Some of these small droplets adhere to the raindrop, thereby *increasing* its mass as it falls. The force on the raindrop is

$$F_{ext} = \frac{dp}{dt} = m\frac{dv}{dt} + v\frac{dm}{dt}$$

Suppose the mass of the raindrop depends on the distance x that it has fallen. Then $m = kx$, where k is a constant, and $dm/dt = kv$. This gives, since $F_{ext} = mg$,

$$mg = m\frac{dv}{dt} + v(kv)$$

Or, dividing by k,

$$xg = x\frac{dv}{dt} + v^2$$

This is a differential equation that has a solution of the form $v = at$, where a is the acceleration and is constant. Take the initial velocity of the raindrop to be zero. (a) Using the proposed solution for v, find the acceleration a. (b) Find the distance the raindrop has fallen in $t = 3.00$ s. (c) Given that $k = 2.00$ g/m, find the mass of the raindrop at $t = 3.00$ s. (For many more intriguing aspects of this problem, see K. S. Krane, *American Journal of Physics,* Vol. 49 (1981), pp. 113–117.)

8.115 •• **CALC** In Section 8.5 we calculated the center of mass by considering objects composed of a *finite* number of point masses or objects that, by symmetry, could be represented by a finite number of point masses. For a solid object whose mass distribution does not allow for a simple determination of the center of mass by symmetry, the sums of Eqs. (8.28) must be generalized to integrals

$$x_{cm} = \frac{1}{M}\int x\,dm \qquad y_{cm} = \frac{1}{M}\int y\,dm$$

where x and y are the coordinates of the small piece of the object that has mass dm. The integration is over the whole of the object.

Consider a thin rod of length L, mass M, and cross-sectional area A. Let the origin of the coordinates be at the left end of the rod and the positive x-axis lie along the rod. (a) If the density $\rho = M/V$ of the object is uniform, perform the integration described above to show that the x-coordinate of the center of mass of the rod is at its geometrical center. (b) If the density of the object varies linearly with x—that is, $\rho = \alpha x$, where α is a positive constant—calculate the x-coordinate of the rod's center of mass.

8.116 •• **CALC** Use the methods of Challenge Problem 8.115 to calculate the x- and y-coordinates of the center of mass of a semicircular metal plate with uniform density ρ and thickness t. Let the radius of the plate be a. The mass of the plate is thus $M = \frac{1}{2}\rho\pi a^2 t$. Use the coordinate system indicated in Fig. P8.116.

Figure **P8.116**

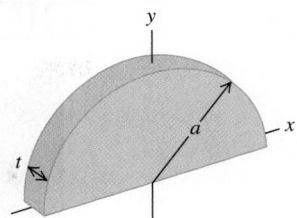

Answers

Chapter Opening Question ?

The two bullets have the same magnitude of momentum $p = mv$ (the product of mass and speed), but the faster, lightweight bullet has twice as much kinetic energy $K = \frac{1}{2}mv^2$. Hence, the light-weight bullet can do twice as much work on the carrot (and twice as much damage) in the process of coming to a halt (see Section 8.1).

Test Your Understanding Questions

8.1 Answer: (v), (i) and (ii) (tied for second place), (iii) and (iv) (tied for third place) We use two interpretations of the impulse of the net force: (1) the net force multiplied by the time that the net force acts, and (2) the change in momentum of the particle on which the net force acts. Which interpretation we use depends on what information we are given. We take the positive x-direction to be to the east. (i) The force is not given, so we use interpretation 2: $J_x = mv_{2x} - mv_{1x} = (1000 \text{ kg})(0) - (1000 \text{ kg})(25 \text{ m/s}) = -25{,}000 \text{ kg} \cdot \text{m/s}$, so the magnitude of the impulse is $25{,}000 \text{ kg} \cdot \text{m/s} = 25{,}000 \text{ N} \cdot \text{s}$. (ii) For the same reason as in (i), we use interpretation 2: $J_x = mv_{2x} - mv_{1x} = (1000 \text{ kg})(0) - (1000 \text{ kg})(25 \text{ m/s}) = -25{,}000 \text{ kg} \cdot \text{m/s}$, and the magnitude of the impulse is again $25{,}000 \text{ kg} \cdot \text{m/s} = 25{,}000 \text{ N} \cdot \text{s}$. (iii) The final velocity is not given, so we use interpretation 1: $J_x = (\Sigma F_x)_{av}(t_2 - t_1) = (2000 \text{ N})(10 \text{ s}) = 20{,}000 \text{ N} \cdot \text{s}$, so the magnitude of the impulse is $20{,}000 \text{ N} \cdot \text{s}$. (iv) For the same reason as in (iii), we use interpretation 1: $J_x = (\Sigma F_x)_{av}(t_2 - t_1) = (-2000 \text{ N})(10 \text{ s}) = -20{,}000 \text{ N} \cdot \text{s}$, so the magnitude of the impulse is $20{,}000 \text{ N} \cdot \text{s}$. (v) The force is not given, so we use interpretation 2: $J_x = mv_{2x} - mv_{1x} = (1000 \text{ kg})(-25 \text{ m/s}) - (1000 \text{ kg})(25 \text{ m/s}) = -50{,}000 \text{ kg} \cdot \text{m/s}$, so the magnitude of the impulse is $50{,}000 \text{ kg} \cdot \text{m/s} = 50{,}000 \text{ N} \cdot \text{s}$.

8.2 Answers: (a) $v_{C2x} > 0, v_{C2y} > 0$, **(b) piece C** There are no external horizontal forces, so the x- and y-components of the total momentum of the system are both conserved. Both components of the total momentum are zero before the spring releases, so they must be zero after the spring releases. Hence,

$$P_x = 0 = m_A v_{A2x} + m_B v_{B2x} + m_C v_{C2x}$$

$$P_y = 0 = m_A v_{A2y} + m_B v_{B2y} + m_C v_{C2y}$$

We are given that $m_A = m_B = m_C$, $v_{A2x} < 0$, $v_{A2y} = 0$, $v_{B2x} = 0$, and $v_{B2y} < 0$. You can solve the above equations to

show that $v_{C2x} = -v_{A2x} > 0$ and $v_{C2y} = -v_{B2y} > 0$, so the velocity components of piece C are both positive. Piece C has speed $\sqrt{v_{C2x}^2 + v_{C2y}^2} = \sqrt{v_{A2x}^2 + v_{B2y}^2}$, which is greater than the speed of either piece A or piece B.

8.3 Answers: (a) elastic, (b) inelastic, (c) completely inelastic In each case gravitational potential energy is converted to kinetic energy as the ball falls, and the collision is between the ball and the ground. In (a) all of the initial energy is converted back to gravitational potential energy, so no kinetic energy is lost in the bounce and the collision is elastic. In (b) there is less gravitational potential energy at the end than at the beginning, so some kinetic energy was lost in the bounce. Hence the collision is inelastic. In (c) the ball loses all the kinetic energy it has to give, the ball and the ground stick together, and the collision is completely inelastic.

8.4 Answer: worse After a collision with a water molecule initially at rest, the speed of the neutron is $|(m_n - m_w)/(m_n + m_w)| = |(1.0 \text{ u} - 18 \text{ u})/(1.0 \text{ u} + 18 \text{ u})| = \frac{17}{19}$ of its initial speed, and its kinetic energy is $\left(\frac{17}{19}\right)^2 = 0.80$ of the initial value. Hence a water molecule is a worse moderator than a carbon atom, for which the corresponding numbers are $\frac{11}{13}$ and $\left(\frac{11}{13}\right)^2 = 0.72$.

8.5 Answer: no If gravity is the only force acting on the system of two fragments, the center of mass will follow the parabolic trajectory of a freely falling object. Once a fragment lands, however, the ground exerts a normal force on that fragment. Hence the net force on the system has changed, and the trajectory of the center of mass changes in response.

8.6 Answers: (a) increasing, (b) decreasing From Eqs. (8.37) and (8.38), the thrust F is equal to $m(dv/dt)$, where m is the rocket's mass and dv/dt is its acceleration. Because m decreases with time, if the thrust F is constant, then the acceleration must increase with time (the same force acts on a smaller mass); if the acceleration dv/dt is constant, then the thrust must decrease with time (a smaller force is all that's needed to accelerate a smaller mass).

Bridging Problem

Answers: **(a)** 1.00 m/s to the right   **(b)** Elastic
(c) 1.93 m/s at $-30.4°$
(d) 2.31 kg · m/s at 149.6°   **(e)** Inelastic
(f) 1.67 m/s in the positive x-direction

ROTATION OF RIGID BODIES

By studying this chapter, you will learn:

- How to describe the rotation of a rigid body in terms of angular coordinate, angular velocity, and angular acceleration.

- How to analyze rigid-body rotation when the angular acceleration is constant.

- How to relate the rotation of a rigid body to the linear velocity and linear acceleration of a point on the body.

- The meaning of a body's moment of inertia about a rotation axis, and how it relates to rotational kinetic energy.

- How to calculate the moment of inertia of various bodies.

? All segments of a rotating wind turbine blade have the same angular velocity. Compared to a given blade segment, how many times greater is the linear speed of a second segment twice as far from the axis of rotation? How many times greater is the radial acceleration?

What do the motions of a compact disc, a Ferris wheel, a circular saw blade, and a ceiling fan have in common? None of these can be represented adequately as a moving *point;* each involves a body that *rotates* about an axis that is stationary in some inertial frame of reference.

Rotation occurs at all scales, from the motions of electrons in atoms to the motions of entire galaxies. We need to develop some general methods for analyzing the motion of a rotating body. In this chapter and the next we consider bodies that have definite size and definite shape, and that in general can have rotational as well as translational motion.

Real-world bodies can be very complicated; the forces that act on them can deform them—stretching, twisting, and squeezing them. We'll neglect these deformations for now and assume that the body has a perfectly definite and unchanging shape and size. We call this idealized model a **rigid body.** This chapter and the next are mostly about rotational motion of a rigid body.

We begin with kinematic language for *describing* rotational motion. Next we look at the kinetic energy of rotation, the key to using energy methods for rotational motion. Then in Chapter 10 we'll develop dynamic principles that relate the forces on a body to its rotational motion.

9.1 Angular Velocity and Acceleration

In analyzing rotational motion, let's think first about a rigid body that rotates about a *fixed axis*—an axis that is at rest in some inertial frame of reference and does not change direction relative to that frame. The rotating rigid body might be a motor shaft, a chunk of beef on a barbecue skewer, or a merry-go-round.

Figure 9.1 shows a rigid body (in this case, the indicator needle of a speedometer) rotating about a fixed axis. The axis passes through point O and is

9.1 A speedometer needle (an example of a rigid body) rotating counterclockwise about a fixed axis.

The angle θ from the $+x$-axis specifies the needle's rotational position.

Direction of needle's rotation

Axis of rotation passes through origin and points out of page.

perpendicular to the plane of the diagram, which we choose to call the *xy*-plane. One way to describe the rotation of this body would be to choose a particular point *P* on the body and to keep track of the *x*- and *y*-coordinates of this point. This isn't a terribly convenient method, since it takes two numbers (the two coordinates *x* and *y*) to specify the rotational position of the body. Instead, we notice that the line *OP* is fixed in the body and rotates with it. The angle θ that this line makes with the +*x*-axis describes the rotational position of the body; we will use this single quantity θ as a *coordinate* for rotation.

The angular coordinate θ of a rigid body rotating around a fixed axis can be positive or negative. If we choose positive angles to be measured counterclockwise from the positive *x*-axis, then the angle θ in Fig. 9.1 is positive. If we instead choose the positive rotation direction to be clockwise, then θ in Fig. 9.1 is negative. When we considered the motion of a particle along a straight line, it was essential to specify the direction of positive displacement along that line; when we discuss rotation around a fixed axis, it's just as essential to specify the direction of positive rotation.

To describe rotational motion, the most natural way to measure the angle θ is not in degrees, but in **radians.** As shown in Fig. 9.2a, one radian (1 rad) is the angle subtended at the center of a circle by an arc with a length equal to the radius of the circle. In Fig. 9.2b an angle θ is subtended by an arc of length *s* on a circle of radius *r*. The value of θ (in radians) is equal to *s* divided by *r*:

$$\theta = \frac{s}{r} \quad \text{or} \quad s = r\theta \qquad (9.1)$$

An angle in radians is the ratio of two lengths, so it is a pure number, without dimensions. If $s = 3.0 \text{ m}$ and $r = 2.0 \text{ m}$, then $\theta = 1.5$, but we will often write this as 1.5 rad to distinguish it from an angle measured in degrees or revolutions.

The circumference of a circle (that is, the arc length all the way around the circle) is 2π times the radius, so there are 2π (about 6.283) radians in one complete revolution (360°). Therefore

$$1 \text{ rad} = \frac{360°}{2\pi} = 57.3°$$

Similarly, $180° = \pi$ rad, $90° = \pi/2$ rad, and so on. If we had insisted on measuring the angle θ in degrees, we would have needed to include an extra factor of $(2\pi/360)$ on the right-hand side of $s = r\theta$ in Eq. (9.1). By measuring angles in radians, we keep the relationship between angle and distance along an arc as simple as possible.

Angular Velocity

The coordinate θ shown in Fig. 9.1 specifies the rotational position of a rigid body at a given instant. We can describe the rotational *motion* of such a rigid body in terms of the rate of change of θ. We'll do this in an analogous way to our description of straight-line motion in Chapter 2. In Fig. 9.3a, a reference line *OP* in a rotating body makes an angle θ_1 with the +*x*-axis at time t_1. At a later time t_2 the angle has changed to θ_2. We define the **average angular velocity** $\omega_{\text{av-}z}$ (the Greek letter omega) of the body in the time interval $\Delta t = t_2 - t_1$ as the ratio of the **angular displacement** $\Delta\theta = \theta_2 - \theta_1$ to Δt:

$$\omega_{\text{av-}z} = \frac{\theta_2 - \theta_1}{t_2 - t_1} = \frac{\Delta\theta}{\Delta t} \qquad (9.2)$$

9.2 Measuring angles in radians.

(a)

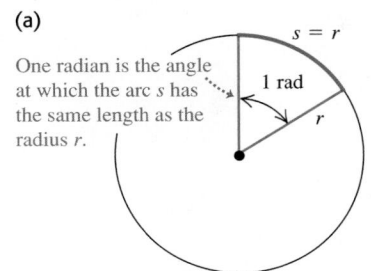

One radian is the angle at which the arc *s* has the same length as the radius *r*.

$s = r$

1 rad

r

(b)

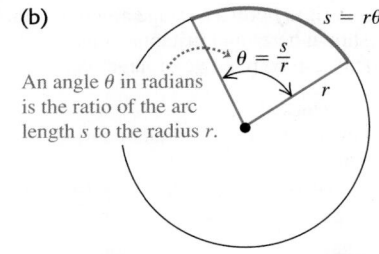

An angle θ in radians is the ratio of the arc length *s* to the radius *r*.

$s = r\theta$

$\theta = \dfrac{s}{r}$

r

9.3 (a) Angular displacement $\Delta\theta$ of a rotating body. (b) Every part of a rotating rigid body has the same average angular velocity $\Delta\theta/\Delta t$.

(a)

Angular displacement $\Delta\theta$ of the rotating needle over a time interval Δt:
$\Delta\theta = \theta_2 - \theta_1$

P at t_2

Direction of rotation

$\Delta\theta$

P at t_1

θ_1 θ_2

O — x

(b)

The subscript z indicates that the body in Fig. 9.3a is rotating about the z-axis, which is perpendicular to the plane of the diagram. The **instantaneous angular velocity** ω_z is the limit of $\omega_{av\text{-}z}$ as Δt approaches zero—that is, the derivative of θ with respect to t:

$$\omega_z = \lim_{\Delta t \to 0} \frac{\Delta\theta}{\Delta t} = \frac{d\theta}{dt} \quad \text{(definition of angular velocity)} \qquad (9.3)$$

When we refer simply to "angular velocity," we mean the instantaneous angular velocity, not the average angular velocity.

The angular velocity ω_z can be positive or negative, depending on the direction in which the rigid body is rotating (Fig. 9.4). The angular *speed* ω, which we will use extensively in Sections 9.3 and 9.4, is the magnitude of angular velocity. Like ordinary (linear) speed v, the angular speed is never negative.

9.4 A rigid body's average angular velocity (shown here) and instantaneous angular velocity can be positive or negative.

Counterclockwise rotation positive:
$\Delta\theta > 0$, so
$\omega_{av\text{-}z} = \Delta\theta/\Delta t > 0$

Clockwise rotation negative:
$\Delta\theta < 0$, so
$\omega_{av\text{-}z} = \Delta\theta/\Delta t < 0$

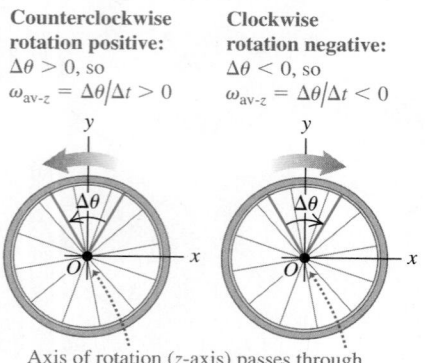

Axis of rotation (z-axis) passes through origin and points out of page.

CAUTION **Angular velocity vs. linear velocity** Keep in mind the distinction between angular velocity ω_z and ordinary velocity, or *linear velocity, v_x* (see Section 2.2). If an object has a velocity v_x, the object as a whole is *moving* along the x-axis. By contrast, if an object has an angular velocity ω_z, then it is *rotating* around the z-axis. We do *not* mean that the object is moving along the z-axis.

Different points on a rotating rigid body move different distances in a given time interval, depending on how far each point lies from the rotation axis. But because the body is rigid, *all* points rotate through the same angle in the same time (Fig. 9.3b). Hence *at any instant, every part of a rotating rigid body has the same angular velocity.* The angular velocity is positive if the body is rotating in the direction of increasing θ and negative if it is rotating in the direction of decreasing θ.

If the angle θ is in radians, the unit of angular velocity is the radian per second (rad/s). Other units, such as the revolution per minute (rev/min or rpm), are often used. Since 1 rev $= 2\pi$ rad, two useful conversions are

$$1 \text{ rev/s} = 2\pi \text{ rad/s} \quad \text{and} \quad 1 \text{ rev/min} = 1 \text{ rpm} = \frac{2\pi}{60} \text{ rad/s}$$

That is, 1 rad/s is about 10 rpm.

Example 9.1 Calculating angular velocity

The angular position θ of a 0.36-m-diameter flywheel is given by

$$\theta = (2.0 \text{ rad/s}^3)t^3$$

(a) Find θ, in radians and in degrees, at $t_1 = 2.0$ s and $t_2 = 5.0$ s.
(b) Find the distance that a particle on the flywheel rim moves over the time interval from $t_1 = 2.0$ s to $t_2 = 5.0$ s. (c) Find the average angular velocity, in rad/s and in rev/min, over that interval. (d) Find the instantaneous angular velocities at $t_1 = 2.0$ s and $t_2 = 5.0$ s.

SOLUTION

IDENTIFY and SET UP: We can find the target variables θ_1 (the angular position at time t_1), θ_2 (the angular position at time t_2), and the angular displacement $\Delta\theta = \theta_2 - \theta_1$ from the given expression. Knowing $\Delta\theta$, we'll find the distance traveled and the average angular velocity between t_1 and t_2 using Eqs. (9.1) and (9.2), respectively. To find the instantaneous angular velocities ω_{1z} (at time t_1) and ω_{2z} (at time t_2), we'll take the derivative of the given equation for θ with respect to time, as in Eq. (9.3).

EXECUTE: (a) We substitute the values of t into the equation for θ:

$$\theta_1 = (2.0 \text{ rad/s}^3)(2.0 \text{ s})^3 = 16 \text{ rad}$$

$$= (16 \text{ rad})\frac{360°}{2\pi \text{ rad}} = 920°$$

$$\theta_2 = (2.0 \text{ rad/s}^3)(5.0 \text{ s})^3 = 250 \text{ rad}$$

$$= (250 \text{ rad})\frac{360°}{2\pi \text{ rad}} = 14{,}000°$$

(b) During the interval from t_1 to t_2 the flywheel's angular displacement is $\Delta\theta = \theta_2 - \theta_1 = 250 \text{ rad} - 16 \text{ rad} = 234 \text{ rad}$.

The radius r is half the diameter, or 0.18 m. To use Eq. (9.1), the angles *must* be expressed in radians:

$$s = r\theta_2 - r\theta_1 = r\Delta\theta = (0.18 \text{ m})(234 \text{ rad}) = 42 \text{ m}$$

We can drop "radians" from the unit for s because θ is a pure, dimensionless number; the distance s is measured in meters, the same as r.

(c) From Eq. (9.2),

$$\omega_{av\text{-}z} = \frac{\theta_2 - \theta_1}{t_2 - t_1} = \frac{250 \text{ rad} - 16 \text{ rad}}{5.0 \text{ s} - 2.0 \text{ s}} = 78 \text{ rad/s}$$

$$= \left(78\frac{\text{rad}}{\text{s}}\right)\left(\frac{1 \text{ rev}}{2\pi \text{ rad}}\right)\left(\frac{60 \text{ s}}{1 \text{ min}}\right) = 740 \text{ rev/min}$$

(d) From Eq. (9.3),

$$\omega_z = \frac{d\theta}{dt} = \frac{d}{dt}[(2.0 \text{ rad/s}^3)t^3] = (2.0 \text{ rad/s}^3)(3t^2)$$

$$= (6.0 \text{ rad/s}^3)t^2$$

At times $t_1 = 2.0$ s and $t_2 = 5.0$ s we have

$$\omega_{1z} = (6.0 \text{ rad/s}^3)(2.0 \text{ s})^2 = 24 \text{ rad/s}$$

$$\omega_{2z} = (6.0 \text{ rad/s}^3)(5.0 \text{ s})^2 = 150 \text{ rad/s}$$

EVALUATE: The angular velocity $\omega_z = (6.0 \text{ rad/s}^3)t^2$ increases with time. Our results are consistent with this; the instantaneous angular velocity at the end of the interval ($\omega_{2z} = 150 \text{ rad/s}$) is greater than at the beginning ($\omega_{1z} = 24 \text{ rad/s}$), and the average angular velocity $\omega_{av\text{-}z} = 78 \text{ rad/s}$ over the interval is intermediate between these two values.

Angular Velocity As a Vector

As we have seen, our notation for the angular velocity ω_z about the z-axis is reminiscent of the notation v_x for the ordinary velocity along the x-axis (see Section 2.2). Just as v_x is the x-component of the velocity vector $\vec{v}$, ω_z is the z-component of an angular velocity *vector* $\vec{\omega}$ directed along the axis of rotation. As Fig. 9.5a shows, the direction of $\vec{\omega}$ is given by the right-hand rule that we used to define the vector

(a)

If you curl the fingers of your right hand in the direction of rotation ...

$\vec{\omega}$ ... your right thumb points in the direction of $\vec{\omega}$.

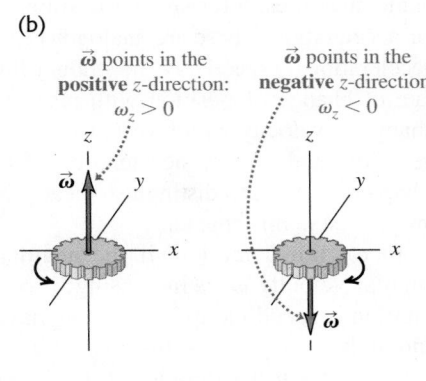

(b)

$\vec{\omega}$ points in the **positive** z-direction: $\omega_z > 0$

$\vec{\omega}$ points in the **negative** z-direction: $\omega_z < 0$

9.5 (a) The right-hand rule for the direction of the angular velocity vector $\vec{\omega}$. Reversing the direction of rotation reverses the direction of $\vec{\omega}$. (b) The sign of ω_z for rotation along the z-axis.

product in Section 1.10. If the rotation is about the z-axis, then $\vec{\boldsymbol{\omega}}$ has only a z-component; this component is positive if $\vec{\boldsymbol{\omega}}$ is along the positive z-axis and negative if $\vec{\boldsymbol{\omega}}$ is along the negative z-axis (Fig. 9.5b).

The vector formulation is especially useful in situations in which the direction of the rotation axis *changes*. We'll examine such situations briefly at the end of Chapter 10. In this chapter, however, we'll consider only situations in which the rotation axis is fixed. Hence throughout this chapter we'll use "angular velocity" to refer to ω_z, the component of the angular velocity vector $\vec{\boldsymbol{\omega}}$ along the axis.

Angular Acceleration

When the angular velocity of a rigid body changes, it has an *angular acceleration*. When you pedal your bicycle harder to make the wheels turn faster or apply the brakes to bring the wheels to a stop, you're giving the wheels an angular acceleration. You also impart an angular acceleration whenever you change the rotation speed of a piece of spinning machinery such as an automobile engine's crankshaft.

If ω_{1z} and ω_{2z} are the instantaneous angular velocities at times t_1 and t_2, we define the **average angular acceleration** $\alpha_{\text{av-}z}$ over the interval $\Delta t = t_2 - t_1$ as the change in angular velocity divided by Δt (Fig. 9.6):

$$\alpha_{\text{av-}z} = \frac{\omega_{2z} - \omega_{1z}}{t_2 - t_1} = \frac{\Delta \omega_z}{\Delta t} \qquad \text{(9.4)}$$

The **instantaneous angular acceleration** α_z is the limit of $\alpha_{\text{av-}z}$ as $\Delta t \to 0$:

$$\alpha_z = \lim_{\Delta t \to 0} \frac{\Delta \omega_z}{\Delta t} = \frac{d\omega_z}{dt} \qquad \text{(definition of angular acceleration)} \quad \text{(9.5)}$$

The usual unit of angular acceleration is the radian per second per second, or rad/s^2. From now on we will use the term "angular acceleration" to mean the instantaneous angular acceleration rather than the average angular acceleration.

Because $\omega_z = d\theta/dt$, we can also express angular acceleration as the second derivative of the angular coordinate:

$$\alpha_z = \frac{d}{dt}\frac{d\theta}{dt} = \frac{d^2\theta}{dt^2} \qquad \text{(9.6)}$$

You have probably noticed that we are using Greek letters for angular kinematic quantities: θ for angular position, ω_z for angular velocity, and α_z for angular acceleration. These are analogous to x for position, v_x for velocity, and a_x for acceleration, respectively, in straight-line motion. In each case, velocity is the rate of change of position with respect to time and acceleration is the rate of change of velocity with respect to time. We will sometimes use the terms "*linear* velocity" and "*linear* acceleration" for the familiar quantities we defined in Chapters 2 and 3 to distinguish clearly between these and the *angular* quantities introduced in this chapter.

In rotational motion, if the angular acceleration α_z is positive, then the angular velocity ω_z is increasing; if α_z is negative, then ω_z is decreasing. The rotation is speeding up if α_z and ω_z have the same sign and slowing down if α_z and ω_z have opposite signs. (These are exactly the same relationships as those between *linear* acceleration a_x and *linear* velocity v_x for straight-line motion; see Section 2.3.)

9.6 Calculating the average angular acceleration of a rotating rigid body.

The average angular acceleration is the change in angular velocity divided by the time interval:

$$\alpha_{\text{av-}z} = \frac{\omega_{2z} - \omega_{1z}}{t_2 - t_1} = \frac{\Delta \omega_z}{\Delta t}$$

ω_{1z} ω_{2z}

At t_1 At t_2

Example 9.2 | **Calculating angular acceleration**

For the flywheel of Example 9.1, (a) find the average angular acceleration between $t_1 = 2.0$ s and $t_2 = 5.0$ s. (b) Find the instantaneous angular accelerations at $t_1 = 2.0$ s and $t_2 = 5.0$ s.

SOLUTION

IDENTIFY and SET UP: We use Eq. (9.4) for the average angular acceleration $\alpha_{av\text{-}z}$ and Eq. (9.5) for the instantaneous angular acceleration α_z.

EXECUTE: (a) From Example 9.1, the values of ω_z at the two times are

$$\omega_{1z} = 24 \text{ rad/s} \qquad \omega_{2z} = 150 \text{ rad/s}$$

From Eq. (9.4), the average angular acceleration is

$$\alpha_{av\text{-}z} = \frac{150 \text{ rad/s} - 24 \text{ rad/s}}{5.0 \text{ s} - 2.0 \text{ s}} = 42 \text{ rad/s}^2$$

(b) From Eq. (9.5), the value of α_z at any time t is

$$\alpha_z = \frac{d\omega_z}{dt} = \frac{d}{dt}[(6.0 \text{ rad/s}^3)(t^2)] = (6.0 \text{ rad/s}^3)(2t)$$
$$= (12 \text{ rad/s}^3)t$$

Hence

$$\alpha_{1z} = (12 \text{ rad/s}^3)(2.0 \text{ s}) = 24 \text{ rad/s}^2$$
$$\alpha_{2z} = (12 \text{ rad/s}^3)(5.0 \text{ s}) = 60 \text{ rad/s}^2$$

EVALUATE: Note that the angular acceleration is *not* constant in this situation. The angular velocity ω_z is always increasing because α_z is always positive. Furthermore, the rate at which the angular velocity increases is itself increasing, since α_z increases with time.

Angular Acceleration As a Vector

Just as we did for angular velocity, it's useful to define an angular acceleration vector $\vec{\alpha}$. Mathematically, $\vec{\alpha}$ is the time derivative of the angular velocity vector $\vec{\omega}$. If the object rotates around the fixed z-axis, then $\vec{\alpha}$ has only a z-component; the quantity α_z is just that component. In this case, $\vec{\alpha}$ is in the same direction as $\vec{\omega}$ if the rotation is speeding up and opposite to $\vec{\omega}$ if the rotation is slowing down (Fig. 9.7).

The angular acceleration vector will be particularly useful in Chapter 10 when we discuss what happens when the rotation axis can change direction. In this chapter, however, the rotation axis will always be fixed and we need use only the z-component α_z.

9.7 When the rotation axis is fixed, the angular acceleration and angular velocity vectors both lie along that axis.

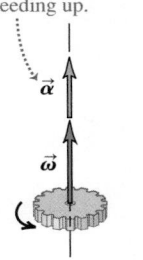

$\vec{\alpha}$ and $\vec{\omega}$ in the **same** direction: Rotation speeding up.

$\vec{\alpha}$ and $\vec{\omega}$ in the **opposite** directions: Rotation slowing down.

Test Your Understanding of Section 9.1 The figure shows a graph of ω_z and α_z versus time for a particular rotating body. (a) During which time intervals is the rotation speeding up? (i) $0 < t < 2$ s; (ii) 2 s $< t < 4$ s; (iii) 4 s $< t < 6$ s. (b) During which time intervals is the rotation slowing down? (i) $0 < t < 2$ s; (ii) 2 s $< t < 4$ s; (iii) 4 s $< t < 6$ s.

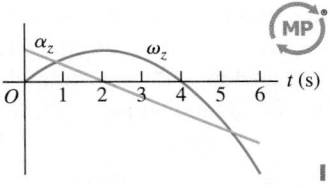

Application **Rotational Motion in Bacteria**
Escherichia coli bacteria (about 2 μm by 0.5 μm) are found in the lower intestines of humans and other warm-blooded animals. The bacteria swim by rotating their long, corkscrew-shaped flagella, which act like the blades of a propeller. Each flagellum is powered by a remarkable protein motor at its base. The motor can rotate the flagellum at angular speeds from 200 to 1000 rev/min (about 20 to 100 rad/s) and can vary its speed to give the flagellum an angular acceleration.

9.2 Rotation with Constant Angular Acceleration

In Chapter 2 we found that straight-line motion is particularly simple when the acceleration is constant. This is also true of rotational motion about a fixed axis. When the angular acceleration is constant, we can derive equations for angular velocity and angular position using exactly the same procedure that we used for straight-line motion in Section 2.4. In fact, the equations we are about to derive are identical to Eqs. (2.8), (2.12), (2.13), and (2.14) if we replace x with θ, v_x with ω_z, and a_x with α_z. We suggest that you review Section 2.4 before continuing.

Let ω_{0z} be the angular velocity of a rigid body at time $t = 0$, and let ω_z be its angular velocity at any later time t. The angular acceleration α_z is constant and equal to the average value for any interval. Using Eq. (9.4) with the interval from 0 to t, we find

$$\alpha_z = \frac{\omega_z - \omega_{0z}}{t - 0} \qquad \text{or}$$

$$\omega_z = \omega_{0z} + \alpha_z t \qquad \text{(constant angular acceleration only)} \qquad (9.7)$$

Flagella

The product $\alpha_z t$ is the total change in ω_z between $t = 0$ and the later time t; the angular velocity ω_z at time t is the sum of the initial value ω_{0z} and this total change. With constant angular acceleration, the angular velocity changes at a uniform rate, so its average value between 0 and t is the average of the initial and final values:

$$\omega_{\text{av-}z} = \frac{\omega_{0z} + \omega_z}{2} \qquad (9.8)$$

We also know that $\omega_{\text{av-}z}$ is the total angular displacement $(\theta - \theta_0)$ divided by the time interval $(t - 0)$:

$$\omega_{\text{av-}z} = \frac{\theta - \theta_0}{t - 0} \qquad (9.9)$$

When we equate Eqs. (9.8) and (9.9) and multiply the result by t, we get

$$\theta - \theta_0 = \tfrac{1}{2}(\omega_{0z} + \omega_z)t \quad \text{(constant angular acceleration only)} \qquad (9.10)$$

To obtain a relationship between θ and t that doesn't contain ω_z, we substitute Eq. (9.7) into Eq. (9.10):

$$\theta - \theta_0 = \tfrac{1}{2}[\omega_{0z} + (\omega_{0z} + \alpha_z t)]t \qquad \text{or}$$

$$\theta = \theta_0 + \omega_{0z}t + \tfrac{1}{2}\alpha_z t^2 \quad \text{(constant angular acceleration only)} \qquad (9.11)$$

That is, if at the initial time $t = 0$ the body is at angular position θ_0 and has angular velocity ω_{0z}, then its angular position θ at any later time t is the sum of three terms: its initial angular position θ_0, plus the rotation $\omega_{0z}t$ it would have if the angular velocity were constant, plus an additional rotation $\tfrac{1}{2}\alpha_z t^2$ caused by the changing angular velocity.

Following the same procedure as for straight-line motion in Section 2.4, we can combine Eqs. (9.7) and (9.11) to obtain a relationship between θ and ω_z that does not contain t. We invite you to work out the details, following the same procedure we used to get Eq. (2.13). (See Exercise 9.12.) In fact, because of the perfect analogy between straight-line and rotational quantities, we can simply take Eq. (2.13) and replace each straight-line quantity by its rotational analog. We get

$$\omega_z^2 = \omega_{0z}^2 + 2\alpha_z(\theta - \theta_0) \quad \text{(constant angular acceleration only)} \qquad (9.12)$$

CAUTION Constant angular acceleration Keep in mind that all of these results are valid *only* when the angular acceleration α_z is *constant*; be careful not to try to apply them to problems in which α_z is *not* constant. Table 9.1 shows the analogy between Eqs. (9.7), (9.10), (9.11), and (9.12) for fixed-axis rotation with constant angular acceleration and the corresponding equations for straight-line motion with constant linear acceleration. ▮

Table 9.1 Comparison of Linear and Angular Motion with Constant Acceleration

Straight-Line Motion with Constant Linear Acceleration		Fixed-Axis Rotation with Constant Angular Acceleration	
$a_x = \text{constant}$		$\alpha_z = \text{constant}$	
$v_x = v_{0x} + a_x t$	(2.8)	$\omega_z = \omega_{0z} + \alpha_z t$	(9.7)
$x = x_0 + v_{0x}t + \tfrac{1}{2}a_x t^2$	(2.12)	$\theta = \theta_0 + \omega_{0z}t + \tfrac{1}{2}\alpha_z t^2$	(9.11)
$v_x^2 = v_{0x}^2 + 2a_x(x - x_0)$	(2.13)	$\omega_z^2 = \omega_{0z}^2 + 2\alpha_z(\theta - \theta_0)$	(9.12)
$x - x_0 = \tfrac{1}{2}(v_{0x} + v_x)t$	(2.14)	$\theta - \theta_0 = \tfrac{1}{2}(\omega_{0z} + \omega_z)t$	(9.10)

Example 9.3 Rotation with constant angular acceleration

You have finished watching a movie on Blu-ray and the disc is slowing to a stop. The disc's angular velocity at $t = 0$ is 27.5 rad/s, and its angular acceleration is a constant -10.0 rad/s². A line PQ on the disc's surface lies along the $+x$-axis at $t = 0$ (Fig. 9.8). (a) What is the disc's angular velocity at $t = 0.300$ s? (b) What angle does the line PQ make with the $+x$-axis at this time?

SOLUTION

IDENTIFY and SET UP: The angular acceleration of the disc is constant, so we can use any of the equations derived in this section (Table 9.1). Our target variables are the angular velocity ω_z and the angular displacement θ at $t = 0.300$ s. Given $\omega_{0z} = 27.5$ rad/s, $\theta_0 = 0$, and $\alpha_z = -10.0$ rad/s², it's easiest to use Eqs. (9.7) and (9.11) to find the target variables.

9.8 A line PQ on a rotating Blu-ray disc at $t = 0$.

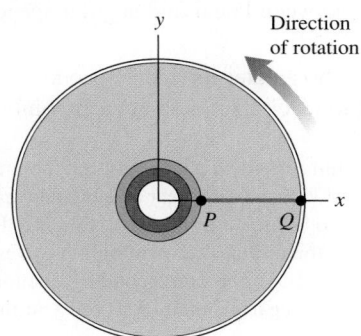

EXECUTE: (a) From Eq. (9.7), at $t = 0.300$ s we have

$$\omega_z = \omega_{0z} + \alpha_z t = 27.5 \text{ rad/s} + (-10.0 \text{ rad/s}^2)(0.300 \text{ s})$$
$$= 24.5 \text{ rad/s}$$

(b) From Eq. (9.11),

$$\theta = \theta_0 + \omega_{0z}t + \tfrac{1}{2}\alpha_z t^2$$

$$= 0 + (27.5 \text{ rad/s})(0.300 \text{ s}) + \tfrac{1}{2}(-10.0 \text{ rad/s}^2)(0.300 \text{ s})^2$$

$$= 7.80 \text{ rad} = 7.80 \text{ rad}\left(\frac{1 \text{ rev}}{2\pi \text{ rad}}\right) = 1.24 \text{ rev}$$

The disc has turned through one complete revolution plus an additional 0.24 revolution—that is, through 360° plus (0.24 rev)(360°/rev) = 87°. Hence the line PQ makes an angle of 87° with the $+x$-axis.

EVALUATE: Our answer to part (a) tells us that the disc's angular velocity has decreased, as it should since $\alpha_z < 0$. We can use our result for ω_z from part (a) with Eq. (9.12) to check our result for θ from part (b). To do so, we solve Eq. (9.12) for θ:

$$\omega_z^2 = \omega_{0z}^2 + 2\alpha_z(\theta - \theta_0)$$

$$\theta = \theta_0 + \left(\frac{\omega_z^2 - \omega_{0z}^2}{2\alpha_z}\right)$$

$$= 0 + \frac{(24.5 \text{ rad/s})^2 - (27.5 \text{ rad/s})^2}{2(-10.0 \text{ rad/s}^2)} = 7.80 \text{ rad}$$

This agrees with our previous result from part (b).

Test Your Understanding of Section 9.2 Suppose the disc in Example 9.3 was initially spinning at twice the rate (55.0 rad/s rather than 27.5 rad/s) and slowed down at twice the rate (-20.0 rad/s² rather than -10.0 rad/s²). (a) Compared to the situation in Example 9.3, how long would it take the disc to come to a stop? (i) the same amount of time; (ii) twice as much time; (iii) 4 times as much time; (iv) $\tfrac{1}{2}$ as much time; (v) $\tfrac{1}{4}$ as much time. (b) Compared to the situation in Example 9.3, through how many revolutions would the disc rotate before coming to a stop? (i) the same number of revolutions; (ii) twice as many revolutions; (iii) 4 times as many revolutions; (iv) $\tfrac{1}{2}$ as many revolutions; (v) $\tfrac{1}{4}$ as many revolutions. ❙

9.3 Relating Linear and Angular Kinematics

How do we find the linear speed and acceleration of a particular point in a rotating rigid body? We need to answer this question to proceed with our study of rotation. For example, to find the kinetic energy of a rotating body, we have to start from $K = \tfrac{1}{2}mv^2$ for a particle, and this requires knowing the speed v for each particle in the body. So it's worthwhile to develop general relationships between the *angular* speed and acceleration of a rigid body rotating about a fixed axis and the *linear* speed and acceleration of a specific point or particle in the body.

Linear Speed in Rigid-Body Rotation

When a rigid body rotates about a fixed axis, every particle in the body moves in a circular path. The circle lies in a plane perpendicular to the axis and is centered on the axis. The speed of a particle is directly proportional to the body's angular

9.9 A rigid body rotating about a fixed axis through point O.

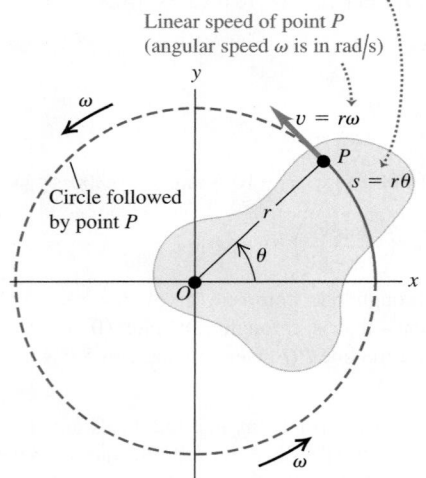

Distance through which point P on the body moves (angle θ is in radians)

Linear speed of point P (angular speed ω is in rad/s)

$v = r\omega$

$s = r\theta$

P

Circle followed by point P

PhET: Ladybug Revolution

velocity; the faster the body rotates, the greater the speed of each particle. In Fig. 9.9, point P is a constant distance r from the axis of rotation, so it moves in a circle of radius r. At any time, the angle θ (in radians) and the arc length s are related by

$$s = r\theta$$

We take the time derivative of this, noting that r is constant for any specific particle, and take the absolute value of both sides:

$$\left|\frac{ds}{dt}\right| = r\left|\frac{d\theta}{dt}\right|$$

Now $|ds/dt|$ is the absolute value of the rate of change of arc length, which is equal to the instantaneous *linear* speed v of the particle. Analogously, $|d\theta/dt|$, the absolute value of the rate of change of the angle, is the instantaneous **angular speed** ω—that is, the magnitude of the instantaneous angular velocity in rad/s. Thus

$$v = r\omega \qquad \text{(relationship between linear and angular speeds)} \qquad (9.13)$$

The farther a point is from the axis, the greater its linear speed. The *direction* of the linear velocity *vector* is tangent to its circular path at each point (Fig. 9.9).

CAUTION **Speed vs. velocity** Keep in mind the distinction between the linear and angular *speeds* v and ω, which appear in Eq. (9.13), and the linear and angular *velocities* v_x and ω_z. The quantities without subscripts, v and ω, are never negative; they are the magnitudes of the vectors $\vec{v}$ and $\vec{\omega}$, respectively, and their values tell you only how fast a particle is moving (v) or how fast a body is rotating (ω). The corresponding quantities with subscripts, v_x and ω_z, can be either positive or negative; their signs tell you the direction of the motion. ▌

Linear Acceleration in Rigid-Body Rotation

We can represent the acceleration of a particle moving in a circle in terms of its centripetal and tangential components, a_{rad} and a_{tan} (Fig. 9.10), as we did in Section 3.4. It would be a good idea to review that section now. We found that the **tangential component of acceleration** a_{tan}, the component parallel to the instantaneous velocity, acts to change the *magnitude* of the particle's velocity (i.e., the speed) and is equal to the rate of change of speed. Taking the derivative of Eq. (9.13), we find

$$a_{\text{tan}} = \frac{dv}{dt} = r\frac{d\omega}{dt} = r\alpha \qquad \text{(tangential acceleration of a point on a rotating body)} \qquad (9.14)$$

9.10 A rigid body whose rotation is speeding up. The acceleration of point P has a component a_{rad} toward the rotation axis (perpendicular to $\vec{v}$) and a component a_{tan} along the circle that point P follows (parallel to $\vec{v}$).

Radial and tangential acceleration components:
• $a_{\text{rad}} = \omega^2 r$ is point P's centripetal acceleration.
• $a_{\text{tan}} = r\alpha$ means that P's rotation is speeding up (the body has angular acceleration).

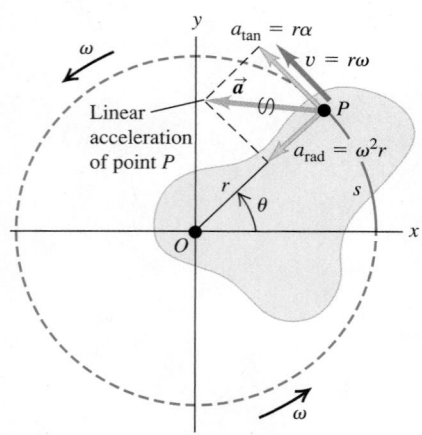

$a_{\text{tan}} = r\alpha$

$v = r\omega$

$\vec{a}$

P

Linear acceleration of point P

$a_{\text{rad}} = \omega^2 r$

This component of a particle's acceleration is always tangent to the circular path of the particle.

The quantity $\alpha = d\omega/dt$ in Eq. (9.14) is the rate of change of the angular *speed*. It is not quite the same as $\alpha_z = d\omega_z/dt$, which is the rate of change of the angular *velocity*. For example, consider a body rotating so that its angular velocity vector points in the $-z$-direction (see Fig. 9.5b). If the body is gaining angular speed at a rate of 10 rad/s per second, then $\alpha = 10$ rad/s². But ω_z is negative and becoming more negative as the rotation gains speed, so $\alpha_z = -10$ rad/s². The rule for rotation about a fixed axis is that α is equal to α_z if ω_z is positive but equal to $-\alpha_z$ if ω_z is negative.

The component of the particle's acceleration directed toward the rotation axis, the **centripetal component of acceleration** a_{rad}, is associated with the **?**

change of *direction* of the particle's velocity. In Section 3.4 we worked out the relationship $a_{rad} = v^2/r$. We can express this in terms of ω by using Eq. (9.13):

$$a_{rad} = \frac{v^2}{r} = \omega^2 r \qquad \text{(centripetal acceleration of a point on a rotating body)} \qquad (9.15)$$

This is true at each instant, *even when ω and v are not constant.* The centripetal component always points toward the axis of rotation.

The vector sum of the centripetal and tangential components of acceleration of a particle in a rotating body is the linear acceleration $\vec{a}$ (Fig. 9.10).

CAUTION **Use angles in radians in all equations** It's important to remember that Eq. (9.1), $s = r\theta$, is valid *only* when θ is measured in radians. The same is true of any equation derived from this, including Eqs. (9.13), (9.14), and (9.15). When you use these equations, you *must* express the angular quantities in radians, not revolutions or degrees (Fig. 9.11). ▌

Equations (9.1), (9.13), and (9.14) also apply to any particle that has the same tangential velocity as a point in a rotating rigid body. For example, when a rope wound around a circular cylinder unwraps without stretching or slipping, its speed and acceleration at any instant are equal to the speed and tangential acceleration of the point at which it is tangent to the cylinder. The same principle holds for situations such as bicycle chains and sprockets, belts and pulleys that turn without slipping, and so on. We will have several opportunities to use these relationships later in this chapter and in Chapter 10. Note that Eq. (9.15) for the centripetal component a_{rad} is applicable to the rope or chain *only* at points that are in contact with the cylinder or sprocket. Other points do not have the same acceleration toward the center of the circle that points on the cylinder or sprocket have.

9.11 Always use radians when relating linear and angular quantities.

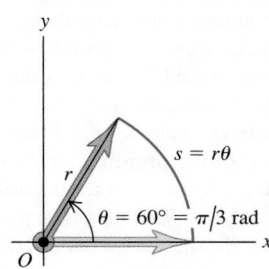

In any equation that relates linear quantities to angular quantities, the angles MUST be expressed in radians ...

RIGHT! ▶ $s = (\pi/3)r$

... never in degrees or revolutions.

WRONG ▶ $s = \cancel{60}r$

Example 9.4 **Throwing a discus**

An athlete whirls a discus in a circle of radius 80.0 cm. At a certain instant, the athlete is rotating at 10.0 rad/s and the angular speed is increasing at 50.0 rad/s². At this instant, find the tangential and centripetal components of the acceleration of the discus and the magnitude of the acceleration.

SOLUTION

IDENTIFY and SET UP: We treat the discus as a particle traveling in a circular path (Fig. 9.12a), so we can use the ideas developed in this section. We are given $r = 0.800$ m, $\omega = 10.0$ rad/s, and $\alpha = 50.0$ rad/s² (Fig. 9.12b). We'll use Eqs. (9.14) and (9.15), respectively, to find the acceleration components a_{tan} and a_{rad}; we'll then find the magnitude a using the Pythagorean theorem.

EXECUTE: From Eqs. (9.14) and (9.15),

$$a_{tan} = r\alpha = (0.800 \text{ m})(50.0 \text{ rad/s}^2) = 40.0 \text{ m/s}^2$$
$$a_{rad} = \omega^2 r = (10.0 \text{ rad/s})^2(0.800 \text{ m}) = 80.0 \text{ m/s}^2$$

Then

$$a = \sqrt{a_{tan}^2 + a_{rad}^2} = 89.4 \text{ m/s}^2$$

EVALUATE: Note that we dropped the unit "radian" from our results for a_{tan}, a_{rad}, and a. We can do this because "radian" is a dimensionless quantity. Can you show that if the angular speed doubles to 20.0 rad/s while α remains the same, the acceleration magnitude a increases to 322 m/s²?

9.12 (a) Whirling a discus in a circle. (b) Our sketch showing the acceleration components for the discus.

(a)

(b)

Example 9.5 **Designing a propeller**

You are designing an airplane propeller that is to turn at 2400 rpm (Fig. 9.13a). The forward airspeed of the plane is to be 75.0 m/s, and the speed of the tips of the propeller blades through the air must not exceed 270 m/s. (This is about 80% of the speed of sound in air. If the speed of the propeller tips were greater than this, they would produce a lot of noise.) (a) What is the maximum possible propeller radius? (b) With this radius, what is the acceleration of the propeller tip?

SOLUTION

IDENTIFY and SET UP: We consider a particle at the tip of the propeller; our target variables are the particle's distance from the axis and its acceleration. The speed of this particle through the air, which cannot exceed 270 m/s, is due to both the propeller's rotation and the forward motion of the airplane. Figure 9.13b shows that the particle's velocity $\vec{v}_{tip}$ is the vector sum of its tangential velocity due to the propeller's rotation of magnitude $v_{tan} = \omega r$, given by Eq. (9.13), and the forward velocity of the airplane of magnitude $v_{plane} = 75.0$ m/s. The propeller rotates in a plane perpendicular to the direction of flight, so $\vec{v}_{tan}$ and $\vec{v}_{plane}$ are perpendicular to each other, and we can use the Pythagorean theorem to obtain an expression for v_{tip} from v_{tan} and v_{plane}. We will then set $v_{tip} = 270$ m/s and solve for the radius r. The angular speed of the propeller is constant, so the acceleration of the propeller tip has only a radial component; we'll find it using Eq. (9.15).

EXECUTE: We first convert ω to rad/s (see Fig. 9.11):

$$\omega = 2400 \text{ rpm} = \left(2400 \frac{\text{rev}}{\text{min}}\right)\left(\frac{2\pi \text{ rad}}{1 \text{ rev}}\right)\left(\frac{1 \text{ min}}{60 \text{ s}}\right)$$
$$= 251 \text{ rad/s}$$

(a) From Fig. 9.13b and Eq. (9.13),

$$v_{tip}{}^2 = v_{plane}{}^2 + v_{tan}{}^2 = v_{plane}{}^2 + r^2\omega^2 \quad \text{so}$$

$$r^2 = \frac{v_{tip}{}^2 - v_{plane}{}^2}{\omega^2} \quad \text{and} \quad r = \frac{\sqrt{v_{tip}{}^2 - v_{plane}{}^2}}{\omega}$$

If $v_{tip} = 270$ m/s, the maximum propeller radius is

$$r = \frac{\sqrt{(270 \text{ m/s})^2 - (75.0 \text{ m/s})^2}}{251 \text{ rad/s}} = 1.03 \text{ m}$$

(b) The centripetal acceleration of the particle is

$$a_{rad} = \omega^2 r = (251 \text{ rad/s})^2(1.03 \text{ m})$$
$$= 6.5 \times 10^4 \text{ m/s}^2 = 6600g$$

The tangential acceleration a_{rad} is zero because the angular speed is constant.

EVALUATE: From $\sum \vec{F} = m\vec{a}$, the propeller must exert a force of 6.5×10^4 N on each kilogram of material at its tip! This is why propellers are made out of tough material, usually aluminum alloy.

9.13 (a) A propeller-driven airplane in flight. (b) Our sketch showing the velocity components for the propeller tip.

Test Your Understanding of Section 9.3 Information is stored on a disc (see Fig. 9.8) in a coded pattern of tiny pits. The pits are arranged in a track that spirals outward toward the rim of the disc. As the disc spins inside a player, the track is scanned at a constant *linear* speed. How must the rotation speed of the disc change as the player's scanning head moves over the track? (i) The rotation speed must increase. (ii) The rotation speed must decrease. (iii) The rotation speed must stay the same.

9.4 Energy in Rotational Motion

A rotating rigid body consists of mass in motion, so it has kinetic energy. As we will see, we can express this kinetic energy in terms of the body's angular speed and a new quantity, called *moment of inertia,* that depends on the body's mass and how the mass is distributed.

To begin, we think of a body as being made up of a large number of particles, with masses $m_1, m_2, \ldots$ at distances $r_1, r_2, \ldots$ from the axis of rotation. We label the particles with the index i: The mass of the ith particle is m_i and its distance from the axis of rotation is r_i. The particles don't necessarily all lie in the same plane, so we specify that r_i is the *perpendicular* distance from the axis to the ith particle.

When a rigid body rotates about a fixed axis, the speed v_i of the ith particle is given by Eq. (9.13), $v_i = r_i\omega$, where ω is the body's angular speed. Different particles have different values of r, but ω is the same for all (otherwise, the body wouldn't be rigid). The kinetic energy of the ith particle can be expressed as

$$\tfrac{1}{2}m_i v_i^2 = \tfrac{1}{2}m_i r_i^2 \omega^2$$

The *total* kinetic energy of the body is the sum of the kinetic energies of all its particles:

$$K = \tfrac{1}{2}m_1 r_1^2 \omega^2 + \tfrac{1}{2}m_2 r_2^2 \omega^2 + \cdots = \sum_i \tfrac{1}{2}m_i r_i^2 \omega^2$$

Taking the common factor $\omega^2/2$ out of this expression, we get

$$K = \tfrac{1}{2}(m_1 r_1^2 + m_2 r_2^2 + \cdots)\omega^2 = \tfrac{1}{2}\left(\sum_i m_i r_i^2\right)\omega^2$$

The quantity in parentheses, obtained by multiplying the mass of each particle by the square of its distance from the axis of rotation and adding these products, is denoted by I and is called the **moment of inertia** of the body for this rotation axis:

$$I = m_1 r_1^2 + m_2 r_2^2 + \cdots = \sum_i m_i r_i^2 \qquad \text{(definition of moment of inertia)} \qquad (9.16)$$

The word "moment" means that I depends on how the body's mass is distributed in space; it has nothing to do with a "moment" of time. For a body with a given rotation axis and a given total mass, the greater the distance from the axis to the particles that make up the body, the greater the moment of inertia. In a rigid body, the distances r_i are all constant and I is independent of how the body rotates around the given axis. The SI unit of moment of inertia is the kilogram-meter^2 $(\text{kg} \cdot \text{m}^2)$.

In terms of moment of inertia I, the **rotational kinetic energy** K of a rigid body is

$$K = \tfrac{1}{2}I\omega^2 \qquad \text{(rotational kinetic energy of a rigid body)} \qquad (9.17)$$

The kinetic energy given by Eq. (9.17) is *not* a new form of energy; it's simply the sum of the kinetic energies of the individual particles that make up the rotating rigid body. To use Eq. (9.17), ω *must* be measured in radians per second, not revolutions or degrees per second, to give K in joules. That's because we used $v_i = r_i\omega$ in our derivation.

Equation (9.17) gives a simple physical interpretation of moment of inertia: *The greater the moment of inertia, the greater the kinetic energy of a rigid body rotating with a given angular speed ω.* We learned in Chapter 6 that the kinetic energy of a body equals the amount of work done to accelerate that body from rest. So the greater a body's moment of inertia, the harder it is to start the body rotating if it's at rest and the harder it is to stop its rotation if it's already rotating (Fig. 9.14). For this reason, I is also called the *rotational inertia*.

The next example shows how *changing* the rotation axis can affect the value of I.

Mastering**PHYSICS**

ActivPhysics 7.7: Rotational Inertia

9.14 An apparatus free to rotate around a vertical axis. To vary the moment of inertia, the two equal-mass cylinders can be locked into different positions on the horizontal shaft.

- Mass close to axis
- Small moment of inertia
- Easy to start apparatus rotating

Rotation axis

- Mass farther from axis
- Greater moment of inertia
- Harder to start apparatus rotating

Rotation axis

Example 9.6 Moments of inertia for different rotation axes

A machine part (Fig. 9.15) consists of three disks linked by lightweight struts. (a) What is this body's moment of inertia about an axis through the center of disk A, perpendicular to the plane of the diagram? (b) What is its moment of inertia about an axis through the centers of disks B and C? (c) What is the body's kinetic energy if it rotates about the axis through A with angular speed ω = 4.0 rad/s?

SOLUTION

IDENTIFY and SET UP: We'll consider the disks as massive particles located at the centers of the disks, and consider the struts as

9.15 An oddly shaped machine part.

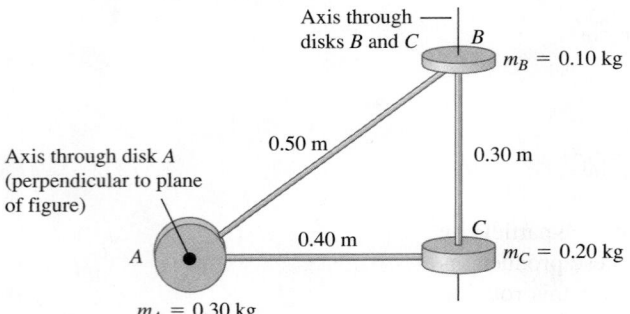

massless. In parts (a) and (b), we'll use Eq. (9.16) to find the moments of inertia. Given the moment of inertia about axis A, we'll use Eq. (9.17) in part (c) to find the rotational kinetic energy.

EXECUTE: (a) The particle at point A lies *on* the axis through A, so its distance r from the axis is zero and it contributes nothing to the moment of inertia. Hence only B and C contribute, and Eq. (9.16) gives

$$I_A = \sum m_i r_i^2 = (0.10 \text{ kg})(0.50 \text{ m})^2 + (0.20 \text{ kg})(0.40 \text{ m})^2$$
$$= 0.057 \text{ kg} \cdot \text{m}^2$$

(b) The particles at B and C both lie on axis BC, so neither particle contributes to the moment of inertia. Hence only A contributes:

$$I_{BC} = \sum m_i r_i^2 = (0.30 \text{ kg})(0.40 \text{ m})^2 = 0.048 \text{ kg} \cdot \text{m}^2$$

(c) From Eq. (9.17),

$$K_A = \tfrac{1}{2} I_A \omega^2 = \tfrac{1}{2}(0.057 \text{ kg} \cdot \text{m}^2)(4.0 \text{ rad/s})^2 = 0.46 \text{ J}$$

EVALUATE: The moment of inertia about axis A is greater than that about axis BC. Hence of the two axes it's easier to make the machine part rotate about axis BC.

Application Moment of Inertia of a Bird's Wing

When a bird flaps its wings, it rotates the wings up and down around the shoulder. A hummingbird has small wings with a small moment of inertia, so the bird can make its wings move rapidly (up to 70 beats per second). By contrast, the Andean condor (*Vultur gryphus*) has immense wings that are hard to move due to their large moment of inertia. Condors flap their wings at about one beat per second on takeoff, but at most times prefer to soar while holding their wings steady.

CAUTION **Moment of inertia depends on the choice of axis** The results of parts (a) and (b) of Example 9.6 show that the moment of inertia of a body depends on the location and orientation of the axis. It's not enough to just say, "The moment of inertia of this body is 0.048 kg · m²." We have to be specific and say, "The moment of inertia of this body *about the axis through B and C* is 0.048 kg · m²." ∎

In Example 9.6 we represented the body as several point masses, and we evaluated the sum in Eq. (9.16) directly. When the body is a *continuous* distribution of matter, such as a solid cylinder or plate, the sum becomes an integral, and we need to use calculus to calculate the moment of inertia. We will give several examples of such calculations in Section 9.6; meanwhile, Table 9.2 gives moments of inertia for several familiar shapes in terms of their masses and dimensions. Each body shown in Table 9.2 is *uniform;* that is, the density has the same value at all points within the solid parts of the body.

CAUTION **Computing the moment of inertia** You may be tempted to try to compute the moment of inertia of a body by assuming that all the mass is concentrated at the center of mass and multiplying the total mass by the square of the distance from the center of mass to the axis. Resist that temptation; it doesn't work! For example, when a uniform thin rod of length L and mass M is pivoted about an axis through one end, perpendicular to the rod, the moment of inertia is $I = ML^2/3$ [case (b) in Table 9.2]. If we took the mass as concentrated at the center, a distance L/2 from the axis, we would obtain the *incorrect* result $I = M(L/2)^2 = ML^2/4$. ∎

Now that we know how to calculate the kinetic energy of a rotating rigid body, we can apply the energy principles of Chapter 7 to rotational motion. Here are some points of strategy and some examples.

Table 9.2 **Moments of Inertia of Various Bodies**

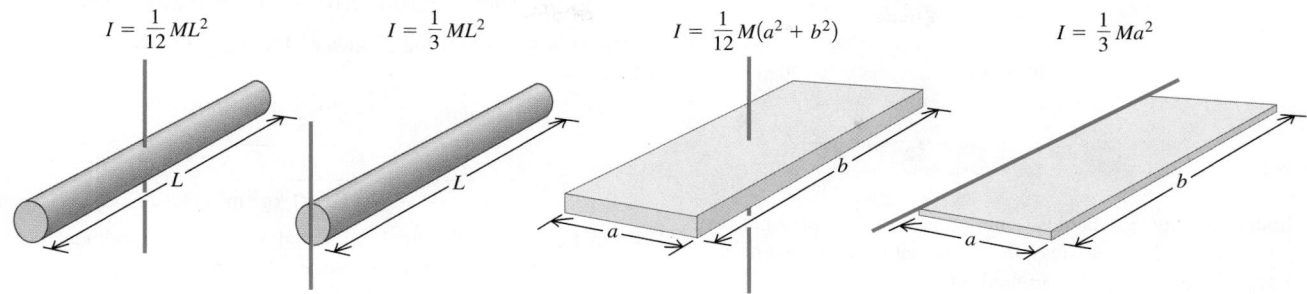

(a) Slender rod, axis through center
$$I = \frac{1}{12}ML^2$$

(b) Slender rod, axis through one end
$$I = \frac{1}{3}ML^2$$

(c) Rectangular plate, axis through center
$$I = \frac{1}{12}M(a^2 + b^2)$$

(d) Thin rectangular plate, axis along edge
$$I = \frac{1}{3}Ma^2$$

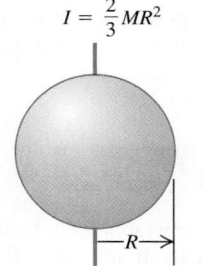

(e) Hollow cylinder
$$I = \frac{1}{2}M(R_1^2 + R_2^2)$$

(f) Solid cylinder
$$I = \frac{1}{2}MR^2$$

(g) Thin-walled hollow cylinder
$$I = MR^2$$

(h) Solid sphere
$$I = \frac{2}{5}MR^2$$

(i) Thin-walled hollow sphere
$$I = \frac{2}{3}MR^2$$

Problem-Solving Strategy 9.1 — Rotational Energy

IDENTIFY *the relevant concepts:* You can use work–energy relationships and conservation of energy to find relationships involving the position and motion of a rigid body rotating around a fixed axis. The energy method is usually not helpful for problems that involve elapsed time. In Chapter 10 we'll see how to approach rotational problems of this kind.

SET UP *the problem* using Problem-Solving Strategy 7.1 (Section 7.1), with the following additions:

5. You can use Eqs. (9.13) and (9.14) in problems involving a rope (or the like) wrapped around a rotating rigid body, if the rope doesn't slip. These equations relate the linear speed and tangential acceleration of a point on the body to the body's angular velocity and angular acceleration. (See Examples 9.7 and 9.8.)

6. Use Table 9.2 to find moments of inertia. Use the parallel-axis theorem, Eq. (9.19) (to be derived in Section 9.5), to find

moments of inertia for rotation about axes parallel to those shown in the table.

EXECUTE *the solution:* Write expressions for the initial and final kinetic and potential energies K_1, K_2, U_1, and U_2 and for the nonconservative work W_{other} (if any), where K_1 and K_2 must now include any rotational kinetic energy $K = \frac{1}{2}I\omega^2$. Substitute these expressions into Eq. (7.14), $K_1 + U_1 + W_{other} = K_2 + U_2$ (if nonconservative work is done), or Eq. (7.11), $K_1 + U_1 = K_2 + U_2$ (if only conservative work is done), and solve for the target variables. It's helpful to draw bar graphs showing the initial and final values of K, U, and $E = K + U$.

EVALUATE *your answer:* Check whether your answer makes physical sense.

Example 9.7 — An unwinding cable I

We wrap a light, nonstretching cable around a solid cylinder of mass 50 kg and diameter 0.120 m, which rotates in frictionless bearings about a stationary horizontal axis (Fig. 9.16). We pull the free end of the cable with a constant 9.0-N force for a distance of 2.0 m; it turns the cylinder as it unwinds without slipping. The cylinder is initially at rest. Find its final angular speed and the final speed of the cable.

SOLUTION

IDENTIFY: We'll solve this problem using energy methods. We'll assume that the cable is massless, so only the cylinder has kinetic energy. There are no changes in gravitational potential energy. There is friction between the cable and the cylinder, but because the cable doesn't slip, there is no motion of the cable relative to the

Continued

9.16 A cable unwinds from a cylinder (side view).

cylinder and no mechanical energy is lost in frictional work. Because the cable is massless, the force that the cable exerts on the cylinder rim is equal to the applied force F.

SET UP: Point 1 is when the cable begins to move. The cylinder starts at rest, so $K_1 = 0$. Point 2 is when the cable has moved a distance $s = 2.0$ m and the cylinder has kinetic energy $K_2 = \frac{1}{2}I\omega^2$. One of our target variables is ω; the other is the speed of the cable at point 2, which is equal to the tangential speed v of the cylinder at that point. We'll use Eq. (9.13) to find v from ω.

EXECUTE: The work done on the cylinder is $W_{\text{other}} = Fs = (9.0 \text{ N})(2.0 \text{ m}) = 18$ J. From Table 9.2 the moment of inertia is

$$I = \tfrac{1}{2}mR^2 = \tfrac{1}{2}(50 \text{ kg})(0.060 \text{ m})^2 = 0.090 \text{ kg} \cdot \text{m}^2$$

(The radius R is half the diameter.) From Eq. (7.14), $K_1 + U_1 + W_{\text{other}} = K_2 + U_2$, so

$$0 + 0 + W_{\text{other}} = \tfrac{1}{2}I\omega^2 + 0$$

$$\omega = \sqrt{\frac{2W_{\text{other}}}{I}} = \sqrt{\frac{2(18 \text{ J})}{0.090 \text{ kg} \cdot \text{m}^2}} = 20 \text{ rad/s}$$

From Eq. (9.13), the final tangential speed of the cylinder, and hence the final speed of the cable, is

$$v = R\omega = (0.060 \text{ m})(20 \text{ rad/s}) = 1.2 \text{ m/s}$$

EVALUATE: If the cable mass is not negligible, some of the 18 J of work would go into the kinetic energy of the cable. Then the cylinder would have less kinetic energy and a lower angular speed than we calculated here.

Example 9.8 An unwinding cable II

We wrap a light, nonstretching cable around a solid cylinder with mass M and radius R. The cylinder rotates with negligible friction about a stationary horizontal axis. We tie the free end of the cable to a block of mass m and release the block from rest at a distance h above the floor. As the block falls, the cable unwinds without stretching or slipping. Find expressions for the speed of the falling block and the angular speed of the cylinder as the block strikes the floor.

SOLUTION

IDENTIFY: As in Example 9.7, the cable doesn't slip and so friction does no work. We assume that the cable is massless, so that the

9.17 Our sketches for this problem.

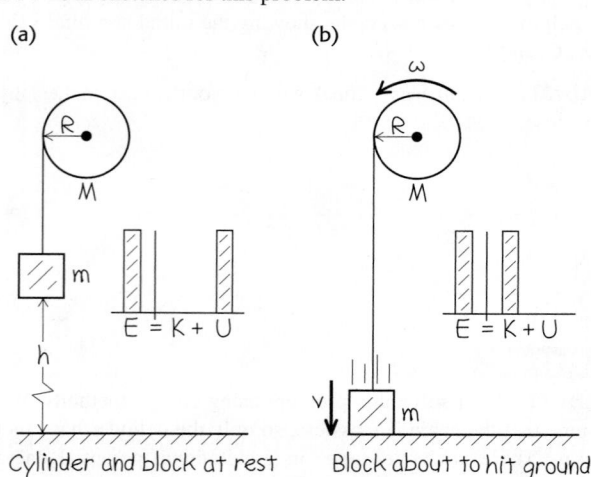

forces it exerts on the cylinder and the block have equal magnitudes. At its upper end the force and displacement are in the same direction, and at its lower end they are in opposite directions, so the cable does no *net* work and $W_{\text{other}} = 0$. Only gravity does work, and mechanical energy is conserved.

SET UP: Figure 9.17a shows the situation before the block begins to fall (point 1). The initial kinetic energy is $K_1 = 0$. We take the gravitational potential energy to be zero when the block is at floor level (point 2), so $U_1 = mgh$ and $U_2 = 0$. (We ignore the gravitational potential energy for the rotating cylinder, since its height doesn't change.) Just before the block hits the floor (Fig. 9.17b), both the block and the cylinder have kinetic energy, so

$$K_2 = \tfrac{1}{2}mv^2 + \tfrac{1}{2}I\omega^2$$

The moment of inertia of the cylinder is $I = \tfrac{1}{2}MR^2$. Also, $v = R\omega$ since the speed of the falling block must be equal to the tangential speed at the outer surface of the cylinder.

EXECUTE: We use our expressions for K_1, U_1, K_2, and U_2 and the relationship $\omega = v/R$ in Eq. (7.4), $K_1 + U_1 = K_2 + U_2$, and solve for v:

$$0 + mgh = \tfrac{1}{2}mv^2 + \tfrac{1}{2}\left(\tfrac{1}{2}MR^2\right)\left(\frac{v}{R}\right)^2 + 0 = \tfrac{1}{2}\left(m + \tfrac{1}{2}M\right)v^2$$

$$v = \sqrt{\frac{2gh}{1 + M/2m}}$$

The final angular speed of the cylinder is $\omega = v/R$.

EVALUATE: When M is much larger than m, v is very small; when M is much smaller than m, v is nearly equal to $\sqrt{2gh}$, the speed of a body that falls freely from height h. Both of these results are as we would expect.

Gravitational Potential Energy for an Extended Body

In Example 9.8 the cable was of negligible mass, so we could ignore its kinetic energy as well as the gravitational potential energy associated with it. If the mass is *not* negligible, we need to know how to calculate the *gravitational potential energy* associated with such an extended body. If the acceleration of gravity g is the same at all points on the body, the gravitational potential energy is the same as though all the mass were concentrated at the center of mass of the body. Suppose we take the y-axis vertically upward. Then for a body with total mass M, the gravitational potential energy U is simply

$$U = Mgy_{cm} \quad \text{(gravitational potential energy for an extended body)} \quad (9.18)$$

where y_{cm} is the y-coordinate of the center of mass. This expression applies to any extended body, whether it is rigid or not (Fig. 9.18).

To prove Eq. (9.18), we again represent the body as a collection of mass elements m_i. The potential energy for element m_i is $m_i g y_i$, so the total potential energy is

$$U = m_1 g y_1 + m_2 g y_2 + \cdots = (m_1 y_1 + m_2 y_2 + \cdots)g$$

But from Eq. (8.28), which defines the coordinates of the center of mass,

$$m_1 y_1 + m_2 y_2 + \cdots = (m_1 + m_2 + \cdots)y_{cm} = My_{cm}$$

where $M = m_1 + m_2 + \cdots$ is the total mass. Combining this with the above expression for U, we find $U = Mgy_{cm}$ in agreement with Eq. (9.18).

We leave the application of Eq. (9.18) to the problems. We'll make use of this relationship in Chapter 10 in the analysis of rigid-body problems in which the axis of rotation moves.

Test Your Understanding of Section 9.4 Suppose the cylinder and block in Example 9.8 have the same mass, so $m = M$. Just before the block strikes the floor, which statement is correct about the relationship between the kinetic energy of the falling block and the rotational kinetic energy of the cylinder? (i) The block has more kinetic energy than the cylinder. (ii) The block has less kinetic energy than the cylinder. (iii) The block and the cylinder have equal amounts of kinetic energy.   ❚

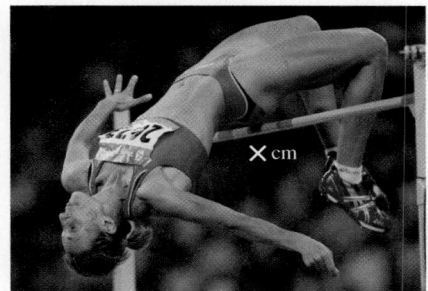

9.18 In a technique called the "Fosbury flop" after its innovator, this athlete arches her body as she passes over the bar in the high jump. As a result, her center of mass actually passes *under* the bar. This technique requires a smaller increase in gravitational potential energy [Eq. (9.18)] than the older method of straddling the bar.

9.5 Parallel-Axis Theorem

We pointed out in Section 9.4 that a body doesn't have just one moment of inertia. In fact, it has infinitely many, because there are infinitely many axes about which it might rotate. But there is a simple relationship between the moment of inertia I_{cm} of a body of mass M about an axis through its center of mass and the moment of inertia I_P about any other axis parallel to the original one but displaced from it by a distance d. This relationship, called the **parallel-axis theorem,** states that

$$I_P = I_{cm} + Md^2 \quad \text{(parallel-axis theorem)} \quad (9.19)$$

To prove this theorem, we consider two axes, both parallel to the z-axis: one through the center of mass and the other through a point P (Fig. 9.19). First we take a very thin slice of the body, parallel to the xy-plane and perpendicular to the z-axis. We take the origin of our coordinate system to be at the center of mass of the body; the coordinates of the center of mass are then $x_{cm} = y_{cm} = z_{cm} = 0$. The axis through the center of mass passes through this thin slice at point O, and the parallel axis passes through point P, whose x- and y-coordinates are (a, b). The distance of this axis from the axis through the center of mass is d, where $d^2 = a^2 + b^2$.

9.19 The mass element m_i has coordinates (x_i, y_i) with respect to an axis of rotation through the center of mass (cm) and coordinates $(x_i - a, y_i - b)$ with respect to the parallel axis through point P.

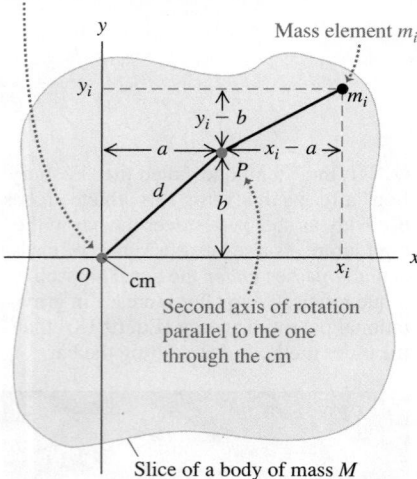

Axis of rotation passing through cm and perpendicular to the plane of the figure

Mass element m_i

Second axis of rotation parallel to the one through the cm

Slice of a body of mass M

We can write an expression for the moment of inertia I_P about the axis through point P. Let m_i be a mass element in our slice, with coordinates (x_i, y_i, z_i). Then the moment of inertia I_{cm} of the slice about the axis through the center of mass (at O) is

$$I_{cm} = \sum_i m_i(x_i^2 + y_i^2)$$

The moment of inertia of the slice about the axis through P is

$$I_P = \sum_i m_i[(x_i - a)^2 + (y_i - b)^2]$$

These expressions don't involve the coordinates z_i measured perpendicular to the slices, so we can extend the sums to include *all* particles in *all* slices. Then I_P becomes the moment of inertia of the *entire* body for an axis through P. We then expand the squared terms and regroup, and obtain

$$I_P = \sum_i m_i(x_i^2 + y_i^2) - 2a\sum_i m_i x_i - 2b\sum_i m_i y_i + (a^2 + b^2)\sum_i m_i$$

The first sum is I_{cm}. From Eq. (8.28), the definition of the center of mass, the second and third sums are proportional to x_{cm} and y_{cm}; these are zero because we have taken our origin to be the center of mass. The final term is d^2 multiplied by the total mass, or Md^2. This completes our proof that $I_P = I_{cm} + Md^2$.

As Eq. (9.19) shows, a rigid body has a lower moment of inertia about an axis through its center of mass than about any other parallel axis. Thus it's easier to start a body rotating if the rotation axis passes through the center of mass. This suggests that it's somehow most natural for a rotating body to rotate about an axis through its center of mass; we'll make this idea more quantitative in Chapter 10.

Example 9.9 **Using the parallel-axis theorem**

A part of a mechanical linkage (Fig. 9.20) has a mass of 3.6 kg. Its moment of inertia I_P about an axis 0.15 m from its center of mass is $I_P = 0.132$ kg $\cdot$ m^2. What is the moment of inertia I_{cm} about a parallel axis through the center of mass?

9.20 Calculating I_{cm} from a measurement of I_P.

0.15 m
cm
Axis through center of mass
P
Axis through P

SOLUTION

IDENTIFY, SET UP, and EXECUTE: We'll determine the target variable I_{cm} using the parallel-axis theorem, Eq. (9.19). Rearranging the equation, we obtain

$$I_{cm} = I_P - Md^2 = 0.132 \text{ kg} \cdot \text{m}^2 - (3.6 \text{ kg})(0.15 \text{ m})^2$$
$$= 0.051 \text{ kg} \cdot \text{m}^2$$

EVALUATE: As we expect, I_{cm} is less than I_P; the moment of inertia for an axis through the center of mass is lower than for any other parallel axis.

Test Your Understanding of Section 9.5 A pool cue is a wooden rod with a uniform composition and tapered with a larger diameter at one end than at the other end. Use the parallel-axis theorem to decide whether a pool cue has a larger moment of inertia (i) for an axis through the thicker end of the rod and perpendicular to the length of the rod, or (ii) for an axis through the thinner end of the rod and perpendicular to the length of the rod. ❙

9.6 Moment-of-Inertia Calculations

If a rigid body is a continuous distribution of mass—like a solid cylinder or a solid sphere—it cannot be represented by a few point masses. In this case the *sum* of masses and distances that defines the moment of inertia [Eq. (9.16)] becomes an *integral*. Imagine dividing the body into elements of mass dm that are very small, so that all points in a particular element are at essentially the same perpendicular distance from the axis of rotation. We call this distance r, as before. Then the moment of inertia is

$$I = \int r^2 \, dm \qquad (9.20)$$

To evaluate the integral, we have to represent r and dm in terms of the same integration variable. When the object is effectively one-dimensional, such as the slender rods (a) and (b) in Table 9.2, we can use a coordinate x along the length and relate dm to an increment dx. For a three-dimensional object it is usually easiest to express dm in terms of an element of volume dV and the *density* ρ of the body. Density is mass per unit volume, $\rho = dm/dV$, so we may also write Eq. (9.20) as

$$I = \int r^2 \rho \, dV$$

This expression tells us that a body's moment of inertia depends on how its density varies within its volume (Fig. 9.21). If the body is uniform in density, then we may take ρ outside the integral:

$$I = \rho \int r^2 \, dV \qquad (9.21)$$

To use this equation, we have to express the volume element dV in terms of the differentials of the integration variables, such as $dV = dx \, dy \, dz$. The element dV must always be chosen so that all points within it are at very nearly the same distance from the axis of rotation. The limits on the integral are determined by the shape and dimensions of the body. For regularly shaped bodies, this integration is often easy to do.

9.21 By measuring small variations in the orbits of satellites, geophysicists can measure the earth's moment of inertia. This tells us how our planet's mass is distributed within its interior. The data show that the earth is far denser at the core than in its outer layers.

Example 9.10 | Hollow or solid cylinder, rotating about axis of symmetry

Figure 9.22 shows a hollow cylinder of uniform mass density ρ with length L, inner radius R_1, and outer radius R_2. (It might be a steel cylinder in a printing press.) Using integration, find its moment of inertia about its axis of symmetry.

9.22 Finding the moment of inertia of a hollow cylinder about its symmetry axis.

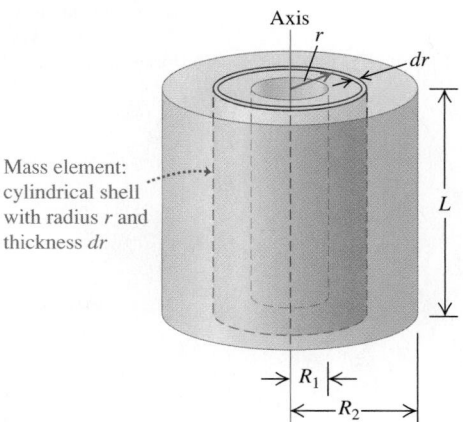

Mass element: cylindrical shell with radius r and thickness dr

SOLUTION

IDENTIFY and SET UP: We choose as a volume element a thin cylindrical shell of radius r, thickness dr, and length L. All parts of this shell are at very nearly the same distance r from the axis. The volume of the shell is very nearly that of a flat sheet with thickness dr, length L, and width $2\pi r$ (the circumference of the shell). The mass of the shell is

$$dm = \rho \, dV = \rho(2\pi r L \, dr)$$

We'll use this expression in Eq. (9.20), integrating from $r = R_1$ to $r = R_2$.

EXECUTE: From Eq. (9.20), the moment of inertia is

$$
\begin{aligned}
I = \int r^2 \, dm &= \int_{R_1}^{R_2} r^2 \rho (2\pi r L \, dr) \\
&= 2\pi \rho L \int_{R_1}^{R_2} r^3 \, dr \\
&= \frac{2\pi \rho L}{4}(R_2{}^4 - R_1{}^4) \\
&= \frac{\pi \rho L}{2}(R_2{}^2 - R_1{}^2)(R_2{}^2 + R_1{}^2)
\end{aligned}
$$

(In the last step we used the identity $a^2 - b^2 = (a - b)(a + b)$.) Let's express this result in terms of the total mass M of the body, which is its density ρ multiplied by the total volume V. The cylinder's volume is

$$V = \pi L (R_2{}^2 - R_1{}^2)$$

so its total mass M is

$$M = \rho V = \pi L \rho (R_2{}^2 - R_1{}^2)$$

Continued

Comparing with the above expression for I, we see that

$$I = \tfrac{1}{2}M(R_1^2 + R_2^2)$$

EVALUATE: Our result agrees with Table 9.2, case (e). If the cylinder is solid, with outer radius $R_2 = R$ and inner radius $R_1 = 0$, its moment of inertia is

$$I = \tfrac{1}{2}MR^2$$

in agreement with case (f). If the cylinder wall is very thin, we have $R_1 \approx R_2 = R$ and the moment of inertia is

$$I = MR^2$$

in agreement with case (g). We could have predicted this last result without calculation; in a thin-walled cylinder, all the mass is at the same distance $r = R$ from the axis, so $I = \int r^2 \, dm = R^2 \int dm = MR^2$.

Example 9.11 | Uniform sphere with radius R, axis through center

Find the moment of inertia of a solid sphere of uniform mass density ρ (like a billiard ball) about an axis through its center.

SOLUTION

IDENTIFY and SET UP: We divide the sphere into thin, solid disks of thickness dx (Fig. 9.23), whose moment of inertia we know from Table 9.2, case (f). We'll integrate over these to find the total moment of inertia.

EXECUTE: The radius and hence the volume and mass of a disk depend on its distance x from the center of the sphere. The radius r of the disk shown in Fig. 9.23 is

$$r = \sqrt{R^2 - x^2}$$

Its volume is

$$dV = \pi r^2 \, dx = \pi(R^2 - x^2) \, dx$$

9.23 Finding the moment of inertia of a sphere about an axis through its center.

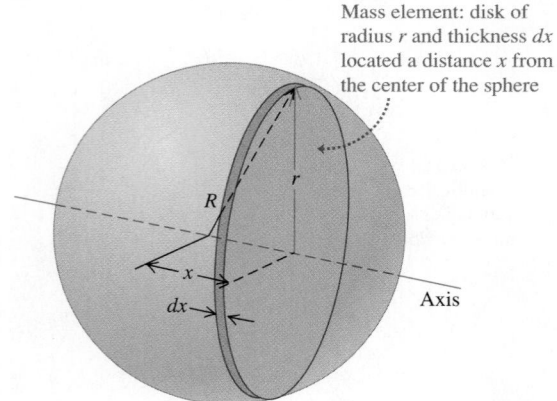

Mass element: disk of radius r and thickness dx located a distance x from the center of the sphere

R

r

x

dx

Axis

and so its mass is

$$dm = \rho \, dV = \pi\rho(R^2 - x^2) \, dx$$

From Table 9.2, case (f), the moment of inertia of a disk of radius r and mass dm is

$$dI = \tfrac{1}{2}r^2 \, dm = \tfrac{1}{2}(R^2 - x^2)[\pi\rho(R^2 - x^2) \, dx]$$
$$= \frac{\pi\rho}{2}(R^2 - x^2)^2 \, dx$$

Integrating this expression from $x = 0$ to $x = R$ gives the moment of inertia of the right hemisphere. The total I for the entire sphere, including both hemispheres, is just twice this:

$$I = (2)\frac{\pi\rho}{2} \int_0^R (R^2 - x^2)^2 \, dx$$

Carrying out the integration, we find

$$I = \frac{8\pi\rho R^5}{15}$$

The volume of the sphere is $V = 4\pi R^3/3$, so in terms of its mass M its density is

$$\rho = \frac{M}{V} = \frac{3M}{4\pi R^3}$$

Hence our expression for I becomes

$$I = \left(\frac{8\pi R^5}{15}\right)\left(\frac{3M}{4\pi R^3}\right) = \tfrac{2}{5}MR^2$$

EVALUATE: This is just as in Table 9.2, case (h). Note that the moment of inertia $I = \tfrac{2}{5}MR^2$ of a solid sphere of mass M and radius R is less than the moment of inertia $I = \tfrac{1}{2}MR^2$ of a solid *cylinder* of the same mass and radius, because more of the sphere's mass is located close to the axis.

Test Your Understanding of Section 9.6 Two hollow cylinders have the same inner and outer radii and the same mass, but they have different lengths. One is made of low-density wood and the other of high-density lead. Which cylinder has the greater moment of inertia around its axis of symmetry? (i) the wood cylinder; (ii) the lead cylinder; (iii) the two moments of inertia are equal.

Rotational kinematics: When a rigid body rotates about a stationary axis (usually called the z-axis), its position is described by an angular coordinate θ. The angular velocity ω_z is the time derivative of θ, and the angular acceleration α_z is the time derivative of ω_z or the second derivative of θ. (See Examples 9.1 and 9.2.) If the angular acceleration is constant, then θ, ω_z, and α_z are related by simple kinematic equations analogous to those for straight-line motion with constant linear acceleration. (See Example 9.3.)

$$\omega_z = \lim_{\Delta t \to 0} \frac{\Delta \theta}{\Delta t} = \frac{d\theta}{dt} \qquad (9.3)$$

$$\alpha_z = \lim_{\Delta t \to 0} \frac{\Delta \omega_z}{\Delta t} = \frac{d\omega_z}{dt} = \frac{d^2\theta}{dt^2}$$
$$(9.5), (9.6)$$

$$\theta = \theta_0 + \omega_{0z}t + \tfrac{1}{2}\alpha_z t^2 \qquad (9.11)$$
(constant α_z only)

$$\theta - \theta_0 = \tfrac{1}{2}(\omega_{0z} + \omega_z)t \qquad (9.10)$$
(constant α_z only)

$$\omega_z = \omega_{0z} + \alpha_z t \qquad (9.7)$$
(constant α_z only)

$$\omega_z^2 = \omega_{0z}^2 + 2\alpha_z(\theta - \theta_0) \qquad (9.12)$$
(constant α_z only)

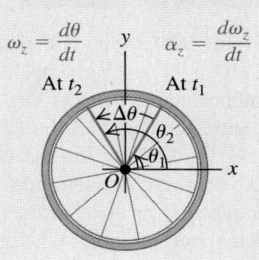

Relating linear and angular kinematics: The angular speed ω of a rigid body is the magnitude of its angular velocity. The rate of change of ω is $\alpha = d\omega/dt$. For a particle in the body a distance r from the rotation axis, the speed v and the components of the acceleration $\vec{a}$ are related to ω and α. (See Examples 9.4 and 9.5.)

$$v = r\omega \qquad (9.13)$$

$$a_{\text{tan}} = \frac{dv}{dt} = r\frac{d\omega}{dt} = r\alpha \qquad (9.14)$$

$$a_{\text{rad}} = \frac{v^2}{r} = \omega^2 r \qquad (9.15)$$

Moment of inertia and rotational kinetic energy: The moment of inertia I of a body about a given axis is a measure of its rotational inertia: The greater the value of I, the more difficult it is to change the state of the body's rotation. The moment of inertia can be expressed as a sum over the particles m_i that make up the body, each of which is at its own perpendicular distance r_i from the axis. The rotational kinetic energy of a rigid body rotating about a fixed axis depends on the angular speed ω and the moment of inertia I for that rotation axis. (See Examples 9.6–9.8.)

$$I = m_1 r_1^2 + m_2 r_2^2 + \cdots$$
$$= \sum_i m_i r_i^2 \qquad (9.16)$$

$$K = \tfrac{1}{2}I\omega^2 \qquad (9.17)$$

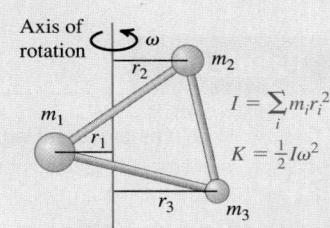

Calculating the moment of inertia: The parallel-axis theorem relates the moments of inertia of a rigid body of mass M about two parallel axes: an axis through the center of mass (moment of inertia I_{cm}) and a parallel axis a distance d from the first axis (moment of inertia I_P). (See Example 9.9.) If the body has a continuous mass distribution, the moment of inertia can be calculated by integration. (See Examples 9.10 and 9.11.)

$$I_P = I_{\text{cm}} + Md^2 \qquad (9.19)$$

BRIDGING PROBLEM **A Rotating, Uniform Thin Rod**

Figure 9.24 shows a slender uniform rod with mass M and length L. It might be a baton held by a twirler in a marching band (less the rubber end caps). (a) Use integration to compute its moment of inertia about an axis through O, at an arbitrary distance h from one end. (b) Initially the rod is at rest. It is given a constant angular acceleration of magnitude α around the axis through O. Find how much work is done on the rod in a time t. (c) At time t, what is the *linear* acceleration of the point on the rod farthest from the axis?

9.24 A thin rod with an axis through O.

SOLUTION GUIDE

See MasteringPhysics® study area for a Video Tutor solution.

IDENTIFY and SET UP

1. Make a list of the target variables for this problem.
2. To calculate the moment of inertia of the rod, you'll have to divide the rod into infinitesimal elements of mass. If an element has length dx, what is the mass of the element? What are the limits of integration?
3. What is the angular speed of the rod at time t? How does the work required to accelerate the rod from rest to this angular speed compare to the rod's kinetic energy at time t?
4. At time t, does the point on the rod farthest from the axis have a centripetal acceleration? A tangential acceleration? Why or why not?

EXECUTE

5. Do the integration required to find the moment of inertia.
6. Use your result from step 5 to calculate the work done in time t to accelerate the rod from rest.
7. Find the linear acceleration components for the point in question at time t. Use these to find the magnitude of the acceleration.

EVALUATE

8. Check your results for the special cases $h = 0$ (the axis passes through one end of the rod) and $h = L/2$ (the axis passes through the middle of the rod). Are these limits consistent with Table 9.2? With the parallel-axis theorem?
9. Is the acceleration magnitude from step 7 constant? Would you expect it to be?

Problems

For instructor-assigned homework, go to www.masteringphysics.com

•, ••, •••: Problems of increasing difficulty. **CP**: Cumulative problems incorporating material from earlier chapters. **CALC**: Problems requiring calculus. **BIO**: Biosciences problems.

DISCUSSION QUESTIONS

Q9.1 Which of the following formulas is valid if the angular acceleration of an object is *not* constant? Explain your reasoning in each case. (a) $v = r\omega$; (b) $a_{\text{tan}} = r\alpha$; (c) $\omega = \omega_0 + \alpha t$; (d) $a_{\text{tan}} = r\omega^2$; (e) $K = \frac{1}{2}I\omega^2$.

Q9.2 A diatomic molecule can be modeled as two point masses, m_1 and m_2, slightly separated (Fig. Q9.2). If the molecule is oriented along the y-axis, it has kinetic energy K when it spins about the x-axis. What will its kinetic energy (in terms of K) be if it spins at the same angular speed about (a) the z-axis and (b) the y-axis?

Figure **Q9.2**

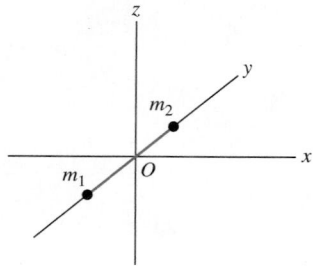

Q9.3 What is the difference between tangential and radial acceleration for a point on a rotating body?

Q9.4 In Fig. Q9.4, all points on the chain have the same linear speed. Is the magnitude of the linear acceleration also the same for all points on the chain? How are the angular accelerations of the two sprockets related? Explain.

Figure **Q9.4**

Q9.5 In Fig. Q9.4, how are the radial accelerations of points at the teeth of the two sprockets related? Explain the reasoning behind your answer.

Q9.6 A flywheel rotates with constant angular velocity. Does a point on its rim have a tangential acceleration? A radial acceleration? Are these accelerations constant in magnitude? In direction? In each case give the reasoning behind your answer.

Q9.7 What is the purpose of the spin cycle of a washing machine? Explain in terms of acceleration components.

Q9.8 Although angular velocity and angular acceleration can be treated as vectors, the angular displacement θ, despite having a magnitude and a direction, cannot. This is because θ does not follow the commutative law of vector addition (Eq. 1.3). Prove this to yourself in the following way: Lay your physics textbook flat on the desk in front of you with the cover side up so you can read the writing on it. Rotate it through 90° about a horizontal axis so that the farthest edge comes toward you. Call this angular displacement θ_1. Then rotate it by 90° about a vertical axis so that the left edge comes toward you. Call this angular displacement θ_2. The spine of the book should now face you, with the writing on it oriented so that you can read it. Now start over again but carry out the two rotations in the reverse order. Do you get a different result? That is, does $\theta_1 + \theta_2$ equal $\theta_2 + \theta_1$? Now repeat this experiment but this time with an angle of 1° rather than 90°. Do you think that the infinitesimal displacement $d\vec{\theta}$ obeys the commutative law of addition and hence qualifies as a vector? If so, how is the direction of $d\vec{\theta}$ related to the direction of $\vec{\omega}$?

Q9.9 Can you think of a body that has the same moment of inertia for all possible axes? If so, give an example, and if not, explain why this is not possible. Can you think of a body that has the same moment of inertia for all axes passing through a certain point? If so, give an example and indicate where the point is located.

Q9.10 To maximize the moment of inertia of a flywheel while minimizing its weight, what shape and distribution of mass should it have? Explain.

Q9.11 How might you determine experimentally the moment of inertia of an irregularly shaped body about a given axis?

Q9.12 A cylindrical body has mass M and radius R. Can the mass be distributed within the body in such a way that its moment of inertia about its axis of symmetry is greater than MR^2? Explain.

Q9.13 Describe how you could use part (b) of Table 9.2 to derive the result in part (d).

Q9.14 A hollow spherical shell of radius R that is rotating about an axis through its center has rotational kinetic energy K. If you want to modify this sphere so that it has three times as much kinetic energy at the same angular speed while keeping the same mass, what should be its radius in terms of R?

Q9.15 For the equations for I given in parts (a) and (b) of Table 9.2 to be valid, must the rod have a circular cross section? Is there any restriction on the size of the cross section for these equations to apply? Explain.

Q9.16 In part (d) of Table 9.2, the thickness of the plate must be much less than a for the expression given for I to apply. But in part (c), the expression given for I applies no matter how thick the plate is. Explain.

Q9.17 Two identical balls, A and B, are each attached to very light string, and each string is wrapped around the rim of a frictionless pulley of mass M. The only difference is that the pulley for ball A is a solid disk, while the one for ball B is a hollow disk, like part (e) in Table 9.2. If both balls are released from rest and fall the same distance, which one will have more kinetic energy, or will they have the same kinetic energy? Explain your reasoning.

Q9.18 An elaborate pulley consists of four identical balls at the ends of spokes extending out from a rotating drum (Fig. Q9.18). A box is connected to a light thin rope wound around the rim of the drum. When it is released from rest, the box acquires a speed V after having fallen a distance d. Now the four balls are moved inward closer to the drum, and the box is again released from rest. After it has fallen a distance d, will its speed be equal to V, greater than V, or less than V? Show or explain why.

Figure **Q9.18**

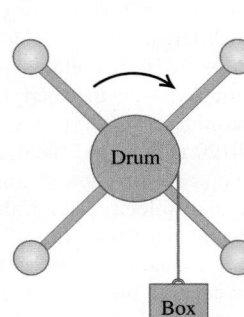

Q9.19 You can use any angular measure—radians, degrees, or revolutions—in some of the equations in Chapter 9, but you can use only radian measure in others. Identify those for which using radians is necessary and those for which it is not, and in each case give the reasoning behind your answer.

Q9.20 When calculating the moment of inertia of an object, can we treat all its mass as if it were concentrated at the center of mass of the object? Justify your answer.

Q9.21 A wheel is rotating about an axis perpendicular to the plane of the wheel and passing through the center of the wheel. The angular speed of the wheel is increasing at a constant rate. Point A is on the rim of the wheel and point B is midway between the rim and center of the wheel. For each of the following quantities, is its magnitude larger at point A or at point B, or is it the same at both points? (a) angular speed; (b) tangential speed; (c) angular acceleration; (d) tangential acceleration; (e) radial acceleration. Justify each of your answers.

Q9.22 Estimate your own moment of inertia about a vertical axis through the center of the top of your head when you are standing up straight with your arms outstretched. Make reasonable approximations and measure or estimate necessary quantities.

EXERCISES

Section 9.1 Angular Velocity and Acceleration

9.1 • (a) What angle in radians is subtended by an arc 1.50 m long on the circumference of a circle of radius 2.50 m? What is this angle in degrees? (b) An arc 14.0 cm long on the circumference of a circle subtends an angle of 128°. What is the radius of the circle? (c) The angle between two radii of a circle with radius 1.50 m is 0.700 rad. What length of arc is intercepted on the circumference of the circle by the two radii?

9.2 • An airplane propeller is rotating at 1900 rpm (rev/min). (a) Compute the propeller's angular velocity in rad/s. (b) How many seconds does it take for the propeller to turn through 35°?

9.3 • **CP CALC** The angular velocity of a flywheel obeys the equation $\omega_z(t) = A + Bt^2$, where t is in seconds and A and B are constants having numerical values 2.75 (for A) and 1.50 (for B). (a) What are the units of A and B if ω_z is in rad/s? (b) What is the angular acceleration of the wheel at (i) $t = 0.00$ and (ii) $t = 5.00$ s? (c) Through what angle does the flywheel turn during the first 2.00 s? (*Hint:* See Section 2.6.)

9.4 •• **CALC** A fan blade rotates with angular velocity given by $\omega_z(t) = \gamma - \beta t^2$, where $\gamma = 5.00$ rad/s and $\beta = 0.800$ rad/s^3. (a) Calculate the angular acceleration as a function of time. (b) Calculate the instantaneous angular acceleration α_z at $t = 3.00$ s and the average angular acceleration $\alpha_{\text{av-}z}$ for the time interval $t = 0$ to $t = 3.00$ s. How do these two quantities compare? If they are different, why are they different?

9.5 •• **CALC** A child is pushing a merry-go-round. The angle through which the merry-go-round has turned varies with time according to $\theta(t) = \gamma t + \beta t^3$, where $\gamma = 0.400$ rad/s and $\beta = 0.0120$ rad/s³. (a) Calculate the angular velocity of the merry-go-round as a function of time. (b) What is the initial value of the angular velocity? (c) Calculate the instantaneous value of the angular velocity ω_z at $t = 5.00$ s and the average angular velocity $\omega_{\text{av-}z}$ for the time interval $t = 0$ to $t = 5.00$ s. Show that $\omega_{\text{av-}z}$ is *not* equal to the average of the instantaneous angular velocities at $t = 0$ and $t = 5.00$ s, and explain why it is not.

9.6 • **CALC** At $t = 0$ the current to a dc electric motor is reversed, resulting in an angular displacement of the motor shaft given by $\theta(t) = (250 \text{ rad/s})t - (20.0 \text{ rad/s}^2)t^2 - (1.50 \text{ rad/s}^3)t^3$. (a) At what time is the angular velocity of the motor shaft zero? (b) Calculate the angular acceleration at the instant that the motor shaft has zero angular velocity. (c) How many revolutions does the motor shaft turn through between the time when the current is reversed and the instant when the angular velocity is zero? (d) How fast was the motor shaft rotating at $t = 0$, when the current was reversed? (e) Calculate the average angular velocity for the time period from $t = 0$ to the time calculated in part (a).

9.7 • **CALC** The angle θ through which a disk drive turns is given by $\theta(t) = a + bt - ct^3$, where a, b, and c are constants, t is in seconds, and θ is in radians. When $t = 0$, $\theta = \pi/4$ rad and the angular velocity is 2.00 rad/s, and when $t = 1.50$ s, the angular acceleration is 1.25 rad/s². (a) Find a, b, and c, including their units. (b) What is the angular acceleration when $\theta = \pi/4$ rad? (c) What are θ and the angular velocity when the angular acceleration is 3.50 rad/s²?

9.8 • A wheel is rotating about an axis that is in the z-direction. The angular velocity ω_z is -6.00 rad/s at $t = 0$, increases linearly with time, and is $+8.00$ rad/s at $t = 7.00$ s. We have taken counterclockwise rotation to be positive. (a) Is the angular acceleration during this time interval positive or negative? (b) During what time interval is the speed of the wheel increasing? Decreasing? (c) What is the angular displacement of the wheel at $t = 7.00$ s?

Section 9.2 Rotation with Constant Angular Acceleration

9.9 • A bicycle wheel has an initial angular velocity of 1.50 rad/s. (a) If its angular acceleration is constant and equal to 0.300 rad/s², what is its angular velocity at $t = 2.50$ s? (b) Through what angle has the wheel turned between $t = 0$ and $t = 2.50$ s?

9.10 •• An electric fan is turned off, and its angular velocity decreases uniformly from 500 rev/min to 200 rev/min in 4.00 s. (a) Find the angular acceleration in rev/s² and the number of revolutions made by the motor in the 4.00-s interval. (b) How many more seconds are required for the fan to come to rest if the angular acceleration remains constant at the value calculated in part (a)?

9.11 •• The rotating blade of a blender turns with constant angular acceleration 1.50 rad/s². (a) How much time does it take to reach an angular velocity of 36.0 rad/s, starting from rest? (b) Through how many revolutions does the blade turn in this time interval?

9.12 • (a) Derive Eq. (9.12) by combining Eqs. (9.7) and (9.11) to eliminate t. (b) The angular velocity of an airplane propeller increases from 12.0 rad/s to 16.0 rad/s while turning through 7.00 rad. What is the angular acceleration in rad/s²?

9.13 •• A turntable rotates with a constant 2.25 rad/s² angular acceleration. After 4.00 s it has rotated through an angle of 60.0 rad. What was the angular velocity of the wheel at the beginning of the 4.00-s interval?

9.14 • A circular saw blade 0.200 m in diameter starts from rest. In 6.00 s it accelerates with constant angular acceleration to an angular velocity of 140 rad/s. Find the angular acceleration and the angle through which the blade has turned.

9.15 •• A high-speed flywheel in a motor is spinning at 500 rpm when a power failure suddenly occurs. The flywheel has mass 40.0 kg and diameter 75.0 cm. The power is off for 30.0 s, and during this time the flywheel slows due to friction in its axle bearings. During the time the power is off, the flywheel makes 200 complete revolutions. (a) At what rate is the flywheel spinning when the power comes back on? (b) How long after the beginning of the power failure would it have taken the flywheel to stop if the power had not come back on, and how many revolutions would the wheel have made during this time?

9.16 •• At $t = 0$ a grinding wheel has an angular velocity of 24.0 rad/s. It has a constant angular acceleration of 30.0 rad/s² until a circuit breaker trips at $t = 2.00$ s. From then on, it turns through 432 rad as it coasts to a stop at constant angular acceleration. (a) Through what total angle did the wheel turn between $t = 0$ and the time it stopped? (b) At what time did it stop? (c) What was its acceleration as it slowed down?

9.17 •• A safety device brings the blade of a power mower from an initial angular speed of ω_1 to rest in 1.00 revolution. At the same constant acceleration, how many revolutions would it take the blade to come to rest from an initial angular speed ω_3 that was three times as great, $\omega_3 = 3\omega_1$?

Section 9.3 Relating Linear and Angular Kinematics

9.18 • In a charming 19th-century hotel, an old-style elevator is connected to a counterweight by a cable that passes over a rotating disk 2.50 m in diameter (Fig. E9.18). The elevator is raised and lowered by turning the disk, and the cable does not slip on the rim of the disk but turns with it. (a) At how many rpm must the disk turn to raise the elevator at 25.0 cm/s? (b) To start the elevator moving, it must be accelerated at $\frac{1}{8}g$. What must be the angular acceleration of the disk, in rad/s²? (c) Through what angle (in radians and degrees) has the disk turned when it has raised the elevator 3.25 m between floors?

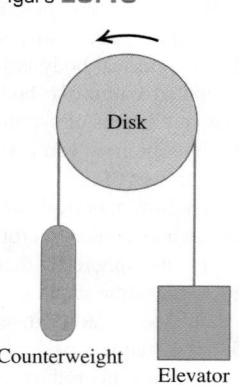

Figure **E9.18**

Disk

Counterweight

Elevator

9.19 • Using astronomical data from Appendix F, along with the fact that the earth spins on its axis once per day, calculate (a) the earth's orbital angular speed (in rad/s) due to its motion around the sun, (b) its angular speed (in rad/s) due to its axial spin, (c) the tangential speed of the earth around the sun (assuming a circular orbit), (d) the tangential speed of a point on the earth's equator due to the planet's axial spin, and (e) the radial and tangential acceleration components of the point in part (d).

9.20 • **Compact Disc.** A compact disc (CD) stores music in a coded pattern of tiny pits 10^{-7} m deep. The pits are arranged in a track that spirals outward toward the rim of the disc; the inner and outer radii of this spiral are 25.0 mm and 58.0 mm, respectively. As the disc spins inside a CD player, the track is scanned at a constant *linear* speed of 1.25 m/s. (a) What is the angular speed of the CD when the innermost part of the track is scanned? The outermost part of the track? (b) The maximum playing time of a CD is 74.0 min. What would be the length of the track on such a maximum-duration CD if it were stretched out in a straight line? (c) What is the average angular acceleration of a maximum-duration CD during its 74.0-min playing time? Take the direction of rotation of the disc to be positive.

9.21 •• A wheel of diameter 40.0 cm starts from rest and rotates with a constant angular acceleration of 3.00 rad/s^2. At the instant the wheel has computed its second revolution, compute the radial acceleration of a point on the rim in two ways: (a) using the relationship $a_{rad} = \omega^2 r$ and (b) from the relationship $a_{rad} = v^2/r$.

9.22 •• You are to design a rotating cylindrical axle to lift 800-N buckets of cement from the ground to a rooftop 78.0 m above the ground. The buckets will be attached to a hook on the free end of a cable that wraps around the rim of the axle; as the axle turns, the buckets will rise. (a) What should the diameter of the axle be in order to raise the buckets at a steady 2.00 cm/s when it is turning at 7.5 rpm? (b) If instead the axle must give the buckets an upward acceleration of 0.400 m/s^2, what should the angular acceleration of the axle be?

9.23 • A flywheel with a radius of 0.300 m starts from rest and accelerates with a constant angular acceleration of 0.600 rad/s^2. Compute the magnitude of the tangential acceleration, the radial acceleration, and the resultant acceleration of a point on its rim (a) at the start; (b) after it has turned through 60.0°; (c) after it has turned through 120.0°.

9.24 •• An electric turntable 0.750 m in diameter is rotating about a fixed axis with an initial angular velocity of 0.250 rev/s and a constant angular acceleration of 0.900 rev/s^2. (a) Compute the angular velocity of the turntable after 0.200 s. (b) Through how many revolutions has the turntable spun in this time interval? (c) What is the tangential speed of a point on the rim of the turntable at $t = 0.200$ s? (d) What is the magnitude of the *resultant* acceleration of a point on the rim at $t = 0.200$ s?

9.25 •• **Centrifuge.** An advertisement claims that a centrifuge takes up only 0.127 m of bench space but can produce a radial acceleration of 3000g at 5000 rev/min. Calculate the required radius of the centrifuge. Is the claim realistic?

9.26 • (a) Derive an equation for the radial acceleration that includes v and ω, but not r. (b) You are designing a merry-go-round for which a point on the rim will have a radial acceleration of 0.500 m/s^2 when the tangential velocity of that point has magnitude 2.00 m/s. What angular velocity is required to achieve these values?

9.27 • **Electric Drill.** According to the shop manual, when drilling a 12.7-mm-diameter hole in wood, plastic, or aluminum, a drill should have a speed of 1250 rev/min. For a 12.7-mm-diameter drill bit turning at a constant 1250 rev/min, find (a) the maximum linear speed of any part of the bit and (b) the maximum radial acceleration of any part of the bit.

9.28 • At $t = 3.00$ s a point on the rim of a 0.200-m-radius wheel has a tangential speed of 50.0 m/s as the wheel slows down with a tangential acceleration of constant magnitude 10.0 m/s^2. (a) Calculate the wheel's constant angular acceleration. (b) Calculate the angular velocities at $t = 3.00$ s and $t = 0$. (c) Through what angle did the wheel turn between $t = 0$ and $t = 3.00$ s? (d) At what time will the radial acceleration equal g?

9.29 • The spin cycles of a washing machine have two angular speeds, 423 rev/min and 640 rev/min. The internal diameter of the drum is 0.470 m. (a) What is the ratio of the maximum radial force on the laundry for the higher angular speed to that for the lower speed? (b) What is the ratio of the maximum tangential speed of the laundry for the higher angular speed to that for the lower speed? (c) Find the laundry's maximum tangential speed and the maximum radial acceleration, in terms of g.

Section 9.4 Energy in Rotational Motion

9.30 • Four small spheres, each of which you can regard as a point of mass 0.200 kg, are arranged in a square 0.400 m on a side and connected by extremely light rods (Fig. E9.30). Find the moment of inertia of the system about an axis (a) through the center of the square, perpendicular to its plane (an axis through point O in the figure); (b) bisecting two opposite sides of the square (an axis along the line AB in the figure); (c) that passes through the centers of the upper left and lower right spheres and through point O.

Figure **E9.30**

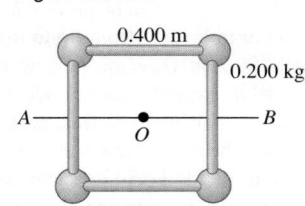

9.31 • Calculate the moment of inertia of each of the following uniform objects about the axes indicated. Consult Table 9.2 as needed. (a) A thin 2.50-kg rod of length 75.0 cm, about an axis perpendicular to it and passing through (i) one end and (ii) its center, and (iii) about an axis parallel to the rod and passing through it. (b) A 3.00-kg sphere 38.0 cm in diameter, about an axis through its center, if the sphere is (i) solid and (ii) a thin-walled hollow shell. (c) An 8.00-kg cylinder, of length 19.5 cm and diameter 12.0 cm, about the central axis of the cylinder, if the cylinder is (i) thin-walled and hollow, and (ii) solid.

9.32 •• Small blocks, each with mass m, are clamped at the ends and at the center of a rod of length L and negligible mass. Compute the moment of inertia of the system about an axis perpendicular to the rod and passing through (a) the center of the rod and (b) a point one-fourth of the length from one end.

9.33 • A uniform bar has two small balls glued to its ends. The bar is 2.00 m long and has mass 4.00 kg, while the balls each have mass 0.500 kg and can be treated as point masses. Find the moment of inertia of this combination about each of the following axes: (a) an axis perpendicular to the bar through its center; (b) an axis perpendicular to the bar through one of the balls; (c) an axis parallel to the bar through both balls; (d) an axis parallel to the bar and 0.500 m from it.

9.34 • A uniform disk of radius R is cut in half so that the remaining half has mass M (Fig. E9.34a). (a) What is the moment of inertia of this half about an axis perpendicular to its plane through point A? (b) Why did your answer in part (a) come out the same as if this were a complete disk of mass M? (c) What would be the moment of inertia of a quarter disk of mass M and radius R about an axis perpendicular to its plane passing through point B (Fig. E9.34b)?

Figure **E9.34**

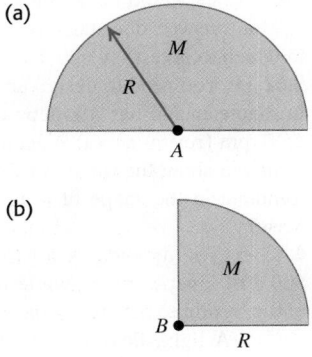

9.35 •• A wagon wheel is constructed as shown in Fig. E9.35. The radius of the wheel is 0.300 m, and the rim has mass 1.40 kg. Each of the eight spokes that lie along a diameter and are 0.300 m long has mass 0.280 kg. What is the moment of inertia of the wheel about an axis through its center and perpendicular to the plane of the wheel? (Use the formulas given in Table 9.2.)

Figure **E9.35**

0.600 m

9.36 •• An airplane propeller is 2.08 m in length (from tip to tip) with mass 117 kg and is rotating at 2400 rpm (rev/min) about an axis through its center. You can model the propeller as a slender rod. (a) What is its rotational kinetic energy? (b) Suppose that, due to weight constraints, you had to reduce the propeller's mass to 75.0% of its original mass, but you still needed to keep the same size and kinetic energy. What would its angular speed have to be, in rpm?

9.37 •• A compound disk of outside diameter 140.0 cm is made up of a uniform solid disk of radius 50.0 cm and area density 3.00 g/cm² surrounded by a concentric ring of inner radius 50.0 cm, outer radius 70.0 cm, and area density 2.00 g/cm². Find the moment of inertia of this object about an axis perpendicular to the plane of the object and passing through its center.

9.38 • A wheel is turning about an axis through its center with constant angular acceleration. Starting from rest, at $t = 0$, the wheel turns through 8.20 revolutions in 12.0 s. At $t = 12.0$ s the kinetic energy of the wheel is 36.0 J. For an axis through its center, what is the moment of inertia of the wheel?

9.39 • A uniform sphere with mass 28.0 kg and radius 0.380 m is rotating at constant angular velocity about a stationary axis that lies along a diameter of the sphere. If the kinetic energy of the sphere is 176 J, what is the tangential velocity of a point on the rim of the sphere?

9.40 •• A hollow spherical shell has mass 8.20 kg and radius 0.220 m. It is initially at rest and then rotates about a stationary axis that lies along a diameter with a constant acceleration of 0.890 rad/s². What is the kinetic energy of the shell after it has turned through 6.00 rev?

9.41 • **Energy from the Moon?** Suppose that some time in the future we decide to tap the moon's rotational energy for use on earth. In additional to the astronomical data in Appendix F, you may need to know that the moon spins on its axis once every 27.3 days. Assume that the moon is uniform throughout. (a) How much total energy could we get from the moon's rotation? (b) The world presently uses about 4.0×10^{20} J of energy per year. If in the future the world uses five times as much energy yearly, for how many years would the moon's rotation provide us energy? In light of your answer, does this seem like a cost-effective energy source in which to invest?

9.42 •• You need to design an industrial turntable that is 60.0 cm in diameter and has a kinetic energy of 0.250 J when turning at 45.0 rpm (rev/min). (a) What must be the moment of inertia of the turntable about the rotation axis? (b) If your workshop makes this turntable in the shape of a uniform solid disk, what must be its mass?

9.43 •• The flywheel of a gasoline engine is required to give up 500 J of kinetic energy while its angular velocity decreases from 650 rev/min to 520 rev/min. What moment of inertia is required?

9.44 • A light, flexible rope is wrapped several times around a *hollow* cylinder, with a weight of 40.0 N and a radius of 0.25 m, that rotates without friction about a fixed horizontal axis. The cylinder is attached to the axle by spokes of a negligible moment of inertia. The cylinder is initially at rest. The free end of the rope is pulled with a constant force P for a distance of 5.00 m, at which point the end of the rope is moving at 6.00 m/s. If the rope does not slip on the cylinder, what is the value of P?

9.45 •• Energy is to be stored in a 70.0-kg flywheel in the shape of a uniform solid disk with radius $R = 1.20$ m. To prevent structural failure of the flywheel, the maximum allowed radial acceleration of a point on its rim is 3500 m/s². What is the maximum kinetic energy that can be stored in the flywheel?

9.46 •• Suppose the solid cylinder in the apparatus described in Example 9.8 (Section 9.4) is replaced by a thin-walled, hollow cylinder with the same mass M and radius R. The cylinder is attached to the axle by spokes of a negligible moment of inertia. (a) Find the speed of the hanging mass m just as it strikes the floor. (b) Use energy concepts to explain why the answer to part (a) is different from the speed found in Example 9.8.

9.47 •• A frictionless pulley has the shape of a uniform solid disk of mass 2.50 kg and radius 20.0 cm. A 1.50-kg stone is attached to a very light wire that is wrapped around the rim of the pulley (Fig. E9.47), and the system is released from rest. (a) How far must the stone fall so that the pulley has 4.50 J of kinetic energy? (b) What percent of the total kinetic energy does the pulley have?

Figure **E9.47**

2.50-kg pulley

1.50-kg stone

9.48 •• A bucket of mass m is tied to a massless cable that is wrapped around the outer rim of a frictionless uniform pulley of radius R, similar to the system shown in Fig. E9.47. In terms of the stated variables, what must be the moment of inertia of the pulley so that it always has half as much kinetic energy as the bucket?

9.49 •• **CP** A thin, light wire is wrapped around the rim of a wheel, as shown in Fig. E9.49. The wheel rotates without friction about a stationary horizontal axis that passes through the center of the wheel. The wheel is a uniform disk with radius $R = 0.280$ m. An object of mass $m = 4.20$ kg is suspended from the free end of the wire. The system is released from rest and the suspended object descends with constant acceleration. If the suspended object moves downward a distance of 3.00 m in 2.00 s, what is the mass of the wheel?

Figure **E9.49**

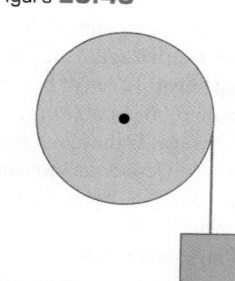

9.50 •• A uniform 2.00-m ladder of mass 9.00 kg is leaning against a vertical wall while making an angle of 53.0° with the floor. A worker pushes the ladder up against the wall until it is vertical. What is the increase in the gravitational potential energy of the ladder?

9.51 •• **How *I* Scales.** If we multiply all the design dimensions of an object by a scaling factor f, its volume and mass will be multiplied by f^3. (a) By what factor will its moment of inertia be multiplied? (b) If a $\frac{1}{48}$-scale model has a rotational kinetic energy of 2.5 J, what will be the kinetic energy for the full-scale object of the same material rotating at the same angular velocity?

9.52 •• A uniform 3.00-kg rope 24.0 m long lies on the ground at the top of a vertical cliff. A mountain climber at the top lets down half of it to help his partner climb up the cliff. What was the change in potential energy of the rope during this maneuver?

Section 9.5 Parallel-Axis Theorem

9.53 •• About what axis will a uniform, balsa-wood sphere have the same moment of inertia as does a thin-walled, hollow, lead sphere of the same mass and radius, with the axis along a diameter?

9.54 •• Find the moment of inertia of a hoop (a thin-walled, hollow ring) with mass M and radius R about an axis perpendicular to the hoop's plane at an edge.

9.55 •• A thin, rectangular sheet of metal has mass M and sides of length a and b. Use the parallel-axis theorem to calculate the moment of inertia of the sheet for an axis that is perpendicular to the plane of the sheet and that passes through one corner of the sheet.

9.56 • (a) For the thin rectangular plate shown in part (d) of Table 9.2, find the moment of inertia about an axis that lies in the plane of the plate, passes through the center of the plate, and is parallel to the axis shown in the figure. (b) Find the moment of inertia of the plate for an axis that lies in the plane of the plate, passes through the center of the plate, and is perpendicular to the axis in part (a).

9.57 •• A thin uniform rod of mass M and length L is bent at its center so that the two segments are now perpendicular to each other. Find its moment of inertia about an axis perpendicular to its plane and passing through (a) the point where the two segments meet and (b) the midpoint of the line connecting its two ends.

Section 9.6 Moment-of-Inertia Calculations

9.58 • CALC Use Eq. (9.20) to calculate the moment of inertia of a slender, uniform rod with mass M and length L about an axis at one end, perpendicular to the rod.

9.59 •• CALC Use Eq. (9.20) to calculate the moment of inertia of a uniform, solid disk with mass M and radius R for an axis perpendicular to the plane of the disk and passing through its center.

9.60 •• CALC A slender rod with length L has a mass per unit length that varies with distance from the left end, where $x = 0$, according to $dm/dx = \gamma x$, where γ has units of kg/m². (a) Calculate the total mass of the rod in terms of γ and L. (b) Use Eq. (9.20) to calculate the moment of inertia of the rod for an axis at the left end, perpendicular to the rod. Use the expression you derived in part (a) to express I in terms of M and L. How does your result compare to that for a uniform rod? Explain this comparison. (c) Repeat part (b) for an axis at the right end of the rod. How do the results for parts (b) and (c) compare? Explain this result.

PROBLEMS

9.61 • CP CALC A flywheel has angular acceleration $\alpha_z(t) = 8.60 \text{ rad/s}^2 - (2.30 \text{ rad/s}^3)t$, where counterclockwise rotation is positive. (a) If the flywheel is at rest at $t = 0$, what is its angular velocity at 5.00 s? (b) Through what angle (in radians) does the flywheel turn in the time interval from $t = 0$ to $t = 5.00$ s?

9.62 •• CALC A uniform disk with radius $R = 0.400$ m and mass 30.0 kg rotates in a horizontal plane on a frictionless vertical axle that passes through the center of the disk. The angle through which the disk has turned varies with time according to $\theta(t) = (1.10 \text{ rad/s})t + (8.60 \text{ rad/s}^2)t^2$. What is the resultant linear acceleration of a point on the rim of the disk at the instant when the disk has turned through 0.100 rev?

9.63 •• CP A circular saw blade with radius 0.120 m starts from rest and turns in a vertical plane with a constant angular acceleration of 3.00 rev/s². After the blade has turned through 155 rev, a small piece of the blade breaks loose from the top of the blade. After the piece breaks loose, it travels with a velocity that is initially horizontal and equal to the tangential velocity of the rim of the blade. The piece travels a vertical distance of 0.820 m to the floor. How far does the piece travel horizontally, from where it broke off the blade until it strikes the floor?

9.64 • CALC A roller in a printing press turns through an angle $\theta(t)$ given by $\theta(t) = \gamma t^2 - \beta t^3$, where $\gamma = 3.20 \text{ rad/s}^2$ and $\beta = 0.500 \text{ rad/s}^3$. (a) Calculate the angular velocity of the roller as a function of time. (b) Calculate the angular acceleration of the roller as a function of time. (c) What is the maximum positive angular velocity, and at what value of t does it occur?

9.65 •• CP CALC A disk of radius 25.0 cm is free to turn about an axle perpendicular to it through its center. It has very thin but strong string wrapped around its rim, and the string is attached to a ball that is pulled tangentially away from the rim of the disk (Fig. P9.65). The pull increases in magnitude and produces an acceleration of the ball that obeys the equation $a(t) = At$, where t is in seconds and A is a constant. The cylinder starts from rest, and at the end of the third second, the ball's acceleration is 1.80 m/s². (a) Find A. (b) Express the angular acceleration of the disk as a function of time. (c) How much time after the disk has begun to turn does it reach an angular speed of 15.0 rad/s? (d) Through what angle has the disk turned just as it reaches 15.0 rad/s? (*Hint:* See Section 2.6.)

Figure **P9.65**

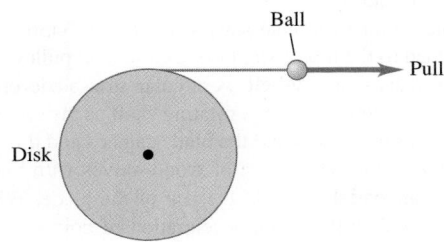

9.66 •• When a toy car is rapidly scooted across the floor, it stores energy in a flywheel. The car has mass 0.180 kg, and its flywheel has moment of inertia $4.00 \times 10^{-5} \text{ kg} \cdot \text{m}^2$. The car is 15.0 cm long. An advertisement claims that the car can travel at a scale speed of up to 700 km/h (440 mi/h). The scale speed is the speed of the toy car multiplied by the ratio of the length of an actual car to the length of the toy. Assume a length of 3.0 m for a real car. (a) For a scale speed of 700 km/h, what is the actual translational speed of the car? (b) If all the kinetic energy that is initially in the flywheel is converted to the translational kinetic energy of the toy, how much energy is originally stored in the flywheel? (c) What initial angular velocity of the flywheel was needed to store the amount of energy calculated in part (b)?

9.67 • A classic 1957 Chevrolet Corvette of mass 1240 kg starts from rest and speeds up with a constant tangential acceleration of 2.00 m/s² on a circular test track of radius 60.0 m. Treat the car as a particle. (a) What is its angular acceleration? (b) What is its angular speed 6.00 s after it starts? (c) What is its radial acceleration at this time? (d) Sketch a view from above showing the circular track, the car, the velocity vector, and the acceleration component vectors 6.00 s after the car starts. (e) What are the magnitudes of the total acceleration and net force for the car at this time? (f) What angle do the total acceleration and net force make with the car's velocity at this time?

9.68 •• Engineers are designing a system by which a falling mass m imparts kinetic energy to a rotating uniform drum to which it is attached by thin, very light wire wrapped around the rim of the drum (Fig. P9.68). There is no appreciable friction in the axle of the drum, and everything starts from rest. This system is being tested on earth, but it is to be used on Mars, where the acceleration due to gravity is 3.71 m/s². In the earth tests, when m is set to 15.0 kg and allowed to fall through 5.00 m, it gives 250.0 J of kinetic energy to the drum. (a) If the system is operated on Mars, through what distance would the 15.0-kg mass have to fall to give the same amount of kinetic energy to the drum? (b) How fast would the 15.0-kg mass be moving on Mars just as the drum gained 250.0 J of kinetic energy?

Figure P9.68

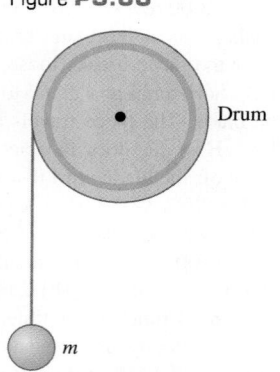

Drum

m

9.69 • A vacuum cleaner belt is looped over a shaft of radius 0.45 cm and a wheel of radius 1.80 cm. The arrangement of the belt, shaft, and wheel is similar to that of the chain and sprockets in Fig. Q9.4. The motor turns the shaft at 60.0 rev/s and the moving belt turns the wheel, which in turn is connected by another shaft to the roller that beats the dirt out of the rug being vacuumed. Assume that the belt doesn't slip on either the shaft or the wheel. (a) What is the speed of a point on the belt? (b) What is the angular velocity of the wheel, in rad/s?

9.70 •• The motor of a table saw is rotating at 3450 rev/min. A pulley attached to the motor shaft drives a second pulley of half the diameter by means of a V-belt. A circular saw blade of diameter 0.208 m is mounted on the same rotating shaft as the second pulley. (a) The operator is careless and the blade catches and throws back a small piece of wood. This piece of wood moves with linear speed equal to the tangential speed of the rim of the blade. What is this speed? (b) Calculate the radial acceleration of points on the outer edge of the blade to see why sawdust doesn't stick to its teeth.

9.71 ••• While riding a multispeed bicycle, the rider can select the radius of the rear sprocket that is fixed to the rear axle. The front sprocket of a bicycle has radius 12.0 cm. If the angular speed of the front sprocket is 0.600 rev/s, what is the radius of the rear sprocket for which the tangential speed of a point on the rim of the rear wheel will be 5.00 m/s? The rear wheel has radius 0.330 m.

9.72 ••• A computer disk drive is turned on starting from rest and has constant angular acceleration. If it took 0.750 s for the drive to make its *second* complete revolution, (a) how long did it take to make the *first* complete revolution, and (b) what is its angular acceleration, in rad/s²?

9.73 • A wheel changes its angular velocity with a constant angular acceleration while rotating about a fixed axis through its center. (a) Show that the change in the magnitude of the radial acceleration during any time interval of a point on the wheel is twice the product of the angular acceleration, the angular displacement, and the perpendicular distance of the point from the axis. (b) The radial acceleration of a point on the wheel that is 0.250 m from the axis changes from 25.0 m/s² to 85.0 m/s² as the wheel rotates through 20.0 rad. Calculate the tangential acceleration of this point. (c) Show that the change in the wheel's kinetic energy during any time interval is the product of the moment of inertia about the axis, the angular

acceleration, and the angular displacement. (d) During the 20.0-rad angular displacement of part (b), the kinetic energy of the wheel increases from 20.0 J to 45.0 J. What is the moment of inertia of the wheel about the rotation axis?

9.74 •• A sphere consists of a solid wooden ball of uniform density 800 kg/m³ and radius 0.30 m and is covered with a thin coating of lead foil with area density 20 kg/m². Calculate the moment of inertia of this sphere about an axis passing through its center.

9.75 ••• It has been argued that power plants should make use of off-peak hours (such as late at night) to generate mechanical energy and store it until it is needed during peak load times, such as the middle of the day. One suggestion has been to store the energy in large flywheels spinning on nearly frictionless ball bearings. Consider a flywheel made of iron (density 7800 kg/m³) in the shape of a 10.0-cm-thick uniform disk. (a) What would the diameter of such a disk need to be if it is to store 10.0 megajoules of kinetic energy when spinning at 90.0 rpm about an axis perpendicular to the disk at its center? (b) What would be the centripetal acceleration of a point on its rim when spinning at this rate?

9.76 •• While redesigning a rocket engine, you want to reduce its weight by replacing a solid spherical part with a hollow spherical shell of the same size. The parts rotate about an axis through their center. You need to make sure that the new part always has the same rotational kinetic energy as the original part had at any given rate of rotation. If the original part had mass M, what must be the mass of the new part?

9.77 • The earth, which is not a uniform sphere, has a moment of inertia of $0.3308MR^2$ about an axis through its north and south poles. It takes the earth 86,164 s to spin once about this axis. Use Appendix F to calculate (a) the earth's kinetic energy due to its rotation about this axis and (b) the earth's kinetic energy due to its orbital motion around the sun. (c) Explain how the value of the earth's moment of inertia tells us that the mass of the earth is concentrated toward the planet's center.

9.78 ••• A uniform, solid disk with mass m and radius R is pivoted about a horizontal axis through its center. A small object of the same mass m is glued to the rim of the disk. If the disk is released from rest with the small object at the end of a horizontal radius, find the angular speed when the small object is directly below the axis.

9.79 •• **CALC** A metal sign for a car dealership is a thin, uniform right triangle with base length b and height h. The sign has mass M. (a) What is the moment of inertia of the sign for rotation about the side of length h? (b) If $M = 5.40$ kg, $b = 1.60$ m, and $h = 1.20$ m, what is the kinetic energy of the sign when it is rotating about an axis along the 1.20-m side at 2.00 rev/s?

9.80 •• **Measuring I.** As an intern with an engineering firm, you are asked to measure the moment of inertia of a large wheel, for rotation about an axis through its center. Since you were a good physics student, you know what to do. You measure the diameter of the wheel to be 0.740 m and find that it weighs 280 N. You mount the wheel, using frictionless bearings, on a horizontal axis through the wheel's center. You wrap a light rope around the wheel and hang an 8.00-kg mass from the free end of the rope, as shown in Fig. 9.17. You release the mass from rest; the mass descends and the wheel turns as the rope unwinds. You find that the mass has speed 5.00 m/s after it has descended 2.00 m. (a) What is the moment of inertia of the wheel for an axis perpendicular to the wheel at its center? (b) Your boss tells you that a larger I is needed. He asks you to design a wheel of the same mass and radius that has $I = 19.0$ kg·m². How do you reply?

9.81 •• **CP** A meter stick with a mass of 0.180 kg is pivoted about one end so it can rotate without friction about a horizontal axis.

The meter stick is held in a horizontal position and released. As it swings through the vertical, calculate (a) the change in gravitational potential energy that has occurred; (b) the angular speed of the stick; (c) the linear speed of the end of the stick opposite the axis. (d) Compare the answer in part (c) to the speed of a particle that has fallen 1.00 m, starting from rest.

9.82 •• Exactly one turn of a flexible rope with mass m is wrapped around a uniform cylinder with mass M and radius R. The cylinder rotates without friction about a horizontal axle along the cylinder axis. One end of the rope is attached to the cylinder. The cylinder starts with angular speed ω_0. After one revolution of the cylinder the rope has unwrapped and, at this instant, hangs vertically down, tangent to the cylinder. Find the angular speed of the cylinder and the linear speed of the lower end of the rope at this time. You can ignore the thickness of the rope. [*Hint:* Use Eq. (9.18).]

9.83 • The pulley in Fig. P9.83 has radius R and a moment of inertia I. The rope does not slip over the pulley, and the pulley spins on a frictionless axle. The coefficient of kinetic friction between block A and the tabletop is μ_k. The system is released from rest, and block B descends. Block A has mass m_A and block B has mass m_B. Use energy methods to calculate the speed of block B as a function of the distance d that it has descended.

Figure **P9.83**

9.84 •• The pulley in Fig. P9.84 has radius 0.160 m and moment of inertia 0.560 kg·m². The rope does not slip on the pulley rim. Use energy methods to calculate the speed of the 4.00-kg block just before it strikes the floor.

Figure **P9.84**

4.00 kg

5.00 m

2.00 kg

9.85 •• You hang a thin hoop with radius R over a nail at the rim of the hoop. You displace it to the side (within the plane of the hoop) through an angle β from its equilibrium position and let it go. What is its angular speed when it returns to its equilibrium position? [*Hint:* Use Eq. (9.18).]

9.86 •• A passenger bus in Zurich, Switzerland, derived its motive power from the energy stored in a large flywheel. The wheel was brought up to speed periodically, when the bus stopped at a station, by an electric motor, which could then be attached to the electric power lines. The flywheel was a solid cylinder with mass 1000 kg and diameter 1.80 m; its top angular speed was 3000 rev/min. (a) At this angular speed, what is the kinetic energy of the flywheel? (b) If the average power required to operate the bus is 1.86×10^4 W, how long could it operate between stops?

9.87 •• Two metal disks, one with radius $R_1 = 2.50$ cm and mass $M_1 = 0.80$ kg and the other with radius $R_2 = 5.00$ cm and mass $M_2 = 1.60$ kg, are welded together and mounted on a frictionless axis through their common center (Fig. P9.87). (a) What is the total moment of inertia of the two disks? (b) A light string is wrapped around the edge of the smaller disk, and a 1.50-kg block is suspended from the free end of the string. If the block is released from rest at a distance of 2.00 m above the floor, what is its speed just before it strikes the floor? (c) Repeat the calculation of part (b), this time with the string wrapped around the edge of the larger disk. In which case is the final speed of the block greater? Explain why this is so.

Figure **P9.87**

R_2

R_1

1.50 kg

9.88 •• A thin, light wire is wrapped around the rim of a wheel, as shown in Fig. E9.49. The wheel rotates about a stationary horizontal axle that passes through the center of the wheel. The wheel has radius 0.180 m and moment of inertia for rotation about the axle of $I = 0.480$ kg·m². A small block with mass 0.340 kg is suspended from the free end of the wire. When the system is released from rest, the block descends with constant acceleration. The bearings in the wheel at the axle are rusty, so friction there does -6.00 J of work as the block descends 3.00 m. What is the magnitude of the angular velocity of the wheel after the block has descended 3.00 m?

9.89 ••• In the system shown in Fig. 9.17, a 12.0-kg mass is released from rest and falls, causing the uniform 10.0-kg cylinder of diameter 30.0 cm to turn about a frictionless axle through its center. How far will the mass have to descend to give the cylinder 480 J of kinetic energy?

9.90 • In Fig. P9.90, the cylinder and pulley turn without friction about stationary horizontal axles that pass through their centers. A light rope is wrapped around the cylinder, passes over the pulley, and has a 3.00-kg box suspended from its free end. There is no slipping between the rope and the pulley surface. The uniform cylinder has mass 5.00 kg and radius 40.0 cm. The pulley is a uniform disk with mass 2.00 kg and radius 20.0 cm. The box is released from rest and descends as the rope unwraps from the cylinder. Find the speed of the box when it has fallen 2.50 m.

Figure **P9.90**

Pulley

Cylinder   Box

9.91 •• A thin, flat, uniform disk has mass M and radius R. A circular hole of radius $R/4$, centered at a point $R/2$ from the disk's center, is then punched in the disk. (a) Find the moment of inertia of the disk with the hole about an axis through the original center of the disk, perpendicular to the plane of the disk. (*Hint:* Find the moment of inertia of the piece punched from the disk.) (b) Find the moment of inertia of the disk with the hole about an axis through the center of the hole, perpendicular to the plane of the disk.

9.92 •• **BIO Human Rotational Energy.** A dancer is spinning at 72 rpm about an axis through her center with her arms outstretched, as shown in Fig. P9.92. From biomedical measurements, the typical distribution of mass in a human body is as follows:

Figure **P9.92**

Head: 7.0%
Arms: 13% (for both)
Trunk and legs: 80.0%

Suppose you are this dancer. Using this information plus length measurements on your own body, calculate (a) your moment of inertia about your spin axis and (b) your rotational kinetic energy. Use the figures in Table 9.2 to model reasonable approximations for the pertinent parts of your body.

9.93 •• **BIO** **The Kinetic Energy of Walking.** If a person of mass M simply moved forward with speed V, his kinetic energy would be $\frac{1}{2}MV^2$. However, in addition to possessing a forward motion, various parts of his body (such as the arms and legs) undergo rotation. Therefore, his total kinetic energy is the sum of the energy from his forward motion plus the rotational kinetic energy of his arms and legs. The purpose of this problem is to see how much this rotational motion contributes to the person's kinetic energy. Biomedical measurements show that the arms and hands together typically make up 13% of a person's mass, while the legs and feet together account for 37%. For a rough (but reasonable) calculation, we can model the arms and legs as thin uniform bars pivoting about the shoulder and hip, respectively. In a brisk walk, the arms and legs each move through an angle of about $\pm 30°$ (a total of 60°) from the vertical in approximately 1 second. We shall assume that they are held straight, rather than being bent, which is not quite true. Let us consider a 75-kg person walking at 5.0 km/h, having arms 70 cm long and legs 90 cm long. (a) What is the average angular velocity of his arms and legs? (b) Using the average angular velocity from part (a), calculate the amount of rotational kinetic energy in this person's arms and legs as he walks. (c) What is the total kinetic energy due to both his forward motion and his rotation? (d) What percentage of his kinetic energy is due to the rotation of his legs and arms?

9.94 •• **BIO** **The Kinetic Energy of Running.** Using Problem 9.93 as a guide, apply it to a person running at 12 km/h, with his arms and legs each swinging through $\pm 30°$ in $\frac{1}{2}$ s. As before, assume that the arms and legs are kept straight.

9.95 •• **Perpendicular-Axis Theorem.** Consider a rigid body that is a thin, plane sheet of arbitrary shape. Take the body to lie in the xy-plane and let the origin O of coordinates be located at any point within or outside the body. Let I_x and I_y be the moments of inertia about the x- and y-axes, and let I_O be the moment of inertia about an axis through O perpendicular to the plane. (a) By considering mass elements m_i with coordinates (x_i, y_i), show that $I_x + I_y = I_O$. This is called the perpendicular-axis theorem. Note that point O does not have to be the center of mass. (b) For a thin washer with mass M and with inner and outer radii R_1 and R_2, use the perpendicular-axis theorem to find the moment of inertia about an axis that is in the plane of the washer and that passes through its center. You may use the information in Table 9.2. (c) Use the perpendicular-axis theorem to show that for a thin, square sheet with mass M and side L, the moment of inertia about *any* axis in the plane of the sheet that passes through the center of the sheet is $\frac{1}{12}ML^2$. You may use the information in Table 9.2.

9.96 ••• A thin, uniform rod is bent into a square of side length a. If the total mass is M, find the moment of inertia about an axis through the center and perpendicular to the plane of the square. (*Hint:* Use the parallel-axis theorem.)

9.97 • **CALC** A cylinder with radius R and mass M has density that increases linearly with distance r from the cylinder axis, $\rho = \alpha r$, where α is a positive constant. (a) Calculate the moment of inertia of the cylinder about a longitudinal axis through its center in terms of M and R. (b) Is your answer greater or smaller than the moment of inertia of a cylinder of the same mass and radius but of uniform density? Explain why this result makes qualitative sense.

9.98 •• **CALC** **Neutron Stars and Supernova Remnants.** The Crab Nebula is a cloud of glowing gas about 10 light-years across, located about 6500 light-years from the earth (Fig. P9.98). It is the remnant of a star that underwent a *supernova explosion,* seen on earth in 1054 A.D. Energy is released by the Crab Nebula at a rate of about 5×10^{31} W, about 10^5 times the rate at which the sun radiates energy. The Crab Nebula obtains its energy from the rotational kinetic energy of a rapidly spinning *neutron star* at its center.

Figure **P9.98**

This object rotates once every 0.0331 s, and this period is increasing by 4.22×10^{-13} s for each second of time that elapses. (a) If the rate at which energy is lost by the neutron star is equal to the rate at which energy is released by the nebula, find the moment of inertia of the neutron star. (b) Theories of supernovae predict that the neutron star in the Crab Nebula has a mass about 1.4 times that of the sun. Modeling the neutron star as a solid uniform sphere, calculate its radius in kilometers. (c) What is the linear speed of a point on the equator of the neutron star? Compare to the speed of light. (d) Assume that the neutron star is uniform and calculate its density. Compare to the density of ordinary rock (3000 kg/m^3) and to the density of an atomic nucleus (about 10^{17} kg/m^3). Justify the statement that a neutron star is essentially a large atomic nucleus.

9.99 •• **CALC** A sphere with radius $R = 0.200$ m has density ρ that decreases with distance r from the center of the sphere according to $\rho = 3.00 \times 10^3$ kg/m$^3 - (9.00 \times 10^3$ kg/m$^4)r$. (a) Calculate the total mass of the sphere. (b) Calculate the moment of inertia of the sphere for an axis along a diameter.

CHALLENGE PROBLEMS

9.100 ••• **CALC** Calculate the moment of inertia of a uniform solid cone about an axis through its center (Fig. P9.100). The cone has mass M and altitude h. The radius of its circular base is R.

9.101 ••• **CALC** On a compact disc (CD), music is coded in a pattern of tiny pits arranged in a track that spirals outward toward the rim of the disc. As the disc spins inside a CD player, the track is scanned at a constant *linear* speed of $v = 1.25$ m/s. Because the radius of the track varies as it spirals outward, the *angular* speed of the disc must change as the CD is played. (See Exercise 9.20.) Let's see what angular acceleration is required to keep v constant. The equation of a spiral is $r(\theta) = r_0 + \beta\theta$, where r_0 is the radius of the spiral at $\theta = 0$ and β is a constant. On a CD, r_0 is the inner radius of the spiral track. If we take the rotation direction of the CD to be positive, β must be positive so that r increases as the disc turns and θ increases. (a) When the disc

Figure **P9.100**

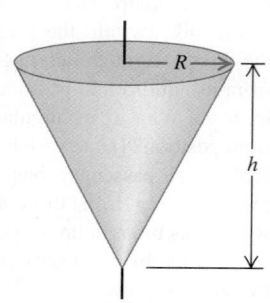

Axis

rotates through a small angle $d\theta$, the distance scanned along the track is $ds = r\,d\theta$. Using the above expression for $r(\theta)$, integrate ds to find the total distance s scanned along the track as a function of the total angle θ through which the disc has rotated. (b) Since the track is scanned at a constant linear speed v, the distance s found in part (a) is equal to vt. Use this to find θ as a function of time. There will be two solutions for θ; choose the positive one, and explain why this is the solution to choose. (c) Use your expression for $\theta(t)$ to find the angular velocity ω_z and the angular acceleration α_z as functions of time. Is α_z constant? (d) On a CD, the inner radius of the track is 25.0 mm, the track radius increases by 1.55 μm per revolution, and the playing time is 74.0 min. Find the values of r_0 and β, and find the total number of revolutions made during the playing time. (e) Using your results from parts (c) and (d), make graphs of ω_z (in rad/s) versus t and α_z (in rad/s^2) versus t between $t = 0$ and $t = 74.0$ min.

Answers

Chapter Opening Question

Both segments of the rigid blade have the same angular speed ω. From Eqs. (9.13) and (9.15), doubling the distance r for the same ω doubles the linear speed $v = r\omega$ and doubles the radial acceleration $a_{\text{rad}} = \omega^2 r$.

Test Your Understanding Questions

9.1 Answers: (a) (i) and (iii), (b) (ii) The rotation is speeding up when the angular velocity and angular acceleration have the same sign, and slowing down when they have opposite signs. Hence it is speeding up for $0 < t < 2$ s (ω_z and α_z are both positive) and for 4 s $< t < 6$ s (ω_z and α_z are both negative), but is slowing down for 2 s $< t < 4$ s (ω_z is positive and α_z is negative). Note that the body is rotating in one direction for $t < 4$ s (ω_z is positive) and in the opposite direction for $t > 4$ s (ω_z is negative).

9.2 Answers: (a) (i), (b) (ii) When the disc comes to rest, $\omega_z = 0$. From Eq. (9.7), the *time* when this occurs is $t = (\omega_z - \omega_{0z})/\alpha_z = -\omega_{0z}/\alpha_z$ (this is a positive time because α_z is negative). If we double the initial angular velocity ω_{0z} and also double the angular acceleration α_z, their ratio is unchanged and the rotation stops in the same amount of time. The *angle* through which the disc rotates is given by Eq. (9.10): $\theta - \theta_0 = \frac{1}{2}(\omega_{0z} + \omega_z)t = \frac{1}{2}\omega_{0z}t$ (since the final angular velocity is $\omega_z = 0$). The initial angular velocity ω_{0z} has been doubled but the time t is the same, so the angular displacement $\theta - \theta_0$ (and hence the number of revolutions) has doubled. You can also come to the same conclusion using Eq. (9.12).

9.3 Answer: (ii) From Eq. (9.13), $v = r\omega$. To maintain a constant linear speed v, the angular speed ω must decrease as the scanning head moves outward (greater r).

9.4 Answer: (i) The kinetic energy in the falling block is $\frac{1}{2}mv^2$, and the kinetic energy in the rotating cylinder is $\frac{1}{2}I\omega^2 = \frac{1}{2}\left(\frac{1}{2}mR^2\right)\left(\frac{v}{R}\right)^2 = \frac{1}{4}mv^2$. Hence the total kinetic energy of the system is $\frac{3}{4}mv^2$, of which two-thirds is in the block and one-third is in the cylinder.

9.5 Answer: (ii) More of the mass of the pool cue is concentrated at the thicker end, so the center of mass is closer to that end. The moment of inertia through a point P at either end is $I_P = I_{\text{cm}} + Md^2$; the thinner end is farther from the center of mass, so the distance d and the moment of inertia I_P are greater for the thinner end.

9.6 Answer: (iii) Our result from Example 9.10 does *not* depend on the cylinder length L. The moment of inertia depends only on the *radial* distribution of mass, not on its distribution along the axis.

Bridging Problem

Answers: (a) $I = \left[\dfrac{M}{L}\left(\dfrac{x^3}{3}\right)\right]_{-h}^{L-h} = \frac{1}{3}M(L^2 - 3Lh + 3h^2)$

(b) $W = \frac{1}{6}M(L^2 - 3Lh + 3h^2)\alpha^2 t^2$

(c) $a = (L - h)\alpha\sqrt{1 + \alpha^2 t^4}$

DYNAMICS OF ROTATIONAL MOTION

LEARNING GOALS

By studying this chapter, you will learn:

- What is meant by the torque produced by a force.

- How the net torque on a body affects the rotational motion of the body.

- How to analyze the motion of a body that both rotates and moves as a whole through space.

- How to solve problems that involve work and power for rotating bodies.

- What is meant by the angular momentum of a particle or of a rigid body.

- How the angular momentum of a system changes with time.

- Why a spinning gyroscope goes through the curious motion called precession.

? If you stand at the north pole, the north star, Polaris, is almost directly overhead, and the other stars appear to trace circles around it. But 5000 years ago a different star, Thuban, was directly above the north pole and was the north star. What caused this change?

We learned in Chapters 4 and 5 that a net force applied to a body gives that body an acceleration. But what does it take to give a body an *angular* acceleration? That is, what does it take to start a stationary body rotating or to bring a spinning body to a halt? A force is required, but it must be applied in a way that gives a twisting or turning action.

In this chapter we will define a new physical quantity, *torque,* that describes the twisting or turning effort of a force. We'll find that the net torque acting on a rigid body determines its angular acceleration, in the same way that the net force on a body determines its linear acceleration. We'll also look at work and power in rotational motion so as to understand such problems as how energy is transmitted by the rotating drive shaft in a car. Finally, we will develop a new conservation principle, *conservation of angular momentum,* that is tremendously useful for understanding the rotational motion of both rigid and nonrigid bodies. We'll finish this chapter by studying *gyroscopes,* rotating devices that seemingly defy common sense and don't fall over when you might think they should—but that actually behave in perfect accordance with the dynamics of rotational motion.

10.1 Torque

We know that forces acting on a body can affect its **translational motion**—that is, the motion of the body as a whole through space. Now we want to learn which aspects of a force determine how effective it is in causing or changing *rotational* motion. The magnitude and direction of the force are important, but so is the point on the body where the force is applied. In Fig. 10.1 a wrench is being used to loosen a tight bolt. Force $\vec{F}_b$, applied near the end of the handle, is more effective than an equal force $\vec{F}_a$ applied near the bolt. Force $\vec{F}_c$ doesn't do any good at all; it's applied at the same point and has the same magnitude as $\vec{F}_b$, but

10.1 Which of these three equal-magnitude forces is most likely to loosen the tight bolt?

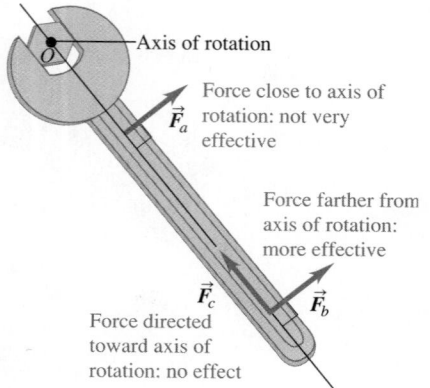

Axis of rotation

Force close to axis of rotation: not very effective
$\vec{F}_a$

Force farther from axis of rotation: more effective

$\vec{F}_c$ $\vec{F}_b$

Force directed toward axis of rotation: no effect

it's directed along the length of the handle. The quantitative measure of the tendency of a force to cause or change a body's rotational motion is called *torque;* we say that $\vec{F}_a$ applies a torque about point O to the wrench in Fig. 10.1, $\vec{F}_b$ applies a greater torque about O, and $\vec{F}_c$ applies zero torque about O.

Figure 10.2 shows three examples of how to calculate torque. The body in the figure can rotate about an axis that is perpendicular to the plane of the figure and passes through point O. Three forces, $\vec{F}_1$, $\vec{F}_2$, and $\vec{F}_3$, act on the body in the plane of the figure. The tendency of the first of these forces, $\vec{F}_1$, to cause a rotation about O depends on its magnitude F_1. It also depends on the *perpendicular* distance l_1 between point O and the **line of action** of the force (that is, the line along which the force vector lies). We call the distance l_1 the **lever arm** (or **moment arm**) of force $\vec{F}_1$ about O. The twisting effort is directly proportional to both F_1 and l_1, so we define the **torque** (or *moment*) of the force $\vec{F}_1$ with respect to O as the product $F_1 l_1$. We use the Greek letter τ (tau) for torque. In general, for a force of magnitude F whose line of action is a perpendicular distance l from O, the torque is

$$\tau = Fl \qquad (10.1)$$

Physicists usually use the term "torque," while engineers usually use "moment" (unless they are talking about a rotating shaft). Both groups use the term "lever arm" or "moment arm" for the distance l.

The lever arm of $\vec{F}_1$ in Fig. 10.2 is the perpendicular distance l_1, and the lever arm of $\vec{F}_2$ is the perpendicular distance l_2. The line of action of $\vec{F}_3$ passes through point O, so the lever arm for $\vec{F}_3$ is zero and its torque with respect to O is zero. In the same way, force $\vec{F}_c$ in Fig. 10.1 has zero torque with respect to point O; $\vec{F}_b$ has a greater torque than $\vec{F}_a$ because its lever arm is greater.

CAUTION **Torque is always measured about a point** Note that torque is *always* defined with reference to a specific point. If we shift the position of this point, the torque of each force may also change. For example, the torque of force $\vec{F}_3$ in Fig. 10.2 is zero with respect to point O, but the torque of $\vec{F}_3$ is *not* zero about point A. It's not enough to refer to "the torque of $\vec{F}$"; you must say "the torque of $\vec{F}$ with respect to point X" or "the torque of $\vec{F}$ about point X." ▌

Force $\vec{F}_1$ in Fig. 10.2 tends to cause *counterclockwise* rotation about O, while $\vec{F}_2$ tends to cause *clockwise* rotation. To distinguish between these two possibilities, we need to choose a positive sense of rotation. With the choice that *counterclockwise torques are positive and clockwise torques are negative*, the torques of $\vec{F}_1$ and $\vec{F}_2$ about O are

$$\tau_1 = +F_1 l_1 \qquad \tau_2 = -F_2 l_2$$

Figure 10.2 shows this choice for the sign of torque. We will often use the symbol ↺ to indicate our choice of the positive sense of rotation.

The SI unit of torque is the newton-meter. In our discussion of work and energy we called this combination the joule. But torque is *not* work or energy, and torque should be expressed in newton-meters, *not* joules.

Figure 10.3 shows a force $\vec{F}$ applied at a point P described by a position vector $\vec{r}$ with respect to the chosen point O. There are three ways to calculate the torque of this force:

1. Find the lever arm l and use $\tau = Fl$.
2. Determine the angle ϕ between the vectors $\vec{r}$ and $\vec{F}$; the lever arm is $r \sin \phi$, so $\tau = rF \sin \phi$.
3. Represent $\vec{F}$ in terms of a radial component F_{rad} along the direction of $\vec{r}$ and a tangential component F_{tan} at right angles, perpendicular to $\vec{r}$. (We call this a tangential component because if the body rotates, the point where the force acts moves in a circle, and this component is tangent to that circle.) Then

10.2 The torque of a force about a point is the product of the force magnitude and the lever arm of the force.

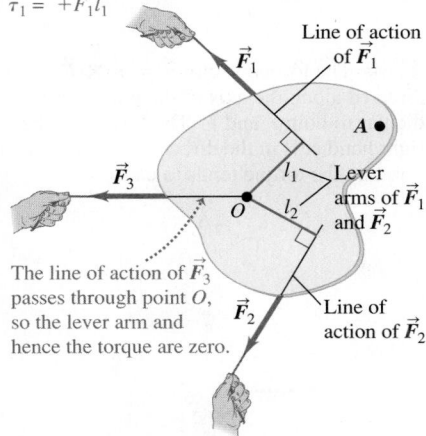

$\vec{F}_1$ tends to cause *counterclockwise* rotation about point O, so its torque is *positive*: $\tau_1 = +F_1 l_1$

The line of action of $\vec{F}_3$ passes through point O, so the lever arm and hence the torque are zero.

$\vec{F}_2$ tends to cause *clockwise* rotation about point O, so its torque is *negative*: $\tau_2 = -F_2 l_2$

10.3 Three ways to calculate the torque of the force $\vec{F}$ about the point O. In this figure, $\vec{r}$ and $\vec{F}$ are in the plane of the page and the torque vector $\vec{\tau}$ points out of the page toward you.

Three ways to calculate torque:
$\tau = Fl = rF \sin \phi = F_{\text{tan}} r$

$F_{\text{tan}} = F \sin \phi$

$F_{\text{rad}} = F \cos \phi$

$\vec{\tau}$ (out of page)

Line of action of $\vec{F}$

$l = r \sin \phi$ = lever arm

$F_{tan} = F \sin\phi$ and $\tau = r(F\sin\phi) = F_{tan}r$. The component F_{rad} produces *no* torque with respect to O because its lever arm with respect to that point is zero (compare to forces $\vec{F}_c$ in Fig. 10.1 and $\vec{F}_3$ in Fig. 10.2).

Summarizing these three expressions for torque, we have

$$\tau = Fl = rF\sin\phi = F_{tan}r \qquad \text{(magnitude of torque)} \qquad (10.2)$$

Torque as a Vector

We saw in Section 9.1 that angular velocity and angular acceleration can be represented as vectors; the same is true for torque. To see how to do this, note that the quantity $rF\sin\phi$ in Eq. (10.2) is the magnitude of the *vector product* $\vec{r} \times \vec{F}$ that we defined in Section 1.10. (You should go back and review that definition.) We now generalize the definition of torque as follows: When a force $\vec{F}$ acts at a point having a position vector $\vec{r}$ with respect to an origin O, as in Fig. 10.3, the torque $\vec{\tau}$ of the force with respect to O is the *vector* quantity

$$\vec{\tau} = \vec{r} \times \vec{F} \qquad \text{(definition of torque vector)} \qquad (10.3)$$

The torque as defined in Eq. (10.2) is just the magnitude of the torque vector $\vec{r} \times \vec{F}$. The direction of $\vec{\tau}$ is perpendicular to both $\vec{r}$ and $\vec{F}$. In particular, if both $\vec{r}$ and $\vec{F}$ lie in a plane perpendicular to the axis of rotation, as in Fig. 10.3, then the torque vector $\vec{\tau} = \vec{r} \times \vec{F}$ is directed along the axis of rotation, with a sense given by the right-hand rule (Fig. 1.29). Figure 10.4 shows the direction relationships.

In diagrams that involve $\vec{r}$, $\vec{F}$, and $\vec{\tau}$, it's common to have one of the vectors oriented perpendicular to the page. (Indeed, by the very nature of the cross product, $\vec{\tau} = \vec{r} \times \vec{F}$ *must* be perpendicular to the plane of the vectors $\vec{r}$ and $\vec{F}$.) We use a dot ($\bullet$) to represent a vector that points out of the page (see Fig. 10.3) and a cross ($\times$) to represent a vector that points into the page.

In the following sections we will usually be concerned with rotation of a body about an axis oriented in a specified constant direction. In that case, only the component of torque along that axis is of interest, and we often call that component the torque with respect to the specified *axis*.

10.4 The torque vector $\vec{\tau} = \vec{r} \times \vec{F}$ is directed along the axis of the bolt, perpendicular to both $\vec{r}$ and $\vec{F}$. The fingers of the right hand curl in the direction of the rotation that the torque tends to cause.

$\vec{\tau}$
$\vec{r}$
$\vec{F}$
(out of page)

If you point the fingers of your right hand in the direction of $\vec{r}$ and then curl them in the direction of $\vec{F}$, your outstretched thumb points in the direction of $\vec{\tau}$.

$\vec{r}$
$\vec{F}$
(out of page)
$\vec{\tau}$

Example 10.1 **Applying a torque**

To loosen a pipe fitting, a weekend plumber slips a piece of scrap pipe (a "cheater") over his wrench handle. He stands on the end of the cheater, applying his full 900-N weight at a point 0.80 m from the center of the fitting (Fig. 10.5a). The wrench handle and cheater make an angle of 19° with the horizontal. Find the magnitude and direction of the torque he applies about the center of the fitting.

10.5 (a) A weekend plumber tries to loosen a pipe fitting by standing on a "cheater." (b) Our vector diagram to find the torque about O.

(a) Diagram of situation

0.80 m

$F = 900$ N

19°

(b) Free-body diagram

Point at which force acts · · · Line of action of force

F_{rad}

Angle ϕ between line of action of force and radial direction

$\phi = 109°$ 71°

F_{tan}

Position vector from point O to the point at which the force acts

$r = 0.80$ m

19°

Point where axis of rotation intersects plane of diagram

O

ℓ

$\vec{\tau}$ (out of page) $\odot$

$F = 900$ N

Lever arm (perpendicular distance from axis of rotation to line of action of force)

SOLUTION

IDENTIFY and SET UP: Figure 10.5b shows the vectors $\vec{r}$ and $\vec{F}$ and the angle between them ($\phi = 109°$). Equation (10.1) or (10.2) will tell us the magnitude of the torque. The right-hand rule with Eq. (10.3), $\vec{\tau} = \vec{r} \times \vec{F}$, will tell us the direction of the torque.

EXECUTE: To use Eq. (10.1), we first calculate the lever arm l. As Fig. 10.5b shows,

$$l = r\sin\phi = (0.80\ \text{m})\sin 109° = (0.80\ \text{m})\sin 71° = 0.76\ \text{m}$$

Then Eq. (10.1) tells us that the magnitude of the torque is

$$\tau = Fl = (900\ \text{N})(0.76\ \text{m}) = 680\ \text{N}\cdot\text{m}$$

We get the same result from Eq. (10.2):

$$\tau = rF\sin\phi = (0.80\ \text{m})(900\ \text{N})(\sin\ 109°) = 680\ \text{N}\cdot\text{m}$$

Alternatively, we can find F_{tan}, the tangential component of $\vec{F}$ that acts perpendicular to $\vec{r}$. Figure 10.5b shows that this component is at an angle of $109° - 90° = 19°$ from $\vec{F}$, so $F_{\text{tan}} = F\sin\phi = F(\cos 19°) = (900\ \text{N})(\cos 19°) = 851\ \text{N}$. Then, from Eq. 10.2,

$$\tau = F_{\text{tan}}r = (851\ \text{N})(0.80\ \text{m}) = 680\ \text{N}\cdot\text{m}$$

Curl the fingers of your right hand from the direction of $\vec{r}$ (in the plane of Fig. 10.5b, to the left and up) into the direction of $\vec{F}$ (straight down). Then your right thumb points out of the plane of the figure: This is the direction of $\vec{\tau}$.

EVALUATE: To check the direction of $\vec{\tau}$, note that the force in Fig. 10.5 tends to produce a counterclockwise rotation about O. If you curl the fingers of your right hand in a counterclockwise direction, the thumb points out of the plane of Fig. 10.5, which is indeed the direction of the torque.

Test Your Understanding of Section 10.1 The figure shows a force P being applied to one end of a lever of length L. What is the magnitude of the torque of this force about point A? (i) $PL\sin\theta$; (ii) $PL\cos\theta$; (iii) $PL\tan\theta$.

10.2 Torque and Angular Acceleration for a Rigid Body

We are now ready to develop the fundamental relationship for the rotational dynamics of a rigid body. We will show that the angular acceleration of a rotating rigid body is directly proportional to the sum of the torque components along the axis of rotation. The proportionality factor is the moment of inertia.

To develop this relationship, we again imagine the body as being made up of a large number of particles. We choose the axis of rotation to be the z-axis; the first particle has mass m_1 and distance r_1 from this axis (Fig. 10.6). The *net force* $\vec{F}_1$ acting on this particle has a component $F_{1,\text{rad}}$ along the radial direction, a component $F_{1,\text{tan}}$ that is tangent to the circle of radius r_1 in which the particle moves as the body rotates, and a component F_{1z} along the axis of rotation. Newton's second law for the tangential component is

$$F_{1,\text{tan}} = m_1 a_{1,\text{tan}} \qquad (10.4)$$

We can express the tangential acceleration of the first particle in terms of the angular acceleration α_z of the body using Eq. (9.14): $a_{1,\text{tan}} = r_1\alpha_z$. Using this relationship and multiplying both sides of Eq. (10.4) by r_1, we obtain

$$F_{1,\text{tan}}r_1 = m_1 r_1^2 \alpha_z \qquad (10.5)$$

From Eq. (10.2), $F_{1,\text{tan}}r_1$ is just the *torque* of the net force with respect to the rotation axis, equal to the component τ_{1z} of the torque vector along the rotation axis. The subscript z is a reminder that the torque affects rotation around the z-axis, in the same way that the subscript on F_{1z} is a reminder that this force affects the motion of particle 1 along the z-axis.

10.6 As a rigid body rotates around the z-axis, a net force $\vec{F}_1$ acts on one particle of the body. Only the force component $F_{1,\text{tan}}$ can affect the rotation, because only $F_{1,\text{tan}}$ exerts a torque about O with a z-component (along the rotation axis).

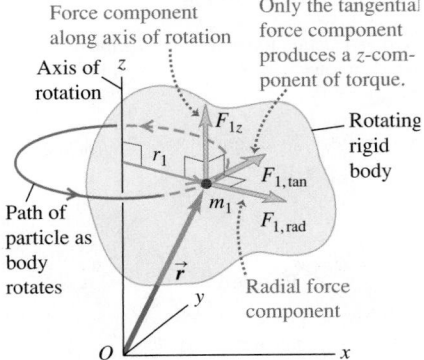

Neither of the components $F_{1,\text{rad}}$ or F_{1z} contributes to the torque about the z-axis, since neither tends to change the particle's rotation about that axis. So $\tau_{1z} = F_{1,\text{tan}}r_1$ is the total torque acting on the particle with respect to the rotation axis. Also, $m_1r_1^2$ is I_1, the moment of inertia of the particle about the rotation axis. Hence we can rewrite Eq. (10.5) as

$$\tau_{1z} = I_1\alpha_z = m_1r_1^2\alpha_z$$

We write an equation like this for every particle in the body and then add all these equations:

$$\tau_{1z} + \tau_{2z} + \cdots = I_1\alpha_z + I_2\alpha_z + \cdots = m_1r_1^2\alpha_z + m_2r_2^2\alpha_z + \cdots$$

or

$$\sum \tau_{iz} = \left(\sum m_ir_i^2\right)\alpha_z \qquad (10.6)$$

The left side of Eq. (10.6) is the sum of all the torques about the rotation axis that act on all the particles. The right side is $I = \sum m_ir_i^2$, the total moment of inertia about the rotation axis, multiplied by the angular acceleration α_z. Note that α_z is the same for every particle because this is a *rigid* body. Thus for the rigid body as a whole, Eq. (10.6) is the *rotational analog of Newton's second law*:

$$\sum \tau_z = I\alpha_z \qquad (10.7)$$

(rotational analog of Newton's second law for a rigid body)

Just as Newton's second law says that the net force on a particle equals the particle's mass times its acceleration, Eq. (10.7) says that the net torque on a rigid body equals the body's moment of inertia about the rotation axis times its angular acceleration (Fig. 10.7).

Note that because our derivation assumed that the angular acceleration α_z is the same for all particles in the body, Eq. (10.7) is valid *only* for *rigid* bodies. Hence this equation doesn't apply to a rotating tank of water or a swirling tornado of air, different parts of which have different angular accelerations. Also note that since our derivation used Eq. (9.14), $a_{\text{tan}} = r\alpha_z$, α_z must be measured in rad/s^2.

The torque on each particle is due to the net force on that particle, which is the vector sum of external and internal forces (see Section 8.2). According to Newton's third law, the *internal* forces that any pair of particles in the rigid body exert on each other are equal and opposite (Fig. 10.8). If these forces act along the line joining the two particles, their lever arms with respect to any axis are also equal. So the torques for each such pair are equal and opposite, and add to zero. Hence *all* the internal torques add to zero, so the sum $\sum \tau_z$ in Eq. (10.7) includes only the torques of the *external* forces.

Often, an important external force acting on a body is its *weight*. This force is not concentrated at a single point; it acts on every particle in the entire body. Nevertheless, it turns out that if $\vec{g}$ has the same value at all points, we always get the correct torque (about any specified axis) if we assume that all the weight is concentrated at the *center of mass* of the body. We will prove this statement in Chapter 11, but meanwhile we will use it for some of the problems in this chapter.

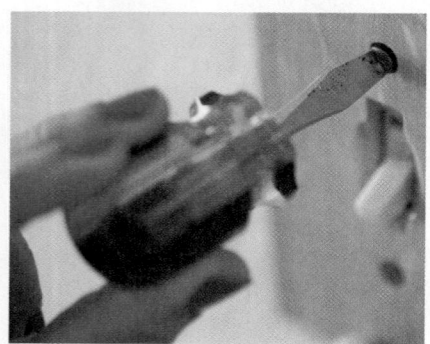

10.7 Loosening or tightening a screw requires giving it an angular acceleration and hence applying a torque. This is made easier by using a screwdriver with a large-radius handle, which provides a large lever arm for the force you apply with your hand.

10.8 Two particles in a rigid body exert equal and opposite forces on each other. If the forces act along the line joining the particles, the lever arms of the forces with respect to an axis through O are the same and the torques due to the two forces are equal and opposite. Only *external* torques affect the body's rotation.

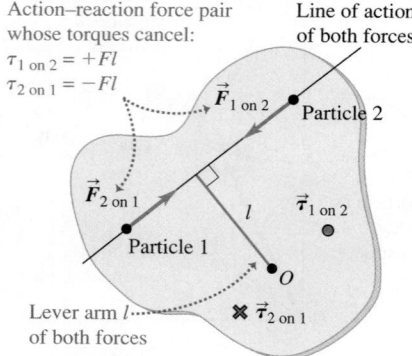

Problem-Solving Strategy 10.1 Rotational Dynamics for Rigid Bodies

Our strategy for solving problems in rotational dynamics is very similar to Problem-Solving Strategy 5.2 for solving problems involving Newton's second law.

IDENTIFY *the relevant concepts:* Equation (10.7), $\sum \tau_z = I\alpha_z$, is useful whenever torques act on a rigid body. Sometimes you can use an energy approach instead, as we did in Section 9.4. However, if the target variable is a force, a torque, an acceleration, an angular acceleration, or an elapsed time, using $\sum \tau_z = I\alpha_z$ is almost always best.

SET UP *the problem* using the following steps:
1. Sketch the situation and identify the body or bodies to be analyzed. Indicate the rotation axis.
2. For each body, draw a free-body diagram that shows the *shape* of each body, including all dimensions and angles that you will need for torque calculations. Label pertinent quantities with algebraic symbols.
3. Choose coordinate axes for each body and indicate a positive sense of rotation (clockwise or counterclockwise) for each rotating body. If you know the sense of α_z, pick that as the positive sense of rotation.

EXECUTE *the solution:*
1. For each body, decide whether it undergoes translational motion, rotational motion, or both. Then apply $\sum \vec{F} = m\vec{a}$ (as in Section 5.2), $\sum \tau_z = I\alpha_z$, or both to the body.
2. Express in algebraic form any *geometrical* relationships between the motions of two or more bodies. An example is a string that unwinds, without slipping, from a pulley or a wheel that rolls without slipping (discussed in Section 10.3). These relationships usually appear as relationships between linear and/or angular accelerations.
3. Ensure that you have as many independent equations as there are unknowns. Solve the equations to find the target variables.

EVALUATE *your answer:* Check that the algebraic signs of your results make sense. As an example, if you are unrolling thread from a spool, your answers should not tell you that the spool is turning in the direction that rolls the thread back on to the spool! Check that any algebraic results are correct for special cases or for extreme values of quantities.

Example 10.2 An unwinding cable I

Figure 10.9a shows the situation analyzed in Example 9.7 using energy methods. What is the cable's acceleration?

SOLUTION

IDENTIFY and SET UP: We can't use the energy method of Section 9.4, which doesn't involve acceleration. Instead we'll apply rotational dynamics to find the angular acceleration of the cylinder (Fig. 10.9b). We'll then find a relationship between the motion of the cable and the motion of the cylinder rim, and use this to find the acceleration of the cable. The cylinder rotates counterclockwise when the cable is pulled, so we take counterclockwise rotation to be positive. The net force on the cylinder must be zero because its center of mass remains at rest. The force F exerted by the cable produces a torque about the rotation axis. The weight (magnitude Mg) and the normal force (magnitude n) exerted by cylinder's bearings produce *no* torque about the rotation axis because they both act along lines through that axis.

EXECUTE: The lever arm of F is equal to the radius $R = 0.060$ m of the cylinder, so the torque is $\tau_z = FR$. (This torque is positive, as it tends to cause a counterclockwise rotation.) From Table 9.2, case (f), the moment of inertia of the cylinder about the rotation axis is $I = \frac{1}{2}MR^2$. Then Eq. (10.7) tells us that

$$\alpha_z = \frac{\tau_z}{I} = \frac{FR}{MR^2/2} = \frac{2F}{MR} = \frac{2(9.0 \text{ N})}{(50 \text{ kg})(0.060 \text{ m})} = 6.0 \text{ rad/s}^2$$

(We can add "rad" to our result because radians are dimensionless.)

To get the linear acceleration of the cable, recall from Section 9.3 that the acceleration of a cable unwinding from a cylinder is the same as the tangential acceleration of a point on the surface of the cylinder where the cable is tangent to it. This tangential acceleration is given by Eq. (9.14):

$$a_{\text{tan}} = R\alpha_z = (0.060 \text{ m})(6.0 \text{ rad/s}^2) = 0.36 \text{ m/s}^2$$

EVALUATE: Can you use this result, together with an equation from Chapter 2, to determine the speed of the cable after it has been pulled 2.0 m? Does your result agree with that of Example 9.7?

10.9 (a) Cylinder and cable. (b) Our free-body diagram for the cylinder.

Example 10.3	An unwinding cable II

In Example 9.8 (Section 9.4), what are the acceleration of the falling block and the tension in the cable?

SOLUTION

IDENTIFY and SET UP: We'll apply translational dynamics to the block and rotational dynamics to the cylinder. As in Example 10.2, we'll relate the linear acceleration of the block (our target variable) to the angular acceleration of the cylinder. Figure 10.10 shows our sketch of the situation and a free-body diagram for each body. We take the positive sense of rotation for the cylinder to be counter-clockwise and the positive direction of the y-coordinate for the block to be downward.

10.10 (a) Our diagram of the situation. (b) Our free-body diagrams for the cylinder and the block. We assume the cable has negligible mass.

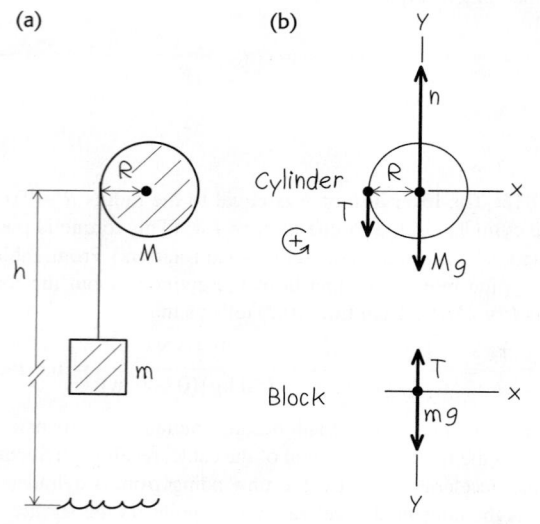

EXECUTE: For the block, Newton's second law gives

$$\sum F_y = mg + (-T) = ma_y$$

For the cylinder, the only torque about its axis is that due to the cable tension T. Hence Eq. (10.7) gives

$$\sum \tau_z = RT = I\alpha_z = \tfrac{1}{2}MR^2\alpha_z$$

As in Example 10.2, the acceleration of the cable is the same as the tangential acceleration of a point on the cylinder rim. From Eq. (9.14), this acceleration is $a_y = a_{tan} = R\alpha_z$. We use this to replace $R\alpha_z$ with a_y in the cylinder equation above, and then divide by R. The result is $T = \tfrac{1}{2}Ma_y$. Now we substitute this expression for T into Newton's second law for the block and solve for the acceleration a_y:

$$mg - \tfrac{1}{2}Ma_y = ma_y$$

$$a_y = \frac{g}{1 + M/2m}$$

To find the cable tension T, we substitute our expression for a_y into the block equation:

$$T = mg - ma_y = mg - m\left(\frac{g}{1 + M/2m}\right) = \frac{mg}{1 + 2m/M}$$

EVALUATE: The acceleration is positive (in the downward direction) and less than g, as it should be, since the cable is holding back the block. The cable tension is *not* equal to the block's weight mg; if it were, the block could not accelerate.

Let's check some particular cases. When M is much larger than m, the tension is nearly equal to mg and the acceleration is correspondingly much less than g. When M is zero, $T = 0$ and $a_y = g$; the object falls freely. If the object starts from rest ($v_{0y} = 0$) a height h above the floor, its y-velocity when it strikes the ground is given by $v_y^2 = v_{0y}^2 + 2a_yh = 2a_yh$, so

$$v_y = \sqrt{2a_yh} = \sqrt{\frac{2gh}{1 + M/2m}}$$

We found this same result from energy considerations in Example 9.8.

Test Your Understanding of Section 10.2 The figure shows a glider of mass m_1 that can slide without friction on a horizontal air track. It is attached to an object of mass m_2 by a massless string. The pulley has radius R and moment of inertia I about its axis of rotation. When released, the hanging object accelerates downward, the glider accelerates to the right, and the string turns the pulley without slipping or stretching. Rank the magnitudes of the following forces that act during the motion, in order from largest to smallest magnitude. (i) the tension force (magnitude T_1) in the horizontal part of the string; (ii) the tension force (magnitude T_2) in the vertical part of the string; (iii) the weight m_2g of the hanging object. ❙

10.3 Rigid-Body Rotation About a Moving Axis

We can extend our analysis of the dynamics of rotational motion to some cases in which the axis of rotation moves. When that happens, the motion of the body is **combined translation and rotation.** The key to understanding such situations is this: Every possible motion of a rigid body can be represented as a combination of *translational motion of the center of mass* and *rotation about an axis through the center of mass.* This is true even when the center of mass accelerates, so that it is not at rest in any inertial frame. Figure 10.11 illustrates this for the motion of a tossed baton: The center of mass of the baton follows a parabolic curve, as though the baton were a particle located at the center of mass. Other examples of combined translational and rotational motions include a ball rolling down a hill and a yo-yo unwinding at the end of a string.

Combined Translation and Rotation: Energy Relationships

It's beyond the scope of this book to prove that the motion of a rigid body can always be divided into translation of the center of mass and rotation about the center of mass. But we can show that this is true for the *kinetic energy* of a rigid body that has both translational and rotational motions. In this case, the body's kinetic energy is the sum of a part $\frac{1}{2}Mv_{cm}^2$ associated with motion of the center of mass and a part $\frac{1}{2}I_{cm}\omega^2$ associated with rotation about an axis through the center of mass:

$$K = \tfrac{1}{2}Mv_{cm}^2 + \tfrac{1}{2}I_{cm}\omega^2 \qquad (10.8)$$

(rigid body with both translation and rotation)

To prove this relationship, we again imagine the rigid body to be made up of particles. Consider a typical particle with mass m_i as shown in Fig. 10.12. The velocity $\vec{v}_i$ of this particle relative to an inertial frame is the vector sum of the velocity $\vec{v}_{cm}$ of the center of mass and the velocity $\vec{v}_i'$ of the particle *relative to the center of mass*:

$$\vec{v}_i = \vec{v}_{cm} + \vec{v}_i' \qquad (10.9)$$

The kinetic energy K_i of this particle in the inertial frame is $\frac{1}{2}m_iv_i^2$, which we can also express as $\frac{1}{2}m_i(\vec{v}_i \cdot \vec{v}_i)$. Substituting Eq. (10.9) into this, we get

$$K_i = \tfrac{1}{2}m_i(\vec{v}_{cm} + \vec{v}_i') \cdot (\vec{v}_{cm} + \vec{v}_i')$$
$$= \tfrac{1}{2}m_i(\vec{v}_{cm} \cdot \vec{v}_{cm} + 2\vec{v}_{cm} \cdot \vec{v}_i' + \vec{v}_i' \cdot \vec{v}_i')$$
$$= \tfrac{1}{2}m_i(v_{cm}^2 + 2\vec{v}_{cm} \cdot \vec{v}_i' + v_i'^2)$$

The total kinetic energy is the sum $\sum K_i$ for all the particles making up the body. Expressing the three terms in this equation as separate sums, we get

$$K = \sum K_i = \sum \left(\tfrac{1}{2}m_iv_{cm}^2\right) + \sum (m_i\vec{v}_{cm} \cdot \vec{v}_i') + \sum \left(\tfrac{1}{2}m_iv_i'^2\right)$$

The first and second terms have common factors that can be taken outside the sum:

$$K = \tfrac{1}{2}\left(\sum m_i\right)v_{cm}^2 + \vec{v}_{cm} \cdot \left(\sum m_i\vec{v}_i'\right) + \sum \left(\tfrac{1}{2}m_iv_i'^2\right) \qquad (10.10)$$

Now comes the reward for our effort. In the first term, $\sum m_i$ is the total mass M. The second term is zero because $\sum m_i\vec{v}_i'$ is M times the velocity of the center of mass *relative to the center of mass,* and this is zero by definition. The last term is the sum of the kinetic energies of the particles computed by using their speeds with respect to the center of mass; this is just the kinetic energy of rotation

10.11 The motion of a rigid body is a combination of translational motion of the center of mass and rotation around the center of mass.

This baton toss can be represented as a combination of ...

... **rotation** about the center of mass ...   ... plus **translation** of the center of mass.

10.12 A rigid body with both translation and rotation.

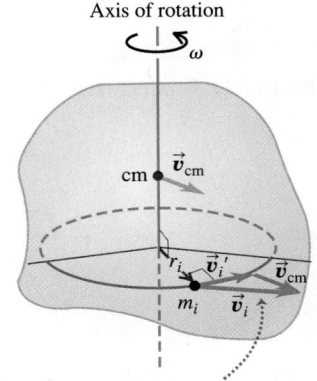

Velocity $\vec{v}_i$ of particle in rotating, translating rigid body = (velocity $\vec{v}_{cm}$ of center of mass) + (particle's velocity $\vec{v}_i'$ relative to center of mass)

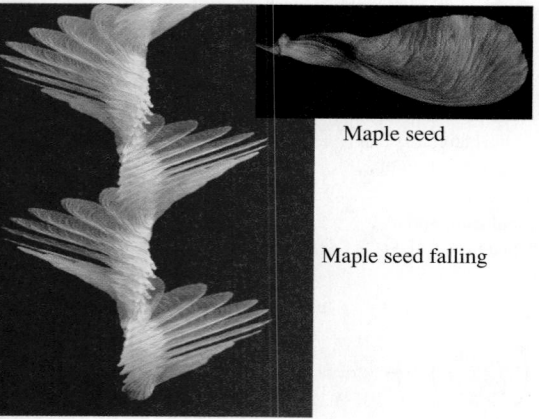

Maple seed

Maple seed falling

Mastering PHYSICS

ActivPhysics 7.11: Race Between a Block and a Disk

around the center of mass. Using the same steps that led to Eq. (9.17) for the rotational kinetic energy of a rigid body, we can write this last term as $\frac{1}{2}I_{cm}\omega^2$, where I_{cm} is the moment of inertia with respect to the axis through the center of mass and ω is the angular speed. So Eq. (10.10) becomes Eq. (10.8):

$$K = \tfrac{1}{2}Mv_{cm}^2 + \tfrac{1}{2}I_{cm}\omega^2$$

Rolling Without Slipping

An important case of combined translation and rotation is **rolling without slipping,** such as the motion of the wheel shown in Fig. 10.13. The wheel is symmetrical, so its center of mass is at its geometric center. We view the motion in an inertial frame of reference in which the surface on which the wheel rolls is at rest. In this frame, the point on the wheel that contacts the surface must be instantaneously *at rest* so that it does not slip. Hence the velocity $\vec{v}_1'$ of the point of contact relative to the center of mass must have the same magnitude but opposite direction as the center-of-mass velocity $\vec{v}_{cm}$. If the radius of the wheel is R and its angular speed about the center of mass is ω, then the magnitude of $\vec{v}_1'$ is $R\omega$; hence we must have

$$v_{cm} = R\omega \quad \text{(condition for rolling without slipping)} \quad (10.11)$$

As Fig. 10.13 shows, the velocity of a point on the wheel is the vector sum of the velocity of the center of mass and the velocity of the point relative to the center of mass. Thus while point 1, the point of contact, is instantaneously at rest, point 3 at the top of the wheel is moving forward *twice as fast* as the center of mass, and points 2 and 4 at the sides have velocities at 45° to the horizontal.

At any instant we can think of the wheel as rotating about an "instantaneous axis" of rotation that passes through the point of contact with the ground. The angular velocity ω is the same for this axis as for an axis through the center of mass; an observer at the center of mass sees the rim make the same number of revolutions per second as does an observer at the rim watching the center of mass spin around him. If we think of the motion of the rolling wheel in Fig. 10.13 in this way, the kinetic energy of the wheel is $K = \frac{1}{2}I_1\omega^2$, where I_1 is the moment of inertia of the wheel about an axis through point 1. But by the parallel-axis theorem, Eq. (9.19), $I_1 = I_{cm} + MR^2$, where M is the total mass of the wheel and I_{cm} is the moment of inertia with respect to an axis through the center of mass. Using Eq. (10.11), the kinetic energy of the wheel is

$$K = \tfrac{1}{2}I_1\omega^2 = \tfrac{1}{2}I_{cm}\omega^2 + \tfrac{1}{2}MR^2\omega^2 = \tfrac{1}{2}I_{cm}\omega^2 + \tfrac{1}{2}Mv_{cm}^2$$

which is the same as Eq. (10.8).

10.13 The motion of a rolling wheel is the sum of the translational motion of the center of mass plus the rotational motion of the wheel around the center of mass.

Translation of center of mass: velocity $\vec{v}_{cm}$

Rotation around center of mass: for rolling without slipping, speed at rim = v_{cm}

Combined motion

Wheel is instantaneously at rest where it contacts the ground.

CAUTION **Rolling without slipping** Note that the relationship $v_{cm} = R\omega$ holds *only* if there is rolling without slipping. When a drag racer first starts to move, the rear tires are spinning very fast even though the racer is hardly moving, so $R\omega$ is greater than v_{cm} (Fig. 10.14). If a driver applies the brakes too heavily so that the car skids, the tires will spin hardly at all and $R\omega$ is less than v_{cm}.

If a rigid body changes height as it moves, we must also consider gravitational potential energy. As we discussed in Section 9.4, the gravitational potential energy associated with any extended body of mass M, rigid or not, is the same as if we replace the body by a particle of mass M located at the body's center of mass. That is,

$$U = Mgy_{cm}$$

10.14 The smoke rising from this drag racer's rear tires shows that the tires are slipping on the road, so v_{cm} is *not* equal to $R\omega$.

Example 10.4 | Speed of a primitive yo-yo

You make a primitive yo-yo by wrapping a massless string around a solid cylinder with mass M and radius R (Fig. 10.15). You hold the free end of the string stationary and release the cylinder from rest. The string unwinds but does not slip or stretch as the cylinder descends and rotates. Using energy considerations, find the speed v_{cm} of the center of mass of the cylinder after it has descended a distance h.

SOLUTION

IDENTIFY and SET UP: The upper end of the string is held fixed, not pulled upward, so your hand does no work on the string–cylinder system. There is friction between the string and the cylinder, but the string doesn't slip so no mechanical energy is lost. Hence we can use conservation of mechanical energy. The initial kinetic energy of the cylinder is $K_1 = 0$, and its final kinetic energy K_2 is given by

10.15 Calculating the speed of a primitive yo-yo.

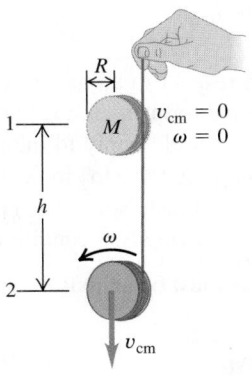

Eq. (10.8); the massless string has no kinetic energy. The moment of inertia is $I = \frac{1}{2}MR^2$, and by Eq. (9.13) $\omega = v_{cm}/R$ because the string doesn't slip. The potential energies are $U_1 = Mgh$ and $U_2 = 0$.

EXECUTE: From Eq. (10.8), the kinetic energy at point 2 is

$$K_2 = \frac{1}{2}Mv_{cm}^2 + \frac{1}{2}\left(\frac{1}{2}MR^2\right)\left(\frac{v_{cm}}{R}\right)^2$$
$$= \frac{3}{4}Mv_{cm}^2$$

The kinetic energy is $1\frac{1}{2}$ times what it would be if the yo-yo were falling at speed v_{cm} without rotating. Two-thirds of the total kinetic energy $\left(\frac{1}{2}Mv_{cm}^2\right)$ is translational and one-third $\left(\frac{1}{4}Mv_{cm}^2\right)$ is rotational. Using conservation of energy,

$$K_1 + U_1 = K_2 + U_2$$
$$0 + Mgh = \frac{3}{4}Mv_{cm}^2 + 0$$
$$v_{cm} = \sqrt{\tfrac{4}{3}gh}$$

EVALUATE: No mechanical energy was lost or gained, so from the energy standpoint the string is merely a way to convert some of the gravitational potential energy (which is released as the cylinder falls) into rotational kinetic energy rather than translational kinetic energy. Because not all of the released energy goes into translation, v_{cm} is less than the speed $\sqrt{2gh}$ of an object dropped from height h with no strings attached.

Example 10.5 | Race of the rolling bodies

In a physics demonstration, an instructor "races" various bodies that roll without slipping from rest down an inclined plane (Fig. 10.16). What shape should a body have to reach the bottom of the incline first?

SOLUTION

IDENTIFY and SET UP: Kinetic friction does no work if the bodies roll without slipping. We can also ignore the effects of *rolling friction,* introduced in Section 5.3, if the bodies and the surface of the

Continued

incline are rigid. (Later in this section we'll explain why this is so.) We can therefore use conservation of energy. Each body starts from rest at the top of an incline with height h, so $K_1 = 0$, $U_1 = Mgh$, and $U_2 = 0$. Equation (10.8) gives the kinetic energy at the bottom of the incline; since the bodies roll without slipping, $\omega = v_{cm}/R$. We can express the moments of inertia of the four round bodies in Table 9.2, cases (f)–(i), as $I_{cm} = cMR^2$, where c is a number less than or equal to 1 that depends on the shape of the body. Our goal is to find the value of c that gives the body the greatest speed v_{cm} after its center of mass has descended a vertical distance h.

10.16 Which body rolls down the incline fastest, and why?

EXECUTE: From conservation of energy,

$$K_1 + U_1 = K_2 + U_2$$

$$0 + Mgh = \tfrac{1}{2}Mv_{cm}^2 + \tfrac{1}{2}cMR^2\left(\frac{v_{cm}}{R}\right)^2 + 0$$

$$Mgh = \tfrac{1}{2}(1 + c)Mv_{cm}^2$$

$$v_{cm} = \sqrt{\frac{2gh}{1 + c}}$$

EVALUATE: For a given value of c, the speed v_{cm} after descending a distance h is *independent* of the body's mass M and radius R. Hence *all* uniform solid cylinders $\left(c = \tfrac{1}{2}\right)$ have the same speed at the bottom, regardless of their mass and radii. The values of c tell us that the order of finish for uniform bodies will be as follows: (1) any solid sphere $\left(c = \tfrac{2}{5}\right)$, (2) any solid cylinder $\left(c = \tfrac{1}{2}\right)$, (3) any thin-walled, hollow sphere $\left(c = \tfrac{2}{3}\right)$, and (4) any thin-walled, hollow cylinder $(c = 1)$. Small-c bodies always beat large-c bodies because less of their kinetic energy is tied up in rotation and so more is available for translation.

Combined Translation and Rotation: Dynamics

We can also analyze the combined translational and rotational motions of a rigid body from the standpoint of dynamics. We showed in Section 8.5 that for a body with total mass M, the acceleration $\vec{a}_{cm}$ of the center of mass is the same as that of a point mass M acted on by all the external forces on the actual body:

$$\sum \vec{F}_{ext} = M\vec{a}_{cm} \qquad (10.12)$$

The rotational motion about the center of mass is described by the rotational analog of Newton's second law, Eq. (10.7):

$$\sum \tau_z = I_{cm}\alpha_z \qquad (10.13)$$

10.17 The axle of a bicycle wheel passes through the wheel's center of mass and is an axis of symmetry. Hence the rotation of the wheel is described by Eq. (10.13), provided the bicycle doesn't turn or tilt to one side (which would change the orientation of the axle).

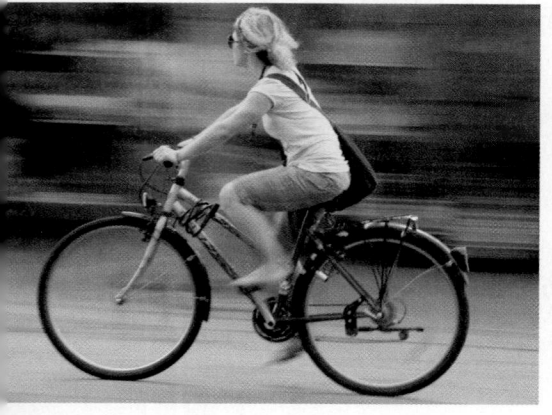

where I_{cm} is the moment of inertia with respect to an axis through the center of mass and the sum $\sum \tau_z$ includes all external torques with respect to this axis. It's not immediately obvious that Eq. (10.13) should apply to the motion of a translating rigid body; after all, our derivation of $\sum \tau_z = I\alpha_z$ in Section 10.2 assumed that the axis of rotation was stationary. But in fact, Eq. (10.13) is valid *even when the axis of rotation moves*, provided the following two conditions are met:

1. The axis through the center of mass must be an axis of symmetry.
2. The axis must not change direction.

These conditions are satisfied for many types of rotation (Fig. 10.17). Note that in general this moving axis of rotation is *not* at rest in an inertial frame of reference.

We can now solve dynamics problems involving a rigid body that undergoes translational and rotational motions at the same time, provided that the rotation axis satisfies the two conditions just mentioned. Problem-Solving Strategy 10.1 (Section 10.2) is equally useful here, and you should review it now. Keep in mind that when a body undergoes translational and rotational motions at the same time, we need two separate equations of motion *for the same body*. One of these, Eq. (10.12), describes the translational motion of the center of mass. The other equation of motion, Eq. (10.13), describes the rotational motion about the axis through the center of mass.

Example 10.6 Acceleration of a primitive yo-yo

For the primitive yo-yo in Example 10.4 (Fig. 10.18a), find the downward acceleration of the cylinder and the tension in the string.

SOLUTION

IDENTIFY and SET UP: Figure 10.18b shows our free-body diagram for the yo-yo, including our choice of positive coordinate directions. Our target variables are $a_{\text{cm-}y}$ and T. We'll use Eq. (10.12) for the

10.18 Dynamics of a primitive yo-yo (see Fig. 10.15).

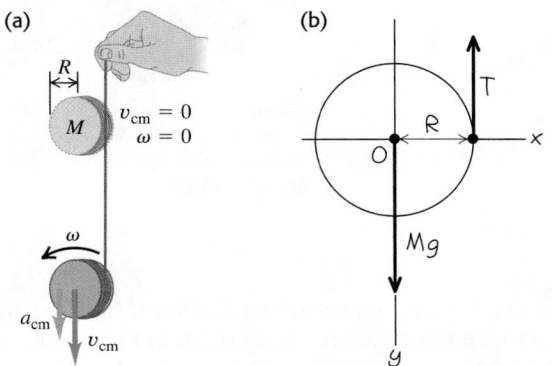

(a)

(b)

translational motion of the center of mass and Eq. (10.13) for rotational motion around the center of mass. We'll also use Eq. (10.11), which says that the string unwinds without slipping. As in Example 10.4, the moment of inertia of the yo-yo for an axis through its center of mass is $I_{\text{cm}} = \frac{1}{2}MR^2$.

EXECUTE: From Eq. (10.12),

$$\sum F_y = Mg + (-T) = Ma_{\text{cm-}y} \qquad (10.14)$$

From Eq. (10.13),

$$\sum \tau_z = TR = I_{\text{cm}}\alpha_z = \tfrac{1}{2}MR^2\alpha_z \qquad (10.15)$$

From Eq. (10.11), $v_{\text{cm-}z} = R\omega_z$; the derivative of this expression with respect to time gives us

$$a_{\text{cm-}y} = R\alpha_z \qquad (10.16)$$

We now use Eq. (10.16) to eliminate α_z from Eq. (10.15) and then solve Eqs. (10.14) and (10.15) simultaneously for T and $a_{\text{cm-}y}$. The results are

$$a_{\text{cm-}y} = \tfrac{2}{3}g \qquad T = \tfrac{1}{3}Mg$$

EVALUATE: The string slows the fall of the yo-yo, but not enough to stop it completely. Hence $a_{\text{cm-}y}$ is less than the free-fall value g and T is less than the yo-yo weight Mg.

Example 10.7 Acceleration of a rolling sphere

A bowling ball rolls without slipping down a ramp, which is inclined at an angle β to the horizontal (Fig. 10.19a). What are the ball's acceleration and the magnitude of the friction force on the ball? Treat the ball as a uniform solid sphere, ignoring the finger holes.

SOLUTION

IDENTIFY and SET UP: The free-body diagram (Fig. 10.19b) shows that only the friction force exerts a torque about the center of mass. Our target variables are the acceleration $a_{\text{cm-}x}$ of the ball's center of mass and the magnitude f of the friction force. (Because

10.19 A bowling ball rolling down a ramp.

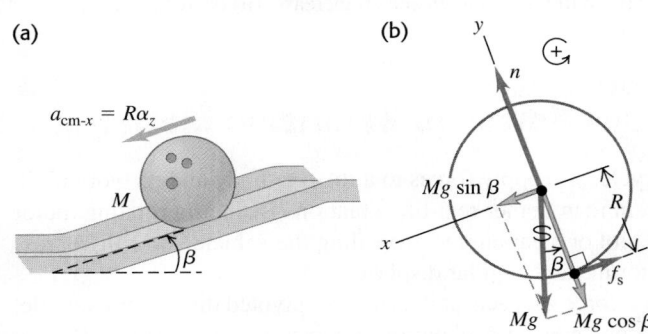

(a)

(b)

the ball does not slip at the instantaneous point of contact with the ramp, this is a *static* friction force; it prevents slipping and gives the ball its angular acceleration.) We use Eqs. (10.12) and (10.13) as in Example 10.6.

EXECUTE: The ball's moment of inertia is $I_{\text{cm}} = \frac{2}{5}MR^2$. The equations of motion are

$$\sum F_x = Mg\sin\beta + (-f) = Ma_{\text{cm-}x} \qquad (10.17)$$

$$\sum \tau_z = fR = I_{\text{cm}}\alpha_z = \left(\tfrac{2}{5}MR^2\right)\alpha_z \qquad (10.18)$$

The ball rolls without slipping, so as in Example 10.6 we use $a_{\text{cm-}x} = R\alpha_z$ to eliminate α_z from Eq. (10.18):

$$fR = \tfrac{2}{5}MRa_{\text{cm-}x}$$

This equation and Eq. (10.17) are two equations for the unknowns $a_{\text{cm-}x}$ and f. We solve Eq. (10.17) for f, substitute that expression into the above equation to eliminate f, and solve for $a_{\text{cm-}x}$:

$$a_{\text{cm-}x} = \tfrac{5}{7}g\sin\beta$$

Finally, we substitute this acceleration into Eq. (10.17) and solve for f:

$$f = \tfrac{2}{7}Mg\sin\beta$$

Continued

EVALUATE: The ball's acceleration is just $\frac{5}{7}$ as large as that of an object *sliding* down the slope without friction. If the ball descends a vertical distance h as it rolls down the ramp, its displacement along the ramp is $h/\sin\beta$. You can show that the speed of the ball at the bottom of the ramp is $v_{cm} = \sqrt{\frac{10}{7}gh}$, the same as our result from Example 10.5 with $c = \frac{2}{5}$.

If the ball were rolling *uphill* without slipping, the force of friction would still be directed uphill as in Fig. 10.19b. Can you see why?

10.20 Rolling down **(a)** a perfectly rigid surface and **(b)** a deformable surface. The deformation in part **(b)** is greatly exaggerated.

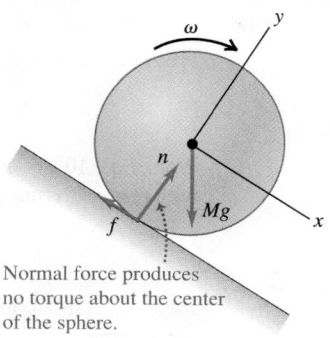

(a) Perfectly rigid sphere rolling on a perfectly rigid surface

Normal force produces no torque about the center of the sphere.

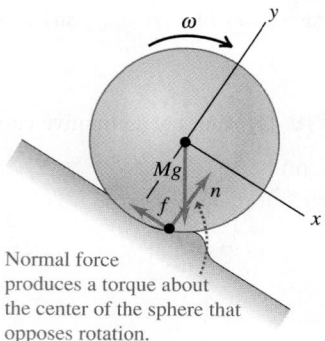

(b) Rigid sphere rolling on a deformable surface

Normal force produces a torque about the center of the sphere that opposes rotation.

Rolling Friction

In Example 10.5 we said that we can ignore rolling friction if both the rolling body and the surface over which it rolls are perfectly rigid. In Fig. 10.20a a perfectly rigid sphere is rolling down a perfectly rigid incline. The line of action of the normal force passes through the center of the sphere, so its torque is zero; there is no sliding at the point of contact, so the friction force does no work. Figure 10.20b shows a more realistic situation, in which the surface "piles up" in front of the sphere and the sphere rides in a shallow trench. Because of these deformations, the contact forces on the sphere no longer act along a single point, but over an area; the forces are concentrated on the front of the sphere as shown. As a result, the normal force now exerts a torque that opposes the rotation. In addition, there is some sliding of the sphere over the surface due to the deformation, causing mechanical energy to be lost. The combination of these two effects is the phenomenon of *rolling friction*. Rolling friction also occurs if the rolling body is deformable, such as an automobile tire. Often the rolling body and the surface are rigid enough that rolling friction can be ignored, as we have assumed in all the examples in this section.

Test Your Understanding of Section 10.3 Suppose the solid cylinder used as a yo-yo in Example 10.6 is replaced by a hollow cylinder of the same mass and radius. (a) Will the acceleration of the yo-yo (i) increase, (ii) decrease, or (iii) remain the same? (b) Will the string tension (i) increase, (ii) decrease, or (iii) remain the same?

10.4 Work and Power in Rotational Motion

When you pedal a bicycle, you apply forces to a rotating body and do work on it. Similar things happen in many other real-life situations, such as a rotating motor shaft driving a power tool or a car engine propelling the vehicle. We can express this work in terms of torque and angular displacement.

Suppose a tangential force $\vec{F}_{tan}$ acts at the rim of a pivoted disk—for example, a child running while pushing on a playground merry-go-round (Fig. 10.21a). The disk rotates through an infinitesimal angle $d\theta$ about a fixed axis during an infinitesimal time interval dt (Fig. 10.21b). The work dW done by the force $\vec{F}_{\text{tan}}$ while a point on the rim moves a distance ds is $dW = F_{\text{tan}} \, ds$. If $d\theta$ is measured in radians, then $ds = R \, d\theta$ and

$$dW = F_{\text{tan}} R \, d\theta$$

Now $F_{\text{tan}} R$ is the *torque* τ_z due to the force $\vec{F}_{\text{tan}}$, so

$$dW = \tau_z \, d\theta \qquad (10.19)$$

The total work W done by the torque during an angular displacement from θ_1 to θ_2 is

$$W = \int_{\theta_1}^{\theta_2} \tau_z \, d\theta \qquad \text{(work done by a torque)} \qquad (10.20)$$

If the torque remains *constant* while the angle changes by a finite amount $\Delta\theta = \theta_2 - \theta_1$, then

$$W = \tau_z(\theta_2 - \theta_1) = \tau_z \Delta\theta \qquad \text{(work done by a constant torque)} \qquad (10.21)$$

The work done by a *constant* torque is the product of torque and the angular displacement. If torque is expressed in newton-meters ($\text{N} \cdot \text{m}$) and angular displacement in radians, the work is in joules. Equation (10.21) is the rotational analog of Eq. (6.1), $W = Fs$, and Eq. (10.20) is the analog of Eq. (6.7), $W = \int F_x \, dx$, for the work done by a force in a straight-line displacement.

If the force in Fig. 10.21 had an axial component (parallel to the rotation axis) or a radial component (directed toward or away from the axis), that component would do no work because the displacement of the point of application has only a tangential component. An axial or radial component of force would also make no contribution to the torque about the axis of rotation. So Eqs. (10.20) and (10.21) are correct for *any* force, no matter what its components.

When a torque does work on a rotating rigid body, the kinetic energy changes by an amount equal to the work done. We can prove this by using exactly the same procedure that we used in Eqs. (6.11) through (6.13) for the translational kinetic energy of a particle. Let τ_z represent the *net* torque on the body so that $\tau_z = I\alpha_z$ from Eq. (10.7), and assume that the body is rigid so that the moment of inertia I is constant. We then transform the integrand in Eq. (10.20) into an integrand with respect to ω_z as follows:

$$\tau_z \, d\theta = (I\alpha_z) \, d\theta = I \frac{d\omega_z}{dt} \, d\theta = I \frac{d\theta}{dt} \, d\omega_z = I\omega_z \, d\omega_z$$

Since τ_z is the net torque, the integral in Eq. (10.20) is the *total* work done on the rotating rigid body. This equation then becomes

$$W_{\text{tot}} = \int_{\omega_1}^{\omega_2} I\omega_z \, d\omega_z = \tfrac{1}{2}I\omega_2^2 - \tfrac{1}{2}I\omega_1^2 \qquad (10.22)$$

The change in the rotational kinetic energy of a *rigid* body equals the work done by forces exerted from outside the body (Fig. 10.22). This equation is analogous to Eq. (6.13), the work–energy theorem for a particle.

What about the *power* associated with work done by a torque acting on a rotating body? When we divide both sides of Eq. (10.19) by the time interval dt during which the angular displacement occurs, we find

$$\frac{dW}{dt} = \tau_z \frac{d\theta}{dt}$$

10.21 A tangential force applied to a rotating body does work.

(a)

Child applies tangential force.

$\vec{F}_{\text{tan}}$

(b) Overhead view of merry-go-round

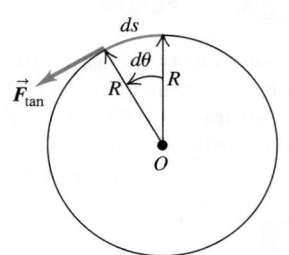

ds

$d\theta$

R

R

$\vec{F}_{\text{tan}}$

O

10.22 The rotational kinetic energy of an airplane propeller is equal to the total work done to set it spinning. When it is spinning at a constant rate, positive work is done on the propeller by the engine and negative work is done on it by air resistance. Hence the net work being done is zero and the kinetic energy remains constant.

But dW/dt is the rate of doing work, or *power P*, and $d\theta/dt$ is angular velocity ω_z, so

$$P = \tau_z\omega_z \tag{10.23}$$

When a torque τ_z (with respect to the axis of rotation) acts on a body that rotates with angular velocity ω_z, its power (rate of doing work) is the product of τ_z and ω_z. This is the analog of the relationship $P = \vec{F}\cdot\vec{v}$ that we developed in Section 6.4 for particle motion.

Example 10.8 Calculating power from torque

An electric motor exerts a constant 10-N · m torque on a grindstone, which has a moment of inertia of 2.0 kg · m² about its shaft. The system starts from rest. Find the work W done by the motor in 8.0 s and the grindstone kinetic energy K at this time. What average power P_{av} is delivered by the motor?

SOLUTION

IDENTIFY and SET UP: The only torque acting is that due to the motor. Since this torque is constant, the grindstone's angular acceleration α_z is constant. We'll use Eq. (10.7) to find α_z, and then use this in the kinematics equations from Section 9.2 to calculate the angle $\Delta\theta$ through which the grindstone rotates in 8.0 s and its final angular velocity ω_z. From these we'll calculate W, K, and P_{av}.

EXECUTE: We have $\sum\tau_z = 10$ N · m and $I = 2.0$ kg · m², so $\sum\tau_z = I\alpha_z$ yields $\alpha_z = 5.0$ rad/s². From Eq. (9.11),

$$\Delta\theta = \tfrac{1}{2}\alpha_z t^2 = \tfrac{1}{2}(5.0 \text{ rad/s}^2)(8.0 \text{ s})^2 = 160 \text{ rad}$$

$$W = \tau_z\Delta\theta = (10 \text{ N}\cdot\text{m})(160 \text{ rad}) = 1600 \text{ J}$$

From Eqs. (9.7) and (9.17),

$$\omega_z = \alpha_z t = (5.0 \text{ rad/s}^2)(8.0 \text{ s}) = 40 \text{ rad/s}$$

$$K = \tfrac{1}{2}I\omega_z^2 = \tfrac{1}{2}(2.0 \text{ kg}\cdot\text{m}^2)(40 \text{ rad/s})^2 = 1600 \text{ J}$$

The average power is the work done divided by the time interval:

$$P_{av} = \frac{1600 \text{ J}}{8.0 \text{ s}} = 200 \text{ J/s} = 200 \text{ W}$$

EVALUATE: The initial kinetic energy was zero, so the work done W must equal the final kinetic energy K [Eq. (10.22)]. This is just as we calculated. We can check our result $P_{av} = 200$ W by considering the *instantaneous* power $P = \tau_z\omega_z$. Because ω_z increases continuously, P increases continuously as well; its value increases from zero at $t = 0$ to $(10$ N · m$)(40$ rad/s$) = 400$ W at $t = 8.0$ s. Both ω_z and P increase *uniformly* with time, so the *average* power is just half this maximum value, or 200 W.

Test Your Understanding of Section 10.4 You apply equal torques to two different cylinders, one of which has a moment of inertia twice as large as the other cylinder. Each cylinder is initially at rest. After one complete rotation, which cylinder has the greater kinetic energy? (i) the cylinder with the larger moment of inertia; (ii) the cylinder with the smaller moment of inertia; (iii) both cylinders have the same kinetic energy.

10.5 Angular Momentum

Every rotational quantity that we have encountered in Chapters 9 and 10 is the analog of some quantity in the translational motion of a particle. The analog of *momentum* of a particle is **angular momentum,** a vector quantity denoted as $\vec{L}$. Its relationship to momentum $\vec{p}$ (which we will often call *linear momentum* for clarity) is exactly the same as the relationship of torque to force, $\vec{\tau} = \vec{r} \times \vec{F}$. For a particle with constant mass m, velocity $\vec{v}$, momentum $\vec{p} = m\vec{v}$, and position vector $\vec{r}$ relative to the origin O of an inertial frame, we define angular momentum $\vec{L}$ as

$$\vec{L} = \vec{r} \times \vec{p} = \vec{r} \times m\vec{v} \qquad \text{(angular momentum of a particle)} \tag{10.24}$$

The value of $\vec{L}$ depends on the choice of origin O, since it involves the particle's position vector relative to O. The units of angular momentum are $kg \cdot m^2/s$.

In Fig. 10.23 a particle moves in the xy-plane; its position vector $\vec{r}$ and momentum $\vec{p} = m\vec{v}$ are shown. The angular momentum vector $\vec{L}$ is perpendicular to the xy-plane. The right-hand rule for vector products shows that its direction is along the $+z$-axis, and its magnitude is

$$L = mvr \sin\phi = mvl \qquad (10.25)$$

where l is the perpendicular distance from the line of $\vec{v}$ to O. This distance plays the role of "lever arm" for the momentum vector.

When a net force $\vec{F}$ acts on a particle, its velocity and momentum change, so its angular momentum may also change. We can show that the *rate of change* of angular momentum is equal to the torque of the net force. We take the time derivative of Eq. (10.24), using the rule for the derivative of a product:

$$\frac{d\vec{L}}{dt} = \left(\frac{d\vec{r}}{dt} \times m\vec{v}\right) + \left(\vec{r} \times m\frac{d\vec{v}}{dt}\right) = (\vec{v} \times m\vec{v}) + (\vec{r} \times m\vec{a})$$

The first term is zero because it contains the vector product of the vector $\vec{v} = d\vec{r}/dt$ with itself. In the second term we replace $m\vec{a}$ with the net force $\vec{F}$, obtaining

$$\frac{d\vec{L}}{dt} = \vec{r} \times \vec{F} = \vec{\tau} \qquad \text{(for a particle acted on by net force } \vec{F}\text{)} \quad (10.26)$$

The rate of change of angular momentum of a particle equals the torque of the net force acting on it. Compare this result to Eq. (8.4), which states that the rate of change $d\vec{p}/dt$ of the *linear* momentum of a particle equals the net force that acts on it.

Angular Momentum of a Rigid Body

We can use Eq. (10.25) to find the total angular momentum of a *rigid body* rotating about the z-axis with angular speed ω. First consider a thin slice of the body lying in the xy-plane (Fig. 10.24). Each particle in the slice moves in a circle centered at the origin, and at each instant its velocity $\vec{v}_i$ is perpendicular to its position vector $\vec{r}_i$, as shown. Hence in Eq. (10.25), $\phi = 90°$ for every particle. A particle with mass m_i at a distance r_i from O has a speed v_i equal to $r_i\omega$. From Eq. (10.25) the magnitude L_i of its angular momentum is

$$L_i = m_i(r_i\omega)r_i = m_i r_i^2 \omega \qquad (10.27)$$

The direction of each particle's angular momentum, as given by the right-hand rule for the vector product, is along the $+z$-axis.

The *total* angular momentum of the slice of the body lying in the xy-plane is the sum $\sum L_i$ of the angular momenta L_i of the particles. Summing Eq. (10.27), we have

$$L = \sum L_i = \left(\sum m_i r_i^2\right)\omega = I\omega$$

where I is the moment of inertia of the slice about the z-axis.

We can do this same calculation for the other slices of the body, all parallel to the xy-plane. For points that do not lie in the xy-plane, a complication arises because the $\vec{r}$ vectors have components in the z-direction as well as the x- and y-directions; this gives the angular momentum of each particle a component perpendicular to the z-axis. But if the z-axis is an axis of symmetry, the perpendicular components for particles on opposite sides of this axis add up to zero (Fig. 10.25). So when a body rotates about an axis of symmetry, its angular momentum vector $\vec{L}$ lies along the symmetry axis, and its magnitude is $L = I\omega$.

10.23 Calculating the angular momentum $\vec{L} = \vec{r} \times m\vec{v} = \vec{r} \times \vec{p}$ of a particle with mass m moving in the xy-plane.

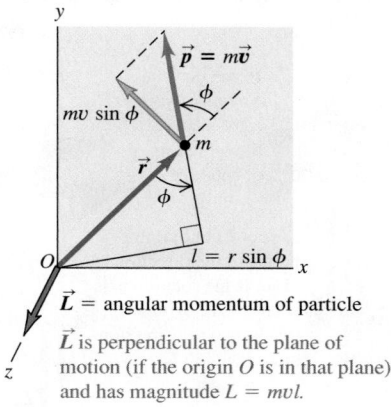

$\vec{L}$ = angular momentum of particle

$\vec{L}$ is perpendicular to the plane of motion (if the origin O is in that plane) and has magnitude $L = mvl$.

10.24 Calculating the angular momentum of a particle of mass m_i in a rigid body rotating at angular speed ω. (Compare Fig. 10.23.)

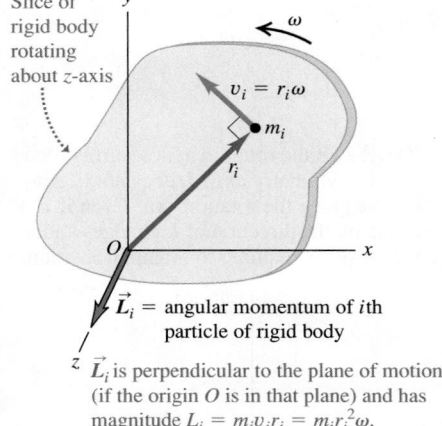

$\vec{L}_i$ = angular momentum of ith particle of rigid body

$\vec{L}_i$ is perpendicular to the plane of motion (if the origin O is in that plane) and has magnitude $L_i = m_i v_i r_i = m_i r_i^2 \omega$.

10.25 Two particles of the same mass located symmetrically on either side of the rotation axis of a rigid body. The angular momentum vectors $\vec{L}_1$ and $\vec{L}_2$ of the two particles do not lie along the rotation axis, but their vector sum $\vec{L}_1 + \vec{L}_2$ does.

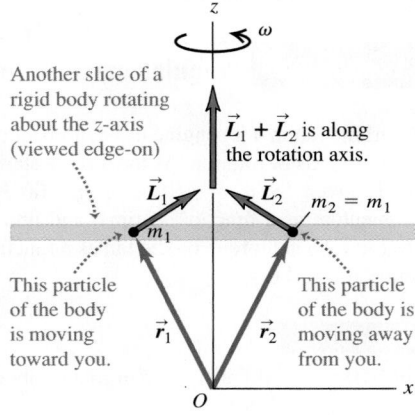

Another slice of a rigid body rotating about the z-axis (viewed edge-on)

$\vec{L}_1 + \vec{L}_2$ is along the rotation axis.

This particle of the body is moving toward you.

This particle of the body is moving away from you.

10.26 For rotation about an axis of symmetry, $\vec{\omega}$ and $\vec{L}$ are parallel and along the axis. The directions of both vectors are given by the right-hand rule (compare Fig. 9.5).

If you curl the fingers of your right hand in the direction of rotation ...

$\vec{\omega}$ $\vec{L}$

... your right thumb points in the direction of $\vec{\omega}$. If the rotation axis is an axis of symmetry, this is also the direction of $\vec{L}$.

$\vec{\omega}$

$\vec{L}$

The angular velocity vector $\vec{\omega}$ also lies along the rotation axis, as we discussed at the end of Section 9.1. Hence for a rigid body rotating around an axis of symmetry, $\vec{L}$ and $\vec{\omega}$ are in the same direction (Fig. 10.26). So we have the *vector* relationship

$$\vec{L} = I\vec{\omega} \quad \text{(for a rigid body rotating around a symmetry axis)} \quad (10.28)$$

From Eq. (10.26) the rate of change of angular momentum of a particle equals the torque of the net force acting on the particle. For any system of particles (including both rigid and nonrigid bodies), the rate of change of the *total* angular momentum equals the sum of the torques of all forces acting on all the particles. The torques of the *internal* forces add to zero if these forces act along the line from one particle to another, as in Fig. 10.8, and so the sum of the torques includes only the torques of the *external* forces. (A similar cancellation occurred in our discussion of center-of-mass motion in Section 8.5.) If the total angular momentum of the system of particles is $\vec{L}$ and the sum of the external torques is $\sum \vec{\tau}$, then

$$\sum \vec{\tau} = \frac{d\vec{L}}{dt} \quad \text{(for any system of particles)} \quad (10.29)$$

Finally, if the system of particles is a rigid body rotating about a symmetry axis (the z-axis), then $L_z = I\omega_z$ and I is constant. If this axis has a fixed direction in space, then the vectors $\vec{L}$ and $\vec{\omega}$ change only in magnitude, not in direction. In that case, $dL_z/dt = I \, d\omega_z/dt = I\alpha_z$, or

$$\sum \tau_z = I\alpha_z$$

which is again our basic relationship for the dynamics of rigid-body rotation. If the body is *not* rigid, I may change, and in that case, L changes even when ω is constant. For a nonrigid body, Eq. (10.29) is still valid, even though Eq. (10.7) is not.

When the axis of rotation is *not* a symmetry axis, the angular momentum is in general *not* parallel to the axis (Fig. 10.27). As the body turns, the angular momentum vector $\vec{L}$ traces out a cone around the rotation axis. Because $\vec{L}$ changes, there must be a net external torque acting on the body even though the angular velocity magnitude ω may be constant. If the body is an unbalanced wheel on a car, this torque is provided by friction in the bearings, which causes the bearings to wear out. "Balancing" a wheel means distributing the mass so that the rotation axis is an axis of symmetry; then $\vec{L}$ points along the rotation axis, and no net torque is required to keep the wheel turning.

In fixed-axis rotation we often use the term "angular momentum of the body" to refer to only the *component* of $\vec{L}$ along the rotation axis of the body (the z-axis in Fig. 10.27), with a positive or negative sign to indicate the sense of rotation just as with angular velocity.

10.27 If the rotation axis of a rigid body is not a symmetry axis, $\vec{L}$ does not in general lie along the rotation axis. Even if $\vec{\omega}$ is constant, the direction of $\vec{L}$ changes and a net torque is required to maintain rotation.

Irregularly shaped body

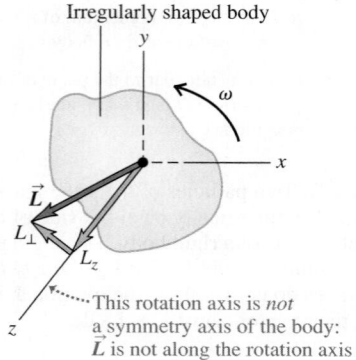

y

ω

x

$\vec{L}$

$L_\perp$

L_z

z

This rotation axis is *not* a symmetry axis of the body: $\vec{L}$ is not along the rotation axis

---

Example 10.9 **Angular momentum and torque**

A turbine fan in a jet engine has a moment of inertia of $2.5 \text{ kg} \cdot \text{m}^2$ about its axis of rotation. As the turbine starts up, its angular velocity is given by $\omega_z = (40 \text{ rad/s}^3)t^2$. (a) Find the fan's angular momentum as a function of time, and find its value at $t = 3.0$ s. (b) Find the net torque on the fan as a function of time, and find its value at $t = 3.0$ s.

SOLUTION

IDENTIFY and SET UP: The fan rotates about its axis of symmetry (the z-axis). Hence the angular momentum vector has only a

z-component L_z, which we can determine from the angular velocity ω_z. Since the direction of angular momentum is constant, the net torque likewise has only a component τ_z along the rotation axis. We'll use Eq. (10.28) to find L_z from ω_z and then use Eq. (10.29) to find τ_z.

EXECUTE: (a) From Eq. (10.28),

$$L_z = I\omega_z = (2.5 \text{ kg} \cdot \text{m}^2)(40 \text{ rad/s}^3)t^2 = (100 \text{ kg} \cdot \text{m}^2/\text{s}^3)t^2$$

(We dropped the dimensionless quantity "rad" from the final expression.) At $t = 3.0$ s, $L_z = 900 \text{ kg} \cdot \text{m}^2/\text{s}$.

(b) From Eq. (10.29),

$$\tau_z = \frac{dL_z}{dt} = (100 \text{ kg} \cdot \text{m}^2/\text{s}^3)(2t) = (200 \text{ kg} \cdot \text{m}^2/\text{s}^3)t$$

At $t = 3.0$ s,

$$\tau_z = (200 \text{ kg} \cdot \text{m}^2/\text{s}^3)(3.0 \text{ s}) = 600 \text{ kg} \cdot \text{m}^2/\text{s}^2 = 600 \text{ N} \cdot \text{m}$$

EVALUATE: As a check on our expression for τ_z, note that the angular acceleration of the turbine is $\alpha_z = d\omega_z/dt = (40 \text{ rad/s}^3)(2t) = (80 \text{ rad/s}^3)t$. Hence from Eq. (10.7), the torque on the fan is $\tau_z = I\alpha_z = (2.5 \text{ kg} \cdot \text{m}^2)(80 \text{ rad/s}^3)t = (200 \text{ kg} \cdot \text{m}^2/\text{s}^3)t$, just as we calculated.

Test Your Understanding of Section 10.5   A ball is attached to one end of a piece of string. You hold the other end of the string and whirl the ball in a circle around your hand. (a) If the ball moves at a constant speed, is its linear momentum $\vec{p}$ constant? Why or why not? (b) Is its angular momentum $\vec{L}$ constant? Why or why not?   ❙

10.28 A falling cat twists different parts of its body in different directions so that it lands feet first. At all times during this process the angular momentum of the cat as a whole remains zero.

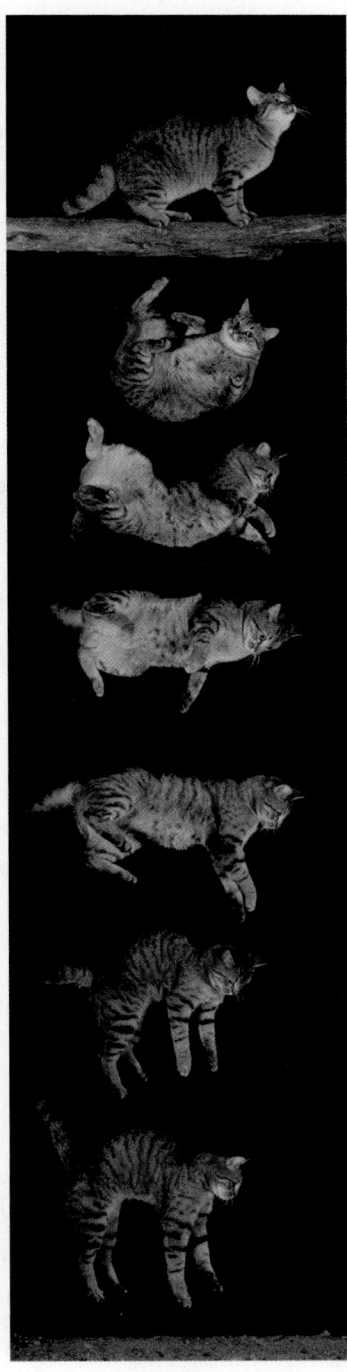

10.6 Conservation of Angular Momentum

We have just seen that angular momentum can be used for an alternative statement of the basic dynamic principle for rotational motion. It also forms the basis for the **principle of conservation of angular momentum.** Like conservation of energy and of linear momentum, this principle is a universal conservation law, valid at all scales from atomic and nuclear systems to the motions of galaxies. This principle follows directly from Eq. (10.29): $\Sigma\vec{\tau} = d\vec{L}/dt$. If $\Sigma\vec{\tau} = 0$, then $d\vec{L}/dt = 0$, and $\vec{L}$ is constant.

> **When the net external torque acting on a system is zero, the total angular momentum of the system is constant (conserved).**

A circus acrobat, a diver, and an ice skater pirouetting on the toe of one skate all take advantage of this principle. Suppose an acrobat has just left a swing with arms and legs extended and rotating counterclockwise about her center of mass. When she pulls her arms and legs in, her moment of inertia I_{cm} with respect to her center of mass changes from a large value I_1 to a much smaller value I_2. The only external force acting on her is her weight, which has no torque with respect to an axis through her center of mass. So her angular momentum $L_z = I_{cm}\omega_z$ remains constant, and her angular velocity ω_z increases as I_{cm} decreases. That is,

$$I_1\omega_{1z} = I_2\omega_{2z} \qquad \text{(zero net external torque)} \qquad (10.30)$$

When a skater or ballerina spins with arms outstretched and then pulls her arms in, her angular velocity increases as her moment of inertia decreases. In each case there is conservation of angular momentum in a system in which the net external torque is zero.

When a system has several parts, the internal forces that the parts exert on one another cause changes in the angular momenta of the parts, but the *total* angular momentum doesn't change. Here's an example. Consider two bodies A and B that interact with each other but not with anything else, such as the astronauts we discussed in Section 8.2 (see Fig. 8.8). Suppose body A exerts a force $\vec{F}_{A \text{ on } B}$ on body B; the corresponding torque (with respect to whatever point we choose) is $\vec{\tau}_{A \text{ on } B}$. According to Eq. (10.29), this torque is equal to the rate of change of angular momentum of B:

$$\vec{\tau}_{A \text{ on } B} = \frac{d\vec{L}_B}{dt}$$

At the same time, body B exerts a force $\vec{F}_{B \text{ on } A}$ on body A, with a corresponding torque $\vec{\tau}_{B \text{ on } A}$, and

$$\vec{\tau}_{B \text{ on } A} = \frac{d\vec{L}_A}{dt}$$

From Newton's third law, $\vec{F}_{B\,on\,A} = -\vec{F}_{A\,on\,B}$. Furthermore, if the forces act along the same line, as in Fig. 10.8, their lever arms with respect to the chosen axis are equal. Thus the *torques* of these two forces are equal and opposite, and $\vec{\tau}_{B\,on\,A} = -\vec{\tau}_{A\,on\,B}$. So if we add the two preceding equations, we find

$$\frac{d\vec{L}_A}{dt} + \frac{d\vec{L}_B}{dt} = 0$$

or, because $\vec{L}_A + \vec{L}_B$ is the *total* angular momentum $\vec{L}$ of the system,

$$\frac{d\vec{L}}{dt} = 0 \qquad \text{(zero net external torque)} \qquad (10.31)$$

That is, the total angular momentum of the system is constant. The torques of the internal forces can transfer angular momentum from one body to the other, but they can't change the *total* angular momentum of the system (Fig. 10.28).

PhET: Torque
ActivPhysics 7.14: Ball Hits Bat

Example 10.10 | Anyone can be a ballerina

A physics professor stands at the center of a frictionless turntable with arms outstretched and a 5.0-kg dumbbell in each hand (Fig. 10.29). He is set rotating about the vertical axis, making one revolution in 2.0 s. Find his final angular velocity if he pulls the dumbbells in to his stomach. His moment of inertia (without the dumbbells) is 3.0 kg·m² with arms outstretched and 2.2 kg·m² with his hands at his stomach. The dumbbells are 1.0 m from the axis initially and 0.20 m at the end.

SOLUTION

IDENTIFY, SET UP, and EXECUTE: No external torques act about the z-axis, so L_z is constant. We'll use Eq. (10.30) to find the final

10.29 Fun with conservation of angular momentum.

Dumbbell   Dumbbell

Professor (not a dumbbell)

ω_1 ω_2

BEFORE   AFTER

angular velocity ω_{2z}. The moment of inertia of the system is $I = I_{prof} + I_{dumbbells}$. We treat each dumbbell as a particle of mass m that contributes mr^2 to $I_{dumbbells}$, where r is the perpendicular distance from the axis to the dumbbell. Initially we have

$$I_1 = 3.0 \text{ kg·m}^2 + 2(5.0 \text{ kg})(1.0 \text{ m})^2 = 13 \text{ kg·m}^2$$

$$\omega_{1z} = \frac{1 \text{ rev}}{2.0 \text{ s}} = 0.50 \text{ rev/s}$$

The final moment of inertia is

$$I_2 = 2.2 \text{ kg·m}^2 + 2(5.0 \text{ kg})(0.20 \text{ m})^2 = 2.6 \text{ kg·m}^2$$

From Eq. (10.30), the final angular velocity is

$$\omega_{2z} = \frac{I_1}{I_2}\omega_{1z} = \frac{13 \text{ kg·m}^2}{2.6 \text{ kg·m}^2}(0.50 \text{ rev/s}) = 2.5 \text{ rev/s} = 5\omega_{1z}$$

Can you see why we didn't have to change "revolutions" to "radians" in this calculation?

EVALUATE: The angular momentum remained constant, but the angular velocity increased by a factor of 5, from $\omega_{1z} = (0.50 \text{ rev/s})(2\pi \text{ rad/rev}) = 3.14 \text{ rad/s}$ to $\omega_{2z} = (2.5 \text{ rev/s})(2\pi \text{ rad/rev}) = 15.7 \text{ rad/s}$. The initial and final kinetic energies are then

$$K_1 = \tfrac{1}{2}I_1\omega_{1z}^2 = \tfrac{1}{2}(13 \text{ kg·m}^2)(3.14 \text{ rad/s})^2 = 64 \text{ J}$$

$$K_2 = \tfrac{1}{2}I_2\omega_{2z}^2 = \tfrac{1}{2}(2.6 \text{ kg·m}^2)(15.7 \text{ rad/s})^2 = 320 \text{ J}$$

The fivefold increase in kinetic energy came from the work that the professor did in pulling his arms and the dumbbells inward.

Example 10.11 | A rotational "collision"

Figure 10.30 shows two disks: an engine flywheel (A) and a clutch plate (B) attached to a transmission shaft. Their moments of inertia are I_A and I_B; initially, they are rotating with constant angular speeds ω_A and ω_B, respectively. We push the disks together with forces acting along the axis, so as not to apply any torque on either disk. The disks rub against each other and eventually reach a common angular speed ω. Derive an expression for ω.

SOLUTION

IDENTIFY, SET UP, and EXECUTE: There are no external torques, so the only torque acting on either disk is the torque applied by the other disk. Hence the total angular momentum of the system of two disks is conserved. At the end they rotate together as one body with total moment of inertia $I = I_A + I_B$ and angular speed ω.

10.30 When the net external torque is zero, angular momentum is conserved.

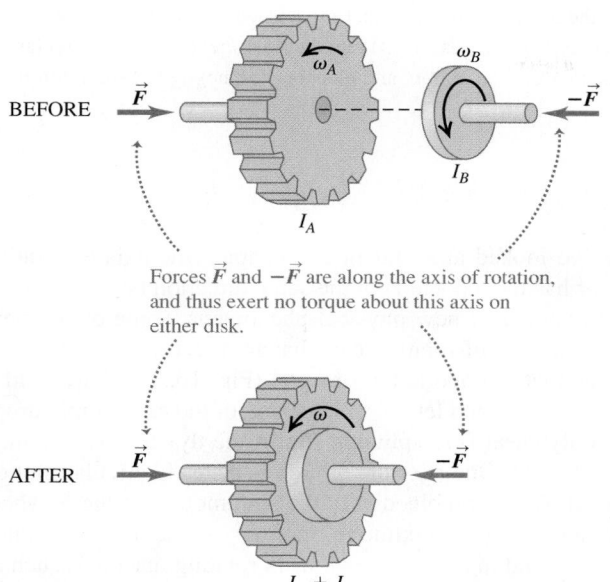

BEFORE

Forces $\vec{F}$ and $-\vec{F}$ are along the axis of rotation, and thus exert no torque about this axis on either disk.

AFTER

Figure 10.30 shows that all angular velocities are in the same direction, so we can regard ω_A, ω_B, and ω as components of angular velocity along the rotation axis. Conservation of angular momentum gives

$$I_A\omega_A + I_B\omega_B = (I_A + I_B)\omega$$
$$\omega = \frac{I_A\omega_A + I_B\omega_B}{I_A + I_B}$$

EVALUATE: This "collision" is analogous to a completely inelastic collision (see Section 8.3). When two objects in translational motion along the same axis collide and stick, the linear momentum of the system is conserved. Here two objects in *rotational* motion around the same axis "collide" and stick, and the *angular* momentum of the system is conserved.

The kinetic energy of a system decreases in a completely inelastic collision. Here kinetic energy is lost because nonconservative (frictional) internal forces act while the two disks rub together. Suppose flywheel A has a mass of 2.0 kg, a radius of 0.20 m, and an initial angular speed of 50 rad/s (about 500 rpm), and clutch plate B has a mass of 4.0 kg, a radius of 0.10 m, and an initial angular speed of 200 rad/s. Can you show that the final kinetic energy is only two-thirds of the initial kinetic energy?

Example 10.12 | **Angular momentum in a crime bust**

A door 1.00 m wide, of mass 15 kg, can rotate freely about a vertical axis through its hinges. A bullet with a mass of 10 g and a speed of 400 m/s strikes the center of the door, in a direction perpendicular to the plane of the door, and embeds itself there. Find the door's angular speed. Is kinetic energy conserved?

SOLUTION

IDENTIFY and SET UP: We consider the door and bullet as a system. There is no external torque about the hinge axis, so angular momentum about this axis is conserved. Figure 10.31 shows our sketch. The initial angular momentum is that of the bullet, as given by Eq. (10.25). The final angular momentum is that of a rigid body

composed of the door and the embedded bullet. We'll equate these quantities and solve for the resulting angular speed ω of the door and bullet.

EXECUTE: From Eq. (10.25), the initial angular momentum of the bullet is

$$L = mvl = (0.010 \text{ kg})(400 \text{ m/s})(0.50 \text{ m}) = 2.0 \text{ kg} \cdot \text{m}^2/\text{s}$$

The final angular momentum is $I\omega$, where $I = I_{\text{door}} + I_{\text{bullet}}$. From Table 9.2, case (d), for a door of width $d = 1.00$ m,

$$I_{\text{door}} = \frac{Md^2}{3} = \frac{(15 \text{ kg})(1.00 \text{ m})^2}{3} = 5.0 \text{ kg} \cdot \text{m}^2$$

The moment of inertia of the bullet (with respect to the axis along the hinges) is

$$I_{\text{bullet}} = ml^2 = (0.010 \text{ kg})(0.50 \text{ m})^2 = 0.0025 \text{ kg} \cdot \text{m}^2$$

Conservation of angular momentum requires that $mvl = I\omega$, or

$$\omega = \frac{mvl}{I} = \frac{2.0 \text{ kg} \cdot \text{m}^2/\text{s}}{5.0 \text{ kg} \cdot \text{m}^2 + 0.0025 \text{ kg} \cdot \text{m}^2} = 0.40 \text{ rad/s}$$

The initial and final kinetic energies are

$$K_1 = \tfrac{1}{2}mv^2 = \tfrac{1}{2}(0.010 \text{ kg})(400 \text{ m/s})^2 = 800 \text{ J}$$

$$K_2 = \tfrac{1}{2}I\omega^2 = \tfrac{1}{2}(5.0025 \text{ kg} \cdot \text{m}^2)(0.40 \text{ rad/s})^2 = 0.40 \text{ J}$$

EVALUATE: The final kinetic energy is only $\frac{1}{2000}$ of the initial value! We did not expect kinetic energy to be conserved: The collision is inelastic because nonconservative friction forces act during the impact. The door's final angular speed is quite slow: At 0.40 rad/s, it takes 3.9 s to swing through 90° ($\pi/2$ radians).

10.31 Our sketch for this problem.

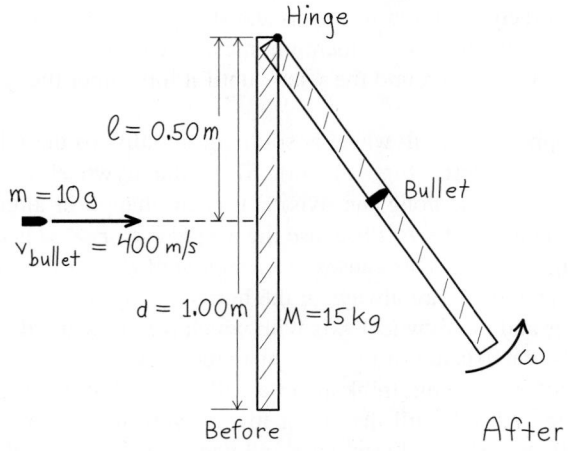

10.32 A gyroscope supported at one end. The horizontal circular motion of the flywheel and axis is called precession. The angular speed of precession is Ω.

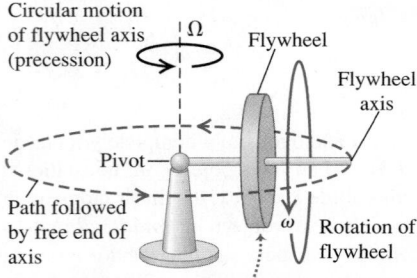

Circular motion of flywheel axis (precession)

Ω

Flywheel

Flywheel axis

Pivot

Path followed by free end of axis

ω Rotation of flywheel

When the flywheel and its axis are stationary, they will fall to the table surface. When the flywheel spins, it and its axis "float" in the air while moving in a circle about the pivot.

10.33 (a) If the flywheel in Fig. 10.32 is initially not spinning, its initial angular momentum is zero (b) In each successive time interval dt, the torque produces a change $d\vec{L} = \vec{\tau}\, dt$ in the angular momentum. The flywheel acquires an angular momentum $\vec{L}$ in the same direction as $\vec{\tau}$, and the flywheel axis falls.

(a) Nonrotating flywheel falls

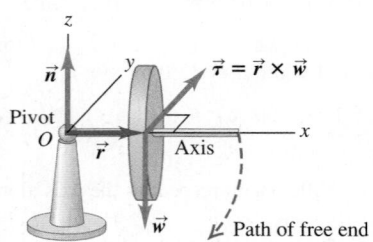

$\vec{n}$

Pivot O

z

y

$\vec{r}$

Axis

$\vec{\tau} = \vec{r} \times \vec{w}$

x

$\vec{w}$

Path of free end

When the flywheel is not rotating, its weight creates a torque around the pivot, causing it to fall along a circular path until its axis rests on the table surface.

(b) View from above as flywheel falls

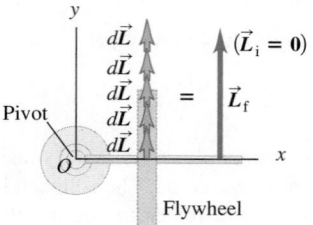

y

$d\vec{L}$
$d\vec{L}$
$d\vec{L}$
$d\vec{L}$
$d\vec{L}$

Pivot

O

Flywheel

$=$

$(\vec{L}_i = 0)$

$\vec{L}_f$

x

In falling, the flywheel rotates about the pivot and thus acquires an angular momentum $\vec{L}$. The *direction* of $\vec{L}$ stays constant.

Test Your Understanding of Section 10.6 If the polar ice caps were to completely melt due to global warming, the melted ice would redistribute itself over the earth. This change would cause the length of the day (the time needed for the earth to rotate once on its axis) to (i) increase; (ii) decrease; (iii) remain the same. (*Hint:* Use angular momentum ideas. Assume that the sun, moon, and planets exert negligibly small torques on the earth.)

10.7 Gyroscopes and Precession

In all the situations we've looked at so far in this chapter, the axis of rotation either has stayed fixed or has moved and kept the same direction (such as rolling without slipping). But a variety of new physical phenomena, some quite unexpected, can occur when the axis of rotation can change direction. For example, consider a toy gyroscope that's supported at one end (Fig. 10.32). If we hold it with the flywheel axis horizontal and let go, the free end of the axis simply drops owing to gravity—*if* the flywheel isn't spinning. But if the flywheel *is* spinning, what happens is quite different. One possible motion is a steady circular motion of the axis in a horizontal plane, combined with the spin motion of the flywheel about the axis. This surprising, nonintuitive motion of the axis is called **precession.** Precession is found in nature as well as in rotating machines such as gyroscopes. As you read these words, the earth itself is precessing; its spin axis (through the north and south poles) slowly changes direction, going through a complete cycle of precession every 26,000 years.

To study this strange phenomenon of precession, we must remember that angular velocity, angular momentum, and torque are all *vector* quantities. In particular, we need the general relationship between the net torque $\sum \vec{\tau}$ that acts on a body and the rate of change of the body's angular momentum $\vec{L}$, given by Eq. (10.29), $\sum \vec{\tau} = d\vec{L}/dt$. Let's first apply this equation to the case in which the flywheel is *not* spinning (Fig. 10.33a). We take the origin O at the pivot and assume that the flywheel is symmetrical, with mass M and moment of inertia I about the flywheel axis. The flywheel axis is initially along the x-axis. The only external forces on the gyroscope are the normal force $\vec{n}$ acting at the pivot (assumed to be frictionless) and the weight $\vec{w}$ of the flywheel that acts at its center of mass, a distance r from the pivot. The normal force has zero torque with respect to the pivot, and the weight has a torque $\vec{\tau}$ in the y-direction, as shown in Fig. 10.33a. Initially, there is no rotation, and the initial angular momentum $\vec{L}_i$ is zero. From Eq. (10.29) the *change* $d\vec{L}$ in angular momentum in a short time interval dt following this is

$$d\vec{L} = \vec{\tau}\, dt \qquad (10.32)$$

This change is in the y-direction because $\vec{\tau}$ is. As each additional time interval dt elapses, the angular momentum changes by additional increments $d\vec{L}$ in the y-direction because the direction of the torque is constant (Fig. 10.33b). The steadily increasing horizontal angular momentum means that the gyroscope rotates downward faster and faster around the y-axis until it hits either the stand or the table on which it sits.

Now let's see what happens if the flywheel *is* spinning initially, so the initial angular momentum $\vec{L}_i$ is not zero (Fig. 10.34a). Since the flywheel rotates around its symmetry axis, $\vec{L}_i$ lies along the axis. But each change in angular momentum $d\vec{L}$ is perpendicular to the axis because the torque $\vec{\tau} = \vec{r} \times \vec{w}$ is perpendicular to the axis (Fig. 10.34b). This causes the *direction* of $\vec{L}$ to change, but not its magnitude. The changes $d\vec{L}$ are always in the horizontal xy-plane, so the angular momentum vector and the flywheel axis with which it moves are always horizontal. In other words, the axis doesn't fall—it just precesses.

If this still seems mystifying to you, think about a ball attached to a string. If the ball is initially at rest and you pull the string toward you, the ball moves toward you also. But if the ball is initially moving and you continuously pull the

(a) Rotating flywheel

When the flywheel is rotating, the system starts with an angular momentum $\vec{L}_i$ parallel to the flywheel's axis of rotation.

(b) View from above

Now the effect of the torque is to cause the angular momentum to precess around the pivot. The gyroscope circles around its pivot without falling.

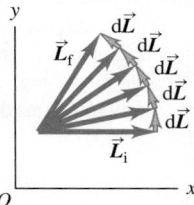

10.34 (a) The flywheel is spinning initially with angular momentum $\vec{L}_i$. The forces (not shown) are the same as those in Fig. 10.33a. (b) Because the initial angular momentum is not zero, each change $d\vec{L} = \vec{\tau}\,dt$ in angular momentum is perpendicular to $\vec{L}$. As a result, the magnitude of $\vec{L}$ remains the same but its direction changes continuously.

string in a direction perpendicular to the ball's motion, the ball moves in a circle around your hand; it does not approach your hand at all. In the first case the ball has zero linear momentum $\vec{p}$ to start with; when you apply a force $\vec{F}$ toward you for a time dt, the ball acquires a momentum $d\vec{p} = \vec{F}\,dt$, which is also toward you. But if the ball already has linear momentum $\vec{p}$, a change in momentum $d\vec{p}$ that's perpendicular to $\vec{p}$ changes the direction of motion, not the speed. Replace $\vec{p}$ with $\vec{L}$ and $\vec{F}$ with $\vec{\tau}$ in this argument, and you'll see that precession is simply the rotational analog of uniform circular motion.

At the instant shown in Fig. 10.34a, the gyroscope has angular momentum $\vec{L}$. A short time interval dt later, the angular momentum is $\vec{L} + d\vec{L}$; the infinitesimal change in angular momentum is $d\vec{L} = \vec{\tau}\,dt$, which is perpendicular to $\vec{L}$. As the vector diagram in Fig. 10.35 shows, this means that the flywheel axis of the gyroscope has turned through a small angle $d\phi$ given by $d\phi = |d\vec{L}|/|\vec{L}|$. The rate at which the axis moves, $d\phi/dt$, is called the **precession angular speed;** denoting this quantity by Ω, we find

10.35 Detailed view of part of Fig. 10.34b.

In a time dt, the angular momentum vector and the flywheel axis (to which it is parallel) precess together through an angle $d\phi$.

$$\Omega = \frac{d\phi}{dt} = \frac{|d\vec{L}|/|\vec{L}|}{dt} = \frac{\tau_z}{L_z} = \frac{wr}{I\omega} \qquad (10.33)$$

Thus the precession angular speed is *inversely* proportional to the angular speed of spin about the axis. A rapidly spinning gyroscope precesses slowly; if friction in its bearings causes the flywheel to slow down, the precession angular speed *increases!* The precession angular speed of the earth is very slow (1 rev/26,000 yr) because its spin angular momentum L_z is large and the torque τ_z, due to the gravitational influences of the moon and sun, is relatively small.

As a gyroscope precesses, its center of mass moves in a circle with radius r in a horizontal plane. Its vertical component of acceleration is zero, so the upward normal force $\vec{n}$ exerted by the pivot must be just equal in magnitude to the weight. The circular motion of the center of mass with angular speed Ω requires a force $\vec{F}$ directed toward the center of the circle, with magnitude $F = M\Omega^2 r$. This force must also be supplied by the pivot.

One key assumption that we made in our analysis of the gyroscope was that the angular momentum vector $\vec{L}$ is associated only with the spin of the flywheel and is purely horizontal. But there will also be a vertical component of angular momentum associated with the precessional motion of the gyroscope. By ignoring this, we've tacitly assumed that the precession is *slow*—that is, that the precession angular speed Ω is very much less than the spin angular speed ω. As Eq. (10.33) shows, a large value of ω automatically gives a small value of Ω, so this approximation is reasonable. When the precession is not slow, additional effects show up, including an up-and-down wobble or *nutation* of the flywheel axis that's superimposed on the precessional motion. You can see nutation occurring in a gyroscope as its spin slows down, so that Ω increases and the vertical component of $\vec{L}$ can no longer be ignored.

Example 10.13 A precessing gyroscope

Figure 10.36a shows a top view of a spinning, cylindrical gyroscope wheel. The pivot is at O, and the mass of the axle is negligible. (a) As seen from above, is the precession clockwise or counterclockwise? (b) If the gyroscope takes 4.0 s for one revolution of precession, what is the angular speed of the wheel?

SOLUTION

IDENTIFY and SET UP: We'll determine the direction of precession using the right-hand rule as in Fig. 10.34, which shows the same kind of gyroscope as Fig. 10.36. We'll use the relationship between precession angular speed Ω and spin angular speed ω, Eq. (10.33), to find ω.

EXECUTE: (a) The right-hand rule shows that $\vec{\omega}$ and $\vec{L}$ are to the left in Fig. 10.36b. The weight $\vec{w}$ points into the page in this top view and acts at the center of mass (denoted by ✖ in the figure). The torque $\vec{\tau} = \vec{r} \times \vec{w}$ is toward the top of the page, so $d\vec{L}/dt$ is

also toward the top of the page. Adding a small $d\vec{L}$ to the initial vector $\vec{L}$ changes the direction of $\vec{L}$ as shown, so the precession is clockwise as seen from above.

(b) Be careful not to confuse ω and Ω! The precession angular speed is $\Omega = (1 \text{ rev})/(4.0 \text{ s}) = (2\pi \text{ rad})/(4.0 \text{ s}) = 1.57 \text{ rad/s}$. The weight is mg, and if the wheel is a solid, uniform cylinder, its moment of inertia about its symmetry axis is $I = \frac{1}{2}mR^2$. From Eq. (10.33),

$$\omega = \frac{wr}{I\Omega} = \frac{mgr}{(mR^2/2)\Omega} = \frac{2gr}{R^2\Omega}$$

$$= \frac{2(9.8 \text{ m/s}^2)(2.0 \times 10^{-2} \text{ m})}{(3.0 \times 10^{-2} \text{ m})^2(1.57 \text{ rad/s})} = 280 \text{ rad/s} = 2600 \text{ rev/min}$$

EVALUATE: The precession angular speed Ω is only about 0.6% of the spin angular speed ω, so this is an example of slow precession.

10.36 In which direction and at what speed does this gyroscope precess?

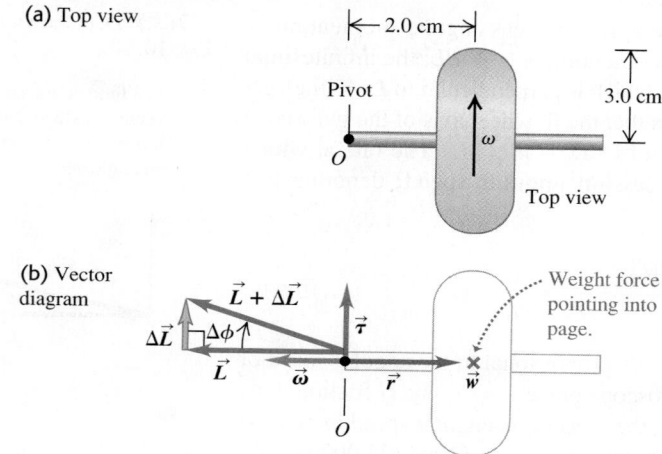

(a) Top view

Pivot

O

2.0 cm

ω

3.0 cm

Top view

(b) Vector diagram

$\vec{L} + \Delta\vec{L}$

$\Delta\vec{L}$

$\Delta\phi$

$\vec{L}$

$\vec{\omega}$

$\vec{\tau}$

$\vec{r}$

✖ $\vec{w}$

Weight force pointing into page.

O

Test Your Understanding of Section 10.7 Suppose the mass of the flywheel in Fig. 10.34 were doubled but all other dimensions and the spin angular speed remained the same. What effect would this change have on the precession angular speed Ω? (i) Ω would increase by a factor of 4; (ii) Ω would double; (iii) Ω would be unaffected; (iv) Ω would be one-half as much; (v) Ω would be one-quarter as much. ❚

(MP)

Torque: When a force $\vec{F}$ acts on a body, the torque of that force with respect to a point O has a magnitude given by the product of the force magnitude F and the lever arm l. More generally, torque is a vector $\vec{\tau}$ equal to the vector product of $\vec{r}$ (the position vector of the point at which the force acts) and $\vec{F}$. (See Example 10.1.)

$$\tau = Fl \quad (10.2)$$

$$\vec{\tau} = \vec{r} \times \vec{F} \quad (10.3)$$

Rotational dynamics: The rotational analog of Newton's second law says that the net torque acting on a body equals the product of the body's moment of inertia and its angular acceleration. (See Examples 10.2 and 10.3.)

$$\sum \tau_z = I\alpha_z \quad (10.7)$$

Combined translation and rotation: If a rigid body is both moving through space and rotating, its motion can be regarded as translational motion of the center of mass plus rotational motion about an axis through the center of mass. Thus the kinetic energy is a sum of translational and rotational kinetic energies. For dynamics, Newton's second law describes the motion of the center of mass, and the rotational equivalent of Newton's second law describes rotation about the center of mass. In the case of rolling without slipping, there is a special relationship between the motion of the center of mass and the rotational motion. (See Examples 10.4–10.7.)

$$K = \tfrac{1}{2}Mv_{cm}^2 + \tfrac{1}{2}I_{cm}\omega^2 \quad (10.8)$$

$$\sum \vec{F}_{ext} = M\vec{a}_{cm} \quad (10.12)$$

$$\sum \tau_z = I_{cm}\alpha_z \quad (10.13)$$

$$v_{cm} = R\omega \quad (10.11)$$
(rolling without slipping)

Work done by a torque: A torque that acts on a rigid body as it rotates does work on that body. The work can be expressed as an integral of the torque. The work–energy theorem says that the total rotational work done on a rigid body is equal to the change in rotational kinetic energy. The power, or rate at which the torque does work, is the product of the torque and the angular velocity (See Example 10.8.)

$$W = \int_{\theta_1}^{\theta_2} \tau_z \, d\theta \quad (10.20)$$

$$W = \tau_z(\theta_2 - \theta_1) = \tau_z \Delta\theta \quad (10.21)$$
(constant torque only)

$$W_{tot} = \tfrac{1}{2}I\omega_2^2 - \tfrac{1}{2}I\omega_1^2 \quad (10.22)$$

$$P = \tau_z \omega_z \quad (10.23)$$

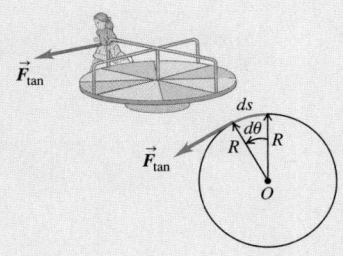

Angular momentum: The angular momentum of a particle with respect to point O is the vector product of the particle's position vector $\vec{r}$ relative to O and its momentum $\vec{p} = m\vec{v}$. When a symmetrical body rotates about a stationary axis of symmetry, its angular momentum is the product of its moment of inertia and its angular velocity vector $\vec{\omega}$. If the body is not symmetrical or the rotation (z) axis is not an axis of symmetry, the component of angular momentum along the rotation axis is $I\omega_z$. (See Example 10.9.)

$$\vec{L} = \vec{r} \times \vec{p} = \vec{r} \times m\vec{v} \quad (10.24)$$
(particle)

$$\vec{L} = I\vec{\omega} \quad (10.28)$$
(rigid body rotating about axis of symmetry)

Rotational dynamics and angular momentum: The net external torque on a system is equal to the rate of change of its angular momentum. If the net external torque on a system is zero, the total angular momentum of the system is constant (conserved). (See Examples 10.10–10.13.)

$$\sum \vec{\tau} = \frac{d\vec{L}}{dt} \quad (10.29)$$

BRIDGING PROBLEM **Billiard Physics**

A cue ball (a uniform solid sphere of mass m and radius R) is at rest on a level pool table. Using a pool cue, you give the ball a sharp, horizontal hit of magnitude F at a height h above the center of the ball (Fig. 10.37). The force of the hit is much greater than the friction force f that the table surface exerts on the ball. The hit lasts for a short time Δt. (a) For what value of h will the ball roll without slipping? (b) If you hit the ball dead center ($h = 0$), the ball will slide across the table for a while, but eventually it will roll without slipping. What will the speed of its center of mass be then?

SOLUTION GUIDE

See MasteringPhysics® study area for a Video Tutor solution.

IDENTIFY and SET UP

1. Draw a free-body diagram for the ball for the situation in part (a), including your choice of coordinate axes. Note that the cue exerts both an impulsive force on the ball and an impulsive torque around the center of mass.
2. The cue force applied for a time Δt gives the ball's center of mass a speed v_{cm}, and the cue torque applied for that same time gives the ball an angular speed ω. What must be the relationship between v_{cm} and ω for the ball to roll without slipping?

10.37

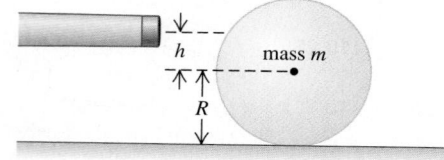

3. Draw two free-body diagrams for the ball in part (b): one showing the forces during the hit and the other showing the forces after the hit but before the ball is rolling without slipping.
4. What is the angular speed of the ball in part (b) just after the hit? While the ball is sliding, does v_{cm} increase or decrease? Does ω increase or decrease? What is the relationship between v_{cm} and ω when the ball is finally rolling without slipping?

EXECUTE

5. In part (a), use the impulse–momentum theorem to find the speed of the ball's center of mass immediately after the hit. Then use the rotational version of the impulse–momentum theorem to find the angular speed immediately after the hit. (*Hint:* To write down the rotational version of the impulse–momentum theorem, remember that the relationship between torque and angular momentum is the same as that between force and linear momentum.)
6. Use your results from step 5 to find the value of h that will cause the ball to roll without slipping immediately after the hit.
7. In part (b), again find the ball's center-of-mass speed and angular speed immediately after the hit. Then write Newton's second law for the translational motion and rotational motion of the ball as it is sliding. Use these equations to write expressions for v_{cm} and ω as functions of the elapsed time t since the hit.
8. Using your results from step 7, find the time t when v_{cm} and ω have the correct relationship for rolling without slipping. Then find the value of v_{cm} at this time.

EVALUATE

9. If you have access to a pool table, test out the results of parts (a) and (b) for yourself!
10. Can you show that if you used a hollow cylinder rather than a solid ball, you would have to hit the top of the cylinder to cause rolling without slipping as in part (a)?

Problems

For instructor-assigned homework, go to www.masteringphysics.com

•, ••, •••: Problems of increasing difficulty. **CP**: Cumulative problems incorporating material from earlier chapters. **CALC**: Problems requiring calculus. **BIO**: Biosciences problems.

DISCUSSION QUESTIONS

Q10.1 When cylinder-head bolts in an automobile engine are tightened, the critical quantity is the *torque* applied to the bolts. Why is the torque more important than the actual *force* applied to the wrench handle?

Q10.2 Can a single force applied to a body change both its translational and rotational motion? Explain.

Q10.3 Suppose you could use wheels of any type in the design of a soapbox-derby racer (an unpowered, four-wheel vehicle that coasts from rest down a hill). To conform to the rules on the total weight of the vehicle and rider, should you design with large massive wheels or small light wheels? Should you use solid wheels or wheels with most of the mass at the rim? Explain.

Q10.4 A four-wheel-drive car is accelerating forward from rest. Show the direction the car's wheels turn and how this causes a friction force due to the pavement that accelerates the car forward.

Q10.5 Serious bicyclists say that if you reduce the weight of a bike, it is more effective if you do so in the wheels rather than in the frame. Why would reducing weight in the wheels make it easier on the bicyclist than reducing the same amount in the frame?

Q10.6 The harder you hit the brakes while driving forward, the more the front end of your car will move down (and the rear end move up). Why? What happens when cars accelerate forward? Why do drag racers not use front-wheel drive only?

Q10.7 When an acrobat walks on a tightrope, she extends her arms straight out from her sides. She does this to make it easier for her to catch herself if she should tip to one side or the other. Explain how this works. [*Hint:* Think about Eq. (10.7).]

Q10.8 When you turn on an electric motor, it takes longer to come up to final speed if a grinding wheel is attached to the shaft. Why?

Q10.9 Experienced cooks can tell whether an egg is raw or hard-boiled by rolling it down a slope (taking care to catch it at the bottom). How is this possible? What are they looking for?

Q10.10 The work done by a force is the product of force and distance. The torque due to a force is the product of force and distance. Does this mean that torque and work are equivalent? Explain.

Q10.11 A valued client brings a treasured ball to your engineering firm, wanting to know whether the ball is solid or hollow. He has tried tapping on it, but that has given insufficient information. Design a simple, inexpensive experiment that you could perform quickly, without injuring the precious ball, to find out whether it is solid or hollow.

Q10.12 You make two versions of the same object out of the same material having uniform density. For one version, all the dimensions are exactly twice as great as for the other one. If the same torque acts on both versions, giving the smaller version angular acceleration α, what will be the angular acceleration of the larger version in terms of α?

Q10.13 Two identical masses are attached to frictionless pulleys by very light strings wrapped around the rim of the pulley and are released from rest. Both pulleys have the same mass and same diameter, but one is solid and the other is a hoop. As the masses fall, in which case is the tension in the string greater, or is it the same in both cases? Justify your answer.

Q10.14 The force of gravity acts on the baton in Fig. 10.11, and forces produce torques that cause a body's angular velocity to change. Why, then, is the angular velocity of the baton in the figure constant?

Q10.15 A certain solid uniform ball reaches a maximum height h_0 when it rolls up a hill without slipping. What maximum height (in terms of h_0) will it reach if you (a) double its diameter, (b) double its mass, (c) double both its diameter and mass, (d) double its angular speed at the bottom of the hill?

Q10.16 A wheel is rolling without slipping on a horizontal surface. In an inertial frame of reference in which the surface is at rest, is there any point on the wheel that has a velocity that is purely vertical? Is there any point that has a horizontal velocity component opposite to the velocity of the center of mass? Explain. Do your answers change if the wheel is slipping as it rolls? Why or why not?

Q10.17 Part of the kinetic energy of a moving automobile is in the rotational motion of its wheels. When the brakes are applied hard on an icy street, the wheels "lock" and the car starts to slide. What becomes of the rotational kinetic energy?

Q10.18 A hoop, a uniform solid cylinder, a spherical shell, and a uniform solid sphere are released from rest at the top of an incline. What is the order in which they arrive at the bottom of the incline? Does it matter whether or not the masses and radii of the objects are all the same? Explain.

Q10.19 A ball is rolling along at speed v without slipping on a horizontal surface when it comes to a hill that rises at a constant angle above the horizontal. In which case will it go higher up the hill: if the hill has enough friction to prevent slipping or if the hill is perfectly smooth? Justify your answers in both cases in terms of energy conservation and in terms of Newton's second law.

Q10.20 You are standing at the center of a large horizontal turntable in a carnival funhouse. The turntable is set rotating on frictionless bearings, and it rotates freely (that is, there is no motor driving the turntable). As you walk toward the edge of the turntable, what happens to the combined angular momentum of you and the turntable? What happens to the rotation speed of the turntable? Explain your answer.

Q10.21 A certain uniform turntable of diameter D_0 has an angular momentum L_0. If you want to redesign it so it retains the same mass but has twice as much angular momentum at the same angular velocity as before, what should be its diameter in terms of D_0?

Q10.22 A point particle travels in a straight line at constant speed, and the closest distance it comes to the origin of coordinates is a distance l. With respect to this origin, does the particle have nonzero angular momentum? As the particle moves along its straight-line path, does its angular momentum with respect to the origin change?

Q10.23 In Example 10.10 (Section 10.6) the angular speed ω changes, and this must mean that there is nonzero angular acceleration. But there is no torque about the rotation axis if the forces the professor applies to the weights are directly, radially inward. Then, by Eq. (10.7), α_z must be zero. Explain what is wrong with this reasoning that leads to this apparent contradiction.

Q10.24 In Example 10.10 (Section 10.6) the rotational kinetic energy of the professor and dumbbells increases. But since there are no external torques, no work is being done to change the rotational kinetic energy. Then, by Eq. (10.22), the kinetic energy must remain the same! Explain what is wrong with this reasoning that leads to this apparent contradiction. Where *does* the extra kinetic energy come from?

Q10.25 As discussed in Section 10.6, the angular momentum of a circus acrobat is conserved as she tumbles through the air. Is her *linear* momentum conserved? Why or why not?

Q10.26 If you stop a spinning raw egg for the shortest possible instant and then release it, the egg will start spinning again. If you do the same to a hard-boiled egg, it will remain stopped. Try it. Explain it.

Q10.27 A helicopter has a large main rotor that rotates in a horizontal plane and provides lift. There is also a small rotor on the tail that rotates in a vertical plane. What is the purpose of the tail rotor? (*Hint:* If there were no tail rotor, what would happen when the pilot changed the angular speed of the main rotor?) Some helicopters have no tail rotor, but instead have two large main rotors that rotate in a horizontal plane. Why is it important that the two main rotors rotate in opposite directions?

Q10.28 In a common design for a gyroscope, the flywheel and flywheel axis are enclosed in a light, spherical frame with the flywheel at the center of the frame. The gyroscope is then balanced on top of a pivot so that the flywheel is directly above the pivot. Does the gyroscope precess if it is released while the flywheel is spinning? Explain.

Q10.29 A gyroscope takes 3.8 s to precess 1.0 revolution about a vertical axis. Two minutes later, it takes only 1.9 s to precess 1.0 revolution. No one has touched the gyroscope. Explain.

Q10.30 A gyroscope is precessing as in Fig. 10.32. What happens if you gently add some weight to the end of the flywheel axis farthest from the pivot?

Q10.31 A bullet spins on its axis as it emerges from a rifle. Explain how this prevents the bullet from tumbling and keeps the streamlined end pointed forward.

EXERCISES

Section 10.1 Torque

10.1 • Calculate the torque (magnitude and direction) about point O due to the force $\vec{F}$ in each of the cases sketched in Fig. E10.1. In each case, the force $\vec{F}$ and the rod both lie in the plane of the page, the rod has length 4.00 m, and the force has magnitude $F = 10.0$ N.

Figure **E10.1**

10.2 • Calculate the net torque about point O for the two forces applied as in Fig. E10.2. The rod and both forces are in the plane of the page.

Figure **E10.2**

10.3 •• A square metal plate 0.180 m on each side is pivoted about an axis through point O at its center and perpendicular to the plate (Fig. E10.3). Calculate the net torque about this axis due to the three forces shown in the figure if the magnitudes of the forces are $F_1 = 18.0$ N, $F_2 = 26.0$ N, and $F_3 = 14.0$ N. The plate and all forces are in the plane of the page.

Figure **E10.3**   Figure **E10.4**

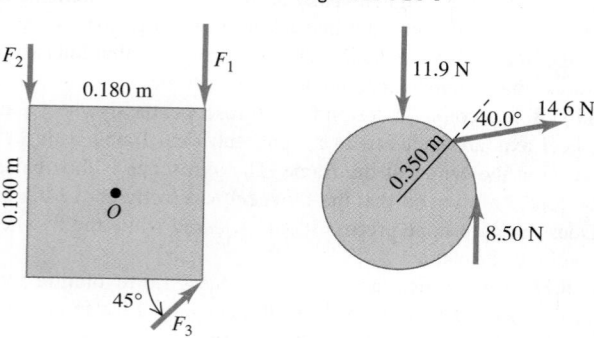

10.4 • Three forces are applied to a wheel of radius 0.350 m, as shown in Fig. E10.4. One force is perpendicular to the rim, one is tangent to it, and the other one makes a 40.0° angle with the radius. What is the net torque on the wheel due to these three forces for an axis perpendicular to the wheel and passing through its center?

10.5 • One force acting on a machine part is $\vec{F} = (-5.00$ N$)\hat{\imath} + (4.00$ N$)\hat{\jmath}$. The vector from the origin to the point where the force is applied is $\vec{r} = (-0.450$ m$)\hat{\imath} + (0.150$ m$)\hat{\jmath}$. (a) In a sketch, show $\vec{r}$, $\vec{F}$, and the origin. (b) Use the right-hand rule to determine the direction of the torque. (c) Calculate the vector torque for an axis at the origin produced by this force. Verify that the direction of the torque is the same as you obtained in part (b).

10.6 • A metal bar is in the xy-plane with one end of the bar at the origin. A force $\vec{F} = (7.00$ N$)\hat{\imath} + (-3.00$ N$)\hat{\jmath}$ is applied to the bar at the point $x = 3.00$ m, $y = 4.00$ m. (a) In terms of unit vectors $\hat{\imath}$ and $\hat{\jmath}$, what is the position vector $\vec{r}$ for the point where the force is applied? (b) What are the magnitude and direction of the torque with respect to the origin produced by $\vec{F}$?

10.7 • In Fig. E10.7, forces $\vec{A}$, $\vec{B}$, $\vec{C}$, and $\vec{D}$ each have magnitude 50 N and act at the same point on the object. (a) What torque (magnitude and direction) does each of these forces exert on the object about point P? (b) What is the total torque about point P?

Figure **E10.7**

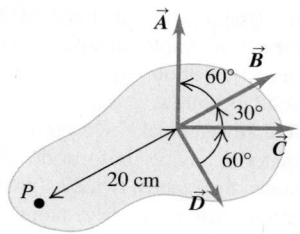

10.8 • A machinist is using a wrench to loosen a nut. The wrench is 25.0 cm long, and he exerts a 17.0-N force at the end of the handle at 37° with the handle (Fig. E10.8). (a) What torque does the machinist exert about the center of the nut? (b) What is the maximum torque he could exert with this force, and how should the force be oriented?

Figure **E10.8**

Section 10.2 Torque and Angular Acceleration for a Rigid Body

10.9 •• The flywheel of an engine has moment of inertia 2.50 kg · m² about its rotation axis. What constant torque is required to bring it up to an angular speed of 400 rev/min in 8.00 s, starting from rest?

10.10 •• A uniform disk with mass 40.0 kg and radius 0.200 m is pivoted at its center about a horizontal, frictionless axle that is stationary. The disk is initially at rest, and then a constant force $F = 30.0$ N is applied tangent to the rim of the disk. (a) What is the magnitude v of the tangential velocity of a point on the rim of the disk after the disk has turned through 0.200 revolution? (b) What is the magnitude a of the resultant acceleration of a point on the rim of the disk after the disk has turned through 0.200 revolution?

10.11 •• A machine part has the shape of a solid uniform sphere of mass 225 g and diameter 3.00 cm. It is spinning about a frictionless axle through its center, but at one point on its equator it is scraping against metal, resulting in a friction force of 0.0200 N at that point. (a) Find its angular acceleration. (b) How long will it take to decrease its rotational speed by 22.5 rad/s?

10.12 • A cord is wrapped around the rim of a solid uniform wheel 0.250 m in radius and of mass 9.20 kg. A steady horizontal pull of 40.0 N to the right is exerted on the cord, pulling it off tangentially from the wheel. The wheel is mounted on frictionless bearings on a horizontal axle through its center. (a) Compute the angular acceleration of the wheel and the acceleration of the part of the cord that has already been pulled off the wheel. (b) Find the magnitude and direction of the force that the axle exerts on the wheel. (c) Which of the answers in parts (a) and (b) would change if the pull were upward instead of horizontal?

10.13 •• **CP** A 2.00-kg textbook rests on a frictionless, horizontal surface. A cord attached to the book passes over a pulley whose diameter is 0.150 m, to a hanging book with mass 3.00 kg. The system is released from rest, and the books are observed to move 1.20 m in 0.800 s. (a) What is the tension in each part of the cord? (b) What is the moment of inertia of the pulley about its rotation axis?

10.14 •• **CP** A stone is suspended from the free end of a wire that is wrapped around the outer rim of a pulley, similar to what is shown in Fig. 10.10. The pulley is a uniform disk with mass 10.0 kg and radius 50.0 cm and turns on frictionless bearings. You measure that the stone travels 12.6 m in the first 3.00 s starting from rest. Find (a) the mass of the stone and (b) the tension in the wire.

10.15 • A wheel rotates without friction about a stationary horizontal axis at the center of the wheel. A constant tangential force equal to 80.0 N is applied to the rim of the wheel. The wheel has radius 0.120 m. Starting from rest, the wheel has an angular speed of 12.0 rev/s after 2.00 s. What is the moment of inertia of the wheel?

10.16 •• **CP** A 15.0-kg bucket of water is suspended by a very light rope wrapped around a solid uniform cylinder 0.300 m in diameter with mass 12.0 kg. The cylinder pivots on a frictionless axle through its center. The bucket is released from rest at the top of a well and falls 10.0 m to the water. (a) What is the tension in the rope while the bucket is falling? (b) With what speed does the bucket strike the water? (c) What is the time of fall? (d) While the bucket is falling, what is the force exerted on the cylinder by the axle?

10.17 •• A 12.0-kg box resting on a horizontal, frictionless surface is attached to a 5.00-kg weight by a thin, light wire that passes over a frictionless pulley (Fig. E10.17). The pulley has the shape of a uniform solid disk of mass 2.00 kg and diameter 0.500 m. After the system is released, find (a) the tension in the wire on both sides of the pulley, (b) the acceleration of the box, and (c) the horizontal and vertical components of the force that the axle exerts on the pulley.

Figure **E10.17**

12.0 kg

5.00 kg

Section 10.3 Rigid-Body Rotation About a Moving Axis

10.18 • **BIO** **Gymnastics.** We can roughly model a gymnastic tumbler as a uniform solid cylinder of mass 75 kg and diameter 1.0 m. If this tumbler rolls forward at 0.50 rev/s, (a) how much total kinetic energy does he have, and (b) what percent of his total kinetic energy is rotational?

10.19 • A 2.20-kg hoop 1.20 m in diameter is rolling to the right without slipping on a horizontal floor at a steady 3.00 rad/s. (a) How fast is its center moving? (b) What is the total kinetic energy of the hoop? (c) Find the velocity vector of each of the following points, as viewed by a person at rest on the ground: (i) the highest point on the hoop; (ii) the lowest point on the hoop; (iii) a point on the right side of the hoop, midway between the top and the bottom. (d) Find the velocity vector for each of the points in part (c), but this time as viewed by someone moving along with the same velocity as the hoop.

10.20 •• A string is wrapped several times around the rim of a small hoop with radius 8.00 cm and mass 0.180 kg. The free end of the string is held in place and the hoop is released from rest (Fig. E10.20). After the hoop has descended 75.0 cm, calculate (a) the angular speed of the rotating hoop and (b) the speed of its center.

Figure **E10.20**

0.0800 m

10.21 • What fraction of the total kinetic energy is rotational for the following objects rolling without slipping on a horizontal surface? (a) a uniform solid cylinder; (b) a uniform sphere; (c) a thin-walled, hollow sphere; (d) a hollow cylinder with outer radius R and inner radius $R/2$.

10.22 •• A hollow, spherical shell with mass 2.00 kg rolls without slipping down a 38.0° slope. (a) Find the acceleration, the friction force, and the minimum coefficient of friction needed to prevent slipping. (b) How would your answers to part (a) change if the mass were doubled to 4.00 kg?

10.23 •• A solid ball is released from rest and slides down a hillside that slopes downward at 65.0° from the horizontal. (a) What minimum value must the coefficient of static friction between the hill and ball surfaces have for no slipping to occur? (b) Would the coefficient of friction calculated in part (a) be sufficient to prevent a hollow ball (such as a soccer ball) from slipping? Justify your answer. (c) In part (a), why did we use the coefficient of static friction and not the coefficient of kinetic friction?

10.24 •• A uniform marble rolls down a symmetrical bowl, starting from rest at the top of the left side. The top of each side is a distance h above the bottom of the bowl. The left half of the bowl is rough enough to cause the marble to roll without slipping, but the right half has no friction because it is coated with oil. (a) How far up the smooth side will the marble go, measured vertically from the bottom? (b) How high would the marble go if both sides were as rough as the left side? (c) How do you account for the fact that the marble goes *higher* with friction on the right side than without friction?

10.25 •• A 392-N wheel comes off a moving truck and rolls without slipping along a highway. At the bottom of a hill it is rotating at 25.0 rad/s. The radius of the wheel is 0.600 m, and its moment of inertia about its rotation axis is $0.800MR^2$. Friction does work on the wheel as it rolls up the hill to a stop, a height h above the bottom of the hill; this work has absolute value 3500 J. Calculate h.

10.26 •• **A Ball Rolling Uphill.** A bowling ball rolls without slipping up a ramp that slopes upward at an angle β to the horizontal (see Example 10.7 in Section 10.3). Treat the ball as a uniform solid sphere, ignoring the finger holes. (a) Draw the free-body diagram for the ball. Explain why the friction force must be directed *uphill*. (b) What is the acceleration of the center of mass of the ball? (c) What minimum coefficient of static friction is needed to prevent slipping?

10.27 •• A thin, light string is wrapped around the outer rim of a uniform hollow cylinder of mass 4.75 kg having inner and outer radii as shown in Fig. E10.27. The cylinder is then released from rest.

Figure **E10.27**

20.0 cm

35.0 cm

(a) How far must the cylinder fall before its center is moving at 6.66 m/s? (b) If you just dropped this cylinder without any string, how fast would its center be moving when it had fallen the distance in part (a)? (c) Why do you get two different answers when the cylinder falls the same distance in both cases?

10.28 ·· A bicycle racer is going downhill at 11.0 m/s when, to his horror, one of his 2.25-kg wheels comes off as he is 75.0 m above the foot of the hill. We can model the wheel as a thin-walled cylinder 85.0 cm in diameter and neglect the small mass of the spokes. (a) How fast is the wheel moving when it reaches the foot of the hill if it rolled without slipping all the way down? (b) How much total kinetic energy does the wheel have when it reaches the bottom of the hill?

10.29 ·· A size-5 soccer ball of diameter 22.6 cm and mass 426 g rolls up a hill without slipping, reaching a maximum height of 5.00 m above the base of the hill. We can model this ball as a thin-walled hollow sphere. (a) At what rate was it rotating at the base of the hill? (b) How much rotational kinetic energy did it have then?

Section 10.4 Work and Power in Rotational Motion

10.30 · An engine delivers 175 hp to an aircraft propeller at 2400 rev/min. (a) How much torque does the aircraft engine provide? (b) How much work does the engine do in one revolution of the propeller?

10.31 · A playground merry-go-round has radius 2.40 m and moment of inertia 2100 kg · m² about a vertical axle through its center, and it turns with negligible friction. (a) A child applies an 18.0-N force tangentially to the edge of the merry-go-round for 15.0 s. If the merry-go-round is initially at rest, what is its angular speed after this 15.0-s interval? (b) How much work did the child do on the merry-go-round? (c) What is the average power supplied by the child?

10.32 ·· An electric motor consumes 9.00 kJ of electrical energy in 1.00 min. If one-third of this energy goes into heat and other forms of internal energy of the motor, with the rest going to the motor output, how much torque will this engine develop if you run it at 2500 rpm?

10.33 · A 1.50-kg grinding wheel is in the form of a solid cylinder of radius 0.100 m. (a) What constant torque will bring it from rest to an angular speed of 1200 rev/min in 2.5 s? (b) Through what angle has it turned during that time? (c) Use Eq. (10.21) to calculate the work done by the torque. (d) What is the grinding wheel's kinetic energy when it is rotating at 1200 rev/min? Compare your answer to the result in part (c).

10.34 ·· An airplane propeller is 2.08 m in length (from tip to tip) and has a mass of 117 kg. When the airplane's engine is first started, it applies a constant torque of 1950 N · m to the propeller, which starts from rest. (a) What is the angular acceleration of the propeller? Model the propeller as a slender rod and see Table 9.2. (b) What is the propeller's angular speed after making 5.00 revolutions? (c) How much work is done by the engine during the first 5.00 revolutions? (d) What is the average power output of the engine during the first 5.00 revolutions? (e) What is the instantaneous power output of the motor at the instant that the propeller has turned through 5.00 revolutions?

10.35 · (a) Compute the torque developed by an industrial motor whose output is 150 kW at an angular speed of 4000 rev/min. (b) A drum with negligible mass, 0.400 m in diameter, is attached to the motor shaft, and the power output of the motor is used to raise a weight hanging from a rope wrapped around the drum. How heavy a weight can the motor lift at constant speed? (c) At what constant speed will the weight rise?

Section 10.5 Angular Momentum

10.36 ·· A woman with mass 50 kg is standing on the rim of a large disk that is rotating at 0.50 rev/s about an axis through its center. The disk has mass 110 kg and radius 4.0 m. Calculate the magnitude of the total angular momentum of the woman–disk system. (Assume that you can treat the woman as a point.)

10.37 · A 2.00-kg rock has a horizontal velocity of magnitude 12.0 m/s when it is at point P in Fig. E10.37. (a) At this instant, what are the magnitude and direction of its angular momentum relative to point O? (b) If the only force acting on the rock is its weight, what is the rate of change (magnitude and direction) of its angular momentum at this instant?

Figure **E10.37**

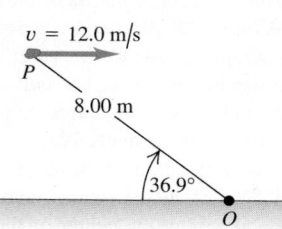

$v = 12.0$ m/s

P

8.00 m

36.9°

O

10.38 ·· (a) Calculate the magnitude of the angular momentum of the earth in a circular orbit around the sun. Is it reasonable to model it as a particle? (b) Calculate the magnitude of the angular momentum of the earth due to its rotation around an axis through the north and south poles, modeling it as a uniform sphere. Consult Appendix E and the astronomical data in Appendix F.

10.39 ·· Find the magnitude of the angular momentum of the second hand on a clock about an axis through the center of the clock face. The clock hand has a length of 15.0 cm and a mass of 6.00 g. Take the second hand to be a slender rod rotating with constant angular velocity about one end.

10.40 ·· **CALC** A hollow, thin-walled sphere of mass 12.0 kg and diameter 48.0 cm is rotating about an axle through its center. The angle (in radians) through which it turns as a function of time (in seconds) is given by $\theta(t) = At^2 + Bt^4$, where A has numerical value 1.50 and B has numerical value 1.10. (a) What are the units of the constants A and B? (b) At the time 3.00 s, find (i) the angular momentum of the sphere and (ii) the net torque on the sphere.

Section 10.6 Conservation of Angular Momentum

10.41 ·· Under some circumstances, a star can collapse into an extremely dense object made mostly of neutrons and called a *neutron star*. The density of a neutron star is roughly 10^{14} times as great as that of ordinary solid matter. Suppose we represent the star as a uniform, solid, rigid sphere, both before and after the collapse. The star's initial radius was 7.0×10^5 km (comparable to our sun); its final radius is 16 km. If the original star rotated once in 30 days, find the angular speed of the neutron star.

10.42 · **CP** A small block on a frictionless, horizontal surface has a mass of 0.0250 kg. It is attached to a massless cord passing through a hole in the surface (Fig. E10.42). The block is originally revolving at a distance of 0.300 m from the hole with an angular speed of 1.75 rad/s. The cord is then pulled from below, shortening the radius of the circle in which the block revolves to 0.150 m. Model the block as a particle. (a) Is the angular momentum of the block conserved? Why or why not? (b) What is the new angular speed? (c) Find the change in kinetic energy of the block. (d) How much work was done in pulling the cord?

Figure **E10.42**

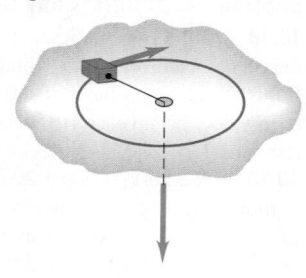

10.43 •• **The Spinning Figure Skater.** The outstretched hands and arms of a figure skater preparing for a spin can be considered a slender rod pivoting about an axis through its center (Fig. E10.43). When the skater's hands and arms are brought in and wrapped around his body to execute the spin, the hands and arms can be considered a thin-walled, hollow cylinder. His hands and arms have a combined mass of 8.0 kg. When outstretched, they span 1.8 m; when wrapped, they form a cylinder of radius 25 cm. The moment of inertia about the rotation axis of the remainder of his body is constant and equal to $0.40 \text{ kg} \cdot \text{m}^2$. If his original angular speed is 0.40 rev/s, what is his final angular speed?

Figure **E10.43**

10.44 •• A diver comes off a board with arms straight up and legs straight down, giving her a moment of inertia about her rotation axis of $18 \text{ kg} \cdot \text{m}^2$. She then tucks into a small ball, decreasing this moment of inertia to $3.6 \text{ kg} \cdot \text{m}^2$. While tucked, she makes two complete revolutions in 1.0 s. If she hadn't tucked at all, how many revolutions would she have made in the 1.5 s from board to water?

10.45 •• A large wooden turntable in the shape of a flat uniform disk has a radius of 2.00 m and a total mass of 120 kg. The turntable is initially rotating at 3.00 rad/s about a vertical axis through its center. Suddenly, a 70.0-kg parachutist makes a soft landing on the turntable at a point near the outer edge. (a) Find the angular speed of the turntable after the parachutist lands. (Assume that you can treat the parachutist as a particle.) (b) Compute the kinetic energy of the system before and after the parachutist lands. Why are these kinetic energies not equal?

10.46 •• A solid wood door 1.00 m wide and 2.00 m high is hinged along one side and has a total mass of 40.0 kg. Initially open and at rest, the door is struck at its center by a handful of sticky mud with mass 0.500 kg, traveling perpendicular to the door at 12.0 m/s just before impact. Find the final angular speed of the door. Does the mud make a significant contribution to the moment of inertia?

10.47 •• A small 10.0-g bug stands at one end of a thin uniform bar that is initially at rest on a smooth horizontal table. The other end of the bar pivots about a nail driven into the table and can rotate freely, without friction. The bar has mass 50.0 g and is 100 cm in length. The bug jumps off in the horizontal direction, perpendicular to the bar, with a speed of 20.0 cm/s relative to the table. (a) What is the angular speed of the bar just after the frisky insect leaps? (b) What is the total kinetic energy of the system just after the bug leaps? (c) Where does this energy come from?

10.48 •• **Asteroid Collision!** Suppose that an asteroid traveling straight toward the center of the earth were to collide with our planet at the equator and bury itself just below the surface. What would have to be the mass of this asteroid, in terms of the earth's mass M, for the day to become 25.0% longer than it presently is as a result of the collision? Assume that the asteroid is very small compared to the earth and that the earth is uniform throughout.

10.49 •• A thin, uniform metal bar, 2.00 m long and weighing 90.0 N, is hanging vertically from the ceiling by a frictionless pivot. Suddenly it is struck 1.50 m below the ceiling by a small 3.00-kg ball, initially traveling horizontally at 10.0 m/s. The ball rebounds in the opposite direction with a speed of 6.00 m/s. (a) Find the angular speed of the bar just after the collision. (b) During the collision, why is the angular momentum conserved but not the linear momentum?

10.50 •• A thin uniform rod has a length of 0.500 m and is rotating in a circle on a frictionless table. The axis of rotation is perpendicular to the length of the rod at one end and is stationary. The rod has an angular velocity of 0.400 rad/s and a moment of inertia about the axis of $3.00 \times 10^{-3} \text{ kg} \cdot \text{m}^2$. A bug initially standing on the rod at the axis of rotation decides to crawl out to the other end of the rod. When the bug has reached the end of the rod and sits there, its tangential speed is 0.160 m/s. The bug can be treated as a point mass. (a) What is the mass of the rod? (b) What is the mass of the bug?

10.51 •• A uniform, 4.5-kg, square, solid wooden gate 1.5 m on each side hangs vertically from a frictionless pivot at the center of its upper edge. A 1.1-kg raven flying horizontally at 5.0 m/s flies into this door at its center and bounces back at 2.0 m/s in the opposite direction. (a) What is the angular speed of the gate just after it is struck by the unfortunate raven? (b) During the collision, why is the angular momentum conserved, but not the linear momentum?

10.52 •• **Sedna.** In November 2003, the now-most-distant-known object in the solar system was discovered by observation with a telescope on Mt. Palomar. This object, known as Sedna, is approximately 1700 km in diameter, takes about 10,500 years to orbit our sun, and reaches a maximum speed of 4.64 km/s. Calculations of its complete path, based on several measurements of its position, indicate that its orbit is highly elliptical, varying from 76 AU to 942 AU in its distance from the sun, where AU is the astronomical unit, which is the average distance of the earth from the sun $(1.50 \times 10^8 \text{ km})$. (a) What is Sedna's minimum speed? (b) At what points in its orbit do its maximum and minimum speeds occur? (c) What is the ratio of Sedna's maximum kinetic energy to its minimum kinetic energy?

Section 10.7 Gyroscopes and Precession

10.53 •• The rotor (flywheel) of a toy gyroscope has mass 0.140 kg. Its moment of inertia about its axis is $1.20 \times 10^{-4} \text{ kg} \cdot \text{m}^2$. The mass of the frame is 0.0250 kg. The gyroscope is supported on a single pivot (Fig. E10.53) with its center of mass a horizontal distance of 4.00 cm from the pivot. The gyroscope is precessing in a horizontal plane at the rate of one revolution in 2.20 s. (a) Find the upward force exerted by the pivot. (b) Find the angular speed with which the rotor is spinning about its axis, expressed in rev/min. (c) Copy the diagram and draw vectors to show the angular momentum of the rotor and the torque acting on it.

Figure **E10.53**

Rotor

4.00 cm

10.54 • **A Gyroscope on the Moon.** A certain gyroscope precesses at a rate of 0.50 rad/s when used on earth. If it were taken to a lunar base, where the acceleration due to gravity is $0.165g$, what would be its precession rate?

10.55 • A gyroscope is precessing about a vertical axis. Describe what happens to the precession angular speed if the following changes in the variables are made, with all other variables remaining the same: (a) the angular speed of the spinning flywheel is doubled; (b) the total weight is doubled; (c) the moment of inertia about the axis of the spinning flywheel is doubled; (d) the distance from the pivot to the center of gravity is doubled. (e) What happens if all four of the variables in parts (a) through (d) are doubled?

10.56 • Stabilization of the Hubble Space Telescope. The Hubble Space Telescope is stabilized to within an angle of about 2-millionths of a degree by means of a series of gyroscopes that spin at 19,200 rpm. Although the structure of these gyroscopes is actually quite complex, we can model each of the gyroscopes as a thin-walled cylinder of mass 2.0 kg and diameter 5.0 cm, spinning about its central axis. How large a torque would it take to cause these gyroscopes to precess through an angle of 1.0×10^{-6} degree during a 5.0-hour exposure of a galaxy?

PROBLEMS

10.57 •• A 50.0-kg grindstone is a solid disk 0.520 m in diameter. You press an ax down on the rim with a normal force of 160 N (Fig. P10.57). The coefficient of kinetic friction between the blade and the stone is 0.60, and there is a constant friction torque of 6.50 N·m between the axle of the stone and its bearings. (a) How much force must be applied tangentially at the end of a crank handle 0.500 m long to bring the stone from rest to 120 rev/min in 9.00 s? (b) After the grindstone attains an angular speed of 120 rev/min, what tangential force at the end of the handle is needed to maintain a constant angular speed of 120 rev/min? (c) How much time does it take the grindstone to come from 120 rev/min to rest if it is acted on by the axle friction alone?

Figure **P10.57**

$m = 50.0$ kg ω

$F = 160$ N

10.58 •• An experimental bicycle wheel is placed on a test stand so that it is free to turn on its axle. If a constant net torque of 7.00 N·m is applied to the tire for 2.00 s, the angular speed of the tire increases from 0 to 100 rev/min. The external torque is then removed, and the wheel is brought to rest by friction in its bearings in 125 s. Compute (a) the moment of inertia of the wheel about the rotation axis; (b) the friction torque; (c) the total number of revolutions made by the wheel in the 125-s time interval.

10.59 ••• A grindstone in the shape of a solid disk with diameter 0.520 m and a mass of 50.0 kg is rotating at 850 rev/min. You press an ax against the rim with a normal force of 160 N (Fig. P10.57), and the grindstone comes to rest in 7.50 s. Find the coefficient of friction between the ax and the grindstone. You can ignore friction in the bearings.

10.60 ••• A uniform, 8.40-kg, spherical shell 50.0 cm in diameter has four small 2.00-kg masses attached to its outer surface and equally spaced around it. This

combination is spinning about an axis running through the center of the sphere and two of the small masses (Fig. P10.60). What friction torque is needed to reduce its angular speed from 75.0 rpm to 50.0 rpm in 30.0 s?

10.61 ••• A solid uniform cylinder with mass 8.25 kg and diameter 15.0 cm is spinning at 220 rpm on a thin, frictionless axle that passes along the cylinder axis. You design a simple friction brake to stop the cylinder by pressing the brake against the outer rim with a normal force. The coefficient of kinetic friction between the brake and rim is 0.333. What must the applied normal force be to bring the cylinder to rest after it has turned through 5.25 revolutions?

10.62 ••• A uniform hollow disk has two pieces of thin, light wire wrapped around its outer rim and is supported from the ceiling (Fig. P10.62). Suddenly one of the wires breaks, and the remaining wire does not slip as the disk rolls down. Use energy conservation to find the speed of the center of this disk after it has fallen a distance of 2.20 m.

Figure **P10.62**

30.0 cm

50.0 cm

10.63 ••• A thin, uniform, 3.80-kg bar, 80.0 cm long, has very small 2.50-kg balls glued on at either end (Fig. P10.63). It is supported horizontally by a thin, horizontal, frictionless axle passing through its center and perpendicular to the bar. Suddenly the right-hand ball becomes detached and falls off, but the other ball remains glued to the bar. (a) Find the angular acceleration of the bar just after the ball falls off. (b) Will the angular acceleration remain constant as the bar continues to swing? If not, will it increase or decrease? (c) Find the angular velocity of the bar just as it swings through its vertical position.

Figure **P10.63**

2.50 kg   2.50 kg

Bar   Axle (seen end-on)

10.64 ••• While exploring a castle, Exena the Exterminator is spotted by a dragon that chases her down a hallway. Exena runs into a room and attempts to swing the heavy door shut before the dragon gets her. The door is initially perpendicular to the wall, so it must be turned through 90° to close. The door is 3.00 m tall and 1.25 m wide, and it weighs 750 N. You can ignore the friction at the hinges. If Exena applies a force of 220 N at the edge of the door and perpendicular to it, how much time does it take her to close the door?

10.65 •• **CALC** You connect a light string to a point on the edge of a uniform vertical disk with radius R and mass M. The disk is free to rotate without friction about a stationary horizontal axis through its center. Initially, the disk is at rest with the string connection at the highest point on the disk. You pull the string with a constant horizontal force $\vec{F}$ until the wheel has made exactly one-quarter revolution about a horizontal axis through its center, and then you let go. (a) Use Eq. (10.20) to find the work done by the string. (b) Use Eq. (6.14) to find the work done by the string. Do you obtain the same result as in part (a)? (c) Find the final angular speed of the disk. (d) Find the maximum tangential acceleration of a point on the disk. (e) Find the maximum radial (centripetal) acceleration of a point on the disk.

10.66 ••• **Balancing Act.** Attached to one end of a long, thin, uniform rod of length L and mass M is a small blob of clay of the same mass M. (a) Locate the position of the center of mass of the system of rod and clay. Note this position on a drawing of the rod.

Figure **P10.60**

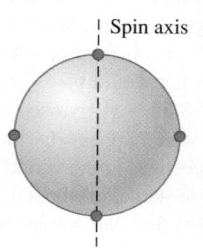

Spin axis

(b) You carefully balance the rod on a frictionless tabletop so that it is standing vertically, with the end without the clay touching the table. If the rod is now tipped so that it is a small angle θ away from the vertical, determine its angular acceleration at this instant. Assume that the end without the clay remains in contact with the tabletop. (*Hint:* See Table 9.2.) (c) You again balance the rod on the frictionless tabletop so that it is standing vertically, but now the end of the rod *with* the clay is touching the table. If the rod is again tipped so that it is a small angle θ away from the vertical, determine its angular acceleration at this instant. Assume that the end with the clay remains in contact with the tabletop. How does this compare to the angular acceleration in part (b)? (d) A pool cue is a tapered wooden rod that is thick at one end and thin at the other. You can easily balance a pool cue vertically on one finger if the thin end is in contact with your finger; this is quite a bit harder to do if the thick end is in contact with your finger. Explain why there is a difference.

10.67 •• **Atwood's Machine.** Figure P10.67 illustrates an Atwood's machine. Find the linear accelerations of blocks *A* and *B*, the angular acceleration of the wheel *C*, and the tension in each side of the cord if there is no slipping between the cord and the surface of the wheel. Let the masses of blocks *A* and *B* be 4.00 kg and 2.00 kg, respectively, the moment of inertia of the wheel about its axis be 0.300 kg·m², and the radius of the wheel be 0.120 m.

Figure **P10.67**

10.68 ••• The mechanism shown in Fig. P10.68 is used to raise a crate of supplies from a ship's hold. The crate has total mass 50 kg. A rope is wrapped around a wooden cylinder that turns on a metal axle. The cylinder has radius 0.25 m and moment of inertia $I = 2.9$ kg·m² about the axle. The crate is suspended from the free end of the rope. One end of the axle pivots on frictionless bearings; a crank handle is attached to the other end. When the crank is turned, the end of the handle rotates about the axle in a vertical circle of radius 0.12 m, the cylinder turns, and the crate is raised. What magnitude of the force $\vec{F}$ applied tangentially to the rotating crank is required to raise the crate with an acceleration of 1.40 m/s²? (You can ignore the mass of the rope as well as the moments of inertia of the axle and the crank.)

Figure **P10.68**

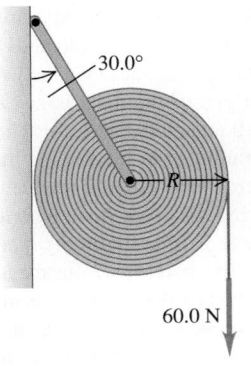

10.69 •• A large 16.0-kg roll of paper with radius $R = 18.0$ cm rests against the wall and is held in place by a bracket attached to a rod through the center of the roll (Fig. P10.69). The rod turns without friction in the bracket, and the moment of inertia of the paper and rod about the axis is 0.260 kg·m². The other end of the bracket is attached by a

Figure **P10.69**

frictionless hinge to the wall such that the bracket makes an angle of 30.0° with the wall. The weight of the bracket is negligible. The coefficient of kinetic friction between the paper and the wall is $\mu_k = 0.25$. A constant vertical force $F = 60.0$ N is applied to the paper, and the paper unrolls. (a) What is the magnitude of the force that the rod exerts on the paper as it unrolls? (b) What is the magnitude of the angular acceleration of the roll?

10.70 •• A block with mass $m = 5.00$ kg slides down a surface inclined 36.9° to the horizontal (Fig. P10.70). The coefficient of kinetic friction is 0.25. A string attached to the block is wrapped around a flywheel on a fixed axis at *O*. The flywheel has mass 25.0 kg and moment of inertia 0.500 kg·m² with respect to the axis of rotation. The string pulls without slipping at a perpendicular distance of 0.200 m from that axis. (a) What is the acceleration of the block down the plane? (b) What is the tension in the string?

Figure **P10.70**

10.71 ••• Two metal disks, one with radius $R_1 = 2.50$ cm and mass $M_1 = 0.80$ kg and the other with radius $R_2 = 5.00$ cm and mass $M_2 = 1.60$ kg, are welded together and mounted on a frictionless axis through their common center, as in Problem 9.87. (a) A light string is wrapped around the edge of the smaller disk, and a 1.50-kg block is suspended from the free end of the string. What is the magnitude of the downward acceleration of the block after it is released? (b) Repeat the calculation of part (a), this time with the string wrapped around the edge of the larger disk. In which case is the acceleration of the block greater? Does your answer make sense?

10.72 •• A lawn roller in the form of a thin-walled, hollow cylinder with mass M is pulled horizontally with a constant horizontal force F applied by a handle attached to the axle. If it rolls without slipping, find the acceleration and the friction force.

10.73 • Two weights are connected by a very light, flexible cord that passes over an 80.0-N frictionless pulley of radius 0.300 m. The pulley is a solid uniform disk and is supported by a hook connected to the ceiling (Fig. P10.73). What force does the ceiling exert on the hook?

Figure **P10.73**

10.74 •• A solid disk is rolling without slipping on a level surface at a constant speed of 3.60 m/s. (a) If the disk rolls up a 30.0° ramp, how far along the ramp will it move before it stops? (b) Explain why your answer in part (a) does not depend on either the mass or the radius of the disk.

10.75 • **The Yo-yo.** A yo-yo is made from two uniform disks, each with mass m and radius R, connected by a light axle of radius b. A light, thin string is wound several times around the axle and then held stationary while the yo-yo is released from rest, dropping as the string unwinds. Find the linear acceleration and angular acceleration of the yo-yo and the tension in the string.

10.76 •• **CP** A thin-walled, hollow spherical shell of mass m and radius r starts from rest and rolls without slipping down the track shown in Fig. P10.76. Points *A* and *B* are on a circular part of the track having radius R. The diameter of the shell is very small compared to h_0 and R, and the work done by rolling friction is negligible. (a) What is the minimum height h_0 for which this shell will make a complete loop-the-loop on the circular part of the track? (b) How hard does the track push on the shell at point B, which is at the same level as the center of the circle? (c) Suppose that the track had no friction and the shell was released from the same height h_0 you found in part (a). Would it make a complete loop-the-loop? How do you know? (d) In part (c), how hard does the track push on the shell at point A, the top of the circle? How hard did it push on the shell in part (a)?

Figure **P10.76**

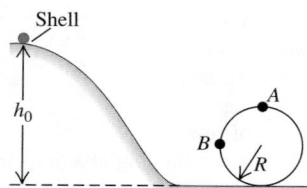

10.77 • Starting from rest, a constant force $F = 100$ N is applied to the free end of a 50-m cable wrapped around the outer rim of a uniform solid cylinder, similar to the situation shown in Fig. 10.9(a). The cylinder has mass 4.00 kg and diameter 30.0 cm and is free to turn about a fixed, frictionless axle through its center. (a) How long does it take to unwrap all the cable, and how fast is the cable moving just as the last bit comes off? (b) Now suppose that the cylinder is replaced by a uniform hoop, with all other quantities remaining unchanged. In this case, would the answers in part (a) be larger or smaller? Explain.

10.78 •• As shown in Fig. E10.20, a string is wrapped several times around the rim of a small hoop with radius 0.0800 m and mass 0.180 kg. The free end of the string is pulled upward in just the right way so that the hoop does not move vertically as the string unwinds. (a) Find the tension in the string as the string unwinds. (b) Find the angular acceleration of the hoop as the string unwinds. (c) Find the upward acceleration of the hand that pulls on the free end of the string. (d) How would your answers be different if the hoop were replaced by a solid disk of the same mass and radius?

10.79 •• A basketball (which can be closely modeled as a hollow spherical shell) rolls down a mountainside into a valley and then up the opposite side, starting from rest at a height H_0 above the bottom. In Fig. P10.79, the rough part of the terrain prevents slipping while the smooth part has no friction. (a) How high, in terms of H_0, will the ball go up the other side? (b) Why doesn't the ball return to height H_0? Has it lost any of its original potential energy?

Figure **P10.79**

10.80 • **CP** A uniform marble rolls without slipping down the path shown in Fig. P10.80, starting from rest. (a) Find the minimum height h required for the marble not to fall into the pit.

(b) The moment of inertia of the marble depends on its radius. Explain why the answer to part (a) does not depend on the radius of the marble. (c) Solve part (a) for a block that slides without friction instead of the rolling marble. How does the minimum h in this case compare to the answer in part (a)?

10.81 •• **Rolling Stones.** A solid, uniform, spherical boulder starts from rest and rolls down a 50.0-m-high hill, as shown in Fig. P10.81. The top half of the hill is rough enough to cause the boulder to roll without slipping, but the lower half is covered with ice and there is no friction. What is the translational speed of the boulder when it reaches the bottom of the hill?

10.82 •• **CP** A solid uniform ball rolls without slipping up a hill, as shown in Fig. P10.82. At the top of the hill, it is moving horizontally, and then it goes over the vertical cliff. (a) How far from the foot of the cliff does the ball land, and how fast is it moving just before it lands? (b) Notice that when the balls lands, it has a greater translational speed than when it was at the bottom of the hill. Does this mean that the ball somehow gained energy? Explain!

10.83 •• A 42.0-cm-diameter wheel, consisting of a rim and six spokes, is constructed from a thin, rigid plastic material having a linear mass density of 25.0 g/cm. This wheel is released from rest at the top of a hill 58.0 m high. (a) How fast is it rolling when it reaches the bottom of the hill? (b) How would your answer change if the linear mass density and the diameter of the wheel were each doubled?

10.84 •• A child rolls a 0.600-kg basketball up a long ramp. The basketball can be considered a thin-walled, hollow sphere. When the child releases the basketball at the bottom of the ramp, it has a speed of 8.0 m/s. When the ball returns to her after rolling up the ramp and then rolling back down, it has a speed of 4.0 m/s. Assume the work done by friction on the basketball is the same when the ball moves up or down the ramp and that the basketball rolls without slipping. Find the maximum vertical height increase of the ball as it rolls up the ramp.

10.85 •• **CP** In a lab experiment you let a uniform ball roll down a curved track. The ball starts from rest and rolls without slipping. While on the track, the ball descends a vertical distance h. The lower end of the track is horizontal and extends over the edge of the lab table; the ball leaves the track traveling horizontally. While free-falling after leaving the track, the ball moves a horizontal distance x and a vertical distance y. (a) Calculate x in terms of h and y, ignoring the work done by friction. (b) Would the answer to part (a) be any different on the moon? (c) Although you do the experiment very carefully, your measured value of x is consistently a bit smaller than the value calculated in part (a). Why? (d) What would x be for the same h and y as in part (a) if you let a silver dollar roll down the track? You can ignore the work done by friction.

Figure **P10.80**

Figure **P10.81**

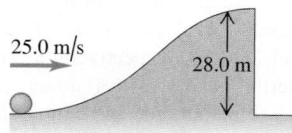

Figure **P10.82**

10.86 •• A uniform drawbridge 8.00 m long is attached to the roadway by a frictionless hinge at one end, and it can be raised by a cable attached to the other end. The bridge is at rest, suspended at 60.0° above the horizontal, when the cable suddenly breaks. (a) Find the angular acceleration of the drawbridge just after the cable breaks. (Gravity behaves as though it all acts at the center of mass.) (b) Could you use the equation $\omega = \omega_0 + \alpha t$ to calculate the angular speed of the drawbridge at a later time? Explain why. (c) What is the angular speed of the drawbridge as it becomes horizontal?

10.87 • A uniform solid cylinder with mass M and radius $2R$ rests on a horizontal tabletop. A string is attached by a yoke to a frictionless axle through the center of the cylinder so that the cylinder can rotate about the axle. The string runs over a disk-shaped pulley with mass M and radius R that is mounted on a frictionless axle through its center. A block of mass M is suspended from the free end of the string (Fig. P10.87). The string doesn't slip over the pulley surface, and the cylinder rolls without slipping on the tabletop. Find the magnitude of the acceleration of the block after the system is released from rest.

Figure **P10.87**

10.88 ••• A uniform, 0.0300-kg rod of length 0.400 m rotates in a horizontal plane about a fixed axis through its center and perpendicular to the rod. Two small rings, each with mass 0.0200 kg, are mounted so that they can slide along the rod. They are initially held by catches at positions 0.0500 m on each side of the center of the rod, and the system is rotating at 30.0 rev/min. With no other changes in the system, the catches are released, and the rings slide outward along the rod and fly off at the ends. (a) What is the angular speed of the system at the instant when the rings reach the ends of the rod? (b) What is the angular speed of the rod after the rings leave it?

10.89 ••• A 5.00-kg ball is dropped from a height of 12.0 m above one end of a uniform bar that pivots at its center. The bar has mass 8.00 kg and is 4.00 m in length. At the other end of the bar sits another 5.00-kg ball, unattached to the bar. The dropped ball sticks to the bar after the collision. How high will the other ball go after the collision?

10.90 •• **Tarzan and Jane in the 21st Century.** Tarzan has foolishly gotten himself into another scrape with the animals and must be rescued once again by Jane. The 60.0-kg Jane starts from rest at a height of 5.00 m in the trees and swings down to the ground using a thin, but very rigid, 30.0-kg vine 8.00 m long. She arrives just in time to snatch the 72.0-kg Tarzan from the jaws of an angry hippopotamus. What is Jane's (and the vine's) angular speed (a) just before she grabs Tarzan and (b) just after she grabs him? (c) How high will Tarzan and Jane go on their first swing after this daring rescue?

10.91 •• A uniform rod of length L rests on a frictionless horizontal surface. The rod pivots about a fixed frictionless axis at one end. The rod is initially at rest. A bullet traveling parallel to the horizontal surface and perpendicular to the rod with speed v strikes the rod at its center and becomes embedded in it. The mass of the bullet is one-fourth the mass of the rod. (a) What is the final angular speed of the rod? (b) What is the ratio of the kinetic energy of the system after the collision to the kinetic energy of the bullet before the collision?

10.92 •• The solid wood door of a gymnasium is 1.00 m wide and 2.00 m high, has total mass 35.0 kg, and is hinged along one side. The door is open and at rest when a stray basketball hits the center of the door head-on, applying an average force of 1500 N to the door for 8.00 ms. Find the angular speed of the door after the impact. [*Hint:* Integrating Eq. (10.29) yields $\Delta L_z = \int_{t_1}^{t_2} (\Sigma \tau_z) dt = (\Sigma \tau_z)_{av} \Delta t$. The quantity $\int_{t_1}^{t_2} (\Sigma \tau_z) dt$ is called the angular impulse.]

10.93 ••• A target in a shooting gallery consists of a vertical square wooden board, 0.250 m on a side and with mass 0.750 kg, that pivots on a horizontal axis along its top edge. The board is struck face-on at its center by a bullet with mass 1.90 g that is traveling at 360 m/s and that remains embedded in the board. (a) What is the angular speed of the board just after the bullet's impact? (b) What maximum height above the equilibrium position does the center of the board reach before starting to swing down again? (c) What minimum bullet speed would be required for the board to swing all the way over after impact?

10.94 •• **Neutron Star Glitches.** Occasionally, a rotating neutron star (see Exercise 10.41) undergoes a sudden and unexpected speedup called a *glitch*. One explanation is that a glitch occurs when the crust of the neutron star settles slightly, decreasing the moment of inertia about the rotation axis. A neutron star with angular speed $\omega_0 = 70.4$ rad/s underwent such a glitch in October 1975 that increased its angular speed to $\omega = \omega_0 + \Delta \omega$, where $\Delta \omega / \omega_0 = 2.01 \times 10^{-6}$. If the radius of the neutron star before the glitch was 11 km, by how much did its radius decrease in the starquake? Assume that the neutron star is a uniform sphere.

10.95 ••• A 500.0-g bird is flying horizontally at 2.25 m/s, not paying much attention, when it suddenly flies into a stationary vertical bar, hitting it 25.0 cm below the top (Fig. P10.95). The bar is uniform, 0.750 m long, has a mass of 1.50 kg, and is hinged at its base. The collision stuns the bird so that it just drops to the ground afterward (but soon recovers to fly happily away). What is the angular velocity of the bar (a) just after it is hit by the bird and (b) just as it reaches the ground?

Figure **P10.95**

10.96 ••• CP A small block with mass 0.250 kg is attached to a string passing through a hole in a frictionless, horizontal surface (see Fig. E10.42). The block is originally revolving in a circle with a radius of 0.800 m about the hole with a tangential speed of 4.00 m/s. The string is then pulled slowly from below, shortening the radius of the circle in which the block revolves. The breaking strength of the string is 30.0 N. What is the radius of the circle when the string breaks?

10.97 • A horizontal plywood disk with mass 7.00 kg and diameter 1.00 m pivots on frictionless bearings about a vertical axis through its center. You attach a circular model-railroad track of negligible mass and average diameter 0.95 m to the disk. A 1.20-kg, battery-driven model train rests on the tracks. To demonstrate conservation of angular momentum, you switch on the train's engine. The train moves counterclockwise, soon attaining a constant speed of 0.600 m/s relative to the tracks. Find the magnitude and direction of the angular velocity of the disk relative to the earth.

10.98 • A 55-kg runner runs around the edge of a horizontal turntable mounted on a vertical, frictionless axis through its center. The runner's velocity relative to the earth has magnitude 2.8 m/s. The turntable is rotating in the opposite direction with an angular velocity of magnitude 0.20 rad/s relative to the earth. The radius of the turntable is 3.0 m, and its moment of inertia about the axis of rotation is 80 kg·m². Find the final angular velocity of the system if the runner comes to rest relative to the turntable. (You can model the runner as a particle.)

10.99 •• **Center of Percussion.** A baseball bat rests on a frictionless, horizontal surface. The bat has a length of 0.900 m, a mass of 0.800 kg, and its center of mass is 0.600 m from the handle end of the bat (Fig. P10.99). The moment of inertia of the bat about its center of mass is 0.0530 kg·m². The bat is struck by a baseball traveling perpendicular to the bat. The impact applies an impulse $J = \int_{t_1}^{t_2} F\, dt$ at a point a distance x from the handle end of the bat. What must x be so that the handle end of the bat remains at rest as the bat begins to move? [*Hint:* Consider the motion of the center of mass and the rotation about the center of mass. Find x so that these two motions combine to give $v = 0$ for the end of the bat just after the collision. Also, note that integration of Eq. (10.29) gives $\Delta L = \int_{t_1}^{t_2}(\Sigma \tau)\, dt$ (see Problem 10.92).] The point on the bat you have located is called the *center of percussion*. Hitting a pitched ball at the center of percussion of the bat minimizes the "sting" the batter experiences on the hands.

Figure **P10.99**

CHALLENGE PROBLEMS

10.100 ••• A uniform ball of radius R rolls without slipping between two rails such that the horizontal distance is d between the two contact points of the rails to the ball. (a) In a sketch, show that at any instant $v_{cm} = \omega\sqrt{R^2 - d^2/4}$. Discuss this expression in the limits $d = 0$ and $d = 2R$. (b) For a uniform ball starting from rest and descending a vertical distance h while rolling without slipping down a ramp, $v_{cm} = \sqrt{10gh/7}$. Replacing the ramp with the two rails, show that

$$v_{cm} = \sqrt{\frac{10gh}{5 + 2/(1 - d^2/4R^2)}}$$

In each case, the work done by friction has been ignored. (c) Which speed in part (b) is smaller? Why? Answer in terms of how the loss of potential energy is shared between the gain in translational and rotational kinetic energies. (d) For which value of the ratio d/R do the two expressions for the speed in part (b) differ by 5.0%? By 0.50%?

10.101 ••• When an object is rolling without slipping, the rolling friction force is much less than the friction force when the object is sliding; a silver dollar will roll on its edge much farther than it will slide on its flat side (see Section 5.3). When an object is rolling without slipping on a horizontal surface, we can approximate the friction force to be zero, so that a_x and α_z are approximately zero and v_x and ω_z are approximately constant. Rolling without slipping means $v_x = r\omega_z$ and $a_x = r\alpha_z$. If an object is set in motion on a surface *without* these equalities, sliding (kinetic) friction will act on the object as it slips until rolling without slipping is established. A solid cylinder with mass M and radius R, rotating with angular speed ω_0 about an axis through its center, is set on a horizontal surface for which the kinetic friction coefficient is μ_k. (a) Draw a free-body diagram for the cylinder on the surface. Think carefully about the direction of the kinetic friction force on the cylinder. Calculate the accelerations a_x of the center of mass and α_z of rotation about the center of mass. (b) The cylinder is initially slipping completely, so initially $\omega_z = \omega_0$ but $v_x = 0$. Rolling without slipping sets in when $v_x = R\omega_z$. Calculate the *distance* the cylinder rolls before slipping stops. (c) Calculate the work done by the friction force on the cylinder as it moves from where it was set down to where it begins to roll without slipping.

10.102 ••• A demonstration gyroscope wheel is constructed by removing the tire from a bicycle wheel 0.650 m in diameter, wrapping lead wire around the rim, and taping it in place. The shaft projects 0.200 m at each side of the wheel, and a woman holds the ends of the shaft in her hands. The mass of the system is 8.00 kg; its entire mass may be assumed to be located at its rim. The shaft is horizontal, and the wheel is spinning about the shaft at 5.00 rev/s. Find the magnitude and direction of the force each hand exerts on the shaft (a) when the shaft is at rest; (b) when the shaft is rotating in a horizontal plane about its center at 0.050 rev/s; (c) when the shaft is rotating in a horizontal plane about its center at 0.300 rev/s. (d) At what rate must the shaft rotate in order that it may be supported at one end only?

10.103 ••• **CP CALC** A block with mass m is revolving with linear speed v_1 in a circle of radius r_1 on a frictionless horizontal surface (see Fig. E10.42). The string is slowly pulled from below until the radius of the circle in which the block is revolving is reduced to r_2. (a) Calculate the tension T in the string as a function of r, the distance of the block from the hole. Your answer will be in terms of the initial velocity v_1 and the radius r_1. (b) Use $W = \int_{r_1}^{r_2} \vec{T}(r) \cdot d\vec{r}$ to calculate the work done by $\vec{T}$ when r changes from r_1 to r_2. (c) Compare the results of part (b) to the change in the kinetic energy of the block.

Answers

Chapter Opening Question ?

The earth precesses like a top due to torques exerted on it by the sun and moon. As a result, its rotation axis (which passes through the earth's north and south poles) slowly changes its orientation relative to the distant stars, taking 26,000 years for a complete cycle of precession. Today the rotation axis points toward Polaris, but 5000 years ago it pointed toward Thuban, and 12,000 years from now it will point toward the bright star Vega.

Test Your Understanding Questions

10.1 Answer: (ii) The force P acts along a vertical line, so the lever arm is the horizontal distance from A to the line of action. This is the horizontal component of the distance L, which is $L\cos\theta$. Hence the magnitude of the torque is the product of the force magnitude P and the lever arm $L\cos\theta$, or $\tau = PL\cos\theta$.

10.2 Answer: (iii), (ii), (i) In order for the hanging object of mass m_2 to accelerate downward, the net force on it must be downward. Hence the magnitude m_2g of the downward weight force must be greater than the magnitude T_2 of the upward tension force. In order for the pulley to have a clockwise angular acceleration, the net torque on the pulley must be clockwise. The tension T_2 tends to rotate the pulley clockwise, while the tension T_1 tends to rotate the pulley counterclockwise. Both tension forces have the same lever arm R, so there is a clockwise torque T_2R and a counterclockwise torque T_1R. In order for the net torque to be clockwise, T_2 must be greater than T_1. Hence $m_2g > T_2 > T_1$.

10.3 Answers: (a) (ii), (b) (i) If you redo the calculation of Example 10.6 with a hollow cylinder (moment of inertia $I_{cm} = MR^2$) instead of a solid cylinder (moment of inertia $I_{cm} = \frac{1}{2}MR^2$), you will find $a_{cm\text{-}y} = \frac{1}{2}g$ and $T = \frac{1}{2}Mg$ (instead of $a_{cm\text{-}y} = \frac{2}{3}g$ and $T = \frac{1}{3}Mg$ for a solid cylinder). Hence the acceleration is less but the tension is greater. You can come to the same conclusion without doing the calculation. The greater moment of inertia means that the hollow cylinder will rotate more slowly and hence will roll downward more slowly. In order to slow the downward motion, a greater upward tension force is needed to oppose the downward force of gravity.

10.4 Answer: (iii) You apply the same torque over the same angular displacement to both cylinders. Hence, by Eq. (10.21), you do the same amount of work to both cylinders and impart the same kinetic energy to both. (The one with the smaller moment of inertia ends up with a greater angular speed, but that isn't what we are asked. Compare Conceptual Example 6.5 in Section 6.2.)

10.5 Answers: (a) no, (b) yes As the ball goes around the circle, the magnitude of $\vec{p} = m\vec{v}$ remains the same (the speed is constant) but its direction changes, so the linear momentum vector isn't constant. But $\vec{L} = \vec{r} \times \vec{p}$ is constant: It has a constant magnitude (the speed and the perpendicular distance from your hand to the ball are both constant) and a constant direction (along the rotation axis, perpendicular to the plane of the ball's motion). The linear momentum changes because there is a net *force* $\vec{F}$ on the ball (toward the center of the circle). The angular momentum remains constant because there is no net *torque;* the vector $\vec{r}$ points from your hand to the ball and the force $\vec{F}$ on the ball is directed toward your hand, so the vector product $\vec{\tau} = \vec{r} \times \vec{F}$ is zero.

10.6 Answer: (i) In the absence of any external torques, the earth's angular momentum $L_z = I\omega_z$ would remain constant. The melted ice would move from the poles toward the equator—that is, away from our planet's rotation axis—and the earth's moment of inertia I would increase slightly. Hence the angular velocity ω_z would decrease slightly and the day would be slightly longer.

10.7 Answer: (iii) Doubling the flywheel mass would double both its moment of inertia I and its weight w, so the ratio I/w would be unchanged. Equation (10.33) shows that the precession angular speed depends on this ratio, so there would be *no* effect on the value of Ω.

Bridging Problem

Answers: (a) $h = \dfrac{2R}{5}$

(b) $\frac{5}{7}$ of the speed it had just after the hit

11 EQUILIBRIUM AND ELASTICITY

LEARNING GOALS

By studying this chapter, you will learn:

- The conditions that must be satisfied for a body or structure to be in equilibrium.

- What is meant by the center of gravity of a body, and how it relates to the body's stability.

- How to solve problems that involve rigid bodies in equilibrium.

- How to analyze situations in which a body is deformed by tension, compression, pressure, or shear.

- What happens when a body is stretched so much that it deforms or breaks.

? This Roman aqueduct uses the principle of the arch to sustain the weight of the structure and the water it carries. Are the blocks that make up the arch being compressed, stretched, or a combination?

We've devoted a good deal of effort to understanding why and how bodies accelerate in response to the forces that act on them. But very often we're interested in making sure that bodies *don't* accelerate. Any building, from a multistory skyscraper to the humblest shed, must be designed so that it won't topple over. Similar concerns arise with a suspension bridge, a ladder leaning against a wall, or a crane hoisting a bucket full of concrete.

A body that can be modeled as a *particle* is in equilibrium whenever the vector sum of the forces acting on it is zero. But for the situations we've just described, that condition isn't enough. If forces act at different points on an extended body, an additional requirement must be satisfied to ensure that the body has no tendency to *rotate:* The sum of the *torques* about any point must be zero. This requirement is based on the principles of rotational dynamics developed in Chapter 10. We can compute the torque due to the weight of a body using the concept of center of gravity, which we introduce in this chapter.

Rigid bodies don't bend, stretch, or squash when forces act on them. But the rigid body is an idealization; all real materials are *elastic* and do deform to some extent. Elastic properties of materials are tremendously important. You want the wings of an airplane to be able to bend a little, but you'd rather not have them break off. The steel frame of an earthquake-resistant building has to be able to flex, but not too much. Many of the necessities of everyday life, from rubber bands to suspension bridges, depend on the elastic properties of materials. In this chapter we'll introduce the concepts of *stress, strain,* and *elastic modulus* and a simple principle called *Hooke's law* that helps us predict what deformations will occur when forces are applied to a real (not perfectly rigid) body.

11.1 Conditions for Equilibrium

We learned in Sections 4.2 and 5.1 that a particle is in *equilibrium*—that is, the particle does not accelerate—in an inertial frame of reference if the vector sum of all the forces acting on the particle is zero, $\sum \vec{F} = 0$. For an *extended* body, the equivalent statement is that the center of mass of the body has zero acceleration if the vector sum of all external forces acting on the body is zero, as discussed in Section 8.5. This is often called the **first condition for equilibrium.** In vector and component forms,

$$\sum \vec{F} = 0$$

$$\sum F_x = 0 \qquad \sum F_y = 0 \qquad \sum F_z = 0 \qquad \begin{array}{l}\text{(first condition}\\ \text{for equilibrium)}\end{array} \qquad (11.1)$$

A second condition for an extended body to be in equilibrium is that the body must have no tendency to *rotate*. This condition is based on the dynamics of rotational motion in exactly the same way that the first condition is based on Newton's first law. A rigid body that, in an inertial frame, is not rotating about a certain point has zero angular momentum about that point. If it is not to start rotating about that point, the rate of change of angular momentum must *also* be zero. From the discussion in Section 10.5, particularly Eq. (10.29), this means that the sum of torques due to all the external forces acting on the body must be zero. A rigid body in equilibrium can't have any tendency to start rotating about *any* point, so the sum of external torques must be zero about any point. This is the **second condition for equilibrium:**

$$\sum \vec{\tau} = 0 \quad \text{about any point} \qquad \text{(second condition for equilibrium)} \qquad (11.2)$$

The sum of the torques due to all external forces acting on the body, with respect to any specified point, must be zero.

In this chapter we will apply the first and second conditions for equilibrium to situations in which a rigid body is at rest (no translation or rotation). Such a body is said to be in **static equilibrium** (Fig. 11.1). But the same conditions apply to a rigid body in uniform *translational* motion (without rotation), such as an airplane in flight with constant speed, direction, and altitude. Such a body is in equilibrium but is not static.

Test Your Understanding of Section 11.1 Which situation satisfies both the first and second conditions for equilibrium? (i) a seagull gliding at a constant angle below the horizontal and at a constant speed; (ii) an automobile crankshaft turning at an increasing angular speed in the engine of a parked car; (iii) a thrown baseball that does not rotate as it sails through the air.

11.2 Center of Gravity

In most equilibrium problems, one of the forces acting on the body is its weight. We need to be able to calculate the *torque* of this force. The weight doesn't act at a single point; it is distributed over the entire body. But we can always calculate the torque due to the body's weight by assuming that the entire force of gravity (weight) is concentrated at a point called the **center of gravity** (abbreviated "cg"). The acceleration due to gravity decreases with altitude; but if we can ignore this variation over the vertical dimension of the body, then the body's center of gravity is identical to its *center of mass* (abbreviated "cm"), which we defined in Section 8.5. We stated this result without proof in Section 10.2, and now we'll prove it.

11.1 To be in static equilibrium, a body at rest must satisfy *both* conditions for equilibrium: It can have no tendency to accelerate as a whole or to start rotating.

(a) This body is in static equilibrium.

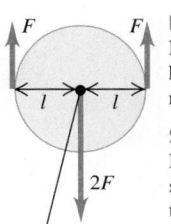

Equilibrium conditions:

First condition satisfied: Net force = 0, so body at rest has no tendency to start moving as a whole.

Second condition satisfied: Net torque about the axis = 0, so body at rest has no tendency to start rotating.

(b) This body has no tendency to accelerate as a whole, but it has a tendency to start rotating.

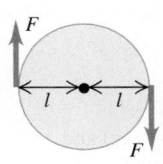

First condition satisfied: Net force = 0, so body at rest has no tendency to start moving as a whole.

Second condition NOT satisfied: There is a net clockwise torque about the axis, so body at rest will start rotating clockwise.

(c) This body has a tendency to accelerate as a whole but no tendency to start rotating.

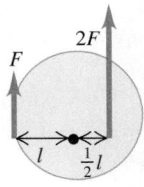

First condition NOT satisfied: There is a net upward force, so body at rest will start moving upward.

Second condition satisfied: Net torque about the axis = 0, so body at rest has no tendency to start rotating.

First let's review the definition of the center of mass. For a collection of particles with masses m_1, m_2,... and coordinates (x_1, y_1, z_1), (x_2, y_2, z_2),..., the coordinates x_{cm}, y_{cm}, and z_{cm} of the center of mass are given by

$$x_{cm} = \frac{m_1 x_1 + m_2 x_2 + m_3 x_3 + \cdots}{m_1 + m_2 + m_3 + \cdots} = \frac{\sum\limits_i m_i x_i}{\sum\limits_i m_i}$$

$$y_{cm} = \frac{m_1 y_1 + m_2 y_2 + m_3 y_3 + \cdots}{m_1 + m_2 + m_3 + \cdots} = \frac{\sum\limits_i m_i y_i}{\sum\limits_i m_i} \quad \text{(center of mass)} \quad (11.3)$$

$$z_{cm} = \frac{m_1 z_1 + m_2 z_2 + m_3 z_3 + \cdots}{m_1 + m_2 + m_3 + \cdots} = \frac{\sum\limits_i m_i z_i}{\sum\limits_i m_i}$$

Also, x_{cm}, y_{cm}, and z_{cm} are the components of the position vector $\vec{r}_{cm}$ of the center of mass, so Eqs. (11.3) are equivalent to the vector equation

$$\vec{r}_{cm} = \frac{m_1 \vec{r}_1 + m_2 \vec{r}_2 + m_3 \vec{r}_3 + \cdots}{m_1 + m_2 + m_3 + \cdots} = \frac{\sum\limits_i m_i \vec{r}_i}{\sum\limits_i m_i} \quad (11.4)$$

11.2 The center of gravity (cg) and center of mass (cm) of an extended body.

The gravitational torque about O on a particle of mass m_i within the body is: $\vec{\tau}_i = \vec{r}_i \times \vec{w}_i$.

If $\vec{g}$ has the same value at all points on the body, the cg is identical to the cm.

The net gravitational torque about O on the entire body can be found by assuming that all the weight acts at the cg: $\vec{\tau} = \vec{r}_{cm} \times \vec{w}$.

Now consider the gravitational torque on a body of arbitrary shape (Fig. 11.2). We assume that the acceleration due to gravity $\vec{g}$ is the same at every point in the body. Every particle in the body experiences a gravitational force, and the total weight of the body is the vector sum of a large number of parallel forces. A typical particle has mass m_i and weight $\vec{w}_i = m_i \vec{g}$. If $\vec{r}_i$ is the position vector of this particle with respect to an arbitrary origin O, then the torque vector $\vec{\tau}_i$ of the weight $\vec{w}_i$ with respect to O is, from Eq. (10.3),

$$\vec{\tau}_i = \vec{r}_i \times \vec{w}_i = \vec{r}_i \times m_i \vec{g}$$

The *total* torque due to the gravitational forces on all the particles is

$$\vec{\tau} = \sum_i \vec{\tau}_i = \vec{r}_1 \times m_1 \vec{g} + \vec{r}_2 \times m_2 \vec{g} + \cdots$$
$$= (m_1 \vec{r}_1 + m_2 \vec{r}_2 + \cdots) \times \vec{g}$$
$$= \left(\sum_i m_i \vec{r}_i \right) \times \vec{g}$$

When we multiply and divide this by the total mass of the body,

$$M = m_1 + m_2 + \cdots = \sum_i m_i$$

we get

$$\vec{\tau} = \frac{m_1 \vec{r}_1 + m_2 \vec{r}_2 + \cdots}{m_1 + m_2 + \cdots} \times M\vec{g} = \frac{\sum\limits_i m_i \vec{r}_i}{\sum\limits_i m_i} \times M\vec{g}$$

The fraction in this equation is just the position vector $\vec{r}_{cm}$ of the center of mass, with components x_{cm}, y_{cm}, and z_{cm}, as given by Eq. (11.4), and $M\vec{g}$ is equal to the total weight $\vec{w}$ of the body. Thus

$$\vec{\tau} = \vec{r}_{cm} \times M\vec{g} = \vec{r}_{cm} \times \vec{w} \quad (11.5)$$

The total gravitational torque, given by Eq. (11.5), is the same as though the total weight $\vec{w}$ were acting on the position $\vec{r}_{cm}$ of the center of mass, which we also call the *center of gravity*. **If $\vec{g}$ has the same value at all points on a body, its center of gravity is identical to its center of mass.** Note, however, that the center of mass is defined independently of any gravitational effect.

While the value of $\vec{g}$ does vary somewhat with elevation, the variation is extremely slight (Fig. 11.3). Hence we will assume throughout this chapter that the center of gravity and center of mass are identical unless explicitly stated otherwise.

Finding and Using the Center of Gravity

We can often use symmetry considerations to locate the center of gravity of a body, just as we did for the center of mass. The center of gravity of a homogeneous sphere, cube, circular sheet, or rectangular plate is at its geometric center. The center of gravity of a right circular cylinder or cone is on its axis of symmetry.

For a body with a more complex shape, we can sometimes locate the center of gravity by thinking of the body as being made of symmetrical pieces. For example, we could approximate the human body as a collection of solid cylinders, with a sphere for the head. Then we can locate the center of gravity of the combination with Eqs. (11.3), letting $m_1, m_2, \ldots$ be the masses of the individual pieces and (x_1, y_1, z_1), $(x_2, y_2, z_2), \ldots$ be the coordinates of their centers of gravity.

When a body acted on by gravity is supported or suspended at a single point, the center of gravity is always at or directly above or below the point of suspension. If it were anywhere else, the weight would have a torque with respect to the point of suspension, and the body could not be in rotational equilibrium. Figure 11.4 shows how to use this fact to determine experimentally the location of the center of gravity of an irregular body.

Using the same reasoning, we can see that a body supported at several points must have its center of gravity somewhere within the area bounded by the supports. This explains why a car can drive on a straight but slanted road if the slant angle is relatively small (Fig. 11.5a) but will tip over if the angle is too steep (Fig. 11.5b). The truck in Fig. 11.5c has a higher center of gravity than the car and will tip over on a shallower incline. When a truck overturns on a highway and blocks traffic for hours, it's the high center of gravity that's to blame.

The lower the center of gravity and the larger the area of support, the more difficult it is to overturn a body. Four-legged animals such as deer and horses have a large area of support bounded by their legs; hence they are naturally stable and need only small feet or hooves. Animals that walk erect on two legs, such as humans and birds, need relatively large feet to give them a reasonable area of

11.3 The acceleration due to gravity at the bottom of the 452-m-tall Petronas Towers in Malaysia is only 0.014% greater than at the top. The center of gravity of the towers is only about 2 cm below the center of mass.

11.4 Finding the center of gravity of an irregularly shaped body—in this case, a coffee mug.

What is the center of gravity of this mug?

(1) Suspend the mug from any point. A vertical line extending down from the point of suspension passes through the center of gravity.

(2) Now suspend the mug from a different point. A vertical line extending down from this point intersects the first line at the center of gravity (which is inside the mug).

Center of gravity

11.5 In (a) the center of gravity is within the area bounded by the supports, and the car is in equilibrium. The car in (b) and the truck in (c) will tip over because their centers of gravity lie outside the area of support.

(a) Center of gravity is over the area of support: car is in equilibrium.

(b) (c) The higher the center of gravity, the smaller the incline needed to tip the vehicle over.

Center of gravity is outside the area of support: vehicle tips over.

support. If a two-legged animal holds its body approximately horizontal, like a chicken or the dinosaur *Tyrannosaurus rex,* it must perform a delicate balancing act as it walks to keep its center of gravity over the foot that is on the ground. A chicken does this by moving its head; *T. rex* probably did it by moving its massive tail.

Example 11.1 Walking the plank

A uniform plank of length $L = 6.0$ m and mass $M = 90$ kg rests on sawhorses separated by $D = 1.5$ m and equidistant from the center of the plank. Cousin Throckmorton wants to stand on the right-hand end of the plank. If the plank is to remain at rest, how massive can Throckmorton be?

SOLUTION

IDENTIFY and SET UP: To just balance, Throckmorton's mass m must be such that the center of gravity of the plank–Throcky system is directly over the right-hand sawhorse (Fig. 11.6). We take the origin at C, the geometric center and center of gravity of the plank, and take the positive x-axis horizontally to the right. Then the centers of gravity of the plank and Throcky are at $x_P = 0$ and $x_T = L/2 = 3.0$ m, respectively, and the right-hand sawhorse is at

11.6 Our sketch for this problem.

$x_S = D/2$. We'll use Eqs. (11.3) to locate the center of gravity x_{cg} of the plank–Throcky system.

EXECUTE: From the first of Eqs. (11.3),

$$x_{cg} = \frac{M(0) + m(L/2)}{M + m} = \frac{m}{M + m}\frac{L}{2}$$

We set $x_{cg} = x_S$ and solve for m:

$$\frac{m}{M + m}\frac{L}{2} = \frac{D}{2}$$

$$mL = (M + m)D$$

$$m = M\frac{D}{L - D} = (90 \text{ kg})\frac{1.5 \text{ m}}{6.0 \text{ m} - 1.5 \text{ m}} = 30 \text{ kg}$$

EVALUATE: As a check, let's repeat the calculation with the origin at the right-hand sawhorse. Now $x_S = 0$, $x_P = -D/2$, and $x_T = (L/2) - (D/2)$, and we require $x_{cg} = x_S = 0$:

$$x_{cg} = \frac{M(-D/2) + m[(L/2) - (D/2)]}{M + m} = 0$$

$$m = \frac{MD/2}{(L/2) - (D/2)} = M\frac{D}{L - D} = 30 \text{ kg}$$

The result doesn't depend on our choice of origin.

A 60-kg adult could stand only halfway between the right-hand sawhorse and the end of the plank. Can you see why?

11.7 At what point will the meter stick with rock attached be in balance?

Rock, mass m Meter stick, mass m

Test Your Understanding of Section 11.2 A rock is attached to the left end of a uniform meter stick that has the same mass as the rock. In order for the combination of rock and meter stick to balance atop the triangular object in Fig. 11.7, how far from the left end of the stick should the triangular object be placed? (i) less than 0.25 m; (ii) 0.25 m; (iii) between 0.25 m and 0.50 m; (iv) 0.50 m; (v) more than 0.50 m.

Mastering PHYSICS

ActivPhysics 7.4: Two Painters on a Beam
ActivPhysics 7.5: Lecturing from a Beam

11.3 Solving Rigid-Body Equilibrium Problems

There are just two key conditions for rigid-body equilibrium: The vector sum of the forces on the body must be zero, and the sum of the torques about any point must be zero. To keep things simple, we'll restrict our attention to situations in which we can treat all forces as acting in a single plane, which we'll call the xy-plane. Then we can ignore the condition $\sum F_z = 0$ in Eqs. (11.1), and in Eq. (11.2) we need consider only the z-components of torque (perpendicular to the plane). The first and second conditions for equilibrium are then

$$\sum F_x = 0 \quad \text{and} \quad \sum F_y = 0 \quad \begin{array}{l}\text{(first condition for equilibrium,}\\ \text{forces in xy-plane)}\end{array}$$

$$\sum \tau_z = 0 \quad \begin{array}{l}\text{(second condition for equilibrium,}\\ \text{forces in xy-plane)}\end{array}$$

(11.6)

CAUTION **Choosing the reference point for calculating torques** In equilibrium problems, the choice of reference point for calculating torques in $\Sigma\tau_z$ is completely arbitrary. But once you make your choice, you must use the *same* point to calculate *all* the torques on a body. Choose the point so as to simplify the calculations as much as possible. █

The challenge is to apply these simple conditions to specific problems. Problem-Solving Strategy 11.1 is very similar to the suggestions given in Section 5.2 for the equilibrium of a particle. You should compare it with Problem-Solving Strategy 10.1 (Section 10.2) for rotational dynamics problems.

Problem-Solving Strategy 11.1   Equilibrium of a Rigid Body

IDENTIFY *the relevant concepts:* The first and second conditions for equilibrium ($\Sigma F_x = 0$, $\Sigma F_y = 0$, and $\Sigma\tau_z = 0$) are applicable to any rigid body that is not accelerating in space and not rotating.

SET UP *the problem* using the following steps:
1. Sketch the physical situation and identify the body in equilibrium to be analyzed. Sketch the body accurately; do *not* represent it as a point. Include dimensions.
2. Draw a free-body diagram showing all forces acting *on* the body. Show the point on the body at which each force acts.
3. Choose coordinate axes and specify their direction. Specify a positive direction of rotation for torques. Represent forces in terms of their components with respect to the chosen axes.
4. Choose a reference point about which to compute torques. Choose wisely; you can eliminate from your torque equation any force whose line of action goes through the point you

choose. The body doesn't actually have to be pivoted about an axis through the reference point.

EXECUTE *the solution* as follows:
1. Write the equations expressing the equilibrium conditions. Remember that $\Sigma F_x = 0$, $\Sigma F_y = 0$, and $\Sigma\tau_z = 0$ are *separate* equations. You can compute the torque of a force by finding the torque of each of its components separately, each with its appropriate lever arm and sign, and adding the results.
2. To obtain as many equations as you have unknowns, you may need to compute torques with respect to two or more reference points; choose them wisely, too.

EVALUATE *your answer:* Check your results by writing $\Sigma\tau_z = 0$ with respect to a different reference point. You should get the same answers.

Example 11.2   Weight distribution for a car

An auto magazine reports that a certain sports car has 53% of its weight on the front wheels and 47% on its rear wheels. (That is, the total normal forces on the front and rear wheels are $0.53w$ and $0.47w$, respectively, where w is the car's weight.) The distance between the axles is 2.46 m. How far in front of the rear axle is the car's center of gravity?

SOLUTION

IDENTIFY and SET UP: We can use the two conditions for equilibrium, Eqs. (11.6), for a car at rest (or traveling in a straight line at constant speed), since the net force and net torque on the car are zero. Figure 11.8 shows our sketch and a free-body diagram, including x- and y-axes and our convention that counterclockwise torques are positive. The weight w acts at the center of gravity. Our target variable is the distance L_{cg}, the lever arm of the weight with respect to the rear axle R, so it is wise to take torques with respect to R. The torque due to the weight is negative because it tends to cause a clockwise rotation about R. The torque due to the upward normal force at the front axle F is positive because it tends to cause a counterclockwise rotation about R.

EXECUTE: The first condition for equilibrium is satisfied (see Fig. 11.8b): $\Sigma F_x = 0$ because there are no x-components of force and $\Sigma F_y = 0$ because $0.47w + 0.53w + (-w) = 0$. We write the torque equation and solve for L_{cg}:

$$\Sigma\tau_R = 0.47w(0) - wL_{cg} + 0.53w(2.46 \text{ m}) = 0$$
$$L_{cg} = 1.30 \text{ m}$$

11.8 Our sketches for this problem.

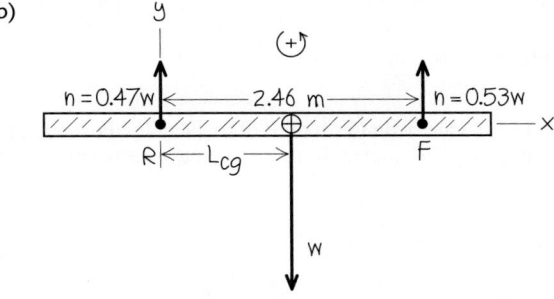

EVALUATE: The center of gravity is between the two supports, as it must be (see Section 11.2). You can check our result by writing the torque equation about the front axle F. You'll find that the center of gravity is 1.16 m behind the front axle, or (2.46 m) − (1.16 m) = 1.30 m in front of the rear axle.

Example 11.3 Will the ladder slip?

Sir Lancelot, who weighs 800 N, is assaulting a castle by climbing a uniform ladder that is 5.0 m long and weighs 180 N (Fig. 11.9a). The bottom of the ladder rests on a ledge and leans across the moat in equilibrium against a frictionless, vertical castle wall. The ladder makes an angle of 53.1° with the horizontal. Lancelot pauses one-third of the way up the ladder. (a) Find the normal and friction forces on the base of the ladder. (b) Find the minimum coefficient of static friction needed to prevent slipping at the base. (c) Find the magnitude and direction of the contact force on the base of the ladder.

SOLUTION

IDENTIFY and SET UP: The ladder–Lancelot system is stationary, so we can use the two conditions for equilibrium to solve part (a). In part (b), we need the relationship among the static friction force, the coefficient of static friction, and the normal force (see Section 5.3). In part (c), the contact force is the vector sum of the normal and friction forces acting at the base of the ladder, found in part (a). Figure 11.9b shows the free-body diagram, with x- and y-directions as shown and with counterclockwise torques taken to be positive. The ladder's center of gravity is at its geometric center. Lancelot's 800-N weight acts at a point one-third of the way up the ladder.

The wall exerts only a normal force n_1 on the top of the ladder. The forces on the base are an upward normal force n_2 and a static friction force f_s, which must point to the right to prevent slipping. The magnitudes n_2 and f_s are the target variables in part (a). From Eq. (5.6), these magnitudes are related by $f_s \leq \mu_s n_2$; the coefficient of static friction μ_s is the target variable in part (b).

EXECUTE: (a) From Eqs. (11.6), the first condition for equilibrium gives

$$\sum F_x = f_s + (-n_1) = 0$$
$$\sum F_y = n_2 + (-800 \text{ N}) + (-180 \text{ N}) = 0$$

These are two equations for the three unknowns n_1, n_2, and f_s. The second equation gives $n_2 = 980$ N. To obtain a third equation, we use the second condition for equilibrium. We take torques about point B, about which n_2 and f_s have no torque. The 53.1° angle creates a 3-4-5 right triangle, so from Fig. 11.9b the lever arm for the ladder's weight is 1.5 m, the lever arm for Lancelot's weight is 1.0 m, and the lever arm for n_1 is 4.0 m. The torque equation for point B is then

$$\sum \tau_B = n_1(4.0 \text{ m}) - (180 \text{ N})(1.5 \text{ m})$$
$$- (800 \text{ N})(1.0 \text{ m}) + n_2(0) + f_s(0) = 0$$

Solving for n_1, we get $n_1 = 268$ N. We substitute this into the $\sum F_x = 0$ equation and get $f_s = 268$ N.

(b) The static friction force f_s cannot exceed $\mu_s n_2$, so the *minimum* coefficient of static friction to prevent slipping is

$$(\mu_s)_{\text{min}} = \frac{f_s}{n_2} = \frac{268 \text{ N}}{980 \text{ N}} = 0.27$$

(c) The components of the contact force $\vec{F}_B$ at the base are the static friction force f_s and the normal force n_2, so

$$\vec{F}_B = f_s \hat{\imath} + n_2 \hat{\jmath} = (268 \text{ N})\hat{\imath} + (980 \text{ N})\hat{\jmath}$$

The magnitude and direction of $\vec{F}_B$ (Fig. 11.9c) are

$$F_B = \sqrt{(268 \text{ N})^2 + (980 \text{ N})^2} = 1020 \text{ N}$$
$$\theta = \arctan \frac{980 \text{ N}}{268 \text{ N}} = 75°$$

EVALUATE: As Fig. 11.9c shows, the contact force $\vec{F}_B$ is *not* directed along the length of the ladder. Can you show that if $\vec{F}_B$ were directed along the ladder, there would be a net counterclockwise torque with respect to the top of the ladder, and equilibrium would be impossible?

As Lancelot climbs higher on the ladder, the lever arm and torque of his weight about B increase. This increases the values of n_1, f_s, and the required friction coefficient $(\mu_s)_{\text{min}}$, so the ladder is more and more likely to slip as he climbs (see Problem 11.10). A simple way to make slipping less likely is to use a larger ladder angle (say, 75° rather than 53.1°). This decreases the lever arms with respect to B of the weights of the ladder and Lancelot and increases the lever arm of n_1, all of which decrease the required friction force.

If we had assumed friction on the wall as well as on the floor, the problem would be impossible to solve by using the equilibrium conditions alone. (Try it!) The difficulty is that it's no longer adequate to treat the body as being perfectly rigid. Another problem of this kind is a four-legged table; there's no way to use the equilibrium conditions alone to find the force on each separate leg.

11.9 (a) Sir Lancelot pauses a third of the way up the ladder, fearing it will slip. (b) Free-body diagram for the system of Sir Lancelot and the ladder. (c) The contact force at B is the superposition of the normal force and the static friction force.

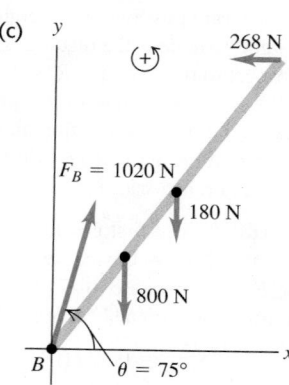

Example 11.4 Equilibrium and pumping iron

Figure 11.10a shows a horizontal human arm lifting a dumbbell. The forearm is in equilibrium under the action of the weight $\vec{w}$ of the dumbbell, the tension $\vec{T}$ in the tendon connected to the biceps muscle, and the force $\vec{E}$ exerted on the forearm by the upper arm at the elbow joint. We neglect the weight of the forearm itself. (For clarity, the point A where the tendon is attached is drawn farther from the elbow than its actual position.) Given the weight w and the angle θ between the tension force and the horizontal, find T and the two components of $\vec{E}$ (three unknown scalar quantities in all).

SOLUTION

IDENTIFY and SET UP: The system is at rest, so we use the conditions for equilibrium. We represent $\vec{T}$ and $\vec{E}$ in terms of their components (Fig. 11.10b). We guess that the directions of E_x and E_y are as shown; the signs of E_x and E_y as given by our solution will tell us the actual directions. Our target variables are T, E_x, and E_y.

EXECUTE: To find T, we take torques about the elbow joint so that the torque equation does not contain E_x, E_y, or T_x:

$$\sum \tau_{\text{elbow}} = Lw - DT_y = 0$$

From this we find

$$T_y = \frac{Lw}{D} \quad \text{and} \quad T = \frac{Lw}{D \sin \theta}$$

To find E_x and E_y, we use the first conditions for equilibrium:

$$\sum F_x = T_x + (-E_x) = 0$$

$$E_x = T_x = T \cos \theta = \frac{Lw}{D \sin \theta} \cos \theta$$

$$= \frac{Lw}{D} \cot \theta = \frac{Lw}{D} \frac{D}{h} = \frac{Lw}{h}$$

$$\sum F_y = T_y + E_y + (-w) = 0$$

$$E_y = w - \frac{Lw}{D} = -\frac{(L-D)w}{D}$$

The negative sign for E_y tells us that it should actually point *down* in Fig. 11.10b.

EVALUATE: We can check our results for E_x and E_y by taking torques about points A and B, about both of which T has zero torque:

$$\sum \tau_A = (L-D)w + DE_y = 0 \quad \text{so} \quad E_y = -\frac{(L-D)w}{D}$$

$$\sum \tau_B = Lw - hE_x = 0 \quad \text{so} \quad E_x = \frac{Lw}{h}$$

As a realistic example, take $w = 200$ N, $D = 0.050$ m, $L = 0.30$ m, and $\theta = 80°$, so that $h = D \tan \theta = (0.050 \text{ m})(5.67) = 0.28$ m. Using our results for T, E_x, and E_y, we find

$$T = \frac{Lw}{D \sin \theta} = \frac{(0.30 \text{ m})(200 \text{ N})}{(0.050 \text{ m})(0.98)} = 1220 \text{ N}$$

$$E_y = -\frac{(L-D)w}{D} = -\frac{(0.30 \text{ m} - 0.050 \text{ m})(200 \text{ N})}{0.050 \text{ m}}$$

$$= -1000 \text{ N}$$

$$E_x = \frac{Lw}{h} = \frac{(0.30 \text{ m})(200 \text{ N})}{0.28 \text{ m}} = 210 \text{ N}$$

The magnitude of the force at the elbow is

$$E = \sqrt{E_x{}^2 + E_y{}^2} = 1020 \text{ N}$$

The large values of T and E suggest that it was reasonable to neglect the weight of the forearm itself, which may be 20 N or so.

11.10 (a) The situation. (b) Our free-body diagram for the forearm. The weight of the forearm is neglected, and the distance D is greatly exaggerated for clarity.

(a)

(b)

We don't know the sign of this component; we draw it positive for convenience.

11.11 What are the tension in the diagonal cable and the force exerted by the hinge at P?

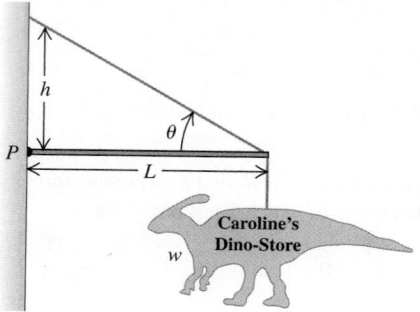

Test Your Understanding of Section 11.3 A metal advertising sign (weight w) for a specialty shop is suspended from the end of a horizontal rod of length L and negligible mass (Fig. 11.11). The rod is supported by a cable at an angle θ from the horizontal and by a hinge at point P. Rank the following force magnitudes in order from greatest to smallest: (i) the weight w of the sign; (ii) the tension in the cable; (iii) the vertical component of force exerted on the rod by the hinge at P.

11.4 Stress, Strain, and Elastic Moduli

The rigid body is a useful idealized model, but the stretching, squeezing, and twisting of real bodies when forces are applied are often too important to ignore. Figure 11.12 shows three examples. We want to study the relationship between the forces and deformations for each case.

For each kind of deformation we will introduce a quantity called **stress** that characterizes the strength of the forces causing the deformation, on a "force per unit area" basis. Another quantity, **strain**, describes the resulting deformation. When the stress and strain are small enough, we often find that the two are directly proportional, and we call the proportionality constant an **elastic modulus.** The harder you pull on something, the more it stretches; the more you squeeze it, the more it compresses. In equation form, this says

$$\frac{\text{Stress}}{\text{Strain}} = \text{Elastic modulus} \qquad \text{(Hooke's law)} \qquad (11.7)$$

The proportionality of stress and strain (under certain conditions) is called **Hooke's law,** after Robert Hooke (1635–1703), a contemporary of Newton. We used one form of Hooke's law in Sections 6.3 and 7.2: The elongation of an ideal spring is proportional to the stretching force. Remember that Hooke's "law" is not really a general law; it is valid over only a limited range. The last section of this chapter discusses what this limited range is.

Tensile and Compressive Stress and Strain

The simplest elastic behavior to understand is the stretching of a bar, rod, or wire when its ends are pulled (Fig. 11.12a). Figure 11.13 shows an object that initially has uniform cross-sectional area A and length l_0. We then apply forces of equal

11.12 Three types of stress. **(a)** Bridge cables under *tensile stress,* being stretched by forces acting at their ends. **(b)** A diver under *bulk stress,* being squeezed from all sides by forces due to water pressure. **(c)** A ribbon under *shear stress,* being deformed and eventually cut by forces exerted by the scissors.

magnitude $F_\perp$ but opposite directions at the ends (this ensures that the object has no tendency to move left or right). We say that the object is in **tension.** We've already talked a lot about tension in ropes and strings; it's the same concept here. The subscript $\perp$ is a reminder that the forces act perpendicular to the cross section.

We define the **tensile stress** at the cross section as the ratio of the force $F_\perp$ to the cross-sectional area A:

$$\text{Tensile stress} = \frac{F_\perp}{A} \qquad (11.8)$$

This is a *scalar* quantity because $F_\perp$ is the *magnitude* of the force. The SI unit of stress is the **pascal** (abbreviated Pa and named for the 17th-century French scientist and philosopher Blaise Pascal). Equation (11.8) shows that 1 pascal equals 1 newton per square meter (N/m^2):

$$1 \text{ pascal} = 1 \text{ Pa} = 1 \text{ N/m}^2$$

In the British system the logical unit of stress would be the pound per square foot, but the pound per square inch $(\text{lb/in.}^2$ or psi) is more commonly used. The conversion factors are

$$1 \text{ psi} = 6895 \text{ Pa} \qquad \text{and} \qquad 1 \text{ Pa} = 1.450 \times 10^{-4} \text{ psi}$$

The units of stress are the same as those of *pressure,* which we will encounter often in later chapters. Air pressure in automobile tires is typically around 3×10^5 Pa = 300 kPa, and steel cables are commonly required to withstand tensile stresses of the order of 10^8 Pa.

The object shown in Fig. 11.13 stretches to a length $l = l_0 + \Delta l$ when under tension. The elongation Δl does not occur only at the ends; every part of the bar stretches in the same proportion. The **tensile strain** of the object is equal to the fractional change in length, which is the ratio of the elongation Δl to the original length l_0:

$$\text{Tensile strain} = \frac{l - l_0}{l_0} = \frac{\Delta l}{l_0} \qquad (11.9)$$

Tensile strain is stretch per unit length. It is a ratio of two lengths, always measured in the same units, and so is a pure (dimensionless) number with no units.

Experiment shows that for a sufficiently small tensile stress, stress and strain are proportional, as in Eq. (11.7). The corresponding elastic modulus is called **Young's modulus,** denoted by Y:

$$Y = \frac{\text{Tensile stress}}{\text{Tensile strain}} = \frac{F_\perp/A}{\Delta l/l_0} = \frac{F_\perp}{A} \frac{l_0}{\Delta l} \qquad \text{(Young's modulus)} \qquad (11.10)$$

Since strain is a pure number, the units of Young's modulus are the same as those of stress: force per unit area. Some typical values are listed in Table 11.1.

11.13 An object in tension. The net force on the object is zero, but the object deforms. The tensile stress (the ratio of the force to the cross-sectional area) produces a tensile strain (the elongation divided by the initial length). The elongation Δl is exaggerated for clarity.

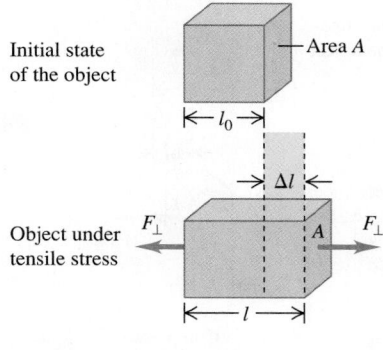

$$\text{Tensile stress} = \frac{F_\perp}{A} \qquad \text{Tensile strain} = \frac{\Delta l}{l_0}$$

Application Young's Modulus of a Tendon

The anterior tibial tendon connects your foot to the large muscle that runs along the side of your shinbone. (You can feel this tendon at the front of your ankle.) Measurements show that this tendon has a Young's modulus of 1.2×10^9 Pa, much less than for the solid materials listed in Table 11.1. Hence this tendon stretches substantially (up to 2.5% of its length) in response to the stresses experienced in walking and running.

Anterior tibial tendon

Table 11.1 Approximate Elastic Moduli

Material	Young's Modulus, Y (Pa)	Bulk Modulus, B (Pa)	Shear Modulus, S (Pa)
Aluminum	7.0×10^{10}	7.5×10^{10}	2.5×10^{10}
Brass	9.0×10^{10}	6.0×10^{10}	3.5×10^{10}
Copper	11×10^{10}	14×10^{10}	4.4×10^{10}
Crown glass	6.0×10^{10}	5.0×10^{10}	2.5×10^{10}
Iron	21×10^{10}	16×10^{10}	7.7×10^{10}
Lead	1.6×10^{10}	4.1×10^{10}	0.6×10^{10}
Nickel	21×10^{10}	17×10^{10}	7.8×10^{10}
Steel	20×10^{10}	16×10^{10}	7.5×10^{10}

11.14 An object in compression. The compressive stress and compressive strain are defined in the same way as tensile stress and strain (see Fig. 11.13), except that Δl now denotes the distance that the object contracts.

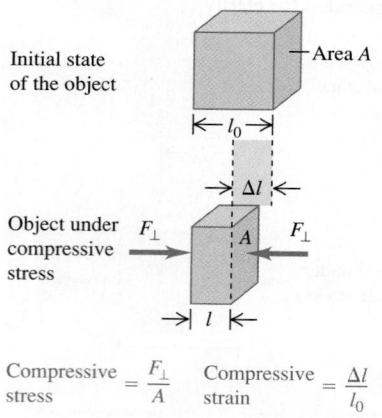

Initial state of the object

Area A

$\leftarrow l_0 \rightarrow$

$\rightarrow \Delta l \leftarrow$

Object under compressive stress

$F_\perp$ A $F_\perp$

$\rightarrow l \leftarrow$

$$\frac{\text{Compressive}}{\text{stress}} = \frac{F_\perp}{A} \qquad \frac{\text{Compressive}}{\text{strain}} = \frac{\Delta l}{l_0}$$

(This table also gives values of two other elastic moduli that we will discuss later in this chapter.) A material with a large value of Y is relatively unstretchable; a large stress is required for a given strain. For example, the value of Y for cast steel (2×10^{11} Pa) is much larger than that for rubber (5×10^8 Pa).

When the forces on the ends of a bar are pushes rather than pulls (Fig. 11.14), the bar is in **compression** and the stress is a **compressive stress.** The **compressive strain** of an object in compression is defined in the same way as the tensile strain, but Δl has the opposite direction. Hooke's law and Eq. (11.10) are valid for compression as well as tension if the compressive stress is not too great. For many materials, Young's modulus has the same value for both tensile and compressive stresses. Composite materials such as concrete and stone are an exception; they can withstand compressive stresses but fail under comparable tensile stresses. Stone was the primary building material used by ancient civilizations such as the Babylonians, Assyrians, and Romans, so their structures had to be designed to avoid tensile stresses. Hence they used arches in doorways and bridges, where the weight of the overlying material compresses the stones of the arch together and does not place them under tension.

In many situations, bodies can experience both tensile and compressive stresses at the same time. As an example, a horizontal beam supported at each end sags under its own weight. As a result, the top of the beam is under compression, while the bottom of the beam is under tension (Fig. 11.15a). To minimize the stress and hence the bending strain, the top and bottom of the beam are given a large cross-sectional area. There is neither compression nor tension along the centerline of the beam, so this part can have a small cross section; this helps to keep the weight of the bar to a minimum and further helps to reduce the stress. The result is an I-beam of the familiar shape used in building construction (Fig. 11.15b).

11.15 (a) A beam supported at both ends is under both compression and tension. (b) The cross-sectional shape of an I-beam minimizes both stress and weight.

(a)

Top of beam is under compression.

Beam's centerline is under neither tension nor compression.

Bottom of beam is under tension.

(b)

The top and bottom of an I-beam are broad to minimize the compressive and tensile stresses.

The beam can be narrow near its centerline, which is under neither compression nor tension.

Example 11.5 | Tensile stress and strain

A steel rod 2.0 m long has a cross-sectional area of 0.30 cm². It is hung by one end from a support, and a 550-kg milling machine is hung from its other end. Determine the stress on the rod and the resulting strain and elongation.

SOLUTION

IDENTIFY, SET UP, and EXECUTE: The rod is under tension, so we can use Eq. (11.8) to find the tensile stress; Eq. (11.9), with the value of Young's modulus Y for steel from Table 11.1, to find the corresponding strain; and Eq. (11.10) to find the elongation Δl:

$$\text{Tensile stress} = \frac{F_\perp}{A} = \frac{(550 \text{ kg})(9.8 \text{ m/s}^2)}{3.0 \times 10^{-5} \text{ m}^2} = 1.8 \times 10^8 \text{ Pa}$$

$$\text{Strain} = \frac{\Delta l}{l_0} = \frac{\text{Stress}}{Y} = \frac{1.8 \times 10^8 \text{ Pa}}{20 \times 10^{10} \text{ Pa}} = 9.0 \times 10^{-4}$$

$$\text{Elongation} = \Delta l = (\text{Strain}) \times l_0$$
$$= (9.0 \times 10^{-4})(2.0 \text{ m}) = 0.0018 \text{ m} = 1.8 \text{ mm}$$

EVALUATE: This small elongation, resulting from a load of over half a ton, is a testament to the stiffness of steel.

Bulk Stress and Strain

When a scuba diver plunges deep into the ocean, the water exerts nearly uniform pressure everywhere on his surface and squeezes him to a slightly smaller volume (see Fig. 11.12b). This is a different situation from the tensile and compressive stresses and strains we have discussed. The stress is now a uniform pressure on all sides, and the resulting deformation is a volume change. We use the terms **bulk stress** (or **volume stress**) and **bulk strain** (or **volume strain**) to describe these quantities.

If an object is immersed in a fluid (liquid or gas) at rest, the fluid exerts a force on any part of the object's surface; this force is *perpendicular* to the surface. (If we tried to make the fluid exert a force parallel to the surface, the fluid would slip sideways to counteract the effort.) The force $F_\perp$ per unit area that the fluid exerts on the surface of an immersed object is called the **pressure** p in the fluid:

$$p = \frac{F_\perp}{A} \qquad \text{(pressure in a fluid)} \qquad (11.11)$$

The pressure in a fluid increases with depth. For example, the pressure of the air is about 21% greater at sea level than in Denver (at an elevation of 1.6 km, or 1.0 mi). If an immersed object is relatively small, however, we can ignore pressure differences due to depth for the purpose of calculating bulk stress. Hence we will treat the pressure as having the same value at all points on an immersed object's surface.

Pressure has the same units as stress; commonly used units include 1 Pa $(=1 \text{ N/m}^2)$ and 1 lb/in.^{2} (1 psi). Also in common use is the **atmosphere,** abbreviated atm. One atmosphere is the approximate average pressure of the earth's atmosphere at sea level:

$$1 \text{ atmosphere} = 1 \text{ atm} = 1.013 \times 10^5 \text{ Pa} = 14.7 \text{ lb/in.}^2$$

CAUTION **Pressure vs. force** Unlike force, pressure has no intrinsic direction: The pressure on the surface of an immersed object is the same no matter how the surface is oriented. Hence pressure is a *scalar* quantity, not a vector quantity. ∎

Pressure plays the role of stress in a volume deformation. The corresponding strain is the fractional change in volume (Fig. 11.16)—that is, the ratio of the volume change ΔV to the original volume V_0:

$$\text{Bulk (volume) strain} = \frac{\Delta V}{V_0} \qquad (11.12)$$

Volume strain is the change in volume per unit volume. Like tensile or compressive strain, it is a pure number, without units.

When Hooke's law is obeyed, an increase in pressure (bulk stress) produces a *proportional* bulk strain (fractional change in volume). The corresponding elastic modulus (ratio of stress to strain) is called the **bulk modulus,** denoted by B. When the pressure on a body changes by a small amount Δp, from p_0 to $p_0 + \Delta p$, and the resulting bulk strain is $\Delta V/V_0$, Hooke's law takes the form

$$B = \frac{\text{Bulk stress}}{\text{Bulk strain}} = -\frac{\Delta p}{\Delta V/V_0} \qquad \text{(bulk modulus)} \qquad (11.13)$$

We include a minus sign in this equation because an *increase* of pressure always causes a *decrease* in volume. In other words, if Δp is positive, ΔV is negative. The bulk modulus B itself is a positive quantity.

For small pressure changes in a solid or a liquid, we consider B to be constant. The bulk modulus of a *gas,* however, depends on the initial pressure p_0. Table 11.1 includes values of the bulk modulus for several solid materials. Its units, force per unit area, are the same as those of pressure (and of tensile or compressive stress).

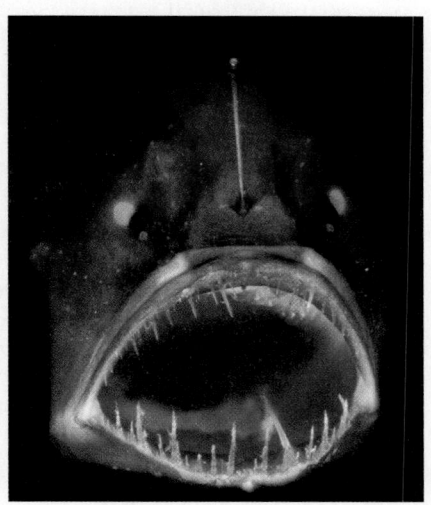

11.16 An object under bulk stress. Without the stress, the cube has volume V_0; when the stress is applied, the cube has a smaller volume V. The volume change ΔV is exaggerated for clarity.

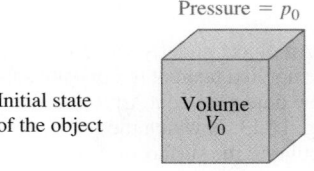

Pressure = p_0

Initial state of the object — Volume V_0

Pressure = $p = p_0 + \Delta p$

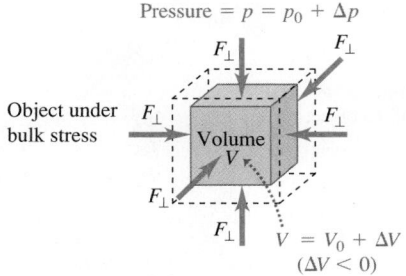

Object under bulk stress — Volume V

$$V = V_0 + \Delta V$$
$$(\Delta V < 0)$$

Bulk stress = Δp Bulk strain = $\dfrac{\Delta V}{V_0}$

Table 11.2 Compressibilities of Liquids

Liquid	Compressibility, k	
	Pa^{-1}	atm^{-1}
Carbon disulfide	93×10^{-11}	94×10^{-6}
Ethyl alcohol	110×10^{-11}	111×10^{-6}
Glycerine	21×10^{-11}	21×10^{-6}
Mercury	3.7×10^{-11}	3.8×10^{-6}
Water	45.8×10^{-11}	46.4×10^{-6}

The reciprocal of the bulk modulus is called the **compressibility** and is denoted by k. From Eq. (11.13),

$$k = \frac{1}{B} = -\frac{\Delta V/V_0}{\Delta p} = -\frac{1}{V_0}\frac{\Delta V}{\Delta p} \quad \text{(compressibility)} \quad (11.14)$$

Compressibility is the fractional decrease in volume, $-\Delta V/V_0$, per unit increase Δp in pressure. The units of compressibility are those of *reciprocal pressure*, Pa^{-1} or atm^{-1}.

Table 11.2 lists the values of compressibility k for several liquids. For example, the compressibility of water is 46.4×10^{-6} atm^{-1}, which means that the volume of water decreases by 46.4 parts per million for each 1-atmosphere increase in pressure. Materials with small bulk modulus and large compressibility are easier to compress.

Example 11.6 Bulk stress and strain

A hydraulic press contains 0.25 m^3 (250 L) of oil. Find the decrease in the volume of the oil when it is subjected to a pressure increase $\Delta p = 1.6 \times 10^7$ Pa (about 160 atm or 2300 psi). The bulk modulus of the oil is $B = 5.0 \times 10^9$ Pa (about 5.0×10^4 atm) and its compressibility is $k = 1/B = 20 \times 10^{-6}$ atm^{-1}.

SOLUTION

IDENTIFY, SET UP, and EXECUTE: This example uses the ideas of bulk stress and strain. We are given both the bulk modulus and the compressibility, and our target variable is ΔV. Solving Eq. (11.13) for ΔV, we find

$$\Delta V = -\frac{V_0\Delta p}{B} = -\frac{(0.25\ m^3)(1.6 \times 10^7\ Pa)}{5.0 \times 10^9\ Pa}$$
$$= -8.0 \times 10^{-4}\ m^3 = -0.80\ L$$

Alternatively, we can use Eq. (11.14) with the approximate unit conversions given above:

$$\Delta V = -kV_0\Delta p = -(20 \times 10^{-6}\ atm^{-1})(0.25\ m^3)(160\ atm)$$
$$= -8.0 \times 10^{-4}\ m^3$$

EVALUATE: The negative value of ΔV means that the volume decreases when the pressure increases. Even though the 160-atm pressure increase is large, the *fractional* change in volume is very small:

$$\frac{\Delta V}{V_0} = \frac{-8.0 \times 10^{-4}\ m^3}{0.25\ m^3} = -0.0032 \quad \text{or} \quad -0.32\%$$

Shear Stress and Strain

The third kind of stress-strain situation is called *shear*. The ribbon in Fig. 11.12c is under **shear stress:** One part of the ribbon is being pushed up while an adjacent part is being pushed down, producing a deformation of the ribbon. Figure 11.17 shows a body being deformed by a shear stress. In the figure, forces of equal magnitude but opposite direction act *tangent* to the surfaces of opposite ends of the object. We define the shear stress as the force $F_\parallel$ acting tangent to the surface divided by the area A on which it acts:

$$\text{Shear stress} = \frac{F_\parallel}{A} \quad (11.15)$$

Shear stress, like the other two types of stress, is a force per unit area.

Figure 11.17 shows that one face of the object under shear stress is displaced by a distance x relative to the opposite face. We define **shear strain** as the ratio of the displacement x to the transverse dimension h:

$$\text{Shear strain} = \frac{x}{h} \quad (11.16)$$

In real-life situations, x is nearly always much smaller than h. Like all strains, shear strain is a dimensionless number; it is a ratio of two lengths.

11.17 An object under shear stress. Forces are applied tangent to opposite surfaces of the object (in contrast to the situation in Fig. 11.13, in which the forces act perpendicular to the surfaces). The deformation x is exaggerated for clarity.

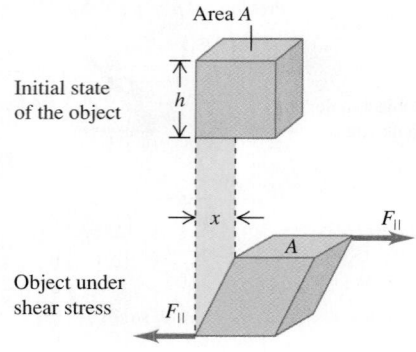

Area A

Initial state of the object

h

Object under shear stress

x

A

$F_\parallel$

$F_\parallel$

Shear stress $= \dfrac{F_\parallel}{A}$ Shear strain $= \dfrac{x}{h}$

If the forces are small enough that Hooke's law is obeyed, the shear strain is *proportional* to the shear stress. The corresponding elastic modulus (ratio of shear stress to shear strain) is called the **shear modulus,** denoted by S:

$$S = \frac{\text{Shear stress}}{\text{Shear strain}} = \frac{F_\parallel/A}{x/h} = \frac{F_\parallel}{A}\frac{h}{x} \quad \text{(shear modulus)} \qquad [11.17]$$

with x and h defined as in Fig. 11.17.

Table 11.1 gives several values of shear modulus. For a given material, S is usually one-third to one-half as large as Young's modulus Y for tensile stress. Keep in mind that the concepts of shear stress, shear strain, and shear modulus apply to *solid* materials only. The reason is that *shear* refers to deforming an object that has a definite shape (see Fig. 11.17). This concept doesn't apply to gases and liquids, which do not have definite shapes.

Example 11.7 Shear stress and strain

Suppose the object in Fig. 11.17 is the brass base plate of an outdoor sculpture that experiences shear forces in an earthquake. The plate is 0.80 m square and 0.50 cm thick. What is the force exerted on each of its edges if the resulting displacement x is 0.16 mm?

SOLUTION

IDENTIFY and SET UP: This example uses the relationship among shear stress, shear strain, and shear modulus. Our target variable is the force $F_\parallel$ exerted parallel to each edge, as shown in Fig. 11.17. We'll find the shear strain using Eq. (11.16), the shear stress using Eq. (11.17), and $F_\parallel$ using Eq. (11.15). Table 11.1 gives the shear modulus of brass. In Fig. 11.17, h represents the 0.80-m length of each side of the plate. The area A in Eq. (11.15) is the product of the 0.80-m length and the 0.50-cm thickness.

EXECUTE: From Eq. (11.16),

$$\text{Shear strain} = \frac{x}{h} = \frac{1.6 \times 10^{-4}\text{ m}}{0.80\text{ m}} = 2.0 \times 10^{-4}$$

From Eq. (11.17),

$$\begin{aligned}\text{Shear stress} &= (\text{Shear strain}) \times S \\ &= (2.0 \times 10^{-4})(3.5 \times 10^{10}\text{ Pa}) = 7.0 \times 10^6\text{ Pa}\end{aligned}$$

Finally, from Eq. (11.15),

$$\begin{aligned}F_\parallel &= (\text{Shear stress}) \times A \\ &= (7.0 \times 10^6\text{ Pa})(0.80\text{ m})(0.0050\text{ m}) = 2.8 \times 10^4\text{ N}\end{aligned}$$

EVALUATE: The shear force supplied by the earthquake is more than 3 tons! The large shear modulus of brass makes it hard to deform. Further, the plate is relatively thick (0.50 cm), so the area A is relatively large and a substantial force $F_\parallel$ is needed to provide the necessary stress $F_\parallel/A$.

Test Your Understanding of Section 11.4 A copper rod of cross-sectional area 0.500 cm^2 and length 1.00 m is elongated by 2.00×10^{-2} mm, and a steel rod of the same cross-sectional area but 0.100 m in length is elongated by 2.00×10^{-3} mm. (a) Which rod has greater tensile *strain*? (i) the copper rod; (ii) the steel rod; (iii) the strain is the same for both. (b) Which rod is under greater tensile *stress*? (i) the copper rod; (ii) the steel rod; (iii) the stress is the same for both. ❚

11.5 Elasticity and Plasticity

Hooke's law—the proportionality of stress and strain in elastic deformations—has a limited range of validity. In the preceding section we used phrases such as "provided that the forces are small enough that Hooke's law is obeyed." Just what *are* the limitations of Hooke's law? We know that if you pull, squeeze, or twist *anything* hard enough, it will bend or break. Can we be more precise than that?

Let's look at tensile stress and strain again. Suppose we plot a graph of stress as a function of strain. If Hooke's law is obeyed, the graph is a straight line with a

11.18 Typical stress-strain diagram for a ductile metal under tension.

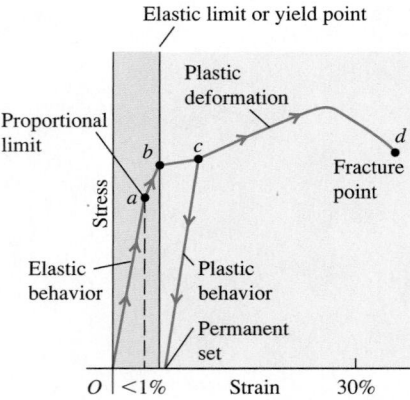

slope equal to Young's modulus. Figure 11.18 shows a typical stress-strain graph for a metal such as copper or soft iron. The strain is shown as the *percent* elongation; the horizontal scale is not uniform beyond the first portion of the curve, up to a strain of less than 1%. The first portion is a straight line, indicating Hooke's law behavior with stress directly proportional to strain. This straight-line portion ends at point *a*; the stress at this point is called the *proportional limit*.

From *a* to *b*, stress and strain are no longer proportional, and Hooke's law is *not* obeyed. If the load is gradually removed, starting at any point between *O* and *b*, the curve is retraced until the material returns to its original length. The deformation is *reversible*, and the forces are conservative; the energy put into the material to cause the deformation is recovered when the stress is removed. In region *Ob* we say that the material shows *elastic behavior*. Point *b*, the end of this region, is called the *yield point*; the stress at the yield point is called the *elastic limit*.

When we increase the stress beyond point *b*, the strain continues to increase. But now when we remove the load at some point beyond *b*, say *c*, the material does not come back to its original length. Instead, it follows the red line in Fig. 11.18. The length at zero stress is now greater than the original length; the material has undergone an irreversible deformation and has acquired what we call a *permanent set*. Further increase of load beyond *c* produces a large increase in strain for a relatively small increase in stress, until a point *d* is reached at which *fracture* takes place. The behavior of the material from *b* to *d* is called *plastic flow* or *plastic deformation*. A plastic deformation is irreversible; when the stress is removed, the material does not return to its original state.

For some materials, such as the one whose properties are graphed in Fig. 11.18, a large amount of plastic deformation takes place between the elastic limit and the fracture point. Such a material is said to be *ductile*. But if fracture occurs soon after the elastic limit is passed, the material is said to be *brittle*. A soft iron wire that can have considerable permanent stretch without breaking is ductile, while a steel piano string that breaks soon after its elastic limit is reached is brittle.

Something very curious can happen when an object is stretched and then allowed to relax. An example is shown in Fig. 11.19, which is a stress-strain curve for vulcanized rubber that has been stretched by more than seven times its original length. The stress is not proportional to the strain, but the behavior is elastic because when the load is removed, the material returns to its original length. However, the material follows *different* curves for increasing and decreasing stress. This is called *elastic hysteresis*. The work done by the material when it returns to its original shape is less than the work required to deform it; there are nonconservative forces associated with internal friction. Rubber with large elastic hysteresis is very useful for absorbing vibrations, such as in engine mounts and shock-absorber bushings for cars.

11.19 Typical stress-strain diagram for vulcanized rubber. The curves are different for increasing and decreasing stress, a phenomenon called elastic hysteresis.

The stress required to cause actual fracture of a material is called the *breaking stress*, the *ultimate strength*, or (for tensile stress) the *tensile strength*. Two materials, such as two types of steel, may have very similar elastic constants but vastly different breaking stresses. Table 11.3 gives typical values of breaking stress for several materials in tension. The conversion factor 6.9×10^8 Pa = 100,000 psi may help put these numbers in perspective. For example, if the breaking stress of a particular steel is 6.9×10^8 Pa, then a bar with a 1-in.^{2} cross section has a breaking strength of 100,000 lb.

Table 11.3 Approximate Breaking Stresses

Material	Breaking Stress (Pa or N/m^2)
Aluminum	2.2×10^8
Brass	4.7×10^8
Glass	10×10^8
Iron	3.0×10^8
Phosphor bronze	5.6×10^8
Steel	$5-20 \times 10^8$

Test Your Understanding of Section 11.5 While parking your car on a crowded street, you accidentally back into a steel post. You pull forward until the car no longer touches the post and then get out to inspect the damage. What does your rear bumper look like if the strain in the impact was (a) less than at the proportional limit; (b) greater than at the proportional limit, but less than at the yield point; (c) greater than at the yield point, but less than at the fracture point; and (d) greater than at the fracture point? ❚

Conditions for equilibrium: For a rigid body to be in equilibrium, two conditions must be satisfied. First, the vector sum of forces must be zero. Second, the sum of torques about any point must be zero. The torque due to the weight of a body can be found by assuming the entire weight is concentrated at the center of gravity, which is at the same point as the center of mass if $\vec{g}$ has the same value at all points. (See Examples 11.1–11.4.)

$$\sum F_x = 0 \qquad \sum F_y = 0 \qquad \sum F_z = 0$$
$$(11.1)$$

$$\sum \vec{\tau} = 0 \quad \text{about } any \text{ point} \qquad (11.2)$$

$$\vec{r}_{cm} = \frac{m_1\vec{r}_1 + m_2\vec{r}_2 + m_3\vec{r}_3 + \cdots}{m_1 + m_2 + m_3 + \cdots}$$
$$(11.4)$$

Stress, strain, and Hooke's law: Hooke's law states that in elastic deformations, stress (force per unit area) is proportional to strain (fractional deformation). The proportionality constant is called the elastic modulus.

$$\frac{\text{Stress}}{\text{Strain}} = \text{Elastic modulus} \qquad (11.7)$$

Tensile and compressive stress: Tensile stress is tensile force per unit area, $F_\perp/A$. Tensile strain is fractional change in length, $\Delta l/l_0$. The elastic modulus is called Young's modulus Y. Compressive stress and strain are defined in the same way. (See Example 11.5.)

$$Y = \frac{\text{Tensile stress}}{\text{Tensile strain}} = \frac{F_\perp/A}{\Delta l/l_0} = \frac{F_\perp}{A}\frac{l_0}{\Delta l}$$
$$(11.10)$$

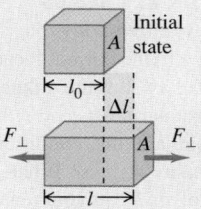

Bulk stress: Pressure in a fluid is force per unit area. Bulk stress is pressure change, Δp, and bulk strain is fractional volume change, $\Delta V/V_0$. The elastic modulus is called the bulk modulus, B. Compressibility, k, is the reciprocal of bulk modulus: $k = 1/B$. (See Example 11.6.)

$$p = \frac{F_\perp}{A} \qquad (11.11)$$

$$B = \frac{\text{Bulk stress}}{\text{Bulk strain}} = -\frac{\Delta p}{\Delta V/V_0} \qquad (11.13)$$

Shear stress: Shear stress is force per unit area, $F_\parallel/A$, for a force applied tangent to a surface. Shear strain is the displacement x of one side divided by the transverse dimension h. The elastic modulus is called the shear modulus, S. (See Example 11.7.)

$$S = \frac{\text{Shear stress}}{\text{Shear strain}} = \frac{F_\parallel/A}{x/h} = \frac{F_\parallel}{A}\frac{h}{x}$$
$$(11.17)$$

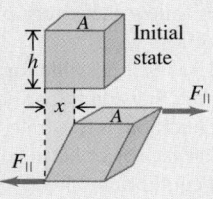

The limits of Hooke's law: The proportional limit is the maximum stress for which stress and strain are proportional. Beyond the proportional limit, Hooke's law is not valid. The elastic limit is the stress beyond which irreversible deformation occurs. The breaking stress, or ultimate strength, is the stress at which the material breaks.

BRIDGING PROBLEM   In Equilibrium and Under Stress

A horizontal, uniform, solid copper rod has an original length l_0, cross-sectional area A, Young's modulus Y, bulk modulus B, shear modulus S, and mass m. It is supported by a frictionless pivot at its right end and by a cable a distance $l_0/4$ from its left end (Fig. 11.20). Both pivot and cable are attached so that they exert their forces uniformly over the rod's cross section. The cable makes an angle θ with the rod and compresses it. (a) Find the tension in the cable. (b) Find the magnitude and direction of the force exerted by the pivot on the right end of the rod. How does this magnitude compare to the cable tension? How does this angle compare to θ? (c) Find the change in length of the rod due to the stresses exerted by the cable and pivot on the rod. (d) By what factor would your answer in part (c) increase if the solid copper rod were twice as long but had the same cross-sectional area?

SOLUTION GUIDE

See MasteringPhysics® study area for a Video Tutor solution.

IDENTIFY and SET UP

1. Draw a free-body diagram for the rod. Be careful to place each force in the correct location.
2. Make a list of the unknown quantities, and decide which are the target variables.
3. What are the conditions that must be met so that the rod remains at rest? What kind of stress (and resulting strain) is involved? Use your answers to select the appropriate equations.

11.20 What are the forces on the rod? What are the stress and strain?

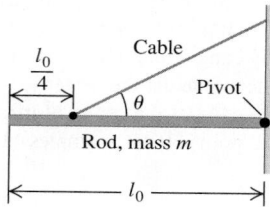

EXECUTE

4. Use your equations to solve for the target variables. (*Hint:* You can make the solution easier by carefully choosing the point around which you calculate torques.)
5. Use your knowledge of trigonometry to decide whether the pivot force or the cable tension has the greater magnitude, as well as to decide whether the angle of the pivot force is greater than, less than, or equal to θ.

EVALUATE

6. Check whether your answers are reasonable. Which force, the cable tension or the pivot force, holds up more of the weight of the rod? Does this make sense?

Problems

For instructor-assigned homework, go to www.masteringphysics.com

•, ••, •••: Problems of increasing difficulty. **CP**: Cumulative problems incorporating material from earlier chapters. **CALC**: Problems requiring calculus. **BIO**: Biosciences problems.

DISCUSSION QUESTIONS

Q11.1 Does a rigid object in uniform rotation about a fixed axis satisfy the first and second conditions for equilibrium? Why? Does it then follow that every particle in this object is in equilibrium? Explain.

Q11.2 (a) Is it possible for an object to be in translational equilibrium (the first condition) but *not* in rotational equilibrium (the second condition)? Illustrate your answer with a simple example. (b) Can an object be in rotational equilibrium yet *not* in translational equilibrium? Justify your answer with a simple example.

Q11.3 Car tires are sometimes "balanced" on a machine that pivots the tire and wheel about the center. Weights are placed around the wheel rim until it does not tip from the horizontal plane. Discuss this procedure in terms of the center of gravity.

Q11.4 Does the center of gravity of a solid body always lie within the material of the body? If not, give a counterexample.

Q11.5 In Section 11.2 we always assumed that the value of g was the same at all points on the body. This is *not* a good approximation if the dimensions of the body are great enough, because the value of g decreases with altitude. If this is taken into account, will the center of gravity of a long, vertical rod be above, below, or at its center of mass? Explain how this can be used to keep the long

axis of an orbiting spacecraft pointed toward the earth. (This would be useful for a weather satellite that must always keep its camera lens trained on the earth.) The moon is not exactly spherical but is somewhat elongated. Explain why this same effect is responsible for keeping the same face of the moon pointed toward the earth at all times.

Q11.6 You are balancing a wrench by suspending it at a single point. Is the equilibrium stable, unstable, or neutral if the point is above, at, or below the wrench's center of gravity? In each case give the reasoning behind your answer. (For rotation, a rigid body is in *stable* equilibrium if a small rotation of the body produces a torque that tends to return the body to equilibrium; it is in *unstable* equilibrium if a small rotation produces a torque that tends to take the body farther from equilibrium; and it is in *neutral* equilibrium if a small rotation produces no torque.)

Q11.7 You can probably stand flatfooted on the floor and then rise up and balance on your tiptoes. Why are you unable do it if your toes are touching the wall of your room? (Try it!)

Q11.8 You freely pivot a horseshoe from a horizontal nail through one of its nail holes. You then hang a long string with a weight at its bottom from the same nail, so that the string hangs vertically in front of the horseshoe without touching it. How do you know that the horseshoe's center of gravity is along the line behind the string? How can you locate the center of gravity by repeating the process at another nail hole? Will the center of gravity be within the solid material of the horseshoe?

Q11.9 An object consists of a ball of weight W glued to the end of a uniform bar also of weight W. If you release it from rest, with the bar horizontal, what will its behavior be as it falls if air resistance is negligible? Will it (a) remain horizontal; (b) rotate about its center of gravity; (c) rotate about the ball; or (d) rotate so that the ball swings downward? Explain your reasoning.

Q11.10 Suppose that the object in Question 11.9 is released from rest with the bar tilted at 60° above the horizontal with the ball at the upper end. As it is falling, will it (a) rotate about its center of gravity until it is horizontal; (b) rotate about its center of gravity until it is vertical with the ball at the bottom; (c) rotate about the ball until it is vertical with the ball at the bottom; or (d) remain at 60° above the horizontal?

Q11.11 Why must a water skier moving with constant velocity lean backward? What determines how far back she must lean? Draw a free-body diagram for the water skier to justify your answers.

Q11.12 In pioneer days, when a Conestoga wagon was stuck in the mud, people would grasp the wheel spokes and try to turn the wheels, rather than simply pushing the wagon. Why?

Q11.13 The mighty Zimbo claims to have leg muscles so strong that he can stand flat on his feet and lean forward to pick up an apple on the floor with his teeth. Should you pay to see him perform, or do you have any suspicions about his claim? Why?

Q11.14 Why is it easier to hold a 10-kg dumbbell in your hand at your side than it is to hold it with your arm extended horizontally?

Q11.15 Certain features of a person, such as height and mass, are fixed (at least over relatively long periods of time). Are the following features also fixed? (a) location of the center of gravity of the body; (b) moment of inertia of the body about an axis through the person's center of mass. Explain your reasoning.

Q11.16 During pregnancy, women often develop back pains from leaning backward while walking. Why do they have to walk this way?

Q11.17 Why is a tapered water glass with a narrow base easier to tip over than a glass with straight sides? Does it matter whether the glass is full or empty?

Q11.18 When a tall, heavy refrigerator is pushed across a rough floor, what factors determine whether it slides or tips?

Q11.19 If a metal wire has its length doubled and its diameter tripled, by what factor does its Young's modulus change?

Q11.20 Why is concrete with steel reinforcing rods embedded in it stronger than plain concrete?

Q11.21 A metal wire of diameter D stretches by 0.100 mm when supporting a weight W. If the same-length wire is used to support a weight three times as heavy, what would its diameter have to be (in terms of D) so it still stretches only 0.100 mm?

Q11.22 Compare the mechanical properties of a steel cable, made by twisting many thin wires together, with the properties of a solid steel rod of the same diameter. What advantages does each have?

Q11.23 The material in human bones and elephant bones is essentially the same, but an elephant has much thicker legs. Explain why, in terms of breaking stress.

Q11.24 There is a small but appreciable amount of elastic hysteresis in the large tendon at the back of a horse's leg. Explain how this can cause damage to the tendon if a horse runs too hard for too long a time.

Q11.25 When rubber mounting blocks are used to absorb machine vibrations through elastic hysteresis, as mentioned in Section 11.5, what becomes of the energy associated with the vibrations?

EXERCISES

Section 11.2 Center of Gravity

11.1 •• A 0.120-kg, 50.0-cm-long uniform bar has a small 0.055-kg mass glued to its left end and a small 0.110-kg mass glued to the other end. The two small masses can each be treated as point masses. You want to balance this system horizontally on a fulcrum placed just under its center of gravity. How far from the left end should the fulcrum be placed?

11.2 •• The center of gravity of a 5.00-kg irregular object is shown in Fig. E11.2. You need to move the center of gravity 2.20 cm to the left by gluing on a 1.50-kg mass, which will then be considered as part of the object. Where should the center of gravity of this additional mass be located?

Figure **E11.2**

Center of gravity

11.3 • A uniform rod is 2.00 m long and has mass 1.80 kg. A 2.40-kg clamp is attached to the rod. How far should the center of gravity of the clamp be from the left-hand end of the rod in order for the center of gravity of the composite object to be 1.20 m from the left-hand end of the rod?

Section 11.3 Solving Rigid-Body Equilibrium Problems

11.4 • A uniform 300-N trapdoor in a floor is hinged at one side. Find the net upward force needed to begin to open it and the total force exerted on the door by the hinges (a) if the upward force is applied at the center and (b) if the upward force is applied at the center of the edge opposite the hinges.

11.5 •• **Raising a Ladder.** A ladder carried by a fire truck is 20.0 m long. The ladder weighs 2800 N and its center of gravity is at its center. The ladder is pivoted at one end (A) about a pin (Fig. E11.5); you can ignore the friction torque at the pin. The ladder is raised into position by a force applied by a hydraulic piston at C. Point C is 8.0 m from A, and the force $\vec{F}$ exerted by the piston makes an angle of 40° with the ladder. What magnitude must $\vec{F}$ have to just lift the ladder off the support bracket at B? Start with a free-body diagram of the ladder.

Figure **E11.5**

11.6 •• Two people are carrying a uniform wooden board that is 3.00 m long and weighs 160 N. If one person applies an upward force equal to 60 N at one end, at what point does the other person lift? Begin with a free-body diagram of the board.

11.7 •• Two people carry a heavy electric motor by placing it on a light board 2.00 m long. One person lifts at one end with a force of 400 N, and the other lifts the opposite end with a force of 600 N.

(a) What is the weight of the motor, and where along the board is its center of gravity located? (b) Suppose the board is not light but weighs 200 N, with its center of gravity at its center, and the two people each exert the same forces as before. What is the weight of the motor in this case, and where is its center of gravity located?

11.8 •• A 60.0-cm, uniform, 50.0-N shelf is supported horizontally by two vertical wires attached to the sloping ceiling (Fig. E11.8). A very small 25.0-N tool is placed on the shelf midway between the points where the wires are attached to it. Find the tension in each wire. Begin by making a free-body diagram of the shelf.

Figure **E11.8**

11.9 •• A 350-N, uniform, 1.50-m bar is suspended horizontally by two vertical cables at each end. Cable A can support a maximum tension of 500.0 N without breaking, and cable B can support up to 400.0 N. You want to place a small weight on this bar. (a) What is the heaviest weight you can put on without breaking either cable, and (b) where should you put this weight?

11.10 •• A uniform ladder 5.0 m long rests against a frictionless, vertical wall with its lower end 3.0 m from the wall. The ladder weighs 160 N. The coefficient of static friction between the foot of the ladder and the ground is 0.40. A man weighing 740 N climbs slowly up the ladder. Start by drawing a free-body diagram of the ladder. (a) What is the maximum frictional force that the ground can exert on the ladder at its lower end? (b) What is the actual frictional force when the man has climbed 1.0 m along the ladder? (c) How far along the ladder can the man climb before the ladder starts to slip?

11.11 • A diving board 3.00 m long is supported at a point 1.00 m from the end, and a diver weighing 500 N stands at the free end (Fig. E11.11). The diving board is of uniform cross section and weighs 280 N. Find (a) the force at the support point and (b) the force at the left-hand end.

Figure **E11.11**

11.12 • A uniform aluminum beam 9.00 m long, weighing 300 N, rests symmetrically on two supports 5.00 m apart (Fig. E11.12). A boy weighing 600 N starts at point A and walks toward the right. (a) In the same diagram construct two graphs showing the upward forces F_A and F_B exerted on the beam at points A and B, as functions of the coordinate x of the boy. Let 1 cm = 100 N vertically, and 1 cm = 1.00 m horizontally. (b) From your diagram, how far beyond point B can the boy walk before the beam tips? (c) How far

Figure **E11.12**

from the right end of the beam should support B be placed so that the boy can walk just to the end of the beam without causing it to tip?

11.13 • Find the tension T in each cable and the magnitude and direction of the force exerted on the strut by the pivot in each of the arrangements in Fig. E11.13. In each case let w be the weight of the suspended crate full of priceless art objects. The strut is uniform and also has weight w. Start each case with a free-body diagram of the strut.

Figure **E11.13**

11.14 • The horizontal beam in Fig. E11.14 weighs 150 N, and its center of gravity is at its center. Find (a) the tension in the cable and (b) the horizontal and vertical components of the force exerted on the beam at the wall.

11.15 •• **BIO** **Push-ups.** To strengthen his arm and chest muscles, an 82-kg athlete who is 2.0 m tall is doing push-ups as shown in Fig. E11.15. His center of mass is 1.15 m from the bottom of his feet, and the centers of

Figure **E11.14**

his palms are 30.0 cm from the top of his head. Find the force that the floor exerts on each of his feet and on each hand, assuming that both feet exert the same force and both palms do likewise. Begin with a free-body diagram of the athlete.

11.16 •• Suppose that you can lift no more than 650 N (around 150 lb) unaided. (a) How much can you lift using a 1.4-m-long wheelbarrow that weighs 80.0 N and whose center of gravity is 0.50 m from the center of the wheel (Fig. E11.16)? The center of gravity of the load carried in

Figure **E11.15**

Figure **E11.16**

the wheelbarrow is also 0.50 m from the center of the wheel. (b) Where does the force come from to enable you to lift more than 650 N using the wheelbarrow?

11.17 •• You take your dog Clea to the vet, and the doctor decides he must locate the little beast's center of gravity. It would be awkward to hang the pooch from the ceiling, so the vet must devise another method. He places Clea's front feet on one scale and her hind feet on another. The front scale reads 157 N, while the rear scale reads 89 N. The vet next measures Clea and finds that her rear feet are 0.95 m behind her front feet. How much does Clea weigh, and where is her center of gravity?

11.18 •• A 15,000-N crane pivots around a friction-free axle at its base and is supported by a cable making a 25° angle with the crane (Fig. E11.18). The crane is 16 m long and is not uniform, its center of gravity being 7.0 m from the axle as measured along the crane. The cable is attached 3.0 m from the upper end of the crane. When the crane is raised to 55° above the horizontal holding an 11,000-N pallet of bricks by a 2.2-m, very light cord, find (a) the tension in the cable and (b) the horizontal and vertical components of the force that the axle exerts on the crane. Start with a free-body diagram of the crane.

Figure **E11.18**

11.19 •• A 3.00-m-long, 240-N, uniform rod at the zoo is held in a horizontal position by two ropes at its ends (Fig. E11.19). The left rope makes an angle of 150° with the rod and the right rope makes an angle θ with the horizontal. A 90-N howler monkey (*Alouatta seniculus*) hangs motionless 0.50 m from the right end of the rod as he carefully studies you. Calculate the tensions in the two ropes and the angle θ. First make a free-body diagram of the rod.

Figure **E11.19**

11.20 •• A nonuniform beam 4.50 m long and weighing 1.00 kN makes an angle of 25.0° below the horizontal. It is held in position by a frictionless pivot at its upper right end and by a cable 3.00 m farther down the beam and perpendicular to it (Fig. E11.20). The center of gravity of the beam is 2.00 m down the beam from the pivot. Lighting equipment exerts a 5.00-kN downward force on the lower left end of the beam. Find the tension T in the cable and the horizontal and vertical components of the force exerted on the beam by the pivot. Start by sketching a free-body diagram of the beam.

Figure **E11.20**

11.21 • **A Couple.** Two forces equal in magnitude and opposite in direction, acting on an object at two different points, form what is called a *couple*. Two antiparallel forces with equal magnitudes $F_1 = F_2 = 8.00$ *N* are applied to a rod as shown in Fig. E11.21. (a) What should the distance *l* between the forces be if they are to provide a net torque of 6.40 N·m about the left end of the rod? (b) Is the sense of this torque clockwise or counterclockwise? (c) Repeat parts (a) and (b) for a pivot at the point on the rod where $\vec{F}_2$ is applied.

Figure **E11.21**

11.22 •• **BIO** **A Good Workout.** You are doing exercises on a Nautilus machine in a gym to strengthen your deltoid (shoulder) muscles. Your arms are raised vertically and can pivot around the shoulder joint, and you grasp the cable of the machine in your hand 64.0 cm from your shoulder joint. The deltoid muscle is attached to the humerus 15.0 cm from the shoulder joint and makes a 12.0° angle with that bone (Fig. E11.22). If you have set the tension in the cable of the machine to 36.0 N on each arm, what is the tension in each deltoid muscle if you simply hold your outstretched arms in place? (*Hint:* Start by making a clear free-body diagram of your arm.)

Figure **E11.22**

11.23 •• **BIO** **Neck Muscles.** A student bends her head at 40.0° from the vertical while intently reading her physics book, pivoting the head around the upper vertebra (point *P* in Fig. E11.23). Her head has a mass of 4.50 kg (which is typical), and its center of mass is 11.0 cm from the pivot point *P*. Her neck muscles are 1.50 cm from point *P*, as measured *perpendicular* to these muscles. The neck itself and the vertebrae are held vertical. (a) Draw a free-body diagram of the student's head. (b) Find the tension in her neck muscles.

Figure **E11.23**

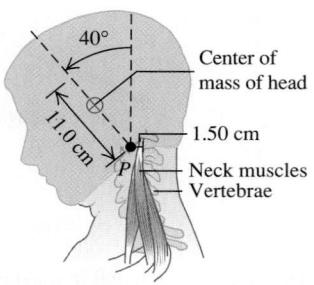

40°

11.0 cm

Center of
mass of head

1.50 cm

P

Neck muscles
Vertebrae

Section 11.4 Stress, Strain, and Elastic Moduli

11.24 • **BIO** **Biceps Muscle.** A relaxed biceps muscle requires a force of 25.0 N for an elongation of 3.0 cm; the same muscle under maximum tension requires a force of 500 N for the same elongation. Find Young's modulus for the muscle tissue under each of these conditions if the muscle is assumed to be a uniform cylinder with length 0.200 m and cross-sectional area 50.0 cm².

11.25 •• A circular steel wire 2.00 m long must stretch no more than 0.25 cm when a tensile force of 400 N is applied to each end of the wire. What minimum diameter is required for the wire?

11.26 •• Two circular rods, one steel and the other copper, are joined end to end. Each rod is 0.750 m long and 1.50 cm in diameter. The combination is subjected to a tensile force with magnitude 4000 N. For each rod, what are (a) the strain and (b) the elongation?

11.27 •• A metal rod that is 4.00 m long and 0.50 cm² in cross-sectional area is found to stretch 0.20 cm under a tension of 5000 N. What is Young's modulus for this metal?

11.28 •• **Stress on a Mountaineer's Rope.** A nylon rope used by mountaineers elongates 1.10 m under the weight of a 65.0-kg climber. If the rope is 45.0 m in length and 7.0 mm in diameter, what is Young's modulus for nylon?

11.29 •• In constructing a large mobile, an artist hangs an aluminum sphere of mass 6.0 kg from a vertical steel wire 0.50 m long and 2.5 × 10⁻³ cm² in cross-sectional area. On the bottom of the sphere he attaches a similar steel wire, from which he hangs a brass cube of mass 10.0 kg. For each wire, compute (a) the tensile strain and (b) the elongation.

11.30 •• A vertical, solid steel post 25 cm in diameter and 2.50 m long is required to support a load of 8000 kg. You can ignore the weight of the post. What are (a) the stress in the post; (b) the strain in the post; and (c) the change in the post's length when the load is applied?

11.31 •• **BIO** **Compression of Human Bone.** The bulk modulus for bone is 15 GPa. (a) If a diver-in-training is put into a pressurized suit, by how much would the pressure have to be raised (in atmospheres) above atmospheric pressure to compress her bones by 0.10% of their original volume? (b) Given that the pressure in the ocean increases by 1.0 × 10⁴ Pa for every meter of depth below the surface, how deep would this diver have to go for her bones to compress by 0.10%? Does it seem that bone compression is a problem she needs to be concerned with when diving?

11.32 • A solid gold bar is pulled up from the hold of the sunken RMS *Titanic*. (a) What happens to its volume as it goes from the pressure at the ship to the lower pressure at the ocean's surface? (b) The pressure difference is proportional to the depth. How many times greater would the volume change have been had the ship been twice as deep? (c) The bulk modulus of lead is one-fourth that of gold. Find the ratio of the volume change of a solid lead bar to that of a gold bar of equal volume for the same pressure change.

11.33 • **BIO** **Downhill Hiking.** During vigorous downhill hiking, the force on the knee cartilage (the medial and lateral meniscus) can be up to eight times body weight. Depending on the angle of descent, this force can cause a large shear force on the cartilage and deform it. The cartilage has an area of about 10 cm² and a shear modulus of 12 MPa. If the hiker plus his pack have a combined mass of 110 kg (not unreasonable), and if the maximum force at impact is 8 times his body weight (which, of course, includes the weight of his pack) at an angle of 12° with the cartilage (Fig. E11.33), through what angle (in degrees) will his knee cartilage be deformed? (Recall that the bone below the cartilage pushes upward with the same force as the downward force.)

Figure **E11.33**

F

12°

Cartilage

12°

F

11.34 •• In the Challenger Deep of the Marianas Trench, the depth of seawater is 10.9 km and the pressure is 1.16 × 10⁸ Pa (about 1.15 × 10³ atm). (a) If a cubic meter of water is taken from the surface to this depth, what is the change in its volume? (Normal atmospheric pressure is about 1.0 × 10⁵ Pa. Assume that k for seawater is the same as the freshwater value given in Table 11.2.) (b) What is the density of seawater at this depth? (At the surface, seawater has a density of 1.03 × 10³ kg/m³.)

11.35 • A specimen of oil having an initial volume of 600 cm³ is subjected to a pressure increase of 3.6 × 10⁶ Pa, and the volume is found to decrease by 0.45 cm³. What is the bulk modulus of the material? The compressibility?

11.36 •• A square steel plate is 10.0 cm on a side and 0.500 cm thick. (a) Find the shear strain that results if a force of magnitude 9.0 × 10⁵ N is applied to each of the four sides, parallel to the side. (b) Find the displacement x in centimeters.

11.37 •• A copper cube measures 6.00 cm on each side. The bottom face is held in place by very strong glue to a flat horizontal surface, while a horizontal force F is applied to the upper face parallel to one of the edges. (Consult Table 11.1.) (a) Show that the glue exerts a force F on the bottom face that is equal but opposite to the force on the top face. (b) How large must F be to cause the cube to deform by 0.250 mm? (c) If the same experiment were performed on a lead cube of the same size as the copper one, by what distance would it deform for the same force as in part (b)?

Figure **E11.38**

11.38 • In lab tests on a 9.25-cm cube of a certain material, a force of 1375 N directed at 8.50° to the cube (Fig. E11.38) causes the cube to deform through an angle of 1.24°. What is the shear modulus of the material?

9.25 cm

9.25 cm

8.50°

1375 N

1375 N

8.50°

Section 11.5 Elasticity and Plasticity

11.39 •• In a materials testing laboratory, a metal wire made from a new alloy is found to break when a tensile force of 90.8 N is applied perpendicular to each end. If the diameter of the wire is 1.84 mm, what is the breaking stress of the alloy?

11.40 • A 4.0-m-long steel wire has a cross-sectional area of 0.050 cm². Its proportional limit has a value of 0.0016 times its Young's modulus (see Table 11.1). Its breaking stress has a value of 0.0065 times its Young's modulus. The wire is fastened at its upper end and hangs vertically. (a) How great a weight can be hung from the wire without exceeding the proportional limit?

(b) How much will the wire stretch under this load? (c) What is the maximum weight that the wire can support?

11.41 •• **CP** A steel cable with cross-sectional area 3.00 cm² has an elastic limit of 2.40×10^8 Pa. Find the maximum upward acceleration that can be given a 1200-kg elevator supported by the cable if the stress is not to exceed one-third of the elastic limit.

11.42 •• A brass wire is to withstand a tensile force of 350 N without breaking. What minimum diameter must the wire have?

PROBLEMS

11.43 ••• A box of negligible mass rests at the left end of a 2.00-m, 25.0-kg plank (Fig. P11.43). The width of the box is 75.0 cm, and sand is to be distributed uniformly throughout it. The center of gravity of the nonuniform plank is 50.0 cm from the right end. What mass of sand should be put into the box so that the plank balances horizontally on a fulcrum placed just below its midpoint?

Figure **P11.43**

11.44 ••• A door 1.00 m wide and 2.00 m high weighs 280 N and is supported by two hinges, one 0.50 m from the top and the other 0.50 m from the bottom. Each hinge supports half the total weight of the door. Assuming that the door's center of gravity is at its center, find the horizontal components of force exerted on the door by each hinge.

11.45 ••• **Mountain Climbing.** Mountaineers often use a rope to lower themselves down the face of a cliff (this is called *rappelling*). They do this with their body nearly horizontal and their feet pushing against the cliff (Fig. P11.45). Suppose that an 82.0-kg climber, who is 1.90 m tall and has a center of gravity 1.1 m from his feet, rappels down a vertical cliff with his body raised 35.0° above the horizontal. He holds the rope 1.40 m from his feet, and it makes a 25.0° angle with the cliff face. (a) What tension does his rope need to support? (b) Find the horizontal and vertical components of the force that the cliff face exerts on the climber's feet. (c) What minimum coefficient of static friction is needed to prevent the climber's feet from slipping on the cliff face if he has one foot at a time against the cliff?

Figure **P11.45**

11.46 • Sir Lancelot rides slowly out of the castle at Camelot and onto the 12.0-m-long drawbridge that passes over the moat (Fig. P11.46). Unbeknownst to him, his enemies have partially severed the vertical cable holding up the front end of the bridge so that it will break under a tension of 5.80×10^3 N. The bridge has mass 200 kg and its center of gravity is at its center. Lancelot, his lance, his armor, and his horse together have a combined mass of 600 kg. Will the cable break before Lancelot reaches the end of the drawbridge? If so, how far from the castle end of the

bridge will the center of gravity of the horse plus rider be when the cable breaks?

Figure **P11.46**

11.47 • Three vertical forces act on an airplane when it is flying at a constant altitude and with a constant velocity. These are the weight of the airplane, an aerodynamic force on the wing of the airplane, and an aerodynamic force on the airplane's horizontal tail. (The aerodynamic forces are exerted by the surrounding air and are reactions to the forces that the wing and tail exert on the air as the airplane flies through it.) For a particular light airplane with a weight of 6700 N, the center of gravity is 0.30 m in front of the point where the wing's vertical aerodynamic force acts and 3.66 m in front of the point where the tail's vertical aerodynamic force acts. Determine the magnitude and direction (upward or downward) of each of the two vertical aerodynamic forces.

11.48 •• A pickup truck has a wheelbase of 3.00 m. Ordinarily, 10,780 N rests on the front wheels and 8820 N on the rear wheels when the truck is parked on a level road. (a) A box weighing 3600 N is now placed on the tailgate, 1.00 m behind the rear axle. How much total weight now rests on the front wheels? On the rear wheels? (b) How much weight would need to be placed on the tailgate to make the front wheels come off the ground?

11.49 •• A uniform, 255-N rod that is 2.00 m long carries a 225-N weight at its right end and an unknown weight W toward the left end (Fig. P11.49). When W is placed 50.0 cm from the left end of the rod, the system just balances horizontally when the fulcrum is located 75.0 cm from the right end. (a) Find W. (b) If W is now moved 25.0 cm to the right, how far and in what direction must the fulcrum be moved to restore balance?

Figure **P11.49**

11.50 •• A uniform, 8.0-m, 1500-kg beam is hinged to a wall and supported by a thin cable attached 2.0 m from the free end of the beam, (Fig. P11.50). The beam is supported at an angle of 30.0° above the horizontal. (a) Draw a free-body diagram of the beam. (b) Find the tension in the cable. (c) How hard does the beam push inward on the wall?

Figure **P11.50**

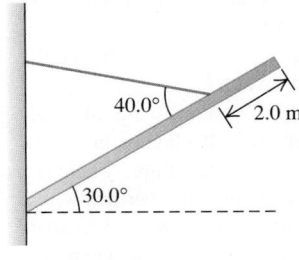

11.51 •• You open a restaurant and hope to entice customers by hanging out a sign (Fig. P11.51). The uniform horizontal beam supporting the sign is 1.50 m long, has a mass of 12.0 kg, and is hinged to the wall. The sign itself is uniform with a mass of 28.0 kg and overall length of 1.20 m. The two wires supporting the sign are each 32.0 cm long, are 90.0 cm apart, and are equally spaced from the middle of the sign. The cable supporting the beam is 2.00 m long. (a) What minimum tension must your cable be able to support without having your sign come crashing down? (b) What minimum vertical force must the hinge be able to support without pulling out of the wall?

Figure **P11.51**

11.52 ••• A claw hammer is used to pull a nail out of a board (Fig. P11.52). The nail is at an angle of 60° to the board, and a force $\vec{F}_1$ of magnitude 400 N applied to the nail is required to pull it from the board. The hammer head contacts the board at point A, which is 0.080 m from where the nail enters the board. A horizontal force $\vec{F}_2$ is applied to the hammer handle at a distance of 0.300 m above the board. What magnitude of force $\vec{F}_2$ is required to apply the required 400-N force (F_1) to the nail? (You can ignore the weight of the hammer.)

Figure **P11.52**

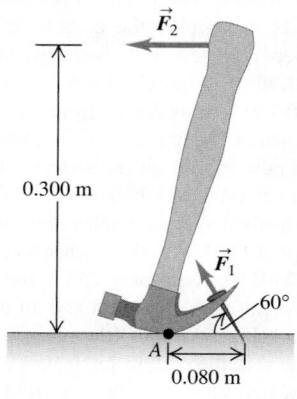

11.53 • End A of the bar AB in Fig. P11.53 rests on a frictionless horizontal surface, and end B is hinged. A horizontal force $\vec{F}$ of magnitude 160 N is exerted on end A. You can ignore the weight of the bar. What are the horizontal and vertical components of the force exerted by the bar on the hinge at B?

Figure **P11.53**

11.54 • A museum of modern art is displaying an irregular 426-N sculpture by hanging it from two thin vertical wires, A and B, that are 1.25 m apart (Fig. P11.54). The center of gravity of this piece of art is located 48.0 cm from its extreme right tip. Find the tension in each wire.

Figure **P11.54**

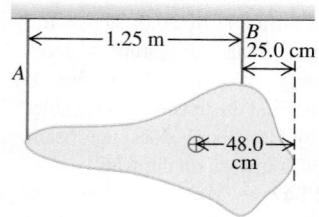

11.55 •• **BIO** **Supporting a Broken Leg.** A therapist tells a 74-kg patient with a broken leg that he must have his leg in a cast suspended horizontally. For minimum discomfort, the leg should be supported by a vertical strap attached at the center of mass of the leg–cast system. (Fig. P11.55). In order to comply with these instructions, the patient consults a table of typical mass distributions and finds that both upper legs (thighs) together typically account for 21.5% of body weight and the center of mass of each thigh is 18.0 cm from the hip joint. The patient also reads that the two lower legs (including the feet) are 14.0% of body weight, with a center of mass 69.0 cm from the hip joint. The cast has a mass of 5.50 kg, and its center of mass is 78.0 cm from the hip joint. How far from the hip joint should the supporting strap be attached to the cast?

Figure **P11.55**

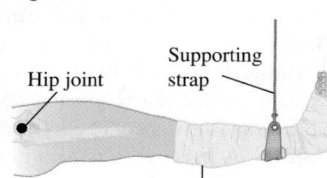

11.56 • **A Truck on a Drawbridge.** A loaded cement mixer drives onto an old drawbridge, where it stalls with its center of gravity three-quarters of the way across the span. The truck driver radios for help, sets the handbrake, and waits. Meanwhile, a boat approaches, so the drawbridge is raised by means of a cable attached to the end opposite the hinge (Fig. P11.56). The drawbridge is 40.0 m long and has a mass of 18,000 kg; its center of gravity is at its midpoint. The cement mixer, with driver, has mass 30,000 kg. When the drawbridge has been raised to an angle of 30° above the horizontal, the cable makes an angle of 70° with the surface of the bridge. (a) What is the tension T in the cable when the drawbridge is held in this position? (b) What are the horizontal and vertical components of the force the hinge exerts on the span?

Figure **P11.56**

11.57 •• **BIO** **Leg Raises.** In a simplified version of the musculature action in leg raises, the abdominal muscles pull on the femur (thigh bone) to raise the leg by pivoting it about one end (Fig. P11.57). When you are lying horizontally, these muscles make an angle of approximately 5° with the femur,

Figure **P11.57**

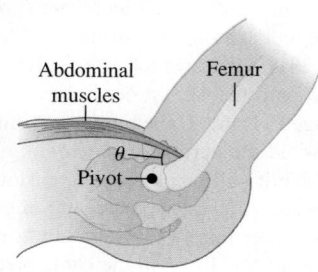

and if you raise your legs, the muscles remain approximately horizontal, so the angle θ increases. We shall assume for simplicity that these muscles attach to the femur in only one place, 10 cm from the hip joint (although, in reality, the situation is more complicated). For a certain 80-kg person having a leg 90 cm long, the mass of the leg is 15 kg and its center of mass is 44 cm from his hip joint as measured along the leg. If the person raises his leg to 60° above the horizontal, the angle between the abdominal muscles and his femur would also be about 60°. (a) With his leg raised to 60°, find the tension in the abdominal muscle on each leg. As usual, begin your solution with a free-body diagram. (b) When is the tension in this muscle greater: when the leg is raised to 60° or when the person just starts to raise it off the ground? Why? (Try this yourself to check your answer.) (c) If the abdominal muscles attached to the femur were perfectly horizontal when a person was lying down, could the person raise his leg? Why or why not?

11.58 • A nonuniform fire escape ladder is 6.0 m long when extended to the icy alley below. It is held at the top by a frictionless pivot, and there is negligible frictional force from the icy surface at the bottom. The ladder weighs 250 N, and its center of gravity is 2.0 m along the ladder from its bottom. A mother and child of total weight 750 N are on the ladder 1.5 m from the pivot. The ladder makes an angle θ with the horizontal. Find the magnitude and direction of (a) the force exerted by the icy alley on the ladder and (b) the force exerted by the ladder on the pivot. (c) Do your answers in parts (a) and (b) depend on the angle θ?

11.59 •• A uniform strut of mass m makes an angle θ with the horizontal. It is supported by a frictionless pivot located at one-third its length from its lower left end and a horizontal rope at its upper right end. A cable and package of total weight w hang from its upper right end. (a) Find the vertical and horizontal components V and H of the pivot's force on the strut as well as the tension T in the rope. (b) If the maximum safe tension in the rope is 700 N and the mass of the strut is 30.0 kg, find the maximum safe weight of the cable and package when the strut makes an angle of 55.0° with the horizontal. (c) For what angle θ can no weight be safely suspended from the right end of the strut?

11.60 • You are asked to design the decorative mobile shown in Fig. P11.60. The strings and rods have negligible weight, and the rods are to hang horizontally. (a) Draw a free-body diagram for each rod. (b) Find the weights of the balls A, B, and C. Find the tensions in the strings S_1, S_2, and S_3. (c) What can you say about the horizontal location of the mobile's center of gravity? Explain.

Figure **P11.60**

11.61 •• A uniform, 7.5-m-long beam weighing 5860 N is hinged to a wall and supported by a thin cable attached 1.5 m from the free end of the beam. The cable runs between the beam and the wall and makes a 40° angle with the beam. What is the tension in the cable when the beam is at an angle of 30° above the horizontal?

11.62 •• **CP** A uniform drawbridge must be held at a 37° angle above the horizontal to allow ships to pass underneath. The drawbridge weighs 45,000 N and is 14.0 m long. A cable is connected 3.5 m from the hinge where the bridge pivots (measured along the bridge) and pulls horizontally on the bridge to hold it in place. (a) What is the tension in the cable? (b) Find the magnitude and direction of the force the hinge exerts on the bridge. (c) If the cable suddenly breaks, what is the magnitude of the angular acceleration of the drawbridge just after the cable breaks? (d) What is the angular speed of the drawbridge as it becomes horizontal?

11.63 •• **BIO** Tendon-Stretching Exercises. As part of an exercise program, a 75-kg person does toe raises in which he raises his entire body weight on the ball of one foot (Fig. P11.63). The Achilles tendon pulls straight upward on the heel bone of his foot. This tendon is 25 cm long and has a cross-sectional area of 78 mm² and a Young's modulus of 1470 MPa. (a) Make a free-body diagram of the person's foot (everything below the ankle joint). You can neglect the weight of the foot. (b) What force does the Achilles tendon exert on the heel during this exercise? Express your answer in newtons and in multiples of his weight. (c) By how many millimeters does the exercise stretch his Achilles tendon?

Figure **P11.63**

11.64 •• (a) In Fig. P11.64 a 6.00-m-long, uniform beam is hanging from a point 1.00 m to the right of its center. The beam weighs 140 N and makes an angle of 30.0° with the vertical. At the right-hand end of the beam a 100.0-N weight is hung; an unknown weight w hangs at the left end. If the system is in equilibrium, what is w? You can ignore the thickness of the beam. (b) If the beam makes, instead, an angle of 45.0° with the vertical, what is w?

Figure **P11.64**

11.65 ••• A uniform, horizontal flagpole 5.00 m long with a weight of 200 N is hinged to a vertical wall at one end. A 600-N stuntwoman hangs from its other end. The flagpole is supported by a guy wire running from its outer end to a point on the wall directly above the pole. (a) If the tension in this wire is not to exceed 1000 N, what is the minimum height above the pole at which it may be fastened to the wall? (b) If the flagpole remains horizontal, by how many newtons would the tension be increased if the wire were fastened 0.50 m below this point?

11.66 • A holiday decoration consists of two shiny glass spheres with masses 0.0240 kg and 0.0360 kg suspended from a uniform rod with mass 0.120 kg and length 1.00 m (Fig. P11.66). The rod is suspended from the ceiling by a vertical cord at each end, so that it is horizontal. Calculate the tension in each of the cords A through F.

Figure **P11.66**

11.67 •• **BIO Downward-Facing Dog.** One yoga exercise, known as the "Downward-Facing Dog," requires stretching your hands straight out above your head and bending down to lean against the floor. This exercise is performed by a 750-N person, as shown in Fig. P11.67. When he bends his body at the hip to a 90° angle between his legs and trunk, his legs, trunk, head, and arms have the dimensions indicated. Furthermore, his legs and feet weigh a total of 277 N, and their center of mass is 41 cm from his hip, measured along his legs. The person's trunk, head, and arms weigh 473 N, and their center of gravity is 65 cm from his hip, measured along the upper body. (a) Find the normal force that the floor exerts on each foot and on each hand, assuming that the person does not favor either hand or either foot. (b) Find the friction force on each foot and on each hand, assuming that it is the same on both feet and on both hands (but not necessarily the same on the feet as on the hands). [*Hint:* First treat his entire body as a system; then isolate his legs (or his upper body).]

Figure **P11.67**

11.68 • When you stretch a wire, rope, or rubber band, it gets thinner as well as longer. When Hooke's law holds, the fractional decrease in width is proportional to the tensile strain. If w_0 is the original width and Δw is the change in width, then $\Delta w/w_0 = -\sigma \Delta l/l_0$, where the minus sign reminds us that width decreases when length increases. The dimensionless constant σ, different for different materials, is called *Poisson's ratio.* (a) If the steel rod of Example 11.5 (Section 11.4) has a circular cross section and a Poisson's ratio of 0.23, what is its change in diameter when the milling machine is hung from it? (b) A cylinder made of nickel (Poisson's ratio = 0.42) has radius 2.0 cm. What tensile force $F_\perp$ must be applied perpendicular to each end of the cylinder to cause its radius to decrease by 0.10 mm? Assume that the breaking stress and proportional limit for the metal are extremely large and are not exceeded.

11.69 • A worker wants to turn over a uniform, 1250-N, rectangular crate by pulling at 53.0° on one of its vertical sides (Fig. P11.69).

The floor is rough enough to prevent the crate from slipping. (a) What pull is needed to just start the crate to tip? (b) How hard does the floor push upward on the crate? (c) Find the friction force on the crate. (d) What is the minimum coefficient of static friction needed to prevent the crate from slipping on the floor?

Figure **P11.69**

11.70 ••• One end of a uniform meter stick is placed against a vertical wall (Fig. P11.70). The other end is held by a lightweight cord that makes an angle θ with the stick. The coefficient of static friction between the end of the meter stick and the wall is 0.40. (a) What is the maximum value the angle θ can have if the stick is to remain in equilibrium? (b) Let the angle θ be 15°. A block of the same weight as the meter stick is suspended from the stick, as shown, at a distance x from the wall. What is the minimum value of x for which the stick will remain in equilibrium? (c) When $\theta = 15°$, how large must the coefficient of static friction be so that the block can be attached 10 cm from the left end of the stick without causing it to slip?

Figure **P11.70**

11.71 •• Two friends are carrying a 200-kg crate up a flight of stairs. The crate is 1.25 m long and 0.500 m high, and its center of gravity is at its center. The stairs make a 45.0° angle with respect to the floor. The crate also is carried at a 45.0° angle, so that its bottom side is parallel to the slope of the stairs (Fig. P11.71). If the force each person applies is vertical, what is the magnitude of each of these forces? Is it better to be the person above or below on the stairs?

Figure **P11.71**

11.72 •• **BIO Forearm.** In the human arm, the forearm and hand pivot about the elbow joint. Consider a simplified model in which the biceps muscle is attached to the forearm 3.80 cm from the elbow joint. Assume that the person's hand and forearm together weigh 15.0 N and that their center of gravity is 15.0 cm from the elbow (not quite halfway to the hand). The forearm is held horizontally at a right angle to the upper arm, with the biceps muscle exerting its force perpendicular to the forearm. (a) Draw a free-body diagram for the forearm, and find the force exerted by the biceps when the hand is empty. (b) Now the person holds a 80.0-N weight in his hand, with the forearm still horizontal. Assume that the center of gravity of this weight is 33.0 cm from the elbow. Construct a free-body diagram for the forearm, and find the force now exerted by the biceps. Explain why the biceps muscle needs to be very strong. (c) Under the conditions of part (b), find the magnitude and direction of the force that the elbow joint exerts on the forearm. (d) While holding the 80.0-N weight, the person raises his forearm until it is at an angle of 53.0° above the horizontal. If the biceps muscle continues to exert its force perpendicular to the forearm, what is this force when the forearm is in this position? Has the force increased or decreased from its value in part (b)? Explain why this is so, and test your answer by actually doing this with your own arm.

11.73 •• **BIO CALC** Refer to the discussion of holding a dumbbell in Example 11.4 (Section 11.3). The maximum weight that can be held in this way is limited by the maximum allowable tendon tension T (determined by the strength of the tendons) and by the distance D from the elbow to where the tendon attaches to the forearm. (a) Let T_{max} represent the maximum value of the tendon tension. Use the results of Example 11.4 to express w_{max} (the maximum weight that can be held) in terms of T_{max}, L, D, and h. Your expression should *not* include the angle θ. (b) The tendons of different primates are attached to the forearm at different values of D. Calculate the derivative of w_{max} with respect to D, and determine whether the derivative is positive or negative. (c) A chimpanzee tendon is attached to the forearm at a point farther from the elbow than for humans. Use this to explain why chimpanzees have stronger arms than humans. (The disadvantage is that chimpanzees have less flexible arms than do humans.)

11.74 •• A uniform, 90.0-N table is 3.6 m long, 1.0 m high, and 1.2 m wide. A 1500-N weight is placed 0.50 m from one end of the table, a distance of 0.60 m from each side of the table. Draw a free-body diagram for the table and find the force that each of the four legs exerts on the floor.

11.75 ••• **Flying Buttress.** (a) A symmetric building has a roof sloping upward at 35.0° above the horizontal on each side. If each side of the uniform roof weighs 10,000 N, find the horizontal force that this roof exerts at the top of the wall, which tends to push out the walls. Which type of building would be more in danger of collapsing: one with tall walls or one with short walls? Explain. (b) As you saw in part (a), tall walls are in danger of collapsing from the weight of the roof. This problem plagued the ancient builders of large structures. A solution used in the great Gothic cathedrals during the 1200s was the flying buttress, a stone support running between the walls and the ground that helped to hold in the walls. A Gothic church has a uniform roof weighing a total of 20,000 N and rising at 40° above the horizontal at each wall. The walls are 40 m tall, and a flying buttress meets each wall 10 m below the base of the roof. What horizontal force must this flying buttress apply to the wall?

11.76 •• You are trying to raise a bicycle wheel of mass m and radius R up over a curb of height h. To do this, you apply a horizontal force $\vec{F}$ (Fig. P11.76). What is the smallest magnitude of the force $\vec{F}$ that will succeed in raising the wheel onto the curb when the force is applied (a) at the center of the wheel and (b) at the top of the wheel? (c) In which case is less force required?

Figure **P11.76**

11.77 • **The Farmyard Gate.** A gate 4.00 m wide and 2.00 m high weighs 500 N. Its center of gravity is at its center, and it is hinged at A and B. To relieve the strain on the top hinge, a

Figure **P11.77**

wire CD is connected as shown in Fig. P11.77. The tension in CD is increased until the horizontal force at hinge A is zero. (a) What is the tension in the wire CD? (b) What is the magnitude of the horizontal component of the force at hinge B? (c) What is the combined vertical force exerted by hinges A and B?

11.78 • If you put a uniform block at the edge of a table, the center of the block must be over the table for the block not to fall off. (a) If you stack two identical blocks at the table edge, the center of the top block must be over the bottom block, and the center of gravity of the two blocks together must be over the table. In terms of the length L of each block, what is the maximum overhang possible (Fig. P11.78)? (b) Repeat part (a) for three identical blocks and for four identical blocks. (c) Is it possible to make a stack of blocks such that the uppermost block is not directly over the table at all? How many blocks would it take to do this? (Try this with your friends using copies of this book.)

Figure **P11.78**

11.79 ••• Two uniform, 75.0-g marbles 2.00 cm in diameter are stacked as shown in Fig. P11.79 in a container that is 3.00 cm wide. (a) Find the force that the container exerts on the marbles at the points of contact A, B, and C. (b) What force does each marble exert on the other?

Figure **P11.79**

11.80 •• Two identical, uniform beams weighing 260 N each are connected at one end by a frictionless hinge. A light horizontal crossbar attached at the midpoints of the beams maintains an angle of 53.0° between the beams. The beams are suspended from the ceiling by vertical wires such that they form a "V," as shown in Fig. P11.80. (a) What force does the crossbar exert on each beam? (b) Is the crossbar under tension or compression? (c) What force (magnitude and direction) does the hinge at point A exert on each beam?

Figure **P11.80**

11.81 • An engineer is designing a conveyor system for loading hay bales into a wagon (Fig. P11.81). Each bale is 0.25 m wide, 0.50 m high, and 0.80 m long (the dimension perpendicular to the plane of the figure), with mass 30.0 kg. The center of

Figure **P11.81**

gravity of each bale is at its geometrical center. The coefficient of static friction between a bale and the conveyor belt is 0.60, and the belt moves with constant speed. (a) The angle β of the conveyor is slowly increased. At some critical angle a bale will tip (if it doesn't slip first), and at some different critical angle it will slip (if it doesn't tip first). Find the two critical angles and determine which happens at the smaller angle. (b) Would the outcome of part (a) be different if the coefficient of friction were 0.40?

11.82 • A weight W is supported by attaching it to a vertical uniform metal pole by a thin cord passing over a pulley having negligible mass and friction. The cord is attached to the pole 40.0 cm below the top and pulls horizontally on it (Fig. P11.82). The pole is pivoted about a hinge at its base, is 1.75 m tall, and weighs 55.0 N. A thin wire connects the top of the pole to a vertical wall. The nail that holds this wire to the wall will

Figure **P11.82**

pull out if an *outward* force greater than 22.0 N acts on it. (a) What is the greatest weight W that can be supported this way without pulling out the nail? (b) What is the *magnitude* of the force that the hinge exerts on the pole?

11.83 •• A garage door is mounted on an overhead rail (Fig. P11.83). The wheels at A and B have rusted so that they do not roll, but rather slide along the track. The coefficient of kinetic friction is 0.52. The distance between the wheels is 2.00 m, and each is 0.50 m from the vertical sides of the door. The door is uniform and

Figure **P11.83**

weighs 950 N. It is pushed to the left at constant speed by a horizontal force $\vec{F}$. (a) If the distance h is 1.60 m, what is the vertical component of the force exerted on each wheel by the track? (b) Find the maximum value h can have without causing one wheel to leave the track.

11.84 •• A horizontal boom is supported at its left end by a frictionless pivot. It is held in place by a cable attached to the right-hand end of the boom. A chain and crate of total weight w hang from somewhere along the boom. The boom's weight w_b cannot be ignored and the boom may or may not be uniform. (a) Show that the tension in the cable is the same whether the cable makes an angle θ or an angle $180° - \theta$ with the horizontal, and that the horizontal force component exerted on the boom by the pivot has equal magnitude but opposite direction for the two angles. (b) Show that the cable cannot be horizontal. (c) Show that the tension in the cable is a minimum when the cable is vertical, pulling upward on the right end of the boom. (d) Show that when the cable is vertical, the force exerted by the pivot on the boom is vertical.

11.85 •• Prior to being placed in its hole, a 5700-N, 9.0-m-long, uniform utility pole makes some nonzero angle with the vertical. A vertical cable attached 2.0 m below its upper end holds it in place while its lower end rests on the ground. (a) Find the tension in the cable and the magnitude and direction of the force exerted by the ground on the pole. (b) Why don't we need to know the angle the pole makes with the vertical, as long as it is not zero?

11.86 ••• **Pyramid Builders.** Ancient pyramid builders are balancing a uniform rectangular slab of stone tipped at an angle θ above the horizontal using a rope (Fig. P11.86). The rope is held by five workers who share the force equally. (a) If $\theta = 20.0°$, what force does each worker exert on the rope? (b) As θ increases, does each worker have to exert more or less force than in part (a), assuming they do not change the angle of the rope? Why? (c) At what angle do the workers need to exert *no force* to balance the slab? What happens if θ exceeds this value?

Figure **P11.86**

11.87 • You hang a floodlamp from the end of a vertical steel wire. The floodlamp stretches the wire 0.18 mm and the stress is proportional to the strain. How much would it have stretched (a) if the wire were twice as long? (b) if the wire had the same length but twice the diameter? (c) for a copper wire of the original length and diameter?

11.88 •• **Hooke's Law for a Wire.** A wire of length l_0 and cross-sectional area A supports a hanging weight W. (a) Show that if the wire obeys Eq. (11.7), it behaves like a spring of force constant AY/l_0, where Y is Young's modulus for the material of which the wire is made. (b) What would the force constant be for a 75.0-cm length of 16-gauge (diameter = 1.291 mm) copper wire? See Table 11.1. (c) What would W have to be to stretch the wire in part (b) by 1.25 mm?

11.89 ••• **CP** A 12.0-kg mass, fastened to the end of an aluminum wire with an unstretched length of 0.50 m, is whirled in a vertical circle with a constant angular speed of 120 rev/min. The cross-sectional area of the wire is 0.014 cm². Calculate the elongation of the wire when the mass is (a) at the lowest point of the path and (b) at the highest point of its path.

11.90 • A metal wire 3.50 m long and 0.70 mm in diameter was given the following test. A load weighing 20 N was originally hung from the wire to keep it taut. The position of the lower end of the wire was read on a scale as load was added.

Added Load (N)	Scale Reading (cm)
0	3.02
10	3.07
20	3.12
30	3.17
40	3.22
50	3.27
60	3.32
70	4.27

(a) Graph these values, plotting the increase in length horizontally and the added load vertically. (b) Calculate the value of Young's modulus. (c) The proportional limit occurred at a scale reading of 3.34 cm. What was the stress at this point?

11.91 ••• A 1.05-m-long rod of negligible weight is supported at its ends by wires A and B of equal length (Fig. P11.91). The cross-sectional area of A is

Figure **P11.91**

2.00 mm² and that of B is 4.00 mm². Young's modulus for wire A is 1.80×10^{11} Pa; that for B is 1.20×10^{11} Pa. At what point along the rod should a weight w be suspended to produce (a) equal stresses in A and B and (b) equal strains in A and B?

11.92 ••• **CP** An amusement park ride consists of airplane-shaped cars attached to steel rods (Fig. P11.92). Each rod has a length of 15.0 m and a cross-sectional area of 8.00 cm². (a) How much is the rod stretched when the ride is at rest? (Assume that each car plus two people seated in it has a total weight of 1900 N.) (b) When operating, the ride has a maximum angular speed of 8.0 rev/min. How much is the rod stretched then?

Figure **P11.92**

11.93 • A brass rod with a length of 1.40 m and a cross-sectional area of 2.00 cm² is fastened end to end to a nickel rod with length L and cross-sectional area 1.00 cm². The compound rod is subjected to equal and opposite pulls of magnitude 4.00×10^4 N at its ends. (a) Find the length L of the nickel rod if the elongations of the two rods are equal. (b) What is the stress in each rod? (c) What is the strain in each rod?

11.94 ••• **CP BIO Stress on the Shin Bone.** The compressive strength of our bones is important in everyday life. Young's modulus for bone is about 1.4×10^{10} Pa. Bone can take only about a 1.0% change in its length before fracturing. (a) What is the maximum force that can be applied to a bone whose minimum cross-sectional area is 3.0 cm²? (This is approximately the cross-sectional area of a tibia, or shin bone, at its narrowest point.) (b) Estimate the maximum height from which a 70-kg man could jump and not fracture the tibia. Take the time between when he first touches the floor and when he has stopped to be 0.030 s, and assume that the stress is distributed equally between his legs.

11.95 ••• A moonshiner produces pure ethanol (ethyl alcohol) late at night and stores it in a stainless steel tank in the form of a cylinder 0.300 m in diameter with a tight-fitting piston at the top. The total volume of the tank is 250 L (0.250 m³). In an attempt to squeeze a little more into the tank, the moonshiner piles 1420 kg of lead bricks on top of the piston. What additional volume of ethanol can the moonshiner squeeze into the tank? (Assume that the wall of the tank is perfectly rigid.)

CHALLENGE PROBLEMS

11.96 ••• Two ladders, 4.00 m and 3.00 m long, are hinged at point A and tied together by a horizontal rope 0.90 m above the floor (Fig. P11.96). The ladders weigh 480 N and 360 N, respectively, and the center of gravity of each is at its center. Assume that

Figure **P11.96**

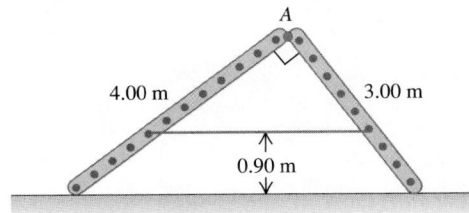

the floor is freshly waxed and frictionless. (a) Find the upward force at the bottom of each ladder. (b) Find the tension in the rope. (c) Find the magnitude of the force one ladder exerts on the other at point A. (d) If an 800-N painter stands at point A, find the tension in the horizontal rope.

11.97 ••• A bookcase weighing 1500 N rests on a horizontal surface for which the coefficient of static friction is $\mu_s = 0.40$. The bookcase is 1.80 m tall and 2.00 m wide; its center of gravity is at its geometrical center. The bookcase rests on four short legs that are each 0.10 m from the edge of the bookcase. A person pulls on a rope attached to an upper corner of the bookcase with a force $\vec{F}$ that makes an angle θ with the bookcase (Fig. P11.97). (a) If $\theta = 90°$, so $\vec{F}$ is horizontal, show that as F is increased from zero, the bookcase will start to slide before it tips, and calculate the magnitude of $\vec{F}$ that will start the bookcase sliding. (b) If $\theta = 0°$, so $\vec{F}$ is vertical, show that the bookcase will tip over rather than slide, and calculate the magnitude of $\vec{F}$ that will cause the bookcase to start to tip. (c) Calculate as a function of θ the magnitude of $\vec{F}$ that will cause the bookcase to start to slide and the magnitude that will cause it to start to tip. What is the smallest value that θ can have so that the bookcase will still start to slide before it starts to tip?

Figure **P11.97**

11.98 ••• **Knocking Over a Post.** One end of a post weighing 400 N and with height h rests on a rough horizontal surface with $\mu_s = 0.30$. The upper end is held by a rope fastened to the surface and making an angle of 36.9° with the post (Fig. P11.98). A horizontal force $\vec{F}$ is exerted on the post as shown. (a) If the force $\vec{F}$ is applied at the midpoint of the post, what is the largest value it can have without causing the post to slip? (b) How large can the force be without causing the post to slip if its point of application is $\frac{6}{10}$ of the way from the ground to the top of the post? (c) Show that if the point of application of the force is too high, the post cannot be made to slip, no matter how great the force. Find the critical height for the point of application.

Figure **P11.98**

11.99 ••• **CALC Minimizing the Tension.** A heavy horizontal girder of length L has several objects suspended from it. It is supported by a frictionless pivot at its left end and a cable of negligible weight that is attached to an I-beam at a point a distance h directly above the girder's center. Where should the other end of the cable be attached to the girder so that the cable's tension is a minimum? (*Hint:* In evaluating and presenting your answer, don't forget that the maximum distance of the point of attachment from the pivot is the length L of the beam.)

11.100 ••• **Bulk Modulus of an Ideal Gas.** The equation of state (the equation relating pressure, volume, and temperature) for an ideal gas is $pV = nRT$, where n and R are constants. (a) Show that if the gas is compressed while the temperature T is held constant, the bulk modulus is equal to the pressure. (b) When an ideal gas is compressed without the transfer of any heat into or out of it, the pressure and volume are related by $pV^{\gamma} = $ constant, where γ is a constant having different values for different gases. Show that, in this case, the bulk modulus is given by $B = \gamma p$.

11.101 ••• **CP** An angler hangs a 4.50-kg fish from a vertical steel wire 1.50 m long and 5.00×10^{-3} cm^2 in cross-sectional area. The upper end of the wire is securely fastened to a support. (a) Calculate the amount the wire is stretched by the hanging fish. The angler now applies a force $\vec{F}$ to the fish, pulling it very slowly downward by 0.500 mm from its equilibrium position. For this downward motion, calculate (b) the work done by gravity; (c) the work done by the force $\vec{F}$; (d) the work done by the force the wire exerts on the fish; and (e) the change in the elastic potential energy (the potential energy associated with the tensile stress in the wire). Compare the answers in parts (d) and (e).

Answers

Chapter Opening Question

Each stone in the arch is under compression, not tension. This is because the forces on the stones tend to push them inward toward the center of the arch and thus squeeze them together. Compared to a solid supporting wall, a wall with arches is just as strong yet much more economical to build.

Test Your Understanding Questions

11.1 Answer: (i) Situation (i) satisfies both equilibrium conditions because the seagull has zero acceleration (so $\Sigma \vec{F} = 0$) and no tendency to start rotating (so $\Sigma \vec{\tau} = 0$). Situation (ii) satisfies the first condition because the crankshaft as a whole does not accelerate through space, but it does not satisfy the second condition; the crankshaft has an angular acceleration, so $\Sigma \vec{\tau}$ is not zero. Situation (iii) satisfies the second condition (there is no tendency to rotate) but not the first one; the baseball accelerates in its flight (due to gravity), so $\Sigma \vec{F}$ is not zero.

11.2 Answer: (ii) In equilibrium, the center of gravity must be at the point of support. Since the rock and meter stick have the same mass and hence the same weight, the center of gravity of the system is midway between their respective centers. The center of gravity of the meter stick alone is 0.50 m from the left end (that is, at the middle of the meter stick), so the center of gravity of the combination of rock and meter stick is 0.25 m from the left end.

11.3 Answer: (ii), (i), (iii) This is the same situation described in Example 11.4, with the rod replacing the forearm, the hinge replacing the elbow, and the cable replacing the tendon. The only difference is that the cable attachment point is at the end of the rod, so the distances D and L are identical. From Example 11.4, the tension is

$$T = \frac{Lw}{L\sin\theta} = \frac{w}{\sin\theta}$$

Since $\sin\theta$ is less than 1, the tension T is greater than the weight w. The vertical component of the force exerted by the hinge is

$$E_y = -\frac{(L - L)w}{L} = 0$$

In this situation, the hinge exerts *no* vertical force. You can see this easily if you calculate torques around the right end of the horizontal rod: The only force that exerts a torque around this point is the vertical component of the hinge force, so this force component must be zero.

11.4 Answers: (a) (iii), (b) (ii) In (a), the copper rod has 10 times the elongation Δl of the steel rod, but it also has 10 times the original length l_0. Hence the tensile strain $\Delta l / l_0$ is the same for both rods. In (b), the stress is equal to Young's modulus Y multiplied by the strain. From Table 11.1, steel has a larger value of Y, so a greater stress is required to produce the same strain.

11.5 In (a) and (b), the bumper will have sprung back to its original shape (although the paint may be scratched). In (c), the bumper will have a permanent dent or deformation. In (d), the bumper will be torn or broken.

Bridging Problem

Answers:

(a) $T = \dfrac{2mg}{3\sin\theta}$

(b) $F = \dfrac{2mg}{3\sin\theta}\sqrt{\cos^2\theta + \frac{1}{4}\sin^2\theta}$, $\phi = \arctan\left(\frac{1}{2}\tan\theta\right)$

(c) $\Delta l = \dfrac{2mgl_0}{3AY\tan\theta}$ (d) 4

FLUID MECHANICS

? This shark must swim constantly to keep from sinking to the bottom of the ocean, yet the orange tropical fish can remain at the same level in the water with little effort. Why is there a difference?

LEARNING GOALS

By studying this chapter, you will learn:

• The meaning of the density of a material and the average density of a body.

• What is meant by the pressure in a fluid, and how it is measured.

• How to calculate the buoyant force that a fluid exerts on a body immersed in it.

• The significance of laminar versus turbulent fluid flow, and how the speed of flow in a tube depends on the tube size.

• How to use Bernoulli's equation to relate pressure and flow speed at different points in certain types of flow.

luids play a vital role in many aspects of everyday life. We drink them, breathe them, swim in them. They circulate through our bodies and control our weather. Airplanes fly through them; ships float in them. A fluid is any substance that can flow; we use the term for both liquids and gases. We usually think of a gas as easily compressed and a liquid as nearly incompressible, although there are exceptional cases.

We begin our study with **fluid statics,** the study of fluids at rest in equilibrium situations. Like other equilibrium situations, it is based on Newton's first and third laws. We will explore the key concepts of density, pressure, and buoyancy. **Fluid dynamics,** the study of fluids in motion, is much more complex; indeed, it is one of the most complex branches of mechanics. Fortunately, we can analyze many important situations using simple idealized models and familiar principles such as Newton's laws and conservation of energy. Even so, we will barely scratch the surface of this broad and interesting topic.

12.1 Density

An important property of any material is its **density,** defined as its mass per unit volume. A homogeneous material such as ice or iron has the same density throughout. We use ρ (the Greek letter rho) for density. If a mass m of homogeneous material has volume V, the density ρ is

$$\rho = \frac{m}{V} \qquad \text{(definition of density)} \qquad (12.1)$$

Two objects made of the same material have the same density even though they may have different masses and different volumes. That's because the *ratio* of mass to volume is the same for both objects (Fig. 12.1).

12.1 Two objects with different masses and different volumes but the same density.

Different mass, same density: Because the wrench and nail are both made of steel, they have the same density (mass per unit volume).

Steel wrench   Steel nail

Table 12.1 Densities of Some Common Substances

Material	Density (kg/m^3)*	Material	Density (kg/m^3)*
Air (1 atm, 20°C)	1.20	Iron, steel	7.8×10^3
Ethanol	0.81×10^3	Brass	8.6×10^3
Benzene	0.90×10^3	Copper	8.9×10^3
Ice	0.92×10^3	Silver	10.5×10^3
Water	1.00×10^3	Lead	11.3×10^3
Seawater	1.03×10^3	Mercury	13.6×10^3
Blood	1.06×10^3	Gold	19.3×10^3
Glycerine	1.26×10^3	Platinum	21.4×10^3
Concrete	2×10^3	White dwarf star	10^{10}
Aluminum	2.7×10^3	Neutron star	10^{18}

*To obtain the densities in grams per cubic centimeter, simply divide by 10^3.

The SI unit of density is the kilogram per cubic meter (1 kg/m^3). The cgs unit, the gram per cubic centimeter (1 g/cm^3), is also widely used:

$$1 \text{ g/cm}^3 = 1000 \text{ kg/m}^3$$

The densities of some common substances at ordinary temperatures are given in Table 12.1. Note the wide range of magnitudes. The densest material found on earth is the metal osmium ($\rho = 22,500 \text{ kg/m}^3$), but its density pales by comparison to the densities of exotic astronomical objects such as white dwarf stars and neutron stars.

The **specific gravity** of a material is the ratio of its density to the density of water at 4.0°C, 1000 kg/m^3; it is a pure number without units. For example, the specific gravity of aluminum is 2.7. "Specific gravity" is a poor term, since it has nothing to do with gravity; "relative density" would have been better.

The density of some materials varies from point to point within the material. One example is the material of the human body, which includes low-density fat (about 940 kg/m^3) and high-density bone (from 1700 to 2500 kg/m^3). Two others are the earth's atmosphere (which is less dense at high altitudes) and oceans (which are denser at greater depths). For these materials, Eq. (12.1) describes the **average density.** In general, the density of a material depends on environmental factors such as temperature and pressure.

Measuring density is an important analytical technique. For example, we can determine the charge condition of a storage battery by measuring the density of its electrolyte, a sulfuric acid solution. As the battery discharges, the sulfuric acid (H_2SO_4) combines with lead in the battery plates to form insoluble lead sulfate ($PbSO_4$), decreasing the concentration of the solution. The density decreases from about $1.30 \times 10^3 \text{ kg/m}^3$ for a fully charged battery to $1.15 \times 10^3 \text{ kg/m}^3$ for a discharged battery.

Another automotive example is permanent-type antifreeze, which is usually a solution of ethylene glycol ($\rho = 1.12 \times 10^3 \text{ kg/m}^3$) and water. The freezing point of the solution depends on the glycol concentration, which can be determined by measuring the specific gravity. Such measurements can be performed by using a device called a hydrometer, which we'll discuss in Section 12.3.

Example 12.1 | **The weight of a roomful of air**

Find the mass and weight of the air at 20°C in a living room with a 4.0 m × 5.0 m floor and a ceiling 3.0 m high, and the mass and weight of an equal volume of water.

SOLUTION

IDENTIFY and SET UP: We assume that the air density is the same throughout the room. (Air is less dense at high elevations than near sea level, but the density varies negligibly over the room's 3.0-m height; see Section 12.2.) We use Eq. (12.1) to relate the mass m_{air} to the room's volume V (which we'll calculate) and the air density ρ_{air} (given in Table 12.1).

EXECUTE: We have $V = (4.0 \text{ m})(5.0 \text{ m})(3.0 \text{ m}) = 60 \text{ m}^3$, so from Eq. (12.1),

$$m_{air} = \rho_{air}V = (1.20 \text{ kg/m}^3)(60 \text{ m}^3) = 72 \text{ kg}$$

$$w_{air} = m_{air}g = (72 \text{ kg})(9.8 \text{ m/s}^2) = 700 \text{ N} = 160 \text{ lb}$$

The mass and weight of an equal volume of water are

$$m_{water} = \rho_{water}V = (1000 \text{ kg/m}^3)(60 \text{ m}^3) = 6.0 \times 10^4 \text{ kg}$$

$$w_{water} = m_{water}g = (6.0 \times 10^4 \text{ kg})(9.8 \text{ m/s}^2)$$

$$= 5.9 \times 10^5 \text{ N} = 1.3 \times 10^5 \text{ lb} = 66 \text{ tons}$$

EVALUATE: A roomful of air weighs about the same as an average adult. Water is nearly a thousand times denser than air, so its mass and weight are larger by the same factor. The weight of a roomful of water would collapse the floor of an ordinary house.

Test Your Understanding of Section 12.1 Rank the following objects in order from highest to lowest average density: (i) mass 4.00 kg, volume $1.60 \times 10^{-3} \text{ m}^3$; (ii) mass 8.00 kg, volume $1.60 \times 10^{-3} \text{ m}^3$; (iii) mass 8.00 kg, volume $3.20 \times 10^{-3} \text{ m}^3$; (iv) mass 2560 kg, volume 0.640 m^3; (v) mass 2560 kg, volume 1.28 m^3.

12.2 Pressure in a Fluid

When a fluid (either liquid or gas) is at rest, it exerts a force perpendicular to any surface in contact with it, such as a container wall or a body immersed in the fluid. This is the force that you feel pressing on your legs when you dangle them in a swimming pool. While the fluid as a whole is at rest, the molecules that make up the fluid are in motion; the force exerted by the fluid is due to molecules colliding with their surroundings.

If we think of an imaginary surface *within* the fluid, the fluid on the two sides of the surface exerts equal and opposite forces on the surface. (Otherwise, the surface would accelerate and the fluid would not remain at rest.) Consider a small surface of area dA centered on a point in the fluid; the normal force exerted by the fluid on each side is $dF_\perp$ (Fig. 12.2). We define the **pressure** p at that point as the normal force per unit area—that is, the ratio of $dF_\perp$ to dA (Fig. 12.3):

$$p = \frac{dF_\perp}{dA} \qquad \text{(definition of pressure)} \qquad (12.2)$$

If the pressure is the same at all points of a finite plane surface with area A, then

$$p = \frac{F_\perp}{A} \qquad (12.3)$$

where $F_\perp$ is the net normal force on one side of the surface. The SI unit of pressure is the **pascal,** where

$$1 \text{ pascal} = 1 \text{ Pa} = 1 \text{ N/m}^2$$

We introduced the pascal in Chapter 11. Two related units, used principally in meteorology, are the *bar,* equal to 10^5 Pa, and the *millibar,* equal to 100 Pa.

Atmospheric pressure p_a is the pressure of the earth's atmosphere, the pressure at the bottom of this sea of air in which we live. This pressure varies with weather changes and with elevation. Normal atmospheric pressure at sea level (an average value) is 1 *atmosphere* (atm), defined to be exactly 101,325 Pa. To four significant figures,

$$(p_a)_{av} = 1 \text{ atm} = 1.013 \times 10^5 \text{ Pa}$$

$$= 1.013 \text{ bar} = 1013 \text{ millibar} = 14.70 \text{ lb/in.}^2$$

12.2 Forces acting on a small surface within a fluid at rest.

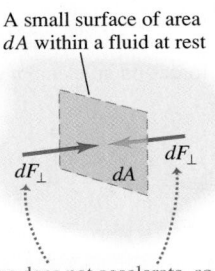

A small surface of area dA within a fluid at rest

The surface does not accelerate, so the surrounding fluid exerts equal normal forces on both sides of it. (The fluid cannot exert any force parallel to the surface, since that would cause the surface to accelerate.)

12.3 The pressure on either side of a surface is force divided by area. Pressure is a scalar with units of newtons per square meter. By contrast, force is a vector with units of newtons.

These surfaces differ in area and orientation ...

... but the pressure on them (force magnitude divided by area) is the same (and is a scalar).

CAUTION **Don't confuse pressure and force** In everyday language the words "pressure" and "force" mean pretty much the same thing. In fluid mechanics, however, these words describe distinct quantities with different characteristics. Fluid pressure acts perpendicular to any surface in the fluid, no matter how that surface is oriented (Fig. 12.3). Hence pressure has no intrinsic direction of its own; it's a scalar. By contrast, force is a vector with a definite direction. Remember, too, that pressure is force per unit area. As Fig. 12.3 shows, a surface with twice the area has twice as much force exerted on it by the fluid, so the pressure is the same. ▌

Example 12.2 **The force of air**

In the room described in Example 12.1, what is the total downward force on the floor due to an air pressure of 1.00 atm?

SOLUTION

IDENTIFY and SET UP: This example uses the relationship among the pressure p of a fluid (air), the area A subjected to that pressure, and the resulting normal force $F_\perp$ the fluid exerts. The pressure is uniform, so we use Eq. (12.3), $F_\perp = pA$, to determine $F_\perp$. The floor is horizontal, so $F_\perp$ is vertical (downward).

EXECUTE: We have $A = (4.0 \text{ m})(5.0 \text{ m}) = 20 \text{ m}^2$, so from Eq. (12.3),

$$F_\perp = pA = (1.013 \times 10^5 \text{ N/m}^2)(20 \text{ m}^2)$$
$$= 2.0 \times 10^6 \text{ N} = 4.6 \times 10^5 \text{ lb} = 230 \text{ tons}$$

EVALUATE: Unlike the water in Example 12.1, $F_\perp$ will not collapse the floor here, because there is an *upward* force of equal magnitude on the floor's underside. If the house has a basement, this upward force is exerted by the air underneath the floor. In this case, if we neglect the thickness of the floor, the *net* force due to air pressure is zero.

Pressure, Depth, and Pascal's Law

12.4 The forces on an element of fluid in equilibrium.

(a)

An element of a fluid at rest with area A and thickness dy

(b)

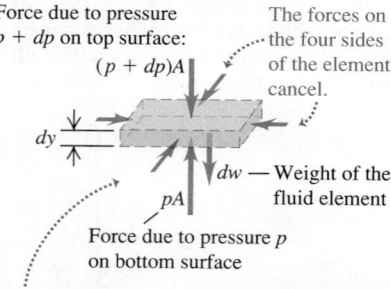

Force due to pressure $p + dp$ on top surface: $(p + dp)A$

The forces on the four sides of the element cancel.

dw — Weight of the fluid element

pA

Force due to pressure p on bottom surface

Because the fluid is in equilibrium, the vector sum of the vertical forces on the fluid element must be zero: $pA - (p + dp)A - dw = 0$.

If the weight of the fluid can be neglected, the pressure in a fluid is the same throughout its volume. We used that approximation in our discussion of bulk stress and strain in Section 11.4. But often the fluid's weight is *not* negligible. Atmospheric pressure is less at high altitude than at sea level, which is why an airplane cabin has to be pressurized when flying at 35,000 feet. When you dive into deep water, your ears tell you that the pressure increases rapidly with increasing depth below the surface.

We can derive a general relationship between the pressure p at any point in a fluid at rest and the elevation y of the point. We'll assume that the density ρ has the same value throughout the fluid (that is, the density is *uniform*), as does the acceleration due to gravity g. If the fluid is in equilibrium, every volume element is in equilibrium. Consider a thin element of fluid with thickness dy (Fig. 12.4a). The bottom and top surfaces each have area A, and they are at elevations y and $y + dy$ above some reference level where $y = 0$. The volume of the fluid element is $dV = A \, dy$, its mass is $dm = \rho \, dV = \rho A \, dy$, and its weight is $dw = dm \, g = \rho g A \, dy$.

What are the other forces on this fluid element (Fig 12.4b)? Let's call the pressure at the bottom surface p; then the total y-component of upward force on this surface is pA. The pressure at the top surface is $p + dp$, and the total y-component of (downward) force on the top surface is $-(p + dp)A$. The fluid element is in equilibrium, so the total y-component of force, including the weight and the forces at the bottom and top surfaces, must be zero:

$$\sum F_y = 0 \quad \text{so} \quad pA - (p + dp)A - \rho g A \, dy = 0$$

When we divide out the area A and rearrange, we get

$$\frac{dp}{dy} = -\rho g \tag{12.4}$$

This equation shows that when y increases, p decreases; that is, as we move upward in the fluid, pressure decreases, as we expect. If p_1 and p_2 are the pressures at elevations y_1 and y_2, respectively, and if ρ and g are constant, then

$$p_2 - p_1 = -\rho g(y_2 - y_1) \quad \text{(pressure in a fluid of uniform density)} \quad (12.5)$$

It's often convenient to express Eq. (12.5) in terms of the *depth* below the surface of a fluid (Fig. 12.5). Take point 1 at any level in the fluid and let p represent the pressure at this point. Take point 2 at the *surface* of the fluid, where the pressure is p_0 (subscript zero for zero depth). The depth of point 1 below the surface is $h = y_2 - y_1$, and Eq. (12.5) becomes

$$p_0 - p = -\rho g(y_2 - y_1) = -\rho gh \quad \text{or}$$

$$p = p_0 + \rho gh \quad \text{(pressure in a fluid of uniform density)} \quad (12.6)$$

The pressure p at a depth h is greater than the pressure p_0 at the surface by an amount ρgh. Note that the pressure is the same at any two points at the same level in the fluid. The *shape* of the container does not matter (Fig. 12.6).

Equation (12.6) shows that if we increase the pressure p_0 at the top surface, possibly by using a piston that fits tightly inside the container to push down on the fluid surface, the pressure p at any depth increases by exactly the same amount. This fact was recognized in 1653 by the French scientist Blaise Pascal (1623–1662) and is called *Pascal's law.*

> **Pascal's law: Pressure applied to an enclosed fluid is transmitted undiminished to every portion of the fluid and the walls of the containing vessel.**

The hydraulic lift shown schematically in Fig. 12.7 illustrates Pascal's law. A piston with small cross-sectional area A_1 exerts a force F_1 on the surface of a liquid such as oil. The applied pressure $p = F_1/A_1$ is transmitted through the connecting pipe to a larger piston of area A_2. The applied pressure is the same in both cylinders, so

$$p = \frac{F_1}{A_1} = \frac{F_2}{A_2} \quad \text{and} \quad F_2 = \frac{A_2}{A_1}F_1 \quad (12.7)$$

The hydraulic lift is a force-multiplying device with a multiplication factor equal to the ratio of the areas of the two pistons. Dentist's chairs, car lifts and jacks, many elevators, and hydraulic brakes all use this principle.

For gases the assumption that the density ρ is uniform is realistic only over short vertical distances. In a room with a ceiling height of 3.0 m filled with air of uniform density 1.2 kg/m³, the difference in pressure between floor and ceiling, given by Eq. (12.6), is

$$\rho gh = (1.2 \text{ kg/m}^3)(9.8 \text{ m/s}^2)(3.0 \text{ m}) = 35 \text{ Pa}$$

or about 0.00035 atm, a very small difference. But between sea level and the summit of Mount Everest (8882 m) the density of air changes by nearly a factor of 3, and in this case we cannot use Eq. (12.6). Liquids, by contrast, are nearly incompressible, and it is usually a very good approximation to regard their density as independent of pressure. A pressure of several hundred atmospheres will cause only a few percent increase in the density of most liquids.

Absolute Pressure and Gauge Pressure

If the pressure inside a car tire is equal to atmospheric pressure, the tire is flat. The pressure has to be *greater* than atmospheric to support the car, so the significant quantity is the *difference* between the inside and outside pressures. When we say that the pressure in a car tire is "32 pounds" (actually 32 lb/in.², equal to 220 kPa or 2.2 × 10⁵ Pa), we mean that it is *greater* than atmospheric pressure

12.5 How pressure varies with depth in a fluid with uniform density.

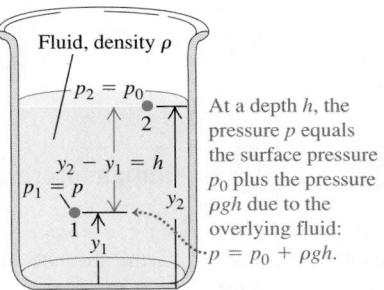

Pressure difference between levels 1 and 2:
$$p_2 - p_1 = -\rho g(y_2 - y_1)$$
The pressure is greater at the lower level.

12.6 Each fluid column has the same height, no matter what its shape.

The pressure at the top of each liquid column is atmospheric pressure, p_0.

The pressure at the bottom of each liquid column has the same value p.

The difference between p and p_0 is ρgh, where h is the distance from the top to the bottom of the liquid column. Hence all columns have the same height.

12.7 The hydraulic lift is an application of Pascal's law. The size of the fluid-filled container is exaggerated for clarity.

A small force is applied to a small piston.

Because the pressure p is the same at all points at a given height in the fluid ...

F_1 F_2

pA_1 pA_2

... a piston of larger area at the same height experiences a larger force.

$(14.7 \text{ lb/in.}^2 \text{ or } 1.01 \times 10^5 \text{ Pa})$ by this amount. The *total* pressure in the tire is then 47 lb/in.^{2} or 320 kPa. The excess pressure above atmospheric pressure is usually called **gauge pressure,** and the total pressure is called **absolute pressure.** Engineers use the abbreviations psig and psia for "pounds per square inch gauge" and "pounds per square inch absolute," respectively. If the pressure is *less* than atmospheric, as in a partial vacuum, the gauge pressure is negative.

Example 12.3 Finding absolute and gauge pressures

Water stands 12.0 m deep in a storage tank whose top is open to the atmosphere. What are the absolute and gauge pressures at the bottom of the tank?

SOLUTION

IDENTIFY and SET UP: Table 11.2 indicates that water is nearly incompressible, so we can treat it as having uniform density. The level of the top of the tank corresponds to point 2 in Fig. 12.5, and the level of the bottom of the tank corresponds to point 1. Our target variable is p in Eq. (12.6). We have $h = 12.0$ m and $p_0 = 1 \text{ atm} = 1.01 \times 10^5$ Pa.

EXECUTE: From Eq. (12.6), the pressures are

absolute:

$$p = p_0 + \rho g h$$
$$= (1.01 \times 10^5 \text{ Pa}) + (1000 \text{ kg/m}^3)(9.80 \text{ m/s}^2)(12.0 \text{ m})$$
$$= 2.19 \times 10^5 \text{ Pa} = 2.16 \text{ atm} = 31.8 \text{ lb/in.}^2$$

gauge: $\quad p - p_0 = (2.19 - 1.01) \times 10^5 \text{ Pa}$
$$= 1.18 \times 10^5 \text{ Pa} = 1.16 \text{ atm} = 17.1 \text{ lb/in.}^2$$

EVALUATE: A pressure gauge at the bottom of such a tank would probably be calibrated to read gauge pressure rather than absolute pressure.

Pressure Gauges

The simplest pressure gauge is the open-tube *manometer* (Fig. 12.8a). The U-shaped tube contains a liquid of density ρ, often mercury or water. The left end of the tube is connected to the container where the pressure p is to be measured, and the right end is open to the atmosphere at pressure $p_0 = p_{\text{atm}}$. The pressure at the bottom of the tube due to the fluid in the left column is $p + \rho g y_1$, and the pressure at the bottom due to the fluid in the right column is $p_{\text{atm}} + \rho g y_2$. These pressures are measured at the same level, so they must be equal:

$$p + \rho g y_1 = p_{\text{atm}} + \rho g y_2$$
$$p - p_{\text{atm}} = \rho g (y_2 - y_1) = \rho g h \tag{12.8}$$

In Eq. (12.8), p is the *absolute pressure,* and the difference $p - p_{\text{atm}}$ between absolute and atmospheric pressure is the gauge pressure. Thus the gauge pressure is proportional to the difference in height $h = y_2 - y_1$ of the liquid columns.

12.8 Two types of pressure gauge.

(a) Open-tube manometer

$p_0 = p_{\text{atm}}$

$h = y_2 - y_1$

Pressure p

y_2

y_1

$p + \rho g y_1$ ⟶ ⟵ $p_{\text{atm}} + \rho g y_2$

The pressure is the same at the bottoms of the two tubes.

(b) Mercury barometer

There is a near-vacuum at the top of the tube.

$p_0 = 0$

The height to which the mercury rises depends on the atmospheric pressure exerted on the mercury in the dish.

$h = y_2 - y_1$

y_2

$p = p_{\text{atm}}$

y_1

Another common pressure gauge is the **mercury barometer.** It consists of a long glass tube, closed at one end, that has been filled with mercury and then inverted in a dish of mercury (Fig. 12.8b). The space above the mercury column contains only mercury vapor; its pressure is negligibly small, so the pressure p_0 at the top of the mercury column is practically zero. From Eq. (12.6),

$$p_{\text{atm}} = p = 0 + \rho g(y_2 - y_1) = \rho g h \qquad (12.9)$$

Thus the mercury barometer reads the atmospheric pressure p_{atm} directly from the height of the mercury column.

Pressures are often described in terms of the height of the corresponding mercury column, as so many "inches of mercury" or "millimeters of mercury" (abbreviated mm Hg). A pressure of 1 mm Hg is called *1 torr,* after Evangelista Torricelli, inventor of the mercury barometer. But these units depend on the density of mercury, which varies with temperature, and on the value of g, which varies with location, so the pascal is the preferred unit of pressure.

Many types of pressure gauges use a flexible sealed tube (Fig. 12.9). A change in the pressure either inside or outside the tube causes a change in its dimensions. This change is detected optically, electrically, or mechanically.

Blood-pressure readings, such as 130/80, give the maximum and minimum gauge pressures in the arteries, measured in mm Hg or torr. Blood pressure varies with vertical position within the body; the standard reference point is the upper arm, level with the heart.

(a)

Changes in the inlet pressure cause the tube to coil or uncoil, which moves the pointer.

Flexible pressure tube

Inlet

Pressure p being measured

(b)

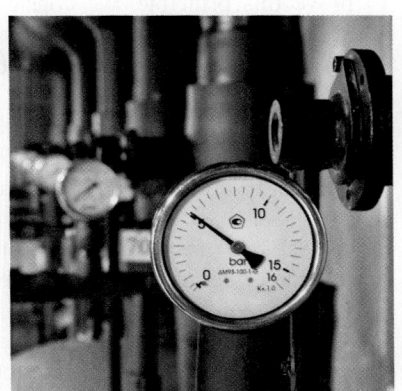

12.9 (a) A Bourdon pressure gauge. When the pressure inside the flexible tube increases, the tube straightens out a little, deflecting the attached pointer. (b) This Bourdon-type pressure gauge is connected to a high-pressure gas line. The gauge pressure shown is just over 5 bars (1 bar = 10^5 Pa).

Example 12.4 | **A tale of two fluids**

A manometer tube is partially filled with water. Oil (which does not mix with water) is poured into the left arm of the tube until the oil–water interface is at the midpoint of the tube as shown. Both arms of the tube are open to the air. Find a relationship between the heights h_{oil} and h_{water}.

SOLUTION

IDENTIFY and SET UP: Figure 12.10 shows our sketch. The relationship between pressure and depth given by Eq. (12.6) applies only to fluids of uniform density; we have two fluids of different densities, so we must write a separate pressure–depth relationship for each. Both fluid columns have pressure p at the bottom (where they are in contact and in equilibrium) and are both at atmospheric pressure p_0 at the top (where both are in contact with and in equilibrium with the air).

EXECUTE: Writing Eq. (12.6) for each fluid gives

$$p = p_0 + \rho_{\text{water}} g h_{\text{water}}$$
$$p = p_0 + \rho_{\text{oil}} g h_{\text{oil}}$$

12.10 Our sketch for this problem.

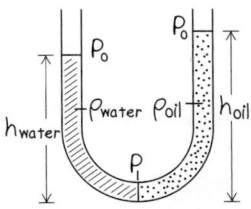

Since the pressure p at the bottom of the tube is the same for both fluids, we set these two expressions equal to each other and solve for h_{oil} in terms of h_{water}. You can show that the result is

$$h_{\text{oil}} = \frac{\rho_{\text{water}}}{\rho_{\text{oil}}} h_{\text{water}}$$

EVALUATE: Water $(\rho_{\text{water}} = 1000 \text{ kg/m}^3)$ is denser than oil $(\rho_{\text{oil}} \approx 850 \text{ kg/m}^3)$, so h_{oil} is greater than h_{water} as Fig. 12.10 shows. It takes a greater height of low-density oil to produce the same pressure p at the bottom of the tube.

Test Your Understanding of Section 12.2 Mercury is less dense at high temperatures than at low temperatures. Suppose you move a mercury barometer from the cold interior of a tightly sealed refrigerator to outdoors on a hot summer day. You find that the column of mercury remains at the same height in the tube. Compared to the air pressure inside the refrigerator, is the air pressure outdoors (i) higher, (ii) lower, or (iii) the same? (Ignore the very small change in the dimensions of the glass tube due to the temperature change.)

12.3 Buoyancy

Buoyancy is a familiar phenomenon: A body immersed in water seems to weigh less than when it is in air. When the body is less dense than the fluid, it floats. The human body usually floats in water, and a helium-filled balloon floats in air.

> **Archimedes's principle:** When a body is completely or partially immersed in a fluid, the fluid exerts an upward force on the body equal to the weight of the fluid displaced by the body.

To prove this principle, we consider an arbitrary element of fluid at rest. In Fig. 12.11a the irregular outline is the surface boundary of this element of fluid. The arrows represent the forces exerted on the boundary surface by the surrounding fluid.

The entire fluid is in equilibrium, so the sum of all the *y*-components of force on this element of fluid is zero. Hence the sum of the *y*-components of the *surface* forces must be an upward force equal in magnitude to the weight *mg* of the fluid inside the surface. Also, the sum of the torques on the element of fluid must be zero, so the line of action of the resultant *y*-component of surface force must pass through the center of gravity of this element of fluid.

Now we remove the fluid inside the surface and replace it with a solid body having exactly the same shape (Fig. 12.11b). The pressure at every point is exactly the same as before. So the total upward force exerted on the body by the fluid is also the same, again equal in magnitude to the weight *mg* of the fluid displaced to make way for the body. We call this upward force the **buoyant force** on the solid body. The line of action of the buoyant force again passes through the center of gravity of the displaced fluid (which doesn't necessarily coincide with the center of gravity of the body).

When a balloon floats in equilibrium in air, its weight (including the gas inside it) must be the same as the weight of the air displaced by the balloon. A fish's flesh is denser than water, yet a fish can float while

12.11 Archimedes's principle.

(a) Arbitrary element of fluid in equilibrium

The forces on the fluid element due to pressure must sum to a buoyant force equal in magnitude to the element's weight.

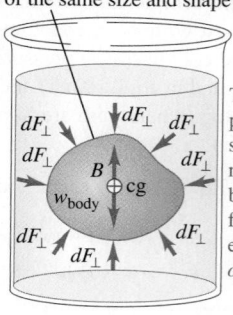

(b) Fluid element replaced with solid body of the same size and shape

The forces due to pressure are the same, so the body must be acted upon by the same buoyant force as the fluid element, *regardless of the body's weight.*

submerged because it has a gas-filled cavity within its body. This makes the fish's *average* density the same as water's, so its net weight is the same as the weight of the water it displaces. A body whose average density is *less* than that of a liquid can float partially submerged at the free upper surface of the liquid. The greater the density of the liquid, the less of the body is submerged. When you swim in seawater (density 1030 kg/m^3), your body floats higher than in fresh water (1000 kg/m^3).

A practical example of buoyancy is the hydrometer, used to measure the density of liquids (Fig. 12.12a). The calibrated float sinks into the fluid until the weight of the fluid it displaces is exactly equal to its own weight. The hydrometer floats *higher* in denser liquids than in less dense liquids, and a scale in the top stem permits direct density readings. Figure 12.12b shows a type of hydrometer that is commonly used to measure the density of battery acid or antifreeze. The bottom of the large tube is immersed in the liquid; the bulb is squeezed to expel air and is then released, like a giant medicine dropper. The liquid rises into the outer tube, and the hydrometer floats in this sample of the liquid.

12.12 Measuring the density of a fluid.

(b) Using a hydrometer to measure the density of battery acid or antifreeze

(a) A simple hydrometer

The depth to which the weighted scale sinks tells you the density of the fluid.

The weight at the bottom makes the scale float upright.

Example 12.5 Buoyancy

A 15.0-kg solid gold statue is raised from the sea bottom (Fig. 12.13a). What is the tension in the hoisting cable (assumed massless) when the statue is (a) at rest and completely underwater and (b) at rest and completely out of the water?

SOLUTION

IDENTIFY and SET UP: In both cases the statue is in equilibrium and experiences three forces: its weight, the cable tension, and a buoyant force equal in magnitude to the weight of the fluid displaced by the statue (seawater in part (a), air in part (b)). Figure 12.13b shows the free-body diagram for the statue. Our target variables are the values of the tension in seawater (T_{sw}) and in air (T_{air}). We are given the mass m_{statue}, and we can calculate the buoyant force in seawater (B_{sw}) and in air (B_{air}) using Archimedes's principle.

EXECUTE: (a) To find B_{sw}, we first find the statue's volume V using the density of gold from Table 12.1:

$$V = \frac{m_{statue}}{\rho_{gold}} = \frac{15.0 \text{ kg}}{19.3 \times 10^3 \text{ kg/m}^3} = 7.77 \times 10^{-4} \text{ m}^3$$

The buoyant force B_{sw} equals the weight of this same volume of seawater. Using Table 12.1 again:

$$B_{sw} = w_{sw} = m_{sw}g = \rho_{sw}Vg$$
$$= (1.03 \times 10^3 \text{ kg/m}^3)(7.77 \times 10^{-4} \text{ m}^3)(9.80 \text{ m/s}^2)$$
$$= 7.84 \text{ N}$$

The statue is at rest, so the net external force acting on it is zero. From Fig. 12.13b,

$$\sum F_y = B_{sw} + T_{sw} + (-m_{statue}g) = 0$$
$$T_{sw} = m_{statue}g - B_{sw} = (15.0 \text{ kg})(9.80 \text{ m/s}^2) - 7.84 \text{ N}$$
$$= 147 \text{ N} - 7.84 \text{ N} = 139 \text{ N}$$

12.13 What is the tension in the cable hoisting the statue?

(a) Immersed statue in equilibrium **(b)** Free-body diagram of statue

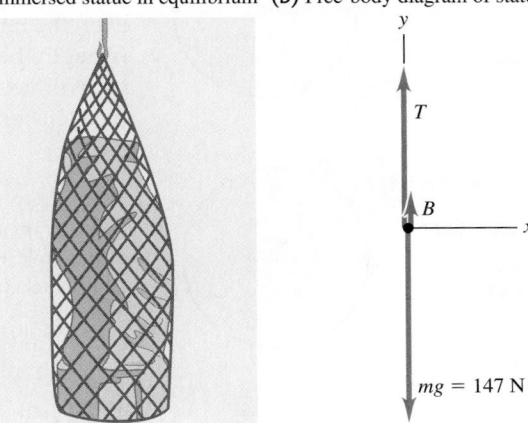

A spring scale attached to the upper end of the cable will indicate a tension 7.84 N less than the statue's actual weight $m_{statue}g = 147$ N.

(b) The density of air is about 1.2 kg/m^3, so the buoyant force of air on the statue is

$$B_{air} = \rho_{air}Vg = (1.2 \text{ kg/m}^3)(7.77 \times 10^{-4} \text{ m}^3)(9.80 \text{ m/s}^2)$$
$$= 9.1 \times 10^{-3} \text{ N}$$

This is negligible compared to the statue's actual weight $m_{statue}g = 147$ N. So within the precision of our data, the tension in the cable with the statue in air is $T_{air} = m_{statue}g = 147$ N.

EVALUATE: Note that the buoyant force is proportional to the density of the *fluid* in which the statue is immersed, *not* the density of

Continued

the statue. The denser the fluid, the greater the buoyant force and the smaller the cable tension. If the fluid had the same density as the statue, the buoyant force would be equal to the statue's weight and the tension would be zero (the cable would go slack). If the fluid were denser than the statue, the tension would be *negative:* The buoyant force would be greater than the statue's weight, and a downward force would be required to keep the statue from rising upward.

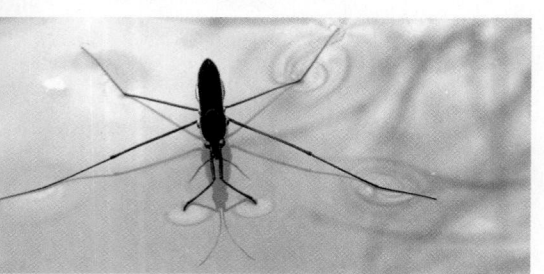

12.14 The surface of the water acts like a membrane under tension, allowing this water strider to literally "walk on water."

Surface Tension

An object less dense than water, such as an air-filled beach ball, floats with part of its volume below the surface. Conversely, a paper clip can rest *atop* a water surface even though its density is several times that of water. This is an example of **surface tension:** The surface of the liquid behaves like a membrane under tension (Fig. 12.14). Surface tension arises because the molecules of the liquid exert attractive forces on each other. There is zero net force on a molecule inside the volume of the liquid, but a surface molecule is drawn into the volume (Fig. 12.15). Thus the liquid tends to minimize its surface area, just as a stretched membrane does.

Surface tension explains why freely falling raindrops are spherical (*not* teardrop-shaped): A sphere has a smaller surface area for its volume than any other shape. It also explains why hot, soapy water is used for washing. To wash clothing thoroughly, water must be forced through the tiny spaces between the fibers (Fig. 12.16). To do so requires increasing the surface area of the water, which is difficult to achieve because of surface tension. The job is made easier by increasing the temperature of the water and adding soap, both of which decrease the surface tension.

Surface tension is important for a millimeter-sized water drop, which has a relatively large surface area for its volume. (A sphere of radius r has surface area $4\pi r^2$ and volume $(4\pi/3)r^3$. The ratio of surface area to volume is $3/r$, which increases with decreasing radius.) For large quantities of liquid, however, the ratio of surface area to volume is relatively small, and surface tension is negligible compared to pressure forces. For the remainder of this chapter, we will consider only fluids in bulk and hence will ignore the effects of surface tension.

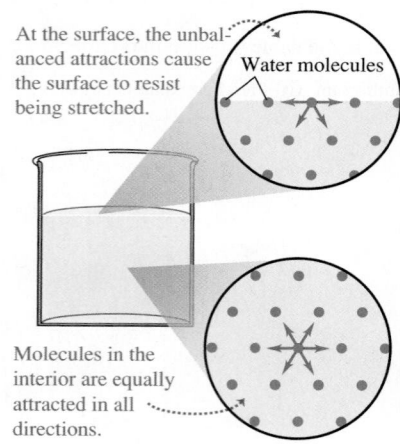

12.15 A molecule at the surface of a liquid is attracted into the bulk liquid, which tends to reduce the liquid's surface area.

Molecules in a liquid are attracted by neighboring molecules.

At the surface, the unbalanced attractions cause the surface to resist being stretched.

Water molecules

Molecules in the interior are equally attracted in all directions.

Test Your Understanding of Section 12.3 You place a container of seawater on a scale and note the reading on the scale. You now suspend the statue of Example 12.5 in the water (Fig. 12.17). How does the scale reading change? (i) It increases by 7.84 N; (ii) it decreases by 7.84 N; (iii) it remains the same; (iv) none of these.

12.4 Fluid Flow

We are now ready to consider *motion* of a fluid. Fluid flow can be extremely complex, as shown by the currents in river rapids or the swirling flames of a campfire. But some situations can be represented by relatively simple idealized models. An **ideal fluid** is a fluid that is *incompressible* (that is, its density cannot change) and has no internal friction (called **viscosity**). Liquids are approximately incompressible in most situations, and we may also treat a gas as incompressible if the pressure differences from one region to another are not too great. Internal friction in a fluid causes shear stresses when two adjacent layers of fluid move relative to each other, as when fluid flows inside a tube or around an obstacle. In some cases we can neglect these shear forces in comparison with forces arising from gravitation and pressure differences.

The path of an individual particle in a moving fluid is called a **flow line.** If the overall flow pattern does not change with time, the flow is called **steady flow.** In

12.16 Surface tension makes it difficult to force water through small crevices. The required water pressure p can be reduced by using hot, soapy water, which has less surface tension.

Water pressure p

Fibers
Air pressure p_0

steady flow, every element passing through a given point follows the same flow line. In this case the "map" of the fluid velocities at various points in space remains constant, although the velocity of a particular particle may change in both magnitude and direction during its motion. A **streamline** is a curve whose tangent at any point is in the direction of the fluid velocity at that point. When the flow pattern changes with time, the streamlines do not coincide with the flow lines. We will consider only steady-flow situations, for which flow lines and streamlines are identical.

The flow lines passing through the edge of an imaginary element of area, such as the area *A* in Fig. 12.18, form a tube called a **flow tube**. From the definition of a flow line, in steady flow no fluid can cross the side walls of a flow tube; the fluids in different flow tubes cannot mix.

Figure 12.19 shows patterns of fluid flow from left to right around three different obstacles. The photographs were made by injecting dye into water flowing between two closely spaced glass plates. These patterns are typical of **laminar flow,** in which adjacent layers of fluid slide smoothly past each other and the flow is steady. (A *lamina* is a thin sheet.) At sufficiently high flow rates, or when boundary surfaces cause abrupt changes in velocity, the flow can become irregular and chaotic. This is called **turbulent flow** (Fig. 12.20). In turbulent flow there is no steady-state pattern; the flow pattern changes continuously.

The Continuity Equation

The mass of a moving fluid doesn't change as it flows. This leads to an important quantitative relationship called the **continuity equation.** Consider a portion of a flow tube between two stationary cross sections with areas A_1 and A_2 (Fig. 12.21). The fluid speeds at these sections are v_1 and v_2, respectively. No fluid flows in or out across the sides of the tube because the fluid velocity is tangent to the wall at every point on the wall. During a small time interval dt, the fluid at A_1 moves a distance $v_1 \, dt$, so a cylinder of fluid with height $v_1 \, dt$ and volume $dV_1 = A_1 v_1 \, dt$ flows into the tube across A_1. During this same interval, a cylinder of volume $dV_2 = A_2 v_2 \, dt$ flows out of the tube across A_2.

Let's first consider the case of an incompressible fluid so that the density ρ has the same value at all points. The mass dm_1 flowing into the tube across A_1 in time dt is $dm_1 = \rho A_1 v_1 \, dt$. Similarly, the mass dm_2 that flows out across A_2 in the same time is $dm_2 = \rho A_2 v_2 \, dt$. In steady flow the total mass in the tube is constant, so $dm_1 = dm_2$ and

$$\rho A_1 v_1 \, dt = \rho A_2 v_2 \, dt \qquad \text{or}$$

$$A_1 v_1 = A_2 v_2 \qquad \text{(continuity equation, incompressible fluid)} \qquad (12.10)$$

12.17 How does the scale reading change when the statue is immersed in water?

12.18 A flow tube bounded by flow lines. In steady flow, fluid cannot cross the walls of a flow tube.

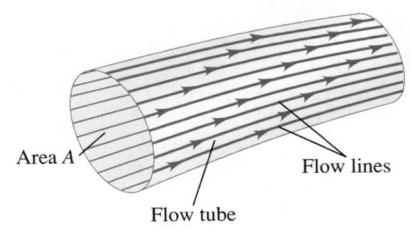

Area *A*

Flow lines

Flow tube

12.19 Laminar flow around obstacles of different shapes.

12.20 The flow of smoke rising from these incense sticks is laminar up to a certain point, and then becomes turbulent.

12.21 A flow tube with changing cross-sectional area. If the fluid is incompressible, the product Av has the same value at all points along the tube.

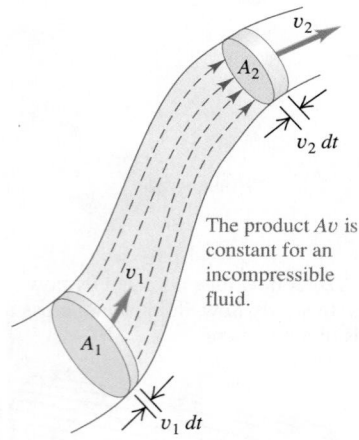

The product Av is constant for an incompressible fluid.

The product Av is the *volume flow rate* dV/dt, the rate at which volume crosses a section of the tube:

$$\frac{dV}{dt} = Av \qquad \text{(volume flow rate)} \qquad (12.11)$$

The *mass* flow rate is the mass flow per unit time through a cross section. This is equal to the density ρ times the volume flow rate dV/dt.

Equation (12.10) shows that the volume flow rate has the same value at all points along any flow tube. When the cross section of a flow tube decreases, the speed increases, and vice versa. A broad, deep part of a river has larger cross section and slower current than a narrow, shallow part, but the volume flow rates are the same in both. This is the essence of the familiar maxim, "Still waters run deep." The stream of water from a faucet narrows as it gains speed during its fall, but dV/dt is the same everywhere along the stream. If a water pipe with 2-cm diameter is connected to a pipe with 1-cm diameter, the flow speed is four times as great in the 1-cm part as in the 2-cm part.

We can generalize Eq. (12.10) for the case in which the fluid is *not* incompressible. If ρ_1 and ρ_2 are the densities at sections 1 and 2, then

$$\rho_1 A_1 v_1 = \rho_2 A_2 v_2 \qquad \text{(continuity equation, compressible fluid)} \qquad (12.12)$$

If the fluid is denser at point 2 than at point 1 ($\rho_2 > \rho_1$), the volume flow rate at point 2 will be less than at point 1 ($A_2 v_2 < A_1 v_1$). We leave the details to you. If the fluid is incompressible so that ρ_1 and ρ_2 are always equal, Eq. (12.12) reduces to Eq. (12.10).

Example 12.6   **Flow of an incompressible fluid**

Incompressible oil of density 850 kg/m³ is pumped through a cylindrical pipe at a rate of 9.5 liters per second. (a) The first section of the pipe has a diameter of 8.0 cm. What is the flow speed of the oil? What is the mass flow rate? (b) The second section of the pipe has a diameter of 4.0 cm. What are the flow speed and mass flow rate in that section?

SOLUTION

IDENTIFY and SET UP: Since the oil is incompressible, the volume flow rate has the *same* value (9.5 L/s) in both sections of pipe. The mass flow rate (the density times the volume flow rate) also has the same value in both sections. (This is just the statement that no fluid is lost or added anywhere along the pipe.) We use the volume flow rate equation, Eq. (12.11), to determine the speed v_1 in the 8.0-cm-diameter section and the continuity equation for incompressible flow, Eq. (12.10), to find the speed v_2 in the 4.0-cm-diameter section.

EXECUTE: (a) From Eq. (12.11) the volume flow rate in the first section is $dV/dt = A_1 v_1$, where A_1 is the cross-sectional area of

the pipe of diameter 8.0 cm and radius 4.0 cm. Hence

$$v_1 = \frac{dV/dt}{A_1} = \frac{(9.5 \text{ L/s})(10^{-3} \text{ m}^3/\text{L})}{\pi(4.0 \times 10^{-2} \text{ m})^2} = 1.9 \text{ m/s}$$

The mass flow rate is $\rho \, dV/dt = (850 \text{ kg/m}^3)(9.5 \times 10^{-3} \text{ m}^3/\text{s}) = 8.1 \text{ kg/s}$.

(b) From the continuity equation, Eq. (12.10),

$$v_2 = \frac{A_1}{A_2} v_1 = \frac{\pi(4.0 \times 10^{-2} \text{ m})^2}{\pi(2.0 \times 10^{-2} \text{ m})^2}(1.9 \text{ m/s}) = 7.6 \text{ m/s} = 4v_1$$

The volume and mass flow rates are the same as in part (a).

EVALUATE: The second section of pipe has one-half the diameter and one-fourth the cross-sectional area of the first section. Hence the speed must be four times greater in the second section, which is just what our result shows.

Test Your Understanding of Section 12.4   A maintenance crew is working on a section of a three-lane highway, leaving only one lane open to traffic. The result is much slower traffic flow (a traffic jam). Do cars on a highway behave like (i) the molecules of an incompressible fluid or (ii) the molecules of a compressible fluid?

(MP)

12.5 Bernoulli's Equation

According to the continuity equation, the speed of fluid flow can vary along the paths of the fluid. The pressure can also vary; it depends on height as in the static situation (see Section 12.2), and it also depends on the speed of flow. We can derive an important relationship called *Bernoulli's equation* that relates the pressure, flow speed, and height for flow of an ideal, incompressible fluid. Bernoulli's equation is an essential tool in analyzing plumbing systems, hydroelectric generating stations, and the flight of airplanes.

The dependence of pressure on speed follows from the continuity equation, Eq. (12.10). When an incompressible fluid flows along a flow tube with varying cross section, its speed *must* change, and so an element of fluid must have an acceleration. If the tube is horizontal, the force that causes this acceleration has to be applied by the surrounding fluid. This means that the pressure *must* be different in regions of different cross section; if it were the same everywhere, the net force on every fluid element would be zero. When a horizontal flow tube narrows and a fluid element speeds up, it must be moving toward a region of lower pressure in order to have a net forward force to accelerate it. If the elevation also changes, this causes an additional pressure difference.

Deriving Bernoulli's Equation

To derive Bernoulli's equation, we apply the work–energy theorem to the fluid in a section of a flow tube. In Fig. 12.22 we consider the element of fluid that at some initial time lies between the two cross sections a and c. The speeds at the lower and upper ends are v_1 and v_2. In a small time interval dt, the fluid that is initially at a moves to b, a distance $ds_1 = v_1\,dt$, and the fluid that is initially at c moves to d, a distance $ds_2 = v_2\,dt$. The cross-sectional areas at the two ends are A_1 and A_2, as shown. The fluid is incompressible; hence by the continuity equation, Eq. (12.10), the volume of fluid dV passing *any* cross section during time dt is the same. That is, $dV = A_1\,ds_1 = A_2\,ds_2$.

Let's compute the *work* done on this fluid element during dt. We assume that there is negligible internal friction in the fluid (i.e., no viscosity), so the only nongravitational forces that do work on the fluid element are due to the pressure of the surrounding fluid. The pressures at the two ends are p_1 and p_2; the force on the cross section at a is p_1A_1, and the force at c is p_2A_2. The net work dW done on the element by the surrounding fluid during this displacement is therefore

$$dW = p_1A_1\,ds_1 - p_2A_2\,ds_2 = (p_1 - p_2)dV \qquad (12.13)$$

The second term has a negative sign because the force at c opposes the displacement of the fluid.

The work dW is due to forces other than the conservative force of gravity, so it equals the change in the total mechanical energy (kinetic energy plus gravitational potential energy) associated with the fluid element. The mechanical energy for the fluid between sections b and c does not change. At the beginning of dt the fluid between a and b has volume $A_1\,ds_1$, mass $\rho A_1\,ds_1$, and kinetic energy $\frac{1}{2}\rho(A_1\,ds_1)v_1^2$. At the end of dt the fluid between c and d has kinetic energy $\frac{1}{2}\rho(A_2\,ds_2)v_2^2$. The net change in kinetic energy dK during time dt is

$$dK = \tfrac{1}{2}\rho\,dV(v_2^2 - v_1^2) \qquad (12.14)$$

What about the change in gravitational potential energy? At the beginning of dt, the potential energy for the mass between a and b is $dm\,gy_1 = \rho\,dV\,gy_1$. At

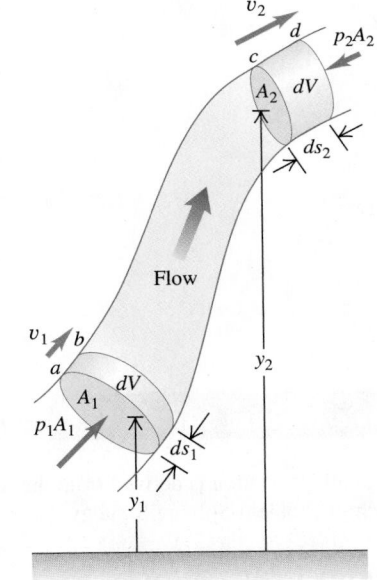

12.22 Deriving Bernoulli's equation. The net work done on a fluid element by the pressure of the surrounding fluid equals the change in the kinetic energy plus the change in the gravitational potential energy.

the end of dt, the potential energy for the mass between c and d is $dm\, gy_2 = \rho\, dV\, gy_2$. The net change in potential energy dU during dt is

$$dU = \rho\, dV\, g(y_2 - y_1) \qquad (12.15)$$

Combining Eqs. (12.13), (12.14), and (12.15) in the energy equation $dW = dK + dU$, we obtain

$$(p_1 - p_2)dV = \tfrac{1}{2}\rho\, dV(v_2^2 - v_1^2) + \rho\, dV\, g(y_2 - y_1)$$
$$p_1 - p_2 = \tfrac{1}{2}\rho(v_2^2 - v_1^2) + \rho g(y_2 - y_1) \qquad (12.16)$$

This is **Bernoulli's equation.** It states that the work done on a unit volume of fluid by the surrounding fluid is equal to the sum of the changes in kinetic and potential energies per unit volume that occur during the flow. We may also interpret Eq. (12.16) in terms of pressures. The first term on the right is the pressure difference associated with the change of speed of the fluid. The second term on the right is the additional pressure difference caused by the weight of the fluid and the difference in elevation of the two ends.

We can also express Eq. (12.16) in a more convenient form as

$$p_1 + \rho g y_1 + \tfrac{1}{2}\rho v_1^2 = p_2 + \rho g y_2 + \tfrac{1}{2}\rho v_2^2 \quad \text{(Bernoulli's equation)} \qquad (12.17)$$

The subscripts 1 and 2 refer to *any* two points along the flow tube, so we can also write

$$p + \rho g y + \tfrac{1}{2}\rho v^2 = \text{constant} \qquad (12.18)$$

Note that when the fluid is *not* moving (so $v_1 = v_2 = 0$), Eq. (12.17) reduces to the pressure relationship we derived for a fluid at rest, Eq. (12.5).

CAUTION **Bernoulli's principle applies only in certain situations** We stress again that Bernoulli's equation is valid for only incompressible, steady flow of a fluid with no internal friction (no viscosity). It's a simple equation that's easy to use; don't let this tempt you to use it in situations in which it doesn't apply! ▌

Problem-Solving Strategy 12.1 | **Bernoulli's Equation**

Bernoulli's equation is derived from the work–energy theorem, so much of Problem-Solving Strategy 7.1 (Section 7.1) is applicable here.

IDENTIFY *the relevant concepts:* Bernoulli's equation is applicable to steady flow of an incompressible fluid that has no internal friction (see Section 12.6). It is generally applicable to flows through large pipes and to flows within bulk fluids (e.g., air flowing around an airplane or water flowing around a fish).

SET UP *the problem* using the following steps:
1. Identify the points 1 and 2 referred to in Bernoulli's equation, Eq. (12.17).
2. Define your coordinate system, particularly the level at which $y = 0$. Take the positive y-direction to be upward.

3. Make lists of the unknown and known quantities in Eq. (12.17). Decide which unknowns are the target variables.

EXECUTE *the solution* as follows: Write Bernoulli's equation and solve for the unknowns. You may need the continuity equation, Eq. (12.10), to get a relationship between the two speeds in terms of cross-sectional areas of pipes or containers. You may also need Eq. (12.11) to find the volume flow rate.

EVALUATE *your answer:* Verify that the results make physical sense. Check that you have used consistent units: In SI units, pressure is in pascals, density in kilograms per cubic meter, and speed in meters per second. Also note that the pressures must be either *all* absolute pressures or *all* gauge pressures.

| Example 12.7 | **Water pressure in the home** |

Water enters a house (Fig. 12.23) through a pipe with an inside diameter of 2.0 cm at an absolute pressure of 4.0×10^5 Pa (about 4 atm). A 1.0-cm-diameter pipe leads to the second-floor bathroom 5.0 m above. When the flow speed at the inlet pipe is 1.5 m/s, find the flow speed, pressure, and volume flow rate in the bathroom.

SOLUTION

IDENTIFY and SET UP: We assume that the water flows at a steady rate. Water is effectively incompressible, so we can use the continuity equation. It's reasonable to ignore internal friction because the pipe has a relatively large diameter, so we can also use Bernoulli's equation. Let points 1 and 2 be at the inlet pipe and at the bathroom, respectively. We are given the pipe diameters at points 1 and 2, from which we calculate the areas A_1 and A_2, as well as the speed $v_1 = 1.5$ m/s and pressure $p_1 = 4.0 \times 10^5$ Pa at the inlet pipe. We take $y_1 = 0$ and $y_2 = 5.0$ m. We find the speed v_2 using the continuity equation and the pressure p_2 using Bernoulli's equation. Knowing v_2, we calculate the volume flow rate $v_2 A_2$.

EXECUTE: From the continuity equation, Eq. (12.10),

$$v_2 = \frac{A_1}{A_2} v_1 = \frac{\pi (1.0 \text{ cm})^2}{\pi (0.50 \text{ cm})^2} (1.5 \text{ m/s}) = 6.0 \text{ m/s}$$

From Bernoulli's equation, Eq. (12.16),

$$p_2 = p_1 - \tfrac{1}{2}\rho(v_2^2 - v_1^2) - \rho g(y_2 - y_1)$$
$$= 4.0 \times 10^5 \text{ Pa}$$
$$- \tfrac{1}{2}(1.0 \times 10^3 \text{ kg/m}^3)(36 \text{ m}^2/\text{s}^2 - 2.25 \text{ m}^2/\text{s}^2)$$
$$- (1.0 \times 10^3 \text{ kg/m}^3)(9.8 \text{ m/s}^2)(5.0 \text{ m})$$
$$= 4.0 \times 10^5 \text{ Pa} - 0.17 \times 10^5 \text{ Pa} - 0.49 \times 10^5 \text{ Pa}$$
$$= 3.3 \times 10^5 \text{ Pa} = 3.3 \text{ atm} = 48 \text{ lb/in.}^2$$

12.23 What is the water pressure in the second-story bathroom of this house?

The volume flow rate is

$$\frac{dV}{dt} = A_2 v_2 = \pi (0.50 \times 10^{-2} \text{ m})^2 (6.0 \text{ m/s})$$
$$= 4.7 \times 10^{-4} \text{ m}^3/\text{s} = 0.47 \text{ L/s}$$

EVALUATE: This is a reasonable flow rate for a bathroom faucet or shower. Note that if the water is turned off, v_1 and v_2 are both zero, the term $\tfrac{1}{2}\rho(v_2^2 - v_1^2)$ in Bernoulli's equation vanishes, and p_2 rises from 3.3×10^5 Pa to 3.5×10^5 Pa.

| Example 12.8 | **Speed of efflux** |

Figure 12.24 shows a gasoline storage tank with cross-sectional area A_1, filled to a depth h. The space above the gasoline contains air at pressure p_0, and the gasoline flows out the bottom of the tank through a short pipe with cross-sectional area A_2. Derive expressions for the flow speed in the pipe and the volume flow rate.

12.24 Calculating the speed of efflux for gasoline flowing out the bottom of a storage tank.

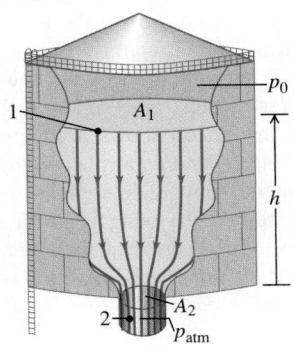

SOLUTION

IDENTIFY and SET UP: We consider the entire volume of moving liquid as a single flow tube of an incompressible fluid with negligible internal friction. Hence, we can use Bernoulli's equation. Points 1 and 2 are at the surface of the gasoline and at the exit pipe, respectively. At point 1 the pressure is p_0, which we assume to be fixed; at point 2 it is atmospheric pressure p_atm. We take $y = 0$ at the exit pipe, so $y_1 = h$ and $y_2 = 0$. Because A_1 is very much larger than A_2, the upper surface of the gasoline will drop very slowly and we can regard v_1 as essentially equal to zero. We find v_2 from Eq. (12.17) and the volume flow rate from Eq. (12.11).

EXECUTE: We apply Bernoulli's equation to points 1 and 2:

$$p_0 + \tfrac{1}{2}\rho v_1^2 + \rho g h = p_\text{atm} + \tfrac{1}{2}\rho v_2^2 + \rho g(0)$$
$$v_2^2 = v_1^2 + 2\left(\frac{p_0 - p_\text{atm}}{\rho}\right) + 2gh$$

Continued

Using $v_1 = 0$, we find

$$v_2 = \sqrt{2\left(\frac{p_0 - p_{atm}}{\rho}\right) + 2gh}$$

From Eq. (12.11), the volume flow rate is $dV/dt = v_2 A_2$.

EVALUATE: The speed v_2, sometimes called the *speed of efflux,* depends on both the pressure difference $(p_0 - p_{atm})$ and the height h of the liquid level in the tank. If the top of the tank is vented to the atmosphere, $p_0 = p_{atm}$ and $p_0 - p_{atm} = 0$. Then

$$v_2 = \sqrt{2gh}$$

That is, the speed of efflux from an opening at a distance h below the top surface of the liquid is the *same* as the speed a body would acquire in falling freely through a height h. This result is called *Torricelli's theorem*. It is valid not only for an opening in the bottom of a container, but also for a hole in a side wall at a depth h below the surface. In this case the volume flow rate is

$$\frac{dV}{dt} = A_2\sqrt{2gh}$$

Example 12.9 The Venturi meter

Figure 12.25 shows a *Venturi meter,* used to measure flow speed in a pipe. Derive an expression for the flow speed v_1 in terms of the cross-sectional areas A_1 and A_2 and the difference in height h of the liquid levels in the two vertical tubes.

SOLUTION

IDENTIFY and SET UP: The flow is steady, and we assume the fluid is incompressible and has negligible internal friction. Hence we can use Bernoulli's equation. We apply that equation to the wide part (point 1) and narrow part (point 2, the *throat*) of the pipe. Equation (12.6) relates h to the pressure difference $p_1 - p_2$.

EXECUTE: Points 1 and 2 have the same vertical coordinate $y_1 = y_2$, so Eq. (12.17) says

$$p_1 + \tfrac{1}{2}\rho v_1^2 = p_2 + \tfrac{1}{2}\rho v_2^2$$

From the continuity equation, $v_2 = (A_1/A_2)v_1$. Substituting this and rearranging, we get

$$p_1 - p_2 = \tfrac{1}{2}\rho v_1^2\left[\left(\frac{A_1}{A_2}\right)^2 - 1\right]$$

12.25 The Venturi meter.

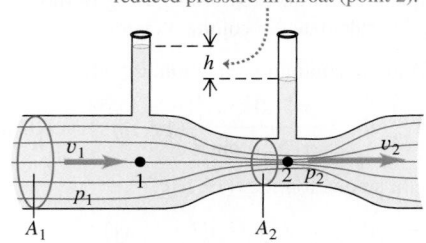

Difference in height results from reduced pressure in throat (point 2).

From Eq. (12.6), the pressure difference $p_1 - p_2$ is also equal to ρgh. Substituting this and solving for v_1, we get

$$v_1 = \sqrt{\frac{2gh}{(A_1/A_2)^2 - 1}}$$

EVALUATE: Because A_1 is greater than A_2, v_2 is greater than v_1 and the pressure p_2 in the throat is *less* than p_1. Those pressure differences produce a net force to the right that makes the fluid speed up as it enters the throat, and a net force to the left that slows it as it leaves.

Conceptual Example 12.10 Lift on an airplane wing

Figure 12.26a shows flow lines around a cross section of an airplane wing. The flow lines crowd together above the wing, corresponding to increased flow speed and reduced pressure, just as in the Venturi throat in Example 12.9. Hence the downward force of the air on the top side of the wing is less than the upward force of the air on the underside of the wing, and there is a net upward force or *lift*. Lift is not simply due to the impulse of air striking the underside of the wing; in fact, the reduced pressure on the upper wing surface makes the greatest contribution to the lift. (This simplified discussion ignores the formation of vortices.)

We can also understand the lift force on the basis of momentum changes. The vector diagram in Fig. 12.26a shows that there is a net *downward* change in the vertical component of momentum of the air flowing past the wing, corresponding to the downward force the wing exerts on the air. The reaction force *on* the wing is *upward*, as we concluded above.

Similar flow patterns and lift forces are found in the vicinity of any humped object in a wind. A moderate wind makes an umbrella "float"; a strong wind can turn it inside out. At high speed, lift can reduce traction on a car's tires; a "spoiler" at the car's tail, shaped like an upside-down wing, provides a compensating downward force.

CAUTION **A misconception about wings** Some discussions of lift claim that air travels faster over the top of a wing because "it has farther to travel." This claim assumes that air molecules that part company at the front of the wing, one traveling over the wing and one under it, must meet again at the wing's trailing edge. Not so! Figure 12.26b shows a computer simulation of parcels of air flowing around an airplane wing. Parcels that are adjacent at the front of the wing do *not* meet at the trailing edge; the flow over the top of the wing is much faster than if the parcels had to meet. In accordance with Bernoulli's equation, this faster speed means that there is even lower pressure above the wing (and hence greater lift) than the "farther-to-travel" claim would suggest.

12.26 Flow around an airplane wing.

(a) Flow lines around an airplane wing

Flow lines are crowded together above the wing, so flow speed is higher there and pressure is lower.

$\vec{p}_i$
$\vec{p}_f$ $\Delta\vec{p}$ (air)

Equivalent explanation: Wing imparts a net downward momentum to the air, so reaction force on airplane is upward.

(b) Computer simulation of air parcels flowing around a wing, showing that air moves much faster over the top than over the bottom.

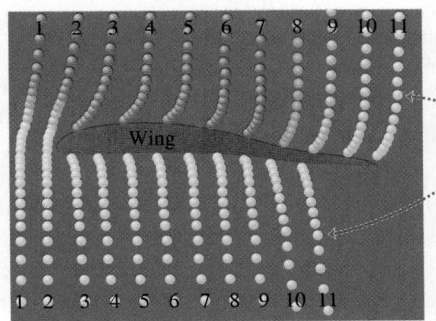

Notice that air particles that are together at the leading edge of the wing do *not* meet up at the trailing edge!

Test Your Understanding of Section 12.5 Which is the most accurate statement of Bernoulli's principle? (i) Fast-moving air causes lower pressure; (ii) lower pressure causes fast-moving air; (iii) both (i) and (ii) are equally accurate.

12.6 Viscosity and Turbulence

In our discussion of fluid flow we assumed that the fluid had no internal friction and that the flow was laminar. While these assumptions are often quite valid, in many important physical situations the effects of viscosity (internal friction) and turbulence (nonlaminar flow) are extremely important. Let's take a brief look at some of these situations.

Viscosity

Viscosity is internal friction in a fluid. Viscous forces oppose the motion of one portion of a fluid relative to another. Viscosity is the reason it takes effort to paddle a canoe through calm water, but it is also the reason the paddle works. Viscous effects are important in the flow of fluids in pipes, the flow of blood, the lubrication of engine parts, and many other situations.

Fluids that flow readily, such as water or gasoline, have smaller viscosities than do "thick" liquids such as honey or motor oil. Viscosities of all fluids are strongly temperature dependent, increasing for gases and decreasing for liquids as the temperature increases (Fig. 12.27). Oils for engine lubrication must flow equally well in cold and warm conditions, and so are designed to have as *little* temperature variation of viscosity as possible.

A viscous fluid always tends to cling to a solid surface in contact with it. There is always a thin *boundary layer* of fluid near the surface, in which the fluid is nearly at rest with respect to the surface. That's why dust particles can cling to a fan blade even when it is rotating rapidly, and why you can't get all the dirt off your car by just squirting a hose at it.

Viscosity has important effects on the flow of liquids through pipes, including the flow of blood in the circulatory system. First think about a fluid with zero viscosity so that we can apply Bernoulli's equation, Eq. (12.17). If the two ends of a long cylindrical pipe are at the same height ($y_1 = y_2$) and the flow speed is the same at both ends (so $v_1 = v_2$), Bernoulli's equation tells us that the pressure is the same at both ends of the pipe. But this result simply isn't true if we take viscosity into account. To see why, consider Fig. 12.28, which shows the flow-speed profile for laminar flow of a viscous fluid in a long cylindrical pipe. Due to viscosity, the speed is *zero* at the pipe walls (to which the fluid clings) and is greatest at the center of the pipe. The motion is like a lot of concentric tubes sliding relative to

12.27 Lava is an example of a viscous fluid. The viscosity decreases with increasing temperature: The hotter the lava, the more easily it can flow.

12.28 Velocity profile for a viscous fluid in a cylindrical pipe.

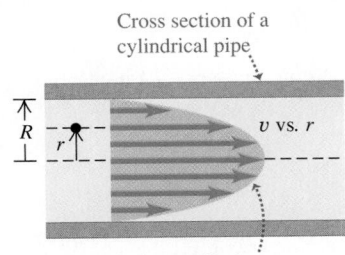

Cross section of a cylindrical pipe

v vs. r

The velocity profile for viscous fluid flowing in the pipe has a parabolic shape.

Application **Listening for Turbulent Flow**
Normal blood flow in the human aorta is laminar, but a small disturbance such as a heart pathology can cause the flow to become turbulent. Turbulence makes noise, which is why listening to blood flow with a stethoscope is a useful diagnostic technique.

one another, with the central tube moving fastest and the outermost tube at rest. Viscous forces between the tubes oppose this sliding, so to keep the flow going we must apply a greater pressure at the back of the flow than at the front. That's why you have to keep squeezing a tube of toothpaste or a packet of ketchup (both viscous fluids) to keep the fluid coming out of its container. Your fingers provide a pressure at the back of the flow that is far greater than the atmospheric pressure at the front of the flow.

The pressure difference required to sustain a given volume flow rate through a cylindrical pipe of length L and radius R turns out to be proportional to L/R^4. If we decrease R by one-half, the required pressure increases by $2^4 = 16$; decreasing R by a factor of 0.90 (a 10% reduction) increases the required pressure difference by a factor of $(1/0.90)^4 = 1.52$ (a 52% increase). This simple relationship explains the connection between a high-cholesterol diet (which tends to narrow the arteries) and high blood pressure. Due to the R^4 dependence, even a small narrowing of the arteries can result in substantially elevated blood pressure and added strain on the heart muscle.

Turbulence

When the speed of a flowing fluid exceeds a certain critical value, the flow is no longer laminar. Instead, the flow pattern becomes extremely irregular and complex, and it changes continuously with time; there is no steady-state pattern. This irregular, chaotic flow is called **turbulence.** Figure 12.20 shows the contrast between laminar and turbulent flow for smoke rising in air. Bernoulli's equation is *not* applicable to regions where there is turbulence because the flow is not steady.

Whether a flow is laminar or turbulent depends in part on the fluid's viscosity. The greater the viscosity, the greater the tendency for the fluid to flow in sheets or lamina and the more likely the flow is to be laminar. (When we discussed Bernoulli's equation in Section 12.5, we assumed that the flow was laminar and that the fluid had zero viscosity. In fact, a *little* viscosity is needed to ensure that the flow is laminar.)

For a fluid of a given viscosity, flow speed is a determining factor for the onset of turbulence. A flow pattern that is stable at low speeds suddenly becomes unstable when a critical speed is reached. Irregularities in the flow pattern can be caused by roughness in the pipe wall, variations in the density of the fluid, and many other factors. At low flow speeds, these disturbances damp out; the flow pattern is *stable* and tends to maintain its laminar nature (Fig. 12.29a). When the critical speed is reached, however, the flow pattern becomes unstable. The disturbances no longer damp out but grow until they destroy the entire laminar-flow pattern (Fig. 12.29b).

12.29 The flow of water from a faucet is (a) laminar at low speeds but (b) turbulent at sufficiently high speeds.

(a)   (b)

Conceptual Example 12.11   **The curve ball**

Does a curve ball *really* curve? Yes, it certainly does, and the reason is turbulence. Figure 12.30a shows a nonspinning ball moving through the air from left to right. The flow lines show that to an observer moving with the ball, the air stream appears to move from right to left. Because of the high speeds that are ordinarily involved (near 35 m/s, or 75 mi/h), there is a region of *turbulent* flow behind the ball.

Figure 12.30b shows a *spinning* ball with "top spin." Layers of air near the ball's surface are pulled around in the direction of the spin by friction between the ball and air and by the air's internal friction (viscosity). Hence air moves relative to the ball's surface more slowly at the top of the ball than at the bottom, and turbulence occurs farther forward on the top side than on the bottom. This asymmetry causes a pressure difference; the average pressure at the top of the ball is now greater than that at the bottom. As Fig. 12.30c shows, the resulting net force deflects the ball downward. "Top spin" is used in tennis to keep a fast serve in the court (Fig. 12.30d).

In baseball, a curve ball spins about a nearly *vertical* axis and the resulting deflection is sideways. In that case, Fig. 12.30c is a *top* view of the situation. A curve ball thrown by a left-handed pitcher spins as shown in Fig. 12.30e and will curve *toward* a right-handed batter, making it harder to hit.

A similar effect occurs with golf balls, which acquire "back spin" from impact with the grooved, slanted club face. Figure 12.30f shows the backspin of a golf ball just after impact. The resulting pressure difference between the top and bottom of the ball causes a *lift* force that keeps the ball in the air longer than would be possible without spin. A well-hit drive appears, from the tee, to "float" or even curve *upward* during the initial portion of its flight. This is a real effect, not an illusion. The dimples on the golf ball play an essential role; the viscosity of air gives a dimpled ball a much longer trajectory than an undimpled one with the same initial velocity and spin.

12.30 (a)–(e) Analyzing the motion of a spinning ball through the air. (f) Stroboscopic photograph of a golf ball being struck by a club. The picture was taken at 1000 flashes per second. The ball rotates about once in eight pictures, corresponding to an angular speed of 125 rev/s, or 7500 rpm.

(a) Motion of air relative to a nonspinning ball

(b) Motion of a spinning ball

This side of the ball moves opposite to the airflow.

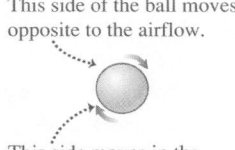

This side moves in the direction of the airflow.

(c) Force generated when a spinning ball moves through air

A moving ball drags the adjacent air with it. So, when air moves past a spinning ball:

On one side, the ball **slows the air**, creating a region of **high pressure**.

On the other side, the ball **speeds the air**, creating a region of **low pressure**.

The resultant force points in the direction of the low-pressure side.

(d) Spin pushing a tennis ball downward

(e) Spin causing a curve ball to be deflected sideways

(f) Backspin of a golf ball

Test Your Understanding of Section 12.6   How much more thumb pressure must a nurse use to administer an injection with a hypodermic needle of inside diameter 0.30 mm compared to one with inside diameter 0.60 mm? Assume that the two needles have the same length and that the volume flow rate is the same in both cases. (i) twice as much; (ii) 4 times as much; (iii) 8 times as much; (iv) 16 times as much; (v) 32 times as much.

Density and pressure: Density is mass per unit volume. If a mass m of homogeneous material has volume V, its density ρ is the ratio m/V. Specific gravity is the ratio of the density of a material to the density of water. (See Example 12.1.)

Pressure is normal force per unit area. Pascal's law states that pressure applied to an enclosed fluid is transmitted undiminished to every portion of the fluid. Absolute pressure is the total pressure in a fluid; gauge pressure is the difference between absolute pressure and atmospheric pressure. The SI unit of pressure is the pascal (Pa): $1 \text{ Pa} = 1 \text{ N/m}^2$. (See Example 12.2.)

$$\rho = \frac{m}{V} \quad (12.1)$$

$$p = \frac{dF_\perp}{dA} \quad (12.2)$$

Small area dA within fluid at rest

Equal normal forces exerted on both sides by surrounding fluid

---

Pressures in a fluid at rest: The pressure difference between points 1 and 2 in a static fluid of uniform density ρ (an incompressible fluid) is proportional to the difference between the elevations y_1 and y_2. If the pressure at the surface of an incompressible liquid at rest is p_0, then the pressure at a depth h is greater by an amount $\rho g h$. (See Examples 12.3 and 12.4.)

$$p_2 - p_1 = -\rho g(y_2 - y_1)$$
(pressure in a fluid of uniform density) $\quad (12.5)$

$$p = p_0 + \rho g h$$
(pressure in a fluid of uniform density) $\quad (12.6)$

Buoyancy: Archimedes's principle states that when a body is immersed in a fluid, the fluid exerts an upward buoyant force on the body equal to the weight of the fluid that the body displaces. (See Example 12.5.)

Fluid element replaced with solid body of the same size and shape

---

Fluid flow: An ideal fluid is incompressible and has no viscosity (no internal friction). A flow line is the path of a fluid particle; a streamline is a curve tangent at each point to the velocity vector at that point. A flow tube is a tube bounded at its sides by flow lines. In laminar flow, layers of fluid slide smoothly past each other. In turbulent flow, there is great disorder and a constantly changing flow pattern.

Conservation of mass in an incompressible fluid is expressed by the continuity equation, which relates the flow speeds v_1 and v_2 for two cross sections A_1 and A_2 in a flow tube. The product Av equals the volume flow rate, dV/dt, the rate at which volume crosses a section of the tube. (See Example 12.6.)

Bernoulli's equation relates the pressure p, flow speed v, and elevation y for any two points, assuming steady flow in an ideal fluid. (See Examples 12.7–12.10.)

$$A_1 v_1 = A_2 v_2$$
(continuity equation, incompressible fluid) $\quad (12.10)$

$$\frac{dV}{dt} = Av$$
(volume flow rate) $\quad (12.11)$

$$p_1 + \rho g y_1 + \tfrac{1}{2}\rho v_1^2 = p_2 + \rho g y_2 + \tfrac{1}{2}\rho v_2^2$$
(Bernoulli's equation) $\quad (12.17)$

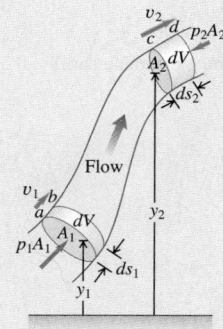

BRIDGING PROBLEM   How Long to Drain?

A large cylindrical tank with diameter D is open to the air at the top. The tank contains water to a height H. A small circular hole with diameter d, where d is very much less than D, is then opened at the bottom of the tank. Ignore any effects of viscosity. (a) Find y, the height of water in the tank a time t after the hole is opened, as a function of t. (b) How long does it take to drain the tank completely? (c) If you double the initial height of water in the tank, by what factor does the time to drain the tank increase?

SOLUTION GUIDE

See MasteringPhysics® study area for a Video Tutor solution.

IDENTIFY and SET UP

1. Draw a sketch of the situation that shows all of the relevant dimensions.
2. Make a list of the unknown quantities, and decide which of these are the target variables.

3. What is the speed at which water flows out of the bottom of the tank? How is this related to the volume flow rate of water out of the tank? How is the volume flow rate related to the rate of change of y?

EXECUTE

4. Use your results from step 3 to write an equation for dy/dt.
5. Your result from step 4 is a relatively simple differential equation. With your knowledge of calculus, you can integrate it to find y as a function of t. (*Hint:* Once you've done the integration, you'll still have to do a little algebra.)
6. Use your result from step 5 to find the time when the tank is empty. How does your result depend on the initial height H?

EVALUATE

7. Check whether your answers are reasonable. A good check is to draw a graph of y versus t. According to your graph, what is the algebraic sign of dy/dt at different times? Does this make sense?

Problems

For instructor-assigned homework, go to www.masteringphysics.com

•, ••, •••: Problems of increasing difficulty. **CP**: Cumulative problems incorporating material from earlier chapters. **CALC**: Problems requiring calculus. **BIO**: Biosciences problems.

DISCUSSION QUESTIONS

Q12.1 A cube of oak wood with very smooth faces normally floats in water. Suppose you submerge it completely and press one face flat against the bottom of a tank so that no water is under that face. Will the block float to the surface? Is there a buoyant force on it? Explain.

Q12.2 A rubber hose is attached to a funnel, and the free end is bent around to point upward. When water is poured into the funnel, it rises in the hose to the same level as in the funnel, even though the funnel has a lot more water in it than the hose does. Why? What supports the extra weight of the water in the funnel?

Q12.3 Comparing Example 12.1 (Section 12.1) and Example 12.2 (Section 12.2), it seems that 700 N of air is exerting a downward force of 2.0×10^6 N on the floor. How is this possible?

Q12.4 Equation (12.7) shows that an area ratio of 100 to 1 can give 100 times more output force than input force. Doesn't this violate conservation of energy? Explain.

Q12.5 You have probably noticed that the lower the tire pressure, the larger the contact area between the tire and the road. Why?

Q12.6 In hot-air ballooning, a large balloon is filled with air heated by a gas burner at the bottom. Why must the air be heated? How does the balloonist control ascent and descent?

Q12.7 In describing the size of a large ship, one uses such expressions as "it displaces 20,000 tons." What does this mean? Can the weight of the ship be obtained from this information?

Q12.8 You drop a solid sphere of aluminum in a bucket of water that sits on the ground. The buoyant force equals the weight of water displaced; this is less than the weight of the sphere, so the sphere sinks to the bottom. If you take the bucket with you on an elevator that accelerates upward, the apparent weight of the water increases and the buoyant force on the sphere increases. Could the

acceleration of the elevator be great enough to make the sphere pop up out of the water? Explain.

Q12.9 A rigid, lighter-than-air dirigible filled with helium cannot continue to rise indefinitely. Why? What determines the maximum height it can attain?

Q12.10 Air pressure decreases with increasing altitude. So why is air near the surface not continuously drawn upward toward the lower-pressure regions above?

Q12.11 The purity of gold can be tested by weighing it in air and in water. How? Do you think you could get away with making a fake gold brick by gold-plating some cheaper material?

Q12.12 During the Great Mississippi Flood of 1993, the levees in St. Louis tended to rupture first at the bottom. Why?

Q12.13 A cargo ship travels from the Atlantic Ocean (salt water) to Lake Ontario (freshwater) via the St. Lawrence River. The ship rides several centimeters lower in the water in Lake Ontario than it did in the ocean. Explain why.

Q12.14 You push a piece of wood under the surface of a swimming pool. After it is completely submerged, you keep pushing it deeper and deeper. As you do this, what will happen to the buoyant force on it? Will the force keep increasing, stay the same, or decrease? Why?

Q12.15 An old question is "Which weighs more, a pound of feathers or a pound of lead?" If the weight in pounds is the gravitational force, will a pound of feathers balance a pound of lead on opposite pans of an equal-arm balance? Explain, taking into account buoyant forces.

Q12.16 Suppose the door of a room makes an airtight but frictionless fit in its frame. Do you think you could open the door if the air pressure on one side were standard atmospheric pressure and the air pressure on the other side differed from standard by 1%? Explain.

Q12.17 At a certain depth in an incompressible liquid, the absolute pressure is p. At twice this depth, will the absolute pressure be equal to $2p$, greater than $2p$, or less than $2p$? Justify your answer.

Q12.18 A piece of iron is glued to the top of a block of wood. When the block is placed in a bucket of water with the iron on top, the block floats. The block is now turned over so that the iron is submerged beneath the wood. Does the block float or sink? Does the water level in the bucket rise, drop, or stay the same? Explain your answers.

Q12.19 You take an empty glass jar and push it into a tank of water with the open mouth of the jar downward, so that the air inside the jar is trapped and cannot get out. If you push the jar deeper into the water, does the buoyant force on the jar stay the same? If not, does it increase or decrease? Explain your answer.

Q12.20 You are floating in a canoe in the middle of a swimming pool. Your friend is at the edge of the pool, carefully noting the level of the water on the side of the pool. You have a bowling ball with you in the canoe. If you carefully drop the bowling ball over the side of the canoe and it sinks to the bottom of the pool, does the water level in the pool rise or fall?

Q12.21 You are floating in a canoe in the middle of a swimming pool. A large bird flies up and lights on your shoulder. Does the water level in the pool rise or fall?

Q12.22 At a certain depth in the incompressible ocean the gauge pressure is p_g. At three times this depth, will the gauge pressure be greater than $3p_g$, equal to $3p_g$, or less than $3p_g$? Justify your answer.

Q12.23 An ice cube floats in a glass of water. When the ice melts, will the water level in the glass rise, fall, or remain unchanged? Explain.

Q12.24 You are told, "Bernoulli's equation tells us that where there is higher fluid speed, there is lower fluid pressure, and vice versa." Is this statement always true, even for an idealized fluid? Explain.

Q12.25 If the velocity at each point in space in steady-state fluid flow is constant, how can a fluid particle accelerate?

Q12.26 In a store-window vacuum cleaner display, a table-tennis ball is suspended in midair in a jet of air blown from the outlet hose of a tank-type vacuum cleaner. The ball bounces around a little but always moves back toward the center of the jet, even if the jet is tilted from the vertical. How does this behavior illustrate Bernoulli's equation?

Q12.27 A tornado consists of a rapidly whirling air vortex. Why is the pressure always much lower in the center than at the outside? How does this condition account for the destructive power of a tornado?

Q12.28 Airports at high elevations have longer runways for take-offs and landings than do airports at sea level. One reason is that aircraft engines develop less power in the thin air well above sea level. What is another reason?

Q12.29 When a smooth-flowing stream of water comes out of a faucet, it narrows as it falls. Explain why this happens.

Q12.30 Identical-size lead and aluminum cubes are suspended at different depths by two wires in a large vat of water (Fig. Q12.30). (a) Which cube experiences a greater buoyant force? (b) For which cube is the tension in the wire greater? (c) Which cube experiences a

Figure **Q12.30**

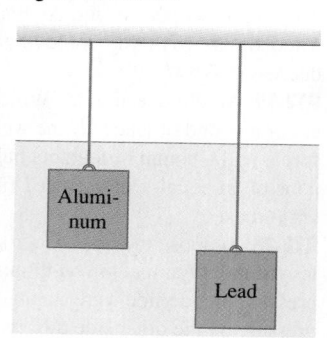

greater force on its lower face? (d) For which cube is the difference in pressure between the upper and lower faces greater?

EXERCISES

Section 12.1 Density

12.1 •• On a part-time job, you are asked to bring a cylindrical iron rod of length 85.8 cm and diameter 2.85 cm from a storage room to a machinist. Will you need a cart? (To answer, calculate the weight of the rod.)

12.2 •• A cube 5.0 cm on each side is made of a metal alloy. After you drill a cylindrical hole 2.0 cm in diameter all the way through and perpendicular to one face, you find that the cube weighs 7.50 N. (a) What is the density of this metal? (b) What did the cube weigh before you drilled the hole in it?

12.3 • You purchase a rectangular piece of metal that has dimensions $5.0 \times 15.0 \times 30.0$ mm and mass 0.0158 kg. The seller tells you that the metal is gold. To check this, you compute the average density of the piece. What value do you get? Were you cheated?

12.4 •• **Gold Brick.** You win the lottery and decide to impress your friends by exhibiting a million-dollar cube of gold. At the time, gold is selling for $426.60 per troy ounce, and 1.0000 troy ounce equals 31.1035 g. How tall would your million-dollar cube be?

12.5 •• A uniform lead sphere and a uniform aluminum sphere have the same mass. What is the ratio of the radius of the aluminum sphere to the radius of the lead sphere?

12.6 • (a) What is the average density of the sun? (b) What is the average density of a neutron star that has the same mass as the sun but a radius of only 20.0 km?

12.7 •• A hollow cylindrical copper pipe is 1.50 m long and has an outside diameter of 3.50 cm and an inside diameter of 2.50 cm. How much does it weigh?

Section 12.2 Pressure in a Fluid

12.8 •• **Black Smokers.** Black smokers are hot volcanic vents that emit smoke deep in the ocean floor. Many of them teem with exotic creatures, and some biologists think that life on earth may have begun around such vents. The vents range in depth from about 1500 m to 3200 m below the surface. What is the gauge pressure at a 3200-m deep vent, assuming that the density of water does not vary? Express your answer in pascals and atmospheres.

12.9 •• **Oceans on Mars.** Scientists have found evidence that Mars may once have had an ocean 0.500 km deep. The acceleration due to gravity on Mars is 3.71 m/s². (a) What would be the gauge pressure at the bottom of such an ocean, assuming it was freshwater? (b) To what depth would you need to go in the earth's ocean to experience the same gauge pressure?

12.10 •• **BIO** (a) Calculate the difference in blood pressure between the feet and top of the head for a person who is 1.65 m tall. (b) Consider a cylindrical segment of a blood vessel 2.00 cm long and 1.50 mm in diameter. What *additional* outward force would such a vessel need to withstand in the person's feet compared to a similar vessel in her head?

12.11 • **BIO** In intravenous feeding, a needle is inserted in a vein in the patient's arm and a tube leads from the needle to a reservoir of fluid (density 1050 kg/m³) located at height h above the arm. The top of the reservoir is open to the air. If the gauge pressure inside the vein is 5980 Pa, what is the minimum value of h that allows fluid to enter the vein? Assume the needle diameter is large enough that you can ignore the viscosity (see Section 12.6) of the fluid.

12.12 • A barrel contains a 0.120-m layer of oil floating on water that is 0.250 m deep. The density of the oil is 600 kg/m^3. (a) What is the gauge pressure at the oil–water interface? (b) What is the gauge pressure at the bottom of the barrel?

12.13 • **BIO** **Standing on Your Head.** (a) What is the *difference* between the pressure of the blood in your brain when you stand on your head and the pressure when you stand on your feet? Assume that you are 1.85 m tall. The density of blood is 1060 kg/m^3. (b) What effect does the increased pressure have on the blood vessels in your brain?

12.14 •• You are designing a diving bell to withstand the pressure of seawater at a depth of 250 m. (a) What is the gauge pressure at this depth? (You can ignore changes in the density of the water with depth.) (b) At this depth, what is the net force due to the water outside and the air inside the bell on a circular glass window 30.0 cm in diameter if the pressure inside the diving bell equals the pressure at the surface of the water? (You can ignore the small variation of pressure over the surface of the window.)

12.15 •• **BIO** **Ear Damage from Diving.** If the force on the tympanic membrane (eardrum) increases by about 1.5 N above the force from atmospheric pressure, the membrane can be damaged. When you go scuba diving in the ocean, below what depth could damage to your eardrum start to occur? The eardrum is typically 8.2 mm in diameter. (Consult Table 12.1.)

12.16 •• The liquid in the open-tube manometer in Fig. 12.8a is mercury, $y_1 = 3.00$ cm, and $y_2 = 7.00$ cm. Atmospheric pressure is 980 millibars. (a) What is the absolute pressure at the bottom of the U-shaped tube? (b) What is the absolute pressure in the open tube at a depth of 4.00 cm below the free surface? (c) What is the absolute pressure of the gas in the container? (d) What is the gauge pressure of the gas in pascals?

12.17 • **BIO** There is a maximum depth at which a diver can breathe through a snorkel tube (Fig. E12.17) because as the depth increases, so does the pressure difference, which tends to collapse the diver's lungs. Since the snorkel connects the air in the lungs to the atmosphere at the surface, the pressure inside the lungs is atmospheric pressure. What is the external–internal pressure difference when the diver's lungs are at a depth of 6.1 m (about 20 ft)? Assume that the diver is in freshwater. (A scuba diver breathing from compressed air tanks can operate at greater depths than can a snorkeler, since the pressure of the air inside the scuba diver's lungs increases to match the external pressure of the water.)

Figure **E12.17**

12.18 •• A tall cylinder with a cross-sectional area 12.0 cm^2 is partially filled with mercury; the surface of the mercury is 5.00 cm above the bottom of the cylinder. Water is slowly poured in on top of the mercury, and the two fluids don't mix. What volume of water must be added to double the gauge pressure at the bottom of the cylinder?

12.19 •• An electrical short cuts off all power to a submersible diving vehicle when it is 30 m below the surface of the ocean. The crew must push out a hatch of area 0.75 m^2 and weight 300 N on the bottom to escape. If the pressure inside is 1.0 atm, what downward force must the crew exert on the hatch to open it?

12.20 •• A closed container is partially filled with water. Initially, the air above the water is at atmospheric pressure (1.01×10^5 Pa) and the gauge pressure at the bottom of the water is 2500 Pa. Then additional air is pumped in, increasing the pressure of the air above the water by 1500 Pa. (a) What is the gauge pressure at the bottom of the water? (b) By how much must the water level in the container be reduced, by drawing some water out through a valve at the bottom of the container, to return the gauge pressure at the bottom of the water to its original value of 2500 Pa? The pressure of the air above the water is maintained at 1500 Pa above atmospheric pressure.

12.21 •• A cylindrical disk of wood weighing 45.0 N and having a diameter of 30.0 cm floats on a cylinder of oil of density 0.850 g/cm^3 (Fig. E12.21). The cylinder of oil is 75.0 cm deep and has a diameter the same as that of the wood. (a) What is the gauge pressure at the top of the oil column? (b) Suppose now that someone puts a weight of 83.0 N on top of the wood, but no oil seeps around the edge of the wood. What is the *change* in pressure at (i) the bottom of the oil and (ii) halfway down in the oil?

Figure **E12.21**

12.22 •• **Exploring Venus.** The surface pressure on Venus is 92 atm, and the acceleration due to gravity there is 0.894g. In a future exploratory mission, an upright cylindrical tank of benzene is sealed at the top but still pressurized at 92 atm just above the benzene. The tank has a diameter of 1.72 m, and the benzene column is 11.50 m tall. Ignore any effects due to the very high temperature on Venus. (a) What total force is exerted on the inside surface of the bottom of the tank? (b) What force does the Venusian atmosphere exert on the outside surface of the bottom of the tank? (c) What total inward force does the atmosphere exert on the vertical walls of the tank?

12.23 •• **Hydraulic Lift I.** For the hydraulic lift shown in Fig. 12.7, what must be the ratio of the diameter of the vessel at the car to the diameter of the vessel where the force F_1 is applied so that a 1520-kg car can be lifted with a force F_1 of just 125 N?

12.24 • **Hydraulic Lift II.** The piston of a hydraulic automobile lift is 0.30 m in diameter. What gauge pressure, in pascals, is required to lift a car with a mass of 1200 kg? Also express this pressure in atmospheres.

Section 12.3 Buoyancy

12.25 • A 950-kg cylindrical can buoy floats vertically in salt water. The diameter of the buoy is 0.900 m. Calculate the additional distance the buoy will sink when a 70.0-kg man stands on top of it.

12.26 •• A slab of ice floats on a freshwater lake. What minimum volume must the slab have for a 45.0-kg woman to be able to stand on it without getting her feet wet?

12.27 •• An ore sample weighs 17.50 N in air. When the sample is suspended by a light cord and totally immersed in water, the tension in the cord is 11.20 N. Find the total volume and the density of the sample.

12.28 •• You are preparing some apparatus for a visit to a newly discovered planet Caasi having oceans of glycerine and a surface acceleration due to gravity of 4.15 m/s^2. If your apparatus floats in the oceans on earth with 25.0% of its volume submerged, what percentage will be submerged in the glycerine oceans of Caasi?

12.29 •• An object of average density ρ floats at the surface of a fluid of density ρ_{fluid}. (a) How must the two densities be related? (b) In view of the answer to part (a), how can steel ships float in water? (c) In terms of ρ and ρ_{fluid}, what fraction of the object is submerged and what fraction is above the fluid? Check that your answers give the correct limiting behavior as $\rho \rightarrow \rho_{fluid}$ and as $\rho \rightarrow 0$. (d) While on board your yacht, your cousin Throckmorton cuts a rectangular piece (dimensions $5.0 \times 4.0 \times 3.0$ cm) out of a life preserver and throws it into the ocean. The piece has a mass of 42 g. As it floats in the ocean, what percentage of its volume is above the surface?

12.30 • A hollow plastic sphere is held below the surface of a freshwater lake by a cord anchored to the bottom of the lake. The sphere has a volume of 0.650 m^3 and the tension in the cord is 900 N. (a) Calculate the buoyant force exerted by the water on the sphere. (b) What is the mass of the sphere? (c) The cord breaks and the sphere rises to the surface. When the sphere comes to rest, what fraction of its volume will be submerged?

12.31 •• A cubical block of wood, 10.0 cm on a side, floats at the interface between oil and water with its lower surface 1.50 cm below the interface (Fig. E12.31). The density of the oil is 790 kg/m^3. (a) What is the gauge pressure at the upper face of the block? (b) What is the gauge pressure at the lower face of the block? (c) What are the mass and density of the block?

Figure **E12.31**

12.32 • A solid aluminum ingot weighs 89 N in air. (a) What is its volume? (b) The ingot is suspended from a rope and totally immersed in water. What is the tension in the rope (the *apparent* weight of the ingot in water)?

12.33 •• A rock is suspended by a light string. When the rock is in air, the tension in the string is 39.2 N. When the rock is totally immersed in water, the tension is 28.4 N. When the rock is totally immersed in an unknown liquid, the tension is 18.6 N. What is the density of the unknown liquid?

Section 12.4 Fluid Flow

12.34 •• Water runs into a fountain, filling all the pipes, at a steady rate of 0.750 m^3/s. (a) How fast will it shoot out of a hole 4.50 cm in diameter? (b) At what speed will it shoot out if the diameter of the hole is three times as large?

12.35 •• A shower head has 20 circular openings, each with radius 1.0 mm. The shower head is connected to a pipe with radius 0.80 cm. If the speed of water in the pipe is 3.0 m/s, what is its speed as it exits the shower-head openings?

12.36 • Water is flowing in a pipe with a varying cross-sectional area, and at all points the water completely fills the pipe. At point 1 the cross-sectional area of the pipe is 0.070 m^2, and the magnitude of the fluid velocity is 3.50 m/s. (a) What is the fluid speed at points in the pipe where the cross-sectional area is (a) 0.105 m^2 and (b) 0.047 m^2? (c) Calculate the volume of water discharged from the open end of the pipe in 1.00 hour.

12.37 • Water is flowing in a pipe with a circular cross section but with varying cross-sectional area, and at all points the water completely fills the pipe. (a) At one point in the pipe the radius is 0.150 m. What is the speed of the water at this point if water is flowing into this pipe at a steady rate of 1.20 m^3/s? (b) At a second point in the

pipe the water speed is 3.80 m/s. What is the radius of the pipe at this point?

12.38 • **Home Repair.** You need to extend a 2.50-inch-diameter pipe, but you have only a 1.00-inch-diameter pipe on hand. You make a fitting to connect these pipes end to end. If the water is flowing at 6.00 cm/s in the wide pipe, how fast will it be flowing through the narrow one?

12.39 • At a point where an irrigation canal having a rectangular cross section is 18.5 m wide and 3.75 m deep, the water flows at 2.50 cm/s. At a point downstream, but on the same level, the canal is 16.5 m wide, but the water flows at 11.0 cm/s. How deep is the canal at this point?

12.40 ••• **BIO Artery Blockage.** A medical technician is trying to determine what percentage of a patient's artery is blocked by plaque. To do this, she measures the blood pressure just before the region of blockage and finds that it is 1.20×10^4 Pa, while in the region of blockage it is 1.15×10^4 Pa. Furthermore, she knows that blood flowing through the normal artery just before the point of blockage is traveling at 30.0 cm/s, and the specific gravity of this patient's blood is 1.06. What percentage of the cross-sectional area of the patient's artery is blocked by the plaque?

Section 12.5 Bernoulli's Equation

12.41 •• A sealed tank containing seawater to a height of 11.0 m also contains air above the water at a gauge pressure of 3.00 atm. Water flows out from the bottom through a small hole. How fast is this water moving?

12.42 • A small circular hole 6.00 mm in diameter is cut in the side of a large water tank, 14.0 m below the water level in the tank. The top of the tank is open to the air. Find (a) the speed of efflux of the water and (b) the volume discharged per second.

12.43 • What gauge pressure is required in the city water mains for a stream from a fire hose connected to the mains to reach a vertical height of 15.0 m? (Assume that the mains have a much larger diameter than the fire hose.)

12.44 •• At one point in a pipeline the water's speed is 3.00 m/s and the gauge pressure is 5.00×10^4 Pa. Find the gauge pressure at a second point in the line, 11.0 m lower than the first, if the pipe diameter at the second point is twice that at the first.

12.45 • At a certain point in a horizontal pipeline, the water's speed is 2.50 m/s and the gauge pressure is 1.80×10^4 Pa. Find the gauge pressure at a second point in the line if the cross-sectional area at the second point is twice that at the first.

12.46 • A soft drink (mostly water) flows in a pipe at a beverage plant with a mass flow rate that would fill 220 0.355-L cans per minute. At point 2 in the pipe, the gauge pressure is 152 kPa and the cross-sectional area is 8.00 cm^2. At point 1, 1.35 m above point 2, the cross-sectional area is 2.00 cm^2. Find the (a) mass flow rate; (b) volume flow rate; (c) flow speeds at points 1 and 2; (d) gauge pressure at point 1.

12.47 •• A golf course sprinkler system discharges water from a horizontal pipe at the rate of 7200 cm^3/s. At one point in the pipe, where the radius is 4.00 cm, the water's absolute pressure is 2.40×10^5 Pa. At a second point in the pipe, the water passes through a constriction where the radius is 2.00 cm. What is the water's absolute pressure as it flows through this constriction?

Section 12.6 Viscosity and Turbulence

12.48 • A pressure difference of 6.00×10^4 Pa is required to maintain a volume flow rate of 0.800 m^3/s for a viscous fluid flowing through a section of cylindrical pipe that has radius 0.210 m.

What pressure difference is required to maintain the same volume flow rate if the radius of the pipe is decreased to 0.0700 m?

12.49 •• **BIO** **Clogged Artery.** Viscous blood is flowing through an artery partially clogged by cholesterol. A surgeon wants to remove enough of the cholesterol to double the flow rate of blood through this artery. If the original diameter of the artery is D, what should be the new diameter (in terms of D) to accomplish this for the same pressure gradient?

PROBLEMS

12.50 •• **CP** The deepest point known in any of the earth's oceans is in the Marianas Trench, 10.92 km deep. (a) Assuming water is incompressible, what is the pressure at this depth? Use the density of seawater. (b) The actual pressure is 1.16×10^8 Pa; your calculated value will be less because the density actually varies with depth. Using the compressibility of water and the actual pressure, find the density of the water at the bottom of the Marianas Trench. What is the percent change in the density of the water?

12.51 ••• In a lecture demonstration, a professor pulls apart two hemispherical steel shells (diameter D) with ease using their attached handles. She then places them together, pumps out the air to an absolute pressure of p, and hands them to a bodybuilder in the back row to pull apart. (a) If atmospheric pressure is p_0, how much force must the bodybuilder exert on each shell? (b) Evaluate your answer for the case $p = 0.025$ atm, $D = 10.0$ cm.

12.52 •• **BIO** **Fish Navigation.** (a) As you can tell by watching them in an aquarium, fish are able to remain at any depth in water with no effort. What does this ability tell you about their density? (b) Fish are able to inflate themselves using a sac (called the *swim bladder*) located under their spinal column. These sacs can be filled with an oxygen–nitrogen mixture that comes from the blood. If a 2.75-kg fish in freshwater inflates itself and increases its volume by 10%, find the *net* force that the *water* exerts on it. (c) What is the net *external* force on it? Does the fish go up or down when it inflates itself?

12.53 ••• **CALC** A swimming pool is 5.0 m long, 4.0 m wide, and 3.0 m deep. Compute the force exerted by the water against (a) the bottom and (b) either end. (*Hint:* Calculate the force on a thin, horizontal strip at a depth h, and integrate this over the end of the pool.) Do not include the force due to air pressure.

12.54 ••• **CP CALC** The upper edge of a gate in a dam runs along the water surface. The gate is 2.00 m high and 4.00 m wide and is hinged along a horizontal line through its center (Fig. P12.54). Calculate the torque about the hinge arising from the force due to the water.

Figure **P12.54**

(*Hint:* Use a procedure similar to that used in Problem 12.53; calculate the torque on a thin, horizontal strip at a depth h and integrate this over the gate.)

12.55 ••• **CP CALC** **Force and Torque on a Dam.** A dam has the shape of a rectangular solid. The side facing the lake has area A and height H. The surface of the freshwater lake behind the dam is at the top of the dam. (a) Show that the net horizontal force exerted by the water on the dam equals $\frac{1}{2}\rho gHA$—that is, the average gauge pressure across the face of the dam times the area (see Problem 12.53). (b) Show that the torque exerted by the water about an axis along the bottom of the dam is $\rho gH^2A/6$. (c) How do the force and torque depend on the size of the lake?

12.56 •• **Ballooning on Mars.** It has been proposed that we could explore Mars using inflated balloons to hover just above the surface. The buoyancy of the atmosphere would keep the balloon aloft. The density of the Martian atmosphere is 0.0154 kg/m^3 (although this varies with temperature). Suppose we construct these balloons of a thin but tough plastic having a density such that each square meter has a mass of 5.00 g. We inflate them with a very light gas whose mass we can neglect. (a) What should be the radius and mass of these balloons so they just hover above the surface of Mars? (b) If we released one of the balloons from part (a) on earth, where the atmospheric density is 1.20 kg/m^3, what would be its initial acceleration assuming it was the same size as on Mars? Would it go up or down? (c) If on Mars these balloons have five times the radius found in part (a), how heavy an instrument package could they carry?

12.57 •• A 0.180-kg cube of ice (frozen water) is floating in glycerine. The glycerine is in a tall cylinder that has inside radius 3.50 cm. The level of the glycerine is well below the top of the cylinder. If the ice completely melts, by what distance does the height of liquid in the cylinder change? Does the level of liquid rise or fall? That is, is the surface of the water above or below the original level of the glycerine before the ice melted?

12.58 •• A narrow, U-shaped glass tube with open ends is filled with 25.0 cm of oil (of specific gravity 0.80) and 25.0 cm of water on opposite sides, with a barrier separating the liquids (Fig. P12.58). (a) Assume that the two liquids do not mix, and find the final heights of the columns of liquid in each side of the tube after the barrier is removed. (b) For the following cases, arrive at your answer by simple physical reasoning, not by calculations: (i) What would be the height on each side if the oil and water had equal densities? (ii) What would the heights be if the oil's density were much less than that of water?

Figure **P12.58**

12.59 • A U-shaped tube open to the air at both ends contains some mercury. A quantity of water is carefully poured into the left arm of the U-shaped tube until the vertical height of the water column is 15.0 cm (Fig. P12.59). (a) What is the gauge pressure at the water–mercury interface? (b) Calculate the vertical distance h from the top of the mercury in the right-hand arm of the tube to the top of the water in the left-hand arm.

Figure **P12.59**

12.60 •• **CALC** **The Great Molasses Flood.** On the afternoon of January 15, 1919, an unusually warm day in Boston, a 17.7-m-high, 27.4-m-diameter cylindrical metal tank used for storing molasses ruptured. Molasses flooded into the streets in a 5-m-deep stream, killing pedestrians and horses and knocking down buildings. The molasses had a density of 1600 kg/m^3. If the tank was full before the accident, what was the total outward force the molasses exerted on its sides? (*Hint:* Consider the outward force on a circular ring of the tank wall of width dy and at a depth y below the surface. Integrate to find the total outward force. Assume that before the tank ruptured, the pressure at the surface of the molasses was equal to the air pressure outside the tank.)

12.61 • An open barge has the dimensions shown in Fig. P12.61. If the barge is made out of 4.0-cm-thick steel plate on each of its four sides and its bottom, what mass of coal can the barge carry in freshwater without sinking? Is there enough room in the barge to hold this amount of coal? (The density of coal is about 1500 kg/m³.)

Figure **P12.61**

40 m, 22 m, 12 m

12.62 ••• A hot-air balloon has a volume of 2200 m³. The balloon fabric (the envelope) weighs 900 N. The basket with gear and full propane tanks weighs 1700 N. If the balloon can barely lift an additional 3200 N of passengers, breakfast, and champagne when the outside air density is 1.23 kg/m³, what is the average density of the heated gases in the envelope?

12.63 •• Advertisements for a certain small car claim that it floats in water. (a) If the car's mass is 900 kg and its interior volume is 3.0 m³, what fraction of the car is immersed when it floats? You can ignore the volume of steel and other materials. (b) Water gradually leaks in and displaces the air in the car. What fraction of the interior volume is filled with water when the car sinks?

12.64 • A single ice cube with mass 9.70 g floats in a glass completely full of 420 cm³ of water. You can ignore the water's surface tension and its variation in density with temperature (as long as it remains a liquid). (a) What volume of water does the ice cube displace? (b) When the ice cube has completely melted, has any water overflowed? If so, how much? If not, explain why this is so. (c) Suppose the water in the glass had been very salty water of density 1050 kg/m³. What volume of salt water would the 9.70-g ice cube displace? (d) Redo part (b) for the freshwater ice cube in the salty water.

12.65 ••• A piece of wood is 0.600 m long, 0.250 m wide, and 0.080 m thick. Its density is 700 kg/m³. What volume of lead must be fastened underneath it to sink the wood in calm water so that its top is just even with the water level? What is the mass of this volume of lead?

12.66 •• A hydrometer consists of a spherical bulb and a cylindrical stem with a cross-sectional area of 0.400 cm² (see Fig. 12.12a). The total volume of bulb and stem is 13.2 cm³. When immersed in water, the hydrometer floats with 8.00 cm of the stem above the water surface. When the hydrometer is immersed in an organic fluid, 3.20 cm of the stem is above the surface. Find the density of the organic fluid. (*Note:* This illustrates the precision of such a hydrometer. Relatively small density differences give rise to relatively large differences in hydrometer readings.)

12.67 •• The densities of air, helium, and hydrogen (at $p = 1.0$ atm and $T = 20°C$) are 1.20 kg/m³, 0.166 kg/m³, and 0.0899 kg/m³, respectively. (a) What is the volume in cubic meters displaced by a hydrogen-filled airship that has a total "lift" of 90.0 kN? (The "lift" is the amount by which the buoyant force exceeds the weight of the gas that fills the airship.) (b) What would be the "lift" if helium were used instead of hydrogen? In view of your answer, why is helium used in modern airships like advertising blimps?

12.68 •• When an open-faced boat has a mass of 5750 kg, including its cargo and passengers, it floats with the water just up to the top of its gunwales (sides) on a freshwater lake. (a) What is the volume of this boat? (b) The captain decides that it is too dangerous to float with his boat on the verge of sinking, so he decides to throw some cargo overboard so that 20% of the boat's volume will be above water. How much mass should he throw out?

12.69 •• **CP** An open cylindrical tank of acid rests at the edge of a table 1.4 m above the floor of the chemistry lab. If this tank springs a small hole in the side at its base, how far from the foot of the table will the acid hit the floor if the acid in the tank is 75 cm deep?

12.70 •• **CP** A firehose must be able to shoot water to the top of a building 28.0 m tall when aimed straight up. Water enters this hose at a steady rate of 0.500 m³/s and shoots out of a round nozzle. (a) What is the maximum diameter this nozzle can have? (b) If the only nozzle available has a diameter twice as great, what is the highest point the water can reach?

12.71 •• **CP** You drill a small hole in the side of a vertical cylindrical water tank that is standing on the ground with its top open to the air. (a) If the water level has a height H, at what height above the base should you drill the hole for the water to reach its greatest distance from the base of the cylinder when it hits the ground? (b) What is the greatest distance the water will reach?

12.72 ••• **CALC** A closed and elevated vertical cylindrical tank with diameter 2.00 m contains water to a depth of 0.800 m. A worker accidently pokes a circular hole with diameter 0.0200 m in the bottom of the tank. As the water drains from the tank, compressed air above the water in the tank maintains a gauge pressure of 5.00×10^3 Pa at the surface of the water. Ignore any effects of viscosity. (a) Just after the hole is made, what is the speed of the water as it emerges from the hole? What is the ratio of this speed to the efflux speed if the top of the tank is open to the air? (b) How much time does it take for all the water to drain from the tank? What is the ratio of this time to the time it takes for the tank to drain if the top of the tank is open to the air?

12.73 •• A block of balsa wood placed in one scale pan of an equal-arm balance is exactly balanced by a 0.115-kg brass mass in the other scale pan. Find the true mass of the balsa wood if its density is 150 kg/m³. Explain why it is accurate to ignore the buoyancy in air of the brass but *not* the buoyancy in air of the balsa wood.

12.74 •• Block A in Fig. P12.74 hangs by a cord from spring balance D and is submerged in a liquid C contained in beaker B. The mass of the beaker is 1.00 kg; the mass of the liquid is 1.80 kg. Balance D reads 3.50 kg, and balance E reads 7.50 kg. The volume of block A is 3.80×10^{-3} m³. (a) What is the density of the liquid? (b) What will each balance read if block A is pulled up out of the liquid?

Figure **P12.74**

12.75 •• A hunk of aluminum is completely covered with a gold shell to form an ingot of weight 45.0 N. When you suspend the ingot from a spring balance and submerge the ingot in water, the balance reads 39.0 N. What is the weight of the gold in the shell?

12.76 •• A plastic ball has radius 12.0 cm and floats in water with 24.0% of its volume submerged. (a) What force must you apply to the ball to hold it at rest totally below the surface of the water? (b) If you let go of the ball, what is its acceleration the instant you release it?

12.77 •• The weight of a king's solid crown is w. When the crown is suspended by a light rope and completely immersed in water, the tension in the rope (the crown's apparent weight) is fw. (a) Prove that the crown's relative density (specific gravity) is $1/(1 - f)$. Discuss the meaning of the limits as f approaches 0 and 1. (b) If the crown is solid gold and weighs 12.9 N in air, what is its apparent weight when completely immersed in water? (c) Repeat part (b) if the crown is solid lead with a very thin gold plating, but still has a weight in air of 12.9 N.

12.78 •• A piece of steel has a weight w, an apparent weight (see Problem 12.77) w_{water} when completely immersed in water, and an apparent weight w_{fluid} when completely immersed in an unknown fluid. (a) Prove that the fluid's density relative to water (specific gravity) is $(w - w_{\text{fluid}})/(w - w_{\text{water}})$. (b) Is this result reasonable for the three cases of w_{fluid} greater than, equal to, or less than w_{water}? (c) The apparent weight of the piece of steel in water of density 1000 kg/m³ is 87.2% of its weight. What percentage of its weight will its apparent weight be in formic acid (density 1220 kg/m³)?

12.79 ••• You cast some metal of density ρ_m in a mold, but you are worried that there might be cavities within the casting. You measure the weight of the casting to be w, and the buoyant force when it is completely surrounded by water to be B. (a) Show that $V_0 = B/(\rho_{\text{water}}g) - w/(\rho_m g)$ is the total volume of any enclosed cavities. (b) If your metal is copper, the casting's weight is 156 N, and the buoyant force is 20 N, what is the total volume of any enclosed cavities in your casting? What fraction is this of the total volume of the casting?

12.80 • A cubical block of wood 0.100 m on a side and with a density of 550 kg/m³ floats in a jar of water. Oil with a density of 750 kg/m³ is poured on the water until the top of the oil layer is 0.035 m below the top of the block. (a) How deep is the oil layer? (b) What is the gauge pressure at the block's lower face?

12.81 •• **Dropping Anchor.** An iron anchor with mass 35.0 kg and density 7860 kg/m³ lies on the deck of a small barge that has vertical sides and floats in a freshwater river. The area of the bottom of the barge is 8.00 m². The anchor is thrown overboard but is suspended above the bottom of the river by a rope; the mass and volume of the rope are small enough to ignore. After the anchor is overboard and the barge has finally stopped bobbing up and down, has the barge risen or sunk down in the water? By what vertical distance?

12.82 •• Assume that crude oil from a supertanker has density 750 kg/m³. The tanker runs aground on a sandbar. To refloat the tanker, its oil cargo is pumped out into steel barrels, each of which has a mass of 15.0 kg when empty and holds 0.120 m³ of oil. You can ignore the volume occupied by the steel from which the barrel is made. (a) If a salvage worker accidentally drops a filled, sealed barrel overboard, will it float or sink in the seawater? (b) If the barrel floats, what fraction of its volume will be above the water surface? If it sinks, what minimum tension would have to be exerted by a rope to haul the barrel up from the ocean floor? (c) Repeat parts (a) and (b) if the density of the oil is 910 kg/m³ and the mass of each empty barrel is 32.0 kg.

12.83 ••• A cubical block of density ρ_B and with sides of length L floats in a liquid of greater density ρ_L. (a) What fraction of the block's volume is above the surface of the liquid? (b) The liquid is denser than water (density ρ_W) and does not mix with it. If water is poured on the surface of the liquid, how deep must the water layer be so that the water surface just rises to the top of the block? Express your answer in terms of L, ρ_B, ρ_L, and ρ_W. (c) Find the depth of the water layer in part (b) if the liquid is mercury, the block is made of iron, and the side length is 10.0 cm.

12.84 •• A barge is in a rectangular lock on a freshwater river. The lock is 60.0 m long and 20.0 m wide, and the steel doors on each end are closed. With the barge floating in the lock, a 2.50×10^6 N load of scrap metal is put onto the barge. The metal has density 9000 kg/m³. (a) When the load of scrap metal, initially on the

bank, is placed onto the barge, what vertical distance does the water in the lock rise? (b) The scrap metal is now pushed overboard into the water. Does the water level in the lock rise, fall, or remain the same? If it rises or falls, by what vertical distance does it change?

12.85 • CP CALC A U-shaped tube with a horizontal portion of length l (Fig. P12.85) contains a liquid. What is the difference in height between the liquid columns in the vertical arms (a) if the tube has an acceleration a toward the right and (b) if the tube is mounted on a horizontal turntable rotating with an angular speed ω with one of the vertical

Figure **P12.85**

arms on the axis of rotation? (c) Explain why the difference in height does not depend on the density of the liquid or on the cross-sectional area of the tube. Would it be the same if the vertical tubes did not have equal cross-sectional areas? Would it be the same if the horizontal portion were tapered from one end to the other? Explain.

12.86 • CP CALC A cylindrical container of an incompressible liquid with density ρ rotates with constant angular speed ω about its axis of symmetry, which we take to be the y-axis (Fig. P12.86). (a) Show that the pressure at a given height within the fluid increases in the radial direction (outward from the axis of rotation) according to $\partial p/\partial r = \rho\omega^2 r$. (b) Integrate this partial differential equation to find the pressure as a

Figure **P12.86**

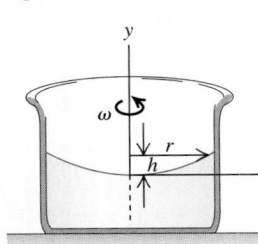

function of distance from the axis of rotation along a horizontal line at $y = 0$. (c) Combine the result of part (b) with Eq. (12.5) to show that the surface of the rotating liquid has a *parabolic* shape; that is, the height of the liquid is given by $h(r) = \omega^2 r^2/2g$. (This technique is used for making parabolic telescope mirrors; liquid glass is rotated and allowed to solidify while rotating.)

12.87 •• CP CALC An incompressible fluid with density ρ is in a horizontal test tube of inner cross-sectional area A. The test tube spins in a horizontal circle in an ultracentrifuge at an angular speed ω. Gravitational forces are negligible. Consider a volume element of the fluid of area A and thickness dr' a distance r' from the rotation axis. The pressure on its inner surface is p and on its outer surface is $p + dp$. (a) Apply Newton's second law to the volume element to show that $dp = \rho\omega^2 r' dr'$. (b) If the surface of the fluid is at a radius r_0 where the pressure is p_0, show that the pressure p at a distance $r \geq r_0$ is $p = p_0 + \rho\omega^2(r^2 - r_0^2)/2$. (c) An object of volume V and density ρ_{ob} has its center of mass at a distance R_{cmob} from the axis. Show that the net horizontal force on the object is $\rho V\omega^2 R_{\text{cm}}$, where R_{cm} is the distance from the axis to the center of mass of the displaced fluid. (d) Explain why the object will move inward if $\rho R_{\text{cm}} > \rho_{\text{ob}} R_{\text{cmob}}$ and outward if $\rho R_{\text{cm}} < \rho_{\text{ob}} R_{\text{cmob}}$. (e) For small objects of uniform density, $R_{\text{cm}} = R_{\text{cmob}}$. What happens to a mixture of small objects of this kind with different densities in an ultracentrifuge?

12.88 ••• CALC Untethered helium balloons, floating in a car that has all the windows rolled up and outside air vents closed, move in the direction of the car's acceleration, but loose balloons filled with air move in the opposite direction. To show why, consider only the horizontal forces acting on the balloons. Let a be the magnitude of the car's forward acceleration. Consider a horizontal tube of air with a cross-sectional area A that extends from the windshield, where $x = 0$ and $p = p_0$, back along the x-axis. Now consider a volume element of thickness dx in this tube. The pressure on its front surface is p and the pressure on its rear surface is $p + dp$. Assume the air has a constant density ρ. (a) Apply Newton's second law to the volume element to show that $dp = \rho a\, dx$. (b) Integrate the result of part (a) to find the pressure at the front surface in terms of a and x. (c) To show that considering ρ constant is reasonable, calculate the pressure difference in atm for a distance as long as 2.5 m and a large acceleration of 5.0 m/s². (d) Show that the net horizontal force on a balloon of volume V is ρVa. (e) For negligible friction forces, show that the acceleration of the balloon (average density ρ_{bal}) is $(\rho/\rho_{bal})a$, so that the acceleration relative to the car is $a_{rel} = [(\rho/\rho_{bal}) - 1]a$. (f) Use the expression for a_{rel} in part (e) to explain the movement of the balloons.

12.89 • **CP** Water stands at a depth H in a large, open tank whose side walls are vertical (Fig. P12.89). A hole is made in one of the walls at a depth h below the water surface. (a) At what distance R from the foot of the wall does the emerging stream strike the floor? (b) How far above the bottom of the tank could a second hole be cut so that the stream emerging from it could have the same range as for the first hole?

Figure **P12.89**

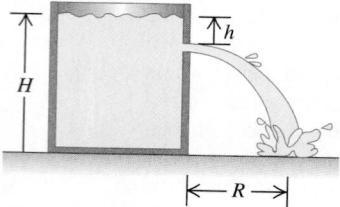

12.90 ••• A cylindrical bucket, open at the top, is 25.0 cm high and 10.0 cm in diameter. A circular hole with a cross-sectional area 1.50 cm² is cut in the center of the bottom of the bucket. Water flows into the bucket from a tube above it at the rate of 2.40×10^{-4} m³/s. How high will the water in the bucket rise?

12.91 • Water flows steadily from an open tank as in Fig. P12.91. The elevation of point 1 is 10.0 m, and the elevation of points 2 and 3 is 2.00 m. The cross-sectional area at point 2 is 0.0480 m²; at point 3 it is 0.0160 m². The area of the tank is very large compared with the cross-sectional area of the pipe. Assuming that Bernoulli's equation applies, compute (a) the discharge rate in cubic meters per second and (b) the gauge pressure at point 2.

Figure **P12.91**

12.92 •• **CP** In 1993 the radius of Hurricane Emily was about 350 km. The wind speed near the center ("eye") of the hurricane, whose radius was about 30 km, reached about 200 km/h. As air swirled in from the rim of the hurricane toward the eye, its angular

momentum remained roughly constant. (a) Estimate the wind speed at the rim of the hurricane. (b) Estimate the pressure difference at the earth's surface between the eye and the rim. (*Hint:* See Table 12.1.) Where is the pressure greater? (c) If the kinetic energy of the swirling air in the eye could be converted completely to gravitational potential energy, how high would the air go? (d) In fact, the air in the eye is lifted to heights of several kilometers. How can you reconcile this with your answer to part (c)?

12.93 •• Two very large open tanks A and F (Fig. P12.93) contain the same liquid. A horizontal pipe BCD, having a constriction at C and open to the air at D, leads out of the bottom of tank A, and a vertical pipe E opens into the constriction at C and dips into the liquid in tank F. Assume streamline flow and no viscosity. If the cross-sectional area at C is one-half the area at D and if D is a distance h_1 below the level of the liquid in A, to what height h_2 will liquid rise in pipe E? Express your answer in terms of h_1.

Figure **P12.93**

12.94 •• The horizontal pipe shown in Fig. P12.94 has a cross-sectional area of 40.0 cm² at the wider portions and 10.0 cm² at the constriction. Water is flowing in the pipe, and the discharge from the pipe is 6.00×10^{-3} m³/s (6.00 L/s). Find (a) the flow speeds at the wide and the narrow portions; (b) the pressure difference between these portions; (c) the difference in height between the mercury columns in the U-shaped tube.

Figure **P12.94**

12.95 • A liquid flowing from a vertical pipe has a definite shape as it flows from the pipe. To get the equation for this shape, assume that the liquid is in free fall once it leaves the pipe. Just as it leaves the pipe, the liquid has speed v_0 and the radius of the stream of liquid is r_0. (a) Find an equation for the speed of the liquid as a function of the distance y it has fallen. Combining this with the equation of continuity, find an expression for the radius of the stream as a function of y. (b) If water flows out of a vertical pipe at a speed of 1.20 m/s, how far below the outlet will the radius be one-half the original radius of the stream?

Challenge Problems

12.96 ••• A rock with mass $m = 3.00$ kg is suspended from the roof of an elevator by a light cord. The rock is totally immersed in a bucket of water that sits on the floor of the elevator, but the rock doesn't touch the bottom or sides of the bucket. (a) When the elevator is at rest, the tension in the cord is 21.0 N. Calculate the volume of the rock. (b) Derive an expression for the tension in the cord when the elevator is accelerating *upward* with an acceleration of magnitude a. Calculate the tension when $a = 2.50$ m/s² upward. (c) Derive an expression for the tension in the cord when the elevator is accelerating *downward* with an acceleration of magnitude *a*. Calculate the tension when $a = 2.50$ m/s^2 downward. (d) What is the tension when the elevator is in free fall with a downward acceleration equal to *g*?

12.97 ••• **CALC** Suppose a piece of styrofoam, $\rho = 180$ kg/m^3, is held completely submerged in water (Fig. P12.97). (a) What is the tension in the cord? Find this using Archimedes's principle. (b) Use $p = p_0 + \rho g h$ to calculate directly the force exerted by the water on the two sloped sides and the bottom of the styrofoam; then show that the vector sum of these forces is the buoyant force.

Figure **P12.97**

12.98 ••• A *siphon,* as shown in Fig. P12.98, is a convenient device for removing liquids from containers. To establish the flow, the tube must be initially filled with fluid. Let the fluid have density ρ, and let the atmospheric pressure be p_{atm}. Assume that the cross-sectional area of the tube is the same at all points along it. (a) If the lower end of the siphon is at a distance *h* below the surface of the liquid in the container, what is the speed of the fluid as it flows out the lower end of the siphon? (Assume that the container has a very large diameter, and ignore any effects of viscosity.) (b) A curious feature of a siphon is that the fluid initially flows "uphill." What is the greatest height *H* that the high point of the tube can have if flow is still to occur?

Figure **P12.98**

Answers

Chapter Opening Question ?

The flesh of both the shark and the tropical fish is denser than seawater, so left to themselves they would sink. However, a tropical fish has a gas-filled body cavity called a swimbladder, so that the *average* density of the fish's body is the same as that of seawater and the fish neither sinks nor rises. Sharks have no such cavity. Hence they must swim constantly to keep from sinking, using their pectoral fins to provide lift much like the wings of an airplane (see Section 12.5).

Test Your Understanding Questions

12.1 Answer: (ii), (iv), (i) and (iii) (tie), (v) In each case the average density equals the mass divided by the volume. Hence we have
(i) $\rho = (4.00 \text{ kg})/(1.60 \times 10^{-3} \text{ m}^3) = 2.50 \times 10^3 \text{ kg/m}^3$;
(ii) $\rho = (8.00 \text{ kg})/(1.60 \times 10^{-3} \text{ m}^3) = 5.00 \times 10^3 \text{ kg/m}^3$;
(iii) $\rho = (8.00 \text{ kg})/(3.20 \times 10^{-3} \text{ m}^3) = 2.50 \times 10^3 \text{ kg/m}^3$;
(iv) $\rho = (2560 \text{ kg})/(0.640 \text{ m}^3) = 4.00 \times 10^3 \text{ kg/m}^3$;
(v) $\rho = (2560 \text{ kg})/(1.28 \text{ m}^3) = 2.00 \times 10^3 \text{ kg/m}^3$. Note that compared to object (i), object (ii) has double the mass but the same volume and so has double the average density. Object (iii) has double the mass and double the volume of object (i), so (i) and (iii) have the same average density. Finally, object (v) has the same mass as object (iv) but double the volume, so (v) has half the average density of (iv).

12.2 Answer: (ii) From Eq. (12.9), the pressure outside the barometer is equal to the product $\rho g h$. When the barometer is taken out of the refrigerator, the density ρ decreases while the height *h* of the mercury column remains the same. Hence the air pressure must be lower outdoors than inside the refrigerator.

12.3 Answer: (i) Consider the water, the statue, and the container together as a system; the total weight of the system does not depend on whether the statue is immersed. The total supporting force, including the tension *T* and the upward force *F* of the scale

on the container (equal to the scale reading), is the same in both cases. But we saw in Example 12.5 that *T* decreases by 7.84 N when the statue is immersed, so the scale reading *F* must *increase* by 7.84 N. An alternative viewpoint is that the water exerts an upward buoyant force of 7.84 N on the statue, so the statue must exert an equal downward force on the water, making the scale reading 7.84 N greater than the weight of water and container.

12.4 Answer: (ii) A highway that narrows from three lanes to one is like a pipe whose cross-sectional area narrows to one-third of its value. If cars behaved like the molecules of an incompressible fluid, then as the cars encountered the one-lane section, the spacing between cars (the "density") would stay the same but the cars would triple their speed. This would keep the "volume flow rate" (number of cars per second passing a point on the highway) the same. In real life cars behave like the molecules of a *compressible* fluid: They end up packed closer (the "density" increases) and fewer cars per second pass a point on the highway (the "volume flow rate" decreases).

12.5 Answer: (ii) Newton's second law tells us that a body accelerates (its velocity changes) in response to a net force. In fluid flow, a pressure difference between two points means that fluid particles moving between those two points experience a force, and this force causes the fluid particles to accelerate and change speed.

12.6 Answer: (iv) The required pressure is proportional to $1/R^4$, where *R* is the inside radius of the needle (half the inside diameter). With the smaller-diameter needle, the pressure is greater by a factor of $[(0.60 \text{ mm})/(0.30 \text{ mm})]^4 = 2^4 = 16$.

Bridging Problem

Answers: (a) $y = H - \left(\dfrac{d}{D}\right)^2 \sqrt{2gH}\, t + \left(\dfrac{d}{D}\right)^4 \dfrac{gt^2}{2}$

(b) $T = \sqrt{\dfrac{2H}{g}} \left(\dfrac{D}{d}\right)^2$ **(c)** $\sqrt{2}$

13

GRAVITATION

LEARNING GOALS

By studying this chapter, you will learn:

- How to calculate the gravitational forces that any two bodies exert on each other.

- How to relate the weight of an object to the general expression for gravitational force.

- How to use and interpret the generalized expression for gravitational potential energy.

- How to relate the speed, orbital period, and mechanical energy of a satellite in a circular orbit.

- The laws that describe the motions of planets, and how to work with these laws.

- What black holes are, how to calculate their properties, and how they are discovered.

? The rings of Saturn are made of countless individual orbiting particles. Do all the ring particles orbit at the same speed, or do the inner particles orbit faster or slower than the outer ones?

Some of the earliest investigations in physical science started with questions that people asked about the night sky. Why doesn't the moon fall to earth? Why do the planets move across the sky? Why doesn't the earth fly off into space rather than remaining in orbit around the sun? The study of gravitation provides the answers to these and many related questions.

As we remarked in Chapter 5, gravitation is one of the four classes of interactions found in nature, and it was the earliest of the four to be studied extensively. Newton discovered in the 17th century that the same interaction that makes an apple fall out of a tree also keeps the planets in their orbits around the sun. This was the beginning of *celestial mechanics,* the study of the dynamics of objects in space. Today, our knowledge of celestial mechanics allows us to determine how to put a satellite into any desired orbit around the earth or to choose just the right trajectory to send a spacecraft to another planet.

In this chapter you will learn the basic law that governs gravitational interactions. This law is *universal:* Gravity acts in the same fundamental way between the earth and your body, between the sun and a planet, and between a planet and one of its moons. We'll apply the law of gravitation to phenomena such as the variation of weight with altitude, the orbits of satellites around the earth, and the orbits of planets around the sun.

13.1 Newton's Law of Gravitation

The example of gravitational attraction that's probably most familiar to you is your *weight,* the force that attracts you toward the earth. During his study of the motions of the planets and of the moon, Newton discovered the fundamental character of the gravitational attraction between *any* two bodies. Along with his three laws of motion, Newton published the **law of gravitation** in 1687. It may be stated as follows:

> **Every particle of matter in the universe attracts every other particle with a force that is directly proportional to the product of the masses of the particles and inversely proportional to the square of the distance between them.**

Translating this into an equation, we have

$$F_g = \frac{Gm_1 m_2}{r^2} \qquad \text{(law of gravitation)} \qquad (13.1)$$

where F_g is the magnitude of the gravitational force on either particle, m_1 and m_2 are their masses, r is the distance between them (Fig. 13.1), and G is a fundamental physical constant called the **gravitational constant.** The numerical value of G depends on the system of units used.

Equation (13.1) tells us that the gravitational force between two particles decreases with increasing distance r: If the distance is doubled, the force is only one-fourth as great, and so on. Although many of the stars in the night sky are far more massive than the sun, they are so far away that their gravitational force on the earth is negligibly small.

CAUTION **Don't confuse g and G** Because the symbols g and G are so similar, it's common to confuse the two very different gravitational quantities that these symbols represent. Lowercase g is the acceleration due to gravity, which relates the weight w of a body to its mass m: $w = mg$. The value of g is different at different locations on the earth's surface and on the surfaces of different planets. By contrast, capital G relates the gravitational force between any two bodies to their masses and the distance between them. We call G a *universal* constant because it has the same value for any two bodies, no matter where in space they are located. In the next section we'll see how the values of g and G are related.

Gravitational forces always act along the line joining the two particles, and they form an action–reaction pair. Even when the masses of the particles are different, the two interaction forces have equal magnitude (Fig. 13.1). The attractive force that your body exerts on the earth has the same magnitude as the force that the earth exerts on you. When you fall from a diving board into a swimming pool, the entire earth rises up to meet you! (You don't notice this because the earth's mass is greater than yours by a factor of about 10^{23}. Hence the earth's acceleration is only 10^{-23} as great as yours.)

Gravitation and Spherically Symmetric Bodies

We have stated the law of gravitation in terms of the interaction between two *particles*. It turns out that the gravitational interaction of any two bodies having *spherically symmetric* mass distributions (such as solid spheres or spherical shells) is the same as though we concentrated all the mass of each at its center, as in Fig. 13.2. Thus, if we model the earth as a spherically symmetric body with mass m_E, the force it exerts on a particle or a spherically symmetric body with mass m, at a distance r between centers, is

$$F_g = \frac{Gm_E m}{r^2} \qquad (13.2)$$

provided that the body lies outside the earth. A force of the same magnitude is exerted *on* the earth by the body. (We will prove these statements in Section 13.6.)

At points *inside* the earth the situation is different. If we could drill a hole to the center of the earth and measure the gravitational force on a body at various depths, we would find that toward the center of the earth the force *decreases,*

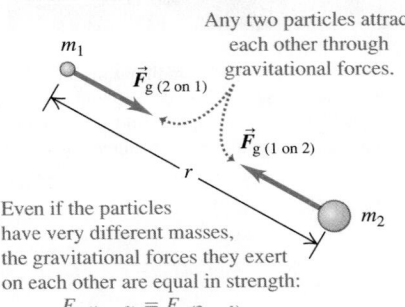

13.1 The gravitational forces between two particles of masses m_1 and m_2.

Any two particles attract each other through gravitational forces.

$\vec{F}_{g\,(2\text{ on }1)}$

$\vec{F}_{g\,(1\text{ on }2)}$

Even if the particles have very different masses, the gravitational forces they exert on each other are equal in strength:
$F_{g\,(1\text{ on }2)} = F_{g\,(2\text{ on }1)}$

13.2 The gravitational effect *outside* any spherically symmetric mass distribution is the same as though all of the mass were concentrated at its center.

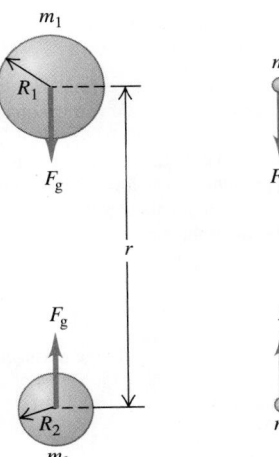

(a) The gravitational force between two spherically symmetric masses m_1 and m_2 ...

(b) ... is the same as if we concentrated all the mass of each sphere at the sphere's center.

13.3 Spherical and nonspherical bodies: the planet Jupiter and one of Jupiter's small moons, Amalthea.

Jupiter's mass is very large (1.90×10^{27} kg), so the mutual gravitational attraction of its parts has pulled it into a nearly spherical shape.

100,000 km

100 km

Amalthea, one of Jupiter's small moons, has a relatively tiny mass (7.17×10^{18} kg, only about 3.8×10^{-9} the mass of Jupiter) and weak mutual gravitation, so it has an irregular shape.

rather than increasing as $1/r^2$. As the body enters the interior of the earth (or other spherical body), some of the earth's mass is on the side of the body opposite from the center and pulls in the opposite direction. Exactly at the center, the earth's gravitational force on the body is zero.

Spherically symmetric bodies are an important case because moons, planets, and stars all tend to be spherical. Since all particles in a body gravitationally attract each other, the particles tend to move to minimize the distance between them. As a result, the body naturally tends to assume a spherical shape, just as a lump of clay forms into a sphere if you squeeze it with equal forces on all sides. This effect is greatly reduced in celestial bodies of low mass, since the gravitational attraction is less, and these bodies tend *not* to be spherical (Fig. 13.3).

Determining the Value of G

To determine the value of the gravitational constant G, we have to *measure* the gravitational force between two bodies of known masses m_1 and m_2 at a known distance r. The force is extremely small for bodies that are small enough to be brought into the laboratory, but it can be measured with an instrument called a *torsion balance,* which Sir Henry Cavendish used in 1798 to determine G.

Figure 13.4 shows a modern version of the Cavendish torsion balance. A light, rigid rod shaped like an inverted T is supported by a very thin, vertical quartz fiber. Two small spheres, each of mass m_1, are mounted at the ends of the horizontal arms of the T. When we bring two large spheres, each of mass m_2, to the positions shown, the attractive gravitational forces twist the T through a small angle. To measure this angle, we shine a beam of light on a mirror fastened to the T. The reflected beam strikes a scale, and as the T twists, the reflected beam moves along the scale.

After calibrating the Cavendish balance, we can measure gravitational forces and thus determine G. The presently accepted value is

$$G = 6.67428(67) \times 10^{-11}\ \mathrm{N \cdot m^2/kg^2}$$

To three significant figures, $G = 6.67 \times 10^{-11}\ \mathrm{N \cdot m^2/kg^2}$. Because $1\ \mathrm{N} = 1\ \mathrm{kg \cdot m/s^2}$, the units of G can also be expressed as $\mathrm{m^3/(kg \cdot s^2)}$.

Gravitational forces combine vectorially. If each of two masses exerts a force on a third, the *total* force on the third mass is the vector sum of the individual forces of the first two. Example 13.3 makes use of this property, which is often called *superposition of forces.*

13.4 The principle of the Cavendish balance, used for determining the value of G. The angle of deflection has been exaggerated here for clarity.

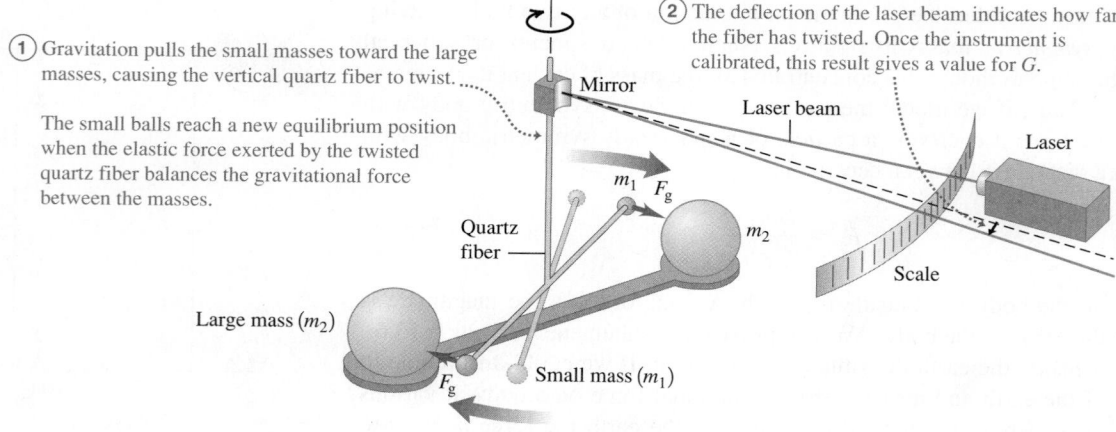

① Gravitation pulls the small masses toward the large masses, causing the vertical quartz fiber to twist.

The small balls reach a new equilibrium position when the elastic force exerted by the twisted quartz fiber balances the gravitational force between the masses.

② The deflection of the laser beam indicates how far the fiber has twisted. Once the instrument is calibrated, this result gives a value for G.

Mirror

Laser beam

Laser

m_1 F_g

m_2

Quartz fiber

Scale

Large mass (m_2)

F_g

Small mass (m_1)

Example 13.1 Calculating gravitational force

The mass m_1 of one of the small spheres of a Cavendish balance is 0.0100 kg, the mass m_2 of the nearest large sphere is 0.500 kg, and the center-to-center distance between them is 0.0500 m. Find the gravitational force F_g on each sphere due to the other.

SOLUTION

IDENTIFY, SET UP, and EXECUTE: Because the spheres are spherically symmetric, we can calculate F_g by treating them as *particles* separated by 0.0500 m, as in Fig. 13.2. Each sphere experiences the same magnitude of force from the other sphere. We use Newton's law of gravitation, Eq. (13.1), to determine F_g:

$$F_g = \frac{(6.67 \times 10^{-11}\,\text{N}\cdot\text{m}^2/\text{kg}^2)(0.0100\,\text{kg})(0.500\,\text{kg})}{(0.0500\,\text{m})^2}$$

$$= 1.33 \times 10^{-10}\,\text{N}$$

EVALUATE: It's remarkable that such a small force could be measured—or even detected—more than 200 years ago. Only a very massive object such as the earth exerts a gravitational force we can feel.

Example 13.2 Acceleration due to gravitational attraction

Suppose the two spheres in Example 13.1 are placed with their centers 0.0500 m apart at a point in space far removed from all other bodies. What is the magnitude of the acceleration of each, relative to an inertial system?

SOLUTION

IDENTIFY, SET UP, and EXECUTE: Each sphere exerts on the other a gravitational force of the same magnitude F_g, which we found in Example 13.1. We can neglect any other forces. The *acceleration* magnitudes a_1 and a_2 are different because the masses are different.

To determine these we'll use Newton's second law:

$$a_1 = \frac{F_g}{m_1} = \frac{1.33 \times 10^{-10}\,\text{N}}{0.0100\,\text{kg}} = 1.33 \times 10^{-8}\,\text{m/s}^2$$

$$a_2 = \frac{F_g}{m_2} = \frac{1.33 \times 10^{-10}\,\text{N}}{0.500\,\text{kg}} = 2.66 \times 10^{-10}\,\text{m/s}^2$$

EVALUATE: The larger sphere has 50 times the mass of the smaller one and hence has $\frac{1}{50}$ the acceleration. These accelerations are *not* constant; the gravitational forces increase as the spheres move toward each other.

Example 13.3 Superposition of gravitational forces

Many stars belong to *systems* of two or more stars held together by their mutual gravitational attraction. Figure 13.5 shows a three-star system at an instant when the stars are at the vertices of a 45° right triangle. Find the total gravitational force exerted on the small star by the two large ones.

SOLUTION

IDENTIFY, SET UP, and EXECUTE: We use the principle of superposition: The total force $\vec{F}$ on the small star is the vector sum of the forces $\vec{F}_1$ and $\vec{F}_2$ due to each large star, as Fig. 13.5 shows. We assume that the stars are spheres as in Fig. 13.2. We first calculate the magnitudes F_1 and F_2 using Eq. (13.1) and then compute the vector sum using components:

$$F_1 = \frac{\left[\begin{array}{c}(6.67 \times 10^{-11}\,\text{N}\cdot\text{m}^2/\text{kg}^2)\\ \times\,(8.00 \times 10^{30}\,\text{kg})(1.00 \times 10^{30}\,\text{kg})\end{array}\right]}{(2.00 \times 10^{12}\,\text{m})^2 + (2.00 \times 10^{12}\,\text{m})^2}$$

$$= 6.67 \times 10^{25}\,\text{N}$$

$$F_2 = \frac{\left[\begin{array}{c}(6.67 \times 10^{-11}\,\text{N}\cdot\text{m}^2/\text{kg}^2)\\ \times\,(8.00 \times 10^{30}\,\text{kg})(1.00 \times 10^{30}\,\text{kg})\end{array}\right]}{(2.00 \times 10^{12}\,\text{m})^2}$$

$$= 1.33 \times 10^{26}\,\text{N}$$

13.5 The total gravitational force on the small star (at O) is the vector sum of the forces exerted on it by the two larger stars. (For comparison, the mass of the sun—a rather ordinary star—is 1.99×10^{30} kg and the earth–sun distance is 1.50×10^{11} m.)

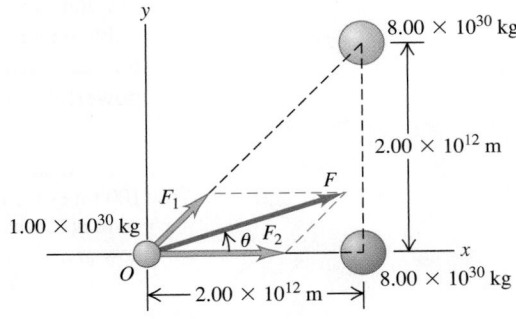

The x- and y-components of these forces are

$$F_{1x} = (6.67 \times 10^{25}\,\text{N})(\cos 45°) = 4.72 \times 10^{25}\,\text{N}$$
$$F_{1y} = (6.67 \times 10^{25}\,\text{N})(\sin 45°) = 4.72 \times 10^{25}\,\text{N}$$
$$F_{2x} = 1.33 \times 10^{26}\,\text{N}$$
$$F_{2y} = 0$$

Continued

The components of the total force $\vec{F}$ on the small star are

$$F_x = F_{1x} + F_{2x} = 1.81 \times 10^{26} \text{ N}$$
$$F_y = F_{1y} + F_{2y} = 4.72 \times 10^{25} \text{ N}$$

The magnitude of $\vec{F}$ and its angle θ (see Fig. 13.5) are

$$F = \sqrt{F_x{}^2 + F_y{}^2} = \sqrt{(1.81 \times 10^{26} \text{ N})^2 + (4.72 \times 10^{25} \text{ N})^2}$$
$$= 1.87 \times 10^{26} \text{ N}$$
$$\theta = \arctan \frac{F_y}{F_x} = \arctan \frac{4.72 \times 10^{25} \text{ N}}{1.81 \times 10^{26} \text{ N}} = 14.6°$$

EVALUATE: While the force magnitude F is tremendous, the magnitude of the resulting acceleration is not: $a = F/m = (1.87 \times 10^{26} \text{ N})/(1.00 \times 10^{30} \text{ kg}) = 1.87 \times 10^{-4} \text{ m/s}^2$. Furthermore, the force $\vec{F}$ is *not* directed toward the center of mass of the two large stars.

13.6 Our solar system is part of a spiral galaxy like this one, which contains roughly 10^{11} stars as well as gas, dust, and other matter. The entire assemblage is held together by the mutual gravitational attraction of all the matter in the galaxy.

Why Gravitational Forces Are Important

Comparing Examples 13.1 and 13.3 shows that gravitational forces are negligible between ordinary household-sized objects, but very substantial between objects that are the size of stars. Indeed, gravitation is *the* most important force on the scale of planets, stars, and galaxies (Fig. 13.6). It is responsible for holding our earth together and for keeping the planets in orbit about the sun. The mutual gravitational attraction between different parts of the sun compresses material at the sun's core to very high densities and temperatures, making it possible for nuclear reactions to take place there. These reactions generate the sun's energy output, which makes it possible for life to exist on earth and for you to read these words.

The gravitational force is so important on the cosmic scale because it acts *at a distance*, without any direct contact between bodies. Electric and magnetic forces have this same remarkable property, but they are less important on astronomical scales because large accumulations of matter are electrically neutral; that is, they contain equal amounts of positive and negative charge. As a result, the electric and magnetic forces between stars or planets are very small or zero. The strong and weak interactions that we discussed in Section 5.5 also act at a distance, but their influence is negligible at distances much greater than the diameter of an atomic nucleus (about 10^{-14} m).

A useful way to describe forces that act at a distance is in terms of a *field*. One body sets up a disturbance or field at all points in space, and the force that acts on a second body at a particular point is its response to the first body's field at that point. There is a field associated with each force that acts at a distance, and so we refer to gravitational fields, electric fields, magnetic fields, and so on. We won't need the field concept for our study of gravitation in this chapter, so we won't discuss it further here. But in later chapters we'll find that the field concept is an extraordinarily powerful tool for describing electric and magnetic interactions.

Test Your Understanding of Section 13.1 The planet Saturn has about 100 times the mass of the earth and is about 10 times farther from the sun than the earth is. Compared to the acceleration of the earth caused by the sun's gravitational pull, how great is the acceleration of Saturn due to the sun's gravitation? (i) 100 times greater; (ii) 10 times greater; (iii) the same; (iv) $\frac{1}{10}$ as great; (v) $\frac{1}{100}$ as great.

Mastering**PHYSICS**

PhET: Lunar Lander

13.2 Weight

We defined the *weight* of a body in Section 4.4 as the attractive gravitational force exerted on it by the earth. We can now broaden our definition:

> **The weight of a body is the total gravitational force exerted on the body by all other bodies in the universe.**

When the body is near the surface of the earth, we can neglect all other gravitational forces and consider the weight as just the earth's gravitational attraction. At the surface of the *moon* we consider a body's weight to be the gravitational attraction of the moon, and so on.

If we again model the earth as a spherically symmetric body with radius R_E and mass m_E, the weight w of a small body of mass m at the earth's surface (a distance R_E from its center) is

$$w = F_g = \frac{Gm_Em}{R_E^2} \qquad \begin{array}{l}\text{(weight of a body of mass } m\\ \text{at the earth's surface)}\end{array} \qquad \text{(13.3)}$$

But we also know from Section 4.4 that the weight w of a body is the force that causes the acceleration g of free fall, so by Newton's second law, $w = mg$. Equating this with Eq. (13.3) and dividing by m, we find

$$g = \frac{Gm_E}{R_E^2} \qquad \text{(acceleration due to gravity at the earth's surface)} \qquad \text{(13.4)}$$

The acceleration due to gravity g is independent of the mass m of the body because m doesn't appear in this equation. We already knew that, but we can now see how it follows from the law of gravitation.

We can *measure* all the quantities in Eq. (13.4) except for m_E, so this relationship allows us to compute the mass of the earth. Solving Eq. (13.4) for m_E and using $R_E = 6380 \text{ km} = 6.38 \times 10^6 \text{ m}$ and $g = 9.80 \text{ m/s}^2$, we find

$$m_E = \frac{gR_E^2}{G} = 5.98 \times 10^{24} \text{ kg}$$

This is very close to the currently accepted value of $5.974 \times 10^{24} \text{ kg}$. Once Cavendish had measured G, he computed the mass of the earth in just this way.

At a point above the earth's surface a distance r from the center of the earth (a distance $r - R_E$ above the surface), the weight of a body is given by Eq. (13.3) with R_E replaced by r:

$$w = F_g = \frac{Gm_Em}{r^2} \qquad \text{(13.5)}$$

The weight of a body decreases inversely with the square of its distance from the earth's center (Fig. 13.7). Figure 13.8 shows how the weight varies with height above the earth for an astronaut who weighs 700 N at the earth's surface.

The *apparent* weight of a body on earth differs slightly from the earth's gravitational force because the earth rotates and is therefore not precisely an inertial frame of reference. We have ignored this effect in our earlier discussion and have assumed that the earth *is* an inertial system. We will return to the effect of the earth's rotation in Section 13.7.

While the earth is an approximately spherically symmetric distribution of mass, it is *not* uniform throughout its volume. To demonstrate this, let's first calculate the average *density,* or mass per unit volume, of the earth. If we assume a spherical earth, the volume is

$$V_E = \tfrac{4}{3}\pi R_E^3 = \tfrac{4}{3}\pi(6.38 \times 10^6 \text{ m})^3 = 1.09 \times 10^{21} \text{ m}^3$$

Application Walking and Running on the Moon
You automatically transition from a walk to a run when the vertical force you exert on the ground—which, by Newton's third law, equals the vertical force the ground exerts on you—exceeds your weight. This transition from walking to running happens at much lower speeds on the moon, where objects weigh only 17% as much as on earth. Hence, the Apollo astronauts found themselves running even when moving relatively slowly during their moon "walks."

13.7 In an airliner at high altitude, you are farther from the center of the earth than when on the ground and hence weigh slightly less. Can you show that at an altitude of 10 km above the surface, you weigh 0.3% less than you do on the ground?

13.8 An astronaut who weighs 700 N at the earth's surface experiences less gravitational attraction when above the surface. The relevant distance r is from the astronaut to the *center* of the earth (*not* from the astronaut to the earth's surface).

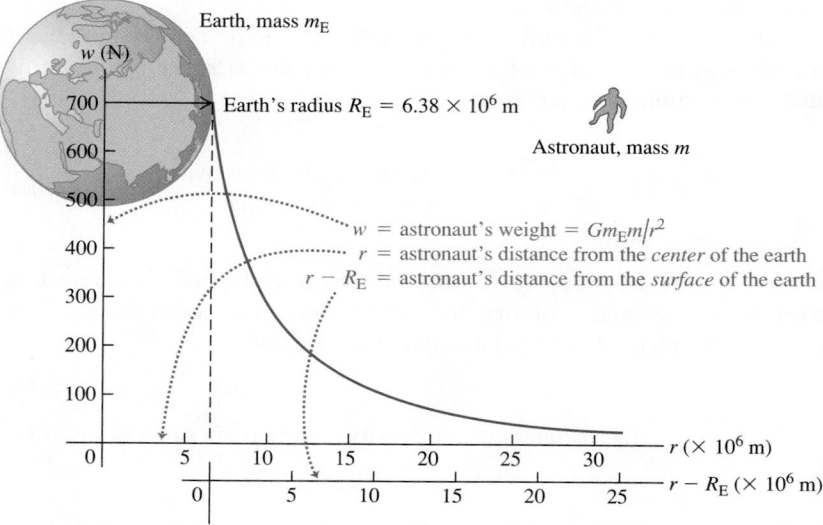

Earth, mass m_E

Earth's radius $R_E = 6.38 \times 10^6$ m

Astronaut, mass m

w = astronaut's weight = $Gm_E m/r^2$
r = astronaut's distance from the *center* of the earth
$r - R_E$ = astronaut's distance from the *surface* of the earth

13.9 The density of the earth decreases with increasing distance from its center.

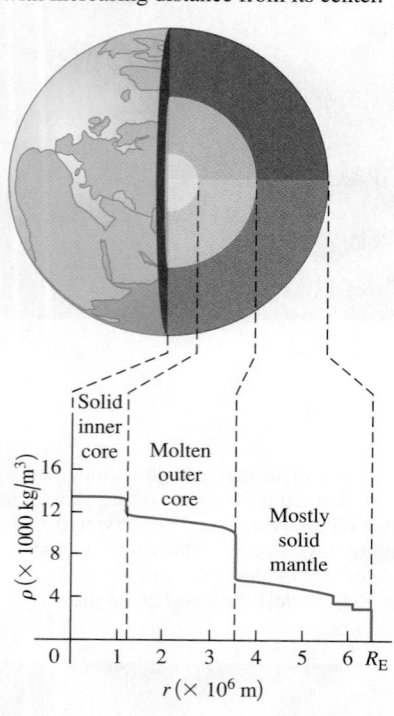

Solid inner core
Molten outer core
Mostly solid mantle

The average density ρ (the Greek letter rho) of the earth is the total mass divided by the total volume:

$$\rho = \frac{m_E}{V_E} = \frac{5.97 \times 10^{24}\ \text{kg}}{1.09 \times 10^{21}\ \text{m}^3}$$

$$= 5500\ \text{kg/m}^3 = 5.5\ \text{g/cm}^3$$

(For comparison, the density of water is 1000 kg/m³ = 1.00 g/cm³.) If the earth were uniform, we would expect rocks near the earth's surface to have this same density. In fact, the density of surface rocks is substantially lower, ranging from about 2000 kg/m³ for sedimentary rocks to about 3300 kg/m³ for basalt. So the earth *cannot* be uniform, and the interior of the earth must be much more dense than the surface in order that the *average* density be 5500 kg/m³. According to geophysical models of the earth's interior, the maximum density at the center is about 13,000 kg/m³. Figure 13.9 is a graph of density as a function of distance from the center.

Example 13.4   **Gravity on Mars**

A robotic lander with an earth weight of 3430 N is sent to Mars, which has radius $R_M = 3.40 \times 10^6$ m and mass $m_M = 6.42 \times 10^{23}$ kg (see Appendix F). Find the weight F_g of the lander on the Martian surface and the acceleration there due to gravity, g_M.

SOLUTION

IDENTIFY and SET UP: To find F_g we use Eq. (13.3), replacing m_E and R_E with m_M and R_M. We determine the lander mass m from the lander's earth weight w and then find g_M from $F_g = mg_M$.

EXECUTE: The lander's earth weight is $w = mg$, so

$$m = \frac{w}{g} = \frac{3430\ \text{N}}{9.80\ \text{m/s}^2} = 350\ \text{kg}$$

The mass is the same no matter where the lander is. From Eq. (13.3), the lander's weight on Mars is

$$F_g = \frac{Gm_M m}{R_M^2}$$

$$= \frac{(6.67 \times 10^{-11}\ \text{N} \cdot \text{m}^2/\text{kg}^2)(6.42 \times 10^{23}\ \text{kg})(350\ \text{kg})}{(3.40 \times 10^6\ \text{m})^2}$$

$$= 1.30 \times 10^3\ \text{N}$$

The acceleration due to gravity on Mars is

$$g_M = \frac{F_g}{m} = \frac{1.30 \times 10^3 \text{ N}}{350 \text{ kg}} = 3.7 \text{ m/s}^2$$

EVALUATE: Even though Mars has just 11% of the earth's mass $(6.42 \times 10^{23} \text{ kg}$ versus $5.98 \times 10^{24} \text{ kg})$, the acceleration due to

gravity g_M (and hence an object's weight F_g) is roughly 40% as large as on earth. That's because g_M is also inversely proportional to the square of the planet's radius, and Mars has only 53% the radius of earth $(3.40 \times 10^6 \text{ m}$ versus $6.38 \times 10^6 \text{ m})$.

You can check our result for g_M by using Eq. (13.4), with appropriate replacements. Do you get the same answer?

Test Your Understanding of Section 13.2 Rank the following hypothetical planets in order from highest to lowest value of g at the surface: (i) mass = 2 times the mass of the earth, radius = 2 times the radius of the earth; (ii) mass = 4 times the mass of the earth, radius = 4 times the radius of the earth; (iii) mass = 4 times the mass of the earth, radius = 2 times the radius of the earth; (iv) mass = 2 times the mass of the earth, radius = 4 times the radius of the earth.

13.3 Gravitational Potential Energy

When we first introduced gravitational potential energy in Section 7.1, we assumed that the gravitational force on a body is constant in magnitude and direction. This led to the expression $U = mgy$. But the earth's gravitational force on a body of mass m at any point outside the earth is given more generally by Eq. (13.2), $F_g = Gm_E m/r^2$, where m_E is the mass of the earth and r is the distance of the body from the earth's center. For problems in which r changes enough that the gravitational force can't be considered constant, we need a more general expression for gravitational potential energy.

To find this expression, we follow the same steps as in Section 7.1. We consider a body of mass m outside the earth, and first compute the work W_{grav} done by the gravitational force when the body moves directly away from or toward the center of the earth from $r = r_1$ to $r = r_2$, as in Fig. 13.10. This work is given by

$$W_{grav} = \int_{r_1}^{r_2} F_r \, dr \tag{13.6}$$

where F_r is the radial component of the gravitational force $\vec{F}$—that is, the component in the direction *outward* from the center of the earth. Because $\vec{F}$ points directly *inward* toward the center of the earth, F_r is negative. It differs from Eq. (13.2), the magnitude of the gravitational force, by a minus sign:

$$F_r = -\frac{Gm_E m}{r^2} \tag{13.7}$$

Substituting Eq. (13.7) into Eq. (13.6), we see that W_{grav} is given by

$$W_{grav} = -Gm_E m \int_{r_1}^{r_2} \frac{dr}{r^2} = \frac{Gm_E m}{r_2} - \frac{Gm_E m}{r_1} \tag{13.8}$$

The path doesn't have to be a straight line; it could also be a curve like the one in Fig. 13.10. By an argument similar to that in Section 7.1, this work depends only on the initial and final values of r, not on the path taken. This also proves that the gravitational force is always *conservative*.

We now define the corresponding potential energy U so that $W_{grav} = U_1 - U_2$, as in Eq. (7.3). Comparing this with Eq. (13.8), we see that the appropriate definition for **gravitational potential energy** is

$$U = -\frac{Gm_E m}{r} \qquad \text{(gravitational potential energy)} \tag{13.9}$$

13.10 Calculating the work done on a body by the gravitational force as the body moves from radial coordinate r_1 to r_2.

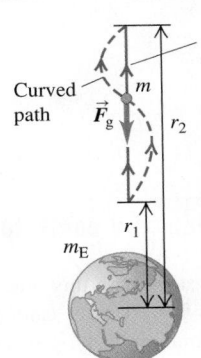

Straight path

Curved path

The gravitational force is conservative: The work done by $\vec{F}_g$ does not depend on the path taken from r_1 to r_2.

13.11 A graph of the gravitational potential energy U for the system of the earth (mass m_E) and an astronaut (mass m) versus the astronaut's distance r from the center of the earth.

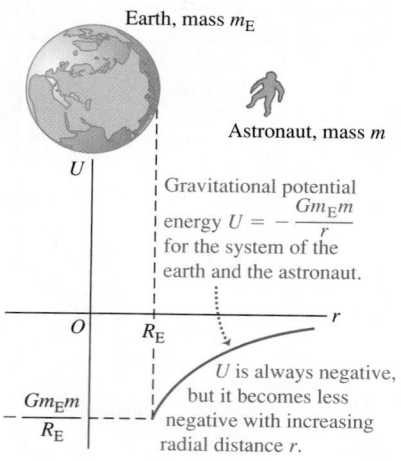

Earth, mass m_E

Astronaut, mass m

Gravitational potential energy $U = -\dfrac{Gm_E m}{r}$ for the system of the earth and the astronaut.

U is always negative, but it becomes less negative with increasing radial distance r.

Figure 13.11 shows how the gravitational potential energy depends on the distance r between the body of mass m and the center of the earth. When the body moves away from the earth, r increases, the gravitational force does negative work, and U increases (i.e., becomes less negative). When the body "falls" toward earth, r decreases, the gravitational work is positive, and the potential energy decreases (i.e., becomes more negative).

You may be troubled by Eq. (13.9) because it states that gravitational potential energy is always negative. But in fact you've seen negative values of U before. In using the formula $U = mgy$ in Section 7.1, we found that U was negative whenever the body of mass m was at a value of y below the arbitrary height we chose to be $y = 0$—that is, whenever the body and the earth were closer together than some certain arbitrary distance. (See, for instance, Example 7.2 in Section 7.1.) In defining U by Eq. (13.9), we have chosen U to be zero when the body of mass m is infinitely far from the earth ($r = \infty$). As the body moves toward the earth, gravitational potential energy decreases and so becomes negative.

If we wanted, we could make $U = 0$ at the surface of the earth, where $r = R_E$, by simply adding the quantity $Gm_E m/R_E$ to Eq. (13.9). This would make U positive when $r > R_E$. We won't do this for two reasons: One, it would make the expression for U more complicated; and two, the added term would not affect the *difference* in potential energy between any two points, which is the only physically significant quantity.

CAUTION Gravitational force vs. gravitational potential energy Be careful not to confuse the expressions for gravitational force, Eq. (13.7), and gravitational potential energy, Eq. (13.9). The force F_r is proportional to $1/r^2$, while potential energy U is proportional to $1/r$. ▮

Armed with Eq. (13.9), we can now use general energy relationships for problems in which the $1/r^2$ behavior of the earth's gravitational force has to be included. If the gravitational force on the body is the only force that does work, the total mechanical energy of the system is constant, or *conserved*. In the following example we'll use this principle to calculate **escape speed,** the speed required for a body to escape completely from a planet.

Example 13.5 **"From the earth to the moon"**

In Jules Verne's 1865 story with this title, three men went to the moon in a shell fired from a giant cannon sunk in the earth in Florida. (a) Find the minimum muzzle speed needed to shoot a shell straight up to a height above the earth equal to the earth's radius R_E. (b) Find the minimum muzzle speed that would allow a shell to escape from the earth completely (the *escape speed*). Neglect air resistance, the earth's rotation, and the gravitational pull of the moon. The earth's radius and mass are $R_E = 6.38 \times 10^6$ m and $m_E = 5.97 \times 10^{24}$ kg.

SOLUTION

IDENTIFY and SET UP: Once the shell leaves the cannon muzzle, only the (conservative) gravitational force does work. Hence we can use conservation of mechanical energy to find the speed at which the shell must leave the muzzle so as to come to a halt (a) at two earth radii from the earth's center and (b) at an infinite distance from earth. The energy-conservation equation is $K_1 + U_1 = K_2 + U_2$, with U given by Eq. (13.9).

Figure 13.12 shows our sketches. Point 1 is at $r_1 = R_E$, where the shell leaves the cannon with speed v_1 (the target variable). Point 2 is where the shell reaches its maximum height; in part

13.12 Our sketches for this problem.

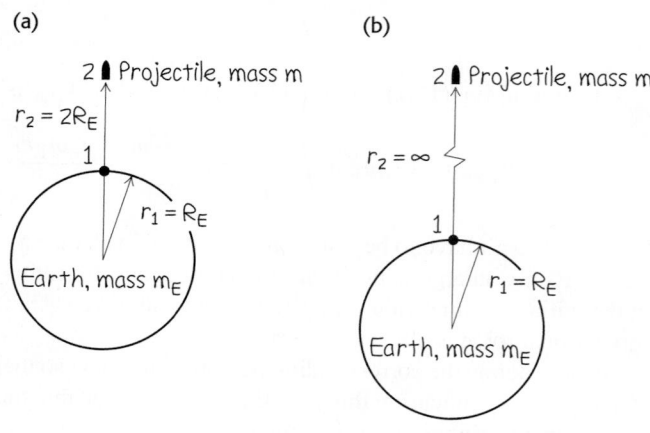

(a) $r_2 = 2R_E$ (Fig. 13.12a), and in part (b) $r_2 = \infty$ (Fig 13.12b). In both cases $v_2 = 0$ and $K_2 = 0$. Let m be the mass of the shell (with passengers).

EXECUTE: (a) We solve the energy-conservation equation for v_1:

$$K_1 + U_1 = K_2 + U_2$$

$$\tfrac{1}{2}mv_1^2 + \left(-\frac{Gm_Em}{R_E}\right) = 0 + \left(-\frac{Gm_Em}{2R_E}\right)$$

$$v_1 = \sqrt{\frac{Gm_E}{R_E}} = \sqrt{\frac{(6.67 \times 10^{-11}\ \text{N}\cdot\text{m}^2/\text{kg}^2)(5.97 \times 10^{24}\ \text{kg})}{6.38 \times 10^6\ \text{m}}}$$

$$= 7900\ \text{m/s}\ (= 28{,}400\ \text{km/h} = 17{,}700\ \text{mi/h})$$

(b) Now $r_2 = \infty$ so $U_2 = 0$ (see Fig. 13.11). Since $K_2 = 0$, the total mechanical energy $K_2 + U_2$ is zero in this case. Again we solve the energy-conservation equation for v_1:

$$\tfrac{1}{2}mv_1^2 + \left(-\frac{Gm_Em}{R_E}\right) = 0 + 0$$

$$v_1 = \sqrt{\frac{2Gm_E}{R_E}}$$

$$= \sqrt{\frac{2(6.67 \times 10^{-11}\ \text{N}\cdot\text{m}^2/\text{kg}^2)(5.97 \times 10^{24}\ \text{kg})}{6.38 \times 10^6\ \text{m}}}$$

$$= 1.12 \times 10^4\ \text{m/s}\ (= 40{,}200\ \text{km/h} = 25{,}000\ \text{mi/h})$$

EVALUATE: Our result in part (b) doesn't depend on the mass of the shell or the direction of launch. A modern spacecraft launched from Florida must attain essentially the speed found in part (b) to escape the earth; however, before launch it's already moving at 410 m/s to the east because of the earth's rotation. Launching to the east takes advantage of this "free" contribution toward escape speed.

To generalize, the initial speed v_1 needed for a body to escape from the surface of a spherical body of mass M and radius R (ignoring air resistance) is $v_1 = \sqrt{2GM/R}$ (escape speed). This equation yields escape speeds of 5.02×10^3 m/s for Mars, 5.95×10^4 m/s for Jupiter, and 6.18×10^5 m/s for the sun.

More on Gravitational Potential Energy

As a final note, let's show that when we are close to the earth's surface, Eq. (13.9) reduces to the familiar $U = mgy$ from Chapter 7. We first rewrite Eq. (13.8) as

$$W_{\text{grav}} = Gm_Em\frac{r_1 - r_2}{r_1r_2}$$

If the body stays close to the earth, then in the denominator we may replace r_1 and r_2 by R_E, the earth's radius, so

$$W_{\text{grav}} = Gm_Em\frac{r_1 - r_2}{R_E^2}$$

According to Eq. (13.4), $g = Gm_E/R_E^2$, so

$$W_{\text{grav}} = mg(r_1 - r_2)$$

If we replace the r's by y's, this is just Eq. (7.1) for the work done by a constant gravitational force. In Section 7.1 we used this equation to derive Eq. (7.2), $U = mgy$, so we may consider Eq. (7.2) for gravitational potential energy to be a special case of the more general Eq. (13.9).

Test Your Understanding of Section 13.3 Is it possible for a planet to have the same surface gravity as the earth (that is, the same value of g at the surface) and yet have a greater escape speed?

13.4 The Motion of Satellites

Artificial satellites orbiting the earth are a familiar part of modern technology (Fig. 13.13). But how do they stay in orbit, and what determines the properties of their orbits? We can use Newton's laws and the law of gravitation to provide the answers. We'll see in the next section that the motion of planets can be analyzed in the same way.

To begin, think back to the discussion of projectile motion in Section 3.3. In Example 3.6 a motorcycle rider rides horizontally off the edge of a cliff, launching himself into a parabolic path that ends on the flat ground at the base of the cliff. If he survives and repeats the experiment with increased launch speed, he will land farther from the starting point. We can imagine him launching himself with great enough speed that the earth's curvature becomes significant. As he falls, the earth curves away beneath him. If he is going fast enough, and if his

13.13 With a length of 13.2 m and a mass of 11,000 kg, the Hubble Space Telescope is among the largest satellites placed in orbit.

13.14 Trajectories of a projectile launched from a great height (ignoring air resistance). Orbits 1 and 2 would be completed as shown if the earth were a point mass at *C*. (This illustration is based on one in Isaac Newton's *Principia*.)

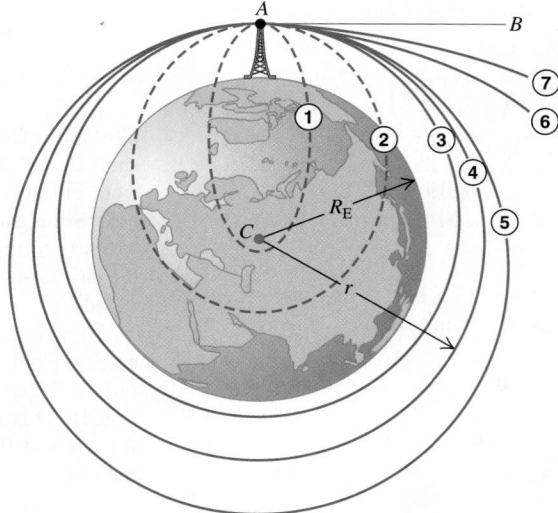

A projectile is launched from *A* toward *B*. Trajectories ① through ⑦ show the effect of increasing initial speed.

13.15 The force $\vec{F}_g$ due to the earth's gravitational attraction provides the centripetal acceleration that keeps a satellite in orbit. Compare to Fig. 5.28.

The satellite is in a circular orbit: Its acceleration $\vec{a}$ is always perpendicular to its velocity $\vec{v}$, so its speed v is constant.

launch point is high enough that he clears the mountaintops, he may be able to go right on around the earth without ever landing.

Figure 13.14 shows a variation on this theme. We launch a projectile from point *A* in the direction *AB*, tangent to the earth's surface. Trajectories 1 through 7 show the effect of increasing the initial speed. In trajectories 3 through 5 the projectile misses the earth and becomes a satellite. If there is no retarding force, the projectile's speed when it returns to point *A* is the same as its initial speed and it repeats its motion indefinitely.

Trajectories 1 through 5 close on themselves and are called **closed orbits.** All closed orbits are ellipses or segments of ellipses; trajectory 4 is a circle, a special case of an ellipse. (We'll discuss the properties of an ellipse in Section 13.5.) Trajectories 6 and 7 are **open orbits.** For these paths the projectile never returns to its starting point but travels ever farther away from the earth.

Satellites: Circular Orbits

A *circular* orbit, like trajectory 4 in Fig. 13.14, is the simplest case. It is also an important case, since many artificial satellites have nearly circular orbits and the orbits of the planets around the sun are also fairly circular. The only force acting on a satellite in circular orbit around the earth is the earth's gravitational attraction, which is directed toward the center of the earth and hence toward the center of the orbit (Fig. 13.15). As we discussed in Section 5.4, this means that the satellite is in *uniform* circular motion and its speed is constant. The satellite isn't falling *toward* the earth; rather, it's constantly falling *around* the earth. In a circular orbit the speed is just right to keep the distance from the satellite to the center of the earth constant.

Let's see how to find the constant speed v of a satellite in a circular orbit. The radius of the orbit is r, measured from the *center* of the earth; the acceleration of the satellite has magnitude $a_{\text{rad}} = v^2/r$ and is always directed toward the center of the circle. By the law of gravitation, the net force (gravitational force) on the satellite of mass m has magnitude $F_g = Gm_{\text{E}}m/r^2$ and is in the same direction as the acceleration. Newton's second law ($\sum\vec{F} = m\vec{a}$) then tells us that

$$\frac{Gm_{\text{E}}m}{r^2} = \frac{mv^2}{r}$$

Solving this for v, we find

$$v = \sqrt{\frac{Gm_{\text{E}}}{r}} \qquad \text{(circular orbit)} \qquad (13.10)$$

This relationship shows that we can't choose the orbit radius r and the speed v independently; for a given radius r, the speed v for a circular orbit is determined.

The satellite's mass m doesn't appear in Eq. (13.10), which shows that the motion of a satellite does not depend on its mass. If we could cut a satellite in half without changing its speed, each half would continue on with the original motion. An astronaut on board a space shuttle is herself a satellite of the earth, held by the earth's gravitational attraction in the same orbit as the shuttle. The astronaut has the same velocity and acceleration as the shuttle, so nothing is pushing her against the floor or walls of the shuttle. She is in a state of *apparent weightlessness,* as in a freely falling elevator; see the discussion following Example 5.9 in Section 5.2. (*True* weightlessness would occur only if the astronaut were infinitely far from any other masses, so that the gravitational force on her would be zero.) Indeed, every part of her body is apparently weightless; she feels nothing pushing her stomach against her intestines or her head against her shoulders (Fig. 13.16).

Apparent weightlessness is not just a feature of circular orbits; it occurs whenever gravity is the only force acting on a spacecraft. Hence it occurs for orbits of any shape, including open orbits such as trajectories 6 and 7 in Fig. 13.14.

We can derive a relationship between the radius r of a circular orbit and the period T, the time for one revolution. The speed v is the distance $2\pi r$ traveled in one revolution, divided by the period:

$$v = \frac{2\pi r}{T} \tag{13.11}$$

To get an expression for T, we solve Eq. (13.11) for T and substitute v from Eq. (13.10):

$$T = \frac{2\pi r}{v} = 2\pi r\sqrt{\frac{r}{Gm_E}} = \frac{2\pi r^{3/2}}{\sqrt{Gm_E}} \quad \text{(circular orbit)} \tag{13.12}$$

Equations (13.10) and (13.12) show that larger orbits correspond to slower speeds and longer periods. As an example, the International Space Station orbits 6800 km from the center of the earth (400 km above the earth's surface) with an orbital speed of 7.7 km/s and an orbital period of 93 minutes. The moon orbits the earth in a much larger orbit of radius 384,000 km, and so has a much slower orbital speed (1.0 km/s) and a much longer orbital period (27.3 days).

It's interesting to compare Eq. (13.10) to the calculation of escape speed in Example 13.5. We see that the escape speed from a spherical body with radius R is $\sqrt{2}$ times greater than the speed of a satellite in a circular orbit at that radius. If our spacecraft is in circular orbit around *any* planet, we have to multiply our speed by a factor of $\sqrt{2}$ to escape to infinity, regardless of the planet's mass.

Since the speed v in a circular orbit is determined by Eq. (13.10) for a given orbit radius r, the total mechanical energy $E = K + U$ is determined as well. Using Eqs. (13.9) and (13.10), we have

$$E = K + U = \tfrac{1}{2}mv^2 + \left(-\frac{Gm_E m}{r}\right) = \tfrac{1}{2}m\left(\frac{Gm_E}{r}\right) - \frac{Gm_E m}{r}$$

$$= -\frac{Gm_E m}{2r} \quad \text{(circular orbit)} \tag{13.13}$$

The total mechanical energy in a circular orbit is negative and equal to one-half the potential energy. Increasing the orbit radius r means increasing the mechanical energy (that is, making E less negative). If the satellite is in a relatively low orbit that encounters the outer fringes of earth's atmosphere, mechanical energy decreases due to negative work done by the force of air resistance; as a result, the orbit radius decreases until the satellite hits the ground or burns up in the atmosphere.

We have talked mostly about earth satellites, but we can apply the same analysis to the circular motion of *any* body under its gravitational attraction to a stationary body. Other examples include the earth's moon and the moons of other worlds (Fig. 13.17).

13.16 These space shuttle astronauts are in a state of apparent weightlessness. Which are right side up and which are upside down?

13.17 The two small satellites of the minor planet Pluto were discovered in 2005. In accordance with Eqs. (13.10) and (13.12), the satellite in the larger orbit has a slower orbital speed and a longer orbital period than the satellite in the smaller orbit.

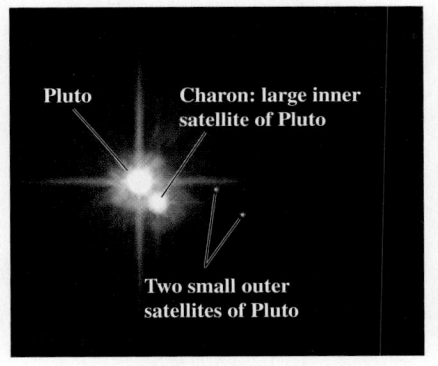

Pluto

Charon: large inner satellite of Pluto

Two small outer satellites of Pluto

Example 13.6 **A satellite orbit**

You wish to put a 1000-kg satellite into a circular orbit 300 km above the earth's surface. (a) What speed, period, and radial acceleration will it have? (b) How much work must be done to the satellite to put it in orbit? (c) How much additional work would have to be done to make the satellite escape the earth? The earth's radius and mass are given in Example 13.5 (Section 13.3).

SOLUTION

IDENTIFY and SET UP: The satellite is in a circular orbit, so we can use the equations derived in this section. In part (a), we first find the radius r of the satellite's orbit from its altitude. We then calculate the speed v and period T using Eqs. (13.10) and (13.12) and the acceleration from $a_{rad} = v^2/r$. In parts (b) and (c), the work required is the difference between the initial and final mechanical energy, which for a circular orbit is given by Eq. (13.13).

EXECUTE: (a) The radius of the satellite's orbit is $r = 6380$ km + 300 km = 6680 km = 6.68×10^6 m. From Eq. (13.10), the orbital speed is

$$v = \sqrt{\frac{Gm_E}{r}} = \sqrt{\frac{(6.67 \times 10^{-11}\,\text{N} \cdot \text{m}^2/\text{kg}^2)(5.97 \times 10^{24}\,\text{kg})}{6.68 \times 10^6\,\text{m}}}$$

$$= 7720\,\text{m/s}$$

We find the orbital period from Eq. (13.12):

$$T = \frac{2\pi r}{v} = \frac{2\pi(6.68 \times 10^6\,\text{m})}{7720\,\text{m/s}} = 5440\,\text{s} = 90.6\,\text{min}$$

Finally, the radial acceleration is

$$a_{rad} = \frac{v^2}{r} = \frac{(7720\,\text{m/s})^2}{6.68 \times 10^6\,\text{m}} = 8.92\,\text{m/s}^2$$

This is the value of g at a height of 300 km above the earth's surface; it is about 10% less than the value of g at the surface.

(b) The work required is the difference between E_2, the total mechanical energy when the satellite is in orbit, and E_1, the total mechanical energy when the satellite was at rest on the launch pad. From Eq. (13.13), the energy in orbit is

$$E_2 = -\frac{Gm_E m}{2r}$$

$$= -\frac{(6.67 \times 10^{-11}\,\text{N} \cdot \text{m}^2/\text{kg}^2)(5.97 \times 10^{24}\,\text{kg})(1000\,\text{kg})}{2(6.68 \times 10^6\,\text{m})}$$

$$= -2.98 \times 10^{10}\,\text{J}$$

The satellite's kinetic energy is zero on the launch pad $(r = R_E)$, so

$$E_1 = K_1 + U_1 = 0 + \left(-\frac{Gm_E m}{R_E}\right)$$

$$= -\frac{(6.67 \times 10^{-11}\,\text{N} \cdot \text{m}^2/\text{kg}^2)(5.97 \times 10^{24}\,\text{kg})(1000\,\text{kg})}{6.38 \times 10^6\,\text{m}}$$

$$= -6.24 \times 10^{10}\,\text{J}$$

Hence the work required is

$$W_{required} = E_2 - E_1 = (-2.98 \times 10^{10}\,\text{J}) - (-6.24 \times 10^{10}\,\text{J})$$

$$= 3.26 \times 10^{10}\,\text{J}$$

(c) We saw in part (b) of Example 13.5 that the minimum total mechanical energy for a satellite to escape to infinity is zero. Here, the total mechanical energy in the circular orbit is $E_2 = -2.98 \times 10^{10}\,\text{J}$; to increase this to zero, an amount of work equal to $2.98 \times 10^{10}\,\text{J}$ would have to be done on the satellite, presumably by rocket engines attached to it.

EVALUATE: In part (b) we ignored the satellite's initial kinetic energy (while it was still on the launch pad) due to the rotation of the earth. How much difference does this make? (See Example 13.5 for useful data.)

Test Your Understanding of Section 13.4 Your personal spacecraft is in a low-altitude circular orbit around the earth. Air resistance from the outer regions of the atmosphere does negative work on the spacecraft, causing the orbital radius to decrease slightly. Does the speed of the spacecraft (i) remain the same, (ii) increase, or (iii) decrease?

(MP)

13.5 Kepler's Laws and the Motion of Planets

The name *planet* comes from a Greek word meaning "wanderer," and indeed the planets continuously change their positions in the sky relative to the background of stars. One of the great intellectual accomplishments of the 16th and 17th centuries was the threefold realization that the earth is also a planet, that all planets orbit the sun, and that the apparent motions of the planets as seen from the earth can be used to precisely determine their orbits.

The first and second of these ideas were published by Nicolaus Copernicus in Poland in 1543. The nature of planetary orbits was deduced between 1601 and 1619 by the German astronomer and mathematician Johannes Kepler, using a voluminous set of precise data on apparent planetary motions compiled by his mentor, the Danish astronomer Tycho Brahe. By trial and error, Kepler discovered three empirical laws that accurately described the motions of the planets:

1. **Each planet moves in an elliptical orbit, with the sun at one focus of the ellipse.**
2. **A line from the sun to a given planet sweeps out equal areas in equal times.**
3. **The periods of the planets are proportional to the $\frac{3}{2}$ powers of the major axis lengths of their orbits.**

Kepler did not know *why* the planets moved in this way. Three generations later, when Newton turned his attention to the motion of the planets, he discovered that each of Kepler's laws can be *derived;* they are consequences of Newton's laws of motion and the law of gravitation. Let's see how each of Kepler's laws arises.

Kepler's First Law

First consider the elliptical orbits described in Kepler's first law. Figure 13.18 shows the geometry of an ellipse. The longest dimension is the *major axis,* with half-length a; this half-length is called the **semi-major axis.** The sum of the distances from S to P and from S' to P is the same for all points on the curve. S and S' are the *foci* (plural of *focus*). The sun is at S, and the planet is at P; we think of them both as points because the size of each is very small in comparison to the distance between them. There is nothing at the other focus S'.

The distance of each focus from the center of the ellipse is ea, where e is a dimensionless number between 0 and 1 called the **eccentricity.** If $e = 0$, the ellipse is a circle. The actual orbits of the planets are fairly circular; their eccentricities range from 0.007 for Venus to 0.206 for Mercury. (The earth's orbit has $e = 0.017$.) The point in the planet's orbit closest to the sun is the *perihelion,* and the point most distant from the sun is the *aphelion.*

Newton was able to show that for a body acted on by an attractive force proportional to $1/r^2$, the only possible closed orbits are a circle or an ellipse; he also showed that open orbits (trajectories 6 and 7 in Fig. 13.14) must be parabolas or hyperbolas. These results can be derived by a straightforward application of Newton's laws and the law of gravitation, together with a lot more differential equations than we're ready for.

Kepler's Second Law

Figure 13.19 shows Kepler's second law. In a small time interval dt, the line from the sun S to the planet P turns through an angle $d\theta$. The area swept out is the colored triangle with height r, base length $r\,d\theta$, and area $dA = \frac{1}{2}r^2\,d\theta$ in Fig. 13.19b. The rate at which area is swept out, dA/dt, is called the *sector velocity:*

$$\frac{dA}{dt} = \frac{1}{2}r^2\frac{d\theta}{dt} \tag{13.14}$$

The essence of Kepler's second law is that the sector velocity has the same value at all points in the orbit. When the planet is close to the sun, r is small and $d\theta/dt$ is large; when the planet is far from the sun, r is large and $d\theta/dt$ is small.

To see how Kepler's second law follows from Newton's laws, we express dA/dt in terms of the velocity vector $\vec{v}$ of the planet P. The component of $\vec{v}$ perpendicular to the radial line is $v_\perp = v\sin\phi$. From Fig. 13.19b the displacement along the direction of $v_\perp$ during time dt is $r\,d\theta$, so we also have $v_\perp = r\,d\theta/dt$. Using this relationship in Eq. (13.14), we find

$$\frac{dA}{dt} = \frac{1}{2}rv\sin\phi \qquad \text{(sector velocity)} \tag{13.15}$$

13.18 Geometry of an ellipse. The sum of the distances SP and $S'P$ is the same for every point on the curve. The sizes of the sun (S) and planet (P) are exaggerated for clarity.

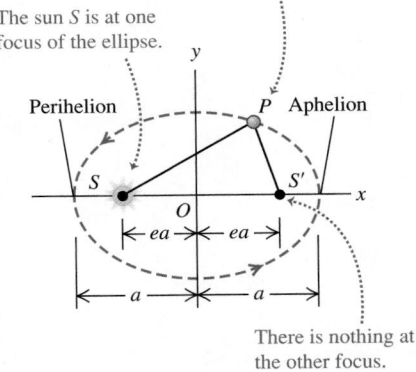

13.19 (a) The planet (P) moves about the sun (S) in an elliptical orbit. (b) In a time dt the line SP sweeps out an area $dA = \frac{1}{2}(r\,d\theta)r = \frac{1}{2}r^2\,d\theta$. (c) The planet's speed varies so that the line SP sweeps out the same area A in a given time t regardless of the planet's position in its orbit.

(a)

(b)

(c)
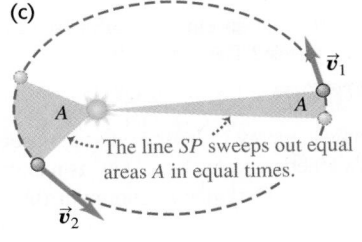

Now $rv\sin\phi$ is the magnitude of the vector product $\vec{r} \times \vec{v}$, which in turn is $1/m$ times the angular momentum $\vec{L} = \vec{r} \times m\vec{v}$ of the planet with respect to the sun. So we have

$$\frac{dA}{dt} = \frac{1}{2m}|\vec{r} \times m\vec{v}| = \frac{L}{2m} \qquad (13.16)$$

Thus Kepler's second law—that sector velocity is constant—means that angular momentum is constant!

It is easy to see why the angular momentum of the planet *must* be constant. According to Eq. (10.26), the rate of change of $\vec{L}$ equals the torque of the gravitational force $\vec{F}$ acting on the planet:

$$\frac{d\vec{L}}{dt} = \vec{\tau} = \vec{r} \times \vec{F}$$

In our situation, $\vec{r}$ is the vector from the sun to the planet, and the force $\vec{F}$ is directed from the planet to the sun. So these vectors always lie along the same line, and their vector product $\vec{r} \times \vec{F}$ is zero. Hence $d\vec{L}/dt = \mathbf{0}$. This conclusion does not depend on the $1/r^2$ behavior of the force; angular momentum is conserved for *any* force that acts always along the line joining the particle to a fixed point. Such a force is called a *central force*. (Kepler's first and third laws are valid *only* for a $1/r^2$ force.)

Conservation of angular momentum also explains why the orbit lies in a plane. The vector $\vec{L} = \vec{r} \times m\vec{v}$ is always perpendicular to the plane of the vectors $\vec{r}$ and $\vec{v}$; since $\vec{L}$ is constant in magnitude *and* direction, $\vec{r}$ and $\vec{v}$ always lie in the same plane, which is just the plane of the planet's orbit.

Kepler's Third Law

We have already derived Kepler's third law for the particular case of circular orbits. Equation (13.12) shows that the period of a satellite or planet in a circular orbit is proportional to the $\frac{3}{2}$ power of the orbit radius. Newton was able to show that this same relationship holds for an *elliptical* orbit, with the orbit radius r replaced by the semi-major axis a:

$$T = \frac{2\pi a^{3/2}}{\sqrt{Gm_S}} \qquad \text{(elliptical orbit around the sun)} \qquad (13.17)$$

Since the planet orbits the sun, not the earth, we have replaced the earth's mass m_E in Eq. (13.12) with the sun's mass m_S. Note that the period does not depend on the eccentricity e. An asteroid in an elongated elliptical orbit with semi-major axis a will have the same orbital period as a planet in a circular orbit of radius a. The key difference is that the asteroid moves at different speeds at different points in its elliptical orbit (Fig. 13.19c), while the planet's speed is constant around its circular orbit.

Application Biological Hazards of Interplanetary Travel

A spacecraft sent from earth to another planet spends most of its journey coasting along an elliptical orbit with the sun at one focus. Rockets are used only at the start and end of the journey, and even the trip to a nearby planet like Mars takes several months. During its journey, the spacecraft is exposed to cosmic rays—radiation that emanates from elsewhere in our galaxy. (On earth we're shielded from this radiation by our planet's magnetic field, as we'll describe in Chapter 27.) This poses no problem for a robotic spacecraft, but would be a severe medical hazard for astronauts undertaking such a voyage.

| **Conceptual Example 13.7** | **Orbital speeds** |

At what point in an elliptical orbit (see Fig. 13.19) does a planet move the fastest? The slowest?

SOLUTION

Mechanical energy is conserved as a planet moves in its orbit. The planet's kinetic energy $K = \frac{1}{2}mv^2$ is maximum when the potential energy $U = -Gm_S m/r$ is minimum (that is, most negative; see Fig. 13.11), which occurs when the sun–planet distance r is a minimum. Hence the speed v is greatest at perihelion. Similarly, K is minimum when r is maximum, so the speed is slowest at aphelion.

Your intuition about falling bodies is helpful here. As the planet falls inward toward the sun, it picks up speed, and its speed is maximum when closest to the sun. The planet slows down as it moves away from the sun, and its speed is minimum at aphelion.

Example 13.8 **Kepler's third law**

The asteroid Pallas has an orbital period of 4.62 years and an orbital eccentricity of 0.233. Find the semi-major axis of its orbit.

SOLUTION

IDENTIFY and SET UP: This example uses Kepler's third law, which relates the period T and the semi-major axis a for an orbiting object (such as an asteroid). We use Eq. (13.17) to determine a; from Appendix F we have $m_S = 1.99 \times 10^{30}$ kg, and a conversion factor from Appendix E gives $T = (4.62 \text{ yr})(3.156 \times 10^7 \text{ s/yr}) = 1.46 \times 10^8$ s. Note that we don't need the value of the eccentricity.

EXECUTE: From Eq. (13.17), $a^{3/2} = [(Gm_S)^{1/2}T]/2\pi$. To solve for a, we raise both sides of this expression to the $\frac{2}{3}$ power and then substitute the values of G, m_S, and T:

$$a = \left(\frac{Gm_S T^2}{4\pi^2} \right)^{1/3} = 4.15 \times 10^{11} \text{ m}$$

(Plug in the numbers yourself to check.)

EVALUATE: Our result is intermediate between the semi-major axes of Mars and Jupiter (see Appendix F). Most known asteroids orbit in an "asteroid belt" between the orbits of these two planets.

Example 13.9 **Comet Halley**

Comet Halley moves in an elongated elliptical orbit around the sun (Fig. 13.20). Its distances from the sun at perihelion and aphelion are 8.75×10^7 km and 5.26×10^9 km, respectively. Find the orbital semi-major axis, eccentricity, and period.

SOLUTION

IDENTIFY and SET UP: We are to find the semi-major axis a, eccentricity e, and orbital period T. We can use Fig. 13.18 to find a and e from the given perihelion and aphelion distances. Knowing a, we can find T from Kepler's third law, Eq. (13.17).

EXECUTE: From Fig. 13.18, the length $2a$ of the major axis equals the sum of the comet–sun distance at perihelion and the comet–sun distance at aphelion. Hence

$$a = \frac{(8.75 \times 10^7 \text{ km}) + (5.26 \times 10^9 \text{ km})}{2} = 2.67 \times 10^9 \text{ km}$$

Figure 13.19 also shows that the comet–sun distance at perihelion is $a - ea = a(1 - e)$. This distance is 8.75×10^7 km, so

$$e = 1 - \frac{8.75 \times 10^7 \text{ km}}{a} = 1 - \frac{8.75 \times 10^7 \text{ km}}{2.67 \times 10^9 \text{ km}} = 0.967$$

From Eq. (13.17), the period is

$$T = \frac{2\pi a^{3/2}}{\sqrt{Gm_S}} = \frac{2\pi (2.67 \times 10^{12} \text{ m})^{3/2}}{\sqrt{(6.67 \times 10^{-11} \text{ N} \cdot \text{m}^2/\text{kg}^2)(1.99 \times 10^{30} \text{ kg})}}$$

$$= 2.38 \times 10^9 \text{ s} = 75.5 \text{ years}$$

EVALUATE: The eccentricity is close to 1, so the orbit is very elongated (see Fig. 13.20a). Comet Halley was at perihelion in early 1986 (Fig. 13.20b); it will next reach perihelion one period later, in 2061.

13.20 (a) The orbit of Comet Halley. (b) Comet Halley as it appeared in 1986. At the heart of the comet is an icy body, called the nucleus, that is about 10 km across. When the comet's orbit carries it close to the sun, the heat of sunlight causes the nucleus to partially evaporate. The evaporated material forms the tail, which can be tens of millions of kilometers long.

(a)

(b)

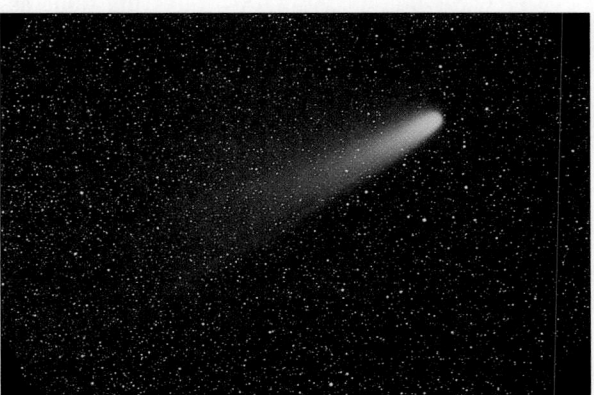

Planetary Motions and the Center of Mass

We have assumed that as a planet or comet orbits the sun, the sun remains absolutely stationary. Of course, this can't be correct; because the sun exerts a

13.21 A star and its planet both orbit about their common center of mass.

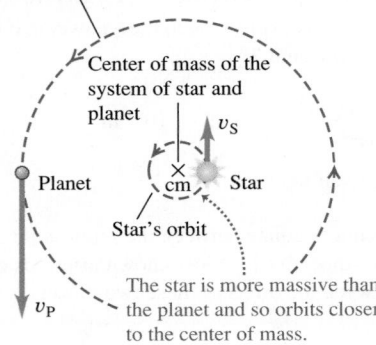

Planet's orbit around the center of mass

Center of mass of the system of star and planet

Planet

Star

Star's orbit

The star is more massive than the planet and so orbits closer to the center of mass.

The planet and star are always on opposite sides of the center of mass.

gravitational force on the planet, the planet exerts a gravitational force on the sun of the same magnitude but opposite direction. In fact, *both* the sun and the planet orbit around their common center of mass (Fig. 13.21). We've made only a small error by ignoring this effect, however; the sun's mass is about 750 times the total mass of all the planets combined, so the center of mass of the solar system is not far from the center of the sun. Remarkably, astronomers have used this effect to detect the presence of planets orbiting other stars. Sensitive telescopes are able to detect the apparent "wobble" of a star as it orbits the common center of mass of the star and an unseen companion planet. (The planets are too faint to observe directly.) By analyzing these "wobbles," astronomers have discovered planets in orbit around hundreds of other stars.

Newton's analysis of planetary motions is used on a daily basis by modern-day astronomers. But the most remarkable result of Newton's work is that the motions of bodies in the heavens obey the *same* laws of motion as do bodies on the earth. This *Newtonian synthesis*, as it has come to be called, is one of the great unifying principles of science. It has had profound effects on the way that humanity looks at the universe—not as a realm of impenetrable mystery, but as a direct extension of our everyday world, subject to scientific study and calculation.

Test Your Understanding of Section 13.5 The orbit of Comet X has a semi-major axis that is four times longer than the semi-major axis of Comet Y. What is the ratio of the orbital period of X to the orbital period of Y? (i) 2; (ii) 4; (iii) 8; (iv) 16; (v) 32; (vi) 64.

13.22 Calculating the gravitational potential energy of interaction between a point mass m outside a spherical shell and a ring on the surface of the shell.

(a) Geometry of the situation

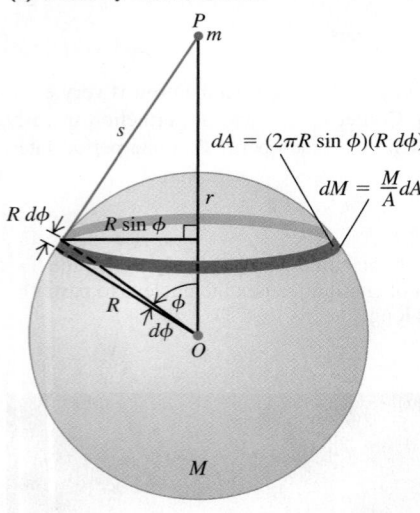

$dA = (2\pi R \sin \phi)(R \, d\phi)$

$dM = \dfrac{M}{A} dA$

(b) The distance s is the hypotenuse of a right triangle with sides $(r - R \cos \phi)$ and $R \sin \phi$.

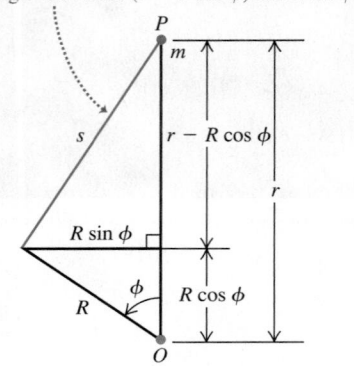

13.6 Spherical Mass Distributions

We have stated without proof that the gravitational interaction between two spherically symmetric mass distributions is the same as though all the mass of each were concentrated at its center. Now we're ready to prove this statement. Newton searched for a proof for several years, and he delayed publication of the law of gravitation until he found one.

Here's our program. Rather than starting with two spherically symmetric masses, we'll tackle the simpler problem of a point mass m interacting with a thin spherical shell with total mass M. We will show that when m is outside the sphere, the *potential energy* associated with this gravitational interaction is the same as though M were all concentrated at the center of the sphere. We learned in Section 7.4 that the force is the negative derivative of the potential energy, so the *force* on m is also the same as for a point mass M. Any spherically symmetric mass distribution can be thought of as being made up of many concentric spherical shells, so our result will also hold for *any* spherically symmetric M.

A Point Mass Outside a Spherical Shell

We start by considering a ring on the surface of the shell (Fig. 13.22a), centered on the line from the center of the shell to m. We do this because all of the particles that make up the ring are the same distance s from the point mass m. From Eq. (13.9) the potential energy of interaction between the earth (mass m_E) and a point mass m, separated by a distance r, is $U = -Gm_E m/r$. By changing notation in this expression, we see that the potential energy of interaction between the point mass m and a particle of mass m_i within the ring is given by

$$U_i = -\frac{Gmm_i}{s}$$

To find the potential energy of interaction between m and the entire ring of mass $dM = \sum_i m_i$, we sum this expression for U_i over all particles in the ring. Calling this potential energy dU, we find

$$dU = \sum_i U_i = \sum_i \left(-\frac{Gmm_i}{s} \right) = -\frac{Gm}{s} \sum_i m_i = -\frac{Gm\,dM}{s} \qquad (13.18)$$

To proceed, we need to know the mass dM of the ring. We can find this with the aid of a little geometry. The radius of the shell is R, so in terms of the angle ϕ shown in the figure, the radius of the ring is $R\sin\phi$, and its circumference is $2\pi R\sin\phi$. The width of the ring is $R\,d\phi$, and its area dA is approximately equal to its width times its circumference:

$$dA = 2\pi R^2 \sin\phi\,d\phi$$

The ratio of the ring mass dM to the total mass M of the shell is equal to the ratio of the area dA of the ring to the total area $A = 4\pi R^2$ of the shell:

$$\frac{dM}{M} = \frac{2\pi R^2 \sin\phi\,d\phi}{4\pi R^2} = \tfrac{1}{2}\sin\phi\,d\phi \qquad (13.19)$$

Now we solve Eq. (13.19) for dM and substitute the result into Eq. (13.18) to find the potential energy of interaction between the point mass m and the ring:

$$dU = -\frac{GMm\sin\phi\,d\phi}{2s} \qquad (13.20)$$

The total potential energy of interaction between the point mass and the *shell* is the integral of Eq. (13.20) over the whole sphere as ϕ varies from 0 to π (*not* 2π!) and s varies from $r - R$ to $r + R$. To carry out the integration, we have to express the integrand in terms of a single variable; we choose s. To express ϕ and $d\phi$ in terms of s, we have to do a little more geometry. Figure 13.22b shows that s is the hypotenuse of a right triangle with sides $(r - R\cos\phi)$ and $R\sin\phi$, so the Pythagorean theorem gives

$$\begin{aligned} s^2 &= (r - R\cos\phi)^2 + (R\sin\phi)^2 \\ &= r^2 - 2rR\cos\phi + R^2 \end{aligned} \qquad (13.21)$$

We take differentials of both sides:

$$2s\,ds = 2rR\sin\phi\,d\phi$$

Next we divide this by $2rR$ and substitute the result into Eq. (13.20):

$$dU = -\frac{GMm}{2s}\frac{s\,ds}{rR} = -\frac{GMm}{2rR}\,ds \qquad (13.22)$$

We can now integrate Eq. (13.22), recalling that s varies from $r - R$ to $r + R$:

$$U = -\frac{GMm}{2rR}\int_{r-R}^{r+R} ds = -\frac{GMm}{2rR}[(r + R) - (r - R)] \qquad (13.23)$$

Finally, we have

$$U = -\frac{GMm}{r} \qquad \text{(point mass } m \text{ outside spherical shell } M) \qquad (13.24)$$

This is equal to the potential energy of two point masses m and M at a distance r. So we have proved that the gravitational potential energy of the spherical shell M and the point mass m at any distance r is the same as though they were point masses. Because the force is given by $F_r = -dU/dr$, the force is also the same.

The Gravitational Force Between Spherical Mass Distributions

Any spherically symmetric mass distribution can be thought of as a combination of concentric spherical shells. Because of the principle of superposition of forces, what is true of one shell is also true of the combination. So we have proved half of what we set out to prove: that the gravitational interaction between any spherically symmetric mass distribution and a point mass is the same as though all the mass of the spherically symmetric distribution were concentrated at its center.

The other half is to prove that *two* spherically symmetric mass distributions interact as though they were both points. That's easier. In Fig. 13.22a the forces the two bodies exert on each other are an action–reaction pair, and they obey Newton's third law. So we have also proved that the force that m exerts *on* the sphere M is the same as though M were a point. But now if we replace m with a spherically symmetric mass distribution centered at m's location, the resulting gravitational force on any part of M is the same as before, and so is the total force. This completes our proof.

A Point Mass Inside a Spherical Shell

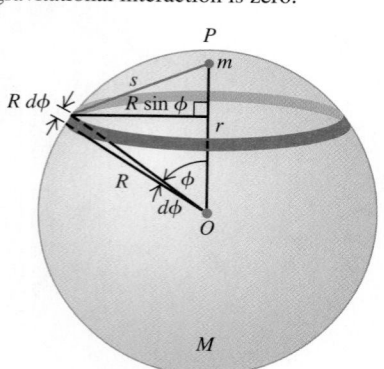

13.23 When a point mass m is *inside* a uniform spherical shell of mass M, the potential energy is the same no matter where inside the shell the point mass is located. The force from the masses' mutual gravitational interaction is zero.

We assumed at the beginning that the point mass m was outside the spherical shell, so our proof is valid only when m is outside a spherically symmetric mass distribution. When m is *inside* a spherical shell, the geometry is as shown in Fig. 13.23. The entire analysis goes just as before; Eqs. (13.18) through (13.22) are still valid. But when we get to Eq. (13.23), the limits of integration have to be changed to $R - r$ and $R + r$. We then have

$$U = -\frac{GMm}{2rR} \int_{R-r}^{R+r} ds = -\frac{GMm}{2rR}[(R + r) - (R - r)] \quad \text{(13.25)}$$

and the final result is

$$U = -\frac{GMm}{R} \quad \text{(point mass } m \text{ inside spherical shell } M\text{)} \quad \text{(13.26)}$$

Compare this result to Eq. (13.24): Instead of having r, the distance between m and the center of M, in the denominator, we have R, the radius of the shell. This means that U in Eq. (13.26) doesn't depend on r and thus has the same value everywhere inside the shell. When m moves around inside the shell, no work is done on it, so the force on m at any point inside the shell must be zero.

More generally, at any point in the interior of any spherically symmetric mass distribution (not necessarily a shell), at a distance r from its center, the gravitational force on a point mass m is the same as though we removed all the mass at points farther than r from the center and concentrated all the remaining mass at the center.

Example 13.10 "Journey to the center of the earth"

Imagine that we drill a hole through the earth along a diameter and drop a mail pouch down the hole. Derive an expression for the gravitational force F_g on the pouch as a function of its distance from the earth's center. Assume that the earth's density is uniform (not a very realistic model; see Fig. 13.9).

SOLUTION

IDENTIFY and SET UP: From the discussion immediately above, the value of F_g at a distance r from the earth's center is determined only by the mass M within a spherical region of radius r

(Fig. 13.24). Hence F_g is the same as if all the mass within radius r were concentrated at the center of the earth. The mass of a uniform sphere is proportional to the volume of the sphere, which is $\frac{4}{3}\pi r^3$ for a sphere of arbitrary radius r and $\frac{4}{3}\pi R_E^3$ for the entire earth.

EXECUTE: The ratio of the mass M of the sphere of radius r to the mass m_E of the earth is

$$\frac{M}{m_E} = \frac{\frac{4}{3}\pi r^3}{\frac{4}{3}\pi R_E^3} = \frac{r^3}{R_E^3} \quad \text{so} \quad M = m_E \frac{r^3}{R_E^3}$$

The magnitude of the gravitational force on m is then

$$F_g = \frac{GMm}{r^2} = \frac{Gm}{r^2}\left(m_E\frac{r^3}{R_E^3}\right) = \frac{Gm_E m}{R_E^3}r$$

EVALUATE: Inside this uniform-density sphere, F_g is *directly proportional* to the distance r from the center, rather than to $1/r^2$ as it is outside the sphere. At the surface $r = R_E$, we have $F_g = Gm_E m/R_E^2$, as we should. In the next chapter we'll learn how to compute the time it would take for the mail pouch to emerge on the other side of the earth.

13.24 A hole through the center of the earth (assumed to be uniform). When an object is a distance r from the center, only the mass inside a sphere of radius r exerts a net gravitational force on it.

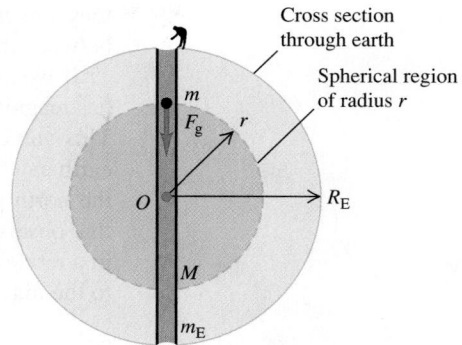

Test Your Understanding of Section 13.6   In the classic 1913 science-fiction novel *At the Earth's Core* by Edgar Rice Burroughs, explorers discover that the earth is a hollow sphere and that an entire civilization lives on the inside of the sphere. Would it be possible to stand and walk on the inner surface of a hollow, nonrotating planet? ❚

13.7 Apparent Weight and the Earth's Rotation

Because the earth rotates on its axis, it is not precisely an inertial frame of reference. For this reason the apparent weight of a body on earth is not precisely equal to the earth's gravitational attraction, which we will call the **true weight** $\vec{w}_0$ of the body. Figure 13.25 is a cutaway view of the earth, showing three observers. Each one holds a spring scale with a body of mass m hanging from it. Each scale applies a tension force $\vec{F}$ to the body hanging from it, and the reading on each scale is the magnitude F of this force. If the observers are unaware of the earth's

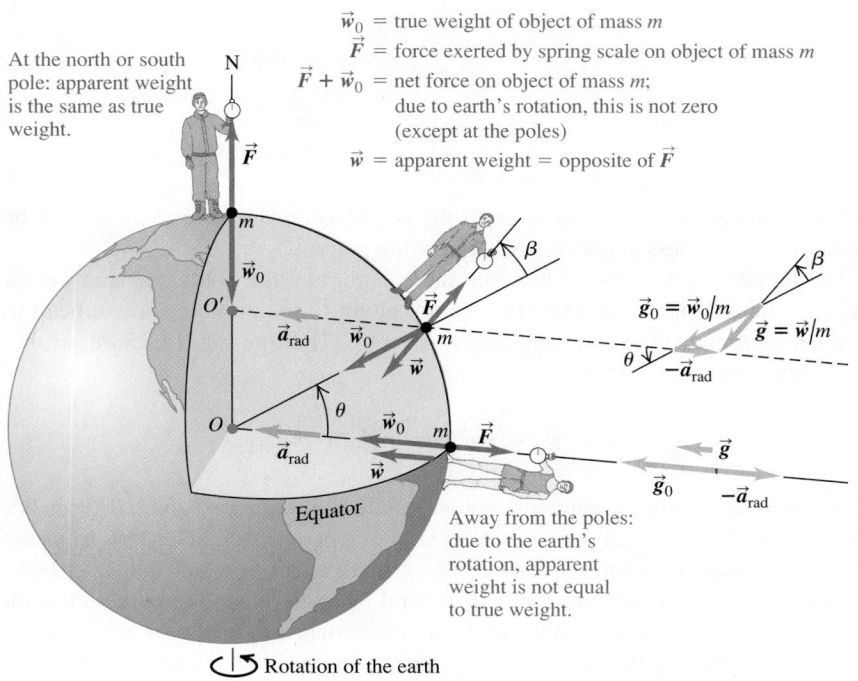

13.25 Except at the poles, the reading for an object being weighed on a scale (the *apparent weight*) is less than the gravitational force of attraction on the object (the *true weight*). The reason is that a net force is needed to provide a centripetal acceleration as the object rotates with the earth. For clarity, the illustration greatly exaggerates the angle β between the true and apparent weight vectors.

rotation, each one *thinks* that the scale reading equals the weight of the body because he thinks the body on his spring scale is in equilibrium. So each observer thinks that the tension $\vec{F}$ must be opposed by an equal and opposite force $\vec{w}$, which we call the **apparent weight.** But if the bodies are rotating with the earth, they are *not* precisely in equilibrium. Our problem is to find the relationship between the apparent weight $\vec{w}$ and the true weight $\vec{w}_0$.

If we assume that the earth is spherically symmetric, then the true weight $\vec{w}_0$ has magnitude $Gm_{\mathrm{E}}m/R_{\mathrm{E}}^2$, where m_{E} and R_{E} are the mass and radius of the earth. This value is the same for all points on the earth's surface. If the center of the earth can be taken as the origin of an inertial coordinate system, then the body at the north pole really *is* in equilibrium in an inertial system, and the reading on that observer's spring scale is equal to w_0. But the body at the equator is moving in a circle of radius R_{E} with speed v, and there must be a net inward force equal to the mass times the centripetal acceleration:

$$w_0 - F = \frac{mv^2}{R_{\mathrm{E}}}$$

So the magnitude of the apparent weight (equal to the magnitude of F) is

$$w = w_0 - \frac{mv^2}{R_{\mathrm{E}}} \qquad \text{(at the equator)} \tag{13.27}$$

If the earth were not rotating, the body when released would have a free-fall acceleration $g_0 = w_0/m$. Since the earth *is* rotating, the falling body's actual acceleration relative to the observer at the equator is $g = w/m$. Dividing Eq. (13.27) by m and using these relationships, we find

$$g = g_0 - \frac{v^2}{R_{\mathrm{E}}} \qquad \text{(at the equator)}$$

To evaluate v^2/R_{E}, we note that in 86,164 s a point on the equator moves a distance equal to the earth's circumference, $2\pi R_{\mathrm{E}} = 2\pi(6.38 \times 10^6\ \text{m})$. (The solar day, 86,400 s, is $\frac{1}{365}$ longer than this because in one day the earth also completes $\frac{1}{365}$ of its orbit around the sun.) Thus we find

$$v = \frac{2\pi(6.38 \times 10^6\ \text{m})}{86,164\ \text{s}} = 465\ \text{m/s}$$

$$\frac{v^2}{R_{\mathrm{E}}} = \frac{(465\ \text{m/s})^2}{6.38 \times 10^6\ \text{m}} = 0.0339\ \text{m/s}^2$$

So for a spherically symmetric earth the acceleration due to gravity should be about 0.03 m/s^2 less at the equator than at the poles.

At locations intermediate between the equator and the poles, the true weight $\vec{w}_0$ and the centripetal acceleration are not along the same line, and we need to write a vector equation corresponding to Eq. (13.27). From Fig. 13.25 we see that the appropriate equation is

$$\vec{w} = \vec{w}_0 - m\vec{a}_{\mathrm{rad}} = m\vec{g}_0 - m\vec{a}_{\mathrm{rad}} \tag{13.28}$$

The difference in the magnitudes of g and g_0 lies between zero and 0.0339 m/s^2. As shown in Fig. 13.25, the *direction* of the apparent weight differs from the direction toward the center of the earth by a small angle β, which is 0.1° or less.

Table 13.1 gives the values of g at several locations, showing variations with latitude. There are also small additional variations due to the lack of perfect spherical symmetry of the earth, local variations in density, and differences in elevation.

Table 13.1 Variations of g with Latitude and Elevation

Station	North Latitude	Elevation (m)	$g(\text{m/s}^2)$
Canal Zone	09°	0	9.78243
Jamaica	18°	0	9.78591
Bermuda	32°	0	9.79806
Denver, CO	40°	1638	9.79609
Pittsburgh, PA	40.5°	235	9.80118
Cambridge, MA	42°	0	9.80398
Greenland	70°	0	9.82534

Test Your Understanding of Section 13.7   Imagine a planet that has the same mass and radius as the earth, but that makes 10 rotations during the time the earth makes one rotation. What would be the difference between the acceleration due to gravity at the planet's equator and the acceleration due to gravity at its poles? (i) 0.00339 m/s^2; (ii) 0.0339 m/s^2; (iii) 0.339 m/s^2; (iv) 3.39 m/s^2.

13.8 Black Holes

The concept of a black hole is one of the most interesting and startling products of modern gravitational theory, yet the basic idea can be understood on the basis of Newtonian principles.

The Escape Speed from a Star

Think first about the properties of our own sun. Its mass $M = 1.99 \times 10^{30}$ kg and radius $R = 6.96 \times 10^8$ m are much larger than those of any planet, but compared to other stars, our sun is not exceptionally massive. You can find the sun's average density ρ in the same way we found the average density of the earth in Section 13.2:

$$\rho = \frac{M}{V} = \frac{M}{\frac{4}{3}\pi R^3} = \frac{1.99 \times 10^{30}\ \text{kg}}{\frac{4}{3}\pi(6.96 \times 10^8\ \text{m})^3} = 1410\ \text{kg/m}^3$$

The sun's temperatures range from 5800 K (about 5500°C or 10,000°F) at the surface up to 1.5×10^7 K (about 2.7×10^7°F) in the interior, so it surely contains no solids or liquids. Yet gravitational attraction pulls the sun's gas atoms together until the sun is, on average, 41% denser than water and about 1200 times as dense as the air we breathe.

Now think about the escape speed for a body at the surface of the sun. In Example 13.5 (Section 13.3) we found that the escape speed from the surface of a spherical mass M with radius R is $v = \sqrt{2GM/R}$. We can relate this to the average density. Substituting $M = \rho V = \rho(\frac{4}{3}\pi R^3)$ into the expression for escape speed gives

$$v = \sqrt{\frac{2GM}{R}} = \sqrt{\frac{8\pi G\rho}{3}}R \qquad (13.29)$$

Using either form of this equation, you can show that the escape speed for a body at the surface of our sun is $v = 6.18 \times 10^5$ m/s (about 2.2 million km/h, or 1.4 million mi/h). This value, roughly $\frac{1}{500}$ the speed of light, is independent of the mass of the escaping body; it depends on only the mass and radius (or average density and radius) of the sun.

Now consider various stars with the same average density ρ and different radii R. Equation (13.29) shows that for a given value of density ρ, the escape speed v is directly proportional to R. In 1783 the Rev. John Mitchell, an amateur astronomer, noted that if a body with the same average density as the sun had about 500 times the radius of the sun, its escape speed would be greater than the speed of light c. With his statement that "all light emitted from such a body would be made to return toward it," Mitchell became the first person to suggest the existence of what we now call a **black hole**—an object that exerts a gravitational force on other bodies but cannot emit any light of its own.

Black Holes, the Schwarzschild Radius, and the Event Horizon

The first expression for escape speed in Eq. (13.29) suggests that a body of mass M will act as a black hole if its radius R is less than or equal to a certain critical radius. How can we determine this critical radius? You might think that you can find the answer by simply setting $v = c$ in Eq. (13.29). As a matter of fact, this does give the correct result, but only because of two compensating errors.

13.26 (a) A body with a radius R greater than the Schwarzschild radius R_S. (b) If the body collapses to a radius smaller than R_S, it is a black hole with an escape speed greater than the speed of light. The surface of the sphere of radius R_S is called the event horizon of the black hole.

(a) When the radius R of a body is greater than the Schwarzschild radius R_S, light can escape from the surface of the body.

(b) If all the mass of the body lies inside radius R_S, the body is a black hole: No light can escape from it.

Gravity acting on the escaping light "red shifts" it to longer wavelengths.

The kinetic energy of light is *not* $mc^2/2$, and the gravitational potential energy near a black hole is *not* given by Eq. (13.9). In 1916, Karl Schwarzschild used Einstein's general theory of relativity (in part a generalization and extension of Newtonian gravitation theory) to derive an expression for the critical radius R_S, now called the **Schwarzschild radius.** The result turns out to be the same as though we had set $v = c$ in Eq. (13.29), so

$$c = \sqrt{\frac{2GM}{R_S}}$$

Solving for the Schwarzschild radius R_S, we find

$$R_S = \frac{2GM}{c^2} \qquad \text{(Schwarzschild radius)} \qquad \text{(13.30)}$$

If a spherical, nonrotating body with mass M has a radius less than R_S, then *nothing* (not even light) can escape from the surface of the body, and the body is a black hole (Fig. 13.26). In this case, any other body within a distance R_S of the center of the black hole is trapped by the gravitational attraction of the black hole and cannot escape from it.

The surface of the sphere with radius R_S surrounding a black hole is called the **event horizon:** Since light can't escape from within that sphere, we can't see events occurring inside. All that an observer outside the event horizon can know about a black hole is its mass (from its gravitational effects on other bodies), its electric charge (from the electric forces it exerts on other charged bodies), and its angular momentum (because a rotating black hole tends to drag space—and everything in that space—around with it). All other information about the body is irretrievably lost when it collapses inside its event horizon.

Example 13.11 Black hole calculations

Astrophysical theory suggests that a burned-out star whose mass is at least three solar masses will collapse under its own gravity to form a black hole. If it does, what is the radius of its event horizon?

SOLUTION

IDENTIFY, SET UP, and EXECUTE: The radius in question is the Schwarzschild radius. We use Eq. (13.30) with a value of M equal to three solar masses, or $M = 3(1.99 \times 10^{30}\,\text{kg}) = 6.0 \times 10^{30}\,\text{kg}$:

$$R_S = \frac{2GM}{c^2} = \frac{2(6.67 \times 10^{-11}\,\text{N} \cdot \text{m}^2/\text{kg}^2)(6.0 \times 10^{30}\,\text{kg})}{(3.00 \times 10^8\,\text{m/s})^2}$$

$$= 8.9 \times 10^3\,\text{m} = 8.9\,\text{km} = 5.5\,\text{mi}$$

EVALUATE: The average density of such an object is

$$\rho = \frac{M}{\frac{4}{3}\pi R^3} = \frac{6.0 \times 10^{30}\ \text{kg}}{\frac{4}{3}\pi(8.9 \times 10^3\ \text{m})^3} = 2.0 \times 10^{18}\ \text{kg/m}^3$$

This is about 10^{15} times as great as the density of familiar matter on earth and is comparable to the densities of atomic nuclei.

In fact, once the body collapses to a radius of R_S, nothing can prevent it from collapsing further. All of the mass ends up being crushed down to a single point called a *singularity* at the center of the event horizon. This point has zero volume and so has *infinite density*.

A Visit to a Black Hole

At points far from a black hole, its gravitational effects are the same as those of any normal body with the same mass. If the sun collapsed to form a black hole, the orbits of the planets would be unaffected. But things get dramatically different close to the black hole. If you decided to become a martyr for science and jump into a black hole, the friends you left behind would notice several odd effects as you moved toward the event horizon, most of them associated with effects of general relativity.

If you carried a radio transmitter to send back your comments on what was happening, your friends would have to retune their receiver continuously to lower and lower frequencies, an effect called the *gravitational red shift*. Consistent with this shift, they would observe that your clocks (electronic or biological) would appear to run more and more slowly, an effect called *time dilation*. In fact, during their lifetimes they would never see you make it to the event horizon.

In your frame of reference, you would make it to the event horizon in a rather short time but in a rather disquieting way. As you fell feet first into the black hole, the gravitational pull on your feet would be greater than that on your head, which would be slightly farther away from the black hole. The *differences* in gravitational force on different parts of your body would be great enough to stretch you along the direction toward the black hole and compress you perpendicular to it. These effects (called *tidal forces*) would rip you to atoms, and then rip your atoms apart, before you reached the event horizon.

Detecting Black Holes

If light cannot escape from a black hole and if black holes are as small as Example 13.11 suggests, how can we know that such things exist? The answer is that any gas or dust near the black hole tends to be pulled into an *accretion disk* that swirls around and into the black hole, rather like a whirlpool (Fig. 13.27). Friction within the accretion disk's material causes it to lose mechanical energy

① Matter is pulled from the ordinary star to form an accretion disk around the black hole.

Ordinary star

② The gas in the accretion disk is compressed and heated to high temperatures, becoming an intense source of x rays.

Black hole

③ Gas in the accretion disk that does not fall into the black hole is ejected in two fast-moving jets.

13.27 A binary star system in which an ordinary star and a black hole orbit each other. The black hole itself cannot be seen, but the x rays from its accretion disk can be detected.

and spiral into the black hole; as it moves inward, it is compressed together. This causes heating of the material, just as air compressed in a bicycle pump gets hotter. Temperatures in excess of 10^6 K can occur in the accretion disk, so hot that the disk emits not just visible light (as do bodies that are "red-hot" or "white-hot") but x rays. Astronomers look for these x rays (emitted by the material *before* it crosses the event horizon) to signal the presence of a black hole. Several promising candidates have been found, and astronomers now express considerable confidence in the existence of black holes.

Black holes in binary star systems like the one depicted in Fig. 13.27 have masses a few times greater than the sun's mass. There is also mounting evidence for the existence of much larger *supermassive black holes*. One example is thought to lie at the center of our Milky Way galaxy, some 26,000 light-years from earth in the direction of the constellation Sagittarius. High-resolution images of the galactic center reveal stars moving at speeds greater than 1500 km/s about an unseen object that lies at the position of a source of radio waves called Sgr A* (Fig. 13.28). By analyzing these motions, astronomers can infer the period T and semi-major axis a of each star's orbit. The mass m_X of the unseen object can then be calculated using Kepler's third law in the form given in Eq. (13.17), with the mass of the sun m_S replaced by m_X:

$$T = \frac{2\pi a^{3/2}}{\sqrt{Gm_X}} \qquad \text{so} \qquad m_X = \frac{4\pi^2 a^3}{GT^2}$$

The conclusion is that the mysterious dark object at the galactic center has a mass of 8.2×10^{36} kg, or 4.1 *million* times the mass of the sun. Yet observations with radio telescopes show that it has a radius no more than 4.4×10^{10} m, about one-third of the distance from the earth to the sun. These observations suggest that this massive, compact object is a black hole with a Schwarzschild radius of 1.1×10^{10} m. Astronomers hope to improve the resolution of their observations so that they can actually see the event horizon of this black hole.

Other lines of research suggest that even larger black holes, in excess of 10^9 times the mass of the sun, lie at the centers of other galaxies. Observational and theoretical studies of black holes of all sizes continue to be an exciting area of research in both physics and astronomy.

13.28 This false-color image shows the motions of stars at the center of our galaxy over a 13-year period. Analyzing these orbits using Kepler's third law indicates that the stars are moving about an unseen object that is some 4.1×10^6 times the mass of the sun. The scale bar indicates a length of 10^{14} m (670 times the distance from the earth to the sun) at the distance of the galactic center.

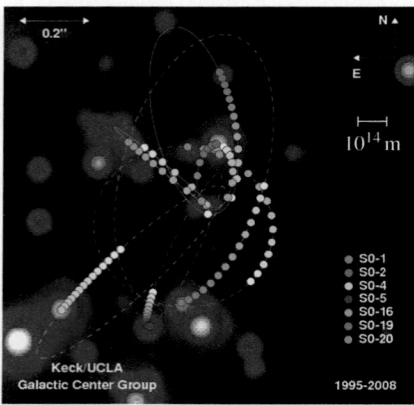

Test Your Understanding of Section 13.8 If the sun somehow collapsed to form a black hole, what effect would this event have on the orbit of the earth? (i) The orbit would shrink; (ii) the orbit would expand; (iii) the orbit would remain the same size.

Newton's law of gravitation: *Any* two bodies with masses m_1 and m_2, a distance r apart, attract each other with forces inversely proportional to r^2. These forces form an action–reaction pair and obey Newton's third law. When two or more bodies exert gravitational forces on a particular body, the total gravitational force on that individual body is the vector sum of the forces exerted by the other bodies. The gravitational interaction between spherical mass distributions, such as planets or stars, is the same as if all the mass of each distribution were concentrated at the center. (See Examples 13.1–13.3 and 13.10.)

$$F_g = \frac{Gm_1m_2}{r^2} \qquad (13.1)$$

Gravitational force, weight, and gravitational potential energy: The weight w of a body is the total gravitational force exerted on it by all other bodies in the universe. Near the surface of the earth (mass m_E and radius R_E), the weight is essentially equal to the gravitational force of the earth alone. The gravitational potential energy U of two masses m and m_E separated by a distance r is inversely proportional to r. The potential energy is never positive; it is zero only when the two bodies are infinitely far apart. (See Examples 13.4 and 13.5.)

$$w = F_g = \frac{Gm_Em}{R_E^2} \qquad (13.3)$$
(weight at earth's surface)

$$g = \frac{Gm_E}{R_E^2} \qquad (13.4)$$
(acceleration due to gravity at earth's surface)

$$U = -\frac{Gm_Em}{r} \qquad (13.9)$$

Orbits: When a satellite moves in a circular orbit, the centripetal acceleration is provided by the gravitational attraction of the earth. Kepler's three laws describe the more general case: an elliptical orbit of a planet around the sun or a satellite around a planet. (See Examples 13.6–13.9.)

$$v = \sqrt{\frac{Gm_E}{r}}$$
(speed in circular orbit) $\qquad (13.10)$

$$T = \frac{2\pi r}{v} = 2\pi r\sqrt{\frac{r}{Gm_E}} = \frac{2\pi r^{3/2}}{\sqrt{Gm_E}}$$
(period in circular orbit) $\qquad (13.12)$

Black holes: If a nonrotating spherical mass distribution with total mass M has a radius less than its Schwarzschild radius R_S, it is called a black hole. The gravitational interaction prevents anything, including light, from escaping from within a sphere with radius R_S. (See Example 13.11.)

$$R_S = \frac{2GM}{c^2} \qquad (13.30)$$
(Schwarzschild radius)

If all of the body is inside its Schwarzschild radius $R_S = 2GM/c^2$, the body is a black hole.

<table>
<tr><td>

BRIDGING PROBLEM   **Speeds in an Elliptical Orbit**

A comet orbits the sun (mass m_S) in an elliptical orbit of semi-major axis a and eccentricity e. (a) Find expressions for the speeds of the comet at perihelion and aphelion. (b) Evaluate these expressions for Comet Halley (see Example 13.9).

SOLUTION GUIDE

See MasteringPhysics® study area for a Video Tutor solution.

IDENTIFY and SET UP

1. Sketch the situation; show all relevant dimensions. Label the perihelion and aphelion.
2. List the unknown quantities, and identify the target variables.
3. Just as for a satellite orbiting the earth, the mechanical energy is conserved for a comet orbiting the sun. (Why?) What other quantity is conserved as the comet moves in its orbit? (*Hint:* See Section 13.5.)

</td><td>

EXECUTE

4. You'll need at least two equations that involve the two unknown speeds, and you'll need expressions for the sun–comet distances at perihelion and aphelion. (*Hint:* See Fig. 13.18.)
5. Solve the equations for your target variables. Compare your expressions: Which speed is lower? Does this make sense?
6. Use your expressions from step 5 to find the perihelion and aphelion speeds for Comet Halley. (*Hint:* See Appendix F.)

EVALUATE

7. Check whether your results make sense for the special case of a circular orbit ($e = 0$).

</td></tr>
</table>

Problems

For instructor-assigned homework, go to www.masteringphysics.com

•, ••, •••: Problems of increasing difficulty. **CP**: Cumulative problems incorporating material from earlier chapters. **CALC**: Problems requiring calculus. **BIO**: Biosciences problems.

DISCUSSION QUESTIONS

Q13.1 A student wrote: "The only reason an apple falls downward to meet the earth instead of the earth rising upward to meet the apple is that the earth is much more massive and so exerts a much greater pull." Please comment.

Q13.2 A planet makes a circular orbit with period T around a star. If it were to orbit, at the same distance, a star with three times the mass of the original star, would the new period (in terms of T) be (a) $3T$, (b) $T\sqrt{3}$, (c) T, (d) $T/\sqrt{3}$, or (e) $T/3$?

Q13.3 If all planets had the same average density, how would the acceleration due to gravity at the surface of a planet depend on its radius?

Q13.4 Is a pound of butter on the earth the same amount as a pound of butter on Mars? What about a kilogram of butter? Explain.

Q13.5 Example 13.2 (Section 13.1) shows that the acceleration of each sphere caused by the gravitational force is inversely proportional to the mass of that sphere. So why does the force of gravity give all masses the same acceleration when they are dropped near the surface of the earth?

Q13.6 When will you attract the sun more: today at noon, or tonight at midnight? Explain.

Q13.7 Since the moon is constantly attracted toward the earth by the gravitational interaction, why doesn't it crash into the earth?

Q13.8 A planet makes a circular orbit with period T around a star. If the planet were to orbit at the same distance around this star, but had three times as much mass, what would the new period (in terms of T) be: (a) $3T$, (b) $T\sqrt{3}$, (c) T, (d) $T/\sqrt{3}$, or (e) $T/3$?

Q13.9 The sun pulls on the moon with a force that is more than twice the magnitude of the force with which the earth attracts the moon. Why, then, doesn't the sun take the moon away from the earth?

Q13.10 As defined in Chapter 7, gravitational potential energy is $U = mgy$ and is positive for a body of mass m above the earth's surface (which is at $y = 0$). But in this chapter, gravitational potential energy is $U = -Gm_E m/r$, which is *negative* for a body of mass m above the earth's surface (which is at $r = R_E$). How can you reconcile these seemingly incompatible descriptions of gravitational potential energy?

Q13.11 A planet is moving at constant speed in a circular orbit around a star. In one complete orbit, what is the net amount of work done on the planet by the star's gravitational force: positive, negative, or zero? What if the planet's orbit is an ellipse, so that the speed is not constant? Explain your answers.

Q13.12 Does the escape speed for an object at the earth's surface depend on the direction in which it is launched? Explain. Does your answer depend on whether or not you include the effects of air resistance?

Q13.13 If a projectile is fired straight up from the earth's surface, what would happen if the total mechanical energy (kinetic plus potential) is (a) less than zero, and (b) greater than zero? In each case, ignore air resistance and the gravitational effects of the sun, the moon, and the other planets.

Q13.14 Discuss whether this statement is correct: "In the absence of air resistance, the trajectory of a projectile thrown near the earth's surface is an *ellipse,* not a parabola."

Q13.15 The earth is closer to the sun in November than in May. In which of these months does it move faster in its orbit? Explain why.

Q13.16 A communications firm wants to place a satellite in orbit so that it is always directly above the earth's 45th parallel (latitude 45° north). This means that the plane of the orbit will not pass through the center of the earth. Is such an orbit possible? Why or why not?

Q13.17 At what point in an elliptical orbit is the acceleration maximum? At what point is it minimum? Justify your answers.

Q13.18 Which takes more fuel: a voyage from the earth to the moon or from the moon to the earth? Explain.

Q13.19 What would Kepler's third law be for circular orbits if an amendment to Newton's law of gravitation made the gravitational force inversely proportional to r^3? Would this change affect Kepler's other two laws? Explain.

Q13.20 In the elliptical orbit of Comet Halley shown in Fig. 13.20a, the sun's gravity is responsible for making the comet fall inward from aphelion to perihelion. But what is responsible for making the comet move from perihelion back outward to aphelion?

Q13.21 Many people believe that orbiting astronauts feel weightless because they are "beyond the pull of the earth's gravity." How far from the earth would a spacecraft have to travel to be truly beyond the earth's gravitational influence? If a spacecraft were really unaffected by the earth's gravity, would it remain in orbit? Explain. What is the real reason astronauts in orbit feel weightless?

Q13.22 As part of their training before going into orbit, astronauts ride in an airliner that is flown along the same parabolic trajectory as a freely falling projectile. Explain why this gives the same experience of apparent weightlessness as being in orbit.

EXERCISES

Section 13.1 Newton's Law of Gravitation

13.1 • What is the ratio of the gravitational pull of the sun on the moon to that of the earth on the moon? (Assume the distance of the moon from the sun can be approximated by the distance of the earth from the sun.) Use the data in Appendix F. Is it more accurate to say that the moon orbits the earth, or that the moon orbits the sun?

13.2 •• **CP Cavendish Experiment.** In the Cavendish balance apparatus shown in Fig. 13.4, suppose that $m_1 = 1.10$ kg, $m_2 = 25.0$ kg, and the rod connecting the m_1 pairs is 30.0 cm long. If, in each pair, m_1 and m_2 are 12.0 cm apart center to center, find (a) the net force and (b) the net torque (about the rotation axis) on the rotating part of the apparatus. (c) Does it seem that the torque in part (b) would be enough to easily rotate the rod? Suggest some ways to improve the sensitivity of this experiment.

13.3 • **Rendezvous in Space!** A couple of astronauts agree to rendezvous in space after hours. Their plan is to let gravity bring them together. One of them has a mass of 65 kg and the other a mass of 72 kg, and they start from rest 20.0 m apart. (a) Make a free-body diagram of each astronaut, and use it to find his or her initial acceleration. As a rough approximation, we can model the astronauts as uniform spheres. (b) If the astronauts' acceleration remained constant, how many days would they have to wait before reaching each other? (Careful! They *both* have acceleration toward each other.) (c) Would their acceleration, in fact, remain constant? If not, would it increase or decrease? Why?

13.4 •• Two uniform spheres, each with mass M and radius R, touch each other. What is the magnitude of their gravitational force of attraction?

13.5 • Two uniform spheres, each of mass 0.260 kg, are fixed at points A and B (Fig. E13.5). Find the magnitude and direction of the initial acceleration of a uniform sphere with mass 0.010 kg if released from rest at

Figure E13.5

point P and acted on only by forces of gravitational attraction of the spheres at A and B.

13.6 •• Find the magnitude and direction of the net gravitational force on mass A due to masses B and C in Fig. E13.6. Each mass is 2.00 kg.

Figure **E13.6**

13.7 • A typical adult human has a mass of about 70 kg. (a) What force does a full moon exert on such a human when it is directly overhead with its center 378,000 km away? (b) Compare this force with the force exerted on the human by the earth.

13.8 •• An 8.00-kg point mass and a 15.0-kg point mass are held in place 50.0 cm apart. A particle of mass m is released from a point between the two masses 20.0 cm from the 8.00-kg mass along the line connecting the two fixed masses. Find the magnitude and direction of the acceleration of the particle.

13.9 •• A particle of mass $3m$ is located 1.00 m from a particle of mass m. (a) Where should you put a third mass M so that the net gravitational force on M due to the two masses is exactly zero? (b) Is the equilibrium of M at this point stable or unstable (i) for points along the line connecting m and $3m$, and (ii) for points along the line passing through M and perpendicular to the line connecting m and $3m$?

13.10 •• The point masses m and $2m$ lie along the x-axis, with m at the origin and $2m$ at $x = L$. A third point mass M is moved along the x-axis. (a) At what point is the net gravitational force on M due to the other two masses equal to zero? (b) Sketch the x-component of the net force on M due to m and $2m$, taking quantities to the right as positive. Include the regions $x < 0$, $0 < x < L$, and $x > L$. Be especially careful to show the behavior of the graph on either side of $x = 0$ and $x = L$.

Section 13.2 Weight

13.11 •• At what distance above the surface of the earth is the acceleration due to the earth's gravity 0.980 m/s² if the acceleration due to gravity at the surface has magnitude 9.80 m/s²?

13.12 • The mass of Venus is 81.5% that of the earth, and its radius is 94.9% that of the earth. (a) Compute the acceleration due to gravity on the surface of Venus from these data. (b) If a rock weighs 75.0 N on earth, what would it weigh at the surface of Venus?

13.13 • Titania, the largest moon of the planet Uranus, has $\frac{1}{8}$ the radius of the earth and $\frac{1}{1700}$ the mass of the earth. (a) What is the acceleration due to gravity at the surface of Titania? (b) What is the average density of Titania? (This is less than the density of rock, which is one piece of evidence that Titania is made primarily of ice.)

13.14 • Rhea, one of Saturn's moons, has a radius of 765 km and an acceleration due to gravity of 0.278 m/s² at its surface. Calculate its mass and average density.

13.15 •• Calculate the earth's gravity force on a 75-kg astronaut who is repairing the Hubble Space Telescope 600 km above the earth's surface, and then compare this value with his weight at the earth's surface. In view of your result, explain why we say astronauts are weightless when they orbit the earth in a satellite such as a space shuttle. Is it because the gravitational pull of the earth is negligibly small?

Section 13.3 Gravitational Potential Energy

13.16 •• **Volcanoes on Io.** Jupiter's moon Io has active volcanoes (in fact, it is the most volcanically active body in the solar system) that eject material as high as 500 km (or even higher) above the surface. Io has a mass of 8.94×10^{22} kg and a radius of 1815 km. Ignore any variation in gravity over the 500-km range of the debris. How high would this material go on earth if it were ejected with the same speed as on Io?

13.17 • Use the results of Example 13.5 (Section 13.3) to calculate the escape speed for a spacecraft (a) from the surface of Mars and (b) from the surface of Jupiter. Use the data in Appendix F. (c) Why is the escape speed for a spacecraft independent of the spacecraft's mass?

13.18 •• Ten days after it was launched toward Mars in December 1998, the *Mars Climate Orbiter* spacecraft (mass 629 kg) was 2.87×10^6 km from the earth and traveling at 1.20×10^4 km/h relative to the earth. At this time, what were (a) the spacecraft's kinetic energy relative to the earth and (b) the potential energy of the earth–spacecraft system?

Section 13.4 The Motion of Satellites

13.19 • For a satellite to be in a circular orbit 780 km above the surface of the earth, (a) what orbital speed must it be given, and (b) what is the period of the orbit (in hours)?

13.20 •• **Aura Mission.** On July 15, 2004, NASA launched the Aura spacecraft to study the earth's climate and atmosphere. This satellite was injected into an orbit 705 km above the earth's surface. Assume a circular orbit. (a) How many hours does it take this satellite to make one orbit? (b) How fast (in km/s) is the Aura spacecraft moving?

13.21 •• Two satellites are in circular orbits around a planet that has radius 9.00×10^6 m. One satellite has mass 68.0 kg, orbital radius 5.00×10^7 m, and orbital speed 4800 m/s. The second satellite has mass 84.0 kg and orbital radius 3.00×10^7 m. What is the orbital speed of this second satellite?

13.22 •• **International Space Station.** The International Space Station makes 15.65 revolutions per day in its orbit around the earth. Assuming a circular orbit, how high is this satellite above the surface of the earth?

13.23 • Deimos, a moon of Mars, is about 12 km in diameter with mass 2.0×10^{15} kg. Suppose you are stranded alone on Deimos and want to play a one-person game of baseball. You would be the pitcher, and you would be the batter! (a) With what speed would you have to throw a baseball so that it would go into a circular orbit just above the surface and return to you so you could hit it? Do you think you could actually throw it at this speed? (b) How long (in hours) after throwing the ball should you be ready to hit it? Would this be an action-packed baseball game?

Section 13.5 Kepler's Laws and the Motion of Planets

13.24 •• **Planet Vulcan.** Suppose that a planet were discovered between the sun and Mercury, with a circular orbit of radius equal to $\frac{2}{3}$ of the average orbit radius of Mercury. What would be the orbital period of such a planet? (Such a planet was once postulated, in part to explain the precession of Mercury's orbit. It was even given the name Vulcan, although we now have no evidence that it actually exists. Mercury's precession has been explained by general relativity.)

13.25 •• The star Rho[1] Cancri is 57 light-years from the earth and has a mass 0.85 times that of our sun. A planet has been detected in a circular orbit around Rho[1] Cancri with an orbital radius equal to 0.11 times the radius of the earth's orbit around the sun. What are (a) the orbital speed and (b) the orbital period of the planet of Rho[1] Cancri?

13.26 •• In March 2006, two small satellites were discovered orbiting Pluto, one at a distance of 48,000 km and the other at 64,000 km. Pluto already was known to have a large satellite Charon, orbiting at 19,600 km with an orbital period of 6.39 days. Assuming that the satellites do not affect each other, find the orbital periods of the two small satellites *without* using the mass of Pluto.

13.27 • (a) Use Fig. 13.18 to show that the sun–planet distance at perihelion is $(1 - e)a$, the sun–planet distance at aphelion is $(1 + e)a$, and therefore the sum of these two distances is $2a$. (b) When the dwarf planet Pluto was at perihelion in 1989, it was almost 100 million km closer to the sun than Neptune. The semimajor axes of the orbits of Pluto and Neptune are 5.92×10^{12} m and 4.50×10^{12} m, respectively, and the eccentricities are 0.248 and 0.010. Find Pluto's closest distance and Neptune's farthest distance from the sun. (c) How many years after being at perihelion in 1989 will Pluto again be at perihelion?

13.28 •• **Hot Jupiters.** In 2004 astronomers reported the discovery of a large Jupiter-sized planet orbiting very close to the star HD 179949 (hence the term "hot Jupiter"). The orbit was just $\frac{1}{9}$ the distance of Mercury from our sun, and it takes the planet only 3.09 days to make one orbit (assumed to be circular). (a) What is the mass of the star? Express your answer in kilograms and as a multiple of our sun's mass. (b) How fast (in km/s) is this planet moving?

13.29 •• **Planets Beyond the Solar System.** On October 15, 2001, a planet was discovered orbiting around the star HD 68988. Its orbital distance was measured to be 10.5 million kilometers from the center of the star, and its orbital period was estimated at 6.3 days. What is the mass of HD 68988? Express your answer in kilograms and in terms of our sun's mass. (Consult Appendix F.)

Section 13.6 Spherical Mass Distributions

13.30 • A uniform, spherical, 1000.0-kg shell has a radius of 5.00 m. (a) Find the gravitational force this shell exerts on a 2.00-kg point mass placed at the following distances from the center of the shell: (i) 5.01 m, (ii) 4.99 m, (iii) 2.72 m. (b) Sketch a qualitative graph of the magnitude of the gravitational force this sphere exerts on a point mass m as a function of the distance r of m from the center of the sphere. Include the region from $r = 0$ to $r \to \infty$.

13.31 •• A uniform, solid, 1000.0-kg sphere has a radius of 5.00 m. (a) Find the gravitational force this sphere exerts on a 2.00-kg point mass placed at the following distances from the center of the sphere: (i) 5.01 m, (ii) 2.50 m. (b) Sketch a qualitative graph of the magnitude of the gravitational force this sphere exerts on a point mass m as a function of the distance r of m from the center of the sphere. Include the region from $r = 0$ to $r \to \infty$.

13.32 • **CALC** A thin, uniform rod has length L and mass M. A small uniform sphere of mass m is placed a distance x from one end of the rod, along the axis of the rod (Fig. E13.32). (a) Calculate

Figure **E13.32**

the gravitational potential energy of the rod–sphere system. Take the potential energy to be zero when the rod and sphere are infinitely far apart. Show that your answer reduces to the expected result when x is much larger than L. (*Hint:* Use the power series expansion for $\ln(1 + x)$ given in Appendix B.) (b) Use $F_x = -dU/dx$ to find the magnitude and direction of the gravitational force exerted on the sphere by the rod (see Section 7.4). Show that your answer reduces to the expected result when x is much larger than L.

13.33 • **CALC** Consider the ring-shaped body of Fig. E13.33. A particle with mass m is placed a distance x from the center of the ring, along the line through the center of the ring and perpendicular to its plane. (a) Calculate the gravitational potential energy U of this system. Take the potential energy to be zero when the two objects are far apart. (b) Show that your answer to part (a) reduces to the expected result when x is much larger than the radius a of the ring. (c) Use $F_x = -dU/dx$ to find the magnitude and direction of the force on the particle (see Section 7.4). (d) Show that your answer to part (c) reduces to the expected result when x is much larger than a. (e) What are the values of U and F_x when $x = 0$? Explain why these results make sense.

Figure **E13.33**

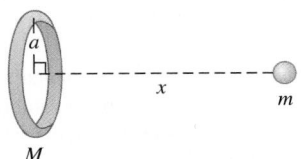

Section 13.7 Apparent Weight and the Earth's Rotation

13.34 •• **A Visit to Santa.** You decide to visit Santa Claus at the north pole to put in a good word about your splendid behavior throughout the year. While there, you notice that the elf Sneezy, when hanging from a rope, produces a tension of 475.0 N in the rope. If Sneezy hangs from a similar rope while delivering presents at the earth's equator, what will the tension in it be? (Recall that the earth is rotating about an axis through its north and south poles.) Consult Appendix F and start with a free-body diagram of Sneezy at the equator.

13.35 • The acceleration due to gravity at the north pole of Neptune is approximately 10.7 m/s². Neptune has mass 1.0×10^{26} kg and radius 2.5×10^4 km and rotates once around its axis in about 16 h. (a) What is the gravitational force on a 5.0-kg object at the north pole of Neptune? (b) What is the apparent weight of this same object at Neptune's equator? (Note that Neptune's "surface" is gaseous, not solid, so it is impossible to stand on it.)

Section 13.8 Black Holes

13.36 •• **Mini Black Holes.** Cosmologists have speculated that black holes the size of a proton could have formed during the early days of the Big Bang when the universe began. If we take the diameter of a proton to be 1.0×10^{-15} m, what would be the mass of a mini black hole?

13.37 •• **At the Galaxy's Core.** Astronomers have observed a small, massive object at the center of our Milky Way galaxy (see Section 13.8). A ring of material orbits this massive object; the ring has a diameter of about 15 light-years and an orbital speed of about 200 km/s. (a) Determine the mass of the object at the center of the Milky Way galaxy. Give your answer both in kilograms and in solar masses (one solar mass is the mass of the sun). (b) Observations of stars, as well as theories of the structure of stars, suggest that it

is impossible for a single star to have a mass of more than about 50 solar masses. Can this massive object be a single, ordinary star? (c) Many astronomers believe that the massive object at the center of the Milky Way galaxy is a black hole. If so, what must the Schwarzschild radius of this black hole be? Would a black hole of this size fit inside the earth's orbit around the sun?

13.38 • (a) Show that a black hole attracts an object of mass m with a force of $mc^2 R_S/(2r^2)$, where r is the distance between the object and the center of the black hole. (b) Calculate the magnitude of the gravitational force exerted by a black hole of Schwarzschild radius 14.0 mm on a 5.00-kg mass 3000 km from it. (c) What is the mass of this black hole?

13.39 • In 2005 astronomers announced the discovery of a large black hole in the galaxy Markarian 766 having clumps of matter orbiting around once every 27 hours and moving at 30,000 km/s. (a) How far are these clumps from the center of the black hole? (b) What is the mass of this black hole, assuming circular orbits? Express your answer in kilograms and as a multiple of our sun's mass. (c) What is the radius of its event horizon?

PROBLEMS

13.40 ••• Four identical masses of 800 kg each are placed at the corners of a square whose side length is 10.0 cm. What is the net gravitational force (magnitude and direction) on one of the masses, due to the other three?

13.41 ••• Neutron stars, such as the one at the center of the Crab Nebula, have about the same mass as our sun but have a *much* smaller diameter. If you weigh 675 N on the earth, what would you weigh at the surface of a neutron star that has the same mass as our sun and a diameter of 20 km?

13.42 ••• **CP Exploring Europa.** There is strong evidence that Europa, a satellite of Jupiter, has a liquid ocean beneath its icy surface. Many scientists think we should land a vehicle there to search for life. Before launching it, we would want to test such a lander under the gravity conditions at the surface of Europa. One way to do this is to put the lander at the end of a rotating arm in an orbiting earth satellite. If the arm is 4.25 m long and pivots about one end, at what angular speed (in rpm) should it spin so that the acceleration of the lander is the same as the acceleration due to gravity at the surface of Europa? The mass of Europa is 4.8×10^{22} kg and its diameter is 3138 km.

13.43 • Three uniform spheres are fixed at the positions shown in Fig. P13.43. (a) What are the magnitude and direction of the force on a 0.0150-kg particle placed at P? (b) If the spheres are in deep outer space and a 0.0150-kg particle is released from rest 300 m from the origin along a line 45° below the $-x$-axis, what will the particle's speed be when it reaches the origin?

Figure **P13.43**

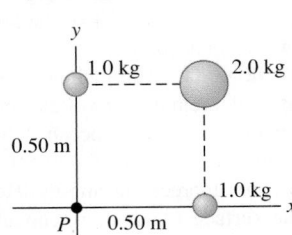

13.44 •• A uniform sphere with mass 60.0 kg is held with its center at the origin, and a second uniform sphere with mass 80.0 kg is held with its center at the point $x = 0$, $y = 3.00$ m. (a) What are the magnitude and direction of the net gravitational force due to these objects on a third uniform sphere with mass 0.500 kg placed at the point $x = 4.00$ m, $y = 0$? (b) Where, other than infinitely far away, could the third sphere be placed such that the net gravitational force acting on it from the other two spheres is equal to zero?

13.45 •• **CP BIO** **Hip Wear on the Moon.** (a) Use data from Appendix F to calculate the acceleration due to gravity on the moon. (b) Calculate the friction force at the hip joint on a 65-kg astronaut carrying a 43-kg instrument pack on the moon if the coefficient of kinetic friction at her hip joint is 0.0050 and the hip joint supports a mass of 85.25 kg. (c) What would be the friction force at the hip joint on earth for this astronaut?

13.46 •• **Mission to Titan.** On December 25, 2004, the *Huygens* probe separated from the *Cassini* spacecraft orbiting Saturn and began a 22-day journey to Saturn's giant moon Titan, on whose surface it landed. Besides the data in Appendix F, it is useful to know that Titan is 1.22×10^6 km from the center of Saturn and has a mass of 1.35×10^{23} kg and a diameter of 5150 km. At what distance from Titan should the gravitational pull of Titan just balance the gravitational pull of Saturn?

13.47 •• The asteroid Toro has a radius of about 5.0 km. Consult Appendix F as necessary. (a) Assuming that the density of Toro is the same as that of the earth (5.5 g/cm^3), find its total mass and find the acceleration due to gravity at its surface. (b) Suppose an object is to be placed in a circular orbit around Toro, with a radius just slightly larger than the asteroid's radius. What is the speed of the object? Could you launch yourself into orbit around Toro by running?

13.48 ••• At a certain instant, the earth, the moon, and a stationary 1250-kg spacecraft lie at the vertices of an equilateral triangle whose sides are 3.84×10^5 km in length. (a) Find the magnitude and direction of the net gravitational force exerted on the spacecraft by the earth and moon. State the direction as an angle measured from a line connecting the earth and the spacecraft. In a sketch, show the earth, the moon, the spacecraft, and the force vector. (b) What is the minimum amount of work that you would have to do to move the spacecraft to a point far from the earth and moon? You can ignore any gravitational effects due to the other planets or the sun.

13.49 ••• **CP** An experiment is performed in deep space with two uniform spheres, one with mass 50.0 kg and the other with mass 100.0 kg. They have equal radii, $r = 0.20$ m. The spheres are released from rest with their centers 40.0 m apart. They accelerate toward each other because of their mutual gravitational attraction. You can ignore all gravitational forces other than that between the two spheres. (a) Explain why linear momentum is conserved. (b) When their centers are 20.0 m apart, find (i) the speed of each sphere and (ii) the magnitude of the relative velocity with which one sphere is approaching the other. (c) How far from the initial position of the center of the 50.0-kg sphere do the surfaces of the two spheres collide?

13.50 •• **CP** **Submarines on Europa.** Some scientists are eager to send a remote-controlled submarine to Jupiter's moon Europa to search for life in its oceans below an icy crust. Europa's mass has been measured to be 4.8×10^{22} kg, its diameter is 3138 km, and it has no appreciable atmosphere. Assume that the layer of ice at the surface is not thick enough to exert substantial force on the water. If the windows of the submarine you are designing are 25.0 cm square and can stand a maximum inward force of 9750 N per window, what is the greatest depth to which this submarine can safely dive?

13.51 • **Geosynchronous Satellites.** Many satellites are moving in a circle in the earth's equatorial plane. They are at such a height above the earth's surface that they always remain above the same point. (a) Find the altitude of these satellites above the earth's surface. (Such an orbit is said to be *geosynchronous*.) (b) Explain, with a sketch, why the radio signals from these satellites cannot directly reach receivers on earth that are north of 81.3° N latitude.

13.52 ••• A landing craft with mass 12,500 kg is in a circular orbit 5.75×10^5 m above the surface of a planet. The period of the orbit is 5800 s. The astronauts in the lander measure the diameter of the planet to be 9.60×10^6 m. The lander sets down at the north pole of the planet. What is the weight of an 85.6-kg astronaut as he steps out onto the planet's surface?

13.53 ••• What is the escape speed from a 300-km-diameter asteroid with a density of 2500 kg/m^3?

13.54 •• (a) Asteroids have average densities of about 2500 kg/m^3 and radii from 470 km down to less than a kilometer. Assuming that the asteroid has a spherically symmetric mass distribution, estimate the radius of the largest asteroid from which you could escape simply by jumping off. (*Hint:* You can estimate your jump speed by relating it to the maximum height that you can jump on earth.) (b) Europa, one of Jupiter's four large moons, has a radius of 1570 km. The acceleration due to gravity at its surface is 1.33 m/s^2. Calculate its average density.

13.55 ••• (a) Suppose you are at the earth's equator and observe a satellite passing directly overhead and moving from west to east in the sky. Exactly 12.0 hours later, you again observe this satellite to be directly overhead. How far above the earth's surface is the satellite's orbit? (b) You observe another satellite directly overhead and traveling east to west. This satellite is again overhead in 12.0 hours. How far is this satellite's orbit above the surface of the earth?

13.56 •• Planet X rotates in the same manner as the earth, around an axis through its north and south poles, and is perfectly spherical. An astronaut who weighs 943.0 N on the earth weighs 915.0 N at the north pole of Planet X and only 850.0 N at its equator. The distance from the north pole to the equator is 18,850 km, measured along the surface of Planet X. (a) How long is the day on Planet X? (b) If a 45,000-kg satellite is placed in a circular orbit 2000 km above the surface of Planet X, what will be its orbital period?

13.57 •• There are two equations from which a change in the gravitational potential energy U of the system of a mass m and the earth can be calculated. One is $U = mgy$ (Eq. 7.2). The other is $U = -Gm_Em/r$ (Eq. 13.9). As shown in Section 13.3, the first equation is correct only if the gravitational force is a constant over the change in height Δy. The second is always correct. Actually, the gravitational force is never exactly constant over any change in height, but if the variation is small, we can ignore it. Consider the difference in U between a mass at the earth's surface and a distance h above it using both equations, and find the value of h for which Eq. (7.2) is in error by 1%. Express this value of h as a fraction of the earth's radius, and also obtain a numerical value for it.

13.58 ••• **CP** Your starship, the *Aimless Wanderer*, lands on the mysterious planet Mongo. As chief scientist-engineer, you make the following measurements: A 2.50-kg stone thrown upward from the ground at 12.0 m/s returns to the ground in 6.00 s; the circumference of Mongo at the equator is 2.00×10^5 km; and there is no appreciable atmosphere on Mongo. The starship commander, Captain Confusion, asks for the following information: (a) What is the mass of Mongo? (b) If the *Aimless Wanderer* goes into a circular orbit 30,000 km above the surface of Mongo, how many hours will it take the ship to complete one orbit?

13.59 •• **CP** An astronaut, whose mission is to go where no one has gone before, lands on a spherical planet in a distant galaxy. As she stands on the surface of the planet, she releases a small rock from rest and finds that it takes the rock 0.480 s to fall 1.90 m. If the radius of the planet is 8.60×10^7 m, what is the mass of the planet?

13.60 •• In Example 13.5 (Section 13.3) we ignored the gravitational effects of the moon on a spacecraft en route from the earth to the moon. In fact, we must include the gravitational potential energy due to the moon as well. For this problem, you can ignore the motion of the earth and moon. (a) If the moon has radius R_M and the distance between the centers of the earth and the moon is R_{EM}, find the total gravitational potential energy of the particle–earth and particle–moon systems when a particle with mass m is between the earth and the moon, and a distance r from the center of the earth. Take the gravitational potential energy to be zero when the objects are far from each other. (b) There is a point along a line between the earth and the moon where the net gravitational force is zero. Use the expression derived in part (a) and numerical values from Appendix F to find the distance of this point from the center of the earth. With what speed must a spacecraft be launched from the surface of the earth just barely to reach this point? (c) If a spacecraft were launched from the earth's surface toward the moon with an initial speed of 11.2 km/s, with what speed would it impact the moon?

13.61 •• Calculate the percent difference between your weight in Sacramento, near sea level, and at the top of Mount Everest, which is 8800 m above sea level.

13.62 •• The 0.100-kg sphere in Fig. P13.62 is released from rest at the position shown in the sketch, with its center 0.400 m from the center of the 5.00-kg mass. Assume that the only forces on the 0.100-kg sphere are the gravitational forces exerted by the other two spheres and that the 5.00-kg and 10.0-kg spheres are held in place at their initial positions. What is the speed of the 0.100-kg sphere when it has moved 0.400 m to the right from its initial position?

Figure **P13.62**

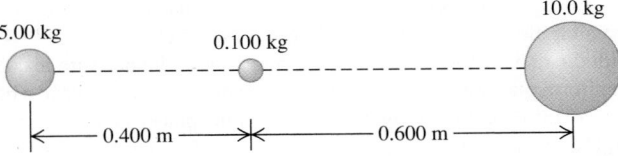

13.63 ••• An unmanned spacecraft is in a circular orbit around the moon, observing the lunar surface from an altitude of 50.0 km (see Appendix F). To the dismay of scientists on earth, an electrical fault causes an on-board thruster to fire, decreasing the speed of the spacecraft by 20.0 m/s. If nothing is done to correct its orbit, with what speed (in km/h) will the spacecraft crash into the lunar surface?

13.64 ••• **Mass of a Comet.** On July 4, 2005, the NASA spacecraft Deep Impact fired a projectile onto the surface of Comet Tempel 1. This comet is about 9.0 km across. Observations of surface debris released by the impact showed that dust with a speed as low as 1.0 m/s was able to escape the comet. (a) Assuming a spherical shape, what is the mass of this comet? (*Hint:* See Example 13.5 in Section 13.3.) (b) How far from the comet's center will this debris be when it has lost (i) 90.0% of its initial kinetic energy at the surface and (ii) all of its kinetic energy at the surface?

13.65 • **Falling Hammer.** A hammer with mass m is dropped from rest from a height h above the earth's surface. This height is not necessarily small compared with the radius R_E of the earth. If you ignore air resistance, derive an expression for the speed v of the hammer when it reaches the surface of the earth. Your expression should involve h, R_E, and m_E, the mass of the earth.

13.66 • (a) Calculate how much work is required to launch a spacecraft of mass m from the surface of the earth (mass m_E, radius R_E) and place it in a circular *low earth orbit*—that is, an orbit whose altitude above the earth's surface is much less than R_E. (As an example, the International Space Station is in low earth orbit at an altitude of about 400 km, much less than $R_E = 6380$ km.) You can ignore the kinetic energy that the spacecraft has on the ground due to the earth's rotation. (b) Calculate the minimum amount of additional work required to move the spacecraft from low earth orbit to a very great distance from the earth. You can ignore the gravitational effects of the sun, the moon, and the other planets. (c) Justify the statement: "In terms of energy, low earth orbit is halfway to the edge of the universe."

13.67 • A spacecraft is to be launched from the surface of the earth so that it will escape from the solar system altogether. (a) Find the speed relative to the center of the earth with which the spacecraft must be launched. Take into consideration the gravitational effects of both the earth and the sun, and include the effects of the earth's orbital speed, but ignore air resistance. (b) The rotation of the earth can help this spacecraft achieve escape speed. Find the speed that the spacecraft must have relative to the earth's *surface* if the spacecraft is launched from Florida at the point shown in Fig. P13.67. The rotation and orbital motions of the earth are in the same direction. The launch facilities in Florida are 28.5° north of the equator. (c) The European Space Agency (ESA) uses launch facilities in French Guiana (immediately north of Brazil), 5.15° north of the equator. What speed relative to the earth's surface would a spacecraft need to escape the solar system if launched from French Guiana?

Figure **P13.67**

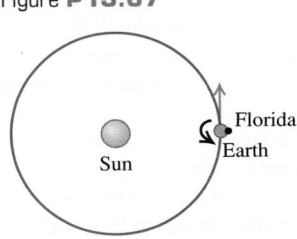

13.68 • **Gravity Inside the Earth.** Find the gravitational force that the earth exerts on a 10.0-kg mass if it is placed at the following locations. Consult Fig. 13.9, and assume a constant density through each of the interior regions (mantle, outer core, inner core), but *not* the same density in each of these regions. Use the graph to estimate the average density for each region: (a) at the surface of the earth; (b) at the outer surface of the molten outer core; (c) at the surface of the solid inner core; (d) at the center of the earth.

13.69 • **Kirkwood Gaps.** Hundreds of thousands of asteroids orbit the sun within the *asteroid belt,* which extends from about 3×10^8 km to about 5×10^8 km from the sun. (a) Find the orbital period (in years) of (i) an asteroid at the inside of the belt and (ii) an asteroid at the outside of the belt. Assume circular orbits. (b) In 1867 the American astronomer Daniel Kirkwood pointed out that several gaps exist in the asteroid belt where relatively few asteroids are found. It is now understood that these *Kirkwood gaps* are caused by the gravitational attraction of Jupiter, the largest planet, which orbits the sun once every 11.86 years. As an example, if an asteroid has an orbital period half that of Jupiter, or 5.93 years, on every other orbit this asteroid would be at its closest to Jupiter and feel a strong attraction toward the planet. This attraction, acting over and over on successive orbits, could sweep asteroids out of the Kirkwood gap. Use this hypothesis to determine the orbital radius for this Kirkwood gap. (c) One of several other Kirkwood gaps appears at a distance from the sun where the orbital period is 0.400 that of Jupiter. Explain why this happens, and find the orbital radius for this Kirkwood gap.

13.70 ••• If a satellite is in a sufficiently low orbit, it will encounter air drag from the earth's atmosphere. Since air drag does negative work (the force of air drag is directed opposite the motion), the mechanical energy will decrease. According to Eq. (13.13), if E decreases (becomes more negative), the radius r of the orbit will decrease. If air drag is relatively small, the satellite can be considered to be in a circular orbit of continually decreasing radius. (a) According to Eq. (13.10), if the radius of a satellite's circular orbit decreases, the satellite's orbital speed v *increases*. How can you reconcile this with the statement that the mechanical energy *decreases*? (*Hint:* Is air drag the only force that does work on the satellite as the orbital radius decreases?) (b) Due to air drag, the radius of a satellite's circular orbit decreases from r to $r - \Delta r$, where the positive quantity Δr is much less than r. The mass of the satellite is m. Show that the increase in orbital speed is $\Delta v = +(\Delta r/2)\sqrt{Gm_E/r^3}$; that the change in kinetic energy is $\Delta K = +(Gm_E\, m/2r^2)\,\Delta r$; that the change in gravitational potential energy is $\Delta U = -2\,\Delta K = -(Gm_E\, m/r^2)\,\Delta r$; and that the amount of work done by the force of air drag is $W = -(Gm_E\, m/2r^2)\,\Delta r$. Interpret these results in light of your comments in part (a). (c) A satellite with mass 3000 kg is initially in a circular orbit 300 km above the earth's surface. Due to air drag, the satellite's altitude decreases to 250 km. Calculate the initial orbital speed; the increase in orbital speed; the initial mechanical energy; the change in kinetic energy; the change in gravitational potential energy; the change in mechanical energy; and the work done by the force of air drag. (d) Eventually a satellite will descend to a low enough altitude in the atmosphere that the satellite burns up and the debris falls to the earth. What becomes of the initial mechanical energy?

13.71 • **Binary Star—Equal Masses.** Two identical stars with mass M orbit around their center of mass. Each orbit is circular and has radius R, so that the two stars are always on opposite sides of the circle. (a) Find the gravitational force of one star on the other. (b) Find the orbital speed of each star and the period of the orbit. (c) How much energy would be required to separate the two stars to infinity?

13.72 •• **CP Binary Star—Different Masses.** Two stars, with masses M_1 and M_2, are in circular orbits around their center of mass. The star with mass M_1 has an orbit of radius R_1; the star with mass M_2 has an orbit of radius R_2. (a) Show that the ratio of the orbital radii of the two stars equals the reciprocal of the ratio of their masses—that is, $R_1/R_2 = M_2/M_1$. (b) Explain why the two stars have the same orbital period, and show that the period T is given by $T = 2\pi(R_1 + R_2)^{3/2}/\sqrt{G(M_1 + M_2)}$. (c) The two stars in a certain binary star system move in circular orbits. The first star, Alpha, has an orbital speed of 36.0 km/s. The second star, Beta, has an orbital speed of 12.0 km/s. The orbital period is 137 d. What are the masses of each of the two stars? (d) One of the best candidates for a black hole is found in the binary system called A0620-0090. The two objects in the binary system are an orange star, V616 Monocerotis, and a compact object believed to be a black hole (see Fig. 13.27). The orbital period of A0620-0090 is 7.75 hours, the mass of V616 Monocerotis is estimated to be 0.67 times the mass of the sun, and the mass of the black hole is estimated to be 3.8 times the mass of the sun. Assuming that the orbits are circular, find the radius of each object's orbit and the orbital speed of each object. Compare these answers to the orbital radius and orbital speed of the earth in its orbit around the sun.

13.73 ••• Comets travel around the sun in elliptical orbits with large eccentricities. If a comet has speed 2.0×10^4 m/s when at a distance of 2.5×10^{11} m from the center of the sun, what is its speed when at a distance of 5.0×10^{10} m?

13.74 •• **CP** An astronaut is standing at the north pole of a newly discovered, spherically symmetric planet of radius R. In his hands he holds a container full of a liquid with mass m and volume V. At the surface of the liquid, the pressure is p_0; at a depth d below the surface, the pressure has a greater value p. From this information, determine the mass of the planet.

13.75 •• **CALC** The earth does not have a uniform density; it is most dense at its center and least dense at its surface. An approximation of its density is $\rho(r) = A - Br$, where $A = 12,700$ kg/m^3 and $B = 1.50 \times 10^{-3}$ kg/m^4. Use $R = 6.37 \times 10^6$ m for the radius of the earth approximated as a sphere. (a) Geological evidence indicates that the densities are 13,100 kg/m^3 and 2400 kg/m^3 at the earth's center and surface, respectively. What values does the linear approximation model give for the densities at these two locations? (b) Imagine dividing the earth into concentric, spherical shells. Each shell has radius r, thickness dr, volume $dV = 4\pi r^2\, dr$, and mass $dm = \rho(r)dV$. By integrating from $r = 0$ to $r = R$, show that the mass of the earth in this model is $M = \frac{4}{3}\pi R^3(A - \frac{3}{4}BR)$. (c) Show that the given values of A and B give the correct mass of the earth to within 0.4%. (d) We saw in Section 13.6 that a uniform spherical shell gives no contribution to g inside it. Show that $g(r) = \frac{4}{3}\pi Gr(A - \frac{3}{4}Br)$ inside the earth in this model. (e) Verify that the expression of part (d) gives $g = 0$ at the center of the earth and $g = 9.85$ m/s^2 at the surface. (f) Show that in this model g does *not* decrease uniformly with depth but rather has a maximum of $4\pi GA^2/9B = 10.01$ m/s^2 at $r = 2A/3B = 5640$ km.

13.76 •• **CP CALC** In Example 13.10 (Section 13.6) we saw that inside a planet of uniform density (not a realistic assumption for the earth) the acceleration due to gravity increases uniformly with distance from the center of the planet. That is, $g(r) = g_s r/R$, where g_s is the acceleration due to gravity at the surface, r is the distance from the center of the planet, and R is the radius of the planet. The interior of the planet can be treated approximately as an incompressible fluid of density ρ. (a) Replace the height y in Eq. (12.4) with the radial coordinate r and integrate to find the pressure inside a uniform planet as a function of r. Let the pressure at the surface be zero. (This means ignoring the pressure of the planet's atmosphere.) (b) Using this model, calculate the pressure at the center of the earth. (Use a value of ρ equal to the average density of the earth, calculated from the mass and radius given in Appendix F.) (c) Geologists estimate the pressure at the center of the earth to be approximately 4×10^{11} Pa. Does this agree with your calculation for the pressure at $r = 0$? What might account for any differences?

13.77 ••• **CP** Consider a spacecraft in an elliptical orbit around the earth. At the low point, or perigee, of its orbit, it is 400 km above the earth's surface; at the high point, or apogee, it is 4000 km above the earth's surface. (a) What is the period of the spacecraft's orbit? (b) Using conservation of angular momentum, find the ratio of the spacecraft's speed at perigee to its speed at apogee. (c) Using conservation of energy, find the speed at perigee and the speed at apogee. (d) It is necessary to have the spacecraft escape from the earth completely. If the spacecraft's rockets are fired at perigee, by how much would the speed have to be increased to achieve this? What if the rockets were fired at apogee? Which point in the orbit is more efficient to use?

13.78 • The planet Uranus has a radius of 25,560 km and a surface acceleration due to gravity of 11.1 m/s^2 at its poles. Its moon Miranda (discovered by Kuiper in 1948) is in a circular orbit about Uranus at an altitude of 104,000 km above the planet's surface. Miranda has a mass of 6.6×10^{19} kg and a radius of 235 km. (a) Calculate the mass of Uranus from the given data. (b) Calculate the magnitude of Miranda's acceleration due to its orbital motion about Uranus. (c) Calculate the acceleration due to Miranda's gravity at the surface of Miranda. (d) Do the answers to parts (b) and (c) mean that an object released 1 m above Miranda's surface on the side toward Uranus will fall *up* relative to Miranda? Explain.

13.79 ••• A 5000-kg spacecraft is in a circular orbit 2000 km above the surface of Mars. How much work must the spacecraft engines perform to move the spacecraft to a circular orbit that is 4000 km above the surface?

13.80 •• One of the brightest comets of the 20th century was Comet Hyakutake, which passed close to the sun in early 1996. The orbital period of this comet is estimated to be about 30,000 years. Find the semi-major axis of this comet's orbit. Compare it to the average sun–Pluto distance and to the distance to Alpha Centauri, the nearest star to the sun, which is 4.3 light-years distant.

13.81 ••• CALC Planets are not uniform inside. Normally, they are densest at the center and have decreasing density outward toward the surface. Model a spherically symmetric planet, with the same radius as the earth, as having a density that decreases linearly with distance from the center. Let the density be $15.0 \times 10^3 \ \text{kg/m}^3$ at the center and $2.0 \times 10^3 \ \text{kg/m}^3$ at the surface. What is the acceleration due to gravity at the surface of this planet?

13.82 •• CALC A uniform wire with mass M and length L is bent into a semicircle. Find the magnitude and direction of the gravitational force this wire exerts on a point with mass m placed at the center of curvature of the semicircle.

13.83 ••• CALC An object in the shape of a thin ring has radius a and mass M. A uniform sphere with mass m and radius R is placed with its center at a distance x to the right of the center of the ring, along a line through the center of the ring, and perpendicular to its plane (see Fig. E13.33). What is the gravitational force that the sphere exerts on the ring-shaped object? Show that your result reduces to the expected result when x is much larger than a.

13.84 ••• CALC A thin, uniform rod has length L and mass M. Calculate the magnitude of the gravitational force the rod exerts on a particle with mass m that is at a point along the axis of the rod a distance x from one end (see Fig. E13.32). Show that your result reduces to the expected result when x is much larger than L.

13.85 • CALC A shaft is drilled from the surface to the center of the earth (see Fig. 13.24). As in Example 13.10 (Section 13.6), make the unrealistic assumption that the density of the earth is uniform. With this approximation, the gravitational force on an object with mass m, that is inside the earth at a distance r from the center, has magnitude $F_g = Gm_E mr/R_E^3$ (as shown in Example 13.10) and points toward the center of the earth. (a) Derive an expression for the gravitational potential energy $U(r)$ of the object–earth system as a function of the object's distance from the center of the earth. Take the potential energy to be zero when the object is at the center of the earth. (b) If an object is released in the shaft at the earth's surface, what speed will it have when it reaches the center of the earth?

CHALLENGE PROBLEMS

13.86 ••• (a) When an object is in a circular orbit of radius r around the earth (mass m_E), the period of the orbit is T, given by Eq. (13.12), and the orbital speed is v, given by Eq. (13.10). Show that when the object is moved into a circular orbit of slightly larger radius $r + \Delta r$, where $\Delta r \ll r$, its new period is $T + \Delta T$ and its new orbital speed is $v - \Delta v$, where Δr, ΔT, and Δv are all positive quantities and

$$\Delta T = \frac{3\pi \, \Delta r}{v} \quad \text{and} \quad \Delta v = \frac{\pi \, \Delta r}{T}$$

[*Hint:* Use the expression $(1 + x)^n \approx 1 + nx$, valid for $|x| \ll 1$.] (b) The International Space Station (ISS) is in a nearly circular orbit at an altitude of 398.00 km above the surface of the earth. A maintenance crew is about to arrive on the space shuttle that is also in a circular orbit in the same orbital plane as the ISS, but with an altitude of 398.10 km. The crew has come to remove a faulty 125-m electrical cable, one end of which is attached to the ISS and the other end of which is floating free in space. The plan is for the shuttle to snag the free end just at the moment that the shuttle, the ISS, and the center of the earth all lie along the same line. The cable will then break free from the ISS when it becomes taut. How long after the free end is caught by the space shuttle will it detach from the ISS? Give your answer in minutes. (c) If the shuttle misses catching the cable, show that the crew must wait a time $t \approx T^2/\Delta T$ before they have a second chance. Find the numerical value of t and explain whether it would be worth the wait.

13.87 ••• **Interplanetary Navigation.** The most efficient way to send a spacecraft from the earth to another planet is by using a *Hohmann transfer orbit* (Fig. P13.87). If the orbits of the departure and destination planets are circular, the Hohmann transfer orbit is an elliptical orbit whose perihelion and aphelion are tangent to the orbits of the two planets. The rockets are fired briefly at the departure planet to put the spacecraft into the transfer orbit; the spacecraft then coasts until it reaches the destination planet. The rockets are then fired again to put the spacecraft into the same orbit about the sun as the destination planet. (a) For a flight from earth to Mars, in what direction must the rockets be fired at the earth and at Mars: in the direction of motion, or opposite the direction of motion? What about for a flight from Mars to the earth? (b) How long does a one-way trip from the the earth to Mars take, between the firings of the rockets? (c) To reach Mars from the earth, the launch must be timed so that Mars will be at the right spot when the spacecraft reaches Mars's orbit around the sun. At launch, what must the angle between a sun–Mars line and a sun–earth line be? Use data from Appendix F.

Figure **P13.87**

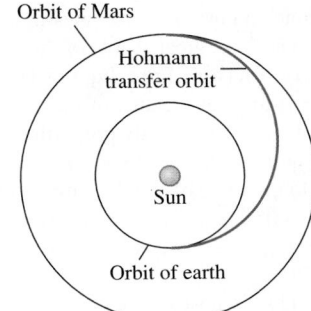

Orbit of Mars

Hohmann transfer orbit

Sun

Orbit of earth

13.88 ••• CP **Tidal Forces near a Black Hole.** An astronaut inside a spacecraft, which protects her from harmful radiation, is orbiting a black hole at a distance of 120 km from its center. The black hole is 5.00 times the mass of the sun and has a Schwarzschild radius of 15.0 km. The astronaut is positioned inside the spaceship such that one of her 0.030-kg ears is 6.0 cm farther from the black hole than the center of mass of the spacecraft and the other ear is 6.0 cm closer. (a) What is the tension between her ears? Would the astronaut find it difficult to keep from being torn apart by the gravitational forces? (Since her whole body orbits with the same angular velocity, one ear is moving too slowly for the radius of its orbit and the other is moving too fast. Hence her head must exert forces on her ears to keep them in their orbits.) (b) Is the center of gravity of her head at the same point as the center of mass? Explain.

13.89 ••• **CALC** Mass M is distributed uniformly over a disk of radius a. Find the gravitational force (magnitude and direction) between this disk-shaped mass and a particle with mass m located a distance x above the center of the disk (Fig. P13.89). Does your result reduce to the correct expression as x becomes very large? (*Hint:* Divide the disk into infinitesimally thin concentric rings, use

Figure **P13.89**

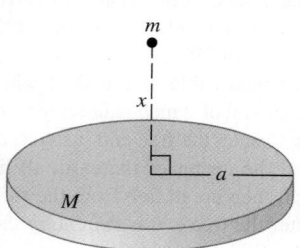

the expression derived in Exercise 13.33 for the gravitational force due to each ring, and integrate to find the total force.)

13.90 ••• **CALC** Mass M is distributed uniformly along a line of length $2L$. A particle with mass m is at a point that is a distance a above the center of the line on its perpendicular bisector (point P in Fig. P13.90). For the gravitational force that the line exerts on the particle, calculate the components perpendicular and parallel to the line. Does your result reduce to the correct expression as a becomes very large?

Figure **P13.90**

Answers

Chapter Opening Question ❓

The smaller the orbital radius r of a satellite, the faster its orbital speed v [see Eq. (13.10)]. Hence a particle near the inner edge of Saturn's rings has a faster speed than a particle near the outer edge of the rings.

Test Your Understanding Questions

13.1 Answer: (v) From Eq. (13.1), the gravitational force of the sun (mass m_1) on a planet (mass m_2) a distance r away has magnitude $F_g = Gm_1m_2/r^2$. Compared to the earth, Saturn has a value of r^2 that is $10^2 = 100$ times greater and a value of m_2 that is also 100 times greater. Hence the *force* that the sun exerts on Saturn has the same magnitude as the force that the sun exerts on earth. The *acceleration* of a planet equals the net force divided by the planet's mass: Since Saturn has 100 times more mass than the earth, its acceleration is $\frac{1}{100}$ as great as that of the earth.

13.2 Answer: (iii), (i), (ii), (iv) From Eq. (13.4), the acceleration due to gravity at the surface of a planet of mass m_P and radius R_P is $g_P = Gm_P/R_P^2$. That is, g_P is directly proportional to the planet's mass and inversely proportional to the square of its radius. It follows that compared to the value of g at the earth's surface, the value of g_P on each planet is (i) $2/2^2 = \frac{1}{2}$ as great; (ii) $4/4^2 = \frac{1}{4}$ as great; (iii) $4/2^2 = 1$ time as great—that is, the same as on earth; and (iv) $2/4^2 = \frac{1}{8}$ as great.

13.3 Answer: yes This is possible because surface gravity and escape speed depend in different ways on the planet's mass m_P and radius R_P: The value of g at the surface is Gm_P/R_P^2, while the escape speed is $\sqrt{2Gm_P/R_P}$. For the planet Saturn, for example, m_P is about 100 times the earth's mass and R_P is about 10 times the earth's radius. The value of g is different than on earth by a factor of $(100)/(10)^2 = 1$ (i.e., it is the same as on earth), while the escape speed is greater by a factor of $\sqrt{100}/10 = 3.2$. It may help to remember that the surface gravity tells you about conditions right next to the planet's surface, while the escape speed (which tells you how fast you must travel to escape to infinity) depends on conditions at *all* points between the planet's surface and infinity.

13.4 Answer: (ii) Equation (13.10) shows that in a smaller-radius orbit, the spacecraft has a faster speed. The negative work

done by air resistance decreases the *total* mechanical energy $E = K + U$; the kinetic energy K increases (becomes more positive), but the gravitational potential energy U decreases (becomes more negative) by a greater amount.

13.5 Answer: (iii) Equation (13.17) shows that the orbital period T is proportional to the $\frac{3}{2}$ power of the semi-major axis a. Hence the orbital period of Comet X is longer than that of Comet Y by a factor of $4^{3/2} = 8$.

13.6 Answer: no Our analysis shows that there is *zero* gravitational force inside a hollow spherical shell. Hence visitors to the interior of a hollow planet would find themselves weightless, and they could not stand or walk on the planet's inner surface.

13.7 Answer: (iv) The discussion following Eq. (13.27) shows that the difference between the acceleration due to gravity at the equator and at the poles is v^2/R_E. Since this planet has the same radius and hence the same circumference as the earth, the speed v at its equator must be 10 times the speed of the earth's equator. Hence v^2/R_E is $10^2 = 100$ times greater than for the earth, or $100(0.0339 \text{ m/s}^2) = 3.39 \text{ m/s}^2$. The acceleration due to gravity at the poles is 9.80 m/s^2, while at the equator it is dramatically less, $9.80 \text{ m/s}^2 - 3.39 \text{ m/s}^2 = 6.41 \text{ m/s}^2$. You can show that if this planet were to rotate 17.0 times faster than the earth, the acceleration due to gravity at the equator would be *zero* and loose objects would fly off the equator's surface!

13.8 Answer: (iii) If the sun collapsed into a black hole (which, according to our understanding of stars, it cannot do), the sun would have the same mass but a much smaller radius. Because the gravitational attraction of the sun on the earth does not depend on the sun's radius, the earth's orbit would be unaffected.

Bridging Problem

Answers: (a) Perihelion: $v_P = \sqrt{\dfrac{Gm_S}{a}\dfrac{(1+e)}{(1-e)}}$

aphelion: $v_A = \sqrt{\dfrac{Gm_S}{a}\dfrac{(1-e)}{(1+e)}}$

(b) $v_P = 54.4 \text{ km/s}$, $v_A = 0.913 \text{ km/s}$

PERIODIC MOTION

14

? Dogs walk with much quicker strides than do humans. Is this primarily because dogs' legs are shorter than human legs, less massive than human legs, or both?

LEARNING GOALS

By studying this chapter, you will learn:

- How to describe oscillations in terms of amplitude, period, frequency, and angular frequency.

- How to do calculations with simple harmonic motion, an important type of oscillation.

- How to use energy concepts to analyze simple harmonic motion.

- How to apply the ideas of simple harmonic motion to different physical situations.

- How to analyze the motions of a simple pendulum.

- What a physical pendulum is, and how to calculate the properties of its motion.

- What determines how rapidly an oscillation dies out.

- How a driving force applied to an oscillator at the right frequency can cause a very large response, or resonance.

Many kinds of motion repeat themselves over and over: the vibration of a quartz crystal in a watch, the swinging pendulum of a grandfather clock, the sound vibrations produced by a clarinet or an organ pipe, and the back-and-forth motion of the pistons in a car engine. This kind of motion, called **periodic motion** or **oscillation,** is the subject of this chapter. Understanding periodic motion will be essential for our later study of waves, sound, alternating electric currents, and light.

A body that undergoes periodic motion always has a stable equilibrium position. When it is moved away from this position and released, a force or torque comes into play to pull it back toward equilibrium. But by the time it gets there, it has picked up some kinetic energy, so it overshoots, stopping somewhere on the other side, and is again pulled back toward equilibrium. Picture a ball rolling back and forth in a round bowl or a pendulum that swings back and forth past its straight-down position.

In this chapter we will concentrate on two simple examples of systems that can undergo periodic motions: spring-mass systems and pendulums. We will also study why oscillations often tend to die out with time and why some oscillations can build up to greater and greater displacements from equilibrium when periodically varying forces act.

14.1 Describing Oscillation

Figure 14.1 shows one of the simplest systems that can have periodic motion. A body with mass *m* rests on a frictionless horizontal guide system, such as a linear air track, so it can move only along the *x*-axis. The body is attached to a spring of negligible mass that can be either stretched or compressed. The left end of the spring is held fixed and the right end is attached to the body. The spring force is the only horizontal force acting on the body; the vertical normal and gravitational forces always add to zero.

14.1 A system that can have periodic motion.

14.2 Model for periodic motion. When the body is displaced from its equilibrium position at $x = 0$, the spring exerts a restoring force back toward the equilibrium position.

(a)

$x > 0$: glider displaced to the right from the equilibrium position.

$F_x < 0$, so $a_x < 0$: stretched spring pulls glider toward equilibrium position.

(b)

$x = 0$: The relaxed spring exerts no force on the glider, so the glider has zero acceleration.

(c)

$x < 0$: glider displaced to the left from the equilibrium position.

$F_x > 0$, so $a_x > 0$: compressed spring pushes glider toward equilibrium position.

It's simplest to define our coordinate system so that the origin O is at the equilibrium position, where the spring is neither stretched nor compressed. Then x is the x-component of the **displacement** of the body from equilibrium and is also the change in the length of the spring. The x-component of the force that the spring exerts on the body is F_x, and the x-component of acceleration a_x is given by $a_x = F_x/m$.

Figure 14.2 shows the body for three different displacements of the spring. Whenever the body is displaced from its equilibrium position, the spring force tends to restore it to the equilibrium position. We call a force with this character a **restoring force.** Oscillation can occur only when there is a restoring force tending to return the system to equilibrium.

Let's analyze how oscillation occurs in this system. If we displace the body to the right to $x = A$ and then let go, the net force and the acceleration are to the left (Fig. 14.2a). The speed increases as the body approaches the equilibrium position O. When the body is at O, the net force acting on it is zero (Fig. 14.2b), but because of its motion it *overshoots* the equilibrium position. On the other side of the equilibrium position the body is still moving to the left, but the net force and the acceleration are to the right (Fig. 14.2c); hence the speed decreases until the body comes to a stop. We will show later that with an ideal spring, the stopping point is at $x = -A$. The body then accelerates to the right, overshoots equilibrium again, and stops at the starting point $x = A$, ready to repeat the whole process. The body is oscillating! If there is no friction or other force to remove mechanical energy from the system, this motion repeats forever; the restoring force perpetually draws the body back toward the equilibrium position, only to have the body overshoot time after time.

In different situations the force may depend on the displacement x from equilibrium in different ways. But oscillation *always* occurs if the force is a *restoring* force that tends to return the system to equilibrium.

Amplitude, Period, Frequency, and Angular Frequency

Here are some terms that we'll use in discussing periodic motions of all kinds:

The **amplitude** of the motion, denoted by A, is the maximum magnitude of displacement from equilibrium—that is, the maximum value of $|x|$. It is always positive. If the spring in Fig. 14.2 is an ideal one, the total overall range of the motion is $2A$. The SI unit of A is the meter. A complete vibration, or **cycle,** is one complete round trip—say, from A to $-A$ and back to A, or from O to A, back through O to $-A$, and back to O. Note that motion from one side to the other (say, $-A$ to A) is a half-cycle, not a whole cycle.

The **period,** T, is the time for one cycle. It is always positive. The SI unit is the second, but it is sometimes expressed as "seconds per cycle."

The **frequency,** f, is the number of cycles in a unit of time. It is always positive. The SI unit of frequency is the hertz:

$$1 \text{ hertz} = 1 \text{ Hz} = 1 \text{ cycle/s} = 1 \text{ s}^{-1}$$

This unit is named in honor of the German physicist Heinrich Hertz (1857–1894), a pioneer in investigating electromagnetic waves.

The **angular frequency,** ω, is 2π times the frequency:

$$\omega = 2\pi f$$

We'll learn shortly why ω is a useful quantity. It represents the rate of change of an angular quantity (not necessarily related to a rotational motion) that is always measured in radians, so its units are rad/s. Since f is in cycle/s, we may regard the number 2π as having units rad/cycle.

From the definitions of period T and frequency f we see that each is the reciprocal of the other:

$$f = \frac{1}{T} \qquad T = \frac{1}{f} \qquad \text{(relationships between frequency and period)} \qquad (14.1)$$

Also, from the definition of ω,

$$\omega = 2\pi f = \frac{2\pi}{T} \qquad \text{(angular frequency)} \qquad (14.2)$$

Example 14.1 **Period, frequency, and angular frequency**

An ultrasonic transducer used for medical diagnosis oscillates at $6.7 \text{ MHz} = 6.7 \times 10^6 \text{ Hz}$. How long does each oscillation take, and what is the angular frequency?

SOLUTION

IDENTIFY and SET UP: The target variables are the period T and the angular frequency ω. We can find these using the given frequency f in Eqs. (14.1) and (14.2).

EXECUTE: From Eqs. (14.1) and (14.2),

$$T = \frac{1}{f} = \frac{1}{6.7 \times 10^6 \text{ Hz}} = 1.5 \times 10^{-7} \text{ s} = 0.15 \ \mu\text{s}$$

$$\omega = 2\pi f = 2\pi(6.7 \times 10^6 \text{ Hz})$$
$$= (2\pi \text{ rad/cycle})(6.7 \times 10^6 \text{ cycle/s})$$
$$= 4.2 \times 10^7 \text{ rad/s}$$

EVALUATE: This is a very rapid vibration, with large f and ω and small T. A slow vibration has small f and ω and large T.

Test Your Understanding of Section 14.1 A body like that shown in Fig. 14.2 oscillates back and forth. For each of the following values of the body's x-velocity v_x and x-acceleration a_x, state whether its displacement x is positive, negative, or zero. (a) $v_x > 0$ and $a_x > 0$; (b) $v_x > 0$ and $a_x < 0$; (c) $v_x < 0$ and $a_x > 0$; (d) $v_x < 0$ and $a_x < 0$; (e) $v_x = 0$ and $a_x < 0$; (f) $v_x > 0$ and $a_x = 0$.

14.2 Simple Harmonic Motion

The simplest kind of oscillation occurs when the restoring force F_x is *directly proportional* to the displacement from equilibrium x. This happens if the spring in Figs. 14.1 and 14.2 is an ideal one that obeys Hooke's law. The constant of proportionality between F_x and x is the force constant k. (You may want to review Hooke's law and the definition of the force constant in Section 6.3.) On either side of the equilibrium position, F_x and x always have opposite signs. In Section 6.3 we represented the force acting *on* a stretched ideal spring as $F_x = kx$. The x-component of force the spring exerts *on the body* is the negative of this, so the x-component of force F_x on the body is

$$F_x = -kx \qquad \text{(restoring force exerted by an ideal spring)} \qquad (14.3)$$

This equation gives the correct magnitude and sign of the force, whether x is positive, negative, or zero (Fig. 14.3). The force constant k is always positive and has units of N/m (a useful alternative set of units is kg/s^2). We are assuming that there is no friction, so Eq. (14.3) gives the *net* force on the body.

When the restoring force is directly proportional to the displacement from equilibrium, as given by Eq. (14.3), the oscillation is called **simple harmonic motion,** abbreviated **SHM.** The acceleration $a_x = d^2x/dt^2 = F_x/m$ of a body in SHM is given by

$$a_x = \frac{d^2x}{dt^2} = -\frac{k}{m}x \qquad \text{(simple harmonic motion)} \qquad (14.4)$$

The minus sign means the acceleration and displacement always have opposite signs. This acceleration is *not* constant, so don't even think of using the constant-acceleration equations from Chapter 2. We'll see shortly how to solve this equation to find the displacement x as a function of time. A body that undergoes simple harmonic motion is called a **harmonic oscillator.**

14.3 An idealized spring exerts a restoring force that obeys Hooke's law, $F_x = -kx$. Oscillation with such a restoring force is called simple harmonic motion.

Restoring force F_x

$x < 0$
$F_x > 0$

Displacement x

O

$x > 0$
$F_x < 0$

The restoring force exerted by an idealized spring is directly proportional to the displacement (Hooke's law, $F_x = -kx$): the graph of F_x versus x is a straight line.

14.4 In most real oscillations Hooke's law applies provided the body doesn't move too far from equilibrium. In such a case small-amplitude oscillations are approximately simple harmonic.

Ideal case: The restoring force obeys Hooke's law ($F_x = -kx$), so the graph of F_x versus x is a straight line.

Restoring force F_x

Typical real case: The restoring force deviates from Hooke's law ...

O

Displacement x

... but $F_x = -kx$ can be a good approximation to the force if the displacement x is sufficiently small.

Why is simple harmonic motion important? Keep in mind that not all periodic motions are simple harmonic; in periodic motion in general, the restoring force depends on displacement in a more complicated way than in Eq. (14.3). But in many systems the restoring force is *approximately* proportional to displacement if the displacement is sufficiently small (Fig. 14.4). That is, if the amplitude is small enough, the oscillations of such systems are approximately simple harmonic and therefore approximately described by Eq. (14.4). Thus we can use SHM as an approximate model for many different periodic motions, such as the vibration of the quartz crystal in a watch, the motion of a tuning fork, the electric current in an alternating-current circuit, and the oscillations of atoms in molecules and solids.

Circular Motion and the Equations of SHM

To explore the properties of simple harmonic motion, we must express the displacement x of the oscillating body as a function of time, $x(t)$. The second derivative of this function, d^2x/dt^2, must be equal to $(-k/m)$ times the function itself, as required by Eq. (14.4). As we mentioned, the formulas for constant acceleration from Section 2.4 are no help because the acceleration changes constantly as the displacement x changes. Instead, we'll find $x(t)$ by noticing a striking similarity between SHM and another form of motion that we've already studied.

Figure 14.5a shows a top view of a horizontal disk of radius A with a ball attached to its rim at point Q. The disk rotates with constant angular speed ω (measured in rad/s), so the ball moves in uniform circular motion. A horizontal light beam shines on the rotating disk and casts a shadow of the ball on a screen. The shadow at point P oscillates back and forth as the ball moves in a circle. We then arrange a body attached to an ideal spring, like the combination shown in Figs. 14.1 and 14.2, so that the body oscillates parallel to the shadow. We will prove that the motion of the body and the motion of the ball's shadow are *identical* if the amplitude of the body's oscillation is equal to the disk radius A, and if the angular frequency $2\pi f$ of the oscillating body is equal to the angular speed ω of the rotating disk. That is, *simple harmonic motion is the projection of uniform circular motion onto a diameter.*

We can verify this remarkable statement by finding the acceleration of the shadow at P and comparing it to the acceleration of a body undergoing SHM, given by Eq. (14.4). The circle in which the ball moves so that its projection matches the motion of the oscillating body is called the **reference circle;** we will call the point Q the *reference point*. We take the reference circle to lie in the

14.5 (a) Relating uniform circular motion and simple harmonic motion. (b) The ball's shadow moves exactly like a body oscillating on an ideal spring.

(a) Apparatus for creating the reference circle

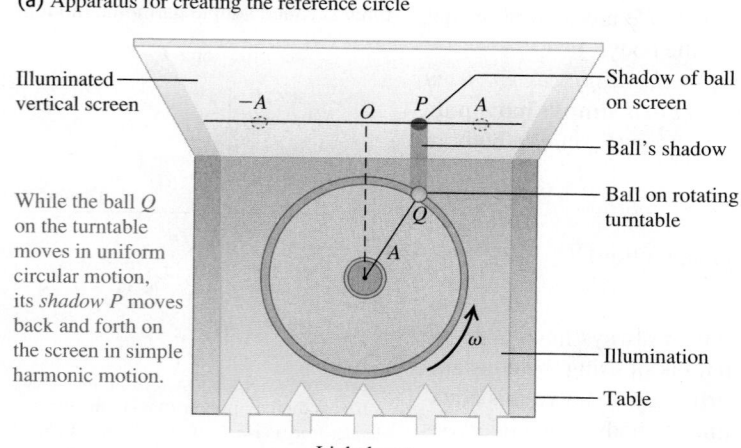

Illuminated vertical screen

Shadow of ball on screen

Ball's shadow

Ball on rotating turntable

While the ball Q on the turntable moves in uniform circular motion, its *shadow P* moves back and forth on the screen in simple harmonic motion.

Illumination

Table

Light beam

(b) An abstract representation of the motion in (a)

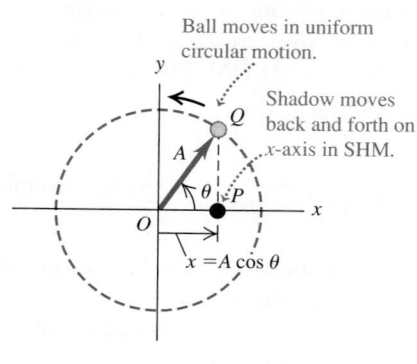

Ball moves in uniform circular motion.

Shadow moves back and forth on x-axis in SHM.

$x = A \cos \theta$

xy-plane, with the origin O at the center of the circle (Fig. 14.5b). At time t the vector OQ from the origin to the reference point Q makes an angle θ with the positive *x*-axis. As the point Q moves around the reference circle with constant angular speed ω, the vector OQ rotates with the same angular speed. Such a rotating vector is called a **phasor.** (This term was in use long before the invention of the Star Trek stun gun with a similar name. The phasor method for analyzing oscillations is useful in many areas of physics. We'll use phasors when we study alternating-current circuits in Chapter 31 and the interference of light in Chapters 35 and 36.)

The *x*-component of the phasor at time t is just the *x*-coordinate of the point Q:

$$x = A\cos\theta \tag{14.5}$$

This is also the *x*-coordinate of the shadow P, which is the *projection* of Q onto the *x*-axis. Hence the *x*-velocity of the shadow P along the *x*-axis is equal to the *x*-component of the velocity vector of point Q (Fig. 14.6a), and the *x*-acceleration of P is equal to the *x*-component of the acceleration vector of Q (Fig. 14.6b). Since point Q is in uniform circular motion, its acceleration vector $\vec{a}_Q$ is always directed toward O. Furthermore, the magnitude of $\vec{a}_Q$ is constant and given by the angular speed squared times the radius of the circle (see Section 9.3):

$$a_Q = \omega^2 A \tag{14.6}$$

Figure 14.6b shows that the *x*-component of $\vec{a}_Q$ is $a_x = -a_Q\cos\theta$. Combining this with Eqs. (14.5) and (14.6), we get that the acceleration of point P is

$$a_x = -a_Q\cos\theta = -\omega^2 A\cos\theta \quad\text{or} \tag{14.7}$$

$$a_x = -\omega^2 x \tag{14.8}$$

The acceleration of point P is directly proportional to the displacement x and always has the opposite sign. These are precisely the hallmarks of simple harmonic motion.

Equation (14.8) is *exactly* the same as Eq. (14.4) for the acceleration of a harmonic oscillator, provided that the angular speed ω of the reference point Q is related to the force constant k and mass m of the oscillating body by

$$\omega^2 = \frac{k}{m} \quad\text{or}\quad \omega = \sqrt{\frac{k}{m}} \tag{14.9}$$

We have been using the same symbol ω for the angular *speed* of the reference point Q and the angular *frequency* of the oscillating point P. The reason is that these quantities are equal! If point Q makes one complete revolution in time T, then point P goes through one complete cycle of oscillation in the same time; hence T is the period of the oscillation. During time T the point Q moves through 2π radians, so its angular speed is $\omega = 2\pi/T$. But this is just the same as Eq. (14.2) for the angular frequency of the point P, which verifies our statement about the two interpretations of ω. This is why we introduced angular frequency in Section 14.1; this quantity makes the connection between oscillation and circular motion. So we reinterpret Eq. (14.9) as an expression for the angular frequency of simple harmonic motion for a body of mass m, acted on by a restoring force with force constant k:

$$\omega = \sqrt{\frac{k}{m}} \quad \text{(simple harmonic motion)} \tag{14.10}$$

When you start a body oscillating in SHM, the value of ω is not yours to choose; it is predetermined by the values of k and m. The units of k are N/m or kg/s^2, so k/m is in $(\text{kg/s}^2)/\text{kg} = \text{s}^{-2}$. When we take the square root in Eq. (14.10), we get s^{-1}, or more properly rad/s because this is an *angular* frequency (recall that a radian is not a true unit).

14.6 The (a) *x*-velocity and (b) *x*-acceleration of the ball's shadow P (see Fig. 14.5) are the *x*-components of the velocity and acceleration vectors, respectively, of the ball Q.

(a) Using the reference circle to determine the *x*-velocity of point P

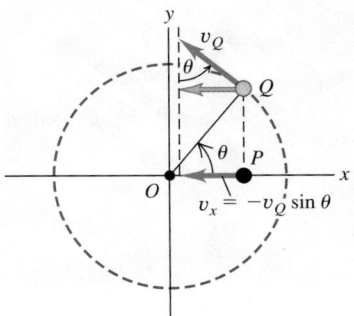

(b) Using the reference circle to determine the *x*-acceleration of point P

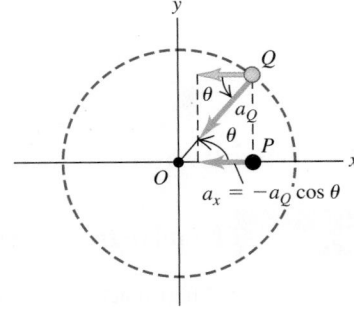

According to Eqs. (14.1) and (14.2), the frequency f and period T are

$$f = \frac{\omega}{2\pi} = \frac{1}{2\pi}\sqrt{\frac{k}{m}} \quad \text{(simple harmonic motion)} \tag{14.11}$$

$$T = \frac{1}{f} = \frac{2\pi}{\omega} = 2\pi\sqrt{\frac{m}{k}} \quad \text{(simple harmonic motion)} \tag{14.12}$$

14.7 The greater the mass m in a tuning fork's tines, the lower the frequency of oscillation $f = (1/2\pi)\sqrt{k/m}$ and the lower the pitch of the sound that the tuning fork produces.

Tines with large mass m: low frequency $f = 128$ Hz

Tines with small mass m: high frequency $f = 4096$ Hz

We see from Eq. (14.12) that a larger mass m, with its greater inertia, will have less acceleration, move more slowly, and take a longer time for a complete cycle (Fig. 14.7). In contrast, a stiffer spring (one with a larger force constant k) exerts a greater force at a given deformation x, causing greater acceleration, higher speeds, and a shorter time T per cycle.

CAUTION **Don't confuse frequency and angular frequency** You can run into trouble if you don't make the distinction between frequency f and angular frequency $\omega = 2\pi f$. Frequency tells you how many cycles of oscillation occur per second, while angular frequency tells you how many radians per second this corresponds to on the reference circle. In solving problems, pay careful attention to whether the goal is to find f or ω.

Period and Amplitude in SHM

Equations (14.11) and (14.12) show that the period and frequency of simple harmonic motion are completely determined by the mass m and the force constant k. *In simple harmonic motion the period and frequency do not depend on the amplitude A.* For given values of m and k, the time of one complete oscillation is the same whether the amplitude is large or small. Equation (14.3) shows why we should expect this. Larger A means that the body reaches larger values of $|x|$ and is subjected to larger restoring forces. This increases the average speed of the body over a complete cycle; this exactly compensates for having to travel a larger distance, so the same total time is involved.

The oscillations of a tuning fork are essentially simple harmonic motion, which means that it always vibrates with the same frequency, independent of amplitude. This is why a tuning fork can be used as a standard for musical pitch. If it were not for this characteristic of simple harmonic motion, it would be impossible to make familiar types of mechanical and electronic clocks run accurately or to play most musical instruments in tune. If you encounter an oscillating body with a period that *does* depend on the amplitude, the oscillation is *not* simple harmonic motion.

Example 14.2 **Angular frequency, frequency, and period in SHM**

A spring is mounted horizontally, with its left end fixed. A spring balance attached to the free end and pulled toward the right (Fig. 14.8a) indicates that the stretching force is proportional to the displacement, and a force of 6.0 N causes a displacement of 0.030 m. We replace the spring balance with a 0.50-kg glider, pull it 0.020 m to the right along a frictionless air track, and release it from rest (Fig. 14.8b). (a) Find the force constant k of the spring. (b) Find the angular frequency ω, frequency f, and period T of the resulting oscillation.

14.8 (a) The force exerted *on* the spring (shown by the vector F) has x-component $F_x = +6.0$ N. The force exerted *by* the spring has x-component $F_x = -6.0$ N. (b) A glider is attached to the same spring and allowed to oscillate.

(a)

$F = 6.0$ N

m

$x = 0$ $x = 0.030$ m

(b)

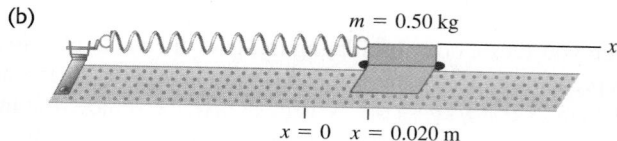

$m = 0.50$ kg

$x = 0$ $x = 0.020$ m

SOLUTION

IDENTIFY and SET UP: Because the spring force (equal in magnitude to the stretching force) is proportional to the displacement, the motion is simple harmonic. We find k using Hooke's law, Eq. (14.3), and ω, f, and T using Eqs. (14.10), (14.11), and (14.12), respectively.

EXECUTE: (a) When $x = 0.030$ m, the force the spring exerts on the spring balance is $F_x = -6.0$ N. From Eq. (14.3),

$$k = -\frac{F_x}{x} = -\frac{-6.0 \text{ N}}{0.030 \text{ m}} = 200 \text{ N/m} = 200 \text{ kg/s}^2$$

(b) From Eq. (14.10), with $m = 0.50$ kg,

$$\omega = \sqrt{\frac{k}{m}} = \sqrt{\frac{200 \text{ kg/s}^2}{0.50 \text{ kg}}} = 20 \text{ rad/s}$$

$$f = \frac{\omega}{2\pi} = \frac{20 \text{ rad/s}}{2\pi \text{ rad/cycle}} = 3.2 \text{ cycle/s} = 3.2 \text{ Hz}$$

$$T = \frac{1}{f} = \frac{1}{3.2 \text{ cycle/s}} = 0.31 \text{ s}$$

EVALUATE: The amplitude of the oscillation is 0.020 m, the distance that we pulled the glider before releasing it. In SHM the angular frequency, frequency, and period are all independent of the amplitude. Note that a period is usually stated in "seconds" rather than "seconds per cycle."

Displacement, Velocity, and Acceleration in SHM

We still need to find the displacement x as a function of time for a harmonic oscillator. Equation (14.4) for a body in simple harmonic motion along the x-axis is identical to Eq. (14.8) for the x-coordinate of the reference point in uniform circular motion with constant angular speed $\omega = \sqrt{k/m}$. Hence Eq. (14.5), $x = A\cos\theta$, describes the x-coordinate for both of these situations. If at $t = 0$ the phasor OQ makes an angle ϕ (the Greek letter phi) with the positive x-axis, then at any later time t this angle is $\theta = \omega t + \phi$. We substitute this into Eq. (14.5) to obtain

$$x = A\cos(\omega t + \phi) \quad \text{(displacement in SHM)} \quad (14.13)$$

where $\omega = \sqrt{k/m}$. Figure 14.9 shows a graph of Eq. (14.13) for the particular case $\phi = 0$. The displacement x is a periodic function of time, as expected for SHM. We could also have written Eq. (14.13) in terms of a sine function rather than a cosine by using the identity $\cos\alpha = \sin(\alpha + \pi/2)$. *In simple harmonic motion the position is a periodic, sinusoidal function of time.* There are many other periodic functions, but none so simple as a sine or cosine function.

The value of the cosine function is always between -1 and 1, so in Eq. (14.13), x is always between $-A$ and A. This confirms that A is the amplitude of the motion.

The period T is the time for one complete cycle of oscillation, as Fig. 14.9 shows. The cosine function repeats itself whenever the quantity in parentheses in Eq. (14.13) increases by 2π radians. Thus, if we start at time $t = 0$, the time T to complete one cycle is given by

$$\omega T = \sqrt{\frac{k}{m}} T = 2\pi \quad \text{or} \quad T = 2\pi\sqrt{\frac{m}{k}}$$

which is just Eq. (14.12). Changing either m or k changes the period of oscillation, as shown in Figs. 14.10a and 14.10b. The period does not depend on the amplitude A (Fig. 14.10c).

PhET: Motion in 2D
ActivPhysics 9.1: Position Graphs and Equations
ActivPhysics 9.2: Describing Vibrational Motion
ActivPhysics 9.5: Age Drops Tarzan

14.9 Graph of x versus t [see Eq. (14.13)] for simple harmonic motion. The case shown has $\phi = 0$.

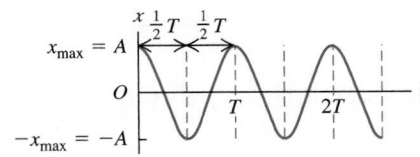

14.10 Variations of simple harmonic motion. All cases shown have $\phi = 0$ [see Eq. (14.13)].

(a) Increasing m; same A and k
Mass m increases from curve 1 to 2 to 3. Increasing m alone increases the period.

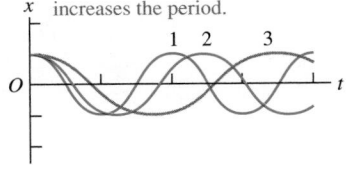

(b) Increasing k; same A and m
Force constant k increases from curve 1 to 2 to 3. Increasing k alone decreases the period.

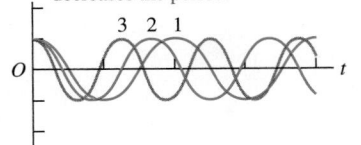

(c) Increasing A; same k and m
Amplitude A increases from curve 1 to 2 to 3. Changing A alone has no effect on the period.

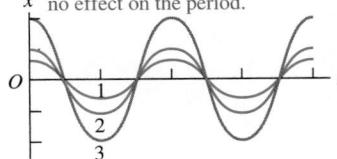

14.11 Variations of SHM: displacement versus time for the same harmonic oscillator with different phase angles ϕ.

These three curves show SHM with the same period T and amplitude A but with different phase angles ϕ.

14.12 Graphs of (a) x versus t, (b) v_x versus t, and (c) a_x versus t for a body in SHM. For the motion depicted in these graphs, $\phi = \pi/3$.

(a) Displacement x as a function of time t

(b) Velocity v_x as a function of time t

The v_x-t graph is shifted by $\frac{1}{4}$ cycle from the x-t graph.

(c) Acceleration a_x as a function of time t

The a_x-t graph is shifted by $\frac{1}{4}$ cycle from the v_x-t graph and by $\frac{1}{2}$ cycle from the x-t graph.

The constant ϕ in Eq. (14.13) is called the **phase angle.** It tells us at what point in the cycle the motion was at $t = 0$ (equivalent to where around the circle the point Q was at $t = 0$). We denote the position at $t = 0$ by x_0. Putting $t = 0$ and $x = x_0$ in Eq. (14.13), we get

$$x_0 = A \cos \phi \qquad (14.14)$$

If $\phi = 0$, then $x_0 = A \cos 0 = A$, and the body starts at its maximum positive displacement. If $\phi = \pi$, then $x_0 = A \cos \pi = -A$, and the particle starts at its maximum *negative* displacement. If $\phi = \pi/2$, then $x_0 = A \cos(\pi/2) = 0$, and the particle is initially at the origin. Figure 14.11 shows the displacement x versus time for three different phase angles.

We find the velocity v_x and acceleration a_x as functions of time for a harmonic oscillator by taking derivatives of Eq. (14.13) with respect to time:

$$v_x = \frac{dx}{dt} = -\omega A \sin(\omega t + \phi) \qquad \text{(velocity in SHM)} \qquad (14.15)$$

$$a_x = \frac{dv_x}{dt} = \frac{d^2x}{dt^2} = -\omega^2 A \cos(\omega t + \phi) \qquad \text{(acceleration in SHM)} \qquad (14.16)$$

The velocity v_x oscillates between $v_{max} = +\omega A$ and $-v_{max} = -\omega A$, and the acceleration a_x oscillates between $a_{max} = +\omega^2 A$ and $-a_{max} = -\omega^2 A$ (Fig. 14.12). Comparing Eq. (14.16) with Eq. (14.13) and recalling that $\omega^2 = k/m$ from Eq. (14.9), we see that

$$a_x = -\omega^2 x = -\frac{k}{m} x$$

which is just Eq. (14.4) for simple harmonic motion. This confirms that Eq. (14.13) for x as a function of time is correct.

We actually derived Eq. (14.16) earlier in a geometrical way by taking the x-component of the acceleration vector of the reference point Q. This was done in Fig. 14.6b and Eq. (14.7) (recall that $\theta = \omega t + \phi$). In the same way, we could have derived Eq. (14.15) by taking the x-component of the velocity vector of Q, as shown in Fig. 14.6b. We'll leave the details for you to work out.

Note that the sinusoidal graph of displacement versus time (Fig. 14.12a) is shifted by one-quarter period from the graph of velocity versus time (Fig. 14.12b) and by one-half period from the graph of acceleration versus time (Fig. 14.12c). Figure 14.13 shows why this is so. When the body is passing through the equilibrium position so that the displacement is zero, the velocity equals either v_{max} or $-v_{max}$ (depending on which way the body is moving) and the acceleration is zero. When the body is at either its maximum positive displacement, $x = +A$, or its maximum negative displacement, $x = -A$, the velocity is zero and the body is instantaneously at rest. At these points, the restoring force $F_x = -kx$ and the acceleration of the body have their maximum magnitudes. At $x = +A$ the acceleration is negative and equal to $-a_{max}$. At $x = -A$ the acceleration is positive: $a_x = +a_{max}$.

If we are given the initial position x_0 and initial velocity v_{0x} for the oscillating body, we can determine the amplitude A and the phase angle ϕ. Here's how to do it. The initial velocity v_{0x} is the velocity at time $t = 0$; putting $v_x = v_{0x}$ and $t = 0$ in Eq. (14.15), we find

$$v_{0x} = -\omega A \sin \phi \qquad (14.17)$$

To find ϕ, we divide Eq. (14.17) by Eq. (14.14). This eliminates A and gives an equation that we can solve for ϕ:

$$\frac{v_{0x}}{x_0} = \frac{-\omega A \sin \phi}{A \cos \phi} = -\omega \tan \phi$$

$$\phi = \arctan\left(-\frac{v_{0x}}{\omega x_0}\right) \quad \text{(phase angle in SHM)} \quad (14.18)$$

It is also easy to find the amplitude A if we are given x_0 and v_{0x}. We'll sketch the derivation, and you can fill in the details. Square Eq. (14.14); then divide Eq. (14.17) by ω, square it, and add to the square of Eq. (14.14). The right side will be $A^2(\sin^2 \phi + \cos^2 \phi)$, which is equal to A^2. The final result is

$$A = \sqrt{x_0^2 + \frac{v_{0x}^2}{\omega^2}} \quad \text{(amplitude in SHM)} \quad (14.19)$$

Note that when the body has both an initial displacement x_0 and a nonzero initial velocity v_{0x}, the amplitude A is *not* equal to the initial displacement. That's reasonable; if you start the body at a positive x_0 but give it a positive velocity v_{0x}, it will go *farther* than x_0 before it turns and comes back.

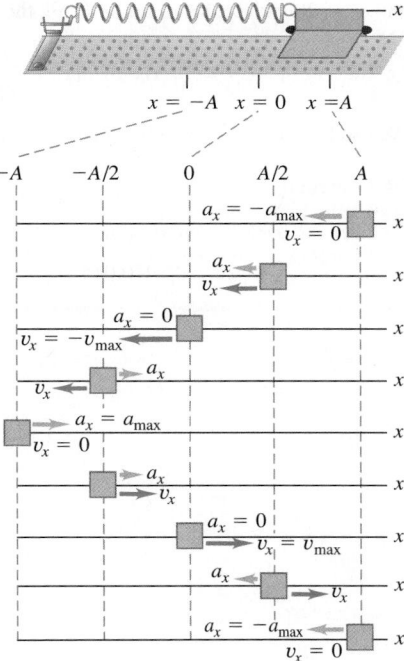

14.13 How x-velocity v_x and x-acceleration a_x vary during one cycle of SHM.

| **Problem-Solving Strategy 14.1** | **Simple Harmonic Motion I: Describing Motion** |

IDENTIFY *the relevant concepts:* An oscillating system undergoes simple harmonic motion (SHM) *only* if the restoring force is directly proportional to the displacement.

SET UP *the problem* using the following steps:
1. Identify the known and unknown quantities, and determine which are the target variables.
2. Distinguish between two kinds of quantities. *Properties of the system* include the mass m, the force constant k, and quantities derived from m and k, such as the period T, frequency f, and angular frequency ω. These are independent of *properties of the motion*, which describe how the system behaves when it is set into motion in a particular way; they include the amplitude A, maximum velocity v_{max}, and phase angle ϕ, and values of x, v_x, and a_x at particular times.
3. If necessary, define an x-axis as in Fig. 14.13, with the equilibrium position at $x = 0$.

EXECUTE *the solution* as follows:
1. Use the equations given in Sections 14.1 and 14.2 to solve for the target variables.
2. To find the values of x, v_x, and a_x at particular times, use Eqs. (14.13), (14.15), and (14.16), respectively. If the initial position x_0 and initial velocity v_{0x} are both given, determine ϕ and A from Eqs. (14.18) and (14.19). If the body has an initial positive displacement x_0 but zero initial velocity ($v_{0x} = 0$), then the amplitude is $A = x_0$ and the phase angle is $\phi = 0$. If it has an initial positive velocity v_{0x} but no initial displacement ($x_0 = 0$), the amplitude is $A = v_{0x}/\omega$ and the phase angle is $\phi = -\pi/2$. Express all phase angles in *radians*.

EVALUATE *your answer:* Make sure that your results are consistent. For example, suppose you used x_0 and v_{0x} to find general expressions for x and v_x at time t. If you substitute $t = 0$ into these expressions, you should get back the given values of x_0 and v_{0x}.

Example 14.3 | **Describing SHM**

We give the glider of Example 14.2 an initial displacement $x_0 = +0.015$ m and an initial velocity $v_{0x} = +0.40$ m/s. (a) Find the period, amplitude, and phase angle of the resulting motion. (b) Write equations for the displacement, velocity, and acceleration as functions of time.

SOLUTION

IDENTIFY and SET UP: As in Example 14.2, the oscillations are SHM. We use equations from this section and the given values $k = 200$ N/m, $m = 0.50$ kg, x_0, and v_{0x} to calculate the target variables A and ϕ and to obtain expressions for x, v_x, and a_x.

Continued

EXECUTE: (a) In SHM the period and angular frequency are *properties of the system* that depend only on k and m, not on the amplitude, and so are the same as in Example 14.2 ($T = 0.31$ s and $\omega = 20$ rad/s). From Eq. (14.19), the amplitude is

$$A = \sqrt{x_0^2 + \frac{v_{0x}^2}{\omega^2}} = \sqrt{(0.015 \text{ m})^2 + \frac{(0.40 \text{ m/s})^2}{(20 \text{ rad/s})^2}} = 0.025 \text{ m}$$

We use Eq. (14.18) to find the phase angle:

$$\phi = \arctan\left(-\frac{v_{0x}}{\omega x_0}\right)$$

$$= \arctan\left(-\frac{0.40 \text{ m/s}}{(20 \text{ rad/s})(0.015 \text{ m})}\right) = -53° = -0.93 \text{ rad}$$

(b) The displacement, velocity, and acceleration at any time are given by Eqs. (14.13), (14.15), and (14.16), respectively. We substitute the values of A, ω, and ϕ into these equations:

$$x = (0.025 \text{ m})\cos\left[(20 \text{ rad/s})t - 0.93 \text{ rad}\right]$$
$$v_x = -(0.50 \text{ m/s})\sin\left[(20 \text{ rad/s})t - 0.93 \text{ rad}\right]$$
$$a_x = -(10 \text{ m/s}^2)\cos\left[(20 \text{ rad/s})t - 0.93 \text{ rad}\right]$$

EVALUATE: You can check the expressions for x and v_x by confirming that if you substitute $t = 0$, they yield $x = x_0 = 0.015$ m and $v_x = v_{0x} = 0.40$ m/s.

Test Your Understanding of Section 14.2 A glider is attached to a spring as shown in Fig. 14.13. If the glider is moved to $x = 0.10$ m and released from rest at time $t = 0$, it will oscillate with amplitude $A = 0.10$ m and phase angle $\phi = 0$. (a) Suppose instead that at $t = 0$ the glider is at $x = 0.10$ m and is moving to the right in Fig. 14.13. In this situation is the amplitude greater than, less than, or equal to 0.10 m? Is the phase angle greater than, less than, or equal to zero? (b) Suppose instead that at $t = 0$ the glider is at $x = 0.10$ m and is moving to the left in Fig. 14.13. In this situation is the amplitude greater than, less than, or equal to 0.10 m? Is the phase angle greater than, less than, or equal to zero?

MasteringPHYSICS

PhET: Masses & Springs
ActivPhysics 9.3: Vibrational Energy
ActivPhysics 9.4: Two Ways to Weigh Young Tarzan
ActivPhysics 9.6: Releasing a Vibrating Skier I
ActivPhysics 9.7: Releasing a Vibrating Skier II
ActivPhysics 9.8: One- and Two-Spring Vibrating Systems
ActivPhysics 9.9: Vibro-Ride

14.3 Energy in Simple Harmonic Motion

We can learn even more about simple harmonic motion by using energy considerations. Take another look at the body oscillating on the end of a spring in Figs. 14.2 and 14.13. We've already noted that the spring force is the only horizontal force on the body. The force exerted by an ideal spring is a conservative force, and the vertical forces do no work, so the total mechanical energy of the system is *conserved*. We also assume that the mass of the spring itself is negligible.

The kinetic energy of the body is $K = \frac{1}{2}mv^2$ and the potential energy of the spring is $U = \frac{1}{2}kx^2$, just as in Section 7.2. (You'll find it helpful to review that section.) There are no nonconservative forces that do work, so the total mechanical energy $E = K + U$ is conserved:

$$E = \tfrac{1}{2}mv_x^2 + \tfrac{1}{2}kx^2 = \text{constant} \qquad (14.20)$$

(Since the motion is one-dimensional, $v^2 = v_x^2$.)

The total mechanical energy E is also directly related to the amplitude A of the motion. When the body reaches the point $x = A$, its maximum displacement from equilibrium, it momentarily stops as it turns back toward the equilibrium position. That is, when $x = A$ (or $-A$), $v_x = 0$. At this point the energy is entirely potential, and $E = \frac{1}{2}kA^2$. Because E is constant, it is equal to $\frac{1}{2}kA^2$ at any other point. Combining this expression with Eq. (14.20), we get

$$E = \tfrac{1}{2}mv_x^2 + \tfrac{1}{2}kx^2 = \tfrac{1}{2}kA^2 = \text{constant} \qquad \begin{array}{l}\text{(total mechanical} \\ \text{energy in SHM)}\end{array} \qquad (14.21)$$

We can verify this equation by substituting x and v_x from Eqs. (14.13) and (14.15) and using $\omega^2 = k/m$ from Eq. (14.9):

$$E = \tfrac{1}{2}mv_x^2 + \tfrac{1}{2}kx^2 = \tfrac{1}{2}m[-\omega A\sin(\omega t + \phi)]^2 + \tfrac{1}{2}k[A\cos(\omega t + \phi)]^2$$
$$= \tfrac{1}{2}kA^2\sin^2(\omega t + \phi) + \tfrac{1}{2}kA^2\cos^2(\omega t + \phi)$$
$$= \tfrac{1}{2}kA^2$$

14.14 Graphs of E, K, and U versus displacement in SHM. The velocity of the body is *not* constant, so these images of the body at equally spaced positions are *not* equally spaced in time.

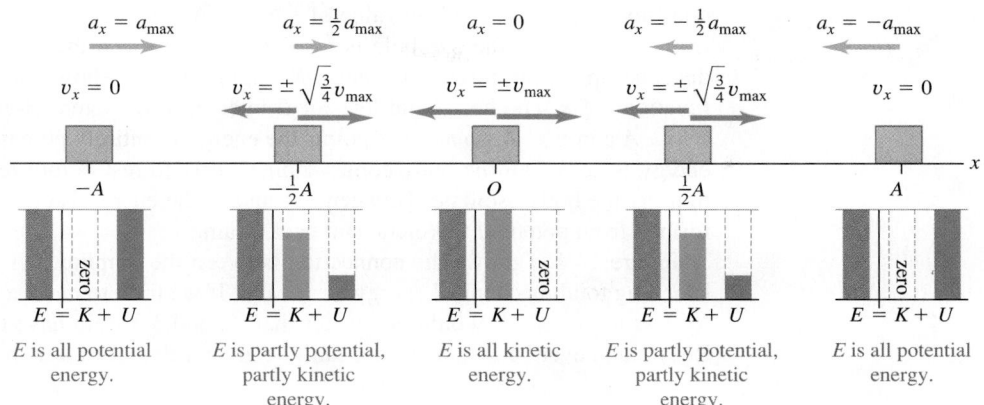

(Recall that $\sin^2\alpha + \cos^2\alpha = 1$.) Hence our expressions for displacement and velocity in SHM are consistent with energy conservation, as they must be.

We can use Eq. (14.21) to solve for the velocity v_x of the body at a given displacement x:

$$v_x = \pm\sqrt{\frac{k}{m}}\sqrt{A^2 - x^2} \qquad (14.22)$$

The $\pm$ sign means that at a given value of x the body can be moving in either direction. For example, when $x = \pm A/2$,

$$v_x = \pm\sqrt{\frac{k}{m}}\sqrt{A^2 - \left(\pm\frac{A}{2}\right)^2} = \pm\sqrt{\frac{3}{4}}\sqrt{\frac{k}{m}}A$$

Equation (14.22) also shows that the *maximum* speed v_{max} occurs at $x = 0$. Using Eq. (14.10), $\omega = \sqrt{k/m}$, we find that

$$v_{max} = \sqrt{\frac{k}{m}}A = \omega A \qquad (14.23)$$

This agrees with Eq. (14.15): v_x oscillates between $-\omega A$ and $+\omega A$.

Interpreting E, K, and U in SHM

Figure 14.14 shows the energy quantities E, K, and U at $x = 0$, $x = \pm A/2$, and $x = \pm A$. Figure 14.15 is a graphical display of Eq. (14.21); energy (kinetic, potential, and total) is plotted vertically and the coordinate x is plotted horizontally.

(a) The potential energy U and total mechanical energy E for a body in SHM as a function of displacement x

The total mechanical energy E is constant.

(b) The same graph as in **(a)**, showing kinetic energy K as well

At $x = \pm A$ the energy is all potential; the kinetic energy is zero.

At $x = 0$ the energy is all kinetic; the potential energy is zero.

At these points the energy is half kinetic and half potential.

14.15 Kinetic energy K, potential energy U, and total mechanical energy E as functions of position for SHM. At each value of x the sum of the values of K and U equals the constant value of E. Can you show that the energy is half kinetic and half potential at $x = \pm\sqrt{\frac{1}{2}}A$?

The parabolic curve in Fig. 14.15a represents the potential energy $U = \frac{1}{2}kx^2$. The horizontal line represents the total mechanical energy E, which is constant and does not vary with x. At any value of x between $-A$ and A, the vertical distance from the x-axis to the parabola is U; since $E = K + U$, the remaining vertical distance up to the horizontal line is K. Figure 14.15b shows both K and U as functions of x. The horizontal line for E intersects the potential-energy curve at $x = -A$ and $x = A$, so at these points the energy is entirely potential, the kinetic energy is zero, and the body comes momentarily to rest before reversing direction. As the body oscillates between $-A$ and A, the energy is continuously transformed from potential to kinetic and back again.

Figure 14.15a shows the connection between the amplitude A and the corresponding total mechanical energy $E = \frac{1}{2}kA^2$. If we tried to make x greater than A (or less than $-A$), U would be greater than E, and K would have to be negative. But K can never be negative, so x can't be greater than A or less than $-A$.

Problem-Solving Strategy 14.2 | **Simple Harmonic Motion II: Energy**

The SHM energy equation, Eq. (14.21), is a useful relationship among velocity, position, and total mechanical energy. If the problem requires you to relate position, velocity, and acceleration without reference to time, consider using Eq. (14.4) (from Newton's second law) or Eq. (14.21) (from energy conservation). Because Eq. (14.21) involves x^2 and v_x^2, you must infer the *signs* of x and v_x from the situation. For instance, if the body is moving from the equilibrium position toward the point of greatest positive displacement, then x is positive and v_x is positive.

Example 14.4 | **Velocity, acceleration, and energy in SHM**

(a) Find the maximum and minimum velocities attained by the oscillating glider of Example 14.2. (b) Find the maximum and minimum accelerations. (c) Find the velocity v_x and acceleration a_x when the glider is halfway from its initial position to the equilibrium position $x = 0$. (d) Find the total energy, potential energy, and kinetic energy at this position.

SOLUTION

IDENTIFY and SET UP: The problem concerns properties of the motion at specified *positions*, not at specified *times*, so we can use the energy relationships of this section. Figure 14.13 shows our choice of x-axis. The maximum displacement from equilibrium is $A = 0.020$ m. We use Eqs. (14.22) and (14.4) to find v_x and a_x for a given x. We then use Eq. (14.21) for given x and v_x to find the total, potential, and kinetic energies E, U, and K.

EXECUTE: (a) From Eq. (14.22), the velocity v_x at any displacement x is

$$v_x = \pm\sqrt{\frac{k}{m}}\sqrt{A^2 - x^2}$$

The glider's maximum *speed* occurs when it is moving through $x = 0$:

$$v_{max} = \sqrt{\frac{k}{m}}\,A = \sqrt{\frac{200\ \text{N/m}}{0.50\ \text{kg}}}(0.020\ \text{m}) = 0.40\ \text{m/s}$$

Its maximum and minimum (most negative) *velocities* are $+0.40$ m/s and -0.40 m/s, which occur when it is moving through $x = 0$ to the right and left, respectively.

(b) From Eq. (14.4), $a_x = -(k/m)x$. The glider's maximum (most positive) acceleration occurs at the most negative value of x, $x = -A$:

$$a_{max} = -\frac{k}{m}(-A) = -\frac{200\ \text{N/m}}{0.50\ \text{kg}}(-0.020\ \text{m}) = 8.0\ \text{m/s}^2$$

The minimum (most negative) acceleration is $a_{min} = -8.0$ m/s^2, which occurs at $x = +A = +0.020$ m.

(c) The point halfway from $x = x_0 = A$ to $x = 0$ is $x = A/2 = 0.010$ m. From Eq. (14.22), at this point

$$v_x = -\sqrt{\frac{200\ \text{N/m}}{0.50\ \text{kg}}}\sqrt{(0.020\ \text{m})^2 - (0.010\ \text{m})^2} = -0.35\ \text{m/s}$$

We choose the negative square root because the glider is moving from $x = A$ toward $x = 0$. From Eq. (14.4),

$$a_x = -\frac{200\ \text{N/m}}{0.50\ \text{kg}}(0.010\ \text{m}) = -4.0\ \text{m/s}^2$$

Figure 14.14 shows the conditions at $x = 0$, $\pm A/2$, and $\pm A$.

(d) The energies are

$$E = \tfrac{1}{2}kA^2 = \tfrac{1}{2}(200\ \text{N/m})(0.020\ \text{m})^2 = 0.040\ \text{J}$$

$$U = \tfrac{1}{2}kx^2 = \tfrac{1}{2}(200\ \text{N/m})(0.010\ \text{m})^2 = 0.010\ \text{J}$$

$$K = \tfrac{1}{2}mv_x^2 = \tfrac{1}{2}(0.50\ \text{kg})(-0.35\ \text{m/s})^2 = 0.030\ \text{J}$$

EVALUATE: At $x = A/2$, the total energy is one-fourth potential energy and three-fourths kinetic energy. You can confirm this by inspecting Fig. 14.15b.

Example 14.5 Energy and momentum in SHM

A block of mass M attached to a horizontal spring with force constant k is moving in SHM with amplitude A_1. As the block passes through its equilibrium position, a lump of putty of mass m is dropped from a small height and sticks to it. (a) Find the new amplitude and period of the motion. (b) Repeat part (a) if the putty is dropped onto the block when it is at one end of its path.

SOLUTION

IDENTIFY and SET UP: The problem involves the motion at a given position, not a given time, so we can use energy methods. Figure 14.16 shows our sketches. Before the putty falls, the mechanical energy of the block–spring system is constant. In part (a), the putty–block collision is completely inelastic: The horizontal component of momentum is conserved, kinetic energy decreases, and the amount of mass that's oscillating increases. After the collision, the mechanical energy remains constant at its new value. In part (b) the oscillating mass also increases, but the block isn't moving when the putty is added; there is effectively no collision at all, and no mechanical energy is lost. We find the amplitude A_2 after each collision from the final energy of the system using Eq. (14.21) and conservation of momentum. The period T_2 after the collision is a *property of the system*, so it is the same in both parts (a) and (b); we find it using Eq. (14.12).

EXECUTE: (a) Before the collision the total mechanical energy of the block and spring is $E_1 = \frac{1}{2}kA_1^2$. The block is at $x = 0$, so $U = 0$ and the energy is purely kinetic (Fig. 14.16a). If we let v_1 be the speed of the block at this point, then $E_1 = \frac{1}{2}kA_1^2 = \frac{1}{2}Mv_1^2$ and

$$v_1 = \sqrt{\frac{k}{M}}A_1$$

During the collision the x-component of momentum of the block–putty system is conserved. (Why?) Just before the collision this component is the sum of Mv_1 (for the block) and zero (for the putty). Just after the collision the block and putty move together with speed v_2, so their combined x-component of momentum is $(M + m)v_2$. From conservation of momentum,

$$Mv_1 + 0 = (M + m)v_2 \quad \text{so} \quad v_2 = \frac{M}{M + m}v_1$$

We assume that the collision lasts a very short time, so that the block and putty are still at the equilibrium position just after the collision. The energy is still purely kinetic but is *less* than before the collision:

$$E_2 = \frac{1}{2}(M + m)v_2^2 = \frac{1}{2}\frac{M^2}{M + m}v_1^2$$

$$= \frac{M}{M + m}\left(\frac{1}{2}Mv_1^2\right) = \left(\frac{M}{M + m}\right)E_1$$

14.16 Our sketches for this problem.

(a)

(b)

Since $E_2 = \frac{1}{2}kA_2^2$, where A_2 is the amplitude after the collision, we have

$$\frac{1}{2}kA_2^2 = \left(\frac{M}{M + m}\right)\frac{1}{2}kA_1^2$$

$$A_2 = A_1\sqrt{\frac{M}{M + m}}$$

From Eq. (14.12), the period of oscillation after the collision is

$$T_2 = 2\pi\sqrt{\frac{M + m}{k}}$$

(b) When the putty falls, the block is instantaneously at rest (Fig. 14.16b). The x-component of momentum is zero both before and after the collision. The block and putty have zero kinetic energy just before and just after the collision. The energy is all potential energy stored in the spring, so adding the putty has *no effect* on the mechanical energy. That is, $E_2 = E_1 = \frac{1}{2}kA_1^2$, and the amplitude is unchanged: $A_2 = A_1$. The period is again $T_2 = 2\pi\sqrt{(M + m)/k}$.

EVALUATE: Energy is lost in part (a) because the putty slides against the moving block during the collision, and energy is dissipated by kinetic friction. No energy is lost in part (b), because there is no sliding during the collision.

Test Your Understanding of Section 14.3 (a) To double the total energy for a mass-spring system oscillating in SHM, by what factor must the amplitude increase? (i) 4; (ii) 2; (iii) $\sqrt{2} = 1.414$; (iv) $\sqrt[4]{2} = 1.189$. (b) By what factor will the frequency change due to this amplitude increase? (i) 4; (ii) 2; (iii) $\sqrt{2} = 1.414$; (iv) $\sqrt[4]{2} = 1.189$; (v) it does not change.

14.4 Applications of Simple Harmonic Motion

So far, we've looked at a grand total of *one* situation in which simple harmonic motion (SHM) occurs: a body attached to an ideal horizontal spring. But SHM can occur in any system in which there is a restoring force that is directly proportional to the displacement from equilibrium, as given by Eq. (14.3), $F_x = -kx$. The restoring force will originate in different ways in different situations, so the force constant k has to be found for each case by examining the net force on the system. Once this is done, it's straightforward to find the angular frequency ω, frequency f, and period T; we just substitute the value of k into Eqs. (14.10), (14.11), and (14.12), respectively. Let's use these ideas to examine several examples of simple harmonic motion.

Vertical SHM

Suppose we hang a spring with force constant k (Fig. 14.17a) and suspend from it a body with mass m. Oscillations will now be vertical; will they still be SHM? In Fig. 14.17b the body hangs at rest, in equilibrium. In this position the spring is stretched an amount Δl just great enough that the spring's upward vertical force $k\,\Delta l$ on the body balances its weight mg:

$$k\,\Delta l = mg$$

Take $x = 0$ to be this equilibrium position and take the positive x-direction to be upward. When the body is a distance x *above* its equilibrium position (Fig. 14.17c), the extension of the spring is $\Delta l - x$. The upward force it exerts on the body is then $k(\Delta l - x)$, and the net x-component of force on the body is

$$F_{\text{net}} = k(\Delta l - x) + (-mg) = -kx$$

that is, a net downward force of magnitude kx. Similarly, when the body is *below* the equilibrium position, there is a net upward force with magnitude kx. In either case there is a restoring force with magnitude kx. If the body is set in vertical motion, it oscillates in SHM with the same angular frequency as though it were horizontal, $\omega = \sqrt{k/m}$. So vertical SHM doesn't differ in any essential way from horizontal SHM. The only real change is that the equilibrium position $x = 0$ no longer corresponds to the point at which the spring is unstretched. The same ideas hold if a body with weight mg is placed atop a compressible spring (Fig. 14.18) and compresses it a distance Δl.

14.17 A body attached to a hanging spring.

(a)

A hanging spring that obeys Hooke's law

$x = 0$

(b) A body is suspended from the spring. It is in equilibrium when the upward force exerted by the stretched spring equals the body's weight.

$F = k\,\Delta l$

mg

(c) If the body is displaced from equilibrium, the net force on the body is proportional to its displacement. The oscillations are SHM.

$\Delta l - x$

$F = k(\Delta l - x)$

mg

Example 14.6 Vertical SHM in an old car

The shock absorbers in an old car with mass 1000 kg are completely worn out. When a 980-N person climbs slowly into the car at its center of gravity, the car sinks 2.8 cm. The car (with the person aboard) hits a bump, and the car starts oscillating up and down in SHM. Model the car and person as a single body on a single spring, and find the period and frequency of the oscillation.

SOLUTION

IDENTIFY and SET UP: The situation is like that shown in Fig. 14.18. The compression of the spring when the person's weight is added tells us the force constant, which we can use to find the period and frequency (the target variables).

EXECUTE: When the force increases by 980 N, the spring compresses an additional 0.028 m, and the x-coordinate of the car

changes by -0.028 m. Hence the effective force constant (including the effect of the entire suspension) is

$$k = -\frac{F_x}{x} = -\frac{980 \text{ N}}{-0.028 \text{ m}} = 3.5 \times 10^4 \text{ kg/s}^2$$

The person's mass is $w/g = (980 \text{ N})/(9.8 \text{ m/s}^2) = 100$ kg. The *total* oscillating mass is $m = 1000$ kg $+ 100$ kg $= 1100$ kg. The period T is

$$T = 2\pi\sqrt{\frac{m}{k}} = 2\pi\sqrt{\frac{1100 \text{ kg}}{3.5 \times 10^4 \text{ kg/s}^2}} = 1.11 \text{ s}$$

The frequency is $f = 1/T = 1/(1.11 \text{ s}) = 0.90$ Hz.

EVALUATE: A persistent oscillation with a period of about 1 second makes for a very unpleasant ride. The purpose of shock absorbers is to make such oscillations die out (see Section 14.7).

Angular SHM

A mechanical watch keeps time based on the oscillations of a balance wheel (Fig. 14.19). The wheel has a moment of inertia I about its axis. A coil spring exerts a restoring torque τ_z that is proportional to the angular displacement θ from the equilibrium position. We write $\tau_z = -\kappa\theta$, where κ (the Greek letter kappa) is a constant called the *torsion constant*. Using the rotational analog of Newton's second law for a rigid body, $\Sigma\tau_z = I\alpha_z = I \, d^2\theta/dt^2$, we can find the equation of motion:

$$-\kappa\theta = I\alpha \quad \text{or} \quad \frac{d^2\theta}{dt^2} = -\frac{\kappa}{I}\theta$$

The form of this equation is exactly the same as Eq. (14.4) for the acceleration in simple harmonic motion, with x replaced by θ and k/m replaced by κ/I. So we are dealing with a form of *angular* simple harmonic motion. The angular frequency ω and frequency f are given by Eqs. (14.10) and (14.11), respectively, with the same replacement:

$$\omega = \sqrt{\frac{\kappa}{I}} \quad \text{and} \quad f = \frac{1}{2\pi}\sqrt{\frac{\kappa}{I}} \quad \text{(angular SHM)} \quad (14.24)$$

The motion is described by the function

$$\theta = \Theta\cos(\omega t + \phi)$$

where Θ (the Greek letter theta) plays the role of an angular amplitude.

It's a good thing that the motion of a balance wheel *is* simple harmonic. If it weren't, the frequency might depend on the amplitude, and the watch would run too fast or too slow as the spring ran down.

Vibrations of Molecules

The following discussion of the vibrations of molecules uses the binomial theorem. If you aren't familiar with this theorem, you should read about it in the appropriate section of a math textbook.

When two atoms are separated from each other by a few atomic diameters, they can exert attractive forces on each other. But if the atoms are so close to each other that their electron shells overlap, the forces between the atoms are repulsive. Between these limits, there can be an equilibrium separation distance at which two atoms form a *molecule*. If these atoms are displaced slightly from equilibrium, they will oscillate.

14.18 If the weight mg compresses the spring a distance Δl, the force constant is $k = mg/\Delta l$ and the angular frequency for vertical SHM is $\omega = \sqrt{k/m}$—the same as if the body were suspended from the spring (see Fig. 14.17).

A body is placed atop the spring. It is in equilibrium when the upward force exerted by the compressed spring equals the body's weight.

14.19 The balance wheel of a mechanical watch. The spring exerts a restoring torque that is proportional to the angular displacement θ, so the motion is angular SHM.

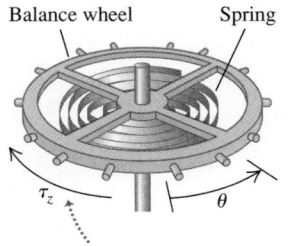

The spring torque τ_z opposes the angular displacement θ.

14.20 (a) Two atoms with centers separated by r. (b) Potential energy U in the van der Waals interaction as a function of r. (c) Force F_r on the right-hand atom as a function of r.

(a) Two-atom system

Distance between atom centers

r

Atoms

F_r = the force exerted by the left-hand atom on the right-hand atom

(b) Potential energy U of the two-atom system as a function of r

$U(r)$ Parabola

Near equilibrium, U can be approximated by a parabola.

R_0

$1.5R_0$

$2R_0$

The equilibrium point is at $r = R_0$ (where U is minimum).

(c) The force F_r as a function of r

$F_r(r)$

Near equilibrium, F_r can be approximated by a straight line.

R_0

$1.5R_0$

$2R_0$

The equilibrium point is at $r = R_0$ (where F_r is zero).

As an example, we'll consider one type of interaction between atoms called the *van der Waals interaction.* Our immediate task here is to study oscillations, so we won't go into the details of how this interaction arises. Let the center of one atom be at the origin and let the center of the other atom be a distance r away (Fig. 14.20a); the equilibrium distance between centers is $r = R_0$. Experiment shows that the van der Waals interaction can be described by the potential-energy function

$$U = U_0 \left[\left(\frac{R_0}{r} \right)^{12} - 2 \left(\frac{R_0}{r} \right)^6 \right] \tag{14.25}$$

where U_0 is a positive constant with units of joules. When the two atoms are very far apart, $U = 0$; when they are separated by the equilibrium distance $r = R_0$, $U = -U_0$. The force on the second atom is the negative derivative of Eq. (14.25):

$$F_r = -\frac{dU}{dr} = U_0 \left[\frac{12R_0^{12}}{r^{13}} - 2\frac{6R_0^6}{r^7} \right] = 12\frac{U_0}{R_0} \left[\left(\frac{R_0}{r} \right)^{13} - \left(\frac{R_0}{r} \right)^7 \right] \tag{14.26}$$

Figures 14.20b and 14.20c plot the potential energy and force, respectively. The force is positive for $r < R_0$ and negative for $r > R_0$, so it is a *restoring* force.

Let's examine the restoring force F_r in Eq. (14.26). We let x represent the displacement from equilibrium:

$$x = r - R_0 \qquad \text{so} \qquad r = R_0 + x$$

In terms of x, the force F_r in Eq. (14.26) becomes

$$F_r = 12\frac{U_0}{R_0} \left[\left(\frac{R_0}{R_0 + x} \right)^{13} - \left(\frac{R_0}{R_0 + x} \right)^7 \right]$$

$$= 12\frac{U_0}{R_0} \left[\frac{1}{(1 + x/R_0)^{13}} - \frac{1}{(1 + x/R_0)^7} \right] \tag{14.27}$$

This looks nothing like Hooke's law, $F_x = -kx$, so we might be tempted to conclude that molecular oscillations cannot be SHM. But let us restrict ourselves to *small-amplitude* oscillations so that the absolute value of the displacement x is small in comparison to R_0 and the absolute value of the ratio x/R_0 is much less than 1. We can then simplify Eq. (14.27) by using the *binomial theorem:*

$$(1 + u)^n = 1 + nu + \frac{n(n-1)}{2!}u^2 + \frac{n(n-1)(n-2)}{3!}u^3 + \cdots \tag{14.28}$$

If $|u|$ is much less than 1, each successive term in Eq. (14.28) is much smaller than the one it follows, and we can safely approximate $(1 + u)^n$ by just the first two terms. In Eq. (14.27), u is replaced by x/R_0 and n equals -13 or -7, so

$$\frac{1}{(1 + x/R_0)^{13}} = (1 + x/R_0)^{-13} \approx 1 + (-13)\frac{x}{R_0}$$

$$\frac{1}{(1 + x/R_0)^{7}} = (1 + x/R_0)^{-7} \approx 1 + (-7)\frac{x}{R_0}$$

$$F_r \approx 12\frac{U_0}{R_0}\left[\left(1 + (-13)\frac{x}{R_0}\right) - \left(1 + (-7)\frac{x}{R_0}\right)\right] = -\left(\frac{72U_0}{R_0^2}\right)x \quad \text{(14.29)}$$

This is just Hooke's law, with force constant $k = 72U_0/R_0^2$. (Note that k has the correct units, J/m^2 or N/m.) So oscillations of molecules bound by the van der Waals interaction can be simple harmonic motion, provided that the amplitude is small in comparison to R_0 so that the approximation $|x/R_0| \ll 1$ used in the derivation of Eq. (14.29) is valid.

You can also use the binomial theorem to show that the potential energy U in Eq. (14.25) can be written as $U \approx \frac{1}{2}kx^2 + C$, where $C = -U_0$ and k is again equal to $72U_0/R_0^2$. Adding a constant to the potential energy has no effect on the physics, so the system of two atoms is fundamentally no different from a mass attached to a horizontal spring for which $U = \frac{1}{2}kx^2$.

Example 14.7 Molecular vibration

Two argon atoms form the molecule Ar_2 as a result of a van der Waals interaction with $U_0 = 1.68 \times 10^{-21}$ J and $R_0 = 3.82 \times 10^{-10}$ m. Find the frequency of small oscillations of one Ar atom about its equilibrium position.

SOLUTION

IDENTIFY and SET UP This is just the situation shown in Fig. 14.20. Because the oscillations are small, we can use Eq. (14.29) to find the force constant k and Eq. (14.11) to find the frequency f of SHM.

EXECUTE: From Eq. (14.29),

$$k = \frac{72U_0}{R_0^2} = \frac{72(1.68 \times 10^{-21}\,\text{J})}{(3.82 \times 10^{-10}\,\text{m})^2} = 0.829\,\text{J/m}^2 = 0.829\,\text{N/m}$$

(This force constant is comparable to that of a loose toy spring like a Slinky™.) From Appendix D, the average atomic mass of argon is $(39.948\,\text{u})(1.66 \times 10^{-27}\,\text{kg/1 u}) = 6.63 \times 10^{-26}$ kg.

From Eq. (14.11), if one atom is fixed and the other oscillates,

$$f = \frac{1}{2\pi}\sqrt{\frac{k}{m}} = \frac{1}{2\pi}\sqrt{\frac{0.829\,\text{N/m}}{6.63 \times 10^{-26}\,\text{kg}}} = 5.63 \times 10^{11}\,\text{Hz}$$

EVALUATE: Our answer for f isn't quite right. If no net external force acts on the molecule, its center of mass (halfway between the atoms) doesn't accelerate, so *both* atoms must oscillate with the same amplitude in opposite directions. It turns out that we can account for this by replacing m with $m/2$ in our expression for f. This makes f larger by a factor of $\sqrt{2}$, so the correct frequency is $f = \sqrt{2}(5.63 \times 10^{11}\,\text{Hz}) = 7.96 \times 10^{11}$ Hz. What's more, on the atomic scale we must use *quantum mechanics* rather than Newtonian mechanics to describe motion; happily, quantum mechanics also yields $f = 7.96 \times 10^{11}$ Hz.

Test Your Understanding of Section 14.4 A block attached to a hanging ideal spring oscillates up and down with a period of 10 s on earth. If you take the block and spring to Mars, where the acceleration due to gravity is only about 40% as large as on earth, what will be the new period of oscillation? (i) 10 s; (ii) more than 10 s; (iii) less than 10 s.

14.5 The Simple Pendulum

A **simple pendulum** is an idealized model consisting of a point mass suspended by a massless, unstretchable string. When the point mass is pulled to one side of its straight-down equilibrium position and released, it oscillates about the equilibrium position. Familiar situations such as a wrecking ball on a crane's cable or a person on a swing (Fig. 14.21a) can be modeled as simple pendulums.

PhET: Pendulum Lab
ActivPhysics 9.10: Pendulum Frequency
ActivPhysics 9.11: Risky Pendulum Walk
ActivPhysics 9.12: Physical Pendulum

14.21 The dynamics of a simple pendulum.

(a) A real pendulum

(b) An idealized simple pendulum

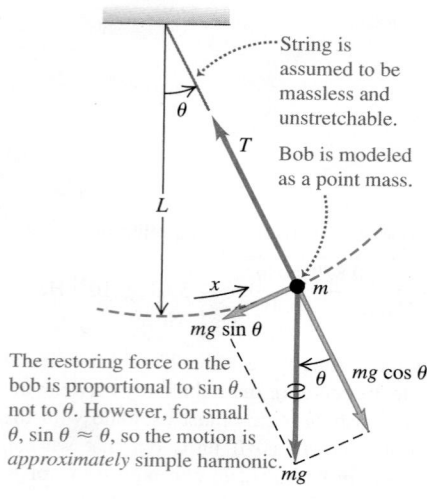

String is assumed to be massless and unstretchable.

Bob is modeled as a point mass.

The restoring force on the bob is proportional to sin θ, not to θ. However, for small θ, sin θ ≈ θ, so the motion is *approximately* simple harmonic.

14.22 For small angular displacements θ, the restoring force $F_\theta = -mg\sin\theta$ on a simple pendulum is approximately equal to $-mg\theta$; that is, it is approximately proportional to the displacement θ. Hence for small angles the oscillations are simple harmonic.

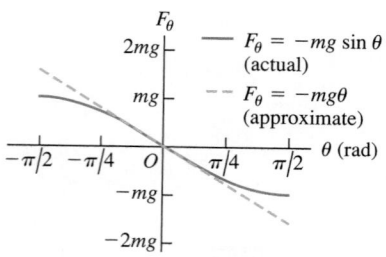

The path of the point mass (sometimes called a pendulum bob) is not a straight line but the arc of a circle with radius L equal to the length of the string (Fig. 14.21b). We use as our coordinate the distance x measured along the arc. If the motion is simple harmonic, the restoring force must be directly proportional to x or (because $x = L\theta$) to θ. Is it?

In Fig. 14.21b we represent the forces on the mass in terms of tangential and radial components. The restoring force F_θ is the tangential component of the net force:

$$F_\theta = -mg\sin\theta \qquad (14.30)$$

The restoring force is provided by gravity; the tension T merely acts to make the point mass move in an arc. The restoring force is proportional *not* to θ but to $\sin\theta$, so the motion is *not* simple harmonic. However, if the angle θ is *small*, $\sin\theta$ is very nearly equal to θ in radians (Fig. 14.22). For example, when $\theta = 0.1$ rad (about 6°), $\sin\theta = 0.0998$, a difference of only 0.2%. With this approximation, Eq. (14.30) becomes

$$F_\theta = -mg\theta = -mg\frac{x}{L} \qquad \text{or}$$

$$F_\theta = -\frac{mg}{L}x \qquad (14.31)$$

The restoring force is then proportional to the coordinate for small displacements, and the force constant is $k = mg/L$. From Eq. (14.10) the angular frequency ω of a simple pendulum with small amplitude is

$$\omega = \sqrt{\frac{k}{m}} = \sqrt{\frac{mg/L}{m}} = \sqrt{\frac{g}{L}} \qquad \text{(simple pendulum, small amplitude)} \qquad (14.32)$$

The corresponding frequency and period relationships are

$$f = \frac{\omega}{2\pi} = \frac{1}{2\pi}\sqrt{\frac{g}{L}} \qquad \text{(simple pendulum, small amplitude)} \qquad (14.33)$$

$$T = \frac{2\pi}{\omega} = \frac{1}{f} = 2\pi\sqrt{\frac{L}{g}} \qquad \text{(simple pendulum, small amplitude)} \qquad (14.34)$$

Note that these expressions do not involve the *mass* of the particle. This is because the restoring force, a component of the particle's weight, is proportional to m. Thus the mass appears on *both* sides of $\Sigma\vec{F} = m\vec{a}$ and cancels out. (This is the same physics that explains why bodies of different masses fall with the same acceleration in a vacuum.) For small oscillations, the period of a pendulum for a given value of g is determined entirely by its length.

The dependence on L and g in Eqs. (14.32) through (14.34) is just what we should expect. A long pendulum has a longer period than a shorter one. Increasing g increases the restoring force, causing the frequency to increase and the period to decrease.

We emphasize again that the motion of a pendulum is only *approximately* simple harmonic. When the amplitude is not small, the departures from simple harmonic motion can be substantial. But how small is "small"? The period can be expressed by an infinite series; when the maximum angular displacement is Θ, the period T is given by

$$T = 2\pi\sqrt{\frac{L}{g}}\left(1 + \frac{1^2}{2^2}\sin^2\frac{\Theta}{2} + \frac{1^2 \cdot 3^2}{2^2 \cdot 4^2}\sin^4\frac{\Theta}{2} + \cdots\right) \qquad (14.35)$$

We can compute the period to any desired degree of precision by taking enough terms in the series. We invite you to check that when $\Theta = 15°$ (on either side of the central position), the true period is longer than that given by the approximate Eq. (14.34) by less than 0.5%.

The usefulness of the pendulum as a timekeeper depends on the period being *very nearly* independent of amplitude, provided that the amplitude is small. Thus, as a pendulum clock runs down and the amplitude of the swings decreases a little, the clock still keeps very nearly correct time.

Example 14.8 | **A simple pendulum**

Find the period and frequency of a simple pendulum 1.000 m long at a location where $g = 9.800$ m/s^2.

SOLUTION

IDENTIFY and SET UP: This is a simple pendulum, so we can use the ideas of this section. We use Eq. (14.34) to determine the pendulum's period T from its length, and Eq. (14.1) to find the frequency f from T.

EXECUTE: From Eqs. (14.34) and (14.1),

$$T = 2\pi\sqrt{\frac{L}{g}} = 2\pi\sqrt{\frac{1.000 \text{ m}}{9.800 \text{ m/s}^2}} = 2.007 \text{ s}$$

$$f = \frac{1}{T} = \frac{1}{2.007 \text{ s}} = 0.4983 \text{ Hz}$$

EVALUATE: The period is almost exactly 2 s. When the metric system was established, the second was *defined* as half the period of a 1-m simple pendulum. This was a poor standard, however, because the value of g varies from place to place. We discussed more modern time standards in Section 1.3.

Test Your Understanding of Section 14.5 When a body oscillating on a horizontal spring passes through its equilibrium position, its acceleration is zero (see Fig. 14.2b). When the bob of an oscillating simple pendulum passes through its equilibrium position, is its acceleration zero?

14.6 The Physical Pendulum

A **physical pendulum** is any *real* pendulum that uses an extended body, as contrasted to the idealized model of the *simple* pendulum with all the mass concentrated at a single point. For small oscillations, analyzing the motion of a real, physical pendulum is almost as easy as for a simple pendulum. Figure 14.23 shows a body of irregular shape pivoted so that it can turn without friction about an axis through point O. In the equilibrium position the center of gravity is directly below the pivot; in the position shown in the figure, the body is displaced from equilibrium by an angle θ, which we use as a coordinate for the system. The distance from O to the center of gravity is d, the moment of inertia of the body about the axis of rotation through O is I, and the total mass is m. When the body is displaced as shown, the weight mg causes a restoring torque

$$\tau_z = -(mg)(d\sin\theta) \qquad (14.36)$$

The negative sign shows that the restoring torque is clockwise when the displacement is counterclockwise, and vice versa.

When the body is released, it oscillates about its equilibrium position. The motion is not simple harmonic because the torque τ_z is proportional to $\sin\theta$ rather than to θ itself. However, if θ is small, we can approximate $\sin\theta$ by θ in radians, just as we did in analyzing the simple pendulum. Then the motion is *approximately* simple harmonic. With this approximation,

$$\tau_z = -(mgd)\theta$$

The equation of motion is $\sum\tau_z = I\alpha_z$, so

$$-(mgd)\theta = I\alpha_z = I\frac{d^2\theta}{dt^2}$$

$$\frac{d^2\theta}{dt^2} = -\frac{mgd}{I}\theta \qquad (14.37)$$

14.23 Dynamics of a physical pendulum.

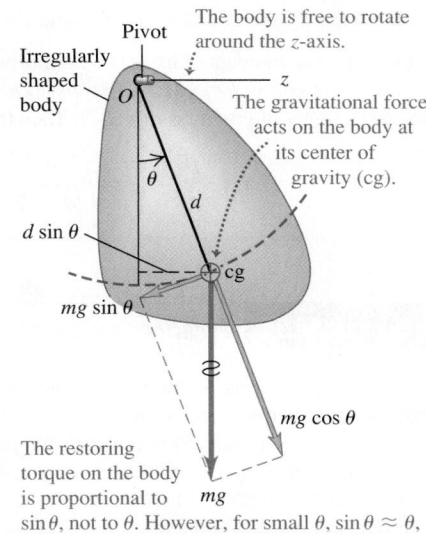

The body is free to rotate around the z-axis.

The gravitational force acts on the body at its center of gravity (cg).

The restoring torque on the body is proportional to $\sin\theta$, not to θ. However, for small θ, $\sin\theta \approx \theta$, so the motion is *approximately* simple harmonic.

Comparing this with Eq. (14.4), we see that the role of (k/m) for the spring-mass system is played here by the quantity (mgd/I). Thus the angular frequency is

$$\omega = \sqrt{\frac{mgd}{I}} \qquad \text{(physical pendulum, small amplitude)} \qquad (14.38)$$

The frequency f is $1/2\pi$ times this, and the period T is

$$T = 2\pi\sqrt{\frac{I}{mgd}} \qquad \text{(physical pendulum, small amplitude)} \qquad (14.39)$$

Equation (14.39) is the basis of a common method for experimentally determining the moment of inertia of a body with a complicated shape. First locate the center of gravity of the body by balancing. Then suspend the body so that it is free to oscillate about an axis, and measure the period T of small-amplitude oscillations. Finally, use Eq. (14.39) to calculate the moment of inertia I of the body about this axis from T, the body's mass m, and the distance d from the axis to the center of gravity (see Exercise 14.53). Biomechanics researchers use this method to find the moments of inertia of an animal's limbs. This information is important for analyzing how an animal walks, as we'll see in the second of the two following examples.

Example 14.9 Physical pendulum versus simple pendulum

If the body in Fig. 14.23 is a uniform rod with length L, pivoted at one end, what is the period of its motion as a pendulum?

SOLUTION

IDENTIFY and SET UP: Our target variable is the oscillation period T of a rod that acts as a physical pendulum. We find the rod's moment of inertia in Table 9.2, and then determine T using Eq. (14.39).

EXECUTE: The moment of inertia of a uniform rod about an axis through one end is $I = \frac{1}{3}ML^2$. The distance from the pivot to the rod's center of gravity is $d = L/2$. Then from Eq. (14.39),

$$T = 2\pi\sqrt{\frac{I}{mgd}} = 2\pi\sqrt{\frac{\frac{1}{3}ML^2}{MgL/2}} = 2\pi\sqrt{\frac{2L}{3g}}$$

EVALUATE: If the rod is a meter stick $(L = 1.00 \text{ m})$ and $g = 9.80 \text{ m/s}^2$, then

$$T = 2\pi\sqrt{\frac{2(1.00 \text{ m})}{3(9.80 \text{ m/s}^2)}} = 1.64 \text{ s}$$

The period is smaller by a factor of $\sqrt{\frac{2}{3}} = 0.816$ than that of a simple pendulum of the same length (see Example 14.8). The rod's moment of inertia around one end, $I = \frac{1}{3}ML^2$, is one-third that of the simple pendulum, and the rod's cg is half as far from the pivot as that of the simple pendulum. You can show that, taken together in Eq. (14.39), these two differences account for the factor $\sqrt{\frac{2}{3}}$ by which the periods differ.

Example 14.10 Tyrannosaurus rex and the physical pendulum

All walking animals, including humans, have a natural walking pace—a number of steps per minute that is more comfortable than a faster or slower pace. Suppose that this pace corresponds to the oscillation of the leg as a physical pendulum. (a) How does this pace depend on the length L of the leg from hip to foot? Treat the leg as a uniform rod pivoted at the hip joint. (b) Fossil evidence shows that *T. rex*, a two-legged dinosaur that lived about 65 million years ago, had a leg length $L = 3.1$ m and a stride length $S = 4.0$ m (the distance from one footprint to the next print of the same foot; see Fig. 14.24). Estimate the walking speed of *T. rex*.

SOLUTION

IDENTIFY and SET UP: Our target variables are (a) the relationship between walking pace and leg length L and (b) the walking speed of *T. rex*. We treat the leg as a physical pendulum, with a period of

14.24 The walking speed of *Tyrannosaurus rex* can be estimated from leg length L and stride length S.

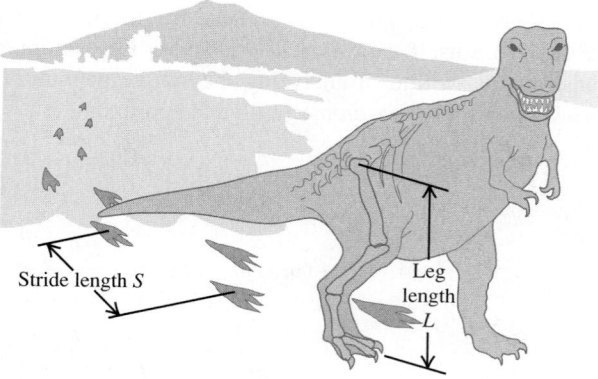

Stride length S

Leg length L

oscillation as found in Example 14.9. We can find the walking speed from the period and the stride length.

EXECUTE: (a) From Example 14.9 the period of oscillation of the leg is $T = 2\pi \sqrt{2L/3g}$, which is proportional to $\sqrt{L}$. Each step takes one-half a period, so the walking pace (in steps per second) is twice the oscillation frequency $f = 1/T$, which is proportional to $1/\sqrt{L}$. The greater the leg length L, the slower the walking pace.

(b) According to our model, *T. rex* traveled one stride length S in a time

$$T = 2\pi\sqrt{\frac{2L}{3g}} = 2\pi\sqrt{\frac{2(3.1 \text{ m})}{3(9.8 \text{ m/s}^2)}} = 2.9 \text{ s}$$

so its walking speed was

$$v = \frac{S}{T} = \frac{4.0 \text{ m}}{2.9 \text{ s}} = 1.4 \text{ m/s} = 5.0 \text{ km/h} = 3.1 \text{ mi/h}$$

This is roughly the walking speed of an adult human.

EVALUATE: A uniform rod isn't a very good model for a leg. The legs of many animals, including both *T. rex* and humans, are tapered; there is more mass between hip and knee than between knee and foot. The center of mass is therefore less than $L/2$ from the hip; a reasonable guess would be about $L/4$. The moment of inertia is therefore *considerably* less than $ML^2/3$—say, $ML^2/15$. Use the analysis of Example 14.9 with these corrections; you'll get a shorter oscillation period and an even greater walking speed for *T. rex*.

Test Your Understanding of Section 14.6 The center of gravity of a simple pendulum of mass m and length L is located at the position of the pendulum bob, a distance L from the pivot point. The center of gravity of a uniform rod of the same mass m and length $2L$ pivoted at one end is also a distance L from the pivot point. How does the period of this uniform rod compare to the period of the simple pendulum? (i) The rod has a longer period; (ii) the rod has a shorter period; (iii) the rod has the same period.

14.7 Damped Oscillations

The idealized oscillating systems we have discussed so far are frictionless. There are no nonconservative forces, the total mechanical energy is constant, and a system set into motion continues oscillating forever with no decrease in amplitude.

Real-world systems always have some dissipative forces, however, and oscillations die out with time unless we replace the dissipated mechanical energy (Fig. 14.25). A mechanical pendulum clock continues to run because potential energy stored in the spring or a hanging weight system replaces the mechanical energy lost due to friction in the pivot and the gears. But eventually the spring runs down or the weights reach the bottom of their travel. Then no more energy is available, and the pendulum swings decrease in amplitude and stop.

The decrease in amplitude caused by dissipative forces is called **damping,** and the corresponding motion is called **damped oscillation.** The simplest case to analyze in detail is a simple harmonic oscillator with a frictional damping force that is directly proportional to the *velocity* of the oscillating body. This behavior occurs in friction involving viscous fluid flow, such as in shock absorbers or sliding between oil-lubricated surfaces. We then have an additional force on the body due to friction, $F_x = -bv_x$, where $v_x = dx/dt$ is the velocity and b is a constant that describes the strength of the damping force. The negative sign shows that the force is always opposite in direction to the velocity. The *net* force on the body is then

$$\sum F_x = -kx - bv_x \tag{14.40}$$

and Newton's second law for the system is

$$-kx - bv_x = ma_x \quad \text{or} \quad -kx - b\frac{dx}{dt} = m\frac{d^2x}{dt^2} \tag{14.41}$$

Equation (14.41) is a differential equation for x; it would be the same as Eq. (14.4), the equation for the acceleration in SHM, except for the added term $-bdx/dt$. Solving this equation is a straightforward problem in differential equations, but we won't go into the details here. If the damping force is relatively small, the motion is described by

$$x = Ae^{-(b/2m)t}\cos(\omega't + \phi) \quad \text{(oscillator with little damping)} \tag{14.42}$$

14.25 A swinging bell left to itself will eventually stop oscillating due to damping forces (air resistance and friction at the point of suspension).

14.26 Graph of displacement versus time for an oscillator with little damping [see Eq. (14.42)] and with phase angle $\phi = 0$. The curves are for two values of the damping constant b.

— $b = 0.1\sqrt{km}$ (weak damping force)
— $b = 0.4\sqrt{km}$ (stronger damping force)

With stronger damping (larger b):
• The amplitude (shown by the dashed curves) decreases more rapidly.
• The period T increases (T_0 = period with zero damping).

14.27 An automobile shock absorber. The viscous fluid causes a damping force that depends on the relative velocity of the two ends of the unit.

Upper cylinder attached to car's frame; moves little.

Piston

Viscous fluid

Lower cylinder attached to axle; moves up and down.

Pushed up

Pushed down

The angular frequency of oscillation ω' is given by

$$\omega' = \sqrt{\frac{k}{m} - \frac{b^2}{4m^2}} \qquad \text{(oscillator with little damping)} \qquad (14.43)$$

You can verify that Eq. (14.42) is a solution of Eq. (14.41) by calculating the first and second derivatives of x, substituting them into Eq. (14.41), and checking whether the left and right sides are equal. This is a straightforward but slightly tedious procedure.

The motion described by Eq. (14.42) differs from the undamped case in two ways. First, the amplitude $Ae^{-(b/2m)t}$ is not constant but decreases with time because of the decreasing exponential factor $e^{-(b/2m)t}$. Figure 14.26 is a graph of Eq. (14.42) for the case $\phi = 0$; it shows that the larger the value of b, the more quickly the amplitude decreases.

Second, the angular frequency ω', given by Eq. (14.43), is no longer equal to $\omega = \sqrt{k/m}$ but is somewhat smaller. It becomes zero when b becomes so large that

$$\frac{k}{m} - \frac{b^2}{4m^2} = 0 \qquad \text{or} \qquad b = 2\sqrt{km} \qquad (14.44)$$

When Eq. (14.44) is satisfied, the condition is called **critical damping.** The system no longer oscillates but returns to its equilibrium position without oscillation when it is displaced and released.

If b is greater than $2\sqrt{km}$, the condition is called **overdamping.** Again there is no oscillation, but the system returns to equilibrium more slowly than with critical damping. For the overdamped case the solutions of Eq. (14.41) have the form

$$x = C_1e^{-a_1t} + C_2e^{-a_2t}$$

where C_1 and C_2 are constants that depend on the initial conditions and a_1 and a_2 are constants determined by m, k, and b.

When b is less than the critical value, as in Eq. (14.42), the condition is called **underdamping.** The system oscillates with steadily decreasing amplitude.

In a vibrating tuning fork or guitar string, it is usually desirable to have as little damping as possible. By contrast, damping plays a beneficial role in the oscillations of an automobile's suspension system. The shock absorbers provide a velocity-dependent damping force so that when the car goes over a bump, it doesn't continue bouncing forever (Fig. 14.27). For optimal passenger comfort, the system should be critically damped or slightly underdamped. Too much damping would be counterproductive; if the suspension is overdamped and the car hits a second bump just after the first one, the springs in the suspension will still be compressed somewhat from the first bump and will not be able to fully absorb the impact.

Energy in Damped Oscillations

In damped oscillations the damping force is nonconservative; the mechanical energy of the system is not constant but decreases continuously, approaching zero after a long time. To derive an expression for the rate of change of energy, we first write an expression for the total mechanical energy E at any instant:

$$E = \tfrac{1}{2}mv_x^2 + \tfrac{1}{2}kx^2$$

To find the rate of change of this quantity, we take its time derivative:

$$\frac{dE}{dt} = mv_x\frac{dv_x}{dt} + kx\frac{dx}{dt}$$

But $dv_x/dt = a_x$ and $dx/dt = v_x$, so

$$\frac{dE}{dt} = v_x(ma_x + kx)$$

From Eq. (14.41), $ma_x + kx = -b\,dx/dt = -bv_x$, so

$$\frac{dE}{dt} = v_x(-bv_x) = -bv_x{}^2 \quad \text{(damped oscillations)} \qquad (14.45)$$

The right side of Eq. (14.45) is **negative** whenever the oscillating body is in motion, whether the x-velocity v_x is positive or negative. This shows that as the body moves, the energy decreases, though not at a uniform rate. The term $-bv_x{}^2 = (-bv_x)v_x$ (force times velocity) is the rate at which the damping force does (negative) work on the system (that is, the damping *power*). This equals the rate of change of the total mechanical energy of the system.

Similar behavior occurs in electric circuits containing inductance, capacitance, and resistance. There is a natural frequency of oscillation, and the resistance plays the role of the damping constant b. We will study these circuits in detail in Chapters 30 and 31.

Test Your Understanding of Section 14.7 An airplane is flying in a straight line at a constant altitude. If a wind gust strikes and raises the nose of the airplane, the nose will bob up and down until the airplane eventually returns to its original attitude. Are these oscillations (i) undamped, (ii) underdamped, (iii) critically damped, or (iv) overdamped?

14.8 Forced Oscillations and Resonance

A damped oscillator left to itself will eventually stop moving altogether. But we can maintain a constant-amplitude oscillation by applying a force that varies with time in a periodic or cyclic way, with a definite period and frequency. As an example, consider your cousin Throckmorton on a playground swing. You can keep him swinging with constant amplitude by giving him a little push once each cycle. We call this additional force a **driving force.**

Damped Oscillation with a Periodic Driving Force

If we apply a periodically varying driving force with angular frequency ω_d to a damped harmonic oscillator, the motion that results is called a **forced oscillation** or a *driven oscillation*. It is different from the motion that occurs when the system is simply displaced from equilibrium and then left alone, in which case the system oscillates with a **natural angular frequency** ω' determined by m, k, and b, as in Eq. (14.43). In a forced oscillation, however, the angular frequency with which the mass oscillates is equal to the driving angular frequency ω_d. This does *not* have to be equal to the angular frequency ω' with which the system would oscillate without a driving force. If you grab the ropes of Throckmorton's swing, you can force the swing to oscillate with any frequency you like.

Suppose we force the oscillator to vibrate with an angular frequency ω_d that is nearly *equal* to the angular frequency ω' it would have with no driving force. What happens? The oscillator is naturally disposed to oscillate at $\omega = \omega'$, so we expect the amplitude of the resulting oscillation to be larger than when the two frequencies are very different. Detailed analysis and experiment show that this is just what happens. The easiest case to analyze is a *sinusoidally* varying force—say, $F(t) = F_{max} \cos \omega_d t$. If we vary the frequency ω_d of the driving force, the amplitude of the resulting forced oscillation varies in an interesting way (Fig. 14.28). When there is very little damping (small b), the amplitude goes through a sharp peak as the driving angular frequency ω_d nears the natural oscillation angular frequency ω'. When the damping is increased (larger b), the peak becomes broader and smaller in height and shifts toward lower frequencies.

We could work out an expression that shows how the amplitude A of the forced oscillation depends on the frequency of a sinusoidal driving force, with

14.28 Graph of the amplitude A of forced oscillation as a function of the angular frequency ω_d of the driving force. The horizontal axis shows the ratio of ω_d to the angular frequency $\omega = \sqrt{k/m}$ of an undamped oscillator. Each curve has a different value of the damping constant b.

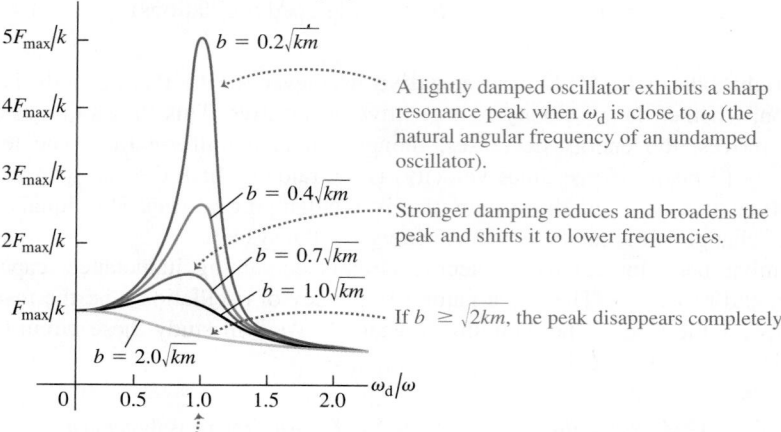

Each curve shows the amplitude A for an oscillator subjected to a driving force at various angular frequencies ω_d. Successive curves from blue to gold represent successively greater damping.

A lightly damped oscillator exhibits a sharp resonance peak when ω_d is close to ω (the natural angular frequency of an undamped oscillator).

Stronger damping reduces and broadens the peak and shifts it to lower frequencies.

If $b \geq \sqrt{2km}$, the peak disappears completely.

Driving frequency ω_d equals natural angular frequency ω of an undamped oscillator.

maximum value F_{max}. That would involve more differential equations than we're ready for, but here is the result:

$$A = \frac{F_{max}}{\sqrt{(k - m\omega_d^2)^2 + b^2\omega_d^2}} \quad \text{(amplitude of a driven oscillator)} \quad (14.46)$$

When $k - m\omega_d^2 = 0$, the first term under the radical is zero, so A has a maximum near $\omega_d = \sqrt{k/m}$. The height of the curve at this point is proportional to $1/b$; the less damping, the higher the peak. At the low-frequency extreme, when $\omega_d = 0$, we get $A = F_{max}/k$. This corresponds to a *constant* force F_{max} and a constant displacement $A = F_{max}/k$ from equilibrium, as we might expect.

Resonance and Its Consequences

The fact that there is an amplitude peak at driving frequencies close to the natural frequency of the system is called **resonance.** Physics is full of examples of resonance; building up the oscillations of a child on a swing by pushing with a frequency equal to the swing's natural frequency is one. A vibrating rattle in a car that occurs only at a certain engine speed or wheel-rotation speed is an all-too-familiar example. Inexpensive loudspeakers often have an annoying boom or buzz when a musical note happens to coincide with the resonant frequency of the speaker cone or the speaker housing. In Chapter 16 we will study other examples of resonance that involve sound. Resonance also occurs in electric circuits, as we will see in Chapter 31; a tuned circuit in a radio or television receiver responds strongly to waves having frequencies near its resonant frequency, and this fact is used to select a particular station and reject the others.

Resonance in mechanical systems can be destructive. A company of soldiers once destroyed a bridge by marching across it in step; the frequency of their steps was close to a natural vibration frequency of the bridge, and the resulting oscillation had large enough amplitude to tear the bridge apart. Ever since, marching soldiers have been ordered to break step before crossing a bridge. Some years ago, vibrations of the engines of a particular airplane had just the right frequency to resonate with the natural frequencies of its wings. Large oscillations built up, and occasionally the wings fell off.

Test Your Understanding of Section 14.8 When driven at a frequency near its natural frequency, an oscillator with very little damping has a much greater response than the same oscillator with more damping. When driven at a frequency that is much higher or lower than the natural frequency, which oscillator will have the greater response: (i) the one with very little damping or (ii) the one with more damping?

Application Canine Resonance
Unlike humans, dogs have no sweat glands and so must pant in order to cool down. The frequency at which a dog pants is very close to the resonant frequency of its respiratory system. This causes the maximum amount of air to move in and out of the dog and so minimizes the effort that the dog must exert to cool itself.

Periodic motion: Periodic motion is motion that repeats itself in a definite cycle. It occurs whenever a body has a stable equilibrium position and a restoring force that acts when it is displaced from equilibrium. Period T is the time for one cycle. Frequency f is the number of cycles per unit time. Angular frequency ω is 2π times the frequency. (See Example 14.1.)

$$f = \frac{1}{T} \qquad T = \frac{1}{f} \qquad (14.1)$$

$$\omega = 2\pi f = \frac{2\pi}{T} \qquad (14.2)$$

Simple harmonic motion: If the restoring force F_x in periodic motion is directly proportional to the displacement x, the motion is called simple harmonic motion (SHM). In many cases this condition is satisfied if the displacement from equilibrium is small. The angular frequency, frequency, and period in SHM do not depend on the amplitude, but only on the mass m and force constant k. The displacement, velocity, and acceleration in SHM are sinusoidal functions of time; the amplitude A and phase angle ϕ of the oscillation are determined by the initial position and velocity of the body. (See Examples 14.2, 14.3, 14.6, and 14.7.)

$$F_x = -kx \qquad (14.3)$$

$$a_x = \frac{F_x}{m} = -\frac{k}{m}x \qquad (14.4)$$

$$\omega = \sqrt{\frac{k}{m}} \qquad (14.10)$$

$$f = \frac{\omega}{2\pi} = \frac{1}{2\pi}\sqrt{\frac{k}{m}} \qquad (14.11)$$

$$T = \frac{1}{f} = 2\pi\sqrt{\frac{m}{k}} \qquad (14.12)$$

$$x = A\cos(\omega t + \phi) \qquad (14.13)$$

Energy in simple harmonic motion: Energy is conserved in SHM. The total energy can be expressed in terms of the force constant k and amplitude A. (See Examples 14.4 and 14.5.)

$$E = \tfrac{1}{2}mv_x^2 + \tfrac{1}{2}kx^2 = \tfrac{1}{2}kA^2 = \text{constant} \qquad (14.21)$$

Angular simple harmonic motion: In angular SHM, the frequency and angular frequency are related to the moment of inertia I and the torsion constant κ.

$$\omega = \sqrt{\frac{\kappa}{I}} \quad \text{and} \quad f = \frac{1}{2\pi}\sqrt{\frac{\kappa}{I}} \qquad (14.24)$$

Spring torque τ_z opposes angular displacement θ.

Simple pendulum: A simple pendulum consists of a point mass m at the end of a massless string of length L. Its motion is approximately simple harmonic for sufficiently small amplitude; the angular frequency, frequency, and period then depend only on g and L, not on the mass or amplitude. (See Example 14.8.)

$$\omega = \sqrt{\frac{g}{L}} \qquad (14.32)$$

$$f = \frac{\omega}{2\pi} = \frac{1}{2\pi}\sqrt{\frac{g}{L}} \qquad (14.33)$$

$$T = \frac{2\pi}{\omega} = \frac{1}{f} = 2\pi\sqrt{\frac{L}{g}} \qquad (14.34)$$

Physical pendulum: A physical pendulum is any body suspended from an axis of rotation. The angular frequency and period for small-amplitude oscillations are independent of amplitude, but depend on the mass m, distance d from the axis of rotation to the center of gravity, and moment of inertia I about the axis. (See Examples 14.9 and 14.10.)

$$\omega = \sqrt{\frac{mgd}{I}} \qquad (14.38)$$

$$T = 2\pi\sqrt{\frac{I}{mgd}} \qquad (14.39)$$

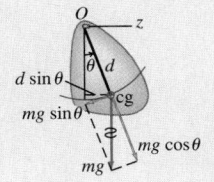

Damped oscillations: When a force $F_x = -bv_x$ proportional to velocity is added to a simple harmonic oscillator, the motion is called a damped oscillation. If $b < 2\sqrt{km}$ (called underdamping), the system oscillates with a decaying amplitude and an angular frequency ω' that is lower than it would be without damping. If $b = 2\sqrt{km}$ (called critical damping) or $b > 2\sqrt{km}$ (called overdamping), when the system is displaced it returns to equilibrium without oscillating.

$$x = Ae^{-(b/2m)t}\cos(\omega' t + \phi) \quad (14.42)$$

$$\omega' = \sqrt{\frac{k}{m} - \frac{b^2}{4m^2}} \quad (14.43)$$

Driven oscillations and resonance: When a sinusoidally varying driving force is added to a damped harmonic oscillator, the resulting motion is called a forced oscillation. The amplitude is a function of the driving frequency ω_d and reaches a peak at a driving frequency close to the natural frequency of the system. This behavior is called resonance.

$$A = \frac{F_{max}}{\sqrt{(k - m\omega_d^2)^2 + b^2\omega_d^2}} \quad (14.46)$$

BRIDGING PROBLEM | **Oscillating and Rolling**

Two uniform, solid cylinders of radius R and total mass M are connected along their common axis by a short, light rod and rest on a horizontal tabletop (Fig. 14.29). A frictionless ring at the center of the rod is attached to a spring with force constant k; the other end of the spring is fixed. The cylinders are pulled to the left a distance x, stretching the spring, and then released from rest. Due to friction between the tabletop and the cylinders, the cylinders roll without slipping as they oscillate. Show that the motion of the center of mass of the cylinders is simple harmonic, and find its period.

14.29

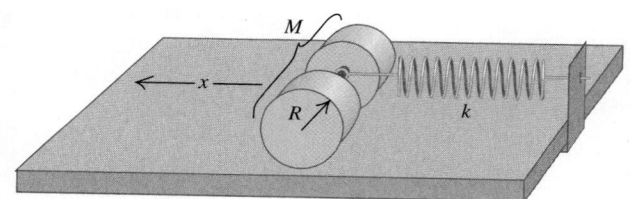

SOLUTION GUIDE

See MasteringPhysics® study area for a Video Tutor solution.

IDENTIFY and SET UP

1. What condition must be satisfied for the motion of the center of mass of the cylinders to be simple harmonic? (*Hint:* See Section 14.2.)
2. Which equations should you use to describe the translational and rotational motions of the cylinders? Which equation should you use to describe the condition that the cylinders roll without slipping? (*Hint:* See Section 10.3.)
3. Sketch the situation and choose a coordinate system. Make a list of the unknown quantities and decide which is the target variable.

EXECUTE

4. Draw a free-body diagram for the cylinders when they are displaced a distance x from equilibrium.
5. Solve the equations to find an expression for the acceleration of the center of mass of the cylinders. What does this expression tell you?
6. Use your result from step 5 to find the period of oscillation of the center of mass of the cylinders.

EVALUATE

7. What would be the period of oscillation if there were no friction and the cylinders didn't roll? Is this period larger or smaller than your result from step 6? Is this reasonable?

•, ••, •••: Problems of increasing difficulty. **CP**: Cumulative problems incorporating material from earlier chapters. **CALC**: Problems requiring calculus. **BIO**: Biosciences problems.

DISCUSSION QUESTIONS

Q14.1 An object is moving with SHM of amplitude A on the end of a spring. If the amplitude is doubled, what happens to the total distance the object travels in one period? What happens to the period? What happens to the maximum speed of the object? Discuss how these answers are related.

Q14.2 Think of several examples in everyday life of motions that are, at least approximately, simple harmonic. In what respects does each differ from SHM?

Q14.3 Does a tuning fork or similar tuning instrument undergo SHM? Why is this a crucial question for musicians?

Q14.4 A box containing a pebble is attached to an ideal horizontal spring and is oscillating on a friction-free air table. When the box has reached its maximum distance from the equilibrium point, the pebble is suddenly lifted out vertically without disturbing the box. Will the following characteristics of the motion increase, decrease, or remain the same in the subsequent motion of the box? Justify each answer. (a) frequency; (b) period; (c) amplitude; (d) the maximum kinetic energy of the box; (e) the maximum speed of the box.

Q14.5 If a uniform spring is cut in half, what is the force constant of each half? Justify your answer. How would the frequency of SHM using a half-spring differ from the frequency using the same mass and the entire spring?

Q14.6 The analysis of SHM in this chapter ignored the mass of the spring. How does the spring's mass change the characteristics of the motion?

Q14.7 Two identical gliders on an air track are connected by an ideal spring. Could such a system undergo SHM? Explain. How would the period compare with that of a single glider attached to a spring whose other end is rigidly attached to a stationary object? Explain.

Q14.8 You are captured by Martians, taken into their ship, and put to sleep. You awake some time later and find yourself locked in a small room with no windows. All the Martians have left you with is your digital watch, your school ring, and your long silver-chain necklace. Explain how you can determine whether you are still on earth or have been transported to Mars.

Q14.9 The system shown in Fig. 14.17 is mounted in an elevator. What happens to the period of the motion (does it increase, decrease, or remain the same) if the elevator (a) accelerates upward at 5.0 m/s^2; (b) moves upward at a steady 5.0 m/s; (c) accelerates downward at 5.0 m/s^2? Justify your answers.

Q14.10 If a pendulum has a period of 2.5 s on earth, what would be its period in a space station orbiting the earth? If a mass hung from a vertical spring has a period of 5.0 s on earth, what would its period be in the space station? Justify each of your answers.

Q14.11 A simple pendulum is mounted in an elevator. What happens to the period of the pendulum (does it increase, decrease, or remain the same) if the elevator (a) accelerates upward at 5.0 m/s^2; (b) moves upward at a steady 5.0 m/s; (c) accelerates downward at 5.0 m/s^2; (d) accelerates downward at 9.8 m/s^2? Justify your answers.

Q14.12 What should you do to the length of the string of a simple pendulum to (a) double its frequency; (b) double its period; (c) double its angular frequency?

Q14.13 If a pendulum clock is taken to a mountaintop, does it gain or lose time, assuming it is correct at a lower elevation? Explain your answer.

Q14.14 When the amplitude of a simple pendulum increases, should its period increase or decrease? Give a qualitative argument; do not rely on Eq. (14.35). Is your argument also valid for a physical pendulum?

Q14.15 Why do short dogs (like Chihuahuas) walk with quicker strides than do tall dogs (like Great Danes)?

Q14.16 At what point in the motion of a simple pendulum is the string tension greatest? Least? In each case give the reasoning behind your answer.

Q14.17 Could a standard of time be based on the period of a certain standard pendulum? What advantages and disadvantages would such a standard have compared to the actual present-day standard discussed in Section 1.3?

Q14.18 For a simple pendulum, clearly distinguish between ω (the angular velocity) and ω (the angular frequency). Which is constant and which is variable?

Q14.19 A glider is attached to a fixed ideal spring and oscillates on a horizontal, friction-free air track. A coin is atop the glider and oscillating with it. At what points in the motion is the friction force on the coin greatest? At what points is it least? Justify your answers.

Q14.20 In designing structures in an earthquake-prone region, how should the natural frequencies of oscillation of a structure relate to typical earthquake frequencies? Why? Should the structure have a large or small amount of damping?

EXERCISES

Section 14.1 Describing Oscillation

14.1 • **BIO** (a) **Music.** When a person sings, his or her vocal cords vibrate in a repetitive pattern that has the same frequency as the note that is sung. If someone sings the note B flat, which has a frequency of 466 Hz, how much time does it take the person's vocal cords to vibrate through one complete cycle, and what is the angular frequency of the cords? (b) **Hearing.** When sound waves strike the eardrum, this membrane vibrates with the same frequency as the sound. The highest pitch that typical humans can hear has a period of 50.0 μs. What are the frequency and angular frequency of the vibrating eardrum for this sound? (c) **Vision.** When light having vibrations with angular frequency ranging from 2.7×10^{15} rad/s to 4.7×10^{15} rad/s strikes the retina of the eye, it stimulates the receptor cells there and is perceived as visible light. What are the limits of the period and frequency of this light? (d) **Ultrasound.** High-frequency sound waves (ultrasound) are used to probe the interior of the body, much as x rays do. To detect small objects such as tumors, a frequency of around 5.0 MHz is used. What are the period and angular frequency of the molecular vibrations caused by this pulse of sound?

14.2 • If an object on a horizontal, frictionless surface is attached to a spring, displaced, and then released, it will oscillate. If it is displaced 0.120 m from its equilibrium position and released with zero initial speed, then after 0.800 s its displacement is found to be

0.120 m on the opposite side, and it has passed the equilibrium position once during this interval. Find (a) the amplitude; (b) the period; (c) the frequency.

14.3 • The tip of a tuning fork goes through 440 complete vibrations in 0.500 s. Find the angular frequency and the period of the motion.

14.4 • The displacement of an oscillating object as a function of time is shown in Fig. E14.4. What are (a) the frequency; (b) the amplitude; (c) the period; (d) the angular frequency of this motion?

Figure **E14.4**

14.5 •• A machine part is undergoing SHM with a frequency of 5.00 Hz and amplitude 1.80 cm. How long does it take the part to go from $x = 0$ to $x = -1.80$ cm?

Section 14.2 Simple Harmonic Motion

14.6 •• In a physics lab, you attach a 0.200-kg air-track glider to the end of an ideal spring of negligible mass and start it oscillating. The elapsed time from when the glider first moves through the equilibrium point to the second time it moves through that point is 2.60 s. Find the spring's force constant.

14.7 • When a body of unknown mass is attached to an ideal spring with force constant 120 N/m, it is found to vibrate with a frequency of 6.00 Hz. Find (a) the period of the motion; (b) the angular frequency; (c) the mass of the body.

14.8 • When a 0.750-kg mass oscillates on an ideal spring, the frequency is 1.33 Hz. What will the frequency be if 0.220 kg are (a) added to the original mass and (b) subtracted from the original mass? Try to solve this problem *without* finding the force constant of the spring.

14.9 •• An object is undergoing SHM with period 0.900 s and amplitude 0.320 m. At $t = 0$ the object is at $x = 0.320$ m and is instantaneously at rest. Calculate the time it takes the object to go (a) from $x = 0.320$ m to $x = 0.160$ m and (b) from $x = 0.160$ m to $x = 0$.

14.10 • A small block is attached to an ideal spring and is moving in SHM on a horizontal, frictionless surface. When the block is at $x = 0.280$ m, the acceleration of the block is -5.30 m/s². What is the frequency of the motion?

14.11 • A 2.00-kg, frictionless block is attached to an ideal spring with force constant 300 N/m. At $t = 0$ the spring is neither stretched nor compressed and the block is moving in the negative direction at 12.0 m/s. Find (a) the amplitude and (b) the phase angle. (c) Write an equation for the position as a function of time.

14.12 •• Repeat Exercise 14.11, but assume that at $t = 0$ the block has velocity -4.00 m/s and displacement $+0.200$ m.

14.13 • The point of the needle of a sewing machine moves in SHM along the x-axis with a frequency of 2.5 Hz. At $t = 0$ its position and velocity components are $+1.1$ cm and -15 cm/s, respectively. (a) Find the acceleration component of the needle at $t = 0$. (b) Write equations giving the position, velocity, and acceleration components of the point as a function of time.

14.14 •• A small block is attached to an ideal spring and is moving in SHM on a horizontal, frictionless surface. When the ampli-

tude of the motion is 0.090 m, it takes the block 2.70 s to travel from $x = 0.090$ m to $x = -0.090$ m. If the amplitude is doubled, to 0.180 m, how long does it take the block to travel (a) from $x = 0.180$ m to $x = -0.180$ m and (b) from $x = 0.090$ m to $x = -0.090$ m?

14.15 • **BIO** **Weighing Astronauts.** This procedure has actually been used to "weigh" astronauts in space. A 42.5-kg chair is attached to a spring and allowed to oscillate. When it is empty, the chair takes 1.30 s to make one complete vibration. But with an astronaut sitting in it, with her feet off the floor, the chair takes 2.54 s for one cycle. What is the mass of the astronaut?

14.16 • A 0.400-kg object undergoing SHM has $a_x = -2.70$ m/s² when $x = 0.300$ m. What is the time for one oscillation?

14.17 • On a frictionless, horizontal air track, a glider oscillates at the end of an ideal spring of force constant 2.50 N/cm. The graph in Fig. E14.17 shows the acceleration of the glider as a function of time. Find (a) the mass of the glider; (b) the maximum displacement of the glider from the equilibrium point; (c) the maximum force the spring exerts on the glider.

Figure **E14.17**

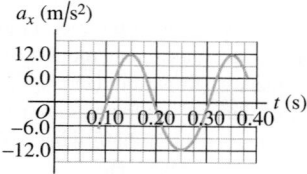

14.18 • A 0.500-kg mass on a spring has velocity as a function of time given by $v_x(t) = -(3.60 \text{ cm/s}) \sin[(4.71 \text{ s}^{-1})t - \pi/2]$. What are (a) the period; (b) the amplitude; (c) the maximum acceleration of the mass; (d) the force constant of the spring?

14.19 • A 1.50-kg mass on a spring has displacement as a function of time given by the equation

$$x(t) = (7.40 \text{ cm}) \cos[(4.16 \text{ s}^{-1})t - 2.42]$$

Find (a) the time for one complete vibration; (b) the force constant of the spring; (c) the maximum speed of the mass; (d) the maximum force on the mass; (e) the position, speed, and acceleration of the mass at $t = 1.00$ s; (f) the force on the mass at that time.

14.20 • **BIO** **Weighing a Virus.** In February 2004, scientists at Purdue University used a highly sensitive technique to measure the mass of a vaccinia virus (the kind used in smallpox vaccine). The procedure involved measuring the frequency of oscillation of a tiny sliver of silicon (just 30 nm long) with a laser, first without the virus and then after the virus had attached itself to the silicon. The difference in mass caused a change in the frequency. We can model such a process as a mass on a spring. (a) Show that the ratio of the frequency with the virus attached (f_{S+V}) to the frequency without the virus (f_S) is given by the formula $\dfrac{f_{S+V}}{f_S} = \dfrac{1}{\sqrt{1 + (m_V/m_S)}}$, where m_V is the mass of the virus and m_S is the mass of the silicon sliver. Notice that it is *not* necessary to know or measure the force constant of the spring. (b) In some data, the silicon sliver has a mass of 2.10×10^{-16} g and a frequency of 2.00×10^{15} Hz without the virus and 2.87×10^{14} Hz with the virus. What is the mass of the virus, in grams and in femtograms?

14.21 •• **CALC** **Jerk.** A guitar string vibrates at a frequency of 440 Hz. A point at its center moves in SHM with an amplitude of

3.0 mm and a phase angle of zero. (a) Write an equation for the position of the center of the string as a function of time. (b) What are the maximum values of the magnitudes of the velocity and acceleration of the center of the string? (c) The derivative of the acceleration with respect to time is a quantity called the *jerk*. Write an equation for the jerk of the center of the string as a function of time, and find the maximum value of the magnitude of the jerk.

Section 14.3 Energy in Simple Harmonic Motion

14.22 •• For the oscillating object in Fig. E14.4, what are (a) its maximum speed and (b) its maximum acceleration?

14.23 • A small block is attached to an ideal spring and is moving in SHM on a horizontal, frictionless surface. The amplitude of the motion is 0.120 m. The maximum speed of the block is 3.90 m/s. What is the maximum magnitude of the acceleration of the block?

14.24 • A small block is attached to an ideal spring and is moving in SHM on a horizontal, frictionless surface. The amplitude of the motion is 0.250 m and the period is 3.20 s. What are the speed and acceleration of the block when $x = 0.160$ m?

14.25 •• A tuning fork labeled 392 Hz has the tip of each of its two prongs vibrating with an amplitude of 0.600 mm. (a) What is the maximum speed of the tip of a prong? (b) A housefly (*Musca domestica*) with mass 0.0270 g is holding onto the tip of one of the prongs. As the prong vibrates, what is the fly's maximum kinetic energy? Assume that the fly's mass has a negligible effect on the frequency of oscillation.

14.26 •• A harmonic oscillator has angular frequency ω and amplitude A. (a) What are the magnitudes of the displacement and velocity when the elastic potential energy is equal to the kinetic energy? (Assume that $U = 0$ at equilibrium.) (b) How often does this occur in each cycle? What is the time between occurrences? (c) At an instant when the displacement is equal to $A/2$, what fraction of the total energy of the system is kinetic and what fraction is potential?

14.27 • A 0.500-kg glider, attached to the end of an ideal spring with force constant $k = 450$ N/m, undergoes SHM with an amplitude of 0.040 m. Compute (a) the maximum speed of the glider; (b) the speed of the glider when it is at $x = -0.015$ m; (c) the magnitude of the maximum acceleration of the glider; (d) the acceleration of the glider at $x = -0.015$ m; (e) the total mechanical energy of the glider at any point in its motion.

14.28 •• A cheerleader waves her pom-pom in SHM with an amplitude of 18.0 cm and a frequency of 0.850 Hz. Find (a) the maximum magnitude of the acceleration and of the velocity; (b) the acceleration and speed when the pom-pom's coordinate is $x = +9.0$ cm; (c) the time required to move from the equilibrium position directly to a point 12.0 cm away. (d) Which of the quantities asked for in parts (a), (b), and (c) can be found using the energy approach used in Section 14.3, and which cannot? Explain.

14.29 • CP For the situation described in part (a) of Example 14.5, what should be the value of the putty mass m so that the amplitude after the collision is one-half the original amplitude? For this value of m, what fraction of the original mechanical energy is converted into heat?

14.30 • A 0.150-kg toy is undergoing SHM on the end of a horizontal spring with force constant $k = 300$ N/m. When the object is 0.0120 m from its equilibrium position, it is observed to have a speed of 0.300 m/s. What are (a) the total energy of the object at any point of its motion; (b) the amplitude of the motion; (c) the maximum speed attained by the object during its motion?

14.31 •• You are watching an object that is moving in SHM. When the object is displaced 0.600 m to the right of its equilibrium position, it has a velocity of 2.20 m/s to the right and an acceleration of 8.40 m/s² to the left. How much farther from this point will the object move before it stops momentarily and then starts to move back to the left?

14.32 •• On a horizontal, frictionless table, an open-topped 5.20-kg box is attached to an ideal horizontal spring having force constant 375 N/m. Inside the box is a 3.44-kg stone. The system is oscillating with an amplitude of 7.50 cm. When the box has reached its maximum speed, the stone is suddenly plucked vertically out of the box without touching the box. Find (a) the period and (b) the amplitude of the resulting motion of the box. (c) Without doing any calculations, is the new period greater or smaller than the original period? How do you know?

14.33 •• A mass is oscillating with amplitude A at the end of a spring. How far (in terms of A) is this mass from the equilibrium position of the spring when the elastic potential energy equals the kinetic energy?

14.34 •• A mass m is attached to a spring of force constant 75 N/m and allowed to oscillate. Figure E14.34 shows a graph of its velocity v_x as a function of time t. Find (a) the period, (b) the frequency, and (c) the angular frequency of this motion. (d) What is the amplitude (in cm), and at what times does the mass reach this position? (e) Find the maximum acceleration of the mass and the times at which it occurs. (f) What is the mass m?

Figure **E14.34**

14.35 • Inside a NASA test vehicle, a 3.50-kg ball is pulled along by a horizontal ideal spring fixed to a friction-free table. The force constant of the spring is 225 N/m. The vehicle has a steady acceleration of 5.00 m/s², and the ball is not oscillating. Suddenly, when the vehicle's speed has reached 45.0 m/s, its engines turn off, thus eliminating its acceleration but not its velocity. Find (a) the amplitude and (b) the frequency of the resulting oscillations of the ball. (c) What will be the ball's maximum speed relative to the vehicle?

Section 14.4 Applications of Simple Harmonic Motion

14.36 • A proud deep-sea fisherman hangs a 65.0-kg fish from an ideal spring having negligible mass. The fish stretches the spring 0.120 m. (a) Find the force constant of the spring. The fish is now pulled down 5.00 cm and released. (b) What is the period of oscillation of the fish? (c) What is the maximum speed it will reach?

14.37 • A 175-g glider on a horizontal, frictionless air track is attached to a fixed ideal spring with force constant 155 N/m. At the instant you make measurements on the glider, it is moving at 0.815 m/s and is 3.00 cm from its equilibrium point. Use *energy conservation* to find (a) the amplitude of the motion and (b) the maximum speed of the glider. (c) What is the angular frequency of the oscillations?

14.38 • A thrill-seeking cat with mass 4.00 kg is attached by a harness to an ideal spring of negligible mass and oscillates vertically in SHM. The amplitude is 0.050 m, and at the highest point of the motion the spring has its natural unstretched length. Calculate the elastic potential energy of the spring (take it to be zero for the unstretched spring), the kinetic energy of the cat, the gravitational potential energy of the system relative to the lowest point of the motion, and the sum of these three energies when the cat is (a) at its highest point; (b) at its lowest point; (c) at its equilibrium position.

14.39 •• A 1.50-kg ball and a 2.00-kg ball are glued together with the lighter one below the heavier one. The upper ball is attached to a vertical ideal spring of force constant 165 N/m, and the system is vibrating vertically with amplitude 15.0 cm. The glue connecting the balls is old and weak, and it suddenly comes loose when the balls are at the lowest position in their motion. (a) Why is the glue more likely to fail at the *lowest* point than at any other point in the motion? (b) Find the amplitude and frequency of the vibrations after the lower ball has come loose.

14.40 •• A uniform, solid metal disk of mass 6.50 kg and diameter 24.0 cm hangs in a horizontal plane, supported at its center by a vertical metal wire. You find that it requires a horizontal force of 4.23 N tangent to the rim of the disk to turn it by 3.34°, thus twisting the wire. You now remove this force and release the disk from rest. (a) What is the torsion constant for the metal wire? (b) What are the frequency and period of the torsional oscillations of the disk? (c) Write the equation of motion for $\theta(t)$ for the disk.

14.41 •• A certain alarm clock ticks four times each second, with each tick representing half a period. The balance wheel consists of a thin rim with radius 0.55 cm, connected to the balance staff by thin spokes of negligible mass. The total mass of the balance wheel is 0.90 g. (a) What is the moment of inertia of the balance wheel about its shaft? (b) What is the torsion constant of the coil spring (Fig. 14.19)?

14.42 • A thin metal disk with mass 2.00×10^{-3} kg and radius 2.20 cm is attached at its center to a long fiber (Fig. E14.42). The disk, when twisted and released, oscillates with a period of 1.00 s. Find the torsion constant of the fiber.

Figure **E14.42**

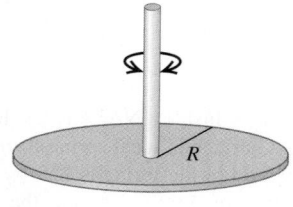

14.43 •• You want to find the moment of inertia of a complicated machine part about an axis through its center of mass. You suspend it from a wire along this axis. The wire has a torsion constant of 0.450 N·m/rad. You twist the part a small amount about this axis and let it go, timing 125 oscillations in 265 s. What is the moment of inertia of the part you want to find?

14.44 •• **CALC** The balance wheel of a watch vibrates with an angular amplitude Θ, angular frequency ω, and phase angle $\phi = 0$. (a) Find expressions for the angular velocity $d\theta/dt$ and angular acceleration $d^2\theta/dt^2$ as functions of time. (b) Find the balance wheel's angular velocity and angular acceleration when its angular displacement is Θ, and when its angular displacement is $\Theta/2$ and θ is decreasing. (*Hint:* Sketch a graph of θ versus t.)

Section 14.5 The Simple Pendulum

14.45 •• You pull a simple pendulum 0.240 m long to the side through an angle of 3.50° and release it. (a) How much time does it take the pendulum bob to reach its highest speed? (b) How much time does it take if the pendulum is released at an angle of 1.75° instead of 3.50°?

14.46 • An 85.0-kg mountain climber plans to swing down, starting from rest, from a ledge using a light rope 6.50 m long. He holds one end of the rope, and the other end is tied higher up on a rock face. Since the ledge is not very far from the rock face, the rope makes a small angle with the vertical. At the lowest point of his swing, he plans to let go and drop a short distance to the ground. (a) How long after he begins his swing will the climber first reach his lowest point? (b) If he missed the first chance to drop off, how long after first beginning his swing will the climber reach his lowest point for the second time?

14.47 • A building in San Francisco has light fixtures consisting of small 2.35-kg bulbs with shades hanging from the ceiling at the end of light, thin cords 1.50 m long. If a minor earthquake occurs, how many swings per second will these fixtures make?

14.48 • **A Pendulum on Mars.** A certain simple pendulum has a period on the earth of 1.60 s. What is its period on the surface of Mars, where $g = 3.71$ m/s²?

14.49 • After landing on an unfamiliar planet, a space explorer constructs a simple pendulum of length 50.0 cm. She finds that the pendulum makes 100 complete swings in 136 s. What is the value of g on this planet?

14.50 •• A small sphere with mass m is attached to a massless rod of length L that is pivoted at the top, forming a simple pendulum. The pendulum is pulled to one side so that the rod is at an angle Θ from the vertical, and released from rest. (a) In a diagram, show the pendulum just after it is released. Draw vectors representing the *forces* acting on the small sphere and the *acceleration* of the sphere. Accuracy counts! At this point, what is the linear acceleration of the sphere? (b) Repeat part (a) for the instant when the pendulum rod is at an angle $\Theta/2$ from the vertical. (c) Repeat part (a) for the instant when the pendulum rod is vertical. At this point, what is the linear speed of the sphere?

14.51 • A simple pendulum 2.00 m long swings through a maximum angle of 30.0° with the vertical. Calculate its period (a) assuming a small amplitude, and (b) using the first three terms of Eq. (14.35). (c) Which of the answers in parts (a) and (b) is more accurate? For the one that is less accurate, by what percent is it in error from the more accurate answer?

Section 14.6 The Physical Pendulum

14.52 •• We want to hang a thin hoop on a horizontal nail and have the hoop make one complete small-angle oscillation each 2.0 s. What must the hoop's radius be?

14.53 • A 1.80-kg connecting rod from a car engine is pivoted about a horizontal knife edge as shown in Fig. E14.53. The center of gravity of the rod was located by balancing and is 0.200 m from the pivot. When the rod is set into small-amplitude oscillation, it makes 100 complete swings in 120 s. Calculate the moment of inertia of the rod about the rotation axis through the pivot.

Figure **E14.53**

$d = 0.200$ m

cg

14.54 •• A 1.80-kg monkey wrench is pivoted 0.250 m from its center of mass and allowed to swing as a physical pendulum. The period for small-angle oscillations is 0.940 s. (a) What is the moment of inertia of the wrench about an axis through the pivot? (b) If the wrench is initially displaced 0.400 rad from its equilibrium position, what is the angular speed of the wrench as it passes through the equilibrium position?

14.55 • Two pendulums have the same dimensions (length L) and total mass (m). Pendulum A is a very small ball swinging at the end of a uniform massless bar. In pendulum B, half the mass is in the ball and half is in the uniform bar. Find the period of each pendulum for small oscillations. Which one takes longer for a swing?

14.56 •• **CP** A holiday ornament in the shape of a hollow sphere with mass $M = 0.015$ kg and radius $R = 0.050$ m is hung from a tree limb by a small loop of wire attached to the surface of the sphere. If the ornament is displaced a small distance and released, it swings back and forth as a physical pendulum with negligible friction. Calculate its period. (*Hint:* Use the parallel-axis theorem to find the moment of inertia of the sphere about the pivot at the tree limb.)

14.57 •• The two pendulums shown in Fig. E14.57 each consist of a uniform solid ball of mass M supported by a rigid massless rod, but the ball for pendulum A is very tiny while the ball for pendulum B is much larger. Find the period of each pendulum for small displacements. Which ball takes longer to complete a swing?

Figure E14.57

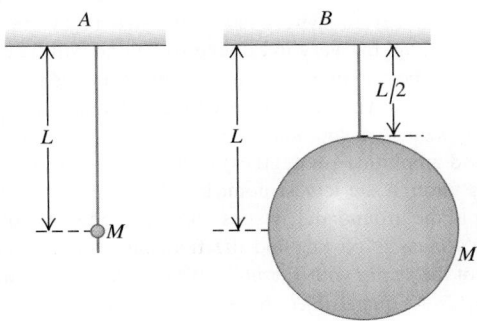

Section 14.7 Damped Oscillations

14.58 • A 2.50-kg rock is attached at the end of a thin, very light rope 1.45 m long. You start it swinging by releasing it when the rope makes an 11° angle with the vertical. You record the observation that it rises only to an angle of 4.5° with the vertical after $10\frac{1}{2}$ swings. (a) How much energy has this system lost during that time? (b) What happened to the "lost" energy? Explain *how* it could have been "lost."

14.59 • An unhappy 0.300-kg rodent, moving on the end of a spring with force constant $k = 2.50$ N/m, is acted on by a damping force $F_x = -bv_x$. (a) If the constant b has the value 0.900 kg/s, what is the frequency of oscillation of the rodent? (b) For what value of the constant b will the motion be critically damped?

14.60 •• A 50.0-g hard-boiled egg moves on the end of a spring with force constant $k = 25.0$ N/m. Its initial displacement is 0.300 m. A damping force $F_x = -bv_x$ acts on the egg, and the amplitude of the motion decreases to 0.100 m in 5.00 s. Calculate the magnitude of the damping constant b.

14.61 •• **CALC** The motion of an underdamped oscillator is described by Eq. (14.42). Let the phase angle ϕ be zero. (a) According to this equation, what is the value of x at $t = 0$? (b) What are the magnitude and direction of the velocity at $t = 0$? What does the result tell you about the slope of the graph of x versus t near $t = 0$? (c) Obtain an expression for the acceleration a_x at $t = 0$. For what value or range of values of the damping constant b (in terms of k and m) is the acceleration at $t = 0$ negative, zero, and positive? Discuss each case in terms of the shape of the graph of x versus t near $t = 0$.

14.62 •• A mass is vibrating at the end of a spring of force constant 225 N/m. Figure E14.62 shows a graph of its position x as a function of time t. (a) At what times is the mass not moving? (b) How much energy did this system originally contain? (c) How much energy did the system lose between $t = 1.0$ s and $t = 4.0$ s? Where did this energy go?

Figure E14.62

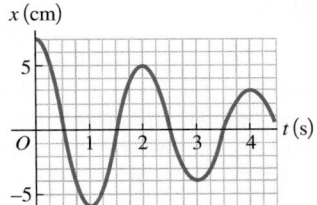

Section 14.8 Forced Oscillations and Resonance

14.63 • A sinusoidally varying driving force is applied to a damped harmonic oscillator. (a) What are the units of the damping constant b? (b) Show that the quantity $\sqrt{km}$ has the same units as b. (c) In terms of F_{max} and k, what is the amplitude for $\omega_d = \sqrt{k/m}$ when (i) $b = 0.2\sqrt{km}$ and (ii) $b = 0.4\sqrt{km}$? Compare your results to Fig. 14.28.

14.64 • A sinusoidally varying driving force is applied to a damped harmonic oscillator of force constant k and mass m. If the damping constant has a value b_1, the amplitude is A_1 when the driving angular frequency equals $\sqrt{k/m}$. In terms of A_1, what is the amplitude for the same driving frequency and the same driving force amplitude F_{max}, if the damping constant is (a) $3b_1$ and (b) $b_1/2$?

PROBLEMS

14.65 •• An object is undergoing SHM with period 1.200 s and amplitude 0.600 m. At $t = 0$ the object is at $x = 0$ and is moving in the negative x-direction. How far is the object from the equilibrium position when $t = 0.480$ s?

14.66 ••• An object is undergoing SHM with period 0.300 s and amplitude 6.00 cm. At $t = 0$ the object is instantaneously at rest at $x = 6.00$ cm. Calculate the time it takes the object to go from $x = 6.00$ cm to $x = -1.50$ cm.

14.67 • **CP SHM in a Car Engine.** The motion of the piston of an automobile engine is approximately simple harmonic. (a) If the stroke of an engine (twice the amplitude) is 0.100 m and the engine runs at 4500 rev/min, compute the acceleration of the piston at the endpoint of its stroke. (b) If the piston has mass 0.450 kg, what net force must be exerted on it at this point? (c) What are the speed and kinetic energy of the piston at the midpoint of its stroke? (d) What average power is required to accelerate the piston from rest to the speed found in part (c)? (e) If the engine runs at 7000 rev/min, what are the answers to parts (b), (c), and (d)?

14.68 • Four passengers with combined mass 250 kg compress the springs of a car with worn-out shock absorbers by 4.00 cm when they get in. Model the car and passengers as a single body on a single ideal spring. If the loaded car has a period of vibration of 1.92 s, what is the period of vibration of the empty car?

14.69 • A glider is oscillating in SHM on an air track with an amplitude A_1. You slow it so that its amplitude is halved. What happens to its (a) period, frequency, and angular frequency;

(b) total mechanical energy; (c) maximum speed; (d) speed at $x = \pm A_1/4$; (e) potential and kinetic energies at $x = \pm A_1/4$?

14.70 ••• **CP** A child with poor table manners is sliding his 250-g dinner plate back and forth in SHM with an amplitude of 0.100 m on a horizontal surface. At a point 0.060 m away from equilibrium, the speed of the plate is 0.400 m/s. (a) What is the period? (b) What is the displacement when the speed is 0.160 m/s? (c) In the center of the dinner plate is a 10.0-g carrot slice. If the carrot slice is just on the verge of slipping at the endpoint of the path, what is the coefficient of static friction between the carrot slice and the plate?

14.71 ••• A 1.50-kg, horizontal, uniform tray is attached to a vertical ideal spring of force constant 185 N/m and a 275-g metal ball is in the tray. The spring is below the tray, so it can oscillate up and down. The tray is then pushed down to point A, which is 15.0 cm below the equilibrium point, and released from rest. (a) How high above point A will the tray be when the metal ball leaves the tray? (*Hint:* This does *not* occur when the ball and tray reach their maximum speeds.) (b) How much time elapses between releasing the system at point A and the ball leaving the tray? (c) How fast is the ball moving just as it leaves the tray?

14.72 •• **CP** A block with mass M rests on a frictionless surface and is connected to a horizontal spring of force constant k. The other end of the spring is attached to a wall (Fig. P14.72). A second block with mass m rests on top of the first block. The coefficient of static friction between the blocks is μ_s. Find the *maximum* amplitude of oscillation such that the top block will not slip on the bottom block.

Figure **P14.72**

14.73 • **CP** A 10.0-kg mass is traveling to the right with a speed of 2.00 m/s on a smooth horizontal surface when it collides with and sticks to a second 10.0-kg mass that is initially at rest but is attached to a light spring with force constant 110.0 N/m. (a) Find the frequency, amplitude, and period of the subsequent oscillations. (b) How long does it take the system to return the first time to the position it had immediately after the collision?

14.74 • **CP** A rocket is accelerating upward at 4.00 m/s^2 from the launchpad on the earth. Inside a small, 1.50-kg ball hangs from the ceiling by a light, 1.10-m wire. If the ball is displaced 8.50° from the vertical and released, find the amplitude and period of the resulting swings of this pendulum.

14.75 ••• An apple weighs 1.00 N. When you hang it from the end of a long spring of force constant 1.50 N/m and negligible mass, it bounces up and down in SHM. If you stop the bouncing and let the apple swing from side to side through a small angle, the frequency of this simple pendulum is half the bounce frequency. (Because the angle is small, the back-and-forth swings do not cause any appreciable change in the length of the spring.) What is the unstretched length of the spring (with the apple removed)?

14.76 ••• **CP SHM of a Floating Object.** An object with height h, mass M, and a uniform cross-sectional area A floats upright in a liquid with density ρ. (a) Calculate the vertical distance from the surface of the liquid to the bottom of the floating object at equilibrium. (b) A downward force with magnitude F is applied to the top of the object. At the new equilibrium position, how much farther below the surface of the liquid is the bottom of the object than it was in part (a)? (Assume that some of the object remains above the surface of the liquid.) (c) Your result in part (b) shows that if the force is suddenly removed, the object will oscillate up and down in SHM. Calculate the period of this motion in terms of the density ρ of the liquid, the mass M, and the cross-sectional area A of the object. You can ignore the damping due to fluid friction (see Section 14.7).

14.77 •• **CP** A 950-kg, cylindrical can buoy floats vertically in salt water. The diameter of the buoy is 0.900 m. (a) Calculate the additional distance the buoy will sink when a 70.0-kg man stands on top of it. (Use the expression derived in part (b) of Problem 14.76.) (b) Calculate the period of the resulting vertical SHM when the man dives off. (Use the expression derived in part (c) of Problem 14.76, and as in that problem, you can ignore the damping due to fluid friction.)

14.78 ••• **CP Tarzan to the Rescue!** Tarzan spies a 35-kg chimpanzee in severe danger, so he swings to the rescue. He adjusts his strong, but very light, vine so that he will first come to rest 4.0 s after beginning his swing, at which time his vine makes a 12° angle with the vertical. (a) How long is Tarzan's vine, assuming that he swings at the bottom end of it? (b) What are the frequency and amplitude (in degrees) of Tarzan's swing? (c) Just as he passes through the lowest point in his swing, Tarzan nabs the chimp from the ground and sweeps him out of the jaws of danger. If Tarzan's mass is 65 kg, find the frequency and amplitude (in degrees) of the swing with Tarzan holding onto the grateful chimp.

14.79 •• **CP** A square object of mass m is constructed of four identical uniform thin sticks, each of length L, attached together. This object is hung on a hook at its upper corner (Fig. P14.79). If it is rotated slightly to the left and then released, at what frequency will it swing back and forth?

Figure **P14.79**

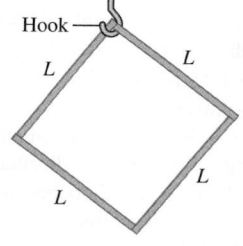

14.80 ••• An object with mass 0.200 kg is acted on by an elastic restoring force with force constant 10.0 N/m. (a) Graph elastic potential energy U as a function of displacement x over a range of x from -0.300 m to $+0.300$ m. On your graph, let 1 cm = 0.05 J vertically and 1 cm = 0.05 m horizontally. The object is set into oscillation with an initial potential energy of 0.140 J and an initial kinetic energy of 0.060 J. Answer the following questions by referring to the graph. (b) What is the amplitude of oscillation? (c) What is the potential energy when the displacement is one-half the amplitude? (d) At what displacement are the kinetic and potential energies equal? (e) What is the value of the phase angle ϕ if the initial velocity is positive and the initial displacement is negative?

14.81 • **CALC** A 2.00-kg bucket containing 10.0 kg of water is hanging from a vertical ideal spring of force constant 125 N/m and oscillating up and down with an amplitude of 3.00 cm. Suddenly the bucket springs a leak in the bottom such that water drops out at a steady rate of 2.00 g/s. When the bucket is half full, find

(a) the period of oscillation and (b) the rate at which the period is changing with respect to time. Is the period getting longer or shorter? (c) What is the shortest period this system can have?

14.82 •• **CP** A hanging wire is 1.80 m long. When a 60.0-kg steel ball is suspended from the wire, the wire stretches by 2.00 mm. If the ball is pulled down a small additional distance and released, at what frequency will it vibrate? Assume that the stress on the wire is less than the proportional limit (see Section 11.5).

14.83 •• A 5.00-kg partridge is suspended from a pear tree by an ideal spring of negligible mass. When the partridge is pulled down 0.100 m below its equilibrium position and released, it vibrates with a period of 4.20 s. (a) What is its speed as it passes through the equilibrium position? (b) What is its acceleration when it is 0.050 m above the equilibrium position? (c) When it is moving upward, how much time is required for it to move from a point 0.050 m below its equilibrium position to a point 0.050 m above it? (d) The motion of the partridge is stopped, and then it is removed from the spring. How much does the spring shorten?

14.84 •• A 0.0200-kg bolt moves with SHM that has an amplitude of 0.240 m and a period of 1.500 s. The displacement of the bolt is +0.240 m when $t = 0$. Compute (a) the displacement of the bolt when $t = 0.500$ s; (b) the magnitude and direction of the force acting on the bolt when $t = 0.500$ s; (c) the minimum time required for the bolt to move from its initial position to the point where $x = -0.180$ m; (d) the speed of the bolt when $x = -0.180$ m.

14.85 •• **CP** **SHM of a Butcher's Scale.** A spring of negligible mass and force constant $k = 400$ N/m is hung vertically, and a 0.200-kg pan is suspended from its lower end. A butcher drops a 2.2-kg steak onto the pan from a height of 0.40 m. The steak makes a totally inelastic collision with the pan and sets the system into vertical SHM. What are (a) the speed of the pan and steak immediately after the collision; (b) the amplitude of the subsequent motion; (c) the period of that motion?

14.86 •• A uniform beam is suspended horizontally by two identical vertical springs that are attached between the ceiling and each end of the beam. The beam has mass 225 kg, and a 175-kg sack of gravel sits on the middle of it. The beam is oscillating in SHM, with an amplitude of 40.0 cm and a frequency of 0.600 cycle/s. (a) The sack of gravel falls off the beam when the beam has its maximum upward displacement. What are the frequency and amplitude of the subsequent SHM of the beam? (b) If the gravel instead falls off when the beam has its maximum speed, what are the frequency and amplitude of the subsequent SHM of the beam?

14.87 ••• **CP** On the planet Newtonia, a simple pendulum having a bob with mass 1.25 kg and a length of 185.0 cm takes 1.42 s, when released from rest, to swing through an angle of 12.5°, where it again has zero speed. The circumference of Newtonia is measured to be 51,400 km. What is the mass of the planet Newtonia?

14.88 •• A 40.0-N force stretches a vertical spring 0.250 m. (a) What mass must be suspended from the spring so that the system will oscillate with a period of 1.00 s? (b) If the amplitude of the motion is 0.050 m and the period is that specified in part (a), where is the object and in what direction is it moving 0.35 s after it has passed the equilibrium position, moving downward? (c) What force (magnitude and direction) does the spring exert on the object when it is 0.030 m below the equilibrium position, moving upward?

14.89 •• **Don't Miss the Boat.** While on a visit to Minnesota ("Land of 10,000 Lakes"), you sign up to take an excursion around one of the larger lakes. When you go to the dock where the 1500-kg boat is tied, you find that the boat is bobbing up and down in the waves, executing simple harmonic motion with amplitude 20 cm. The boat takes 3.5 s to make one complete up-and-down cycle.

When the boat is at its highest point, its deck is at the same height as the stationary dock. As you watch the boat bob up and down, you (mass 60 kg) begin to feel a bit woozy, due in part to the previous night's dinner of lutefisk. As a result, you refuse to board the boat unless the level of the boat's deck is within 10 cm of the dock level. How much time do you have to board the boat comfortably during each cycle of up-and-down motion?

14.90 • **CP** An interesting, though highly impractical example of oscillation is the motion of an object dropped down a hole that extends from one side of the earth, through its center, to the other side. With the assumption (not realistic) that the earth is a sphere of uniform density, prove that the motion is simple harmonic and find the period. [*Note:* The gravitational force on the object as a function of the object's distance r from the center of the earth was derived in Example 13.10 (Section 13.6). The motion is simple harmonic if the acceleration a_x and the displacement from equilibrium x are related by Eq. (14.8), and the period is then $T = 2\pi/\omega$.]

14.91 ••• **CP** A rifle bullet with mass 8.00 g and initial horizontal velocity 280 m/s strikes and embeds itself in a block with mass 0.992 kg that rests on a frictionless surface and is attached to one end of an ideal spring. The other end of the spring is attached to the wall. The impact compresses the spring a maximum distance of 18.0 cm. After the impact, the block moves in SHM. Calculate the period of this motion.

14.92 •• **CP CALC** For a certain oscillator the net force on the body with mass m is given by $F_x = -cx^3$. (a) What is the potential energy function for this oscillator if we take $U = 0$ at $x = 0$? (b) One-quarter of a period is the time for the body to move from $x = 0$ to $x = A$. Calculate this time and hence the period. [*Hint:* Begin with Eq. (14.20), modified to include the potential-energy function you found in part (a), and solve for the velocity v_x as a function of x. Then replace v_x with dx/dt. Separate the variable by writing all factors containing x on one side and all factors containing t on the other side so that each side can be integrated. In the x-integral make the change of variable $u = x/A$. The resulting integral can be evaluated by numerical methods on a computer and has the value $\int_0^1 du/\sqrt{1 - u^4} = 1.31$.] (c) According to the result you obtained in part (b), does the period depend on the amplitude A of the motion? Are the oscillations simple harmonic?

14.93 • **CP CALC** An approximation for the potential energy of a KCl molecule is $U = A[(R_0^7/8r^8) - 1/r]$, where $R_0 = 2.67 \times 10^{-10}$ m, $A = 2.31 \times 10^{-28}$ J·m, and r is the distance between the two atoms. Using this approximation: (a) Show that the radial component of the force on each atom is $F_r = A[(R_0^7/r^9) - 1/r^2]$. (b) Show that R_0 is the equilibrium separation. (c) Find the minimum potential energy. (d) Use $r = R_0 + x$ and the first two terms of the binomial theorem (Eq. 14.28) to show that $F_r \approx -(7A/R_0^3)x$, so that the molecule's force constant is $k = 7A/R_0^3$. (e) With both the K and Cl atoms vibrating in opposite directions on opposite sides of the molecule's center of mass, $m_1 m_2/(m_1 + m_2) = 3.06 \times 10^{-26}$ kg is the mass to use in calculating the frequency. Calculate the frequency of small-amplitude vibrations.

14.94 ••• **CP** Two uniform solid spheres, each with mass $M = 0.800$ kg and radius $R = 0.0800$ m, are connected by a short, light rod that is along a diameter of each sphere and are at rest on a horizontal tabletop. A spring with force constant $k = 160$ N/m has one end attached to the wall and the other end attached to a frictionless ring that passes over the rod at the center of mass of the spheres, which is midway between the centers of the two spheres. The spheres are each pulled the same distance from the wall, stretching the spring, and released. There is sufficient friction between the tabletop and the spheres for the spheres to roll without slipping as they move back and forth on the end of the spring. Show that the motion of the center of mass of the spheres is simple harmonic and calculate the period.

14.95 • **CP** In Fig. P14.95 the upper ball is released from rest, collides with the stationary lower ball, and sticks to it. The strings are both 50.0 cm long. The upper ball has mass 2.00 kg, and it is initially 10.0 cm higher than the lower ball, which has mass 3.00 kg. Find the frequency and maximum angular displacement of the motion after the collision.

Figure **P14.95**

14.96 •• **CP BIO** *T. rex.* Model the leg of the *T. rex* in Example 14.10 (Section 14.6) as two uniform rods, each 1.55 m long, joined rigidly end to end. Let the lower rod have mass M and the upper rod mass $2M$. The composite object is pivoted about the top of the upper rod. Compute the oscillation period of this object for small-amplitude oscillations. Compare your result to that of Example 14.10.

14.97 •• **CALC** A slender, uniform, metal rod with mass M is pivoted without friction about an axis through its midpoint and perpendicular to the rod. A horizontal spring with force constant k is attached to the lower end of the rod, with the other end of the spring attached to a rigid support. If the rod is displaced by a small angle Θ from the vertical (Fig. P14.97) and released, show that it moves in angular SHM and calculate the period. (*Hint:* Assume that the angle Θ is small enough for the approximations $\sin\Theta \approx \Theta$ and $\cos\Theta \approx 1$ to be valid. The motion is simple harmonic if $d^2\theta/dt^2 = -\omega^2\theta$, and the period is then $T = 2\pi/\omega$.)

Figure **P14.97**

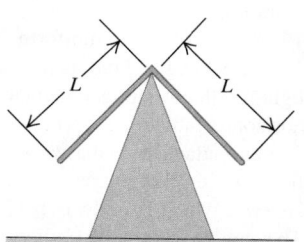

14.98 •• **The Silently Ringing Bell Problem.** A large bell is hung from a wooden beam so it can swing back and forth with negligible friction. The center of mass of the bell is 0.60 m below the pivot, the bell has mass 34.0 kg, and the moment of inertia of the bell about an axis at the pivot is 18.0 kg·m². The clapper is a small, 1.8-kg mass attached to one end of a slender rod that has length L and negligible mass. The other end of the rod is attached to the inside of the bell so it can swing freely about the same axis as the bell. What should be the length L of the clapper rod for the bell to ring silently—that is, for the period of oscillation for the bell to equal that for the clapper?

14.99 ••• Two identical thin rods, each with mass m and length L, are joined at right angles to form an L-shaped object. This object is balanced on top of a sharp edge (Fig. P14.99). If the L-shaped object is deflected slightly, it oscillates. Find the frequency of oscillation.

Figure **P14.99**

14.100 • **CP CALC** A uniform rod of length L oscillates through small angles about a point a distance x from its center. (a) Prove that its angular frequency is $\sqrt{gx/[(L^2/12) + x^2]}$. (b) Show that its maximum angular frequency occurs when $x = L/\sqrt{12}$. (c) What is the length of the rod if the maximum angular frequency is 2π rad/s?

CHALLENGE PROBLEMS

14.101 ••• **The Effective Force Constant of Two Springs.** Two springs with the same unstretched length but different force constants k_1 and k_2 are attached to a block with mass m on a level, frictionless surface. Calculate the effective force constant k_{eff} in each of the three cases (a), (b), and (c) depicted in Fig. P14.101. (The effective force constant is defined by $\Sigma F_x = -k_{\text{eff}}x$.) (d) An object with mass m, suspended from a uniform spring with a force constant k, vibrates with a frequency f_1. When the spring is cut in half and the same object is suspended from one of the halves, the frequency is f_2. What is the ratio f_2/f_1?

Figure **P14.101**

14.102 ••• Two springs, each with unstretched length 0.200 m but with different force constants k_1 and k_2, are attached to opposite ends of a block with mass m on a level, frictionless surface. The outer ends of the springs are now attached to two pins P_1 and P_2, 0.100 m from the original positions of the ends of the springs (Fig. P14.102). Let $k_1 = 2.00$ N/m, $k_2 = 6.00$ N/m, and $m = 0.100$ kg. (a) Find the length of each spring when the block is in its new equilibrium position after the springs have been attached to the pins. (b) Find the period of vibration of the block if it is slightly displaced from its new equilibrium position and released.

Figure **P14.102**

14.103 ••• **CALC** **A Spring with Mass.** The preceding problems in this chapter have assumed that the springs had negligible mass. But of course no spring is completely massless. To find the effect of the spring's mass, consider a spring with mass M, equilibrium length L_0, and spring constant k. When stretched or compressed to a length L, the potential energy is $\frac{1}{2}kx^2$, where $x = L - L_0$. (a) Consider a spring, as described above, that has one end fixed and the other end moving with speed v. Assume that the speed of points along the length of the spring varies linearly with distance l from the fixed end. Assume also that the mass M of the spring is distributed uniformly along the length of the spring. Calculate the kinetic energy of the spring in terms of M and v. (*Hint:* Divide the spring into pieces of length dl; find the speed of each piece in terms of l, v, and L; find the mass of each piece in terms of dl, M, and L; and integrate from 0 to L. The result is *not* $\frac{1}{2}Mv^2$, since not all of the spring moves with the same speed.) (b) Take the time derivative of the conservation of energy equation, Eq. (14.21), for a mass m moving on the end of a *massless* spring. By comparing your results to Eq. (14.8), which defines ω, show that the angular frequency of oscillation is $\omega = \sqrt{k/m}$. (c) Apply the procedure of part (b) to obtain the angular frequency of oscillation ω of the spring considered in part (a). If the *effective mass M'* of the spring is defined by $\omega = \sqrt{k/M'}$, what is M' in terms of M?

Answers

Chapter Opening Question ?

The length of the leg is more important. The back-and-forth motion of a leg during walking is like a physical pendulum, for which the oscillation period is $T = 2\pi\sqrt{I/mgd}$ [see Eq. (14.39)]. In this expression I is the moment of inertia of the pendulum, m is its mass, and d is the distance from the rotation axis to the pendulum center of mass. The moment of inertia I is proportional to the mass m, so the mass cancels out of this expression for the period T. Hence only the dimensions of the leg matter. (See Examples 14.9 and 14.10.)

Test Your Understanding Questions

14.1 Answers: (a) $x < 0$, **(b)** $x > 0$, **(c)** $x < 0$, **(d)** $x > 0$, **(e)** $x > 0$, **(f)** $x = 0$ Figure 14.2 shows that the net x-component of force F_x and the x-acceleration a_x are both positive when $x < 0$ (so the body is displaced to the left and the spring is compressed), while F_x and a_x are both negative when $x > 0$ (so the body is displaced to the right and the spring is stretched). Hence x and a_x always have *opposite* signs. This is true whether the object is moving to the right ($v_x > 0$), to the left ($v_x < 0$), or not at all ($v_x = 0$), since the force exerted by the spring depends only on whether it is compressed or stretched and by what distance. This explains the answers to (a) through (e). If the acceleration is zero as in (f), the net force must also be zero and so the spring must be relaxed; hence $x = 0$.

14.2 Answers: (a) $A > 0.10$ m, $\phi < 0$; **(b)** $A > 0.10$ m, $\phi > 0$ In both situations the initial ($t = 0$) x-velocity v_{0x} is nonzero, so from Eq. (14.19) the amplitude $A = \sqrt{x_0^2 + (v_{0x}^2/\omega^2)}$ is greater than the initial x-coordinate $x_0 = 0.10$ m. From Eq. (14.18) the phase angle is $\phi = \arctan(-v_{0x}/\omega x_0)$, which is positive if the quantity $-v_{0x}/\omega x_0$ (the argument of the arctangent function) is positive and negative if $-v_{0x}/\omega x_0$ is negative. In part (a) x_0 and v_{0x} are both positive, so $-v_{0x}/\omega x_0 < 0$ and $\phi < 0$. In part (b) x_0 is positive and v_{0x} is negative, so $-v_{0x}/\omega x_0 > 0$ and $\phi > 0$.

14.3 Answers: (a) (iii), (b) (v) To increase the total energy $E = \frac{1}{2}kA^2$ by a factor of 2, the amplitude A must increase by a factor of $\sqrt{2}$. Because the motion is SHM, changing the amplitude has no effect on the frequency.

14.4 Answer: (i) The oscillation period of a body of mass m attached to a hanging spring of force constant k is given by $T = 2\pi\sqrt{m/k}$, the same expression as for a body attached to a horizontal spring. Neither m nor k changes when the apparatus is taken to Mars, so the period is unchanged. The only difference is that in equilibrium, the spring will stretch a shorter distance on Mars than on earth due to the weaker gravity.

14.5 Answer: no Just as for an object oscillating on a spring, at the equilibrium position the *speed* of the pendulum bob is instantaneously not changing (this is where the speed is maximum, so its derivative at this time is zero). But the *direction* of motion is changing because the pendulum bob follows a circular path. Hence the bob must have a component of acceleration perpendicular to the path and toward the center of the circle (see Section 3.4). To cause this acceleration at the equilibrium position when the string is vertical, the upward tension force at this position must be greater than the weight of the bob. This causes a net upward force on the bob and an upward acceleration toward the center of the circular path.

14.6 Answer: (i) The period of a physical pendulum is given by Eq. (14.39), $T = 2\pi\sqrt{I/mgd}$. The distance $d = L$ from the pivot to the center of gravity is the same for both the rod and the simple pendulum, as is the mass m. This means that for any displacement angle θ the same restoring torque acts on both the rod and the simple pendulum. However, the rod has a greater moment of inertia: $I_{rod} = \frac{1}{3}m(2L)^2 = \frac{4}{3}mL^2$ and $I_{simple} = mL^2$ (all the mass of the pendulum is a distance L from the pivot). Hence the rod has a longer period.

14.7 Answer: (ii) The oscillations are underdamped with a decreasing amplitude on each cycle of oscillation, like those graphed in Fig. 14.26. If the oscillations were undamped, they would continue indefinitely with the same amplitude. If they were critically damped or overdamped, the nose would not bob up and down but would return smoothly to the original equilibrium attitude without overshooting.

14.8 Answer: (i) Figure 14.28 shows that the curve of amplitude versus driving frequency moves upward at *all* frequencies as the value of the damping constant b is decreased. Hence for fixed values of k and m, the oscillator with the least damping (smallest value of b) will have the greatest response at any driving frequency.

Bridging Problem

Answer: $T = 2\pi\sqrt{3M/2k}$

15 MECHANICAL WAVES

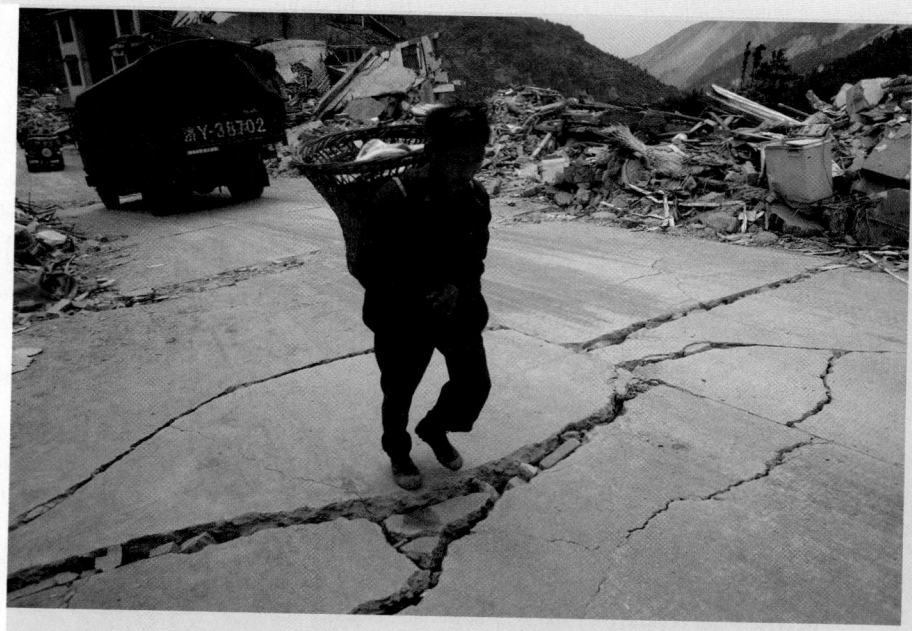

? When an earthquake strikes, the news of the event travels through the body of the earth in the form of seismic waves. Which aspects of a seismic wave determine how much power is carried by the wave?

LEARNING GOALS

By studying this chapter, you will learn:

- What is meant by a mechanical wave, and the different varieties of mechanical waves.

- How to use the relationship among speed, frequency, and wavelength for a periodic wave.

- How to interpret and use the mathematical expression for a sinusoidal periodic wave.

- How to calculate the speed of waves on a rope or string.

- How to calculate the rate at which a mechanical wave transports energy.

- What happens when mechanical waves overlap and interfere.

- The properties of standing waves on a string, and how to analyze these waves.

- How stringed instruments produce sounds of specific frequencies.

Ripples on a pond, musical sounds, seismic tremors triggered by an earthquake—all these are *wave* phenomena. Waves can occur whenever a system is disturbed from equilibrium and when the disturbance can travel, or *propagate,* from one region of the system to another. As a wave propagates, it carries energy. The energy in light waves from the sun warms the surface of our planet; the energy in seismic waves can crack our planet's crust.

This chapter and the next are about mechanical waves—waves that travel within some material called a *medium.* (Chapter 16 is concerned with sound, an important type of mechanical wave.) We'll begin this chapter by deriving the basic equations for describing waves, including the important special case of *sinusoidal* waves in which the wave pattern is a repeating sine or cosine function. To help us understand waves in general, we'll look at the simple case of waves that travel on a stretched string or rope.

Waves on a string play an important role in music. When a musician strums a guitar or bows a violin, she makes waves that travel in opposite directions along the instrument's strings. What happens when these oppositely directed waves overlap is called *interference.* We'll discover that sinusoidal waves can occur on a guitar or violin string only for certain special frequencies, called *normal-mode frequencies,* determined by the properties of the string. The normal-mode frequencies of a stringed instrument determine the pitch of the musical sounds that the instrument produces. (In the next chapter we'll find that interference also helps explain the pitches of *wind* instruments such as flutes and pipe organs.)

Not all waves are mechanical in nature. *Electromagnetic* waves—including light, radio waves, infrared and ultraviolet radiation, and x rays—can propagate even in empty space, where there is *no* medium. We'll explore these and other nonmechanical waves in later chapters.

15.1 Types of Mechanical Waves

A **mechanical wave** is a disturbance that travels through some material or substance called the **medium** for the wave. As the wave travels through the medium, the particles that make up the medium undergo displacements of various kinds, depending on the nature of the wave.

Figure 15.1 shows three varieties of mechanical waves. In Fig. 15.1a the medium is a string or rope under tension. If we give the left end a small upward shake or wiggle, the wiggle travels along the length of the string. Successive sections of string go through the same motion that we gave to the end, but at successively later times. Because the displacements of the medium are perpendicular or *transverse* to the direction of travel of the wave along the medium, this is called a **transverse wave.**

In Fig. 15.1b the medium is a liquid or gas in a tube with a rigid wall at the right end and a movable piston at the left end. If we give the piston a single back-and-forth motion, displacement and pressure fluctuations travel down the length of the medium. This time the motions of the particles of the medium are back and forth along the *same* direction that the wave travels. We call this a **longitudinal wave.**

In Fig. 15.1c the medium is a liquid in a channel, such as water in an irrigation ditch or canal. When we move the flat board at the left end forward and back once, a wave disturbance travels down the length of the channel. In this case the displacements of the water have *both* longitudinal and transverse components.

Each of these systems has an equilibrium state. For the stretched string it is the state in which the system is at rest, stretched out along a straight line. For the fluid in a tube it is a state in which the fluid is at rest with uniform pressure. And for the liquid in a trough it is a smooth, level water surface. In each case the wave motion is a disturbance from the equilibrium state that travels from one region of the medium to another. And in each case there are forces that tend to restore the system to its equilibrium position when it is displaced, just as the force of gravity tends to pull a pendulum toward its straight-down equilibrium position when it is displaced.

Application Waves on a Snake's Body
A snake moves itself along the ground by producing waves that travel backward along its body from its head to its tail. The waves remain stationary with respect to the ground as they push against the ground, so the snake moves forward.

Mastering**PHYSICS**

ActivPhysics 10.1: Properties of Mechanical Waves

15.1 Three ways to make a wave that moves to the right. **(a)** The hand moves the string up and then returns, producing a transverse wave. **(b)** The piston moves to the right, compressing the gas or liquid, and then returns, producing a longitudinal wave. **(c)** The board moves to the right and then returns, producing a combination of longitudinal and transverse waves.

(a) Transverse wave on a string

Motion of the wave

Particles of the string

As the wave passes, each particle of the string moves up and then down, *transversely* to the motion of the wave itself.

(b) Longitudinal wave in a fluid

Particles of the fluid

As the wave passes, each particle of the fluid moves forward and then back, *parallel* to the motion of the wave itself.

(c) Waves on the surface of a liquid

Surface particles of the liquid

As the wave passes, each particle of the liquid surface moves in a circle.

15.2 "Doing the wave" at a sports stadium is an example of a mechanical wave: The disturbance propagates through the crowd, but there is no transport of matter (none of the spectators moves from one seat to another).

These examples have three things in common. First, in each case the disturbance travels or *propagates* with a definite speed through the medium. This speed is called the speed of propagation, or simply the **wave speed.** Its value is determined in each case by the mechanical properties of the medium. We will use the symbol v for wave speed. (The wave speed is *not* the same as the speed with which particles move when they are disturbed by the wave. We'll return to this point in Section 15.3.) Second, the medium itself does not travel through space; its individual particles undergo back-and-forth or up-and-down motions around their equilibrium positions. The overall pattern of the wave disturbance is what travels. Third, to set any of these systems into motion, we have to put in energy by doing mechanical work on the system. The wave motion transports this energy from one region of the medium to another. *Waves transport energy, but not matter, from one region to another* (Fig. 15.2).

Test Your Understanding of Section 15.1 What type of wave is "the wave" shown in Fig. 15.2? (i) transverse; (ii) longitudinal; (iii) a combination of transverse and longitudinal.

15.2 Periodic Waves

The transverse wave on a stretched string in Fig. 15.1a is an example of a *wave pulse*. The hand shakes the string up and down just once, exerting a transverse force on it as it does so. The result is a single "wiggle," or pulse, that travels along the length of the string. The tension in the string restores its straight-line shape once the pulse has passed.

A more interesting situation develops when we give the free end of the string a repetitive, or *periodic*, motion. (You may want to review the discussion of periodic motion in Chapter 14 before going ahead.) Then each particle in the string also undergoes periodic motion as the wave propagates, and we have a **periodic wave.**

Periodic Transverse Waves

In particular, suppose we move the string up and down with *simple harmonic motion* (SHM) with amplitude A, frequency f, angular frequency $\omega = 2\pi f$, and period $T = 1/f = 2\pi/\omega$. Figure 15.3 shows one way to do this. The wave that results is a symmetrical sequence of *crests* and *troughs*. As we will see, periodic

15.3 A block of mass m attached to a spring undergoes simple harmonic motion, producing a sinusoidal wave that travels to the right on the string. (In a real-life system a driving force would have to be applied to the block to replace the energy carried away by the wave.)

The SHM of the spring and mass generates a sinusoidal wave in the string. Each particle in the string exhibits the same harmonic motion as the spring and mass; the amplitude of the wave is the amplitude of this motion.

waves with simple harmonic motion are particularly easy to analyze; we call them **sinusoidal waves.** It also turns out that *any* periodic wave can be represented as a combination of sinusoidal waves. So this particular kind of wave motion is worth special attention.

In Fig. 15.3 the wave that advances along the string is a *continuous succession* of transverse sinusoidal disturbances. Figure 15.4 shows the shape of a part of the string near the left end at time intervals of $\frac{1}{8}$ of a period, for a total time of one period. The wave shape advances steadily toward the right, as indicated by the highlighted area. As the wave moves, any point on the string (any of the red dots, for example) oscillates up and down about its equilibrium position with simple harmonic motion. *When a sinusoidal wave passes through a medium, every particle in the medium undergoes simple harmonic motion with the same frequency.*

CAUTION **Wave motion vs. particle motion** Be very careful to distinguish between the motion of the *transverse wave* along the string and the motion of a *particle* of the string. The wave moves with constant speed v *along* the length of the string, while the motion of the particle is simple harmonic and *transverse* (perpendicular) to the length of the string. ▮

For a periodic wave, the shape of the string at any instant is a repeating pattern. The length of one complete wave pattern is the distance from one crest to the next, or from one trough to the next, or from any point to the corresponding point on the next repetition of the wave shape. We call this distance the **wavelength** of the wave, denoted by λ (the Greek letter lambda). The wave pattern travels with constant speed v and advances a distance of one wavelength λ in a time interval of one period T. So the wave speed v is given by $v = \lambda/T$ or, because $f = 1/T$,

$$v = \lambda f \qquad \text{(periodic wave)} \qquad (15.1)$$

The speed of propagation equals the product of wavelength and frequency. The frequency is a property of the *entire* periodic wave because all points on the string oscillate with the same frequency f.

Waves on a string propagate in just one dimension (in Fig. 15.4, along the x-axis). But the ideas of frequency, wavelength, and amplitude apply equally well to waves that propagate in two or three dimensions. Figure 15.5 shows a wave propagating in two dimensions on the surface of a tank of water. As with waves on a string, the wavelength is the distance from one crest to the next, and the amplitude is the height of a crest above the equilibrium level.

In many important situations including waves on a string, the wave speed v is determined entirely by the mechanical properties of the medium. In this case, increasing f causes λ to decrease so that the product $v = \lambda f$ remains the same, and waves of *all* frequencies propagate with the same wave speed. In this chapter we will consider *only* waves of this kind. (In later chapters we will study the propagation of light waves in matter for which the wave speed depends on frequency; this turns out to be the reason prisms break white light into a spectrum and raindrops create a rainbow.)

Periodic Longitudinal Waves

To understand the mechanics of a periodic *longitudinal* wave, we consider a long tube filled with a fluid, with a piston at the left end as in Fig. 15.1b. If we push the piston in, we compress the fluid near the piston, increasing the pressure in this

15.4 A sinusoidal transverse wave traveling to the right along a string. The vertical scale is exaggerated.

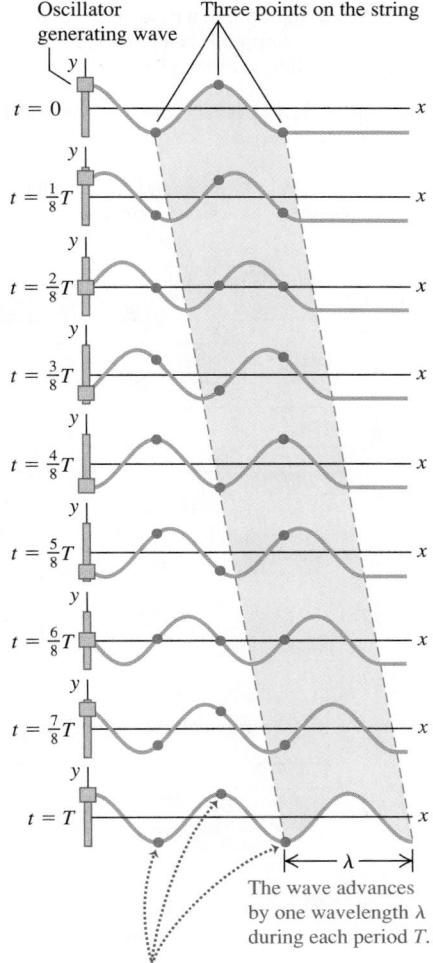

The string is shown at time intervals of $\frac{1}{8}$ period for a total of one period T. The highlighting shows the motion of one wavelength of the wave.

The wave advances by one wavelength λ during each period T.

Each point moves up and down in place. Particles one wavelength apart move in phase with each other.

15.5 A series of drops falling into water produces a periodic wave that spreads radially outward. The wave crests and troughs are concentric circles. The wavelength λ is the radial distance between adjacent crests or adjacent troughs.

15.6 Using an oscillating piston to make a sinusoidal longitudinal wave in a fluid.

Forward motion of the plunger creates a compression (a zone of high density); backward motion creates a rarefaction (a zone of low density).

Wavelength λ is the distance between corresponding points on successive cycles.

15.7 A sinusoidal longitudinal wave traveling to the right in a fluid. The wave has the same amplitude A and period T as the oscillation of the piston.

Longitudinal waves are shown at intervals of $\frac{1}{8}T$ for one period T.

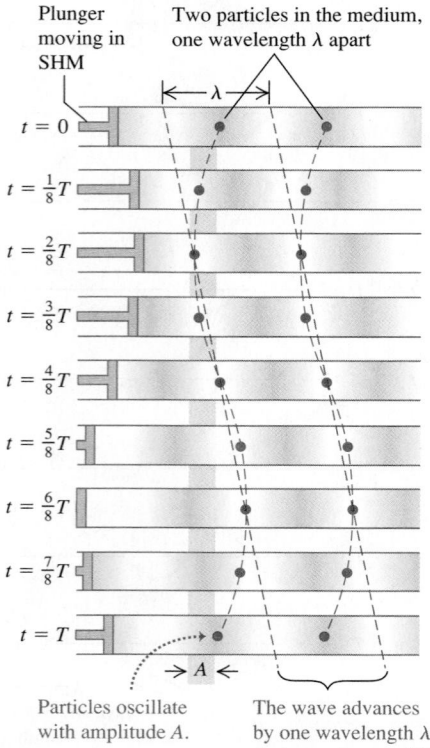

Particles oscillate with amplitude A.

The wave advances by one wavelength λ during each period T.

region. This region then pushes against the neighboring region of fluid, and so on, and a wave pulse moves along the tube.

Now suppose we move the piston back and forth with simple harmonic motion, along a line parallel to the axis of the tube (Fig. 15.6). This motion forms regions in the fluid where the pressure and density are greater or less than the equilibrium values. We call a region of increased density a *compression;* a region of reduced density is a *rarefaction.* Figure 15.6 shows compressions as darkly shaded areas and rarefactions as lightly shaded areas. The wavelength is the distance from one compression to the next or from one rarefaction to the next.

Figure 15.7 shows the wave propagating in the fluid-filled tube at time intervals of $\frac{1}{8}$ of a period, for a total time of one period. The pattern of compressions and rarefactions moves steadily to the right, just like the pattern of crests and troughs in a sinusoidal transverse wave (compare Fig. 15.4). Each particle in the fluid oscillates in SHM parallel to the direction of wave propagation (that is, left and right) with the same amplitude A and period T as the piston. The particles shown by the two red dots in Fig. 15.7 are one wavelength apart, and so oscillate in phase with each other.

Just like the sinusoidal transverse wave shown in Fig. 15.4, in one period T the longitudinal wave in Fig. 15.7 travels one wavelength λ to the right. Hence the fundamental equation $v = \lambda f$ holds for longitudinal waves as well as for transverse waves, and indeed for *all* types of periodic waves. Just as for transverse waves, in this chapter and the next we will consider only situations in which the speed of longitudinal waves does not depend on the frequency.

| Example 15.1 | **Wavelength of a musical sound** |

Sound waves are longitudinal waves in air. The speed of sound depends on temperature; at 20°C it is 344 m/s (1130 ft/s). What is the wavelength of a sound wave in air at 20°C if the frequency is 262 Hz (the approximate frequency of middle C on a piano)?

SOLUTION

IDENTIFY and SET UP: This problem involves Eq. (15.1), $v = \lambda f$, which relates wave speed v, wavelength λ, and frequency f for a periodic wave. The target variable is the wavelength λ. We are given $v = 344$ m/s and $f = 262$ Hz $= 262$ s^{-1}.

EXECUTE: We solve Eq. (15.1) for λ:

$$\lambda = \frac{v}{f} = \frac{344 \text{ m/s}}{262 \text{ Hz}} = \frac{344 \text{ m/s}}{262 \text{ s}^{-1}} = 1.31 \text{ m}$$

EVALUATE: The speed v of sound waves does *not* depend on the frequency. Hence $\lambda = v/f$ says that wavelength changes in inverse proportion to frequency. As an example, high (soprano) C is two octaves above middle C. Each octave corresponds to a factor of 2 in frequency, so the frequency of high C is four times that of middle C: $f = 4(262 \text{ Hz}) = 1048$ Hz. Hence the *wavelength* of high C is *one-fourth* as large: $\lambda = (1.31 \text{ m})/4 = 0.328$ m.

Test Your Understanding of Section 15.2 If you double the wavelength of a wave on a particular string, what happens to the wave speed v and the frequency f? (i) v doubles and f is unchanged; (ii) v is unchanged and f doubles; (iii) v becomes one-half as great and f is unchanged; (iv) v is unchanged and f becomes one-half as great; (v) none of these.

15.3 Mathematical Description of a Wave

Many characteristics of periodic waves can be described by using the concepts of wave speed, amplitude, period, frequency, and wavelength. Often, though, we need a more detailed description of the positions and motions of individual particles of the medium at particular times during wave propagation.

As a specific example, let's look at waves on a stretched string. If we ignore the sag of the string due to gravity, the equilibrium position of the string is along a straight line. We take this to be the x-axis of a coordinate system. Waves on a string are *transverse;* during wave motion a particle with equilibrium position x is displaced some distance y in the direction perpendicular to the x-axis. The value of y depends on which particle we are talking about (that is, y depends on x) and also on the time t when we look at it. Thus y is a *function* of both x and t; $y = y(x, t)$. We call $y(x, t)$ the **wave function** that describes the wave. If we know this function for a particular wave motion, we can use it to find the displacement (from equilibrium) of any particle at any time. From this we can find the velocity and acceleration of any particle, the shape of the string, and anything else we want to know about the behavior of the string at any time.

Wave Function for a Sinusoidal Wave

Let's see how to determine the form of the wave function for a sinusoidal wave. Suppose a sinusoidal wave travels from left to right (the direction of increasing x) along the string, as in Fig. 15.8. Every particle of the string oscillates with simple harmonic motion with the same amplitude and frequency. But the oscillations of particles at different points on the string are *not* all in step with each other. The particle at point B in Fig. 15.8 is at its maximum positive value of y at $t = 0$ and returns to $y = 0$ at $t = \frac{2}{8}T$; these same events occur for a particle at point A or point C at $t = \frac{4}{8}T$ and $t = \frac{6}{8}T$, exactly one half-period later. For any two particles of the string, the motion of the particle on the right (in terms of the wave, the "downstream" particle) lags behind the motion of the particle on the left by an amount proportional to the distance between the particles.

Hence the cyclic motions of various points on the string are out of step with each other by various fractions of a cycle. We call these differences *phase differences,* and we say that the *phase* of the motion is different for different points. For example, if one point has its maximum positive displacement at the same time that another has its maximum negative displacement, the two are a half-cycle out of phase. (This is the case for points A and B, or points B and C.)

Suppose that the displacement of a particle at the left end of the string ($x = 0$), where the wave originates, is given by

$$y(x = 0, t) = A\cos\omega t = A\cos 2\pi ft \qquad (15.2)$$

That is, the particle oscillates in simple harmonic motion with amplitude A, frequency f, and angular frequency $\omega = 2\pi f$. The notation $y(x = 0, t)$ reminds us that the motion of this particle is a special case of the wave function $y(x, t)$ that describes the entire wave. At $t = 0$ the particle at $x = 0$ is at its maximum positive displacement ($y = A$) and is instantaneously at rest (because the value of y is a maximum).

The wave disturbance travels from $x = 0$ to some point x to the right of the origin in an amount of time given by x/v, where v is the wave speed. So the motion of point x at time t is the same as the motion of point $x = 0$ at the earlier time $t - x/v$. Hence we can find the displacement of point x at time t by simply

15.8 Tracking the oscillations of three points on a string as a sinusoidal wave propagates along it.

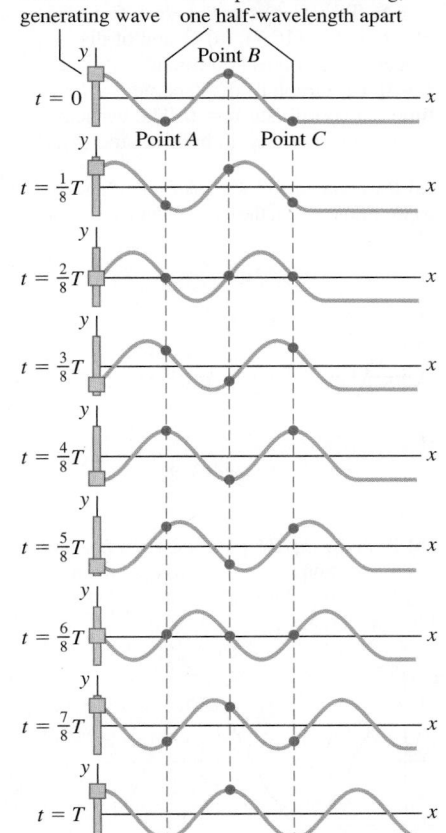

The string is shown at time intervals of $\frac{1}{8}$ period for a total of one period T.

replacing t in Eq. (15.2) by $(t - x/v)$. When we do that, we find the following expression for the wave function:

$$y(x, t) = A \cos\left[\omega\left(t - \frac{x}{v}\right)\right]$$

Because $\cos(-\theta) = \cos\theta$, we can rewrite the wave function as

$$y(x, t) = A \cos\left[\omega\left(\frac{x}{v} - t\right)\right] = A \cos\left[2\pi f\left(\frac{x}{v} - t\right)\right] \quad \begin{array}{l}\text{(sinusoidal wave} \\ \text{moving in} \\ \text{+}x\text{-direction)}\end{array} \quad (15.3)$$

The displacement $y(x, t)$ is a function of both the location x of the point and the time t. We could make Eq. (15.3) more general by allowing for different values of the phase angle, as we did for simple harmonic motion in Section 14.2, but for now we omit this.

We can rewrite the wave function given by Eq. (15.3) in several different but useful forms. We can express it in terms of the period $T = 1/f$ and the wavelength $\lambda = v/f$:

$$y(x, t) = A \cos\left[2\pi\left(\frac{x}{\lambda} - \frac{t}{T}\right)\right] \quad \begin{array}{l}\text{(sinusoidal wave moving} \\ \text{in +}x\text{-direction)}\end{array} \quad (15.4)$$

It's convenient to define a quantity k, called the **wave number:**

$$k = \frac{2\pi}{\lambda} \quad \text{(wave number)} \quad (15.5)$$

Substituting $\lambda = 2\pi/k$ and $f = \omega/2\pi$ into the wavelength–frequency relationship $v = \lambda f$ gives

$$\omega = vk \quad \text{(periodic wave)} \quad (15.6)$$

We can then rewrite Eq. (15.4) as

$$y(x, t) = A \cos(kx - \omega t) \quad \begin{array}{l}\text{(sinusoidal wave moving} \\ \text{in +}x\text{-direction)}\end{array} \quad (15.7)$$

Which of these various forms for the wave function $y(x, t)$ we use in any specific problem is a matter of convenience. Note that ω has units rad/s, so for unit consistency in Eqs. (15.6) and (15.7) the wave number k must have the units rad/m. (Some physicists define the wave number as $1/\lambda$ rather than $2\pi/\lambda$. When reading other texts, be sure to determine how this term is defined.)

Graphing the Wave Function

Figure 15.9a graphs the wave function $y(x, t)$ as a function of x for a specific time t. This graph gives the displacement y of a particle from its equilibrium position as a function of the coordinate x of the particle. If the wave is a transverse wave on a string, the graph in Fig. 15.9a represents the shape of the string at that instant, like a flash photograph of the string. In particular, at time $t = 0$,

$$y(x, t = 0) = A \cos kx = A \cos 2\pi \frac{x}{\lambda}$$

Figure 15.9b is a graph of the wave function versus time t for a specific coordinate x. This graph gives the displacement y of the particle at that coordinate as a function of time; that is, it describes the motion of that particle. In particular, at the position $x = 0$,

$$y(x = 0, t) = A \cos(-\omega t) = A \cos \omega t = A \cos 2\pi \frac{t}{T}$$

This is consistent with our original statement about the motion at $x = 0$, Eq. (15.2).

15.9 Two graphs of the wave function $y(x, t)$ in Eq. (15.7). **(a)** Graph of displacement y versus coordinate x at time $t = 0$. **(b)** Graph of displacement y versus time t at coordinate $x = 0$. The vertical scale is exaggerated in both (a) and (b).

(a) If we use Eq. (15.7) to plot y as a function of x for time $t = 0$, the curve shows the *shape* of the string at $t = 0$.

(b) If we use Eq. (15.7) to plot y as a function of t for position $x = 0$, the curve shows the *displacement* y of the particle at $x = 0$ as a function of time.

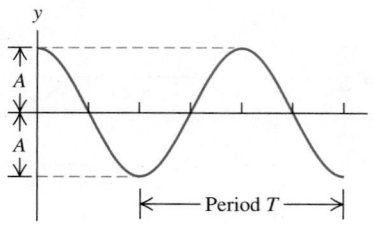

CAUTION **Wave graphs** Although they may look the same at first glance, Figs. 15.9a and 15.9b are *not* identical. Figure 15.9a is a picture of the shape of the string at $t = 0$, while Fig. 15.9b is a graph of the displacement y of a particle at $x = 0$ as a function of time. ▮

More on the Wave Function

We can modify Eqs. (15.3) through (15.7) to represent a wave traveling in the *negative x*-direction. In this case the displacement of point x at time t is the same as the motion of point $x = 0$ at the *later* time $(t + x/v)$, so in Eq. (15.2) we replace t by $(t + x/v)$. For a wave traveling in the negative x-direction,

$$y(x, t) = A \cos\left[2\pi f\left(\frac{x}{v} + t\right)\right] = A \cos\left[2\pi\left(\frac{x}{\lambda} + \frac{t}{T}\right)\right] = A \cos(kx + \omega t) \quad (15.8)$$

(sinusoidal wave moving in $-x$-direction)

In the expression $y(x, t) = A \cos(kx \pm \omega t)$ for a wave traveling in the $-x$- or $+x$-direction, the quantity $(kx \pm \omega t)$ is called the **phase.** It plays the role of an angular quantity (always measured in radians) in Eq. (15.7) or (15.8), and its value for any values of x and t determines what part of the sinusoidal cycle is occurring at a particular point and time. For a crest (where $y = A$ and the cosine function has the value 1), the phase could be 0, 2π, 4π, and so on; for a trough (where $y = -A$ and the cosine has the value -1), it could be π, 3π, 5π, and so on.

The wave speed is the speed with which we have to move along with the wave to keep alongside a point of a given phase, such as a particular crest of a wave on a string. For a wave traveling in the $+x$-direction, that means $kx - \omega t = $ constant. Taking the derivative with respect to t, we find $k \, dx/dt = \omega$, or

$$\frac{dx}{dt} = \frac{\omega}{k}$$

Comparing this with Eq. (15.6), we see that dx/dt is equal to the speed v of the wave. Because of this relationship, v is sometimes called the *phase velocity* of the wave. (*Phase speed* would be a better term.)

Problem-Solving Strategy 15.1   Mechanical Waves

IDENTIFY *the relevant concepts:* As always, identify the target variables; these may include mathematical *expressions* (for example, the wave function for a given situation). Note that wave problems fall into two categories. *Kinematics* problems, concerned with describing wave motion, involve wave speed v, wavelength λ (or wave number k), frequency f (or angular frequency ω), and amplitude A. They may also involve the position, velocity, and acceleration of individual particles in the medium. *Dynamics* problems also use concepts from Newton's laws. Later in this chapter we'll encounter problems that involve the relationship of wave speed to the mechanical properties of the medium.

SET UP *the problem* using the following steps:
1. List the given quantities. Sketch graphs of y versus x (like Fig. 15.9a) and of y versus t (like Fig. 15.9b), and label them with known values.
2. Identify useful equations. These may include Eq. (15.1) ($v = \lambda f$), Eq. (15.6) ($\omega = vk$), and Eqs. (15.3), (15.4), and

(15.7), which express the wave function in various forms. From the wave function, you can find the value of y at any point (value of x) and at any time t.
3. If you need to determine the wave speed v and don't know both λ and f, you may be able to use a relationship between v and the mechanical properties of the system. (In the next section we'll develop this relationship for waves on a string.)

EXECUTE *the solution:* Solve for the unknown quantities using the equations you've identified. To determine the wave function from Eq. (15.3), (15.4), or (15.7), you must know A and any two of v, λ, and f (or v, k, and ω).

EVALUATE *your answer:* Confirm that the values of v, f, and λ (or v, ω, and k) agree with the relationships given in Eq. (15.1) or (15.6). If you've calculated the wave function, check one or more special cases for which you can predict the results.

Example 15.2 **Wave on a clothesline**

Cousin Throckmorton holds one end of the clothesline taut and wiggles it up and down sinusoidally with frequency 2.00 Hz and amplitude 0.075 m. The wave speed on the clothesline is $v = 12.0$ m/s. At $t = 0$ Throcky's end has maximum positive displacement and is instantaneously at rest. Assume that no wave bounces back from the far end. (a) Find the wave amplitude A, angular frequency ω, period T, wavelength λ, and wave number k. (b) Write a wave function describing the wave. (c) Write equations for the displacement, as a function of time, of Throcky's end of the clothesline and of a point 3.00 m from that end.

SOLUTION

IDENTIFY and SET UP: This is a kinematics problem about the clothesline's wave motion. Throcky produces a sinusoidal wave that propagates along the clothesline, so we can use all of the expressions of this section. In part (a) our target variables are A, ω, T, λ, and k. We use the relationships $\omega = 2\pi f$, $f = 1/T$, $v = \lambda f$, and $k = 2\pi/\lambda$. In parts (b) and (c) our target "variables" are expressions for displacement, which we'll obtain from an appropriate equation for the wave function. We take the positive x-direction to be the direction in which the wave propagates, so either Eq. (15.4) or (15.7) will yield the desired expression. A photograph of the clothesline at time $t = 0$ would look like Fig. 15.9a, with the maximum displacement at $x = 0$ (the end that Throcky holds).

EXECUTE: (a) The wave amplitude and frequency are the same as for the oscillations of Throcky's end of the clothesline, $A = 0.075$ m and $f = 2.00$ Hz. Hence

$$\omega = 2\pi f = \left(2\pi \frac{\text{rad}}{\text{cycle}}\right)\left(2.00 \frac{\text{cycles}}{\text{s}}\right)$$

$$= 4.00\pi \text{ rad/s} = 12.6 \text{ rad/s}$$

The period is $T = 1/f = 0.500$ s, and from Eq. (15.1),

$$\lambda = \frac{v}{f} = \frac{12.0 \text{ m/s}}{2.00 \text{ s}^{-1}} = 6.00 \text{ m}$$

We find the wave number from Eq. (15.5) or (15.6):

$$k = \frac{2\pi}{\lambda} = \frac{2\pi \text{ rad}}{6.00 \text{ m}} = 1.05 \text{ rad/m}$$

or

$$k = \frac{\omega}{v} = \frac{4.00\pi \text{ rad/s}}{12.0 \text{ m/s}} = 1.05 \text{ rad/m}$$

(b) We write the wave function using Eq. (15.4) and the values of A, T, and λ from part (a):

$$y(x,t) = A\cos 2\pi\left(\frac{x}{\lambda} - \frac{t}{T}\right)$$

$$= (0.075 \text{ m})\cos 2\pi\left(\frac{x}{6.00 \text{ m}} - \frac{t}{0.500 \text{ s}}\right)$$

$$= (0.075 \text{ m})\cos[(1.05 \text{ rad/m})x - (12.6 \text{ rad/s})t]$$

We can also get this same expression from Eq. (15.7) by using the values of ω and k from part (a).

(c) We can find the displacement as a function of time at $x = 0$ and $x = +3.00$ m by substituting these values into the wave function from part (b):

$$y(x = 0, t) = (0.075 \text{ m})\cos 2\pi\left(\frac{0}{6.00 \text{ m}} - \frac{t}{0.500 \text{ s}}\right)$$

$$= (0.075 \text{ m})\cos(12.6 \text{ rad/s})t$$

$$y(x = +3.00 \text{ m}, t) = (0.075 \text{ m})\cos 2\pi\left(\frac{3.00 \text{ m}}{6.00 \text{ m}} - \frac{t}{0.500 \text{ s}}\right)$$

$$= (0.075 \text{ m})\cos[\pi - (12.6 \text{ rad/s})t]$$

$$= -(0.075 \text{ m})\cos(12.6 \text{ rad/s})t$$

EVALUATE: In part (b), the quantity $(1.05 \text{ rad/m})x - (12.6 \text{ rad/s})t$ is the *phase* of a point x on the string at time t. The two points in part (c) oscillate in SHM with the same frequency and amplitude, but their oscillations differ in phase by $(1.05 \text{ rad/m})(3.00 \text{ m}) = 3.15$ rad $= \pi$ radians—that is, one-half cycle—because the points are separated by one half-wavelength: $\lambda/2 = (6.00 \text{ m})/2 = 3.00$ m. Thus, while a graph of y versus t for the point at $x = 0$ is a cosine curve (like Fig. 15.9b), a graph of y versus t for the point $x = 3.00$ m is a *negative* cosine curve (the same as a cosine curve shifted by one half-cycle).

Using the expression for $y(x = 0, t)$ in part (c), can you show that the end of the string at $x = 0$ is instantaneously at rest at $t = 0$, as stated at the beginning of this example? (*Hint:* Calculate the y-velocity at this point by taking the derivative of y with respect to t.)

Particle Velocity and Acceleration in a Sinusoidal Wave

From the wave function we can get an expression for the transverse velocity of any *particle* in a transverse wave. We call this v_y to distinguish it from the wave propagation speed v. To find the transverse velocity v_y at a particular point x, we take the derivative of the wave function $y(x, t)$ with respect to t, keeping x constant. If the wave function is

$$y(x, t) = A\cos(kx - \omega t)$$

then

$$v_y(x, t) = \frac{\partial y(x, t)}{\partial t} = \omega A \sin(kx - \omega t) \qquad (15.9)$$

The ∂ in this expression is a modified d, used to remind us that $y(x, t)$ is a function of *two* variables and that we are allowing only one (t) to vary. The other (x) is constant because we are looking at a particular point on the string. This derivative is called a *partial derivative*. If you haven't reached this point yet in your study of calculus, don't fret; it's a simple idea.

Equation (15.9) shows that the transverse velocity of a particle varies with time, as we expect for simple harmonic motion. The maximum particle speed is ωA; this can be greater than, less than, or equal to the wave speed v, depending on the amplitude and frequency of the wave.

The *acceleration* of any particle is the *second* partial derivative of $y(x, t)$ with respect to t:

$$a_y(x, t) = \frac{\partial^2 y(x, t)}{\partial t^2} = -\omega^2 A \cos(kx - \omega t) = -\omega^2 y(x, t) \quad (15.10)$$

The acceleration of a particle equals $-\omega^2$ times its displacement, which is the result we obtained in Section 14.2 for simple harmonic motion.

We can also compute partial derivatives of $y(x, t)$ with respect to x, holding t constant. The first derivative $\partial y(x, t)/\partial x$ is the *slope* of the string at point x and at time t. The second partial derivative with respect to x is the *curvature* of the string:

$$\frac{\partial^2 y(x, t)}{\partial x^2} = -k^2 A \cos(kx - \omega t) = -k^2 y(x, t) \quad (15.11)$$

From Eqs. (15.10) and (15.11) and the relationship $\omega = vk$ we see that

$$\frac{\partial^2 y(x, t)/\partial t^2}{\partial^2 y(x, t)/\partial x^2} = \frac{\omega^2}{k^2} = v^2 \quad \text{and}$$

$$\frac{\partial^2 y(x, t)}{\partial x^2} = \frac{1}{v^2} \frac{\partial^2 y(x, t)}{\partial t^2} \quad \text{(wave equation)} \quad (15.12)$$

We've derived Eq. (15.12) for a wave traveling in the positive x-direction. You can use the same steps to show that the wave function for a sinusoidal wave propagating in the *negative* x-direction, $y(x, t) = A \cos(kx + \omega t)$, also satisfies this equation.

Equation (15.12), called the **wave equation,** is one of the most important equations in all of physics. Whenever it occurs, we know that a disturbance can propagate as a wave along the x-axis with wave speed v. The disturbance need not be a sinusoidal wave; we'll see in the next section that *any* wave on a string obeys Eq. (15.12), whether the wave is periodic or not (see also Problem 15.65). In Chapter 32 we will find that electric and magnetic fields satisfy the wave equation; the wave speed will turn out to be the speed of light, which will lead us to the conclusion that light is an electromagnetic wave.

Figure 15.10a shows the transverse velocity v_y and transverse acceleration a_y, given by Eqs. (15.9) and (15.10), for several points on a string as a sinusoidal wave passes along it. Note that at points where the string has an upward curvature $(\partial^2 y/\partial x^2 > 0)$, the acceleration of that point is positive $(a_y = \partial^2 y/\partial t^2 > 0)$; this follows from the wave equation, Eq. (15.12). For the same reason the acceleration is negative $(a_y = \partial^2 y/\partial t^2 < 0)$ at points where the string has a downward curvature $(\partial^2 y/\partial x^2 < 0)$, and the acceleration is zero $(a_y = \partial^2 y/\partial t^2 = 0)$ at points of inflection where the curvature is zero $(\partial^2 y/\partial x^2 = 0)$. We emphasize again that v_y and a_y are the *transverse* velocity and acceleration of points on the string; these points move along the y-direction, not along the propagation direction of the

15.10 (a) Another view of the wave at $t = 0$ in Fig. 15.9a. The vectors show the transverse velocity v_y and transverse acceleration a_y at several points on the string. (b) From $t = 0$ to $t = 0.05T$, a particle at point 1 is displaced to point 1′, a particle at point 2 is displaced to point 2′, and so on.

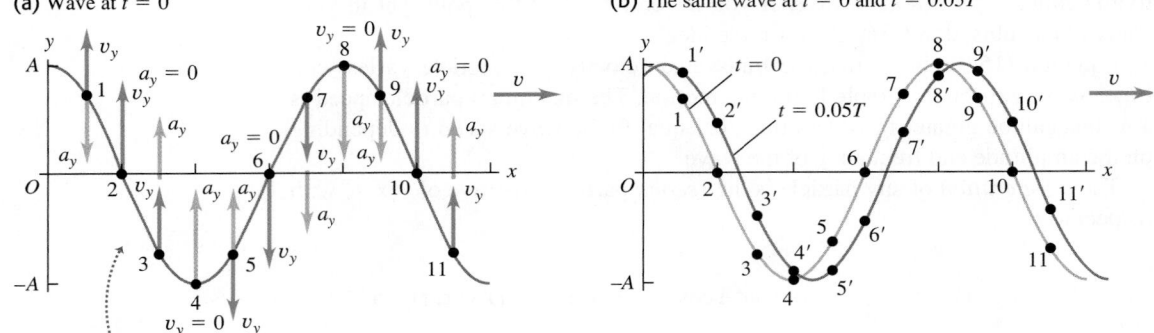

(a) Wave at $t = 0$

(b) The same wave at $t = 0$ and $t = 0.05T$

- Acceleration a_y at each point on the string is proportional to displacement y at that point.
- Acceleration is upward where string curves upward, downward where string curves downward.

wave. Figure 15.10b shows the transverse motions of several points on the string.

The concept of wave function is equally useful with *longitudinal* waves. The quantity y still measures the displacement of a particle of the medium from its equilibrium position; the difference is that for a longitudinal wave, this displacement is *parallel* to the x-axis instead of perpendicular to it. We'll discuss longitudinal waves in detail in Chapter 16.

Test Your Understanding of Section 15.3 Figure 15.8 shows a sinusoidal wave of period T on a string at times $0, \frac{1}{8}T, \frac{2}{8}T, \frac{3}{8}T, \frac{4}{8}T, \frac{5}{8}T, \frac{6}{8}T, \frac{7}{8}T$, and T. (a) At which time is point A on the string moving upward with maximum speed? (b) At which time does point B on the string have the greatest upward acceleration? (c) At which time does point C on the string have a downward acceleration but an upward velocity?

15.4 Speed of a Transverse Wave

One of the key properties of any wave is the wave *speed*. Light waves in air have a much greater speed of propagation than do sound waves in air (3.00×10^8 m/s versus 344 m/s); that's why you see the flash from a bolt of lightning before you hear the clap of thunder. In this section we'll see what determines the speed of propagation of one particular kind of wave: transverse waves on a string. The speed of these waves is important to understand because it is an essential part of analyzing stringed musical instruments, as we'll discuss later in this chapter. Furthermore, the speeds of many kinds of mechanical waves turn out to have the same basic mathematical expression as does the speed of waves on a string.

The physical quantities that determine the speed of transverse waves on a string are the *tension* in the string and its *mass per unit length* (also called *linear mass density*). We might guess that increasing the tension should increase the restoring forces that tend to straighten the string when it is disturbed, thus increasing the wave speed. We might also guess that increasing the mass should make the motion more sluggish and decrease the speed. Both these guesses turn out to be right. We'll develop the exact relationship among wave speed, tension, and mass per unit length by two different methods. The first is simple in concept and considers a specific wave shape; the second is more general but also more formal. Choose whichever you like better.

15.11 Propagation of a transverse wave on a string.

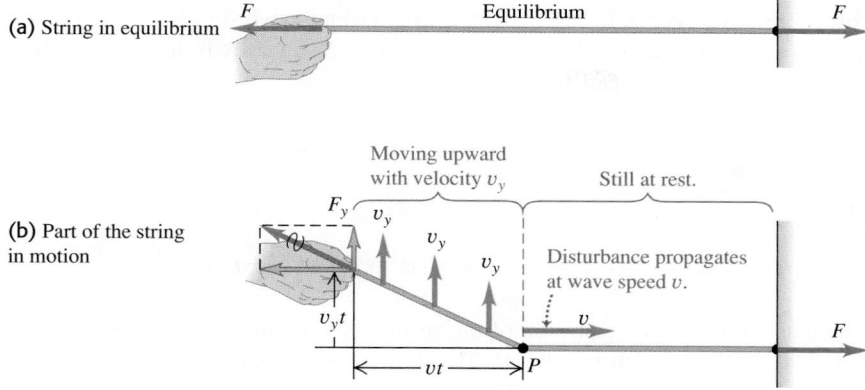

(a) String in equilibrium

F Equilibrium F

(b) Part of the string in motion

Moving upward with velocity v_y Still at rest.

F_y v_y

v_y

v_y

Disturbance propagates at wave speed v.

v_yt

v

F

$\longleftarrow vt \longrightarrow$ P

Wave Speed on a String: First Method

We consider a perfectly flexible string (Fig. 15.11). In the equilibrium position the tension is F and the linear mass density (mass per unit length) is μ. (When portions of the string are displaced from equilibrium, the mass per unit length decreases a little, and the tension increases a little.) We ignore the weight of the string so that when the string is at rest in the equilibrium position, the string forms a perfectly straight line as in Fig. 15.11a.

Starting at time $t = 0$, we apply a constant upward force F_y at the left end of the string. We might expect that the end would move with constant acceleration; that would happen if the force were applied to a *point* mass. But here the effect of the force F_y is to set successively more and more mass in motion. The wave travels with constant speed v, so the division point P between moving and nonmoving portions moves with the same constant speed v (Fig. 15.11b).

Figure 15.11b shows that all particles in the moving portion of the string move upward with constant *velocity* v_y, not constant acceleration. To see why this is so, we note that the *impulse* of the force F_y up to time t is F_yt. According to the impulse–momentum theorem (see Section 8.1), the impulse is equal to the change in the total transverse component of momentum $(mv_y - 0)$ of the moving part of the string. Because the system started with *no* transverse momentum, this is equal to the total momentum at time t:

$$F_yt = mv_y$$

The total momentum thus must increase proportionately with time. But since the division point P moves with constant speed, the length of string that is in motion and hence the total mass m in motion are also proportional to the time t that the force has been acting. So the *change* of momentum must be associated entirely with the increasing amount of mass in motion, not with an increasing velocity of an individual mass element. That is, mv_y changes because m, not v_y, changes.

At time t, the left end of the string has moved up a distance v_yt, and the boundary point P has advanced a distance vt. The total force at the left end of the string has components F and F_y. Why F? There is no motion in the direction along the length of the string, so there is no unbalanced horizontal force. Therefore F, the magnitude of the horizontal component, does not change when the string is displaced. In the displaced position the tension is $(F^2 + F_y^2)^{1/2}$ (greater than F), and the string stretches somewhat.

To derive an expression for the wave speed v, we again apply the impulse–momentum theorem to the portion of the string in motion at time t—that is, the portion to the left of P in Fig. 15.11b. The transverse *impulse* (transverse force times time) is equal to the change of transverse *momentum* of the moving portion (mass times transverse component of velocity). The impulse of the transverse force F_y in time t is $F_y t$. In Fig. 15.11b the right triangle whose vertex is at P, with sides $v_y t$ and vt, is similar to the right triangle whose vertex is at the position of the hand, with sides F_y and F. Hence

$$\frac{F_y}{F} = \frac{v_y t}{vt} \qquad F_y = F\frac{v_y}{v}$$

and

$$\text{Transverse impulse} = F_y t = F\frac{v_y}{v}t$$

The mass of the moving portion of the string is the product of the mass per unit length μ and the length vt, or μvt. The transverse momentum is the product of this mass and the transverse velocity v_y:

$$\text{Transverse momentum} = (\mu vt)v_y$$

We note again that the momentum increases with time *not* because mass is moving faster, as was usually the case in Chapter 8, but because *more mass* is brought into motion. But the impulse of the force F_y is still equal to the total change in momentum of the system. Applying this relationship, we obtain

$$F\frac{v_y}{v}t = \mu vt v_y$$

Solving this for v, we find

$$v = \sqrt{\frac{F}{\mu}} \qquad \text{(speed of a transverse wave on a string)} \qquad (15.13)$$

Equation (15.13) confirms our prediction that the wave speed v should increase when the tension F increases but decrease when the mass per unit length μ increases (Fig. 15.12).

Note that v_y does not appear in Eq. (15.13); thus the wave speed doesn't depend on v_y. Our calculation considered only a very special kind of pulse, but we can consider *any* shape of wave disturbance as a series of pulses with different values of v_y. So even though we derived Eq. (15.13) for a special case, it is valid for *any* transverse wave motion on a string, including the sinusoidal and other periodic waves we discussed in Section 15.3. Note also that the wave speed doesn't depend on the amplitude or frequency of the wave, in accordance with our assumptions in Section 15.3.

Wave Speed on a String: Second Method

Here is an alternative derivation of Eq. (15.13). If you aren't comfortable with partial derivatives, it can be omitted. We apply Newton's second law, $\sum \vec{F} = m\vec{a}$, to a small segment of string whose length in the equilibrium position is Δx (Fig. 15.13). The mass of the segment is $m = \mu \, \Delta x$; the forces at the ends are represented in terms of their x- and y-components. The x-components have equal magnitude F and add to zero because the motion is transverse and there is no component of acceleration in the x-direction. To obtain F_{1y} and F_{2y}, we note that the ratio F_{1y}/F is equal in magnitude to the *slope* of the string at point x and that F_{2y}/F is equal to the slope at point $x + \Delta x$. Taking proper account of signs, we find

$$\frac{F_{1y}}{F} = -\left(\frac{\partial y}{\partial x}\right)_x \qquad \frac{F_{2y}}{F} = \left(\frac{\partial y}{\partial x}\right)_{x+\Delta x} \qquad (15.14)$$

15.12 These cables have a relatively large amount of mass per unit length (μ) and a low tension (F). If the cables are disturbed—say, by a bird landing on them—transverse waves will travel along them at a slow speed $v = \sqrt{F/\mu}$.

15.13 Free-body diagram for a segment of string. The force at each end of the string is tangent to the string at the point of application.

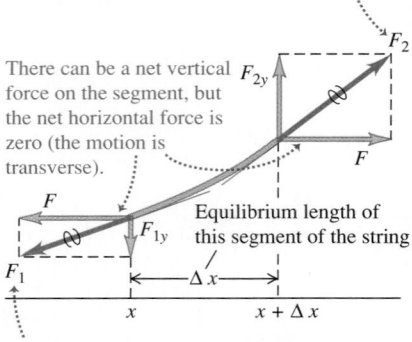

The string to the right of the segment (not shown) exerts a force $\vec{F}_2$ on the segment.

There can be a net vertical force on the segment, but the net horizontal force is zero (the motion is transverse).

Equilibrium length of this segment of the string

The string to the left of the segment (not shown) exerts a force $\vec{F}_1$ on the segment.

The notation reminds us that the derivatives are evaluated at points x and $x + \Delta x$, respectively. From Eq. (15.14) we find that the net y-component of force is

$$F_y = F_{1y} + F_{2y} = F\left[\left(\frac{\partial y}{\partial x}\right)_{x+\Delta x} - \left(\frac{\partial y}{\partial x}\right)_x\right] \qquad (15.15)$$

We now equate F_y from Eq. (15.15) to the mass $\mu\,\Delta x$ times the y-component of acceleration $\partial^2 y/\partial t^2$. We obtain

$$F\left[\left(\frac{\partial y}{\partial x}\right)_{x+\Delta x} - \left(\frac{\partial y}{\partial x}\right)_x\right] = \mu\,\Delta x\frac{\partial^2 y}{\partial t^2} \qquad (15.16)$$

or, dividing by $F\Delta x$,

$$\frac{\left(\dfrac{\partial y}{\partial x}\right)_{x+\Delta x} - \left(\dfrac{\partial y}{\partial x}\right)_x}{\Delta x} = \frac{\mu}{F}\frac{\partial^2 y}{\partial t^2} \qquad (15.17)$$

We now take the limit as $\Delta x \to 0$. In this limit, the left side of Eq. (15.17) becomes the derivative of $\partial y/\partial x$ with respect to x (at constant t)—that is, the *second* (partial) derivative of y with respect to x:

$$\frac{\partial^2 y}{\partial x^2} = \frac{\mu}{F}\frac{\partial^2 y}{\partial t^2} \qquad (15.18)$$

Now, finally, comes the punch line of our story. Equation (15.18) has exactly the same form as the *wave equation,* Eq. (15.12), that we derived at the end of Section 15.3. That equation and Eq. (15.18) describe the very same wave motion, so they must be identical. Comparing the two equations, we see that for this to be so, we must have

$$v = \sqrt{\frac{F}{\mu}} \qquad (15.19)$$

which is the same expression as Eq. (15.13).

In going through this derivation, we didn't make any special assumptions about the shape of the wave. Since our derivation led us to rediscover Eq. (15.12), the wave equation, we conclude that the wave equation is valid for waves on a string that have *any* shape.

The Speed of Mechanical Waves

Equation (15.13) or (15.19) gives the wave speed for only the special case of mechanical waves on a stretched string or rope. Remarkably, it turns out that for many types of mechanical waves, including waves on a string, the expression for wave speed has the same general form:

$$v = \sqrt{\frac{\text{Restoring force returning the system to equilibrium}}{\text{Inertia resisting the return to equilibrium}}}$$

To interpret this expression, let's look at the now-familiar case of waves on a string. The tension F in the string plays the role of the restoring force; it tends to bring the string back to its undisturbed, equilibrium configuration. The mass of the string—or, more properly, the linear mass density μ—provides the inertia that prevents the string from returning instantaneously to equilibrium. Hence we have $v = \sqrt{F/\mu}$ for the speed of waves on a string.

In Chapter 16 we'll see a similar expression for the speed of sound waves in a gas. Roughly speaking, the gas pressure provides the force that tends to return the gas to its undisturbed state when a sound wave passes through. The inertia is provided by the density, or mass per unit volume, of the gas.

| Example 15.3 | Calculating wave speed |

One end of a 2.00-kg rope is tied to a support at the top of a mine shaft 80.0 m deep (Fig. 15.14). The rope is stretched taut by a 20.0-kg box of rocks attached at the bottom. (a) The geologist at the bottom of the shaft signals to a colleague at the top by jerking the rope sideways. What is the speed of a transverse wave on the rope? (b) If a point on the rope is in transverse SHM with $f = 2.00$ Hz, how many cycles of the wave are there in the rope's length?

SOLUTION

IDENTIFY and SET UP: In part (a) we can find the wave speed (our target variable) using the *dynamic* relationship $v = \sqrt{F/\mu}$

15.14 Sending signals along a vertical rope using transverse waves.

$m_{\text{rope}} = 2.00$ kg

80.0 m

$m_{\text{samples}} = 20.0$ kg

[Eq. (15.13)]. In part (b) we find the wavelength from the *kinematic* relationship $v = f\lambda$; from that we can find the target variable, the number of wavelengths that fit into the rope's 80.0-m length. We'll assume that the rope is massless (even though its weight is 10% that of the box), so that the box alone provides the tension in the rope.

EXECUTE: (a) The tension in the rope due to the box is

$$F = m_{\text{box}}g = (20.0 \text{ kg})(9.80 \text{ m/s}^2) = 196 \text{ N}$$

and the rope's linear mass density is

$$\mu = \frac{m_{\text{rope}}}{L} = \frac{2.00 \text{ kg}}{80.0 \text{ m}} = 0.0250 \text{ kg/m}$$

Hence, from Eq. (15.13), the wave speed is

$$v = \sqrt{\frac{F}{\mu}} = \sqrt{\frac{196 \text{ N}}{0.0250 \text{ kg/m}}} = 88.5 \text{ m/s}$$

(b) From Eq. (15.1), the wavelength is

$$\lambda = \frac{v}{f} = \frac{88.5 \text{ m/s}}{2.00 \text{ s}^{-1}} = 44.3 \text{ m}$$

There are $(80.0 \text{ m})/(44.3 \text{ m}) = 1.81$ wavelengths (that is, cycles of the wave) in the rope.

EVALUATE: Because of the rope's weight, its tension is greater at the top than at the bottom. Hence both the wave speed and the wavelength increase as a wave travels up the rope. If you take account of this, can you verify that the wave speed at the top of the rope is 92.9 m/s?

Test Your Understanding of Section 15.4 The six strings of a guitar are the same length and under nearly the same tension, but they have different thicknesses. On which string do waves travel the fastest? (i) the thickest string; (ii) the thinnest string; (iii) the wave speed is the same on all strings.

Application Surface Waves and the Swimming Speed of Ducks
When a duck swims, it necessarily produces waves on the surface of the water. The faster the duck swims, the larger the wave amplitude and the more power the duck must supply to produce these waves. The maximum power available from their leg muscles limits the maximum swimming speed of ducks to only about 0.7 m/s (2.5 km/h = 1.6 mi/h).

15.5 Energy in Wave Motion

Every wave motion has *energy* associated with it. The energy we receive from sunlight and the destructive effects of ocean surf and earthquakes bear this out. To produce any of the wave motions we have discussed in this chapter, we have to apply a force to a portion of the wave medium; the point where the force is applied moves, so we do *work* on the system. As the wave propagates, each portion of the medium exerts a force and does work on the adjoining portion. In this way a wave can transport energy from one region of space to another.

As an example of energy considerations in wave motion, let's look again at transverse waves on a string. How is energy transferred from one portion of string to another? Picture a wave traveling from left to right (the positive x-direction) on the string, and consider a particular point a on the string (Fig. 15.15a). The string to the left of point a exerts a force on the string to the right of it, and vice versa. In Fig. 15.15b the string to the left of a has been removed, and the force it exerts at a is represented by the components F and F_y, as we did in Figs. 15.11 and

15.13. We note again that F_y/F is equal to the negative of the *slope* of the string at a, which is also given by $\partial y/\partial x$. Putting these together, we have

$$F_y(x, t) = -F\frac{\partial y(x, t)}{\partial x} \qquad (15.20)$$

We need the negative sign because F_y is negative when the slope is positive. We write the vertical force as $F_y(x, t)$ as a reminder that its value may be different at different points along the string and at different times.

When point a moves in the y-direction, the force F_y does *work* on this point and therefore transfers energy into the part of the string to the right of a. The corresponding power P (rate of doing work) at the point a is the transverse force $F_y(x, t)$ at a times the transverse velocity $v_y(x, t) = \partial y(x, t)/\partial t$ of that point:

$$P(x, t) = F_y(x, t)v_y(x, t) = -F\frac{\partial y(x, t)}{\partial x}\frac{\partial y(x, t)}{\partial t} \qquad (15.21)$$

This power is the *instantaneous* rate at which energy is transferred along the string. Its value depends on the position x on the string and on the time t. Note that energy is being transferred only at points where the string has a nonzero slope ($\partial y/\partial x$ is nonzero), so that there is a transverse component of the tension force, and where the string has a nonzero transverse velocity ($\partial y/\partial t$ is nonzero) so that the transverse force can do work.

Equation (15.21) is valid for *any* wave on a string, sinusoidal or not. For a sinusoidal wave with wave function given by Eq. (15.7), we have

$$y(x, t) = A\cos(kx - \omega t)$$
$$\frac{\partial y(x, t)}{\partial x} = -kA\sin(kx - \omega t)$$
$$\frac{\partial y(x, t)}{\partial t} = \omega A\sin(kx - \omega t)$$
$$P(x, t) = Fk\omega A^2\sin^2(kx - \omega t) \qquad (15.22)$$

By using the relationships $\omega = vk$ and $v^2 = F/\mu$, we can also express Eq. (15.22) in the alternative form

$$P(x, t) = \sqrt{\mu F}\,\omega^2 A^2\sin^2(kx - \omega t) \qquad (15.23)$$

The $\sin^2$ function is never negative, so the instantaneous power in a sinusoidal wave is either positive (so that energy flows in the positive x-direction) or zero (at points where there is no energy transfer). Energy is never transferred in the direction opposite to the direction of wave propagation (Fig. 15.16).

The maximum value of the instantaneous power $P(x, t)$ occurs when the $\sin^2$ function has the value unity:

$$P_{\text{max}} = \sqrt{\mu F}\,\omega^2 A^2 \qquad (15.24)$$

To obtain the *average* power from Eq. (15.23), we note that the *average* value of the $\sin^2$ function, averaged over any whole number of cycles, is $\frac{1}{2}$. Hence the average power is

$$P_{\text{av}} = \tfrac{1}{2}\sqrt{\mu F}\,\omega^2 A^2 \quad \text{(average power, sinusoidal wave on a string)} \quad (15.25)$$

The average power is just one-half of the maximum instantaneous power (see Fig. 15.16).

The average rate of energy transfer is proportional to the square of the amplitude and to the square of the frequency. This proportionality is a general result for mechanical waves of all types, including seismic waves (see the photo that opens this chapter). For a mechanical wave, the rate of energy transfer

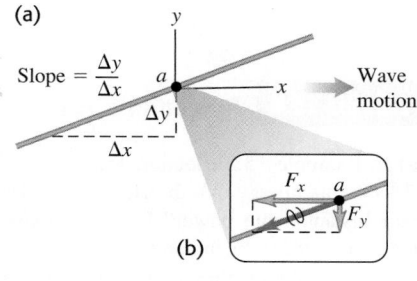

15.15 (a) Point a on a string carrying a wave from left to right. (b) The components of the force exerted on the part of the string to the right of point a by the part of the string to the left of point a.

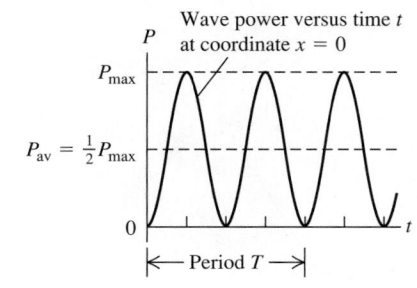

15.16 The instantaneous power $P(x, t)$ in a sinusoidal wave as given by Eq. (15.23), shown as a function of time at coordinate $x = 0$. The power is never negative, which means that energy never flows opposite to the direction of wave propagation.

quadruples if the frequency is doubled (for the same amplitude) or if the amplitude is doubled (for the same frequency).

Electromagnetic waves turn out to be a bit different. While the average rate of energy transfer in an electromagnetic wave is proportional to the square of the amplitude, just as for mechanical waves, it is independent of the value of ω.

Example 15.4   Power in a wave

(a) In Example 15.2 (Section 15.3), at what maximum rate does Throcky put energy into the clothesline? That is, what is his maximum instantaneous power? The linear mass density of the clothesline is $\mu = 0.250$ kg/m, and Throcky applies tension $F = 36.0$ N. (b) What is his average power? (c) As Throcky tires, the amplitude decreases. What is the average power when the amplitude is 7.50 mm?

SOLUTION

IDENTIFY and SET UP: In part (a) our target variable is the *maximum instantaneous* power P_{max}, while in parts (b) and (c) it is the *average* power. For part (a) we'll use Eq. (15.24), and for parts (b) and (c) we'll use Eq. (15.25); Example 15.2 gives us all the needed quantities.

EXECUTE: (a) From Eq. (15.24),

$$P_{max} = \sqrt{\mu F \omega^2 A^2}$$
$$= \sqrt{(0.250 \text{ kg/m})(36.0 \text{ N})(4.00\pi \text{ rad/s})^2(0.075 \text{ m})^2}$$
$$= 2.66 \text{ W}$$

(b) From Eqs. (15.24) and (15.25), the average power is one-half of the maximum instantaneous power, so

$$P_{av} = \tfrac{1}{2}P_{max} = \tfrac{1}{2}(2.66 \text{ W}) = 1.33 \text{ W}$$

(c) The new amplitude is $\frac{1}{10}$ of the value we used in parts (a) and (b). From Eq. (15.25), the average power is proportional to A^2, so the new average power is

$$P_{av} = \left(\tfrac{1}{10}\right)^2(1.33 \text{ W}) = 0.0133 \text{ W} = 13.3 \text{ mW}$$

EVALUATE: Equation (15.23) shows that P_{max} occurs when $\sin^2(kx - \omega t) = 1$. At any given position x, this happens twice per period of the wave—once when the sine function is equal to $+1$, and once when it's equal to -1. The *minimum* instantaneous power is zero; this occurs when $\sin^2(kx - \omega t) = 0$, which also happens twice per period.

Can you confirm that the given values of μ and F give the wave speed mentioned in Example 15.2?

Wave Intensity

Waves on a string carry energy in just one dimension of space (along the direction of the string). But other types of waves, including sound waves in air and seismic waves in the body of the earth, carry energy across all three dimensions of space. For waves that travel in three dimensions, we define the **intensity** (denoted by I) to be *the time average rate at which energy is transported by the wave, per unit area,* across a surface perpendicular to the direction of propagation. That is, intensity I is average power per unit area. It is usually measured in watts per square meter (W/m^2).

If waves spread out equally in all directions from a source, the intensity at a distance r from the source is inversely proportional to r^2 (Fig. 15.17). This follows directly from energy conservation. If the power output of the source is P, then the average intensity I_1 through a sphere with radius r_1 and surface area $4\pi r_1^2$ is

$$I_1 = \frac{P}{4\pi r_1^2}$$

The average intensity I_2 through a sphere with a different radius r_2 is given by a similar expression. If no energy is absorbed between the two spheres, the power P must be the same for both, and

$$4\pi r_1^2 I_1 = 4\pi r_2^2 I_2$$

$$\frac{I_1}{I_2} = \frac{r_2^2}{r_1^2} \qquad \text{(inverse-square law for intensity)} \qquad (15.26)$$

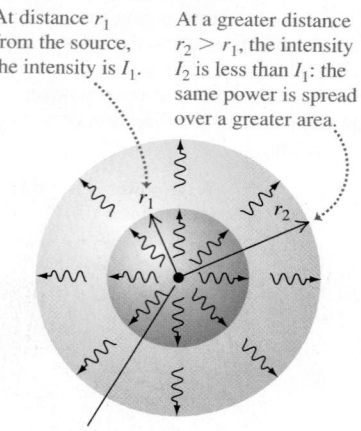

15.17 The greater the distance from a wave source, the greater the area over which the wave power is distributed and the smaller the wave intensity.

At distance r_1 from the source, the intensity is I_1.

At a greater distance $r_2 > r_1$, the intensity I_2 is less than I_1: the same power is spread over a greater area.

Source of waves

The intensity I at any distance r is therefore inversely proportional to r^2. This relationship is called the *inverse-square law* for intensity.

Example 15.5   The inverse-square law

A siren on a tall pole radiates sound waves uniformly in all directions. At a distance of 15.0 m from the siren, the sound intensity is 0.250 W/m^2. At what distance is the intensity 0.010 W/m^2?

SOLUTION

IDENTIFY and SET UP: Because sound is radiated uniformly in all directions, we can use the inverse-square law, Eq. (15.26). At $r_1 = 15.0$ m the intensity is $I_1 = 0.250$ W/m^2, and the target variable is the distance r_2 at which the intensity is $I_2 = 0.010$ W/m^2.

EXECUTE: We solve Eq. (15.26) for r_2:

$$r_2 = r_1 \sqrt{\frac{I_1}{I_2}} = (15.0 \text{ m}) \sqrt{\frac{0.250 \text{ W/m}^2}{0.010 \text{ W/m}^2}} = 75.0 \text{ m}$$

EVALUATE: As a check on our answer, note that r_2 is five times greater than r_1. By the inverse-square law, the intensity I_2 should be $1/5^2 = 1/25$ as great as I_1, and indeed it is.

By using the inverse-square law, we've assumed that the sound waves travel in straight lines away from the siren. A more realistic solution, which is beyond our scope, would account for the reflection of sound waves from the ground.

Test Your Understanding of Section 15.5 Four identical strings each carry a sinusoidal wave of frequency 10 Hz. The string tension and wave amplitude are different for different strings. Rank the following strings in order from highest to lowest value of the average wave power: (i) tension 10 N, amplitude 1.0 mm; (ii) tension 40 N, amplitude 1.0 mm; (iii) tension 10 N, amplitude 4.0 mm; (iv) tension 20 N, amplitude 2.0 mm.

15.6 Wave Interference, Boundary Conditions, and Superposition

Up to this point we've been discussing waves that propagate continuously in the same direction. But when a wave strikes the boundaries of its medium, all or part of the wave is *reflected*. When you yell at a building wall or a cliff face some distance away, the sound wave is reflected from the rigid surface and you hear an echo. When you flip the end of a rope whose far end is tied to a rigid support, a pulse travels the length of the rope and is reflected back to you. In both cases, the initial and reflected waves overlap in the same region of the medium. This overlapping of waves is called **interference.** (In general, the term "interference" refers to what happens when two or more waves pass through the same region at the same time.)

As a simple example of wave reflections and the role of the boundary of a wave medium, let's look again at transverse waves on a stretched string. What happens when a wave pulse or a sinusoidal wave arrives at the *end* of the string?

If the end is fastened to a rigid support, it is a *fixed* end that cannot move. The arriving wave exerts a force on the support; the reaction to this force, exerted *by* the support *on* the string, "kicks back" on the string and sets up a *reflected* pulse or wave traveling in the reverse direction. Figure 15.18 is a series of photographs showing the reflection of a pulse at the fixed end of a long coiled spring. The reflected pulse moves in the opposite direction from the initial, or *incident,* pulse, and its displacement is also opposite. Figure 15.19a illustrates this situation for a wave pulse on a string.

The opposite situation from an end that is held stationary is a *free* end, one that is perfectly free to move in the direction perpendicular to the length of the string. For example, the string might be tied to a light ring that slides on a frictionless rod perpendicular to the string, as in Fig. 15.19b. The ring and rod maintain the tension but exert no transverse force. When a wave arrives at this free end, the ring slides along the rod. The ring reaches a maximum displacement, and both it and the string come momentarily to rest, as in drawing 4 in Fig. 15.19b. But the string is now stretched, giving increased tension, so the free end of the string is

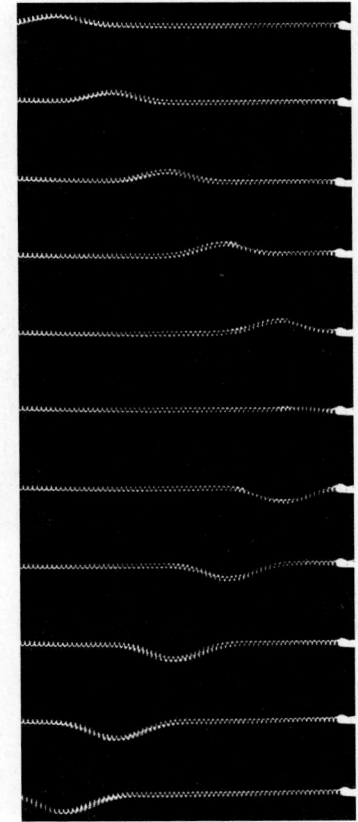

15.18 A series of images of a wave pulse, equally spaced in time from top to bottom. The pulse starts at the left in the top image, travels to the right, and is reflected from the fixed end at the right.

15.19 Reflection of a wave pulse **(a)** at a fixed end of a string and **(b)** at a free end. Time increases from top to bottom in each figure.

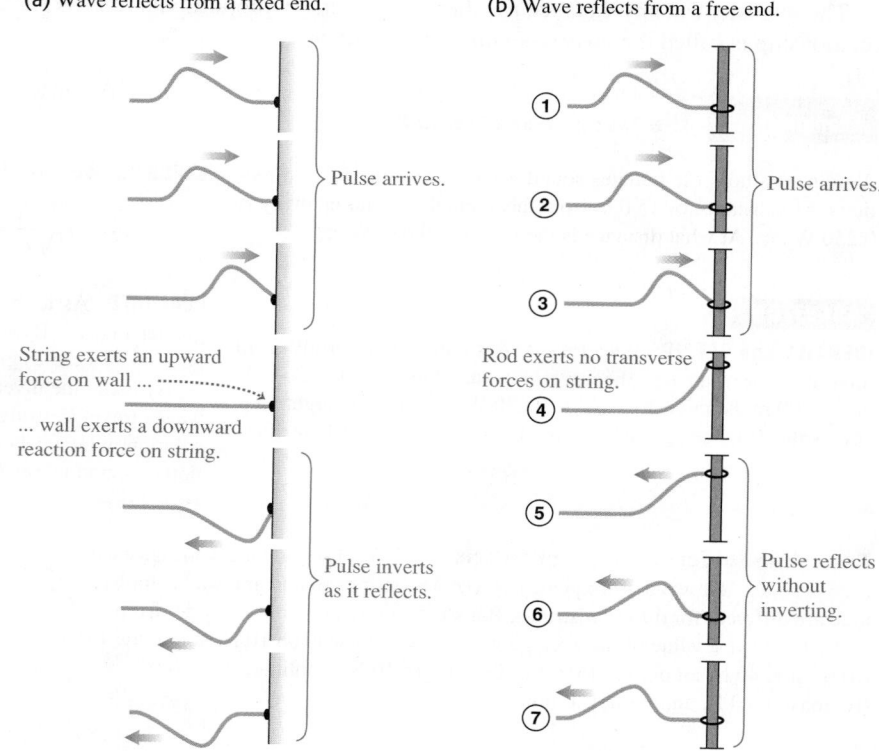

(a) Wave reflects from a fixed end.

Pulse arrives.

String exerts an upward force on wall ...

... wall exerts a downward reaction force on string.

Pulse inverts as it reflects.

(b) Wave reflects from a free end.

Pulse arrives.

Rod exerts no transverse forces on string.

Pulse reflects without inverting.

15.20 Overlap of two wave pulses—one right side up, one inverted—traveling in opposite directions. Time increases from top to bottom.

As the pulses overlap, the displacement of the string at any point is the algebraic sum of the displacements due to the individual pulses.

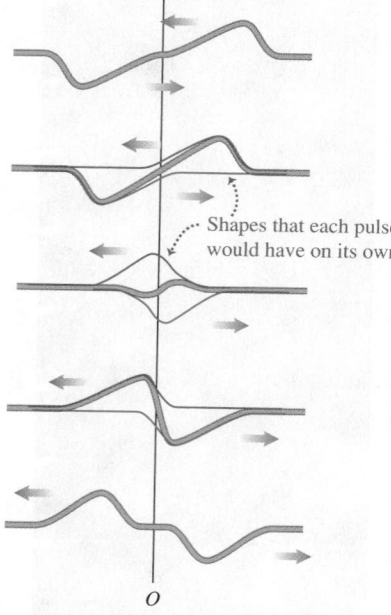

Shapes that each pulse would have on its own

O

pulled back down, and again a reflected pulse is produced (drawing 7). As for a fixed end, the reflected pulse moves in the opposite direction from the initial pulse, but now the direction of the displacement is the same as for the initial pulse. The conditions at the end of the string, such as a rigid support or the complete absence of transverse force, are called **boundary conditions.**

The formation of the reflected pulse is similar to the overlap of two pulses traveling in opposite directions. Figure 15.20 shows two pulses with the same shape, one inverted with respect to the other, traveling in opposite directions. As the pulses overlap and pass each other, the total displacement of the string is the *algebraic sum* of the displacements at that point in the individual pulses. Because these two pulses have the same shape, the total displacement at point O in the middle of the figure is zero at all times. Thus the motion of the left half of the string would be the same if we cut the string at point O, threw away the right side, and held the end at O fixed. The two pulses on the left side then correspond to the incident and reflected pulses, combining so that the total displacement at O is *always* zero. For this to occur, the reflected pulse must be inverted relative to the incident pulse.

Figure 15.21 shows two pulses with the same shape, traveling in opposite directions but *not* inverted relative to each other. The displacement at point O in the middle of the figure is not zero, but the slope of the string at this point is always zero. According to Eq. (15.20), this corresponds to the absence of any transverse force at this point. In this case the motion of the left half of the string would be the same as if we cut the string at point O and attached the end to a frictionless sliding ring (Fig. 15.19b) that maintains tension without exerting any transverse force. In other words, this situation corresponds to reflection of a pulse at a free end of a string at point O. In this case the reflected pulse is *not* inverted.

The Principle of Superposition

Combining the displacements of the separate pulses at each point to obtain the actual displacement is an example of the **principle of superposition:** When two waves overlap, the actual displacement of any point on the string at any time is obtained by adding the displacement the point would have if only the first wave were present and the displacement it would have if only the second wave were present. In other words, the wave function $y(x, t)$ that describes the resulting motion in this situation is obtained by *adding* the two wave functions for the two separate waves:

$$y(x, t) = y_1(x, t) + y_2(x, t) \quad \text{(principle of superposition)} \quad (15.27)$$

Mathematically, this additive property of wave functions follows from the form of the wave equation, Eq. (15.12) or (15.18), which every physically possible wave function must satisfy. Specifically, the wave equation is *linear;* that is, it contains the function $y(x, t)$ only to the first power (there are no terms involving $y(x, t)^2$, $y(x, t)^{1/2}$, etc.). As a result, if any two functions $y_1(x, t)$ and $y_2(x, t)$ satisfy the wave equation separately, their sum $y_1(x, t) + y_2(x, t)$ also satisfies it and is therefore a physically possible motion. Because this principle depends on the linearity of the wave equation and the corresponding linear-combination property of its solutions, it is also called the *principle of linear superposition.* For some physical systems, such as a medium that does not obey Hooke's law, the wave equation is *not* linear; this principle does not hold for such systems.

The principle of superposition is of central importance in all types of waves. When a friend talks to you while you are listening to music, you can distinguish the sound of speech and the sound of music from each other. This is precisely because the total sound wave reaching your ears is the algebraic sum of the wave produced by your friend's voice and the wave produced by the speakers of your stereo. If two sound waves did *not* combine in this simple linear way, the sound you would hear in this situation would be a hopeless jumble. Superposition also applies to electromagnetic waves (such as light) and many other types of waves.

Test Your Understanding of Section 15.6 Figure 15.22 shows two wave pulses with different shapes traveling in different directions along a string. Make a series of sketches like Fig. 15.21 showing the shape of the string as the two pulses approach, overlap, and then pass each other.

15.21 Overlap of two wave pulses—both right side up—traveling in opposite directions. Time increases from top to bottom. Compare to Fig. 15.20.

O

15.22 Two wave pulses with different shapes.

15.7 Standing Waves on a String

We have talked about the reflection of a wave *pulse* on a string when it arrives at a boundary point (either a fixed end or a free end). Now let's look at what happens when a *sinusoidal* wave is reflected by a fixed end of a string. We'll again approach the problem by considering the superposition of two waves propagating through the string, one representing the original or incident wave and the other representing the wave reflected at the fixed end.

Figure 15.23 shows a string that is fixed at its left end. Its right end is moved up and down in simple harmonic motion to produce a wave that travels to the left; the wave reflected from the fixed end travels to the right. The resulting motion when the two waves combine no longer looks like two waves traveling in opposite directions. The string appears to be subdivided into a number of segments, as in the time-exposure photographs of Figs. 15.23a, 15.23b, 15.23c, and 15.23d. Figure 15.23e shows two instantaneous shapes of the string in Fig. 15.23b. Let's compare this behavior with the waves we studied in Sections 15.1 through 15.5. In a wave that travels along the string, the amplitude is constant and the wave pattern moves with a speed equal to the wave speed. Here, instead, the wave pattern remains in the same position along the string and its amplitude

15.23 (a)–(d) Time exposures of standing waves in a stretched string. From (a) to (d), the frequency of oscillation of the right-hand end increases and the wavelength of the standing wave decreases. (e) The extremes of the motion of the standing wave in part (b), with nodes at the center and at the ends. The right-hand end of the string moves very little compared to the antinodes and so is essentially a node.

(a) String is one-half wavelength long.

(b) String is one wavelength long.

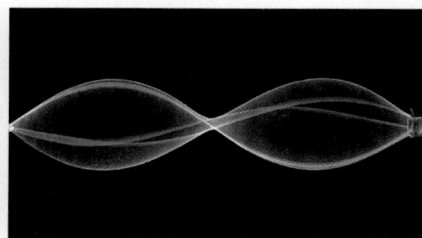

(c) String is one and a half wavelengths long.

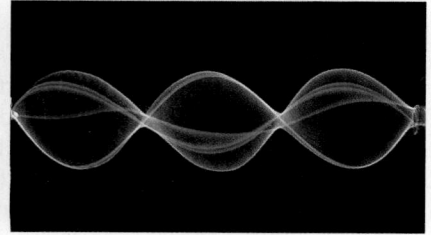

(d) String is two wavelengths long.

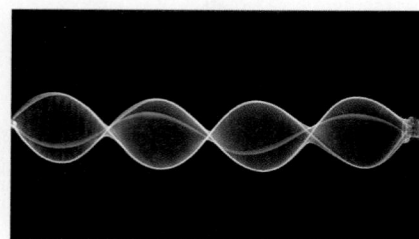

(e) The shape of the string in (b) at two different instants

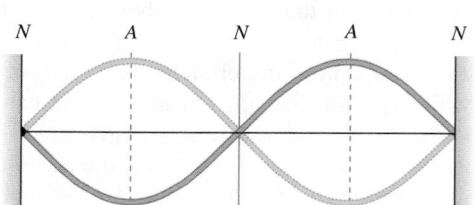

N = **nodes:** points at which the string never moves

A = **antinodes:** points at which the amplitude of string motion is greatest

fluctuates. There are particular points called **nodes** (labeled N in Fig. 15.23e) that never move at all. Midway between the nodes are points called **antinodes** (labeled A in Fig. 15.23e) where the amplitude of motion is greatest. Because the wave pattern doesn't appear to be moving in either direction along the string, it is called a **standing wave.** (To emphasize the difference, a wave that *does* move along the string is called a **traveling wave.**)

The principle of superposition explains how the incident and reflected waves combine to form a standing wave. In Fig. 15.24 the red curves show a wave traveling to the left. The blue curves show a wave traveling to the right with the same propagation speed, wavelength, and amplitude. The waves are shown at nine instants, $\frac{1}{16}$ of a period apart. At each point along the string, we add the displacements (the values of y) for the two separate waves; the result is the total wave on the string, shown in brown.

At certain instants, such as $t = \frac{1}{4}T$, the two wave patterns are exactly in phase with each other, and the shape of the string is a sine curve with twice the amplitude of either individual wave. At other instants, such as $t = \frac{1}{2}T$, the two waves are exactly out of phase with each other, and the total wave at that instant is zero. The resultant displacement is *always* zero at those places marked N at the bottom of Fig. 15.24. These are the *nodes.* At a node the displacements of the two waves in red and blue are always equal and opposite and cancel each other out. This cancellation is called **destructive interference.** Midway between the nodes are the points of *greatest* amplitude, or the *antinodes,* marked A. At the antinodes the displacements of the two waves in red and blue are always identical, giving a large resultant displacement; this phenomenon is called **constructive interference.** We can see from the figure that the distance between successive nodes or between successive antinodes is one half-wavelength, or $\lambda/2$.

We can derive a wave function for the standing wave of Fig. 15.24 by adding the wave functions $y_1(x, t)$ and $y_2(x, t)$ for two waves with equal amplitude, period, and wavelength traveling in opposite directions. Here $y_1(x, t)$ (the red curves in Fig. 15.24) represents an incoming, or *incident,* wave traveling to the

15.24 Formation of a standing wave. A wave traveling to the left (red curves) combines with a wave traveling to the right (blue curves) to form a standing wave (brown curves).

Equilibrium position of string is along the horizontal x-axis.

$t = \frac{1}{16}T$

$t = \frac{1}{8}T$

$t = \frac{3}{16}T$

$t = \frac{1}{4}T$

At this instant the waves coincide, so they add to give maximum string displacement.

$t = \frac{5}{16}T$

$t = \frac{3}{8}T$

$t = \frac{7}{16}T$

$t = \frac{1}{2}T$

At this instant the waves exactly cancel, so the string displacement is zero.

$t = \frac{9}{16}T$

N A N A N A N A N A N A N

left along the $+x$-axis, arriving at the point $x = 0$ and being reflected; $y_2(x, t)$ (the blue curves in Fig. 15.24) represents the *reflected* wave traveling to the right from $x = 0$. We noted in Section 15.6 that the wave reflected from a fixed end of a string is inverted, so we give a negative sign to one of the waves:

$$y_1(x, t) = -A\cos(kx + \omega t) \quad \text{(incident wave traveling to the left)}$$
$$y_2(x, t) = A\cos(kx - \omega t) \quad \text{(reflected wave traveling to the right)}$$

Note also that the change in sign corresponds to a shift in *phase* of 180° or π radians. At $x = 0$ the motion from the reflected wave is $A\cos\omega t$ and the motion from the incident wave is $-A\cos\omega t$, which we can also write as $A\cos(\omega t + \pi)$. From Eq. (15.27), the wave function for the standing wave is the sum of the individual wave functions:

$$y(x, t) = y_1(x, t) + y_2(x, t) = A[-\cos(kx + \omega t) + \cos(kx - \omega t)]$$

We can rewrite each of the cosine terms by using the identities for the cosine of the sum and difference of two angles: $\cos(a \mp b) = \cos a \cos b \mp \sin a \sin b$.

Applying these and combining terms, we obtain the wave function for the standing wave:

$$y(x, t) = y_1(x, t) + y_2(x, t) = (2A \sin kx) \sin \omega t \quad \text{or}$$

$$y(x, t) = (A_{SW} \sin kx) \sin \omega t \quad \begin{pmatrix} \text{standing wave on a} \\ \text{string, fixed end at } x = 0 \end{pmatrix} \quad (15.28)$$

The standing-wave amplitude A_{SW} is twice the amplitude A of either of the original traveling waves:

$$A_{SW} = 2A$$

Equation (15.28) has two factors: a function of x and a function of t. The factor $A_{SW} \sin kx$ shows that at each instant the shape of the string is a sine curve. But unlike a wave traveling along a string, the wave shape stays in the same position, oscillating up and down as described by the $\sin \omega t$ factor. This behavior is shown graphically by the brown curves in Fig. 15.24. Each point in the string still undergoes simple harmonic motion, but all the points between any successive pair of nodes oscillate *in phase*. This is in contrast to the phase differences between oscillations of adjacent points that we see with a wave traveling in one direction.

We can use Eq. (15.28) to find the positions of the nodes; these are the points for which $\sin kx = 0$, so the displacement is *always* zero. This occurs when $kx = 0, \pi, 2\pi, 3\pi, \ldots$, or, using $k = 2\pi/\lambda$,

$$x = 0, \frac{\pi}{k}, \frac{2\pi}{k}, \frac{3\pi}{k}, \ldots$$

$$= 0, \frac{\lambda}{2}, \frac{2\lambda}{2}, \frac{3\lambda}{2}, \ldots \quad \begin{pmatrix} \text{nodes of a standing wave on} \\ \text{a string, fixed end at } x = 0 \end{pmatrix} \quad (15.29)$$

In particular, there is a node at $x = 0$, as there should be, since this point is a fixed end of the string.

A standing wave, unlike a traveling wave, *does not* transfer energy from one end to the other. The two waves that form it would individually carry equal amounts of power in opposite directions. There is a local flow of energy from each node to the adjacent antinodes and back, but the *average* rate of energy transfer is zero at every point. If you evaluate the wave power given by Eq. (15.21) using the wave function of Eq. (15.28), you will find that the average power is zero.

Problem-Solving Strategy 15.2 **Standing Waves**

IDENTIFY *the relevant concepts:* Identify the target variables. Then determine whether the problem is purely *kinematic* (involving only such quantities as wave speed v, wavelength λ, and frequency f) or whether *dynamic* properties of the medium (such as F and μ for transverse waves on a string) are also involved.

SET UP *the problem* using the following steps:
1. Sketch the shape of the standing wave at a particular instant. This will help you visualize the nodes (label them N) and antinodes (A). The distance between adjacent nodes (or antinodes) is $\lambda/2$; the distance between a node and the adjacent antinode is $\lambda/4$.
2. Choose the equations you'll use. The wave function for the standing wave, like Eq. (15.28), is often useful.

3. You can determine the wave speed if you know λ and f (or, equivalently, $k = 2\pi/\lambda$ and $\omega = 2\pi f$) or if you know the relevant properties of the medium (for a string, F and μ).

EXECUTE *the solution:* Solve for the target variables. Once you've found the wave function, you can find the displacement y at any point x and at any time t. You can find the velocity and acceleration of a particle in the medium by taking the first and second partial derivatives of y with respect to time.

EVALUATE *your answer:* Compare your numerical answers with your sketch. Check that the wave function satisfies the boundary conditions (for example, the displacement should be zero at a fixed end).

Example 15.6   Standing waves on a guitar string

A guitar string lies along the x-axis when in equilibrium. The end of the string at $x = 0$ (the bridge of the guitar) is fixed. A sinusoidal wave with amplitude $A = 0.750$ mm $= 7.50 \times 10^{-4}$ m and frequency $f = 440$ Hz, corresponding to the red curves in Fig. 15.24, travels along the string in the $-x$-direction at 143 m/s. It is reflected from the fixed end, and the superposition of the incident and reflected waves forms a standing wave. (a) Find the equation giving the displacement of a point on the string as a function of position and time. (b) Locate the nodes. (c) Find the amplitude of the standing wave and the maximum transverse velocity and acceleration.

SOLUTION

IDENTIFY and SET UP: This is a *kinematics* problem (see Problem-Solving Strategy 15.1 in Section 15.3). The target variables are: in part (a), the wave function of the standing wave; in part (b), the locations of the nodes; and in part (c), the maximum displacement y, transverse velocity v_y, and transverse acceleration a_y. Since there is a fixed end at $x = 0$, we can use Eqs. (15.28) and (15.29) to describe this standing wave. We will need the relationships $\omega = 2\pi f$, $v = \omega/k$, and $v = \lambda f$.

EXECUTE: (a) The standing-wave amplitude is $A_{SW} = 2A = 1.50 \times 10^{-3}$ m (twice the amplitude of either the incident or reflected wave). The angular frequency and wave number are

$$\omega = 2\pi f = (2\pi \text{ rad})(440 \text{ s}^{-1}) = 2760 \text{ rad/s}$$

$$k = \frac{\omega}{v} = \frac{2760 \text{ rad/s}}{143 \text{ m/s}} = 19.3 \text{ rad/m}$$

Equation (15.28) then gives

$$y(x, t) = (A_{SW} \sin kx) \sin \omega t$$
$$= [(1.50 \times 10^{-3} \text{ m}) \sin(19.3 \text{ rad/m})x] \sin(2760 \text{ rad/s})t$$

(b) From Eq. (15.29), the positions of the nodes are $x = 0, \lambda/2,$ $\lambda, 3\lambda/2, \ldots$. The wavelength is $\lambda = v/f = (143 \text{ m/s})/(440 \text{ Hz})$

$= 0.325$ m, so the nodes are at $x = 0$, 0.163 m, 0.325 m, 0.488 m, ...

(c) From the expression for $y(x, t)$ in part (a), the maximum displacement from equilibrium is $A_{SW} = 1.50 \times 10^{-3}$ m $= 1.50$ mm. This occurs at the *antinodes*, which are midway between adjacent nodes (that is, at $x = 0.081$ m, 0.244 m, 0.406 m, ...).

For a particle on the string at any point x, the transverse (y-) velocity is

$$v_y(x, t) = \frac{\partial y(x, t)}{\partial t}$$
$$= [(1.50 \times 10^{-3} \text{ m}) \sin(19.3 \text{ rad/m})x]$$
$$\times [(2760 \text{ rad/s}) \cos(2760 \text{ rad/s})t]$$
$$= [(4.15 \text{ m/s}) \sin(19.3 \text{ rad/m})x] \cos(2760 \text{ rad/s})t$$

At an antinode, $\sin(19.3 \text{ rad/m})x = \pm 1$ and the transverse velocity varies between $+4.15$ m/s and -4.15 m/s. As is always the case in SHM, the maximum velocity occurs when the particle is passing through the equilibrium position ($y = 0$).

The transverse acceleration $a_y(x, t)$ is the *second* partial derivative of $y(x, t)$ with respect to time. You can show that

$$a_y(x, t) = \frac{\partial v_y(x, t)}{\partial t} = \frac{\partial^2 y(x, t)}{\partial t^2}$$
$$= [(-1.15 \times 10^4 \text{ m/s}^2) \sin(19.3 \text{ rad/m})x]$$
$$\times \sin(2760 \text{ rad/s})t$$

At the antinodes, the transverse acceleration varies between $+1.15 \times 10^4$ m/s^2 and -1.15×10^4 m/s^2.

EVALUATE: The maximum transverse velocity at an antinode is quite respectable (about 15 km/h, or 9.3 mi/h). But the maximum transverse acceleration is tremendous, 1170 times the acceleration due to gravity! Guitar strings are actually fixed at *both* ends; we'll see the consequences of this in the next section.

Test Your Understanding of Section 15.7 Suppose the frequency of the standing wave in Example 15.6 were doubled from 440 Hz to 880 Hz. Would all of the nodes for $f = 440$ Hz also be nodes for $f = 880$ Hz? If so, would there be additional nodes for $f = 880$ Hz? If not, which nodes are absent for $f = 880$ Hz?

15.8 Normal Modes of a String

When we described standing waves on a string rigidly held at one end, as in Fig. 15.23, we made no assumptions about the length of the string or about what was happening at the other end. Let's now consider a string of a definite length L, rigidly held at *both* ends. Such strings are found in many musical instruments, including pianos, violins, and guitars. When a guitar string is plucked, a wave is produced in the string; this wave is reflected and re-reflected from the ends of the string, making a standing wave. This standing wave on the string in turn produces a sound wave in the air, with a frequency determined by the properties of the string. This is what makes stringed instruments so useful in making music.

To understand these properties of standing waves on a string fixed at both ends, let's first examine what happens when we set up a sinusoidal wave on such a string. The standing wave that results must have a node at *both* ends of the string. We saw in the preceding section that adjacent nodes are one half-wavelength

PhET: Fourier: Making Waves
PhET: Waves on a String
ActivPhysics 10.4: Standing Waves on Strings
ActivPhysics 10.5: Tuning a Stringed Instrument: Standing Waves
ActivPhysics 10.6: String Mass and Standing Waves

$(\lambda/2)$ apart, so the length of the string must be $\lambda/2$, or $2(\lambda/2)$, or $3(\lambda/2)$, or in general some integer number of half-wavelengths:

$$L = n\frac{\lambda}{2} \quad (n = 1, 2, 3, \dots) \quad \text{(string fixed at both ends)} \quad (15.30)$$

That is, if a string with length L is fixed at both ends, a standing wave can exist only if its wavelength satisfies Eq. (15.30).

Solving this equation for λ and labeling the possible values of λ as λ_n, we find

$$\lambda_n = \frac{2L}{n} \quad (n = 1, 2, 3, \dots) \quad \text{(string fixed at both ends)} \quad (15.31)$$

Waves can exist on the string if the wavelength is *not* equal to one of these values, but there cannot be a steady wave pattern with nodes and antinodes, and the total wave cannot be a standing wave. Equation (15.31) is illustrated by the standing waves shown in Figs. 15.23a, 15.23b, 15.23c, and 15.23d; these represent $n = 1, 2, 3,$ and 4, respectively.

Corresponding to the series of possible standing-wave wavelengths λ_n is a series of possible standing-wave frequencies f_n, each related to its corresponding wavelength by $f_n = v/\lambda_n$. The smallest frequency f_1 corresponds to the largest wavelength (the $n = 1$ case), $\lambda_1 = 2L$:

$$f_1 = \frac{v}{2L} \quad \text{(string fixed at both ends)} \quad (15.32)$$

This is called the **fundamental frequency.** The other standing-wave frequencies are $f_2 = 2v/2L$, $f_3 = 3v/2L$, and so on. These are all integer multiples of the fundamental frequency f_1, such as $2f_1$, $3f_1$, $4f_1$, and so on, and we can express *all* the frequencies as

$$f_n = n\frac{v}{2L} = nf_1 \quad (n = 1, 2, 3, \dots) \quad \text{(string fixed at both ends)} \quad (15.33)$$

These frequencies are called **harmonics,** and the series is called a **harmonic series.** Musicians sometimes call f_2, f_3, and so on **overtones;** f_2 is the second harmonic or the first overtone, f_3 is the third harmonic or the second overtone, and so on. The first harmonic is the same as the fundamental frequency (Fig. 15.25).

For a string with fixed ends at $x = 0$ and $x = L$, the wave function $y(x, t)$ of the nth standing wave is given by Eq. (15.28) (which satisfies the condition that there is a node at $x = 0$), with $\omega = \omega_n = 2\pi f_n$ and $k = k_n = 2\pi/\lambda_n$:

$$y_n(x, t) = A_{SW} \sin k_n x \sin \omega_n t \quad (15.34)$$

You can easily show that this wave function has nodes at both $x = 0$ and $x = L$, as it must.

A **normal mode** of an oscillating system is a motion in which all particles of the system move sinusoidally with the same frequency. For a system made up of a string of length L fixed at both ends, each of the wavelengths given by Eq. (15.31) corresponds to a possible normal-mode pattern and frequency. There are infinitely many normal modes, each with its characteristic frequency and vibration pattern. Figure 15.26 shows the first four normal-mode patterns and their associated frequencies and wavelengths; these correspond to Eq. (15.34) with $n = 1, 2, 3,$ and 4. By contrast, a harmonic oscillator, which has only one oscillating particle, has only one normal mode and one characteristic frequency. The string fixed at both ends has infinitely many normal modes because it is made up of a very large (effectively infinite) number of particles. More complicated oscillating systems also have infinite numbers of normal modes, though with more complex normal-mode patterns than a string (Fig. 15.27).

15.25 Each string of a violin naturally oscillates at one or more of its harmonic frequencies, producing sound waves in the air with the same frequencies.

15.26 The first four normal modes of a string fixed at both ends. (Compare these to the photographs in Fig. 15.23.)

(a) $n = 1$: fundamental frequency, f_1

N   A   N

$\frac{\lambda}{2} = L$

(b) $n = 2$: second harmonic, f_2 (first overtone)

N   A   N   A   N

$2\frac{\lambda}{2} = L$

(c) $n = 3$: third harmonic, f_3 (second overtone)

N   A   N   A   N   A   N

$3\frac{\lambda}{2} = L$

(d) $n = 4$: fourth harmonic, f_4 (third overtone)

N   A   N   A   N   A   N   A   N

$4\frac{\lambda}{2} = L$

Complex Standing Waves

If we could displace a string so that its shape is the same as one of the normal-mode patterns and then release it, it would vibrate with the frequency of that mode. Such a vibrating string would displace the surrounding air with the same frequency, producing a traveling sinusoidal sound wave that your ears would perceive as a pure tone. But when a string is struck (as in a piano) or plucked (as is done to guitar strings), the shape of the displaced string is *not* as simple as one of the patterns in Fig. 15.26. The fundamental as well as many overtones are present in the resulting vibration. This motion is therefore a combination or *superposition* of many normal modes. Several simple harmonic motions of different frequencies are present simultaneously, and the displacement of any point on the string is the sum (or superposition) of the displacements associated with the individual modes. The sound produced by the vibrating string is likewise a superposition of traveling sinusoidal sound waves, which you perceive as a rich, complex tone with the fundamental frequency f_1. The standing wave on the string and the traveling sound wave in the air have similar **harmonic content** (the extent to which frequencies higher than the fundamental are present). The harmonic content depends on how the string is initially set into motion. If you pluck the strings of an acoustic guitar in the normal location over the sound hole, the sound that you hear has a different harmonic content than if you pluck the strings next to the fixed end on the guitar body.

It is possible to represent every possible motion of the string as some superposition of normal-mode motions. Finding this representation for a given vibration pattern is called *harmonic analysis*. The sum of sinusoidal functions that represents a complex wave is called a *Fourier series*. Figure 15.28 shows how a standing wave that is produced by plucking a guitar string of length L at a point $L/4$ from one end can be represented as a combination of sinusoidal functions.

Standing Waves and String Instruments

As we have seen, the fundamental frequency of a vibrating string is $f_1 = v/2L$. The speed v of waves on the string is determined by Eq. (15.13), $v = \sqrt{F/\mu}$. Combining these equations, we find

$$f_1 = \frac{1}{2L}\sqrt{\frac{F}{\mu}} \qquad \text{(string fixed at both ends)} \qquad (15.35)$$

This is also the fundamental frequency of the sound wave created in the surrounding air by the vibrating string. Familiar musical instruments show how f_1 depends on the properties of the string. The inverse dependence of frequency on length L is illustrated by the long strings of the bass (low-frequency) section of the piano or the bass viol compared with the shorter strings of the treble section of the piano or the violin (Fig. 15.29). The pitch of a violin or guitar is usually

15.27 Astronomers have discovered that the sun oscillates in several different normal modes. This computer simulation shows one such mode.

Cross section of the sun's interior

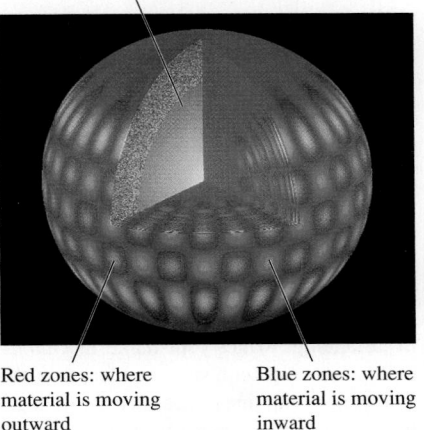

Red zones: where material is moving outward

Blue zones: where material is moving inward

Mastering**PHYSICS**

ActivPhysics 10.10: Complex Waves: Fourier Analysis

15.28 When a guitar string is plucked (pulled into a triangular shape) and released, a standing wave results. The standing wave is well represented (except at the sharp maximum point) by the sum of just three sinusoidal functions. Including additional sinusoidal functions further improves the representation.

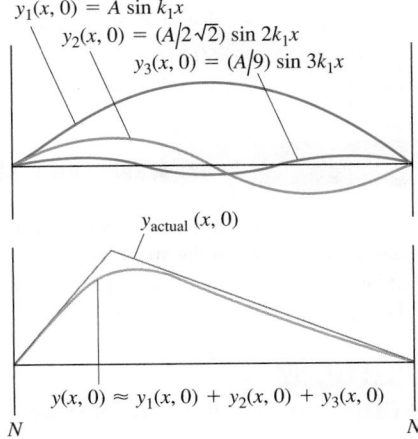

$y_1(x, 0) = A \sin k_1 x$

$y_2(x, 0) = (A/2\sqrt{2}) \sin 2k_1 x$

$y_3(x, 0) = (A/9) \sin 3k_1 x$

$y_{\text{actual}}(x, 0)$

$y(x, 0) \approx y_1(x, 0) + y_2(x, 0) + y_3(x, 0)$

Bass viol

Cello

Viola

Violin

Bass viol

Cello

Viola

Violin

15.29 Comparing the range of a concert grand piano to the ranges of a bass viol, a cello, a viola, and a violin. In all cases, longer strings produce bass notes and shorter strings produce treble notes.

varied by pressing a string against the fingerboard with the fingers to change the length L of the vibrating portion of the string. Increasing the tension F increases the wave speed v and thus increases the frequency (and the pitch). All string instruments are "tuned" to the correct frequencies by varying the tension; you tighten the string to raise the pitch. Finally, increasing the mass per unit length μ decreases the wave speed and thus the frequency. The lower notes on a steel guitar are produced by thicker strings, and one reason for winding the bass strings of a piano with wire is to obtain the desired low frequency from a relatively short string.

Wind instruments such as saxophones and trombones also have normal modes. As for stringed instruments, the frequencies of these normal modes determine the pitch of the musical tones that these instruments produce. We'll discuss these instruments and many other aspects of sound in Chapter 16.

Example 15.7 | **A giant bass viol**

In an attempt to get your name in *Guinness World Records,* you build a bass viol with strings of length 5.00 m between fixed points. One string, with linear mass density 40.0 g/m, is tuned to a 20.0-Hz fundamental frequency (the lowest frequency that the human ear can hear). Calculate (a) the tension of this string, (b) the frequency and wavelength on the string of the second harmonic, and (c) the frequency and wavelength on the string of the second overtone.

SOLUTION

IDENTIFY and SET UP: In part (a) the target variable is the string tension F; we'll use Eq. (15.35), which relates F to the known values $f_1 = 20.0$ Hz, $L = 5.00$ m, and $\mu = 40.0$ g/m. In parts (b) and (c) the target variables are the frequency and wavelength of a given harmonic and a given overtone. We determine these from the given length of the string and the fundamental frequency, using Eqs. (15.31) and (15.33).

EXECUTE: (a) We solve Eq. (15.35) for F:

$$F = 4\mu L^2 f_1{}^2 = 4(40.0 \times 10^{-3} \text{ kg/m})(5.00 \text{ m})^2(20.0 \text{ s}^{-1})^2$$
$$= 1600 \text{ N} = 360 \text{ lb}$$

(b) From Eqs. (15.33) and (15.31), the frequency and wavelength of the second harmonic ($n = 2$) are

$$f_2 = 2f_1 = 2(20.0 \text{ Hz}) = 40.0 \text{ Hz}$$
$$\lambda_2 = \frac{2L}{2} = \frac{2(5.00 \text{ m})}{2} = 5.00 \text{ m}$$

(c) The second overtone is the "second tone over" (above) the fundamental—that is, $n = 3$. Its frequency and wavelength are

$$f_3 = 3f_1 = 3(20.0 \text{ Hz}) = 60.0 \text{ Hz}$$
$$\lambda_3 = \frac{2L}{3} = \frac{2(5.00 \text{ m})}{3} = 3.33 \text{ m}$$

EVALUATE: The string tension in a real bass viol is typically a few hundred newtons; the tension in part (a) is a bit higher than that. The wavelengths in parts (b) and (c) are equal to the length of the string and two-thirds the length of the string, respectively, which agrees with the drawings of standing waves in Fig. 15.26.

Example 15.8 | **From waves on a string to sound waves in air**

What are the frequency and wavelength of the sound waves produced in the air when the string in Example 15.7 is vibrating at its fundamental frequency? The speed of sound in air at 20°C is 344 m/s.

SOLUTION

IDENTIFY and SET UP: Our target variables are the frequency and wavelength for the *sound wave* produced by the bass viol string. The frequency of the sound wave is the same as the fundamental frequency f_1 of the standing wave, because the string forces the surrounding air to vibrate at the same frequency. The wavelength of the sound wave is $\lambda_{1(\text{sound})} = v_{\text{sound}}/f_1$.

EXECUTE: We have $f = f_1 = 20.0$ Hz, so

$$\lambda_{1(\text{sound})} = \frac{v_{\text{sound}}}{f_1} = \frac{344 \text{ m/s}}{20.0 \text{ Hz}} = 17.2 \text{ m}$$

EVALUATE: In Example 15.7, the wavelength of the fundamental on the string was $\lambda_{1(\text{string})} = 2L = 2(5.00 \text{ m}) = 10.0$ m. Here $\lambda_{1(\text{sound})} = 17.2$ m is greater than that by the factor of $17.2/10.0 = 1.72$. This is as it should be: Because the frequencies of the sound wave and the standing wave are equal, $\lambda = v/f$ says that the wavelengths in air and on the string are in the same ratio as the corresponding wave speeds; here $v_{\text{sound}} = 344$ m/s is greater than $v_{\text{string}} = (10.0 \text{ m})(20.0 \text{ Hz}) = 200$ m/s by just the factor 1.72.

Test Your Understanding of Section 15.8 While a guitar string is vibrating, you gently touch the midpoint of the string to ensure that the string does not vibrate at that point. Which normal modes *cannot* be present on the string while you are touching it in this way?

Waves and their properties: A wave is any disturbance that propagates from one region to another. A mechanical wave travels within some material called the medium. The wave speed v depends on the type of wave and the properties of the medium.

In a periodic wave, the motion of each point of the medium is periodic with frequency f and period T. The wavelength λ is the distance over which the wave pattern repeats, and the amplitude A is the maximum displacement of a particle in the medium. The product of λ and f equals the wave speed. A sinusoidal wave is a special periodic wave in which each point moves in simple harmonic motion. (See Example 15.1.)

$$v = \lambda f \qquad (15.1)$$

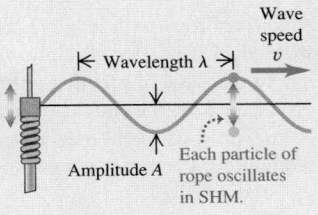

Wave functions and wave dynamics: The wave function $y(x, t)$ describes the displacements of individual particles in the medium. Equations (15.3), (15.4), and (15.7) give the wave equation for a sinusoidal wave traveling in the $+x$-direction. If the wave is moving in the $-x$-direction, the minus signs in the cosine functions are replaced by plus signs. (See Example 15.2.)

The wave function obeys a partial differential equation called the wave equation, Eq. (15.12).

The speed of transverse waves on a string depends on the tension F and mass per unit length μ. (See Example 15.3.)

$$y(x, t) = A \cos\left[\omega\left(\frac{x}{v} - t \right) \right]$$
$$= A \cos 2\pi f \left(\frac{x}{v} - t \right) \qquad (15.3)$$
$$y(x, t) = A \cos 2\pi \left(\frac{x}{\lambda} - \frac{t}{T} \right) \qquad (15.4)$$
$$y(x, t) = A \cos(kx - \omega t) \qquad (15.7)$$
where $k = 2\pi/\lambda$ and $\omega = 2\pi f = vk$

$$\frac{\partial^2 y(x, t)}{\partial x^2} = \frac{1}{v^2} \frac{\partial^2 y(x, t)}{\partial t^2} \qquad (15.12)$$

$$v = \sqrt{\frac{F}{\mu}} \qquad \text{(waves on a string)} \quad (15.13)$$

Wave power: Wave motion conveys energy from one region to another. For a sinusoidal mechanical wave, the average power P_{av} is proportional to the square of the wave amplitude and the square of the frequency. For waves that spread out in three dimensions, the wave intensity I is inversely proportional to the square of the distance from the source. (See Examples 15.4 and 15.5.)

$$P_{av} = \frac{1}{2} \sqrt{\mu F} \, \omega^2 A^2 \qquad (15.25)$$
(average power, sinusoidal wave)

$$\frac{I_1}{I_2} = \frac{r_2^2}{r_1^2} \qquad (15.26)$$
(inverse-square law for intensity)

Wave superposition: A wave reflects when it reaches a boundary of its medium. At any point where two or more waves overlap, the total displacement is the sum of the displacements of the individual waves (principle of superposition).

$$y(x, t) = y_1(x, t) + y_2(x, t) \qquad (15.27)$$
(principle of superposition)

Standing waves on a string: When a sinusoidal wave is reflected from a fixed or free end of a stretched string, the incident and reflected waves combine to form a standing sinusoidal wave with nodes and antinodes. Adjacent nodes are spaced a distance $\lambda/2$ apart, as are adjacent antinodes. (See Example 15.6.)

When both ends of a string with length L are held fixed, standing waves can occur only when L is an integer multiple of $\lambda/2$. Each frequency with its associated vibration pattern is called a normal mode. (See Examples 15.7 and 15.8.)

$$y(x, t) = (A_{SW} \sin kx) \sin \omega t \qquad (15.28)$$
(standing wave on a string, fixed end at $x = 0$)

$$f_n = n \frac{v}{2L} = n f_1 \ (n = 1, 2, 3, \dots) \qquad (15.33)$$

$$f_1 = \frac{1}{2L} \sqrt{\frac{F}{\mu}} \qquad (15.35)$$
(string fixed at both ends)

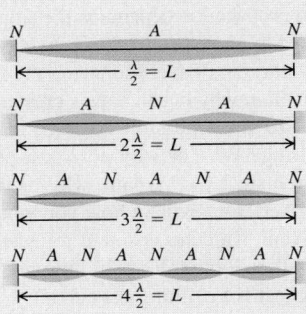

BRIDGING PROBLEM | Waves on a Rotating Rope

A uniform rope with length L and mass m is held at one end and whirled in a horizontal circle with angular velocity ω. You can ignore the force of gravity on the rope. (a) At a point on the rope a distance r from the end that is held, what is the tension F? (b) What is the speed of transverse waves at this point? (c) Find the time required for a transverse wave to travel from one end of the rope to the other.

SOLUTION GUIDE

See MasteringPhysics® Study Area for a Video Tutor solution.

IDENTIFY and SET UP

1. Draw a sketch of the situation and label the distances r and L. The tension in the rope will be different at different values of r. Do you see why? Where on the rope do you expect the tension to be greatest? Where do you expect it will be least?

2. Where on the rope do you expect the wave speed to be greatest? Where do you expect it will be least?

3. Think about the portion of the rope that is farther out than r from the end that is held. What forces act on this portion? (Remember that you can ignore gravity.) What is the mass of this portion? How far is its center of mass from the rotation axis?

4. Make a list of the unknown quantities and decide which are your target variables.

EXECUTE

5. Draw a free-body diagram for the portion of the rope that is farther out than r from the end that is held.

6. Use your free-body diagram to help you determine the tension in the rope at distance r.

7. Use your result from step 6 to find the wave speed at distance r.

8. Use your result from step 7 to find the time for a wave to travel from one end to the other. (*Hint:* The wave speed is $v = dr/dt$, so the time for the wave to travel a distance dr along the rope is $dt = dr/v$. Integrate this to find the total time. See Appendix B.)

EVALUATE

9. Do your results for parts (a) and (b) agree with your expectations from steps 1 and 2? Are the units correct?

10. Check your result for part (a) by considering the net force on a small segment of the rope at distance r with length dr and mass $dm = (m/L)dr$. [*Hint:* The tension forces on this segment are $F(r)$ on one side and $F(r + dr)$ on the other side. You will get an equation for dF/dr that you can integrate to find F as a function of r.]

Problems

For instructor-assigned homework, go to www.masteringphysics.com

•, ••, •••: Problems of increasing difficulty. **CP**: Cumulative problems incorporating material from earlier chapters. **CALC**: Problems requiring calculus. **BIO**: Biosciences problems.

DISCUSSION QUESTIONS

Q15.1 Two waves travel on the same string. Is it possible for them to have (a) different frequencies; (b) different wavelengths; (c) different speeds; (d) different amplitudes; (e) the same frequency but different wavelengths? Explain your reasoning.

Q15.2 Under a tension F, it takes 2.00 s for a pulse to travel the length of a taut wire. What tension is required (in terms of F) for the pulse to take 6.00 s instead?

Q15.3 What kinds of energy are associated with waves on a stretched string? How could you detect such energy experimentally?

Q15.4 The amplitude of a wave decreases gradually as the wave travels down a long, stretched string. What happens to the energy of the wave when this happens?

Q15.5 For the wave motions discussed in this chapter, does the speed of propagation depend on the amplitude? What makes you say this?

Q15.6 The speed of ocean waves depends on the depth of the water; the deeper the water, the faster the wave travels. Use this to explain why ocean waves crest and "break" as they near the shore.

Q15.7 Is it possible to have a longitudinal wave on a stretched string? Why or why not? Is it possible to have a transverse wave on a steel rod? Again, why or why not? If your answer is yes in either case, explain how you would create such a wave.

Q15.8 An echo is sound reflected from a distant object, such as a wall or a cliff. Explain how you can determine how far away the object is by timing the echo.

Q15.9 Why do you see lightning before you hear the thunder? A familiar rule of thumb is to start counting slowly, once per second, when you see the lightning; when you hear the thunder, divide the number you have reached by 3 to obtain your distance from the lightning in kilometers (or divide by 5 to obtain your distance in miles). Why does this work, or does it?

Q15.10 For transverse waves on a string, is the wave speed the same as the speed of any part of the string? Explain the difference between these two speeds. Which one is constant?

Q15.11 Children make toy telephones by sticking each end of a long string through a hole in the bottom of a paper cup and knotting it so it will not pull out. When the spring is pulled taut, sound can be transmitted from one cup to the other. How does this work? Why is the transmitted sound louder than the sound traveling through air for the same distance?

Q15.12 The four strings on a violin have different thicknesses, but are all under approximately the same tension. Do waves travel faster on the thick strings or the thin strings? Why? How does the fundamental vibration frequency compare for the thick versus the thin strings?

Q15.13 A sinusoidal wave can be described by a cosine function, which is negative just as often as positive. So why isn't the average power delivered by this wave zero?

Q15.14 Two strings of different mass per unit length μ_1 and μ_2 are tied together and stretched with a tension F. A wave travels along the string and passes the discontinuity in μ. Which of the following wave properties will be the same on both sides of the discontinuity, and which ones will change? speed of the wave; frequency; wavelength. Explain the physical reasoning behind each of your answers.

Q15.15 A long rope with mass m is suspended from the ceiling and hangs vertically. A wave pulse is produced at the lower end of the rope, and the pulse travels up the rope. Does the speed of the wave pulse change as it moves up the rope, and if so, does it increase or decrease?

Q15.16 In a transverse wave on a string, the motion of the string is perpendicular to the length of the string. How, then, is it possible for energy to move along the length of the string?

Q15.17 Both wave intensity and gravitation obey inverse-square laws. Do they do so for the same reason? Discuss the reason for each of these inverse-square laws as well as you can.

Q15.18 Energy can be transferred along a string by wave motion. However, in a standing wave on a string, no energy can ever be transferred past a node. Why not?

Q15.19 Can a standing wave be produced on a string by superposing two waves traveling in opposite directions with the same frequency but different amplitudes? Why or why not? Can a standing wave be produced by superposing two waves traveling in opposite directions with different frequencies but the same amplitude? Why or why not?

Q15.20 If you stretch a rubber band and pluck it, you hear a (somewhat) musical tone. How does the frequency of this tone change as you stretch the rubber band further? (Try it!) Does this agree with Eq. (15.35) for a string fixed at both ends? Explain.

Q15.21 A musical interval of an *octave* corresponds to a factor of 2 in frequency. By what factor must the tension in a guitar or violin string be increased to raise its pitch one octave? To raise it two octaves? Explain your reasoning. Is there any danger in attempting these changes in pitch?

Q15.22 By touching a string lightly at its center while bowing, a violinist can produce a note exactly one octave above the note to which the string is tuned—that is, a note with exactly twice the frequency. Why is this possible?

Q15.23 As we discussed in Section 15.1, water waves are a combination of longitudinal and transverse waves. Defend the following statement: "When water waves hit a vertical wall, the wall is a node of the longitudinal displacement but an antinode of the transverse displacement."

Q15.24 Violins are short instruments, while cellos and basses are long. In terms of the frequency of the waves they produce, explain why this is so.

Q15.25 What is the purpose of the frets on a guitar? In terms of frequency of the vibration of the strings, explain their use.

EXERCISES

Section 15.2 Periodic Waves

15.1 • The speed of sound in air at 20°C is 344 m/s. (a) What is the wavelength of a sound wave with a frequency of 784 Hz, corresponding to the note G_5 on a piano, and how many milliseconds does each vibration take? (b) What is the wavelength of a sound wave one octave higher than the note in part (a)?

15.2 • **BIO Audible Sound.** Provided the amplitude is sufficiently great, the human ear can respond to longitudinal waves over a range of frequencies from about 20.0 Hz to about 20.0 kHz. (a) If you were to mark the beginning of each complete wave pattern with a red dot for the long-wavelength sound and a blue dot

for the short-wavelength sound, how far apart would the red dots be, and how far apart would the blue dots be? (b) In reality would adjacent dots in each set be far enough apart for you to easily measure their separation with a meter stick? (c) Suppose you repeated part (a) in water, where sound travels at 1480 m/s. How far apart would the dots be in each set? Could you readily measure their separation with a meter stick?

15.3 • **Tsunami!** On December 26, 2004, a great earthquake occurred off the coast of Sumatra and triggered immense waves (tsunami) that killed some 200,000 people. Satellites observing these waves from space measured 800 km from one wave crest to the next and a period between waves of 1.0 hour. What was the speed of these waves in m/s and in km/h? Does your answer help you understand why the waves caused such devastation?

15.4 • **BIO Ultrasound Imaging.** Sound having frequencies above the range of human hearing (about 20,000 Hz) is called *ultrasound.* Waves above this frequency can be used to penetrate the body and to produce images by reflecting from surfaces. In a typical ultrasound scan, the waves travel through body tissue with a speed of 1500 m/s. For a good, detailed image, the wavelength should be no more than 1.0 mm. What frequency sound is required for a good scan?

15.5 • **BIO** (a) **Audible wavelengths.** The range of audible frequencies is from about 20 Hz to 20,000 Hz. What is the range of the wavelengths of audible sound in air? (b) **Visible light.** The range of visible light extends from 400 nm to 700 nm. What is the range of visible frequencies of light? (c) **Brain surgery.** Surgeons can remove brain tumors by using a cavitron ultrasonic surgical aspirator, which produces sound waves of frequency 23 kHz. What is the wavelength of these waves in air? (d) **Sound in the body.** What would be the wavelength of the sound in part (c) in bodily fluids in which the speed of sound is 1480 m/s but the frequency is unchanged?

15.6 •• A fisherman notices that his boat is moving up and down periodically, owing to waves on the surface of the water. It takes 2.5 s for the boat to travel from its highest point to its lowest, a total distance of 0.62 m. The fisherman sees that the wave crests are spaced 6.0 m apart. (a) How fast are the waves traveling? (b) What is the amplitude of each wave? (c) If the total vertical distance traveled by the boat were 0.30 m but the other data remained the same, how would the answers to parts (a) and (b) be affected?

Section 15.3 Mathematical Description of a Wave

15.7 • Transverse waves on a string have wave speed 8.00 m/s, amplitude 0.0700 m, and wavelength 0.320 m. The waves travel in the $-x$-direction, and at $t = 0$ the $x = 0$ end of the string has its maximum upward displacement. (a) Find the frequency, period, and wave number of these waves. (b) Write a wave function describing the wave. (c) Find the transverse displacement of a particle at $x = 0.360$ m at time $t = 0.150$ s. (d) How much time must elapse from the instant in part (c) until the particle at $x = 0.360$ m next has maximum upward displacement?

15.8 • A certain transverse wave is described by

$$y(x, t) = (6.50 \text{ mm}) \cos 2\pi \left(\frac{x}{28.0 \text{ cm}} - \frac{t}{0.0360 \text{ s}} \right)$$

Determine the wave's (a) amplitude; (b) wavelength; (c) frequency; (d) speed of propagation; (e) direction of propagation.

15.9 • **CALC** Which of the following wave functions satisfies the wave equation, Eq. (15.12)? (a) $y(x, t) = A \cos(kx + \omega t)$; (b) $y(x, t) = A \sin(kx + \omega t)$; (c) $y(x, t) = A(\cos kx + \cos \omega t)$. (d) For the wave of part (b), write the equations for the transverse velocity and transverse acceleration of a particle at point x.

15.10 • A water wave traveling in a straight line on a lake is described by the equation

$$y(x, t) = (3.75 \text{ cm}) \cos(0.450 \text{ cm}^{-1} x + 5.40 \text{ s}^{-1} t)$$

where y is the displacement perpendicular to the undisturbed surface of the lake. (a) How much time does it take for one complete wave pattern to go past a fisherman in a boat at anchor, and what horizontal distance does the wave crest travel in that time? (b) What are the wave number and the number of waves per second that pass the fisherman? (c) How fast does a wave crest travel past the fisherman, and what is the maximum speed of his cork floater as the wave causes it to bob up and down?

15.11 • A sinusoidal wave is propagating along a stretched string that lies along the x-axis. The displacement of the string as a function of time is graphed in Fig. E15.11 for particles at $x = 0$ and at $x = 0.0900$ m. (a) What is the amplitude of the

Figure **E15.11**

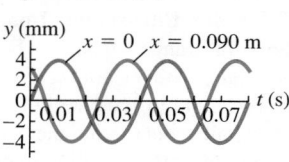

wave? (b) What is the period of the wave? (c) You are told that the two points $x = 0$ and $x = 0.0900$ m are within one wavelength of each other. If the wave is moving in the $+x$-direction, determine the wavelength and the wave speed. (d) If instead the wave is moving in the $-x$-direction, determine the wavelength and the wave speed. (e) Would it be possible to determine definitively the wavelength in parts (c) and (d) if you were not told that the two points were within one wavelength of each other? Why or why not?

15.12 •• **CALC** **Speed of Propagation vs. Particle Speed.** (a) Show that Eq. (15.3) may be written as

$$y(x, t) = A \cos\left[\frac{2\pi}{\lambda}(x - vt)\right]$$

(b) Use $y(x, t)$ to find an expression for the transverse velocity v_y of a particle in the string on which the wave travels. (c) Find the maximum speed of a particle of the string. Under what circumstances is this equal to the propagation speed v? Less than v? Greater than v?

15.13 •• A transverse wave on a string has amplitude 0.300 cm, wavelength 12.0 cm, and speed 6.00 cm/s. It is represented by $y(x, t)$ as given in Exercise 15.12. (a) At time $t = 0$, compute y at 1.5-cm intervals of x (that is, at $x = 0$, $x = 1.5$ cm, $x = 3.0$ cm, and so on) from $x = 0$ to $x = 12.0$ cm. Graph the results. This is the shape of the string at time $t = 0$. (b) Repeat the calculations for the same values of x at times $t = 0.400$ s and $t = 0.800$ s. Graph the shape of the string at these instants. In what direction is the wave traveling?

15.14 • A wave on a string is described by $y(x, t) = A \cos(kx - \omega t)$. (a) Graph y, v_y, and a_y as functions of x for time $t = 0$. (b) Consider the following points on the string: (i) $x = 0$; (ii) $x = \pi/4k$; (iii) $x = \pi/2k$; (iv) $x = 3\pi/4k$; (v) $x = \pi/k$; (vi) $x = 5\pi/4k$; (vii) $x = 3\pi/2k$; (viii) $x = 7\pi/4k$. For a particle at each of these points at $t = 0$, describe in words whether the particle is moving and in what direction, and whether the particle is speeding up, slowing down, or instantaneously not accelerating.

Section 15.4 Speed of a Transverse Wave

15.15 • One end of a horizontal rope is attached to a prong of an electrically driven tuning fork that vibrates the rope transversely at 120 Hz. The other end passes over a pulley and supports a 1.50-kg mass. The linear mass density of the rope is 0.0550 kg/m.

(a) What is the speed of a transverse wave on the rope? (b) What is the wavelength? (c) How would your answers to parts (a) and (b) change if the mass were increased to 3.00 kg?

15.16 • With what tension must a rope with length 2.50 m and mass 0.120 kg be stretched for transverse waves of frequency 40.0 Hz to have a wavelength of 0.750 m?

15.17 •• The upper end of a 3.80-m-long steel wire is fastened to the ceiling, and a 54.0-kg object is suspended from the lower end of the wire. You observe that it takes a transverse pulse 0.0492 s to travel from the bottom to the top of the wire. What is the mass of the wire?

15.18 •• A 1.50-m string of weight 0.0125 N is tied to the ceiling at its upper end, and the lower end supports a weight W. Neglect the very small variation in tension along the length of the string that is produced by the weight of the string. When you pluck the string slightly, the waves traveling up the string obey the equation

$$y(x, t) = (8.50 \text{ mm}) \cos(172 \text{ m}^{-1} x - 4830 \text{ s}^{-1} t)$$

Assume that the tension of the string is constant and equal to W. (a) How much time does it take a pulse to travel the full length of the string? (b) What is the weight W? (c) How many wavelengths are on the string at any instant of time? (d) What is the equation for waves traveling *down* the string?

15.19 • A thin, 75.0-cm wire has a mass of 16.5 g. One end is tied to a nail, and the other end is attached to a screw that can be adjusted to vary the tension in the wire. (a) To what tension (in newtons) must you adjust the screw so that a transverse wave of wavelength 3.33 cm makes 875 vibrations per second? (b) How fast would this wave travel?

15.20 • **Weighty Rope.** If in Example 15.3 (Section 15.4) we do *not* neglect the weight of the rope, what is the wave speed (a) at the bottom of the rope; (b) at the middle of the rope; (c) at the top of the rope?

15.21 • A simple harmonic oscillator at the point $x = 0$ generates a wave on a rope. The oscillator operates at a frequency of 40.0 Hz and with an amplitude of 3.00 cm. The rope has a linear mass density of 50.0 g/m and is stretched with a tension of 5.00 N. (a) Determine the speed of the wave. (b) Find the wavelength. (c) Write the wave function $y(x, t)$ for the wave. Assume that the oscillator has its maximum upward displacement at time $t = 0$. (d) Find the maximum transverse acceleration of points on the rope. (e) In the discussion of transverse waves in this chapter, the force of gravity was ignored. Is that a reasonable approximation for this wave? Explain.

Section 15.5 Energy in Wave Motion

15.22 •• A piano wire with mass 3.00 g and length 80.0 cm is stretched with a tension of 25.0 N. A wave with frequency 120.0 Hz and amplitude 1.6 mm travels along the wire. (a) Calculate the average power carried by the wave. (b) What happens to the average power if the wave amplitude is halved?

15.23 • A horizontal wire is stretched with a tension of 94.0 N, and the speed of transverse waves for the wire is 492 m/s. What must the amplitude of a traveling wave of frequency 69.0 Hz be in order for the average power carried by the wave to be 0.365 W?

15.24 •• A light wire is tightly stretched with tension F. Transverse traveling waves of amplitude A and wavelength λ_1 carry average power $P_{av,1} = 0.400$ W. If the wavelength of the waves is doubled, so $\lambda_2 = 2\lambda_1$, while the tension F and amplitude A are not altered, what then is the average power $P_{av,2}$ carried by the waves?

15.25 •• A jet plane at takeoff can produce sound of intensity 10.0 W/m² at 30.0 m away. But you prefer the tranquil sound of normal conversation, which is 1.0 μW/m^2. Assume that the plane behaves like a point source of sound. (a) What is the closest distance you should live from the airport runway to preserve your peace of mind? (b) What intensity from the jet does your friend experience if she lives twice as far from the runway as you do? (c) What power of sound does the jet produce at takeoff?

15.26 •• **Threshold of Pain.** You are investigating the report of a UFO landing in an isolated portion of New Mexico, and you encounter a strange object that is radiating sound waves uniformly in all directions. Assume that the sound comes from a point source and that you can ignore reflections. You are slowly walking toward the source. When you are 7.5 m from it, you measure its intensity to be 0.11 W/m^2. An intensity of 1.0 W/m^2 is often used as the "threshold of pain." How much closer to the source can you move before the sound intensity reaches this threshold?

15.27 • **Energy Output.** By measurement you determine that sound waves are spreading out equally in all directions from a point source and that the intensity is 0.026 W/m^2 at a distance of 4.3 m from the source. (a) What is the intensity at a distance of 3.1 m from the source? (b) How much sound energy does the source emit in one hour if its power output remains constant?

15.28 • A fellow student with a mathematical bent tells you that the wave function of a traveling wave on a thin rope is $y(x, t) = 2.30$ mm $\cos[(6.98 \text{ rad/m})x + (742 \text{ rad/s})t]$. Being more practical, you measure the rope to have a length of 1.35 m and a mass of 0.00338 kg. You are then asked to determine the following: (a) amplitude; (b) frequency; (c) wavelength; (d) wave speed; (e) direction the wave is traveling; (f) tension in the rope; (g) average power transmitted by the wave.

15.29 • At a distance of 7.00×10^{12} m from a star, the intensity of the radiation from the star is 15.4 W/m^2. Assuming that the star radiates uniformly in all directions, what is the total power output of the star?

Section 15.6 Wave Interference, Boundary Conditions, and Superposition

15.30 • **Reflection.** A wave pulse on a string has the dimensions shown in Fig. E15.30 at $t = 0$. The wave speed is 40 cm/s. (a) If point O is a fixed end, draw the total wave on the string at $t = 15$ ms, 20 ms, 25 ms, 30 ms, 35 ms, 40 ms, and 45 ms. (b) Repeat part (a) for the case in which point O is a free end.

Figure **E15.30**

Figure **E15.31**

15.31 • **Reflection.** A wave pulse on a string has the dimensions shown in Fig. E15.31 at $t = 0$. The wave speed is 5.0 m/s. (a) If point O is a fixed end, draw the total wave on the string at $t = 1.0$ ms, 2.0 ms, 3.0 ms, 4.0 ms, 5.0 ms, 6.0 ms, and 7.0 ms. (b) Repeat part (a) for the case in which point O is a free end.

15.32 • **Interference of Triangular Pulses.** Two triangular wave pulses are traveling toward each other on a stretched string as shown in Fig. E15.32. Each pulse is identical to the other and travels at 2.00 cm/s. The leading edges of the pulses are 1.00 cm apart at $t = 0$. Sketch the shape of the string at $t = 0.250$ s, $t = 0.500$ s, $t = 0.750$ s, $t = 1.000$ s, and $t = 1.250$ s.

Figure **E15.32**

15.33 • Suppose that the left-traveling pulse in Exercise 15.32 is *below* the level of the unstretched string instead of above it. Make the same sketches that you did in that exercise.

15.34 •• Two pulses are moving in opposite directions at 1.0 cm/s on a taut string, as shown in Fig. E15.34. Each square is 1.0 cm. Sketch the shape of the string at the end of (a) 6.0 s; (b) 7.0 s; (c) 8.0 s.

Figure **E15.34**

15.35 •• **Interference of Rectangular Pulses.** Figure E15.35 shows two rectangular wave pulses on a stretched string traveling toward each other. Each pulse is traveling with a speed of 1.00 mm/s and has the height and width shown in the figure. If the leading edges of the pulses are 8.00 mm apart at $t = 0$, sketch the shape of the string at $t = 4.00$ s, $t = 6.00$ s, and $t = 10.0$ s.

Figure **E15.35**

Section 15.7 Standing Waves on a String
Section 15.8 Normal Modes of a String

15.36 •• **CALC** Adjacent antinodes of a standing wave on a string are 15.0 cm apart. A particle at an antinode oscillates in simple harmonic motion with amplitude 0.850 cm and period 0.0750 s. The string lies along the $+x$-axis and is fixed at $x = 0$. (a) How far apart are the adjacent nodes? (b) What are the wavelength, amplitude, and speed of the two traveling waves that form this pattern? (c) Find the maximum and minimum transverse speeds of a point at an antinode. (d) What is the shortest distance along the string between a node and an antinode?

15.37 • Standing waves on a wire are described by Eq. (15.28), with $A_{SW} = 2.50$ mm, $\omega = 942$ rad/s, and $k = 0.750\pi$ rad/m. The left end of the wire is at $x = 0$. At what distances from the left end are (a) the nodes of the standing wave and (b) the antinodes of the standing wave?

15.38 • **CALC Wave Equation and Standing Waves.** (a) Prove by direct substitution that $y(x, t) = (A_{SW} \sin kx) \sin \omega t$ is a solution of the wave equation, Eq. (15.12), for $v = \omega/k$. (b) Explain why the relationship $v = \omega/k$ for *traveling* waves also applies to *standing* waves.

15.39 • CALC Let $y_1(x, t) = A \cos(k_1 x - \omega_1 t)$ and $y_2(x, t) = A \cos(k_2 x - \omega_2 t)$ be two solutions to the wave equation, Eq. (15.12), for the same v. Show that $y(x, t) = y_1(x, t) + y_2(x, t)$ is also a solution to the wave equation.

15.40 • A 1.50-m-long rope is stretched between two supports with a tension that makes the speed of transverse waves 48.0 m/s. What are the wavelength and frequency of (a) the fundamental; (b) the second overtone; (c) the fourth harmonic?

15.41 • A wire with mass 40.0 g is stretched so that its ends are tied down at points 80.0 cm apart. The wire vibrates in its fundamental mode with frequency 60.0 Hz and with an amplitude at the antinodes of 0.300 cm. (a) What is the speed of propagation of transverse waves in the wire? (b) Compute the tension in the wire. (c) Find the maximum transverse velocity and acceleration of particles in the wire.

15.42 • A piano tuner stretches a steel piano wire with a tension of 800 N. The steel wire is 0.400 m long and has a mass of 3.00 g. (a) What is the frequency of its fundamental mode of vibration? (b) What is the number of the highest harmonic that could be heard by a person who is capable of hearing frequencies up to 10,000 Hz?

15.43 • CALC A thin, taut string tied at both ends and oscillating in its third harmonic has its shape described by the equation $y(x, t) = (5.60 \text{ cm}) \sin[(0.0340 \text{ rad/cm})x] \sin[(50.0 \text{ rad/s})t]$, where the origin is at the left end of the string, the x-axis is along the string, and the y-axis is perpendicular to the string. (a) Draw a sketch that shows the standing-wave pattern. (b) Find the amplitude of the two traveling waves that make up this standing wave. (c) What is the length of the string? (d) Find the wavelength, frequency, period, and speed of the traveling waves. (e) Find the maximum transverse speed of a point on the string. (f) What would be the equation $y(x, t)$ for this string if it were vibrating in its eighth harmonic?

15.44 • The wave function of a standing wave is $y(x, t) = 4.44 \text{ mm} \sin[(32.5 \text{ rad/m})x] \sin[(754 \text{ rad/s})t]$. For the two traveling waves that make up this standing wave, find the (a) amplitude; (b) wavelength; (c) frequency; (d) wave speed; (e) wave functions. (f) From the information given, can you determine which harmonic this is? Explain.

15.45 •• Consider again the rope and traveling wave of Exercise 15.28. Assume that the ends of the rope are held fixed and that this traveling wave and the reflected wave are traveling in the opposite direction. (a) What is the wave function $y(x, t)$ for the standing wave that is produced? (b) In which harmonic is the standing wave oscillating? (c) What is the frequency of the fundamental oscillation?

15.46 •• One string of a certain musical instrument is 75.0 cm long and has a mass of 8.75 g. It is being played in a room where the speed of sound is 344 m/s. (a) To what tension must you adjust the string so that, when vibrating in its second overtone, it produces sound of wavelength 0.765 m? (Assume that the breaking stress of the wire is very large and isn't exceeded.) (b) What frequency sound does this string produce in its fundamental mode of vibration?

15.47 • The portion of the string of a certain musical instrument between the bridge and upper end of the finger board (that part of the string that is free to vibrate) is 60.0 cm long, and this length of the string has mass 2.00 g. The string sounds an A_4 note (440 Hz) when played. (a) Where must the player put a finger (what distance x from the bridge) to play a D_5 note (587 Hz)? (See Fig. E15.47.)

For both the A_4 and D_5 notes, the string vibrates in its fundamental mode. (b) Without retuning, is it possible to play a G_4 note (392 Hz) on this string? Why or why not?

Figure **E15.47**

15.48 •• (a) A horizontal string tied at both ends is vibrating in its fundamental mode. The traveling waves have speed v, frequency f, amplitude A, and wavelength λ. Calculate the maximum transverse velocity and maximum transverse acceleration of points located at (i) $x = \lambda/2$, (ii) $x = \lambda/4$, and (iii) $x = \lambda/8$ from the left-hand end of the string. (b) At each of the points in part (a), what is the amplitude of the motion? (c) At each of the points in part (a), how much time does it take the string to go from its largest upward displacement to its largest downward displacement?

15.49 • Guitar String. One of the 63.5-cm-long strings of an ordinary guitar is tuned to produce the note B_3 (frequency 245 Hz) when vibrating in its fundamental mode. (a) Find the speed of transverse waves on this string. (b) If the tension in this string is increased by 1.0%, what will be the new fundamental frequency of the string? (c) If the speed of sound in the surrounding air is 344 m/s, find the frequency and wavelength of the sound wave produced in the air by the vibration of the B_3 string. How do these compare to the frequency and wavelength of the standing wave on the string?

15.50 • Waves on a Stick. A flexible stick 2.0 m long is not fixed in any way and is free to vibrate. Make clear drawings of this stick vibrating in its first three harmonics, and then use your drawings to find the wavelengths of each of these harmonics. (*Hint*: Should the ends be nodes or antinodes?)

PROBLEMS

15.51 • CALC A transverse sine wave with an amplitude of 2.50 mm and a wavelength of 1.80 m travels from left to right along a long, horizontal, stretched string with a speed of 36.0 m/s. Take the origin at the left end of the undisturbed string. At time $t = 0$ the left end of the string has its maximum upward displacement. (a) What are the frequency, angular frequency, and wave number of the wave? (b) What is the function $y(x, t)$ that describes the wave? (c) What is $y(t)$ for a particle at the left end of the string? (d) What is $y(t)$ for a particle 1.35 m to the right of the origin? (e) What is the maximum magnitude of transverse velocity of any particle of the string? (f) Find the transverse displacement and the transverse velocity of a particle 1.35 m to the right of the origin at time $t = 0.0625$ s.

15.52 • A transverse wave on a rope is given by

$$y(x, t) = (0.750 \text{ cm}) \cos \pi[(0.400 \text{ cm}^{-1})x + (250 \text{ s}^{-1})t]$$

(a) Find the amplitude, period, frequency, wavelength, and speed of propagation. (b) Sketch the shape of the rope at these values of t: 0, 0.0005 s, 0.0010 s. (c) Is the wave traveling in the $+x$- or $-x$-direction? (d) The mass per unit length of the rope is 0.0500 kg/m. Find the tension. (e) Find the average power of this wave.

15.53 •• Three pieces of string, each of length L, are joined together end to end, to make a combined string of length $3L$. The first piece of string has mass per unit length μ_1, the second piece has mass per unit length $\mu_2 = 4\mu_1$, and the third piece has mass per unit length $\mu_3 = \mu_1/4$. (a) If the combined string is under tension F, how much time does it take a transverse wave to travel the entire length $3L$? Give your answer in terms of L, F, and μ_1. (b) Does your answer to part (a) depend on the order in which the three pieces are joined together? Explain.

15.54 ·· **CP** A 1750-N irregular beam is hanging horizontally by its ends from the ceiling by two vertical wires (A and B), each 1.25 m long and weighing 0.360 N. The center of gravity of this beam is one-third of the way along the beam from the end where wire A is attached. If you pluck both strings at the same time at the beam, what is the time delay between the arrival of the two pulses at the ceiling? Which pulse arrives first? (Neglect the effect of the weight of the wires on the tension in the wires.)

15.55 · **CALC** **Ant Joy Ride.** You place your pet ant Klyde (mass m) on top of a horizontal, stretched rope, where he holds on tightly. The rope has mass M and length L and is under tension F. You start a sinusoidal transverse wave of wavelength λ and amplitude A propagating along the rope. The motion of the rope is in a vertical plane. Klyde's mass is so small that his presence has no effect on the propagation of the wave. (a) What is Klyde's top speed as he oscillates up and down? (b) Klyde enjoys the ride and begs for more. You decide to double his top speed by changing the tension while keeping the wavelength and amplitude the same. Should the tension be increased or decreased, and by what factor?

15.56 ·· **Weightless Ant.** An ant with mass m is standing peacefully on top of a horizontal, stretched rope. The rope has mass per unit length μ and is under tension F. Without warning, Cousin Throckmorton starts a sinusoidal transverse wave of wavelength λ propagating along the rope. The motion of the rope is in a vertical plane. What minimum wave amplitude will make the ant become momentarily weightless? Assume that m is so small that the presence of the ant has no effect on the propagation of the wave.

15.57 · **CP** When a transverse sinusoidal wave is present on a string, the particles of the string undergo SHM. This is the same motion as that of a mass m attached to an ideal spring of force constant k', for which the angular frequency of oscillation was found in Chapter 14 to be $\omega = \sqrt{k'/m}$. Consider a string with tension F and mass per unit length μ, along which is propagating a sinusoidal wave with amplitude A and wavelength λ. (a) Find the "force constant" k' of the restoring force that acts on a short segment of the string of length Δx (where $\Delta x \ll \lambda$). (b) How does the "force constant" calculated in part (b) depend on F, μ, A, and λ? Explain the physical reasons this should be so.

15.58 ·· **Music.** You are designing a two-string instrument with metal strings 35.0 cm long, as shown in Fig. P15.58. Both strings are under the *same tension.* String S_1 has a mass of 8.00 g and produces the note middle C (frequency 262 Hz) in its fundamental mode. (a) What should be the tension in the string? (b) What should be the mass of string S_2 so that it will produce A-sharp (frequency 466 Hz) as its fundamental? (c) To extend the range of your instrument, you include a fret located just under the strings but not normally touching them. How far from the upper end should you put this fret so that when you press S_1 tightly against it, this string will produce C-sharp (frequency 277 Hz) in its fundamental? That is, what is x in the figure?

Figure **P15.58**

(d) If you press S_2 against the fret, what frequency of sound will it produce in its fundamental?

15.59 ··· **CP** The lower end of a uniform bar of mass 45.0 kg is attached to a wall by a frictionless hinge. The bar is held by a horizontal wire attached at its upper end so that the bar makes an angle of 30.0° with the wall. The wire has length 0.330 m and mass 0.0920 kg. What is the frequency of the fundamental standing wave for transverse waves on the wire?

15.60 ··· **CP** You are exploring a newly discovered planet. The radius of the planet is 7.20×10^7 m. You suspend a lead weight from the lower end of a light string that is 4.00 m long and has mass 0.0280 kg. You measure that it takes 0.0600 s for a transverse pulse to travel from the lower end to the upper end of the string. On earth, for the same string and lead weight, it takes 0.0390 s for a transverse pulse to travel the length of the string. The weight of the string is small enough that its effect on the tension in the string can be neglected. Assuming that the mass of the planet is distributed with spherical symmetry, what is its mass?

15.61 ·· For a string stretched between two supports, two successive standing-wave frequencies are 525 Hz and 630 Hz. There are other standing-wave frequencies lower than 525 Hz and higher than 630 Hz. If the speed of transverse waves on the string is 384 m/s, what is the length of the string? Assume that the mass of the wire is small enough for its effect on the tension in the wire to be neglected.

15.62 ··· **CP** A 5.00-m, 0.732-kg wire is used to support two uniform 235-N posts of equal length (Fig. P15.62). Assume that the wire is essentially horizontal and that the speed of sound is 344 m/s. A strong wind is blowing, causing the wire to vibrate in its 5th overtone. What are the frequency and wavelength of the sound this wire produces?

Figure **P15.62**

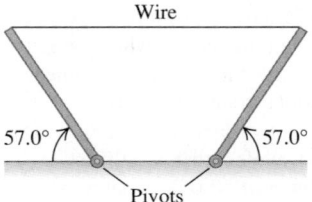

15.63 ··· **CP** A 1.80-m-long uniform bar that weighs 536 N is suspended in a horizontal position by two vertical wires that are attached to the ceiling. One wire is aluminum and the other is copper. The aluminum wire is attached to the left-hand end of the bar, and the copper wire is attached 0.40 m to the left of the right-hand end. Each wire has length 0.600 m and a circular cross section with radius 0.280 mm. What is the fundamental frequency of transverse standing waves for each wire?

15.64 ·· A continuous succession of sinusoidal wave pulses are produced at one end of a very long string and travel along the length of the string. The wave has frequency 70.0 Hz, amplitude 5.00 mm, and wavelength 0.600 m. (a) How long does it take the wave to travel a distance of 8.00 m along the length of the string? (b) How long does it take a point on the string to travel a distance of 8.00 m, once the wave train has reached the point and set it into motion? (c) In parts (a) and (b), how does the time change if the amplitude is doubled?

15.65 ··· **CALC** **Waves of Arbitrary Shape.** (a) Explain why *any* wave described by a function of the form $y(x, t) = f(x - vt)$ moves in the $+x$-direction with speed v. (b) Show that $y(x, t) = f(x - vt)$ satisfies the wave equation, no matter what the functional form of f. To do this, write $y(x, t) = f(u)$, where

$u = x - vt$. Then, to take partial derivatives of $y(x, t)$, use the chain rule:

$$\frac{\partial y(x, t)}{\partial t} = \frac{df(u)}{du}\frac{\partial u}{\partial t} = \frac{df(u)}{du}(-v)$$

$$\frac{\partial y(x, t)}{\partial t} = \frac{df(u)}{du}\frac{\partial u}{\partial x} = \frac{df(u)}{du}$$

(c) A wave pulse is described by the function $y(x, t) = De^{-(Bx - Ct)^2}$, where B, C, and D are all positive constants. What is the speed of this wave?

15.66 ••• **CP** A vertical, 1.20-m length of 18-gauge (diameter of 1.024 mm) copper wire has a 100.0-N ball hanging from it. (a) What is the wavelength of the third harmonic for this wire? (b) A 500.0-N ball now *replaces* the original ball. What is the change in the wavelength of the third harmonic caused by replacing the light ball with the heavy one? (*Hint:* See Table 11.1 for Young's modulus.)

15.67 • (a) Show that Eq. (15.25) can also be written as $P_{av} = \frac{1}{2}Fk\omega A^2$, where k is the wave number of the wave. (b) If the tension F in the string is quadrupled while the amplitude A is kept the same, how must k and ω each change to keep the average power constant? [*Hint:* Recall Eq. (15.6).]

15.68 ••• **CALC** Equation (15.7) for a sinusoidal wave can be made more general by including a phase angle ϕ, where $0 \le \phi \le 2\pi$ (in radians). Then the wave function $y(x, t)$ becomes

$$y(x, t) = A\cos(kx - \omega t + \phi)$$

(a) Sketch the wave as a function of x at $t = 0$ for $\phi = 0$, $\phi = \pi/4$, $\phi = \pi/2$, $\phi = 3\pi/4$, and $\phi = 3\pi/2$. (b) Calculate the transverse velocity $v_y = \partial y/\partial t$. (c) At $t = 0$, a particle on the string at $x = 0$ has displacement $y = A/\sqrt{2}$. Is this enough information to determine the value of ϕ? In addition, if you are told that a particle at $x = 0$ is moving toward $y = 0$ at $t = 0$, what is the value of ϕ? (d) Explain in general what you must know about the wave's behavior at a given instant to determine the value of ϕ.

15.69 ••• A sinusoidal transverse wave travels on a string. The string has length 8.00 m and mass 6.00 g. The wave speed is 30.0 m/s, and the wavelength is 0.200 m. (a) If the wave is to have an average power of 50.0 W, what must be the amplitude of the wave? (b) For this same string, if the amplitude and wavelength are the same as in part (a), what is the average power for the wave if the tension is increased such that the wave speed is doubled?

15.70 ••• **CALC** **Energy in a Triangular Pulse.** A triangular wave pulse on a taut string travels in the positive x-direction with speed v. The tension in the string is F, and the linear mass density of the string is μ. At $t = 0$, the shape of the pulse is given by

$$y(x, 0) = \begin{cases} 0 & \text{if } x < -L \\ h(L + x)/L & \text{for } -L < x < 0 \\ h(L - x)/L & \text{for } 0 < x < L \\ 0 & \text{for } x > L \end{cases}$$

(a) Draw the pulse at $t = 0$. (b) Determine the wave function $y(x, t)$ at all times t. (c) Find the instantaneous power in the wave. Show that the power is zero except for $-L < (x - vt) < L$ and that in this interval the power is constant. Find the value of this constant power.

15.71 ••• **CALC** **Instantaneous Power in a Wave.** (a) Graph $y(x, t)$ as given by Eq. (15.7) as a function of x for a given time t (say, $t = 0$). On the same axes, make a graph of the instantaneous power $P(x, t)$ as given by Eq. (15.23). (b) Explain the connection between the slope of the graph of $y(x, t)$ versus x and the value of $P(x, t)$. In particular, explain what is happening at points where $P = 0$, where there is no instantaneous energy transfer. (c) The quantity $P(x, t)$ always has the same sign. What does this imply about the direction of energy flow? (d) Consider a wave moving in the $-x$-direction, for which $y(x, t) = A\cos(kx + \omega t)$. Calculate $P(x, t)$ for this wave, and make a graph of $y(x, t)$ and $P(x, t)$ as functions of x for a given time t (say, $t = 0$). What differences arise from reversing the direction of the wave?

15.72 •• A vibrating string 50.0 cm long is under a tension of 1.00 N. The results from five successive stroboscopic pictures are shown in Fig. P15.72. The strobe rate is set at 5000 flashes per minute, and observations reveal that the maximum displacement occurred at flashes 1 and 5 with no other maxima in between. (a) Find the period, frequency, and wavelength for the traveling waves on this string. (b) In what normal mode (harmonic) is the string vibrating? (c) What is the speed of the traveling waves on the string? (d) How fast is point P moving when the string is in (i) position 1 and (ii) position 3? (e) What is the mass of this string? (See Section 15.3.)

Figure **P15.72**

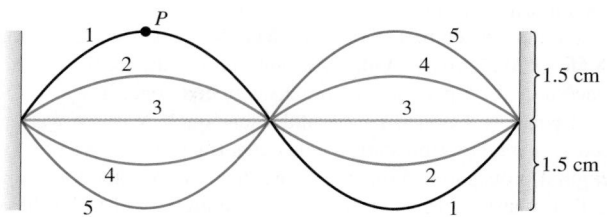

15.73 • **Clothesline Nodes.** Cousin Throckmorton is once again playing with the clothesline in Example 15.2 (Section 15.3). One end of the clothesline is attached to a vertical post. Throcky holds the other end loosely in his hand, so that the speed of waves on the clothesline is a relatively slow 0.720 m/s. He finds several frequencies at which he can oscillate his end of the clothesline so that a light clothespin 45.0 cm from the post doesn't move. What are these frequencies?

15.74 ••• **CALC** A guitar string is vibrating in its fundamental mode, with nodes at each end. The length of the segment of the string that is free to vibrate is 0.386 m. The maximum transverse acceleration of a point at the middle of the segment is 8.40×10^3 m/s^2 and the maximum transverse velocity is 3.80 m/s. (a) What is the amplitude of this standing wave? (b) What is the wave speed for the transverse traveling waves on this string?

15.75 •• **CALC** A string that lies along the $+x$-axis has a free end at $x = 0$. (a) By using steps similar to those used to derive Eq. (15.28), show that an incident traveling wave $y_1(x, t) = A\cos(kx + \omega t)$ gives rise to a standing wave $y(x, t) = 2A\cos\omega t\cos kx$. (b) Show that the standing wave has an antinode at its free end ($x = 0$). (c) Find the maximum displacement, maximum speed, and maximum acceleration of the free end of the string.

15.76 •• A string with both ends held fixed is vibrating in its third harmonic. The waves have a speed of 192 m/s and a frequency of 240 Hz. The amplitude of the standing wave at an antinode is 0.400 cm. (a) Calculate the amplitude at points on the string a distance of (i) 40.0 cm; (ii) 20.0 cm; and (iii) 10.0 cm from the left end of the string. (b) At each point in part (a), how much time does it take the string to go from its largest upward displacement to its largest downward displacement? (c) Calculate the maximum transverse velocity and the maximum transverse acceleration of the string at each of the points in part (a).

15.77 ••• A uniform cylindrical steel wire, 55.0 cm long and 1.14 mm in diameter, is fixed at both ends. To what tension must it be adjusted so that, when vibrating in its first overtone, it produces the note D-sharp of frequency 311 Hz? Assume that it stretches an insignificant amount. (*Hint:* See Table 12.1.)

15.78 • **Holding Up Under Stress.** A string or rope will break apart if it is placed under too much tensile stress [Eq. (11.8)]. Thicker ropes can withstand more tension without breaking because the thicker the rope, the greater the cross-sectional area and the smaller the stress. One type of steel has density 7800 kg/m^3 and will break if the tensile stress exceeds 7.0×10^8 N/m^2. You want to make a guitar string from 4.0 g of this type of steel. In use, the guitar string must be able to withstand a tension of 900 N without breaking. Your job is the following: (a) Determine the maximum length and minimum radius the string can have. (b) Determine the highest possible fundamental frequency of standing waves on this string, if the entire length of the string is free to vibrate.

15.79 ••• **Combining Standing Waves.** A guitar string of length L is plucked in such a way that the total wave produced is the sum of the fundamental and the second harmonic. That is, the standing wave is given by

$$y(x, t) = y_1(x, t) + y_2(x, t)$$

where

$$y_1(x, t) = C \sin \omega_1 t \sin k_1 x$$
$$y_2(x, t) = C \sin \omega_2 t \sin k_2 x$$

with $\omega_1 = v k_1$ and $\omega_2 = v k_2$. (a) At what values of x are the nodes of y_1? (b) At what values of x are the nodes of y_2? (c) Graph the total wave at $t = 0$, $t = \frac{1}{8} f_1$, $t = \frac{1}{4} f_1$, $t = \frac{3}{8} f_1$, and $t = \frac{1}{2} f_1$. (d) Does the sum of the two standing waves y_1 and y_2 produce a standing wave? Explain.

15.80 •• **CP** When a massive aluminum sculpture is hung from a steel wire, the fundamental frequency for transverse standing waves on the wire is 250.0 Hz. The sculpture (but not the wire) is then completely submerged in water. (a) What is the new fundamental frequency? (*Hint:* See Table 12.1.) (b) Why is it a good approximation to treat the wire as being fixed at both ends?

15.81 ••• **CP** A large rock that weighs 164.0 N is suspended from the lower end of a thin wire that is 3.00 m long. The density of the rock is 3200 kg/m^3. The mass of the wire is small enough that its effect on the tension in the wire can be neglected. The upper end of the wire is held fixed. When the rock is in air, the fundamental frequency for transverse standing waves on the wire is 42.0 Hz. When the rock is totally submerged in a liquid, with the top of the rock just below the surface, the fundamental frequency for the wire is 28.0 Hz. What is the density of the liquid?

15.82 •• **Tuning an Instrument.** A musician tunes the C-string of her instrument to a fundamental frequency of 65.4 Hz. The vibrating portion of the string is 0.600 m long and has a mass of 14.4 g. (a) With what tension must the musician stretch it? (b) What percent increase in tension is needed to increase the frequency from 65.4 Hz to 73.4 Hz, corresponding to a rise in pitch from C to D?

15.83 ••• One type of steel has a density of 7.8×10^3 kg/m^3 and a breaking stress of 7.0×10^8 N/m^2. A cylindrical guitar string is to be made of 4.00 g of this steel. (a) What are the length and radius of the longest and thinnest string that can be placed under a tension of 900 N without breaking? (b) What is the highest fundamental frequency that this string could have?

CHALLENGE PROBLEMS

15.84 ••• **CP CALC** A deep-sea diver is suspended beneath the surface of Loch Ness by a 100-m-long cable that is attached to a boat on the surface (Fig. P15.84). The diver and his suit have a total mass of 120 kg and a volume of 0.0800 m^3. The cable has a diameter of 2.00 cm and a linear mass density of $\mu = 1.10$ kg/m. The diver thinks he sees something moving in the murky depths and jerks the end of the cable back and forth to send transverse waves up the cable as a signal to his compan-

Figure **P15.84**

ions in the boat. (a) What is the tension in the cable at its lower end, where it is attached to the diver? Do not forget to include the buoyant force that the water (density 1000 kg/m^3) exerts on him. (b) Calculate the tension in the cable a distance x above the diver. The buoyant force on the cable must be included in your calculation. (c) The speed of transverse waves on the cable is given by $v = \sqrt{F/\mu}$ (Eq. 15.13). The speed therefore varies along the cable, since the tension is not constant. (This expression neglects the damping force that the water exerts on the moving cable.) Integrate to find the time required for the first signal to reach the surface.

15.85 ••• **CALC** (a) Show that for a wave on a string, the kinetic energy *per unit length of string* is

$$u_k(x, t) = \frac{1}{2} \mu v_y^2(x, t) = \frac{1}{2} \mu \left(\frac{\partial y(x, t)}{\partial t} \right)^2$$

where μ is the mass per unit length. (b) Calculate $u_k(x, t)$ for a sinusoidal wave given by Eq. (15.7). (c) There is also elastic potential energy in the string, associated with the work required to deform and stretch the string. Consider a short segment of string at position x that has unstretched length Δx, as in Fig. 15.13. Ignoring the (small) curvature of the segment, its slope is $\partial y(x, t)/\partial x$. Assume that the displacement of the string from equilibrium is small, so that $\partial y/\partial x$ has a magnitude much less than unity. Show that the stretched length of the segment is approximately

$$\Delta x \left[1 + \frac{1}{2} \left(\frac{\partial y(x, t)}{\partial x} \right)^2 \right]$$

(*Hint:* Use the relationship $\sqrt{1 + u} \approx 1 + \frac{1}{2} u$, valid for $|u| \ll 1$.) (d) The potential energy stored in the segment equals the work done by the string tension F (which acts along the string) to stretch the segment from its unstretched length Δx to the length calculated in part (c). Calculate this work and show that the potential energy *per unit length of string* is

$$u_p(x, t) = \frac{1}{2} F \left(\frac{\partial y(x, t)}{\partial x} \right)^2$$

(e) Calculate $u_p(x, t)$ for a sinusoidal wave given by Eq. (15.7). (f) Show that $u_k(x, t), = u_p(x, t)$, for all x and t. (g) Show $y(x, t)$, $u_k(x, t)$, and $u_p(x, t)$ as functions of x for $t = 0$ in one graph with all three functions on the same axes. Explain why u_k and u_p are maximum where y is zero, and vice versa. (h) Show that the instantaneous power in the wave, given by Eq. (15.22), is equal to the total energy per unit length multiplied by the wave speed v. Explain why this result is reasonable.

Answers

Chapter Opening Question ?

The power of a mechanical wave depends on its frequency and amplitude [see Eq. (15.25)].

Test Your Understanding Questions

15.1 Answer: (i) The "wave" travels horizontally from one spectator to the next along each row of the stadium, but the displacement of each spectator is vertically upward. Since the displacement is perpendicular to the direction in which the wave travels, the wave is transverse.

15.2 Answer: (iv) The speed of waves on a string, v, does not depend on the wavelength. We can rewrite the relationship $v = \lambda f$ as $f = v/\lambda$, which tells us that if the wavelength λ doubles, the frequency f becomes one-half as great.

15.3 Answers: (a) $\frac{2}{8}T$, **(b)** $\frac{4}{8}T$, **(c)** $\frac{5}{8}T$ Since the wave is sinusoidal, each point on the string oscillates in simple harmonic motion (SHM). Hence we can apply all of the ideas from Chapter 14 about SHM to the wave depicted in Fig. 15.8. (a) A particle in SHM has its maximum speed when it is passing through the equilibrium position ($y = 0$ in Fig. 15.8). The particle at point A is moving upward through this position at $t = \frac{2}{8}T$. (b) In vertical SHM the greatest *upward* acceleration occurs when a particle is at its maximum *downward* displacement. This occurs for the particle at point B at $t = \frac{4}{8}T$. (c) A particle in vertical SHM has a *downward* acceleration when its displacement is *upward*. The particle at C has an upward displacement and is moving downward at $t = \frac{5}{8}T$.

15.4 Answer: (ii) The relationship $v = \sqrt{F/\mu}$ [Eq. (15.13)] says that the wave speed is greatest on the string with the smallest linear mass density. This is the thinnest string, which has the smallest amount of mass m and hence the smallest linear mass density $\mu = m/L$ (all strings are the same length).

15.5 Answer: (iii), (iv), (ii), (i) Equation (15.25) says that the average power in a sinusoidal wave on a string is $P_{av} = \frac{1}{2}\sqrt{\mu F}\omega^2 A^2$. All four strings are identical, so all have the same mass, the same length, and the same linear mass density μ. The frequency f is the same for each wave, as is the angular frequency $\omega = 2\pi f$. Hence the average wave power for each string is proportional to the square root of the string tension F and the square of the amplitude A. Compared to string (i), the average power in each string is (ii) $\sqrt{4} = 2$ times greater; (iii) $4^2 = 16$ times greater; and (iv) $\sqrt{2}(2)^2 = 4\sqrt{2}$ times greater.

15.6 Answer:

15.7 Answers: yes, yes Doubling the frequency makes the wavelength half as large. Hence the spacing between nodes (equal to $\lambda/2$) is also half as large. There are nodes at all of the previous positions, but there is also a new node between every pair of old nodes.

15.8 Answers: $n = 1, 3, 5, \ldots$ When you touch the string at its center, you are demanding that there be a node at the center. Hence only standing waves with a node at $x = L/2$ are allowed. From Figure 15.26 you can see that the normal modes $n = 1, 3, 5, \ldots$ cannot be present.

Bridging Problem

Answers: (a) $F(r) = \dfrac{m\omega^2}{2L}(L^2 - r^2)$

(b) $v(r) = \omega\sqrt{\dfrac{L^2 - r^2}{2}}$

(c) $\dfrac{\pi}{\omega\sqrt{2}}$

SOUND AND HEARING

? Most people like to listen to music, but hardly anyone likes to listen to noise. What is the physical difference between musical sound and noise?

LEARNING GOALS

By studying this chapter, you will learn:

- How to describe a sound wave in terms of either particle displacements or pressure fluctuations.

- How to calculate the speed of sound waves in different materials.

- How to calculate the intensity of a sound wave.

- What determines the particular frequencies of sound produced by an organ or a flute.

- How resonance occurs in musical instruments.

- What happens when sound waves from different sources overlap.

- How to describe what happens when two sound waves of slightly different frequencies are combined.

- Why the pitch of a siren changes as it moves past you.

Of all the mechanical waves that occur in nature, the most important in our everyday lives are longitudinal waves in a medium—usually air—called *sound* waves. The reason is that the human ear is tremendously sensitive and can detect sound waves even of very low intensity. Besides their use in spoken communication, our ears allow us to pick up a myriad of cues about our environment, from the welcome sound of a meal being prepared to the warning sound of an approaching car. The ability to hear an unseen nocturnal predator was essential to the survival of our ancestors, so it is no exaggeration to say that we humans owe our existence to our highly evolved sense of hearing.

Up to this point we have described mechanical waves primarily in terms of displacement; however, a description of sound waves in terms of *pressure* fluctuations is often more appropriate, largely because the ear is primarily sensitive to changes in pressure. We'll study the relationships among displacement, pressure fluctuation, and intensity and the connections between these quantities and human sound perception.

When a source of sound or a listener moves through the air, the listener may hear a frequency different from the one emitted by the source. This is the Doppler effect, which has important applications in medicine and technology.

16.1 Sound Waves

The most general definition of **sound** is a longitudinal wave in a medium. Our main concern in this chapter is with sound waves in air, but sound can travel through any gas, liquid, or solid. You may be all too familiar with the propagation of sound through a solid if your neighbor's stereo speakers are right next to your wall.

The simplest sound waves are sinusoidal waves, which have definite frequency, amplitude, and wavelength. The human ear is sensitive to waves in the frequency range from about 20 to 20,000 Hz, called the **audible range,** but we also use the

16.1 A sinusoidal longitudinal wave traveling to the right in a fluid. (Compare to Fig. 15.7.)

Longitudinal waves are shown at intervals of $\frac{1}{8}T$ for one period T.

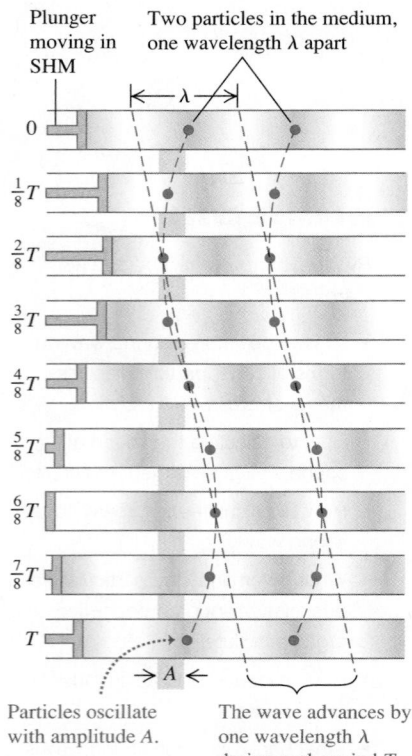

Particles oscillate with amplitude A.

The wave advances by one wavelength λ during each period T.

16.2 As a sound wave propagates along the x-axis, the left and right ends undergo different displacements y_1 and y_2.

Undisturbed cylinder of fluid has cross-sectional area S, length Δx, and volume $S\Delta x$.

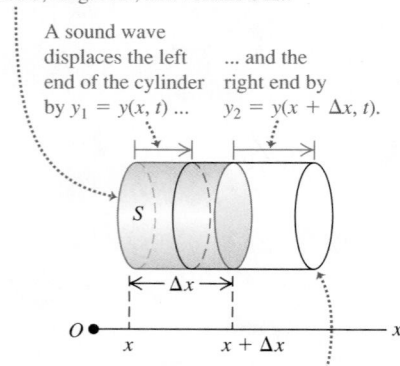

The change in volume of the disturbed cylinder of fluid is $S(y_2 - y_1)$.

term "sound" for similar waves with frequencies above (**ultrasonic**) and below (**infrasonic**) the range of human hearing.

Sound waves usually travel out in all directions from the source of sound, with an amplitude that depends on the direction and distance from the source. We'll return to this point in the next section. For now, we concentrate on the idealized case of a sound wave that propagates in the positive x-direction only. As we discussed in Section 15.3, such a wave is described by a wave function $y(x, t)$, which gives the instantaneous displacement y of a particle in the medium at position x at time t. If the wave is sinusoidal, we can express it using Eq. (15.7):

$$y(x, t) = A\cos(kx - \omega t) \quad \begin{array}{l}\text{(sound wave propagating}\\ \text{in the } +x\text{-direction)}\end{array} \quad (16.1)$$

Remember that in a longitudinal wave the displacements are *parallel* to the direction of travel of the wave, so distances x and y are measured parallel to each other, not perpendicular as in a transverse wave. The amplitude A is the maximum displacement of a particle in the medium from its equilibrium position (Fig. 16.1). Hence A is also called the **displacement amplitude.**

Sound Waves As Pressure Fluctuations

Sound waves may also be described in terms of variations of *pressure* at various points. In a sinusoidal sound wave in air, the pressure fluctuates above and below atmospheric pressure p_a in a sinusoidal variation with the same frequency as the motions of the air particles. The human ear operates by sensing such pressure variations. A sound wave entering the ear canal exerts a fluctuating pressure on one side of the eardrum; the air on the other side of the eardrum, vented to the outside by the Eustachian tube, is at atmospheric pressure. The pressure difference on the two sides of the eardrum sets it into motion. Microphones and similar devices also usually sense pressure differences, not displacements, so it is very useful to develop a relationship between these two descriptions.

Let $p(x, t)$ be the instantaneous pressure fluctuation in a sound wave at any point x at time t. That is, $p(x, t)$ is the amount by which the pressure *differs* from normal atmospheric pressure p_a. Think of $p(x, t)$ as the *gauge pressure* defined in Section 12.2; it can be either positive or negative. The *absolute* pressure at a point is then $p_a + p(x, t)$.

To see the connection between the pressure fluctuation $p(x, t)$ and the displacement $y(x, t)$ in a sound wave propagating in the $+x$-direction, consider an imaginary cylinder of a wave medium (gas, liquid, or solid) with cross-sectional area S and axis along the direction of propagation (Fig. 16.2). When no sound wave is present, the cylinder has length Δx and volume $V = S\Delta x$, as shown by the shaded volume in Fig. 16.2. When a wave is present, at time t the end of the cylinder that is initially at x is displaced by $y_1 = y(x, t)$, and the end that is initially at $x + \Delta x$ is displaced by $y_2 = y(x + \Delta x, t)$; this is shown by the red lines. If $y_2 > y_1$, as shown in Fig. 16.2, the cylinder's volume has increased, which causes a decrease in pressure. If $y_2 < y_1$, the cylinder's volume has decreased and the pressure has increased. If $y_2 = y_1$, the cylinder is simply shifted to the left or right; there is no volume change and no pressure fluctuation. The pressure fluctuation depends on the *difference* between the displacements at neighboring points in the medium.

Quantitatively, the change in volume ΔV of the cylinder is

$$\Delta V = S(y_2 - y_1) = S[y(x + \Delta x, t) - y(x, t)]$$

In the limit as $\Delta x \to 0$, the fractional change in volume dV/V (volume change divided by original volume) is

$$\frac{dV}{V} = \lim_{\Delta x \to 0} \frac{S[y(x + \Delta x, t) - y(x, t)]}{S\Delta x} = \frac{\partial y(x, t)}{\partial x} \quad (16.2)$$

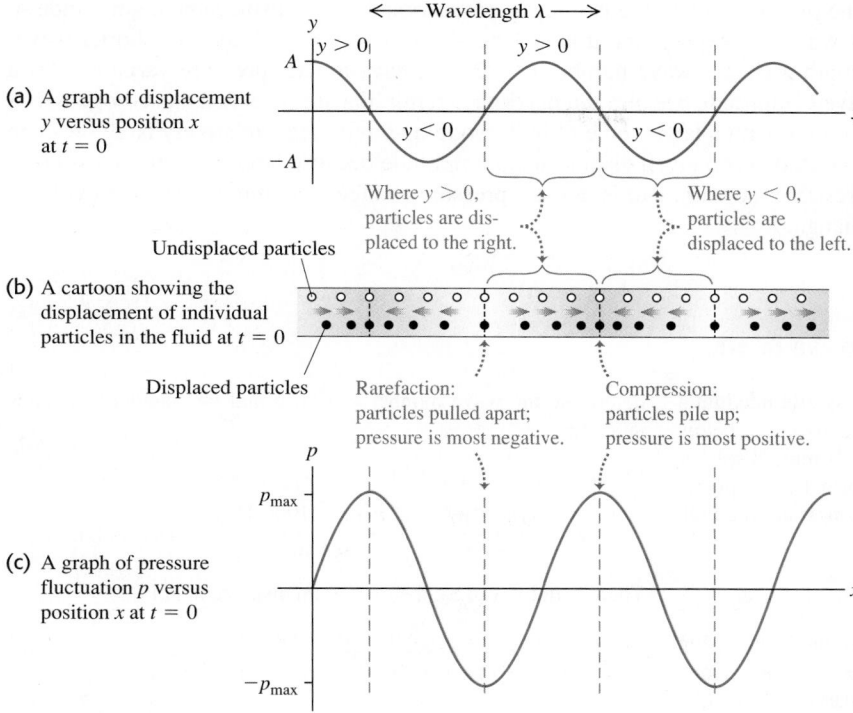

(a) A graph of displacement y versus position x at $t = 0$

(b) A cartoon showing the displacement of individual particles in the fluid at $t = 0$

(c) A graph of pressure fluctuation p versus position x at $t = 0$

Where $y > 0$, particles are displaced to the right.

Where $y < 0$, particles are displaced to the left.

Rarefaction: particles pulled apart; pressure is most negative.

Compression: particles pile up; pressure is most positive.

The fractional volume change is related to the pressure fluctuation by the bulk modulus B, which by definition [Eq. (11.13)] is $B = -p(x, t)/(dV/V)$ (see Section 11.4). Solving for $p(x, t)$, we have

$$p(x, t) = -B \frac{\partial y(x, t)}{\partial x} \qquad (16.3)$$

The negative sign arises because when $\partial y(x, t)/\partial x$ is positive, the displacement is greater at $x + \Delta x$ than at x, corresponding to an increase in volume and a *decrease* in pressure.

When we evaluate $\partial y(x, t)/\partial x$ for the sinusoidal wave of Eq. (16.1), we find

$$p(x, t) = BkA \sin(kx - \omega t) \qquad (16.4)$$

Figure 16.3 shows $y(x, t)$ and $p(x, t)$ for a sinusoidal sound wave at $t = 0$. It also shows how individual particles of the wave are displaced at this time. While $y(x, t)$ and $p(x, t)$ describe the same wave, these two functions are one-quarter cycle out of phase: At any time, the displacement is greatest where the pressure fluctuation is zero, and vice versa. In particular, note that the compressions (points of greatest pressure and density) and rarefactions (points of lowest pressure and density) are points of *zero* displacement.

CAUTION **Graphs of a sound wave** Keep in mind that the graphs in Fig. 16.3 show the wave at only *one* instant of time. Because the wave is propagating in the $+x$-direction, as time goes by the wave patterns in the functions $y(x, t)$ and $p(x, t)$ move to the right at the wave speed $v = \omega/k$. Hence the positions of the compressions and rarefactions also move to the right at this same speed. The particles, by contrast, simply oscillate back and forth in simple harmonic motion as shown in Fig. 16.1. ▌

Equation (16.4) shows that the quantity BkA represents the maximum pressure fluctuation. We call this the **pressure amplitude,** denoted by p_{max}:

$$p_{max} = BkA \qquad \text{(sinusoidal sound wave)} \qquad (16.5)$$

The pressure amplitude is directly proportional to the displacement amplitude A, as we might expect, and it also depends on wavelength. Waves of shorter wavelength λ (larger wave number $k = 2\pi/\lambda$) have greater pressure variations for a given amplitude because the maxima and minima are squeezed closer together. A medium with a large value of bulk modulus B requires a relatively large pressure amplitude for a given displacement amplitude because large B means a less compressible medium; that is, greater pressure change is required for a given volume change.

Example 16.1 Amplitude of a sound wave in air

In a sinusoidal sound wave of moderate loudness, the maximum pressure variations are about 3.0×10^{-2} Pa above and below atmospheric pressure. Find the corresponding maximum displacement if the frequency is 1000 Hz. In air at normal atmospheric pressure and density, the speed of sound is 344 m/s and the bulk modulus is 1.42×10^5 Pa.

SOLUTION

IDENTIFY and SET UP: This problem involves the relationship between two ways of describing a sound wave: in terms of displacement and in terms of pressure. The target variable is the displacement amplitude A. We are given the pressure amplitude p_{max}, wave speed v, frequency f, and bulk modulus B. Equation (16.5) relates the target variable A to p_{max}. We use $\omega = vk$ [Eq. (15.6)] to determine the wave number k from v and the angular frequency $\omega = 2\pi f$.

EXECUTE: From Eq. (15.6),

$$k = \frac{\omega}{v} = \frac{2\pi f}{v} = \frac{(2\pi \text{ rad})(1000 \text{ Hz})}{344 \text{ m/s}} = 18.3 \text{ rad/m}$$

Then from Eq. (16.5), the maximum displacement is

$$A = \frac{p_{max}}{Bk} = \frac{3.0 \times 10^{-2} \text{ Pa}}{(1.42 \times 10^5 \text{ Pa})(18.3 \text{ rad/m})} = 1.2 \times 10^{-8} \text{ m}$$

EVALUATE: This displacement amplitude is only about $\frac{1}{100}$ the size of a human cell. The ear actually senses pressure fluctuations; it detects these minuscule displacements only indirectly.

Example 16.2 Amplitude of a sound wave in the inner ear

A sound wave that enters the human ear sets the eardrum into oscillation, which in turn causes oscillation of the *ossicles,* a chain of three tiny bones in the middle ear (Fig. 16.4). The ossicles transmit this oscillation to the fluid (mostly water) in the inner ear; there the fluid motion disturbs hair cells that send nerve impulses to the brain with information about the sound. The area of the moving part of the eardrum is about 43 mm², and that of the stapes (the smallest of the ossicles) where it connects to the inner ear is about 3.2 mm². For the sound in Example 16.1, determine (a) the pressure amplitude and (b) the displacement amplitude of the wave in the fluid of the inner ear, in which the speed of sound is 1500 m/s.

SOLUTION

IDENTIFY and SET UP: Although the sound wave here travels in liquid rather than air, the same principles and relationships among the properties of the wave apply. We can neglect the mass of the tiny ossicles (about 58 mg = 5.8×10^{-5} kg), so the force they exert on the inner-ear fluid is the same as that exerted on the eardrum and ossicles by the incident sound wave. (In Chapters 4 and 5 we used the same idea to say that the tension is the same at either end of a massless rope.) Hence the pressure amplitude in the inner ear, $p_{max \text{ (inner ear)}}$, is greater than in the outside air, $p_{max \text{ (air)}}$, because the same force is exerted on a smaller area (the area of the stapes versus the area of the eardrum). Given $p_{max \text{ (inner ear)}}$, we find the displacement amplitude $A_{inner ear}$ using Eq. (16.5).

16.4 The anatomy of the human ear. The middle ear is the size of a small marble; the ossicles (incus, malleus, and stapes) are the smallest bones in the human body.

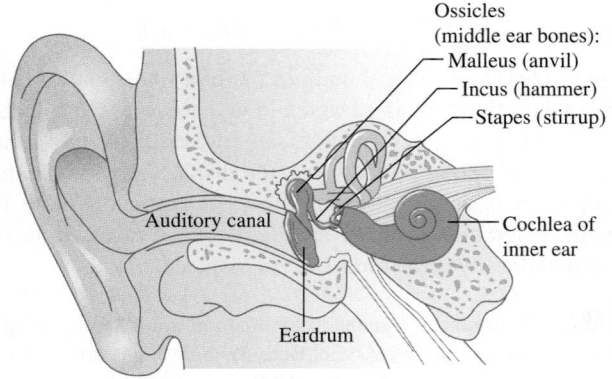

EXECUTE: (a) From the area of the eardrum and the pressure amplitude in air found in Example 16.1, the maximum force exerted by the sound wave in air on the eardrum is $F_{max} = p_{max \text{ (air)}} S_{eardrum}$. Then

$$p_{max \text{ (inner ear)}} = \frac{F_{max}}{S_{stapes}} = p_{max \text{ (air)}} \frac{S_{eardrum}}{S_{stapes}}$$

$$= (3.0 \times 10^{-2} \text{ Pa}) \frac{43 \text{ mm}^2}{3.2 \text{ mm}^2} = 0.40 \text{ Pa}$$

(b) To find the maximum displacement $A_{\text{inner ear}}$, we use $A = p_{\text{max}}/Bk$ as in Example 16.1. The inner-ear fluid is mostly water, which has a much greater bulk modulus B than air. From Table 11.2 the compressibility of water (unfortunately also called k) is 45.8×10^{-11} Pa^{-1}, so $B_{\text{fluid}} = 1/(45.8 \times 10^{-11}$ Pa$^{-1}) = 2.18 \times 10^9$ Pa.

The wave in the inner ear has the same angular frequency ω as the wave in the air because the air, eardrum, ossicles, and inner-ear fluid all oscillate together (see Example 15.8 in Section 15.8). But because the wave speed v is greater in the inner ear than in the air (1500 m/s versus 344 m/s), the wave number $k = \omega/v$ is smaller. Using the value of ω from Example 16.1,

$$k_{\text{inner ear}} = \frac{\omega}{v_{\text{inner ear}}} = \frac{(2\pi \text{ rad})(1000 \text{ Hz})}{1500 \text{ m/s}} = 4.2 \text{ rad/m}$$

Putting everything together, we have

$$A_{\text{inner ear}} = \frac{p_{\text{max (inner ear)}}}{B_{\text{fluid}} k_{\text{inner ear}}} = \frac{0.40 \text{ Pa}}{(2.18 \times 10^9 \text{ Pa})(4.2 \text{ rad/m})}$$
$$= 4.4 \times 10^{-11} \text{ m}$$

EVALUATE: In part (a) we see that the ossicles increase the pressure amplitude by a factor of $(43 \text{ mm}^2)/(3.2 \text{ mm}^2) = 13$. This amplification helps give the human ear its great sensitivity.

The displacement amplitude in the inner ear is even smaller than in the air. But *pressure* variations within the inner-ear fluid are what set the hair cells into motion, so what matters is that the pressure amplitude is larger in the inner ear than in the air.

Perception of Sound Waves

The physical characteristics of a sound wave are directly related to the perception of that sound by a listener. For a given frequency, the greater the pressure amplitude of a sinusoidal sound wave, the greater the perceived **loudness.** The relationship between pressure amplitude and loudness is not a simple one, and it varies from one person to another. One important factor is that the ear is not equally sensitive to all frequencies in the audible range. A sound at one frequency may seem louder than one of equal pressure amplitude at a different frequency. At 1000 Hz the minimum pressure amplitude that can be perceived with normal hearing is about 3×10^{-5} Pa; to produce the same loudness at 200 Hz or 15,000 Hz requires about 3×10^{-4} Pa. Perceived loudness also depends on the health of the ear. A loss of sensitivity at the high-frequency end usually happens naturally with age but can be further aggravated by excessive noise levels.

The frequency of a sound wave is the primary factor in determining the **pitch** of a sound, the quality that lets us classify the sound as "high" or "low." The higher the frequency of a sound (within the audible range), the higher the pitch that a listener will perceive. Pressure amplitude also plays a role in determining pitch. When a listener compares two sinusoidal sound waves with the same frequency but different pressure amplitudes, the one with the greater pressure amplitude is usually perceived as louder but also as slightly lower in pitch.

Musical sounds have wave functions that are more complicated than a simple sine function. The pressure fluctuation in the sound wave produced by a clarinet is shown in Fig. 16.5a. The pattern is so complex because the column of air in a wind instrument like a clarinet vibrates at a fundamental frequency and at many harmonics at the same time. (In Section 15.8, we described this same behavior for a string that has been plucked, bowed, or struck. We'll examine the physics of wind instruments in Section 16.5.) The sound wave produced in the surrounding air has a similar amount of each harmonic—that is, a similar *harmonic content*. Figure 16.5b shows the harmonic content of the sound of a clarinet. The mathematical process of translating a pressure–time graph like Fig. 16.5a into a graph of harmonic content like Fig. 16.5b is called *Fourier analysis*.

Two tones produced by different instruments might have the same fundamental frequency (and thus the same pitch) but sound different because of different harmonic content. The difference in sound is called *tone color, quality,* or **timbre** and is often described in subjective terms such as reedy, golden, round, mellow, and tinny. A tone that is rich in harmonics, like the clarinet tone in Figs. 16.5a and 16.5b, usually sounds thin and "stringy" or "reedy," while a tone containing mostly a fundamental, like the alto recorder tone in Figs. 16.5c and 16.5d, is more mellow and flutelike. The same principle applies to the human voice, which is another example of a wind instrument; the vowels "a" and "e" sound different because of differences in harmonic content.

Application Hearing Loss from Amplified Sound
Due to exposure to highly amplified music, many young popular musicians have suffered permanent ear damage and have hearing typical of persons 65 years of age. Headphones for personal music players used at high volume pose similar threats to hearing. Be careful!

16.5 Different representations of the sound of (a), (b) a clarinet and (c), (d) an alto recorder. (Graphs adapted from R.E. Berg and D.G. Stork, *The Physics of Sound,* Prentice-Hall, 1982.)

Pressure fluctuation versus time for a clarinet with fundamental frequency $f_1 = 233$ Hz

$T = 4.29$ ms

(a)

(b) Harmonic content of the sound in (a)

O $5f_1$ $10f_1$ $20f_1$ $30f_1$ $40f_1$

Pressure fluctuation versus time for an alto recorder with fundamental frequency $f_1 = 523$ Hz

$T = 1.91$ ms

(c)

(d) Harmonic content of the sound in (c)

O $5f_1$ $10f_1$ $20f_1$ $30f_1$ $40f_1$

Another factor in determining tone quality is the behavior at the beginning (*attack*) and end (*decay*) of a tone. A piano tone begins with a thump and then dies away gradually. A harpsichord tone, in addition to having different harmonic content, begins much more quickly with a click, and the higher harmonics begin before the lower ones. When the key is released, the sound also dies away much more rapidly with a harpsichord than with a piano. Similar effects are present in other musical instruments. With wind and string instruments the player has considerable control over the attack and decay of the tone, and these characteristics help to define the unique characteristics of each instrument.

Unlike the tones made by musical instruments or the vowels in human speech, **noise** is a combination of *all* frequencies, not just frequencies that are integer multiples of a fundamental frequency. (An extreme case is "white noise," which contains equal amounts of all frequencies across the audible range.) Examples include the sound of the wind and the hissing sound you make in saying the consonant "s."

Test Your Understanding of Section 16.1 You use an electronic signal generator to produce a sinusoidal sound wave in air. You then increase the frequency of the wave from 100 Hz to 400 Hz while keeping the pressure amplitude constant. What effect does this have on the displacement amplitude of the sound wave? (i) It becomes four times greater; (ii) it becomes twice as great; (iii) it is unchanged; (iv) it becomes $\frac{1}{2}$ as great; (v) it becomes $\frac{1}{4}$ as great.

16.2 Speed of Sound Waves

We found in Section 15.4 that the speed v of a transverse wave on a string depends on the string tension F and the linear mass density μ: $v = \sqrt{F/\mu}$. What, we may ask, is the corresponding expression for the speed of sound waves in a gas or liquid? On what properties of the medium does the speed depend?

We can make an educated guess about these questions by remembering a claim that we made in Section 15.4: For mechanical waves in general, the expression for the wave speed is of the form

$$v = \sqrt{\frac{\text{Restoring force returning the system to equilibrium}}{\text{Inertia resisting the return to equilibrium}}}$$

A sound wave in a bulk fluid causes compressions and rarefactions of the fluid, so the restoring-force term in the above expression must be related to how easy or difficult it is to compress the fluid. This is precisely what the bulk modulus B of the medium tells us. According to Newton's second law, inertia is related to mass. The "massiveness" of a bulk fluid is described by its density, or mass per unit volume, ρ. (The corresponding quantity for a string is the mass per unit length, μ.) Hence we expect that the speed of sound waves should be of the form $v = \sqrt{B/\rho}$.

To check our guess, we'll derive the speed of sound waves in a fluid in a pipe. This is a situation of some importance, since all musical wind instruments are fundamentally pipes in which a longitudinal wave (sound) propagates in a fluid (air) (Fig. 16.6). Human speech works on the same principle; sound waves propagate in your vocal tract, which is basically an air-filled pipe connected to the lungs at one end (your larynx) and to the outside air at the other end (your mouth). The steps in our derivation are completely parallel to those we used in Section 15.4 to find the speed of transverse waves, so you'll find it useful to review that section.

Speed of Sound in a Fluid

Figure 16.7 shows a fluid (either liquid or gas) with density ρ in a pipe with cross-sectional area A. In the equilibrium state, the fluid is under a uniform pressure p. In Fig. 16.7a the fluid is at rest. We take the x-axis along the length of the pipe. This is also the direction in which we make a longitudinal wave propagate, so the displacement y is also measured along the pipe, just as in Section 16.1 (see Fig. 16.2).

At time $t = 0$ we start the piston at the left end moving toward the right with constant speed v_y. This initiates a wave motion that travels to the right along the length of the pipe, in which successive sections of fluid begin to move and become compressed at successively later times.

Figure 16.7b shows the fluid at time t. All portions of fluid to the left of point P are moving to the right with speed v_y, and all portions to the right of P are still at rest. The boundary between the moving and stationary portions travels to the right with a speed equal to the speed of propagation or wave speed v. At time t the piston has moved a distance v_yt, and the boundary has advanced a distance vt. As with a transverse disturbance in a string, we can compute the speed of propagation from the impulse–momentum theorem.

The quantity of fluid set in motion in time t is the amount that originally occupied a section of the cylinder with length vt, cross-sectional area A, and volume vtA. The mass of this fluid is ρvtA, and its longitudinal momentum (that is, momentum along the length of the pipe) is

$$\text{Longitudinal momentum} = (\rho vtA)v_y$$

Next we compute the increase of pressure, Δp, in the moving fluid. The original volume of the moving fluid, Avt, has decreased by an amount Av_yt. From the definition of the bulk modulus B, Eq. (11.13) in Section 11.5,

$$B = \frac{-\text{Pressure change}}{\text{Fractional volume change}} = \frac{-\Delta p}{-Av_yt/Avt}$$

$$\Delta p = B\frac{v_y}{v}$$

The pressure in the moving fluid is $p + \Delta p$ and the force exerted on it by the piston is $(p + \Delta p)A$. The net force on the moving fluid (see Fig. 16.7b) is ΔpA, and the longitudinal impulse is

$$\text{Longitudinal impulse} = \Delta pAt = B\frac{v_y}{v}At$$

Because the fluid was at rest at time $t = 0$, the change in momentum up to time t is equal to the momentum at that time. Applying the impulse–momentum theorem (see Section 8.1), we find

$$B\frac{v_y}{v}At = \rho vtAv_y \qquad (16.6)$$

When we solve this expression for v, we get

$$v = \sqrt{\frac{B}{\rho}} \qquad \text{(speed of a longitudinal wave in a fluid)} \qquad (16.7)$$

which agrees with our educated guess. Thus the speed of propagation of a longitudinal pulse in a fluid depends only on the bulk modulus B and the density ρ of the medium.

While we derived Eq. (16.7) for waves in a pipe, it also applies to longitudinal waves in a bulk fluid. Thus the speed of sound waves traveling in air or water is determined by this equation.

Speed of Sound in a Solid

When a longitudinal wave propagates in a *solid* rod or bar, the situation is somewhat different. The rod expands sideways slightly when it is compressed

16.6 When a wind instrument like this French horn is played, sound waves propagate through the air within the instrument's pipes. The properties of the sound that emerges from the large bell depend on the speed of these waves.

16.7 A sound wave propagating in a fluid confined to a tube. **(a)** Fluid in equilibrium. **(b)** A time t after the piston begins moving to the right at speed v_y, the fluid between the piston and point P is in motion. The speed of sound waves is v.

longitudinally, while a fluid in a pipe with constant cross section cannot move sideways. Using the same kind of reasoning that led us to Eq. (16.7), we can show that the speed of a longitudinal pulse in the rod is given by

$$v = \sqrt{\frac{Y}{\rho}} \quad \text{(speed of a longitudinal wave in a solid rod)} \quad (16.8)$$

where Y is Young's modulus, defined in Section 11.4.

CAUTION **Solid rods vs. bulk solids** Equation (16.8) applies only to a rod or bar whose sides are free to bulge and shrink a little as the wave travels. It does not apply to longitudinal waves in a *bulk* solid, since in these materials, sideways motion in any element of material is prevented by the surrounding material. The speed of longitudinal waves in a bulk solid depends on the density, the bulk modulus, and the *shear* modulus; a full discussion is beyond the scope of this book. ▌

As with the derivation for a transverse wave on a string, Eqs. (16.7) and (16.8) are valid for sinusoidal and other periodic waves, not just for the special case discussed here.

Table 16.1 lists the speed of sound in several bulk materials. Sound waves travel more slowly in lead than in aluminum or steel because lead has a lower bulk modulus and shear modulus and a higher density.

Table 16.1 Speed of Sound in Various Bulk Materials

Material	Speed of Sound (m/s)
Gases	
Air (20°C)	344
Helium (20°C)	999
Hydrogen (20°C)	1330
Liquids	
Liquid helium (4 K)	211
Mercury (20°C)	1451
Water (0°C)	1402
Water (20°C)	1482
Water (100°C)	1543
Solids	
Aluminum	6420
Lead	1960
Steel	5941

Example 16.3 Wavelength of sonar waves

A ship uses a sonar system (Fig. 16.8) to locate underwater objects. Find the speed of sound waves in water using Eq. (16.7), and find the wavelength of a 262-Hz wave.

SOLUTION

IDENTIFY and SET UP: Our target variables are the speed and wavelength of a sound wave in water. In Eq. (16.7), we use the density of water, $\rho = 1.00 \times 10^3 \text{ kg/m}^3$, and the bulk modulus of water, which we find from the compressibility (see Table 11.2). Given the speed and the frequency $f = 262$ Hz, we find the wavelength from $v = f\lambda$.

EXECUTE: In Example 16.2, we used Table 11.2 to find $B = 2.18 \times 10^9$ Pa. Then

$$v = \sqrt{\frac{B}{\rho}} = \sqrt{\frac{2.18 \times 10^9 \text{ Pa}}{1.00 \times 10^3 \text{ kg/m}^3}} = 1480 \text{ m/s}$$

and

$$\lambda = \frac{v}{f} = \frac{1480 \text{ m/s}}{262 \text{ s}^{-1}} = 5.65 \text{ m}$$

EVALUATE: The calculated value of v agrees well with the value in Table 16.1. Water is denser than air (ρ is larger) but is also much

16.8 A sonar system uses underwater sound waves to detect and locate submerged objects.

more incompressible (B is much larger), and so the speed $v = \sqrt{B/\rho}$ is greater than the 344-m/s speed of sound in air at ordinary temperatures. The relationship $\lambda = v/f$ then says that a sound wave in water must have a longer wavelength than a wave of the same frequency in air. Indeed, we found in Example 15.1 (Section 15.2) that a 262-Hz sound wave in air has a wavelength of only 1.31 m.

Dolphins emit high-frequency sound waves (typically 100,000 Hz) and use the echoes for guidance and for hunting. The corresponding wavelength in water is 1.48 cm. With this high-frequency "sonar" system they can sense objects that are roughly as small as the wavelength (but not much smaller). *Ultrasonic imaging* is a medical technique that uses exactly the same physical principle; sound waves of very high frequency and very short wavelength, called *ultrasound*, are scanned over the human body, and the "echoes" from interior organs are used to create an image. With ultrasound of frequency 5 MHz = 5 × 10⁶ Hz, the wavelength in water (the primary constituent of the body) is 0.3 mm, and features as small as this can be discerned in the image. Ultrasound is used for the study of heart-valve action, detection of tumors, and prenatal examinations (Fig. 16.9). Ultrasound is more sensitive than x rays in distinguishing various kinds of tissues and does not have the radiation hazards associated with x rays.

Speed of Sound in a Gas

Most of the sound waves that we encounter on a daily basis propagate in air. To use Eq. (16.7) to find the speed of sound waves in air, we must keep in mind that the bulk modulus of a gas depends on the pressure of the gas: The greater the pressure applied to a gas to compress it, the more it resists further compression and hence the greater the bulk modulus. (That's why specific values of the bulk modulus for gases are not given in Table 11.1.) The expression for the bulk modulus of a gas for use in Eq. (16.7) is

$$B = \gamma p_0 \qquad (16.9)$$

where p_0 is the equilibrium pressure of the gas. The quantity γ (the Greek letter gamma) is called the *ratio of heat capacities*. It is a dimensionless number that characterizes the thermal properties of the gas. (We'll learn more about this quantity in Chapter 19.) As an example, the ratio of heat capacities for air is $\gamma = 1.40$. At normal atmospheric pressure $p_0 = 1.013 \times 10^5$ Pa, so $B = (1.40)(1.013 \times 10^5 \text{ Pa}) = 1.42 \times 10^5$ Pa. This value is minuscule compared to the bulk modulus of a typical solid (see Table 11.1), which is approximately 10^{10} to 10^{11} Pa. This shouldn't be surprising: It's simply a statement that air is far easier to compress than steel.

The density ρ of a gas also depends on the pressure, which in turn depends on the temperature. It turns out that the ratio B/ρ for a given type of ideal gas does *not* depend on the pressure at all, only the temperature. From Eq. (16.7), this means that the speed of sound in a gas is fundamentally a function of temperature T:

$$v = \sqrt{\frac{\gamma RT}{M}} \qquad \text{(speed of sound in an ideal gas)} \qquad (16.10)$$

This expression incorporates several quantities that you may recognize from your study of ideal gases in chemistry and that we will study in Chapters 17, 18, and 19. The temperature T is the *absolute* temperature in kelvins (K), equal to the Celsius temperature plus 273.15; thus 20.00°C corresponds to $T = 293.15$ K. The quantity M is the *molar mass,* or mass per mole of the substance of which the gas is composed. The *gas constant* R has the same value for all gases. The current best numerical value of R is

$$R = 8.314472(15) \text{ J/mol} \cdot \text{K}$$

which for practical calculations we can write as 8.314 J/mol · K.

For any particular gas, γ, R, and M are constants, and the wave speed is proportional to the square root of the absolute temperature. We will see in Chapter 18 that Eq. (16.10) is almost identical to the expression for the average speed of molecules in an ideal gas. This shows that sound speeds and molecular speeds are closely related.

16.9 This three-dimensional image of a fetus in the womb was made using a sequence of ultrasound scans. Each individual scan reveals a two-dimensional "slice" through the fetus; many such slices were then combined digitally.

Example 16.4 | **Speed of sound in air**

Find the speed of sound in air at $T = 20°C$, and find the range of wavelengths in air to which the human ear (which can hear frequencies in the range of 20–20,000 Hz) is sensitive. The mean molar mass for air (a mixture of mostly nitrogen and oxygen) is $M = 28.8 \times 10^{-3}$ kg/mol and the ratio of heat capacities is $\gamma = 1.40$.

Continued

SOLUTION

IDENTIFY and SET UP: We use Eq. (16.10) to find the sound speed from γ, T, and M, and we use $v = f\lambda$ to find the wavelengths corresponding to the frequency limits. Note that in Eq. (16.10) temperature T *must* be expressed in kelvins, not Celsius degrees.

EXECUTE: At $T = 20°C = 293$ K, we find

$$v = \sqrt{\frac{\gamma RT}{M}} = \sqrt{\frac{(1.40)(8.314 \text{ J/mol} \cdot \text{K})(293 \text{ K})}{28.8 \times 10^{-3} \text{ kg/mol}}} = 344 \text{ m/s}$$

Using this value of v in $\lambda = v/f$, we find that at 20°C the frequency $f = 20$ Hz corresponds to $\lambda = 17$ m and $f = 20{,}000$ Hz to $\lambda = 1.7$ cm.

EVALUATE: Our calculated value of v agrees with the measured sound speed at $T = 20°C$ to within 0.3%.

In this discussion we have treated a gas as a continuous medium. A gas is actually composed of molecules in random motion, separated by distances that are large in comparison with their diameters. The vibrations that constitute a wave in a gas are superposed on the random thermal motion. At atmospheric pressure, a molecule travels an average distance of about 10^{-7} m between collisions, while the displacement amplitude of a faint sound may be only 10^{-9} m. We can think of a gas with a sound wave passing through as being comparable to a swarm of bees; the swarm as a whole oscillates slightly while individual insects move about through the swarm, apparently at random.

Test Your Understanding of Section 16.2   Mercury is 13.6 times denser than water. Based on Table 16.1, at 20°C which of these liquids has the greater bulk modulus? (i) mercury; (ii) water; (iii) both are about the same; (iv) not enough information is given to decide.

16.3 Sound Intensity

Traveling sound waves, like all other traveling waves, transfer energy from one region of space to another. We saw in Section 15.5 that a useful way to describe the energy carried by a sound wave is through the *wave intensity I*, equal to the time average rate at which energy is transported per unit area across a surface perpendicular to the direction of propagation. Let's see how to express the intensity of a sound wave in terms of the displacement amplitude A or pressure amplitude p_{max}.

Intensity and Displacement Amplitude

For simplicity, let us consider a sound wave propagating in the $+x$-direction so that we can use our expressions from Section 16.1 for the displacement $y(x, t)$ and pressure fluctuation $p(x, t)$—Eqs. (16.1) and (16.4), respectively. In Section 6.4 we saw that power equals the product of force and velocity [see Eq. (6.18)]. So the power per unit area in this sound wave equals the product of $p(x, t)$ (force per unit area) and the *particle* velocity $v_y(x, t)$. The particle velocity $v_y(x, t)$ is the velocity at time t of that portion of the wave medium at coordinate x. Using Eqs. (16.1) and (16.4), we find

$$v_y(x, t) = \frac{\partial y(x, t)}{\partial t} = \omega A \sin(kx - \omega t)$$

$$p(x, t)v_y(x, t) = [BkA\sin(kx - \omega t)][\omega A\sin(kx - \omega t)]$$

$$= B\omega kA^2 \sin^2(kx - \omega t)$$

CAUTION **Wave velocity vs. particle velocity** Remember that the velocity of the wave as a whole is *not* the same as the particle velocity. While the wave continues to move in the direction of propagation, individual particles in the wave medium merely slosh back and forth, as shown in Fig. 16.1. Furthermore, the maximum speed of a particle of the medium can be very different from the wave speed.

The intensity is, by definition, the time average value of $p(x, t)v_y(x, t)$. For any value of x the average value of the function $\sin^2(kx - \omega t)$ over one period $T = 2\pi/\omega$ is $\frac{1}{2}$, so

$$I = \frac{1}{2}B\omega kA^2 \tag{16.11}$$

By using the relationships $\omega = vk$ and $v^2 = B/\rho$, we can transform Eq. (16.11) into the form

$$I = \frac{1}{2}\sqrt{\rho B}\,\omega^2 A^2 \qquad \text{(intensity of a sinusoidal sound wave)} \tag{16.12}$$

This equation shows why in a stereo system, a low-frequency woofer has to vibrate with much larger amplitude than a high-frequency tweeter to produce the same sound intensity.

Intensity and Pressure Amplitude

It is usually more useful to express I in terms of the pressure amplitude p_{max}. Using Eq. (16.5) and the relationship $\omega = vk$, we find

$$I = \frac{\omega p_{max}^2}{2Bk} = \frac{vp_{max}^2}{2B} \tag{16.13}$$

By using the wave speed relationship $v^2 = B/\rho$, we can also write Eq. (16.13) in the alternative forms

$$I = \frac{p_{max}^2}{2\rho v} = \frac{p_{max}^2}{2\sqrt{\rho B}} \qquad \text{(intensity of a sinusoidal sound wave)} \tag{16.14}$$

You should verify these expressions. Comparison of Eqs. (16.12) and (16.14) shows that sinusoidal sound waves of the same intensity but different frequency have different displacement amplitudes A but the *same* pressure amplitude p_{max}. This is another reason it is usually more convenient to describe a sound wave in terms of pressure fluctuations, not displacement.

The *total* average power carried across a surface by a sound wave equals the product of the intensity at the surface and the surface area, if the intensity over the surface is uniform. The average total sound power emitted by a person speaking in an ordinary conversational tone is about 10^{-5} W, while a loud shout corresponds to about 3×10^{-2} W. If all the residents of New York City were to talk at the same time, the total sound power would be about 100 W, equivalent to the electric power requirement of a medium-sized light bulb. On the other hand, the power required to fill a large auditorium or stadium with loud sound is considerable (see Example 16.7.)

If the sound source emits waves in all directions equally, the intensity decreases with increasing distance r from the source according to the inverse-square law: The intensity is proportional to $1/r^2$. We discussed this law and its consequences in Section 15.5. If the sound goes predominantly in one direction, the inverse-square law does not apply and the intensity decreases with distance more slowly than $1/r^2$ (Fig. 16.10).

16.10 By cupping your hands like this, you direct the sound waves emerging from your mouth so that they don't propagate to the sides. Hence the intensity decreases with distance more slowly than the inverse-square law would predict, and you can be heard at greater distances.

The inverse-square relationship also does not apply indoors because sound energy can reach a listener by reflection from the walls and ceiling. Indeed, part of the architect's job in designing an auditorium is to tailor these reflections so that the intensity is as nearly uniform as possible over the entire auditorium.

Problem-Solving Strategy 16.1   **Sound Intensity**

IDENTIFY *the relevant concepts:* The relationships between the intensity and amplitude of a sound wave are straightforward. Other quantities are involved in these relationships, however, so it's particularly important to decide which is your target variable.

SET UP *the problem* using the following steps:
1. Sort the physical quantities into categories. Wave properties include the displacement and pressure amplitudes A and p_{max} and the frequency f, which can be determined from the angular frequency ω, the wave number k, or the wavelength λ. These quantities are related through the wave speed v, which is determined by properties of the medium (B and ρ for a liquid, and γ, T, and M for a gas).

2. List the given quantities and identify the target variables. Find relationships that take you where you want to go.

EXECUTE *the solution*: Use your selected equations to solve for the target variables. Express the temperature in kelvins (Celsius temperature plus 273.15) to calculate the speed of sound in a gas.

EVALUATE *your answer:* If possible, use an alternative relationship to check your results.

Example 16.5   **Intensity of a sound wave in air**

Find the intensity of the sound wave in Example 16.1, with $p_{max} = 3.0 \times 10^{-2}$ Pa. Assume the temperature is 20°C so that the density of air is $\rho = 1.20$ kg/m^3 and the speed of sound is $v = 344$ m/s.

SOLUTION

IDENTIFY and SET UP: Our target variable is the intensity I of the sound wave. We are given the pressure amplitude p_{max} of the wave as well as the density ρ and wave speed v for the medium. We can determine I from p_{max}, ρ, and v using Eq. (16.14).

EXECUTE: From Eq. (16.14),

$$I = \frac{p_{max}^2}{2\rho v} = \frac{(3.0 \times 10^{-2} \text{ Pa})^2}{2(1.20 \text{ kg/m}^3)(344 \text{ m/s})}$$
$$= 1.1 \times 10^{-6} \text{ J/(s} \cdot \text{m}^2) = 1.1 \times 10^{-6} \text{ W/m}^2$$

EVALUATE: This seems like a very low intensity, but it is well within the range of sound intensities encountered on a daily basis. A very loud sound wave at the threshold of pain has a pressure amplitude of about 30 Pa and an intensity of about 1 W/m^2. The pressure amplitude of the faintest sound wave that can be heard is about 3×10^{-5} Pa, and the corresponding intensity is about 10^{-12} W/m^2. (Try these values of p_{max} in Eq. (16.14) to check that the corresponding intensities are as we have stated.)

Example 16.6   **Same intensity, different frequencies**

What are the pressure and displacement amplitudes of a 20-Hz sound wave with the same intensity as the 1000-Hz sound wave of Examples 16.1 and 16.5?

SOLUTION

IDENTIFY and SET UP: In Examples 16.1 and 16.5 we found that for a 1000-Hz sound wave with $p_{max} = 3.0 \times 10^{-2}$ Pa, $A = 1.2 \times 10^{-8}$ m and $I = 1.1 \times 10^{-6}$ W/m^2. Our target variables are p_{max} and A for a 20-Hz sound wave of the same intensity I. We can find these using Eqs. (16.14) and (16.12), respectively.

EXECUTE: We can rearrange Eqs. (16.14) and (16.12) as $p_{max}^2 = 2I\sqrt{\rho B}$ and $\omega^2 A^2 = 2I/\sqrt{\rho B}$, respectively. These tell us that for a given sound intensity I in a given medium (constant ρ and B), the

quantities p_{max} and ωA (or, equivalently, fA) are both *constants* that don't depend on frequency. From the first result we immediately have $p_{max} = 3.0 \times 10^{-2}$ Pa for $f = 20$ Hz, the same as for $f = 1000$ Hz. If write the second result as $f_{20}A_{20} = f_{1000}A_{1000}$, we have

$$A_{20} = \left(\frac{f_{1000}}{f_{20}}\right)A_{1000}$$
$$= \left(\frac{1000 \text{ Hz}}{20 \text{ Hz}}\right)(1.2 \times 10^{-8} \text{ m}) = 6.0 \times 10^{-7} \text{ m} = 0.60 \text{ } \mu\text{m}$$

EVALUATE: Our result reinforces the idea that pressure amplitude is a more convenient description of a sound wave and its intensity than displacement amplitude.

"Play it loud!"

For an outdoor concert we want the sound intensity to be 1 W/m^2 at a distance of 20 m from the speaker array. If the sound intensity is uniform in all directions, what is the required acoustic power output of the array?

SOLUTION

IDENTIFY, SET UP, and EXECUTE: This example uses the definition of sound intensity as power per unit area. The total power is the target variable; the area in question is a hemisphere centered on the speaker array. We assume that the speakers are on the ground and

that none of the acoustic power is directed into the ground, so the acoustic power is uniform over a hemisphere 20 m in radius. The surface area of this hemisphere is $(\frac{1}{2})(4\pi)(20 \text{ m})^2$, or about 2500 m^2. The required power is the product of this area and the intensity: $(1 \text{ W/m}^2)(2500 \text{ m}^2) = 2500 \text{ W} = 2.5 \text{ kW}$.

EVALUATE: The electrical power input to the speaker would need to be considerably greater than 2.5 kW, because speaker efficiency is not very high (typically a few percent for ordinary speakers, and up to 25% for horn-type speakers).

The Decibel Scale

Because the ear is sensitive over a broad range of intensities, a *logarithmic* intensity scale is usually used. The **sound intensity level** β of a sound wave is defined by the equation

$$\beta = (10 \text{ dB}) \log \frac{I}{I_0} \quad \text{(definition of sound intensity level)} \quad (16.15)$$

In this equation, I_0 is a reference intensity, chosen to be 10^{-12} W/m^2, approximately the threshold of human hearing at 1000 Hz. Recall that "log" means the logarithm to base 10. Sound intensity levels are expressed in **decibels,** abbreviated dB. A decibel is $\frac{1}{10}$ of a *bel,* a unit named for Alexander Graham Bell (the inventor of the telephone). The bel is inconveniently large for most purposes, and the decibel is the usual unit of sound intensity level.

If the intensity of a sound wave equals I_0 or 10^{-12} W/m^2, its sound intensity level is 0 dB. An intensity of 1 W/m^2 corresponds to 120 dB. Table 16.2 gives the sound intensity levels in decibels of some familiar sounds. You can use Eq. (16.15) to check the value of sound intensity level β given for each intensity in the table.

Because the ear is not equally sensitive to all frequencies in the audible range, some sound-level meters weight the various frequencies unequally. One such scheme leads to the so-called dBA scale; this scale deemphasizes the low and very high frequencies, where the ear is less sensitive than at midrange frequencies.

Table 16.2 Sound Intensity Levels from Various Sources (Representative Values)

Source or Description of Sound	Sound Intensity Level, β (dB)	Intensity, I (W/m²)
Military jet aircraft 30 m away	140	10^2
Threshold of pain	120	1
Riveter	95	3.2×10^{-3}
Elevated train	90	10^{-3}
Busy street traffic	70	10^{-5}
Ordinary conversation	65	3.2×10^{-6}
Quiet automobile	50	10^{-7}
Quiet radio in home	40	10^{-8}
Average whisper	20	10^{-10}
Rustle of leaves	10	10^{-11}
Threshold of hearing at 1000 Hz	0	10^{-12}

Example 16.8 Temporary—or permanent—hearing loss

A 10-min exposure to 120-dB sound will temporarily shift your threshold of hearing at 1000 Hz from 0 dB up to 28 dB. Ten years of exposure to 92-dB sound will cause a *permanent* shift to 28 dB. What sound intensities correspond to 28 dB and 92 dB?

SOLUTION

IDENTIFY and SET UP: We are given two sound intensity levels β; our target variables are the corresponding intensities. We can solve Eq. (16.15) to find the intensity I that corresponds to each value of β.

EXECUTE: We solve Eq. (16.15) for I by dividing both sides by 10 dB and using the relationship $10^{\log x} = x$:

$$I = I_0 10^{\beta/(10 \text{ dB})}$$

For $\beta = 28$ dB and $\beta = 92$ dB, the exponents are $\beta/(10 \text{ dB}) = 2.8$ and 9.2, respectively, so that

$$I_{28 \text{ dB}} = (10^{-12} \text{ W/m}^2)10^{2.8} = 6.3 \times 10^{-10} \text{ W/m}^2$$
$$I_{92 \text{ dB}} = (10^{-12} \text{ W/m}^2)10^{9.2} = 1.6 \times 10^{-3} \text{ W/m}^2$$

EVALUATE: If your answers are a factor of 10 too large, you may have entered 10×10^{-12} in your calculator instead of 1×10^{-12}. Be careful!

Example 16.9 A bird sings in a meadow

Consider an idealized bird (treated as a point source) that emits constant sound power, with intensity obeying the inverse-square law (Fig. 16.11). If you move twice the distance from the bird, by how many decibels does the sound intensity level drop?

SOLUTION

IDENTIFY and SET UP: The decibel scale is logarithmic, so the *difference* between two sound intensity levels (the target variable) corresponds to the *ratio* of the corresponding intensities, which is determined by the inverse-square law. We label the two points P_1 and P_2 (Fig. 16.11). We use Eq. (16.15), the definition of sound intensity level, at each point. We use Eq. (15.26), the inverse-square law, to relate the intensities at the two points.

EXECUTE: The difference $\beta_2 - \beta_1$ between any two sound intensity levels is related to the corresponding intensities by

$$\beta_2 - \beta_1 = (10 \text{ dB})\left(\log\frac{I_2}{I_0} - \log\frac{I_1}{I_0}\right)$$
$$= (10 \text{ dB})[(\log I_2 - \log I_0) - (\log I_1 - \log I_0)]$$
$$= (10 \text{ dB})\log\frac{I_2}{I_1}$$

For this inverse-square-law source, Eq. (15.26) yields $I_2/I_1 = r_1^2/r_2^2 = \frac{1}{4}$, so

$$\beta_2 - \beta_1 = (10 \text{ dB})\log\frac{I_1}{I_2} = (10 \text{ dB})\log\tfrac{1}{4} = -6.0 \text{ dB}$$

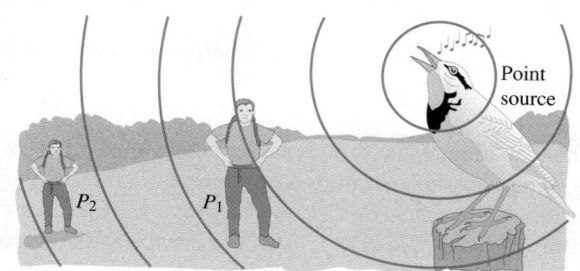

16.11 When you double your distance from a point source of sound, by how much does the sound intensity level decrease?

Point source

P_2 P_1

EVALUATE: Our result is negative, which tells us (correctly) that the sound intensity level is less at P_2 than at P_1. The 6-dB difference doesn't depend on the sound intensity level at P_1; *any* doubling of the distance from an inverse-square-law source reduces the sound intensity level by 6 dB.

Note that the perceived *loudness* of a sound is not directly proportional to its intensity. For example, most people interpret an increase of 8 dB to 10 dB in sound intensity level (corresponding to increasing intensity by a factor of 6 to 10) as a doubling of loudness.

Test Your Understanding of Section 16.3 You double the intensity of a sound wave in air while leaving the frequency unchanged. (The pressure, density, and temperature of the air remain unchanged as well.) What effect does this have on the displacement amplitude, pressure amplitude, bulk modulus, sound speed, and sound intensity level?

16.4 Standing Sound Waves and Normal Modes

When longitudinal (sound) waves propagate in a fluid in a pipe with finite length, the waves are reflected from the ends in the same way that transverse waves on a string are reflected at its ends. The superposition of the waves traveling in opposite directions again forms a standing wave. Just as for transverse standing waves on a string (see Section 15.7), standing sound waves (normal modes) in a pipe can

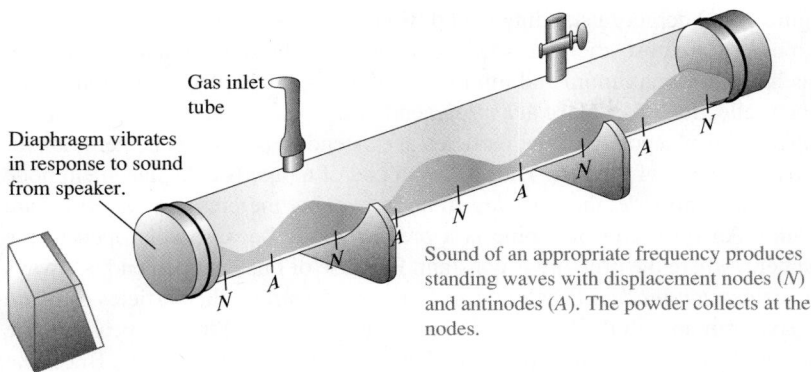

Gas inlet tube

Diaphragm vibrates in response to sound from speaker.

Speaker

Sound of an appropriate frequency produces standing waves with displacement nodes (*N*) and antinodes (*A*). The powder collects at the nodes.

16.12 Demonstrating standing sound waves using a Kundt's tube. The blue shading represents the density of the gas at an instant when the gas pressure at the displacement nodes is a maximum or a minimum.

be used to create sound waves in the surrounding air. This is the operating principle of the human voice as well as many musical instruments, including woodwinds, brasses, and pipe organs.

Transverse waves on a string, including standing waves, are usually described only in terms of the displacement of the string. But, as we have seen, sound waves in a fluid may be described either in terms of the displacement of the fluid or in terms of the pressure variation in the fluid. To avoid confusion, we'll use the terms **displacement node** and **displacement antinode** to refer to points where particles of the fluid have zero displacement and maximum displacement, respectively.

We can demonstrate standing sound waves in a column of gas using an apparatus called a Kundt's tube (Fig. 16.12). A horizontal glass tube a meter or so long is closed at one end and has a flexible diaphragm at the other end that can transmit vibrations. A nearby loudspeaker is driven by an audio oscillator and amplifier; this produces sound waves that force the diaphragm to vibrate sinusoidally with a frequency that we can vary. The sound waves within the tube are reflected at the other, closed end of the tube. We spread a small amount of light powder uniformly along the bottom of the tube. As we vary the frequency of the sound, we pass through frequencies at which the amplitude of the standing waves becomes large enough for the powder to be swept along the tube at those points where the gas is in motion. The powder therefore collects at the displacement nodes (where the gas is not moving). Adjacent nodes are separated by a distance equal to $\lambda/2$, and we can measure this distance. Given the wavelength, we can use this experiment to determine the wave speed: We read the frequency f from the oscillator dial, and we can then calculate the speed v of the waves from the relationship $v = \lambda f$.

Figure 16.13 shows the motions of nine different particles within a gas-filled tube in which there is a standing sound wave. A particle at a displacement node (*N*) does not move, while a particle at a displacement antinode (*A*) oscillates with maximum amplitude. Note that particles on opposite sides of a displacement node vibrate in opposite phase. When these particles approach each other, the gas between them is compressed and the pressure rises; when they recede from each other, there is an expansion and the pressure drops. Hence at a displacement *node* the gas undergoes the maximum amount of compression and expansion, and the variations in pressure and density above and below the average have their maximum value. By contrast, particles on opposite sides of a displacement *antinode* vibrate *in phase;* the distance between the particles is nearly constant, and there is *no* variation in pressure or density at a displacement antinode.

We use the term **pressure node** to describe a point in a standing sound wave at which the pressure and density do not vary and the term **pressure antinode** to describe a point at which the variations in pressure and density are greatest. Using these terms, we can summarize our observations about standing sound waves as follows:

A pressure node is always a displacement antinode, and a pressure antinode is always a displacement node.

16.13 In a standing sound wave, a displacement node *N* is a pressure antinode (a point where the pressure fluctuates the most) and a displacement antinode *A* is a pressure node (a point where the pressure does not fluctuate at all).

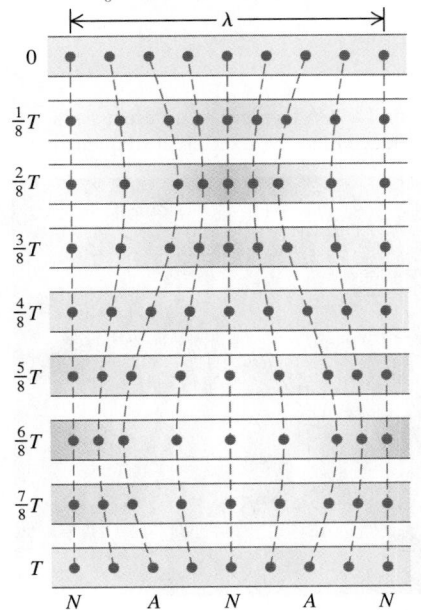

A standing wave shown at intervals of $\frac{1}{8}T$ for one period T

λ

0

$\frac{1}{8}T$

$\frac{2}{8}T$

$\frac{3}{8}T$

$\frac{4}{8}T$

$\frac{5}{8}T$

$\frac{6}{8}T$

$\frac{7}{8}T$

T

$N \qquad A \qquad N \qquad A \qquad N$

N = a displacement node = a pressure antinode
A = a displacement antinode = a pressure node

Figure 16.12 depicts a standing sound wave at an instant at which the pressure variations are greatest; the blue shading shows that the density and pressure of the gas have their maximum and minimum values at the displacement nodes.

When reflection takes place at a *closed* end of a pipe (an end with a rigid barrier or plug), the displacement of the particles at this end must always be zero, analogous to a fixed end of a string. Thus a closed end of a pipe is a displacement node and a pressure antinode; the particles do not move, but the pressure variations are maximum. An *open* end of a pipe is a pressure node because it is open to the atmosphere, where the pressure is constant. Because of this, an open end is always a displacement *antinode,* in analogy to a free end of a string; the particles oscillate with maximum amplitude, but the pressure does not vary. (Strictly speaking, the pressure node actually occurs somewhat beyond an open end of a pipe. But if the diameter of the pipe is small in comparison to the wavelength, which is true for most musical instruments, this effect can safely be neglected.) Thus longitudinal waves in a column of fluid are reflected at the closed and open ends of a pipe in the same way that transverse waves in a string are reflected at fixed and free ends, respectively.

Conceptual Example 16.10 **The sound of silence**

A directional loudspeaker directs a sound wave of wavelength λ at a wall (Fig. 16.14). At what distances from the wall could you stand and hear no sound at all?

SOLUTION

Your ear detects pressure variations in the air; you will therefore hear no sound if your ear is at a *pressure node,* which is a displacement antinode. The wall is at a displacement node; the distance from any node to an adjacent antinode is $\lambda/4$, and the distance from one antinode to the next is $\lambda/2$ (Fig. 16.14). Hence the displacement antinodes (pressure nodes), at which no sound will be heard, are at distances $d = \lambda/4$, $d = \lambda/4 + \lambda/2 = 3\lambda/4$, $d = 3\lambda/4 + \lambda/2 = 5\lambda/4, \ldots$ from the wall. If the loudspeaker is not highly directional, this effect is hard to notice because of reflections of sound waves from the floor, ceiling, and other walls.

16.14 When a sound wave is directed at a wall, it interferes with the reflected wave to create a standing wave. The N's and A's are *displacement* nodes and antinodes.

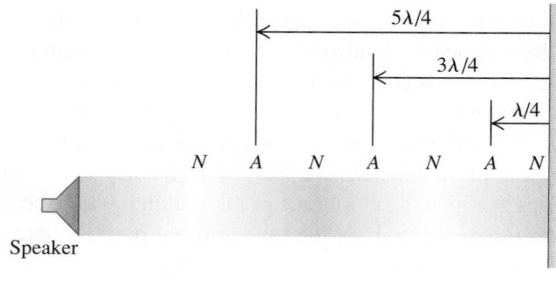

Organ Pipes and Wind Instruments

16.15 Organ pipes of different sizes produce tones with different frequencies.

The most important application of standing sound waves is the production of musical tones by wind instruments. Organ pipes are one of the simplest examples (Fig. 16.15). Air is supplied by a blower, at a gauge pressure typically of the order of 10^3 Pa $(10^{-2}$ atm$)$, to the bottom end of the pipe (Fig. 16.16). A stream of air emerges from the narrow opening at the edge of the horizontal surface and is directed against the top edge of the opening, which is called the *mouth* of the pipe. The column of air in the pipe is set into vibration, and there is a series of possible normal modes, just as with the stretched string. The mouth always acts as an open end; thus it is a pressure node and a displacement antinode. The other end of the pipe (at the top in Fig. 16.16) may be either open or closed.

In Fig. 16.17, both ends of the pipe are open, so both ends are pressure nodes and displacement antinodes. An organ pipe that is open at both ends is called an *open pipe.* The fundamental frequency f_1 corresponds to a standing-wave pattern with a displacement antinode at each end and a displacement node in the middle (Fig. 16.17a). The distance between adjacent antinodes is always equal to one half-wavelength, and in this case that is equal to the length L of the pipe; $\lambda/2 = L$. The corresponding frequency, obtained from the relationship $f = v/\lambda$, is

$$f_1 = \frac{v}{2L} \quad \text{(open pipe)} \tag{16.16}$$

Figures 16.17b and 16.17c show the second and third harmonics (first and second overtones); their vibration patterns have two and three displacement nodes, respectively. For these, a half-wavelength is equal to $L/2$ and $L/3$, respectively, and the frequencies are twice and three times the fundamental, respectively. That is, $f_2 = 2f_1$ and $f_3 = 3f_1$. For *every* normal mode of an open pipe the length L must be an integer number of half-wavelengths, and the possible wavelengths λ_n are given by

$$L = n\frac{\lambda_n}{2} \quad \text{or} \quad \lambda_n = \frac{2L}{n} \quad (n = 1, 2, 3, \ldots) \quad \text{(open pipe)} \tag{16.17}$$

The corresponding frequencies f_n are given by $f_n = v/\lambda_n$, so all the normal-mode frequencies for a pipe that is open at both ends are given by

$$f_n = \frac{nv}{2L} \quad (n = 1, 2, 3, \ldots) \quad \text{(open pipe)} \tag{16.18}$$

The value $n = 1$ corresponds to the fundamental frequency, $n = 2$ to the second harmonic (or first overtone), and so on. Alternatively, we can say

$$f_n = nf_1 \quad (n = 1, 2, 3, \ldots) \quad \text{(open pipe)} \tag{16.19}$$

with f_1 given by Eq. (16.16).

Figure 16.18 shows a pipe that is open at the left end but closed at the right end. This is called a *stopped pipe*. The left (open) end is a displacement antinode (pressure node), but the right (closed) end is a displacement node (pressure antinode). The distance between a node and the adjacent antinode is always one quarter-wavelength. Figure 16.18a shows the lowest-frequency mode; the length

16.16 Cross sections of an organ pipe at two instants one half-period apart. The N's and A's are *displacement* nodes and antinodes; as the blue shading shows, these are points of maximum pressure variation and zero pressure variation, respectively.

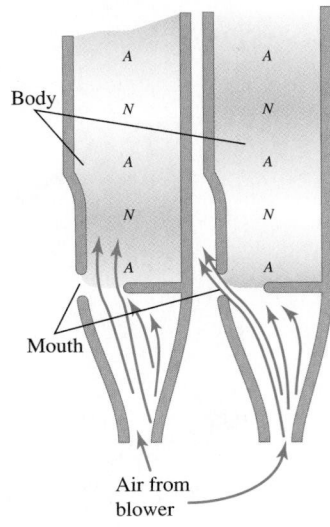

16.17 A cross section of an open pipe showing the first three normal modes. The shading indicates the pressure variations. The red curves are graphs of the displacement along the pipe axis at two instants separated in time by one half-period. The N's and A's are the *displacement* nodes and antinodes; interchange these to show the *pressure* nodes and antinodes.

(a) Fundamental: $f_1 = \dfrac{v}{2L}$

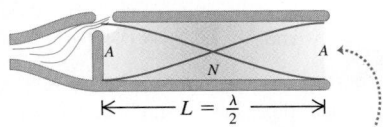

Open end is always a displacement antinode.

(b) Second harmonic: $f_2 = 2\dfrac{v}{2L} = 2f_1$

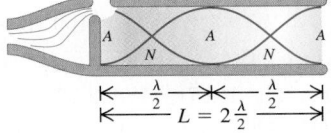

(c) Third harmonic: $f_3 = 3\dfrac{v}{2L} = 3f_1$

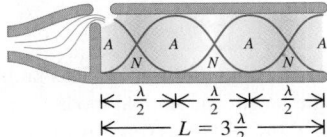

16.18 A cross section of a stopped pipe showing the first three normal modes as well as the *displacement* nodes and antinodes. Only odd harmonics are possible.

(a) Fundamental: $f_1 = \dfrac{v}{4L}$

Closed end is always a displacement node.

(b) Third harmonic: $f_3 = 3\dfrac{v}{4L} = 3f_1$

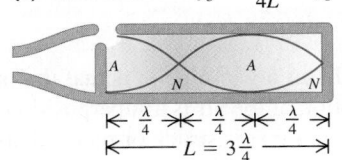

(c) Fifth harmonic: $f_5 = 5\dfrac{v}{4L} = 5f_1$

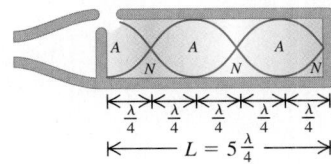

of the pipe is a quarter-wavelength ($L = \lambda_1/4$). The fundamental frequency is $f_1 = v/\lambda_1$, or

$$f_1 = \frac{v}{4L} \qquad \text{(stopped pipe)} \qquad (16.20)$$

This is one-half the fundamental frequency for an *open* pipe of the same length. In musical language, the *pitch* of a closed pipe is one octave lower (a factor of 2 in frequency) than that of an open pipe of the same length. Figure 16.18b shows the next mode, for which the length of the pipe is *three-quarters* of a wavelength, corresponding to a frequency $3f_1$. For Fig. 16.18c, $L = 5\lambda/4$ and the frequency is $5f_1$. The possible wavelengths are given by

$$L = n\frac{\lambda_n}{4} \qquad \text{or} \qquad \lambda_n = \frac{4L}{n} \quad (n = 1, 3, 5, \ldots) \quad \text{(stopped pipe)} \quad (16.21)$$

The normal-mode frequencies are given by $f_n = v/\lambda_n$, or

$$f_n = \frac{nv}{4L} \qquad (n = 1, 3, 5, \ldots) \qquad \text{(stopped pipe)} \qquad (16.22)$$

or

$$f_n = nf_1 \qquad (n = 1, 3, 5, \ldots) \qquad \text{(stopped pipe)} \qquad (16.23)$$

with f_1 given by Eq. (16.20). We see that the second, fourth, and all *even* harmonics are missing. In a pipe that is closed at one end, the fundamental frequency is $f_1 = v/4L$, and only the odd harmonics in the series ($3f_1, 5f_1, \ldots$) are possible.

A final possibility is a pipe that is closed at *both* ends, with displacement nodes and pressure antinodes at both ends. This wouldn't be of much use as a musical instrument because there would be no way for the vibrations to get out of the pipe.

Example 16.11 | **A tale of two pipes**

On a day when the speed of sound is 345 m/s, the fundamental frequency of a particular stopped organ pipe is 220 Hz. (a) How long is this pipe? (b) The second *overtone* of this pipe has the same wavelength as the third *harmonic* of an *open* pipe. How long is the open pipe?

SOLUTION

IDENTIFY and SET UP: This problem uses the relationship between the length and normal-mode frequencies of open pipes (Fig. 16.17) and stopped pipes (Fig. 16.18). In part (a), we determine the length of the stopped pipe from Eq. (16.22). In part (b), we must determine the length of an open pipe, for which Eq. (16.18) gives the frequencies.

EXECUTE: (a) For a stopped pipe $f_1 = v/4L$, so

$$L_{\text{stopped}} = \frac{v}{4f_1} = \frac{345 \text{ m/s}}{4(220 \text{ s}^{-1})} = 0.392 \text{ m}$$

(b) The frequency of the second overtone of a stopped pipe (the *third* possible frequency) is $f_5 = 5f_1 = 5(220 \text{ Hz}) = 1100 \text{ Hz}$. If the wavelengths for the two pipes are the same, the frequencies are also the same. Hence the frequency of the third harmonic of the open pipe, which is at $3f_1 = 3(v/2L)$, equals 1100 Hz. Then

$$1100 \text{ Hz} = 3\left(\frac{345 \text{ m/s}}{2L_{\text{open}}}\right) \qquad \text{and} \qquad L_{\text{open}} = 0.470 \text{ m}$$

EVALUATE: The 0.392-m stopped pipe has a fundamental frequency of 220 Hz; the *longer* (0.470-m) open pipe has a *higher* fundamental frequency, $(1100 \text{ Hz})/3 = 367 \text{ Hz}$. This is not a contradiction, as you can see if you compare Figs. 16.17a and 16.18a.

In an organ pipe in actual use, several modes are always present at once; the motion of the air is a superposition of these modes. This situation is analogous to a string that is struck or plucked, as in Fig. 15.28. Just as for a vibrating string, a complex standing wave in the pipe produces a traveling sound wave in the surrounding air with a harmonic content similar to that of the standing wave. A very narrow pipe produces a sound wave rich in higher harmonics, which we hear as a thin and "stringy" tone; a fatter pipe produces mostly the fundamental mode, heard as a softer, more flutelike tone. The harmonic content also depends on the shape of the pipe's mouth.

We have talked about organ pipes, but this discussion is also applicable to other wind instruments. The flute and the recorder are directly analogous. The most significant difference is that those instruments have holes along the pipe. Opening and closing the holes with the fingers changes the effective length L of the air column and thus changes the pitch. Any individual organ pipe, by comparison, can play only a single note. The flute and recorder behave as *open* pipes, while the clarinet acts as a *stopped* pipe (closed at the reed end, open at the bell).

Equations (16.18) and (16.22) show that the frequencies of any wind instrument are proportional to the speed of sound v in the air column inside the instrument. As Eq. (16.10) shows, v depends on temperature; it increases when temperature increases. Thus the pitch of all wind instruments rises with increasing temperature. An organ that has some of its pipes at one temperature and others at a different temperature is bound to sound out of tune.

Test Your Understanding of Section 16.4   If you connect a hose to one end of a metal pipe and blow compressed air into it, the pipe produces a musical tone. If instead you blow compressed helium into the pipe at the same pressure and temperature, will the pipe produce (i) the same tone, (ii) a higher-pitch tone, or (iii) a lower-pitch tone?

16.5 Resonance and Sound

Many mechanical systems have normal modes of oscillation. As we have seen, these include columns of air (as in an organ pipe) and stretched strings (as in a guitar; see Section 15.8). In each mode, every particle of the system oscillates with simple harmonic motion at the same frequency as the mode. Air columns and stretched strings have an infinite series of normal modes, but the basic concept is closely related to the simple harmonic oscillator, discussed in Chapter 14, which has only a single normal mode (that is, only one frequency at which it oscillates after being disturbed).

Suppose we apply a periodically varying force to a system that can oscillate. The system is then forced to oscillate with a frequency equal to the frequency of the applied force (called the *driving frequency*). This motion is called a *forced oscillation*. We talked about forced oscillations of the harmonic oscillator in Section 14.8, and we suggest that you review that discussion. In particular, we described the phenomenon of mechanical **resonance**. A simple example of resonance is pushing Cousin Throckmorton on a swing. The swing is a pendulum; it has only a single normal mode, with a frequency determined by its length. If we push the swing periodically with this frequency, we can build up the amplitude of the motion. But if we push with a very different frequency, the swing hardly moves at all.

Resonance also occurs when a periodically varying force is applied to a system with many normal modes. An example is shown in Fig. 16.19a. An open organ pipe is placed next to a loudspeaker that is driven by an amplifier and emits pure sinusoidal sound waves of frequency f, which can be varied by adjusting the amplifier. The air in the pipe is forced to vibrate with the same frequency f as the *driving force* provided by the loudspeaker. In general the amplitude of this motion is relatively small, and the air inside the pipe will not move in any of the normal-mode patterns shown in Fig. 16.17. But if the frequency f of the force is close to one of the normal-mode frequencies, the air in the pipe moves in the normal-mode pattern for that frequency, and the amplitude can become quite large. Figure 16.19b shows the amplitude of oscillation of the air

16.19 (a) The air in an open pipe is forced to oscillate at the same frequency as the sinusoidal sound waves coming from the loudspeaker. (b) The resonance curve of the open pipe graphs the amplitude of the standing sound wave in the pipe as a function of the driving frequency.

(a)

Speaker emits frequency f Open organ pipe

• Air in pipe oscillates at the same frequency f emitted by the speaker.
• Wave amplitude A depends on frequency.

Amplifier

(b) Resonance curve: graph of amplitude A versus driving frequency f. Peaks occur at normal-mode frequencies of the pipe: f_1, $f_2 = 2f_1$, $f_3 = 3f_1$, ...

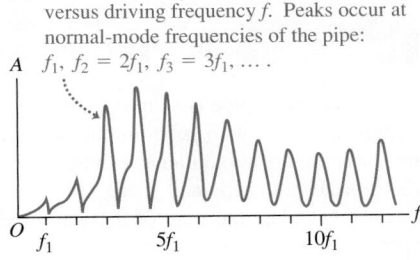

The auditory canal of the human ear (see Fig. 16.4) is an air-filled pipe open at one end and closed at the other (eardrum) end. The canal is about 2.5 cm = 0.025 m long, so it has a resonance at its fundamental frequency $f_1 = v/4L = (344 \text{ m/s})/[4(0.025 \text{ m})] = 3440$ Hz. The resonance means that a sound at this frequency produces a strong oscillation of the eardrum. That's why your ear is most sensitive to sounds near 3440 Hz.

16.20 The frequency of the sound from this trumpet exactly matches one of the normal-mode frequencies of the goblet. The resonant vibrations of the goblet have such large amplitude that the goblet tears itself apart.

in the pipe as a function of the driving frequency f. The shape of this graph is called the **resonance curve** of the pipe; it has peaks where f equals the normal-mode frequencies of the pipe. The detailed shape of the resonance curve depends on the geometry of the pipe.

If the frequency of the force is precisely *equal* to a normal-mode frequency, the system is in resonance, and the amplitude of the forced oscillation is maximum. If there were no friction or other energy-dissipating mechanism, a driving force at a normal-mode frequency would continue to add energy to the system, and the amplitude would increase indefinitely. In such an idealized case the peaks in the resonance curve of Fig. 16.19b would be infinitely high. But in any real system there is always some dissipation of energy, or damping, as we discussed in Section 14.8; the amplitude of oscillation in resonance may be large, but it cannot be infinite.

The "sound of the ocean" you hear when you put your ear next to a large seashell is due to resonance. The noise of the outside air moving past the seashell is a mixture of sound waves of almost all audible frequencies, which forces the air inside the seashell to oscillate. The seashell behaves like an organ pipe, with a set of normal-mode frequencies; hence the inside air oscillates most strongly at those frequencies, producing the seashell's characteristic sound. To hear a similar phenomenon, uncap a full bottle of your favorite beverage and blow across the open top. The noise is provided by your breath blowing across the top, and the "organ pipe" is the column of air inside the bottle above the surface of the liquid. If you take a drink and repeat the experiment, you will hear a lower tone because the "pipe" is longer and the normal-mode frequencies are lower.

Resonance also occurs when a stretched string is forced to oscillate (see Section 15.8). Suppose that one end of a stretched string is held fixed while the other is given a transverse sinusoidal motion with small amplitude, setting up standing waves. If the frequency of the driving mechanism is *not* equal to one of the normal-mode frequencies of the string, the amplitude at the antinodes is fairly small. However, if the frequency is equal to any one of the normal-mode frequencies, the string is in resonance, and the amplitude at the antinodes is very much larger than that at the driven end. The driven end is not precisely a node, but it lies much closer to a node than to an antinode when the string is in resonance. The photographs in Fig. 15.23 were made this way, with the left end of the string fixed and the right end oscillating vertically with small amplitude; large-amplitude standing waves resulted when the frequency of oscillation of the right end was equal to the fundamental frequency or to one of the first three overtones.

It is easy to demonstrate resonance with a piano. Push down the damper pedal (the right-hand pedal) so that the dampers are lifted and the strings are free to vibrate, and then sing a steady tone into the piano. When you stop singing, the piano seems to continue to sing the same note. The sound waves from your voice excite vibrations in the strings that have natural frequencies close to the frequencies (fundamental and harmonics) present in the note you sang.

A more spectacular example is a singer breaking a wine glass with her amplified voice. A good-quality wine glass has normal-mode frequencies that you can hear by tapping it. If the singer emits a loud note with a frequency corresponding exactly to one of these normal-mode frequencies, large-amplitude oscillations can build up and break the glass (Fig. 16.20).

Example 16.12 **An organ–guitar duet**

A stopped organ pipe is sounded near a guitar, causing one of the strings to vibrate with large amplitude. We vary the string tension until we find the maximum amplitude. The string is 80% as long as the pipe. If both pipe and string vibrate at their fundamental frequency, calculate the ratio of the wave speed on the string to the speed of sound in air.

SOLUTION

IDENTIFY and SET UP: The large response of the string is an example of resonance. It occurs because the organ pipe and the guitar string have the same fundamental frequency. Letting the subscripts a and s stand for the air in the pipe and the string, respectively, the condition for resonance is $f_{1a} = f_{1s}$. Equation (16.20) gives the fundamental frequency for a stopped pipe, and Eq. (15.32) gives the fundamental frequency for a guitar string held at both ends. These expressions involve the wave speed in air (v_a) and on the string (v_s) and the lengths of the pipe and string. We are given that $L_s = 0.80L_a$; our target variable is the ratio v_s/v_a.

EXECUTE: From Eqs. (16.20) and (15.32), $f_{1a} = v_a/4L_a$ and $f_{1s} = v_s/2L_s$. These frequencies are equal, so

$$\frac{v_a}{4L_a} = \frac{v_s}{2L_s}$$

Substituting $L_s = 0.80L_a$ and rearranging, we get $v_s/v_a = 0.40$.

EVALUATE: As an example, if the speed of sound in air is 345 m/s, the wave speed on the string is $(0.40)(345 \text{ m/s}) = 138$ m/s. Note that while the standing waves in the pipe and on the string have the same frequency, they have different *wavelengths* $\lambda = v/f$ because the two media have different wave speeds v. Which standing wave has the greater wavelength?

Test Your Understanding of Section 16.5 A stopped organ pipe of length L has a fundamental frequency of 220 Hz. For which of the following organ pipes will there be a resonance if a tuning fork of frequency 660 Hz is sounded next to the pipe? (There may be more than one correct answer.) (i) a stopped organ pipe of length L; (ii) a stopped organ pipe of length $2L$; (iii) an open organ pipe of length L; (iii) an open organ pipe of length $2L$.

16.6 Interference of Waves

Wave phenomena that occur when two or more waves overlap in the same region of space are grouped under the heading *interference*. As we have seen, standing waves are a simple example of an interference effect: Two waves traveling in opposite directions in a medium combine to produce a standing wave pattern with nodes and antinodes that do not move.

Figure 16.21 shows an example of another type of interference that involves waves that spread out in space. Two speakers, driven in phase by the same amplifier, emit identical sinusoidal sound waves with the same constant frequency. We place a microphone at point P in the figure, equidistant from the speakers. Wave crests emitted from the two speakers at the same time travel equal distances and arrive at point P at the same time; hence the waves arrive in phase, and there is constructive interference. The total wave amplitude at P is twice the amplitude from each individual wave, and we can measure this combined amplitude with the microphone.

Now let's move the microphone to point Q, where the distances from the two speakers to the microphone differ by a half-wavelength. Then the two waves arrive a half-cycle out of step, or *out of phase;* a positive crest from one speaker arrives at the same time as a negative crest from the other. Destructive interference takes place, and the amplitude measured by the microphone is much *smaller* than when only one speaker is present. If the amplitudes from the two speakers are equal, the two waves cancel each other out completely at point Q, and the total amplitude there is zero.

CAUTION **Interference and traveling waves** Although this situation bears some resemblance to standing waves in a pipe, the total wave in Fig. 16.21 is a *traveling* wave, not a standing wave. To see why, recall that in a standing wave there is no net flow of energy in any direction. By contrast, in Fig. 16.21 there is an overall flow of energy from the speakers into the surrounding air; this is characteristic of a traveling wave. The interference between the waves from the two speakers simply causes the energy flow to be *channeled* into certain directions (for example, toward P) and away from other directions (for example, away from Q). You can see another difference between Fig. 16.21 and a standing wave by considering a point, such as Q, where destructive interference occurs. Such a point is *both* a displacement node *and* a pressure node because there is no wave at all at this point. Compare this to a standing wave, in which a pressure node is a displacement antinode, and vice versa.

16.21 Two speakers driven by the same amplifier. Constructive interference occurs at point P, and destructive interference occurs at point Q.

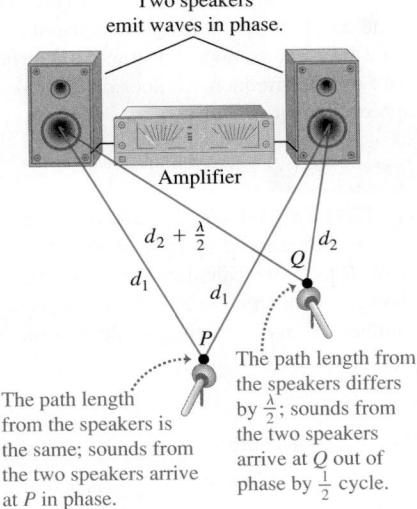

Two speakers emit waves in phase.

Amplifier

$d_2 + \frac{\lambda}{2}$

d_2

Q

d_1

d_1

P

The path length from the speakers is the same; sounds from the two speakers arrive at P in phase.

The path length from the speakers differs by $\frac{\lambda}{2}$; sounds from the two speakers arrive at Q out of phase by $\frac{1}{2}$ cycle.

Constructive interference occurs wherever the distances traveled by the two waves differ by a whole number of wavelengths, $0, \lambda, 2\lambda, 3\lambda, \ldots$; in all these cases the waves arrive at the microphone in phase (Fig. 16.22a). If the distances from the two speakers to the microphone differ by any half-integer number of wavelengths, $\lambda/2, 3\lambda/2, 5\lambda/2, \ldots$, the waves arrive at the microphone out of phase and there will be destructive interference (Fig. 16.22b). In this case, little or no sound energy flows toward the microphone directly in front of the speakers. The energy is instead directed to the sides, where constructive interference occurs.

16.22 Two speakers driven by the same amplifier, emitting waves in phase. Only the waves directed toward the microphone are shown, and they are separated for clarity. **(a)** Constructive interference occurs when the path difference is 0, $\lambda, 2\lambda, 3\lambda, \ldots$. **(b)** Destructive interference occurs when the path difference is $\lambda/2, 3\lambda/2, 5\lambda/2, \ldots$.

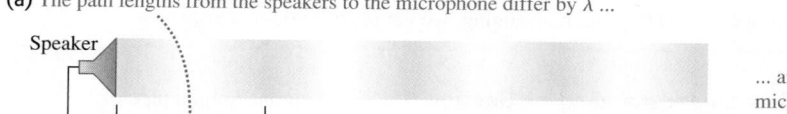

(a) The path lengths from the speakers to the microphone differ by λ ...

Speaker

λ

... so there is constructive interference ...

... and the microphone detects a loud sound.

Speaker

Amplifier

(b) The path lengths from the speakers to the microphone differ by $\frac{\lambda}{2}$ ...

Speaker

$\lambda/2$

... so there is destructive interference ...

... and the microphone detects little or no sound.

Speaker

Amplifier

Example 16.13 Loudspeaker interference

Two small loudspeakers, A and B (Fig. 16.23), are driven by the same amplifier and emit pure sinusoidal waves in phase. (a) For what frequencies does constructive interference occur at point P? (b) For what frequencies does destructive interference occur? The speed of sound is 350 m/s.

SOLUTION

IDENTIFY and SET UP: The nature of the interference at P depends on the difference d in path lengths from point A to P and from point B to P. We calculate the path lengths using the Pythagorean theorem. Constructive interference occurs when d equals a whole number of wavelengths, while destructive interference occurs

when d is a half-integer number of wavelengths. To find the corresponding frequencies, we use $v = f\lambda$.

EXECUTE: The distance from A to P is $[(2.00 \text{ m})^2 + (4.00 \text{ m})^2]^{1/2} = 4.47$ m, and the distance from B to P is $[(1.00 \text{ m})^2 + (4.00 \text{ m})^2]^{1/2} = 4.12$ m. The path difference is $d = 4.47 \text{ m} - 4.12 \text{ m} = 0.35$ m.

(a) Constructive interference occurs when $d = 0, \lambda, 2\lambda, \ldots$ or $d = 0, v/f, 2v/f, \ldots = nv/f$. So the possible frequencies are

$$f_n = \frac{nv}{d} = n\frac{350 \text{ m/s}}{0.35 \text{ m}} \quad (n = 1, 2, 3, \ldots)$$

$$= 1000 \text{ Hz}, 2000 \text{ Hz}, 3000 \text{ Hz}, \ldots$$

(b) Destructive interference occurs when $d = \lambda/2, 3\lambda/2, 5\lambda/2, \ldots$ or $d = v/2f, 3v/2f, 5v/2f, \ldots$. The possible frequencies are

$$f_n = \frac{nv}{2d} = n\frac{350 \text{ m/s}}{2(0.35 \text{ m})} \quad (n = 1, 3, 5, \ldots)$$

$$= 500 \text{ Hz}, 1500 \text{ Hz}, 2500 \text{ Hz}, \ldots$$

EVALUATE: As we increase the frequency, the sound at point P alternates between large and small (near zero) amplitudes, with maxima and minima at the frequencies given above. This effect may not be strong in an ordinary room because of reflections from the walls, floor, and ceiling. It is stronger outdoors and best in an anechoic chamber, which has walls that absorb almost all sound and thereby eliminate reflections.

16.23 What sort of interference occurs at P?

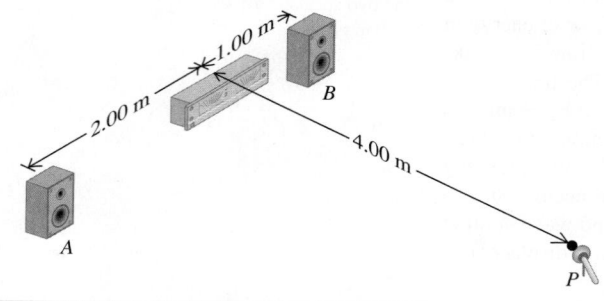

1.00 m

2.00 m

B

4.00 m

A

P

Interference effects are used to control noise from very loud sound sources such as gas-turbine power plants or jet engine test cells. The idea is to use additional sound sources that in some regions of space interfere destructively with the unwanted sound and cancel it out. Microphones in the controlled area feed signals back to the sound sources, which are continuously adjusted for optimum cancellation of noise in the controlled area.

Test Your Understanding of Section 16.6 Suppose that speaker *A* in Fig. 16.23 emits a sinusoidal sound wave of frequency 500 Hz and speaker *B* emits a sinusoidal sound wave of frequency 1000 Hz. What sort of interference will there be between these two waves? (i) constructive interference at various points, including point *P*, and destructive interference at various other points; (ii) destructive interference at various points, including point *P*, and constructive interference at various points; (iii) neither (i) nor (ii). ❚

16.7 Beats

Mastering**PHYSICS**

ActivPhysics 10.7: Beats and Beat Frequency

In Section 16.6 we talked about *interference* effects that occur when two different waves with the same frequency overlap in the same region of space. Now let's look at what happens when we have two waves with equal amplitude but slightly different frequencies. This occurs, for example, when two tuning forks with slightly different frequencies are sounded together, or when two organ pipes that are supposed to have exactly the same frequency are slightly "out of tune."

Consider a particular point in space where the two waves overlap. The displacements of the individual waves at this point are plotted as functions of time in Fig. 16.24a. The total length of the time axis represents 1 second, and the frequencies are 16 Hz (blue graph) and 18 Hz (red graph). Applying the principle of superposition, we add the two displacements at each instant of time to find the total displacement at that time. The result is the graph of Fig. 16.24b. At certain times the two waves are in phase; their maxima coincide and their amplitudes add. But because of their slightly different frequencies, the two waves cannot be in phase at all times. Indeed, at certain times (like $t = 0.50$ s in Fig. 16.24) the two waves are exactly *out* of phase. The two waves then cancel each other, and the total amplitude is zero.

The resultant wave in Fig. 16.24b looks like a single sinusoidal wave with a varying amplitude that goes from a maximum to zero and back. In this example the amplitude goes through two maxima and two minima in 1 second, so the frequency of this amplitude variation is 2 Hz. The amplitude variation causes variations of loudness called **beats,** and the frequency with which the loudness varies is called the **beat frequency.** In this example the beat frequency is the *difference*

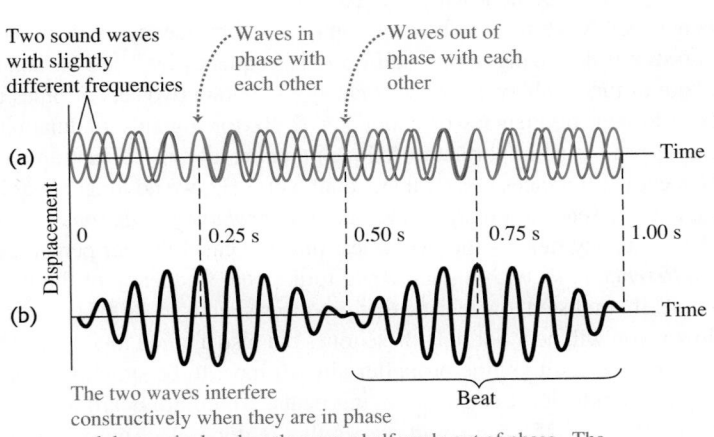

16.24 Beats are fluctuations in amplitude produced by two sound waves of slightly different frequency, here 16 Hz and 18 Hz. (a) Individual waves. (b) Resultant wave formed by superposition of the two waves. The beat frequency is 18 Hz − 16 Hz = 2 Hz.

Two sound waves with slightly different frequencies

Waves in phase with each other

Waves out of phase with each other

(a) ———— Time

Displacement

0   0.25 s   0.50 s   0.75 s   1.00 s

(b) ———— Time

The two waves interfere constructively when they are in phase and destructively when they are a half-cycle out of phase. The resultant wave rises and falls in intensity, forming beats.

Beat

of the two frequencies. If the beat frequency is a few hertz, we hear it as a waver or pulsation in the tone.

We can prove that the beat frequency is *always* the difference of the two frequencies f_a and f_b. Suppose f_a is larger than f_b; the corresponding periods are T_a and T_b, with $T_a < T_b$. If the two waves start out in phase at time $t = 0$, they are again in phase when the first wave has gone through exactly one more cycle than the second. This happens at a value of t equal to T_{beat}, the *period* of the beat. Let n be the number of cycles of the first wave in time T_{beat}; then the number of cycles of the second wave in the same time is $(n - 1)$, and we have the relationships

$$T_{\text{beat}} = nT_a \qquad \text{and} \qquad T_{\text{beat}} = (n - 1)T_b$$

Eliminating n between these two equations, we find

$$T_{\text{beat}} = \frac{T_a T_b}{T_b - T_a}$$

The reciprocal of the beat period is the beat *frequency,* $f_{\text{beat}} = 1/T_{\text{beat}}$, so

$$f_{\text{beat}} = \frac{T_b - T_a}{T_a T_b} = \frac{1}{T_a} - \frac{1}{T_b}$$

and finally

$$f_{\text{beat}} = f_a - f_b \qquad \text{(beat frequency)} \tag{16.24}$$

As claimed, the beat frequency is the difference of the two frequencies. In using Eq. (16.24), remember that f_a is the higher frequency.

An alternative way to derive Eq. (16.24) is to write functions to describe the curves in Fig. 16.24a and then add them. Suppose that at a certain position the two waves are given by $y_a(t) = A \sin 2\pi f_a t$ and $y_b(t) = -A \sin 2\pi f_b t$. We use the trigonometric identity

$$\sin a - \sin b = 2 \sin \tfrac{1}{2}(a - b) \cos \tfrac{1}{2}(a + b)$$

We can then express the total wave $y(t) = y_a(t) + y_b(t)$ as

$$y_a(t) + y_b(t) = \left[2A \sin \tfrac{1}{2}(2\pi)(f_a - f_b)t \right] \cos \tfrac{1}{2}(2\pi)(f_a + f_b)t$$

The amplitude factor (the quantity in brackets) varies slowly with frequency $\tfrac{1}{2}(f_a - f_b)$. The cosine factor varies with a frequency equal to the *average* frequency $\tfrac{1}{2}(f_a + f_b)$. The *square* of the amplitude factor, which is proportional to the intensity that the ear hears, goes through two maxima and two minima per cycle. So the beat frequency f_{beat} that is heard is twice the quantity $\tfrac{1}{2}(f_a - f_b)$, or just $f_a - f_b$, in agreement with Eq. (16.24).

Beats between two tones can be heard up to a beat frequency of about 6 or 7 Hz. Two piano strings or two organ pipes differing in frequency by 2 or 3 Hz sound wavery and "out of tune," although some organ stops contain two sets of pipes deliberately tuned to beat frequencies of about 1 to 2 Hz for a gently undulating effect. Listening for beats is an important technique in tuning all musical instruments.

At frequency differences greater than about 6 or 7 Hz, we no longer hear individual beats, and the sensation merges into one of *consonance* or *dissonance,* depending on the frequency ratio of the two tones. In some cases the ear perceives a tone called a *difference tone,* with a pitch equal to the beat frequency of the two tones. For example, if you listen to a whistle that produces sounds at 1800 Hz and 1900 Hz when blown, you will hear not only these tones but also a much lower 100-Hz tone.

The engines on multiengine propeller aircraft have to be synchronized so that the propeller sounds don't cause annoying beats, which are heard as loud throbbing sounds (Fig. 16.25). On some planes this is done electronically; on others the pilot does it by ear, just like tuning a piano.

16.25 If the two propellers on this airplane are not precisely synchronized, the pilots, passengers, and listeners on the ground will hear beats.

Test Your Understanding of Section 16.7   One tuning fork vibrates at 440 Hz, while a second tuning fork vibrates at an unknown frequency. When both tuning forks are sounded simultaneously, you hear a tone that rises and falls in intensity three times per second. What is the frequency of the second tuning fork? (i) 434 Hz; (ii) 437 Hz; (iii) 443 Hz; (iv) 446 Hz; (v) either 434 Hz or 446 Hz; (vi) either 437 Hz or 443 Hz. ❚

16.8 The Doppler Effect

You've probably noticed that when a car approaches you with its horn sounding, the pitch seems to drop as the car passes. This phenomenon, first described by the 19th-century Austrian scientist Christian Doppler, is called the **Doppler effect.** When a source of sound and a listener are in motion relative to each other, the frequency of the sound heard by the listener is not the same as the source frequency. A similar effect occurs for light and radio waves; we'll return to this later in this section.

To analyze the Doppler effect for sound, we'll work out a relationship between the frequency shift and the velocities of source and listener relative to the medium (usually air) through which the sound waves propagate. To keep things simple, we consider only the special case in which the velocities of both source and listener lie along the line joining them. Let v_S and v_L be the velocity components along this line for the source and the listener, respectively, relative to the medium. We choose the positive direction for both v_S and v_L to be the direction from the listener L to the source S. The speed of sound relative to the medium, v, is always considered positive.

Moving Listener and Stationary Source

Let's think first about a listener L moving with velocity v_L toward a stationary source S (Fig. 16.26). The source emits a sound wave with frequency f_S and wavelength $\lambda = v/f_S$. The figure shows four wave crests, separated by equal distances λ. The wave crests approaching the moving listener have a speed of propagation *relative to the listener* of $(v + v_L)$. So the frequency f_L with which the crests arrive at the listener's position (that is, the frequency the listener hears) is

$$f_L = \frac{v + v_L}{\lambda} = \frac{v + v_L}{v/f_S} \qquad (16.25)$$

or

$$f_L = \left(\frac{v + v_L}{v}\right)f_S = \left(1 + \frac{v_L}{v}\right)f_S \qquad \begin{array}{l}\text{(moving listener,}\\ \text{stationary source)}\end{array} \quad (16.26)$$

MasteringPHYSICS

ActivPhysics 10.8: Doppler Effect: Conceptual Introduction
ActivPhysics 10.9: Doppler Effect: Problems

16.26 A listener moving toward a stationary source hears a frequency that is higher than the source frequency. This is because the relative speed of listener and wave is greater than the wave speed v.

- Velocity of listener (L) = v_L
- Velocity of source (S) = 0 (at rest)
- Speed of sound wave = v
- Positive direction: from listener to source

16.27 Wave crests emitted by a moving source are crowded together in front of the source (to the right of this source) and stretched out behind it (to the left of this source).

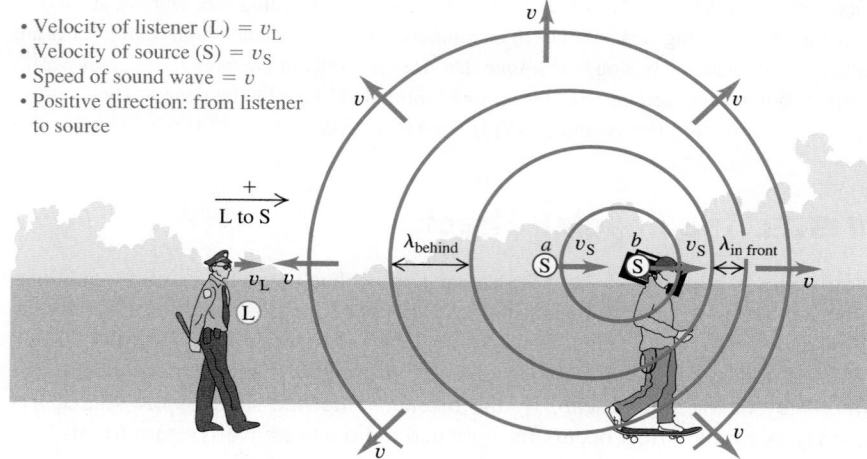

- Velocity of listener (L) = v_L
- Velocity of source (S) = v_S
- Speed of sound wave = v
- Positive direction: from listener to source

So a listener moving toward a source ($v_L > 0$), as in Fig. 16.26, hears a higher frequency (higher pitch) than does a stationary listener. A listener moving away from the source ($v_L < 0$) hears a lower frequency (lower pitch).

Moving Source and Moving Listener

Now suppose the source is also moving, with velocity v_S (Fig. 16.27). The wave speed relative to the wave medium (air) is still v; it is determined by the properties of the medium and is not changed by the motion of the source. But the wavelength is no longer equal to v/f_S. Here's why. The time for emission of one cycle of the wave is the period $T = 1/f_S$. During this time, the wave travels a distance $vT = v/f_S$ and the source moves a distance $v_S T = v_S/f_S$. The wavelength is the distance between successive wave crests, and this is determined by the *relative* displacement of source and wave. As Fig. 16.27 shows, this is different in front of and behind the source. In the region to the right of the source in Fig. 16.27 (that is, in front of the source), the wavelength is

$$\lambda_{\text{in front}} = \frac{v}{f_S} - \frac{v_S}{f_S} = \frac{v - v_S}{f_S} \qquad \begin{array}{l}\text{(wavelength in front} \\ \text{of a moving source)}\end{array} \qquad (16.27)$$

In the region to the left of the source (that is, behind the source), it is

$$\lambda_{\text{behind}} = \frac{v + v_S}{f_S} \qquad \begin{array}{l}\text{(wavelength behind} \\ \text{a moving source)}\end{array} \qquad (16.28)$$

The waves in front of and behind the source are compressed and stretched out, respectively, by the motion of the source.

To find the frequency heard by the listener behind the source, we substitute Eq. (16.28) into the first form of Eq. (16.25):

$$f_L = \frac{v + v_L}{\lambda_{\text{behind}}} = \frac{v + v_L}{(v + v_S)/f_S}$$

$$f_L = \frac{v + v_L}{v + v_S} f_S \qquad \begin{array}{l}\text{(Doppler effect, moving source} \\ \text{and moving listener)}\end{array} \qquad (16.29)$$

This expresses the frequency f_L heard by the listener in terms of the frequency f_S of the source.

Although we derived it for the particular situation shown in Fig. 16.27, Eq. (16.29) includes *all* possibilities for motion of source and listener (relative to the medium) along the line joining them. If the listener happens to be at rest in the medium, v_L is zero. When both source and listener are at rest or have the same velocity relative to the medium, $v_L = v_S$ and $f_L = f_S$. Whenever the direction of the source or listener velocity is opposite to the direction from the listener toward the source (which we have defined as positive), the corresponding velocity to be used in Eq. (16.29) is negative.

As an example, the frequency heard by a listener at rest $(v_L = 0)$ is $f_L = [v/(v + v_S)]f_S$. If the source is moving toward the listener (in the negative direction), then $v_S < 0$, $f_L > f_S$, and the listener hears a higher frequency than that emitted by the source. If instead the source is moving away from the listener (in the positive direction), then $v_S > 0$, $f_L < f_S$, and the listener hears a lower frequency. This explains the change in pitch that you hear from the siren of an ambulance as it passes you (Fig. 16.28).

16.28 The Doppler effect explains why the siren on a fire engine or ambulance has a high pitch $(f_L > f_S)$ when it is approaching you $(v_S < 0)$ and a low pitch $(f_L < f_S)$ when it is moving away $(v_S > 0)$.

Problem-Solving Strategy 16.2 | Doppler Effect

IDENTIFY *the relevant concepts:* The Doppler effect occurs whenever the source of waves, the wave detector (listener), or both are in motion.

SET UP *the problem* using the following steps:
1. Establish a coordinate system, with the positive direction from the listener toward the source. Carefully determine the signs of all relevant velocities. A velocity in the direction from the listener toward the source is positive; a velocity in the opposite direction is negative. All velocities must be measured relative to the air in which the sound travels.
2. Use consistent subscripts to identify the various quantities: S for source and L for listener.
3. Identify which unknown quantities are the target variables.

EXECUTE *the solution* as follows:
1. Use Eq. (16.29) to relate the frequencies at the source and the listener, the sound speed, and the velocities of the source and

the listener according to the sign convention of step 1. If the source is moving, you can find the wavelength measured by the listener using Eq. (16.27) or (16.28).
2. When a wave is reflected from a stationary or moving surface, solve the problem in two steps. In the first, the surface is the "listener"; the frequency with which the wave crests arrive at the surface is f_L. In the second, the surface is the "source," emitting waves with this same frequency f_L. Finally, determine the frequency heard by a listener detecting this new wave.

EVALUATE *your answer:* Is the *direction* of the frequency shift reasonable? If the source and the listener are moving toward each other, $f_L > f_S$; if they are moving apart, $f_L < f_S$. If the source and the listener have no relative motion, $f_L = f_S$.

Example 16.14 | Doppler effect I: Wavelengths

A police car's siren emits a sinusoidal wave with frequency $f_S = 300$ Hz. The speed of sound is 340 m/s and the air is still. (a) Find the wavelength of the waves if the siren is at rest. (b) Find the wavelengths of the waves in front of and behind the siren if it is moving at 30 m/s.

SOLUTION

IDENTIFY and SET UP: In part (a) there is no Doppler effect because neither source nor listener is moving with respect to the air; $v = \lambda f$ gives the wavelength. Figure 16.29 shows the situation in part (b): The source is in motion, so we find the wavelengths using Eqs. (16.27) and (16.28) for the Doppler effect.

EXECUTE: (a) When the source is at rest,

$$\lambda = \frac{v}{f_S} = \frac{340 \text{ m/s}}{300 \text{ Hz}} = 1.13 \text{ m}$$

16.29 Our sketch for this problem.

Police car

$\lambda_{\text{behind}} = ?$ $v_S = 30$ m/s $\lambda_{\text{in front}} = ?$

(b) From Eq. (16.27), in front of the siren

$$\lambda_{\text{in front}} = \frac{v - v_S}{f_S} = \frac{340 \text{ m/s} - 30 \text{ m/s}}{300 \text{ Hz}} = 1.03 \text{ m}$$

From Eq. (16.28), behind the siren

$$\lambda_{\text{behind}} = \frac{v + v_S}{f_S} = \frac{340 \text{ m/s} + 30 \text{ m/s}}{300 \text{ Hz}} = 1.23 \text{ m}$$

EVALUATE: The wavelength is shorter in front of the siren and longer behind it, as we expect.

Example 16.15 Doppler effect II: Frequencies

If a listener L is at rest and the siren in Example 16.14 is moving away from L at 30 m/s, what frequency does the listener hear?

SOLUTION

IDENTIFY and SET UP: Our target variable is the frequency f_L heard by a listener behind the moving source. Figure 16.30 shows the situation. We have $v_L = 0$ and $v_S = +30$ m/s (positive, since the velocity of the source is in the direction from listener to source).

EXECUTE: From Eq. (16.29),

$$f_L = \frac{v}{v + v_S} f_S = \frac{340 \text{ m/s}}{340 \text{ m/s} + 30 \text{ m/s}} (300 \text{ Hz}) = 276 \text{ Hz}$$

EVALUATE: The source and listener are moving apart, so $f_L < f_S$. Here's a check on our numerical result. From Example 16.14, the

16.30 Our sketch for this problem.

wavelength behind the source (where the listener in Fig. 16.30 is located) is 1.23 m. The wave speed relative to the stationary listener is $v = 340$ m/s even though the source is moving, so

$$f_L = \frac{v}{\lambda} = \frac{340 \text{ m/s}}{1.23 \text{ m}} = 276 \text{ Hz}$$

Example 16.16 Doppler effect III: A moving listener

If the siren is at rest and the listener is moving away from it at 30 m/s, what frequency does the listener hear?

SOLUTION

IDENTIFY and SET UP: Again our target variable is f_L, but now L is in motion and S is at rest. Figure 16.31 shows the situation. The velocity of the listener is $v_L = -30$ m/s (negative, since the motion is in the direction from source to listener).

EXECUTE: From Eq. (16.29),

$$f_L = \frac{v + v_L}{v} f_S = \frac{340 \text{ m/s} + (-30 \text{ m/s})}{340 \text{ m/s}} (300 \text{ Hz}) = 274 \text{ Hz}$$

16.31 Our sketch for this problem.

EVALUATE: Again the source and listener are moving apart, so $f_L < f_S$. Note that the *relative velocity* of source and listener is the same as in Example 16.15, but the Doppler shift is different because v_S and v_L are different.

Example 16.17 Doppler effect IV: Moving source, moving listener

The siren is moving away from the listener with a speed of 45 m/s relative to the air, and the listener is moving toward the siren with a speed of 15 m/s relative to the air. What frequency does the listener hear?

SOLUTION

IDENTIFY and SET UP: Now *both* L and S are in motion. Again our target variable is f_L. Both the source velocity $v_S = +45$ m/s and the listener's velocity $v_L = +15$ m/s are positive because both velocities are in the direction from listener to source.

EXECUTE: From Eq. (16.29),

$$f_L = \frac{v + v_L}{v + v_S} f_S = \frac{340 \text{ m/s} + 15 \text{ m/s}}{340 \text{ m/s} + 45 \text{ m/s}} (300 \text{ Hz}) = 277 \text{ Hz}$$

16.32 Our sketch for this problem.

Listener
$f_L = ?$
$v_L = 15$ m/s
L to S
+
Police car
$v_S = 45$ m/s

EVALUATE: As in Examples 16.15 and 16.16, the source and listener again move away from each other at 30 m/s, so again $f_L < f_S$. But f_L is different in all three cases because the Doppler effect for sound depends on how the source and listener are moving relative to the *air*, not simply on how they move relative to each other.

Example 16.18 Doppler effect V: A double Doppler shift

The police car is moving toward a warehouse at 30 m/s. What frequency does the driver hear reflected from the warehouse?

SOLUTION

IDENTIFY: In this situation there are *two* Doppler shifts (Fig. 16.33). In the first shift, the warehouse is the stationary "listener."

16.33 Two stages of the sound wave's motion from the police car to the warehouse and back to the police car.

(a) Sound travels from police car's siren (source S) to warehouse ("listener" L).

(b) Reflected sound travels from warehouse (source S) to police car (listener L).

The frequency of sound reaching the warehouse, which we call f_W, is greater than 300 Hz because the source is approaching. In the second shift, the warehouse acts as a source of sound with frequency f_W, and the listener is the driver of the police car; she hears a frequency greater than f_W because she is approaching the source.

SET UP: To determine f_W, we use Eq. (16.29) with f_L replaced by f_W. For this part of the problem, $v_L = v_W = 0$ (the warehouse is at rest) and $v_S = -30$ m/s (the siren is moving in the negative direction from source to listener).

To determine the frequency heard by the driver (our target variable), we again use Eq. (16.29) but now with f_S replaced by f_W. For this second part of the problem, $v_S = 0$ because the stationary warehouse is the source and the velocity of the listener (the driver) is $v_L = +30$ m/s. (The listener's velocity is positive because it is in the direction from listener to source.)

EXECUTE: The frequency reaching the warehouse is

$$f_W = \frac{v}{v + v_S} f_S = \frac{340 \text{ m/s}}{340 \text{ m/s} + (-30 \text{ m/s})}(300 \text{ Hz}) = 329 \text{ Hz}$$

Then the frequency heard by the driver is

$$f_L = \frac{v + v_L}{v} f_W = \frac{340 \text{ m/s} + 30 \text{ m/s}}{340 \text{ m/s}}(329 \text{ Hz}) = 358 \text{ Hz}$$

EVALUATE: Because there are two Doppler shifts, the reflected sound heard by the driver has an even higher frequency than the sound heard by a stationary listener in the warehouse.

Doppler Effect for Electromagnetic Waves

In the Doppler effect for sound, the velocities v_L and v_S are always measured relative to the *air* or whatever medium we are considering. There is also a Doppler effect for *electromagnetic* waves in empty space, such as light waves or radio waves. In this case there is no medium that we can use as a reference to measure velocities, and all that matters is the *relative* velocity of source and receiver. (By contrast, the Doppler effect for sound does not depend simply on this relative velocity, as discussed in Example 16.17.)

To derive the expression for the Doppler frequency shift for light, we have to use the special theory of relativity. We will discuss this in Chapter 37, but for now we quote the result without derivation. The wave speed is the speed of light, usually denoted by c, and it is the same for both source and receiver. In the frame of reference in which the receiver is at rest, the source is moving away from the receiver with velocity v. (If the source is *approaching* the receiver, v is negative.) The source frequency is again f_S. The frequency f_R measured by the receiver R (the frequency of arrival of the waves at the receiver) is then

$$f_R = \sqrt{\frac{c - v}{c + v}} f_S \qquad \text{(Doppler effect for light)} \qquad (16.30)$$

When v is positive, the source is moving directly *away* from the receiver and f_R is always *less* than f_S; when v is negative, the source is moving directly *toward* the receiver and f_R is *greater* than f_S. The qualitative effect is the same as for sound, but the quantitative relationship is different.

A familiar application of the Doppler effect for radio waves is the radar device mounted on the side window of a police car to check other cars' speeds. The electromagnetic wave emitted by the device is reflected from a moving car, which acts as a moving source, and the wave reflected back to the device is Doppler-shifted in frequency. The transmitted and reflected signals are combined to produce beats, and the speed can be computed from the frequency of the beats. Similar techniques ("Doppler radar") are used to measure wind velocities in the atmosphere.

16.34 Change of velocity component along the line of sight of a satellite passing a tracking station. The frequency received at the tracking station changes from high to low as the satellite passes overhead.

The Doppler effect is also used to track satellites and other space vehicles. In Fig. 16.34 a satellite emits a radio signal with constant frequency f_S. As the satellite orbits past, it first approaches and then moves away from the receiver; the frequency f_R of the signal received on earth changes from a value greater than f_S to a value less than f_S as the satellite passes overhead.

Test Your Understanding of Section 16.8 You are at an outdoor concert with a wind blowing at 10 m/s from the performers toward you. Is the sound you hear Doppler-shifted? If so, is it shifted to lower or higher frequencies? ❙

16.9 Shock Waves

You may have experienced "sonic booms" caused by an airplane flying overhead faster than the speed of sound. We can see qualitatively why this happens from Fig. 16.35. Let v_S denote the *speed* of the airplane relative to the air, so that it is always positive. The motion of the airplane through the air produces sound; if v_S is less than the speed of sound v, the waves in front of the airplane are crowded together with a wavelength given by Eq. (16.27):

$$\lambda_{\text{in front}} = \frac{v - v_S}{f_S}$$

As the speed v_S of the airplane approaches the speed of sound v, the wavelength approaches zero and the wave crests pile up on each other (Fig. 16.35a). The airplane must exert a large force to compress the air in front of it; by Newton's third law, the air exerts an equally large force back on the airplane. Hence there is a large increase in aerodynamic drag (air resistance) as the airplane approaches the speed of sound, a phenomenon known as the "sound barrier."

When v_S is greater in magnitude than v, the source of sound is **supersonic,** and Eqs. (16.27) and (16.29) for the Doppler effect no longer describe the sound wave in front of the source. Figure 16.35b shows a cross section of what happens. As the airplane moves, it displaces the surrounding air and produces sound. A series of wave crests is emitted from the nose of the airplane; each spreads out in a circle centered at the position of the airplane when it emitted the crest. After a time t the crest emitted from point S_1 has spread to a circle with radius vt, and the airplane has moved a greater distance v_St to position S_2. You can see that the circular crests interfere constructively at points along the blue line that makes an angle α with

16.35 Wave crests around a sound source S moving (a) slightly slower than the speed of sound v and (b) faster than the sound speed v. (c) This photograph shows a T-38 jet airplane moving at 1.1 times the speed of sound. Separate shock waves are produced by the nose, wings, and tail. The angles of these waves vary because the air speeds up and slows down as it moves around the airplane, so the relative speed v_S of the airplane and air is different for shock waves produced at different points.

(a) Sound source S (airplane) moving at nearly the speed of sound

(b) Sound source moving faster than the speed of sound

(c) Shock waves around a supersonic airplane

the direction of the airplane velocity, leading to a very-large-amplitude wave crest along this line. This large-amplitude crest is called a **shock wave** (Fig. 16.35c).

From the right triangle in Fig. 16.35b we can see that the angle α is given by

$$\sin\alpha = \frac{vt}{v_S t} = \frac{v}{v_S} \quad \text{(shock wave)} \tag{16.31}$$

In this relationship, v_S is the *speed* of the source (the magnitude of its velocity) relative to the air and is always positive. The ratio v_S/v is called the **Mach number.** It is greater than unity for all supersonic speeds, and $\sin\alpha$ in Eq. (16.31) is the reciprocal of the Mach number. The first person to break the sound barrier was Capt. Chuck Yeager of the U.S. Air Force, flying the Bell X-1 at Mach 1.06 on October 14, 1947 (Fig. 16.36).

Shock waves are actually three-dimensional; a shock wave forms a *cone* around the direction of motion of the source. If the source (possibly a supersonic jet airplane or a rifle bullet) moves with constant velocity, the angle α is constant, and the shock-wave cone moves along with the source. It's the arrival of this shock wave that causes the sonic boom you hear after a supersonic airplane has passed by. The larger the airplane, the stronger the sonic boom; the shock wave produced at ground level by the (now retired) Concorde supersonic airliner flying at 12,000 m (40,000 ft) caused a sudden jump in air pressure of about 20 Pa. In front of the shock-wave cone, there is no sound. Inside the cone a stationary listener hears the Doppler-shifted sound of the airplane moving away.

CAUTION Shock waves We emphasize that a shock wave is produced *continuously* by any object that moves through the air at supersonic speed, not only at the instant that it "breaks the sound barrier." The sound waves that combine to form the shock wave, as in Fig. 16.35b, are created by the motion of the object itself, not by any sound source that the object may carry. The cracking noises of a bullet and of the tip of a circus whip are due to their supersonic motion. A supersonic jet airplane may have very loud engines, but these do not cause the shock wave. Indeed, a space shuttle makes a very loud sonic boom when coming in for a landing; its engines are out of fuel at this point, so it is a supersonic glider.

Shock waves have applications outside of aviation. They are used to break up kidney stones and gallstones without invasive surgery, using a technique with the impressive name *extracorporeal shock-wave lithotripsy*. A shock wave produced outside the body is focused by a reflector or acoustic lens so that as much of it as possible converges on the stone. When the resulting stresses in the stone exceed its tensile strength, it breaks into small pieces and can be eliminated. This technique requires accurate determination of the location of the stone, which may be done using ultrasonic imaging techniques (see Fig. 16.9).

16.36 The first supersonic airplane, the Bell X-1, was shaped much like a 50-caliber bullet—which was known to be able to travel faster than sound.

Example 16.19 Sonic boom from a supersonic airplane

An airplane is flying at Mach 1.75 at an altitude of 8000 m, where the speed of sound is 320 m/s. How long after the plane passes directly overhead will you hear the sonic boom?

SOLUTION

IDENTIFY and SET UP: The shock wave forms a cone trailing backward from the airplane, so the problem is really asking for how much time elapses from when the airplane flies overhead to when the shock wave reaches you at point L (Fig. 16.37). During the time t (our target variable) since the airplane traveling at speed

v_S passed overhead, it has traveled a distance $v_S t$. Equation (16.31) gives the shock cone angle α; we use trigonometry to solve for t.

EXECUTE: From Eq. (16.31) the angle α of the shock cone is

$$\alpha = \arcsin\frac{1}{1.75} = 34.8°$$

The speed of the plane is the speed of sound multiplied by the Mach number:

$$v_S = (1.75)(320 \text{ m/s}) = 560 \text{ m/s}$$

Continued

16.37 You hear a sonic boom when the shock wave reaches you at L (*not* just when the plane breaks the sound barrier). A listener to the right of L has not yet heard the sonic boom but will shortly; a listener to the left of L has already heard the sonic boom.

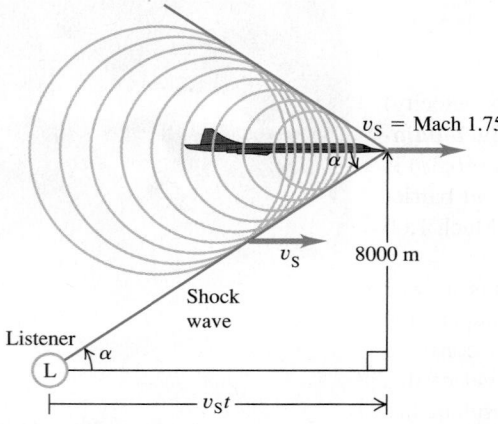

From Fig. 16.37 we have

$$\tan \alpha = \frac{8000 \text{ m}}{v_S t}$$

$$t = \frac{8000 \text{ m}}{(560 \text{ m/s})(\tan 34.8°)} = 20.5 \text{ s}$$

EVALUATE: You hear the boom 20.5 s after the airplane passes overhead, at which time it has traveled $(560 \text{ m/s})(20.5 \text{ s}) = 11.5$ km since it passed overhead. We have assumed that the speed of sound is the same at all altitudes, so that $\alpha = \arcsin v/v_S$ is constant and the shock wave forms a perfect cone. In fact, the speed of sound decreases with increasing altitude. How would this affect the value of t?

Test Your Understanding of Section 16.9 What would you hear if you were directly behind (to the left of) the supersonic airplane in Fig. 16.37? (i) a sonic boom; (ii) the sound of the airplane, Doppler-shifted to higher frequencies; (iii) the sound of the airplane, Doppler-shifted to lower frequencies; (iv) nothing. ❚

Sound waves: Sound consists of longitudinal waves in a medium. A sinusoidal sound wave is characterized by its frequency f and wavelength λ (or angular frequency ω and wave number k) and by its displacement amplitude A. The pressure amplitude p_{max} is directly proportional to the displacement amplitude, the wave number, and the bulk modulus B of the wave medium. (See Examples 16.1 and 16.2.)

The speed of a sound wave in a fluid depends on the bulk modulus B and density ρ. If the fluid is an ideal gas, the speed can be expressed in terms of the temperature T, molar mass M, and ratio of heat capacities γ of the gas. The speed of longitudinal waves in a solid rod depends on the density and Young's modulus Y. (See Examples 16.3 and 16.4.)

$$p_{max} = BkA \quad \text{(16.5)}$$
(sinusoidal sound wave)

$$v = \sqrt{\frac{B}{\rho}} \quad \text{(16.7)}$$
(longitudinal wave in a fluid)

$$v = \sqrt{\frac{\gamma RT}{M}} \quad \text{(16.10)}$$
(sound wave in an ideal gas)

$$v = \sqrt{\frac{Y}{\rho}} \quad \text{(16.8)}$$
(longitudinal wave in a solid rod)

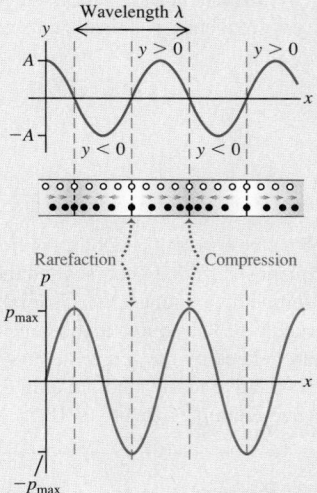

Intensity and sound intensity level: The intensity I of a sound wave is the time average rate at which energy is transported by the wave, per unit area. For a sinusoidal wave, the intensity can be expressed in terms of the displacement amplitude A or the pressure amplitude p_{max}. (See Examples 16.5–16.7.)

The sound intensity level β of a sound wave is a logarithmic measure of its intensity. It is measured relative to I_0, an arbitrary intensity defined to be 10^{-12} W/m². Sound intensity levels are expressed in decibels (dB). (See Examples 16.8 and 16.9.)

$$I = \frac{1}{2}\sqrt{\rho B}\,\omega^2 A^2 = \frac{p_{max}^2}{2\rho v}$$

$$= \frac{p_{max}^2}{2\sqrt{\rho B}} \quad \text{(16.12), (16.14)}$$
(intensity of a sinusoidal sound wave)

$$\beta = (10\text{ dB})\log\frac{I}{I_0} \quad \text{(16.15)}$$
(definition of sound intensity level)

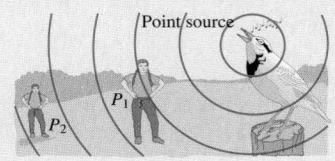

Standing sound waves: Standing sound waves can be set up in a pipe or tube. A closed end is a displacement node and a pressure antinode; an open end is a displacement antinode and a pressure node. For a pipe of length L open at both ends, the normal-mode frequencies are integer multiples of the sound speed divided by $2L$. For a stopped pipe (one that is open at only one end), the normal-mode frequencies are the odd multiples of the sound speed divided by $4L$. (See Examples 16.10 and 16.11.)

A pipe or other system with normal-mode frequencies can be driven to oscillate at any frequency. A maximum response, or resonance, occurs if the driving frequency is close to one of the normal-mode frequencies of the system. (See Example 16.12.)

$$f_n = \frac{nv}{2L} \quad (n = 1, 2, 3, \ldots) \quad \text{(16.18)}$$
(open pipe)

$$f_n = \frac{nv}{4L} \quad (n = 1, 3, 5, \ldots) \quad \text{(16.22)}$$
(stopped pipe)

Interference: When two or more waves overlap in the same region of space, the resulting effects are called interference. The resulting amplitude can be either larger or smaller than the amplitude of each individual wave, depending on whether the waves are in phase (constructive interference) or out of phase (destructive interference). (See Example 16.13.)

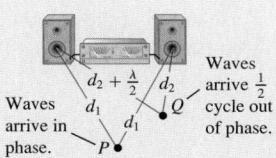

Beats: Beats are heard when two tones with slightly different frequencies f_a and f_b are sounded together. The beat frequency f_{beat} is the difference between f_a and f_b.

$$f_{beat} = f_a - f_b \qquad (16.24)$$
(beat frequency)

Doppler effect: The Doppler effect for sound is the frequency shift that occurs when there is motion of a source of sound, a listener, or both, relative to the medium. The source and listener frequencies f_S and f_L are related by the source and listener velocities v_S and v_L relative to the medium and to the speed of sound v. (See Examples 16.14–16.18.)

$$f_L = \frac{v + v_L}{v + v_S} f_S \qquad (16.29)$$
(Doppler effect, moving source and moving listener)

Shock waves: A sound source moving with a speed v_S greater than the speed of sound v creates a shock wave. The wave front is a cone with angle α. (See Example 16.19.)

$$\sin \alpha = \frac{v}{v_S} \qquad \text{(shock wave)} \qquad (16.31)$$

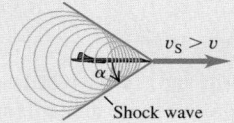

BRIDGING PROBLEM   Loudspeaker Interference

Loudspeakers A and B are 7.00 m apart and vibrate in phase at 172 Hz. They radiate sound uniformly in all directions. Their acoustic power outputs are 8.00×10^{-4} W and 6.00×10^{-5} W, respectively. The air temperature is 20°C. (a) Determine the difference in phase of the two signals at a point C along the line joining A and B, 3.00 m from B and 4.00 m from A. (b) Determine the intensity and sound intensity level at C from speaker A alone (with B turned off) and from speaker B alone (with A turned off). (c) Determine the intensity and sound intensity level at C from both speakers together.

SOLUTION GUIDE

See MasteringPhysics® study area for a Video Tutor solution.

IDENTIFY and SET UP

1. Sketch the situation and label the distances between A, B, and C.
2. Choose the equations that relate power, distance from the source, intensity, pressure amplitude, and sound intensity level.
3. Decide how you will determine the phase difference in part (a). Once you have found the phase difference, how can you use it to find the amplitude of the combined wave at C due to both sources?

4. List the unknown quantities for each part of the problem and identify your target variables.

EXECUTE

5. Determine the phase difference at point C.
6. Find the intensity, sound intensity level, and pressure amplitude at C due to each speaker alone.
7. Use your results from steps 5 and 6 to find the pressure amplitude at C due to both loudspeakers together.
8. Use your result from step 7 to find the intensity and sound intensity level at C due to both loudspeakers together.

EVALUATE

9. How do your results from part (c) for intensity and sound intensity level at C compare to those from part (b)? Does this make sense?
10. What result would you have gotten in part (c) if you had (incorrectly) combined the *intensities* from A and B directly, rather than (correctly) combining the *pressure amplitudes* as you did in step 7?

Problems

For instructor-assigned homework, go to www.masteringphysics.com

•, ••, •••: Problems of increasing difficulty. **CP**: Cumulative problems incorporating material from earlier chapters. **CALC**: Problems requiring calculus. **BIO**: Biosciences problems.

DISCUSSION QUESTIONS

Q16.1 When sound travels from air into water, does the frequency of the wave change? The speed? The wavelength? Explain your reasoning.

Q16.2 The hero of a western movie listens for an oncoming train by putting his ear to the track. Why does this method give an earlier warning of the approach of a train than just listening in the usual way?

Q16.3 Would you expect the pitch (or frequency) of an organ pipe to increase or decrease with increasing temperature? Explain.

Q16.4 In most modern wind instruments the pitch is changed by using keys or valves to change the length of the vibrating air column. The bugle, however, has no valves or keys, yet it can play many notes. How might this be possible? Are there restrictions on what notes a bugle can play?

Q16.5 Symphonic musicians always "warm up" their wind instruments by blowing into them before a performance. What purpose does this serve?

Q16.6 In a popular and amusing science demonstration, a person inhales helium and then his voice becomes high and squeaky. Why does this happen? (*Warning:* Inhaling too much helium can cause unconsciousness or death.)

Q16.7 Lane dividers on highways sometimes have regularly spaced ridges or ripples. When the tires of a moving car roll along such a divider, a musical note is produced. Why? Explain how this phenomenon could be used to measure the car's speed.

Q16.8 The tone quality of an acoustic guitar is different when the strings are plucked near the bridge (the lower end of the strings) than when they are plucked near the sound hole (close to the center of the strings). Why?

Q16.9 Which has a more direct influence on the loudness of a sound wave: the *displacement* amplitude or the *pressure* amplitude? Explain your reasoning.

Q16.10 If the pressure amplitude of a sound wave is halved, by what factor does the intensity of the wave decrease? By what factor must the pressure amplitude of a sound wave be increased in order to increase the intensity by a factor of 16? Explain.

Q16.11 Does the sound intensity level β obey the inverse-square law? Why?

Q16.12 A small fraction of the energy in a sound wave is absorbed by the air through which the sound passes. How does this modify the inverse-square relationship between intensity and distance from the source? Explain your reasoning.

Q16.13 A wire under tension and vibrating in its first overtone produces sound of wavelength λ. What is the new wavelength of the sound (in terms of λ) if the tension is doubled?

Q16.14 A small metal band is slipped onto one of the tines of a tuning fork. As this band is moved closer and closer to the end of the tine, what effect does this have on the wavelength and frequency of the sound the tine produces? Why?

Q16.15 An organist in a cathedral plays a loud chord and then releases the keys. The sound persists for a few seconds and gradually dies away. Why does it persist? What happens to the sound energy when the sound dies away?

Q16.16 Two vibrating tuning forks have identical frequencies, but one is stationary and the other is mounted at the rim of a rotating platform. What does a listener hear? Explain.

Q16.17 A large church has part of the organ in the front of the church and part in the back. A person walking rapidly down the aisle while both segments are playing at once reports that the two segments sound out of tune. Why?

Q16.18 A sound source and a listener are both at rest on the earth, but a strong wind is blowing from the source toward the listener. Is there a Doppler effect? Why or why not?

Q16.19 Can you think of circumstances in which a Doppler effect would be observed for surface waves in water? For elastic waves propagating in a body of water deep below the surface? If so, describe the circumstances and explain your reasoning. If not, explain why not.

Q16.20 Stars other than our sun normally appear featureless when viewed through telescopes. Yet astronomers can readily use the light from these stars to determine that they are rotating and even measure the speed of their surface. How do you think they can do this?

Q16.21 If you wait at a railroad crossing as a train approaches and passes, you hear a Doppler shift in its sound. But if you listen closely, you hear that the change in frequency is continuous; it does not suddenly go from one high frequency to another low frequency. Instead the frequency *smoothly* (but rather quickly) changes from high to low as the train passes. Why does this smooth change occur?

Q16.22 In case 1, a source of sound approaches a stationary observer at speed v. In case 2, the observer moves toward the stationary source at the same speed v. If the source is always producing the same frequency sound, will the observer hear the same frequency in both cases, since the relative speed is the same each time? Why or why not?

Q16.23 Does an aircraft make a sonic boom only at the instant its speed exceeds Mach 1? Explain your reasoning.

Q16.24 If you are riding in a supersonic aircraft, what do you hear? Explain your reasoning. In particular, do you hear a continuous sonic boom? Why or why not?

Q16.25 A jet airplane is flying at a constant altitude at a steady speed v_S greater than the speed of sound. Describe what observers at points A, B, and C hear at the instant shown in Fig. Q16.25, when the shock wave has just reached point B. Explain your reasoning.

Figure **Q16.25**

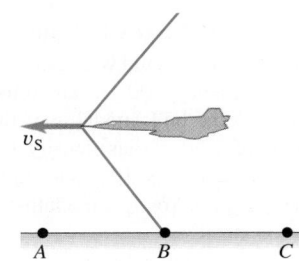

EXERCISES

Unless indicated otherwise, assume the speed of sound in air to be $v = 344$ m/s.

Section 16.1 Sound Waves

16.1 • Example 16.1 (Section 16.1) showed that for sound waves in air with frequency 1000 Hz, a displacement amplitude of

1.2×10^{-8} m produces a pressure amplitude of 3.0×10^{-2} Pa. (a) What is the wavelength of these waves? (b) For 1000-Hz waves in air, what displacement amplitude would be needed for the pressure amplitude to be at the pain threshold, which is 30 Pa? (c) For what wavelength and frequency will waves with a displacement amplitude of 1.2×10^{-8} m produce a pressure amplitude of 1.5×10^{-3} Pa?

16.2 • Example 16.1 (Section 16.1) showed that for sound waves in air with frequency 1000 Hz, a displacement amplitude of 1.2×10^{-8} m produces a pressure amplitude of 3.0×10^{-2} Pa. Water at 20°C has a bulk modulus of 2.2×10^{9} Pa, and the speed of sound in water at this temperature is 1480 m/s. For 1000-Hz sound waves in 20°C water, what displacement amplitude is produced if the pressure amplitude is 3.0×10^{-2} Pa? Explain why your answer is much less than 1.2×10^{-8} m.

16.3 • Consider a sound wave in air that has displacement amplitude 0.0200 mm. Calculate the pressure amplitude for frequencies of (a) 150 Hz; (b) 1500 Hz; (c) 15,000 Hz. In each case compare the result to the pain threshold, which is 30 Pa.

16.4 • A loud factory machine produces sound having a displacement amplitude of 1.00 μm, but the frequency of this sound can be adjusted. In order to prevent ear damage to the workers, the maximum pressure amplitude of the sound waves is limited to 10.0 Pa. Under the conditions of this factory, the bulk modulus of air is 1.42×10^{5} Pa. What is the highest-frequency sound to which this machine can be adjusted without exceeding the prescribed limit? Is this frequency audible to the workers?

16.5 • **BIO** **Ultrasound and Infrasound.** (a) **Whale communication.** Blue whales apparently communicate with each other using sound of frequency 17 Hz, which can be heard nearly 1000 km away in the ocean. What is the wavelength of such a sound in seawater, where the speed of sound is 1531 m/s? (b) **Dolphin clicks.** One type of sound that dolphins emit is a sharp click of wavelength 1.5 cm in the ocean. What is the frequency of such clicks? (c) **Dog whistles.** One brand of dog whistles claims a frequency of 25 kHz for its product. What is the wavelength of this sound? (d) **Bats.** While bats emit a wide variety of sounds, one type emits pulses of sound having a frequency between 39 kHz and 78 kHz. What is the range of wavelengths of this sound? (e) **Sonograms.** Ultrasound is used to view the interior of the body, much as x rays are utilized. For sharp imagery, the wavelength of the sound should be around one-fourth (or less) the size of the objects to be viewed. Approximately what frequency of sound is needed to produce a clear image of a tumor that is 1.0 mm across if the speed of sound in the tissue is 1550 m/s?

Section 16.2 Speed of Sound Waves

16.6 • (a) In a liquid with density 1300 kg/m^3, longitudinal waves with frequency 400 Hz are found to have wavelength 8.00 m. Calculate the bulk modulus of the liquid. (b) A metal bar with a length of 1.50 m has density 6400 kg/m^3. Longitudinal sound waves take 3.90×10^{-4} s to travel from one end of the bar to the other. What is Young's modulus for this metal?

16.7 • A submerged scuba diver hears the sound of a boat horn directly above her on the surface of the lake. At the same time, a friend on dry land 22.0 m from the boat also hears the horn (Fig. E16.7). The horn is 1.2 m above the surface of the water.

What is the distance (labeled by "?" in Fig. E16.7) from the horn to the diver? Both air and water are at 20°C.

16.8 • At a temperature of 27.0°C, what is the speed of longitudinal waves in (a) hydrogen (molar mass 2.02 g/mol); (b) helium (molar mass 4.00 g/mol); (c) argon (molar mass 39.9 g/mol)? See Table 19.1 for values of γ. (d) Compare your answers for parts (a), (b), and (c) with the speed in air at the same temperature.

16.9 • An oscillator vibrating at 1250 Hz produces a sound wave that travels through an ideal gas at 325 m/s when the gas temperature is 22.0°C. For a certain experiment, you need to have the same oscillator produce sound of wavelength 28.5 cm in this gas. What should the gas temperature be to achieve this wavelength?

16.10 •• **CALC** (a) Show that the fractional change in the speed of sound (dv/v) due to a very small temperature change dT is given by $dv/v = \frac{1}{2}dT/T$. (*Hint:* Start with Eq. 16.10.) (b) The speed of sound in air at 20°C is found to be 344 m/s. Use the result in part (a) to find the change in the speed of sound for a 1.0°C change in air temperature.

16.11 •• An 80.0-m-long brass rod is struck at one end. A person at the other end hears two sounds as a result of two longitudinal waves, one traveling in the metal rod and the other traveling in the air. What is the time interval between the two sounds? (The speed of sound in air is 344 m/s; relevant information about brass can be found in Table 11.1 and Table 12.1.)

16.12 •• What must be the stress (F/A) in a stretched wire of a material whose Young's modulus is Y for the speed of longitudinal waves to equal 30 times the speed of transverse waves?

Section 16.3 Sound Intensity

16.13 •• **BIO** **Energy Delivered to the Ear.** Sound is detected when a sound wave causes the tympanic membrane (the eardrum) to vibrate. Typically, the diameter of this membrane is about 8.4 mm in humans. (a) How much energy is delivered to the eardrum each second when someone whispers (20 dB) a secret in your ear? (b) To comprehend how sensitive the ear is to very small amounts of energy, calculate how fast a typical 2.0-mg mosquito would have to fly (in mm/s) to have this amount of kinetic energy.

16.14 • Use information from Table 16.2 to answer the following questions about sound in air. At 20°C the bulk modulus for air is 1.42×10^{5} Pa and its density is 1.20 kg/m^3. At this temperature, what are the pressure amplitude (in Pa and atm) and the displacement amplitude (in m and nm) (a) for the softest sound a person can normally hear at 1000 Hz and (b) for the sound from a riveter at the same frequency? (c) How much energy per second does each wave deliver to a square 5.00 mm on a side?

16.15 •• **Longitudinal Waves in Different Fluids.** (a) A longitudinal wave propagating in a water-filled pipe has intensity 3.00×10^{-6} W/m^2 and frequency 3400 Hz. Find the amplitude A and wavelength λ of the wave. Water has density 1000 kg/m^3 and bulk modulus 2.18×10^{9} Pa. (b) If the pipe is filled with air at pressure 1.00×10^{5} Pa and density 1.20 kg/m^3, what will be the amplitude A and wavelength λ of a longitudinal wave with the same intensity and frequency as in part (a)? (c) In which fluid is the amplitude larger, water or air? What is the ratio of the two amplitudes? Why is this ratio so different from 1.00?

16.16 •• **BIO** **Human Hearing.** A fan at a rock concert is 30 m from the stage, and at this point the sound intensity level is 110 dB. (a) How much energy is transferred to her eardrums each second? (b) How fast would a 2.0-mg mosquito have to fly (in mm/s) to have this much kinetic energy? Compare the mosquito's speed with that found for the whisper in part (a) of Exercise 16.13.

Figure **E16.7**

22.0 m

?

16.17 • A sound wave in air at 20°C has a frequency of 150 Hz and a displacement amplitude of 5.00×10^{-3} mm. For this sound wave calculate the (a) pressure amplitude (in Pa); (b) intensity (in W/m^2); (c) sound intensity level (in decibels).

16.18 •• You live on a busy street, but as a music lover, you want to reduce the traffic noise. (a) If you install special sound-reflecting windows that reduce the sound intensity level (in dB) by 30 dB, by what fraction have you lowered the sound intensity (in W/m^2)? (b) If, instead, you reduce the intensity by half, what change (in dB) do you make in the sound intensity level?

16.19 • BIO For a person with normal hearing, the faintest sound that can be heard at a frequency of 400 Hz has a pressure amplitude of about 6.0×10^{-5} Pa. Calculate the (a) intensity; (b) sound intensity level; (c) displacement amplitude of this sound wave at 20°C.

16.20 •• The intensity due to a number of independent sound sources is the sum of the individual intensities. (a) When four quadruplets cry simultaneously, how many decibels greater is the sound intensity level than when a single one cries? (b) To increase the sound intensity level again by the same number of decibels as in part (a), how many more crying babies are required?

16.21 • CP A baby's mouth is 30 cm from her father's ear and 1.50 m from her mother's ear. What is the difference between the sound intensity levels heard by the father and by the mother?

16.22 •• The Sacramento City Council adopted a law to reduce the allowed sound intensity level of the much-despised leaf blowers from their current level of about 95 dB to 70 dB. With the new law, what is the ratio of the new allowed intensity to the previously allowed intensity?

16.23 •• CP At point A, 3.0 m from a small source of sound that is emitting uniformly in all directions, the sound intensity level is 53 dB. (a) What is the intensity of the sound at A? (b) How far from the source must you go so that the intensity is one-fourth of what it was at A? (c) How far must you go so that the sound intensity level is one-fourth of what it was at A? (d) Does intensity obey the inverse-square law? What about sound intensity level?

16.24 •• (a) If two sounds differ by 5.00 dB, find the ratio of the intensity of the louder sound to that of the softer one. (b) If one sound is 100 times as intense as another, by how much do they differ in sound intensity level (in decibels)? (c) If you increase the volume of your stereo so that the intensity doubles, by how much does the sound intensity level increase?

Section 16.4 Standing Sound Waves and Normal Modes

16.25 • Standing sound waves are produced in a pipe that is 1.20 m long. For the fundamental and first two overtones, determine the locations along the pipe (measured from the left end) of the displacement nodes and the pressure nodes if (a) the pipe is open at both ends and (b) the pipe is closed at the left end and open at the right end.

16.26 • The fundamental frequency of a pipe that is open at both ends is 594 Hz. (a) How long is this pipe? If one end is now closed, find (b) the wavelength and (c) the frequency of the new fundamental.

16.27 • BIO The Human Voice. The human vocal tract is a pipe that extends about 17 cm from the lips to the vocal folds (also called "vocal cords") near the middle of your throat. The vocal folds behave rather like the reed of a clarinet, and the vocal tract acts like a stopped pipe. Estimate the first three standing-wave frequencies of the vocal tract. Use $v = 344$ m/s. (The answers are only an estimate, since the position of lips and tongue affects the motion of air in the vocal tract.)

16.28 •• BIO The Vocal Tract. Many opera singers (and some pop singers) have a range of about $2\frac{1}{2}$ octaves or even greater. Suppose a soprano's range extends from A below middle C (frequency 220 Hz) up to E^b-flat above high C (frequency 1244 Hz). Although the vocal tract is quite complicated, we can model it as a resonating air column, like an organ pipe, that is open at the top and closed at the bottom. The column extends from the mouth down to the diaphragm in the chest cavity, and we can also assume that the lowest note is the fundamental. How long is this column of air if $v = 354$ m/s? Does your result seem reasonable, on the basis of observations of your own body?

16.29 •• A certain pipe produces a fundamental frequency of 262 Hz in air. (a) If the pipe is filled with helium at the same temperature, what fundamental frequency does it produce? (The molar mass of air is 28.8 g/mol, and the molar mass of helium is 4.00 g/mol.) (b) Does your answer to part (a) depend on whether the pipe is open or stopped? Why or why not?

16.30 • Singing in the Shower. A pipe closed at both ends can have standing waves inside of it, but you normally don't hear them because little of the sound can get out. But you *can* hear them if you are *inside* the pipe, such as someone singing in the shower. (a) Show that the wavelengths of standing waves in a pipe of length L that is closed at both ends are $\lambda_n = 2L/n$ and the frequencies are given by $f_n = nv/2L = nf_1$, where $n = 1, 2, 3, \ldots$. (b) Modeling it as a pipe, find the frequency of the fundamental and the first two overtones for a shower 2.50 m tall. Are these frequencies audible?

Section 16.5 Resonance and Sound

16.31 • You blow across the open mouth of an empty test tube and produce the fundamental standing wave of the air column inside the test tube. The speed of sound in air is 344 m/s and the test tube acts as a stopped pipe. (a) If the length of the air column in the test tube is 14.0 cm, what is the frequency of this standing wave? (b) What is the frequency of the fundamental standing wave in the air column if the test tube is half filled with water?

16.32 •• CP You have a stopped pipe of adjustable length close to a taut 85.0-cm, 7.25-g wire under a tension of 4110 N. You want to adjust the length of the pipe so that, when it produces sound at its fundamental frequency, this sound causes the wire to vibrate in its second *overtone* with very large amplitude. How long should the pipe be?

Section 16.6 Interference of Waves

16.33 • Two loudspeakers, A and B (Fig. E16.33), are driven by the same amplifier and emit sinusoidal waves in phase. Speaker B is 2.00 m to the right of speaker A. Consider point Q along the extension of the line connecting the speakers, 1.00 m to the right of speaker B. Both speakers emit sound waves that travel directly from the speaker to point Q. (a) What is the lowest frequency for which *constructive* interference occurs at point Q? (b) What is the lowest frequency for which *destructive* interference occurs at point Q?

Figure **E16.33**

16.34 •• Two loudspeakers, A and B (see Fig. E16.33), are driven by the same amplifier and emit sinusoidal waves in phase. Speaker B is 2.00 m to the right of speaker A. The frequency of the sound waves produced by the loudspeakers is 206 Hz. Consider point P between the speakers and along the line connecting them, a distance x to the right of speaker A. Both speakers emit sound waves that travel directly from the speaker to point P. (a) For what values of x will *destructive* interference occur at point P? (b) For what values of x will *constructive* interference occur at point P? (c) Interference effects like those in parts (a) and (b) are almost never a factor in listening to home stereo equipment. Why not?

16.35 •• Two loudspeakers, A and B, are driven by the same amplifier and emit sinusoidal waves in phase. Speaker B is 12.0 m to the right of speaker A. The frequency of the waves emitted by each speaker is 688 Hz. You are standing between the speakers, along the line connecting them, and are at a point of constructive interference. How far must you walk toward speaker B to move to a point of destructive interference?

16.36 • Two loudspeakers, A and B, are driven by the same amplifier and emit sinusoidal waves in phase. The frequency of the waves emitted by each speaker is 172 Hz. You are 8.00 m from A. What is the closest you can be to B and be at a point of destructive interference?

16.37 • Two loudspeakers, A and B, are driven by the same amplifier and emit sinusoidal waves in phase. The frequency of the waves emitted by each speaker is 860 Hz. Point P is 12.0 m from A and 13.4 m from B. Is the interference at P constructive or destructive? Give the reasoning behind your answer.

16.38 •• Two small stereo speakers are driven in step by the same variable-frequency oscillator. Their sound is picked up by a microphone arranged as shown in Fig. E16.38. For what frequencies does their sound at the speakers produce (a) constructive interference and (b) destructive interference?

Figure **E16.38**

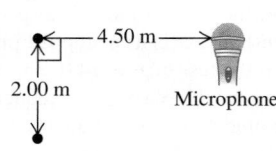

Section 16.7 Beats

16.39 •• **Tuning a Violin.** A violinist is tuning her instrument to concert A (440 Hz). She plays the note while listening to an electronically generated tone of exactly that frequency and hears a beat of frequency 3 Hz, which increases to 4 Hz when she tightens her violin string slightly. (a) What was the frequency of the note played by her violin when she heard the 3-Hz beat? (b) To get her violin perfectly tuned to concert A, should she tighten or loosen her string from what it was when she heard the 3-Hz beat?

16.40 •• Two guitarists attempt to play the same note of wavelength 6.50 cm at the same time, but one of the instruments is slightly out of tune and plays a note of wavelength 6.52 cm instead. What is the frequency of the beat these musicians hear when they play together?

16.41 •• Two organ pipes, open at one end but closed at the other, are each 1.14 m long. One is now lengthened by 2.00 cm. Find the frequency of the beat they produce when playing together in their fundamental.

16.42 •• **Adjusting Airplane Motors.** The motors that drive airplane propellers are, in some cases, tuned by using beats. The whirring motor produces a sound wave having the same frequency as the propeller. (a) If one single-bladed propeller is turning at 575 rpm and you hear a 2.0-Hz beat when you run the second propeller, what are the two possible frequencies (in rpm) of the second propeller? (b) Suppose you increase the speed of the second propeller slightly and find that the beat frequency changes to 2.1 Hz. In part (a), which of the two answers was the correct one for the frequency of the second single-bladed propeller? How do you know?

Section 16.8 The Doppler Effect

16.43 ••• On the planet Arrakis a male ornithoid is flying toward his mate at 25.0 m/s while singing at a frequency of 1200 Hz. If the stationary female hears a tone of 1240 Hz, what is the speed of sound in the atmosphere of Arrakis?

16.44 •• In Example 16.18 (Section 16.8), suppose the police car is moving away from the warehouse at 20 m/s. What frequency does the driver of the police car hear reflected from the warehouse?

16.45 • Two train whistles, A and B, each have a frequency of 392 Hz. A is stationary and B is moving toward the right (away from A) at a speed of 35.0 m/s. A listener is between the two whistles and is moving toward the right with a speed of 15.0 m/s (Fig. E16.45). No wind is blowing. (a) What is the frequency from A as heard by the listener? (b) What is the frequency from B as heard by the listener? (c) What is the beat frequency detected by the listener?

Figure **E16.45**

16.46 • A railroad train is traveling at 25.0 m/s in still air. The frequency of the note emitted by the locomotive whistle is 400 Hz. What is the wavelength of the sound waves (a) in front of the locomotive and (b) behind the locomotive? What is the frequency of the sound heard by a stationary listener (c) in front of the locomotive and (d) behind the locomotive?

16.47 • A swimming duck paddles the water with its feet once every 1.6 s, producing surface waves with this period. The duck is moving at constant speed in a pond where the speed of surface waves is 0.32 m/s, and the crests of the waves ahead of the duck are spaced 0.12 m apart. (a) What is the duck's speed? (b) How far apart are the crests behind the duck?

16.48 • **Moving Source vs. Moving Listener.** (a) A sound source producing 1.00-kHz waves moves toward a stationary listener at one-half the speed of sound. What frequency will the listener hear? (b) Suppose instead that the source is stationary and the listener moves toward the source at one-half the speed of sound. What frequency does the listener hear? How does your answer compare to that in part (a)? Explain on physical grounds why the two answers differ.

16.49 • A car alarm is emitting sound waves of frequency 520 Hz. You are on a motorcycle, traveling directly away from the car. How fast must you be traveling if you detect a frequency of 490 Hz?

16.50 • A railroad train is traveling at 30.0 m/s in still air. The frequency of the note emitted by the train whistle is 262 Hz. What frequency is heard by a passenger on a train moving in the opposite direction to the first at 18.0 m/s and (a) approaching the first and (b) receding from the first?

16.51 • Two swift canaries fly toward each other, each moving at 15.0 m/s relative to the ground, each warbling a note of frequency 1750 Hz. (a) What frequency note does each bird hear from the other one? (b) What wavelength will each canary measure for the note from the other one?

16.52 •• The siren of a fire engine that is driving northward at 30.0 m/s emits a sound of frequency 2000 Hz. A truck in front of this fire engine is moving northward at 20.0 m/s. (a) What is the frequency of the siren's sound that the fire engine's driver hears reflected from the back of the truck? (b) What wavelength would this driver measure for these reflected sound waves?

16.53 •• How fast (as a percentage of light speed) would a star have to be moving so that the frequency of the light we receive from it is 10.0% higher than the frequency of the light it is emitting? Would it be moving away from us or toward us? (Assume it is moving either directly away from us or directly toward us.)

16.54 • **Extrasolar Planets.** In the not-too-distant future, it should be possible to detect the presence of planets moving around other stars by measuring the Doppler shift in the infrared light they emit. If a planet is going around its star at 50.00 km/s while emitting infrared light of frequency 3.330×10^{14} Hz, what frequency light will be received from this planet when it is moving directly away from us? (*Note:* Infrared light is light having wavelengths longer than those of visible light.)

Section 16.9 Shock Waves

16.55 •• A jet plane flies overhead at Mach 1.70 and at a constant altitude of 950 m. (a) What is the angle α of the shock-wave cone? (b) How much time after the plane passes directly overhead do you hear the sonic boom? Neglect the variation of the speed of sound with altitude.

16.56 • The shock-wave cone created by the space shuttle at one instant during its reentry into the atmosphere makes an angle of 58.0° with its direction of motion. The speed of sound at this altitude is 331 m/s. (a) What is the Mach number of the shuttle at this instant, and (b) how fast (in m/s and in mi/h) is it traveling relative to the atmosphere? (c) What would be its Mach number and the angle of its shock-wave cone if it flew at the same speed but at low altitude where the speed of sound is 344 m/s?

PROBLEMS

16.57 ••• **CP** Two identical taut strings under the same tension F produce a note of the same fundamental frequency f_0. The tension in one of them is now increased by a very small amount ΔF. (a) If they are played together in their fundamental, show that the frequency of the beat produced is $f_{\text{beat}} = f_0(\Delta F/2F)$. (b) Two identical violin strings, when in tune and stretched with the same tension, have a fundamental frequency of 440.0 Hz. One of the strings is retuned by increasing its tension. When this is done, 1.5 beats per second are heard when both strings are plucked simultaneously at their centers. By what percentage was the string tension changed?

16.58 •• **CALC** (a) Defend the following statement: "In a sinusoidal sound wave, the pressure variation given by Eq. (16.4) is greatest where the displacement given by Eq. (16.1) is zero." (b) For a sinusoidal sound wave given by Eq. (16.1) with amplitude $A = 10.0\ \mu\text{m}$ and wavelength $\lambda = 0.250$ m, graph the displacement y and pressure fluctuation p as functions of x at time $t = 0$. Show at least two wavelengths of the wave on your graphs. (c) The displacement y in a *non*sinusoidal sound wave is shown in Fig. P16.58 as a function of x for $t = 0$. Draw a graph showing the pressure fluctuation p in this wave as a function of x at $t = 0$. This sound wave has the same 10.0-μm amplitude as the wave in part (b). Does it have the same pressure amplitude? Why or why not? (d) Is the statement in part (a) necessarily true if the sound wave is *not* sinusoidal? Explain your reasoning.

Figure **P16.58**

16.59 •• A soprano and a bass are singing a duet. While the soprano sings an A-sharp at 932 Hz, the bass sings an A-sharp but three octaves lower. In this concert hall, the density of air is 1.20 kg/m³ and its bulk modulus is 1.42×10^5 Pa. In order for their notes to have the same sound intensity level, what must be (a) the ratio of the pressure amplitude of the bass to that of the soprano and (b) the ratio of the displacement amplitude of the bass to that of the soprano? (c) What displacement amplitude (in m and in nm) does the soprano produce to sing her A-sharp at 72.0 dB?

16.60 •• **CP** The sound from a trumpet radiates uniformly in all directions in 20°C air. At a distance of 5.00 m from the trumpet the sound intensity level is 52.0 dB. The frequency is 587 Hz. (a) What is the pressure amplitude at this distance? (b) What is the displacement amplitude? (c) At what distance is the sound intensity level 30.0 dB?

16.61 ••• **A Thermometer.** Suppose you have a tube of length L containing a gas whose temperature you want to take, but you cannot get inside the tube. One end is closed, and the other end is open but a small speaker producing sound of variable frequency is at that end. You gradually increase the frequency of the speaker until the sound from the tube first becomes very loud. With further increase of the frequency, the loudness decreases but then gets very loud again at still higher frequencies. Call f_0 the lowest frequency at which the sound is very loud. (a) Show that the absolute temperature of this gas is given by $T = 16ML^2f_0{}^2/\gamma R$, where M is the molar mass of the gas, γ is the ratio of its heat capacities, and R is the ideal gas constant. (b) At what frequency above f_0 will the sound from the tube next reach a maximum in loudness? (c) How could you determine the speed of sound in this tube at temperature T?

16.62 •• **CP** A uniform 165-N bar is supported horizontally by two identical wires A and B (Fig. P16.62). A small 185-N cube of lead is placed three-fourths of the way from A to B. The wires are each 75.0 cm long and have a mass of 5.50 g.

Figure **P16.62**

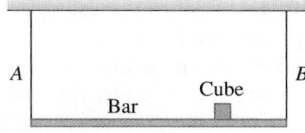

If both of them are simultaneously plucked at the center, what is the frequency of the beats that they will produce when vibrating in their fundamental?

16.63 • **CP** A person is playing a small flute 10.75 cm long, open at one end and closed at the other, near a taut string having a fundamental frequency of 600.0 Hz. If the speed of sound is 344.0 m/s, for which harmonics of the flute will the string resonate? In each case, which harmonic of the string is in resonance?

16.64 ••• **CP** **A New Musical Instrument.** You have designed a new musical instrument of very simple construction. Your design consists of a metal tube with length L and diameter $L/10$. You have stretched a string of mass per unit length μ across the open end of the tube. The other end of the tube is closed. To produce the musical effect you're looking for, you want the frequency of the third-harmonic standing wave on the string to be the same as the fundamental frequency for sound waves in the air column in the tube. The speed of sound waves in this air column is v_s. (a) What must be the tension of the string to produce the desired effect? (b) What happens to the sound produced by the instrument if the tension is changed to twice the value calculated in part (a)? (c) For the tension calculated in part (a), what other harmonics of the string, if any, are in resonance with standing waves in the air column?

16.65 • An organ pipe has two successive harmonics with frequencies 1372 and 1764 Hz. (a) Is this an open or a stopped pipe? Explain. (b) What two harmonics are these? (c) What is the length of the pipe?

16.66 • **Longitudinal Standing Waves in a Solid.** Longitudinal standing waves can be produced in a solid rod by holding it at some point between the fingers of one hand and stroking it with the other hand. The rod oscillates with antinodes at both ends. (a) Why are the ends antinodes and not nodes? (b) The fundamental frequency can be obtained by stroking the rod while it is held at its center. Explain why this is the *only* place to hold the rod to obtain the fundamental. (c) Calculate the fundamental frequency of a steel rod of length 1.50 m (see Table 16.1). (d) What is the next possible standing-wave frequency of this rod? Where should the rod be held to excite a standing wave of this frequency?

16.67 •• A long tube contains air at a pressure of 1.00 atm and a temperature of 77.0°C. The tube is open at one end and closed at the other by a movable piston. A tuning fork near the open end is vibrating with a frequency of 500 Hz. Resonance is produced when the piston is at distances 18.0, 55.5, and 93.0 cm from the open end. (a) From these measurements, what is the speed of sound in air at 77.0°C? (b) From the result of part (a), what is the value of γ? (c) These data show that a displacement antinode is slightly outside of the open end of the tube. How far outside is it?

16.68 ••• The frequency of the note F_4 is 349 Hz. (a) If an organ pipe is open at one end and closed at the other, what length must it have for its fundamental mode to produce this note at 20.0°C? (b) At what air temperature will the frequency be 370 Hz, corresponding to a rise in pitch from F to F-sharp? (Ignore the change in length of the pipe due to the temperature change.)

16.69 • A standing wave with a frequency of 1100 Hz in a column of methane (CH_4) at 20.0°C produces nodes that are 0.200 m apart. What is the value of γ for methane? (The molar mass of methane is 16.0 g/mol.)

16.70 •• Two identical loudspeakers are located at points A and B, 2.00 m apart. The loudspeakers are driven by the same amplifier and produce sound waves with a frequency of 784 Hz. Take the speed of sound in air to be 344 m/s. A small microphone is moved out from point B along a line perpendicular to the line connecting A and B (line BC in Fig. P16.70). (a) At what distances from B will there be *destructive* interference? (b) At what distances from B will there be *constructive* interference? (c) If the frequency is made low enough, there will be no positions along the line BC at which destructive interference occurs. How low must the frequency be for this to be the case?

Figure **P16.70**

16.71 • **Wagnerian Opera.** A man marries a great Wagnerian soprano but, alas, he discovers he cannot stand Wagnerian opera. In order to save his eardrums, the unhappy man decides he must silence his larklike wife for good. His plan is to tie her to the front of his car and send car and soprano speeding toward a brick wall. This soprano is quite shrewd, however, having studied physics in her student days at the music conservatory. She realizes that this wall has a resonant frequency of 600 Hz, which means that if a continuous sound wave of this frequency hits the wall, it will fall down, and she will be saved to sing more Isoldes. The car is heading toward the wall at a high speed of 30 m/s. (a) At what frequency must the soprano sing so that the wall will crumble? (b) What frequency will the soprano hear reflected from the wall just before it crumbles?

16.72 •• A bat flies toward a wall, emitting a steady sound of frequency 1.70 kHz. This bat hears its own sound plus the sound reflected by the wall. How fast should the bat fly in order to hear a beat frequency of 10.0 Hz?

16.73 •• **CP** A person leaning over a 125-m-deep well accidentally drops a siren emitting sound of frequency 2500 Hz. Just before this siren hits the bottom of the well, find the frequency and wavelength of the sound the person hears (a) coming directly from the siren and (b) reflected off the bottom of the well. (c) What beat frequency does this person perceive?

16.74 ••• **BIO** **Ultrasound in Medicine.** A 2.00-MHz sound wave travels through a pregnant woman's abdomen and is reflected from the fetal heart wall of her unborn baby. The heart wall is moving toward the sound receiver as the heart beats. The reflected sound is then mixed with the transmitted sound, and 72 beats per second are detected. The speed of sound in body tissue is 1500 m/s. Calculate the speed of the fetal heart wall at the instant this measurement is made.

16.75 •• The sound source of a ship's sonar system operates at a frequency of 22.0 kHz. The speed of sound in water (assumed to be at a uniform 20°C) is 1482 m/s. (a) What is the wavelength of the waves emitted by the source? (b) What is the difference in frequency between the directly radiated waves and the waves reflected from a whale traveling directly toward the ship at 4.95 m/s? The ship is at rest in the water.

16.76 • **CP** A police siren of frequency f_{siren} is attached to a vibrating platform. The platform and siren oscillate up and down in simple harmonic motion with amplitude A_p and frequency f_p. (a) Find the maximum and minimum sound frequencies that you would hear at a position directly above the siren. (b) At what point in the motion of the platform is the maximum frequency heard? The minimum frequency? Explain.

16.77 ••• **BIO** Horseshoe bats (genus *Rhinolophus*) emit sounds from their nostrils and then listen to the frequency of the sound reflected from their prey to determine the prey's speed. (The "horseshoe" that gives the bat its name is a depression around the nostrils that acts like a focusing mirror, so that the bat emits sound in a narrow beam like a flashlight.) A *Rhinolophus* flying at speed v_{bat} emits sound of frequency f_{bat}; the sound it hears reflected from an insect flying toward it has a higher frequency f_{refl}. (a) Show that the speed of the insect is

$$v_{insect} = v\left[\frac{f_{refl}(v - v_{bat}) - f_{bat}(v + v_{bat})}{f_{refl}(v - v_{bat}) + f_{bat}(v + v_{bat})}\right]$$

where v is the speed of sound. (b) If $f_{bat} = 80.7$ kHz, $f_{refl} = 83.5$ kHz, and $v_{bat} = 3.9$ m/s, calculate the speed of the insect.

16.78 •• (a) Show that Eq. (16.30) can be written as

$$f_R = f_S\left(1 - \frac{v}{c}\right)^{1/2}\left(1 + \frac{v}{c}\right)^{-1/2}$$

(b) Use the binomial theorem to show that if $v \ll c$, this is approximately equal to

$$f_R = f_S\left(1 - \frac{v}{c}\right)$$

(c) A pilotless reconnaissance aircraft emits a radio signal with a frequency of 243 MHz. It is flying directly toward a test engineer on the ground. The engineer detects beats between the received signal and a local signal also of frequency 243 MHz. The beat frequency is 46.0 Hz. What is the speed of the aircraft? (Radio waves travel at the speed of light, $c = 3.00 \times 10^8$ m/s.)

16.79 •• **Supernova!**   The gas cloud known as the Crab Nebula can be seen with even a small telescope. It is the remnant of a *supernova*, a cataclysmic explosion of a star. The explosion was seen on the earth on July 4, 1054 C.E. The streamers glow with the characteristic red color of heated hydrogen gas. In a laboratory on the earth, heated hydrogen produces red light with frequency 4.568×10^{14} Hz; the red light received from streamers in the Crab Nebula pointed toward the earth has frequency 4.586×10^{14} Hz. (a) Estimate the speed with which the outer edges of the Crab Nebula are expanding. Assume that the speed of the center of the nebula relative to the earth is negligible. (You may use the formulas derived in Problem 16.78. The speed of light is 3.00×10^8 m/s.) (b) Assuming that the expansion speed has been constant since the supernova explosion, estimate the diameter of the Crab Nebula. Give your answer in meters and in light-years. (c) The angular diameter of the Crab Nebula as seen from earth is about 5 arc minutes (1 arc minute = $\frac{1}{60}$ degree). Estimate the distance (in light-years) to the Crab Nebula, and estimate the year in which the supernova explosion actually took place.

16.80 •• **CP** A turntable 1.50 m in diameter rotates at 75 rpm. Two speakers, each giving off sound of wavelength 31.3 cm, are attached to the rim of the table at opposite ends of a diameter. A listener stands in front of the turntable. (a) What is the greatest beat frequency the listener will receive from this system? (b) Will the listener be able to distinguish individual beats?

16.81 •• A woman stands at rest in front of a large, smooth wall. She holds a vibrating tuning fork of frequency f_0 directly in front of her (between her and the wall). (a) The woman now runs toward the wall with speed v_W. She detects beats due to the interference between the sound waves reaching her directly from the fork and those reaching her after being reflected from the wall. How many beats per second will she detect? (*Note:* If the beat frequency is too large, the woman may have to use some instrumentation other than

her ears to detect and count the beats.) (b) If the woman instead runs away from the wall, holding the tuning fork at her back so it is between her and the wall, how many beats per second will she detect?

16.82 •• On a clear day you see a jet plane flying overhead. From the apparent size of the plane, you determine that it is flying at a constant altitude h. You hear the sonic boom at time T after the plane passes directly overhead. Show that if the speed of sound v is the same at all altitudes, the speed of the plane is

$$v_S = \frac{hv}{\sqrt{h^2 - v^2T^2}}$$

(*Hint:* Trigonometric identities will be useful.)

CHALLENGE PROBLEMS

16.83 ••• **CALC** Figure P16.83 shows the pressure fluctuation p of a nonsinusoidal sound wave as a function of x for $t = 0$. The wave is traveling in the $+x$-direction. (a) Graph the pressure fluctuation p as a function of t for $x = 0$. Show at least two cycles of oscillation. (b) Graph the displacement y in this sound wave as a function of x at $t = 0$. At $x = 0$, the displacement at $t = 0$ is zero. Show at least two wavelengths of the wave. (c) Graph the displacement y as a function of t for $x = 0$. Show at least two cycles of oscillation. (d) Calculate the maximum velocity and the maximum acceleration of an element of the air through which this sound wave is traveling. (e) Describe how the cone of a loudspeaker must move as a function of time to produce the sound wave in this problem.

Figure **P16.83**

16.84 ••• **CP Longitudinal Waves on a Spring.**   A long spring such as a Slinky™ is often used to demonstrate longitudinal waves. (a) Show that if a spring that obeys Hooke's law has mass m, length L, and force constant k', the speed of longitudinal waves on the spring is $v = L\sqrt{k'/m}$. (see Section 16.2). (b) Evaluate v for a spring with $m = 0.250$ kg, $L = 2.00$ m, and $k' = 1.50$ N/m.

Answers

Chapter Opening Question ?

Both musical sound and noise are made up of a combination of sinusoidal sound waves. The difference is that the frequencies of the sine waves in musical sound are all integer multiples of a fundamental frequency, while *all* frequencies are present in noise.

Test Your Understanding Questions

16.1 Answer: (v) From Eq. (16.5), the displacement amplitude is $A = p_{max}/Bk$. The pressure amplitude p_{max} and bulk modulus B remain the same, but the frequency f increases by a factor of 4. Hence the wave number $k = \omega/v = 2\pi f/v$ also increases by a factor of 4. Since A is inversely proportional to k, the displacement amplitude becomes $\frac{1}{4}$ as great. In other words, at higher frequency

a smaller maximum displacement is required to produce the same maximum pressure fluctuation.

16.2 Answer: (i) From Eq. (16.7), the speed of longitudinal waves (sound) in a fluid is $v = \sqrt{B/\rho}$. We can rewrite this to give an expression for the bulk modulus B in terms of the fluid density ρ and the sound speed v: $B = \rho v^2$. At 20°C the speed of sound in mercury is slightly less than in water (1451 m/s versus 1482 m/s), but the density of mercury is greater than that of water by a large factor (13.6). Hence the bulk modulus of mercury is greater than that of water by a factor of $(13.6)(1451/1482)^2 = 13.0$.

16.3 Answer: A and p_{max} increase by a factor of $\sqrt{2}$, B and v are unchanged, β increases by 3.0 dB Equations (16.9) and (16.10) show that the bulk modulus B and sound speed v remain the same because the physical properties of the air are unchanged. From Eqs. (16.12) and (16.14), the intensity is proportional to the square of the displacement amplitude or the square of the pressure amplitude. Hence doubling the intensity means that A and p_{max} both increase by a factor of $\sqrt{2}$. Example 16.9 shows that *multiplying* the intensity by a factor of $2 (I_2/I_1 = 2)$ corresponds to *adding* to the sound intensity level by $(10\text{ dB}) \log (I_2/I_1) = (10\text{ dB}) \log 2 = 3.0$ dB.

16.4 Answer: (ii) Helium is less dense and has a lower molar mass than air, so sound travels faster in helium than in air. The normal-mode frequencies for a pipe are proportional to the sound speed v, so the frequency and hence the pitch increase when the air in the pipe is replaced with helium.

16.5 Answer: (i) and (iv) There will be a resonance if 660 Hz is one of the pipe's normal-mode frequencies. A stopped organ pipe has normal-mode frequencies that are odd multiples of its fundamental frequency [see Eq. (16.22) and Fig. 16.18]. Hence pipe (i), which has fundamental frequency 220 Hz, also has a normal-mode frequency of $3 (220\text{ Hz}) = 660$ Hz. Pipe (ii) has twice the length of pipe (i); from Eq. (16.20), the fundamental frequency of a stopped pipe is inversely proportional to the length, so pipe (ii) has a fundamental frequency of $\left(\frac{1}{2}\right)(220\text{ Hz}) = 110$ Hz. Its other normal-mode frequencies are 330 Hz, 550 Hz, 770 Hz, . . . , so a 660-Hz tuning fork will not cause resonance. Pipe (iii) is an open pipe of the same length as pipe (i), so its fundamental frequency is twice as great as for pipe (i) [compare Eqs. (16.16) and (16.20)], or $2 (220\text{ Hz}) = 440$ Hz. Its other normal-mode frequencies are integer multiples of the fundamental frequency [see Eq. (16.19)], or 880 Hz, 1320 Hz, . . . , none of which match the 660-Hz frequency of the tuning fork. Pipe (iv) is also an open pipe but with twice the length of pipe (iii) [see Eq. (16.18)], so its normal-mode frequencies are one-half those of pipe (iii): 220 Hz, 440 Hz, 660 Hz, . . . , so the third harmonic will resonate with the tuning fork.

16.6 Answer: (iii) Constructive and destructive interference between two waves can occur only if the two waves have the same frequency. In this case the frequencies are different, so there are no points where the two waves always reinforce each other (constructive interference) or always cancel each other (destructive interference).

16.7 Answer: (vi) The beat frequency is 3 Hz, so the difference between the two tuning fork frequencies is also 3 Hz. Hence the second tuning fork vibrates at a frequency of either 443 Hz or 437 Hz. You can distinguish between the two possibilities by comparing the pitches of the two tuning forks sounded one at a time: The frequency is 437 Hz if the second tuning fork has a lower pitch and 443 Hz if it has a higher pitch.

16.8 Answer: no The air (the medium for sound waves) is moving from the source toward the listener. Hence, relative to the air, both the source and the listener are moving in the direction from listener to source. So both velocities are positive and $v_S = v_L = +10$ m/s. The equality of these two velocities means that the numerator and the denominator in Eq. (16.29) are the same, so $f_L = f_S$ and there is *no* Doppler shift.

16.9 Answer: (iii) Figure 16.37 shows that there are sound waves inside the cone of the shock wave. Behind the airplane the wave crests are spread apart, just as they are behind the moving source in Fig. 16.27. Hence the waves that reach you have an increased wavelength and a lower frequency.

Bridging Problem

Answers: (a) $180° = \pi$ rad

(b) A alone: $I = 3.98 \times 10^{-6}$ W/m^2, $\beta = 66.0$ dB; B alone: $I = 5.31 \times 10^{-7}$ W/m^2, $\beta = 57.2$ dB

(c) $I = 1.60 \times 10^{-6}$ W/m^2, $\beta = 62.1$ dB

TEMPERATURE AND HEAT

? At a steelworks, molten iron is heated to 1500° Celsius to remove impurities. Is it accurate to say that the molten iron contains heat?

LEARNING GOALS

By studying this chapter, you will learn:

- The meaning of thermal equilibrium, and what thermometers really measure.

- How different types of thermometers function.

- The physics behind the absolute, or Kelvin, temperature scale.

- How the dimensions of an object change as a result of a temperature change.

- The meaning of heat, and how it differs from temperature.

- How to do calculations that involve heat flow, temperature changes, and changes of phase.

- How heat is transferred by conduction, convection, and radiation.

Whether it's a sweltering summer day or a frozen midwinter night, your body needs to be kept at a nearly constant temperature. It has effective temperature-control mechanisms, but sometimes it needs help. On a hot day you wear less clothing to improve heat transfer from your body to the air and for better cooling by evaporation of perspiration. You drink cold beverages and may sit near a fan or in an air-conditioned room. On a cold day you wear more clothes or stay indoors where it's warm. When you're outside, you keep active and drink hot liquids to stay warm. The concepts in this chapter will help you understand the basic physics of keeping warm or cool.

The terms "temperature" and "heat" are often used interchangeably in everyday language. In physics, however, these two terms have very different meanings. In this chapter we'll define temperature in terms of how it's measured and see how temperature changes affect the dimensions of objects. We'll see that heat refers to energy transfer caused by temperature differences and learn how to calculate and control such energy transfers.

Our emphasis in this chapter is on the concepts of temperature and heat as they relate to *macroscopic* objects such as cylinders of gas, ice cubes, and the human body. In Chapter 18 we'll look at these same concepts from a *microscopic* viewpoint in terms of the behavior of individual atoms and molecules. These two chapters lay the groundwork for the subject of **thermodynamics,** the study of energy transformations involving heat, mechanical work, and other aspects of energy and how these transformations relate to the properties of matter. Thermodynamics forms an indispensable part of the foundation of physics, chemistry, and the life sciences, and its applications turn up in such places as car engines, refrigerators, biochemical processes, and the structure of stars. We'll explore the key ideas of thermodynamics in Chapters 19 and 20.

17.1 Temperature and Thermal Equilibrium

The concept of **temperature** is rooted in qualitative ideas of "hot" and "cold" based on our sense of touch. A body that feels hot usually has a higher temperature than a similar body that feels cold. That's pretty vague, and the senses can be deceived. But many properties of matter that we can *measure* depend on temperature. The length of a metal rod, steam pressure in a boiler, the ability of a wire to conduct an electric current, and the color of a very hot glowing object—all these depend on temperature.

Temperature is also related to the kinetic energies of the molecules of a material. In general this relationship is fairly complex, so it's not a good place to start in *defining* temperature. In Chapter 18 we will look at the relationship between temperature and the energy of molecular motion for an ideal gas. It is important to understand, however, that temperature and heat can be defined independently of any detailed molecular picture. In this section we'll develop a *macroscopic* definition of temperature.

To use temperature as a measure of hotness or coldness, we need to construct a temperature scale. To do this, we can use any measurable property of a system that varies with its "hotness" or "coldness." Figure 17.1a shows a familiar system that is used to measure temperature. When the system becomes hotter, the colored liquid (usually mercury or ethanol) expands and rises in the tube, and the value of *L* increases. Another simple system is a quantity of gas in a constant-volume container (Fig. 17.1b). The pressure *p*, measured by the gauge, increases or decreases as the gas becomes hotter or colder. A third example is the electrical resistance *R* of a conducting wire, which also varies when the wire becomes hotter or colder. Each of these properties gives us a number (*L*, *p*, or *R*) that varies with hotness and coldness, so each property can be used to make a **thermometer.**

To measure the temperature of a body, you place the thermometer in contact with the body. If you want to know the temperature of a cup of hot coffee, you stick the thermometer in the coffee; as the two interact, the thermometer becomes hotter and the coffee cools off a little. After the thermometer settles down to a steady value, you read the temperature. The system has reached an *equilibrium* condition, in which the interaction between the thermometer and the coffee causes no further change in the system. We call this a state of **thermal equilibrium.**

If two systems are separated by an insulating material or **insulator** such as wood, plastic foam, or fiberglass, they influence each other more slowly. Camping coolers are made with insulating materials to delay the ice and cold food inside from warming up and attaining thermal equilibrium with the hot summer air outside. An *ideal insulator* is a material that permits no interaction at all between the two systems. It prevents the systems from attaining thermal equilibrium if they aren't in thermal equilibrium at the start. An ideal insulator is just that, an idealization; real insulators, like those in camping coolers, aren't ideal, so the contents of the cooler will warm up eventually.

The Zeroth Law of Thermodynamics

We can discover an important property of thermal equilibrium by considering three systems, *A*, *B*, and *C*, that initially are not in thermal equilibrium (Fig. 17.2). We surround them with an ideal insulating box so that they cannot interact with anything except each other. We separate systems *A* and *B* with an ideal insulating wall (the green slab in Fig. 17.2a), but we let system *C* interact with both systems *A* and *B*. This interaction is shown in the figure by a yellow slab representing a thermal **conductor,** a material that *permits* thermal interactions through it. We wait until thermal equilibrium is attained; then *A* and *B* are each in thermal equilibrium with *C*. But are they in thermal equilibrium *with each other?*

To find out, we separate system *C* from systems *A* and *B* with an ideal insulating wall (Fig. 17.2b), and then we replace the insulating wall between *A* and *B*

17.1 Two devices for measuring temperature.

(a) Changes in temperature cause the liquid's volume to change.

Thick glass wall

Capillary of small volume

L

Zero level

Liquid (mercury or ethanol)

Thin glass wall

(b) Changes in temperature cause the pressure of the gas to change.

p

Container of gas at constant volume

with a *conducting* wall that lets *A* and *B* interact. What happens? Experiment shows that *nothing* happens; there are no additional changes to *A* or *B*. We conclude:

> **If *C* is initially in thermal equilibrium with both *A* and *B*, then *A* and *B* are also in thermal equilibrium with each other. This result is called the zeroth law of thermodynamics.**

(The importance of this law was recognized only after the first, second, and third laws of thermodynamics had been named. Since it is fundamental to all of them, the name "zeroth" seemed appropriate.)

Now suppose system *C* is a thermometer, such as the liquid-in-tube system of Fig. 17.1a. In Fig. 17.2a the thermometer *C* is in contact with both *A* and *B*. In thermal equilibrium, when the thermometer reading reaches a stable value, the thermometer measures the temperature of both *A* and *B*; hence *A* and *B* both have the *same* temperature. Experiment shows that thermal equilibrium isn't affected by adding or removing insulators, so the reading of thermometer *C* wouldn't change if it were in contact only with *A* or only with *B*. We conclude:

> **Two systems are in thermal equilibrium if and only if they have the same temperature.**

This is what makes a thermometer useful; a thermometer actually measures *its own* temperature, but when a thermometer is in thermal equilibrium with another body, the temperatures must be equal. When the temperatures of two systems are different, they *cannot* be in thermal equilibrium.

Test Your Understanding of Section 17.1 You put a thermometer in a pot of hot water and record the reading. What temperature have you recorded? (i) the temperature of the water; (ii) the temperature of the thermometer; (iii) an equal average of the temperatures of the water and thermometer; (iv) a weighted average of the temperatures of the water and thermometer, with more emphasis on the temperature of the water; (v) a weighted average of the water and thermometer, with more emphasis on the temperature of the thermometer.

17.2 Thermometers and Temperature Scales

To make the liquid-in-tube device shown in Fig. 17.1a into a useful thermometer, we need to mark a scale on the tube wall with numbers on it. These numbers are arbitrary, and historically many different schemes have been used. Suppose we label the thermometer's liquid level at the freezing temperature of pure water "zero" and the level at the boiling temperature "100," and divide the distance between these two points into 100 equal intervals called *degrees*. The result is the **Celsius temperature scale** (formerly called the *centigrade* scale in English-speaking countries). The Celsius temperature for a state colder than freezing water is a negative number. The Celsius scale is used, both in everyday life and in science and industry, almost everywhere in the world.

Another common type of thermometer uses a *bimetallic strip,* made by bonding strips of two different metals together (Fig. 17.3a). When the temperature of the composite strip increases, one metal expands more than the other and the strip bends (Fig. 17.3b). This strip is usually formed into a spiral, with the outer end anchored to the thermometer case and the inner end attached to a pointer (Fig. 17.3c). The pointer rotates in response to temperature changes.

In a *resistance thermometer* the changing electrical resistance of a coil of fine wire, a carbon cylinder, or a germanium crystal is measured. Resistance thermometers are usually more precise than most other types.

17.2 The zeroth law of thermodynamics.

(a) If systems *A* and *B* are each in thermal equilibrium with system *C* ...

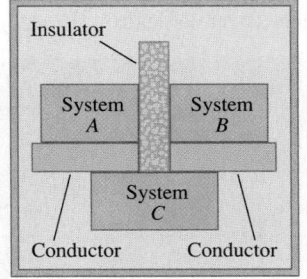

(b) ... then systems *A* and *B* are in thermal equilibrium with each other.

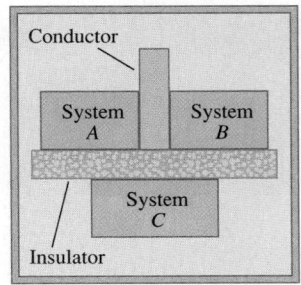

17.3 Use of a bimetallic strip as a thermometer.

(a) A bimetallic strip

(b) The strip bends when its temperature is raised.

When heated, metal 2 expands more than metal 1.

(c) A bimetallic strip used in a thermometer

17.4 A temporal artery thermometer measures infrared radiation from the skin that overlies one of the important arteries in the head. Although the thermometer cover touches the skin, the infrared detector inside the cover does not.

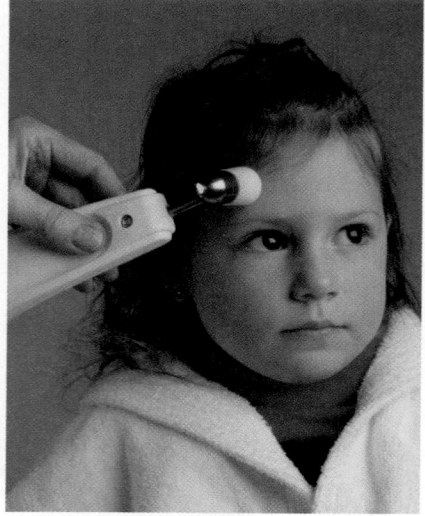

Some thermometers work by detecting the amount of infrared radiation emitted by an object. (We'll see in Section 17.7 that *all* objects emit electromagnetic radiation, including infrared, as a consequence of their temperature.) A modern example is a *temporal artery thermometer* (Fig. 17.4). A nurse runs this over a patient's forehead in the vicinity of the temporal artery, and an infrared sensor in the thermometer measures the radiation from the skin. Tests show that this device gives more accurate values of body temperature than do oral or ear thermometers.

In the **Fahrenheit temperature scale,** still used in everyday life in the United States, the freezing temperature of water is 32°F (thirty-two degrees Fahrenheit) and the boiling temperature is 212°F, both at standard atmospheric pressure. There are 180 degrees between freezing and boiling, compared to 100 on the Celsius scale, so one Fahrenheit degree represents only $\frac{100}{180}$, or $\frac{5}{9}$, as great a temperature change as one Celsius degree.

To convert temperatures from Celsius to Fahrenheit, note that a Celsius temperature T_C is the number of Celsius degrees above freezing; the number of Fahrenheit degrees above freezing is $\frac{9}{5}$ of this. But freezing on the Fahrenheit scale is at 32°F, so to obtain the actual Fahrenheit temperature T_F, multiply the Celsius value by $\frac{9}{5}$ and then add 32°. Symbolically,

$$T_F = \tfrac{9}{5}T_C + 32° \qquad (17.1)$$

To convert Fahrenheit to Celsius, solve this equation for T_C:

$$T_C = \tfrac{5}{9}(T_F - 32°) \qquad (17.2)$$

In words, subtract 32° to get the number of Fahrenheit degrees above freezing, and then multiply by $\frac{5}{9}$ to obtain the number of Celsius degrees above freezing—that is, the Celsius temperature.

We don't recommend memorizing Eqs. (17.1) and (17.2). Instead, try to understand the reasoning that led to them so that you can derive them on the spot when you need them, checking your reasoning with the relationship 100°C = 212°F.

It is useful to distinguish between an actual temperature and a temperature *interval* (a difference or change in temperature). An actual temperature of 20° is stated as 20°C (twenty degrees Celsius), and a temperature *interval* of 10° is 10 C° (ten Celsius degrees). A beaker of water heated from 20°C to 30°C undergoes a temperature change of 10 C°.

Test Your Understanding of Section 17.2 Which of the following types of thermometers have to be in thermal equilibrium with the object being measured in order to give accurate readings? (i) a bimetallic strip; (ii) a resistance thermometer; (iii) a temporal artery thermometer; (iv) both (i) and (ii); (v) all of (i), (ii), and (iii).

Application **Mammalian Body Temperatures**
Most mammals maintain body temperatures in the range from 36°C to 40°C (309 K to 313 K). A high metabolic rate warms the animal from within, and insulation (such as fur, feathers, and body fat) slows heat loss.

17.3 Gas Thermometers and the Kelvin Scale

When we calibrate two thermometers, such as a liquid-in-tube system and a resistance thermometer, so that they agree at 0°C and 100°C, they may not agree exactly at intermediate temperatures. Any temperature scale defined in this way always depends somewhat on the specific properties of the material used. Ideally, we would like to define a temperature scale that *doesn't* depend on the properties of a particular material. To establish a truly material-independent scale, we first need to develop some principles of thermodynamics. We'll return to this fundamental problem in Chapter 20. Here we'll discuss a thermometer that comes close to the ideal, the *gas thermometer.*

17.5 (a) Using a constant-volume gas thermometer to measure temperature. (b) The greater the amount of gas in the thermometer, the higher the graph of pressure p versus temperature T.

(a) A constant-volume gas thermometer

(b) Graphs of pressure versus temperature at constant volume for three different types and quantities of gas

Plots of pressure as a function of temperature for gas thermometers containing different types and quantities of gas

Dashed lines show the plots extrapolated to zero pressure.

The extrapolated plots all reach zero pressure at the same temperature: $-273.15°C$.

The principle of a gas thermometer is that the pressure of a gas at constant volume increases with temperature. A quantity of gas is placed in a constant-volume container (Fig. 17.5a), and its pressure is measured by one of the devices described in Section 12.2. To calibrate a constant-volume gas thermometer, we measure the pressure at two temperatures, say 0°C and 100°C, plot these points on a graph, and draw a straight line between them. Then we can read from the graph the temperature corresponding to any other pressure. Figure 17.5b shows the results of three such experiments, each using a different type and quantity of gas.

By extrapolating this graph, we see that there is a hypothetical temperature, $-273.15°C$, at which the absolute pressure of the gas would become zero. We might expect that this temperature would be different for different gases, but it turns out to be the *same* for many different gases (at least in the limit of very low gas density). We can't actually observe this zero-pressure condition. Gases liquefy and solidify at very low temperatures, and the proportionality of pressure to temperature no longer holds.

We use this extrapolated zero-pressure temperature as the basis for a temperature scale with its zero at this temperature. This is the **Kelvin temperature scale,** named for the British physicist Lord Kelvin (1824–1907). The units are the same size as those on the Celsius scale, but the zero is shifted so that $0 \text{ K} = -273.15°C$ and $273.15 \text{ K} = 0°C$; that is,

$$T_\text{K} = T_\text{C} + 273.15 \qquad (17.3)$$

Figure 17.5b shows both the Celsius and Kelvin scales. A common room temperature, 20°C (= 68°F), is 20 + 273.15, or about 293 K.

CAUTION **Never say "degrees kelvin"** In SI nomenclature, "degree" is not used with the Kelvin scale; the temperature mentioned above is read "293 kelvins," not "degrees kelvin" (Fig. 17.6). We capitalize Kelvin when it refers to the temperature scale; however, the *unit* of temperature is the *kelvin,* which is not capitalized (but is nonetheless abbreviated as a capital K). ▮

17.6 Correct and incorrect uses of the Kelvin scale.

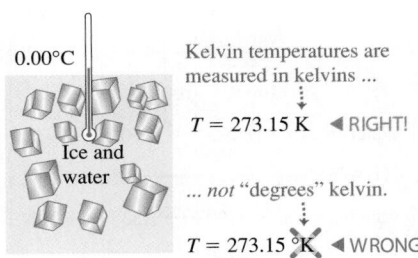

0.00°C

Ice and water

Kelvin temperatures are measured in kelvins ...

$T = 273.15 \text{ K}$ ◀ RIGHT!

... *not* "degrees" kelvin.

$T = 273.15 °\text{K}$ ◀ WRONG

You place a small piece of ice in your mouth. Eventually, the water all converts from ice at $T_1 = 32.00°F$ to body temperature, $T_2 = 98.60°F$. Express these temperatures in both Celsius degrees and kelvins, and find $\Delta T = T_2 - T_1$ in both cases.

SOLUTION

IDENTIFY and SET UP: Our target variables are stated above. We convert Fahrenheit temperatures to Celsius using Eq. (17.2), and Celsius temperatures to Kelvin using Eq. (17.3).

EXECUTE: From Eq. (17.2), $T_1 = 0.00°C$ and $T_2 = 37.00°C$; then $\Delta T = T_2 - T_1 = 37.00\ C°$. To get the Kelvin temperatures, just add 273.15 to each Celsius temperature: $T_1 = 273.15\ K$ and $T_2 = 310.15\ K$. The temperature difference is $\Delta T = T_2 - T_1 = 37.00\ K$.

EVALUATE: The Celsius and Kelvin scales have different zero points but the same size degrees. Therefore *any* temperature difference ΔT is the *same* on the Celsius and Kelvin scales. However, ΔT is *not* the same on the Fahrenheit scale; here, for example, $\Delta T = 66.60\ F°$.

The Kelvin Scale and Absolute Temperature

The Celsius scale has two fixed points: the normal freezing and boiling temperatures of water. But we can define the Kelvin scale using a gas thermometer with only a single reference temperature. We define the ratio of any two temperatures T_1 and T_2 on the Kelvin scale as the ratio of the corresponding gas-thermometer pressures p_1 and p_2:

$$\frac{T_2}{T_1} = \frac{p_2}{p_1} \qquad \text{(constant-volume gas thermometer, } T \text{ in kelvins)} \quad (17.4)$$

The pressure p is directly proportional to the Kelvin temperature, as shown in Fig. 17.5b. To complete the definition of T, we need only specify the Kelvin temperature of a single specific state. For reasons of precision and reproducibility, the state chosen is the *triple point* of water. This is the unique combination of temperature and pressure at which solid water (ice), liquid water, and water vapor can all coexist. It occurs at a temperature of 0.01°C and a water-vapor pressure of 610 Pa (about 0.006 atm). (This is the pressure of the *water;* it has nothing to do directly with the gas pressure in the *thermometer.*) The triple-point temperature T_{triple} of water is *defined* to have the value $T_{triple} = 273.16\ K$, corresponding to 0.01°C. From Eq. (17.4), if p_{triple} is the pressure in a gas thermometer at temperature T_{triple} and p is the pressure at some other temperature T, then T is given on the Kelvin scale by

$$T = T_{triple}\frac{p}{p_{triple}} = (273.16\ K)\frac{p}{p_{triple}} \qquad (17.5)$$

Low-pressure gas thermometers using various gases are found to agree very closely, but they are large, bulky, and very slow to come to thermal equilibrium. They are used principally to establish high-precision standards and to calibrate other thermometers.

Figure 17.7 shows the relationships among the three temperature scales we have discussed. The Kelvin scale is called an **absolute temperature scale,** and its zero point ($T = 0\ K = -273.15°C$, the temperature at which $p = 0$ in Eq. (17.5)) is called **absolute zero.** At absolute zero a system of molecules (such as a quantity of a gas, a liquid, or a solid) has its *minimum* possible total energy (kinetic plus potential); because of quantum effects, however, it is *not* correct to say that all molecular motion ceases at absolute zero. To define more completely what we mean by absolute zero, we need to use the thermodynamic principles developed in the next several chapters. We will return to this concept in Chapter 20.

17.7 Relationships among Kelvin (K), Celsius (C), and Fahrenheit (F) temperature scales. Temperatures have been rounded off to the nearest degree.

	K	C	F
Water boils	373	100°	212°
	100 K	100 C°	180 F°
Water freezes	273	0°	32°
CO_2 solidifies	195	−78°	−109°
Oxygen liquefies	90	−183°	−298°
Absolute zero	0	−273°	−460°

Test Your Understanding of Section 17.3 Rank the following temperatures from highest to lowest: (i) 0.00°C; (ii) 0.00°F; (iii) 260.00 K; (iv) 77.00 K; (v) −180.00°C.

17.4 Thermal Expansion

Most materials expand when their temperatures increase. Rising temperatures make the liquid expand in a liquid-in-tube thermometer (Fig. 17.1a) and bend bimetallic strips (Fig. 17.3b). The decks of bridges need special joints and supports to allow for expansion. A completely filled and tightly capped bottle of water cracks when it is heated, but you can loosen a metal jar lid by running hot water over it. These are all examples of *thermal expansion*.

Linear Expansion

Suppose a rod of material has a length L_0 at some initial temperature T_0. When the temperature changes by ΔT, the length changes by ΔL. Experiments show that if ΔT is not too large (say, less than 100 C° or so), ΔL is *directly proportional* to ΔT (Fig. 17.8a). If two rods made of the same material have the same temperature change, but one is twice as long as the other, then the *change* in its length is also twice as great. Therefore ΔL must also be proportional to L_0 (Fig. 17.8b). Introducing a proportionality constant α (which is different for different materials), we may express these relationships in an equation:

$$\Delta L = \alpha L_0 \Delta T \qquad \text{(linear thermal expansion)} \qquad (17.6)$$

If a body has length L_0 at temperature T_0, then its length L at a temperature $T = T_0 + \Delta T$ is

$$L = L_0 + \Delta L = L_0 + \alpha L_0 \Delta T = L_0(1 + \alpha \Delta T) \qquad (17.7)$$

The constant α, which describes the thermal expansion properties of a particular material, is called the **coefficient of linear expansion.** The units of α are K^{-1} or $(C°)^{-1}$. (Remember that a temperature *interval* is the same in the Kelvin and Celsius scales.) For many materials, every linear dimension changes according to Eq. (17.6) or (17.7). Thus L could be the thickness of a rod, the side length of a square sheet, or the diameter of a hole. Some materials, such as wood or single crystals, expand differently in different directions. We won't consider this complication.

We can understand thermal expansion qualitatively on a molecular basis. Picture the interatomic forces in a solid as springs, as in Fig. 17.9a. (We explored the analogy between spring forces and interatomic forces in Section 14.4.) Each atom vibrates about its equilibrium position. When the temperature increases, the energy and amplitude of the vibration also increase. The interatomic spring forces are not symmetrical about the equilibrium position; they usually behave like a spring that is easier to stretch than to compress. As a result, when the amplitude of vibration increases, the *average* distance between atoms also increases (Fig. 17.9b). As the atoms get farther apart, every dimension increases.

17.8 How the length of a rod changes with a change in temperature. (Length changes are exaggerated for clarity.)

(a) For moderate temperature changes, ΔL is directly proportional to ΔT.

(b) ΔL is also directly proportional to L_0.

(a) A model of the forces between neighboring atoms in a solid

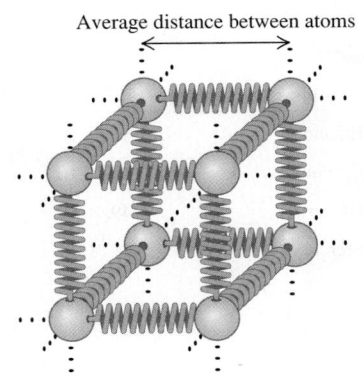

(b) A graph of the "spring" potential energy $U(x)$

As energy increases from E_1 to E_2 to E_3, average distance between atoms increases.

17.9 **(a)** We can model atoms in a solid as being held together by "springs" that are easier to stretch than to compress. **(b)** A graph of the "spring" potential energy $U(x)$ versus distance x between neighboring atoms is *not* symmetrical (compare Fig. 14.20b). As the energy increases and the atoms oscillate with greater amplitude, the average distance increases.

17.10 When an object undergoes thermal expansion, any holes in the object expand as well. (The expansion is exaggerated.)

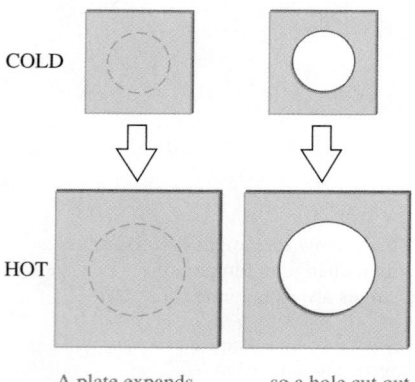

COLD

HOT

A plate expands when heated . . .

. . . so a hole cut out of the plate must expand, too.

17.11 When this SR-71 aircraft is sitting on the ground, its wing panels fit together so loosely that fuel leaks out of the wings onto the ground. But once it is in flight at over three times the speed of sound, air friction heats the panels so much that they expand to make a perfect fit. (In-flight refueling makes up for the lost fuel.)

CAUTION **Heating an object with a hole** If a solid object has a hole in it, what happens to the size of the hole when the temperature of the object increases? A common misconception is that if the object expands, the hole will shrink because material expands into the hole. But the truth of the matter is that if the object expands, the hole will expand too (Fig. 17.10); as we stated above, *every* linear dimension of an object changes in the same way when the temperature changes. If you're not convinced, think of the atoms in Fig. 17.9a as outlining a cubical hole. When the object expands, the atoms move apart and the hole increases in size. The only situation in which a "hole" will fill in due to thermal expansion is when two separate objects expand and close the gap between them (Fig. 17.11). ▮

The direct proportionality expressed by Eq. (17.6) is not exact; it is *approximately* correct only for sufficiently small temperature changes. For a given material, α varies somewhat with the initial temperature T_0 and the size of the temperature interval. We'll ignore this complication here, however. Average values of α for several materials are listed in Table 17.1. Within the precision of these values we don't need to worry whether T_0 is 0°C or 20°C or some other temperature. Note that typical values of α are very small; even for a temperature change of 100 C°, the fractional length change $\Delta L/L_0$ is only of the order of $\frac{1}{1000}$ for the metals in the table.

Volume Expansion

Increasing temperature usually causes increases in *volume* for both solid and liquid materials. Just as with linear expansion, experiments show that if the temperature change ΔT is not too great (less than 100 C° or so), the increase in volume ΔV is approximately proportional to both the temperature change ΔT and the initial volume V_0:

$$\Delta V = \beta V_0 \Delta T \qquad \text{(volume thermal expansion)} \qquad (17.8)$$

The constant β characterizes the volume expansion properties of a particular material; it is called the **coefficient of volume expansion.** The units of β are K^{-1} or $(\text{C}°)^{-1}$. As with linear expansion, β varies somewhat with temperature, and Eq. (17.8) is an approximate relationship that is valid only for small temperature changes. For many substances, β decreases at low temperatures. Several values of β in the neighborhood of room temperature are listed in Table 17.2. Note that the values for liquids are generally much larger than those for solids.

For solid materials there is a simple relationship between the volume expansion coefficient β and the linear expansion coefficient α. To derive this relationship, we consider a cube of material with side length L and volume $V = L^3$. At the initial temperature the values are L_0 and V_0. When the temperature increases by dT, the side length increases by dL and the volume increases by an amount dV given by

$$dV = \frac{dV}{dL} dL = 3L^2 \, dL$$

Table 17.1 Coefficients of Linear Expansion

Material	$\alpha\,[\text{K}^{-1} \text{ or } (\text{C}°)^{-1}]$
Aluminum	2.4×10^{-5}
Brass	2.0×10^{-5}
Copper	1.7×10^{-5}
Glass	$0.4\text{--}0.9 \times 10^{-5}$
Invar (nickel–iron alloy)	0.09×10^{-5}
Quartz (fused)	0.04×10^{-5}
Steel	1.2×10^{-5}

Table 17.2 Coefficients of Volume Expansion

Solids	$\beta\,[\text{K}^{-1} \text{ or } (\text{C}°)^{-1}]$	Liquids	$\beta\,[\text{K}^{-1} \text{ or } (\text{C}°)^{-1}]$
Aluminum	7.2×10^{-5}	Ethanol	75×10^{-5}
Brass	6.0×10^{-5}	Carbon disulfide	115×10^{-5}
Copper	5.1×10^{-5}	Glycerin	49×10^{-5}
Glass	$1.2\text{--}2.7 \times 10^{-5}$	Mercury	18×10^{-5}
Invar	0.27×10^{-5}		
Quartz (fused)	0.12×10^{-5}		
Steel	3.6×10^{-5}		

Now we replace L and V by the initial values L_0 and V_0. From Eq. (17.6), dL is

$$dL = \alpha L_0 \, dT$$

Since $V_0 = L_0^3$, this means that dV can also be expressed as

$$dV = 3L_0^2 \alpha L_0 \, dT = 3\alpha V_0 \, dT$$

This is consistent with the infinitesimal form of Eq. (17.8), $dV = \beta V_0 \, dT$, only if

$$\beta = 3\alpha \qquad (17.9)$$

You should check this relationship for some of the materials listed in Tables 17.1 and 17.2.

Problem-Solving Strategy 17.1 | **Thermal Expansion**

IDENTIFY *the relevant concepts:* Decide whether the problem involves changes in length (linear thermal expansion) or in volume (volume thermal expansion).

SET UP *the problem* using the following steps:
1. List the known and unknown quantities and identify the target variables.
2. Choose Eq. (17.6) for linear expansion and Eq. (17.8) for volume expansion.

EXECUTE *the solution* as follows:
1. Solve for the target variables. If you are given an initial temperature T_0 and must find a final temperature T corresponding to a given length or volume change, find ΔT and calculate $T = T_0 + \Delta T$. Remember that the size of a hole in a material varies with temperature just as any other linear dimension, and that the volume of a hole (such as the interior of a container) varies just as that of the corresponding solid shape.
2. Maintain unit consistency. Both L_0 and ΔL (or V_0 and ΔV) must have the same units. If you use a value of α or β in K^{-1} or $(\mathrm{C}°)^{-1}$, then ΔT must be in either kelvins or Celsius degrees; from Example 17.1, the two scales are equivalent *for temperature differences*.

EVALUATE *your answer:* Check whether your results make sense.

Example 17.2 | **Length change due to temperature change**

A surveyor uses a steel measuring tape that is exactly 50.000 m long at a temperature of 20°C. The markings on the tape are calibrated for this temperature. (a) What is the length of the tape when the temperature is 35°C? (b) When it is 35°C, the surveyor uses the tape to measure a distance. The value that she reads off the tape is 35.794 m. What is the actual distance?

SOLUTION

IDENTIFY and SET UP: This problem concerns the linear expansion of a measuring tape. We are given the tape's initial length $L_0 = 50.000$ m at $T_0 = 20$°C. In part (a) we use Eq. (17.6) to find the change ΔL in the tape's length at $T = 35$°C, and use Eq. (17.7) to find L. (Table 17.1 gives the value of α for steel.) Since the tape expands, at 35°C the distance between two successive meter marks is greater than 1 m. Hence the actual distance in part (b) is *larger* than the distance read off the tape by a factor equal to the ratio of the tape's length L at 35°C to its length L_0 at 20°C.

EXECUTE: (a) The temperature change is $\Delta T = T - T_0 = 15$ C°; from Eqs. (17.6) and (17.7),

$$\Delta L = \alpha L_0 \, \Delta T = (1.2 \times 10^{-5} \text{ K}^{-1})(50 \text{ m})(15 \text{ K})$$
$$= 9.0 \times 10^{-3} \text{ m} = 9.0 \text{ mm}$$
$$L = L_0 + \Delta L = 50.000 \text{ m} + 0.009 \text{ m} = 50.009 \text{ m}$$

(b) Our result from part (a) shows that at 35°C, the slightly expanded tape reads a distance of 50.000 m when the true distance is 50.009 m. We can rewrite the algebra of part (a) as $L = L_0(1 + \alpha \, \Delta T)$; at 35°C, *any* true distance will be greater than the reading by the factor $50.009/50.000 = 1 + \alpha \, \Delta T = 1 + 1.8 \times 10^{-4}$. The true distance is therefore

$$(1 + 1.8 \times 10^{-4})(35.794 \text{ m}) = 35.800 \text{ m}$$

EVALUATE: Note that in part (a) we needed only two of the five significant figures of L_0 to compute ΔL to the same number of decimal places as L_0. Our result shows that metals expand very little under moderate temperature changes. However, even the small difference 0.009 m = 9 mm found in part (b) between the scale reading and the true distance can be important in precision work.

Example 17.3 | **Volume change due to temperature change**

A 200-cm³ glass flask is filled to the brim with mercury at 20°C. How much mercury overflows when the temperature of the system is raised to 100°C? The coefficient of *linear* expansion of the glass is 0.40×10^{-5} K^{-1}.

Continued

SOLUTION

IDENTIFY and SET UP: This problem involves the volume expansion of the glass and of the mercury. The amount of overflow depends on the *difference* between the volume changes ΔV for these two materials, both given by Eq. (17.8). The mercury will overflow if its coefficient of volume expansion β (given in Table 17.2) is greater than that of glass, which we find from Eq. (17.9) using the given value of α.

EXECUTE: From Table 17.2, $\beta_{Hg} = 18 \times 10^{-5} \, K^{-1}$. That is indeed greater than β_{glass}: From Eq. (17.9), $\beta_{glass} = 3\alpha_{glass} = 3(0.40 \times 10^{-5} \, K^{-1}) = 1.2 \times 10^{-5} \, K^{-1}$. The volume overflow is then

$$\Delta V_{Hg} - \Delta V_{glass} = \beta_{Hg} V_0 \Delta T - \beta_{glass} V_0 \Delta T$$
$$= V_0 \Delta T (\beta_{Hg} - \beta_{glass})$$
$$= (200 \, cm^3)(80 \, C°)(18 \times 10^{-5} - 1.2 \times 10^{-5})$$
$$= 2.7 \, cm^3$$

EVALUATE: This is basically how a mercury-in-glass thermometer works; the column of mercury inside a sealed tube rises as T increases because mercury expands faster than glass.

As Tables 17.1 and 17.2 show, glass has smaller coefficients of expansion α and β than do most metals. This is why you can use hot water to loosen a metal lid on a glass jar; the metal expands more than the glass does.

Thermal Expansion of Water

Water, in the temperature range from 0°C to 4°C, *decreases* in volume with increasing temperature. In this range its coefficient of volume expansion is *negative*. Above 4°C, water expands when heated (Fig. 17.12). Hence water has its greatest density at 4°C. Water also expands when it freezes, which is why ice humps up in the middle of the compartments in an ice cube tray. By contrast, most materials contract when they freeze.

This anomalous behavior of water has an important effect on plant and animal life in lakes. A lake cools from the surface down; above 4°C, the cooled water at the surface flows to the bottom because of its greater density. But when the surface temperature drops below 4°C, the water near the surface is less dense than the warmer water below. Hence the downward flow ceases, and the water near the surface remains colder than that at the bottom. As the surface freezes, the ice floats because it is less dense than water. The water at the bottom remains at 4°C until nearly the entire lake is frozen. If water behaved like most substances, contracting continuously on cooling and freezing, lakes would freeze from the bottom up. Circulation due to density differences would continuously carry warmer water to the surface for efficient cooling, and lakes would freeze solid much more easily. This would destroy all plant and animal life that cannot withstand freezing. If water did not have this special property, the evolution of life would have taken a very different course.

17.12 The volume of 1 gram of water in the temperature range from 0°C to 100°C. By 100°C the volume has increased to 1.034 cm³. If the coefficient of volume expansion were constant, the curve would be a straight line.

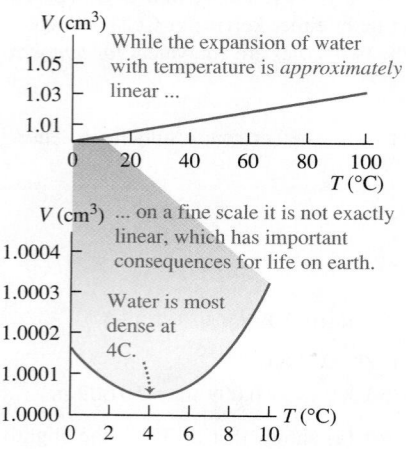

Thermal Stress

If we clamp the ends of a rod rigidly to prevent expansion or contraction and then change the temperature, **thermal stresses** develop. The rod would like to expand or contract, but the clamps won't let it. The resulting stresses may become large enough to strain the rod irreversibly or even break it. (You may want to review the discussion of stress and strain in Section 11.4).

Engineers must account for thermal stress when designing structures. Concrete highways and bridge decks usually have gaps between sections, filled with a flexible material or bridged by interlocking teeth (Fig. 17.13), to permit expansion and contraction of the concrete. Long steam pipes have expansion joints or U-shaped sections to prevent buckling or stretching with temperature changes. If one end of a steel bridge is rigidly fastened to its abutment, the other end usually rests on rollers.

To calculate the thermal stress in a clamped rod, we compute the amount the rod *would* expand (or contract) if not held and then find the stress needed to compress (or stretch) it back to its original length. Suppose that a rod with length L_0 and cross-sectional area A is held at constant length while the temperature is reduced (negative ΔT), causing a tensile stress. The fractional change in length if the rod were free to contract would be

$$\left(\frac{\Delta L}{L_0}\right)_{thermal} = \alpha \, \Delta T \tag{17.10}$$

17.13 Expansion joints on bridges are needed to accommodate changes in length that result from thermal expansion.

Both ΔL and ΔT are negative. The tension must increase by an amount F that is just enough to produce an equal and opposite fractional change in length $(\Delta L/L_0)_{tension}$. From the definition of Young's modulus, Eq. (11.10),

$$Y = \frac{F/A}{\Delta L/L_0} \quad \text{so} \quad \left(\frac{\Delta L}{L_0}\right)_{tension} = \frac{F}{AY} \qquad (17.11)$$

If the length is to be constant, the *total* fractional change in length must be zero. From Eqs. (17.10) and (17.11), this means that

$$\left(\frac{\Delta L}{L_0}\right)_{thermal} + \left(\frac{\Delta L}{L_0}\right)_{tension} = \alpha \, \Delta T + \frac{F}{AY} = 0$$

Solving for the tensile stress F/A required to keep the rod's length constant, we find

$$\frac{F}{A} = -Y\alpha \, \Delta T \qquad \text{(thermal stress)} \qquad (17.12)$$

For a decrease in temperature, ΔT is negative, so F and F/A are positive; this means that a *tensile* force and stress are needed to maintain the length. If ΔT is positive, F and F/A are negative, and the required force and stress are *compressive*.

If there are temperature differences within a body, nonuniform expansion or contraction will result and thermal stresses can be induced. You can break a glass bowl by pouring very hot water into it; the thermal stress between the hot and cold parts of the bowl exceeds the breaking stress of the glass, causing cracks. The same phenomenon makes ice cubes crack when dropped into warm water. Heat-resistant glasses such as Pyrex™ have exceptionally low expansion coefficients and high strength.

Example 17.4 Thermal stress

An aluminum cylinder 10 cm long, with a cross-sectional area of 20 cm², is used as a spacer between two steel walls. At 17.2°C it just slips between the walls. Calculate the stress in the cylinder and the total force it exerts on each wall when it warms to 22.3°C, assuming that the walls are perfectly rigid and a constant distance apart.

SOLUTION

IDENTIFY and SET UP: Figure 17.14 shows our sketch of the situation. Our target variables are the thermal stress F/A in the cylinder, whose cross-sectional area A is given, and the associated force F it

17.14 Our sketch for this problem.

exerts on the walls. We use Eq. (17.12) to relate F/A to the temperature change ΔT, and from that calculate F. (The length of the cylinder is irrelevant.) We find Young's modulus Y_{Al} and the coefficient of linear expansion α_{Al} from Tables 11.1 and 17.1, respectively.

EXECUTE: We have $Y_{Al} = 7.0 \times 10^{10}$ Pa and $\alpha_{Al} = 2.4 \times 10^{-5}$ K^{-1}, and $\Delta T = 22.3°C - 17.2°C = 5.1$ C° $= 5.1$ K. From Eq. (17.12), the stress is

$$\frac{F}{A} = -Y_{Al}\alpha_{Al}\Delta T$$
$$= -(7.0 \times 10^{10} \text{ Pa})(2.4 \times 10^{-5} \text{ K}^{-1})(5.1 \text{ K})$$
$$= -8.6 \times 10^6 \text{ Pa} = -1200 \text{ lb/in.}^2$$

The total force is the cross-sectional area times the stress:

$$F = A\left(\frac{F}{A}\right) = (20 \times 10^{-4} \text{ m}^2)(-8.6 \times 10^6 \text{ Pa})$$
$$= -1.7 \times 10^4 \text{ N} = 1.9 \text{ tons}$$

EVALUATE: The stress on the cylinder and the force it exerts on each wall are immense. Such thermal stresses must be accounted for in engineering.

Test Your Understanding of Section 17.4 In the bimetallic strip shown in Fig. 17.3a, metal 1 is copper. Which of the following materials could be used for metal 2? (There may be more than one correct answer). (i) steel; (ii) brass; (iii) aluminum. ❚

17.5 Quantity of Heat

17.15 The same temperature change of the same system may be accomplished by (a) doing work on it or (b) adding heat to it.

(a) Raising the temperature of water by doing work on it

The water warms as the paddle does work on it; the temperature rise is proportional to the amount of work done.

(b) Raising the temperature of water by direct heating

Direct heating can produce the same temperature change as doing work on the water.

When you put a cold spoon into a cup of hot coffee, the spoon warms up and the coffee cools down as they approach thermal equilibrium. The interaction that causes these temperature changes is fundamentally a transfer of *energy* from one substance to another. Energy transfer that takes place solely because of a temperature difference is called *heat flow* or *heat transfer,* and energy transferred in this way is called **heat.**

An understanding of the relationship between heat and other forms of energy emerged during the 18th and 19th centuries. Sir James Joule (1818–1889) studied how water can be warmed by vigorous stirring with a paddle wheel (Fig. 17.15a). The paddle wheel adds energy to the water by doing *work* on it, and Joule found that *the temperature rise is directly proportional to the amount of work done.* The same temperature change can also be caused by putting the water in contact with some hotter body (Fig. 17.15b); hence this interaction must also involve an energy exchange. We will explore the relationship between heat and mechanical energy in Chapters 19 and 20.

CAUTION **Temperature vs. heat** It is absolutely essential for you to distinguish between *temperature* and *heat.* Temperature depends on the physical state of a material and is a quantitative description of its hotness or coldness. In physics the term "heat" always refers to energy in transit from one body or system to another because of a temperature difference, never to the amount of energy contained within a particular system. We can change the temperature of a body by adding heat to it or taking heat away, or by adding or subtracting energy in other ways, such as mechanical work (Fig. 17.15a). If we cut a body in half, each half has the same temperature as the whole; but to raise the temperature of each half by a given interval, we add *half* as much heat as for the whole. ▮

We can define a *unit* of quantity of heat based on temperature changes of some specific material. The **calorie** (abbreviated cal) is defined as *the amount of heat required to raise the temperature of 1 gram of water from 14.5°C to 15.5°C.* The kilocalorie (kcal), equal to 1000 cal, is also used; a food-value calorie is actually a kilocalorie (Fig. 17.16). A corresponding unit of heat using Fahrenheit degrees and British units is the **British thermal unit,** or Btu. One Btu is the quantity of heat required to raise the temperature of 1 pound (weight) of water 1 F° from 63°F to 64°F.

Because heat is energy in transit, there must be a definite relationship between these units and the familiar mechanical energy units such as the joule. Experiments similar in concept to Joule's have shown that

$$1 \text{ cal} = 4.186 \text{ J}$$
$$1 \text{ kcal} = 1000 \text{ cal} = 4186 \text{ J}$$
$$1 \text{ Btu} = 778 \text{ ft} \cdot \text{lb} = 252 \text{ cal} = 1055 \text{ J}$$

The calorie is not a fundamental SI unit. The International Committee on Weights and Measures recommends using the joule as the basic unit of energy in all forms, including heat. We will follow that recommendation in this book.

Specific Heat

17.16 The word "energy" is of Greek origin. This label on a can of Greek coffee shows that 100 milliliters of prepared coffee have an energy content (ενέργεια) of 9.6 kilojoules or 2.3 kilocalories.

We use the symbol Q for quantity of heat. When it is associated with an infinitesimal temperature change dT, we call it dQ. The quantity of heat Q required to increase the temperature of a mass m of a certain material from T_1 to T_2 is found to be approximately proportional to the temperature change $\Delta T = T_2 - T_1$. It is also proportional to the mass m of material. When you're heating water to make tea, you need twice as much heat for two cups as for one if the temperature change is the same. The quantity of heat needed also depends on the nature of the material; raising the temperature of 1 kilogram of water by 1 C° requires 4190 J of heat, but only 910 J is needed to raise the temperature of 1 kilogram of aluminum by 1 C°.

Putting all these relationships together, we have

$$Q = mc\,\Delta T \quad \text{(heat required for temperature change } \Delta T \text{ of mass } m\text{)} \quad (17.13)$$

where c is a quantity, different for different materials, called the **specific heat** of the material. For an infinitesimal temperature change dT and corresponding quantity of heat dQ,

$$dQ = mc\,dT \qquad (17.14)$$

$$c = \frac{1}{m}\frac{dQ}{dT} \quad \text{(specific heat)} \qquad (17.15)$$

In Eqs. (17.13), (17.14), and (17.15), Q (or dQ) and ΔT (or dT) can be either positive or negative. When they are positive, heat enters the body and its temperature increases; when they are negative, heat leaves the body and its temperature decreases.

CAUTION **The definition of heat** Remember that dQ does not represent a change in the amount of heat *contained* in a body; this is a meaningless concept. Heat is always energy *in transit* as a result of a temperature difference. There is no such thing as "the amount of heat in a body." ▌

The specific heat of water is approximately

$$4190 \text{ J/kg}\cdot\text{K} \qquad 1 \text{ cal/g}\cdot\text{C}° \qquad \text{or} \qquad 1 \text{ Btu/lb}\cdot\text{F}°$$

The specific heat of a material always depends somewhat on the initial temperature and the temperature interval. Figure 17.17 shows this dependence for water. In the problems and examples in this chapter we will usually ignore this small variation.

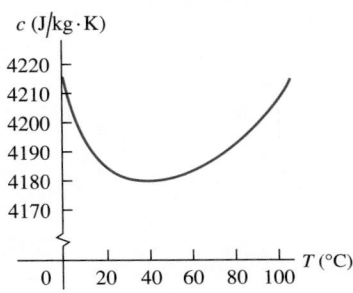

17.17 Specific heat of water as a function of temperature. The value of c varies by less than 1% between 0°C and 100°C.

Example 17.5 **Feed a cold, starve a fever**

During a bout with the flu an 80-kg man ran a fever of 39.0°C (102.2°F) instead of the normal body temperature of 37.0°C (98.6°F). Assuming that the human body is mostly water, how much heat is required to raise his temperature by that amount?

SOLUTION

IDENTIFY and SET UP: This problem uses the relationship among heat (the target variable), mass, specific heat, and temperature change. We use Eq. (17.13) to determine the required heat Q, with $m = 80$ kg, $c = 4190$ J/kg·K (for water), and $\Delta T = 39.0°C - 37.0°C = 2.0$ C° = 2.0 K.

EXECUTE: From Eq. (17.13),

$$Q = mc\,\Delta T = (80 \text{ kg})(4190 \text{ J/kg}\cdot\text{K})(2.0 \text{ K}) = 6.7 \times 10^5 \text{ J}$$

EVALUATE: This corresponds to 160 kcal. In fact, the specific heat of the human body is about 3480 J/kg·K, 83% that of water, because protein, fat, and minerals have lower specific heats. Hence a more accurate answer is $Q = 5.6 \times 10^5$ J = 133 kcal. Either result shows us that were it not for the body's temperature-regulating systems, taking in energy in the form of food would produce measurable changes in body temperature. (The elevated temperature of a person with the flu results from the body's extra activity in response to infection.)

Example 17.6 **Overheating electronics**

You are designing an electronic circuit element made of 23 mg of silicon. The electric current through it adds energy at the rate of 7.4 mW = 7.4×10^{-3} J/s. If your design doesn't allow any heat transfer out of the element, at what rate does its temperature increase? The specific heat of silicon is 705 J/kg·K.

SOLUTION

IDENTIFY and SET UP: The energy added to the circuit element gives rise to a temperature increase, just as if heat were flowing into the element at the rate $dQ/dt = 7.4 \times 10^{-3}$ J/s. Our target variable is the rate of temperature change dT/dt. We can use Eq. (17.14),

which relates infinitesimal temperature changes dT to the corresponding heat dQ, to obtain an expression for dQ/dt in terms of dT/dt.

EXECUTE: We divide both sides of Eq. (17.14) by dt and rearrange:

$$\frac{dT}{dt} = \frac{dQ/dt}{mc} = \frac{7.4 \times 10^{-3} \text{ J/s}}{(23 \times 10^{-6} \text{ kg})(705 \text{ J/kg}\cdot\text{K})} = 0.46 \text{ K/s}$$

EVALUATE: At this rate of temperature rise (27 K/min), the circuit element would soon self-destruct. Heat transfer is an important design consideration in electronic circuit elements.

Molar Heat Capacity

Sometimes it's more convenient to describe a quantity of substance in terms of the number of *moles n* rather than the *mass m* of material. Recall from your study of chemistry that a mole of any pure substance always contains the same number of molecules. (We will discuss this point in more detail in Chapter 18.) The *molar mass* of any substance, denoted by M, is the mass per mole. (The quantity M is sometimes called *molecular weight,* but *molar mass* is preferable; the quantity depends on the mass of a molecule, not its weight.) For example, the molar mass of water is 18.0 g/mol $= 18.0 \times 10^{-3}$ kg/mol; 1 mole of water has a mass of 18.0 g $= 0.0180$ kg. The total mass m of material is equal to the mass per mole M times the number of moles n:

$$m = nM \tag{17.16}$$

Replacing the mass m in Eq. (17.13) by the product nM, we find

$$Q = nMc\,\Delta T \tag{17.17}$$

The product Mc is called the **molar heat capacity** (or *molar specific heat*) and is denoted by C (capitalized). With this notation we rewrite Eq. (17.17) as

$$Q = nC\,\Delta T \quad \text{(heat required for temperature change of } n \text{ moles)} \tag{17.18}$$

Comparing to Eq. (17.15), we can express the molar heat capacity C (heat per mole per temperature change) in terms of the specific heat c (heat per mass per temperature change) and the molar mass M (mass per mole):

$$C = \frac{1}{n}\frac{dQ}{dT} = Mc \quad \text{(molar heat capacity)} \tag{17.19}$$

For example, the molar heat capacity of water is

$$C = Mc = (0.0180 \text{ kg/mol})(4190 \text{ J/kg} \cdot \text{K}) = 75.4 \text{ J/mol} \cdot \text{K}$$

Values of specific heat and molar heat capacity for several substances are given in Table 17.3. Note the remarkably large specific heat for water (Fig. 17.18).

CAUTION **The meaning of "heat capacity"** The term "heat capacity" is unfortunate because it gives the erroneous impression that a body *contains* a certain amount of heat. Remember, heat is energy in transit to or from a body, not the energy residing in the body. ▮

17.18 Water has a much higher specific heat than the glass or metals used to make cookware. This helps explain why it takes several minutes to boil water on a stove, even though the pot or kettle reaches a high temperature very quickly.

Table 17.3 Approximate Specific Heats and Molar Heat Capacities (Constant Pressure)

Substance	Specific Heat, c (J/kg · K)	Molar Mass, M (kg/mol)	Molar Heat Capacity, C (J/mol · K)
Aluminum	910	0.0270	24.6
Beryllium	1970	0.00901	17.7
Copper	390	0.0635	24.8
Ethanol	2428	0.0461	111.9
Ethylene glycol	2386	0.0620	148.0
Ice (near 0°C)	2100	0.0180	37.8
Iron	470	0.0559	26.3
Lead	130	0.207	26.9
Marble ($CaCO_3$)	879	0.100	87.9
Mercury	138	0.201	27.7
Salt (NaCl)	879	0.0585	51.4
Silver	234	0.108	25.3
Water (liquid)	4190	0.0180	75.4

Precise measurements of specific heats and molar heat capacities require great experimental skill. Usually, a measured quantity of energy is supplied by an electric current in a heater wire wound around the specimen. The temperature change ΔT is measured with a resistance thermometer or thermocouple embedded in the specimen. This sounds simple, but great care is needed to avoid or compensate for unwanted heat transfer between the sample and its surroundings. Measurements for solid materials are usually made at constant atmospheric pressure; the corresponding values are called the *specific heat* and *molar heat capacity at constant pressure,* denoted by c_p and C_p. For a gas it is usually easier to keep the substance in a container with constant *volume;* the corresponding values are called the *specific heat* and *molar heat capacity at constant volume,* denoted by c_V and C_V. For a given substance, C_V and C_p are different. If the system can expand while heat is added, there is additional energy exchange through the performance of *work* by the system on its surroundings. If the volume is constant, the system does no work. For gases the difference between C_p and C_V is substantial. We will study heat capacities of gases in detail in Section 19.7.

The last column of Table 17.3 shows something interesting. The molar heat capacities for most elemental solids are about the same: about 25 J/mol · K. This correlation, named the *rule of Dulong and Petit* (for its discoverers), forms the basis for a very important idea. The number of atoms in 1 mole is the same for all elemental substances. This means that on a *per atom* basis, about the same amount of heat is required to raise the temperature of each of these elements by a given amount, even though the *masses* of the atoms are very different. The heat required for a given temperature increase depends only on *how many* atoms the sample contains, not on the mass of an individual atom. We will see the reason the rule of Dulong and Petit works so well when we study the molecular basis of heat capacities in greater detail in Chapter 18.

Test Your Understanding of Section 17.5 You wish to raise the temperature of each of the following samples from 20°C to 21°C. Rank these in order of the amount of heat needed to do this, from highest to lowest. (i) 1 kilogram of mercury; (ii) 1 kilogram of ethanol; (iii) 1 mole of mercury; (iv) 1 mole of ethanol.

17.6 Calorimetry and Phase Changes

Calorimetry means "measuring heat." We have discussed the energy transfer (heat) involved in temperature changes. Heat is also involved in *phase changes,* such as the melting of ice or boiling of water. Once we understand these additional heat relationships, we can analyze a variety of problems involving quantity of heat.

Phase Changes

We use the term **phase** to describe a specific state of matter, such as a solid, liquid, or gas. The compound H_2O exists in the *solid phase* as ice, in the *liquid phase* as water, and in the *gaseous phase* as steam. (These are also referred to as **states of matter:** the solid state, the liquid state, and the gaseous state.) A transition from one phase to another is called a **phase change** or *phase transition.* For any given pressure a phase change takes place at a definite temperature, usually accompanied by absorption or emission of heat and a change of volume and density.

A familiar example of a phase change is the melting of ice. When we add heat to ice at 0°C and normal atmospheric pressure, the temperature of the ice *does not* increase. Instead, some of it melts to form liquid water. If we add the heat slowly, to maintain the system very close to thermal equilibrium, the temperature remains at 0°C until all the ice is melted (Fig. 17.19). The effect of adding heat to this system is not to raise its temperature but to change its *phase* from solid to liquid.

To change 1 kg of ice at 0°C to 1 kg of liquid water at 0°C and normal atmospheric pressure requires 3.34×10^5 J of heat. The heat required per unit mass is

17.19 The surrounding air is at room temperature, but this ice–water mixture remains at 0°C until all of the ice has melted and the phase change is complete.

called the **heat of fusion** (or sometimes *latent heat of fusion*), denoted by L_f. For water at normal atmospheric pressure the heat of fusion is

$$L_f = 3.34 \times 10^5 \, \text{J/kg} = 79.6 \, \text{cal/g} = 143 \, \text{Btu/lb}$$

More generally, to melt a mass m of material that has a heat of fusion L_f requires a quantity of heat Q given by

$$Q = mL_f$$

This process is *reversible*. To freeze liquid water to ice at 0°C, we have to *remove* heat; the magnitude is the same, but in this case, Q is negative because heat is removed rather than added. To cover both possibilities and to include other kinds of phase changes, we write

$$Q = \pm mL \quad \text{(heat transfer in a phase change)} \quad (17.20)$$

The plus sign (heat entering) is used when the material melts; the minus sign (heat leaving) is used when it freezes. The heat of fusion is different for different materials, and it also varies somewhat with pressure.

For any given material at any given pressure, the freezing temperature is the same as the melting temperature. At this unique temperature the liquid and solid phases (liquid water and ice, for example) can coexist in a condition called **phase equilibrium.**

We can go through this whole story again for *boiling* or *evaporation,* a phase transition between liquid and gaseous phases. The corresponding heat (per unit mass) is called the **heat of vaporization** L_v. At normal atmospheric pressure the heat of vaporization L_v for water is

$$L_v = 2.256 \times 10^6 \, \text{J/kg} = 539 \, \text{cal/g} = 970 \, \text{Btu/lb}$$

That is, it takes 2.256×10^6 J to change 1 kg of liquid water at 100°C to 1 kg of water vapor at 100°C. By comparison, to raise the temperature of 1 kg of water from 0°C to 100°C requires $Q = mc \, \Delta T = (1.00 \, \text{kg})(4190 \, \text{J/kg} \cdot \text{C}°) \times (100 \, \text{C}°) = 4.19 \times 10^5$ J, less than one-fifth as much heat as is required for vaporization at 100°C. This agrees with everyday kitchen experience; a pot of water may reach boiling temperature in a few minutes, but it takes a much longer time to completely evaporate all the water away.

Like melting, boiling is a reversible transition. When heat is removed from a gas at the boiling temperature, the gas returns to the liquid phase, or *condenses,* giving up to its surroundings the same quantity of heat (heat of vaporization) that was needed to vaporize it. At a given pressure the boiling and condensation temperatures are always the same; at this temperature the liquid and gaseous phases can coexist in phase equilibrium.

Both L_v and the boiling temperature of a material depend on pressure. Water boils at a lower temperature (about 95°C) in Denver than in Pittsburgh because Denver is at higher elevation and the average atmospheric pressure is lower. The heat of vaporization is somewhat greater at this lower pressure, about 2.27×10^6 J/kg.

Table 17.4 lists heats of fusion and vaporization for some materials and their melting and boiling temperatures at normal atmospheric pressure. Very few *elements* have melting temperatures in the vicinity of ordinary room temperatures; one of the few is the metal gallium, shown in Fig. 17.20.

Figure 17.21 shows how the temperature varies when we add heat continuously to a specimen of ice with an initial temperature below 0°C (point *a*). The temperature rises until we reach the melting point (point *b*). As more heat is added, the temperature remains constant until all the ice has melted (point *c*). Then the temperature rises again until the boiling temperature is reached (point *d*). At that point the temperature again is constant until all the water is transformed into the vapor phase (point *e*). If the rate of heat input is constant, the line for the solid phase (ice) has a steeper slope than does the line for the liquid phase (water). Do you see why? (See Table 17.3.)

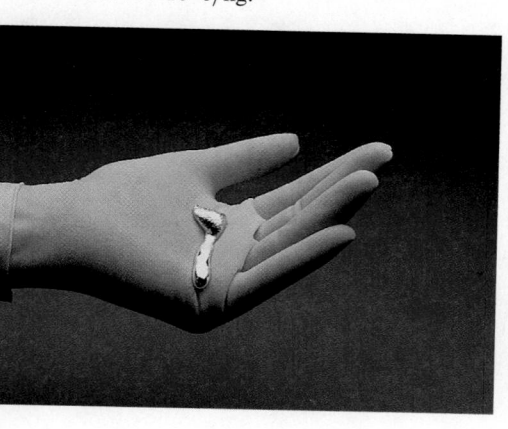

17.20 The metal gallium, shown here melting in a person's hand, is one of the few elements that melt in the vicinity of room temperature. Its melting temperature is 29.8°C, and its heat of fusion is 8.04×10^4 J/kg.

Table 17.4 Heats of Fusion and Vaporization

Substance	Normal Melting Point		Heat of Fusion, L_f (J/kg)	Normal Boiling Point		Heat of Vaporization, L_v (J/kg)
	K	**°C**		**K**	**°C**	
Helium	*	*	*	4.216	−268.93	20.9×10^3
Hydrogen	13.84	−259.31	58.6×10^3	20.26	−252.89	452×10^3
Nitrogen	63.18	−209.97	25.5×10^3	77.34	−195.8	201×10^3
Oxygen	54.36	−218.79	13.8×10^3	90.18	−183.0	213×10^3
Ethanol	159	−114	104.2×10^3	351	78	854×10^3
Mercury	234	−39	11.8×10^3	630	357	272×10^3
Water	273.15	0.00	334×10^3	373.15	100.00	2256×10^3
Sulfur	392	119	38.1×10^3	717.75	444.60	326×10^3
Lead	600.5	327.3	24.5×10^3	2023	1750	871×10^3
Antimony	903.65	630.50	165×10^3	1713	1440	561×10^3
Silver	1233.95	960.80	88.3×10^3	2466	2193	2336×10^3
Gold	1336.15	1063.00	64.5×10^3	2933	2660	1578×10^3
Copper	1356	1083	134×10^3	1460	1187	5069×10^3

*A pressure in excess of 25 atmospheres is required to make helium solidify. At 1 atmosphere pressure, helium remains a liquid down to absolute zero.

A substance can sometimes change directly from the solid to the gaseous phase. This process is called *sublimation,* and the solid is said to *sublime.* The corresponding heat is called the *heat of sublimation, L_s.* Liquid carbon dioxide cannot exist at a pressure lower than about 5×10^5 Pa (about 5 atm), and "dry ice" (solid carbon dioxide) sublimes at atmospheric pressure. Sublimation of water from frozen food causes freezer burn. The reverse process, a phase change from gas to solid, occurs when frost forms on cold bodies such as refrigerator cooling coils.

Very pure water can be cooled several degrees below the freezing temperature without freezing; the resulting unstable state is described as *supercooled.* When a small ice crystal is dropped in or the water is agitated, it crystallizes within a second or less. Supercooled water *vapor* condenses quickly into fog droplets when a disturbance, such as dust particles or ionizing radiation, is introduced. This principle is used in "seeding" clouds, which often contain supercooled water vapor, to cause condensation and rain.

A liquid can sometimes be *superheated* above its normal boiling temperature. Any small disturbance such as agitation causes local boiling with bubble formation.

Steam heating systems for buildings use a boiling–condensing process to transfer heat from the furnace to the radiators. Each kilogram of water that is

Phase of water changes. During these periods, temperature stays constant and the phase change proceeds as heat is added: $Q = +mL$.

Ice melts to liquid water at 0°C.
Liquid water vaporizes to steam at 100°C.
Boiling point
Melting point
Ice warms. Liquid water warms. Steam warms.

Temperature of water changes. During these periods, temperature rises as heat is added: $Q = mc\,\Delta T$.

17.21 Graph of temperature versus time for a specimen of water initially in the solid phase (ice). Heat is added to the specimen at a constant rate. The temperature remains constant during each change of phase, provided that the pressure remains constant.

17.22 The water may be warm and it may be a hot day, but these children will feel cold when they first step out of the swimming pool. That's because as water evaporates from their skin, it removes the heat of vaporization from their bodies. To stay warm, they will need to dry off immediately.

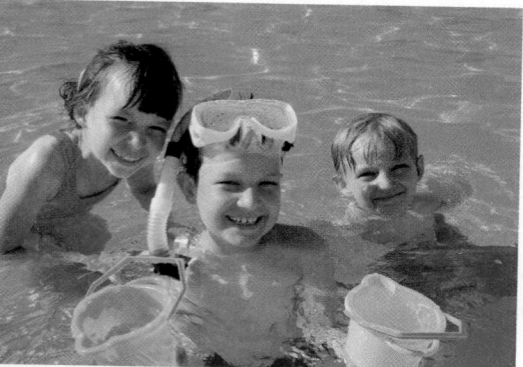

turned to steam in the boiler absorbs over 2×10^6 J (the heat of vaporization L_v of water) from the boiler and gives it up when it condenses in the radiators. Boiling–condensing processes are also used in refrigerators, air conditioners, and heat pumps. We will discuss these systems in Chapter 20.

The temperature-control mechanisms of many warm-blooded animals make use of heat of vaporization, removing heat from the body by using it to evaporate water from the tongue (panting) or from the skin (sweating). Evaporative cooling enables humans to maintain normal body temperature in hot, dry desert climates where the air temperature may reach 55°C (about 130°F). The skin temperature may be as much as 30°C cooler than the surrounding air. Under these conditions a normal person may perspire several liters per day, and this lost water must be replaced. Old-time desert rats (such as one of the authors) state that in the desert, any canteen that holds less than a gallon should be viewed as a toy! Evaporative cooling also explains why you feel cold when you first step out of a swimming pool (Fig. 17.22).

Evaporative cooling is also used to cool buildings in hot, dry climates and to condense and recirculate "used" steam in coal-fired or nuclear-powered electric-generating plants. That's what goes on in the large, tapered concrete towers that you see at such plants.

Chemical reactions such as combustion are analogous to phase changes in that they involve definite quantities of heat. Complete combustion of 1 gram of gasoline produces about 46,000 J or about 11,000 cal, so the **heat of combustion** L_c of gasoline is

$$L_c = 46,000 \text{ J/g} = 4.6 \times 10^7 \text{ J/kg}$$

Energy values of foods are defined similarly. When we say that a gram of peanut butter "contains 6 calories," we mean that 6 kcal of heat (6000 cal or 25,000 J) is released when the carbon and hydrogen atoms in the peanut butter react with oxygen (with the help of enzymes) and are completely converted to CO_2 and H_2O. Not all of this energy is directly useful for mechanical work. We will study the *efficiency* of energy utilization in Chapter 20.

Heat Calculations

Let's look at some examples of calorimetry calculations (calculations with heat). The basic principle is very simple: When heat flow occurs between two bodies that are isolated from their surroundings, the amount of heat lost by one body must equal the amount gained by the other. Heat is energy in transit, so this principle is really just conservation of energy. Calorimetry, dealing entirely with one conserved quantity, is in many ways the simplest of all physical theories!

Problem-Solving Strategy 17.2 **Calorimetry Problems**

IDENTIFY *the relevant concepts:* When heat flow occurs between two or more bodies that are isolated from their surroundings, the *algebraic sum* of the quantities of heat transferred to all the bodies is zero. We take a quantity of heat *added* to a body as *positive* and a quantity *leaving* a body as *negative.*

SET UP *the problem* using the following steps:
1. Identify the objects that exchange heat.
2. Each object may undergo a temperature change only, a phase change at constant temperature, or both. Use Eq. (17.13) for the heat transferred in a temperature change and Eq. (17.20) for the heat transferred in a phase change.
3. Consult Table 17.3 for values of specific heat or molar heat capacity and Table 17.4 for heats of fusion or vaporization.
4. List the known and unknown quantities and identify the target variables.

EXECUTE *the solution* as follows:
1. Use Eq. (17.13) and/or Eq. (17.20) and the energy-conservation relation $\Sigma Q = 0$ to solve for the target variables. Ensure that you use the correct algebraic signs for Q and ΔT terms, and that you correctly write $\Delta T = T_{final} - T_{initial}$ and not the reverse.
2. If a phase change occurs, you may not know in advance whether all, or only part, of the material undergoes a phase change. Make a reasonable guess; if that leads to an unreasonable result (such as a final temperature higher or lower than any initial temperature), the guess was wrong. Try again!

EVALUATE *your answer:* Double-check your calculations, and ensure that the results are physically sensible.

Example 17.7 A temperature change with no phase change

A camper pours 0.300 kg of coffee, initially in a pot at 70.0°C, into a 0.120-kg aluminum cup initially at 20.0°C. What is the equilibrium temperature? Assume that coffee has the same specific heat as water and that no heat is exchanged with the surroundings.

SOLUTION

IDENTIFY and SET UP: The target variable is the common final temperature T of the cup and coffee. No phase changes occur, so we need only Eq. (17.13). With subscripts C for coffee, W for water, and Al for aluminum, we have $T_{0C} = 70.0°$ and $T_{0Al} = 20.0°$; Table 17.3 gives $c_W = 4190 \text{ J/kg} \cdot \text{K}$ and $c_{Al} = 910 \text{ J/kg} \cdot \text{K}$.

EXECUTE: The (negative) heat gained by the coffee is $Q_C = m_C c_W \Delta T_C$. The (positive) heat gained by the cup is $Q_{Al} = m_{Al} c_{Al} \Delta T_{Al}$. We set $Q_C + Q_{Al} = 0$ (see Problem-Solving Strategy 17.2) and substitute $\Delta T_C = T - T_{0C}$ and $\Delta T_{Al} = T - T_{0Al}$:

$$Q_C + Q_{Al} = m_C c_W \Delta T_C + m_{Al} c_{Al} \Delta T_{Al} = 0$$
$$m_C c_W (T - T_{0C}) + m_{Al} c_{Al}(T - T_{0Al}) = 0$$

Then we solve this expression for the final temperature T. A little algebra gives

$$T = \frac{m_C c_W T_{0C} + m_{Al} c_{Al} T_{0Al}}{m_C c_W + m_{Al} c_{Al}} = 66.0°C$$

EVALUATE: The final temperature is much closer to the initial temperature of the coffee than to that of the cup; water has a much higher specific heat than aluminum, and we have more than twice as much mass of water. We can also find the quantities of heat by substituting the value $T = 66.0°C$ back into the original equations. We find $Q_C = -5.0 \times 10^3 \text{ J}$ and $Q_{Al} = +5.0 \times 10^3 \text{ J}$. As expected, Q_C is negative: The coffee loses heat to the cup.

Example 17.8 Changes in both temperature and phase

A glass contains 0.25 kg of Omni-Cola (mostly water) initially at 25°C. How much ice, initially at −20°C, must you add to obtain a final temperature of 0°C with all the ice melted? Neglect the heat capacity of the glass.

SOLUTION

IDENTIFY and SET UP: The Omni-Cola and ice exchange heat. The cola undergoes a temperature change; the ice undergoes both a temperature change and a phase change from solid to liquid. We use subscripts C for cola, I for ice, and W for water. The target variable is the mass of ice, m_I. We use Eq. (17.13) to obtain an expression for the amount of heat involved in cooling the drink to $T = 0°C$ and warming the ice to $T = 0°C$, and Eq. (17.20) to obtain an expression for the heat required to melt the ice at 0°C. We have $T_{0C} = 25°C$ and $T_{0I} = -20°C$, Table 17.3 gives $c_W = 4190 \text{ J/kg} \cdot \text{K}$ and $c_I = 2100 \text{ J/kg} \cdot \text{K}$, and Table 17.4 gives $L_f = 3.34 \times 10^5 \text{ J/kg}$.

EXECUTE: From Eq. (17.13), the (negative) heat gained by the Omni-Cola is $Q_C = m_C c_W \Delta T_C$. The (positive) heat gained by the ice in warming is $Q_1 = m_I c_I \Delta T_I$. The (positive) heat required to melt the ice is $Q_2 = m_I L_f$. We set $Q_C + Q_1 + Q_2 = 0$, insert $\Delta T_C = T - T_{0C}$ and $\Delta T_I = T - T_{0I}$, and solve for m_I:

$$m_C c_W \Delta T_C + m_I c_I \Delta T_I + m_I L_f = 0$$
$$m_C c_W (T - T_{0C}) + m_I c_I (T - T_{0I}) + m_I L_f = 0$$
$$m_I[c_I(T - T_{0I}) + L_f] = -m_C c_W(T - T_{0C})$$
$$m_I = m_C \frac{c_W(T_{0C} - T)}{c_I(T - T_{0I}) + L_f}$$

Substituting numerical values, we find that $m_I = 0.070 \text{ kg} = 70 \text{ g}$.

EVALUATE: Three or four medium-size ice cubes would make about 70 g, which seems reasonable given the 250 g of Omni-Cola to be cooled.

Example 17.9 What's cooking?

A hot copper pot of mass 2.0 kg (including its copper lid) is at a temperature of 150°C. You pour 0.10 kg of cool water at 25°C into the pot, then quickly replace the lid so no steam can escape. Find the final temperature of the pot and its contents, and determine the phase of the water (liquid, gas, or a mixture). Assume that no heat is lost to the surroundings.

SOLUTION

IDENTIFY and SET UP: The water and the pot exchange heat. Three outcomes are possible: (1) No water boils, and the final temperature T is less than 100°C; (2) some water boils, giving a mixture of water and steam at 100°C; or (3) all the water boils, giving 0.10 kg of steam at 100°C or greater. We use Eq. (17.13) for the heat transferred in a temperature change and Eq. (17.20) for the heat transferred in a phase change.

EXECUTE: First consider case (1), which parallels Example 17.8 exactly. The equation that states that the heat flow into the water equals the heat flow out of the pot is

$$Q_W + Q_{Cu} = m_W c_W(T - T_{0W}) + m_{Cu} c_{Cu}(T - T_{0Cu}) = 0$$

Here we use subscripts W for water and Cu for copper, with $m_W = 0.10 \text{ kg}$, $m_{Cu} = 2.0 \text{ kg}$, $T_{0W} = 25°C$, and $T_{0Cu} = 150°C$. From Table 17.3, $c_W = 4190 \text{ J/kg} \cdot \text{K}$ and $c_{Cu} = 390 \text{ J/kg} \cdot \text{K}$. Solving for the final temperature T and substituting these values, we get

$$T = \frac{m_W c_W T_{0W} + m_{Cu} c_{Cu} T_{0Cu}}{m_W c_W + m_{Cu} c_{Cu}} = 106°C$$

But this is above the boiling point of water, which contradicts our assumption that no water boils! So at least some of the water boils.

Continued

So consider case (2), in which the final temperature is $T = 100°C$ and some unknown fraction x of the water boils, where (if this case is correct) x is greater than zero and less than or equal to 1. The (positive) amount of heat needed to vaporize this water is xm_WL_v. The energy-conservation condition $Q_W + Q_{Cu} = 0$ is then

$$m_Wc_W(100°C - T_{0W}) + xm_WL_v + m_{Cu}c_{Cu}(100°C - T_{0Cu}) = 0$$

We solve for the target variable x:

$$x = \frac{-m_{Cu}c_{Cu}(100°C - T_{0Cu}) - m_Wc_W(100°C - T_{0W})}{m_WL_v}$$

With $L_v = 2.256 \times 10^6$ J from Table 17.4, this yields $x = 0.034$. We conclude that the final temperature of the water and copper is $100°C$ and that $0.034(0.10$ kg$) = 0.0034$ kg $= 3.4$ g of the water is converted to steam at $100°C$.

EVALUATE: Had x turned out to be greater than 1, case (3) would have held; all the water would have vaporized, and the final temperature would have been greater than $100°C$. Can you show that this would have been the case if we had originally poured less than 15 g of $25°C$ water into the pot?

Example 17.10 Combustion, temperature change, and phase change

In a particular camp stove, only 30% of the energy released in burning gasoline goes to heating the water in a pot on the stove. How much gasoline must we burn to heat 1.00 L (1.00 kg) of water from 20°C to 100°C and boil away 0.25 kg of it?

SOLUTION

IDENTIFY and SET UP: All of the water undergoes a temperature change and part of it undergoes a phase change, from liquid to gas. We determine the heat required to cause both of these changes, and then use the 30% combustion efficiency to determine the amount of gasoline that must be burned (the target variable). We use Eqs. (17.13) and (17.20) and the idea of heat of combustion.

EXECUTE: To raise the temperature of the water from 20°C to 100°C requires

$$Q_1 = mc\,\Delta T = (1.00 \text{ kg})(4190 \text{ J/kg} \cdot \text{K})(80 \text{ K})$$
$$= 3.35 \times 10^5 \text{ J}$$

To boil 0.25 kg of water at 100°C requires

$$Q_2 = mL_v = (0.25 \text{ kg})(2.256 \times 10^6 \text{ J/kg}) = 5.64 \times 10^5 \text{ J}$$

The total energy needed is $Q_1 + Q_2 = 8.99 \times 10^5$ J. This is 30% = 0.30 of the total heat of combustion, which is therefore $(8.99 \times 10^5 \text{ J})/0.30 = 3.00 \times 10^6$ J. As we mentioned earlier, the combustion of 1 g of gasoline releases 46,000 J, so the mass of gasoline required is $(3.00 \times 10^6 \text{ J})/(46,000 \text{ J/g}) = 65$ g, or a volume of about 0.09 L of gasoline.

EVALUATE: This result suggests the tremendous amount of energy released in burning even a small quantity of gasoline. Another 123 g of gasoline would be required to boil away the remaining water; can you prove this?

Test Your Understanding of Section 17.6 You take a block of ice at 0°C and add heat to it at a steady rate. It takes a time t to completely convert the block of ice to steam at 100°C. What do you have at time $t/2$? (i) all ice at 0°C; (ii) a mixture of ice and water at 0°C; (iii) water at a temperature between 0°C and 100°C; (iv) a mixture of water and steam at 100°C.

PhET: The Greenhouse Effect

17.7 Mechanisms of Heat Transfer

We have talked about *conductors* and *insulators,* materials that permit or prevent heat transfer between bodies. Now let's look in more detail at *rates* of energy transfer. In the kitchen you use a metal or glass pot for good heat transfer from the stove to whatever you're cooking, but your refrigerator is insulated with a material that *prevents* heat from flowing into the food inside the refrigerator. How do we describe the difference between these two materials?

The three mechanisms of heat transfer are conduction, convection, and radiation. *Conduction* occurs within a body or between two bodies in contact. *Convection* depends on motion of mass from one region of space to another. *Radiation* is heat transfer by electromagnetic radiation, such as sunshine, with no need for matter to be present in the space between bodies.

Conduction

If you hold one end of a copper rod and place the other end in a flame, the end you are holding gets hotter and hotter, even though it is not in direct contact with the flame. Heat reaches the cooler end by **conduction** through the material.

On the atomic level, the atoms in the hotter regions have more kinetic energy, on the average, than their cooler neighbors. They jostle their neighbors, giving them some of their energy. The neighbors jostle *their* neighbors, and so on through the material. The atoms themselves do not move from one region of material to another, but their energy does.

Most metals also use another, more effective mechanism to conduct heat. Within the metal, some electrons can leave their parent atoms and wander through the crystal lattice. These "free" electrons can rapidly carry energy from the hotter to the cooler regions of the metal, so metals are generally good conductors of heat. A metal rod at 20°C feels colder than a piece of wood at 20°C because heat can flow more easily from your hand into the metal. The presence of "free" electrons also causes most metals to be good electrical conductors.

Heat transfer occurs only between regions that are at different temperatures, and the direction of heat flow is always from higher to lower temperature. Figure 17.23a shows a rod of conducting material with cross-sectional area A and length L. The left end of the rod is kept at a temperature T_H and the right end at a lower temperature T_C, so heat flows from left to right. The sides of the rod are covered by an ideal insulator, so no heat transfer occurs at the sides.

When a quantity of heat dQ is transferred through the rod in a time dt, the rate of heat flow is dQ/dt. We call this rate the **heat current,** denoted by H. That is, $H = dQ/dt$. Experiments show that the heat current is proportional to the cross-sectional area A of the rod (Fig. 17.23b) and to the temperature difference $(T_H - T_C)$ and is inversely proportional to the rod length L (Fig. 17.23c). Introducing a proportionality constant k called the **thermal conductivity** of the material, we have

$$H = \frac{dQ}{dt} = kA\frac{T_H - T_C}{L} \qquad \text{(heat current in conduction)} \qquad (17.21)$$

The quantity $(T_H - T_C)/L$ is the temperature difference *per unit length;* it is called the magnitude of the **temperature gradient.** The numerical value of k depends on the material of the rod. Materials with large k are good conductors of heat; materials with small k are poor conductors, or insulators. Equation (17.21) also gives the heat current through a slab or through *any* homogeneous body with uniform cross section A perpendicular to the direction of flow; L is the length of the heat-flow path.

The units of heat current H are units of energy per time, or power; the SI unit of heat current is the watt ($1 \text{ W} = 1 \text{ J/s}$). We can find the units of k by solving Eq. (17.21) for k; you can show that the SI units are $\text{W/m} \cdot \text{K}$. Some numerical values of k are given in Table 17.5.

The thermal conductivity of "dead" (that is, nonmoving) air is very small. A wool sweater keeps you warm because it traps air between the fibers. In fact, many insulating materials such as Styrofoam and fiberglass are mostly dead air.

If the temperature varies in a nonuniform way along the length of the conducting rod, we introduce a coordinate x along the length and generalize the temperature gradient to be dT/dx. The corresponding generalization of Eq. (17.21) is

$$H = \frac{dQ}{dt} = -kA\frac{dT}{dx} \qquad (17.22)$$

The negative sign shows that heat always flows in the direction of *decreasing* temperature.

17.23 Steady-state heat flow due to conduction in a uniform rod.

(a) Heat current H

(b) Doubling the cross-sectional area of the conductor doubles the heat current (H is proportional to A).

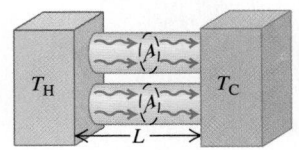

(c) Doubling the length of the conductor halves the heat current (H is inversely proportional to L).

Table 17.5 Thermal Conductivities

Substance	k $(\text{W/m} \cdot \text{K})$
Metals	
Aluminum	205.0
Brass	109.0
Copper	385.0
Lead	34.7
Mercury	8.3
Silver	406.0
Steel	50.2
Solids (representative values)	
Brick, insulating	0.15
Brick, red	0.6
Concrete	0.8
Cork	0.04
Felt	0.04
Fiberglass	0.04
Glass	0.8
Ice	1.6
Rock wool	0.04
Styrofoam	0.027
Wood	0.12–0.04
Gases	
Air	0.024
Argon	0.016
Helium	0.14
Hydrogen	0.14
Oxygen	0.023

Application **Fur Versus Blubber**
The fur of an arctic fox is a good thermal insulator because it traps air, which has a low thermal conductivity k. (The value $k = 0.04$ W/m·K for fur is higher than for air, $k = 0.024$ W/m·K, because fur also includes solid hairs.) The layer of fat beneath a bowhead whale's skin, called blubber, has six times the thermal conductivity of fur ($k = 0.24$ W/m·K). So a 6-cm thickness of blubber ($L = 6$ cm) is required to give the same insulation as 1 cm of fur.

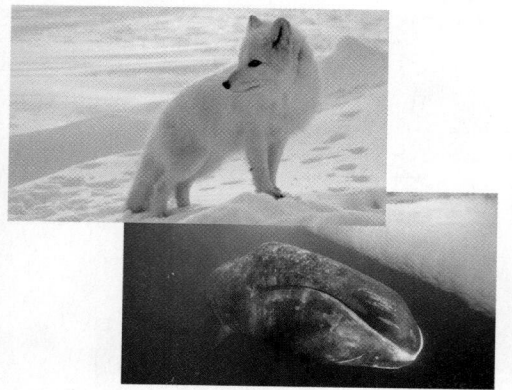

For thermal insulation in buildings, engineers use the concept of **thermal resistance,** denoted by R. The thermal resistance R of a slab of material with area A is defined so that the heat current H through the slab is

$$H = \frac{A(T_H - T_C)}{R} \qquad (17.23)$$

where T_H and T_C are the temperatures on the two sides of the slab. Comparing this with Eq. (17.21), we see that R is given by

$$R = \frac{L}{k} \qquad (17.24)$$

where L is the thickness of the slab. The SI unit of R is 1 m²·K/W. In the units used for commercial insulating materials in the United States, H is expressed in Btu/h, A is in ft², and $T_H - T_C$ in F°. (1 Btu/h = 0.293 W.) The units of R are then ft²·F°·h/Btu, though values of R are usually quoted without units; a 6-inch-thick layer of fiberglass has an R value of 19 (that is, $R = 19$ ft²·F°·h/Btu), a 2-inch-thick slab of polyurethane foam has an R value of 12, and so on. Doubling the thickness doubles the R value. Common practice in new construction in severe northern climates is to specify R values of around 30 for exterior walls and ceilings. When the insulating material is in layers, such as a plastered wall, fiberglass insulation, and wood exterior siding, the R values are additive. Do you see why? (See Problem 17.108.)

Problem-Solving Strategy 17.3 **Heat Conduction**

IDENTIFY *the relevant concepts:* Heat conduction occurs whenever two objects at different temperatures are placed in contact.

SET UP *the problem* using the following steps:
1. Identify the direction of heat flow (from hot to cold). In Eq. (17.21), L is measured along this direction, and A is an area perpendicular to this direction. You can often approximate an irregular-shaped container with uniform wall thickness as a flat slab with the same thickness and total wall area.

2. List the known and unknown quantities and identify the target variable.

EXECUTE *the solution* as follows:
1. If heat flows through a single object, use Eq. (17.21) to solve for the target variable.
2. If the heat flows through two different materials in succession (in *series*), the temperature T at the interface between them is intermediate between T_H and T_C, so that the temperature differences across the two materials are $(T_H - T)$ and $(T - T_C)$. In steady-state heat flow, the same heat must pass through both materials, so the heat current H must be the *same* in both materials.

3. If heat flows through two or more *parallel* paths, then the total heat current H is the sum of the currents H_1, H_2, ... for the separate paths. An example is heat flow from inside a room to outside, both through the glass in a window and through the surrounding wall. In parallel heat flow the temperature difference is the same for each path, but L, A, and k may be different for each path.
4. Use consistent units. If k is expressed in W/m·K, for example, use distances in meters, heat in joules, and T in kelvins.

EVALUATE *your answer:* Are the results physically reasonable?

Example 17.11 **Conduction into a picnic cooler**

A Styrofoam cooler (Fig. 17.24a) has total wall area (including the lid) of 0.80 m² and wall thickness 2.0 cm. It is filled with ice, water, and cans of Omni-Cola, all at 0°C. What is the rate of heat flow into the cooler if the temperature of the outside wall is 30°C? How much ice melts in 3 hours?

SOLUTION

IDENTIFY and SET UP: The target variables are the heat current H and the mass m of ice melted. We use Eq. (17.21) to determine H and Eq. (17.20) to determine m.

EXECUTE: We assume that the total heat flow is the same as it would be through a flat Styrofoam slab of area 0.80 m² and thickness 2.0 cm = 0.020 m (Fig. 17.24b). We find k from Table 17.5. From Eq. (17.21),

$$H = kA\frac{T_H - T_C}{L} = (0.027 \text{ W/m·K})(0.80 \text{ m}^2)\frac{30°C - 0°C}{0.020 \text{ m}}$$
$$= 32.4 \text{ W} = 32.4 \text{ J/s}$$

The total heat flow is $Q = Ht$, with $t = 3$ h $= 10,800$ s. From Table 17.4, the heat of fusion of ice is $L_f = 3.34 \times 10^5$ J/kg, so from Eq. (17.20) the mass of ice that melts is

$$m = \frac{Q}{L_f} = \frac{(32.4 \text{ J/s})(10,800 \text{ s})}{3.34 \times 10^5 \text{ J/kg}} = 1.0 \text{ kg}$$

EVALUATE: The low heat current is a result of the low thermal conductivity of Styrofoam.

17.24 Conduction of heat across the walls of a Styrofoam cooler.

(a) A cooler at the beach

(b) Our sketch for this problem

Example 17.12 | **Conduction through two bars I**

A steel bar 10.0 cm long is welded end to end to a copper bar 20.0 cm long. Each bar has a square cross section, 2.00 cm on a side. The free end of the steel bar is kept at 100°C by placing it in contact with steam, and the free end of the copper bar is kept at 0°C by placing it in contact with ice. Both bars are perfectly insulated on their sides. Find the steady-state temperature at the junction of the two bars and the total rate of heat flow through the bars.

17.25 Our sketch for this problem.

SOLUTION

IDENTIFY and SET UP: Figure 17.25 shows the situation. The heat currents in these end-to-end bars must be the same (see Problem-Solving Strategy 17.3). We are given "hot" and "cold" temperatures $T_H = 100°C$ and $T_C = 0°C$. With subscripts S for steel and Cu for copper, we write Eq. (17.21) separately for the heat currents H_S and H_{Cu} and set the resulting expressions equal to each other.

EXECUTE: Setting $H_S = H_{Cu}$, we have from Eq. (17.21)

$$H_S = k_S A \frac{T_H - T}{L_S} = H_{Cu} = k_{Cu} A \frac{T - T_C}{L_{Cu}}$$

We divide out the equal cross-sectional areas A and solve for T:

$$T = \frac{\dfrac{k_S}{L_S}T_H + \dfrac{k_{Cu}}{L_{Cu}}T_C}{\left(\dfrac{k_S}{L_S} + \dfrac{k_{Cu}}{L_{Cu}}\right)}$$

Substituting $L_S = 10.0$ cm and $L_{Cu} = 20.0$ cm, the given values of T_H and T_C, and the values of k_S and k_{Cu} from Table 17.5, we find $T = 20.7°C$.

We can find the total heat current by substituting this value of T into either the expression for H_S or the one for H_{Cu}:

$$H_S = (50.2 \text{ W/m} \cdot \text{K})(0.0200 \text{ m})^2 \frac{100°C - 20.7°C}{0.100 \text{ m}}$$
$$= 15.9 \text{ W}$$

$$H_{Cu} = (385 \text{ W/m} \cdot \text{K})(0.0200 \text{ m})^2 \frac{20.7°C}{0.200 \text{ m}} = 15.9 \text{ W}$$

EVALUATE: Even though the steel bar is shorter, the temperature drop across it is much greater (from 100°C to 20.7°C) than across the copper bar (from 20.7°C to 0°C). That's because steel is a much poorer conductor than copper.

Example 17.13 | **Conduction through two bars II**

Suppose the two bars of Example 17.12 are separated. One end of each bar is kept at 100°C and the other end of each bar is kept at 0°C. What is the *total* heat current in the two bars?

17.26 Our sketch for this problem.

SOLUTION

IDENTIFY and SET UP: Figure 17.26 shows the situation. For each bar, $T_H - T_C = 100°C - 0°C = 100$ K. The total heat current is the sum of the currents in the two bars, $H_S + H_{Cu}$.

Continued

EXECUTE: We write the heat currents for the two rods individually, and then add them to get the total heat current:

$$H = H_S + H_{Cu} = k_S A \frac{T_H - T_C}{L_S} + k_{Cu} A \frac{T_H - T_C}{L_{Cu}}$$

$$= (50.2 \text{ W/m} \cdot \text{K})(0.0200 \text{ m})^2 \frac{100 \text{ K}}{0.100 \text{ m}}$$

$$+ (385 \text{ W/m} \cdot \text{K})(0.0200 \text{ m})^2 \frac{100 \text{ K}}{0.200 \text{ m}}$$

$$= 20.1 \text{ W} + 77.0 \text{ W} = 97.1 \text{ W}$$

EVALUATE: The heat flow in the copper bar is much greater than that in the steel bar, even though it is longer, because the thermal conductivity of copper is much larger. The total heat flow is greater than in Example 17.12 because the total cross section for heat flow is greater and because the full 100-K temperature difference appears across each bar.

Convection

Convection is the transfer of heat by mass motion of a fluid from one region of space to another. Familiar examples include hot-air and hot-water home heating systems, the cooling system of an automobile engine, and the flow of blood in the body. If the fluid is circulated by a blower or pump, the process is called *forced convection;* if the flow is caused by differences in density due to thermal expansion, such as hot air rising, the process is called *natural convection* or *free convection* (Fig. 17.27).

17.27 A heating element in the tip of this submerged tube warms the surrounding water, producing a complex pattern of free convection.

Free convection in the atmosphere plays a dominant role in determining the daily weather, and convection in the oceans is an important global heat-transfer mechanism. On a smaller scale, soaring hawks and glider pilots make use of thermal updrafts from the warm earth. The most important mechanism for heat transfer within the human body (needed to maintain nearly constant temperature in various environments) is *forced* convection of blood, with the heart serving as the pump.

Convective heat transfer is a very complex process, and there is no simple equation to describe it. Here are a few experimental facts:

1. The heat current due to convection is directly proportional to the surface area. This is the reason for the large surface areas of radiators and cooling fins.
2. The viscosity of fluids slows natural convection near a stationary surface, giving a surface film that on a vertical surface typically has about the same insulating value as 1.3 cm of plywood (*R* value = 0.7). Forced convection decreases the thickness of this film, increasing the rate of heat transfer. This is the reason for the "wind-chill factor"; you get cold faster in a cold wind than in still air with the same temperature.
3. The heat current due to convection is found to be approximately proportional to the $\frac{5}{4}$ power of the temperature difference between the surface and the main body of fluid.

Radiation

Radiation is the transfer of heat by electromagnetic waves such as visible light, infrared, and ultraviolet radiation. Everyone has felt the warmth of the sun's radiation and the intense heat from a charcoal grill or the glowing coals in a fireplace. Most of the heat from these very hot bodies reaches you not by conduction or convection in the intervening air but by *radiation*. This heat transfer would occur even if there were nothing but vacuum between you and the source of heat.

Every body, even at ordinary temperatures, emits energy in the form of electromagnetic radiation. Around 20°C, nearly all the energy is carried by infrared waves with wavelengths much longer than those of visible light (see Figs. 17.4 and 17.28). As the temperature rises, the wavelengths shift to shorter values. At 800°C, a body emits enough visible radiation to appear "red-hot," although even at this temperature most of the energy is carried by infrared waves. At 3000°C,

17.28 This false-color infrared photograph reveals radiation emitted by various parts of the man's body. The strongest emission (colored red) comes from the warmest areas, while there is very little emission from the bottle of cold beverage.

the temperature of an incandescent lamp filament, the radiation contains enough visible light that the body appears "white-hot."

The rate of energy radiation from a surface is proportional to the surface area A and to the fourth power of the absolute (Kelvin) temperature T. The rate also depends on the nature of the surface; this dependence is described by a quantity e called the **emissivity**. A dimensionless number between 0 and 1, e represents the ratio of the rate of radiation from a particular surface to the rate of radiation from an equal area of an ideal radiating surface at the same temperature. Emissivity also depends somewhat on temperature. Thus the heat current $H = dQ/dt$ due to radiation from a surface area A with emissivity e at absolute temperature T can be expressed as

$$H = Ae\sigma T^4 \qquad \text{(heat current in radiation)} \qquad (17.25)$$

where σ is a fundamental physical constant called the **Stefan–Boltzmann constant**. This relationship is called the **Stefan–Boltzmann law** in honor of its late-19th-century discoverers. The current best numerical value of σ is

$$\sigma = 5.670400(40) \times 10^{-8} \text{ W/m}^2 \cdot \text{K}^4$$

We invite you to check unit consistency in Eq. (17.25). Emissivity (e) is often larger for dark surfaces than for light ones. The emissivity of a smooth copper surface is about 0.3, but e for a dull black surface can be close to unity.

Example 17.14   Heat transfer by radiation

A thin, square steel plate, 10 cm on a side, is heated in a blacksmith's forge to 800°C. If the emissivity is 0.60, what is the total rate of radiation of energy from the plate?

SOLUTION

IDENTIFY and SET UP: The target variable is H, the rate of emission of energy from the plate's two surfaces. We use Eq. (17.25) to calculate H.

EXECUTE: The total surface area is $2(0.10 \text{ m})^2 = 0.020 \text{ m}^2$, and $T = 800°C = 1073$ K. Then Eq. (17.25) gives

$$
\begin{aligned}
H &= Ae\sigma T^4 \\
&= (0.020 \text{ m}^2)(0.60)(5.67 \times 10^{-8} \text{ W/m}^2 \cdot \text{K}^4)(1073 \text{ K})^4 \\
&= 900 \text{ W}
\end{aligned}
$$

EVALUATE: The nearby blacksmith will easily feel the heat radiated from this plate.

Radiation and Absorption

While a body at absolute temperature T is radiating, its surroundings at temperature T_s are also radiating, and the body *absorbs* some of this radiation. If it is in thermal equilibrium with its surroundings, $T = T_s$ and the rates of radiation and absorption must be equal. For this to be true, the rate of absorption must be given in general by $H = Ae\sigma T_s^4$. Then the *net* rate of radiation from a body at temperature T with surroundings at temperature T_s is

$$H_{\text{net}} = Ae\sigma T^4 - Ae\sigma T_s^4 = Ae\sigma(T^4 - T_s^4) \qquad (17.26)$$

In this equation a positive value of H means a net heat flow *out of* the body. Equation (17.26) shows that for radiation, as for conduction and convection, the heat current depends on the temperature *difference* between two bodies.

Example 17.15   Radiation from the human body

What is the total rate of radiation of energy from a human body with surface area 1.20 m^2 and surface temperature 30°C = 303 K? If the surroundings are at a temperature of 20°C, what is the *net* rate of radiative heat loss from the body? The emissivity of the human body is very close to unity, irrespective of skin pigmentation.

SOLUTION

IDENTIFY and SET UP: We must consider both the radiation that the body emits and the radiation that it absorbs from its surroundings. Equation (17.25) gives the rate of radiation of energy from the body, and Eq. (17.26) gives the net rate of heat loss.

Continued

EXECUTE: Taking $e = 1$ in Eq. (17.25), we find that the body radiates at a rate

$$H = Ae\sigma T^4$$

$$= (1.20 \text{ m}^2)(1)(5.67 \times 10^{-8} \text{ W/m}^2 \cdot \text{K}^4)(303 \text{ K})^4 = 574 \text{ W}$$

This loss is partly offset by absorption of radiation, which depends on the temperature of the surroundings. From Eq. (17.26), the *net* rate of radiative energy transfer is

$$H_{\text{net}} = Ae\sigma(T^4 - T_s^{\,4})$$

$$= (1.20 \text{ m}^2)(1)(5.67 \times 10^{-8} \text{ W/m}^2 \cdot \text{K}^4)[(303 \text{ K})^4$$

$$- (293 \text{ K})^4] = 72 \text{ W}$$

EVALUATE: The value of H_{net} is positive because the body is losing heat to its colder surroundings.

Applications of Radiation

Heat transfer by radiation is important in some surprising places. A premature baby in an incubator can be cooled dangerously by radiation if the walls of the incubator happen to be cold, even when the *air* in the incubator is warm. Some incubators regulate the air temperature by measuring the baby's skin temperature.

A body that is a good absorber must also be a good emitter. An ideal radiator, with an emissivity of unity, is also an ideal absorber, absorbing *all* of the radiation that strikes it. Such an ideal surface is called an ideal black body or simply a **blackbody.** Conversely, an ideal *reflector,* which absorbs *no* radiation at all, is also a very ineffective radiator.

This is the reason for the silver coatings on vacuum ("Thermos") bottles, invented by Sir James Dewar (1842–1923). A vacuum bottle has double glass walls. The air is pumped out of the spaces between the walls; this eliminates nearly all heat transfer by conduction and convection. The silver coating on the walls reflects most of the radiation from the contents back into the container, and the wall itself is a very poor emitter. Thus a vacuum bottle can keep coffee or soup hot for several hours. The Dewar flask, used to store very cold liquefied gases, is exactly the same in principle.

Radiation, Climate, and Climate Change

Our planet constantly absorbs radiation coming from the sun. In thermal equilibrium, the rate at which our planet absorbs solar radiation must equal the rate at which it emits radiation into space. The presence of an atmosphere on our planet has a significant effect on this equilibrium.

Most of the radiation emitted by the sun (which has a surface temperature of 5800 K) is in the visible part of the spectrum, to which our atmosphere is transparent. But the average surface temperature of the earth is only 287 K (14°C). Hence most of the radiation that our planet emits into space is infrared radiation, just like the radiation from the person shown in Fig. 17.28. However, our atmosphere is *not* completely transparent to infrared radiation. This is because our atmosphere contains carbon dioxide (CO_2), which is its fourth most abundant constituent (after nitrogen, oxygen, and argon). Molecules of CO_2 in the atmosphere have the property that they *absorb* some of the infrared radiation coming upward from the surface. They then re-radiate the absorbed energy, but some of the re-radiated energy is directed back down toward the surface instead of escaping into space. In order to maintain thermal equilibrium, the earth's surface must compensate for this by increasing its temperature T and hence its total rate of radiating energy (which is proportional to T^4). This phenomenon, called the **greenhouse effect,** makes our planet's surface temperature about 33°C higher than it would be if there were no atmospheric CO_2. If CO_2 were absent, the earth's average surface temperature would be below the freezing point of water, and life as we know it would be impossible.

While atmospheric CO_2 has a beneficial effect, too much of it can have extremely negative consequences. Measurements of air trapped in ancient Antarctic ice show that over the past 650,000 years CO_2 has constituted less than 300 parts per million of our atmosphere. Since the beginning of the industrial age, however, the burning of fossil fuels such as coal and petroleum has elevated the atmospheric CO_2 concentration to unprecedented levels (Fig. 17.29a). As a consequence, since the 1950s the global average surface temperature has increased by 0.6°C and the earth has experienced the hottest years ever recorded (Fig. 17.29b). If we continue to consume fossil fuels at the same rate, by 2050 the atmospheric CO_2 concentration will reach 600 parts per million, well off the scale of Fig. 17.29a. The resulting temperature increase will have dramatic effects on climate around the world. In the polar regions massive quantities of ice will melt and run from solid land to the sea, thus raising ocean levels worldwide and threatening the homes and lives of hundreds of millions of people who live near the coast. Coping with these threats is one of the greatest challenges facing 21st-century civilization.

17.29 (a) The concentration of atmospheric CO_2 has increased by 22% since continuous measurements began in 1958. (The yearly variations are due to increased intake of CO_2 by plants in spring and summer.) (b) The increase in global average temperature since the beginning of the industrial era is a result of the increase in CO_2 concentration.

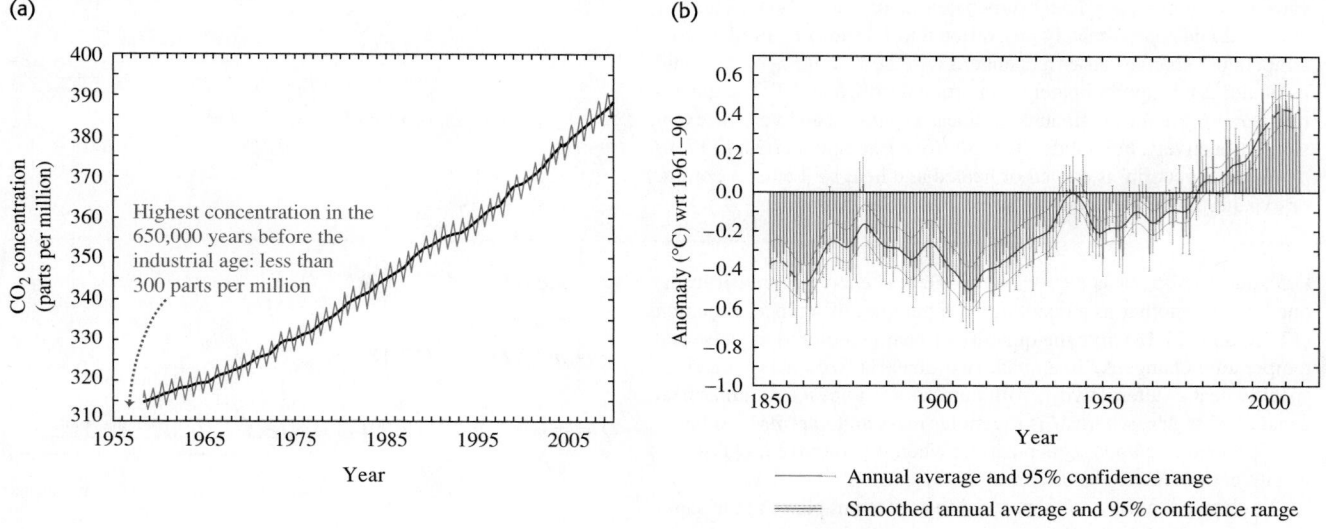

(a)

(b)

Test Your Understanding of Section 17.7   A room has one wall made of concrete, one wall made of copper, and one wall made of steel. All of the walls are the same size and at the same temperature of 20°C. Which wall feels coldest to the touch? (i) the concrete wall; (ii) the copper wall; (iii) the steel wall; (iv) all three walls feel equally cold.

Temperature and temperature scales: Two bodies in thermal equilibrium must have the same temperature. A conducting material between two bodies permits them to interact and come to thermal equilibrium; an insulating material impedes this interaction.

The Celsius and Fahrenheit temperature scales are based on the freezing ($0°C = 32°F$) and boiling ($100°C = 212°F$) temperatures of water. One Celsius degree equals $\frac{9}{5}$ Fahrenheit degrees. (See Example 17.1.)

The Kelvin scale has its zero at the extrapolated zero-pressure temperature for a gas thermometer, $-273.15°C = 0$ K. In the gas-thermometer scale, the ratio of two temperatures T_1 and T_2 is defined to be equal to the ratio of the two corresponding gas-thermometer pressures p_1 and p_2.

$$T_F = \tfrac{9}{5}T_C + 32° \quad (17.1)$$

$$T_C = \tfrac{5}{9}(T_F - 32°) \quad (17.2)$$

$$T_K = T_C + 273.15 \quad (17.3)$$

$$\frac{T_2}{T_1} = \frac{p_2}{p_1} \quad (17.4)$$

If systems A and B are each in thermal equilibrium with system C …

… then systems A and B are in thermal equilibrium with each other.

Thermal expansion and thermal stress: A temperature change ΔT causes a change in any linear dimension L_0 of a solid body. The change ΔL is approximately proportional to L_0 and ΔT. Similarly, a temperature change causes a change ΔV in the volume V_0 of any solid or liquid; ΔV is approximately proportional to V_0 and ΔT. The quantities α and β are the coefficients of linear expansion and volume expansion, respectively. For solids, $\beta = 3\alpha$. (See Examples 17.2 and 17.3.)

When a material is cooled or heated and held so it cannot contract or expand, it is under a tensile stress F/A. (See Example 17.4.)

$$\Delta L = \alpha L_0 \, \Delta T \quad (17.6)$$

$$\Delta V = \beta V_0 \, \Delta T \quad (17.8)$$

$$\frac{F}{A} = -Y\alpha \, \Delta T \quad (17.12)$$

$$L = L_0 + \Delta L$$
$$= L_0(1 + \alpha \, \Delta T)$$

Heat, phase changes, and calorimetry: Heat is energy in transit from one body to another as a result of a temperature difference. Equations (17.13) and (17.18) give the quantity of heat Q required to cause a temperature change ΔT in a quantity of material with mass m and specific heat c (alternatively, with number of moles n and molar heat capacity $C = Mc$, where M is the molar mass and $m = nM$). When heat is added to a body, Q is positive; when it is removed, Q is negative. (See Examples 17.5 and 17.6.)

To change a mass m of a material to a different phase at the same temperature (such as liquid to vapor), a quantity of heat given by Eq. (17.20) must be added or subtracted. Here L is the heat of fusion, vaporization, or sublimation.

In an isolated system whose parts interact by heat exchange, the algebraic sum of the Q's for all parts of the system must be zero. (See Examples 17.7–17.10.)

$$Q = mc \, \Delta T \quad (17.13)$$

$$Q = nC \, \Delta T \quad (17.18)$$

$$Q = \pm mL \quad (17.20)$$

Phase changes, temperature is constant:
$Q = +mL$

Temperature rises, phase does not change:
$Q = mc\,\Delta T$

Conduction, convection, and radiation: Conduction is the transfer of heat within materials without bulk motion of the materials. The heat current H depends on the area A through which the heat flows, the length L of the heat-flow path, the temperature difference $(T_H - T_C)$, and the thermal conductivity k of the material. (See Examples 17.11–17.13.)

Convection is a complex heat-transfer process that involves mass motion from one region to another.

Radiation is energy transfer through electromagnetic radiation. The radiation heat current H depends on the surface area A, the emissivity e of the surface (a pure number between 0 and 1), and the Kelvin temperature T. Here σ is the Stefan–Boltzmann constant. The *net* radiation heat current H_{net} from a body at temperature T to its surroundings at temperature T_s depends on both T and T_s. (See Examples 17.14 and 17.15.)

$$H = \frac{dQ}{dt} = kA\frac{T_H - T_C}{L} \quad (17.21)$$

$$H = Ae\sigma T^4 \quad (17.25)$$

$$H_{net} = Ae\sigma(T^4 - T_s^4) \quad (17.26)$$

Heat current H

Heat current $H = kA\dfrac{T_H - T_C}{L}$

BRIDGING PROBLEM   Steady-State Heat Flow: Radiation and Conduction

One end of a solid cylindrical copper rod 0.200 m long and 0.0250 m in radius is inserted into a large block of solid hydrogen at its melting temperature, 13.84 K. The other end is blackened and exposed to thermal radiation from surrounding walls at 500.0 K. The sides of the rod are insulated, so no energy is lost or gained except at the ends of the rod. (a) When equilibrium is reached, what is the temperature of the blackened end? The thermal conductivity of copper at temperatures near 20 K is 1670 Wm · K. (b) At what rate (in kg/h) does the solid hydrogen melt?

SOLUTION GUIDE

See MasteringPhysics® study area for a Video Tutor solution.

IDENTIFY and SET UP

1. Draw a sketch of the situation, showing all relevant dimensions.
2. List the known and unknown quantities, and identify the target variables.
3. In order for the rod to be in equilibrium, how must the radiation heat current from the walls into the blackened end of the rod compare to the conduction heat current from this end to the

other end and into the solid hydrogen? Use your answers to select the appropriate equations for part (a).
4. How does the heat current from the rod into the hydrogen determine the rate at which the hydrogen melts? (*Hint:* See Table 17.4.) Use your answer to select the appropriate equations for part (b).

EXECUTE

5. Solve for the temperature of the blackened end of the rod. (*Hint:* Since copper is an excellent conductor of heat at low temperature, you can assume that the temperature of the blackened end is only slightly higher than 13.84 K.)
6. Use your result from step 5 to find the rate at which the hydrogen melts.

EVALUATE

7. Is your result from step 5 consistent with the hint given in that step?
8. How would your results from steps 5 and 6 be affected if the rod had twice the radius?

Problems

For instructor-assigned homework, go to www.masteringphysics.com

•, ••, •••: Problems of increasing difficulty. **CP**: Cumulative problems incorporating material from earlier chapters. **CALC**: Problems requiring calculus. **BIO**: Biosciences problems.

DISCUSSION QUESTIONS

Q17.1 Explain why it would not make sense to use a full-size glass thermometer to measure the temperature of a thimbleful of hot water.

Q17.2 If you heat the air inside a rigid, sealed container until its Kelvin temperature doubles, the air pressure in the container will also double. Is the same thing true if you double the Celsius temperature of the air in the container? Explain.

Q17.3 Many automobile engines have cast-iron cylinders and aluminum pistons. What kinds of problems could occur if the engine gets too hot? (The coefficient of volume expansion of cast iron is approximately the same as that of steel.)

Q17.4 Why do frozen water pipes burst? Would a mercury thermometer break if the temperature went below the freezing temperature of mercury? Why or why not?

Q17.5 Two bodies made of the same material have the same external dimensions and appearance, but one is solid and the other is hollow. When their temperature is increased, is the overall volume expansion the same or different? Why?

Q17.6 The inside of an oven is at a temperature of 200°C (392°F). You can put your hand in the oven without injury as long as you don't touch anything. But since the air inside the oven is also at 200°C, why isn't your hand burned just the same?

Q17.7 A newspaper article about the weather states that "the temperature of a body measures how much heat the body contains." Is this description correct? Why or why not?

Q17.8 To raise the temperature of an object, must you add heat to it? If you add heat to an object, must you raise its temperature? Explain.

Q17.9 A student asserts that a suitable unit for specific heat is $1 \ m^2/s^2 \cdot C°$. Is she correct? Why or why not?

Q17.10 In some household air conditioners used in dry climates, air is cooled by blowing it through a water-soaked filter, evaporating some of the water. How does this cool the air? Would such a system work well in a high-humidity climate? Why or why not?

Q17.11 The units of specific heat c are J/kg · K, but the units of heat of fusion L_f or heat of vaporization L_v are simply J/kg. Why do the units of L_f and L_v not include a factor of $(K)^{-1}$ to account for a temperature change?

Q17.12 Why is a hot, humid day in the tropics generally more uncomfortable for human beings than a hot, dry day in the desert?

Q17.13 A piece of aluminum foil used to wrap a potato for baking in a hot oven can usually be handled safely within a few seconds after the potato is removed from the oven. The same is not true of the potato, however! Give two reasons for this difference.

Q17.14 Desert travelers sometimes keep water in a canvas bag. Some water seeps through the bag and evaporates. How does this cool the water inside the bag?

Q17.15 When you first step out of the shower, you feel cold. But as soon as you are dry you feel warmer, even though the room temperature does not change. Why?

Q17.16 The climate of regions adjacent to large bodies of water (like the Pacific and Atlantic coasts) usually features a narrower range of temperature than the climate of regions far from large bodies of water (like the prairies). Why?

Q17.17 When water is placed in ice-cube trays in a freezer, why doesn't the water freeze all at once when the temperature has reached 0°C? In fact, the water freezes first in a layer adjacent to the sides of the tray. Why?

Q17.18 Before giving you an injection, a physician swabs your arm with isopropyl alcohol at room temperature. Why does this make your arm feel cold? (*Hint:* The reason is *not* the fear of the injection! The boiling point of isopropyl alcohol is 82.4°C.)

Q17.19 A cold block of metal feels colder than a block of wood at the same temperature. Why? A *hot* block of metal feels hotter than a block of wood at the same temperature. Again, why? Is there any temperature at which the two blocks feel equally hot or cold? What temperature is this?

Q17.20 A person pours a cup of hot coffee, intending to drink it five minutes later. To keep the coffee as hot as possible, should she put cream in it now or wait until just before she drinks it? Explain.

Q17.21 When a freshly baked apple pie has just been removed from the oven, the crust and filling are both at the same temperature. Yet if you sample the pie, the filling will burn your tongue but the crust will not. Why is there a difference? (*Hint:* The filling is moist while the crust is dry.)

Q17.22 Old-time kitchen lore suggests that things cook better (evenly and without burning) in heavy cast-iron pots. What desirable characteristics do such pots have?

Q17.23 In coastal regions in the winter, the temperature over the land is generally colder than the temperature over the nearby ocean; in the summer, the reverse is usually true. Explain. (*Hint:* The specific heat of soil is only 0.2–0.8 times as great as that of water.)

Q17.24 It is well known that a potato bakes faster if a large nail is stuck through it. Why? Does an aluminum nail work better than a steel one? Why or why not? (*Note:* Don't try this in a microwave oven!) There is also a gadget on the market to hasten the roasting of meat; it consists of a hollow metal tube containing a wick and some water. This is claimed to work much better than a solid metal rod. How does it work?

Q17.25 Glider pilots in the Midwest know that thermal updrafts are likely to occur in the vicinity of freshly plowed fields. Why?

Q17.26 Some folks claim that ice cubes freeze faster if the trays are filled with hot water, because hot water cools off faster than cold water. What do you think?

Q17.27 We're lucky that the earth isn't in thermal equilibrium with the sun (which has a surface temperature of 5800 K). But why aren't the two bodies in thermal equilibrium?

Q17.28 When energy shortages occur, magazine articles sometimes urge us to keep our homes at a constant temperature day and night to conserve fuel. They argue that when we turn down the heat at night, the walls, ceilings, and other areas cool off and must be reheated in the morning. So if we keep the temperature constant, these parts of the house will not cool off and will not have to be reheated. Does this argument make sense? Would we really save energy by following this advice?

EXERCISES

Section 17.2 Thermometers and Temperature Scales

17.1 • Convert the following Celsius temperatures to Fahrenheit: (a) −62.8°C, the lowest temperature ever recorded in North America (February 3, 1947, Snag, Yukon); (b) 56.7°C, the highest temperature ever recorded in the United States (July 10, 1913, Death Valley, California); (c) 31.1°C, the world's highest average annual temperature (Lugh Ferrandi, Somalia).

17.2 • **BIO** **Temperatures in Biomedicine.** (a) **Normal body temperature.** The average normal body temperature measured in the mouth is 310 K. What would Celsius and Fahrenheit thermometers read for this temperature? (b) **Elevated body temperature.** During very vigorous exercise, the body's temperature can go as high as 40°C. What would Kelvin and Fahrenheit thermometers read for this temperature? (c) **Temperature difference in the body.** The surface temperature of the body is normally about 7 C° lower than the internal temperature. Express this temperature difference in kelvins and in Fahrenheit degrees. (d) **Blood storage.** Blood stored at 4.0°C lasts safely for about 3 weeks, whereas blood stored at −160°C lasts for 5 years. Express both temperatures on the Fahrenheit and Kelvin scales. (e) **Heat stroke.** If the body's temperature is above 105°F for a prolonged period, heat stroke can result. Express this temperature on the Celsius and Kelvin scales.

17.3 • (a) On January 22, 1943, the temperature in Spearfish, South Dakota, rose from −4.0°F to 45.0°F in just 2 minutes. What was the temperature change in Celsius degrees? (b) The temperature in Browning, Montana, was 44.0°F on January 23, 1916. The next day the temperature plummeted to −56°F. What was the temperature change in Celsius degrees?

Section 17.3 Gas Thermometers and the Kelvin Scale

17.4 • (a) Calculate the one temperature at which Fahrenheit and Celsius thermometers agree with each other. (b) Calculate the one temperature at which Fahrenheit and Kelvin thermometers agree with each other.

17.5 •• You put a bottle of soft drink in a refrigerator and leave it until its temperature has dropped 10.0 K. What is its temperature change in (a) F° and (b) C°?

17.6 • Convert the following Kelvin temperatures to the Celsius and Fahrenheit scales: (a) the midday temperature at the surface of the moon (400 K); (b) the temperature at the tops of the clouds in the atmosphere of Saturn (95 K); (c) the temperature at the center of the sun (1.55×10^7 K).

17.7 • The pressure of a gas at the triple point of water is 1.35 atm. If its volume remains unchanged, what will its pressure be at the temperature at which CO_2 solidifies?

17.8 •• A gas thermometer registers an absolute pressure corresponding to 325 mm of mercury when in contact with water at the triple point. What pressure does it read when in contact with water at the normal boiling point?

17.9 •• **A Constant-Volume Gas Thermometer.** An experimenter using a gas thermometer found the pressure at the triple point of water (0.01°C) to be 4.80×10^4 Pa and the pressure at the normal boiling point (100°C) to be 6.50×10^4 Pa. (a) Assuming that the pressure varies linearly with temperature, use these two data points to find the Celsius temperature at which the gas pressure would be zero (that is, find the Celsius temperature of absolute zero). (b) Does the gas in this thermometer obey Eq. (17.4) precisely? If that equation were precisely obeyed and the pressure at 100°C were 6.50×10^4 Pa, what pressure would the experimenter have measured at 0.01°C? (As we will learn in Section 18.1, Eq. (17.4) is accurate only for gases at very low density.)

17.10 • Like the Kelvin scale, the *Rankine scale* is an absolute temperature scale: Absolute zero is zero degrees Rankine (0°R). However, the units of this scale are the same size as those of the Fahrenheit scale rather than the Celsius scale. What is the numerical value of the triple-point temperature of water on the Rankine scale?

Section 17.4 Thermal Expansion

17.11 • The Humber Bridge in England has the world's longest single span, 1410 m. Calculate the change in length of the steel deck of the span when the temperature increases from −5.0°C to 18.0°C.

17.12 • One of the tallest buildings in the world is the Taipei 101 in Taiwan, at a height of 1671 feet. Assume that this height was measured on a cool spring day when the temperature was 15.5°C. You could use the building as a sort of giant thermometer on a hot summer day by carefully measuring its height. Suppose you do this and discover that the Taipei 101 is 0.471 foot taller than its official height. What is the temperature, assuming that the building is in thermal equilibrium with the air and that its entire frame is made of steel?

17.13 • A U.S. penny has a diameter of 1.9000 cm at 20.0°C. The coin is made of a metal alloy (mostly zinc) for which the coefficient of linear expansion is 2.6×10^{-5} K^{-1}. What would its diameter be on a hot day in Death Valley (48.0°C)? On a cold night in the mountains of Greenland (−53°C)?

17.14 • **Ensuring a Tight Fit.** Aluminum rivets used in airplane construction are made slightly larger than the rivet holes and cooled by "dry ice" (solid CO_2) before being driven. If the diameter of a hole is 4.500 mm, what should be the diameter of a rivet at 23.0°C if its diameter is to equal that of the hole when the rivet is cooled to −78.0°C, the temperature of dry ice? Assume that the expansion coefficient remains constant at the value given in Table 17.1.

17.15 •• The outer diameter of a glass jar and the inner diameter of its iron lid are both 725 mm at room temperature (20.0°C). What will be the size of the difference in these diameters if the lid is briefly held under hot water until its temperature rises to 50.0°C, without changing the temperature of the glass?

17.16 •• A geodesic dome constructed with an aluminum framework is a nearly perfect hemisphere; its diameter measures 55.0 m on a winter day at a temperature of −15°C. How much more interior space does the dome have in the summer, when the temperature is 35°C?

17.17 •• A copper cylinder is initially at 20.0°C. At what temperature will its volume be 0.150% larger than it is at 20.0°C?

17.18 •• A steel tank is completely filled with 2.80 m^3 of ethanol when both the tank and the ethanol are at a temperature of 32.0°C. When the tank and its contents have cooled to 18.0°C, what additional volume of ethanol can be put into the tank?

17.19 •• A glass flask whose volume is 1000.00 cm^3 at 0.0°C is completely filled with mercury at this temperature. When flask and mercury are warmed to 55.0°C, 8.95 cm^3 of mercury overflow. If the coefficient of volume expansion of mercury is 18.0×10^{-5} K^{-1}, compute the coefficient of volume expansion of the glass.

17.20 •• (a) If an area measured on the surface of a solid body is A_0 at some initial temperature and then changes by ΔA when the temperature changes by ΔT, show that

$$\Delta A = (2\alpha)A_0 \Delta T$$

where α is the coefficient of linear expansion. (b) A circular sheet of aluminum is 55.0 cm in diameter at 15.0°C. By how much does the area of one side of the sheet change when the temperature increases to 27.5°C?

17.21 •• A machinist bores a hole of diameter 1.35 cm in a steel plate at a temperature of 25.0°C. What is the cross-sectional area of the hole (a) at 25.0°C and (b) when the temperature of the plate is increased to 175°C? Assume that the coefficient of linear expansion remains constant over this temperature range. (*Hint:* See Exercise 17.20.)

17.22 •• As a new mechanical engineer for Engines Inc., you have been assigned to design brass pistons to slide inside steel cylinders. The engines in which these pistons will be used will operate between 20.0°C and 150.0°C. Assume that the coefficients of expansion are constant over this temperature range. (a) If the piston just fits inside the chamber at 20.0°C, will the engines be able to run at higher temperatures? Explain. (b) If the cylindrical pistons are 25.000 cm in diameter at 20.0°C, what should be the minimum diameter of the cylinders at that temperature so the pistons will operate at 150.0°C?

17.23 • (a) A wire that is 1.50 m long at 20.0°C is found to increase in length by 1.90 cm when warmed to 420.0°C. Compute its average coefficient of linear expansion for this temperature range. (b) The wire is stretched just taut (zero tension) at 420.0°C. Find the stress in the wire if it is cooled to 20.0°C without being allowed to contract. Young's modulus for the wire is 2.0×10^{11} Pa.

17.24 •• A brass rod is 185 cm long and 1.60 cm in diameter. What force must be applied to each end of the rod to prevent it from contracting when it is cooled from 120.0°C to 10.0°C?

17.25 •• Steel train rails are laid in 12.0-m-long segments placed end to end. The rails are laid on a winter day when their temperature is −2.0°C. (a) How much space must be left between adjacent rails if they are just to touch on a summer day when their temperature is 33.0°C? (b) If the rails are originally laid in contact, what is the stress in them on a summer day when their temperature is 33.0°C?

Section 17.5 Quantity of Heat

17.26 • In an effort to stay awake for an all-night study session, a student makes a cup of coffee by first placing a 200-W electric immersion heater in 0.320 kg of water. (a) How much heat must be added to the water to raise its temperature from 20.0°C to 80.0°C? (b) How much time is required? Assume that all of the heater's power goes into heating the water.

17.27 •• An aluminum tea kettle with mass 1.50 kg and containing 1.80 kg of water is placed on a stove. If no heat is lost to the surroundings, how much heat must be added to raise the temperature from 20.0°C to 85.0°C?

17.28 • **BIO** **Heat Loss During Breathing.** In very cold weather a significant mechanism for heat loss by the human body is energy expended in warming the air taken into the lungs with each breath. (a) On a cold winter day when the temperature is −20°C, what amount of heat is needed to warm to body temperature (37°C) the 0.50 L of air exchanged with each breath? Assume that the specific heat of air is 1020 J/kg · K and that 1.0 L of air has mass 1.3×10^{-3} kg. (b) How much heat is lost per hour if the respiration rate is 20 breaths per minute?

17.29 • You are given a sample of metal and asked to determine its specific heat. You weigh the sample and find that its weight is 28.4 N. You carefully add 1.25×10^4 J of heat energy to the sample and find that its temperature rises 18.0 C°. What is the sample's specific heat?

17.30 •• **On-Demand Water Heaters.** Conventional hot-water heaters consist of a tank of water maintained at a fixed temperature. The hot water is to be used when needed. The drawbacks are that energy is wasted because the tank loses heat when it is not in use and that you can run out of hot water if you use too much. Some utility companies are encouraging the use of *on-demand* water heaters (also known as *flash heaters*), which consist of heating units to heat the water as you use it. No water tank is involved, so no heat is wasted. A typical household shower flow rate is 2.5 gal/min

(9.46 L/min) with the tap water being heated from 50°F (10°C) to 120°F (49°C) by the on-demand heater. What rate of heat input (either electrical or from gas) is required to operate such a unit, assuming that all the heat goes into the water?

17.31 • **BIO** While running, a 70-kg student generates thermal energy at a rate of 1200 W. For the runner to maintain a constant body temperature of 37°C, this energy must be removed by perspiration or other mechanisms. If these mechanisms failed and the heat could not flow out of the student's body, for what amount of time could a student run before irreversible body damage occurred? (*Note:* Protein structures in the body are irreversibly damaged if body temperature rises to 44°C or higher. The specific heat of a typical human body is 3480 J/kg·K, slightly less than that of water. The difference is due to the presence of protein, fat, and minerals, which have lower specific heats.)

17.32 • **CP** While painting the top of an antenna 225 m in height, a worker accidentally lets a 1.00-L water bottle fall from his lunchbox. The bottle lands in some bushes at ground level and does not break. If a quantity of heat equal to the magnitude of the change in mechanical energy of the water goes into the water, what is its increase in temperature?

17.33 •• **CP** A crate of fruit with mass 35.0 kg and specific heat 3650 J/kg·K slides down a ramp inclined at 36.9° below the horizontal. The ramp is 8.00 m long. (a) If the crate was at rest at the top of the incline and has a speed of 2.50 m/s at the bottom, how much work was done on the crate by friction? (b) If an amount of heat equal to the magnitude of the work done by friction goes into the crate of fruit and the fruit reaches a uniform final temperature, what is its temperature change?

17.34 • **CP** A 25,000-kg subway train initially traveling at 15.5 m/s slows to a stop in a station and then stays there long enough for its brakes to cool. The station's dimensions are 65.0 m long by 20.0 m wide by 12.0 m high. Assuming all the work done by the brakes in stopping the train is transferred as heat uniformly to all the air in the station, by how much does the air temperature in the station rise? Take the density of the air to be 1.20 kg/m³ and its specific heat to be 1020 J/kg·K.

17.35 • **CP** A nail driven into a board increases in temperature. If we assume that 60% of the kinetic energy delivered by a 1.80-kg hammer with a speed of 7.80 m/s is transformed into heat that flows into the nail and does not flow out, what is the temperature increase of an 8.00-g aluminum nail after it is struck ten times?

17.36 • A technician measures the specific heat of an unidentified liquid by immersing an electrical resistor in it. Electrical energy is converted to heat transferred to the liquid for 120 s at a constant rate of 65.0 W. The mass of the liquid is 0.780 kg, and its temperature increases from 18.55°C to 22.54°C. (a) Find the average specific heat of the liquid in this temperature range. Assume that negligible heat is transferred to the container that holds the liquid and that no heat is lost to the surroundings. (b) Suppose that in this experiment heat transfer from the liquid to the container or surroundings cannot be ignored. Is the result calculated in part (a) an *overestimate* or an *underestimate* of the average specific heat? Explain.

17.37 •• **CP** A 15.0-g bullet traveling horizontally at 865 m/s passes through a tank containing 13.5 kg of water and emerges with a speed of 534 m/s. What is the maximum temperature increase that the water could have as a result of this event?

Section 17.6 Calorimetry and Phase Changes

17.38 •• As a physicist, you put heat into a 500.0-g solid sample at the rate of 10.0 kJ/min, while recording its temperature as a function of time. You plot your data and obtain the graph shown in Fig. E17.38. (a) What is the latent heat of fusion for this solid? (b) What are the specific heats of the liquid and solid states of the material?

Figure **E17.38**

17.39 •• A 500.0-g chunk of an unknown metal, which has been in boiling water for several minutes, is quickly dropped into an insulating Styrofoam beaker containing 1.00 kg of water at room temperature (20.0°C). After waiting and gently stirring for 5.00 minutes, you observe that the water's temperature has reached a constant value of 22.0°C. (a) Assuming that the Styrofoam absorbs a negligibly small amount of heat and that no heat was lost to the surroundings, what is the specific heat of the metal? (b) Which is more useful for storing thermal energy: this metal or an equal weight of water? Explain. (c) What if the heat absorbed by the Styrofoam actually is not negligible. How would the specific heat you calculated in part (a) be in error? Would it be too large, too small, or still correct? Explain.

17.40 • **BIO** **Treatment for a Stroke.** One suggested treatment for a person who has suffered a stroke is immersion in an ice-water bath at 0°C to lower the body temperature, which prevents damage to the brain. In one set of tests, patients were cooled until their internal temperature reached 32.0°C. To treat a 70.0-kg patient, what is the minimum amount of ice (at 0°C) you need in the bath so that its temperature remains at 0°C? The specific heat of the human body is 3480 J/kg·C°, and recall that normal body temperature is 37.0°C.

17.41 •• A copper pot with a mass of 0.500 kg contains 0.170 kg of water, and both are at a temperature of 20.0°C. A 0.250-kg block of iron at 85.0°C is dropped into the pot. Find the final temperature of the system, assuming no heat loss to the surroundings.

17.42 •• **BIO** **Bicycling on a Warm Day.** If the air temperature is the same as the temperature of your skin (about 30°C), your body cannot get rid of heat by transferring it to the air. In that case, it gets rid of the heat by evaporating water (sweat). During bicycling, a typical 70-kg person's body produces energy at a rate of about 500 W due to metabolism, 80% of which is converted to heat. (a) How many kilograms of water must the person's body evaporate in an hour to get rid of this heat? The heat of vaporization of water at body temperature is 2.42×10^6 J/kg. (b) The evaporated water must, of course, be replenished, or the person will dehydrate. How many 750-mL bottles of water must the bicyclist drink per hour to replenish the lost water? (Recall that the mass of a liter of water is 1.0 kg.)

17.43 •• **BIO** **Overheating.** (a) By how much would the body temperature of the bicyclist in the preceeding problem increase in an hour if he were unable to get rid of the excess heat? (b) Is this temperature increase large enough to be serious? To find out, how high a fever would it be equivalent to, in °F? (Recall that the normal internal body temperature is 98.6°F and the specific heat of the body is 3480 J/kg · C°.)

17.44 • In a container of negligible mass, 0.200 kg of ice at an initial temperature of −40.0°C is mixed with a mass m of water that has an initial temperature of 80.0°C. No heat is lost to the surroundings. If the final temperature of the system is 20.0°C, what is the mass m of the water that was initially at 80.0°C?

17.45 • A 6.00-kg piece of solid copper metal at an initial temperature T is placed with 2.00 kg of ice that is initially at −20.0°C. The ice is in an insulated container of negligible mass and no heat is exchanged with the surroundings. After thermal equilibrium is reached, there is 1.20 kg of ice and 0.80 kg of liquid water. What was the initial temperature of the piece of copper?

17.46 • BIO Before going in for his annual physical, a 70.0-kg man whose body temperature is 37.0°C consumes an entire 0.355-L can of a soft drink (mostly water) at 12.0°C. (a) What will his body temperature be after equilibrium is attained? Ignore any heating by the man's metabolism. The specific heat of the man's body is 3480 J/kg · K. (b) Is the change in his body temperature great enough to be measured by a medical thermometer?

17.47 •• BIO In the situation described in Exercise 17.46, the man's metabolism will eventually return the temperature of his body (and of the soft drink that he consumed) to 37.0°C. If his body releases energy at a rate of 7.00×10^3 kJ/day (the *basal metabolic rate,* or BMR), how long does this take? Assume that all of the released energy goes into raising the temperature.

17.48 •• An ice-cube tray of negligible mass contains 0.350 kg of water at 18.0°C. How much heat must be removed to cool the water to 0.00°C and freeze it? Express your answer in joules, calories, and Btu.

17.49 • How much heat is required to convert 12.0 g of ice at −10.0°C to steam at 100.0°C? Express your answer in joules, calories, and Btu.

17.50 •• An open container holds 0.550 kg of ice at −15.0°C. The mass of the container can be ignored. Heat is supplied to the container at the constant rate of 800.0 J/min for 500.0 min. (a) After how many minutes does the ice *start* to melt? (b) After how many minutes, from the time when the heating is first started, does the temperature begin to rise above 0.0°C? (c) Plot a curve showing the temperature as a function of the elapsed time.

17.51 • CP What must the initial speed of a lead bullet be at a temperature of 25.0°C so that the heat developed when it is brought to rest will be just sufficient to melt it? Assume that all the initial mechanical energy of the bullet is converted to heat and that no heat flows from the bullet to its surroundings. (Typical rifles have muzzle speeds that exceed the speed of sound in air, which is 347 m/s at 25.0°C.)

17.52 •• BIO **Steam Burns Versus Water Burns.** What is the amount of heat input to your skin when it receives the heat released (a) by 25.0 g of steam initially at 100.0°C, when it is cooled to skin temperature (34.0°C)? (b) By 25.0 g of water initially at 100.0°C, when it is cooled to 34.0°C? (c) What does this tell you about the relative severity of steam and hot water burns?

17.53 • BIO **"The Ship of the Desert."** Camels require very little water because they are able to tolerate relatively large changes in their body temperature. While humans keep their body temperatures constant to within one or two Celsius degrees, a dehydrated camel permits its body temperature to drop to 34.0°C overnight and rise to 40.0°C during the day. To see how effective this mechanism is for saving water, calculate how many liters of water a 400-kg

camel would have to drink if it attempted to keep its body temperature at a constant 34.0°C by evaporation of sweat during the day (12 hours) instead of letting it rise to 40.0°C. (*Note:* The specific heat of a camel or other mammal is about the same as that of a typical human, 3480 J/kg · K. The heat of vaporization of water at 34°C is 2.42×10^6 J/kg.)

17.54 • BIO Evaporation of sweat is an important mechanism for temperature control in some warm-blooded animals. (a) What mass of water must evaporate from the skin of a 70.0-kg man to cool his body 1.00 C°? The heat of vaporization of water at body temperature (37°C) is 2.42×10^6 J/kg. The specific heat of a typical human body is 3480 J/kg · K (see Exercise 17.31). (b) What volume of water must the man drink to replenish the evaporated water? Compare to the volume of a soft-drink can (355 cm³).

17.55 •• CP An asteroid with a diameter of 10 km and a mass of 2.60×10^{15} kg impacts the earth at a speed of 32.0 km/s, landing in the Pacific Ocean. If 1.00% of the asteroid's kinetic energy goes to boiling the ocean water (assume an initial water temperature of 10.0°C), what mass of water will be boiled away by the collision? (For comparison, the mass of water contained in Lake Superior is about 2×10^{15} kg.)

17.56 • A laboratory technician drops a 0.0850-kg sample of unknown solid material, at a temperature of 100.0°C, into a calorimeter. The calorimeter can, initially at 19.0°C, is made of 0.150 kg of copper and contains 0.200 kg of water. The final temperature of the calorimeter can and contents is 26.1°C. Compute the specific heat of the sample.

17.57 •• An insulated beaker with negligible mass contains 0.250 kg of water at a temperature of 75.0°C. How many kilograms of ice at a temperature of −20.0°C must be dropped into the water to make the final temperature of the system 40.0°C?

17.58 •• A glass vial containing a 16.0-g sample of an enzyme is cooled in an ice bath. The bath contains water and 0.120 kg of ice. The sample has specific heat 2250 J/kg · K; the glass vial has mass 6.00 g and specific heat 2800 J/kg · K. How much ice melts in cooling the enzyme sample from room temperature (19.5°C) to the temperature of the ice bath?

17.59 • A 4.00-kg silver ingot is taken from a furnace, where its temperature is 750.0°C, and placed on a large block of ice at 0.0°C. Assuming that all the heat given up by the silver is used to melt the ice, how much ice is melted?

17.60 •• A copper calorimeter can with mass 0.100 kg contains 0.160 kg of water and 0.0180 kg of ice in thermal equilibrium at atmospheric pressure. If 0.750 kg of lead at a temperature of 255°C is dropped into the calorimeter can, what is the final temperature? Assume that no heat is lost to the surroundings.

17.61 •• A vessel whose walls are thermally insulated contains 2.40 kg of water and 0.450 kg of ice, all at a temperature of 0.0°C. The outlet of a tube leading from a boiler in which water is boiling at atmospheric pressure is inserted into the water. How many grams of steam must condense inside the vessel (also at atmospheric pressure) to raise the temperature of the system to 28.0°C? You can ignore the heat transferred to the container.

Section 17.7 Mechanisms of Heat Transfer

17.62 •• Two rods, one made of brass and the other made of copper, are joined end to end. The length of the brass section is 0.200 m and the length of the copper section is 0.800 m. Each segment has cross-sectional area 0.00500 m². The free end of the brass segment is in boiling water and the free end of the copper segment is in an ice and water mixture, in both cases under normal atmospheric pressure. The sides of the rods are insulated so there is no heat loss to the surroundings. (a) What is the temperature of the point where the brass and copper segments are joined? (b) What mass of ice is melted in 5.00 min by the heat conducted by the composite rod?

17.63 • Suppose that the rod in Fig. 17.23a is made of copper, is 45.0 cm long, and has a cross-sectional area of 1.25 cm². Let $T_H = 100.0°C$ and $T_C = 0.0°C$. (a) What is the final steady-state temperature gradient along the rod? (b) What is the heat current in the rod in the final steady state? (c) What is the final steady-state temperature at a point in the rod 12.0 cm from its left end?

17.64 •• One end of an insulated metal rod is maintained at 100.0°C, and the other end is maintained at 0.00°C by an ice–water mixture. The rod is 60.0 cm long and has a cross-sectional area of 1.25 cm². The heat conducted by the rod melts 8.50 g of ice in 10.0 min. Find the thermal conductivity k of the metal.

17.65 •• A carpenter builds an exterior house wall with a layer of wood 3.0 cm thick on the outside and a layer of Styrofoam insulation 2.2 cm thick on the inside wall surface. The wood has $k = 0.080$ W/m·K, and the Styrofoam has $k = 0.010$ W/m·K. The interior surface temperature is 19.0°C, and the exterior surface temperature is −10.0°C. (a) What is the temperature at the plane where the wood meets the Styrofoam? (b) What is the rate of heat flow per square meter through this wall?

17.66 • An electric kitchen range has a total wall area of 1.40 m² and is insulated with a layer of fiberglass 4.00 cm thick. The inside surface of the fiberglass has a temperature of 175°C, and its outside surface is at 35.0°C. The fiberglass has a thermal conductivity of 0.040 W/m·K. (a) What is the heat current through the insulation, assuming it may be treated as a flat slab with an area of 1.40 m²? (b) What electric-power input to the heating element is required to maintain this temperature?

17.67 • **BIO** **Conduction Through the Skin.** The blood plays an important role in removing heat from the body by bringing this heat directly to the surface where it can radiate away. Nevertheless, this heat must still travel through the skin before it can radiate away. We shall assume that the blood is brought to the bottom layer of skin at a temperature of 37.0°C and that the outer surface of the skin is at 30.0°C. Skin varies in thickness from 0.50 mm to a few millimeters on the palms and soles, so we shall assume an average thickness of 0.75 mm. A 165-lb, 6-ft-tall person has a surface area of about 2.0 m² and loses heat at a net rate of 75 W while resting. On the basis of our assumptions, what is the thermal conductivity of this person's skin?

17.68 • A long rod, insulated to prevent heat loss along its sides, is in perfect thermal contact with boiling water (at atmospheric pressure) at one end and with an ice–water mixture at the other (Fig. E17.68). The rod consists of a 1.00-m section of copper (one end in boiling water) joined end to end to a length L_2 of steel (one end in the ice–water mixture). Both sections of the rod have cross-sectional areas of 4.00 cm². The temperature of the copper–steel junction is 65.0°C after a steady state has been set up. (a) How much heat per second flows from the boiling water to the ice–water mixture? (b) What is the length L_2 of the steel section?

Figure **E17.68**

17.69 • A pot with a steel bottom 8.50 mm thick rests on a hot stove. The area of the bottom of the pot is 0.150 m². The water inside the pot is at 100.0°C, and 0.390 kg are evaporated every 3.00 min. Find the temperature of the lower surface of the pot, which is in contact with the stove.

17.70 •• You are asked to design a cylindrical steel rod 50.0 cm long, with a circular cross section, that will conduct 150.0 J/s from a furnace at 400.0°C to a container of boiling water under 1 atmosphere. What must the rod's diameter be?

17.71 •• A picture window has dimensions of 1.40 m × 2.50 m and is made of glass 5.20 mm thick. On a winter day, the outside temperature is −20.0°C, while the inside temperature is a comfortable 19.5°C. (a) At what rate is heat being lost through the window by conduction? (b) At what rate would heat be lost through the window if you covered it with a 0.750-mm-thick layer of paper (thermal conductivity 0.0500 W/m·K)?

17.72 • What is the rate of energy radiation per unit area of a blackbody at a temperature of (a) 273 K and (b) 2730 K?

17.73 • **Size of a Light-Bulb Filament.** The operating temperature of a tungsten filament in an incandescent light bulb is 2450 K, and its emissivity is 0.350. Find the surface area of the filament of a 150-W bulb if all the electrical energy consumed by the bulb is radiated by the filament as electromagnetic waves. (Only a fraction of the radiation appears as visible light.)

17.74 •• The emissivity of tungsten is 0.350. A tungsten sphere with radius 1.50 cm is suspended within a large evacuated enclosure whose walls are at 290.0 K. What power input is required to maintain the sphere at a temperature of 3000.0 K if heat conduction along the supports is neglected?

17.75 • **The Sizes of Stars.** The hot glowing surfaces of stars emit energy in the form of electromagnetic radiation. It is a good approximation to assume $e = 1$ for these surfaces. Find the radii of the following stars (assumed to be spherical): (a) Rigel, the bright blue star in the constellation Orion, which radiates energy at a rate of 2.7×10^{32} W and has surface temperature 11,000 K; (b) Procyon B (visible only using a telescope), which radiates energy at a rate of 2.1×10^{23} W and has surface temperature 10,000 K. (c) Compare your answers to the radius of the earth, the radius of the sun, and the distance between the earth and the sun. (Rigel is an example of a *supergiant* star, and Procyon B is an example of a *white dwarf* star.)

PROBLEMS

17.76 •• Suppose that a steel hoop could be constructed to fit just around the earth's equator at a temperature of 20.0°C. What would be the thickness of space between the hoop and the earth if the temperature of the hoop were increased by 0.500 C°?

17.77 ••• You propose a new temperature scale with temperatures given in °M. You define 0.0°M to be the normal melting point of mercury and 100.0° to be the normal boiling point of mercury. (a) What is the normal boiling point of water in °M? (b) A temperature change of 10.0 M° corresponds to how many C°?

17.78 • **CP, CALC** A 250-kg weight is hanging from the ceiling by a thin copper wire. In its fundamental mode, this wire vibrates at the frequency of concert A (440 Hz). You then increase the temperature of the wire by 40 C°. (a) By how much will the fundamental frequency change? Will it increase or decrease? (b) By what percentage will the speed of a wave on the wire change? (c) By what percentage will the wavelength of the fundamental standing wave change? Will it increase or decrease?

17.79 ••• You are making pesto for your pasta and have a cylindrical measuring cup 10.0 cm high made of ordinary glass $[\beta = 2.7 \times 10^{-5} \, (C°)^{-1}]$ that is filled with olive oil $[\beta = 6.8 \times 10^{-4} \, (C°)^{-1}]$ to a height of 2.00 mm below the top of the cup. Initially, the cup and oil are at room temperature (22.0°C). You get a phone call and forget about the olive oil, which you inadvertently leave on the hot stove. The cup and oil heat up slowly and have a common temperature. At what temperature will the olive oil start to spill out of the cup?

17.80 •• A surveyor's 30.0-m steel tape is correct at a temperature of 20.0°C. The distance between two points, as measured by this tape on a day when its temperature is 5.00°C, is 25.970 m. What is the true distance between the points?

17.81 •• **CP** A Foucault pendulum consists of a brass sphere with a diameter of 35.0 cm suspended from a steel cable 10.5 m long (both measurements made at 20.0°C). Due to a design oversight, the swinging sphere clears the floor by a distance of only 2.00 mm when the temperature is 20.0°C. At what temperature will the sphere begin to brush the floor?

17.82 •• You pour 108 cm³ of ethanol, at a temperature of −10.0°C, into a graduated cylinder initially at 20.0°C, filling it to the very top. The cylinder is made of glass with a specific heat of 840 J/kg·K and a coefficient of volume expansion of $1.2 \times 10^{-5} \, K^{-1}$; its mass is 0.110 kg. The mass of the ethanol is 0.0873 kg. (a) What will be the final temperature of the ethanol, once thermal equilibrium is reached? (b) How much ethanol will overflow the cylinder before thermal equilibrium is reached?

17.83 •• A metal rod that is 30.0 cm long expands by 0.0650 cm when its temperature is raised from 0.0°C to 100.0°C. A rod of a different metal and of the same length expands by 0.0350 cm for the same rise in temperature. A third rod, also 30.0 cm long, is made up of pieces of each of the above metals placed end to end and expands 0.0580 cm between 0.0°C and 100.0°C. Find the length of each portion of the composite rod.

17.84 •• On a cool (4.0°C) Saturday morning, a pilot fills the fuel tanks of her Pitts S-2C (a two-seat aerobatic airplane) to their full capacity of 106.0 L. Before flying on Sunday morning, when the temperature is again 4.0°C, she checks the fuel level and finds only 103.4 L of gasoline in the tanks. She realizes that it was hot on Saturday afternoon, and that thermal expansion of the gasoline caused the missing fuel to empty out of the tank's vent. (a) What was the maximum temperature (in °C) reached by the fuel and the tank on Saturday afternoon? The coefficient of volume expansion of gasoline is $9.5 \times 10^{-4} \, K^{-1}$, and the tank is made of aluminum. (b) In order to have the maximum amount of fuel available for flight, when should the pilot have filled the fuel tanks?

17.85 ••• (a) Equation (17.12) gives the stress required to keep the length of a rod constant as its temperature changes. Show that if the length is permitted to change by an amount ΔL when its temperature changes by ΔT, the stress is equal to

$$\frac{F}{A} = Y\left(\frac{\Delta L}{L_0} - \alpha \Delta T\right)$$

where F is the tension on the rod, L_0 is the original length of the rod, A its cross-sectional area, α its coefficient of linear expansion, and Y its Young's modulus. (b) A heavy brass bar has projections at its ends, as in Fig. P17.85. Two fine steel wires, fastened between the pro-

Figure **P17.85**

Steel wires   Brass

jections, are just taut (zero tension) when the whole system is at 20°C. What is the tensile stress in the steel wires when the temperature of the system is raised to 140°C? Make any simplifying assumptions you think are justified, but state what they are.

17.86 •• **CP** A metal wire, with density ρ and Young's modulus Y, is stretched between rigid supports. At temperature T, the speed of a transverse wave is found to be v_1. When the temperature is increased to $T + \Delta T$, the speed decreases to $v_2 < v_1$. Determine the coefficient of linear expansion of the wire.

17.87 •• **CP** **Out of Tune.** The B-string of a guitar is made of steel (density 7800 kg/m³), is 63.5 cm long, and has diameter 0.406 mm. The fundamental frequency is $f = 247.0$ Hz. (a) Find the string tension. (b) If the tension F is changed by a small amount ΔF, the frequency f changes by a small amount Δf. Show that

$$\frac{\Delta f}{f} = \frac{\Delta F}{2F}$$

(c) The string is tuned to a fundamental frequency of 247.0 Hz when its temperature is 18.5°C. Strenuous playing can make the temperature of the string rise, changing its vibration frequency. Find Δf if the temperature of the string rises to 29.5°C. The steel string has a Young's modulus of 2.00×10^{11} Pa and a coefficient of linear expansion of $1.20 \times 10^{-5} \, (C°)^{-1}$. Assume that the temperature of the body of the guitar remains constant. Will the vibration frequency rise or fall?

17.88 ••• A steel rod 0.450 m long and an aluminum rod 0.250 m long, both with the same diameter, are placed end to end between rigid supports with no initial stress in the rods. The temperature of the rods is now raised by 60.0 C°. What is the stress in each rod? (*Hint:* The length of the combined rod remains the same, but the lengths of the individual rods do not. See Problem 17.85.)

17.89 •• A steel ring with a 2.5000-in. inside diameter at 20.0°C is to be warmed and slipped over a brass shaft with a 2.5020-in. outside diameter at 20.0°C. (a) To what temperature should the ring be warmed? (b) If the ring and the shaft together are cooled by some means such as liquid air, at what temperature will the ring just slip off the shaft?

17.90 •• **Bulk Stress Due to a Temperature Increase.** (a) Prove that, if an object under pressure has its temperature raised but is not allowed to expand, the increase in pressure is

$$\Delta p = B\beta \Delta T$$

where the bulk modulus B and the average coefficient of volume expansion β are both assumed positive and constant. (b) What pressure is necessary to prevent a steel block from expanding when its temperature is increased from 20.0°C to 35.0°C?

17.91 •• A liquid is enclosed in a metal cylinder that is provided with a piston of the same metal. The system is originally at a pressure of 1.00 atm (1.013×10^5 Pa) and at a temperature of 30.0°C. The piston is forced down until the pressure on the liquid is increased by 50.0 atm, and then clamped in this position. Find the new temperature at which the pressure of the liquid is again 1.00 atm. Assume that the cylinder is sufficiently strong so that its volume is not altered by changes in pressure, but only by changes in temperature. Use the result derived in Problem 17.90. (*Hint:* See Section 11.4.)
Compressibility of liquid: $k = 8.50 \times 10^{-10} \, Pa^{-1}$
Coefficient of volume expansion of liquid: $\beta = 4.80 \times 10^{-4} \, K^{-1}$
Coefficient of volume expansion of metal: $\beta = 3.90 \times 10^{-5} \, K^{-1}$

17.92 •• You cool a 100.0-g slug of red-hot iron (temperature 745°C) by dropping it into an insulated cup of negligible mass containing 85.0 g of water at 20.0°C. Assuming no heat exchange with the surroundings, (a) what is the final temperature of the water and (b) what is the final mass of the iron and the remaining water?

17.93 • **CP Spacecraft Reentry.** A spacecraft made of aluminum circles the earth at a speed of 7700 m/s. (a) Find the ratio of its kinetic energy to the energy required to raise its temperature from 0°C to 600°C. (The melting point of aluminum is 660°C. Assume a constant specific heat of 910 J/kg·K.) (b) Discuss the bearing of your answer on the problem of the reentry of a manned space vehicle into the earth's atmosphere.

17.94 • **CP** A capstan is a rotating drum or cylinder over which a rope or cord slides in order to provide a great amplification of the rope's tension while keeping both ends free (Fig. P17.94). Since the added tension in the rope is due to friction, the capstan generates thermal energy. (a) If the difference in tension between the two ends of the rope is 520.0 N and the capstan has a diameter of 10.0 cm and turns once in 0.900 s, find the rate at which thermal energy is generated. Why does the number of turns not matter? (b) If the capstan is made of iron and has mass 6.00 kg, at what rate does its temperature rise? Assume that the temperature in the capstan is uniform and that all the thermal energy generated flows into it.

Figure **P17.94**

17.95 •• **CALC Debye's T^3 Law.** At very low temperatures the molar heat capacity of rock salt varies with temperature according to Debye's T^3 law:

$$C = k\frac{T^3}{\Theta^3}$$

where $k = 1940$ J/mol·K and $\Theta = 281$ K. (a) How much heat is required to raise the temperature of 1.50 mol of rock salt from 10.0 K to 40.0 K? (*Hint:* Use Eq. (17.18) in the form $dQ = nC\, dT$ and integrate.) (b) What is the average molar heat capacity in this range? (c) What is the true molar heat capacity at 40.0 K?

17.96 •• **CP** A person of mass 70.0 kg is sitting in the bathtub. The bathtub is 190.0 cm by 80.0 cm; before the person got in, the water was 16.0 cm deep. The water is at a temperature of 37.0°C. Suppose that the water were to cool down spontaneously to form ice at 0.0°C, and that all the energy released was used to launch the hapless bather vertically into the air. How high would the bather go? (As you will see in Chapter 20, this event is allowed by energy conservation but is prohibited by the second law of thermodynamics.)

17.97 • **Hot Air in a Physics Lecture.** (a) A typical student listening attentively to a physics lecture has a heat output of 100 W. How much heat energy does a class of 90 physics students release into a lecture hall over the course of a 50-min lecture? (b) Assume that all the heat energy in part (a) is transferred to the 3200 m³ of air in the room. The air has specific heat 1020 J/kg·K and density 1.20 kg/m³. If none of the heat escapes and the air conditioning system is off, how much will the temperature of the air in the room rise during the 50-min lecture? (c) If the class is taking an exam,

the heat output per student rises to 280 W. What is the temperature rise during 50 min in this case?

17.98 ••• **CALC** The molar heat capacity of a certain substance varies with temperature according to the empirical equation

$$C = 29.5 \text{ J/mol·K} + (8.20 \times 10^{-3} \text{ J/mol·K}^2)T$$

How much heat is necessary to change the temperature of 3.00 mol of this substance from 27°C to 227°C? (*Hint:* Use Eq. (17.18) in the form $dQ = nC\, dT$ and integrate.)

17.99 ••• For your cabin in the wilderness, you decide to build a primitive refrigerator out of Styrofoam, planning to keep the interior cool with a block of ice that has an initial mass of 24.0 kg. The box has dimensions of 0.500 m × 0.800 m × 0.500 m. Water from melting ice collects in the bottom of the box. Suppose the ice block is at 0.00°C and the outside temperature is 21.0°C. If the top of the empty box is never opened and you want the interior of the box to remain at 5.00°C for exactly one week, until all the ice melts, what must be the thickness of the Styrofoam?

17.100 •• **Hot Water Versus Steam Heating.** In a household hot-water heating system, water is delivered to the radiators at 70.0°C (158.0°F) and leaves at 28.0°C (82.4°F). The system is to be replaced by a steam system in which steam at atmospheric pressure condenses in the radiators and the condensed steam leaves the radiators at 35.0°C (95.0°F). How many kilograms of steam will supply the same heat as was supplied by 1.00 kg of hot water in the first system?

17.101 ••• A copper calorimeter can with mass 0.446 kg contains 0.0950 kg of ice. The system is initially at 0.0°C. (a) If 0.0350 kg of steam at 100.0°C and 1.00 atm pressure is added to the can, what is the final temperature of the calorimeter can and its contents? (b) At the final temperature, how many kilograms are there of ice, how many of liquid water, and how many of steam?

17.102 • A Styrofoam bucket of negligible mass contains 1.75 kg of water and 0.450 kg of ice. More ice, from a refrigerator at −15.0°C, is added to the mixture in the bucket, and when thermal equilibrium has been reached, the total mass of ice in the bucket is 0.868 kg. Assuming no heat exchange with the surroundings, what mass of ice was added?

17.103 ••• In a container of negligible mass, 0.0400 kg of steam at 100°C and atmospheric pressure is added to 0.200 kg of water at 50.0°C. (a) If no heat is lost to the surroundings, what is the final temperature of the system? (b) At the final temperature, how many kilograms are there of steam and how many of liquid water?

17.104 •• **BIO Mammal Insulation.** Animals in cold climates often depend on *two* layers of insulation: a layer of body fat (of thermal conductivity 0.20 W/m·K) surrounded by a layer of air trapped inside fur or down. We can model a black bear (*Ursus americanus*) as a sphere 1.5 m in diameter having a layer of fat 4.0 cm thick. (Actually, the thickness varies with the season, but we are interested in hibernation, when the fat layer is thickest.) In studies of bear hibernation, it was found that the outer surface layer of the fur is at 2.7°C during hibernation, while the inner surface of the fat layer is at 31.0°C. (a) What is the temperature at the fat–inner fur boundary? (b) How thick should the air layer (contained within the fur) be so that the bear loses heat at a rate of 50.0 W?

17.105 ••• A worker pours 1.250 kg of molten lead at a temperature of 327.3°C into 0.5000 kg of water at a temperature of 75.00°C in an insulated bucket of negligible mass. Assuming no heat loss to the surroundings, calculate the mass of lead and water remaining in the bucket when the materials have reached thermal equilibrium.

17.106 •• One experimental method of measuring an insulating material's thermal conductivity is to construct a box of the material and measure the power input to an electric heater inside the box that maintains the interior at a measured temperature above the outside surface. Suppose that in such an apparatus a power input of 180 W is required to keep the interior surface of the box 65.0 C° (about 120 F°) above the temperature of the outer surface. The total area of the box is 2.18 m², and the wall thickness is 3.90 cm. Find the thermal conductivity of the material in SI units.

17.107 •• **Effect of a Window in a Door.** A carpenter builds a solid wood door with dimensions 2.00 m × 0.95 m × 5.0 cm. Its thermal conductivity is $k = 0.120$ W/m·K. The air films on the inner and outer surfaces of the door have the same combined thermal resistance as an additional 1.8-cm thickness of solid wood. The inside air temperature is 20.0°C, and the outside air temperature is −8.0°C. (a) What is the rate of heat flow through the door? (b) By what factor is the heat flow increased if a window 0.500 m on a side is inserted in the door? The glass is 0.450 cm thick, and the glass has a thermal conductivity of 0.80 W/m·K. The air films on the two sides of the glass have a total thermal resistance that is the same as an additional 12.0 cm of glass.

17.108 • A wood ceiling with thermal resistance R_1 is covered with a layer of insulation with thermal resistance R_2. Prove that the effective thermal resistance of the combination is $R = R_1 + R_2$.

17.109 •• Compute the ratio of the rate of heat loss through a single-pane window with area 0.15 m² to that for a double-pane window with the same area. The glass of a single pane is 4.2 mm thick, and the air space between the two panes of the double-pane window is 7.0 mm thick. The glass has thermal conductivity 0.80 W/m·K. The air films on the room and outdoor surfaces of either window have a combined thermal resistance of 0.15 m²·K/W.

17.110 • Rods of copper, brass, and steel are welded together to form a Y-shaped figure. The cross-sectional area of each rod is 2.00 cm². The free end of the copper rod is maintained at 100.0°C, and the free ends of the brass and steel rods at 0.0°C. Assume there is no heat loss from the surfaces of the rods. The lengths of the rods are: copper, 13.0 cm; brass, 18.0 cm; steel, 24.0 cm. (a) What is the temperature of the junction point? (b) What is the heat current in each of the three rods?

17.111 ••• **CALC** **Time Needed for a Lake to Freeze Over.** (a) When the air temperature is below 0°C, the water at the surface of a lake freezes to form an ice sheet. Why doesn't freezing occur throughout the entire volume of the lake? (b) Show that the thickness of the ice sheet formed on the surface of a lake is proportional to the square root of the time if the heat of fusion of the water freezing on the underside of the ice sheet is conducted through the sheet. (c) Assuming that the upper surface of the ice sheet is at −10°C and the bottom surface is at 0°C, calculate the time it will take to form an ice sheet 25 cm thick. (d) If the lake in part (c) is uniformly 40 m deep, how long would it take to freeze all the water in the lake? Is this likely to occur?

17.112 •• A rod is initially at a uniform temperature of 0°C throughout. One end is kept at 0°C, and the other is brought into contact with a steam bath at 100°C. The surface of the rod is insulated so that heat can flow only lengthwise along the rod. The cross-sectional area of the rod is 2.50 cm², its length is 120 cm, its thermal conductivity is 380 W/m·K, its density is 1.00×10^4 kg/m³, and its specific heat is 520 J/kg·K. Consider a short cylindrical element of the rod 1.00 cm in length. (a) If the temperature gradient at the cooler end of this element is 140 C°/m, how many joules of heat energy flow across this end per second? (b) If the average temperature of the element is

increasing at the rate of what is the temperature gradient at the other end of the element?

17.113 •• A rustic cabin has a floor area of 3.50 m × 3.00 m. Its walls, which are 2.50 m tall, are made of wood (thermal conductivity 0.0600 W/m·K) 1.80 cm thick and are further insulated with 1.50 cm of a synthetic material. When the outside temperature is 2.00°C, it is found necessary to heat the room at a rate of 1.25 kW to maintain its temperature at 19.0°C. Calculate the thermal conductivity of the insulating material. Neglect the heat lost through the ceiling and floor. Assume the inner and outer surfaces of the wall have the same temperature as the air inside and outside the cabin.

17.114 • The rate at which radiant energy from the sun reaches the earth's upper atmosphere is about 1.50 kW/m². The distance from the earth to the sun is 1.50×10^{11} m, and the radius of the sun is 6.96×10^8 m. (a) What is the rate of radiation of energy per unit area from the sun's surface? (b) If the sun radiates as an ideal blackbody, what is the temperature of its surface?

17.115 ••• **A Thermos for Liquid Helium.** A physicist uses a cylindrical metal can 0.250 m high and 0.090 m in diameter to store liquid helium at 4.22 K; at that temperature the heat of vaporization of helium is 2.09×10^4 J/kg. Completely surrounding the metal can are walls maintained at the temperature of liquid nitrogen, 77.3 K, with vacuum between the can and the surrounding walls. How much helium is lost per hour? The emissivity of the metal can is 0.200. The only heat transfer between the metal can and the surrounding walls is by radiation.

17.116 •• **BIO** **Basal Metabolic Rate.** The basal metabolic rate is the rate at which energy is produced in the body when a person is at rest. A 75-kg (165-lb) person of height 1.83 m (6 ft) has a body surface area of approximately 2.0 m². (a) What is the net amount of heat this person could radiate per second into a room at 18°C (about 65°F) if his skin's surface temperature is 30°C? (At such temperatures, nearly all the heat is infrared radiation, for which the body's emissivity is 1.0, regardless of the amount of pigment.) (b) Normally, 80% of the energy produced by metabolism goes into heat, while the rest goes into things like pumping blood and repairing cells. Also normally, a person at rest can get rid of this excess heat just through radiation. Use your answer to part (a) to find this person's basal metabolic rate.

17.117 •• **BIO** **Jogging in the Heat of the Day.** You have probably seen people jogging in extremely hot weather and wondered Why? As we shall see, there are good reasons not to do this! When jogging strenuously, an average runner of mass 68 kg and surface area 1.85 m² produces energy at a rate of up to 1300 W, 80% of which is converted to heat. The jogger radiates heat, but actually absorbs more from the hot air than he radiates away. At such high levels of activity, the skin's temperature can be elevated to around 33°C instead of the usual 30°C. (We shall neglect conduction, which would bring even more heat into his body.) The only way for the body to get rid of this extra heat is by evaporating water (sweating). (a) How much heat per second is produced just by the act of jogging? (b) How much *net* heat per second does the runner gain just from radiation if the air temperature is 40.0°C (104°F)? (Remember that he radiates out, but the environment radiates back in.) (c) What is the *total* amount of excess heat this runner's body must get rid of per second? (d) How much water must the jogger's body evaporate every minute due to his activity? The heat of vaporization of water at body temperature is 2.42×10^6 J/kg. (e) How many 750-mL bottles of water must he drink after (or preferably before!) jogging for a half hour? Recall that a liter of water has a mass of 1.0 kg.

17.118 •• **BIO Overheating While Jogging.** (a) If the jogger in the preceding problem were not able to get rid of the excess heat, by how much would his body temperature increase above the normal 37°C in a half hour of jogging? The specific heat for a human is about 3500 J/kg·K. (b) How high a fever (in °F) would this temperature increase be equivalent to? Is the increase large enough to be of concern? (Recall that normal body temperature is 98.6°F.)

17.119 •• An engineer is developing an electric water heater to provide a continuous supply of hot water. One trial design is shown in Fig. P17.119. Water is flowing at the rate of 0.500 kg/min, the inlet thermometer registers 18.0°C, the voltmeter reads 120 V, and the ammeter reads 15.0 A [corresponding to a power input of (120 V) × (15.0 A) = 1800 W]. (a) When a steady state is finally reached, what is the reading of the outlet thermometer? (b) Why is it unnecessary to take into account the heat capacity *mc* of the apparatus itself?

Figure **P17.119**

17.120 • **Food Intake of a Hamster.** The energy output of an animal engaged in an activity is called the basal metabolic rate (BMR) and is a measure of the conversion of food energy into other forms of energy. A simple calorimeter to measure the BMR consists of an insulated box with a thermometer to measure the temperature of the air. The air has density 1.20 kg/m^3 and specific heat 1020 J/kg·K. A 50.0-g hamster is placed in a calorimeter that contains 0.0500 m^3 of air at room temperature. (a) When the hamster is running in a wheel, the temperature of the air in the calorimeter rises 1.60 C° per hour. How much heat does the running hamster generate in an hour? Assume that all this heat goes into the air in the calorimeter. You can ignore the heat that goes into the walls of the box and into the thermometer, and assume that no heat is lost to the surroundings. (b) Assuming that the hamster converts seed into heat with an efficiency of 10% and that hamster seed has a food energy value of 24 J/g, how many grams of seed must the hamster eat per hour to supply this energy?

17.121 •• The icecaps of Greenland and Antarctica contain about 1.75% of the total water (by mass) on the earth's surface; the oceans contain about 97.5%, and the other 0.75% is mainly groundwater. Suppose the icecaps, currently at an average temperature of about −30°C, somehow slid into the ocean and melted. What would be the resulting temperature decrease of the ocean? Assume that the average temperature of ocean water is currently 5.00°C.

17.122 •• **Why Do the Seasons Lag?** In the northern hemisphere, June 21 (the summer solstice) is both the longest day of the year and the day on which the sun's rays strike the earth most vertically, hence delivering the greatest amount of heat to the surface. Yet the hottest summer weather usually occurs about a month or so later. Let us see why this is the case. Because of the large specific heat of water, the oceans are slower to warm up than the land (and also slower to cool off in winter). In addition to perusing pertinent information in the tables included in this book, it is useful to know

that approximately two-thirds of the earth's surface is ocean composed of salt water having a specific heat of 3890 J/kg·K and that the oceans, on the average, are 4000 m deep. Typically, an average of 1050 W/m^2 of solar energy falls on the earth's surface, and the oceans absorb essentially all of the light that strikes them. However, most of that light is absorbed in the upper 100 m of the surface. Depths below that do not change temperature seasonally. Assume that the sunlight falls on the surface for only 12 hours per day and that the ocean retains all the heat it absorbs. What will be the rise in temperature of the upper 100 m of the oceans during the month following the summer solstice? Does this seem to be large enough to be perceptible?

CHALLENGE PROBLEMS

17.123 ••• **CALC** Suppose that both ends of the rod in Fig. 17.23a are kept at a temperature of 0°C, and that the initial temperature distribution along the rod is given by $T = (100°C) \sin \pi x/L$, where x is measured from the left end of the rod. Let the rod be copper, with length $L = 0.100$ m and cross-sectional area 1.00 cm^2. (a) Show the initial temperature distribution in a diagram. (b) What is the final temperature distribution after a very long time has elapsed? (c) Sketch curves that you think would represent the temperature distribution at intermediate times. (d) What is the initial temperature gradient at the ends of the rod? (e) What is the initial heat current from the ends of the rod into the bodies making contact with its ends? (f) What is the initial heat current at the center of the rod? Explain. What is the heat current at this point at any later time? (g) What is the value of the *thermal diffusivity* $k/\rho c$ for copper, and in what unit is it expressed? (Here k is the thermal conductivity, $\rho = 8.9 \times 10^3$ kg/m^3 is the density, and c is the specific heat.) (h) What is the initial time rate of change of temperature at the center of the rod? (i) How much time would be required for the center of the rod to reach its final temperature if the temperature continued to decrease at this rate? (This time is called the *relaxation time* of the rod.) (j) From the graphs in part (c), would you expect the magnitude of the rate of temperature change at the midpoint to remain constant, increase, or decrease as a function of time? (k) What is the initial rate of change of temperature at a point in the rod 2.5 cm from its left end?

17.124 ••• **CALC** (a) A spherical shell has inner and outer radii a and b, respectively, and the temperatures at the inner and outer surfaces are T_2 and T_1. The thermal conductivity of the material of which the shell is made is k. Derive an equation for the total heat current through the shell. (b) Derive an equation for the temperature variation within the shell in part (a); that is, calculate T as a function of r, the distance from the center of the shell. (c) A hollow cylinder has length L, inner radius a, and outer radius b, and the temperatures at the inner and outer surfaces are T_2 and T_1. (The cylinder could represent an insulated hot-water pipe, for example.) The thermal conductivity of the material of which the cylinder is made is k. Derive an equation for the total heat current through the walls of the cylinder. (d) For the cylinder of part (c), derive an equation for the temperature variation inside the cylinder walls. (e) For the spherical shell of part (a) and the hollow cylinder of part (c), show that the equation for the total heat current in each case reduces to Eq. (17.21) for linear heat flow when the shell or cylinder is very thin.

17.125 ••• A steam pipe with a radius of 2.00 cm, carrying steam at 140°C, is surrounded by a cylindrical jacket with inner and outer radii 2.00 cm and 4.00 cm and made of a type of cork with thermal conductivity 4.00×10^{-2} W/m·K. This in turn is surrounded by a cylindrical jacket made of a brand of Styrofoam with thermal conductivity 1.00×10^{-2} W/m·K and having inner and outer radii 4.00 cm and 6.00 cm (Fig. P17.125). The outer surface of the Styrofoam is in contact with air at 15°C. Assume that this outer surface has a temperature of 15°C. (a) What is the temperature at a radius of 4.00 cm, where the two insulating layers meet? (b) What is the total rate of transfer of heat out of a 2.00-m length of pipe? (*Hint:* Use the expression derived in part (c) of Challenge Problem 17.124.)

Figure **P17.125**

17.126 ••• **CP Temperature Change in a Clock.** A pendulum clock is designed to tick off one second on each side-to-side swing of the pendulum (two ticks per complete period). (a) Will a pendulum clock gain time in hot weather and lose it in cold, or the reverse? Explain your reasoning. (b) A particular pendulum clock keeps correct time at 20.0°C. The pendulum shaft is steel, and its mass can be ignored compared with that of the bob. What is the fractional change in the length of the shaft when it is cooled to 10.0°C? (c) How many seconds per day will the clock gain or lose at 10.0°C? (d) How closely must the temperature be controlled if the clock is not to gain or lose more than 1.00 s a day? Does the answer depend on the period of the pendulum?

17.127 ••• **BIO A Walk in the Sun.** Consider a poor lost soul walking at 5 km/h on a hot day in the desert, wearing only a bathing suit. This person's skin temperature tends to rise due to four mechanisms: (i) energy is generated by metabolic reactions in the body at a rate of 280 W, and almost all of this energy is converted to heat that flows to the skin; (ii) heat is delivered to the skin by convection from the outside air at a rate equal to $k'A_{skin}(T_{air} - T_{skin})$, where k' is 54 J/h·C°·m², the exposed skin area A_{skin} is 1.5 m², the air temperature T_{air} is 47°C, and the skin temperature T_{skin} is 36°C; (iii) the skin absorbs radiant energy from the sun at a rate of 1400 W/m²; (iv) the skin absorbs radiant energy from the environment, which has temperature 47°C. (a) Calculate the net rate (in watts) at which the person's skin is heated by all four of these mechanisms. Assume that the emissivity of the skin is $e = 1$ and that the skin temperature is initially 36°C. Which mechanism is the most important? (b) At what rate (in L/h) must perspiration evaporate from this person's skin to maintain a constant skin temperature? (The heat of vaporization of water at 36°C is 2.42×10^6 J/kg.) (c) Suppose instead the person is protected by light-colored clothing ($e \approx 0$) so that the exposed skin area is only 0.45 m². What rate of perspiration is required now? Discuss the usefulness of the traditional clothing worn by desert peoples.

Answers

Chapter Opening Question

No. By "heat" we mean energy that is in transit from one body to another as a result of temperature difference between the bodies. Bodies do not *contain* heat.

Test Your Understanding Questions

17.1 Answer: (ii) A liquid-in-tube thermometer actually measures its own temperature. If the thermometer stays in the hot water long enough, it will come to thermal equilibrium with the water and its temperature will be the same as that of the water.

17.2 Answer: (iv) Both a bimetallic strip and a resistance thermometer measure their own temperature. For this to be equal to the temperature of the object being measured, the thermometer and object must be in contact and in thermal equilibrium. A temporal artery thermometer detects the infrared radiation from a person's skin, so there is no need for the detector and skin to be at the same temperature.

17.3 Answer: (i), (iii), (ii), (v), (iv) To compare these temperatures, convert them all to the Kelvin scale. For (i), the Kelvin temperature is $T_K = T_C + 273.15 = 0.00 + 273.15 = 273.15$ K; for (ii), $T_C = \frac{5}{9}(T_F - 32°) = \frac{5}{9}(0.00° - 32°) = -17.78°$C and $T_K = T_C + 273.15 = -17.78 + 273.15 = 255.37$ K; for (iii), $T_K = 260.00$ K; for (iv), $T_K = 77.00$ K; and for (v), $T_K = T_C + 273.15 = -180.00 + 273.15 = 93.15$ K.

17.4 Answer: (ii) and (iii) Metal 2 must expand more than metal 1 when heated and so must have a larger coefficient of linear expansion α. From Table 17.1, brass and aluminum have larger values of α than copper, but steel does not.

17.5 Answer: (ii), (i), (iv), (iii) For (i) and (ii), the relevant quantity is the specific heat c of the substance, which is the amount of heat required to raise the temperature of 1 *kilogram* of that substance by 1 K (1 C°). From Table 17.3, these values are (i) 138 J for mercury and (ii) 2428 J for ethanol. For (iii) and (iv) we need the molar heat capacity C, which is the amount of heat required to raise the temperature of 1 *mole* of that substance by 1 C°. Again from Table 17.3, these values are (iii) 27.7 J for mercury and (iv) 111.9 J for ethanol. (The ratio of molar heat capacities is different from the ratio of the specific heats because a mole of mercury and a mole of ethanol have different masses.)

17.6 Answer: (iv) In time t the system goes from point b to point e in Fig. 17.21. According to this figure, at time $t/2$ (halfway along the horizontal axis from b to e), the system is at 100°C and is still boiling; that is, it is a mixture of liquid and gas. This says that most of the heat added goes into boiling the water.

17.7 Answer: (ii) When you touch one of the walls, heat flows from your hand to the lower-temperature wall. The more rapidly heat flows from your hand, the colder you will feel. Equation (17.21) shows that the rate of heat flow is proportional to the thermal conductivity k. From Table 17.5, copper has a much higher thermal conductivity (385.0 W/m·K) than steel (50.2 W/m·K) or concrete (0.8 W/m·K), and so the copper wall feels the coldest.

Bridging Problem

Answers: (a) 14.26 K (b) 0.427 kg/h

18 THERMAL PROPERTIES OF MATTER

LEARNING GOALS

By studying this chapter, you will learn:

- How to relate the pressure, volume, and temperature of a gas.

- How the interactions between the molecules of a substance determine the properties of the substance.

- How the pressure and temperature of a gas are related to the kinetic energy of its molecules.

- How the heat capacities of a gas reveal whether its molecules are rotating or vibrating.

- What determines whether a substance is a gas, a liquid, or a solid.

? The higher the temperature of a gas, the greater the average kinetic energy of its molecules. How much faster are molecules moving in the air above a frying pan (100°C) than in the surrounding kitchen air (25°C)?

The kitchen is a great place to learn about how the properties of matter depend on temperature. When you boil water in a tea kettle, the increase in temperature produces steam that whistles out of the spout at high pressure. If you forget to poke holes in a potato before baking it, the high-pressure steam produced inside the potato can cause it to explode messily. Water vapor in the air can condense into droplets of liquid on the sides of a glass of ice water; if the glass is just out of the freezer, frost will form on the sides as water vapor changes to a solid.

These examples show the relationships among the large-scale or *macroscopic* properties of a substance, such as pressure, volume, temperature, and mass. But we can also describe a substance using a *microscopic* perspective. This means investigating small-scale quantities such as the masses, speeds, kinetic energies, and momenta of the individual molecules that make up a substance.

The macroscopic and microscopic descriptions are intimately related. For example, the (microscopic) forces that occur when air molecules strike a solid surface (such as your skin) cause (macroscopic) atmospheric pressure. To produce standard atmospheric pressure of 1.01×10^5 Pa, 10^{32} molecules strike your skin every day with an average speed of over 1700 km/h (1000 mi/h)!

In this chapter we'll begin our study of the thermal properties of matter by looking at some macroscopic aspects of matter in general. We'll pay special attention to the *ideal gas,* one of the simplest types of matter to understand. Using our knowledge of momentum and kinetic energy, we'll relate the macroscopic properties of an ideal gas to the microscopic behavior of its individual molecules. We'll also use microscopic ideas to understand the heat capacities of both gases and solids. Finally, we'll take a look at the various phases of matter—gas, liquid, and solid—and the conditions under which each occurs.

18.1 Equations of State

The conditions in which a particular material exists are described by physical quantities such as pressure, volume, temperature, and amount of substance. For example, a tank of oxygen in a welding outfit has a pressure gauge and a label stating its volume. We could add a thermometer and place the tank on a scale to determine its mass. These variables describe the *state* of the material and are called **state variables.**

The volume V of a substance is usually determined by its pressure p, temperature T, and amount of substance, described by the mass m_{total} or number of moles n. (We are calling the total mass of a substance m_{total} because later in the chapter we will use m for the mass of one molecule.) Ordinarily, we can't change one of these variables without causing a change in another. When the tank of oxygen gets hotter, the pressure increases. If the tank gets too hot, it explodes.

In a few cases the relationship among p, V, T, and m (or n) is simple enough that we can express it as an equation called the **equation of state.** When it's too complicated for that, we can use graphs or numerical tables. Even then, the relationship among the variables still exists; we call it an equation of state even when we don't know the actual equation.

Here's a simple (though approximate) equation of state for a solid material. The temperature coefficient of volume expansion β (see Section 17.4) is the fractional volume change $\Delta V/V_0$ per unit temperature change, and the compressibility k (see Section 11.4) is the negative of the fractional volume change $\Delta V/V_0$ per unit pressure change. If a certain amount of material has volume V_0 when the pressure is p_0 and the temperature is T_0, the volume V at slightly differing pressure p and temperature T is approximately

$$V = V_0[1 + \beta(T - T_0) - k(p - p_0)] \qquad (18.1)$$

(There is a negative sign in front of the term $k(p - p_0)$ because an *increase* in pressure causes a *decrease* in volume.)

The Ideal-Gas Equation

Another simple equation of state is the one for an *ideal gas.* Figure 18.1 shows an experimental setup to study the behavior of a gas. The cylinder has a movable piston to vary the volume, the temperature can be varied by heating, and we can pump any desired amount of any gas into the cylinder. We then measure the pressure, volume, temperature, and amount of gas. Note that *pressure* refers both to the force per unit area exerted by the cylinder on the gas and to the force per unit area exerted by the gas on the cylinder; by Newton's third law, these must be equal.

It is usually easiest to describe the amount of gas in terms of the number of moles n, rather than the mass. (We did this when we defined molar heat capacity in Section 17.5.) The **molar mass** M of a compound (sometimes called *molecular weight*) is the mass per mole, and the total mass m_{total} of a given quantity of that compound is the number of moles n times the mass per mole M:

$$m_{\text{total}} = nM \qquad \text{(total mass, number of moles, and molar mass)} \qquad (18.2)$$

Hence if we know the number of moles of gas in the cylinder, we can determine the mass of gas using Eq. (18.2).

Measurements of the behavior of various gases lead to three conclusions:

1. The volume V is proportional to the number of moles n. If we double the number of moles, keeping pressure and temperature constant, the volume doubles.

ActivPhysics 8.4: State Variables and Ideal Gas Law

18.1 A hypothetical setup for studying the behavior of gases. By heating the gas, varying the volume with a movable piston, and adding more gas, we can control the gas pressure p, volume V, temperature T, and number of moles n.

2. The volume varies *inversely* with the absolute pressure p. If we double the pressure while holding the temperature T and number of moles n constant, the gas compresses to one-half of its initial volume. In other words, $pV = $ constant when n and T are constant.

3. The pressure is proportional to the *absolute* temperature. If we double the absolute temperature, keeping the volume and number of moles constant, the pressure doubles. In other words, $p = $ (constant)T when n and V are constant.

18.2 The ideal-gas equation $pV = nRT$ gives a good description of the air inside an inflated vehicle tire, where the pressure is about 3 atmospheres and the temperature is much too high for nitrogen or oxygen to liquefy. As the tire warms (T increases), the volume V changes only slightly but the pressure p increases.

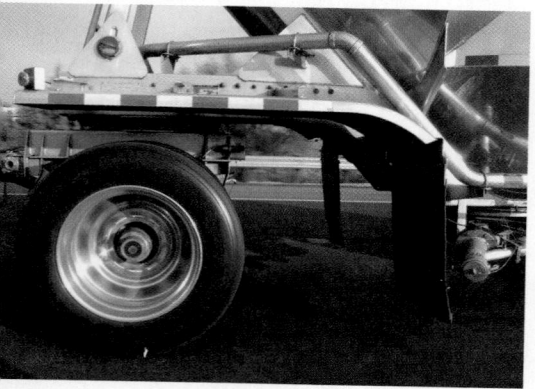

These three relationships can be combined neatly into a single equation, called the **ideal-gas equation:**

$$pV = nRT \qquad \text{(ideal-gas equation)} \tag{18.3}$$

where R is a proportionality constant. An **ideal gas** is one for which Eq. (18.3) holds precisely for *all* pressures and temperatures. This is an idealized model; it works best at very low pressures and high temperatures, when the gas molecules are far apart and in rapid motion. It is reasonably good (within a few percent) at moderate pressures (such as a few atmospheres) and at temperatures well above those at which the gas liquefies (Fig. 18.2).

We might expect that the constant R in the ideal-gas equation would have different values for different gases, but it turns out to have the same value for *all* gases, at least at sufficiently high temperature and low pressure. It is called the **gas constant** (or *ideal-gas constant*). The numerical value of R depends on the units of p, V, and T. In SI units, in which the unit of p is Pa $(1 \text{ Pa} = 1 \text{ N/m}^2)$ and the unit of V is m^3, the current best numerical value of R is

$$R = 8.314472(15) \text{ J/mol} \cdot \text{K}$$

or $R = 8.314 \text{ J/mol} \cdot \text{K}$ to four significant figures. Note that the units of pressure times volume are the same as the units of work or energy (for example, N/m^2 times m^3); that's why R has units of energy per mole per unit of absolute temperature. In chemical calculations, volumes are often expressed in liters (L) and pressures in atmospheres (atm). In this system, to four significant figures,

$$R = 0.08206 \frac{\text{L} \cdot \text{atm}}{\text{mol} \cdot \text{K}}$$

Application Respiration and the Ideal-Gas Equation

To breathe, you rely on the ideal-gas equation $pV = nRT$. Contraction of the dome-shaped diaphragm muscle increases the volume V of the thoracic cavity (which encloses the lungs), decreasing its pressure p. The lowered pressure causes the lungs to expand and fill with air. (The temperature T is kept constant.) When you exhale, the diaphragm relaxes, allowing the lungs to contract and expel the air.

We can express the ideal-gas equation, Eq. (18.3), in terms of the mass m_{total} of gas, using $m_{\text{total}} = nM$ from Eq. (18.2):

$$pV = \frac{m_{\text{total}}}{M}RT \tag{18.4}$$

From this we can get an expression for the density $\rho = m_{\text{total}}/V$ of the gas:

$$\rho = \frac{pM}{RT} \tag{18.5}$$

CAUTION **Density vs. pressure** When using Eq. (18.5), be certain that you distinguish between the Greek letter ρ (rho) for density and the letter p for pressure. ▮

Inhalation: Diaphragm contracts; lungs expand

Exhalation: Diaphragm relaxes; lungs contract

Diaphragm

For a *constant mass* (or constant number of moles) of an ideal gas the product nR is constant, so the quantity pV/T is also constant. If the subscripts 1 and 2 refer to any two states of the same mass of a gas, then

$$\frac{p_1 V_1}{T_1} = \frac{p_2 V_2}{T_2} = \text{constant} \qquad \text{(ideal gas, constant mass)} \tag{18.6}$$

Notice that you don't need the value of R to use this equation.

We used the proportionality of pressure to absolute temperature in Chapter 17 to define a temperature scale in terms of pressure in a constant-volume gas thermometer. That may make it seem that the pressure–temperature relationship in the ideal-gas equation, Eq. (18.3), is just a result of the way we define temperature. But the equation also tells us what happens when we change the volume or the amount of substance. Also, we'll see in Chapter 20 that the gas-thermometer scale corresponds closely to a temperature scale that does *not* depend on the properties of any particular material. For now, consider Eq. (18.6) as being based on this genuinely material-independent temperature scale.

Problem-Solving Strategy 18.1   Ideal Gases

IDENTIFY *the relevant concepts:* Unless the problem states otherwise, you can use the ideal-gas equation to find quantities related to the state of a gas, such as pressure p, volume V, temperature T, and/or number of moles n.

SET UP *the problem* using the following steps:
1. List the known and unknown quantities. Identify the target variables.
2. If the problem concerns only one state of the system, use Eq. (18.3), $pV = nRT$.
3. Use Eq. (18.5), $\rho = pM/RT$, as an alternative to Eq. (18.3) if the problem involves the density ρ rather than n and V.
4. In problems that concern two states (call them 1 and 2) of the same amount of gas, if all but one of the six quantities p_1, p_2, V_1, V_2, T_1, and T_2 are known, use Eq. (18.6), $p_1 V_1 / T_1 = p_2 V_2 / T_2 = $ constant. Otherwise, use Eq. (18.3) or Eq. (18.5) as appropriate.

EXECUTE *the solution* as follows:
1. Use consistent units. (SI units are entirely consistent.) The problem statement may make one system of units more con-

venient than others. Make appropriate unit conversions, such as from atmospheres to pascals or from liters to cubic meters.
2. You may have to convert between mass m_{total} and number of moles n, using $m_{total} = Mn$, where M is the molar mass. If you use Eq. (18.4), you *must* use the same mass units for m_{total} and M. So if M is in grams per mole (the usual units for molar mass), then m_{total} must also be in grams. To use m_{total} in kilograms, you must convert M to kg/mol. For example, the molar mass of oxygen is 32 g/mol or 32×10^{-3} kg/mol.
3. Remember that in the ideal-gas equations, T is always an *absolute* (Kelvin) temperature and p is always an absolute (not gauge) pressure.
4. Solve for the target variables.

EVALUATE *your answer:* Do your results make physical sense? Use benchmarks, such as the result of Example 18.1 below that a mole of an ideal gas at 1 atmosphere pressure occupies a volume of 22.4 liters.

Example 18.1   Volume of an ideal gas at STP

What is the volume of a container that holds exactly 1 mole of an ideal gas at *standard temperature and pressure* (STP), defined as $T = 0°C = 273.15$ K and $p = 1$ atm $= 1.013 \times 10^5$ Pa?

SOLUTION

IDENTIFY and SET UP: This problem involves the properties of a single state of an ideal gas, so we use Eq. (18.3). We are given the pressure p, temperature T, and number of moles n; our target variable is the corresponding volume V.

EXECUTE: From Eq. (18.3), using R in J/mol $\cdot$ K,

$$V = \frac{nRT}{p} = \frac{(1 \text{ mol})(8.314 \text{ J/mol} \cdot \text{K})(273.15 \text{ K})}{1.013 \times 10^5 \text{ Pa}}$$

$$= 0.0224 \text{ m}^3 = 22.4 \text{ L}$$

EVALUATE: At STP, 1 mole of an ideal gas occupies 22.4 L. This is the volume of a cube 0.282 m (11.1 in.) on a side, or of a sphere 0.350 m (13.8 in.) in diameter.

Example 18.2   Compressing gas in an automobile engine

In an automobile engine, a mixture of air and vaporized gasoline is compressed in the cylinders before being ignited. A typical engine has a compression ratio of 9.00 to 1; that is, the gas in the cylinders is compressed to $\frac{1}{9.00}$ of its original volume (Fig. 18.3). The intake

and exhaust valves are closed during the compression, so the quantity of gas is constant. What is the final temperature of the compressed gas if its initial temperature is 27°C and the initial and final pressures are 1.00 atm and 21.7 atm, respectively?

Continued

SOLUTION

IDENTIFY and SET UP: We must compare two states of the same quantity of ideal gas, so we use Eq. (18.6). In the uncompressed state, $p_1 = 1.00$ atm and $T_1 = 27°C = 300$ K. In the compressed state 2, $p_2 = 21.7$ atm. The cylinder volumes are not given, but we have $V_1 = 9.00V_2$. The temperature T_2 of the compressed gas is the target variable.

EXECUTE: We solve Eq. (18.6) for T_2:

$$T_2 = T_1 \frac{p_2 V_2}{p_1 V_1} = (300 \text{ K}) \frac{(21.7 \text{ atm})V_2}{(1.00 \text{ atm})(9.00V_2)} = 723 \text{ K} = 450°C$$

EVALUATE: This is the temperature of the air–gasoline mixture *before* the mixture is ignited; when burning starts, the temperature becomes higher still.

18.3 Cutaway of an automobile engine. While the air–gasoline mixture is being compressed prior to ignition, the intake and exhaust valves are both in the closed (up) position.

Example 18.3 **Mass of air in a scuba tank**

An "empty" aluminum scuba tank contains 11.0 L of air at 21°C and 1 atm. When the tank is filled rapidly from a compressor, the air temperature is 42°C and the gauge pressure is 2.10×10^7 Pa. What mass of air was added? (Air is about 78% nitrogen, 21% oxygen, and 1% miscellaneous; its average molar mass is 28.8 g/mol = 28.8×10^{-3} kg/mol.)

SOLUTION

IDENTIFY and SET UP: Our target variable is the difference $m_2 - m_1$ between the masses present at the end (state 2) and at the beginning (state 1). We are given the molar mass M of air, so we can use Eq. (18.2) to find the target variable if we know the number of moles present in states 1 and 2. We determine n_1 and n_2 by applying Eq. (18.3) to each state individually.

EXECUTE: We convert temperatures to the Kelvin scale by adding 273 and convert the pressure to absolute by adding 1.013×10^5 Pa.

The tank's volume is hardly affected by the increased temperature and pressure, so $V_2 = V_1$. From Eq. (18.3), the numbers of moles in the empty tank (n_1) and the full tank (n_2) are

$$n_1 = \frac{p_1 V_1}{RT_1} = \frac{(1.013 \times 10^5 \text{ Pa})(11.0 \times 10^{-3} \text{ m}^3)}{(8.314 \text{ J/mol} \cdot \text{K})(294 \text{ K})} = 0.46 \text{ mol}$$

$$n_2 = \frac{p_2 V_2}{RT_2} = \frac{(2.11 \times 10^7 \text{ Pa})(11.0 \times 10^{-3} \text{ m}^3)}{(8.314 \text{ J/mol} \cdot \text{K})(315 \text{ K})} = 88.6 \text{ mol}$$

We added $n_2 - n_1 = 88.6$ mol $- 0.46$ mol $= 88.1$ mol to the tank. From Eq. (18.2), the added mass is $M(n_2 - n_1) = (28.8 \times 10^{-3} \text{ kg/mol})(88.1 \text{ mol}) = 2.54$ kg.

EVALUATE: The added mass is not insubstantial: You could certainly use a scale to determine whether the tank was empty or full.

Example 18.4 **Variation of atmospheric pressure with elevation**

Find the variation of atmospheric pressure with elevation in the earth's atmosphere. Assume that at all elevations, $T = 0°C$ and $g = 9.80$ m/s².

SOLUTION

IDENTIFY and SET UP: As the elevation y increases, both the atmospheric pressure p and the density ρ decrease. Hence we have *two* unknown functions of y; to solve for them, we need two independent equations. One is the ideal-gas equation, Eq. (18.5), which is expressed in terms of p and ρ. The other is Eq. (12.4), the relationship that we found in Section 12.2 among p, ρ, and y in a fluid in equilibrium: $dp/dy = -\rho g$. We are told to assume that g and T are the same at all elevations; we also assume that the atmosphere has the same chemical composition, and hence the same molar mass M, at all heights. We combine the two equations and solve for $p(y)$.

EXECUTE: We substitute $\rho = pM/RT$ into $dp/dy = -\rho g$, separate variables, and integrate, letting p_1 be the pressure at elevation y_1 and p_2 be the pressure at y_2:

$$\frac{dp}{dy} = -\frac{pM}{RT}g$$

$$\int_{p_1}^{p_2} \frac{dp}{p} = -\frac{Mg}{RT} \int_{y_1}^{y_2} dy$$

$$\ln \frac{p_2}{p_1} = -\frac{Mg}{RT}(y_2 - y_1)$$

$$\frac{p_2}{p_1} = e^{-Mg(y_2 - y_1)/RT}$$

Now let $y_1 = 0$ be at sea level and let the pressure at that point be $p_0 = 1.013 \times 10^5$ Pa. Then the pressure p at any height y is

$$p = p_0 e^{-Mgy/RT}$$

EVALUATE: According to our calculation, the pressure decreases exponentially with elevation. The graph in Fig. 18.4 shows that the slope dp/dy becomes less negative with greater elevation. That result makes sense, since $dp/dy = -\rho g$ and the density also

18.4 The variation of atmospheric pressure p with elevation y, assuming a constant temperature T.

decreases with elevation. At the summit of Mount Everest, where $y = 8863$ m,

$$\frac{Mgy}{RT} = \frac{(28.8 \times 10^{-3}\ \text{kg/mol})(9.80\ \text{m/s}^2)(8863\ \text{m})}{(8.314\ \text{J/mol} \cdot \text{K})(273\ \text{K})} = 1.10$$

$$p = (1.013 \times 10^5\ \text{Pa})e^{-1.10} = 0.337 \times 10^5\ \text{Pa} = 0.33\ \text{atm}$$

The assumption of constant temperature isn't realistic, and g decreases a little with increasing elevation (see Challenge Problem 18.92). Even so, this example shows why mountaineers need to carry oxygen on Mount Everest. It also shows why jet airliners, which typically fly at altitudes of 8000 to 12,000 m, *must* have pressurized cabins for passenger comfort and health.

The van der Waals Equation

The ideal-gas equation, Eq. (18.3), can be obtained from a simple molecular model that ignores the volumes of the molecules themselves and the attractive forces between them (Fig. 18.5a). We'll examine that model in Section 18.3. Meanwhile, we mention another equation of state, the **van der Waals equation,** that makes approximate corrections for these two omissions (Fig. 18.5b). This equation was developed by the 19th-century Dutch physicist J. D. van der Waals; the interaction between atoms that we discussed in Section 14.4 was named the *van der Waals interaction* after him. The van der Waals equation is

$$\left(p + \frac{an^2}{V^2}\right)(V - nb) = nRT \qquad [18.7]$$

The constants a and b are empirical constants, different for different gases. Roughly speaking, b represents the volume of a mole of molecules; the total volume of the molecules is then nb, and the volume remaining in which the molecules can move is $V - nb$. The constant a depends on the attractive intermolecular forces, which reduce the pressure of the gas for given values of n, V, and T by *pulling* the molecules together as they *push* on the walls of the container. The decrease in pressure is proportional to the number of molecules per unit volume in a layer near the wall (which are exerting the pressure on the wall) and is also proportional to the number per unit volume in the next layer beyond the wall (which are doing the attracting). Hence the decrease in pressure due to intermolecular forces is proportional to n^2/V^2.

When n/V is small (that is, when the gas is *dilute*), the average distance between molecules is large, the corrections in the van der Waals equation become insignificant, and Eq. (18.7) reduces to the ideal-gas equation. As an example, for carbon dioxide gas (CO_2) the constants in the van der Waals equation are $a = 0.364\ \text{J} \cdot \text{m}^3/\text{mol}^2$ and $b = 4.27 \times 10^{-5}\ \text{m}^3/\text{mol}$. We found in Example 18.1 that 1 mole of an ideal gas at $T = 0°\text{C} = 273.15$ K and $p = 1$ atm $= 1.013 \times 10^5$ Pa occupies a volume $V = 0.0224\ \text{m}^3$; according to Eq. (18.7),

(a) An idealized model of a gas

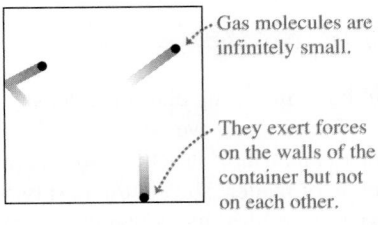

Gas molecules are infinitely small.

They exert forces on the walls of the container but not on each other.

(b) A more realistic model of a gas

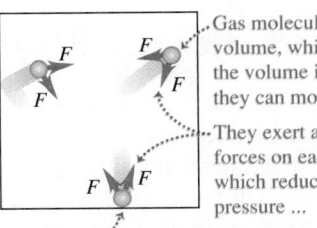

Gas molecules have volume, which reduces the volume in which they can move.

They exert attractive forces on each other, which reduces the pressure ...

... and they exert forces on the container's walls.

18.5 A gas as modeled by **(a)** the ideal-gas equation and **(b)** the van der Waals equation.

1 mole of CO_2 occupying this volume at this temperature would be at a pressure 532 Pa less than 1 atm, a difference of only 0.5% from the ideal-gas value.

pV-Diagrams

We could in principle represent the p-V-T relationship graphically as a *surface* in a three-dimensional space with coordinates p, V, and T. This representation sometimes helps us grasp the overall behavior of the substance, but ordinary two-dimensional graphs are usually more convenient. One of the most useful of these is a set of graphs of pressure as a function of volume, each for a particular constant temperature. Such a diagram is called a ***pV*-diagram.** Each curve, representing behavior at a specific temperature, is called an **isotherm,** or a *pV-isotherm.*

Figure 18.6 shows pV-isotherms for a constant amount of an ideal gas. Since $p = nRT/V$ from Eq. (18.3), along an isotherm (constant T) the pressure p is inversely proportional to the volume V and the isotherms are hyperbolic curves.

Figure 18.7 shows a pV-diagram for a material that *does not* obey the ideal-gas equation. At temperatures below T_c the isotherms develop flat regions in which we can compress the material (that is, reduce the volume V) without increasing the pressure p. Observation shows that the gas is *condensing* from the vapor (gas) to the liquid phase. The flat parts of the isotherms in the shaded area of Fig. 18.7 represent conditions of liquid-vapor *phase equilibrium.* As the volume decreases, more and more material goes from vapor to liquid, but the pressure does not change. (To keep the temperature constant during condensation, we have to remove the heat of vaporization, discussed in Section 17.6.)

When we compress such a gas at a constant temperature T_2 in Fig. 18.7, it is vapor until point a is reached. Then it begins to liquefy; as the volume decreases further, more material liquefies, and *both* the pressure and the temperature remain constant. At point b, all the material is in the liquid state. After this, any further compression requires a very rapid rise of pressure, because liquids are in general much less compressible than gases. At a lower constant temperature T_1, similar behavior occurs, but the condensation begins at lower pressure and greater volume than at the constant temperature T_2. At temperatures greater than T_c, *no* phase transition occurs as the material is compressed; at the highest temperatures, such as T_4, the curves resemble the ideal-gas curves of Fig. 18.6. We call T_c the *critical temperature* for this material. In Section 18.6 we'll discuss what happens to the phase of the gas above the critical temperature.

We will use pV-diagrams often in the next two chapters. We will show that the *area* under a pV-curve (whether or not it is an isotherm) represents the *work* done by the system during a volume change. This work, in turn, is directly related to heat transfer and changes in the *internal energy* of the system.

18.6 Isotherms, or constant-temperature curves, for a constant amount of an ideal gas. The highest temperature is T_4; the lowest is T_1. This is a graphical representation of the ideal-gas equation of state.

Each curve represents pressure as a function of volume for an ideal gas at a single temperature.

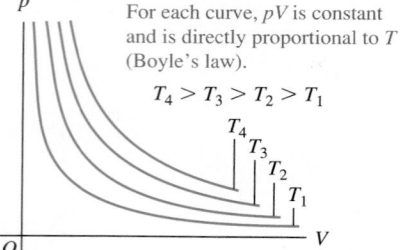

For each curve, pV is constant and is directly proportional to T (Boyle's law).

$$T_4 > T_3 > T_2 > T_1$$

18.7 A pV-diagram for a nonideal gas, showing isotherms for temperatures above and below the critical temperature T_c. The liquid–vapor equilibrium region is shown as a green shaded area. At still lower temperatures the material might undergo phase transitions from liquid to solid or from gas to solid; these are not shown in this diagram.

$$T_4 > T_3 > T_c > T_2 > T_1$$

Above the critical temperature T_c there is no liquid–vapor phase transition.

Below T_c, the material condenses to liquid as it is compressed.

Liquid–vapor phase equilibrium region

Test Your Understanding of Section 18.1 Rank the following ideal gases in order from highest to lowest number of moles: (i) pressure 1 atm, volume 1 L, and temperature 300 K; (ii) pressure 2 atm, volume 1 L, and temperature 300 K; (iii) pressure 1 atm, volume 2 L, and temperature 300 K; (iv) pressure 1 atm, volume 1 L, and temperature 600 K; (v) pressure 2 atm, volume 1 L, and temperature 600 K.

18.2 Molecular Properties of Matter

We have studied several properties of matter in bulk, including elasticity, density, surface tension, heat capacities, and equations of state. Now we want to look in more detail at the relationship of bulk behavior to *molecular* structure. We begin with a general discussion of the molecular structure of matter. Then in the next two sections we develop the kinetic-molecular model of an ideal gas, obtaining from this molecular model the equation of state and an expression for heat capacity.

Molecules and Intermolecular Forces

Any specific chemical compound is made up of identical **molecules.** The smallest molecules contain one atom each and are of the order of 10^{-10} m in size; the largest contain many atoms and are at least 10,000 times larger. In gases the molecules move nearly independently; in liquids and solids they are held together by intermolecular forces. These forces arise from interactions among the electrically charged particles that make up the molecules. Gravitational forces between molecules are negligible in comparison with electrical forces.

The interaction of two *point* electric charges is described by a force (repulsive for like charges, attractive for unlike charges) with a magnitude proportional to $1/r^2$, where r is the distance between the points. We will study this relationship, called *Coulomb's law,* in Chapter 21. Molecules are *not* point charges but complex structures containing both positive and negative charge, and their interactions are more complex. The force between molecules in a gas varies with the distance r between molecules somewhat as shown in Fig. 18.8, where a positive F_r corresponds to a repulsive force and a negative F_r to an attractive force. When molecules are far apart, the intermolecular forces are very small and usually attractive. As a gas is compressed and its molecules are brought closer together, the attractive forces increase. The intermolecular force becomes zero at an equilibrium spacing r_0, corresponding roughly to the spacing between molecules in the liquid and solid states. In liquids and solids, relatively large pressures are needed to compress the substance appreciably. This shows that at molecular distances slightly *less* than the equilibrium spacing, the forces become *repulsive* and relatively large.

Figure 18.8 also shows the potential energy as a function of r. This function has a *minimum* at r_0, where the force is zero. The two curves are related by $F_r(r) = -dU/dr$, as we showed in Section 7.4. Such a potential-energy function is often called a **potential well.** A molecule at rest at a distance r_0 from a second molecule would need an additional energy $|U_0|$, the "depth" of the potential well, to "escape" to an indefinitely large value of r.

Molecules are always in motion; their kinetic energies usually increase with temperature. At very low temperatures the average kinetic energy of a molecule may be much *less* than the depth of the potential well. The molecules then condense into the liquid or solid phase with average intermolecular spacings of about r_0. But at higher temperatures the average kinetic energy becomes larger than the depth $|U_0|$ of the potential well. Molecules can then escape the intermolecular force and become free to move independently, as in the gaseous phase of matter.

In *solids,* molecules vibrate about more or less fixed points. In a crystalline solid these points are arranged in a *crystal lattice.* Figure 18.9 shows the cubic crystal structure of sodium chloride, and Fig. 18.10 shows a scanning tunneling microscope image of individual silicon atoms on the surface of a crystal.

The vibration of molecules in a solid about their equilibrium positions may be nearly simple harmonic if the potential well is approximately parabolic in shape at distances close to r_0. (We discussed this kind of simple harmonic motion in Section 14.4.) But if the potential-energy curve rises more gradually for $r > r_0$ than for $r < r_0$, as in Fig. 18.8, the average position shifts to larger r with increasing amplitude. As we pointed out in Section 17.4, this is the basis of thermal expansion.

In a *liquid,* the intermolecular distances are usually only slightly greater than in the solid phase of the same substance, but the molecules have much greater freedom of movement. Liquids show regularity of structure only in the immediate neighborhood of a few molecules.

The molecules of a *gas* are usually widely separated and so have only very small attractive forces. A gas molecule moves in a straight line until it collides with another molecule or with a wall of the container. In molecular terms, an *ideal gas* is a gas whose molecules exert *no* attractive forces on each other (see Fig. 18.5a) and therefore have no *potential* energy.

18.8 How the force between molecules and their potential energy of interaction depend on their separation r.

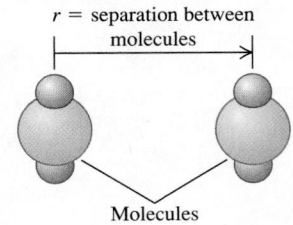

r = separation between molecules

Molecules

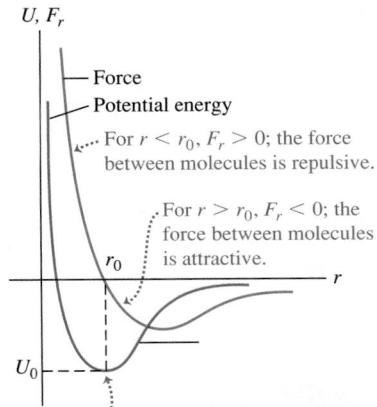

U, F_r

Force

Potential energy

For $r < r_0$, $F_r > 0$; the force between molecules is repulsive.

For $r > r_0$, $F_r < 0$; the force between molecules is attractive.

r_0

r

U_0

At a separation $r = r_0$, the potential energy of the two molecules is minimum and the force between the molecules is zero.

18.9 Schematic representation of the cubic crystal structure of sodium chloride (ordinary salt).

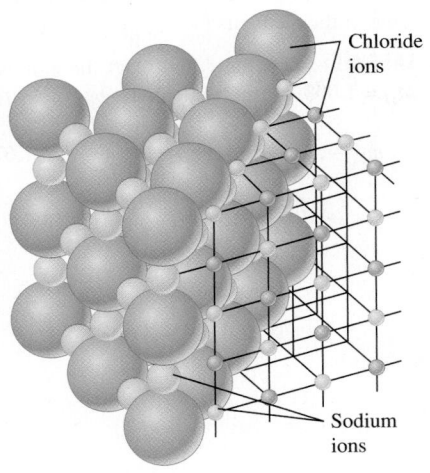

Chloride ions

Sodium ions

18.10 A scanning tunneling microscope image of the surface of a silicon crystal. The area shown is only 9.0 nm $(9.0 \times 10^{-9}$ m) across. Each blue "bead" is an individual silicon atom; you can clearly see how these atoms are arranged in a (nearly) perfect array of hexagons.

At low temperatures, most common substances are in the solid phase. As the temperature rises, a substance melts and then vaporizes. From a molecular point of view, these transitions are in the direction of increasing molecular kinetic energy. Thus temperature and molecular kinetic energy are closely related.

Moles and Avogadro's Number

We have used the mole as a measure of quantity of substance. One **mole** of any pure chemical element or compound contains a definite number of molecules, the same number for all elements and compounds. The official SI definition is:

> One mole is the amount of substance that contains as many elementary entities as there are atoms in 0.012 kilogram of carbon-12.

In our discussion, the "elementary entities" are molecules. (In a monatomic substance such as carbon or helium, each molecule is a single atom.) Atoms of a given element may occur in any of several isotopes, which are chemically identical but have different atomic masses; "carbon-12" is a specific isotope of carbon.

The number of molecules in a mole is called **Avogadro's number,** denoted by N_A. The current best numerical value of N_A is

$$N_A = 6.02214179(30) \times 10^{23} \text{ molecules/mol (Avogadro's number)}$$

The *molar mass M* of a compound is the mass of 1 mole. It is equal to the mass m of a single molecule multiplied by Avogadro's number:

$$M = N_A m \qquad \begin{array}{l}\text{(molar mass, Avogadro's number,}\\ \text{and mass of a molecule)}\end{array} \qquad (18.8)$$

When the molecule consists of a single atom, the term *atomic mass* is often used instead of molar mass or molecular weight.

Example 18.5 | **Atomic and molecular mass**

Find the mass of a single hydrogen atom and of a single oxygen molecule.

SOLUTION

IDENTIFY and SET UP: This problem involves the relationship between the mass of a molecule or atom (our target variable) and the corresponding molar mass M. We use Eq. (18.8) in the form $m = M/N_A$ and the values of the atomic masses from the periodic table of the elements (see Appendix D).

EXECUTE: For atomic hydrogen the atomic mass (molar mass) is $M_H = 1.008$ g/mol, so the mass m_H of a single hydrogen atom is

$$m_H = \frac{1.008 \text{ g/mol}}{6.022 \times 10^{23} \text{ atoms/mol}} = 1.674 \times 10^{-24} \text{ g/atom}$$

For oxygen the atomic mass is 16.0 g/mol, so for the diatomic (two-atom) oxygen molecule the molar mass is 32.0 g/mol. Then the mass of a single oxygen molecule is

$$m_{O_2} = \frac{32.0 \text{ g/mol}}{6.022 \times 10^{23} \text{ molecules/mol}} = 53.1 \times 10^{-24} \text{ g/molecule}$$

EVALUATE: We note that the values in Appendix D are for the *average* atomic masses of a natural sample of each element. Such a sample may contain several *isotopes* of the element, each with a different atomic mass. Natural samples of hydrogen and oxygen are almost entirely made up of just one isotope.

Test Your Understanding of Section 18.2 Suppose you could adjust the value of r_0 for the molecules of a certain chemical compound (Fig. 18.8) by turning a dial. If you doubled the value of r_0, the density of the solid form of this compound would become (i) twice as great; (ii) four times as great; (iii) eight times as great; (iv) $\frac{1}{2}$ as great; (v) $\frac{1}{4}$ as great; (vi) $\frac{1}{8}$ as great.

18.3 Kinetic-Molecular Model of an Ideal Gas

Mastering**PHYSICS**

PhET: Balloons & Buoyancy
PhET: Friction
PhET: Gas Properties
ActivPhysics 8.1: Characteristics of a Gas

The goal of any molecular theory of matter is to understand the *macroscopic* properties of matter in terms of its atomic or molecular structure and behavior. Once we have this understanding, we can design materials to have specific desired properties. Theories have led to the development of high-strength steels, semiconductor materials for electronic devices, and countless other materials essential to contemporary technology.

In this and the following sections we will consider a simple molecular model of an ideal gas. This *kinetic-molecular model* represents the gas as a large number of particles bouncing around in a closed container. In this section we use the kinetic-molecular model to understand how the ideal-gas equation of state, Eq. (18.3), is related to Newton's laws. In the following section we use the kinetic-molecular model to predict the molar heat capacity of an ideal gas. We'll go on to elaborate the model to include "particles" that are not points but have a finite size.

Our discussion of the kinetic-molecular model has several steps, and you may need to go over them several times. Don't get discouraged!

Here are the assumptions of our model:

1. A container with volume V contains a very large number N of identical molecules, each with mass m.
2. The molecules behave as point particles that are small compared to the size of the container and to the average distance between molecules.
3. The molecules are in constant motion. Each molecule collides occasionally with a wall of the container. These collisions are perfectly elastic.
4. The container walls are rigid and infinitely massive and do not move.

CAUTION **Molecules vs. moles** Make sure you don't confuse N, the number of *molecules* in the gas, with n, the number of *moles*. The number of molecules is equal to the number of moles multiplied by Avogadro's number: $N = nN_A$. ▮

Collisions and Gas Pressure

During collisions the molecules exert *forces* on the walls of the container; this is the origin of the *pressure* that the gas exerts. In a typical collision (Fig. 18.11) the velocity component parallel to the wall is unchanged, and the component perpendicular to the wall reverses direction but does not change in magnitude.

Our program is first to determine the *number* of collisions that occur per unit time for a certain area A of wall. Then we find the total momentum change associated with these collisions and the force needed to cause this momentum change. From this we can determine the pressure, which is force per unit area, and compare the result to the ideal-gas equation. We'll find a direct connection between the temperature of the gas and the kinetic energy of the gas molecules.

To begin, we will assume that all molecules in the gas have the same *magnitude* of x-velocity, $|v_x|$. This isn't right, but making this temporary assumption helps to clarify the basic ideas. We will show later that this assumption isn't really necessary.

As shown in Fig. 18.11, for each collision the x-component of velocity changes from $-|v_x|$ to $+|v_x|$. So the x-component of momentum changes from $-m|v_x|$ to $+m|v_x|$, and the *change* in the x-component of momentum is $m|v_x| - (-m|v_x|) = 2m|v_x|$.

If a molecule is going to collide with a given wall area A during a small time interval dt, then at the beginning of dt it must be within a distance $|v_x|\,dt$ from the wall (Fig. 18.12) and it must be headed toward the wall. So the number of molecules that collide with A during dt is equal to the number of molecules within a cylinder with base area A and length $|v_x|\,dt$ that have their x-velocity aimed

18.11 Elastic collision of a molecule with an idealized container wall.

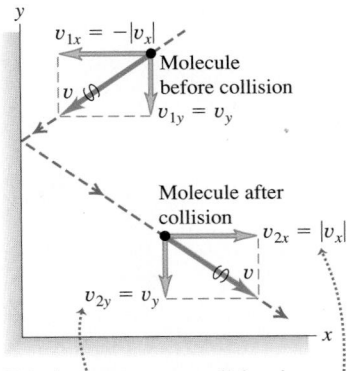

- Velocity component parallel to the wall (y-component) does not change.
- Velocity component perpendicular to the wall (x-component) reverses direction.
- Speed v does not change.

18.12 For a molecule to strike the wall in area A during a time interval dt, the molecule must be headed for the wall and be within the shaded cylinder of length $|v_x|\,dt$ at the beginning of the interval.

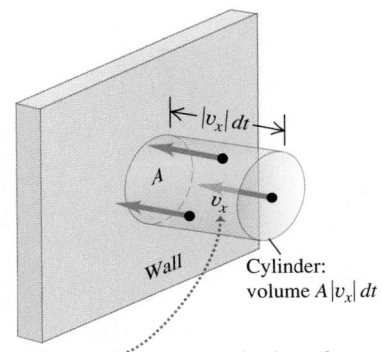

All molecules are assumed to have the same magnitude $|v_x|$ of x-velocity.

toward the wall. The volume of such a cylinder is $A|v_x|\,dt$. Assuming that the number of molecules per unit volume (N/V) is uniform, the *number* of molecules in this cylinder is $(N/V)(A|v_x|\,dt)$. On the average, half of these molecules are moving toward the wall and half are moving away from it. So the number of collisions with A during dt is

$$\tfrac{1}{2}\left(\frac{N}{V}\right)(A|v_x|\,dt)$$

For the system of all molecules in the gas, the total momentum change dP_x during dt is the *number* of collisions multiplied by $2m|v_x|$:

$$dP_x = \tfrac{1}{2}\left(\frac{N}{V}\right)(A|v_x|\,dt)(2m|v_x|) = \frac{NAmv_x^2\,dt}{V} \tag{18.9}$$

(We are using capital P for total momentum and small p for pressure. Be careful!) We wrote v_x^2 rather than $|v_x|^2$ in the final expression because the square of the absolute value of a number is equal to the square of that number. The *rate* of change of momentum component P_x is

$$\frac{dP_x}{dt} = \frac{NAmv_x^2}{V} \tag{18.10}$$

According to Newton's second law, this rate of change of momentum equals the force exerted by the wall area A on the gas molecules. From Newton's *third* law this is equal and opposite to the force exerted *on* the wall *by* the molecules. Pressure p is the magnitude of the force exerted on the wall per unit area, and we obtain

$$p = \frac{F}{A} = \frac{Nmv_x^2}{V} \tag{18.11}$$

The pressure exerted by the gas depends on the number of molecules per volume (N/V), the mass m per molecule, and the speed of the molecules.

Pressure and Molecular Kinetic Energies

We mentioned that $|v_x|$ is really *not* the same for all the molecules. But we could have sorted the molecules into groups having the same $|v_x|$ within each group, then added up the resulting contributions to the pressure. The net effect of all this is just to replace v_x^2 in Eq. (18.11) by the *average* value of v_x^2, which we denote by $(v_x^2)_{av}$. We can relate $(v_x^2)_{av}$ to the *speeds* of the molecules. The speed v of a molecule is related to the velocity components v_x, v_y, and v_z by

$$v^2 = v_x^2 + v_y^2 + v_z^2$$

We can average this relation over all molecules:

$$(v^2)_{av} = (v_x^2)_{av} + (v_y^2)_{av} + (v_z^2)_{av}$$

But there is no real difference in our model between the x-, y-, and z-directions. (Molecular speeds are very fast in a typical gas, so the effects of gravity are negligibly small.) It follows that $(v_x^2)_{av}$, $(v_y^2)_{av}$, and $(v_z^2)_{av}$ must all be *equal*. Hence $(v^2)_{av}$ is equal to $3(v_x^2)_{av}$ and

$$(v_x^2)_{av} = \tfrac{1}{3}(v^2)_{av}$$

so Eq. (18.11) becomes

$$pV = \tfrac{1}{3}Nm(v^2)_{av} = \tfrac{2}{3}N\left[\tfrac{1}{2}m(v^2)_{av}\right] \tag{18.12}$$

We notice that $\tfrac{1}{2}m(v^2)_{av}$ is the average translational kinetic energy of a single molecule. The product of this and the total number of molecules N equals the total random kinetic energy K_{tr} of translational motion of all the molecules. (The notation K_{tr} reminds us that this is the energy of *translational* motion. There may also be energies associated with molecular rotation and vibration.) The product pV equals two-thirds of the total translational kinetic energy:

$$pV = \tfrac{2}{3}K_{tr} \qquad (18.13)$$

Now we compare this with the ideal-gas equation,

$$pV = nRT$$

which is based on experimental studies of gas behavior. For the two equations to agree, we must have

$$K_{tr} = \tfrac{3}{2}nRT \qquad \text{(average translational kinetic energy of } n \text{ moles of ideal gas)} \qquad (18.14)$$

This remarkably simple result shows that K_{tr} is *directly proportional* to the absolute temperature T (Fig. 18.13).

The average translational kinetic energy of a single molecule is the total translational kinetic energy K_{tr} of all molecules divided by the number of molecules, N:

$$\frac{K_{tr}}{N} = \tfrac{1}{2}m(v^2)_{av} = \frac{3nRT}{2N}$$

Also, the total number of molecules N is the number of moles n multiplied by Avogadro's number N_A, so

$$N = nN_A \qquad \frac{n}{N} = \frac{1}{N_A}$$

and

$$\frac{K_{tr}}{N} = \tfrac{1}{2}m(v^2)_{av} = \tfrac{3}{2}\left(\frac{R}{N_A}\right)T \qquad (18.15)$$

The ratio R/N_A occurs frequently in molecular theory. It is called the **Boltzmann constant**, k:

$$k = \frac{R}{N_A} = \frac{8.314 \text{ J/mol} \cdot \text{K}}{6.022 \times 10^{23} \text{ molecules/mol}}$$

$$= 1.381 \times 10^{-23} \text{ J/molecule} \cdot \text{K}$$

(The current best numerical value of k is $1.3806504(24) \times 10^{-23}$ J/molecule $\cdot$ K). In terms of k we can rewrite Eq. (18.15) as

$$\tfrac{1}{2}m(v^2)_{av} = \tfrac{3}{2}kT \qquad \text{(average translational kinetic energy of a gas molecule)} \qquad (18.16)$$

This shows that the average translational kinetic energy *per molecule* depends only on the temperature, not on the pressure, volume, or kind of molecule. We can obtain the average translational kinetic energy *per mole* by multiplying Eq. (18.16) by Avogadro's number and using the relation $M = N_A m$:

$$N_A \tfrac{1}{2}m(v^2)_{av} = \tfrac{1}{2}M(v^2)_{av} = \tfrac{3}{2}RT \qquad \text{(average translational kinetic energy per mole of gas)} \qquad (18.17)$$

The translational kinetic energy of a mole of an ideal gas depends only on T.

18.13 Summer air (top) is warmer than winter air (bottom); that is, the average translational kinetic energy of air molecules is greater in summer.

Finally, it is sometimes convenient to rewrite the ideal-gas equation on a molecular basis. We use $N = N_A n$ and $R = N_A k$ to obtain this alternative form:

$$pV = NkT \tag{18.18}$$

This shows that we can think of the Boltzmann constant k as a gas constant on a "per-molecule" basis instead of the usual "per-mole" basis for R.

Molecular Speeds

From Eqs. (18.16) and (18.17) we can obtain expressions for the square root of $(v^2)_{av}$, called the **root-mean-square speed** (or **rms speed**) v_{rms}:

$$v_{rms} = \sqrt{(v^2)_{av}} = \sqrt{\frac{3kT}{m}} = \sqrt{\frac{3RT}{M}} \quad \begin{array}{l}\text{(root-mean-square speed} \\ \text{of a gas molecule)}\end{array} \tag{18.19}$$

It might seem more natural to characterize molecular speeds by their *average* value rather than by v_{rms}, but we see that v_{rms} follows more directly from Eqs. (18.16) and (18.17). To compute the rms speed, we square each molecular speed, add, divide by the number of molecules, and take the square root; v_{rms} is the *root* of the *mean* of the *squares*. Example 18.7 illustrates this procedure.

Equations (18.16) and (18.19) show that at a given temperature T, gas molecules of different mass m have the same average kinetic energy but different root-mean-square speeds. On average, the nitrogen molecules ($M = 28$ g/mol) in the air around you are moving faster than are the oxygen molecules ($M = 32$ g/mol). Hydrogen molecules ($M = 2$ g/mol) are fastest of all; this is why there is hardly any hydrogen in the earth's atmosphere, despite its being the most common element in the universe (Fig. 18.14). A sizable fraction of any H_2 molecules in the atmosphere would have speeds greater than the earth's escape speed of 1.12×10^4 m/s (calculated in Example 13.5 in Section 13.3) and would escape into space. The heavier, slower-moving gases cannot escape so easily, which is why they predominate in our atmosphere.

The assumption that individual molecules undergo perfectly elastic collisions with the container wall is actually a little too simple. More detailed investigation has shown that in most cases, molecules actually adhere to the wall for a short time and then leave again with speeds that are characteristic of the temperature *of the wall*. However, the gas and the wall are ordinarily in thermal equilibrium and have the same temperature. So there is no net energy transfer between gas and wall, and this discovery does not alter the validity of our conclusions.

18.14 While hydrogen is a desirable fuel for vehicles, it is only a trace constituent of our atmosphere (0.00005% by volume). Hence hydrogen fuel has to be generated by electrolysis of water, which is itself an energy-intensive process.

Problem-Solving Strategy 18.2 | **Kinetic-Molecular Theory**

IDENTIFY *the relevant concepts:* Use the results of the kinetic-molecular model to relate the macroscopic properties of a gas, such as temperature and pressure, to microscopic properties, such as molecular speeds.

SET UP *the problem* using the following steps:
1. List knowns and unknowns; identify the target variables.
2. Choose appropriate equation(s) from among Eqs. (18.14), (18.16), and (18.19).

EXECUTE *the solution* as follows: Maintain consistency in units. Note especially the following:
1. The usual units for molar mass M are grams per mole; these units are often omitted in tables. In equations such as Eq. (18.19), when you use SI units you must express M in kilograms per

mole. For example, for oxygen $M_{O_2} = 32$ g/mol $= 32 \times 10^{-3}$ kg/mol.
2. Are you working on a "per-molecule" basis (with m, N, and k) or a "per-mole" basis (with M, n, and R)? To check units, think of N as having units of "molecules"; then m has units of mass per molecule, and k has units of joules per molecule per kelvin. Similarly, n has units of moles; then M has units of mass per mole and R has units of joules per mole per kelvin.
3. Remember that T is always *absolute* (Kelvin) temperature.

EVALUATE *your answer:* Are your answers reasonable? Here's a benchmark: Typical molecular speeds at room temperature are several hundred meters per second.

Example 18.6 **Molecular kinetic energy and v_{rms}**

(a) What is the average translational kinetic energy of an ideal-gas molecule at 27°C? (b) What is the total random translational kinetic energy of the molecules in 1 mole of this gas? (c) What is the root-mean-square speed of oxygen molecules at this temperature?

SOLUTION

IDENTIFY and SET UP: This problem involves the translational kinetic energy of an ideal gas on a per-molecule and per-mole basis, as well as the root-mean-square molecular speed v_{rms}. We are given $T = 27°C = 300$ K and $n = 1$ mol; we use the molecular mass m for oxygen. We use Eq. (18.16) to determine the average kinetic energy of a molecule, Eq. (18.14) to find the total molecular kinetic energy K_{tr} of 1 mole, and Eq. (18.19) to find v_{rms}.

EXECUTE: (a) From Eq. (18.16),

$$\tfrac{1}{2}m(v^2)_{av} = \tfrac{3}{2}kT = \tfrac{3}{2}(1.38 \times 10^{-23}\text{ J/K})(300\text{ K})$$
$$= 6.21 \times 10^{-21}\text{ J}$$

(b) From Eq. (18.14), the kinetic energy of one mole is

$$K_{tr} = \tfrac{3}{2}nRT = \tfrac{3}{2}(1\text{ mol})(8.314\text{ J/mol·K})(300\text{ K}) = 3740\text{ J}$$

(c) We found the mass per molecule m and molar mass M of molecular oxygen in Example 18.5. Using Eq. (18.19), we can calculate v_{rms} in two ways:

$$v_{rms} = \sqrt{\frac{3kT}{m}} = \sqrt{\frac{3(1.38 \times 10^{-23}\text{ J/K})(300\text{ K})}{5.31 \times 10^{-26}\text{ kg}}}$$
$$= 484\text{ m/s} = 1740\text{ km/h} = 1080\text{ mi/h}$$

$$v_{rms} = \sqrt{\frac{3RT}{M}} = \sqrt{\frac{3(8.314\text{ J/mol·K})(300\text{ K})}{32.0 \times 10^{-3}\text{ kg/mol}}} = 484\text{ m/s}$$

EVALUATE: The answer in part (a) does not depend on the mass of the molecule. We can check our result in part (b) by noting that the translational kinetic energy per mole must be equal to the product of the average translational kinetic energy per molecule from part (a) and Avogadro's number N_A: $K_{tr} = (6.022 \times 10^{23}$ molecules) $(6.21 \times 10^{-21}$ J/molecule$) = 3740$ J.

Example 18.7 **Calculating rms and average speeds**

Five gas molecules chosen at random are found to have speeds of 500, 600, 700, 800, and 900 m/s. What is the rms speed? What is the *average* speed?

SOLUTION

IDENTIFY and SET UP: We use the definitions of the root mean square and the average of a collection of numbers. To find v_{rms}, we square each speed, find the average (mean) of the squares, and take the square root of the result. We find v_{av} as usual.

EXECUTE: The average value of v^2 and the resulting v_{rms} for the five molecules are

$$(v^2)_{av} = \frac{500^2 + 600^2 + 700^2 + 800^2 + 900^2}{5}\text{ m}^2/\text{s}^2$$
$$= 5.10 \times 10^5\text{ m}^2/\text{s}^2$$

$$v_{rms} = \sqrt{(v^2)_{av}} = 714\text{ m/s}$$

The average speed v_{av} is

$$v_{av} = \frac{500 + 600 + 700 + 800 + 900}{5}\text{ m/s} = 700\text{ m/s}$$

EVALUATE: In general v_{rms} and v_{av} are *not* the same. Roughly speaking, v_{rms} gives greater weight to the higher speeds than does v_{av}.

Collisions Between Molecules

We have ignored the possibility that two gas molecules might collide. If they are really points, they *never* collide. But consider a more realistic model in which the molecules are rigid spheres with radius r. How often do they collide with other molecules? How far do they travel, on average, between collisions? We can get approximate answers from the following rather primitive model.

Consider N spherical molecules with radius r in a volume V. Suppose only one molecule is moving. When it collides with another molecule, the distance between centers is $2r$. Suppose we draw a cylinder with radius $2r$, with its axis parallel to the velocity of the molecule (Fig. 18.15). The moving molecule collides with any other molecule whose center is inside this cylinder. In a short time dt a molecule with speed v travels a distance $v\,dt$; during this time it collides with any molecule that is in the cylindrical volume of radius $2r$ and length $v\,dt$. The volume of the cylinder is $4\pi r^2 v\,dt$. There are N/V molecules per unit volume, so the number dN with centers in this cylinder is

$$dN = 4\pi r^2 v\,dt\,N/V$$

18.15 In a time dt a molecule with radius r will collide with any other molecule within a cylindrical volume of radius $2r$ and length $v\,dt$.

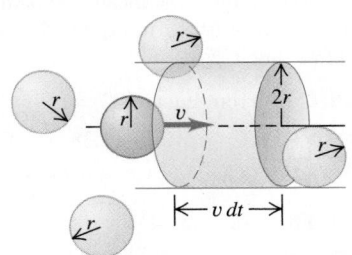

Thus the number of collisions *per unit time* is

$$\frac{dN}{dt} = \frac{4\pi r^2 v N}{V}$$

This result assumes that only one molecule is moving. The analysis is quite a bit more involved when all the molecules move at once. It turns out that in this case the collisions are more frequent, and the above equation has to be multiplied by a factor of $\sqrt{2}$:

$$\frac{dN}{dt} = \frac{4\pi \sqrt{2}\, r^2 v N}{V}$$

The average time t_{mean} between collisions, called the *mean free time,* is the reciprocal of this expression:

$$t_{mean} = \frac{V}{4\pi \sqrt{2}\, r^2 v N} \tag{18.20}$$

The average distance traveled between collisions is called the **mean free path,** denoted by λ (the Greek letter lambda). In our simple model, this is just the molecule's speed v multiplied by t_{mean}:

18.16 If you try to walk through a crowd, your mean free path—the distance you can travel on average without running into another person—depends on how large the people are and how closely they are spaced.

$$\lambda = v t_{mean} = \frac{V}{4\pi \sqrt{2}\, r^2 N} \quad \text{(mean free path of a gas molecule)} \tag{18.21}$$

The mean free path is inversely proportional to the number of molecules per unit volume (N/V) and inversely proportional to the cross-sectional area πr^2 of a molecule; the more molecules there are and the larger the molecule, the shorter the mean distance between collisions (Fig. 18.16). Note that the mean free path *does not* depend on the speed of the molecule.

We can express Eq. (18.21) in terms of macroscopic properties of the gas, using the ideal-gas equation in the form of Eq. (18.18), $pV = NkT$. We find

$$\lambda = \frac{kT}{4\pi \sqrt{2}\, r^2 p} \tag{18.22}$$

If the temperature is increased at constant pressure, the gas expands, the average distance between molecules increases, and λ increases. If the pressure is increased at constant temperature, the gas compresses and λ decreases.

Example 18.8 Calculating mean free path

(a) Estimate the mean free path of a molecule of air at 27°C and 1 atm. Model the molecules as spheres with radius $r = 2.0 \times 10^{-10}$ m. (b) Estimate the mean free time of an oxygen molecule with $v = v_{rms}$ at 27°C and 1 atm.

SOLUTION

IDENTIFY and SET UP: This problem uses the concepts of mean free path and mean free time (our target variables). We use Eq. (18.22) to determine the mean free path λ. We then use the basic relationship $\lambda = v t_{mean}$ in Eq. (18.21), with $v = v_{rms}$, to find the mean free time t_{mean}.

EXECUTE: (a) From Eq. (18.22),

$$\lambda = \frac{kT}{4\pi \sqrt{2} r^2 p} = \frac{(1.38 \times 10^{-23} \text{ J/K})(300 \text{ K})}{4\pi \sqrt{2}(2.0 \times 10^{-10} \text{ m})^2 (1.01 \times 10^5 \text{ Pa})}$$

$$= 5.8 \times 10^{-8} \text{ m}$$

(b) From Example 18.6, for oxygen at 27°C the root-mean-square speed is $v_{rms} = 484$ m/s, so the mean free time for a molecule with this speed is

$$t_{mean} = \frac{\lambda}{v} = \frac{5.8 \times 10^{-8} \text{ m}}{484 \text{ m/s}} = 1.2 \times 10^{-10} \text{ s}$$

This molecule undergoes about 10^{10} collisions per second!

EVALUATE: Note that from Eqs. (18.21) and (18.22) the mean free *path* doesn't depend on the molecule's speed, but the mean free *time* does. Slower molecules have a longer average time interval t_{mean} between collisions than do fast ones, but the average *distance* λ between collisions is the same no matter what the molecule's speed. Our answer to part (a) says that the molecule doesn't go far between collisions, but the mean free path is still several hundred times the molecular radius r.

Test Your Understanding of Section 18.3 Rank the following gases in order from (a) highest to lowest rms speed of molecules and (b) highest to lowest average translational kinetic energy of a molecule: (i) oxygen ($M = 32.0$ g/mol) at 300 K; (ii) nitrogen ($M = 28.0$ g/mol) at 300 K; (iii) oxygen at 330 K; (iv) nitrogen at 330 K.

18.4 Heat Capacities

When we introduced the concept of heat capacity in Section 17.5, we talked about ways to *measure* the specific heat or molar heat capacity of a particular material. Now we'll see how to *predict* these on theoretical grounds.

Heat Capacities of Gases

The basis of our analysis is that heat is *energy* in transit. When we add heat to a substance, we are increasing its molecular energy. In this discussion the volume of the gas will remain constant so that we don't have to worry about energy transfer through mechanical work. If we were to let the gas expand, it would do work by pushing on moving walls of its container, and this additional energy transfer would have to be included in our calculations. We'll return to this more general case in Chapter 19. For now, with the volume held constant, we are concerned with C_V, the molar heat capacity *at constant volume*.

In the simple kinetic-molecular model of Section 18.3 the molecular energy consists only of the translational kinetic energy K_{tr} of the pointlike molecules. This energy is directly proportional to the absolute temperature T, as shown by Eq. (18.14), $K_{tr} = \frac{3}{2}nRT$. When the temperature changes by a small amount dT, the corresponding change in kinetic energy is

$$dK_{tr} = \tfrac{3}{2} nR \, dT \qquad (18.23)$$

From the definition of molar heat capacity at constant volume, C_V (see Section 17.5), we also have

$$dQ = nC_V \, dT \qquad (18.24)$$

where dQ is the heat input needed for a temperature change dT. Now if K_{tr} represents the total molecular energy, as we have assumed, then dQ and dK_{tr} must be *equal* (Fig. 18.17). From Eqs. (18.23) and (18.24), this says

$$nC_V \, dT = \tfrac{3}{2} nR \, dT$$

$$C_V = \tfrac{3}{2}R \qquad \text{(ideal gas of point particles)} \qquad (18.25)$$

This surprisingly simple result says that the molar heat capacity at constant volume is $3R/2$ for *any* gas whose molecules can be represented as points.

Does Eq. (18.25) agree with measured values of molar heat capacities? In SI units, Eq. (18.25) gives

$$C_V = \tfrac{3}{2}(8.314 \text{ J/mol} \cdot \text{K}) = 12.47 \text{ J/mol} \cdot \text{K}$$

For comparison, Table 18.1 gives measured values of C_V for several gases. We see that for *monatomic* gases our prediction is right on the money, but that it is way off for diatomic and polyatomic gases.

This comparison tells us that our point-molecule model is good enough for monatomic gases but that for diatomic and polyatomic molecules we need something more sophisticated. For example, we can picture a diatomic molecule as

18.17 (a) A fixed volume V of a monatomic ideal gas. (b) When an amount of heat dQ is added to the gas, the total translational kinetic energy increases by $dK_{tr} = dQ$ and the temperature increases by $dT = dQ/nC_V$.

(a)

(b)

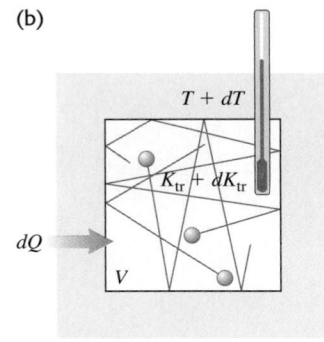

Table 18.1 Molar Heat Capacities of Gases

Type of Gas	Gas	C_V (J/mol · K)
Monatomic	He	12.47
	Ar	12.47
Diatomic	H_2	20.42
	N_2	20.76
	O_2	20.85
	CO	20.85
Polyatomic	CO_2	28.46
	SO_2	31.39
	H_2S	25.95

18.18 Motions of a diatomic molecule.

(a) Translational motion. The molecule moves as a whole; its velocity may be described as the *x*-, *y*-, and *z*-velocity components of its center of mass.

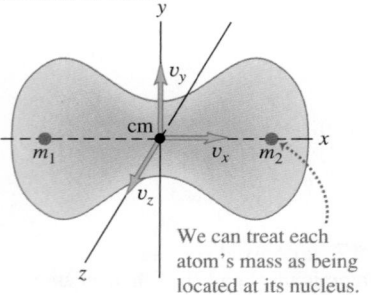

We can treat each atom's mass as being located at its nucleus.

(b) Rotational motion. The molecule rotates about its center of mass. This molecule has two independent axes of rotation.

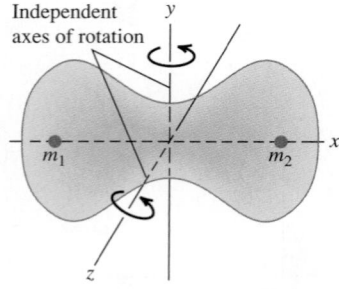

Independent axes of rotation

(c) Vibrational motion. The molecule oscillates as though the nuclei were connected by a spring.

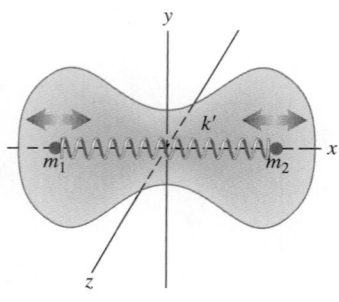

two point masses, like a little elastic dumbbell, with an interaction force between the atoms of the kind shown in Fig. 18.8. Such a molecule can have additional kinetic energy associated with *rotation* about axes through its center of mass. The atoms may also have *vibrating* motion along the line joining them, with additional kinetic and potential energies. Figure 18.18 shows these possibilities.

When heat flows into a *monatomic* gas at constant volume, *all* of the added energy goes into an increase in random *translational* molecular kinetic energy. Equation (18.23) shows that this gives rise to an increase in temperature. But when the temperature is increased by the same amount in a *diatomic* or *polyatomic* gas, additional heat is needed to supply the increased rotational and vibrational energies. Thus polyatomic gases have *larger* molar heat capacities than monatomic gases, as Table 18.1 shows.

But how do we know how much energy is associated with each additional kind of motion of a complex molecule, compared to the translational kinetic energy? The new principle that we need is called the principle of **equipartition of energy.** It can be derived from sophisticated statistical-mechanics considerations; that derivation is beyond our scope, and we will treat the principle as an axiom.

The principle of equipartition of energy states that each velocity component (either linear or angular) has, on average, an associated kinetic energy per molecule of $\frac{1}{2}kT$, or one-half the product of the Boltzmann constant and the absolute temperature. The number of velocity components needed to describe the motion of a molecule completely is called the number of **degrees of freedom.** For a monatomic gas, there are three degrees of freedom (for the velocity components v_x, v_y, and v_z); this gives a total average kinetic energy per molecule of $3(\frac{1}{2}kT)$, consistent with Eq. (18.16).

For a *diatomic* molecule there are two possible axes of rotation, perpendicular to each other and to the molecule's axis. (We don't include rotation about the molecule's own axis because in ordinary collisions there is no way for this rotational motion to change.) If we assign five degrees of freedom to a diatomic molecule, the average total kinetic energy per molecule is $\frac{5}{2}kT$ instead of $\frac{3}{2}kT$. The total kinetic energy of n moles is $K_{\text{total}} = nN_A(\frac{5}{2}kT) = \frac{5}{2}n(kN_A)T = \frac{5}{2}nRT$, and the molar heat capacity (at constant volume) is

$$C_V = \frac{5}{2}R \qquad \text{(diatomic gas, including rotation)} \qquad (18.26)$$

In SI units,

$$C_V = \frac{5}{2}(8.314 \text{ J/mol} \cdot \text{K}) = 20.79 \text{ J/mol} \cdot \text{K}$$

This agrees within a few percent with the measured values for diatomic gases given in Table 18.1.

Vibrational motion can also contribute to the heat capacities of gases. Molecular bonds are not rigid; they can stretch and bend, and the resulting vibrations lead to additional degrees of freedom and additional energies. For most diatomic gases, however, vibrational motion does *not* contribute appreciably to heat capacity. The reason for this is a little subtle and involves some concepts of quantum mechanics. Briefly, vibrational energy can change only in finite steps. If the energy change of the first step is much larger than the energy possessed by most molecules, then nearly all the molecules remain in the minimum-energy state of motion. In that case, changing the temperature does not change their average vibrational energy appreciably, and the vibrational degrees of freedom are said to be "frozen out." In more complex molecules the gaps between permitted energy levels are sometimes much smaller, and then vibration *does* contribute to heat capacity. The rotational energy of a molecule also changes by finite steps, but they are usually much smaller; the "freezing out" of rotational degrees of freedom occurs only in rare instances, such as for the hydrogen molecule below about 100 K.

In Table 18.1 the large values of C_V for some polyatomic molecules show the contributions of vibrational energy. In addition, a molecule with three or more atoms that are not in a straight line has three, not two, rotational degrees of freedom.

From this discussion we expect heat capacities to be temperature-dependent, generally increasing with increasing temperature. Figure 18.19 is a graph of the temperature dependence of C_V for hydrogen gas (H_2), showing the temperatures at which the rotational and vibrational energies begin to contribute.

Heat Capacities of Solids

We can carry out a similar heat-capacity analysis for a crystalline solid. Consider a crystal consisting of N identical atoms (a *monatomic solid*). Each atom is bound to an equilibrium position by interatomic forces. The elasticity of solid materials shows us that these forces must permit stretching and bending of the bonds. We can think of a crystal as an array of atoms connected by little springs (Fig. 18.20). Each atom can *vibrate* about its equilibrium position.

Each atom has three degrees of freedom, corresponding to its three components of velocity. According to the equipartition principle, each atom has an average kinetic energy of $\frac{1}{2}kT$ for each degree of freedom. In addition, each atom has *potential* energy associated with the elastic deformation. For a simple harmonic oscillator (discussed in Chapter 14) it is not hard to show that the average kinetic energy of an atom is *equal* to its average potential energy. In our model of a crystal, each atom is essentially a three-dimensional harmonic oscillator; it can be shown that the equality of average kinetic and potential energies also holds here, provided that the "spring" forces obey Hooke's law.

Thus we expect each atom to have an average kinetic energy $\frac{3}{2}kT$ and an average potential energy $\frac{3}{2}kT$, or an average total energy $3kT$ per atom. If the crystal contains N atoms or n moles, its total energy is

$$E_{\text{total}} = 3NkT = 3nRT \qquad (18.27)$$

From this we conclude that the molar heat capacity of a crystal should be

$$C_V = 3R \qquad \text{(ideal monatomic solid)} \qquad (18.28)$$

In SI units,

$$C_V = (3)(8.314 \text{ J/mol} \cdot \text{K}) = 24.9 \text{ J/mol} \cdot \text{K}$$

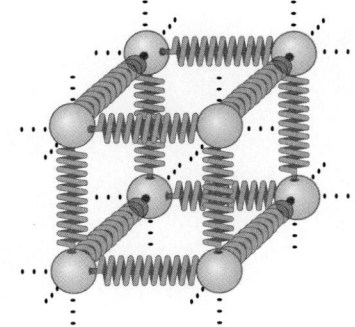

18.20 To visualize the forces between neighboring atoms in a crystal, envision every atom as being attached to its neighbors by springs.

This is the **rule of Dulong and Petit,** which we encountered as an *empirical* finding in Section 17.5: Elemental solids all have molar heat capacities of about 25 J/mol · K. Now we have *derived* this rule from kinetic theory. The agreement is only approximate, to be sure, but considering the very simple nature of our model, it is quite significant.

At low temperatures, the heat capacities of most solids *decrease* with decreasing temperature (Fig. 18.21) for the same reason that vibrational degrees of freedom of molecules are frozen out at low temperatures. At very low temperatures the quantity kT is much *smaller* than the smallest energy step the vibrating atoms can take. Hence most of the atoms remain in their lowest energy states because the next higher energy level is out of reach. The average vibrational energy per atom is then *less* than $3kT$, and the heat capacity per molecule is *less* than $3k$. At higher temperatures when kT is *large* in comparison to the minimum energy step, the equipartition principle holds, and the total heat capacity is $3k$ per molecule or $3R$ per mole as the Dulong and Petit rule predicts. Quantitative understanding of the temperature variation of heat capacities was one of the triumphs of quantum mechanics during its initial development in the 1920s.

Test Your Understanding of Section 18.4 A cylinder with a fixed volume contains hydrogen gas (H_2) at 25 K. You then add heat to the gas at a constant rate until its temperature reaches 500 K. Does the temperature of the gas increase at a constant rate? Why or why not? If not, does the temperature increase most rapidly near the beginning or near the end of this process? ❙

18.21 Experimental values of C_V for lead, aluminum, silicon, and diamond. At high temperatures, C_V for each solid approaches about $3R$, in agreement with the rule of Dulong and Petit. At low temperatures, C_V is much less than $3R$.

18.5 Molecular Speeds

As we mentioned in Section 18.3, the molecules in a gas don't all have the same speed. Figure 18.22 shows one experimental scheme for measuring the distribution of molecular speeds. A substance is vaporized in a hot oven; molecules of the vapor escape through an aperture in the oven wall and into a vacuum chamber. A series of slits blocks all molecules except those in a narrow beam, which is aimed at a pair of rotating disks. A molecule passing through the slit in the first disk is blocked by the second disk unless it arrives just as the slit in the second disk is lined up with the beam. The disks function as a speed selector that passes only molecules within a certain narrow speed range. This range can be varied by changing the disk rotation speed, and we can measure how many molecules lie within each of various speed ranges.

To describe the results of such measurements, we define a function $f(v)$ called a *distribution function*. If we observe a total of N molecules, the number dN having speeds in the range between v and $v + dv$ is given by

$$dN = Nf(v)\,dv \qquad (18.29)$$

18.22 A molecule with a speed v passes through the slit in the first rotating disk. When the molecule reaches the second rotating disk, the disks have rotated through the offset angle θ. If $v = \omega x/\theta$, the molecule passes through the slit in the second rotating disk and reaches the detector.

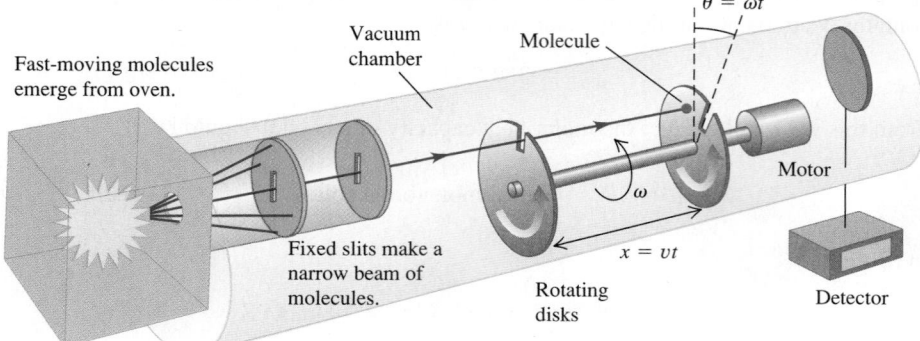

We can also say that the *probability* that a randomly chosen molecule will have a speed in the interval v to $v + dv$ is $f(v)\,dv$. Hence $f(v)$ is the probability per unit speed *interval;* it is *not* equal to the probability that a molecule has speed exactly equal to v. Since a probability is a pure number, $f(v)$ has units of reciprocal speed (s/m).

Figure 18.23a shows distribution functions for three different temperatures. At each temperature the height of the curve for any value of v is proportional to the number of molecules with speeds near v. The peak of the curve represents the *most probable speed* v_{mp} for the corresponding temperature. As the temperature increases, the average molecular kinetic energy increases, and so the peak of $f(v)$ shifts to higher and higher speeds.

Figure 18.23b shows that the area under a curve between any two values of v represents the fraction of all the molecules having speeds in that range. Every molecule must have *some* value of v, so the integral of $f(v)$ over all v must be unity for any T.

If we know $f(v)$, we can calculate the most probable speed v_{mp}, the average speed v_{av}, and the rms speed v_{rms}. To find v_{mp}, we simply find the point where $df/dv = 0$; this gives the value of the speed where the curve has its peak. To find v_{av}, we take the number $Nf(v)\,dv$ having speeds in each interval dv, multiply each number by the corresponding speed v, add all these products (by integrating over all v from zero to infinity), and finally divide by N. That is,

$$v_{av} = \int_0^\infty v f(v)\,dv \qquad (18.30)$$

The rms speed is obtained similarly; the average of v^2 is given by

$$(v^2)_{av} = \int_0^\infty v^2 f(v)\,dv \qquad (18.31)$$

and v_{rms} is the square root of this.

The Maxwell–Boltzmann Distribution

The function $f(v)$ describing the actual distribution of molecular speeds is called the **Maxwell–Boltzmann distribution.** It can be derived from statistical-mechanics considerations, but that derivation is beyond our scope. Here is the result:

$$f(v) = 4\pi\left(\frac{m}{2\pi kT}\right)^{3/2} v^2 e^{-mv^2/2kT} \qquad \begin{array}{l}\text{(Maxwell–Boltzmann} \\ \text{distribution)}\end{array} \qquad (18.32)$$

We can also express this function in terms of the translational kinetic energy of a molecule, which we denote by ϵ; that is, $\epsilon = \frac{1}{2}mv^2$. We invite you to verify that when this is substituted into Eq. (18.32), the result is

$$f(v) = \frac{8\pi}{m}\left(\frac{m}{2\pi kT}\right)^{3/2} \epsilon e^{-\epsilon/kT} \qquad (18.33)$$

This form shows that the exponent in the Maxwell–Boltzmann distribution function is $-\epsilon/kT$ and that the shape of the curve is determined by the relative magnitude of ϵ and kT at any point. We leave it to you (see Exercise 18.48) to prove that the *peak* of each curve occurs where $\epsilon = kT$, corresponding to a most probable speed v_{mp} given by

$$v_{mp} = \sqrt{\frac{2kT}{m}} \qquad (18.34)$$

To find the average speed, we substitute Eq. (18.32) into Eq. (18.30) and carry out the integration, making a change of variable $v^2 = x$ and then integrating by parts. The result is

$$v_{av} = \sqrt{\frac{8kT}{\pi m}} \qquad (18.35)$$

18.23 (a) Curves of the Maxwell–Boltzmann distribution function $f(v)$ for three temperatures. (b) The shaded areas under the curve represent the fractions of molecules within certain speed ranges. The most probable speed v_{mp} for a given temperature is at the peak of the curve.

(a)

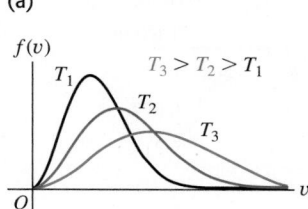

As temperature increases:
• the curve flattens.
• the maximum shifts to higher speeds.

(b)

Fraction of molecules with speeds from v_1 to v_2

Fraction of molecules with speeds greater than v_A

Mastering**PHYSICS**

ActivPhysics 8.2: Maxwell-Boltzmann Distribution—Conceptual Analysis
ActivPhysics 8.3: Maxwell-Boltzmann Distribution—Quantitative Analysis

Finally, to find the rms speed, we substitute Eq. (18.32) into Eq. (18.31). Evaluating the resulting integral takes some mathematical acrobatics, but we can find it in a table of integrals. The result is

$$v_{rms} = \sqrt{\frac{3kT}{m}} \qquad (18.36)$$

This result agrees with Eq. (18.19); it *must* agree if the Maxwell–Boltzmann distribution is to be consistent with the equipartition principle and our other kinetic-theory calculations.

Table 18.2 shows the fraction of all the molecules in an ideal gas that have speeds *less than* various multiples of v_{rms}. These numbers were obtained by numerical integration; they are the same for all ideal gases.

The distribution of molecular speeds in liquids is similar, although not identical, to that for gases. We can understand the vapor pressure of a liquid and the phenomenon of boiling on this basis. Suppose a molecule must have a speed at least as great as v_A in Fig. 18.23b to escape from the surface of a liquid into the adjacent vapor. The number of such molecules, represented by the area under the "tail" of each curve (to the right of v_A), increases rapidly with temperature. Thus the rate at which molecules can escape is strongly temperature-dependent. This process is balanced by another one in which molecules in the vapor phase collide inelastically with the surface and are trapped back into the liquid phase. The number of molecules suffering this fate per unit time is proportional to the pressure in the vapor phase. Phase equilibrium between liquid and vapor occurs when these two competing processes proceed at exactly the same rate. So if the molecular speed distributions are known for various temperatures, we can make a theoretical prediction of vapor pressure as a function of temperature. When liquid evaporates, it's the high-speed molecules that escape from the surface. The ones that are left have less energy on average; this gives us a molecular view of evaporative cooling.

Rates of chemical reactions are often strongly temperature-dependent, and the reason is contained in the Maxwell–Boltzmann distribution. When two reacting molecules collide, the reaction can occur only when the molecules are close enough for the electric-charge distributions of their electrons to interact strongly. This requires a minimum energy, called the *activation energy,* and thus a certain minimum molecular speed. Figure 18.23a shows that the number of molecules in the high-speed tail of the curve increases rapidly with temperature. Thus we expect the rate of any reaction that depends on an activation energy to increase rapidly with temperature.

Table 18.2 Fractions of Molecules in an Ideal Gas with Speeds Less than Various Multiples of v/v_{rms}

v/v_{rms}	Fraction
0.20	0.011
0.40	0.077
0.60	0.218
0.80	0.411
1.00	0.608
1.20	0.771
1.40	0.882
1.60	0.947
1.80	0.979
2.00	0.993

Application Activation Energy and Moth Activity

This hawkmoth of genus *Manduca* cannot fly if the temperature of its muscles is below 29°C. The reason is that the enzyme-catalyzed reactions that power aerobic metabolism and enable muscle action require a minimum molecular energy (activation energy). Just like the molecules in an ideal gas, at low temperatures very few of the molecules involved in these reactions have high energy. As the temperature increases, more molecules have the required minimum energy and the reactions take place at a greater rate. Above 29°C, enough power is generated to allow the hawkmoth to fly.

Test Your Understanding of Section 18.5 A quantity of gas containing N molecules has a speed distribution function $f(v)$. How many molecules have speeds between v_1 and $v_2 > v_1$? (i) $\int_0^{v_2} f(v)\, dv - \int_0^{v_1} f(v)\, dv$; (ii) $N[\int_0^{v_2} f(v)\, dv - \int_0^{v_1} f(v)\, dv]$; (iii) $\int_0^{v_1} f(v)\, dv - \int_0^{v_2} f(v)\, dv$; (iv) $N[\int_0^{v_1} f(v)\, dv - \int_0^{v_2} f(v)\, dv]$; (v) none of these.

18.6 Phases of Matter

An ideal gas is the simplest system to analyze from a molecular viewpoint because we ignore the interactions between molecules. But those interactions are the very thing that makes matter condense into the liquid and solid phases under some conditions. So it's not surprising that theoretical analysis of liquid and solid structure and behavior is a lot more complicated than that for gases. We won't try to go far here with a microscopic picture, but we can talk in general about phases of matter, phase equilibrium, and phase transitions.

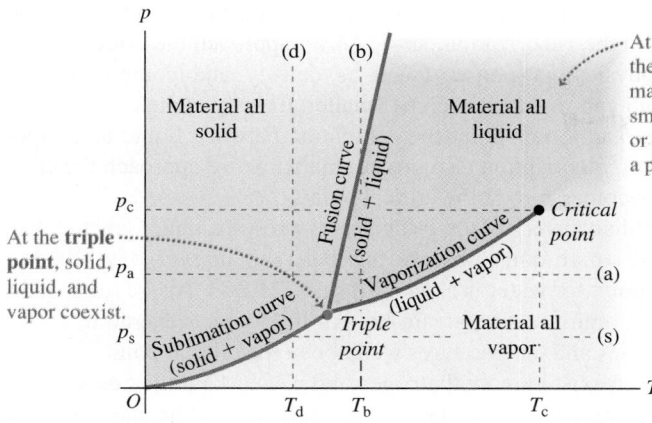

18.24 A typical pT phase diagram, showing regions of temperature and pressure at which the various phases exist and where phase changes occur.

In Section 17.6 we learned that each phase is stable only in certain ranges of temperature and pressure. A transition from one phase to another ordinarily requires **phase equilibrium** between the two phases, and for a given pressure this occurs at only one specific temperature. We can represent these conditions on a graph with axes p and T, called a **phase diagram;** Fig. 18.24 shows an example. Each point on the diagram represents a pair of values of p and T.

Only a single phase can exist at each point in Fig. 18.24, except for points on the solid lines, where two phases can coexist in phase equilibrium. The fusion curve separates the solid and liquid areas and represents possible conditions of solid-liquid phase equilibrium. The vaporization curve separates the liquid and vapor areas, and the sublimation curve separates the solid and vapor areas. All three curves meet at the **triple point,** the only condition under which all three phases can coexist (Fig. 18.25). In Section 17.3 we used the triple-point temperature of water to define the Kelvin temperature scale. Table 18.3 gives triple-point data for several substances.

If we add heat to a substance at a constant pressure p_a, it goes through a series of states represented by the horizontal line (a) in Fig. 18.24. The melting and boiling temperatures at this pressure are the temperatures at which the line intersects the fusion and vaporization curves, respectively. When the pressure is p_s, constant-pressure heating transforms a substance from solid directly to vapor. This process is called *sublimation;* the intersection of line (s) with the sublimation curve gives the temperature T_s at which it occurs for a pressure p_s. At any pressure less than the triple-point pressure, no liquid phase is possible. The triple-point pressure for carbon dioxide is 5.1 atm. At normal atmospheric pressure, solid carbon dioxide ("dry ice") undergoes sublimation; there is no liquid phase at this pressure.

Line (b) in Fig. 18.24 represents compression at a constant temperature T_b. The material passes from vapor to liquid and then to solid at the points where line (b) crosses the vaporization curve and fusion curve, respectively. Line (d) shows constant-temperature compression at a lower temperature T_d; the material passes from vapor to solid at the point where line (d) crosses the sublimation curve.

We saw in the pV-diagram of Fig. 18.7 that a liquid-vapor phase transition occurs only when the temperature and pressure are less than those at the point lying at the top of the green shaded area labeled "Liquid-vapor phase equilibrium region." This point corresponds to the endpoint at the top of the vaporization curve in Fig. 18.24. It is called the **critical point,** and the corresponding values of p and T are called the critical pressure and temperature, p_c and T_c. A gas at a pressure *above* the critical pressure does not separate into two phases when it is cooled at constant pressure (along a horizontal line above the critical point in Fig. 18.24). Instead, its properties change gradually and continuously from those we ordinarily associate with a gas (low density, large compressibility) to those of a liquid (high density, small compressibility) *without a phase transition.*

18.25 Atmospheric pressure on earth is higher than the triple-point pressure of water (see line (a) in Fig. 18.24). Depending on the temperature, water can exist as a vapor (in the atmosphere), as a liquid (in the ocean), or as a solid (like the iceberg shown here).

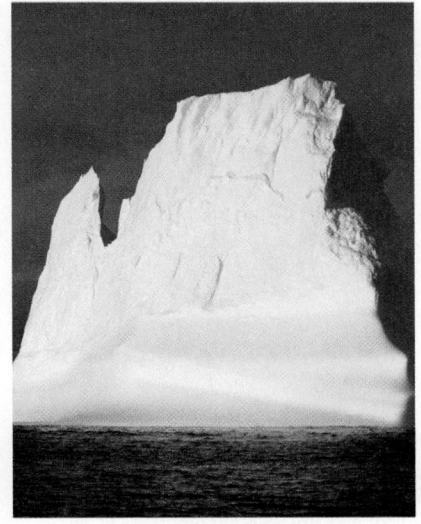

Table 18.3 Triple-Point Data

Substance	Temperature (K)	Pressure (Pa)
Hydrogen	13.80	0.0704×10^5
Deuterium	18.63	0.171×10^5
Neon	24.56	0.432×10^5
Nitrogen	63.18	0.125×10^5
Oxygen	54.36	0.00152×10^5
Ammonia	195.40	0.0607×10^5
Carbon dioxide	216.55	5.17×10^5
Sulfur dioxide	197.68	0.00167×10^5
Water	273.16	0.00610×10^5

You can understand this by thinking about liquid-phase transitions at successively higher points on the vaporization curve. As we approach the critical point, the *differences* in physical properties (such as density and compressibility) between the liquid and vapor phases become smaller. Exactly *at* the critical point they all become zero, and at this point the distinction between liquid and vapor disappears. The heat of vaporization also grows smaller as we approach the critical point, and it too becomes zero at the critical point.

For nearly all familiar materials the critical pressures are much greater than atmospheric pressure, so we don't observe this behavior in everyday life. For example, the critical point for water is at 647.4 K and 221.2×10^5 Pa (about 218 atm or 3210 psi). But high-pressure steam boilers in electric generating plants regularly run at pressures and temperatures well above the critical point.

Many substances can exist in more than one solid phase. A familiar example is carbon, which exists as noncrystalline soot and crystalline graphite and diamond. Water is another example; at least eight types of ice, differing in crystal structure and physical properties, have been observed at very high pressures.

pVT-Surfaces

We remarked in Section 18.1 that the equation of state of any material can be represented graphically as a surface in a three-dimensional space with coordinates p, V, and T. Visualizing such a surface can add to our understanding of the behavior of materials at various temperatures and pressures. Figure 18.26 shows a typical *pVT*-surface. The light lines represent *pV*-isotherms; projecting them onto the *pV*-plane gives a diagram similar to Fig. 18.7. The *pV*-isotherms represent contour lines on the *pVT*-surface, just as contour lines on a topographic map represent the elevation (the third dimension) at each point. The projections of the edges of the surface onto the *pT*-plane give the *pT* phase diagram of Fig. 18.24.

Line *abcdef* in Fig. 18.26 represents constant-pressure heating, with melting along *bc* and vaporization along *de*. Note the volume changes that occur as T increases along this line. Line *ghjklm* corresponds to an isothermal (constant temperature) compression, with liquefaction along *hj* and solidification along *kl*. Between these, segments *gh* and *jk* represent isothermal compression with increase in pressure; the pressure increases are much greater in the liquid region

18.26 A *pVT*-surface for a substance that expands on melting. Projections of the boundaries on the surface onto the *pT*- and *pV*-planes are also shown.

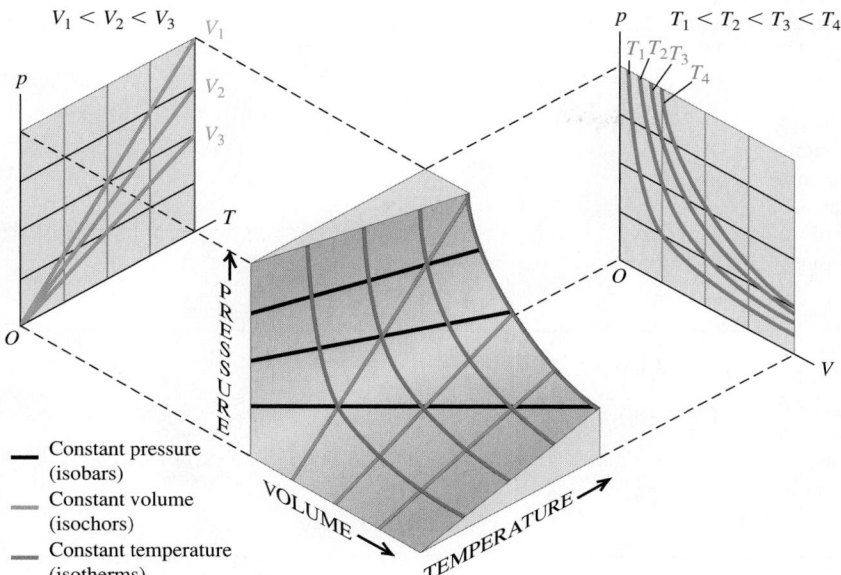

- Constant pressure (isobars)
- Constant volume (isochors)
- Constant temperature (isotherms)

18.27 A pVT-surface for an ideal gas. At the left, each red line corresponds to a certain constant volume; at the right, each green line corresponds to a certain constant temperature.

jk and the solid region lm than in the vapor region gh. Finally, line $nopq$ represents isothermal solidification directly from vapor, as in the formation of snowflakes or frost.

Figure 18.27 shows the much simpler pVT-surface for a substance that obeys the ideal-gas equation of state under all conditions. The projections of the constant-temperature curves onto the pV-plane correspond to the curves of Fig. 18.6, and the projections of the constant-volume curves onto the pT-plane show that pressure is directly proportional to absolute temperature.

Test Your Understanding of Section 18.6 The average atmospheric pressure on Mars is 6.0×10^2 Pa. Could there be lakes or rivers of liquid water on Mars today? What about in the past, when the atmospheric pressure is thought to have been substantially greater than today?

Equations of state: The pressure p, volume V, and absolute temperature T of a given quantity of a substance are related by an equation of state. This relationship applies only for equilibrium states, in which p and T are uniform throughout the system. The ideal-gas equation of state, Eq. (18.3), involves the number of moles n and a constant R that is the same for all gases. (See Examples 18.1–18.4.)

$$pV = nRT \qquad (18.3)$$

Molecular properties of matter: The molar mass M of a pure substance is the mass per mole. The mass m_{total} of a quantity of substance equals M multiplied by the number of moles n. Avogadro's number N_A is the number of molecules in a mole. The mass m of an individual molecule is M divided by N_A. (See Example 18.5.)

$$m_{total} = nM \qquad (18.2)$$

$$M = N_A m \qquad (18.8)$$

Kinetic-molecular model of an ideal gas: In an ideal gas, the total translational kinetic energy of the gas as a whole (K_{tr}) and the average translational kinetic energy per molecule $[\frac{1}{2}m(v^2)_{av}]$ are proportional to the absolute temperature T, and the root-mean-square speed of molecules is proportional to the square root of T. These expressions involve the Boltzmann constant $k = R/N_A$. (See Examples 18.6 and 18.7.) The mean free path λ of molecules in an ideal gas depends on the number of molecules per volume (N/V) and the molecular radius r. (See Example 18.8.)

$$K_{tr} = \tfrac{3}{2}nRT \qquad (18.14)$$

$$\tfrac{1}{2}m(v^2)_{av} = \tfrac{3}{2}kT \qquad (18.16)$$

$$v_{rms} = \sqrt{(v^2)_{av}} = \sqrt{\frac{3kT}{m}} \qquad (18.19)$$

$$\lambda = v t_{mean} = \frac{V}{4\pi\sqrt{2}\, r^2 N} \qquad (18.21)$$

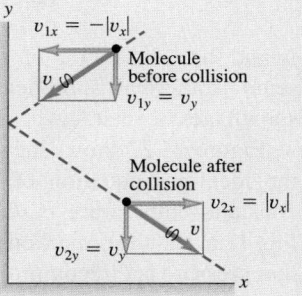

Heat capacities: The molar heat capacity at constant volume C_V is a simple multiple of the gas constant R for certain idealized cases: an ideal monatomic gas [Eq. (18.25)]; an ideal diatomic gas including rotational energy [Eq. (18.26)]; and an ideal monatomic solid [Eq. (18.28)]. Many real systems are approximated well by these idealizations.

$$C_V = \tfrac{3}{2}R \quad \text{(monatomic gas)} \qquad (18.25)$$

$$C_V = \tfrac{5}{2}R \quad \text{(diatomic gas)} \qquad (18.26)$$

$$C_V = 3R \quad \text{(monatomic solid)} \qquad (18.28)$$

Molecular speeds: The speeds of molecules in an ideal gas are distributed according to the Maxwell–Boltzmann distribution $f(v)$. The quantity $f(v)\, dv$ describes what fraction of the molecules have speeds between v and $v + dv$.

$$f(v) = 4\pi\left(\frac{m}{2\pi kT}\right)^{3/2} v^2 e^{-mv^2/2kT}$$
$$(18.32)$$

Phases of matter: Ordinary matter exists in the solid, liquid, and gas phases. A phase diagram shows conditions under which two phases can coexist in phase equilibrium. All three phases can coexist at the triple point. The vaporization curve ends at the critical point, above which the distinction between the liquid and gas phases disappears.

BRIDGING PROBLEM Gas on Jupiter's Moon Europa

An astronaut visiting Jupiter's satellite Europa leaves a canister of 1.20 mol of nitrogen gas (28.0 g/mol) at 25.0°C on the satellite's surface. Europa has no significant atmosphere, and the acceleration due to gravity at its surface is 1.30 m/s². The canister springs a leak, allowing molecules to escape from a small hole. (a) What is the maximum height (in km) above Europa's surface that is reached by a nitrogen molecule whose speed equals the rms speed? Assume that the molecule is shot straight up out of the hole in the canister, and ignore the variation in *g* with altitude. (b) The escape speed from Europa is 2025 m/s. Can any of the nitrogen molecules escape from Europa and into space?

SOLUTION GUIDE

See MasteringPhysics® study area for a Video Tutor solution.

IDENTIFY and SET UP

1. Draw a sketch of the situation, showing all relevant dimensions.
2. Make a list of the unknown quantities, and decide which are the target variables.

3. How will you find the rms speed of the nitrogen molecules? What principle will you use to find the maximum height that a molecule with this speed can reach?
4. Does the rms speed of molecules in an ideal gas represent the maximum speed of the molecules? If not, what is the maximum speed?

EXECUTE

5. Solve for the rms speed. Use this to calculate the maximum height that a molecule with this speed can reach.
6. Use your result from step 5 to answer the question in part (b).

EVALUATE

7. Do your results depend on the amount of gas in the container? Why or why not?
8. How would your results from steps 5 and 6 be affected if the gas cylinder were instead left on Jupiter's satellite Ganymede, which has higher surface gravity than Europa and a higher escape speed? Like Europa, Ganymede has no significant atmosphere.

Problems

For instructor-assigned homework, go to www.masteringphysics.com

•, ••, •••: Problems of increasing difficulty. **CP**: Cumulative problems incorporating material from earlier chapters. **CALC**: Problems requiring calculus. **BIO**: Biosciences problems.

DISCUSSION QUESTIONS

Q18.1 Section 18.1 states that ordinarily, pressure, volume, and temperature cannot change individually without one affecting the others. Yet when a liquid evaporates, its volume changes, even though its pressure and temperature are constant. Is this inconsistent? Why or why not?

Q18.2 In the ideal-gas equation, could an equivalent Celsius temperature be used instead of the Kelvin one if an appropriate numerical value of the constant *R* is used? Why or why not?

Q18.3 On a chilly morning you can "see your breath." Can you really? What are you actually seeing? Does this phenomenon depend on the temperature of the air, the humidity, or both? Explain.

Q18.4 When a car is driven some distance, the air pressure in the tires increases. Why? Should you let out some air to reduce the pressure? Why or why not?

Q18.5 The coolant in an automobile radiator is kept at a pressure higher than atmospheric pressure. Why is this desirable? The radiator cap will release coolant when the gauge pressure of the coolant reaches a certain value, typically 15 lb/in.² or so. Why not just seal the system completely?

Q18.6 Unwrapped food placed in a freezer experiences dehydration, known as "freezer burn." Why?

Q18.7 "Freeze-drying" food involves the same process as "freezer burn," referred to in Discussion Question Q18.6. For freeze-drying, the food is usually frozen first, and then placed in a vacuum chamber and irradiated with infrared radiation. What is the purpose of the vacuum? The radiation? What advantages might freeze-drying have in comparison to ordinary drying?

Q18.8 A group of students drove from their university (near sea level) up into the mountains for a skiing weekend. Upon arriving at the slopes, they discovered that the bags of potato chips they had brought for snacks had all burst open. What caused this to happen?

Q18.9 How does evaporation of perspiration from your skin cool your body?

Q18.10 A rigid, perfectly insulated container has a membrane dividing its volume in half. One side contains a gas at an absolute temperature T_0 and pressure p_0, while the other half is completely empty. Suddenly a small hole develops in the membrane, allowing the gas to leak out into the other half until it eventually occupies twice its original volume. In terms of T_0 and p_0, what will be the new temperature and pressure of the gas when it is distributed equally in both halves of the container? Explain your reasoning.

Q18.11 (a) Which has more atoms: a kilogram of hydrogen or a kilogram of lead? Which has more mass? (b) Which has more atoms: a mole of hydrogen or a mole of lead? Which has more mass? Explain your reasoning.

Q18.12 Use the concepts of the kinetic-molecular model to explain: (a) why the pressure of a gas in a rigid container increases as heat is added to the gas and (b) why the pressure of a gas increases as we compress it, even if we do not change its temperature.

Q18.13 The proportions of various gases in the earth's atmosphere change somewhat with altitude. Would you expect the proportion of oxygen at high altitude to be greater or less than at sea level compared to the proportion of nitrogen? Why?

Q18.14 Comment on the following statement: *When two gases are mixed, if they are to be in thermal equilibrium, they must have the*

same average molecular speed. Is the statement correct? Why or why not?

Q18.15 The kinetic-molecular model contains a hidden assumption about the temperature of the container walls. What is this assumption? What would happen if this assumption were not valid?

Q18.16 The temperature of an ideal gas is directly proportional to the average kinetic energy of its molecules. If a container of ideal gas is moving past you at 2000 m/s, is the temperature of the gas higher than if the container was at rest? Explain your reasoning.

Q18.17 If the pressure of an ideal monatomic gas is increased while the number of moles is kept constant, what happens to the average translational kinetic energy of one atom of the gas? Is it possible to change *both* the volume and the pressure of an ideal gas and keep the average translational kinetic energy of the atoms constant? Explain.

Q18.18 In deriving the ideal-gas equation from the kinetic-molecular model, we ignored potential energy due to the earth's gravity. Is this omission justified? Why or why not?

Q18.19 The derivation of the ideal-gas equation included the assumption that the number of molecules is very large, so that we could compute the average force due to many collisions. However, the ideal-gas equation holds accurately only at low pressures, where the molecules are few and far between. Is this inconsistent? Why or why not?

Q18.20 A gas storage tank has a small leak. The pressure in the tank drops more quickly if the gas is hydrogen or helium than if it is oxygen. Why?

Q18.21 Consider two specimens of ideal gas at the same temperature. Specimen A has the same total mass as specimen B, but the molecules in specimen A have greater molar mass than they do in specimen B. In which specimen is the total kinetic energy of the gas greater? Does your answer depend on the molecular structure of the gases? Why or why not?

Q18.22 The temperature of an ideal monatomic gas is increased from 25°C to 50°C. Does the average translational kinetic energy of each gas atom double? Explain. If your answer is no, what would the final temperature be if the average translational kinetic energy was doubled?

Q18.23 If the root-mean-square speed of the atoms of an ideal gas is to be doubled, by what factor must the Kelvin temperature of the gas be increased? Explain.

Q18.24 (a) If you apply the same amount of heat to 1.00 mol of an ideal monatomic gas and 1.00 mol of an ideal diatomic gas, which one (if any) will increase more in temperature? (b) Physically, *why* do diatomic gases have a greater molar heat capacity than monatomic gases?

Q18.25 The discussion in Section 18.4 concluded that all ideal diatomic gases have the same heat capacity C_V. Does this mean that it takes the same amount of heat to raise the temperature of 1.0 g of each one by 1.0 K? Explain your reasoning.

Q18.26 In a gas that contains *N* molecules, is it accurate to say that the number of molecules with speed v is equal to $f(v)$? Is it accurate to say that this number is given by $Nf(v)$? Explain your answers.

Q18.27 Imagine a special air filter placed in a window of a house. The tiny holes in the filter allow only air molecules moving faster than a certain speed to exit the house, and allow only air molecules moving slower than that speed to enter the house from outside. What effect would this filter have on the temperature inside the house? (It turns out that the second law of thermodynamics—which we will discuss in Chapter 20—tells us that such a wonderful air filter would be impossible to make.)

Q18.28 A beaker of water at room temperature is placed in an enclosure, and the air pressure in the enclosure is slowly reduced. When the air pressure is reduced sufficiently, the water begins to boil. The temperature of the water does not rise when it boils; in fact, the temperature *drops* slightly. Explain these phenomena.

Q18.29 Ice is slippery to walk on, and especially slippery if you wear ice skates. What does this tell you about how the melting temperature of ice depends on pressure? Explain.

Q18.30 Hydrothermal vents are openings in the ocean floor that discharge very hot water. The water emerging from one such vent off the Oregon coast, 2400 m below the surface, has a temperature of 279°C. Despite its high temperature, the water doesn't boil. Why not?

Q18.31 The dark areas on the moon's surface are called *maria,* Latin for "seas," and were once thought to be bodies of water. In fact, the maria are not "seas" at all, but plains of solidified lava. Given that there is no atmosphere on the moon, how can you explain the absence of liquid water on the moon's surface?

Q18.32 In addition to the normal cooking directions printed on the back of a box of rice, there are also "high-altitude directions." The only difference is that the "high-altitude directions" suggest increasing the cooking time and using a greater volume of boiling water in which to cook the rice. Why should the directions depend on the altitude in this way?

EXERCISES

Section 18.1 Equations of State

18.1 • A 20.0-L tank contains 4.86×10^{-4} kg of helium at 18.0°C. The molar mass of helium is 4.00 g/mol. (a) How many moles of helium are in the tank? (b) What is the pressure in the tank, in pascals and in atmospheres?

18.2 •• Helium gas with a volume of 2.60 L, under a pressure of 0.180 atm and at a temperature of 41.0°C, is warmed until both pressure and volume are doubled. (a) What is the final temperature? (b) How many grams of helium are there? The molar mass of helium is 4.00 g/mol.

18.3 • A cylindrical tank has a tight-fitting piston that allows the volume of the tank to be changed. The tank originally contains 0.110 m³ of air at a pressure of 0.355 atm. The piston is slowly pulled out until the volume of the gas is increased to 0.390 m³. If the temperature remains constant, what is the final value of the pressure?

18.4 • A 3.00-L tank contains air at 3.00 atm and 20.0°C. The tank is sealed and cooled until the pressure is 1.00 atm. (a) What is the temperature then in degrees Celsius? Assume that the volume of the tank is constant. (b) If the temperature is kept at the value found in part (a) and the gas is compressed, what is the volume when the pressure again becomes 3.00 atm?

18.5 • **Planetary Atmospheres.** (a) Calculate the density of the atmosphere at the surface of Mars (where the pressure is 650 Pa and the temperature is typically 253 K, with a CO_2 atmosphere), Venus (with an average temperature of 730 K and pressure of 92 atm, with a CO_2 atmosphere), and Saturn's moon Titan (where the pressure is 1.5 atm and the temperature is −178°C, with a N_2 atmosphere). (b) Compare each of these densities with that of the earth's atmosphere, which is 1.20 kg/m³. Consult the periodic chart in Appendix D to determine molar masses.

18.6 •• You have several identical balloons. You experimentally determine that a balloon will break if its volume exceeds 0.900 L. The pressure of the gas inside the balloon equals air pressure (1.00 atm). (a) If the air inside the balloon is at a constant temperature of

22.0°C and behaves as an ideal gas, what mass of air can you blow into one of the balloons before it bursts? (b) Repeat part (a) if the gas is helium rather than air.

18.7 •• A Jaguar XK8 convertible has an eight-cylinder engine. At the beginning of its compression stroke, one of the cylinders contains 499 cm^3 of air at atmospheric pressure (1.01×10^5 Pa) and a temperature of 27.0°C. At the end of the stroke, the air has been compressed to a volume of 46.2 cm^3 and the gauge pressure has increased to 2.72×10^6 Pa. Compute the final temperature.

18.8 •• A welder using a tank of volume 0.0750 m^3 fills it with oxygen (molar mass 32.0 g/mol) at a gauge pressure of 3.00×10^5 Pa and temperature of 37.0°C. The tank has a small leak, and in time some of the oxygen leaks out. On a day when the temperature is 22.0°C, the gauge pressure of the oxygen in the tank is 1.80×10^5 Pa. Find (a) the initial mass of oxygen and (b) the mass of oxygen that has leaked out.

18.9 •• A large cylindrical tank contains 0.750 m^3 of nitrogen gas at 27°C and 7.50×10^3 Pa (absolute pressure). The tank has a tight-fitting piston that allows the volume to be changed. What will be the pressure if the volume is decreased to 0.480 m^3 and the temperature is increased to 157°C?

18.10 • An empty cylindrical canister 1.50 m long and 90.0 cm in diameter is to be filled with pure oxygen at 22.0°C to store in a space station. To hold as much gas as possible, the absolute pressure of the oxygen will be 21.0 atm. The molar mass of oxygen is 32.0 g/mol. (a) How many moles of oxygen does this canister hold? (b) For someone lifting this canister, by how many kilograms does this gas increase the mass to be lifted?

18.11 • The gas inside a balloon will always have a pressure nearly equal to atmospheric pressure, since that is the pressure applied to the outside of the balloon. You fill a balloon with helium (a nearly ideal gas) to a volume of 0.600 L at a temperature of 19.0°C. What is the volume of the balloon if you cool it to the boiling point of liquid nitrogen (77.3 K)?

18.12 • **Deviations from the Ideal-Gas Equation.** For carbon dioxide gas (CO_2), the constants in the van der Waals equation are $a = 0.364$ J $\cdot$ m^3/mol^2 and $b = 4.27 \times 10^{-5}$ m^3/mol. (a) If 1.00 mol of CO_2 gas at 350 K is confined to a volume of 400 cm^3, find the pressure of the gas using the ideal-gas equation and the van der Waals equation. (b) Which equation gives a lower pressure? Why? What is the percentage difference of the van der Waals equation result from the ideal-gas equation result? (c) The gas is kept at the same temperature as it expands to a volume of 4000 cm^3. Repeat the calculations of parts (a) and (b). (d) Explain how your calculations show that the van der Waals equation is equivalent to the ideal-gas equation if n/V is small.

18.13 •• If a certain amount of ideal gas occupies a volume V at STP on earth, what would be its volume (in terms of V) on Venus, where the temperature is 1003°C and the pressure is 92 atm?

18.14 • A diver observes a bubble of air rising from the bottom of a lake (where the absolute pressure is 3.50 atm) to the surface (where the pressure is 1.00 atm). The temperature at the bottom is 4.0°C, and the temperature at the surface is 23.0°C. (a) What is the ratio of the volume of the bubble as it reaches the surface to its volume at the bottom? (b) Would it be safe for the diver to hold his breath while ascending from the bottom of the lake to the surface? Why or why not?

18.15 • A metal tank with volume 3.10 L will burst if the absolute pressure of the gas it contains exceeds 100 atm. (a) If 11.0 mol of an ideal gas is put into the tank at a temperature of 23.0°C, to what temperature can the gas be warmed before the tank ruptures? You can ignore the thermal expansion of the tank. (b) Based on your answer to part (a), is it reasonable to ignore the thermal expansion of the tank? Explain.

18.16 • Three moles of an ideal gas are in a rigid cubical box with sides of length 0.200 m. (a) What is the force that the gas exerts on each of the six sides of the box when the gas temperature is 20.0°C? (b) What is the force when the temperature of the gas is increased to 100.0°C?

18.17 • With the assumptions of Example 18.4 (Section 18.1), at what altitude above sea level is air pressure 90% of the pressure at sea level?

18.18 • Make the same assumptions as in Example 18.4 (Section 18.1). How does the percentage decrease in air pressure in going from sea level to an altitude of 100 m compare to that when going from sea level to an altitude of 1000 m? If your second answer is not 10 times your first answer, explain why.

18.19 •• (a) Calculate the mass of nitrogen present in a volume of 3000 cm^3 if the temperature of the gas is 22.0°C and the absolute pressure of 2.00×10^{-13} atm is a partial vacuum easily obtained in laboratories. (b) What is the density (in kg/m^3) of the N_2?

18.20 •• With the assumption that the air temperature is a uniform 0.0°C (as in Example 18.4), what is the density of the air at an altitude of 1.00 km as a percentage of the density at the surface?

18.21 • At an altitude of 11,000 m (a typical cruising altitude for a jet airliner), the air temperature is −56.5°C and the air density is 0.364 kg/m^3. What is the pressure of the atmosphere at that altitude? (*Note:* The temperature at this altitude is not the same as at the surface of the earth, so the calculation of Example 18.4 in Section 18.1 doesn't apply.)

Section 18.2 Molecular Properties of Matter

18.22 • A large organic molecule has a mass of 1.41×10^{-21} kg. What is the molar mass of this compound?

18.23 •• Suppose you inherit 3.00 mol of gold from your uncle (an eccentric chemist) at a time when this metal is selling for $14.75 per gram. Consult the periodic table in Appendix D and Table 12.1. (a) To the nearest dollar, what is this gold worth? (b) If you have your gold formed into a spherical nugget, what is its diameter?

18.24 •• Modern vacuum pumps make it easy to attain pressures of the order of 10^{-13} atm in the laboratory. Consider a volume of air and treat the air as an ideal gas. (a) At a pressure of 9.00×10^{-14} atm and an ordinary temperature of 300.0 K, how many molecules are present in a volume of 1.00 cm^3? (b) How many molecules would be present at the same temperature but at 1.00 atm instead?

18.25 •• The Lagoon Nebula (Fig. E18.25) is a cloud of hydrogen gas located 3900 light-years from the earth. The cloud is about 45 light-years in diameter and glows because of its high temperature of 7500 K. (The gas is raised to this temperature by the stars that

Figure **E18.25**

lie within the nebula.) The cloud is also very thin; there are only 80 molecules per cubic centimeter. (a) Find the gas pressure (in atmospheres) in the Lagoon Nebula. Compare it to the laboratory pressure referred to in Exercise 18.24. (b) Science-fiction films sometimes show starships being buffeted by turbulence as they fly through gas clouds such as the Lagoon Nebula. Does this seem realistic? Why or why not?

18.26 •• In a gas at standard conditions, what is the length of the side of a cube that contains a number of molecules equal to the population of the earth (about 6×10^9 people)?

18.27 • How many moles are in a 1.00-kg bottle of water? How many molecules? The molar mass of water is 18.0 g/mol.

18.28 •• **How Close Together Are Gas Molecules?** Consider an ideal gas at 27°C and 1.00 atm pressure. To get some idea how close these molecules are to each other, on the average, imagine them to be uniformly spaced, with each molecule at the center of a small cube. (a) What is the length of an edge of each cube if adjacent cubes touch but do not overlap? (b) How does this distance compare with the diameter of a typical molecule? (c) How does their separation compare with the spacing of atoms in solids, which typically are about 0.3 nm apart?

18.29 •• Consider 5.00 mol of liquid water. (a) What volume is occupied by this amount of water? The molar mass of water is 18.0 g/mol. (b) Imagine the molecules to be, on average, uniformly spaced, with each molecule at the center of a small cube. What is the length of an edge of each small cube if adjacent cubes touch but don't overlap? (c) How does this distance compare with the diameter of a molecule?

Section 18.3 Kinetic-Molecular Model of an Ideal Gas

18.30 • A flask contains a mixture of neon (Ne), krypton (Kr), and radon (Rn) gases. Compare (a) the average kinetic energies of the three types of atoms and (b) the root-mean-square speeds. (*Hint:* The periodic table in Appendix D shows the molar mass (in g/mol) of each element under the chemical symbol for that element.)

18.31 • **Gaseous Diffusion of Uranium.** (a) A process called *gaseous diffusion* is often used to separate isotopes of uranium—that is, atoms of the elements that have different masses, such as ^{235}U and ^{238}U. The only gaseous compound of uranium at ordinary temperatures is uranium hexafluoride, UF_6. Speculate on how $^{235}UF_6$ and $^{238}UF_6$ molecules might be separated by diffusion. (b) The molar masses for $^{235}UF_6$ and $^{238}UF_6$ molecules are 0.349 kg/mol and 0.352 kg/mol, respectively. If uranium hexafluoride acts as an ideal gas, what is the ratio of the root-mean-square speed of $^{235}UF_6$ molecules to that of $^{238}UF_6$ molecules if the temperature is uniform?

18.32 • The ideas of average and root-mean-square value can be applied to any distribution. A class of 150 students had the following scores on a 100-point quiz:

Score	Number of Students
10	11
20	12
30	24
40	15
50	19
60	10
70	12
80	20
90	17
100	10

(a) Find the average score for the class. (b) Find the root-mean-square score for the class.

18.33 • We have two equal-size boxes, A and B. Each box contains gas that behaves as an ideal gas. We insert a thermometer into each box and find that the gas in box A is at a temperature of 50°C while the gas in box B is at 10°C. This is all we know about the gas in the boxes. Which of the following statements *must* be true? Which *could* be true? (a) The pressure in A is higher than in B. (b) There are more molecules in A than in B. (c) A and B do not contain the same type of gas. (d) The molecules in A have more average kinetic energy per molecule than those in B. (e) The molecules in A are moving faster than those in B. Explain the reasoning behind your answers.

18.34 • A container with volume 1.48 L is initially evacuated. Then it is filled with 0.226 g of N_2. Assume that the pressure of the gas is low enough for the gas to obey the ideal-gas law to a high degree of accuracy. If the root-mean-square speed of the gas molecules is 182 m/s, what is the pressure of the gas?

18.35 •• (a) A deuteron, ^2_1H, is the nucleus of a hydrogen isotope and consists of one proton and one neutron. The plasma of deuterons in a nuclear fusion reactor must be heated to about 300 million K. What is the rms speed of the deuterons? Is this a significant fraction of the speed of light ($c = 3.0 \times 10^8$ m/s)? (b) What would the temperature of the plasma be if the deuterons had an rms speed equal to $0.10c$?

18.36 • **Martian Climate.** The atmosphere of Mars is mostly CO_2 (molar mass 44.0 g/mol) under a pressure of 650 Pa, which we shall assume remains constant. In many places the temperature varies from 0.0°C in summer to −100°C in winter. Over the course of a Martian year, what are the ranges of (a) the rms speeds of the CO_2 molecules and (b) the density (in mol/m³) of the atmosphere?

18.37 •• (a) Oxygen (O_2) has a molar mass of 32.0 g/mol. What is the average translational kinetic energy of an oxygen molecule at a temperature of 300 K? (b) What is the average value of the square of its speed? (c) What is the root-mean-square speed? (d) What is the momentum of an oxygen molecule traveling at this speed? (e) Suppose an oxygen molecule traveling at this speed bounces back and forth between opposite sides of a cubical vessel 0.10 m on a side. What is the average force the molecule exerts on one of the walls of the container? (Assume that the molecule's velocity is perpendicular to the two sides that it strikes.) (f) What is the average force per unit area? (g) How many oxygen molecules traveling at this speed are necessary to produce an average pressure of 1 atm? (h) Compute the number of oxygen molecules that are actually contained in a vessel of this size at 300 K and atmospheric pressure. (i) Your answer for part (h) should be three times as large as the answer for part (g). Where does this discrepancy arise?

18.38 •• Calculate the mean free path of air molecules at a pressure of 3.50×10^{-13} atm and a temperature of 300 K. (This pressure is readily attainable in the laboratory; see Exercise 18.24.) As in Example 18.8, model the air molecules as spheres of radius 2.0×10^{-10} m.

18.39 •• At what temperature is the root-mean-square speed of nitrogen molecules equal to the root-mean-square speed of hydrogen molecules at 20.0°C? (*Hint:* The periodic table in Appendix D shows the molar mass (in g/mol) of each element under the chemical symbol for that element. The molar mass of H_2 is twice the molar mass of hydrogen atoms, and similarly for N_2.)

18.40 • Smoke particles in the air typically have masses of the order of 10^{-16} kg. The Brownian motion (rapid, irregular movement) of these particles, resulting from collisions with air molecules, can be observed with a microscope. (a) Find the root-mean-square speed of Brownian motion for a particle with a mass of 3.00×10^{-16} kg in air at 300 K. (b) Would the root-mean-square speed be different if the particle were in hydrogen gas at the same temperature? Explain.

Section 18.4 Heat Capacities

18.41 • (a) How much heat does it take to increase the temperature of 2.50 mol of a diatomic ideal gas by 50.0 K near room temperature if the gas is held at constant volume? (b) What is the answer to the question in part (a) if the gas is monatomic rather than diatomic?

18.42 •• Perfectly rigid containers each hold n moles of ideal gas, one being hydrogen (H_2) and other being neon (Ne). If it takes 300 J of heat to increase the temperature of the hydrogen by 2.50°C, by how many degrees will the same amount of heat raise the temperature of the neon?

18.43 •• (a) Compute the specific heat at constant volume of nitrogen (N_2) gas, and compare it with the specific heat of liquid water. The molar mass of N_2 is 28.0 g/mol. (b) You warm 1.00 kg of water at a constant volume of 1.00 L from 20.0°C to 30.0°C in a kettle. For the same amount of heat, how many kilograms of 20.0°C air would you be able to warm to 30.0°C? What volume (in liters) would this air occupy at 20.0°C and a pressure of 1.00 atm? Make the simplifying assumption that air is 100% N_2.

18.44 •• (a) Calculate the specific heat at constant volume of water vapor, assuming the nonlinear triatomic molecule has three translational and three rotational degrees of freedom and that vibrational motion does not contribute. The molar mass of water is 18.0 g/mol. (b) The actual specific heat of water vapor at low pressures is about 2000 J/kg·K. Compare this with your calculation and comment on the actual role of vibrational motion.

18.45 •• (a) Use Eq. 18.28 to calculate the specific heat at constant volume of aluminum in units of J/kg·K. Consult the periodic table in Appendix D. (b) Compare the answer in part (a) with the value given in Table 17.3. Try to explain any disagreement between these two values.

Section 18.5 Molecular Speeds

18.46 • For a gas of nitrogen molecules (N_2), what must the temperature be if 94.7% of all the molecules have speeds less than (a) 1500 m/s; (b) 1000 m/s; (c) 500 m/s? Use Table 18.2. The molar mass of N_2 is 28.0 g/mol.

18.47 • For diatomic carbon dioxide gas (CO_2, molar mass 44.0 g/mol) at $T = 300$ K, calculate (a) the most probable speed v_{mp}; (b) the average speed v_{av}; (c) the root-mean-square speed v_{rms}.

18.48 • **CALC** Prove that $f(v)$ as given by Eq. (18.33) is maximum for $\epsilon = kT$. Use this result to obtain Eq. (18.34).

Section 18.6 Phases of Matter

18.49 • Solid water (ice) is slowly warmed from a very low temperature. (a) What minimum external pressure p_1 must be applied to the solid if a melting phase transition is to be observed? Describe the sequence of phase transitions that occur if the applied pressure p is such that $p < p_1$. (b) Above a certain maximum pressure p_2, no boiling transition is observed. What is this pressure? Describe the sequence of phase transitions that occur if $p_1 < p < p_2$.

18.50 • Puffy cumulus clouds, which are made of water droplets, occur at lower altitudes in the atmosphere. Wispy cirrus clouds, which are made of ice crystals, occur only at higher altitudes. Find the altitude y (measured from sea level) above which only cirrus clouds can occur. On a typical day and at altitudes less than 11 km, the temperature at an altitude y is given by $T = T_0 - \alpha y$, where $T_0 = 15.0$°C and $\alpha = 6.0$ C°/1000 m.

18.51 • The atmosphere of the planet Mars is 95.3% carbon dioxide (CO_2) and about 0.03% water vapor. The atmospheric pressure is only about 600 Pa, and the surface temperature varies from −30°C to −100°C. The polar ice caps contain both CO_2 ice and water ice. Could there be *liquid* CO_2 on the surface of Mars? Could there be liquid water? Why or why not?

18.52 • A physics lecture room has a volume of 216 m³. (a) For a pressure of 1.00 atm and a temperature of 27.0°C, use the ideal-gas law to estimate the number of air molecules in the room. Assume all the air is N_2. (b) Calculate the particle density—that is, the number of N_2 molecules per cubic centimeter. (c) Calculate the mass of the air in the room.

PROBLEMS

18.53 •• **CP BIO** **The Effect of Altitude on the Lungs.** (a) Calculate the *change* in air pressure you will experience if you climb a 1000-m mountain, assuming that the temperature and air density do not change over this distance and that they were 22°C and 1.2 kg/m³, respectively, at the bottom of the mountain. (Note that the result of Example 18.4 doesn't apply, since the expression derived in that example accounts for the variation of air density with altitude and we are told to ignore that in this problem.) (b) If you took a 0.50-L breath at the foot of the mountain and managed to hold it until you reached the top, what would be the volume of this breath when you exhaled it there?

18.54 •• **CP BIO** **The Bends.** If deep-sea divers rise to the surface too quickly, nitrogen bubbles in their blood can expand and prove fatal. This phenomenon is known as the *bends*. If a scuba diver rises quickly from a depth of 25 m in Lake Michigan (which is fresh water), what will be the volume at the surface of an N_2 bubble that occupied 1.0 mm³ in his blood at the lower depth? Does it seem that this difference is large enough to be a problem? (Assume that the pressure difference is due only to the changing water pressure, not to any temperature difference, an assumption that is reasonable, since we are warm-blooded creatures.)

18.55 ••• **CP** A hot-air balloon stays aloft because hot air at atmospheric pressure is less dense than cooler air at the same pressure. If the volume of the balloon is 500.0 m³ and the surrounding air is at 15.0°C, what must the temperature of the air in the balloon be for it to lift a total load of 290 kg (in addition to the mass of the hot air)? The density of air at 15.0°C and atmospheric pressure is 1.23 kg/m³.

18.56 •• (a) Use Eq. (18.1) to estimate the change in the volume of a solid steel sphere of volume 11 L when the temperature and pressure increase from 21°C and 1.013×10^5 Pa to 42°C and 2.10×10^7 Pa. (*Hint:* Consult Chapters 11 and 17 to determine the values of β and k.) (b) In Example 18.3 the change in volume of an 11-L steel scuba tank was ignored. Was this a good approximation? Explain.

18.57 ••• A cylinder 1.00 m tall with inside diameter 0.120 m is used to hold propane gas (molar mass 44.1 g/mol) for use in a barbecue. It is initially filled with gas until the gauge pressure is 1.30×10^6 Pa and the temperature is 22.0°C. The temperature of the gas remains constant as it is partially emptied out of the tank, until the gauge pressure is 2.50×10^5 Pa. Calculate the mass of propane that has been used.

18.58 • **CP** During a test dive in 1939, prior to being accepted by the U.S. Navy, the submarine *Squalus* sank at a point where the depth of water was 73.0 m. The temperature at the surface was 27.0°C, and at the bottom it was 7.0°C. The density of seawater is 1030 kg/m³. (a) A diving bell was used to rescue 33 trapped crewmen from the *Squalus*. The diving bell was in the form of a circular cylinder 2.30 m high, open at the bottom and closed at the top. When the diving bell was lowered to the bottom of the sea, to what height did water rise within the diving bell? (*Hint:* You may ignore the relatively small variation in water pressure between the bottom of the bell and the surface of the water within the bell.) (b) At what gauge pressure must compressed air have been supplied to the bell while on the bottom to expel all the water from it?

18.59 • **Atmosphere of Titan.** Titan, the largest satellite of Saturn, has a thick nitrogen atmosphere. At its surface, the pressure is 1.5 earth-atmospheres and the temperature is 94 K. (a) What is the surface temperature in °C? (b) Calculate the surface density in Titan's atmosphere in molecules per cubic meter. (c) Compare the density of Titan's surface atmosphere to the density of earth's atmosphere at 22°C. Which body has denser atmosphere?

18.60 • **Pressure on Venus.** At the surface of Venus the average temperature is a balmy 460°C due to the greenhouse effect (global warming!), the pressure is 92 earth-atmospheres, and the acceleration due to gravity is 0.894g_{earth}. The atmosphere is nearly all CO_2 (molar mass 44.0 g/mol) and the temperature remains remarkably constant. We shall assume that the temperature does not change at all with altitude. (a) What is the atmospheric pressure 1.00 km above the surface of Venus? Express your answer in Venus-atmospheres and earth-atmospheres. (b) What is the root-mean-square speed of the CO_2 molecules at the surface of Venus and at an altitude of 1.00 km?

18.61 •• An automobile tire has a volume of 0.0150 m³ on a cold day when the temperature of the air in the tire is 5.0°C and atmospheric pressure is 1.02 atm. Under these conditions the gauge pressure is measured to be 1.70 atm (about 25 lb/in.²). After the car is driven on the highway for 30 min, the temperature of the air in the tires has risen to 45.0°C and the volume has risen to 0.0159 m³. What then is the gauge pressure?

18.62 •• A flask with a volume of 1.50 L, provided with a stopcock, contains ethane gas (C_2H_6) at 300 K and atmospheric pressure (1.013×10^5 Pa). The molar mass of ethane is 30.1 g/mol. The system is warmed to a temperature of 490 K, with the stopcock open to the atmosphere. The stopcock is then closed, and the flask is cooled to its original temperature. (a) What is the final pressure of the ethane in the flask? (b) How many grams of ethane remain in the flask?

18.63 •• **CP** A balloon whose volume is 750 m³ is to be filled with hydrogen at atmospheric pressure (1.01×10^5 Pa). (a) If the hydrogen is stored in cylinders with volumes of 1.90 m³ at a gauge pressure of 1.20×10^6 Pa, how many cylinders are required? Assume that the temperature of the hydrogen remains constant. (b) What is the total weight (in addition to the weight of the gas) that can be supported by the balloon if the gas in the balloon and the surrounding air are both at 15.0°C? The molar mass of hydrogen (H_2) is 2.02 g/mol. The density of air at 15.0°C and atmospheric pressure is 1.23 kg/m³. See Chapter 12 for a discussion of buoyancy. (c) What weight could be supported if the balloon were filled with helium (molar mass 4.00 g/mol) instead of hydrogen, again at 15.0°C?

18.64 •• A vertical cylindrical tank contains 1.80 mol of an ideal gas under a pressure of 0.500 atm at 20.0°C. The round part of the tank has a radius of 10.0 cm, and the gas is supporting a piston that

can move up and down in the cylinder without friction. There is a vacuum above the piston. (a) What is the mass of this piston? (b) How tall is the column of gas that is supporting the piston?

18.65 •• **CP** A large tank of water has a hose connected to it, as shown in Fig. P18.65. The tank is sealed at the top and has compressed air between the water surface and the top. When the water height h has the value 3.50 m, the absolute pressure p of the compressed air is 4.20×10^5 Pa. Assume that the air above the water expands at constant temperature, and take the atmospheric pressure to be 1.00×10^5 Pa. (a) What is the speed with which water flows out of the hose when $h = 3.50$ m? (b) As water flows out of the tank, h decreases. Calculate the speed of flow for $h = 3.00$ m and for $h = 2.00$ m. (c) At what value of h does the flow stop?

Figure **P18.65**

18.66 •• **BIO** A person at rest inhales 0.50 L of air with each breath at a pressure of 1.00 atm and a temperature of 20.0°C. The inhaled air is 21.0% oxygen. (a) How many oxygen molecules does this person inhale with each breath? (b) Suppose this person is now resting at an elevation of 2000 m but the temperature is still 20.0°C. Assuming that the oxygen percentage and volume per inhalation are the same as stated above, how many oxygen molecules does this person now inhale with each breath? (c) Given that the body still requires the same number of oxygen molecules per second as at sea level to maintain its functions, explain why some people report "shortness of breath" at high elevations.

18.67 •• **BIO** **How Many Atoms Are You?** Estimate the number of atoms in the body of a 50-kg physics student. Note that the human body is mostly water, which has molar mass 18.0 g/mol, and that each water molecule contains three atoms.

18.68 • The size of an oxygen molecule is about 2.0×10^{-10} m. Make a rough estimate of the pressure at which the finite volume of the molecules should cause noticeable deviations from ideal-gas behavior at ordinary temperatures ($T = 300$ K).

18.69 •• You have two identical containers, one containing gas A and the other gas B. The masses of these molecules are $m_A = 3.34 \times 10^{-27}$ kg and $m_B = 5.34 \times 10^{-26}$ kg. Both gases are under the same pressure and are at 10.0°C. (a) Which molecules (A or B) have greater translational kinetic energy per molecule and rms speeds? (b) Now you want to raise the temperature of only one of these containers so that both gases will have the same rms speed. For which gas should you raise the temperature? (c) At what temperature will you accomplish your goal? (d) Once you have accomplished your goal, which molecules (A or B) now have greater average translational kinetic energy per molecule?

18.70 • **Insect Collisions.** A cubical cage 1.25 m on each side contains 2500 angry bees, each flying randomly at 1.10 m/s. We can model these insects as spheres 1.50 cm in diameter. On the average, (a) how far does a typical bee travel between collisions, (b) what is the average time between collisions, and (c) how many collisions per second does a bee make?

18.71 •• You blow up a spherical balloon to a diameter of 50.0 cm until the absolute pressure inside is 1.25 atm and the temperature is 22.0°C. Assume that all the gas is N_2, of molar mass 28.0 g/mol. (a) Find the mass of a single N_2 molecule. (b) How much translational kinetic energy does an average N_2 molecule have? (c) How many N_2 molecules are in this balloon? (d) What is the *total* translational kinetic energy of all the molecules in the balloon?

18.72 • **CP** (a) Compute the increase in gravitational potential energy for a nitrogen molecule (molar mass 28.0 g/mol) for an increase in elevation of 400 m near the earth's surface. (b) At what temperature is this equal to the average kinetic energy of a nitrogen molecule? (c) Is it possible that a nitrogen molecule near sea level where $T = 15.0°C$ could rise to an altitude of 400 m? Is it likely that it could do so without hitting any other molecules along the way? Explain.

18.73 •• **CP, CALC** **The Lennard-Jones Potential.** A commonly used potential-energy function for the interaction of two molecules (see Fig. 18.8) is the Lennard-Jones 6-12 potential:

$$U(r) = U_0 \left[\left(\frac{R_0}{r} \right)^{12} - 2 \left(\frac{R_0}{r} \right)^6 \right]$$

where r is the distance between the centers of the molecules and U_0 and R_0 are positive constants. The corresponding force $F(r)$ is given in Eq. (14.26). (a) Graph $U(r)$ and $F(r)$ versus r. (b) Let r_1 be the value of r at which $U(r) = 0$, and let r_2 be the value of r at which $F(r) = 0$. Show the locations of r_1 and r_2 on your graphs of $U(r)$ and $F(r)$. Which of these values represents the equilibrium separation between the molecules? (c) Find the values of r_1 and r_2 in terms of R_0, and find the ratio r_1/r_2. (d) If the molecules are located a distance r_2 apart [as calculated in part (c)], how much work must be done to pull them apart so that $r \to \infty$?

18.74 • (a) What is the total random translational kinetic energy of 5.00 L of hydrogen gas (molar mass 2.016 g/mol) with pressure 1.01×10^5 Pa and temperature 300 K? (*Hint:* Use the procedure of Problem 18.71 as a guide.) (b) If the tank containing the gas is placed on a swift jet moving at 300.0 m/s, by what percentage is the *total* kinetic energy of the gas increased? (c) Since the kinetic energy of the gas molecules is greater when it is on the jet, does this mean that its temperature has gone up? Explain.

18.75 • The speed of propagation of a sound wave in air at 27°C is about 350 m/s. Calculate, for comparison, (a) v_{rms} for nitrogen molecules and (b) the rms value of v_x at this temperature. The molar mass of nitrogen (N_2) is 28.0 g/mol.

18.76 • **Hydrogen on the Sun.** The surface of the sun has a temperature of about 5800 K and consists largely of hydrogen atoms. (a) Find the rms speed of a hydrogen atom at this temperature. (The mass of a single hydrogen atom is 1.67×10^{-27} kg.) (b) The escape speed for a particle to leave the gravitational influence of the sun is given by $(2GM/R)^{1/2}$, where M is the sun's mass, R its radius, and G the gravitational constant (see Example 13.5 of Section 13.3). Use the data in Appendix F to calculate this escape speed. (c) Can appreciable quantities of hydrogen escape from the sun? Can *any* hydrogen escape? Explain.

18.77 •• **CP** (a) Show that a projectile with mass m can "escape" from the surface of a planet if it is launched vertically upward with a kinetic energy greater than mgR_p, where g is the acceleration due to gravity at the planet's surface and R_p is the planet's radius. Ignore air resistance. (See Problem 18.76.) (b) If the planet in question is the earth, at what temperature does the average translational kinetic energy of a nitrogen molecule (molar mass 28.0 g/mol) equal that required to escape? What about a hydrogen molecule (molar mass 2.02 g/mol)? (c) Repeat part (b) for the moon, for which $g = 1.63$ m/s^2 and $R_p = 1740$ km. (d) While the earth and the moon have similar average surface temperatures, the moon has essentially no atmosphere. Use your results from parts (b) and (c) to explain why.

18.78 • **Planetary Atmospheres.** (a) The temperature near the top of Jupiter's multicolored cloud layer is about 140 K. The temperature at the top of the earth's troposphere, at an altitude of about 20 km, is about 220 K. Calculate the rms speed of hydrogen molecules in both these environments. Give your answers in m/s and as a fraction of the escape speed from the respective planet (see Problem 18.76). (b) Hydrogen gas (H_2) is a rare element in the earth's atmosphere. In the atmosphere of Jupiter, by contrast, 89% of all molecules are H_2. Explain why, using your results from part (a). (c) Suppose an astronomer claims to have discovered an oxygen (O_2) atmosphere on the asteroid Ceres. How likely is this? Ceres has a mass equal to 0.014 times the mass of the moon, a density of 2400 kg/m^3, and a surface temperature of about 200 K.

18.79 •• (a) For what mass of molecule or particle is v_{rms} equal to 1.00 mm/s at 300 K? (b) If the particle is an ice crystal, how many molecules does it contain? The molar mass of water is 18.0 g/mol. (c) Calculate the diameter of the particle if it is a spherical piece of ice. Would it be visible to the naked eye?

18.80 •• In describing the heat capacities of solids in Section 18.4, we stated that the potential energy $U = \frac{1}{2}kx^2$ of a harmonic oscillator averaged over one period of the motion is equal to the kinetic energy $K = \frac{1}{2}mv^2$ averaged over one period. Prove this result using Eqs. (14.13) and (14.15) for the position and velocity of a simple harmonic oscillator. For simplicity, assume that the initial position and velocity make the phase angle ϕ equal to zero. (*Hint:* Use the trigonometric identities $\cos^2(\theta) = [1 + \cos(2\theta)]/2$ and $\sin^2(\theta) = [1 - \cos(2\theta)]/2$. What is the average value of $\cos(2\omega t)$ over one period?)

18.81 •• It is possible to make crystalline solids that are only one layer of atoms thick. Such "two-dimensional" crystals can be created by depositing atoms on a very flat surface. (a) If the atoms in such a two-dimensional crystal can move only within the plane of the crystal, what will be its molar heat capacity near room temperature? Give your answer as a multiple of R and in J/mol·K. (b) At very low temperatures, will the molar heat capacity of a two-dimensional crystal be greater than, less than, or equal to the result you found in part (a)? Explain why.

18.82 •• (a) Calculate the total *rotational* kinetic energy of the molecules in 1.00 mol of a diatomic gas at 300 K. (b) Calculate the moment of inertia of an oxygen molecule (O_2) for rotation about either the y- or z-axis shown in Fig. 18.18b. Treat the molecule as two massive points (representing the oxygen atoms) separated by a distance of 1.21×10^{-10} m. The molar mass of oxygen *atoms* is 16.0 g/mol. (c) Find the rms angular velocity of rotation of an oxygen molecule about either the y- or z-axis shown in Fig. 18.18b. How does your answer compare to the angular velocity of a typical piece of rapidly rotating machinery (10,000 rev/min)?

18.83 • For each polyatomic gas in Table 18.1, compute the value of the molar heat capacity at constant volume, C_V, on the assumption that there is no vibrational energy. Compare with the measured values in the table, and compute the fraction of the total heat capacity that is due to vibration for each of the three gases. (*Note:* CO_2 is linear; SO_2 and H_2S are not. Recall that a linear polyatomic molecule has two rotational degrees of freedom, and a nonlinear molecule has three.)

18.84 •• **CALC** (a) Show that $\int_0^\infty f(v)\, dv = 1$, where $f(v)$ is the Maxwell–Boltzmann distribution of Eq. (18.32). (b) In terms of the physical definition of $f(v)$, explain why the integral in part (a) *must* have this value.

18.85 •• **CALC** Calculate the integral in Eq. (18.31), $\int_0^\infty v^2 f(v)\, dv$, and compare this result to $(v^2)_{av}$ as given by Eq. (18.16). (*Hint:* You may use the tabulated integral

$$\int_0^\infty x^{2n} e^{-\alpha x^2}\, dx = \frac{1 \cdot 3 \cdot 5 \cdots (2n-1)}{2^{n+1} \alpha^n} \sqrt{\frac{\pi}{\alpha}}$$

where n is a positive integer and α is a positive constant.)

18.86 ·· **CALC** Calculate the integral in Eq. (18.30), $\int_0^\infty v f(v) \, dv$, and compare this result to v_{av} as given by Eq. (18.35). (*Hint:* Make the change of variable $v^2 = x$ and use the tabulated integral

$$\int_0^\infty x^n e^{-\alpha x} \, dx = \frac{n!}{\alpha^{n+1}}$$

where n is a positive integer and α is a positive constant.)

18.87 ·· **CALC** (a) Explain why in a gas of N molecules, the number of molecules having speeds in the *finite* interval v to $v + \Delta v$ is $\Delta N = N \int_v^{v+\Delta v} f(v) \, dv$. (b) If Δv is small, then $f(v)$ is approximately constant over the interval and $\Delta N \approx N f(v) \Delta v$. For oxygen gas ($O_2$, molar mass 32.0 g/mol) at $T = 300$ K, use this approximation to calculate the number of molecules with speeds within $\Delta v = 20$ m/s of v_{mp}. Express your answer as a multiple of N. (c) Repeat part (b) for speeds within $\Delta v = 20$ m/s of $7v_{mp}$. (d) Repeat parts (b) and (c) for a temperature of 600 K. (e) Repeat parts (b) and (c) for a temperature of 150 K. (f) What do your results tell you about the shape of the distribution as a function of temperature? Do your conclusions agree with what is shown in Fig. 18.23?

18.88 · **Meteorology.** The *vapor pressure* is the pressure of the vapor phase of a substance when it is in equilibrium with the solid or liquid phase of the substance. The *relative humidity* is the partial pressure of water vapor in the air divided by the vapor pressure of water at that same temperature, expressed as a percentage. The air is saturated when the humidity is 100%. (a) The vapor pressure of water at 20.0°C is 2.34×10^3 Pa. If the air temperature is 20.0°C and the relative humidity is 60%, what is the partial pressure of water vapor in the atmosphere (that is, the pressure due to water vapor alone)? (b) Under the conditions of part (a), what is the mass of water in 1.00 m³ of air? (The molar mass of water is 18.0 g/mol. Assume that water vapor can be treated as an ideal gas.)

18.89 · **The Dew Point.** The vapor pressure of water (see Problem 18.88) decreases as the temperature decreases. If the amount of water vapor in the air is kept constant as the air is cooled, a temperature is reached, called the *dew point*, at which the partial pressure and vapor pressure coincide and the vapor is saturated. If the air is cooled further, vapor condenses to liquid until the partial pressure again equals the vapor pressure at that temperature. The temperature in a room is 30.0°C. A meteorologist cools a metal can by gradually adding cold water. When the can temperature reaches 16.0°C, water droplets form on its outside surface. What is the relative humidity of the 30.0°C air in the room? The table lists the vapor pressure of water at various temperatures:

Temperature (°C)	Vapor Pressure (Pa)
10.0	1.23×10^3
12.0	1.40×10^3
14.0	1.60×10^3
16.0	1.81×10^3
18.0	2.06×10^3
20.0	2.34×10^3
22.0	2.65×10^3
24.0	2.99×10^3
26.0	3.36×10^3
28.0	3.78×10^3
30.0	4.25×10^3

18.90 ··· **Altitude at Which Clouds Form.** On a spring day in the midwestern United States, the air temperature at the surface is 28.0°C. Puffy cumulus clouds form at an altitude where the air temperature equals the dew point (see Problem 18.89). If the air temperature decreases with altitude at a rate of 0.6 C°/100 m, at approximately what height above the ground will clouds form if the relative humidity at the surface is 35% and 80%? (*Hint:* Use the table in Problem 18.89.)

CHALLENGE PROBLEMS

18.91 ··· **CP Dark Nebulae and the Interstellar Medium.** The dark area in Fig. P18.91 that appears devoid of stars is a *dark nebula*, a cold gas cloud in interstellar space that contains enough material to block out light from the stars behind it. A typical dark nebula is about 20 light-years in diameter and contains about 50 hydrogen atoms per cubic centimeter (monatomic hydrogen, *not* H_2) at a temperature of about 20 K. (A light-year is the distance light travels in vacuum in one year and is equal to 9.46×10^{15} m.) (a) Estimate the mean free path for a hydrogen atom in a dark nebula. The radius of a hydrogen atom is 5.0×10^{-11} m. (b) Estimate the rms speed of a hydrogen atom and the mean free time (the average time between collisions for a given atom). Based on this result, do you think that atomic collisions, such as those leading to H_2 molecule formation, are very important in determining the composition of the nebula? (c) Estimate the pressure inside a dark nebula. (d) Compare the rms speed of a hydrogen atom to the escape speed at the surface of the nebula (assumed spherical). If the space around the nebula were a vacuum, would such a cloud be stable or would it tend to evaporate? (e) The stability of dark nebulae is explained by the presence of the *interstellar medium* (ISM), an even thinner gas that permeates space and in which the dark nebulae are embedded. Show that for dark nebulae to be in equilibrium with the ISM, the numbers of atoms per volume (N/V) and the temperatures (T) of dark nebulae and the ISM must be related by

$$\frac{(N/V)_{nebula}}{(N/V)_{ISM}} = \frac{T_{ISM}}{T_{nebula}}$$

(f) In the vicinity of the sun, the ISM contains about 1 hydrogen atom per 200 cm³. Estimate the temperature of the ISM in the vicinity of the sun. Compare to the temperature of the sun's surface, about 5800 K. Would a spacecraft coasting through interstellar space burn up? Why or why not?

Figure **P18.91**

18.92 ··· **CALC Earth's Atmosphere.** In the *troposphere,* the part of the atmosphere that extends from earth's surface to an altitude of about 11 km, the temperature is not uniform but decreases with increasing elevation. (a) Show that if the temperature variation is approximated by the linear relationship

$$T = T_0 - \alpha y$$

where T_0 is the temperature at the earth's surface and T is the temperature at height y, the pressure p at height y is given by

$$\ln\left(\frac{p}{p_0}\right) = \frac{Mg}{R\alpha} \ln\left(\frac{T_0 - \alpha y}{T_0}\right)$$

where p_0 is the pressure at the earth's surface and M is the molar mass for air. The coefficient α is called the lapse rate of temperature. It varies with atmospheric conditions, but an average value is about 0.6 C°/100 m. (b) Show that the above result reduces to the result of Example 18.4 (Section 18.1) in the limit that $\alpha \to 0$. (c) With $\alpha = 0.6$ C°/100 m, calculate p for $y = 8863$ m and compare your answer to the result of Example 18.4. Take $T_0 = 288$ K and $p_0 = 1.00$ atm.

18.93 ••• In Example 18.7 (Section 18.3) we saw that $v_{rms} > v_{av}$. It is not difficult to show that this is *always* the case. (The only exception is when the particles have the same speed, in which case $v_{rms} = v_{av}$.) (a) For two particles with speeds v_1 and v_2, show that $v_{rms} \geq v_{av}$, regardless of the numerical values of v_1 and v_2. Then show that $v_{rms} > v_{av}$ if $v_1 \neq v_2$. (b) Suppose that for a collection of N particles you know that $v_{rms} > v_{av}$. Another particle, with speed u, is added to the collection of particles. If the new rms and average speeds are denoted as v'_{rms} and v'_{av}, show that

$$v'_{rms} = \sqrt{\frac{N v_{rms}^2 + u^2}{N + 1}} \quad \text{and} \quad v'_{av} = \frac{N v_{av} + u}{N + 1}$$

(c) Use the expressions in part (b) to show that $v'_{rms} > v'_{av}$ regardless of the numerical value of u. (d) Explain why your results for (a) and (c) together show that $v_{rms} > v_{av}$ for any collection of particles if the particles do not all have the same speed.

Answers

Chapter Opening Question ?

From Eq. (18.19), the root-mean-square speed of a gas molecule is proportional to the square root of the absolute temperature T. The temperature range we're considering is from $(25 + 273.15)$ K $=$ 298 K to $(100 + 273.15)$ K $= 373$ K. Hence the speeds increase by a factor of $\sqrt{(373 \text{ K})/(298 \text{ K})} = 1.12$; that is, there is a 12% increase. While 100°C feels far warmer than 25°C, the difference in molecular speeds is relatively small.

Test Your Understanding Questions

18.1 Answer: (ii) and (iii) (tie), (i) and (v) (tie), (iv) We can rewrite the ideal-gas equation, Eq. (18.3), as $n = pV/RT$. This tells us that the number of moles n is proportional to the pressure and volume and inversely proportional to the absolute temperature. Hence, compared to (i), the number of moles in each case is (ii) $(2)(1)/(1) = 2$ times as much, (iii) $(1)(2)/(1) = 2$ times as much, (iv) $(1)(1)/(2) = \frac{1}{2}$ as much, and (v) $(2)(1)/(2) = 1$ time as much (that is, equal).

18.2 Answer: (vi) The value of r_0 determines the equilibrium separation of the molecules in the solid phase, so doubling r_0 means that the separation doubles as well. Hence a solid cube of this compound might grow from 1 cm on a side to 2 cm on a side. The volume would then be $2^3 = 8$ times larger, and the density (mass divided by volume) would be $\frac{1}{8}$ as great.

18.3 Answers: (a) (iv), (ii), (iii), (i); (b) (iii) and (iv) (tie), (i) and (ii) (tie) (a) Equation (18.19) tells us that $v_{rms} = \sqrt{3RT/M}$, so the rms speed is proportional to the square root of the ratio of absolute temperature T to molar mass M. Compared to (i) oxygen at 300 K, v_{rms} in the other cases is (ii) $\sqrt{(32.0 \text{ g/mol})/(28.0 \text{ g/mol})} = 1.07$ times faster, (iii) $\sqrt{(330 \text{ K})/(300 \text{ K})} = 1.05$ times faster, and (iv) $\sqrt{(330 \text{ K})(32.0 \text{ g/mol})/(300 \text{ K})(28.0 \text{ g/mol})} = 1.12$ times faster. (b) From Eq. (18.16), the average translational kinetic energy per molecule is $\frac{1}{2}m(v^2)_{av} = \frac{3}{2}kT$, which is directly proportional to T and independent of M. We have $T = 300$ K for cases (i) and (ii) and $T = 330$ K for cases (iii) and (iv), so $\frac{1}{2}m(v^2)_{av}$ has equal values for cases (iii) and (iv) and equal (but smaller) values for cases (i) and (ii).

18.4 Answers: no, near the beginning Adding a small amount of heat dQ to the gas changes the temperature by dT, where $dQ = nC_V dT$ from Eq. (18.24). Figure 18.19 shows that C_V for H_2 varies with temperature between 25 K and 500 K, so a given amount of heat gives rise to different amounts of temperature change during the process. Hence the temperature will *not* increase at a constant rate. The temperature change $dT = dQ/nC_V$ is inversely proportional to C_V, so the temperature increases most rapidly at the beginning of the process when the temperature is lowest and C_V is smallest (see Fig. 18.19).

18.5 Answer: (ii) Figure 18.23b shows that the *fraction* of molecules with speeds between v_1 and v_2 equals the area under the curve of $f(v)$ versus v from $v = v_1$ to $v = v_2$. This is equal to the integral $\int_{v_1}^{v_2} f(v)\,dv$, which in turn is equal to the difference between the integrals $\int_0^{v_2} f(v)\,dv$ (the fraction of molecules with speeds from 0 to v_2) and $\int_0^{v_1} f(v)\,dv$ (the fraction of molecules with speeds from 0 to the slower speed v_1). The *number* of molecules with speeds from v_1 to v_2 equals the fraction of molecules in this speed range multiplied by N, the total number of molecules.

18.6 Answers: no, yes The triple-point pressure of water from Table 18.3 is 6.10×10^2 Pa. The present-day pressure on Mars is just less than this value, corresponding to the line labeled p_s in Fig. 18.24. Hence liquid water cannot exist on the present-day Martian surface, and there are no rivers or lakes. Planetary scientists conclude that liquid water could have and almost certainly did exist on Mars in the past, when the atmosphere was thicker.

Bridging Problem

Answers: (a) 102 km   **(b)** yes

19 THE FIRST LAW OF THERMODYNAMICS

? A steam locomotive operates using the first law of thermodynamics: Water is heated and boils, and the expanding steam does work to propel the locomotive. Would it be possible for the steam to propel the locomotive by doing work as it *condenses?*

Every time you drive a car, turn on an air conditioner, or cook a meal, you reap the practical benefits of *thermodynamics,* the study of relationships involving heat, mechanical work, and other aspects of energy and energy transfer. For example, in a car engine heat is generated by the chemical reaction of oxygen and vaporized gasoline in the engine's cylinders. The heated gas pushes on the pistons within the cylinders, doing mechanical work that is used to propel the car. This is an example of a *thermodynamic process.*

The first law of thermodynamics, central to the understanding of such processes, is an extension of the principle of conservation of energy. It broadens this principle to include energy exchange by both heat transfer and mechanical work and introduces the concept of the *internal energy* of a system. Conservation of energy plays a vital role in every area of physical science, and the first law has extremely broad usefulness. To state energy relationships precisely, we need the concept of a *thermodynamic system.* We'll discuss *heat* and *work* as two means of transferring energy into or out of such a system.

19.1 Thermodynamic Systems

We have studied energy transfer through mechanical work (Chapter 6) and through heat transfer (Chapters 17 and 18). Now we are ready to combine and generalize these principles.

We always talk about energy transfer to or from some specific *system.* The system might be a mechanical device, a biological organism, or a specified quantity of material, such as the refrigerant in an air conditioner or steam expanding in a turbine. In general, a **thermodynamic system** is any collection of objects that is convenient to regard as a unit, and that may have the potential to exchange energy with its surroundings. A familiar example is a quantity of popcorn kernels in a pot with a lid. When the pot is placed on a stove, energy is added to the popcorn

19.1 The popcorn in the pot is a thermodynamic system. In the thermodynamic process shown here, heat is added to the system, and the system does work on its surroundings to lift the lid of the pot.

by conduction of heat. As the popcorn pops and expands, it does work as it exerts an upward force on the lid and moves it through a displacement (Fig. 19.1). The *state* of the popcorn changes in this process, since the volume, temperature, and pressure of the popcorn all change as it pops. A process such as this one, in which there are changes in the state of a thermodynamic system, is called a **thermodynamic process.**

In mechanics we used the concept of *system* with free-body diagrams and with conservation of energy and momentum. For *thermodynamic* systems, as for all others, it is essential to define clearly at the start exactly what is and is not included in the system. Only then can we describe unambiguously the energy transfers into and out of that system. For instance, in our popcorn example we defined the system to include the popcorn but not the pot, lid, or stove.

Thermodynamics has its roots in many practical problems other than popping popcorn (Fig. 19.2). The gasoline engine in an automobile, the jet engines in an airplane, and the rocket engines in a launch vehicle use the heat of combustion of their fuel to perform mechanical work in propelling the vehicle. Muscle tissue in living organisms metabolizes chemical energy in food and performs mechanical work on the organism's surroundings. A steam engine or steam turbine uses the heat of combustion of coal or other fuel to perform mechanical work such as driving an electric generator or pulling a train.

Signs for Heat and Work in Thermodynamics

We describe the energy relationships in any thermodynamic process in terms of the quantity of heat Q added *to* the system and the work W done *by* the system. Both Q and W may be positive, negative, or zero (Fig. 19.3). A positive value of Q represents heat flow *into* the system, with a corresponding input of energy to it; negative Q represents heat flow *out of* the system. A positive value of W represents work done *by* the system against its surroundings, such as work done by an expanding gas, and hence corresponds to energy *leaving* the system. Negative W, such as work done during compression of a gas in which work is done *on the gas* by its surroundings, represents energy *entering* the system. We will use these conventions consistently in the examples in this chapter and the next.

CAUTION   **Be careful with the sign of work** W Note that our sign rule for work is *opposite* to the one we used in mechanics, in which we always spoke of the work done by the forces acting *on* a body. In thermodynamics it is usually more convenient to call W the work done *by* the system so that when a system expands, the pressure, volume change, and work are all positive. Take care to use the sign rules for work and heat consistently! ∎

Test Your Understanding of Section 19.1   In Example 17.8 (Section 17.6), what is the sign of Q for the coffee? For the aluminum cup? If a block slides along a horizontal surface with friction, what is the sign of W for the block?   ∎

19.2 Work Done During Volume Changes

A simple but common example of a thermodynamic system is a quantity of gas enclosed in a cylinder with a movable piston. Internal-combustion engines, steam engines, and compressors in refrigerators and air conditioners all use some version of such a system. In the next several sections we will use the gas-in-cylinder system to explore several kinds of processes involving energy transformations.

We'll use a microscopic viewpoint, based on the kinetic and potential energies of individual molecules in a material, to develop intuition about thermodynamic quantities. But it is important to understand that the central principles of thermodynamics can be treated in a completely *macroscopic* way, without reference to microscopic models. Indeed, part of the great power and generality of thermodynamics is that it does *not* depend on details of the structure of matter.

19.2 (a) A rocket engine uses the heat of combustion of its fuel to do work propelling the launch vehicle. (b) Humans and other biological organisms are more complicated systems than we can analyze fully in this book, but the same basic principles of thermodynamics apply to them.

(a)   (b)

19.3 A thermodynamic system may exchange energy with its surroundings (environment) by means of heat, work, or both. Note the sign conventions for Q and W.

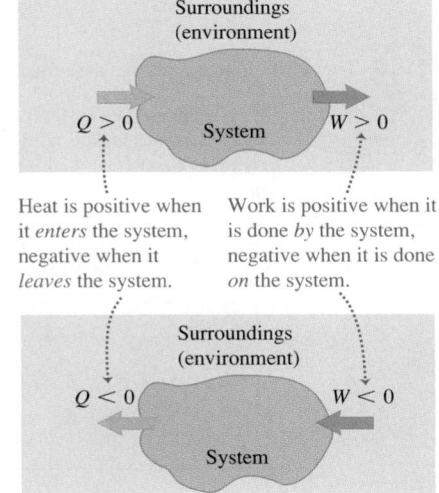

19.4 A molecule striking a piston (a) does positive work if the piston is moving away from the molecule and (b) does negative work if the piston is moving toward the molecule. Hence a gas does positive work when it expands as in (a) but does negative work when it compresses as in (b).

(a)

Piston moves away from molecule during collision.

Molecule bounces off piston.

v_{before}

Motion of piston

v_{after}

Molecule loses kinetic energy, does positive work on piston.

(b)

Piston moves toward molecule during collision.

v_{before}

Motion of piston

v_{after}

Molecule gains kinetic energy, does negative work on piston.

19.5 The infinitesimal work done by the system during the small expansion dx is $dW = pA\, dx$.

System

dx

pA

A

Force that system exerts on piston

First we consider the *work* done by the system during a volume change. When a gas expands, it pushes outward on its boundary surfaces as they move outward. Hence an expanding gas always does positive work. The same thing is true of any solid or fluid material that expands under pressure, such as the popcorn in Fig. 19.1.

We can understand the work done by a gas in a volume change by considering the molecules that make up the gas. When one such molecule collides with a stationary surface, it exerts a momentary force on the wall but does no work because the wall does not move. But if the surface is moving, like a piston in a gasoline engine, the molecule *does* do work on the surface during the collision. If the piston in Fig. 19.4a moves to the right, so that the volume of the gas increases, the molecules that strike the piston exert a force through a distance and do *positive* work on the piston. If the piston moves toward the left as in Fig. 19.4b, so the volume of the gas decreases, then positive work is done *on* the molecule during the collision. Hence the gas molecules do *negative* work on the piston.

Figure 19.5 shows a system whose volume can change (a gas, liquid, or solid) in a cylinder with a movable piston. Suppose that the cylinder has cross-sectional area A and that the pressure exerted by the system at the piston face is p. The total force F exerted by the system on the piston is $F = pA$. When the piston moves out an infinitesimal distance dx, the work dW done by this force is

$$dW = F\, dx = pA\, dx$$

But

$$A\, dx = dV$$

where dV is the infinitesimal change of volume of the system. Thus we can express the work done by the system in this infinitesimal volume change as

$$dW = p\, dV \tag{19.1}$$

In a finite change of volume from V_1 to V_2,

$$W = \int_{V_1}^{V_2} p\, dV \quad \text{(work done in a volume change)} \tag{19.2}$$

In general, the pressure of the system may vary during the volume change. For example, this is the case in the cylinders of an automobile engine as the pistons move back and forth. To evaluate the integral in Eq. (19.2), we have to know how the pressure varies as a function of volume. We can represent this relationship as a graph of p as a function of V (a pV-diagram, described at the end of Section 18.1). Figure 19.6 a shows a simple example. In this figure, Eq. (19.2) is represented

19.6 The work done equals the area under the curve on a pV-diagram.

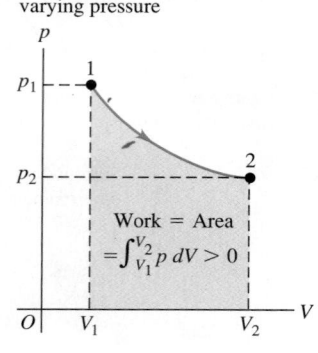

(a) pV-diagram for a system undergoing an expansion with varying pressure

p

p_1 1

p_2 2

Work = Area
$= \int_{V_1}^{V_2} p\, dV > 0$

O V_1 V_2 V

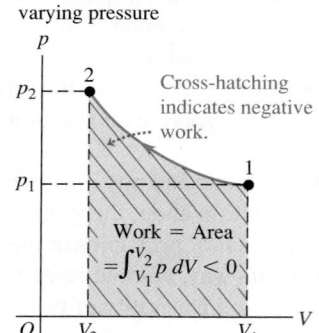

(b) pV-diagram for a system undergoing a compression with varying pressure

p

p_2 2

Cross-hatching indicates negative work.

p_1 1

Work = Area
$= \int_{V_1}^{V_2} p\, dV < 0$

O V_2 V_1 V

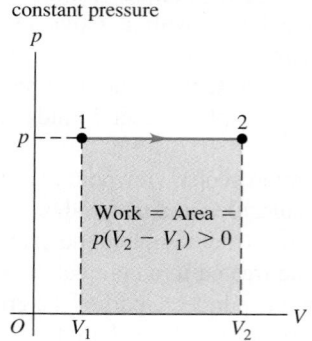

(c) pV-diagram for a system undergoing an expansion with constant pressure

p

p 1   2

Work = Area =
$p(V_2 - V_1) > 0$

O V_1 V_2 V

graphically as the *area* under the curve of p versus V between the limits V_1 and V_2. (In Section 6.3 we used a similar interpretation of the work done by a force F as the area under the curve of F versus x between the limits x_1 and x_2.)

According to the rule we stated in Section 19.1, work is *positive* when a system *expands*. In an expansion from state 1 to state 2 in Fig. 19.6a, the area under the curve and the work are positive. A *compression* from 1 to 2 in Fig. 19.6b gives a *negative* area; when a system is compressed, its volume decreases and it does *negative* work on its surroundings (see also Fig. 19.4b).

CAUTION **Be careful with subscripts 1 and 2** When using Eq. (19.2), always remember that V_1 is the *initial* volume and V_2 is the *final* volume. That's why the labels 1 and 2 are reversed in Fig. 19.6b compared to Fig. 19.6a, even though both processes move between the same two thermodynamic states. ▮

If the pressure p remains constant while the volume changes from V_1 to V_2 (Fig. 19.6c), the work done by the system is

$$W = p(V_2 - V_1) \qquad \text{(work done in a volume change at constant pressure)} \qquad (19.3)$$

In any process in which the volume is *constant,* the system does no work because there is no displacement.

Mastering**PHYSICS**

ActivPhysics 8.5: Work Done By a Gas

Example 19.1 **Isothermal expansion of an ideal gas**

As an ideal gas undergoes an *isothermal* (constant-temperature) expansion at temperature T, its volume changes from V_1 to V_2. How much work does the gas do?

SOLUTION

IDENTIFY and SET UP: The ideal-gas equation, Eq. (18.3), tells us that if the temperature T of n moles of an ideal gas is constant, the quantity $pV = nRT$ is also constant: p and V are inversely related. If V changes, p changes as well, so we *cannot* use Eq. (19.3) to calculate the work done. Instead we must use Eq. (19.2). To evaluate the integral in Eq. (19.2) we must know p as a function of V; for this we use Eq. (18.3).

EXECUTE: From Eq. (18.3),

$$p = \frac{nRT}{V}$$

We substitute this into the integral of Eq. (19.2), take the constant factor nRT outside, and evaluate the integral:

$$W = \int_{V_1}^{V_2} p \, dV$$

$$= nRT \int_{V_1}^{V_2} \frac{dV}{V} = nRT \ln\frac{V_2}{V_1} \quad \text{(ideal gas, isothermal process)}$$

We can rewrite this expression for W in terms of p_1 and p_2. Because $pV = nRT$ is constant,

$$p_1 V_1 = p_2 V_2 \qquad \text{or} \qquad \frac{V_2}{V_1} = \frac{p_1}{p_2}$$

so

$$W = nRT \ln\frac{p_1}{p_2} \quad \text{(ideal gas, isothermal process)}$$

EVALUATE: We check our result by noting that in an expansion $V_2 > V_1$ and the ratio V_2/V_1 is greater than 1. The logarithm of a number greater than 1 is positive, so $W > 0$, as it should be. As an additional check, look at our second expression for W: In an isothermal expansion the volume increases and the pressure drops, so $p_2 < p_1$, the ratio $p_1/p_2 > 1$, and $W = nRT \ln(p_1/p_2)$ is again positive.

These results also apply to an isothermal *compression* of a gas, for which $V_2 < V_1$ and $p_2 > p_1$.

Test Your Understanding of Section 19.2 A quantity of ideal gas undergoes an expansion that increases its volume from V_1 to $V_2 = 2V_1$. The final pressure of the gas is p_2. Does the gas do more work on its surroundings if the expansion is at constant *pressure* or at constant *temperature*? (i) constant pressure; (ii) constant temperature; (iii) the same amount of work is done in both cases; (iv) not enough information is given to decide.

MP

19.3 Paths Between Thermodynamic States

We've seen that if a thermodynamic process involves a change in volume, the system undergoing the process does work (either positive or negative) on its surroundings. Heat also flows into or out of the system during the process if there is a temperature difference between the system and its surroundings. Let's now examine how the work done by and the heat added to the system during a thermodynamic process depend on the details of how the process takes place.

Work Done in a Thermodynamic Process

When a thermodynamic system changes from an initial state to a final state, it passes through a series of intermediate states. We call this series of states a **path.** There are always infinitely many different possibilities for these intermediate states. When they are all equilibrium states, the path can be plotted on a pV-diagram (Fig. 19.7a). Point 1 represents an initial state with pressure p_1 and volume V_1, and point 2 represents a final state with pressure p_2 and volume V_2. To pass from state 1 to state 2, we could keep the pressure constant at p_1 while the system expands to volume V_2 (point 3 in Fig. 19.7b), then reduce the pressure to p_2 (probably by decreasing the temperature) while keeping the volume constant at V_2 (to point 2 on the diagram). The work done by the system during this process is the area under the line $1 \rightarrow 3$; no work is done during the constant-volume process $3 \rightarrow 2$. Or the system might traverse the path $1 \rightarrow 4 \rightarrow 2$ (Fig. 19.7c); in that case the work is the area under the line $4 \rightarrow 2$, since no work is done during the constant-volume process $1 \rightarrow 4$. The smooth curve from 1 to 2 is another possibility (Fig. 19.7d), and the work for this path is different from that for either of the other paths.

We conclude that *the work done by the system depends not only on the initial and final states, but also on the intermediate states—that is, on the path.* Furthermore, we can take the system through a series of states forming a closed loop, such as $1 \rightarrow 3 \rightarrow 2 \rightarrow 4 \rightarrow 1$. In this case the final state is the same as the initial state, but the total work done by the system is *not* zero. (In fact, it is represented on the graph by the area enclosed by the loop; can you prove that? See Exercise 19.7.) It follows that it doesn't make sense to talk about the amount of work *contained in* a system. In a particular state, a system may have definite values of the state coordinates p, V, and T, but it wouldn't make sense to say that it has a definite value of W.

Heat Added in a Thermodynamic Process

Like work, the *heat* added to a thermodynamic system when it undergoes a change of state depends on the path from the initial state to the final state. Here's an example. Suppose we want to change the volume of a certain quantity of an ideal gas from 2.0 L to 5.0 L while keeping the temperature constant at $T = 300$ K. Figure 19.8 shows two different ways in which we can do this. In Fig. 19.8a the gas is contained in a cylinder with a piston, with an initial volume of 2.0 L. We let the gas expand slowly, supplying heat from the electric heater to keep the temperature at 300 K. After expanding in this slow, controlled, isothermal manner, the gas reaches its final volume of 5.0 L; it absorbs a definite amount of heat in the process.

Figure 19.8b shows a different process leading to the same final state. The container is surrounded by insulating walls and is divided by a thin, breakable partition into two compartments. The lower part has volume 2.0 L and the upper part has volume 3.0 L. In the lower compartment we place the same amount of the same gas as in Fig. 19.8a, again at $T = 300$ K. The initial state is the same as before. Now we break the partition; the gas undergoes a rapid, uncontrolled expansion, with no heat passing through the insulating walls. The final volume is 5.0 L, the same as in Fig. 19.8a. The gas does no work during this expansion

19.7 The work done by a system during a transition between two states depends on the path chosen.

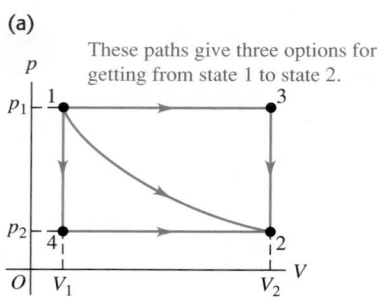

(a)

These paths give three options for getting from state 1 to state 2.

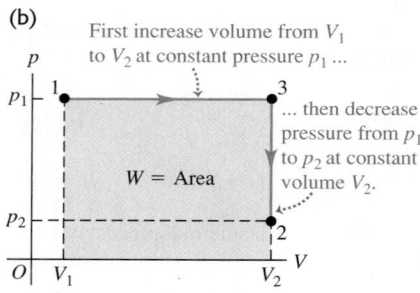

(b)

First increase volume from V_1 to V_2 at constant pressure p_1 ...

... then decrease pressure from p_1 to p_2 at constant volume V_2.

$W = $ Area

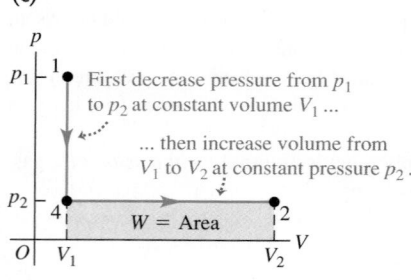

(c)

First decrease pressure from p_1 to p_2 at constant volume V_1 ...

... then increase volume from V_1 to V_2 at constant pressure p_2.

$W = $ Area

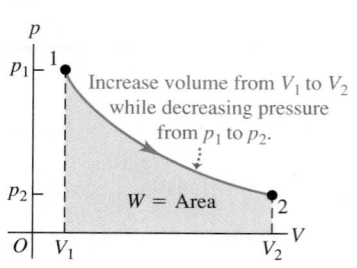

(d)

Increase volume from V_1 to V_2 while decreasing pressure from p_1 to p_2.

$W = $ Area

because it doesn't push against anything that moves. This uncontrolled expansion of a gas into vacuum is called a **free expansion;** we will discuss it further in Section 19.6.

Experiments have shown that when an ideal gas undergoes a free expansion, there is no temperature change. Therefore the final state of the gas is the same as in Fig. 19.8a. The intermediate states (pressures and volumes) during the transition from state 1 to state 2 are entirely different in the two cases; Figs. 19.8a and 19.8b represent *two different paths* connecting the *same states* 1 and 2. For the path in Fig. 19.8b, *no* heat is transferred into the system, and the system does no work. Like work, *heat depends not only on the initial and final states but also on the path.*

Because of this path dependence, it would not make sense to say that a system "contains" a certain quantity of heat. To see this, suppose we assign an arbitrary value to the "heat in a body" in some reference state. Then presumably the "heat in the body" in some other state would equal the heat in the reference state plus the heat added when the body goes to the second state. But that's ambiguous, as we have just seen; the heat added depends on the *path* we take from the reference state to the second state. We are forced to conclude that there is *no* consistent way to define "heat in a body"; it is not a useful concept.

While it doesn't make sense to talk about "work in a body" or "heat in a body," it *does* make sense to speak of the amount of *internal energy* in a body. This important concept is our next topic.

Test Your Understanding of Section 19.3 The system described in Fig. 19.7a undergoes four different thermodynamic processes. Each process is represented in a pV-diagram as a straight line from the initial state to the final state. (These processes are different from those shown in the pV-diagrams of Fig. 19.7.) Rank the processes in order of the amount of work done by the system, from the most positive to the most negative. (i) $1 \rightarrow 2$; (ii) $2 \rightarrow 1$; (iii) $3 \rightarrow 4$; (iv) $4 \rightarrow 3$. ▮

19.4 Internal Energy and the First Law of Thermodynamics

Internal energy is one of the most important concepts in thermodynamics. In Section 7.3, when we discussed energy changes for a body sliding with friction, we stated that warming a body increased its internal energy and that cooling the body decreased its internal energy. But what *is* internal energy? We can look at it in various ways; let's start with one based on the ideas of mechanics. Matter consists of atoms and molecules, and these are made up of particles having kinetic and potential energies. We *tentatively* define the **internal energy** of a system as the sum of the kinetic energies of all of its constituent particles, plus the sum of all the potential energies of interaction among these particles.

> **CAUTION** **Is it internal?** Note that internal energy does *not* include potential energy arising from the interaction between the system and its surroundings. If the system is a glass of water, placing it on a high shelf increases the gravitational potential energy arising from the interaction between the glass and the earth. But this has no effect on the interaction between the molecules of the water, and so the internal energy of the water does not change. ▮

We use the symbol U for internal energy. (We used this same symbol in our study of mechanics to represent potential energy. You may have to remind yourself occasionally that U has a different meaning in thermodynamics.) During a change of state of the system, the internal energy may change from an initial value U_1 to a final value U_2. We denote the change in internal energy as $\Delta U = U_2 - U_1$.

19.8 (a) Slow, controlled isothermal expansion of a gas from an initial state 1 to a final state 2 with the same temperature but lower pressure. (b) Rapid, uncontrolled expansion of the same gas starting at the same state 1 and ending at the same state 2.

(a) System does work on piston; hot plate adds heat to system ($W > 0$ and $Q > 0$).

(b) System does no work; no heat enters or leaves system ($W = 0$ and $Q = 0$).

Mastering**PHYSICS**

ActivPhysics 8.6: Heat, Internal Energy, and First Law of Thermodynamics

19.9 In a thermodynamic process, the internal energy U of a system may (a) increase ($\Delta U > 0$), (b) decrease ($\Delta U < 0$), or (c) remain the same ($\Delta U = 0$).

(a) More heat is added to system than system does work: Internal energy of system increases.

Surroundings (environment)

$Q = 150$ J $W = 100$ J

System

$\Delta U = Q - W = +50$ J

(b) More heat flows out of system than work is done: Internal energy of system decreases.

Surroundings (environment)

$Q = -150$ J $W = -100$ J

System

$\Delta U = Q - W = -50$ J

(c) Heat added to system equals work done by system: Internal energy of system unchanged.

Surroundings (environment)

$Q = 150$ J $W = 150$ J

System

$\Delta U = Q - W = 0$

Application The First Law of Exercise Thermodynamics
Your body is a thermodynamic system. When you exercise, your body does work (such as the work done to lift your body as a whole in a push-up). Hence $W > 0$. Your body also warms up during exercise; by perspiration and other means the body rids itself of this heat, so $Q < 0$. Since Q is negative and W is positive, $\Delta U = Q - W < 0$ and the body's internal energy decreases. That's why exercise helps you lose weight: It uses up some of the internal energy stored in your body in the form of fat.

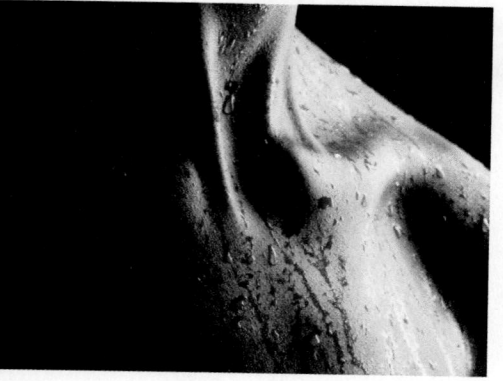

When we add a quantity of heat Q to a system and the system does no work during the process (so $W = 0$), the internal energy increases by an amount equal to Q; that is, $\Delta U = Q$. When a system does work W by expanding against its surroundings and no heat is added during the process, energy leaves the system and the internal energy decreases: W is positive, Q is zero, and $\Delta U = -W$. When *both* heat transfer and work occur, the *total* change in internal energy is

$$U_2 - U_1 = \Delta U = Q - W \quad \text{(first law of thermodynamics)} \quad (19.4)$$

We can rearrange this to the form

$$Q = \Delta U + W \quad (19.5)$$

The message of Eq. (19.5) is that in general, when heat Q is added to a system, some of this added energy remains within the system, changing its internal energy by an amount ΔU; the remainder leaves the system again as the system does work W against its surroundings. Because W and Q may be positive, negative, or zero, ΔU can be positive, negative, or zero for different processes (Fig. 19.9).

Equation (19.4) or (19.5) is the **first law of thermodynamics.** It is a generalization of the principle of conservation of energy to include energy transfer through heat as well as mechanical work. As you will see in later chapters, this principle can be extended to ever-broader classes of phenomena by identifying additional forms of energy and energy transfer. In every situation in which it seems that the total energy in all known forms is not conserved, it has been possible to identify a new form of energy such that the total energy, including the new form, *is* conserved. There is energy associated with electric fields, with magnetic fields, and, according to the theory of relativity, even with mass itself.

Understanding the First Law of Thermodynamics

At the beginning of this discussion we tentatively defined internal energy in terms of microscopic kinetic and potential energies. This has drawbacks, however. Actually *calculating* internal energy in this way for any real system would be hopelessly complicated. Furthermore, this definition isn't an *operational* one because it doesn't describe how to determine internal energy from physical quantities that we can measure directly.

So let's look at internal energy in another way. Starting over, we define the *change* in internal energy ΔU during any change of a system as the quantity given by Eq. (19.4), $\Delta U = Q - W$. This *is* an operational definition because we can measure Q and W. It does not define U itself, only ΔU. This is not a shortcoming because we can *define* the internal energy of a system to have a specified value in some reference state, and then use Eq. (19.4) to define the internal energy in any other state. This is analogous to our treatment of potential energy in Chapter 7, in which we arbitrarily defined the potential energy of a mechanical system to be zero at a certain position.

This new definition trades one difficulty for another. If we define ΔU by Eq. (19.4), then when the system goes from state 1 to state 2 by two different paths, how do we know that ΔU is the same for the two paths? We have already seen that Q and W are, in general, *not* the same for different paths. If ΔU, which equals $Q - W$, is also path dependent, then ΔU is ambiguous. If so, the concept of internal energy of a system is subject to the same criticism as the erroneous concept of quantity of heat in a system, as we discussed at the end of Section 19.3.

The only way to answer this question is through *experiment*. For various materials we measure Q and W for various changes of state and various paths to learn whether ΔU is or is not path dependent. The results of many such investigations are clear and unambiguous: While Q and W depend on the path, $\Delta U = Q - W$ is independent of path. *The change in internal energy of a system during any thermodynamic process depends only on the initial and final states, not on the path leading from one to the other.*

Experiment, then, is the ultimate justification for believing that a thermodynamic system in a specific state has a unique internal energy that depends only on that state. An equivalent statement is that the internal energy U of a system is a function of the state coordinates p, V, and T (actually, any two of these, since the three variables are related by the equation of state).

To say that the first law of thermodynamics, given by Eq. (19.4) or (19.5), represents conservation of energy for thermodynamic processes is correct, as far as it goes. But an important *additional* aspect of the first law is the fact that internal energy depends only on the state of a system (Fig. 19.10). In changes of state, the change in internal energy is independent of the path.

All this may seem a little abstract if you are satisfied to think of internal energy as microscopic mechanical energy. There's nothing wrong with that view, and we will make use of it at various times during our discussion. But in the interest of precise *operational* definitions, internal energy, like heat, can and must be defined in a way that is independent of the detailed microscopic structure of the material.

Cyclic Processes and Isolated Systems

Two special cases of the first law of thermodynamics are worth mentioning. A process that eventually returns a system to its initial state is called a *cyclic process*. For such a process, the final state is the same as the initial state, and so the *total* internal energy change must be zero. Then

$$U_2 = U_1 \quad \text{and} \quad Q = W$$

If a net quantity of work W is done by the system during this process, an equal amount of energy must have flowed into the system as heat Q. But there is no reason either Q or W individually has to be zero (Fig. 19.11).

Another special case occurs in an *isolated system,* one that does no work on its surroundings and has no heat flow to or from its surroundings. For any process taking place in an isolated system,

$$W = Q = 0$$

and therefore

$$U_2 = U_1 = \Delta U = 0$$

In other words, *the internal energy of an isolated system is constant.*

19.10 The internal energy of a cup of coffee depends on just its thermodynamic state—how much water and ground coffee it contains, and what its temperature is. It does not depend on the history of how the coffee was prepared—that is, the thermodynamic path that led to its current state.

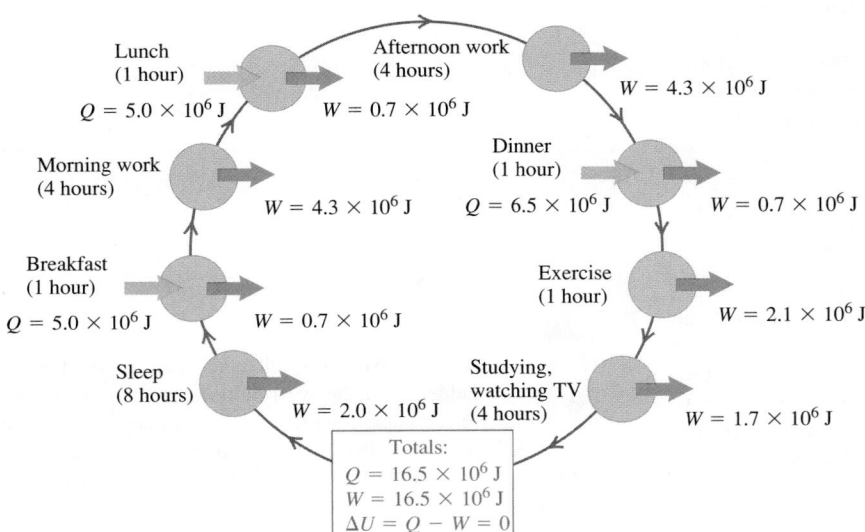

Lunch
(1 hour)

$Q = 5.0 \times 10^6$ J

Afternoon work
(4 hours)

$W = 0.7 \times 10^6$ J

$W = 4.3 \times 10^6$ J

Morning work
(4 hours)

$W = 4.3 \times 10^6$ J

Dinner
(1 hour)

$Q = 6.5 \times 10^6$ J

$W = 0.7 \times 10^6$ J

Breakfast
(1 hour)

$Q = 5.0 \times 10^6$ J

$W = 0.7 \times 10^6$ J

Exercise
(1 hour)

$W = 2.1 \times 10^6$ J

Sleep
(8 hours)

$W = 2.0 \times 10^6$ J

Studying,
watching TV
(4 hours)

$W = 1.7 \times 10^6$ J

Totals:
$Q = 16.5 \times 10^6$ J
$W = 16.5 \times 10^6$ J
$\Delta U = Q - W = 0$

19.11 Every day, your body (a thermodynamic system) goes through a cyclic thermodynamic process like this one. Heat Q is added by metabolizing food, and your body does work W in breathing, walking, and other activities. If you return to the same state at the end of the day, $Q = W$ and the net change in your internal energy is zero.

Problem-Solving Strategy 19.1 The First Law of Thermodynamics

IDENTIFY *the relevant concepts:* The first law of thermodynamics is the statement of the law of conservation of energy in its most general form. You can apply it to *any* thermodynamic process in which the internal energy of a system changes, heat flows into or out of the system, and/or work is done by or on the system.

SET UP *the problem* using the following steps:
1. Define the thermodynamic system to be considered.
2. If the thermodynamic process has more than one step, identify the initial and final states for each step.
3. List the known and unknown quantities and identify the target variables.
4. Confirm that you have enough equations. You can apply the first law, $\Delta U = Q - W$, just once to each step in a thermodynamic process, so you will often need additional equations. These may include Eq. (19.2), $W = \int_{V_1}^{V_2} p \, dV$, which gives the work W done in a volume change, and the equation of state of the material that makes up the thermodynamic system (for an ideal gas, $pV = nRT$).

EXECUTE *the solution* as follows:
1. Be sure to use consistent units. If p is in Pa and V in m³, then W is in joules. If a heat capacity is given in terms of calories,

convert it to joules. When you use $n = m_{total}/M$ to relate total mass m_{total} to number of moles n, remember that if m_{total} is in kilograms, M must be in *kilograms* per mole; M is usually tabulated in *grams* per mole.
2. The internal energy change ΔU in any thermodynamic process or series of processes is independent of the path, whether the substance is an ideal gas or not. If you can calculate ΔU for *any* path between given initial and final states, you know ΔU for *every possible path* between those states; you can then relate the various energy quantities for any of those other paths.
3. In a process comprising several steps, tabulate Q, W, and ΔU for each step, with one line per step and with the Q's, W's, and ΔU's forming columns (see Example 19.4). You can apply the first law to each line, and you can add each column and apply the first law to the sums. Do you see why?
4. Using steps 1–3, solve for the target variables.

EVALUATE *your answer:* Check your results for reasonableness. Ensure that each of your answers has the correct algebraic sign. A positive Q means that heat flows *into* the system; a negative Q means that heat flows *out of* the system. A positive W means that work is done *by* the system on its environment; a negative W means that work is done *on* the system by its environment.

Example 19.2 Working off your dessert

You propose to climb several flights of stairs to work off the energy you took in by eating a 900-calorie hot fudge sundae. How high must you climb? Assume that your mass is 60.0 kg.

SOLUTION

IDENTIFY and SET UP: The thermodynamic system is your body. You climb the stairs to make the final state of the system the same as the initial state (no fatter, no leaner). There is therefore no net change in internal energy: $\Delta U = 0$. Eating the hot fudge sundae corresponds to a heat flow into your body, and you do work climbing the stairs. We can relate these quantities using the first law of thermodynamics. We are given that $Q = 900$ food calories (900 kcal) of heat flow into your body. The work you must do to raise your mass m a height h is $W = mgh$; our target variable is h.

EXECUTE: From the first law of thermodynamics, $\Delta U = 0 = Q - W$, so $W = mgh = Q$. Hence you must climb to height $h = Q/mg$. First convert units: $Q = (900 \text{ kcal})(4186 \text{ J}/1 \text{ kcal}) = 3.77 \times 10^6$ J. Then

$$h = \frac{Q}{mg} = \frac{3.77 \times 10^6 \text{ J}}{(60.0 \text{ kg})(9.80 \text{ m/s}^2)} = 6410 \text{ m}$$

EVALUATE: We have unrealistically assumed 100% efficiency in the conversion of food energy into mechanical work. You would in fact have to climb considerably *less* than 6140 m (about 21,000 ft).

Example 19.3 A cyclic process

Figure 19.12 shows a *pV*-diagram for a *cyclic* process in which the initial and final states of some thermodynamic system are the same. As shown, the state of the system starts at point *a* and proceeds counterclockwise in the *pV*-diagram to point *b*, then back to *a*; the total work is $W = -500$ J. (a) Why is the work negative? (b) Find the change in internal energy and the heat added during this process.

SOLUTION

IDENTIFY and SET UP: We must relate the change in internal energy, the heat added, and the work done in a thermodynamic process. Hence we can apply the first law of thermodynamics. The process is cyclic, and it has two steps: $a \rightarrow b$ via the lower curve in Fig. 19.12 and $b \rightarrow a$ via the upper curve. We are asked only about the *entire* cyclic process $a \rightarrow b \rightarrow a$.

19.12 The net work done by the system in the process *aba* is −500 J. What would it have been if the process had proceeded clockwise in this *pV*-diagram?

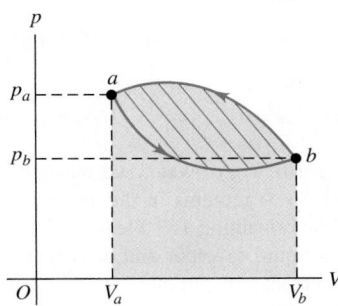

EXECUTE: (a) The work done in any step equals the area under the curve in the *pV*-diagram, with the area taken as positive if $V_2 > V_1$

and negative if $V_2 < V_1$; this rule yields the signs that result from the actual integrations in Eq. (19.2), $W = \int_{V_1}^{V_2} p\, dV$. The area under the lower curve $a \to b$ is therefore positive, but it is smaller than the absolute value of the (negative) area under the upper curve $b \to a$. Therefore the net area (the area enclosed by the path, shown with red stripes) and the net work W are negative. In other words, 500 J more work is done *on* the system than *by* the system in the complete process.

(b) In any cyclic process, $\Delta U = 0$, so $Q = W$. Here, that means $Q = -500$ J; that is, 500 J of heat flows *out of* the system.

EVALUATE:

In cyclic processes, the total work is positive if the process goes clockwise around the *pV*-diagram representing the cycle, and negative if the process goes counterclockwise (as here).

Example 19.4 **Comparing thermodynamic processes**

The *pV*-diagram of Fig. 19.13 shows a series of thermodynamic processes. In process *ab*, 150 J of heat is added to the system; in process *bd*, 600 J of heat is added. Find (a) the internal energy change in process *ab*; (b) the internal energy change in process *abd* (shown in light blue); and (c) the total heat added in process *acd* (shown in dark blue).

SOLUTION

IDENTIFY and SET UP: In each process we use $\Delta U = Q - W$ to determine the desired quantity. We are given $Q_{ab} = +150$ J and $Q_{bd} = +600$ J (both values are positive because heat is *added* to the system). Our target variables are (a) ΔU_{ab}, (b) ΔU_{abd}, and (c) Q_{acd}.

EXECUTE: (a) No volume change occurs during process *ab*, so the system does no work: $W_{ab} = 0$ and so $\Delta U_{ab} = Q_{ab} = 150$ J.

(b) Process *bd* is an expansion at constant pressure, so from Eq. (19.3),

$$W_{bd} = p(V_2 - V_1)$$
$$= (8.0 \times 10^4 \text{ Pa})(5.0 \times 10^{-3} \text{ m}^3 - 2.0 \times 10^{-3} \text{ m}^3)$$
$$= 240 \text{ J}$$

The total work for the two-step process *abd* is then

$$W_{abd} = W_{ab} + W_{bd} = 0 + 240 \text{ J} = 240 \text{ J}$$

and the total heat is

$$Q_{abd} = Q_{ab} + Q_{bd} = 150 \text{ J} + 600 \text{ J} = 750 \text{ J}$$

Applying Eq. (19.4) to *abd*, we then have

$$\Delta U_{abd} = Q_{abd} - W_{abd} = 750 \text{ J} - 240 \text{ J} = 510 \text{ J}$$

(c) Because ΔU is *independent of the path* from *a* to *d*, the internal energy change is the same for path *acd* as for path *abd*:

$$\Delta U_{acd} = \Delta U_{abd} = 510 \text{ J}$$

19.13 A *pV*-diagram showing the various thermodynamic processes.

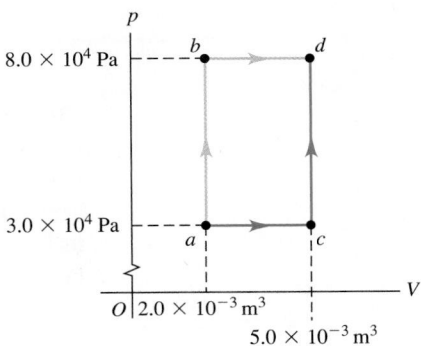

The total work for path *acd* is

$$W_{acd} = W_{ac} + W_{cd} = p(V_2 - V_1) + 0$$
$$= (3.0 \times 10^4 \text{ Pa})(5.0 \times 10^{-3} \text{ m}^3 - 2.0 \times 10^{-3} \text{ m}^3)$$
$$= 90 \text{ J}$$

Now we apply Eq. (19.5) to process *acd*:

$$Q_{acd} = \Delta U_{acd} + W_{acd} = 510 \text{ J} + 90 \text{ J} = 600 \text{ J}$$

We tabulate the quantities above:

Step	Q	W	$\Delta U = Q - W$	Step	Q	W	$\Delta U = Q - W$
ab	150 J	0 J	150 J	ac	?	90 J	?
bd	600 J	240 J	360 J	cd	?	0 J	?
abd	750 J	240 J	510 J	acd	600 J	90 J	510 J

EVALUATE: Be sure that you understand how each entry in the table above was determined. Although ΔU is the same (510 J) for *abd* and *acd*, W (240 J versus 90 J) and Q (750 J versus 600 J) are quite different. Although we couldn't find Q or ΔU for processes *ac* and *cd*, we could analyze the composite process *acd* by comparing it with process *abd*, which has the same initial and final states and for which we have more information.

Example 19.5 **Thermodynamics of boiling water**

One gram of water (1 cm^3) becomes 1671 cm^3 of steam when boiled at a constant pressure of 1 atm (1.013×10^5 Pa). The heat of vaporization at this pressure is $L_v = 2.256 \times 10^6$ J/kg. Compute (a) the work done by the water when it vaporizes and (b) its increase in internal energy.

SOLUTION

IDENTIFY and SET UP: The heat added causes the system (water) to change phase from liquid to vapor. We can analyze this process using the first law of thermodynamics, which holds for thermodynamic processes of *all* kinds. The water is boiled at constant pressure, so we can use Eq. (19.3) to calculate the work W done by the vaporizing water as it expands. We are given the mass of water and the heat of vaporization, so we can use Eq. (17.20), $Q = mL_v$, to calculate the heat Q added to the water. We can then find the internal energy change using Eq. (19.4), $\Delta U = Q - W$.

EXECUTE: (a) From Eq. (19.3), the water does work

$$W = p(V_2 - V_1)$$
$$= (1.013 \times 10^5 \text{ Pa})(1671 \times 10^{-6} \text{ m}^3 - 1 \times 10^{-6} \text{ m}^3)$$
$$= 169 \text{ J}$$

(b) From Eq. (17.20), the heat added to the water is

$$Q = mL_v = (10^{-3} \text{ kg})(2.256 \times 10^6 \text{ J/kg}) = 2256 \text{ J}$$

Then from Eq. (19.4),

$$\Delta U = Q - W = 2256 \text{ J} - 169 \text{ J} = 2087 \text{ J}$$

EVALUATE: To vaporize 1 g of water, we must add 2256 J of heat, most of which (2087 J) remains in the system as an increase in internal energy. The remaining 169 J leaves the system as the system expands from liquid to vapor and does work against the surroundings. (The increase in internal energy is associated mostly with the attractive intermolecular forces that hold the molecules together in the liquid state. The associated potential energies are greater after work has been done to pull the molecules apart, forming the vapor state. It's like increasing gravitational potential energy by pulling an elevator farther from the center of the earth.)

Infinitesimal Changes of State

In the preceding examples the initial and final states differ by a finite amount. Later we will consider *infinitesimal* changes of state in which a small amount of heat dQ is added to the system, the system does a small amount of work dW, and its internal energy changes by an amount dU. For such a process we state the first law in differential form as

$$dU = dQ - dW \qquad \text{(first law of thermodynamics, infinitesimal process)} \qquad (19.6)$$

For the systems we will discuss, the work dW is given by $dW = p\,dV$, so we can also state the first law as

$$dU = dQ - p\,dV \qquad (19.7)$$

Test Your Understanding of Section 19.4 Rank the following thermodynamic processes according to the change in internal energy in each process, from most positive to most negative. (i) As you do 250 J of work on a system, it transfers 250 J of heat to its surroundings; (ii) as you do 250 J of work on a system, it absorbs 250 J of heat from its surroundings; (iii) as a system does 250 J of work on you, it transfers 250 J of heat to its surroundings; (iv) as a system does 250 J of work on you, it absorbs 250 J of heat from its surroundings.

ActivPhysics 8.4: State Variables and Ideal Gas Law

19.5 Kinds of Thermodynamic Processes

In this section we describe four specific kinds of thermodynamic processes that occur often in practical situations. These can be summarized briefly as "no heat transfer" or *adiabatic,* "constant volume" or *isochoric,* "constant pressure" or *isobaric,* and "constant temperature" or *isothermal.* For some of these processes we can use a simplified form of the first law of thermodynamics.

Adiabatic Process

An **adiabatic process** (pronounced "ay-dee-ah-*bat*-ic") is defined as one with no heat transfer into or out of a system; $Q = 0$. We can prevent heat flow either by surrounding the system with thermally insulating material or by carrying out the process so quickly that there is not enough time for appreciable heat flow. From the first law we find that for every adiabatic process,

$$U_2 - U_1 = \Delta U = -W \qquad \text{(adiabatic process)} \qquad (19.8)$$

When a system expands adiabatically, W is positive (the system does work on its surroundings), so ΔU is negative and the internal energy decreases. When a system is *compressed* adiabatically, W is negative (work is done on the system by its surroundings) and U increases. In many (but not all) systems an increase of internal energy is accompanied by a rise in temperature, and a decrease in internal energy by a drop in temperature (Fig. 19.14).

The compression stroke in an internal-combustion engine is an approximately adiabatic process. The temperature rises as the air–fuel mixture in the cylinder is compressed. The expansion of the burned fuel during the power stroke is also an approximately adiabatic expansion with a drop in temperature. In Section 19.8 we'll consider adiabatic processes in an ideal gas.

Isochoric Process

An **isochoric process** (pronounced "eye-so-*kor*-ic") is a *constant-volume* process. When the volume of a thermodynamic system is constant, it does no work on its surroundings. Then $W = 0$ and

$$U_2 - U_1 = \Delta U = Q \qquad \text{(isochoric process)} \qquad (19.9)$$

In an isochoric process, all the energy added as heat remains in the system as an increase in internal energy. Heating a gas in a closed constant-volume container is an example of an isochoric process. The processes *ab* and *cd* in Example 19.4 are also examples of isochoric processes. (Note that there are types of work that do not involve a volume change. For example, we can do work on a fluid by stirring it. In some literature, "isochoric" is used to mean that no work of any kind is done.)

Isobaric Process

An **isobaric process** (pronounced "eye-so-*bear*-ic") is a *constant-pressure* process. In general, none of the three quantities ΔU, Q, and W is zero in an isobaric process, but calculating W is easy nonetheless. From Eq. (19.3),

$$W = p(V_2 - V_1) \qquad \text{(isobaric process)} \qquad (19.10)$$

Example 19.5 concerns an isobaric process, boiling water at constant pressure (Fig. 19.15).

Isothermal Process

An **isothermal process** is a *constant-temperature* process. For a process to be isothermal, any heat flow into or out of the system must occur slowly enough that thermal equilibrium is maintained. In general, none of the quantities ΔU, Q, or W is zero in an isothermal process.

In some special cases the internal energy of a system depends *only* on its temperature, not on its pressure or volume. The most familiar system having this special property is an ideal gas, as we'll discuss in the next section. For such systems, if the temperature is constant, the internal energy is also constant; $\Delta U = 0$ and $Q = W$. That is, any energy entering the system as heat Q must leave it again as work W done by the system. Example 19.1, involving an ideal gas, is an example of an isothermal process in which U is also constant. For most systems other than ideal gases, the internal energy depends on pressure as well as temperature, so U may vary even when T is constant.

19.14 When the cork is popped on a bottle of champagne, the pressurized gases inside the bottle expand rapidly and do work on the outside air ($W > 0$). There is no time for the gases to exchange heat with their surroundings, so the expansion is adiabatic ($Q = 0$). Hence the internal energy of the expanding gases decreases ($\Delta U = -W < 0$) and their temperature drops. This makes water vapor condense and form a miniature cloud.

19.15 Most cooking involves isobaric processes. That's because the air pressure above a saucepan or frying pan, or inside a microwave oven, remains essentially constant while the food is being heated.

19.16 Four different processes for a constant amount of an ideal gas, all starting at state a. For the adiabatic process, $Q = 0$; for the isochoric process, $W = 0$; and for the isothermal process, $\Delta U = 0$. The temperature increases only during the isobaric expansion.

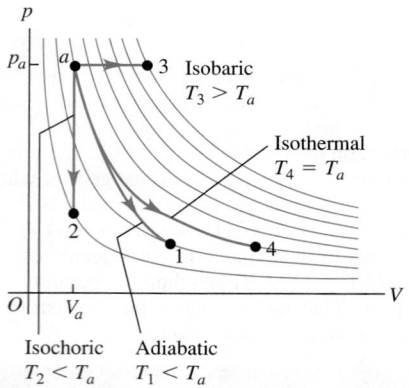

19.17 The partition is broken (or removed) to start the free expansion of gas into the vacuum region.

Figure 19.16 shows a pV-diagram for these four processes for a constant amount of an ideal gas. The path followed in an adiabatic process (a to 1) is called an **adiabat.** A vertical line (constant volume) is an **isochor,** a horizontal line (constant pressure) is an **isobar,** and a curve of constant temperature (shown as light blue lines in Fig. 19.16) is an **isotherm.**

Test Your Understanding of Section 19.5 Which of the processes in Fig. 19.7 are isochoric? Which are isobaric? Is it possible to tell if any of the processes are isothermal or adiabatic? ▌

19.6 Internal Energy of an Ideal Gas

We now show that for an ideal gas, the internal energy U depends only on temperature, not on pressure or volume. Let's think again about the free-expansion experiment described in Section 19.3. A thermally insulated container with rigid walls is divided into two compartments by a partition (Fig. 19.17). One compartment has a quantity of an ideal gas and the other is evacuated.

When the partition is removed or broken, the gas expands to fill both parts of the container. The gas does no work on its surroundings because the walls of the container don't move, and there is no heat flow through the insulation. So both Q and W are zero and the internal energy U is constant. This is true of any substance, whether it is an ideal gas or not.

Does the *temperature* change during a free expansion? Suppose it *does* change, while the internal energy stays the same. In that case we have to conclude that the internal energy depends on both the temperature and the volume or on both the temperature and the pressure, but certainly not on the temperature alone. But if T is constant during a free expansion, for which we know that U is constant even though both p and V change, then we have to conclude that U depends only on T, not on p or V.

Many experiments have shown that when a low-density gas undergoes a free expansion, its temperature *does not* change. Such a gas is essentially an ideal gas. The conclusion is:

> **The internal energy of an ideal gas depends only on its temperature, not on its pressure or volume.**

This property, in addition to the ideal-gas equation of state, is part of the ideal-gas model. Make sure you understand that U depends only on T for an ideal gas, for we will make frequent use of this fact.

For nonideal gases, some temperature change occurs during free expansions, even though the internal energy is constant. This shows that the internal energy cannot depend *only* on temperature; it must depend on pressure as well. From the microscopic viewpoint, in which internal energy U is the sum of the kinetic and potential energies for all the particles that make up the system, this is not surprising. Nonideal gases usually have attractive intermolecular forces, and when molecules move farther apart, the associated potential energies increase. If the total internal energy is constant, the kinetic energies must decrease. Temperature is directly related to molecular *kinetic* energy, and for such a gas a free expansion is usually accompanied by a *drop* in temperature.

Test Your Understanding of Section 19.6 Is the internal energy of a solid likely to be independent of its volume, as is the case for an ideal gas? Explain your reasoning. (*Hint:* See Fig. 18.20.) ▌

19.7 Heat Capacities of an Ideal Gas

We defined specific heat and molar heat capacity in Section 17.5. We also remarked at the end of that section that the specific heat or molar heat capacity of a substance depends on the conditions under which the heat is added. It is usually easiest to measure the heat capacity of a gas in a closed container under constant-volume conditions. The corresponding heat capacity is the **molar heat capacity at constant volume,** denoted by C_V. Heat capacity measurements for solids and liquids are usually carried out in the atmosphere under constant atmospheric pressure, and we call the corresponding heat capacity the **molar heat capacity at constant pressure,** C_p. If neither p nor V is constant, we have an infinite number of possible heat capacities.

Let's consider C_V and C_p for an ideal gas. To measure C_V, we raise the temperature of an ideal gas in a rigid container with constant volume, neglecting its thermal expansion (Fig. 19.18a). To measure C_p, we let the gas expand just enough to keep the pressure constant as the temperature rises (Fig. 19.18b).

Why should these two molar heat capacities be different? The answer lies in the first law of thermodynamics. In a constant-volume temperature increase, the system does no work, and the change in internal energy ΔU equals the heat added Q. In a constant-pressure temperature increase, on the other hand, the volume *must* increase; otherwise, the pressure (given by the ideal-gas equation of state, $p = nRT/V$) could not remain constant. As the material expands, it does an amount of work W. According to the first law,

$$Q = \Delta U + W \qquad (19.11)$$

For a given temperature increase, the internal energy change ΔU of an ideal gas has the same value no matter what the process (remember that the internal energy of an ideal gas depends only on temperature, not on pressure or volume). Equation (19.11) then shows that the heat input for a constant-pressure process must be *greater* than that for a constant-volume process because additional energy must be supplied to account for the work W done during the expansion. So C_p is greater than C_V for an ideal gas. The pV-diagram in Fig. 19.19 shows this relationship. For air, C_p is 40% greater than C_V.

For a very few substances (one of which is water between 0°C and 4°C) the volume *decreases* during heating. In this case, W is negative, the heat input is *less* than in the constant-volume case, and C_p is *less* than C_V.

Relating C_p and C_V for an Ideal Gas

We can derive a simple relationship between C_p and C_V for an ideal gas. First consider the constant-*volume* process. We place n moles of an ideal gas at temperature T in a constant-volume container. We place it in thermal contact with a hotter body; an infinitesimal quantity of heat dQ flows into the gas, and its temperature increases by an infinitesimal amount dT. By the definition of C_V, the molar heat capacity at constant volume,

$$dQ = nC_V\,dT \qquad (19.12)$$

The pressure increases during this process, but the gas does no work ($dW = 0$) because the volume is constant. The first law in differential form, Eq. (19.6), is $dQ = dU + dW$. Since $dW = 0$, $dQ = dU$ and Eq. (19.12) can also be written as

$$dU = nC_V\,dT \qquad (19.13)$$

Now consider a constant-*pressure* process with the same temperature change dT. We place the same gas in a cylinder with a piston that we can allow to move just enough to maintain constant pressure, as shown in Fig. 19.18b. Again we bring the system into contact with a hotter body. As heat flows into

19.18 Measuring the molar heat capacity of an ideal gas **(a)** at constant volume and **(b)** at constant pressure.

(a) Constant volume: $dQ = nC_V\,dT$

(b) Constant pressure: $dQ = nC_p\,dT$

19.19 Raising the temperature of an ideal gas from T_1 to T_2 by a constant-volume or a constant-pressure process. For an ideal gas, U depends only on T, so ΔU is the same for both processes. But for the constant-pressure process, more heat Q must be added to both increase U and do work W. Hence $C_p > C_V$.

ActivPhysics 8.7: Heat Capacity
ActivPhysics 8.8: Isochoric Process
ActivPhysics 8.9: Isobaric Process
ActivPhysics 8.10: Isothermal Process

the gas, it expands at constant pressure and does work. By the definition of C_p, the molar heat capacity at constant pressure, the amount of heat dQ entering the gas is

$$dQ = nC_p\, dT \tag{19.14}$$

The work dW done by the gas in this constant-pressure process is

$$dW = p\, dV$$

We can also express dW in terms of the temperature change dT by using the ideal-gas equation of state, $pV = nRT$. Because p is constant, the change in V is proportional to the change in T:

$$dW = p\, dV = nR\, dT \tag{19.15}$$

Now we substitute Eqs. (19.14) and (19.15) into the first law, $dQ = dU + dW$. We obtain

$$nC_p\, dT = dU + nR\, dT \tag{19.16}$$

Now here comes the crux of the calculation. The internal energy change dU for the constant-pressure process is again given by Eq. (19.13), $dU = nC_V\, dT$, *even though now the volume is not constant.* Why is this so? Recall the discussion of Section 19.6; one of the special properties of an ideal gas is that its internal energy depends *only* on temperature. Thus the *change* in internal energy during any process must be determined only by the temperature change. If Eq. (19.13) is valid for an ideal gas for one particular kind of process, it must be valid for an ideal gas for *every* kind of process with the same dT. So we may replace dU in Eq. (19.16) by $nC_V\, dT$:

$$nC_p\, dT = nC_V\, dT + nR\, dT$$

When we divide each term by the common factor $n\, dT$, we get

$$C_p = C_V + R \qquad \text{(molar heat capacities of an ideal gas)} \tag{19.17}$$

As we predicted, the molar heat capacity of an ideal gas at constant pressure is *greater* than the molar heat capacity at constant volume; the difference is the gas constant R. (Of course, R must be expressed in the same units as C_p and C_V, such as J/mol · K.)

We have used the ideal-gas model to derive Eq. (19.17), but it turns out to be obeyed to within a few percent by many real gases at moderate pressures. Measured values of C_p and C_V are given in Table 19.1 for several real gases at low pressures; the difference in most cases is approximately $R = 8.314$ J/mol · K.

The table also shows that the molar heat capacity of a gas is related to its molecular structure, as we discussed in Section 18.4. In fact, the first two columns of Table 19.1 are the same as Table 18.1.

Table 19.1 Molar Heat Capacities of Gases at Low Pressure

Type of Gas	Gas	C_V (J/mol · K)	C_p (J/mol · K)	$C_p - C_V$ (J/mol · K)	$\gamma = C_p/C_V$
Monatomic	He	12.47	20.78	8.31	1.67
	Ar	12.47	20.78	8.31	1.67
Diatomic	H_2	20.42	28.74	8.32	1.41
	N_2	20.76	29.07	8.31	1.40
	O_2	20.85	29.17	8.31	1.40
	CO	20.85	29.16	8.31	1.40
Polyatomic	CO_2	28.46	36.94	8.48	1.30
	SO_2	31.39	40.37	8.98	1.29
	H_2S	25.95	34.60	8.65	1.33

The Ratio of Heat Capacities

The last column of Table 19.1 lists the values of the dimensionless **ratio of heat capacities,** C_p/C_V, denoted by γ (the Greek letter gamma):

$$\gamma = \frac{C_p}{C_V} \qquad \text{(ratio of heat capacities)} \qquad (19.18)$$

(This is sometimes called the "ratio of specific heats.") For gases, C_p is always greater than C_V and γ is always greater than unity. This quantity plays an important role in *adiabatic* processes for an ideal gas, which we will study in the next section.

We can use our kinetic-theory discussion of the molar heat capacity of an ideal gas (see Section 18.4) to predict values of γ. As an example, an ideal monatomic gas has $C_V = \frac{3}{2}R$. From Eq. (19.17),

$$C_p = C_V + R = \tfrac{3}{2}R + R = \tfrac{5}{2}R$$

so

$$\gamma = \frac{C_p}{C_V} = \frac{\frac{5}{2}R}{\frac{3}{2}R} = \tfrac{5}{3} = 1.67$$

As Table 19.1 shows, this agrees well with values of γ computed from measured heat capacities. For most diatomic gases near room temperature, $C_V = \frac{5}{2}R$, $C_p = C_V + R = \frac{7}{2}R$, and

$$\gamma = \frac{C_p}{C_V} = \frac{\frac{7}{2}R}{\frac{5}{2}R} = \tfrac{7}{5} = 1.40$$

also in good agreement with measured values.

Here's a final reminder: For an ideal gas the internal energy change in *any* process is given by $\Delta U = nC_V \Delta T$, *whether the volume is constant or not.* This relationship, which comes in handy in the following example, holds for other substances *only* when the volume is constant.

Example 19.6   Cooling your room

A typical dorm room or bedroom contains about 2500 moles of air. Find the change in the internal energy of this much air when it is cooled from 35.0°C to 26.0°C at a constant pressure of 1.00 atm. Treat the air as an ideal gas with $\gamma = 1.400$.

SOLUTION

IDENTIFY and SET UP: Our target variable is the change in the internal energy ΔU of an ideal gas in a constant-pressure process. We are given the number of moles, the temperature change, and the value of γ for air. We use Eq. (19.13), $\Delta U = nC_V \Delta T$, which gives the internal energy change for an ideal gas in *any* process, *whether the volume is constant or not.* [See the discussion following Eq. (19.16).] We use Eqs. (19.17) and (19.18) to find C_V.

EXECUTE: From Eqs. (19.17) and (19.18),

$$\gamma = \frac{C_p}{C_V} = \frac{C_V + R}{C_V} = 1 + \frac{R}{C_V}$$

$$C_V = \frac{R}{\gamma - 1} = \frac{8.314 \text{ J/mol} \cdot \text{K}}{1.400 - 1} = 20.79 \text{ J/mol} \cdot \text{K}$$

Then from Eq. (19.13),

$$\begin{aligned}
\Delta U &= nC_V \Delta T \\
&= (2500 \text{ mol})(20.79 \text{ J/mol} \cdot \text{K})(26.0°\text{C} - 35.0°\text{C}) \\
&= -4.68 \times 10^5 \text{ J}
\end{aligned}$$

EVALUATE: To cool 2500 moles of air from 35.0°C to 26.0°C, a room air conditioner must extract this much internal energy from the air and transfer it to the air outside. In Chapter 20 we'll discuss how this is done.

Test Your Understanding of Section 19.7 You want to cool a storage cylinder containing 10 moles of compressed gas from 30°C to 20°C. For which kind of gas would this be easiest? (i) a monatomic gas; (ii) a diatomic gas; (iii) a polyatomic gas; (iv) it would be equally easy for all of these.

19.8 Adiabatic Processes for an Ideal Gas

An adiabatic process, defined in Section 19.5, is a process in which no heat transfer takes place between a system and its surroundings. Zero heat transfer is an idealization, but a process is approximately adiabatic if the system is well insulated or if the process takes place so quickly that there is not enough time for appreciable heat flow to occur.

In an adiabatic process, $Q = 0$, so from the first law, $\Delta U = -W$. An adiabatic process for an ideal gas is shown in the pV-diagram of Fig. 19.20. As the gas expands from volume V_a to V_b, it does positive work, so its internal energy decreases and its temperature drops. If point a, representing the initial state, lies on an isotherm at temperature $T + dT$, then point b for the final state is on a different isotherm at a lower temperature T. For an ideal gas an adiabatic curve (adiabat) at any point is always *steeper* than the isotherm passing through the same point. For an adiabatic *compression* from V_b to V_a the situation is reversed and the temperature rises.

The air in the output hoses of air compressors used in gasoline stations, in paint-spraying equipment, and to fill scuba tanks is always warmer than the air entering the compressor; this is because the compression is rapid and hence approximately adiabatic. Adiabatic *cooling* occurs when you open a bottle of your favorite carbonated beverage. The gas just above the beverage surface expands rapidly in a nearly adiabatic process; the temperature of the gas drops so much that water vapor in the gas condenses, forming a miniature cloud (see Fig. 19.14).

CAUTION **"Heating" and "cooling" without heat** Keep in mind that when we talk about "adiabatic heating" and "adiabatic cooling," we really mean "raising the temperature" and "lowering the temperature," respectively. In an adiabatic process, the temperature change is due to work done by or on the system; there is *no* heat flow at all.

19.20 A pV-diagram of an adiabatic ($Q = 0$) process for an ideal gas. As the gas expands from V_a to V_b, it does positive work W on its environment, its internal energy decreases ($\Delta U = -W < 0$), and its temperature drops from $T + dT$ to T. (An adiabatic process is also shown in Fig. 19.16.)

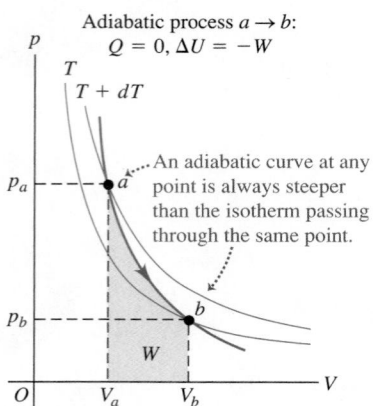

Adiabatic process $a \rightarrow b$:
$Q = 0$, $\Delta U = -W$

An adiabatic curve at any point is always steeper than the isotherm passing through the same point.

Mastering PHYSICS

ActivPhysics 8.11: Adiabatic Process

Adiabatic Ideal Gas: Relating V, T, and p

We can derive a relationship between volume and temperature changes for an infinitesimal adiabatic process in an ideal gas. Equation (19.13) gives the internal energy change dU for *any* process for an ideal gas, adiabatic or not, so we have $dU = nC_V \, dT$. Also, the work done by the gas during the process is given by $dW = p \, dV$. Then, since $dU = -dW$ for an adiabatic process, we have

$$nC_V \, dT = -p \, dV \qquad (19.19)$$

To obtain a relationship containing only the volume V and temperature T, we eliminate p using the ideal-gas equation in the form $p = nRT/V$. Substituting this into Eq. (19.19) and rearranging, we get

$$nC_V \, dT = -\frac{nRT}{V} \, dV$$

$$\frac{dT}{T} + \frac{R}{C_V} \frac{dV}{V} = 0$$

The coefficient R/C_V can be expressed in terms of $\gamma = C_p/C_V$. We have

$$\frac{R}{C_V} = \frac{C_p - C_V}{C_V} = \frac{C_p}{C_V} - 1 = \gamma - 1$$

$$\frac{dT}{T} + (\gamma - 1)\frac{dV}{V} = 0 \qquad (19.20)$$

Because γ is always greater than unity for a gas, $(\gamma - 1)$ is always positive. This means that in Eq. (19.20), dV and dT always have opposite signs. An adiabatic *expansion* of an ideal gas $(dV > 0)$ always occurs with a *drop* in temperature $(dT < 0)$, and an adiabatic *compression* $(dV < 0)$ always occurs with a *rise* in temperature $(dT > 0)$; this confirms our earlier prediction.

For finite changes in temperature and volume we integrate Eq. (19.20), obtaining

$$\ln T + (\gamma - 1) \ln V = \text{constant}$$

$$\ln T + \ln V^{\gamma-1} = \text{constant}$$

$$\ln (TV^{\gamma-1}) = \text{constant}$$

and finally,

$$TV^{\gamma-1} = \text{constant} \qquad (19.21)$$

Thus for an initial state (T_1, V_1) and a final state (T_2, V_2),

$$T_1 V_1^{\gamma-1} = T_2 V_2^{\gamma-1} \quad \text{(adiabatic process, ideal gas)} \qquad (19.22)$$

Because we have used the ideal-gas equation in our derivation of Eqs. (19.21) and (19.22), the T's must always be *absolute* (Kelvin) temperatures.

We can also convert Eq. (19.21) into a relationship between pressure and volume by eliminating T, using the ideal-gas equation in the form $T = pV/nR$. Substituting this into Eq. (19.21), we find

$$\frac{pV}{nR}V^{\gamma-1} = \text{constant}$$

or, because n and R are constant,

$$pV^{\gamma} = \text{constant} \qquad (19.23)$$

For an initial state (p_1, V_1) and a final state (p_2, V_2), Eq. (19.23) becomes

$$p_1 V_1^{\gamma} = p_2 V_2^{\gamma} \quad \text{(adiabatic process, ideal gas)} \qquad (19.24)$$

We can also calculate the *work* done by an ideal gas during an adiabatic process. We know that $Q = 0$ and $W = -\Delta U$ for *any* adiabatic process. For an ideal gas, $\Delta U = nC_V(T_2 - T_1)$. If the number of moles n and the initial and final temperatures T_1 and T_2 are known, we have simply

$$W = nC_V(T_1 - T_2) \quad \text{(adiabatic process, ideal gas)} \qquad (19.25)$$

We may also use $pV = nRT$ in this equation to obtain

$$W = \frac{C_V}{R}(p_1 V_1 - p_2 V_2) = \frac{1}{\gamma - 1}(p_1 V_1 - p_2 V_2) \quad \begin{matrix}\text{(adiabatic process,}\\ \text{ideal gas)}\end{matrix} \qquad (19.26)$$

(We used the result $C_V = R/(\gamma - 1)$ from Example 19.6.) If the process is an expansion, the temperature drops, T_1 is greater than T_2, $p_1 V_1$ is greater than $p_2 V_2$, and the work is *positive,* as we should expect. If the process is a compression, the work is negative.

Throughout this analysis of adiabatic processes we have used the ideal-gas equation of state, which is valid only for *equilibrium* states. Strictly speaking, our results are valid only for a process that is fast enough to prevent appreciable heat exchange with the surroundings (so that $Q = 0$ and the process is adiabatic), yet slow enough that the system does not depart very much from thermal and mechanical equilibrium. Even when these conditions are not strictly satisfied, though, Eqs. (19.22), (19.24), and (19.26) give useful approximate results.

Example 19.7 **Adiabatic compression in a diesel engine**

The compression ratio of a diesel engine is 15.0 to 1; that is, air in a cylinder is compressed to $\frac{1}{(15.0)}$ of its initial volume (Fig. 19.21). (a) If the initial pressure is 1.01×10^5 Pa and the initial temperature is 27°C (300 K), find the final pressure and the temperature after adiabatic compression. (b) How much work does the gas do during the compression if the initial volume of the cylinder is $1.00 \text{ L} = 1.00 \times 10^{-3} \text{ m}^3$? Use the values $C_V = 20.8 \text{ J/mol} \cdot \text{K}$ and $\gamma = 1.400$ for air.

SOLUTION

IDENTIFY and SET UP: This problem involves the adiabatic compression of an ideal gas, so we can use the ideas of this section. In part (a) we are given the initial pressure and temperature $p_1 = 1.01 \times 10^5$ Pa and $T_1 = 300$ K; the ratio of initial and final volumes is $V_1/V_2 = 15.0$. We use Eq. (19.22) to find the final temperature T_2 and Eq. (19.24) to find the final pressure p_2. In part (b) our target variable is W, the work done *by* the gas during the adiabatic compression. We use Eq. (19.26) to calculate W.

19.21 Adiabatic compression of air in a cylinder of a diesel engine.

Initial volume

Maximum compression

$V_2 = \dfrac{1}{(15.0)} V_1$

V_1

EXECUTE: (a) From Eqs. (19.22) and (19.24),

$$T_2 = T_1\left(\frac{V_1}{V_2}\right)^{\gamma-1} = (300 \text{ K})(15.0)^{0.40} = 886 \text{ K} = 613°\text{C}$$

$$p_2 = p_1\left(\frac{V_1}{V_2}\right)^{\gamma} = (1.01 \times 10^5 \text{ Pa})(15.0)^{1.40}$$

$$= 44.8 \times 10^5 \text{ Pa} = 44 \text{ atm}$$

(b) From Eq. (19.26), the work done is

$$W = \frac{1}{\gamma - 1}(p_1V_1 - p_2V_2)$$

Using $V_1/V_2 = 15.0$, this becomes

$$W = \frac{1}{1.400 - 1}\left[\begin{array}{c} (1.01 \times 10^5 \text{ Pa})(1.00 \times 10^{-3} \text{ m}^3) \\ - (44.8 \times 10^5 \text{ Pa})\left(\dfrac{1.00 \times 10^{-3} \text{ m}^3}{15.0}\right) \end{array} \right]$$

$$= -494 \text{ J}$$

EVALUATE: If the compression had been isothermal, the final pressure would have been 15.0 atm. Because the temperature also increases during an adiabatic compression, the final pressure is much greater. When fuel is injected into the cylinders near the end of the compression stroke, the high temperature of the air attained during compression causes the fuel to ignite spontaneously without the need for spark plugs.

We can check our result in part (b) using Eq. (19.25). The number of moles of gas in the cylinder is

$$n = \frac{p_1V_1}{RT_1} = \frac{(1.01 \times 10^5 \text{ Pa})(1.00 \times 10^{-3} \text{ m}^3)}{(8.314 \text{ J/mol} \cdot \text{K})(300 \text{ K})} = 0.0405 \text{ mol}$$

Then Eq. (19.25) gives

$$W = nC_V(T_1 - T_2)$$
$$= (0.0405 \text{ mol})(20.8 \text{ J/mol} \cdot \text{K})(300 \text{ K} - 886 \text{ K})$$
$$= -494 \text{ J}$$

The work is negative because the gas is compressed.

Test Your Understanding of Section 19.8 You have four samples of ideal gas, each of which contains the same number of moles of gas and has the same initial temperature, volume, and pressure. You compress each sample to one-half of its initial volume. Rank the four samples in order from highest to lowest value of the final pressure. (i) a monatomic gas compressed isothermally; (ii) a monatomic gas compressed adiabatically; (iii) a diatomic gas compressed isothermally; (iv) a diatomic gas compressed adiabatically.

Heat and work in thermodynamic processes: A thermodynamic system has the potential to exchange energy with its surroundings by heat transfer or by mechanical work. When a system at pressure p changes volume from V_1 to V_2, it does an amount of work W given by the integral of p with respect to volume. If the pressure is constant, the work done is equal to p times the change in volume. A negative value of W means that work is done on the system. (See Example 19.1.)

In any thermodynamic process, the heat added to the system and the work done by the system depend not only on the initial and final states, but also on the path (the series of intermediate states through which the system passes).

$$W = \int_{V_1}^{V_2} p \, dV \qquad (19.2)$$

$$W = p(V_2 - V_1) \qquad (19.3)$$
(constant pressure only)

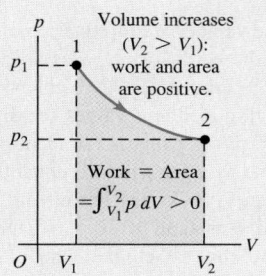

The first law of thermodynamics: The first law of thermodynamics states that when heat Q is added to a system while the system does work W, the internal energy U changes by an amount equal to $Q - W$. This law can also be expressed for an infinitesimal process. (See Examples 19.2, 19.3, and 19.5.)

The internal energy of any thermodynamic system depends only on its state. The change in internal energy in any process depends only on the initial and final states, not on the path. The internal energy of an isolated system is constant. (See Example 19.4.)

$$\Delta U = Q - W \qquad (19.4)$$

$$dU = dQ - dW \qquad (19.6)$$
(infinitesimal process)

Important kinds of thermodynamic processes:

- Adiabatic process: No heat transfer into or out of a system; $Q = 0$.
- Isochoric process: Constant volume; $W = 0$.
- Isobaric process: Constant pressure; $W = p(V_2 - V_1)$.
- Isothermal process: Constant temperature.

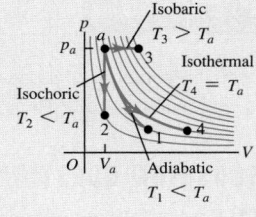

Thermodynamics of ideal gases: The internal energy of an ideal gas depends only on its temperature, not on its pressure or volume. For other substances the internal energy generally depends on both pressure and temperature.

The molar heat capacities C_V and C_p of an ideal gas differ by R, the ideal-gas constant. The dimensionless ratio of heat capacities, C_p/C_V, is denoted by γ. (See Example 19.6.)

$$C_p = C_V + R \qquad (19.17)$$

$$\gamma = \frac{C_p}{C_V} \qquad (19.18)$$

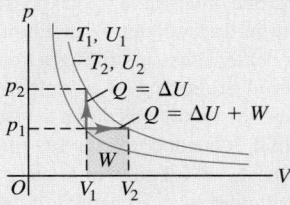

Adiabatic processes in ideal gases: For an adiabatic process for an ideal gas, the quantities $TV^{\gamma-1}$ and pV^{γ} are constant. The work done by an ideal gas during an adiabatic expansion can be expressed in terms of the initial and final values of temperature, or in terms of the initial and final values of pressure and volume. (See Example 19.7.)

$$W = nC_V(T_1 - T_2)$$

$$= \frac{C_V}{R}(p_1 V_1 - p_2 V_2) \qquad (19.25)$$

$$= \frac{1}{\gamma - 1}(p_1 V_1 - p_2 V_2) \qquad (19.26)$$

| BRIDGING PROBLEM | Work Done by a Van Der Waals Gas |

The van der Waals equation of state, an approximate representation of the behavior of gases at high pressure, is given by Eq. (18.7): $[p + (an^2/V^2)](V - nb) = nRT$, where a and b are constants having different values for different gases. (In the special case of $a = b = 0$, this is the ideal-gas equation.) (a) Calculate the work done by a gas with this equation of state in an isothermal expansion from V_1 to V_2. (b) For ethane gas (C_2H_6), $a = 0.554$ J·m³/mol² and $b = 6.38 \times 10^{-5}$ m³/mol. Calculate the work W done by 1.80 mol of ethane when it expands from 2.00×10^{-3} m³ to 4.00×10^{-3} m³ at a constant temperature of 300 K. Do the calculation using (i) the van der Waals equation of state and (ii) the ideal-gas equation of state. (c) For which equation of state is W larger? Why should this be so?

| SOLUTION GUIDE |

See MasteringPhysics® study area for a Video Tutor solution.

IDENTIFY and SET UP

1. Review the discussion of the van der Waals equation of state in Section 18.1. What is the significance of the quantities a and b?
2. Decide how to find the work done by an expanding gas whose pressure p does not depend on V in the same way as for an ideal gas. (*Hint:* See Section 19.2.)
3. How will you find the work done by an expanding ideal gas?

EXECUTE

4. Find the general expression for the work done by a van der Waals gas as it expands from volume V_1 to volume V_2. (*Hint:* If you set $a = b = 0$ in your result, it should reduce to the expression for the work done by an expanding ideal gas.)
5. Use your result from step 4 to solve part (b) for ethane treated as a van der Waals gas.
6. Use the formula you chose in step 3 to solve part (b) for ethane treated as an ideal gas.

EVALUATE

7. Is the difference between W for the two equations of state large enough to be significant?
8. Does the term with a in the van der Waals equation of state increase or decrease the amount of work done? What about the term with b? Which one is more important for the ethane in this problem?

| Problems | For instructor-assigned homework, go to www.masteringphysics.com |

•, ••, •••: Problems of increasing difficulty. **CP**: Cumulative problems incorporating material from earlier chapters. **CALC**: Problems requiring calculus. **BIO**: Biosciences problems.

DISCUSSION QUESTIONS

Q19.1 For the following processes, is the work done by the system (defined as the expanding or contracting gas) on the environment positive or negative? (a) expansion of the burned gasoline–air mixture in the cylinder of an automobile engine; (b) opening a bottle of champagne; (c) filling a scuba tank with compressed air; (d) partial crumpling of a sealed, empty water bottle, as you drive from the mountains down to sea level.

Q19.2 It is not correct to say that a body contains a certain amount of heat, yet a body can transfer heat to another body. How can a body give away something it does not have in the first place?

Q19.3 In which situation must you do more work: inflating a balloon at sea level or inflating the same balloon to the same volume at the summit of Mt. McKinley? Explain in terms of pressure and volume change.

Q19.4 If you are told the initial and final states of a system and the associated change in internal energy, can you determine whether the internal energy change was due to work or to heat transfer? Explain.

Q19.5 Discuss the application of the first law of thermodynamics to a mountaineer who eats food, gets warm and perspires a lot during a climb, and does a lot of mechanical work in raising herself to the summit. The mountaineer also gets warm during the descent. Is the source of this energy the same as the source during the ascent?

Q19.6 When ice melts at 0°C, its volume decreases. Is the internal energy change greater than, less than, or equal to the heat added? How can you tell?

Q19.7 You hold an inflated balloon over a hot-air vent in your house and watch it slowly expand. You then remove it and let it cool back to room temperature. During the expansion, which was larger: the heat added to the balloon or the work done by the air inside it? Explain. (Assume that air is an ideal gas.) Once the balloon has returned to room temperature, how does the net heat gained or lost by the air inside it compare to the net work done on or by the surrounding air?

Q19.8 You bake chocolate chip cookies and put them, still warm, in a container with a loose (not airtight) lid. What kind of process does the air inside the container undergo as the cookies gradually cool to room temperature (isothermal, isochoric, adiabatic, isobaric, or some combination)? Explain your answer.

Q19.9 Imagine a gas made up entirely of negatively charged electrons. Like charges repel, so the electrons exert repulsive forces on each other. Would you expect that the temperature of such a gas would rise, fall, or stay the same in a free expansion? Why?

Q19.10 There are a few materials that contract when their temperature is increased, such as water between 0°C and 4°C. Would you expect C_p for such materials to be greater or less than C_V? Explain?

Q19.11 When you blow on the back of your hand with your mouth wide open, your breath feels warm. But if you partially close your mouth to form an "o" and then blow on your hand, your breath feels cool. Why?

Q19.12 An ideal gas expands while the pressure is kept constant. During this process, does heat flow into the gas or out of the gas? Justify your answer.

Q19.13 A liquid is irregularly stirred in a well-insulated container and thereby undergoes a rise in temperature. Regard the liquid as the system. Has heat been transferred? How can you tell? Has work been done? How can you tell? Why is it important that the stirring is irregular? What is the sign of ΔU? How can you tell?

Q19.14 When you use a hand pump to inflate the tires of your bicycle, the pump gets warm after a while. Why? What happens to the temperature of the air in the pump as you compress it? Why does this happen? When you raise the pump handle to draw outside air into the pump, what happens to the temperature of the air taken in? Again, why does this happen?

Q19.15 In the carburetor of an aircraft or automobile engine, air flows through a relatively small aperture and then expands. In cool, foggy weather, ice sometimes forms in this aperture even though the outside air temperature is above freezing. Why?

Q19.16 On a sunny day, large "bubbles" of air form on the sun-warmed earth, gradually expand, and finally break free to rise through the atmosphere. Soaring birds and glider pilots are fond of using these "thermals" to gain altitude easily. This expansion is essentially an adiabatic process. Why?

Q19.17 The prevailing winds on the Hawaiian island of Kauai blow from the northeast. The winds cool as they go up the slope of Mt. Waialeale (elevation 1523 m), causing water vapor to condense and rain to fall. There is much more precipitation at the summit than at the base of the mountain. In fact, Mt. Waialeale is the rainiest spot on earth, averaging 11.7 m of rainfall a year. But what makes the winds cool?

Q19.18 Applying the same considerations as in Question Q19.17, explain why the island of Niihau, a few kilometers to the southwest of Kauai, is almost a desert and farms there need to be irrigated.

Q19.19 In a constant-volume process, $dU = nC_V\,dT$. But in a constant-pressure process, it is *not* true that $dU = nC_p\,dT$. Why not?

Q19.20 When a gas surrounded by air is compressed adiabatically, its temperature rises even though there is no heat input to the gas. Where does the energy come from to raise the temperature?

Q19.21 When a gas expands adiabatically, it does work on its surroundings. But if there is no heat input to the gas, where does the energy come from to do the work?

Q19.22 The gas used in separating the two uranium isotopes ^{235}U and ^{238}U has the formula UF_6. If you added heat at equal rates to a mole of UF_6 gas and a mole of H_2 gas, which one's temperature would you expect to rise faster? Explain.

EXERCISES

Section 19.2 Work Done During Volume Changes and Section 19.3 Paths Between Thermodynamic States

19.1 •• Two moles of an ideal gas are heated at constant pressure from $T = 27°C$ to $T = 107°C$. (a) Draw a pV-diagram for this process. (b) Calculate the work done by the gas.

19.2 • Six moles of an ideal gas are in a cylinder fitted at one end with a movable piston. The initial temperature of the gas is 27.0°C and the pressure is constant. As part of a machine design project, calculate the final temperature of the gas after it has done 2.40×10^3 J of work.

19.3 •• **CALC** Two moles of an ideal gas are compressed in a cylinder at a constant temperature of 65.0°C until the original pressure has tripled. (a) Sketch a pV-diagram for this process. (b) Calculate the amount of work done.

19.4 •• **BIO** **Work Done by the Lungs.** The graph in Fig. E19.4 shows a pV-diagram of the air in a human lung when a person is inhaling and then exhaling a deep breath. Such graphs, obtained in clinical practice, are normally somewhat curved, but we have modeled one as a set of straight lines of the same general shape. (*Important:* The

Figure **E19.4**

pressure shown is the *gauge* pressure, *not* the absolute pressure.) (a) How many joules of *net* work does this person's lung do during one complete breath? (b) The process illustrated here is somewhat different from those we have been studying, because the pressure change is due to changes in the amount of gas in the lung, not to temperature changes. (Think of your own breathing. Your lungs do not expand because they've gotten hot.) If the temperature of the air in the lung remains a reasonable 20°C, what is the maximum number of moles in this person's lung during a breath?

19.5 •• **CALC** During the time 0.305 mol of an ideal gas undergoes an isothermal compression at 22.0°C, 468 J of work is done on it by the surroundings. (a) If the final pressure is 1.76 atm, what was the initial pressure? (b) Sketch a pV-diagram for the process.

19.6 •• A gas undergoes two processes. In the first, the volume remains constant at 0.200 m³ and the pressure increases from 2.00×10^5 Pa to 5.00×10^5 Pa. The second process is a compression to a volume of 0.120 m³ at a constant pressure of 5.00×10^5 Pa. (a) In a pV-diagram, show both processes. (b) Find the total work done by the gas during both processes.

19.7 • **Work Done in a Cyclic Process.** (a) In Fig. 19.7a, consider the closed loop $1 \rightarrow 3 \rightarrow 2 \rightarrow 4 \rightarrow 1$. This is a *cyclic* process in which the initial and final states are the same. Find the total work done by the system in this cyclic process, and show that it is equal to the area enclosed by the loop. (b) How is the work done for the process in part (a) related to the work done if the loop is traversed in the opposite direction, $1 \rightarrow 4 \rightarrow 2 \rightarrow 3 \rightarrow 1$? Explain.

Section 19.4 Internal Energy and the First Law of Thermodynamics

19.8 •• Figure E19.8 shows a pV-diagram for an ideal gas in which its absolute temperature at b is one-fourth of its absolute temperature at a. (a) What volume does this gas occupy at point b? (b) How many joules of work was done by or on the gas in this process?

Figure **E19.8**

Was it done by or on the gas? (c) Did the internal energy of the gas increase or decrease from a to b? How do you know? (d) Did heat enter or leave the gas from a to b? How do you know?

19.9 • A gas in a cylinder expands from a volume of 0.110 m³ to 0.320 m³. Heat flows into the gas just rapidly enough to keep the pressure constant at 1.65×10^5 Pa during the expansion. The total heat added is 1.15×10^5 J. (a) Find the work done by the gas. (b) Find the change in internal energy of the gas. (c) Does it matter whether the gas is ideal? Why or why not?

19.10 •• Five moles of an ideal monatomic gas with an initial temperature of 127°C expand and, in the process, absorb 1200 J of heat and do 2100 J of work. What is the final temperature of the gas?

19.11 •• The process *abc* shown in the *pV*-diagram in Fig. E19.11 involves 0.0175 mole of an ideal gas. (a) What was the lowest temperature the gas reached in this process? Where did it occur? (b) How much work was done by or on the gas from *a* to *b*? From *b* to *c*? (c) If 215 J of heat was put into the gas during *abc*, how many of those joules went into internal energy?

Figure **E19.11**

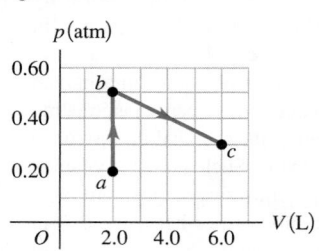

19.12 • A gas in a cylinder is held at a constant pressure of 1.80×10^5 Pa and is cooled and compressed from 1.70 m³ to 1.20 m³. The internal energy of the gas decreases by 1.40×10^5 J. (a) Find the work done by the gas. (b) Find the absolute value $|Q|$ of the heat flow into or out of the gas, and state the direction of the heat flow. (c) Does it matter whether the gas is ideal? Why or why not?

19.13 • **BIO Doughnuts: Breakfast of Champions!** A typical doughnut contains 2.0 g of protein, 17.0 g of carbohydrates, and 7.0 g of fat. The average food energy values of these substances are 4.0 kcal/g for protein and carbohydrates and 9.0 kcal/g for fat. (a) During heavy exercise, an average person uses energy at a rate of 510 kcal/h. How long would you have to exercise to "work off" one doughnut? (b) If the energy in the doughnut could somehow be converted into the kinetic energy of your body as a whole, how fast could you move after eating the doughnut? Take your mass to be 60 kg, and express your answer in m/s and in km/h.

19.14 • **Boiling Water at High Pressure.** When water is boiled at a pressure of 2.00 atm, the heat of vaporization is 2.20×10^6 J/kg and the boiling point is 120°C. At this pressure, 1.00 kg of water has a volume of 1.00×10^{-3} m³, and 1.00 kg of steam has a volume of 0.824 m³. (a) Compute the work done when 1.00 kg of steam is formed at this temperature. (b) Compute the increase in internal energy of the water.

19.15 • An ideal gas is taken from *a* to *b* on the *pV*-diagram shown in Fig. E19.15. During this process, 700 J of heat is added and the pressure doubles. (a) How much work is done by or on the gas? Explain. (b) How does the temperature of the gas at *a* compare to its temperature at *b*? Be specific. (c) How does the internal energy of the gas at *a* compare to the internal energy at *b*? Again, be specific and explain.

Figure **E19.15**

19.16 • A system is taken from state *a* to state *b* along the three paths shown in Fig. E19.16. (a) Along which path is the work done by the system the greatest? The least? (b) If $U_b > U_a$, along which path is

Figure **E19.16**

the absolute value $|Q|$ of the heat transfer the greatest? For this path, is heat absorbed or liberated by the system?

19.17 • A thermodynamic system undergoes a cyclic process as shown in Fig. E19.17. The cycle consists of two closed loops: I and II. (a) Over one complete cycle, does the system do positive or negative work? (b) In each of loops I and II, is the net work done by the system positive or negative? (c) Over one complete cycle, does heat flow into or out of the system? (d) In each of loops I and II, does heat flow into or out of the system?

Figure **E19.17**

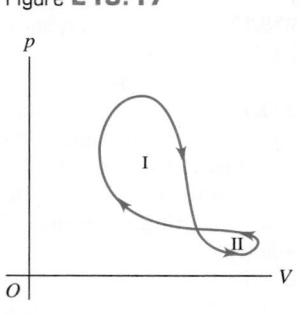

Section 19.5 Kinds of Thermodynamic Processes, Section 19.6 Internal Energy of an Ideal Gas, and Section 19.7 Heat Capacities of an Ideal Gas

19.18 • During an isothermal compression of an ideal gas, 335 J of heat must be removed from the gas to maintain constant temperature. How much work is done by the gas during the process?

19.19 • A cylinder contains 0.250 mol of carbon dioxide (CO_2) gas at a temperature of 27.0°C. The cylinder is provided with a frictionless piston, which maintains a constant pressure of 1.00 atm on the gas. The gas is heated until its temperature increases to 127.0°C. Assume that the CO_2 may be treated as an ideal gas. (a) Draw a *pV*-diagram for this process. (b) How much work is done by the gas in this process? (c) On what is this work done? (d) What is the change in internal energy of the gas? (e) How much heat was supplied to the gas? (f) How much work would have been done if the pressure had been 0.50 atm?

19.20 • A cylinder contains 0.0100 mol of helium at $T = 27.0°C$. (a) How much heat is needed to raise the temperature to 67.0°C while keeping the volume constant? Draw a *pV*-diagram for this process. (b) If instead the pressure of the helium is kept constant, how much heat is needed to raise the temperature from 27.0°C to 67.0°C? Draw a *pV*-diagram for this process. (c) What accounts for the difference between your answers to parts (a) and (b)? In which case is more heat required? What becomes of the additional heat? (d) If the gas is ideal, what is the change in its internal energy in part (a)? In part (b)? How do the two answers compare? Why?

19.21 • In an experiment to simulate conditions inside an automobile engine, 0.185 mol of air at a temperature of 780 K and a pressure of 3.00×10^6 Pa is contained in a cylinder of volume 40.0 cm³. Then 645 J of heat is transferred to the cylinder. (a) If the volume of the cylinder is constant while the heat is added, what is the final temperature of the air? Assume that the air is essentially nitrogen gas, and use the data in Table 19.1 even though the pressure is not low. Draw a *pV*-diagram for this process. (b) If instead the volume of the cylinder is allowed to increase while the pressure remains constant, find the final temperature of the air. Draw a *pV*-diagram for this process.

19.22 •• When a quantity of monatomic ideal gas expands at a constant pressure of 4.00×10^4 Pa, the volume of the gas increases from 2.00×10^{-3} m³ to 8.00×10^{-3} m³. What is the change in the internal energy of the gas?

19.23 • Heat Q flows into a monatomic ideal gas, and the volume increases while the pressure is kept constant. What fraction of the heat energy is used to do the expansion work of the gas?

19.24 • Three moles of an ideal monatomic gas expands at a constant pressure of 2.50 atm; the volume of the gas changes from $3.20 \times 10^{-2} \, m^3$ to $4.50 \times 10^{-2} \, m^3$. (a) Calculate the initial and final temperatures of the gas. (b) Calculate the amount of work the gas does in expanding. (c) Calculate the amount of heat added to the gas. (d) Calculate the change in internal energy of the gas.

19.25 • A cylinder with a movable piston contains 3.00 mol of N_2 gas (assumed to behave like an ideal gas). (a) The N_2 is heated at constant volume until 1557 J of heat have been added. Calculate the change in temperature. (b) Suppose the same amount of heat is added to the N_2, but this time the gas is allowed to expand while remaining at constant pressure. Calculate the temperature change. (c) In which case, (a) or (b), is the final internal energy of the N_2 higher? How do you know? What accounts for the difference between the two cases?

19.26 • Propane gas (C_3H_8) behaves like an ideal gas with $\gamma = 1.127$. Determine the molar heat capacity at constant volume and the molar heat capacity at constant pressure.

19.27 • **CALC** The temperature of 0.150 mol of an ideal gas is held constant at 77.0°C while its volume is reduced to 25.0% of its initial volume. The initial pressure of the gas is 1.25 atm. (a) Determine the work done by the gas. (b) What is the change in its internal energy? (c) Does the gas exchange heat with its surroundings? If so, how much? Does the gas absorb or liberate heat?

19.28 • An experimenter adds 970 J of heat to 1.75 mol of an ideal gas to heat it from 10.0°C to 25.0°C at constant pressure. The gas does +223 J of work during the expansion. (a) Calculate the change in internal energy of the gas. (b) Calculate γ for the gas.

Section 19.8 Adiabatic Processes for an Ideal Gas

19.29 • A monatomic ideal gas that is initially at a pressure of 1.50×10^5 Pa and has a volume of $0.0800 \, m^3$ is compressed adiabatically to a volume of $0.0400 \, m^3$. (a) What is the final pressure? (b) How much work is done by the gas? (c) What is the ratio of the final temperature of the gas to its initial temperature? Is the gas heated or cooled by this compression?

19.30 • In an adiabatic process for an ideal gas, the pressure decreases. In this process does the internal energy of the gas increase or decrease? Explain your reasoning.

19.31 •• Two moles of carbon monoxide (CO) start at a pressure of 1.2 atm and a volume of 30 liters. The gas is then compressed adiabatically to $\frac{1}{3}$ this volume. Assume that the gas may be treated as ideal. What is the change in the internal energy of the gas? Does the internal energy increase or decrease? Does the temperature of the gas increase or decrease during this process? Explain.

19.32 • The engine of a Ferrari F355 F1 sports car takes in air at 20.0°C and 1.00 atm and compresses it adiabatically to 0.0900 times the original volume. The air may be treated as an ideal gas with $\gamma = 1.40$. (a) Draw a pV-diagram for this process. (b) Find the final temperature and pressure.

19.33 • During an adiabatic expansion the temperature of 0.450 mol of argon (Ar) drops from 50.0°C to 10.0°C. The argon may be treated as an ideal gas. (a) Draw a pV-diagram for this process. (b) How much work does the gas do? (c) What is the change in internal energy of the gas?

19.34 •• A player bounces a basketball on the floor, compressing it to 80.0% of its original volume. The air (assume it is essentially N_2 gas) inside the ball is originally at a temperature of 20.0°C and a pressure of 2.00 atm. The ball's inside diameter is 23.9 cm. (a) What temperature does the air in the ball reach at its maximum compression? Assume the compression is adiabatic and treat the gas as ideal. (b) By how much does the internal energy of the air change between the ball's original state and its maximum compression?

19.35 •• On a warm summer day, a large mass of air (atmospheric pressure 1.01×10^5 Pa) is heated by the ground to a temperature of 26.0°C and then begins to rise through the cooler surrounding air. (This can be treated approximately as an adiabatic process; why?) Calculate the temperature of the air mass when it has risen to a level at which atmospheric pressure is only 0.850×10^5 Pa. Assume that air is an ideal gas, with $\gamma = 1.40$. (This rate of cooling for dry, rising air, corresponding to roughly 1°C per 100 m of altitude, is called the *dry adiabatic lapse rate*.)

19.36 • A cylinder contains 0.100 mol of an ideal monatomic gas. Initially the gas is at a pressure of 1.00×10^5 Pa and occupies a volume of $2.50 \times 10^{-3} \, m^3$. (a) Find the initial temperature of the gas in kelvins. (b) If the gas is allowed to expand to twice the initial volume, find the final temperature (in kelvins) and pressure of the gas if the expansion is (i) isothermal; (ii) isobaric; (iii) adiabatic.

PROBLEMS

19.37 •• One mole of ideal gas is slowly compressed to one-third of its original volume. In this compression, the work done on the gas has magnitude 600 J. For the gas, $C_p = 7R/2$. (a) If the process is isothermal, what is the heat flow Q for the gas? Does heat flow into or out of the gas? (b) If the process is isobaric, what is the change in internal energy of the gas? Does the internal energy increase or decrease?

19.38 •• **CALC** Figure P19.38 shows the pV-diagram for an isothermal expansion of 1.50 mol of an ideal gas, at a temperature of 15.0°C. (a) What is the change in internal energy of the gas? Explain. (b) Calculate the work done by (or on) the gas and the heat absorbed (or released) by the gas during the expansion.

Figure **P19.38**

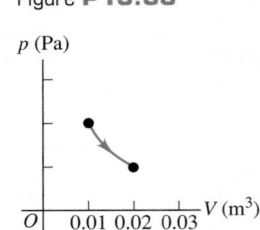

19.39 •• A quantity of air is taken from state a to state b along a path that is a straight line in the pV-diagram (Fig. P19.39). (a) In this process, does the temperature of the gas increase, decrease, or stay the same? Explain. (b) If $V_a = 0.0700 \, m^3$, $V_b = 0.1100 \, m^3$, $p_a = 1.00 \times 10^5$ Pa, and $p_b = 1.40 \times 10^5$ Pa, what is the work W done by the gas in this process? Assume that the gas may be treated as ideal.

Figure **P19.39**

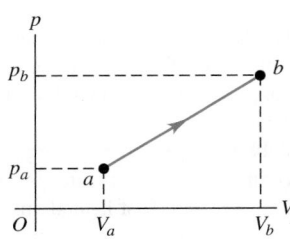

19.40 • One-half mole of an ideal gas is taken from state a to state c, as shown in Fig. P19.40. (a) Calculate the final temperature of the gas. (b) Calculate the work done on (or by) the gas as it moves from state a to state c. (c) Does heat leave the system or enter the system during this process? How much heat? Explain.

Figure **P19.40**

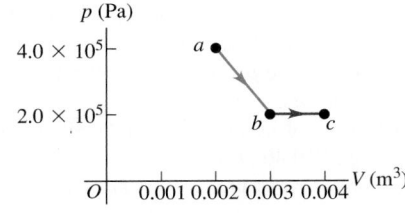

19.41 •• When a system is taken from state a to state b in Fig. P19.41 along the path acb, 90.0 J of heat flows into the system and 60.0 J of work is done by the system. (a) How much heat flows into the system along path adb if the work done by the system is 15.0 J? (b) When the system is returned from b to a along the curved path, the absolute value of the work done by the system is 35.0 J. Does the system absorb or liberate heat? How much heat? (c) If $U_a = 0$ and $U_d = 8.0$ J, find the heat absorbed in the processes ad and db.

Figure **P19.41**

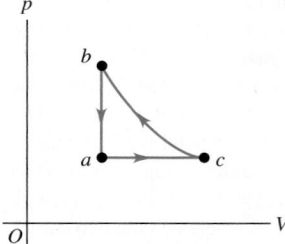

19.42 • A thermodynamic system is taken from state a to state c in Fig. P19.42 along either path abc or path adc. Along path abc, the work W done by the system is 450 J. Along path adc, W is 120 J. The internal energies of each of the four states shown in the figure are $U_a = 150$ J, $U_b = 240$ J, $U_c = 680$ J, and $U_d = 330$ J. Calculate the heat flow Q for each of the four processes ab, bc, ad, and dc. In each process, does the system absorb or liberate heat?

Figure **P19.42**

19.43 • A volume of air (assumed to be an ideal gas) is first cooled without changing its volume and then expanded without changing its pressure, as shown by the path abc in Fig. P19.43. (a) How does the final temperature of the gas compare with its initial temperature? (b) How much heat does the air exchange with its surroundings during the process abc? Does the air absorb heat or release heat during this process? Explain. (c) If the air instead expands from state a to state c by the straight-line path shown, how much heat does it exchange with its surroundings?

Figure **P19.43**

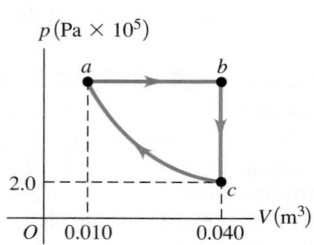

19.44 • Three moles of argon gas (assumed to be an ideal gas) originally at a pressure of 1.50×10^4 Pa and a volume of 0.0280 m^3 are first heated and expanded at constant pressure to a volume of 0.0435 m^3, then heated at constant volume until the pressure reaches 3.50×10^4 Pa, then cooled and compressed at constant pressure until the volume is again 0.0280 m^3, and finally cooled at constant volume until the pressure drops to its original value of 1.50×10^4 Pa. (a) Draw the pV-diagram for this cycle. (b) Calculate the total work done by (or on) the gas during the cycle. (c) Calculate the net heat exchanged with the surroundings. Does the gas gain or lose heat overall?

19.45 •• Two moles of an ideal monatomic gas go through the cycle abc. For the complete cycle, 800 J of heat flows out of the gas. Process ab is at constant pressure, and process bc is at constant volume. States a and b have temperatures $T_a = 200$ K and $T_b = 300$ K. (a) Sketch the pV-diagram for the cycle. (b) What is the work W for the process ca?

19.46 •• Three moles of an ideal gas are taken around the cycle acb shown in Fig. P19.46. For this gas, $C_p = 29.1$ J/mol · K. Process ac is at constant pressure, process ba is at constant volume, and process cb is adiabatic. The temperatures of the gas in states a, c, and b are $T_a = 300$ K, $T_c = 492$ K, and $T_b = 600$ K. Calculate the total work W for the cycle.

Figure **P19.46**

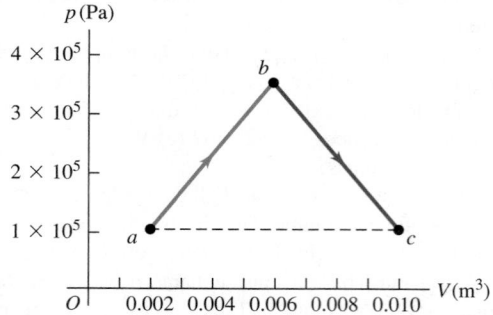

Figure **P19.47**

19.47 •• Figure P19.47 shows a pV-diagram for 0.0040 mole of ideal H_2 gas. The temperature of the gas does not change during segment bc. (a) What volume does this gas occupy at point c? (b) Find the temperature of the gas at points a, b, and c. (c) How much heat went into or out of the gas during segments ab, ca, and bc? Indicate whether the heat has gone into or out of the gas. (d) Find the change in the internal energy of this hydrogen during segments ab, bc, and ca. Indicate whether the internal energy increased or decreased during each of these segments.

19.48 •• The graph in Fig. P19.48 shows a pV-diagram for 3.25 moles of ideal helium (He) gas. Part ca of this process is isothermal. (a) Find the pressure of the He at point a. (b) Find the temperature of the He at points a, b, and c. (c) How much heat entered or left the He during segments ab, bc, and ca? In each segment, did the heat enter or leave? (d) By how much did the internal energy of the He change from a to b, from b to c, and from c to a? Indicate whether this energy increased or decreased.

Figure **P19.48**

19.49 •• (a) One-third of a mole of He gas is taken along the path abc shown as the solid line in Fig. P19.49. Assume that the gas may be treated as ideal. How much heat is transferred into or out of the gas? (b) If the gas instead went from state a to state c along the horizontal dashed line in Fig. P19.49, how much heat would be transferred into or out of the gas? (c) How does Q in part (b) compare with Q in part (a)? Explain.

Figure **P19.49**

19.50 •• Two moles of helium are initially at a temperature of 27.0°C and occupy a volume of 0.0300 m^3. The helium first expands at constant pressure until its volume has doubled. Then it expands adiabatically until the temperature returns to its initial value. Assume that the helium can be treated as an ideal gas. (a) Draw a diagram of the process in the pV-plane. (b) What is the total heat supplied to the helium in the process? (c) What is the total change in internal energy of the helium? (d) What is the total work done by the helium? (e) What is the final volume of the helium?

19.51 ••• Starting with 2.50 mol of N_2 gas (assumed to be ideal) in a cylinder at 1.00 atm and 20.0°C, a chemist first heats the gas at constant volume, adding 1.52×10^4 J of heat, then continues heating and allows the gas to expand at constant pressure to twice its original volume. (a) Calculate the final temperature of the gas. (b) Calculate the amount of work done by the gas. (c) Calculate the amount of heat added to the gas while it was expanding. (d) Calculate the change in internal energy of the gas for the whole process.

19.52 •• Nitrogen gas in an expandable container is cooled from 50.0°C to 10.0°C with the pressure held constant at 3.00×10^5 Pa. The total heat liberated by the gas is 2.50×10^4 J. Assume that the gas may be treated as ideal. (a) Find the number of moles of gas. (b) Find the change in internal energy of the gas. (c) Find the work done by the gas. (d) How much heat would be liberated by the gas for the same temperature change if the volume were constant?

19.53 • In a certain process, 2.15×10^5 J of heat is liberated by a system, and at the same time the system contracts under a constant external pressure of 9.50×10^5 Pa. The internal energy of the system is the same at the beginning and end of the process. Find the change in volume of the system. (The system is *not* an ideal gas.)

19.54 • **CALC** A cylinder with a frictionless, movable piston like that shown in Fig. 19.5 contains a quantity of helium gas. Initially the gas is at a pressure of 1.00×10^5 Pa, has a temperature of 300 K, and occupies a volume of 1.50 L. The gas then undergoes two processes. In the first, the gas is heated and the piston is allowed to move to keep the temperature equal to 300 K. This continues until the pressure reaches 2.50×10^4 Pa. In the second process, the gas is compressed at constant pressure until it returns to its original volume of 1.50 L. Assume that the gas may be treated as ideal. (a) In a pV-diagram, show both processes. (b) Find the volume of the gas at the end of the first process, and find the pressure and temperature at the end of the second process. (c) Find the total work done by the gas during both processes. (d) What would you have to do to the gas to return it to its original pressure and temperature?

19.55 •• **CP A Thermodynamic Process in a Liquid.** A chemical engineer is studying the properties of liquid methanol (CH_3OH). She uses a steel cylinder with a cross-sectional area of 0.0200 m² and containing 1.20×10^{-2} m³ of methanol. The cylinder is equipped with a tightly fitting piston that supports a load of 3.00×10^4 N. The temperature of the system is increased from 20.0°C to 50.0°C. For methanol, the coefficient of volume expansion is 1.20×10^{-3} K^{-1}, the density is 791 kg/m³, and the specific heat at constant pressure is $c_p = 2.51 \times 10^3$ J/kg·K. You can ignore the expansion of the steel cylinder. Find (a) the increase in volume of the methanol; (b) the mechanical work done by the methanol against the 3.00×10^4 N force; (c) the amount of heat added to the methanol; (d) the change in internal energy of the methanol. (e) Based on your results, explain whether there is any substantial difference between the specific heats c_p (at constant pressure) and c_V (at constant volume) for methanol under these conditions.

19.56 • **CP A Thermodynamic Process in a Solid.** A cube of copper 2.00 cm on a side is suspended by a string. (The physical properties of copper are given in Tables 14.1, 17.2, and 17.3.) The cube is heated with a burner from 20.0°C to 90.0°C. The air surrounding the cube is at atmospheric pressure (1.01×10^5 Pa). Find (a) the increase in volume of the cube; (b) the mechanical work done by the cube to expand against the pressure of the surrounding air; (c) the amount of heat added to the cube; (d) the change in internal energy of the cube. (e) Based on your results, explain whether there is any substantial difference between the specific heats c_p (at constant pressure) and c_V (at constant volume) for copper under these conditions.

19.57 • **BIO A Thermodynamic Process in an Insect.** The African bombardier beetle (*Stenaptinus insignis*) can emit a jet of defensive spray from the movable tip of its abdomen (Fig. P19.57). The beetle's body has reservoirs of two different chemicals; when the beetle is disturbed, these chemicals are combined in a reaction chamber, producing a compound that is warmed from 20°C to 100°C by the heat of reaction. The high pressure produced allows the compound to be sprayed out at speeds up to 19 m/s (68 km/h), scaring away predators of all kinds. (The beetle shown in the figure is 2 cm long.) Calculate the heat of reaction of the two chemicals (in J/kg). Assume that the specific heat of the two chemicals and the spray is the same as that of water, 4.19×10^3 J/kg·K, and that the initial temperature of the chemicals is 20°C.

Figure **P19.57**

19.58 ••• **High-Altitude Research.** A large research balloon containing 2.00×10^3 m³ of helium gas at 1.00 atm and a temperature of 15.0°C rises rapidly from ground level to an altitude at which the atmospheric pressure is only 0.900 atm (Fig. P19.58). Assume the helium behaves like an ideal gas and the balloon's ascent is too rapid to permit much heat exchange with the surrounding air. (a) Calculate the volume of the gas at the higher altitude. (b) Calculate the temperature of the gas at the higher altitude. (c) What is the change in internal energy of the helium as the balloon rises to the higher altitude?

Figure **P19.58**

19.59 ••• **Chinook.** During certain seasons strong winds called chinooks blow from the west across the eastern slopes of the Rockies and downhill into Denver and nearby areas. Although the mountains are cool, the wind in Denver is very hot; within a few minutes after the chinook wind arrives, the temperature can climb 20 C° ("chinook" is a Native American word meaning "snow eater"). Similar winds occur in the Alps (called foehns) and in southern California (called Santa Anas). (a) Explain why the temperature of the chinook wind rises as it descends the slopes. Why is it important that the wind be fast moving? (b) Suppose a strong wind is blowing toward Denver (elevation 1630 m) from Grays Peak (80 km west of Denver, at an elevation of 4350 m), where the air pressure is

5.60×10^4 Pa and the air temperature is $-15.0°C$. The temperature and pressure in Denver before the wind arrives are $2.0°C$ and 8.12×10^4 Pa. By how many Celsius degrees will the temperature in Denver rise when the chinook arrives?

19.60 •• A certain ideal gas has molar heat capacity at constant volume C_V. A sample of this gas initially occupies a volume V_0 at pressure p_0 and absolute temperature T_0. The gas expands isobarically to a volume $2V_0$ and then expands further adiabatically to a final volume $4V_0$. (a) Draw a pV-diagram for this sequence of processes. (b) Compute the total work done by the gas for this sequence of processes. (c) Find the final temperature of the gas. (d) Find the absolute value $|Q|$ of the total heat flow into or out of the gas for this sequence of processes, and state the direction of heat flow.

19.61 ••• An air pump has a cylinder 0.250 m long with a movable piston. The pump is used to compress air from the atmosphere (at absolute pressure 1.01×10^5 Pa) into a very large tank at 4.20×10^5 Pa gauge pressure. (For air, $C_V = 20.8$ J/mol·K.) (a) The piston begins the compression stroke at the open end of the cylinder. How far down the length of the cylinder has the piston moved when air first begins to flow from the cylinder into the tank? Assume that the compression is adiabatic. (b) If the air is taken into the pump at $27.0°C$, what is the temperature of the compressed air? (c) How much work does the pump do in putting 20.0 mol of air into the tank?

19.62 •• **Engine Turbochargers and Intercoolers.** The power output of an automobile engine is directly proportional to the mass of air that can be forced into the volume of the engine's cylinders to react chemically with gasoline. Many cars have a *turbocharger,* which compresses the air before it enters the engine, giving a greater mass of air per volume. This rapid, essentially adiabatic compression also heats the air. To compress it further, the air then passes through an *intercooler* in which the air exchanges heat with its surroundings at essentially constant pressure. The air is then drawn into the cylinders. In a typical installation, air is taken into the turbocharger at atmospheric pressure (1.01×10^5 Pa), density $\rho = 1.23$ kg/m³, and temperature $15.0°C$. It is compressed adiabatically to 1.45×10^5 Pa. In the intercooler, the air is cooled to the original temperature of $15.0°C$ at a constant pressure of 1.45×10^5 Pa. (a) Draw a pV-diagram for this sequence of processes. (b) If the volume of one of the engine's cylinders is 575 cm³, what mass of air exiting from the intercooler will fill the cylinder at 1.45×10^5 Pa? Compared to the power output of an engine that takes in air at 1.01×10^5 Pa at $15.0°C$, what percentage increase in power is obtained by using the turbocharger and intercooler? (c) If the intercooler is not used, what mass of air exiting from the turbocharger will fill the cylinder at 1.45×10^5 Pa? Compared to the power output of an engine that takes in air at 1.01×10^5 Pa at $15.0°C$, what percentage increase in power is obtained by using the turbocharger alone?

19.63 • A monatomic ideal gas expands slowly to twice its original volume, doing 300 J of work in the process. Find the heat added to the gas and the change in internal energy of the gas if the process is (a) isothermal; (b) adiabatic; (c) isobaric.

19.64 •• **CALC** A cylinder with a piston contains 0.250 mol of oxygen at 2.40×10^5 Pa and 355 K. The oxygen may be treated as an ideal gas. The gas first expands isobarically to twice its original volume. It is then compressed isothermally back to its original volume, and finally it is cooled isochorically to its original pressure. (a) Show the series of processes on a pV-diagram. (b) Compute the temperature during the isothermal compression. (c) Compute the maximum pressure. (d) Compute the total work done by the piston on the gas during the series of processes.

19.65 • Use the conditions and processes of Problem 19.64 to compute (a) the work done by the gas, the heat added to it, and its internal energy change during the initial expansion; (b) the work done, the heat added, and the internal energy change during the final cooling; (c) the internal energy change during the isothermal compression.

19.66 •• **CALC** A cylinder with a piston contains 0.150 mol of nitrogen at 1.80×10^5 Pa and 300 K. The nitrogen may be treated as an ideal gas. The gas is first compressed isobarically to half its original volume. It then expands adiabatically back to its original volume, and finally it is heated isochorically to its original pressure. (a) Show the series of processes in a pV-diagram. (b) Compute the temperatures at the beginning and end of the adiabatic expansion. (c) Compute the minimum pressure.

19.67 • Use the conditions and processes of Problem 19.66 to compute (a) the work done by the gas, the heat added to it, and its internal energy change during the initial compression; (b) the work done by the gas, the heat added to it, and its internal energy change during the adiabatic expansion; (c) the work done, the heat added, and the internal energy change during the final heating.

19.68 • **Comparing Thermodynamic Processes.** In a cylinder, 1.20 mol of an ideal monatomic gas, initially at 3.60×10^5 Pa and 300 K, expands until its volume triples. Compute the work done by the gas if the expansion is (a) isothermal; (b) adiabatic; (c) isobaric. (d) Show each process in a pV-diagram. In which case is the absolute value of the work done by the gas greatest? Least? (e) In which case is the absolute value of the heat transfer greatest? Least? (f) In which case is the absolute value of the change in internal energy of the gas greatest? Least?

CHALLENGE PROBLEMS

19.69 ••• **CP Oscillations of a Piston.** A vertical cylinder of radius r contains a quantity of ideal gas and is fitted with a piston with mass m that is free to move (Fig. P19.69). The piston and the walls of the cylinder are frictionless, and the entire cylinder is placed in a constant-temperature bath. The outside air pressure is p_0. In equilibrium, the piston sits at a height h above the bottom of the cylinder. (a) Find the absolute pressure of the gas trapped below the piston when in equilibrium. (b) The piston is pulled up by a small distance and released. Find the net force acting on the piston when its base is a distance $h + y$ above the bottom of the cylinder, where y is much less than h. (c) After the piston is displaced from equilibrium and released, it oscillates up and down. Find the frequency of these small oscillations. If the displacement is not small, are the oscillations simple harmonic? How can you tell?

Figure **P19.69**

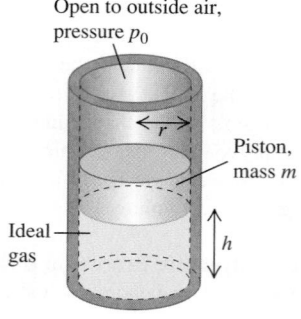

Open to outside air, pressure p_0

Piston, mass m

Ideal gas

r

h

Answers

Chapter Opening Question ?

No. The work done by a gas as its volume changes from V_1 to V_2 is equal to the integral $\int p\, dV$ between those two volume limits. If the volume of the gas contracts, the final volume V_2 is less than the initial volume V_1 and the gas does negative work. Propelling the locomotive requires that the gas do positive work, so the gas doesn't contribute to propulsion while contracting.

Test Your Understanding Questions

19.1 Answers: negative, positive, positive Heat flows out of the coffee, so $Q_{coffee} < 0$; heat flows into the aluminum cup, so $Q_{aluminum} > 0$. In mechanics, we would say that negative work is done *on* the block, since the surface exerts a force on the block that opposes the block's motion. But in thermodynamics we use the opposite convention and say that $W > 0$, which means that positive work is done *by* the block on the surface.

19.2 Answer: (ii) The work done in an expansion is represented by the area under the curve of pressure p versus volume V. In an isothermal expansion the pressure decreases as the volume increases, so the pV-diagram looks like Fig. 19.6a and the work done equals the shaded area under the blue curve from point 1 to point 2. If, however, the expansion is at constant pressure, the curve of p versus V would be the same as the dashed horizontal line at pressure p_2 in Fig. 19.6a. The area under this dashed line is smaller than the area under the blue curve for an isothermal expansion, so less work is done in the constant-pressure expansion than in the isothermal expansion.

19.3 Answer: (i) and (iv) (tie), (ii) and (iii) (tie) The accompanying figure shows the pV-diagrams for each of the four processes. The trapezoidal area under the curve, and hence the absolute value of the work, is the same in all four cases. In cases (i) and (iv) the volume increases, so the system does positive work as it expands against its surroundings. In cases (ii) and (iii) the volume decreases, so the system does negative work (shown by cross-hatching) as the surroundings push inward on it.

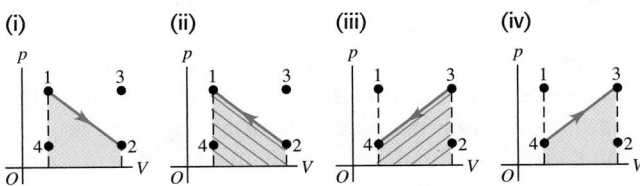

19.4 Answer: (ii), (i) and (iv) (tie), (iii) In the expression $\Delta U = Q - W$, Q is the heat *added* to the system and W is the work done *by* the system. If heat is transferred from the system to its surroundings, Q is negative; if work is done on the system, W is negative. Hence we have (i) $Q = -250$ J, $W = -250$ J, $\Delta U = -250$ J $- (-250$ J$) = 0$; (ii) $Q = 250$ J, $W = -250$ J, $\Delta U = 250$ J $- (-250$ J$) = 500$ J; (iii) $Q = -250$ J, $W = 250$ J, $\Delta U = -250$ J $- 250$ J $= -500$ J; and (iv) $Q = 250$ J, $W = 250$ J, $\Delta U = 250$ J $- 250$ J $= 0$.

19.5 Answers: $1 \to 4$ and $3 \to 2$ are isochoric; $1 \to 3$ and $4 \to 2$ are isobaric; no In a pV-diagram like those shown in Fig. 19.7, isochoric processes are represented by vertical lines (lines of constant volume) and isobaric processes are represented by horizontal lines (lines of constant pressure). The process $1 \to 2$ in Fig. 19.7 is shown as a curved line, which superficially resembles the adiabatic and isothermal processes for an ideal gas in Fig. 19.16. Without more information we can't tell whether process $1 \to 2$ is isothermal, adiabatic, or neither.

19.6 Answer: no Using the model of a solid in Fig. 18.20, we can see that the internal energy of a solid *does* depend on its volume. Compressing the solid means compressing the "springs" between the atoms, thereby increasing their stored potential energy and hence the internal energy of the solid.

19.7 Answer: (i) For a given number of moles n and a given temperature change ΔT, the amount of heat that must be transferred out of a fixed volume of air is $Q = nC_V \Delta T$. Hence the amount of heat transfer required is least for the gas with the smallest value of C_V. From Table 19.1, C_V is smallest for monatomic gases.

19.8 Answer: (ii), (iv), (i) and (iii) (tie) Samples (i) and (iii) are compressed isothermally, so $pV =$ constant. The volume of each sample decreases to one-half of its initial value, so the final pressure is twice the initial pressure. Samples (ii) and (iv) are compressed adiabatically, so $pV^\gamma =$ constant and the pressure increases by a factor of 2^γ. Sample (ii) is a monatomic gas for which $\gamma = \frac{5}{3}$, so its final pressure is $2^{5/3} = 3.17$ times greater than the initial pressure. Sample (iv) is a diatomic gas for which $\gamma = \frac{7}{5}$, so its final pressure is greater than the initial pressure by a factor of $2^{7/5} = 2.64$.

Bridging Problem

Answers: (a) $W = nRT \ln\left[\dfrac{V_2 - nb}{V_1 - nb}\right] + an^2\left[\dfrac{1}{V_2} - \dfrac{1}{V_1}\right]$

(b) (i) $W = 2.80 \times 10^3$ J, (ii) $W = 3.11 \times 10^3$ J

(c) Ideal gas, for which there is no attraction between molecules

20 THE SECOND LAW OF THERMODYNAMICS

LEARNING GOALS

By studying this chapter, you will learn:

• What determines whether a thermo-dynamic process is reversible or irreversible.

• What a heat engine is, and how to calculate its efficiency.

• The physics of internal-combustion engines.

• How refrigerators and heat engines are related, and how to analyze the performance of a refrigerator.

• How the second law of thermo-dynamics sets limits on the efficiency of engines and the performance of refrigerators.

• How to do calculations involving the idealized Carnot cycle for engines and refrigerators.

• What is meant by entropy, and how to use this concept to analyze thermodynamic processes.

? The second law of thermodynamics tells us that heat naturally flows from a hot body (such as molten lava, shown here flowing into the ocean in Hawaii) to a cold one (such as ocean water, which is heated to make steam). Is it *ever* possible for heat to flow from a cold body to a hot one?

Many thermodynamic processes proceed naturally in one direction but not the opposite. For example, heat by itself always flows from a hot body to a cooler body, never the reverse. Heat flow from a cool body to a hot body would not violate the first law of thermodynamics; energy would be conserved. But it doesn't happen in nature. Why not? As another example, note that it is easy to convert mechanical energy completely into heat; this happens every time we use a car's brakes to stop it. In the reverse direction, there are plenty of devices that convert heat *partially* into mechanical energy. (An automobile engine is an example.) But no one has ever managed to build a machine that converts heat *completely* into mechanical energy. Again, why not?

The answer to both of these questions has to do with the *directions* of thermodynamic processes and is called the *second law of thermodynamics*. This law places fundamental limitations on the efficiency of an engine or a power plant. It also places limitations on the minimum energy input needed to operate a refrigerator. So the second law is directly relevant for many important practical problems.

We can also state the second law in terms of the concept of *entropy,* a quantitative measure of the degree of disorder or randomness of a system. The idea of entropy helps explain why ink mixed with water never spontaneously unmixes and why we never observe a host of other seemingly possible processes.

20.1 Directions of Thermodynamic Processes

Thermodynamic processes that occur in nature are all **irreversible processes.** These are processes that proceed spontaneously in one direction but not the other (Fig. 20.1a). The flow of heat from a hot body to a cooler body is irreversible, as is the free expansion of a gas discussed in Sections 19.3 and 19.6. Sliding a book across a table converts mechanical energy into heat by friction;

(a) A block of ice melts *irreversibly* when we place it in a hot (70°C) metal box.

Heat flows from the box into the ice and water, never the reverse.

(b) A block of ice at 0°C can be melted *reversibly* if we put it in a 0°C metal box.

By infinitesimally raising or lowering the temperature of the box, we can make heat flow into the ice to melt it or make heat flow out of the water to refreeze it.

20.1 Reversible and irreversible processes.

this process is irreversible, for no one has ever observed the reverse process (in which a book initially at rest on the table would spontaneously start moving and the table and book would cool down). Our main topic for this chapter is the *second law of thermodynamics,* which determines the preferred direction for such processes.

Despite this preferred direction for every natural process, we can think of a class of idealized processes that *would* be reversible. A system that undergoes such an idealized **reversible process** is always very close to being in thermodynamic equilibrium within itself and with its surroundings. Any change of state that takes place can then be reversed by making only an infinitesimal change in the conditions of the system. For example, we can reverse heat flow between two bodies whose temperatures differ only infinitesimally by making only a very small change in one temperature or the other (Fig. 20.1b).

Reversible processes are thus **equilibrium processes,** with the system always in thermodynamic equilibrium. Of course, if a system were *truly* in thermodynamic equilibrium, no change of state would take place. Heat would not flow into or out of a system with truly uniform temperature throughout, and a system that is truly in mechanical equilibrium would not expand and do work against its surroundings. A reversible process is an idealization that can never be precisely attained in the real world. But by making the temperature gradients and the pressure differences in the substance very small, we can keep the system very close to equilibrium states and make the process nearly reversible.

By contrast, heat flow with finite temperature difference, free expansion of a gas, and conversion of work to heat by friction are all *irreversible* processes; no small change in conditions could make any of them go the other way. They are also all *nonequilibrium* processes, in that the system is not in thermodynamic equilibrium at any point until the end of the process.

Disorder and Thermodynamic Processes

There is a relationship between the direction of a process and the *disorder* or *randomness* of the resulting state. For example, imagine a thousand names written on file cards and arranged in alphabetical order. Throw the alphabetized stack of cards into the air, and they will likely come down in a random, disordered state. In the free expansion of a gas discussed in Sections 19.3 and 19.6, the air is more disordered after it has expanded into the entire box than when it was confined in one side, just as your clothes are more disordered when scattered all over your floor than when confined to your closet.

Similarly, macroscopic kinetic energy is energy associated with organized, coordinated motions of many molecules, but heat transfer involves changes in energy of random, disordered molecular motion. Therefore conversion of mechanical energy into heat involves an increase of randomness or disorder.

In the following sections we will introduce the second law of thermodynamics by considering two broad classes of devices: *heat engines,* which are partly successful in converting heat into work, and *refrigerators,* which are partly successful in transporting heat from cooler to hotter bodies.

Test Your Understanding of Section 20.1   Your left and right hands are normally at the same temperature, just like the metal box and ice in Fig. 20.1b. Is rubbing your hands together to warm them (i) a reversible process or (ii) an irreversible process? ∎

MasteringPHYSICS

ActivPhysics 8.12: Cyclic Process—Strategies
ActivPhysics 8.13: Cyclic Process—Problems

20.2 All motorized vehicles other than purely electric vehicles use heat engines for propulsion. (Hybrid vehicles use their internal-combustion engine to help charge the batteries for the electric motor.)

20.2 Heat Engines

The essence of our technological society is the ability to use sources of energy other than muscle power. Sometimes, mechanical energy is directly available; water power and wind power are examples. But most of our energy comes from the burning of fossil fuels (coal, oil, and gas) and from nuclear reactions. They supply energy that is transferred as *heat.* This is directly useful for heating buildings, for cooking, and for chemical processing, but to operate a machine or propel a vehicle, we need *mechanical* energy.

Thus it's important to know how to take heat from a source and convert as much of it as possible into mechanical energy or work. This is what happens in gasoline engines in automobiles, jet engines in airplanes, steam turbines in electric power plants, and many other systems. Closely related processes occur in the animal kingdom; food energy is "burned" (that is, carbohydrates combine with oxygen to yield water, carbon dioxide, and energy) and partly converted to mechanical energy as an animal's muscles do work on its surroundings.

Any device that transforms heat partly into work or mechanical energy is called a **heat engine** (Fig. 20.2). Usually, a quantity of matter inside the engine undergoes inflow and outflow of heat, expansion and compression, and sometimes change of phase. We call this matter the **working substance** of the engine. In internal-combustion engines, such as those used in automobiles, the working substance is a mixture of air and fuel; in a steam turbine it is water.

The simplest kind of engine to analyze is one in which the working substance undergoes a **cyclic process,** a sequence of processes that eventually leaves the substance in the same state in which it started. In a steam turbine the water is recycled and used over and over. Internal-combustion engines do not use the same air over and over, but we can still analyze them in terms of cyclic processes that approximate their actual operation.

Hot and Cold Reservoirs

All heat engines *absorb* heat from a source at a relatively high temperature, perform some mechanical work, and *discard* or *reject* some heat at a lower temperature. As far as the engine is concerned, the discarded heat is wasted. In internal-combustion engines the waste heat is that discarded in the hot exhaust gases and the cooling system; in a steam turbine it is the heat that must flow out of the used steam to condense and recycle the water.

When a system is carried through a cyclic process, its initial and final internal energies are equal. For any cyclic process, the first law of thermodynamics requires that

$$U_2 - U_1 = 0 = Q - W \quad \text{so} \quad Q = W$$

That is, the net heat flowing into the engine in a cyclic process equals the net work done by the engine.

When we analyze heat engines, it helps to think of two bodies with which the working substance of the engine can interact. One of these, called the *hot reservoir,* represents the heat source; it can give the working substance large amounts of heat at a constant temperature T_H without appreciably changing its own temperature. The other body, called the *cold reservoir,* can absorb large amounts of discarded heat from the engine at a constant lower temperature T_C. In a steam-turbine system the flames and hot gases in the boiler are the hot reservoir, and the cold water and air used to condense and cool the used steam are the cold reservoir.

We denote the quantities of heat transferred from the hot and cold reservoirs as Q_H and Q_C, respectively. A quantity of heat Q is positive when heat is transferred *into* the working substance and is negative when heat leaves the working substance. Thus in a heat engine, Q_H is positive but Q_C is negative, representing heat *leaving* the working substance. This sign convention is consistent with the rules we stated in Section 19.1; we will continue to use those rules here. For clarity, we'll often state the relationships in terms of the absolute values of the Q's and W's because absolute values are always positive.

Energy-Flow Diagrams and Efficiency

We can represent the energy transformations in a heat engine by the *energy-flow diagram* of Fig. 20.3. The engine itself is represented by the circle. The amount of heat Q_H supplied to the engine by the hot reservoir is proportional to the width of the incoming "pipeline" at the top of the diagram. The width of the outgoing pipeline at the bottom is proportional to the magnitude $|Q_C|$ of the heat rejected in the exhaust. The branch line to the right represents the portion of the heat supplied that the engine converts to mechanical work, W.

When an engine repeats the same cycle over and over, Q_H and Q_C represent the quantities of heat absorbed and rejected by the engine *during one cycle; Q_H is positive, and Q_C is negative. The *net heat Q absorbed per cycle is

$$Q = Q_H + Q_C = |Q_H| - |Q_C| \qquad (20.1)$$

The useful output of the engine is the net work W done by the working substance. From the first law,

$$W = Q = Q_H + Q_C = |Q_H| - |Q_C| \qquad (20.2)$$

Ideally, we would like to convert *all* the heat Q_H into work; in that case we would have $Q_H = W$ and $Q_C = 0$. Experience shows that this is impossible; there is always some heat wasted, and Q_C *is never zero*. We define the **thermal efficiency** of an engine, denoted by e, as the quotient

$$e = \frac{W}{Q_H} \qquad (20.3)$$

The thermal efficiency e represents the fraction of Q_H that *is* converted to work. To put it another way, e is what you get divided by what you pay for. This is always less than unity, an all-too-familiar experience! In terms of the flow diagram of Fig. 20.3, the most efficient engine is one for which the branch pipeline representing the work output is as wide as possible and the exhaust pipeline representing the heat thrown away is as narrow as possible.

When we substitute the two expressions for W given by Eq. (20.2) into Eq. (20.3), we get the following equivalent expressions for e:

$$e = \frac{W}{Q_H} = 1 + \frac{Q_C}{Q_H} = 1 - \left|\frac{Q_C}{Q_H}\right| \quad \text{(thermal efficiency of an engine)} \quad (20.4)$$

Note that e is a quotient of two energy quantities and thus is a pure number, without units. Of course, we must always express W, Q_H, and Q_C in the same units.

20.3 Schematic energy-flow diagram for a heat engine.

$$W = Q_H + Q_C$$
$$= |Q_H| - |Q_C|$$

Application **Biological Efficiency**
Although a biological organism is not a heat engine, the concept of efficiency still applies: Here e is the ratio of the work done to the energy that was used to do that work. To exercise on a stationary bike, your body must first convert the chemical-bond energy in glucose to chemical-bond energy in ATP (adenosine triphosphate), then convert energy from ATP into motion of your leg muscles, and finally convert muscular motion into motion of the pedals. The overall efficiency of this entire process is only about 25%. The remaining 75% of the energy liberated from glucose goes into heating your body.

Problem-Solving Strategy 20.1 Heat Engines

Problems involving heat engines are, fundamentally, problems in the first law of thermodynamics. You should review Problem-Solving Strategy 19.1 (Section 19.4).

IDENTIFY *the relevant concepts:* A heat engine is any device that converts heat partially to work, as shown schematically in Fig. 20.3. We will see in Section 20.4 that a refrigerator is essentially a heat engine running in reverse, so many of the same concepts apply.

SET UP *the problem* as suggested in Problem-Solving Strategy 19.1. Use Eq. (20.4) if the thermal efficiency of the engine is relevant. Sketch an energy-flow diagram like Fig. 20.3.

EXECUTE *the solution* as follows:
1. Be careful with the sign conventions for W and the various Q's. W is positive when the system expands and does work; W is

negative when the system is compressed and work is done on it. Each Q is positive if it represents heat entering the system and is negative if it represents heat leaving the system. When you know that a quantity is negative, such as Q_C in the above discussion, it sometimes helps to write it as $Q_C = -|Q_C|$.
2. Power is work per unit time ($P = W/t$), and rate of heat transfer (heat current) H is heat transfer per unit time ($H = Q/t$). In problems involving these concepts it helps to ask, "What is W or Q in one second (or one hour)?"
3. Keeping steps 1 and 2 in mind, solve for the target variables.

EVALUATE *your answer:* Use the first law of thermodynamics to check your results. Pay particular attention to algebraic signs.

Example 20.1 Analyzing a heat engine

A gasoline truck engine takes in 10,000 J of heat and delivers 2000 J of mechanical work per cycle. The heat is obtained by burning gasoline with heat of combustion $L_c = 5.0 \times 10^4$ J/g. (a) What is the thermal efficiency of this engine? (b) How much heat is discarded in each cycle? (c) If the engine goes through 25 cycles per second, what is its power output in watts? In horsepower? (d) How much gasoline is burned in each cycle? (e) How much gasoline is burned per second? Per hour?

SOLUTION

IDENTIFY and SET UP: This problem concerns a heat engine, so we can use the ideas of this section. Figure 20.4 is our energy-flow diagram for one cycle. In each cycle the engine does $W = 2000$ J of work and takes in heat $Q_H = 10,000$ J. We use Eq. (20.4), in the form $e = W/Q_H$, to find the thermal efficiency. We use Eq. (20.2) to find the amount of heat Q_C rejected per cycle. The heat of combustion tells us how much gasoline must be burned per cycle and hence per unit time. The power output is the time rate at which the work W is done.

20.4 Our sketch for this problem.

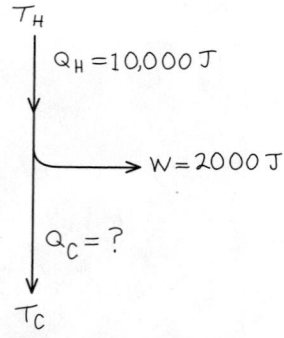

EXECUTE: (a) From Eq. (20.4), the thermal efficiency is

$$e = \frac{W}{Q_H} = \frac{2000 \text{ J}}{10,000 \text{ J}} = 0.20 = 20\%$$

(b) From Eq. (20.2), $W = Q_H + Q_C$, so

$$Q_C = W - Q_H = 2000 \text{ J} - 10,000 \text{ J} = -8000 \text{ J}$$

That is, 8000 J of heat leaves the engine during each cycle.

(c) The power P equals the work per cycle multiplied by the number of cycles per second:

$$P = (2000 \text{ J/cycle})(25 \text{ cycles/s}) = 50,000 \text{ W} = 50 \text{ kW}$$

$$= (50,000 \text{ W})\frac{1 \text{ hp}}{746 \text{ W}} = 67 \text{ hp}$$

(d) Let m be the mass of gasoline burned during each cycle. Then $Q_H = mL_c$ and

$$m = \frac{Q_H}{L_c} = \frac{10,000 \text{ J}}{5.0 \times 10^4 \text{ J/g}} = 0.20 \text{ g}$$

(e) The mass of gasoline burned per second equals the mass per cycle multiplied by the number of cycles per second:

$$(0.20 \text{ g/cycle})(25 \text{ cycles/s}) = 5.0 \text{ g/s}$$

The mass burned per hour is

$$(5.0 \text{ g/s})\frac{3600 \text{ s}}{1 \text{ h}} = 18,000 \text{ g/h} = 18 \text{ kg/h}$$

EVALUATE: An efficiency of 20% is fairly typical for cars and trucks if W includes only the work delivered to the wheels. We can check the mass burned per hour by expressing it in miles per gallon ("mileage"). The density of gasoline is about 0.70 g/cm³, so this is about 25,700 cm³, 25.7 L, or 6.8 gallons of gasoline per hour. If the truck is traveling at 55 mi/h (88 km/h), this represents fuel consumption of 8.1 miles/gallon (3.4 km/L). This is a fairly typical mileage for large trucks.

Test Your Understanding of Section 20.2   Rank the following heat engines in order from highest to lowest thermal efficiency. (i) an engine that in one cycle absorbs 5000 J of heat and rejects 4500 J of heat; (ii) an engine that in one cycle absorbs 25,000 J of heat and does 2000 J of work; (iii) an engine that in one cycle does 400 J of work and rejects 2800 J of heat.

20.3 Internal-Combustion Engines

The gasoline engine, used in automobiles and many other types of machinery, is a familiar example of a heat engine. Let's look at its thermal efficiency. Figure 20.5 shows the operation of one type of gasoline engine. First a mixture of air and gasoline vapor flows into a cylinder through an open intake valve while the piston descends, increasing the volume of the cylinder from a minimum of V (when the piston is all the way up) to a maximum of rV (when it is all the way down). The quantity r is called the **compression ratio;** for present-day automobile engines its value is typically 8 to 10. At the end of this *intake stroke,* the intake valve closes and the mixture is compressed, approximately adiabatically, to volume V during the *compression stroke*. The mixture is then ignited by the spark plug, and the heated gas expands, approximately adiabatically, back to volume rV, pushing on the piston and doing work; this is the *power stroke*. Finally, the exhaust valve opens, and the combustion products are pushed out (during the *exhaust stroke*), leaving the cylinder ready for the next intake stroke.

The Otto Cycle

Figure 20.6 is a *pV*-diagram for an idealized model of the thermodynamic processes in a gasoline engine. This model is called the **Otto cycle.** At point a the gasoline–air mixture has entered the cylinder. The mixture is compressed adiabatically to point b and is then ignited. Heat Q_H is added to the system by the burning gasoline along line bc, and the power stroke is the adiabatic expansion to d. The gas is cooled to the temperature of the outside air along line da; during this process, heat $|Q_C|$ is rejected. This gas leaves the engine as exhaust and does not enter the engine again. But since an equivalent amount of gasoline and air enters, we may consider the process to be cyclic.

20.5 Cycle of a four-stroke internal-combustion engine.

Intake stroke: Piston moves down, causing a partial vacuum in cylinder; gasoline–air mixture enters through intake valve.

Compression stroke: Intake valve closes; mixture is compressed as piston moves up.

Ignition: Spark plug ignites mixture.

Power stroke: Hot burned mixture expands, pushing piston down.

Exhaust stroke: Exhaust valve opens; piston moves up, expelling exhaust and leaving cylinder ready for next intake stroke.

20.6 The pV-diagram for the Otto cycle, an idealized model of the thermodynamic processes in a gasoline engine.

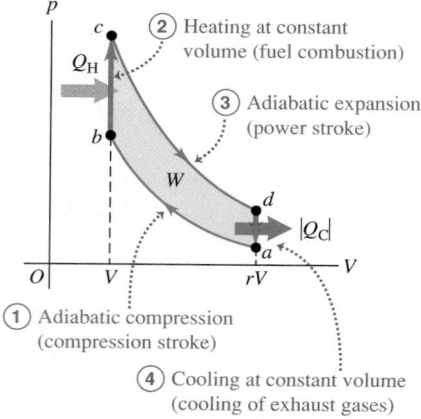

Otto cycle

(1) Adiabatic compression (compression stroke)

(2) Heating at constant volume (fuel combustion)

(3) Adiabatic expansion (power stroke)

(4) Cooling at constant volume (cooling of exhaust gases)

We can calculate the efficiency of this idealized cycle. Processes bc and da are constant-volume, so the heats Q_H and Q_C are related simply to the temperatures:

$$Q_H = nC_V(T_c - T_b) > 0$$
$$Q_C = nC_V(T_a - T_d) < 0$$

The thermal efficiency is given by Eq. (20.4). Inserting the above expressions and cancelling out the common factor nC_V, we find

$$e = \frac{Q_H + Q_C}{Q_H} = \frac{T_c - T_b + T_a - T_d}{T_c - T_b} \tag{20.5}$$

To simplify this further, we use the temperature–volume relationship for adiabatic processes for an ideal gas, Eq. (19.22). For the two adiabatic processes ab and cd,

$$T_a(rV)^{\gamma-1} = T_bV^{\gamma-1} \quad \text{and} \quad T_d(rV)^{\gamma-1} = T_cV^{\gamma-1}$$

We divide each of these equations by the common factor $V^{\gamma-1}$ and substitute the resulting expressions for T_b and T_c back into Eq. (20.5). The result is

$$e = \frac{T_dr^{\gamma-1} - T_ar^{\gamma-1} + T_a - T_d}{T_dr^{\gamma-1} - T_ar^{\gamma-1}} = \frac{(T_d - T_a)(r^{\gamma-1} - 1)}{(T_d - T_a)r^{\gamma-1}}$$

Dividing out the common factor $(T_d - T_a)$, we get

$$e = 1 - \frac{1}{r^{\gamma-1}} \qquad \text{(thermal efficiency in Otto cycle)} \tag{20.6}$$

The thermal efficiency given by Eq. (20.6) is always less than unity, even for this idealized model. With $r = 8$ and $\gamma = 1.4$ (the value for air) the theoretical efficiency is $e = 0.56$, or 56%. The efficiency can be increased by increasing r. However, this also increases the temperature at the end of the adiabatic compression of the air–fuel mixture. If the temperature is too high, the mixture explodes spontaneously during compression instead of burning evenly after the spark plug ignites it. This is called *pre-ignition* or *detonation;* it causes a knocking sound and can damage the engine. The octane rating of a gasoline is a measure of its antiknock qualities. The maximum practical compression ratio for high-octane, or "premium," gasoline is about 10 to 13.

The Otto cycle is a highly idealized model. It assumes that the mixture behaves as an ideal gas; it neglects friction, turbulence, loss of heat to cylinder walls, and many other effects that reduce the efficiency of an engine. Efficiencies of real gasoline engines are typically around 35%.

The Diesel Cycle

The Diesel engine is similar in operation to the gasoline engine. The most important difference is that there is no fuel in the cylinder at the beginning of the compression stroke. A little before the beginning of the power stroke, the injectors start to inject fuel directly into the cylinder, just fast enough to keep the pressure approximately constant during the first part of the power stroke. Because of the high temperature developed during the adiabatic compression, the fuel ignites spontaneously as it is injected; no spark plugs are needed.

Figure 20.7 shows the idealized **Diesel cycle.** Starting at point a, air is compressed adiabatically to point b, heated at constant pressure to point c, expanded adiabatically to point d, and cooled at constant volume to point a. Because there is no fuel in the cylinder during most of the compression stroke, pre-ignition cannot occur, and the compression ratio r can be much higher than for a gasoline engine. This improves efficiency and ensures reliable ignition when the fuel is injected (because of the high temperature reached during the

20.7 The pV-diagram for the idealized Diesel cycle.

Diesel cycle

(1) Adiabatic compression (compression stroke)

(2) Fuel ignition, heating at constant pressure (fuel combustion). This is a significant difference between the Diesel and Otto cycles.

(3) Adiabatic expansion (power stroke)

(4) Cooling at constant volume (cooling of exhaust gases)

adiabatic compression). Values of r of 15 to 20 are typical; with these values and $\gamma = 1.4$, the theoretical efficiency of the idealized Diesel cycle is about 0.65 to 0.70. As with the Otto cycle, the efficiency of any actual engine is substantially less than this. While Diesel engines are very efficient, they must be built to much tighter tolerances than gasoline engines and the fuel-injection system requires careful maintenance.

Test Your Understanding of Section 20.3   For an Otto-cycle engine with cylinders of a fixed size and a fixed compression ratio, which of the following aspects of the pV-diagram in Fig. 20.6 would change if you doubled the amount of fuel burned per cycle? (There may be more than one correct answer.) (i) the vertical distance between points b and c; (ii) the vertical distance between points a and d; (iii) the horizontal distance between points b and a.

20.4 Refrigerators

We can think of a **refrigerator** as a heat engine operating in reverse. A heat engine takes heat from a hot place and gives off heat to a colder place. A refrigerator does the opposite; it takes heat from a cold place (the inside of the refrigerator) and gives it off to a warmer place (usually the air in the room where the refrigerator is located). A heat engine has a net *output* of mechanical work; the refrigerator requires a net *input* of mechanical work. Using the sign conventions from Section 20.2, for a refrigerator Q_C is positive but both W and Q_H are negative; hence $|W| = -W$ and $|Q_H| = -Q_H$.

Figure 20.8 shows an energy-flow diagram for a refrigerator. From the first law for a cyclic process,

$$Q_H + Q_C - W = 0 \quad \text{or} \quad -Q_H = Q_C - W$$

or, because both Q_H and W are negative,

$$|Q_H| = Q_C + |W| \qquad (20.7)$$

Thus, as the diagram shows, the heat $|Q_H|$ leaving the working substance and given to the hot reservoir is always *greater* than the heat Q_C taken from the cold reservoir. Note that the absolute-value relationship

$$|Q_H| = |Q_C| + |W| \qquad (20.8)$$

is valid for both heat engines and refrigerators.

From an economic point of view, the best refrigeration cycle is one that removes the greatest amount of heat $|Q_C|$ from the inside of the refrigerator for the least expenditure of mechanical work, $|W|$. The relevant ratio is therefore $|Q_C|/|W|$; the larger this ratio, the better the refrigerator. We call this ratio the **coefficient of performance**, denoted by K. From Eq. (20.8), $|W| = |Q_H| - |Q_C|$, so

$$K = \frac{|Q_C|}{|W|} = \frac{|Q_C|}{|Q_H| - |Q_C|} \qquad \begin{array}{l}\text{(coefficient of performance} \\ \text{of a refrigerator)}\end{array} \qquad (20.9)$$

As always, we measure Q_H, Q_C, and W all in the same energy units; K is then a dimensionless number.

Practical Refrigerators

The principles of the common refrigeration cycle are shown schematically in Fig. 20.9a. The fluid "circuit" contains a refrigerant fluid (the working substance). The left side of the circuit (including the cooling coils inside the refrigerator) is at low temperature and low pressure; the right side (including the condenser coils outside the refrigerator) is at high temperature and high pressure. Ordinarily, both sides contain liquid and vapor in phase equilibrium.

20.8 Schematic energy-flow diagram of a refrigerator.

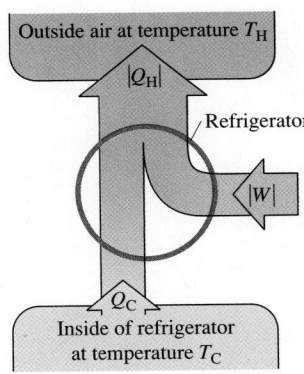

20.9 (a) Principle of the mechanical refrigeration cycle. (b) How the key elements are arranged in a practical refrigerator.

(a)

(b)

The compressor takes in fluid, compresses it adiabatically, and delivers it to the condenser coil at high pressure. The fluid temperature is then higher than that of the air surrounding the condenser, so the refrigerant gives off heat $|Q_H|$ and partially condenses to liquid. The fluid then expands adiabatically into the evaporator at a rate controlled by the expansion valve. As the fluid expands, it cools considerably, enough that the fluid in the evaporator coil is colder than its surroundings. It absorbs heat $|Q_C|$ from its surroundings, cooling them and partially vaporizing. The fluid then enters the compressor to begin another cycle. The compressor, usually driven by an electric motor (Fig. 20.9b), requires energy input and does work $|W|$ *on* the working substance during each cycle.

An air conditioner operates on exactly the same principle. In this case the refrigerator box becomes a room or an entire building. The evaporator coils are inside, the condenser is outside, and fans circulate air through these (Fig. 20.10). In large installations the condenser coils are often cooled by water. For air conditioners the quantities of greatest practical importance are the *rate* of heat removal (the heat current H from the region being cooled) and the *power* input $P = W/t$

20.10 An air conditioner works on the same principle as a refrigerator.

to the compressor. If heat $|Q_C|$ is removed in time t, then $H = |Q_C|/t$. Then we can express the coefficient of performance as

$$K = \frac{|Q_C|}{|W|} = \frac{Ht}{Pt} = \frac{H}{P}$$

Typical room air conditioners have heat removal rates H of 5000 to 10,000 Btu/h, or about 1500–3000 W, and require electric power input of about 600 to 1200 W. Typical coefficients of performance are about 3; the actual values depend on the inside and outside temperatures.

A variation on this theme is the **heat pump,** used to heat buildings by cooling the outside air. It functions like a refrigerator turned inside out. The evaporator coils are outside, where they take heat from cold air, and the condenser coils are inside, where they give off heat to the warmer air. With proper design, the heat $|Q_H|$ delivered to the inside per cycle can be considerably greater than the work $|W|$ required to get it there.

Work is *always* needed to transfer heat from a colder to a hotter body. Heat flows spontaneously from hotter to colder, and to reverse this flow requires the addition of work from the outside. Experience shows that it is impossible to make a refrigerator that transports heat from a colder body to a hotter body without the addition of work. If no work were needed, the coefficient of performance would be infinite. We call such a device a *workless refrigerator;* it is a mythical beast, like the unicorn and the free lunch.

Test Your Understanding of Section 20.4 Can you cool your house by leaving the refrigerator door open?

20.5 The Second Law of Thermodynamics

Experimental evidence suggests strongly that it is *impossible* to build a heat engine that converts heat completely to work—that is, an engine with 100% thermal efficiency. This impossibility is the basis of one statement of the **second law of thermodynamics,** as follows:

> **It is impossible for any system to undergo a process in which it absorbs heat from a reservoir at a single temperature and converts the heat completely into mechanical work, with the system ending in the same state in which it began.**

We will call this the "engine" statement of the second law. (It is also known to physicists as the *Kelvin–Planck statement* of this law.)

The basis of the second law of thermodynamics is the difference between the nature of internal energy and that of macroscopic mechanical energy. In a moving body the molecules have random motion, but superimposed on this is a coordinated motion of every molecule in the direction of the body's velocity. The kinetic energy associated with this *coordinated* macroscopic motion is what we call the kinetic energy of the moving body. The kinetic and potential energies associated with the *random* motion constitute the internal energy.

When a body sliding on a surface comes to rest as a result of friction, the organized motion of the body is converted to random motion of molecules in the body and in the surface. Since we cannot control the motions of individual molecules, we cannot convert this random motion completely back to organized motion. We can convert *part* of it, and this is what a heat engine does.

If the second law were *not* true, we could power an automobile or run a power plant by cooling the surrounding air. Neither of these impossibilities violates the *first* law of thermodynamics. The second law, therefore, is not a deduction from the first but stands by itself as a separate law of nature. The first law denies the possibility of creating or destroying energy; the second law limits the *availability* of energy and the ways in which it can be used and converted.

Restating the Second Law

Our analysis of refrigerators in Section 20.4 forms the basis for an alternative statement of the second law of thermodynamics. Heat flows spontaneously from hotter to colder bodies, never the reverse. A refrigerator does take heat from a colder to a hotter body, but its operation requires an input of mechanical energy or work. Generalizing this observation, we state:

> **It is impossible for any process to have as its sole result the transfer of heat from a cooler to a hotter body.**

We'll call this the "refrigerator" statement of the second law. (It is also known as the *Clausius statement.*) It may not seem to be very closely related to the "engine" statement. In fact, though, the two statements are completely equivalent. For example, if we could build a workless refrigerator, violating the second or "refrigerator" statement of the second law, we could use it in conjunction with a heat engine, pumping the heat rejected by the engine back to the hot reservoir to be reused. This composite machine (Fig. 20.11a) would violate the "engine" statement of the second law because its net effect would be to take a net quantity of heat $Q_H - |Q_C|$ from the hot reservoir and convert it completely to work W.

Alternatively, if we could make an engine with 100% thermal efficiency, in violation of the first statement, we could run it using heat from the hot reservoir and use the work output to drive a refrigerator that pumps heat from the cold reservoir to the hot (Fig. 20.11b). This composite device would violate the "refrigerator" statement because its net effect would be to take heat Q_C from the

20.11 Energy-flow diagrams showing that the two forms of the second law are equivalent.

(a) The "engine" statement of the second law of thermodynamics

If a workless refrigerator were possible, it could be used in conjunction with an ordinary heat engine to form a 100%-efficient engine, converting heat $Q_H - |Q_C|$ completely to work.

(b) The "refrigerator" statement of the second law of thermodynamics

If a 100%-efficient engine were possible, it could be used in conjunction with an ordinary refrigerator to form a workless refrigerator, transferring heat Q_C from the cold to the hot reservoir with no input of work.

cold reservoir and deliver it to the hot reservoir without requiring any input of work. Thus any device that violates one form of the second law can be used to make a device that violates the other form. If violations of the first form are impossible, so are violations of the second!

The conversion of work to heat and the heat flow from hot to cold across a finite temperature gradient are *irreversible* processes. The "engine" and "refrigerator" statements of the second law state that these processes can be only partially reversed. We could cite other examples. Gases naturally flow from a region of high pressure to a region of low pressure; gases and miscible liquids left by themselves always tend to mix, not to unmix. The second law of thermodynamics is an expression of the inherent one-way aspect of these and many other irreversible processes. Energy conversion is an essential aspect of all plant and animal life and of human technology, so the second law of thermodynamics is of fundamental importance.

Test Your Understanding of Section 20.5 Would a 100%-efficient engine (Fig. 20.11a) violate the *first* law of thermodynamics? What about a workless refrigerator (Fig. 20.11b)? ❙

20.6 The Carnot Cycle

According to the second law, no heat engine can have 100% efficiency. How great an efficiency *can* an engine have, given two heat reservoirs at temperatures T_H and T_C? This question was answered in 1824 by the French engineer Sadi Carnot (1796–1832), who developed a hypothetical, idealized heat engine that has the maximum possible efficiency consistent with the second law. The cycle of this engine is called the **Carnot cycle.**

To understand the rationale of the Carnot cycle, we return to *reversibility* and its relationship to directions of thermodynamic processes. Conversion of work to heat is an irreversible process; the purpose of a heat engine is a *partial* reversal of this process, the conversion of heat to work with as great an efficiency as possible. For maximum heat-engine efficiency, therefore, *we must avoid all irreversible processes* (Fig. 20.12).

Heat flow through a finite temperature drop is an irreversible process. Therefore, during heat transfer in the Carnot cycle there must be *no* finite temperature difference. When the engine takes heat from the hot reservoir at temperature T_H, the working substance of the engine must also be at T_H; otherwise, irreversible heat flow would occur. Similarly, when the engine discards heat to the cold reservoir at T_C, the engine itself must be at T_C. That is, every process that involves heat transfer must be *isothermal* at either T_H or T_C.

Conversely, in any process in which the temperature of the working substance of the engine is intermediate between T_H and T_C, there must be *no* heat transfer between the engine and either reservoir because such heat transfer could not be reversible. Therefore any process in which the temperature T of the working substance changes must be *adiabatic*.

The bottom line is that every process in our idealized cycle must be either isothermal or adiabatic. In addition, thermal and mechanical equilibrium must be maintained at all times so that each process is completely reversible.

Steps of the Carnot Cycle

The Carnot cycle consists of two reversible isothermal and two reversible adiabatic processes. Figure 20.13 shows a Carnot cycle using as its working substance an ideal gas in a cylinder with a piston. It consists of the following steps:

1. The gas expands isothermally at temperature T_H, absorbing heat Q_H (*ab*).
2. It expands adiabatically until its temperature drops to T_C (*bc*).
3. It is compressed isothermally at T_C, rejecting heat $|Q_C|$ (*cd*).
4. It is compressed adiabatically back to its initial state at temperature T_H (*da*).

20.12 The temperature of the firebox of a steam engine is much higher than the temperature of water in the boiler, so heat flows irreversibly from firebox to water. Carnot's quest to understand the efficiency of steam engines led him to the idea that an ideal engine would involve only *reversible* processes.

20.13 The Carnot cycle for an ideal gas. The light blue lines in the pV-diagram are isotherms (curves of constant temperature) and the dark blue lines are adiabats (curves of zero heat flow).

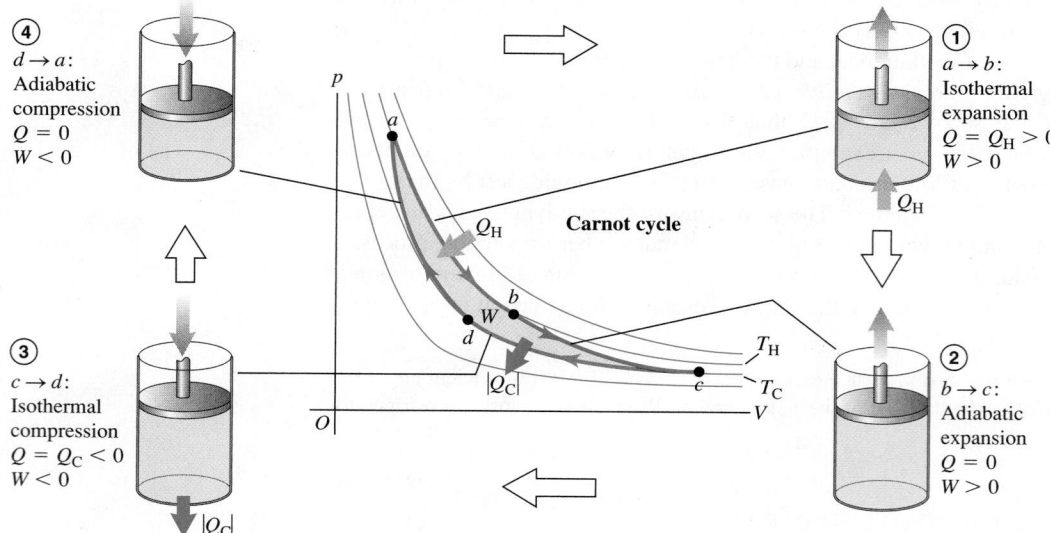

We can calculate the thermal efficiency e of a Carnot engine in the special case shown in Fig. 20.13 in which the working substance is an *ideal gas*. To carry out this calculation, we will first find the ratio Q_C/Q_H of the quantities of heat transferred in the two isothermal processes and then use Eq. (20.4) to find e.

For an ideal gas the internal energy U depends only on temperature and is thus constant in any isothermal process. For the isothermal expansion ab, $\Delta U_{ab} = 0$ and Q_H is equal to the work W_{ab} done by the gas during its isothermal expansion at temperature T_H. We calculated this work in Example 19.1 (Section 19.2); using that result, we have

$$Q_H = W_{ab} = nRT_H \ln \frac{V_b}{V_a} \qquad (20.10)$$

Similarly,

$$Q_C = W_{cd} = nRT_C \ln \frac{V_d}{V_c} = -nRT_C \ln \frac{V_c}{V_d} \qquad (20.11)$$

Because V_d is less than V_c, Q_C is negative ($Q_C = -|Q_C|$); heat flows out of the gas during the isothermal compression at temperature T_C.

The ratio of the two quantities of heat is thus

$$\frac{Q_C}{Q_H} = -\left(\frac{T_C}{T_H}\right) \frac{\ln (V_c/V_d)}{\ln (V_b/V_a)} \qquad (20.12)$$

This can be simplified further by use of the temperature–volume relationship for an adiabatic process, Eq. (19.22). We find for the two adiabatic processes:

$$T_H V_b^{\gamma-1} = T_C V_c^{\gamma-1} \qquad \text{and} \qquad T_H V_a^{\gamma-1} = T_C V_d^{\gamma-1}$$

Dividing the first of these by the second, we find

$$\frac{V_b^{\gamma-1}}{V_a^{\gamma-1}} = \frac{V_c^{\gamma-1}}{V_d^{\gamma-1}} \qquad \text{and} \qquad \frac{V_b}{V_a} = \frac{V_c}{V_d}$$

Thus the two logarithms in Eq. (20.12) are equal, and that equation reduces to

$$\frac{Q_C}{Q_H} = -\frac{T_C}{T_H} \quad \text{or} \quad \frac{|Q_C|}{|Q_H|} = \frac{T_C}{T_H} \qquad \text{(heat transfer in a Carnot engine)} \quad \text{(20.13)}$$

The ratio of the heat rejected at T_C to the heat absorbed at T_H is just equal to the ratio T_C/T_H. Then from Eq. (20.4) the efficiency of the Carnot engine is

$$e_{\text{Carnot}} = 1 - \frac{T_C}{T_H} = \frac{T_H - T_C}{T_H} \qquad \text{(efficiency of a Carnot engine)} \quad \text{(20.14)}$$

This simple result says that the efficiency of a Carnot engine depends only on the temperatures of the two heat reservoirs. The efficiency is large when the temperature *difference* is large, and it is very small when the temperatures are nearly equal. The efficiency can never be exactly unity unless $T_C = 0$; we'll see later that this, too, is impossible.

CAUTION **Use Kelvin temperature in Carnot calculations** In all calculations involving the Carnot cycle, you must make sure that you use *absolute* (Kelvin) temperatures only. That's because Eqs. (20.10) through (20.14) come from the ideal-gas equation $pV = nRT$, in which T is absolute temperature. ▮

Example 20.2   **Analyzing a Carnot engine I**

A Carnot engine takes 2000 J of heat from a reservoir at 500 K, does some work, and discards some heat to a reservoir at 350 K. How much work does it do, how much heat is discarded, and what is its efficiency?

SOLUTION

IDENTIFY and SET UP: This problem involves a Carnot engine, so we can use the ideas of this section and those of Section 20.2 (which apply to heat engines of all kinds). Figure 20.14 shows the energy-flow diagram. We have $Q_H = 2000$ J, $T_H = 500$ K, and $T_C = 350$ K. We use Eq. (20.13) to find Q_C, and then use the first law of thermodynamics as given by Eq. (20.2) to find W. We find the efficiency e from T_C and T_H using Eq. (20.14).

EXECUTE: From Eq. (20.13),

$$Q_C = -Q_H \frac{T_C}{T_H} = -(2000 \text{ J})\frac{350 \text{ K}}{500 \text{ K}} = -1400 \text{ J}$$

Then from Eq. (20.2), the work done is

$$W = Q_H + Q_C = 2000 \text{ J} + (-1400 \text{ J}) = 600 \text{ J}$$

From Eq. (20.14), the thermal efficiency is

$$e = 1 - \frac{T_C}{T_H} = 1 - \frac{350 \text{ K}}{500 \text{ K}} = 0.30 = 30\%$$

EVALUATE: The negative sign of Q_C is correct: It shows that 1400 J of heat flows *out* of the engine and into the cold reservoir. We can check our result for e by using the basic definition of thermal efficiency, Eq. (20.3):

$$e = \frac{W}{Q_H} = \frac{600 \text{ J}}{2000 \text{ J}} = 0.30 = 30\%$$

20.14 Our sketch for this problem.

$T_H = 500$ K
$Q_H = 2000$ J
$W = ?$
$e = ?$
$Q_C = ?$
$T_C = 350$ K

Example 20.3   **Analyzing a Carnot engine II**

Suppose 0.200 mol of an ideal diatomic gas ($\gamma = 1.40$) undergoes a Carnot cycle between 227°C and 27°C, starting at $p_a = 10.0 \times 10^5$ Pa at point a in the pV-diagram of Fig. 20.13. The volume doubles during the isothermal expansion step $a \rightarrow b$. (a) Find the pressure and volume at points a, b, c, and d. (b) Find Q, W, and ΔU for each step and for the entire cycle. (c) Find the efficiency directly from the results of part (b), and compare with the value calculated from Eq. (20.14).

Continued

SOLUTION

IDENTIFY and SET UP: This problem involves the properties of the Carnot cycle and those of an ideal gas. We are given the number of moles n and the pressure and temperature at point a (which is at the higher of the two reservoir temperatures); we can find the volume at a using the ideal-gas equation $pV = nRT$. We then find the pressure and volume at points b, c, and d from the known doubling of volume in step $a \to b$, from equations given in this section, and from $pV = nRT$. In each step we use Eqs. (20.10) and (20.11) to find the heat flow and work done and Eq. (19.13) to find the internal energy change.

EXECUTE: (a) With $T_H = (227 + 273.15)$ K $= 500$ K and $T_C = (27 + 273.15)$ K $= 300$ K, $pV = nRT$ yields

$$V_a = \frac{nRT_H}{p_a} = \frac{(0.200 \text{ mol})(8.314 \text{ J/mol}\cdot\text{K})(500 \text{ K})}{10.0 \times 10^5 \text{ Pa}}$$
$$= 8.31 \times 10^{-4} \text{ m}^3$$

The volume doubles during the isothermal expansion $a \to b$:

$$V_b = 2V_a = 2(8.31 \times 10^{-4} \text{ m}^3) = 16.6 \times 10^{-4} \text{ m}^3$$

Because the expansion $a \to b$ is isothermal, $p_a V_a = p_b V_b$, so

$$p_b = \frac{p_a V_a}{V_b} = 5.00 \times 10^5 \text{ Pa}$$

For the adiabatic expansion $b \to c$, we use the equation $T_H V_b^{\gamma-1} = T_C V_c^{\gamma-1}$ that follows Eq. (20.12) as well as the ideal-gas equation:

$$V_c = V_b \left(\frac{T_H}{T_C}\right)^{1/(\gamma-1)} = (16.6 \times 10^{-4} \text{ m}^3)\left(\frac{500 \text{ K}}{300 \text{ K}}\right)^{2.5}$$
$$= 59.6 \times 10^{-4} \text{ m}^3$$

$$p_c = \frac{nRT_C}{V_c} = \frac{(0.200 \text{ mol})(8.314 \text{ J/mol}\cdot\text{K})(300 \text{ K})}{59.6 \times 10^{-4} \text{ m}^3}$$
$$= 0.837 \times 10^5 \text{ Pa}$$

For the adiabatic compression $d \to a$ we have $T_C V_d^{\gamma-1} = T_H V_a^{\gamma-1}$ and so

$$V_d = V_a \left(\frac{T_H}{T_C}\right)^{1/(\gamma-1)} = (8.31 \times 10^{-4} \text{ m}^3)\left(\frac{500 \text{ K}}{300 \text{ K}}\right)^{2.5}$$
$$= 29.8 \times 10^{-4} \text{ m}^3$$

$$p_d = \frac{nRT_C}{V_d} = \frac{(0.200 \text{ mol})(8.314 \text{ J/mol}\cdot\text{K})(300 \text{ K})}{29.8 \times 10^{-4} \text{ m}^3}$$
$$= 1.67 \times 10^5 \text{ Pa}$$

(b) For the isothermal expansion $a \to b$, $\Delta U_{ab} = 0$. From Eq. (20.10),

$$W_{ab} = Q_H = nRT_H \ln \frac{V_b}{V_a}$$
$$= (0.200 \text{ mol})(8.314 \text{ J/mol}\cdot\text{K})(500 \text{ K})(\ln 2) = 576 \text{ J}$$

For the adiabatic expansion $b \to c$, $Q_{bc} = 0$. From the first law of thermodynamics, $\Delta U_{bc} = Q_{bc} - W_{bc} = -W_{bc}$; the work W_{bc} done by the gas in this adiabatic expansion equals the negative of the change in internal energy of the gas. From Eq. (19.13) we have $\Delta U = nC_V \Delta T$, where $\Delta T = T_C - T_H$. Using $C_V = 20.8 \text{ J/mol}\cdot\text{K}$ for an ideal diatomic gas, we find

$$W_{bc} = -\Delta U_{bc} = -nC_V(T_C - T_H) = nC_V(T_H - T_C)$$
$$= (0.200 \text{ mol})(20.8 \text{ J/mol}\cdot\text{K})(500 \text{ K} - 300 \text{ K}) = 832 \text{ J}$$

For the isothermal compression $c \to d$, $\Delta U_{cd} = 0$; Eq. (20.11) gives

$$W_{cd} = Q_C = nRT_C \ln \frac{V_d}{V_c}$$
$$= (0.200 \text{ mol})(8.314 \text{ J/mol}\cdot\text{K})(300 \text{ K})\left(\ln \frac{29.8 \times 10^{-4} \text{ m}^3}{59.6 \times 10^{-4} \text{ m}^3}\right)$$
$$= -346 \text{ J}$$

For the adiabatic compression $d \to a$, $Q_{da} = 0$ and

$$W_{da} = -\Delta U_{da} = -nC_V(T_H - T_C) = nC_V(T_C - T_H)$$
$$= (0.200 \text{ mol})(20.8 \text{ J/mol}\cdot\text{K})(300 \text{ K} - 500 \text{ K}) = -832 \text{ J}$$

We can tabulate these results as follows:

Process	Q	W	ΔU
$a \to b$	576 J	576 J	0
$b \to c$	0	832 J	−832 J
$c \to d$	−346 J	−346 J	0
$d \to a$	0	−832 J	832 J
Total	230 J	230 J	0

(c) From the above table, $Q_H = 576$ J and the total work is 230 J. Thus

$$e = \frac{W}{Q_H} = \frac{230 \text{ J}}{576 \text{ J}} = 0.40 = 40\%$$

We can compare this to the result from Eq. (20.14),

$$e = \frac{T_H - T_C}{T_H} = \frac{500 \text{ K} - 300 \text{ K}}{500 \text{ K}} = 0.40 = 40\%$$

EVALUATE: The table in part (b) shows that for the entire cycle $Q = W$ and $\Delta U = 0$, just as we would expect: In a complete cycle, the *net* heat input is used to do work, and there is zero net change in the internal energy of the system. Note also that the quantities of work in the two adiabatic processes are negatives of each other. Can you show from the analysis leading to Eq. (20.13) that this must *always* be the case in a Carnot cycle?

The Carnot Refrigerator

Because each step in the Carnot cycle is reversible, the *entire cycle* may be reversed, converting the engine into a refrigerator. The coefficient of performance of the Carnot refrigerator is obtained by combining the general definition of K, Eq. (20.9), with Eq. (20.13) for the Carnot cycle. We first rewrite Eq. (20.9) as

$$K = \frac{|Q_C|}{|Q_H| - |Q_C|} = \frac{|Q_C|/|Q_H|}{1 - |Q_C|/|Q_H|}$$

Then we substitute Eq. (20.13), $|Q_C|/|Q_H| = T_C/T_H$, into this expression. The result is

$$K_{\text{Carnot}} = \frac{T_C}{T_H - T_C} \qquad \begin{array}{l}\text{(coefficient of performance}\\\text{of a Carnot refrigerator)}\end{array} \qquad (20.15)$$

When the temperature difference $T_H - T_C$ is small, K is much larger than unity; in this case a lot of heat can be "pumped" from the lower to the higher temperature with only a little expenditure of work. But the greater the temperature difference, the smaller the value of K and the more work is required to transfer a given quantity of heat.

Example 20.4 | **Analyzing a Carnot refrigerator**

If the cycle described in Example 20.3 is run backward as a refrigerator, what is its coefficient of performance?

SOLUTION

IDENTIFY and SET UP: This problem uses the ideas of Section 20.3 (for refrigerators in general) and the above discussion of Carnot refrigerators. Equation (20.9) gives the coefficient of performance K of *any* refrigerator in terms of the heat Q_C extracted from the cold reservoir per cycle and the work W that must be done per cycle.

EXECUTE: In Example 20.3 we found that in one cycle the Carnot engine rejects heat $Q_C = -346$ J to the cold reservoir and does work $W = 230$ J. When run in reverse as a refrigerator, the system

extracts heat $Q_C = +346$ J from the cold reservoir while requiring a work input of $W = -230$ J. From Eq. (20.9),

$$K = \frac{|Q_C|}{|W|} = \frac{346 \text{ J}}{230 \text{ J}} = 1.50$$

Because this is a Carnot cycle, we can also use Eq. (20.15):

$$K = \frac{T_C}{T_H - T_C} = \frac{300 \text{ K}}{500 \text{ K} - 300 \text{ K}} = 1.50$$

EVALUATE: Equations (20.14) and (20.15) show that e and K for a Carnot cycle depend only on T_H and T_C, and we don't need to calculate Q and W. For cycles containing irreversible processes, however, these two equations are not valid, and more detailed calculations are necessary.

The Carnot Cycle and the Second Law

We can prove that **no engine can be more efficient than a Carnot engine operating between the same two temperatures.** The key to the proof is the above observation that since each step in the Carnot cycle is reversible, the *entire cycle* may be reversed. Run backward, the engine becomes a refrigerator. Suppose we have an engine that is more efficient than a Carnot engine (Fig. 20.15). Let the Carnot engine, run backward as a refrigerator by negative work $-|W|$, take

20.15 Proving that the Carnot engine has the highest possible efficiency. A "superefficient" engine (more efficient than a Carnot engine) combined with a Carnot refrigerator could convert heat completely into work with no net heat transfer to the cold reservoir. This would violate the second law of thermodynamics.

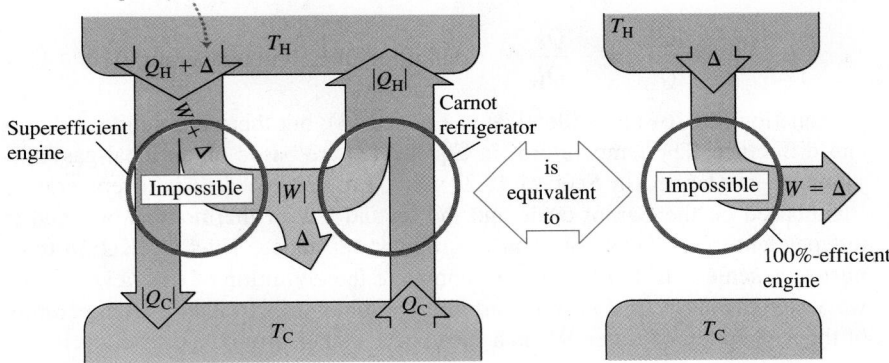

If a superefficient engine were possible, it could be used in conjunction with a Carnot refrigerator to convert the heat Δ completely to work, with no net transfer to the cold reservoir.

in heat Q_C from the cold reservoir and expel heat $|Q_H|$ to the hot reservoir. The superefficient engine expels heat $|Q_C|$, but to do this, it takes in a greater amount of heat $Q_H + \Delta$. Its work output is then $W + \Delta$, and the net effect of the two machines together is to take a quantity of heat Δ and convert it completely into work. This violates the engine statement of the second law. We could construct a similar argument that a superefficient engine could be used to violate the refrigerator statement of the second law. Note that we don't have to assume that the superefficient engine is reversible. In a similar way we can show that *no refrigerator can have a greater coefficient of performance than a Carnot refrigerator operating between the same two temperatures.*

Thus the statement that no engine can be more efficient than a Carnot engine is yet another equivalent statement of the second law of thermodynamics. It also follows directly that **all Carnot engines operating between the same two temperatures have the same efficiency, irrespective of the nature of the working substance.** Although we derived Eq. (20.14) for a Carnot engine using an ideal gas as its working substance, it is in fact valid for *any* Carnot engine, no matter what its working substance.

Equation (20.14), the expression for the efficiency of a Carnot engine, sets an upper limit to the efficiency of a real engine such as a steam turbine. To maximize this upper limit and the actual efficiency of the real engine, the designer must make the intake temperature T_H as high as possible and the exhaust temperature T_C as low as possible (Fig. 20.16).

The exhaust temperature cannot be lower than the lowest temperature available for cooling the exhaust. For a steam turbine at an electric power plant, T_C may be the temperature of river or lake water; then we want the boiler temperature T_H to be as high as possible. The vapor pressures of all liquids increase rapidly with temperature, so we are limited by the mechanical strength of the boiler. At 500°C the vapor pressure of water is about 240×10^5 Pa (235 atm); this is about the maximum practical pressure in large present-day steam boilers.

20.16 To maximize efficiency, the temperatures inside a jet engine are made as high as possible. Exotic ceramic materials are used that can withstand temperatures in excess of 1000°C without melting or becoming soft.

The Kelvin Temperature Scale

In Chapter 17 we expressed the need for a temperature scale that doesn't depend on the properties of any particular material. We can now use the Carnot cycle to define such a scale. The thermal efficiency of a Carnot engine operating between two heat reservoirs at temperatures T_H and T_C is independent of the nature of the working substance and depends only on the temperatures. From Eq. (20.4), this thermal efficiency is

$$e = \frac{Q_H + Q_C}{Q_H} = 1 + \frac{Q_C}{Q_H}$$

Therefore the ratio Q_C/Q_H is the same for *all* Carnot engines operating between two given temperatures T_H and T_C.

Kelvin proposed that we *define* the ratio of the temperatures, T_C/T_H, to be equal to the magnitude of the ratio Q_C/Q_H of the quantities of heat absorbed and rejected:

$$\frac{T_C}{T_H} = \frac{|Q_C|}{|Q_H|} = -\frac{Q_C}{Q_H} \qquad \text{(definition of Kelvin temperature)} \quad \text{(20.16)}$$

Equation (20.16) looks identical to Eq. (20.13), but there is a subtle and crucial difference. The temperatures in Eq. (20.13) are based on an ideal-gas thermometer, as defined in Section 17.3, while Eq. (20.16) *defines* a temperature scale based on the Carnot cycle and the second law of thermodynamics and is independent of the behavior of any particular substance. Thus the **Kelvin temperature scale** is truly *absolute*. To complete the definition of the Kelvin scale, we assign, as in Section 17.3, the arbitrary value of 273.16 K to the temperature of the triple point of water. When a substance is taken around a Carnot cycle, the ratio of the heats absorbed and rejected, $|Q_H|/|Q_C|$, is equal to the ratio of the temperatures of the reservoirs *as expressed on the gas-thermometer scale* defined in Section 17.3. Since the triple point of water is chosen to be 273.16 K in both scales, it follows that *the Kelvin and ideal-gas scales are identical.*

The zero point on the Kelvin scale is called **absolute zero.** At absolute zero a system has its *minimum* possible total internal energy (kinetic plus potential). Because of quantum effects, however, it is *not* true that at $T = 0$, all molecular motion ceases. There are theoretical reasons for believing that absolute zero cannot be attained experimentally, although temperatures below 10^{-7} K have been achieved. The more closely we approach absolute zero, the more difficult it is to get closer. One statement of the *third law of thermodynamics* is that it is impossible to reach absolute zero in a finite number of thermodynamic steps.

Test Your Understanding of Section 20.6 An inventor looking for financial support comes to you with an idea for a gasoline engine that runs on a novel type of thermodynamic cycle. His design is made entirely of copper and is air-cooled. He claims that the engine will be 85% efficient. Should you invest in this marvelous new engine? (*Hint:* See Table 17.4.) ❙

20.7 Entropy

The second law of thermodynamics, as we have stated it, is not an equation or a quantitative relationship but rather a statement of *impossibility.* However, the second law *can* be stated as a quantitative relationship with the concept of *entropy,* the subject of this section.

We have talked about several processes that proceed naturally in the direction of increasing disorder. Irreversible heat flow increases disorder because the molecules are initially sorted into hotter and cooler regions; this sorting is lost when the system comes to thermal equilibrium. Adding heat to a body increases its disorder because it increases average molecular speeds and therefore the randomness of molecular motion. Free expansion of a gas increases its disorder because the molecules have greater randomness of position after the expansion than before. Figure 20.17 shows another process in which disorder increases.

20.17 When firecrackers explode, disorder increases: The neatly packaged chemicals within each firecracker are dispersed in all directions, and the stored chemical energy is converted to random kinetic energy of the fragments.

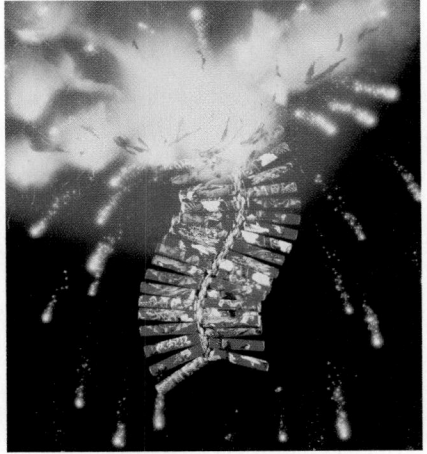

Entropy and Disorder

Entropy provides a *quantitative* measure of disorder. To introduce this concept, let's consider an infinitesimal isothermal expansion of an ideal gas. We add heat dQ and let the gas expand just enough to keep the temperature constant. Because the internal energy of an ideal gas depends only on its temperature, the internal energy is also constant; thus from the first law, the work dW done by the gas is equal to the heat dQ added. That is,

$$dQ = dW = p\,dV = \frac{nRT}{V}\,dV \quad \text{so} \quad \frac{dV}{V} = \frac{dQ}{nRT}$$

The gas is more disordered after the expansion than before: The molecules are moving in a larger volume and have more randomness of position. Thus the fractional volume change dV/V is a measure of the increase in disorder, and the above equation shows that it is proportional to the quantity dQ/T. We introduce the symbol S for the entropy of the system, and we define the infinitesimal entropy change dS during an infinitesimal reversible process at absolute temperature T as

$$dS = \frac{dQ}{T} \quad \text{(infinitesimal reversible process)} \qquad (20.17)$$

If a total amount of heat Q is added during a reversible isothermal process at absolute temperature T, the total entropy change $\Delta S = S_2 - S_1$ is given by

$$\Delta S = S_2 - S_1 = \frac{Q}{T} \quad \text{(reversible isothermal process)} \qquad (20.18)$$

Entropy has units of energy divided by temperature; the SI unit of entropy is 1 J/K.

We can see how the quotient Q/T is related to the increase in disorder. Higher temperature means greater randomness of motion. If the substance is initially cold, with little molecular motion, adding heat Q causes a substantial fractional increase in molecular motion and randomness. But if the substance is already hot, the same quantity of heat adds relatively little to the greater molecular motion already present. So Q/T is an appropriate characterization of the increase in randomness or disorder when heat flows into a system.

Example 20.5 Entropy change in melting

What is the change of entropy of 1 kg of ice that is melted reversibly at 0°C and converted to water at 0°C? The heat of fusion of water is $L_f = 3.34 \times 10^5$ J/kg.

SOLUTION

IDENTIFY and SET UP: The melting occurs at a constant temperature $T = 0°C = 273$ K, so this is an *isothermal* reversible process. We can calculate the added heat Q required to melt the ice, then calculate the entropy change ΔS using Eq. (20.18).

EXECUTE: The heat needed to melt the ice is $Q = mL_f = 3.34 \times 10^5$ J. Then from Eq. (20.18),

$$\Delta S = S_2 - S_1 = \frac{Q}{T} = \frac{3.34 \times 10^5 \text{ J}}{273 \text{ K}} = 1.22 \times 10^3 \text{ J/K}$$

EVALUATE: This entropy increase corresponds to the increase in disorder when the water molecules go from the highly ordered state of a crystalline solid to the much more disordered state of a liquid. In *any* isothermal reversible process, the entropy change equals the heat transferred divided by the absolute temperature. When we refreeze the water, Q has the opposite sign, and the entropy change is $\Delta S = -1.22 \times 10^3$ J/K. The water molecules rearrange themselves into a crystal to form ice, so disorder and entropy both decrease.

Entropy in Reversible Processes

We can generalize the definition of entropy change to include *any* reversible process leading from one state to another, whether it is isothermal or not. We represent the process as a series of infinitesimal reversible steps. During a typical step, an infinitesimal quantity of heat dQ is added to the system at absolute temperature T. Then we sum (integrate) the quotients dQ/T for the entire process; that is,

$$\Delta S = \int_1^2 \frac{dQ}{T} \qquad \text{(entropy change in a reversible process)} \quad (20.19)$$

The limits 1 and 2 refer to the initial and final states.

Because entropy is a measure of the disorder of a system in any specific state, it must depend only on the current state of the system, not on its past history. (We will verify this later.) When a system proceeds from an initial state with entropy S_1 to a final state with entropy S_2, the change in entropy $\Delta S = S_2 - S_1$ defined by Eq. (20.19) does not depend on the path leading from the initial to the final state but is the same for *all possible* processes leading from state 1 to state 2. Thus the entropy of a system must also have a definite value for any given state of the system. *Internal energy,* introduced in Chapter 19, also has this property, although entropy and internal energy are very different quantities.

Since entropy is a function only of the state of a system, we can also compute entropy changes in *irreversible* (nonequilibrium) processes for which Eqs. (20.17) and (20.19) are not applicable. We simply invent a path connecting the given initial and final states that *does* consist entirely of reversible equilibrium processes and compute the total entropy change for that path. It is not the actual path, but the entropy change must be the same as for the actual path.

As with internal energy, the above discussion does not tell us how to calculate entropy itself, but only the change in entropy in any given process. Just as with internal energy, we may arbitrarily assign a value to the entropy of a system in a specified reference state and then calculate the entropy of any other state with reference to this.

Example 20.6 **Entropy change in a temperature change**

One kilogram of water at 0°C is heated to 100°C. Compute its change in entropy. Assume that the specific heat of water is constant at 4190 J/kg · K over this temperature range.

SOLUTION

IDENTIFY and SET UP: The entropy change of the water depends only on the initial and final states of the system, no matter whether the process is reversible or irreversible. We can imagine a reversible process in which the water temperature is increased in a sequence of infinitesimal steps dT. We can use Eq. (20.19) to integrate over all these steps and calculate the entropy change for such a reversible process. (Heating the water on a stove whose cooking surface is maintained at 100°C would be an irreversible process. The entropy change would be the same, however.)

EXECUTE: From Eq. (17.14) the heat required to carry out each infinitesimal step is $dQ = mc \, dT$. Substituting this into Eq. (20.19) and integrating, we find

$$\Delta S = S_2 - S_1 = \int_1^2 \frac{dQ}{T} = \int_{T_1}^{T_2} mc \frac{dT}{T} = mc \ln \frac{T_2}{T_1}$$

$$= (1.00 \text{ kg})(4190 \text{ J/kg} \cdot \text{K}) \left(\ln \frac{373 \text{ K}}{273 \text{ K}} \right)$$

$$= 1.31 \times 10^3 \text{ J/K}$$

EVALUATE: The entropy change is positive, as it must be for a process in which the system absorbs heat. Our assumption about the specific heat is a pretty good one, since c for water increases by only 1% between 0°C and 100°C.

CAUTION **When $\Delta S = Q/T$ can (and cannot) be used** In solving this problem you might be tempted to avoid doing an integral by using the simpler expression in Eq. (20.18), $\Delta S = Q/T$. This would be incorrect, however, because Eq. (20.18) is applicable only to *isothermal* processes, and the initial and final temperatures in our example are *not* the same. The *only* correct way to find the entropy change in a process with different initial and final temperatures is to use Eq. (20.19). ∎

Conceptual Example 20.7 **Entropy change in a reversible adiabatic process**

A gas expands adiabatically and reversibly. What is its change in entropy?

SOLUTION

In an adiabatic process, no heat enters or leaves the system. Hence $dQ = 0$ and there is *no* change in entropy in this reversible process: $\Delta S = 0$. Every *reversible* adiabatic process is a constant-entropy process. (That's why such processes are also called *isentropic* processes.) The increase in disorder resulting from the gas occupying a greater volume is exactly balanced by the decrease in disorder associated with the lowered temperature and reduced molecular speeds.

Example 20.8 **Entropy change in a free expansion**

A partition divides a thermally insulated box into two compartments, each of volume V (Fig. 20.18). Initially, one compartment contains n moles of an ideal gas at temperature T, and the other compartment is evacuated. We break the partition and the gas expands, filling both compartments. What is the entropy change in this free-expansion process?

SOLUTION

IDENTIFY and SET UP: For this process, $Q = 0$, $W = 0$, $\Delta U = 0$, and therefore (because the system is an ideal gas) $\Delta T = 0$. We might think that the entropy change is zero because there is no heat exchange. But Eq. (20.19) can be used to calculate entropy changes for *reversible* processes only; this free expansion is *not* reversible, and there *is* an entropy change. As we mentioned at the beginning of this section, entropy increases in a free expansion because the positions of the molecules are more random than before the expansion. To calculate ΔS, we recall that the entropy change depends only on the initial and final states. We can devise a

20.18 (a, b) Free expansion of an insulated ideal gas. (c) The free-expansion process doesn't pass through equilibrium states from a to b. However, the entropy change $S_b - S_a$ can be calculated by using the isothermal path shown or *any* reversible path from a to b.

reversible process having the same endpoints as this free expansion, and in general we can then use Eq. (20.19) to calculate its entropy change, which will be the same as for the free expansion.

Continued

An appropriate reversible process is an *isothermal* expansion from V to $2V$ at temperature T, which allows us to use the simpler Eq. (20.18) to calculate ΔS. The gas does work W during this expansion, so an equal amount of heat Q must be supplied to keep the internal energy constant.

EXECUTE: We saw in Example 19.1 that the work done by n moles of ideal gas in an isothermal expansion from V_1 to V_2 is $W = nRT \ln(V_2/V_1)$. With $V_1 = V$ and $V_2 = 2V$, we have

$$Q = W = nRT \ln \frac{2V}{V} = nRT \ln 2$$

From Eq. (20.18), the entropy change is

$$\Delta S = \frac{Q}{T} = nR \ln 2$$

EVALUATE: For 1 mole, $\Delta S = (1 \text{ mol})(8.314 \text{ J/mol} \cdot \text{K})(\ln 2) = 5.76 \text{ J/K}$. The entropy change is positive, as we predicted. The factor $(\ln 2)$ in our answer is a result of the volume having increased by a factor of 2, from V to $2V$. Can you show that if the volume increases in a free expansion from V to xV, where x is an arbitrary number, the entropy change is $\Delta S = nR \ln x$?

Example 20.9   Entropy and the Carnot cycle

For the Carnot engine in Example 20.2 (Section 20.6), what is the total entropy change during one cycle?

SOLUTION

IDENTIFY and SET UP: All four steps in the Carnot cycle (see Fig. 20.13) are reversible, so we can use our expressions for the entropy change ΔS in a reversible process. We find ΔS for each step and add them to get ΔS for the complete cycle.

EXECUTE: There is no entropy change during the adiabatic expansion $b \to c$ or the adiabatic compression $d \to a$. During the isothermal expansion $a \to b$ at $T_H = 500$ K, the engine takes in 2000 J of heat, and from Eq. (20.18),

$$\Delta S_H = \frac{Q_H}{T_H} = \frac{2000 \text{ J}}{500 \text{ K}} = 4.0 \text{ J/K}$$

During the isothermal compression $c \to d$ at $T_C = 350$ K, the engine gives off 1400 J of heat, and

$$\Delta S_C = \frac{Q_C}{T_C} = \frac{-1400 \text{ J}}{350 \text{ K}} = -4.0 \text{ J/K}$$

The total entropy change in the engine during one cycle is $\Delta S_{tot} = \Delta S_H + \Delta S_C = 4.0 \text{ J/K} + (-4.0 \text{ J/K}) = 0$.

EVALUATE: The result $\Delta S_{total} = 0$ tells us that when the Carnot engine completes a cycle, it has the same entropy as it did at the beginning of the cycle. We'll explore this result in the next subsection.

What is the total entropy change of the engine's *environment* during this cycle? During the reversible isothermal expansion $a \to b$, the hot (500 K) reservoir gives off 2000 J of heat, so its entropy change is $(-2000 \text{ J})/(500 \text{ K}) = -4.0 \text{ J/K}$. During the reversible isothermal compression $c \to d$, the cold (350 K) reservoir absorbs 1400 J of heat, so its entropy change is $(+1400 \text{ J})/(350 \text{ K}) = +4.0 \text{ J/K}$. Thus the hot and cold reservoirs each have an entropy change, but the sum of these changes—that is, the total entropy change of the system's environment—is zero.

These results apply to the special case of the Carnot cycle, for which *all* of the processes are reversible. In this case we find that the total entropy change of the system and the environment together is zero. We will see that if the cycle includes irreversible processes (as is the case for the Otto and Diesel cycles of Section 20.3), the total entropy change of the system and the environment *cannot* be zero, but rather must be positive.

Entropy in Cyclic Processes

Example 20.9 showed that the total entropy change for a cycle of a particular Carnot engine, which uses an ideal gas as its working substance, is zero. This result follows directly from Eq. (20.13), which we can rewrite as

$$\frac{Q_H}{T_H} + \frac{Q_C}{T_C} = 0 \tag{20.20}$$

The quotient Q_H/T_H equals ΔS_H, the entropy change of the engine that occurs at $T = T_H$. Likewise, Q_C/T_C equals ΔS_C, the (negative) entropy change of the engine that occurs at $T = T_C$. Hence Eq. (20.20) says that $\Delta S_H + \Delta S_C = 0$; that is, there is zero net entropy change in one cycle.

What about Carnot engines that use a different working substance? According to the second law, *any* Carnot engine operating between given temperatures T_H and T_C has the same efficiency $e = 1 - T_C/T_H$ [Eq. (20.14)]. Combining this expression for e with Eq. (20.4), $e = 1 + Q_C/Q_H$, just reproduces Eq. (20.20). So Eq. (20.20) is valid for any Carnot engine working between these temperatures, whether its working substance is an ideal gas or not. We conclude that *the total entropy change in one cycle of any Carnot engine is zero.*

20.19 (a) A reversible cyclic process for an ideal gas is shown as a red closed path on a pV-diagram. Several ideal-gas isotherms are shown in blue. (b) We can approximate the path in (a) by a series of long, thin Carnot cycles; one of these is highlighted in gold. The total entropy change is zero for each Carnot cycle and for the actual cyclic process. (c) The entropy change between points a and b is independent of the path.

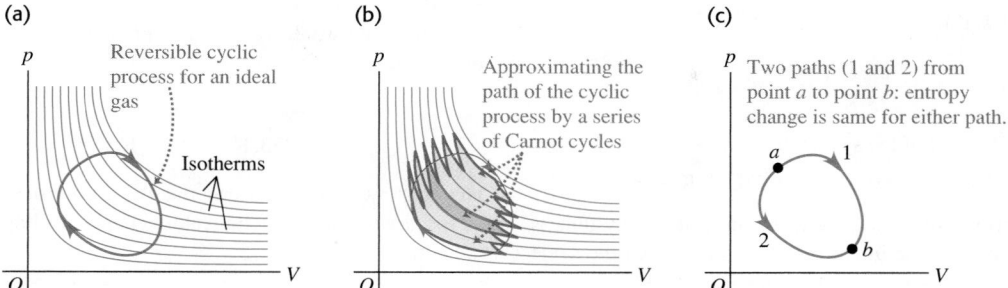

This result can be generalized to show that the total entropy change during *any* reversible cyclic process is zero. A reversible cyclic process appears on a pV-diagram as a closed path (Fig. 20.19a). We can approximate such a path as closely as we like by a sequence of isothermal and adiabatic processes forming parts of many long, thin Carnot cycles (Fig. 20.19b). The total entropy change for the full cycle is the sum of the entropy changes for each small Carnot cycle, each of which is zero. So **the total entropy change during *any* reversible cycle is zero:**

$$\int \frac{dQ}{T} = 0 \qquad \text{(reversible cyclic process)} \qquad (20.21)$$

It follows that when a system undergoes a reversible process leading from any state a to any other state b, *the entropy change of the system is independent of the path* (Fig. 20.19c). If the entropy change for path 1 were different from the change for path 2, the system could be taken along path 1 and then backward along path 2 to the starting point, with a nonzero net change in entropy. This would violate the conclusion that the total entropy change in such a cyclic process must be zero. Because the entropy change in such processes is independent of path, we conclude that in any given state, the system has a definite value of entropy that depends only on the state, not on the processes that led to that state.

Entropy in Irreversible Processes

In an idealized, reversible process involving only equilibrium states, the total entropy change of the system and its surroundings is zero. But all *irreversible* processes involve an increase in entropy. Unlike energy, *entropy is not a conserved quantity.* The entropy of an isolated system *can* change, but as we shall see, it can never decrease. The free expansion of a gas, described in Example 20.8, is an irreversible process in an isolated system in which there is an entropy increase.

Application **Entropy Changes in a Living Organism**
When a kitten or other growing animal eats, it takes ordered chemical energy from the food and uses it to make new cells that are even more highly ordered. This process alone lowers entropy. But most of the energy in the food is either excreted in the animal's feces or used to generate heat, processes that lead to a large increase in entropy. So while the entropy of the animal alone decreases, the *total* entropy of animal plus food *increases.*

Example 20.10 **Entropy change in an irreversible process**

Suppose 1.00 kg of water at 100°C is placed in thermal contact with 1.00 kg of water at 0°C. What is the total change in entropy? Assume that the specific heat of water is constant at 4190 J/kg · K over this temperature range.

SOLUTION

IDENTIFY and SET UP: This process involves irreversible heat flow because of the temperature differences. There are equal masses of 0°C water and 100°C water, so the final temperature is the average of these two temperatures: 50°C = 323 K. Although the processes are irreversible, we can calculate the entropy changes for the (initially) hot water and the (initially) cold water by assuming that the process occurs reversibly. As in Example 20.6, we must use Eq. (20.19) to calculate ΔS for each substance because the temperatures are not constant.

Continued

EXECUTE: The entropy changes of the hot water (subscript H) and the cold water (subscript C) are

$$\Delta S_H = mc \int_{T_1}^{T_2} \frac{dT}{T} = (1.00 \text{ kg})(4190 \text{ J/kg} \cdot \text{K}) \int_{373 \text{ K}}^{323 \text{ K}} \frac{dT}{T}$$

$$= (4190 \text{ J/K})\left(\ln\frac{323 \text{ K}}{373 \text{ K}}\right) = -603 \text{ J/K}$$

$$\Delta S_C = (4190 \text{ J/K})\left(\ln\frac{323 \text{ K}}{273 \text{ K}}\right) = +705 \text{ J/K}$$

The *total* entropy change of the system is

$$\Delta S_{tot} = \Delta S_H + \Delta S_C = (-603 \text{ J/K}) + 705 \text{ J/K} = +102 \text{ J/K}$$

EVALUATE: An irreversible heat flow in an isolated system is accompanied by an increase in entropy. We could reach the same

end state by mixing the hot and cold water, which is also an irreversible process; the total entropy change, which depends only on the initial and final states of the system, would again be 102 J/K.

Note that the entropy of the system increases *continuously* as the two quantities of water come to equilibrium. For example, the first 4190 J of heat transferred cools the hot water to 99°C and warms the cold water to 1°C. The net change in entropy for this step is approximately

$$\Delta S = \frac{-4190 \text{ J}}{373 \text{ K}} + \frac{4190 \text{ J}}{273 \text{ K}} = +4.1 \text{ J/K}.$$

Can you show in a similar way that the net entropy change is positive for *any* one-degree temperature change leading to the equilibrium condition?

Entropy and the Second Law

The results of Example 20.10 about the flow of heat from a higher to a lower temperature are characteristic of *all* natural (that is, irreversible) processes. When we include the entropy changes of all the systems taking part in the process, the increases in entropy are always greater than the decreases. In the special case of a *reversible* process, the increases and decreases are equal. Hence we can state the general principle: **When all systems taking part in a process are included, the entropy either remains constant or increases.** In other words: **No process is possible in which the total entropy decreases, when all systems taking part in the process are included.** This is an alternative statement of the second law of thermodynamics in terms of entropy. Thus it is equivalent to the "engine" and "refrigerator" statements discussed earlier. Figure 20.20 shows a specific example of this general principle.

The increase of entropy in every natural, irreversible process measures the increase of disorder or randomness in the universe associated with that process. Consider again the example of mixing hot and cold water (Example 20.10). We *might* have used the hot and cold water as the high- and low-temperature reservoirs of a heat engine. While removing heat from the hot water and giving heat to the cold water, we could have obtained some mechanical work. But once the hot and cold water have been mixed and have come to a uniform temperature, this opportunity to convert heat to mechanical work is lost irretrievably. The lukewarm water will never *unmix* itself and separate into hotter and colder portions. No decrease in *energy* occurs when the hot and cold water are mixed. What has been lost is the *opportunity* to convert part of the heat from the hot water into mechanical work. Hence when entropy increases, energy becomes less *available*, and the universe becomes more random or "run down."

20.20 The mixing of colored ink and water starts from a state of relative order (low entropy) in which each fluid is separate and distinct from the other. The final state after mixing is more disordered (has greater entropy). Spontaneous unmixing of the ink and water, a process in which there would be a net decrease in entropy, is never observed.

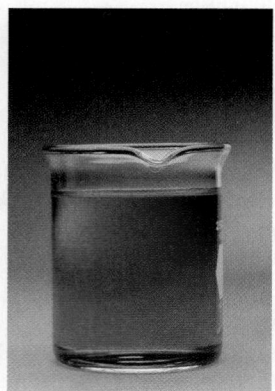

Test Your Understanding of Section 20.7   Suppose 2.00 kg of water at 50°C spontaneously changes temperature, so that half of the water cools to 0°C while the other half spontaneously warms to 100°C. (All of the water remains liquid, so it doesn't freeze or boil.) What would be the entropy change of the water? Is this process possible? (*Hint:* See Example 20.10.)   ❚

20.8 Microscopic Interpretation of Entropy

We described in Section 19.4 how the internal energy of a system could be calculated, at least in principle, by adding up all the kinetic energies of its constituent particles and all the potential energies of interaction among the particles. This is called a *microscopic* calculation of the internal energy. We can also make a microscopic calculation of the entropy S of a system. Unlike energy, however, entropy is not something that belongs to each individual particle or pair of particles in the system. Rather, entropy is a measure of the disorder of the system as a whole. To see how to calculate entropy microscopically, we first have to introduce the idea of *macroscopic* and *microscopic states.*

Suppose you toss N identical coins on the floor, and half of them show heads and half show tails. This is a description of the large-scale or **macroscopic state** of the system of N coins. A description of the **microscopic state** of the system includes information about each individual coin: Coin 1 was heads, coin 2 was tails, coin 3 was tails, and so on. There can be many microscopic states that correspond to the same macroscopic description. For instance, with $N = 4$ coins there are six possible states in which half are heads and half are tails (Fig. 20.21). The number of microscopic states grows rapidly with increasing N; for $N = 100$ there are $2^{100} = 1.27 \times 10^{30}$ microscopic states, of which 1.01×10^{29} are half heads and half tails.

The least probable outcomes of the coin toss are the states that are either all heads or all tails. It is certainly possible that you could throw 100 heads in a row, but don't bet on it; the probability of doing this is only 1 in 1.27×10^{30}. The most probable outcome of tossing N coins is that half are heads and half are tails. The reason is that this *macroscopic* state has the greatest number of corresponding *microscopic* states, as Fig. 20.21 shows.

To make the connection to the concept of entropy, note that N coins that are all heads constitute a completely ordered macroscopic state; the description "all heads" completely specifies the state of each one of the N coins. The same is true if the coins are all tails. But the macroscopic description "half heads, half tails" by itself tells you very little about the state (heads or tails) of each individual coin. We say that the system is *disordered* because we know so little about its microscopic state. Compared to the state "all heads" or "all tails," the state "half heads, half tails" has a much greater number of possible microscopic states, much greater disorder, and hence much greater entropy (which is a quantitative measure of disorder).

Now instead of N coins, consider a mole of an ideal gas containing Avogadro's number of molecules. The macroscopic state of this gas is given by its pressure p, volume V, and temperature T; a description of the microscopic state involves stating the position and velocity for each molecule in the gas. At a given pressure, volume, and temperature, the gas may be in any one of an astronomically large number of microscopic states, depending on the positions and velocities of its 6.02×10^{23} molecules. If the gas undergoes a free expansion into a greater volume, the range of possible positions increases, as does the number of possible microscopic states. The system becomes more disordered, and the entropy increases as calculated in Example 20.8 (Section 20.7).

We can draw the following general conclusion: **For any system, the most probable macroscopic state is the one with the greatest number of corresponding microscopic states, which is also the macroscopic state with the greatest disorder and the greatest entropy.**

20.21 All possible microscopic states of four coins. There can be several possible microscopic states for each macroscopic state.

Calculating Entropy: Microscopic States

Let w represent the number of possible microscopic states for a given macroscopic state. (For the four coins shown in Fig. 20.21 the state of four heads has $w = 1$, the state of three heads and one tails has $w = 4$, and so on.) Then the entropy S of a macroscopic state can be shown to be given by

$$S = k\ln w \qquad \text{(microscopic expression for entropy)} \qquad (20.22)$$

where $k = R/N_A$ is the Boltzmann constant (gas constant per molecule) introduced in Section 18.3. As Eq. (20.22) shows, increasing the number of possible microscopic states w increases the entropy S.

What matters in a thermodynamic process is not the absolute entropy S but the *difference* in entropy between the initial and final states. Hence an equally valid and useful definition would be $S = k\ln w + C$, where C is a constant, since C cancels in any calculation of an entropy difference between two states. But it's convenient to set this constant equal to zero and use Eq. (20.22). With this choice, since the smallest possible value of w is unity, the smallest possible value of S for any system is $k\ln 1 = 0$. Entropy can *never* be negative.

In practice, calculating w is a difficult task, so Eq. (20.22) is typically used only to calculate the absolute entropy S of certain special systems. But we can use this relationship to calculate *differences* in entropy between one state and another. Consider a system that undergoes a thermodynamic process that takes it from macroscopic state 1, for which there are w_1 possible microscopic states, to macroscopic state 2, with w_2 associated microscopic states. The change in entropy in this process is

$$\Delta S = S_2 - S_1 = k\ln w_2 - k\ln w_1 = k\ln\frac{w_2}{w_1} \qquad (20.23)$$

The *difference* in entropy between the two macroscopic states depends on the *ratio* of the numbers of possible microscopic states.

As the following example shows, using Eq. (20.23) to calculate a change in entropy from one macroscopic state to another gives the same results as considering a reversible process connecting those two states and using Eq. (20.19).

Example 20.11 — A microscopic calculation of entropy change

Use Eq. (20.23) to calculate the entropy change in the free expansion of n moles of gas at temperature T described in Example 20.8 (Fig. 20.22).

SOLUTION

IDENTIFY and SET UP: We are asked to calculate the entropy change using the number of microstates in the initial and final macroscopic states (Figs. 20.22a and b). When the partition is broken, no work is done, so the velocities of the molecules are unaffected. But each molecule now has twice as much volume in which it can move and hence has twice the number of possible positions. This is all we need to calculate the entropy change using Eq. (20.23).

EXECUTE: Let w_1 be the number of microscopic states of the system as a whole when the gas occupies volume V (Fig. 20.22a). The number of molecules is $N = nN_A$, and each of these N molecules has twice as many possible states after the partition is broken. Hence the number w_2 of microscopic states when the gas occupies volume $2V$ (Fig. 20.22b) is greater by a factor of 2^N; that is, $w_2 = 2^N w_1$. The change in entropy in this process is

$$\Delta S = k\ln\frac{w_2}{w_1} = k\ln\frac{2^N w_1}{w_1} = k\ln 2^N = Nk\ln 2$$

Since $N = nN_A$ and $k = R/N_A$, this becomes

$$\Delta S = (nN_A)(R/N_A)\ln 2 = nR\ln 2$$

EVALUATE: We found the same result as in Example 20.8, but without any reference to the thermodynamic path taken.

20.22 In a free expansion of N molecules in which the volume doubles, the number of possible microscopic states increases by 2^N.

(a) Gas occupies volume V; number of microstates $= w_1$.

(b) Gas occupies volume $2V$; number of microstates $= w_2 = 2^N w_1$.

Microscopic States and the Second Law

The relationship between entropy and the number of microscopic states gives us new insight into the entropy statement of the second law of thermodynamics: that the entropy of a closed system can never decrease. From Eq. (20.22) this means that a closed system can never spontaneously undergo a process that decreases the number of possible microscopic states.

An example of such a forbidden process would be if all of the air in your room spontaneously moved to one half of the room, leaving a vacuum in the other half. Such a "free compression" would be the reverse of the free expansion of Examples 20.8 and 20.11. This would decrease the number of possible microscopic states by a factor of 2^N. Strictly speaking, this process is not impossible! The probability of finding a given molecule in one half of the room is $\frac{1}{2}$, so the probability of finding all of the molecules in one half of the room at once is $(\frac{1}{2})^N$. (This is exactly the same as the probability of having a tossed coin come up heads N times in a row.) This probability is *not* zero. But lest you worry about suddenly finding yourself gasping for breath in the evacuated half of your room, consider that a typical room might hold 1000 moles of air, and so $N = 1000N_A = 6.02 \times 10^{26}$ molecules. The probability of all the molecules being in the same half of the room is therefore $(\frac{1}{2})^{6.02 \times 10^{26}}$. Expressed as a decimal, this number has more than 10^{26} zeros to the right of the decimal point!

Because the probability of such a "free compression" taking place is so vanishingly small, it has almost certainly never occurred anywhere in the universe since the beginning of time. We conclude that for all practical purposes the second law of thermodynamics is never violated.

Test Your Understanding of Section 20.8 A quantity of N molecules of an ideal gas initially occupies volume V. The gas then expands to volume $2V$. The number of microscopic states of the gas increases in this expansion. Under which of the following circumstances will this number increase the most? (i) if the expansion is reversible and isothermal; (ii) if the expansion is reversible and adiabatic; (iii) the number will change by the same amount for both circumstances. |

Reversible and irreversible processes: A reversible process is one whose direction can be reversed by an infinitesimal change in the conditions of the process, and in which the system is always in or very close to thermal equilibrium. All other thermodynamic processes are irreversible.

Heat engines: A heat engine takes heat Q_H from a source, converts part of it to work W, and discards the remainder $|Q_C|$ at a lower temperature. The thermal efficiency e of a heat engine measures how much of the absorbed heat is converted to work. (See Example 20.1.)

$$e = \frac{W}{Q_H} = 1 + \frac{Q_C}{Q_H} = 1 - \left|\frac{Q_C}{Q_H}\right| \quad (20.4)$$

The Otto cycle: A gasoline engine operating on the Otto cycle has a theoretical maximum thermal efficiency e that depends on the compression ratio r and the ratio of heat capacities γ of the working substance.

$$e = 1 - \frac{1}{r^{\gamma-1}} \quad (20.6)$$

Refrigerators: A refrigerator takes heat Q_C from a colder place, has a work input $|W|$, and discards heat $|Q_H|$ at a warmer place. The effectiveness of the refrigerator is given by its coefficient of performance K.

$$K = \frac{|Q_C|}{|W|} = \frac{|Q_C|}{|Q_H| - |Q_C|} \quad (20.9)$$

The second law of thermodynamics: The second law of thermodynamics describes the directionality of natural thermodynamic processes. It can be stated in several equivalent forms. The *engine* statement is that no cyclic process can convert heat completely into work. The *refrigerator* statement is that no cyclic process can transfer heat from a colder place to a hotter place with no input of mechanical work.

The Carnot cycle: The Carnot cycle operates between two heat reservoirs at temperatures T_H and T_C and uses only reversible processes. Its thermal efficiency depends only on T_H and T_C. An additional equivalent statement of the second law is that no engine operating between the same two temperatures can be more efficient than a Carnot engine. (See Examples 20.2 and 20.3.)

A Carnot engine run backward is a Carnot refrigerator. Its coefficient of performance depends only on T_H and T_C. Another form of the second law states that no refrigerator operating between the same two temperatures can have a larger coefficient of performance than a Carnot refrigerator. (See Example 20.4.)

$$e_{Carnot} = 1 - \frac{T_C}{T_H} = \frac{T_H - T_C}{T_H} \quad (20.14)$$

$$K_{Carnot} = \frac{T_C}{T_H - T_C} \quad (20.15)$$

Entropy: Entropy is a quantitative measure of the disorder of a system. The entropy change in any reversible process depends on the amount of heat flow and the absolute temperature T. Entropy depends only on the state of the system, and the change in entropy between given initial and final states is the same for all processes leading from one state to the other. This fact can be used to find the entropy change in an irreversible process. (See Examples 20.5–20.10.)

$$\Delta S = \int_1^2 \frac{dQ}{T} \quad (20.19)$$
(reversible process)

An important statement of the second law of thermo-dynamics is that the entropy of an isolated system may increase but can never decrease. When a system interacts with its surroundings, the total entropy change of system and surroundings can never decrease. When the interaction involves only reversible processes, the total entropy is constant and $\Delta S = 0$; when there is any irreversible process, the total entropy increases and $\Delta S > 0$.

Entropy and microscopic states: When a system is in a particular macroscopic state, the particles that make up the system may be in any of w possible microscopic states. The greater the number w, the greater the entropy. (See Example 20.11.)

$$S = k \ln w \qquad (20.22)$$

w microstates $2^N w$ microstates

BRIDGING PROBLEM   Entropy Changes: Cold Ice in Hot Water

An insulated container of negligible mass holds 0.600 kg of water at 45.0°C. You put a 0.0500-kg ice cube at −15.0°C in the water. (a) Calculate the final temperature of the water once the ice has melted. (b) Calculate the change in entropy of the system.

SOLUTION GUIDE

See MasteringPhysics® study area for a Video Tutor solution.

IDENTIFY and SET UP

1. Make a list of the known and unknown quantities, and identify the target variables.
2. How will you find the final temperature of the ice–water mixture? How will you decide whether or not all the ice melts?
3. Once you find the final temperature of the mixture, how will you determine the changes in entropy of (i) the ice initially at −15.0°C and (ii) the water initially at 45.0°C?

EXECUTE

4. Use the methods of Chapter 17 to calculate the final temperature T. (*Hint:* First assume that all of the ice melts, then write

an equation which says that the heat that flows into the ice equals the heat that flows out of the water. If your assumption is correct, the final temperature that you calculate will be greater than 0°C. If your assumption is incorrect, the final temperature will be 0°C or less, which means that some ice remains. You'll then need to redo the calculation to account for this.)

5. Use your result from step 4 to calculate the entropy changes of the ice and the water. (*Hint:* You must include the heat flow associated with temperature changes, as in Example 20.6, as well as the heat flow associated with the change of phase.)
6. Find the total change in entropy of the system.

EVALUATE

7. Do the signs of the entropy changes make sense? Why or why not?

Problems

For instructor-assigned homework, go to www.masteringphysics.com

•, ••, •••: Problems of increasing difficulty. **CP**: Cumulative problems incorporating material from earlier chapters. **CALC**: Problems requiring calculus. **BIO**: Biosciences problems.

DISCUSSION QUESTIONS

Q20.1 A pot is half-filled with water, and a lid is placed on it, forming a tight seal so that no water vapor can escape. The pot is heated on a stove, forming water vapor inside the pot. The heat is then turned off and the water vapor condenses back to liquid. Is this cycle reversible or irreversible? Why?

Q20.2 Give two examples of reversible processes and two examples of irreversible processes in purely mechanical systems, such as blocks sliding on planes, springs, pulleys, and strings. Explain what makes each process reversible or irreversible.

Q20.3 What irreversible processes occur in a gasoline engine? Why are they irreversible?

Q20.4 Suppose you try to cool the kitchen of your house by leaving the refrigerator door open. What happens? Why? Would the result be the same if you left open a picnic cooler full of ice? Explain the reason for any differences.

Q20.5 A member of the U.S. Congress proposed a scheme to produce energy as follows. Water molecules (H_2O) are to be broken apart to produce hydrogen and oxygen. The hydrogen is then burned (that is, combined with oxygen), releasing energy in the process. The only product of this combustion is water, so there is no pollution. In light of the second law of thermodynamics, what do you think of this energy-producing scheme?

Q20.6 Is it a violation of the second law of thermodynamics to convert mechanical energy completely into heat? To convert heat completely into work? Explain your answers.

Q20.7 Imagine a special air filter placed in a window of a house. The tiny holes in the filter allow only air molecules moving faster than a certain speed to exit the house, and allow only air molecules moving slower than that speed to enter the house from outside. Explain why such an air filter would cool the house, and why the second law of thermodynamics makes building such a filter an impossible task.

Q20.8 An electric motor has its shaft coupled to that of an electric generator. The motor drives the generator, and some current from the generator is used to run the motor. The excess current is used to light a home. What is wrong with this scheme?

Q20.9 When a wet cloth is hung up in a hot wind in the desert, it is cooled by evaporation to a temperature that may be 20 C° or so below that of the air. Discuss this process in light of the second law of thermodynamics.

Q20.10 Compare the pV-diagram for the Otto cycle in Fig. 20.6 with the diagram for the Carnot heat engine in Fig. 20.13. Explain some of the important differences between the two cycles.

Q20.11 If no real engine can be as efficient as a Carnot engine operating between the same two temperatures, what is the point of developing and using Eq. (20.14)?

Q20.12 The efficiency of heat engines is high when the temperature difference between the hot and cold reservoirs is large. Refrigerators, on the other hand, work better when the temperature difference is small. Thinking of the mechanical refrigeration cycle shown in Fig. 20.9, explain in physical terms why it takes less work to remove heat from the working substance if the two reservoirs (the inside of the refrigerator and the outside air) are at nearly the same temperature, than if the outside air is much warmer than the interior of the refrigerator.

Q20.13 What would be the efficiency of a Carnot engine operating with $T_H = T_C$? What would be the efficiency if $T_C = 0$ K and T_H were any temperature above 0 K? Interpret your answers.

Q20.14 Real heat engines, like the gasoline engine in a car, always have some friction between their moving parts, although lubricants keep the friction to a minimum. Would a heat engine with completely frictionless parts be 100% efficient? Why or why not? Does the answer depend on whether or not the engine runs on the Carnot cycle? Again, why or why not?

Q20.15 Does a refrigerator full of food consume more power if the room temperature is 20°C than if it is 15°C? Or is the power consumption the same? Explain your reasoning.

Q20.16 In Example 20.4, a Carnot refrigerator requires a work input of only 230 J to extract 346 J of heat from the cold reservoir. Doesn't this discrepancy imply a violation of the law of conservation of energy? Explain why or why not.

Q20.17 Explain why each of the following processes is an example of increasing disorder or randomness: mixing hot and cold water; free expansion of a gas; irreversible heat flow; developing heat by mechanical friction. Are entropy increases involved in all of these? Why or why not?

Q20.18 The free expansion of a gas is an adiabatic process and so no heat is transferred. No work is done, so the internal energy does not change. Thus, $Q/T = 0$, yet the disorder of the system and thus its entropy have increased after the expansion. Why does Eq. (20.19) not apply to this situation?

Q20.19 Are the earth and sun in thermal equilibrium? Are there entropy changes associated with the transmission of energy from the sun to the earth? Does radiation differ from other modes of heat transfer with respect to entropy changes? Explain your reasoning.

Q20.20 Discuss the entropy changes involved in the preparation and consumption of a hot fudge sundae.

Q20.21 If you run a movie film backward, it is as if the direction of time were reversed. In the time-reversed movie, would you see processes that violate conservation of energy? Conservation of linear momentum? Would you see processes that violate the second law of thermodynamics? In each case, if law-breaking processes could occur, give some examples.

Q20.22 **BIO** Some critics of biological evolution claim that it violates the second law of thermodynamics, since evolution involves simple life forms developing into more complex and more highly ordered organisms. Explain why this is not a valid argument against evolution.

Q20.23 **BIO** A growing plant creates a highly complex and organized structure out of simple materials such as air, water, and trace minerals. Does this violate the second law of thermodynamics? Why or why not? What is the plant's ultimate source of energy? Explain your reasoning.

EXERCISES

Section 20.2 Heat Engines

20.1 • A diesel engine performs 2200 J of mechanical work and discards 4300 J of heat each cycle. (a) How much heat must be supplied to the engine in each cycle? (b) What is the thermal efficiency of the engine?

20.2 • An aircraft engine takes in 9000 J of heat and discards 6400 J each cycle. (a) What is the mechanical work output of the engine during one cycle? (b) What is the thermal efficiency of the engine?

20.3 • **A Gasoline Engine.** A gasoline engine takes in 1.61×10^4 J of heat and delivers 3700 J of work per cycle. The heat is obtained by burning gasoline with a heat of combustion of 4.60×10^4 J/g. (a) What is the thermal efficiency? (b) How much heat is discarded in each cycle? (c) What mass of fuel is burned in each cycle? (d) If the engine goes through 60.0 cycles per second, what is its power output in kilowatts? In horsepower?

20.4 • A gasoline engine has a power output of 180 kW (about 241 hp). Its thermal efficiency is 28.0%. (a) How much heat must be supplied to the engine per second? (b) How much heat is discarded by the engine per second?

20.5 •• The pV-diagram in Fig. E20.5 shows a cycle of a heat engine that uses 0.250 mole of an ideal gas having $\gamma = 1.40$. The curved part ab

Figure **E20.5**

of the cycle is adiabatic. (a) Find the pressure of the gas at point a. (b) How much heat enters this gas per cycle, and where does it happen? (c) How much heat leaves this gas in a cycle, and where does it occur? (d) How much work does this engine do in a cycle? (e) What is the thermal efficiency of the engine?

Section 20.3 Internal-Combustion Engines

20.6 • (a) Calculate the theoretical efficiency for an Otto-cycle engine with $\gamma = 1.40$ and $r = 9.50$. (b) If this engine takes in 10,000 J of heat from burning its fuel, how much heat does it discard to the outside air?

20.7 •• The Otto-cycle engine in a Mercedes-Benz SLK230 has a compression ratio of 8.8. (a) What is the ideal efficiency of the engine? Use $\gamma = 1.40$. (b) The engine in a Dodge Viper GT2 has a slightly higher compression ratio of 9.6. How much increase in the ideal efficiency results from this increase in the compression ratio?

Section 20.4 Refrigerators

20.8 • The coefficient of performance $K = H/P$ is a dimensionless quantity. Its value is independent of the units used for H and P, as long as the same units, such as watts, are used for both quantities. However, it is common practice to express H in Btu/h and P in watts. When these mixed units are used, the ratio H/P is called the energy efficiency rating (EER). If a room air conditioner has a coefficient of performance $K = 3.0$, what is its EER?

20.9 • A refrigerator has a coefficient of performance of 2.10. In each cycle it absorbs 3.40×10^4 J of heat from the cold reservoir. (a) How much mechanical energy is required each cycle to operate the refrigerator? (b) During each cycle, how much heat is discarded to the high-temperature reservoir?

20.10 • A room air conditioner has a coefficient of performance of 2.9 on a hot day and uses 850 W of electrical power. (a) How many joules of heat does the air conditioner remove from the room in one minute? (b) How many joules of heat does the air conditioner deliver to the hot outside air in one minute? (c) Explain why your answers to parts (a) and (b) are not the same.

20.11 •• A refrigerator has a coefficient of performance of 2.25, runs on an input of 95 W of electrical power, and keeps its inside compartment at 5°C. If you put a dozen 1.0-L plastic bottles of water at 31°C into this refrigerator, how long will it take for them to be cooled down to 5°C? (Ignore any heat that leaves the plastic.)

20.12 •• A freezer has a coefficient of performance of 2.40. The freezer is to convert 1.80 kg of water at 25.0°C to 1.80 kg of ice at −5.0°C in one hour. (a) What amount of heat must be removed from the water at 25.0°C to convert it to ice at −5.0°C? (b) How much electrical energy is consumed by the freezer during this hour? (c) How much wasted heat is delivered to the room in which the freezer sits?

Section 20.6 The Carnot Cycle

20.13 • A Carnot engine whose high-temperature reservoir is at 620 K takes in 550 J of heat at this temperature in each cycle and gives up 335 J to the low-temperature reservoir. (a) How much mechanical work does the engine perform during each cycle? (b) What is the temperature of the low-temperature reservoir? (c) What is the thermal efficiency of the cycle?

20.14 • A Carnot engine is operated between two heat reservoirs at temperatures of 520 K and 300 K. (a) If the engine receives 6.45 kJ of heat energy from the reservoir at 520 K in each cycle, how many joules per cycle does it discard to the reservoir at 300 K? (b) How much mechanical work is performed by the engine during each cycle? (c) What is the thermal efficiency of the engine?

20.15 • A Carnot engine has an efficiency of 59% and performs 2.5×10^4 J of work in each cycle. (a) How much heat does the engine extract from its heat source in each cycle? (b) Suppose the engine exhausts heat at room temperature (20.0°C). What is the temperature of its heat source?

20.16 •• An ice-making machine operates in a Carnot cycle. It takes heat from water at 0.0°C and rejects heat to a room at 24.0°C. Suppose that 85.0 kg of water at 0.0°C are converted to ice at 0.0°C. (a) How much heat is discharged into the room? (b) How much energy must be supplied to the device?

20.17 • A Carnot refrigerator is operated between two heat reservoirs at temperatures of 320 K and 270 K. (a) If in each cycle the refrigerator receives 415 J of heat energy from the reservoir at 270 K, how many joules of heat energy does it deliver to the reservoir at 320 K? (b) If the refrigerator completes 165 cycles each minute, what power input is required to operate it? (c) What is the coefficient of performance of the refrigerator?

20.18 •• A certain brand of freezer is advertised to use 730 kW · h of energy per year. (a) Assuming the freezer operates for 5 hours each day, how much power does it require while operating? (b) If the freezer keeps its interior at a temperature of −5.0°C in a 20.0°C room, what is its theoretical maximum performance coefficient? (c) What is the theoretical maximum amount of ice this freezer could make in an hour, starting with water at 20.0°C?

20.19 •• A Carnot heat engine has a thermal efficiency of 0.600, and the temperature of its hot reservoir is 800 K. If 3000 J of heat is rejected to the cold reservoir in one cycle, what is the work output of the engine during one cycle?

20.20 •• A Carnot heat engine uses a hot reservoir consisting of a large amount of boiling water and a cold reservoir consisting of a large tub of ice and water. In 5 minutes of operation, the heat rejected by the engine melts 0.0400 kg of ice. During this time, how much work W is performed by the engine?

20.21 •• You design an engine that takes in 1.50×10^4 J of heat at 650 K in each cycle and rejects heat at a temperature of 350 K. The engine completes 240 cycles in 1 minute. What is the theoretical maximum power output of your engine, in horsepower?

Section 20.7 Entropy

20.22 • A 4.50-kg block of ice at 0.00°C falls into the ocean and melts. The average temperature of the ocean is 3.50°C, including all the deep water. By how much does the melting of this ice change the entropy of the world? Does it make it larger or smaller? (*Hint:* Do you think that the ocean will change temperature appreciably as the ice melts?)

20.23 • A sophomore with nothing better to do adds heat to 0.350 kg of ice at 0.0°C until it is all melted. (a) What is the change in entropy of the water? (b) The source of heat is a very massive body at a temperature of 25.0°C. What is the change in entropy of this body? (c) What is the total change in entropy of the water and the heat source?

20.24 • **CALC** You decide to take a nice hot bath but discover that your thoughtless roommate has used up most of the hot water. You fill the tub with 270 kg of 30.0°C water and attempt to warm it further by pouring in 5.00 kg of boiling water from the stove. (a) Is this a reversible or an irreversible process? Use physical reasoning to explain. (b) Calculate the final temperature of the bath water. (c) Calculate the net change in entropy of the system (bath water + boiling water), assuming no heat exchange with the air or the tub itself.

20.25 •• A 15.0-kg block of ice at 0.0°C melts to liquid water at 0.0°C inside a large room that has a temperature of 20.0°C. Treat the ice and the room as an isolated system, and assume that the room is large enough for its temperature change to be ignored. (a) Is the melting of the ice reversible or irreversible? Explain, using simple physical reasoning without resorting to any equations. (b) Calculate the net entropy change of the system during this process. Explain whether or not this result is consistent with your answer to part (a).

20.26 •• **CALC** You make tea with 0.250 kg of 85.0°C water and let it cool to room temperature (20.0°C) before drinking it. (a) Calculate the entropy change of the water while it cools. (b) The cooling process is essentially isothermal for the air in your kitchen. Calculate the change in entropy of the air while the tea cools, assuming that all the heat lost by the water goes into the air. What is the total entropy change of the system tea + air?

20.27 • Three moles of an ideal gas undergo a reversible isothermal compression at 20.0°C. During this compression, 1850 J of work is done on the gas. What is the change of entropy of the gas?

20.28 •• What is the change in entropy of 0.130 kg of helium gas at the normal boiling point of helium when it all condenses isothermally to 1.00 L of liquid helium? (*Hint:* See Table 17.4 in Section 17.6.)

20.29 • (a) Calculate the change in entropy when 1.00 kg of water at 100°C is vaporized and converted to steam at 100°C (see Table 17.4). (b) Compare your answer to the change in entropy when 1.00 kg of ice is melted at 0°C, calculated in Example 20.5 (Section 20.7). Is the change in entropy greater for melting or for vaporization? Interpret your answer using the idea that entropy is a measure of the randomness of a system.

20.30 • (a) Calculate the change in entropy when 1.00 mol of water (molecular mass 18.0 g/mol) at 100°C evaporates to form water vapor at 100°C. (b) Repeat the calculation of part (a) for 1.00 mol of liquid nitrogen, 1.00 mol of silver, and 1.00 mol of mercury when each is vaporized at its normal boiling point. (See Table 17.4 for the heats of vaporization, and Appendix D for the molar masses. Note that the nitrogen molecule is N_2.) (c) Your results in parts (a) and (b) should be in relatively close agreement. (This is called the *rule of Drepez and Trouton.*) Explain why this should be so, using the idea that entropy is a measure of the randomness of a system.

20.31 •• A 10.0-L gas tank containing 3.20 moles of ideal He gas at 20.0°C is placed inside a completely evacuated, insulated bell jar of volume 35.0 L. A small hole in the tank allows the He to leak out into the jar until the gas reaches a final equilibrium state with no more leakage. (a) What is the change in entropy of this system due to the leaking of the gas? (b) Is the process reversible or irreversible? How do you know?

Section 20.8 Microscopic Interpretation of Entropy

20.32 • A box is separated by a partition into two parts of equal volume. The left side of the box contains 500 molecules of nitrogen gas; the right side contains 100 molecules of oxygen gas. The two gases are at the same temperature. The partition is punctured, and equilibrium is eventually attained. Assume that the volume of the box is large enough for each gas to undergo a free expansion and not change temperature. (a) On average, how many molecules of each type will there be in either half of the box? (b) What is the change in entropy of the system when the partition is punctured? (c) What is the probability that the molecules will be found in the same distribution as they were before the partition was punctured—that is, 500 nitrogen molecules in the left half and 100 oxygen molecules in the right half?

20.33 • **CALC** Two moles of an ideal gas occupy a volume V. The gas expands isothermally and reversibly to a volume 3V. (a) Is the velocity distribution changed by the isothermal expansion? Explain. (b) Use Eq. (20.23) to calculate the change in entropy of the gas. (c) Use Eq. (20.18) to calculate the change in entropy of the gas. Compare this result to that obtained in part (b).

20.34 • **CALC** A lonely party balloon with a volume of 2.40 L and containing 0.100 mol of air is left behind to drift in the temporarily uninhabited and depressurized International Space Station. Sunlight coming through a porthole heats and explodes the balloon, causing the air in it to undergo a free expansion into the empty station, whose total volume is 425 m³. Calculate the entropy change of the air during the expansion.

PROBLEMS

20.35 •• **CP** An ideal Carnot engine operates between 500°C and 100°C with a heat input of 250 J per cycle. (a) How much heat is delivered to the cold reservoir in each cycle? (b) What minimum number of cycles is necessary for the engine to lift a 500-kg rock through a height of 100 m?

20.36 • You are designing a Carnot engine that has 2 mol of CO_2 as its working substance; the gas may be treated as ideal. The gas is to have a maximum temperature of 527°C and a maximum pressure of 5.00 atm. With a heat input of 400 J per cycle, you want 300 J of useful work. (a) Find the temperature of the cold reservoir. (b) For how many cycles must this engine run to melt completely a 10.0-kg block of ice originally at 0.0°C, using only the heat rejected by the engine?

20.37 •• **CP** A certain heat engine operating on a Carnot cycle absorbs 150 J of heat per cycle at its hot reservoir at 135°C and has a thermal efficiency of 22.0%. (a) How much work does this engine do per cycle? (b) How much heat does the engine waste each cycle? (c) What is the temperature of the cold reservoir? (d) By how much does the engine change the entropy of the world each cycle? (e) What mass of water could this engine pump per cycle from a well 35.0 m deep?

20.38 •• **BIO** **Entropy of Metabolism.** An average sleeping person metabolizes at a rate of about 80 W by digesting food or burning fat. Typically, 20% of this energy goes into bodily functions, such as cell repair, pumping blood, and other uses of mechanical energy, while the rest goes to heat. Most people get rid of all this excess heat by transferring it (by conduction and the flow of blood) to the surface of the body, where it is radiated away. The normal internal temperature of the body (where the metabolism takes place) is 37°C, and the skin is typically 7 C° cooler. By how much does the person's entropy change per second due to this heat transfer?

20.39 •• **BIO** **Entropy Change from Digesting Fat.** Digesting fat produces 9.3 food calories per gram of fat, and typically 80% of this energy goes to heat when metabolized. (One food calorie is 1000 calories and therefore equals 4186 J.) The body then moves all this heat to the surface by a combination of thermal conductivity and motion of the blood. The internal temperature of the body (where digestion occurs) is normally 37°C, and the surface is usually about 30°C. By how much do the digestion and metabolism of a 2.50-g pat of butter change your body's entropy? Does it increase or decrease?

20.40 • A heat engine takes 0.350 mol of a diatomic ideal gas around the cycle shown in the *pV*-diagram of Fig. P20.40. Process 1 → 2 is at constant volume, process 2 → 3 is adiabatic, and process $3 \rightarrow 1$ is at a constant pressure of 1.00 atm. The value of γ for this gas is 1.40. (a) Find the pressure and volume at points 1, 2, and 3. (b) Calculate Q, W, and ΔU for each of the three processes. (c) Find the net work done by the gas in the cycle. (d) Find the net heat flow into the engine in one cycle. (e) What is the thermal efficiency of the engine? How does this compare to the efficiency of a Carnot-cycle engine operating between the same minimum and maximum temperatures T_1 and T_2?

Figure **P20.40**

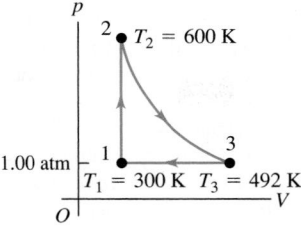

20.41 •• **CALC** You build a heat engine that takes 1.00 mol of an ideal diatomic gas through the cycle shown in Fig. P20.41. (a) Show that segment ab is an isothermal compression. (b) During which segment(s) of the cycle is heat absorbed by the gas? During which segment(s) is heat rejected?

Figure **P20.41**

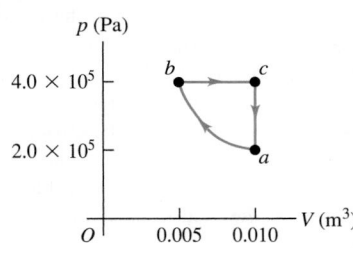

How do you know? (c) Calculate the temperature at points a, b, and c. (d) Calculate the net heat exchanged with the surroundings and the net work done by the engine in one cycle. (e) Calculate the thermal efficiency of the engine.

20.42 • **Heat Pump.** A heat pump is a heat engine run in reverse. In winter it pumps heat from the cold air outside into the warmer air inside the building, maintaining the building at a comfortable temperature. In summer it pumps heat from the cooler air inside the building to the warmer air outside, acting as an air conditioner. (a) If the outside temperature in winter is $-5.0°C$ and the inside temperature is $17.0°C$, how many joules of heat will the heat pump deliver to the inside for each joule of electrical energy used to run the unit, assuming an ideal Carnot cycle? (b) Suppose you have the option of using electrical resistance heating rather than a heat pump. How much electrical energy would you need in order to deliver the same amount of heat to the inside of the house as in part (a)? Consider a Carnot heat pump delivering heat to the inside of a house to maintain it at $68°F$. Show that the heat pump delivers less heat for each joule of electrical energy used to operate the unit as the outside temperature decreases. Notice that this behavior is opposite to the dependence of the efficiency of a Carnot heat engine on the difference in the reservoir temperatures. Explain why this is so.

20.43 • **CALC** A heat engine operates using the cycle shown in Fig. P20.43. The working substance is 2.00 mol of helium gas, which reaches a maximum temperature of $327°C$. Assume the helium can be treated as an ideal gas. Process bc is isothermal. The pressure in states a and c is 1.00×10^5 Pa, and the pressure in state b is 3.00×10^5 Pa. (a) How much heat enters the gas and how much

Figure **P20.43**

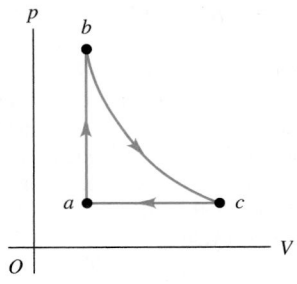

leaves the gas each cycle? (b) How much work does the engine do each cycle, and what is its efficiency? (c) Compare this engine's efficiency with the maximum possible efficiency attainable with the hot and cold reservoirs used by this cycle.

20.44 • **CP** As a budding mechanical engineer, you are called upon to design a Carnot engine that has 2.00 mol of a monatomic ideal gas as its working substance and operates from a high-temperature reservoir at $500°C$. The engine is to lift a 15.0-kg weight 2.00 m per cycle, using 500 J of heat input. The gas in the engine chamber can have a minimum volume of 5.00 L during the cycle. (a) Draw a pV-diagram for this cycle. Show in your diagram where heat enters and leaves the gas. (b) What must be the temperature of the cold reservoir? (c) What is the thermal efficiency of the engine? (d) How much heat energy does this engine waste per cycle? (e) What is the maximum pressure that the gas chamber will have to withstand?

20.45 ••• An experimental power plant at the Natural Energy Laboratory of Hawaii generates electricity from the temperature gradient of the ocean. The surface and deep-water temperatures are $27°C$ and $6°C$, respectively. (a) What is the maximum theoretical efficiency of this power plant? (b) If the power plant is to produce 210 kW of power, at what rate must heat be extracted from the warm water? At what rate must heat be absorbed by the cold water? Assume the maximum theoretical efficiency. (c) The cold water that enters the plant leaves it at a temperature of $10°C$. What must be the flow rate of cold water through the system? Give your answer in kg/h and in L/h.

20.46 •• What is the thermal efficiency of an engine that operates by taking n moles of diatomic ideal gas through the cycle $1 \rightarrow 2 \rightarrow 3 \rightarrow 4 \rightarrow 1$ shown in Fig. P20.46?

Figure **P20.46**

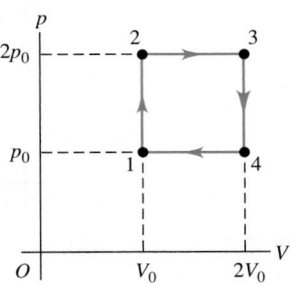

20.47 • **CALC** A cylinder contains oxygen at a pressure of 2.00 atm. The volume is 4.00 L, and the temperature is 300 K. Assume that the oxygen may be treated as an ideal gas. The oxygen is carried through the following processes:

(i) Heated at constant pressure from the initial state (state 1) to state 2, which has $T = 450$ K.
(ii) Cooled at constant volume to 250 K (state 3).
(iii) Compressed at constant temperature to a volume of 4.00 L (state 4).
(iv) Heated at constant volume to 300 K, which takes the system back to state 1.

(a) Show these four processes in a pV-diagram, giving the numerical values of p and V in each of the four states. (b) Calculate Q and W for each of the four processes. (c) Calculate the net work done by the oxygen in the complete cycle. (d) What is the efficiency of this device as a heat engine? How does this compare to the efficiency of a Carnot-cycle engine operating between the same minimum and maximum temperatures of 250 K and 450 K?

20.48 •• **CP BIO Human Entropy.** A person who has skin of surface area 1.85 m^2 and temperature $30.0°C$ is resting in an insulated room where the ambient air temperature is $20.0°C$. In this state, a person gets rid of excess heat by radiation. By how much does the person change the entropy of the air in this room each second?

(Recall that the room radiates back into the person and that the emissivity of the skin is 1.00.)

20.49 •• **CP** **BIO** **A Human Engine.** You decide to use your body as a Carnot heat engine. The operating gas is in a tube with one end in your mouth (where the temperature is 37.0°C) and the other end at the surface of your skin, at 30.0°C. (a) What is the maximum efficiency of such a heat engine? Would it be a very useful engine? (b) Suppose you want to use this human engine to lift a 2.50-kg box from the floor to a tabletop 1.20 m above the floor. How much must you increase the gravitational potential energy, and how much heat input is needed to accomplish this? (c) If your favorite candy bar has 350 food calories (1 food calorie = 4186 J) and 80% of the food energy goes into heat, how many of these candy bars must you eat to lift the box in this way?

20.50 •• **CP** **Entropy Change Due to the Sun.** Our sun radiates from a surface at 5800 K (with an emissivity of 1.0) into the near-vacuum of space, which is at a temperature of 3 K. (a) By how much does our sun change the entropy of the universe every second? (Consult Appendix F.) (b) Is the process reversible or irreversible? Is your answer to part (a) consistent with this conclusion? Explain.

20.51 • A monatomic ideal gas is taken around the cycle shown in Fig. P20.51 in the direction shown in the figure. The path for process $c \to a$ is a straight line in the pV-diagram. (a) Calculate Q, W, and ΔU for each process $a \to b$, $b \to c$, and $c \to a$. (b) What are Q, W, and ΔU for one complete cycle? (c) What is the efficiency of the cycle?

Figure **P20.51**

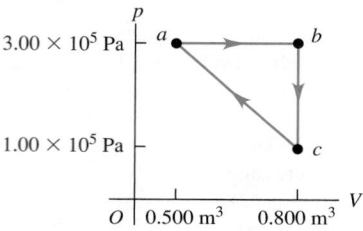

20.52 •• **CALC** **A Stirling-Cycle Engine.** The *Stirling cycle* is similar to the Otto cycle, except that the compression and expansion of the gas are done at constant temperature, not adiabatically as in the Otto cycle. The Stirling cycle is used in *external* combustion engines (in fact, burning fuel is not necessary; *any* way of producing a temperature difference will do—solar, geothermal, ocean temperature gradient, etc.), which means that the gas inside the cylinder is not used in the combustion process. Heat is supplied by burning fuel steadily outside the cylinder, instead of explosively inside the cylinder as in the Otto cycle. For this reason Stirling-cycle engines are quieter than Otto-cycle engines, since there are no intake and exhaust valves (a major source of engine noise). While small Stirling engines are used for a variety of purposes, Stirling engines for automobiles have not been successful because they are larger, heavier, and more expensive than conventional automobile engines. In the cycle, the working fluid goes through the following sequence of steps (Fig. P20.52):

Figure **P20.52**

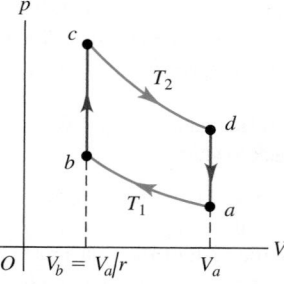

(i) Compressed isothermally at temperature T_1 from the initial state a to state b, with a compression ratio r.
(ii) Heated at constant volume to state c at temperature T_2.
(iii) Expanded isothermally at T_2 to state d.
(iv) Cooled at constant volume back to the initial state a.

Assume that the working fluid is n moles of an ideal gas (for which C_V is independent of temperature). (a) Calculate Q, W, and ΔU for each of the processes $a \to b$, $b \to c$, $c \to d$, and $d \to a$. (b) In the Stirling cycle, the heat transfers in the processes $b \to c$ and $d \to a$ do not involve external heat sources but rather use *regeneration:* The same substance that transfers heat to the gas inside the cylinder in the process $b \to c$ also absorbs heat back from the gas in the process $d \to a$. Hence the heat transfers $Q_{b \to c}$ and $Q_{d \to a}$ do not play a role in determining the efficiency of the engine. Explain this last statement by comparing the expressions for $Q_{b \to c}$ and $Q_{d \to a}$ calculated in part (a). (c) Calculate the efficiency of a Stirling-cycle engine in terms of the temperatures T_1 and T_2. How does this compare to the efficiency of a Carnot-cycle engine operating between these same two temperatures? (Historically, the Stirling cycle was devised before the Carnot cycle.) Does this result violate the second law of thermodynamics? Explain. Unfortunately, actual Stirling-cycle engines cannot achieve this efficiency due to problems with the heat-transfer processes and pressure losses in the engine.

20.53 •• A Carnot engine operates between two heat reservoirs at temperatures T_H and T_C. An inventor proposes to increase the efficiency by running one engine between T_H and an intermediate temperature T' and a second engine between T' and T_C, using as input the heat expelled by the first engine. Compute the efficiency of this composite system, and compare it to that of the original engine.

20.54 ••• A typical coal-fired power plant generates 1000 MW of usable power at an overall thermal efficiency of 40%. (a) What is the rate of heat input to the plant? (b) The plant burns anthracite coal, which has a heat of combustion of 2.65×10^7 J/kg. How much coal does the plant use per day, if it operates continuously? (c) At what rate is heat ejected into the cool reservoir, which is the nearby river? (d) The river's temperature is 18.0°C before it reaches the power plant and 18.5°C after it has received the plant's waste heat. Calculate the river's flow rate, in cubic meters per second. (e) By how much does the river's entropy increase each second?

20.55 • **Automotive Thermodynamics.** A Volkswagen Passat has a six-cylinder Otto-cycle engine with compression ratio $r = 10.6$. The diameter of each cylinder, called the *bore* of the engine, is 82.5 mm. The distance that the piston moves during the compression in Fig. 20.5, called the *stroke* of the engine, is 86.4 mm. The initial pressure of the air–fuel mixture (at point a in Fig. 20.6) is 8.50×10^4 Pa, and the initial temperature is 300 K (the same as the outside air). Assume that 200 J of heat is added to each cylinder in each cycle by the burning gasoline, and that the gas has $C_V = 20.5$ J/mol·K and $\gamma = 1.40$. (a) Calculate the total work done in one cycle in each cylinder of the engine, and the heat released when the gas is cooled to the temperature of the outside air. (b) Calculate the volume of the air–fuel mixture at point a in the cycle. (c) Calculate the pressure, volume, and temperature of the gas at points b, c, and d in the cycle. In a pV-diagram, show the numerical values of p, V, and T for each of the four states. (d) Compare the efficiency of this engine with the efficiency of a Carnot-cycle engine operating between the same maximum and minimum temperatures.

20.56 • An air conditioner operates on 800 W of power and has a performance coefficient of 2.80 with a room temperature of 21.0°C and an outside temperature of 35.0°C. (a) Calculate the rate of heat removal for this unit. (b) Calculate the rate at which heat is discharged to the outside air. (c) Calculate the total entropy change in the room if the air conditioner runs for 1 hour. Calculate the total entropy change in the outside air for the same time period. (d) What is the net change in entropy for the system (room + outside air)?

20.57 •• **CALC** **Unavailable Energy.** The discussion of entropy and the second law that follows Example 20.10 (Section 20.7) says that the increase in entropy in an irreversible process is associated with energy becoming less available. Consider a Carnot cycle that uses a low-temperature reservoir with Kelvin temperature T_c. This is a true reservoir—that is, large enough not to change temperature when it accepts heat from the engine. Let the engine accept heat from an object of temperature T', where $T' > T_c$. The object is of finite size, so it cools as heat is extracted from it. The engine continues to operate until $T' = T_c$. (a) Show that the total magnitude of heat rejected to the low-temperature reservoir is $T_c|\Delta S_h|$, where ΔS_h is the change in entropy of the high-temperature reservoir. (b) Apply the result of part (a) to 1.00 kg of water initially at a temperature of 373 K as the heat source for the engine and $T_c = 273$ K. How much total mechanical work can be performed by the engine until it stops? (c) Repeat part (b) for 2.00 kg of water at 323 K. (d) Compare the amount of work that can be obtained from the energy in the water of Example 20.10 before and after it is mixed. Discuss whether your result shows that energy has become less available.

20.58 ••• **CP** The maximum power that can be extracted by a wind turbine from an air stream is approximately

$$P = kd^2v^3$$

where d is the blade diameter, v is the wind speed, and the constant $k = 0.5$ W·s³/m⁵. (a) Explain the dependence of P on d and on v by considering a cylinder of air that passes over the turbine blades in time t (Fig. P20.58). This cylinder has diameter d, length $L = vt$, and density ρ. (b) The Mod-5B wind turbine at Kahaku on the Hawaiian island of Oahu has a blade diameter of 97 m (slightly longer than a football field) and sits atop a 58-m tower. It can produce 3.2 MW of electric power. Assuming 25% efficiency, what wind speed is required to produce this amount of power? Give your answer in m/s and in km/h. (c) Commercial wind turbines are commonly located in or downwind of mountain passes. Why?

Figure **P20.58**

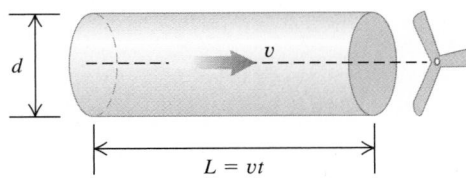

20.59 •• **CALC** (a) For the Otto cycle shown in Fig. 20.6, calculate the changes in entropy of the gas in each of the constant-volume processes $b \to c$ and $d \to a$ in terms of the temperatures T_a, T_b, T_c, and T_d and the number of moles n and the heat capacity C_V of the gas. (b) What is the total entropy change in the engine during one cycle? (*Hint:* Use the relationships between T_a and T_b and between T_d and T_c.) (c) The processes $b \to c$ and $d \to a$ occur irreversibly in a real Otto engine. Explain how can this be reconciled with your result in part (b).

20.60 •• **CALC** **A TS-Diagram.** (a) Graph a Carnot cycle, plotting Kelvin temperature vertically and entropy horizontally. This is called a temperature–entropy diagram, or *TS*-diagram. (b) Show that the area under any curve representing a reversible path in a temperature–entropy diagram represents the heat absorbed by the system. (c) Derive from your diagram the expression for the thermal efficiency of a Carnot cycle. (d) Draw a temperature–entropy diagram for the Stirling cycle described in Problem 20.52. Use this diagram to relate the efficiencies of the Carnot and Stirling cycles.

20.61 • A physics student immerses one end of a copper rod in boiling water at 100°C and the other end in an ice–water mixture at 0°C. The sides of the rod are insulated. After steady-state conditions have been achieved in the rod, 0.120 kg of ice melts in a certain time interval. For this time interval, find (a) the entropy change of the boiling water; (b) the entropy change of the ice–water mixture; (c) the entropy change of the copper rod; (d) the total entropy change of the entire system.

20.62 •• **CALC** To heat 1 cup of water (250 cm³) to make coffee, you place an electric heating element in the cup. As the water temperature increases from 20°C to 78°C, the temperature of the heating element remains at a constant 120°C. Calculate the change in entropy of (a) the water; (b) the heating element; (c) the system of water and heating element. (Make the same assumption about the specific heat of water as in Example 20.10 in Section 20.7, and ignore the heat that flows into the ceramic coffee cup itself.) (d) Is this process reversible or irreversible? Explain.

20.63 •• **CALC** An object of mass m_1, specific heat c_1, and temperature T_1 is placed in contact with a second object of mass m_2, specific heat c_2, and temperature $T_2 > T_1$. As a result, the temperature of the first object increases to T and the temperature of the second object decreases to T'. (a) Show that the entropy increase of the system is

$$\Delta S = m_1 c_1 \ln\frac{T}{T_1} + m_2 c_2 \ln\frac{T'}{T_2}$$

and show that energy conservation requires that

$$m_1 c_1 (T - T_1) = m_2 c_2 (T_2 - T')$$

(b) Show that the entropy change ΔS, considered as a function of T, is a *maximum* if $T = T'$, which is just the condition of thermodynamic equilibrium. (c) Discuss the result of part (b) in terms of the idea of entropy as a measure of disorder.

CHALLENGE PROBLEM

20.64 ••• Consider a Diesel cycle that starts (at point a in Fig. 20.7) with air at temperature T_a. The air may be treated as an ideal gas. (a) If the temperature at point c is T_c, derive an expression for the efficiency of the cycle in terms of the compression ratio r. (b) What is the efficiency if $T_a = 300$ K, $T_c = 950$ K, $\gamma = 1.40$, and $r = 21.0$?

Answers

Chapter Opening Question ?

Yes. That's what a refrigerator does: It makes heat flow from the cold interior of the refrigerator to the warm outside. The second law of thermodynamics says that heat cannot *spontaneously* flow from a cold body to a hot one. A refrigerator has a motor that does work on the system to *force* the heat to flow in that way.

Test Your Understanding Questions

20.1 Answer: (ii) Like sliding a book across a table, rubbing your hands together uses friction to convert mechanical energy into heat. The (impossible) reverse process would involve your hands spontaneously getting colder, with the released energy forcing your hands to move rhythmically back and forth!

20.2 Answer: (iii), (i), (ii) From Eq. (20.4) the efficiency is $e = W/Q_H$, and from Eq. (20.2) $W = Q_H + Q_C = |Q_H| - |Q_C|$. For engine (i) $Q_H = 5000\,J$ and $Q_C = -4500\,J$, so $W = 5000\,J + (-4500\,J) = 500\,J$ and $e = (500\,J)/(5000\,H) = 0.100$. For engine (ii) $Q_H = 25{,}000\,J$ and $W = 2000\,J$, so $e = (2000\,J)/(25{,}000\,J) = 0.080$. For engine (iii) $W = 400\,J$ and $Q_C = -2800\,J$, so $Q_H = W - Q_C = 400\,J - (-2800\,J) = 3200\,J$ and $e = (400\,J)/(3200\,J) = 0.125$.

20.3 Answers: (i), (ii) Doubling the amount of fuel burned per cycle means that Q_H is doubled, so the resulting pressure increase from b to c in Fig. 20.6 is greater. The compression ratio and hence the efficiency remain the same, so $|Q_C|$ (the amount of heat rejected to the environment) must increase by the same factor as Q_H. Hence the pressure drop from d to a in Fig. 20.6 is also greater. The volume V and the compression ratio r don't change, so the horizontal dimensions of the pV-diagram don't change.

20.4 Answer: no A refrigerator uses an input of work to transfer heat from one system (the refrigerator's interior) to another system (its exterior, which includes the house in which the refrigerator is installed). If the door is open, these two systems are really the *same* system and will eventually come to the same temperature. By the first law of thermodynamics, all of the work input to the refrigerator motor will be converted into heat and the temperature in your house will actually *increase*. To cool the house you need a system that will transfer heat from it to the outside world, such as an air conditioner or heat pump.

20.5 Answers: no, no Both the 100%-efficient engine of Fig. 20.11a and the workless refrigerator of Fig. 20.11b return to the same state at the end of a cycle as at the beginning, so the net change in internal energy of each system is zero ($\Delta U = 0$). For the 100%-efficient engine, the net heat flow into the engine equals the net work done, so $Q = W$, $Q - W = 0$, and the first law ($\Delta U = Q - W$) is obeyed. For the workless refrigerator, no net work is done (so $W = 0$) and as much heat flows into it as out (so $Q = 0$), so again $Q - W = 0$ and $\Delta U = Q - W$ in accordance with the first law. It is the *second* law of thermodynamics that tells us that both the 100%-efficient engine and the workless refrigerator are impossible.

20.6 Answer: no The efficiency can be no better than that of a Carnot engine running between the same two temperature limits, $e_{Carnot} = 1 - (T_C/T_H)$ [Eq. (20.14)]. The temperature T_C of the cold reservoir for this air-cooled engine is about 300 K (ambient temperature), and the temperature T_H of the hot reservoir cannot exceed the melting point of copper, 1356 K (see Table 17.4). Hence the maximum possible Carnot efficiency is $e = 1 - (300\,K)/(1356\,K) = 0.78$, or 78%. The temperature of any real engine would be less than this, so it would be impossible for the inventor's engine to attain 85% efficiency. You should invest your money elsewhere.

20.7 Answers: $-102\,J/K$, no The process described is exactly the opposite of the process used in Example 20.10. The result violates the second law of thermodynamics, which states that the entropy of an isolated system cannot decrease.

20.8 Answer: (i) For case (i), we saw in Example 20.8 (Section 20.7) that for an ideal gas, the entropy change in a free expansion is the same as in an isothermal expansion. From Eq. (20.23), this implies that the ratio of the number of microscopic states after and before the expansion, w_2/w_1, is also the same for these two cases. From Example 20.11, $w_2/w_1 = 2^N$, so the number of microscopic states increases by a factor 2^N. For case (ii), in a reversible expansion the entropy change is $\Delta S = \int dQ/T = 0$; if the expansion is adiabatic there is no heat flow, so $\Delta S = 0$. From Eq. (20.23), $w_2/w_1 = 1$ and there is *no* change in the number of microscopic states. The difference is that in an adiabatic expansion the temperature drops and the molecules move more slowly, so they have fewer microscopic states available to them than in an isothermal expansion.

Bridging Problem

Answers: (a) 34.83°C **(b)** +12.1 J/K

ELECTRIC CHARGE AND ELECTRIC FIELD

? Water makes life possible: The cells of your body could not function without water in which to dissolve essential biological molecules. What electrical properties of water make it such a good solvent?

In Chapter 5 we mentioned the four kinds of fundamental forces. To this point the only one of these forces that we have examined in any detail is gravity. Now we are ready to examine the force of *electromagnetism,* which encompasses both electricity and magnetism. Electromagnetic phenomena will occupy our attention for most of the remainder of this book.

Electromagnetic interactions involve particles that have a property called *electric charge,* an attribute that is as fundamental as mass. Just as objects with mass are accelerated by gravitational forces, so electrically charged objects are accelerated by electric forces. The shock you feel when you scuff your shoes across a carpet and then reach for a metal doorknob is due to charged particles leaping between your finger and the doorknob. Electric currents are simply streams of charged particles flowing within wires in response to electric forces. Even the forces that hold atoms together to form solid matter, and that keep the atoms of solid objects from passing through each other, are fundamentally due to electric interactions between the charged particles within atoms.

We begin our study of electromagnetism in this chapter by examining the nature of electric charge. We'll find that charge is quantized and obeys a conservation principle. When charges are at rest in our frame of reference, they exert *electrostatic* forces on each other. These forces are of tremendous importance in chemistry and biology and have many technological applications. Electrostatic forces are governed by a simple relationship known as *Coulomb's law* and are most conveniently described by using the concept of *electric field.* In later chapters we'll expand our discussion to include electric charges in motion. This will lead us to an understanding of magnetism and, remarkably, of the nature of light.

While the key ideas of electromagnetism are conceptually simple, applying them to practical problems will make use of many of your mathematical skills, especially your knowledge of geometry and integral calculus. For this reason you may find this chapter and those that follow to be more mathematically demanding than earlier chapters. The reward for your extra effort will be a deeper understanding of principles that are at the heart of modern physics and technology.

21.1 Electric Charge

The ancient Greeks discovered as early as 600 B.C. that after they rubbed amber with wool, the amber could attract other objects. Today we say that the amber has acquired a net **electric charge,** or has become *charged*. The word "electric" is derived from the Greek word *elektron,* meaning amber. When you scuff your shoes across a nylon carpet, you become electrically charged, and you can charge a comb by passing it through dry hair.

Plastic rods and fur (real or fake) are particularly good for demonstrating **electrostatics,** the interactions between electric charges that are at rest (or nearly so). After we charge both plastic rods in Fig. 21.1a by rubbing them with the piece of fur, we find that the rods repel each other.

When we rub glass rods with silk, the glass rods also become charged and repel each other (Fig. 21.1b). But a charged plastic rod *attracts* a charged glass rod; furthermore, the plastic rod and the fur attract each other, and the glass rod and the silk attract each other (Fig. 21.1c).

These experiments and many others like them have shown that there are exactly two kinds of electric charge: the kind on the plastic rod rubbed with fur and the kind on the glass rod rubbed with silk. Benjamin Franklin (1706–1790) suggested calling these two kinds of charge *negative* and *positive,* respectively, and these names are still used. The plastic rod and the silk have negative charge; the glass rod and the fur have positive charge.

> **Two positive charges or two negative charges repel each other. A positive charge and a negative charge attract each other.**

CAUTION **Electric attraction and repulsion** The attraction and repulsion of two charged objects are sometimes summarized as "Like charges repel, and opposite charges attract." But keep in mind that the phrase "like charges" does *not* mean that the two charges are exactly identical, only that both charges have the same algebraic *sign* (both positive or both negative). "Opposite charges" means that both objects have an electric charge, and those charges have different signs (one positive and the other negative). ▌

21.1 Experiments in electrostatics. **(a)** Negatively charged objects repel each other. **(b)** Positively charged objects repel each other. **(c)** Positively charged objects and negatively charged objects attract each other.

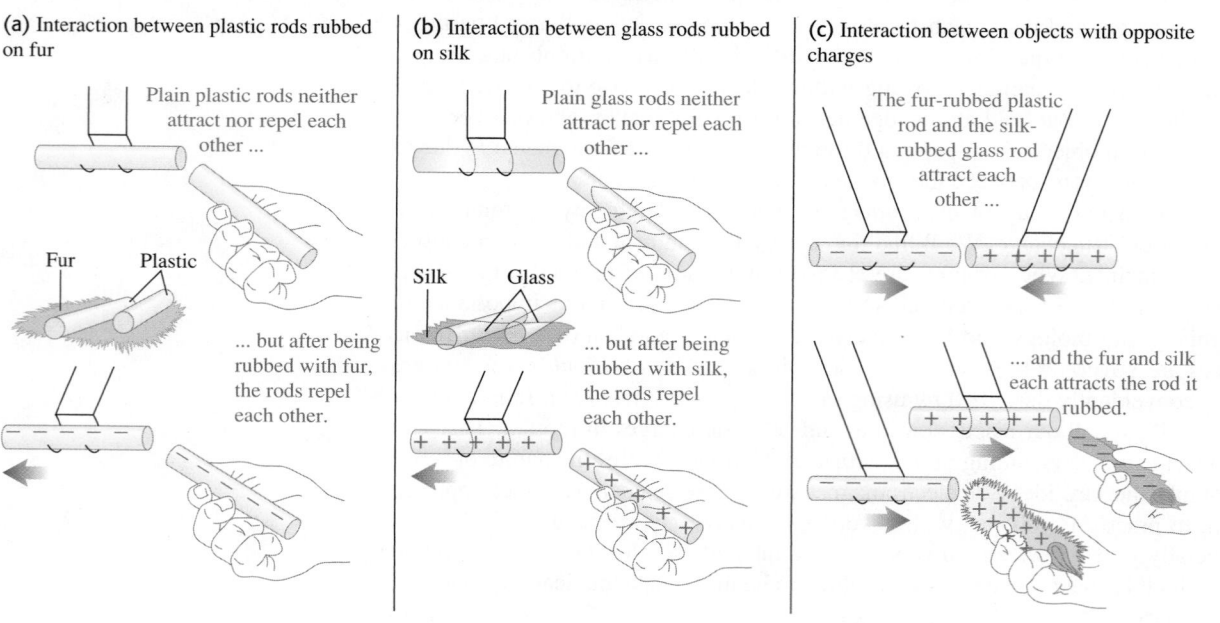

(a) Interaction between plastic rods rubbed on fur

Plain plastic rods neither attract nor repel each other ...

Fur   Plastic

... but after being rubbed with fur, the rods repel each other.

(b) Interaction between glass rods rubbed on silk

Plain glass rods neither attract nor repel each other ...

Silk   Glass

... but after being rubbed with silk, the rods repel each other.

(c) Interaction between objects with opposite charges

The fur-rubbed plastic rod and the silk-rubbed glass rod attract each other ...

... and the fur and silk each attracts the rod it rubbed.

21.2 Schematic diagram of the operation of a laser printer.

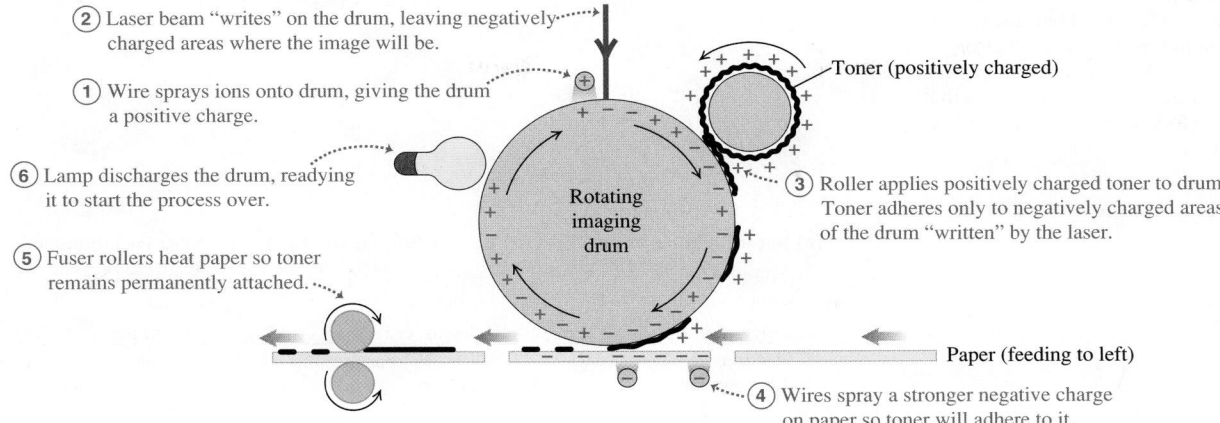

② Laser beam "writes" on the drum, leaving negatively charged areas where the image will be.

① Wire sprays ions onto drum, giving the drum a positive charge.

⑥ Lamp discharges the drum, readying it to start the process over.

⑤ Fuser rollers heat paper so toner remains permanently attached.

Toner (positively charged)

Rotating imaging drum

③ Roller applies positively charged toner to drum. Toner adheres only to negatively charged areas of the drum "written" by the laser.

Paper (feeding to left)

④ Wires spray a stronger negative charge on paper so toner will adhere to it.

One application of forces between charged bodies is in a laser printer (Fig. 21.2). The printer's light-sensitive imaging drum is given a positive charge. As the drum rotates, a laser beam shines on selected areas of the drum, leaving those areas with a *negative* charge. Positively charged particles of toner adhere only to the areas of the drum "written" by the laser. When a piece of paper is placed in contact with the drum, the toner particles stick to the paper and form an image.

Electric Charge and the Structure of Matter

When you charge a rod by rubbing it with fur or silk as in Fig. 21.1, there is no visible change in the appearance of the rod. What, then, actually happens to the rod when you charge it? To answer this question, we must look more closely at the structure of atoms, the building blocks of ordinary matter.

The structure of atoms can be described in terms of three particles: the negatively charged **electron,** the positively charged **proton,** and the uncharged **neutron** (Fig. 21.3). The proton and neutron are combinations of other entities called *quarks,* which have charges of $\pm\frac{1}{3}$ and $\pm\frac{2}{3}$ times the electron charge. Isolated quarks have not been observed, and there are theoretical reasons to believe that it is impossible in principle to observe a quark in isolation.

The protons and neutrons in an atom make up a small, very dense core called the **nucleus,** with dimensions of the order of 10^{-15} m. Surrounding the nucleus are the electrons, extending out to distances of the order of 10^{-10} m from the nucleus. If an atom were a few kilometers across, its nucleus would be the size of a tennis ball. The negatively charged electrons are held within the atom by the attractive electric forces exerted on them by the positively charged nucleus. (The protons and neutrons are held within stable atomic nuclei by an attractive interaction, called the *strong nuclear force,* that overcomes the electric repulsion of the protons. The strong nuclear force has a short range, and its effects do not extend far beyond the nucleus.)

The masses of the individual particles, to the precision that they are presently known, are

$$\text{Mass of electron} = m_e = 9.10938215(45) \times 10^{-31} \text{ kg}$$

$$\text{Mass of proton} = m_p = 1.672621637(83) \times 10^{-27} \text{ kg}$$

$$\text{Mass of neutron} = m_n = 1.674927211(84) \times 10^{-27} \text{ kg}$$

The numbers in parentheses are the uncertainties in the last two digits. Note that the masses of the proton and neutron are nearly equal and are roughly 2000 times

21.3 The structure of an atom. The particular atom depicted here is lithium (see Fig. 21.4a).

Atom

Most of the atom's volume is occupied sparsely by electrons.

$\leftarrow \sim 10^{-10}$ m $\rightarrow$

Nucleus

Tiny compared with the rest of the atom, the nucleus contains over 99.9% of the atom's mass.

$\sim 10^{-15}$ m

⊕ **Proton:** Positive charge
Mass = 1.673×10^{-27} kg

Neutron: No charge
Mass = 1.675×10^{-27} kg

⊖ **Electron:** Negative charge
Mass = 9.109×10^{-31} kg

The charges of the electron and proton are equal in magnitude.

21.4 (a) A neutral atom has as many electrons as it does protons. (b) A positive ion has a deficit of electrons. (c) A negative ion has an excess of electrons. (The electron "shells" are a schematic representation of the actual electron distribution, a diffuse cloud many times larger than the nucleus.)

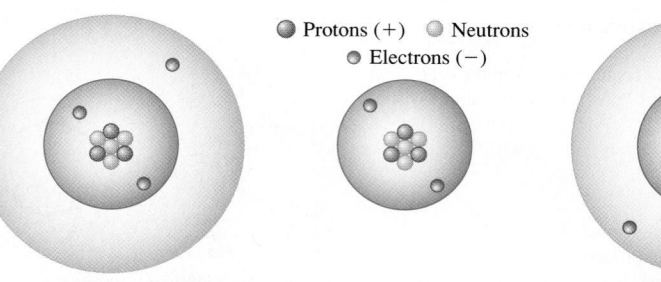

● Protons (+)   ○ Neutrons
⊙ Electrons (−)

(a) Neutral lithium atom (Li):
3 protons (3+)
4 neutrons
3 electrons (3−)
Electrons equal protons:
Zero net charge

(b) Positive lithium ion (Li⁺):
3 protons (3+)
4 neutrons
2 electrons (2−)
Fewer electrons than protons:
Positive net charge

(c) Negative lithium ion (Li⁻):
3 protons (3+)
4 neutrons
4 electrons (4−)
More electrons than protons:
Negative net charge

the mass of the electron. Over 99.9% of the mass of any atom is concentrated in its nucleus.

The negative charge of the electron has (within experimental error) *exactly* the same magnitude as the positive charge of the proton. In a neutral atom the number of electrons equals the number of protons in the nucleus, and the net electric charge (the algebraic sum of all the charges) is exactly zero (Fig. 21.4a). The number of protons or electrons in a neutral atom of an element is called the **atomic number** of the element. If one or more electrons are removed from an atom, what remains is called a **positive ion** (Fig. 21.4b). A **negative ion** is an atom that has *gained* one or more electrons (Fig. 21.4c). This gain or loss of electrons is called **ionization.**

When the total number of protons in a macroscopic body equals the total number of electrons, the total charge is zero and the body as a whole is electrically neutral. To give a body an excess negative charge, we may either *add negative* charges to a neutral body or *remove positive* charges from that body. Similarly, we can create an excess positive charge by either *adding positive* charge or *removing negative* charge. In most cases, negatively charged (and highly mobile) electrons are added or removed, and a "positively charged body" is one that has lost some of its normal complement of electrons. When we speak of the charge of a body, we always mean its *net* charge. The net charge is always a very small fraction (typically no more than 10^{-12}) of the total positive charge or negative charge in the body.

Electric Charge Is Conserved

Implicit in the foregoing discussion are two very important principles. First is the **principle of conservation of charge:**

> **The algebraic sum of all the electric charges in any closed system is constant.**

If we rub together a plastic rod and a piece of fur, both initially uncharged, the rod acquires a negative charge (since it takes electrons from the fur) and the fur acquires a positive charge of the *same* magnitude (since it has lost as many electrons as the rod has gained). Hence the total electric charge on the two bodies together does not change. In any charging process, charge is not created or destroyed; it is merely *transferred* from one body to another.

Conservation of charge is thought to be a *universal* conservation law. No experimental evidence for any violation of this principle has ever been observed. Even in high-energy interactions in which particles are created and destroyed, such as the creation of electron–positron pairs, the total charge of any closed system is exactly constant.

The second important principle is:

The magnitude of charge of the electron or proton is a natural unit of charge.

Every observable amount of electric charge is always an integer multiple of this basic unit. We say that charge is *quantized*. A familiar example of quantization is money. When you pay cash for an item in a store, you have to do it in one-cent increments. Cash can't be divided into amounts smaller than one cent, and electric charge can't be divided into amounts smaller than the charge of one electron or proton. (The quark charges, $\pm\frac{1}{3}$ and $\pm\frac{2}{3}$ of the electron charge, are probably not observable as isolated charges.) Thus the charge on any macroscopic body is always either zero or an integer multiple (negative or positive) of the electron charge.

Understanding the electric nature of matter gives us insight into many aspects of the physical world (Fig. 21.5). The chemical bonds that hold atoms together to form molecules are due to electric interactions between the atoms. They include the strong ionic bonds that hold sodium and chlorine atoms together to make table salt and the relatively weak bonds between the strands of DNA that record your body's genetic code. The normal force exerted on you by the chair in which you're sitting arises from electric forces between charged particles in the atoms of your seat and in the atoms of your chair. The tension force in a stretched string and the adhesive force of glue are likewise due to the electric interactions of atoms.

Test Your Understanding of Section 21.1 (a) Strictly speaking, does the plastic rod in Fig. 21.1 weigh more, less, or the same after rubbing it with fur? (b) What about the glass rod after rubbing it with silk? What about (c) the fur and (d) the silk? ❙

21.2 Conductors, Insulators, and Induced Charges

Some materials permit electric charge to move easily from one region of the material to another, while others do not. For example, Fig. 21.6a shows a copper wire supported by a nylon thread. Suppose you touch one end of the wire to a charged plastic rod and attach the other end to a metal ball that is initially uncharged; you then remove the charged rod and the wire. When you bring another charged body up close to the ball (Figs. 21.6b and 21.6c), the ball is attracted or repelled, showing that the ball has become electrically charged. Electric charge has been transferred through the copper wire between the ball and the surface of the plastic rod.

The copper wire is called a **conductor** of electricity. If you repeat the experiment using a rubber band or nylon thread in place of the wire, you find that *no* charge is transferred to the ball. These materials are called **insulators.** Conductors permit the easy movement of charge through them, while insulators do not. (The supporting nylon threads shown in Fig. 21.6 are insulators, which prevents charge from leaving the metal ball and copper wire.)

As an example, carpet fibers on a dry day are good insulators. As you walk across a carpet, the rubbing of your shoes against the fibers causes charge to build up on you, and this charge remains on you because it can't flow through the insulating fibers. If you then touch a conducting object such as a doorknob, a rapid charge transfer takes place between your finger and the doorknob, and you feel a shock. One way to prevent this is to wind some of the carpet fibers around conducting cores so that any charge that builds up on you can be transferred harmlessly to the carpet. Another solution is to coat the carpet fibers with an anti-static layer that does not easily transfer electrons to or from your shoes; this prevents any charge from building up on you in the first place.

21.5 Most of the forces on this water skier are electric. Electric interactions between adjacent molecules give rise to the force of the water on the ski, the tension in the tow rope, and the resistance of the air on the skier's body. Electric interactions also hold the atoms of the skier's body together. Only one wholly nonelectric force acts on the skier: the force of gravity.

MasteringPHYSICS

PhET: Balloons and Static Electricity
PhET: John Travoltage

21.6 Copper is a good conductor of electricity; nylon is a good insulator. **(a)** The copper wire conducts charge between the metal ball and the charged plastic rod to charge the ball negatively. Afterward, the metal ball is **(b)** repelled by a negatively charged plastic rod and **(c)** attracted to a positively charged glass rod.

(a)

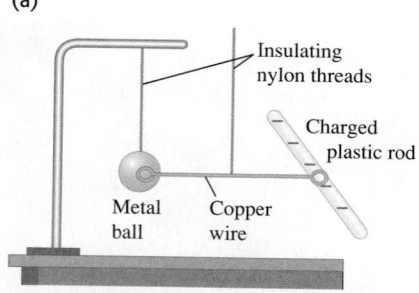

The wire conducts charge from the negatively charged plastic rod to the metal ball.

(b)

(c)

Most metals are good conductors, while most nonmetals are insulators. Within a solid metal such as copper, one or more outer electrons in each atom become detached and can move freely throughout the material, just as the molecules of a gas can move through the spaces between the grains in a bucket of sand. The other electrons remain bound to the positively charged nuclei, which themselves are bound in nearly fixed positions within the material. In an insulator there are no, or very few, free electrons, and electric charge cannot move freely through the material. Some materials called *semiconductors* are intermediate in their properties between good conductors and good insulators.

Charging by Induction

We can charge a metal ball using a copper wire and an electrically charged plastic rod, as in Fig. 21.6a. In this process, some of the excess electrons on the rod are transferred from it to the ball, leaving the rod with a smaller negative charge. But there is a different technique in which the plastic rod can give another body a charge of *opposite* sign without losing any of its own charge. This process is called charging by **induction.**

Figure 21.7 shows an example of charging by induction. An uncharged metal ball is supported on an insulating stand (Fig. 21.7a). When you bring a negatively charged rod near it, without actually touching it (Fig. 21.7b), the free electrons in the metal ball are repelled by the excess electrons on the rod, and they shift toward the right, away from the rod. They cannot escape from the ball because the supporting stand and the surrounding air are insulators. So we get excess negative charge at the right surface of the ball and a deficiency of negative charge (that is, a net positive charge) at the left surface. These excess charges are called **induced charges.**

Not all of the free electrons move to the right surface of the ball. As soon as any induced charge develops, it exerts forces toward the *left* on the other free electrons. These electrons are repelled by the negative induced charge on the right and attracted toward the positive induced charge on the left. The system reaches an equilibrium state in which the force toward the right on an electron, due to the charged rod, is just balanced by the force toward the left due to the induced charge. If we remove the charged rod, the free electrons shift back to the left, and the original neutral condition is restored.

What happens if, while the plastic rod is nearby, you touch one end of a conducting wire to the right surface of the ball and the other end to the earth (Fig. 21.7c)? The earth is a conductor, and it is so large that it can act as a practically infinite source of extra electrons or sink of unwanted electrons. Some of the negative charge flows through the wire to the earth. Now suppose you disconnect the wire (Fig. 21.7d) and then remove the rod (Fig. 21.7e); a net positive charge is left on the ball. The charge on the negatively charged rod has not changed during this process. The earth acquires a negative charge that is equal in magnitude to the induced positive charge remaining on the ball.

21.7 Charging a metal ball by induction.

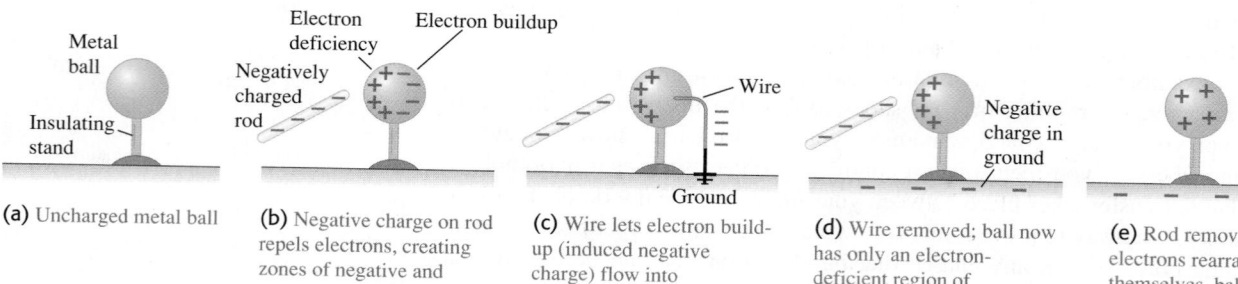

(a) Uncharged metal ball

(b) Negative charge on rod repels electrons, creating zones of negative and positive **induced charge.**

(c) Wire lets electron build-up (induced negative charge) flow into ground.

(d) Wire removed; ball now has only an electron-deficient region of positive charge.

(e) Rod removed; electrons rearrange themselves, ball has overall electron deficiency (net positive charge).

21.8 The charges within the molecules of an insulating material can shift slightly. As a result, a comb with either sign of charge attracts a neutral insulator. By Newton's third law the neutral insulator exerts an equal-magnitude attractive force on the comb.

(a) A charged comb picking up uncharged pieces of plastic

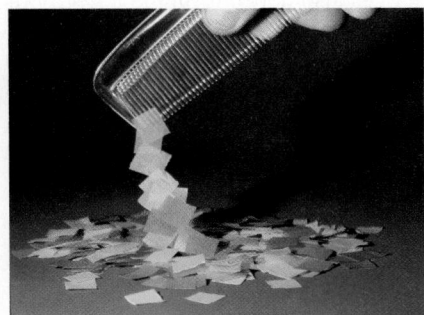

(b) How a negatively charged comb attracts an insulator

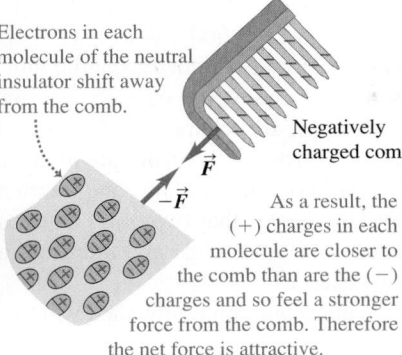

Electrons in each molecule of the neutral insulator shift away from the comb.

Negatively charged comb

$\vec{F}$

$-\vec{F}$

As a result, the (+) charges in each molecule are closer to the comb than are the (−) charges and so feel a stronger force from the comb. Therefore the net force is attractive.

(c) How a positively charged comb attracts an insulator

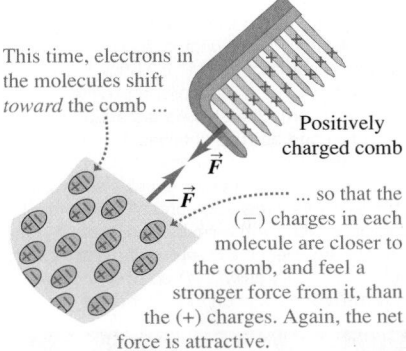

This time, electrons in the molecules shift *toward* the comb ...

Positively charged comb

$\vec{F}$

$-\vec{F}$

... so that the (−) charges in each molecule are closer to the comb, and feel a stronger force from it, than the (+) charges. Again, the net force is attractive.

Electric Forces on Uncharged Objects

Finally, we note that a charged body can exert forces even on objects that are *not* charged themselves. If you rub a balloon on the rug and then hold the balloon against the ceiling, it sticks, even though the ceiling has no net electric charge. After you electrify a comb by running it through your hair, you can pick up uncharged bits of paper or plastic with the comb (Fig. 21.8a). How is this possible?

This interaction is an induced-charge effect. Even in an insulator, electric charges can shift back and forth a little when there is charge nearby. This is shown in Fig. 21.8b; the negatively charged plastic comb causes a slight shifting of charge within the molecules of the neutral insulator, an effect called *polarization*. The positive and negative charges in the material are present in equal amounts, but the positive charges are closer to the plastic comb and so feel an attraction that is stronger than the repulsion felt by the negative charges, giving a net attractive force. (In Section 21.3 we will study how electric forces depend on distance.) Note that a neutral insulator is also attracted to a *positively* charged comb (Fig. 21.8c). Now the charges in the insulator shift in the opposite direction; the negative charges in the insulator are closer to the comb and feel an attractive force that is stronger than the repulsion felt by the positive charges in the insulator. Hence a charged object of *either* sign exerts an attractive force on an uncharged insulator. Figure 21.9 shows an industrial application of this effect.

Test Your Understanding of Section 21.2   You have two lightweight metal spheres, each hanging from an insulating nylon thread. One of the spheres has a net negative charge, while the other sphere has no net charge. (a) If the spheres are close together but do not touch, will they (i) attract each other, (ii) repel each other, or (iii) exert no force on each other? (b) You now allow the two spheres to touch. Once they have touched, will the two spheres (i) attract each other, (ii) repel each other, or (iii) exert no force on each other? ❙

21.3 Coulomb's Law

Charles Augustin de Coulomb (1736–1806) studied the interaction forces of charged particles in detail in 1784. He used a torsion balance (Fig. 21.10a) similar to the one used 13 years later by Cavendish to study the much weaker gravitational interaction, as we discussed in Section 13.1. For **point charges,** charged

21.9 The electrostatic painting process (compare Figs. 21.7b and 21.7c). A metal object to be painted is connected to the earth ("ground"), and the paint droplets are given an electric charge as they exit the sprayer nozzle. Induced charges of the opposite sign appear in the object as the droplets approach, just as in Fig. 21.7b, and they attract the droplets to the surface. This process minimizes overspray from clouds of stray paint particles and gives a particularly smooth finish.

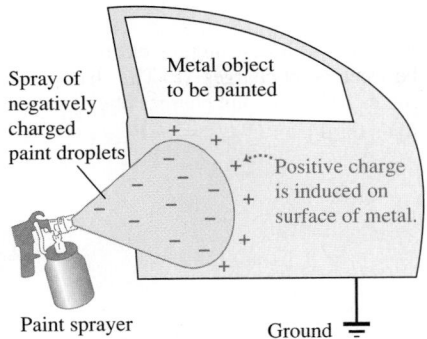

Spray of negatively charged paint droplets

Metal object to be painted

Positive charge is induced on surface of metal.

Paint sprayer

Ground

Mastering**PHYSICS**

ActivPhysics 11.1: Electric Force: Coulomb's Law
ActivPhysics 11.2: Electric Force: Superposition Principle
ActivPhysics 11.3: Electric Force: Superposition (Quantitative)

Application Electric Forces, Sweat, and Cystic Fibrosis
One way to test for the genetic disease cystic fibrosis (CF) is by measuring the salt content of a person's sweat. Sweat is a mixture of water and ions, including the sodium (Na^+) and chloride (Cl^-) ions that make up ordinary salt (NaCl). When sweat is secreted by epithelial cells, some of the Cl^- ions flow from the sweat back into these cells (a process called reabsorption). The electric attraction between negative and positive charges pulls Na^+ ions along with the Cl^-. Water molecules cannot flow back into the epithelial cells, so sweat on the skin has a low salt content. However, in persons with CF the reabsorption of Cl^- ions is blocked. Hence the sweat of persons with CF is unusually salty, with up to four times the normal concentration of Cl^- and Na^+.

bodies that are very small in comparison with the distance r between them, Coulomb found that the electric force is proportional to $1/r^2$. That is, when the distance r doubles, the force decreases to one-quarter of its initial value; when the distance is halved, the force increases to four times its initial value.

The electric force between two point charges also depends on the quantity of charge on each body, which we will denote by q or Q. To explore this dependence, Coulomb divided a charge into two equal parts by placing a small charged spherical conductor into contact with an identical but uncharged sphere; by symmetry, the charge is shared equally between the two spheres. (Note the essential role of the principle of conservation of charge in this procedure.) Thus he could obtain one-half, one-quarter, and so on, of any initial charge. He found that the forces that two point charges q_1 and q_2 exert on each other are proportional to each charge and therefore are proportional to the *product* q_1q_2 of the two charges.

Thus Coulomb established what we now call **Coulomb's law:**

> **The magnitude of the electric force between two point charges is directly proportional to the product of the charges and inversely proportional to the square of the distance between them.**

In mathematical terms, the magnitude F of the force that each of two point charges q_1 and q_2 a distance r apart exerts on the other can be expressed as

$$F = k\frac{|q_1q_2|}{r^2} \tag{21.1}$$

where k is a proportionality constant whose numerical value depends on the system of units used. The absolute value bars are used in Eq. (21.1) because the charges q_1 and q_2 can be either positive or negative, while the force magnitude F is always positive.

The directions of the forces the two charges exert on each other are always along the line joining them. When the charges q_1 and q_2 have the same sign, either both positive or both negative, the forces are repulsive; when the charges have opposite signs, the forces are attractive (Fig. 21.10b). The two forces obey Newton's third law; they are always equal in magnitude and opposite in direction, even when the charges are not equal in magnitude.

21.10 (a) Measuring the electric force between point charges. (b) The electric forces between point charges obey Newton's third law: $\vec{F}_{1\text{ on }2} = -\vec{F}_{2\text{ on }1}$.

(a) A torsion balance of the type used by Coulomb to measure the electric force

Torsion fiber

The negatively charged ball attracts the positively charged one; the positive ball moves until the elastic forces in the torsion fiber balance the electrostatic attraction.

Charged pith balls

Scale

(b) Interactions between point charges

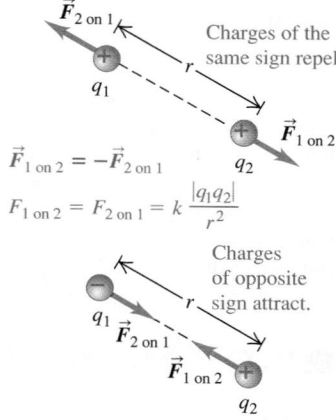

$\vec{F}_{2\text{ on }1}$

q_1

Charges of the same sign repel.

r

$\vec{F}_{1\text{ on }2}$

q_2

$\vec{F}_{1\text{ on }2} = -\vec{F}_{2\text{ on }1}$

$F_{1\text{ on }2} = F_{2\text{ on }1} = k\dfrac{|q_1q_2|}{r^2}$

q_1

$\vec{F}_{2\text{ on }1}$

Charges of opposite sign attract.

r

$\vec{F}_{1\text{ on }2}$

q_2

The proportionality of the electric force to $1/r^2$ has been verified with great precision. There is no reason to suspect that the exponent is different from precisely 2. Thus the form of Eq. (21.1) is the same as that of the law of gravitation. But electric and gravitational interactions are two distinct classes of phenomena. Electric interactions depend on electric charges and can be either attractive or repulsive, while gravitational interactions depend on mass and are always attractive (because there is no such thing as negative mass).

Fundamental Electric Constants

The value of the proportionality constant k in Coulomb's law depends on the system of units used. In our study of electricity and magnetism we will use SI units exclusively. The SI electric units include most of the familiar units such as the volt, the ampere, the ohm, and the watt. (There is *no* British system of electric units.) The SI unit of electric charge is called one **coulomb** (1 C). In SI units the constant k in Eq. (21.1) is

$$k = 8.987551787 \times 10^9 \text{ N} \cdot \text{m}^2/\text{C}^2 \cong 8.988 \times 10^9 \text{ N} \cdot \text{m}^2/\text{C}^2$$

The value of k is known to such a large number of significant figures because this value is closely related to the speed of light in vacuum. (We will show this in Chapter 32 when we study electromagnetic radiation.) As we discussed in Section 1.3, this speed is *defined* to be exactly $c = 2.99792458 \times 10^8$ m/s. The numerical value of k is defined in terms of c to be precisely

$$k = (10^{-7} \text{ N} \cdot \text{s}^2/\text{C}^2)c^2$$

You should check this expression to confirm that k has the right units.

In principle we can measure the electric force F between two equal charges q at a measured distance r and use Coulomb's law to determine the charge. Thus we could regard the value of k as an operational definition of the coulomb. For reasons of experimental precision it is better to define the coulomb instead in terms of a unit of electric *current* (charge per unit time), the *ampere,* equal to 1 coulomb per second. We will return to this definition in Chapter 28.

In SI units we usually write the constant k in Eq. (21.1) as $1/4\pi\epsilon_0$, where ϵ_0 ("epsilon-nought" or "epsilon-zero") is another constant. This appears to complicate matters, but it actually simplifies many formulas that we will encounter in later chapters. From now on, we will usually write Coulomb's law as

$$F = \frac{1}{4\pi\epsilon_0} \frac{|q_1 q_2|}{r^2} \qquad \text{(Coulomb's law: force between two point charges)} \qquad \text{(21.2)}$$

The constants in Eq. (21.2) are approximately

$$\epsilon_0 = 8.854 \times 10^{-12} \text{ C}^2/\text{N} \cdot \text{m}^2 \quad \text{and} \quad \frac{1}{4\pi\epsilon_0} = k = 8.988 \times 10^9 \text{ N} \cdot \text{m}^2/\text{C}^2$$

In examples and problems we will often use the approximate value

$$\frac{1}{4\pi\epsilon_0} = 9.0 \times 10^9 \text{ N} \cdot \text{m}^2/\text{C}^2$$

which is within about 0.1% of the correct value.

As we mentioned in Section 21.1, the most fundamental unit of charge is the magnitude of the charge of an electron or a proton, which is denoted by e. The most precise value available as of the writing of this book is

$$e = 1.602176487(40) \times 10^{-19} \text{ C}$$

One coulomb represents the negative of the total charge of about 6×10^{18} electrons. For comparison, a copper cube 1 cm on a side contains about 2.4×10^{24}

electrons. About 10^{19} electrons pass through the glowing filament of a flashlight bulb every second.

In electrostatics problems (that is, problems that involve charges at rest), it's very unusual to encounter charges as large as 1 coulomb. Two 1-C charges separated by 1 m would exert forces on each other of magnitude 9×10^9 N (about 1 million tons)! The total charge of all the electrons in a copper one-cent coin is even greater, about 1.4×10^5 C, which shows that we can't disturb electric neutrality very much without using enormous forces. More typical values of charge range from about 10^{-9} to about 10^{-6} C. The microcoulomb (1 μC $= 10^{-6}$ C) and the nanocoulomb (1 nC $= 10^{-9}$ C) are often used as practical units of charge.

Example 21.1   Electric force versus gravitational force

An α particle (the nucleus of a helium atom) has mass $m = 6.64 \times 10^{-27}$ kg and charge $q = +2e = 3.2 \times 10^{-19}$ C. Compare the magnitude of the electric repulsion between two α ("alpha") particles with that of the gravitational attraction between them.

SOLUTION

IDENTIFY and SET UP: This problem involves Newton's law for the gravitational force F_g between particles (see Section 13.1) and Coulomb's law for the electric force F_e between point charges. To compare these forces, we make our target variable the ratio F_e/F_g. We use Eq. (21.2) for F_e and Eq. (13.1) for F_g.

21.11 Our sketch for this problem.

EXECUTE: Figure 21.11 shows our sketch. From Eqs. (21.2) and (13.1),

$$F_e = \frac{1}{4\pi\epsilon_0}\frac{q^2}{r^2} \qquad F_g = G\frac{m^2}{r^2}$$

These are both inverse-square forces, so the r^2 factors cancel when we take the ratio:

$$\frac{F_e}{F_g} = \frac{1}{4\pi\epsilon_0 G}\frac{q^2}{m^2}$$

$$= \frac{9.0 \times 10^9 \text{ N} \cdot \text{m}^2/\text{C}^2}{6.67 \times 10^{-11} \text{ N} \cdot \text{m}^2/\text{kg}^2}\frac{(3.2 \times 10^{-19} \text{ C})^2}{(6.64 \times 10^{-27} \text{ kg})^2}$$

$$= 3.1 \times 10^{35}$$

EVALUATE: This astonishingly large number shows that the gravitational force in this situation is completely negligible in comparison to the electric force. This is always true for interactions of atomic and subnuclear particles. But within objects the size of a person or a planet, the positive and negative charges are nearly equal in magnitude, and the net electric force is usually much *smaller* than the gravitational force.

Superposition of Forces

Coulomb's law as we have stated it describes only the interaction of two *point* charges. Experiments show that when two charges exert forces simultaneously on a third charge, the total force acting on that charge is the *vector sum* of the forces that the two charges would exert individually. This important property, called the **principle of superposition of forces,** holds for any number of charges. By using this principle, we can apply Coulomb's law to *any* collection of charges. Two of the examples at the end of this section use the superposition principle.

Strictly speaking, Coulomb's law as we have stated it should be used only for point charges *in a vacuum.* If matter is present in the space between the charges, the net force acting on each charge is altered because charges are induced in the molecules of the intervening material. We will describe this effect later. As a practical matter, though, we can use Coulomb's law unaltered for point charges in air. At normal atmospheric pressure, the presence of air changes the electric force from its vacuum value by only about one part in 2000.

Problem-Solving Strategy 21.1 **Coulomb's Law**

IDENTIFY *the relevant concepts:* Coulomb's law describes the electric force between charged particles.

SET UP *the problem* using the following steps:

1. Sketch the locations of the charged particles and label each particle with its charge.
2. If the charges do not all lie on a single line, set up an *xy*-coordinate system.
3. The problem will ask you to find the electric force on one or more particles. Identify which these are.

EXECUTE *the solution* as follows:

1. For each particle that exerts an electric force on a given particle of interest, use Eq. (21.2) to calculate the magnitude of that force.
2. Using those magnitudes, sketch a free-body diagram showing the electric force vectors acting on each particle of interest. The force exerted by particle 1 on particle 2 points from particle 2 toward particle 1 if the charges have opposite signs, but points from particle 2 directly away from particle 1 if the charges have the same sign.
3. Use the principle of superposition to calculate the total electric force—a *vector* sum—on each particle of interest. (Review the

vector algebra in Sections 1.7 through 1.9. The method of components is often helpful.)

4. Use consistent units; SI units are completely consistent. With $1/4\pi\epsilon_0 = 9.0 \times 10^9 \ \mathrm{N \cdot m^2/C^2}$, distances must be in meters, charges in coulombs, and forces in newtons.
5. Some examples and problems in this and later chapters involve *continuous* distributions of charge along a line, over a surface, or throughout a volume. In these cases the vector sum in step 3 becomes a vector *integral*. We divide the charge distribution into infinitesimal pieces, use Coulomb's law for each piece, and integrate to find the vector sum. Sometimes this can be done without actual integration.
6. Exploit any symmetries in the charge distribution to simplify your problem solving. For example, two identical charges q exert zero net electric force on a charge Q midway between them, because the forces on Q have equal magnitude and opposite direction.

EVALUATE *your answer:* Check whether your numerical results are reasonable. Confirm that the direction of the net electric force agrees with the principle that charges of the same sign repel and charges of opposite sign attract.

Example 21.2 **Force between two point charges**

Two point charges, $q_1 = +25$ nC and $q_2 = -75$ nC, are separated by a distance $r = 3.0$ cm (Fig. 21.12a). Find the magnitude and direction of the electric force (a) that q_1 exerts on q_2 and (b) that q_2 exerts on q_1.

SOLUTION

IDENTIFY and SET UP: This problem asks for the electric forces that two charges exert on each other. We use Coulomb's law, Eq. (21.2), to calculate the magnitudes of the forces. The signs of the charges will determine the directions of the forces.

EXECUTE: (a) After converting the units of r to meters and the units of q_1 and q_2 to coulombs, Eq. (21.2) gives us

$$F_{1 \text{ on } 2} = \frac{1}{4\pi\epsilon_0} \frac{|q_1 q_2|}{r^2}$$

$$= (9.0 \times 10^9 \ \mathrm{N \cdot m^2/C^2}) \frac{|(+25 \times 10^{-9} \ \mathrm{C})(-75 \times 10^{-9} \ \mathrm{C})|}{(0.030 \ \mathrm{m})^2}$$

$$= 0.019 \ \mathrm{N}$$

The charges have opposite signs, so the force is attractive (to the left in Fig. 21.12b); that is, the force that acts on q_2 is directed toward q_1 along the line joining the two charges.

21.12 What force does q_1 exert on q_2, and what force does q_2 exert on q_1? Gravitational forces are negligible.

(a) The two charges   **(b)** Free-body diagram   **(c)** Free-body diagram
                              for charge q_2 for charge q_1

(b) Proceeding as in part (a), we have

$$F_{1 \text{ on } 2} = \frac{1}{4\pi\epsilon_0} \frac{|q_2 q_1|}{r^2} = F_{2 \text{ on } 1} = 0.019 \ \mathrm{N}$$

The attractive force that acts on q_1 is to the right, toward q_2 (Fig. 21.12c).

EVALUATE: Newton's third law applies to the electric force. Even though the charges have different magnitudes, the magnitude of the force that q_2 exerts on q_1 is the same as the magnitude of the force that q_1 exerts on q_2, and these two forces are in opposite directions.

Example 21.3 **Vector addition of electric forces on a line**

Two point charges are located on the *x*-axis of a coordinate system: $q_1 = 1.0$ nC is at $x = +2.0$ cm, and $q_2 = -3.0$ nC is at $x = +4.0$ cm. What is the total electric force exerted by q_1 and q_2 on a charge $q_3 = 5.0$ nC at $x = 0$?

SOLUTION

IDENTIFY and SET UP: Figure 21.13a shows the situation. To find the total force on q_3, our target variable, we find the vector sum of the two electric forces on it.

Continued

EXECUTE: Figure 21.13b is a free-body diagram for q_3, which is repelled by q_1 (which has the same sign) and attracted to q_2 (which has the opposite sign): $\vec{F}_{1\text{ on }3}$ is in the $-x$-direction and $\vec{F}_{2\text{ on }3}$ is in the $+x$-direction. After unit conversions, we have from Eq. (21.2)

$$F_{1\text{ on }3} = \frac{1}{4\pi\epsilon_0}\frac{|q_1 q_3|}{r_{13}^2}$$

$$= (9.0 \times 10^9\ \text{N}\cdot\text{m}^2/\text{C}^2)\frac{(1.0 \times 10^{-9}\ \text{C})(5.0 \times 10^{-9}\ \text{C})}{(0.020\ \text{m})^2}$$

$$= 1.12 \times 10^{-4}\ \text{N} = 112\ \mu\text{N}$$

In the same way you can show that $F_{2\text{ on }3} = 84\ \mu\text{N}$. We thus have $\vec{F}_{1\text{ on }3} = (-112\ \mu\text{N})\hat{\imath}$ and $\vec{F}_{2\text{ on }3} = (84\ \mu\text{N})\hat{\imath}$. The net force on q_3 is

$$\vec{F}_3 = \vec{F}_{1\text{ on }3} + \vec{F}_{2\text{ on }3} = (-112\ \mu\text{N})\hat{\imath} + (84\ \mu\text{N})\hat{\imath} = (-28\ \mu\text{N})\hat{\imath}$$

21.13 Our sketches for this problem.

(a) Our diagram of the situation   **(b)** Free-body diagram for q_3

EVALUATE: As a check, note that the magnitude of q_2 is three times that of q_1, but q_2 is twice as far from q_3 as q_1. Equation (21.2) then says that $F_{2\text{ on }3}$ must be $3/2^2 = 3/4 = 0.75$ as large as $F_{1\text{ on }3}$. This agrees with our calculated values: $F_{2\text{ on }3}/F_{1\text{ on }3} = (84\ \mu\text{N})/(112\ \mu\text{N}) = 0.75$. Because $F_{2\text{ on }3}$ is the weaker force, the direction of the net force is that of $\vec{F}_{1\text{ on }3}$—that is, in the negative x-direction.

Example 21.4   Vector addition of electric forces in a plane

Two equal positive charges $q_1 = q_2 = 2.0\ \mu\text{C}$ are located at $x = 0$, $y = 0.30$ m and $x = 0$, $y = -0.30$ m, respectively. What are the magnitude and direction of the total electric force that q_1 and q_2 exert on a third charge $Q = 4.0\ \mu\text{C}$ at $x = 0.40$ m, $y = 0$?

SOLUTION

IDENTIFY and SET UP: As in Example 21.3, we must compute the force that each charge exerts on Q and then find the vector sum of those forces. Figure 21.14 shows the situation. Since the three charges do not all lie on a line, the best way to calculate the forces is to use components.

21.14 Our sketch for this problem.

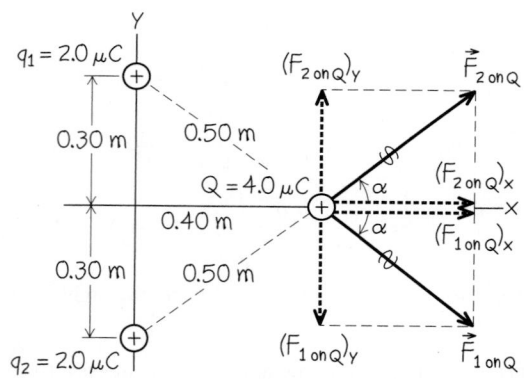

EXECUTE: Figure 21.14 shows the forces $\vec{F}_{1\text{ on }Q}$ and $\vec{F}_{2\text{ on }Q}$ due to the identical charges q_1 and q_2, which are at equal distances from Q. From Coulomb's law, *both* forces have magnitude

$$F_{1\text{ or }2\text{ on }Q} = (9.0 \times 10^9\ \text{N}\cdot\text{m}^2/\text{C}^2)$$

$$\times\frac{(4.0 \times 10^{-6}\ \text{C})(2.0 \times 10^{-6}\ \text{C})}{(0.50\ \text{m})^2} = 0.29\ \text{N}$$

The x-components of the two forces are equal:

$$(F_{1\text{ or }2\text{ on }Q})_x = (F_{1\text{ or }2\text{ on }Q})\cos\alpha = (0.29\ \text{N})\frac{0.40\ \text{m}}{0.50\ \text{m}} = 0.23\ \text{N}$$

From symmetry we see that the y-components of the two forces are equal and opposite. Hence their sum is zero and the total force $\vec{F}$ on Q has only an x-component $F_x = 0.23\ \text{N} + 0.23\ \text{N} = 0.46\ \text{N}$. The total force on Q is in the $+x$-direction, with magnitude 0.46 N.

EVALUATE: The total force on Q points neither directly away from q_1 nor directly away from q_2. Rather, this direction is a compromise that points away from the *system* of charges q_1 and q_2. Can you see that the total force would *not* be in the $+x$-direction if q_1 and q_2 were not equal or if the geometrical arrangement of the changes were not so symmetric?

Test Your Understanding of Section 21.3 Suppose that charge q_2 in Example 21.4 were $-2.0\ \mu\text{C}$. In this case, the total electric force on Q would be (i) in the positive x-direction; (ii) in the negative x-direction; (iii) in the positive y-direction; (iv) in the negative y-direction; (v) zero; (vi) none of these.

21.4 Electric Field and Electric Forces

When two electrically charged particles in empty space interact, how does each one know the other is there? We can begin to answer this question, and at the same time reformulate Coulomb's law in a very useful way, by using the concept of *electric field*.

Electric Field

To introduce this concept, let's look at the mutual repulsion of two positively charged bodies A and B (Fig. 21.15a). Suppose B has charge q_0, and let $\vec{F}_0$ be the electric force of A on B. One way to think about this force is as an "action-at-a-distance" force—that is, as a force that acts across empty space without needing any matter (such as a push rod or a rope) to transmit it through the intervening space. (Gravity can also be thought of as an "action-at-a-distance" force.) But a more fruitful way to visualize the repulsion between A and B is as a two-stage process. We first envision that body A, as a result of the charge that it carries, somehow *modifies the properties of the space around it*. Then body B, as a result of the charge that *it* carries, senses how space has been modified at its position. The response of body B is to experience the force $\vec{F}_0$.

To elaborate how this two-stage process occurs, we first consider body A by itself: We remove body B and label its former position as point P (Fig. 21.15b). We say that the charged body A produces or causes an **electric field** at point P (and at all other points in the neighborhood). This electric field is present at P even if there is no charge at P; it is a consequence of the charge on body A only. If a point charge q_0 is then placed at point P, it experiences the force $\vec{F}_0$. We take the point of view that this force is exerted on q_0 *by the field* at P (Fig. 21.15c). Thus the electric field is the intermediary through which A communicates its presence to q_0. Because the point charge q_0 would experience a force at *any* point in the neighborhood of A, the electric field that A produces exists at all points in the region around A.

We can likewise say that the point charge q_0 produces an electric field in the space around it and that this electric field exerts the force $-\vec{F}_0$ on body A. For each force (the force of A on q_0 and the force of q_0 on A), one charge sets up an electric field that exerts a force on the second charge. We emphasize that this is an *interaction* between *two* charged bodies. A single charge produces an electric field in the surrounding space, but this electric field cannot exert a net force on the charge that created it; as we discussed in Section 4.3, a body cannot exert a net force on itself. (If this wasn't true, you would be able to lift yourself to the ceiling by pulling up on your belt!)

> **The electric force on a charged body is exerted by the electric field created by *other* charged bodies.**

To find out experimentally whether there is an electric field at a particular point, we place a small charged body, which we call a **test charge,** at the point (Fig. 21.15c). If the test charge experiences an electric force, then there is an electric field at that point. This field is produced by charges other than q_0.

Force is a vector quantity, so electric field is also a vector quantity. (Note the use of vector signs as well as boldface letters and plus, minus, and equals signs in the following discussion.) We define the *electric field* $\vec{E}$ at a point as the electric force $\vec{F}_0$ experienced by a test charge q_0 at the point, divided by the charge q_0. That is, the electric field at a certain point is equal to the *electric force per unit charge* experienced by a charge at that point:

$$\vec{E} = \frac{\vec{F}_0}{q_0} \quad \begin{array}{l}\text{(definition of electric field as electric} \\ \text{force per unit charge)}\end{array} \qquad (21.3)$$

In SI units, in which the unit of force is 1 N and the unit of charge is 1 C, the unit of electric field magnitude is 1 newton per coulomb (1 N/C).

If the field $\vec{E}$ at a certain point is known, rearranging Eq. (21.3) gives the force $\vec{F}_0$ experienced by a point charge q_0 placed at that point. This force is just equal to the electric field $\vec{E}$ produced at that point by charges other than q_0, multiplied by the charge q_0:

$$\vec{F}_0 = q_0\vec{E} \quad \begin{array}{l}\text{(force exerted on a point charge } q_0 \\ \text{by an electric field } \vec{E}\text{)}\end{array} \qquad (21.4)$$

21.15 A charged body creates an electric field in the space around it.

(a) *A and B exert electric forces on each other.*

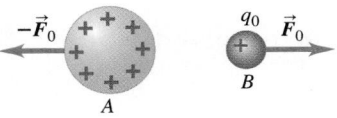

(b) Remove body B ...

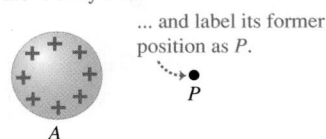

... and label its former position as P.

(c) Body A sets up an electric field $\vec{E}$ at point P.

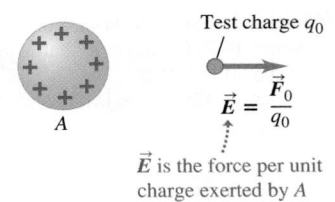

Test charge q_0

$$\vec{E} = \frac{\vec{F}_0}{q_0}$$

$\vec{E}$ is the force per unit charge exerted by A on a test charge at P.

Application Sharks and the "Sixth Sense"

Sharks have the ability to locate prey (such as flounder and other bottom-dwelling fish) that are completely hidden beneath the sand at the bottom of the ocean. They do this by sensing the weak electric fields produced by muscle contractions in their prey. Sharks derive their sensitivity to electric fields (a "sixth sense") from jelly-filled canals in their bodies. These canals end in pores on the shark's skin (shown in this photograph). An electric field as weak as 5×10^{-7} N/C causes charge flow within the canals and triggers a signal in the shark's nervous system. Because the shark has canals with different orientations, it can measure different components of the electric-field vector and hence determine the direction of the field.

21.16 The force $\vec{F}_0 = q_0\vec{E}$ exerted on a point charge q_0 placed in an electric field $\vec{E}$.

The force on a positive test charge q_0 points in the direction of the electric field.

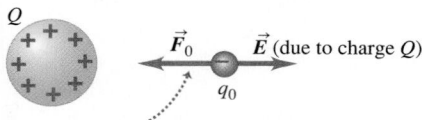

The force on a negative test charge q_0 points opposite to the electric field.

Mastering**PHYSICS**

ActivPhysics 11.4: Electric Field: Point Charge
ActivPhysics 11.9: Motion of a Charge in an Electric Field: Introduction
ActivPhysics 11.10: Motion in an Electric Field: Problems

The charge q_0 can be either positive or negative. If q_0 is *positive*, the force $\vec{F}_0$ experienced by the charge is the same direction as $\vec{E}$; if q_0 is *negative*, $\vec{F}_0$ and $\vec{E}$ are in opposite directions (Fig. 21.16).

While the electric field concept may be new to you, the basic idea—that one body sets up a field in the space around it and a second body responds to that field—is one that you've actually used before. Compare Eq. (21.4) to the familiar expression for the gravitational force $\vec{F}_g$ that the earth exerts on a mass m_0:

$$\vec{F}_g = m_0\vec{g} \tag{21.5}$$

In this expression, $\vec{g}$ is the acceleration due to gravity. If we divide both sides of Eq. (21.5) by the mass m_0, we obtain

$$\vec{g} = \frac{\vec{F}_g}{m_0}$$

Thus $\vec{g}$ can be regarded as the gravitational force per unit mass. By analogy to Eq. (21.3), we can interpret $\vec{g}$ as the *gravitational field*. Thus we treat the gravitational interaction between the earth and the mass m_0 as a two-stage process: The earth sets up a gravitational field $\vec{g}$ in the space around it, and this gravitational field exerts a force given by Eq. (21.5) on the mass m_0 (which we can regard as a *test mass*). The gravitational field $\vec{g}$, or gravitational force per unit mass, is a useful concept because it does not depend on the mass of the body on which the gravitational force is exerted; likewise, the electric field $\vec{E}$, or electric force per unit charge, is useful because it does not depend on the charge of the body on which the electric force is exerted.

> **CAUTION** $\vec{F}_0 = q_0\vec{E}_0$ **is for *point* test charges only** The electric force experienced by a test charge q_0 can vary from point to point, so the electric field can also be different at different points. For this reason, Eq. (21.4) can be used only to find the electric force on a *point* charge. If a charged body is large enough in size, the electric field $\vec{E}$ may be noticeably different in magnitude and direction at different points on the body, and calculating the net electric force on the body can become rather complicated. ▮

Electric Field of a Point Charge

If the source distribution is a point charge q, it is easy to find the electric field that it produces. We call the location of the charge the **source point,** and we call the point P where we are determining the field the **field point.** It is also useful to introduce a *unit vector* $\hat{r}$ that points along the line from source point to field point (Fig. 21.17a). This unit vector is equal to the displacement vector $\vec{r}$ from the source point to the field point, divided by the distance $r = |\vec{r}|$ between these two points; that is, $\hat{r} = \vec{r}/r$. If we place a small test charge q_0 at the field point P, at a

21.17 The electric field $\vec{E}$ produced at point P by an isolated point charge q at S. Note that in both (b) and (c), $\vec{E}$ is *produced* by q [see Eq. (21.7)] but *acts* on the charge q_0 at point P [see Eq. (21.4)].

(a)

Unit vector $\hat{r}$ points from source point S to field point P.

(b)

At each point P, the electric field set up by an isolated *positive* point charge q points directly *away* from the charge in the *same* direction as $\hat{r}$.

(c)

At each point P, the electric field set up by an isolated *negative* point charge q points directly *toward* the charge in the *opposite* direction from $\hat{r}$.

distance r from the source point, the magnitude F_0 of the force is given by Coulomb's law, Eq. (21.2):

$$F_0 = \frac{1}{4\pi\epsilon_0} \frac{|qq_0|}{r^2}$$

From Eq. (21.3) the magnitude E of the electric field at P is

$$E = \frac{1}{4\pi\epsilon_0} \frac{|q|}{r^2} \qquad \text{(magnitude of electric field of a point charge)} \qquad (21.6)$$

Using the unit vector $\hat{r}$, we can write a *vector* equation that gives both the magnitude and direction of the electric field $\vec{E}$:

$$\vec{E} = \frac{1}{4\pi\epsilon_0} \frac{q}{r^2} \hat{r} \qquad \text{(electric field of a point charge)} \qquad (21.7)$$

By definition, the electric field of a point charge always points *away from* a positive charge (that is, in the same direction as $\hat{r}$; see Fig. 21.17b) but *toward* a negative charge (that is, in the direction opposite $\hat{r}$; see Fig. 21.17c).

We have emphasized calculating the electric field $\vec{E}$ at a certain point. But since $\vec{E}$ can vary from point to point, it is not a single vector quantity but rather an *infinite* set of vector quantities, one associated with each point in space. This is an example of a **vector field.** Figure 21.18 shows a number of the field vectors produced by a positive or negative point charge. If we use a rectangular (x, y, z) coordinate system, each component of $\vec{E}$ at any point is in general a function of the coordinates (x, y, z) of the point. We can represent the functions as $E_x(x, y, z)$, $E_y(x, y, z)$, and $E_z(x, y, z)$. Vector fields are an important part of the language of physics, not just in electricity and magnetism. One everyday example of a vector field is the velocity $\vec{v}$ of wind currents; the magnitude and direction of $\vec{v}$, and hence its vector components, vary from point to point in the atmosphere.

In some situations the magnitude and direction of the field (and hence its vector components) have the same values everywhere throughout a certain region; we then say that the field is *uniform* in this region. An important example of this is the electric field inside a *conductor.* If there is an electric field within a conductor, the field exerts a force on every charge in the conductor, giving the free charges a net motion. By definition an electrostatic situation is one in which the charges have *no* net motion. We conclude that *in electrostatics the electric field at every point within the material of a conductor must be zero.* (Note that we are not saying that the field is necessarily zero in a *hole* inside a conductor.)

In summary, our description of electric interactions has two parts. First, a given charge distribution acts as a source of electric field. Second, the electric field exerts a force on any charge that is present in the field. Our analysis often has two corresponding steps: first, calculating the field caused by a source charge distribution; second, looking at the effect of the field in terms of force and motion. The second step often involves Newton's laws as well as the principles of electric interactions. In the next section we show how to calculate fields caused by various source distributions, but first here are three examples of calculating the field due to a point charge and of finding the force on a charge due to a given field $\vec{E}$.

21.18 A point charge q produces an electric field $\vec{E}$ at *all* points in space. The field strength decreases with increasing distance.

(a) The field produced by a positive point charge points *away from* the charge.

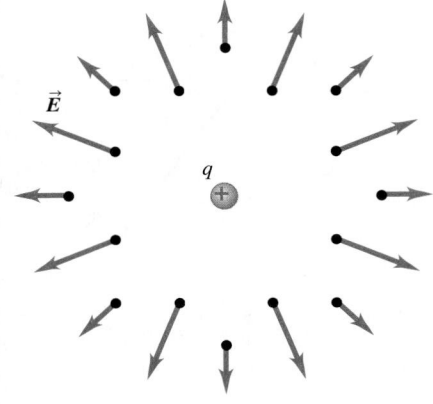

(b) The field produced by a negative point charge points *toward* the charge.

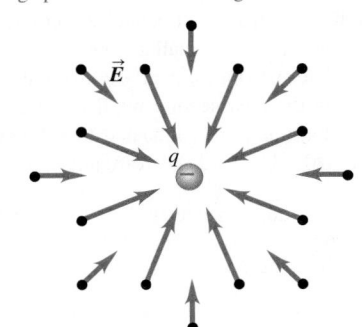

Example 21.5 | **Electric-field magnitude for a point charge**

What is the magnitude of the electric field $\vec{E}$ at a field point 2.0 m from a point charge $q = 4.0$ nC?

SOLUTION

IDENTIFY and SET UP: This problem concerns the electric field due to a point charge. We are given the magnitude of the charge

Continued

and the distance from the charge to the field point, so we use Eq. (21.6) to calculate the field magnitude E.

EXECUTE: From Eq. (21.6),

$$E = \frac{1}{4\pi\epsilon_0}\frac{|q|}{r^2} = (9.0 \times 10^9 \text{ N} \cdot \text{m}^2/\text{C}^2)\frac{4.0 \times 10^{-9} \text{ C}}{(2.0 \text{ m})^2}$$

$$= 9.0 \text{ N/C}$$

EVALUATE: Our result $E = 9.0$ N/C means that if we placed a 1.0-C charge at a point 2.0 m from q, it would experience a 9.0-N force. The force on a 2.0-C charge at that point would be $(2.0 \text{ C})(9.0 \text{ N/C}) = 18$ N, and so on.

Example 21.6 Electric-field vector for a point charge

A point charge $q = -8.0$ nC is located at the origin. Find the electric-field vector at the field point $x = 1.2$ m, $y = -1.6$ m.

SOLUTION

IDENTIFY and SET UP: We must find the electric-field *vector* $\vec{E}$ due to a point charge. Figure 21.19 shows the situation. We use Eq. (21.7); to do this, we must find the distance r from the source point S (the position of the charge q, which in this example is at the origin O) to the field point P, and we must obtain an expression for the unit vector $\hat{r} = \vec{r}/r$ that points from S to P.

EXECUTE: The distance from S to P is

$$r = \sqrt{x^2 + y^2} = \sqrt{(1.2 \text{ m})^2 + (-1.6 \text{ m})^2} = 2.0 \text{ m}$$

The unit vector $\hat{r}$ is then

$$\hat{r} = \frac{\vec{r}}{r} = \frac{x\hat{i} + y\hat{j}}{r}$$

$$= \frac{(1.2 \text{ m})\hat{i} + (-1.6 \text{ m})\hat{j}}{2.0 \text{ m}} = 0.60\hat{i} - 0.80\hat{j}$$

Then, from Eq. (21.7),

$$\vec{E} = \frac{1}{4\pi\epsilon_0}\frac{q}{r^2}\hat{r}$$

$$= (9.0 \times 10^9 \text{ N} \cdot \text{m}^2/\text{C}^2)\frac{(-8.0 \times 10^{-9} \text{ C})}{(2.0 \text{ m})^2}(0.60\hat{i} - 0.80\hat{j})$$

$$= (-11 \text{ N/C})\hat{i} + (14 \text{ N/C})\hat{j}$$

EVALUATE: Since q is negative, $\vec{E}$ points from the field point to the charge (the source point), in the direction opposite to $\hat{r}$ (compare Fig. 21.17c). We leave the calculation of the magnitude and direction of $\vec{E}$ to you (see Exercise 21.36).

21.19 Our sketch for this problem.

Example 21.7 Electron in a uniform field

When the terminals of a battery are connected to two parallel conducting plates with a small gap between them, the resulting charges on the plates produce a nearly uniform electric field $\vec{E}$ between the plates. (In the next section we'll see why this is.) If the plates are 1.0 cm apart and are connected to a 100-volt battery as shown in Fig. 21.20, the field is vertically upward and has magnitude $E = 1.00 \times 10^4$ N/C. (a) If an electron (charge $-e = -1.60 \times 10^{-19}$ C, mass $m = 9.11 \times 10^{-31}$ kg) is released from rest at the upper plate, what is its acceleration? (b) What speed and kinetic energy does it acquire while traveling 1.0 cm to the lower plate? (c) How long does it take to travel this distance?

SOLUTION

IDENTIFY and SET UP: This example involves the relationship between electric field and electric force. It also involves the relationship between force and acceleration, the definition of kinetic energy, and the kinematic relationships among acceleration, distance, velocity, and time. Figure 21.20 shows our coordinate system. We are given the electric field, so we use Eq. (21.4) to find the force on the electron and Newton's second law to find its acceleration. Because the field is uniform, the force is constant and we can use the constant-acceleration formulas from Chapter 2 to find the electron's velocity and travel time. We find the kinetic energy using $K = \frac{1}{2}mv^2$.

21.20 A uniform electric field between two parallel conducting plates connected to a 100-volt battery. (The separation of the plates is exaggerated in this figure relative to the dimensions of the plates.)

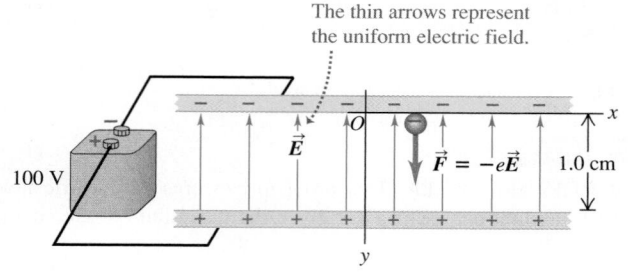

The thin arrows represent the uniform electric field.

EXECUTE: (a) Although $\vec{E}$ is upward (in the $+y$-direction), $\vec{F}$ is downward (because the electron's charge is negative) and so F_y is negative. Because F_y is constant, the electron's acceleration is constant:

$$a_y = \frac{F_y}{m} = \frac{-eE}{m} = \frac{(-1.60 \times 10^{-19} \text{ C})(1.00 \times 10^4 \text{ N/C})}{9.11 \times 10^{-31} \text{ kg}}$$

$$= -1.76 \times 10^{15} \text{ m/s}^2$$

(b) The electron starts from rest, so its motion is in the y-direction only (the direction of the acceleration). We can find the electron's speed at any position y using the constant-acceleration Eq. (2.13), $v_y^2 = v_{0y}^2 + 2a_y(y - y_0)$. We have $v_{0y} = 0$ and $y_0 = 0$, so at $y = -1.0$ cm $= -1.0 \times 10^{-2}$ m we have

$$|v_y| = \sqrt{2a_y y} = \sqrt{2(-1.76 \times 10^{15} \text{ m/s}^2)(-1.0 \times 10^{-2} \text{ m})}$$

$$= 5.9 \times 10^6 \text{ m/s}$$

The velocity is downward, so $v_y = -5.9 \times 10^6$ m/s. The electron's kinetic energy is

$$K = \tfrac{1}{2}mv^2 = \tfrac{1}{2}(9.11 \times 10^{-31} \text{ kg})(5.9 \times 10^6 \text{ m/s})^2$$

$$= 1.6 \times 10^{-17} \text{ J}$$

(c) From Eq. (2.8) for constant acceleration, $v_y = v_{0y} + a_y t$,

$$t = \frac{v_y - v_{0y}}{a_y} = \frac{(-5.9 \times 10^6 \text{ m/s}) - (0 \text{ m/s})}{-1.76 \times 10^{15} \text{ m/s}^2}$$

$$= 3.4 \times 10^{-9} \text{ s}$$

EVALUATE: Our results show that in problems concerning subatomic particles such as electrons, many quantities—including acceleration, speed, kinetic energy, and time—will have *very* different values from those typical of everyday objects such as baseballs and automobiles.

Test Your Understanding of Section 21.4 (a) A negative point charge moves along a straight-line path directly toward a stationary positive point charge. Which aspect(s) of the electric force on the negative point charge will remain constant as it moves? (i) magnitude; (ii) direction; (iii) both magnitude and direction; (iv) neither magnitude nor direction. (b) A negative point charge moves along a circular orbit around a positive point charge. Which aspect(s) of the electric force on the negative point charge will remain constant as it moves? (i) magnitude; (ii) direction; (iii) both magnitude and direction; (iv) neither magnitude nor direction.

21.5 Electric-Field Calculations

Equation (21.7) gives the electric field caused by a single point charge. But in most realistic situations that involve electric fields and forces, we encounter charge that is *distributed* over space. The charged plastic and glass rods in Fig. 21.1 have electric charge distributed over their surfaces, as does the imaging drum of a laser printer (Fig. 21.2). In this section we'll learn to calculate electric fields caused by various distributions of electric charge. Calculations of this kind are of tremendous importance for technological applications of electric forces. To determine the trajectories of atomic nuclei in an accelerator for cancer radiotherapy or of charged particles in a semiconductor electronic device, you have to know the detailed nature of the electric field acting on the charges.

The Superposition of Electric Fields

To find the field caused by a charge distribution, we imagine the distribution to be made up of many point charges $q_1, q_2, q_3, \ldots$. (This is actually quite a realistic description, since we have seen that charge is carried by electrons and protons that are so small as to be almost pointlike.) At any given point P, each point charge produces its own electric field $\vec{E}_1, \vec{E}_2, \vec{E}_3, \ldots$, so a test charge q_0 placed at P experiences a force $\vec{F}_1 = q_0\vec{E}_1$ from charge q_1, a force $\vec{F}_2 = q_0\vec{E}_2$ from charge q_2, and so on. From the principle of superposition of forces discussed in Section 21.3, the *total* force $\vec{F}_0$ that the charge distribution exerts on q_0 is the vector sum of these individual forces:

$$\vec{F}_0 = \vec{F}_1 + \vec{F}_2 + \vec{F}_3 + \cdots = q_0\vec{E}_1 + q_0\vec{E}_2 + q_0\vec{E}_3 + \cdots$$

The combined effect of all the charges in the distribution is described by the *total* electric field $\vec{E}$ at point P. From the definition of electric field, Eq. (21.3), this is

$$\vec{E} = \frac{\vec{F}_0}{q_0} = \vec{E}_1 + \vec{E}_2 + \vec{E}_3 + \cdots$$

Mastering**PHYSICS**

ActivPhysics 11.5: Electric Field Due to a Dipole
ActivPhysics 11.6: Electric Field: Problems

21.21 Illustrating the principle of superposition of electric fields.

The total electric field $\vec{E}$ at point P is the vector sum of $\vec{E}_1$ and $\vec{E}_2$.

The total electric field at P is the vector sum of the fields at P due to each point charge in the charge distribution (Fig. 21.21). This is the **principle of superposition of electric fields.**

When charge is distributed along a line, over a surface, or through a volume, a few additional terms are useful. For a line charge distribution (such as a long, thin, charged plastic rod), we use λ (the Greek letter lambda) to represent the **linear charge density** (charge per unit length, measured in C/m). When charge is distributed over a surface (such as the surface of the imaging drum of a laser printer), we use σ (sigma) to represent the **surface charge density** (charge per unit area, measured in C/m²). And when charge is distributed through a volume, we use ρ (rho) to represent the **volume charge density** (charge per unit volume, C/m³).

Some of the calculations in the following examples may look fairly intricate. After you've worked through the examples one step at a time, the process will seem less formidable. We will use many of the calculational techniques in these examples in Chapter 28 to calculate the *magnetic* fields caused by charges in motion.

Problem-Solving Strategy 21.2 | **Electric-Field Calculations**

IDENTIFY *the relevant concepts:* Use the principle of superposition to calculate the electric field due to a discrete or continuous charge distribution.

SET UP *the problem* using the following steps:
1. Make a drawing showing the locations of the charges and your choice of coordinate axes.
2. On your drawing, indicate the position of the *field point P* (the point at which you want to calculate the electric field $\vec{E}$).

EXECUTE *the solution* as follows:
1. Use consistent units. Distances must be in meters and charge must be in coulombs. If you are given centimeters or nanocoulombs, don't forget to convert.
2. Distinguish between the source point S and the field point P. The field produced by a point charge always points from S to P if the charge is positive, and from P to S if the charge is negative.
3. Use *vector* addition when applying the principle of superposition; review the treatment of vector addition in Chapter 1 if necessary.

4. Simplify your calculations by exploiting any symmetries in the charge distribution.
5. If the charge distribution is continuous, define a small element of charge that can be considered as a point, find its electric field at P, and find a way to add the fields of all the charge elements by doing an integral. Usually it is easiest to do this for each component of $\vec{E}$ separately, so you may need to evaluate more than one integral. Ensure that the limits on your integrals are correct; especially when the situation has symmetry, don't count a charge twice.

EVALUATE *your answer:* Check that the direction of $\vec{E}$ is reasonable. If your result for the electric-field magnitude E is a function of position (say, the coordinate x), check your result in any limits for which you know what the magnitude should be. When possible, check your answer by calculating it in a different way.

Example 21.8 | **Field of an electric dipole**

Point charges $q_1 = +12$ nC and $q_2 = -12$ nC are 0.100 m apart (Fig. 21.22). (Such pairs of point charges with equal magnitude and opposite sign are called *electric dipoles.*) Compute the electric field caused by q_1, the field caused by q_2, and the total field (a) at point a; (b) at point b; and (c) at point c.

SOLUTION

IDENTIFY and SET UP: We must find the total electric field at various points due to two point charges. We use the principle of superposition: $\vec{E} = \vec{E}_1 + \vec{E}_2$. Figure 21.22 shows the coordinate system and the locations of the field points a, b, and c.

EXECUTE: At each field point, $\vec{E}$ depends on $\vec{E}_1$ and $\vec{E}_2$ there; we first calculate the magnitudes E_1 and E_2 at each field point. At a the magnitude of the field $\vec{E}_{1a}$ caused by q_1 is

$$E_{1a} = \frac{1}{4\pi\epsilon_0}\frac{|q_1|}{r^2} = (9.0 \times 10^9 \text{ N}\cdot\text{m}^2/\text{C}^2)\frac{12 \times 10^{-9} \text{ C}}{(0.060 \text{ m})^2}$$

$$= 3.0 \times 10^4 \text{ N/C}$$

We calculate the other field magnitudes in a similar way. The results are

$$E_{1a} = 3.0 \times 10^4 \text{ N/C} \qquad E_{1b} = 6.8 \times 10^4 \text{ N/C}$$
$$E_{1c} = 6.39 \times 10^3 \text{ N/C}$$

$$E_{2a} = 6.8 \times 10^4 \text{ N/C} \qquad E_{2b} = 0.55 \times 10^4 \text{ N/C}$$
$$E_{2c} = E_{1c} = 6.39 \times 10^3 \text{ N/C}$$

The *directions* of the corresponding fields are in all cases *away* from the positive charge q_1 and *toward* the negative charge q_2.

21.22 Electric field at three points, a, b, and c, set up by charges q_1 and q_2, which form an electric dipole.

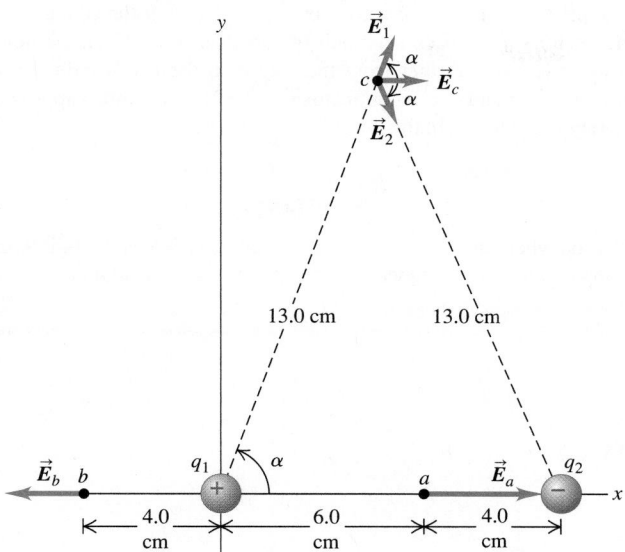

(a) At a, $\vec{E}_{1a}$ and $\vec{E}_{2a}$ are both directed to the right, so

$$\vec{E}_a = E_{1a}\hat{\imath} + E_{2a}\hat{\imath} = (9.8 \times 10^4 \text{ N/C})\hat{\imath}$$

(b) At b, $\vec{E}_{1b}$ is directed to the left and $\vec{E}_{2b}$ is directed to the right, so

$$\vec{E}_b = -E_{1b}\hat{\imath} + E_{2b}\hat{\imath} = (-6.2 \times 10^4 \text{ N/C})\hat{\imath}$$

(c) Figure 21.22 shows the directions of $\vec{E}_1$ and $\vec{E}_2$ at c. Both vectors have the same x-component:

$$E_{1cx} = E_{2cx} = E_{1c}\cos\alpha = (6.39 \times 10^3 \text{ N/C})\left(\tfrac{5}{13}\right)$$
$$= 2.46 \times 10^3 \text{ N/C}$$

From symmetry, E_{1y} and E_{2y} are equal and opposite, so their sum is zero. Hence

$$\vec{E}_c = 2(2.46 \times 10^3 \text{ N/C})\hat{\imath} = (4.9 \times 10^3 \text{ N/C})\hat{\imath}$$

EVALUATE: We can also find $\vec{E}_c$ using Eq. (21.7) for the field of a point charge. The displacement vector $\vec{r}_1$ from q_1 to point c is $\vec{r}_1 = r\cos\alpha\,\hat{\imath} + r\sin\alpha\,\hat{\jmath}$. Hence the unit vector that points from q_1 to point c is $\hat{r}_1 = \vec{r}_1/r = \cos\alpha\,\hat{\imath} + \sin\alpha\,\hat{\jmath}$. By symmetry, the unit vector that points from q_2 to point c has the opposite x-component but the same y-component: $\hat{r}_2 = -\cos\alpha\,\hat{\imath} + \sin\alpha\,\hat{\jmath}$. We can now use Eq. (21.7) to write the fields $\vec{E}_{1c}$ and $\vec{E}_{2c}$ at c in vector form, then find their sum. Since $q_2 = -q_1$ and the distance r to c is the same for both charges,

$$\vec{E}_c = \vec{E}_{1c} + \vec{E}_{2c} = \frac{1}{4\pi\epsilon_0}\frac{q_1}{r^2}\hat{r}_1 + \frac{1}{4\pi\epsilon_0}\frac{q_2}{r^2}\hat{r}_2$$

$$= \frac{1}{4\pi\epsilon_0 r^2}(q_1\hat{r}_1 + q_2\hat{r}_2) = \frac{q_1}{4\pi\epsilon_0 r^2}(\hat{r}_1 - \hat{r}_2)$$

$$= \frac{1}{4\pi\epsilon_0}\frac{q_1}{r^2}(2\cos\alpha\,\hat{\imath})$$

$$= 2(9.0 \times 10^9 \text{ N} \cdot \text{m}^2/\text{C}^2)\frac{12 \times 10^{-9} \text{ C}}{(0.13 \text{ m})^2}\left(\tfrac{5}{13}\right)\hat{\imath}$$

$$= (4.9 \times 10^3 \text{ N/C})\hat{\imath}$$

This is the same as we calculated in part (c).

Example 21.9   **Field of a ring of charge**

Charge Q is uniformly distributed around a conducting ring of radius a (Fig. 21.23). Find the electric field at a point P on the ring axis at a distance x from its center.

SOLUTION

IDENTIFY and SET UP: This is a problem in the superposition of electric fields. Each bit of charge around the ring produces an electric field at an arbitrary point on the x-axis; our target variable is the total field at this point due to all such bits of charge.

21.23 Calculating the electric field on the axis of a ring of charge. In this figure, the charge is assumed to be positive.

EXECUTE: We divide the ring into infinitesimal segments ds as shown in Fig. 21.23. In terms of the linear charge density $\lambda = Q/2\pi a$, the charge in a segment of length ds is $dQ = \lambda\,ds$. Consider two identical segments, one as shown in the figure at $y = a$ and another halfway around the ring at $y = -a$. From Example 21.4, we see that the net force $d\vec{F}$ they exert on a point test charge at P, and thus their net field $d\vec{E}$, are directed along the x-axis. The same is true for any such pair of segments around the ring, so the *net* field at P is along the x-axis: $\vec{E} = E_x\hat{\imath}$.

To calculate E_x, note that the square of the distance r from a single ring segment to the point P is $r^2 = x^2 + a^2$. Hence the magnitude of this segment's contribution $d\vec{E}$ to the electric field at P is

$$dE = \frac{1}{4\pi\epsilon_0}\frac{dQ}{x^2 + a^2}$$

The x-component of this field is $dE_x = dE\cos\alpha$. We know $dQ = \lambda\,ds$ and Fig. 21.23 shows that $\cos\alpha = x/r = x/(x^2 + a^2)^{1/2}$, so

$$dE_x = dE\cos\alpha = \frac{1}{4\pi\epsilon_0}\frac{dQ}{x^2 + a^2}\frac{x}{\sqrt{x^2 + a^2}}$$

$$= \frac{1}{4\pi\epsilon_0}\frac{\lambda x}{(x^2 + a^2)^{3/2}}ds$$

Continued

To find E_x we integrate this expression over the entire ring—that is, for s from 0 to $2\pi a$ (the circumference of the ring). The integrand has the same value for all points on the ring, so it can be taken outside the integral. Hence we get

$$E_x = \int dE_x = \frac{1}{4\pi\epsilon_0} \frac{\lambda x}{(x^2 + a^2)^{3/2}} \int_0^{2\pi a} ds$$

$$= \frac{1}{4\pi\epsilon_0} \frac{\lambda x}{(x^2 + a^2)^{3/2}}(2\pi a)$$

$$\vec{E} = E_x \hat{\imath} = \frac{1}{4\pi\epsilon_0} \frac{Qx}{(x^2 + a^2)^{3/2}} \hat{\imath} \qquad (21.8)$$

EVALUATE: Equation (21.8) shows that $\vec{E} = 0$ at the center of the ring ($x = 0$). This makes sense; charges on opposite sides of the ring push in opposite directions on a test charge at the center, and the vector sum of each such pair of forces is zero. When the field point P is much farther from the ring than the ring's radius, we have $x \gg a$ and the denominator in Eq. (21.8) becomes approximately equal to x^3. In this limit the electric field at P is

$$\vec{E} = \frac{1}{4\pi\epsilon_0} \frac{Q}{x^2} \hat{\imath}$$

That is, when the ring is so far away that its radius is negligible in comparison to the distance x, its field is the same as that of a point charge.

Example 21.10 Field of a charged line segment

Positive charge Q is distributed uniformly along the y-axis between $y = -a$ and $y = +a$. Find the electric field at point P on the x-axis at a distance x from the origin.

SOLUTION

IDENTIFY and SET UP: Figure 21.24 shows the situation. As in Example 21.9, we must find the electric field due to a continuous distribution of charge. Our target variable is an expression for the electric field at P as a function of x. The x-axis is a perpendicular bisector of the segment, so we can use a symmetry argument.

EXECUTE: We divide the line charge of length $2a$ into infinitesimal segments of length dy. The linear charge density is $\lambda = Q/2a$, and the charge in a segment is $dQ = \lambda\,dy = (Q/2a)dy$. The distance r from a segment at height y to the field point P is $r = (x^2 + y^2)^{1/2}$, so the magnitude of the field at P due to the segment at height y is

$$dE = \frac{1}{4\pi\epsilon_0} \frac{dQ}{r^2} = \frac{1}{4\pi\epsilon_0} \frac{Q}{2a} \frac{dy}{(x^2 + y^2)}$$

Figure 21.24 shows that the x- and y-components of this field are $dE_x = dE \cos\alpha$ and $dE_y = -dE \sin\alpha$, where $\cos\alpha = x/r$ and $\sin\alpha = y/r$. Hence

$$dE_x = \frac{1}{4\pi\epsilon_0} \frac{Q}{2a} \frac{x\,dy}{(x^2 + y^2)^{3/2}}$$

$$dE_y = \frac{1}{4\pi\epsilon_0} \frac{Q}{2a} \frac{y\,dy}{(x^2 + y^2)^{3/2}}$$

To find the total field at P, we must sum the fields from all segments along the line—that is, we must integrate from $y = -a$ to $y = +a$. You should work out the details of the integration (a table of integrals will help). The results are

$$E_x = \frac{1}{4\pi\epsilon_0} \frac{Q}{2a} \int_{-a}^{+a} \frac{x\,dy}{(x^2 + y^2)^{3/2}} = \frac{Q}{4\pi\epsilon_0} \frac{1}{x\sqrt{x^2 + a^2}}$$

$$E_y = \frac{1}{4\pi\epsilon_0} \frac{Q}{2a} \int_{-a}^{+a} \frac{y\,dy}{(x^2 + y^2)^{3/2}} = 0$$

or, in vector form,

$$\vec{E} = \frac{1}{4\pi\epsilon_0} \frac{Q}{x\sqrt{x^2 + a^2}} \hat{\imath} \qquad (21.9)$$

21.24 Our sketch for this problem.

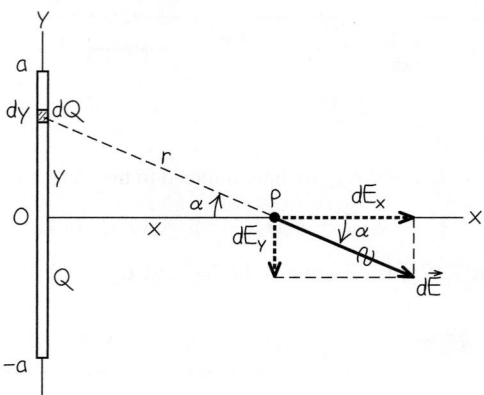

$\vec{E}$ points away from the line of charge if λ is positive and toward the line of charge if λ is negative.

EVALUATE: Using a symmetry argument as in Example 21.9, we could have guessed that E_y would be zero; if we place a positive test charge at P, the upper half of the line of charge pushes downward on it, and the lower half pushes up with equal magnitude. Symmetry also tells us that the upper and lower halves of the segment contribute equally to the total field at P.

If the segment is very *short* (or the field point is very far from the segment) so that $x \gg a$, we can neglect a in the denominator of Eq. (21.9). Then the field becomes that of a point charge, just as in Example 21.9:

$$\vec{E} = \frac{1}{4\pi\epsilon_0} \frac{Q}{x^2} \hat{\imath}$$

To see what happens if the segment is very *long* (or the field point is very close to it) so that $a \gg x$, we first rewrite Eq. (21.9) slightly:

$$\vec{E} = \frac{1}{2\pi\epsilon_0} \frac{\lambda}{x\sqrt{(x^2/a^2) + 1}} \hat{\imath} \qquad (21.10)$$

In the limit $a \gg x$ we can neglect x^2/a^2 in the denominator of Eq. (21.10), so

$$\vec{E} = \frac{\lambda}{2\pi\epsilon_0 x} \hat{\imath}$$

This is the field of an *infinitely long* line of charge. At any point P at a perpendicular distance r from the line in *any* direction, $\vec{E}$ has magnitude

$$E = \frac{\lambda}{2\pi\epsilon_0 r} \qquad \text{(infinite line of charge)}$$

Note that this field is proportional to $1/r$ rather than to $1/r^2$ as for a point charge.

There's really no such thing in nature as an infinite line of charge. But when the field point is close enough to the line, there's very little difference between the result for an infinite line and the real-life finite case. For example, if the distance r of the field point from the center of the line is 1% of the length of the line, the value of E differs from the infinite-length value by less than 0.02%.

Example 21.11   Field of a uniformly charged disk

A nonconducting disk of radius R has a uniform positive surface charge density σ. Find the electric field at a point along the axis of the disk a distance x from its center. Assume that x is positive.

SOLUTION

IDENTIFY and SET UP: Figure 21.25 shows the situation. We represent the charge distribution as a collection of concentric rings of charge dQ. In Example 21.9 we obtained Eq. (21.8) for the field on the axis of a single uniformly charged ring, so all we need do here is integrate the contributions of our rings.

EXECUTE: A typical ring has charge dQ, inner radius r, and outer radius $r + dr$. Its area is approximately equal to its width dr times its circumference $2\pi r$, or $dA = 2\pi r\, dr$. The charge per unit area is $\sigma = dQ/dA$, so the charge of the ring is $dQ = \sigma\, dA = 2\pi\sigma r\, dr$. We use dQ in place of Q in Eq. (21.8), the expression for the field due to a ring that we found in Example 21.9, and replace the ring radius a with r. Then the field component dE_x at point P due to this ring is

$$dE_x = \frac{1}{4\pi\epsilon_0} \frac{2\pi\sigma r x\, dr}{(x^2 + r^2)^{3/2}}$$

21.25 Our sketch for this problem.

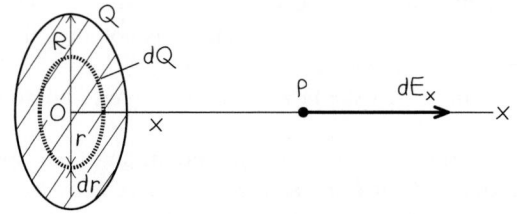

To find the total field due to all the rings, we integrate dE_x over r from $r = 0$ to $r = R$ (*not* from $-R$ to R):

$$E_x = \int_0^R \frac{1}{4\pi\epsilon_0} \frac{(2\pi\sigma r\, dr)x}{(x^2 + r^2)^{3/2}} = \frac{\sigma x}{4\epsilon_0} \int_0^R \frac{2r\, dr}{(x^2 + r^2)^{3/2}}$$

You can evaluate this integral by making the substitution $t = x^2 + r^2$ (which yields $dt = 2r\, dr$); you can work out the details. The result is

$$E_x = \frac{\sigma x}{2\epsilon_0}\left[-\frac{1}{\sqrt{x^2 + R^2}} + \frac{1}{x} \right]$$

$$= \frac{\sigma}{2\epsilon_0}\left[1 - \frac{1}{\sqrt{(R^2/x^2) + 1}} \right] \qquad (21.11)$$

EVALUATE: If the disk is very large (or if we are very close to it), so that $R \gg x$, the term $1/\sqrt{(R^2/x^2) + 1}$ in Eq. (21.11) is very much less than 1. Then Eq. (21.11) becomes

$$E = \frac{\sigma}{2\epsilon_0} \qquad (21.12)$$

Our final result does not contain the distance x from the plane. Hence the electric field produced by an *infinite* plane sheet of charge is *independent of the distance from the sheet*. The field direction is everywhere perpendicular to the sheet, away from it. There is no such thing as an infinite sheet of charge, but if the dimensions of the sheet are much larger than the distance x of the field point P from the sheet, the field is very nearly given by Eq. (21.12).

If P is to the *left* of the plane ($x < 0$), the result is the same except that the direction of $\vec{E}$ is to the left instead of the right. If the surface charge density is negative, the directions of the fields on both sides of the plane are toward it rather than away from it.

Example 21.12   Field of two oppositely charged infinite sheets

Two infinite plane sheets with uniform surface charge densities $+\sigma$ and $-\sigma$ are placed parallel to each other with separation d (Fig. 21.26). Find the electric field between the sheets, above the upper sheet, and below the lower sheet.

SOLUTION

IDENTIFY and SET UP: Equation (21.12) gives the electric field due to a single infinite plane sheet of charge. To find the field due to *two* such sheets, we combine the fields using the principle of superposition (Fig. 21.26).

21.26 Finding the electric field due to two oppositely charged infinite sheets. The sheets are seen edge-on; only a portion of the infinite sheets can be shown!

Continued

EXECUTE: From Eq. (21.12), both $\vec{E}_1$ and $\vec{E}_2$ have the same magnitude at all points, independent of distance from either sheet:

$$E_1 = E_2 = \frac{\sigma}{2\epsilon_0}$$

From Example 21.11, $\vec{E}_1$ is everywhere directed away from sheet 1, and $\vec{E}_2$ is everywhere directed toward sheet 2.

Between the sheets, $\vec{E}_1$ and $\vec{E}_2$ reinforce each other; above the upper sheet and below the lower sheet, they cancel each other. Thus the total field is

$$\vec{E} = \vec{E}_1 + \vec{E}_2 = \begin{cases} 0 & \text{above the upper sheet} \\ \dfrac{\sigma}{\epsilon_0}\hat{\jmath} & \text{between the sheets} \\ 0 & \text{below the lower sheet} \end{cases}$$

EVALUATE: Because we considered the sheets to be infinite, our result does not depend on the separation d. Our result shows that the field between oppositely charged plates is essentially uniform if the plate separation is much smaller than the dimensions of the plates. We actually used this result in Example 21.7 (Section 21.4).

CAUTION **Electric fields are not "flows"** You may have thought that the field $\vec{E}_1$ of sheet 1 would be unable to "penetrate" sheet 2, and that field $\vec{E}_2$ caused by sheet 2 would be unable to "penetrate" sheet 1. You might conclude this if you think of the electric field as some kind of physical substance that "flows" into or out of charges. But in fact there is no such substance, and the electric fields $\vec{E}_1$ and $\vec{E}_2$ depend only on the individual charge distributions that create them. The *total* field at every point is just the vector sum of $\vec{E}_1$ and $\vec{E}_2$.

Test Your Understanding of Section 21.5 Suppose that the line of charge in Fig. 21.25 (Example 21.11) had charge $+Q$ distributed uniformly between $y = 0$ and $y = +a$ and had charge $-Q$ distributed uniformly between $y = 0$ and $y = -a$. In this situation, the electric field at P would be (i) in the positive x-direction; (ii) in the negative x-direction; (iii) in the positive y-direction; (iv) in the negative y-direction; (v) zero; (vi) none of these.

21.6 Electric Field Lines

The concept of an electric field can be a little elusive because you can't see an electric field directly. Electric field *lines* can be a big help for visualizing electric fields and making them seem more real. An **electric field line** is an imaginary line or curve drawn through a region of space so that its tangent at any point is in the direction of the electric-field vector at that point. Figure 21.27 shows the basic idea. (We used a similar concept in our discussion of fluid flow in Section 12.5. A *streamline* is a line or curve whose tangent at any point is in the direction of the velocity of the fluid at that point. However, the similarity between electric field lines and fluid streamlines is a mathematical one only; there is nothing "flowing" in an electric field.) The English scientist Michael Faraday (1791–1867) first introduced the concept of field lines. He called them "lines of force," but the term "field lines" is preferable.

Electric field lines show the direction of $\vec{E}$ at each point, and their spacing gives a general idea of the *magnitude* of $\vec{E}$ at each point. Where $\vec{E}$ is strong, we draw lines bunched closely together; where $\vec{E}$ is weaker, they are farther apart. At any particular point, the electric field has a unique direction, so only one field line can pass through each point of the field. In other words, *field lines never intersect.*

Figure 21.28 shows some of the electric field lines in a plane containing (a) a single positive charge; (b) two equal-magnitude charges, one positive and one negative (a dipole); and (c) two equal positive charges. Diagrams such as these are sometimes called *field maps;* they are cross sections of the actual three-dimensional patterns. The direction of the total electric field at every point in each diagram is along the tangent to the electric field line passing through the point. Arrowheads indicate the direction of the $\vec{E}$-field vector along each field line. The actual field vectors have been drawn at several points in each pattern. Notice that in general, the magnitude of the electric field is different at different points on a given field line; a field line is *not* a curve of constant electric-field magnitude!

Figure 21.28 shows that field lines are directed *away* from positive charges (since close to a positive point charge, $\vec{E}$ points away from the charge) and

21.27 The direction of the electric field at any point is tangent to the field line through that point.

PhET: Charges and Fields
PhET: Electric Field of Dreams
PhET: Electric Field Hockey

21.28 Electric field lines for three different charge distributions. In general, the magnitude of $\vec{E}$ is different at different points along a given field line.

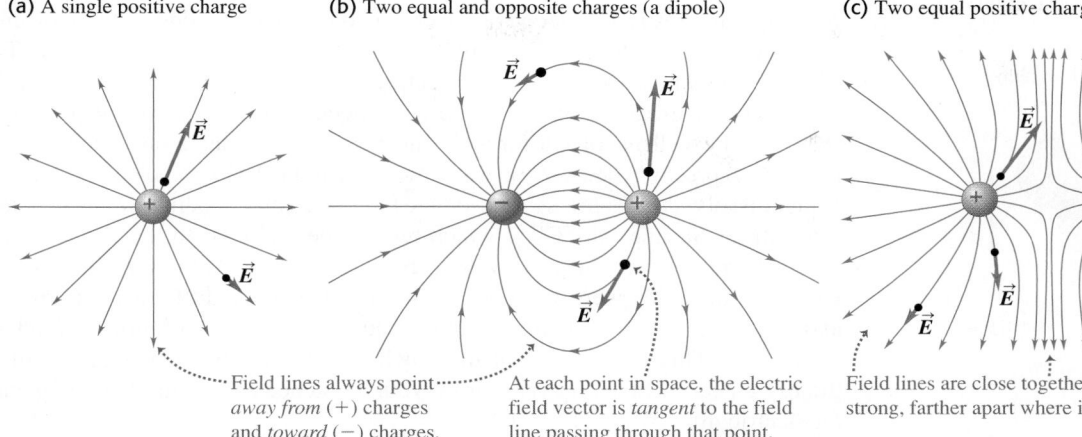

(a) A single positive charge

(b) Two equal and opposite charges (a dipole)

(c) Two equal positive charges

······ Field lines always point *away from* (+) charges and *toward* (−) charges.

At each point in space, the electric field vector is *tangent* to the field line passing through that point.

Field lines are close together where the field is strong, farther apart where it is weaker.

toward negative charges (since close to a negative point charge, $\vec{E}$ points toward the charge). In regions where the field magnitude is large, such as between the positive and negative charges in Fig. 21.28b, the field lines are drawn close together. In regions where the field magnitude is small, such as between the two positive charges in Fig. 21.28c, the lines are widely separated. In a *uniform* field, the field lines are straight, parallel, and uniformly spaced, as in Fig. 21.20.

Figure 21.29 is a view from above of a demonstration setup for visualizing electric field lines. In the arrangement shown here, the tips of two positively charged wires are inserted in a container of insulating liquid, and some grass seeds are floated on the liquid. The grass seeds are electrically neutral insulators, but the electric field of the two charged wires causes *polarization* of the grass seeds; there is a slight shifting of the positive and negative charges within the molecules of each seed, like that shown in Fig. 21.8. The positively charged end of each grass seed is pulled in the direction of $\vec{E}$ and the negatively charged end is pulled opposite $\vec{E}$. Hence the long axis of each grass seed tends to orient parallel to the electric field, in the direction of the field line that passes through the position of the seed (Fig. 21.29b).

CAUTION **Electric field lines are not the same as trajectories** It's a common misconception that if a charged particle of charge q is in motion where there is an electric field, the particle must move along an electric field line. Because $\vec{E}$ at any point is tangent to the field line that passes through that point, it is indeed true that the *force* $\vec{F} = q\vec{E}$ on the particle, and hence the particle's acceleration, are tangent to the field line. But we learned in Chapter 3 that when a particle moves on a curved path, its acceleration *cannot* be tangent to the path. So in general, the trajectory of a charged particle is *not* the same as a field line. ▮

Test Your Understanding of Section 21.6 Suppose the electric field lines in a region of space are straight lines. If a charged particle is released from rest in that region, will the trajectory of the particle be along a field line? ▮

21.7 Electric Dipoles

An **electric dipole** is a pair of point charges with equal magnitude and opposite sign (a positive charge q and a negative charge $-q$) separated by a distance d. We introduced electric dipoles in Example 21.8 (Section 21.5); the concept is worth exploring further because many physical systems, from molecules to TV antennas, can be described as electric dipoles. We will also use this concept extensively in our discussion of dielectrics in Chapter 24.

21.29 (a) Electric field lines produced by two equal point charges. The pattern is formed by grass seeds floating on a liquid above two charged wires. Compare this pattern with Fig. 21.28c. (b) The electric field causes polarization of the grass seeds, which in turn causes the seeds to align with the field.

(a)

(b)

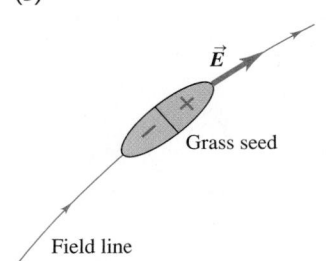

21.30 (a) A water molecule is an example of an electric dipole. (b) Each test tube contains a solution of a different substance in water. The large electric dipole moment of water makes it an excellent solvent.

(a) A water molecule, showing positive charge as red and negative charge as blue

The electric dipole moment $\vec{p}$ is directed from the negative end to the positive end of the molecule.

(b) Various substances dissolved in water

21.31 The net force on this electric dipole is zero, but there is a torque directed into the page that tends to rotate the dipole clockwise.

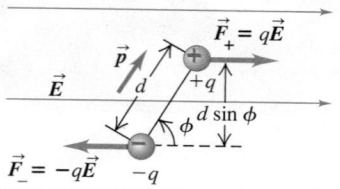

Figure 21.30a shows a molecule of water (H_2O), which in many ways behaves like an electric dipole. The water molecule as a whole is electrically neutral, but the chemical bonds within the molecule cause a displacement of charge; the result is a net negative charge on the oxygen end of the molecule and a net positive charge on the hydrogen end, forming an electric dipole. The effect is equivalent to shifting one electron only about 4×10^{-11} m (about the radius of a hydrogen atom), but the consequences of this shift are profound. Water is an excellent solvent for ionic substances such as table salt (sodium chloride, NaCl) precisely because the water molecule is an electric dipole (Fig. 21.30b). When dissolved in water, salt dissociates into a positive sodium ion (Na^+) and a negative chlorine ion (Cl^-), which tend to be attracted to the negative and positive ends, respectively, of water molecules; this holds the ions in solution. If water molecules were not electric dipoles, water would be a poor solvent, and almost all of the chemistry that occurs in aqueous solutions would be impossible. This includes all of the biochemical reactions that occur in all of the life on earth. In a very real sense, your existence as a living being depends on electric dipoles!

We examine two questions about electric dipoles. First, what forces and torques does an electric dipole experience when placed in an external electric field (that is, a field set up by charges outside the dipole)? Second, what electric field does an electric dipole itself produce?

Force and Torque on an Electric Dipole

To start with the first question, let's place an electric dipole in a *uniform* external electric field $\vec{E}$, as shown in Fig. 21.31. The forces $\vec{F}_+$ and $\vec{F}_-$ on the two charges both have magnitude qE, but their directions are opposite, and they add to zero. *The net force on an electric dipole in a uniform external electric field is zero.*

However, the two forces don't act along the same line, so their *torques* don't add to zero. We calculate torques with respect to the center of the dipole. Let the angle between the electric field $\vec{E}$ and the dipole axis be ϕ; then the lever arm for both $\vec{F}_+$ and $\vec{F}_-$ is $(d/2) \sin \phi$. The torque of $\vec{F}_+$ and the torque of $\vec{F}_-$ both have the same magnitude of $(qE)(d/2) \sin \phi$, and both torques tend to rotate the dipole clockwise (that is, $\vec{\tau}$ is directed into the page in Fig. 21.31). Hence the magnitude of the net torque is twice the magnitude of either individual torque:

$$\tau = (qE)(d \sin \phi) \tag{21.13}$$

where $d \sin \phi$ is the perpendicular distance between the lines of action of the two forces.

The product of the charge q and the separation d is the magnitude of a quantity called the **electric dipole moment,** denoted by p:

$$p = qd \quad \text{(magnitude of electric dipole moment)} \tag{21.14}$$

The units of p are charge times distance (C · m). For example, the magnitude of the electric dipole moment of a water molecule is $p = 6.13 \times 10^{-30}$ C · m.

CAUTION **The symbol p has multiple meanings** Be careful not to confuse dipole moment with momentum or pressure. There aren't as many letters in the alphabet as there are physical quantities, so some letters are used several times. The context usually makes it clear what we mean, but be careful. ▮

We further define the electric dipole moment to be a *vector* quantity $\vec{p}$. The magnitude of $\vec{p}$ is given by Eq. (21.14), and its direction is along the dipole axis from the negative charge to the positive charge as shown in Fig. 21.31.

In terms of p, Eq. (21.13) for the magnitude τ of the torque exerted by the field becomes

$$\tau = pE \sin\phi \qquad \text{(magnitude of the torque on an electric dipole)} \qquad (21.15)$$

Since the angle ϕ in Fig. 21.31 is the angle between the directions of the vectors $\vec{p}$ and $\vec{E}$, this is reminiscent of the expression for the magnitude of the *vector product* discussed in Section 1.10. (You may want to review that discussion.) Hence we can write the torque on the dipole in vector form as

$$\vec{\tau} = \vec{p} \times \vec{E} \qquad \text{(torque on an electric dipole, in vector form)} \qquad (21.16)$$

You can use the right-hand rule for the vector product to verify that in the situation shown in Fig. 21.31, $\vec{\tau}$ is directed into the page. The torque is greatest when $\vec{p}$ and $\vec{E}$ are perpendicular and is zero when they are parallel or antiparallel. The torque always tends to turn $\vec{p}$ to line it up with $\vec{E}$. The position $\phi = 0$, with $\vec{p}$ parallel to $\vec{E}$, is a position of stable equilibrium, and the position $\phi = \pi$, with $\vec{p}$ and $\vec{E}$ antiparallel, is a position of unstable equilibrium. The polarization of a grass seed in the apparatus of Fig. 21.29b gives it an electric dipole moment; the torque exerted by $\vec{E}$ then causes the seed to align with $\vec{E}$ and hence with the field lines.

Potential Energy of an Electric Dipole

When a dipole changes direction in an electric field, the electric-field torque does *work* on it, with a corresponding change in potential energy. The work dW done by a torque τ during an infinitesimal displacement $d\phi$ is given by Eq. (10.19): $dW = \tau\, d\phi$. Because the torque is in the direction of decreasing ϕ, we must write the torque as $\tau = -pE \sin\phi$, and

$$dW = \tau\, d\phi = -pE \sin\phi\, d\phi$$

In a finite displacement from ϕ_1 to ϕ_2 the total work done on the dipole is

$$W = \int_{\phi_1}^{\phi_2} (-pE \sin\phi)\, d\phi$$

$$= pE \cos\phi_2 - pE \cos\phi_1$$

The work is the negative of the change of potential energy, just as in Chapter 7: $W = U_1 - U_2$. So a suitable definition of potential energy U for this system is

$$U(\phi) = -pE \cos\phi \qquad (21.17)$$

In this expression we recognize the *scalar product* $\vec{p} \cdot \vec{E} = pE \cos\phi$, so we can also write

$$U = -\vec{p} \cdot \vec{E} \qquad \text{(potential energy for a dipole in an electric field)} \qquad (21.18)$$

The potential energy has its minimum (most negative) value $U = -pE$ at the stable equilibrium position, where $\phi = 0$ and $\vec{p}$ is parallel to $\vec{E}$. The potential energy is maximum when $\phi = \pi$ and $\vec{p}$ is antiparallel to $\vec{E}$; then $U = +pE$. At $\phi = \pi/2$, where $\vec{p}$ is perpendicular to $\vec{E}$, U is zero. We could define U differently so that it is zero at some other orientation of $\vec{p}$, but our definition is simplest.

Equation (21.18) gives us another way to look at the effect shown in Fig. 21.29. The electric field $\vec{E}$ gives each grass seed an electric dipole moment, and the grass seed then aligns itself with $\vec{E}$ to minimize the potential energy.

MasteringPHYSICS

PhET: Microwaves

Example 21.13 **Force and torque on an electric dipole**

Figure 21.32a shows an electric dipole in a uniform electric field of magnitude 5.0×10^5 N/C that is directed parallel to the plane of the figure. The charges are $\pm 1.6 \times 10^{-19}$ C; both lie in the plane and are separated by 0.125 nm $= 0.125 \times 10^{-9}$ m. Find (a) the net force exerted by the field on the dipole; (b) the magnitude and

Continued

21.32 (a) An electric dipole. (b) Directions of the electric dipole moment, electric field, and torque ($\vec{\tau}$ points out of the page).

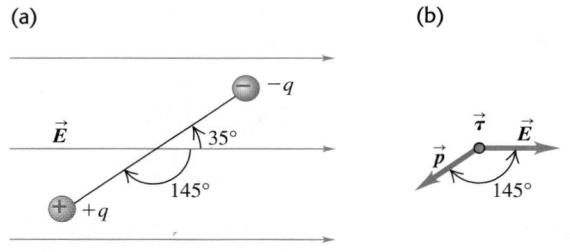

(a)   (b)

direction of the electric dipole moment; (c) the magnitude and direction of the torque; (d) the potential energy of the system in the position shown.

SOLUTION

IDENTIFY and SET UP: This problem uses the ideas of this section about an electric dipole placed in an electric field. We use the relationship $\vec{F} = q\vec{E}$ for each point charge to find the force on the dipole as a whole. Equation (21.14) gives the dipole moment, Eq. (21.16) gives the torque on the dipole, and Eq. (21.18) gives the potential energy of the system.

EXECUTE: (a) The field is uniform, so the forces on the two charges are equal and opposite. Hence the total force on the dipole is zero.

(b) The magnitude p of the electric dipole moment $\vec{p}$ is

$$p = qd = (1.6 \times 10^{-19}\ \text{C})(0.125 \times 10^{-9}\ \text{m})$$
$$= 2.0 \times 10^{-29}\ \text{C} \cdot \text{m}$$

The direction of $\vec{p}$ is from the negative to the positive charge, 145° clockwise from the electric-field direction (Fig. 21.32b).

(c) The magnitude of the torque is

$$\tau = pE\sin\phi = (2.0 \times 10^{-29}\ \text{C})(5.0 \times 10^{5}\ \text{N/C})(\sin 145°)$$
$$= 5.7 \times 10^{-24}\ \text{N} \cdot \text{m}$$

From the right-hand rule for vector products (see Section 1.10), the direction of the torque $\vec{\tau} = \vec{p} \times \vec{E}$ is out of the page. This corresponds to a counterclockwise torque that tends to align $\vec{p}$ with $\vec{E}$.

(d) The potential energy

$$U = -pE\cos\phi$$
$$= -(2.0 \times 10^{-29}\ \text{C} \cdot \text{m})(5.0 \times 10^{5}\ \text{N/C})(\cos 145°)$$
$$= 8.2 \times 10^{-24}\ \text{J}$$

EVALUATE: The charge magnitude, the distance between the charges, the dipole moment, and the potential energy are all very small, but are all typical of molecules.

In this discussion we have assumed that $\vec{E}$ is uniform, so there is no net force on the dipole. If $\vec{E}$ is not uniform, the forces at the ends may not cancel completely, and the net force may not be zero. Thus a body with zero net charge but an electric dipole moment can experience a net force in a nonuniform electric field. As we mentioned in Section 21.1, an uncharged body can be polarized by an electric field, giving rise to a separation of charge and an electric dipole moment. This is how uncharged bodies can experience electrostatic forces (see Fig. 21.8).

Field of an Electric Dipole

Now let's think of an electric dipole as a *source* of electric field. What does the field look like? The general shape of things is shown by the field map of Fig. 21.28b. At each point in the pattern the total $\vec{E}$ field is the vector sum of the fields from the two individual charges, as in Example 21.8 (Section 21.5). Try drawing diagrams showing this vector sum for several points.

To get quantitative information about the field of an electric dipole, we have to do some calculating, as illustrated in the next example. Notice the use of the principle of superposition of electric fields to add up the contributions to the field of the individual charges. Also notice that we need to use approximation techniques even for the relatively simple case of a field due to two charges. Field calculations often become very complicated, and computer analysis is typically used to determine the field due to an arbitrary charge distribution.

Example 21.14   **Field of an electric dipole, revisited**

An electric dipole is centered at the origin, with $\vec{p}$ in the direction of the $+y$-axis (Fig. 21.33). Derive an approximate expression for the electric field at a point P on the y-axis for which y is much larger than d. To do this, use the binomial expansion $(1 + x)^n \cong 1 + nx + n(n - 1)x^2/2 + \cdots$ (valid for the case $|x| < 1$).

SOLUTION

IDENTIFY and SET UP: We use the principle of superposition: The total electric field is the vector sum of the field produced by the positive charge and the field produced by the negative charge. At the field point P shown in Fig. 21.33, the field $\vec{E}_+$ of the positive charge has a positive (upward) y-component and the field $\vec{E}_-$ of

21.33 Finding the electric field of an electric dipole at a point on its axis.

the negative charge has a negative (downward) y-component. We add these components to find the total field and then apply the approximation that y is much greater than d.

EXECUTE: The total y-component E_y of electric field from the two charges is

$$E_y = \frac{q}{4\pi\epsilon_0}\left[\frac{1}{(y - d/2)^2} - \frac{1}{(y + d/2)^2}\right]$$

$$= \frac{q}{4\pi\epsilon_0 y^2}\left[\left(1 - \frac{d}{2y}\right)^{-2} - \left(1 + \frac{d}{2y}\right)^{-2}\right]$$

We used this same approach in Example 21.8 (Section 21.5). Now the approximation: When we are far from the dipole compared to its size, so $y \gg d$, we have $d/2y \ll 1$. With $n = -2$ and with $d/2y$ replacing x in the binomial expansion, we keep only the first two terms (the terms we discard are much smaller). We then have

$$\left(1 - \frac{d}{2y}\right)^{-2} \cong 1 + \frac{d}{y} \quad \text{and} \quad \left(1 + \frac{d}{2y}\right)^{-2} \cong 1 - \frac{d}{y}$$

Hence E_y is given approximately by

$$E_y \cong \frac{q}{4\pi\epsilon_0 y^2}\left[1 + \frac{d}{y} - \left(1 - \frac{d}{y}\right)\right] = \frac{qd}{2\pi\epsilon_0 y^3} = \frac{p}{2\pi\epsilon_0 y^3}$$

EVALUATE: An alternative route to this result is to put the fractions in the first expression for E_y over a common denominator, add, and then approximate the denominator $(y - d/2)^2(y + d/2)^2$ as y^4. We leave the details to you (see Exercise 21.60).

For points P off the coordinate axes, the expressions are more complicated, but at *all* points far away from the dipole (in any direction) the field drops off as $1/r^3$. We can compare this with the $1/r^2$ behavior of a point charge, the $1/r$ behavior of a long line charge, and the independence of r for a large sheet of charge. There are charge distributions for which the field drops off even more quickly. At large distances, the field of an *electric quadrupole,* which consists of two equal dipoles with opposite orientation, separated by a small distance, drops off as $1/r^4$.

Test Your Understanding of Section 21.7   An electric dipole is placed in a region of uniform electric field $\vec{E}$, with the electric dipole moment $\vec{p}$, pointing in the direction opposite to $\vec{E}$. Is the dipole (i) in stable equilibrium, (ii) in unstable equilibrium, or (iii) neither? (*Hint:* You many want to review Section 7.5.)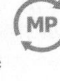

Electric charge, conductors, and insulators: The fundamental quantity in electrostatics is electric charge. There are two kinds of charge, positive and negative. Charges of the same sign repel each other; charges of opposite sign attract. Charge is conserved; the total charge in an isolated system is constant.

All ordinary matter is made of protons, neutrons, and electrons. The positive protons and electrically neutral neutrons in the nucleus of an atom are bound together by the nuclear force; the negative electrons surround the nucleus at distances much greater than the nuclear size. Electric interactions are chiefly responsible for the structure of atoms, molecules, and solids.

Conductors are materials in which charge moves easily; in insulators, charge does not move easily. Most metals are good conductors; most nonmetals are insulators.

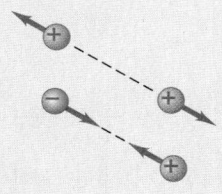

Coulomb's law: For charges q_1 and q_2 separated by a distance r, the magnitude of the electric force on either charge is proportional to the product q_1q_2 and inversely proportional to r^2. The force on each charge is along the line joining the two charges—repulsive if q_1 and q_2 have the same sign, attractive if they have opposite signs. In SI units the unit of electric charge is the coulomb, abbreviated C. (See Examples 21.1 and 21.2.)

When two or more charges each exert a force on a charge, the total force on that charge is the vector sum of the forces exerted by the individual charges. (See Examples 21.3 and 21.4.)

$$F = \frac{1}{4\pi\epsilon_0}\frac{|q_1q_2|}{r^2} \qquad (21.2)$$

$$\frac{1}{4\pi\epsilon_0} = 8.988 \times 10^9 \ \text{N}\cdot\text{m}^2/\text{C}^2$$

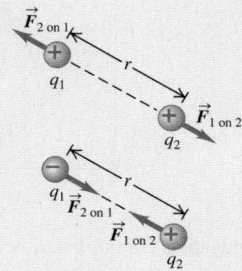

Electric field: Electric field $\vec{E}$, a vector quantity, is the force per unit charge exerted on a test charge at any point. The electric field produced by a point charge is directed radially away from or toward the charge. (See Examples 21.5–21.7.)

$$\vec{E} = \frac{\vec{F}_0}{q_0} \qquad (21.3)$$

$$\vec{E} = \frac{1}{4\pi\epsilon_0}\frac{q}{r^2}\hat{r} \qquad (21.7)$$

Superposition of electric fields: The electric field $\vec{E}$ of any combination of charges is the vector sum of the fields caused by the individual charges. To calculate the electric field caused by a continuous distribution of charge, divide the distribution into small elements, calculate the field caused by each element, and then carry out the vector sum, usually by integrating. Charge distributions are described by linear charge density λ, surface charge density σ, and volume charge density ρ. (See Examples 21.8–21.12.)

Electric field lines: Field lines provide a graphical representation of electric fields. At any point on a field line, the tangent to the line is in the direction of $\vec{E}$ at that point. The number of lines per unit area (perpendicular to their direction) is proportional to the magnitude of $\vec{E}$ at the point.

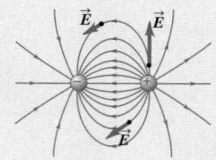

Electric dipoles: An electric dipole is a pair of electric charges of equal magnitude q but opposite sign, separated by a distance d. The electric dipole moment $\vec{p}$ has magnitude $p = qd$. The direction of $\vec{p}$ is from negative toward positive charge. An electric dipole in an electric field $\vec{E}$ experiences a torque $\vec{\tau}$ equal to the vector product of $\vec{p}$ and $\vec{E}$. The magnitude of the torque depends on the angle ϕ between $\vec{p}$ and $\vec{E}$. The potential energy U for an electric dipole in an electric field also depends on the relative orientation of $\vec{p}$ and $\vec{E}$. (See Examples 21.13 and 21.14.)

$$\tau = pE\sin\phi \qquad (21.15)$$

$$\vec{\tau} = \vec{p} \times \vec{E} \qquad (21.16)$$

$$U = -\vec{p}\cdot\vec{E} \qquad (21.18)$$

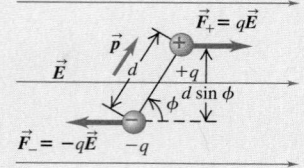

BRIDGING PROBLEM   Calculating Electric Field: Half a Ring of Charge

Positive charge Q is uniformly distributed around a semicircle of radius a as shown in Fig. 21.34. Find the magnitude and direction of the resulting electric field at point P, the center of curvature of the semicircle.

SOLUTION GUIDE

See MasteringPhysics® study area for a Video Tutor solution.

IDENTIFY and SET UP

1. The target variables are the components of the electric field at P.
2. Divide the semicircle into infinitesimal segments, each of which is a short circular arc of radius a and angle $d\theta$. What is the length of such a segment? How much charge is on a segment?

21.34

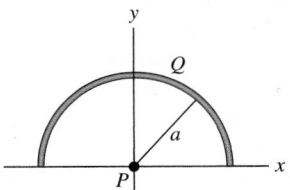

3. Consider an infinitesimal segment located at an angular position θ on the semicircle, measured from the lower right corner of the semicircle at $x = a$, $y = 0$. (Thus $\theta = \pi/2$ at $x = 0$, $y = a$ and $\theta = \pi$ at $x = -a$, $y = 0$.) What are the x- and y-components of the electric field at P (dE_x and dE_y) produced by just this segment?

EXECUTE

4. Integrate your expressions for dE_x and dE_y from $\theta = 0$ to $\theta = \pi$. The results will be the x-component and y-component of the electric field at P.
5. Use your results from step 4 to find the magnitude and direction of the field at P.

EVALUATE

6. Does your result for the electric-field magnitude have the correct units?
7. Explain how you could have found the x-component of the electric field using a symmetry argument.
8. What would be the electric field at P if the semicircle were extended to a full circle centered at P?

Problems

For instructor-assigned homework, go to www.masteringphysics.com

•, ••, •••: Problems of increasing difficulty. **CP**: Cumulative problems incorporating material from earlier chapters. **CALC**: Problems requiring calculus. **BIO**: Biosciences problems.

DISCUSSION QUESTIONS

Q21.1 If you peel two strips of transparent tape off the same roll and immediately let them hang near each other, they will repel each other. If you then stick the sticky side of one to the shiny side of the other and rip them apart, they will attract each other. Give a plausible explanation, involving transfer of electrons between the strips of tape, for this sequence of events.

Q21.2 Two metal spheres are hanging from nylon threads. When you bring the spheres close to each other, they tend to attract. Based on this information alone, discuss all the possible ways that the spheres could be charged. Is it possible that after the spheres touch, they will cling together? Explain.

Q21.3 The electric force between two charged particles becomes weaker with increasing distance. Suppose instead that the electric force were *independent* of distance. In this case, would a charged comb still cause a neutral insulator to become polarized as in Fig. 21.8? Why or why not? Would the neutral insulator still be attracted to the comb? Again, why or why not?

Q21.4 Your clothing tends to cling together after going through the dryer. Why? Would you expect more or less clinging if all your clothing were made of the same material (say, cotton) than if you dried different kinds of clothing together? Again, why? (You may want to experiment with your next load of laundry.)

Q21.5 An uncharged metal sphere hangs from a nylon thread. When a positively charged glass rod is brought close to the metal sphere, the sphere is drawn toward the rod. But if the sphere touches the rod, it suddenly flies away from the rod. Explain why the sphere is first attracted and then repelled.

Q21.6 The free electrons in a metal are gravitationally attracted toward the earth. Why, then, don't they all settle to the bottom of the conductor, like sediment settling to the bottom of a river?

Q21.7 • Figure Q21.7 shows some of the electric field lines due to three point charges arranged along the vertical axis. All three charges have the same magnitude. (a) What are the signs of the three charges? Explain your reasoning. (b) At what point(s) is the magnitude of the electric field the smallest? Explain your reasoning. Explain how the fields produced by each individual point charge combine to give a small net field at this point or points.

Figure **Q21.7**

Q21.8 Good electrical conductors, such as metals, are typically good conductors of heat; electrical insulators, such as wood, are typically poor conductors of heat. Explain why there should be a relationship between electrical conduction and heat conduction in these materials.

Q21.9 • Suppose the charge shown in Fig. 21.28a is fixed in position. A small, positively charged particle is then placed at some point in the figure and released. Will the trajectory of the particle follow an electric field line? Why or why not? Suppose instead that the particle is placed at some point in Fig. 21.28b and released (the positive and negative charges shown in the figure are fixed in position). Will its trajectory follow an electric field line? Again, why or why not? Explain any differences between your answers for the two different situations.

Q21.10 Two identical metal objects are mounted on insulating stands. Describe how you could place charges of opposite sign but exactly equal magnitude on the two objects.

Q21.11 You can use plastic food wrap to cover a container by stretching the material across the top and pressing the overhanging material against the sides. What makes it stick? (*Hint:* The answer involves the electric force.) Does the food wrap stick to itself with equal tenacity? Why or why not? Does it work with metallic containers? Again, why or why not?

Q21.12 If you walk across a nylon rug and then touch a large metal object such as a doorknob, you may get a spark and a shock. Why does this tend to happen more on dry days than on humid days? (*Hint:* See Fig. 21.30.) Why are you less likely to get a shock if you touch a *small* metal object, such as a paper clip?

Q21.13 You have a negatively charged object. How can you use it to place a net negative charge on an insulated metal sphere? To place a net positive charge on the sphere?

Q21.14 When two point charges of equal mass and charge are released on a frictionless table, each has an initial acceleration a_0. If instead you keep one fixed and release the other one, what will be its initial acceleration: a_0, $2a_0$, or $a_0/2$? Explain.

Q21.15 A point charge of mass m and charge Q and another point charge of mass m but charge $2Q$ are released on a frictionless table. If the charge Q has an initial acceleration a_0, what will be the acceleration of $2Q$: a_0, $2a_0$, $4a_0$, $a_0/2$, or $a_0/4$? Explain.

Q21.16 A proton is placed in a uniform electric field and then released. Then an electron is placed at this same point and released. Do these two particles experience the same force? The same acceleration? Do they move in the same direction when released?

Q21.17 In Example 21.1 (Section 21.3) we saw that the electric force between two α particles is of the order of 10^{35} times as strong as the gravitational force. So why do we readily feel the gravity of the earth but no electrical force from it?

Q21.18 What similarities do electrical forces have with gravitational forces? What are the most significant differences?

Q21.19 Two irregular objects A and B carry charges of opposite sign. Figure Q21.19 shows the electric field lines near each of these objects. (a) Which object is positive, A or B? How do you know? (b) Where is the electric field stronger, close to A or close to B? How do you know?

Figure **Q21.19**

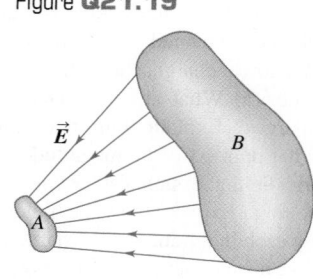

Q21.20 Atomic nuclei are made of protons and neutrons. This shows that there must be another kind of interaction in addition to gravitational and electric forces. Explain.

Q21.21 Sufficiently strong electric fields can cause atoms to become positively ionized—that is, to lose one or more electrons. Explain how this can happen. What determines how strong the field must be to make this happen?

Q21.22 The electric fields at point P due to the positive charges q_1 and q_2 are shown in Fig. Q21.22. Does the fact that they cross each other violate the statement in Section 21.6 that electric field lines never cross? Explain.

Figure **Q21.22**

Q21.23 The air temperature and the velocity of the air have different values at different places in the earth's atmosphere. Is the air velocity a vector field? Why or why not? Is the air temperature a vector field? Again, why or why not?

EXERCISES

Section 21.3 Coulomb's Law

21.1 •• Excess electrons are placed on a small lead sphere with mass 8.00 g so that its net charge is -3.20×10^{-9} C. (a) Find the number of excess electrons on the sphere. (b) How many excess electrons are there per lead atom? The atomic number of lead is 82, and its atomic mass is 207 g/mol.

21.2 • Lightning occurs when there is a flow of electric charge (principally electrons) between the ground and a thundercloud. The maximum rate of charge flow in a lightning bolt is about 20,000 C/s; this lasts for 100 μs or less. How much charge flows between the ground and the cloud in this time? How many electrons flow during this time?

21.3 •• **BIO** Estimate how many electrons there are in your body. Make any assumptions you feel are necessary, but clearly state what they are. (*Hint:* Most of the atoms in your body have equal numbers of electrons, protons, and neutrons.) What is the combined charge of all these electrons?

21.4 • **Particles in a Gold Ring.** You have a pure (24-karat) gold ring with mass 17.7 g. Gold has an atomic mass of 197 g/mol and an atomic number of 79. (a) How many protons are in the ring, and what is their total positive charge? (b) If the ring carries no net charge, how many electrons are in it?

21.5 • **BIO Signal Propagation in Neurons.** *Neurons* are components of the nervous system of the body that transmit signals as electrical impulses travel along their length. These impulses propagate when charge suddenly rushes into and then out of a part of the neuron called an *axon*. Measurements have shown that, during the inflow part of this cycle, approximately 5.6×10^{11} Na$^+$ (sodium ions) per meter, each with charge $+e$, enter the axon. How many coulombs of charge enter a 1.5-cm length of the axon during this process?

21.6 • Two small spheres spaced 20.0 cm apart have equal charge. How many excess electrons must be present on each sphere if the magnitude of the force of repulsion between them is 4.57×10^{-21} N?

21.7 •• An average human weighs about 650 N. If two such generic humans each carried 1.0 coulomb of excess charge, one positive and one negative, how far apart would they have to be for the electric attraction between them to equal their 650-N weight?

21.8 •• Two small aluminum spheres, each having mass 0.0250 kg, are separated by 80.0 cm. (a) How many electrons does each sphere contain? (The atomic mass of aluminum is 26.982 g/mol, and its atomic number is 13.) (b) How many electrons would have to be removed from one sphere and added to the other to cause an attractive force between the spheres of magnitude 1.00×10^4 N (roughly 1 ton)? Assume that the spheres may be treated as point charges. (c) What fraction of all the electrons in each sphere does this represent?

21.9 •• Two small plastic spheres are given positive electrical charges. When they are 15.0 cm apart, the repulsive force between them has magnitude 0.220 N. What is the charge on each sphere (a) if the two charges are equal and (b) if one sphere has four times the charge of the other?

21.10 •• **What If We Were Not Neutral?**   A 75-kg person holds out his arms so that his hands are 1.7 m apart. Typically, a person's hand makes up about 1.0% of his or her body weight. For round numbers, we shall assume that all the weight of each hand is due to the calcium in the bones, and we shall treat the hands as point charges. One mole of Ca contains 40.18 g, and each atom has 20 protons and 20 electrons. Suppose that only 1.0% of the positive charges in each hand were unbalanced by negative charge. (a) How many Ca atoms does each hand contain? (b) How many coulombs of unbalanced charge does each hand contain? (c) What force would the person's arms have to exert on his hands to prevent them from flying off? Does it seem likely that his arms are capable of exerting such a force?

21.11 •• Two very small 8.55-g spheres, 15.0 cm apart from center to center, are charged by adding equal numbers of electrons to each of them. Disregarding all other forces, how many electrons would you have to add to each sphere so that the two spheres will accelerate at $25.0g$ when released? Which way will they accelerate?

21.12 •• **Just How Strong Is the Electric Force?**   Suppose you had two small boxes, each containing 1.0 g of protons. (a) If one were placed on the moon by an astronaut and the other were left on the earth, and if they were connected by a very light (and very long!) string, what would be the tension in the string? Express your answer in newtons and in pounds. Do you need to take into account the gravitational forces of the earth and moon on the protons? Why? (b) What gravitational force would each box of protons exert on the other box?

21.13 • In an experiment in space, one proton is held fixed and another proton is released from rest a distance of 2.50 mm away. (a) What is the initial acceleration of the proton after it is released? (b) Sketch qualitative (no numbers!) acceleration–time and velocity–time graphs of the released proton's motion.

21.14 • A negative charge of $-0.550\ \mu C$ exerts an upward 0.200-N force on an unknown charge 0.300 m directly below it. (a) What is the unknown charge (magnitude and sign)? (b) What are the magnitude and direction of the force that the unknown charge exerts on the $-0.550\text{-}\mu C$ charge?

21.15 •• Three point charges are arranged on a line. Charge $q_3 = +5.00$ nC and is at the origin. Charge $q_2 = -3.00$ nC and is at $x = +4.00$ cm. Charge q_1 is at $x = +2.00$ cm. What is q_1 (magnitude and sign) if the net force on q_3 is zero?

21.16 •• In Example 21.4, suppose the point charge on the y-axis at $y = -0.30$ m has negative charge $-2.0\ \mu C$, and the other charges remain the same. Find the magnitude and direction of the net force on Q. How does your answer differ from that in Example 21.4? Explain the differences.

21.17 •• In Example 21.3, calculate the net force on charge q_1.

21.18 •• In Example 21.4, what is the net force (magnitude and direction) on charge q_1 exerted by the other two charges?

21.19 •• Three point charges are arranged along the x-axis. Charge $q_1 = +3.00\ \mu C$ is at the origin, and charge $q_2 = -5.00\ \mu C$ is at $x = 0.200$ m. Charge $q_3 = -8.00\ \mu C$. Where is q_3 located if the net force on q_1 is 7.00 N in the $-x$-direction?

21.20 •• Repeat Exercise 21.19 for $q_3 = +8.00\ \mu C$.

21.21 •• Two point charges are located on the y-axis as follows: charge $q_1 = -1.50$ nC at $y = -0.600$ m, and charge $q_2 = +3.20$ nC at the origin ($y = 0$). What is the total force (magnitude and direction) exerted by these two charges on a third charge $q_3 = +5.00$ nC located at $y = -0.400$ m?

21.22 •• Two point charges are placed on the x-axis as follows: Charge $q_1 = +4.00$ nC is located at $x = 0.200$ m, and charge $q_2 = +5.00$ nC is at $x = -0.300$ m. What are the magnitude and direction of the total force exerted by these two charges on a negative point charge $q_3 = -6.00$ nC that is placed at the origin?

21.23 •• **BIO Base Pairing in DNA, I.**   The two sides of the DNA double helix are connected by pairs of bases (adenine, thymine, cytosine, and guanine). Because of the geometric shape of these molecules, adenine bonds with thymine and cytosine bonds with guanine. Figure E21.23 shows the thymine–adenine bond. Each charge shown is $\pm e$, and the H—N distance is 0.110 nm. (a) Calculate the *net* force that thymine exerts on adenine. Is it attractive or repulsive? To keep the calculations fairly simple, yet reasonable, consider only the forces due to the O—H—N and the N—H—N combinations, assuming that these two combinations are parallel to each other. Remember, however, that in the O—H—N set, the O^- exerts a force on both the H^+ and the N^-, and likewise along the N—H—N set. (b) Calculate the force on the electron in the hydrogen atom, which is 0.0529 nm from the proton. Then compare the strength of the bonding force of the electron in hydrogen with the bonding force of the adenine–thymine molecules.

Figure **E21.23**

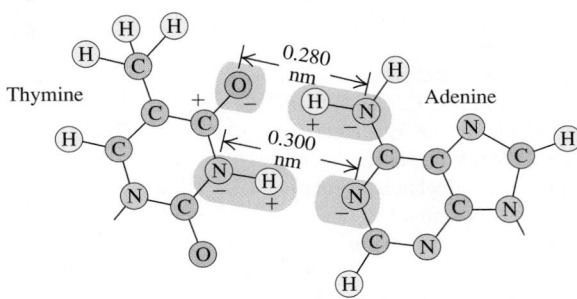

21.24 •• **BIO Base Pairing in DNA, II.**   Refer to Exercise 21.23. Figure E21.24 shows the bonding of the cytosine and guanine molecules. The O—H and H—N distances are each 0.110 nm. In this case, assume that the bonding is due only to the forces along the O—H—O, N—H—N, and O—H—N combinations, and assume also that these three combinations are parallel to each other. Calculate the *net* force that cytosine exerts on guanine due to the preceding three combinations. Is this force attractive or repulsive?

Figure **E21.24**

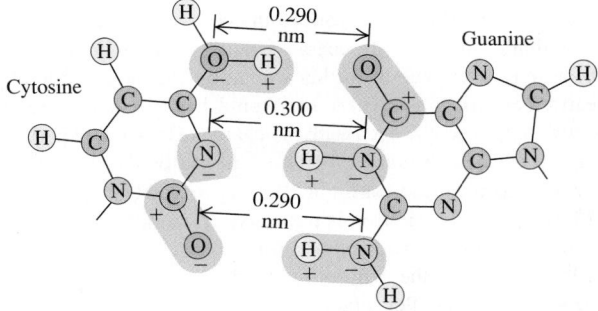

Section 21.4 Electric Field and Electric Forces

21.25 • **CP** A proton is placed in a uniform electric field of 2.75×10^3 N/C. Calculate: (a) the magnitude of the electric force felt by the proton; (b) the proton's acceleration; (c) the proton's speed after 1.00 μs in the field, assuming it starts from rest.

21.26 • A particle has charge -3.00 nC. (a) Find the magnitude and direction of the electric field due to this particle at a point 0.250 m directly above it. (b) At what distance from this particle does its electric field have a magnitude of 12.0 N/C?

21.27 • **CP** A proton is traveling horizontally to the right at 4.50×10^6 m/s. (a) Find the magnitude and direction of the weakest electric field that can bring the proton uniformly to rest over a distance of 3.20 cm. (b) How much time does it take the proton to stop after entering the field? (c) What minimum field (magnitude and direction) would be needed to stop an electron under the conditions of part (a)?

21.28 • **CP** An electron is released from rest in a uniform electric field. The electron accelerates vertically upward, traveling 4.50 m in the first 3.00 μs after it is released. (a) What are the magnitude and direction of the electric field? (b) Are we justified in ignoring the effects of gravity? Justify your answer quantitatively.

21.29 •• (a) What must the charge (sign and magnitude) of a 1.45-g particle be for it to remain stationary when placed in a downward-directed electric field of magnitude 650 N/C? (b) What is the magnitude of an electric field in which the electric force on a proton is equal in magnitude to its weight?

21.30 •• A point charge is placed at each corner of a square with side length a. The charges all have the same magnitude q. Two of the charges are positive and two are negative, as shown in Fig. E21.30. What is the direction of the net electric field at the center of the square due to the four charges, and what is its magnitude in terms of q and a?

Figure **E21.30**

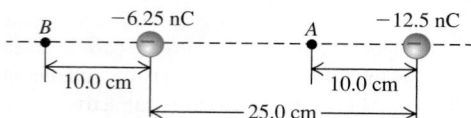

21.31 • Two point charges are separated by 25.0 cm (Fig. E21.31). Find the net electric field these charges produce at (a) point A and (b) point B. (c) What would be the magnitude and direction of the electric force this combination of charges would produce on a proton at A?

Figure **E21.31**

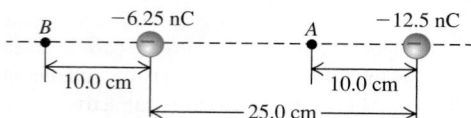

21.32 •• **Electric Field of the Earth.** The earth has a net electric charge that causes a field at points near its surface equal to 150 N/C and directed in toward the center of the earth. (a) What magnitude and sign of charge would a 60-kg human have to acquire to overcome his or her weight by the force exerted by the earth's electric field? (b) What would be the force of repulsion between two people each with the charge calculated in part (a) and separated by a distance of 100 m? Is use of the earth's electric field a feasible means of flight? Why or why not?

21.33 •• **CP** An electron is projected with an initial speed $v_0 = 1.60 \times 10^6$ m/s into the uniform field between the parallel plates in Fig. E21.33. Assume that the field between the plates is uniform and directed vertically downward, and that the field outside the plates is zero. The electron enters the field at a point

Figure **E21.33**

midway between the plates. (a) If the electron just misses the upper plate as it emerges from the field, find the magnitude of the electric field. (b) Suppose that in Fig. E21.33 the electron is replaced by a proton with the same initial speed v_0. Would the proton hit one of the plates? If the proton would not hit one of the plates, what would be the magnitude and direction of its vertical displacement as it exits the region between the plates? (c) Compare the paths traveled by the electron and the proton and explain the differences. (d) Discuss whether it is reasonable to ignore the effects of gravity for each particle.

21.34 •• Point charge $q_1 = -5.00$ nC is at the origin and point charge $q_2 = +3.00$ nC is on the x-axis at $x = 3.00$ cm. Point P is on the y-axis at $y = 4.00$ cm. (a) Calculate the electric fields $\vec{E}_1$ and $\vec{E}_2$ at point P due to the charges q_1 and q_2. Express your results in terms of unit vectors (see Example 21.6). (b) Use the results of part (a) to obtain the resultant field at P, expressed in unit vector form.

21.35 •• **CP** In Exercise 21.33, what is the speed of the electron as it emerges from the field?

21.36 • (a) Calculate the magnitude and direction (relative to the $+x$-axis) of the electric field in Example 21.6. (b) A -2.5-nC point charge is placed at point P in Fig. 21.19. Find the magnitude and direction of (i) the force that the -8.0-nC charge at the origin exerts on this charge and (ii) the force that this charge exerts on the -8.0-nC charge at the origin.

21.37 •• If two electrons are each 1.50×10^{-10} m from a proton, as shown in Fig. E21.37, find the magnitude and direction of the net electric force they will exert on the proton.

Figure **E21.37**

21.38 •• **CP** A uniform electric field exists in the region between two oppositely charged plane parallel plates. A proton is released from rest at the surface of the positively charged plate and strikes the surface of the opposite plate, 1.60 cm distant from the first, in a time interval of 1.50×10^{-6} s. (a) Find the magnitude of the electric field. (b) Find the speed of the proton when it strikes the negatively charged plate.

21.39 • A point charge is at the origin. With this point charge as the source point, what is the unit vector $\hat{r}$ in the direction of (a) the field point at $x = 0$, $y = -1.35$ m; (b) the field point at $x = 12.0$ cm, $y = 12.0$ cm; (c) the field point at $x = -1.10$ m, $y = 2.60$ m? Express your results in terms of the unit vectors $\hat{i}$ and $\hat{j}$.

21.40 •• A $+8.75$-μC point charge is glued down on a horizontal frictionless table. It is tied to a -6.50-μC point charge by a light, nonconducting 2.50-cm wire. A uniform electric field of magnitude 1.85×10^8 N/C is directed parallel to the wire, as shown in Fig. E21.40. (a) Find the tension in the wire. (b) What would the tension be if both charges were negative?

Figure **E21.40**

21.41 •• (a) An electron is moving east in a uniform electric field of 1.50 N/C directed to the west. At point A, the velocity of the electron is 4.50×10^5 m/s toward the east. What is the speed of the electron when it reaches point B, 0.375 m east of point A? (b) A proton is moving in the uniform electric field of part (a). At point A, the velocity of the proton is 1.90×10^4 m/s, east. What is the speed of the proton at point B?

Section 21.5 Electric-Field Calculations

21.42 • Two point charges Q and $+q$ (where q is positive) produce the net electric field shown at point P in Fig. E21.42. The field points parallel to the line connecting the two charges. (a) What can you conclude about the sign and magnitude of Q? Explain your reasoning. (b) If the lower charge were negative instead, would it be possible for the field to have the direction shown in the figure? Explain your reasoning.

Figure **E21.42**

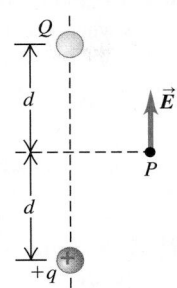

21.43 •• Two positive point charges q are placed on the x-axis, one at $x = a$ and one at $x = -a$. (a) Find the magnitude and direction of the electric field at $x = 0$. (b) Derive an expression for the electric field at points on the x-axis. Use your result to graph the x-component of the electric field as a function of x, for values of x between $-4a$ and $+4a$.

21.44 • The two charges q_1 and q_2 shown in Fig. E21.44 have equal magnitudes. What is the direction of the net electric field due to these two charges at points A (midway between the charges), B, and C if (a) both charges are negative, (b) both charges are positive, (c) q_1 is positive and q_2 is negative.

Figure **E21.44**

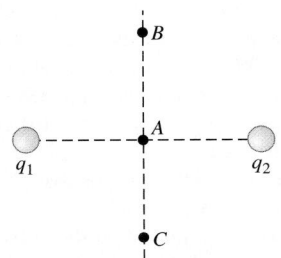

21.45 • A $+2.00$-nC point charge is at the origin, and a second -5.00-nC point charge is on the x-axis at $x = 0.800$ m. (a) Find the electric field (magnitude and direction) at each of the following points on the x-axis: (i) $x = 0.200$ m; (ii) $x = 1.20$ m; (iii) $x = -0.200$ m. (b) Find the net electric force that the two charges would exert on an electron placed at each point in part (a).

21.46 •• Repeat Exercise 21.44, but now let $q_1 = -4.00$ nC.

21.47 • Three negative point charges lie along a line as shown in Fig. E21.47. Find the magnitude and direction of the electric field this combination of charges produces at point P, which lies 6.00 cm from the -2.00-μC charge measured perpendicular to the line connecting the three charges.

Figure **E21.47**

21.48 •• **BIO** **Electric Field of Axons.** A nerve signal is transmitted through a neuron when an excess of Na^+ ions suddenly enters the axon, a long cylindrical part of the neuron. Axons are approximately 10.0 μm in diameter, and measurements show that about 5.6×10^{11} Na^+ ions per meter (each of charge $+e$) enter during this process. Although the axon is a long cylinder, the charge does not all enter everywhere at the same time. A plausible model would be a series of point charges moving along the axon. Let us look at a 0.10-mm length of the axon and model it as a point charge. (a) If the charge that enters each meter of the axon gets distributed uniformly along it, how many coulombs of charge enter a 0.10-mm length of the axon? (b) What electric field (magnitude and direction) does the sudden influx of charge produce at the surface of the body if the axon is 5.00 cm below the skin? (c) Certain sharks can respond to electric fields as weak as 1.0 μN/C.

How far from this segment of axon could a shark be and still detect its electric field?

21.49 • In a rectangular coordinate system a positive point charge $q = 6.00 \times 10^{-9}$ C is placed at the point $x = +0.150$ m, $y = 0$, and an identical point charge is placed at $x = -0.150$ m, $y = 0$. Find the x- and y-components, the magnitude, and the direction of the electric field at the following points: (a) the origin; (b) $x = 0.300$ m, $y = 0$; (c) $x = 0.150$ m, $y = -0.400$ m; (d) $x = 0$, $y = 0.200$ m.

21.50 •• A point charge $q_1 = -4.00$ nC is at the point $x = 0.600$ m, $y = 0.800$ m, and a second point charge $q_2 = +6.00$ nC is at the point $x = 0.600$ m, $y = 0$. Calculate the magnitude and direction of the net electric field at the origin due to these two point charges.

21.51 •• Repeat Exercise 21.49 for the case where the point charge at $x = +0.150$ m, $y = 0$ is positive and the other is negative, each with magnitude 6.00×10^{-9} C.

21.52 •• A very long, straight wire has charge per unit length 1.50×10^{-10} C/m. At what distance from the wire is the electric-field magnitude equal to 2.50 N/C?

21.53 • A ring-shaped conductor with radius $a = 2.50$ cm has a total positive charge $Q = +0.125$ nC uniformly distributed around it, as shown in Fig. 21.23. The center of the ring is at the origin of coordinates O. (a) What is the electric field (magnitude and direction) at point P, which is on the x-axis at $x = 40.0$ cm? (b) A point charge $q = -2.50$ μC is placed at the point P described in part (a). What are the magnitude and direction of the force exerted *by* the charge q *on* the ring?

21.54 •• A straight, nonconducting plastic wire 8.50 cm long carries a charge density of $+175$ nC/m distributed uniformly along its length. It is lying on a horizontal tabletop. (a) Find the magnitude and direction of the electric field this wire produces at a point 6.00 cm directly above its midpoint. (b) If the wire is now bent into a circle lying flat on the table, find the magnitude and direction of the electric field it produces at a point 6.00 cm directly above its center.

21.55 •• A charge of -6.50 nC is spread uniformly over the surface of one face of a nonconducting disk of radius 1.25 cm. (a) Find the magnitude and direction of the electric field this disk produces at a point P on the axis of the disk a distance of 2.00 cm from its center. (b) Suppose that the charge were all pushed away from the center and distributed uniformly on the outer rim of the disk. Find the magnitude and direction of the electric field at point P. (c) If the charge is all brought to the center of the disk, find the magnitude and direction of the electric field at point P. (d) Why is the field in part (a) stronger than the field in part (b)? Why is the field in part (c) the strongest of the three fields?

Section 21.7 Electric Dipoles

21.56 • The ammonia molecule (NH_3) has a dipole moment of 5.0×10^{-30} C·m. Ammonia molecules in the gas phase are placed in a uniform electric field $\vec{E}$ with magnitude 1.6×10^6 N/C. (a) What is the change in electric potential energy when the dipole moment of a molecule changes its orientation with respect to $\vec{E}$ from parallel to perpendicular? (b) At what absolute temperature T is the average translational kinetic energy $\frac{3}{2}kT$ of a molecule equal to the change in potential energy calculated in part (a)? (*Note:* Above this temperature, thermal agitation prevents the dipoles from aligning with the electric field.)

21.57 • Point charges $q_1 = -4.5$ nC and $q_2 = +4.5$ nC are separated by 3.1 mm, forming an electric dipole. (a) Find the electric dipole moment (magnitude and direction). (b) The charges are in a uniform electric field whose direction makes an angle of 36.9° with the line connecting the charges. What is the magnitude of this field if the torque exerted on the dipole has magnitude 7.2×10^{-9} N·m?

21.58 • The dipole moment of the water molecule (H_2O) is 6.17×10^{-30} C·m. Consider a water molecule located at the origin whose dipole moment $\vec{p}$ points in the $+x$-direction. A chlorine ion (Cl^-), of charge -1.60×10^{-19} C, is located at $x = 3.00 \times 10^{-9}$ m. Find the magnitude and direction of the electric force that the water molecule exerts on the chlorine ion. Is this force attractive or repulsive? Assume that x is much larger than the separation d between the charges in the dipole, so that the approximate expression for the electric field along the dipole axis derived in Example 21.14 can be used.

21.59 • **Torque on a Dipole.** An electric dipole with dipole moment $\vec{p}$ is in a uniform electric field $\vec{E}$. (a) Find the orientations of the dipole for which the torque on the dipole is zero. (b) Which of the orientations in part (a) is stable, and which is unstable? (*Hint:* Consider a small displacement away from the equilibrium position and see what happens.) (c) Show that for the stable orientation in part (b), the dipole's own electric field tends to oppose the external field.

21.60 •• Consider the electric dipole of Example 21.14. (a) Derive an expression for the magnitude of the electric field produced by the dipole at a point on the x-axis in Fig. 21.33. What is the direction of this electric field? (b) How does the electric field at points on the x-axis depend on x when x is very large?

21.61 • Three charges are at the corners of an isosceles triangle as shown in Fig. E21.61. The ± 5.00-μC charges form a dipole. (a) Find the force (magnitude and direction) the -10.00-μC charge exerts on the dipole. (b) For an axis perpendicular to the line connecting the ± 5.00-μC charges at the midpoint of this line, find the torque (magnitude and direction) exerted on the dipole by the -10.00-μC charge.

Figure **E21.61**

21.62 • A dipole consisting of charges $\pm e$, 220 nm apart, is placed between two very large (essentially infinite) sheets carrying equal but opposite charge densities of 125 $\mu C/m^2$. (a) What is the maximum potential energy this dipole can have due to the sheets, and how should it be oriented relative to the sheets to attain this value? (b) What is the maximum torque the sheets can exert on the dipole, and how should it be oriented relative to the sheets to attain this value? (c) What net force do the two sheets exert on the dipole?

PROBLEMS

21.63 ••• Four identical charges Q are placed at the corners of a square of side L. (a) In a free-body diagram, show all of the forces that act on one of the charges. (b) Find the magnitude and direction of the total force exerted on one charge by the other three charges.

21.64 ••• Two charges, one of 2.50 μC and the other of -3.50 μC, are placed on the x-axis, one at the origin and the other at $x = 0.600$ m, as shown in Fig. P21.64. Find the position on the x-axis where the net force on a small charge $+q$ would be zero.

Figure **P21.64**

21.65 •• Three point charges are arranged along the x-axis. Charge $q_1 = -4.50$ nC is located at $x = 0.200$ m, and charge $q_2 = +2.50$ nC is at $x = -0.300$ m. A positive point charge q_3 is located at the origin. (a) What must the value of q_3 be for the net force on this point charge to have magnitude 4.00 μN? (b) What is the direction of the net force on q_3? (c) Where along the x-axis can q_3 be placed and the net force on it be zero, other than the trivial answers of $x = +\infty$ and $x = -\infty$?

21.66 •• A charge $q_1 = +5.00$ nC is placed at the origin of an xy-coordinate system, and a charge $q_2 = -2.00$ nC is placed on the positive x-axis at $x = 4.00$ cm. (a) If a third charge $q_3 = +6.00$ nC is now placed at the point $x = 4.00$ cm, $y = 3.00$ cm, find the x- and y-components of the total force exerted on this charge by the other two. (b) Find the magnitude and direction of this force.

21.67 •• **CP** Two positive point charges Q are held fixed on the x-axis at $x = a$ and $x = -a$. A third positive point charge q, with mass m, is placed on the x-axis away from the origin at a coordinate x such that $|x| \ll a$. The charge q, which is free to move along the x-axis, is then released. (a) Find the frequency of oscillation of the charge q. (*Hint:* Review the definition of simple harmonic motion in Section 14.2. Use the binomial expansion $(1 + z)^n = 1 + nz + n(n-1)z^2/2 + \cdots$, valid for the case $|z| < 1$.) (b) Suppose instead that the charge q were placed on the y-axis at a coordinate y such that $|y| \ll a$, and then released. If this charge is free to move anywhere in the xy-plane, what will happen to it? Explain your answer.

21.68 •• **CP** Two identical spheres with mass m are hung from silk threads of length L, as shown in Fig. P21.68. Each sphere has the same charge, so $q_1 = q_2 = q$. The radius of each sphere is very small compared to the distance between the spheres, so they may be treated as point charges. Show that if the angle θ is small, the equilibrium separation d between the spheres is $d = (q^2L/2\pi\epsilon_0 mg)^{1/3}$. (*Hint:* If θ is small, then $\tan\theta \cong \sin\theta$.)

Figure **P21.68**

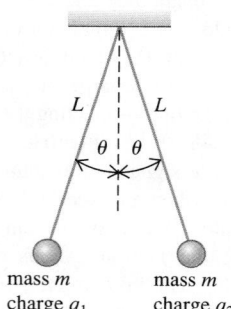

mass m mass m
charge q_1 charge q_2

21.69 ••• **CP** Two small spheres with mass $m = 15.0$ g are hung by silk threads of length $L = 1.20$ m from a common point (Fig. P21.68). When the spheres are given equal quantities of negative charge, so that $q_1 = q_2 = q$, each thread hangs at $\theta = 25.0°$ from the vertical. (a) Draw a diagram showing the forces on each sphere. Treat the spheres as point charges. (b) Find the magnitude of q. (c) Both threads are now shortened to length $L = 0.600$ m, while the charges q_1 and q_2 remain unchanged. What new angle will each thread make with the vertical? (*Hint:* This part of the problem can be solved numerically by using trial values for θ and adjusting the values of θ until a self-consistent answer is obtained.)

21.70 •• **CP** Two identical spheres are each attached to silk threads of length $L = 0.500$ m and hung from a common point (Fig. P21.68). Each sphere has mass $m = 8.00$ g. The radius of each sphere is very small compared to the distance between the spheres, so they may be treated as point charges. One sphere is given positive charge q_1, and the other a different positive charge q_2; this causes the spheres to separate so that when the spheres are in equilibrium, each thread makes an angle $\theta = 20.0°$ with the vertical. (a) Draw a free-body diagram for each sphere when in equilibrium, and label all the forces that act on each sphere. (b) Determine the magnitude of the electrostatic force that acts on each sphere, and determine the tension in each thread. (c) Based on the information you have been given, what can you say about the magnitudes of q_1 and q_2? Explain your answers. (d) A small wire is now connected between the spheres, allowing charge to be transferred from one sphere to the other until the two spheres have equal charges; the wire is then removed. Each thread now makes an angle of $30.0°$ with the vertical. Determine the original charges. (*Hint:* The total charge on the pair of spheres is conserved.)

21.71 •• Sodium chloride (NaCl, ordinary table salt) is made up of positive sodium ions (Na^+) and negative chloride ions (Cl^-). (a) If a point charge with the same charge and mass as all the Na^+ ions in 0.100 mol of NaCl is 2.00 cm from a point charge with the same charge and mass as all the Cl^- ions, what is the magnitude of the attractive force between these two point charges? (b) If the positive point charge in part (a) is held in place and the negative point charge is released from rest, what is its initial acceleration? (See Appendix D for atomic masses.) (c) Does it seem reasonable that the ions in NaCl could be separated in this way? Why or why not? (In fact, when sodium chloride dissolves in water, it breaks up into Na^+ and Cl^- ions. However, in this situation there are additional electric forces exerted by the water molecules on the ions.)

21.72 •• A -5.00-nC point charge is on the x-axis at $x = 1.20$ m. A second point charge Q is on the x-axis at -0.600 m. What must be the sign and magnitude of Q for the resultant electric field at the origin to be (a) 45.0 N/C in the $+x$-direction, (b) 45.0 N/C in the $-x$-direction?

21.73 •• **CP** A small 12.3-g plastic ball is tied to a very light 28.6-cm string that is attached to the vertical wall of a room (Fig. P21.73). A uniform horizontal electric field exists in this room. When the ball has been given an excess charge of $-1.11\ \mu C$, you observe that it remains suspended, with the string making an angle of $17.4°$ with the wall. Find the magnitude and direction of the electric field in the room.

Figure **P21.73**

17.4°

21.74 •• **CP** At $t = 0$ a very small object with mass 0.400 mg and charge $+9.00\ \mu C$ is traveling at 125 m/s in the $-x$-direction. The charge is moving in a uniform electric field that is in the $+y$-direction and that has magnitude $E = 895$ N/C. The gravitational force on the particle can be neglected. How far is the particle from the origin at $t = 7.00$ ms?

21.75 •• Two particles having charges $q_1 = 0.500$ nC and $q_2 = 8.00$ nC are separated by a distance of 1.20 m. At what point along the line connecting the two charges is the total electric field due to the two charges equal to zero?

21.76 ••• Two point charges q_1 and q_2 are held in place 4.50 cm apart. Another point charge $Q = -1.75\ \mu C$ of mass 5.00 g is initially located 3.00 cm from each of these charges (Fig. P21.76) and released from rest. You observe that the initial acceleration of Q is 324 m/s² upward, parallel to the line connecting the two point charges. Find q_1 and q_2.

Figure **P21.76**

q_1

$\vec{a}$ 3.00 cm

Q 4.50 cm

3.00 cm

q_2

21.77 • Three identical point charges q are placed at each of three corners of a square of side L. Find the magnitude and direction of the net force on a point charge $-3q$ placed (a) at the center of the square and (b) at the vacant corner of the square. In each case, draw a free-body diagram showing the forces exerted on the $-3q$ charge by each of the other three charges.

21.78 ••• Three point charges are placed on the y-axis: a charge q at $y = a$, a charge $-2q$ at the origin, and a charge q at $y = -a$. Such an arrangement is called an electric quadrupole. (a) Find the magnitude and direction of the electric field at points on the positive x-axis. (b) Use the binomial expansion to find an approximate expression for the electric field valid for $x \gg a$. Contrast this behavior to that of the electric field of a point charge and that of the electric field of a dipole.

21.79 •• **CP** **Strength of the Electric Force.** Imagine two 1.0-g bags of protons, one at the earth's north pole and the other at the south pole. (a) How many protons are in each bag? (b) Calculate the gravitational attraction and the electrical repulsion that each bag exerts on the other. (c) Are the forces in part (b) large enough for you to feel if you were holding one of the bags?

21.80 • **Electric Force Within the Nucleus.** Typical dimensions of atomic nuclei are of the order of 10^{-15} m (1 fm). (a) If two protons in a nucleus are 2.0 fm apart, find the magnitude of the electric force each one exerts on the other. Express the answer in newtons and in pounds. Would this force be large enough for a person to feel? (b) Since the protons repel each other so strongly, why don't they shoot out of the nucleus?

21.81 •• **If Atoms Were Not Neutral . . .** Because the charges on the electron and proton have the same absolute value, atoms are electrically neutral. Suppose this were not precisely true, and the absolute value of the charge of the electron were less than the charge of the proton by 0.00100%. (a) Estimate what the net charge of this textbook would be under these circumstances. Make any assumptions you feel are justified, but state clearly what they are. (*Hint:* Most of the atoms in this textbook have equal numbers of electrons, protons, and neutrons.) (b) What would be the magnitude of the electric force between two textbooks placed 5.0 m apart? Would this force be attractive or repulsive? Estimate what the acceleration of each book would be if the books were 5.0 m apart and there were no non-electric forces on them. (c) Discuss how the fact that ordinary matter is stable shows that the absolute values of the charges on the electron and proton must be identical to a *very* high level of accuracy.

21.82 ••• **CP** Two tiny spheres of mass 6.80 mg carry charges of equal magnitude,

Figure **P21.82**

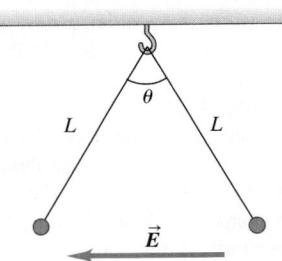

θ

L L

$\vec{E}$

72.0 nC, but opposite sign. They are tied to the same ceiling hook by light strings of length 0.530 m. When a horizontal uniform electric field E that is directed to the left is turned on, the spheres hang at rest with the angle θ between the strings equal to 50.0° (Fig. P21.82). (a) Which ball (the one on the right or the one on the left) has positive charge? (b) What is the magnitude E of the field?

21.83 ·· CP Consider a model of a hydrogen atom in which an electron is in a circular orbit of radius $r = 5.29 \times 10^{-11}$ m around a stationary proton. What is the speed of the electron in its orbit?

21.84 ·· CP A small sphere with mass 9.00 μg and charge -4.30 μC is moving in a circular orbit around a stationary sphere that has charge $+7.50$ μC. If the speed of the small sphere is 5.90×10^3 m/s, what is the radius of its orbit? Treat the spheres as point charges and ignore gravity.

21.85 ·· Two small copper spheres each have radius 1.00 mm. (a) How many atoms does each sphere contain? (b) Assume that each copper atom contains 29 protons and 29 electrons. We know that electrons and protons have charges of exactly the same magnitude, but let's explore the effect of small differences (see also Problem 21.81). If the charge of a proton is $+e$ and the magnitude of the charge of an electron is 0.100% smaller, what is the net charge of each sphere and what force would one sphere exert on the other if they were separated by 1.00 m?

21.86 ··· CP Operation of an Inkjet Printer. In an inkjet printer, letters are built up by squirting drops of ink at the paper from a rapidly moving nozzle. The ink drops, which have a mass of 1.4×10^{-8} g each, leave the nozzle and travel toward the paper at 20 m/s, passing through a charging unit that gives each drop a positive charge q by removing some electrons from it. The drops then pass between parallel deflecting plates 2.0 cm long where there is a uniform vertical electric field with magnitude 8.0×10^4 N/C. If a drop is to be deflected 0.30 mm by the time it reaches the end of the deflection plates, what magnitude of charge must be given to the drop?

21.87 ·· CP A proton is projected into a uniform electric field that points vertically upward and has magnitude E. The initial velocity of the proton has a magnitude v_0 and is directed at an angle α below the horizontal. (a) Find the maximum distance h_{max} that the proton descends vertically below its initial elevation. You can ignore gravitational forces. (b) After what horizontal distance d does the proton return to its original elevation? (c) Sketch the trajectory of the proton. (d) Find the numerical values of h_{max} and d if $E = 500$ N/C, $v_0 = 4.00 \times 10^5$ m/s, and $\alpha = 30.0°$.

21.88 · A negative point charge $q_1 = -4.00$ nC is on the x-axis at $x = 0.60$ m. A second point charge q_2 is on the x-axis at $x = -1.20$ m. What must the sign and magnitude of q_2 be for the net electric field at the origin to be (a) 50.0 N/C in the $+x$-direction and (b) 50.0 N/C in the $-x$-direction?

21.89 ·· CALC Positive charge Q is distributed uniformly along the x-axis from $x = 0$ to $x = a$. A positive point charge q is located on the positive x-axis at $x = a + r$, a distance r to the right of the end of Q (Fig. P21.89). (a) Calculate the x- and y-components of the electric field produced by the charge distribution Q at points on the positive x-axis where $x > a$. (b) Calculate the force (magnitude and direction) that the charge distribution Q exerts on q. (c) Show that if $r \gg a$, the magnitude of the force in part (b) is approximately $Qq/4\pi\epsilon_0 r^2$. Explain why this result is obtained.

Figure **P21.89**

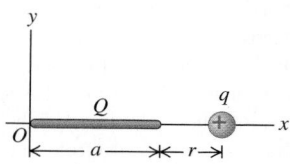

21.90 ·· CALC Positive charge Q is distributed uniformly along the positive y-axis between $y = 0$ and $y = a$. A negative point charge $-q$ lies on the positive x-axis, a distance x from the origin (Fig. P21.90). (a) Calculate the x- and y-components of the electric field produced by the charge distribution Q at points on the positive x-axis. (b) Calculate the x- and y-components of the force that the charge distribution Q exerts on q. (c) Show that if $x \gg a$, $F_x \cong -Qq/4\pi\epsilon_0 x^2$ and $F_y \cong +Qqa/8\pi\epsilon_0 x^3$. Explain why this result is obtained.

Figure **P21.90**

21.91 ·· A charged line like that shown in Fig. 21.24 extends from $y = 2.50$ cm to $y = -2.50$ cm. The total charge distributed uniformly along the line is -7.00 nC. (a) Find the electric field (magnitude and direction) on the x-axis at $x = 10.0$ cm. (b) Is the magnitude of the electric field you calculated in part (a) larger or smaller than the electric field 10.0 cm from a point charge that has the same total charge as this finite line of charge? In terms of the approximation used to derive $E = Q/4\pi\epsilon_0 x^2$ for a point charge from Eq. (21.9), explain why this is so. (c) At what distance x does the result for the finite line of charge differ by 1.0% from that for the point charge?

21.92 · CP A Parallel Universe. Imagine a parallel universe in which the electric force has the same properties as in our universe but there is no gravity. In this parallel universe, the sun carries charge Q, the earth carries charge $-Q$, and the electric attraction between them keeps the earth in orbit. The earth in the parallel universe has the same mass, the same orbital radius, and the same orbital period as in our universe. Calculate the value of Q. (Consult Appendix F as needed.)

21.93 ··· A uniformly charged disk like the disk in Fig. 21.25 has radius 2.50 cm and carries a total charge of 7.0×10^{-12} C. (a) Find the electric field (magnitude and direction) on the x-axis at $x = 20.0$ cm. (b) Show that for $x \gg R$, Eq. (21.11) becomes $E = Q/4\pi\epsilon_0 x^2$, where Q is the total charge on the disk. (c) Is the magnitude of the electric field you calculated in part (a) larger or smaller than the electric field 20.0 cm from a point charge that has the same total charge as this disk? In terms of the approximation used in part (b) to derive $E = Q/4\pi\epsilon_0 x^2$ for a point charge from Eq. (21.11), explain why this is so. (d) What is the percent difference between the electric fields produced by the finite disk and by a point charge with the same charge at $x = 20.0$ cm and at $x = 10.0$ cm?

21.94 ·· BIO Electrophoresis. Electrophoresis is a process used by biologists to separate different biological molecules (such as proteins) from each other according to their ratio of charge to size. The materials to be separated are in a viscous solution that produces a drag force F_D proportional to the size and speed of the molecule. We can express this relationship as $F_D = KRv$, where R is the radius of the molecule (modeled as being spherical), v is its speed, and K is a constant that depends on the viscosity of the

Figure **P21.94**

solution. The solution is placed in an external electric field E so that the electric force on a particle of charge q is $F = qE$. (a) Show that when the electric field is adjusted so that the two forces (electric and viscous drag) just balance, the ratio of q to R is Kv/E. (b) Show that if we leave the electric field on for a time T, the distance x that the molecule moves during that time is $x = (ET/k)(q/R)$. (c) Suppose you have a sample containing three different biological molecules for which the molecular ratio q/R for material 2 is twice that of material 1 and the ratio for material 3 is three times that of material 1. Show that the distances migrated by these molecules after the same amount of time are $x_2 = 2x_1$ and $x_3 = 3x_1$. In other words, material 2 travels twice as far as material 1, and material 3 travels three times as far as material 1. Therefore, we have separated these molecules according to their ratio of charge to size. In practice, this process can be carried out in a special gel or paper, along which the biological molecules migrate. (Fig. P21.94). The process can be rather slow, requiring several hours for separations of just a centimeter or so.

21.95 • **CALC** Positive charge $+Q$ is distributed uniformly along the $+x$-axis from $x = 0$ to $x = a$. Negative charge $-Q$ is distributed uniformly along the $-x$-axis from $x = 0$ to $x = -a$. (a) A positive point charge q lies on the positive y-axis, a distance y from the origin. Find the force (magnitude and direction) that the positive and negative charge distributions together exert on q. Show that this force is proportional to y^{-3} for $y \gg a$. (b) Suppose instead that the positive point charge q lies on the positive x-axis, a distance $x > a$ from the origin. Find the force (magnitude and direction) that the charge distribution exerts on q. Show that this force is proportional to x^{-3} for $x \gg a$.

21.96 •• **CP** A small sphere with mass m carries a positive charge q and is attached to one end of a silk fiber of length L. The other end of the fiber is attached to a large vertical insulating sheet that has a positive surface charge density σ. Show that when the sphere is in equilibrium, the fiber makes an angle equal to arctan $(q\sigma/2mg\epsilon_0)$ with the vertical sheet.

21.97 •• **CALC** Negative charge $-Q$ is distributed uniformly around a quarter-circle of radius a that lies in the first quadrant, with the center of curvature at the origin. Find the x- and y-components of the net electric field at the origin.

21.98 •• **CALC** A semicircle of radius a is in the first and second quadrants, with the center of curvature at the origin. Positive charge $+Q$ is distributed uniformly around the left half of the semicircle, and negative charge $-Q$ is distributed uniformly around the right half of the semicircle (Fig. P21.98). What are the magnitude and direction of the net electric field at the origin produced by this distribution of charge?

Figure **P21.98**

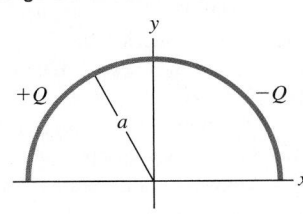

21.99 •• Two 1.20-m nonconducting wires meet at a right angle. One segment carries $+2.50 \mu C$ of charge distributed uniformly along its length, and the other carries $-2.50 \mu C$ distributed uniformly along it, as shown in Fig. P21.99.

Figure **P21.99**

(a) Find the magnitude and direction of the electric field these wires produce at point P, which is 60.0 cm from each wire. (b) If an electron is released at P, what are the magnitude and direction of the net force that these wires exert on it?

21.100 • Two very large parallel sheets are 5.00 cm apart. Sheet A carries a uniform surface charge density of $-9.50 \mu C/m^2$, and sheet B, which is to the right of A, carries a uniform charge density of $-11.6 \mu C/m^2$. Assume the sheets are large enough to be treated as infinite. Find the magnitude and direction of the net electric field these sheets produce at a point (a) 4.00 cm to the right of sheet A; (b) 4.00 cm to the left of sheet A; (c) 4.00 cm to the right of sheet B.

21.101 • Repeat Problem 21.100 for the case where sheet B is positive.

21.102 • Two very large horizontal sheets are 4.25 cm apart and carry equal but opposite uniform surface charge densities of magnitude σ. You want to use these sheets to hold stationary in the region between them an oil droplet of mass 324 μg that carries an excess of five electrons. Assuming that the drop is in vacuum, (a) which way should the electric field between the plates point, and (b) what should σ be?

21.103 •• An infinite sheet with positive charge per unit area σ lies in the xy-plane. A second infinite sheet with negative charge per unit area $-\sigma$ lies in the yz-plane. Find the net electric field at all points that do not lie in either of these planes. Express your answer in terms of the unit vectors $\hat{\imath}, \hat{\jmath}$, and $\hat{k}$.

21.104 •• **CP** A thin disk with a circular hole at its center, called an *annulus*, has inner radius R_1 and outer radius R_2 (Fig. P21.104). The disk has a uniform positive surface charge density σ on its surface. (a) Determine the total electric charge on the annulus. (b) The annulus lies in the yz-plane, with its center at the origin. For an arbitrary point on the x-axis (the axis of the annulus),

Figure **P21.104**

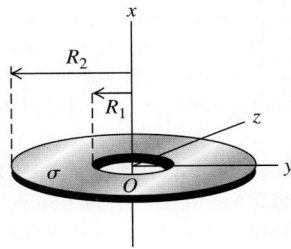

find the magnitude and direction of the electric field $\vec{E}$. Consider points both above and below the annulus in Fig. P21.104. (c) Show that at points on the x-axis that are sufficiently close to the origin, the magnitude of the electric field is approximately proportional to the distance between the center of the annulus and the point. How close is "sufficiently close"? (d) A point particle with mass m and negative charge $-q$ is free to move along the x-axis (but cannot move off the axis). The particle is originally placed at rest at $x = 0.01R_1$ and released. Find the frequency of oscillation of the particle. (*Hint*: Review Section 14.2. The annulus is held stationary.)

CHALLENGE PROBLEMS

21.105 ••• Three charges are placed as shown in Fig. P21.105. The magnitude of q_1 is 2.00 μC, but its sign and the value of the charge q_2 are not known. Charge q_3 is $+4.00 \mu C$, and the net force $\vec{F}$ on q_3 is entirely in the negative x-direction. (a) Considering the different possible signs of q_1, there are four possible force diagrams representing the forces $\vec{F}_1$ and $\vec{F}_2$ that q_1 and q_2 exert on q_3. Sketch these four possible force configurations.

Figure **P21.105**

(b) Using the sketches from part (a) and the direction of $\vec{F}$, deduce the signs of the charges q_1 and q_2. (c) Calculate the magnitude of q_2. (d) Determine F, the magnitude of the net force on q_3.

21.106 ••• Two charges are placed as shown in Fig. P21.106. The magnitude of q_1 is 3.00 μC, but its sign and the value of the charge q_2 are not known. The direction of the net electric field $\vec{E}$ at point P is entirely in the negative y-direction. (a) Considering the different possible signs of q_1 and q_2, there are four possible diagrams that could represent the electric fields $\vec{E}_1$ and $\vec{E}_2$ produced by q_1 and q_2. Sketch the four possible electric-field configurations. (b) Using the sketches from part (a) and the direction of $\vec{E}$, deduce the signs of q_1 and q_2. (c) Determine the magnitude of $\vec{E}$.

Figure **P21.106**

P

5.0 cm

12.0 cm

$\vec{E}$

q_1 13.0 cm q_2

21.107 ••• **CALC** Two thin rods of length L lie along the x-axis, one between $x = a/2$ and $x = a/2 + L$ and the other between $x = -a/2$ and $x = -a/2 - L$. Each rod has positive charge Q distributed uniformly along its length. (a) Calculate the electric field produced by the second rod at points along the positive x-axis. (b) Show that the magnitude of the force that one rod exerts on the other is

$$F = \frac{Q^2}{4\pi\epsilon_0 L^2} \ln\left[\frac{(a + L)^2}{a(a + 2L)}\right]$$

(c) Show that if $a \gg L$, the magnitude of this force reduces to $F = Q^2/4\pi\epsilon_0 a^2$. (*Hint:* Use the expansion $\ln(1 + z) = z - z^2/2 + z^3/3 - \cdots$, valid for $|z| \ll 1$. Carry *all* expansions to at least order L^2/a^2.) Interpret this result.

Answers

Chapter Opening Question ?

Water molecules have a permanent electric dipole moment: One end of the molecule has a positive charge and the other end has a negative charge. These ends attract negative and positive ions, respectively, holding the ions apart in solution. Water is less effective as a solvent for materials whose molecules do not ionize (called *nonionic* substances), such as oils.

Test Your Understanding Questions

21.1 Answers: (a) the plastic rod weighs more, (b) the glass rod weighs less, (c) the fur weighs less, (d) the silk weighs more The plastic rod gets a negative charge by taking electrons from the fur, so the rod weighs a little more and the fur weighs a little less after the rubbing. By contrast, the glass rod gets a positive charge by giving electrons to the silk. Hence, after they are rubbed together, the glass rod weighs a little less and the silk weighs a little more. The weight change is *very* small: The number of electrons transferred is a small fraction of a mole, and a mole of electrons has a mass of only $(6.02 \times 10^{23}$ electrons$)(9.11 \times 10^{-31}$ kg/electron$) = 5.48 \times 10^{-7}$ kg = 0.548 milligram!

21.2 Answers: (a) (i), (b) (ii) Before the two spheres touch, the negatively charged sphere exerts a repulsive force on the electrons in the other sphere, causing zones of positive and negative induced charge (see Fig. 21.7b). The positive zone is closer to the negatively charged sphere than the negative zone, so there is a net force of attraction that pulls the spheres together, like the comb and insulator in Fig. 21.8b. Once the two metal spheres touch, some of the excess electrons on the negatively charged sphere will flow onto the other sphere (because metals are conductors). Then both spheres will have a net negative charge and will repel each other.

21.3 Answer: (iv) The force exerted by q_1 on Q is still as in Example 21.4. The magnitude of the force exerted by q_2 on Q is still equal to $F_{1\text{ on }Q}$, but the direction of the force is now *toward* q_2 at an angle α below the x-axis. Hence the x-components of the two forces cancel while the (negative) y-components add together, and the total electric force is in the negative y-direction.

21.4 Answers: (a) (ii), (b) (i) The electric field $\vec{E}$ produced by a positive point charge points directly away from the charge (see Fig. 21.18a) and has a magnitude that depends on the distance r from the charge to the field point. Hence a second, negative point charge $q < 0$ will feel a force $\vec{F} = q\vec{E}$ that points directly toward the positive charge and has a magnitude that depends on the distance r between the two charges. If the negative charge moves directly toward the positive charge, the direction of the force remains the same but the force magnitude increases as the distance r decreases. If the negative charge moves in a circle around the positive charge, the force magnitude stays the same (because the distance r is constant) but the force direction changes.

21.5 Answer: (iv) Think of a pair of segments of length dy, one at coordinate $y > 0$ and the other at coordinate $-y < 0$. The upper segment has a positive charge and produces an electric field $d\vec{E}$ at P that points away from the segment, so this $d\vec{E}$ has a positive x-component and a negative y-component, like the vector $d\vec{E}$ in Fig. 21.24. The lower segment has the same amount of negative charge. It produces a $d\vec{E}$ that has the same magnitude but points *toward* the lower segment, so it has a negative x-component and a negative y-component. By symmetry, the two x-components are equal but opposite, so they cancel. Thus the total electric field has only a negative y-component.

21.6 Answer: yes If the field lines are straight, $\vec{E}$ must point in the same direction throughout the region. Hence the force $\vec{F} = q\vec{E}$ on a particle of charge q is always in the same direction. A particle released from rest accelerates in a straight line in the direction of $\vec{F}$, and so its trajectory is a straight line along a field line.

21.7 Answer: (ii) Equations (21.17) and (21.18) tell us that the potential energy for a dipole in an electric field is $U = -\vec{p} \cdot \vec{E} = -pE\cos\phi$, where ϕ is the angle between the directions of $\vec{p}$ and $\vec{E}$. If $\vec{p}$ and $\vec{E}$ point in opposite directions, so that $\phi = 180°$, we have $\cos\phi = -1$ and $U = +pE$. This is the maximum value that U can have. From our discussion of energy diagrams in Section 7.5, it follows that this is a situation of unstable equilibrium.

Bridging Problem

Answer: $E = 2kQ/\pi a^2$ in the $-y$-direction

GAUSS'S LAW

? This child acquires an electric charge by touching the charged metal sphere. The charged hairs on the child's head repel and stand out. If the child stands *inside* a large, charged metal sphere, will her hair stand on end?

LEARNING GOALS

By studying this chapter, you will learn:

- How you can determine the amount of charge within a closed surface by examining the electric field on the surface.

- What is meant by electric flux, and how to calculate it.

- How Gauss's law relates the electric flux through a closed surface to the charge enclosed by the surface.

- How to use Gauss's law to calculate the electric field due to a symmetric charge distribution.

- Where the charge is located on a charged conductor.

Often, there are both an easy way and a hard way to do a job; the easy way may involve nothing more than using the right tools. In physics, an important tool for simplifying problems is the *symmetry properties* of systems. Many physical systems have symmetry; for example, a cylindrical body doesn't look any different after you've rotated it around its axis, and a charged metal sphere looks just the same after you've turned it about any axis through its center.

Gauss's law is part of the key to using symmetry considerations to simplify electric-field calculations. For example, the field of a straight-line or plane-sheet charge distribution, which we derived in Section 21.5 using some fairly strenuous integrations, can be obtained in a few lines with the help of Gauss's law. But Gauss's law is more than just a way to make certain calculations easier. Indeed, it is a fundamental statement about the relationship between electric charges and electric fields. Among other things, Gauss's law can help us understand how electric charge distributes itself over conducting bodies.

Here's what Gauss's law is all about. Given any general distribution of charge, we surround it with an imaginary surface that encloses the charge. Then we look at the electric field at various points on this imaginary surface. Gauss's law is a relationship between the field at *all* the points on the surface and the total charge enclosed within the surface. This may sound like a rather indirect way of expressing things, but it turns out to be a tremendously useful relationship. Above and beyond its use as a calculational tool, Gauss's law can help us gain deeper insights into electric fields. We will make use of these insights repeatedly in the next several chapters as we pursue our study of electromagnetism.

22.1 Charge and Electric Flux

In Chapter 21 we asked the question, "Given a charge distribution, what is the electric field produced by that distribution at a point *P*?" We saw that the answer could be found by representing the distribution as an assembly of point charges,

The discussion of Gauss's law in this section is based on and inspired by the innovative ideas of Ruth W. Chabay and Bruce A. Sherwood in *Electric and Magnetic Interactions* (John Wiley & Sons, 1994).

22.1 How can you measure the charge inside a box without opening it?

(a) A box containing an unknown amount of charge

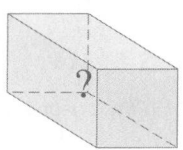

(b) Using a test charge outside the box to probe the amount of charge inside the box

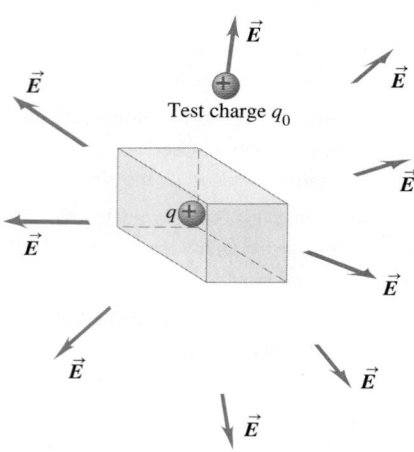

each of which produces an electric field $\vec{E}$ given by Eq. (21.7). The total field at P is then the vector sum of the fields due to all the point charges.

But there is an alternative relationship between charge distributions and electric fields. To discover this relationship, let's stand the question of Chapter 21 on its head and ask, "If the electric field pattern is known in a given region, what can we determine about the charge distribution in that region?"

Here's an example. Consider the box shown in Fig. 22.1a, which may or may not contain electric charge. We'll imagine that the box is made of a material that has no effect on any electric fields; it's of the same breed as the massless rope and the frictionless incline. Better still, let the box represent an *imaginary* surface that may or may not enclose some charge. We'll refer to the box as a **closed surface** because it completely encloses a volume. How can you determine how much (if any) electric charge lies within the box?

Knowing that a charge distribution produces an electric field and that an electric field exerts a force on a test charge, you move a test charge q_0 around the vicinity of the box. By measuring the force $\vec{F}$ experienced by the test charge at different positions, you make a three-dimensional map of the electric field $\vec{E} = \vec{F}/q_0$ outside the box. In the case shown in Fig. 22.1b, the map turns out to be the same as that of the electric field produced by a positive point charge (Fig. 21.28a). From the details of the map, you can find the exact value of the point charge inside the box.

To determine the contents of the box, we actually need to measure $\vec{E}$ only on the *surface* of the box. In Fig. 22.2a there is a single *positive* point charge inside the box, and in Fig. 22.2b there are two such charges. The field patterns on the surfaces of the boxes are different in detail, but in each case the electric field points *out* of the box. Figures 22.2c and 22.2d show cases with one and two *negative* point charges, respectively, inside the box. Again, the details of $\vec{E}$ are different for the two cases, but the electric field points *into* each box.

Electric Flux and Enclosed Charge

In Section 21.4 we mentioned the analogy between electric-field vectors and the velocity vectors of a fluid in motion. This analogy can be helpful, even though an electric field does not actually "flow." Using this analogy, in Figs. 22.2a and 22.2b, in which the electric field vectors point out of the surface, we say that there is an *outward* **electric flux**. (The word "flux" comes from a Latin word meaning "flow.") In Figs. 22.2c and 22.2d the $\vec{E}$ vectors point into the surface, and the electric flux is *inward*.

Figure 22.2 suggests a simple relationship: Positive charge inside the box goes with an outward electric flux through the box's surface, and negative charge inside goes with an inward electric flux. What happens if there is *zero* charge

22.2 The electric field on the surface of boxes containing **(a)** a single positive point charge, **(b)** two positive point charges, **(c)** a single negative point charge, or **(d)** two negative point charges.

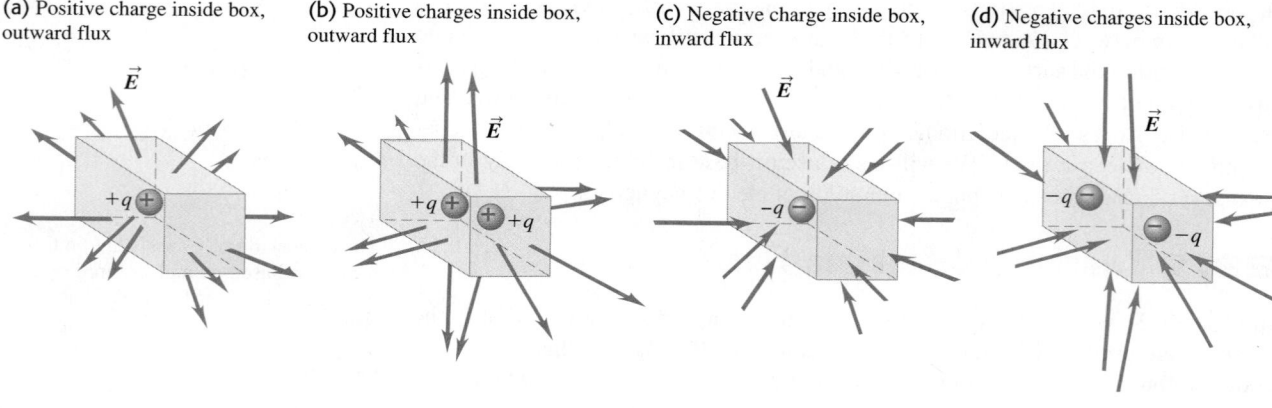

(a) Positive charge inside box, outward flux

(b) Positive charges inside box, outward flux

(c) Negative charge inside box, inward flux

(d) Negative charges inside box, inward flux

inside the box? In Fig. 22.3a the box is empty and $\vec{E} = 0$ everywhere, so there is no electric flux into or out of the box. In Fig. 22.3b, one positive and one negative point charge of equal magnitude are enclosed within the box, so the *net* charge inside the box is zero. There is an electric field, but it "flows into" the box on half of its surface and "flows out of" the box on the other half. Hence there is no *net* electric flux into or out of the box.

The box is again empty in Fig. 22.3c. However, there is charge present *outside* the box; the box has been placed with one end parallel to a uniformly charged infinite sheet, which produces a uniform electric field perpendicular to the sheet (as we learned in Example 21.11 of Section 21.5). On one end of the box, $\vec{E}$ points into the box; on the opposite end, $\vec{E}$ points out of the box; and on the sides, $\vec{E}$ is parallel to the surface and so points neither into nor out of the box. As in Fig. 22.3b, the inward electric flux on one part of the box exactly compensates for the outward electric flux on the other part. So in all of the cases shown in Fig. 22.3, there is no *net* electric flux through the surface of the box, and no *net* charge is enclosed in the box.

Figures 22.2 and 22.3 demonstrate a connection between the *sign* (positive, negative, or zero) of the *net* charge enclosed by a closed surface and the sense (outward, inward, or none) of the net electric flux through the surface. There is also a connection between the *magnitude* of the net charge inside the closed surface and the *strength* of the net "flow" of $\vec{E}$ over the surface. In both Figs. 22.4a and 22.4b there is a single point charge inside the box, but in Fig. 22.4b the magnitude of the charge is twice as great, and so $\vec{E}$ is everywhere twice as great in magnitude as in Fig. 22.4a. If we keep in mind the fluid-flow analogy, this means that the net outward electric flux is also twice as great in Fig. 22.4b as in Fig. 22.4a. This suggests that the net electric flux through the surface of the box is *directly proportional* to the magnitude of the net charge enclosed by the box.

This conclusion is independent of the size of the box. In Fig. 22.4c the point charge $+q$ is enclosed by a box with twice the linear dimensions of the box in Fig. 22.4a. The magnitude of the electric field of a point charge decreases with distance according to $1/r^2$, so the average magnitude of $\vec{E}$ on each face of the large box in Fig. 22.4c is just $\frac{1}{4}$ of the average magnitude on the corresponding face in Fig. 22.4a. But each face of the large box has exactly four times the area of the corresponding face of the small box. Hence the outward electric flux is the *same* for the two boxes if we *define* electric flux as follows: For each face of the box, take the product of the average perpendicular component of $\vec{E}$ and the area of that face; then add up the results from all faces of the box. With this definition the net electric flux due to a single point charge inside the box is independent of the size of the box and depends only on the net charge inside the box.

22.3 Three cases in which there is zero *net* charge inside a box and no net electric flux through the surface of the box. (a) An empty box with $\vec{E} = 0$. (b) A box containing one positive and one equal-magnitude negative point charge. (c) An empty box immersed in a uniform electric field.

(a) No charge inside box, zero flux

(b) Zero *net* charge inside box, inward flux cancels outward flux.

(c) No charge inside box, inward flux cancels outward flux.

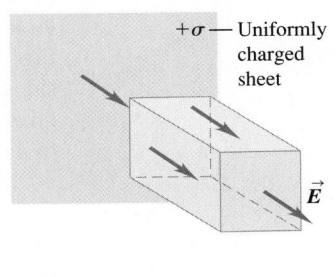

22.4 (a) A box enclosing a positive point charge $+q$. (b) Doubling the charge causes the magnitude of $\vec{E}$ to double, and it doubles the electric flux through the surface. (c) If the charge stays the same but the dimensions of the box are doubled, the flux stays the same. The magnitude of $\vec{E}$ on the surface decreases by a factor of $\frac{1}{4}$, but the area through which $\vec{E}$ "flows" increases by a factor of 4.

(a) A box containing a charge

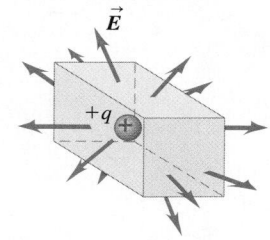

(b) Doubling the enclosed charge doubles the flux.

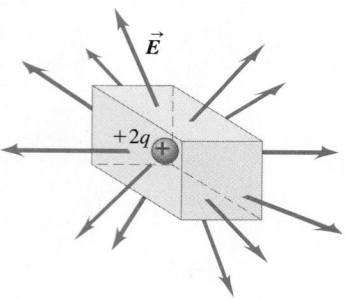

(c) Doubling the box dimensions *does not change* the flux.

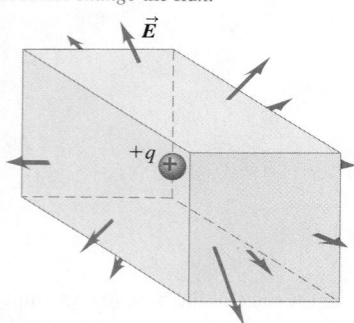

To summarize, for the special cases of a closed surface in the shape of a rectangular box and charge distributions made up of point charges or infinite charged sheets, we have found:

1. Whether there is a net outward or inward electric flux through a closed surface depends on the sign of the enclosed charge.
2. Charges *outside* the surface do not give a net electric flux through the surface.
3. The net electric flux is directly proportional to the net amount of charge enclosed within the surface but is otherwise independent of the size of the closed surface.

These observations are a qualitative statement of *Gauss's law*.

Do these observations hold true for other kinds of charge distributions and for closed surfaces of arbitrary shape? The answer to these questions will prove to be yes. But to explain why this is so, we need a precise mathematical statement of what we mean by electric flux. We develop this in the next section.

Test Your Understanding of Section 22.1 If all of the dimensions of the box in Fig. 22.2a are increased by a factor of 3, what effect will this change have on the electric flux through the box? (i) The flux will be $3^2 = 9$ times greater; (ii) the flux will be 3 times greater; (iii) the flux will be unchanged; (iv) the flux will be $\frac{1}{3}$ as great; (v) the flux will be $\left(\frac{1}{3}\right)^2 = \frac{1}{9}$ as great; (vi) not enough information is given to decide.

22.2 Calculating Electric Flux

In the preceding section we introduced the concept of *electric flux*. We used this to give a rough qualitative statement of Gauss's law: The net electric flux through a closed surface is directly proportional to the net charge inside that surface. To be able to make full use of this law, we need to know how to *calculate* electric flux. To do this, let's again make use of the analogy between an electric field $\vec{E}$ and the field of velocity vectors $\vec{v}$ in a flowing fluid. (Again, keep in mind that this is only an analogy; an electric field is *not* a flow.)

Flux: Fluid-Flow Analogy

Figure 22.5 shows a fluid flowing steadily from left to right. Let's examine the volume flow rate dV/dt (in, say, cubic meters per second) through the wire rectangle with area A. When the area is perpendicular to the flow velocity $\vec{v}$ (Fig. 22.5a) and the flow velocity is the same at all points in the fluid, the volume flow rate dV/dt is the area A multiplied by the flow speed v:

$$\frac{dV}{dt} = vA$$

When the rectangle is tilted at an angle ϕ (Fig. 22.5b) so that its face is not perpendicular to $\vec{v}$, the area that counts is the silhouette area that we see when we look in the direction of $\vec{v}$. This area, which is outlined in red and labeled $A_\perp$ in Fig. 22.5b, is the *projection* of the area A onto a surface perpendicular to $\vec{v}$. Two sides of the projected rectangle have the same length as the original one, but the other two are foreshortened by a factor of $\cos\phi$, so the projected area $A_\perp$ is equal to $A\cos\phi$. Then the volume flow rate through A is

$$\frac{dV}{dt} = vA\cos\phi$$

If $\phi = 90°$, $dV/dt = 0$; the wire rectangle is edge-on to the flow, and no fluid passes through the rectangle.

Also, $v\cos\phi$ is the component of the vector $\vec{v}$ perpendicular to the plane of the area A. Calling this component $v_\perp$, we can rewrite the volume flow rate as

$$\frac{dV}{dt} = v_\perp A$$

We can express the volume flow rate more compactly by using the concept of *vector area $\vec{A}$*, a vector quantity with magnitude A and a direction perpendicular to the plane of the area we are describing. The vector area $\vec{A}$ describes both the size of an area and its orientation in space. In terms of $\vec{A}$, we can write the volume flow rate of fluid through the rectangle in Fig. 22.5b as a scalar (dot) product:

$$\frac{dV}{dt} = \vec{v}\cdot\vec{A}$$

Flux of a Uniform Electric Field

Using the analogy between electric field and fluid flow, we now define electric flux in the same way as we have just defined the volume flow rate of a fluid; we simply replace the fluid velocity $\vec{v}$ by the electric field $\vec{E}$. The symbol that we use for electric flux is Φ_E (the capital Greek letter phi; the subscript E is a reminder that this is *electric* flux). Consider first a flat area A perpendicular to a uniform electric field $\vec{E}$ (Fig. 22.6a). We define the electric flux through this area to be the product of the field magnitude E and the area A:

$$\Phi_E = EA$$

Roughly speaking, we can picture Φ_E in terms of the field lines passing through A. Increasing the area means that more lines of $\vec{E}$ pass through the area, increasing the flux; a stronger field means more closely spaced lines of $\vec{E}$ and therefore more lines per unit area, so again the flux increases.

If the area A is flat but not perpendicular to the field $\vec{E}$, then fewer field lines pass through it. In this case the area that counts is the silhouette area that we see when looking in the direction of $\vec{E}$. This is the area $A_\perp$ in Fig. 22.6b and is equal to $A\cos\phi$ (compare to Fig. 22.5b). We generalize our definition of electric flux for a uniform electric field to

$$\Phi_E = EA\cos\phi \quad \text{(electric flux for uniform $\vec{E}$, flat surface)} \quad (22.1)$$

Since $E\cos\phi$ is the component of $\vec{E}$ perpendicular to the area, we can rewrite Eq. (22.1) as

$$\Phi_E = E_\perp A \quad \text{(electric flux for uniform $\vec{E}$, flat surface)} \quad (22.2)$$

In terms of the vector area $\vec{A}$ perpendicular to the area, we can write the electric flux as the scalar product of $\vec{E}$ and $\vec{A}$:

$$\Phi_E = \vec{E}\cdot\vec{A} \quad \text{(electric flux for uniform $\vec{E}$, flat surface)} \quad (22.3)$$

Equations (22.1), (22.2), and (22.3) express the electric flux for a *flat* surface and a *uniform* electric field in different but equivalent ways. The SI unit for electric flux is $1\ \text{N}\cdot\text{m}^2/\text{C}$. Note that if the area is edge-on to the field, $\vec{E}$ and $\vec{A}$ are perpendicular and the flux is zero (Fig. 22.6c).

We can represent the direction of a vector area $\vec{A}$ by using a *unit vector $\hat{n}$* perpendicular to the area; $\hat{n}$ stands for "normal." Then

$$\vec{A} = A\hat{n} \quad (22.4)$$

A surface has two sides, so there are two possible directions for $\hat{n}$ and $\vec{A}$. We must always specify which direction we choose. In Section 22.1 we related the charge inside a *closed* surface to the electric flux through the surface. With a closed surface we will always choose the direction of $\hat{n}$ to be *outward*, and we

22.5 The volume flow rate of fluid through the wire rectangle (a) is vA when the area of the rectangle is perpendicular to $\vec{v}$ and (b) is $vA\cos\phi$ when the rectangle is tilted at an angle ϕ.

(a) A wire rectangle in a fluid

(b) The wire rectangle tilted by an angle ϕ

Application Flux Through a Basking Shark's Mouth
Unlike aggressive carnivorous sharks such as great whites, a basking shark feeds passively on plankton in the water that passes through the shark's gills as it swims. To survive on these tiny organisms requires a huge flux of water through a basking shark's immense mouth, which can be up to a meter across. The water flux—the product of the shark's speed through the water and the area of its mouth—can be up to $0.5\ \text{m}^3/\text{s}$ (500 liters per second, or almost 5×10^5 gallons per hour). In a similar way, the flux of electric field through a surface depends on the magnitude of the field and the area of the surface (as well as the relative orientation of the field and surface).

22.6 A flat surface in a uniform electric field. The electric flux Φ_E through the surface equals the scalar product of the electric field $\vec{E}$ and the area vector $\vec{A}$.

(a) Surface is face-on to electric field:
- $\vec{E}$ and $\vec{A}$ are parallel (the angle between $\vec{E}$ and $\vec{A}$ is $\phi = 0$).
- The flux $\Phi_E = \vec{E} \cdot \vec{A} = EA$.

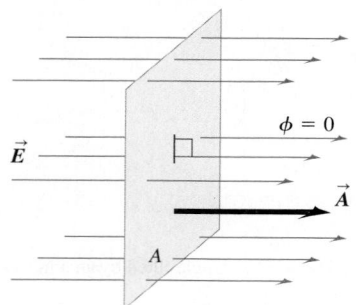

(b) Surface is tilted from a face-on orientation by an angle ϕ:
- The angle between $\vec{E}$ and $\vec{A}$ is ϕ.
- The flux $\Phi_E = \vec{E} \cdot \vec{A} = EA \cos \phi$.

(c) Surface is edge-on to electric field:
- $\vec{E}$ and $\vec{A}$ are perpendicular (the angle between $\vec{E}$ and $\vec{A}$ is $\phi = 90°$).
- The flux $\Phi_E = \vec{E} \cdot \vec{A} = EA \cos 90° = 0$.

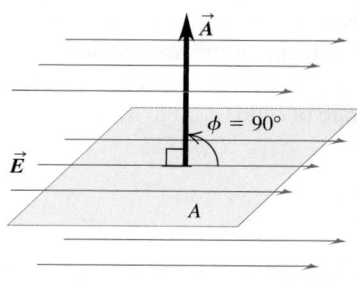

will speak of the flux *out of* a closed surface. Thus what we called "outward electric flux" in Section 22.1 corresponds to a *positive* value of Φ_E, and what we called "inward electric flux" corresponds to a *negative* value of Φ_E.

Flux of a Nonuniform Electric Field

What happens if the electric field $\vec{E}$ isn't uniform but varies from point to point over the area A? Or what if A is part of a curved surface? Then we divide A into many small elements dA, each of which has a unit vector $\hat{n}$ perpendicular to it and a vector area $d\vec{A} = \hat{n}\,dA$. We calculate the electric flux through each element and integrate the results to obtain the total flux:

$$\Phi_E = \int E \cos \phi \, dA = \int E_\perp \, dA = \int \vec{E} \cdot d\vec{A} \qquad \begin{array}{l}\text{(general definition} \\ \text{of electric flux)}\end{array} \qquad (22.5)$$

We call this integral the **surface integral** of the component $E_\perp$ over the area, or the surface integral of $\vec{E} \cdot d\vec{A}$. In specific problems, one form of the integral is sometimes more convenient than another. Example 22.3 at the end of this section illustrates the use of Eq. (22.5).

In Eq. (22.5) the electric flux $\int E_\perp \, dA$ is equal to the *average* value of the perpendicular component of the electric field, multiplied by the area of the surface. This is the same definition of electric flux that we were led to in Section 22.1, now expressed more mathematically. In the next section we will see the connection between the total electric flux through *any* closed surface, no matter what its shape, and the amount of charge enclosed within that surface.

Example 22.1 | **Electric flux through a disk**

A disk of radius 0.10 m is oriented with its normal unit vector $\hat{n}$ at 30° to a uniform electric field $\vec{E}$ of magnitude 2.0×10^3 N/C (Fig. 22.7). (Since this isn't a closed surface, it has no "inside" or "outside." That's why we have to specify the direction of $\hat{n}$ in the figure.) (a) What is the electric flux through the disk? (b) What is the flux through the disk if it is turned so that $\hat{n}$ is perpendicular to $\vec{E}$? (c) What is the flux through the disk if $\hat{n}$ is parallel to $\vec{E}$?

22.7 The electric flux Φ_E through a disk depends on the angle between its normal $\hat{n}$ and the electric field $\vec{E}$.

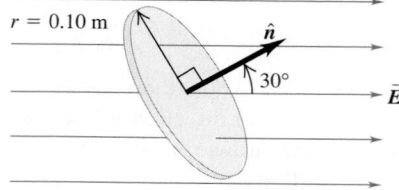

SOLUTION

IDENTIFY and SET UP: This problem is about a flat surface in a uniform electric field, so we can apply the ideas of this section. We calculate the electric flux using Eq. (22.1).

EXECUTE: (a) The area is $A = \pi(0.10 \text{ m})^2 = 0.0314 \text{ m}^2$ and the angle between $\vec{E}$ and $\vec{A} = A\hat{n}$ is $\phi = 30°$, so from Eq. (22.1),

$$\Phi_E = EA\cos\phi = (2.0 \times 10^3 \text{ N/C})(0.0314 \text{ m}^2)(\cos 30°)$$
$$= 54 \text{ N} \cdot \text{m}^2/\text{C}$$

(b) The normal to the disk is now perpendicular to $\vec{E}$, so $\phi = 90°$, $\cos\phi = 0$, and $\Phi_E = 0$.

(c) The normal to the disk is parallel to $\vec{E}$, so $\phi = 0$ and $\cos\phi = 1$:

$$\Phi_E = EA\cos\phi = (2.0 \times 10^3 \text{ N/C})(0.0314 \text{ m}^2)(1)$$
$$= 63 \text{ N} \cdot \text{m}^2/\text{C}$$

EVALUATE: As a check on our results, note that our answer to part (b) is smaller than that to part (a), which is in turn smaller than that to part (c). Is all this as it should be?

Example 22.2 | **Electric flux through a cube**

An imaginary cubical surface of side L is in a region of uniform electric field $\vec{E}$. Find the electric flux through each face of the cube and the total flux through the cube when (a) it is oriented with two of its faces perpendicular to $\vec{E}$ (Fig. 22.8a) and (b) the cube is turned by an angle θ about a vertical axis (Fig. 22.8b).

SOLUTION

IDENTIFY and SET UP: Since $\vec{E}$ is uniform and each of the six faces of the cube is flat, we find the flux Φ_{Ei} through each face using Eqs. (22.3) and (22.4). The total flux through the cube is the sum of the six individual fluxes.

EXECUTE: (a) Figure 22.8a shows the unit vectors $\hat{n}_1$ through $\hat{n}_6$ for each face; each unit vector points *outward* from the cube's closed surface. The angle between $\vec{E}$ and $\hat{n}_1$ is 180°, the angle between $\vec{E}$ and $\hat{n}_2$ is 0°, and the angle between $\vec{E}$ and each of the other four unit vectors is 90°. Each face of the cube has area L^2, so the fluxes through the faces are

$$\Phi_{E1} = \vec{E} \cdot \hat{n}_1 A = EL^2\cos 180° = -EL^2$$
$$\Phi_{E2} = \vec{E} \cdot \hat{n}_2 A = EL^2\cos 0° = +EL^2$$
$$\Phi_{E3} = \Phi_{E4} = \Phi_{E5} = \Phi_{E6} = EL^2\cos 90° = 0$$

The flux is negative on face 1, where $\vec{E}$ is directed into the cube, and positive on face 2, where $\vec{E}$ is directed out of the cube. The total flux through the cube is

$$\Phi_E = \Phi_{E1} + \Phi_{E2} + \Phi_{E3} + \Phi_{E4} + \Phi_{E5} + \Phi_{E6}$$
$$= -EL^2 + EL^2 + 0 + 0 + 0 + 0 = 0$$

(b) The field $\vec{E}$ is directed into faces 1 and 3, so the fluxes through them are negative; $\vec{E}$ is directed out of faces 2 and 4, so the fluxes through them are positive. We find

$$\Phi_{E1} = \vec{E} \cdot \hat{n}_1 A = EL^2\cos(180° - \theta) = -EL^2\cos\theta$$
$$\Phi_{E2} = \vec{E} \cdot \hat{n}_2 A = +EL^2\cos\theta$$
$$\Phi_{E3} = \vec{E} \cdot \hat{n}_3 A = EL^2\cos(90° + \theta) = -EL^2\sin\theta$$
$$\Phi_{E4} = \vec{E} \cdot \hat{n}_4 A = EL^2\cos(90° - \theta) = +EL^2\sin\theta$$
$$\Phi_{E5} = \Phi_{E6} = EL^2\cos 90° = 0$$

The total flux $\Phi_E = \Phi_{E1} + \Phi_{E2} + \Phi_{E3} + \Phi_{E4} + \Phi_{E5} + \Phi_{E6}$ through the surface of the cube is again zero.

EVALUATE: We came to the same conclusion in our discussion of Fig. 22.3c: There is zero net flux of a uniform electric field through a closed surface that contains no electric charge.

22.8 Electric flux of a uniform field $\vec{E}$ through a cubical box of side L in two orientations.

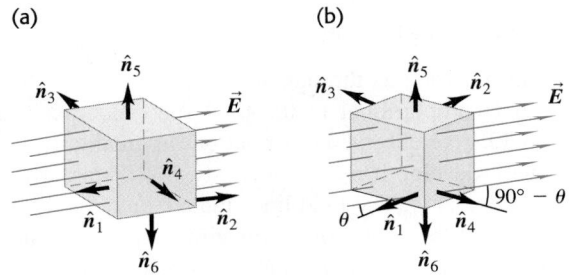

(a)   (b)

Example 22.3 | **Electric flux through a sphere**

A point charge $q = +3.0 \mu\text{C}$ is surrounded by an imaginary sphere of radius $r = 0.20$ m centered on the charge (Fig. 22.9). Find the resulting electric flux through the sphere.

SOLUTION

IDENTIFY and SET UP: The surface is not flat and the electric field is not uniform, so to calculate the electric flux (our target variable)

we must use the general definition, Eq. (22.5). We use Eq. (22.5) to calculate the electric flux (our target variable). Because the sphere is centered on the point charge, at any point on the spherical surface, $\vec{E}$ is directed out of the sphere perpendicular to the surface. The positive direction for both $\hat{n}$ and $E_\perp$ is outward, so $E_\perp = E$ and the flux through a surface element dA is $\vec{E} \cdot d\vec{A} = E\,dA$. This greatly simplifies the integral in Eq. (22.5).

Continued

22.9 Electric flux through a sphere centered on a point charge.

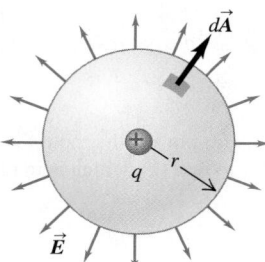

EXECUTE: We must evaluate the integral of Eq. (22.5), $\Phi_E = \int E\,dA$. At any point on the sphere of radius r the electric field has the same magnitude $E = q/4\pi\epsilon_0 r^2$. Hence E can be taken outside the integral, which becomes $\Phi_E = E \int dA = EA$, where A is the

area of the spherical surface: $A = 4\pi r^2$. Hence the total flux through the sphere is

$$\Phi_E = EA = \frac{q}{4\pi\epsilon_0 r^2}4\pi r^2 = \frac{q}{\epsilon_0}$$

$$= \frac{3.0 \times 10^{-6}\,\text{C}}{8.85 \times 10^{-12}\,\text{C}^2/\text{N}\cdot\text{m}^2} = 3.4 \times 10^5\,\text{N}\cdot\text{m}^2/\text{C}$$

EVALUATE: The radius r of the sphere cancels out of the result for Φ_E. We would have obtained the same flux with a sphere of radius 2.0 m or 200 m. We came to essentially the same conclusion in our discussion of Fig. 22.4 in Section 22.1, where we considered rectangular closed surfaces of two different sizes enclosing a point charge. There we found that the flux of $\vec{E}$ was independent of the size of the surface; the same result holds true for a spherical surface. Indeed, the flux through *any* surface enclosing a single point charge is independent of the shape or size of the surface, as we'll soon see.

Test Your Understanding of Section 22.2 Rank the following surfaces in order from most positive to most negative electric flux. (i) a flat rectangular surface with vector area $\vec{A} = (6.0\,\text{m}^2)\hat{\imath}$ in a uniform electric field $\vec{E} = (4.0\,\text{N/C})\hat{\jmath}$; (ii) a flat circular surface with vector area $\vec{A} = (3.0\,\text{m}^2)\hat{\jmath}$ in a uniform electric field $\vec{E} = (4.0\,\text{N/C})\hat{\imath} + (2.0\,\text{N/C})\hat{\jmath}$; (iii) a flat square surface with vector area $\vec{A} = (3.0\,\text{m}^2)\hat{\imath} + (7.0\,\text{m}^2)\hat{\jmath}$ in a uniform electric field $\vec{E} = (4.0\,\text{N/C})\hat{\imath} - (2.0\,\text{N/C})\hat{\jmath}$; (iv) a flat oval surface with vector area $\vec{A} = (3.0\,\text{m}^2)\hat{\imath} - (7.0\,\text{m}^2)\hat{\jmath}$ in a uniform electric field $\vec{E} = (4.0\,\text{N/C})\hat{\imath} - (2.0\,\text{N/C})\hat{\jmath}$.

22.3 Gauss's Law

22.10 Carl Friedrich Gauss helped develop several branches of mathematics, including differential geometry, real analysis, and number theory. The "bell curve" of statistics is one of his inventions. Gauss also made state-of-the-art investigations of the earth's magnetism and calculated the orbit of the first asteroid to be discovered.

Gauss's law is an alternative to Coulomb's law. While completely equivalent to Coulomb's law, Gauss's law provides a different way to express the relationship between electric charge and electric field. It was formulated by Carl Friedrich Gauss (1777–1855), one of the greatest mathematicians of all time (Fig. 22.10).

Point Charge Inside a Spherical Surface

Gauss's law states that the total electric flux through any closed surface (a surface enclosing a definite volume) is proportional to the total (net) electric charge inside the surface. In Section 22.1 we observed this relationship qualitatively for certain special cases; now we'll develop it more rigorously. We'll start with the field of a single positive point charge q. The field lines radiate out equally in all directions. We place this charge at the center of an imaginary spherical surface with radius R. The magnitude E of the electric field at every point on the surface is given by

$$E = \frac{1}{4\pi\epsilon_0}\frac{q}{R^2}$$

At each point on the surface, $\vec{E}$ is perpendicular to the surface, and its magnitude is the same at every point, just as in Example 22.3 (Section 22.2). The total electric flux is the product of the field magnitude E and the total area $A = 4\pi R^2$ of the sphere:

$$\Phi_E = EA = \frac{1}{4\pi\epsilon_0}\frac{q}{R^2}(4\pi R^2) = \frac{q}{\epsilon_0} \tag{22.6}$$

The flux is independent of the radius R of the sphere. It depends only on the charge q enclosed by the sphere.

We can also interpret this result in terms of field lines. Figure 22.11 shows two spheres with radii R and $2R$ centered on the point charge q. Every field line that passes through the smaller sphere also passes through the larger sphere, so the total flux through each sphere is the same.

What is true of the entire sphere is also true of any portion of its surface. In Fig. 22.11 an area dA is outlined on the sphere of radius R and then projected onto the sphere of radius $2R$ by drawing lines from the center through points on the boundary of dA. The area projected on the larger sphere is clearly $4\,dA$. But since the electric field due to a point charge is inversely proportional to r^2, the field magnitude is $\frac{1}{4}$ as great on the sphere of radius $2R$ as on the sphere of radius R. Hence the electric flux is the same for both areas and is independent of the radius of the sphere.

Point Charge Inside a Nonspherical Surface

This projection technique shows us how to extend this discussion to nonspherical surfaces. Instead of a second sphere, let us surround the sphere of radius R by a surface of irregular shape, as in Fig. 22.12a. Consider a small element of area dA on the irregular surface; we note that this area is *larger* than the corresponding element on a spherical surface at the same distance from q. If a normal to dA makes an angle ϕ with a radial line from q, two sides of the area projected onto the spherical surface are foreshortened by a factor $\cos\phi$ (Fig. 22.12b). The other two sides are unchanged. Thus the electric flux through the spherical surface element is equal to the flux $E\,dA\cos\phi$ through the corresponding irregular surface element.

We can divide the entire irregular surface into elements dA, compute the electric flux $E\,dA\cos\phi$ for each, and sum the results by integrating, as in Eq. (22.5). Each of the area elements projects onto a corresponding spherical surface element. Thus the *total* electric flux through the irregular surface, given by any of the forms of Eq. (22.5), must be the same as the total flux through a sphere, which Eq. (22.6) shows is equal to q/ϵ_0. Thus, for the irregular surface,

$$\Phi_E = \oint \vec{E} \cdot d\vec{A} = \frac{q}{\epsilon_0} \qquad (22.7)$$

Equation (22.7) holds for a surface of *any* shape or size, provided only that it is a *closed* surface enclosing the charge q. The circle on the integral sign reminds us that the integral is always taken over a *closed* surface.

The area elements $d\vec{A}$ and the corresponding unit vectors $\hat{n}$ always point *out of* the volume enclosed by the surface. The electric flux is then positive in areas

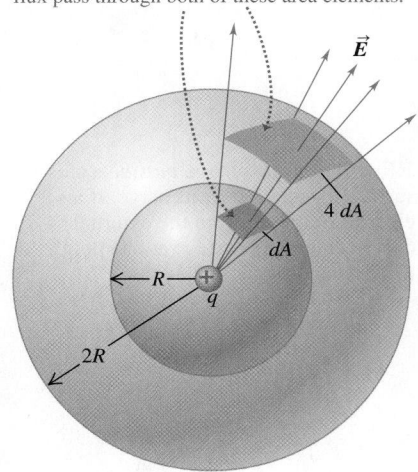

22.11 Projection of an element of area dA of a sphere of radius R onto a concentric sphere of radius $2R$. The projection multiplies each linear dimension by 2, so the area element on the larger sphere is $4\,dA$.

The same number of field lines and the same flux pass through both of these area elements.

(a) The outward normal to the surface makes an angle ϕ with the direction of $\vec{E}$.

(b)

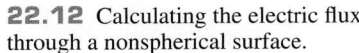

The projection of the area element dA onto the spherical surface is $dA\cos\phi$.

22.12 Calculating the electric flux through a nonspherical surface.

where the electric field points out of the surface and negative where it points inward. Also, $E_\perp$ is positive at points where $\vec{E}$ points out of the surface and negative at points where $\vec{E}$ points into the surface.

If the point charge in Fig. 22.12 is negative, the $\vec{E}$ field is directed radially *inward;* the angle ϕ is then greater than 90°, its cosine is negative, and the integral in Eq. (22.7) is negative. But since q is also negative, Eq. (22.7) still holds.

For a closed surface enclosing *no* charge,

$$\Phi_E = \oint \vec{E} \cdot d\vec{A} = 0$$

This is the mathematical statement that when a region contains no charge, any field lines caused by charges *outside* the region that enter on one side must leave again on the other side. (In Section 22.1 we came to the same conclusion by considering the special case of a rectangular box in a uniform field.) Figure 22.13 illustrates this point. *Electric field lines can begin or end inside a region of space only when there is charge in that region.*

General Form of Gauss's Law

Now comes the final step in obtaining the general form of Gauss's law. Suppose the surface encloses not just one point charge q but several charges q_1, q_2, $q_3, \ldots$. The total (resultant) electric field $\vec{E}$ at any point is the vector sum of the $\vec{E}$ fields of the individual charges. Let Q_{encl} be the *total* charge enclosed by the surface: $Q_{\text{encl}} = q_1 + q_2 + q_3 + \cdots$. Also let $\vec{E}$ be the *total* field at the position of the surface area element $d\vec{A}$, and let $E_\perp$ be its component perpendicular to the plane of that element (that is, parallel to $d\vec{A}$). Then we can write an equation like Eq. (22.7) for each charge and its corresponding field and add the results. When we do, we obtain the general statement of Gauss's law:

$$\Phi_E = \oint \vec{E} \cdot d\vec{A} = \frac{Q_{\text{encl}}}{\epsilon_0} \qquad \text{(Gauss's law)} \qquad (22.8)$$

> **The total electric flux through a closed surface is equal to the total (net) electric charge inside the surface, divided by ϵ_0.**

CAUTION Gaussian surfaces are imaginary Remember that the closed surface in Gauss's law is *imaginary;* there need not be any material object at the position of the surface. We often refer to a closed surface used in Gauss's law as a **Gaussian surface.**

Using the definition of Q_{encl} and the various ways to express electric flux given in Eq. (22.5), we can express Gauss's law in the following equivalent forms:

$$\Phi_E = \oint E\cos\phi\, dA = \oint E_\perp\, dA = \oint \vec{E} \cdot d\vec{A} = \frac{Q_{\text{encl}}}{\epsilon_0} \quad \begin{array}{l}\text{(various forms}\\\text{of Gauss's law)}\end{array} \quad (22.9)$$

As in Eq. (22.5), the various forms of the integral all express the same thing, the total electric flux through the Gaussian surface, in different terms. One form is sometimes more convenient than another.

As an example, Fig. 22.14a shows a spherical Gaussian surface of radius r around a positive point charge $+q$. The electric field points out of the Gaussian surface, so at every point on the surface $\vec{E}$ is in the same direction as $d\vec{A}$, $\phi = 0$, and $E_\perp$ is equal to the field magnitude $E = q/4\pi\epsilon_0 r^2$. Since E is the same at all points

22.13 A point charge *outside* a closed surface that encloses no charge. If an electric field line from the external charge enters the surface at one point, it must leave at another.

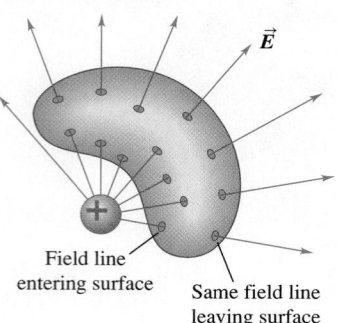

Field line entering surface

$\vec{E}$

Same field line leaving surface

(a) Gaussian surface around positive charge: positive (outward) flux

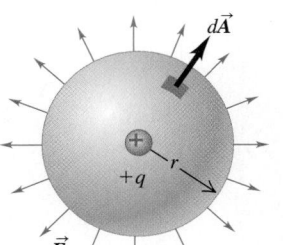

(b) Gaussian surface around negative charge: negative (inward) flux

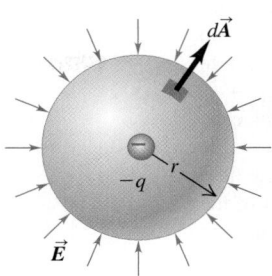

22.14 Spherical Gaussian surfaces around **(a)** a positive point charge and **(b)** a negative point charge.

on the surface, we can take it outside the integral in Eq. (22.9). Then the remaining integral is $\int dA = A = 4\pi r^2$, the area of the sphere. Hence Eq. (22.9) becomes

$$\Phi_E = \oint E_\perp \, dA = \oint \left(\frac{q}{4\pi\epsilon_0 r^2}\right) dA = \frac{q}{4\pi\epsilon_0 r^2} \oint dA = \frac{q}{4\pi\epsilon_0 r^2} 4\pi r^2 = \frac{q}{\epsilon_0}$$

The enclosed charge Q_{encl} is just the charge $+q$, so this agrees with Gauss's law. If the Gaussian surface encloses a *negative* point charge as in Fig. 22.14b, then $\vec{E}$ points *into* the surface at each point in the direction opposite $d\vec{A}$. Then $\phi = 180°$ and $E_\perp$ is equal to the negative of the field magnitude: $E_\perp = -E = -|-q|/4\pi\epsilon_0 r^2 = -q/4\pi\epsilon_0 r^2$. Equation (22.9) then becomes

$$\Phi_E = \oint E_\perp \, dA = \oint \left(\frac{-q}{4\pi\epsilon_0 r^2}\right) dA = \frac{-q}{4\pi\epsilon_0 r^2} \oint dA = \frac{-q}{4\pi\epsilon_0 r^2} 4\pi r^2 = \frac{-q}{\epsilon_0}$$

This again agrees with Gauss's law because the enclosed charge in Fig. 22.14b is $Q_{encl} = -q$.

In Eqs. (22.8) and (22.9), Q_{encl} is always the algebraic sum of all the positive and negative charges enclosed by the Gaussian surface, and $\vec{E}$ is the *total* field at each point on the surface. Also note that in general, this field is caused partly by charges inside the surface and partly by charges outside. But as Fig. 22.13 shows, the outside charges do *not* contribute to the total (net) flux through the surface. So Eqs. (22.8) and (22.9) are correct even when there are charges outside the surface that contribute to the electric field at the surface. When $Q_{encl} = 0$, the total flux through the Gaussian surface must be zero, even though some areas may have positive flux and others may have negative flux (see Fig. 22.3b).

Gauss's law is the definitive answer to the question we posed at the beginning of Section 22.1: "If the electric field pattern is known in a given region, what can we determine about the charge distribution in that region?" It provides a relationship between the electric field on a closed surface and the charge distribution within that surface. But in some cases we can use Gauss's law to answer the reverse question: "If the charge distribution is known, what can we determine about the electric field that the charge distribution produces?" Gauss's law may seem like an unappealing way to address this question, since it may look as though evaluating the integral in Eq. (22.8) is a hopeless task. Sometimes it is, but other times it is surprisingly easy. Here's an example in which *no* integration is involved at all; we'll work out several more examples in the next section.

Conceptual Example 22.4 **Electric flux and enclosed charge**

Figure 22.15 shows the field produced by two point charges $+q$ and $-q$ (an electric dipole). Find the electric flux through each of the closed surfaces A, B, C, and D.

SOLUTION

Gauss's law, Eq. (22.8), says that the total electric flux through a closed surface is equal to the total enclosed charge divided by ϵ_0. In

Continued

Fig. 22.15, surface A (shown in red) encloses the positive charge, so $Q_{encl} = +q$; surface B (in blue) encloses the negative charge, so $Q_{encl} = -q$; surface C (in purple) encloses *both* charges, so $Q_{encl} = +q + (-q) = 0$; and surface D (in yellow) encloses no charges, so $Q_{encl} = 0$. Hence, without having to do any integration, we have $\Phi_{EA} = +q/\epsilon_0$, $\Phi_{EB} = -q/\epsilon_0$, and $\Phi_{EC} = \Phi_{ED} = 0$. These results depend only on the charges enclosed within each Gaussian surface, not on the precise shapes of the surfaces.

We can draw similar conclusions by examining the electric field lines. All the field lines that cross surface A are directed out of the surface, so the flux through A must be positive. Similarly, the flux through B must be negative since all of the field lines that cross that surface point inward. For both surface C and surface D, there are as many field lines pointing into the surface as there are field lines pointing outward, so the flux through each of these surfaces is zero.

22.15 The net number of field lines leaving a closed surface is proportional to the total charge enclosed by that surface.

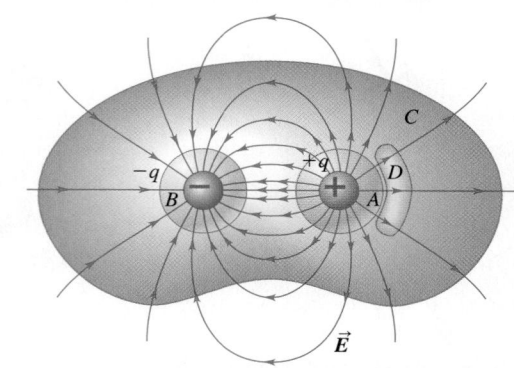

22.16 Five Gaussian surfaces and six point charges.

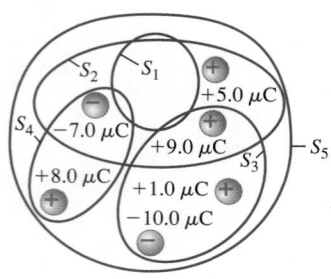

Test Your Understanding of Section 22.3 Figure 22.16 shows six point charges that all lie in the same plane. Five Gaussian surfaces—S_1, S_2, S_3, S_4, and S_5—each enclose part of this plane, and Fig. 22.16 shows the intersection of each surface with the plane. Rank these five surfaces in order of the electric flux through them, from most positive to most negative.

22.4 Applications of Gauss's Law

Gauss's law is valid for *any* distribution of charges and for *any* closed surface. Gauss's law can be used in two ways. If we know the charge distribution, and if it has enough symmetry to let us evaluate the integral in Gauss's law, we can find the field. Or if we know the field, we can use Gauss's law to find the charge distribution, such as charges on conducting surfaces.

In this section we present examples of both kinds of applications. As you study them, watch for the role played by the symmetry properties of each system. We will use Gauss's law to calculate the electric fields caused by several simple charge distributions; the results are collected in a table in the chapter summary.

In practical problems we often encounter situations in which we want to know the electric field caused by a charge distribution on a conductor. These calculations are aided by the following remarkable fact: *When excess charge is placed on a solid conductor and is at rest, it resides entirely on the surface, not in the interior of the material.* (By *excess* we mean charges other than the ions and free electrons that make up the neutral conductor.) Here's the proof. We know from Section 21.4 that in an electrostatic situation (with all charges at rest) the electric field $\vec{E}$ at every point in the interior of a conducting material is zero. If $\vec{E}$ were *not* zero, the excess charges would move. Suppose we construct a Gaussian surface inside the conductor, such as surface A in Fig. 22.17. Because $\vec{E} = 0$ everywhere on this surface, Gauss's law requires that the net charge inside the surface is zero. Now imagine shrinking the surface like a collapsing balloon until it encloses a region so small that we may consider it as a point P; then the charge at that point must be zero. We can do this anywhere inside the conductor, so *there can be no excess charge at any point within a solid conductor; any excess charge must reside on the conductor's surface.* (This result is for a *solid* conductor. In the next section we'll discuss what can happen if the conductor has cavities in its interior.) We will make use of this fact frequently in the examples that follow.

22.17 Under electrostatic conditions (charges not in motion), any excess charge on a solid conductor resides entirely on the conductor's surface.

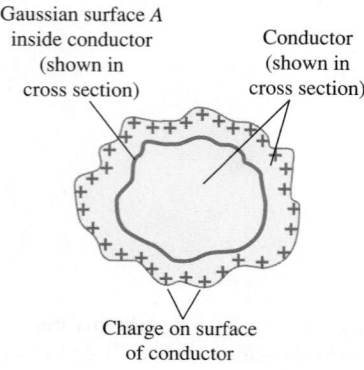

Problem-Solving Strategy 22.1   Gauss's Law

IDENTIFY *the relevant concepts:* Gauss's law is most useful when the charge distribution has spherical, cylindrical, or planar symmetry. In these cases the symmetry determines the direction of $\vec{E}$. Then Gauss's law yields the magnitude of $\vec{E}$ if we are given the charge distribution, and vice versa. In either case, begin the analysis by asking the question: What is the symmetry?

SET UP *the problem* using the following steps:
1. List the known and unknown quantities and identify the target variable.
2. Select the appropriate closed, imaginary Gaussian surface. For spherical symmetry, use a concentric spherical surface. For cylindrical symmetry, use a coaxial cylindrical surface with flat ends perpendicular to the axis of symmetry (like a soup can). For planar symmetry, use a cylindrical surface (like a tuna can) with its flat ends parallel to the plane.

EXECUTE *the solution* as follows:
1. Determine the appropriate size and placement of your Gaussian surface. To evaluate the field magnitude at a particular point, the surface must include that point. It may help to place one end of a can-shaped surface within a conductor, where $\vec{E}$ and therefore Φ are zero, or to place its ends equidistant from a charged plane.
2. Evaluate the integral $\oint E_\perp \, dA$ in Eq. (22.9). In this equation $E_\perp$ is the perpendicular component of the *total* electric field at each point on the Gaussian surface. A well-chosen Gaussian surface should make integration trivial or unnecessary. If the surface comprises several separate surfaces, such as the sides and ends

of a cylinder, the integral $\oint E_\perp \, dA$ over the entire closed surface is the sum of the integrals $\int E_\perp \, dA$ over the separate surfaces. Consider points 3–6 as you work.
3. If $\vec{E}$ is *perpendicular* (normal) at every point to a surface with area A, if it points *outward* from the interior of the surface, and if it has the same *magnitude* at every point on the surface, then $E_\perp = E =$ constant, and $\int E_\perp \, dA$ over that surface is equal to EA. (If $\vec{E}$ is inward, then $E_\perp = -E$ and $\int E_\perp \, dA = -EA$.) This should be the case for part or all of your Gaussian surface. If $\vec{E}$ is tangent to a surface at every point, then $E_\perp = 0$ and the integral over that surface is zero. This may be the case for parts of a cylindrical Gaussian surface. If $\vec{E} = 0$ at every point on a surface, the integral is zero.
4. Even when there is *no* charge within a Gaussian surface, the field at any given point on the surface is not necessarily zero. In that case, however, the total electric flux through the surface is always zero.
5. The flux integral $\oint E_\perp \, dA$ can be approximated as the difference between the numbers of electric lines of force leaving and entering the Gaussian surface. In this sense the flux gives the sign of the enclosed charge, but is only proportional to it; zero flux corresponds to zero enclosed charge.
6. Once you have evaluated $\oint E_\perp \, dA$, use Eq. (22.9) to solve for your target variable.

EVALUATE *your answer:* If your result is a *function* that describes how the magnitude of the electric field varies with position, ensure that it makes sense.

Example 22.5   Field of a charged conducting sphere

We place a total positive charge q on a solid conducting sphere with radius R (Fig. 22.18). Find $\vec{E}$ at any point inside or outside the sphere.

22.18 Calculating the electric field of a conducting sphere with positive charge q. Outside the sphere, the field is the same as if all of the charge were concentrated at the center of the sphere.

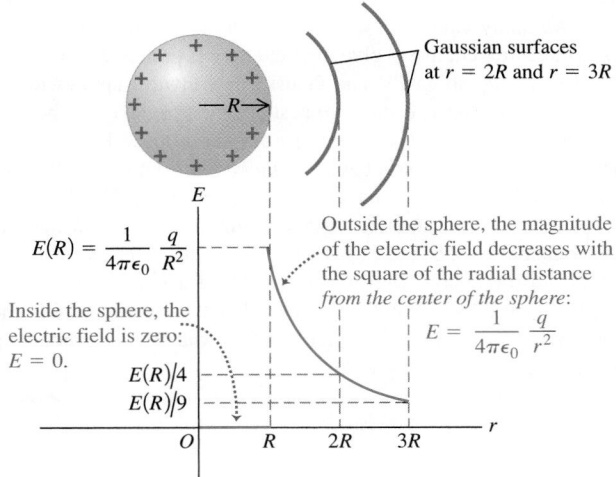

Inside the sphere, the electric field is zero: $E = 0$.

$$E(R) = \frac{1}{4\pi\epsilon_0} \frac{q}{R^2}$$

Outside the sphere, the magnitude of the electric field decreases with the square of the radial distance *from the center of the sphere*:

$$E = \frac{1}{4\pi\epsilon_0} \frac{q}{r^2}$$

SOLUTION

IDENTIFY and SET UP: As we discussed earlier in this section, all of the charge must be on the surface of the sphere. The charge is free to move on the conductor, and there is no preferred position on the surface; the charge is therefore distributed *uniformly* over the surface, and the system is spherically symmetric. To exploit this symmetry, we take as our Gaussian surface a sphere of radius r centered on the conductor. We can calculate the field inside or outside the conductor by taking $r < R$ or $r > R$, respectively. In either case, the point at which we want to calculate $\vec{E}$ lies on the Gaussian surface.

EXECUTE: The spherical symmetry means that the direction of the electric field must be radial; that's because there is no preferred direction parallel to the surface, so $\vec{E}$ can have no component parallel to the surface. There is also no preferred orientation of the sphere, so the field magnitude E can depend only on the distance r from the center and must have the same value at all points on the Gaussian surface.

For $r > R$ the entire conductor is within the Gaussian surface, so the enclosed charge is q. The area of the Gaussian surface is $4\pi r^2$, and $\vec{E}$ is uniform over the surface and perpendicular to it at each point. The flux integral $\oint E_\perp \, dA$ is then just $E(4\pi r^2)$, and Eq. (22.8) gives

Continued

$$E(4\pi r^2) = \frac{q}{\epsilon_0} \quad \text{and}$$

$$E = \frac{1}{4\pi\epsilon_0} \frac{q}{r^2} \quad \text{(outside a charged conducting sphere)}$$

This expression is the same as that for a point charge; outside the charged sphere, its field is the same as though the entire charge were concentrated at its center. Just outside the surface of the sphere, where $r = R$,

$$E = \frac{1}{4\pi\epsilon_0} \frac{q}{R^2} \quad \text{(at the surface of a charged conducting sphere)}$$

CAUTION **Flux can be positive or negative** Remember that we have chosen the charge q to be *positive*. If the charge is negative, the electric field is radially *inward* instead of radially outward, and the electric flux through the Gaussian surface is negative. The electric-field magnitudes outside and at the surface of the sphere are given by the same expressions as above, except that q denotes the *magnitude* (absolute value) of the charge. ▮

For $r < R$ we again have $E(4\pi r^2) = Q_{\text{encl}}/\epsilon_0$. But now our Gaussian surface (which lies entirely within the conductor)

encloses *no* charge, so $Q_{\text{encl}} = 0$. The electric field inside the conductor is therefore zero.

EVALUATE: We already knew that $\vec{E} = 0$ inside a solid conductor (whether spherical or not) when the charges are at rest. Figure 22.18 shows E as a function of the distance r from the center of the sphere. Note that in the limit as $R \rightarrow 0$, the sphere becomes a point charge; there is then only an "outside," and the field is everywhere given by $E = q/4\pi\epsilon_0 r^2$. Thus we have deduced Coulomb's law from Gauss's law. (In Section 22.3 we deduced Gauss's law from Coulomb's law; the two laws are equivalent.)

We can also use this method for a conducting spherical *shell* (a spherical conductor with a concentric spherical hole inside) if there is no charge inside the hole. We use a spherical Gaussian surface with radius r less than the radius of the hole. If there *were* a field inside the hole, it would have to be radial and spherically symmetric as before, so $E = Q_{\text{encl}}/4\pi\epsilon_0 r^2$. But now there is no enclosed charge, so $Q_{\text{encl}} = 0$ and $E = 0$ inside the hole.

Can you use this same technique to find the electric field in the region between a charged sphere and a concentric hollow conducting sphere that surrounds it?

Example 22.6   Field of a uniform line charge

Electric charge is distributed uniformly along an infinitely long, thin wire. The charge per unit length is λ (assumed positive). Find the electric field using Gauss's law.

SOLUTION

IDENTIFY and SET UP: We found in Example 21.10 (Section 21.5) that the field $\vec{E}$ of a uniformly charged, infinite wire is radially outward if λ is positive and radially inward if λ is negative, and that the field magnitude E depends only on the radial distance from the wire. This suggests that we use a *cylindrical* Gaussian surface, of radius r and arbitrary length l, coaxial with the wire and with its ends perpendicular to the wire (Fig. 22.19).

EXECUTE: The flux through the flat ends of our Gaussian surface is zero because the radial electric field is parallel to these ends, and so $\vec{E} \cdot \hat{n} = 0$. On the cylindrical part of our surface we have $\vec{E} \cdot \hat{n} = E_\perp = E$ everywhere. (If λ were negative, we would have

22.19 A coaxial cylindrical Gaussian surface is used to find the electric field outside an infinitely long, charged wire.

$\vec{E} \cdot \hat{n} = E_\perp = -E$ everywhere.) The area of the cylindrical surface is $2\pi r l$, so the flux through it—and hence the total flux Φ_E through the Gaussian surface—is $EA = 2\pi r l E$. The total enclosed charge is $Q_{\text{encl}} = \lambda l$, and so from Gauss's law, Eq. (22.8),

$$\Phi_E = 2\pi r l E = \frac{\lambda l}{\epsilon_0} \quad \text{and}$$

$$E = \frac{1}{2\pi\epsilon_0} \frac{\lambda}{r} \quad \text{(field of an infinite line of charge)}$$

We found this same result in Example 21.10 with *much* more effort.

If λ is *negative*, $\vec{E}$ is directed radially inward, and in the above expression for E we must interpret λ as the absolute value of the charge per unit length.

EVALUATE: We saw in Example 21.10 that the *entire* charge on the wire contributes to the field at any point, and yet we consider only that part of the charge $Q_{\text{encl}} = \lambda l$ within the Gaussian surface when we apply Gauss's law. There's nothing inconsistent here; it takes the entire charge to give the field the properties that allow us to calculate Φ_E so easily, and Gauss's law always applies to the enclosed charge only. If the wire is short, the symmetry of the infinite wire is lost, and E is not uniform over a coaxial, cylindrical Gaussian surface. Gauss's law then *cannot* be used to find Φ_E; we must solve the problem the hard way, as in Example 21.10.

We can use the Gaussian surface in Fig. 22.19 to show that the field outside a long, uniformly charged cylinder is the same as though all the charge were concentrated on a line along its axis (see Problem 22.42). We can also calculate the electric field in the space between a charged cylinder and a coaxial hollow conducting cylinder surrounding it (see Problem 22.39).

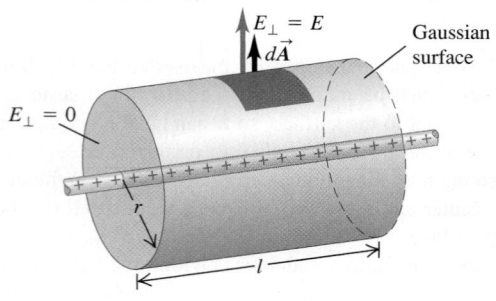

Example 22.7   Field of an infinite plane sheet of charge

Use Gauss's law to find the electric field caused by a thin, flat, infinite sheet with a uniform positive surface charge density σ.

SOLUTION

IDENTIFY and SET UP: In Example 21.11 (Section 21.5) we found that the field $\vec{E}$ of a uniformly charged infinite sheet is normal to the sheet, and that its magnitude is independent of the distance from the sheet. To take advantage of these symmetry properties, we use a cylindrical Gaussian surface with ends of area A and with its axis perpendicular to the sheet of charge (Fig. 22.20).

22.20 A cylindrical Gaussian surface is used to find the field of an infinite plane sheet of charge.

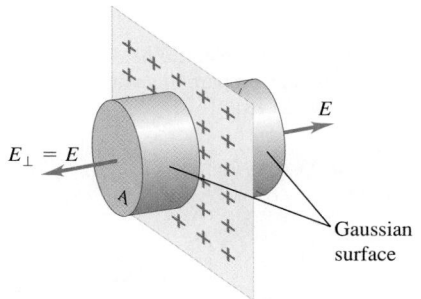

EXECUTE: The flux through the cylindrical part of our Gaussian surface is zero because $\vec{E} \cdot \hat{n} = 0$ everywhere. The flux through each flat end of the surface is $+EA$ because $\vec{E} \cdot \hat{n} = E_\perp = E$ everywhere, so the total flux through both ends—and hence the total flux Φ_E through the Gaussian surface—is $+2EA$. The total enclosed charge is $Q_{encl} = \sigma A$, and so from Gauss's law,

$$2EA = \frac{\sigma A}{\epsilon_0} \quad \text{and}$$

$$E = \frac{\sigma}{2\epsilon_0} \quad \text{(field of an infinite sheet of charge)}$$

In Example 21.11 we found this same result using a much more complex calculation.

If σ is negative, $\vec{E}$ is directed *toward* the sheet, the flux through the Gaussian surface in Fig. 22.20 is negative, and σ in the expression $E = \sigma/2\epsilon_0$ denotes the magnitude (absolute value) of the charge density.

EVALUATE: Again we see that, given favorable symmetry, we can deduce electric fields using Gauss's law much more easily than using Coulomb's law.

Example 22.8   Field between oppositely charged parallel conducting plates

Two large plane parallel conducting plates are given charges of equal magnitude and opposite sign; the surface charge densities are $+\sigma$ and $-\sigma$. Find the electric field in the region between the plates.

SOLUTION

IDENTIFY and SET UP: Figure 22.21a shows the field. Because opposite charges attract, most of the charge accumulates at the opposing faces of the plates. A small amount of charge resides on the *outer* surfaces of the plates, and there is some spreading or "fringing" of the field at the edges. But if the plates are very large in comparison to the distance between them, the amount of charge on the outer surfaces is negligibly small, and the fringing can be neglected except near the edges. In this case we can assume that the field is uniform in the interior region between the plates, as in Fig. 22.21b, and that the charges are distributed uniformly over the opposing surfaces. To exploit this symmetry, we can use the shaded Gaussian surfaces S_1, S_2, S_3, and S_4. These surfaces are cylinders with flat ends of area A; one end of each surface lies *within* a plate.

22.21 Electric field between oppositely charged parallel plates.

(a) Realistic drawing

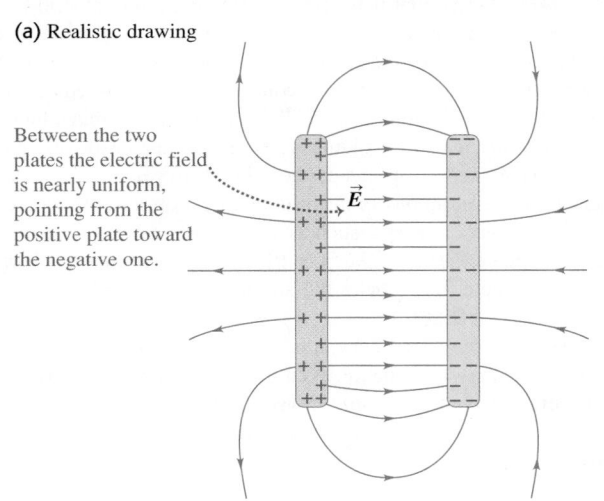

Between the two plates the electric field is nearly uniform, pointing from the positive plate toward the negative one.

(b) Idealized model

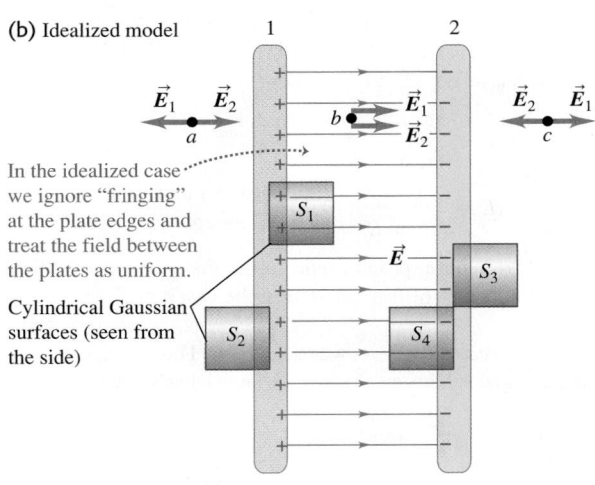

In the idealized case we ignore "fringing" at the plate edges and treat the field between the plates as uniform.

Cylindrical Gaussian surfaces (seen from the side)

Continued

EXECUTE: The left-hand end of surface S_1 is within the positive plate 1. Since the field is zero within the volume of any solid conductor under electrostatic conditions, there is no electric flux through this end. The electric field between the plates is perpendicular to the right-hand end, so on that end, $E_\perp$ is equal to E and the flux is EA; this is positive, since $\vec{E}$ is directed out of the Gaussian surface. There is no flux through the side walls of the cylinder, since these walls are parallel to $\vec{E}$. So the total flux integral in Gauss's law is EA. The net charge enclosed by the cylinder is σA, so Eq. (22.8) yields $EA = \sigma A / \epsilon_0$; we then have

$$E = \frac{\sigma}{\epsilon_0} \quad \text{(field between oppositely charged conducting plates)}$$

The field is uniform and perpendicular to the plates, and its magnitude is independent of the distance from either plate. The Gaussian surface S_4 yields the same result. Surfaces S_2 and S_3 yield $E = 0$ to the left of plate 1 and to the right of plate 2, respectively. We leave these calculations to you (see Exercise 22.29).

EVALUATE: We obtained the same results in Example 21.11 by using the principle of superposition of electric fields. The fields due to the two sheets of charge (one on each plate) are $\vec{E}_1$ and $\vec{E}_2$; from Example 22.7, both of these have magnitude $\sigma/2\epsilon_0$. The total electric field at any point is the vector sum $\vec{E} = \vec{E}_1 + \vec{E}_2$. At points a and c in Fig. 22.21b, $\vec{E}_1$ and $\vec{E}_2$ point in opposite directions, and their sum is zero. At point b, $\vec{E}_1$ and $\vec{E}_2$ are in the same direction; their sum has magnitude $E = \sigma/\epsilon_0$, just as we found above using Gauss's law.

Example 22.9 Field of a uniformly charged sphere

Positive electric charge Q is distributed uniformly *throughout the volume* of an *insulating* sphere with radius R. Find the magnitude of the electric field at a point P a distance r from the center of the sphere.

SOLUTION

IDENTIFY and SET UP: As in Example 22.5, the system is spherically symmetric. Hence we can use the conclusions of that example about the direction and magnitude of $\vec{E}$. To make use of the spherical symmetry, we choose as our Gaussian surface a sphere with radius r, concentric with the charge distribution.

EXECUTE: From symmetry, the direction of $\vec{E}$ is radial at every point on the Gaussian surface, so $E_\perp = E$ and the field magnitude E is the same at every point on the surface. Hence the total electric flux through the Gaussian surface is the product of E and the total area of the surface $A = 4\pi r^2$—that is, $\Phi_E = 4\pi r^2 E$.

The amount of charge enclosed within the Gaussian surface depends on r. To find E *inside* the sphere, we choose $r < R$. The volume charge density ρ is the charge Q divided by the volume of the entire charged sphere of radius R:

$$\rho = \frac{Q}{4\pi R^3/3}$$

The volume V_{encl} enclosed by the Gaussian surface is $\frac{4}{3}\pi r^3$, so the total charge Q_{encl} enclosed by that surface is

$$Q_{\text{encl}} = \rho V_{\text{encl}} = \left(\frac{Q}{4\pi R^3/3}\right)\left(\tfrac{4}{3}\pi r^3\right) = Q\frac{r^3}{R^3}$$

Then Gauss's law, Eq. (22.8), becomes

$$4\pi r^2 E = \frac{Q}{\epsilon_0}\frac{r^3}{R^3} \quad \text{or}$$

$$E = \frac{1}{4\pi\epsilon_0}\frac{Qr}{R^3} \quad \begin{array}{l}\text{(field inside a uniformly}\\\text{charged sphere)}\end{array}$$

The field magnitude is proportional to the distance r of the field point from the center of the sphere (see the graph of E versus r in Fig. 22.22).

To find E *outside* the sphere, we take $r > R$. This surface encloses the entire charged sphere, so $Q_{\text{encl}} = Q$, and Gauss's law gives

$$4\pi r^2 E = \frac{Q}{\epsilon_0} \quad \text{or}$$

$$E = \frac{1}{4\pi\epsilon_0}\frac{Q}{r^2} \quad \begin{array}{l}\text{(field outside a uniformly}\\\text{charged sphere)}\end{array}$$

22.22 The magnitude of the electric field of a uniformly charged insulating sphere. Compare this with the field for a conducting sphere (see Fig. 22.18).

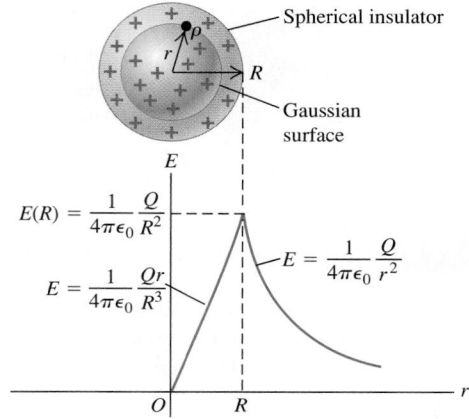

The field outside *any* spherically symmetric charged body varies as $1/r^2$, as though the entire charge were concentrated at the center. This is graphed in Fig. 22.22.

If the charge is *negative*, $\vec{E}$ is radially *inward* and in the expressions for E we interpret Q as the absolute value of the charge.

EVALUATE: Notice that if we set $r = R$ in either expression for E, we get the same result $E = Q/4\pi\epsilon_0 R^2$ for the magnitude of the field at the surface of the sphere. This is because the magnitude E is a *continuous* function of r. By contrast, for the charged conducting sphere of Example 22.5 the electric-field magnitude is *discontinuous* at $r = R$ (it jumps from $E = 0$ just inside the sphere to $E = Q/4\pi\epsilon_0 R^2$ just outside the sphere). In general, the electric field $\vec{E}$ is discontinuous in magnitude, direction, or both wherever there is a *sheet* of charge, such as at the surface of a charged conducting sphere (Example 22.5), at the surface of an infinite charged sheet (Example 22.7), or at the surface of a charged conducting plate (Example 22.8).

The approach used here can be applied to *any* spherically symmetric distribution of charge, even if it is not radially uniform, as it was here. Such charge distributions occur within many atoms and atomic nuclei, so Gauss's law is useful in atomic and nuclear physics.

Example 22.10 Charge on a hollow sphere

A thin-walled, hollow sphere of radius 0.250 m has an unknown charge distributed uniformly over its surface. At a distance of 0.300 m from the center of the sphere, the electric field points radially inward and has magnitude 1.80×10^2 N/C. How much charge is on the sphere?

SOLUTION

IDENTIFY and SET UP: The charge distribution is spherically symmetric. As in Examples 22.5 and 22.9, it follows that the electric field is radial everywhere and its magnitude is a function only of the radial distance r from the center of the sphere. We use a spherical Gaussian surface that is concentric with the charge distribution and has radius $r = 0.300$ m. Our target variable is $Q_{encl} = q$.

EXECUTE: The charge distribution is the same as if the charge were on the surface of a 0.250-m-radius conducting sphere. Hence we can borrow the results of Example 22.5. We note that the electric field here is directed toward the sphere, so that q must be *negative*. Furthermore, the electric field is directed into the Gaussian surface, so that $E_\perp = -E$ and $\Phi_E = \oint E_\perp \, dA = -E(4\pi r^2)$.

By Gauss's law, the flux is equal to the charge q on the sphere (all of which is enclosed by the Gaussian surface) divided by ϵ_0. Solving for q, we find

$$q = -E(4\pi\epsilon_0 r^2) = -(1.80 \times 10^2 \text{ N/C})(4\pi)$$
$$\times (8.854 \times 10^{-12} \text{ C}^2/\text{N} \cdot \text{m}^2)(0.300 \text{ m})^2$$
$$= -1.80 \times 10^{-9} \text{ C} = -1.80 \text{ nC}$$

EVALUATE: To determine the charge, we had to know the electric field at *all* points on the Gaussian surface so that we could calculate the flux integral. This was possible here because the charge distribution is highly symmetric. If the charge distribution is irregular or lacks symmetry, Gauss's law is not very useful for calculating the charge distribution from the field, or vice versa.

Test Your Understanding of Section 22.4 You place a known amount of charge Q on the irregularly shaped conductor shown in Fig. 22.17. If you know the size and shape of the conductor, can you use Gauss's law to calculate the electric field at an arbitrary position outside the conductor? ▮

22.5 Charges on Conductors

We have learned that in an electrostatic situation (in which there is no net motion of charge) the electric field at every point within a conductor is zero and that any excess charge on a solid conductor is located entirely on its surface (Fig. 22.23a). But what if there is a *cavity* inside the conductor (Fig. 22.23b)? If there is no charge within the cavity, we can use a Gaussian surface such as A (which lies completely within the material of the conductor) to show that the *net* charge on the *surface of the cavity* must be zero, because $\vec{E} = 0$ everywhere on the Gaussian surface. In fact, we can prove in this situation that there can't be any charge *anywhere* on the cavity surface. We will postpone detailed proof of this statement until Chapter 23.

Suppose we place a small body with a charge q inside a cavity within a conductor (Fig. 22.23c). The conductor is uncharged and is insulated from the charge q. Again $\vec{E} = 0$ everywhere on surface A, so according to Gauss's law the *total* charge inside this surface must be zero. Therefore there must be a charge $-q$ distributed on the surface of the cavity, drawn there by the charge q inside the cavity. The *total* charge on the conductor must remain zero, so a charge $+q$ must appear

---

22.23 Finding the electric field within a charged conductor.

(a) Solid conductor with charge q_C	(b) The same conductor with an internal cavity	(c) An isolated charge q placed in the cavity

The charge q_C resides entirely on the surface of the conductor. The situation is electrostatic, so $\vec{E} = 0$ within the conductor.

Because $\vec{E} = 0$ at all points within the conductor, the electric field at all points on the Gaussian surface must be zero.

For $\vec{E}$ to be zero at all points on the Gaussian surface, the surface of the cavity must have a total charge $-q$.

either on its outer surface or inside the material. But we showed that in an electrostatic situation there can't be any excess charge within the material of a conductor. So we conclude that the charge $+q$ must appear on the outer surface. By the same reasoning, if the conductor originally had a charge q_C, then the total charge on the outer surface must be $q_C + q$ after the charge q is inserted into the cavity.

Conceptual Example 22.11 | **A conductor with a cavity**

A solid conductor with a cavity carries a total charge of $+7$ nC. Within the cavity, insulated from the conductor, is a point charge of -5 nC. How much charge is on each surface (inner and outer) of the conductor?

22.24 Our sketch for this problem. There is zero electric field inside the bulk conductor and hence zero flux through the Gaussian surface shown, so the charge on the cavity wall must be the opposite of the point charge.

SOLUTION

Figure 22.24 shows the situation. If the charge in the cavity is $q = -5$ nC, the charge on the inner cavity surface must be $-q = -(-5 \text{ nC}) = +5$ nC. The conductor carries a *total* charge of $+7$ nC, none of which is in the interior of the material. If $+5$ nC is on the inner surface of the cavity, then there must be $(+7 \text{ nC}) - (+5 \text{ nC}) = +2$ nC on the outer surface of the conductor.

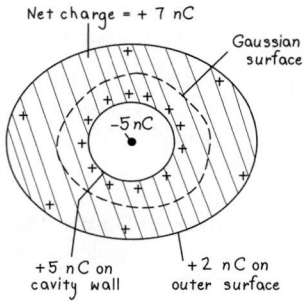

Net charge = + 7 nC

Gaussian surface

−5 nC

+5 nC on cavity wall

+2 nC on outer surface

Testing Gauss's Law Experimentally

We can now consider a historic experiment, shown in Fig. 22.25. We mount a conducting container on an insulating stand. The container is initially uncharged. Then we hang a charged metal ball from an insulating thread (Fig. 22.25a), lower it into the container, and put the lid on (Fig. 22.25b). Charges are induced on the walls of the container, as shown. But now we let the ball *touch* the inner wall (Fig. 22.25c). The surface of the ball becomes part of the cavity surface. The situation is now the same as Fig. 22.23b; if Gauss's law is correct, the net charge on the cavity surface must be zero. Thus the ball must lose all its charge. Finally, we pull the ball out; we find that it has indeed lost all its charge.

This experiment was performed in the 19th century by the English scientist Michael Faraday, using a metal icepail with a lid, and it is called **Faraday's icepail experiment**. The result confirms the validity of Gauss's law and therefore of

22.25 (a) A charged conducting ball suspended by an insulating thread outside a conducting container on an insulating stand. (b) The ball is lowered into the container, and the lid is put on. (c) The ball is touched to the inner surface of the container.

(a) Insulating thread — Charged conducting ball

Metal container

Insulating stand

(b) Metal lid

Charged ball induces charges on the interior and exterior of the container.

(c) Metal lid

Once the ball touches the container, it is part of the interior surface; all the charge moves to the container's exterior.

22.26 Cutaway view of the essential parts of a Van de Graaff electrostatic generator. The electron sink at the bottom draws electrons from the belt, giving it a positive charge; at the top the belt attracts electrons away from the conducting shell, giving the shell a positive charge.

Coulomb's law. Faraday's result was significant because Coulomb's experimental method, using a torsion balance and dividing of charges, was not very precise; it is very difficult to confirm the $1/r^2$ dependence of the electrostatic force by direct force measurements. By contrast, experiments like Faraday's test the validity of Gauss's law, and therefore of Coulomb's law, with much greater precision. Modern versions of this experiment have shown that the exponent 2 in the $1/r^2$ of Coulomb's law does not differ from precisely 2 by more than 10^{-16}. So there is no reason to believe it is anything other than exactly 2.

The same principle behind Faraday's icepail experiment is used in a *Van de Graaff electrostatic generator* (Fig. 22.26). A charged belt continuously carries charge to the inside of a conducting shell. By Gauss's law, there can never be any charge on the inner surface of this shell, so the charge is immediately carried away to the outside surface of the shell. As a result, the charge on the shell and the electric field around it can become very large very rapidly. The Van de Graaff generator is used as an accelerator of charged particles and for physics demonstrations.

This principle also forms the basis for *electrostatic shielding*. Suppose we have a very sensitive electronic instrument that we want to protect from stray electric fields that might cause erroneous measurements. We surround the instrument with a conducting box, or we line the walls, floor, and ceiling of the room with a conducting material such as sheet copper. The external electric field redistributes the free electrons in the conductor, leaving a net positive charge on the outer surface in some regions and a net negative charge in others (Fig. 22.27). This charge distribution causes an additional electric field such that the *total* field at every point inside the box is zero, as Gauss's law says it must be. The charge distribution on the box also alters the shapes of the field lines near the box, as the figure shows. Such a setup is often called a *Faraday cage*. The same physics tells

22.27 (a) A conducting box (a Faraday cage) immersed in a uniform electric field. The field of the induced charges on the box combines with the uniform field to give zero total field inside the box. (b) This person is inside a Faraday cage, and so is protected from the powerful electric discharge.

(a)

Field pushes electrons toward left side.   Net positive charge remains on right side.

$\vec{E}$ $\vec{E} = 0$ $\vec{E}$

Field perpendicular to conductor surface

(b)

you that one of the safest places to be in a lightning storm is inside an automobile; if the car is struck by lightning, the charge tends to remain on the metal skin of the vehicle, and little or no electric field is produced inside the passenger compartment.

Field at the Surface of a Conductor

Finally, we note that there is a direct relationship between the $\vec{E}$ field at a point just outside any conductor and the surface charge density σ at that point. In general, σ varies from point to point on the surface. We will show in Chapter 23 that at any such point, the direction of $\vec{E}$ is always *perpendicular* to the surface. (You can see this effect in Fig. 22.27a.)

To find a relationship between σ at any point on the surface and the perpendicular component of the electric field at that point, we construct a Gaussian surface in the form of a small cylinder (Fig. 22.28). One end face, with area A, lies within the conductor and the other lies just outside. The electric field is zero at all points within the conductor. Outside the conductor the component of $\vec{E}$ perpendicular to the side walls of the cylinder is zero, and over the end face the perpendicular component is equal to $E_\perp$. (If σ is positive, the electric field points out of the conductor and $E_\perp$ is positive; if σ is negative, the field points inward and $E_\perp$ is negative.) Hence the total flux through the surface is $E_\perp A$. The charge enclosed within the Gaussian surface is σA, so from Gauss's law,

22.28 The field just outside a charged conductor is perpendicular to the surface, and its perpendicular component $E_\perp$ is equal to σ/ϵ_0.

Outer surface of charged conductor

A

$E_\perp = E$

$E_\perp = 0$

Gaussian surface

$E = 0$

A

$$E_\perp A = \frac{\sigma A}{\epsilon_0} \quad \text{and} \quad E_\perp = \frac{\sigma}{\epsilon_0} \quad \begin{array}{l}\text{(field at the surface} \\ \text{of a conductor)}\end{array} \quad (22.10)$$

We can check this with the results we have obtained for spherical, cylindrical, and plane surfaces.

We showed in Example 22.8 that the field magnitude between two infinite flat oppositely charged conducting plates also equals σ/ϵ_0. In this case the field magnitude E is the same at *all* distances from the plates, but in all other cases E decreases with increasing distance from the surface.

Conceptual Example 22.12 **Field at the surface of a conducting sphere**

Verify Eq. (22.10) for a conducting sphere with radius R and total charge q.

SOLUTION

In Example 22.5 (Section 22.4) we showed that the electric field just outside the surface is

$$E = \frac{1}{4\pi\epsilon_0}\frac{q}{R^2}$$

The surface charge density is uniform and equal to q divided by the surface area of the sphere:

$$\sigma = \frac{q}{4\pi R^2}$$

Comparing these two expressions, we see that $E = \sigma/\epsilon_0$, which verifies Eq. (22.10).

Example 22.13 **Electric field of the earth**

The earth (a conductor) has a net electric charge. The resulting electric field near the surface has an average value of about 150 N/C, directed toward the center of the earth. (a) What is the corresponding surface charge density? (b) What is the *total* surface charge of the earth?

SOLUTION

IDENTIFY and SET UP: We are given the electric-field magnitude at the surface of the conducting earth. We can calculate the surface charge density σ using Eq. (22.10). The total charge Q on the earth's surface is then the product of σ and the earth's surface area.

EXECUTE: (a) The direction of the field means that σ is negative (corresponding to $\vec{E}$ being directed *into* the surface, so $E_\perp$ is negative). From Eq. (22.10),

$$\sigma = \epsilon_0 E_\perp = (8.85 \times 10^{-12}\ \text{C}^2/\text{N}\cdot\text{m}^2)(-150\ \text{N/C})$$
$$= -1.33 \times 10^{-9}\ \text{C/m}^2 = -1.33\ \text{nC/m}^2$$

(b) The earth's surface area is $4\pi R_\text{E}^2$, where $R_\text{E} = 6.38 \times 10^6$ m is the radius of the earth (see Appendix F). The total charge Q is the product $4\pi R_\text{E}^2 \sigma$, or

$$Q = 4\pi(6.38 \times 10^6\ \text{m})^2(-1.33 \times 10^{-9}\ \text{C/m}^2)$$
$$= -6.8 \times 10^5\ \text{C} = -680\ \text{kC}$$

EVALUATE: You can check our result in part (b) using the result of Example 22.5. Solving for Q, we find

$$Q = 4\pi\epsilon_0 R^2 E_\perp$$
$$= \frac{1}{9.0 \times 10^9\ \text{N}\cdot\text{m}^2/\text{C}^2}(6.38 \times 10^6\ \text{m})^2\,(-150\ \text{N/C})$$
$$= -6.8 \times 10^5\ \text{C}$$

One electron has a charge of -1.60×10^{-19} C. Hence this much excess negative electric charge corresponds to there being $(-6.8 \times 10^5\ \text{C})/(-1.60 \times 10^{-19}\ \text{C}) = 4.2 \times 10^{24}$ excess electrons on the earth, or about 7 moles of excess electrons. This is compensated by an equal *deficiency* of electrons in the earth's upper atmosphere, so the combination of the earth and its atmosphere is electrically neutral.

Test Your Understanding of Section 22.5 A hollow conducting sphere has no net charge. There is a positive point charge q at the center of the spherical cavity within the sphere. You connect a conducting wire from the outside of the sphere to ground. Will you measure an electric field outside the sphere? ❙

Electric flux: Electric flux is a measure of the "flow" of electric field through a surface. It is equal to the product of an area element and the perpendicular component of $\vec{E}$, integrated over a surface. (See Examples 22.1–22.3.)

$$\Phi_E = \int E \cos\phi \, dA$$

$$= \int E_\perp \, dA = \int \vec{E} \cdot d\vec{A} \quad (22.5)$$

Gauss's law: Gauss's law states that the total electric flux through a closed surface, which can be written as the surface integral of the component of $\vec{E}$ normal to the surface, equals a constant times the total charge Q_{encl} enclosed by the surface. Gauss's law is logically equivalent to Coulomb's law, but its use greatly simplifies problems with a high degree of symmetry. (See Examples 22.4–22.10.)

When excess charge is placed on a conductor and is at rest, it resides entirely on the surface, and $\vec{E} = 0$ everywhere in the material of the conductor. (See Examples 22.11–22.13.)

$$\Phi_E = \oint E \cos\phi \, dA$$

$$= \oint E_\perp \, dA = \oint \vec{E} \cdot d\vec{A}$$

$$= \frac{Q_{encl}}{\epsilon_0} \quad (22.8), (22.9)$$

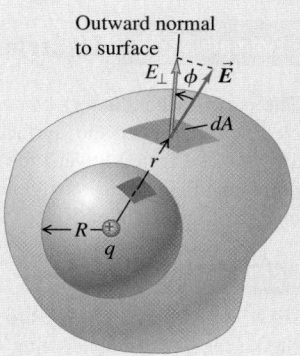

Electric field of various symmetric charge distributions: The following table lists electric fields caused by several symmetric charge distributions. In the table, q, Q, λ, and σ refer to the *magnitudes* of the quantities.

Charge Distribution	Point in Electric Field	Electric Field Magnitude
Single point charge q	Distance r from q	$E = \dfrac{1}{4\pi\epsilon_0} \dfrac{q}{r^2}$
Charge q on surface of conducting sphere with radius R	Outside sphere, $r > R$	$E = \dfrac{1}{4\pi\epsilon_0} \dfrac{q}{r^2}$
	Inside sphere, $r < R$	$E = 0$
Infinite wire, charge per unit length λ	Distance r from wire	$E = \dfrac{1}{2\pi\epsilon_0} \dfrac{\lambda}{r}$
Infinite conducting cylinder with radius R, charge per unit length λ	Outside cylinder, $r > R$	$E = \dfrac{1}{2\pi\epsilon_0} \dfrac{\lambda}{r}$
	Inside cylinder, $r < R$	$E = 0$
Solid insulating sphere with radius R, charge Q distributed uniformly throughout volume	Outside sphere, $r > R$	$E = \dfrac{1}{4\pi\epsilon_0} \dfrac{Q}{r^2}$
	Inside sphere, $r < R$	$E = \dfrac{1}{4\pi\epsilon_0} \dfrac{Qr}{R^3}$
Infinite sheet of charge with uniform charge per unit area σ	Any point	$E = \dfrac{\sigma}{2\epsilon_0}$
Two oppositely charged conducting plates with surface charge densities $+\sigma$ and $-\sigma$	Any point between plates	$E = \dfrac{\sigma}{\epsilon_0}$
Charged conductor	Just outside the conductor	$E = \dfrac{\sigma}{\epsilon_0}$

BRIDGING PROBLEM   Electric Field Inside a Hydrogen Atom

A hydrogen atom is made up of a proton of charge $+Q = 1.60 \times 10^{-19}$ C and an electron of charge $-Q = -1.60 \times 10^{-19}$ C. The proton may be regarded as a point charge at $r = 0$, the center of the atom. The motion of the electron causes its charge to be "smeared out" into a spherical distribution around the proton, so that the electron is equivalent to a charge per unit volume of $\rho(r) = -(Q/\pi a_0^3)e^{-2r/a_0}$, where $a_0 = 5.29 \times 10^{-11}$ m is called the *Bohr radius*. (a) Find the total amount of the hydrogen atom's charge that is enclosed within a sphere with radius r centered on the proton. (b) Find the electric field (magnitude and direction) caused by the charge of the hydrogen atom as a function of r. (c) Make a graph as a function of r of the ratio of the electric-field magnitude E to the magnitude of the field due to the proton alone.

SOLUTION GUIDE

See MasteringPhysics® study area for a Video Tutor solution.

IDENTIFY and SET UP

1. The charge distribution in this problem is spherically symmetric, just as in Example 22.9, so you can solve it using Gauss's law.
2. The charge within a sphere of radius r includes the proton charge $+Q$ plus the portion of the electron charge distribution that lies within the sphere. The difference from Example 22.9 is that the electron charge distribution is *not* uniform, so the charge enclosed within a sphere of radius r is *not* simply the charge density multiplied by the volume $4\pi r^3/3$ of the sphere. Instead, you'll have to do an integral.

3. Consider a thin spherical shell centered on the proton, with radius r' and infinitesimal thickness dr'. Since the shell is so thin, every point within the shell is at essentially the same radius from the proton. Hence the amount of electron charge within this shell *is* equal to the electron charge density $\rho(r')$ at this radius multiplied by the volume dV of the shell. What is dV in terms of r'?
4. The total electron charge within a radius r equals the integral of $\rho(r')dV$ from $r' = 0$ to $r' = r$. Set up this integral (but don't solve it yet), and use it to write an expression for the total charge (including the proton) within a sphere of radius r.

EXECUTE

5. Integrate your expression from step 4 to find the charge within radius r. *Hint:* Integrate by substitution: Change the integration variable from r' to $x = 2r'/a_0$. You can calculate the integral $\int x^2 e^{-x}\, dx$ using integration by parts, or you can look it up in a table of integrals or on the World Wide Web.
6. Use Gauss's law and your results from step 5 to find the electric field at a distance r from the proton.
7. Find the ratio referred to in part (c) and graph it versus r. (You'll actually find it simplest to graph this function versus the quantity r/a_0.)

EVALUATE

8. How do your results for the enclosed charge and the electric-field magnitude behave in the limit $r \to 0$? In the limit $r \to \infty$? Explain your results.

Problems

For instructor-assigned homework, go to www.masteringphysics.com

•, ••, •••: Problems of increasing difficulty. **CP**: Cumulative problems incorporating material from earlier chapters. **CALC**: Problems requiring calculus. **BIO**: Biosciences problems.

DISCUSSION QUESTIONS

Q22.1 A rubber balloon has a single point charge in its interior. Does the electric flux through the balloon depend on whether or not it is fully inflated? Explain your reasoning.

Q22.2 Suppose that in Fig. 22.15 both charges were positive. What would be the fluxes through each of the four surfaces in the example?

Q22.3 In Fig. 22.15, suppose a third point charge were placed outside the purple Gaussian surface C. Would this affect the electric flux through any of the surfaces A, B, C, or D in the figure? Why or why not?

Q22.4 A certain region of space bounded by an imaginary closed surface contains no charge. Is the electric field always zero everywhere on the surface? If not, under what circumstances is it zero on the surface?

Q22.5 A spherical Gaussian surface encloses a point charge q. If the point charge is moved from the center of the sphere to a point away from the center, does the electric field at a point on the surface change? Does the total flux through the Gaussian surface change? Explain.

Q22.6 You find a sealed box on your doorstep. You suspect that the box contains several charged metal spheres packed in insulating material. How can you determine the total net charge inside the box without opening the box? Or isn't this possible?

Q22.7 A solid copper sphere has a net positive charge. The charge is distributed uniformly over the surface of the sphere, and the electric field inside the sphere is zero. Then a negative point charge outside the sphere is brought close to the surface of the sphere. Is all the net charge on the sphere still on its surface? If so, is this charge still distributed uniformly over the surface? If it is not uniform, how is it distributed? Is the electric field inside the sphere still zero? In each case justify your answers.

Q22.8 If the electric field of a point charge were proportional to $1/r^3$ instead of $1/r^2$, would Gauss's law still be valid? Explain your reasoning. (*Hint:* Consider a spherical Gaussian surface centered on a single point charge.)

Q22.9 In a conductor, one or more electrons from each atom are free to roam throughout the volume of the conductor. Does this contradict the statement that any excess charge on a solid conductor must reside on its surface? Why or why not?

Q22.10 You charge up the van de Graaff generator shown in Fig. 22.26, and then bring an identical but uncharged hollow conducting sphere near it, without letting the two spheres touch. Sketch the distribution of charges on the second sphere. What is the net flux through the second sphere? What is the electric field inside the second sphere?

Q22.11 A lightning rod is a rounded copper rod mounted on top of a building and welded to a heavy copper cable running down into the ground. Lightning rods are used to protect houses and barns from lightning; the lightning current runs through the copper rather than through the building. Why? Why should the end of the rod be rounded?

Q22.12 A solid conductor has a cavity in its interior. Would the presence of a point charge inside the cavity affect the electric field outside the conductor? Why or why not? Would the presence of a point charge outside the conductor affect the electric field inside the cavity? Again, why or why not?

Q22.13 Explain this statement: "In a static situation, the electric field at the surface of a conductor can have no component parallel to the surface because this would violate the condition that the charges on the surface are at rest." Would this same statement be valid for the electric field at the surface of an *insulator*? Explain your answer and the reason for any differences between the cases of a conductor and an insulator.

Q22.14 In a certain region of space, the electric field $\vec{E}$ is uniform. (a) Use Gauss's law to prove that this region of space must be electrically neutral; that is, the volume charge density ρ must be zero. (b) Is the converse true? That is, in a region of space where there is no charge, must $\vec{E}$ be uniform? Explain.

Q22.15 (a) In a certain region of space, the volume charge density ρ has a uniform positive value. Can $\vec{E}$ be uniform in this region? Explain. (b) Suppose that in this region of uniform positive ρ there is a "bubble" within which $\rho = 0$. Can $\vec{E}$ be uniform within this bubble? Explain.

EXERCISES

Section 22.2 Calculating Electric Flux

22.1 • A flat sheet of paper of area 0.250 m^2 is oriented so that the normal to the sheet is at an angle of 60° to a uniform electric field of magnitude 14 N/C. (a) Find the magnitude of the electric flux through the sheet. (b) Does the answer to part (a) depend on the shape of the sheet? Why or why not? (c) For what angle ϕ between the normal to the sheet and the electric field is the magnitude of the flux through the sheet (i) largest and (ii) smallest? Explain your answers.

22.2 •• A flat sheet is in the shape of a rectangle with sides of lengths 0.400 m and 0.600 m. The sheet is immersed in a uniform electric field of magnitude 75.0 N/C that is directed at 20° from the plane of the sheet (Fig. E22.2). Find the magnitude of the electric flux through the sheet.

Figure **E22.2**

22.3 • You measure an electric field of 1.25×10^6 N/C at a distance of 0.150 m from a point charge. There is no other source of electric field in the region other than this point charge. (a) What is the electric flux through the surface of a sphere that has this charge

at its center and that has radius 0.150 m? (b) What is the magnitude of this charge?

22.4 • It was shown in Example 21.11 (Section 21.5) that the electric field due to an infinite line of charge is perpendicular to the line and has magnitude $E = \lambda/2\pi\epsilon_0 r$. Consider an imaginary cylinder with radius $r = 0.250$ m and length $l = 0.400$ m that has an infinite line of positive charge running along its axis. The charge per unit length on the line is $\lambda = 3.00 \ \mu C/m$. (a) What is the electric flux through the cylinder due to this infinite line of charge? (b) What is the flux through the cylinder if its radius is increased to $r = 0.500$ m? (c) What is the flux through the cylinder if its length is increased to $l = 0.800$ m?

22.5 •• A hemispherical surface with radius r in a region of uniform electric field $\vec{E}$ has its axis aligned parallel to the direction of the field. Calculate the flux through the surface.

22.6 • The cube in Fig. E22.6 has sides of length L =10.0 cm. The electric field is uniform, has magnitude $E = 4.00 \times 10^3$ N/C, and is parallel to the xy-plane at an angle of 53.1° measured from the $+x$-axis toward the $+y$-axis. (a) What is the electric flux through each of the six cube faces S_1, S_2, S_3, S_4, S_5, and S_6? (b) What is the total electric flux through all faces of the cube?

Figure **E22.6**

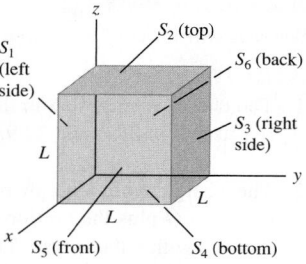

Section 22.3 Gauss's Law

22.7 • **BIO** As discussed in Section 22.5, human nerve cells have a net negative charge and the material in the interior of the cell is a good conductor. If a cell has a net charge of -8.65 pC, what are the magnitude and direction (inward or outward) of the net flux through the cell boundary?

22.8 • The three small spheres shown in Fig. E22.8 carry charges $q_1 = 4.00$ nC, $q_2 = -7.80$ nC, and $q_3 = 2.40$ nC. Find the net electric flux through each of the following closed surfaces shown in cross section in the figure: (a) S_1; (b) S_2; (c) S_3; (d) S_4; (e) S_5. (f) Do your answers to parts (a)–(e) depend on how the charge is distributed over each small sphere? Why or why not?

Figure **E22.8**

22.9 •• A charged paint is spread in a very thin uniform layer over the surface of a plastic sphere of diameter 12.0 cm, giving it a charge of $-35.0 \ \mu C$. Find the electric field (a) just inside the paint layer; (b) just outside the paint layer; (c) 5.00 cm outside the surface of the paint layer.

22.10 • A point charge $q_1 = 4.00$ nC is located on the x-axis at $x = 2.00$ m, and a second point charge $q_2 = -6.00$ nC is on the y-axis at $y = 1.00$ m. What is the total electric flux due to these two point charges through a spherical surface centered at the origin and with radius (a) 0.500 m, (b) 1.50 m, (c) 2.50 m?

22.11 • A 6.20-μC point charge is at the center of a cube with sides of length 0.500 m. (a) What is the electric flux through one of the six faces of the cube? (b) How would your answer to part (a) change if the sides were 0.250 m long? Explain.

22.12 • **Electric Fields in an Atom.** The nuclei of large atoms, such as uranium, with 92 protons, can be modeled as spherically symmetric spheres of charge. The radius of the uranium nucleus is approximately 7.4×10^{-15} m. (a) What is the electric field this nucleus produces just outside its surface? (b) What magnitude of electric field does it produce at the distance of the electrons, which is about 1.0×10^{-10} m? (c) The electrons can be modeled as forming a uniform shell of negative charge. What net electric field do they produce at the location of the nucleus?

22.13 • A point charge of $+5.00$ μC is located on the x-axis at $x = 4.00$ m, next to a spherical surface of radius 3.00 m centered at the origin. (a) Calculate the magnitude of the electric field at $x = 3.00$ m. (b) Calculate the magnitude of the electric field at $x = -3.00$ m. (c) According to Gauss's law, the net flux through the sphere is zero because it contains no charge. Yet the field due to the external charge is much stronger on the near side of the sphere (i.e., at $x = 3.00$ m) than on the far side (at $x = -3.00$ m). How, then, can the flux into the sphere (on the near side) equal the flux out of it (on the far side)? Explain. A sketch will help.

Section 22.4 Applications of Gauss's Law and
Section 22.5 Charges on Conductors

22.14 •• A solid metal sphere with radius 0.450 m carries a net charge of 0.250 nC. Find the magnitude of the electric field (a) at a point 0.100 m outside the surface of the sphere and (b) at a point inside the sphere, 0.100 m below the surface.

22.15 •• Two very long uniform lines of charge are parallel and are separated by 0.300 m. Each line of charge has charge per unit length $+5.20$ μC/m. What magnitude of force does one line of charge exert on a 0.0500-m section of the other line of charge?

22.16 • Some planetary scientists have suggested that the planet Mars has an electric field somewhat similar to that of the earth, producing a net electric flux of 3.63×10^{16} N$\cdot$m^2/C at the planet's surface, directed toward the center of the planet. Calculate: (a) the total electric charge on the planet; (b) the electric field at the planet's surface (refer to the astronomical data inside the back cover); (c) the charge density on Mars, assuming all the charge is uniformly distributed over the planet's surface.

22.17 •• How many excess electrons must be added to an isolated spherical conductor 32.0 cm in diameter to produce an electric field of 1150 N/C just outside the surface?

22.18 •• The electric field 0.400 m from a very long uniform line of charge is 840 N/C. How much charge is contained in a 2.00-cm section of the line?

22.19 •• A very long uniform line of charge has charge per unit length 4.80 μC/m and lies along the x-axis. A second long uniform line of charge has charge per unit length -2.40 μC/m and is parallel to the x-axis at $y = 0.400$ m. What is the net electric field (magnitude and direction) at the following points on the y-axis: (a) $y = 0.200$ m and (b) $y = 0.600$ m?

22.20 • (a) At a distance of 0.200 cm from the center of a charged conducting sphere with radius 0.100 cm, the electric field is 480 N/C. What is the electric field 0.600 cm from the center of the sphere? (b) At a distance of 0.200 cm from the axis of a very long charged conducting cylinder with radius 0.100 cm, the electric field is 480 N/C. What is the electric field 0.600 cm from the axis of the cylinder? (c) At a distance of 0.200 cm from a large uniform sheet of charge, the electric field is 480 N/C. What is the electric field 1.20 cm from the sheet?

22.21 •• A hollow, conducting sphere with an outer radius of 0.250 m and an inner radius of 0.200 m has a uniform surface charge density of $+6.37 \times 10^{-6}$ C/m^2. A charge of -0.500 μC is now introduced into the cavity inside the sphere. (a) What is the new charge density on the outside of the sphere? (b) Calculate the strength of the electric field just outside the sphere. (c) What is the electric flux through a spherical surface just inside the inner surface of the sphere?

22.22 •• A point charge of -2.00 μC is located in the center of a spherical cavity of radius 6.50 cm inside an insulating charged solid. The charge density in the solid is $\rho = 7.35 \times 10^{-4}$ C/m^3. Calculate the electric field inside the solid at a distance of 9.50 cm from the center of the cavity.

22.23 •• The electric field at a distance of 0.145 m from the surface of a solid insulating sphere with radius 0.355 m is 1750 N/C. (a) Assuming the sphere's charge is uniformly distributed, what is the charge density inside it? (b) Calculate the electric field inside the sphere at a distance of 0.200 m from the center.

22.24 •• CP A very small object with mass 8.20×10^{-9} kg and positive charge 6.50×10^{-9} C is projected directly toward a very large insulating sheet of positive charge that has uniform surface charge density 5.90×10^{-8} C/m^2. The object is initially 0.400 m from the sheet. What initial speed must the object have in order for its closest distance of approach to the sheet to be 0.100 m?

22.25 •• CP At time $t = 0$ a proton is a distance of 0.360 m from a very large insulating sheet of charge and is moving parallel to the sheet with speed 9.70×10^2 m/s. The sheet has uniform surface charge density 2.34×10^{-9} C/m^2. What is the speed of the proton at $t = 5.00 \times 10^{-8}$ s?

22.26 •• CP An electron is released from rest at a distance of 0.300 m from a large insulating sheet of charge that has uniform surface charge density $+2.90 \times 10^{-12}$ C/m^2. (a) How much work is done on the electron by the electric field of the sheet as the electron moves from its initial position to a point 0.050 m from the sheet? (b) What is the speed of the electron when it is 0.050 m from the sheet?

22.27 ••• CP CALC An insulating sphere of radius $R = 0.160$ m has uniform charge density $\rho = +7.20 \times 10^{-9}$ C/m^3. A small object that can be treated as a point charge is released from rest just outside the surface of the sphere. The small object has positive charge $q = 3.40 \times 10^{-6}$ C. How much work does the electric field of the sphere do on the object as the object moves to a point very far from the sphere?

22.28 • A conductor with an inner cavity, like that shown in Fig. 22.23c, carries a total charge of $+5.00$ nC. The charge within the cavity, insulated from the conductor, is -6.00 nC. How much charge is on (a) the inner surface of the conductor and (b) the outer surface of the conductor?

22.29 • Apply Gauss's law to the Gaussian surfaces S_2, S_3, and S_4 in Fig. 22.21b to calculate the electric field between and outside the plates.

22.30 • A square insulating sheet 80.0 cm on a side is held horizontally. The sheet has 7.50 nC of charge spread uniformly over its area. (a) Calculate the electric field at a point 0.100 mm above the center of the sheet. (b) Estimate the electric field at a point 100 m above the center of the sheet. (c) Would the answers to parts (a) and (b) be different if the sheet were made of a conducting material? Why or why not?

22.31 • An infinitely long cylindrical conductor has radius R and uniform surface charge density σ. (a) In terms of σ and R, what is the charge per unit length λ for the cylinder? (b) In terms of σ, what is the magnitude of the electric field produced by the charged cylinder at a distance $r > R$ from its axis? (c) Express the result of part (b) in terms of λ and show that the electric field outside the cylinder is the same as if all the charge were on the axis. Compare your result to the result for a line of charge in Example 22.6 (Section 22.4).

22.32 • Two very large, nonconducting plastic sheets, each 10.0 cm thick, carry uniform charge densities σ_1, σ_2, σ_3, and σ_4 on their surfaces, as shown in Fig. E22.32. These surface charge densities have the values $\sigma_1 = -6.00 \ \mu C/m^2$, $\sigma_2 = +5.00 \ \mu C/m^2$, $\sigma_3 = +2.00 \ \mu C/m^2$, and $\sigma_4 = +4.00 \ \mu C/m^2$. Use Gauss's law to find the magnitude and direction of the electric field at the following points, far from

Figure E22.32

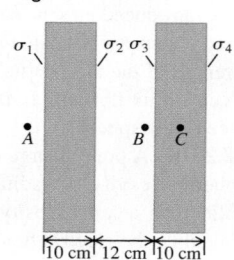

the edges of these sheets: (a) point A, 5.00 cm from the left face of the left-hand sheet; (b) point B, 1.25 cm from the inner surface of the right-hand sheet; (c) point C, in the middle of the right-hand sheet.

22.33 • A negative charge $-Q$ is placed inside the cavity of a hollow metal solid. The outside of the solid is grounded by connecting a conducting wire between it and the earth. (a) Is there any excess charge induced on the inner surface of the piece of metal? If so, find its sign and magnitude. (b) Is there any excess charge on the outside of the piece of metal? Why or why not? (c) Is there an electric field in the cavity? Explain. (d) Is there an electric field within the metal? Why or why not? Is there an electric field outside the piece of metal? Explain why or why not. (e) Would someone outside the solid measure an electric field due to the charge $-Q$? Is it reasonable to say that the grounded conductor has *shielded* the region from the effects of the charge $-Q$? In principle, could the same thing be done for gravity? Why or why not?

PROBLEMS

22.34 •• A cube has sides of length $L = 0.300$ m. It is placed with one corner at the origin as shown in Fig. E22.6. The electric field is not uniform but is given by $\vec{E} = (-5.00 \ N/C \cdot m)x\hat{\imath} + (3.00 \ N/C \cdot m)z\hat{k}$. (a) Find the electric flux through each of the six cube faces $S_1, S_2, S_3, S_4, S_5,$ and S_6. (b) Find the total electric charge inside the cube.

22.35 • The electric field $\vec{E}$ in Fig. P22.35 is everywhere parallel to the x-axis, so the components E_y and E_z are zero. The x-component of the field E_x depends on x but not on y and z. At points in the yz-plane (where $x = 0$), $E_x = 125$ N/C. (a) What is the electric flux through surface I in Fig. P22.35? (b) What is the electric flux through surface II? (c) The volume shown in the figure is a small section of a

Figure P22.35

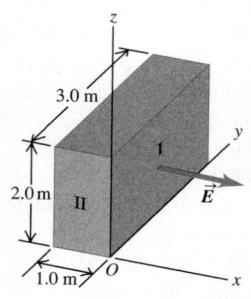

very large insulating slab 1.0 m thick. If there is a total charge of -24.0 nC within the volume shown, what are the magnitude and direction of $\vec{E}$ at the face opposite surface I? (d) Is the electric field produced only by charges within the slab, or is the field also due to charges outside the slab? How can you tell?

22.36 •• **CALC** In a region of space there is an electric field $\vec{E}$ that is in the z-direction and that has magnitude $E = (964 \ N/(C \cdot m))x$. Find the flux for this field through a square in the xy-plane at $z = 0$ and with side length 0.350 m. One side of the square is along the $+x$-axis and another side is along the $+y$-axis.

22.37 •• The electric field $\vec{E}_1$ at one face of a parallelepiped is uniform over the entire face and is directed out of the face. At the opposite face, the electric field $\vec{E}_2$ is also uniform over the entire face and is directed into that face (Fig. P22.37). The two faces in question are inclined at 30.0° from the horizontal, while $\vec{E}_1$ and $\vec{E}_2$ are both horizontal;

Figure P22.37

$\vec{E}_1$ has a magnitude of 2.50×10^4 N/C, and $\vec{E}_2$ has a magnitude of 7.00×10^4 N/C. (a) Assuming that no other electric field lines cross the surfaces of the parallelepiped, determine the net charge contained within. (b) Is the electric field produced only by the charges within the parallelepiped, or is the field also due to charges outside the parallelepiped? How can you tell?

22.38 • A long line carrying a uniform linear charge density $+50.0 \ \mu C/m$ runs parallel to and 10.0 cm from the surface of a large, flat plastic sheet that has a uniform surface charge density of $-100 \ \mu C/m^2$ on one side. Find the location of all points where an α particle would feel no force due to this arrangement of charged objects.

22.39 • **The Coaxial Cable.** A long coaxial cable consists of an inner cylindrical conductor with radius a and an outer coaxial cylinder with inner radius b and outer radius c. The outer cylinder is mounted on insulating supports and has no net charge. The inner cylinder has a uniform positive charge per unit length λ. Calculate the electric field (a) at any point between the cylinders a distance r from the axis and (b) at any point outside the outer cylinder. (c) Graph the magnitude of the electric field as a function of the distance r from the axis of the cable, from $r = 0$ to $r = 2c$. (d) Find the charge per unit length on the inner surface and on the outer surface of the outer cylinder.

22.40 • A very long conducting tube (hollow cylinder) has inner radius a and outer radius b. It carries charge per unit length $+\alpha$, where α is a positive constant with units of C/m. A line of charge lies along the axis of the tube. The line of charge has charge per unit length $+\alpha$. (a) Calculate the electric field in terms of α and the distance r from the axis of the tube for (i) $r < a$; (ii) $a < r < b$; (iii) $r > b$. Show your results in a graph of E as a function of r. (b) What is the charge per unit length on (i) the inner surface of the tube and (ii) the outer surface of the tube?

22.41 • Repeat Problem 22.40, but now let the conducting tube have charge per unit length $-\alpha$. As in Problem 22.40, the line of charge has charge per unit length $+\alpha$.

22.42 • A very long, solid cylinder with radius R has positive charge uniformly distributed throughout it, with charge per unit volume ρ. (a) Derive the expression for the electric field inside the volume at a distance r from the axis of the cylinder in terms of the charge density ρ. (b) What is the electric field at a point outside the volume in terms of the charge per unit length λ in the cylinder? (c) Compare the answers to parts (a) and (b) for $r = R$. (d) Graph the electric-field magnitude as a function of r from $r = 0$ to $r = 3R$.

22.43 •• **CP** A small sphere with a mass of 4.00×10^{-6} kg and carrying a charge of 5.00×10^{-8} C hangs from a thread near a very large, charged insulating

Figure P22.43

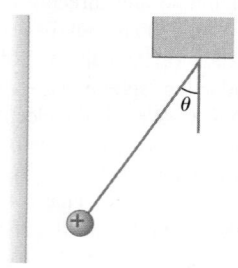

sheet, as shown in Fig. P22.43. The charge density on the surface of the sheet is uniform and equal to 2.50×10^{-9} C/m². Find the angle of the thread.

22.44 • **A Sphere in a Sphere.** A solid conducting sphere carrying charge q has radius a. It is inside a concentric hollow conducting sphere with inner radius b and outer radius c. The hollow sphere has no net charge. (a) Derive expressions for the electric-field magnitude in terms of the distance r from the center for the regions $r < a$, $a < r < b$, $b < r < c$, and $r > c$. (b) Graph the magnitude of the electric field as a function of r from $r = 0$ to $r = 2c$. (c) What is the charge on the inner surface of the hollow sphere? (d) On the outer surface? (e) Represent the charge of the small sphere by four plus signs. Sketch the field lines of the system within a spherical volume of radius $2c$.

22.45 • A solid conducting sphere with radius R that carries positive charge Q is concentric with a very thin insulating shell of radius $2R$ that also carries charge Q. The charge Q is distributed uniformly over the insulating shell. (a) Find the electric field (magnitude and direction) in each of the regions $0 < r < R$, $R < r < 2R$, and $r > 2R$. (b) Graph the electric-field magnitude as a function of r.

22.46 • A conducting spherical shell with inner radius a and outer radius b has a positive point charge Q located at its center. The total charge on the shell is $-3Q$, and it is insulated from its surroundings (Fig. P22.46). (a) Derive expressions for the electric-field magnitude in terms of the distance r from the center for the regions $r < a$, $a < r < b$, and $r > b$. (b) What is the surface charge density on the inner surface of the conducting shell? (c) What is the surface charge density on the outer surface of the conducting shell? (d) Sketch the electric field lines and the location of all charges. (e) Graph the electric-field magnitude as a function of r.

Figure **P22.46**

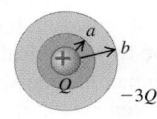

22.47 • **Concentric Spherical Shells.** A small conducting spherical shell with inner radius a and outer radius b is concentric with a larger conducting spherical shell with inner radius c and outer radius d (Fig. P22.47). The inner shell has total charge $+2q$, and the outer shell has charge $+4q$. (a) Calculate the electric field (magnitude and direction) in terms of q and the distance r from the common center of the two shells for (i) $r < a$; (ii) $a < r < b$; (iii) $b < r < c$; (iv) $c < r < d$; (v) $r > d$. Show your results in a graph of the radial component of $\vec{E}$ as a function of r. (b) What is the total charge on the (i) inner surface of the small shell; (ii) outer surface of the small shell; (iii) inner surface of the large shell; (iv) outer surface of the large shell?

Figure **P22.47**

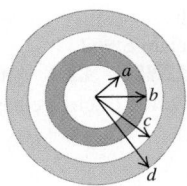

22.48 • Repeat Problem 22.47, but now let the outer shell have charge $-2q$. As in Problem 22.47, the inner shell has charge $+2q$.

22.49 • Repeat Problem 22.47, but now let the outer shell have charge $-4q$. As in Problem 22.47, the inner shell has charge $+2q$.

22.50 • A solid conducting sphere with radius R carries a positive total charge Q. The sphere is surrounded by an insulating shell with inner radius R and outer radius $2R$. The insulating shell has a uniform charge density ρ. (a) Find the value of ρ so that the net charge of the entire system is zero. (b) If ρ has the value found in part (a), find the electric field (magnitude and direction) in each of the regions $0 < r < R$, $R < r < 2R$, and $r > 2R$. Show your results in a graph of the radial component of $\vec{E}$ as a function of r. (c) As a general rule, the electric field is discontinuous only at locations where there is a thin sheet of charge. Explain how your results in part (b) agree with this rule.

22.51 • Negative charge $-Q$ is distributed uniformly over the surface of a thin spherical insulating shell with radius R. Calculate the force (magnitude and direction) that the shell exerts on a positive point charge q located (a) a distance $r > R$ from the center of the shell (outside the shell) and (b) a distance $r < R$ from the center of the shell (inside the shell).

22.52 •• (a) How many excess electrons must be distributed uniformly within the volume of an isolated plastic sphere 30.0 cm in diameter to produce an electric field of 1390 N/C just outside the surface of the sphere? (b) What is the electric field at a point 10.0 cm outside the surface of the sphere?

22.53 ••• **CALC** An insulating hollow sphere has inner radius a and outer radius b. Within the insulating material the volume charge density is given by $\rho(r) = \frac{\alpha}{r}$, where α is a positive constant. (a) In terms of α and a, what is the magnitude of the electric field at a distance r from the center of the shell, where $a < r < b$? (b) A point charge q is placed at the center of the hollow space, at $r = 0$. In terms of α and a, what value must q have (sign and magnitude) in order for the electric field to be constant in the region $a < r < b$, and what then is the value of the constant field in this region?

22.54 •• **CP Thomson's Model of the Atom.** In the early years of the 20th century, a leading model of the structure of the atom was that of the English physicist J. J. Thomson (the discoverer of the electron). In Thomson's model, an atom consisted of a sphere of positively charged material in which were embedded negatively charged electrons, like chocolate chips in a ball of cookie dough. Consider such an atom consisting of one electron with mass m and charge $-e$, which may be regarded as a point charge, and a uniformly charged sphere of charge $+e$ and radius R. (a) Explain why the equilibrium position of the electron is at the center of the nucleus. (b) In Thomson's model, it was assumed that the positive material provided little or no resistance to the motion of the electron. If the electron is displaced from equilibrium by a distance less than R, show that the resulting motion of the electron will be simple harmonic, and calculate the frequency of oscillation. (*Hint:* Review the definition of simple harmonic motion in Section 14.2. If it can be shown that the net force on the electron is of this form, then it follows that the motion is simple harmonic. Conversely, if the net force on the electron does not follow this form, the motion is not simple harmonic.) (c) By Thomson's time, it was known that excited atoms emit light waves of only certain frequencies. In his model, the frequency of emitted light is the same as the oscillation frequency of the electron or electrons in the atom. What would the radius of a Thomson-model atom have to be for it to produce red light of frequency 4.57×10^{14} Hz? Compare your answer to the radii of real atoms, which are of the order of 10^{-10} m (see Appendix F for data about the electron). (d) If the electron were displaced from equilibrium by a distance greater than R, would the electron oscillate? Would its motion be simple harmonic? Explain your reasoning. (*Historical note:* In 1910, the atomic nucleus was discovered, proving the Thomson model to be incorrect. An atom's positive charge is not spread over its volume as Thomson supposed, but is concentrated in the tiny nucleus of radius 10^{-14} to 10^{-15} m.)

22.55 • **Thomson's Model of the Atom, Continued.** Using Thomson's (outdated) model of the atom described in Problem 22.54, consider an atom consisting of two electrons, each of charge $-e$, embedded in a sphere of charge $+2e$ and radius R. In

Figure **P22.55**

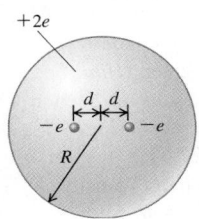

equilibrium, each electron is a distance d from the center of the atom (Fig. P22.55). Find the distance d in terms of the other properties of the atom.

22.56 • **A Uniformly Charged Slab.** A slab of insulating material has thickness $2d$ and is oriented so that its faces are parallel to the yz-plane and given by the planes $x = d$ and $x = -d$. The y- and z-dimensions of the slab are very large compared to d and may be treated as essentially infinite. The slab has a uniform positive charge density ρ. (a) Explain why the electric field due to the slab is zero at the center of the slab ($x = 0$). (b) Using Gauss's law, find the electric field due to the slab (magnitude and direction) at all points in space.

22.57 • **CALC** **A Nonuniformly Charged Slab.** Repeat Problem 22.56, but now let the charge density of the slab be given by $\rho(x) = \rho_0(x/d)^2$, where ρ_0 is a positive constant.

22.58 • **CALC** A nonuniform, but spherically symmetric, distribution of charge has a charge density $\rho(r)$ given as follows:

$$\rho(r) = \rho_0(1 - 4r/3R) \qquad \text{for } r \leq R$$
$$\rho(r) = 0 \qquad \text{for } r \geq R$$

where ρ_0 is a positive constant. (a) Find the total charge contained in the charge distribution. (b) Obtain an expression for the electric field in the region $r \geq R$. (c) Obtain an expression for the electric field in the region $r \leq R$. (d) Graph the electric-field magnitude E as a function of r. (e) Find the value of r at which the electric field is maximum, and find the value of that maximum field.

22.59 • **CP CALC** **Gauss's Law for Gravitation.** The gravitational force between two point masses separated by a distance r is proportional to $1/r^2$, just like the electric force between two point charges. Because of this similarity between gravitational and electric interactions, there is also a Gauss's law for gravitation. (a) Let $\vec{g}$ be the acceleration due to gravity caused by a point mass m at the origin, so that $\vec{g} = -(Gm/r^2)\hat{r}$. Consider a spherical Gaussian surface with radius r centered on this point mass, and show that the flux of $\vec{g}$ through this surface is given by

$$\oint \vec{g} \cdot d\vec{A} = -4\pi Gm$$

(b) By following the same logical steps used in Section 22.3 to obtain Gauss's law for the electric field, show that the flux of $\vec{g}$ through *any* closed surface is given by

$$\oint \vec{g} \cdot d\vec{A} = -4\pi GM_{\text{encl}}$$

where M_{encl} is the total mass enclosed within the closed surface.

22.60 • **CP** **Applying Gauss's Law for Gravitation.** Using Gauss's law for gravitation (derived in part (b) of Problem 22.59), show that the following statements are true: (a) For any spherically symmetric mass distribution with total mass M, the acceleration due to gravity outside the distribution is the same as though all the mass were concentrated at the center. (*Hint:* See Example 22.5 in Section 22.4.) (b) At any point inside a spherically symmetric shell of mass, the acceleration due to gravity is zero. (*Hint:* See Example 22.5.) (c) If we could drill a hole through a spherically symmetric planet to its center, and if the density were uniform, we would find that the magnitude of $\vec{g}$ is directly proportional to the distance r from the center. (*Hint:* See Example 22.9 in Section 22.4.) We proved these results in Section 13.6 using some fairly strenuous analysis; the proofs using Gauss's law for gravitation are *much* easier.

22.61 • (a) An insulating sphere with radius a has a uniform charge density ρ. The sphere is not centered at the origin but at $\vec{r} = \vec{b}$. Show that the electric field inside the sphere is given by

$\vec{E} = \rho(\vec{r} - \vec{b})/3\epsilon_0$. (b) An insulating sphere of radius R has a spherical hole of radius a located within its volume and centered a distance b from the center of the sphere, where $a < b < R$ (a cross section of the sphere is shown in Fig. P22.61). The solid part of the sphere has a uniform volume charge density ρ. Find the magnitude and direction of the electric field $\vec{E}$ inside the hole, and show that $\vec{E}$ is uniform over the entire hole. [*Hint:* Use the principle of superposition and the result of part (a).]

Figure **P22.61**

Charge density ρ

22.62 • A very long, solid insulating cylinder with radius R has a cylindrical hole with radius a bored along its entire length. The axis of the hole is a distance b from the axis of the cylinder, where $a < b < R$ (Fig. P22.62). The solid material of the cylinder has a uniform volume charge density ρ. Find the magnitude and direction of the electric field $\vec{E}$ inside the hole, and show that $\vec{E}$ is uniform over the entire hole. (*Hint:* See Problem 22.61.)

Figure **P22.62**

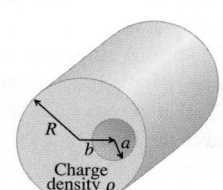
Charge density ρ

22.63 • Positive charge Q is distributed uniformly over each of two spherical volumes with radius R. One sphere of charge is centered at the origin and the other at $x = 2R$ (Fig. P22.63). Find the magnitude and direction of the net electric field due to these two distributions of charge at the following points on the x-axis: (a) $x = 0$; (b) $x = R/2$; (c) $x = R$; (d) $x = 3R$.

Figure **P22.63**

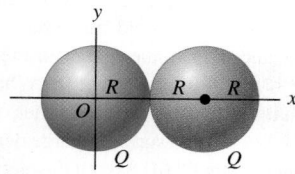

22.64 • Repeat Problem 22.63, but now let the left-hand sphere have positive charge Q and let the right-hand sphere have negative charge $-Q$.

22.65 •• **CALC** A nonuniform, but spherically symmetric, distribution of charge has a charge density $\rho(r)$ given as follows:

$$\rho(r) = \rho_0(1 - r/R) \qquad \text{for } r \leq R$$
$$\rho(r) = 0 \qquad \text{for } r \geq R$$

where $\rho_0 = 3Q/\pi R^3$ is a positive constant. (a) Show that the total charge contained in the charge distribution is Q. (b) Show that the electric field in the region $r \geq R$ is identical to that produced by a point charge Q at $r = 0$. (c) Obtain an expression for the electric field in the region $r \leq R$. (d) Graph the electric-field magnitude E as a function of r. (e) Find the value of r at which the electric field is maximum, and find the value of that maximum field.

CHALLENGE PROBLEMS

22.66 ••• **CP CALC** A region in space contains a total positive charge Q that is distributed spherically such that the volume charge density $\rho(r)$ is given by

$$\rho(r) = \alpha \qquad \text{for } r \leq R/2$$
$$\rho(r) = 2\alpha(1 - r/R) \qquad \text{for } R/2 \leq r \leq R$$
$$\rho(r) = 0 \qquad \text{for } r \geq R$$

Here α is a positive constant having units of C/m^3. (a) Determine α in terms of Q and R. (b) Using Gauss's law, derive an expression for the magnitude of $\vec{E}$ as a function of r. Do this separately for all three regions. Express your answers in terms of the total charge Q. Be sure to check that your results agree on the boundaries of the regions. (c) What fraction of the total charge is contained within the region $r \leq R/2$? (d) If an electron with charge $q' = -e$ is oscillating back and forth about $r = 0$ (the center of the distribution) with an amplitude less than $R/2$, show that the motion is simple harmonic. (*Hint:* Review the discussion of simple harmonic motion in Section 14.2. If, and only if, the net force on the electron is proportional to its displacement from equilibrium, then the motion is simple harmonic.) (e) What is the period of the motion in part (d)? (f) If the amplitude of the motion described in part (e) is greater than $R/2$, is the motion still simple harmonic? Why or why not?

22.67 ••• **CP CALC** A region in space contains a total positive charge Q that is distributed spherically such that the volume charge density $\rho(r)$ is given by

$$\rho(r) = 3\alpha r/(2R) \qquad \text{for } r \leq R/2$$
$$\rho(r) = \alpha[1 - (r/R)^2] \qquad \text{for } R/2 \leq r \leq R$$
$$\rho(r) = 0 \qquad \text{for } r \geq R$$

Here α is a positive constant having units of C/m^3. (a) Determine α in terms of Q and R. (b) Using Gauss's law, derive an expression for the magnitude of the electric field as a function of r. Do this separately for all three regions. Express your answers in terms of the total charge Q. (c) What fraction of the total charge is contained within the region $R/2 \leq r \leq R$? (d) What is the magnitude of $\vec{E}$ at $r = R/2$? (e) If an electron with charge $q' = -e$ is released from rest at any point in any of the three regions, the resulting motion will be oscillatory but not simple harmonic. Why? (See Challenge Problem 22.66.)

Answers

Chapter Opening Question ?

No. The electric field inside a cavity within a conductor is zero, so there is no electric effect on the child. (See Section 22.5.)

Test Your Understanding Questions

22.1 Answer: (iii) Each part of the surface of the box will be three times farther from the charge $+q$, so the electric field will be $\left(\frac{1}{3}\right)^2 = \frac{1}{9}$ as strong. But the area of the box will increase by a factor of $3^2 = 9$. Hence the electric flux will be multiplied by a factor of $\left(\frac{1}{9}\right)(9) = 1$. In other words, the flux will be unchanged.

22.2 Answer: (iv), (ii), (i), (iii) In each case the electric field is uniform, so the flux is $\Phi_E = \vec{E} \cdot \vec{A}$. We use the relationships for the scalar products of unit vectors: $\hat{\imath} \cdot \hat{\imath} = \hat{\jmath} \cdot \hat{\jmath} = 1$, $\hat{\imath} \cdot \hat{\jmath} = 0$. In case (i) we have $\Phi_E = (4.0 \text{ N/C})(6.0 \text{ m}^2)\hat{\imath} \cdot \hat{\jmath} = 0$ (the electric field and vector area are perpendicular, so there is zero flux). In case (ii) we have $\Phi_E [(4.0 \text{ N/C})\hat{\imath} + (2.0 \text{ N/C})\hat{\jmath}] \cdot (3.0 \text{ m}^2)\hat{\jmath} = (2.0 \text{ N/C}) \cdot (3.0 \text{ m}^2) = 6.0 \text{ N} \cdot \text{m}^2/\text{C}$. Similarly, in case (iii) we have $\Phi_E = [(4.0 \text{ N/C})\hat{\imath} - (2.0 \text{ N/C})\hat{\jmath}] \cdot [(3.0 \text{ m}^2)\hat{\imath} + (7.0 \text{ m}^2)\hat{\jmath}] = (4.0 \text{ N/C})(3.0 \text{ m}^2) - (2.0 \text{ N/C})(7.0 \text{ m}^2) = -2 \text{ N} \cdot \text{m}^2/\text{C}$, and in case (iv) we have $\Phi_E = [(4.0 \text{ N/C})\hat{\imath} - (2.0 \text{ N/C})\hat{\jmath}] \cdot [(3.0 \text{ m}^2)\hat{\imath} - (7.0 \text{ m}^2)\hat{\jmath}] = (4.0 \text{ N/C})(3.0 \text{ m}^2) + (2.0 \text{ N/C}) \cdot (7.0 \text{ m}^2) = 26 \text{ N} \cdot \text{m}^2/\text{C}$.

22.3 Answer: S_2, S_5, S_4, S_1 **and** S_3 **(tie)** Gauss's law tells us that the flux through a closed surface is proportional to the amount of charge enclosed within that surface. So an ordering of these surfaces by their fluxes is the same as an ordering by the amount of enclosed charge. Surface S_1 encloses no charge, surface S_2 encloses $9.0 \ \mu\text{C} + 5.0 \ \mu\text{C} + (-7.0 \ \mu\text{C}) = 7.0 \ \mu\text{C}$, surface S_3 encloses $9.0 \ \mu\text{C} + 1.0 \ \mu\text{C} + (-10.0 \ \mu\text{C}) = 0$, surface S_4 encloses $8.0 \ \mu\text{C} + (-7.0 \ \mu\text{C}) = 1.0 \ \mu\text{C}$, and surface S_5 encloses $8.0 \ \mu\text{C} + (-7.0 \ \mu\text{C}) + (-10.0 \ \mu\text{C}) + (1.0 \ \mu\text{C}) + (9.0 \ \mu\text{C}) + (5.0 \ \mu\text{C}) = 6.0 \ \mu\text{C}$.

22.4 Answer: no You might be tempted to draw a Gaussian surface that is an enlarged version of the conductor, with the same shape and placed so that it completely encloses the conductor.

While you know the flux through this Gaussian surface (by Gauss's law, it's $\Phi_E = Q/\epsilon_0$), the direction of the electric field need not be perpendicular to the surface and the magnitude of the field need not be the same at all points on the surface. It's not possible to do the flux integral $\oint E_\perp \, dA$, and we can't calculate the electric field. Gauss's law is useful for calculating the electric field only when the charge distribution is *highly* symmetric.

22.5 Answer: no Before you connect the wire to the sphere, the presence of the point charge will induce a charge $-q$ on the inner surface of the hollow sphere and a charge q on the outer surface (the net charge on the sphere is zero). There will be an electric field outside the sphere due to the charge on the outer surface. Once you touch the conducting wire to the sphere, however, electrons will flow from ground to the outer surface of the sphere to neutralize the charge there (see Fig. 21.7c). As a result the sphere will have no charge on its outer surface and no electric field outside.

Bridging Problem

Answers: (a) $Q(r) = Qe^{-2r/a_0}[2(r/a_0)^2 + 2(r/a_0) + 1]$

(b) $E = \dfrac{kQe^{-2r/a_0}}{r^2}[2(r/a_0)^2 + 2(r/a_0) + 1]$

(c)

23 ELECTRIC POTENTIAL

LEARNING GOALS

By studying this chapter, you will learn:

- How to calculate the electric potential energy of a collection of charges.

- The meaning and significance of electric potential.

- How to calculate the electric potential that a collection of charges produces at a point in space.

- How to use equipotential surfaces to visualize how the electric potential varies in space.

- How to use electric potential to calculate the electric field.

? In one type of welding, electric charge flows between the welding tool and the metal pieces that are to be joined together. This produces a glowing arc whose high temperature fuses the pieces together. Why must the tool be held close to the pieces being welded?

This chapter is about energy associated with electrical interactions. Every time you turn on a light, listen to an MP3 player, or talk on a mobile phone, you are using electrical energy, an indispensable ingredient of our technological society. In Chapters 6 and 7 we introduced the concepts of *work* and *energy* in the context of mechanics; now we'll combine these concepts with what we've learned about electric charge, electric forces, and electric fields. Just as we found for many problems in mechanics, using energy ideas makes it easier to solve a variety of problems in electricity.

When a charged particle moves in an electric field, the field exerts a force that can do *work* on the particle. This work can always be expressed in terms of electric potential energy. Just as gravitational potential energy depends on the height of a mass above the earth's surface, electric potential energy depends on the position of the charged particle in the electric field. We'll describe electric potential energy using a new concept called *electric potential,* or simply *potential.* In circuits, a difference in potential from one point to another is often called *voltage.* The concepts of potential and voltage are crucial to understanding how electric circuits work and have equally important applications to electron beams used in cancer radiotherapy, high-energy particle accelerators, and many other devices.

23.1 Electric Potential Energy

The concepts of work, potential energy, and conservation of energy proved to be extremely useful in our study of mechanics. In this section we'll show that these concepts are just as useful for understanding and analyzing electrical interactions.

Let's begin by reviewing three essential points from Chapters 6 and 7. First, when a force $\vec{F}$ acts on a particle that moves from point a to point b, the work $W_{a\to b}$ done by the force is given by a *line integral*:

$$W_{a\to b} = \int_a^b \vec{F} \cdot d\vec{l} = \int_a^b F\cos\phi \, dl \quad \text{(work done by a force)} \qquad (23.1)$$

where $d\vec{l}$ is an infinitesimal displacement along the particle's path and ϕ is the angle between $\vec{F}$ and $d\vec{l}$ at each point along the path.

Second, if the force $\vec{F}$ is *conservative*, as we defined the term in Section 7.3, the work done by $\vec{F}$ can always be expressed in terms of a **potential energy** U. When the particle moves from a point where the potential energy is U_a to a point where it is U_b, the change in potential energy is $\Delta U = U_b - U_a$ and the work $W_{a\to b}$ done by the force is

$$W_{a\to b} = U_a - U_b = -(U_b - U_a) = -\Delta U \quad \begin{array}{l}\text{(work done by a}\\\text{conservative force)}\end{array} \qquad (23.2)$$

When $W_{a\to b}$ is positive, U_a is greater than U_b, ΔU is negative, and the potential energy *decreases*. That's what happens when a baseball falls from a high point (a) to a lower point (b) under the influence of the earth's gravity; the force of gravity does positive work, and the gravitational potential energy decreases (Fig. 23.1). When a tossed ball is moving upward, the gravitational force does negative work during the ascent, and the potential energy increases.

Third, the work–energy theorem says that the change in kinetic energy $\Delta K = K_b - K_a$ during a displacement equals the *total* work done on the particle. If only conservative forces do work, then Eq. (23.2) gives the total work, and $K_b - K_a = -(U_b - U_a)$. We usually write this as

$$K_a + U_a = K_b + U_b \qquad (23.3)$$

That is, the total mechanical energy (kinetic plus potential) is *conserved* under these circumstances.

23.1 The work done on a baseball moving in a uniform gravitational field.

Object moving in a uniform gravitational field

$\vec{w} = m\vec{g}$

The work done by the gravitational force is the same for any path from a to b:
$W_{a\to b} = -\Delta U = mgh$

Electric Potential Energy in a Uniform Field

Let's look at an electrical example of these basic concepts. In Fig. 23.2 a pair of charged parallel metal plates sets up a uniform, downward electric field with magnitude E. The field exerts a downward force with magnitude $F = q_0E$ on a positive test charge q_0. As the charge moves downward a distance d from point a to point b, the force on the test charge is constant and independent of its location. So the work done by the electric field is the product of the force magnitude and the component of displacement in the (downward) direction of the force:

$$W_{a\to b} = Fd = q_0Ed \qquad (23.4)$$

This work is positive, since the force is in the same direction as the net displacement of the test charge.

The y-component of the electric force, $F_y = -q_0E$, is constant, and there is no x- or z-component. This is exactly analogous to the gravitational force on a mass m near the earth's surface; for this force, there is a constant y-component $F_y = -mg$ and the x- and z-components are zero. Because of this analogy, we can conclude that the force exerted on q_0 by the uniform electric field in Fig. 23.2 is *conservative*, just as is the gravitational force. This means that the work $W_{a\to b}$ done by the field is independent of the path the particle takes from a to b. We can represent this work with a *potential-energy* function U, just as we did for gravitational potential energy

23.2 The work done on a point charge moving in a uniform electric field. Compare with Fig. 23.1.

Point charge moving in a uniform electric field

$\vec{E}$

q_0

$\vec{F} = q_0\vec{E}$

The work done by the electric force is the same for any path from a to b:
$W_{a\to b} = -\Delta U = q_0Ed$

23.3 A positive charge moving (a) in the direction of the electric field $\vec{E}$ and (b) in the direction opposite $\vec{E}$.

(a) Positive charge moves in the direction of $\vec{E}$:
• Field does *positive* work on charge.
• *U decreases.*

(b) Positive charge moves opposite $\vec{E}$:
• Field does *negative* work on charge.
• *U increases.*

in Section 7.1. The potential energy for the gravitational force $F_y = -mg$ was $U = mgy$; hence the potential energy for the electric force $F_y = -q_0E$ is

$$U = q_0Ey \tag{23.5}$$

When the test charge moves from height y_a to height y_b, the work done on the charge by the field is given by

$$W_{a \to b} = -\Delta U = -(U_b - U_a) = -(q_0Ey_b - q_0Ey_a) = q_0E(y_a - y_b) \tag{23.6}$$

When y_a is greater than y_b (Fig. 23.3a), the positive test charge q_0 moves downward, in the same direction as $\vec{E}$; the displacement is in the same direction as the force $\vec{F} = q_0\vec{E}$, so the field does positive work and U decreases. [In particular, if $y_a - y_b = d$ as in Fig. 23.2, Eq. (23.6) gives $W_{a \to b} = q_0Ed$, in agreement with Eq. (23.4).] When y_a is less than y_b (Fig. 23.3b), the positive test charge q_0 moves upward, in the opposite direction to $\vec{E}$; the displacement is opposite the force, the field does negative work, and U increases.

If the test charge q_0 is *negative,* the potential energy increases when it moves with the field and decreases when it moves against the field (Fig. 23.4).

Whether the test charge is positive or negative, the following general rules apply: *U increases* if the test charge q_0 moves in the direction *opposite* the electric force $\vec{F} = q_0\vec{E}$ (Figs. 23.3b and 23.4a); *U decreases* if q_0 moves in the *same*

23.4 A negative charge moving (a) in the direction of the electric field $\vec{E}$ and (b) in the direction opposite $\vec{E}$. Compare with Fig. 23.3.

(a) Negative charge moves in the direction of $\vec{E}$:
• Field does *negative* work on charge.
• *U increases.*

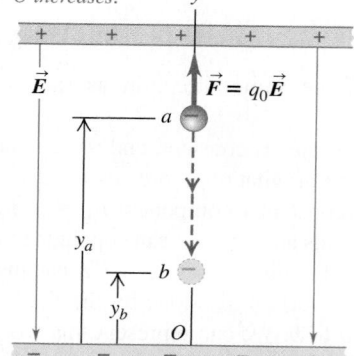

(b) Negative charge moves opposite $\vec{E}$:
• Field does *positive* work on charge.
• *U decreases.*

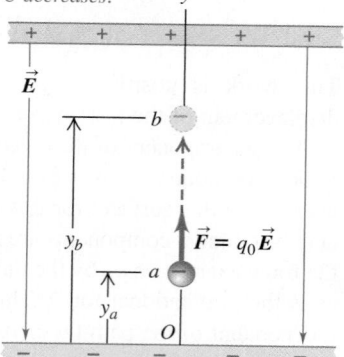

direction as $\vec{F} = q_0\vec{E}$ (Figs. 23.3a and 23.4b). This is the same behavior as for gravitational potential energy, which increases if a mass m moves upward (opposite the direction of the gravitational force) and decreases if m moves downward (in the same direction as the gravitational force).

> **CAUTION** **Electric potential energy** The relationship between electric potential energy change and motion in an electric field is an important one that we'll use often, but that takes some effort to truly understand. Take the time to carefully study the preceding paragraph as well as Figs. 23.3 and 23.4. Doing so now will help you tremendously later!

Electric Potential Energy of Two Point Charges

The idea of electric potential energy isn't restricted to the special case of a uniform electric field. Indeed, we can apply this concept to a point charge in *any* electric field caused by a static charge distribution. Recall from Chapter 21 that we can represent any charge distribution as a collection of point charges. Therefore it's useful to calculate the work done on a test charge q_0 moving in the electric field caused by a single, stationary point charge q.

We'll consider first a displacement along the *radial* line in Fig. 23.5. The force on q_0 is given by Coulomb's law, and its radial component is

$$F_r = \frac{1}{4\pi\epsilon_0} \frac{qq_0}{r^2} \tag{23.7}$$

If q and q_0 have the same sign ($+$ or $-$) the force is repulsive and F_r is positive; if the two charges have opposite signs, the force is attractive and F_r is negative. The force is *not* constant during the displacement, and we have to integrate to calculate the work $W_{a\to b}$ done on q_0 by this force as q_0 moves from a to b:

$$W_{a\to b} = \int_{r_a}^{r_b} F_r\,dr = \int_{r_a}^{r_b} \frac{1}{4\pi\epsilon_0} \frac{qq_0}{r^2}\,dr = \frac{qq_0}{4\pi\epsilon_0}\left(\frac{1}{r_a} - \frac{1}{r_b}\right) \tag{23.8}$$

The work done by the electric force for this particular path depends only on the endpoints.

Now let's consider a more general displacement (Fig. 23.6) in which a and b do *not* lie on the same radial line. From Eq. (23.1) the work done on q_0 during this displacement is given by

$$W_{a\to b} = \int_{r_a}^{r_b} F\cos\phi\,dl = \int_{r_a}^{r_b} \frac{1}{4\pi\epsilon_0} \frac{qq_0}{r^2}\cos\phi\,dl$$

But Fig. 23.6 shows that $\cos\phi\,dl = dr$. That is, the work done during a small displacement $d\vec{l}$ depends only on the change dr in the distance r between the charges, which is the *radial component* of the displacement. Thus Eq. (23.8) is valid even for this more general displacement; the work done on q_0 by the electric field $\vec{E}$ produced by q depends only on r_a and r_b, not on the details of the path. Also, if q_0 returns to its starting point a by a different path, the total work done in the round-trip displacement is zero (the integral in Eq. (23.8) is from r_a back to r_a). These are the needed characteristics for a conservative force, as we defined it in Section 7.3. Thus the force on q_0 is a *conservative* force.

We see that Eqs. (23.2) and (23.8) are consistent if we define the potential energy to be $U_a = qq_0/4\pi\epsilon_0 r_a$ when q_0 is a distance r_a from q, and to be $U_b = qq_0/4\pi\epsilon_0 r_b$ when q_0 is a distance r_b from q. Thus the potential energy U when the test charge q_0 is at *any* distance r from charge q is

$$U = \frac{1}{4\pi\epsilon_0} \frac{qq_0}{r} \quad \begin{array}{l}\text{(electric potential energy of}\\ \text{two point charges } q \text{ and } q_0)\end{array} \tag{23.9}$$

23.5 Test charge q_0 moves along a straight line extending radially from charge q. As it moves from a to b, the distance varies from r_a to r_b.

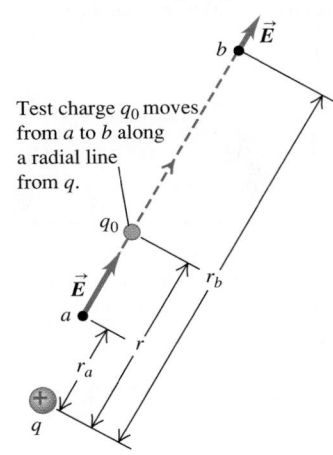

23.6 The work done on charge q_0 by the electric field of charge q does not depend on the path taken, but only on the distances r_a and r_b.

Test charge q_0 moves from a to b along an arbitrary path.

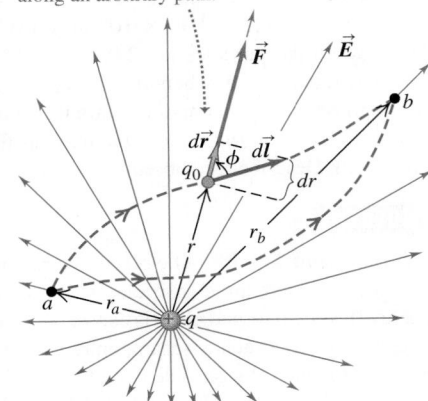

23.7 Graphs of the potential energy U of two point charges q and q_0 versus their separation r.

(a) q and q_0 have the same sign.

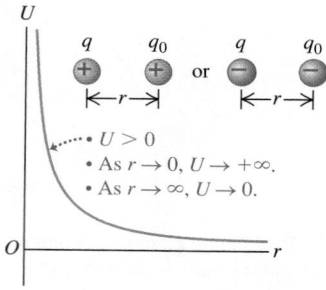

- $U > 0$
- As $r \to 0$, $U \to +\infty$.
- As $r \to \infty$, $U \to 0$.

(b) q and q_0 have opposite signs.

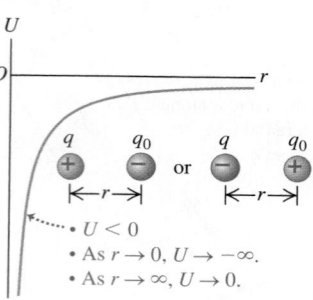

- $U < 0$
- As $r \to 0$, $U \to -\infty$.
- As $r \to \infty$, $U \to 0$.

Equation (23.9) is valid no matter what the signs of the charges q and q_0. The potential energy is positive if the charges q and q_0 have the same sign (Fig. 23.7a) and negative if they have opposite signs (Fig. 23.7b).

CAUTION **Electric potential energy vs. electric force** Don't confuse Eq. (23.9) for the potential energy of two point charges with the similar expression in Eq. (23.7) for the radial component of the electric force that one charge exerts on the other. Potential energy U is proportional to $1/r$, while the force component F_r is proportional to $1/r^2$. ▮

Potential energy is always defined relative to some reference point where $U = 0$. In Eq. (23.9), U is zero when q and q_0 are infinitely far apart and $r = \infty$. Therefore U represents the work that would be done on the test charge q_0 by the field of q if q_0 moved from an initial distance r to infinity. If q and q_0 have the same sign, the interaction is repulsive, this work is positive, and U is positive at any finite separation (Fig. 23.7a). If the charges have opposite signs, the interaction is attractive, the work done is negative, and U is negative (Fig. 23.7b).

We emphasize that the potential energy U given by Eq. (23.9) is a *shared* property of the two charges. If the distance between q and q_0 is changed from r_a to r_b, the change in potential energy is the same whether q is held fixed and q_0 is moved or q_0 is held fixed and q is moved. For this reason, we never use the phrase "the electric potential energy *of* a point charge." (Likewise, if a mass m is at a height h above the earth's surface, the gravitational potential energy is a shared property of the mass m and the earth. We emphasized this in Sections 7.1 and 13.3.)

Equation (23.9) also holds if the charge q_0 is outside a spherically symmetric charge *distribution* with total charge q; the distance r is from q_0 to the center of the distribution. That's because Gauss's law tells us that the electric field outside such a distribution is the same as if all of its charge q were concentrated at its center (see Example 22.9 in Section 22.4).

Example 23.1 Conservation of energy with electric forces

A positron (the electron's antiparticle) has mass 9.11×10^{-31} kg and charge $q_0 = +e = +1.60 \times 10^{-19}$ C. Suppose a positron moves in the vicinity of an α (alpha) particle, which has charge $q = +2e = 3.20 \times 10^{-19}$ C and mass 6.64×10^{-27} kg. The α particle's mass is more than 7000 times that of the positron, so we assume that the α particle remains at rest. When the positron is 1.00×10^{-10} m from the α particle, it is moving directly away from the α particle at 3.00×10^6 m/s. (a) What is the positron's speed when the particles are 2.00×10^{-10} m apart? (b) What is the positron's speed when it is very far from the α particle? (c) Suppose the initial conditions are the same but the moving particle is an electron (with the same mass as the positron but charge $q_0 = -e$). Describe the subsequent motion.

SOLUTION

IDENTIFY and SET UP: The electric force between a positron (or an electron) and an α particle is conservative, so mechanical energy (kinetic plus potential) is conserved. Equation (23.9) gives the potential energy U at any separation r: The potential-energy function for parts (a) and (b) looks like that of Fig. 23.7a, and the function for part (c) looks like that of Fig. 23.7b. We are given the positron speed $v_a = 3.00 \times 10^6$ m/s when the separation between the particles is $r_a = 1.00 \times 10^{-10}$ m. In parts (a) and (b) we use Eqs. (23.3) and (23.9) to find the speed for $r = r_b = 2.00 \times 10^{-10}$ m and $r = r_c \to \infty$, respectively. In part (c) we replace the positron with an electron and reconsider the problem.

EXECUTE: (a) Both particles have positive charge, so the positron speeds up as it moves away from the α particle. From the energy-conservation equation, Eq. (23.3), the final kinetic energy is

$$K_b = \tfrac{1}{2}mv_b^2 = K_a + U_a - U_b$$

In this expression,

$$K_a = \tfrac{1}{2}mv_a^2 = \tfrac{1}{2}(9.11 \times 10^{-31} \text{ kg})(3.00 \times 10^6 \text{ m/s})^2$$
$$= 4.10 \times 10^{-18} \text{ J}$$

$$U_a = \frac{1}{4\pi\epsilon_0}\frac{qq_0}{r_a} = (9.0 \times 10^9 \text{ N} \cdot \text{m}^2/\text{C}^2)$$
$$\times \frac{(3.20 \times 10^{-19} \text{ C})(1.60 \times 10^{-19} \text{ C})}{1.00 \times 10^{-10} \text{ m}}$$
$$= 4.61 \times 10^{-18} \text{ J}$$

$$U_b = \frac{1}{4\pi\epsilon_0}\frac{qq_0}{r_b} = 2.30 \times 10^{-18} \text{ J}$$

Hence the positron kinetic energy and speed at $r = r_b = 2.00 \times 10^{-10}$ m are

$$K_b = \tfrac{1}{2}mv_b^2 = 4.10 \times 10^{-18} \text{ J} + 4.61 \times 10^{-18} \text{ J} - 2.30 \times 10^{-18} \text{ J}$$
$$= 6.41 \times 10^{-18} \text{ J}$$

$$v_b = \sqrt{\frac{2K_b}{m}} = \sqrt{\frac{2(6.41 \times 10^{-18} \text{ J})}{9.11 \times 10^{-31} \text{ kg}}} = 3.8 \times 10^6 \text{ m/s}$$

(b) When the positron and α particle are very far apart so that $r = r_c \to \infty$, the final potential energy U_c approaches zero. Again from energy conservation, the final kinetic energy and speed of the positron in this case are

$$K_c = K_a + U_a - U_c = 4.10 \times 10^{-18}\,\text{J} + 4.61 \times 10^{-18}\,\text{J} - 0$$
$$= 8.71 \times 10^{-18}\,\text{J}$$
$$v_c = \sqrt{\frac{2K_c}{m}} = \sqrt{\frac{2(8.71 \times 10^{-18}\,\text{J})}{9.11 \times 10^{-31}\,\text{kg}}} = 4.4 \times 10^6\,\text{m/s}$$

(c) The electron and α particle have opposite charges, so the force is attractive and the electron slows down as it moves away. Changing the moving particle's sign from $+e$ to $-e$ means that the initial potential energy is now $U_a = -4.61 \times 10^{-18}$ J, which makes the total mechanical energy *negative*:

$$K_a + U_a = (4.10 \times 10^{-18}\,\text{J}) - (4.61 \times 10^{-18}\,\text{J})$$
$$= -0.51 \times 10^{-18}\,\text{J}$$

The total mechanical energy would have to be positive for the electron to move infinitely far away from the α particle. Like a rock thrown upward at low speed from the earth's surface, it will reach a maximum separation $r = r_d$ from the α particle before reversing direction. At this point its speed and its kinetic energy K_d are zero, so at separation r_d we have

$$U_d = K_a + U_a - K_d = (-0.51 \times 10^{-18}\,\text{J}) - 0$$
$$U_d = \frac{1}{4\pi\epsilon_0}\frac{qq_0}{r_d} = -0.51 \times 10^{-18}\,\text{J}$$
$$r_d = \frac{1}{U_d}\frac{qq_0}{4\pi\epsilon_0}$$
$$= \frac{(9.0 \times 10^9\,\text{N}\cdot\text{m}^2/\text{C}^2)}{-0.51 \times 10^{-18}\,\text{J}}(3.20 \times 10^{-19}\,\text{C})(-1.60 \times 10^{-19}\,\text{C})$$
$$= 9.0 \times 10^{-10}\,\text{m}$$

For $r_b = 2.00 \times 10^{-10}$ m we have $U_b = -2.30 \times 10^{-18}$ J, so the electron kinetic energy and speed at this point are

$$K_b = \tfrac{1}{2}mv_b^2 = 4.10 \times 10^{-18}\,\text{J} + (-4.61 \times 10^{-18}\,\text{J})$$
$$- (-2.30 \times 10^{-18}\,\text{J}) = 1.79 \times 10^{-18}\,\text{J}$$
$$v_b = \sqrt{\frac{2K_b}{m}} = \sqrt{\frac{2(1.79 \times 10^{-18}\,\text{J})}{9.11 \times 10^{-31}\,\text{kg}}} = 2.0 \times 10^6\,\text{m/s}$$

EVALUATE: Both particles behave as expected as they move away from the α particle: The positron speeds up, and the electron slows down and eventually turns around. How fast would an electron have to be moving at $r_a = 1.00 \times 10^{-10}$ m to travel infinitely far from the α particle? (*Hint:* See Example 13.4 in Section 13.3.)

Electric Potential Energy with Several Point Charges

Suppose the electric field $\vec{E}$ in which charge q_0 moves is caused by *several* point charges $q_1, q_2, q_3, \ldots$ at distances $r_1, r_2, r_3, \ldots$ from q_0, as in Fig. 23.8. For example, q_0 could be a positive ion moving in the presence of other ions (Fig. 23.9). The total electric field at each point is the *vector sum* of the fields due to the individual charges, and the total work done on q_0 during any displacement is the sum of the contributions from the individual charges. From Eq. (23.9) we conclude that the potential energy associated with the test charge q_0 at point a in Fig. 23.8 is the *algebraic* sum (*not* a vector sum):

23.8 The potential energy associated with a charge q_0 at point a depends on the other charges q_1, q_2, and q_3 and on their distances r_1, r_2, and r_3 from point a.

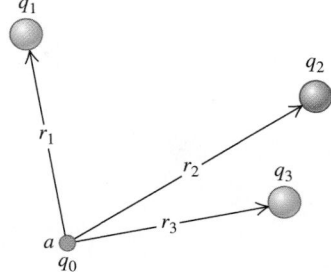

$$U = \frac{q_0}{4\pi\epsilon_0}\left(\frac{q_1}{r_1} + \frac{q_2}{r_2} + \frac{q_3}{r_3} + \cdots\right) = \frac{q_0}{4\pi\epsilon_0}\sum_i \frac{q_i}{r_i} \qquad \begin{array}{l}\text{(point charge } q_0 \\ \text{and collection} \qquad (23.10) \\ \text{of charges } q_i)\end{array}$$

When q_0 is at a different point b, the potential energy is given by the same expression, but $r_1, r_2, \ldots$ are the distances from $q_1, q_2, \ldots$ to point b. The work done on charge q_0 when it moves from a to b along any path is equal to the difference $U_a - U_b$ between the potential energies when q_0 is at a and at b.

We can represent *any* charge distribution as a collection of point charges, so Eq. (23.10) shows that we can always find a potential-energy function for *any* static electric field. It follows that **for every electric field due to a static charge distribution, the force exerted by that field is conservative.**

Equations (23.9) and (23.10) define U to be zero when all the distances $r_1, r_2, \ldots$ are infinite—that is, when the test charge q_0 is very far away from all the charges that produce the field. As with any potential-energy function, the point where $U = 0$ is arbitrary; we can always add a constant to make U equal zero at any point we choose. In electrostatics problems it's usually simplest to choose this point to be at infinity. When we analyze electric circuits in Chapters 25 and 26, other choices will be more convenient.

23.9 This ion engine for spacecraft uses electric forces to eject a stream of positive xenon ions (Xe^+) at speeds in excess of 30 km/s. The thrust produced is very low (about 0.09 newton) but can be maintained continuously for days, in contrast to chemical rockets, which produce a large thrust for a short time (see Fig. 8.33). Such ion engines have been used for maneuvering interplanetary spacecraft.

Equation (23.10) gives the potential energy associated with the presence of the test charge q_0 in the $\vec{E}$ field produced by $q_1, q_2, q_3, \ldots$. But there is also potential energy involved in assembling these charges. If we start with charges $q_1, q_2, q_3, \ldots$ all separated from each other by infinite distances and then bring them together so that the distance between q_i and q_j is r_{ij}, the *total* potential energy U is the sum of the potential energies of interaction for each pair of charges. We can write this as

$$U = \frac{1}{4\pi\epsilon_0} \sum_{i<j} \frac{q_i q_j}{r_{ij}} \qquad (23.11)$$

This sum extends over all *pairs* of charges; we don't let $i = j$ (because that would be an interaction of a charge with itself), and we include only terms with $i < j$ to make sure that we count each pair only once. Thus, to account for the interaction between q_3 and q_4, we include a term with $i = 3$ and $j = 4$ but not a term with $i = 4$ and $j = 3$.

Interpreting Electric Potential Energy

As a final comment, here are two viewpoints on electric potential energy. We have defined it in terms of the work done *by the electric field* on a charged particle moving in the field, just as in Chapter 7 we defined potential energy in terms of the work done by gravity or by a spring. When a particle moves from point a to point b, the work done on it by the electric field is $W_{a\rightarrow b} = U_a - U_b$. Thus the potential-energy difference $U_a - U_b$ equals *the work that is done by the electric force when the particle moves from a to b*. When U_a is greater than U_b, the field does positive work on the particle as it "falls" from a point of higher potential energy (a) to a point of lower potential energy (b).

An alternative but equivalent viewpoint is to consider how much work we would have to do to "raise" a particle from a point b where the potential energy is U_b to a point a where it has a greater value U_a (pushing two positive charges closer together, for example). To move the particle slowly (so as not to give it any kinetic energy), we need to exert an additional external force $\vec{F}_{ext}$ that is equal and opposite to the electric-field force and does positive work. The potential-energy difference $U_a - U_b$ is then defined as *the work that must be done by an external force to move the particle slowly from b to a against the electric force*. Because $\vec{F}_{ext}$ is the negative of the electric-field force and the displacement is in the opposite direction, this definition of the potential difference $U_a - U_b$ is equivalent to that given above. This alternative viewpoint also works if U_a is less than U_b, corresponding to "lowering" the particle; an example is moving two positive charges away from each other. In this case, $U_a - U_b$ is again equal to the work done by the external force, but now this work is negative.

We will use both of these viewpoints in the next section to interpret what is meant by electric *potential*, or potential energy per unit charge.

Example 23.2   A system of point charges

Two point charges are located on the x-axis, $q_1 = -e$ at $x = 0$ and $q_2 = +e$ at $x = a$. (a) Find the work that must be done by an external force to bring a third point charge $q_3 = +e$ from infinity to $x = 2a$. (b) Find the total potential energy of the system of three charges.

SOLUTION

IDENTIFY and SET UP: Figure 23.10 shows the final arrangement of the three charges. In part (a) we need to find the work W that must be done on q_3 by an external force $\vec{F}_{ext}$ to bring q_3 in from

infinity to $x = 2a$. We do this by using Eq. (23.10) to find the potential energy associated with q_3 in the presence of q_1 and q_2. In part (b) we use Eq. (23.11), the expression for the potential energy of a collection of point charges, to find the total potential energy of the system.

23.10 Our sketch of the situation after the third charge has been brought in from infinity.

$$q_1 = -e \qquad q_2 = +e \qquad q_3 = +e$$
$$\underset{x=0}{\ominus} \rule{1.5cm}{0.4pt} \underset{x=a}{\oplus} \rule{1.5cm}{0.4pt} \underset{x=2a}{\oplus} \rule{0.5cm}{0.4pt} x$$

EXECUTE: (a) The work W equals the difference between (i) the potential energy U associated with q_3 when it is at $x = 2a$ and (ii) the potential energy when it is infinitely far away. The second of these is zero, so the work required is equal to U. The distances between the charges are $r_{13} = 2a$ and $r_{23} = a$, so from Eq. (23.10),

$$W = U = \frac{q_3}{4\pi\epsilon_0}\left(\frac{q_1}{r_{13}} + \frac{q_2}{r_{23}}\right) = \frac{+e}{4\pi\epsilon_0}\left(\frac{-e}{2a} + \frac{+e}{a}\right) = \frac{+e^2}{8\pi\epsilon_0 a}$$

This is positive, just as we should expect. If we bring q_3 in from infinity along the $+x$-axis, it is attracted by q_1 but is repelled more strongly by q_2. Hence we must do positive work to push q_3 to the position at $x = 2a$.

(b) From Eq. (23.11), the total potential energy of the three-charge system is

$$U = \frac{1}{4\pi\epsilon_0}\sum_{i<j}\frac{q_i q_j}{r_{ij}} = \frac{1}{4\pi\epsilon_0}\left(\frac{q_1 q_2}{r_{12}} + \frac{q_1 q_3}{r_{13}} + \frac{q_2 q_3}{r_{23}}\right)$$

$$= \frac{1}{4\pi\epsilon_0}\left[\frac{(-e)(e)}{a} + \frac{(-e)(e)}{2a} + \frac{(e)(e)}{a}\right] = \frac{-e^2}{8\pi\epsilon_0 a}$$

EVALUATE: Our negative result in part (b) means that the system has lower potential energy than it would if the three charges were infinitely far apart. An external force would have to do *negative* work to bring the three charges from infinity to assemble this entire arrangement and would have to do *positive* work to move the three charges back to infinity.

Test Your Understanding of Section 23.1   Consider the system of three point charges in Example 21.4 (Section 21.3) and shown in Fig. 21.14. (a) What is the sign of the total potential energy of this system? (i) positive; (ii) negative; (iii) zero. (b) What is the sign of the total amount of work you would have to do to move these charges infinitely far from each other? (i) positive; (ii) negative; (iii) zero. ❙

23.2 Electric Potential

In Section 23.1 we looked at the potential energy U associated with a test charge q_0 in an electric field. Now we want to describe this potential energy on a "per unit charge" basis, just as electric field describes the force per unit charge on a charged particle in the field. This leads us to the concept of *electric potential*, often called simply *potential*. This concept is very useful in calculations involving energies of charged particles. It also facilitates many electric-field calculations because electric potential is closely related to the electric field $\vec{E}$. When we need to determine an electric field, it is often easier to determine the potential first and then find the field from it.

Potential is *potential energy per unit charge*. We define the potential V at any point in an electric field as the potential energy U *per unit charge* associated with a test charge q_0 at that point:

$$V = \frac{U}{q_0} \quad \text{or} \quad U = q_0 V \qquad (23.12)$$

Potential energy and charge are both scalars, so potential is a scalar. From Eq. (23.12) its units are the units of energy divided by those of charge. The SI unit of potential, called one **volt** (1 V) in honor of the Italian electrical experimenter Alessandro Volta (1745–1827), equals 1 joule per coulomb:

$$1\ \text{V} = 1\ \text{volt} = 1\ \text{J/C} = 1\ \text{joule/coulomb}$$

Let's put Eq. (23.2), which equates the work done by the electric force during a displacement from a to b to the quantity $-\Delta U = -(U_b - U_a)$, on a "work per unit charge" basis. We divide this equation by q_0, obtaining

$$\frac{W_{a\to b}}{q_0} = -\frac{\Delta U}{q_0} = -\left(\frac{U_b}{q_0} - \frac{U_a}{q_0}\right) = -(V_b - V_a) = V_a - V_b \qquad (23.13)$$

where $V_a = U_a/q_0$ is the potential energy per unit charge at point a and similarly for V_b. We call V_a and V_b the *potential at point a* and *potential at point b*, respectively. Thus the work done per unit charge by the electric force when a charged body moves from a to b is equal to the potential at a minus the potential at b.

MasteringPHYSICS

PhET: Charges and Fields
ActivPhysics 11.13: Electrical Potential Energy and Potential

23.11 The voltage of this battery equals the difference in potential $V_{ab} = V_a - V_b$ between its positive terminal (point a) and its negative terminal (point b).

Point a (positive terminal)

$V_{ab} = 1.5$ volts

Point b (negative terminal)

The difference $V_a - V_b$ is called the *potential of a with respect to b*; we sometimes abbreviate this difference as $V_{ab} = V_a - V_b$ (note the order of the subscripts). This is often called the potential difference between a and b, but that's ambiguous unless we specify which is the reference point. In electric circuits, which we will analyze in later chapters, the potential difference between two points is often called **voltage** (Fig. 23.11). Equation (23.13) then states: V_{ab}, **the potential of a with respect to b, equals the work done by the electric force when a UNIT charge moves from a to b.**

Another way to interpret the potential difference V_{ab} in Eq. (23.13) is to use the alternative viewpoint mentioned at the end of Section 23.1. In that viewpoint, $U_a - U_b$ is the amount of work that must be done by an *external* force to move a particle of charge q_0 slowly from b to a against the electric force. The work that must be done *per unit charge* by the external force is then $(U_a - U_b)/q_0 = V_a - V_b = V_{ab}$. In other words: V_{ab}, **the potential of a with respect to b, equals the work that must be done to move a UNIT charge slowly from b to a against the electric force.**

An instrument that measures the difference of potential between two points is called a *voltmeter*. (In Chapter 26 we'll discuss how these devices work.) Voltmeters that can measure a potential difference of 1 μV are common, and sensitivities down to 10^{-12} V can be attained.

Calculating Electric Potential

To find the potential V due to a single point charge q, we divide Eq. (23.9) by q_0:

$$V = \frac{U}{q_0} = \frac{1}{4\pi\epsilon_0}\frac{q}{r} \quad \text{(potential due to a point charge)} \qquad (23.14)$$

where r is the distance from the point charge q to the point at which the potential is evaluated. If q is positive, the potential that it produces is positive at all points; if q is negative, it produces a potential that is negative everywhere. In either case, V is equal to zero at $r = \infty$, an infinite distance from the point charge. Note that potential, like electric field, is independent of the test charge q_0 that we use to define it.

Similarly, we divide Eq. (23.10) by q_0 to find the potential due to a collection of point charges:

$$V = \frac{U}{q_0} = \frac{1}{4\pi\epsilon_0}\sum_i \frac{q_i}{r_i} \quad \begin{array}{l}\text{(potential due to a collection}\\\text{of point charge)}\end{array} \qquad (23.15)$$

In this expression, r_i is the distance from the ith charge, q_i, to the point at which V is evaluated. Just as the electric field due to a collection of point charges is the *vector* sum of the fields produced by each charge, the electric potential due to a collection of point charges is the *scalar* sum of the potentials due to each charge. When we have a continuous distribution of charge along a line, over a surface, or through a volume, we divide the charge into elements dq, and the sum in Eq. (23.15) becomes an integral:

$$V = \frac{1}{4\pi\epsilon_0}\int\frac{dq}{r} \quad \begin{array}{l}\text{(potential due to a continuous}\\\text{distribution of charge)}\end{array} \qquad (23.16)$$

where r is the distance from the charge element dq to the field point where we are finding V. We'll work out several examples of such cases. The potential defined by Eqs. (23.15) and (23.16) is zero at points that are infinitely far away from *all* the charges. Later we'll encounter cases in which the charge distribution itself

Application **Electrocardiography**
The electrodes used in an electrocardiogram—EKG or ECG for short—measure the potential differences (typically no greater than 1 mV = 10^{-3} V) between different parts of the patient's skin. These are indicative of the potential differences between regions of the heart, and so provide a sensitive way to detect any abnormalities in the electrical activity that drives cardiac function.

extends to infinity. We'll find that in such cases we cannot set $V = 0$ at infinity, and we'll need to exercise care in using and interpreting Eqs. (23.15) and (23.16).

CAUTION **What is electric potential?** Before getting too involved in the details of how to calculate electric potential, you should stop and remind yourself what potential is. The electric *potential* at a certain point is the potential energy that would be associated with a *unit* charge placed at that point. That's why potential is measured in joules per coulomb, or volts. Keep in mind, too, that there doesn't have to be a charge at a given point for a potential V to exist at that point. (In the same way, an electric field can exist at a given point even if there's no charge there to respond to it.) ▮

Finding Electric Potential from Electric Field

When we are given a collection of point charges, Eq. (23.15) is usually the easiest way to calculate the potential V. But in some problems in which the electric field is known or can be found easily, it is easier to determine V from $\vec{E}$. The force $\vec{F}$ on a test charge q_0 can be written as $\vec{F} = q_0\vec{E}$, so from Eq. (23.1) the work done by the electric force as the test charge moves from a to b is given by

$$W_{a \to b} = \int_a^b \vec{F} \cdot d\vec{l} = \int_a^b q_0\vec{E} \cdot d\vec{l}$$

If we divide this by q_0 and compare the result with Eq. (23.13), we find

$$V_a - V_b = \int_a^b \vec{E} \cdot d\vec{l} = \int_a^b E\cos\phi\, dl \qquad \begin{array}{l}\text{(potential difference}\\ \text{as an integral of } \vec{E})\end{array} \qquad \text{(23.17)}$$

The value of $V_a - V_b$ is independent of the path taken from a to b, just as the value of $W_{a \to b}$ is independent of the path. To interpret Eq. (23.17), remember that $\vec{E}$ is the electric force per unit charge on a test charge. If the line integral $\int_a^b \vec{E} \cdot d\vec{l}$ is positive, the electric field does positive work on a positive test charge as it moves from a to b. In this case the electric potential energy decreases as the test charge moves, so the potential energy per unit charge decreases as well; hence V_b is less than V_a and $V_a - V_b$ is positive.

As an illustration, consider a positive point charge (Fig. 23.12a). The electric field is directed away from the charge, and $V = q/4\pi\epsilon_0 r$ is positive at any finite distance from the charge. If you move away from the charge, in the direction of $\vec{E}$, you move toward lower values of V; if you move toward the charge, in the direction opposite $\vec{E}$, you move toward greater values of V. For the negative point charge in Fig. 23.12b, $\vec{E}$ is directed toward the charge and $V = q/4\pi\epsilon_0 r$ is negative at any finite distance from the charge. In this case, if you move toward the charge, you are moving in the direction of $\vec{E}$ and in the direction of decreasing (more negative) V. Moving away from the charge, in the direction opposite $\vec{E}$, moves you toward increasing (less negative) values of V. The general rule, valid for *any* electric field, is: Moving *with* the direction of $\vec{E}$ means moving in the direction of *decreasing* V, and moving *against* the direction of $\vec{E}$ means moving in the direction of *increasing* V.

Also, a positive test charge q_0 experiences an electric force in the direction of $\vec{E}$, toward lower values of V; a negative test charge experiences a force opposite $\vec{E}$, toward higher values of V. Thus a positive charge tends to "fall" from a high-potential region to a lower-potential region. The opposite is true for a negative charge.

Notice that Eq. (23.17) can be rewritten as

$$V_a - V_b = -\int_b^a \vec{E} \cdot d\vec{l} \qquad \text{(23.18)}$$

This has a negative sign compared to the integral in Eq. (23.17), and the limits are reversed; hence Eqs. (23.17) and (23.18) are equivalent. But Eq. (23.18) has a slightly different interpretation. To move a unit charge slowly against the electric

23.12 If you move in the direction of $\vec{E}$, electric potential V decreases; if you move in the direction opposite $\vec{E}$, V increases.

(a) A positive point charge

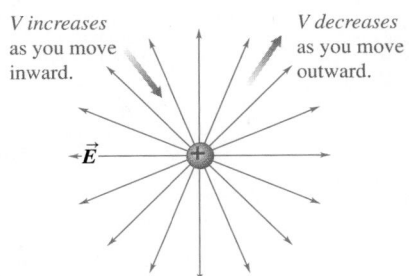

V increases as you move inward.

V decreases as you move outward.

(b) A negative point charge

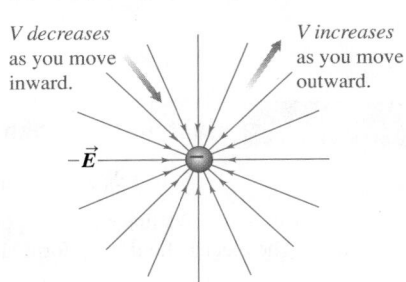

V decreases as you move inward.

V increases as you move outward.

force, we must apply an *external* force per unit charge equal to $-\vec{E}$, equal and opposite to the electric force per unit charge $\vec{E}$. Equation (23.18) says that $V_a - V_b = V_{ab}$, the potential of a with respect to b, equals the work done per unit charge by this external force to move a unit charge from b to a. This is the same alternative interpretation we discussed under Eq. (23.13).

Equations (23.17) and (23.18) show that the unit of potential difference (1 V) is equal to the unit of electric field (1 N/C) multiplied by the unit of distance (1 m). Hence the unit of electric field can be expressed as 1 *volt per meter* (1 V/m), as well as 1 N/C:

$$1 \text{ V/m} = 1 \text{ volt/meter} = 1 \text{ N/C} = 1 \text{ newton/coulomb}$$

In practice, the volt per meter is the usual unit of electric-field magnitude.

Electron Volts

The magnitude e of the electron charge can be used to define a unit of energy that is useful in many calculations with atomic and nuclear systems. When a particle with charge q moves from a point where the potential is V_b to a point where it is V_a, the change in the potential energy U is

$$U_a - U_b = q(V_a - V_b) = qV_{ab}$$

If the charge q equals the magnitude e of the electron charge, 1.602×10^{-19} C, and the potential difference is $V_{ab} = 1$ V, the change in energy is

$$U_a - U_b = (1.602 \times 10^{-19} \text{ C})(1 \text{ V}) = 1.602 \times 10^{-19} \text{ J}$$

This quantity of energy is defined to be 1 **electron volt** (1 eV):

$$1 \text{ eV} = 1.602 \times 10^{-19} \text{ J}$$

The multiples meV, keV, MeV, GeV, and TeV are often used.

CAUTION **Electron volts vs. volts** Remember that the electron volt is a unit of energy, *not* a unit of potential or potential difference! ▌

When a particle with charge e moves through a potential difference of 1 volt, the change in potential *energy* is 1 eV. If the charge is some multiple of e—say Ne—the change in potential energy in electron volts is N times the potential difference in volts. For example, when an alpha particle, which has charge $2e$, moves between two points with a potential difference of 1000 V, the change in potential energy is $2(1000 \text{ eV}) = 2000$ eV. To confirm this, we write

$$U_a - U_b = qV_{ab} = (2e)(1000 \text{ V}) = (2)(1.602 \times 10^{-19} \text{ C})(1000 \text{ V})$$
$$= 3.204 \times 10^{-16} \text{ J} = 2000 \text{ eV}$$

Although we have defined the electron volt in terms of *potential* energy, we can use it for *any* form of energy, such as the kinetic energy of a moving particle. When we speak of a "one-million-electron-volt proton," we mean a proton with a kinetic energy of one million electron volts (1 MeV), equal to $(10^6)(1.602 \times 10^{-19} \text{ J}) = 1.602 \times 10^{-13}$ J. The Large Hadron Collider near Geneva, Switzerland, is designed to accelerate protons to a kinetic energy of 7 TeV (7×10^{12} eV).

Example 23.3 **Electric force and electric potential**

A proton (charge $+e = 1.602 \times 10^{-19}$ C) moves a distance $d = 0.50$ m in a straight line between points a and b in a linear accelerator. The electric field is uniform along this line, with mag- nitude $E = 1.5 \times 10^7 \text{ V/m} = 1.5 \times 10^7 \text{ N/C}$ in the direction from a to b. Determine (a) the force on the proton; (b) the work done on it by the field; (c) the potential difference $V_a - V_b$.

SOLUTION

IDENTIFY and SET UP: This problem uses the relationship between electric field and electric force. It also uses the relationship among force, work, and potential-energy difference. We are given the electric field, so it is straightforward to find the electric force on the proton. Calculating the work is also straightforward because $\vec{E}$ is uniform, so the force on the proton is constant. Once the work is known, we find $V_a - V_b$ using Eq. (23.13).

EXECUTE: (a) The force on the proton is in the same direction as the electric field, and its magnitude is

$$F = qE = (1.602 \times 10^{-19} \text{ C})(1.5 \times 10^7 \text{ N/C})$$
$$= 2.4 \times 10^{-12} \text{ N}$$

(b) The force is constant and in the same direction as the displacement, so the work done on the proton is

$$W_{a \to b} = Fd = (2.4 \times 10^{-12} \text{ N})(0.50 \text{ m}) = 1.2 \times 10^{-12} \text{ J}$$

$$= (1.2 \times 10^{-12} \text{ J}) \frac{1 \text{ eV}}{1.602 \times 10^{-19} \text{ J}}$$

$$= 7.5 \times 10^6 \text{ eV} = 7.5 \text{ MeV}$$

(c) From Eq. (23.13) the potential difference is the work per unit charge, which is

$$V_a - V_b = \frac{W_{a \to b}}{q} = \frac{1.2 \times 10^{-12} \text{ J}}{1.602 \times 10^{-19} \text{ C}}$$
$$= 7.5 \times 10^6 \text{ J/C} = 7.5 \times 10^6 \text{ V}$$
$$= 7.5 \text{ MV}$$

We can get this same result even more easily by remembering that 1 electron volt equals 1 volt multiplied by the charge e. The work done is 7.5×10^6 eV and the charge is e, so the potential difference is $(7.5 \times 10^6 \text{ eV})/e = 7.5 \times 10^6$ V.

EVALUATE: We can check our result in part (c) by using Eq. (23.17) or Eq. (23.18). The angle ϕ between the constant field $\vec{E}$ and the displacement is zero, so Eq. (23.17) becomes

$$V_a - V_b = \int_a^b E \cos \phi \, dl = \int_a^b E \, dl = E \int_a^b dl$$

The integral of dl from a to b is just the distance d, so we again find

$$V_a - V_b = Ed = (1.5 \times 10^7 \text{ V/m})(0.50 \text{ m}) = 7.5 \times 10^6 \text{ V}$$

Example 23.4   Potential due to two point charges

An electric dipole consists of point charges $q_1 = +12$ nC and $q_2 = -12$ nC placed 10.0 cm apart (Fig. 23.13). Compute the electric potentials at points a, b, and c.

SOLUTION

IDENTIFY and SET UP: This is the same arrangement as in Example 21.8, in which we calculated the electric *field* at each point by doing a *vector* sum. Here our target variable is the electric *potential* V at three points, which we find by doing the *algebraic* sum in Eq. (23.15).

EXECUTE: At point a we have $r_1 = 0.060$ m and $r_2 = 0.040$ m, so Eq. (23.15) becomes

$$V_a = \frac{1}{4\pi\epsilon_0} \sum_i \frac{q_i}{r_i} = \frac{1}{4\pi\epsilon_0} \frac{q_1}{r_1} + \frac{1}{4\pi\epsilon_0} \frac{q_2}{r_2}$$

$$= (9.0 \times 10^9 \text{ N} \cdot \text{m}^2/\text{C}^2) \frac{12 \times 10^{-9} \text{ C}}{0.060 \text{ m}}$$

$$+ (9.0 \times 10^9 \text{ N} \cdot \text{m}^2/\text{C}^2) \frac{(-12 \times 10^{-9} \text{ C})}{0.040 \text{ m}}$$

$$= 1800 \text{ N} \cdot \text{m/C} + (-2700 \text{ N} \cdot \text{m/C})$$

$$= 1800 \text{ V} + (-2700 \text{ V}) = -900 \text{ V}$$

In a similar way you can show that the potential at point b (where $r_1 = 0.040$ m and $r_2 = 0.140$ m) is $V_b = 1930$ V and that the potential at point c (where $r_1 = r_2 = 0.130$ m) is $V_c = 0$.

EVALUATE: Let's confirm that these results make sense. Point a is closer to the -12-nC charge than to the $+12$-nC charge, so the potential at a is negative. The potential is positive at point b, which

23.13 What are the potentials at points a, b, and c due to this electric dipole?

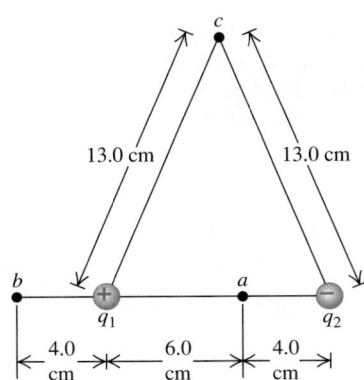

is closer to the $+12$-nC charge than the -12-nC charge. Finally, point c is equidistant from the $+12$-nC charge and the -12-nC charge, so the potential there is zero. (The potential is also equal to zero at a point infinitely far from both charges.)

Comparing this example with Example 21.8 shows that it's much easier to calculate electric potential (a scalar) than electric field (a vector). We'll take advantage of this simplification whenever possible.

Example 23.5 Potential and potential energy

Compute the potential energy associated with a +4.0-nC point charge if it is placed at points a, b, and c in Fig. 23.13.

SOLUTION

IDENTIFY and SET UP: The potential energy U associated with a point charge q at a location where the electric potential is V is $U = qV$. We use the values of V from Example 23.4.

EXECUTE: At the three points we find

$$U_a = qV_a = (4.0 \times 10^{-9}\,\text{C})(-900\,\text{J/C}) = -3.6 \times 10^{-6}\,\text{J}$$
$$U_b = qV_b = (4.0 \times 10^{-9}\,\text{C})(1930\,\text{J/C}) = 7.7 \times 10^{-6}\,\text{J}$$
$$U_c = qV_c = 0$$

All of these values correspond to U and V being zero at infinity.

EVALUATE: Note that *zero* net work is done on the 4.0-nC charge if it moves from point c to infinity *by any path*. In particular, let the path be along the perpendicular bisector of the line joining the other two charges q_1 and q_2 in Fig. 23.13. As shown in Example 21.8 (Section 21.5), at points on the bisector, the direction of $\vec{E}$ is perpendicular to the bisector. Hence the force on the 4.0-nC charge is perpendicular to the path, and no work is done in any displacement along it.

Example 23.6 Finding potential by integration

By integrating the electric field as in Eq. (23.17), find the potential at a distance r from a point charge q.

SOLUTION

IDENTIFY and SET UP: We let point a in Eq. (23.17) be at distance r and let point b be at infinity (Fig. 23.14). As usual, we choose the potential to be zero at an infinite distance from the charge q.

EXECUTE: To carry out the integral, we can choose any path we like between points a and b. The most convenient path is a radial line as shown in Fig. 23.14, so that $d\vec{l}$ is in the radial direction and has magnitude dr. Writing $d\vec{l} = \hat{r}\,dr$, we have from Eq. (23.17)

$$V - 0 = V = \int_r^\infty \vec{E} \cdot d\vec{l}$$
$$= \int_r^\infty \frac{1}{4\pi\epsilon_0} \frac{q}{r^2} \hat{r} \cdot \hat{r}\,dr = \int_r^\infty \frac{q}{4\pi\epsilon_0 r^2}\,dr$$
$$= -\frac{q}{4\pi\epsilon_0 r}\Big|_r^\infty = 0 - \left(-\frac{q}{4\pi\epsilon_0 r}\right)$$
$$V = \frac{q}{4\pi\epsilon_0 r}$$

23.14 Calculating the potential by integrating $\vec{E}$ for a single point charge.

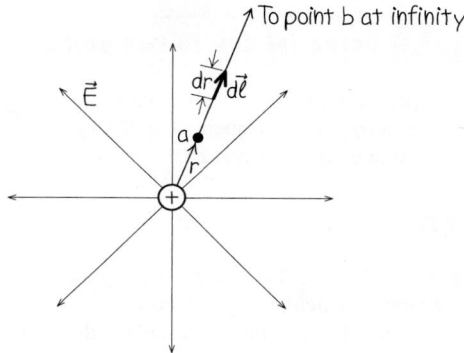

EVALUATE: Our result agrees with Eq. (23.14) and is correct for positive or negative q.

Example 23.7 Moving through a potential difference

In Fig. 23.15 a dust particle with mass $m = 5.0 \times 10^{-9}\,\text{kg}$ $= 5.0\,\mu\text{g}$ and charge $q_0 = 2.0\,\text{nC}$ starts from rest and moves in a straight line from point a to point b. What is its speed v at point b?

SOLUTION

IDENTIFY and SET UP: Only the conservative electric force acts on the particle, so mechanical energy is conserved: $K_a + U_a = K_b + U_b$. We get the potential energies U from the

23.15 The particle moves from point a to point b; its acceleration is not constant.

corresponding potentials V using Eq. (23.12): $U_a = q_0V_a$ and $U_b = q_0V_b$.

EXECUTE: We have $K_a = 0$ and $K_b = \frac{1}{2}mv^2$. We substitute these and our expressions for U_a and U_b into the energy-conservation equation, then solve for v. We find

$$0 + q_0 V_a = \frac{1}{2}mv^2 + q_0 V_b$$

$$v = \sqrt{\frac{2q_0(V_a - V_b)}{m}}$$

We calculate the potentials using Eq. (23.15), $V = q/4\pi\epsilon_0 r$:

$$V_a = (9.0 \times 10^9 \, \text{N} \cdot \text{m}^2/\text{C}^2)\left(\frac{3.0 \times 10^{-9} \, \text{C}}{0.010 \, \text{m}}\right.$$

$$\left. + \frac{(-3.0 \times 10^{-9} \, \text{C})}{0.020 \, \text{m}}\right)$$

$$= 1350 \, \text{V}$$

$$V_b = (9.0 \times 10^9 \, \text{N} \cdot \text{m}^2/\text{C}^2)\left(\frac{3.0 \times 10^{-9} \, \text{C}}{0.020 \, \text{m}}\right.$$

$$\left. + \frac{(-3.0 \times 10^{-9} \, \text{C})}{0.010 \, \text{m}}\right)$$

$$= -1350 \, \text{V}$$

$$V_a - V_b = (1350 \, \text{V}) - (-1350 \, \text{V}) = 2700 \, \text{V}$$

Finally,

$$v = \sqrt{\frac{2(2.0 \times 10^{-9} \, \text{C})(2700 \, \text{V})}{5.0 \times 10^{-9} \, \text{kg}}} = 46 \, \text{m/s}$$

EVALUATE: Our result makes sense: The positive test charge speeds up as it moves away from the positive charge and toward the negative charge. To check unit consistency in the final line of the calculation, note that $1 \, \text{V} = 1 \, \text{J/C}$, so the numerator under the radical has units of J or $\text{kg} \cdot \text{m}^2/\text{s}^2$.

Test Your Understanding of Section 23.2 If the electric *potential* at a certain point is zero, does the electric *field* at that point have to be zero? (*Hint:* Consider point c in Example 23.4 and Example 21.8.) ❙

23.3 Calculating Electric Potential

When calculating the potential due to a charge distribution, we usually follow one of two routes. If we know the charge distribution, we can use Eq. (23.15) or (23.16). Or if we know how the electric field depends on position, we can use Eq. (23.17), defining the potential to be zero at some convenient place. Some problems require a combination of these approaches.

As you read through these examples, compare them with the related examples of calculating electric *field* in Section 21.5. You'll see how much easier it is to calculate scalar electric potentials than vector electric fields. The moral is clear: Whenever possible, solve problems using an energy approach (using electric potential and electric potential energy) rather than a dynamics approach (using electric fields and electric forces).

Problem-Solving Strategy 23.1 **Calculating Electric Potential**

IDENTIFY *the relevant concepts:* Remember that electric potential is *potential energy per unit charge.*

SET UP *the problem* using the following steps:
1. Make a drawing showing the locations and values of the charges (which may be point charges or a continuous distribution of charge) and your choice of coordinate axes.
2. Indicate on your drawing the position of the point at which you want to calculate the electric potential V. Sometimes this position will be an arbitrary one (say, a point a distance r from the center of a charged sphere).

EXECUTE *the solution* as follows:
1. To find the potential due to a collection of point charges, use Eq. (23.15). If you are given a continuous charge distribution, devise a way to divide it into infinitesimal elements and use Eq. (23.16). Carry out the integration, using appropriate limits to include the entire charge distribution.
2. If you are given the electric field, or if you can find it using any of the methods presented in Chapter 21 or 22, it may be

easier to find the potential difference between points a and b using Eq. (23.17) or (23.18). When appropriate, make use of your freedom to define V to be zero at some convenient place, and choose this place to be point b. (For point charges, this will usually be at infinity. For other distributions of charge—especially those that themselves extend to infinity—it may be necessary to define V_b to be zero at some finite distance from the charge distribution.) Then the potential at any other point, say a, can by found from Eq. (23.17) or (23.18) with $V_b = 0$.
3. Although potential V is a *scalar* quantity, you may have to use components of the vectors $\vec{E}$ and $d\vec{l}$ when you use Eq. (23.17) or (23.18) to calculate V.

EVALUATE *your answer:* Check whether your answer agrees with your intuition. If your result gives V as a function of position, graph the function to see whether it makes sense. If you know the electric field, you can make a rough check of your result for V by verifying that V decreases if you move in the direction of $\vec{E}$.

Example 23.8 A charged conducting sphere

A solid conducting sphere of radius R has a total charge q. Find the electric potential everywhere, both outside and inside the sphere.

SOLUTION

IDENTIFY and SET UP: In Example 22.5 (Section 22.4) we used Gauss's law to find the electric *field* at all points for this charge distribution. We can use that result to determine the corresponding potential.

EXECUTE: From Example 22.5, the field *outside* the sphere is the same as if the sphere were removed and replaced by a point charge q. We take $V = 0$ at infinity, as we did for a point charge. Then the potential at a point outside the sphere at a distance r from its center is the same as that due to a point charge q at the center:

$$V = \frac{1}{4\pi\epsilon_0}\frac{q}{r}$$

The potential at the surface of the sphere is $V_{surface} = q/4\pi\epsilon_0 R$.

Inside the sphere, $\vec{E}$ is zero everywhere. Hence no work is done on a test charge that moves from any point to any other point inside the sphere. This means that the potential is the same at every point inside the sphere and is equal to its value $q/4\pi\epsilon_0 R$ at the surface.

EVALUATE: Figure 23.16 shows the field and potential for a positive charge q. In this case the electric field points radially away from the sphere. As you move away from the sphere, in the direction of $\vec{E}$, V decreases (as it should).

23.16 Electric-field magnitude E and potential V at points inside and outside a positively charged spherical conductor.

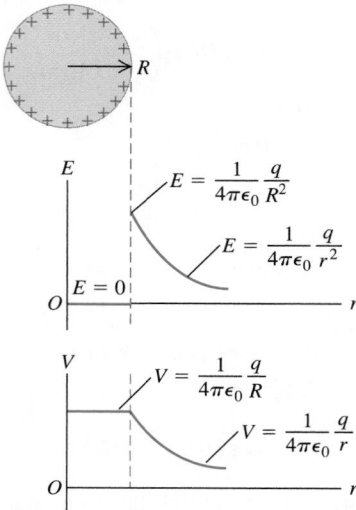

Ionization and Corona Discharge

The results of Example 23.8 have numerous practical consequences. One consequence relates to the maximum potential to which a conductor in air can be raised. This potential is limited because air molecules become *ionized*, and air becomes a conductor, at an electric-field magnitude of about 3×10^6 V/m. Assume for the moment that q is positive. When we compare the expressions in Example 23.8 for the potential $V_{surface}$ and field magnitude $E_{surface}$ at the surface of a charged conducting sphere, we note that $V_{surface} = E_{surface}R$. Thus, if E_m represents the electric-field magnitude at which air becomes conductive (known as the *dielectric strength* of air), then the maximum potential V_m to which a spherical conductor can be raised is

$$V_m = RE_m$$

For a conducting sphere 1 cm in radius in air, $V_m = (10^{-2}\text{ m})(3 \times 10^6\text{ V/m}) = 30,000$ V. No amount of "charging" could raise the potential of a conducting sphere of this size in air higher than about 30,000 V; attempting to raise the potential further by adding extra charge would cause the surrounding air to become ionized and conductive, and the extra added charge would leak into the air.

To attain even higher potentials, high-voltage machines such as Van de Graaff generators use spherical terminals with very large radii (see Fig. 22.26 and the photograph that opens Chapter 22). For example, a terminal of radius $R = 2$ m has a maximum potential $V_m = (2\text{ m})(3 \times 10^6\text{ V/m}) = 6 \times 10^6\text{ V} = 6$ MV.

Our result in Example 23.8 also explains what happens with a charged conductor with a very *small* radius of curvature, such as a sharp point or thin wire. Because the maximum potential is proportional to the radius, even relatively small potentials applied to sharp points in air produce sufficiently high fields just outside the point to ionize the surrounding air, making it become a conductor. The resulting current and its associated glow (visible in a dark room) are called *corona*. Laser printers and photocopying machines use corona from fine wires to spray charge on the imaging drum (see Fig. 21.2).

A large-radius conductor is used in situations where it's important to *prevent* corona. An example is the metal ball at the end of a car radio antenna, which prevents the static that would be caused by corona. Another example is the blunt end of a metal lightning rod (Fig. 23.17). If there is an excess charge in the atmosphere, as happens during thunderstorms, a substantial charge of the opposite sign can build up on this blunt end. As a result, when the atmospheric charge is discharged through a lightning bolt, it tends to be attracted to the charged lightning rod rather than to other nearby structures that could be damaged. (A conducting wire connecting the lightning rod to the ground then allows the acquired charge to dissipate harmlessly.) A lightning rod with a sharp end would allow less charge buildup and hence would be less effective.

23.17 The metal mast at the top of the Empire State Building acts as a lightning rod. It is struck by lightning as many as 500 times each year.

Example 23.9 Oppositely charged parallel plates

Find the potential at any height y between the two oppositely charged parallel plates discussed in Section 23.1 (Fig. 23.18).

SOLUTION

IDENTIFY and SET UP: We discussed this situation in Section 23.1. From Eq. (23.5), we know the electric *potential energy* U for a test charge q_0 is $U = q_0Ey$. (We set $y = 0$ and $U = 0$ at the bottom plate.) We use Eq. (23.12), $U = q_0V$, to find the electric *potential* V as a function of y.

EXECUTE: The potential $V(y)$ at coordinate y is the potential energy per unit charge:

$$V(y) = \frac{U(y)}{q_0} = \frac{q_0Ey}{q_0} = Ey$$

The potential decreases as we move in the direction of $\vec{E}$ from the upper to the lower plate. At point a, where $y = d$ and $V(y) = V_a$,

$$V_a - V_b = Ed \quad \text{and} \quad E = \frac{V_a - V_b}{d} = \frac{V_{ab}}{d}$$

where V_{ab} is the potential of the positive plate with respect to the negative plate. That is, the electric field equals the potential difference between the plates divided by the distance between them. For a given potential difference V_{ab}, the smaller the distance d between the two plates, the greater the magnitude E of the electric field. (This relationship between E and V_{ab} holds *only* for the planar geometry

23.18 The charged parallel plates from Fig. 23.2.

we have described. It does *not* work for situations such as concentric cylinders or spheres in which the electric field is not uniform.)

EVALUATE: Our result shows that $V = 0$ at the bottom plate (at $y = 0$). This is consistent with our choice that $U = q_0V = 0$ for a test charge placed at the bottom plate.

CAUTION **"Zero potential" is arbitrary** You might think that if a conducting body has zero potential, it must necessarily also have zero net charge. But that just isn't so! As an example, the plate at $y = 0$ in Fig. 23.18 has zero potential ($V = 0$) but has a nonzero charge per unit area $-\sigma$. There's nothing particularly special about the place where potential is zero; we can *define* this place to be wherever we want it to be. ∎

Example 23.10 An infinite line charge or charged conducting cylinder

Find the potential at a distance r from a very long line of charge with linear charge density (charge per unit length) λ.

SOLUTION

IDENTIFY and SET UP: In both Example 21.10 (Section 21.5) and Example 22.6 (Section 22.4) we found that the electric field at a

radial distance r from a long straight-line charge (Fig. 23.19a) has only a radial component given by $E_r = \lambda/2\pi\epsilon_0r$. We use this expression to find the potential by integrating $\vec{E}$ as in Eq. (23.17).

EXECUTE: Since the field has only a radial component, we have $\vec{E} \cdot d\vec{l} = E_r\,dr$. Hence from Eq. (23.17) the potential of any point a

Continued

with respect to any other point b, at radial distances r_a and r_b from the line of charge, is

$$V_a - V_b = \int_a^b \vec{E} \cdot d\vec{l} = \int_a^b E_r \, dr = \frac{\lambda}{2\pi\epsilon_0} \int_{r_a}^{r_b} \frac{dr}{r} = \frac{\lambda}{2\pi\epsilon_0} \ln\frac{r_b}{r_a}$$

If we take point b at infinity and set $V_b = 0$, we find that V_a is *infinite* for any finite distance r_a from the line charge: $V_a = (\lambda/2\pi\epsilon_0) \ln(\infty/r_a) = \infty$. This is *not* a useful way to define V for this problem! The difficulty is that the charge distribution itself extends to infinity.

Instead, as recommended in Problem-Solving Strategy 23.1, we set $V_b = 0$ at point b at an arbitrary but *finite* radial distance r_0. Then the potential $V = V_a$ at point a at a radial distance r is given by $V - 0 = (\lambda/2\pi\epsilon_0) \ln(r_0/r)$, or

$$V = \frac{\lambda}{2\pi\epsilon_0} \ln\frac{r_0}{r}$$

EVALUATE: According to our result, if λ is positive, then V decreases as r increases. This is as it should be: V decreases as we move in the direction of $\vec{E}$.

From Example 22.6, the expression for E_r with which we started also applies outside a long, charged conducting cylinder with charge per unit length λ (Fig. 23.19b). Hence our result also gives the potential for such a cylinder, but only for values

23.19 Electric field outside **(a)** a long, positively charged wire and **(b)** a long, positively charged cylinder.

(a)

(b)

of r (the distance from the cylinder axis) equal to or greater than the radius R of the cylinder. If we choose r_0 to be the cylinder radius R, so that $V = 0$ when $r = R$, then at any point for which $r > R$,

$$V = \frac{\lambda}{2\pi\epsilon_0} \ln\frac{R}{r}$$

Inside the cylinder, $\vec{E} = 0$, and V has the same value (zero) as on the cylinder's surface.

Example 23.11 A ring of charge

Electric charge Q is distributed uniformly around a thin ring of radius a (Fig. 23.20). Find the potential at a point P on the ring axis at a distance x from the center of the ring.

SOLUTION

IDENTIFY and SET UP: We divide the ring into infinitesimal segments and use Eq. (23.16) to find V. All parts of the ring (and therefore all elements of the charge distribution) are at the same distance from P.

EXECUTE: Figure 23.20 shows that the distance from each charge element dq to P is $r = \sqrt{x^2 + a^2}$. Hence we can take the factor $1/r$ outside the integral in Eq. (23.16), and

$$V = \frac{1}{4\pi\epsilon_0} \int \frac{dq}{r} = \frac{1}{4\pi\epsilon_0} \frac{1}{\sqrt{x^2 + a^2}} \int dq = \frac{1}{4\pi\epsilon_0} \frac{Q}{\sqrt{x^2 + a^2}}$$

EVALUATE: When x is much larger than a, our expression for V becomes approximately $V = Q/4\pi\epsilon_0 x$, which is the potential at a distance x from a point charge Q. Very far away from a charged

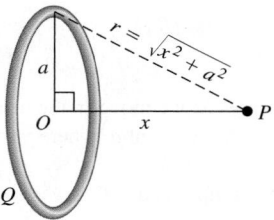

23.20 All the charge in a ring of charge Q is the same distance r from a point P on the ring axis.

ring, its electric potential looks like that of a point charge. We drew a similar conclusion about the electric *field* of a ring in Example 21.9 (Section 21.5).

We know the electric field at all points along the x-axis from Example 21.9 (Section 21.5), so we can also find V along this axis by integrating $\vec{E} \cdot d\vec{l}$ as in Eq. (23.17).

Example 23.12 Potential of a line of charge

Positive electric charge Q is distributed uniformly along a line of length $2a$ lying along the y-axis between $y = -a$ and $y = +a$ (Fig. 23.21). Find the electric potential at a point P on the x-axis at a distance x from the origin.

SOLUTION

IDENTIFY and SET UP: This is the same situation as in Example 21.10 (Section 21.5), where we found an expression for the electric field $\vec{E}$ at an arbitrary point on the x-axis. We can find V at point P by integrating over the charge distribution using Eq. (23.16). Unlike the situation in Example 23.11, each charge element dQ is a *different* distance from point P, so the integration will take a little more effort.

EXECUTE: As in Example 21.10, the element of charge dQ corresponding to an element of length dy on the rod is $dQ = (Q/2a)dy$. The distance from dQ to P is $\sqrt{x^2 + y^2}$, so the contribution dV that the charge element makes to the potential at P is

$$dV = \frac{1}{4\pi\epsilon_0}\frac{Q}{2a}\frac{dy}{\sqrt{x^2 + y^2}}$$

To find the potential at P due to the entire rod, we integrate dV over the length of the rod from $y = -a$ to $y = a$:

$$V = \frac{1}{4\pi\epsilon_0}\frac{Q}{2a}\int_{-a}^{a}\frac{dy}{\sqrt{x^2 + y^2}}$$

You can look up the integral in a table. The final result is

$$V = \frac{1}{4\pi\epsilon_0}\frac{Q}{2a}\ln\left(\frac{\sqrt{a^2 + x^2} + a}{\sqrt{a^2 + x^2} - a}\right)$$

23.21 Our sketch for this problem.

EVALUATE: We can check our result by letting x approach infinity. In this limit the point P is infinitely far from all of the charge, so we expect V to approach zero; you can verify that it does.

We know the electric field at all points along the x-axis from Example 21.10. We invite you to use this information to find V along this axis by integrating $\vec{E}$ as in Eq. (23.17).

Test Your Understanding of Section 23.3 If the electric *field* at a certain point is zero, does the electric *potential* at that point have to be zero? (*Hint:* Consider the center of the ring in Example 23.11 and Example 21.9.) ∎

23.4 Equipotential Surfaces

Field lines (see Section 21.6) help us visualize electric fields. In a similar way, the potential at various points in an electric field can be represented graphically by *equipotential surfaces*. These use the same fundamental idea as topographic maps like those used by hikers and mountain climbers (Fig. 23.22). On a topographic map, contour lines are drawn through points that are all at the same elevation. Any number of these could be drawn, but typically only a few contour lines are shown at equal spacings of elevation. If a mass m is moved over the terrain along such a contour line, the gravitational potential energy mgy does not change because the elevation y is constant. Thus contour lines on a topographic map are really curves of constant gravitational potential energy. Contour lines are close together where the terrain is steep and there are large changes in elevation over a small horizontal distance; the contour lines are farther apart where the terrain is gently sloping. A ball allowed to roll downhill will experience the greatest downhill gravitational force where contour lines are closest together.

By analogy to contour lines on a topographic map, an **equipotential surface** is a three-dimensional surface on which the *electric potential* V is the same at every point. If a test charge q_0 is moved from point to point on such a surface, the *electric potential energy* q_0V remains constant. In a region where an electric field is present, we can construct an equipotential surface through any point. In diagrams we usually show only a few representative equipotentials, often with equal potential differences between adjacent surfaces. No point can be at two different potentials, so equipotential surfaces for different potentials can never touch or intersect.

23.22 Contour lines on a topographic map are curves of constant elevation and hence of constant gravitational potential energy.

Equipotential Surfaces and Field Lines

Because potential energy does not change as a test charge moves over an equipotential surface, the electric field can do no work on such a charge. It follows that $\vec{E}$ must be perpendicular to the surface at every point so that the electric force $q_0\vec{E}$ is always perpendicular to the displacement of a charge moving on the surface.

23.23 Cross sections of equipotential surfaces (blue lines) and electric field lines (red lines) for assemblies of point charges. There are equal potential differences between adjacent surfaces. Compare these diagrams to those in Fig. 21.28, which showed only the electric field lines.

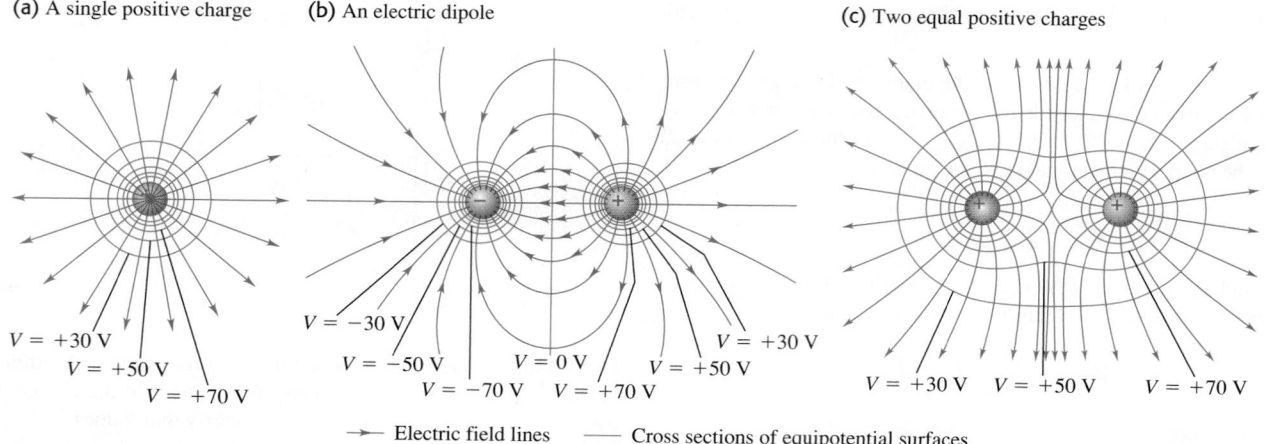

(a) A single positive charge

$V = +30$ V
$V = +50$ V
$V = +70$ V

(b) An electric dipole

$V = -30$ V
$V = -50$ V $V = 0$ V $V = +30$ V
$V = -70$ V $V = +70$ V $V = +50$ V

(c) Two equal positive charges

$V = +30$ V $V = +50$ V $V = +70$ V

→— Electric field lines   —— Cross sections of equipotential surfaces

Field lines and equipotential surfaces are always mutually perpendicular. In general, field lines are curves, and equipotentials are curved surfaces. For the special case of a *uniform* field, in which the field lines are straight, parallel, and equally spaced, the equipotentials are parallel *planes* perpendicular to the field lines.

Figure 23.23 shows three arrangements of charges. The field lines in the plane of the charges are represented by red lines, and the intersections of the equipotential surfaces with this plane (that is, cross sections of these surfaces) are shown as blue lines. The actual equipotential surfaces are three-dimensional. At each crossing of an equipotential and a field line, the two are perpendicular.

In Fig. 23.23 we have drawn equipotentials so that there are equal potential differences between adjacent surfaces. In regions where the magnitude of $\vec{E}$ is large, the equipotential surfaces are close together because the field does a relatively large amount of work on a test charge in a relatively small displacement. This is the case near the point charge in Fig. 23.23a or between the two point charges in Fig. 23.23b; note that in these regions the field lines are also closer together. This is directly analogous to the downhill force of gravity being greatest in regions on a topographic map where contour lines are close together. Conversely, in regions where the field is weaker, the equipotential surfaces are farther apart; this happens at larger radii in Fig. 23.23a, to the left of the negative charge or the right of the positive charge in Fig. 23.23b, and at greater distances from both charges in Fig. 23.23c. (It may appear that two equipotential surfaces intersect at the center of Fig. 23.23c, in violation of the rule that this can never happen. In fact this is a single figure-8–shaped equipotential surface.)

CAUTION *E* **need not be constant over an equipotential surface** On a given equipotential surface, the potential *V* has the same value at every point. In general, however, the electric-field magnitude *E* is *not* the same at all points on an equipotential surface. For instance, on the equipotential surface labeled "*V* = −30 V" in Fig. 23.23b, the magnitude *E* is less to the left of the negative charge than it is between the two charges. On the figure-8–shaped equipotential surface in Fig. 23.23c, *E* = 0 at the middle point halfway between the two charges; at any other point on this surface, *E* is nonzero. ▌

Equipotentials and Conductors

Here's an important statement about equipotential surfaces: **When all charges are at rest, the surface of a conductor is always an equipotential surface.**

Since the electric field $\vec{E}$ is always perpendicular to an equipotential surface, we can prove this statement by proving that **when all charges are at rest, the electric field just outside a conductor must be perpendicular to the surface at every point** (Fig. 23.24). We know that $\vec{E} = 0$ everywhere inside the conductor; otherwise, charges would move. In particular, at any point just inside the surface the component of $\vec{E}$ tangent to the surface is zero. It follows that the tangential component of $\vec{E}$ is also zero just *outside* the surface. If it were not, a charge could move around a rectangular path partly inside and partly outside (Fig. 23.25) and return to its starting point with a net amount of work having been done on it. This would violate the conservative nature of electrostatic fields, so the tangential component of $\vec{E}$ just outside the surface must be zero at every point on the surface. Thus $\vec{E}$ is perpendicular to the surface at each point, proving our statement.

It also follows that **when all charges are at rest, the entire solid volume of a conductor is at the same potential.** Equation (23.17) states that the potential difference between two points a and b within the conductor's solid volume, $V_a - V_b$, is equal to the line integral $\int_a^b \vec{E} \cdot d\vec{l}$ of the electric field from a to b. Since $\vec{E} = 0$ everywhere inside the conductor, the integral is guaranteed to be zero for any two such points a and b. Hence the potential is the same for any two points within the solid volume of the conductor. We describe this by saying that the solid volume of the conductor is an *equipotential volume.*

Finally, we can now prove a theorem that we quoted without proof in Section 22.5. The theorem is as follows: In an electrostatic situation, if a conductor contains a cavity and if no charge is present inside the cavity, then there can be no net charge *anywhere* on the surface of the cavity. This means that if you're inside a charged conducting box, you can safely touch any point on the inside walls of the box without being shocked. To prove this theorem, we first prove that *every point in the cavity is at the same potential.* In Fig. 23.26 the conducting surface A of the cavity is an equipotential surface, as we have just proved. Suppose point P in the cavity is at a different potential; then we can construct a different equipotential surface B including point P.

Now consider a Gaussian surface, shown in Fig. 23.26, between the two equipotential surfaces. Because of the relationship between $\vec{E}$ and the equipotentials, we know that the field at every point between the equipotentials is from A toward B, or else at every point it is from B toward A, depending on which equipotential surface is at higher potential. In either case the flux through this Gaussian surface is certainly not zero. But then Gauss's law says that the charge enclosed by the Gaussian surface cannot be zero. This contradicts our initial assumption that there is *no* charge in the cavity. So the potential at P *cannot* be different from that at the cavity wall.

The entire region of the cavity must therefore be at the same potential. But for this to be true, *the electric field inside the cavity must be zero everywhere.* Finally, Gauss's law shows that the electric field at any point on the surface of a conductor is proportional to the surface charge density σ at that point. We conclude that *the surface charge density on the wall of the cavity is zero at every point.* This chain of reasoning may seem tortuous, but it is worth careful study.

CAUTION **Equipotential surfaces vs. Gaussian surfaces** Don't confuse equipotential surfaces with the Gaussian surfaces we encountered in Chapter 22. Gaussian surfaces have relevance only when we are using Gauss's law, and we can choose *any* Gaussian surface that's convenient. We are *not* free to choose the shape of equipotential surfaces; the shape is determined by the charge distribution. ▮

Test Your Understanding of Section 23.4 Would the shapes of the equipotential surfaces in Fig. 23.23 change if the sign of each charge were reversed? ▮

23.24 When charges are at rest, a conducting surface is always an equipotential surface. Field lines are perpendicular to a conducting surface.

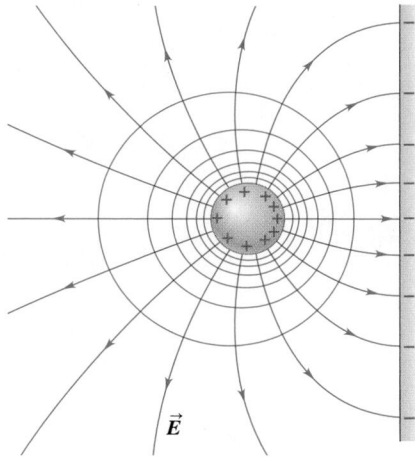

—— Cross sections of equipotential surfaces
⟶ Electric field lines

23.25 At all points on the surface of a conductor, the electric field must be perpendicular to the surface. If $\vec{E}$ had a tangential component, a net amount of work would be done on a test charge by moving it around a loop as shown here—which is impossible because the electric force is conservative.

An impossible electric field
If the electric field just outside a conductor had a tangential component $E_\parallel$, a charge could move in a loop with net work done.

23.26 A cavity in a conductor. If the cavity contains no charge, every point in the cavity is at the same potential, the electric field is zero everywhere in the cavity, and there is no charge anywhere on the surface of the cavity.

Cross section of equipotential surface through P

Gaussian surface (in cross section)

Surface of cavity

Conductor

23.5 Potential Gradient

Electric field and potential are closely related. Equation (23.17), restated here, expresses one aspect of that relationship:

$$V_a - V_b = \int_a^b \vec{E} \cdot d\vec{l}$$

If we know $\vec{E}$ at various points, we can use this equation to calculate potential differences. In this section we show how to turn this around; if we know the potential V at various points, we can use it to determine $\vec{E}$. Regarding V as a function of the coordinates (x, y, z) of a point in space, we will show that the components of $\vec{E}$ are related to the *partial derivatives* of V with respect to x, y, and z.

In Eq. (23.17), $V_a - V_b$ is the potential of a with respect to b—that is, the change of potential encountered on a trip from b to a. We can write this as

$$V_a - V_b = \int_b^a dV = -\int_a^b dV$$

where dV is the infinitesimal change of potential accompanying an infinitesimal element $d\vec{l}$ of the path from b to a. Comparing to Eq. (23.17), we have

$$-\int_a^b dV = \int_a^b \vec{E} \cdot d\vec{l}$$

These two integrals must be equal for *any* pair of limits a and b, and for this to be true the *integrands* must be equal. Thus, for *any* infinitesimal displacement $d\vec{l}$,

$$-dV = \vec{E} \cdot d\vec{l}$$

To interpret this expression, we write $\vec{E}$ and $d\vec{l}$ in terms of their components: $\vec{E} = \hat{\imath} E_x + \hat{\jmath} E_y + \hat{k} E_z$ and $d\vec{l} = \hat{\imath} dx + \hat{\jmath} dy + \hat{k} dz$. Then we have

$$-dV = E_x dx + E_y dy + E_z dz$$

Suppose the displacement is parallel to the x-axis, so $dy = dz = 0$. Then $-dV = E_x dx$ or $E_x = -(dV/dx)_{y,\,z\text{ constant}}$, where the subscript reminds us that only x varies in the derivative; recall that V is in general a function of x, y, and z. But this is just what is meant by the partial derivative $\partial V / \partial x$. The y- and z-components of $\vec{E}$ are related to the corresponding derivatives of V in the same way, so we have

$$E_x = -\frac{\partial V}{\partial x} \qquad E_y = -\frac{\partial V}{\partial y} \qquad E_z = -\frac{\partial V}{\partial z} \qquad \begin{array}{l}\text{(components of } \vec{E} \\ \text{in terms of } V)\end{array} \qquad (23.19)$$

This is consistent with the units of electric field being V/m. In terms of unit vectors we can write $\vec{E}$ as

$$\vec{E} = -\left(\hat{\imath} \frac{\partial V}{\partial x} + \hat{\jmath} \frac{\partial V}{\partial y} + \hat{k} \frac{\partial V}{\partial z} \right) \qquad (\vec{E} \text{ in terms of } V) \qquad (23.20)$$

In vector notation the following operation is called the **gradient** of the function f:

$$\vec{\nabla} f = \left(\hat{\imath} \frac{\partial}{\partial x} + \hat{\jmath} \frac{\partial}{\partial y} + \hat{k} \frac{\partial}{\partial z} \right) f \qquad (23.21)$$

The operator denoted by the symbol $\vec{\nabla}$ is called "grad" or "del." Thus in vector notation,

$$\vec{E} = -\vec{\nabla} V \qquad (23.22)$$

This is read "$\vec{E}$ is the negative of the gradient of V" or "$\vec{E}$ equals negative grad V." The quantity $\vec{\nabla} V$ is called the *potential gradient*.

Application **Potential Gradient Across a Cell Membrane**

The interior of a human cell is at a lower electric potential V than the exterior. (The potential difference when the cell is inactive is about –70 mV in neurons and about –95 mV in skeletal muscle cells.) Hence there is a potential gradient $\vec{\nabla} V$ that points from the *interior* to the *exterior* of the cell membrane, and an electric field $\vec{E} = -\vec{\nabla} V$ that points from the *exterior* to the *interior*. This field affects how ions flow into or out of the cell through special channels in the membrane.

- Calcium (Ca)
- Potassium (K)
- Sodium (Na)

At each point, the potential gradient points in the direction in which V *increases* most rapidly with a change in position. Hence at each point the direction of $\vec{E}$ is the direction in which V *decreases* most rapidly and is always perpendicular to the equipotential surface through the point. This agrees with our observation in Section 23.2 that moving in the direction of the electric field means moving in the direction of decreasing potential.

Equation (23.22) doesn't depend on the particular choice of the zero point for V. If we were to change the zero point, the effect would be to change V at every point by the same amount; the derivatives of V would be the same.

If $\vec{E}$ is radial with respect to a point or an axis and r is the distance from the point or the axis, the relationship corresponding to Eqs. (23.19) is

$$E_r = -\frac{\partial V}{\partial r} \qquad \text{(radial electric field)} \qquad (23.23)$$

Often we can compute the electric field caused by a charge distribution in either of two ways: directly, by adding the $\vec{E}$ fields of point charges, or by first calculating the potential and then taking its gradient to find the field. The second method is often easier because potential is a *scalar* quantity, requiring at worst the integration of a scalar function. Electric field is a *vector* quantity, requiring computation of components for each element of charge and a separate integration for each component. Thus, quite apart from its fundamental significance, potential offers a very useful computational technique in field calculations. Below, we present two examples in which a knowledge of V is used to find the electric field.

We stress once more that if we know $\vec{E}$ as a function of position, we can calculate V using Eq. (23.17) or (23.18), and if we know V as a function of position, we can calculate $\vec{E}$ using Eq. (23.19), (23.20), or (23.23). Deriving V from $\vec{E}$ requires integration, and deriving $\vec{E}$ from V requires differentiation.

Example 23.13   Potential and field of a point charge

From Eq. (23.14) the potential at a radial distance r from a point charge q is $V = q/4\pi\epsilon_0 r$. Find the vector electric field from this expression for V.

SOLUTION

IDENTIFY and SET UP: This problem uses the general relationship between the electric potential as a function of position and the electric-field vector. By symmetry, the electric field here has only a radial component E_r. We use Eq. (23.23) to find this component.

EXECUTE: From Eq. (23.23),

$$E_r = -\frac{\partial V}{\partial r} = -\frac{\partial}{\partial r}\left(\frac{1}{4\pi\epsilon_0}\frac{q}{r}\right) = \frac{1}{4\pi\epsilon_0}\frac{q}{r^2}$$

so the vector electric field is

$$\vec{E} = \hat{r}E_r = \frac{1}{4\pi\epsilon_0}\frac{q}{r^2}\hat{r}$$

EVALUATE: Our result agrees with Eq. (21.7), as it must.

An alternative approach is to ignore the radial symmetry, write the radial distance as $r = \sqrt{x^2 + y^2 + z^2}$, and take the derivatives of V with respect to x, y, and z as in Eq. (23.20). We find

$$\frac{\partial V}{\partial x} = \frac{\partial}{\partial x}\left(\frac{1}{4\pi\epsilon_0}\frac{q}{\sqrt{x^2 + y^2 + z^2}}\right) = -\frac{1}{4\pi\epsilon_0}\frac{qx}{(x^2 + y^2 + z^2)^{3/2}}$$

$$= -\frac{qx}{4\pi\epsilon_0 r^3}$$

and similarly

$$\frac{\partial V}{\partial y} = -\frac{qy}{4\pi\epsilon_0 r^3} \qquad \frac{\partial V}{\partial z} = -\frac{qz}{4\pi\epsilon_0 r^3}$$

Then from Eq. (23.20),

$$\vec{E} = -\left[\hat{i}\left(-\frac{qx}{4\pi\epsilon_0 r^3}\right) + \hat{j}\left(-\frac{qy}{4\pi\epsilon_0 r^3}\right) + \hat{k}\left(-\frac{qz}{4\pi\epsilon_0 r^3}\right)\right]$$

$$= \frac{1}{4\pi\epsilon_0}\frac{q}{r^2}\left(\frac{x\hat{i} + y\hat{j} + z\hat{k}}{r}\right) = \frac{1}{4\pi\epsilon_0}\frac{q}{r^2}\hat{r}$$

This approach gives us the same answer, but with more effort. Clearly it's best to exploit the symmetry of the charge distribution whenever possible.

Example 23.14 | Potential and field of a ring of charge

In Example 23.11 (Section 23.3) we found that for a ring of charge with radius a and total charge Q, the potential at a point P on the ring's symmetry axis a distance x from the center is

$$V = \frac{1}{4\pi\epsilon_0} \frac{Q}{\sqrt{x^2 + a^2}}$$

Find the electric field at P.

SOLUTION

IDENTIFY and SET UP: Figure 23.20 shows the situation. We are given V as a function of x along the x-axis, and we wish to find the electric field at a point on this axis. From the symmetry of the charge distribution, the electric field along the symmetry (x-) axis of the ring can have only an x-component. We find it using the first of Eqs. (23.19).

EXECUTE: The x-component of the electric field is

$$E_x = -\frac{\partial V}{\partial x} = \frac{1}{4\pi\epsilon_0} \frac{Qx}{(x^2 + a^2)^{3/2}}$$

EVALUATE: This agrees with our result in Example 21.9.

CAUTION Don't use expressions where they don't apply In this example, V is not a function of y or z on the ring axis, so that $\partial V/\partial y = \partial V/\partial z = 0$ and $E_y = E_z = 0$. But that does not mean that it's true *everywhere*; our expressions for V and E_x are valid *only on the ring axis*. If we had an expression for V valid at *all* points in space, we could use it to find the components of $\vec{E}$ at any point using Eqs. (23.19). ▮

Test Your Understanding of Section 23.5 In a certain region of space the potential is given by $V = A + Bx + Cy^3 + Dxy$, where A, B, C, and D are positive constants. Which of these statements about the electric field $\vec{E}$ in this region of space is correct? (There may be more than one correct answer.) (i) Increasing the value of A will increase the value of $\vec{E}$ at all points; (ii) increasing the value of A will decrease the value of $\vec{E}$ at all points; (iii) $\vec{E}$ has no z-component; (iv) the electric field is zero at the origin ($x = 0$, $y = 0$, $z = 0$). ▮

Electric potential energy: The electric force caused by any collection of charges at rest is a conservative force. The work W done by the electric force on a charged particle moving in an electric field can be represented by the change in a potential-energy function U.

The electric potential energy for two point charges q and q_0 depends on their separation r. The electric potential energy for a charge q_0 in the presence of a collection of charges q_1, q_2, q_3 depends on the distance from q_0 to each of these other charges. (See Examples 23.1 and 23.2.)

$$W_{a \to b} = U_a - U_b \qquad (23.2)$$

$$U = \frac{1}{4\pi\epsilon_0} \frac{qq_0}{r} \qquad (23.9)$$
(two point charges)

$$U = \frac{q_0}{4\pi\epsilon_0}\left(\frac{q_1}{r_1} + \frac{q_2}{r_2} + \frac{q_3}{r_3} + \cdots\right)$$
$$= \frac{q_0}{4\pi\epsilon_0}\sum_i \frac{q_i}{r_i} \qquad (23.10)$$
(q_0 in presence of other point charges)

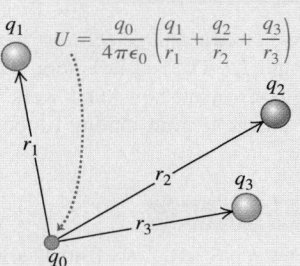

Electric potential: Potential, denoted by V, is potential energy per unit charge. The potential difference between two points equals the amount of work that would be required to move a unit positive test charge between those points. The potential V due to a quantity of charge can be calculated by summing (if the charge is a collection of point charges) or by integrating (if the charge is a distribution). (See Examples 23.3, 23.4, 23.5, 23.7, 23.11, and 23.12.)

The potential difference between two points a and b, also called the potential of a with respect to b, is given by the line integral of $\vec{E}$. The potential at a given point can be found by first finding $\vec{E}$ and then carrying out this integral. (See Examples 23.6, 23.8, 23.9, and 23.10.)

$$V = \frac{U}{q_0} = \frac{1}{4\pi\epsilon_0}\frac{q}{r} \qquad (23.14)$$
(due to a point charge)

$$V = \frac{U}{q_0} = \frac{1}{4\pi\epsilon_0}\sum_i \frac{q_i}{r_i} \qquad (23.15)$$
(due to a collection of point charges)

$$V = \frac{1}{4\pi\epsilon_0}\int \frac{dq}{r} \qquad (23.16)$$
(due to a charge distribution)

$$V_a - V_b = \int_a^b \vec{E} \cdot d\vec{l} = \int_a^b E\cos\phi\, dl \qquad (23.17)$$

Equipotential surfaces: An equipotential surface is a surface on which the potential has the same value at every point. At a point where a field line crosses an equipotential surface, the two are perpendicular. When all charges are at rest, the surface of a conductor is always an equipotential surface and all points in the interior of a conductor are at the same potential. When a cavity within a conductor contains no charge, the entire cavity is an equipotential region and there is no surface charge anywhere on the surface of the cavity.

Finding electric field from electric potential: If the potential V is known as a function of the coordinates x, y, and z, the components of electric field $\vec{E}$ at any point are given by partial derivatives of V. (See Examples 23.13 and 23.14.)

$$E_x = -\frac{\partial V}{\partial x} \quad E_y = -\frac{\partial V}{\partial y} \quad E_z = -\frac{\partial V}{\partial z} \qquad (23.19)$$

$$\vec{E} = -\left(\hat{\imath}\frac{\partial V}{\partial x} + \hat{\jmath}\frac{\partial V}{\partial y} + \hat{k}\frac{\partial V}{\partial z}\right) \qquad (23.20)$$
(vector form)

BRIDGING PROBLEM **A Point Charge and a Line of Charge**

Positive electric charge Q is distributed uniformly along a thin rod of length $2a$. The rod lies along the x-axis between $x = -a$ and $x = +a$. Calculate how much work you must do to bring a positive point charge q from infinity to the point $x = +L$ on the x-axis, where $L > a$.

SOLUTION GUIDE

See MasteringPhysics® study area for a Video Tutor solution. (MP)

IDENTIFY and SET UP

1. In this problem you must first calculate the potential V at $x = +L$ due to the charged rod. You can then find the change in potential energy involved in bringing the point charge q from infinity (where $V = 0$) to $x = +L$.

2. To find V, divide the rod into infinitesimal segments of length dx'. How much charge is on such a segment? Consider one such segment located at $x = x'$, where $-a \leq x' \leq a$. What is the potential dV at $x = +L$ due to this segment?

3. The total potential at $x = +L$ is the integral of dV, including contributions from all of the segments for x' from $-a$ to $+a$. Set up this integral.

EXECUTE

4. Integrate your expression from step 3 to find the potential V at $x = +L$. A simple, standard substitution will do the trick; use a table of integrals only as a last resort.

5. Use your result from step 4 to find the potential energy for a point charge q placed at $x = +L$.

6. Use your result from step 5 to find the work you must do to bring the point charge from infinity to $x = +L$.

EVALUATE

7. What does your result from step 5 become in the limit $a \rightarrow 0$? Does this make sense?

8. Suppose the point charge q were negative rather than positive. How would this affect your result in step 4? In step 5?

Problems

For instructor-assigned homework, go to www.masteringphysics.com (MP)

•, ••, •••: Problems of increasing difficulty. **CP**: Cumulative problems incorporating material from earlier chapters. **CALC**: Problems requiring calculus. **BIO**: Biosciences problems.

DISCUSSION QUESTIONS

Q23.1 A student asked, "Since electrical potential is always proportional to potential energy, why bother with the concept of potential at all?" How would you respond?

Q23.2 The potential (relative to a point at infinity) midway between two charges of equal magnitude and opposite sign is zero. Is it possible to bring a test charge from infinity to this midpoint in such a way that no work is done in any part of the displacement? If so, describe how it can be done. If it is not possible, explain why.

Q23.3 Is it possible to have an arrangement of two point charges separated by a finite distance such that the electric potential energy of the arrangement is the same as if the two charges were infinitely far apart? Why or why not? What if there are three charges? Explain your reasoning.

Q23.4 Since potential can have any value you want depending on the choice of the reference level of zero potential, how does a voltmeter know what to read when you connect it between two points?

Q23.5 If $\vec{E}$ is zero everywhere along a certain path that leads from point A to point B, what is the potential difference between those two points? Does this mean that $\vec{E}$ is zero everywhere along *any* path from A to B? Explain.

Q23.6 If $\vec{E}$ is zero throughout a certain region of space, is the potential necessarily also zero in this region? Why or why not? If not, what *can* be said about the potential?

Q23.7 If you carry out the integral of the electric field $\int \vec{E} \cdot d\vec{l}$ for a *closed* path like that shown in Fig. Q23.7, the integral will *always* be equal to zero, independent of the shape of the

Figure **Q23.7**

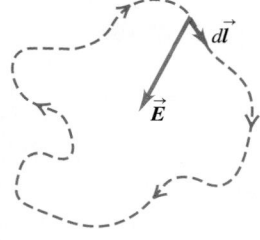

path and independent of where charges may be located relative to the path. Explain why.

Q23.8 The potential difference between the two terminals of an AA battery (used in flashlights and portable stereos) is 1.5 V. If two AA batteries are placed end to end with the positive terminal of one battery touching the negative terminal of the other, what is the potential difference between the terminals at the exposed ends of the combination? What if the two positive terminals are touching each other? Explain your reasoning.

Q23.9 It is easy to produce a potential difference of several thousand volts between your body and the floor by scuffing your shoes across a nylon carpet. When you touch a metal doorknob, you get a mild shock. Yet contact with a power line of comparable voltage would probably be fatal. Why is there a difference?

Q23.10 If the electric potential at a single point is known, can $\vec{E}$ at that point be determined? If so, how? If not, why not?

Q23.11 Because electric field lines and equipotential surfaces are always perpendicular, two equipotential surfaces can never cross; if they did, the direction of $\vec{E}$ would be ambiguous at the crossing points. Yet two equipotential surfaces appear to cross at the center of Fig. 23.23c. Explain why there is no ambiguity about the direction of $\vec{E}$ in this particular case.

Q23.12 A uniform electric field is directed due east. Point B is 2.00 m west of point A, point C is 2.00 m east of point A, and point D is 2.00 m south of A. For each point, B, C, and D, is the potential at that point larger, smaller, or the same as at point A? Give the reasoning behind your answers.

Q23.13 We often say that if point A is at a higher potential than point B, A is at positive potential and B is at negative potential. Does it necessarily follow that a point at positive potential is positively charged, or that a point at negative potential is negatively charged? Illustrate your answers with clear, simple examples.

Q23.14 A conducting sphere is to be charged by bringing in positive charge a little at a time until the total charge is Q. The total work required for this process is alleged to be proportional to Q^2. Is this correct? Why or why not?

Q23.15 Three pairs of parallel metal plates (A, B, and C) are connected as shown in Fig. Q23.15, and a battery maintains a potential of 1.5 V across ab. What can you say about the potential difference across each pair of plates? Why?

Figure **Q23.15**

Q23.16 A conducting sphere is placed between two charged parallel plates such as those shown in Fig. 23.2. Does the electric field inside the sphere depend on precisely where between the plates the sphere is placed? What about the electric potential inside the sphere? Do the answers to these questions depend on whether or not there is a net charge on the sphere? Explain your reasoning.

Q23.17 A conductor that carries a net charge Q has a hollow, empty cavity in its interior. Does the potential vary from point to point within the material of the conductor? What about within the cavity? How does the potential inside the cavity compare to the potential within the material of the conductor?

Q23.18 A high-voltage dc power line falls on a car, so the entire metal body of the car is at a potential of 10,000 V with respect to the ground. What happens to the occupants (a) when they are sitting in the car and (b) when they step out of the car? Explain your reasoning.

Q23.19 When a thunderstorm is approaching, sailors at sea sometimes observe a phenomenon called "St. Elmo's fire," a bluish flickering light at the tips of masts. What causes this? Why does it occur at the tips of masts? Why is the effect most pronounced when the masts are wet? (*Hint:* Seawater is a good conductor of electricity.)

Q23.20 A positive point charge is placed near a very large conducting plane. A professor of physics asserted that the field caused by this configuration is the same as would be obtained by removing the plane and placing a negative point charge of equal magnitude in the mirror-image position behind the initial position of the plane. Is this correct? Why or why not? (*Hint:* Inspect Fig. 23.23b.)

Q23.21 In electronics it is customary to define the potential of ground (thinking of the earth as a large conductor) as zero. Is this consistent with the fact that the earth has a net electric charge that is not zero? (Refer to Exercise 21.32.)

EXERCISES

Section 23.1 Electric Potential Energy

23.1 •• A point charge $q_1 = +2.40\ \mu\text{C}$ is held stationary at the origin. A second point charge $q_2 = -4.30\ \mu\text{C}$ moves from the point $x = 0.150$ m, $y = 0$ to the point $x = 0.250$ m, $y = 0.250$ m. How much work is done by the electric force on q_2?

23.2 • A point charge q_1 is held stationary at the origin. A second charge q_2 is placed at point a, and the electric potential energy of the pair of charges is $+5.4 \times 10^{-8}$ J. When the second charge is moved to point b, the electric force on the charge does -1.9×10^{-8} J of work. What is the electric potential energy of the pair of charges when the second charge is at point b?

23.3 •• **Energy of the Nucleus.** How much work is needed to assemble an atomic nucleus containing three protons (such as Be) if we model it as an equilateral triangle of side 2.00×10^{-15} m

with a proton at each vertex? Assume the protons started from very far away.

23.4 •• (a) How much work would it take to push two protons very slowly from a separation of 2.00×10^{-10} m (a typical atomic distance) to 3.00×10^{-15} m (a typical nuclear distance)? (b) If the protons are both released from rest at the closer distance in part (a), how fast are they moving when they reach their original separation?

23.5 •• A small metal sphere, carrying a net charge of $q_1 = -2.80\ \mu\text{C}$, is held in a stationary position by insulating supports. A second small metal sphere, with a net charge of $q_2 = -7.80\ \mu\text{C}$ and mass 1.50 g, is projected toward q_1.

Figure **E23.5**

When the two spheres are 0.800 m apart, q_2 is moving toward q_1 with speed 22.0 m/s (Fig. E23.5). Assume that the two spheres can be treated as point charges. You can ignore the force of gravity. (a) What is the speed of q_2 when the spheres are 0.400 m apart? (b) How close does q_2 get to q_1?

23.6 •• **BIO Energy of DNA Base Pairing, I.** (See Exercise 21.23.) (a) Calculate the electric potential energy of the adenine–thymine bond, using the same combinations of molecules ($O-H-N$ and $N-H-N$) as in Exercise 21.23. (b) Compare this energy with the potential energy of the proton–electron pair in the hydrogen atom.

23.7 •• **BIO Energy of DNA Base Pairing, II.** (See Exercise 21.24.) Calculate the electric potential energy of the guanine–cytosine bond, using the same combinations of molecules ($O-H-O$, $N-H-N$, and $O-H-N$) as in Exercise 21.24.

23.8 •• Three equal 1.20-μC point charges are placed at the corners of an equilateral triangle whose sides are 0.500 m long. What is the potential energy of the system? (Take as zero the potential energy of the three charges when they are infinitely far apart.)

23.9 •• Two protons are released from rest when they are 0.750 nm apart. (a) What is the maximum speed they will reach? When does this speed occur? (b) What is the maximum acceleration they will achieve? When does this acceleration occur?

23.10 •• Four electrons are located at the corners of a square 10.0 nm on a side, with an alpha particle at its midpoint. How much work is needed to move the alpha particle to the midpoint of one of the sides of the square?

23.11 •• Three point charges, which initially are infinitely far apart, are placed at the corners of an equilateral triangle with sides d. Two of the point charges are identical and have charge q. If zero net work is required to place the three charges at the corners of the triangle, what must the value of the third charge be?

23.12 •• Starting from a separation of several meters, two protons are aimed directly toward each other by a cyclotron accelerator with speeds of 1000 km/s, measured relative to the earth. Find the maximum electrical force that these protons will exert on each other.

Section 23.2 Electric Potential

23.13 • A small particle has charge $-5.00\ \mu\text{C}$ and mass 2.00×10^{-4} kg. It moves from point A, where the electric potential is $V_A = +200$ V, to point B, where the electric potential is $V_B = +800$ V. The electric force is the only force acting on the particle. The particle has speed 5.00 m/s at point A. What is its speed at point B? Is it moving faster or slower at B than at A? Explain.

23.14 • A particle with a charge of $+4.20$ nC is in a uniform electric field $\vec{E}$ directed to the left. It is released from rest and moves to the left; after it has moved 6.00 cm, its kinetic energy is found to be $+1.50 \times 10^{-6}$ J. (a) What work was done by the electric force? (b) What is the potential of the starting point with respect to the end point? (c) What is the magnitude of $\vec{E}$?

23.15 • A charge of 28.0 nC is placed in a uniform electric field that is directed vertically upward and has a magnitude of 4.00×10^4 V/m. What work is done by the electric force when the charge moves (a) 0.450 m to the right; (b) 0.670 m upward; (c) 2.60 m at an angle of $45.0°$ downward from the horizontal?

23.16 • Two stationary point charges $+3.00$ nC and $+2.00$ nC are separated by a distance of 50.0 cm. An electron is released from rest at a point midway between the two charges and moves along the line connecting the two charges. What is the speed of the electron when it is 10.0 cm from the $+3.00$-nC charge?

23.17 •• Point charges $q_1 = +2.00$ μC and $q_2 = -2.00$ μC are placed at adjacent corners of a square for which the length of each side is 3.00 cm. Point a is at the center of the square, and point b is at the empty corner closest to q_2. Take the electric potential to be zero at a distance far from both charges. (a) What is the electric potential at point a due to q_1 and q_2? (b) What is the electric potential at point b? (c) A point charge $q_3 = -5.00$ μC moves from point a to point b. How much work is done on q_3 by the electric forces exerted by q_1 and q_2? Is this work positive or negative?

23.18 • Two charges of equal magnitude Q are held a distance d apart. Consider only points on the line passing through both charges. (a) If the two charges have the same sign, find the location of all points (if there are any) at which (i) the potential (relative to infinity) is zero (is the electric field zero at these points?), and (ii) the electric field is zero (is the potential zero at these points?). (b) Repeat part (a) for two charges having opposite signs.

23.19 • Two point charges $q_1 = +2.40$ nC and $q_2 = -6.50$ nC are 0.100 m apart. Point A is midway between them; point B is 0.080 m from q_1 and 0.060 m from q_2 (Fig. E23.19). Take the electric potential to be zero at infinity. Find (a) the potential at point A; (b) the potential at point B; (c) the work done by the electric field on a charge of 2.50 nC that travels from point B to point A.

Figure **E23.19**

23.20 • A positive charge $+q$ is located at the point $x = 0$, $y = -a$, and a negative charge $-q$ is located at the point $x = 0$, $y = +a$. (a) Derive an expression for the potential V at points on the y-axis as a function of the coordinate y. Take V to be zero at an infinite distance from the charges. (b) Graph V at points on the y-axis as a function of y over the range from $y = -4a$ to $y = +4a$. (c) Show that for $y > a$, the potential at a point on the positive y-axis is given by $V = -(1/4\pi\epsilon_0)2qa/y^2$. (d) What are the answers to parts (a) and (c) if the two charges are interchanged so that $+q$ is at $y = +a$ and $-q$ is at $y = -a$?

23.21 •• A positive charge q is fixed at the point $x = 0$, $y = 0$, and a negative charge $-2q$ is fixed at the point $x = a$, $y = 0$. (a) Show the positions of the charges in a diagram. (b) Derive an expression for the potential V at points on the x-axis as a function of the coordinate x. Take V to be zero at an infinite distance from the charges. (c) At which positions on the x-axis is $V = 0$? (d) Graph V at points on the x-axis as a function of x in the range from $x = -2a$ to $x = +2a$. (e) What does the answer to part (b) become when $x \gg a$? Explain why this result is obtained.

23.22 •• Consider the arrangement of point charges described in Exercise 23.21. (a) Derive an expression for the potential V at points on the y-axis as a function of the coordinate y. Take V to be zero at an infinite distance from the charges. (b) At which positions on the y-axis is $V = 0$? (c) Graph V at points on the y-axis as a function of y in the range from $y = -2a$ to $y = +2a$. (d) What does the answer to part (a) become when $y > a$? Explain why this result is obtained.

23.23 •• (a) An electron is to be accelerated from 3.00×10^6 m/s to 8.00×10^6 m/s. Through what potential difference must the electron pass to accomplish this? (b) Through what potential difference must the electron pass if it is to be slowed from 8.00×10^6 m/s to a halt?

23.24 • At a certain distance from a point charge, the potential and electric-field magnitude due to that charge are 4.98 V and 12.0 V/m, respectively. (Take the potential to be zero at infinity.) (a) What is the distance to the point charge? (b) What is the magnitude of the charge? (c) Is the electric field directed toward or away from the point charge?

23.25 • A uniform electric field has magnitude E and is directed in the negative x-direction. The potential difference between point a (at $x = 0.60$ m) and point b (at $x = 0.90$ m) is 240 V. (a) Which point, a or b, is at the higher potential? (b) Calculate the value of E. (c) A negative point charge $q = -0.200$ μC is moved from b to a. Calculate the work done on the point charge by the electric field.

23.26 • For each of the following arrangements of two point charges, find all the points along the line passing through both charges for which the electric potential V is zero (take $V = 0$ infinitely far from the charges) and for which the electric field E is zero: (a) charges $+Q$ and $+2Q$ separated by a distance d, and (b) charges $-Q$ and $+2Q$ separated by a distance d. (c) Are both V and E zero at the same places? Explain.

Section 23.3 Calculating Electric Potential

23.27 •• A thin spherical shell with radius $R_1 = 3.00$ cm is concentric with a larger thin spherical shell with radius $R_2 = 5.00$ cm. Both shells are made of insulating material. The smaller shell has charge $q_1 = +6.00$ nC distributed uniformly over its surface, and the larger shell has charge $q_2 = -9.00$ nC distributed uniformly over its surface. Take the electric potential to be zero at an infinite distance from both shells. (a) What is the electric potential due to the two shells at the following distance from their common center: (i) $r = 0$; (ii) $r = 4.00$ cm; (iii) $r = 6.00$ cm? (b) What is the magnitude of the potential difference between the surfaces of the two shells? Which shell is at higher potential: the inner shell or the outer shell?

23.28 • A total electric charge of 3.50 nC is distributed uniformly over the surface of a metal sphere with a radius of 24.0 cm. If the potential is zero at a point at infinity, find the value of the potential at the following distances from the center of the sphere: (a) 48.0 cm; (b) 24.0 cm; (c) 12.0 cm.

23.29 •• A uniformly charged, thin ring has radius 15.0 cm and total charge $+24.0$ nC. An electron is placed on the ring's axis a distance 30.0 cm from the center of the ring and is constrained to stay on the axis of the ring. The electron is then released from rest. (a) Describe the subsequent motion of the electron. (b) Find the speed of the electron when it reaches the center of the ring.

23.30 •• An infinitely long line of charge has linear charge density 5.00×10^{-12} C/m. A proton (mass 1.67×10^{-27} kg, charge $+1.60 \times 10^{-19}$ C) is 18.0 cm from the line and moving directly toward the line at 1.50×10^3 m/s. (a) Calculate the proton's initial kinetic energy. (b) How close does the proton get to the line of charge?

23.31 • A very long wire carries a uniform linear charge density λ. Using a voltmeter to measure potential difference, you find that when one probe of the meter is placed 2.50 cm from the wire and the other probe is 1.00 cm farther from the wire, the meter reads 575 V. (a) What is λ? (b) If you now place one probe at 3.50 cm from the wire and the other probe 1.00 cm farther away, will the voltmeter read 575 V? If not, will it read more or less than 575 V? Why? (c) If you place both probes 3.50 cm from the wire but 17.0 cm from each other, what will the voltmeter read?

23.32 •• A very long insulating cylinder of charge of radius 2.50 cm carries a uniform linear density of 15.0 nC/m. If you put one probe of a voltmeter at the surface, how far from the surface must the other probe be placed so that the voltmeter reads 175 V?

23.33 •• A very long insulating cylindrical shell of radius 6.00 cm carries charge of linear density 8.50 μC/m spread uniformly over its outer surface. What would a voltmeter read if it were connected between (a) the surface of the cylinder and a point 4.00 cm above the surface, and (b) the surface and a point 1.00 cm from the central axis of the cylinder?

23.34 • A ring of diameter 8.00 cm is fixed in place and carries a charge of +5.00 μC uniformly spread over its circumference. (a) How much work does it take to move a tiny +3.00-μC charged ball of mass 1.50 g from very far away to the center of the ring? (b) Is it necessary to take a path along the axis of the ring? Why? (c) If the ball is slightly displaced from the center of the ring, what will it do and what is the maximum speed it will reach?

23.35 •• A very small sphere with positive charge $q = +8.00 \ \mu$C is released from rest at a point 1.50 cm from a very long line of uniform linear charge density $\lambda = +3.00 \ \mu$C/m. What is the kinetic energy of the sphere when it is 4.50 cm from the line of charge if the only force on it is the force exerted by the line of charge?

23.36 • Charge $Q = 5.00 \ \mu$C is distributed uniformly over the volume of an insulating sphere that has radius $R = 12.0$ cm. A small sphere with charge $q = +3.00 \ \mu$C and mass 6.00×10^{-5} kg is projected toward the center of the large sphere from an initial large distance. The large sphere is held at a fixed position and the small sphere can be treated as a point charge. What minimum speed must the small sphere have in order to come within 8.00 cm of the surface of the large sphere?

23.37 • **BIO** **Axons.** Neurons are the basic units of the nervous system. They contain long tubular structures called *axons* that propagate electrical signals away from the ends of the neurons. The axon contains a solution of potassium (K^+) ions and large negative organic ions. The axon membrane prevents the large ions from leaking out, but the smaller K^+ ions are able to penetrate the membrane to some degree (Fig. E23.37). This leaves an excess negative charge on the inner surface of the axon membrane and an excess positive charge on the outer surface, resulting in a potential difference across the membrane that prevents further K^+ ions from leaking out. Measurements show that this potential difference is typically about 70 mV. The thickness of the axon membrane itself varies from about 5 to 10 nm, so we'll use an average of 7.5 nm. We can model the membrane as a large sheet having equal and opposite charge densities on its faces. (a) Find the electric field inside the axon membrane, assuming (not too realistically) that it is filled with air. Which way does it point: into or out of the axon?

Figure **E23.37**

(b) Which is at a higher potential: the inside surface or the outside surface of the axon membrane?

23.38 • **CP** Two large, parallel conducting plates carrying opposite charges of equal magnitude are separated by 2.20 cm. (a) If the surface charge density for each plate has magnitude 47.0 nC/m^2, what is the magnitude of $\vec{E}$ in the region between the plates? (b) What is the potential difference between the two plates? (c) If the separation between the plates is doubled while the surface charge density is kept constant at the value in part (a), what happens to the magnitude of the electric field and to the potential difference?

23.39 • Two large, parallel, metal plates carry opposite charges of equal magnitude. They are separated by 45.0 mm, and the potential difference between them is 360 V. (a) What is the magnitude of the electric field (assumed to be uniform) in the region between the plates? (b) What is the magnitude of the force this field exerts on a particle with charge +2.40 nC? (c) Use the results of part (b) to compute the work done by the field on the particle as it moves from the higher-potential plate to the lower. (d) Compare the result of part (c) to the change of potential energy of the same charge, computed from the electric potential.

23.40 • **BIO** **Electrical Sensitivity of Sharks.** Certain sharks can detect an electric field as weak as 1.0 μV/m. To grasp how weak this field is, if you wanted to produce it between two parallel metal plates by connecting an ordinary 1.5-V AA battery across these plates, how far apart would the plates have to be?

23.41 •• (a) Show that V for a spherical shell of radius R, that has charge q distributed uniformly over its surface, is the same as V for a solid conductor with radius R and charge q. (b) You rub an inflated balloon on the carpet and it acquires a potential that is 1560 V lower than its potential before it became charged. If the charge is uniformly distributed over the surface of the balloon and if the radius of the balloon is 15 cm, what is the net charge on the balloon? (c) In light of its 1200-V potential difference relative to you, do you think this balloon is dangerous? Explain.

23.42 •• (a) How much excess charge must be placed on a copper sphere 25.0 cm in diameter so that the potential of its center, relative to infinity, is 1.50 kV? (b) What is the potential of the sphere's surface relative to infinity?

23.43 • The electric field at the surface of a charged, solid, copper sphere with radius 0.200 m is 3800 N/C, directed toward the center of the sphere. What is the potential at the center of the sphere, if we take the potential to be zero infinitely far from the sphere?

Section 23.4 Equipotential Surfaces and
Section 23.5 Potential Gradient

23.44 • A very large plastic sheet carries a uniform charge density of −6.00 nC/m^2 on one face. (a) As you move away from the sheet along a line perpendicular to it, does the potential increase or decrease? How do you know, without doing any calculations? Does your answer depend on where you choose the reference point for potential? (b) Find the spacing between equipotential surfaces that differ from each other by 1.00 V. What type of surfaces are these?

23.45 • **CALC** In a certain region of space, the electric potential is $V(x, y, z) = Axy - Bx^2 + Cy$, where A, B, and C are positive constants. (a) Calculate the x-, y-, and z-components of the electric field. (b) At which points is the electric field equal to zero?

23.46 • **CALC** In a certain region of space the electric potential is given by $V = +Ax^2y - Bxy^2$, where $A = 5.00$ V/m^3 and $B = 8.00$ V/m^3. Calculate the magnitude and direction of the electric field at the point in the region that has coordinates $x = 2.00$ m, $y = 0.400$ m, and $z = 0$.

23.47 •• **CALC** A metal sphere with radius r_a is supported on an insulating stand at the center of a hollow, metal, spherical shell with radius r_b. There is charge $+q$ on the inner sphere and charge $-q$ on the outer spherical shell. (a) Calculate the potential $V(r)$ for (i) $r < r_a$; (ii) $r_a < r < r_b$; (iii) $r > r_b$. (*Hint:* The net potential is the sum of the potentials due to the individual spheres.) Take V to be zero when r is infinite. (b) Show that the potential of the inner sphere with respect to the outer is

$$V_{ab} = \frac{q}{4\pi\epsilon_0}\left(\frac{1}{r_a} - \frac{1}{r_b}\right)$$

(c) Use Eq. (23.23) and the result from part (a) to show that the electric field at any point between the spheres has magnitude

$$E(r) = \frac{V_{ab}}{(1/r_a - 1/r_b)}\frac{1}{r^2}$$

(d) Use Eq. (23.23) and the result from part (a) to find the electric field at a point outside the larger sphere at a distance r from the center, where $r > r_b$. (e) Suppose the charge on the outer sphere is not $-q$ but a negative charge of different magnitude, say $-Q$. Show that the answers for parts (b) and (c) are the same as before but the answer for part (d) is different.

23.48 • A metal sphere with radius $r_a = 1.20$ cm is supported on an insulating stand at the center of a hollow, metal, spherical shell with radius $r_b = 9.60$ cm. Charge $+q$ is put on the inner sphere and charge $-q$ on the outer spherical shell. The magnitude of q is chosen to make the potential difference between the spheres 500 V, with the inner sphere at higher potential. (a) Use the result of Exercise 23.47(b) to calculate q. (b) With the help of the result of Exercise 23.47(a), sketch the equipotential surfaces that correspond to 500, 400, 300, 200, 100, and 0 V. (c) In your sketch, show the electric field lines. Are the electric field lines and equipotential surfaces mutually perpendicular? Are the equipotential surfaces closer together when the magnitude of $\vec{E}$ is largest?

23.49 • A very long cylinder of radius 2.00 cm carries a uniform charge density of 1.50 nC/m. (a) Describe the shape of the equipotential surfaces for this cylinder. (b) Taking the reference level for the zero of potential to be the surface of the cylinder, find the radius of equipotential surfaces having potentials of 10.0 V, 20.0 V, and 30.0 V. (c) Are the equipotential surfaces equally spaced? If not, do they get closer together or farther apart as r increases?

PROBLEMS

23.50 • **CP** A point charge $q_1 = +5.00\ \mu C$ is held fixed in space. From a horizontal distance of 6.00 cm, a small sphere with mass 4.00×10^{-3} kg and charge $q_2 = +2.00\ \mu C$ is fired toward the fixed charge with an initial speed of 40.0 m/s. Gravity can be neglected. What is the acceleration of the sphere at the instant when its speed is 25.0 m/s?

23.51 ••• A point charge $q_1 = 4.00$ nC is placed at the origin, and a second point charge $q_2 = -3.00$ nC is placed on the x-axis at $x = +20.0$ cm. A third point charge $q_3 = 2.00$ nC is to be placed on the x-axis between q_1 and q_2. (Take as zero the potential energy of the three charges when they are infinitely far apart.) (a) What is the potential energy of the system of the three charges if q_3 is placed at $x = +10.0$ cm? (b) Where should q_3 be placed to make the potential energy of the system equal to zero?

23.52 ••• A small sphere with mass 5.00×10^{-7} kg and charge $+3.00\ \mu C$ is released from rest a distance of 0.400 m above a large horizontal insulating sheet of charge that has uniform surface charge density $\sigma = +8.00\ pC/m^2$. Using energy methods, calculate the speed of the sphere when it is 0.100 m above the sheet of charge?

23.53 •• **Determining the Size of the Nucleus.** When radium-226 decays radioactively, it emits an alpha particle (the nucleus of helium), and the end product is radon-222. We can model this decay by thinking of the radium-226 as consisting of an alpha particle emitted from the surface of the spherically symmetric radon-222 nucleus, and we can treat the alpha particle as a point charge. The energy of the alpha particle has been measured in the laboratory and has been found to be 4.79 MeV when the alpha particle is essentially infinitely far from the nucleus. Since radon is much heavier than the alpha particle, we can assume that there is no appreciable recoil of the radon after the decay. The radon nucleus contains 86 protons, while the alpha particle has 2 protons and the radium nucleus has 88 protons. (a) What was the electric potential energy of the alpha–radon combination just before the decay, in MeV and in joules? (b) Use your result from part (a) to calculate the radius of the radon nucleus.

23.54 •• **CP** A proton and an alpha particle are released from rest when they are 0.225 nm apart. The alpha particle (a helium nucleus) has essentially four times the mass and two times the charge of a proton. Find the maximum *speed* and maximum *acceleration* of each of these particles. When do these maxima occur: just following the release of the particles or after a very long time?

23.55 • A particle with charge $+7.60$ nC is in a uniform electric field directed to the left. Another force, in addition to the electric force, acts on the particle so that when it is released from rest, it moves to the right. After it has moved 8.00 cm, the additional force has done 6.50×10^{-5} J of work and the particle has 4.35×10^{-5} J of kinetic energy. (a) What work was done by the electric force? (b) What is the potential of the starting point with respect to the end point? (c) What is the magnitude of the electric field?

23.56 • **CP** In the *Bohr model* of the hydrogen atom, a single electron revolves around a single proton in a circle of radius r. Assume that the proton remains at rest. (a) By equating the electric force to the electron mass times its acceleration, derive an expression for the electron's speed. (b) Obtain an expression for the electron's kinetic energy, and show that its magnitude is just half that of the electric potential energy. (c) Obtain an expression for the total energy, and evaluate it using $r = 5.29 \times 10^{-11}$ m. Give your numerical result in joules and in electron volts.

23.57 •• **CALC** A vacuum tube diode consists of concentric cylindrical electrodes, the negative cathode and the positive anode. Because of the accumulation of charge near the cathode, the electric potential between the electrodes is not a linear function of the position, even with planar geometry, but is given by

$$V(x) = Cx^{4/3}$$

where x is the distance from the cathode and C is a constant, characteristic of a particular diode and operating conditions. Assume that the distance between the cathode and anode is 13.0 mm and the potential difference between electrodes is 240 V. (a) Determine the value of C. (b) Obtain a formula for the electric field between the electrodes as a function of x. (c) Determine the force on an electron when the electron is halfway between the electrodes.

23.58 •• Two oppositely charged, identical insulating spheres, each 50.0 cm in diameter and carrying a uniform charge of magnitude 250 μC, are placed 1.00 m apart center to center (Fig. P23.58). (a) If a voltmeter is connected between the nearest points (a and b) on their surfaces, what will it read? (b) Which point, a or b, is at the higher potential? How can you know this without any calculations?

Figure **P23.58**

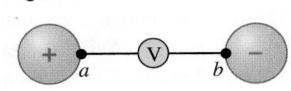

23.59 •• **An Ionic Crystal.** Figure P23.59 shows eight point charges arranged at the corners of a cube with sides of length d. The values of the charges are $+q$ and $-q$, as shown. This is a model of one cell of a cubic ionic crystal. In sodium chloride (NaCl), for instance, the positive ions are Na$^+$ and the negative ions are Cl$^-$. (a) Calculate the potential energy U of this arrangement. (Take as zero the potential energy of the eight charges when they are infinitely far apart.) (b) In part (a), you should have found that $U < 0$. Explain the relationship between this result and the observation that such ionic crystals exist in nature.

Figure **P23.59**

23.60 • (a) Calculate the potential energy of a system of two small spheres, one carrying a charge of 2.00 μC and the other a charge of -3.50 μC, with their centers separated by a distance of 0.250 m. Assume zero potential energy when the charges are infinitely separated. (b) Suppose that one of the spheres is held in place and the other sphere, which has a mass of 1.50 g, is shot away from it. What minimum initial speed would the moving sphere need in order to escape completely from the attraction of the fixed sphere? (To escape, the moving sphere would have to reach a velocity of zero when it was infinitely distant from the fixed sphere.)

23.61 •• **The H$_2^+$ Ion.** The H$_2^+$ ion is composed of two protons, each of charge $+e = 1.60 \times 10^{-19}$ C, and an electron of charge $-e$ and mass 9.11×10^{-31} kg. The separation between the protons is 1.07×10^{-10} m. The protons and the electron may be treated as point charges. (a) Suppose the electron is located at the point midway between the two protons. What is the potential energy of the interaction between the electron and the two protons? (Do not include the potential energy due to the interaction between the two protons.) (b) Suppose the electron in part (a) has a velocity of magnitude 1.50×10^6 m/s in a direction along the perpendicular bisector of the line connecting the two protons. How far from the point midway between the two protons can the electron move? Because the masses of the protons are much greater than the electron mass, the motions of the protons are very slow and can be ignored. (*Note:* A realistic description of the electron motion requires the use of quantum mechanics, not Newtonian mechanics.)

23.62 •• **CP** A small sphere with mass 1.50 g hangs by a thread between two parallel vertical plates 5.00 cm apart (Fig. P23.62). The plates are insulating and have uniform

Figure **P23.62**

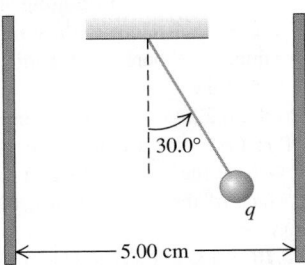

surface charge densities $+\sigma$ and $-\sigma$. The charge on the sphere is $q = 8.90 \times 10^{-6}$ C. What potential difference between the plates will cause the thread to assume an angle of 30.0° with the vertical?

23.63 • **CALC Coaxial Cylinders.** A long metal cylinder with radius a is supported on an insulating stand on the axis of a long, hollow, metal tube with radius b. The positive charge per unit length on the inner cylinder is λ, and there is an equal negative charge per unit length on the outer cylinder. (a) Calculate the potential $V(r)$ for (i) $r < a$; (ii) $a < r < b$; (iii) $r > b$. (*Hint:* The net potential is the sum of the potentials due to the individual conductors.) Take $V = 0$ at $r = b$. (b) Show that the potential of the inner cylinder with respect to the outer is

$$V_{ab} = \frac{\lambda}{2\pi\epsilon_0}\ln\frac{b}{a}$$

(c) Use Eq. (23.23) and the result from part (a) to show that the electric field at any point between the cylinders has magnitude

$$E(r) = \frac{V_{ab}}{\ln(b/a)}\frac{1}{r}$$

(d) What is the potential difference between the two cylinders if the outer cylinder has no net charge?

23.64 •• A *Geiger counter* detects radiation such as alpha particles by using the fact that the radiation ionizes the air along its path. A thin wire lies on the axis of a hollow metal cylinder and is insulated from it (Fig. P23.64). A large potential difference is established between the wire and the outer cylinder, with the wire at higher potential; this sets up a strong electric field directed radially outward. When ionizing radiation enters the device, it ionizes a few air molecules. The free electrons produced are accelerated by the electric field toward the wire and, on the way there, ionize many more air molecules. Thus a current pulse is produced that can be detected by appropriate electronic circuitry and converted to an audible "click." Suppose the radius of the central wire is 145 μm and the radius of the hollow cylinder is 1.80 cm. What potential difference between the wire and the cylinder produces an electric field of 2.00×10^4 V/m at a distance of 1.20 cm from the axis of the wire? (The wire and cylinder are both very long in comparison to their radii, so the results of Problem 23.63 apply.)

Figure **P23.64**

Radiation

Free electron

Counter

23.65 • **CP Deflection in a CRT.** Cathode-ray tubes (CRTs) are often found in oscilloscopes and computer monitors. In Fig. P23.65 an electron with an initial speed of 6.50×10^6 m/s is projected along the axis midway between the deflection plates of a cathode-ray tube. The potential difference between the two plates is 22.0 V and the lower plate is the one at higher potential. (a) What is the force (magnitude and direction) on the electron when it is between the plates? (b) What is the acceleration of the electron (magnitude and direction) when acted on by the force in part (a)? (c) How far below the axis has the electron moved when it reaches the end of the plates? (d) At what angle with the axis is it moving as it leaves the plates? (e) How far below the axis will it strike the fluorescent screen S?

Figure **P23.65**

23.66 •• **CP** **Deflecting Plates of an Oscilloscope.** The vertical deflecting plates of a typical classroom oscilloscope are a pair of parallel square metal plates carrying equal but opposite charges. Typical dimensions are about 3.0 cm on a side, with a separation of about 5.0 mm. The potential difference between the plates is 25.0 V. The plates are close enough that we can ignore fringing at the ends. Under these conditions: (a) how much charge is on each plate, and (b) how strong is the electric field between the plates? (c) If an electron is ejected at rest from the negative plate, how fast is it moving when it reaches the positive plate?

23.67 •• *Electrostatic precipitators* use electric forces to remove pollutant particles from smoke, in particular in the smokestacks of coal-burning power plants. One form of precipitator consists of a vertical, hollow, metal cylinder with a thin wire, insulated from the cylinder, running along its axis (Fig. P23.67). A large potential difference is established between the wire and the outer cylinder, with the wire at lower potential. This sets up a strong radial electric field directed inward. The field produces a region of ionized air near the wire. Smoke enters the precipitator at the bottom, ash and dust in it pick up electrons, and the charged pollutants are accelerated toward the outer cylinder wall by the electric field. Suppose the radius of the central wire is 90.0 μm, the radius of the cylinder is 14.0 cm, and a potential difference of 50.0 kV is established between the wire and the cylinder. Also assume that the wire and cylinder are both very long in comparison to the cylinder radius, so the results of Problem 23.63 apply. (a) What is the magnitude of the electric field midway between the wire and the cylinder wall? (b) What magnitude of charge must a 30.0-μg ash particle have if the electric field computed in part (a) is to exert a force ten times the weight of the particle?

Figure **P23.67**

23.68 •• **CALC** A disk with radius R has uniform surface charge density σ. (a) By regarding the disk as a series of thin concentric rings, calculate the electric potential V at a point on the disk's axis a distance x from the center of the disk. Assume that the potential is zero at infinity. (*Hint:* Use the result of Example 23.11 in Section 23.3.) (b) Calculate $-\partial V/\partial x$. Show that the result agrees with the expression for E_x calculated in Example 21.11 (Section 21.5).

23.69 •• **CALC** (a) From the expression for E obtained in Problem 22.42, find the expressions for the electric potential V as a function of r, both inside and outside the cylinder. Let $V = 0$ at the surface of the cylinder. In each case, express your result in terms of the charge per unit length λ of the charge distribution. (b) Graph V and E as functions of r from $r = 0$ to $r = 3R$.

23.70 • **CALC** A thin insulating rod is bent into a semicircular arc of radius a, and a total electric charge Q is distributed uniformly

along the rod. Calculate the potential at the center of curvature of the arc if the potential is assumed to be zero at infinity.

23.71 ••• **CALC** **Self-Energy of a Sphere of Charge.** A solid sphere of radius R contains a total charge Q distributed uniformly throughout its volume. Find the energy needed to assemble this charge by bringing infinitesimal charges from far away. This energy is called the "self-energy" of the charge distribution. (*Hint:* After you have assembled a charge q in a sphere of radius r, how much energy would it take to add a spherical shell of thickness dr having charge dq? Then integrate to get the total energy.)

23.72 •• **CALC** (a) From the expression for E obtained in Example 22.9 (Section 22.4), find the expression for the electric potential V as a function of r both inside and outside the uniformly charged sphere. Assume that $V = 0$ at infinity. (b) Graph V and E as functions of r from $r = 0$ to $r = 3R$.

23.73 •• Charge $Q = +4.00 \ \mu$C is distributed uniformly over the volume of an insulating sphere that has radius $R = 5.00$ cm. What is the potential difference between the center of the sphere and the surface of the sphere?

23.74 • An insulating spherical shell with inner radius 25.0 cm and outer radius 60.0 cm carries a charge of $+150.0 \ \mu$C uniformly distributed over its outer surface (see Exercise 23.41). Point a is at the center of the shell, point b is on the inner surface, and point c is on the outer surface. (a) What will a voltmeter read if it is connected between the following points: (i) a and b; (ii) b and c; (iii) c and infinity; (iv) a and c? (b) Which is at higher potential: (i) a or b; (ii) b or c; (iii) a or c? (c) Which, if any, of the answers would change sign if the charge were $-150 \ \mu$C?

23.75 •• Exercise 23.41 shows that, outside a spherical shell with uniform surface charge, the potential is the same as if all the charge were concentrated into a point charge at the center of the sphere. (a) Use this result to show that for two uniformly charged insulating shells, the force they exert on each other and their mutual electrical energy are the same as if all the charge were concentrated at their centers. (*Hint:* See Section 13.6.) (b) Does this same result hold for solid insulating spheres, with charge distributed uniformly throughout their volume? (c) Does this same result hold for the force between two charged conducting shells? Between two charged solid conductors? Explain.

23.76 •• **CP** Two plastic spheres, each carrying charge uniformly distributed throughout its interior, are initially placed in contact and then released. One sphere is 60.0 cm in diameter, has mass 50.0 g, and contains $-10.0 \ \mu$C of charge. The other sphere is 40.0 cm in diameter, has mass 150.0 g, and contains $-30.0 \ \mu$C of charge. Find the maximum acceleration and the maximum speed achieved by each sphere (relative to the fixed point of their initial location in space), assuming that no other forces are acting on them. (*Hint:* The uniformly distributed charges behave as though they were concentrated at the centers of the two spheres.)

23.77 • **CALC** Use the electric field calculated in Problem 22.45 to calculate the potential difference between the solid conducting sphere and the thin insulating shell.

23.78 • **CALC** Consider a solid conducting sphere inside a hollow conducting sphere, with radii and charges specified in Problem 22.44. Take $V = 0$ as $r \rightarrow \infty$. Use the electric field calculated in Problem 22.44 to calculate the potential V at the following values of r: (a) $r = c$ (at the outer surface of the hollow sphere); (b) $r = b$ (at the inner surface of the hollow sphere); (c) $r = a$ (at the surface of the solid sphere); (d) $r = 0$ (at the center of the solid sphere).

23.79 • **CALC** Electric charge is distributed uniformly along a thin rod of length a, with total charge Q. Take the potential to be zero at infinity. Find the potential at the following points (Fig. P23.79): (a) point P, a distance x to the right of the rod, and (b) point R, a distance y above the right-hand end of the rod. (c) In parts (a) and (b), what does your result reduce to as x or y becomes much larger than a?

Figure **P23.79**

23.80 • (a) If a spherical raindrop of radius 0.650 mm carries a charge of -3.60 pC uniformly distributed over its volume, what is the potential at its surface? (Take the potential to be zero at an infinite distance from the raindrop.) (b) Two identical raindrops, each with radius and charge specified in part (a), collide and merge into one larger raindrop. What is the radius of this larger drop, and what is the potential at its surface, if its charge is uniformly distributed over its volume?

23.81 •• Two metal spheres of different sizes are charged such that the electric potential is the same at the surface of each. Sphere A has a radius three times that of sphere B. Let Q_A and Q_B be the charges on the two spheres, and let E_A and E_B be the electric-field magnitudes at the surfaces of the two spheres. What are (a) the ratio Q_B/Q_A and (b) the ratio E_B/E_A?

23.82 • An alpha particle with kinetic energy 11.0 MeV makes a head-on collision with a lead nucleus at rest. What is the distance of closest approach of the two particles? (Assume that the lead nucleus remains stationary and that it may be treated as a point charge. The atomic number of lead is 82. The alpha particle is a helium nucleus, with atomic number 2.)

23.83 • A metal sphere with radius R_1 has a charge Q_1. Take the electric potential to be zero at an infinite distance from the sphere. (a) What are the electric field and electric potential at the surface of the sphere? This sphere is now connected by a long, thin conducting wire to another sphere of radius R_2 that is several meters from the first sphere. Before the connection is made, this second sphere is uncharged. After electrostatic equilibrium has been reached, what are (b) the total charge on each sphere; (c) the electric potential at the surface of each sphere; (d) the electric field at the surface of each sphere? Assume that the amount of charge on the wire is much less than the charge on each sphere.

23.84 ••• **CALC** Use the charge distribution and electric field calculated in Problem 22.65. (a) Show that for $r \geq R$ the potential is identical to that produced by a point charge Q. (Take the potential to be zero at infinity.) (b) Obtain an expression for the electric potential valid in the region $r \leq R$.

23.85 •• **CP Nuclear Fusion in the Sun.** The source of the sun's energy is a sequence of nuclear reactions that occur in its core. The first of these reactions involves the collision of two protons, which fuse together to form a heavier nucleus and release energy. For this process, called *nuclear fusion,* to occur, the two protons must first approach until their surfaces are essentially in contact. (a) Assume both protons are moving with the same speed and they collide head-on. If the radius of the proton is 1.2×10^{-15} m, what is the minimum speed that will allow fusion to occur? The charge distribution within a proton is spherically symmetric, so the electric field and potential outside a proton are the same as if it were a point charge. The mass of the proton is 1.67×10^{-27} kg. (b) Another nuclear fusion reaction that occurs in the sun's core involves a collision between two helium nuclei, each of which has 2.99 times the mass of the proton, charge $+2e$, and radius 1.7×10^{-15} m. Assuming the same collision geometry as in part (a), what minimum speed is required for this fusion reaction to take place if the nuclei must approach a center-to-center

distance of about 3.5×10^{-15} m? As for the proton, the charge of the helium nucleus is uniformly distributed throughout its volume. (c) In Section 18.3 it was shown that the average translational kinetic energy of a particle with mass m in a gas at absolute temperature T is $\frac{3}{2}kT$, where k is the Boltzmann constant (given in Appendix F). For two protons with kinetic energy equal to this average value to be able to undergo the process described in part (a), what absolute temperature is required? What absolute temperature is required for two average helium nuclei to be able to undergo the process described in part (b)? (At these temperatures, atoms are completely ionized, so nuclei and electrons move separately.) (d) The temperature in the sun's core is about 1.5×10^7 K. How does this compare to the temperatures calculated in part (c)? How can the reactions described in parts (a) and (b) occur at all in the interior of the sun? (*Hint:* See the discussion of the distribution of molecular speeds in Section 18.5.)

23.86 • **CALC** The electric potential V in a region of space is given by

$$V(x, y, z) = A(x^2 - 3y^2 + z^2)$$

where A is a constant. (a) Derive an expression for the electric field $\vec{E}$ at any point in this region. (b) The work done by the field when a $1.50\text{-}\mu C$ test charge moves from the point $(x, y, z) = (0, 0, 0.250 \text{ m})$ to the origin is measured to be 6.00×10^{-5} J. Determine A. (c) Determine the electric field at the point $(0, 0, 0.250 \text{ m})$. (d) Show that in every plane parallel to the xz-plane the equipotential contours are circles. (e) What is the radius of the equipotential contour corresponding to $V = 1280$ V and $y = 2.00$ m?

23.87 •• **Nuclear Fission.** The unstable nucleus of uranium-236 can be regarded as a uniformly charged sphere of charge $Q = +92e$ and radius $R = 7.4 \times 10^{-15}$ m. In nuclear fission, this can divide into two smaller nuclei, each with half the charge and half the volume of the original uranium-236

Figure **P23.87**

$Q = +92e$

BEFORE

$Q = +46e$ ⇩ $Q = +46e$

AFTER

nucleus. This is one of the reactions that occurred in the nuclear weapon that exploded over Hiroshima, Japan, in August 1945. (a) Find the radii of the two "daughter" nuclei of charge $+46e$. (b) In a simple model for the fission process, immediately after the uranium-236 nucleus has undergone fission, the "daughter" nuclei are at rest and just touching, as shown in Fig. P23.87. Calculate the kinetic energy that each of the "daughter" nuclei will have when they are very far apart. (c) In this model the sum of the kinetic energies of the two "daughter" nuclei, calculated in part (b), is the energy released by the fission of one uranium-236 nucleus. Calculate the energy released by the fission of 10.0 kg of uranium-236. The atomic mass of uranium-236 is 236 u, where 1 u = 1 atomic mass unit = 1.66×10^{-24} kg. Express your answer both in joules and in kilotons of TNT (1 kiloton of TNT releases 4.18×10^{12} J when it explodes). (d) In terms of this model, discuss why an atomic bomb could just as well be called an "electric bomb."

CHALLENGE PROBLEMS

23.88 ••• **CP CALC** In a certain region, a charge distribution exists that is spherically symmetric but nonuniform. That is, the volume charge density $\rho(r)$ depends on the distance r from the center of the distribution but not on the spherical polar angles θ and ϕ. The electric potential $V(r)$ due to this charge distribution is

$$V(r) = \begin{cases} \dfrac{\rho_0 a^2}{18\epsilon_0}\left[1 - 3\left(\dfrac{r}{a}\right)^2 + 2\left(\dfrac{r}{a}\right)^3\right] & \text{for } r \le a \\ 0 & \text{for } r \ge a \end{cases}$$

where ρ_0 is a constant having units of C/m^3 and a is a constant having units of meters. (a) Derive expressions for $\vec{E}$ for the regions $r \le a$ and $r \ge a$. [*Hint:* Use Eq. (23.23).] Explain why $\vec{E}$ has only a radial component. (b) Derive an expression for $\rho(r)$ in each of the two regions $r \le a$ and $r \ge a$. [*Hint:* Use Gauss's law for two spherical shells, one of radius r and the other of radius $r + dr$. The charge contained in the infinitesimal spherical shell of radius dr is $dq = 4\pi r^2 \rho(r)\, dr$.] (c) Show that the net charge contained in the volume of a sphere of radius greater than or equal to a is zero. [*Hint:* Integrate the expressions derived in part (b) for $\rho(r)$ over a spherical volume of radius greater than or equal to a.] Is this result consistent with the electric field for $r > a$ that you calculated in part (a)?

23.89 ••• **CP** In experiments in which atomic nuclei collide, head-on collisions like that described in Problem 23.82 do happen, but "near misses" are more common. Suppose the alpha particle in Problem 23.82 was not "aimed" at the center of the lead nucleus, but had an initial nonzero angular momentum (with respect to the stationary lead nucleus) of magnitude $L = p_0 b$, where p_0 is the magnitude of the initial momentum of the alpha particle and $b = 1.00 \times 10^{-12}$ m. What is the distance of closest approach? Repeat for $b = 1.00 \times 10^{-13}$ m and $b = 1.00 \times 10^{-14}$ m.

23.90 ••• **CALC** A hollow, thin-walled insulating cylinder of radius R and length L (like the cardboard tube in a roll of toilet paper) has charge Q uniformly distributed over its surface. (a) Calculate the electric potential at all points along the axis of the tube. Take the origin to be at the center of the tube, and take the potential to be zero at infinity. (b) Show that if $L \ll R$, the result of part (a) reduces to the potential on the axis of a ring of charge of radius R. (See Example 23.11 in Section 23.3.) (c) Use the result of part (a) to find the electric field at all points along the axis of the tube.

23.91 ••• **The Millikan Oil-Drop Experiment.** The charge of an electron was first measured by the American physicist Robert Millikan during 1909–1913. In his experiment, oil is sprayed in very fine drops (around 10^{-4} mm in diameter) into the space between two parallel horizontal plates separated by a distance d. A potential difference V_{AB} is maintained between the parallel plates, causing a downward electric field between them. Some of the oil drops acquire a negative charge because of frictional effects or because of ionization of the surrounding air by x rays or radioactivity. The drops are observed through a microscope. (a) Show that an oil drop of radius r at rest between the plates will remain at rest if the magnitude of its charge is

$$q = \frac{4\pi}{3}\frac{\rho r^3 g d}{V_{AB}}$$

where ρ is the density of the oil. (Ignore the buoyant force of the air.) By adjusting V_{AB} to keep a given drop at rest, the charge on that drop can be determined, provided its radius is known. (b) Millikan's oil drops were much too small to measure their radii directly. Instead, Millikan determined r by cutting off the electric field and measuring the *terminal speed* v_t of the drop as it fell. (We discussed the concept of terminal speed in Section 5.3.) The viscous force F on a sphere of radius r moving with speed v through a fluid with viscosity η is given by Stokes's law: $F = 6\pi\eta r v$. When the drop is falling at v_t, the viscous force just balances the weight $w = mg$ of the drop. Show that the magnitude of the charge on the drop is

$$q = 18\pi \frac{d}{V_{AB}}\sqrt{\frac{\eta^3 v_t^3}{2\rho g}}$$

Within the limits of their experimental error, every one of the thousands of drops that Millikan and his coworkers measured had a charge equal to some small integer multiple of a basic charge e. That is, they found drops with charges of $\pm 2e$, $\pm 5e$, and so on, but none with values such as $0.76e$ or $2.49e$. A drop with charge $-e$ has acquired one extra electron; if its charge is $-2e$, it has acquired two extra electrons, and so on. (c) A charged oil drop in a Millikan oil-drop apparatus is observed to fall 1.00 mm at constant speed in 39.3 s if $V_{AB} = 0$. The same drop can be held at rest between two plates separated by 1.00 mm if $V_{AB} = 9.16$ V. How many excess electrons has the drop acquired, and what is the radius of the drop? The viscosity of air is 1.81×10^{-5} N $\cdot$ s/m², and the density of the oil is 824 kg/m³.

23.92 •• **CP** Two point charges are moving to the right along the x-axis. Point charge 1 has charge $q_1 = 2.00\ \mu C$, mass $m_1 = 6.00 \times 10^{-5}$ kg, and speed v_1. Point charge 2 is to the right of q_1 and has charge $q_2 = -5.00\ \mu C$, mass $m_2 = 3.00 \times 10^{-5}$ kg, and speed v_2. At a particular instant, the charges are separated by a distance of 9.00 mm and have speeds $v_1 = 400$ m/s and $v_2 = 1300$ m/s. The only forces on the particles are the forces they exert on each other. (a) Determine the speed v_{cm} of the center of mass of the system. (b) The *relative energy* E_{rel} of the system is defined as the total energy minus the kinetic energy contributed by the motion of the center of mass:

$$E_{rel} = E - \tfrac{1}{2}(m_1 + m_2)v_{cm}^2$$

where $E = \tfrac{1}{2}m_1 v_1^2 + \tfrac{1}{2}m_2 v_2^2 + q_1 q_2/4\pi\epsilon_0 r$ is the total energy of the system and r is the distance between the charges. Show that $E_{rel} = \tfrac{1}{2}\mu v^2 + q_1 q_2/4\pi\epsilon_0 r$, where $\mu = m_1 m_2/(m_1 + m_2)$ is called the *reduced mass* of the system and $v = v_2 - v_1$ is the relative speed of the moving particles. (c) For the numerical values given above, calculate the numerical value of E_{rel}. (d) Based on the result of part (c), for the conditions given above, will the particles escape from one another? Explain. (e) If the particles do escape, what will be their final relative speed when $r \to \infty$? If the particles do not escape, what will be their distance of maximum separation? That is, what will be the value of r when $v = 0$? (f) Repeat parts (c)–(e) for $v_1 = 400$ m/s and $v_2 = 1800$ m/s when the separation is 9.00 mm.

Answers

Chapter Opening Question ?

A large, constant potential difference V_{ab} is maintained between the welding tool (a) and the metal pieces to be welded (b). From Example 23.9 (Section 23.3) the electric field between two conductors separated by a distance d has magnitude $E = V_{ab}/d$. Hence d must be small in order for the field magnitude E to be large enough to ionize the gas between the conductors a and b (see Section 23.3) and produce an arc through this gas.

Test Your Understanding Questions

23.1 Answers: (a) (i), (b) (ii) The three charges q_1, q_2, and q_3 are all positive, so all three of the terms in the sum in Eq. (23.11)— q_1q_2/r_{12}, q_1q_3/r_{13}, and q_2q_3/r_{23}—are positive. Hence the total electric potential energy U is positive. This means that it would take positive work to bring the three charges from infinity to the positions shown in Fig. 21.14, and hence *negative* work to move the three charges from these positions back to infinity.

23.2 Answer: no If $V = 0$ at a certain point, $\vec{E}$ does *not* have to be zero at that point. An example is point c in Figs. 21.23 and 23.13, for which there is an electric field in the $+x$-direction (see Example 21.9 in Section 21.5) even though $V = 0$ (see Example 23.4). This isn't a surprising result because V and $\vec{E}$ are quite different quantities: V is the net amount of work required to bring a unit charge from infinity to the point in question, whereas $\vec{E}$ is the electric force that acts on a unit charge when it arrives at that point.

23.3 Answer: no If $\vec{E} = 0$ at a certain point, V does *not* have to be zero at that point. An example is point O at the center of the charged ring in Figs. 21.23 and 23.21. From Example 21.9 (Section 21.5), the electric field is zero at O because the electric-field contributions from different parts of the ring completely cancel. From Example 23.11, however, the potential at O is *not* zero: This point corresponds to $x = 0$, so $V = (1/4\pi\epsilon_0)(Q/a)$. This value of V corresponds to the work that would have to be done to move a unit positive test charge along a path from infinity to point O; it is nonzero because the charged ring repels the test charge, so positive work must be done to move the test charge toward the ring.

23.4 Answer: no If the positive charges in Fig. 23.23 were replaced by negative charges, and vice versa, the equipotential surfaces would be the same but the sign of the potential would be reversed. For example, the surfaces in Fig. 23.23b with potential $V = +30$ V and $V = -50$ V would have potential $V = -30$ V and $V = +50$ V, respectively.

23.5 Answer: (iii) From Eqs. (23.19), the components of the electric field are $E_x = -\partial V/\partial x = B + Dy$, $E_y = -\partial V/\partial y = 3Cy^2 + Dx$, and $E_z = -\partial V/\partial z = 0$. The value of A has no effect, which means that we can add a constant to the electric potential at all points without changing $\vec{E}$ or the potential difference between two points. The potential does not depend on z, so the z-component of $\vec{E}$ is zero. Note that at the origin the electric field is not zero because it has a nonzero x-component: $E_x = B$, $E_y = 0$, $E_z = 0$.

Bridging Problem

Answer: $\dfrac{qQ}{8\pi\epsilon_0 a} \ln\left(\dfrac{L+a}{L-a}\right)$

24 CAPACITANCE AND DIELECTRICS

? The energy used in a camera's flash unit is stored in a capacitor, which consists of two closely spaced conductors that carry opposite charges. If the amount of charge on the conductors is doubled, by what factor does the stored energy increase?

LEARNING GOALS

By studying this chapter, you will learn:

- The nature of capacitors, and how to calculate a quantity that measures their ability to store charge.

- How to analyze capacitors connected in a network.

- How to calculate the amount of energy stored in a capacitor.

- What dielectrics are, and how they make capacitors more effective.

When you set an old-fashioned spring mousetrap or pull back the string of an archer's bow, you are storing mechanical energy as elastic potential energy. A capacitor is a device that stores *electric* potential energy and electric charge. To make a capacitor, just insulate two conductors from each other. To store energy in this device, transfer charge from one conductor to the other so that one has a negative charge and the other has an equal amount of positive charge. Work must be done to move the charges through the resulting potential difference between the conductors, and the work done is stored as electric potential energy.

Capacitors have a tremendous number of practical applications in devices such as electronic flash units for photography, pulsed lasers, air bag sensors for cars, and radio and television receivers. We'll encounter many of these applications in later chapters (particularly Chapter 31, in which we'll see the crucial role played by capacitors in the alternating-current circuits that pervade our technological society). In this chapter, however, our emphasis is on the fundamental properties of capacitors. For a particular capacitor, the ratio of the charge on each conductor to the potential difference between the conductors is a constant, called the *capacitance*. The capacitance depends on the sizes and shapes of the conductors and on the insulating material (if any) between them. Compared to the case in which there is only vacuum between the conductors, the capacitance increases when an insulating material (a *dielectric*) is present. This happens because a redistribution of charge, called *polarization*, takes place within the insulating material. Studying polarization will give us added insight into the electrical properties of matter.

Capacitors also give us a new way to think about electric potential energy. The energy stored in a charged capacitor is related to the electric field in the space between the conductors. We will see that electric potential energy can be regarded as being stored *in the field itself*. The idea that the electric field is itself a storehouse of energy is at the heart of the theory of electromagnetic waves and our modern understanding of the nature of light, to be discussed in Chapter 32.

24.1 Capacitors and Capacitance

Any two conductors separated by an insulator (or a vacuum) form a **capacitor** (Fig. 24.1). In most practical applications, each conductor initially has zero net charge and electrons are transferred from one conductor to the other; this is called *charging* the capacitor. Then the two conductors have charges with equal magnitude and opposite sign, and the *net* charge on the capacitor as a whole remains zero. We will assume throughout this chapter that this is the case. When we say that a capacitor has charge Q, or that a charge Q is *stored* on the capacitor, we mean that the conductor at higher potential has charge $+Q$ and the conductor at lower potential has charge $-Q$ (assuming that Q is positive). Keep this in mind in the following discussion and examples.

In circuit diagrams a capacitor is represented by either of these symbols:

In either symbol the vertical lines (straight or curved) represent the conductors and the horizontal lines represent wires connected to either conductor. One common way to charge a capacitor is to connect these two wires to opposite terminals of a battery. Once the charges Q and $-Q$ are established on the conductors, the battery is disconnected. This gives a fixed *potential difference* V_{ab} between the conductors (that is, the potential of the positively charged conductor a with respect to the negatively charged conductor b) that is just equal to the voltage of the battery.

The electric field at any point in the region between the conductors is proportional to the magnitude Q of charge on each conductor. It follows that the potential difference V_{ab} between the conductors is also proportional to Q. If we double the magnitude of charge on each conductor, the charge density at each point doubles, the electric field at each point doubles, and the potential difference between conductors doubles; however, the *ratio* of charge to potential difference does not change. This ratio is called the **capacitance** C of the capacitor:

$$C = \frac{Q}{V_{ab}} \qquad \text{(definition of capacitance)} \qquad (24.1)$$

The SI unit of capacitance is called one **farad** (1 F), in honor of the 19th-century English physicist Michael Faraday. From Eq. (24.1), one farad is equal to one *coulomb per volt* (1 C/V):

$$1 \text{ F} = 1 \text{ farad} = 1 \text{ C/V} = 1 \text{ coulomb/volt}$$

CAUTION **Capacitance vs. coulombs** Don't confuse the symbol C for capacitance (which is always in italics) with the abbreviation C for coulombs (which is never italicized). ▮

The greater the capacitance C of a capacitor, the greater the magnitude Q of charge on either conductor for a given potential difference V_{ab} and hence the greater the amount of stored energy. (Remember that potential is potential energy per unit charge.) Thus *capacitance is a measure of the ability of a capacitor to store energy.* We will see that the value of the capacitance depends only on the shapes and sizes of the conductors and on the nature of the insulating material between them. (The above remarks about capacitance being independent of Q and V_{ab} do not apply to certain special types of insulating materials. We won't discuss these materials in this book, however.)

Calculating Capacitance: Capacitors in Vacuum

We can calculate the capacitance C of a given capacitor by finding the potential difference V_{ab} between the conductors for a given magnitude of charge Q and

24.1 Any two conductors a and b insulated from each other form a capacitor.

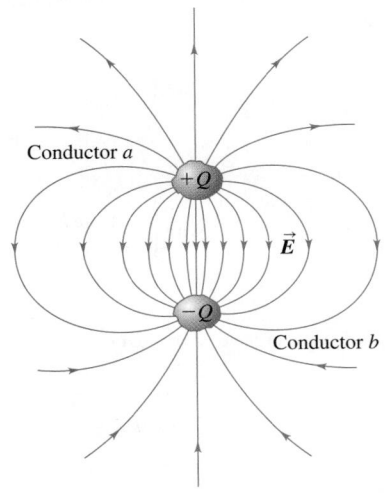

Conductor a

$\vec{E}$

Conductor b

Mastering**PHYSICS**

ActivPhysics 11.11.6: Electric Potential: Qualitative Introduction
ActivPhysics 11.12.1 and 11.12.3: Electric Potential, Field, and Force

24.2 A charged parallel-plate capacitor.

(a) Arrangement of the capacitor plates

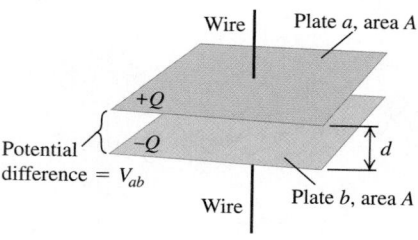

Potential difference = V_{ab}

(b) Side view of the electric field $\vec{E}$

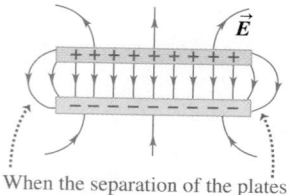

When the separation of the plates is small compared to their size, the fringing of the field is slight.

24.3 Inside a condenser microphone is a capacitor with one rigid plate and one flexible plate. The two plates are kept at a constant potential difference V_{ab}. Sound waves cause the flexible plate to move back and forth, varying the capacitance C and causing charge to flow to and from the capacitor in accordance with the relationship $C = Q/V_{ab}$. Thus a sound wave is converted to a charge flow that can be amplified and recorded digitally.

then using Eq. (24.1). For now we'll consider only *capacitors in vacuum;* that is, we'll assume that the conductors that make up the capacitor are separated by empty space.

The simplest form of capacitor consists of two parallel conducting plates, each with area A, separated by a distance d that is small in comparison with their dimensions (Fig. 24.2a). When the plates are charged, the electric field is almost completely localized in the region between the plates (Fig. 24.2b). As we discussed in Example 22.8 (Section 22.4), the field between such plates is essentially *uniform,* and the charges on the plates are uniformly distributed over their opposing surfaces. We call this arrangement a **parallel-plate capacitor.**

We worked out the electric-field magnitude E for this arrangement in Example 21.12 (Section 21.5) using the principle of superposition of electric fields and again in Example 22.8 (Section 22.4) using Gauss's law. It would be a good idea to review those examples. We found that $E = \sigma/\epsilon_0$, where σ is the magnitude (absolute value) of the surface charge density on each plate. This is equal to the magnitude of the total charge Q on each plate divided by the area A of the plate, or $\sigma = Q/A$, so the field magnitude E can be expressed as

$$E = \frac{\sigma}{\epsilon_0} = \frac{Q}{\epsilon_0 A}$$

The field is uniform and the distance between the plates is d, so the potential difference (voltage) between the two plates is

$$V_{ab} = Ed = \frac{1}{\epsilon_0} \frac{Qd}{A}$$

From this we see that the capacitance C of a parallel-plate capacitor in vacuum is

$$C = \frac{Q}{V_{ab}} = \epsilon_0 \frac{A}{d} \qquad \text{(capacitance of a parallel-plate capacitor in vacuum)} \qquad (24.2)$$

The capacitance depends only on the geometry of the capacitor; it is directly proportional to the area A of each plate and inversely proportional to their separation d. The quantities A and d are constants for a given capacitor, and ϵ_0 is a universal constant. Thus in vacuum the capacitance C is a constant independent of the charge on the capacitor or the potential difference between the plates. If one of the capacitor plates is flexible, the capacitance C changes as the plate separation d changes. This is the operating principle of a condenser microphone (Fig. 24.3).

When matter is present between the plates, its properties affect the capacitance. We will return to this topic in Section 24.4. Meanwhile, we remark that if the space contains air at atmospheric pressure instead of vacuum, the capacitance differs from the prediction of Eq. (24.2) by less than 0.06%.

In Eq. (24.2), if A is in square meters and d in meters, C is in farads. The units of ϵ_0 are $C^2/N \cdot m^2$, so we see that

$$1\ F = 1\ C^2/N \cdot m = 1\ C^2/J$$

Because $1\ V = 1\ J/C$ (energy per unit charge), this is consistent with our definition $1\ F = 1\ C/V$. Finally, the units of ϵ_0 can be expressed as $1\ C^2/N \cdot m^2 = 1\ F/m$, so

$$\epsilon_0 = 8.85 \times 10^{-12}\ F/m$$

This relationship is useful in capacitance calculations, and it also helps us to verify that Eq. (24.2) is dimensionally consistent.

One farad is a very large capacitance, as the following example shows. In many applications the most convenient units of capacitance are the *microfarad*

(1 μF $= 10^{-6}$ F) and the *picofarad* (1 pF $= 10^{-12}$ F). For example, the flash unit in a point-and-shoot camera uses a capacitor of a few hundred microfarads (Fig. 24.4), while capacitances in a radio tuning circuit are typically from 10 to 100 picofarads.

For *any* capacitor in vacuum, the capacitance C depends only on the shapes, dimensions, and separation of the conductors that make up the capacitor. If the conductor shapes are more complex than those of the parallel-plate capacitor, the expression for capacitance is more complicated than in Eq. (24.2). In the following examples we show how to calculate C for two other conductor geometries.

24.4 A commercial capacitor is labeled with the value of its capacitance. For these capacitors, $C = 2200$ μF, 1000 μF, and 470 μF.

Example 24.1   **Size of a 1-F capacitor**

The parallel plates of a 1.0-F capacitor are 1.0 mm apart. What is their area?

SOLUTION

IDENTIFY and SET UP: This problem uses the relationship among the capacitance C, plate separation d, and plate area A (our target variable) for a parallel-plate capacitor. We solve Eq. (24.2) for A.

EXECUTE: From Eq. (24.2),

$$A = \frac{Cd}{\epsilon_0} = \frac{(1.0\ \text{F})(1.0 \times 10^{-3}\ \text{m})}{8.85 \times 10^{-12}\ \text{F/m}} = 1.1 \times 10^{8}\ \text{m}^2$$

EVALUATE: This corresponds to a square about 10 km (about 6 miles) on a side! The volume of such a capacitor would be at least $Ad = 1.1 \times 10^5$ m^3, equivalent to that of a cube about 50 m on a side. In fact, it's possible to make 1-F capacitors a few *centimeters* on a side. The trick is to have an appropriate substance between the plates rather than a vacuum, so that (among other things) the plate separation d can greatly reduced. We'll explore this further in Section 24.4.

Example 24.2   **Properties of a parallel-plate capacitor**

The plates of a parallel-plate capacitor in vacuum are 5.00 mm apart and 2.00 m^2 in area. A 10.0-kV potential difference is applied across the capacitor. Compute (a) the capacitance; (b) the charge on each plate; and (c) the magnitude of the electric field between the plates.

SOLUTION

IDENTIFY and SET UP: We are given the plate area A, the plate spacing d, and the potential difference $V_{ab} = 1.00 \times 10^4$ V for this parallel-plate capacitor. Our target variables are the capacitance C, the charge Q on each plate, and the electric-field magnitude E. We use Eq. (24.2) to calculate C and then use Eq. (24.1) and V_{ab} to find Q. We use $E = Q/\epsilon_0 A$ to find E.

EXECUTE: (a) From Eq. (24.2),

$$C = \epsilon_0 \frac{A}{d} = (8.85 \times 10^{-12}\ \text{F/m})\frac{(2.00\ \text{m}^2)}{5.00 \times 10^{-3}\ \text{m}}$$
$$= 3.54 \times 10^{-9}\ \text{F} = 0.00354\ \mu\text{F}$$

(b) The charge on the capacitor is

$$Q = CV_{ab} = (3.54 \times 10^{-9}\ \text{C/V})(1.00 \times 10^4\ \text{V})$$
$$= 3.54 \times 10^{-5}\ \text{C} = 35.4\ \mu\text{C}$$

The plate at higher potential has charge $+35.4$ μC, and the other plate has charge -35.4 μC.

(c) The electric-field magnitude is

$$E = \frac{\sigma}{\epsilon_0} = \frac{Q}{\epsilon_0 A} = \frac{3.54 \times 10^{-5}\ \text{C}}{(8.85 \times 10^{-12}\ \text{C}^2/\text{N} \cdot \text{m}^2)(2.00\ \text{m}^2)}$$
$$= 2.00 \times 10^6\ \text{N/C}$$

EVALUATE: We can also find E by recalling that the electric field is equal in magnitude to the potential gradient [Eq. (23.22)]. The field between the plates is uniform, so

$$E = \frac{V_{ab}}{d} = \frac{1.00 \times 10^4\ \text{V}}{5.00 \times 10^{-3}\ \text{m}} = 2.00 \times 10^6\ \text{V/m}$$

(Remember that 1 N/C $=$ 1 V/m.)

Example 24.3 **A spherical capacitor**

Two concentric spherical conducting shells are separated by vacuum (Fig. 24.5). The inner shell has total charge $+Q$ and outer radius r_a, and the outer shell has charge $-Q$ and inner radius r_b. Find the capacitance of this spherical capacitor.

SOLUTION

IDENTIFY and SET UP: By definition, the capacitance C is the magnitude Q of the charge on either sphere divided by the potential difference V_{ab} between the spheres. We first find V_{ab}, and then use Eq. (24.1) to find the capacitance $C = Q/V_{ab}$.

EXECUTE: Using a Gaussian surface such as that shown in Fig. 24.5, we found in Example 22.5 (Section 22.4) that the charge on a conducting sphere produces zero field *inside* the sphere, so the outer sphere makes no contribution to the field between the spheres. Therefore the electric field *and* the electric potential

24.5 A spherical capacitor.

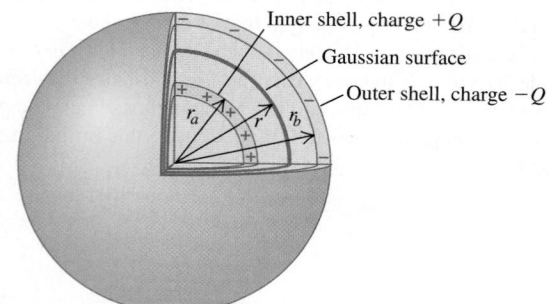

- Inner shell, charge $+Q$
- Gaussian surface
- Outer shell, charge $-Q$

between the shells are the same as those outside a charged conducting sphere with charge $+Q$. We considered that problem in Example 23.8 (Section 23.3), so the same result applies here: The potential at any point between the spheres is $V = Q/4\pi\epsilon_0 r$. Hence the potential of the inner (positive) conductor at $r = r_a$ with respect to that of the outer (negative) conductor at $r = r_b$ is

$$V_{ab} = V_a - V_b = \frac{Q}{4\pi\epsilon_0 r_a} - \frac{Q}{4\pi\epsilon_0 r_b}$$

$$= \frac{Q}{4\pi\epsilon_0}\left(\frac{1}{r_a} - \frac{1}{r_b}\right) = \frac{Q}{4\pi\epsilon_0}\frac{r_b - r_a}{r_a r_b}$$

The capacitance is then

$$C = \frac{Q}{V_{ab}} = 4\pi\epsilon_0 \frac{r_a r_b}{r_b - r_a}$$

As an example, if $r_a = 9.5$ cm and $r_b = 10.5$ cm,

$$C = 4\pi(8.85 \times 10^{-12}\ \text{F/m})\frac{(0.095\ \text{m})(0.105\ \text{m})}{0.010\ \text{m}}$$

$$= 1.1 \times 10^{-10}\ \text{F} = 110\ \text{pF}$$

EVALUATE: We can relate our expression for C to that for a parallel-plate capacitor. The quantity $4\pi r_a r_b$ is intermediate between the areas $4\pi r_a^2$ and $4\pi r_b^2$ of the two spheres; in fact, it's the *geometric mean* of these two areas, which we can denote by A_{gm}. The distance between spheres is $d = r_b - r_a$, so we can write $C = 4\pi\epsilon_0 r_a r_b/(r_b - r_a) = \epsilon_0 A_{gm}/d$. This has the same form as for parallel plates: $C = \epsilon_0 A/d$. If the distance between spheres is very small in comparison to their radii, their capacitance is the same as that of parallel plates with the same area and spacing.

Example 24.4 **A cylindrical capacitor**

Two long, coaxial cylindrical conductors are separated by vacuum (Fig. 24.6). The inner cylinder has radius r_a and linear charge density $+\lambda$. The outer cylinder has inner radius r_b and linear charge density $-\lambda$. Find the capacitance per unit length for this capacitor.

SOLUTION

IDENTIFY and SET UP: As in Example 24.3, we use the definition of capacitance, $C = Q/V_{ab}$. We use the result of Example 23.10

24.6 A long cylindrical capacitor. The linear charge density λ is assumed to be positive in this figure. The magnitude of charge in a length L of either cylinder is λL.

(Section 23.3) to find the potential difference V_{ab} between the cylinders, and find the charge Q on a length L of the cylinders from the linear charge density. We then find the corresponding capacitance C using Eq. (24.1). Our target variable is this capacitance divided by L.

EXECUTE: As in Example 24.3, the potential V between the cylinders is not affected by the presence of the charged outer cylinder. Hence our result in Example 23.10 for the potential outside a charged conducting cylinder also holds in this example for potential in the space between the cylinders:

$$V = \frac{\lambda}{2\pi\epsilon_0}\ln\frac{r_0}{r}$$

Here r_0 is the arbitrary, *finite* radius at which $V = 0$. We take $r_0 = r_b$, the radius of the inner surface of the outer cylinder. Then the potential at the outer surface of the inner cylinder (at which $r = r_a$) is just the potential V_{ab} of the inner (positive) cylinder a with respect to the outer (negative) cylinder b:

$$V_{ab} = \frac{\lambda}{2\pi\epsilon_0}\ln\frac{r_b}{r_a}$$

If λ is positive as in Fig. 24.6, then V_{ab} is positive as well: The inner cylinder is at higher potential than the outer.

The total charge Q in a length L is $Q = \lambda L$, so from Eq. (24.1) the capacitance C of a length L is

$$C = \frac{Q}{V_{ab}} = \frac{\lambda L}{\dfrac{\lambda}{2\pi\epsilon_0}\ln\dfrac{r_b}{r_a}} = \frac{2\pi\epsilon_0 L}{\ln(r_b/r_a)}$$

The capacitance per unit length is

$$\frac{C}{L} = \frac{2\pi\epsilon_0}{\ln(r_b/r_a)}$$

Substituting $\epsilon_0 = 8.85 \times 10^{-12}$ F/m $= 8.85$ pF/m, we get

$$\frac{C}{L} = \frac{55.6 \text{ pF/m}}{\ln(r_b/r_a)}$$

EVALUATE: The capacitance of coaxial cylinders is determined entirely by their dimensions, just as for parallel-plate and spherical capacitors. Ordinary coaxial cables are made like this but with an insulating material instead of vacuum between the conductors. A typical cable used for connecting a television to a cable TV feed has a capacitance per unit length of 69 pF/m.

Test Your Understanding of Section 24.1 A capacitor has vacuum in the space between the conductors. If you double the amount of charge on each conductor, what happens to the capacitance? (i) It increases; (ii) it decreases; (iii) it remains the same; (iv) the answer depends on the size or shape of the conductors.

24.7 An assortment of commercially available capacitors.

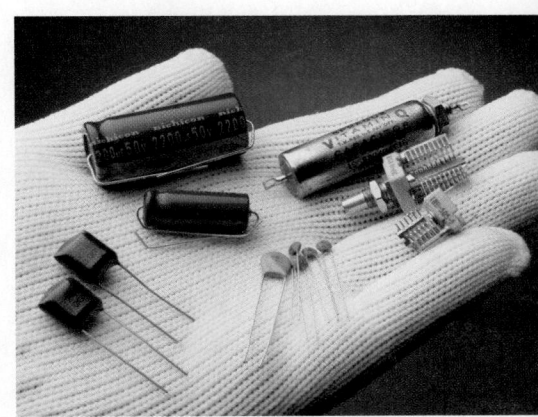

24.2 Capacitors in Series and Parallel

Capacitors are manufactured with certain standard capacitances and working voltages (Fig. 24.7). However, these standard values may not be the ones you actually need in a particular application. You can obtain the values you need by combining capacitors; many combinations are possible, but the simplest combinations are a series connection and a parallel connection.

Capacitors in Series

Figure 24.8a is a schematic diagram of a **series connection.** Two capacitors are connected in series (one after the other) by conducting wires between points a and b. Both capacitors are initially uncharged. When a constant positive potential difference V_{ab} is applied between points a and b, the capacitors become charged; the figure shows that the charge on *all* conducting plates has the same magnitude. To see why, note first that the top plate of C_1 acquires a positive charge Q. The electric field of this positive charge pulls negative charge up to the bottom plate of C_1 until all of the field lines that begin on the top plate end on the bottom plate. This requires that the bottom plate have charge $-Q$. These negative charges had to come from the top plate of C_2, which becomes positively charged with charge $+Q$. This positive charge then pulls negative charge $-Q$ from the connection at point b onto the bottom plate of C_2. The total charge on the lower plate of C_1 and the upper plate of C_2 together must always be zero because these plates aren't connected to anything except each other. Thus *in a series connection the magnitude of charge on all plates is the same.*

Referring to Fig. 24.8a, we can write the potential differences between points a and c, c and b, and a and b as

$$V_{ac} = V_1 = \frac{Q}{C_1} \qquad V_{cb} = V_2 = \frac{Q}{C_2}$$

$$V_{ab} = V = V_1 + V_2 = Q\left(\frac{1}{C_1} + \frac{1}{C_2}\right)$$

and so

$$\frac{V}{Q} = \frac{1}{C_1} + \frac{1}{C_2} \tag{24.3}$$

Following a common convention, we use the symbols V_1, V_2, and V to denote the potential *differences* V_{ac} (across the first capacitor), V_{cb} (across the second capacitor), and V_{ab} (across the entire combination of capacitors), respectively.

24.8 A series connection of two capacitors.

(a) Two capacitors in series

Capacitors in series:
• The capacitors have the same charge Q.
• Their potential differences add:
 $V_{ac} + V_{cb} = V_{ab}$.

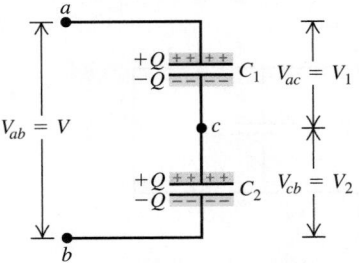

(b) The equivalent single capacitor

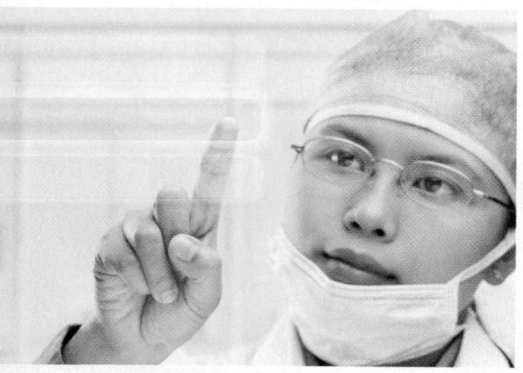

The **equivalent capacitance** C_{eq} of the series combination is defined as the capacitance of a *single* capacitor for which the charge Q is the same as for the combination, when the potential difference V is the same. In other words, the combination can be replaced by an *equivalent capacitor* of capacitance C_{eq}. For such a capacitor, shown in Fig. 24.8b,

$$C_{eq} = \frac{Q}{V} \quad \text{or} \quad \frac{1}{C_{eq}} = \frac{V}{Q} \qquad (24.4)$$

Combining Eqs. (24.3) and (24.4), we find

$$\frac{1}{C_{eq}} = \frac{1}{C_1} + \frac{1}{C_2}$$

We can extend this analysis to any number of capacitors in series. We find the following result for the *reciprocal* of the equivalent capacitance:

$$\frac{1}{C_{eq}} = \frac{1}{C_1} + \frac{1}{C_2} + \frac{1}{C_3} + \cdots \qquad \text{(capacitors in series)} \qquad (24.5)$$

The reciprocal of the equivalent capacitance of a series combination equals the sum of the reciprocals of the individual capacitances. In a series connection the equivalent capacitance is always *less than* any individual capacitance.

CAUTION **Capacitors in series** The magnitude of charge is the same on all plates of all the capacitors in a series combination; however, the potential differences of the individual capacitors are *not* the same unless their individual capacitances are the same. The potential differences of the individual capacitors add to give the total potential difference across the series combination: $V_{total} = V_1 + V_2 + V_3 + \cdots$. ∎

Capacitors in Parallel

The arrangement shown in Fig. 24.9a is called a **parallel connection.** Two capacitors are connected in parallel between points a and b. In this case the upper plates of the two capacitors are connected by conducting wires to form an equipotential surface, and the lower plates form another. Hence *in a parallel connection the potential difference for all individual capacitors is the same* and is equal to $V_{ab} = V$. The charges Q_1 and Q_2 are not necessarily equal, however, since charges can reach each capacitor independently from the source (such as a battery) of the voltage V_{ab}. The charges are

$$Q_1 = C_1 V \quad \text{and} \quad Q_2 = C_2 V$$

The *total* charge Q of the combination, and thus the total charge on the equivalent capacitor, is

$$Q = Q_1 + Q_2 = (C_1 + C_2)V$$

so

$$\frac{Q}{V} = C_1 + C_2 \qquad (24.6)$$

The parallel combination is equivalent to a single capacitor with the same total charge $Q = Q_1 + Q_2$ and potential difference V as the combination (Fig. 24.9b). The equivalent capacitance of the combination, C_{eq}, is the same as the capacitance Q/V of this single equivalent capacitor. So from Eq. (24.6),

$$C_{eq} = C_1 + C_2$$

24.9 A parallel connection of two capacitors.

(a) Two capacitors in parallel

Capacitors in parallel:
- The capacitors have the same potential V.
- The charge on each capacitor depends on its capacitance: $Q_1 = C_1 V$, $Q_2 = C_2 V$.

(b) The equivalent single capacitor

Charge is the sum of the individual charges:
$Q = Q_1 + Q_2$
Equivalent capacitance:
$C_{eq} = C_1 + C_2$

In the same way we can show that for any number of capacitors in parallel,

$$C_{eq} = C_1 + C_2 + C_3 + \cdots \qquad \text{(capacitors in parallel)} \qquad (24.7)$$

The equivalent capacitance of a parallel combination equals the *sum* of the individual capacitances. In a parallel connection the equivalent capacitance is always *greater than* any individual capacitance.

CAUTION   **Capacitors in parallel** The potential differences are the same for all the capacitors in a parallel combination; however, the charges on individual capacitors are *not* the same unless their individual capacitances are the same. The charges on the individual capacitors add to give the total charge on the parallel combination: $Q_{total} = Q_1 + Q_2 + Q_3 + \cdots$. [Compare these statements to those in the "Caution" paragraph following Eq. (24.5).] ▮

Problem-Solving Strategy 24.1 **Equivalent Capacitance**

IDENTIFY *the relevant concepts:* The concept of equivalent capacitance is useful whenever two or more capacitors are connected.

SET UP *the problem* using the following steps:
1. Make a drawing of the capacitor arrangement.
2. Identify all groups of capacitors that are connected in series or in parallel.
3. Keep in mind that when we say a capacitor "has charge Q," we mean that the plate at higher potential has charge $+Q$ and the other plate has charge $-Q$.

EXECUTE *the solution* as follows:
1. Use Eq. (24.5) to find the equivalent capacitance of capacitors connected in series, as in Fig. 24.8. Such capacitors each have the *same charge* if they were uncharged before they were connected; that charge is the same as that on the equivalent capacitor. The potential difference across the combination is the sum of the potential differences across the individual capacitors.

2. Use Eq. (24.7) to find the equivalent capacitance of capacitors connected in parallel, as in Fig. 24.9. Such capacitors all have the *same potential difference* across them; that potential difference is the same as that across the equivalent capacitor. The total charge on the combination is the sum of the charges on the individual capacitors.

3. After replacing all the series or parallel groups you initially identified, you may find that more such groups reveal themselves. Replace those groups using the same procedure as above until no more replacements are possible. If you then need to find the charge or potential difference for an individual original capacitor, you may have to retrace your steps.

EVALUATE *your answer:* Check whether your result makes sense. If the capacitors are connected in series, the equivalent capacitance C_{eq} must be *smaller* than any of the individual capacitances. If the capacitors are connected in parallel, C_{eq} must be *greater* than any of the individual capacitances.

Example 24.5   **Capacitors in series and in parallel**

In Figs. 24.8 and 24.9, let $C_1 = 6.0 \ \mu F$, $C_2 = 3.0 \ \mu F$, and $V_{ab} = 18 \ V$. Find the equivalent capacitance and the charge and potential difference for each capacitor when the capacitors are connected (a) in series (see Fig. 24.8) and (b) in parallel (see Fig. 24.9).

SOLUTION

IDENTIFY and SET UP: In both parts of this example a target variable is the equivalent capacitance C_{eq}, which is given by Eq. (24.5) for the series combination in part (a) and by Eq. (24.7) for the parallel combination in part (b). In each part we find the charge and potential difference using the definition of capacitance, Eq. (24.1), and the rules outlined in Problem-Solving Strategy 24.1.

EXECUTE: (a) From Eq. (24.5) for a series combination,

$$\frac{1}{C_{eq}} = \frac{1}{C_1} + \frac{1}{C_2} = \frac{1}{6.0 \ \mu F} + \frac{1}{3.0 \ \mu F} \qquad C_{eq} = 2.0 \ \mu F$$

The charge Q on each capacitor in series is the same as that on the equivalent capacitor:

$$Q = C_{eq}V = (2.0 \ \mu F)(18 \ V) = 36 \ \mu C$$

The potential difference across each capacitor is inversely proportional to its capacitance:

$$V_{ac} = V_1 = \frac{Q}{C_1} = \frac{36 \ \mu C}{6.0 \ \mu F} = 6.0 \ V$$

$$V_{cb} = V_2 = \frac{Q}{C_2} = \frac{36 \ \mu C}{3.0 \ \mu F} = 12.0 \ V$$

(b) From Eq. (24.7) for a parallel combination,

$$C_{eq} = C_1 + C_2 = 6.0 \ \mu F + 3.0 \ \mu F = 9.0 \ \mu F$$

The potential difference across each of the capacitors is the same as that across the equivalent capacitor, 18 V. The charge on each capacitor is directly proportional to its capacitance:

$$Q_1 = C_1V = (6.0 \ \mu F)(18 \ V) = 108 \ \mu C$$
$$Q_2 = C_2V = (3.0 \ \mu F)(18 \ V) = 54 \ \mu C$$

EVALUATE: As expected, the equivalent capacitance C_{eq} for the series combination in part (a) is less than either C_1 or C_2, while

Continued

that for the parallel combination in part (b) is greater than either C_1 or C_2. For two capacitors in series, as in part (a), the charge is the same on either capacitor and the *larger* potential difference appears across the capacitor with the *smaller* capacitance. Furthermore, the sum of the potential differences across the individual capacitors in series equals the potential difference across the

equivalent capacitor: $V_{ac} + V_{cb} = V_{ab} = 18$ V. By contrast, for two capacitors in parallel, as in part (b), each capacitor has the same potential difference and the *larger* charge appears on the capacitor with the *larger* capacitance. Can you show that the total charge $Q_1 + Q_2$ on the parallel combination is equal to the charge $Q = C_{eq}V$ on the equivalent capacitor?

Example 24.6 A capacitor network

Find the equivalent capacitance of the five-capacitor network shown in Fig. 24.10a.

SOLUTION

IDENTIFY and SET UP: These capacitors are neither all in series nor all in parallel. We can, however, identify portions of the arrangement that *are* either in series or parallel. We combine these as described in Problem-Solving Strategy 24.1 to find the net equivalent capacitance, using Eq. (24.5) for series connections and Eq. (24.7) for parallel connections.

EXECUTE: The caption of Fig. 24.10 outlines our procedure. We first use Eq. (24.5) to replace the 12-μF and 6-μF series combination by its equivalent capacitance C':

$$\frac{1}{C'} = \frac{1}{12\ \mu\text{F}} + \frac{1}{6\ \mu\text{F}} \qquad C' = 4\ \mu\text{F}$$

This gives us the equivalent combination of Fig. 24.10b. Now we see three capacitors in parallel, and we use Eq. (24.7) to replace them with their equivalent capacitance C'':

$$C'' = 3\ \mu\text{F} + 11\ \mu\text{F} + 4\ \mu\text{F} = 18\ \mu\text{F}$$

This gives us the equivalent combination of Fig. 24.10c, which has two capacitors in series. We use Eq. (24.5) to replace them with their equivalent capacitance C_{eq}, which is our target variable (Fig. 24.10d):

$$\frac{1}{C_{eq}} = \frac{1}{18\ \mu\text{F}} + \frac{1}{9\ \mu\text{F}} \qquad C_{eq} = 6\ \mu\text{F}$$

EVALUATE: If the potential difference across the entire network in Fig. 24.10a is $V_{ab} = 9.0$ V, the net charge on the network is $Q = C_{eq}V_{ab} = (6\ \mu\text{F})(9.0\ \text{V}) = 54\ \mu\text{C}$. Can you find the charge on, and the voltage across, each of the five individual capacitors?

24.10 (a) A capacitor network between points a and b. (b) The 12-μF and 6-μF capacitors in series in (a) are replaced by an equivalent 4-μF capacitor. (c) The 3-μF, 11-μF, and 4-μF capacitors in parallel in (b) are replaced by an equivalent 18-μF capacitor. (d) Finally, the 18-μF and 9-μF capacitors in series in (c) are replaced by an equivalent 6-μF capacitor.

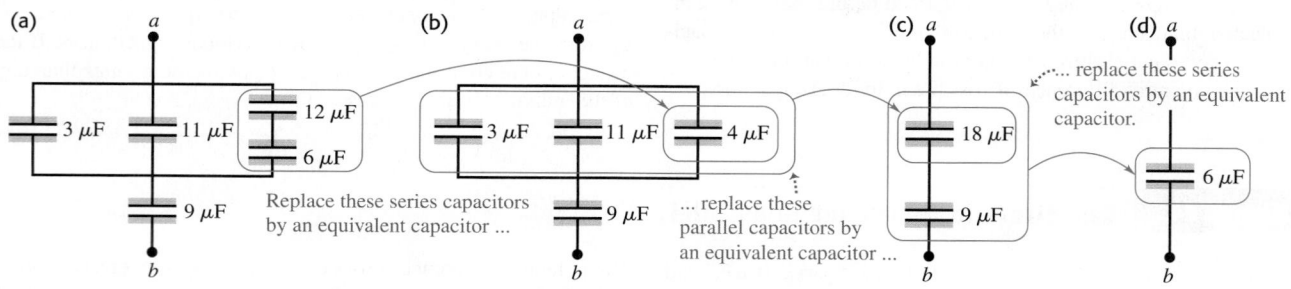

Test Your Understanding of Section 24.2 You want to connect a 4-μF capacitor and an 8-μF capacitor. (a) With which type of connection will the 4-μF capacitor have a greater potential difference across it than the 8-μF capacitor? (i) series; (ii) parallel; (iii) either series or parallel; (iv) neither series nor parallel. (b) With which type of connection will the 4-μF capacitor have a greater charge than the 8-μF capacitor? (i) series; (ii) parallel; (iii) either series or parallel; (iv) neither series nor parallel.

24.3 Energy Storage in Capacitors and Electric-Field Energy

Many of the most important applications of capacitors depend on their ability to store energy. The electric potential energy stored in a charged capacitor is just equal to the amount of work required to charge it—that is, to separate opposite charges and place them on different conductors. When the capacitor is discharged, this stored energy is recovered as work done by electrical forces.

We can calculate the potential energy U of a charged capacitor by calculating the work W required to charge it. Suppose that when we are done charging the capacitor, the final charge is Q and the final potential difference is V. From Eq. (24.1) these quantities are related by

$$V = \frac{Q}{C}$$

Let q and v be the charge and potential difference, respectively, at an intermediate stage during the charging process; then $v = q/C$. At this stage the work dW required to transfer an additional element of charge dq is

$$dW = v \, dq = \frac{q \, dq}{C}$$

The total work W needed to increase the capacitor charge q from zero to a final value Q is

$$W = \int_0^W dW = \frac{1}{C} \int_0^Q q \, dq = \frac{Q^2}{2C} \quad \text{(work to charge a capacitor)} \quad (24.8)$$

This is also equal to the total work done by the electric field on the charge when the capacitor discharges. Then q *decreases* from an initial value Q to zero as the elements of charge dq "fall" through potential differences v that vary from V down to zero.

If we define the potential energy of an *uncharged* capacitor to be zero, then W in Eq. (24.8) is equal to the potential energy U of the charged capacitor. The final stored charge is $Q = CV$, so we can express U (which is equal to W) as

$$U = \frac{Q^2}{2C} = \tfrac{1}{2}CV^2 = \tfrac{1}{2}QV \qquad \begin{array}{l}\text{(potential energy stored} \\ \text{in a capacitor)}\end{array} \quad (24.9)$$

When Q is in coulombs, C in farads (coulombs per volt), and V in volts (joules per coulomb), U is in joules.

The last form of Eq. (24.9), $U = \tfrac{1}{2}QV$, shows that the total work W required to charge the capacitor is equal to the total charge Q multiplied by the *average* potential difference $\tfrac{1}{2}V$ during the charging process.

The expression $U = \tfrac{1}{2}(Q^2/C)$ in Eq. (24.9) shows that a charged capacitor is the electrical analog of a stretched spring with elastic potential energy $U = \tfrac{1}{2}kx^2$. The charge Q is analogous to the elongation x, and the *reciprocal* of the capacitance, $1/C$, is analogous to the force constant k. The energy supplied to a capacitor in the charging process is analogous to the work we do on a spring when we stretch it.

Equations (24.8) and (24.9) tell us that capacitance measures the ability of a capacitor to store both energy and charge. If a capacitor is charged by connecting it to a battery or other source that provides a fixed potential difference V, then increasing the value of C gives a greater charge $Q = CV$ and a greater amount of stored energy $U = \tfrac{1}{2}CV^2$. If instead the goal is to transfer a given quantity of charge Q from one conductor to another, Eq. (24.8) shows that the work W required is inversely proportional to C; the greater the capacitance, the easier it is to give a capacitor a fixed amount of charge.

Applications of Capacitors: Energy Storage

Most practical applications of capacitors take advantage of their ability to store and release energy. In electronic flash units used by photographers, the energy stored in a capacitor (see Fig. 24.4) is released by depressing the camera's shutter button. This provides a conducting path from one capacitor plate to the other through the flash tube. Once this path is established, the stored energy is rapidly converted into a brief but intense flash of light. An extreme example of the same principle is the Z machine at Sandia National Laboratories in New Mexico,

24.11 The Z machine uses a large number of capacitors in parallel to give a tremendous equivalent capacitance C (see Section 24.2). Hence a large amount of energy $U = \frac{1}{2}CV^2$ can be stored with even a modest potential difference V. The arcs shown here are produced when the capacitors discharge their energy into a target, which is no larger than a spool of thread. This heats the target to a temperature higher than 2×10^9 K.

which is used in experiments in controlled nuclear fusion (Fig. 24.11). A bank of charged capacitors releases more than a million joules of energy in just a few billionths of a second. For that brief space of time, the power output of the Z machine is 2.9×10^{14} W, or about 80 times the power output of all the electric power plants on earth combined!

In other applications, the energy is released more slowly. Springs in the suspension of an automobile help smooth out the ride by absorbing the energy from sudden jolts and releasing that energy gradually; in an analogous way, a capacitor in an electronic circuit can smooth out unwanted variations in voltage due to power surges. We'll discuss these circuits in detail in Chapter 26.

Electric-Field Energy

We can charge a capacitor by moving electrons directly from one plate to another. This requires doing work against the electric field between the plates. Thus we can think of the energy as being stored *in the field* in the region between the plates. To develop this relationship, let's find the energy *per unit volume* in the space between the plates of a parallel-plate capacitor with plate area A and separation d. We call this the **energy density,** denoted by u. From Eq. (24.9) the total stored potential energy is $\frac{1}{2}CV^2$ and the volume between the plates is just Ad; hence the energy density is

$$u = \text{Energy density} = \frac{\frac{1}{2}CV^2}{Ad} \qquad (24.10)$$

From Eq. (24.2) the capacitance C is given by $C = \epsilon_0 A/d$. The potential difference V is related to the electric-field magnitude E by $V = Ed$. If we use these expressions in Eq. (24.10), the geometric factors A and d cancel, and we find

$$u = \frac{1}{2}\epsilon_0 E^2 \qquad \text{(electric energy density in a vacuum)} \qquad (24.11)$$

Although we have derived this relationship only for a parallel-plate capacitor, it turns out to be valid for any capacitor in vacuum and indeed *for any electric field configuration in vacuum.* This result has an interesting implication. We think of vacuum as space with no matter in it, but vacuum can nevertheless have electric fields and therefore energy. Thus "empty" space need not be truly empty after all. We will use this idea and Eq. (24.11) in Chapter 32 in connection with the energy transported by electromagnetic waves.

> **CAUTION** **Electric-field energy is electric potential energy** It's a common misconception that electric-field energy is a new kind of energy, different from the electric potential energy described before. This is *not* the case; it is simply a different way of interpreting electric potential energy. We can regard the energy of a given system of charges as being a shared property of all the charges, or we can think of the energy as being a property of the electric field that the charges create. Either interpretation leads to the same value of the potential energy. ▮

Example 24.7 | **Transferring charge and energy between capacitors**

We connect a capacitor $C_1 = 8.0\ \mu\text{F}$ to a power supply, charge it to a potential difference $V_0 = 120$ V, and disconnect the power supply (Fig. 24.12). Switch S is open. (a) What is the charge Q_0 on C_1? (b) What is the energy stored in C_1? (c) Capacitor $C_2 = 4.0\ \mu\text{F}$ is initially uncharged. We close switch S. After charge no longer flows, what is the potential difference across each capacitor, and what is the charge on each capacitor? (d) What is the final energy of the system?

24.12 When the switch S is closed, the charged capacitor C_1 is connected to an uncharged capacitor C_2. The center part of the switch is an insulating handle; charge can flow only between the two upper terminals and between the two lower terminals.

SOLUTION

IDENTIFY and SET UP: In parts (a) and (b) we find the charge Q_0 and stored energy U_{initial} for the single charged capacitor C_1 using Eqs. (24.1) and (24.9), respectively. After we close switch S, one wire connects the upper plates of the two capacitors and another wire connects the lower plates; the capacitors are now connected in parallel. In part (c) we use the character of the parallel connection to determine how Q_0 is shared between the two capacitors. In part (d) we again use Eq. (24.9) to find the energy stored in capacitors C_1 and C_2; the energy of the system is the sum of these values.

EXECUTE: (a) The initial charge Q_0 on C_1 is

$$Q_0 = C_1V_0 = (8.0\ \mu\text{F})(120\ \text{V}) = 960\ \mu\text{C}$$

(b) The energy initially stored in C_1 is

$$U_{\text{initial}} = \tfrac{1}{2}Q_0V_0 = \tfrac{1}{2}(960 \times 10^{-6}\ \text{C})(120\ \text{V}) = 0.058\ \text{J}$$

(c) When we close the switch, the positive charge Q_0 is distributed over the upper plates of both capacitors and the negative charge $-Q_0$ is distributed over the lower plates. Let Q_1 and Q_2 be the magnitudes of the final charges on the capacitors. Conservation of charge requires that $Q_1 + Q_2 = Q_0$. The potential difference V between the plates is the same for both capacitors because they are connected in parallel, so the charges are $Q_1 = C_1V$ and $Q_2 = C_2V$. We now have three independent equations relating the three unknowns Q_1, Q_2, and V. Solving these, we find

$$V = \frac{Q_0}{C_1 + C_2} = \frac{960\ \mu\text{C}}{8.0\ \mu\text{F} + 4.0\ \mu\text{F}} = 80\ \text{V}$$

$$Q_1 = 640\ \mu\text{C} \qquad Q_2 = 320\ \mu\text{C}$$

(d) The final energy of the system is

$$U_{\text{final}} = \tfrac{1}{2}Q_1V + \tfrac{1}{2}Q_2V = \tfrac{1}{2}Q_0V$$

$$= \tfrac{1}{2}(960 \times 10^{-6}\ \text{C})(80\ \text{V}) = 0.038\ \text{J}$$

EVALUATE: The final energy is less than the initial energy; the difference was converted to energy of some other form. The conductors become a little warmer because of their resistance, and some energy is radiated as electromagnetic waves. We'll study the behavior of capacitors in more detail in Chapters 26 and 31.

Example 24.8 Electric-field energy

(a) What is the magnitude of the electric field required to store 1.00 J of electric potential energy in a volume of 1.00 m³ in vacuum? (b) If the field magnitude is 10 times larger than that, how much energy is stored per cubic meter?

SOLUTION

IDENTIFY and SET UP: We use the relationship between the electric-field magnitude E and the energy density u. In part (a) we use the given information to find u; then we use Eq. (24.11) to find the corresponding value of E. In part (b), Eq. (24.11) tells us how u varies with E.

EXECUTE: (a) The desired energy density is $u = 1.00\ \text{J/m}^3$. Then from Eq. (24.11),

$$E = \sqrt{\frac{2u}{\epsilon_0}} = \sqrt{\frac{2(1.00\ \text{J/m}^3)}{8.85 \times 10^{-12}\ \text{C}^2/\text{N}\cdot\text{m}^2}}$$

$$= 4.75 \times 10^5\ \text{N/C} = 4.75 \times 10^5\ \text{V/m}$$

(b) Equation (24.11) shows that u is proportional to E^2. If E increases by a factor of 10, u increases by a factor of $10^2 = 100$, so the energy density becomes $u = 100\ \text{J/m}^3$.

EVALUATE: Dry air can sustain an electric field of about $3 \times 10^6\ \text{V/m}$ without experiencing *dielectric breakdown*, which we will discuss in Section 24.4. There we will see that field magnitudes in practical insulators can be as great as this or even larger.

Example 24.9 Two ways to calculate energy stored in a capacitor

The spherical capacitor described in Example 24.3 (Section 24.1) has charges $+Q$ and $-Q$ on its inner and outer conductors. Find the electric potential energy stored in the capacitor (a) by using the capacitance C found in Example 24.3 and (b) by integrating the electric-field energy density u.

SOLUTION

IDENTIFY and SET UP: We can determine the energy U stored in a capacitor in two ways: in terms of the work done to put the charges on the two conductors, and in terms of the energy in the electric field between the conductors. The descriptions are equivalent, so they must give us the same result. In Example 24.3 we found the capacitance C and the field magnitude E in the space between the conductors. (The electric field is zero inside the inner sphere and is also zero outside the inner surface of the outer sphere, because a Gaussian surface with radius $r < r_a$ or $r > r_b$ encloses zero net charge. Hence the energy density is nonzero only in the space between the spheres, $r_a < r < r_b$.) In part (a) we use Eq. (24.9) to find U. In part (b) we use Eq. (24.11) to find u, which we integrate over the volume between the spheres to find U.

EXECUTE: (a) From Example 24.3, the spherical capacitor has capacitance

$$C = 4\pi\epsilon_0 \frac{r_ar_b}{r_b - r_a}$$

where r_a and r_b are the radii of the inner and outer conducting spheres, respectively. From Eq. (24.9) the energy stored in this capacitor is

$$U = \frac{Q^2}{2C} = \frac{Q^2}{8\pi\epsilon_0}\frac{r_b - r_a}{r_ar_b}$$

Continued

(b) The electric field in the region $r_a < r < r_b$ between the two conducting spheres has magnitude $E = Q/4\pi\epsilon_0 r^2$. The energy density in this region is

$$u = \tfrac{1}{2}\epsilon_0 E^2 = \tfrac{1}{2}\epsilon_0 \left(\frac{Q}{4\pi\epsilon_0 r^2}\right)^2 = \frac{Q^2}{32\pi^2\epsilon_0 r^4}$$

The energy density is *not* uniform; it decreases rapidly with increasing distance from the center of the capacitor. To find the total electric-field energy, we integrate u (the energy per unit volume) over the region $r_a < r < r_b$. We divide this region into spherical shells of radius r, surface area $4\pi r^2$, thickness dr, and volume $dV = 4\pi r^2\, dr$. Then

$$U = \int u\, dV = \int_{r_a}^{r_b}\left(\frac{Q^2}{32\pi^2\epsilon_0 r^4}\right)4\pi r^2\, dr$$

$$= \frac{Q^2}{8\pi\epsilon_0}\int_{r_a}^{r_b}\frac{dr}{r^2} = \frac{Q^2}{8\pi\epsilon_0}\left(-\frac{1}{r_b} + \frac{1}{r_a}\right)$$

$$= \frac{Q^2}{8\pi\epsilon_0}\frac{r_b - r_a}{r_a r_b}$$

EVALUATE: Electric potential energy can be regarded as being associated with either the *charges,* as in part (a), or the *field,* as in part (b); the calculated amount of stored energy is the same in either case.

Test Your Understanding of Section 24.3 You want to connect a 4-μF capacitor and an 8-μF capacitor. With which type of connection will the 4-μF capacitor have a greater amount of *stored energy* than the 8-μF capacitor? (i) series; (ii) parallel; (iii) either series or parallel; (iv) neither series nor parallel.

24.4 Dielectrics

24.13 A common type of capacitor uses dielectric sheets to separate the conductors.

Conductor (metal foil)

Conductor (metal foil)

Dielectric (plastic sheets)

Most capacitors have a nonconducting material, or **dielectric,** between their conducting plates. A common type of capacitor uses long strips of metal foil for the plates, separated by strips of plastic sheet such as Mylar. A sandwich of these materials is rolled up, forming a unit that can provide a capacitance of several microfarads in a compact package (Fig. 24.13).

Placing a solid dielectric between the plates of a capacitor serves three functions. First, it solves the mechanical problem of maintaining two large metal sheets at a very small separation without actual contact.

Second, using a dielectric increases the maximum possible potential difference between the capacitor plates. As we described in Section 23.3, any insulating material, when subjected to a sufficiently large electric field, experiences a partial ionization that permits conduction through it. This is called **dielectric breakdown.** Many dielectric materials can tolerate stronger electric fields without breakdown than can air. Thus using a dielectric allows a capacitor to sustain a higher potential difference V and so store greater amounts of charge and energy.

Third, the capacitance of a capacitor of given dimensions is *greater* when there is a dielectric material between the plates than when there is vacuum. We can demonstrate this effect with the aid of a sensitive *electrometer,* a device that measures the potential difference between two conductors without letting any appreciable charge flow from one to the other. Figure 24.14a shows an electrometer connected across a charged capacitor, with magnitude of charge Q on each plate and potential difference V_0. When we insert an uncharged sheet of dielectric, such as glass, paraffin, or polystyrene, between the plates, experiment shows that the potential difference *decreases* to a smaller value V (Fig. 24.14b). When we remove the dielectric, the potential difference returns to its original value V_0, showing that the original charges on the plates have not changed.

The original capacitance C_0 is given by $C_0 = Q/V_0$, and the capacitance C with the dielectric present is $C = Q/V$. The charge Q is the same in both cases, and V is less than V_0, so we conclude that the capacitance C with the dielectric present is *greater* than C_0. When the space between plates is completely filled by the dielectric, the ratio of C to C_0 (equal to the ratio of V_0 to V) is called the **dielectric constant** of the material, K:

$$K = \frac{C}{C_0} \qquad \text{(definition of dielectric constant)} \qquad (24.12)$$

When the charge is constant, $Q = C_0 V_0 = CV$ and $C/C_0 = V_0/V$. In this case, Eq. (24.12) can be rewritten as

$$V = \frac{V_0}{K} \quad \text{(when } Q \text{ is constant)} \quad (24.13)$$

With the dielectric present, the potential difference for a given charge Q is *reduced* by a factor K.

The dielectric constant K is a pure number. Because C is always greater than C_0, K is always greater than unity. Some representative values of K are given in Table 24.1. For vacuum, $K = 1$ by definition. For air at ordinary temperatures and pressures, K is about 1.0006; this is so nearly equal to 1 that for most purposes an air capacitor is equivalent to one in vacuum. Note that while water has a very large value of K, it is usually not a very practical dielectric for use in capacitors. The reason is that while pure water is a very poor conductor, it is also an excellent ionic solvent. Any ions that are dissolved in the water will cause charge to flow between the capacitor plates, so the capacitor discharges.

Table 24.1 Values of Dielectric Constant K at 20°C

Material	K	Material	K
Vacuum	1	Polyvinyl chloride	3.18
Air (1 atm)	1.00059	Plexiglas®	3.40
Air (100 atm)	1.0548	Glass	5–10
Teflon	2.1	Neoprene	6.70
Polyethylene	2.25	Germanium	16
Benzene	2.28	Glycerin	42.5
Mica	3–6	Water	80.4
Mylar	3.1	Strontium titanate	310

No real dielectric is a perfect insulator. Hence there is always some *leakage current* between the charged plates of a capacitor with a dielectric. We tacitly ignored this effect in Section 24.2 when we derived expressions for the equivalent capacitances of capacitors in series, Eq. (24.5), and in parallel, Eq. (24.7). But if a leakage current flows for a long enough time to substantially change the charges from the values we used to derive Eqs. (24.5) and (24.7), those equations may no longer be accurate.

Induced Charge and Polarization

When a dielectric material is inserted between the plates while the charge is kept constant, the potential difference between the plates decreases by a factor K. Therefore the electric field between the plates must decrease by the same factor. If E_0 is the vacuum value and E is the value with the dielectric, then

$$E = \frac{E_0}{K} \quad \text{(when } Q \text{ is constant)} \quad (24.14)$$

Since the electric-field magnitude is smaller when the dielectric is present, the surface charge density (which causes the field) must be smaller as well. The surface charge on the conducting plates does not change, but an *induced* charge of the opposite sign appears on each surface of the dielectric (Fig. 24.15). The dielectric was originally electrically neutral and is still neutral; the induced surface charges arise as a result of *redistribution* of positive and negative charge within the dielectric material, a phenomenon called **polarization.** We first encountered polarization in Section 21.2, and we suggest that you reread the discussion of Fig. 21.8. We will assume that the induced surface charge is *directly proportional* to the electric-field magnitude E in the material; this is indeed the case for many common dielectrics. (This direct proportionality is analogous to

24.14 Effect of a dielectric between the plates of a parallel-plate capacitor. **(a)** With a given charge, the potential difference is V_0. **(b)** With the same charge but with a dielectric between the plates, the potential difference V is smaller than V_0.

(a)

Vacuum

Q $-Q$

V_0

Electrometer (measures potential difference across plates)

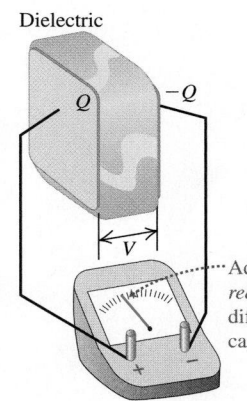

(b)

Dielectric

Q $-Q$

V

Adding the dielectric *reduces* the potential difference across the capacitor.

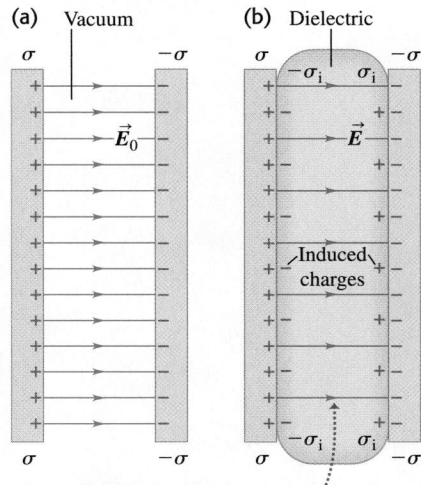

24.15 Electric field lines with **(a)** vacuum between the plates and **(b)** dielectric between the plates.

(a) Vacuum

σ $-\sigma$

$\vec{E}_0$

σ $-\sigma$

(b) Dielectric

σ $-\sigma$

$-\sigma_i$ σ_i

$\vec{E}$

Induced charges

$-\sigma_i$ σ_i

σ $-\sigma$

For a given charge density σ, the induced charges on the dielectric's surfaces reduce the electric field between the plates.

Hooke's law for a spring.) In that case, K is a constant for any particular material. When the electric field is very strong or if the dielectric is made of certain crystalline materials, the relationship between induced charge and the electric field can be more complex; we won't consider such cases here.

We can derive a relationship between this induced surface charge and the charge on the plates. Let's denote the magnitude of the charge per unit area induced on the surfaces of the dielectric (the induced surface charge density) by σ_i. The magnitude of the surface charge density on the capacitor plates is σ, as usual. Then the *net* surface charge on each side of the capacitor has magnitude $(\sigma - \sigma_i)$, as shown in Fig. 24.15b. As we found in Example 21.12 (Section 21.5) and in Example 22.8 (Section 22.4), the field between the plates is related to the net surface charge density by $E = \sigma_{\text{net}}/\epsilon_0$. Without and with the dielectric, respectively, we have

$$E_0 = \frac{\sigma}{\epsilon_0} \qquad E = \frac{\sigma - \sigma_i}{\epsilon_0} \tag{24.15}$$

Using these expressions in Eq. (24.14) and rearranging the result, we find

$$\sigma_i = \sigma\left(1 - \frac{1}{K}\right) \qquad \text{(induced surface charge density)} \tag{24.16}$$

This equation shows that when K is very large, σ_i is nearly as large as σ. In this case, σ_i nearly cancels σ, and the field and potential difference are much smaller than their values in vacuum.

The product $K\epsilon_0$ is called the **permittivity** of the dielectric, denoted by ϵ:

$$\epsilon = K\epsilon_0 \qquad \text{(definition of permittivity)} \tag{24.17}$$

In terms of ϵ we can express the electric field within the dielectric as

$$E = \frac{\sigma}{\epsilon} \tag{24.18}$$

The capacitance when the dielectric is present is given by

$$C = KC_0 = K\epsilon_0\frac{A}{d} = \epsilon\frac{A}{d} \qquad \begin{array}{l}\text{(parallel-plate capacitor,}\\ \text{dielectric between plates)}\end{array} \tag{24.19}$$

We can repeat the derivation of Eq. (24.11) for the energy density u in an electric field for the case in which a dielectric is present. The result is

$$u = \tfrac{1}{2}K\epsilon_0E^2 = \tfrac{1}{2}\epsilon E^2 \qquad \text{(electric energy density in a dielectric)} \tag{24.20}$$

In empty space, where $K = 1$, $\epsilon = \epsilon_0$ and Eqs. (24.19) and (24.20) reduce to Eqs. (24.2) and (24.11), respectively, for a parallel-plate capacitor in vacuum. For this reason, ϵ_0 is sometimes called the "permittivity of free space" or the "permittivity of vacuum." Because K is a pure number, ϵ and ϵ_0 have the same units, $C^2/N \cdot m^2$ or F/m.

Equation (24.19) shows that extremely high capacitances can be obtained with plates that have a large surface area A and are separated by a small distance d by a dielectric with a large value of K. In an *electrolytic double-layer capacitor,* tiny carbon granules adhere to each plate: The value of A is the combined surface area of the granules, which can be tremendous. The plates with granules attached are separated by a very thin dielectric sheet. A capacitor of this kind can have a capacitance of 5000 farads yet fit in the palm of your hand (compare Example 24.1 in Section 24.1).

Several practical devices make use of the way in which a capacitor responds to a change in dielectric constant. One example is an electric stud finder, used by home repair workers to locate metal studs hidden behind a wall's surface. It consists of a metal plate with associated circuitry. The plate acts as one half of a capacitor, with the wall acting as the other half. If the stud finder moves over a metal stud, the effective dielectric constant for the capacitor changes, changing the capacitance and triggering a signal.

Problem-Solving Strategy 24.2 | Dielectrics

IDENTIFY *the relevant concepts:* The relationships in this section are useful whenever there is an electric field in a dielectric, such as a dielectric between charged capacitor plates. Typically you must relate the potential difference V_{ab} between the plates, the electric field magnitude E in the capacitor, the charge density σ on the capacitor plates, and the induced charge density σ_i on the surfaces of the capacitor.

SET UP *the problem* using the following steps:
1. Make a drawing of the situation.
2. Identify the target variables, and choose which equations from this section will help you solve for those variables.

EXECUTE *the solution* as follows:
1. In problems such as the next example, it is easy to get lost in a blizzard of formulas. Ask yourself at each step what kind of

quantity each symbol represents. For example, distinguish clearly between charges and charge densities, and between electric fields and electric potential differences.
2. Check for consistency of units. Distances must be in meters. A microfarad is 10^{-6} farad, and so on. Don't confuse the numerical value of ϵ_0 with the value of $1/4\pi\epsilon_0$. Electric-field magnitude can be expressed in both N/C and V/m. The units of ϵ_0 are $C^2/N \cdot m^2$ or F/m.

EVALUATE *your answer:* With a dielectric present, (a) the capacitance is greater than without a dielectric; (b) for a given charge on the capacitor, the electric field and potential difference are less than without a dielectric; and (c) the magnitude of the induced surface charge density σ_i on the dielectric is less than that of the charge density σ on the capacitor plates.

Example 24.10 | A capacitor with and without a dielectric

Suppose the parallel plates in Fig. 24.15 each have an area of 2000 cm² (2.00×10^{-1} m²) and are 1.00 cm (1.00×10^{-2} m) apart. We connect the capacitor to a power supply, charge it to a potential difference $V_0 = 3.00$ kV, and disconnect the power supply. We then insert a sheet of insulating plastic material between the plates, completely filling the space between them. We find that the potential difference decreases to 1.00 kV while the charge on each capacitor plate remains constant. Find (a) the original capacitance C_0; (b) the magnitude of charge Q on each plate; (c) the capacitance C after the dielectric is inserted; (d) the dielectric constant K of the dielectric; (e) the permittivity ϵ of the dielectric; (f) the magnitude of the induced charge Q_i on each face of the dielectric; (g) the original electric field E_0 between the plates; and (h) the electric field E after the dielectric is inserted.

SOLUTION

IDENTIFY and SET UP: This problem uses most of the relationships we have discussed for capacitors and dielectrics. (Energy relationships are treated in Example 24.11.) Most of the target variables can be obtained in several ways. The methods used below are a sample; we encourage you to think of others and compare your results.

EXECUTE: (a) With vacuum between the plates, we use Eq. (24.19) with $K = 1$:

$$C_0 = \epsilon_0 \frac{A}{d} = (8.85 \times 10^{-12} \text{ F/m}) \frac{2.00 \times 10^{-1} \text{ m}^2}{1.00 \times 10^{-2} \text{ m}}$$

$$= 1.77 \times 10^{-10} \text{ F} = 177 \text{ pF}$$

(b) From the definition of capacitance, Eq. (24.1),

$$Q = C_0 V_0 = (1.77 \times 10^{-10} \text{ F})(3.00 \times 10^3 \text{ V})$$

$$= 5.31 \times 10^{-7} \text{ C} = 0.531 \text{ } \mu\text{C}$$

(c) When the dielectric is inserted, Q is unchanged but the potential difference decreases to $V = 1.00$ kV. Hence from Eq. (24.1), the new capacitance is

$$C = \frac{Q}{V} = \frac{5.31 \times 10^{-7} \text{ C}}{1.00 \times 10^3 \text{ V}} = 5.31 \times 10^{-10} \text{ F} = 531 \text{ pF}$$

(d) From Eq. (24.12), the dielectric constant is

$$K = \frac{C}{C_0} = \frac{5.31 \times 10^{-10} \text{ F}}{1.77 \times 10^{-10} \text{ F}} = \frac{531 \text{ pF}}{177 \text{ pF}} = 3.00$$

Alternatively, from Eq. (24.13),

$$K = \frac{V_0}{V} = \frac{3000 \text{ V}}{1000 \text{ V}} = 3.00$$

(e) Using K from part (d) in Eq. (24.17), the permittivity is

$$\epsilon = K\epsilon_0 = (3.00)(8.85 \times 10^{-12} \text{ C}^2/\text{N} \cdot \text{m}^2)$$

$$= 2.66 \times 10^{-11} \text{ C}^2/\text{N} \cdot \text{m}^2$$

(f) Multiplying both sides of Eq. (24.16) by the plate area A gives the induced charge $Q_i = \sigma_i A$ in terms of the charge $Q = \sigma A$ on each plate:

$$Q_i = Q\left(1 - \frac{1}{K}\right) = (5.31 \times 10^{-7} \text{ C})\left(1 - \frac{1}{3.00}\right)$$

$$= 3.54 \times 10^{-7} \text{ C}$$

Continued

(g) Since the electric field between the plates is uniform, its magnitude is the potential difference divided by the plate separation:

$$E_0 = \frac{V_0}{d} = \frac{3000 \text{ V}}{1.00 \times 10^{-2} \text{ m}} = 3.00 \times 10^5 \text{ V/m}$$

(h) After the dielectric is inserted,

$$E = \frac{V}{d} = \frac{1000 \text{ V}}{1.00 \times 10^{-2} \text{ m}} = 1.00 \times 10^5 \text{ V/m}$$

or, from Eq. (24.18),

$$E = \frac{\sigma}{\epsilon} = \frac{Q}{\epsilon A} = \frac{5.31 \times 10^{-7} \text{ C}}{(2.66 \times 10^{-11} \text{ C}^2/\text{N} \cdot \text{m}^2)(2.00 \times 10^{-1} \text{ m}^2)}$$

$$= 1.00 \times 10^5 \text{ V/m}$$

or, from Eq. (24.15),

$$E = \frac{\sigma - \sigma_i}{\epsilon_0} = \frac{Q - Q_i}{\epsilon_0 A}$$

$$= \frac{(5.31 - 3.54) \times 10^{-7} \text{ C}}{(8.85 \times 10^{-12} \text{ C}^2/\text{N} \cdot \text{m}^2)(2.00 \times 10^{-1} \text{ m}^2)}$$

$$= 1.00 \times 10^5 \text{ V/m}$$

or, from Eq. (24.14),

$$E = \frac{E_0}{K} = \frac{3.00 \times 10^5 \text{ V/m}}{3.00} = 1.00 \times 10^5 \text{ V/m}$$

EVALUATE: Inserting the dielectric increased the capacitance by a factor of $K = 3.00$ and reduced the electric field between the plates by a factor of $1/K = 1/3.00$. It did so by developing induced charges on the faces of the dielectric of magnitude $Q(1 - 1/K) = Q(1 - 1/3.00) = 0.667Q$.

Example 24.11   **Energy storage with and without a dielectric**

Find the energy stored in the electric field of the capacitor in Example 24.10 and the energy density, both before and after the dielectric sheet is inserted.

SOLUTION

IDENTIFY and SET UP: We now consider the ideas of energy stored in a capacitor and of electric-field energy density. We use Eq. (24.9) to find the stored energy and Eq. (24.20) to find the energy density.

EXECUTE: From Eq. (24.9), the stored energies U_0 and U without and with the dielectric in place are

$$U_0 = \tfrac{1}{2}C_0 V_0^2 = \tfrac{1}{2}(1.77 \times 10^{-10} \text{ F})(3000 \text{ V})^2 = 7.97 \times 10^{-4} \text{ J}$$

$$U = \tfrac{1}{2}CV^2 = \tfrac{1}{2}(5.31 \times 10^{-10} \text{ F})(1000 \text{ V})^2 = 2.66 \times 10^{-4} \text{ J}$$

The final energy is one-third of the original energy.

Equation (24.20) gives the energy densities without and with the dielectric:

$$u_0 = \tfrac{1}{2}\epsilon_0 E_0^2 = \tfrac{1}{2}(8.85 \times 10^{-12} \text{ C}^2/\text{N} \cdot \text{m}^2)(3.00 \times 10^5 \text{ N/C})^2$$

$$= 0.398 \text{ J/m}^3$$

$$u = \tfrac{1}{2}\epsilon E^2 = \tfrac{1}{2}(2.66 \times 10^{-11} \text{ C}^2/\text{N} \cdot \text{m}^2)(1.00 \times 10^5 \text{ N/C})^2$$

$$= 0.133 \text{ J/m}^3$$

The energy density with the dielectric is one-third of the original energy density.

EVALUATE: We can check our answer for u_0 by noting that the volume between the plates is $V = (0.200 \text{ m}^2)(0.0100 \text{ m}) =$ 0.00200 m^3. Since the electric field between the plates is uniform, u_0 is uniform as well and the energy density is just the stored energy divided by the volume:

$$u_0 = \frac{U_0}{V} = \frac{7.97 \times 10^{-4} \text{ J}}{0.00200 \text{ m}^3} = 0.398 \text{ J/m}^3$$

This agrees with our earlier answer. You can use the same approach to check our result for u.

In general, when a dielectric is inserted into a capacitor while the charge on each plate remains the same, the permittivity ϵ increases by a factor of K (the dielectric constant), and the electric field E and the energy density $u = \tfrac{1}{2}\epsilon E^2$ decrease by a factor of $1/K$. Where does the energy go? The answer lies in the fringing field at the edges of a real parallel-plate capacitor. As Fig. 24.16 shows, that field tends to pull the dielectric into the space between the plates, doing work on it as it does so. We could attach a spring to the left end of the dielectric in Fig. 24.16 and use this force to stretch the spring. Because work is done by the field, the field energy density decreases.

24.16 The fringing field at the edges of the capacitor exerts forces $\vec{F}_{-i}$ and $\vec{F}_{+i}$ on the negative and positive induced surface charges of a dielectric, pulling the dielectric into the capacitor.

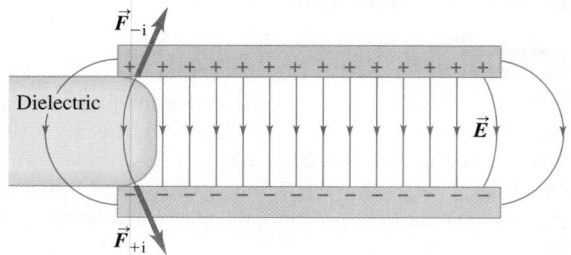

Dielectric Breakdown

We mentioned earlier that when a dielectric is subjected to a sufficiently strong electric field, *dielectric breakdown* takes place and the dielectric becomes a conductor. This occurs when the electric field is so strong that electrons are ripped loose from their molecules and crash into other molecules, liberating even more electrons. This avalanche of moving charge forms a spark or arc discharge. Lightning is a dramatic example of dielectric breakdown in air.

Because of dielectric breakdown, capacitors always have maximum voltage ratings. When a capacitor is subjected to excessive voltage, an arc may form through a layer of dielectric, burning or melting a hole in it. This arc creates a conducting path (a short circuit) between the conductors. If a conducting path remains after the arc is extinguished, the device is rendered permanently useless as a capacitor.

The maximum electric-field magnitude that a material can withstand without the occurrence of breakdown is called its **dielectric strength.** This quantity is affected significantly by temperature, trace impurities, small irregularities in the metal electrodes, and other factors that are difficult to control. For this reason we can give only approximate figures for dielectric strengths. The dielectric strength of dry air is about 3×10^6 V/m. Table 24.2 lists the dielectric strengths of a few common insulating materials. Note that the values are all substantially greater than the value for air. For example, a layer of polycarbonate 0.01 mm thick (about the smallest practical thickness) has 10 times the dielectric strength of air and can withstand a maximum voltage of about $(3 \times 10^7 \text{ V/m})(1 \times 10^{-5} \text{ m}) = 300$ V.

Table 24.2 Dielectric Constant and Dielectric Strength of Some Insulating Materials

Material	Dielectric Constant, K	Dielectric Strength, E_m (V/m)
Polycarbonate	2.8	3×10^7
Polyester	3.3	6×10^7
Polypropylene	2.2	7×10^7
Polystyrene	2.6	2×10^7
Pyrex glass	4.7	1×10^7

Test Your Understanding of Section 24.4 The space between the plates of an isolated parallel-plate capacitor is filled by a slab of dielectric with dielectric constant K. The two plates of the capacitor have charges Q and $-Q$. You pull out the dielectric slab. If the charges do not change, how does the energy in the capacitor change when you remove the slab? (i) It increases; (ii) it decreases; (iii) it remains the same.

24.5 Molecular Model of Induced Charge

In Section 24.4 we discussed induced surface charges on a dielectric in an electric field. Now let's look at how these surface charges can arise. If the material were a *conductor,* the answer would be simple. Conductors contain charge that is free to move, and when an electric field is present, some of the charge redistributes itself on the surface so that there is no electric field inside the conductor. But an ideal dielectric has *no* charges that are free to move, so how can a surface charge occur?

To understand this, we have to look again at rearrangement of charge at the *molecular* level. Some molecules, such as H_2O and N_2O, have equal amounts of positive and negative charges but a lopsided distribution, with excess positive charge concentrated on one side of the molecule and negative charge on the other. As we described in Section 21.7, such an arrangement is called an *electric dipole,* and the molecule is called a *polar molecule.* When no electric field is present in a gas or liquid with polar molecules, the molecules are oriented randomly (Fig. 24.17a). When they are placed in an electric field, however, they tend to orient themselves as in Fig. 24.17b, as a result of the electric-field torques described in Section 21.7. Because of thermal agitation, the alignment of the molecules with $\vec{E}$ is not perfect.

Application Dielectric Cell Membrane

The membrane of a living cell behaves like a dielectric between the plates of a capacitor. The membrane is made of two sheets of lipid molecules, with their water-insoluble ends in the middle and their water-soluble ends (shown in red) on the surfaces of the membrane. The conductive fluids on either side of the membrane (water with negative ions inside the cell, water with positive ions outside) act as charged capacitor plates, and the nonconducting membrane acts as a dielectric with K of about 10. The potential difference V across the membrane is about 0.07 V and the membrane thickness d is about 7×10^{-9} m, so the electric field $E = V/d$ in the membrane is about 10^7 V/m—close to the dielectric strength of the membrane. If the membrane were made of air, V and E would be larger by a factor of $K \approx 10$ and dielectric breakdown would occur.

24.17 Polar molecules (a) without and (b) with an applied electric field $\vec{E}$.

(a)

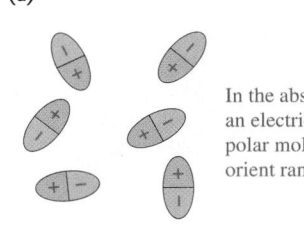

In the absence of an electric field, polar molecules orient randomly.

(b)

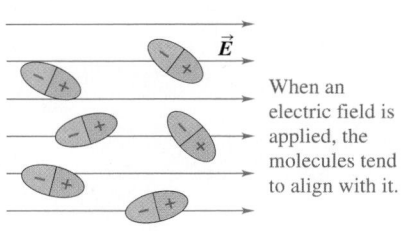

When an electric field is applied, the molecules tend to align with it.

24.18 Nonpolar molecules (a) without and (b) with an applied electric field $\vec{E}$.

(a)

In the absence of an electric field, nonpolar molecules are not electric dipoles.

(b)

$\vec{E}$

An electric field causes the molecules' positive and negative charges to separate slightly, making the molecule effectively polar.

Even a molecule that is *not* ordinarily polar *becomes* a dipole when it is placed in an electric field because the field pushes the positive charges in the molecules in the direction of the field and pushes the negative charges in the opposite direction. This causes a redistribution of charge within the molecule (Fig. 24.18). Such dipoles are called *induced* dipoles.

With either polar or nonpolar molecules, the redistribution of charge caused by the field leads to the formation of a layer of charge on each surface of the dielectric material (Fig. 24.19). These layers are the surface charges described in Section 24.4; their surface charge density is denoted by σ_i. The charges are *not* free to move indefinitely, as they would be in a conductor, because each charge is bound to a molecule. They are in fact called **bound charges** to distinguish them from the **free charges** that are added to and removed from the conducting capacitor plates. In the interior of the material the net charge per unit volume remains zero. As we have seen, this redistribution of charge is called *polarization,* and we say that the material is *polarized.*

The four parts of Fig. 24.20 show the behavior of a slab of dielectric when it is inserted in the field between a pair of oppositely charged capacitor plates. Figure 24.20a shows the original field. Figure 24.20b is the situation after the dielectric has been inserted but before any rearrangement of charges has occurred. Figure 24.20c shows by thinner arrows the additional field set up in the dielectric by its induced surface charges. This field is *opposite* to the original field, but it is not great enough to cancel the original field completely because the charges in the dielectric are not free to move indefinitely. The resultant field

24.19 Polarization of a dielectric in an electric field $\vec{E}$ gives rise to thin layers of bound charges on the surfaces, creating surface charge densities σ_i and $-\sigma_i$. The sizes of the molecules are greatly exaggerated for clarity.

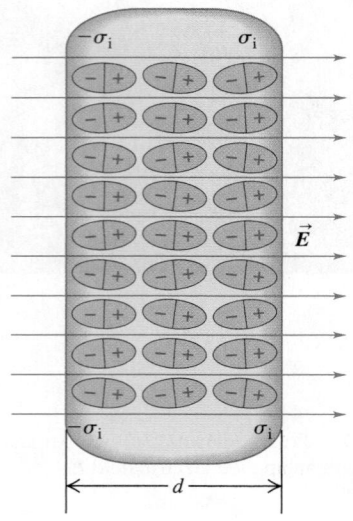

24.20 (a) Electric field of magnitude E_0 between two charged plates. (b) Introduction of a dielectric of dielectric constant K. (c) The induced surface charges and their field. (d) Resultant field of magnitude E_0/K.

(a) No dielectric

(b) Dielectric just inserted

(c) Induced charges create electric field

(d) Resultant field

Original electric field

Weaker field in dielectric due to induced (bound) charges

in the dielectric, shown in Fig. 24.20d, is therefore decreased in magnitude. In the field-line representation, some of the field lines leaving the positive plate go through the dielectric, while others terminate on the induced charges on the faces of the dielectric.

As we discussed in Section 21.2, polarization is also the reason a charged body, such as an electrified plastic rod, can exert a force on an *uncharged* body such as a bit of paper or a pith ball. Figure 24.21 shows an uncharged dielectric sphere B in the radial field of a positively charged body A. The induced positive charges on B experience a force toward the right, while the force on the induced negative charges is toward the left. The negative charges are closer to A, and thus are in a stronger field, than are the positive charges. The force toward the left is stronger than that toward the right, and B is attracted toward A, even though its net charge is zero. The attraction occurs whether the sign of A's charge is positive or negative (see Fig. 21.8). Furthermore, the effect is not limited to dielectrics; an uncharged conducting body would be attracted in the same way.

Test Your Understanding of Section 24.5 A parallel-plate capacitor has charges Q and $-Q$ on its two plates. A dielectric slab with $K = 3$ is then inserted into the space between the plates as shown in Fig. 24.20. Rank the following electric-field magnitudes in order from largest to smallest. (i) the field before the slab is inserted; (ii) the resultant field after the slab is inserted; (iii) the field due to the bound charges. ❙

24.6 Gauss's Law in Dielectrics

We can extend the analysis of Section 24.4 to reformulate Gauss's law in a form that is particularly useful for dielectrics. Figure 24.22 is a close-up view of the left capacitor plate and left surface of the dielectric in Fig. 24.15b. Let's apply Gauss's law to the rectangular box shown in cross section by the purple line; the surface area of the left and right sides is A. The left side is embedded in the conductor that forms the left capacitor plate, and so the electric field everywhere on that surface is zero. The right side is embedded in the dielectric, where the electric field has magnitude E, and $E_\perp = 0$ everywhere on the other four sides. The total charge enclosed, including both the charge on the capacitor plate and the induced charge on the dielectric surface, is $Q_{\text{encl}} = (\sigma - \sigma_i)A$, so Gauss's law gives

$$EA = \frac{(\sigma - \sigma_i)A}{\epsilon_0} \tag{24.21}$$

This equation is not very illuminating as it stands because it relates two unknown quantities: E inside the dielectric and the induced surface charge density σ_i. But now we can use Eq. (24.16), developed for this same situation, to simplify this equation by eliminating σ_i. Equation (24.16) is

$$\sigma_i = \sigma\left(1 - \frac{1}{K}\right) \quad \text{or} \quad \sigma - \sigma_i = \frac{\sigma}{K}$$

Combining this with Eq. (24.21), we get

$$EA = \frac{\sigma A}{K\epsilon_0} \quad \text{or} \quad KEA = \frac{\sigma A}{\epsilon_0} \tag{24.22}$$

Equation (24.22) says that the flux of $K\vec{E}$, not $\vec{E}$, through the Gaussian surface in Fig. 24.22 is equal to the enclosed *free* charge σA divided by ϵ_0. It turns out that for *any* Gaussian surface, whenever the induced charge is proportional to the electric field in the material, we can rewrite Gauss's law as

$$\oint K\vec{E} \cdot d\vec{A} = \frac{Q_{\text{encl-free}}}{\epsilon_0} \quad \text{(Gauss's law in a dielectric)} \tag{24.23}$$

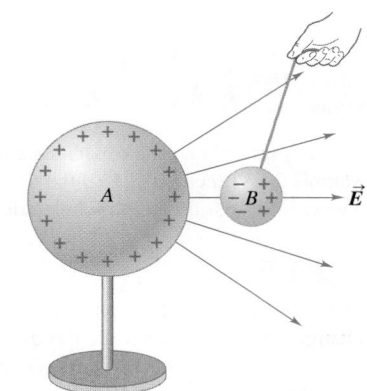

24.21 A neutral sphere B in the radial electric field of a positively charged sphere A is attracted to the charge because of polarization.

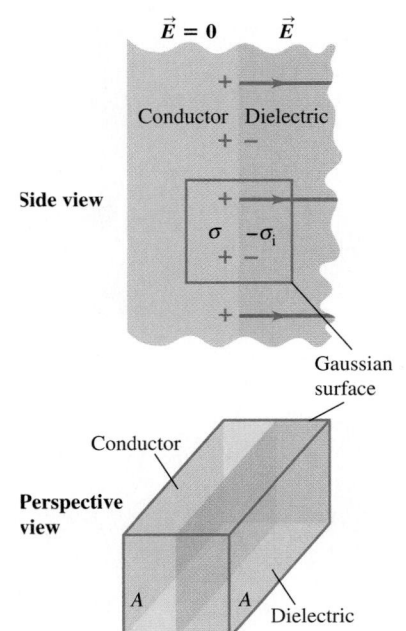

24.22 Gauss's law with a dielectric. This figure shows a close-up of the left-hand capacitor plate in Fig. 24.15b. The Gaussian surface is a rectangular box that lies half in the conductor and half in the dielectric.

where $Q_{\text{encl-free}}$ is the total *free* charge (not bound charge) enclosed by the Gaussian surface. The significance of these results is that the right sides contain only the *free* charge on the conductor, not the bound (induced) charge. In fact, although we have not proved it, Eq. (24.23) remains valid even when different parts of the Gaussian surface are embedded in dielectrics having different values of K, provided that the value of K in each dielectric is independent of the electric field (usually the case for electric fields that are not too strong) and that we use the appropriate value of K for each point on the Gaussian surface.

Example 24.12   A spherical capacitor with dielectric

Use Gauss's law to find the capacitance of the spherical capacitor of Example 24.3 (Section 24.1) if the volume between the shells is filled with an insulating oil with dielectric constant K.

SOLUTION

IDENTIFY and SET UP: The spherical symmetry of the problem is not changed by the presence of the dielectric, so as in Example 24.3, we use a concentric spherical Gaussian surface of radius r between the shells. Since a dielectric is present, we use Gauss's law in the form of Eq. (24.23).

EXECUTE: From Eq. (24.23),

$$\oint K\vec{E} \cdot d\vec{A} = \oint KE\, dA = KE \oint dA = (KE)(4\pi r^2) = \frac{Q}{\epsilon_0}$$

$$E = \frac{Q}{4\pi K\epsilon_0 r^2} = \frac{Q}{4\pi\epsilon r^2}$$

where $\epsilon = K\epsilon_0$. Compared to the case in which there is vacuum between the shells, the electric field is reduced by a factor of $1/K$. The potential difference V_{ab} between the shells is reduced by the same factor, and so the capacitance $C = Q/V_{ab}$ is *increased* by a factor of K, just as for a parallel-plate capacitor when a dielectric is inserted. Using the result of Example 24.3, we find that the capacitance with the dielectric is

$$C = \frac{4\pi K\epsilon_0 r_a r_b}{r_b - r_a} = \frac{4\pi\epsilon r_a r_b}{r_b - r_a}$$

EVALUATE: If the dielectric fills the volume between the two conductors, the capacitance is just K times the value with no dielectric. The result is more complicated if the dielectric only partially fills this volume (see Challenge Problem 24.78).

Test Your Understanding of Section 24.6   A single point charge q is imbedded in a dielectric of dielectric constant K. At a point inside the dielectric a distance r from the point charge, what is the magnitude of the electric field? (i) $q/4\pi\epsilon_0 r^2$; (ii) $Kq/4\pi\epsilon_0 r^2$; (iii) $q/4\pi K\epsilon_0 r^2$; (iv) none of these. ❚

Capacitors and capacitance: A capacitor is any pair of conductors separated by an insulating material. When the capacitor is charged, there are charges of equal magnitude Q and opposite sign on the two conductors, and the potential V_{ab} of the positively charged conductor with respect to the negatively charged conductor is proportional to Q. The capacitance C is defined as the ratio of Q to V_{ab}. The SI unit of capacitance is the farad (F): $1\ \text{F} = 1\ \text{C/V}$.

A parallel-plate capacitor consists of two parallel conducting plates, each with area A, separated by a distance d. If they are separated by vacuum, the capacitance depends only on A and d. For other geometries, the capacitance can be found by using the definition $C = Q/V_{ab}$. (See Examples 24.1–24.4.)

$$C = \frac{Q}{V_{ab}} \quad (24.1)$$

$$C = \frac{Q}{V_{ab}} = \epsilon_0 \frac{A}{d} \quad (24.2)$$

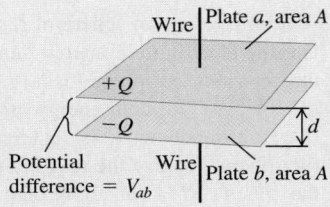

Capacitors in series and parallel: When capacitors with capacitances $C_1, C_2, C_3, \ldots$ are connected in series, the reciprocal of the equivalent capacitance C_{eq} equals the sum of the reciprocals of the individual capacitances. When capacitors are connected in parallel, the equivalent capacitance C_{eq} equals the sum of the individual capacitances. (See Examples 24.5 and 24.6.)

$$\frac{1}{C_{eq}} = \frac{1}{C_1} + \frac{1}{C_2} + \frac{1}{C_3} + \cdots \quad (24.5)$$
(capacitors in series)

$$C_{eq} = C_1 + C_2 + C_3 + \cdots \quad (24.7)$$
(capacitors in parallel)

Energy in a capacitor: The energy U required to charge a capacitor C to a potential difference V and a charge Q is equal to the energy stored in the capacitor. This energy can be thought of as residing in the electric field between the conductors; the energy density u (energy per unit volume) is proportional to the square of the electric-field magnitude. (See Examples 24.7–24.9.)

$$U = \frac{Q^2}{2C} = \frac{1}{2}CV^2 = \frac{1}{2}QV \quad (24.9)$$

$$u = \frac{1}{2}\epsilon_0 E^2 \quad (24.11)$$

Dielectrics: When the space between the conductors is filled with a dielectric material, the capacitance increases by a factor K, called the dielectric constant of the material. The quantity $\epsilon = K\epsilon_0$ is called the permittivity of the dielectric. For a fixed amount of charge on the capacitor plates, induced charges on the surface of the dielectric decrease the electric field and potential difference between the plates by the same factor K. The surface charge results from polarization, a microscopic rearrangement of charge in the dielectric. (See Example 24.10.)

Under sufficiently strong fields, dielectrics become conductors, a situation called dielectric breakdown. The maximum field that a material can withstand without breakdown is called its dielectric strength.

In a dielectric, the expression for the energy density is the same as in vacuum but with ϵ_0 replaced by $\epsilon = K\epsilon$. (See Example 24.11.)

Gauss's law in a dielectric has almost the same form as in vacuum, with two key differences: $\vec{E}$ is replaced by $K\vec{E}$ and Q_{encl} is replaced by $Q_{encl\text{-}free}$, which includes only the free charge (not bound charge) enclosed by the Gaussian surface. (See Example 24.12.)

$$C = KC_0 = K\epsilon_0 \frac{A}{d} = \epsilon \frac{A}{d} \quad (24.19)$$
(parallel-plate capacitor filled with dielectric)

$$u = \frac{1}{2}K\epsilon_0 E^2 = \frac{1}{2}\epsilon E^2 \quad (24.20)$$

$$\oint K\vec{E} \cdot d\vec{A} = \frac{Q_{encl\text{-}free}}{\epsilon_0} \quad (24.23)$$

Dielectric between plates

BRIDGING PROBLEM | **Electric-Field Energy and Capacitance of a Conducting Sphere**

A solid conducting sphere of radius R carries a charge Q. Calculate the electric-field energy density at a point a distance r from the center of the sphere for (a) $r < R$ and (b) $r > R$. (c) Calculate the total electric-field energy associated with the charged sphere. (d) How much work is required to assemble the charge Q on the sphere? (e) Use the result of part (c) to find the capacitance of the sphere. (You can think of the second conductor as a hollow conducting shell of infinite radius.)

SOLUTION GUIDE

See MasteringPhysics® study area for a Video Tutor solution

IDENTIFY and SET UP

1. You know the electric field for this situation at all values of r from Example 22.5 (Section 22.4). You'll use this to find the electric-field energy density u and the *total* electric-field energy U. You can then find the capacitance from the relationship $U = Q^2/2C$.
2. To find U, consider a spherical shell of radius r and thickness dr that has volume $dV = 4\pi r^2 dr$. The energy stored in this volume

is $u\, dV$, and the total energy is the integral of $u\, dV$ from $r = 0$ to $r \to \infty$. Set up this integral.

EXECUTE

3. Find u for $r < R$ and for $r > R$.
4. Substitute your results from step 3 into the expression from step 2. Then calculate the integral to find the total electric-field energy U.
5. Use your understanding of the energy stored in a charge distribution to find the work required to assemble the charge Q.
6. Find the capacitance of the sphere.

EVALUATE

7. Where is the electric-field energy density greatest? Where is it least?
8. How would the results be affected if the solid sphere were replaced by a hollow conducting sphere of the same radius R?
9. You can find the potential difference between the sphere and infinity from $C = Q/V$. Does this agree with the result of Example 23.8 (Section 23.3)?

Problems

For instructor-assigned homework, go to www.masteringphysics.com

•, ••, •••: Problems of increasing difficulty. **CP**: Cumulative problems incorporating material from earlier chapters. **CALC**: Problems requiring calculus. **BIO**: Biosciences problems.

DISCUSSION QUESTIONS

Q24.1 Equation (24.2) shows that the capacitance of a parallel-plate capacitor becomes larger as the plate separation d decreases. However, there is a practical limit to how small d can be made, which places limits on how large C can be. Explain what sets the limit on d. (*Hint:* What happens to the magnitude of the electric field as $d \to 0$?)

Q24.2 Suppose several different parallel-plate capacitors are charged up by a constant-voltage source. Thinking of the actual movement and position of the charges on an atomic level, why does it make sense that the capacitances are proportional to the surface areas of the plates? Why does it make sense that the capacitances are *inversely* proportional to the distance between the plates?

Q24.3 Suppose the two plates of a capacitor have different areas. When the capacitor is charged by connecting it to a battery, do the charges on the two plates have equal magnitude, or may they be different? Explain your reasoning.

Q24.4 At the Fermi National Accelerator Laboratory (Fermilab) in Illinois, protons are accelerated around a ring 2 km in radius to speeds that approach that of light. The energy for this is stored in capacitors the size of a house. When these capacitors are being charged, they make a very loud creaking sound. What is the origin of this sound?

Q24.5 In the parallel-plate capacitor of Fig. 24.2, suppose the plates are pulled apart so that the separation d is much larger than the size of the plates. (a) Is it still accurate to say that the electric field between the plates is uniform? Why or why not? (b) In the sit-

uation shown in Fig. 24.2, the potential difference between the plates is $V_{ab} = Qd/\epsilon_0 A$. If the plates are pulled apart as described above, is V_{ab} more or less than this formula would indicate? Explain your reasoning. (c) With the plates pulled apart as described above, is the capacitance more than, less than, or the same as that given by Eq. (24.2)? Explain your reasoning.

Q24.6 A parallel-plate capacitor is charged by being connected to a battery and is kept connected to the battery. The separation between the plates is then doubled. How does the electric field change? The charge on the plates? The total energy? Explain your reasoning.

Q24.7 A parallel-plate capacitor is charged by being connected to a battery and is then disconnected from the battery. The separation between the plates is then doubled. How does the electric field change? The potential difference? The total energy? Explain your reasoning.

Q24.8 Two parallel-plate capacitors, identical except that one has twice the plate separation of the other, are charged by the same voltage source. Which capacitor has a stronger electric field between the plates? Which capacitor has a greater charge? Which has greater energy density? Explain your reasoning.

Q24.9 The charged plates of a capacitor attract each other, so to pull the plates farther apart requires work by some external force. What becomes of the energy added by this work? Explain your reasoning.

Q24.10 The two plates of a capacitor are given charges $\pm Q$. The capacitor is then disconnected from the charging device so that the charges on the plates can't change, and the capacitor is immersed in a tank of oil. Does the electric field between the plates increase, decrease, or stay the same? Explain your reasoning. How can this field be measured?

Q24.11 As shown in Table 24.1, water has a very large dielectric constant $K = 80.4$. Why do you think water is not commonly used as a dielectric in capacitors?

Q24.12 Is dielectric strength the same thing as dielectric constant? Explain any differences between the two quantities. Is there a simple relationship between dielectric strength and dielectric constant (see Table 24.2)?

Q24.13 A capacitor made of aluminum foil strips separated by Mylar film was subjected to excessive voltage, and the resulting dielectric breakdown melted holes in the Mylar. After this, the capacitance was found to be about the same as before, but the breakdown voltage was much less. Why?

Q24.14 Suppose you bring a slab of dielectric close to the gap between the plates of a charged capacitor, preparing to slide it between the plates. What force will you feel? What does this force tell you about the energy stored between the plates once the dielectric is in place, compared to before the dielectric is in place?

Q24.15 The freshness of fish can be measured by placing a fish between the plates of a capacitor and measuring the capacitance. How does this work? (*Hint:* As time passes, the fish dries out. See Table 24.1.)

Q24.16 *Electrolytic* capacitors use as their dielectric an extremely thin layer of nonconducting oxide between a metal plate and a conducting solution. Discuss the advantage of such a capacitor over one constructed using a solid dielectric between the metal plates.

Q24.17 In terms of the dielectric constant K, what happens to the electric flux through the Gaussian surface shown in Fig. 24.22 when the dielectric is inserted into the previously empty space between the plates? Explain.

Q24.18 A parallel-plate capacitor is connected to a power supply that maintains a fixed potential difference between the plates. (a) If a sheet of dielectric is then slid between the plates, what happens to (i) the electric field between the plates, (ii) the magnitude of charge on each plate, and (iii) the energy stored in the capacitor? (b) Now suppose that before the dielectric is inserted, the charged capacitor is disconnected from the power supply. In this case, what happens to (i) the electric field between the plates, (ii) the magnitude of charge on each plate, and (iii) the energy stored in the capacitor? Explain any differences between the two situations.

Q24.19 Liquid dielectrics that have polar molecules (such as water) always have dielectric constants that decrease with increasing temperature. Why?

Q24.20 A conductor is an extreme case of a dielectric, since if an electric field is applied to a conductor, charges are free to move within the conductor to set up "induced charges." What is the dielectric constant of a perfect conductor? Is it $K = 0$, $K \rightarrow \infty$, or something in between? Explain your reasoning.

EXERCISES

Section 24.1 Capacitors and Capacitance

24.1 • The plates of a parallel-plate capacitor are 2.50 mm apart, and each carries a charge of magnitude 80.0 nC. The plates are in vacuum. The electric field between the plates has a magnitude of 4.00×10^6 V/m. (a) What is the potential difference between the plates? (b) What is the area of each plate? (c) What is the capacitance?

24.2 • The plates of a parallel-plate capacitor are 3.28 mm apart, and each has an area of 12.2 cm². Each plate carries a charge of magnitude 4.35×10^{-8} C. The plates are in vacuum. (a) What is the capacitance? (b) What is the potential difference between the plates? (c) What is the magnitude of the electric field between the plates?

24.3 • A parallel-plate air capacitor of capacitance 245 pF has a charge of magnitude 0.148 μC on each plate. The plates are 0.328 mm apart. (a) What is the potential difference between the plates? (b) What is the area of each plate? (c) What is the electric-field magnitude between the plates? (d) What is the surface charge density on each plate?

24.4 •• **Capacitance of an Oscilloscope.** Oscilloscopes have parallel metal plates inside them to deflect the electron beam. These plates are called the *deflecting plates*. Typically, they are squares 3.0 cm on a side and separated by 5.0 mm, with vacuum in between. What is the capacitance of these deflecting plates and hence of the oscilloscope? (*Note:* This capacitance can sometimes have an effect on the circuit you are trying to study and must be taken into consideration in your calculations.)

24.5 • A 10.0-μF parallel-plate capacitor with circular plates is connected to a 12.0-V battery. (a) What is the charge on each plate? (b) How much charge would be on the plates if their separation were doubled while the capacitor remained connected to the battery? (c) How much charge would be on the plates if the capacitor were connected to the 12.0-V battery after the radius of each plate was doubled without changing their separation?

24.6 • A 10.0-μF parallel-plate capacitor is connected to a 12.0-V battery. After the capacitor is fully charged, the battery is disconnected without loss of any of the charge on the plates. (a) A voltmeter is connected across the two plates without discharging them. What does it read? (b) What would the voltmeter read if (i) the plate separation were doubled; (ii) the radius of each plate were doubled but their separation was unchanged?

24.7 •• How far apart would parallel pennies have to be to make a 1.00-pF capacitor? Does your answer suggest that you are justified in treating these pennies as infinite sheets? Explain.

24.8 • A 5.00-pF, parallel-plate, air-filled capacitor with circular plates is to be used in a circuit in which it will be subjected to potentials of up to 1.00×10^2 V. The electric field between the plates is to be no greater than 1.00×10^4 N/C. As a budding electrical engineer for Live-Wire Electronics, your tasks are to (a) design the capacitor by finding what its physical dimensions and separation must be; (b) find the maximum charge these plates can hold.

24.9 • A parallel-plate air capacitor is to store charge of magnitude 240.0 pC on each plate when the potential difference between the plates is 42.0 V. (a) If the area of each plate is 6.80 cm², what is the separation between the plates? (b) If the separation between the two plates is double the value calculated in part (a), what potential difference is required for the capacitor to store charge of magnitude 240.0 pC on each plate?

24.10 • A cylindrical capacitor consists of a solid inner conducting core with radius 0.250 cm, surrounded by an outer hollow conducting tube. The two conductors are separated by air, and the length of the cylinder is 12.0 cm. The capacitance is 36.7 pF. (a) Calculate the inner radius of the hollow tube. (b) When the capacitor is charged to 125 V, what is the charge per unit length λ on the capacitor?

24.11 • A capacitor is made from two hollow, coaxial, iron cylinders, one inside the other. The inner cylinder is negatively charged and the outer is positively charged; the magnitude of the charge on each is 10.0 pC. The inner cylinder has radius 0.50 mm, the outer one has radius 5.00 mm, and the length of each cylinder is 18.0 cm. (a) What is the capacitance? (b) What applied potential difference is necessary to produce these charges on the cylinders?

24.12 •• A cylindrical capacitor has an inner conductor of radius 1.5 mm and an outer conductor of radius 3.5 mm. The two conductors are separated by vacuum, and the entire capacitor is 2.8 m long. (a) What is the capacitance per unit length? (b) The potential of the inner conductor is 350 mV higher than that of the outer conductor. Find the charge (magnitude and sign) on both conductors.

24.13 •• A spherical capacitor contains a charge of 3.30 nC when connected to a potential difference of 220 V. If its plates are separated by vacuum and the inner radius of the outer shell is 4.00 cm, calculate: (a) the capacitance; (b) the radius of the inner sphere; (c) the electric field just outside the surface of the inner sphere.

24.14 • A spherical capacitor is formed from two concentric, spherical, conducting shells separated by vacuum. The inner sphere has radius 15.0 cm and the capacitance is 116 pF. (a) What is the radius of the outer sphere? (b) If the potential difference between the two spheres is 220 V, what is the magnitude of charge on each sphere?

Section 24.2 Capacitors in Series and Parallel

24.15 • **BIO Electric Eels.** Electric eels and electric fish generate large potential differences that are used to stun enemies and prey. These potentials are produced by cells that each can generate 0.10 V. We can plausibly model such cells as charged capacitors. (a) How should these cells be connected (in series or in parallel) to produce a total potential of more than 0.10 V? (b) Using the connection in part (a), how many cells must be connected together to produce the 500-V surge of the electric eel?

24.16 • For the system of capacitors shown in Fig. E24.16, find the equivalent capacitance (a) between b and c, and (b) between a and c.

24.17 • In Fig. E24.17, each capacitor has $C = 4.00 \ \mu F$ and $V_{ab} = +28.0$ V. Calculate (a) the charge on each capacitor; (b) the potential difference across each capacitor; (c) the potential difference between points a and d.

24.18 • In Fig. 24.8a, let $C_1 = 3.00 \ \mu F$, $C_2 = 5.00 \ \mu F$, and $V_{ab} = +52.0$ V. Calculate (a) the charge on each capacitor and (b) the potential difference across each capacitor.

24.19 • In Fig. 24.9a, let $C_1 = 3.00 \ \mu F$, $C_2 = 5.00 \ \mu F$, and $V_{ab} = +52.0$ V. Calculate (a) the charge on each capacitor and (b) the potential difference across each capacitor.

24.20 • In Fig. E24.20, $C_1 = 6.00 \ \mu F$, $C_2 = 3.00 \ \mu F$, and $C_3 = 5.00 \ \mu F$. The capacitor network is connected to an applied potential V_{ab}. After the charges on the capacitors have reached their final values, the charge on C_2 is 40.0 μC. (a) What are the charges on capacitors C_1 and C_3? (b) What is the applied voltage V_{ab}?

24.21 •• For the system of capacitors shown in Fig. E24.21, a potential difference of 25 V is maintained across ab. (a) What is the equivalent capacitance of this system between a and b? (b) How much charge is stored by this system? (c) How much charge does the 6.5-nF capacitor store? (d) What is the potential difference across the 7.5-nF capacitor?

24.22 • Figure E24.22 shows a system of four capacitors, where the potential difference across ab is 50.0 V. (a) Find the equivalent capacitance of this system between a and b. (b) How much charge is stored by this combination of capacitors? (c) How much charge is stored in each of the 10.0-μF and the 9.0-μF capacitors?

24.23 •• Suppose the 3-μF capacitor in Fig. 24.10a were removed and replaced by a different one, and that this changed the equivalent capacitance between points a and b to 8 μF. What would be the capacitance of the replacement capacitor?

Section 24.3 Energy Storage in Capacitors and Electric-Field Energy

24.24 • A parallel-plate air capacitor has a capacitance of 920 pF. The charge on each plate is 2.55 μC. (a) What is the potential difference between the plates? (b) If the charge is kept constant, what will be the potential difference between the plates if the separation is doubled? (c) How much work is required to double the separation?

24.25 • A 5.80-μF, parallel-plate, air capacitor has a plate separation of 5.00 mm and is charged to a potential difference of 400 V. Calculate the energy density in the region between the plates, in units of J/m^3.

24.26 • An air capacitor is made from two flat parallel plates 1.50 mm apart. The magnitude of charge on each plate is 0.0180 μC when the potential difference is 200 V. (a) What is the capacitance? (b) What is the area of each plate? (c) What maximum voltage can be applied without dielectric breakdown? (Dielectric breakdown for air occurs at an electric-field strength of 3.0×10^6 V/m.) (d) When the charge is 0.0180 μC, what total energy is stored?

24.27 • A parallel-plate vacuum capacitor with plate area A and separation x has charges $+Q$ and $-Q$ on its plates. The capacitor is disconnected from the source of charge, so the charge on each plate remains fixed. (a) What is the total energy stored in the capacitor? (b) The plates are pulled apart an additional distance dx. What is the change in the stored energy? (c) If F is the force with

Figure **E24.16**

Figure **E24.17**

Figure **E24.20**

Figure **E24.21**

Figure **E24.22**

which the plates attract each other, then the change in the stored energy must equal the work $dW = Fdx$ done in pulling the plates apart. Find an expression for F. (d) Explain why F is *not* equal to QE, where E is the electric field between the plates.

24.28 •• A parallel-plate vacuum capacitor has 8.38 J of energy stored in it. The separation between the plates is 2.30 mm. If the separation is decreased to 1.15 mm, what is the energy stored (a) if the capacitor is disconnected from the potential source so the charge on the plates remains constant, and (b) if the capacitor remains connected to the potential source so the potential difference between the plates remains constant?

24.29 • You have two identical capacitors and an external potential source. (a) Compare the total energy stored in the capacitors when they are connected to the applied potential in series and in parallel. (b) Compare the maximum amount of charge stored in each case. (c) Energy storage in a capacitor can be limited by the maximum electric field between the plates. What is the ratio of the electric field for the series and parallel combinations?

24.30 • For the capacitor network shown in Fig. E24.30, the potential difference across ab is 36 V. Find (a) the total charge stored in this network; (b) the charge on each capacitor; (c) the total energy stored in the network; (d) the energy stored in each capacitor; (e) the potential differences across each capacitor.

Figure **E24.30**

150 nF   120 nF

$a \bullet$ —||—||— $\bullet b$

24.31 • For the capacitor network shown in Fig. E24.31, the potential difference across ab is 220 V. Find (a) the total charge stored in this network; (b) the charge on each capacitor; (c) the total energy stored in the network; (d) the energy stored in each capacitor; (e) the potential difference across each capacitor.

Figure **E24.31**

35 nF

$a \bullet$ —( )— $\bullet b$

75 nF

24.32 • A 0.350-m-long cylindrical capacitor consists of a solid conducting core with a radius of 1.20 mm and an outer hollow conducting tube with an inner radius of 2.00 mm. The two conductors are separated by air and charged to a potential difference of 6.00 V. Calculate (a) the charge per length for the capacitor; (b) the total charge on the capacitor; (c) the capacitance; (d) the energy stored in the capacitor when fully charged.

24.33 • A cylindrical air capacitor of length 15.0 m stores 3.20×10^{-9} J of energy when the potential difference between the two conductors is 4.00 V. (a) Calculate the magnitude of the charge on each conductor. (b) Calculate the ratio of the radii of the inner and outer conductors.

24.34 •• A capacitor is formed from two concentric spherical conducting shells separated by vacuum. The inner sphere has radius 12.5 cm, and the outer sphere has radius 14.8 cm. A potential difference of 120 V is applied to the capacitor. (a) What is the energy density at $r = 12.6$ cm, just outside the inner sphere? (b) What is the energy density at $r = 14.7$ cm, just inside the outer sphere? (c) For a parallel-plate capacitor the energy density is uniform in the region between the plates, except near the edges of the plates. Is this also true for a spherical capacitor?

Section 24.4 Dielectrics

24.35 • A 12.5-μF capacitor is connected to a power supply that keeps a constant potential difference of 24.0 V across the plates. A piece of material having a dielectric constant of 3.75 is placed between the plates, completely filling the space between them. (a) How much energy is stored in the capacitor before and after the dielectric is inserted? (b) By how much did the energy change during the insertion? Did it increase or decrease?

24.36 • A parallel-plate capacitor has capacitance $C_0 = 5.00$ pF when there is air between the plates. The separation between the plates is 1.50 mm. (a) What is the maximum magnitude of charge Q that can be placed on each plate if the electric field in the region between the plates is not to exceed 3.00×10^4 V/m? (b) A dielectric with $K = 2.70$ is inserted between the plates of the capacitor, completely filling the volume between the plates. Now what is the maximum magnitude of charge on each plate if the electric field between the plates is not to exceed 3.00×10^4 V/m?

24.37 • Two parallel plates have equal and opposite charges. When the space between the plates is evacuated, the electric field is $E = 3.20 \times 10^5$ V/m. When the space is filled with dielectric, the electric field is $E = 2.50 \times 10^5$ V/m. (a) What is the charge density on each surface of the dielectric? (b) What is the dielectric constant?

24.38 • A budding electronics hobbyist wants to make a simple 1.0-nF capacitor for tuning her crystal radio, using two sheets of aluminum foil as plates, with a few sheets of paper between them as a dielectric. The paper has a dielectric constant of 3.0, and the thickness of one sheet of it is 0.20 mm. (a) If the sheets of paper measure 22 × 28 cm and she cuts the aluminum foil to the same dimensions, how many sheets of paper should she use between her plates to get the proper capacitance? (b) Suppose for convenience she wants to use a single sheet of posterboard, with the same dielectric constant but a thickness of 12.0 mm, instead of the paper. What area of aluminum foil will she need for her plates to get her 1.0 nF of capacitance? (c) Suppose she goes high-tech and finds a sheet of Teflon of the same thickness as the posterboard to use as a dielectric. Will she need a larger or smaller area of Teflon than of posterboard? Explain.

24.39 • The dielectric to be used in a parallel-plate capacitor has a dielectric constant of 3.60 and a dielectric strength of 1.60×10^7 V/m. The capacitor is to have a capacitance of 1.25×10^{-9} F and must be able to withstand a maximum potential difference of 5500 V. What is the minimum area the plates of the capacitor may have?

24.40 •• **BIO** **Potential in Human Cells.** Some cell walls in the human body have a layer of negative charge on the inside surface and a layer of positive charge of equal magnitude on the outside surface. Suppose that the charge density on either surface is $\pm 0.50 \times 10^{-3}$ C/m^2, the cell wall is 5.0 nm thick, and the cell-wall material is air. (a) Find the magnitude of $\vec{E}$ in the wall between the two layers of charge. (b) Find the potential difference between the inside and the outside of the cell. Which is at the higher potential? (c) A typical cell in the human body has a volume of 10^{-16} m^3. Estimate the total electric-field energy stored in the wall of a cell of this size. (*Hint:* Assume that the cell is spherical, and calculate the volume of the cell wall.) (d) In reality, the cell wall is made up, not of air, but of tissue with a dielectric constant of 5.4. Repeat parts (a) and (b) in this case.

24.41 • A capacitor has parallel plates of area 12 cm^2 separated by 2.0 mm. The space between the plates is filled with polystyrene (see Table 24.2). (a) Find the permittivity of polystyrene. (b) Find the maximum permissible voltage across the capacitor to avoid dielectric breakdown. (c) When the voltage equals the value found in part (b), find the surface charge density on each plate and the induced surface charge density on the surface of the dielectric.

24.42 • A constant potential difference of 12 V is maintained between the terminals of a 0.25-μF, parallel-plate, air capacitor. (a) A sheet of Mylar is inserted between the plates of the capacitor, completely filling the space between the plates. When this is done, how much additional charge flows onto the positive plate of the capacitor (see Table 24.1)? (b) What is the total induced charge on either face of the Mylar sheet? (c) What effect does the Mylar sheet have on the electric field between the plates? Explain how you can reconcile this with the increase in charge on the plates, which acts to *increase* the electric field.

24.43 • When a 360-nF air capacitor $(1\ \text{nF} = 10^{-9}\ \text{F})$ is connected to a power supply, the energy stored in the capacitor is 1.85×10^{-5} J. While the capacitor is kept connected to the power supply, a slab of dielectric is inserted that completely fills the space between the plates. This increases the stored energy by 2.32×10^{-5} J. (a) What is the potential difference between the capacitor plates? (b) What is the dielectric constant of the slab?

24.44 • A parallel-plate capacitor has capacitance $C = 12.5$ pF when the volume between the plates is filled with air. The plates are circular, with radius 3.00 cm. The capacitor is connected to a battery, and a charge of magnitude 25.0 pC goes onto each plate. With the capacitor still connected to the battery, a slab of dielectric is inserted between the plates, completely filling the space between the plates. After the dielectric has been inserted, the charge on each plate has magnitude 45.0 pC. (a) What is the dielectric constant K of the dielectric? (b) What is the potential difference between the plates before and after the dielectric has been inserted? (c) What is the electric field at a point midway between the plates before and after the dielectric has been inserted?

Section 24.6 Gauss's Law in Dielectrics

24.45 • A parallel-plate capacitor has the volume between its plates filled with plastic with dielectric constant K. The magnitude of the charge on each plate is Q. Each plate has area A, and the distance between the plates is d. (a) Use Gauss's law as stated in Eq. (24.23) to calculate the magnitude of the electric field in the dielectric. (b) Use the electric field determined in part (a) to calculate the potential difference between the two plates. (c) Use the result of part (b) to determine the capacitance of the capacitor. Compare your result to Eq. (24.12).

24.46 • A parallel-plate capacitor has plates with area 0.0225 m^2 separated by 1.00 mm of Teflon. (a) Calculate the charge on the plates when they are charged to a potential difference of 12.0 V. (b) Use Gauss's law (Eq. 24.23) to calculate the electric field inside the Teflon. (c) Use Gauss's law to calculate the electric field if the voltage source is disconnected and the Teflon is removed.

PROBLEMS

24.47 • Electronic flash units for cameras contain a capacitor for storing the energy used to produce the flash. In one such unit, the flash lasts for $\frac{1}{675}$ s with an average light power output of 2.70×10^5 W. (a) If the conversion of electrical energy to light is 95% efficient (the rest of the energy goes to thermal energy), how much energy must be stored in the capacitor for one flash? (b) The capacitor has a potential difference between its plates of 125 V when the stored energy equals the value calculated in part (a). What is the capacitance?

24.48 • A parallel-plate air capacitor is made by using two plates 16 cm square, spaced 3.7 mm apart. It is connected to a 12-V battery. (a) What is the capacitance? (b) What is the charge on each

plate? (c) What is the electric field between the plates? (d) What is the energy stored in the capacitor? (e) If the battery is disconnected and then the plates are pulled apart to a separation of 7.4 mm, what are the answers to parts (a)–(d)?

24.49 •• Suppose the battery in Problem 24.48 remains connected while the plates are pulled apart. What are the answers then to parts (a)–(d) after the plates have been pulled apart?

24.50 ••• **BIO** **Cell Membranes.** Cell membranes (the walled enclosure around a cell) are typically about 7.5 nm thick. They are partially permeable to allow charged material to pass in and out, as needed. Equal but opposite charge densities build up on the inside and outside faces of such a membrane, and these charges prevent additional charges from passing through the cell wall. We can model a cell membrane as a parallel-plate capacitor, with the membrane itself containing proteins embedded in an organic material to give the membrane a dielectric constant of about 10. (See Fig. P24.50.) (a) What is the capacitance per square centimeter of such a cell wall? (b) In its normal resting state, a cell has a potential difference of 85 mV across its membrane. What is the electric field inside this membrane?

Figure **P24.50**

24.51 • A capacitor is made from two hollow, coaxial copper cylinders, one inside the other. There is air in the space between the cylinders. The inner cylinder has net positive charge and the outer cylinder has net negative charge. The inner cylinder has radius 2.50 mm, the outer cylinder has radius 3.10 mm, and the length of each cylinder is 36.0 cm. If the potential difference between the surfaces of the two cylinders is 80.0 V, what is the magnitude of the electric field at a point between the two cylinders that is a distance of 2.80 mm from their common axis and midway between the ends of the cylinders?

24.52 ••• In one type of computer keyboard, each key holds a small metal plate that serves as one plate of a parallel-plate, air-filled capacitor. When the key is depressed, the plate separation decreases and the capacitance increases. Electronic circuitry detects the change in capacitance and thus detects that the key has been pressed. In one particular keyboard, the area of each metal plate is 42.0 mm^2, and the separation between the plates is 0.700 mm before the key is depressed. (a) Calculate the capacitance before the key is depressed. (b) If the circuitry can detect a change in capacitance of 0.250 pF, how far must the key be depressed before the circuitry detects its depression?

24.53 •• A 20.0-μF capacitor is charged to a potential difference of 800 V. The terminals of the charged capacitor are then connected to those of an uncharged 10.0-μF capacitor. Compute (a) the original charge of the system, (b) the final potential difference across each capacitor, (c) the final energy of the system, and (d) the decrease in energy when the capacitors are connected.

24.54 •• In Fig. 24.9a, let $C_1 = 9.0\ \mu$F, $C_2 = 4.0\ \mu$F, and $V_{ab} = 36$ V. Suppose the charged capacitors are disconnected from the source and from each other, and then reconnected to each other with plates of *opposite* sign together. By how much does the energy of the system decrease?

24.55 • For the capacitor network shown in Fig. P24.55, the potential difference across ab is 12.0 V. Find (a) the total energy stored in this network and (b) the energy stored in the 4.80-μF capacitor.

Figure **P24.55**

24.56 •• Several 0.25-μF capacitors are available. The voltage across each is not to exceed 600 V. You need to make a capacitor with capacitance 0.25 μF to be connected across a potential difference of 960 V. (a) Show in a diagram how an equivalent capacitor with the desired properties can be obtained. (b) No dielectric is a perfect insulator that would not permit the flow of any charge through its volume. Suppose that the dielectric in one of the capacitors in your diagram is a moderately good conductor. What will happen in this case when your combination of capacitors is connected across the 960-V potential difference?

24.57 • In Fig. P24.57, $C_1 = C_5 = 8.4$ μF and $C_2 = C_3 = C_4 = 4.2$ μF. The applied potential is $V_{ab} = 220$ V. (a) What is the equivalent capacitance of the network between points a and b? (b) Calculate the charge on each capacitor and the potential difference across each capacitor.

Figure **P24.57**

24.58 •• You are working on an electronics project requiring a variety of capacitors, but you have only a large supply of 100-nF capacitors available. Show how you can connect these capacitors to produce each of the following equivalent capacitances: (a) 50 nF; (b) 450 nF; (c) 25 nF; (d) 75 nF.

24.59 •• In Fig. E24.20, $C_1 = 3.00$ μF and $V_{ab} = 150$ V. The charge on capacitor C_1 is 150 μC and the charge on C_3 is 450 μC. What are the values of the capacitances of C_2 and C_3?

24.60 • The capacitors in Fig. P24.60 are initially uncharged and are connected, as in the diagram, with switch S open. The applied potential difference is $V_{ab} = +210$ V. (a) What is the potential difference V_{cd}? (b) What is the potential difference across each capacitor after switch S is closed? (c) How much charge flowed through the switch when it was closed?

Figure **P24.60**

24.61 •• Three capacitors having capacitances of 8.4, 8.4, and 4.2 μF are connected in series across a 36-V potential difference. (a) What is the charge on the 4.2-μF capacitor? (b) What is the total energy stored in all three capacitors? (c) The capacitors are disconnected from the potential difference without allowing them to discharge. They are then reconnected in parallel with each other, with the positively charged plates connected together. What is the voltage across each capacitor in the parallel combination? (d) What is the total energy now stored in the capacitors?

24.62 • **Capacitance of a Thundercloud.** The charge center of a thundercloud, drifting 3.0 km above the earth's surface, contains 20 C of negative charge. Assuming the charge center has a radius of 1.0 km, and modeling the charge center and the earth's surface as parallel plates, calculate: (a) the capacitance of the system; (b)

the potential difference between charge center and ground; (c) the average strength of the electric field between cloud and ground; (d) the electrical energy stored in the system.

24.63 •• In Fig. P24.63, each capacitance C_1 is 6.9 μF, and each capacitance C_2 is 4.6 μF. (a) Compute the equivalent capacitance of the network between points a and b. (b) Compute the charge on each of the three capacitors nearest a and b when $V_{ab} = 420$ V. (c) With 420 V across a and b, compute V_{cd}.

Figure **P24.63**

24.64 • Each combination of capacitors between points a and b in Fig. P24.64 is first connected across a 120-V battery, charging the combination to 120 V. These combinations are then connected to make the circuits shown. When the switch S is thrown, a surge of charge for the discharging capacitors flows to trigger the signal device. How much charge flows through the signal device in each case?

Figure **P24.64**

(a)

(b)

24.65 • A parallel-plate capacitor with only air between the plates is charged by connecting it to a battery. The capacitor is then disconnected from the battery, without any of the charge leaving the plates. (a) A voltmeter reads 45.0 V when placed across the capacitor. When a dielectric is inserted between the plates, completely filling the space, the voltmeter reads 11.5 V. What is the dielectric constant of this material? (b) What will the voltmeter read if the dielectric is now pulled partway out so it fills only one-third of the space between the plates?

24.66 •• An air capacitor is made by using two flat plates, each with area A, separated by a distance d. Then a metal slab having thickness a (less than d) and the same shape and size as the plates is inserted between them, parallel to the plates and not touching either plate (Fig. P24.66). (a) What is the capacitance of this arrangement? (b) Express the capacitance as a multiple of the capacitance C_0 when the metal slab is not present. (c) Discuss what happens to the capacitance in the limits $a \to 0$ and $a \to d$.

Figure **P24.66**

24.67 •• **Capacitance of the Earth.** Consider a spherical capacitor with one conductor being a solid conducting sphere of radius R and the other conductor being at infinity. (a) Use Eq. (24.1) and what you know about the potential at the surface of a conducting sphere with charge Q to derive an expression for the capacitance of the charged sphere. (b) Use your result in part (a) to calculate the capacitance of the earth. The earth is a good conductor and has a radius of 6380 km. Compare your results to the capacitance of typical capacitors used in electronic circuits, which ranges from 10 pF to 100 pF.

24.68 •• A potential difference $V_{ab} = 48.0$ V is applied across the capacitor network of Fig. E24.17. If $C_1 = C_2 = 4.00$ μF and

$C_4 = 8.00 \ \mu F$, what must the capacitance C_3 be if the network is to store 2.90×10^{-3} J of electrical energy?

24.69 • Earth-Ionosphere Capacitance. The earth can be considered as a single-conductor capacitor (see Problem 24.67). It can also be considered in combination with a charged layer of the atmosphere, the ionosphere, as a spherical capacitor with two plates, the surface of the earth being the negative plate. The ionosphere is at a level of about 70 km, and the potential difference between earth and ionosphere is about 350,000 V. Calculate: (a) the capacitance of this system; (b) the total charge on the capacitor; (c) the energy stored in the system.

24.70 • CALC The inner cylinder of a long, cylindrical capacitor has radius r_a and linear charge density $+\lambda$. It is surrounded by a coaxial cylindrical conducting shell with inner radius r_b and linear charge density $-\lambda$ (see Fig. 24.6). (a) What is the energy density in the region between the conductors at a distance r from the axis? (b) Integrate the energy density calculated in part (a) over the volume between the conductors in a length L of the capacitor to obtain the total electric-field energy per unit length. (c) Use Eq. (24.9) and the capacitance per unit length calculated in Example 24.4 (Section 24.1) to calculate U/L. Does your result agree with that obtained in part (b)?

24.71 •• CP A capacitor has a potential difference of 2.25×10^3 V between its plates. A short aluminum wire with initial temperature $23.0°C$ is connected between the plates of the capacitor and all the energy stored in the capacitor goes into heating the wire. The wire has mass 12.0 g. If no heat is lost to the surroundings and the final temperature of the wire is $34.2°C$, what is the capacitance of the capacitor?

24.72 •• A parallel-plate capacitor is made from two plates 12.0 cm on each side and 4.50 mm apart. Half of the space between these plates contains only air, but the other half is filled with Plexiglas® of dielectric constant 3.40 (Fig. P24.72). An 18.0-V battery is connected across the plates. (a) What is the capacitance of this combination? (*Hint:* Can you think of this capacitor as equivalent to two capacitors in parallel?) (b) How much energy is stored in the capacitor? (c) If we remove the Plexiglas® but change nothing else, how much energy will be stored in the capacitor?

Figure **P24.72**

Plexiglas®   Air

24.73 •• A parallel-plate capacitor has square plates that are 8.00 cm on each side and 3.80 mm apart. The space between the plates is completely filled with two square slabs of dielectric, each 8.00 cm on a side and 1.90 mm thick. One slab is pyrex glass and the other is polystyrene. If the potential difference between the plates is 86.0 V, how much electrical energy is stored in the capacitor?

24.74 •• A fuel gauge uses a capacitor to determine the height of the fuel in a tank. The effective dielectric constant K_{eff} changes from a value of 1 when the tank is empty to a value of K, the dielectric constant of the fuel, when the tank is full. The appropriate electronic circuitry can determine the effective dielectric constant of the combined air and fuel between the capacitor plates. Each of the two rectangular plates has a width

Figure **P24.74**

w and a length L (Fig. P24.74). The height of the fuel between the plates is h. You can ignore any fringing effects. (a) Derive an expression for K_{eff} as a function of h. (b) What is the effective dielectric constant for a tank $\frac{1}{4}$ full, $\frac{1}{2}$ full, and $\frac{3}{4}$ full if the fuel is gasoline $(K = 1.95)$? (c) Repeat part (b) for methanol $(K = 33.0)$. (d) For which fuel is this fuel gauge more practical?

24.75 •• Three square metal plates A, B, and C, each 12.0 cm on a side and 1.50 mm thick, are arranged as in Fig. P24.75. The plates are separated by sheets of paper 0.45 mm thick and with dielectric constant 4.2. The outer plates are connected together and connected to point b. The inner plate is connected to point a. (a) Copy the diagram and show by plus and minus signs the charge distribution on the plates when point a is maintained at a positive potential relative to point b. (b) What is the capacitance between points a and b?

Figure **P24.75**

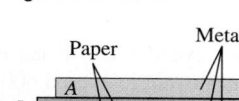

Paper   Metal

CHALLENGE PROBLEMS

24.76 ••• CP The parallel-plate air capacitor in Fig. P24.76 consists of two horizontal conducting plates of equal area A. The bottom plate rests on a fixed support, and the top plate is suspended by four springs with spring constant k, positioned at each of the four corners of the top plate as shown in the figure. When uncharged, the plates are separated by a distance z_0. A battery is connected to the plates and produces a potential difference V between them. This causes the plate separation to decrease to z. Neglect any fringing effects. (a) Show that the electrostatic force between the charged plates has a magnitude $\epsilon_0 A V^2 / 2z^2$. (*Hint:* See Exercise 24.27.) (b) Obtain an expression that relates the plate separation z to the potential difference V. The resulting equation will be cubic in z. (c) Given the values $A = 0.300 \ \text{m}^2$, $z_0 = 1.20$ mm, $k = 25.0$ N/m, and $V = 120$ V, find the two values of z for which the top plate will be in equilibrium. (*Hint:* You can solve the cubic equation by plugging a trial value of z into the equation and then adjusting your guess until the equation is satisfied to three significant figures. Locating the roots of the cubic equation graphically can help you pick starting values of z for this trial-and-error procedure. One root of the cubic equation has a nonphysical negative value.) (d) For each of the two values of z found in part (c), is the equilibrium stable or unstable? For stable equilibrium a small displacement of the object will give rise to a net force tending to return the object to the equilibrium position. For unstable equilibrium a small displacement gives rise to a net force that takes the object farther away from equilibrium.

Figure **P24.76**

24.77 ••• Two square conducting plates with sides of length L are separated by a distance D. A dielectric slab with constant K with dimensions $L \times L \times D$ is inserted a distance x into the space between the plates, as shown in Fig. P24.77. (a) Find the capacitance

C of this system. (b) Suppose that the capacitor is connected to a battery that maintains a constant potential difference V between the plates. If the dielectric slab is inserted an additional distance dx into the space between the plates, show that the change in stored energy is

$$dU = +\frac{(K-1)\epsilon_0 V^2 L}{2D}dx$$

(c) Suppose that before the slab is moved by dx, the plates are disconnected from the battery, so that the charges on the plates remain constant. Determine the magnitude of the charge on each plate, and then show that when the slab is moved dx farther into the space between the plates, the stored energy changes by an amount that is the *negative* of the expression for dU given in part (b). (d) If F is the force exerted on the slab by the charges on the plates, then dU should equal the work done *against* this force to move the slab a distance dx. Thus $dU = -F\,dx$. Show that applying this expression to the result of part (b) suggests that the electric force on the slab pushes it *out* of the capacitor, while the result of part (c) suggests that the force

Figure **P24.77**

Dielectric slab, constant K

pulls the slab *into* the capacitor. (e) Figure 24.16 shows that the force in fact pulls the slab into the capacitor. Explain why the result of part (b) gives an incorrect answer for the direction of this force, and calculate the magnitude of the force. (This method does not require knowledge of the nature of the fringing field.)

24.78 ••• An isolated spherical capacitor has charge $+Q$ on its inner conductor (radius r_a) and charge $-Q$ on its outer conductor (radius r_b). Half of the volume between the two conductors is then filled with a liquid dielectric of constant K, as shown in cross section in Fig. P24.78. (a) Find the capacitance of the half-filled capacitor. (b) Find the magnitude of $\vec{E}$ in the volume between the two conductors as a function of the distance r from the center of the capacitor. Give answers for both the upper and lower halves of this volume. (c) Find the surface density of free charge on the upper and lower halves of the inner and outer conductors. (d) Find the surface density of bound charge on the inner ($r = r_a$) and outer ($r = r_b$) surfaces of the dielectric. (e) What is the surface density of bound charge on the flat surface of the dielectric? Explain.

Figure **P24.78**

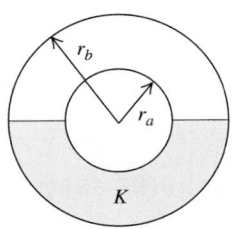

Answers

Chapter Opening Question ?

Equation (24.9) shows that the energy stored in a capacitor with capacitance C and charge Q is $U = Q^2/2C$. If the charge Q is doubled, the stored energy increases by a factor of $2^2 = 4$. Note that if the value of Q is too great, the electric-field magnitude inside the capacitor will exceed the dielectric strength of the material between the plates and dielectric breakdown will occur (see Section 24.4). This puts a practical limit on the amount of energy that can be stored.

Test Your Understanding Questions

24.1 Answer: (iii) The capacitance does not depend on the value of the charge Q. Doubling the value of Q causes the potential difference V_{ab} to double, so the capacitance $C = Q/V_{ab}$ remains the same. These statements are true no matter what the geometry of the capacitor.

24.2 Answers: (a) (i), (b) (iv) In a series connection the two capacitors carry the same charge Q but have different potential differences $V_{ab} = Q/C$; the capacitor with the smaller capacitance C has the greater potential difference. In a parallel connection the two capacitors have the same potential difference V_{ab} but carry different charges $Q = CV_{ab}$; the capacitor with the larger capacitance C has the greater charge. Hence a 4-μF capacitor will have a greater potential difference than an 8-μF capacitor if the two are connected in series. The 4-μF capacitor cannot carry more charge than the 8-μF capacitor no matter how they are connected: In a series connection they will carry the same charge, and in a parallel connection the 8-μF capacitor will carry more charge.

24.3 Answer: (i) Capacitors connected in series carry the same charge Q. To compare the amount of energy stored, we use the expression $U = Q^2/2C$ from Eq. (24.9); it shows that the capacitor with the *smaller* capacitance ($C = 4\,\mu$F) has more stored

energy in a series combination. By contrast, capacitors in parallel have the same potential difference V, so to compare them we use $U = \frac{1}{2}CV^2$ from Eq. (24.9). It shows that in a parallel combination, the capacitor with the *larger* capacitance ($C = 8\,\mu$F) has more stored energy. (If we had instead used $U = \frac{1}{2}CV^2$ to analyze the series combination, we would have to account for the different potential differences across the two capacitors. Likewise, using $U = Q^2/2C$ to study the parallel combination would require us to account for the different charges on the capacitors.)

24.4 Answer: (i) Here Q remains the same, so we use $U = Q^2/2C$ from Eq. (24.9) for the stored energy. Removing the dielectric lowers the capacitance by a factor of $1/K$; since U is inversely proportional to C, the stored energy *increases* by a factor of K. It takes work to pull the dielectric slab out of the capacitor because the fringing field tries to pull the slab back in (Fig. 24.16). The work that you do goes into the energy stored in the capacitor.

24.5 Answer: (i), (iii), (ii) Equation (24.14) says that if E_0 is the initial electric-field magnitude (before the dielectric slab is inserted), then the resultant field magnitude after the slab is inserted is $E_0/K = E_0/3$. The magnitude of the resultant field equals the difference between the initial field magnitude and the magnitude E_i of the field due to the bound charges (see Fig. 24.20). Hence $E_0 - E_i = E_0/3$ and $E_i = 2E_0/3$.

24.6 Answer: (iii) Equation (24.23) shows that this situation is the same as an isolated point charge in vacuum but with $\vec{E}$ replaced by $K\vec{E}$. Hence KE at the point of interest is equal to $q/4\pi\epsilon_0 r^2$, and so $E = q/4\pi K\epsilon_0 r^2$. As in Example 24.12, filling the space with a dielectric reduces the electric field by a factor of $1/K$.

Bridging Problem

Answers: (a) 0 **(b)** $Q^2/32\pi^2\epsilon_0 r^4$ **(c)** $Q^2/8\pi\epsilon_0 R$ **(d)** $Q^2/8\pi\epsilon_0 R$ **(e)** $C = 4\pi\epsilon_0 R$

25 CURRENT, RESISTANCE, AND ELECTROMOTIVE FORCE

LEARNING GOALS

By studying this chapter, you will learn:

- The meaning of electric current, and how charges move in a conductor.

- What is meant by the resistivity and conductivity of a substance.

- How to calculate the resistance of a conductor from its dimensions and its resistivity.

- How an electromotive force (emf) makes it possible for current to flow in a circuit.

- How to do calculations involving energy and power in circuits.

? In a flashlight, is the amount of current that flows out of the bulb less than, greater than, or equal to the amount of current that flows into the bulb?

In the past four chapters we studied the interactions of electric charges *at rest;* now we're ready to study charges *in motion.* An *electric current* consists of charges in motion from one region to another. If the charges follow a conducting path that forms a closed loop, the path is called an *electric circuit.*

Fundamentally, electric circuits are a means for conveying *energy* from one place to another. As charged particles move within a circuit, electric potential energy is transferred from a source (such as a battery or generator) to a device in which that energy is either stored or converted to another form: into sound in a stereo system or into heat and light in a toaster or light bulb. Electric circuits are useful because they allow energy to be transported without any moving parts (other than the moving charged particles themselves). They are at the heart of flashlights, computers, radio and television transmitters and receivers, and household and industrial power distribution systems. Your nervous system is a specialized electric circuit that carries vital signals from one part of your body to another.

In Chapter 26 we will see how to analyze electric circuits and will examine some practical applications of circuits. Before we can do so, however, you must understand the basic properties of electric currents. These properties are the subject of this chapter. We'll begin by describing the nature of electric conductors and considering how they are affected by temperature. We'll learn why a short, fat, cold copper wire is a better conductor than a long, skinny, hot steel wire. We'll study the properties of batteries and see how they cause current and energy transfer in a circuit. In this analysis we will use the concepts of current, potential difference (or voltage), resistance, and electromotive force. Finally, we'll look at electric current in a material from a microscopic viewpoint.

25.1 Current

A **current** is any motion of charge from one region to another. In this section we'll discuss currents in conducting materials. The vast majority of technological applications of charges in motion involve currents of this kind.

In electrostatic situations (discussed in Chapters 21 through 24) the electric field is zero everywhere within the conductor, and there is *no* current. However, this does not mean that all charges within the conductor are at rest. In an ordinary metal such as copper or alumium, some of the electrons are free to move within the conducting material. These free electrons move randomly in all directions, somewhat like the molecules of a gas but with much greater speeds, of the order of 10^6 m/s. The electrons nonetheless do not escape from the conducting material, because they are attracted to the positive ions of the material. The motion of the electrons is random, so there is no *net* flow of charge in any direction and hence no current.

Now consider what happens if a constant, steady electric field $\vec{E}$ is established inside a conductor. (We'll see later how this can be done.) A charged particle (such as a free electron) inside the conducting material is then subjected to a steady force $\vec{F} = q\vec{E}$. If the charged particle were moving in *vacuum,* this steady force would cause a steady acceleration in the direction of $\vec{F}$, and after a time the charged particle would be moving in that direction at high speed. But a charged particle moving in a *conductor* undergoes frequent collisions with the massive, nearly stationary ions of the material. In each such collision the particle's direction of motion undergoes a random change. The net effect of the electric field $\vec{E}$ is that in addition to the random motion of the charged particles within the conductor, there is also a very slow net motion or *drift* of the moving charged particles as a group in the direction of the electric force $\vec{F} = q\vec{E}$ (Fig. 25.1). This motion is described in terms of the **drift velocity** $\vec{v}_d$ of the particles. As a result, there is a net current in the conductor.

While the random motion of the electrons has a very fast average speed of about 10^6 m/s, the drift speed is very slow, often on the order of 10^{-4} m/s. Given that the electrons move so slowly, you may wonder why the light comes on immediately when you turn on the switch of a flashlight. The reason is that the electric field is set up in the wire with a speed approaching the speed of light, and electrons start to move all along the wire at very nearly the same time. The time that it takes any individual electron to get from the switch to the light bulb isn't really relevant. A good analogy is a group of soldiers standing at attention when the sergeant orders them to start marching; the order reaches the soldiers' ears at the speed of sound, which is much faster than their marching speed, so all the soldiers start to march essentially in unison.

The Direction of Current Flow

The drift of moving charges through a conductor can be interpreted in terms of work and energy. The electric field $\vec{E}$ does work on the moving charges. The resulting kinetic energy is transferred to the material of the conductor by means of collisions with the ions, which vibrate about their equilibrium positions in the crystalline structure of the conductor. This energy transfer increases the average vibrational energy of the ions and therefore the temperature of the material. Thus much of the work done by the electric field goes into heating the conductor, *not* into making the moving charges move ever faster and faster. This heating is sometimes useful, as in an electric toaster, but in many situations is simply an unavoidable by-product of current flow.

In different current-carrying materials, the charges of the moving particles may be positive or negative. In metals the moving charges are always (negative) electrons, while in an ionized gas (plasma) or an ionic solution the moving charges may include both electrons and positively charged ions. In a semiconductor

25.1 If there is no electric field inside a conductor, an electron moves randomly from point P_1 to point P_2 in a time Δt. If an electric field $\vec{E}$ is present, the electric force $\vec{F} = q\vec{E}$ imposes a small drift (greatly exaggerated here) that takes the electron to point P'_2, a distance $v_d \Delta t$ from P_2 in the direction of the force.

Conductor without internal $\vec{E}$ field

Path of electron without $\vec{E}$ field. Electron moves randomly.

Path of electron with $\vec{E}$ field. The motion is mostly random, but ...

P_1

P_2 $\quad P'_2$

$v_d \Delta t$

...the $\vec{E}$ field results in a net displacement along the wire.

Conductor with internal $\vec{E}$ field

$\vec{E}$ $\quad \vec{F} = q\vec{E}$ $\quad \vec{E}$

An electron has a negative charge q, so the force on it due to the $\vec{E}$ field is in the direction opposite to $\vec{E}$.

25.2 The same current can be produced by (a) positive charges moving in the direction of the electric field $\vec{E}$ or (b) the same number of negative charges moving at the same speed in the direction opposite to $\vec{E}$.

(a)

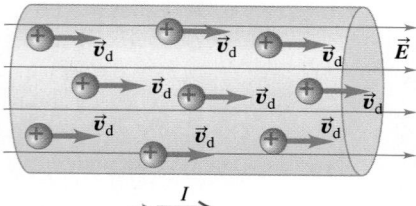

A **conventional current** is treated as a flow of positive charges, regardless of whether the free charges in the conductor are positive, negative, or both.

(b)

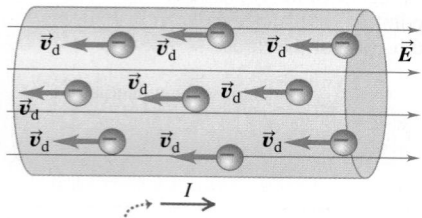

In a metallic conductor, the moving charges are electrons — but the *current* still points in the direction positive charges would flow.

25.3 The current I is the time rate of charge transfer through the cross-sectional area A. The random component of each moving charged particle's motion averages to zero, and the current is in the same direction as $\vec{E}$ whether the moving charges are positive (as shown here) or negative (see Fig. 25.2b).

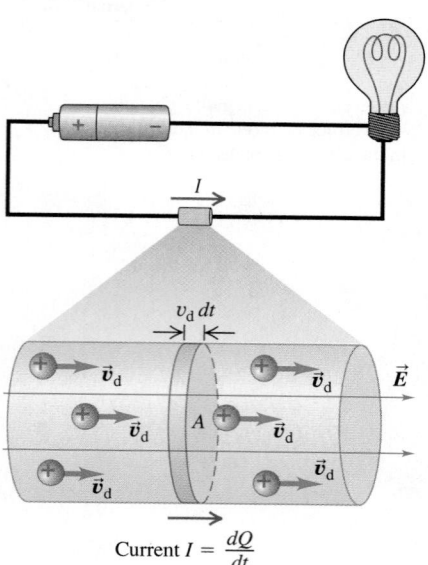

Current $I = \dfrac{dQ}{dt}$

material such as germanium or silicon, conduction is partly by electrons and partly by motion of *vacancies,* also known as *holes;* these are sites of missing electrons and act like positive charges.

Figure 25.2 shows segments of two different current-carrying materials. In Fig. 25.2a the moving charges are positive, the electric force is in the same direction as $\vec{E}$, and the drift velocity $\vec{v}_d$ is from left to right. In Fig. 25.2b the charges are negative, the electric force is opposite to $\vec{E}$, and the drift velocity $\vec{v}_d$ is from right to left. In both cases there is a net flow of positive charge from left to right, and positive charges end up to the right of negative ones. We *define* the current, denoted by I, to be in the direction in which there is a flow of *positive* charge. Thus we describe currents as though they consisted entirely of positive charge flow, even in cases in which we know that the actual current is due to electrons. Hence the current is to the right in both Figs. 25.2a and 25.2b. This choice or convention for the direction of current flow is called **conventional current.** While the direction of the conventional current is *not* necessarily the same as the direction in which charged particles are actually moving, we'll find that the sign of the moving charges is of little importance in analyzing electric circuits.

Figure 25.3 shows a segment of a conductor in which a current is flowing. We consider the moving charges to be *positive,* so they are moving in the same direction as the current. We define the current through the cross-sectional area A to be *the net charge flowing through the area per unit time.* Thus, if a net charge dQ flows through an area in a time dt, the current I through the area is

$$I = \frac{dQ}{dt} \qquad \text{(definition of current)} \qquad (25.1)$$

CAUTION **Current is not a vector** Although we refer to the *direction* of a current, current as defined by Eq. (25.1) is *not* a vector quantity. In a current-carrying wire, the current is always along the length of the wire, regardless of whether the wire is straight or curved. No single vector could describe motion along a curved path, which is why current is not a vector. We'll usually describe the direction of current either in words (as in "the current flows clockwise around the circuit") or by choosing a current to be positive if it flows in one direction along a conductor and negative if it flows in the other direction. ▮

The SI unit of current is the **ampere;** one ampere is defined to be *one coulomb per second* (1 A = 1 C/s). This unit is named in honor of the French scientist André Marie Ampère (1775–1836). When an ordinary flashlight (D-cell size) is turned on, the current in the flashlight is about 0.5 to 1 A; the current in the wires of a car engine's starter motor is around 200 A. Currents in radio and television circuits are usually expressed in *milliamperes* (1 mA = 10^{-3} A) or *microamperes* (1 μA = 10^{-6} A), and currents in computer circuits are expressed in *nanoamperes* (1 nA = 10^{-9} A) or *picoamperes* (1 pA = 10^{-12} A).

Current, Drift Velocity, and Current Density

We can express current in terms of the drift velocity of the moving charges. Let's consider again the situation of Fig. 25.3 of a conductor with cross-sectional area A and an electric field $\vec{E}$ directed from left to right. To begin with, we'll assume that the free charges in the conductor are positive; then the drift velocity is in the same direction as the field.

Suppose there are n moving charged particles per unit volume. We call n the **concentration** of particles; its SI unit is m^{-3}. Assume that all the particles move with the same drift velocity with magnitude v_d. In a time interval dt, each particle moves a distance $v_d\, dt$. The particles that flow out of the right end of the shaded cylinder with length $v_d\, dt$ during dt are the particles that were within this cylinder at the beginning of the interval dt. The volume of the cylinder is $Av_d\, dt$, and the number of particles within it is $nAv_d\, dt$. If each particle has a charge q, the charge dQ that flows out of the end of the cylinder during time dt is

$$dQ = q(nAv_d\,dt) = nqv_dA\,dt$$

and the current is

$$I = \frac{dQ}{dt} = nqv_dA$$

The current *per unit cross-sectional area* is called the **current density** J:

$$J = \frac{I}{A} = nqv_d$$

The units of current density are amperes per square meter (A/m^2).

If the moving charges are negative rather than positive, as in Fig. 25.2b, the drift velocity is opposite to $\vec{E}$. But the *current* is still in the same direction as $\vec{E}$ at each point in the conductor. Hence the current I and current density J don't depend on the sign of the charge, and so in the above expressions for I and J we replace the charge q by its absolute value $|q|$:

$$I = \frac{dQ}{dt} = n|q|v_dA \quad \text{(general expression for current)} \quad (25.2)$$

$$J = \frac{I}{A} = n|q|v_d \quad \text{(general expression for current density)} \quad (25.3)$$

The current in a conductor is the product of the concentration of moving charged particles, the magnitude of charge of each such particle, the magnitude of the drift velocity, and the cross-sectional area of the conductor.

We can also define a *vector* current density $\vec{J}$ that includes the direction of the drift velocity:

$$\vec{J} = nq\vec{v}_d \quad \text{(vector current density)} \quad (25.4)$$

There are *no* absolute value signs in Eq. (25.4). If q is positive, $\vec{v}_d$ is in the same direction as $\vec{E}$; if q is negative, $\vec{v}_d$ is opposite to $\vec{E}$. In either case, $\vec{J}$ is in the same direction as $\vec{E}$. Equation (25.3) gives the *magnitude* J of the vector current density $\vec{J}$.

CAUTION **Current density vs. current** Note that current density $\vec{J}$ is a vector, but current I is not. The difference is that the current density $\vec{J}$ describes how charges flow at a certain point, and the vector's direction tells you about the direction of the flow at that point. By contrast, the current I describes how charges flow through an extended object such as a wire. For example, I has the same value at all points in the circuit of Fig. 25.3, but $\vec{J}$ does not: The current density is directed downward in the left-hand side of the loop and upward in the right-hand side. The magnitude of $\vec{J}$ can also vary around a circuit. In Fig. 25.3 the current density magnitude $J = I/A$ is less in the battery (which has a large cross-sectional area A) than in the wires (which have a small cross-sectional area). ▮

In general, a conductor may contain several different kinds of moving charged particles having charges $q_1, q_2, \ldots$, concentrations $n_1, n_2, \ldots$, and drift velocities with magnitudes $v_{d1}, v_{d2}, \ldots$. An example is current flow in an ionic solution (Fig. 25.4). In a sodium chloride solution, current can be carried by both positive sodium ions and negative chlorine ions; the total current I is found by adding up the currents due to each kind of charged particle, using Eq. (25.2). Likewise, the total vector current density $\vec{J}$ is found by using Eq. (25.4) for each kind of charged particle and adding the results.

We will see in Section 25.4 that it is possible to have a current that is *steady* (that is, one that is constant in time) only if the conducting material forms a

25.4 Part of the electric circuit that includes this light bulb passes through a beaker with a solution of sodium chloride. The current in the solution is carried by both positive charges (Na^+ ions) and negative charges (Cl^- ions).

closed loop, called a *complete circuit*. In such a steady situation, the total charge in every segment of the conductor is constant. Hence the rate of flow of charge *out* at one end of a segment at any instant equals the rate of flow of charge *in* at the other end of the segment, and *the current is the same at all cross sections of the circuit.* We'll make use of this observation when we analyze electric circuits later in this chapter.

In many simple circuits, such as flashlights or cordless electric drills, the direction of the current is always the same; this is called *direct current*. But home appliances such as toasters, refrigerators, and televisions use *alternating current,* in which the current continuously changes direction. In this chapter we'll consider direct current only. Alternating current has many special features worthy of detailed study, which we'll examine in Chapter 31.

Example 25.1   **Current density and drift velocity in a wire**

An 18-gauge copper wire (the size usually used for lamp cords), with a diameter of 1.02 mm, carries a constant current of 1.67 A to a 200-W lamp. The free-electron density in the wire is 8.5×10^{28} per cubic meter. Find (a) the current density and (b) the drift speed.

SOLUTION

IDENTIFY and SET UP: This problem uses the relationships among current I, current density J, and drift speed v_d. We are given I and the wire diameter d, so we use Eq. (25.3) to find J. We use Eq. (25.3) again to find v_d from J and the known electron density n.

EXECUTE: (a) The cross-sectional area is

$$A = \frac{\pi d^2}{4} = \frac{\pi (1.02 \times 10^{-3} \text{ m})^2}{4} = 8.17 \times 10^{-7} \text{ m}^2$$

The magnitude of the current density is then

$$J = \frac{I}{A} = \frac{1.67 \text{ A}}{8.17 \times 10^{-7} \text{ m}^2} = 2.04 \times 10^6 \text{ A/m}^2$$

(b) From Eq. (25.3) for the drift velocity magnitude v_d, we find

$$v_d = \frac{J}{n|q|} = \frac{2.04 \times 10^6 \text{ A/m}^2}{(8.5 \times 10^{28} \text{ m}^{-3})|-1.60 \times 10^{-19} \text{ C}|}$$

$$= 1.5 \times 10^{-4} \text{ m/s} = 0.15 \text{ mm/s}$$

EVALUATE: At this speed an electron would require 6700 s (almost 2 h) to travel 1 m along this wire. The speeds of random motion of the electrons are roughly 10^6 m/s, around 10^{10} times the drift speed. Picture the electrons as bouncing around frantically, with a very slow drift!

Test Your Understanding of Section 25.1   Suppose we replaced the wire in Example 25.1 with 12-gauge copper wire, which has twice the diameter of 18-gauge wire. If the current remains the same, what effect would this have on the magnitude of the drift velocity v_d? (i) none—v_d would be unchanged; (ii) v_d would be twice as great; (iii) v_d would be four times greater; (iv) v_d would be half as great; (v) v_d would be one-fourth as great.

25.2 Resistivity

The current density $\vec{J}$ in a conductor depends on the electric field $\vec{E}$ and on the properties of the material. In general, this dependence can be quite complex. But for some materials, especially metals, at a given temperature, $\vec{J}$ is nearly *directly proportional* to $\vec{E}$, and the ratio of the magnitudes of E and J is constant. This relationship, called Ohm's law, was discovered in 1826 by the German physicist Georg Simon Ohm (1787–1854). The word "law" should actually be in quotation marks, since **Ohm's law,** like the ideal-gas equation and Hooke's law, is an *idealized model* that describes the behavior of some materials quite well but is not a general description of *all* matter. In the following discussion we'll assume that Ohm's law is valid, even though there are many situations in which it is not. The situation is comparable to our representation of the behavior of the static and kinetic friction forces; we treated these friction forces as being directly proportional to the normal force, even though we knew that this was at best an approximate description.

Table 25.1 Resistivities at Room Temperature (20°C)

Substance		ρ ($\Omega \cdot$ m)	Substance		ρ ($\Omega \cdot$ m)
Conductors			**Semiconductors**		
Metals	Silver	1.47×10^{-8}		Pure carbon (graphite)	3.5×10^{-5}
	Copper	1.72×10^{-8}		Pure germanium	0.60
	Gold	2.44×10^{-8}		Pure silicon	2300
	Aluminum	2.75×10^{-8}	**Insulators**		
	Tungsten	5.25×10^{-8}		Amber	5×10^{14}
	Steel	20×10^{-8}		Glass	10^{10}–10^{14}
	Lead	22×10^{-8}		Lucite	$>10^{13}$
	Mercury	95×10^{-8}		Mica	10^{11}–10^{15}
Alloys	Manganin (Cu 84%, Mn 12%, Ni 4%)	44×10^{-8}		Quartz (fused)	75×10^{16}
	Constantan (Cu 60%, Ni 40%)	49×10^{-8}		Sulfur	10^{15}
	Nichrome	100×10^{-8}		Teflon	$>10^{13}$
				Wood	10^{8}–10^{11}

We define the **resistivity** ρ of a material as the ratio of the magnitudes of electric field and current density:

$$\rho = \frac{E}{J} \quad \text{(definition of resistivity)} \qquad (25.5)$$

The greater the resistivity, the greater the field needed to cause a given current density, or the smaller the current density caused by a given field. From Eq. (25.5) the units of ρ are $(V/m)/(A/m^2) = V \cdot m/A$. As we will discuss in the next section, 1 V/A is called one *ohm* (1 Ω; we use the Greek letter Ω, or omega, which is alliterative with "ohm"). So the SI units for ρ are $\Omega \cdot m$ (ohm-meters). Table 25.1 lists some representative values of resistivity. A perfect conductor would have zero resistivity, and a perfect insulator would have an infinite resistivity. Metals and alloys have the smallest resistivities and are the best conductors. The resistivities of insulators are greater than those of the metals by an enormous factor, on the order of 10^{22}.

The reciprocal of resistivity is **conductivity.** Its units are $(\Omega \cdot m)^{-1}$. Good conductors of electricity have larger conductivity than insulators. Conductivity is the direct electrical analog of thermal conductivity. Comparing Table 25.1 with Table 17.5 (Thermal Conductivities), we note that good electrical conductors, such as metals, are usually also good conductors of heat. Poor electrical conductors, such as ceramic and plastic materials, are also poor thermal conductors. In a metal the free electrons that carry charge in electrical conduction also provide the principal mechanism for heat conduction, so we should expect a correlation between electrical and thermal conductivity. Because of the enormous difference in conductivity between electrical conductors and insulators, it is easy to confine electric currents to well-defined paths or circuits (Fig. 25.5). The variation in *thermal* conductivity is much less, only a factor of 10^3 or so, and it is usually impossible to confine heat currents to that extent.

Semiconductors have resistivities intermediate between those of metals and those of insulators. These materials are important because of the way their resistivities are affected by temperature and by small amounts of impurities.

A material that obeys Ohm's law reasonably well is called an *ohmic* conductor or a *linear* conductor. For such materials, at a given temperature, ρ is a *constant* that does not depend on the value of E. Many materials show substantial departures from Ohm's-law behavior; they are *nonohmic,* or *nonlinear.* In these materials, J depends on E in a more complicated manner.

Analogies with fluid flow can be a big help in developing intuition about electric current and circuits. For example, in the making of wine or maple syrup, the product is sometimes filtered to remove sediments. A pump forces the fluid through the filter under pressure; if the flow rate (analogous to J) is proportional to the pressure difference between the upstream and downstream sides (analogous to E), the behavior is analogous to Ohm's law.

25.5 The copper "wires," or traces, on this circuit board are printed directly onto the surface of the dark-colored insulating board. Even though the traces are very close to each other (only about a millimeter apart), the board has such a high resistivity (and low conductivity) that no current can flow between the traces.

Conducting paths
(traces)

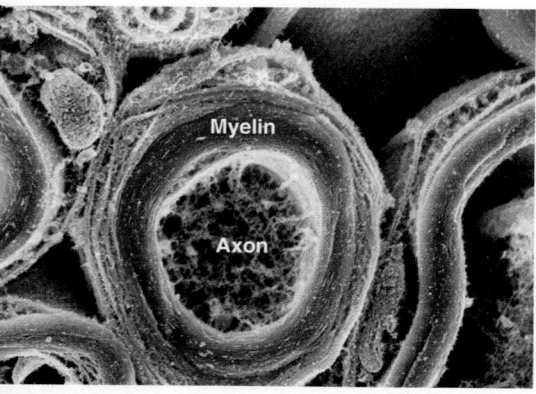

Resistivity and Temperature

The resistivity of a *metallic* conductor nearly always increases with increasing temperature, as shown in Fig. 25.6a. As temperature increases, the ions of the conductor vibrate with greater amplitude, making it more likely that a moving electron will collide with an ion as in Fig. 25.1; this impedes the drift of electrons through the conductor and hence reduces the current. Over a small temperature range (up to 100 C° or so), the resistivity of a metal can be represented approximately by the equation

$$\rho(T) = \rho_0[1 + \alpha(T - T_0)]$$

(temperature dependence of resistivity) (25.6)

where ρ_0 is the resistivity at a reference temperature T_0 (often taken as 0°C or 20°C) and $\rho(T)$ is the resistivity at temperature T, which may be higher or lower than T_0. The factor α is called the **temperature coefficient of resistivity.** Some representative values are given in Table 25.2. The resistivity of the alloy manganin is practically independent of temperature.

Table 25.2 Temperature Coefficients of Resistivity (Approximate Values Near Room Temperature)

Material	$\alpha\,[(°C)^{-1}]$	Material	$\alpha\,[(°C)^{-1}]$
Aluminum	0.0039	Lead	0.0043
Brass	0.0020	Manganin	0.00000
Carbon (graphite)	−0.0005	Mercury	0.00088
Constantan	0.00001	Nichrome	0.0004
Copper	0.00393	Silver	0.0038
Iron	0.0050	Tungsten	0.0045

25.6 Variation of resistivity ρ with absolute temperature T for (a) a normal metal, (b) a semiconductor, and (c) a superconductor. In (a) the linear approximation to ρ as a function of T is shown as a green line; the approximation agrees exactly at $T = T_0$, where $\rho = \rho_0$.

(a)

(b)

(c)

The resistivity of graphite (a nonmetal) *decreases* with increasing temperature, since at higher temperatures, more electrons are "shaken loose" from the atoms and become mobile; hence the temperature coefficient of resistivity of graphite is negative. This same behavior occurs for semiconductors (Fig. 25.6b). Measuring the resistivity of a small semiconductor crystal is therefore a sensitive measure of temperature; this is the principle of a type of thermometer called a *thermistor.*

Some materials, including several metallic alloys and oxides, show a phenomenon called *superconductivity.* As the temperature decreases, the resistivity at first decreases smoothly, like that of any metal. But then at a certain critical temperature T_c a phase transition occurs and the resistivity suddenly drops to zero, as shown in Fig. 25.6c. Once a current has been established in a superconducting ring, it continues indefinitely without the presence of any driving field.

Superconductivity was discovered in 1911 by the Dutch physicist Heike Kamerlingh Onnes (1853–1926). He discovered that at very low temperatures, below 4.2 K, the resistivity of mercury suddenly dropped to zero. For the next 75 years, the highest T_c attained was about 20 K. This meant that superconductivity occurred only when the material was cooled using expensive liquid helium, with a boiling-point temperature of 4.2 K, or explosive liquid hydrogen, with a boiling point of 20.3 K. But in 1986 Karl Müller and Johannes Bednorz discovered an oxide of barium, lanthanum, and copper with a T_c of nearly 40 K, and the race was on to develop "high-temperature" superconducting materials.

By 1987 a complex oxide of yttrium, copper, and barium had been found that has a value of T_c well above the 77 K boiling temperature of liquid nitrogen, a refrigerant that is both inexpensive and safe. The current (2010) record for T_c at atmospheric pressure is 138 K, and materials that are superconductors at room temperature may become a reality. The implications of these discoveries for power-distribution systems, computer design, and transportation are enormous. Meanwhile, superconducting electromagnets cooled by liquid helium are used in particle accelerators and some experimental magnetic-levitation railroads.

Superconductors have other exotic properties that require an understanding of magnetism to explore; we will discuss these further in Chapter 29.

Test Your Understanding of Section 25.2 You maintain a constant electric field inside a piece of semiconductor while lowering the semiconductor's temperature. What happens to the current density in the semiconductor? (i) It increases; (ii) it decreases; (iii) it remains the same.

25.3 Resistance

For a conductor with resistivity ρ, the current density $\vec{J}$ at a point where the electric field is $\vec{E}$ is given by Eq. (25.5), which we can write as

$$\vec{E} = \rho\vec{J} \qquad (25.7)$$

When Ohm's law is obeyed, ρ is constant and independent of the magnitude of the electric field, so $\vec{E}$ is directly proportional to $\vec{J}$. Often, however, we are more interested in the total current in a conductor than in $\vec{J}$ and more interested in the potential difference between the ends of the conductor than in $\vec{E}$. This is so largely because current and potential difference are much easier to measure than are $\vec{J}$ and $\vec{E}$.

Suppose our conductor is a wire with uniform cross-sectional area A and length L, as shown in Fig. 25.7. Let V be the potential difference between the higher-potential and lower-potential ends of the conductor, so that V is positive. The *direction* of the current is always from the higher-potential end to the lower-potential end. That's because current in a conductor flows in the direction of $\vec{E}$, no matter what the sign of the moving charges (Fig. 25.2), and because $\vec{E}$ points in the direction of *decreasing* electric potential (see Section 23.2). As the current flows through the potential difference, electric potential energy is lost; this energy is transferred to the ions of the conducting material during collisions.

We can also relate the *value* of the current I to the potential difference between the ends of the conductor. If the magnitudes of the current density $\vec{J}$ and the electric field $\vec{E}$ are uniform throughout the conductor, the total current I is given by $I = JA$, and the potential difference V between the ends is $V = EL$. When we solve these equations for J and E, respectively, and substitute the results in Eq. (25.7), we obtain

$$\frac{V}{L} = \frac{\rho I}{A} \quad \text{or} \quad V = \frac{\rho L}{A}I \qquad (25.8)$$

This shows that when ρ is constant, the total current I is proportional to the potential difference V.

The ratio of V to I for a particular conductor is called its **resistance** R:

$$R = \frac{V}{I} \qquad (25.9)$$

Comparing this definition of R to Eq. (25.8), we see that the resistance R of a particular conductor is related to the resistivity ρ of its material by

$$R = \frac{\rho L}{A} \qquad \text{(relationship between resistance and resistivity)} \qquad (25.10)$$

If ρ is constant, as is the case for ohmic materials, then so is R.

The equation

$$V = IR \qquad \text{(relationship among voltage, current, and resistance)} \qquad (25.11)$$

is often called Ohm's law, but it is important to understand that the real content of Ohm's law is the direct proportionality (for some materials) of V to I or of J to E.

25.7 A conductor with uniform cross section. The current density is uniform over any cross section, and the electric field is constant along the length.

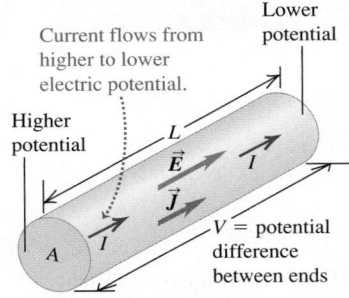

Equation (25.9) or (25.11) *defines* resistance R for *any* conductor, whether or not it obeys Ohm's law, but only when R is constant can we correctly call this relationship Ohm's law.

Interpreting Resistance

Equation (25.10) shows that the resistance of a wire or other conductor of uniform cross section is directly proportional to its length and inversely proportional to its cross-sectional area. It is also proportional to the resistivity of the material of which the conductor is made.

The flowing-fluid analogy is again useful. In analogy to Eq. (25.10), a narrow water hose offers more resistance to flow than a fat one, and a long hose has more resistance than a short one (Fig. 25.8). We can increase the resistance to flow by stuffing the hose with cotton or sand; this corresponds to increasing the resistivity. The flow rate is approximately proportional to the pressure difference between the ends. Flow rate is analogous to current, and pressure difference is analogous to potential difference ("voltage"). Let's not stretch this analogy too far, though; the water flow rate in a pipe is usually *not* proportional to its cross-sectional area (see Section 14.6).

The SI unit of resistance is the **ohm,** equal to one volt per ampere ($1\,\Omega = 1\,\text{V/A}$). The *kilohm* ($1\,\text{k}\Omega = 10^3\,\Omega$) and the *megohm* ($1\,\text{M}\Omega = 10^6\,\Omega$) are also in common use. A 100-m length of 12-gauge copper wire, the size usually used in household wiring, has a resistance at room temperature of about $0.5\,\Omega$. A 100-W, 120-V light bulb has a resistance (at operating temperature) of $140\,\Omega$. If the same current I flows in both the copper wire and the light bulb, the potential difference $V = IR$ is much greater across the light bulb, and much more potential energy is lost per charge in the light bulb. This lost energy is converted by the light bulb filament into light and heat. You don't want your household wiring to glow white-hot, so its resistance is kept low by using wire of low resistivity and large cross-sectional area.

Because the resistivity of a material varies with temperature, the resistance of a specific conductor also varies with temperature. For temperature ranges that are not too great, this variation is approximately a linear relationship, analogous to Eq. (25.6):

$$R(T) = R_0[1 + \alpha(T - T_0)] \qquad (25.12)$$

In this equation, $R(T)$ is the resistance at temperature T and R_0 is the resistance at temperature T_0, often taken to be 0°C or 20°C. The *temperature coefficient of resistance* α is the same constant that appears in Eq. (25.6) if the dimensions L and A in Eq. (25.10) do not change appreciably with temperature; this is indeed the case for most conducting materials (see Problem 25.67). Within the limits of validity of Eq. (25.12), the *change* in resistance resulting from a temperature change $T - T_0$ is given by $R_0\alpha(T - T_0)$.

A circuit device made to have a specific value of resistance between its ends is called a **resistor.** Resistors in the range 0.01 to $10^7\,\Omega$ can be bought off the shelf. Individual resistors used in electronic circuitry are often cylindrical, a few millimeters in diameter and length, with wires coming out of the ends. The resistance may be marked with a standard code using three or four color bands near one end (Fig. 25.9), according to the scheme shown in Table 25.3. The first two bands (starting with the band nearest an end) are digits, and the third is a power-of-10 multiplier, as shown in Fig. 25.9. For example, green–violet–red means $57 \times 10^2\,\Omega$, or $5.7\,\text{k}\Omega$. The fourth band, if present, indicates the precision (tolerance) of the value; no band means $\pm 20\%$, a silver band $\pm 10\%$, and a gold band $\pm 5\%$. Another important characteristic of a resistor is the maximum *power* it can dissipate without damage. We'll return to this point in Section 25.5.

25.8 A long fire hose offers substantial resistance to water flow. To make water pass through the hose rapidly, the upstream end of the hose must be at much higher pressure than the end where the water emerges. In an analogous way, there must be a large potential difference between the ends of a long wire in order to cause a substantial electric current through the wire.

25.9 This resistor has a resistance of $5.7\,\text{k}\Omega$ with a precision (tolerance) of $\pm 10\%$.

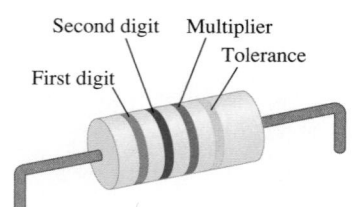

Second digit   Multiplier
First digit   Tolerance

Table 25.3 Color Codes for Resistors

Color	Value as Digit	Value as Multiplier
Black	0	1
Brown	1	10
Red	2	10^2
Orange	3	10^3
Yellow	4	10^4
Green	5	10^5
Blue	6	10^6
Violet	7	10^7
Gray	8	10^8
White	9	10^9

25.10 Current–voltage relationships for two devices. Only for a resistor that obeys Ohm's law as in **(a)** is current I proportional to voltage V.

(a)

Ohmic resistor (e.g., typical metal wire): At a given temperature, current is proportional to voltage.

Slope $= \dfrac{1}{R}$

(b)

Semiconductor diode: a nonohmic resistor

In the direction of positive current and voltage, I increases nonlinearly with V.

In the direction of negative current and voltage, little current flows.

For a resistor that obeys Ohm's law, a graph of current as a function of potential difference (voltage) is a straight line (Fig. 25.10a). The slope of the line is $1/R$. If the sign of the potential difference changes, so does the sign of the current produced; in Fig. 25.7 this corresponds to interchanging the higher- and lower-potential ends of the conductor, so the electric field, current density, and current all reverse direction. In devices that do not obey Ohm's law, the relationship of voltage to current may not be a direct proportion, and it may be different for the two directions of current. Figure 25.10b shows the behavior of a semiconductor *diode*, a device used to convert alternating current to direct current and to perform a wide variety of logic functions in computer circuitry. For positive potentials V of the anode (one of two terminals of the diode) with respect to the cathode (the other terminal), I increases exponentially with increasing V; for negative potentials the current is extremely small. Thus a positive V causes a current to flow in the positive direction, but a potential difference of the other sign causes little or no current. Hence a diode acts like a one-way valve in a circuit.

Example 25.2   **Electric field, potential difference, and resistance in a wire**

The 18-gauge copper wire of Example 25.1 has a cross-sectional area of $8.20 \times 10^{-7}\,\text{m}^2$. It carries a current of 1.67 A. Find (a) the electric-field magnitude in the wire; (b) the potential difference between two points in the wire 50.0 m apart; (c) the resistance of a 50.0-m length of this wire.

SOLUTION

IDENTIFY and SET UP: We are given the cross-sectional area A and current I. Our target variables are the electric-field magnitude E, potential difference V, and resistance R. The current density is $J = I/A$. We find E from Eq. (25.5), $E = \rho J$ (Table 25.1 gives the resistivity ρ for copper). The potential difference is then the product of E and the length of the wire. We can use either Eq. (25.10) or Eq. (25.11) to find R.

EXECUTE: (a) From Table 25.1, $\rho = 1.72 \times 10^{-8}\,\Omega \cdot \text{m}$. Hence, using Eq. (25.5),

$$E = \rho J = \frac{\rho I}{A} = \frac{(1.72 \times 10^{-8}\,\Omega \cdot \text{m})(1.67\,\text{A})}{8.20 \times 10^{-7}\,\text{m}^2}$$

$$= 0.0350\,\text{V/m}$$

(b) The potential difference is

$$V = EL = (0.0350\,\text{V/m})(50.0\,\text{m}) = 1.75\,\text{V}$$

(c) From Eq. (25.10) the resistance of 50.0 m of this wire is

$$R = \frac{\rho L}{A} = \frac{(1.72 \times 10^{-8}\,\Omega \cdot \text{m})(50.0\,\text{m})}{8.20 \times 10^{-7}\,\text{m}^2} = 1.05\,\Omega$$

Alternatively, we can find R using Eq. (25.11):

$$R = \frac{V}{I} = \frac{1.75\,\text{V}}{1.67\,\text{A}} = 1.05\,\Omega$$

EVALUATE: We emphasize that the resistance of the wire is *defined* to be the ratio of voltage to current. If the wire is made of nonohmic material, then R is different for different values of V but is always given by $R = V/I$. Resistance is also always given by $R = \rho L/A$; if the material is nonohmic, ρ is not constant but depends on E (or, equivalently, on $V = EL$).

Example 25.3 Temperature dependence of resistance

Suppose the resistance of a copper wire is 1.05 Ω at 20°C. Find the resistance at 0°C and 100°C.

SOLUTION

IDENTIFY and SET UP: We are given the resistance $R_0 = 1.05\ \Omega$ at a reference temperature $T_0 = 20°C$. We use Eq. (25.12) to find the resistances at $T = 0°C$ and $T = 100°C$ (our target variables), taking the temperature coefficient of resistivity from Table 25.2.

EXECUTE: From Table 25.2, $\alpha = 0.00393\ (C°)^{-1}$ for copper. Then from Eq. (25.12),

$$R = R_0[1 + \alpha(T - T_0)]$$
$$= (1.05\ \Omega)\{1 + [0.00393\ (C°)^{-1}][0°C - 20°C]\}$$
$$= 0.97\ \Omega\ \text{at}\ T = 0°C$$

$$R = (1.05\ \Omega)\{1 + [0.00393\ (C°)^{-1}][100°C - 20°C]\}$$
$$= 1.38\ \Omega\ \text{at}\ T = 100°C$$

EVALUATE: The resistance at 100°C is greater than that at 0°C by a factor of $(1.38\ \Omega)/(0.97\ \Omega) = 1.42$: Raising the temperature of copper wire from 0°C to 100°C increases its resistance by 42%. From Eq. (25.11), $V = IR$, this means that 42% more voltage is required to produce the same current at 100°C than at 0°C. Designers of electric circuits that must operate over a wide temperature range must take this substantial effect into account.

Test Your Understanding of Section 25.3 Suppose you increase the voltage across the copper wire in Examples 25.2 and 25.3. The increased voltage causes more current to flow, which makes the temperature of the wire increase. (The same thing happens to the coils of an electric oven or a toaster when a voltage is applied to them. We'll explore this issue in more depth in Section 25.5.) If you double the voltage across the wire, the current in the wire increases. By what factor does it increase? (i) 2; (ii) greater than 2; (iii) less than 2.

25.4 Electromotive Force and Circuits

25.11 If an electric field is produced inside a conductor that is *not* part of a complete circuit, current flows for only a very short time.

(a) An electric field E_1 produced inside an isolated conductor causes a current.

(b) The current causes charge to build up at the ends.

The charge buildup produces an opposing field $\vec{E}_2$, thus reducing the current.

(c) After a very short time $\vec{E}_2$ has the same magnitude as $\vec{E}_1$; then the total field is $\vec{E}_{\text{total}} = 0$ and the current stops completely.

For a conductor to have a steady current, it must be part of a path that forms a closed loop or **complete circuit.** Here's why. If you establish an electric field $\vec{E}_1$ inside an isolated conductor with resistivity ρ that is *not* part of a complete circuit, a current begins to flow with current density $\vec{J} = \vec{E}_1/\rho$ (Fig. 25.11a). As a result a net positive charge quickly accumulates at one end of the conductor and a net negative charge accumulates at the other end (Fig. 25.11b). These charges themselves produce an electric field $\vec{E}_2$ in the direction opposite to $\vec{E}_1$, causing the total electric field and hence the current to decrease. Within a very small fraction of a second, enough charge builds up on the conductor ends that the total electric field $\vec{E} = \vec{E}_1 + \vec{E}_2 = 0$ inside the conductor. Then $\vec{J} = 0$ as well, and the current stops altogether (Fig. 25.11c). So there can be no steady motion of charge in such an *incomplete* circuit.

To see how to maintain a steady current in a *complete* circuit, we recall a basic fact about electric potential energy: If a charge q goes around a complete circuit and returns to its starting point, the potential energy must be the same at the end of the round trip as at the beginning. As described in Section 25.3, there is always a *decrease* in potential energy when charges move through an ordinary conducting material with resistance. So there must be some part of the circuit in which the potential energy *increases.*

The problem is analogous to an ornamental water fountain that recycles its water. The water pours out of openings at the top, cascades down over the terraces and spouts (moving in the direction of decreasing gravitational potential energy), and collects in a basin in the bottom. A pump then lifts it back to the top (increasing the potential energy) for another trip. Without the pump, the water would just fall to the bottom and stay there.

Electromotive Force

In an electric circuit there must be a device somewhere in the loop that acts like the water pump in a water fountain (Fig. 25.12). In this device a charge travels "uphill," from lower to higher potential energy, even though the electrostatic force is trying to push it from higher to lower potential energy. The direction of current in such a device is from lower to higher potential, just the opposite of what happens in an ordinary conductor. The influence that makes current flow from lower to higher potential is called **electromotive force** (abbreviated **emf** and pronounced "ee-em-eff"). This is a poor term because emf is *not* a force but an energy-per-unit-charge quantity, like potential. The SI unit of emf is the same as that for potential, the volt ($1\ \text{V} = 1\ \text{J/C}$). A typical flashlight battery has an emf of 1.5 V; this means that the battery does 1.5 J of work on every coulomb of charge that passes through it. We'll use the symbol $\mathcal{E}$ (a script capital E) for emf.

Every complete circuit with a steady current must include some device that provides emf. Such a device is called a **source of emf.** Batteries, electric generators, solar cells, thermocouples, and fuel cells are all examples of sources of emf. All such devices convert energy of some form (mechanical, chemical, thermal, and so on) into electric potential energy and transfer it into the circuit to which the device is connected. An *ideal* source of emf maintains a constant potential difference between its terminals, independent of the current through it. We define electromotive force quantitatively as the magnitude of this potential difference. As we will see, such an ideal source is a mythical beast, like the frictionless plane and the massless rope. We will discuss later how real-life sources of emf differ in their behavior from this idealized model.

Figure 25.13 is a schematic diagram of an ideal source of emf that maintains a potential difference between conductors a and b, called the *terminals* of the device. Terminal a, marked +, is maintained at *higher* potential than terminal b, marked −. Associated with this potential difference is an electric field $\vec{E}$ in the region around the terminals, both inside and outside the source. The electric field inside the device is directed from a to b, as shown. A charge q within the source experiences an electric force $\vec{F}_e = q\vec{E}$. But the source also provides an additional influence, which we represent as a nonelectrostatic force $\vec{F}_n$. This force, operating inside the device, pushes charge from b to a in an "uphill" direction against the electric force $\vec{F}_e$. Thus $\vec{F}_n$ maintains the potential difference between the terminals. If $\vec{F}_n$ were not present, charge would flow between the terminals until the potential difference was zero. The origin of the additional influence $\vec{F}_n$ depends on the kind of source. In a generator it results from magnetic-field forces on moving charges. In a battery or fuel cell it is associated with diffusion processes and varying electrolyte concentrations resulting from chemical reactions. In an electrostatic machine such as a Van de Graaff generator (see Fig. 22.26), an actual mechanical force is applied by a moving belt or wheel.

If a positive charge q is moved from b to a inside the source, the nonelectrostatic force $\vec{F}_n$ does a positive amount of work $W_n = q\mathcal{E}$ on the charge. This displacement is *opposite* to the electrostatic force $\vec{F}_e$, so the potential energy associated with the charge *increases* by an amount equal to qV_{ab}, where $V_{ab} = V_a - V_b$ is the (positive) potential of point a with respect to point b. For the ideal source of emf that we've described, $\vec{F}_e$ and $\vec{F}_n$ are equal in magnitude but opposite in direction, so the total work done on the charge q is zero; there is an increase in potential energy but *no* change in the kinetic energy of the charge. It's like lifting a book from the floor to a high shelf at constant speed. The increase in potential energy is just equal to the nonelectrostatic work W_n, so $q\mathcal{E} = qV_{ab}$, or

$$V_{ab} = \mathcal{E} \qquad \text{(ideal source of emf)} \qquad (25.13)$$

Now let's make a complete circuit by connecting a wire with resistance R to the terminals of a source (Fig. 25.14). The potential difference between terminals a and b sets up an electric field within the wire; this causes current to flow around the loop from a toward b, from higher to lower potential. Where the wire bends, equal amounts of positive and negative charge persist on the "inside" and "outside"

25.12 Just as a water fountain requires a pump, an electric circuit requires a source of electromotive force to sustain a steady current.

25.13 Schematic diagram of a source of emf in an "open-circuit" situation. The electric-field force $\vec{F}_e = q\vec{E}$ and the nonelectrostatic force $\vec{F}_n$ are shown for a positive charge q.

When the emf source is not part of a closed circuit, $F_n = F_e$ and there is no net motion of charge between the terminals.

PhET: Battery Voltage
PhET: Signal Circuit
ActivPhysics 12.1: DC Series Circuits (Qualitative)

25.14 Schematic diagram of an ideal source of emf in a complete circuit. The electric-field force $\vec{F}_e = q\vec{E}$ and the non-electrostatic force $\vec{F}_n$ are shown for a positive charge q. The current is in the direction from a to b in the external circuit and from b to a within the source.

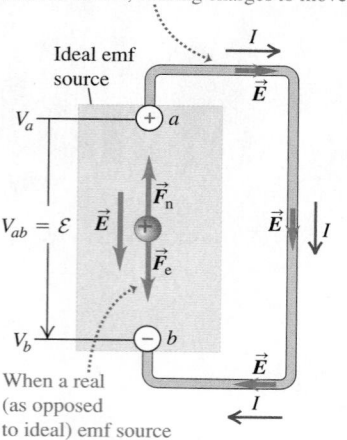

Potential across terminals creates electric field in circuit, causing charges to move.

Ideal emf source

V_a $+$ a

$\vec{E}$

I

$\vec{F}_n$

$V_{ab} = \mathcal{E}$ $\vec{E}$ $+$ $\vec{E}$ I

$\vec{F}_e$

V_b $-$ b

$\vec{E}$

I

When a real (as opposed to ideal) emf source is connected to a circuit, V_{ab} and thus F_e fall, so that $F_n > F_e$ and $\vec{F}_n$ does work on the charges.

Application **Danger: Electric Ray!**
Electric rays deliver electric shocks to stun their prey and to discourage predators. (In ancient Rome, physicians practiced a primitive form of electroconvulsive therapy by placing electric rays on their patients to cure headaches and gout.) The shocks are produced by specialized flattened cells called electroplaques. Such a cell moves ions across membranes to produce an emf of about 0.05 V. Thousands of electroplaques are stacked on top of each other, so their emfs add to a total of as much as 200 V. These stacks make up more than half of an electric ray's body mass. A ray can use these to deliver an impressive current of up to 30 A for a few milliseconds.

of the bend. These charges exert the forces that cause the current to follow the bends in the wire.

From Eq. (25.11) the potential difference between the ends of the wire in Fig. 25.14 is given by $V_{ab} = IR$. Combining with Eq. (25.13), we have

$$\mathcal{E} = V_{ab} = IR \qquad \text{(ideal source of emf)} \qquad (25.14)$$

That is, when a positive charge q flows around the circuit, the potential *rise* $\mathcal{E}$ as it passes through the ideal source is numerically equal to the potential *drop* $V_{ab} = IR$ as it passes through the remainder of the circuit. Once $\mathcal{E}$ and R are known, this relationship determines the current in the circuit.

CAUTION **Current is not "used up" in a circuit** It's a common misconception that in a closed circuit, current is something that squirts out of the positive terminal of a battery and is consumed or "used up" by the time it reaches the negative terminal. In fact the current is the *same* at every point in a simple loop circuit like that in Fig. 25.14, even if the thickness of the wires is different at different points in the circuit. This happens because charge is conserved (that is, it can be neither created nor destroyed) and because charge cannot accumulate in the circuit devices we have described. If charge did accumulate, the potential differences would change with time. It's like the flow of water in an ornamental fountain; water flows out of the top of the fountain at the same rate at which it reaches the bottom, no matter what the dimensions of the fountain. None of the water is "used up" along the way!

Internal Resistance

Real sources of emf in a circuit don't behave in exactly the way we have described; the potential difference across a real source in a circuit is *not* equal to the emf as in Eq. (25.14). The reason is that charge moving through the material of any real source encounters *resistance*. We call this the **internal resistance** of the source, denoted by r. If this resistance behaves according to Ohm's law, r is constant and independent of the current I. As the current moves through r, it experiences an associated drop in potential equal to Ir. Thus, when a current is flowing through a source from the negative terminal b to the positive terminal a, the potential difference V_{ab} between the terminals is

$$V_{ab} = \mathcal{E} - Ir \qquad \text{(terminal voltage, source with internal resistance)} \qquad (25.15)$$

The potential V_{ab}, called the **terminal voltage,** is less than the emf $\mathcal{E}$ because of the term Ir representing the potential drop across the internal resistance r. Expressed another way, the increase in potential energy qV_{ab} as a charge q moves from b to a within the source is now less than the work $q\mathcal{E}$ done by the nonelectrostatic force $\vec{F}_n$, since some potential energy is lost in traversing the internal resistance.

A 1.5-V battery has an emf of 1.5 V, but the terminal voltage V_{ab} of the battery is equal to 1.5 V only if no current is flowing through it so that $I = 0$ in Eq. (25.15). If the battery is part of a complete circuit through which current is flowing, the terminal voltage will be less than 1.5 V. *For a real source of emf, the terminal voltage equals the emf only if no current is flowing through the source* (Fig. 25.15). Thus we can describe the behavior of a source in terms of two properties: an emf $\mathcal{E}$, which supplies a constant potential difference independent of current, in series with an internal resistance r.

The current in the external circuit connected to the source terminals a and b is still determined by $V_{ab} = IR$. Combining this with Eq. (25.15), we find

$$\mathcal{E} - Ir = IR \quad \text{or} \quad I = \frac{\mathcal{E}}{R + r} \qquad \text{(current, source with internal resistance)} \qquad (25.16)$$

That is, the current equals the source emf divided by the *total* circuit resistance $(R + r)$.

> **CAUTION** **A battery is not a "current source"** You might have thought that a battery or other source of emf always produces the same current, no matter what circuit it's used in. Equation (25.16) shows that this isn't so! The greater the resistance R of the external circuit, the less current the source will produce. It's analogous to pushing an object through a thick, viscous liquid such as oil or molasses; if you exert a certain steady push (emf), you can move a small object at high speed (small R, large I) or a large object at low speed (large R, small I).

Symbols for Circuit Diagrams

An important part of analyzing any electric circuit is drawing a schematic *circuit diagram*. Table 25.4 shows the usual symbols used in circuit diagrams. We will use these symbols extensively in this chapter and the next. We usually assume that the wires that connect the various elements of the circuit have negligible resistance; from Eq. (25.11), $V = IR$, the potential difference between the ends of such a wire is zero.

Table 25.4 includes two *meters* that are used to measure the properties of circuits. Idealized meters do not disturb the circuit in which they are connected. A **voltmeter,** introduced in Section 23.2, measures the potential difference between its terminals; an idealized voltmeter has infinitely large resistance and measures potential difference without having any current diverted through it. An ammeter measures the current passing through it; an idealized **ammeter** has zero resistance and has no potential difference between its terminals. Because meters act as part of the circuit in which they are connected, these properties are important to remember.

25.15 The emf of this battery—that is, the terminal voltage when it's not connected to anything—is 12 V. But because the battery has internal resistance, the terminal voltage of the battery is less than 12 V when it is supplying current to a light bulb.

Table 25.4 **Symbols for Circuit Diagrams**

———————	Conductor with negligible resistance
R resistor symbol	Resistor
$+\|\mathcal{E}$	Source of emf (longer vertical line always represents the positive terminal, usually the terminal with higher potential)
$\mathcal{E}\|+$ or $+\|\mathcal{E}$	Source of emf with internal resistance r (r can be placed on either side)
(V)	Voltmeter (measures potential difference between its terminals)
(A)	Ammeter (measures current through it)

Conceptual Example 25.4 **A source in an open circuit**

Figure 25.16 shows a source (a battery) with emf $\mathcal{E} = 12$ V and internal resistance $r = 2\ \Omega$. (For comparison, the internal resistance of a commercial 12-V lead storage battery is only a few thousandths of an ohm.) The wires to the left of a and to the right of the ammeter A are not connected to anything. Determine the respective readings V_{ab} and I of the idealized voltmeter V and the idealized ammeter A.

25.16 A source of emf in an open circuit.

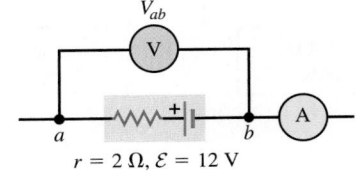

$$r = 2\ \Omega, \mathcal{E} = 12\text{ V}$$

Continued

SOLUTION

There is *zero* current because there is no complete circuit. (Our idealized voltmeter has an infinitely large resistance, so no current flows through it.) Hence the ammeter reads $I = 0$. Because there is no current through the battery, there is no potential difference across

its internal resistance. From Eq. (25.15) with $I = 0$, the potential difference V_{ab} across the battery terminals is equal to the emf. So the voltmeter reads $V_{ab} = \mathcal{E} = 12$ V. The terminal voltage of a real, nonideal source equals the emf *only* if there is no current flowing through the source, as in this example.

Example 25.5   A source in a complete circuit

We add a 4-Ω resistor to the battery in Conceptual Example 25.4, forming a complete circuit (Fig. 25.17). What are the voltmeter and ammeter readings V_{ab} and I now?

SOLUTION

IDENTIFY and SET UP: Our target variables are the current I through the circuit $aa'b'b$ and the potential difference V_{ab}. We first find I using Eq. (25.16). To find V_{ab}, we can use either Eq. (25.11) or Eq. (25.15).

25.17 A source of emf in a complete circuit.

EXECUTE: The ideal ammeter has zero resistance, so the total resistance external to the source is $R = 4$ Ω. From Eq. (25.16), the current through the circuit $aa'b'b$ is then

$$I = \frac{\mathcal{E}}{R + r} = \frac{12 \text{ V}}{4 \text{ }\Omega + 2 \text{ }\Omega} = 2 \text{ A}$$

Our idealized conducting wires and the idealized ammeter have zero resistance, so there is no potential difference between points a and a' or between points b and b'; that is, $V_{ab} = V_{a'b'}$. We find V_{ab} by considering a and b as the terminals of the resistor: From Ohm's law, Eq. (25.11), we then have

$$V_{a'b'} = IR = (2 \text{ A})(4 \text{ }\Omega) = 8 \text{ V}$$

Alternatively, we can consider a and b as the terminals of the source. Then, from Eq. (25.15),

$$V_{ab} = \mathcal{E} - Ir = 12 \text{ V} - (2 \text{ A})(2 \text{ }\Omega) = 8 \text{ V}$$

Either way, we see that the voltmeter reading is 8 V.

EVALUATE: With current flowing through the source, the terminal voltage V_{ab} is less than the emf $\mathcal{E}$. The smaller the internal resistance r, the less the difference between V_{ab} and $\mathcal{E}$.

Conceptual Example 25.6   Using voltmeters and ammeters

We move the voltmeter and ammeter in Example 25.5 to different positions in the circuit. What are the readings of the ideal voltmeter and ammeter in the situations shown in (a) Fig. 25.18a and (b) Fig. 25.18b?

SOLUTION

(a) The voltmeter now measures the potential difference between points a' and b'. As in Example 25.5, $V_{ab} = V_{a'b'}$, so the voltmeter reads the same as in Example 25.5: $V_{a'b'} = 8$ V.

25.18 Different placements of a voltmeter and an ammeter in a complete circuit.

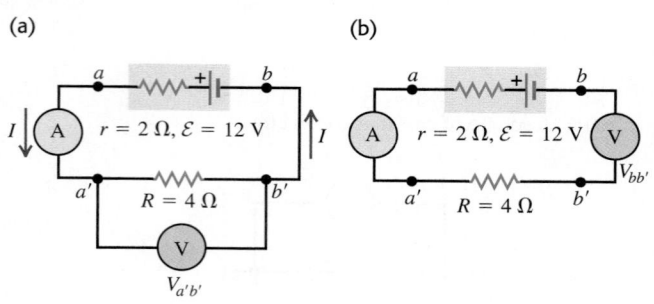

CAUTION **Current in a simple loop** As charges move through a resistor, there is a decrease in electric potential energy, but there is *no* change in the current. *The current in a simple loop is the same at every point;* it is not "used up" as it moves through a resistor. Hence the ammeter in Fig. 25.17 ("downstream" of the 4-Ω resistor) and the ammeter in Fig. 25.18b ("upstream" of the resistor) both read $I = 2$ A.

(b) There is no current through the ideal voltmeter because it has infinitely large resistance. Since the voltmeter is now part of the circuit, there is no current at all in the circuit, and the ammeter reads $I = 0$.

The voltmeter measures the potential difference $V_{bb'}$ between points b and b'. Since $I = 0$, the potential difference across the resistor is $V_{a'b'} = IR = 0$, and the potential difference between the ends a and a' of the idealized ammeter is also zero. So $V_{bb'}$ is equal to V_{ab}, the terminal voltage of the source. As in Conceptual Example 25.4, there is no current, so the terminal voltage equals the emf, and the voltmeter reading is $V_{ab} = \mathcal{E} = 12$ V.

This example shows that ammeters and voltmeters are circuit elements, too. Moving the voltmeter from the position in Fig. 25.18a to that in Fig. 25.18b makes large changes in the current and potential differences in the circuit. If you want to measure the potential difference between two points in a circuit without disturbing the circuit, use a voltmeter as in Fig. 25.17 or 25.18a, *not* as in Fig. 25.18b.

Example 25.7   A source with a short circuit

In the circuit of Example 25.5 we replace the 4-Ω resistor with a zero-resistance conductor. What are the meter readings now?

SOLUTION

IDENTIFY and SET UP: Figure 25.19 shows the new circuit. Our target variables are again I and V_{ab}. There is now a zero-resistance path between points a and b, through the lower loop, so the potential difference between these points must be zero.

25.19 Our sketch for this problem.

EXECUTE: We must have $V_{ab} = IR = I(0) = 0$, no matter what the current. We can therefore find the current I from Eq. (25.15):

$$V_{ab} = \mathcal{E} - Ir = 0$$
$$I = \frac{\mathcal{E}}{r} = \frac{12 \text{ V}}{2 \text{ }\Omega} = 6 \text{ A}$$

EVALUATE: The current has a different value than in Example 25.5, even though the same battery is used; the current depends on both the internal resistance r and the resistance of the external circuit.

The situation here is called a *short circuit*. The external-circuit resistance is zero, because terminals of the battery are connected directly to each other. The short-circuit current is equal to the emf $\mathcal{E}$ divided by the internal resistance r. *Warning:* Short circuits can be dangerous! An automobile battery or a household power line has very small internal resistance (much less than in these examples), and the short-circuit current can be great enough to melt a small wire or cause a storage battery to explode.

Potential Changes Around a Circuit

The net change in potential energy for a charge q making a round trip around a complete circuit must be zero. Hence the net change in *potential* around the circuit must also be zero; in other words, the algebraic sum of the potential differences and emfs around the loop is zero. We can see this by rewriting Eq. (25.16) in the form

$$\mathcal{E} - Ir - IR = 0$$

A potential gain of $\mathcal{E}$ is associated with the emf, and potential drops of Ir and IR are associated with the internal resistance of the source and the external circuit, respectively. Figure 25.20 is a graph showing how the potential varies as we go around the complete circuit of Fig. 25.17. The horizontal axis doesn't necessarily represent actual distances, but rather various points in the loop. If we take the potential to be zero at the negative terminal of the battery, then we have a rise $\mathcal{E}$ and a drop Ir in the battery and an additional drop IR in the external resistor, and as we finish our trip around the loop, the potential is back where it started.

In this section we have considered only situations in which the resistances are ohmic. If the circuit includes a nonlinear device such as a diode (see Fig. 25.10b), Eq. (25.16) is still valid but cannot be solved algebraically because R is not a constant. In such a situation, the current I can be found by using numerical techniques.

Finally, we remark that Eq. (25.15) is not always an adequate representation of the behavior of a source. The emf may not be constant, and what we have described as an internal resistance may actually be a more complex voltage–current relationship that doesn't obey Ohm's law. Nevertheless, the concept of internal resistance frequently provides an adequate description of batteries, generators, and other energy converters. The principal difference between a fresh flashlight battery and an old one is not in the emf, which decreases only slightly with use, but in the internal resistance, which may increase from less than an ohm when the battery is fresh to as much as 1000 Ω or more after long use. Similarly, a car battery can deliver less current to the starter motor on a cold morning than when the battery is warm, not because the emf is appreciably less but because the internal resistance increases with decreasing temperature.

25.20 Potential rises and drops in a circuit.

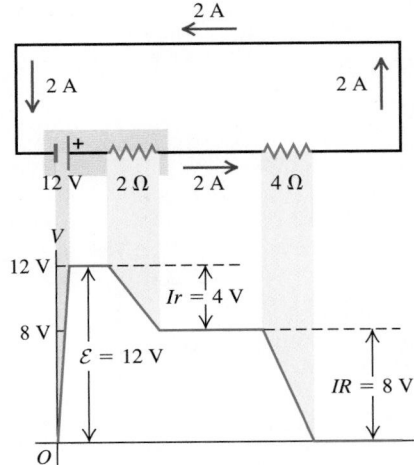

Test Your Understanding of Section 25.4 Rank the following circuits in order from highest to lowest current. (i) a 1.4-Ω resistor connected to a 1.5-V battery that has an internal resistance of 0.10 Ω; (ii) a 1.8-Ω resistor connected to a 4.0-V battery that has a terminal voltage of 3.6 V but an unknown internal resistance; (iii) an unknown resistor connected to a 12.0-V battery that has an internal resistance of 0.20 Ω and a terminal voltage of 11.0 V.

25.5 Energy and Power in Electric Circuits

25.21 The power input to the circuit element between *a* and *b* is $P = (V_a - V_b)I = V_{ab}I$.

Let's now look at some energy and power relationships in electric circuits. The box in Fig. 25.21 represents a circuit element with potential difference $V_a - V_b = V_{ab}$ between its terminals and current I passing through it in the direction from *a* toward *b*. This element might be a resistor, a battery, or something else; the details don't matter. As charge passes through the circuit element, the electric field does work on the charge. In a source of emf, additional work is done by the force $\vec{F}_n$ that we mentioned in Section 25.4.

As an amount of charge q passes through the circuit element, there is a change in potential energy equal to qV_{ab}. For example, if $q > 0$ and $V_{ab} = V_a - V_b$ is positive, potential energy decreases as the charge "falls" from potential V_a to lower potential V_b. The moving charges don't gain *kinetic* energy, because the current (the rate of charge flow) out of the circuit element must be the same as the current into the element. Instead, the quantity qV_{ab} represents energy transferred into the circuit element. This situation occurs in the coils of a toaster or electric oven, in which electrical energy is converted to thermal energy.

If the potential at *a* is lower than at *b*, then V_{ab} is negative and there is a net transfer of energy *out* of the circuit element. The element then acts as a source, delivering electrical energy into the circuit to which it is attached. This is the usual situation for a battery, which converts chemical energy into electrical energy and delivers it to the external circuit. Thus qV_{ab} can denote either a quantity of energy delivered to a circuit element or a quantity of energy extracted from that element.

In electric circuits we are most often interested in the *rate* at which energy is either delivered to or extracted from a circuit element. If the current through the element is I, then in a time interval dt an amount of charge $dQ = I \, dt$ passes through the element. The potential energy change for this amount of charge is $V_{ab} \, dQ = V_{ab} I \, dt$. Dividing this expression by dt, we obtain the *rate* at which energy is transferred either into or out of the circuit element. The time rate of energy transfer is *power*, denoted by P, so we write

$$P = V_{ab}I \qquad \text{(rate at which energy is delivered to or extracted from a circuit element)} \qquad (25.17)$$

The unit of V_{ab} is one volt, or one joule per coulomb, and the unit of I is one ampere, or one coulomb per second. Hence the unit of $P = V_{ab}I$ is one watt, as it should be:

$$(1 \text{ J/C})(1 \text{ C/s}) = 1 \text{ J/s} = 1 \text{ W}$$

Let's consider a few special cases.

Power Input to a Pure Resistance

If the circuit element in Fig. 25.21 is a resistor, the potential difference is $V_{ab} = IR$. From Eq. (25.17) the electrical power delivered to the resistor by the circuit is

$$P = V_{ab}I = I^2R = \frac{V_{ab}^2}{R} \qquad \text{(power delivered to a resistor)} \qquad (25.18)$$

In this case the potential at *a* (where the current enters the resistor) is always higher than that at *b* (where the current exits). Current enters the higher-potential terminal of the device, and Eq. (25.18) represents the rate of transfer of electric potential energy *into* the circuit element.

What becomes of this energy? The moving charges collide with atoms in the resistor and transfer some of their energy to these atoms, increasing the internal energy of the material. Either the temperature of the resistor increases or there is a flow of heat out of it, or both. In any of these cases we say that energy is *dissipated* in the resistor at a rate I^2R. Every resistor has a *power rating,* the maximum power the device can dissipate without becoming overheated and damaged. Some devices, such as electric heaters, are designed to get hot and transfer heat to their surroundings. But if the power rating is exceeded, even such a device may melt or even explode.

Power Output of a Source

The upper rectangle in Fig. 25.22a represents a source with emf $\mathcal{E}$ and internal resistance *r*, connected by ideal (resistanceless) conductors to an external circuit represented by the lower box. This could describe a car battery connected to one of the car's headlights (Fig. 25.22b). Point *a* is at higher potential than point *b*, so $V_a > V_b$ and V_{ab} is positive. Note that the current *I* is *leaving* the source at the higher-potential terminal (rather than entering there). Energy is being delivered to the external circuit, at a rate given by Eq. (25.17):

$$P = V_{ab}I$$

For a source that can be described by an emf $\mathcal{E}$ and an internal resistance *r*, we may use Eq. (25.15):

$$V_{ab} = \mathcal{E} - Ir$$

Multiplying this equation by *I*, we find

$$P = V_{ab}I = \mathcal{E}I - I^2r \qquad (25.19)$$

What do the terms $\mathcal{E}I$ and I^2r mean? In Section 25.4 we defined the emf $\mathcal{E}$ as the work per unit charge performed on the charges by the nonelectrostatic force as the charges are pushed "uphill" from *b* to *a* in the source. In a time *dt*, a charge $dQ = I\,dt$ flows through the source; the work done on it by this nonelectrostatic force is $\mathcal{E}\,dQ = \mathcal{E}I\,dt$. Thus $\mathcal{E}I$ is the *rate* at which work is done on the circulating charges by whatever agency causes the nonelectrostatic force in the source. This term represents the rate of conversion of nonelectrical energy to electrical energy within the source. The term I^2r is the rate at which electrical energy is *dissipated* in the internal resistance of the source. The difference $\mathcal{E}I - I^2r$ is the *net* electrical power output of the source—that is, the rate at which the source delivers electrical energy to the remainder of the circuit.

Power Input to a Source

Suppose that the lower rectangle in Fig. 25.22a is itself a source, with an emf *larger* than that of the upper source and with its emf opposite to that of the upper source. Figure 25.23 shows a practical example, an automobile battery (the upper circuit element) being charged by the car's alternator (the lower element). The current *I* in the circuit is then *opposite* to that shown in Fig. 25.22; the lower source is pushing current backward through the upper source. Because of this reversal of current, instead of Eq. (25.15) we have for the upper source

$$V_{ab} = \mathcal{E} + Ir$$

and instead of Eq. (25.19), we have

$$P = V_{ab}I = \mathcal{E}I + I^2r \qquad (25.20)$$

25.22 Energy conversion in a simple circuit.

(a) Diagrammatic circuit

- The emf source converts nonelectrical to electrical energy at a rate $\mathcal{E}I$.
- Its internal resistance *dissipates* energy at a rate I^2r.
- The difference $\mathcal{E}I - I^2r$ is its power output.

emf source with internal resistance *r*

External circuit

(b) A real circuit of the type shown in (a)

Battery

Headlight

25.23 When two sources are connected in a simple loop circuit, the source with the larger emf delivers energy to the other source.

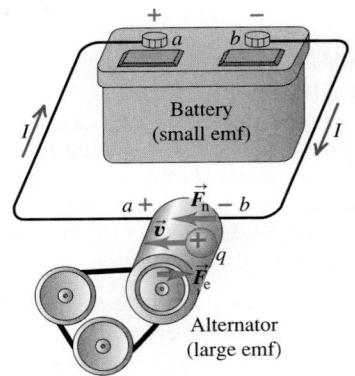

Battery (small emf)

Alternator (large emf)

Work is being done *on*, rather than *by*, the agent that causes the nonelectrostatic force in the upper source. There is a conversion of electrical energy into nonelectrical energy in the upper source at a rate $\mathcal{E}I$. The term I^2r in Eq. (25.20) is again the rate of dissipation of energy in the internal resistance of the upper source, and the sum $\mathcal{E}I + I^2r$ is the total electrical power *input* to the upper source. This is what happens when a rechargeable battery (a storage battery) is connected to a charger. The charger supplies electrical energy to the battery; part of it is converted to chemical energy, to be reconverted later, and the remainder is dissipated (wasted) in the battery's internal resistance, warming the battery and causing a heat flow out of it. If you have a power tool or laptop computer with a rechargeable battery, you may have noticed that it gets warm while it is charging.

Problem-Solving Strategy 25.1 **Power and Energy in Circuits**

IDENTIFY *the relevant concepts:* The ideas of electric power input and output can be applied to any electric circuit. Many problems will ask you to explicitly consider power or energy.

SET UP *the problem* using the following steps:
1. Make a drawing of the circuit.
2. Identify the circuit elements, including sources of emf and resistors. We will introduce other circuit elements later, including capacitors (Chapter 26) and inductors (Chapter 30).
3. Identify the target variables. Typically they will be the power input or output for each circuit element, or the total amount of energy put into or taken out of a circuit element in a given time.

EXECUTE *the solution* as follows:
1. A source of emf $\mathcal{E}$ delivers power $\mathcal{E}I$ into a circuit when current I flows through the source in the direction from $-$ to $+$. (For example, energy is converted from chemical energy in a battery, or from mechanical energy in a generator.) In this case there is a *positive* power output to the circuit or, equivalently, a *negative* power input to the source.
2. A source of emf takes power $\mathcal{E}I$ from a circuit when current passes through the source from $+$ to $-$. (This occurs in charging a storage battery, when electrical energy is converted to chemical energy.) In this case there is a *negative* power output

to the circuit or, equivalently, a *positive* power input to the source.
3. There is always a *positive* power input to a resistor through which current flows, irrespective of the direction of current flow. This process removes energy from the circuit, converting it to heat at the rate $VI = I^2R = V^2/R$, where V is the potential difference across the resistor.
4. Just as in item 3, there always is a positive power input to the internal resistance r of a source through which current flows, irrespective of the direction of current flow. This process likewise removes energy from the circuit, converting it into heat at the rate I^2r.
5. If the power into or out of a circuit element is constant, the energy delivered to or extracted from that element is the product of power and elapsed time. (In Chapter 26 we will encounter situations in which the power is not constant. In such cases, calculating the total energy requires an integral over the relevant time interval.)

EVALUATE *your answer:* Check your results; in particular, check that energy is conserved. This conservation can be expressed in either of two forms: "net power input = net power output" or "the algebraic sum of the power inputs to the circuit elements is zero."

Example 25.8 **Power input and output in a complete circuit**

For the circuit that we analyzed in Example 25.5, find the rates of energy conversion (chemical to electrical) and energy dissipation in the battery, the rate of energy dissipation in the 4-Ω resistor, and the battery's net power output.

SOLUTION

IDENTIFY and SET UP: Figure 25.24 shows the circuit, gives values of quantities known from Example 25.5, and indicates how we find the target variables. We use Eq. (25.19) to find the battery's net power output, the rate of chemical-to-electrical energy conversion, and the rate of energy dissipation in the battery's internal resistance. We use Eq. (25.18) to find the power delivered to (and dissipated in) the 4-Ω resistor.

EXECUTE: From the first term in Eq. (25.19), the rate of energy conversion in the battery is

$$\mathcal{E}I = (12 \text{ V})(2 \text{ A}) = 24 \text{ W}$$

25.24 Our sketch for this problem.

From the second term in Eq. (25.19), the rate of dissipation of energy in the battery is

$$I^2r = (2 \text{ A})^2(2 \text{ }\Omega) = 8 \text{ W}$$

The *net* electrical power output of the battery is the difference between these: $\mathcal{E}I - I^2r = 16$ W. From Eq. (25.18), the electrical power input to, and the equal rate of dissipation of electrical energy in, the 4-Ω resistor are

$$V_{a'b'}I = (8\text{ V})(2\text{ A}) = 16\text{ W}\quad\text{and}$$
$$I^2R = (2\text{ A})^2(4\text{ }\Omega) = 16\text{ W}$$

EVALUATE: The rate $V_{a'b'}I$ at which energy is supplied to the 4-Ω resistor equals the rate I^2R at which energy is dissipated there. This is also equal to the battery's net power output: $P = V_{ab}I = (8\text{ V})(2\text{ A}) = 16$ W. In summary, the rate at which the source of emf supplies energy is $\mathcal{E}I = 24$ W, of which $I^2r = 8$ W is dissipated in the battery's internal resistor and $I^2R = 16$ W is dissipated in the external resistor.

Example 25.9 Increasing the resistance

Suppose we replace the external 4-Ω resistor in Fig. 25.24 with an 8-Ω resistor. How does this affect the electrical power dissipated in this resistor?

SOLUTION

IDENTIFY and SET UP: Our target variable is the power dissipated in the resistor to which the battery is connected. The situation is the same as in Example 25.8, but with a higher external resistance R.

EXECUTE: According to Eq. (25.18), the power dissipated in the resistor is $P = I^2R$. You might conclude that making the resistance R twice as great as in Example 25.8 should also make the power twice as great, or $2(16\text{ W}) = 32$ W. If instead you used the formula $P = V_{ab}^2/R$, you might conclude that the power should be one-half as great as in the preceding example, or $(16\text{ W})/2 = 8$ W. Which answer is correct?

In fact, *both* of these answers are *incorrect*. The first is wrong because changing the resistance R also changes the current in the circuit (remember, a source of emf does *not* generate the same current in all situations). The second answer is wrong because the potential difference V_{ab} across the resistor changes when the current changes. To get the correct answer, we first find the current just as we did in Example 25.5:

$$I = \frac{\mathcal{E}}{R + r} = \frac{12\text{ V}}{8\text{ }\Omega + 2\text{ }\Omega} = 1.2\text{ A}$$

The greater resistance causes the current to decrease. The potential difference across the resistor is

$$V_{ab} = IR = (1.2\text{ A})(8\text{ }\Omega) = 9.6\text{ V}$$

which is greater than that with the 4-Ω resistor. We can then find the power dissipated in the resistor in either of two ways:

$$P = I^2R = (1.2\text{ A})^2(8\text{ }\Omega) = 12\text{ W}\quad\text{or}$$
$$P = \frac{V_{ab}^2}{R} = \frac{(9.6\text{ V})^2}{8\text{ }\Omega} = 12\text{ W}$$

EVALUATE: Increasing the resistance R causes a *reduction* in the power input to the resistor. In the expression $P = I^2R$ the decrease in current is more important than the increase in resistance; in the expression $P = V_{ab}^2/R$ the increase in resistance is more important than the increase in V_{ab}. This same principle applies to ordinary light bulbs; a 50-W light bulb has a greater resistance than does a 100-W light bulb.

Can you show that replacing the 4-Ω resistor with an 8-Ω resistor decreases both the rate of energy conversion (chemical to electrical) in the battery and the rate of energy dissipation in the battery?

Example 25.10 Power in a short circuit

For the short-circuit situation of Example 25.7, find the rates of energy conversion and energy dissipation in the battery and the net power output of the battery.

SOLUTION

IDENTIFY and SET UP: Our target variables are again the power inputs and outputs associated with the battery. Figure 25.25 shows

25.25 Our sketch for this problem.

the circuit. This is the same situation as in Example 25.8, but now the external resistance R is zero.

EXECUTE: We found in Example 25.7 that the current in this situation is $I = 6$ A. From Eq. (25.19), the rate of energy conversion (chemical to electrical) in the battery is then

$$\mathcal{E}I = (12\text{ V})(6\text{ A}) = 72\text{ W}$$

and the rate of dissipation of energy in the battery is

$$I^2r = (6\text{ A})^2(2\text{ }\Omega) = 72\text{ W}$$

The net power output of the source is $\mathcal{E}I - I^2r = 0$. We get this same result from the expression $P = V_{ab}I$, because the terminal voltage V_{ab} of the source is zero.

EVALUATE: With ideal wires and an ideal ammeter, so that $R = 0$, *all* of the converted energy from the source is dissipated within the source. This is why a short-circuited battery is quickly ruined and may explode.

25.6 Theory of Metallic Conduction

25.26 Random motions of an electron in a metallic crystal **(a)** with zero electric field and **(b)** with an electric field that causes drift. The curvatures of the paths are greatly exaggerated.

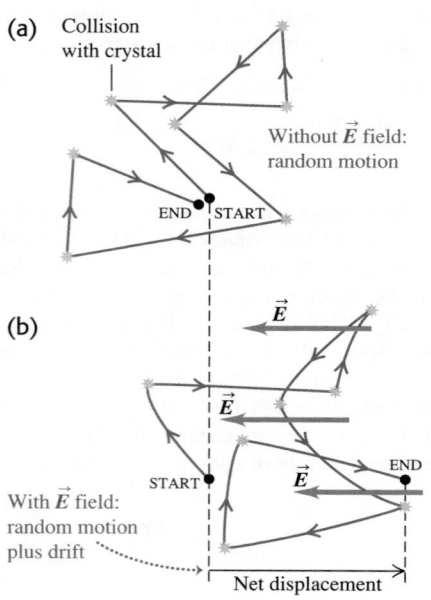

(a)   Collision with crystal

Without $\vec{E}$ field: random motion

END START

(b)

$\vec{E}$

$\vec{E}$

START $\vec{E}$ END

With $\vec{E}$ field: random motion plus drift

Net displacement

Mastering**PHYSICS**

PhET: Conductivity

25.27 The motion of a ball rolling down an inclined plane and bouncing off pegs in its path is analogous to the motion of an electron in a metallic conductor with an electric field present.

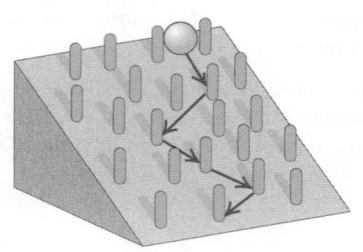

We can gain additional insight into electrical conduction by looking at the microscopic origin of conductivity. We'll consider a very simple model that treats the electrons as classical particles and ignores their quantum-mechanical behavior in solids. Using this model, we'll derive an expression for the resistivity of a metal. Even though this model is not entirely correct, it will still help you to develop an intuitive idea of the microscopic basis of conduction.

In the simplest microscopic model of conduction in a metal, each atom in the metallic crystal gives up one or more of its outer electrons. These electrons are then free to move through the crystal, colliding at intervals with the stationary positive ions. The motion of the electrons is analogous to the motion of molecules of a gas moving through a porous bed of sand.

If there is no electric field, the electrons move in straight lines between collisions, the directions of their velocities are random, and on average they never get anywhere (Fig. 25.26a). But if an electric field is present, the paths curve slightly because of the acceleration caused by electric-field forces. Figure 25.26b shows a few paths of an electron in an electric field directed from right to left. As we mentioned in Section 25.1, the average speed of random motion is of the order of 10^6 m/s, while the average drift speed is *much* slower, of the order of 10^{-4} m/s. The average time between collisions is called the **mean free time,** denoted by τ. Figure 25.27 shows a mechanical analog of this electron motion.

We would like to derive from this model an expression for the resistivity ρ of a material, defined by Eq. (25.5):

$$\rho = \frac{E}{J} \tag{25.21}$$

where E and J are the magnitudes of electric field and current density, respectively. The current density $\vec{J}$ is in turn given by Eq. (25.4):

$$\vec{J} = nq\vec{v}_d \tag{25.22}$$

where n is the number of free electrons per unit volume, $q = -e$ is the charge of each, and $\vec{v}_d$ is their average drift velocity.

We need to relate the drift velocity $\vec{v}_d$ to the electric field $\vec{E}$. The value of $\vec{v}_d$ is determined by a steady-state condition in which, on average, the velocity *gains* of the charges due to the force of the $\vec{E}$ field are just balanced by the velocity *losses* due to collisions. To clarify this process, let's imagine turning on the two effects one at a time. Suppose that before time $t = 0$ there is no field. The electron motion is then completely random. A typical electron has velocity $\vec{v}_0$ at time $t = 0$, and the value of $\vec{v}_0$ averaged over many electrons (that is, the initial velocity of an average electron) is zero: $(\vec{v}_0)_{av} = 0$. Then at time $t = 0$ we turn on a constant electric field $\vec{E}$. The field exerts a force $\vec{F} = q\vec{E}$ on each charge, and this causes an acceleration $\vec{a}$ in the direction of the force, given by

$$\vec{a} = \frac{\vec{F}}{m} = \frac{q\vec{E}}{m}$$

where m is the electron mass. *Every* electron has this acceleration.

We wait for a time τ, the average time between collisions, and then "turn on" the collisions. An electron that has velocity $\vec{v}_0$ at time $t = 0$ has a velocity at time $t = \tau$ equal to

$$\vec{v} = \vec{v}_0 + \vec{a}\tau$$

The velocity $\vec{v}_{av}$ of an *average* electron at this time is the sum of the averages of the two terms on the right. As we have pointed out, the initial velocity $\vec{v}_0$ is zero for an average electron, so

$$\vec{v}_{av} = \vec{a}\tau = \frac{q\tau}{m}\vec{E} \qquad (25.23)$$

After time $t = \tau$, the tendency of the collisions to decrease the velocity of an average electron (by means of randomizing collisions) just balances the tendency of the $\vec{E}$ field to increase this velocity. Thus the velocity of an average electron, given by Eq. (25.23), is maintained over time and is equal to the drift velocity $\vec{v}_d$:

$$\vec{v}_d = \frac{q\tau}{m}\vec{E}$$

Now we substitute this equation for the drift velocity $\vec{v}_d$ into Eq. (25.22):

$$\vec{J} = nq\vec{v}_d = \frac{nq^2\tau}{m}\vec{E}$$

Comparing this with Eq. (25.21), which we can rewrite as $\vec{J} = \vec{E}/\rho$, and substituting $q = -e$ for an electron, we see that the resistivity ρ is given by

$$\rho = \frac{m}{ne^2\tau} \qquad (25.24)$$

If n and τ are independent of $\vec{E}$, then the resistivity is independent of $\vec{E}$ and the conducting material obeys Ohm's law.

Turning the interactions on one at a time may seem artificial. But the derivation would come out the same if each electron had its own clock and the $t = 0$ times were different for different electrons. If τ is the average time between collisions, then $\vec{v}_d$ is still the average electron drift velocity, even though the motions of the various electrons aren't actually correlated in the way we postulated.

What about the temperature dependence of resistivity? In a perfect crystal with no atoms out of place, a correct quantum-mechanical analysis would let the free electrons move through the crystal with no collisions at all. But the atoms vibrate about their equilibrium positions. As the temperature increases, the amplitudes of these vibrations increase, collisions become more frequent, and the mean free time τ decreases. So this theory predicts that the resistivity of a metal increases with temperature. In a superconductor, roughly speaking, there are no inelastic collisions, τ is infinite, and the resistivity ρ is zero.

In a pure semiconductor such as silicon or germanium, the number of charge carriers per unit volume, n, is not constant but increases very rapidly with increasing temperature. This increase in n far outweighs the decrease in the mean free time, and in a semiconductor the resistivity always decreases rapidly with increasing temperature. At low temperatures, n is very small, and the resistivity becomes so large that the material can be considered an insulator.

Electrons gain energy between collisions through the work done on them by the electric field. During collisions they transfer some of this energy to the atoms of the material of the conductor. This leads to an increase in the material's internal energy and temperature; that's why wires carrying current get warm. If the electric field in the material is large enough, an electron can gain enough energy between collisions to knock off electrons that are normally bound to atoms in the material. These can then knock off more electrons, and so on, leading to an avalanche of current. This is the basis of dielectric breakdown in insulators (see Section 24.4).

Example 25.11 **Mean free time in copper**

Calculate the mean free time between collisions in copper at room temperature.

SOLUTION

IDENTIFY and SET UP: We can obtain an expression for mean free time τ in terms of n, ρ, e, and m by rearranging Eq. (25.24). From Example 25.1 and Table 25.1, for copper $n = 8.5 \times 10^{28}$ m^{-3} and $\rho = 1.72 \times 10^{-8}$ $\Omega \cdot$m. In addition, $e = 1.60 \times 10^{-19}$ C and $m = 9.11 \times 10^{-31}$ kg for electrons.

EXECUTE: From Eq. (25.24), we get

$$\tau = \frac{m}{ne^2\rho}$$

$$= \frac{9.11 \times 10^{-31} \text{ kg}}{(8.5 \times 10^{28} \text{ m}^{-3})(1.60 \times 10^{-19} \text{ C})^2(1.72 \times 10^{-8} \ \Omega \cdot \text{m})}$$

$$= 2.4 \times 10^{-14} \text{ s}$$

EVALUATE: The mean free time is the average time between collisions for a given electron. Taking the reciprocal of this time, we find that each electron averages $1/\tau = 4.2 \times 10^{13}$ collisions per second!

Test Your Understanding of Section 25.6 Which of the following factors will, if increased, make it more difficult to produce a certain amount of current in a conductor? (There may be more than one correct answer.) (i) the mass of the moving charged particles in the conductor; (ii) the number of moving charged particles per cubic meter; (iii) the amount of charge on each moving particle; (iv) the average time between collisions for a typical moving charged particle. ∣

Current and current density: Current is the amount of charge flowing through a specified area, per unit time. The SI unit of current is the ampere (1 A = 1 C/s). The current I through an area A depends on the concentration n and charge q of the charge carriers, as well as on the magnitude of their drift velocity $\vec{v}_d$. The current density is current per unit cross-sectional area. Current is usually described in terms of a flow of positive charge, even when the charges are actually negative or of both signs. (See Example 25.1.)

$$I = \frac{dQ}{dt} = n|q|v_d A \quad (25.2)$$

$$\vec{J} = nq\vec{v}_d \quad (25.4)$$

Resistivity: The resistivity ρ of a material is the ratio of the magnitudes of electric field and current density. Good conductors have small resistivity; good insulators have large resistivity. Ohm's law, obeyed approximately by many materials, states that ρ is a constant independent of the value of E. Resistivity usually increases with temperature; for small temperature changes this variation is represented approximately by Eq. (25.6), where α is the temperature coefficient of resistivity.

$$\rho = \frac{E}{J} \quad (25.5)$$

$$\rho(T) = \rho_0[1 + \alpha(T - T_0)] \quad (25.6)$$

Metal: ρ increases with increasing T.

Resistors: The potential difference V across a sample of material that obeys Ohm's law is proportional to the current I through the sample. The ratio $V/I = R$ is the resistance of the sample. The SI unit of resistance is the ohm (1 Ω = 1 V/A). The resistance of a cylindrical conductor is related to its resistivity ρ, length L, and cross-sectional area A. (See Examples 25.2 and 25.3.)

$$V = IR \quad (25.11)$$

$$R = \frac{\rho L}{A} \quad (25.10)$$

Circuits and emf: A complete circuit has a continuous current-carrying path. A complete circuit carrying a steady current must contain a source of electromotive force (emf) $\mathcal{E}$. The SI unit of electromotive force is the volt (1 V). Every real source of emf has some internal resistance r, so its terminal potential difference V_{ab} depends on current. (See Examples 25.4–25.7.)

$$V_{ab} = \mathcal{E} - Ir \quad (25.15)$$
(source with internal resistance)

Energy and power in circuits: A circuit element with a potential difference $V_a - V_b = V_{ab}$ and a current I puts energy into a circuit if the current direction is from lower to higher potential in the device, and it takes energy out of the circuit if the current is opposite. The power P equals the product of the potential difference and the current. A resistor always takes electrical energy out of a circuit. (See Examples 25.8–25.10.)

$$P = V_{ab}I \quad (25.17)$$
(general circuit element)

$$P = V_{ab}I = I^2 R = \frac{V_{ab}^2}{R} \quad (25.18)$$
(power into a resistor)

Conduction in metals: The microscopic basis of conduction in metals is the motion of electrons that move freely through the metallic crystal, bumping into ion cores in the crystal. In a crude classical model of this motion, the resistivity of the material can be related to the electron mass, charge, speed of random motion, density, and mean free time between collisions. (See Example 25.11.)

BRIDGING PROBLEM   Resistivity, Temperature, and Power

A toaster using a Nichrome heating element operates on 120 V. When it is switched on at 20°C, the heating element carries an initial current of 1.35 A. A few seconds later the current reaches the steady value of 1.23 A. (a) What is the final temperature of the element? The average value of the temperature coefficient of resistivity for Nichrome over the relevant temperature range is 4.5×10^{-4} $(C°)^{-1}$. (b) What is the power dissipated in the heating element initially and when the current reaches 1.23 A?

SOLUTION GUIDE

See MasteringPhysics® study area for a Video Tutor solution.

IDENTIFY and SET UP

1. A heating element acts as a resistor that converts electrical energy into thermal energy. The resistivity ρ of Nichrome depends on temperature, and hence so does the resistance $R = \rho L/A$ of the heating element and the current $I = V/R$ that it carries.
2. We are given $V = 120$ V and the initial and final values of I. Select an equation that will allow you to find the initial and

final values of resistance, and an equation that relates resistance to temperature [the target variable in part (a)].
3. The power P dissipated in the heating element depends on I and V. Select an equation that will allow you to calculate the initial and final values of P.

EXECUTE

4. Combine your equations from step 2 to give a relationship between the initial and final values of I and the initial and final temperatures (20°C and T_{final}).
5. Solve your expression from step 4 for T_{final}.
6. Use your equation from step 3 to find the initial and final powers.

EVALUATE

7. Is the final temperature greater than or less than 20°C? Does this make sense?
8. Is the final resistance greater than or less than the initial resistance? Again, does this make sense?
9. Is the final power greater than or less than the initial power? Does this agree with your observations in step 8?

Problems

For instructor-assigned homework, go to www.masteringphysics.com

•, ••, •••: Problems of increasing difficulty. **CP**: Cumulative problems incorporating material from earlier chapters. **CALC**: Problems requiring calculus. **BIO**: Biosciences problems.

DISCUSSION QUESTIONS

Q25.1 The definition of resistivity ($\rho = E/J$) implies that an electric field exists inside a conductor. Yet we saw in Chapter 21 that there can be no electric field inside a conductor. Is there a contradiction here? Explain.

Q25.2 A cylindrical rod has resistance R. If we triple its length and diameter, what is its resistance, in terms of R?

Q25.3 A cylindrical rod has resistivity ρ. If we triple its length and diameter, what is its resistivity, in terms of ρ?

Q25.4 Two copper wires with different diameters are joined end to end. If a current flows in the wire combination, what happens to electrons when they move from the larger-diameter wire into the smaller-diameter wire? Does their drift speed increase, decrease, or stay the same? If the drift speed changes, what is the force that causes the change? Explain your reasoning.

Q25.5 When is a 1.5-V AAA battery *not* actually a 1.5-V battery? That is, when do its terminals provide a potential difference of less than 1.5 V?

Q25.6 Can the potential difference between the terminals of a battery ever be opposite in direction to the emf? If it can, give an example. If it cannot, explain why not.

Q25.7 A rule of thumb used to determine the internal resistance of a source is that it is the open-circuit voltage divided by the short-circuit current. Is this correct? Why or why not?

Q25.8 Batteries are always labeled with their emf; for instance, an AA flashlight battery is labeled "1.5 volts." Would it also be appropriate to put a label on batteries stating how much current they provide? Why or why not?

Q25.9 We have seen that a coulomb is an enormous amount of charge; it is virtually impossible to place a charge of 1 C on an object. Yet, a current of 10 A, 10 C/s, is quite reasonable. Explain this apparent discrepancy.

Q25.10 Electrons in an electric circuit pass through a resistor. The wire on either side of the resistor has the same diameter. (a) How does the drift speed of the electrons before entering the resistor compare to the speed after leaving the resistor? Explain your reasoning. (b) How does the potential energy for an electron before entering the resistor compare to the potential energy after leaving the resistor? Explain your reasoning.

Q25.11 Current causes the temperature of a real resistor to increase. Why? What effect does this heating have on the resistance? Explain.

Q25.12 Which of the graphs in Fig. Q25.12 best illustrates the current I in a real resistor as a function of the potential difference V across it? Explain. (*Hint*: See Discussion Question Q25.11.)

Figure **Q25.12**

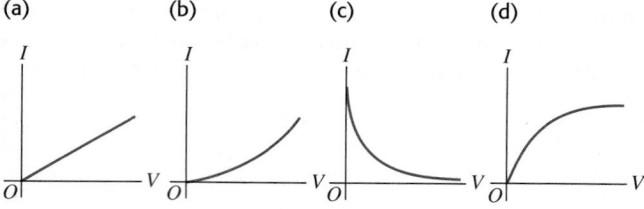

Q25.13 Why does an electric light bulb nearly always burn out just as you turn on the light, almost never while the light is shining?

Q25.14 A light bulb glows because it has resistance. The brightness of a light bulb increases with the electrical power dissipated in the bulb. (a) In the circuit shown in Fig. Q25.14a, the two bulbs A and B are identical. Compared to bulb A, does bulb B glow more brightly, just as brightly, or less brightly? Explain your reasoning. (b) Bulb B is removed from the circuit and the circuit is completed as shown in Fig. Q25.14b. Compared to the brightness of bulb A in Fig. Q25.14a, does bulb A now glow more brightly, just as brightly, or less brightly? Explain your reasoning.

Figure **Q25.14**

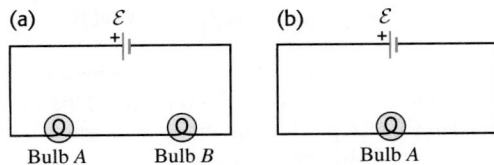

Q25.15 (See Discussion Question Q25.14.) An ideal ammeter A is placed in a circuit with a battery and a light bulb as shown in Fig. Q25.15a, and the ammeter reading is noted. The circuit is then reconnected as in Fig. Q25.15b, so that the positions of the ammeter and light bulb are reversed. (a) How does the ammeter reading in the situation shown in Fig. Q25.15a compare to the reading in the situation shown in Fig. Q25.15b? Explain your reasoning. (b) In which situation does the light bulb glow more brightly? Explain your reasoning.

Figure **Q25.15**

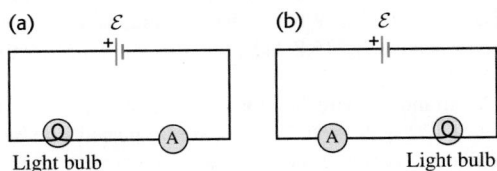

Q25.16 (See Discussion Question Q25.14.) Will a light bulb glow more brightly when it is connected to a battery as shown in Fig. Q25.16a, in which an ideal ammeter A is placed in the circuit, or when it is connected as shown in Fig. 25.16b, in which an ideal voltmeter V is placed in the circuit? Explain your reasoning.

Figure **Q25.16**

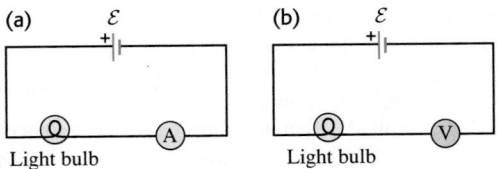

Q25.17 The energy that can be extracted from a storage battery is always less than the energy that goes into it while it is being charged. Why?

Q25.18 Eight flashlight batteries in series have an emf of about 12 V, similar to that of a car battery. Could they be used to start a car with a dead battery? Why or why not?

Q25.19 Small aircraft often have 24-V electrical systems rather than the 12-V systems in automobiles, even though the electrical power requirements are roughly the same in both applications. The explanation given by aircraft designers is that a 24-V system weighs less than a 12-V system because thinner wires can be used. Explain why this is so.

Q25.20 Long-distance, electric-power, transmission lines always operate at very high voltage, sometimes as much as 750 kV. What are the advantages of such high voltages? What are the disadvantages?

Q25.21 Ordinary household electric lines in North America usually operate at 120 V. Why is this a desirable voltage, rather than a value considerably larger or smaller? On the other hand, automobiles usually have 12-V electrical systems. Why is this a desirable voltage?

Q25.22 A fuse is a device designed to break a circuit, usually by melting when the current exceeds a certain value. What characteristics should the material of the fuse have?

Q25.23 High-voltage power supplies are sometimes designed intentionally to have rather large internal resistance as a safety precaution. Why is such a power supply with a large internal resistance safer than a supply with the same voltage but lower internal resistance?

Q25.24 The text states that good thermal conductors are also good electrical conductors. If so, why don't the cords used to connect toasters, irons, and similar heat-producing appliances get hot by conduction of heat from the heating element?

EXERCISES

Section 25.1 Current

25.1 • **Lightning Strikes.** During lightning strikes from a cloud to the ground, currents as high as 25,000 A can occur and last for about 40 μs. How much charge is transferred from the cloud to the earth during such a strike?

25.2 • A silver wire 2.6 mm in diameter transfers a charge of 420 C in 80 min. Silver contains 5.8×10^{28} free electrons per cubic meter. (a) What is the current in the wire? (b) What is the magnitude of the drift velocity of the electrons in the wire?

25.3 • A 5.00-A current runs through a 12-gauge copper wire (diameter 2.05 mm) and through a light bulb. Copper has 8.5×10^{28} free electrons per cubic meter. (a) How many electrons pass through the light bulb each second? (b) What is the current density in the wire? (c) At what speed does a typical electron pass by any given point in the wire? (d) If you were to use wire of twice the diameter, which of the above answers would change? Would they increase or decrease?

25.4 • An 18-gauge copper wire (diameter 1.02 mm) carries a current with a current density of 1.50×10^6 A/m². The density of free electrons for copper is 8.5×10^{28} electrons per cubic meter. Calculate (a) the current in the wire and (b) the drift velocity of electrons in the wire.

25.5 •• Copper has 8.5×10^{28} free electrons per cubic meter. A 71.0-cm length of 12-gauge copper wire that is 2.05 mm in diameter carries 4.85 A of current. (a) How much time does it take for an electron to travel the length of the wire? (b) Repeat part (a) for 6-gauge copper wire (diameter 4.12 mm) of the same length that carries the same current. (c) Generally speaking, how does changing the diameter of a wire that carries a given amount of current affect the drift velocity of the electrons in the wire?

25.6 • Consider the 18-gauge wire in Example 25.1. How many atoms are in 1.00 m³ of copper? With the density of free electrons given in the example, how many free electrons are there per copper atom?

25.7 • **CALC** The current in a wire varies with time according to the relationship $I = 55$ A $- (0.65$ A/s$^2)t^2$. (a) How many coulombs of charge pass a cross section of the wire in the time interval between $t = 0$ and $t = 8.0$ s? (b) What constant current would transport the same charge in the same time interval?

25.8 • Current passes through a solution of sodium chloride. In 1.00 s, 2.68×10^{16} Na$^+$ ions arrive at the negative electrode and 3.92×10^{16} Cl$^-$ ions arrive at the positive electrode. (a) What is the current passing between the electrodes? (b) What is the direction of the current?

25.9 • **BIO Transmission of Nerve Impulses.** Nerve cells transmit electric signals through their long tubular axons. These signals propagate due to a sudden rush of Na$^+$ ions, each with charge $+e$, into the axon. Measurements have revealed that typically about 5.6×10^{11} Na$^+$ ions enter each meter of the axon during a time of 10 ms. What is the current during this inflow of charge in a meter of axon?

Section 25.2 Resistivity and Section 25.3 Resistance

25.10 • (a) At room temperature what is the strength of the electric field in a 12-gauge copper wire (diameter 2.05 mm) that is needed to cause a 2.75-A current to flow? (b) What field would be needed if the wire were made of silver instead?

25.11 •• A 1.50-m cylindrical rod of diameter 0.500 cm is connected to a power supply that maintains a constant potential difference of 15.0 V across its ends, while an ammeter measures the current through it. You observe that at room temperature (20.0°C) the ammeter reads 18.5 A, while at 92.0°C it reads 17.2 A. You can ignore any thermal expansion of the rod. Find (a) the resistivity at 20.0°C and (b) the temperature coefficient of resistivity at 20°C for the material of the rod.

25.12 • A copper wire has a square cross section 2.3 mm on a side. The wire is 4.0 m long and carries a current of 3.6 A. The density of free electrons is 8.5×10^{28}/m^3. Find the magnitudes of (a) the current density in the wire and (b) the electric field in the wire. (c) How much time is required for an electron to travel the length of the wire?

25.13 • A 14-gauge copper wire of diameter 1.628 mm carries a current of 12.5 mA. (a) What is the potential difference across a 2.00-m length of the wire? (b) What would the potential difference in part (a) be if the wire were silver instead of copper, but all else were the same?

25.14 •• A wire 6.50 m long with diameter of 2.05 mm has a resistance of 0.0290 Ω. What material is the wire most likely made of?

25.15 •• A cylindrical tungsten filament 15.0 cm long with a diameter of 1.00 mm is to be used in a machine for which the temperature will range from room temperature (20°C) up to 120°C. It will carry a current of 12.5 A at all temperatures (consult Tables 25.1 and 25.2). (a) What will be the maximum electric field in this filament, and (b) what will be its resistance with that field? (c) What will be the maximum potential drop over the full length of the filament?

25.16 •• A ductile metal wire has resistance R. What will be the resistance of this wire in terms of R if it is stretched to three times its original length, assuming that the density and resistivity of the material do not change when the wire is stretched? (*Hint:* The amount of metal does not change, so stretching out the wire will affect its cross-sectional area.)

25.17 • In household wiring, copper wire 2.05 mm in diameter is often used. Find the resistance of a 24.0-m length of this wire.

25.18 •• What diameter must a copper wire have if its resistance is to be the same as that of an equal length of aluminum wire with diameter 3.26 mm?

25.19 • You need to produce a set of cylindrical copper wires 3.50 m long that will have a resistance of 0.125 Ω each. What will be the mass of each of these wires?

25.20 • A tightly coiled spring having 75 coils, each 3.50 cm in diameter, is made of insulated metal wire 3.25 mm in diameter. An ohmmeter connected across its opposite ends reads 1.74 Ω. What is the resistivity of the metal?

25.21 • An aluminum cube has sides of length 1.80 m. What is the resistance between two opposite faces of the cube?

25.22 • You apply a potential difference of 4.50 V between the ends of a wire that is 2.50 m in length and 0.654 mm in radius. The resulting current through the wire is 17.6 A. What is the resistivity of the wire?

25.23 • A current-carrying gold wire has diameter 0.84 mm. The electric field in the wire is 0.49 V/m. What are (a) the current carried by the wire; (b) the potential difference between two points in the wire 6.4 m apart; (c) the resistance of a 6.4-m length of this wire?

25.24 • A hollow aluminum cylinder is 2.50 m long and has an inner radius of 3.20 cm and an outer radius of 4.60 cm. Treat each surface (inner, outer, and the two end faces) as an equipotential surface. At room temperature, what will an ohmmeter read if it is connected between (a) the opposite faces and (b) the inner and outer surfaces?

25.25 • (a) What is the resistance of a Nichrome wire at 0.0°C if its resistance is 100.00 Ω at 11.5°C? (b) What is the resistance of a carbon rod at 25.8°C if its resistance is 0.0160 Ω at 0.0°C?

25.26 • A carbon resistor is to be used as a thermometer. On a winter day when the temperature is 4.0°C, the resistance of the carbon resistor is 217.3 Ω. What is the temperature on a spring day when the resistance is 215.8 Ω? (Take the reference temperature T_0 to be 4.0°C.)

25.27 • A strand of wire has resistance 5.60 $\mu\Omega$. Find the net resistance of 120 such strands if they are (a) placed side by side to form a cable of the same length as a single strand, and (b) connected end to end to form a wire 120 times as long as a single strand.

Section 25.4 Electromotive Force and Circuits

25.28 • Consider the circuit shown in Fig. E25.28. The terminal voltage of the 24.0-V battery is 21.2 V. What are (a) the internal resistance r of the battery and (b) the resistance R of the circuit resistor?

Figure **E25.28**

25.29 • A copper transmission cable 100 km long and 10.0 cm in diameter carries a current of 125 A. (a) What is the potential drop across the cable? (b) How much electrical energy is dissipated as thermal energy every hour?

Figure **E25.30**

25.30 • An idealized ammeter is connected to a battery as shown in Fig. E25.30. Find (a) the reading of the ammeter, (b) the current through the 4.00-Ω resistor, (c) the terminal voltage of the battery.

25.31 • An ideal voltmeter V is connected to a 2.0-Ω resistor and a battery with emf 5.0 V and internal resistance 0.5 Ω as shown in Fig. E25.31. (a) What is the current in the 2.0-Ω resistor? (b) What is the terminal voltage of the battery? (c) What is the reading on the voltmeter? Explain your answers.

Figure E25.31

25.32 • The circuit shown in Fig. E25.32 contains two batteries, each with an emf and an internal resistance, and two resistors. Find (a) the current in the circuit (magnitude *and* direction); (b) the terminal voltage V_{ab} of the 16.0-V battery; (c) the potential difference V_{ac} of point *a* with respect to point *c*. (d) Using Fig. 25.20 as a model, graph the potential rises and drops in this circuit.

Figure E25.32

25.33 • When switch S in Fig. E25.33 is open, the voltmeter V of the battery reads 3.08 V. When the switch is closed, the voltmeter reading drops to 2.97 V, and the ammeter A reads 1.65 A. Find the emf, the internal resistance of the battery, and the circuit resistance R. Assume that the two meters are ideal, so they don't affect the circuit.

Figure E25.33

25.34 • In the circuit of Fig. E25.32, the 5.0-Ω resistor is removed and replaced by a resistor of unknown resistance R. When this is done, an ideal voltmeter connected across the points *b* and *c* reads 1.9 V. Find (a) the current in the circuit and (b) the resistance R. (c) Graph the potential rises and drops in this circuit (see Fig. 25.20).

25.35 • In the circuit shown in Fig. E25.32, the 16.0-V battery is removed and reinserted with the opposite polarity, so that its negative terminal is now next to point *a*. Find (a) the current in the circuit (magnitude *and* direction); (b) the terminal voltage V_{ba} of the 16.0-V battery; (c) the potential difference V_{ac} of point *a* with respect to point *c*. (d) Graph the potential rises and drops in this circuit (see Fig. 25.20).

25.36 • The following measurements were made on a Thyrite resistor:

I (A)	0.50	1.00	2.00	4.00
V_{ab} (V)	2.55	3.11	3.77	4.58

(a) Graph V_{ab} as a function of *I*. (b) Does Thyrite obey Ohm's law? How can you tell? (c) Graph the resistance $R = V_{ab}/I$ as a function of *I*.

25.37 • The following measurements of current and potential difference were made on a resistor constructed of Nichrome wire:

I (A)	0.50	1.00	2.00	4.00
V_{ab} (V)	1.94	3.88	7.76	15.52

(a) Graph V_{ab} as a function of *I*. (b) Does Nichrome obey Ohm's law? How can you tell? (c) What is the resistance of the resistor in ohms?

25.38 •• The circuit shown in Fig. E25.38 contains two batteries, each with an emf and an internal resistance, and two resistors. Find (a) the current in the circuit (magnitude and direction) and (b) the terminal voltage V_{ab} of the 16.0-V battery.

Figure E25.38

Section 25.5 Energy and Power in Electric Circuits

25.39 • **Light Bulbs.** The power rating of a light bulb (such as a 100-W bulb) is the power it dissipates when connected across a 120-V potential difference. What is the resistance of (a) a 100-W bulb and (b) a 60-W bulb? (c) How much current does each bulb draw in normal use?

25.40 • If a "75-W" bulb (see Problem 25.39) is connected across a 220-V potential difference (as is used in Europe), how much power does it dissipate?

25.41 • **European Light Bulb.** In Europe the standard voltage in homes is 220 V instead of the 120 V used in the United States. Therefore a "100-W" European bulb would be intended for use with a 220-V potential difference (see Problem 25.40). (a) If you bring a "100-W" European bulb home to the United States, what should be its U.S. power rating? (b) How much current will the 100-W European bulb draw in normal use in the United States?

25.42 • A battery-powered global positioning system (GPS) receiver operating on 9.0 V draws a current of 0.13 A. How much electrical energy does it consume during 1.5 h?

25.43 • Consider a resistor with length L, uniform cross-sectional area A, and uniform resistivity ρ that is carrying a current with uniform current density J. Use Eq. (25.18) to find the electrical power dissipated per unit volume, p. Express your result in terms of (a) E and J; (b) J and ρ; (c) E and ρ.

25.44 • **BIO Electric Eels.** Electric eels generate electric pulses along their skin that can be used to stun an enemy when they come into contact with it. Tests have shown that these pulses can be up to 500 V and produce currents of 80 mA (or even larger). A typical pulse lasts for 10 ms. What power and how much energy are delivered to the unfortunate enemy with a single pulse, assuming a steady current?

25.45 • **BIO Treatment of Heart Failure.** A heart defibrillator is used to enable the heart to start beating if it has stopped. This is done by passing a large current of 12 A through the body at 25 V for a very short time, usually about 3.0 ms. (a) What power does the defibrillator deliver to the body, and (b) how much energy is transferred?

25.46 • Consider the circuit of Fig. E25.32. (a) What is the total rate at which electrical energy is dissipated in the 5.0-Ω and 9.0-Ω resistors? (b) What is the power output of the 16.0-V battery? (c) At what rate is electrical energy being converted to other forms in the 8.0-V battery? (d) Show that the power output of the 16.0-V battery equals the overall rate of dissipation of electrical energy in the rest of the circuit.

25.47 •• The capacity of a storage battery, such as those used in automobile electrical systems, is rated in ampere-hours (A·h). A 50-A·h battery can supply a current of 50 A for 1.0 h, or 25 A for 2.0 h, and so on. (a) What total energy can be supplied by a 12-V, 60-A·h battery if its internal resistance is negligible? (b) What volume (in liters) of gasoline has a total heat of combustion equal to the energy obtained in part (a)? (See Section 17.6; the density of gasoline is 900 kg/m^3.) (c) If a generator with an average electrical power output of 0.45 kW is connected to the battery, how much time will be required for it to charge the battery fully?

25.48 • In the circuit analyzed in Example 25.8 the 4.0-Ω resistor is replaced by a 8.0-Ω resistor, as in Example 25.9. (a) Calculate the rate of conversion of chemical energy to electrical energy in the battery. How does your answer compare to the result calculated in Example 25.8? (b) Calculate the rate of electrical energy dissipation in the internal resistance of the battery. How does your answer compare to the result calculated in Example 25.8? (c) Use the results of parts (a) and (b) to calculate the net power output of the battery. How does your result compare to the electrical power dissipated in the 8.0-Ω resistor as calculated for this circuit in Example 25.9?

25.49 •• A 25.0-Ω bulb is connected across the terminals of a 12.0-V battery having 3.50 Ω of internal resistance. What percentage of the power of the battery is dissipated across the internal resistance and hence is not available to the bulb?

25.50 • An idealized voltmeter is connected across the terminals of a 15.0-V battery, and a 75.0-Ω appliance is also connected across its terminals. If the voltmeter reads 11.3 V: (a) how much power is being dissipated by the appliance, and (b) what is the internal resistance of the battery?

25.51 • In the circuit in Fig. E25.51, find (a) the rate of conversion of internal (chemical) energy to electrical energy within the battery; (b) the rate of dissipation of electrical energy in the battery; (c) the rate of dissipation of electrical energy in the external resistor.

Figure **E25.51**

25.52 •• A typical small flashlight contains two batteries, each having an emf of 1.5 V, connected in series with a bulb having resistance 17 Ω. (a) If the internal resistance of the batteries is negligible, what power is delivered to the bulb? (b) If the batteries last for 5.0 h, what is the total energy delivered to the bulb? (c) The resistance of real batteries increases as they run down. If the initial internal resistance is negligible, what is the combined internal resistance of both batteries when the power to the bulb has decreased to half its initial value? (Assume that the resistance of the bulb is constant. Actually, it will change somewhat when the current through the filament changes, because this changes the temperature of the filament and hence the resistivity of the filament wire.)

25.53 • A "540-W" electric heater is designed to operate from 120-V lines. (a) What is its resistance? (b) What current does it draw? (c) If the line voltage drops to 110 V, what power does the heater take? (Assume that the resistance is constant. Actually, it will change because of the change in temperature.) (d) The heater coils are metallic, so that the resistance of the heater decreases with decreasing temperature. If the change of resistance with temperature is taken into account, will the electrical power consumed by the heater be larger or smaller than what you calculated in part (c)? Explain.

Section 25.6 Theory of Metallic Conduction

25.54 •• Pure silicon contains approximately 1.0×10^{16} free electrons per cubic meter. (a) Referring to Table 25.1, calculate the mean free time τ for silicon at room temperature. (b) Your answer in part (a) is much greater than the mean free time for copper given in Example 25.11. Why, then, does pure silicon have such a high resistivity compared to copper?

PROBLEMS

25.55 • An electrical conductor designed to carry large currents has a circular cross section 2.50 mm in diameter and is 14.0 m long. The resistance between its ends is 0.104 Ω. (a) What is the resistivity of the material? (b) If the electric-field magnitude in the conductor is 1.28 V/m, what is the total current? (c) If the material has 8.5×10^{28} free electrons per cubic meter, find the average drift speed under the conditions of part (b).

25.56 •• A plastic tube 25.0 m long and 3.00 cm in diameter is dipped into a silver solution, depositing a layer of silver 0.100 mm thick uniformly over the outer surface of the tube. If this coated tube is then connected across a 12.0-V battery, what will be the current?

25.57 •• On your first day at work as an electrical technician, you are asked to determine the resistance per meter of a long piece of wire. The company you work for is poorly equipped. You find a battery, a voltmeter, and an ammeter, but no meter for directly measuring resistance (an ohmmeter). You put the leads from the voltmeter across the terminals of the battery, and the meter reads 12.6 V. You cut off a 20.0-m length of wire and connect it to the battery, with an ammeter in series with it to measure the current in the wire. The ammeter reads 7.00 A. You then cut off a 40.0-m length of wire and connect it to the battery, again with the ammeter in series to measure the current. The ammeter reads 4.20 A. Even though the equipment you have available to you is limited, your boss assures you of its high quality: The ammeter has very small resistance, and the voltmeter has very large resistance. What is the resistance of 1 meter of wire?

25.58 • A 2.0-mm length of wire is made by welding the end of a 120-cm-long silver wire to the end of an 80-cm-long copper wire. Each piece of wire is 0.60 mm in diameter. The wire is at room temperature, so the resistivities are as given in Table 25.1. A potential difference of 5.0 V is maintained between the ends of the 2.0-m composite wire. (a) What is the current in the copper section? (b) What is the current in the silver section? (c) What is the magnitude of $\vec{E}$ in the copper? (d) What is the magnitude of $\vec{E}$ in the silver? (e) What is the potential difference between the ends of the silver section of wire?

25.59 • A 3.00-m length of copper wire at 20°C has a 1.20-m-long section with diameter 1.60 mm and a 1.80-m-long section with diameter 0.80 mm. There is a current of 2.5 mA in the 1.60-mm-diameter section. (a) What is the current in the 0.80-mm-diameter section? (b) What is the magnitude of $\vec{E}$ in the 1.60-mm-diameter section? (c) What is the magnitude of $\vec{E}$ in the 0.80-mm-diameter section? (d) What is the potential difference between the ends of the 3.00-m length of wire?

25.60 • **Critical Current Density in Superconductors.** One problem with some of the newer high-temperature superconductors is getting a large enough current density for practical use without causing the resistance to reappear. The maximum current density for which the material will remain a superconductor is called the critical current density of the material. In 1987, IBM research labs had produced thin films with critical current densities of 1.0×10^5 A/cm^2. (a) How much current could an 18-gauge wire (see Example 25.1 in Section 25.1) of this material carry and still remain superconducting? (b) Researchers are trying to develop superconductors with critical current densities of 1.0×10^6 A/cm^2. What diameter cylindrical wire of such a material would be needed to carry 1000 A without losing its superconductivity?

25.61 •• **CP** A Nichrome heating element that has resistance 28.0 Ω is connected to a battery that has emf 96.0 V and internal resistance 1.2 Ω. An aluminum cup with mass 0.130 kg contains 0.200 kg of water. The heating element is placed in the water and the electrical energy dissipated in the resistance of the heating element all goes into the cup and water. The element itself has very small mass. How much time does it take for the temperature of the cup and water to rise from 21.2°C to 34.5°C? (The change of the resistance of the Nichrome due to its temperature change can be neglected.)

25.62 •• A resistor with resistance R is connected to a battery that has emf 12.0 V and internal resistance $r = 0.40$ Ω. For what two values of R will the power dissipated in the resistor be 80.0 W?

25.63 •• **CP BIO Struck by Lightning.** Lightning strikes can involve currents as high as 25,000 A that last for about 40 μs. If a person is struck by a bolt of lightning with these properties, the current will pass through his body. We shall assume that his mass is 75 kg, that he is wet (after all, he is in a rainstorm) and therefore has a resistance of 1.0 kΩ, and that his body is all water (which is reasonable for a rough, but plausible, approximation). (a) By how many degrees Celsius would this lightning bolt increase the temperature of 75 kg of water? (b) Given that the internal body temperature is about 37°C, would the person's temperature actually increase that much? Why not? What would happen first?

25.64 •• In the Bohr model of the hydrogen atom, the electron makes 6.0×10^{15} rev/s around the nucleus. What is the average current at a point on the orbit of the electron?

25.65 • **CALC** A material of resistivity ρ is formed into a solid, truncated cone of height h and radii r_1 and r_2 at either end (Fig. P25.65). (a) Calculate the resistance of the cone between the two flat end faces. (*Hint:* Imagine slicing the cone into very many thin disks, and calculate the resistance of one such disk.) (b) Show that your result agrees with Eq. (25.10) when $r_1 = r_2$.

Figure **P25.65**

25.66 • **CALC** The region between two concentric conducting spheres with radii a and b is filled with a conducting material with resistivity ρ. (a) Show that the resistance between the spheres is given by

$$R = \frac{\rho}{4\pi}\left(\frac{1}{a} - \frac{1}{b}\right)$$

(b) Derive an expression for the current density as a function of radius, in terms of the potential difference V_{ab} between the spheres. (c) Show that the result in part (a) reduces to Eq. (25.10) when the separation $L = b - a$ between the spheres is small.

25.67 ••• The temperature coefficient of resistance α in Eq. (25.12) equals the temperature coefficient of resistivity α in Eq. (25.6) only if the coefficient of thermal expansion is small. A cylindrical column of mercury is in a vertical glass tube. At 20°C, the length of the mercury column is 12.0 cm. The diameter of the mercury column is 1.6 mm and doesn't change with temperature because glass has a small coefficient of thermal expansion. The coefficient of volume expansion of the mercury is given in Table 17.2, its resistivity at 20°C is given in Table 25.1, and its temperature coefficient of resistivity is given in Table 25.2. (a) At 20°C, what is the resistance between the ends of the mercury column? (b) The mercury column is heated to 60°C. What is the change in its resistivity? (c) What is the change in its length? Explain why the coefficient of volume expansion, rather than the coefficient of linear expansion, determines the change in length. (d) What is the change in its resistance? (*Hint:* Since the percentage changes in ρ and L are small, you may find it helpful to derive from Eq. (25.10) an

equation for ΔR in terms of $\Delta\rho$ and ΔL.) (e) What is the temperature coefficient of resistance α for the mercury column, as defined in Eq. (25.12)? How does this value compare with the temperature coefficient of resistivity? Is the effect of the change in length important?

25.68 • (a) What is the potential difference V_{ad} in the circuit of Fig. P25.68? (b) What is the terminal voltage of the 4.00-V battery? (c) A battery with emf 10.30 V and internal resistance 0.50 Ω is inserted in the circuit at d, with its negative terminal connected to the negative terminal of the 8.00-V battery. What is the difference of potential V_{bc} between the terminals of the 4.00-V battery now?

Figure **P25.68**

25.69 • The potential difference across the terminals of a battery is 8.40 V when there is a current of 1.50 A in the battery from the negative to the positive terminal. When the current is 3.50 A in the reverse direction, the potential difference becomes 10.20 V. (a) What is the internal resistance of the battery? (b) What is the emf of the battery?

25.70 •• **BIO** A person with body resistance between his hands of 10 kΩ accidentally grasps the terminals of a 14-kV power supply. (a) If the internal resistance of the power supply is 2000 Ω, what is the current through the person's body? (b) What is the power dissipated in his body? (c) If the power supply is to be made safe by increasing its internal resistance, what should the internal resistance be for the maximum current in the above situation to be 1.00 mA or less?

25.71 • **BIO** The average bulk resistivity of the human body (apart from surface resistance of the skin) is about 5.0 Ω · m. The conducting path between the hands can be represented approximately as a cylinder 1.6 m long and 0.10 m in diameter. The skin resistance can be made negligible by soaking the hands in salt water. (a) What is the resistance between the hands if the skin resistance is negligible? (b) What potential difference between the hands is needed for a lethal shock current of 100 mA? (Note that your result shows that small potential differences produce dangerous currents when the skin is damp.) (c) With the current in part (b), what power is dissipated in the body?

25.72 • A typical cost for electric power is $0.120 per kilowatt-hour. (a) Some people leave their porch light on all the time. What is the yearly cost to keep a 75-W bulb burning day and night? (b) Suppose your refrigerator uses 400 W of power when it's running, and it runs 8 hours a day. What is the yearly cost of operating your refrigerator?

25.73 •• A 12.6-V car battery with negligible internal resistance is connected to a series combination of a 3.2-Ω resistor that obeys Ohm's law and a thermistor that does not obey Ohm's law but instead has a current–voltage relationship $V = \alpha I + \beta I^2$, with $\alpha = 3.8$ Ω and $\beta = 1.3$ Ω/A. What is the current through the 3.2-Ω resistor?

25.74 •• A cylindrical copper cable 1.50 km long is connected across a 220.0-V potential difference. (a) What should be its diameter so that it produces heat at a rate of 75.0 W? (b) What is the electric field inside the cable under these conditions?

25.75 •• **A Nonideal Ammeter.** Unlike the idealized ammeter described in Section 25.4, any real ammeter has a nonzero resistance. (a) An ammeter with resistance R_A is connected in series with a resistor R and a battery of emf $\mathcal{E}$ and internal resistance r. The current measured by the ammeter is I_A. Find the current through the circuit if the ammeter is removed so that the battery and the resistor form a complete circuit. Express your answer in terms of I_A, r, R_A, and R. The more "ideal" the ammeter, the smaller the difference between this current and the current I_A. (b) If $R = 3.80 \ \Omega$, $\mathcal{E} = 7.50 \ \text{V}$, and $r = 0.45 \ \Omega$, find the maximum value of the ammeter resistance R_A so that I_A is within 1.0% of the current in the circuit when the ammeter is absent. (c) Explain why your answer in part (b) represents a *maximum* value.

25.76 • **CALC** A 1.50-m cylinder of radius 1.10 cm is made of a complicated mixture of materials. Its resistivity depends on the distance x from the left end and obeys the formula $\rho(x) = a + bx^2$, where a and b are constants. At the left end, the resistivity is $2.25 \times 10^{-8} \ \Omega \cdot \text{m}$, while at the right end it is $8.50 \times 10^{-8} \ \Omega \cdot \text{m}$. (a) What is the resistance of this rod? (b) What is the electric field at its midpoint if it carries a 1.75-A current? (c) If we cut the rod into two 75.0-cm halves, what is the resistance of each half?

25.77 •• According to the U.S. National Electrical Code, copper wire used for interior wiring of houses, hotels, office buildings, and industrial plants is permitted to carry no more than a specified maximum amount of current. The table below shows the maximum current I_{max} for several common sizes of wire with varnished cambric insulation. The "wire gauge" is a standard used to describe the diameter of wires. Note that the larger the diameter of the wire, the *smaller* the wire gauge.

Wire gauge	Diameter (cm)	I_{max} (A)
14	0.163	18
12	0.205	25
10	0.259	30
8	0.326	40
6	0.412	60
5	0.462	65
4	0.519	85

(a) What considerations determine the maximum current-carrying capacity of household wiring? (b) A total of 4200 W of power is to be supplied through the wires of a house to the household electrical appliances. If the potential difference across the group of appliances is 120 V, determine the gauge of the thinnest permissible wire that can be used. (c) Suppose the wire used in this house is of the gauge found in part (b) and has total length 42.0 m. At what rate is energy dissipated in the wires? (d) The house is built in a community where the consumer cost of electric energy is $0.11 per kilowatt-hour. If the house were built with wire of the next larger diameter than that found in part (b), what would be the savings in electricity costs in one year? Assume that the appliances are kept on for an average of 12 hours a day.

25.78 •• **Compact Fluorescent Bulbs.** Compact fluorescent bulbs are much more efficient at producing light than are ordinary incandescent bulbs. They initially cost much more, but they last far longer and use much less electricity. According to one study of these bulbs, a compact bulb that produces as much light as a 100-W incandescent bulb uses only 23 W of power. The compact bulb lasts 10,000 hours, on the average, and costs $11.00, whereas the incandescent bulb costs only $0.75, but lasts just 750 hours. The study assumed that electricity costs $0.080 per kilowatt-hour and that the bulbs are on for 4.0 h per day. (a) What is the total cost (including the price of the bulbs) to run each bulb for 3.0 years? (b) How much do you save over 3.0 years if you use a compact fluorescent bulb instead of an incandescent bulb? (c) What is the resistance of a "100-W" fluorescent bulb? (Remember, it actually uses only 23 W of power and operates across 120 V.)

25.79 • In the circuit of Fig. P25.79, find (a) the current through the 8.0-Ω resistor and (b) the total rate of dissipation of electrical energy in the 8.0-Ω resistor and in the internal resistance of the batteries. (c) In one of the batteries, chemical energy is being converted into electrical energy. In which one is this happening, and at what rate? (d) In one of the batteries, electrical energy is being converted into chemical energy. In which one is this happening, and at what rate? (e) Show that the overall rate of production of electrical energy equals the overall rate of consumption of electrical energy in the circuit.

Figure **P25.79**

$\mathcal{E}_1 = 12.0 \ \text{V} \quad r_1 = 1.0 \ \Omega$

$R = 8.0 \ \Omega$

$\mathcal{E}_2 = 8.0 \ \text{V} \quad r_2 = 1.0 \ \Omega$

25.80 • A lightning bolt strikes one end of a steel lightning rod, producing a 15,000-A current burst that lasts for 65 μs. The rod is 2.0 m long and 1.8 cm in diameter, and its other end is connected to the ground by 35 m of 8.0-mm-diameter copper wire. (a) Find the potential difference between the top of the steel rod and the lower end of the copper wire during the current burst. (b) Find the total energy deposited in the rod and wire by the current burst.

25.81 • A 12.0-V battery has an internal resistance of 0.24 Ω and a capacity of 50.0 A·h (see Exercise 25.47). The battery is charged by passing a 10-A current through it for 5.0 h. (a) What is the terminal voltage during charging? (b) What total electrical energy is supplied to the battery during charging? (c) What electrical energy is dissipated in the internal resistance during charging? (d) The battery is now completely discharged through a resistor, again with a constant current of 10 A. What is the external circuit resistance? (e) What total electrical energy is supplied to the external resistor? (f) What total electrical energy is dissipated in the internal resistance? (g) Why are the answers to parts (b) and (e) not the same?

25.82 • Repeat Problem 25.81 with charge and discharge currents of 30 A. The charging and discharging times will now be 1.7 h rather than 5.0 h. What differences in performance do you see?

25.83 •• **CP** Consider the circuit shown in Fig. P25.83. The emf source has negligible internal resistance. The resistors have resistances $R_1 = 6.00 \ \Omega$ and $R_2 = 4.00 \ \Omega$. The capacitor has capacitance $C = 9.00 \ \mu\text{F}$. When the capacitor is fully charged, the magnitude of the charge on its plates is $Q = 36.0 \ \mu\text{C}$. Calculate the emf $\mathcal{E}$.

Figure **P25.83**

25.84 •• **CP** Consider the circuit shown in Fig. P25.84. The battery has emf 60.0 V and negligible internal resistance. $R_2 = 2.00 \ \Omega$, $C_1 = 3.00 \ \mu\text{F}$, and $C_2 = 6.00 \ \mu\text{F}$. After the capacitors have attained their final charges, the charge on C_1 is $Q_1 = 18.0 \ \mu\text{C}$. (a) What is the final charge on C_2? (b) What is the resistance R_1?

Figure **P25.84**

CHALLENGE PROBLEMS

25.85 ••• The *Tolman-Stewart experiment* in 1916 demonstrated that the free charges in a metal have negative charge and provided a quantitative measurement of their charge-to-mass ratio,

Figure **P25.85**

$|q|/m$. The experiment consisted of abruptly stopping a rapidly rotating spool of wire and measuring the potential difference that this produced between the ends of the wire. In a simplified model of this experiment, consider a metal rod of length L that is given a uniform acceleration $\vec{a}$ to the right. Initially the free charges in the metal lag behind the rod's motion, thus setting up an electric field $\vec{E}$ in the rod. In the steady state this field exerts a force on the free charges that makes them accelerate along with the rod. (a) Apply $\Sigma \vec{F} = m\vec{a}$ to the free charges to obtain an expression for $|q|/m$ in terms of the magnitudes of the induced electric field $\vec{E}$ and the acceleration $\vec{a}$. (b) If all the free charges in the metal rod have the same acceleration, the electric field $\vec{E}$ is the same at all points in the rod. Use this fact to rewrite the expression for $|q|/m$ in terms of the potential V_{bc} between the ends of the rod (Fig. P25.85). (c) If the free charges have negative charge, which end of the rod, b or c, is at higher potential? (d) If the rod is 0.50 m long and the free charges are electrons (charge $q = -1.60 \times 10^{-19}$ C, mass 9.11×10^{-31} kg), what magnitude of acceleration is required to produce a potential difference of 1.0 mV between the ends of the rod? (e) Discuss why the actual experiment used a rotating spool of thin wire rather than a moving bar as in our simplified analysis.

25.86 ••• **CALC** A source with emf $\mathcal{E}$ and internal resistance r is connected to an external circuit. (a) Show that the power output of the source is maximum when the current in the circuit is one-half the short-circuit current of the source. (b) If the external circuit consists of a resistance R, show that the power output is maximum when $R = r$ and that the maximum power is $\mathcal{E}^2/4r$.

25.87 ••• **CALC** The resistivity of a semiconductor can be modified by adding different amounts of impurities. A rod of semiconducting material of length L and cross-sectional area A lies along the x-axis between $x = 0$ and $x = L$. The material obeys Ohm's law, and its resistivity varies along the rod according to $\rho(x) = \rho_0 \exp(-x/L)$. The end of the rod at $x = 0$ is at a potential V_0 greater than the end at $x = L$. (a) Find the total resistance of the rod and the current in the rod. (b) Find the electric-field magnitude $E(x)$ in the rod as a function of x. (c) Find the electric potential $V(x)$ in the rod as a function of x. (d) Graph the functions $\rho(x)$, $E(x)$, and $V(x)$ for values of x between $x = 0$ and $x = L$.

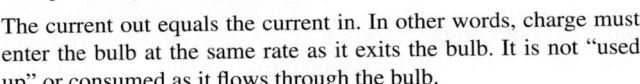

Chapter Opening Question ?

The current out equals the current in. In other words, charge must enter the bulb at the same rate as it exits the bulb. It is not "used up" or consumed as it flows through the bulb.

Test Your Understanding Questions

25.1 Answer: (v) Doubling the diameter increases the cross-sectional area A by a factor of 4. Hence the current-density magnitude $J = I/A$ is reduced to $\frac{1}{4}$ of the value in Example 25.1, and the magnitude of the drift velocity $v_d = J/n|q|$ is reduced by the same factor. The new magnitude is $v_d = (0.15 \text{ mm/s})/4 = 0.038 \text{ mm/s}$. This behavior is the same as that of an incompressible fluid, which slows down when it moves from a narrow pipe to a broader one (see Section 14.4).

25.2 Answer: (ii) Figure 25.6b shows that the resistivity ρ of a semiconductor increases as the temperature decreases. From Eq. (25.5), the magnitude of the current density is $J = E/\rho$, so the current density decreases as the temperature drops and the resistivity increases.

25.3 Answer: (iii) Solving Eq. (25.11) for the current shows that $I = V/R$. If the resistance R of the wire remained the same, doubling the voltage V would make the current I double as well. However, we saw in Example 25.3 that the resistance is *not* constant: As the current increases and the temperature increases, R increases as well. Thus doubling the voltage produces a current that is *less* than double the original current. An ohmic conductor is one for which $R = V/I$ has the same value no matter what the voltage, so the wire is *nonohmic*. (In many practical problems the temperature change of the wire is so small that it can be ignored, so we can safely regard the wire as being ohmic. We do so in almost all examples in this book.)

25.4 Answer: (iii), (ii), (i) For circuit (i), we find the current from Eq. (25.16): $I = \mathcal{E}/(R + r) = (1.5 \text{ V})/(1.4 \ \Omega + 0.10 \ \Omega) = 1.0 \text{ A}$. For circuit (ii), we note that the terminal voltage $v_{ab} = 3.6 \text{ V}$ equals the voltage IR across the 1.8-Ω resistor: $V_{ab} = IR$, so $I = V_{ab}/R = (3.6 \text{ V})/(1.8 \ \Omega) = 2.0 \text{ A}$. For circuit (iii), we use Eq. (25.15) for the terminal voltage: $V_{ab} = \mathcal{E} - Ir$, so $I = (\mathcal{E} - V_{ab})/r = (12.0 \text{ V} - 11.0 \text{ V})/(0.20 \ \Omega) = 5.0 \text{ A}$.

25.5 Answer: (iii), (ii), (i) These are the same circuits that we analyzed in Test Your Understanding of Section 25.4. In each case the net power output of the battery is $P = V_{ab}I$, where V_{ab} is the battery terminal voltage. For circuit (i), we found that $I = 1.0 \text{ A}$, so $V_{ab} = \mathcal{E} - Ir = 1.5 \text{ V} - (1.0 \text{ A})(0.10 \ \Omega) = 1.4 \text{ V}$, so $P = (1.4 \text{ V})(1.0 \text{ A}) = 1.4 \text{ W}$. For circuit (ii), we have $V_{ab} = 3.6 \text{ V}$ and found that $I = 2.0 \text{ A}$, so $P = (3.6 \text{ V})(2.0 \text{ A}) = 7.2 \text{ W}$. For circuit (iii), we have $V_{ab} = 11.0 \text{ V}$ and found that $I = 5.0 \text{ A}$, so $P = (11.0 \text{ V})(5.0 \text{ A}) = 55 \text{ A}$.

25.6 Answer: (i) The difficulty of producing a certain amount of current increases as the resistivity ρ increases. From Eq. (25.24), $\rho = m/ne^2\tau$, so increasing the mass m will increase the resistivity. That's because a more massive charged particle will respond more sluggishly to an applied electric field and hence drift more slowly. To produce the same current, a greater electric field would be needed. (Increasing n, e, or τ would decrease the resistivity and make it easier to produce a given current.)

Bridging Problem

Answers: (a) 237°C **(b)** 162 W initially, 148 W at 1.23 A

26 DIRECT-CURRENT CIRCUITS

? In a complex circuit like the one on this circuit board, is it possible to connect several resistors with different resistances so that they all have the same potential difference? If so, will the current be the same through all of the resistors?

I f you look inside your TV, your computer, or under the hood of a car, you will find circuits of much greater complexity than the simple circuits we studied in Chapter 25. Whether connected by wires or integrated in a semiconductor chip, these circuits often include several sources, resistors, and other circuit elements interconnected in a *network*.

In this chapter we study general methods for analyzing such networks, including how to find voltages and currents of circuit elements. We'll learn how to determine the equivalent resistance for several resistors connected in series or in parallel. For more general networks we need two rules called *Kirchhoff's rules*. One is based on the principle of conservation of charge applied to a junction; the other is derived from energy conservation for a charge moving around a closed loop. We'll discuss instruments for measuring various electrical quantities. We'll also look at a circuit containing resistance and capacitance, in which the current varies with time.

Our principal concern in this chapter is with **direct-current** (dc) circuits, in which the direction of the current does not change with time. Flashlights and automobile wiring systems are examples of direct-current circuits. Household electrical power is supplied in the form of **alternating current** (ac), in which the current oscillates back and forth. The same principles for analyzing networks apply to both kinds of circuits, and we conclude this chapter with a look at household wiring systems. We'll discuss alternating-current circuits in detail in Chapter 31.

26.1 Resistors in Series and Parallel

Resistors turn up in all kinds of circuits, ranging from hair dryers and space heaters to circuits that limit or divide current or reduce or divide a voltage. Such circuits often contain several resistors, so it's appropriate to consider *combinations* of resistors. A simple example is a string of light bulbs used for holiday decorations;

each bulb acts as a resistor, and from a circuit-analysis perspective the string of bulbs is simply a combination of resistors.

Suppose we have three resistors with resistances R_1, R_2, and R_3. Figure 26.1 shows four different ways in which they might be connected between points a and b. When several circuit elements such as resistors, batteries, and motors are connected in sequence as in Fig. 26.1a, with only a single current path between the points, we say that they are connected in **series.** We studied *capacitors* in series in Section 24.2; we found that, because of conservation of charge, capacitors in series all have the same charge if they are initially uncharged. In circuits we're often more interested in the *current,* which is charge flow per unit time.

The resistors in Fig. 26.1b are said to be connected in **parallel** between points a and b. Each resistor provides an alternative path between the points. For circuit elements that are connected in parallel, the *potential difference* is the same across each element. We studied capacitors in parallel in Section 24.2.

In Fig. 26.1c, resistors R_2 and R_3 are in parallel, and this combination is in series with R_1. In Fig. 26.1d, R_2 and R_3 are in series, and this combination is in parallel with R_1.

For any combination of resistors we can always find a *single* resistor that could replace the combination and result in the same total current and potential difference. For example, a string of holiday light bulbs could be replaced by a single, appropriately chosen light bulb that would draw the same current and have the same potential difference between its terminals as the original string of bulbs. The resistance of this single resistor is called the **equivalent resistance** of the combination. If any one of the networks in Fig. 26.1 were replaced by its equivalent resistance R_{eq}, we could write

$$V_{ab} = IR_{eq} \quad \text{or} \quad R_{eq} = \frac{V_{ab}}{I}$$

where V_{ab} is the potential difference between terminals a and b of the network and I is the current at point a or b. To compute an equivalent resistance, we assume a potential difference V_{ab} across the actual network, compute the corresponding current I, and take the ratio V_{ab}/I.

Resistors in Series

We can derive general equations for the equivalent resistance of a series or parallel combination of resistors. If the resistors are in *series,* as in Fig. 26.1a, the current I must be the same in all of them. (As we discussed in Section 25.4, current is *not* "used up" as it passes through a circuit.) Applying $V = IR$ to each resistor, we have

$$V_{ax} = IR_1 \qquad V_{xy} = IR_2 \qquad V_{yb} = IR_3$$

The potential differences across each resistor need not be the same (except for the special case in which all three resistances are equal). The potential difference V_{ab} across the entire combination is the sum of these individual potential differences:

$$V_{ab} = V_{ax} + V_{xy} + V_{yb} = I(R_1 + R_2 + R_3)$$

and so

$$\frac{V_{ab}}{I} = R_1 + R_2 + R_3$$

The ratio V_{ab}/I is, by definition, the equivalent resistance R_{eq}. Therefore

$$R_{eq} = R_1 + R_2 + R_3$$

It is easy to generalize this to any number of resistors:

$$R_{eq} = R_1 + R_2 + R_3 + \cdots \quad \text{(resistors in series)} \qquad (26.1)$$

26.1 Four different ways of connecting three resistors.

(a) R_1, R_2, and R_3 in series

(b) R_1, R_2, and R_3 in parallel

(c) R_1 in series with parallel combination of R_2 and R_3

(d) R_1 in parallel with series combination of R_2 and R_3

The equivalent resistance of *any number* of resistors in series equals the sum of their individual resistances.

The equivalent resistance is *greater than* any individual resistance.

Let's compare this result with Eq. (24.5) for *capacitors* in series. Resistors in series add directly because the voltage across each is directly proportional to its resistance and to the common current. Capacitors in series add reciprocally because the voltage across each is directly proportional to the common charge but *inversely* proportional to the individual capacitance.

Resistors in Parallel

If the resistors are in *parallel,* as in Fig. 26.1b, the current through each resistor need not be the same. But the potential difference between the terminals of each resistor must be the same and equal to V_{ab} (Fig. 26.2). (Remember that the potential difference between any two points does not depend on the path taken between the points.) Let's call the currents in the three resistors I_1, I_2, and I_3. Then from $I = V/R$,

$$I_1 = \frac{V_{ab}}{R_1} \qquad I_2 = \frac{V_{ab}}{R_2} \qquad I_3 = \frac{V_{ab}}{R_3}$$

In general, the current is different through each resistor. Because charge is not accumulating or draining out of point a, the total current I must equal the sum of the three currents in the resistors:

$$I = I_1 + I_2 + I_3 = V_{ab}\left(\frac{1}{R_1} + \frac{1}{R_2} + \frac{1}{R_3}\right) \quad \text{or}$$

$$\frac{I}{V_{ab}} = \frac{1}{R_1} + \frac{1}{R_2} + \frac{1}{R_3}$$

But by the definition of the equivalent resistance R_{eq}, $I/V_{ab} = 1/R_{eq}$, so

$$\frac{1}{R_{eq}} = \frac{1}{R_1} + \frac{1}{R_2} + \frac{1}{R_3}$$

Again it is easy to generalize to *any number* of resistors in parallel:

$$\frac{1}{R_{eq}} = \frac{1}{R_1} + \frac{1}{R_2} + \frac{1}{R_3} + \cdots \qquad \text{(resistors in parallel)} \qquad \text{(26.2)}$$

For *any number* of resistors in parallel, the *reciprocal* of the equivalent resistance equals the *sum of the reciprocals* of their individual resistances.

The equivalent resistance is always *less than* any individual resistance.

Compare this with Eq. (24.7) for *capacitors* in parallel. Resistors in parallel add reciprocally because the current in each is proportional to the common voltage across them and *inversely* proportional to the resistance of each. Capacitors in parallel add directly because the charge on each is proportional to the common voltage across them and *directly* proportional to the capacitance of each.

For the special case of *two* resistors in parallel,

$$\frac{1}{R_{eq}} = \frac{1}{R_1} + \frac{1}{R_2} = \frac{R_1 + R_2}{R_1 R_2} \quad \text{and}$$

$$R_{eq} = \frac{R_1 R_2}{R_1 + R_2} \qquad \text{(two resistors in parallel)} \qquad \text{(26.3)}$$

26.2 A car's headlights and taillights are connected in parallel. Hence each light is exposed to the full potential difference supplied by the car's electrical system, giving maximum brightness. Another advantage is that if one headlight or taillight burns out, the other one keeps shining (see Example 26.2).

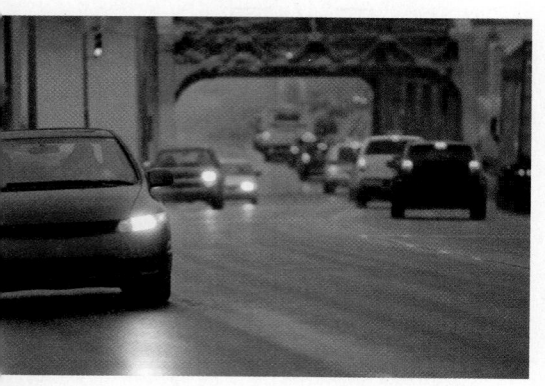

MasteringPHYSICS
ActivPhysics 12.2: DC Parallel Circuits

Because $V_{ab} = I_1 R_1 = I_2 R_2$, it follows that

$$\frac{I_1}{I_2} = \frac{R_2}{R_1} \quad \text{(two resistors in parallel)} \qquad (26.4)$$

This shows that the currents carried by two resistors in parallel are *inversely proportional* to their resistances. More current goes through the path of least resistance.

Problem-Solving Strategy 26.1 **Resistors in Series and Parallel**

IDENTIFY *the relevant concepts:* As in Fig. 26.1, many resistor networks are made up of resistors in series, in parallel, or a combination thereof. Such networks can be replaced by a single equivalent resistor. The logic is similar to that of Problem-Solving Strategy 24.1 for networks of capacitors.

SET UP *the problem* using the following steps:
1. Make a drawing of the resistor network.
2. Identify groups of resistors connected in series or parallel.
3. Identify the target variables. They could include the equivalent resistance of the network, the potential difference across each resistor, or the current through each resistor.

EXECUTE *the solution* as follows:
1. Use Eq. (26.1) or (26.2), respectively, to find the equivalent resistance for series or parallel combinations.
2. If the network is more complex, try reducing it to series and parallel combinations. For example, in Fig. 26.1c we first replace the parallel combination of R_2 and R_3 with its equivalent resistance;

this then forms a series combination with R_1. In Fig. 26.1d, the combination of R_2 and R_3 in series forms a parallel combination with R_1.
3. Keep in mind that the total potential difference across resistors connected in series is the sum of the individual potential differences. The potential difference across resistors connected in parallel is the same for every resistor and equals the potential difference across the combination.
4. The current through resistors connected in series is the same through every resistor and equals the current through the combination. The total current through resistors connected in parallel is the sum of the currents through the individual resistors.

EVALUATE *your answer:* Check whether your results are consistent. The equivalent resistance of resistors connected in series should be greater than that of any individual resistor; that of resistors in parallel should be less than that of any individual resistor.

Example 26.1 **Equivalent resistance**

Find the equivalent resistance of the network in Fig. 26.3a and the current in each resistor. The source of emf has negligible internal resistance.

SOLUTION

IDENTIFY and SET UP: This network of three resistors is a *combination* of series and parallel resistances, as in Fig. 26.1c. We determine

26.3 Steps in reducing a combination of resistors to a single equivalent resistor and finding the current in each resistor.

Continued

the equivalent resistance of the parallel 6-Ω and 3-Ω resistors, and then that of their series combination with the 4-Ω resistor: This is the equivalent resistance R_{eq} of the network as a whole. We then find the current in the emf, which is the same as that in the 4-Ω resistor. The potential difference is the same across each of the parallel 6-Ω and 3-Ω resistors; we use this to determine how the current is divided between these.

EXECUTE: Figures 26.3b and 26.3c show successive steps in reducing the network to a single equivalent resistance R_{eq}. From Eq. (26.2), the 6-Ω and 3-Ω resistors in parallel in Fig. 26.3a are equivalent to the single 2-Ω resistor in Fig. 26.3b:

$$\frac{1}{R_{6\,\Omega + 3\,\Omega}} = \frac{1}{6\ \Omega} + \frac{1}{3\ \Omega} = \frac{1}{2\ \Omega}$$

[Equation (26.3) gives the same result.] From Eq. (26.1) the series combination of this 2-Ω resistor with the 4-Ω resistor is equivalent to the single 6-Ω resistor in Fig. 26.3c.

We reverse these steps to find the current in each resistor of the original network. In the circuit shown in Fig. 26.3d (identical to Fig. 26.3c), the current is $I = V_{ab}/R = (18\ \text{V})/(6\ \Omega) = 3\ \text{A}$. So the current in the 4-Ω and 2-Ω resistors in Fig. 26.3e (identical to Fig. 26.3b) is also 3 A. The potential difference V_{cb} across the 2-Ω resistor is therefore $V_{cb} = IR = (3\ \text{A})(2\ \Omega) = 6\ \text{V}$. This potential difference must also be 6 V in Fig. 26.3f (identical to Fig. 26.3a). From $I = V_{cb}/R$, the currents in the 6-Ω and 3-Ω resistors in Fig. 26.3f are respectively $(6\ \text{V})/(6\ \Omega) = 1\ \text{A}$ and $(6\ \text{V})/(3\ \Omega) = 2\ \text{A}$.

EVALUATE: Note that for the two resistors in parallel between points c and b in Fig. 26.3f, there is twice as much current through the 3-Ω resistor as through the 6-Ω resistor; more current goes through the path of least resistance, in accordance with Eq. (26.4). Note also that the total current through these two resistors is 3 A, the same as it is through the 4-Ω resistor between points a and c.

Example 26.2 Series versus parallel combinations

Two identical light bulbs, each with resistance $R = 2\ \Omega$, are connected to a source with $\mathcal{E} = 8\ \text{V}$ and negligible internal resistance. Find the current through each bulb, the potential difference across each bulb, and the power delivered to each bulb and to the entire network if the bulbs are connected (a) in series and (b) in parallel. (c) Suppose one of the bulbs burns out; that is, its filament breaks and current can no longer flow through it. What happens to the other bulb in the series case? In the parallel case?

SOLUTION

IDENTIFY and SET UP: The light bulbs are just resistors in simple series and parallel connections (Figs. 26.4a and 26.4b). Once we find the current I through each bulb, we can find the power delivered to each bulb using Eq. (25.18), $P = I^2R = V^2/R$.

EXECUTE: (a) From Eq. (26.1) the equivalent resistance of the two bulbs between points a and c in Fig. 26.4a is $R_{eq} = 2R = 2(2\ \Omega) = 4\ \Omega$. In series, the current is the same through each bulb:

$$I = \frac{V_{ac}}{R_{eq}} = \frac{8\ \text{V}}{4\ \Omega} = 2\ \text{A}$$

Since the bulbs have the same resistance, the potential difference is the same across each bulb:

$$V_{ab} = V_{bc} = IR = (2\ \text{A})(2\ \Omega) = 4\ \text{V}$$

From Eq. (25.18), the power delivered to each bulb is

$$P = I^2R = (2\ \text{A})^2(2\ \Omega) = 8\ \text{W} \qquad \text{or}$$

$$P = \frac{V_{ab}^2}{R} = \frac{V_{bc}^2}{R} = \frac{(4\ \text{V})^2}{2\ \Omega} = 8\ \text{W}$$

The total power delivered to both bulbs is $P_{tot} = 2P = 16\ \text{W}$.

(b) If the bulbs are in parallel, as in Fig. 26.4b, the potential difference V_{de} across each bulb is the same and equal to 8 V, the terminal voltage of the source. Hence the current through each light bulb is

$$I = \frac{V_{de}}{R} = \frac{8\ \text{V}}{2\ \Omega} = 4\ \text{A}$$

and the power delivered to each bulb is

$$P = I^2R = (4\ \text{A})^2(2\ \Omega) = 32\ \text{W} \qquad \text{or}$$

$$P = \frac{V_{de}^2}{R} = \frac{(8\ \text{V})^2}{2\ \Omega} = 32\ \text{W}$$

Both the potential difference across each bulb and the current through each bulb are twice as great as in the series case. Hence the power delivered to each bulb is *four* times greater, and each bulb is brighter.

The total power delivered to the parallel network is $P_{total} = 2P = 64\ \text{W}$, four times greater than in the series case. The

26.4 Our sketches for this problem.

(a) Light bulbs in series

(b) Light bulbs in parallel

26.5 When connected to the same source, two light bulbs in series (shown at top) draw less power and glow less brightly than when they are in parallel (shown at bottom).

increased power compared to the series case isn't obtained "for free"; energy is extracted from the source four times more rapidly in the parallel case than in the series case. If the source is a battery, it will be used up four times as fast.

(c) In the series case the same current flows through both bulbs. If one bulb burns out, there will be no current in the circuit, and neither bulb will glow.

In the parallel case the potential difference across either bulb is unchanged if a bulb burns out. The current through the functional bulb and the power delivered to it are unchanged.

EVALUATE: Our calculation isn't completely accurate, because the resistance $R = V/I$ of real light bulbs depends on the potential difference V across the bulb. That's because the filament resistance increases with increasing operating temperature and therefore with increasing V. But bulbs connected in series across a source do in fact glow less brightly than when connected in parallel across the same source (Fig. 26.5).

Test Your Understanding of Section 26.1 Suppose all three of the resistors shown in Fig. 26.1 have the same resistance, so $R_1 = R_2 = R_3 = R$. Rank the four arrangements shown in parts (a)–(d) of Fig. 26.1 in order of their equivalent resistance, from highest to lowest.

26.2 Kirchhoff's Rules

Many practical resistor networks cannot be reduced to simple series-parallel combinations. Figure 26.6a shows a dc power supply with emf $\mathcal{E}_1$ charging a battery with a smaller emf $\mathcal{E}_2$ and feeding current to a light bulb with resistance R. Figure 26.6b is a "bridge" circuit, used in many different types of measurement and control systems. (Problem 26.81 describes one important application of a "bridge" circuit.) To compute the currents in these networks, we'll use the techniques developed by the German physicist Gustav Robert Kirchhoff (1824–1887).

First, here are two terms that we will use often. A **junction** in a circuit is a point where three or more conductors meet. A **loop** is any closed conducting path. In Fig. 26.6a points a and b are junctions, but points c and d are not; in Fig. 26.6b the points a, b, c, and d are junctions, but points e and f are not. The blue lines in Figs. 26.6a and 26.6b show some possible loops in these circuits.

Kirchhoff's rules are the following two statements:

Kirchhoff's junction rule: *The algebraic sum of the currents into any junction is zero.* That is,

$$\sum I = 0 \quad \text{(junction rule, valid at any junction)} \quad \text{(26.5)}$$

Kirchhoff's loop rule: *The algebraic sum of the potential differences in any loop,* including those associated with emfs and those of resistive elements, *must equal zero.* That is,

$$\sum V = 0 \quad \text{(loop rule, valid for any closed loop)} \quad \text{(26.6)}$$

26.6 Two networks that cannot be reduced to simple series-parallel combinations of resistors.

(a)

(b)

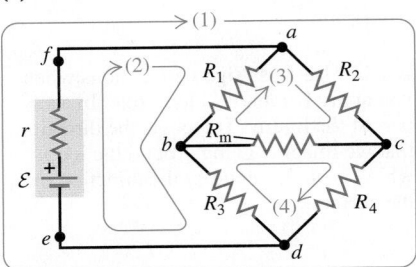

26.7 Kirchhoff's junction rule states that as much current flows into a junction as flows out of it.

(a) Kirchhoff's junction rule

Junction

$I_1 \rightarrow$ $\leftarrow I_2$

The current leaving a junction equals the current entering it.

$I_1 + I_2 \downarrow$

(b) Water-pipe analogy

The flow rate of water leaving the pipe equals the flow rate entering it.

The junction rule is based on *conservation of electric charge.* No charge can accumulate at a junction, so the total charge entering the junction per unit time must equal the total charge leaving per unit time (Fig. 26.7a). Charge per unit time is current, so if we consider the currents entering a junction to be positive and those leaving to be negative, the algebraic sum of currents into a junction must be zero. It's like a T branch in a water pipe (Fig. 26.7b); if you have a total of 1 liter per minute coming in the two pipes, you can't have 3 liters per minute going out the third pipe. We may as well confess that we used the junction rule (without saying so) in Section 26.1 in the derivation of Eq. (26.2) for resistors in parallel.

The loop rule is a statement that the electrostatic force is *conservative.* Suppose we go around a loop, measuring potential differences across successive circuit elements as we go. When we return to the starting point, we must find that the *algebraic sum* of these differences is zero; otherwise, we could not say that the potential at this point has a definite value.

Sign Conventions for the Loop Rule

In applying the loop rule, we need some sign conventions. Problem-Solving Strategy 26.2 describes in detail how to use these, but here's a quick overview. We first assume a direction for the current in each branch of the circuit and mark it on a diagram of the circuit. Then, starting at any point in the circuit, we imagine traveling around a loop, adding emfs and *IR* terms as we come to them. When we travel through a source in the direction from − to +, the emf is considered to be *positive;* when we travel from + to −, the emf is considered to be *negative* (Fig. 26.8a). When we travel through a resistor in the *same* direction as the assumed current, the *IR* term is *negative* because the current goes in the direction of decreasing potential. When we travel through a resistor in the direction *opposite* to the assumed current, the *IR* term is *positive* because this represents a rise of potential (Fig. 26.8b).

Kirchhoff's two rules are all we need to solve a wide variety of network problems. Usually, some of the emfs, currents, and resistances are known, and others are unknown. We must always obtain from Kirchhoff's rules a number of independent equations equal to the number of unknowns so that we can solve the equations simultaneously. Often the hardest part of the solution is not understanding the basic principles but keeping track of algebraic signs!

26.8 Use these sign conventions when you apply Kirchhoff's loop rule. In each part of the figure "Travel" is the direction that we imagine going around the loop, which is not necessarily the direction of the current.

(a) Sign conventions for emfs

$+\mathcal{E}$: Travel direction from − to +:

— Travel →

$\mathcal{E}$

$-\mathcal{E}$: Travel direction from + to −:

← Travel —

$\mathcal{E}$

(b) Sign conventions for resistors

$+IR$: Travel *opposite to* current direction:

— Travel →

R

$-IR$: Travel *in* current direction:

← Travel —

R

Problem-Solving Strategy 26.2 Kirchhoff's Rules

IDENTIFY *the relevant concepts:* Kirchhoff's rules are useful for analyzing any electric circuit.

SET UP *the problem* using the following steps:
1. Draw a circuit diagram, leaving room to label all quantities, known and unknown. Indicate an assumed direction for each unknown current and emf. (Kirchhoff's rules will yield the magnitudes *and directions* of unknown currents and emfs. If the actual direction of a quantity is opposite to your assumption, the resulting quantity will have a negative sign.)
2. As you label currents, it helpful to use Kirchhoff's junction rule, as in Fig. 26.9, so as to express the currents in terms of as few quantities as possible.
3. Identify the target variables.

EXECUTE *the solution* as follows:
1. Choose any loop in the network and choose a direction (clockwise or counterclockwise) to travel around the loop as you apply Kirchhoff's loop rule. The direction need not be the same as any assumed current direction.

2. Travel around the loop in the chosen direction, adding potential differences algebraically as you cross them. Use the sign conventions of Fig. 26.8.
3. Equate the sum obtained in step 2 to zero in accordance with the loop rule.
4. If you need more independent equations, choose another loop and repeat steps 1–3; continue until you have as many independent equations as unknowns or until every circuit element has been included in at least one loop.
5. Solve the equations simultaneously to determine the unknowns.
6. You can use the loop-rule bookkeeping system to find the potential V_{ab} of any point *a* with respect to any other point *b*. Start at *b* and add the potential changes you encounter in going from *b* to *a*, using the same sign rules as in step 2. The algebraic sum of these changes is $V_{ab} = V_a - V_b$.

EVALUATE *your answer:* Check all the steps in your algebra. Apply steps 1 and 2 to a loop you have not yet considered; if the sum of potential drops isn't zero, you've made an error somewhere.

26.9 Applying the junction rule to point *a* reduces the number of unknown currents from three to two.

(a) Three unknown currents: I_1, I_2, I_3

(b) Applying the junction rule to point *a* eliminates I_3.

Example 26.3 A single-loop circuit

The circuit shown in Fig. 26.10a contains two batteries, each with an emf and an internal resistance, and two resistors. Find (a) the current in the circuit, (b) the potential difference V_{ab}, and (c) the power output of the emf of each battery.

SOLUTION

IDENTIFY and SET UP: There are no junctions in this single-loop circuit, so we don't need Kirchhoff's junction rule. To apply Kirchhoff's loop rule, we first assume a direction for the current; let's assume a counterclockwise direction as shown in Fig. 26.10a.

EXECUTE: (a) Starting at *a* and traveling counterclockwise with the current, we add potential increases and decreases and equate the sum to zero as in Eq. (26.6):

$$-I(4\ \Omega) - 4\ \text{V} - I(7\ \Omega) + 12\ \text{V} - I(2\ \Omega) - I(3\ \Omega) = 0$$

Collecting like terms and solving for I, we find

$$8\ \text{V} = I(16\ \Omega) \quad \text{and} \quad I = 0.5\ \text{A}$$

The positive result for I shows that our assumed current direction is correct.

(b) To find V_{ab}, the potential at *a* with respect to *b*, we start at *b* and add potential changes as we go toward *a*. There are two paths from *b* to *a*; taking the lower one, we find

$$V_{ab} = (0.5\ \text{A})(7\ \Omega) + 4\ \text{V} + (0.5\ \text{A})(4\ \Omega)$$
$$= 9.5\ \text{V}$$

Point *a* is at 9.5 V higher potential than *b*. All the terms in this sum, including the IR terms, are positive because each represents an *increase* in potential as we go from *b* to *a*. Taking the upper path, we find

$$V_{ab} = 12\ \text{V} - (0.5\ \text{A})(2\ \Omega) - (0.5\ \text{A})(3\ \Omega)$$
$$= 9.5\ \text{V}$$

Here the IR terms are negative because our path goes in the direction of the current, with potential decreases through the resistors. The results for V_{ab} are the same for both paths, as they must be in order for the total potential change around the loop to be zero.

Continued

(c) The power outputs of the emf of the 12-V and 4-V batteries are

$$P_{12V} = \mathcal{E}I = (12 \text{ V})(0.5 \text{ A}) = 6 \text{ W}$$

$$P_{4V} = \mathcal{E}I = (-4 \text{ V})(0.5 \text{ A}) = -2 \text{ W}$$

The negative sign in $\mathcal{E}$ for the 4-V battery appears because the current actually runs from the higher-potential side of the battery to the lower-potential side. The negative value of P means that we are *storing* energy in that battery; the 12-V battery is *recharging* it (if it is in fact rechargeable; otherwise, we're destroying it).

EVALUATE: By applying the expression $P = I^2R$ to each of the four resistors in Fig. 26.10a, you can show that the total power dissipated in all four resistors is 4 W. Of the 6 W provided by the emf of the 12-V battery, 2 W goes into storing energy in the 4-V battery and 4 W is dissipated in the resistances.

The circuit shown in Fig. 26.10a is much like that used when a fully charged 12-V storage battery (in a car with its engine running) is used to "jump-start" a car with a run-down battery (Fig. 26.10b). The run-down battery is slightly recharged in the process. The 3-Ω and 7-Ω resistors in Fig. 26.10a represent the resistances of the jumper cables and of the conducting path through the automobile with the run-down battery. (The values of the resistances in actual automobiles and jumper cables are considerably lower.)

26.10 (a) In this example we travel around the loop in the same direction as the assumed current, so all the *IR* terms are negative. The potential decreases as we travel from $+$ to $-$ through the bottom emf but increases as we travel from $-$ to $+$ through the top emf. (b) A real-life example of a circuit of this kind.

Example 26.4 Charging a battery

In the circuit shown in Fig. 26.11, a 12-V power supply with unknown internal resistance r is connected to a run-down rechargeable battery with unknown emf $\mathcal{E}$ and internal resistance 1 Ω and to an indicator light bulb of resistance 3 Ω carrying a current of 2 A. The current through the run-down battery is 1 A in the direction shown. Find r, $\mathcal{E}$, and the current I through the power supply.

SOLUTION

IDENTIFY and SET UP: This circuit has more than one loop, so we must apply both the junction and loop rules. We assume the direction of the current through the 12-V power supply, and the polarity of the run-down battery, to be as shown in Fig. 26.11. There are three target variables, so we need three equations.

26.11 In this circuit a power supply charges a run-down battery and lights a bulb. An assumption has been made about the polarity of the emf $\mathcal{E}$ of the run-down battery. Is this assumption correct?

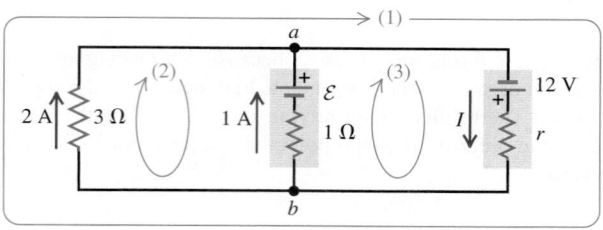

EXECUTE: We apply the junction rule, Eq. (26.5), to point *a*:

$$-I + 1 \text{ A} + 2 \text{ A} = 0 \quad \text{so} \quad I = 3 \text{ A}$$

To determine r, we apply the loop rule, Eq. (26.6), to the large, outer loop (1):

$$12 \text{ V} - (3 \text{ A})r - (2 \text{ A})(3 \text{ }\Omega) = 0 \quad \text{so} \quad r = 2 \text{ }\Omega$$

To determine $\mathcal{E}$, we apply the loop rule to the left-hand loop (2):

$$-\mathcal{E} + (1 \text{ A})(1 \text{ }\Omega) - (2 \text{ A})(3 \text{ }\Omega) = 0 \quad \text{so} \quad \mathcal{E} = -5 \text{ V}$$

The negative value for $\mathcal{E}$ shows that the actual polarity of this emf is opposite to that shown in Fig. 26.11. As in Example 26.3, the battery is being recharged.

EVALUATE: Try applying the junction rule at point *b* instead of point *a*, and try applying the loop rule by traveling counterclockwise rather than clockwise around loop (1). You'll get the same results for I and r. We can check our result for $\mathcal{E}$ by using the right-hand loop (3):

$$12 \text{ V} - (3 \text{ A})(2 \text{ }\Omega) - (1 \text{ A})(1 \text{ }\Omega) + \mathcal{E} = 0$$

which again gives us $\mathcal{E} = -5 \text{ V}$.

As an additional check, we note that $V_{ba} = V_b - V_a$ equals the voltage across the 3-Ω resistance, which is $(2 \text{ A})(3 \text{ }\Omega) = 6 \text{ V}$. Going from *a* to *b* by the top branch, we encounter potential differences $+12 \text{ V} - (3 \text{ A})(2 \text{ }\Omega) = +6 \text{ V}$, and going by the middle branch, we find $-(-5 \text{ V}) + (1 \text{ A})(1 \text{ }\Omega) = +6 \text{ V}$. The three ways of getting V_{ba} give the same results.

Example 26.5   Power in a battery-charging circuit

In the circuit of Example 26.4 (shown in Fig. 26.11), find the power delivered by the 12-V power supply and by the battery being recharged, and find the power dissipated in each resistor.

SOLUTION

IDENTIFY and SET UP: We use the results of Section 25.5, in which we found that the power delivered *from* an emf to a circuit is $\mathcal{E}I$ and the power delivered *to* a resistor from a circuit is $V_{ab}I = I^2R$. We know the values of all relevant quantities from Example 26.4.

EXECUTE: The power output P_s from the emf of the power supply is

$$P_{\text{supply}} = \mathcal{E}_{\text{supply}}I_{\text{supply}} = (12\ \text{V})(3\ \text{A}) = 36\ \text{W}$$

The power dissipated in the power supply's internal resistance r is

$$P_{r-\text{supply}} = I_{\text{supply}}^2 r_{\text{supply}} = (3\ \text{A})^2(2\ \Omega) = 18\ \text{W}$$

so the power supply's *net* power output is $P_{\text{net}} = 36\ \text{W} - 18\ \text{W} = 18\ \text{W}$. Alternatively, from Example 26.4 the terminal voltage of the battery is $V_{ba} = 6\ \text{V}$, so the net power output is

$$P_{\text{net}} = V_{ba}I_{\text{supply}} = (6\ \text{V})(3\ \text{A}) = 18\ \text{W}$$

The power output of the emf $\mathcal{E}$ of the battery being charged is

$$P_{\text{emf}} = \mathcal{E}I_{\text{battery}} = (-5\ \text{V})(1\ \text{A}) = -5\ \text{W}$$

This is negative because the 1-A current runs through the battery from the higher-potential side to the lower-potential side. (As we mentioned in Example 26.4, the polarity assumed for this battery in Fig. 26.11 was wrong.) We are storing energy in the battery as we charge it. Additional power is dissipated in the battery's internal resistance; this power is

$$P_{r\text{-battery}} = I_{\text{battery}}^2 r_{\text{battery}} = (1\ \text{A})^2(1\ \Omega) = 1\ \text{W}$$

The total power input to the battery is thus $1\ \text{W} + |-5\ \text{W}| = 6\ \text{W}$. Of this, 5 W represents useful energy stored in the battery; the remainder is wasted in its internal resistance.

The power dissipated in the light bulb is

$$P_{\text{bulb}} = I_{\text{bulb}}^2 R_{\text{bulb}} = (2\ \text{A})^2(3\ \Omega) = 12\ \text{W}$$

EVALUATE: As a check, note that all of the power from the supply is accounted for. Of the 18 W of net power from the power supply, 5 W goes to recharge the battery, 1 W is dissipated in the battery's internal resistance, and 12 W is dissipated in the light bulb.

Example 26.6   A complex network

Figure 26.12 shows a "bridge" circuit of the type described at the beginning of this section (see Fig. 26.6b). Find the current in each resistor and the equivalent resistance of the network of five resistors.

SOLUTION

IDENTIFY and SET UP: This network is neither a series combination nor a parallel combination. Hence we must use Kirchhoff's rules to find the values of the target variables. There are five unknown currents, but by applying the junction rule to junctions a and b, we can represent them in terms of three unknown currents I_1, I_2, and I_3, as shown in Fig. 26.12.

26.12 A network circuit with several resistors.

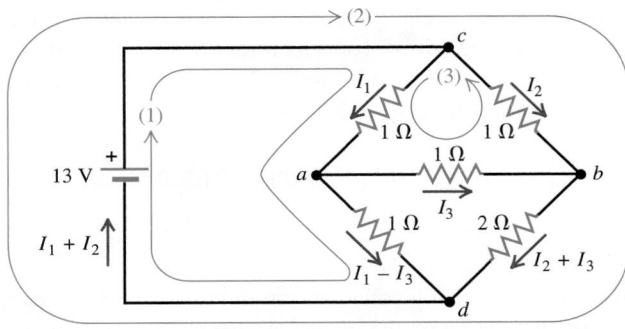

EXECUTE: We apply the loop rule to the three loops shown:

$$13\ \text{V} - I_1(1\ \Omega) - (I_1 - I_3)(1\ \Omega) = 0 \quad (1)$$
$$-I_2(1\ \Omega) - (I_2 + I_3)(2\ \Omega) + 13\ \text{V} = 0 \quad (2)$$
$$-I_1(1\ \Omega) - I_3(1\ \Omega) + I_2(1\ \Omega) = 0 \quad (3)$$

One way to solve these simultaneous equations is to solve Eq. (3) for I_2, obtaining $I_2 = I_1 + I_3$, and then substitute this expression into Eq. (2) to eliminate I_2. We then have

$$13\ \text{V} = I_1(2\ \Omega) - I_3(1\ \Omega) \quad (1')$$
$$13\ \text{V} = I_1(3\ \Omega) + I_3(5\ \Omega) \quad (2')$$

Now we can eliminate I_3 by multiplying Eq. (1') by 5 and adding the two equations. We obtain

$$78\ \text{V} = I_1(13\ \Omega) \qquad I_1 = 6\ \text{A}$$

We substitute this result into Eq. (1') to obtain $I_3 = -1\ \text{A}$, and from Eq. (3) we find $I_2 = 5\ \text{A}$. The negative value of I_3 tells us that its direction is opposite to the direction we assumed.

The total current through the network is $I_1 + I_2 = 11\ \text{A}$, and the potential drop across it is equal to the battery emf, 13 V. The equivalent resistance of the network is therefore

$$R_{\text{eq}} = \frac{13\ \text{V}}{11\ \text{A}} = 1.2\ \Omega$$

EVALUATE: You can check our results for I_1, I_2, and I_3 by substituting them back into Eqs. (1)–(3). What do you find?

Example 26.7 | **A potential difference in a complex network**

In the circuit of Example 26.6 (Fig. 26.12), find the potential difference V_{ab}.

SOLUTION

IDENTIFY and SET UP: Our target variable $V_{ab} = V_a - V_b$ is the potential at point a with respect to point b. To find it, we start at point b and follow a path to point a, adding potential rises and drops as we go. We can follow any of several paths from b to a; the result must be the same for all such paths, which gives us a way to check our result.

EXECUTE: The simplest path is through the center 1-Ω resistor. In Example 26.6 we found $I_3 = -1$ A, showing that the actual current direction through this resistor is from right to left. Thus, as we go from b to a, there is a *drop* of potential with magnitude $|I_3|R = (1 \text{ A})(1 \ \Omega) = 1$ V. Hence $V_{ab} = -1$ V, and the potential at a is 1 V less than at point b.

EVALUATE: To check our result, let's try a path from b to a that goes through the lower two resistors. The currents through these are

$$I_2 + I_3 = 5 \text{ A} + (-1 \text{ A}) = 4 \text{ A} \quad \text{and}$$
$$I_1 - I_3 = 6 \text{ A} - (-1 \text{ A}) = 7 \text{ A}$$

and so

$$V_{ab} = -(4 \text{ A})(2 \ \Omega) + (7 \text{ A})(1 \ \Omega) = -1 \text{ V}$$

You can confirm this result using some other paths from b to a.

26.13 This ammeter (top) and voltmeter (bottom) are both d'Arsonval galvanometers. The difference has to do with their internal connections (see Fig. 26.15).

Test Your Understanding of Section 26.2 Subtract Eq. (1) from Eq. (2) in Example 26.6. To which loop in Fig. 26.12 does this equation correspond? Would this equation have simplified the solution of Example 26.6?

26.3 Electrical Measuring Instruments

We've been talking about potential difference, current, and resistance for two chapters, so it's about time we said something about how to *measure* these quantities. Many common devices, including car instrument panels, battery chargers, and inexpensive electrical instruments, measure potential difference (voltage), current, or resistance using a **d'Arsonval galvanometer** (Fig. 26.13). In the following discussion we'll often call it just a *meter*. A pivoted coil of fine wire is placed in the magnetic field of a permanent magnet (Fig. 26.14). Attached to the coil is a spring, similar to the hairspring on the balance wheel of a watch. In the equilibrium position, with no current in the coil, the pointer is at zero. When there is a current in the coil, the magnetic field exerts a torque on the coil that is proportional to the current. (We'll discuss this magnetic interaction in detail in Chapter 27.) As the coil turns, the spring exerts a restoring torque that is proportional to the angular displacement.

Thus the angular deflection of the coil and pointer is directly proportional to the coil current, and the device can be calibrated to measure current. The maximum deflection, typically 90° or so, is called *full-scale deflection*. The essential electrical characteristics of the meter are the current I_{fs} required for full-scale deflection (typically on the order of 10 μA to 10 mA) and the resistance R_c of the coil (typically on the order of 10 to 1000 Ω).

The meter deflection is proportional to the *current* in the coil. If the coil obeys Ohm's law, the current is proportional to the *potential difference* between the terminals of the coil, and the deflection is also proportional to this potential difference. For example, consider a meter whose coil has a resistance $R_c = 20.0 \ \Omega$ and that deflects full scale when the current in its coil is $I_{fs} = 1.00$ mA. The corresponding potential difference for full-scale deflection is

$$V = I_{fs}R_c = (1.00 \times 10^{-3} \text{ A})(20.0 \ \Omega) = 0.0200 \text{ V}$$

26.14 A d'Arsonval galvanometer, showing a pivoted coil with attached pointer, a permanent magnet supplying a magnetic field that is uniform in magnitude, and a spring to provide restoring torque, which opposes magnetic-field torque.

Magnetic-field torque tends to push pointer away from zero.

Spring torque tends to push pointer toward zero.

Spring

Magnetic field

Permanent magnet

Soft-iron core

Pivoted coil

Ammeters

A current-measuring instrument is usually called an **ammeter** (or milliammeter, microammeter, and so forth, depending on the range). *An ammeter always measures the current passing through it.* An *ideal* ammeter, discussed in Section 25.4, would have *zero* resistance, so including it in a branch of a circuit would not affect the current in that branch. Real ammeters always have some finite resistance, but it is always desirable for an ammeter to have as little resistance as possible.

We can adapt any meter to measure currents that are larger than its full-scale reading by connecting a resistor in parallel with it (Fig. 26.15a) so that some of the current bypasses the meter coil. The parallel resistor is called a **shunt resistor** or simply a *shunt,* denoted as R_{sh}.

Suppose we want to make a meter with full-scale current I_{fs} and coil resistance R_c into an ammeter with full-scale reading I_a. To determine the shunt resistance R_{sh} needed, note that at full-scale deflection the total current through the parallel combination is I_a, the current through the coil of the meter is I_{fs}, and the current through the shunt is the difference $I_a - I_{\text{fs}}$. The potential difference V_{ab} is the same for both paths, so

$$I_{\text{fs}}R_c = (I_a - I_{\text{fs}})R_{\text{sh}} \qquad \text{(for an ammeter)} \qquad (26.7)$$

26.15 Using the same meter to measure (a) current and (b) voltage.

(a) Moving-coil ammeter

(b) Moving-coil voltmeter

Example 26.8 | Designing an ammeter

What shunt resistance is required to make the 1.00-mA, 20.0-Ω meter described above into an ammeter with a range of 0 to 50.0 mA?

SOLUTION

IDENTIFY and SET UP: Since the meter is being used as an ammeter, its internal connections are as shown in Fig. 26.15a. Our target variable is the shunt resistance R_{sh}, which we will find using Eq. (26.7). The ammeter must handle a maximum current $I_a = 50.0 \times 10^{-3}$ A. The coil resistance is $R_c = 20.0$ Ω, and the meter shows full-scale deflection when the current through the coil is $I_{\text{fs}} = 1.00 \times 10^{-3}$ A.

EXECUTE: Solving Eq. (26.7) for R_{sh}, we find

$$R_{\text{sh}} = \frac{I_{\text{fs}}R_c}{I_a - I_{\text{fs}}} = \frac{(1.00 \times 10^{-3}\ \text{A})(20.0\ \Omega)}{50.0 \times 10^{-3}\ \text{A} - 1.00 \times 10^{-3}\ \text{A}}$$
$$= 0.408\ \Omega$$

EVALUATE: It's useful to consider the equivalent resistance R_{eq} of the ammeter as a whole. From Eq. (26.2),

$$R_{\text{eq}} = \left(\frac{1}{R_c} + \frac{1}{R_{\text{sh}}}\right)^{-1} = \left(\frac{1}{20.0\ \Omega} + \frac{1}{0.408\ \Omega}\right)^{-1}$$
$$= 0.400\ \Omega$$

The shunt resistance is so small in comparison to the coil resistance that the equivalent resistance is very nearly equal to the shunt resistance. The result is an ammeter with a low equivalent resistance and the desired 0–50.0-mA range. At full-scale deflection, $I = I_a = 50.0$ mA, the current through the galvanometer is 1.00 mA, the current through the shunt resistor is 49.0 mA, and $V_{ab} = 0.0200$ V. If the current I is *less* than 50.0 mA, the coil current and the deflection are proportionally less.

Voltmeters

This same basic meter may also be used to measure potential difference or *voltage*. A voltage-measuring device is called a **voltmeter**. A voltmeter always measures the potential difference between two points, and its terminals must be connected to these points. (Example 25.6 in Section 25.4 described what can happen if a voltmeter is connected incorrectly.) As we discussed in Section 25.4, an ideal voltmeter would have *infinite* resistance, so connecting it between two points in a circuit would not alter any of the currents. Real voltmeters always have finite resistance, but a voltmeter should have large enough resistance that connecting it in a circuit does not change the other currents appreciably.

For the meter described in Example 26.8 the voltage across the meter coil at full-scale deflection is only $I_{\text{fs}}R_c = (1.00 \times 10^{-3}\ \text{A})(20.0\ \Omega) = 0.0200$ V. We can extend this range by connecting a resistor R_s in *series* with the coil (Fig. 26.15b). Then only a fraction of the total potential difference appears across the coil itself, and the remainder appears across R_s. For a voltmeter with full-scale reading V_V, we need a series resistor R_s in Fig. 26.15b such that

$$V_V = I_{\text{fs}}(R_c + R_s) \qquad \text{(for a voltmeter)} \qquad (26.8)$$

Application Electromyography
A fine needle containing two electrodes is being inserted into a muscle in this patient's hand. By using a sensitive voltmeter to measure the potential difference between these electrodes, a physician can probe the muscle's electrical activity. This is an important technique for diagnosing neurological and neuromuscular diseases.

Example 26.9 **Designing a voltmeter**

What series resistance is required to make the 1.00-mA, 20.0-Ω meter described above into a voltmeter with a range of 0 to 10.0 V?

SOLUTION

IDENTIFY and SET UP: Since this meter is being used as a voltmeter, its internal connections are as shown in Fig. 26.15b. Our target variable is the series resistance R_s. The maximum allowable voltage across the voltmeter is $V_V = 10.0$ V. We want this to occur when the current through the coil is $I_{fs} = 1.00 \times 10^{-3}$ A. Our target variable is the series resistance R_s, which we find using Eq. (26.8).

EXECUTE: From Eq. (26.8),

$$R_s = \frac{V_V}{I_{fs}} - R_c = \frac{10.0 \text{ V}}{0.00100 \text{ A}} - 20.0 \ \Omega = 9980 \ \Omega$$

EVALUATE: At full-scale deflection, $V_{ab} = 10.0$ V, the voltage across the meter is 0.0200 V, the voltage across R_s is 9.98 V, and the current through the voltmeter is 0.00100 A. Most of the voltage appears across the series resistor. The meter's equivalent resistance is a desirably high $R_{eq} = 20.0 \ \Omega + 9980 \ \Omega = 10,000 \ \Omega$. Such a meter is called a "1000 ohms-per-volt" meter, referring to the ratio of resistance to full-scale deflection. In normal operation the current through the circuit element being measured (I in Fig. 26.15b) is much greater than 0.00100 A, and the resistance between points a and b in the circuit is much less than 10,000 Ω. The voltmeter draws off only a small fraction of the current and thus disturbs the circuit being measured only slightly.

Mastering PHYSICS

ActivPhysics 12.4: Using Ammeters and Voltmeters

Ammeters and Voltmeters in Combination

A voltmeter and an ammeter can be used together to measure *resistance* and *power.* The resistance R of a resistor equals the potential difference V_{ab} between its terminals divided by the current I; that is, $R = V_{ab}/I$. The power input P to any circuit element is the product of the potential difference across it and the current through it: $P = V_{ab}I$. In principle, the most straightforward way to measure R or P is to measure V_{ab} and I simultaneously.

With practical ammeters and voltmeters this isn't quite as simple as it seems. In Fig. 26.16a, ammeter A reads the current I in the resistor R. Voltmeter V, however, reads the *sum* of the potential difference V_{ab} across the resistor and the potential difference V_{bc} across the ammeter. If we transfer the voltmeter terminal from c to b, as in Fig. 26.16b, then the voltmeter reads the potential difference V_{ab} correctly, but the ammeter now reads the *sum* of the current I in the resistor and the current I_V in the voltmeter. Either way, we have to correct the reading of one instrument or the other unless the corrections are small enough to be negligible.

26.16 Ammeter–voltmeter method for measuring resistance.

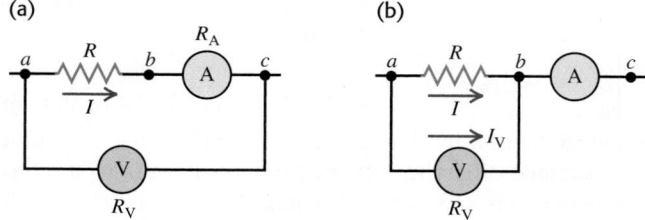

(a)   (b)

Example 26.10 **Measuring resistance I**

The voltmeter in the circuit of Fig. 26.16a reads 12.0 V and the ammeter reads 0.100 A. The meter resistances are $R_V = 10,000 \ \Omega$ (for the voltmeter) and $R_A = 2.00 \ \Omega$ (for the ammeter). What are the resistance R and the power dissipated in the resistor?

SOLUTION

IDENTIFY and SET UP: The ammeter reads the current $I = 0.100$ A through the resistor, and the voltmeter reads the potential difference

between a and c. If the ammeter were *ideal* (that is, if $R_A = 0$), there would be zero potential difference between b and c, the voltmeter reading $V = 12.0$ V would be equal to the potential difference V_{ab} across the resistor, and the resistance would simply be equal to $R = V/I = (12.0 \text{ V})/(0.100 \text{ A}) = 120 \ \Omega$. The ammeter is *not* ideal, however (its resistance is $R_A = 2.00 \ \Omega$), so the voltmeter reading V is actually the sum of the potential differences V_{bc} (across the ammeter) and V_{ab} (across the resistor). We use Ohm's law to find the voltage V_{bc} from the known current and ammeter resistance. Then we solve for V_{ab} and the resistance R. Given these, we are able to calculate the power P into the resistor.

EXECUTE: From Ohm's law, $V_{bc} = IR_A = (0.100 \text{ A})(2.00 \text{ }\Omega) = 0.200 \text{ V}$ and $V_{ab} = IR$. The sum of these is $V = 12.0 \text{ V}$, so the potential difference across the resistor is $V_{ab} = V - V_{bc} = (12.0 \text{ V}) - (0.200 \text{ V}) = 11.8 \text{ V}$. Hence the resistance is

$$R = \frac{V_{ab}}{I} = \frac{11.8 \text{ V}}{0.100 \text{ A}} = 118 \text{ }\Omega$$

The power dissipated in this resistor is

$$P = V_{ab}I = (11.8 \text{ V})(0.100 \text{ A}) = 1.18 \text{ W}$$

EVALUATE: You can confirm this result for the power by using the alternative formula $P = I^2R$. Do you get the same answer?

Example 26.11   Measuring resistance II

Suppose the meters of Example 26.10 are connected to a different resistor as shown in Fig. 26.16b, and the readings obtained on the meters are the same as in Example 26.10. What is the value of this new resistance R, and what is the power dissipated in the resistor?

SOLUTION

IDENTIFY and SET UP: In Example 26.10 the ammeter read the actual current through the resistor, but the voltmeter reading was not the same as the potential difference across the resistor. Now the situation is reversed: The voltmeter reading $V = 12.0 \text{ V}$ shows the actual potential difference V_{ab} across the resistor, but the ammeter reading $I_A = 0.100 \text{ A}$ is *not* equal to the current I through the resistor. Applying the junction rule at b in Fig. 26.16b shows that $I_A = I + I_V$, where I_V is the current through the voltmeter. We find I_V from the given values of V and the voltmeter resistance R_V, and we use this value to find the resistor current I. We then determine the resistance R from I and the voltmeter reading, and calculate the power as in Example 26.10.

EXECUTE: We have $I_V = V/R_V = (12.0 \text{ V})/(10,000 \text{ }\Omega) = 1.20 \text{ mA}$. The actual current I in the resistor is $I = I_A - I_V = 0.100 \text{ A} - 0.0012 \text{ A} = 0.0988 \text{ A}$, and the resistance is

$$R = \frac{V_{ab}}{I} = \frac{12.0 \text{ V}}{0.0988 \text{ A}} = 121 \text{ }\Omega$$

The power dissipated in the resistor is

$$P = V_{ab}I = (12.0 \text{ V})(0.0988 \text{ A}) = 1.19 \text{ W}$$

EVALUATE: Had the meters been ideal, our results would have been $R = 12.0 \text{ V}/0.100 \text{ A} = 120 \text{ }\Omega$ and $P = VI = (12.0 \text{ V}) \times (0.100 \text{ A}) = 1.2 \text{ W}$ both here and in Example 26.10. The actual (correct) results are not too different in either case. That's because the ammeter and voltmeter are nearly ideal: Compared with the resistance R under test, the ammeter resistance R_A is very small and the voltmeter resistance R_V is very large. Under these conditions, treating the meters as ideal yields pretty good results; accurate work requires calculations as in these two examples.

Ohmmeters

An alternative method for measuring resistance is to use a d'Arsonval meter in an arrangement called an **ohmmeter.** It consists of a meter, a resistor, and a source (often a flashlight battery) connected in series (Fig. 26.17). The resistance R to be measured is connected between terminals x and y.

The series resistance R_s is variable; it is adjusted so that when terminals x and y are short-circuited (that is, when $R = 0$), the meter deflects full scale. When nothing is connected to terminals x and y, so that the circuit between x and y is *open* (that is, when $R \to \infty$), there is no current and hence no deflection. For any intermediate value of R the meter deflection depends on the value of R, and the meter scale can be calibrated to read the resistance R directly. Larger currents correspond to smaller resistances, so this scale reads backward compared to the scale showing the current.

In situations in which high precision is required, instruments containing d'Arsonval meters have been supplanted by electronic instruments with direct digital readouts. Digital voltmeters can be made with extremely high internal resistance, of the order of 100 MΩ. Figure 26.18 shows a digital *multimeter,* an instrument that can measure voltage, current, or resistance over a wide range.

26.17 Ohmmeter circuit. The resistor R_s has a variable resistance, as is indicated by the arrow through the resistor symbol. To use the ohmmeter, first connect x directly to y and adjust R_s until the meter reads zero. Then connect x and y across the resistor R and read the scale.

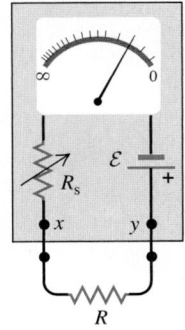

The Potentiometer

The *potentiometer* is an instrument that can be used to measure the emf of a source without drawing any current from the source; it also has a number of other useful applications. Essentially, it balances an unknown potential difference against an adjustable, measurable potential difference.

26.18 This digital multimeter can be used as a voltmeter (red arc), ammeter (yellow arc), or ohmmeter (green arc).

26.19 A potentiometer.

(a) Potentiometer circuit

(b) Circuit symbol for potentiometer (variable resistor)

The principle of the potentiometer is shown schematically in Fig. 26.19a. A resistance wire ab of total resistance R_{ab} is permanently connected to the terminals of a source of known emf $\mathcal{E}_1$. A sliding contact c is connected through the galvanometer G to a second source whose emf $\mathcal{E}_2$ is to be measured. As contact c is moved along the resistance wire, the resistance R_{cb} between points c and b varies; if the resistance wire is uniform, R_{cb} is proportional to the length of wire between c and b. To determine the value of $\mathcal{E}_2$, contact c is moved until a position is found at which the galvanometer shows no deflection; this corresponds to zero current passing through $\mathcal{E}_2$. With $I_2 = 0$, Kirchhoff's loop rule gives

$$\mathcal{E}_2 = IR_{cb}$$

With $I_2 = 0$, the current I produced by the emf $\mathcal{E}_1$ has the same value no matter what the value of the emf $\mathcal{E}_2$. We calibrate the device by replacing $\mathcal{E}_2$ by a source of known emf; then any unknown emf $\mathcal{E}_2$ can be found by measuring the length of wire cb for which $I_2 = 0$. Note that for this to work, V_{ab} must be greater than $\mathcal{E}_2$.

The term *potentiometer* is also used for any variable resistor, usually having a circular resistance element and a sliding contact controlled by a rotating shaft and knob. The circuit symbol for a potentiometer is shown in Fig. 26.19b.

Test Your Understanding of Section 26.3 You want to measure the current through and the potential difference across the 2-Ω resistor shown in Fig. 26.12 (Example 26.6 in Section 26.2). (a) How should you connect an ammeter and a voltmeter to do this? (i) ammeter and voltmeter both in series with the 2-Ω resistor; (ii) ammeter in series with the 2-Ω resistor and voltmeter connected between points b and d; (iii) ammeter connected between points b and d and voltmeter in series with the 2-Ω resistor; (iv) ammeter and voltmeter both connected between points b and d. (b) What resistances should these meters have? (i) Ammeter and voltmeter resistances should both be much greater than 2 Ω; (ii) ammeter resistance should be much greater than 2 Ω and voltmeter resistance should be much less than 2 Ω; (iii) ammeter resistance should be much less than 2 Ω and voltmeter resistance should be much greater than 2 Ω; (iv) ammeter and voltmeter resistances should both be much less than 2 Ω.

26.4 *R-C* Circuits

In the circuits we have analyzed up to this point, we have assumed that all the emfs and resistances are *constant* (time independent) so that all the potentials, currents, and powers are also independent of time. But in the simple act of charging or discharging a capacitor we find a situation in which the currents, voltages, and powers *do* change with time.

Many devices incorporate circuits in which a capacitor is alternately charged and discharged. These include flashing traffic lights, automobile turn signals, and electronic flash units. Understanding what happens in such circuits is thus of great practical importance.

Charging a Capacitor

Figure 26.20 shows a simple circuit for charging a capacitor. A circuit such as this that has a resistor and a capacitor in series is called an ***R-C* circuit.** We idealize the battery (or power supply) to have a constant emf $\mathcal{E}$ and zero internal resistance ($r = 0$), and we neglect the resistance of all the connecting conductors.

We begin with the capacitor initially uncharged (Fig. 26.20a); then at some initial time $t = 0$ we close the switch, completing the circuit and permitting current around the loop to begin charging the capacitor (Fig. 26.20b). For all practical purposes, the current begins at the same instant in every conducting part of the circuit, and at each instant the current is the same in every part.

CAUTION **Lowercase means time-varying** Up to this point we have been working with constant potential differences (voltages), currents, and charges, and we have used *capital* letters *V*, *I*, and *Q*, respectively, to denote these quantities. To distinguish between quantities that vary with time and those that are constant, we will use *lowercase* letters v, i, and q for time-varying voltages, currents, and charges, respectively. We suggest that you follow this same convention in your own work. ▮

Because the capacitor in Fig. 26.20 is initially uncharged, the potential difference v_{bc} across it is zero at $t = 0$. At this time, from Kirchhoff's loop law, the voltage v_{ab} across the resistor R is equal to the battery emf $\mathcal{E}$. The initial ($t = 0$) current through the resistor, which we will call I_0, is given by Ohm's law: $I_0 = v_{ab}/R = \mathcal{E}/R$.

As the capacitor charges, its voltage v_{bc} increases and the potential difference v_{ab} across the resistor decreases, corresponding to a decrease in current. The sum of these two voltages is constant and equal to $\mathcal{E}$. After a long time the capacitor becomes fully charged, the current decreases to zero, and the potential difference v_{ab} across the resistor becomes zero. Then the entire battery emf $\mathcal{E}$ appears across the capacitor and $v_{bc} = \mathcal{E}$.

Let q represent the charge on the capacitor and i the current in the circuit at some time t after the switch has been closed. We choose the positive direction for the current to correspond to positive charge flowing onto the left-hand capacitor plate, as in Fig. 26.20b. The instantaneous potential differences v_{ab} and v_{bc} are

$$v_{ab} = iR \qquad v_{bc} = \frac{q}{C}$$

Using these in Kirchhoff's loop rule, we find

$$\mathcal{E} - iR - \frac{q}{C} = 0 \qquad (26.9)$$

The potential drops by an amount iR as we travel from a to b and by q/C as we travel from b to c. Solving Eq. (26.9) for i, we find

$$i = \frac{\mathcal{E}}{R} - \frac{q}{RC} \qquad (26.10)$$

At time $t = 0$, when the switch is first closed, the capacitor is uncharged, and so $q = 0$. Substituting $q = 0$ into Eq. (26.10), we find that the *initial* current I_0 is given by $I_0 = \mathcal{E}/R$, as we have already noted. If the capacitor were not in the circuit, the last term in Eq. (26.10) would not be present; then the current would be *constant* and equal to $\mathcal{E}/R$.

As the charge q increases, the term q/RC becomes larger and the capacitor charge approaches its final value, which we will call Q_f. The current decreases and eventually becomes zero. When $i = 0$, Eq. (26.10) gives

$$\frac{\mathcal{E}}{R} = \frac{Q_f}{RC} \qquad Q_f = C\mathcal{E} \qquad (26.11)$$

Note that the final charge Q_f does not depend on R.

Figure 26.21 shows the current and capacitor charge as functions of time. At the instant the switch is closed ($t = 0$), the current jumps from zero to its initial value $I_0 = \mathcal{E}/R$; after that, it gradually approaches zero. The capacitor charge starts at zero and gradually approaches the final value given by Eq. (26.11), $Q_f = C\mathcal{E}$.

We can derive general expressions for the charge q and current i as functions of time. With our choice of the positive direction for current (Fig. 26.20b), i equals the rate at which positive charge arrives at the left-hand (positive)

26.20 Charging a capacitor. (a) Just before the switch is closed, the charge q is zero. (b) When the switch closes (at $t = 0$), the current jumps from zero to $\mathcal{E}/R$. As time passes, q approaches Q_f and the current i approaches zero.

(a) Capacitor initially uncharged

(b) Charging the capacitor

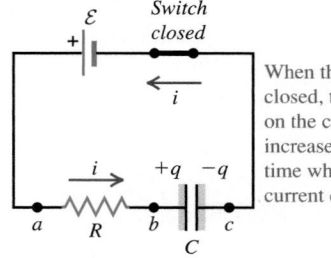

When the switch is closed, the charge on the capacitor increases over time while the current decreases.

26.21 Current i and capacitor charge q as functions of time for the circuit of Fig. 26.20. The initial current is I_0 and the initial capacitor charge is zero. The current asymptotically approaches zero, and the capacitor charge asymptotically approaches a final value of Q_f.

(a) Graph of current versus time for a charging capacitor

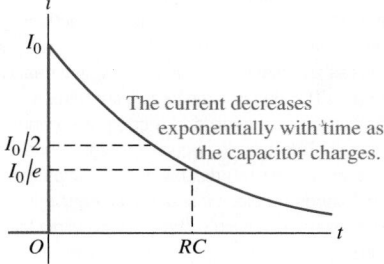

The current decreases exponentially with time as the capacitor charges.

(b) Graph of capacitor charge versus time for a charging capacitor

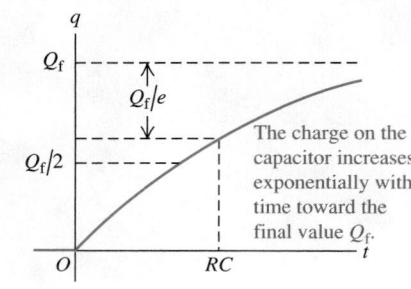

The charge on the capacitor increases exponentially with time toward the final value Q_f.

plate of the capacitor, so $i = dq/dt$. Making this substitution in Eq. (26.10), we have

$$\frac{dq}{dt} = \frac{\mathcal{E}}{R} - \frac{q}{RC} = -\frac{1}{RC}(q - C\mathcal{E})$$

We can rearrange this to

$$\frac{dq}{q - C\mathcal{E}} = -\frac{dt}{RC}$$

and then integrate both sides. We change the integration variables to q' and t' so that we can use q and t for the upper limits. The lower limits are $q' = 0$ and $t' = 0$:

$$\int_0^q \frac{dq'}{q' - C\mathcal{E}} = -\int_0^t \frac{dt'}{RC}$$

When we carry out the integration, we get

$$\ln\left(\frac{q - C\mathcal{E}}{-C\mathcal{E}}\right) = -\frac{t}{RC}$$

Exponentiating both sides (that is, taking the inverse logarithm) and solving for q, we find

$$\frac{q - C\mathcal{E}}{-C\mathcal{E}} = e^{-t/RC}$$

$$q = C\mathcal{E}(1 - e^{-t/RC}) = Q_f(1 - e^{-t/RC}) \qquad \text{(R-C circuit, charging capacitor)} \qquad (26.12)$$

The instantaneous current i is just the time derivative of Eq. (26.12):

$$i = \frac{dq}{dt} = \frac{\mathcal{E}}{R}e^{-t/RC} = I_0 e^{-t/RC} \qquad \text{(R-C circuit, charging capacitor)} \qquad (26.13)$$

The charge and current are both *exponential* functions of time. Figure 26.21a is a graph of Eq. (26.13) and Fig. 26.21b is a graph of Eq. (26.12).

Time Constant

After a time equal to RC, the current in the R-C circuit has decreased to $1/e$ (about 0.368) of its initial value. At this time, the capacitor charge has reached $(1 - 1/e) = 0.632$ of its final value $Q_f = C\mathcal{E}$. The product RC is therefore a measure of how quickly the capacitor charges. We call RC the **time constant,** or the **relaxation time,** of the circuit, denoted by τ:

$$\tau = RC \qquad \text{(time constant for R-C circuit)} \qquad (26.14)$$

When τ is small, the capacitor charges quickly; when it is larger, the charging takes more time. If the resistance is small, it's easier for current to flow, and the capacitor charges more quickly. If R is in ohms and C in farads, τ is in seconds.

In Fig. 26.21a the horizontal axis is an *asymptote* for the curve. Strictly speaking, i never becomes exactly zero. But the longer we wait, the closer it gets. After a time equal to $10RC$, the current has decreased to 0.000045 of its initial value. Similarly, the curve in Fig. 26.21b approaches the horizontal dashed line labeled Q_f as an asymptote. The charge q never attains exactly this value, but after a time equal to $10RC$, the difference between q and Q_f is only 0.000045 of Q_f. We invite you to verify that the product RC has units of time.

Application Pacemakers and Capacitors

This x-ray image shows a pacemaker implanted in a patient with a malfunctioning sinoatrial node, the part of the heart that generates the electrical signal to trigger heartbeats. The pacemaker circuit contains a battery, a capacitor, and a computer-controlled switch. To maintain regular beating, once per second the switch discharges the capacitor and sends an electrical pulse along the lead to the heart. The switch then flips to allow the capacitor to recharge for the next pulse.

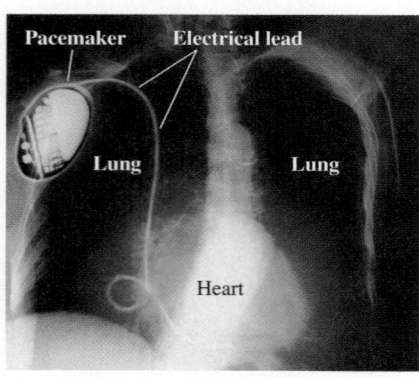

Discharging a Capacitor

Now suppose that after the capacitor in Fig. 26.21b has acquired a charge Q_0, we remove the battery from our *R-C* circuit and connect points *a* and *c* to an open switch (Fig. 26.22a). We then close the switch and at the same instant reset our stopwatch to $t = 0$; at that time, $q = Q_0$. The capacitor then *discharges* through the resistor, and its charge eventually decreases to zero.

Again let *i* and *q* represent the time-varying current and charge at some instant after the connection is made. In Fig. 26.22b we make the same choice of the positive direction for current as in Fig. 26.20b. Then Kirchhoff's loop rule gives Eq. (26.10) but with $\mathcal{E} = 0$; that is,

$$i = \frac{dq}{dt} = -\frac{q}{RC} \tag{26.15}$$

The current *i* is now negative; this is because positive charge *q* is leaving the left-hand capacitor plate in Fig. 26.22b, so the current is in the direction opposite to that shown in the figure. At time $t = 0$, when $q = Q_0$, the initial current is $I_0 = -Q_0/RC$.

To find *q* as a function of time, we rearrange Eq. (26.15), again change the names of the variables to q' and t', and integrate. This time the limits for q' are Q_0 to *q*. We get

$$\int_{Q_0}^{q} \frac{dq'}{q'} = -\frac{1}{RC} \int_0^t dt'$$

$$\ln \frac{q}{Q_0} = -\frac{t}{RC}$$

$$q = Q_0 e^{-t/RC} \quad \text{(R-C circuit, discharging capacitor)} \tag{26.16}$$

The instantaneous current *i* is the derivative of this with respect to time:

$$i = \frac{dq}{dt} = -\frac{Q_0}{RC} e^{-t/RC} = I_0 e^{-t/RC} \quad \begin{matrix}\text{(R-C circuit,}\\\text{discharging capacitor)}\end{matrix} \tag{26.17}$$

We graph the current and the charge in Fig. 26.23; both quantities approach zero exponentially with time. Comparing these results with Eqs. (26.12) and (26.13), we note that the expressions for the current are identical, apart from the sign of I_0. The capacitor charge approaches zero asymptotically in Eq. (26.16), while the *difference* between *q* and *Q* approaches zero asymptotically in Eq. (26.12).

Energy considerations give us additional insight into the behavior of an *R-C* circuit. While the capacitor is charging, the instantaneous rate at which the battery delivers energy to the circuit is $P = \mathcal{E}i$. The instantaneous rate at which electrical energy is dissipated in the resistor is i^2R, and the rate at which energy is stored in the capacitor is $iv_{bc} = iq/C$. Multiplying Eq. (26.9) by *i*, we find

$$\mathcal{E}i = i^2R + \frac{iq}{C} \tag{26.18}$$

This means that of the power $\mathcal{E}i$ supplied by the battery, part (i^2R) is dissipated in the resistor and part (iq/C) is stored in the capacitor.

The *total* energy supplied by the battery during charging of the capacitor equals the battery emf $\mathcal{E}$ multiplied by the total charge Q_f, or $\mathcal{E}Q_f$. The total energy stored in the capacitor, from Eq. (24.9), is $Q_f\mathcal{E}/2$. Thus, of the energy supplied by the battery, *exactly half* is stored in the capacitor, and the other half is dissipated in the resistor. This half-and-half division of energy doesn't depend on *C*, *R*, or $\mathcal{E}$. You can verify this result by taking the integral over time of each of the power quantities in Eq. (26.18) (see Problem 26.88).

26.22 Discharging a capacitor. **(a)** Before the switch is closed at time $t = 0$, the capacitor charge is Q_0 and the current is zero. **(b)** At time *t* after the switch is closed, the capacitor charge is *q* and the current is *i*. The actual current direction is opposite to the direction shown; *i* is negative. After a long time, *q* and *i* both approach zero.

(a) Capacitor initially charged

(b) Discharging the capacitor

When the switch is closed, the charge on the capacitor and the current both decrease over time.

26.23 Current *i* and capacitor charge *q* as functions of time for the circuit of Fig. 26.22. The initial current is I_0 and the initial capacitor charge is Q_0. Both *i* and *q* asymptotically approach zero.

(a) Graph of current versus time for a discharging capacitor

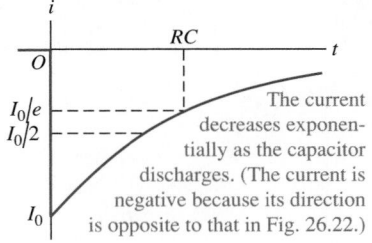

The current decreases exponentially as the capacitor discharges. (The current is negative because its direction is opposite to that in Fig. 26.22.)

(b) Graph of capacitor charge versus time for a discharging capacitor

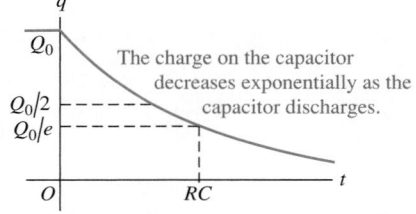

The charge on the capacitor decreases exponentially as the capacitor discharges.

Example 26.12 Charging a capacitor

A 10-MΩ resistor is connected in series with a 1.0-μF capacitor and a battery with emf 12.0 V. Before the switch is closed at time $t = 0$, the capacitor is uncharged. (a) What is the time constant? (b) What fraction of the final charge Q_f is on the capacitor at $t = 46$ s? (c) What fraction of the initial current I_0 is still flowing at $t = 46$ s?

SOLUTION

IDENTIFY and SET UP: This is the same situation as shown in Fig. 26.20, with $R = 10$ MΩ, $C = 1.0$ μF, and $\mathcal{E} = 12.0$ V. The charge q and current i vary with time as shown in Fig. 26.21. Our target variables are (a) the time constant τ, (b) the ratio q/Q_f at $t = 46$ s, and (c) the ratio i/I_0 at $t = 46$ s. Equation (26.14) gives τ. For a capacitor being charged, Eq. (26.12) gives q and Eq. (26.13) gives i.

EXECUTE: (a) From Eq. (26.14),

$$\tau = RC = (10 \times 10^6 \ \Omega)(1.0 \times 10^{-6} \ \text{F}) = 10 \ \text{s}$$

(b) From Eq. (26.12),

$$\frac{q}{Q_f} = 1 - e^{-t/RC} = 1 - e^{-(46 \ \text{s})/(10 \ \text{s})} = 0.99$$

(c) From Eq. (26.13),

$$\frac{i}{I_0} = e^{-t/RC} = e^{-(46 \ \text{s})/(10 \ \text{s})} = 0.010$$

EVALUATE: After 4.6 time constants the capacitor is 99% charged and the charging current has decreased to 1.0% of its initial value. The circuit would charge more rapidly if we reduced the time constant by using a smaller resistance.

Example 26.13 Discharging a capacitor

The resistor and capacitor of Example 26.12 are reconnected as shown in Fig. 26.22. The capacitor has an initial charge of 5.0 μC and is discharged by closing the switch at $t = 0$. (a) At what time will the charge be equal to 0.50 μC? (b) What is the current at this time?

SOLUTION

IDENTIFY and SET UP: Now the capacitor is being discharged, so q and i vary with time as in Fig. 26.23, with $Q_0 = 5.0 \times 10^{-6}$ C. Again we have $RC = \tau = 10$ s. Our target variables are (a) the value of t at which $q = 0.50$ μC and (b) the value of i at this time. We first solve Eq. (26.16) for t, and then solve Eq. (26.17) for i.

EXECUTE: (a) Solving Eq. (26.16) for the time t gives

$$t = -RC \ln\frac{q}{Q_0} = -(10 \ \text{s}) \ln\frac{0.50 \ \mu\text{C}}{5.0 \ \mu\text{C}} = 23 \ \text{s} = 2.3\tau$$

(b) From Eq. (26.17), with $Q_0 = 5.0$ μC $= 5.0 \times 10^{-6}$ C,

$$i = -\frac{Q_0}{RC}e^{-t/RC} = -\frac{5.0 \times 10^{-6} \ \text{C}}{10 \ \text{s}}e^{-2.3} = -5.0 \times 10^{-8} \ \text{A}$$

EVALUATE: The current in part (b) is negative because i has the opposite sign when the capacitor is discharging than when it is charging. Note that we could have avoided evaluating $e^{-t/RC}$ by noticing that at the time in question, $q = 0.10Q_0$; from Eq. (26.16) this means that $e^{-t/RC} = 0.10$.

Test Your Understanding of Section 26.4 The energy stored in a capacitor is equal to $q^2/2C$. When a capacitor is discharged, what fraction of the initial energy remains after an elapsed time of one time constant? (i) $1/e$; (ii) $1/e^2$; (iii) $1 - 1/e$; (iv) $(1 - 1/e)^2$; (v) answer depends on how much energy was stored initially.

26.5 Power Distribution Systems

We conclude this chapter with a brief discussion of practical household and automotive electric-power distribution systems. Automobiles use direct-current (dc) systems, while nearly all household, commercial, and industrial systems use alternating current (ac) because of the ease of stepping voltage up and down with transformers. Most of the same basic wiring concepts apply to both. We'll talk about alternating-current circuits in greater detail in Chapter 31.

The various lamps, motors, and other appliances to be operated are always connected in *parallel* to the power source (the wires from the power company for houses, or from the battery and alternator for a car). If appliances were connected in series, shutting one appliance off would shut them all off (see Example 26.2 in Section 26.1). Figure 26.24 shows the basic idea of house wiring. One side of the "line," as the pair of conductors is called, is called the *neutral* side; it is always connected to

26.24 Schematic diagram of part of a house wiring system. Only two branch circuits are shown; an actual system might have four to thirty branch circuits. Lamps and appliances may be plugged into the outlets. The grounding wires, which normally carry no current, are not shown.

"ground" at the entrance panel. For houses, *ground* is an actual electrode driven into the earth (which is usually a good conductor) or sometimes connected to the household water pipes. Electricians speak of the "hot" side and the "neutral" side of the line. Most modern house wiring systems have *two* hot lines with opposite polarity with respect to the neutral. We'll return to this detail later.

Household voltage is nominally 120 V in the United States and Canada, and often 240 V in Europe. (For alternating current, which varies sinusoidally with time, these numbers represent the *root-mean-square* voltage, which is $1/\sqrt{2}$ times the peak voltage. We'll discuss this further in Section 31.1.) The amount of current *I* drawn by a given device is determined by its power input *P*, given by Eq. (25.17): $P = VI$. Hence $I = P/V$. For example, the current in a 100-W light bulb is

$$I = \frac{P}{V} = \frac{100 \text{ W}}{120 \text{ V}} = 0.83 \text{ A}$$

The power input to this bulb is actually determined by its resistance *R*. Using Eq. (25.18), which states that $P = VI = I^2R = V^2/R$ for a resistor, the resistance of this bulb at operating temperature is

$$R = \frac{V}{I} = \frac{120 \text{ V}}{0.83 \text{ A}} = 144 \ \Omega \qquad \text{or} \qquad R = \frac{V^2}{P} = \frac{(120 \text{ V})^2}{100 \text{ W}} = 144 \ \Omega$$

Similarly, a 1500-W waffle iron draws a current of $(1500 \text{ W})/(120 \text{ V}) = 12.5$ A and has a resistance, at operating temperature, of 9.6 Ω. Because of the temperature dependence of resistivity, the resistances of these devices are considerably less when they are cold. If you measure the resistance of a 100-W light bulb with an ohmmeter (whose small current causes very little temperature rise), you will probably get a value of about 10 Ω. When a light bulb is turned on, this low resistance causes an initial surge of current until the filament heats up. That's why a light bulb that's ready to burn out nearly always does so just when you turn it on.

Circuit Overloads and Short Circuits

The maximum current available from an individual circuit is limited by the resistance of the wires. As we discussed in Section 25.5, the I^2R power loss in the wires causes them to become hot, and in extreme cases this can cause a fire or melt the wires. Ordinary lighting and outlet wiring in houses usually uses 12-gauge wire. This has a diameter of 2.05 mm and can carry a maximum current of 20 A safely (without overheating). Larger-diameter wires of the same length have lower resistance [see Eq. (25.10)]. Hence 8-gauge (3.26 mm) or 6-gauge (4.11 mm) are used for high-current appliances such as clothes dryers, and 2-gauge (6.54 mm) or larger is used for the main power lines entering a house.

26.25 (a) Excess current will melt the thin wire of lead–tin alloy that runs along the length of a fuse, inside the transparent housing. (b) The switch on this circuit breaker will flip if the maximum allowable current is exceeded.

(a)

(b)

Protection against overloading and overheating of circuits is provided by fuses or circuit breakers. A *fuse* contains a link of lead–tin alloy with a very low melting temperature; the link melts and breaks the circuit when its rated current is exceeded (Fig. 26.25a). A *circuit breaker* is an electromechanical device that performs the same function, using an electromagnet or a bimetallic strip to "trip" the breaker and interrupt the circuit when the current exceeds a specified value (Fig. 26.25b). Circuit breakers have the advantage that they can be reset after they are tripped, while a blown fuse must be replaced.

If your system has fuses and you plug too many high-current appliances into the same outlet, the fuse blows. *Do not* replace the fuse with one of larger rating; if you do, you risk overheating the wires and starting a fire. The only safe solution is to distribute the appliances among several circuits. Modern kitchens often have three or four separate 20-A circuits.

Contact between the hot and neutral sides of the line causes a *short circuit.* Such a situation, which can be caused by faulty insulation or by any of a variety of mechanical malfunctions, provides a very low-resistance current path, permitting a very large current that would quickly melt the wires and ignite their insulation if the current were not interrupted by a fuse or circuit breaker (see Example 25.10 in Section 25.5). An equally dangerous situation is a broken wire that interrupts the current path, creating an *open circuit.* This is hazardous because of the sparking that can occur at the point of intermittent contact.

In approved wiring practice, a fuse or breaker is placed *only* in the hot side of the line, never in the neutral side. Otherwise, if a short circuit should develop because of faulty insulation or other malfunction, the ground-side fuse could blow. The hot side would still be live and would pose a shock hazard if you touched the live conductor and a grounded object such as a water pipe. For similar reasons the wall switch for a light fixture is always in the hot side of the line, never the neutral side.

Further protection against shock hazard is provided by a third conductor called the *grounding wire,* included in all present-day wiring. This conductor corresponds to the long round or U-shaped prong of the three-prong connector plug on an appliance or power tool. It is connected to the neutral side of the line at the entrance panel. The grounding wire normally carries no current, but it connects the metal case or frame of the device to ground. If a conductor on the hot side of the line accidentally contacts the frame or case, the grounding conductor provides a current path, and the fuse blows. Without the ground wire, the frame could become "live"—that is, at a potential 120 V above ground. Then if you touched it and a water pipe (or even a damp basement floor) at the same time, you could get a dangerous shock (Fig. 26.26). In some situations, especially outlets located outdoors or near a sink or other water pipes, a special kind of circuit breaker called a *ground-fault interrupter* (GFI or GFCI) is used. This device senses the difference in current between the hot and neutral conductors (which is normally zero) and trips when this difference exceeds some very small value, typically 5 mA.

Household and Automotive Wiring

Most modern household wiring systems actually use a slight elaboration of the system described above. The power company provides *three* conductors. One is neutral; the other two are both at 120 V with respect to the neutral but with opposite polarity, giving a voltage between them of 240 V. The power company calls this a *three-wire line,* in contrast to the 120-V two-wire (plus ground wire) line described above. With a three-wire line, 120-V lamps and appliances can be connected between neutral and either hot conductor, and high-power devices requiring 240 V, such as electric ranges and clothes dryers, are connected between the two hot lines.

All of the above discussion can be applied directly to automobile wiring. The voltage is about 13 V (direct current); the power is supplied by the battery and by

(a) Two-prong plug   (b) Three-prong plug

26.26 (a) If a malfunctioning electric drill is connected to a wall socket via a two-prong plug, a person may receive a shock. (b) When the drill malfunctions when connected via a three-prong plug, a person touching it receives no shock, because electric charge flows through the ground wire (shown in green) to the third prong and into the ground rather than into the person's body. If the ground current is appreciable, the fuse blows.

the alternator, which charges the battery when the engine is running. The neutral side of each circuit is connected to the body and frame of the vehicle. For this low voltage a separate grounding conductor is not required for safety. The fuse or circuit breaker arrangement is the same in principle as in household wiring. Because of the lower voltage (less energy per charge), more current (a greater number of charges per second) is required for the same power; a 100-W headlight bulb requires a current of about $(100 \text{ W})/(13 \text{ V}) = 8 \text{ A}$.

Although we spoke of *power* in the above discussion, what we buy from the power company is *energy*. Power is energy transferred per unit time, so energy is average power multiplied by time. The usual unit of energy sold by the power company is the kilowatt-hour $(1 \text{ kW} \cdot \text{h})$:

$$1 \text{ kW} \cdot \text{h} = (10^3 \text{ W})(3600 \text{ s}) = 3.6 \times 10^6 \text{ W} \cdot \text{s} = 3.6 \times 10^6 \text{ J}$$

In the United States, one kilowatt-hour typically costs 8 to 27 cents, depending on the location and quantity of energy purchased. To operate a 1500-W (1.5-kW) waffle iron continuously for 1 hour requires $1.5 \text{ kW} \cdot \text{h}$ of energy; at 10 cents per kilowatt-hour, the energy cost is 15 cents. The cost of operating any lamp or appliance for a specified time can be calculated in the same way if the power rating is known. However, many electric cooking utensils (including waffle irons) cycle on and off to maintain a constant temperature, so the average power may be less than the power rating marked on the device.

Example 26.14   **A kitchen circuit**

An 1800-W toaster, a 1.3-kW electric frying pan, and a 100-W lamp are plugged into the same 20-A, 120-V circuit. (a) What current is drawn by each device, and what is the resistance of each device? (b) Will this combination trip the circuit breaker?

SOLUTION

IDENTIFY and SET UP: When plugged into the same circuit, the three devices are connected in parallel, so the voltage across each appliance is $V = 120 \text{ V}$. We find the current I drawn by each device using the relationship $P = VI$, where P is the power input of the device. To find the resistance R of each device we use the relationship $P = V^2/R$.

EXECUTE: (a) To simplify the calculation of current and resistance, we note that $I = P/V$ and $R = V^2/P$. Hence

$$I_{\text{toaster}} = \frac{1800 \text{ W}}{120 \text{ V}} = 15 \text{ A} \qquad R_{\text{toaster}} = \frac{(120 \text{ V})^2}{1800 \text{ W}} = 8 \text{ }\Omega$$

$$I_{\text{frying pan}} = \frac{1300 \text{ W}}{120 \text{ V}} = 11 \text{ A} \qquad R_{\text{frying pan}} = \frac{(120 \text{ V})^2}{1300 \text{ W}} = 11 \text{ }\Omega$$

$$I_{\text{lamp}} = \frac{100 \text{ W}}{120 \text{ V}} = 0.83 \text{ A} \qquad R_{\text{lamp}} = \frac{(120 \text{ V})^2}{100 \text{ W}} = 144 \text{ }\Omega$$

For constant voltage the device with the *least* resistance (in this case the toaster) draws the most current and receives the most power.

Continued

(b) The total current through the line is the sum of the currents drawn by the three devices:

$$I = I_{\text{toaster}} + I_{\text{frying pan}} + I_{\text{lamp}}$$
$$= 15 \text{ A} + 11 \text{ A} + 0.83 \text{ A} = 27 \text{ A}$$

This exceeds the 20-A rating of the line, and the circuit breaker will indeed trip.

EVALUATE: We could also find the total current by using $I = P/V$ and dividing the total power P delivered to all three devices by the voltage:

$$I = \frac{P_{\text{toaster}} + P_{\text{frying pan}} + P_{\text{lamp}}}{V}$$

$$= \frac{1800 \text{ W} + 1300 \text{ W} + 100 \text{ W}}{120 \text{ V}} = 27 \text{ A}$$

A third way to determine I is to use $I = V/R_{\text{eq}}$, where R_{eq} is the equivalent resistance of the three devices in parallel:

$$I = \frac{V}{R_{\text{eq}}} = (120 \text{ V})\left(\frac{1}{8 \ \Omega} + \frac{1}{11 \ \Omega} + \frac{1}{144 \ \Omega}\right) = 27 \text{ A}$$

Appliances with such current demands are common, so modern kitchens have more than one 20-A circuit. To keep currents safely below 20 A, the toaster and frying pan should be plugged into different circuits.

Test Your Understanding of Section 26.5 To prevent the circuit breaker in Example 26.14 from blowing, a home electrician replaces the circuit breaker with one rated at 40 A. Is this a reasonable thing to do? ❙

Resistors in series and parallel: When several resistors $R_1, R_2, R_3, \ldots$ are connected in series, the equivalent resistance R_{eq} is the sum of the individual resistances. The same *current* flows through all the resistors in a series connection. When several resistors are connected in parallel, the reciprocal of the equivalent resistance R_{eq} is the sum of the reciprocals of the individual resistances. All resistors in a parallel connection have the same *potential difference* between their terminals. (See Examples 26.1 and 26.2.)

$$R_{eq} = R_1 + R_2 + R_3 + \cdots \quad \text{(26.1)}$$
(resistors in series)

$$\frac{1}{R_{eq}} = \frac{1}{R_1} + \frac{1}{R_2} + \frac{1}{R_3} + \cdots \quad \text{(26.2)}$$
(resistors in parallel)

Kirchhoff's rules: Kirchhoff's junction rule is based on conservation of charge. It states that the algebraic sum of the currents into any junction must be zero. Kirchhoff's loop rule is based on conservation of energy and the conservative nature of electrostatic fields. It states that the algebraic sum of potential differences around any loop must be zero. Careful use of consistent sign rules is essential in applying Kirchhoff's rules. (See Examples 26.3–26.7.)

$$\sum I = 0 \quad \text{(junction rule)} \quad \text{(26.5)}$$
$$\sum V = 0 \quad \text{(loop rule)} \quad \text{(26.6)}$$

Electrical measuring instruments: In a d'Arsonval galvanometer, the deflection is proportional to the current in the coil. For a larger current range, a shunt resistor is added, so some of the current bypasses the meter coil. Such an instrument is called an ammeter. If the coil and any additional series resistance included obey Ohm's law, the meter can also be calibrated to read potential difference or voltage. The instrument is then called a voltmeter. A good ammeter has very low resistance; a good voltmeter has very high resistance. (See Examples 26.8–26.11.)

R-C circuits: When a capacitor is charged by a battery in series with a resistor, the current and capacitor charge are not constant. The charge approaches its final value asymptotically and the current approaches zero asymptotically. The charge and current in the circuit are given by Eqs. (26.12) and (26.13). After a time $\tau = RC$, the charge has approached within $1/e$ of its final value. This time is called the time constant or relaxation time of the circuit. When the capacitor discharges, the charge and current are given as functions of time by Eqs. (26.16) and (26.17). The time constant is the same for charging and discharging. (See Examples 26.12 and 26.13.)

Capacitor charging:
$$q = C\mathcal{E}\left(1 - e^{-t/RC}\right)$$
$$= Q_f\left(1 - e^{-t/RC}\right) \quad \text{(26.12)}$$
$$i = \frac{dq}{dt} = \frac{\mathcal{E}}{R}e^{-t/RC}$$
$$= I_0 e^{-t/RC} \quad \text{(26.13)}$$

Capacitor discharging:
$$q = Q_0 e^{-t/RC} \quad \text{(26.16)}$$
$$i = \frac{dq}{dt} = -\frac{Q_0}{RC}e^{-t/RC}$$
$$= I_0 e^{-t/RC} \quad \text{(26.17)}$$

Household wiring: In household wiring systems, the various electrical devices are connected in parallel across the power line, which consists of a pair of conductors, one "hot" and the other "neutral." An additional "ground" wire is included for safety. The maximum permissible current in a circuit is determined by the size of the wires and the maximum temperature they can tolerate. Protection against excessive current and the resulting fire hazard is provided by fuses or circuit breakers. (See Example 26.14.)

BRIDGING PROBLEM — Two Capacitors and Two Resistors

A 2.40-μF capacitor and a 3.60-μF capacitor are connected in series. (a) A charge of 5.20 mC is placed on each capacitor. What is the energy stored in the capacitors? (b) A 655-Ω resistor is connected to the terminals of the capacitor combination, and a voltmeter with resistance $4.58 \times 10^4 \ \Omega$ is connected across the resistor. What is the rate of change of the energy stored in the capacitors just after the connection is made? (c) How long after the connection is made has the energy stored in the capacitors decreased to $1/e$ of its initial value? (d) At the instant calculated in part (c), what is the rate of change of the energy stored in the capacitors?

SOLUTION GUIDE

See MasteringPhysics® study area for a Video Tutor solution.

IDENTIFY and SET UP

1. The two capacitors act as a single equivalent capacitor (see Section 24.2), and the resistor and voltmeter act as a single equivalent resistor. Select equations that will allow you to calculate the values of these equivalent circuit elements.

2. Equation (24.9) gives the energy stored in a capacitor. Equations (26.16) and (26.17) give the capacitor charge and current as functions of time. Use these to set up the solutions to the various parts of this problem. (*Hint:* The rate at which energy is lost by the capacitors equals the rate at which energy is dissipated in the resistances.)

EXECUTE

3. Find the stored energy at $t = 0$.
4. Find the rate of change of the stored energy at $t = 0$.
5. Find the value of t at which the stored energy has $1/e$ of the value you found in step 3.
6. Find the rate of change of the stored energy at the time you found in step 5.

EVALUATE

7. Check your results from steps 4 and 6 by calculating the rate of change in a different way. (*Hint:* The rate of change of the stored energy U is dU/dt.)

Problems

For instructor-assigned homework, go to www.masteringphysics.com

•, ••, •••: Problems of increasing difficulty. **CP**: Cumulative problems incorporating material from earlier chapters. **CALC**: Problems requiring calculus. **BIO**: Biosciences problems.

DISCUSSION QUESTIONS

Q26.1 In which 120-V light bulb does the filament have greater resistance: a 60-W bulb or a 120-W bulb? If the two bulbs are connected to a 120-V line in series, through which bulb will there be the greater voltage drop? What if they are connected in parallel? Explain your reasoning.

Q26.2 Two 120-V light bulbs, one 25-W and one 200-W, were connected in series across a 240-V line. It seemed like a good idea at the time, but one bulb burned out almost immediately. Which one burned out, and why?

Q26.3 You connect a number of identical light bulbs to a flashlight battery. (a) What happens to the brightness of each bulb as more and more bulbs are added to the circuit if you connect them (i) in series and (ii) in parallel? (b) Will the battery last longer if the bulbs are in series or in parallel? Explain your reasoning.

Q26.4 In the circuit shown in Fig. Q26.4, three identical light bulbs are connected to a flashlight battery. How do the brightnesses of the bulbs compare? Which light bulb has the greatest current passing through it? Which light bulb has the greatest potential difference between its terminals? What happens if bulb A is unscrewed? Bulb B? Bulb C? Explain your reasoning.

Figure **Q26.4**

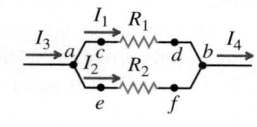

Q26.5 If two resistors R_1 and R_2 ($R_2 > R_1$) are connected in series as shown in Fig. Q26.5, which of the following must be true? In each case justify your answer. (a) $I_1 = I_2 = I_3$. (b) The current is greater in R_1 than in R_2. (c) The electrical power consumption is the same for both resistors. (d) The electrical power consumption is greater in R_2 than in R_1.

Figure **Q26.5**

(e) The potential drop is the same across both resistors. (f) The potential at point a is the same as at point c. (g) The potential at point b is lower than at point c. (h) The potential at point c is lower than at point b.

Q26.6 If two resistors R_1 and R_2 ($R_2 > R_1$) are connected in parallel as shown in Fig. Q26.6, which of the following must be true? In each case justify your answer. (a) $I_1 = I_2$. (b) $I_3 = I_4$. (c) The current is greater in R_1 than in R_2. (d) The rate of electrical energy consumption is the same for both resistors. (e) The rate of electrical energy consumption is greater in R_2 than in R_1. (f) $V_{cd} = V_{ef} = V_{ab}$. (g) Point c is at higher potential than point d. (h) Point f is at higher potential than point e. (i) Point c is at higher potential than point e.

Figure **Q26.6**

Q26.7 Why do the lights on a car become dimmer when the starter is operated?

Q26.8 A resistor consists of three identical metal strips connected as shown in Fig. Q26.8. If one of the strips is cut out, does the ammeter reading increase, decrease, or stay the same? Why?

Figure **Q26.8**

Q26.9 A light bulb is connected in the circuit shown in Fig. Q26.9. If we close the switch S, does the bulb's brightness increase, decrease, or remain the same? Explain why.

Figure **Q26.9**

Q26.10 A real battery, having nonnegligible internal resistance, is connected across a light bulb as shown in Fig. Q26.10. When the switch S is closed, what happens to the brightness of the bulb? Why?

Figure **Q26.10**

Q26.11 If the battery in Discussion Question Q26.10 is ideal with no internal resistance, what will happen to the brightness of the bulb when S is closed? Why?

Q26.12 For the circuit shown in Fig. Q26.12 what happens to the brightness of the bulbs when the switch S is closed if the battery (a) has no internal resistance and (b) has nonnegligible internal resistance? Explain why.

Figure **Q26.12**

Q26.13 Is it possible to connect resistors together in a way that cannot be reduced to some combination of series and parallel combinations? If so, give examples. If not, state why not.

Q26.14 The direction of current in a battery can be reversed by connecting it to a second battery of greater emf with the positive terminals of the two batteries together. When the direction of current is reversed in a battery, does its emf also reverse? Why or why not?

Q26.15 In a two-cell flashlight, the batteries are usually connected in series. Why not connect them in parallel? What possible advantage could there be in connecting several identical batteries in parallel?

Q26.16 The greater the diameter of the wire used in household wiring, the greater the maximum current that can safely be carried by the wire. Why is this? Does the maximum permissible current depend on the length of the wire? Does it depend on what the wire is made of? Explain your reasoning.

Q26.17 The emf of a flashlight battery is roughly constant with time, but its internal resistance increases with age and use. What sort of meter should be used to test the freshness of a battery?

Q26.18 Is it possible to have a circuit in which the potential difference across the terminals of a battery in the circuit is zero? If so, give an example. If not, explain why not.

Q26.19 Verify that the time constant RC has units of time.

Q26.20 For very large resistances it is easy to construct R-C circuits that have time constants of several seconds or minutes. How might this fact be used to measure very large resistances, those that are too large to measure by more conventional means?

Q26.21 Whan a capacitor, battery, and resistor are connected in series, does the resistor affect the maximum charge stored on the capacitor? Why or why not? What purpose does the resistor serve?

EXERCISES

Section 26.1 Resistors in Series and Parallel

26.1 •• A uniform wire of resistance R is cut into three equal lengths. One of these is formed into a circle and connected between the other two (Fig. E26.1). What is the resistance between the opposite ends a and b?

Figure **E26.1**

26.2 •• A machine part has a resistor X protruding from an opening in the side. This resistor is connected to three other resistors, as shown in Fig. E26.2. An ohmmeter connected across a and b reads 2.00 Ω. What is the resistance of X?

Figure **E26.2**

26.3 • A resistor with $R_1 = 25.0\ \Omega$ is connected to a battery that has negligible internal resistance and electrical energy is dissipated by R_1 at a rate of 36.0 W. If a second resistor with $R_2 = 15.0\ \Omega$ is connected in series with R_1, what is the total rate at which electrical energy is dissipated by the two resistors?

26.4 • A 32-Ω resistor and a 20-Ω resistor are connected in parallel, and the combination is connected across a 240-V dc line. (a) What is the resistance of the parallel combination? (b) What is the total current through the parallel combination? (c) What is the current through each resistor?

26.5 • A triangular array of resistors is shown in Fig. E26.5. What current will this array draw from a 35.0-V battery having negligible internal resistance if we connect it across (a) ab; (b) bc; (c) ac? (d) If the battery has an internal resistance of 3.00 Ω, what current will the array draw if the battery is connected across bc?

Figure **E26.5**

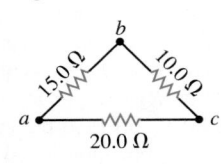

26.6 •• For the circuit shown in Fig. E26.6 both meters are idealized, the battery has no appreciable internal resistance, and the ammeter reads 1.25 A. (a) What does the voltmeter read? (b) What is the emf $\mathcal{E}$ of the battery?

Figure **E26.6**

26.7 •• For the circuit shown in Fig. E26.7 find the reading of the idealized ammeter if the battery has an internal resistance of 3.26 Ω.

Figure **E26.7**

26.8 • Three resistors having resistances of 1.60 Ω, 2.40 Ω, and 4.80 Ω are connected in parallel to a 28.0-V battery that has negligible internal resistance. Find (a) the equivalent resistance of the combination; (b) the current in each resistor; (c) the total current through the battery; (d) the voltage across each resistor; (e) the power dissipated in each resistor. (f) Which resistor dissipates the most power: the one with the greatest resistance or the least resistance? Explain why this should be.

26.9 • Now the three resistors of Exercise 26.8 are connected in series to the same battery. Answer the same questions for this situation.

26.10 •• **Power Rating of a Resistor.** The *power rating* of a resistor is the maximum power the resistor can safely dissipate without too great a rise in temperature and hence damage to the resistor. (a) If the power rating of a 15-kΩ resistor is 5.0 W, what is the maximum allowable potential difference across the terminals of the resistor? (b) A 9.0-kΩ resistor is to be connected across a 120-V potential difference. What power rating is required? (c) A 100.0-Ω and a 150.0-Ω resistor, both rated at 2.00 W, are connected in series across a variable potential difference. What is the greatest this potential difference can be without overheating either resistor, and what is the rate of heat generated in each resistor under these conditions?

26.11 • In Fig. E26.11, $R_1 = 3.00\ \Omega$, $R_2 = 6.00\ \Omega$, and $R_3 = 5.00\ \Omega$. The battery has negligible internal resistance. The current I_2 through R_2 is 4.00 A. (a) What are the currents I_1 and I_3? (b) What is the emf of the battery?

Figure **E26.11**

26.12 •• In Fig. E26.11 the battery has emf 25.0 V and negligible internal resistance. $R_1 = 5.00\ \Omega$. The current through R_1 is 1.50 A and the current through $R_3 = 4.50$ A. What are the resistances R_2 and R_3?

26.13 • Compute the equivalent resistance of the network in Fig. E26.13, and find the current in each resistor. The battery has negligible internal resistance.

Figure **E26.13**

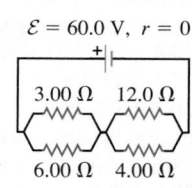

$\mathcal{E} = 60.0$ V, $r = 0$

3.00 Ω 12.0 Ω

6.00 Ω 4.00 Ω

26.14 • Compute the equivalent resistance of the network in Fig. E26.14, and find the current in each resistor. The battery has negligible internal resistance.

Figure **E26.14**

$\mathcal{E} = 48.0$ V, $r = 0$

1.00 Ω 3.00 Ω

7.00 Ω 5.00 Ω

26.15 • In the circuit of Fig. E26.15, each resistor represents a light bulb. Let $R_1 = R_2 = R_3 = R_4 = 4.50\ \Omega$ and $\mathcal{E} = 9.00$ V. (a) Find the current in each bulb. (b) Find the power dissipated in each bulb. Which bulb or bulbs glow the brightest? (c) Bulb R_4 is now removed from the circuit, leaving a break in the wire at its position. Now what is the current in each of the remaining bulbs R_1, R_2, and R_3? (d) With bulb R_4 removed, what is the power dissipated in each of the remaining bulbs? (e) Which light bulb(s) glow brighter as a result of removing R_4? Which bulb(s) glow less brightly? Discuss why there are different effects on different bulbs.

Figure **E26.15**

26.16 • Consider the circuit shown in Fig. E26.16. The current through the 6.00-Ω resistor is 4.00 A, in the direction shown. What are the currents through the 25.0-Ω and 20.0-Ω resistors?

Figure **E26.16**

4.00 A 6.00 Ω
25.0 Ω
8.00 Ω
20.0 Ω
$\mathcal{E}$

26.17 • In the circuit shown in Fig. E26.17, the voltage across the 2.00-Ω resistor is 12.0 V. What are the emf of the battery and the current through the 6.00-Ω resistor?

Figure **E26.17**

$\mathcal{E}$
1.00 Ω 2.00 Ω
6.00 Ω

26.18 • **A Three-Way Light Bulb.** A three-way light bulb has three brightness settings (low, medium, and high) but only two filaments. (a) A particular three-way light bulb connected across a 120-V line can dissipate 60 W, 120 W, or 180 W. Describe how the two filaments are arranged in the bulb, and calculate the resistance of each filament. (b) Suppose the filament with the higher resistance burns out. How much power will the bulb dissipate on each of the three brightness settings? What will be the brightness (low, medium, or high) on each setting? (c) Repeat part (b) for the situation in which the filament with the lower resistance burns out.

26.19 •• **Working Late!** You are working late in your electronics shop and find that you need various resistors for a project. But alas, all you have is a big box of 10.0-Ω resistors. Show how you can make each of the following equivalent resistances by a combination of your 10.0-Ω resistors: (a) 35 Ω, (b) 1.0 Ω, (c) 3.33 Ω, (d) 7.5 Ω.

26.20 • In the circuit shown in Fig. E26.20, the rate at which R_1 is dissipating electrical energy is 20.0 W. (a) Find R_1 and R_2. (b) What is the emf of the battery? (c) Find the current through both R_2 and the 10.0-Ω resistor. (d) Calculate the total electrical power consumption in all the resistors and the electrical power delivered

Figure **E26.20**

3.50 A

$\mathcal{E}$ 10.0 Ω R_2 R_1

2.00 A

by the battery. Show that your results are consistent with conservation of energy.

26.21 • **Light Bulbs in Series and in Parallel.** Two light bulbs have resistances of 400 Ω and 800 Ω. If the two light bulbs are connected in series across a 120-V line, find (a) the current through each bulb; (b) the power dissipated in each bulb; (c) the total power dissipated in both bulbs. The two light bulbs are now connected in parallel across the 120-V line. Find (d) the current through each bulb; (e) the power dissipated in each bulb; (f) the total power dissipated in both bulbs. (g) In each situation, which of the two bulbs glows the brightest? (h) In which situation is there a greater total light output from both bulbs combined?

26.22 • **Light Bulbs in Series.** A 60-W, 120-V light bulb and a 200-W, 120-V light bulb are connected in series across a 240-V line. Assume that the resistance of each bulb does not vary with current. (*Note:* This description of a light bulb gives the power it dissipates when connected to the stated potential difference; that is, a 25-W, 120-V light bulb dissipates 25 W when connected to a 120-V line.) (a) Find the current through the bulbs. (b) Find the power dissipated in each bulb. (c) One bulb burns out very quickly. Which one? Why?

26.23 •• **CP** In the circuit in Fig. E26.23, a 20.0-Ω resistor is inside 100 g of pure water that is surrounded by insulating styrofoam. If the water is initially at 10.0°C, how long will it take for its temperature to rise to 58.0°C?

Figure **E26.23**

10.0 Ω 10.0 Ω
20.0 Ω 10.0 Ω 10.0 Ω
Water 5.0 Ω 5.0 Ω
30.0 V 5.0 Ω

Section 26.2 Kirchhoff's Rules

26.24 •• The batteries shown in the circuit in Fig. E26.24 have negligibly small internal resistances. Find the current through (a) the 30.0-Ω resistor; (b) the 20.0-Ω resistor; (c) the 10.0-V battery.

Figure **E26.24**

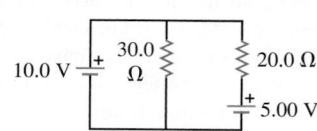

10.0 V 30.0 Ω 20.0 Ω
5.00 V

26.25 • In the circuit shown in Fig. E26.25 find (a) the current in resistor R; (b) the resistance R; (c) the unknown emf $\mathcal{E}$. (d) If the circuit is broken at point x, what is the current in resistor R?

26.26 • Find the emfs $\mathcal{E}_1$ and $\mathcal{E}_2$ in the circuit of Fig. E26.26, and find the potential difference of point b relative to point a.

Figure **E26.25**

28.0 V R
$\mathcal{E}$ 4.00 A
x 6.00 Ω
6.00 A
3.00 Ω

Figure **E26.26**

1.00 Ω 20.0 V
6.00 Ω
1.00 A
4.00 Ω
1.00 Ω $\mathcal{E}_1$
a
2.00 A
1.00 Ω $\mathcal{E}_2$
2.00 Ω
b

26.27 • In the circuit shown in Fig. E26.27, find (a) the current in the 3.00-Ω resistor; (b) the unknown emfs $\mathcal{E}_1$ and $\mathcal{E}_2$; (c) the resistance R. Note that three currents are given.

Figure **E26.27**

26.28 •• In the circuit shown in Fig. E26.28, find (a) the current in each branch and (b) the potential difference V_{ab} of point a relative to point b.

Figure **E26.28**

26.29 • The 10.00-V battery in Fig. E26.28 is removed from the circuit and reinserted with the opposite polarity, so that its positive terminal is now next to point a. The rest of the circuit is as shown in the figure. Find (a) the current in each branch and (b) the potential difference V_{ab} of point a relative to point b.

26.30 • The 5.00-V battery in Fig. E26.28 is removed from the circuit and replaced by a 20.00-V battery, with its negative terminal next to point b. The rest of the circuit is as shown in the figure. Find (a) the current in each branch and (b) the potential difference V_{ab} of point a relative to point b.

26.31 •• In the circuit shown in Fig. E26.31 the batteries have negligible internal resistance and the meters are both idealized. With the switch S open, the voltmeter reads 15.0 V. (a) Find the emf $\mathcal{E}$ of the battery. (b) What will the ammeter read when the switch is closed?

Figure **E26.31**

26.32 •• In the circuit shown in Fig. E26.32 both batteries have insignificant internal resistance and the idealized ammeter reads 1.50 A in the direction shown. Find the emf $\mathcal{E}$ of the battery. Is the polarity shown correct?

Figure **E26.32**

26.33 • In the circuit shown in Fig. E26.33 all meters are idealized and the batteries have no appreciable internal resistance. (a) Find the reading of the voltmeter with the switch S open. Which point is at a higher potential: a or b? (b) With the switch closed, find the reading of the voltmeter and the ammeter. Which way (up or down) does the current flow through the switch?

Figure **E26.33**

26.34 •• In the circuit shown in Fig. E26.34, the 6.0-Ω resistor is consuming energy at a rate of 24 J/s when the current through it flows as shown. (a) Find the current through the ammeter A. (b) What are the polarity and emf $\mathcal{E}$ of the battery, assuming it has negligible internal resistance?

Figure **E26.34**

Section 26.3 Electrical Measuring Instruments

26.35 • The resistance of a galvanometer coil is 25.0 Ω, and the current required for full-scale deflection is 500 μA. (a) Show in a diagram how to convert the galvanometer to an ammeter reading 20.0 mA full scale, and compute the shunt resistance. (b) Show how to convert the galvanometer to a voltmeter reading 500 mV full scale, and compute the series resistance.

26.36 • The resistance of the coil of a pivoted-coil galvanometer is 9.36 Ω, and a current of 0.0224 A causes it to deflect full scale. We want to convert this galvanometer to an ammeter reading 20.0 A full scale. The only shunt available has a resistance of 0.0250 Ω. What resistance R must be connected in series with the coil (Fig. E26.36)?

Figure **E26.36**

26.37 • A circuit consists of a series combination of 6.00-kΩ and 5.00-kΩ resistors connected across a 50.0-V battery having negligible internal resistance. You want to measure the true potential difference (that is, the potential difference without the meter present) across the 5.00-kΩ resistor using a voltmeter having an internal resistance of 10.0 kΩ. (a) What potential difference does the voltmeter measure across the 5.00-kΩ resistor? (b) What is the *true* potential difference across this resistor when the meter is not present? (c) By what percentage is the voltmeter reading in error from the true potential difference?

26.38 • A galvanometer having a resistance of 25.0 Ω has a 1.00-Ω shunt resistance installed to convert it to an ammeter. It is then used to measure the current in a circuit consisting of a 15.0-Ω resistor connected across the terminals of a 25.0-V battery having no appreciable internal resistance. (a) What current does the ammeter measure? (b) What should be the *true* current in the circuit (that is, the current without the ammeter present)? (c) By what percentage is the ammeter reading in error from the *true* current?

26.39 • In the ohmmeter in Fig. E26.39 M is a 2.50-mA meter of resistance 65.0 Ω. (A 2.50-mA meter deflects full scale when the current through it is 2.50 mA.) The battery B has an emf of 1.52 V and negligible internal resistance. R is chosen so that when the terminals a and b are shorted ($R_x = 0$), the meter reads full scale. When a and b are open ($R_x = \infty$), the meter reads zero. (a) What is the resistance of the resistor R? (b) What current indicates a resistance R_x of 200 Ω? (c) What values of R_x correspond to meter deflections of $\frac{1}{4}$, $\frac{1}{2}$, and $\frac{3}{4}$ of full scale if the deflection is proportional to the current through the galvanometer?

Figure **E26.39**

Section 26.4 *R-C* Circuits

26.40 • A 4.60-μF capacitor that is initially uncharged is connected in series with a 7.50-kΩ resistor and an emf source with $\mathcal{E} = 245$ V and negligible internal resistance. Just after the circuit is completed, what are (a) the voltage drop across the capacitor;

(b) the voltage drop across the resistor; (c) the charge on the capacitor; (d) the current through the resistor? (e) A long time after the circuit is completed (after many time constants) what are the values of the quantities in parts (a)–(d)?

26.41 • A capacitor is charged to a potential of 12.0 V and is then connected to a voltmeter having an internal resistance of 3.40 MΩ. After a time of 4.00 s the voltmeter reads 3.0 V. What are (a) the capacitance and (b) the time constant of the circuit?

26.42 • A 12.4-μF capacitor is connected through a 0.895-MΩ resistor to a constant potential difference of 60.0 V. (a) Compute the charge on the capacitor at the following times after the connections are made: 0, 5.0 s, 10.0 s, 20.0 s, and 100.0 s. (b) Compute the charging currents at the same instants. (c) Graph the results of parts (a) and (b) for t between 0 and 20 s.

26.43 •• CP In the circuit shown in Fig. E26.43 both capacitors are initially charged to 45.0 V. (a) How long after closing the switch S will the potential across each capacitor be reduced to 10.0 V, and (b) what will be the current at that time?

Figure **E26.43**

26.44 • A resistor and a capacitor are connected in series to an emf source. The time constant for the circuit is 0.870 s. (a) A second capacitor, identical to the first, is added in series. What is the time constant for this new circuit? (b) In the original circuit a second capacitor, identical to the first, is connected in parallel with the first capacitor. What is the time constant for this new circuit?

26.45 • An emf source with $\mathcal{E} = 120$ V, a resistor with $R = 80.0 \ \Omega$, and a capacitor with $C = 4.00 \ \mu$F are connected in series. As the capacitor charges, when the current in the resistor is 0.900 A, what is the magnitude of the charge on each plate of the capacitor?

26.46 • A 1.50-μF capacitor is charging through a 12.0-Ω resistor using a 10.0-V battery. What will be the current when the capacitor has acquired $\frac{1}{4}$ of its maximum charge? Will it be $\frac{1}{4}$ of the maximum current?

26.47 •• CP In the circuit shown in Fig. E26.47 each capacitor initially has a charge of magnitude 3.50 nC on its plates. After the switch S is closed, what will be the current in the circuit at the instant that the capacitors have lost 80.0% of their initial stored energy?

Figure **E26.47**

26.48 • A 12.0-μF capacitor is charged to a potential of 50.0 V and then discharged through a 175-Ω resistor. How long does it take the capacitor to lose (a) half of its charge and (b) half of its stored energy?

26.49 • In the circuit in Fig. E26.49 the capacitors are all initially uncharged, the battery has no internal resistance, and the ammeter is idealized. Find the reading of the ammeter (a) just after the switch S is closed and (b) after the switch has been closed for a very long time.

Figure **E26.49**

26.50 • In the circuit shown in Fig. E26.50, $C = 5.90 \ \mu$F, $\mathcal{E} = 28.0$ V, and the emf has negligible resistance. Initially the capacitor is uncharged and the switch S is in position 1. The switch is then moved to position 2, so that the capacitor begins to charge. (a) What will be the charge on the capacitor a long time after the switch is moved to position 2? (b) After the switch has been in position 2 for 3.00 ms, the charge on the capacitor is measured to be 110 μC. What is the value of the resistance R? (c) How long after the switch is moved to position 2 will the charge on the capacitor be equal to 99.0% of the final value found in part (a)?

Figure **E26.50**

26.51 • A capacitor with $C = 1.50 \times 10^{-5}$ F is connected as shown in Fig. E26.50 with a resistor with $R = 980 \ \Omega$ and an emf source with $\mathcal{E} = 18.0$ V and negligible internal resistance. Initially the capacitor is uncharged and the switch S is in position 1. The switch is then moved to position 2, so that the capacitor begins to charge. After the switch has been in position 2 for 10.0 ms, the switch is moved back to position 1 so that the capacitor begins to discharge. (a) Compute the charge on the capacitor just *before* the switch is thrown from position 2 back to position 1. (b) Compute the voltage drops across the resistor and across the capacitor at the instant described in part (a). (c) Compute the voltage drops across the resistor and across the capacitor just *after* the switch is thrown from position 2 back to position 1. (d) Compute the charge on the capacitor 10.0 ms after the switch is thrown from position 2 back to position 1.

Section 26.5 Power Distribution Systems

26.52 • The heating element of an electric dryer is rated at 4.1 kW when connected to a 240-V line. (a) What is the current in the heating element? Is 12-gauge wire large enough to supply this current? (b) What is the resistance of the dryer's heating element at its operating temperature? (c) At 11 cents per kWh, how much does it cost per hour to operate the dryer?

26.53 • A 1500-W electric heater is plugged into the outlet of a 120-V circuit that has a 20-A circuit breaker. You plug an electric hair dryer into the same outlet. The hair dryer has power settings of 600 W, 900 W, 1200 W, and 1500 W. You start with the hair dryer on the 600-W setting and increase the power setting until the circuit breaker trips. What power setting caused the breaker to trip?

26.54 • CP The heating element of an electric stove consists of a heater wire embedded within an electrically insulating material, which in turn is inside a metal casing. The heater wire has a resistance of 20 Ω at room temperature (23.0°C) and a temperature coefficient of resistivity $\alpha = 2.8 \times 10^{-3} (\text{C}°)^{-1}$. The heating element operates from a 120-V line. (a) When the heating element is first turned on, what current does it draw and what electrical power does it dissipate? (b) When the heating element has reached an operating temperature of 280°C (536°F), what current does it draw and what electrical power does it dissipate?

PROBLEMS

26.55 •• In Fig. P26.55, the battery has negligible internal resistance and $\mathcal{E} = 48.0$ V. $R_1 = R_2 = 4.00 \ \Omega$ and $R_4 = 3.00 \ \Omega$. What must the resistance R_3 be for the resistor network to dissipate electrical energy at a rate of 295 W?

Figure **P26.55**

26.56 • A 400-Ω, 2.4-W resistor is needed, but only several 400-Ω, 1.2-W resistors are available (see Exercise 26.10). (a) What two different combinations of the available units give the required resistance and power rating? (b) For each of the resistor networks from part (a), what power is dissipated in each resistor when 2.4 W is dissipated by the combination?

26.57 • **CP** A 20.0-m-long cable consists of a solid-inner, cylindrical, nickel core 10.0 cm in diameter surrounded by a solid-outer cylindrical shell of copper 10.0 cm in inside diameter and 20.0 cm in outside diameter. The resistivity of nickel is $7.8 \times 10^{-8} \, \Omega \cdot m$. (a) What is the resistance of this cable? (b) If we think of this cable as a single material, what is its equivalent resistivity?

26.58 • Two identical 3.00-Ω wires are laid side by side and soldered together so they touch each other for half of their lengths. What is the equivalent resistance of this combination?

26.59 • The two identical light bulbs in Example 26.2 (Section 26.1) are connected in parallel to a different source, one with $\mathcal{E} = 8.0$ V and internal resistance 0.8 Ω. Each light bulb has a resistance $R = 2.0 \, \Omega$ (assumed independent of the current through the bulb). (a) Find the current through each bulb, the potential difference across each bulb, and the power delivered to each bulb. (b) Suppose one of the bulbs burns out, so that its filament breaks and current no longer flows through it. Find the power delivered to the remaining bulb. Does the remaining bulb glow more or less brightly after the other bulb burns out than before?

26.60 •• Each of the three resistors in Fig. P26.60 has a resistance of 2.4 Ω and can dissipate a maximum of 48 W without becoming excessively heated. What is the maximum power the circuit can dissipate?

Figure **P26.60**

26.61 • If an ohmmeter is connected between points *a* and *b* in each of the circuits shown in Fig. P26.61, what will it read?

Figure **P26.61**

(a) *a* 100.0 Ω *b*

50.0 Ω
75.0 Ω 25.0 Ω
40.0 Ω 50.0 Ω

(b) 7.00 Ω *a* 10.0 Ω *b*

60.0 Ω
20.0 Ω 30.0 Ω
45.0 Ω

26.62 •• **CP** For the circuit shown in Fig. P26.62 a 20.0-Ω resistor is embedded in a large block of ice at 0.00°C, and the battery has negligible internal resistance. At what rate (in g/s) is this circuit melting the ice? (The latent heat of fusion for ice is 3.34×10^5 J/kg.)

Figure **P26.62**

Figure **P26.63**

5.00 Ω   8.00 Ω
I_2 I_1
12.00   1.00   1.00   9.00
V   Ω I_3 Ω   V
10.00 Ω

26.63 • Calculate the three currents I_1, I_2, and I_3 indicated in the circuit diagram shown in Fig. P26.63.

26.64 ••• What must the emf $\mathcal{E}$ in Fig. P26.64 be in order for the current through the 7.00-Ω resistor to be 1.80 A? Each emf source has negligible internal resistance.

Figure **P26.64**

24.0 V $\mathcal{E}$
3.00 Ω   2.00 Ω   7.00 Ω

Figure **P26.65**

20.0 V   5.00 Ω
2.00 Ω   4.00 Ω   36.0 V
14.0 V

26.65 • Find the current through each of the three resistors of the circuit shown in Fig. P26.65. The emf sources have negligible internal resistance.

26.66 • (a) Find the current through the battery and each resistor in the circuit shown in Fig. P26.66. (b) What is the equivalent resistance of the resistor network?

Figure **P26.66**

$R_1 = 1.00 \, \Omega$ $R_2 = 2.00 \, \Omega$
$R_3 = 1.00 \, \Omega$
14.0 V
$R_4 = 2.00 \, \Omega$ $R_5 = 1.00 \, \Omega$

26.67 •• (a) Find the potential of point *a* with respect to point *b* in Fig. P26.67. (b) If points *a* and *b* are connected by a wire with negligible resistance, find the current in the 12.0-V battery.

Figure **P26.67**

1.00 Ω 12.0 V
2.00 Ω   *a*   *b* 1.00 Ω 10.0 V   1.00 Ω
2.00 Ω   1.00 Ω 8.0 V 3.00 Ω   2.00 Ω

26.68 •• Consider the circuit shown in Fig. P26.68. (a) What must the emf $\mathcal{E}$ of the battery be in order for a current of 2.00 A to flow through the 5.00-V battery as shown? Is the polarity of the battery correct as shown? (b) How long does it take for 60.0 J of thermal energy to be produced in the 10.0-Ω resistor?

Figure **P26.68**

10.0 Ω   20.0 Ω
60.0 Ω
2.00 A   60.0 Ω
30.0 Ω   5.00 Ω
5.0 V
5.0 Ω   15.0 Ω $\mathcal{E}$ 10.0 V
20.0 Ω

26.69 •• **CP** A 1.00-km cable having a cross-sectional area of 0.500 cm² is to be constructed out of equal lengths of copper and aluminum. This could be accomplished either by making a 0.50-km cable of each one and welding them together end to end or by making two parallel 1.00-km cables, one of each metal (Fig. P26.69). Calculate the resistance of the 1.00-km cable for both designs to see which one provides the least resistance.

Figure **P26.69**

26.70 ••• In the circuit shown in Fig. P26.70 all the resistors are rated at a maximum power of 2.00 W. What is the maximum emf $\mathcal{E}$ that the battery can have without burning up any of the resistors?

Figure **P26.70**

26.71 • In the circuit shown in Fig. P26.71, the current in the 20.0-V battery is 5.00 A in the direction shown and the voltage across the 8.00-Ω resistor is 16.0 V, with the lower end of the resistor at higher potential. Find (a) the emf (including its polarity) of the battery X; (b) the current I through the 200.0-V battery (including its direction); (c) the resistance R.

Figure **P26.71**

26.72 •• Three identical resistors are connected in series. When a certain potential difference is applied across the combination, the total power dissipated is 36 W. What power would be dissipated if the three resistors were connected in parallel across the same potential difference?

26.73 • A resistor R_1 consumes electrical power P_1 when connected to an emf $\mathcal{E}$. When resistor R_2 is connected to the same emf, it consumes electrical power P_2. In terms of P_1 and P_2, what is the total electrical power consumed when they are both connected to this emf source (a) in parallel and (b) in series?

26.74 • The capacitor in Fig. P26.74 is initially uncharged. The switch is closed at $t = 0$. (a) Immediately after the switch is closed, what is the current through each resistor? (b) What is the final charge on the capacitor?

Figure **P26.74**

$R_1 = 8.00 \ \Omega$

$+ | \mathcal{E} = 42.0 \ \text{V}$
$R_2 =$
$6.00 \ \Omega$

$R_3 = 3.00 \ \Omega$

$C = 4.00 \ \mu\text{F}$

26.75 •• A 2.00-μF capacitor that is initially uncharged is connected in series with a 6.00-kΩ resistor and an emf source with $\mathcal{E} = 90.0$ V and negligible internal resistance. The circuit is completed at $t = 0$. (a) Just after the circuit is completed, what is the rate at which electrical energy is being dissipated in the resistor? (b) At what value of t is the rate at which electrical energy is being dissipated in the resistor equal to the rate at which electrical energy is being stored in the capacitor? (c) At the time calculated in part (b), what is the rate at which electrical energy is being dissipated in the resistor?

26.76 •• A 6.00-μF capacitor that is initially uncharged is connected in series with a 5.00-Ω resistor and an emf source with $\mathcal{E} = 50.0$ V and negligible internal resistance. At the instant when the resistor is dissipating electrical energy at a rate of 250 W, how much energy has been stored in the capacitor?

26.77 • Figure P26.77 employs a convention often used in circuit diagrams. The battery (or other power supply) is not shown explicitly. It is understood that the point at the top, labeled "36.0 V," is connected to the positive terminal of a 36.0-V battery having negligible internal resistance, and that the "ground" symbol at the bottom is connected to the negative terminal of the battery. The circuit is completed through the battery, even though it is not shown on the diagram. (a) What is the potential difference V_{ab}, the potential of point a relative to point b, when the switch S is open? (b) What is the current through switch S when it is closed? (c) What is the equivalent resistance when switch S is closed?

Figure **P26.77**

$V = 36.0$ V

$6.00 \ \Omega$ 3.00 Ω 3.00 Ω

a ——S—— b

$3.00 \ \Omega$ 6.00 Ω

26.78 • (See Problem 26.77.) (a) What is the potential of point a with respect to point b in Fig. P26.78 when switch S is open? (b) Which point, a or b, is at the higher potential? (c) What is the final potential of point b with respect to ground when switch S is closed? (d) How much does the charge on each capacitor change when S is closed?

Figure **P26.78**

$V = 18.0$ V

$6.00 \ \Omega$ 6.00 μF

a ——S—— b

$3.00 \ \mu$F   3.00 Ω

26.79 • Point a in Fig. P26.79 is maintained at a constant potential of 400 V above ground. (See Problem 26.77.) (a) What is the reading of a voltmeter with the proper range and with resistance $5.00 \times 10^4 \ \Omega$ when connected between point b and ground? (b) What is the reading of a voltmeter with resistance $5.00 \times 10^6 \ \Omega$? (c) What is the reading of a voltmeter with infinite resistance?

Figure **P26.79**

100 kΩ 200 kΩ

a ——— b ———

26.80 •• A 150-V voltmeter has a resistance of 30,000 Ω. When connected in series with a large resistance R across a 110-V line, the meter reads 74 V. Find the resistance R.

26.81 •• **The Wheatstone Bridge.** The circuit shown in Fig. P26.81, called a *Wheatstone bridge,* is used to determine the value of an unknown resistor X by comparison with three resistors M, N, and P whose resistances can be varied. For each setting, the resistance of each resistor is precisely known. With switches K_1 and K_2 closed,

Figure **P26.81**

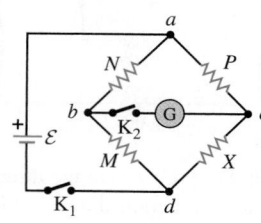

these resistors are varied until the current in the galvanometer G is zero; the bridge is then said to be *balanced.* (a) Show that under this condition the unknown resistance is given by $X = MP/N$. (This method permits very high precision in comparing resistors.) (b) If the galvanometer G shows zero deflection when $M = 850.0\ \Omega$, $N = 15.00\ \Omega$, and $P = 33.48\ \Omega$, what is the unknown resistance X?

26.82 • A 2.36-μF capacitor that is initially uncharged is connected in series with a 5.86-Ω resistor and an emf source with $\mathcal{E} = 120$ V and negligible internal resistance. (a) Just after the connection is made, what are (i) the rate at which electrical energy is being dissipated in the resistor; (ii) the rate at which the electrical energy stored in the capacitor is increasing; (iii) the electrical power output of the source? How do the answers to parts (i), (ii), and (iii) compare? (b) Answer the same questions as in part (a) at a long time after the connection is made. (c) Answer the same questions as in part (a) at the instant when the charge on the capacitor is one-half its final value.

26.83 • A 224-Ω resistor and a 589-Ω resistor are connected in series across a 90.0-V line. (a) What is the voltage across each resistor? (b) A voltmeter connected across the 224-Ω resistor reads 23.8 V. Find the voltmeter resistance. (c) Find the reading of the same voltmeter if it is connected across the 589-Ω resistor. (d) The readings on this voltmeter are lower than the "true" voltages (that is, without the voltmeter present). Would it be possible to design a voltmeter that gave readings *higher* than the "true" voltages? Explain.

26.84 • A resistor with $R = 850\ \Omega$ is connected to the plates of a charged capacitor with capacitance $C = 4.62\ \mu$F. Just before the connection is made, the charge on the capacitor is 6.90 mC. (a) What is the energy initially stored in the capacitor? (b) What is the electrical power dissipated in the resistor just after the connection is made? (c) What is the electrical power dissipated in the resistor at the instant when the energy stored in the capacitor has decreased to half the value calculated in part (a)?

26.85 • A capacitor that is initially uncharged is connected in series with a resistor and an emf source with $\mathcal{E} = 110$ V and negligible internal resistance. Just after the circuit is completed, the current through the resistor is 6.5×10^{-5} A. The time constant for the circuit is 5.2 s. What are the resistance of the resistor and the capacitance of the capacitor?

26.86 ••• An R-C circuit has a time constant RC. (a) If the circuit is discharging, how long will it take for its stored energy to be reduced to $1/e$ of its initial value? (b) If it is charging, how long will it take for the stored energy to reach $1/e$ of its maximum value?

26.87 • Strictly speaking, Eq. (26.16) implies that an *infinite* amount of time is required to discharge a capacitor completely. Yet for practical purposes, a capacitor may be considered to be fully discharged after a finite length of time. To be specific, consider a capacitor with capacitance C connected to a resistor R to be fully discharged if its charge q differs from zero by no more than the charge of one electron. (a) Calculate the time required to reach this state if $C = 0.920\ \mu$F, $R = 670$ kΩ, and $Q_0 = 7.00\ \mu$C. How many time constants is this? (b) For a given Q_0, is the time required to reach this state always the same number of time constants, independent of the values of C and R? Why or why not?

26.88 • **CALC** The current in a charging capacitor is given by Eq. (26.13). (a) The instantaneous power supplied by the battery is $\mathcal{E}i$. Integrate this to find the total energy supplied by the battery. (b) The instantaneous power dissipated in the resistor is i^2R. Integrate this to find the total energy dissipated in the resistor. (c) Find the final energy stored in the capacitor, and show that this equals the total energy supplied by the battery less the energy dissipated in

the resistor, as obtained in parts (a) and (b). (d) What fraction of the energy supplied by the battery is stored in the capacitor? How does this fraction depend on R?

26.89 ••• **CALC** (a) Using Eq. (26.17) for the current in a discharging capacitor, derive an expression for the instantaneous power $P = i^2R$ dissipated in the resistor. (b) Integrate the expression for P to find the total energy dissipated in the resistor, and show that this is equal to the total energy initially stored in the capacitor.

CHALLENGE PROBLEMS

26.90 ••• **A Capacitor Burglar Alarm.** The capacitance of a capacitor can be affected by dielectric material that, although not inside the capacitor, is near enough to the capacitor to be polarized by the fringing electric field that exists near a charged capacitor. This effect is usually of the order of picofarads (pF), but it can be used with appropriate electronic circuitry to detect a change in the dielectric material surrounding the capacitor. Such a dielectric material might be the human body, and the effect described above might be used in the design of a burglar alarm. Consider the simplified circuit shown in Fig. P26.90. The voltage source has emf $\mathcal{E} = 1000$ V, and the capacitor has capacitance $C = 10.0$ pF. The electronic circuitry for detecting the current, represented as an ammeter in the diagram, has negligible resistance and is capable of detecting a current that persists at a level of at least 1.00 μA for at least 200 μs after the capacitance has changed abruptly from C to C'. The burglar alarm is designed to be activated if the capacitance changes by 10%. (a) Determine the charge on the 10.0-pF capacitor when it is fully charged. (b) If the capacitor is fully charged before the intruder is detected, assuming that the time taken for the capacitance to change by 10% is short enough to be ignored, derive an equation that expresses the current through the resistor R as a function of the time t since the capacitance has changed. (c) Determine the range of values of the resistance R that will meet the design specifications of the burglar alarm. What happens if R is too small? Too large? (*Hint:* You will not be able to solve this part analytically but must use numerical methods. Express R as a logarithmic function of R plus known quantities. Use a trial value of R and calculate from the expression a new value. Continue to do this until the input and output values of R agree to within three significant figures.)

Figure **P26.90**

26.91 ••• **An Infinite Network.** As shown in Fig. P26.91, a network of resistors of resistances R_1 and R_2 extends to infinity toward the right. Prove that the total resistance R_T of the infinite network is equal to

$$R_T = R_1 + \sqrt{R_1^2 + 2R_1R_2}$$

Figure **P26.91**

(*Hint:* Since the network is infinite, the resistance of the network to the right of points c and d is also equal to R_T.)

26.92 ••• Suppose a resistor R lies along each edge of a cube (12 resistors in all) with connections at the corners. Find the equivalent resistance between two diagonally opposite corners of the cube (points a and b in Fig. P26.92).

Figure **P26.92**

26.93 ••• **BIO Attenuator Chains and Axons.** The infinite network of resistors shown in Fig. P26.91 is known as an *attenuator chain,* since this chain of resistors causes the potential difference between the upper and lower wires to decrease, or attenuate, along the length of the chain. (a) Show that if the potential difference between the points a and b in Fig. 26.91 is V_{ab}, then the potential difference between points c and d is $V_{cd} = V_{ab}/(1 + \beta)$, where $\beta = 2R_1(R_T + R_2)/R_T R_2$ and R_T, the total resistance of the network, is given in Challenge Problem 26.91. (See the hint given in that problem.) (b) If the potential difference between terminals a and b at the left end of the infinite network is V_0, show that the potential difference between the upper and lower wires n segments from the left end is $V_n = V_0/(1 + \beta)^n$. If $R_1 = R_2$, how many segments are needed to decrease the potential difference V_n to less than 1.0% of V_0? (c) An infinite attenuator chain provides a model of the propagation of a voltage pulse along a nerve fiber, or axon. Each segment of the network in Fig. P26.91 represents a short segment of the axon of length Δx. The resistors R_1 represent the resistance of the fluid inside and outside the membrane wall of the axon. The resistance of the membrane to current flowing through the wall is represented by R_2. For an axon segment of length $\Delta x = 1.0\ \mu m$, $R_1 = 6.4 \times 10^3\ \Omega$ and $R_2 = 8.0 \times 10^8\ \Omega$ (the membrane wall

is a good insulator). Calculate the total resistance R_T and β for an infinitely long axon. (This is a good approximation, since the length of an axon is much greater than its width; the largest axons in the human nervous system are longer than 1 m but only about 10^{-7} m in radius.) (d) By what fraction does the potential difference between the inside and outside of the axon decrease over a distance of 2.0 mm? (e) The attenuation of the potential difference calculated in part (d) shows that the axon cannot simply be a passive, current-carrying electrical cable; the potential difference must periodically be reinforced along the axon's length. This reinforcement mechanism is slow, so a signal propagates along the axon at only about 30 m/s. In situations where faster response is required, axons are covered with a segmented sheath of fatty myelin. The segments are about 2 mm long, separated by gaps called the *nodes of Ranvier.* The myelin increases the resistance of a 1.0-μm-long segment of the membrane to $R_2 = 3.3 \times 10^{12}\ \Omega$. For such a myelinated axon, by what fraction does the potential difference between the inside and outside of the axon decrease over the distance from one node of Ranvier to the next? This smaller attenuation means the propagation speed is increased.

Answers

Chapter Opening Question **?**

The potential difference V is the same across resistors connected in parallel. However, there is a different current I through each resistor if the resistances R are different: $I = V/R$.

Test Your Understanding Questions

26.1 Answer: (a), (c), (d), (b) Here's why: The three resistors in Fig. 26.1a are in series, so $R_{eq} = R + R + R = 3R$. In Fig. 26.1b the three resistors are in parallel, so $1/R_{eq} = 1/R + 1/R + 1/R = 3/R$ and $R_{eq} = R/3$. In Fig. 26.1c the second and third resistors are in parallel, so their equivalent resistance R_{23} is given by $1/R_{23} = 1/R + 1/R = 2/R$; hence $R_{23} = R/2$. This combination is in series with the first resistor, so the three resistors together have equivalent resistance $R_{eq} = R + R/2 = 3R/2$. In Fig. 26.1d the second and third resistors are in series, so their equivalent resistance is $R_{23} = R + R = 2R$. This combination is in parallel with the first resistor, so the equivalent resistance of the three-resistor combination is given by $1/R_{eq} = 1/R + 1/2R = 3/2R$. Hence $R_{eq} = 2R/3$.

26.2 Answer: loop *cbdac* Equation (2) minus Eq. (1) gives $-I_2(1\ \Omega) - (I_2 + I_3)(2\ \Omega) + (I_1 - I_3)(1\ \Omega) + I_1(1\ \Omega) = 0$. We can obtain this equation by applying the loop rule around the path from c to b to d to a to c in Fig. 26.12. This isn't a new equa-

tion, so it would not have helped with the solution of Example 26.6.

26.3 Answers: (a) (ii), (b) (iii) An ammeter must always be placed in series with the circuit element of interest, and a voltmeter must always be placed in parallel. Ideally the ammeter would have zero resistance and the voltmeter would have infinite resistance so that their presence would have no effect on either the resistor current or the voltage. Neither of these idealizations is possible, but the ammeter resistance should be much less than 2 Ω and the voltmeter resistance should be much greater than 2 Ω.

26.4 Answer: (ii) After one time constant, $t = RC$ and the initial charge Q_0 has decreased to $Q_0 e^{-t/RC} = Q_0 e^{-RC/RC} = Q_0 e^{-1} = Q_0/e$. Hence the stored energy has decreased from $Q_0^2/2C$ to $(Q_0/e)^2/2C = Q_0^2/2Ce^2$, a fraction $1/e^2 = 0.135$ of its initial value. This result doesn't depend on the initial value of the energy.

26.5 Answer: no This is a very dangerous thing to do. The circuit breaker will allow currents up to 40 A, double the rated value of the wiring. The amount of power $P = I^2 R$ dissipated in a section of wire can therefore be up to four times the rated value, so the wires could get very warm and start a fire.

Bridging Problem

Answers: (a) 9.39 J **(b)** 2.02×10^4 W **(c)** 4.65×10^{-4} s **(d)** 7.43×10^3 W

MAGNETIC FIELD AND MAGNETIC FORCES

? Magnetic resonance imaging (MRI) makes it possible to see details of soft tissue (such as in the foot shown here) that aren't visible in x-ray images. Yet soft tissue isn't a magnetic material (it's not attracted to a magnet). How does MRI work?

Everybody uses magnetic forces. They are at the heart of electric motors, microwave ovens, loudspeakers, computer printers, and disk drives. The most familiar examples of magnetism are permanent magnets, which attract unmagnetized iron objects and can also attract or repel other magnets. A compass needle aligning itself with the earth's magnetism is an example of this interaction. But the *fundamental* nature of magnetism is the interaction of moving electric charges. Unlike electric forces, which act on electric charges whether they are moving or not, magnetic forces act only on *moving* charges.

We saw in Chapter 21 that the electric force arises in two stages: (1) a charge produces an electric field in the space around it, and (2) a second charge responds to this field. Magnetic forces also arise in two stages. First, a *moving* charge or a collection of moving charges (that is, an electric current) produces a *magnetic* field. Next, a second current or moving charge responds to this magnetic field, and so experiences a magnetic force.

In this chapter we study the second stage in the magnetic interaction—that is, how moving charges and currents *respond* to magnetic fields. In particular, we will see how to calculate magnetic forces and torques, and we will discover why magnets can pick up iron objects like paper clips. In Chapter 28 we will complete our picture of the magnetic interaction by examining how moving charges and currents *produce* magnetic fields.

27.1 Magnetism

Magnetic phenomena were first observed at least 2500 years ago in fragments of magnetized iron ore found near the ancient city of Magnesia (now Manisa, in western Turkey). These fragments were examples of what are now called **permanent magnets;** you probably have several permanent magnets on your refrigerator

27.1 (a) Two bar magnets attract when opposite poles (N and S, or S and N) are next to each other. (b) The bar magnets repel when like poles (N and N, or S and S) are next to each other.

(a) Opposite poles attract.

(b) Like poles repel.

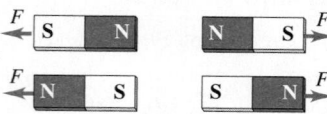

27.2 (a) Either pole of a bar magnet attracts an unmagnetized object that contains iron, such as a nail. (b) A real-life example of this effect.

(a)

(b)

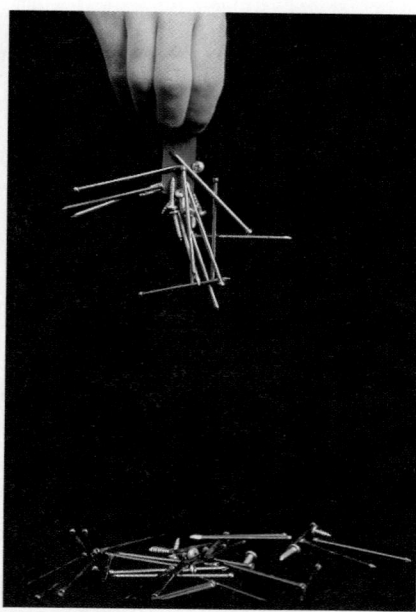

door at home. Permanent magnets were found to exert forces on each other as well as on pieces of iron that were not magnetized. It was discovered that when an iron rod is brought in contact with a natural magnet, the rod also becomes magnetized. When such a rod is floated on water or suspended by a string from its center, it tends to line itself up in a north-south direction. The needle of an ordinary compass is just such a piece of magnetized iron.

Before the relationship of magnetic interactions to moving charges was understood, the interactions of permanent magnets and compass needles were described in terms of *magnetic poles*. If a bar-shaped permanent magnet, or *bar magnet,* is free to rotate, one end points north. This end is called a *north pole* or *N pole;* the other end is a *south pole* or *S pole.* Opposite poles attract each other, and like poles repel each other (Fig. 27.1). An object that contains iron but is not itself magnetized (that is, it shows no tendency to point north or south) is attracted by *either* pole of a permanent magnet (Fig. 27.2). This is the attraction that acts between a magnet and the unmagnetized steel door of a refrigerator. By analogy to electric interactions, we describe the interactions in Figs. 27.1 and 27.2 by saying that a bar magnet sets up a *magnetic field* in the space around it and a second body responds to that field. A compass needle tends to align with the magnetic field at the needle's position.

The earth itself is a magnet. Its north geographic pole is close to a magnetic *south* pole, which is why the north pole of a compass needle points north. The earth's magnetic axis is not quite parallel to its geographic axis (the axis of rotation), so a compass reading deviates somewhat from geographic north. This deviation, which varies with location, is called *magnetic declination* or *magnetic variation.* Also, the magnetic field is not horizontal at most points on the earth's surface; its angle up or down is called *magnetic inclination.* At the magnetic poles the magnetic field is vertical.

Figure 27.3 is a sketch of the earth's magnetic field. The lines, called *magnetic field lines,* show the direction that a compass would point at each location; they are discussed in detail in Section 27.3. The direction of the field at any point can be defined as the direction of the force that the field would exert on a magnetic

27.3 A sketch of the earth's magnetic field. The field, which is caused by currents in the earth's molten core, changes with time; geologic evidence shows that it reverses direction entirely at irregular intervals of 10^4 to 10^6 years.

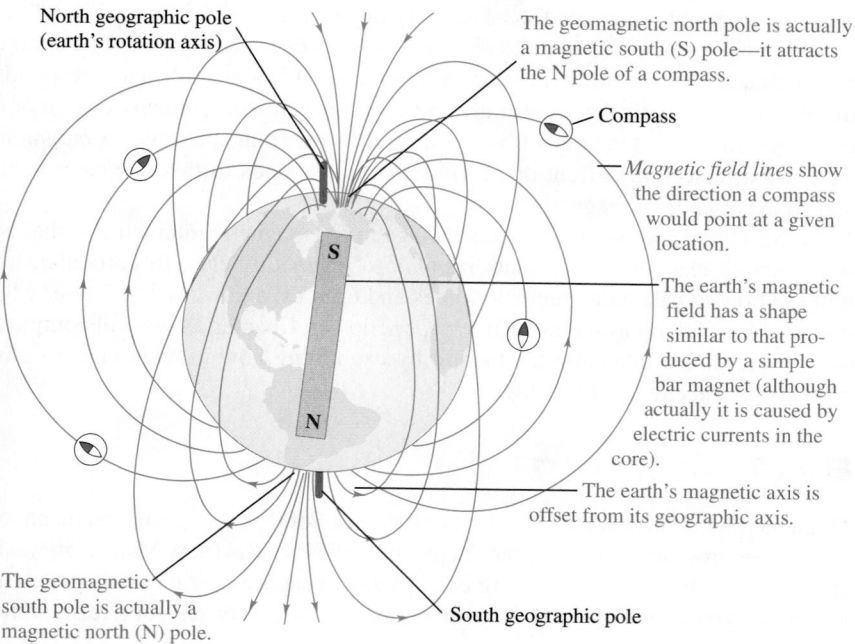

north pole. In Section 27.2 we'll describe a more fundamental way to define the direction and magnitude of a magnetic field.

Magnetic Poles Versus Electric Charge

The concept of magnetic poles may appear similar to that of electric charge, and north and south poles may seem analogous to positive and negative charge. But the analogy can be misleading. While isolated positive and negative charges exist, there is *no* experimental evidence that a single isolated magnetic pole exists; poles always appear in pairs. If a bar magnet is broken in two, each broken end becomes a pole (Fig. 27.4). The existence of an isolated magnetic pole, or **magnetic monopole,** would have sweeping implications for theoretical physics. Extensive searches for magnetic monopoles have been carried out, but so far without success.

The first evidence of the relationship of magnetism to moving charges was discovered in 1820 by the Danish scientist Hans Christian Oersted. He found that a compass needle was deflected by a current-carrying wire, as shown in Fig. 27.5 Similar investigations were carried out in France by André Ampère. A few years later, Michael Faraday in England and Joseph Henry in the United States discovered that moving a magnet near a conducting loop can cause a current in the loop. We now know that the magnetic forces between two bodies shown in Figs. 27.1 and 27.2 are fundamentally due to interactions between moving electrons in the atoms of the bodies. (There are also *electric* interactions between the two bodies, but these are far weaker than the magnetic interactions because the two bodies are electrically neutral.) Inside a magnetized body such as a permanent magnet, there is a *coordinated* motion of certain of the atomic electrons; in an unmagnetized body these motions are not coordinated. (We'll describe these motions further in Section 27.7, and see how the interactions shown in Figs. 27.1 and 27.2 come about.)

Electric and magnetic interactions prove to be intimately connected. Over the next several chapters we will develop the unifying principles of electromagnetism, culminating in the expression of these principles in *Maxwell's equations.* These equations represent the synthesis of electromagnetism, just as Newton's laws of motion are the synthesis of mechanics, and like Newton's laws they represent a towering achievement of the human intellect.

Test Your Understanding of Section 27.1 Suppose you cut off the part of the compass needle shown in Fig. 27.5a that is painted gray. You discard this part, drill a hole in the remaining red part, and place the red part on the pivot at the center of the compass. Will the red part still swing east and west when a current is applied as in Fig. 27.5b? ❙

27.2 Magnetic Field

To introduce the concept of magnetic field properly, let's review our formulation of *electric* interactions in Chapter 21, where we introduced the concept of *electric* field. We represented electric interactions in two steps:

1. A distribution of electric charge at rest creates an electric field $\vec{E}$ in the surrounding space.
2. The electric field exerts a force $\vec{F} = q\vec{E}$ on any other charge q that is present in the field.

We can describe magnetic interactions in a similar way:

1. A moving charge or a current creates a **magnetic field** in the surrounding space (in addition to its *electric* field).
2. The magnetic field exerts a force $\vec{F}$ on any other moving charge or current that is present in the field.

27.4 Breaking a bar magnet. Each piece has a north and south pole, even if the pieces are different sizes. (The smaller the piece, the weaker its magnetism.)

In contrast to electric charges, magnetic poles always come in pairs and can't be isolated.

Breaking a magnet in two ...

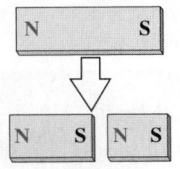

... yields two magnets, not two isolated poles.

27.5 In Oersted's experiment, a compass is placed directly over a horizontal wire (here viewed from above). When the compass is placed directly under the wire, the compass deflection is reversed.

(a)

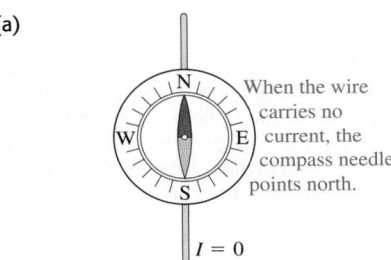

When the wire carries no current, the compass needle points north.

$I = 0$

(b)

When the wire carries a current, the compass needle deflects. The direction of deflection depends on the direction of the current.

27.6 The magnetic force $\vec{F}$ acting on a positive charge q moving with velocity $\vec{v}$ is perpendicular to both $\vec{v}$ and the magnetic field $\vec{B}$. For given values of the speed v and magnetic field strength B, the force is greatest when $\vec{v}$ and $\vec{B}$ are perpendicular.

(a)

A charge moving **parallel** to a magnetic field experiences **zero magnetic force.**

(b)

A charge moving at an angle ϕ to a magnetic field experiences a magnetic force with magnitude $F = |q|v_\perp B = |q|vB \sin \phi$.

$\vec{F}$ is perpendicular to the plane containing $\vec{v}$ and $\vec{B}$.

(c)

A charge moving **perpendicular** to a magnetic field experiences a maximal magnetic force with magnitude $F_{max} = qvB$.

In this chapter we'll concentrate on the *second* aspect of the interaction: Given the presence of a magnetic field, what force does it exert on a moving charge or a current? In Chapter 28 we will come back to the problem of how magnetic fields are *created* by moving charges and currents.

Like electric field, magnetic field is a *vector field*—that is, a vector quantity associated with each point in space. We will use the symbol $\vec{B}$ for magnetic field. At any position the direction of $\vec{B}$ is defined as the direction in which the north pole of a compass needle tends to point. The arrows in Fig. 27.3 suggest the direction of the earth's magnetic field; for any magnet, $\vec{B}$ points out of its north pole and into its south pole.

Magnetic Forces on Moving Charges

There are four key characteristics of the magnetic force on a moving charge. First, its magnitude is proportional to the magnitude of the charge. If a 1-μC charge and a 2-μC charge move through a given magnetic field with the same velocity, experiments show that the force on the 2-μC charge is twice as great as the force on the 1-μC charge. Second, the magnitude of the force is also proportional to the magnitude, or "strength," of the field; if we double the magnitude of the field (for example, by using two identical bar magnets instead of one) without changing the charge or its velocity, the force doubles.

A third characteristic is that the magnetic force depends on the particle's velocity. This is quite different from the electric-field force, which is the same whether the charge is moving or not. A charged particle at rest experiences *no* magnetic force. And fourth, we find by experiment that the magnetic force $\vec{F}$ *does not* have the same direction as the magnetic field $\vec{B}$ but instead is always *perpendicular* to both $\vec{B}$ and the velocity $\vec{v}$. The magnitude F of the force is found to be proportional to the component of $\vec{v}$ perpendicular to the field; when that component is zero (that is, when $\vec{v}$ and $\vec{B}$ are parallel or antiparallel), the force is zero.

Figure 27.6 shows these relationships. The direction of $\vec{F}$ is always perpendicular to the plane containing $\vec{v}$ and $\vec{B}$. Its magnitude is given by

$$F = |q|v_\perp B = |q|vB \sin \phi \qquad (27.1)$$

where $|q|$ is the magnitude of the charge and ϕ is the angle measured from the direction of $\vec{v}$ to the direction of $\vec{B}$, as shown in the figure.

This description does not specify the direction of $\vec{F}$ completely; there are always two directions, opposite to each other, that are both perpendicular to the plane of $\vec{v}$ and $\vec{B}$. To complete the description, we use the same right-hand rule that we used to define the vector product in Section 1.10. (It would be a good idea to review that section before you go on.) Draw the vectors $\vec{v}$ and $\vec{B}$ with their tails together, as in Fig. 27.7a. Imagine turning $\vec{v}$ until it points in the direction of $\vec{B}$ (turning through the smaller of the two possible angles). Wrap the fingers of your right hand around the line perpendicular to the plane of $\vec{v}$ and $\vec{B}$ so that they curl around with the sense of rotation from $\vec{v}$ to $\vec{B}$. Your thumb then points in the direction of the force $\vec{F}$ on a *positive* charge. (Alternatively, the direction of the force $\vec{F}$ on a positive charge is the direction in which a right-hand-thread screw would advance if turned the same way.)

This discussion shows that the force on a charge q moving with velocity $\vec{v}$ in a magnetic field $\vec{B}$ is given, both in magnitude and in direction, by

$$\vec{F} = q\vec{v} \times \vec{B} \qquad \text{(magnetic force on a moving charged particle)} \quad (27.2)$$

This is the first of several vector products we will encounter in our study of magnetic-field relationships. It's important to note that Eq. (27.2) was *not* deduced theoretically; it is an observation based on *experiment*.

27.7 Finding the direction of the magnetic force on a moving charged particle.

(a)

Right-hand rule for the direction of magnetic force on a **positive** charge moving in a magnetic field:

① Place the $\vec{v}$ and $\vec{B}$ vectors tail to tail.

② Imagine turning $\vec{v}$ toward $\vec{B}$ in the $\vec{v}$-$\vec{B}$ plane (through the smaller angle).

③ The force acts along a line perpendicular to the $\vec{v}$-$\vec{B}$ plane. Curl the fingers of your *right hand* around this line in the same direction you rotated $\vec{v}$. Your thumb now points in the direction the force acts.

(b)

If the charge is negative, the direction of the force is *opposite* to that given by the right-hand rule.

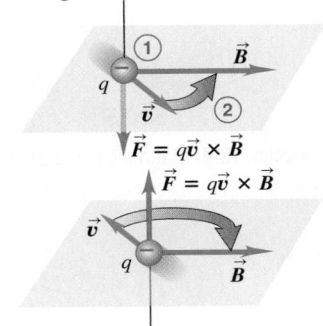

Equation (27.2) is valid for both positive and negative charges. When q is negative, the direction of the force $\vec{F}$ is opposite to that of $\vec{v} \times \vec{B}$ (Fig. 27.7b). If two charges with equal magnitude and opposite sign move in the same $\vec{B}$ field with the same velocity (Fig. 27.8), the forces have equal magnitude and opposite direction. Figures 27.6, 27.7, and 27.8 show several examples of the relationships of the directions of $\vec{F}$, $\vec{v}$, and $\vec{B}$ for both positive and negative charges. Be sure you understand the relationships shown in these figures.

Equation (27.1) gives the magnitude of the magnetic force $\vec{F}$ in Eq. (27.2). We can express this magnitude in a different but equivalent way. Since ϕ is the angle between the directions of vectors $\vec{v}$ and $\vec{B}$, we may interpret $B \sin \phi$ as the component of $\vec{B}$ perpendicular to $\vec{v}$—that is, $B_\perp$. With this notation the force magnitude is

$$F = |q|vB_\perp \qquad (27.3)$$

This form is sometimes more convenient, especially in problems involving *currents* rather than individual particles. We will discuss forces on currents later in this chapter.

From Eq. (27.1) the *units* of B must be the same as the units of F/qv. Therefore the SI unit of B is equivalent to $1 \ \text{N} \cdot \text{s}/\text{C} \cdot \text{m}$, or, since one ampere is one coulomb per second $(1 \ \text{A} = 1 \ \text{C/s})$, $1 \ \text{N}/\text{A} \cdot \text{m}$. This unit is called the **tesla** (abbreviated T), in honor of Nikola Tesla (1856–1943), the prominent Serbian-American scientist and inventor:

$$1 \ \text{tesla} = 1 \ \text{T} = 1 \ \text{N}/\text{A} \cdot \text{m}$$

Another unit of B, the **gauss** $(1 \ \text{G} = 10^{-4} \ \text{T})$, is also in common use.

The magnetic field of the earth is of the order of 10^{-4} T or 1 G. Magnetic fields of the order of 10 T occur in the interior of atoms and are important in the analysis of atomic spectra. The largest steady magnetic field that can be produced at present in the laboratory is about 45 T. Some pulsed-current electromagnets can produce fields of the order of 120 T for millisecond time intervals.

Measuring Magnetic Fields with Test Charges

To explore an unknown magnetic field, we can measure the magnitude and direction of the force on a *moving* test charge and then use Eq. (27.2) to determine $\vec{B}$. The electron beam in a cathode-ray tube, such as that in an older television set (not a flat screen), is a convenient device for this. The electron gun shoots out a narrow beam of electrons at a known speed. If there is no force to deflect the beam, it strikes the center of the screen.

27.8 Two charges of the same magnitude but opposite sign moving with the same velocity in the same magnetic field. The magnetic forces on the charges are equal in magnitude but opposite in direction.

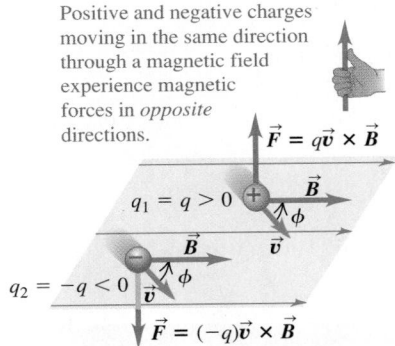

Application **Magnetic Fields of the Body**
All living cells are electrically active, and the feeble electric currents within the body produce weak but measurable magnetic fields. The fields produced by skeletal muscles have magnitudes less than 10^{-10} T, about one-millionth as strong as the earth's magnetic field. The brain produces magnetic fields that are far weaker, only about 10^{-12} T.

If a magnetic field is present, in general the electron beam is deflected. But if the beam is parallel or antiparallel to the field, then $\phi = 0$ or π in Eq. (27.1) and $F = 0$; there is no force and hence no deflection. If we find that the electron beam is not deflected when its direction is parallel to a certain axis as in Fig. 27.9a, the $\vec{B}$ vector must point either up or down along that axis.

If we then turn the tube 90° (Fig. 27.9b), $\phi = \pi/2$ in Eq. (27.1) and the magnetic force is maximum; the beam is deflected in a direction perpendicular to the plane of $\vec{B}$ and $\vec{v}$. The direction and magnitude of the deflection determine the direction and magnitude of $\vec{B}$. We can perform additional experiments in which the angle between $\vec{B}$ and $\vec{v}$ is between zero and 90° to confirm Eq. (27.1). We note that the electron has a negative charge; the force in Fig. 27.9b is opposite in direction to the force on a positive charge.

When a charged particle moves through a region of space where *both* electric and magnetic fields are present, both fields exert forces on the particle. The total force $\vec{F}$ is the vector sum of the electric and magnetic forces:

$$\vec{F} = q(\vec{E} + \vec{v} \times \vec{B}) \tag{27.4}$$

27.9 Determining the direction of a magnetic field using a cathode-ray tube. Because electrons have a negative charge, the magnetic force $\vec{F} = q\vec{v} \times \vec{B}$ in part (b) points opposite to the direction given by the right-hand rule (see Fig. 27.7b).

(a) If the tube axis is parallel to the y-axis, the beam is undeflected, so $\vec{B}$ is in either the +y- or the −y-direction.

Electron beam

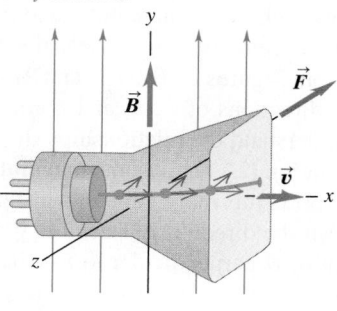

(b) If the tube axis is parallel to the x-axis, the beam is deflected in the −z-direction, so $\vec{B}$ is in the +y-direction.

Problem-Solving Strategy 27.1 | Magnetic Forces

IDENTIFY *the relevant concepts:* The equation $\vec{F} = q\vec{v} \times \vec{B}$ allows you to determine the magnetic force on a moving charged particle.

SET UP *the problem* using the following steps:
1. Draw the velocity $\vec{v}$ and magnetic field $\vec{B}$ with their tails together so that you can visualize the plane that contains them.
2. Determine the angle ϕ between $\vec{v}$ and $\vec{B}$.
3. Identify the target variables.

EXECUTE *the solution* as follows:
1. Express the magnetic force using Eq. (27.2), $\vec{F} = q\vec{v} \times \vec{B}$. Equation (27.1) gives the magnitude of the force, $F = qvB\sin\phi$.

2. Remember that $\vec{F}$ is perpendicular to the plane containing $\vec{v}$ and $\vec{B}$. The right-hand rule (see Fig. 27.7) gives the direction of $\vec{v} \times \vec{B}$. If q is negative, $\vec{F}$ is *opposite* to $\vec{v} \times \vec{B}$.

EVALUATE *your answer:* Whenever possible, solve the problem in two ways to confirm that the results agree. Do it directly from the geometric definition of the vector product. Then find the components of the vectors in some convenient coordinate system and calculate the vector product from the components. Verify that the results agree.

Example 27.1 | Magnetic force on a proton

A beam of protons ($q = 1.6 \times 10^{-19}$ C) moves at 3.0×10^5 m/s through a uniform 2.0-T magnetic field directed along the positive z-axis, as in Fig. 27.10. The velocity of each proton lies in the xz-plane and is directed at 30° to the +z-axis. Find the force on a proton.

SOLUTION

IDENTIFY and SET UP: This problem uses the expression $\vec{F} = q\vec{v} \times \vec{B}$ for the magnetic force $\vec{F}$ on a moving charged particle. The target variable is $\vec{F}$.

EXECUTE: The charge is positive, so the force is in the same direction as the vector product $\vec{v} \times \vec{B}$. From the right-hand rule, this direction is along the negative y-axis. The magnitude of the force, from Eq. (27.1), is

27.10 Directions of $\vec{v}$ and $\vec{B}$ for a proton in a magnetic field.

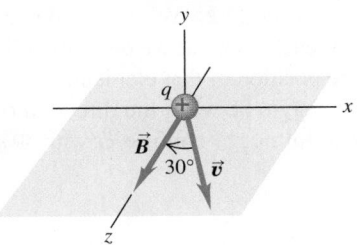

$$F = qvB \sin \phi$$
$$= (1.6 \times 10^{-19} \text{ C})(3.0 \times 10^5 \text{ m/s})(2.0 \text{ T})(\sin 30°)$$
$$= 4.8 \times 10^{-14} \text{ N}$$

EVALUATE: We check our result by evaluating the force using vector language and Eq. (27.2). We have

$$\vec{v} = (3.0 \times 10^5 \text{ m/s})(\sin 30°)\hat{\imath} + (3.0 \times 10^5 \text{ m/s})(\cos 30°)\hat{k}$$
$$\vec{B} = (2.0 \text{ T})\hat{k}$$
$$\vec{F} = q\vec{v} \times \vec{B}$$
$$= (1.6 \times 10^{-19} \text{ C})(3.0 \times 10^5 \text{ m/s})(2.0 \text{ T})$$
$$\times (\sin 30°\hat{\imath} + \cos 30°\hat{k}) \times \hat{k}$$
$$= (-4.8 \times 10^{-14} \text{ N})\hat{\jmath}$$

(Recall that $\hat{\imath} \times \hat{k} = -\hat{\jmath}$ and $\hat{k} \times \hat{k} = \mathbf{0}$.) We again find that the force is in the negative y-direction with magnitude 4.8×10^{-14} N.

If the beam consists of *electrons* rather than protons, the charge is negative ($q = -1.6 \times 10^{-19}$ C) and the direction of the force is reversed. The force is now directed along the *positive* y-axis, but the magnitude is the same as before, $F = 4.8 \times 10^{-14}$ N.

Test Your Understanding of Section 27.2 The figure at right shows a uniform magnetic field $\vec{B}$ directed into the plane of the paper (shown by the blue $\times$'s). A particle with a negative charge moves in the plane. Which of the three paths—1, 2, or 3—does the particle follow?

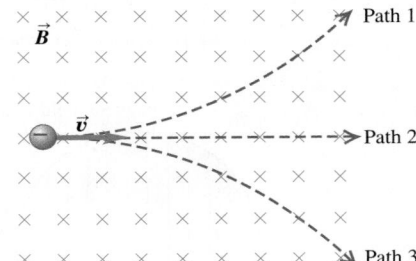

27.3 Magnetic Field Lines and Magnetic Flux

We can represent any magnetic field by **magnetic field lines,** just as we did for the earth's magnetic field in Fig. 27.3. The idea is the same as for the electric field lines we introduced in Section 21.6. We draw the lines so that the line through any point is tangent to the magnetic field vector $\vec{B}$ at that point (Fig. 27.11). Just as with electric field lines, we draw only a few representative lines; otherwise, the lines would fill up all of space. Where adjacent field lines are close together, the field magnitude is large; where these field lines are far apart, the field magnitude is small. Also, because the direction of $\vec{B}$ at each point is unique, field lines never intersect.

CAUTION **Magnetic field lines are not "lines of force"** Magnetic field lines are sometimes called "magnetic lines of force," but that's not a good name for them; unlike electric field lines, they *do not* point in the direction of the force on a charge (Fig. 27.12). Equation (27.2) shows that the force on a moving charged particle is always perpendicular to the magnetic field, and hence to the magnetic field line that passes through the particle's position. The direction of the force depends on the particle's velocity and the sign of its charge, so just looking at magnetic field lines cannot in itself tell you the direction of the force on an arbitrary moving charged particle. Magnetic field lines *do* have the direction that a compass needle would point at each location; this may help you to visualize them.

Figures 27.11 and 27.13 show magnetic field lines produced by several common sources of magnetic field. In the gap between the poles of the magnet shown in Fig. 27.13a, the field lines are approximately straight, parallel, and equally spaced, showing that the magnetic field in this region is approximately *uniform* (that is, constant in magnitude and direction).

27.11 The magnetic field lines of a permanent magnet. Note that the field lines pass through the interior of the magnet.

At each point, the field line is tangent to the magnetic field vector $\vec{B}$.

The more densely the field lines are packed, the stronger the field is at that point.

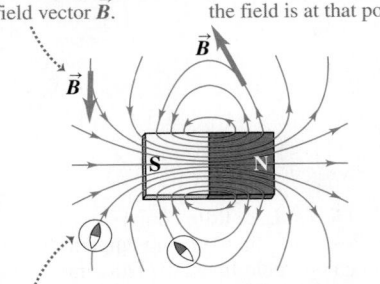

At each point, the field lines point in the same direction a compass would . . .

. . . therefore, magnetic field lines point *away from* N poles and *toward* S poles.

27.12 Magnetic field lines are *not* "lines of force."

Magnetic field lines are *not* "lines of force." The force on a charged particle is not along the direction of a field line.

The direction of the magnetic force depends on the velocity $\vec{v}$, as expressed by the magnetic force law $\vec{F} = q\vec{v} \times \vec{B}$.

Because magnetic-field patterns are three-dimensional, it's often necessary to draw magnetic field lines that point into or out of the plane of a drawing. To do this we use a dot (·) to represent a vector directed out of the plane and a cross (×) to represent a vector directed into the plane (Fig. 27.13b). To remember these, think of a dot as the head of an arrow coming directly toward you, and think of a cross as the feathers of an arrow flying directly away from you.

Iron filings, like compass needles, tend to align with magnetic field lines. Hence they provide an easy way to visualize field lines (Fig. 27.14).

Magnetic Flux and Gauss's Law for Magnetism

We define the **magnetic flux** Φ_B through a surface just as we defined electric flux in connection with Gauss's law in Section 22.2. We can divide any surface into elements of area dA (Fig. 27.15). For each element we determine $B_\perp$, the component of $\vec{B}$ normal to the surface at the position of that element, as shown. From the figure, $B_\perp = B\cos\phi$, where ϕ is the angle between the direction of $\vec{B}$ and a line perpendicular to the surface. (Be careful not to confuse ϕ with Φ_B.) In general,

27.13 Magnetic field lines produced by some common sources of magnetic field.

(a) Magnetic field of a C-shaped magnet

Between flat, parallel magnetic poles, the magnetic field is nearly uniform.

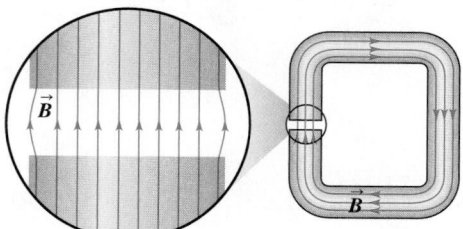

(b) Magnetic field of a straight current-carrying wire

To represent a field coming out of or going into the plane of the paper, we use dots and crosses, respectively.

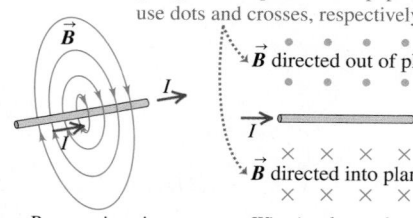

$\vec{B}$ directed out of plane

$\vec{B}$ directed into plane

Perspective view   *Wire in plane of paper*

(c) Magnetic fields of a current-carrying loop and a current-carrying coil (solenoid)

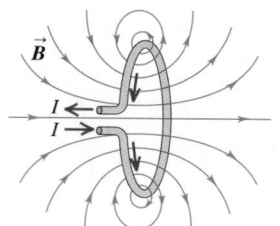

Notice that the field of the loop and, especially, that of the coil look like the field of a bar magnet (see Fig. 27.11).

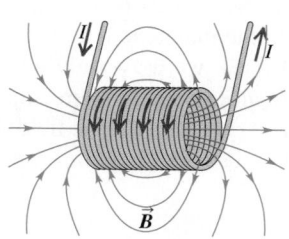

27.14 (a) Like little compass needles, iron filings line up tangent to magnetic field lines. (b) Drawing of the field lines for the situation shown in (a).

(a)

(b)

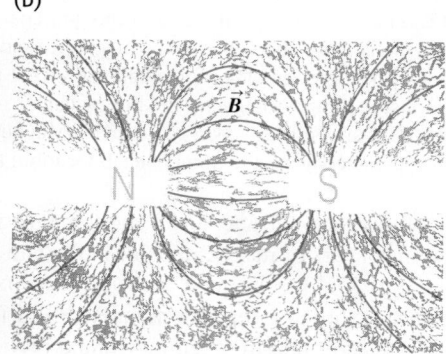

this component varies from point to point on the surface. We define the magnetic flux $d\Phi_B$ through this area as

$$d\Phi_B = B_\perp \, dA = B\cos\phi \, dA = \vec{B} \cdot d\vec{A} \qquad \text{(27.5)}$$

The *total* magnetic flux through the surface is the sum of the contributions from the individual area elements:

$$\Phi_B = \int B_\perp \, dA = \int B\cos\phi \, dA = \int \vec{B} \cdot d\vec{A} \qquad \begin{array}{l}\text{(magnetic flux}\\ \text{through a surface)}\end{array} \quad \text{(27.6)}$$

(This equation uses the concepts of vector area and surface integral that we introduced in Section 22.2; you may want to review that discussion.)

Magnetic flux is a *scalar* quantity. If $\vec{B}$ is uniform over a plane surface with total area A, then $B_\perp$ and ϕ are the same at all points on the surface, and

$$\Phi_B = B_\perp A = BA\cos\phi \qquad \text{(27.7)}$$

If $\vec{B}$ happens to be perpendicular to the surface, then $\cos\phi = 1$ and Eq. (27.7) reduces to $\Phi_B = BA$. We will use the concept of magnetic flux extensively during our study of electromagnetic induction in Chapter 29.

The SI unit of magnetic flux is equal to the unit of magnetic field (1 T) times the unit of area (1 m^2). This unit is called the **weber** (1 Wb), in honor of the German physicist Wilhelm Weber (1804–1891):

$$1 \text{ Wb} = 1 \text{ T} \cdot \text{m}^2$$

Also, $1 \text{ T} = 1 \text{ N/A} \cdot \text{m}$, so

$$1 \text{ Wb} = 1 \text{ T} \cdot \text{m}^2 = 1 \text{ N} \cdot \text{m/A}$$

In Gauss's law the total *electric* flux through a closed surface is proportional to the total electric charge enclosed by the surface. For example, if the closed surface encloses an electric dipole, the total electric flux is zero because the total charge is zero. (You may want to review Section 22.3 on Gauss's law.) By analogy, if there were such a thing as a single magnetic charge (magnetic monopole), the total *magnetic* flux through a closed surface would be proportional to the total magnetic charge enclosed. But we have mentioned that no magnetic monopole has ever been observed, despite intensive searches. We conclude:

The total magnetic flux through a closed surface is always zero.

Symbolically,

$$\oint \vec{B} \cdot d\vec{A} = 0 \qquad \text{(magnetic flux through any closed surface)} \quad \text{(27.8)}$$

This equation is sometimes called *Gauss's law for magnetism.* You can verify it by examining Figs. 27.11 and 27.13; if you draw a closed surface anywhere in any of the field maps shown in those figures, you will see that every field line that enters the surface also exits from it; the net flux through the surface is zero. It also follows from Eq. (27.8) that magnetic field lines always form closed loops.

CAUTION **Magnetic field lines have no ends** Unlike electric field lines that begin and end on electric charges, magnetic field lines *never* have end points; such a point would indicate the presence of a monopole. You might be tempted to draw magnetic field lines that begin at the north pole of a magnet and end at a south pole. But as Fig. 27.11 shows, the field lines of a magnet actually continue through the interior of the magnet. Like all other magnetic field lines, they form closed loops. ▐

27.15 The magnetic flux through an area element dA is defined to be $d\Phi_B = B_\perp \, dA$.

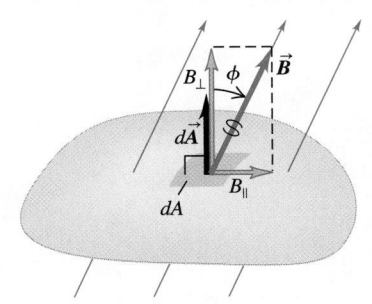

MasteringPHYSICS

PhET: Magnet and Compass
PhET: Magnets and Electromagnets

For Gauss's law, which always deals with *closed* surfaces, the vector area element $d\vec{A}$ in Eq. (27.6) always points *out of* the surface. However, some applications of *magnetic* flux involve an *open* surface with a boundary line; there is then an ambiguity of sign in Eq. (27.6) because of the two possible choices of direction for $d\vec{A}$. In these cases we choose one of the two sides of the surface to be the "positive" side and use that choice consistently.

If the element of area dA in Eq. (27.5) is at right angles to the field lines, then $B_\perp = B$; calling the area $dA_\perp$, we have

$$B = \frac{d\Phi_B}{dA_\perp} \qquad (27.9)$$

That is, the magnitude of magnetic field is equal to *flux per unit area* across an area at right angles to the magnetic field. For this reason, magnetic field $\vec{B}$ is sometimes called **magnetic flux density.**

Example 27.2 | **Magnetic flux calculations**

Figure 27.16a is a perspective view of a flat surface with area 3.0 cm^2 in a uniform magnetic field $\vec{B}$. The magnetic flux through this surface is +0.90 mWb. Find the magnitude of the magnetic field and the direction of the area vector $\vec{A}$.

27.16 (a) A flat area A in a uniform magnetic field $\vec{B}$. (b) The area vector $\vec{A}$ makes a 60° angle with $\vec{B}$. (If we had chosen $\vec{A}$ to point in the opposite direction, ϕ would have been 120° and the magnetic flux Φ_B would have been negative.)

(a) Perspective view

(b) Our sketch of the problem (edge-on view)

SOLUTION

IDENTIFY and SET UP: Our target variables are the field magnitude B and the direction of the area vector. Because $\vec{B}$ is uniform, B and ϕ are the same at all points on the surface. Hence we can use Eq. (27.7), $\Phi_B = BA\cos\phi$.

EXECUTE: The area A is 3.0×10^{-4} m^2; the direction of $\vec{A}$ is perpendicular to the surface, so ϕ could be either 60° or 120°. But Φ_B, B, and A are all positive, so $\cos\phi$ must also be positive. This rules out 120°, so $\phi = 60°$ (Fig. 27.16b). Hence we find

$$B = \frac{\Phi_B}{A\cos\phi} = \frac{0.90 \times 10^{-3}\ \text{Wb}}{(3.0 \times 10^{-4}\ \text{m}^2)(\cos 60°)} = 6.0\ \text{T}$$

EVALUATE: In many problems we are asked to calculate the flux of a given magnetic field through a given area. This example is somewhat different: It tests your understanding of the definition of magnetic flux.

Test Your Understanding of Section 27.3 Imagine moving along the axis of the current-carrying loop in Fig. 27.13c, starting at a point well to the left of the loop and ending at a point well to the right of the loop. (a) How would the magnetic field strength vary as you moved along this path? (i) It would be the same at all points along the path; (ii) it would increase and then decrease; (iii) it would decrease and then increase. (b) Would the magnetic field direction vary as you moved along the path?

27.4 Motion of Charged Particles in a Magnetic Field

When a charged particle moves in a magnetic field, it is acted on by the magnetic force given by Eq. (27.2), and the motion is determined by Newton's laws. Figure 27.17a shows a simple example. A particle with positive charge q is at point O, moving with velocity $\vec{v}$ in a uniform magnetic field $\vec{B}$ directed into the plane of the figure. The vectors $\vec{v}$ and $\vec{B}$ are perpendicular, so the magnetic force $\vec{F} = q\vec{v} \times \vec{B}$ has magnitude $F = qvB$ and a direction as shown in the figure. The force is *always* perpendicular to $\vec{v}$, so it cannot change the *magnitude* of the velocity, only its direction. To put it differently, the magnetic force never has a component parallel to the particle's motion, so the magnetic force can never do *work* on the particle. This is true even if the magnetic field is not uniform.

Motion of a charged particle under the action of a magnetic field alone is always motion with constant speed.

Using this principle, we see that in the situation shown in Fig. 27.17a the magnitudes of both $\vec{F}$ and $\vec{v}$ are constant. At points such as P and S the directions of force and velocity have changed as shown, but their magnitudes are the same. The particle therefore moves under the influence of a constant-magnitude force that is always at right angles to the velocity of the particle. Comparing the discussion of circular motion in Sections 3.4 and 5.4, we see that the particle's path is a *circle,* traced out with constant speed v. The centripetal acceleration is v^2/R and only the magnetic force acts, so from Newton's second law,

$$F = |q|vB = m\frac{v^2}{R} \tag{27.10}$$

where m is the mass of the particle. Solving Eq. (27.10) for the radius R of the circular path, we find

$$R = \frac{mv}{|q|B} \quad \text{(radius of a circular orbit in a magnetic field)} \tag{27.11}$$

We can also write this as $R = p/|q|B$, where $p = mv$ is the magnitude of the particle's momentum. If the charge q is negative, the particle moves *clockwise* around the orbit in Fig. 27.17a.

The angular speed ω of the particle can be found from Eq. (9.13), $v = R\omega$. Combining this with Eq. (27.11), we get

$$\omega = \frac{v}{R} = v\frac{|q|B}{mv} = \frac{|q|B}{m} \tag{27.12}$$

The number of revolutions per unit time is $f = \omega/2\pi$. This frequency f is independent of the radius R of the path. It is called the **cyclotron frequency;** in a particle accelerator called a *cyclotron,* particles moving in nearly circular paths are given a boost twice each revolution, increasing their energy and their orbital radii but not their angular speed or frequency. Similarly, one type of *magnetron,* a common source of microwave radiation for microwave ovens and radar systems, emits radiation with a frequency equal to the frequency of circular motion of electrons in a vacuum chamber between the poles of a magnet.

If the direction of the initial velocity is *not* perpendicular to the field, the velocity *component* parallel to the field is constant because there is no force parallel to the field. Then the particle moves in a helix (Fig. 27.18). The radius of the helix is given by Eq. (27.11), where v is now the component of velocity perpendicular to the $\vec{B}$ field.

Motion of a charged particle in a nonuniform magnetic field is more complex. Figure 27.19 shows a field produced by two circular coils separated by some distance. Particles near either coil experience a magnetic force toward the center of the region; particles with appropriate speeds spiral repeatedly from one end of the region to the other and back. Because charged particles can be trapped in such a magnetic field, it is called a *magnetic bottle.* This technique is used to confine very hot plasmas with temperatures of the order of 10^6 K. In a similar way the earth's nonuniform magnetic field traps charged particles coming from the sun in doughnut-shaped regions around the earth, as shown in Fig. 27.20. These regions, called the *Van Allen radiation belts,* were discovered in 1958 using data obtained by instruments aboard the Explorer I satellite.

27.17 A charged particle moves in a plane perpendicular to a uniform magnetic field $\vec{B}$.

(a) The orbit of a charged particle in a uniform magnetic field

A charge moving at right angles to a uniform $\vec{B}$ field moves in a circle at constant speed because $\vec{F}$ and $\vec{v}$ are always perpendicular to each other.

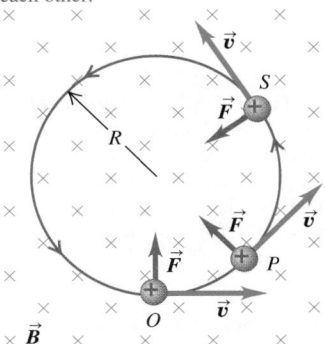

(b) An electron beam (seen as a white arc) curving in a magnetic field

27.18 The general case of a charged particle moving in a uniform magnetic field $\vec{B}$. The magnetic field does no work on the particle, so its speed and kinetic energy remain constant.

This particle's motion has components both parallel ($v_\parallel$) and perpendicular ($v_\perp$) to the magnetic field, so it moves in a helical path.

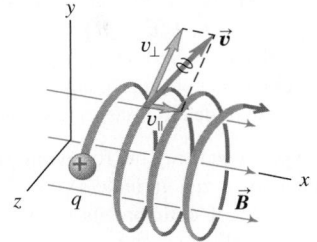

27.19 A magnetic bottle. Particles near either end of the region experience a magnetic force toward the center of the region. This is one way of containing an ionized gas that has a temperature of the order of 10^6 K, which would vaporize any material container.

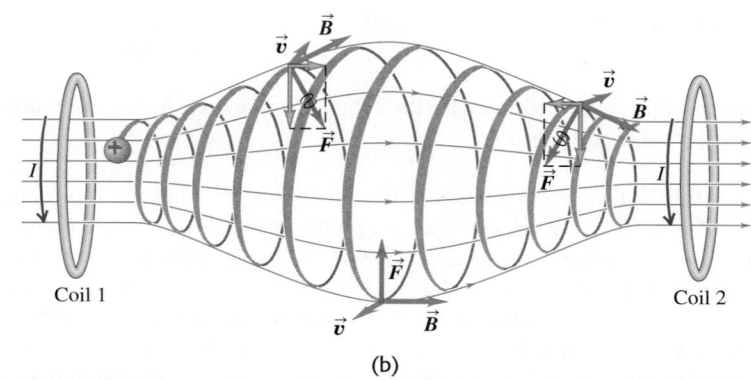

Coil 1   Coil 2

27.20 (a) The Van Allen radiation belts around the earth. Near the poles, charged particles from these belts can enter the atmosphere, producing the aurora borealis ("northern lights") and aurora australis ("southern lights"). (b) A photograph of the aurora borealis.

(a)   (b)

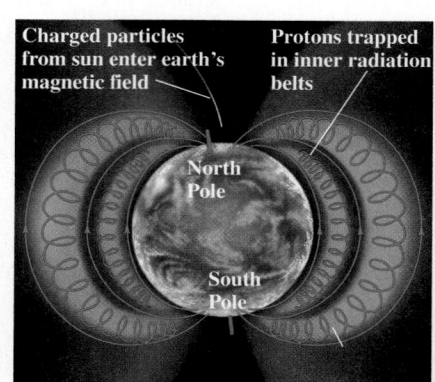

Charged particles from sun enter earth's magnetic field

Protons trapped in inner radiation belts

North Pole

South Pole

27.21 This bubble chamber image shows the result of a high-energy gamma ray (which does not leave a track) that collides with an electron in a hydrogen atom. This electron flies off to the right at high speed. Some of the energy in the collision is transformed into a second electron and a positron (a positively charged electron). A magnetic field is directed into the plane of the image, which makes the positive and negative particles curve off in different directions.

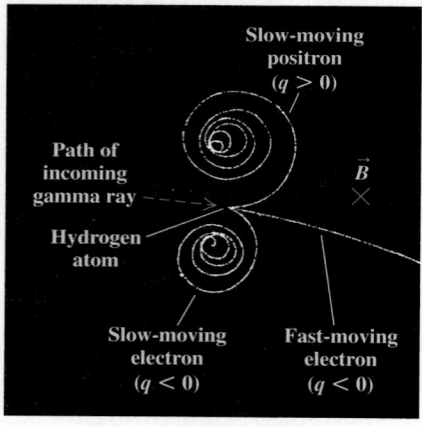

Slow-moving positron ($q > 0$)

Path of incoming gamma ray

$\vec{B}$

Hydrogen atom

Slow-moving electron ($q < 0$)

Fast-moving electron ($q < 0$)

Magnetic forces on charged particles play an important role in studies of elementary particles. Figure 27.21 shows a chamber filled with liquid hydrogen and with a magnetic field directed into the plane of the photograph. A high-energy gamma ray dislodges an electron from a hydrogen atom, sending it off at high speed and creating a visible track in the liquid hydrogen. The track shows the electron curving downward due to the magnetic force. The energy of the collision also produces another electron and a *positron* (a positively charged electron). Because of their opposite charges, the trajectories of the electron and the positron curve in opposite directions. As these particles plow through the liquid hydrogen, they collide with other charged particles, losing energy and speed. As a result, the radius of curvature decreases as suggested by Eq. (27.11). (The electron's speed is comparable to the speed of light, so Eq. (27.11) isn't directly applicable here.) Similar experiments allow physicists to determine the mass and charge of newly discovered particles.

Problem-Solving Strategy 27.2 **Motion in Magnetic Fields**

(MP)

IDENTIFY *the relevant concepts:* In analyzing the motion of a charged particle in electric and magnetic fields, you will apply Newton's second law of motion, $\sum \vec{F} = m\vec{a}$, with the net force given by $\sum \vec{F} = q(\vec{E} + \vec{v} \times \vec{B})$. Often other forces such as gravity can be neglected. Many of the problems are similar to the trajectory and circular-motion problems in Sections 3.3, 3.4, and 5.4; it would be a good idea to review those sections.

SET UP *the problem* using the following steps:
1. Determine the target variable(s).
2. Often the use of components is the most efficient approach. Choose a coordinate system and then express all vector quanti-

ties (including , and) in terms of their components in this system.

EXECUTE *the solution* as follows:
1. If the particle moves perpendicular to a uniform magnetic field, the trajectory is a circle with a radius and angular speed given by Eqs. (27.11) and (27.12), respectively.
2. If your calculation involves a more complex trajectory, use $\sum \vec{F} = m\vec{a}$ in component form: $\sum F_x = ma_x$, and so forth. This approach is particularly useful when both electric and magnetic fields are present.

EVALUATE *your answer:* Check whether your results are reasonable.

Example 27.3   Electron motion in a magnetron

A magnetron in a microwave oven emits electromagnetic waves with frequency f = 2450 MHz. What magnetic field strength is required for electrons to move in circular paths with this frequency?

SOLUTION

IDENTIFY and SET UP: The problem refers to circular motion as shown in Fig. 27.17a. We use Eq. (27.12) to solve for the field magnitude B.

EXECUTE: The angular speed that corresponds to the frequency f is $\omega = 2\pi f = (2\pi)(2450 \times 10^6 \text{ s}^{-1}) = 1.54 \times 10^{10} \text{ s}^{-1}$. Then from Eq. (27.12),

$$B = \frac{m\omega}{|q|} = \frac{(9.11 \times 10^{-31} \text{ kg})(1.54 \times 10^{10} \text{ s}^{-1})}{1.60 \times 10^{-19} \text{ C}} = 0.0877 \text{ T}$$

EVALUATE: This is a moderate field strength, easily produced with a permanent magnet. Incidentally, 2450-MHz electromagnetic waves are useful for heating and cooking food because they are strongly absorbed by water molecules.

Example 27.4   Helical particle motion in a magnetic field

In a situation like that shown in Fig. 27.18, the charged particle is a proton ($q = 1.60 \times 10^{-19}$ C, $m = 1.67 \times 10^{-27}$ kg) and the uniform, 0.500-T magnetic field is directed along the x-axis. At $t = 0$ the proton has velocity components $v_x = 1.50 \times 10^5$ m/s, $v_y = 0$, and $v_z = 2.00 \times 10^5$ m/s. Only the magnetic force acts on the proton. (a) At $t = 0$, find the force on the proton and its acceleration. (b) Find the radius of the resulting helical path, the angular speed of the proton, and the *pitch* of the helix (the distance traveled along the helix axis per revolution).

SOLUTION

IDENTIFY and SET UP: The magnetic force is $\vec{F} = q\vec{v} \times \vec{B}$; New-ton's second law gives the resulting acceleration. Because $\vec{F}$ is per-pendicular to $\vec{v}$, the proton's speed does not change. Hence Eq. (27.11) gives the radius of the helical trajectory if we replace v with the velocity component perpendicular to $\vec{B}$. Equation (27.12) gives the angular speed ω, which yields the time T for one revolu-tion (the *period*). Given the velocity component parallel to the magnetic field, we can then determine the pitch.

EXECUTE: (a) With $\vec{B} = B\hat{\imath}$ and $\vec{v} = v_x\hat{\imath} + v_z\hat{k}$, Eq. (27.2) yields

$$\vec{F} = q\vec{v} \times \vec{B} = q(v_x\hat{\imath} + v_z\hat{k}) \times B\hat{\imath} = qv_zB\hat{\jmath}$$
$$= (1.60 \times 10^{-19} \text{ C})(2.00 \times 10^5 \text{ m/s})(0.500 \text{ T})\hat{\jmath}$$
$$= (1.60 \times 10^{-14} \text{ N})\hat{\jmath}$$

(Recall that that $\hat{\imath} \times \hat{\imath} = 0$ and $\hat{k} \times \hat{\imath} = \hat{\jmath}$.) The resulting acceler-ation is

$$\vec{a} = \frac{\vec{F}}{m} = \frac{1.60 \times 10^{-14} \text{ N}}{1.67 \times 10^{-27} \text{ kg}}\hat{\jmath} = (9.58 \times 10^{12} \text{ m/s}^2)\hat{\jmath}$$

(b) Since $v_y = 0$, the component of velocity perpendicular to $\vec{B}$ is v_z; then from Eq. (27.11),

$$R = \frac{mv_z}{|q|B} = \frac{(1.67 \times 10^{-27} \text{ kg})(2.00 \times 10^5 \text{ m/s})}{(1.60 \times 10^{-19} \text{ C})(0.500 \text{ T})}$$
$$= 4.18 \times 10^{-3} \text{ m} = 4.18 \text{ mm}$$

From Eq. (27.12) the angular speed is

$$\omega = \frac{|q|B}{m} = \frac{(1.60 \times 10^{-19} \text{ C})(0.500 \text{ T})}{1.67 \times 10^{-27} \text{ kg}} = 4.79 \times 10^7 \text{ rad/s}$$

The period is $T = 2\pi/\omega = 2\pi/(4.79 \times 10^7 \text{ s}^{-1}) = 1.31 \times 10^{-7}$ s. The pitch is the distance traveled along the x-axis in this time, or

$$v_xT = (1.50 \times 10^5 \text{ m/s})(1.31 \times 10^{-7} \text{ s})$$
$$= 0.0197 \text{ m} = 19.7 \text{ mm}$$

EVALUATE: Although the magnetic force has a tiny magnitude, it produces an immense acceleration because the proton mass is so small. Note that the pitch of the helix is almost five times greater than the radius R, so this helix is much more "stretched out" than that shown in Fig. 27.18.

Test Your Understanding of Section 27.4   (a) If you double the speed of the charged particle in Fig. 27.17a while keeping the magnetic field the same (as well as the charge and the mass), how does this affect the radius of the trajec-tory? (i) The radius is unchanged; (ii) the radius is twice as large; (iii) the radius is four times as large; (iv) the radius is $\frac{1}{2}$ as large; (v) the radius is $\frac{1}{4}$ as large. (b) How does this affect the time required for one complete circular orbit? (i) The time is unchanged; (ii) the time is twice as long; (iii) the time is four times as long; (iv) the time is $\frac{1}{2}$ as long; (v) the time is $\frac{1}{4}$ as long.

27.5 Applications of Motion of Charged Particles

This section describes several applications of the principles introduced in this chapter. Study them carefully, watching for applications of Problem-Solving Strategy 27.2 (Section 27.4).

Velocity Selector

In a beam of charged particles produced by a heated cathode or a radioactive material, not all particles move with the same speed. Many applications, however, require a beam in which all the particle speeds are the same. Particles of a specific speed can be selected from the beam using an arrangement of electric and magnetic fields called a *velocity selector*. In Fig. 27.22a a charged particle with mass m, charge q, and speed v enters a region of space where the electric and magnetic fields are perpendicular to the particle's velocity and to each other. The electric field $\vec{E}$ is to the left, and the magnetic field $\vec{B}$ is into the plane of the figure. If q is positive, the electric force is to the left, with magnitude qE, and the magnetic force is to the right, with magnitude qvB. For given field magnitudes E and B, for a particular value of v the electric and magnetic forces will be equal in magnitude; the total force is then zero, and the particle travels in a straight line with constant velocity. For zero total force, $\sum F_y = 0$, we need $-qE + qvB = 0$; solving for the speed v for which there is no deflection, we find

$$v = \frac{E}{B} \qquad (27.13)$$

Only particles with speeds equal to E/B can pass through without being deflected by the fields (Fig. 27.22b). By adjusting E and B appropriately, we can select particles having a particular speed for use in other experiments. Because q divides out in Eq. (27.13), a velocity selector for positively charged particles also works for electrons or other negatively charged particles.

Thomson's e/m Experiment

In one of the landmark experiments in physics at the end of the 19th century, J. J. Thomson (1856–1940) used the idea just described to measure the ratio of charge to mass for the electron. For this experiment, carried out in 1897 at the Cavendish Laboratory in Cambridge, England, Thomson used the apparatus shown in Fig. 27.23. In a highly evacuated glass container, electrons from the hot cathode are accelerated and formed into a beam by a potential difference V between the two anodes A and A'. The speed v of the electrons is determined by the accelerating

27.22 (a) A velocity selector for charged particles uses perpendicular $\vec{E}$ and $\vec{B}$ fields. Only charged particles with $v = E/B$ move through undeflected. (b) The electric and magnetic forces on a positive charge. The forces are reversed if the charge is negative.

(a) Schematic diagram of velocity selector

Source of charged particles

By the right-hand rule, the force of the $\vec{B}$ field on the charge points to the right.

The force of the $\vec{E}$ field on the charge points to the left.

For a negative charge, the directions of *both* forces are reversed.

(b) Free-body diagram for a positive particle

$F_E = qE \qquad F_B = qvB$

Only if a charged particle has $v = E/B$ do the electric and magnetic forces cancel. All other particles are deflected.

27.23 Thomson's apparatus for measuring the ratio e/m for the electron.

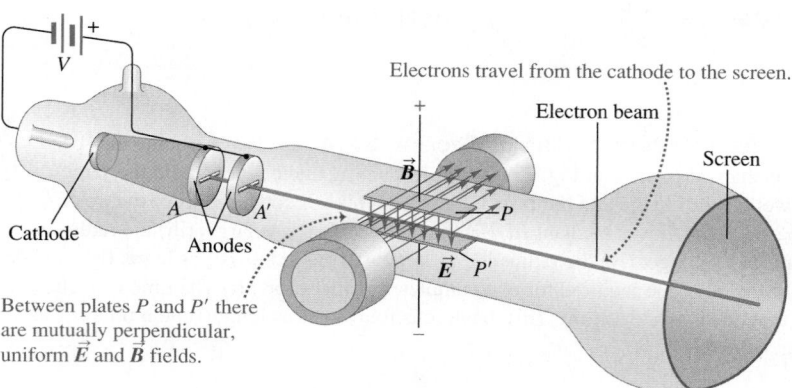

Electrons travel from the cathode to the screen.

Electron beam

Screen

Cathode

Anodes

A A'

$\vec{B}$

P

$\vec{E}$ P'

Between plates P and P' there are mutually perpendicular, uniform $\vec{E}$ and $\vec{B}$ fields.

potential V. The gained kinetic energy $\frac{1}{2}mv^2$ equals the lost electric potential energy eV, where e is the magnitude of the electron charge:

$$\tfrac{1}{2}mv^2 = eV \quad \text{or} \quad v = \sqrt{\frac{2eV}{m}} \qquad (27.14)$$

The electrons pass between the plates P and P' and strike the screen at the end of the tube, which is coated with a material that fluoresces (glows) at the point of impact. The electrons pass straight through the plates when Eq. (27.13) is satisfied; combining this with Eq. (27.14), we get

$$\frac{E}{B} = \sqrt{\frac{2eV}{m}} \quad \text{so} \quad \frac{e}{m} = \frac{E^2}{2VB^2} \qquad (27.15)$$

All the quantities on the right side can be measured, so the ratio e/m of charge to mass can be determined. It is *not* possible to measure e or m separately by this method, only their ratio.

The most significant aspect of Thomson's e/m measurements was that he found a *single value* for this quantity. It did not depend on the cathode material, the residual gas in the tube, or anything else about the experiment. This independence showed that the particles in the beam, which we now call electrons, are a common constituent of all matter. Thus Thomson is credited with the first discovery of a subatomic particle, the electron.

The most precise value of e/m available as of this writing is

$$e/m = 1.758820150(44) \times 10^{11} \text{ C/kg}$$

In this expression, (44) indicates the likely uncertainty in the last two digits, 50.

Fifteen years after Thomson's experiments, the American physicist Robert Millikan succeeded in measuring the charge of the electron precisely (see Challenge Problem 23.91). This value, together with the value of e/m, enables us to determine the *mass* of the electron. The most precise value available at present is

$$m = 9.10938215(45) \times 10^{-31} \text{ kg}$$

Mass Spectrometers

Techniques similar to Thomson's e/m experiment can be used to measure masses of ions and thus measure atomic and molecular masses. In 1919, Francis Aston (1877–1945), a student of Thomson's, built the first of a family of instruments called **mass spectrometers.** A variation built by Bainbridge is shown in Fig. 27.24. Positive ions from a source pass through the slits S_1 and S_2, forming a narrow beam. Then the ions pass through a velocity selector with crossed $\vec{E}$ and $\vec{B}$ fields, as we have described, to block all ions except those with speeds v equal to E/B. Finally, the ions pass into a region with a magnetic field $\vec{B}'$ perpendicular to the figure, where they move in circular arcs with radius R determined by Eq. (27.11): $R = mv/qB'$. Ions with different masses strike the detector (in Bainbridge's design, a photographic plate) at different points, and the values of R can be measured. We assume that each ion has lost one electron, so the net charge of each ion is just $+e$. With everything known in this equation except m, we can compute the mass m of the ion.

One of the earliest results from this work was the discovery that neon has two species of atoms, with atomic masses 20 and 22 g/mol. We now call these species **isotopes** of the element. Later experiments have shown that many elements have several isotopes, atoms that are identical in their chemical behavior but different in mass owing to differing numbers of neutrons in their nuclei. This is just one of the many applications of mass spectrometers in chemistry and physics.

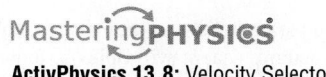

Mastering**PHYSICS**

ActivPhysics 13.8: Velocity Selector

27.24 Bainbridge's mass spectrometer utilizes a velocity selector to produce particles with uniform speed v. In the region of magnetic field B', particles with greater mass ($m_2 > m_1$) travel in paths with larger radius ($R_2 > R_1$).

Velocity selector selects particles with speed v.

Particle detector

Magnetic field separates particles by mass; the greater a particle's mass, the larger is the radius of its path.

Example 27.5 **An e/m demonstration experiment**

You set out to reproduce Thomson's e/m experiment with an accelerating potential of 150 V and a deflecting electric field of magnitude 6.0×10^6 N/C. (a) At what fraction of the speed of light do the electrons move? (b) What magnetic-field magnitude will yield zero beam deflection? (c) With this magnetic field, how will the electron beam behave if you increase the accelerating potential above 150 V?

SOLUTION

IDENTIFY and SET UP: This is the situation shown in Fig. 27.23. We use Eq. (27.14) to determine the electron speed and Eq. (27.13) to determine the required magnetic field B.

EXECUTE: (a) From Eq. (27.14), the electron speed v is

$$v = \sqrt{2(e/m)V} = \sqrt{2(1.76 \times 10^{11} \text{ C/kg})(150 \text{ V})}$$
$$= 7.27 \times 10^6 \text{ m/s} = 0.024c$$

(b) From Eq. (27.13), the required field strength is

$$B = \frac{E}{v} = \frac{6.00 \times 10^6 \text{ N/C}}{7.27 \times 10^6 \text{ m/s}} = 0.83 \text{ T}$$

(c) Increasing the accelerating potential V increases the electron speed v. In Fig. 27.23 this doesn't change the upward electric force eE, but it increases the downward magnetic force evB. Therefore the electron beam will turn *downward* and will hit the end of the tube below the undeflected position.

EVALUATE: The required magnetic field is relatively large because the electrons are moving fairly rapidly (2.4% of the speed of light). If the maximum available magnetic field is less than 0.83 T, the electric field strength E would have to be reduced to maintain the desired ratio E/B in Eq. (27.15).

Example 27.6 **Finding leaks in a vacuum system**

There is almost no helium in ordinary air, so helium sprayed near a leak in a vacuum system will quickly show up in the output of a vacuum pump connected to such a system. You are designing a leak detector that uses a mass spectrometer to detect He^+ ions (charge $+e = +1.60 \times 10^{-19}$ C, mass 6.65×10^{-27} kg). The ions emerge from the velocity selector with a speed of 1.00×10^5 m/s. They are curved in a semicircular path by a magnetic field B' and are detected at a distance of 10.16 cm from the slit S_3 in Fig. 27.24. Calculate the magnitude of the magnetic field B'.

SOLUTION

IDENTIFY and SET UP: After it passes through the slit, the ion follows a circular path as described in Section 27.4 (see Fig. 27.17). We solve Eq. (27.11) for B'.

EXECUTE: The distance given is the *diameter* of the semicircular path shown in Fig. 27.24, so the radius is $R = \frac{1}{2}(10.16 \times 10^{-2}$ m). From Eq. (27.11), $R = mv/qB'$, we get

$$B' = \frac{mv}{qR} = \frac{(6.65 \times 10^{-27} \text{ kg})(1.00 \times 10^5 \text{ m/s})}{(1.60 \times 10^{-19} \text{ C})(5.08 \times 10^{-2} \text{ m})}$$
$$= 0.0818 \text{ T}$$

EVALUATE: Helium leak detectors are widely used with high-vacuum systems. Our result shows that only a small magnetic field is required, so leak detectors can be relatively compact.

ActivPhysics 13.7: Mass Spectrometer

Test Your Understanding of Section 27.5 In Example 27.6 He^+ ions with charge $+e$ move at 1.00×10^5 m/s in a straight line through a velocity selector. Suppose the He^+ ions were replaced with He^{2+} ions, in which both electrons have been removed from the helium atom and the ion charge is $+2e$. At what speed must the He^{2+} ions travel through the same velocity selector in order to move in a straight line? (i) about 4.00×10^5 m/s; (ii) about 2.00×10^5 m/s; (iii) 1.00×10^5 m/s; (iv) about 0.50×10^5 m/s; (v) about 0.25×10^5 m/s.

27.25 Forces on a moving positive charge in a current-carrying conductor.

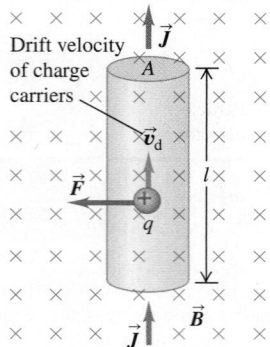

Drift velocity of charge carriers

27.6 Magnetic Force on a Current-Carrying Conductor

What makes an electric motor work? Within the motor are conductors that carry currents (that is, whose charges are in motion), as well as magnets that exert forces on the moving charges. Hence there is a magnetic force along the length of each current-carrying conductor, and these forces make the motor turn. The moving-coil galvanometer that we described in Section 26.3 also uses magnetic forces on conductors.

We can compute the force on a current-carrying conductor starting with the magnetic force $\vec{F} = q\vec{v} \times \vec{B}$ on a single moving charge. Figure 27.25 shows a straight segment of a conducting wire, with length l and cross-sectional area A; the current is from bottom to top. The wire is in a uniform magnetic field $\vec{B}$, perpendicular to the plane of the diagram and directed *into* the plane. Let's assume first that the moving charges are positive. Later we'll see what happens when they are negative.

The drift velocity $\vec{v}_d$ is upward, perpendicular to $\vec{B}$. The average force on each charge is $\vec{F} = q\vec{v}_d \times \vec{B}$, directed to the left as shown in the figure; since $\vec{v}_d$ and $\vec{B}$ are perpendicular, the magnitude of the force is $F = qv_dB$.

We can derive an expression for the *total* force on all the moving charges in a length l of conductor with cross-sectional area A using the same language we used in Eqs. (25.2) and (25.3) of Section 25.1. The number of charges per unit volume is n; a segment of conductor with length l has volume Al and contains a number of charges equal to nAl. The total force $\vec{F}$ on *all* the moving charges in this segment has magnitude

$$F = (nAl)(qv_dB) = (nqv_dA)(lB) \quad (27.16)$$

From Eq. (25.3) the current density is $J = nqv_d$. The product JA is the total current I, so we can rewrite Eq. (27.16) as

$$F = IlB \quad (27.17)$$

If the $\vec{B}$ field is not perpendicular to the wire but makes an angle ϕ with it, we handle the situation the same way we did in Section 27.2 for a single charge. Only the component of $\vec{B}$ perpendicular to the wire (and to the drift velocities of the charges) exerts a force; this component is $B_\perp = B\sin\phi$. The magnetic force on the wire segment is then

$$F = IlB_\perp = IlB\sin\phi \quad (27.18)$$

The force is always perpendicular to both the conductor and the field, with the direction determined by the same right-hand rule we used for a moving positive charge (Fig. 27.26). Hence this force can be expressed as a vector product, just like the force on a single moving charge. We represent the segment of wire with a vector $\vec{l}$ along the wire in the direction of the current; then the force $\vec{F}$ on this segment is

$$\boxed{\vec{F} = I\vec{l} \times \vec{B}} \quad \text{(magnetic force on a straight wire segment)} \quad (27.19)$$

Figure 27.27 illustrates the directions of $\vec{B}$, $\vec{l}$, and $\vec{F}$ for several cases.

If the conductor is not straight, we can divide it into infinitesimal segments $d\vec{l}$. The force $d\vec{F}$ on each segment is

$$\boxed{d\vec{F} = I\,d\vec{l} \times \vec{B}} \quad \text{(magnetic force on an infinitesimal wire section)} \quad (27.20)$$

Then we can integrate this expression along the wire to find the total force on a conductor of any shape. The integral is a *line integral*, the same mathematical operation we have used to define work (Section 6.3) and electric potential (Section 23.2).

CAUTION **Current is not a vector** Recall from Section 25.1 that the current I is not a vector. The direction of current flow is described by $d\vec{l}$, not I. If the conductor is curved, the current I is the same at all points along its length, but $d\vec{l}$ changes direction so that it is always tangent to the conductor. ▮

Finally, what happens when the moving charges are negative, such as electrons in a metal? Then in Fig. 27.25 an upward current corresponds to a downward drift velocity. But because q is now negative, the direction of the force $\vec{F}$ is the same as before. Thus Eqs. (27.17) through (27.20) are valid for *both* positive and negative charges and even when *both* signs of charge are present at once. This happens in some semiconductor materials and in ionic solutions.

A common application of the magnetic forces on a current-carrying wire is found in loudspeakers (Fig. 27.28). The radial magnetic field created by the permanent

27.26 A straight wire segment of length $\vec{l}$ carries a current I in the direction of $\vec{l}$. The magnetic force on this segment is perpendicular to both $\vec{l}$ and the magnetic field $\vec{B}$.

Force $\vec{F}$ on a straight wire carrying a positive current and oriented at an angle ϕ to a magnetic field $\vec{B}$:

• Magnitude is $F = IlB_\perp = IlB\sin\phi$.
• Direction of $\vec{F}$ is given by the right-hand rule.

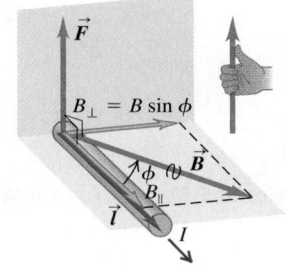

27.27 Magnetic field $\vec{B}$, length $\vec{l}$, and force $\vec{F}$ vectors for a straight wire carrying a current I.

(a)

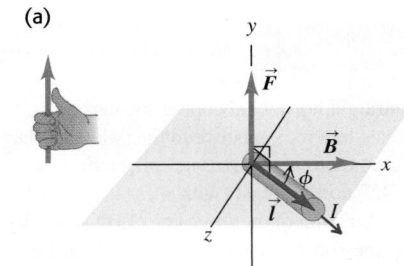

(b)

Reversing $\vec{B}$ reverses the force direction.

(c)

Reversing the current [relative to (b)] reverses the force direction.

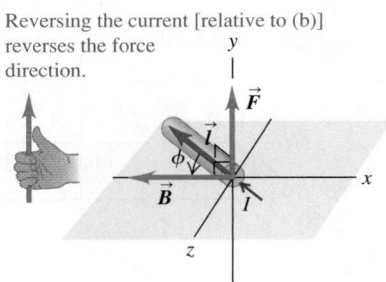

27.28 (a) Components of a loudspeaker. (b) The permanent magnet creates a magnetic field that exerts forces on the current in the voice coil; for a current I in the direction shown, the force is to the right. If the electric current in the voice coil oscillates, the speaker cone attached to the voice coil oscillates at the same frequency.

(a)

(b)

MasteringPHYSICS

ActivPhysics 13.5: Magnetic Force on Wire

magnet exerts a force on the voice coil that is proportional to the current in the coil; the direction of the force is either to the left or to the right, depending on the direction of the current. The signal from the amplifier causes the current to oscillate in direction and magnitude. The coil and the speaker cone to which it is attached respond by oscillating with an amplitude proportional to the amplitude of the current in the coil. Turning up the volume knob on the amplifier increases the current amplitude and hence the amplitudes of the cone's oscillation and of the sound wave produced by the moving cone.

Example 27.7 | **Magnetic force on a straight conductor**

A straight horizontal copper rod carries a current of 50.0 A from west to east in a region between the poles of a large electromagnet. In this region there is a horizontal magnetic field toward the northeast (that is, 45° north of east) with magnitude 1.20 T. (a) Find the magnitude and direction of the force on a 1.00-m section of rod. (b) While keeping the rod horizontal, how should it be oriented to maximize the magnitude of the force? What is the force magnitude in this case?

SOLUTION

IDENTIFY and SET UP: Figure 27.29 shows the situation. This is a straight wire segment in a uniform magnetic field, as in Fig. 27.26. Our target variables are the force $\vec{F}$ on the segment and the angle ϕ for which the force magnitude F is greatest. We find the magnitude of the magnetic force using Eq. (27.18) and the direction from the right-hand rule.

EXECUTE: (a) The angle ϕ between the directions of current and field is 45°. From Eq. (27.18) we obtain

27.29 Our sketch of the copper rod as seen from overhead.

$F = IlB\sin\phi = (50.0 \text{ A})(1.00 \text{ m})(1.20 \text{ T})(\sin 45°) = 42.4 \text{ N}$

The direction of the force is perpendicular to the plane of the current and the field, both of which lie in the horizontal plane. Thus the force must be vertical; the right-hand rule shows that it is vertically *upward* (out of the plane of the figure).

(b) From $F = IlB\sin\phi$, F is maximum for $\phi = 90°$, so that $\vec{l}$ and $\vec{B}$ are perpendicular. To keep $\vec{F} = I\vec{l} \times \vec{B}$ upward, we rotate the rod *clockwise* by 45° from its orientation in Fig. 27.29, so that the current runs toward the southeast. Then $F = IlB = (50.0 \text{ A})(1.00 \text{ m})(1.20 \text{ T}) = 60.0 \text{ N}$.

EVALUATE: You can check the result in part (a) by using Eq. (27.19) to calculate the force vector. If we use a coordinate system with the x-axis pointing east, the y-axis north, and the z-axis upward, we have $\vec{l} = (1.00 \text{ m})\hat{\imath}$, $\vec{B} = (1.20 \text{ T})[(\cos 45°)\hat{\imath} + (\sin 45°)\hat{\jmath}]$, and

$$\vec{F} = I\vec{l} \times \vec{B}$$
$$= (50 \text{ A})(1.00 \text{ m})\hat{\imath} \times (1.20 \text{ T})[(\cos 45°)\hat{\imath} + (\sin 45°)\hat{\jmath}]$$
$$= (42.4 \text{ N})\hat{k}$$

Note that the maximum upward force of 60.0 N can hold the conductor in midair against the force of gravity—that is, *magnetically levitate* the conductor—if its weight is 60.0 N and its mass is $m = w/g = (60.0 \text{ N})/(9.8 \text{ m/s}^2) = 6.12 \text{ kg}$. Magnetic levitation is used in some high-speed trains to suspend the train over the tracks. Eliminating rolling friction in this way allows the train to achieve speeds of over 400 km/h.

Example 27.8 Magnetic force on a curved conductor

In Fig. 27.30 the magnetic field $\vec{B}$ is uniform and perpendicular to the plane of the figure, pointing out of the page. The conductor, carrying current I to the left, has three segments: (1) a straight segment with length L perpendicular to the plane of the figure, (2) a semicircle with radius R, and (3) another straight segment with length L parallel to the x-axis. Find the total magnetic force on this conductor.

SOLUTION

IDENTIFY and SET UP: The magnetic field $\vec{B} = B\hat{k}$ is uniform, so we find the forces $\vec{F}_1$ and $\vec{F}_3$ on the straight segments (1) and (3) using Eq. (27.19). We divide the curved segment (2) into infinitesimal straight segments and find the corresponding force $d\vec{F}_2$ on each straight segment using Eq. (27.20). We then integrate to find $\vec{F}_2$. The total magnetic force on the conductor is then $\vec{F} = \vec{F}_1 + \vec{F}_2 + \vec{F}_3$.

27.30 What is the total magnetic force on the conductor?

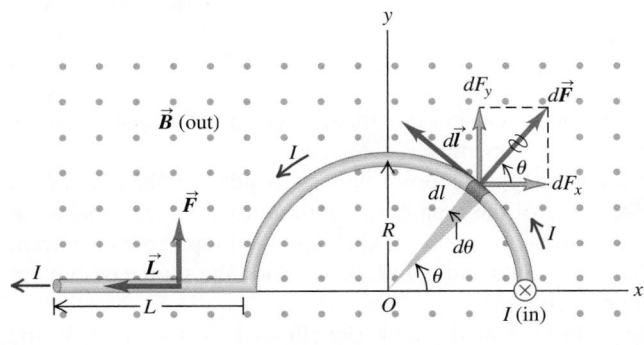

EXECUTE: For segment (1), $\vec{L} = -L\hat{k}$. Hence from Eq. (27.19), $\vec{F}_1 = I\vec{L} \times \vec{B} = 0$. For segment (3), $\vec{L} = -L\hat{i}$, so $\vec{F}_3 = I\vec{L} \times \vec{B} = I(-L\hat{i}) \times (B\hat{k}) = ILB\hat{j}$.

For the curved segment (2), Fig. 27.20 shows a segment $d\vec{l}$ with length $dl = R\,d\theta$, at angle θ. The right-hand rule shows that the direction of $d\vec{l} \times \vec{B}$ is radially outward from the center; make sure you can verify this. Because $d\vec{l}$ and $\vec{B}$ are perpendicular, the magnitude dF_2 of the force on the segment $d\vec{l}$ is just $dF_2 = I\,dl\,B = I(R\,d\theta)B$. The components of the force on this segment are

$$dF_{2x} = IR\,d\theta\,B\cos\theta \qquad dF_{2y} = IR\,d\theta\,B\sin\theta$$

To find the components of the total force, we integrate these expressions with respect to θ from $\theta = 0$ to $\theta = \pi$ to take in the whole semicircle. The results are

$$F_{2x} = IRB \int_0^\pi \cos\theta\,d\theta = 0$$

$$F_{2y} = IRB \int_0^\pi \sin\theta\,d\theta = 2IRB$$

Hence $\vec{F}_2 = 2IRB\hat{j}$. Finally, adding the forces on all three segments, we find that the total force is in the positive y-direction:

$$\vec{F} = \vec{F}_1 + \vec{F}_2 + \vec{F}_3 = 0 + 2IRB\hat{j} + ILB\hat{j} = IB(2R + L)\hat{j}$$

EVALUATE: We could have predicted from symmetry that the x-component of $\vec{F}_2$ would be zero: On the right half of the semicircle the x-component of the force is positive (to the right) and on the left half it is negative (to the left); the positive and negative contributions to the integral cancel. The result is that $\vec{F}_2$ is the force that would be exerted if we replaced the semicircle with a *straight* segment of length $2R$ along the x-axis. Do you see why?

Test Your Understanding of Section 27.6 The figure at right shows a top view of two conducting rails on which a conducting bar can slide. A uniform magnetic field is directed perpendicular to the plane of the figure as shown. A battery is to be connected to the two rails so that when the switch is closed, current will flow through the bar and cause a magnetic force to push the bar to the right. In which orientation, A or B, should the battery be placed in the circuit?

27.7 Force and Torque on a Current Loop

Current-carrying conductors usually form closed loops, so it is worthwhile to use the results of Section 27.6 to find the *total* magnetic force and torque on a conductor in the form of a loop. Many practical devices make use of the magnetic force or torque on a conducting loop, including loudspeakers (see Fig. 27.28) and galvanometers (see Section 26.3). Hence the results of this section are of substantial practical importance. These results will also help us understand the behavior of bar magnets described in Section 27.1.

As an example, let's look at a rectangular current loop in a uniform magnetic field. We can represent the loop as a series of straight line segments. We will find

Mastering PHYSICS

ActivPhysics 13.6: Magnetic Torque on a Loop

27.31 Finding the torque on a current-carrying loop in a uniform magnetic field.

(a)

The two pairs of forces acting on the loop cancel, so no net force acts on the loop.

However, the forces on the a sides of the loop ($\vec{F}$ and $-\vec{F}$) produce a torque $\tau = (IBa)(b\sin\phi)$ on the loop.

ϕ is the angle between a vector normal to the loop and the magnetic field.

(b)

The torque is maximal when $\phi = 90°$ (so $\vec{B}$ is in the plane of the loop).

x (direction normal to loop)

(c)

z (direction normal to loop)

The torque is zero when $\phi = 0°$ (as shown here) or $\phi = 180°$. In both cases, $\vec{B}$ is perpendicular to the plane of the loop.

The loop is in stable equilibrium when $\phi = 0$; it is in unstable equilibrium when $\phi = 180°$.

that the total *force* on the loop is zero but that there can be a net *torque* acting on the loop, with some interesting properties.

Figure 27.31a shows a rectangular loop of wire with side lengths a and b. A line perpendicular to the plane of the loop (i.e., a *normal* to the plane) makes an angle ϕ with the direction of the magnetic field $\vec{B}$, and the loop carries a current I. The wires leading the current into and out of the loop and the source of emf are omitted to keep the diagram simple.

The force $\vec{F}$ on the right side of the loop (length a) is to the right, in the $+x$-direction as shown. On this side, $\vec{B}$ is perpendicular to the current direction, and the force on this side has magnitude

$$F = IaB \qquad (27.21)$$

A force $-\vec{F}$ with the same magnitude but opposite direction acts on the opposite side of the loop, as shown in the figure.

The sides with length b make an angle $(90° - \phi)$ with the direction of $\vec{B}$. The forces on these sides are the vectors $\vec{F}'$ and $-\vec{F}'$; their magnitude F' is given by

$$F' = IbB\sin(90° - \phi) = IbB\cos\phi$$

The lines of action of both forces lie along the y-axis.

The *total* force on the loop is zero because the forces on opposite sides cancel out in pairs.

> **The net force on a current loop in a uniform magnetic field is zero. However, the net torque is not in general equal to zero.**

(You may find it helpful at this point to review the discussion of torque in Section 10.1.) The two forces $\vec{F}'$ and $-\vec{F}'$ in Fig. 27.31a lie along the same line and so give rise to zero net torque with respect to any point. The two forces $\vec{F}$ and $-\vec{F}$ lie along different lines, and each gives rise to a torque about the y-axis. According to the right-hand rule for determining the direction of torques, the vector torques due to $\vec{F}$ and $-\vec{F}$ are both in the $+y$-direction; hence the net vector torque $\vec{\tau}$ is in the $+y$-direction as well. The moment arm for each of these forces (equal to the perpendicular distance from the rotation axis to the line of action of the force) is $(b/2)\sin\phi$, so the torque due to each force has magnitude $F(b/2)\sin\phi$. If we use Eq. (27.21) for F, the magnitude of the net torque is

$$\tau = 2F(b/2)\sin\phi = (IBa)(b\sin\phi) \tag{27.22}$$

The torque is greatest when $\phi = 90°$, $\vec{B}$ is in the plane of the loop, and the normal to this plane is perpendicular to $\vec{B}$ (Fig. 27.31b). The torque is zero when ϕ is $0°$ or $180°$ and the normal to the loop is parallel or antiparallel to the field (Fig. 27.31c). The value $\phi = 0°$ is a stable equilibrium position because the torque is zero there, and when the loop is rotated slightly from this position, the resulting torque tends to rotate it back toward $\phi = 0°$. The position $\phi = 180°$ is an *unstable* equilibrium position; if displaced slightly from this position, the loop tends to move farther away from $\phi = 180°$. Figure 27.31 shows rotation about the *y*-axis, but because the net force on the loop is zero, Eq. (27.22) for the torque is valid for *any* choice of axis.

The area A of the loop is equal to ab, so we can rewrite Eq. (27.22) as

$$\tau = IBA\sin\phi \qquad \text{(magnitude of torque on a current loop)} \tag{27.23}$$

The product IA is called the **magnetic dipole moment** or **magnetic moment** of the loop, for which we use the symbol μ (the Greek letter mu):

$$\mu = IA \tag{27.24}$$

It is analogous to the electric dipole moment introduced in Section 21.7. In terms of μ, the magnitude of the torque on a current loop is

$$\tau = \mu B\sin\phi \tag{27.25}$$

where ϕ is the angle between the normal to the loop (the direction of the vector area $\vec{A}$) and $\vec{B}$. The torque tends to rotate the loop in the direction of *decreasing* ϕ—that is, toward its stable equilibrium position in which the loop lies in the *xy*-plane perpendicular to the direction of the field $\vec{B}$ (Fig. 27.31c). A current loop, or any other body that experiences a magnetic torque given by Eq. (27.25), is also called a **magnetic dipole.**

Magnetic Torque: Vector Form

We can also define a vector magnetic moment $\vec{\mu}$ with magnitude IA: This is shown in Fig. 27.31. The direction of $\vec{\mu}$ is defined to be perpendicular to the plane of the loop, with a sense determined by a right-hand rule, as shown in Fig. 27.32. Wrap the fingers of your right hand around the perimeter of the loop in the direction of the current. Then extend your thumb so that it is perpendicular to the plane of the loop; its direction is the direction of $\vec{\mu}$ (and of the vector area $\vec{A}$ of the loop). The torque is greatest when $\vec{\mu}$ and $\vec{B}$ are perpendicular and is zero when they are parallel or antiparallel. In the stable equilibrium position, $\vec{\mu}$ and $\vec{B}$ are parallel.

Finally, we can express this interaction in terms of the torque vector $\vec{\tau}$, which we used for *electric*-dipole interactions in Section 21.7. From Eq. (27.25) the magnitude of $\vec{\tau}$ is equal to the magnitude of $\vec{\mu} \times \vec{B}$, and reference to Fig. 27.31 shows that the directions are also the same. So we have

$$\vec{\tau} = \vec{\mu} \times \vec{B} \qquad \text{(vector torque on a current loop)} \tag{27.26}$$

This result is directly analogous to the result we found in Section 21.7 for the torque exerted by an *electric* field $\vec{E}$ on an *electric* dipole with dipole moment $\vec{p}$.

Potential Energy for a Magnetic Dipole

When a magnetic dipole changes orientation in a magnetic field, the field does work on it. In an infinitesimal angular displacement $d\phi$, the work dW is given by

27.32 The right-hand rule determines the direction of the magnetic moment of a current-carrying loop. This is also the direction of the loop's area vector $\vec{A}$; $\vec{\mu} = I\vec{A}$ is a vector equation.

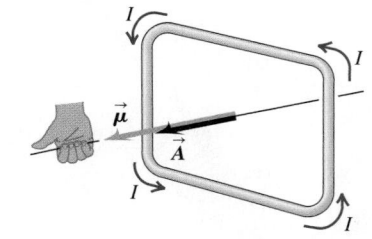

$\tau \, d\phi$, and there is a corresponding change in potential energy. As the above discussion suggests, the potential energy is least when $\vec{\mu}$ and $\vec{B}$ are parallel and greatest when they are antiparallel. To find an expression for the potential energy U as a function of orientation, we can make use of the beautiful symmetry between the electric and magnetic dipole interactions. The torque on an *electric* dipole in an *electric* field is $\vec{\tau} = \vec{p} \times \vec{E}$; we found in Section 21.7 that the corresponding potential energy is $U = -\vec{p} \cdot \vec{E}$. The torque on a *magnetic* dipole in a *magnetic* field is $\vec{\tau} = \vec{\mu} \times \vec{B}$, so we can conclude immediately that the corresponding potential energy is

$$U = -\vec{\mu} \cdot \vec{B} = -\mu B \cos \phi \quad \text{(potential energy for a magnetic dipole)} \quad (27.27)$$

With this definition, U is zero when the magnetic dipole moment is perpendicular to the magnetic field.

Magnetic Torque: Loops and Coils

Although we have derived Eqs. (27.21) through (27.27) for a rectangular current loop, all these relationships are valid for a plane loop of any shape at all. Any planar loop may be approximated as closely as we wish by a very large number of rectangular loops, as shown in Fig. 27.33. If these loops all carry equal currents in the same clockwise sense, then the forces and torques on the sides of two loops adjacent to each other cancel, and the only forces and torques that do not cancel are due to currents around the boundary. Thus all the above relationships are valid for a plane current loop of any shape, with the magnetic moment $\vec{\mu}$ given by $\vec{\mu} = I\vec{A}$.

We can also generalize this whole formulation to a coil consisting of N planar loops close together; the effect is simply to multiply each force, the magnetic moment, the torque, and the potential energy by a factor of N.

An arrangement of particular interest is the **solenoid,** a helical winding of wire, such as a coil wound on a circular cylinder (Fig. 27.34). If the windings are closely spaced, the solenoid can be approximated by a number of circular loops lying in planes at right angles to its long axis. The total torque on a solenoid in a magnetic field is simply the sum of the torques on the individual turns. For a solenoid with N turns in a uniform field B, the magnetic moment is $\mu = NIA$ and

$$\tau = NIAB \sin \phi \quad (27.28)$$

where ϕ is the angle between the axis of the solenoid and the direction of the field. The magnetic moment vector $\vec{\mu}$ is along the solenoid axis. The torque is greatest when the solenoid axis is perpendicular to the magnetic field and zero when they are parallel. The effect of this torque is to tend to rotate the solenoid into a position where its axis is parallel to the field. Solenoids are also useful as *sources* of magnetic field, as we'll discuss in Chapter 28.

The d'Arsonval galvanometer, described in Section 26.3, makes use of a magnetic torque on a coil carrying a current. As Fig. 26.14 shows, the magnetic field is not uniform but is *radial,* so the side thrusts on the coil are always perpendicular to its plane. Thus the angle ϕ in Eq. (27.28) is always 90°, and the magnetic torque is directly proportional to the current, no matter what the orientation of the coil. A restoring torque proportional to the angular displacement of the coil is provided by two hairsprings, which also serve as current leads to the coil. When current is supplied to the coil, it rotates along with its attached pointer until the restoring spring torque just balances the magnetic torque. Thus the pointer deflection is proportional to the current.

An important medical application of the torque on a magnetic dipole is *magnetic resonance imaging* (MRI). A patient is placed in a magnetic field of about 1.5 T, more than 10^4 times stronger than the earth's field. The nucleus of each hydrogen atom in the tissue to be imaged has a magnetic dipole moment,

27.33 The collection of rectangles exactly matches the irregular plane loop in the limit as the number of rectangles approaches infinity and the width of each rectangle approaches zero.

A planar current loop of any shape can be approximated by a set of rectangular loops.

27.34 The torque $\vec{\tau} = \vec{\mu} \times \vec{B}$ on this solenoid in a uniform magnetic field is directed straight into the page. An actual solenoid has many more turns, wrapped closely together.

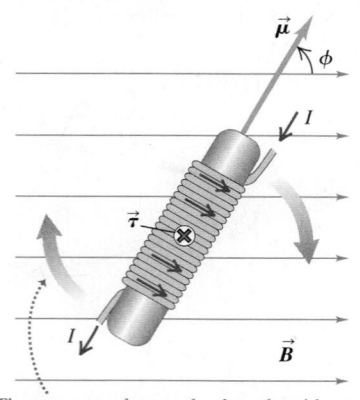

The torque tends to make the solenoid rotate clockwise in the plane of the page, aligning magnetic moment $\vec{\mu}$ with field $\vec{B}$.

which experiences a torque that aligns it with the applied field. The tissue is then illuminated with radio waves of just the right frequency to flip these magnetic moments out of alignment. The extent to which these radio waves are absorbed in the tissue is proportional to the amount of hydrogen present. Hence hydrogen-rich soft tissue looks quite different from hydrogen-deficient bone, which makes MRI ideal for analyzing details in soft tissue that cannot be seen in x-ray images (see the image that opens this chapter).

Example 27.9   Magnetic torque on a circular coil

A circular coil 0.0500 m in radius, with 30 turns of wire, lies in a horizontal plane. It carries a counterclockwise (as viewed from above) current of 5.00 A. The coil is in a uniform 1.20-T magnetic field directed toward the right. Find the magnitudes of the magnetic moment and the torque on the coil.

SOLUTION

IDENTIFY and SET UP: This problem uses the definition of magnetic moment and the expression for the torque on a magnetic dipole in a magnetic field. Figure 27.35 shows the situation. Equation (27.24) gives the magnitude μ of the magnetic moment of a single turn of wire; for N turns, the magnetic moment is N times greater. Equation (27.25) gives the magnitude τ of the torque.

EXECUTE: The area of the coil is $A = \pi r^2$. From Eq. (27.24), the total magnetic moment of all 30 turns is

$$\mu_{\text{total}} = NIA = 30(5.00 \text{ A})\pi(0.0500 \text{ m})^2 = 1.18 \text{ A} \cdot \text{m}^2$$

The angle ϕ between the direction of $\vec{B}$ and the direction of $\vec{\mu}$ (which is along the normal to the plane of the coil) is 90°. From Eq. (27.25), the torque on the coil is

$$\tau = \mu_{\text{total}}B \sin \phi = (1.18 \text{ A} \cdot \text{m}^2)(1.20 \text{ T})(\sin 90°)$$
$$= 1.41 \text{ N} \cdot \text{m}$$

EVALUATE: The torque tends to rotate the right side of the coil down and the left side up, into a position where the normal to its plane is parallel to $\vec{B}$.

27.35 Our sketch for this problem.

Example 27.10   Potential energy for a coil in a magnetic field

If the coil in Example 27.9 rotates from its initial orientation to one in which its magnetic moment $\vec{\mu}$ is parallel to $\vec{B}$, what is the change in potential energy?

SOLUTION

IDENTIFY and SET UP: Equation (27.27) gives the potential energy for each orientation. The initial position is as shown in Fig. 27.35, with $\phi_1 = 90°$. In the final orientation, the coil has been rotated 90° clockwise so that $\vec{\mu}$ and $\vec{B}$ are parallel, so the angle between these vectors is $\phi_2 = 0$.

EXECUTE: From Eq. (27.27), the potential energy change is

$$\Delta U = U_2 - U_1 = -\mu B \cos \phi_2 - (-\mu B \cos \phi_1)$$
$$= -\mu B(\cos \phi_2 - \cos \phi_1)$$
$$= -(1.18 \text{ A} \cdot \text{m}^2)(1.20 \text{ T})(\cos 0° - \cos 90°) = -1.41 \text{ J}$$

EVALUATE: The potential energy decreases because the rotation is in the direction of the magnetic torque that we found in Example 27.9.

Magnetic Dipole in a Nonuniform Magnetic Field

We have seen that a current loop (that is, a magnetic dipole) experiences zero net force in a uniform magnetic field. Figure 27.36 shows two current loops in the *nonuniform* $\vec{B}$ field of a bar magnet; in both cases the net force on the loop is *not* zero. In Fig. 27.36a the magnetic moment $\vec{\mu}$ is in the direction opposite to the field, and the force $d\vec{F} = I \, d\vec{l} \times \vec{B}$ on a segment of the loop has both a radial component and a component to the right. When these forces are summed to find the net force $\vec{F}$ on the loop, the radial components cancel so that the net force is to the right, away from the magnet. Note that in this case the force is toward the region where the field lines are farther apart and the field magnitude B is less. The polarity of the bar magnet is reversed in Fig. 27.36b, so $\vec{\mu}$ and $\vec{B}$ are parallel;

27.36 Forces on current loops in a nonuniform $\vec{B}$ field. In each case the axis of the bar magnet is perpendicular to the plane of the loop and passes through the center of the loop.

(a) Net force on this coil is away from north pole of magnet.

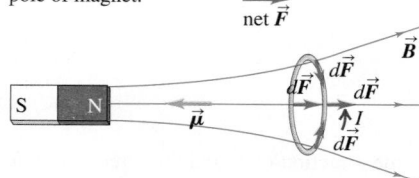

(b) Net force on same coil is toward south pole of magnet.

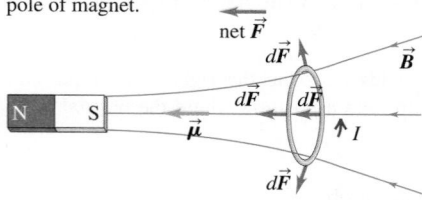

27.37 (a) An unmagnetized piece of iron. (Only a few representative atomic moments are shown.) (b) A magnetized piece of iron (bar magnet). The net magnetic moment of the bar magnet points from its south pole to its north pole. (c) A bar magnet in a magnetic field.

(a) Unmagnetized iron: magnetic moments are oriented randomly.

(b) In a bar magnet, the magnetic moments are aligned.

(c) A magnetic field creates a torque on the bar magnet that tends to align its dipole moment with the $\vec{B}$ field.

now the net force on the loop is to the left, toward the region of greater field magnitude near the magnet. Later in this section we'll use these observations to explain why bar magnets can pick up unmagnetized iron objects.

Magnetic Dipoles and How Magnets Work

The behavior of a solenoid in a magnetic field (see Fig. 27.34) resembles that of a bar magnet or compass needle; if free to turn, both the solenoid and the magnet orient themselves with their axes parallel to the $\vec{B}$ field. In both cases this is due to the interaction of moving electric charges with a magnetic field; the difference is that in a bar magnet the motion of charge occurs on the microscopic scale of the atom.

Think of an electron as being like a spinning ball of charge. In this analogy the circulation of charge around the spin axis is like a current loop, and so the electron has a net magnetic moment. (This analogy, while helpful, is inexact; an electron isn't really a spinning sphere. A full explanation of the origin of an electron's magnetic moment involves quantum mechanics, which is beyond our scope here.) In an iron atom a substantial fraction of the electron magnetic moments align with each other, and the atom has a nonzero magnetic moment. (By contrast, the atoms of most elements have little or no net magnetic moment.) In an unmagnetized piece of iron there is no overall alignment of the magnetic moments of the atoms; their vector sum is zero, and the net magnetic moment is zero (Fig. 27.37a). But in an iron bar magnet the magnetic moments of many of the atoms are parallel, and there is a substantial net magnetic moment $\vec{\mu}$ (Fig. 27.37b). If the magnet is placed in a magnetic field $\vec{B}$, the field exerts a torque given by Eq. (27.26) that tends to align $\vec{\mu}$ with $\vec{B}$ (Fig. 27.37c). A bar magnet tends to align with a $\vec{B}$ field so that a line from the south pole to the north pole of the magnet is in the direction of $\vec{B}$; hence the real significance of a magnet's north and south poles is that they represent the head and tail, respectively, of the magnet's dipole moment $\vec{\mu}$.

The torque experienced by a current loop in a magnetic field also explains how an unmagnetized iron object like that in Fig. 27.37a becomes magnetized. If an unmagnetized iron paper clip is placed next to a powerful magnet, the magnetic moments of the paper clip's atoms tend to align with the $\vec{B}$ field of the magnet. When the paper clip is removed, its atomic dipoles tend to remain aligned, and the paper clip has a net magnetic moment. The paper clip can be demagnetized by being dropped on the floor or heated; the added internal energy jostles and re-randomizes the atomic dipoles.

The magnetic-dipole picture of a bar magnet explains the attractive and repulsive forces between bar magnets shown in Fig. 27.1. The magnetic moment $\vec{\mu}$ of a bar magnet points from its south pole to its north pole, so the current loops in Figs. 27.36a and 27.36b are both equivalent to a magnet with its north pole on the left. Hence the situation in Fig. 27.36a is equivalent to two bar magnets with their north poles next to each other; the resultant force is repulsive, just as in Fig. 27.1b. In Fig. 27.36b we again have the equivalent of two bar magnets end to end, but with the south pole of the left-hand magnet next to the north pole of the right-hand magnet. The resultant force is attractive, as in Fig. 27.1a.

Finally, we can explain how a magnet can attract an unmagnetized iron object (see Fig. 27.2). It's a two-step process. First, the atomic magnetic moments of the iron tend to align with the $\vec{B}$ field of the magnet, so the iron acquires a net magnetic dipole moment $\vec{\mu}$ parallel to the field. Second, the nonuniform field of the magnet attracts the magnetic dipole. Figure 27.38a shows an example. The north pole of the magnet is closer to the nail (which contains iron), and the magnetic dipole produced in the nail is equivalent to a loop with a current that circulates in a direction opposite to that shown in Fig. 27.36a. Hence the net magnetic force on the nail is opposite to the force on the loop in Fig. 27.36a, and the nail is attracted toward the magnet. Changing the polarity of the magnet, as in Fig. 27.38b, reverses the directions of both $\vec{B}$ and $\vec{\mu}$. The situation is now equivalent to that shown in Fig. 27.36b; like the loop in that figure, the nail is attracted toward the magnet. Hence a previously unmagnetized object containing iron is attracted to *either* pole of a magnet. By contrast, objects made of brass, aluminum, or wood hardly respond at all to a magnet; the atomic magnetic dipoles of these materials, if present at all, have less tendency to align with an external field.

Our discussion of how magnets and pieces of iron interact has just scratched the surface of a diverse subject known as *magnetic properties of materials*. We'll discuss these properties in more depth in Section 28.8.

Test Your Understanding of Section 27.7 Figure 27.13c depicts the magnetic field lines due to a circular current-carrying loop. (a) What is the direction of the magnetic moment of this loop? (b) Which side of the loop is equivalent to the north pole of a magnet, and which side is equivalent to the south pole? ∎

27.8 The Direct-Current Motor

Electric motors play an important role in contemporary society. In a motor a magnetic torque acts on a current-carrying conductor, and electric energy is converted to mechanical energy. As an example, let's look at a simple type of direct-current (dc) motor, shown in Fig. 27.39.

The moving part of the motor is the *rotor,* a length of wire formed into an open-ended loop and free to rotate about an axis. The ends of the rotor wires are attached to circular conducting segments that form a *commutator.* In Fig. 27.39a, each of the two commutator segments makes contact with one of the terminals, or *brushes,* of an external circuit that includes a source of emf. This causes a current to flow into the rotor on one side, shown in red, and out of the rotor on the other side, shown in blue. Hence the rotor is a current loop with a magnetic moment $\vec{\mu}$. The rotor lies between opposing poles of a permanent magnet, so there is a magnetic field $\vec{B}$ that exerts a torque $\vec{\tau} = \vec{\mu} \times \vec{B}$ on the rotor. For the rotor orientation shown in Fig. 27.39a the torque causes the rotor to turn counterclockwise, in the direction that will align $\vec{\mu}$ with $\vec{B}$.

In Fig. 27.39b the rotor has rotated by 90° from its orientation in Fig. 27.39a. If the current through the rotor were constant, the rotor would now be in its equilibrium orientation; it would simply oscillate around this orientation. But here's where the commutator comes into play; each brush is now in contact with *both* segments of the commutator. There is no potential difference between the

27.38 A bar magnet attracts an unmagnetized iron nail in two steps. First, the $\vec{B}$ field of the bar magnet gives rise to a net magnetic moment in the nail. Second, because the field of the bar magnet is not uniform, this magnetic dipole is attracted toward the magnet. The attraction is the same whether the nail is closer to (a) the magnet's north pole or (b) the magnet's south pole.

(a)

(b)
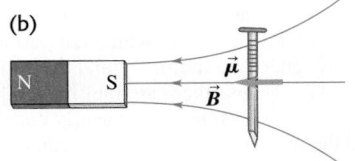

27.39 Schematic diagram of a simple dc motor. The rotor is a wire loop that is free to rotate about an axis; the rotor ends are attached to the two curved conductors that form the commutator. (The rotor halves are colored red and blue for clarity.) The commutator segments are insulated from one another.

(a) Brushes are aligned with commutator segments.
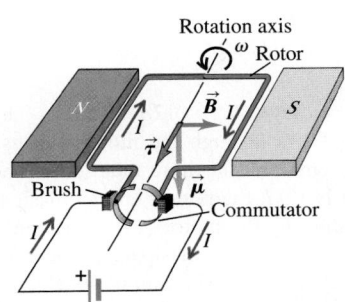
• Current flows into the red side of the rotor and out of the blue side.
• Therefore the magnetic torque causes the rotor to spin counterclockwise.

(b) Rotor has turned 90°.
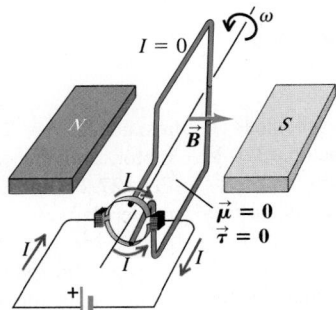
• Each brush is in contact with both commutator segments, so the current bypasses the rotor altogether.
• No magnetic torque acts on the rotor.

(c) Rotor has turned 180°.
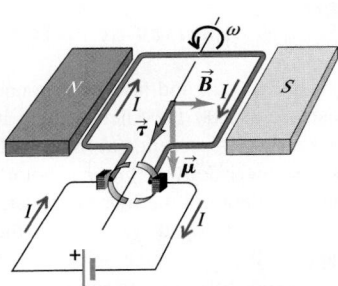
• The brushes are again aligned with commutator segments. This time the current flows into the blue side of the rotor and out of the red side.
• Therefore the magnetic torque again causes the rotor to spin counterclockwise.

commutators, so at this instant no current flows through the rotor, and the magnetic moment is zero. The rotor continues to rotate counterclockwise because of its inertia, and current again flows through the rotor as in Fig. 27.39c. But now current enters on the *blue* side of the rotor and exits on the *red* side, just the opposite of the situation in Fig. 27.39a. While the direction of the current has reversed with respect to the rotor, the rotor itself has rotated 180° and the magnetic moment $\vec{\mu}$ is in the same direction with respect to the magnetic field. Hence the magnetic torque $\vec{\tau}$ is in the same direction in Fig. 27.39c as in Fig. 27.39a. Thanks to the commutator, the current reverses after every 180° of rotation, so the torque is always in the direction to rotate the rotor counterclockwise. When the motor has come "up to speed," the average magnetic torque is just balanced by an opposing torque due to air resistance, friction in the rotor bearings, and friction between the commutator and brushes.

The simple motor shown in Fig. 27.39 has only a single turn of wire in its rotor. In practical motors, the rotor has many turns; this increases the magnetic moment and the torque so that the motor can spin larger loads. The torque can also be increased by using a stronger magnetic field, which is why many motor designs use electromagnets instead of a permanent magnet. Another drawback of the simple design in Fig. 27.39 is that the magnitude of the torque rises and falls as the rotor spins. This can be remedied by having the rotor include several independent coils of wire oriented at different angles (Fig. 27.40).

Power for Electric Motors

Because a motor converts electric energy to mechanical energy or work, it requires electric energy input. If the potential difference between its terminals is V_{ab} and the current is I, then the power input is $P = V_{ab}I$. Even if the motor coils have negligible resistance, there must be a potential difference between the terminals if P is to be different from zero. This potential difference results principally from magnetic forces exerted on the currents in the conductors of the rotor as they rotate through the magnetic field. The associated electromotive force $\mathcal{E}$ is called an *induced* emf; it is also called a *back* emf because its sense is opposite to that of the current. In Chapter 29 we will study induced emfs resulting from motion of conductors in magnetic fields.

In a *series* motor the rotor is connected in series with the electromagnet that produces the magnetic field; in a *shunt* motor they are connected in parallel. In a series motor with internal resistance r, V_{ab} is greater than $\mathcal{E}$, and the difference is the potential drop Ir across the internal resistance. That is,

$$V_{ab} = \mathcal{E} + Ir \qquad (27.29)$$

Because the magnetic force is proportional to velocity, $\mathcal{E}$ is *not* constant but is proportional to the speed of rotation of the rotor.

27.40 This motor from a computer disk drive has 12 current-carrying coils. They interact with permanent magnets on the turntable (not shown) to make the turntable rotate. (This design is the reverse of the design in Fig. 27.39, in which the permanent magnets are stationary and the coil rotates.) Because there are multiple coils, the magnetic torque is very nearly constant and the turntable spins at a very constant rate.

Coils

Example 27.11 | **A series dc motor**

A dc motor with its rotor and field coils connected in series has an internal resistance of 2.00 Ω. When running at full load on a 120-V line, it draws a 4.00-A current. (a) What is the emf in the rotor? (b) What is the power delivered to the motor? (c) What is the rate of dissipation of energy in the internal resistance? (d) What is the mechanical power developed? (e) What is the motor's efficiency? (f) What happens if the machine being driven by the motor jams, so that the rotor suddenly stops turning?

SOLUTION

IDENTIFY and SET UP: This problem uses the ideas of power and potential drop in a series dc motor. We are given the internal resistance $r = 2.00$ Ω, the voltage $V_{ab} = 120$ V across the motor, and the current $I = 4.00$ A through the motor. We use Eq. (27.29) to determine the emf $\mathcal{E}$ from these quantities. The power delivered to the motor is $V_{ab}I$, the rate of energy dissipation is I^2r, and the power output by the motor is the difference between the power input and the power dissipated. The efficiency e is the ratio of mechanical power output to electric power input.

EXECUTE: (a) From Eq. (27.29), $V_{ab} = \mathcal{E} + Ir$, we have

$$120 \text{ V} = \mathcal{E} + (4.00 \text{ A})(2.00 \text{ Ω}) \qquad \text{and so} \qquad \mathcal{E} = 112 \text{ V}$$

(b) The power delivered to the motor from the source is

$$P_{\text{input}} = V_{ab}I = (120 \text{ V})(4.00 \text{ A}) = 480 \text{ W}$$

(c) The power dissipated in the resistance r is

$$P_{\text{dissipated}} = I^2r = (4.00 \text{ A})^2(2.00 \text{ }\Omega) = 32 \text{ W}$$

(d) The mechanical power output is the electric power input minus the rate of dissipation of energy in the motor's resistance (assuming that there are no other power losses):

$$P_{\text{output}} = P_{\text{input}} - P_{\text{dissipated}} = 480 \text{ W} - 32 \text{ W} = 448 \text{ W}$$

(e) The efficiency e is the ratio of mechanical power output to electric power input:

$$e = \frac{P_{\text{output}}}{P_{\text{input}}} = \frac{448 \text{ W}}{480 \text{ W}} = 0.93 = 93\%$$

(f) With the rotor stalled, the back emf $\mathcal{E}$ (which is proportional to rotor speed) goes to zero. From Eq. (27.29) the current becomes

$$I = \frac{V_{ab}}{r} = \frac{120 \text{ V}}{2.00 \text{ }\Omega} = 60 \text{ A}$$

and the power dissipated in the resistance r becomes

$$P_{\text{dissipated}} = I^2r = (60 \text{ A})^2(2.00 \text{ }\Omega) = 7200 \text{ W}$$

EVALUATE: If this massive overload doesn't blow a fuse or trip a circuit breaker, the coils will quickly melt. When the motor is first turned on, there's a momentary surge of current until the motor picks up speed. This surge causes greater-than-usual voltage drops ($V = IR$) in the power lines supplying the current. Similar effects are responsible for the momentary dimming of lights in a house when an air conditioner or dishwasher motor starts.

Test Your Understanding of Section 27.8   In the circuit shown in Fig. 27.39, you add a switch in series with the source of emf so that the current can be turned on and off. When you close the switch and allow current to flow, will the rotor begin to turn no matter what its original orientation?   ❙

27.9 The Hall Effect

The reality of the forces acting on the moving charges in a conductor in a magnetic field is strikingly demonstrated by the *Hall effect,* an effect analogous to the transverse deflection of an electron beam in a magnetic field in vacuum. (The effect was discovered by the American physicist Edwin Hall in 1879 while he was still a graduate student.) To describe this effect, let's consider a conductor in the form of a flat strip, as shown in Fig. 27.41. The current is in the direction of the $+x$-axis and there is a uniform magnetic field $\vec{B}$ perpendicular to the plane of the strip, in the $+y$-direction. The drift velocity of the moving charges (charge magnitude $|q|$) has magnitude v_{d}. Figure 27.41a shows the case of negative charges, such as electrons in a metal, and Fig. 27.41b shows positive charges. In both cases the magnetic force is upward, just as the magnetic force on a conductor is the same whether the moving charges are positive or negative. In either case a moving charge is driven toward the *upper* edge of the strip by the magnetic force $F_z = |q|v_{\text{d}}B$.

If the charge carriers are electrons, as in Fig. 27.41a, an excess negative charge accumulates at the upper edge of the strip, leaving an excess positive charge at its lower edge. This accumulation continues until the resulting transverse electrostatic field $\vec{E}_{\text{e}}$ becomes large enough to cause a force (magnitude $|q|E_{\text{e}}$) that is equal and opposite to the magnetic force (magnitude $|q|v_{\text{d}}B$). After that, there is no longer any net transverse force to deflect the moving charges. This electric field causes a transverse potential difference between opposite edges of the strip, called the *Hall voltage* or the *Hall emf.* The polarity depends on whether the moving charges are positive or negative. Experiments show that for metals the upper edge of the strip in Fig. 27.41a *does* become negatively charged, showing that the charge carriers in a metal are indeed negative electrons.

However, if the charge carriers are *positive,* as in Fig. 27.41b, then *positive* charge accumulates at the upper edge, and the potential difference is *opposite* to the situation with negative charges. Soon after the discovery of the Hall effect in 1879, it was observed that some materials, particularly some *semiconductors,* show a Hall emf opposite to that of the metals, as if their charge carriers were positively charged. We now know that these materials conduct by a process known as *hole conduction.* Within such a material there are locations, called *holes,* that would normally be occupied by an electron but are actually empty. A missing negative charge is equivalent to a positive charge. When an electron moves in one direction to fill a hole, it leaves another hole behind it. The hole migrates in the direction opposite to that of the electron.

27.41 Forces on charge carriers in a conductor in a magnetic field.

(a) Negative charge carriers (electrons)

The charge carriers are pushed toward the top of the strip ...

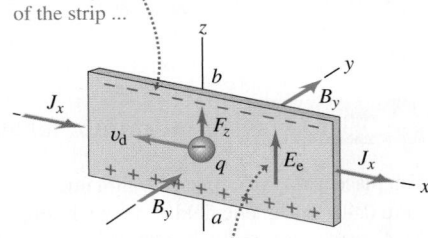

... so point a is at a higher potential than point b.

(b) Positive charge carriers

The charge carriers are again pushed toward the top of the strip ...

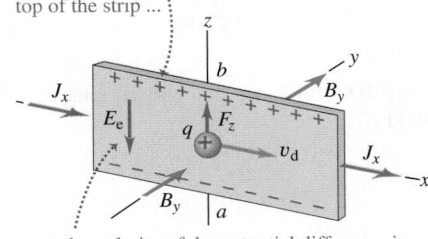

... so the polarity of the potential difference is opposite to that for negative charge carriers.

In terms of the coordinate axes in Fig. 27.41b, the electrostatic field $\vec{E}_e$ for the positive-q case is in the $-z$-direction; its z-component E_z is negative. The magnetic field is in the $+y$-direction, and we write it as B_y. The magnetic force (in the $+z$-direction) is qv_dB_y. The current density J_x is in the $+x$-direction. In the steady state, when the forces qE_z and qv_dB_y are equal in magnitude and opposite in direction,

$$qE_z + qv_dB_y = 0 \qquad \text{or} \qquad E_z = -v_dB_y$$

This confirms that when q is positive, E_z is negative. The current density J_x is

$$J_x = nqv_d$$

Eliminating v_d between these equations, we find

$$nq = \frac{-J_xB_y}{E_z} \qquad \text{(Hall effect)} \qquad (27.30)$$

Note that this result (as well as the entire derivation) is valid for both positive and negative q. When q is negative, E_z is positive, and conversely.

We can measure J_x, B_y, and E_z, so we can compute the product nq. In both metals and semiconductors, q is equal in magnitude to the electron charge, so the Hall effect permits a direct measurement of n, the concentration of current-carrying charges in the material. The *sign* of the charges is determined by the polarity of the Hall emf, as we have described.

The Hall effect can also be used for a direct measurement of electron drift speed v_d in metals. As we saw in Chapter 25, these speeds are very small, often of the order of 1 mm/s or less. If we move the entire conductor in the opposite direction to the current with a speed equal to the drift speed, then the electrons are at rest with respect to the magnetic field, and the Hall emf disappears. Thus the conductor speed needed to make the Hall emf vanish is equal to the drift speed.

Example 27.12   A Hall-effect measurement

You place a strip of copper, 2.0 mm thick and 1.50 cm wide, in a uniform 0.40-T magnetic field as shown in Fig. 27.41a. When you run a 75-A current in the $+x$-direction, you find that the potential at the bottom of the slab is $0.81\,\mu$V higher than at the top. From this measurement, determine the concentration of mobile electrons in copper.

SOLUTION

IDENTIFY and SET UP: This problem describes a Hall-effect experiment. We use Eq. (27.30) to determine the mobile electron concentration n.

EXECUTE: First we find the current density J_x and the electric field E_z:

$$J_x = \frac{I}{A} = \frac{75\ \text{A}}{(2.0 \times 10^{-3}\ \text{m})(1.50 \times 10^{-2}\ \text{m})} = 2.5 \times 10^6\ \text{A/m}^2$$

$$E_z = \frac{V}{d} = \frac{0.81 \times 10^{-6}\ \text{V}}{1.5 \times 10^{-2}\ \text{m}} = 5.4 \times 10^{-5}\ \text{V/m}$$

Then, from Eq. (27.30),

$$n = \frac{-J_xB_y}{qE_z} = \frac{-(2.5 \times 10^6\ \text{A/m}^2)(0.40\ \text{T})}{(-1.60 \times 10^{-19}\ \text{C})(5.4 \times 10^{-5}\ \text{V/m})}$$

$$= 11.6 \times 10^{28}\ \text{m}^{-3}$$

EVALUATE: The actual value of n for copper is $8.5 \times 10^{28}\ \text{m}^{-3}$. The difference shows that our simple model of the Hall effect, which ignores quantum effects and electron interactions with the ions, must be used with caution. This example also shows that with good conductors, the Hall emf is very small even with large current densities. In practice, Hall-effect devices for magnetic-field measurements use semiconductor materials, for which moderate current densities give much larger Hall emfs.

Test Your Understanding of Section 27.9 A copper wire of square cross section is oriented vertically. The four sides of the wire face north, south, east, and west. There is a uniform magnetic field directed from east to west, and the wire carries current downward. Which side of the wire is at the highest electric potential? (i) north side; (ii) south side; (iii) east side; (iv) west side.

Magnetic forces: Magnetic interactions are fundamentally interactions between moving charged particles. These interactions are described by the vector magnetic field, denoted by $\vec{B}$. A particle with charge q moving with velocity $\vec{v}$ in a magnetic field $\vec{B}$ experiences a force $\vec{F}$ that is perpendicular to both $\vec{v}$ and $\vec{B}$. The SI unit of magnetic field is the tesla ($1\text{ T} = 1\text{ N/A} \cdot \text{m}$). (See Example 27.1.)

$$\vec{F} = q\vec{v} \times \vec{B} \qquad (27.2)$$

Magnetic field lines and flux: A magnetic field can be represented graphically by magnetic field lines. At each point a magnetic field line is tangent to the direction of $\vec{B}$ at that point. Where field lines are close together the field magnitude is large, and vice versa. Magnetic flux Φ_B through an area is defined in an analogous way to electric flux. The SI unit of magnetic flux is the weber ($1\text{ Wb} = 1\text{ T} \cdot \text{m}^2$). The net magnetic flux through any closed surface is zero (Gauss's law for magnetism). As a result, magnetic field lines always close on themselves. (See Example 27.2.)

$$\Phi_B = \int B_\perp \, dA$$

$$= \int B \cos\phi \, dA \qquad (27.6)$$

$$= \int \vec{B} \cdot d\vec{A}$$

$$\oint \vec{B} \cdot d\vec{A} = 0 \quad \text{(closed surface)} \qquad (27.8)$$

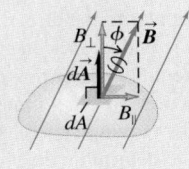

Motion in a magnetic field: The magnetic force is always perpendicular to $\vec{v}$; a particle moving under the action of a magnetic field alone moves with constant speed. In a uniform field, a particle with initial velocity perpendicular to the field moves in a circle with radius R that depends on the magnetic field strength B and the particle mass m, speed v, and charge q. (See Examples 27.3 and 27.4.)

Crossed electric and magnetic fields can be used as a velocity selector. The electric and magnetic forces exactly cancel when $v = E/B$. (See Examples 27.5 and 27.6.)

$$R = \frac{mv}{|q|B} \qquad (27.11)$$

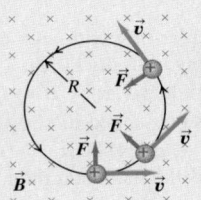

Magnetic force on a conductor: A straight segment of a conductor carrying current I in a uniform magnetic field $\vec{B}$ experiences a force $\vec{F}$ that is perpendicular to both $\vec{B}$ and the vector $\vec{l}$, which points in the direction of the current and has magnitude equal to the length of the segment. A similar relationship gives the force $d\vec{F}$ on an infinitesimal current-carrying segment $d\vec{l}$. (See Examples 27.7 and 27.8.)

$$\vec{F} = I\vec{l} \times \vec{B} \qquad (27.19)$$

$$d\vec{F} = I \, d\vec{l} \times \vec{B} \qquad (27.20)$$

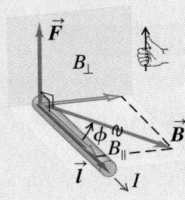

Magnetic torque: A current loop with area A and current I in a uniform magnetic field $\vec{B}$ experiences no net magnetic force, but does experience a magnetic torque of magnitude τ. The vector torque $\vec{\tau}$ can be expressed in terms of the magnetic moment $\vec{\mu} = I\vec{A}$ of the loop, as can the potential energy U of a magnetic moment in a magnetic field $\vec{B}$. The magnetic moment of a loop depends only on the current and the area; it is independent of the shape of the loop. (See Examples 27.9 and 27.10.)

$$\tau = IBA \sin\phi \qquad (27.23)$$

$$\vec{\tau} = \vec{\mu} \times \vec{B} \qquad (27.26)$$

$$U = -\vec{\mu} \cdot \vec{B} = -\mu B \cos\phi \qquad (27.27)$$

Electric motors: In a dc motor a magnetic field exerts a torque on a current in the rotor. Motion of the rotor through the magnetic field causes an induced emf called a back emf. For a series motor, in which the rotor coil is in parallel with coils that produce the magnetic field, the terminal voltage is the sum of the back emf and the drop Ir across the internal resistance. (See Example 27.11.)

The Hall effect: The Hall effect is a potential difference perpendicular to the direction of current in a conductor, when the conductor is placed in a magnetic field. The Hall potential is determined by the requirement that the associated electric field must just balance the magnetic force on a moving charge. Hall-effect measurements can be used to determine the sign of charge carriers and their concentration n. (See Example 27.12.)

$$nq = \frac{-J_x B_y}{E_z} \qquad (27.30)$$

BRIDGING PROBLEM   Magnetic Torque on a Current-Carrying Ring

A circular ring with area 4.45 cm² is carrying a current of 12.5 A. The ring, initially at rest, is immersed in a region of uniform magnetic field given by $\vec{B} = (1.15 \times 10^{-2}\ \text{T})(12\hat{\imath} + 3\hat{\jmath} - 4\hat{k})$. The ring is positioned initially such that its magnetic moment is given by $\vec{\mu}_i = \mu(-0.800\hat{\imath} + 0.600\hat{\jmath})$, where μ is the (positive) magnitude of the magnetic moment. (a) Find the initial magnetic torque on the ring. (b) The ring (which is free to rotate around one diameter) is released and turns through an angle of 90.0°, at which point its magnetic moment is given by $\vec{\mu}_f = -\mu\hat{k}$. Determine the decrease in potential energy. (c) If the moment of inertia of the ring about a diameter is $8.50 \times 10^{-7}\ \text{kg} \cdot \text{m}^2$, determine the angular speed of the ring as it passes through the second position.

SOLUTION GUIDE

See MasteringPhysics® study area for a Video Tutor solution.

IDENTIFY and SET UP

1. The current-carrying ring acts as a magnetic dipole, so you can use the equations for a magnetic dipole in a uniform magnetic field.

2. There are no nonconservative forces acting on the ring as it rotates, so the sum of its rotational kinetic energy (discussed in Section 9.4) and the potential energy is conserved.

EXECUTE

3. Use the vector expression for the torque on a magnetic dipole to find the answer to part (a). (*Hint:* You may want to review Section 1.10.)

4. Find the change in potential energy from the first orientation of the ring to the second orientation.

5. Use your result from step 4 to find the rotational kinetic energy of the ring when it is in the second orientation.

6. Use your result from step 5 to find the ring's angular speed when it is in the second orientation.

EVALUATE

7. If the ring were free to rotate around *any* diameter, in what direction would the magnetic moment point when the ring is in a state of stable equilibrium?

Problems

For instructor-assigned homework, go to www.masteringphysics.com

•, ••, •••: Problems of increasing difficulty. **CP**: Cumulative problems incorporating material from earlier chapters. **CALC**: Problems requiring calculus. **BIO**: Biosciences problems.

DISCUSSION QUESTIONS

Q27.1 Can a charged particle move through a magnetic field without experiencing any force? If so, how? If not, why not?

Q27.2 At any point in space, the electric field $\vec{E}$ is defined to be in the direction of the electric force on a positively charged particle at that point. Why don't we similarly define the magnetic field $\vec{B}$ to be in the direction of the magnetic force on a moving, positively charged particle?

Q27.3 Section 27.2 describes a procedure for finding the direction of the magnetic force using your right hand. If you use the same procedure, but with your left hand, will you get the correct direction for the force? Explain.

Q27.4 The magnetic force on a moving charged particle is always perpendicular to the magnetic field $\vec{B}$. Is the trajectory of a moving charged particle always perpendicular to the magnetic field lines? Explain your reasoning.

Q27.5 A charged particle is fired into a cubical region of space where there is a uniform magnetic field. Outside this region, there is no magnetic field. Is it possible that the particle will remain inside the cubical region? Why or why not?

Q27.6 If the magnetic force does no work on a charged particle, how can it have any effect on the particle's motion? Are there other examples of forces that do no work but have a significant effect on a particle's motion?

Q27.7 A charged particle moves through a region of space with constant velocity (magnitude and direction). If the external magnetic field is zero in this region, can you conclude that the external electric field in the region is also zero? Explain. (By "external" we mean fields other than those produced by the charged particle.) If the external electric field is zero in the region, can you conclude that the external magnetic field in the region is also zero?

Q27.8 How might a loop of wire carrying a current be used as a compass? Could such a compass distinguish between north and south? Why or why not?

Q27.9 How could the direction of a magnetic field be determined by making only *qualitative* observations of the magnetic force on a straight wire carrying a current?

Q27.10 A loose, floppy loop of wire is carrying current I. The loop of wire is placed on a horizontal table in a uniform magnetic field $\vec{B}$ perpendicular to the plane of the table. This causes the loop of wire to expand into a circular shape while still lying on the table. In a diagram, show all possible orientations of the current I and magnetic field $\vec{B}$ that could cause this to occur. Explain your reasoning.

Q27.11 Several charges enter a uniform magnetic field directed into the page. (a) What path would a positive charge q moving with a velocity of magnitude v follow through the field? (b) What path would a positive charge q moving with a velocity of magnitude $2v$ follow through the field? (c) What path would a negative charge $-q$ moving with a velocity of magnitude v follow through the field? (d) What path would a neutral particle follow through the field?

Q27.12 Each of the lettered points at the corners of the cube in Fig. Q27.12 represents a positive charge q moving with a velocity of magnitude v in the direction indicated. The region in the figure is in a uniform magnetic field $\vec{B}$, parallel to the x-axis and directed toward the right. Which charges experience a force due to $\vec{B}$? What is the direction of the force on each charge?

Figure **Q27.12**

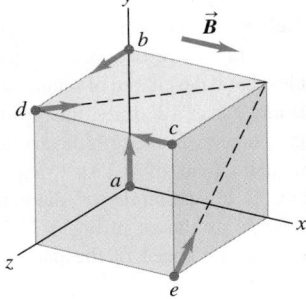

Q27.13 A student claims that if lightning strikes a metal flagpole, the force exerted by the earth's magnetic field on the current in the pole can be large enough to bend it. Typical lightning currents are of the order of 10^4 to 10^5 A. Is the student's opinion justified? Explain your reasoning.

Q27.14 Could an accelerator be built in which *all* the forces on the particles, for steering and for increasing speed, are magnetic forces? Why or why not?

Q27.15 An ordinary loudspeaker such as that shown in Fig. 27.28 should not be placed next to a computer monitor or TV screen. Why not?

Q27.16 The magnetic force acting on a charged particle can never do work because at every instant the force is perpendicular to the velocity. The torque exerted by a magnetic field can do work on a current loop when the loop rotates. Explain how these seemingly contradictory statements can be reconciled.

Q27.17 If an emf is produced in a dc motor, would it be possible to use the motor somehow as a generator or source, taking power out of it rather than putting power into it? How might this be done?

Q27.18 When the polarity of the voltage applied to a dc motor is reversed, the direction of motion does *not* reverse. Why not? How *could* the direction of motion be reversed?

Q27.19 In a Hall-effect experiment, is it possible that *no* transverse potential difference will be observed? Under what circumstances might this happen?

Q27.20 Hall-effect voltages are much greater for relatively poor conductors (such as germanium) than for good conductors (such as copper), for comparable currents, fields, and dimensions. Why?

EXERCISES

Section 27.2 Magnetic Field

27.1 • A particle with a charge of -1.24×10^{-8} C is moving with instantaneous velocity $\vec{v} = (4.19 \times 10^4 \text{ m/s})\hat{\imath} + (-3.85 \times 10^4 \text{ m/s})\hat{\jmath}$. What is the force exerted on this particle by a magnetic field (a) $\vec{B} = (1.40 \text{ T})\hat{\imath}$ and (b) $\vec{B} = (1.40 \text{ T})\hat{k}$?

27.2 • A particle of mass 0.195 g carries a charge of -2.50×10^{-8} C. The particle is given an initial horizontal velocity that is due north and has magnitude 4.00×10^4 m/s. What are the magnitude and direction of the minimum magnetic field that will keep the particle moving in the earth's gravitational field in the same horizontal, northward direction?

27.3 • In a 1.25-T magnetic field directed vertically upward, a particle having a charge of magnitude 8.50 μC and initially moving northward at 4.75 km/s is deflected toward the east. (a) What is the sign of the charge of this particle? Make a sketch to illustrate how you found your answer. (b) Find the magnetic force on the particle.

27.4 • A particle with mass 1.81×10^{-3} kg and a charge of 1.22×10^{-8} C has, at a given instant, a velocity $\vec{v} = (3.00 \times 10^4 \text{ m/s})\hat{\jmath}$. What are the magnitude and direction of the particle's acceleration produced by a uniform magnetic field $\vec{B} = (1.63 \text{ T})\hat{\imath} + (0.980 \text{ T})\hat{\jmath}$?

27.5 • An electron experiences a magnetic force of magnitude 4.60×10^{-15} N when moving at an angle of $60.0°$ with respect to a magnetic field of magnitude 3.50×10^{-3} T. Find the speed of the electron.

27.6 • An electron moves at 2.50×10^6 m/s through a region in which there is a magnetic field of unspecified direction and magnitude 7.40×10^{-2} T. (a) What are the largest and smallest possible magnitudes of the acceleration of the electron due to the magnetic field? (b) If the actual acceleration of the electron is one-fourth of the largest magnitude in part (a), what is the angle between the electron velocity and the magnetic field?

27.7 •• **CP** A particle with charge 7.80 μC is moving with velocity $\vec{v} = -(3.80 \times 10^3 \text{ m/s})\hat{\jmath}$. The magnetic force on the particle is measured to be $\vec{F} = +(7.60 \times 10^{-3} \text{ N})\hat{\imath} - (5.20 \times 10^{-3} \text{ N})\hat{k}$. (a) Calculate all the components of the magnetic field you can from this information. (b) Are there components of the magnetic field that are not determined by the measurement of the force? Explain. (c) Calculate the scalar product $\vec{B} \cdot \vec{F}$. What is the angle between $\vec{B}$ and $\vec{F}$?

27.8 •• **CP** A particle with charge -5.60 nC is moving in a uniform magnetic field $\vec{B} = -(1.25 \text{ T})\hat{k}$. The magnetic force on the particle is measured to be $\vec{F} = -(3.40 \times 10^{-7}\text{ N})\hat{\imath} + (7.40 \times 10^{-7}\text{ N})\hat{\jmath}$. (a) Calculate all the components of the velocity of the particle that you can from this information. (b) Are there components of the velocity that are not determined by the measurement of the force? Explain. (c) Calculate the scalar product $\vec{v} \cdot \vec{F}$. What is the angle between $\vec{v}$ and $\vec{F}$?

27.9 •• A group of particles is traveling in a magnetic field of unknown magnitude and direction. You observe that a proton moving at 1.50 km/s in the $+x$-direction experiences a force of 2.25×10^{-16} N in the $+y$-direction, and an electron moving at 4.75 km/s in the $-z$-direction experiences a force of 8.50×10^{-16} N in the $+y$-direction. (a) What are the magnitude and direction of the magnetic field? (b) What are the magnitude and direction of the magnetic force on an electron moving in the $-y$-direction at 3.20 km/s?

Section 27.3 Magnetic Field Lines and Magnetic Flux

27.10 • A flat, square surface with side length 3.40 cm is in the xy-plane at $z = 0$. Calculate the magnitude of the flux through this surface produced by a magnetic field $\vec{B} = (0.200\text{ T})\hat{\imath} + (0.300\text{ T})\hat{\jmath} - (0.500\text{ T})\hat{k}$.

27.11 • A circular area with a radius of 6.50 cm lies in the xy-plane. What is the magnitude of the magnetic flux through this circle due to a uniform magnetic field $B = 0.230$ T (a) in the $+z$-direction; (b) at an angle of 53.1° from the $+z$-direction; (c) in the $+y$-direction?

27.12 • A horizontal rectangular surface has dimensions 2.80 cm by 3.20 cm and is in a uniform magnetic field that is directed at an angle of 30.0° above the horizontal. What must the magnitude of the magnetic field be in order to produce a flux of 4.20×10^{-4} Wb through the surface?

27.13 •• An open plastic soda bottle with an opening diameter of 2.5 cm is placed on a table. A uniform 1.75-T magnetic field directed upward and oriented 25° from vertical encompasses the bottle. What is the total magnetic flux through the plastic of the soda bottle?

27.14 •• The magnetic field $\vec{B}$ in a certain region is 0.128 T, and its direction is that of the $+z$-axis in Fig. E27.14. (a) What is the magnetic flux across the surface *abcd* in the figure? (b) What is the magnetic flux across the surface *befc*? (c) What is the magnetic flux across the surface *aefd*? (d) What is the net flux through all five surfaces that enclose the shaded volume?

Figure **E27.14**

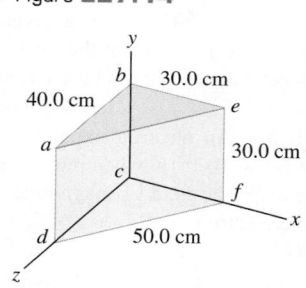

Section 27.4 Motion of Charged Particles in a Magnetic Field

27.15 •• An electron at point A in Fig. E27.15 has a speed v_0 of 1.41×10^6 m/s. Find (a) the magnitude and direction of the magnetic field that will cause the electron to follow the semicircular path from A to B, and (b) the time required for the electron to move from A to B.

Figure **E27.15**

27.16 •• Repeat Exercise 27.15 for the case in which the particle is a proton rather than an electron.

27.17 • CP A 150-g ball containing 4.00×10^8 excess electrons is dropped into a 125-m vertical shaft. At the bottom of the shaft, the ball suddenly enters a uniform horizontal magnetic field that has magnitude 0.250 T and direction from east to west. If air resistance is negligibly small, find the magnitude and direction of the force that this magnetic field exerts on the ball just as it enters the field.

27.18 • An alpha particle (a He nucleus, containing two protons and two neutrons and having a mass of 6.64×10^{-27} kg) traveling horizontally at 35.6 km/s enters a uniform, vertical, 1.10-T magnetic field. (a) What is the diameter of the path followed by this alpha particle? (b) What effect does the magnetic field have on the speed of the particle? (c) What are the magnitude and direction of the acceleration of the alpha particle while it is in the magnetic field? (d) Explain why the speed of the particle does not change even though an unbalanced external force acts on it.

27.19 • CP A particle with charge 6.40×10^{-19} C travels in a circular orbit with radius 4.68 mm due to the force exerted on it by a magnetic field with magnitude 1.65 T and perpendicular to the orbit. (a) What is the magnitude of the linear momentum $\vec{p}$ of the particle? (b) What is the magnitude of the angular momentum $\vec{L}$ of the particle?

27.20 • (a) An ^{16}O nucleus (charge $+8e$) moving horizontally from west to east with a speed of 500 km/s experiences a magnetic force of 0.00320 nN vertically downward. Find the magnitude and direction of the weakest magnetic field required to produce this force. Explain how this same force could be caused by a larger magnetic field. (b) An electron moves in a uniform, horizontal, 2.10-T magnetic field that is toward the west. What must the magnitude and direction of the minimum velocity of the electron be so that the magnetic force on it will be 4.60 pN, vertically upward? Explain how the velocity could be greater than this minimum value and the force still have this same magnitude and direction.

27.21 • A deuteron (the nucleus of an isotope of hydrogen) has a mass of 3.34×10^{-27} kg and a charge of $+e$. The deuteron travels in a circular path with a radius of 6.96 mm in a magnetic field with magnitude 2.50 T. (a) Find the speed of the deuteron. (b) Find the time required for it to make half a revolution. (c) Through what potential difference would the deuteron have to be accelerated to acquire this speed?

27.22 • In an experiment with cosmic rays, a vertical beam of particles that have charge of magnitude $3e$ and mass 12 times the proton mass enters a uniform horizontal magnetic field of 0.250 T and is bent in a semicircle of diameter 95.0 cm, as shown in Fig. E27.22. (a) Find the speed of the particles and the sign of their charge. (b) Is it reasonable to ignore the gravity force on the particles? (c) How does the speed of the particles as they enter the field compare to their speed as they exit the field?

Figure **E27.22**

27.23 • A physicist wishes to produce electromagnetic waves of frequency 3.0 THz (1 THz = 1 terahertz = 10^{12} Hz) using a magnetron (see Example 27.3). (a) What magnetic field would be required? Compare this field with the strongest constant magnetic fields yet produced on earth, about 45 T. (b) Would there be any advantage to using protons instead of electrons in the magnetron? Why or why not?

27.24 •• A beam of protons traveling at 1.20 km/s enters a uniform magnetic field, traveling perpendicular to the field. The beam exits the magnetic field, leaving the field in a direction perpendicular to its original direction (Fig. E27.24). The beam travels a distance of 1.18 cm *while in the field*. What is the magnitude of the magnetic field?

Figure **E27.24**

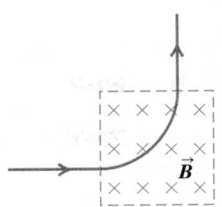

27.25 • An electron in the beam of a TV picture tube is accelerated by a potential difference of 2.00 kV. Then it passes through a region of transverse magnetic field, where it moves in a circular arc with radius 0.180 m. What is the magnitude of the field?

27.26 • A singly charged ion of ^{7}Li (an isotope of lithium) has a mass of 1.16×10^{-26} kg. It is accelerated through a potential difference of 220 V and then enters a magnetic field with magnitude 0.723 T perpendicular to the path of the ion. What is the radius of the ion's path in the magnetic field?

27.27 •• A proton ($q = 1.60 \times 10^{-19}$ C, $m = 1.67 \times 10^{-27}$ kg) moves in a uniform magnetic field $\vec{B} = (0.500 \text{ T})\hat{\imath}$. At $t = 0$ the proton has velocity components $v_x = 1.50 \times 10^5$ m/s, $v_y = 0$, and $v_z = 2.00 \times 10^5$ m/s (see Example 27.4). (a) What are the magnitude and direction of the magnetic force acting on the proton? In addition to the magnetic field there is a uniform electric field in the $+x$-direction, $\vec{E} = (+2.00 \times 10^4 \text{ V/m})\hat{\imath}$. (b) Will the proton have a component of acceleration in the direction of the electric field? (c) Describe the path of the proton. Does the electric field affect the radius of the helix? Explain. (d) At $t = T/2$, where T is the period of the circular motion of the proton, what is the x-component of the displacement of the proton from its position at $t = 0$?

Section 27.5 Applications of Motion of Charged Particles

27.28 • (a) What is the speed of a beam of electrons when the simultaneous influence of an electric field of 1.56×10^4 V/m and a magnetic field of 4.62×10^{-3} T, with both fields normal to the beam and to each other, produces no deflection of the electrons? (b) In a diagram, show the relative orientation of the vectors $\vec{v}$, $\vec{E}$, and $\vec{B}$. (c) When the electric field is removed, what is the radius of the electron orbit? What is the period of the orbit?

27.29 • In designing a velocity selector that uses uniform perpendicular electric and magnetic fields, you want to select positive ions of charge $+5e$ that are traveling perpendicular to the fields at 8.75 km/s. The magnetic field available to you has a magnitude of 0.550 T. (a) What magnitude of electric field do you need? (b) Show how the two fields should be oriented relative to each other and to the velocity of the ions. (c) Will your velocity selector also allow the following ions (having the same velocity as the $+5e$ ions) to pass through undeflected: (i) negative ions of charge $-5e$, (ii) positive ions of charge different from $+5e$?

27.30 • **Crossed $\vec{E}$ and $\vec{B}$ Fields.** A particle with initial velocity $\vec{v}_0 = (5.85 \times 10^3 \text{ m/s})\hat{\jmath}$ enters a region of uniform electric and magnetic fields. The magnetic field in the region is $\vec{B} = -(1.35 \text{ T})\hat{k}$. Calculate the magnitude and direction of the electric field in the region if the particle is to pass through undeflected, for a particle of charge (a) $+0.640$ nC and (b) -0.320 nC. You can ignore the weight of the particle.

27.31 •• A 150-V battery is connected across two parallel metal plates of area 28.5 cm^2 and separation 8.20 mm. A beam of alpha particles (charge $+2e$, mass 6.64×10^{-27} kg) is accelerated from rest through a potential difference of 1.75 kV and enters the region between the plates perpendicular to the electric field, as shown in Fig. E27.31. What magnitude and direction of magnetic field are needed so that the alpha particles emerge undeflected from between the plates?

Figure **E27.31**

27.32 • A singly ionized (one electron removed) ^{40}K atom passes through a velocity selector consisting of uniform perpendicular electric and magnetic fields. The selector is adjusted to allow ions having a speed of 4.50 km/s to pass through undeflected when the magnetic field is 0.0250 T. The ions next enter a second uniform magnetic field (B') oriented at right angles to their velocity. ^{40}K contains 19 protons and 21 neutrons and has a mass of 6.64×10^{-26} kg. (a) What is the magnitude of the electric field in the velocity selector? (b) What must be the magnitude of B' so that the ions will be bent into a semicircle of radius 12.5 cm?

27.33 • Singly ionized (one electron removed) atoms are accelerated and then passed through a velocity selector consisting of perpendicular electric and magnetic fields. The electric field is 155 V/m and the magnetic field is 0.0315 T. The ions next enter a uniform magnetic field of magnitude 0.0175 T that is oriented perpendicular to their velocity. (a) How fast are the ions moving when they emerge from the velocity selector? (b) If the radius of the path of the ions in the second magnetic field is 17.5 cm, what is their mass?

27.34 • In the Bainbridge mass spectrometer (see Fig. 27.24), the magnetic-field magnitude in the velocity selector is 0.650 T, and ions having a speed of 1.82×10^6 m/s pass through undeflected. (a) What is the electric-field magnitude in the velocity selector? (b) If the separation of the plates is 5.20 mm, what is the potential difference between plates P and P'?

27.35 •• **BIO Ancient Meat Eating.** The amount of meat in prehistoric diets can be determined by measuring the ratio of the isotopes nitrogen-15 to nitrogen-14 in bone from human remains. Carnivores concentrate ^{15}N, so this ratio tells archaeologists how much meat was consumed by ancient people. Use the spectrometer of Exercise 27.34 to find the separation of the ^{14}N and ^{15}N isotopes at the detector. The measured masses of these isotopes are 2.32×10^{-26} kg (^{14}N) and 2.49×10^{-26} kg (^{15}N).

Section 27.6 Magnetic Force on a Current-Carrying Conductor

27.36 • A straight, 2.5-m wire carries a typical household current of 1.5 A (in one direction) at a location where the earth's magnetic field is 0.55 gauss from south to north. Find the magnitude and direction of the force that our planet's magnetic field exerts on this wire if is oriented so that the current in it is running (a) from west to east, (b) vertically upward, (c) from north to south. (d) Is the magnetic force ever large enough to cause significant effects under normal household conditions?

27.37 • A straight, 2.00-m, 150-g wire carries a current in a region where the earth's magnetic field is horizontal with a magnitude of 0.55 gauss. (a) What is the minimum value of the current in this wire so that its weight is completely supported by the magnetic force due to earth's field, assuming that no other forces except gravity act on it? Does it seem likely that such a wire could support this size of current? (b) Show how the wire would have to be oriented relative to the earth's magnetic field to be supported in this way.

27.38 •• An electromagnet produces a magnetic field of 0.550 T in a cylindrical region of radius 2.50 cm between its poles. A straight wire carrying a current of 10.8 A passes through the center of this region and is perpendicular to both the axis of the cylindrical region and the magnetic field. What magnitude of force is exerted on the wire?

27.39 •• A long wire carrying 4.50 A of current makes two 90° bends, as shown in Fig. E27.39. The bent part of the wire passes through a uniform 0.240-T magnetic field directed as shown in the figure and confined to a limited region of space. Find the magnitude and direction of the force that the magnetic field exerts on the wire.

Figure **E27.39**

27.40 • A straight, vertical wire carries a current of 1.20 A downward in a region between the poles of a large superconducting electromagnet, where the magnetic field has magnitude $B = 0.588$ T and is horizontal. What are the magnitude and direction of the magnetic force on a 1.00-cm section of the wire that is in this uniform magnetic field, if the magnetic field direction is (a) east; (b) south; (c) 30.0° south of west?

27.41 • A thin, 50.0-cm-long metal bar with mass 750 g rests on, but is not attached to, two metallic supports in a uniform 0.450-T magnetic field, as shown in Fig. E27.41. A battery and a 25.0-Ω resistor in series are connected to the supports. (a) What is the highest voltage the battery can have without breaking the circuit at the supports? (b) The battery voltage has the maximum value calculated in part (a). If the resistor suddenly gets partially short-circuited, decreasing its resistance to 2.0 Ω, find the initial acceleration of the bar.

Figure **E27.41**

27.42 • **Magnetic Balance.** The circuit shown in Fig. E27.42 is used to make a magnetic balance to weigh objects. The mass m to be measured is hung from the center of the bar that is in a uniform magnetic field of 1.50 T, directed into the plane of the figure. The battery voltage can be adjusted to vary the current in the circuit. The horizontal bar is 60.0 cm long and is made of extremely light-weight material. It is connected to the battery by thin vertical wires that can support no appreciable tension; all the weight of the suspended mass m is supported by the magnetic force on the bar. A resistor with $R = 5.00$ Ω is in series with the bar; the resistance of the rest of the circuit is much less than this. (a) Which point, a or b, should be the positive terminal of the battery? (b) If the maximum terminal voltage of the battery is 175 V, what is the greatest mass m that this instrument can measure?

Figure **E27.42**

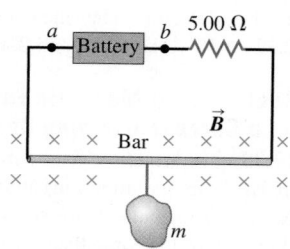

27.43 • Consider the conductor and current in Example 27.8, but now let the magnetic field be parallel to the x-axis. (a) What are the magnitude and direction of the total magnetic force on the conductor? (b) In Example 27.8, the total force is the same as if we replaced the semicircle with a straight segment along the x-axis. Is

that still true when the magnetic field is in this different direction? Can you explain why, or why not?

Section 27.7 Force and Torque on a Current Loop

27.44 •• The plane of a 5.0 cm × 8.0 cm rectangular loop of wire is parallel to a 0.19-T magnetic field. The loop carries a current of 6.2 A. (a) What torque acts on the loop? (b) What is the magnetic moment of the loop? (c) What is the maximum torque that can be obtained with the same total length of wire carrying the same current in this magnetic field?

27.45 • The 20.0 cm × 35.0 cm rectangular circuit shown in Fig. E27.45 is hinged along side ab. It carries a clockwise 5.00-A current and is located in a uniform 1.20-T magnetic field oriented perpendicular to two of its sides, as shown. (a) Draw a clear diagram showing the direction of the force that the magnetic field exerts on each segment of the circuit (ab, bc, etc.). (b) Of the four forces you drew in part (a), decide which ones exert a torque about the hinge ab. Then calculate only those forces that exert this torque. (c) Use your results from part (b) to calculate the torque that the magnetic field exerts on the circuit about the hinge axis ab.

Figure **E27.45**

27.46 • A rectangular coil of wire, 22.0 cm by 35.0 cm and carrying a current of 1.40 A, is oriented with the plane of its loop perpendicular to a uniform 1.50-T magnetic field, as shown in Fig. E27.46. (a) Calculate the net force and torque that the magnetic field exerts on the coil. (b) The coil is rotated through a 30.0° angle about the axis shown, with the left side coming out of the plane of the figure and the right side going into the plane. Calculate the net force and torque that the magnetic field now exerts on the coil. (*Hint:* In order to help visualize this three-dimensional problem, make a careful drawing of the coil as viewed along the rotation axis.)

Figure **E27.46**

27.47 • **CP** A uniform rectangular coil of total mass 212 g and dimensions 0.500 m × 1.00 m is oriented with its plane parallel to a uniform 3.00-T magnetic field (Fig. E27.47). A current of 2.00 A is suddenly started in the coil. (a) About which axis (A_1 or A_2) will the coil begin to rotate? Why? (b) Find the initial angular acceleration of the coil just after the current is started.

Figure **E27.47**

27.48 • A circular coil with area A and N turns is free to rotate about a diameter that coincides with the x-axis. Current I is circulating in the coil. There is a uniform magnetic field $\vec{B}$ in the positive y-direction. Calculate the magnitude and direction of the torque $\vec{\tau}$

Figure **E27.48**

and the value of the potential energy U, as given in Eq. (27.27), when the coil is oriented as shown in parts (a) through (d) of Fig. E27.48.

27.49 •• A coil with magnetic moment 1.45 A·m^2 is oriented initially with its magnetic moment antiparallel to a uniform 0.835-T magnetic field. What is the change in potential energy of the coil when it is rotated 180° so that its magnetic moment is parallel to the field?

Section 27.8 The Direct-Current Motor

27.50 • A dc motor with its rotor and field coils connected in series has an internal resistance of 3.2 Ω. When the motor is running at full load on a 120-V line, the emf in the rotor is 105 V. (a) What is the current drawn by the motor from the line? (b) What is the power delivered to the motor? (c) What is the mechanical power developed by the motor?

27.51 •• In a shunt-wound dc motor with the field coils and rotor connected in parallel (Fig. E27.51), the resistance R_f of the field coils is 106 Ω, and the resistance R_r of the rotor is 5.9 Ω. When a potential difference of 120 V is applied to the

Figure **E27.51**

brushes and the motor is running at full speed delivering mechanical power, the current supplied to it is 4.82 A. (a) What is the current in the field coils? (b) What is the current in the rotor? (c) What is the induced emf developed by the motor? (d) How much mechanical power is developed by this motor?

27.52 • A shunt-wound dc motor with the field coils and rotor connected in parallel (see Fig. E27.51) operates from a 120-V dc power line. The resistance of the field windings, R_f, is 218 Ω. The resistance of the rotor, R_r, is 5.9 Ω. When the motor is running, the rotor develops an emf $\mathcal{E}$. The motor draws a current of 4.82 A from the line. Friction losses amount to 45.0 W. Compute (a) the field current; (b) the rotor current; (c) the emf $\mathcal{E}$; (d) the rate of development of thermal energy in the field windings; (e) the rate of development of thermal energy in the rotor; (f) the power input to the motor; (g) the efficiency of the motor.

Section 27.9 The Hall Effect

27.53 • Figure E27.53 shows a portion of a silver ribbon with $z_1 = 11.8$ mm and $y_1 = 0.23$ mm, carrying a current of 120 A in the +x-direction. The ribbon lies in a uniform magnetic field, in the y-direction, with magnitude 0.95 T. Apply the simplified model of the Hall effect presented in Section 27.9. If

Figure **E27.53**

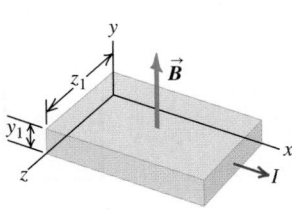

there are 5.85×10^{28} free electrons per cubic meter, find (a) the magnitude of the drift velocity of the electrons in the x-direction; (b) the magnitude and direction of the electric field in the z-direction due to the Hall effect; (c) the Hall emf.

27.54 • Let Fig. E27.53 represent a strip of an unknown metal of the same dimensions as those of the silver ribbon in Exercise 27.53. When the magnetic field is 2.29 T and the current is 78.0 A, the Hall emf is found to be 131 μV. What does the simplified model of the Hall effect presented in Section 27.9 give for the density of free electrons in the unknown metal?

PROBLEMS

27.55 • When a particle of charge $q > 0$ moves with a velocity of $\vec{v}_1$ at 45.0° from the +x-axis in the xy-plane, a uniform magnetic field exerts a force $\vec{F}_1$ along the $-z$-axis (Fig. P27.55). When the same particle moves with a velocity $\vec{v}_2$ with the same magnitude as $\vec{v}_1$ but along the +z-axis, a force $\vec{F}_2$ of magnitude F_2 is exerted on it along the +x-axis. (a) What are the magnitude (in terms of q, v_1, and F_2) and direction of the magnetic field? (b) What is the magnitude of $\vec{F}_1$ in terms of F_2?

Figure **P27.55**

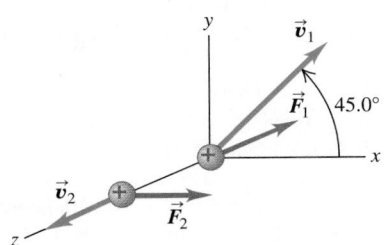

27.56 • A particle with charge 9.45×10^{-8} C is moving in a region where there is a uniform magnetic field of 0.650 T in the +x-direction. At a particular instant of time the velocity of the particle has components $v_x = -1.68 \times 10^4$ m/s, $v_y = -3.11 \times 10^4$ m/s, and $v_z = 5.85 \times 10^4$ m/s. What are the components of the force on the particle at this time?

27.57 ••• **CP Fusion Reactor.** If two deuterium nuclei (charge $+e$, mass 3.34×10^{-27} kg) get close enough together, the attraction of the strong nuclear force will fuse them to make an isotope of helium, releasing vast amounts of energy. The range of this force is about 10^{-15} m. This is the principle behind the fusion reactor. The deuterium nuclei are moving much too fast to be contained by physical walls, so they are confined magnetically. (a) How fast would two nuclei have to move so that in a head-on collision they would get close enough to fuse? (Assume their speeds are equal. Treat the nuclei as point charges, and assume that a separation of 1.0×10^{-15} is required for fusion.) (b) What strength magnetic field is needed to make deuterium nuclei with this speed travel in a circle of diameter 2.50 m?

27.58 •• **Magnetic Moment of the Hydrogen Atom.** In the Bohr model of the hydrogen atom (see Section 38.5), in the lowest energy state the electron orbits the proton at a speed of 2.2×10^6 m/s in a circular orbit of radius 5.3×10^{-11} m. (a) What is the orbital period of the electron? (b) If the orbiting electron is considered to be a current loop, what is the current I? (c) What is the magnetic moment of the atom due to the motion of the electron?

27.59 •• You wish to hit a target from several meters away with a charged coin having a mass of 4.25 g and a charge of +2500 μC. The coin is given an initial velocity of 12.8 m/s, and a downward, uniform electric field with field strength 27.5 N/C exists throughout the region. If you aim directly at the target and fire the coin horizontally, what magnitude and direction of uniform magnetic field are needed in the region for the coin to hit the target?

27.60 • A cyclotron is to accelerate protons to an energy of 5.4 MeV. The superconducting electromagnet of the cyclotron produces a 2.9-T magnetic field perpendicular to the proton orbits. (a) When the protons have achieved a kinetic energy of 2.7 MeV, what is the radius of their circular orbit and what is their angular speed? (b) Repeat part (a) when the protons have achieved their final kinetic energy of 5.4 MeV.

27.61 • The magnetic poles of a small cyclotron produce a magnetic field with magnitude 0.85 T. The poles have a radius of 0.40 m, which is the maximum radius of the orbits of the accelerated particles. (a) What is the maximum energy to which protons ($q = 1.60 \times 10^{-19}$ C, $m = 1.67 \times 10^{-27}$ kg) can be accelerated by this cyclotron? Give your answer in electron volts and in joules. (b) What is the time for one revolution of a proton orbiting at this maximum radius? (c) What would the magnetic-field magnitude have to be for the maximum energy to which a proton can be accelerated to be twice that calculated in part (a)? (d) For $B = 0.85$ T, what is the maximum energy to which alpha particles ($q = 3.20 \times 10^{-19}$ C, $m = 6.65 \times 10^{-27}$ kg) can be accelerated by this cyclotron? How does this compare to the maximum energy for protons?

27.62 •• A particle with charge q is moving with speed v in the $-y$-direction. It is moving in a uniform magnetic field $\vec{B} = B_x\hat{i} + B_y\hat{j} + B_z\hat{k}$. (a) What are the components of the force $\vec{F}$ exerted on the particle by the magnetic field? (b) If $q > 0$, what must the signs of the components of $\vec{B}$ be if the components of $\vec{F}$ are all nonnegative? (c) If $q < 0$ and $B_x = B_y = B_z > 0$, find the direction of $\vec{F}$ and find the magnitude of $\vec{F}$ in terms of $|q|$, v, and B_x.

27.63 •• A particle with negative charge q and mass $m = 2.58 \times 10^{-15}$ kg is traveling through a region containing a uniform magnetic field $\vec{B} = -(0.120 \text{ T})\hat{k}$. At a particular instant of time the velocity of the particle is $\vec{v} = (1.05 \times 10^6 \text{ m/s})(-3\hat{i} + 4\hat{j} + 12\hat{k})$ and the force $\vec{F}$ on the particle has a magnitude of 2.45 N. (a) Determine the charge q. (b) Determine the acceleration $\vec{a}$ of the particle. (c) Explain why the path of the particle is a helix, and determine the radius of curvature R of the circular component of the helical path. (d) Determine the cyclotron frequency of the particle. (e) Although helical motion is not periodic in the full sense of the word, the x- and y-coordinates do vary in a periodic way. If the coordinates of the particle at $t = 0$ are $(x, y, z) = (R, 0, 0)$, determine its coordinates at a time $t = 2T$, where T is the period of the motion in the xy-plane.

27.64 •• **BIO Medical Uses of Cyclotrons.** The largest cyclotron in the United States is the *Tevatron* at Fermilab, near Chicago, Illinois. It is called a Tevatron because it can accelerate particles to energies in the TeV range: 1 tera-eV $= 10^{12}$ eV. Its circumference is 6.4 km, and it currently can produce a maximum energy of 2.0 TeV. In a certain medical experiment, protons will be accelerated to energies of 1.25 MeV and aimed at a tumor to destroy its cells. (a) How fast are these protons moving when they hit the tumor? (b) How strong must the magnetic field be to bend the protons in the circle indicated?

27.65 • A magnetic field exerts a torque τ on a round current-carrying loop of wire. What will be the torque on this loop (in terms of τ) if its diameter is tripled?

27.66 •• A particle of charge $q > 0$ is moving at speed v in the $+z$-direction through a region of uniform magnetic field $\vec{B}$. The magnetic force on the particle is $\vec{F} = F_0(3\hat{i} + 4\hat{j})$, where F_0 is a positive constant. (a) Determine the components B_x, B_y, and B_z, or at least as many of the three components as is possible from the information given. (b) If it is given in addition that the magnetic field has magnitude $6F_0/qv$, determine as much as you can about the remaining components of $\vec{B}$.

27.67 •• Suppose the electric field between the plates in Fig. 27.24 is 1.88×10^4 V/m and the magnetic field in both regions is 0.682 T. If the source contains the three isotopes of krypton, ^{82}Kr, ^{84}Kr, and ^{86}Kr, and the ions are singly charged, find the distance between the lines formed by the three isotopes on the particle

detector. Assume the atomic masses of the isotopes (in atomic mass units) are equal to their mass numbers, 82, 84, and 86. (One atomic mass unit = 1 u = 1.66×10^{-27} kg.)

27.68 •• **Mass Spectrograph.** A mass spectrograph is used to measure the masses of ions, or to separate ions of different masses (see Section 27.5). In one design for such an instrument, ions with mass m and charge q are accelerated through a potential difference V. They then enter a uniform magnetic field that is perpendicular to their velocity, and they are deflected in a semicircular path of radius R. A detector measures where the ions complete the semicircle and from this it is easy to calculate R. (a) Derive the equation for calculating the mass of the ion from measurements of B, V, R, and q. (b) What potential difference V is needed so that singly ionized ^{12}C atoms will have $R = 50.0$ cm in a 0.150-T magnetic field? (c) Suppose the beam consists of a mixture of ^{12}C and ^{14}C ions. If v and B have the same values as in part (b), calculate the separation of these two isotopes at the detector. Do you think that this beam separation is sufficient for the two ions to be distinguished? (Make the assumption described in Problem 27.67 for the masses of the ions.)

27.69 •• A straight piece of conducting wire with mass M and length L is placed on a frictionless incline tilted at an angle θ from the horizontal (Fig. P27.69). There is a uniform, vertical magnetic field $\vec{B}$ at all points (produced by an arrangement of magnets not shown in the figure). To keep the wire from sliding down the incline, a voltage source is attached to the ends of the wire. When just the right amount of current flows through the wire, the wire remains at rest. Determine the magnitude and direction of the current in the wire that will cause the wire to remain at rest. Copy the figure and draw the direction of the current on your copy. In addition, show in a free-body diagram all the forces that act on the wire.

Figure **P27.69**

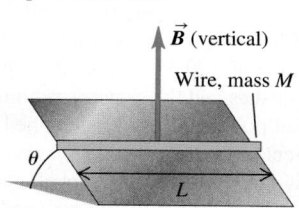

27.70 •• **CP** A 2.60-N metal bar, 1.50 m long and having a resistance of 10.0 Ω, rests horizontally on conducting wires connecting it to the circuit shown in Fig. P27.70. The bar is in a uniform, horizontal, 1.60-T magnetic field and is not attached to the wires in the circuit. What is the acceleration of the bar just after the switch S is closed?

Figure **P27.70**

27.71 •• **Using Gauss's Law for Magnetism.** In a certain region of space, the magnetic field $\vec{B}$ is not uniform. The magnetic field has both a z-component and a component that points radially away from or toward the z-axis. The z-component is given by $B_z(z) = \beta z$, where β is a positive constant. The radial component B_r depends only on r, the radial distance from the z-axis. (a) Use Gauss's law for magnetism, Eq. (27.8), to find the radial component B_r as a function of r. (*Hint:* Try a cylindrical Gaussian surface of radius r concentric with the z-axis, with one end at $z = 0$ and the other at $z = L$.) (b) Sketch the magnetic field lines.

27.72 •• **CP** A plastic circular loop has radius R, and a positive charge q is distributed uniformly around the circumference of the loop. The loop is then rotated around its central axis, perpendicular to the plane of the loop, with angular speed ω. If the loop is in a region where there is a uniform magnetic field $\vec{B}$ directed parallel to the plane of the loop, calculate the magnitude of the magnetic torque on the loop.

27.73 •• **BIO** **Determining Diet.** One method for determining the amount of corn in early Native American diets is the *stable isotope ratio analysis* (SIRA) technique. As corn photosynthesizes, it concentrates the isotope carbon-13, whereas most other plants concentrate carbon-12. Overreliance on corn consumption can then be correlated with certain diseases, because corn lacks the essential amino acid lysine. Archaeologists use a mass spectrometer to separate the ^{12}C and ^{13}C isotopes in samples of human remains. Suppose you use a velocity selector to obtain singly ionized (missing one electron) atoms of speed 8.50 km/s, and you want to bend them within a uniform magnetic field in a semicircle of diameter 25.0 cm for the ^{12}C. The measured masses of these isotopes are 1.99×10^{-26} kg (^{12}C) and 2.16×10^{-26} kg (^{13}C). (a) What strength of magnetic field is required? (b) What is the diameter of the ^{13}C semicircle? (c) What is the separation of the ^{12}C and ^{13}C ions at the detector at the end of the semicircle? Is this distance large enough to be easily observed?

27.74 •• **CP** **An Electromagnetic Rail Gun.** A conducting bar with mass m and length L slides over horizontal rails that are connected to a voltage source. The voltage source maintains a constant current I in the rails and bar, and a constant, uniform, vertical magnetic field $\vec{B}$ fills the region between the rails (Fig. P27.74). (a) Find the magnitude and direction of the net force on the conducting bar. Ignore friction, air resistance, and electrical resistance. (b) If the bar has mass m, find the distance d that the bar must move along the rails from rest to attain speed v. (c) It has been suggested that rail guns based on this principle could accelerate payloads into earth orbit or beyond. Find the distance the bar must travel along the rails if it is to reach the escape speed for the earth (11.2 km/s). Let $B = 0.80$ T, $I = 2.0 \times 10^3$ A, $m = 25$ kg, and $L = 50$ cm. For simplicity assume the net force on the object is equal to the magnetic force, as in parts (a) and (b), even though gravity plays an important role in an actual launch in space.

Figure **P27.74**

27.75 • A long wire carrying a 6.00-A current reverses direction by means of two right-angle bends, as shown in Fig. P27.75. The part of the wire where the bend occurs is in a magnetic field of 0.666 T confined to the circular region of diameter 75 cm, as shown. Find the magnitude and direction of the net force that the magnetic field exerts on this wire.

Figure **P27.75**

27.76 • A wire 25.0 cm long lies along the z-axis and carries a current of 7.40 A in the $+z$-direction. The magnetic field is uniform and has components $B_x = -0.242$ T, $B_y = -0.985$ T, and $B_z = -0.336$ T. (a) Find the components of the magnetic force on the wire. (b) What is the magnitude of the net magnetic force on the wire?

27.77 •• **CP** The rectangular loop of wire shown in Fig. P27.77 has a mass of 0.15 g per centimeter of length and is pivoted about side ab on a frictionless axis. The current in the wire is 8.2 A in the direction shown. Find the magnitude and direction of the magnetic field parallel to the y-axis that will cause the loop to swing up until its plane makes an angle of 30.0° with the yz-plane.

Figure **P27.77**

27.78 •• The rectangular loop shown in Fig. P27.78 is pivoted about the y-axis and carries a current of 15.0 A in the direction indicated. (a) If the loop is in a uniform magnetic field with magnitude 0.48 T in the $+x$-direction, find the magnitude and direction of the torque required to hold the loop in the position shown. (b) Repeat part (a) for the case in which the field is in the $-z$-direction. (c) For each of the above magnetic fields, what torque would be required if the loop were pivoted about an axis through its center, parallel to the y-axis?

Figure **P27.78**

27.79 •• **CP CALC** A thin, uniform rod with negligible mass and length 0.200 m is attached to the floor by a frictionless hinge at point P (Fig. P27.79). A horizontal spring with force constant $k = 4.80$ N/m connects the other end of the rod to a vertical wall. The rod is in a uniform magnetic field $B = 0.340$ T directed into the plane of the figure. There is current $I = 6.50$ A in the rod, in the direction shown. (a) Calculate the torque due to the magnetic force on the rod, for an axis at P. Is it correct to take the total magnetic force to act at the center of gravity of the rod when calculating the torque? Explain. (b) When the rod is in equilibrium and makes an angle of 53.0° with the floor, is the spring stretched or compressed? (c) How much energy is stored in the spring when the rod is in equilibrium?

Figure **P27.79**

27.80 •• The loop of wire shown in Fig. P27.80 forms a right triangle and carries a current $I = 5.00$ A in the direction shown. The loop is in a uniform magnetic field that has magnitude $B = 3.00$ T and the same direction as the current in side PQ of the loop. (a) Find the force exerted by the magnetic field on each side of the triangle. If the force is not zero, specify its direction. (b) What is the net force on the loop? (c) The loop is pivoted about an axis that lies along side *PR*. Use the forces calculated in part (a) to calculate the torque on each side of the loop (see Problem 27.79). (d) What is the magnitude of the net torque on the loop? Calculate the net torque from the torques calculated in part (c) and also from Eq. (27.28). Do these two results agree? (e) Is the net torque directed to rotate point *Q* into the plane of the figure or out of the plane of the figure?

Figure **P27.80**

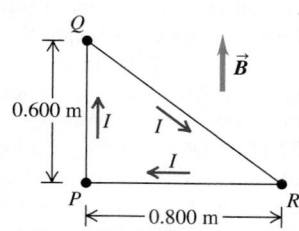

27.81 •• **CP** A uniform, 458-g metal bar 75.0 cm long carries a current *I* in a uniform, horizontal 1.25-T magnetic field as shown in Fig. P27.81. The directions of *I* and $\vec{B}$ are shown in the figure. The bar is free to rotate about a frictionless hinge at point *b*. The other end of the bar rests on a conducting support at point *a* but is not attached there. The bar rests at an angle of 60.0° above the horizontal. What is the largest value the current *I* can have without breaking the electrical contact at *a*? (See Problem 27.77.)

Figure **P27.81**

27.82 •• **Paleoclimate.** Climatologists can determine the past temperature of the earth by comparing the ratio of the isotope oxygen-18 to the isotope oxygen-16 in air trapped in ancient ice sheets, such as those in Greenland. In one method for separating these isotopes, a sample containing both of them is first singly ionized (one electron is removed) and then accelerated from rest through a potential difference *V*. This beam then enters a magnetic field *B* at right angles to the field and is bent into a quarter-circle. A particle detector at the end of the path measures the amount of each isotope. (a) Show that the separation Δr of the two isotopes at the detector is given by

$$\Delta r = \frac{\sqrt{2eV}}{eB}\left(\sqrt{m_{18}} - \sqrt{m_{16}}\right)$$

where m_{16} and m_{18} are the masses of the two oxygen isotopes. (b) The measured masses of the two isotopes are 2.66×10^{-26} kg (^{16}O) and 2.99×10^{-26} kg (^{18}O). If the magnetic field is 0.050 T, what must be the accelerating potential *V* so that these two isotopes will be separated by 4.00 cm at the detector?

27.83 •• **CALC A Voice Coil.** It was shown in Section 27.7 that the net force on a current loop in a *uniform* magnetic field is zero. The magnetic force on the voice coil of a loudspeaker (see Fig. 27.28) is nonzero because the magnetic field at the coil is not uniform. A voice coil in a loudspeaker has 50 turns of wire and a diameter of 1.56 cm, and the current in the coil is 0.950 A. Assume that the magnetic field at each point of the coil has a constant magnitude of 0.220 T and is directed at an angle of 60.0° outward from the normal to the plane of the coil (Fig. P27.83). Let the axis of the coil be in the *y*-direction. The current in the coil is in the direction shown (counterclockwise as viewed from a point above the coil on the *y*-axis). Calculate the magnitude and direction of the net magnetic force on the coil.

Figure **P27.83**

27.84 •• **Quark Model of the Neutron.** The neutron is a particle with zero charge. Nonetheless, it has a nonzero magnetic moment with *z*-component 9.66×10^{-27} A·m². This can be explained by the internal structure of the neutron. A substantial body of evidence indicates that a neutron is composed of three fundamental particles called *quarks*: an "up" (*u*) quark, of charge $+2e/3$, and two "down" (*d*) quarks, each of charge $-e/3$. The combination of the three quarks produces a net charge of $2e/3 - e/3 - e/3 = 0$. If the quarks are in motion, they can produce a nonzero magnetic moment. As a very simple model, suppose the *u* quark moves in a counterclockwise circular path and the *d* quarks move in a clockwise circular path, all of radius *r* and all with the same speed *v* (Fig. P27.84). (a) Determine the current due to the circulation of the *u* quark. (b) Determine the magnitude of the magnetic moment due to the circulating *u* quark. (c) Determine the magnitude of the magnetic moment of the three-quark system. (Be careful to use the correct magnetic moment directions.) (d) With what speed *v* must the quarks move if this model is to reproduce the magnetic moment of the neutron? Use $r = 1.20 \times 10^{-15}$ m (the radius of the neutron) for the radius of the orbits.

Figure **P27.84**

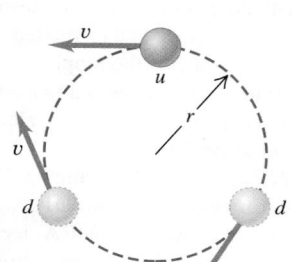

27.85 •• **CALC Force on a Current Loop in a Nonuniform Magnetic Field.** It was shown in Section 27.7 that the net force on a current loop in a *uniform* magnetic field is zero. But what if $\vec{B}$ is *not* uniform? Figure P27.85 shows a square loop of wire that lies in the *xy*-plane. The loop has corners at $(0, 0)$, $(0, L)$, $(L, 0)$, and (L, L) and carries a constant current *I* in the clockwise direction. The magnetic field has no *x*-component but has both *y*- and *z*-components: $\vec{B} = (B_0z/L)\hat{j} + (B_0y/L)\hat{k}$, where B_0 is a positive constant. (a) Sketch the magnetic field lines in the *yz*-plane. (b) Find the magnitude and direction of the magnetic force exerted on each of the sides of the loop by integrating Eq. (27.20). (c) Find the magnitude and direction of the net magnetic force on the loop.

Figure **P27.85**

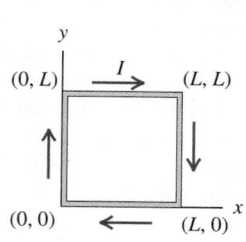

27.86 • **CALC Torque on a Current Loop in a Nonuniform Magnetic Field.** In Section 27.7 the expression for the torque on a current loop was derived assuming that the magnetic field $\vec{B}$ was uniform. But what if $\vec{B}$ is *not* uniform? Figure P27.85 shows a square loop of wire that lies in the *xy*-plane. The loop has corners at $(0, 0)$, $(0, L)$, $(L, 0)$, and (L, L) and carries a constant current *I* in the clockwise direction. The magnetic field has no *z*-component but has both *x*- and *y*-components: $\vec{B} = (B_0y/L)\hat{i} + (B_0x/L)\hat{j}$, where B_0 is a positive constant. (a) Sketch the magnetic field lines in the *xy*-plane. (b) Find the magnitude and direction of the magnetic force exerted on each of the sides of the loop by integrating Eq. (27.20). (c) If the loop is free to rotate about the *x*-axis, find the magnitude and direction of the magnetic torque on the loop. (d) Repeat part (c) for the case in which the loop is free to rotate about the *y*-axis. (e) Is Eq. (27.26), $\vec{\tau} = \vec{\mu} \times \vec{B}$, an appropriate description of the torque on this loop? Why or why not?

27.87 •• **CP** An insulated wire with mass $m = 5.40 \times 10^{-5}$ kg is bent into the shape of an inverted U such that the horizontal part has a length $l = 15.0$ cm. The bent ends of the wire are partially immersed in two pools of mercury, with 2.5 cm of each end below the mercury's surface. The entire structure is in a region containing a uniform 0.00650-T magnetic field directed into the page (Fig. P27.87). An electrical connection from the mercury pools is made through the ends of the wires. The mercury pools are connected to a 1.50-V battery and a switch S. When switch S is closed, the wire jumps 35.0 cm into the air, measured from its initial position. (a) Determine the speed v of the wire as it leaves the mercury. (b) Assuming that the current I through the wire was constant from the time the switch was closed until the wire left the mercury, determine I. (c) Ignoring the resistance of the mercury and the circuit wires, determine the resistance of the moving wire.

Figure **P27.87**

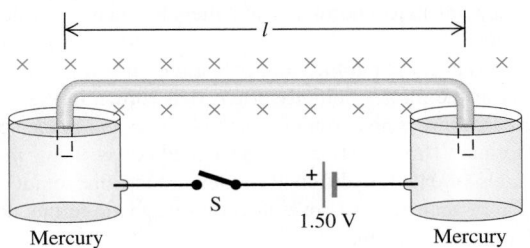

27.88 •• A circular loop of wire with area A lies in the xy-plane. As viewed along the z-axis looking in the $-z$-direction toward the origin, a current I is circulating clockwise around the loop. The torque produced by an external magnetic field $\vec{B}$ is given by $\vec{\tau} = D(4\hat{\imath} - 3\hat{\jmath})$, where D is a positive constant, and for this orientation of the loop the magnetic potential energy $U = -\vec{\mu} \cdot \vec{B}$ is negative. The magnitude of the magnetic field is $B_0 = 13D/IA$. (a) Determine the vector magnetic moment of the current loop. (b) Determine the components B_x, B_y, and B_z of $\vec{B}$.

CHALLENGE PROBLEMS

27.89 ••• A particle with charge 2.15 μC and mass 3.20 × 10^{-11} kg is initially traveling in the $+y$-direction with a speed $v_0 = 1.45 \times 10^5$ m/s. It then enters a region containing a uniform magnetic field that is directed into, and perpendicular to, the page in Fig. P27.89. The magnitude of the field is 0.420 T. The

Figure **P27.89**

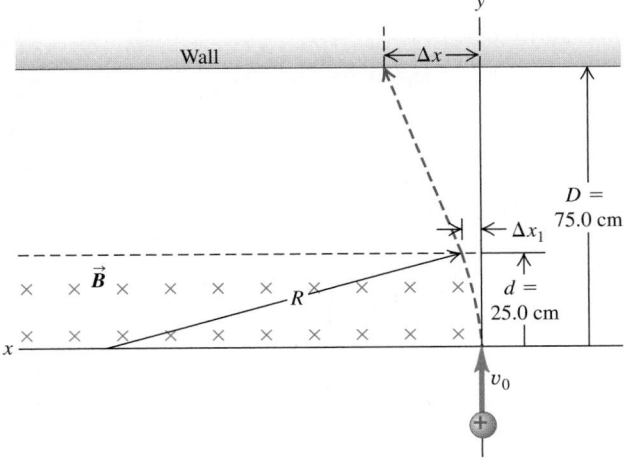

region extends a distance of 25.0 cm along the initial direction of travel; 75.0 cm from the point of entry into the magnetic field region is a wall. The length of the field-free region is thus 50.0 cm. When the charged particle enters the magnetic field, it follows a curved path whose radius of curvature is R. It then leaves the magnetic field after a time t_1, having been deflected a distance Δx_1. The particle then travels in the field-free region and strikes the wall after undergoing a total deflection Δx. (a) Determine the radius R of the curved part of the path. (b) Determine t_1, the time the particle spends in the magnetic field. (c) Determine Δx_1, the horizontal deflection at the point of exit from the field. (d) Determine Δx, the total horizontal deflection.

27.90 ••• **The Electromagnetic Pump.** Magnetic forces acting on conducting fluids provide a convenient means of pumping these fluids. For example, this method can be used to pump blood without the damage to the cells that can be caused by a mechanical pump. A horizontal tube with rectangular cross section (height h, width w) is placed at right angles to a uniform magnetic field with magnitude B so that a length l is in the field (Fig. P27.90). The tube is filled with a conducting liquid, and an electric current of density J is maintained in the third mutually perpendicular direction. (a) Show that the difference of pressure between a point in the liquid on a vertical plane through ab and a point in the liquid on another vertical plane through cd, under conditions in which the liquid is prevented from flowing, is $\Delta p = JlB$. (b) What current density is needed to provide a pressure difference of 1.00 atm between these two points if $B = 2.20$ T and $l = 35.0$ mm?

Figure **P27.90**

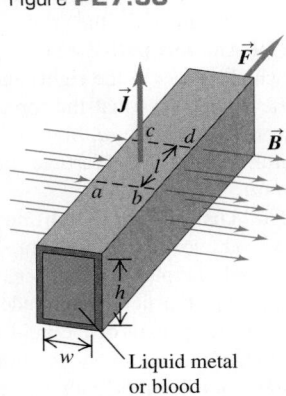

Liquid metal or blood

27.91 ••• **CP A Cycloidal Path.** A particle with mass m and positive charge q starts from rest at the origin shown in Fig. P27.91. There is a uniform electric field $\vec{E}$ in the $+y$-direction and a uniform magnetic field $\vec{B}$ directed out of the page. It is shown in more advanced books that the path is a *cycloid* whose radius of curvature at the top points is twice the y-coordinate at that level. (a) Explain why the path has this general shape and why it is repetitive. (b) Prove that the speed at any point is equal to $\sqrt{2qEy/m}$. (*Hint:* Use energy conservation.) (c) Applying Newton's second law at the top point and taking as given that the radius of curvature here equals $2y$, prove that the speed at this point is $2E/B$.

Figure **P27.91**

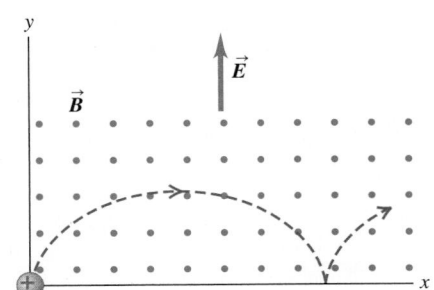

Answers

Chapter Opening Question ?

In MRI the nuclei of hydrogen atoms within soft tissue act like miniature current loops whose magnetic moments align with an applied field. See Section 27.7 for details.

Test Your Understanding Questions

27.1 Answer: yes When a magnet is cut apart, each part has a north and south pole (see Fig. 27.4). Hence the small red part behaves much like the original, full-sized compass needle.

27.2 Answer: path 3 Applying the right-hand rule to the vectors $\vec{v}$ (which points to the right) and $\vec{B}$ (which points into the plane of the figure) says that the force $\vec{F} = q\vec{v} \times \vec{B}$ on a *positive* charge would point *upward*. Since the charge is *negative,* the force points *downward* and the particle follows a trajectory that curves downward.

27.3 Answers: (a) (ii), (b) no The magnitude of $\vec{B}$ would increase as you moved to the right, reaching a maximum as you pass through the plane of the loop. As you moved beyond the plane of the loop, the field magnitude would decrease. You can tell this from the spacing of the field lines: The closer the field lines, the stronger the field. The direction of the field would be to the right at all points along the path, since the path is along a field line and the direction of $\vec{B}$ at any point is tangent to the field line through that point.

27.4 Answers: (a) (ii), (b) (i) The radius of the orbit as given by Eq. (27.11) is directly proportional to the speed, so doubling the particle speed causes the radius to double as well. The particle has twice as far to travel to complete one orbit but is traveling at double the speed, so the time for one orbit is unchanged. This result also follows from Eq. (27.12), which states that the angular speed ω is independent of the linear speed v. Hence the time per orbit, $T = 2\pi/\omega$, likewise does not depend on v.

27.5 Answer: (iii) From Eq. (27.13), the speed $v = E/B$ at which particles travel straight through the velocity selector does not depend on the magnitude or sign of the charge or the mass of the particle. All that is required is that the particles (in this case, ions) have a nonzero charge.

27.6 Answer: A This orientation will cause current to flow clockwise around the circuit and hence through the conducting bar in the direction from the top to the bottom of the figure. From the right-hand rule, the magnetic force $\vec{F} = I\vec{l} \times \vec{B}$ on the bar will then point to the right.

27.7 Answers: (a) to the right; (b) north pole on the right, south pole on the left If you wrap the fingers of your right hand around the coil in the direction of the current, your right thumb points to the right (perpendicular to the plane of the coil). This is the direction of the magnetic moment $\vec{\mu}$. The magnetic moment points from the south pole to the north pole, so the right side of the loop is equivalent to a north pole and the left side is equivalent to a south pole.

27.8 Answer: no The rotor will not begin to turn when the switch is closed if the rotor is initially oriented as shown in Fig. 27.39b. In this case there is no current through the rotor and hence no magnetic torque. This situation can be remedied by using multiple rotor coils oriented at different angles around the rotation axis. With this arrangement, there is always a magnetic torque no matter what the orientation.

27.9 Answer: (ii) The mobile charge carriers in copper are negatively charged electrons, which move upward through the wire to give a downward current. From the right-hand rule, the force on a positively charged particle moving upward in a westward-pointing magnetic field would be to the south; hence the force on a negatively charged particle is to the north. The result is an excess of negative charge on the north side of the wire, leaving an excess of positive charge—and hence a higher electric potential—on the south side.

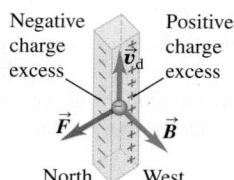

Bridging Problem

Answers: (a) $\tau_x = -1.54 \times 10^{-4}\,\text{N} \cdot \text{m}$,
$\tau_y = -2.05 \times 10^{-4}\,\text{N} \cdot \text{m}$,
$\tau_z = -6.14 \times 10^{-4}\,\text{N} \cdot \text{m}$
(b) $-7.55 \times 10^{-4}\,\text{J}$ (c) 42.1 rad/s

SOURCES OF MAGNETIC FIELD

? The immense cylinder in this photograph is actually a current-carrying coil, or solenoid, that generates a uniform magnetic field in its interior as part of an experiment at CERN, the European Organization for Nuclear Research. If two such solenoids were joined end to end, how much stronger would the magnetic field become?

LEARNING GOALS

By studying this chapter, you will learn:

- The nature of the magnetic field produced by a single moving charged particle.

- How to describe the magnetic field produced by an element of a current-carrying conductor.

- How to calculate the magnetic field produced by a long, straight, current-carrying wire.

- Why wires carrying current in the same direction attract, while wires carrying opposing currents repel.

- How to calculate the magnetic field produced by a current-carrying wire bent into a circle.

- What Ampere's law is, and what it tells us about magnetic fields.

- How to use Ampere's law to calculate the magnetic field of symmetric current distributions.

In Chapter 27 we studied the forces exerted on moving charges and on current-carrying conductors in a magnetic field. We didn't worry about how the magnetic field got there; we simply took its existence as a given fact. But how are magnetic fields *created*? We know that both permanent magnets and electric currents in electromagnets create magnetic fields. In this chapter we will study these sources of magnetic field in detail.

We've learned that a charge creates an electric field and that an electric field exerts a force on a charge. But a *magnetic* field exerts a force only on a *moving* charge. Is it also true that a charge *creates* a magnetic field only when the charge is moving? In a word, yes.

Our analysis will begin with the magnetic field created by a single moving point charge. We can use this analysis to determine the field created by a small segment of a current-carrying conductor. Once we can do that, we can in principle find the magnetic field produced by *any* shape of conductor.

Then we will introduce Ampere's law, which plays a role in magnetism analogous to the role of Gauss's law in electrostatics. Ampere's law lets us exploit symmetry properties in relating magnetic fields to their sources.

Moving charged particles within atoms respond to magnetic fields and can also act as sources of magnetic field. We'll use these ideas to understand how certain magnetic materials can be used to intensify magnetic fields as well as why some materials such as iron act as permanent magnets.

28.1 Magnetic Field of a Moving Charge

Let's start with the basics, the magnetic field of a single point charge q moving with a constant velocity $\vec{v}$. In practical applications, such as the solenoid shown in the photo that opens this chapter, magnetic fields are produced by tremendous numbers of charged particles moving together in a current. But once we understand how to calculate the magnetic field due to a single point charge, it's a small leap to calculate the field due to a current-carrying wire or collection of wires.

As we did for electric fields, we call the location of the moving charge at a given instant the **source point** and the point P where we want to find the field the **field point.** In Section 21.4 we found that at a field point a distance r from a point charge q, the magnitude of the *electric* field $\vec{E}$ caused by the charge is proportional to the charge magnitude $|q|$ and to $1/r^2$, and the direction of $\vec{E}$ (for positive q) is along the line from source point to field point. The corresponding relationship for the *magnetic* field $\vec{B}$ of a point charge q moving with constant velocity has some similarities and some interesting differences.

Experiments show that the magnitude of $\vec{B}$ is also proportional to $|q|$ and to $1/r^2$. But the *direction* of $\vec{B}$ is *not* along the line from source point to field point. Instead, $\vec{B}$ is perpendicular to the plane containing this line and the particle's velocity vector $\vec{v}$, as shown in Fig. 28.1. Furthermore, the field *magnitude B* is also proportional to the particle's speed v and to the sine of the angle ϕ. Thus the magnetic field magnitude at point P is given by

$$B = \frac{\mu_0}{4\pi} \frac{|q| v \sin\phi}{r^2} \qquad (28.1)$$

where $\mu_0/4\pi$ is a proportionality constant (μ_0 is read as "mu-nought" or "mu-sub-zero"). The reason for writing the constant in this particular way will emerge shortly. We did something similar with Coulomb's law in Section 21.3.

Moving Charge: Vector Magnetic Field

We can incorporate both the magnitude and direction of $\vec{B}$ into a single vector equation using the vector product. To avoid having to say "the direction from the source q to the field point P" over and over, we introduce a *unit* vector $\hat{r}$ ("r-hat") that points from the source point to the field point. (We used $\hat{r}$ for the same purpose in Section 21.4.) This unit vector is equal to the vector $\vec{r}$ from the source to the field point divided by its magnitude: $\hat{r} = \vec{r}/r$. Then the $\vec{B}$ field of a moving point charge is

$$\vec{B} = \frac{\mu_0}{4\pi} \frac{q\vec{v} \times \hat{r}}{r^2} \qquad \begin{array}{l}\text{(magnetic field of a point charge} \\ \text{with constant velocity)}\end{array} \qquad (28.2)$$

Figure 28.1 shows the relationship of $\hat{r}$ to P and also shows the magnetic field $\vec{B}$ at several points in the vicinity of the charge. At all points along a line through the charge parallel to the velocity $\vec{v}$, the field is zero because $\sin\phi = 0$ at all such points. At any distance r from q, $\vec{B}$ has its greatest magnitude at points lying in the plane perpendicular to $\vec{v}$, because there $\phi = 90°$ and $\sin\phi = 1$. If q is negative, the directions of $\vec{B}$ are opposite to those shown in Fig. 28.1.

Moving Charge: Magnetic Field Lines

A point charge in motion also produces an *electric* field, with field lines that radiate outward from a positive charge. The *magnetic* field lines are completely different. For a point charge moving with velocity $\vec{v}$, the magnetic field lines are *circles* centered on the line of $\vec{v}$ and lying in planes perpendicular to this line. The field-line directions for a positive charge are given by the following *right-hand rule,* one of several that we will encounter in this chapter: Grasp the velocity vector $\vec{v}$ with your right hand so that your right thumb points in the direction of $\vec{v}$; your fingers then curl around the line of $\vec{v}$ in the same sense as the magnetic field lines, assuming q is positive. Figure 28.1a shows parts of a few field lines; Fig. 28.1b shows some field lines in a plane through q, perpendicular to $\vec{v}$. If the point charge is negative, the directions of the field and field lines are the opposite of those shown in Fig. 28.1.

28.1 (a) Magnetic-field vectors due to a moving positive point charge q. At each point, $\vec{B}$ is perpendicular to the plane of $\vec{r}$ and $\vec{v}$, and its magnitude is proportional to the sine of the angle between them. (b) Magnetic field lines in a plane containing a moving positive charge.

(a) Perspective view

Right-hand rule for the magnetic field due to a positive charge moving at constant velocity: Point the thumb of your right hand in the direction of the velocity. Your fingers now curl around the charge in the direction of the magnetic field lines. (If the charge is negative, the field lines are in the opposite direction.)

For these field points, $\vec{r}$ and $\vec{v}$ both lie in the beige plane, and $\vec{B}$ is perpendicular to this plane.

For these field points, $\vec{r}$ and $\vec{v}$ both lie in the gold plane, and $\vec{B}$ is perpendicular to this plane.

(b) View from behind the charge

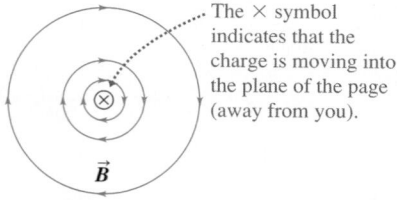

The $\times$ symbol indicates that the charge is moving into the plane of the page (away from you).

Equations (28.1) and (28.2) describe the $\vec{B}$ field of a point charge moving with *constant* velocity. If the charge *accelerates,* the field can be much more complicated. We won't need these more complicated results for our purposes. (The moving charged particles that make up a current in a wire accelerate at points where the wire bends and the direction of $\vec{v}$ changes. But because the magnitude v_d of the drift velocity in a conductor is typically very small, the centripetal acceleration v_d^2/r is so small that we can ignore its effects.)

As we discussed in Section 27.2, the unit of B is one tesla (1 T):

$$1\,\text{T} = 1\,\text{N} \cdot \text{s/C} \cdot \text{m} = 1\,\text{N/A} \cdot \text{m}$$

Using this with Eq. (28.1) or (28.2), we find that the units of the constant μ_0 are

$$1\,\text{N} \cdot \text{s}^2/\text{C}^2 = 1\,\text{N/A}^2 = 1\,\text{Wb/A} \cdot \text{m} = 1\,\text{T} \cdot \text{m/A}$$

In SI units the numerical value of μ_0 is exactly $4\pi \times 10^{-7}$. Thus

$$\mu_0 = 4\pi \times 10^{-7}\,\text{N} \cdot \text{s}^2/\text{C}^2 = 4\pi \times 10^{-7}\,\text{Wb/A} \cdot \text{m}$$
$$= 4\pi \times 10^{-7}\,\text{T} \cdot \text{m/A} \qquad (28.3)$$

It may seem incredible that μ_0 has *exactly* this numerical value! In fact this is a *defined* value that arises from the definition of the ampere, as we'll discuss in Section 28.4.

We mentioned in Section 21.3 that the constant $1/4\pi\epsilon_0$ in Coulomb's law is related to the speed of light c:

$$k = \frac{1}{4\pi\epsilon_0} = (10^{-7}\,\text{N} \cdot \text{s}^2/\text{C}^2)c^2$$

When we study electromagnetic waves in Chapter 32, we will find that their speed of propagation in vacuum, which is equal to the speed of light c, is given by

$$c^2 = \frac{1}{\epsilon_0\mu_0} \qquad (28.4)$$

If we solve the equation $k = 1/4\pi\epsilon_0$ for ϵ_0, substitute the resulting expression into Eq. (28.4), and solve for μ_0, we indeed get the value of μ_0 stated above. This discussion is a little premature, but it may give you a hint that electric and magnetic fields are intimately related to the nature of light.

Example 28.1 Forces between two moving protons

Two protons move parallel to the x-axis in opposite directions (Fig. 28.2) at the same speed v (small compared to the speed of light c). At the instant shown, find the electric and magnetic forces on the upper proton and compare their magnitudes.

SOLUTION

IDENTIFY and SET UP: Coulomb's law [Eq. (21.2)] gives the electric force F_E on the upper proton. The magnetic force law [Eq. (27.2)] gives the magnetic force on the upper proton; to use it, we must first use Eq. (28.2) to find the magnetic field that the lower proton produces at the position of the upper proton. The unit vector from the lower proton (the source) to the position of the upper proton is $\hat{r} = \hat{\jmath}$.

EXECUTE: From Coulomb's law, the magnitude of the electric force on the upper proton is

$$F_E = \frac{1}{4\pi\epsilon_0}\frac{q^2}{r^2}$$

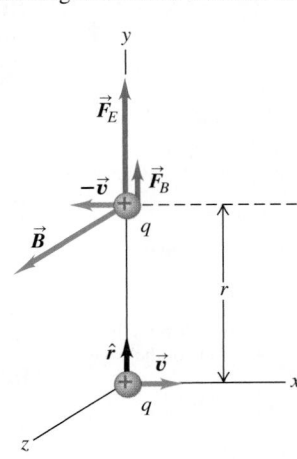

28.2 Electric and magnetic forces between two moving protons.

Continued

The forces are repulsive, and the force on the upper proton is vertically upward (in the +y-direction).

The velocity of the lower proton is $\vec{v} = v\hat{\imath}$. From the right-hand rule for the cross product $\vec{v} \times \hat{r}$ in Eq. (28.2), the $\vec{B}$ field due to the lower proton at the position of the upper proton is in the +z-direction (see Fig. 28.2). From Eq. (28.2), the field is

$$\vec{B} = \frac{\mu_0}{4\pi} \frac{q(v\hat{\imath}) \times \hat{\jmath}}{r^2} = \frac{\mu_0}{4\pi} \frac{qv}{r^2} \hat{k}$$

The velocity of the upper proton is $-\vec{v} = -v\hat{\imath}$, so the magnetic force on it is

$$\vec{F}_B = q(-\vec{v}) \times \vec{B} = q(-v\hat{\imath}) \times \frac{\mu_0}{4\pi} \frac{qv}{r^2} \hat{k} = \frac{\mu_0}{4\pi} \frac{q^2v^2}{r^2} \hat{\jmath}$$

The magnetic interaction in this situation is also repulsive. The ratio of the force magnitudes is

$$\frac{F_B}{F_E} = \frac{\mu_0 q^2 v^2 / 4\pi r^2}{q^2 / 4\pi\epsilon_0 r^2} = \frac{\mu_0 v^2}{1/\epsilon_0} = \epsilon_0 \mu_0 v^2$$

With the relationship $\epsilon_0\mu_0 = 1/c^2$, Eq. (28.4), this becomes

$$\frac{F_B}{F_E} = \frac{v^2}{c^2}$$

When v is small in comparison to the speed of light, the magnetic force is much smaller than the electric force.

EVALUATE: We have described the velocities, fields, and forces as they are measured by an observer who is stationary in the coordinate system of Fig. 28.2. In a coordinate system that moves with one of the charges, one of the velocities would be zero, so there would be *no* magnetic force. The explanation of this apparent paradox provided one of the paths that led to the special theory of relativity.

Test Your Understanding of Section 28.1 (a) If two protons are traveling parallel to each other in the *same* direction and at the same speed, is the magnetic force between them (i) attractive or (ii) repulsive? (b) Is the net force between them (i) attractive, (ii) repulsive, or (iii) zero? (Assume that the protons' speed is much slower than the speed of light.)

28.3 (a) Magnetic-field vectors due to a current element $d\vec{l}$. (b) Magnetic field lines in a plane containing the current element $d\vec{l}$. Compare this figure to Fig. 28.1 for the field of a moving point charge.

(a) Perspective view

Right-hand rule for the magnetic field due to a current element: Point the thumb of your right hand in the direction of the current. Your fingers now curl around the current element in the direction of the magnetic field lines.

For these field points, $\vec{r}$ and $d\vec{l}$ both lie in the beige plane, and $d\vec{B}$ is perpendicular to this plane.

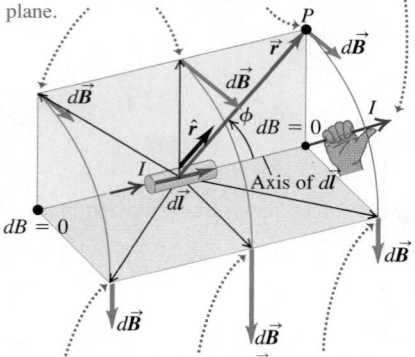

For these field points, $\vec{r}$ and $d\vec{l}$ both lie in the gold plane, and $d\vec{B}$ is perpendicular to this plane.

(b) View along the axis of the current element

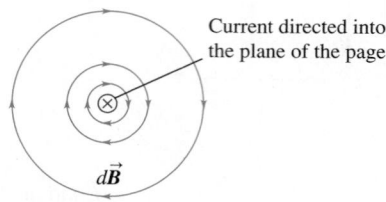

Current directed into the plane of the page

28.2 Magnetic Field of a Current Element

Just as for the electric field, there is a **principle of superposition of magnetic fields:**

> **The total magnetic field caused by several moving charges is the vector sum of the fields caused by the individual charges.**

We can use this principle with the results of Section 28.1 to find the magnetic field produced by a current in a conductor.

We begin by calculating the magnetic field caused by a short segment $d\vec{l}$ of a current-carrying conductor, as shown in Fig. 28.3a. The volume of the segment is $A\, dl$, where A is the cross-sectional area of the conductor. If there are n moving charged particles per unit volume, each of charge q, the total moving charge dQ in the segment is

$$dQ = nqA\, dl$$

The moving charges in this segment are equivalent to a single charge dQ, traveling with a velocity equal to the *drift* velocity $\vec{v}_d$. (Magnetic fields due to the *random* motions of the charges will, on average, cancel out at every point.) From Eq. (28.1) the magnitude of the resulting field $d\vec{B}$ at any field point P is

$$dB = \frac{\mu_0}{4\pi} \frac{|dQ|v_d \sin\phi}{r^2} = \frac{\mu_0}{4\pi} \frac{n|q|v_d A\, dl \sin\phi}{r^2}$$

But from Eq. (25.2), $n|q|v_d A$ equals the current I in the element. So

$$dB = \frac{\mu_0}{4\pi} \frac{I\, dl \sin\phi}{r^2} \tag{28.5}$$

Current Element: Vector Magnetic Field

In vector form, using the unit vector $\hat{r}$ as in Section 28.1, we have

$$d\vec{B} = \frac{\mu_0}{4\pi} \frac{I\, d\vec{l} \times \hat{r}}{r^2} \qquad \text{(magnetic field of a current element)} \tag{28.6}$$

where $d\vec{l}$ is a vector with length dl, in the same direction as the current in the conductor.

Equations (28.5) and (28.6) are called the **law of Biot and Savart** (pronounced "Bee-oh" and "Suh-var"). We can use this law to find the total magnetic field $\vec{B}$ at any point in space due to the current in a complete circuit. To do this, we integrate Eq. (28.6) over all segments $d\vec{l}$ that carry current; symbolically,

$$\vec{B} = \frac{\mu_0}{4\pi} \int \frac{I\, d\vec{l} \times \hat{r}}{r^2} \qquad (28.7)$$

In the following sections we will carry out this vector integration for several examples.

Current Element: Magnetic Field Lines

As Fig. 28.3 shows, the field vectors $d\vec{B}$ and the magnetic field lines of a current element are exactly like those set up by a positive charge dQ moving in the direction of the drift velocity $\vec{v}_d$. The field lines are circles in planes perpendicular to $d\vec{l}$ and centered on the line of $d\vec{l}$. Their directions are given by the same right-hand rule that we introduced for point charges in Section 28.1.

We can't verify Eq. (28.5) or (28.6) directly because we can never experiment with an isolated segment of a current-carrying circuit. What we measure experimentally is the *total* $\vec{B}$ for a complete circuit. But we can still verify these equations indirectly by calculating $\vec{B}$ for various current configurations using Eq. (28.7) and comparing the results with experimental measurements.

If matter is present in the space around a current-carrying conductor, the field at a field point P in its vicinity will have an additional contribution resulting from the *magnetization* of the material. We'll return to this point in Section 28.8. However, unless the material is iron or some other ferromagnetic material, the additional field is small and is usually negligible. Additional complications arise if time-varying electric or magnetic fields are present or if the material is a superconductor; we'll return to these topics later.

Problem-Solving Strategy 28.1 | **Magnetic-Field Calculations**

IDENTIFY *the relevant concepts:* The Biot–Savart law [Eqs. (28.5) and (28.6)] allows you to calculate the magnetic field at a field point P due to a current-carrying wire of any shape. The idea is to calculate the field element $d\vec{B}$ at P due to a representative current element in the wire and integrate all such field elements to find the field $\vec{B}$ at P.

SET UP *the problem* using the following steps:
1. Make a diagram showing a representative current element and the field point P.
2. Draw the current element $d\vec{l}$, being careful that it points in the direction of the current.
3. Draw the unit vector $\hat{r}$ directed *from* the current element (the source point) to P.
4. Identify the target variable (usually $\vec{B}$).

EXECUTE *the solution* as follows:
1. Use Eq. (28.5) or (28.6) to express the magnetic field $d\vec{B}$ at P from the representative current element.
2. Add up all the $d\vec{B}$'s to find the total field at point P. In some situations the $d\vec{B}$'s at point P have the same direction for all the current elements; then the magnitude of the total $\vec{B}$ field is the

sum of the magnitudes of the $d\vec{B}$'s. But often the $d\vec{B}$'s have different directions for different current elements. Then you have to set up a coordinate system and represent each $d\vec{B}$ in terms of its components. The integral for the total $\vec{B}$ is then expressed in terms of an integral for each component.
3. Sometimes you can use the symmetry of the situation to prove that one component of $\vec{B}$ must vanish. Always be alert for ways to use symmetry to simplify the problem.
4. Look for ways to use the principle of superposition of magnetic fields. Later in this chapter we'll determine the fields produced by certain simple conductor shapes; if you encounter a conductor of a complex shape that can be represented as a combination of these simple shapes, you can use superposition to find the field of the complex shape. Examples include a rectangular loop and a semicircle with straight line segments on both sides.

EVALUATE *your answer:* Often your answer will be a mathematical expression for $\vec{B}$ as a function of the position of the field point. Check the answer by examining its behavior in as many limits as you can.

| Example 28.2 | **Magnetic field of a current segment** |

A copper wire carries a steady 125-A current to an electroplating tank (Fig. 28.4). Find the magnetic field due to a 1.0-cm segment of this wire at a point 1.2 m away from it, if the point is (a) point P_1, straight out to the side of the segment, and (b) point P_2, in the xy-plane and on a line at 30° to the segment.

SOLUTION

IDENTIFY and SET UP: Although Eqs. (28.5) and (28.6) apply only to infinitesimal current elements, we may use either of them here because the segment length is much less than the distance to the field point. The current element is shown in red in Fig. 28.4 and points in the $-x$-direction (the direction of the current), so $d\vec{l} = dl(-\hat{\imath})$. The unit vector $\hat{r}$ for each field point is directed from the current element toward that point: $\hat{r}$ is in the $+y$-direction for point P_1 and at an angle of 30° above the $-x$-direction for point P_2.

28.4 Finding the magnetic field at two points due to a 1.0-cm segment of current-carrying wire (not shown to scale).

EXECUTE: (a) At point P_1, $\hat{r} = \hat{\jmath}$, so

$$\vec{B} = \frac{\mu_0}{4\pi}\frac{I\,d\vec{l}\times\hat{r}}{r^2} = \frac{\mu_0}{4\pi}\frac{I\,dl(-\hat{\imath})\times\hat{\jmath}}{r^2} = -\frac{\mu_0}{4\pi}\frac{I\,dl}{r^2}\hat{k}$$

$$= -(10^{-7}\,\text{T}\cdot\text{m/A})\frac{(125\,\text{A})(1.0\times10^{-2}\,\text{m})}{(1.2\,\text{m})^2}\hat{k}$$

$$= -(8.7\times10^{-8}\,\text{T})\hat{k}$$

The direction of $\vec{B}$ at P_1 is into the xy-plane of Fig. 28.4.

(b) At P_2, the unit vector is $\hat{r} = (-\cos30°)\hat{\imath} + (\sin30°)\hat{\jmath}$. From Eq. (28.6),

$$\vec{B} = \frac{\mu_0}{4\pi}\frac{I\,d\vec{l}\times\hat{r}}{r^2} = \frac{\mu_0}{4\pi}\frac{I\,dl(-\hat{\imath})\times(-\cos30°\hat{\imath}+\sin30°\hat{\jmath})}{r^2}$$

$$= -\frac{\mu_0 I}{4\pi}\frac{dl\sin30°}{r^2}\hat{k}$$

$$= -(10^{-7}\,\text{T}\cdot\text{m/A})\frac{(125\,\text{A})(1.0\times10^{-2}\,\text{m})(\sin30°)}{(1.2\,\text{m})^2}\hat{k}$$

$$= -(4.3\times10^{-8}\,\text{T})\hat{k}$$

The direction of $\vec{B}$ at P_2 is also into the xy-plane of Fig. 28.4.

EVALUATE: We can check our results for the direction of $\vec{B}$ by comparing them with Fig. 28.3. The xy-plane in Fig. 28.4 corresponds to the beige plane in Fig. 28.3, but here the direction of the current and hence of $d\vec{l}$ is the reverse of that shown in Fig. 28.3. Hence the direction of the magnetic field is reversed as well. Hence the field at points in the xy-plane in Fig. 28.4 must point *into*, not out of, that plane. This is just what we concluded above.

Test Your Understanding of Section 28.2 An infinitesimal current element located at the origin ($x = y = z = 0$) carries current I in the positive y-direction. Rank the following locations in order of the strength of the magnetic field that the current element produces at that location, from largest to smallest value. (i) $x = L$, $y = 0$, $z = 0$; (ii) $x = 0$, $y = L$, $z = 0$; (iii) $x = 0$, $y = 0$, $z = L$; (iv) $x = L/\sqrt{2}$, $y = L/\sqrt{2}$, $z = 0$.

28.5 Magnetic field produced by a straight current-carrying conductor of length $2a$.

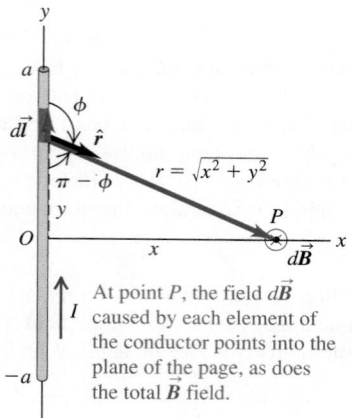

At point P, the field $d\vec{B}$ caused by each element of the conductor points into the plane of the page, as does the total $\vec{B}$ field.

28.3 Magnetic Field of a Straight Current-Carrying Conductor

Let's use the law of Biot and Savart to find the magnetic field produced by a straight current-carrying conductor. This result is useful because straight conducting wires are found in essentially all electric and electronic devices. Figure 28.5 shows such a conductor with length $2a$ carrying a current I. We will find $\vec{B}$ at a point a distance x from the conductor on its perpendicular bisector.

We first use the law of Biot and Savart, Eq. (28.5), to find the field $d\vec{B}$ caused by the element of conductor of length $dl = dy$ shown in Fig. 28.5. From the figure, $r = \sqrt{x^2 + y^2}$ and $\sin\phi = \sin(\pi - \phi) = x/\sqrt{x^2 + y^2}$. The right-hand rule for the vector product $d\vec{l} \times \hat{r}$ shows that the *direction* of $d\vec{B}$ is into the plane of the figure, perpendicular to the plane; furthermore, the directions of the $d\vec{B}$'s from *all* elements of the conductor are the same. Thus in integrating Eq. (28.7), we can just add the *magnitudes* of the $d\vec{B}$'s, a significant simplification.

Putting the pieces together, we find that the magnitude of the total $\vec{B}$ field is

$$B = \frac{\mu_0 I}{4\pi} \int_{-a}^{a} \frac{x\, dy}{(x^2 + y^2)^{3/2}}$$

We can integrate this by trigonometric substitution or by using an integral table:

$$B = \frac{\mu_0 I}{4\pi} \frac{2a}{x\sqrt{x^2 + a^2}} \qquad (28.8)$$

When the length $2a$ of the conductor is very great in comparison to its distance x from point P, we can consider it to be infinitely long. When a is much larger than x, $\sqrt{x^2 + a^2}$ is approximately equal to a; hence in the limit $a \to \infty$, Eq. (28.8) becomes

$$B = \frac{\mu_0 I}{2\pi x}$$

The physical situation has axial symmetry about the y-axis. Hence $\vec{B}$ must have the same *magnitude* at all points on a circle centered on the conductor and lying in a plane perpendicular to it, and the *direction* of $\vec{B}$ must be everywhere tangent to such a circle (Fig. 28.6). Thus, at all points on a circle of radius r around the conductor, the magnitude B is

$$B = \frac{\mu_0 I}{2\pi r} \qquad \text{(near a long, straight, current-carrying conductor)} \quad (28.9)$$

The geometry of this problem is similar to that of Example 21.10 (Section 21.5), in which we solved the problem of the *electric* field caused by an infinite line of charge. The same integral appears in both problems, and the field magnitudes in both problems are proportional to $1/r$. But the lines of $\vec{B}$ in the magnetic problem have completely different shapes than the lines of $\vec{E}$ in the analogous electrical problem. Electric field lines radiate outward from a positive line charge distribution (inward for negative charges). By contrast, magnetic field lines *encircle* the current that acts as their source. Electric field lines due to charges begin and end at those charges, but magnetic field lines always form closed loops and *never* have end points, irrespective of the shape of the current-carrying conductor that sets up the field. As we discussed in Section 27.3, this is a consequence of Gauss's law for magnetism, which states that the total magnetic flux through *any* closed surface is always zero:

$$\oint \vec{B} \cdot d\vec{A} = 0 \qquad \text{(magnetic flux through any closed surface)} \quad (28.10)$$

Any magnetic field line that enters a closed surface must also emerge from that surface.

28.6 Magnetic field around a long, straight, current-carrying conductor. The field lines are circles, with directions determined by the right-hand rule.

Right-hand rule for the magnetic field around a current-carrying wire: Point the thumb of your right hand in the direction of the current. Your fingers now curl around the wire in the direction of the magnetic field lines.

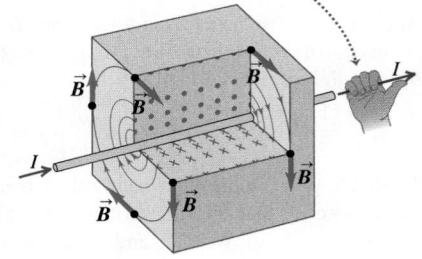

Mastering**PHYSICS**

ActivPhysics 13.1: Magnetic Field of a Wire

Example 28.3   Magnetic field of a single wire

A long, straight conductor carries a 1.0-A current. At what distance from the axis of the conductor does the resulting magnetic field have magnitude $B = 0.5 \times 10^{-4}$ T (about that of the earth's magnetic field in Pittsburgh)?

SOLUTION

IDENTIFY and SET UP: The length of a "long" conductor is much greater than the distance from the conductor to the field point. Hence we can use the ideas of this section. The geometry is the same as that of Fig. 28.6, so we use Eq. (28.9). All of the quantities in this equation are known except the target variable, the distance r.

EXECUTE: We solve Eq. (28.9) for r:

$$r = \frac{\mu_0 I}{2\pi B} = \frac{(4\pi \times 10^{-7}\, \text{T} \cdot \text{m/A})(1.0\, \text{A})}{(2\pi)(0.5 \times 10^{-4}\, \text{T})}$$

$$= 4 \times 10^{-3}\, \text{m} = 4\, \text{mm}$$

EVALUATE: As we saw in Example 26.14, currents of an ampere or more are typical of those found in the wiring of home appliances. This example shows that the magnetic fields produced by these appliances are very weak even very close to the wire; the fields are proportional to $1/r$, so they become even weaker at greater distances.

Example 28.4 | Magnetic field of two wires

Figure 28.7a is an end-on view of two long, straight, parallel wires perpendicular to the *xy*-plane, each carrying a current *I* but in opposite directions. (a) Find $\vec{B}$ at points P_1, P_2, and P_3. (b) Find an expression for $\vec{B}$ at any point on the *x*-axis to the right of wire 2.

SOLUTION

IDENTIFY and SET UP: We can find the magnetic fields $\vec{B}_1$ and $\vec{B}_2$ due to wires 1 and 2 at each point using the ideas of this section. By the superposition principle, the magnetic field at each point is then $\vec{B} = \vec{B}_1 + \vec{B}_2$. We use Eq. (28.9) to find the magnitudes B_1 and B_2 of these fields and the right-hand rule to find the corresponding directions. Figure 28.7a shows $\vec{B}_1$, $\vec{B}_2$, and $\vec{B} = \vec{B}_{total}$ at each point; you should confirm that the directions and relative magnitudes shown are correct. Figure 28.7b shows some of the magnetic field lines due to this two-wire system.

EXECUTE: (a) Since point P_1 is a distance $2d$ from wire 1 and a distance $4d$ from wire 2, $B_1 = \mu_0 I / 2\pi(2d) = \mu_0 I / 4\pi d$ and $B_2 = \mu_0 I / 2\pi(4d) = \mu_0 I / 8\pi d$. The right-hand rule shows that $\vec{B}_1$ is in the negative *y*-direction and $\vec{B}_2$ is in the positive *y*-direction, so

$$\vec{B}_{total} = \vec{B}_1 + \vec{B}_2 = -\frac{\mu_0 I}{4\pi d}\hat{j} + \frac{\mu_0 I}{8\pi d}\hat{j} = -\frac{\mu_0 I}{8\pi d}\hat{j} \quad \text{(point } P_1\text{)}$$

At point P_2, a distance d from both wires, $\vec{B}_1$ and $\vec{B}_2$ are both in the positive *y*-direction, and both have the same magnitude $B_1 = B_2 = \mu_0 I / 2\pi d$. Hence

$$\vec{B}_{total} = \vec{B}_1 + \vec{B}_2 = \frac{\mu_0 I}{2\pi d}\hat{j} + \frac{\mu_0 I}{2\pi d}\hat{j} = \frac{\mu_0 I}{\pi d}\hat{j} \quad \text{(point } P_2\text{)}$$

Finally, at point P_3 the right-hand rule shows that $\vec{B}_1$ is in the positive *y*-direction and $\vec{B}_2$ is in the negative *y*-direction. This point is a distance $3d$ from wire 1 and a distance d from wire 2, so $B_1 = \mu_0 I / 2\pi(3d) = \mu_0 I / 6\pi d$ and $B_2 = \mu_0 I / 2\pi d$. The total field at P_3 is

$$\vec{B}_{total} = \vec{B}_1 + \vec{B}_2 = \frac{\mu_0 I}{6\pi d}\hat{j} - \frac{\mu_0 I}{2\pi d}\hat{j} = -\frac{\mu_0 I}{3\pi d}\hat{j} \quad \text{(point } P_3\text{)}$$

The same technique can be used to find $\vec{B}_{total}$ at any point; for points off the *x*-axis, caution must be taken in vector addition, since $\vec{B}_1$ and $\vec{B}_2$ need no longer be simply parallel or antiparallel.

(b) At any point on the *x*-axis to the right of wire 2 (that is, for $x > d$), $\vec{B}_1$ and $\vec{B}_2$ are in the same directions as at P_3. Such a point is a distance $x + d$ from wire 1 and a distance $x - d$ from wire 2, so the total field is

$$\vec{B}_{total} = \vec{B}_1 + \vec{B}_2 = \frac{\mu_0 I}{2\pi(x+d)}\hat{j} - \frac{\mu_0 I}{2\pi(x-d)}\hat{j}$$

$$= -\frac{\mu_0 I d}{\pi(x^2 - d^2)}\hat{j}$$

where we combined the two terms using a common denominator.

EVALUATE: Consider our result from part (b) at a point very far from the wires, so that *x* is much larger than *d*. Then the d^2 term in the denominator can be neglected, and the magnitude of the total field is approximately $B_{total} = \mu_0 I d / \pi x^2$. For a single wire, Eq. (28.9) shows that the magnetic field decreases with distance in proportion to $1/x$; for two wires carrying opposite currents, $\vec{B}_1$ and $\vec{B}_2$ partially cancel each other, and so B_{total} decreases more rapidly, in proportion to $1/x^2$. This effect is used in communication systems such as telephone or computer networks. The wiring is arranged so that a conductor carrying a signal in one direction and the conductor carrying the return signal are side by side, as in Fig. 28.7a, or twisted around each other (Fig. 28.8). As a result, the magnetic field due to these signals *outside* the conductors is very small, making it less likely to exert unwanted forces on other information-carrying currents.

28.8 Computer cables, or cables for audio-video equipment, create little or no magnetic field. This is because within each cable, closely spaced wires carry current in both directions along the length of the cable. The magnetic fields from these opposing currents cancel each other.

28.7 (a) Two long, straight conductors carrying equal currents in opposite directions. The conductors are seen end-on. (b) Map of the magnetic field produced by the two conductors. The field lines are closest together between the conductors, where the field is strongest.

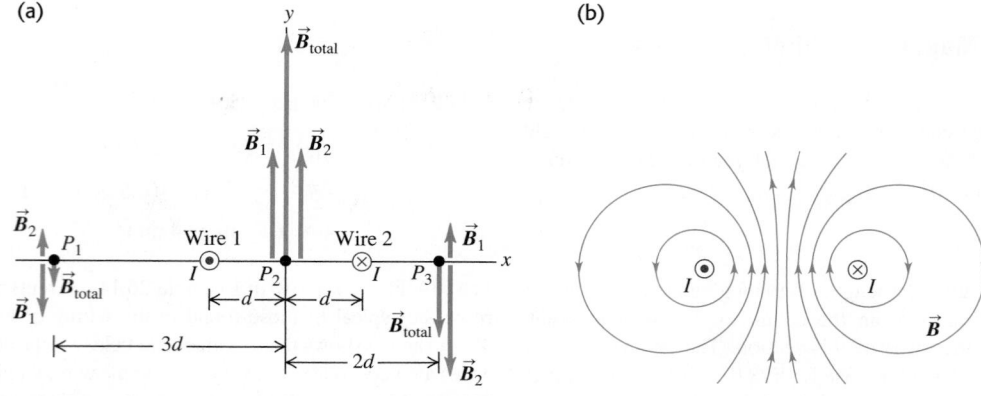

Test Your Understanding of Section 28.3   The figure at right shows a circuit that lies on a horizontal table. A compass is placed on top of the circuit as shown. A battery is to be connected to the circuit so that when the switch is closed, the compass needle deflects counterclockwise. In which orientation, A or B, should the battery be placed in the circuit?

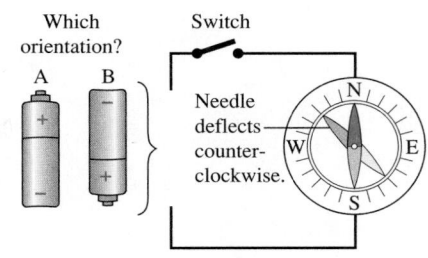

28.4 Force Between Parallel Conductors

In Example 28.4 (Section 28.3) we showed how to use the principle of superposition of magnetic fields to find the total field due to two long current-carrying conductors. Another important aspect of this configuration is the *interaction force* between the conductors. This force plays a role in many practical situations in which current-carrying wires are close to each other. Figure 28.9 shows segments of two long, straight, parallel conductors separated by a distance r and carrying currents I and I' in the same direction. Each conductor lies in the magnetic field set up by the other, so each experiences a force. The figure shows some of the field lines set up by the current in the lower conductor.

From Eq. (28.9) the lower conductor produces a $\vec{B}$ field that, at the position of the upper conductor, has magnitude

$$B = \frac{\mu_0 I}{2\pi r}$$

From Eq. (27.19) the force that this field exerts on a length L of the upper conductor is $\vec{F} = I'\vec{L} \times \vec{B}$, where the vector $\vec{L}$ is in the direction of the current I' and has magnitude L. Since $\vec{B}$ is perpendicular to the length of the conductor and hence to $\vec{L}$, the magnitude of this force is

$$F = I'LB = \frac{\mu_0 I I' L}{2\pi r}$$

and the force *per unit length* F/L is

$$\frac{F}{L} = \frac{\mu_0 I I'}{2\pi r} \qquad \text{(two long, parallel, current-carrying conductors)} \qquad \text{(28.11)}$$

Applying the right-hand rule to $\vec{F} = I'\vec{L} \times \vec{B}$ shows that the force on the upper conductor is directed *downward*.

The current in the *upper* conductor also sets up a field at the position of the *lower* one. Two successive applications of the right-hand rule for vector products (one to find the direction of the $\vec{B}$ field due to the upper conductor, as in Section 28.2, and one to find the direction of the force that this field exerts on the lower conductor, as in Section 27.6) show that the force on the lower conductor is *upward*. Thus *two parallel conductors carrying current in the same direction attract each other.* If the direction of either current is reversed, the forces also reverse. *Parallel conductors carrying currents in opposite directions repel each other.*

Magnetic Forces and Defining the Ampere

The attraction or repulsion between two straight, parallel, current-carrying conductors is the basis of the official SI definition of the **ampere:**

> *One ampere* is that unvarying current that, if present in each of two parallel conductors of infinite length and one meter apart in empty space, causes each conductor to experience a force of exactly 2×10^{-7} newtons per meter of length.

From Eq. (28.11) you can see that this definition of the ampere is what leads us to choose the value of $4\pi \times 10^{-7}$ T·m/A for μ_0. It also forms the basis of the SI

28.9 Parallel conductors carrying currents in the same direction attract each other. The diagrams show how the magnetic field $\vec{B}$ caused by the current in the lower conductor exerts a force $\vec{F}$ on the upper conductor.

The magnetic field of the lower wire exerts an attractive force on the upper wire. By the same token, the upper wire attracts the lower one.

If the wires had currents in *opposite* directions, they would *repel* each other.

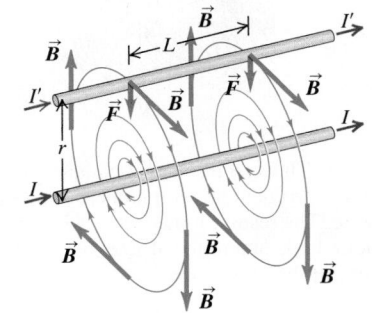

definition of the coulomb, which is the amount of charge transferred in one second by a current of one ampere.

This is an *operational definition;* it gives us an actual experimental procedure for measuring current and defining a unit of current. For high-precision standardization of the ampere, coils of wire are used instead of straight wires, and their separation is only a few centimeters. Even more precise measurements of the standardized ampere are possible using a version of the Hall effect (see Section 27.9).

Mutual forces of attraction exist not only between *wires* carrying currents in the same direction, but also between the longitudinal elements of a single current-carrying conductor. If the conductor is a liquid or an ionized gas (a plasma), these forces result in a constriction of the conductor. This is called the *pinch effect.* The high temperature produced by the pinch effect in a plasma has been used in one technique to bring about nuclear fusion.

Example 28.5 Forces between parallel wires

Two straight, parallel, superconducting wires 4.5 mm apart carry equal currents of 15,000 A in opposite directions. What force, per unit length, does each wire exert on the other?

SOLUTION

IDENTIFY and SET UP: Figure 28.10 shows the situation. We find F/L, the magnetic force per unit length of wire, using Eq. (28.11).

28.10 Our sketch for this problem.

$I = 15,000 \, A$

L

$r = 4.5 \, mm$

$I' = 15,000 \, A$

EXECUTE: The conductors *repel* each other because the currents are in opposite directions. From Eq. (28.11) the force per unit length is

$$\frac{F}{L} = \frac{\mu_0 II'}{2\pi r} = \frac{(4\pi \times 10^{-7} \, \text{T} \cdot \text{m/A})(15,000 \, \text{A})^2}{(2\pi)(4.5 \times 10^{-3} \, \text{m})}$$

$$= 1.0 \times 10^4 \, \text{N/m}$$

EVALUATE: This is a large force, more than one ton per meter. Currents and separations of this magnitude are used in superconducting electromagnets in particle accelerators, and mechanical stress analysis is a crucial part of the design process.

28.11 This electromagnet contains a current-carrying coil with numerous turns of wire. The resulting magnetic field can pick up large quantities of steel bars and other iron-bearing items.

Test Your Understanding of Section 28.4 A solenoid is a wire wound into a helical coil. The figure at left shows a solenoid that carries a current I. (a) Is the *magnetic* force that one turn of the coil exerts on an adjacent turn (i) attractive, (ii) repulsive, or (iii) zero? (b) Is the *electric* force that one turn of the coil exerts on an adjacent turn (i) attractive, (ii) repulsive, or (iii) zero? (c) Is the *magnetic* force between opposite sides of the same turn of the coil (i) attractive, (ii) repulsive, or (iii) zero? (d) Is the *electric* force between opposite sides of the same turn of the coil (i) attractive, (ii) repulsive, or (iii) zero?

28.5 Magnetic Field of a Circular Current Loop

If you look inside a doorbell, a transformer, an electric motor, or an electromagnet (Fig. 28.11), you will find coils of wire with a large number of turns, spaced so closely that each turn is very nearly a planar circular loop. A current in such a coil is used to establish a magnetic field. So it is worthwhile to derive an expression for the magnetic field produced by a single circular conducting loop carrying a current or by N closely spaced circular loops forming a coil. In Section 27.7 we considered the force and torque on such a current loop placed in an external magnetic field produced by other currents; we are now about to find the magnetic field produced by the loop itself.

Figure 28.12 shows a circular conductor with radius a. A current I is led into and out of the loop through two long, straight wires side by side; the currents in these straight wires are in opposite directions, and their magnetic fields very nearly cancel each other (see Example 28.4 in Section 28.3).

We can use the law of Biot and Savart, Eq. (28.5) or (28.6), to find the magnetic field at a point P on the axis of the loop, at a distance x from the center. As the figure shows, $d\vec{l}$ and $\hat{r}$ are perpendicular, and the direction of the field $d\vec{B}$ caused by this particular element $d\vec{l}$ lies in the xy-plane. Since $r^2 = x^2 + a^2$, the magnitude dB of the field due to element $d\vec{l}$ is

$$dB = \frac{\mu_0 I}{4\pi} \frac{dl}{(x^2 + a^2)} \qquad (28.12)$$

The components of the vector $d\vec{B}$ are

$$dB_x = dB\cos\theta = \frac{\mu_0 I}{4\pi} \frac{dl}{(x^2 + a^2)} \frac{a}{(x^2 + a^2)^{1/2}} \qquad (28.13)$$

$$dB_y = dB\sin\theta = \frac{\mu_0 I}{4\pi} \frac{dl}{(x^2 + a^2)} \frac{x}{(x^2 + a^2)^{1/2}} \qquad (28.14)$$

The *total* field $\vec{B}$ at P has only an x-component (it is perpendicular to the plane of the loop). Here's why: For every element $d\vec{l}$ there is a corresponding element on the opposite side of the loop, with opposite direction. These two elements give equal contributions to the x-component of $d\vec{B}$, given by Eq. (28.13), but *opposite* components perpendicular to the x-axis. Thus all the perpendicular components cancel and only the x-components survive.

To obtain the x-component of the total field $\vec{B}$, we integrate Eq. (28.13), including all the $d\vec{l}$'s around the loop. Everything in this expression except dl is constant and can be taken outside the integral, and we have

$$B_x = \int \frac{\mu_0 I}{4\pi} \frac{a \, dl}{(x^2 + a^2)^{3/2}} = \frac{\mu_0 I a}{4\pi(x^2 + a^2)^{3/2}} \int dl$$

The integral of dl is just the circumference of the circle, $\int dl = 2\pi a$, and we finally get

$$B_x = \frac{\mu_0 I a^2}{2(x^2 + a^2)^{3/2}} \qquad \text{(on the axis of a circular loop)} \qquad (28.15)$$

The *direction* of the magnetic field on the axis of a current-carrying loop is given by a right-hand rule. If you curl the fingers of your right hand around the loop in the direction of the current, your right thumb points in the direction of the field (Fig. 28.13).

Magnetic Field on the Axis of a Coil

Now suppose that instead of the single loop in Fig. 28.12 we have a coil consisting of N loops, all with the same radius. The loops are closely spaced so that the plane of each loop is essentially the same distance x from the field point P. Then the total field is N times the field of a single loop:

$$B_x = \frac{\mu_0 N I a^2}{2(x^2 + a^2)^{3/2}} \qquad \text{(on the axis of } N \text{ circular loops)} \qquad (28.16)$$

The factor N in Eq. (28.16) is the reason coils of wire, not single loops, are used to produce strong magnetic fields; for a desired field strength, using a single loop might require a current I so great as to exceed the rating of the loop's wire.

Figure 28.14 shows a graph of B_x as a function of x. The maximum value of the field is at $x = 0$, the center of the loop or coil:

$$B_x = \frac{\mu_0 N I}{2a} \qquad \text{(at the center of } N \text{ circular loops)} \qquad (28.17)$$

As we go out along the axis, the field decreases in magnitude.

28.12 Magnetic field on the axis of a circular loop. The current in the segment $d\vec{l}$ causes the field $d\vec{B}$, which lies in the xy-plane. The currents in other $d\vec{l}$'s cause $d\vec{B}$'s with different components perpendicular to the x-axis; these components add to zero. The x-components of the $d\vec{B}$'s combine to give the total $\vec{B}$ field at point P.

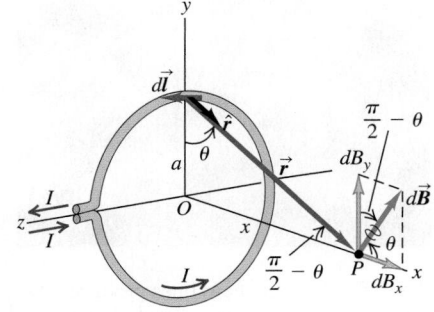

Mastering**PHYSICS**

PhET: Faraday's Electromagnetic Lab
PhET: Magnets and Electromagnets
ActivPhysics 13.2: Magnetic Field of a Loop

28.13 The right-hand rule for the direction of the magnetic field produced on the axis of a current-carrying coil.

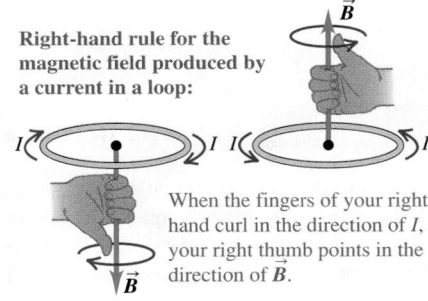

Right-hand rule for the magnetic field produced by a current in a loop:

When the fingers of your right hand curl in the direction of I, your right thumb points in the direction of $\vec{B}$.

28.14 Graph of the magnetic field along the axis of a circular coil with N turns. When x is much larger than a, the field magnitude decreases approximately as $1/x^3$.

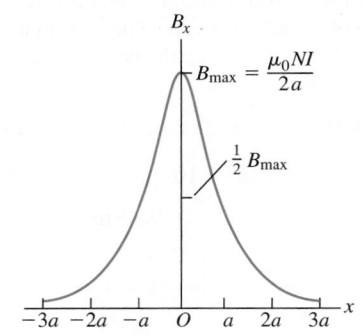

Application **Magnetic Fields for MRI**
The diagnostic technique called MRI, or magnetic resonance imaging (see Section 27.7), requires a magnetic field of about 1.5 T. In a typical MRI scan, the patient lies inside a coil that produces the intense field. The currents required are very high, so the coils are bathed in liquid helium at a temperature of 4.2 K to keep them from overheating.

28.15 Magnetic field lines produced by the current in a circular loop. At points on the axis the $\vec{B}$ field has the same direction as the magnetic moment of the loop.

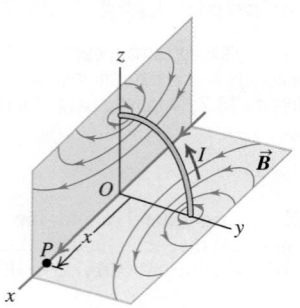

In Section 27.7 we defined the *magnetic dipole moment* μ (or *magnetic moment*) of a current-carrying loop to be equal to IA, where A is the cross-sectional area of the loop. If there are N loops, the total magnetic moment is NIA. The circular loop in Fig. 28.12 has area $A = \pi a^2$, so the magnetic moment of a single loop is $\mu = I\pi a^2$; for N loops, $\mu = NI\pi a^2$. Substituting these results into Eqs. (28.15) and (28.16), we find that both of these expressions can be written as

$$B_x = \frac{\mu_0 \mu}{2\pi(x^2 + a^2)^{3/2}} \qquad \text{(on the axis of any number of circular loops)} \qquad (28.18)$$

We described a magnetic dipole in Section 27.7 in terms of its response to a magnetic field produced by currents outside the dipole. But a magnetic dipole is also a *source* of magnetic field; Eq. (28.18) describes the magnetic field *produced* by a magnetic dipole for points along the dipole axis. This field is directly proportional to the magnetic dipole moment μ. Note that the field along the x-axis is in the same direction as the vector magnetic moment $\vec{\mu}$; this is true on both the positive and negative x-axis.

CAUTION **Magnetic field of a coil** Equations (28.15), (28.16), and (28.18) are valid only on the *axis* of a loop or coil. Don't attempt to apply these equations at other points! ▮

Figure 28.15 shows some of the magnetic field lines surrounding a circular current loop (magnetic dipole) in planes through the axis. The directions of the field lines are given by the same right-hand rule as for a long, straight conductor. Grab the conductor with your right hand, with your thumb in the direction of the current; your fingers curl around in the same direction as the field lines. The field lines for the circular current loop are closed curves that encircle the conductor; they are *not* circles, however.

Example 28.6 **Magnetic field of a coil**

A coil consisting of 100 circular loops with radius 0.60 m carries a 5.0-A current. (a) Find the magnetic field at a point along the axis of the coil, 0.80 m from the center. (b) Along the axis, at what distance from the center of the coil is the field magnitude $\frac{1}{8}$ as great as it is at the center?

SOLUTION

IDENTIFY and SET UP: This problem concerns the magnetic field magnitude B along the axis of a current-carrying coil, so we can use the ideas of this section, and in particular Eq. (28.16). We are given $N = 100$, $I = 5.0$ A, and $a = 0.60$ m. In part (a) our target variable is B_x at a given value of x. In part (b) the target variable is the value of x at which the field has $\frac{1}{8}$ of the magnitude that it has at the origin.

EXECUTE: (a) Using $x = 0.80$ m, from Eq. (28.16) we have

$$B_x = \frac{(4\pi \times 10^{-7}\,\text{T}\cdot\text{m/A})(100)(5.0\,\text{A})(0.60\,\text{m})^2}{2[(0.80\,\text{m})^2 + (0.60\,\text{m})^2]^{3/2}}$$

$$= 1.1 \times 10^{-4}\,\text{T}$$

(b) Considering Eq. (28.16), we want to find a value of x such that

$$\frac{1}{(x^2 + a^2)^{3/2}} = \frac{1}{8}\frac{1}{(0^2 + a^2)^{3/2}}$$

To solve this for x, we take the reciprocal of the whole thing and then take the 2/3 power of both sides; the result is

$$x = \pm\sqrt{3}a = \pm 1.04\,\text{m}$$

EVALUATE: We check our answer in part (a) by finding the coil's magnetic moment and substituting the result into Eq. (28.18):

$$\mu = NI\pi a^2 = (100)(5.0\,\text{A})\pi(0.60\,\text{m})^2 = 5.7 \times 10^2\,\text{A}\cdot\text{m}^2$$

$$B_x = \frac{(4\pi \times 10^{-7}\,\text{T}\cdot\text{m/A})(5.7 \times 10^2\,\text{A}\cdot\text{m}^2)}{2\pi[(0.80\,\text{m})^2 + (0.60\,\text{m})^2]^{3/2}} = 1.1 \times 10^{-4}\,\text{T}$$

The magnetic moment μ is relatively large, yet it produces a rather small field, comparable to that of the earth. This illustrates how difficult it is to produce strong fields of 1 T or more.

Test Your Understanding of Section 28.5   Figure 28.12 shows the magnetic field $d\vec{B}$ produced at point P by a segment $d\vec{l}$ that lies on the positive y-axis (at the top of the loop). This field has components $dB_x > 0$, $dB_y > 0$, $dB_z = 0$. (a) What are the signs of the components of the field $d\vec{B}$ produced at P by a segment $d\vec{l}$ on the negative y-axis (at the bottom of the loop)? (i) $dB_x > 0$, $dB_y > 0$, $dB_z = 0$; (ii) $dB_x > 0$, $dB_y < 0$, $dB_z = 0$; (iii) $dB_x < 0$, $dB_y > 0$, $dB_z = 0$; (iv) $dB_x < 0$, $dB_y < 0$, $dB_z = 0$; (v) none of these. (b) What are the signs of the components of the field $d\vec{B}$ produced at P by a segment $d\vec{l}$ on the negative z-axis (at the right-hand side of the loop)? (i) $dB_x > 0$, $dB_y > 0$, $dB_z = 0$; (ii) $dB_x > 0$, $dB_y < 0$, $dB_z = 0$; (iii) $dB_x < 0$, $dB_y > 0$, $dB_z = 0$; (iv) $dB_x < 0$, $dB_y < 0$, $dB_z = 0$; (v) none of these. |

28.6 Ampere's Law

So far our calculations of the magnetic field due to a current have involved finding the infinitesimal field $d\vec{B}$ due to a current element and then summing all the $d\vec{B}$'s to find the total field. This approach is directly analogous to our *electric-field* calculations in Chapter 21.

For the electric-field problem we found that in situations with a highly symmetric charge distribution, it was often easier to use Gauss's law to find $\vec{E}$. There is likewise a law that allows us to more easily find the *magnetic* fields caused by highly symmetric *current* distributions. But the law that allows us to do this, called *Ampere's law,* is rather different in character from Gauss's law.

Gauss's law for electric fields involves the flux of $\vec{E}$ through a closed surface; it states that this flux is equal to the total charge enclosed within the surface, divided by the constant ϵ_0. Thus this law relates electric fields and charge distributions. By contrast, Gauss's law for *magnetic* fields, Eq. (28.10), is *not* a relationship between magnetic fields and current distributions; it states that the flux of $\vec{B}$ through *any* closed surface is always zero, whether or not there are currents within the surface. So Gauss's law for $\vec{B}$ can't be used to determine the magnetic field produced by a particular current distribution.

Ampere's law is formulated not in terms of magnetic flux, but rather in terms of the *line integral* of $\vec{B}$ around a closed path, denoted by

$$\oint \vec{B} \cdot d\vec{l}$$

We used line integrals to define work in Chapter 6 and to calculate electric potential in Chapter 23. To evaluate this integral, we divide the path into infinitesimal segments $d\vec{l}$, calculate the scalar product of $\vec{B} \cdot d\vec{l}$ for each segment, and sum these products. In general, $\vec{B}$ varies from point to point, and we must use the value of $\vec{B}$ at the location of each $d\vec{l}$. An alternative notation is $\oint B_\parallel \, dl$, where $B_\parallel$ is the component of $\vec{B}$ parallel to $d\vec{l}$ at each point. The circle on the integral sign indicates that this integral is always computed for a *closed* path, one whose beginning and end points are the same.

Ampere's Law for a Long, Straight Conductor

To introduce the basic idea of Ampere's law, let's consider again the magnetic field caused by a long, straight conductor carrying a current I. We found in Section 28.3 that the field at a distance r from the conductor has magnitude

$$B = \frac{\mu_0 I}{2\pi r}$$

The magnetic field lines are circles centered on the conductor. Let's take the line integral of $\vec{B}$ around one such circle with radius r, as in Fig. 28.16a. At every point on the circle, $\vec{B}$ and $d\vec{l}$ are parallel, and so $\vec{B} \cdot d\vec{l} = B \, dl$; since r is constant around the circle, B is constant as well. Alternatively, we can say that $B_\parallel$ is constant and equal to B at every point on the circle. Hence we can take B

28.16 Three integration paths for the line integral of $\vec{B}$ in the vicinity of a long, straight conductor carrying current I *out* of the plane of the page (as indicated by the circle with a dot). The conductor is seen end-on.

(a) Integration path is a circle centered on the conductor; integration goes around the circle counterclockwise.

Result: $\oint \vec{B} \cdot d\vec{l} = \mu_0 I$

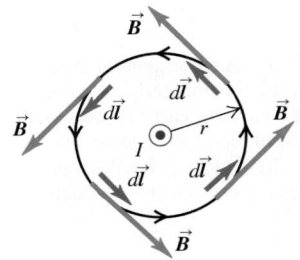

(b) Same integration path as in (a), but integration goes around the circle clockwise.

Result: $\oint \vec{B} \cdot d\vec{l} = -\mu_0 I$

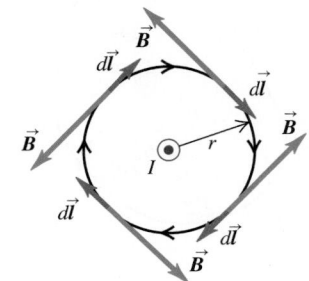

(c) An integration path that does not enclose the conductor

Result: $\oint B \cdot dl = 0$

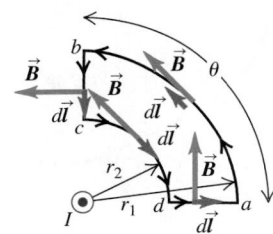

outside of the integral. The remaining integral is just the circumference of the circle, so

$$\oint \vec{B} \cdot d\vec{l} = \oint B_{\parallel} \, dl = B \oint dl = \frac{\mu_0 I}{2\pi r}(2\pi r) = \mu_0 I$$

The line integral is thus independent of the radius of the circle and is equal to μ_0 multiplied by the current passing through the area bounded by the circle.

In Fig. 28.16b the situation is the same, but the integration path now goes around the circle in the opposite direction. Now $\vec{B}$ and $d\vec{l}$ are antiparallel, so $\vec{B} \cdot d\vec{l} = -B \, dl$ and the line integral equals $-\mu_0 I$. We get the same result if the integration path is the same as in Fig. 28.16a, but the direction of the current is reversed. Thus $\oint \vec{B} \cdot d\vec{l}$ equals μ_0 multiplied by the current passing through the area bounded by the integration path, with a positive or negative sign depending on the direction of the current relative to the direction of integration.

There's a simple rule for the sign of the current; you won't be surprised to learn that it uses your right hand. Curl the fingers of your right hand around the integration path so that they curl in the direction of integration (that is, the direction that you use to evaluate $\oint \vec{B} \cdot d\vec{l}$). Then your right thumb indicates the positive current direction. Currents that pass through the integration path in this direction are positive; those in the opposite direction are negative. Using this rule, you should be able to convince yourself that the current is positive in Fig. 28.16a and negative in Fig. 28.16b. Here's another way to say the same thing: Looking at the surface bounded by the integration path, integrate counterclockwise around the path as in Fig. 28.16a. Currents moving toward you through the surface are positive, and those going away from you are negative.

An integration path that does *not* enclose the conductor is used in Fig. 28.16c. Along the circular arc ab of radius r_1, $\vec{B}$ and $d\vec{l}$ are parallel, and $B_{\parallel} = B_1 = \mu_0 I/2\pi r_1$; along the circular arc cd of radius r_2, $\vec{B}$ and $d\vec{l}$ are antiparallel and $B_{\parallel} = -B_2 = -\mu_0 I/2\pi r_2$. The $\vec{B}$ field is perpendicular to $d\vec{l}$ at each point on the straight sections bc and da, so $B_{\parallel} = 0$ and these sections contribute zero to the line integral. The total line integral is then

$$\oint \vec{B} \cdot d\vec{l} = \oint B_{\parallel} \, dl = B_1 \int_a^b dl + (0) \int_b^c dl + (-B_2) \int_c^d dl + (0) \int_d^a dl$$

$$= \frac{\mu_0 I}{2\pi r_1}(r_1 \theta) + 0 - \frac{\mu_0 I}{2\pi r_2}(r_2 \theta) + 0 = 0$$

The magnitude of $\vec{B}$ is greater on arc cd than on arc ab, but the arc length is less, so the contributions from the two arcs exactly cancel. Even though there is a magnetic field everywhere along the integration path, the line integral $\oint \vec{B} \cdot d\vec{l}$ is zero if there is no current passing through the area bounded by the path.

We can also derive these results for more general integration paths, such as the one in Fig. 28.17a. At the position of the line element $d\vec{l}$, the angle between $d\vec{l}$ and $\vec{B}$ is ϕ, and

$$\vec{B} \cdot d\vec{l} = B \, dl \cos \phi$$

From the figure, $dl \cos \phi = r \, d\theta$, where $d\theta$ is the angle subtended by $d\vec{l}$ at the position of the conductor and r is the distance of $d\vec{l}$ from the conductor. Thus

$$\oint \vec{B} \cdot d\vec{l} = \oint \frac{\mu_0 I}{2\pi r}(r \, d\theta) = \frac{\mu_0 I}{2\pi} \oint d\theta$$

But $\oint d\theta$ is just equal to 2π, the total angle swept out by the radial line from the conductor to $d\vec{l}$ during a complete trip around the path. So we get

28.17 (a) A more general integration path for the line integral of $\vec{B}$ around a long, straight conductor carrying current I *out* of the plane of the page. The conductor is seen end-on. (b) A more general integration path that does not enclose the conductor.

(a)

(b)

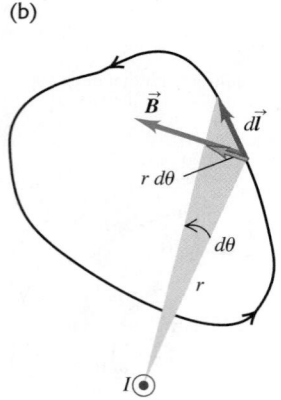

$$\oint \vec{B} \cdot d\vec{l} = \mu_0 I \qquad (28.19)$$

This result doesn't depend on the shape of the path or on the position of the wire inside it. If the current in the wire is opposite to that shown, the integral has the opposite sign. But if the path doesn't enclose the wire (Fig. 28.17b), then the net change in θ during the trip around the integration path is zero; $\oint d\theta$ is zero instead of 2π and the line integral is zero.

Ampere's Law: General Statement

Equation (28.19) is almost, but not quite, the general statement of Ampere's law. To generalize it even further, suppose *several* long, straight conductors pass through the surface bounded by the integration path. The total magnetic field $\vec{B}$ at any point on the path is the vector sum of the fields produced by the individual conductors. Thus the line integral of the total $\vec{B}$ equals μ_0 times the *algebraic sum* of the currents. In calculating this sum, we use the sign rule for currents described above. If the integration path does not enclose a particular wire, the line integral of the $\vec{B}$ field of that wire is zero, because the angle θ for that wire sweeps through a net change of zero rather than 2π during the integration. Any conductors present that are not enclosed by a particular path may still contribute to the value of $\vec{B}$ at every point, but the *line integrals* of their fields around the path are zero.

Thus we can replace I in Eq. (28.19) with I_{encl}, the algebraic sum of the currents *enclosed* or *linked* by the integration path, with the sum evaluated by using the sign rule just described (Fig. 28.18). Our statement of **Ampere's law** is then

$$\oint \vec{B} \cdot d\vec{l} = \mu_0 I_{encl} \qquad \text{(Ampere's law)} \qquad (28.20)$$

While we have derived Ampere's law only for the special case of the field of several long, straight, parallel conductors, Eq. (28.20) is in fact valid for conductors and paths of *any* shape. The general derivation is no different in principle from what we have presented, but the geometry is more complicated.

If $\oint \vec{B} \cdot d\vec{l} = 0$, it *does not* necessarily mean that $\vec{B} = 0$ everywhere along the path, only that the total current through an area bounded by the path is zero. In Figs. 28.16c and 28.17b, the integration paths enclose no current at all; in Fig. 28.19 there are positive and negative currents of equal magnitude through the area enclosed by the path. In both cases, $I_{encl} = 0$ and the line integral is zero.

> **CAUTION** **Line integrals of electric and magnetic fields** In Chapter 23 we saw that the line integral of the electrostatic field $\vec{E}$ around any closed path is equal to zero; this is a statement that the electrostatic force $\vec{F} = q\vec{E}$ on a point charge q is conservative, so this force does zero work on a charge that moves around a closed path that returns to the starting point. You might think that the value of the line integral $\oint \vec{B} \cdot d\vec{l}$ is similarly related to the question of whether the *magnetic* force is conservative. This isn't the case at all. Remember that the magnetic force $\vec{F} = q\vec{v} \times \vec{B}$ on a moving charged particle is always *perpendicular* to $\vec{B}$, so $\oint \vec{B} \cdot d\vec{l}$ is *not* related to the work done by the magnetic force; as stated in Ampere's law, this integral is related only to the total current through a surface bounded by the integration path. In fact, the magnetic force on a moving charged particle is *not* conservative. A conservative force depends only on the position of the body on which the force is exerted, but the magnetic force on a moving charged particle also depends on the *velocity* of the particle.

Equation (28.20) turns out to be valid only if the currents are steady and if no magnetic materials or time-varying electric fields are present. In Chapter 29 we will see how to generalize Ampere's law for time-varying fields.

28.18 Ampere's law.

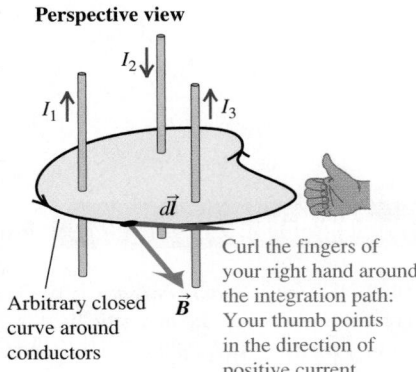

Perspective view

Arbitrary closed curve around conductors

Curl the fingers of your right hand around the integration path: Your thumb points in the direction of positive current.

Top view

Plane of curve

$I_{encl} = I_1 - I_2 + I_3$

Ampere's law: If we calculate the line integral of the magnetic field around a closed curve, the result equals μ_0 times the total enclosed current: $\oint \vec{B} \cdot d\vec{l} = \mu_0 I_{encl}$.

28.19 Two long, straight conductors carrying equal currents in opposite directions. The conductors are seen end-on, and the integration path is counterclockwise. The line integral $\oint \vec{B} \cdot d\vec{l}$ gets zero contribution from the upper and lower segments, a positive contribution from the left segment, and a negative contribution from the right segment; the net integral is zero.

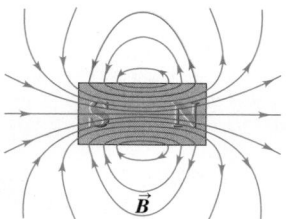

$\vec{B}$

Test Your Understanding of Section 28.6 The figure at left shows magnetic field lines through the center of a permanent magnet. The magnet is not connected to a source of emf. One of the field lines is colored red. What can you conclude about the currents inside the permanent magnet within the region enclosed by this field line? (i) There are no currents inside the magnet; (ii) there are currents directed out of the plane of the page; (iii) there are currents directed into the plane of the page; (iv) not enough information is given to decide.

28.7 Applications of Ampere's Law

Ampere's law is useful when we can exploit the symmetry of a situation to evaluate the line integral of $\vec{B}$. Several examples are given below. Problem-Solving Strategy 28.2 is directly analogous to Problem-Solving Strategy 22.1 (Section 22.4) for applications of Gauss's law; we suggest you review that strategy now and compare the two methods.

Problem-Solving Strategy 28.2 Ampere's Law

IDENTIFY *the relevant concepts:* Like Gauss's law, Ampere's law is most useful when the magnetic field is highly symmetric. In the form $\oint \vec{B} \cdot d\vec{l} = \mu_0 I_{\text{encl}}$, it can yield the magnitude of $\vec{B}$ as a function of position if we are given the magnitude and direction of the field-generating electric current.

SET UP *the problem* using the following steps:
1. Determine the target variable(s). Usually one will be the magnitude of the $\vec{B}$ field as a function of position.
2. Select the integration path you will use with Ampere's law. If you want to determine the magnetic field at a certain point, then the path must pass through that point. The integration path doesn't have to be any actual physical boundary. Usually it is a purely geometric curve; it may be in empty space, embedded in a solid body, or some of each. The integration path has to have enough *symmetry* to make evaluation of the integral possible. Ideally the path will be tangent to $\vec{B}$ in regions of interest; elsewhere the path should be perpendicular to $\vec{B}$ or should run through regions in which $\vec{B} = 0$.

EXECUTE *the solution* as follows:
1. Carry out the integral $\oint \vec{B} \cdot d\vec{l}$ along the chosen path. If $\vec{B}$ is tangent to all or some portion of the path and has the same magnitude B at every point, then its line integral is the product of B and the length of that portion of the path. If $\vec{B}$ is perpendicular to some portion of the path, or if $\vec{B} = 0$, that portion makes no contribution to the integral.
2. In the integral $\oint \vec{B} \cdot d\vec{l}$, $\vec{B}$ is the *total* magnetic field at each point on the path; it can be caused by currents enclosed *or not enclosed* by the path. If *no* net current is enclosed by the path, the field at points on the path need not be zero, but the integral $\oint \vec{B} \cdot d\vec{l}$ is always zero.
3. Determine the current I_{encl} enclosed by the integration path. A right-hand rule gives the sign of this current: If you curl the fingers of your right hand so that they follow the path in the direction of integration, then your right thumb points in the direction of positive current. If $\vec{B}$ is tangent to the path everywhere and I_{encl} is positive, the direction of $\vec{B}$ is the same as the direction of integration. If instead I_{encl} is negative, $\vec{B}$ is in the direction opposite to that of the integration.
4. Use Ampere's law $\oint \vec{B} \cdot d\vec{l} = \mu_0 I$ to solve for the target variable.

EVALUATE *your answer:* If your result is an expression for the field magnitude as a function of position, check it by examining how the expression behaves in different limits.

Example 28.7 Field of a long, straight, current-carrying conductor

In Section 28.6 we derived Ampere's law using Eq. (28.9) for the field $\vec{B}$ of a long, straight, current-carrying conductor. Reverse this process, and use Ampere's law to find $\vec{B}$ for this situation.

SOLUTION

IDENTIFY and SET UP: The situation has cylindrical symmetry, so in Ampere's law we take our integration path to be a circle with radius r centered on the conductor and lying in a plane perpendicular to it, as in Fig. 28.16a. The field $\vec{B}$ is everywhere tangent to this circle and has the same magnitude B everywhere on the circle.

EXECUTE: With our choice of integration path, Ampere's law [Eq. (28.20)] becomes

$$\oint \vec{B} \cdot d\vec{l} = \oint B_{\parallel} \, dl = B(2\pi r) = \mu_0 I$$

Equation (28.9), $B = \mu_0 I / 2\pi r$, follows immediately.

Ampere's law determines the direction of $\vec{B}$ as well as its magnitude. Since we chose to go counterclockwise around the integration path, the positive direction for current is out of the plane of Fig. 28.16a; this is the same as the actual current direction in the figure, so I is positive and the integral $\oint \vec{B} \cdot d\vec{l}$ is also positive. Since the $d\vec{l}$'s run counterclockwise, the direction of $\vec{B}$ must be counterclockwise as well, as shown in Fig. 28.16a.

EVALUATE: Our results are consistent with those in Section 28.6.

Example 28.8 Field of a long cylindrical conductor

A cylindrical conductor with radius R carries a current I (Fig. 28.20). The current is uniformly distributed over the cross-sectional area of the conductor. Find the magnetic field as a function of the distance r from the conductor axis for points both inside $(r < R)$ and outside $(r > R)$ the conductor.

SOLUTION

IDENTIFY and SET UP: As in Example 28.7, the current distribution has cylindrical symmetry, and the magnetic field lines must be circles concentric with the conductor axis. To find the magnetic field inside and outside the conductor, we choose circular integration paths with radii $r < R$ and $r > R$, respectively (see Fig. 28.20).

EXECUTE: In either case the field $\vec{B}$ has the same magnitude at every point on the circular integration path and is tangent to the path. Thus the magnitude of the line integral is simply $B(2\pi r)$. To find the current I_{encl} enclosed by a circular integration path inside the conductor $(r < R,)$, note that the current density (current per unit area) is $J = I/\pi R^2$, so $I_{encl} = J(\pi r^2) = Ir^2/R^2$. Hence Ampere's law gives $B(2\pi r) = \mu_0 Ir^2/R^2$, or

28.20 To find the magnetic field at radius $r < R$, we apply Ampere's law to the circle enclosing the gray area. The current through the red area is $(r^2/R^2)I$. To find the magnetic field at radius $r > R$, we apply Ampere's law to the circle enclosing the entire conductor.

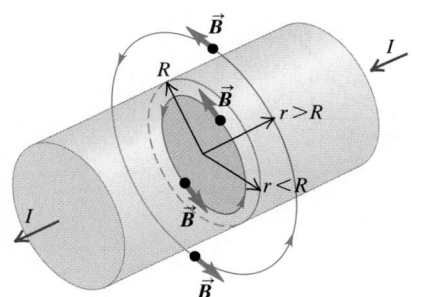

$$B = \frac{\mu_0 I}{2\pi}\frac{r}{R^2} \qquad \begin{array}{l}\text{(inside the conductor,} \\ r < R)\end{array} \qquad (28.21)$$

A circular integration path outside the conductor encloses the total current in the conductor, so $I_{encl} = I$. Applying Ampere's law gives the same equation as in Example 28.7, with the same result for B:

$$B = \frac{\mu_0 I}{2\pi r} \qquad \begin{array}{l}\text{(outside the conductor,} \\ r > R)\end{array} \qquad (28.22)$$

Outside the conductor, the magnetic field is the same as that of a long, straight conductor carrying current I, independent of the radius R over which the current is distributed. Indeed, the magnetic field outside *any* cylindrically symmetric current distribution is the same as if the entire current were concentrated along the axis of the distribution. This is analogous to the results of Examples 22.5 and 22.9 (Section 22.4), in which we found that the *electric* field outside a spherically symmetric *charged* body is the same as though the entire charge were concentrated at the center.

EVALUATE: Note that at the surface of the conductor $(r = R)$, Eqs. (28.21) and (28.22) agree, as they must. Figure 28.21 shows a graph of B as a function of r.

28.21 Magnitude of the magnetic field inside and outside a long, straight cylindrical conductor with radius R carrying a current I.

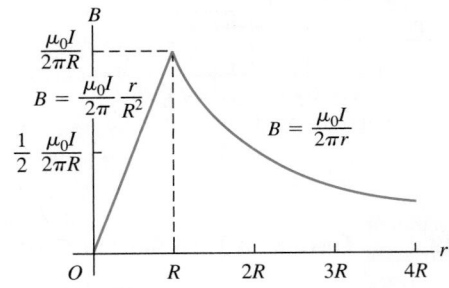

Example 28.9 Field of a solenoid

A solenoid consists of a helical winding of wire on a cylinder, usually circular in cross section. There can be thousands of closely spaced turns (often in several layers), each of which can be regarded as a circular loop. For simplicity, Fig. 28.22 shows a solenoid with only a few turns. All turns carry the same current I, and the total $\vec{B}$ field at every point is the vector sum of the fields caused by the individual turns. The figure shows field lines in the xy- and xz-planes. We draw field lines that are uniformly spaced at the center of the solenoid. Exact calculations show that for a long, closely wound solenoid, half of these field lines emerge from the ends and half "leak out" through the windings between the center and the end, as the figure suggests.

If the solenoid is long in comparison with its cross-sectional diameter and the coils are tightly wound, the field inside the solenoid near its midpoint is very nearly uniform over the cross section and parallel to the axis; the *external* field near the midpoint is very small.

28.22 Magnetic field lines produced by the current in a solenoid. For clarity, only a few turns are shown.

Continued

Use Ampere's law to find the field at or near the center of such a solenoid if it has n turns per unit length and carries current I.

SOLUTION

IDENTIFY and SET UP: We assume that $\vec{B}$ is uniform inside the solenoid and zero outside. Figure 28.23 shows the situation and our chosen integration path, rectangle $abcd$. Side ab, with length L, is parallel to the axis of the solenoid. Sides bc and da are taken to be very long so that side cd is far from the solenoid; then the field at side cd is negligibly small.

EXECUTE: Along side ab, $\vec{B}$ is parallel to the path and is constant. Our Ampere's-law integration takes us along side ab in the same direction as $\vec{B}$, so here $B_{\parallel} = +B$ and

$$\int_a^b \vec{B} \cdot d\vec{l} = BL$$

Along sides bc and da, $\vec{B}$ is perpendicular to the path and so $B_{\parallel} = 0$; along side cd, $\vec{B} = 0$ and so $B_{\parallel} = 0$. Around the entire closed path, then, we have $\oint \vec{B} \cdot d\vec{l} = BL$.

In a length L there are nL turns, each of which passes once through $abcd$ carrying current I. Hence the total current enclosed by the rectangle is $I_{\text{encl}} = nLI$. The integral $\oint \vec{B} \cdot d\vec{l}$ is

positive, so from Ampere's law I_{encl} must be positive as well. This means that the current passing through the surface bounded by the integration path must be in the direction shown in Fig. 28.23. Ampere's law then gives $BL = \mu_0 nLI$, or

$$B = \mu_0 nI \qquad \text{(solenoid)} \qquad (28.23)$$

Side ab need not lie on the axis of the solenoid, so this result demonstrates that the field is uniform over the entire cross section at the center of the solenoid's length.

EVALUATE: Note that the *direction* of $\vec{B}$ inside the solenoid is in the same direction as the solenoid's vector magnetic moment $\vec{\mu}$, as we found in Section 28.5 for a single current-carrying loop.

For points along the axis, the field is strongest at the center of the solenoid and drops off near the ends. For a solenoid very long in comparison to its diameter, the field magnitude at each end is exactly half that at the center. This is approximately the case even for a relatively short solenoid, as Fig. 28.24 shows.

28.23 Our sketch for this problem.

Central part of solenoid

28.24 Magnitude of the magnetic field at points along the axis of a solenoid with length $4a$, equal to four times its radius a. The field magnitude at each end is about half its value at the center. (Compare with Fig. 28.14 for the field of N circular loops.)

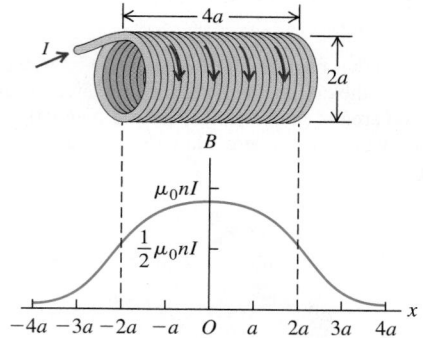

Example 28.10   Field of a toroidal solenoid

Figure 28.25a shows a doughnut-shaped **toroidal solenoid**, tightly wound with N turns of wire carrying a current I. (In a practical solenoid the turns would be much more closely spaced than they are in the figure.) Find the magnetic field at all points.

SOLUTION

IDENTIFY and SET UP: Ignoring the slight pitch of the helical windings, we can consider each turn of a tightly wound toroidal solenoid as a loop lying in a plane perpendicular to the large, circular axis of the toroid. The symmetry of the situation then tells us that the magnetic field lines must be circles concentric with the toroid axis. Therefore we choose circular integration paths (of which Fig. 28.25b shows three) for use with Ampere's law, so that the field $\vec{B}$ (if any) is tangent to each path at all points along the path.

EXECUTE: Along each path, $\oint \vec{B} \cdot d\vec{l}$ equals the product of B and the path circumference $l = 2\pi r$. The total current enclosed by path 1 is zero, so from Ampere's law the field $\vec{B} = 0$ everywhere on this path.

Each turn of the winding passes *twice* through the area bounded by path 3, carrying equal currents in opposite directions. The *net*

28.25 (a) A toroidal solenoid. For clarity, only a few turns of the winding are shown. (b) Integration paths (black circles) used to compute the magnetic field $\vec{B}$ set up by the current (shown as dots and crosses).

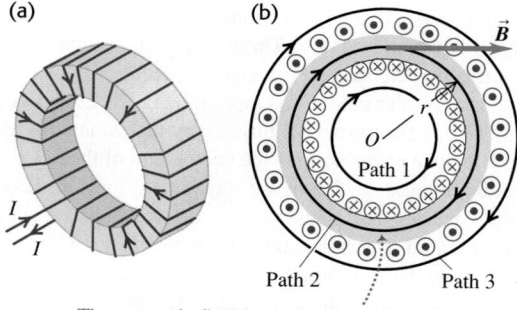

The magnetic field is confined almost entirely to the space enclosed by the windings (in blue).

current enclosed is therefore zero, and hence $\vec{B} = 0$ at all points on this path as well. We conclude that *the field of an ideal toroidal* solenoid is confined to the space enclosed by the windings. We can think of such a solenoid as a tightly wound, straight solenoid that has been bent into a circle.

For path 2, we have $\oint \vec{B} \cdot d\vec{l} = 2\pi rB$. Each turn of the winding passes *once* through the area bounded by this path, so $I_{encl} = NI$. We note that I_{encl} is positive for the clockwise direction of integration in Fig. 28.25b, so $\vec{B}$ is in the direction shown. Ampere's law then says that $2\pi rB = \mu_0 NI$, so

$$B = \frac{\mu_0 NI}{2\pi r} \quad \text{(toroidal solenoid)} \quad (28.24)$$

EVALUATE: Equation (28.24) indicates that B is *not* uniform over the interior of the core, because different points in the interior are difference distances r from the toroid axis. However, if the radial extent of the core is small in comparison to r, the variation is slight. In that case, considering that $2\pi r$ is the circumferential length of the toroid and that $N/2\pi r$ is the number of turns per unit length n, the field may be written as $B = \mu_0 nI$, just as it is at the center of a long, *straight* solenoid.

In a real toroidal solenoid the turns are not precisely circular loops but rather segments of a bent helix. As a result, the external field is not exactly zero. To estimate its magnitude, we imagine Fig. 28.25a as being *very* roughly equivalent, for points outside the torus, to a *single-turn* circular loop with radius r. At the center of such a loop, Eq. (28.17) gives $B = \mu_0 I/2r$; this is smaller than the field inside the solenoid by the factor N/π.

The equations we have derived for the field in a closely wound straight or toroidal solenoid are strictly correct only for windings in *vacuum*. For most practical purposes, however, they can be used for windings in air or on a core of any nonmagnetic, nonsuperconducting material. In the next section we will show how these equations are modified if the core is a magnetic material.

Test Your Understanding of Section 28.7 Consider a conducting wire that runs along the central axis of a hollow conducting cylinder. Such an arrangement, called a *coaxial cable,* has many applications in telecommunications. (The cable that connects a television set to a local cable provider is an example of a coaxial cable.) In such a cable a current I runs in one direction along the hollow conducting cylinder and is spread uniformly over the cylinder's cross-sectional area. An equal current runs in the opposite direction along the central wire. How does the magnitude B of the magnetic field outside such a cable depend on the distance r from the central axis of the cable? (i) B is proportional to $1/r$; (ii) B is proportional to $1/r^2$; (iii) B is zero at all points outside the cable.

Hollow conducting cylinder   Insulator   Central wire

28.8 Magnetic Materials

In discussing how currents cause magnetic fields, we have assumed that the conductors are surrounded by vacuum. But the coils in transformers, motors, generators, and electromagnets nearly always have iron cores to increase the magnetic field and confine it to desired regions. Permanent magnets, magnetic recording tapes, and computer disks depend directly on the magnetic properties of materials; when you store information on a computer disk, you are actually setting up an array of microscopic permanent magnets on the disk. So it is worthwhile to examine some aspects of the magnetic properties of materials. After describing the atomic origins of magnetic properties, we will discuss three broad classes of magnetic behavior that occur in materials; these are called *paramagnetism, diamagnetism,* and *ferromagnetism.*

The Bohr Magneton

As we discussed briefly in Section 27.7, the atoms that make up all matter contain moving electrons, and these electrons form microscopic current loops that produce magnetic fields of their own. In many materials these currents are randomly oriented and cause no net magnetic field. But in some materials an external field (a field produced by currents outside the material) can cause these loops to become oriented preferentially with the field, so their magnetic fields *add* to the external field. We then say that the material is *magnetized.*

Let's look at how these microscopic currents come about. Figure 28.26 shows a primitive model of an electron in an atom. We picture the electron (mass m, charge $-e$) as moving in a circular orbit with radius r and speed v. This moving charge is equivalent to a current loop. In Section 27.7 we found that a current loop with area A and current I has a magnetic dipole moment μ given by $\mu = IA$; for the orbiting electron the area of the loop is $A = \pi r^2$. To find the current

28.26 An electron moving with speed v in a circular orbit of radius r has an angular momentum $\vec{L}$ and an oppositely directed orbital magnetic dipole moment $\vec{\mu}$. It also has a spin angular momentum and an oppositely directed spin magnetic dipole moment.

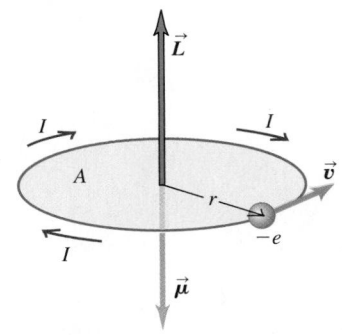

associated with the electron, we note that the orbital period T (the time for the electron to make one complete orbit) is the orbit circumference divided by the electron speed: $T = 2\pi r/v$. The equivalent current I is the total charge passing any point on the orbit per unit time, which is just the magnitude e of the electron charge divided by the orbital period T:

$$I = \frac{e}{T} = \frac{ev}{2\pi r}$$

The magnetic moment $\mu = IA$ is then

$$\mu = \frac{ev}{2\pi r}(\pi r^2) = \frac{evr}{2} \qquad (28.25)$$

It is useful to express μ in terms of the *angular momentum* L of the electron. For a particle moving in a circular path, the magnitude of angular momentum equals the magnitude of momentum mv multiplied by the radius r—that is, $L = mvr$ (see Section 10.5). Comparing this with Eq. (28.25), we can write

$$\mu = \frac{e}{2m}L \qquad (28.26)$$

Equation (28.26) is useful in this discussion because atomic angular momentum is *quantized;* its component in a particular direction is always an integer multiple of $h/2\pi$, where h is a fundamental physical constant called *Planck's constant.* The numerical value of h is

$$h = 6.626 \times 10^{-34}\,\text{J} \cdot \text{s}$$

The quantity $h/2\pi$ thus represents a fundamental unit of angular momentum in atomic systems, just as e is a fundamental unit of charge. Associated with the quantization of $\vec{L}$ is a fundamental uncertainty in the *direction* of $\vec{L}$ and therefore of $\vec{\mu}$. In the following discussion, when we speak of the magnitude of a magnetic moment, a more precise statement would be "maximum component in a given direction." Thus, to say that a magnetic moment $\vec{\mu}$ is aligned with a magnetic field $\vec{B}$ really means that $\vec{\mu}$ has its maximum possible component in the direction of $\vec{B}$; such components are always quantized.

Equation (28.26) shows that associated with the fundamental unit of angular momentum is a corresponding fundamental unit of magnetic moment. If $L = h/2\pi$, then

$$\mu = \frac{e}{2m}\left(\frac{h}{2\pi}\right) = \frac{eh}{4\pi m} \qquad (28.27)$$

This quantity is called the **Bohr magneton,** denoted by μ_B. Its numerical value is

$$\mu_B = 9.274 \times 10^{-24}\,\text{A} \cdot \text{m}^2 = 9.274 \times 10^{-24}\,\text{J/T}$$

You should verify that these two sets of units are consistent. The second set is useful when we compute the potential energy $U = -\vec{\mu} \cdot \vec{B}$ for a magnetic moment in a magnetic field.

Electrons also have an intrinsic angular momentum, called *spin,* that is not related to orbital motion but that can be pictured in a classical model as spinning on an axis. This angular momentum also has an associated magnetic moment, and its magnitude turns out to be almost exactly one Bohr magneton. (Effects having to do with quantization of the electromagnetic field cause the spin magnetic moment to be about 1.001 μ_B.)

Paramagnetism

In an atom, most of the various orbital and spin magnetic moments of the electrons add up to zero. However, in some cases the atom has a net magnetic moment that is of the order of μ_B. When such a material is placed in a magnetic field, the field exerts a torque on each magnetic moment, as given by Eq. (27.26): $\vec{\tau} = \vec{\mu} \times \vec{B}$. These torques tend to align the magnetic moments with the field, as we discussed in Section 27.7. In this position, the directions of the current loops are such as to *add* to the externally applied magnetic field.

We saw in Section 28.5 that the $\vec{B}$ field produced by a current loop is proportional to the loop's magnetic dipole moment. In the same way, the additional $\vec{B}$ field produced by microscopic electron current loops is proportional to the total magnetic moment $\vec{\mu}_{\text{total}}$ per unit volume V in the material. We call this vector quantity the **magnetization** of the material, denoted by $\vec{M}$:

$$\vec{M} = \frac{\vec{\mu}_{\text{total}}}{V} \qquad (28.28)$$

The additional magnetic field due to magnetization of the material turns out to be equal simply to $\mu_0 \vec{M}$, where μ_0 is the same constant that appears in the law of Biot and Savart and Ampere's law. When such a material completely surrounds a current-carrying conductor, the total magnetic field $\vec{B}$ in the material is

$$\vec{B} = \vec{B}_0 + \mu_0 \vec{M} \qquad (28.29)$$

where $\vec{B}_0$ is the field caused by the current in the conductor.

To check that the units in Eq. (28.29) are consistent, note that magnetization $\vec{M}$ is magnetic moment per unit volume. The units of magnetic moment are current times area $(\text{A} \cdot \text{m}^2)$, so the units of magnetization are $(\text{A} \cdot \text{m}^2)/\text{m}^3 = \text{A}/\text{m}$. From Section 28.1, the units of the constant μ_0 are $\text{T} \cdot \text{m}/\text{A}$. So the units of $\mu_0 \vec{M}$ are the same as the units of $\vec{B}$: $(\text{T} \cdot \text{m}/\text{A})(\text{A}/\text{m}) = \text{T}$.

A material showing the behavior just described is said to be **paramagnetic.** The result is that the magnetic field at any point in such a material is greater by a dimensionless factor K_m, called the **relative permeability** of the material, than it would be if the material were replaced by vacuum. The value of K_m is different for different materials; for common paramagnetic solids and liquids at room temperature, K_m typically ranges from 1.00001 to 1.003.

All of the equations in this chapter that relate magnetic fields to their sources can be adapted to the situation in which the current-carrying conductor is embedded in a paramagnetic material. All that need be done is to replace μ_0 by $K_m \mu_0$. This product is usually denoted as μ and is called the **permeability** of the material:

$$\mu = K_m \mu_0 \qquad (28.30)$$

CAUTION Two meanings of the symbol μ Equation (28.30) involves some really dangerous notation because we have also used μ for magnetic dipole moment. It's customary to use μ for both quantities, but beware: From now on, every time you see a μ, make sure you know whether it is permeability or magnetic moment. You can usually tell from the context.

The amount by which the relative permeability differs from unity is called the **magnetic susceptibility,** denoted by χ_m:

$$\chi_m = K_m - 1 \qquad (28.31)$$

Both K_m and χ_m are dimensionless quantities. Table 28.1 lists values of magnetic susceptibility for several materials. For example, for aluminum, $\chi_m = 2.2 \times 10^{-5}$ and $K_m = 1.000022$. The first group of materials in the table are paramagnetic; we'll discuss the second group of materials, which are called *diamagnetic,* very shortly.

The tendency of atomic magnetic moments to align themselves parallel to the magnetic field (where the potential energy is minimum) is opposed by random thermal motion, which tends to randomize their orientations. For this reason, paramagnetic susceptibility always decreases with increasing temperature.

Table 28.1 Magnetic Susceptibilities of Paramagnetic and Diamagnetic Materials at $T = 20°\text{C}$

Material	$\chi_m = K_m - 1 \, (\times 10^{-5})$
Paramagnetic	
Iron ammonium alum	66
Uranium	40
Platinum	26
Aluminum	2.2
Sodium	0.72
Oxygen gas	0.19
Diamagnetic	
Bismuth	−16.6
Mercury	−2.9
Silver	−2.6
Carbon (diamond)	−2.1
Lead	−1.8
Sodium chloride	−1.4
Copper	−1.0

In many cases it is inversely proportional to the absolute temperature T, and the magnetization M can be expressed as

$$M = C\frac{B}{T} \tag{28.32}$$

This relationship is called *Curie's law,* after its discoverer, Pierre Curie (1859–1906). The quantity C is a constant, different for different materials, called the *Curie constant.*

As we described in Section 27.7, a body with atomic magnetic dipoles is attracted to the poles of a magnet. In most paramagnetic substances this attraction is very weak due to thermal randomization of the atomic magnetic moments. But at very low temperatures the thermal effects are reduced, the magnetization increases in accordance with Curie's law, and the attractive forces are greater.

Example 28.11 Magnetic dipoles in a paramagnetic material

Nitric oxide (NO) is a paramagnetic compound. The magnetic moment of each NO molecule has a maximum component in any direction of about one Bohr magneton. Compare the interaction energy of such magnetic moments in a 1.5-T magnetic field with the average translational kinetic energy of molecules at 300 K.

SOLUTION

IDENTIFY and SET UP: This problem involves the energy of a magnetic moment in a magnetic field and the average thermal kinetic energy. We have Eq. (27.27), $U = -\vec{\mu} \cdot \vec{B}$, for the interaction energy of a magnetic moment $\vec{\mu}$ with a $\vec{B}$ field, and Eq. (18.16), $K = \frac{3}{2}kT$, for the average translational kinetic energy of a molecule at temperature T.

EXECUTE: We can write $U = -\mu_\parallel B$, where $\mu_\parallel$ is the component of the magnetic moment $\vec{\mu}$ in the direction of the $\vec{B}$ field. Here the maximum value of $\mu_\parallel$ is about μ_B, so

$$|U|_{max} \approx \mu_B B = (9.27 \times 10^{-24}\,\text{J/T})(1.5\,\text{T})$$
$$= 1.4 \times 10^{-23}\,\text{J} = 8.7 \times 10^{-5}\,\text{eV}$$

The average translational kinetic energy K is

$$K = \frac{3}{2}kT = \frac{3}{2}(1.38 \times 10^{-23}\,\text{J/K})(300\,\text{K})$$
$$= 6.2 \times 10^{-21}\,\text{J} = 0.039\,\text{eV}$$

EVALUATE: At 300 K the magnetic interaction energy is only about 0.2% of the thermal kinetic energy, so we expect only a slight degree of alignment. This is why paramagnetic susceptibilities at ordinary temperature are usually very small.

Diamagnetism

In some materials the total magnetic moment of all the atomic current loops is zero when no magnetic field is present. But even these materials have magnetic effects because an external field alters electron motions within the atoms, causing additional current loops and induced magnetic dipoles comparable to the induced *electric* dipoles we studied in Section 28.5. In this case the additional field caused by these current loops is always *opposite* in direction to that of the external field. (This behavior is explained by Faraday's law of induction, which we will study in Chapter 29. An induced current always tends to cancel the field change that caused it.)

Such materials are said to be **diamagnetic.** They always have negative susceptibility, as shown in Table 28.1, and relative permeability K_m slightly *less* than unity, typically of the order of 0.99990 to 0.99999 for solids and liquids. Diamagnetic susceptibilities are very nearly temperature independent.

Ferromagnetism

There is a third class of materials, called **ferromagnetic** materials, that includes iron, nickel, cobalt, and many alloys containing these elements. In these materials, strong interactions between atomic magnetic moments cause them to line up parallel to each other in regions called **magnetic domains,** even when no external field is present. Figure 28.27 shows an example of magnetic domain structure. Within each domain, nearly all of the atomic magnetic moments are parallel.

When there is no externally applied field, the domain magnetizations are randomly oriented. But when a field $\vec{B}_0$ (caused by external currents) is present, the domains tend to orient themselves parallel to the field. The domain boundaries also shift; the domains that are magnetized in the field direction grow, and those that are magnetized in other directions shrink. Because the total magnetic moment of a domain may be many thousands of Bohr magnetons, the torques that tend to align the domains with an external field are much stronger than occur with paramagnetic materials. The relative permeability K_m is *much* larger than unity, typically of the order of 1000 to 100,000. As a result, an object made of a ferromagnetic material such as iron is strongly magnetized by the field from a permanent magnet and is attracted to the magnet (see Fig. 27.38). A paramagnetic material such as aluminum is also attracted to a permanent magnet, but K_m for paramagnetic materials is so much smaller for such a material than for ferromagnetic materials that the attraction is very weak. Thus a magnet can pick up iron nails, but not aluminum cans.

As the external field is increased, a point is eventually reached at which nearly *all* the magnetic moments in the ferromagnetic material are aligned parallel to the external field. This condition is called *saturation magnetization;* after it is reached, further increase in the external field causes no increase in magnetization or in the additional field caused by the magnetization.

Figure 28.28 shows a "magnetization curve," a graph of magnetization M as a function of external magnetic field B_0, for soft iron. An alternative description of this behavior is that K_m is not constant but decreases as B_0 increases. (Paramagnetic materials also show saturation at sufficiently strong fields. But the magnetic fields required are so large that departures from a linear relationship between M and B_0 in these materials can be observed only at very low temperatures, 1 K or so.)

For many ferromagnetic materials the relationship of magnetization to external magnetic field is different when the external field is increasing from when it is decreasing. Figure 28.29a shows this relationship for such a material. When the material is magnetized to saturation and then the external field is reduced to zero, some magnetization remains. This behavior is characteristic of permanent magnets, which retain most of their saturation magnetization when the magnetizing field is removed. To reduce the magnetization to zero requires a magnetic field in the reverse direction.

This behavior is called **hysteresis,** and the curves in Fig. 28.29 are called *hysteresis loops.* Magnetizing and demagnetizing a material that has hysteresis involve the dissipation of energy, and the temperature of the material increases during such a process.

28.27 In this drawing adapted from a magnified photo, the arrows show the directions of magnetization in the domains of a single crystal of nickel. Domains that are magnetized in the direction of an applied magnetic field grow larger.

(a) No field

(b) Weak field

(c) Stronger field
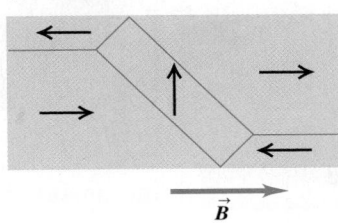

28.28 A magnetization curve for a ferromagnetic material. The magnetization M approaches its saturation value M_{sat} as the magnetic field B_0 (caused by external currents) becomes large.

28.29 Hysteresis loops. The materials of both (a) and (b) remain strongly magnetized when B_0 is reduced to zero. Since (a) is also hard to demagnetize, it would be good for permanent magnets. Since (b) magnetizes and demagnetizes more easily, it could be used as a computer memory material. The material of (c) would be useful for transformers and other alternating-current devices where zero hysteresis would be optimal.

100 μm

Ferromagnetic materials are widely used in electromagnets, transformer cores, and motors and generators, in which it is desirable to have as large a magnetic field as possible for a given current. Because hysteresis dissipates energy, materials that are used in these applications should usually have as narrow a hysteresis loop as possible. Soft iron is often used; it has high permeability without appreciable hysteresis. For permanent magnets a broad hysteresis loop is usually desirable, with large zero-field magnetization and large reverse field needed to demagnetize. Many kinds of steel and many alloys, such as Alnico, are commonly used for permanent magnets. The remaining magnetic field in such a material, after it has been magnetized to near saturation, is typically of the order of 1 T, corresponding to a remaining magnetization $M = B/\mu_0$ of about 800,000 A/m.

Example 28.12 A ferromagnetic material

A cube-shaped permanent magnet is made of a ferromagnetic material with a magnetization M of about 8×10^5 A/m. The side length is 2 cm. (a) Find the magnetic dipole moment of the magnet. (b) Estimate the magnetic field due to the magnet at a point 10 cm from the magnet along its axis.

SOLUTION

IDENTIFY and SET UP: This problem uses the relationship between magnetization M and magnetic dipole moment μ_{total} and the idea that a magnetic dipole produces a magnetic field. We find μ_{total} using Eq. (28.28). To estimate the field, we approximate the magnet as a current loop with this same magnetic moment and use Eq. (28.18).

EXECUTE: (a) From Eq. (28.28),

$$\mu_{\text{total}} = MV = (8 \times 10^5 \text{ A/m})(2 \times 10^{-2} \text{ m})^3 = 6 \text{ A} \cdot \text{m}^2$$

(b) From Eq. (28.18), the magnetic field on the axis of a current loop with magnetic moment μ_{total} is

$$B = \frac{\mu_0 \mu_{\text{total}}}{2\pi(x^2 + a^2)^{3/2}}$$

where x is the distance from the loop and a is its radius. We can use this expression here if we take a to refer to the size of the permanent magnet. Strictly speaking, there are complications because our magnet does not have the same geometry as a circular current loop. But because $x = 10$ cm is fairly large in comparison to the 2-cm size of the magnet, the term a^2 is negligible in comparison to x^2 and can be ignored. So

$$B \approx \frac{\mu_0 \mu_{\text{total}}}{2\pi x^3} = \frac{(4\pi \times 10^{-7} \text{ T} \cdot \text{m/A})(6 \text{ A} \cdot \text{m}^2)}{2\pi(0.1 \text{ m})^3}$$
$$= 1 \times 10^{-3} \text{ T} = 10 \text{ G}$$

which is about ten times stronger than the earth's magnetic field.

EVALUATE: We calculated B at a point *outside* the magnetic material and therefore used μ_0, not the permeability μ of the magnetic material, in our calculation. You would substitute permeability μ for μ_0 if you were calculating B *inside* a material with relative permeability K_m, for which $\mu = K_m\mu_0$.

Test Your Understanding of Section 28.8 Which of the following materials are attracted to a magnet? (i) sodium; (ii) bismuth; (iii) lead; (iv) uranium. ❙

Magnetic field of a moving charge: The magnetic field $\vec{B}$ created by a charge q moving with velocity $\vec{v}$ depends on the distance r from the source point (the location of q) to the field point (where $\vec{B}$ is measured). The $\vec{B}$ field is perpendicular to $\vec{v}$ and to $\hat{r}$, the unit vector directed from the source point to the field point. The principle of superposition of magnetic fields states that the total $\vec{B}$ field produced by several moving charges is the vector sum of the fields produced by the individual charges. (See Example 28.1.)

$$\vec{B} = \frac{\mu_0}{4\pi} \frac{q\vec{v} \times \hat{r}}{r^2} \qquad (28.2)$$

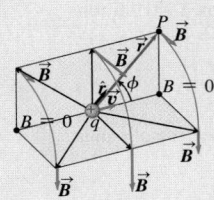

Magnetic field of a current-carrying conductor: The law of Biot and Savart gives the magnetic field $d\vec{B}$ created by an element $d\vec{l}$ of a conductor carrying current I. The field $d\vec{B}$ is perpendicular to both $d\vec{l}$ and $\hat{r}$, the unit vector from the element to the field point. The $\vec{B}$ field created by a finite current-carrying conductor is the integral of $d\vec{B}$ over the length of the conductor. (See Example 28.2.)

$$d\vec{B} = \frac{\mu_0}{4\pi} \frac{I\, d\vec{l} \times \hat{r}}{r^2} \qquad (28.6)$$

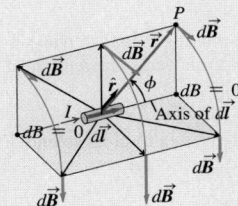

Magnetic field of a long, straight, current-carrying conductor: The magnetic field $\vec{B}$ at a distance r from a long, straight conductor carrying a current I has a magnitude that is inversely proportional to r. The magnetic field lines are circles coaxial with the wire, with directions given by the right-hand rule. (See Examples 28.3 and 28.4.)

$$B = \frac{\mu_0 I}{2\pi r} \qquad (28.9)$$

Magnetic force between current-carrying conductors: Two long, parallel, current-carrying conductors attract if the currents are in the same direction and repel if the currents are in opposite directions. The magnetic force per unit length between the conductors depends on their currents I and I' and their separation r. The definition of the ampere is based on this relationship. (See Example 28.5.)

$$\frac{F}{L} = \frac{\mu_0 II'}{2\pi r} \qquad (28.11)$$

Magnetic field of a current loop: The law of Biot and Savart allows us to calculate the magnetic field produced along the axis of a circular conducting loop of radius a carrying current I. The field depends on the distance x along the axis from the center of the loop to the field point. If there are N loops, the field is multiplied by N. At the center of the loop, $x = 0$. (See Example 28.6.)

$$B_x = \frac{\mu_0 I a^2}{2(x^2 + a^2)^{3/2}} \qquad (28.15)$$
(circular loop)

$$B_x = \frac{\mu_0 N I}{2a} \qquad (28.17)$$
(center of N circular loops)

Ampere's law: Ampere's law states that the line integral of $\vec{B}$ around any closed path equals μ_0 times the net current through the area enclosed by the path. The positive sense of current is determined by a right-hand rule. (See Examples 28.7–28.10.)

$$\oint \vec{B} \cdot d\vec{l} = \mu_0 I_{\text{encl}} \qquad (28.20)$$

Magnetic fields due to current distributions: The table lists magnetic fields caused by several current distributions. In each case the conductor is carrying current I.

Current Distribution	Point in Magnetic Field	Magnetic-Field Magnitude
Long, straight conductor	Distance r from conductor	$B = \dfrac{\mu_0 I}{2\pi r}$
Circular loop of radius a	On axis of loop	$B = \dfrac{\mu_0 I a^2}{2(x^2 + a^2)^{3/2}}$
	At center of loop	$B = \dfrac{\mu_0 I}{2a}$ (for N loops, multiply these expressions by N)
Long cylindrical conductor of radius R	Inside conductor, $r < R$	$B = \dfrac{\mu_0 I}{2\pi}\dfrac{r}{R^2}$
	Outside conductor, $r > R$	$B = \dfrac{\mu_0 I}{2\pi r}$
Long, closely wound solenoid with n turns per unit length, near its midpoint	Inside solenoid, near center	$B = \mu_0 n I$
	Outside solenoid	$B \approx 0$
Tightly wound toroidal solenoid (toroid) with N turns	Within the space enclosed by the windings, distance r from symmetry axis	$B = \dfrac{\mu_0 N I}{2\pi r}$
	Outside the space enclosed by the windings	$B \approx 0$

Magnetic materials: When magnetic materials are present, the magnetization of the material causes an additional contribution to $\vec{B}$. For paramagnetic and diamagnetic materials, μ_0 is replaced in magnetic-field expressions by $\mu = K_m \mu_0$, where μ is the permeability of the material and K_m is its relative permeability. The magnetic susceptibility χ_m is defined as $\chi_m = K_m - 1$. Magnetic susceptibilities for paramagnetic materials are small positive quantities; those for diamagnetic materials are small negative quantities. For ferromagnetic materials, K_m is much larger than unity and is not constant. Some ferromagnetic materials are permanent magnets, retaining their magnetization even after the external magnetic field is removed. (See Examples 28.11 and 28.12.)

BRIDGING PROBLEM   **Magnetic Field of a Charged, Rotating Dielectric Disk**

A thin dielectric disk with radius a has a total charge $+Q$ distributed uniformly over its surface. It rotates n times per second about an axis perpendicular to the surface of the disk and passing through its center. Find the magnetic field at the center of the disk.

SOLUTION GUIDE

See MasteringPhysics® study area for a Video Tutor solution. **(MP)**

IDENTIFY and SET UP

1. Think of the rotating disk as a series of concentric rotating rings. Each ring acts as a circular current loop that produces a magnetic field at the center of the disk.
2. Use the results of Section 28.5 to find the magnetic field due to a single ring. Then integrate over all rings to find the total field.

EXECUTE

3. Find the charge on a ring with inner radius r and outer radius $r + dr$.

4. How long does it take the charge found in step 3 to make a complete trip around the rotating ring? Use this to find the current of the rotating ring.
5. Use a result from Section 28.5 to determine the magnetic field that this ring produces at the center of the disk.
6. Integrate your result from step 5 to find the total magnetic field from all rings with radii from $r = 0$ to $r = a$.

EVALUATE

7. Does your answer have the correct units?
8. Suppose all of the charge were concentrated at the rim of the disk (at $r = a$). Would this increase or decrease the field at the center of the disk?

Problems

•, ••, •••: Problems of increasing difficulty. **CP**: Cumulative problems incorporating material from earlier chapters. **CALC**: Problems requiring calculus. **BIO**: Biosciences problems.

DISCUSSION QUESTIONS

Q28.1 A topic of current interest in physics research is the search (thus far unsuccessful) for an isolated magnetic pole, or magnetic *monopole*. If such an entity were found, how could it be recognized? What would its properties be?

Q28.2 Streams of charged particles emitted from the sun during periods of solar activity create a disturbance in the earth's magnetic field. How does this happen?

Q28.3 The text discussed the magnetic field of an infinitely long, straight conductor carrying a current. Of course, there is no such thing as an infinitely long *anything*. How do you decide whether a particular wire is long enough to be considered infinite?

Q28.4 Two parallel conductors carrying current in the same direction attract each other. If they are permitted to move toward each other, the forces of attraction do work. From where does the energy come? Does this contradict the assertion in Chapter 27 that magnetic forces on moving charges do no work? Explain.

Q28.5 Pairs of conductors carrying current into or out of the power-supply components of electronic equipment are sometimes twisted together to reduce magnetic-field effects. Why does this help?

Q28.6 Suppose you have three long, parallel wires arranged so that in cross section they are at the corners of an equilateral triangle. Is there any way to arrange the currents so that all three wires attract each other? So that all three wires repel each other? Explain.

Q28.7 In deriving the force on one of the long, current-carrying conductors in Section 28.4, why did we use the magnetic field due to only one of the conductors? That is, why didn't we use the *total* magnetic field due to *both* conductors?

Q28.8 Two concentric, coplanar, circular loops of wire of different diameter carry currents in the same direction. Describe the nature of the force exerted on the inner loop by the outer loop and on the outer loop by the inner loop.

Q28.9 A current was sent through a helical coil spring. The spring contracted, as though it had been compressed. Why?

Q28.10 What are the relative advantages and disadvantages of Ampere's law and the law of Biot and Savart for practical calculations of magnetic fields?

Q28.11 Magnetic field lines never have a beginning or an end. Use this to explain why it is reasonable for the field of a toroidal solenoid to be confined entirely to its interior, while a straight solenoid *must* have some field outside.

Q28.12 If the magnitude of the magnetic field a distance R from a very long, straight, current-carrying wire is B, at what distance from the wire will the field have magnitude $3B$?

Q28.13 Two very long, parallel wires carry equal currents in opposite directions. (a) Is there any place that their magnetic fields completely cancel? If so, where? If not, why not? (b) How would the answer to part (a) change if the currents were in the same direction?

Q28.14 In the circuit shown in Fig. Q28.14, when switch S is suddenly closed, the wire L is pulled toward the lower wire carrying current I. Which (a or b) is the positive terminal of the battery? How do you know?

Figure **Q28.14**

Q28.15 A metal ring carries a current that causes a magnetic field B_0 at the center of the ring and a field B at point P a distance x from the center along the axis of the ring. If the radius of the ring is doubled, find the magnetic field at the center. Will the field at point P change by the same factor? Why?

Q28.16 Why should the permeability of a paramagnetic material be expected to decrease with increasing temperature?

Q28.17 If a magnet is suspended over a container of liquid air, it attracts droplets to its poles. The droplets contain only liquid oxygen; even though nitrogen is the primary constituent of air, it is not attracted to the magnet. Explain what this tells you about the magnetic susceptibilities of oxygen and nitrogen, and explain why a magnet in ordinary, room-temperature air doesn't attract molecules of oxygen *gas* to its poles.

Q28.18 What features of atomic structure determine whether an element is diamagnetic or paramagnetic? Explain.

Q28.19 The magnetic susceptibility of paramagnetic materials is quite strongly temperature dependent, but that of diamagnetic materials is nearly independent of temperature. Why the difference?

Q28.20 A cylinder of iron is placed so that it is free to rotate around its axis. Initially the cylinder is at rest, and a magnetic field is applied to the cylinder so that it is magnetized in a direction parallel to its axis. If the direction of the *external* field is suddenly reversed, the direction of magnetization will also reverse and the cylinder will begin rotating around its axis. (This is called the *Einstein–de Haas effect*.) Explain why the cylinder begins to rotate.

Q28.21 The discussion of magnetic forces on current loops in Section 27.7 commented that no net force is exerted on a complete loop in a uniform magnetic field, only a torque. Yet magnetized materials that contain atomic current loops certainly *do* experience net forces in magnetic fields. How is this discrepancy resolved?

Q28.22 Show that the units $A \cdot m^2$ and J/T for the Bohr magneton are equivalent.

EXERCISES

Section 26.1 Magnetic Field of a Moving Charge

28.1 •• A $+6.00$-μC point charge is moving at a constant 8.00×10^6 m/s in the $+y$-direction, relative to a reference frame. At the instant when the point charge is at the origin of this reference frame, what is the magnetic-field vector $\vec{B}$ it produces at the following points: (a) $x = 0.500$ m, $y = 0$, $z = 0$; (b) $x = 0$, $y = -0.500$ m, $z = 0$; (c) $x = 0$, $y = 0$, $z = +0.500$ m; (d) $x = 0$, $y = -0.500$ m, $z = +0.500$ m?

28.2 • **Fields Within the Atom.** In the Bohr model of the hydrogen atom, the electron moves in a circular orbit of radius 5.3×10^{-11} m with a speed of 2.2×10^6 m/s. If we are viewing the atom in such a way that the electron's orbit is in the plane of the paper with the electron moving clockwise, find the magnitude and direction of the electric and magnetic fields that the electron produces at the location of the nucleus (treated as a point).

28.3 • An electron moves at $0.100c$ as shown in Fig. E28.3. Find the magnitude and direction of the magnetic field this electon produces at the following points, each 2.00 μm from the electron: (a) points A and B; (b) point C; (c) point D.

Figure **E28.3**

28.4 •• An alpha particle (charge $+2e$) and an electron move in opposite directions from the same point, each with the speed of 2.50×10^5 m/s (Fig. E28.4). Find the magnitude and direction of the total magnetic field these charges produce at point P, which is 1.75 nm from each of them.

Figure **E28.4**

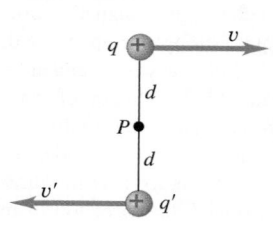

28.5 • A -4.80-μC charge is moving at a constant speed of 6.80×10^5 m/s in the $+x$-direction relative to a reference frame. At the instant when the point charge is at the origin, what is the magnetic-field vector it produces at the following points: (a) $x = 0.500$ m, $y = 0$, $z = 0$; (b) $x = 0$, $y = 0.500$ m, $z = 0$; (c) $x = 0.500$ m, $y = 0.500$ m, $z = 0$; (d) $x = 0$, $y = 0$, $z = 0.500$ m?

28.6 • Positive point charges $q = +8.00\ \mu$C and $q' = +3.00\ \mu$C are moving relative to an observer at point P, as shown in Fig. E28.6. The distance d is 0.120 m, $v = 4.50 \times 10^6$ m/s, and $v' = 9.00 \times 10^6$ m/s. (a) When the two charges are at the locations shown in the figure, what are the magnitude and direction of the net magnetic field they produce at point P? (b) What are the magnitude and direction of the electric and magnetic forces that each charge exerts on the other, and what is the ratio of the magnitude of the electric force to the magnitude of the magnetic force? (c) If the direction of $\vec{v}'$ is reversed, so both charges are moving in the same direction, what are the magnitude and direction of the magnetic forces that the two charges exert on each other?

Figure **E28.6**

28.7 •• Figure E28.6 shows two point charges, q and q', moving relative to an observer at point P. Suppose that the lower charge is actually *negative*, with $q' = -q$. (a) Find the magnetic field (magnitude and direction) produced by the two charges at point P if (i) $v' = v/2$; (ii) $v' = v$; (iii) $v' = 2v$. (b) Find the direction of the magnetic force that q exerts on q', and find the direction of the magnetic force that q' exerts on q. (c) If $v = v' = 3.00 \times 10^5$ m/s, what is the ratio of the magnitude of the magnetic force acting on each charge to that of the Coulomb force acting on each charge?

28.8 •• An electron and a proton are each moving at 845 km/s in perpendicular paths as shown in Fig. E28.8. At the instant when they are at the positions shown in the figure, find the magnitude and direction of (a) the total magnetic field they produce at the origin; (b) the magnetic field the electron produces at the location of the proton; (c) the total electric force and the total magnetic force that the electron exerts on the proton.

Figure **E28.8**

28.9 • A negative charge $q = -3.60 \times 10^{-6}$ C is located at the origin and has velocity $\vec{v} = (7.50 \times 10^4$ m/s$)\hat{\imath} + (-4.90 \times 10^4$ m/s$)\hat{\jmath}$. At this instant what are the magnitude and direction of

the magnetic field produced by this charge at the point $x = 0.200$ m, $y = -0.300$ m, $z = 0$?

Section 28.2 Magnetic Field of a Current Element

28.10 • A short current element $d\vec{l} = (0.500$ mm$)\hat{\jmath}$ carries a current of 8.20 A in the same direction as $d\vec{l}$. Point P is located at $\vec{r} = (-0.730$ m$)\hat{\imath} + (0.390$ m$)\hat{k}$. Use unit vectors to express the magnetic field at P produced by this current element.

28.11 • A straight wire carries a 10.0-A current (Fig. E28.11). $ABCD$ is a rectangle with point D in the middle of a 1.10-mm segment of the wire and point C in the wire. Find the magnitude and direction of the magnetic field due to this segment at (a) point A; (b) point B; (c) point C.

Figure **E28.11**

28.12 • A long, straight wire, carrying a current of 200 A, runs through a cubical wooden box, entering and leaving through holes in the centers of opposite faces (Fig. E28.12). The length of each side of the box is 20.0 cm. Consider an element dl of the wire 0.100 cm long at the center of the box. Compute the magnitude dB of

Figure **E28.12**

the magnetic field produced by this element at the points a, b, c, d, and e in Fig. E28.12. Points a, c, and d are at the centers of the faces of the cube; point b is at the midpoint of one edge; and point e is at a corner. Copy the figure and show the directions and relative magnitudes of the field vectors. (*Note:* Assume that the length dl is small in comparison to the distances from the current element to the points where the magnetic field is to be calculated.)

28.13 •• A long, straight wire lies along the z-axis and carries a 4.00-A current in the $+z$-direction. Find the magnetic field (magnitude and direction) produced at the following points by a 0.500-mm segment of the wire centered at the origin: (a) $x = 2.00$ m, $y = 0$, $z = 0$; (b) $x = 0$, $y = 2.00$ m, $z = 0$; (c) $x = 2.00$ m, $y = 2.00$ m, $z = 0$; (d) $x = 0$, $y = 0$, $z = 2.00$ m.

28.14 •• Two parallel wires are 5.00 cm apart and carry currents in opposite directions, as shown in Fig. E28.14. Find the magnitude and direction of the magnetic field at point P due to two 1.50-mm segments of wire that are opposite each other and each 8.00 cm from P.

Figure **E28.14**

28.15 • A wire carrying a 28.0-A current bends through a right angle. Consider two 2.00-mm segments of wire, each 3.00 cm from the bend (Fig. E28.15). Find the magnitude and direction of the magnetic field these two segments produce at point P, which is midway between them.

Figure **E28.15**

28.16 •• A square wire loop 10.0 cm on each side carries a clockwise current of 15.0 A. Find the magnitude and direction of the magnetic field at its center due to the four 1.20-mm wire segments at the midpoint of each side.

Section 28.3 Magnetic Field of a Straight Current-Carrying Conductor

28.17 • **The Magnetic Field from a Lightning Bolt.** Lightning bolts can carry currents up to approximately 20 kA. We can model such a current as the equivalent of a very long, straight wire. (a) If you were unfortunate enough to be 5.0 m away from such a lightning bolt, how large a magnetic field would you experience? (b) How does this field compare to one you would experience by being 5.0 cm from a long, straight household current of 10 A?

28.18 • A very long, straight horizontal wire carries a current such that 3.50×10^{18} electrons per second pass any given point going from west to east. What are the magnitude and direction of the magnetic field this wire produces at a point 4.00 cm directly above it?

28.19 • **BIO Currents in the Heart.** The body contains many small currents caused by the motion of ions in the organs and cells. Measurements of the magnetic field around the chest due to currents in the heart give values of about 10 μG. Although the actual currents are rather complicated, we can gain a rough understanding of their magnitude if we model them as a long, straight wire. If the surface of the chest is 5.0 cm from this current, how large is the current in the heart?

28.20 • **BIO Bacteria Navigation.** Certain bacteria (such as *Aquaspirillum magnetotacticum*) tend to swim toward the earth's geographic north pole because they contain tiny particles, called magnetosomes, that are sensitive to a magnetic field. If a transmission line carrying 100 A is laid underwater, at what range of distances would the magnetic field from this line be great enough to interfere with the migration of these bacteria? (Assume that a field less than 5 percent of the earth's field would have little effect on the bacteria. Take the earth's field to be 5.0×10^{-5} T and ignore the effects of the seawater.)

28.21 • (a) How large a current would a very long, straight wire have to carry so that the magnetic field 2.00 cm from the wire is equal to 1.00 G (comparable to the earth's northward-pointing magnetic field)? (b) If the wire is horizontal with the current running from east to west, at what locations would the magnetic field of the wire point in the same direction as the horizontal component of the earth's magnetic field? (c) Repeat part (b) except the wire is vertical with the current going upward.

28.22 • Two long, straight wires, one above the other, are separated by a distance $2a$ and are parallel to the x-axis. Let the $+y$-axis be in the plane of the wires in the direction from the lower wire to the upper wire. Each wire carries current I in the $+x$-direction. What are the magnitude and direction of the net magnetic field of the two wires at a point in the plane of the wires (a) midway between them; (b) at a distance a above the upper wire; (c) at a distance a below the lower wire?

28.23 •• A long, straight wire lies along the y-axis and carries a current $I = 8.00$ A in the $-y$-direction (Fig. E28.23). In addition to the magnetic field due to the current in the wire, a uniform magnetic field $\vec{B}_0$ with magnitude 1.50×10^{-6} T is in the $+x$-direction What is the total field (magnitude and direction) at the following points in the xz-plane: (a) $x = 0$, $z = 1.00$ m; (b) $x = 1.00$ m, $z = 0$; (c) $x = 0$, $z = -0.25$ m?

Figure **E28.23**

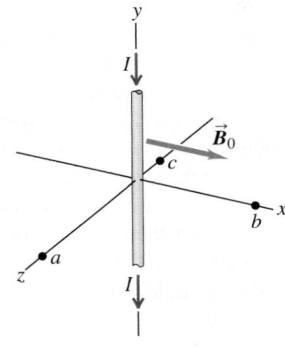

28.24 •• **BIO EMF.** Currents in dc transmission lines can be 100 A or more. Some people have expressed concern that the electromagnetic fields (EMFs) from such lines near their homes could cause health dangers. For a line with current 150 A and at a height of 8.0 m above the ground, what magnetic field does the line produce at ground level? Express your answer in teslas and as a percent of the earth's magnetic field, which is 0.50 gauss. Does this seem to be cause for worry?

28.25 • Two long, straight, parallel wires, 10.0 cm apart, carry equal 4.00-A currents in the same direction, as shown in Fig. E28.25. Find the magnitude and direction of the magnetic field at (a) point P_1, midway between the wires; (b) point P_2, 25.0 cm to the right of P_1; (c) point P_3, 20.0 cm directly above P_1.

Figure **E28.25**

28.26 •• A rectangular loop with dimensions 4.20 cm by 9.50 cm carries current I. The current in the loop produces a magnetic field at the center of the loop that has magnitude 5.50×10^{-5} T and direction away from you as you view the plane of the loop. What are the magnitude and direction (clockwise or counterclockwise) of the current in the loop?

28.27 • Four, long, parallel power lines each carry 100-A currents. A cross-sectional diagram of these lines is a square, 20.0 cm on each side. For each of the three cases shown in Fig. E28.27, calculate the magnetic field at the center of the square.

Figure **E28.27**

⊗	⊗	⊗	⊙	⊗	⊗
⊗	⊗	⊙	⊗	⊙	⊙
(a)		(b)		(c)	

28.28 • Four very long, current-carrying wires in the same plane intersect to form a square 40.0 cm on each side, as shown in Fig. E28.28. Find the magnitude and direction of the current I so that the magnetic field at the center of the square is zero.

Figure **E28.28**   Figure **E28.29**

28.29 •• Two insulated wires perpendicular to each other in the same plane carry currents as shown in Fig. E28.29. Find the magnitude of the *net* magnetic field these wires produce at points P and Q if the 10.0 A-current is (a) to the right or (b) to the left.

Section 28.4 Force Between Parallel Conductors

28.30 • Three parallel wires each carry current I in the directions shown in Fig. E28.30. If the separation between adjacent wires is d, calculate the magnitude and direction of the net magnetic force per unit length on each wire.

Figure **E28.30**

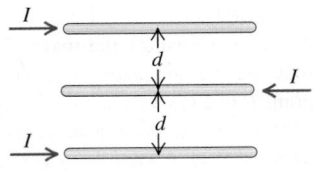

28.31 • Two long, parallel wires are separated by a distance of 0.400 m (Fig. E28.31). The currents I_1 and I_2 have the directions shown. (a) Calculate the magnitude of the force exerted by each wire on a 1.20-m length of the other. Is the force attractive or repulsive? (b) Each current is doubled, so that I_1 becomes 10.0 A and I_2 becomes 4.00 A. Now what is the magnitude of the force that each wire exerts on a 1.20-m length of the other?

Figure **E28.31**

28.32 • Two long, parallel wires are separated by a distance of 2.50 cm. The force per unit length that each wire exerts on the other is 4.00×10^{-5} N/m, and the wires repel each other. The current in one wire is 0.600 A. (a) What is the current in the second wire? (b) Are the two currents in the same direction or in opposite directions?

28.33 • Lamp Cord Wires. The wires in a household lamp cord are typically 3.0 mm apart center to center and carry equal currents in opposite directions. If the cord carries current to a 100-W light bulb connected across a 120-V potential difference, what force per meter does each wire of the cord exert on the other? Is the force attractive or repulsive? Is this force large enough so it should be considered in the design of the lamp cord? (Model the lamp cord as a very long straight wire.)

28.34 • A long, horizontal wire AB rests on the surface of a table and carries a current I. Horizontal wire CD is vertically above wire AB and is free to slide up and down on the two vertical metal guides C and D (Fig. E28.34). Wire CD is connected through the sliding contacts to another wire that also carries a current I, opposite in direction to the current in wire AB. The mass per unit length of the wire CD is λ. To what equilibrium height h will the wire CD rise, assuming that the magnetic force on it is due entirely to the current in the wire AB?

Figure **E28.34**

Section 28.5 Magnetic Field of a Circular Current Loop

28.35 • BIO Currents in the Brain. The magnetic field around the head has been measured to be approximately 3.0×10^{-8} G. Although the currents that cause this field are quite complicated, we can get a rough estimate of their size by modeling them as a single circular current loop 16 cm (the width of a typical head) in diameter. What is the current needed to produce such a field at the center of the loop?

28.36 • Calculate the magnitude and direction of the magnetic field at point P due to the current in the semicircular section of wire shown in Fig. E28.36. (*Hint:* Does the current in the long, straight section of the wire produce any field at P?)

Figure **E28.36**

28.37 •• Calculate the magnitude of the magnetic field at point P of Fig. E28.37 in terms of R, I_1, and I_2. What does your expression give when $I_1 = I_2$?

Figure **E28.37**

28.38 •• A closely wound, circular coil with radius 2.40 cm has 800 turns. (a) What must the current in the coil be if the magnetic field at the center of the coil is 0.0580 T? (b) At what distance x from the center of the coil, on the axis of the coil, is the magnetic field half its value at the center?

28.39 •• A closely wound, circular coil with a diameter of 4.00 cm has 600 turns and carries a current of 0.500 A. What is the magnitude of the magnetic field (a) at the center of the coil and (b) at a point on the axis of the coil 8.00 cm from its center?

28.40 •• A closely wound coil has a radius of 6.00 cm and carries a current of 2.50 A. How many turns must it have if, at a point on the coil axis 6.00 cm from the center of the coil, the magnetic field is 6.39×10^{-4} T?

28.41 •• Two concentric circular loops of wire lie on a tabletop, one inside the other. The inner wire has a diameter of 20.0 cm and carries a clockwise current of 12.0 A, as viewed from above, and the outer wire has a diameter of 30.0 cm. What must be the magnitude and direction (as viewed from above) of the current in the outer wire so that the net magnetic field due to this combination of wires is zero at the common center of the wires?

Section 28.6 Ampere's Law

28.42 • Figure E28.42 shows, in cross section, several conductors that carry currents through the plane of the figure. The currents have the magnitudes $I_1 = 4.0$ A, $I_2 = 6.0$ A, and $I_3 = 2.0$ A, and the directions shown. Four paths, labeled a through d, are shown. What is the line integral $\oint \vec{B} \cdot d\vec{l}$ for each path? Each integral involves going around the path in the counterclockwise direction. Explain your answers.

Figure **E28.42**

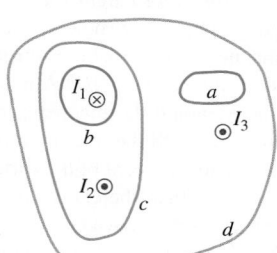

28.43 • A closed curve encircles several conductors. The line integral $\oint \vec{B} \cdot d\vec{l}$ around this curve is 3.83×10^{-4} T·m. (a) What is the net current in the conductors? (b) If you were to integrate around the curve in the opposite direction, what would be the value of the line integral? Explain.

Section 28.7 Applications of Ampere's Law

28.44 •• As a new electrical technician, you are designing a large solenoid to produce a uniform 0.150-T magnetic field near the center of the solenoid. You have enough wire for 4000 circular turns. This solenoid must be 1.40 m long and 2.80 cm in diameter. What current will you need to produce the necessary field?

28.45 • Coaxial Cable. A solid conductor with radius a is supported by insulating disks on the axis of a conducting tube with inner radius b and outer radius c (Fig. E28.45). The central conductor and tube carry equal currents I in opposite directions. The currents are distributed uniformly over the cross sections of each conductor. Derive an expression for the magnitude of the magnetic field (a) at points outside the central, solid conductor but inside the tube ($a < r < b$) and (b) at points outside the tube ($r > c$).

Figure **E28.45**

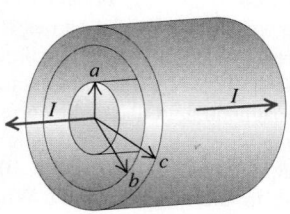

28.46 • Repeat Exercise 28.45 for the case in which the current in the central, solid conductor is I_1, the current in the tube is I_2, and these currents are in the same direction rather than in opposite directions.

28.47 • A long, straight, cylindrical wire of radius R carries a current uniformly distributed over its cross section. At what locations is the magnetic field produced by this current equal to half of its largest value? Consider points inside and outside the wire.

28.48 •• A 15.0-cm-long solenoid with radius 0.750 cm is closely wound with 600 turns of wire. The current in the windings is 8.00 A. Compute the magnetic field at a point near the center of the solenoid.

28.49 •• A solenoid is designed to produce a magnetic field of 0.0270 T at its center. It has radius 1.40 cm and length 40.0 cm, and the wire can carry a maximum current of 12.0 A. (a) What minimum number of turns per unit length must the solenoid have? (b) What total length of wire is required?

28.50 • A toroidal solenoid has an inner radius of 12.0 cm and an outer radius of 15.0 cm. It carries a current of 1.50 A. How many equally spaced turns must it have so that it will produce a magnetic field of 3.75 mT at points within the coils 14.0 cm from its center?

28.51 • A magnetic field of 37.2 T has been achieved at the MIT Francis Bitter National Magnetic Laboratory. Find the current needed to achieve such a field (a) 2.00 cm from a long, straight wire; (b) at the center of a circular coil of radius 42.0 cm that has 100 turns; (c) near the center of a solenoid with radius 2.40 cm, length 32.0 cm, and 40,000 turns.

28.52 • A toroidal solenoid (see Example 28.10) has inner radius $r_1 = 15.0$ cm and outer radius $r_2 = 18.0$ cm. The solenoid has 250 turns and carries a current of 8.50 A. What is the magnitude of the magnetic field at the following distances from the center of the torus: (a) 12.0 cm; (b) 16.0 cm; (c) 20.0 cm?

28.53 •• A wooden ring whose mean diameter is 14.0 cm is wound with a closely spaced toroidal winding of 600 turns. Compute the magnitude of the magnetic field at the center of the cross section of the windings when the current in the windings is 0.650 A.

Section 28.8 Magnetic Materials

28.54 •• A toroidal solenoid with 400 turns of wire and a mean radius of 6.0 cm carries a current of 0.25 A. The relative permeability of the core is 80. (a) What is the magnetic field in the core? (b) What part of the magnetic field is due to atomic currents?

28.55 • A toroidal solenoid with 500 turns is wound on a ring with a mean radius of 2.90 cm. Find the current in the winding that is required to set up a magnetic field of 0.350 T in the ring (a) if the ring is made of annealed iron ($K_m = 1400$) and (b) if the ring is made of silicon steel ($K_m = 5200$).

28.56 • The current in the windings of a toroidal solenoid is 2.400 A. There are 500 turns, and the mean radius is 25.00 cm. The toroidal solenoid is filled with a magnetic material. The magnetic field inside the windings is found to be 1.940 T. Calculate (a) the relative permeability and (b) the magnetic susceptibility of the material that fills the toroid.

28.57 • A long solenoid with 60 turns of wire per centimeter carries a current of 0.15 A. The wire that makes up the solenoid is wrapped around a solid core of silicon steel ($K_m = 5200$). (The wire of the solenoid is jacketed with an insulator so that none of the current flows into the core.) (a) For a point inside the core, find the magnitudes of (i) the magnetic field $\vec{B}_0$ due to the solenoid current; (ii) the magnetization $\vec{M}$; (iii) the total magnetic field $\vec{B}$. (b) In a sketch of the solenoid and core, show the directions of the vectors $\vec{B}$, $\vec{B}_0$, and $\vec{M}$ inside the core.

28.58 • When a certain paramagnetic material is placed in an external magnetic field of 1.5000 T, the field inside the material is measured to be 1.5023 T. Find (a) the relative permeability and (b) the magnetic permeability of this material.

PROBLEMS

28.59 •• A pair of point charges, $q = +8.00$ μC and $q' = -5.00$ μC, are moving as shown in Fig. P28.59 with speeds $v = 9.00 \times 10^4$ m/s and $v' = 6.50 \times 10^4$ m/s. When the charges are at the locations shown in the figure, what are the magnitude and direction of (a) the magnetic field produced at the origin and (b) the magnetic force that q' exerts on q?

Figure **P28.59**

28.60 •• At a particular instant, charge $q_1 = +4.80 \times 10^{-6}$ C is at the point (0, 0.250 m, 0) and has velocity $\vec{v}_1 = (9.20 \times 10^5 \text{ m/s})\hat{\imath}$. Charge $q_2 = -2.90 \times 10^{-6}$ C is at the point (0.150 m, 0, 0) and has velocity $\vec{v}_2 = (-5.30 \times 10^5 \text{ m/s})\hat{\jmath}$. At this instant, what are the magnitude and direction of the magnetic force that q_1 exerts on q_2?

28.61 ••• Two long, parallel transmission lines, 40.0 cm apart, carry 25.0-A and 75.0-A currents. Find all locations where the net magnetic field of the two wires is zero if these currents are in (a) the same direction and (b) the opposite direction.

28.62 • A long, straight wire carries a current of 5.20 A. An electron is traveling in the vicinity of the wire. At the instant when the electron is 4.50 cm from the wire and traveling with a speed of 6.00×10^4 m/s directly toward the wire, what are the magnitude and direction (relative to the direction of the current) of the force that the magnetic field of the current exerts on the electron?

28.63 • CP A long, straight wire carries a 13.0-A current. An electron is fired parallel to this wire with a velocity of 250 km/s in the same direction as the current, 2.00 cm from the wire. (a) Find the magnitude and direction of the electron's initial acceleration. (b) What should be the magnitude and direction of a uniform electric field that will allow the electron to continue to travel parallel to the wire? (c) Is it necessary to include the effects of gravity? Justify your answer.

28.64 • Two very long, straight wires carry currents as shown in Fig. P28.64. For each case, find all locations where the net magnetic field is zero.

Figure **P28.64**

(a)   (b)   (c)

28.65 •• CP Two identical circular, wire loops 40.0 cm in diameter each carry a current of 3.80 A in the same direction. These loops are parallel to each other and are 25.0 cm apart. Line ab is normal to the plane of the loops and passes through their centers.

A proton is fired at 2400 km/s perpendicular to line ab from a point midway between the centers of the loops. Find the magnitude of the magnetic force these loops exert on the proton just after it is fired.

28.66 • A negative point charge $q = -7.20$ mC is moving in a reference frame. When the point charge is at the origin, the magnetic field it produces at the point $x = 25.0$ cm, $y = 0$, $z = 0$ is $\vec{B} = (6.00\ \mu\text{T})\hat{\jmath}$, and its speed is 800 m/s. (a) What are the x-, y-, and z-components of the velocity $\vec{v}_0$ of the charge? (b) At this same instant, what is the magnitude of the magnetic field that the charge produces at the point $x = 0$, $y = 25.0$ cm, $z = 0$?

28.67 • Two long, straight, parallel wires are 1.00 m apart (Fig. P28.67). The wire on the left carries a current I_1 of 6.00 A into the plane of the paper. (a) What must the magnitude and direction of the current I_2 be for the net field at point P to be zero? (b) Then what are the magnitude and direction of the net field at Q? (c) Then what is the magnitude of the net field at S?

Figure **P28.67**

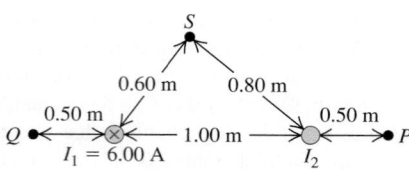

28.68 •• Figure P28.68 shows an end view of two long, parallel wires perpendicular to the xy-plane, each carrying a current I but in opposite directions. (a) Copy the diagram, and draw vectors to show the $\vec{B}$ field of each wire and the net $\vec{B}$ field at point P. (b) Derive the expression for the magnitude of $\vec{B}$ at any point on the x-axis in terms of the x-coordinate of the point. What is the direction of $\vec{B}$? (c) Graph the magnitude of $\vec{B}$ at points on the x-axis. (d) At what value of x is the magnitude of $\vec{B}$ a maximum? (e) What is the magnitude of $\vec{B}$ when $x \gg a$?

Figure **P28.68**

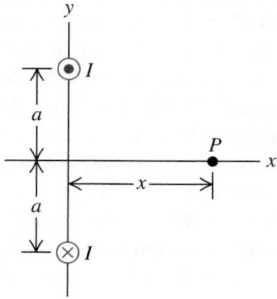

28.69 • Refer to the situation in Problem 28.68. Suppose that a third long, straight wire, parallel to the other two, passes through point P (see Fig. P28.68) and that each wire carries a current $I = 6.00$ A. Let $a = 40.0$ cm and $x = 60.0$ cm. Find the magnitude and direction of the force per unit length on the third wire, (a) if the current in it is directed into the plane of the figure, and (b) if the current in it is directed out of the plane of the figure.

28.70 •• **CP** A pair of long, rigid metal rods, each of length L, lie parallel to each other on a perfectly smooth table. Their ends are connected by identical, very light conducting springs of force constant k (Fig. P28.70) and negligible unstretched length. If a current I runs through this circuit, the springs will stretch. At what separation will the rods remain at rest? Assume that k is large enough so that the separation of the rods will be much less than L.

Figure **P28.70**

28.71 ••• **CP** Two long, parallel wires hang by 4.00-cm-long cords from a common axis (Fig. P28.71). The wires have a mass per unit length of 0.0125 kg/m and carry the same current in opposite directions. What is the current in each wire if the cords hang at an angle of 6.00° with the vertical?

Figure **P28.71**

28.72 • The long, straight wire AB shown in Fig. P28.72 carries a current of 14.0 A. The rectangular loop whose long edges are parallel to the wire carries a current of 5.00 A. Find the magnitude and direction of the net force exerted on the loop by the magnetic field of the wire.

Figure **P28.72**

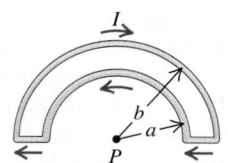

28.73 •• **CP** A flat, round iron ring 5.00 cm in diameter has a current running through it that produces a magnetic field of 75.4 μT at its center. This ring is placed in a uniform external magnetic field of 0.375 T. What is the maximum torque the external field can exert on the ring? Show how the ring should be oriented relative to the field for the torque to have its maximum value.

28.74 • The wire semicircles shown in Fig. P28.74 have radii a and b. Calculate the net magnetic field (magnitude and direction) that the current in the wires produces at point P.

Figure **P28.74**

28.75 • **CALC** Helmholtz Coils. Figure 28.75 is a sectional view of two circular coils with radius a, each wound with N turns of wire carrying a current I, circulating in the same direction in both coils. The coils are separated by a distance a equal to their radii. In this configuration the coils are called Helmholtz coils; they produce a very uniform magnetic field in the region between them. (a) Derive the expression for the magnitude B of the magnetic field at a point on the axis a distance x to the right of point P, which is midway between the coils. (b) Graph B versus x for $x = 0$ to $x = a/2$. Compare this graph to one for the magnetic field due to the right-hand coil alone. (c) From part (a), obtain an expression for the magnitude of the magnetic field at point P. (d) Calculate the magnitude of the magnetic field at P if $N = 300$ turns, $I = 6.00$ A, and $a = 8.00$ cm. (e) Calculate dB/dx and d^2B/dx^2 at $P(x = 0)$. Discuss how your results show that the field is very uniform in the vicinity of P.

Figure **P28.75**

Figure **P28.76**

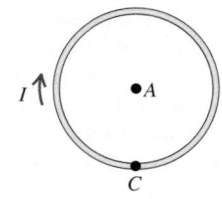

28.76 • A circular wire of diameter D lies on a horizontal table and carries a current I. In Fig. P28.76 point A marks the center of the circle and point C is on its rim. (a) Find the magnitude and direction of the magnetic field at point A. (b) The wire is now unwrapped so it is straight, centered on point C, and perpendicular to the line AC, but the same current is maintained in it. Now find the magnetic field at point A. (c) Which field is greater: the one in part (a) or in part (b)? By what factor? Why is this result physically reasonable?

28.77 • **CALC** A long, straight wire with a circular cross section of radius R carries a current I. Assume that the current density is not constant across the cross section of the wire, but rather varies as $J = \alpha r$, where α is a constant. (a) By the requirement that J integrated over the cross section of the wire gives the total current I, calculate the constant α in terms of I and R. (b) Use Ampere's law to calculate the magnetic field $B(r)$ for (i) $r \leq R$ and (ii) $r \geq R$. Express your answers in terms of I.

28.78 • **CALC** The wire shown in Fig. P28.78 is infinitely long and carries a current I. Calculate the magnitude and direction of the magnetic field that this current produces at point P.

Figure **P28.78**

28.79 • A conductor is made in the form of a hollow cylinder with inner and outer radii a and b, respectively. It carries a current I uniformly distributed over its cross section. Derive expressions for the magnitude of the magnetic field in the regions (a) $r < a$; (b) $a < r < b$; (c) $r > b$.

28.80 • A circular loop has radius R and carries current I_2 in a clockwise direction (Fig. P28.80). The center of the loop is a distance D above a long, straight wire. What are the magnitude and direction of the current I_1 in the wire if the magnetic field at the center of the loop is zero?

Figure **P28.80**

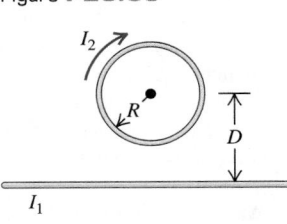

28.81 • **CALC** A long, straight, solid cylinder, oriented with its axis in the z-direction, carries a current whose current density is $\vec{J}$. The current density, although symmetric about the cylinder axis, is not constant but varies according to the relationship

$$\vec{J} = \frac{2I_0}{\pi a^2}\left[1 - \left(\frac{r}{a}\right)^2\right]\hat{k} \quad \text{for } r \leq a$$
$$= 0 \quad \text{for } r \geq a$$

where a is the radius of the cylinder, r is the radial distance from the cylinder axis, and I_0 is a constant having units of amperes. (a) Show that I_0 is the total current passing through the entire cross section of the wire. (b) Using Ampere's law, derive an expression for the magnitude of the magnetic field $\vec{B}$ in the region $r \geq a$. (c) Obtain an expression for the current I contained in a circular cross section of radius $r \leq a$ and centered at the cylinder axis. (d) Using Ampere's law, derive an expression for the magnitude of the magnetic field $\vec{B}$ in the region $r \leq a$. How do your results in parts (b) and (d) compare for $r = a$?

28.82 • A long, straight, solid cylinder, oriented with its axis in the z-direction, carries a current whose current density is $\vec{J}$. The current density, although symmetric about the cylinder axis, is not constant and varies according to the relationship

$$\vec{J} = \left(\frac{b}{r}\right)e^{(r-a)/\delta}\hat{k} \quad \text{for } r \leq a$$
$$= 0 \quad \text{for } r \geq a$$

where the radius of the cylinder is $a = 5.00$ cm, r is the radial distance from the cylinder axis, b is a constant equal to 600 A/m, and δ is a constant equal to 2.50 cm. (a) Let I_0 be the total current passing

through the entire cross section of the wire. Obtain an expression for I_0 in terms of b, δ, and a. Evaluate your expression to obtain a numerical value for I_0. (b) Using Ampere's law, derive an expression for the magnetic field $\vec{B}$ in the region $r \geq a$. Express your answer in terms of I_0 rather than b. (c) Obtain an expression for the current I contained in a circular cross section of radius $r \leq a$ and centered at the cylinder axis. Express your answer in terms of I_0 rather than b. (d) Using Ampere's law, derive an expression for the magnetic field $\vec{B}$ in the region $r \leq a$. (e) Evaluate the magnitude of the magnetic field at $r = \delta$, $r = a$, and $r = 2a$.

28.83 • **An Infinite Current Sheet.** Long, straight conductors with square cross sections and each carrying current I are laid side by side to form an infinite current sheet (Fig. P28.83). The conductors lie in the xy-plane, are parallel to the y-axis, and carry current in the $+y$-direction There are n conductors per unit length measured along the x-axis. (a) What are the magnitude and direction of the magnetic field a distance a below the current sheet? (b) What are the magnitude and direction of the magnetic field a distance a above the current sheet?

Figure **P28.83**

28.84 • Long, straight conductors with square cross section, each carrying current I, are laid side by side to form an infinite current sheet with current directed out of the plane of the page (Fig. P28.84). A second infinite current sheet is a distance d below the first and is parallel to it. The second sheet carries current into the plane of the page. Each sheet has n conductors per unit length. (Refer to Problem 28.83.) Calculate the magnitude and direction of the net magnetic field at (a) point P (above the upper sheet); (b) point R (midway between the two sheets); (c) point S (below the lower sheet).

Figure **P28.84**

28.85 • **CP** A piece of iron has magnetization $M = 6.50 \times 10^4$ A/m. Find the average magnetic dipole moment *per atom* in this piece of iron. Express your answer both in A $\cdot$ m^2 and in Bohr magnetons. The density of iron is given in Table 14.1, and the atomic mass of iron (in grams per mole) is given in Appendix D. The chemical symbol for iron is Fe.

CHALLENGE PROBLEMS

28.86 ••• A wide, long, insulating belt has a uniform positive charge per unit area σ on its upper surface. Rollers at each end move the belt to the right at a constant speed v. Calculate the magnitude and direction of the magnetic field produced by the moving belt at a point just above its surface. (*Hint:* At points near the surface and far from its edges or ends, the moving belt can be considered to be an infinite current sheet like that in Problem 28.83.)

28.87 ••• **CP** Two long, straight conducting wires with linear mass density λ are suspended from cords so that they are each horizontal, parallel to each other, and a distance d apart. The back ends of the wires are connected to each other by a slack, low-resistance connecting wire. A charged capacitor (capacitance C) is now added to the system; the positive plate of the capacitor (initial charge $+Q_0$) is connected to the front end of one of the wires, and the negative plate of the capacitor (initial charge $-Q_0$) is connected to the front end of the other wire (Fig. P28.87). Both of these connections are also made by slack, low-resistance wires. When the connection is made, the wires are pushed aside by the repulsive force between the wires, and each wire has an initial horizontal velocity of magnitude v_0. Assume that the time constant for the capacitor to discharge is negligible compared to the time it takes for any appreciable displacement in the position of the wires to occur. (a) Show that the initial speed v_0 of either wire is given by

$$v_0 = \frac{\mu_0 Q_0^2}{4\pi \lambda RCd}$$

Figure **P28.87**

where R is the total resistance of the circuit. (b) To what height h will each wire rise as a result of the circuit connection?

28.88 ••• **CALC** A wire in the shape of a semicircle with radius a is oriented in the yz-plane with its center of curvature at the origin (Fig. P28.88). If the current in the wire is I, calculate the magnetic-field components produced at point P, a distance x out along the x-axis. (*Note:* Do not forget the contribution from the straight wire at the bottom of the semicircle that runs from $z = -a$ to $z = +a$. You may use the fact that the fields of the two antiparallel currents at $z > a$ cancel, but you must explain *why* they cancel.)

Figure **P28.88**

Answers

Chapter Opening Question ?

There would be *no* change in the magnetic field strength. From Example 28.9 (Section 28.7), the field inside a solenoid has magnitude $B = \mu_0 nI$, where n is the number of turns of wire per unit length. Joining two solenoids end to end doubles both the number of turns and the length, so the number of turns per unit length is unchanged.

Test Your Understanding Questions

28.1 Answers: (a) (i), (b) (ii) The situation is the same as shown in Fig. 28.2 except that the upper proton has velocity $\vec{v}$ rather than $-\vec{v}$. The magnetic field due to the lower proton is the same as shown in Fig. 28.2, but the direction of the magnetic force $\vec{F} = q\vec{v} \times \vec{B}$ on the upper proton is reversed. Hence the magnetic force is attractive. Since the speed v is small compared to c, the magnetic force is much smaller in magnitude than the repulsive electric force and the net force is still repulsive.

28.2 Answer: (i) and (iii) (tie), (iv), (ii) From Eq. (28.5), the magnitude of the field dB due to a current element of length dl carrying current I is $dB = (\mu/4\pi)(I\,dl\sin\phi/r^2)$. In this expression r is the distance from the element to the field point, and ϕ is the angle between the direction of the current and a vector from the current element to the field point. All four points are the same distance $r = L$ from the current element, so the value of dB is proportional to the value of $\sin\phi$. For the four points the angle is (i) $\phi = 90°$, (ii) $\phi = 0$, (iii) $\phi = 90°$, and (iv) $\phi = 45°$, so the values of $\sin\phi$ are (i) 1, (ii) 0, (iii) 1, and (iv) $1/\sqrt{2}$.

28.3 Answer: A This orientation will cause current to flow clockwise around the circuit. Hence current will flow south through the wire that lies under the compass. From the right-hand rule for the magnetic field produced by a long, straight, current-carrying conductor, this will produce a magnetic field that points to the left at the position of the compass (which lies atop the wire). The combination of the northward magnetic field of the earth and the westward field produced by the current gives a net magnetic field to the northwest, so the compass needle will swing counterclockwise to align with this field.

28.4 Answers: (a) (i), (b) (iii), (c) (ii), (d) (iii) Current flows in the same direction in adjacent turns of the coil, so the magnetic forces between these turns are attractive. Current flows in opposite directions on opposite sides of the same turn, so the magnetic forces between these sides are repulsive. Thus the magnetic forces on the

solenoid turns squeeze them together in the direction along its axis but push them apart radially. The *electric* forces are zero because the wire is electrically neutral, with as much positive charge as there is negative charge.

28.5 Answers: (a) (ii), (b) (v) The vector $d\vec{B}$ is in the direction of $d\vec{l} \times \hat{r}$. For a segment on the negative y-axis, $d\vec{l} = -\hat{k}\,dl$ points in the negative z-direction and $\vec{r} = x\hat{i} + a\hat{j}$. Hence $d\vec{l} \times \vec{r} = (a\,dl)\hat{i} - (x\,dl)\hat{j}$, which has a positive x-component, a negative y-component, and zero z-component. For a segment on the negative z-axis, $d\vec{l} = \hat{j}\,dl$ points in the positive y-direction and $\vec{r} = x\hat{i} + a\hat{k}$. Hence $d\vec{l} \times \vec{r} = 1(a\,dl)\hat{i} - 1(x\,dl)\hat{k}$, which has a positive x-component, zero y-component, and a negative z-component.

28.6 Answer: (ii) Imagine carrying out the integral $\oint \vec{B} \cdot d\vec{l}$ along an integration path that goes counterclockwise around the red magnetic field line. At each point along the path the magnetic field $\vec{B}$ and the infinitesimal segment $d\vec{l}$ are both tangent to the path, so $\vec{B} \cdot d\vec{l}$ is positive at each point and the integral $\oint \vec{B} \cdot d\vec{l}$ is likewise positive. It follows from Ampere's law $\oint \vec{B} \cdot d\vec{l} = \mu_0 I_{encl}$ and the right-hand rule that the integration path encloses a current directed out of the plane of the page. There are no currents in the empty space outside the magnet, so there must be currents inside the magnet (see Section 28.8).

28.7 Answer: (iii) By symmetry, any $\vec{B}$ field outside the cable must circulate around the cable, with circular field lines like those surrounding the solid cylindrical conductor in Fig. 28.20. Choose an integration path like the one shown in Fig. 28.20 with radius $r > R$, so that the path completely encloses the cable. As in Example 28.8, the integral $\oint \vec{B} \cdot d\vec{l}$ for this path has magnitude $B(2\pi r)$. From Ampere's law this is equal to $\mu_0 I_{encl}$. The net enclosed current I_{encl} is zero because it includes two currents of equal magnitude but opposite direction: one in the central wire and one in the hollow cylinder. Hence $B(2\pi r) = 0$, and so $B = 0$ for any value of r outside the cable. (The field is nonzero *inside* the cable; see Exercise 28.45.)

28.8 Answer: (i), (iv) Sodium and uranium are paramagnetic materials and hence are attracted to a magnet, while bismuth and lead are diamagnetic materials that are repelled by a magnet. (See Table 28.1.)

Bridging Problem

Answer: $B = \dfrac{\mu_0 nQ}{a}$

ELECTROMAGNETIC INDUCTION

? When a credit card is "swiped" through a card reader, the information coded in a magnetic pattern on the back of the card is transmitted to the cardholder's bank. Why is it necessary to swipe the card rather than holding it motionless in the card reader's slot?

LEARNING GOALS

By studying this chapter, you will learn:

- The experimental evidence that a changing magnetic field induces an emf.

- How Faraday's law relates the induced emf in a loop to the change in magnetic flux through the loop.

- How to determine the direction of an induced emf.

- How to calculate the emf induced in a conductor moving through a magnetic field.

- How a changing magnetic flux generates an electric field that is very different from that produced by an arrangement of charges.

- The four fundamental equations that completely describe both electricity and magnetism.

Almost every modern device or machine, from a computer to a washing machine to a power drill, has electric circuits at its heart. We learned in Chapter 25 that an electromotive force (emf) is required for a current to flow in a circuit; in Chapters 25 and 26 we almost always took the source of emf to be a battery. But for the vast majority of electric devices that are used in industry and in the home (including any device that you plug into a wall socket), the source of emf is *not* a battery but an electric generating station. Such a station produces electric energy by converting other forms of energy: gravitational potential energy at a hydroelectric plant, chemical energy in a coal- or oil-fired plant, nuclear energy at a nuclear plant. But how is this energy conversion done?

The answer is a phenomenon known as *electromagnetic induction:* If the magnetic flux through a circuit changes, an emf and a current are induced in the circuit. In a power-generating station, magnets move relative to coils of wire to produce a changing magnetic flux in the coils and hence an emf. Other key components of electric power systems, such as transformers, also depend on magnetically induced emfs.

The central principle of electromagnetic induction, and the keystone of this chapter, is *Faraday's law.* This law relates induced emf to changing magnetic flux in any loop, including a closed circuit. We also discuss Lenz's law, which helps us to predict the directions of induced emfs and currents. These principles will allow us to understand electrical energy-conversion devices such as motors, generators, and transformers.

Electromagnetic induction tells us that a time-varying magnetic field can act as a source of electric field. We will also see how a time-varying *electric* field can act as a source of *magnetic* field. These remarkable results form part of a neat package of formulas, called *Maxwell's equations,* that describe the behavior of electric and magnetic fields in *any* situation. Maxwell's equations pave the way toward an understanding of electromagnetic waves, the topic of Chapter 32.

29.1 Induction Experiments

During the 1830s, several pioneering experiments with magnetically induced emf were carried out in England by Michael Faraday and in the United States by Joseph Henry (1797–1878), later the first director of the Smithsonian Institution. Figure 29.1 shows several examples. In Fig. 29.1a, a coil of wire is connected to a galvanometer. When the nearby magnet is stationary, the meter shows no current. This isn't surprising; there is no source of emf in the circuit. But when we *move* the magnet either toward or away from the coil, the meter shows current in the circuit, but *only* while the magnet is moving (Fig. 29.1b). If we keep the magnet stationary and move the coil, we again detect a current during the motion. We call this an **induced current,** and the corresponding emf required to cause this current is called an **induced emf.**

In Fig. 29.1c we replace the magnet with a second coil connected to a battery. When the second coil is stationary, there is no current in the first coil. However, when we move the second coil toward or away from the first or move the first toward or away from the second, there is current in the first coil, but again *only* while one coil is moving relative to the other.

Finally, using the two-coil setup in Fig. 29.1d, we keep both coils stationary and vary the current in the second coil, either by opening and closing the switch or by changing the resistance of the second coil with the switch closed (perhaps by changing the second coil's temperature). We find that as we open or close the switch, there is a momentary current pulse in the first circuit. When we vary the resistance (and thus the current) in the second coil, there is an induced current in the first circuit, but only while the current in the second circuit is changing.

To explore further the common elements in these observations, let's consider a more detailed series of experiments (Fig. 29.2). We connect a coil of wire to a galvanometer and then place the coil between the poles of an electromagnet whose magnetic field we can vary. Here's what we observe:

1. When there is no current in the electromagnet, so that $\vec{B} = 0$, the galvanometer shows no current.
2. When the electromagnet is turned on, there is a momentary current through the meter as $\vec{B}$ increases.
3. When $\vec{B}$ levels off at a steady value, the current drops to zero, no matter how large $\vec{B}$ is.
4. With the coil in a horizontal plane, we squeeze it so as to decrease the cross-sectional area of the coil. The meter detects current only *during* the

29.1 Demonstrating the phenomenon of induced current.

(a) A stationary magnet does NOT induce a current in a coil.

All these actions DO induce a current in the coil. What do they have in common?*

(b) Moving the magnet toward or away from the coil

(c) Moving a second, current-carrying coil toward or away from the coil

(d) Varying the current in the second coil (by closing or opening a switch)

Meter shows zero current.

Meter shows induced current.

*They cause the magnetic field through the coil to *change*.

deformation, not before or after. When we increase the area to return the coil to its original shape, there is current in the opposite direction, but only while the area of the coil is changing.

5. If we rotate the coil a few degrees about a horizontal axis, the meter detects current during the rotation, in the same direction as when we decreased the area. When we rotate the coil back, there is a current in the opposite direction during this rotation.

6. If we jerk the coil out of the magnetic field, there is a current during the motion, in the same direction as when we decreased the area.

7. If we decrease the number of turns in the coil by unwinding one or more turns, there is a current during the unwinding, in the same direction as when we decreased the area. If we wind more turns onto the coil, there is a current in the opposite direction during the winding.

8. When the magnet is turned off, there is a momentary current in the direction opposite to the current when it was turned on.

9. The faster we carry out any of these changes, the greater the current.

10. If all these experiments are repeated with a coil that has the same shape but different material and different resistance, the current in each case is inversely proportional to the total circuit resistance. This shows that the induced emfs that are causing the current do not depend on the material of the coil but only on its shape and the magnetic field.

29.2 A coil in a magnetic field. When the $\vec{B}$ field is constant and the shape, location, and orientation of the coil do not change, no current is induced in the coil. A current is induced when any of these factors change.

The common element in all these experiments is changing *magnetic flux* Φ_B through the coil connected to the galvanometer. In each case the flux changes either because the magnetic field changes with time or because the coil is moving through a nonuniform magnetic field. Faraday's law of induction, the subject of the next section, states that in all of these situations the induced emf is proportional to the *rate of change* of magnetic flux Φ_B through the coil. The *direction* of the induced emf depends on whether the flux is increasing or decreasing. If the flux is constant, there is no induced emf.

Induced emfs are not mere laboratory curiosities but have a tremendous number of practical applications. If you are reading these words indoors, you are making use of induced emfs right now! At the power plant that supplies your neighborhood, an electric generator produces an emf by varying the magnetic flux through coils of wire. (In the next section we'll see in detail how this is done.) This emf supplies the voltage between the terminals of the wall sockets in your home, and this voltage supplies the power to your reading lamp. Indeed, any appliance that you plug into a wall socket makes use of induced emfs.

Magnetically induced emfs, just like the emfs discussed in Section 25.4, are the result of *nonelectrostatic* forces. We have to distinguish carefully between the electrostatic electric fields produced by charges (according to Coulomb's law) and the nonelectrostatic electric fields produced by changing magnetic fields. We'll return to this distinction later in this chapter and the next.

29.2 Faraday's Law

The common element in all induction effects is changing magnetic flux through a circuit. Before stating the simple physical law that summarizes all of the kinds of experiments described in Section 29.1, let's first review the concept of magnetic flux Φ_B (which we introduced in Section 27.3). For an infinitesimal-area element $d\vec{A}$ in a magnetic field $\vec{B}$ (Fig. 29.3), the magnetic flux $d\Phi_B$ through the area is

$$d\Phi_B = \vec{B} \cdot d\vec{A} = B_\perp \, dA = B \, dA \cos\phi$$

where $B_\perp$ is the component of $\vec{B}$ perpendicular to the surface of the area element and ϕ is the angle between $\vec{B}$ and $d\vec{A}$. (As in Chapter 27, be careful to distinguish

29.3 Calculating the magnetic flux through an area element.

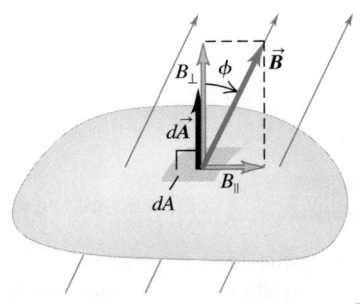

Magnetic flux through element of area $d\vec{A}$:
$$d\Phi_B = \vec{B} \cdot d\vec{A} = B_\perp \, dA = B \, dA \cos\phi$$

29.4 Calculating the flux of a uniform magnetic field through a flat area. (Compare to Fig. 22.6, which shows the rules for calculating the flux of a uniform *electric* field.)

Surface is face-on to magnetic field:
- $\vec{B}$ and $\vec{A}$ are parallel (the angle between $\vec{B}$ and $\vec{A}$ is $\phi = 0$).
- The magnetic flux $\Phi_B = \vec{B} \cdot \vec{A} = BA$.

Surface is tilted from a face-on orientation by an angle ϕ:
- The angle between $\vec{B}$ and $\vec{A}$ is ϕ.
- The magnetic flux $\Phi_B = \vec{B} \cdot \vec{A} = BA \cos \phi$.

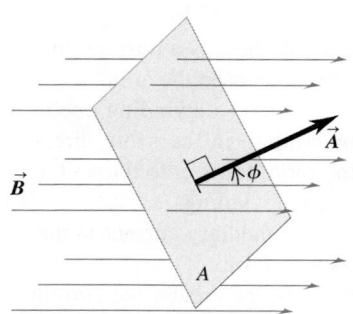

Surface is edge-on to magnetic field:
- $\vec{B}$ and $\vec{A}$ are perpendicular (the angle between $\vec{B}$ and $\vec{A}$ is $\phi = 90°$).
- The magnetic flux $\Phi_B = \vec{B} \cdot \vec{A} = BA \cos 90° = 0$.

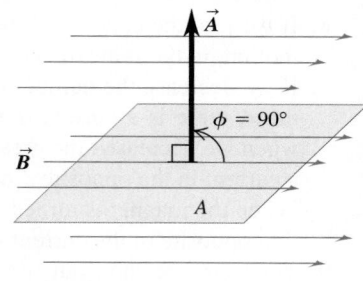

between two quantities named "phi," ϕ and Φ_B.) The total magnetic flux Φ_B through a finite area is the integral of this expression over the area:

$$\Phi_B = \int \vec{B} \cdot d\vec{A} = \int B \, dA \cos \phi \qquad (29.1)$$

If $\vec{B}$ is uniform over a flat area $\vec{A}$, then

$$\Phi_B = \vec{B} \cdot \vec{A} = BA \cos \phi \qquad (29.2)$$

Figure 29.4 reviews the rules for using Eq. (29.2).

CAUTION **Choosing the direction of $d\vec{A}$ or $\vec{A}$** In Eqs. (29.1) and (29.2) we have to be careful to define the direction of the vector area $d\vec{A}$ or $\vec{A}$ unambiguously. There are always two directions perpendicular to any given area, and the sign of the magnetic flux through the area depends on which one we choose to be positive. For example, in Fig. 29.3 we chose $d\vec{A}$ to point upward so ϕ is less than 90° and $\vec{B} \cdot d\vec{A}$ is positive. We could have chosen instead to have $d\vec{A}$ point downward, in which case ϕ would have been greater than 90° and $\vec{B} \cdot d\vec{A}$ would have been negative. Either choice is equally good, but once we make a choice we must stick with it. ▮

Faraday's law of induction states:

> **The induced emf in a closed loop equals the negative of the time rate of change of magnetic flux through the loop.**

In symbols, Faraday's law is

$$\mathcal{E} = -\frac{d\Phi_B}{dt} \qquad \text{(Faraday's law of induction)} \qquad (29.3)$$

To understand the negative sign, we have to introduce a sign convention for the induced emf $\mathcal{E}$. But first let's look at a simple example of this law in action.

Example 29.1 **Emf and current induced in a loop**

The magnetic field between the poles of the electromagnet in Fig. 29.5 is uniform at any time, but its magnitude is increasing at the rate of 0.020 T/s. The area of the conducting loop in the field is 120 cm², and the total circuit resistance, including the meter, is 5.0 Ω. (a) Find the induced emf and the induced current in the circuit. (b) If the loop is replaced by one made of an insulator, what effect does this have on the induced emf and induced current?

SOLUTION

IDENTIFY and SET UP: The magnetic flux Φ_B through the loop changes as the magnetic field changes. Hence there will be an induced emf $\mathcal{E}$ and an induced current I in the loop. We calculate Φ_B using Eq. (29.2), then find $\mathcal{E}$ using Faraday's law. Finally, we calculate I using $\mathcal{E} = IR$, where R is the total resistance of the circuit that includes the loop.

EXECUTE: (a) The area vector $\vec{A}$ for the loop is perpendicular to the plane of the loop; we take $\vec{A}$ to be vertically upward. Then $\vec{A}$ and $\vec{B}$ are parallel, and because $\vec{B}$ is uniform the magnetic flux through the loop is $\Phi_B = \vec{B} \cdot \vec{A} = BA\cos 0 = BA$. The area $A = 0.012 \text{ m}^2$ is constant, so the rate of change of magnetic flux is

29.5 A stationary conducting loop in an increasing magnetic field.

$$\frac{d\Phi_B}{dt} = \frac{d(BA)}{dt} = \frac{dB}{dt}A = (0.020 \text{ T/s})(0.012 \text{ m}^2)$$
$$= 2.4 \times 10^{-4} \text{ V} = 0.24 \text{ mV}$$

This, apart from a sign that we haven't discussed yet, is the induced emf $\mathcal{E}$. The corresponding induced current is

$$I = \frac{\mathcal{E}}{R} = \frac{2.4 \times 10^{-4} \text{ V}}{5.0 \text{ }\Omega} = 4.8 \times 10^{-5} \text{A} = 0.048 \text{ mA}$$

(b) By changing to an insulating loop, we've made the resistance of the loop very high. Faraday's law, Eq. (29.3), does not involve the resistance of the circuit in any way, so the induced *emf* does not change. But the *current* will be smaller, as given by the equation $I = \mathcal{E}/R$. If the loop is made of a perfect insulator with infinite resistance, the induced current is zero. This situation is analogous to an isolated battery whose terminals aren't connected to anything: An emf is present, but no current flows.

EVALUATE: Let's verify unit consistency in this calculation. One way to do this is to note that the magnetic-force relationship $\vec{F} = q\vec{v} \times \vec{B}$ implies that the units of $\vec{B}$ are the units of force divided by the units of (charge times velocity): $1 \text{ T} = (1 \text{ N})/(1 \text{ C} \cdot \text{m/s})$. The units of magnetic flux are then $(1 \text{ T})(1 \text{ m}^2) = 1 \text{ N} \cdot \text{s} \cdot \text{m/C}$, and the rate of change of magnetic flux is $1 \text{ N} \cdot \text{m/C} = 1 \text{ J/C} = 1 \text{ V}$. Thus the unit of $d\Phi_B/dt$ is the volt, as required by Eq. (29.3). Also recall that the unit of magnetic flux is the weber (Wb): $1 \text{ T} \cdot \text{m}^2 = 1 \text{ Wb}$, so $1 \text{ V} = 1 \text{ Wb/s}$.

Direction of Induced emf

We can find the direction of an induced emf or current by using Eq. (29.3) together with some simple sign rules. Here's the procedure:

1. Define a positive direction for the vector area $\vec{A}$.
2. From the directions of $\vec{A}$ and the magnetic field $\vec{B}$, determine the sign of the magnetic flux Φ_B and its rate of change $d\Phi_B/dt$. Figure 29.6 shows several examples.
3. Determine the sign of the induced emf or current. If the flux is increasing, so $d\Phi_B/dt$ is positive, then the induced emf or current is negative; if the flux is decreasing, $d\Phi_B/dt$ is negative and the induced emf or current is positive.
4. Finally, determine the direction of the induced emf or current using your right hand. Curl the fingers of your right hand around the $\vec{A}$ vector, with your right thumb in the direction of $\vec{A}$. If the induced emf or current in the circuit is *positive,* it is in the same direction as your curled fingers; if the induced emf or current is *negative,* it is in the opposite direction.

In Example 29.1, in which $\vec{A}$ is upward, a positive $\mathcal{E}$ would be directed counterclockwise around the loop, as seen from above. Both $\vec{A}$ and $\vec{B}$ are upward in this example, so Φ_B is positive; the magnitude B is increasing, so $d\Phi_B/dt$ is positive. Hence by Eq. (29.3), $\mathcal{E}$ in Example 29.1 is *negative*. Its actual direction is thus *clockwise* around the loop, as seen from above.

If the loop in Fig. 29.5 is a conductor, an induced current results from this emf; this current is also clockwise, as Fig. 29.5 shows. This induced current produces an additional magnetic field through the loop, and the right-hand rule described in Section 28.6 shows that this field is *opposite* in direction to the increasing field produced by the electromagnet. This is an example of a general rule called *Lenz's law,* which says that any induction effect tends to oppose the change that caused it; in this case the change is the increase in the flux of the

Application **Exploring the Brain with Induced emfs**
Transcranial magnetic stimulation (TMS) is a technique for studying the function of various parts of the brain. A coil held to the subject's head carries a varying electric current, and so produces a varying magnetic field. This field causes an induced emf, and that triggers electric activity in the region of the brain underneath the coil. By observing how the TMS subject responds (for instance, which muscles move as a result of stimulating a certain part of the brain), a physician can test for various neurological conditions.

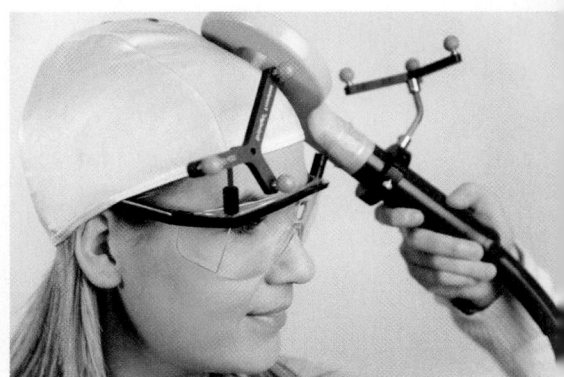

29.6 The magnetic flux is becoming **(a)** more positive, **(b)** less positive, **(c)** more negative, and **(d)** less negative. Therefore Φ_B is increasing in **(a)** and **(d)** and decreasing in **(b)** and **(c)**. In **(a)** and **(d)** the emfs are negative (they are opposite to the direction of the curled fingers of your right hand when your right thumb points along $\vec{A}$). In **(b)** and **(c)** the emfs are positive (in the same direction as the curled fingers).

(a)

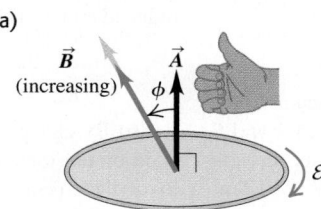

- Flux is positive ($\Phi_B > 0$) ...
- ... and becoming more positive ($d\Phi_B/dt > 0$).
- Induced emf is negative ($\mathcal{E} < 0$).

(b)

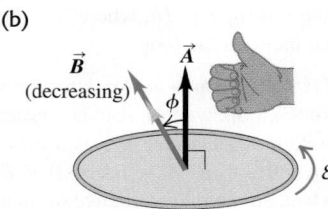

- Flux is positive ($\Phi_B > 0$) ...
- ... and becoming less positive ($d\Phi_B/dt < 0$).
- Induced emf is positive ($\mathcal{E} > 0$).

(c)

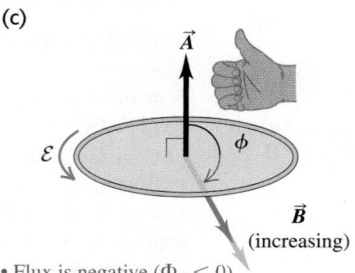

- Flux is negative ($\Phi_B < 0$) ...
- ... and becoming more negative ($d\Phi_B/dt < 0$).
- Induced emf is positive ($\mathcal{E} > 0$).

(d)

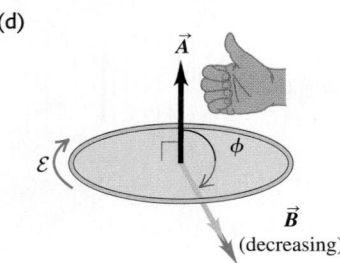

- Flux is negative ($\Phi_B < 0$) ...
- ... and becoming less negative ($d\Phi_B/dt > 0$).
- Induced emf is negative ($\mathcal{E} < 0$).

electromagnet's field through the loop. (We'll study this law in detail in the next section.)

You should check out the signs of the induced emfs and currents for the list of experiments in Section 29.1. For example, when the loop in Fig. 29.2 is in a constant field and we tilt it or squeeze it to *decrease* the flux through it, the induced emf and current are counterclockwise, as seen from above.

> **CAUTION** **Induced emfs are caused by changes in flux** Since magnetic flux plays a central role in Faraday's law, it's tempting to think that *flux* is the cause of induced emf and that an induced emf will appear in a circuit whenever there is a magnetic field in the region bordered by the circuit. But Eq. (29.3) shows that only a *change* in flux through a circuit, not flux itself, can induce an emf in a circuit. If the flux through a circuit has a constant value, whether positive, negative, or zero, there is no induced emf.

If we have a coil with N identical turns, and if the flux varies at the same rate through each turn, the *total* rate of change through all the turns is N times as large as for a single turn. If Φ_B is the flux through each turn, the total emf in a coil with N turns is

$$\mathcal{E} = -N\frac{d\Phi_B}{dt} \tag{29.4}$$

As we discussed in this chapter's introduction, induced emfs play an essential role in the generation of electric power for commercial use. Several of the following examples explore different methods of producing emfs by the motion of a conductor relative to a magnetic field, giving rise to a changing flux through a circuit.

MasteringPHYSICS

PhET: Faraday's Electromagnetic Lab
PhET: Faraday's Law
PhET: Generator

Problem-Solving Strategy 29.1   Faraday's Law

IDENTIFY *the relevant concepts:* Faraday's law applies when there is a changing magnetic flux. To use the law, identify an area through which there is a flux of magnetic field. This will usually be the area enclosed by a loop made of a conducting material (though not always—see part (b) of Example 29.1). Identify the target variables.

SET UP *the problem* using the following steps:
1. Faraday's law relates the induced emf to the rate of change of magnetic flux. To calculate this rate of change, you first have to understand what is making the flux change. Is the conductor moving? Is it changing orientation? Is the magnetic field changing? Remember that it's not the flux itself that counts, but its *rate of change*.
2. The area vector $\vec{A}$ (or $d\vec{A}$) must be perpendicular to the plane of the area. You always have two choices of its direction; for example, if the area is in a horizontal plane, $\vec{A}$ could point up or

down. Choose a direction and use it consistently throughout the problem.

EXECUTE *the solution* as follows:
1. Calculate the magnetic flux using Eq. (29.2) if $\vec{B}$ is uniform over the area of the loop or Eq. (29.1) if it isn't uniform. Remember the direction you chose for the area vector.
2. Calculate the induced emf using Eq. (29.3) or (if your conductor has N turns in a coil) Eq. (29.4). Apply the sign rule (described just after Example 29.1) to determine the positive direction of emf.
3. If the circuit resistance is known, you can calculate the magnitude of the induced current I using $\mathcal{E} = IR$.

EVALUATE *your answer:* Check your results for the proper units, and double-check that you have properly implemented the sign rules for magnetic flux and induced emf.

Example 29.2   Magnitude and direction of an induced emf

A 500-loop circular wire coil with radius 4.00 cm is placed between the poles of a large electromagnet. The magnetic field is uniform and makes an angle of 60° with the plane of the coil; it decreases at 0.200 T/s. What are the magnitude and direction of the induced emf?

SOLUTION

IDENTIFY and SET UP: Our target variable is the emf induced by a varying magnetic flux through the coil. The flux varies because the magnetic field decreases in amplitude. We choose the area vector $\vec{A}$ to be in the direction shown in Fig. 29.7. With this choice, the geometry is similar to that of Fig. 29.6b. That figure will help us determine the direction of the induced emf.

29.7 Our sketch for this problem.

EXECUTE: The magnetic field is uniform over the loop, so we can calculate the flux using Eq. (29.2): $\Phi_B = BA\cos\phi$, where $\phi = 30°$. In this expression, the only quantity that changes with time is the magnitude B of the field, so $d\Phi_B/dt = (dB/dt)A\cos\phi$.

CAUTION **Remember how ϕ is defined** You may have been tempted to say that $\phi = 60°$ in this problem. If so, remember that ϕ is the angle between $\vec{A}$ and $\vec{B}$, *not* the angle between $\vec{B}$ and the plane of the loop. ▮

From Eq. (29.4), the induced emf in the coil of $N = 500$ turns is

$$\mathcal{E} = -N\frac{d\Phi_B}{dt} = N\frac{dB}{dt}A\cos\phi$$
$$= 500(-0.200 \text{ T/s})\pi(0.0400 \text{ m})^2(\cos 30°) = 0.435 \text{ V}$$

The positive answer means that when you point your right thumb in the direction of the area vector $\vec{A}$ (30° below the magnetic field $\vec{B}$ in Fig. 29.7), the positive direction for $\mathcal{E}$ is in the direction of the curled fingers of your right hand. If you viewed the coil from the left in Fig. 29.7 and looked in the direction of $\vec{A}$, the emf would be clockwise.

EVALUATE: If the ends of the wire are connected, the direction of current in the coil is in the same direction as the emf—that is, clockwise as seen from the left side of the coil. A clockwise current increases the magnetic flux through the coil, and therefore tends to oppose the decrease in total flux. This is an example of Lenz's law, which we'll discuss in Section 29.3.

Example 29.3   Generator I: A simple alternator

Figure 29.8a shows a simple *alternator,* a device that generates an emf. A rectangular loop is rotated with constant angular speed ω about the axis shown. The magnetic field $\vec{B}$ is uniform and constant. At time $t = 0$, $\phi = 0$. Determine the induced emf.

SOLUTION

IDENTIFY and SET UP: The magnetic field $\vec{B}$ and the area A of the loop are both constant, but the flux through the loop varies because the loop rotates and so the angle ϕ between $\vec{B}$ and the area vector

Continued

29.8 (a) Schematic diagram of an alternator. A conducting loop rotates in a magnetic field, producing an emf. Connections from each end of the loop to the external circuit are made by means of that end's slip ring. The system is shown at the time when the angle $\phi = \omega t = 90°$. (b) Graph of the flux through the loop and the resulting emf between terminals a and b, along with the corresponding positions of the loop during one complete rotation.

$\vec{A}$ changes (Fig. 29.8a). Because the angular speed is constant and $\phi = 0$ at $t = 0$, the angle as a function of time is given by $\phi = \omega t$.

EXECUTE: The magnetic field is uniform over the loop, so the magnetic flux is $\Phi_B = BA \cos \phi = BA \cos \omega t$. Hence, by Faraday's law [Eq. (29.3)] the induced emf is

$$\mathcal{E} = -\frac{d\Phi_B}{dt} = -\frac{d}{dt}(BA \cos \omega t) = \omega BA \sin \omega t$$

EVALUATE: The induced emf $\mathcal{E}$ varies sinusoidally with time (Fig. 29.8b). When the plane of the loop is perpendicular to $\vec{B}$ ($\phi = 0$ or 180°), Φ_B reaches its maximum and minimum values. At these times, its instantaneous rate of change is zero and $\mathcal{E}$ is zero. Conversely, $\mathcal{E}$ reaches its maximum and minimum values when the plane of the loop is parallel to $\vec{B}$ ($\phi = 90°$ or 270°) and Φ_B is changing most rapidly. We note that the induced emf does not depend on the *shape* of the loop, but only on its area.

We can use the alternator as a source of emf in an external circuit by using two *slip rings* that rotate with the loop, as shown in Fig. 29.8a. The rings slide against stationary contacts called *brushes*, which are connected to the output terminals a and b. Since the emf varies sinusoidally, the current that results in the circuit is an *alternating* current that also varies sinusoidally in magnitude and direction. The amplitude of the emf can be increased by increasing the rotation speed, the field magnitude, or the loop area or by using N loops instead of one, as in Eq. (29.4).

Alternators are used in automobiles to generate the currents in the ignition, the lights, and the entertainment system. The arrangement is a little different than in this example; rather than having a rotating loop in a magnetic field, the loop stays fixed and an electromagnet rotates. (The rotation is provided by a mechanical connection between the alternator and the engine.) But the result is the same; the flux through the loop varies sinusoidally, producing a sinusoidally varying emf. Larger alternators of this same type are used in electric power plants (Fig. 29.9).

29.9 A commercial alternator uses many loops of wire wound around a barrel-like structure called an armature. The armature and wire remain stationary while electromagnets rotate on a shaft (not shown) through the center of the armature. The resulting induced emf is far larger than would be possible with a single loop of wire.

Example 29.4 Generator II: A DC generator and back emf in a motor

The alternator in Example 29.3 produces a sinusoidally varying emf and hence an alternating current. Figure 29.10a shows a *direct-current* (dc) *generator* that produces an emf that always has the same sign. The arrangement of split rings, called a *commutator*, reverses the connections to the external circuit at angular positions at which the emf reverses. Figure 29.10b shows the resulting emf. Commercial dc generators have a large number of coils and commutator segments, smoothing out the bumps in the emf so that the terminal voltage is not only one-directional but also practically constant. This brush-and-commutator arrangement is the same as that in the direct-current motor discussed in Section 27.8. The

motor's *back emf* is just the emf induced by the changing magnetic flux through its rotating coil. Consider a motor with a square, 500-turn coil 10.0 cm on a side. If the magnetic field has magnitude 0.200 T, at what rotation speed is the *average* back emf of the motor equal to 112 V?

SOLUTION

IDENTIFY and SET UP: As far as the rotating loop is concerned, the situation is the same as in Example 29.3 except that we now have N turns of wire. Without the commutator, the emf would alternate between positive and negative values and have an average value of zero (see Fig. 29.8b). With the commutator, the emf is never negative and its average value is positive (Fig. 29.10b). We'll use our result from Example 29.3 to obtain an expression for this average value and solve it for the rotational speed ω.

EXECUTE: Comparison of Figs. 29.8b and 29.10b shows that the back emf of the motor is just N times the absolute value of the emf found for an alternator in Example 29.3, as in Eq. (29.4): $|\mathcal{E}| = N\omega BA |\sin \omega t|$. To find the *average* back emf, we must replace $|\sin \omega t|$ by its average value. We find this by integrating $|\sin \omega t|$ over half a cycle, from $t = 0$ to $t = T/2 = \pi/\omega$, and dividing by the elapsed time π/ω. During this half cycle, the sine function is positive, so $|\sin \omega t| = \sin \omega t$, and we find

$$(|\sin \omega t|)_{av} = \frac{\int_0^{\pi/\omega} \sin \omega t \, dt}{\pi/\omega} = \frac{2}{\pi}$$

The average back emf is then

$$\mathcal{E}_{av} = \frac{2N\omega BA}{\pi}$$

This confirms that the back emf is proportional to the rotation speed ω, as we stated without proof in Section 27.8. Solving for ω, we obtain

$$\omega = \frac{\pi \mathcal{E}_{av}}{2NBA}$$
$$= \frac{\pi(112 \text{ V})}{2(500)(0.200 \text{ T})(0.100 \text{ m})^2} = 176 \text{ rad/s}$$

(We used the unit relationships $1 \text{ V} = 1 \text{ Wb/s} = 1 \text{ T} \cdot \text{m}^2/\text{s}$ from Example 29.1.)

EVALUATE: The average back emf is directly proportional to ω. Hence the slower the rotation speed, the less the back emf and the greater the possibility of burning out the motor, as we described in Example 27.11 (Section 27.8).

29.10 (a) Schematic diagram of a dc generator, using a split-ring commutator. The ring halves are attached to the loop and rotate with it. (b) Graph of the resulting induced emf between terminals a and b. Compare to Fig. 29.8b.

Example 29.5 **Generator III: The slidewire generator**

Figure 29.11 shows a U-shaped conductor in a uniform magnetic field $\vec{B}$ perpendicular to the plane of the figure and directed *into* the page. We lay a metal rod (the "slidewire") with length L across the two arms of the conductor, forming a circuit, and move it to the right with constant velocity $\vec{v}$. This induces an emf and a current, which is why this device is called a *slidewire generator*. Find the magnitude and direction of the resulting induced emf.

SOLUTION

IDENTIFY and SET UP: The magnetic flux changes because the area of the loop—bounded on the right by the moving rod—is

29.11 A slidewire generator. The magnetic field $\vec{B}$ and the vector area $\vec{A}$ are both directed into the figure. The increase in magnetic flux (caused by an increase in area) induces the emf and current.

increasing. Our target variable is the emf $\mathcal{E}$ induced in this expanding loop. The magnetic field is uniform over the area of the loop, so we can find the flux using $\Phi_B = BA \cos \phi$. We choose the area vector $\vec{A}$ to point straight into the page, in the same direction as $\vec{B}$. With this choice a positive emf will be one that is directed clockwise around the loop. (You can check this with the right-hand rule: Using your right hand, point your thumb into the page and curl your fingers as in Fig. 29.6.)

EXECUTE: Since $\vec{B}$ and $\vec{A}$ point in the same direction, the angle $\phi = 0$ and $\Phi_B = BA$. The magnetic field magnitude B is constant, so the induced emf is

$$\mathcal{E} = -\frac{d\Phi_B}{dt} = -B\frac{dA}{dt}$$

To calculate dA/dt, note that in a time dt the sliding rod moves a distance $v \, dt$ (Fig. 29.11) and the loop area increases by an amount $dA = Lv \, dt$. Hence the induced emf is

$$\mathcal{E} = -B\frac{Lv \, dt}{dt} = -BLv$$

The minus sign tells us that the emf is directed *counterclockwise* around the loop. The induced current is also counterclockwise, as shown in the figure.

Continued

EVALUATE: The emf of a slidewire generator is constant if $\vec{v}$ is constant. Hence the slidewire generator is a *direct-current* generator. It's not a very practical device because the rod eventually moves beyond the U-shaped conductor and loses contact, after which the current stops.

Example 29.6 | Work and power in the slidewire generator

In the slidewire generator of Example 29.5, energy is dissipated in the circuit owing to its resistance. Let the resistance of the circuit (made up of the moving slidewire and the U-shaped conductor that connects the ends of the slidewire) at a given point in the slidewire's motion be R. Find the rate at which energy is dissipated in the circuit and the rate at which work must be done to move the rod through the magnetic field.

SOLUTION

IDENTIFY and SET UP: Our target variables are the *rates* at which energy is dissipated and at which work is done. Energy is dissipated in the circuit at the rate $P_{\text{dissipated}} = I^2R$. The current I in the circuit equals $|\mathcal{E}|/R$; we found an expression for the induced emf $\mathcal{E}$ in this circuit in Example 29.5. There is a magnetic force $\vec{F} = I\vec{L} \times \vec{B}$ on the rod, where $\vec{L}$ points along the rod in the direction of the current. Figure 29.12 shows that this force is opposite to the rod velocity $\vec{v}$; to maintain the motion, whoever is pushing the rod must apply a force of equal magnitude in the direction of $\vec{v}$. This force does work at the rate $P_{\text{applied}} = Fv$.

29.12 The magnetic force $\vec{F} = I\vec{L} \times \vec{B}$ that acts on the rod due to the induced current is to the left, opposite to $\vec{v}$.

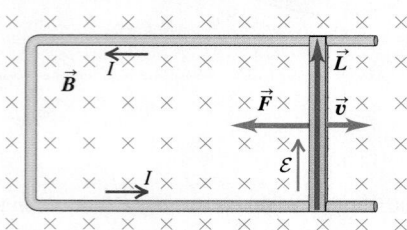

EXECUTE: First we'll calculate $P_{\text{dissipated}}$. From Example 29.5, $\mathcal{E} = -BLv$, so the current in the rod is $I = |\mathcal{E}|/R = Blv/R$. Hence

$$P_{\text{dissipated}} = I^2R = \left(\frac{BLv}{R}\right)^2 R = \frac{B^2L^2v^2}{R}$$

To calculate P_{applied}, we first calculate the magnitude of $\vec{F} = I\vec{L} \times \vec{B}$. Since $\vec{L}$ and $\vec{B}$ are perpendicular, this magnitude is

$$F = ILB = \frac{BLv}{R}LB = \frac{B^2L^2v}{R}$$

The applied force has the same magnitude and does work at the rate

$$P_{\text{applied}} = Fv = \frac{B^2L^2v^2}{R}$$

EVALUATE: The rate at which work is done is exactly *equal* to the rate at which energy is dissipated in the resistance.

CAUTION **You can't violate energy conservation** You might think that reversing the direction of $\vec{B}$ or of $\vec{v}$ might make it possible to have the magnetic force $\vec{F} = I\vec{L} \times \vec{B}$ be in the *same* direction as $\vec{v}$. This would be a pretty neat trick. Once the rod was moving, the changing magnetic flux would induce an emf and a current, and the magnetic force on the rod would make it move even faster, increasing the emf and current; this would go on until the rod was moving at tremendous speed and producing electric power at a prodigious rate. If this seems too good to be true, not to mention a violation of energy conservation, that's because it is. Reversing $\vec{B}$ also reverses the sign of the induced emf and current and hence the direction of $\vec{L}$, so the magnetic force still opposes the motion of the rod; a similar result holds true if we reverse $\vec{v}$. ▮

Generators As Energy Converters

Example 29.6 shows that the slidewire generator doesn't produce electric energy out of nowhere; the energy is supplied by whatever body exerts the force that keeps the rod moving. All that the generator does is *convert* that energy into a different form. The equality between the rate at which *mechanical* energy is supplied to a generator and the rate at which *electric* energy is generated holds for all types of generators. This is true in particular for the alternator described in Example 29.3. (We are neglecting the effects of friction in the bearings of an alternator or between the rod and the U-shaped conductor of a slidewire generator. If these are included, the conservation of energy demands that the energy lost to friction is not available for conversion to electric energy. In real generators the friction is kept to a minimum to keep the energy-conversion process as efficient as possible.)

In Chapter 27 we stated that the magnetic force on moving charges can never do work. But you might think that the magnetic force $\vec{F} = I\vec{L} \times \vec{B}$ in Example 29.6 *is* doing (negative) work on the current-carrying rod as it moves, contradicting our earlier statement. In fact, the work done by the magnetic force is actually zero. The moving charges that make up the current in the rod in Fig. 29.12 have a vertical component of velocity, causing a horizontal component of force on these charges. As a result, there is a horizontal displacement of charge within the rod, the left side acquiring a net positive charge and the right side a net negative charge. The result is a horizontal component of electric field, perpendicular to the length of the rod (analogous to the Hall effect, described in Section 27.9). It is this field, in the direction of motion of the rod, that does work on the mobile charges in the rod and hence indirectly on the atoms making up the rod.

Test Your Understanding of Section 29.2   The figure at right shows a wire coil being squeezed in a uniform magnetic field. (a) While the coil is being squeezed, is the induced emf in the coil (i) clockwise, (ii) counterclockwise, or (iii) zero? (b) Once the coil has reached its final squeezed shape, is the induced emf in the coil (i) clockwise, (ii) counterclockwise, or (iii) zero?

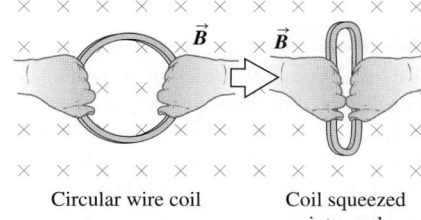

Circular wire coil   Coil squeezed into oval

29.3 Lenz's Law

Lenz's law is a convenient alternative method for determining the direction of an induced current or emf. Lenz's law is not an independent principle; it can be derived from Faraday's law. It always gives the same results as the sign rules we introduced in connection with Faraday's law, but it is often easier to use. Lenz's law also helps us gain intuitive understanding of various induction effects and of the role of energy conservation. H. F. E. Lenz (1804–1865) was a Russian scientist who duplicated independently many of the discoveries of Faraday and Henry. **Lenz's law** states:

> **The direction of any magnetic induction effect is such as to oppose the cause of the effect.**

The "cause" may be changing flux through a stationary circuit due to a varying magnetic field, changing flux due to motion of the conductors that make up the circuit, or any combination. If the flux in a stationary circuit changes, as in Examples 29.1 and 29.2, the induced current sets up a magnetic field of its own. Within the area bounded by the circuit, this field is *opposite* to the original field if the original field is *increasing* but is in the *same* direction as the original field if the latter is *decreasing*. That is, the induced current opposes the *change in flux* through the circuit (*not* the flux itself).

If the flux change is due to motion of the conductors, as in Examples 29.3 through 29.6, the direction of the induced current in the moving conductor is such that the direction of the magnetic-field force on the conductor is opposite in direction to its motion. Thus the motion of the conductor, which caused the induced current, is opposed. We saw this explicitly for the slidewire generator in Example 29.6. In all these cases the induced current tries to preserve the *status quo* by opposing motion or a change of flux.

Lenz's law is also directly related to energy conservation. If the induced current in Example 29.6 were in the direction opposite to that given by Lenz's law, the magnetic force on the rod would accelerate it to ever-increasing speed with no external energy source, even though electric energy is being dissipated in the circuit. This would be a clear violation of energy conservation and doesn't happen in nature.

Conceptual Example 29.7   **Lenz's law and the slidewire generator**

In Fig. 29.11, the induced current in the loop causes an additional magnetic field in the area bounded by the loop. The direction of the induced current is counterclockwise, so from the discussion of Section 28.2, this additional magnetic field is directed *out of* the plane of the figure. That direction is opposite that of the original magnetic field, so it tends to cancel the effect of that field. This is just what Lenz's law predicts.

Conceptual Example 29.8 | **Lenz's law and the direction of induced current**

In Fig. 29.13 there is a uniform magnetic field $\vec{B}$ through the coil. The magnitude of the field is increasing, so there is an induced emf. Use Lenz's law to determine the direction of the resulting induced current.

SOLUTION

This situation is the same as in Example 29.1 (Section 29.2). By Lenz's law the induced current must produce a magnetic field $\vec{B}_{induced}$ inside the coil that is downward, opposing the change in flux. From the right-hand rule we described in Section 28.5 for the direction of the magnetic field produced by a circular loop, $\vec{B}_{induced}$ will be in the desired direction if the induced current flows as shown in Fig. 29.13.

Figure 29.14 shows several applications of Lenz's law to the similar situation of a magnet moving near a conducting loop. In each case, the induced current produces a magnetic field whose

29.13 The induced current due to the change in $\vec{B}$ is clockwise, as seen from above the loop. The added field $\vec{B}_{induced}$ that it causes is downward, opposing the change in the upward field $\vec{B}$.

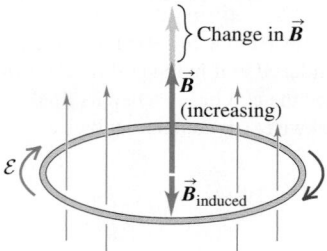

direction opposes the change in flux through the loop due to the magnet's motion.

29.14 Directions of induced currents as a bar magnet moves along the axis of a conducting loop. If the bar magnet is stationary, there is no induced current.

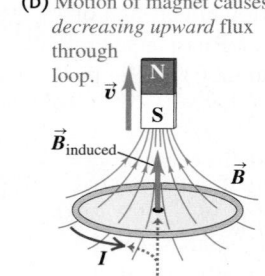

(a) Motion of magnet causes *increasing downward* flux through loop.

(b) Motion of magnet causes *decreasing upward* flux through loop.

The induced magnetic field is *upward* to oppose the flux change. To produce this induced field, the induced current must be *counterclockwise* as seen from above the loop.

(c) Motion of magnet causes *decreasing downward* flux through loop.

(d) Motion of magnet causes *increasing upward* flux through loop.

The induced magnetic field is *downward* to oppose the flux change. To produce this induced field, the induced current must be *clockwise* as seen from above the loop.

Lenz's Law and the Response to Flux Changes

Since an induced current always opposes any change in magnetic flux through a circuit, how is it possible for the flux to change at all? The answer is that Lenz's law gives only the *direction* of an induced current; the *magnitude* of the current depends on the resistance of the circuit. The greater the circuit resistance, the less the induced current that appears to oppose any change in flux and the easier it is for a flux change to take effect. If the loop in Fig. 29.14 were made out of wood (an insulator), there would be almost no induced current in response to changes in the flux through the loop.

Conversely, the less the circuit resistance, the greater the induced current and the more difficult it is to change the flux through the circuit. If the loop in Fig. 29.14 is a good conductor, an induced current flows as long as the magnet moves relative to the loop. Once the magnet and loop are no longer in relative motion, the induced current very quickly decreases to zero because of the nonzero resistance in the loop.

The extreme case occurs when the resistance of the circuit is *zero*. Then the induced current in Fig. 29.14 will continue to flow even after the induced emf has disappeared—that is, even after the magnet has stopped moving relative to the loop. Thanks to this *persistent current,* it turns out that the flux through the loop is exactly the same as it was before the magnet started to move, so the flux through a loop of zero resistance *never* changes. Exotic materials called *superconductors* do indeed have zero resistance; we discuss these further in Section 29.8.

Test Your Understanding of Section 29.3 (a) Suppose the magnet in Fig. 29.14a were stationary and the loop of wire moved upward. Would the induced current in the loop be (i) in the same direction as shown in Fig. 29.14a, (ii) in the direction opposite to that shown in Fig. 29.14a, or (iii) zero? (b) Suppose the magnet and loop of wire in Fig. 29.14a both moved downward at the same velocity. Would the induced current in the loop be (i) in the same direction as shown in Fig. 29.14a, (ii) in the direction opposite to that shown in Fig. 29.14a, or (iii) zero?

29.4 Motional Electromotive Force

We've seen several situations in which a conductor moves in a magnetic field, as in the generators discussed in Examples 29.3 through 29.6. We can gain additional insight into the origin of the induced emf in these situations by considering the magnetic forces on mobile charges in the conductor. Figure 29.15a shows the same moving rod that we discussed in Example 29.5, separated for the moment from the U-shaped conductor. The magnetic field $\vec{B}$ is uniform and directed into the page, and we move the rod to the right at a constant velocity $\vec{v}$. A charged particle q in the rod then experiences a magnetic force $\vec{F} = q\vec{v} \times \vec{B}$ with magnitude $F = |q|vB$. We'll assume in the following discussion that q is positive; in that case the direction of this force is upward along the rod, from b toward a.

This magnetic force causes the free charges in the rod to move, creating an excess of positive charge at the upper end a and negative charge at the lower end b. This in turn creates an electric field $\vec{E}$ within the rod, in the direction from a toward b (opposite to the magnetic force). Charge continues to accumulate at the ends of the rod until $\vec{E}$ becomes large enough for the downward electric force (with magnitude qE) to cancel exactly the *upward* magnetic force (with magnitude qvB). Then $qE = qvB$ and the charges are in equilibrium.

The magnitude of the potential difference $V_{ab} = V_a - V_b$ is equal to the electric-field magnitude E multiplied by the length L of the rod. From the above discussion, $E = vB$, so

$$V_{ab} = EL = vBL \qquad (29.5)$$

with point a at higher potential than point b.

Now suppose the moving rod slides along a stationary U-shaped conductor, forming a complete circuit (Fig. 29.15b). No *magnetic* force acts on the charges in the stationary U-shaped conductor, but the charge that was near points a and b redistributes itself along the stationary conductor, creating an *electric* field within it. This field establishes a current in the direction shown. The moving rod has become a source of electromotive force; within it, charge moves from lower to higher potential, and in the remainder of the circuit, charge moves from higher to lower potential. We call this emf a **motional electromotive force,** denoted by $\mathcal{E}$. From the above discussion, the magnitude of this emf is

$$\mathcal{E} = vBL \qquad \text{(motional emf; length and velocity perpendicular to uniform } \vec{B}) \qquad (29.6)$$

corresponding to a force per unit charge of magnitude vB acting for a distance L along the moving rod. If the total circuit resistance of the U-shaped conductor and the sliding rod is R, the induced current I in the circuit is given by $vBL = IR$. This is the same result we obtained in Section 29.2 using Faraday's law, and indeed motional emf is a particular case of Faraday's law, one of the several examples described in Section 29.2.

The emf associated with the moving rod in Fig. 29.15 is analogous to that of a battery with its positive terminal at a and its negative terminal at b, although the origins of the two emfs are quite different. In each case a nonelectrostatic force acts on the charges in the device, in the direction from b to a, and the emf is the work per unit charge done by this force when a charge moves from b to a in the device. When the device is connected to an external circuit, the direction of

29.15 A conducting rod moving in a uniform magnetic field. **(a)** The rod, the velocity, and the field are mutually perpendicular. **(b)** Direction of induced current in the circuit.

(a) Isolated moving rod

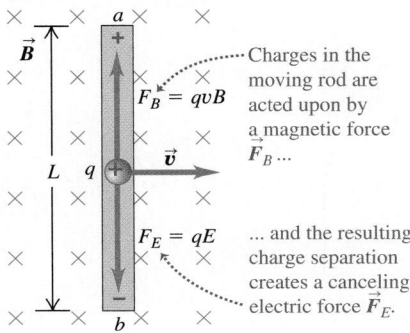

Charges in the moving rod are acted upon by a magnetic force $\vec{F}_B$...

... and the resulting charge separation creates a canceling electric force $\vec{F}_E$.

(b) Rod connected to stationary conductor

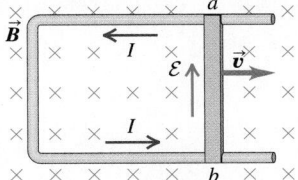

The motional emf $\mathcal{E}$ in the moving rod creates an electric field in the stationary conductor.

ActivPhysics 13.10: Motional EMF

current is from *b* to *a* in the device and from *a* to *b* in the external circuit. While we have discussed motional emf in terms of a closed circuit like that in Fig. 29.15b, a motional emf is also present in the isolated moving rod in Fig. 29.15a, in the same way that a battery has an emf even when it's not part of a circuit.

The direction of the induced emf in Fig. 29.15 can be deduced by using Lenz's law, even if (as in Fig. 29.15a) the conductor does not form a complete circuit. In this case we can mentally complete the circuit between the ends of the conductor and use Lenz's law to determine the direction of the current. From this we can deduce the polarity of the ends of the open-circuit conductor. The direction from the − end to the + end within the conductor is the direction the current would have if the circuit were complete.

You should verify that if we express *v* in meters per second, *B* in teslas, and *L* in meters, then $\mathcal{E}$ is in volts. (Recall that 1 V = 1 J/C.)

Motional emf: General Form

We can generalize the concept of motional emf for a conductor with *any* shape, moving in any magnetic field, uniform or not (assuming that the magnetic field at each point does not vary with time). For an element $d\vec{l}$ of the conductor, the contribution $d\mathcal{E}$ to the emf is the magnitude dl multiplied by the component of $\vec{v} \times \vec{B}$ (the magnetic force per unit charge) parallel to $d\vec{l}$; that is,

$$d\mathcal{E} = (\vec{v} \times \vec{B}) \cdot d\vec{l}$$

For any closed conducting loop, the total emf is

$$\mathcal{E} = \oint (\vec{v} \times \vec{B}) \cdot d\vec{l} \qquad \text{(motional emf; closed conducting loop)} \quad \text{(29.7)}$$

This expression looks very different from our original statement of Faraday's law, Eq. (29.3), which stated that $\mathcal{E} = -d\Phi_B/dt$. In fact, though, the two statements are equivalent. It can be shown that the rate of change of magnetic flux through a moving conducting loop is always given by the negative of the expression in Eq. (29.7). Thus this equation gives us an alternative formulation of Faraday's law. This alternative is often more convenient than the original one in problems with *moving* conductors. But when we have *stationary* conductors in changing magnetic fields, Eq. (29.7) *cannot* be used; in this case, $\mathcal{E} = -d\Phi_B/dt$ is the only correct way to express Faraday's law.

Example 29.9 **Motional emf in the slidewire generator**

Suppose the moving rod in Fig. 29.15b is 0.10 m long, the velocity *v* is 2.5 m/s, the total resistance of the loop is 0.030 Ω, and *B* is 0.60 T. Find the motional emf, the induced current, and the force acting on the rod.

SOLUTION

IDENTIFY and SET UP: The first target variable is the motional emf $\mathcal{E}$ due to the rod's motion, which we'll find using Eq. (29.6). We'll find the current from the values of $\mathcal{E}$ and the resistance *R*. The force on the rod is a *magnetic* force, exerted by $\vec{B}$ on the current in the rod; we'll find this force using $\vec{F} = I\vec{L} \times \vec{B}$.

EXECUTE: From Eq. (29.6) the motional emf is

$$\mathcal{E} = vBL = (2.5 \text{ m/s})(0.60 \text{ T})(0.10 \text{ m}) = 0.15 \text{ V}$$

The induced current in the loop is

$$I = \frac{\mathcal{E}}{R} = \frac{0.15 \text{ V}}{0.030 \text{ }\Omega} = 5.0 \text{ A}$$

In the expression for the magnetic force, $\vec{F} = I\vec{L} \times \vec{B}$, the vector $\vec{L}$ points in the same direction as the induced current in the rod (from *b* to *a* in Fig. 29.15). Applying the right-hand rule for vector products shows that this force is directed *opposite* to the rod's motion. Since $\vec{L}$ and $\vec{B}$ are perpendicular, the magnetic force has magnitude

$$F = ILB = (5.0 \text{ A})(0.10 \text{ m})(0.60 \text{ T}) = 0.30 \text{ N}$$

EVALUATE: We can check our answer for the direction of $\vec{F}$ by using Lenz's law. If we take the area vector $\vec{A}$ to point into the plane of the loop, the magnetic flux is positive and increasing as the rod moves to the right and increases the area of the loop. Lenz's law tells us that a force appears to oppose this increase in flux. Hence the force on the rod is to the left, opposite its motion.

Example 29.10 **The Faraday disk dynamo**

Figure 29.16 shows a conducting disk with radius R that lies in the xy-plane and rotates with constant angular velocity ω about the z-axis. The disk is in a uniform, constant $\vec{B}$ field in the z-direction. Find the induced emf between the center and the rim of the disk.

SOLUTION

IDENTIFY and SET UP: A motional emf arises because the conducting disk moves relative to $\vec{B}$. The complication is that different

29.16 A conducting disk with radius R rotating at an angular speed ω in a magnetic field $\vec{B}$. The emf is induced along radial lines of the disk and is applied to an external circuit through the two sliding contacts labeled b.

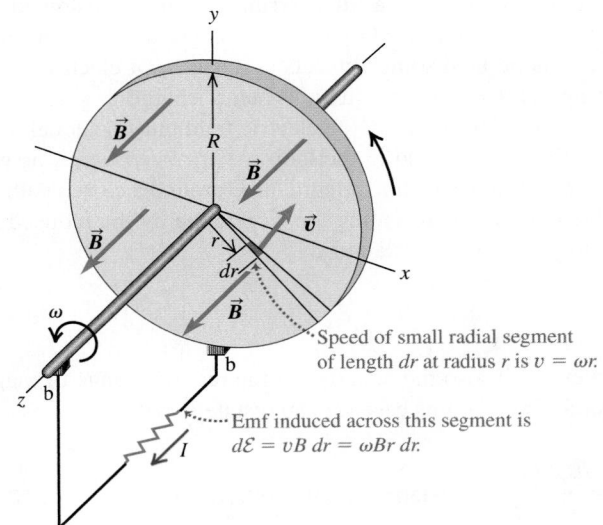

Speed of small radial segment of length dr at radius r is $v = \omega r$.

Emf induced across this segment is
$d\mathcal{E} = vB\,dr = \omega Br\,dr$.

parts of the disk move at different speeds v, depending on their distance from the rotation axis. We'll address this by considering small segments of the disk and integrating their contributions to determine our target variable, the emf between the center and the rim. Consider the small segment of the disk shown in red in Fig. 29.16 and labeled by its velocity vector $\vec{v}$. The magnetic force per unit charge on this segment is $\vec{v} \times \vec{B}$, which points radially outward from the center of the disk. Hence the induced emf tends to make a current flow radially outward, which tells us that the moving conducting path to think about here is a straight line from the center to the rim. We can find the emf from each small disk segment along this line using $d\mathcal{E} = (\vec{v} \times \vec{B}) \cdot d\vec{l}$ and then integrate to find the total emf.

EXECUTE: The length vector $d\vec{l}$ (of length dr) associated with the segment points radially outward, in the same direction as $\vec{v} \times \vec{B}$. The vectors $\vec{v}$ and $\vec{B}$ are perpendicular, and the magnitude of $\vec{v}$ is $v = \omega r$. The emf from the segment is then $d\mathcal{E} = \omega Br\,dr$. The total emf is the integral of $d\mathcal{E}$ from the center $(r = 0)$ to the rim $(r = R)$:

$$\mathcal{E} = \int_0^R \omega Br\,dr = \tfrac{1}{2}\omega BR^2$$

EVALUATE: We can use this device as a source of emf in a circuit by completing the circuit through two stationary brushes (labeled b in the figure) that contact the disk and its conducting shaft as shown. Such a disk is called a *Faraday disk dynamo* or a *homopolar generator*. Unlike the alternator in Example 29.3, the Faraday disk dynamo is a direct-current generator; it produces an emf that is constant in time. Can you use Lenz's law to show that for the direction of rotation in Fig. 29.16, the current in the external circuit must be in the direction shown?

Test Your Understanding of Section 29.4 The earth's magnetic field points toward (magnetic) north. For simplicity, assume that the field has no vertical component (as is the case near the earth's equator). (a) If you hold a metal rod in your hand and walk toward the east, how should you orient the rod to get the maximum motional emf between its ends? (i) east-west; (ii) north-south; (iii) up-down; (iv) you get the same motional emf with all of these orientations. (b) How should you hold it to get *zero* emf as you walk toward the east? (i) east-west; (ii) north-south; (iii) up-down; (iv) none of these. (c) In which direction should you travel so that the motional emf across the rod is zero no matter how the rod is oriented? (i) west; (ii) north; (iii) south; (iv) straight up; (v) straight down.

29.5 Induced Electric Fields

When a conductor moves in a magnetic field, we can understand the induced emf on the basis of magnetic forces on charges in the conductor, as described in Section 29.4. But an induced emf also occurs when there is a changing flux through a stationary conductor. What is it that pushes the charges around the circuit in this type of situation?

As an example, let's consider the situation shown in Fig. 29.17. A long, thin solenoid with cross-sectional area A and n turns per unit length is encircled at its center by a circular conducting loop. The galvanometer G measures the current in the loop. A current I in the winding of the solenoid sets up a magnetic field $\vec{B}$ along the solenoid axis, as shown, with magnitude B as calculated in Example 28.9 (Section 28.7): $B = \mu_0 nI$, where n is the number of turns per unit length.

29.17 (a) The windings of a long sole-noid carry a current I that is increasing at a rate dI/dt. The magnetic flux in the sole-noid is increasing at a rate $d\Phi_B/dt$, and this changing flux passes through a wire loop. An emf $\mathcal{E} = -d\Phi_B/dt$ is induced in the loop, inducing a current I' that is measured by the galvanometer G. (b) Cross-sectional view.

(a)

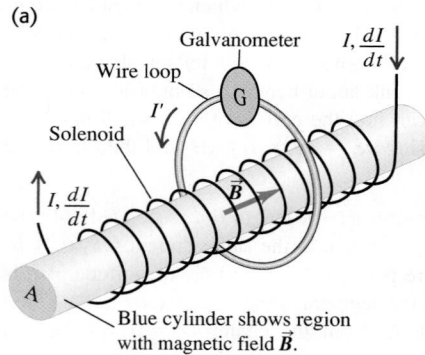

Galvanometer $I, \dfrac{dI}{dt}$

Wire loop

I'

G

Solenoid

$I, \dfrac{dI}{dt}$

$\vec{B}$

A

Blue cylinder shows region with magnetic field $\vec{B}$.

(b)

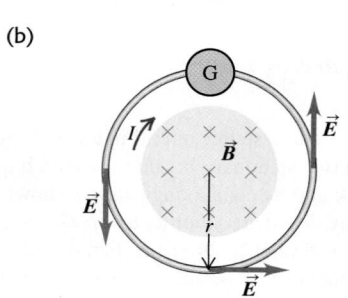

G

I $\times$ $\times$ $\times$

$\vec{B}$

$\times$ $\times$ $\times$

$\vec{E}$

$\vec{E}$

$\times$ $\times$ $\times$

r

$\vec{E}$

$\vec{E}$

If we neglect the small field outside the solenoid and take the area vector $\vec{A}$ to point in the same direction as $\vec{B}$, then the magnetic flux Φ_B through the loop is

$$\Phi_B = BA = \mu_0 nIA$$

When the solenoid current I changes with time, the magnetic flux Φ_B also changes, and according to Faraday's law the induced emf in the loop is given by

$$\mathcal{E} = -\frac{d\Phi_B}{dt} = -\mu_0 nA \frac{dI}{dt} \tag{29.8}$$

If the total resistance of the loop is R, the induced current in the loop, which we may call I', is $I' = \mathcal{E}/R$.

But what *force* makes the charges move around the wire loop? It can't be a magnetic force because the loop isn't even *in* a magnetic field. We are forced to conclude that there has to be an **induced electric field** in the conductor *caused by the changing magnetic flux*. This may be a little jarring; we are accustomed to thinking about electric field as being caused by electric charges, and now we are saying that a changing magnetic field somehow acts as a source of electric field. Furthermore, it's a strange sort of electric field. When a charge q goes once around the loop, the total work done on it by the electric field must be equal to q times the emf $\mathcal{E}$. That is, the electric field in the loop *is not conservative*, as we used the term in Chapter 23, because the line integral of $\vec{E}$ around a closed path is not zero. Indeed, this line integral, representing the work done by the induced $\vec{E}$ field per unit charge, is equal to the induced emf $\mathcal{E}$:

$$\oint \vec{E} \cdot d\vec{l} = \mathcal{E} \tag{29.9}$$

From Faraday's law the emf $\mathcal{E}$ is also the negative of the rate of change of mag-netic flux through the loop. Thus for this case we can restate Faraday's law as

$$\oint \vec{E} \cdot d\vec{l} = -\frac{d\Phi_B}{dt} \quad \text{(stationary integration path)} \tag{29.10}$$

Note that Faraday's law is *always* true in the form $\mathcal{E} = -d\Phi_B/dt$; the form given in Eq. (29.10) is valid *only* if the path around which we integrate is *stationary*.

As an example of a situation to which Eq. (29.10) can be applied, consider the stationary circular loop in Fig. 29.17b, which we take to have radius r. Because of cylindrical symmetry, the electric field $\vec{E}$ has the same magnitude at every point on the circle and is tangent to it at each point. (Symmetry would also permit the field to be *radial,* but then Gauss's law would require the presence of a net charge inside the circle, and there is none.) The line integral in Eq. (29.10) becomes simply the magnitude E times the circumference $2\pi r$ of the loop, $\oint \vec{E} \cdot d\vec{l} = 2\pi rE$, and Eq. (29.10) gives

$$E = \frac{1}{2\pi r} \left| \frac{d\Phi_B}{dt} \right| \tag{29.11}$$

The directions of $\vec{E}$ at points on the loop are shown in Fig. 29.17b. We know that $\vec{E}$ has to have the direction shown when $\vec{B}$ in the solenoid is increasing, because $\oint \vec{E} \cdot d\vec{l}$ has to be negative when $d\Phi_B/dt$ is positive. The same approach can be used to find the induced electric field *inside* the solenoid when the solenoid $\vec{B}$ field is changing; we leave the details to you (see Exercise 29.35).

Nonelectrostatic Electric Fields

Now let's summarize what we've learned. Faraday's law, Eq. (29.3), is valid for two rather different situations. In one, an emf is induced by magnetic forces on charges when a conductor moves through a magnetic field. In the other, a time-varying magnetic field induces an electric field in a stationary conductor and hence induces an emf; in fact, the $\vec{E}$ field is induced even when no conductor is present. This $\vec{E}$ field differs from an electro*static* field in an important way. It is *nonconservative;* the line integral $\oint \vec{E} \cdot d\vec{l}$ around a closed path is not zero, and when a charge moves around a closed path, the field does a nonzero amount of work on it. It follows that for such a field the concept of *potential* has no meaning. We call such a field a **nonelectrostatic field.** In contrast, an *electrostatic* field is *always* conservative, as we discussed in Section 23.1, and always has an associated potential function. Despite this difference, the fundamental effect of *any* electric field is to exert a force $\vec{F} = q\vec{E}$ on a charge q. This relationship is valid whether $\vec{E}$ is a conservative field produced by a charge distribution or a nonconservative field caused by changing magnetic flux.

So a changing magnetic field acts as a source of electric field of a sort that we *cannot* produce with any static charge distribution. This may seem strange, but it's the way nature behaves. What's more, we'll see in Section 29.7 that a changing *electric* field acts as a source of *magnetic* field. We'll explore this symmetry between the two fields in greater detail in our study of electromagnetic waves in Chapter 32.

If any doubt remains in your mind about the reality of magnetically induced electric fields, consider a few of the many practical applications (Fig. 29.18). Pickups in electric guitars use currents induced in stationary pickup coils by the vibration of nearby ferromagnetic strings. Alternators in most cars use rotating magnets to induce currents in stationary coils. Whether we realize it or not, magnetically induced electric fields play an important role in everyday life.

29.18 Applications of induced electric fields. (a) Data are stored on a computer hard disk in a pattern of magnetized areas on the surface of the disk. To read these data, a coil on a movable arm is placed next to the spinning disk. The coil experiences a changing magnetic flux, inducing a current whose characteristics depend on the pattern coded on the disk. (b) This hybrid automobile has both a gasoline engine and an electric motor. As the car comes to a halt, the spinning wheels run the motor backward so that it acts as a generator. The resulting induced current is used to recharge the car's batteries. (c) The rotating crankshaft of a piston-engine airplane spins a magnet, inducing an emf in an adjacent coil and generating the spark that ignites fuel in the engine cylinders. This keeps the engine running even if the airplane's other electrical systems fail.

(a)

(b)

(c)

Example 29.11 | **Induced electric fields**

Suppose the long solenoid in Fig. 29.17a has 500 turns per meter and cross-sectional area 4.0 cm². The current in its windings is increasing at 100 A/s. (a) Find the magnitude of the induced emf in the wire loop outside the solenoid. (b) Find the magnitude of the induced electric field within the loop if its radius is 2.0 cm.

SOLUTION

IDENTIFY and SET UP: As in Fig. 29.17b, the increasing magnetic field inside the solenoid causes a change in the magnetic flux through the wire loop and hence induces an electric field $\vec{E}$ around the loop. Our target variables are the induced emf $\mathcal{E}$ and the electric-field magnitude E. We use Eq. (29.8) to determine the emf.

Determining the field magnitude E is simplified because the loop and the solenoid share the same central axis. Hence, by symmetry, the electric field is tangent to the loop and has the same magnitude all the way around its circumference. We can therefore use Eq. (29.9) to find E.

EXECUTE: (a) From Eq. (29.8), the induced emf is

$$\mathcal{E} = -\frac{d\Phi_B}{dt} = -\mu_0 n A \frac{dI}{dt}$$
$$= -(4\pi \times 10^{-7}\,\text{Wb/A} \cdot \text{m})(500\,\text{turns/m})$$
$$\times (4.0 \times 10^{-4}\,\text{m}^2)(100\,\text{A/s})$$
$$= -25 \times 10^{-6}\,\text{Wb/s} = -25 \times 10^{-6}\,\text{V} = -25\,\mu\text{V}$$

Continued

(b) By symmetry the line integral $\oint \vec{E} \cdot d\vec{l}$ has absolute value $2\pi rE$ no matter which direction we integrate around the loop. This is equal to the absolute value of the emf, so

$$E = \frac{|\mathcal{E}|}{2\pi r} = \frac{25 \times 10^{-6}\,\text{V}}{2\pi(2.0 \times 10^{-2}\,\text{m})} = 2.0 \times 10^{-4}\,\text{V/m}$$

EVALUATE: In Fig. 29.17b the magnetic flux *into* the plane of the figure is increasing. According to the right-hand rule for induced emf (illustrated in Fig. 29.6), a positive emf would be clockwise around the loop; the negative sign of $\mathcal{E}$ shows that the emf is in the counterclockwise direction. Can you also show this using Lenz's law?

Test Your Understanding of Section 29.5 If you wiggle a magnet back and forth in your hand, are you generating an electric field? If so, is this electric field conservative?

29.6 Eddy Currents

29.19 Eddy currents induced in a rotating metal disk.

(a) Metal disk rotating through a magnetic field

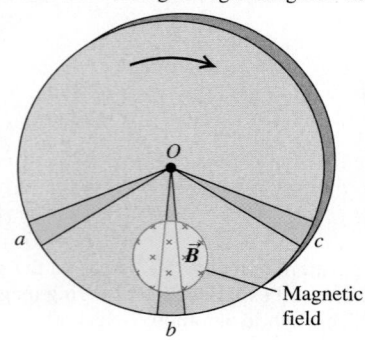

(b) Resulting eddy currents and braking force

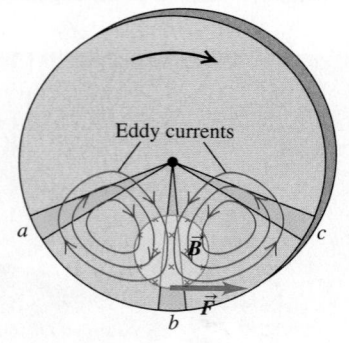

In the examples of induction effects that we have studied, the induced currents have been confined to well-defined paths in conductors and other components forming a circuit. However, many pieces of electrical equipment contain masses of metal moving in magnetic fields or located in changing magnetic fields. In situations like these we can have induced currents that circulate throughout the volume of a material. Because their flow patterns resemble swirling eddies in a river, we call these **eddy currents.**

As an example, consider a metallic disk rotating in a magnetic field perpendicular to the plane of the disk but confined to a limited portion of the disk's area, as shown in Fig. 29.19a. Sector *Ob* is moving across the field and has an emf induced in it. Sectors *Oa* and *Oc* are not in the field, but they provide return conducting paths for charges displaced along *Ob* to return from *b* to *O*. The result is a circulation of eddy currents in the disk, somewhat as sketched in Fig. 29.19b.

We can use Lenz's law to decide on the direction of the induced current in the neighborhood of sector *Ob*. This current must experience a magnetic force $\vec{F} = I\vec{L} \times \vec{B}$ that *opposes* the rotation of the disk, and so this force must be to the right in Fig. 29.19b. Since $\vec{B}$ is directed into the plane of the disk, the current and hence $\vec{L}$ have downward components. The return currents lie outside the field, so they do not experience magnetic forces. The interaction between the eddy currents and the field causes a braking action on the disk. Such effects can be used to stop the rotation of a circular saw quickly when the power is turned off. Some sensitive balances use this effect to damp out vibrations. Eddy current braking is used on some electrically powered rapid-transit vehicles. Electromagnets mounted in the cars induce eddy currents in the rails; the resulting magnetic fields cause braking forces on the electromagnets and thus on the cars.

Eddy currents have many other practical uses. The shiny metal disk in the electric power company's meter outside your house rotates as a result of eddy currents. These currents are induced in the disk by magnetic fields caused by sinusoidally varying currents in a coil. In induction furnaces, eddy currents are used to heat materials in completely sealed containers for processes in which it is essential to avoid the slightest contamination of the materials. The metal detectors used at airport security checkpoints (Fig. 29.20a) operate by detecting eddy currents induced in metallic objects. Similar devices (Fig. 29.20b) are used to find buried treasure such as bottlecaps and lost pennies.

Eddy currents also have undesirable effects. In an alternating-current transformer, coils wrapped around an iron core carry a sinusoidally varying current. The resulting eddy currents in the core waste energy through I^2R heating and themselves set up an unwanted opposing emf in the coils. To minimize these effects, the core is designed so that the paths for eddy currents are as narrow as possible. We'll describe how this is done when we discuss transformers in detail in Section 31.6.

(a) (b)

Transmitting coil

Receiver coil

Pulse of current

Eddy currents

29.20 (a) A metal detector at an airport security checkpoint generates an alternating magnetic field $\vec{B}_0$. This induces eddy currents in a conducting object carried through the detector. The eddy currents in turn produce an alternating magnetic field $\vec{B}'$, and this field induces a current in the detector's receiver coil. (b) Portable metal detectors work on the same principle.

Test Your Understanding of Section 29.6 Suppose that the magnetic field in Fig. 29.19 were directed out of the plane of the figure and the disk were rotating counterclockwise. Compared to the directions of the force $\vec{F}$ and the eddy currents shown in Fig. 29.19b, what would the new directions be? (i) The force $\vec{F}$ and the eddy currents would both be in the same direction; (ii) the force $\vec{F}$ would be in the same direction, but the eddy currents would be in the opposite direction; (iii) the force $\vec{F}$ would be in the opposite direction, but the eddy currents would be in the same direction; (iv) the force $\vec{F}$ and the eddy currents would be in the opposite directions. ‖

29.7 Displacement Current and Maxwell's Equations

We have seen that a varying magnetic field gives rise to an induced electric field. In one of the more remarkable examples of the symmetry of nature, it turns out that a varying *electric* field gives rise to a *magnetic* field. This effect is of tremendous importance, for it turns out to explain the existence of radio waves, gamma rays, and visible light, as well as all other forms of electromagnetic waves.

Generalizing Ampere's Law

To see the origin of the relationship between varying electric fields and magnetic fields, let's return to Ampere's law as given in Section 28.6, Eq. (28.20):

$$\oint \vec{B} \cdot d\vec{l} = \mu_0 I_{encl}$$

The problem with Ampere's law in this form is that it is *incomplete*. To see why, let's consider the process of charging a capacitor (Fig. 29.21). Conducting wires lead current i_C into one plate and out of the other; the charge Q increases, and the electric field $\vec{E}$ between the plates increases. The notation i_C indicates *conduction* current to distinguish it from another kind of current we are about to encounter, called *displacement* current i_D. We use lowercase i's and v's to denote instantaneous values of currents and potential differences, respectively, that may vary with time.

Let's apply Ampere's law to the circular path shown. The integral $\oint \vec{B} \cdot d\vec{l}$ around this path equals $\mu_0 I_{encl}$. For the plane circular area bounded by the circle, I_{encl} is just the current i_C in the left conductor. But the surface that bulges out to the right is bounded by the same circle, and the current through that surface is zero. So $\oint \vec{B} \cdot d\vec{l}$ is equal to $\mu_0 i_C$, and at the same time it is equal to zero! This is a clear contradiction.

But something else is happening on the bulged-out surface. As the capacitor charges, the electric field $\vec{E}$ and the electric *flux* Φ_E through the surface are increasing. We can determine their rates of change in terms of the charge and

Application **Eddy Currents Help Power Io's Volcanoes**
Jupiter's moon Io is slightly larger than the earth's moon. It moves at more than 60,000 km/h through Jupiter's intense magnetic field (about ten times stronger than the earth's field), which sets up strong eddy currents within Io that dissipate energy at a rate of 10^{12} W. This dissipated energy helps to heat Io's interior and causes volcanic eruptions on its surface, as shown in the lower close-up image. (Gravitational effects from Jupiter cause even more heating.)

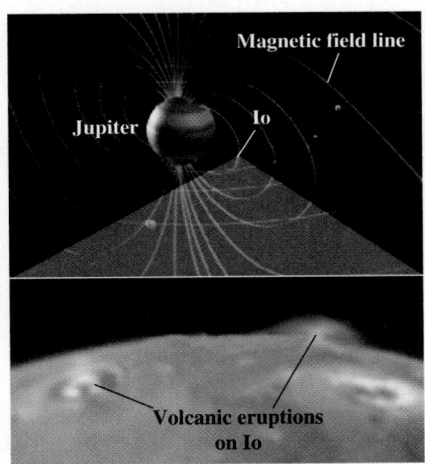

Magnetic field line

Jupiter Io

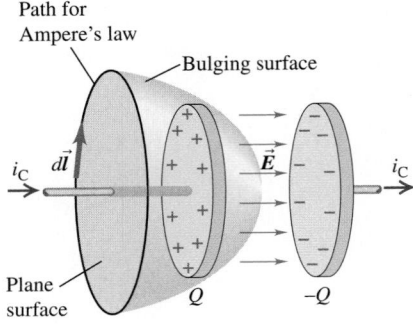

Volcanic eruptions on Io

29.21 Parallel-plate capacitor being charged. The conduction current through the plane surface is i_C, but there is no conduction current through the surface that bulges out to pass between the plates. The two surfaces have a common boundary, so this difference in I_{encl} leads to an apparent contradiction in applying Ampere's law.

Path for Ampere's law

Bulging surface

i_C $d\vec{l}$ $\vec{E}$ i_C

Plane surface

Q $-Q$

current. The instantaneous charge is $q = Cv$, where C is the capacitance and v is the instantaneous potential difference. For a parallel-plate capacitor, $C = \epsilon_0 A/d$, where A is the plate area and d is the spacing. The potential difference v between plates is $v = Ed$, where E is the electric-field magnitude between plates. (We neglect fringing and assume that $\vec{E}$ is uniform in the region between the plates.) If this region is filled with a material with permittivity ϵ, we replace ϵ_0 by ϵ everywhere; we'll use ϵ in the following discussion.

Substituting these expressions for C and v into $q = Cv$, we can express the capacitor charge q as

$$q = Cv = \frac{\epsilon A}{d}(Ed) = \epsilon EA = \epsilon \Phi_E \qquad (29.12)$$

where $\Phi_E = EA$ is the electric flux through the surface.

As the capacitor charges, the rate of change of q is the conduction current, $i_C = dq/dt$. Taking the derivative of Eq. (29.12) with respect to time, we get

$$i_C = \frac{dq}{dt} = \epsilon \frac{d\Phi_E}{dt} \qquad (29.13)$$

Now, stretching our imagination a little, we invent a fictitious **displacement current** i_D in the region between the plates, defined as

$$i_D = \epsilon \frac{d\Phi_E}{dt} \qquad \text{(displacement current)} \qquad (29.14)$$

That is, we imagine that the changing flux through the curved surface in Fig. 29.21 is somehow equivalent, in Ampere's law, to a conduction current through that surface. We include this fictitious current, along with the real conduction current i_C, in Ampere's law:

$$\oint \vec{B} \cdot d\vec{l} = \mu_0 (i_C + i_D)_{\text{encl}} \qquad \text{(generalized Ampere's law)} \qquad (29.15)$$

Ampere's law in this form is obeyed no matter which surface we use in Fig. 29.21. For the flat surface, i_D is zero; for the curved surface, i_C is zero; and i_C for the flat surface equals i_D for the curved surface. Equation (29.15) remains valid in a magnetic material, provided that the magnetization is proportional to the external field and we replace μ_0 by μ.

The fictitious current i_D was invented in 1865 by the Scottish physicist James Clerk Maxwell (1831–1879), who called it displacement current. There is a corresponding *displacement current density* $j_D = i_D/A$; using $\Phi_E = EA$ and dividing Eq. (29.14) by A, we find

$$j_D = \epsilon \frac{dE}{dt} \qquad (29.16)$$

We have pulled the concept out of thin air, as Maxwell did, but we see that it enables us to save Ampere's law in situations such as that in Fig. 29.21.

Another benefit of displacement current is that it lets us generalize Kirchhoff's junction rule, discussed in Section 26.2. Considering the left plate of the capacitor, we have conduction current into it but none out of it. But when we include the displacement current, we have conduction current coming in one side and an equal displacement current coming out the other side. With this generalized meaning of the term "current," we can speak of current going *through* the capacitor.

The Reality of Displacement Current

You might well ask at this point whether displacement current has any real physical significance or whether it is just a ruse to satisfy Ampere's law and Kirchhoff's junction rule. Here's a fundamental experiment that helps to answer that question. We take a plane circular area between the capacitor plates (Fig. 29.22). If displacement current really plays the role in Ampere's law that we have claimed, then there ought to be a magnetic field in the region between the plates while the capacitor is charging. We can use our generalized Ampere's law, including displacement current, to predict what this field should be.

To be specific, let's picture round capacitor plates with radius R. To find the magnetic field at a point in the region between the plates at a distance r from the axis, we apply Ampere's law to a circle of radius r passing through the point, with $r < R$. This circle passes through points a and b in Fig. 29.22. The total current enclosed by the circle is j_D times its area, or $(i_D/\pi R^2)(\pi r^2)$. The integral $\oint \vec{B} \cdot d\vec{l}$ in Ampere's law is just B times the circumference $2\pi r$ of the circle, and because $i_D = i_C$ for the charging capacitor, Ampere's law becomes

$$\oint \vec{B} \cdot d\vec{l} = 2\pi r B = \mu_0 \frac{r^2}{R^2} i_C \qquad \text{or}$$

$$B = \frac{\mu_0}{2\pi} \frac{r}{R^2} i_C \qquad (29.17)$$

This result predicts that in the region between the plates $\vec{B}$ is zero at the axis and increases linearly with distance from the axis. A similar calculation shows that *outside* the region between the plates (that is, for $r > R$), $\vec{B}$ is the same as though the wire were continuous and the plates not present at all.

When we *measure* the magnetic field in this region, we find that it really is there and that it behaves just as Eq. (29.17) predicts. This confirms directly the role of displacement current as a source of magnetic field. It is now established beyond reasonable doubt that displacement current, far from being just an artifice, is a fundamental fact of nature. Maxwell's discovery was the bold step of an extraordinary genius.

Maxwell's Equations of Electromagnetism

We are now in a position to wrap up in a single package *all* of the relationships between electric and magnetic fields and their sources. This package consists of four equations, called **Maxwell's equations.** Maxwell did not discover all of these equations single-handedly (though he did develop the concept of displacement current). But he did put them together and recognized their significance, particularly in predicting the existence of electromagnetic waves.

For now we'll state Maxwell's equations in their simplest form, for the case in which we have charges and currents in otherwise empty space. In Chapter 32 we'll discuss how to modify these equations if a dielectric or a magnetic material is present.

Two of Maxwell's equations involve an integral of $\vec{E}$ or $\vec{B}$ over a closed surface. The first is simply Gauss's law for electric fields, Eq. (22.8), which states that the surface integral of $E_\perp$ over any closed surface equals $1/\epsilon_0$ times the total charge Q_{encl} enclosed within the surface:

$$\oint \vec{E} \cdot d\vec{A} = \frac{Q_{encl}}{\epsilon_0} \qquad \text{(Gauss's law for } \vec{E}\text{)} \qquad (29.18)$$

The second is the analogous relationship for *magnetic* fields, Eq. (27.8), which states that the surface integral of $B_\perp$ over any closed surface is always zero:

$$\oint \vec{B} \cdot d\vec{A} = 0 \qquad \text{(Gauss's law for } \vec{B}\text{)} \qquad (29.19)$$

This statement means, among other things, that there are no magnetic monopoles (single magnetic charges) to act as sources of magnetic field.

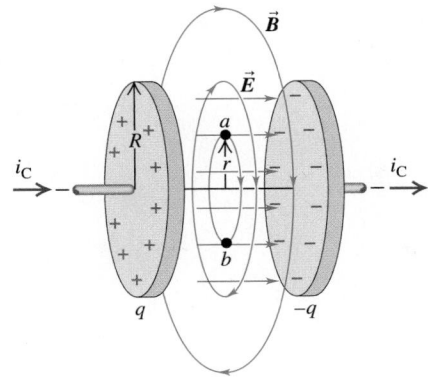

29.22 A capacitor being charged by a current i_C has a displacement current equal to i_C between the plates, with displacement-current density $j_D = \epsilon \, dE/dt$. This can be regarded as the source of the magnetic field between the plates.

The third equation is Ampere's law including displacement current. This states that both conduction current i_C and displacement current $\epsilon_0 d\Phi_E/dt$, where Φ_E is electric flux, act as sources of magnetic field:

$$\oint \vec{B} \cdot d\vec{l} = \mu_0 \left(i_C + \epsilon_0 \frac{d\Phi_E}{dt} \right)_{encl} \quad \text{(Ampere's law)} \qquad (29.20)$$

The fourth and final equation is Faraday's law. It states that a changing magnetic field or magnetic flux induces an electric field:

$$\oint \vec{E} \cdot d\vec{l} = -\frac{d\Phi_B}{dt} \quad \text{(Faraday's law)} \qquad (29.21)$$

If there is a changing magnetic flux, the line integral in Eq. (29.21) is not zero, which shows that the $\vec{E}$ field produced by a changing magnetic flux is not conservative. Recall that this line integral must be carried out over a *stationary* closed path.

It's worthwhile to look more carefully at the electric field $\vec{E}$ and its role in Maxwell's equations. In general, the total $\vec{E}$ field at a point in space can be the superposition of an electrostatic field $\vec{E}_c$ caused by a distribution of charges at rest and a magnetically induced, nonelectrostatic field $\vec{E}_n$. (The subscript c stands for Coulomb or conservative; the subscript n stands for non-Coulomb, nonelectrostatic, or nonconservative.) That is,

$$\vec{E} = \vec{E}_c + \vec{E}_n$$

The electrostatic part $\vec{E}_c$ is *always* conservative, so $\oint \vec{E}_c \cdot d\vec{l} = 0$. This conservative part of the field does not contribute to the integral in Faraday's law, so we can take $\vec{E}$ in Eq. (29.21) to be the *total* electric field $\vec{E}$, including both the part $\vec{E}_c$ due to charges and the magnetically induced part $\vec{E}_n$. Similarly, the nonconservative part $\vec{E}_n$ of the $\vec{E}$ field does not contribute to the integral in Gauss's law, because this part of the field is not caused by static charges. Hence $\oint \vec{E}_n \cdot d\vec{A}$ is always zero. We conclude that in all the Maxwell equations, $\vec{E}$ is the total electric field; these equations don't distinguish between conservative and nonconservative fields.

Symmetry in Maxwell's Equations

There is a remarkable symmetry in Maxwell's four equations. In empty space where there is no charge, the first two equations (Eqs. (29.18) and (29.19)) are identical in form, one containing $\vec{E}$ and the other containing $\vec{B}$. When we compare the second two equations, Eq. (29.20) says that a changing electric flux creates a magnetic field, and Eq. (29.21) says that a changing magnetic flux creates an electric field. In empty space, where there is no conduction current, $i_C = 0$ and the two equations have the same form, apart from a numerical constant and a negative sign, with the roles of $\vec{E}$ and $\vec{B}$ exchanged in the two equations.

We can rewrite Eqs. (29.20) and (29.21) in a different but equivalent form by introducing the definitions of electric and magnetic flux, $\Phi_E = \int \vec{E} \cdot d\vec{A}$ and $\Phi_B = \int \vec{B} \cdot d\vec{A}$, respectively. In empty space, where there is no charge or conduction current, $i_C = 0$ and $Q_{encl} = 0$, and we have

$$\oint \vec{B} \cdot d\vec{l} = \epsilon_0 \mu_0 \frac{d}{dt} \int \vec{E} \cdot d\vec{A} \qquad (29.22)$$

$$\oint \vec{E} \cdot d\vec{l} = -\frac{d}{dt} \int \vec{B} \cdot d\vec{A} \qquad (29.23)$$

Again we notice the symmetry between the roles of $\vec{E}$ and $\vec{B}$ in these expressions.

The most remarkable feature of these equations is that a time-varying field of *either* kind induces a field of the other kind in neighboring regions of space. Maxwell recognized that these relationships predict the existence of electromagnetic disturbances consisting of time-varying electric and magnetic fields that travel or *propagate* from one region of space to another, even if no matter is present in the intervening space. Such disturbances, called *electromagnetic waves,* provide the physical basis for light, radio and television waves, infrared, ultraviolet, x rays, and the rest of the electromagnetic spectrum. We will return to this vitally important topic in Chapter 32.

Although it may not be obvious, *all* the basic relationships between fields and their sources are contained in Maxwell's equations. We can derive Coulomb's law from Gauss's law, we can derive the law of Biot and Savart from Ampere's law, and so on. When we add the equation that defines the $\vec{E}$ and $\vec{B}$ fields in terms of the forces that they exert on a charge q, namely,

$$\vec{F} = q(\vec{E} + \vec{v} \times \vec{B}) \qquad (29.24)$$

we have *all* the fundamental relationships of electromagnetism!

Finally, we note that Maxwell's equations would have even greater symmetry between the $\vec{E}$ and $\vec{B}$ fields if single magnetic charges (magnetic monopoles) existed. The right side of Eq. (29.19) would contain the total *magnetic* charge enclosed by the surface, and the right side of Eq. (29.21) would include a magnetic monopole current term. Perhaps you can begin to see why some physicists wish that magnetic monopoles existed; they would help to perfect the mathematical poetry of Maxwell's equations.

The discovery that electromagnetism can be wrapped up so neatly and elegantly is a very satisfying one. In conciseness and generality, Maxwell's equations are in the same league with Newton's laws of motion and the laws of thermodynamics. Indeed, a major goal of science is learning how to express very broad and general relationships in a concise and compact form. Maxwell's synthesis of electromagnetism stands as a towering intellectual achievement, comparable to the Newtonian synthesis we described at the end of Section 13.5 and to the development of relativity and quantum mechanics in the 20th century.

Test Your Understanding of Section 29.7 (a) Which of Maxwell's equations explains how a credit card reader works? (b) Which one describes how a wire carrying a steady current generates a magnetic field? ❙

29.8 Superconductivity

The most familiar property of a superconductor is the sudden disappearance of all electrical resistance when the material is cooled below a temperature called the *critical temperature,* denoted by T_c. We discussed this behavior and the circumstances of its discovery in Section 25.2. But superconductivity is far more than just the absence of measurable resistance. As we'll see in this section, superconductors also have extraordinary *magnetic* properties.

The first hint of unusual magnetic properties was the discovery that for any superconducting material the critical temperature T_c changes when the material is placed in an externally produced magnetic field $\vec{B}_0$. Figure 29.23 shows this dependence for mercury, the first element in which superconductivity was observed. As the external field magnitude B_0 increases, the superconducting transition occurs at lower and lower temperature. When B_0 is greater than 0.0412 T, *no* superconducting transition occurs. The minimum magnitude of magnetic field that is needed to eliminate superconductivity at a temperature below T_c is called the *critical field,* denoted by B_c.

The Meissner Effect

Another aspect of the magnetic behavior of superconductors appears if we place a homogeneous sphere of a superconducting material in a uniform applied magnetic field $\vec{B}_0$ at a temperature T greater than T_c. The material is then in the normal phase, not the superconducting phase (Fig. 29.24a). Now we lower the temperature until the superconducting transition occurs. (We assume that the magnitude of $\vec{B}_0$ is not large enough to prevent the phase transition.) What happens to the field?

29.23 Phase diagram for pure mercury, showing the critical magnetic field B_c and its dependence on temperature. Superconductivity is impossible above the critical temperature T_c. The curves for other superconducting materials are similar but with different numerical values.

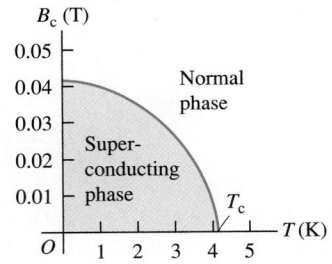

29.24 A superconducting material (a) above the critical temperature and (b), (c) below the critical temperature.

(a) Superconducting material in an external magnetic field $\vec{B}_0$ at $T > T_c$.

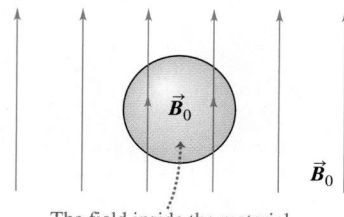

The field inside the material is very nearly equal to $\vec{B}_0$.

(b) The temperature is lowered to $T < T_c$, so the material becomes superconducting.

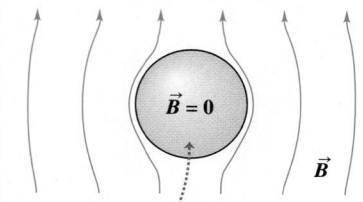

Magnetic flux is expelled from the material, and the field inside it is zero (Meissner effect).

(c) When the external field is turned off at $T < T_c$, the field is zero everywhere.

There is no change in magnetic flux in the material.

29.25 A superconductor (the black slab) exerts a repulsive force on a magnet (the metallic cylinder), supporting the magnet in midair.

Measurements of the field outside the sphere show that the field lines become distorted as in Fig. 29.24b. There is no longer any field inside the material, except possibly in a very thin surface layer a hundred or so atoms thick. If a coil is wrapped around the sphere, the emf induced in the coil shows that during the superconducting transition the magnetic flux through the coil decreases from its initial value to zero; this is consistent with the absence of field inside the material. Finally, if the field is now turned off while the material is still in its superconducting phase, no emf is induced in the coil, and measurements show no field outside the sphere (Fig. 29.24c).

We conclude that during a superconducting transition in the presence of the field $\vec{B}_0$, all of the magnetic flux is expelled from the bulk of the sphere, and the magnetic flux Φ_B through the coil becomes zero. This expulsion of magnetic flux is called the *Meissner effect*. As Fig. 29.24b shows, this expulsion crowds the magnetic field lines closer together to the side of the sphere, increasing $\vec{B}$ there.

Superconductor Levitation and Other Applications

The diamagnetic nature of a superconductor has some interesting *mechanical* consequences. A paramagnetic or ferromagnetic material is attracted by a permanent magnet because the magnetic dipoles in the material align with the nonuniform magnetic field of the permanent magnet. (We discussed this in Section 27.7.) For a diamagnetic material the magnetization is in the opposite sense, and a diamagnetic material is *repelled* by a permanent magnet. By Newton's third law the magnet is also repelled by the diamagnetic material. Figure 29.25 shows the repulsion between a specimen of a high-temperature superconductor and a magnet; the magnet is supported ("levitated") by this repulsive magnetic force.

The behavior we have described is characteristic of what are called *type-I superconductors*. There is another class of superconducting materials called *type-II superconductors*. When such a material in the superconducting phase is placed in a magnetic field, the bulk of the material remains superconducting, but thin filaments of material, running parallel to the field, may return to the normal phase. Currents circulate around the boundaries of these filaments, and there *is* magnetic flux inside them. Type-II superconductors are used for electromagnets because they usually have much larger values of B_c than do type-I materials, permitting much larger magnetic fields without destroying the superconducting state. Type-II superconductors have *two* critical magnetic fields: The first, B_{c1}, is the field at which magnetic flux begins to enter the material, forming the filaments just described, and the second, B_{c2}, is the field at which the material becomes normal.

Many important and exciting applications of superconductors are under development. Superconducting electromagnets have been used in research laboratories for several years. Their advantages compared to conventional electromagnets include greater efficiency, compactness, and greater field magnitudes. Once a current is established in the coil of a superconducting electromagnet, no additional power input is required because there is no resistive energy loss. The coils can also be made more compact because there is no need to provide channels for the circulation of cooling fluids. Superconducting magnets routinely attain steady fields of the order of 10 T, much larger than the maximum fields that are available with ordinary electromagnets.

Superconductors are attractive for long-distance electric power transmission and for energy-conversion devices, including generators, motors, and transformers. Very sensitive measurements of magnetic fields can be made with superconducting quantum interference devices (SQUIDs), which can detect changes in magnetic flux of less than 10^{-14} Wb; these devices have applications in medicine, geology, and other fields. The number of potential uses for superconductors has increased greatly since the discovery in 1987 of high-temperature superconductors. These materials have critical temperatures that are above the temperature of liquid nitrogen (about 77 K) and so are comparatively easy to attain. Development of practical applications of superconductor science promises to be an exciting chapter in contemporary technology.

CHAPTER 29 SUMMARY

Faraday's law: Faraday's law states that the induced emf in a closed loop equals the negative of the time rate of change of magnetic flux through the loop. This relationship is valid whether the flux change is caused by a changing magnetic field, motion of the loop, or both. (See Examples 29.1–29.6.)

$$\mathcal{E} = -\frac{d\Phi_B}{dt} \qquad (29.3)$$

The magnet's motion causes a *changing* magnetic field through the coil, inducing a current in the coil.

Lenz's law: Lenz's law states that an induced current or emf always tends to oppose or cancel out the change that caused it. Lenz's law can be derived from Faraday's law and is often easier to use. (See Examples 29.7 and 29.8.)

Change in $\vec{B}$
$\vec{B}$ (increasing)

Motional emf: If a conductor moves in a magnetic field, a motional emf is induced. (See Examples 29.9 and 29.10.)

$$\mathcal{E} = vBL \qquad (29.6)$$
(conductor with length L moves in uniform $\vec{B}$ field, $\vec{L}$ and $\vec{v}$ both perpendicular to $\vec{B}$ and to each other)

$$\mathcal{E} = \oint (\vec{v} \times \vec{B}) \cdot d\vec{l} \qquad (29.7)$$

(all or part of a closed loop moves in a $\vec{B}$ field)

Induced electric fields: When an emf is induced by a changing magnetic flux through a stationary conductor, there is an induced electric field $\vec{E}$ of nonelectrostatic origin. This field is nonconservative and cannot be associated with a potential. (See Example 29.11.)

$$\oint \vec{E} \cdot d\vec{l} = -\frac{d\Phi_B}{dt} \qquad (29.10)$$

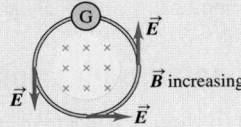

$\vec{B}$ increasing

Displacement current and Maxwell's equations: A time-varying electric field generates a displacement current i_D, which acts as a source of magnetic field in exactly the same way as conduction current. The relationships between electric and magnetic fields and their sources can be stated compactly in four equations, called Maxwell's equations. Together they form a complete basis for the relationship of $\vec{E}$ and $\vec{B}$ fields to their sources.

$$i_D = \epsilon \frac{d\Phi_E}{dt} \qquad (29.14)$$
(displacement current)

$$\oint \vec{E} \cdot d\vec{A} = \frac{Q_{encl}}{\epsilon_0} \qquad (29.18)$$
(Gauss's law for $\vec{E}$ fields)

$$\oint \vec{B} \cdot d\vec{A} = 0 \qquad (29.19)$$
(Gauss's law for $\vec{B}$ fields)

$$\oint \vec{B} \cdot d\vec{l} = \mu_0 \left(i_C + \epsilon_0 \frac{d\Phi_E}{dt} \right)_{encl} \qquad (29.20)$$
(Ampere's law including displacement current)

$$\oint \vec{E} \cdot d\vec{l} = -\frac{d\Phi_B}{dt} \qquad (29.21)$$
(Faraday's law)

BRIDGING PROBLEM A Falling Square Loop

A vertically oriented square loop of copper wire falls from rest in a region in which the field $\vec{B}$ is horizontal, uniform, and perpendicular to the plane of the loop, into a field-free region. The side length of the loop is s and the wire diameter is d. The resistivity of copper is ρ_R and the density of copper is ρ_m. If the loop reaches its terminal speed while its upper segment is still in the magnetic-field region, find an expression for the terminal speed.

SOLUTION GUIDE

See MasteringPhysics® study area for a Video Tutor solution.

IDENTIFY and SET UP

1. The motion of the loop through the magnetic field induces an emf and a current in the loop. The field then gives rise to a magnetic force on this current that opposes the downward force of gravity.
2. Consider the case in which the entire loop is in the magnetic-field region. Is there an induced emf in this case? If so, what is its direction?

3. Consider the case in which only the upper segment of the loop is in the magnetic-field region. Is there an induced emf in this case? If so, what is its direction?
4. For the case in which there is an induced emf and hence an induced current, what is the direction of the magnetic force on each of the four sides of the loop? What is the direction of the *net* magnetic force on the loop?

EXECUTE

5. For the case in which the loop is falling at speed v and there is an induced emf, find (i) the emf, (ii) the induced current, and (iii) the magnetic force on the loop in terms of its resistance R.
6. Find R and the mass of the loop in terms of the given information about the loop.
7. Use your results from steps 5 and 6 to find an expression for the terminal speed.

EVALUATE

8. How does the terminal speed depend on the magnetic-field magnitude B? Explain why this makes sense.

Problems

For instructor-assigned homework, go to www.masteringphysics.com

•, ••, •••: Problems of increasing difficulty. **CP:** Cumulative problems incorporating material from earlier chapters. **CALC:** Problems requiring calculus. **BIO:** Biosciences problems.

DISCUSSION QUESTIONS

Q29.1 A sheet of copper is placed between the poles of an electromagnet with the magnetic field perpendicular to the sheet. When the sheet is pulled out, a considerable force is required, and the force required increases with speed. Explain.

Q29.2 In Fig. 29.8, if the angular speed ω of the loop is doubled, then the frequency with which the induced current changes direction doubles, and the maximum emf also doubles. Why? Does the torque required to turn the loop change? Explain.

Q29.3 Two circular loops lie side by side in the same plane. One is connected to a source that supplies an increasing current; the other is a simple closed ring. Is the induced current in the ring in the same direction as the current in the loop connected to the source, or opposite? What if the current in the first loop is decreasing? Explain.

Q29.4 For Eq. (29.6), show that if v is in meters per second, B in teslas, and L in meters, then the units of the right-hand side of the equation are joules per coulomb or volts (the correct SI units for $\mathcal{E}$).

Q29.5 A long, straight conductor passes through the center of a metal ring, perpendicular to its plane. If the current in the conductor increases, is a current induced in the ring? Explain.

Q29.6 A student asserted that if a permanent magnet is dropped down a vertical copper pipe, it eventually reaches a terminal velocity even if there is no air resistance. Why should this be? Or should it?

Q29.7 An airplane is in level flight over Antarctica, where the magnetic field of the earth is mostly directed upward away from the ground. As viewed by a passenger facing toward the front of the plane, is the left or the right wingtip at higher potential? Does your answer depend on the direction the plane is flying?

Q29.8 Consider the situation in Exercise 29.19. In part (a), find the direction of the force that the large circuit exerts on the small one. Explain how this result is consistent with Lenz's law.

Q29.9 A metal rectangle is close to a long, straight, current-carrying wire, with two of its sides parallel to the wire. If the current in the long wire is decreasing, is the rectangle repelled by or attracted to the wire? Explain why this result is consistent with Lenz's law.

Q29.10 A square conducting loop is in a region of uniform, constant magnetic field. Can the loop be rotated about an axis along one side and no emf be induced in the loop? Discuss, in terms of the orientation of the rotation axis relative to the magnetic-field direction.

Q29.11 Example 29.6 discusses the external force that must be applied to the slidewire to move it at constant speed. If there were a break in the left-hand end of the U-shaped conductor, how much force would be needed to move the slidewire at constant speed? As in the example, you can ignore friction.

Q29.12 In the situation shown in Fig. 29.17, would it be appropriate to ask how much *energy* an electron gains during a complete trip around the wire loop with current I'? Would it be appropriate to ask what *potential difference* the electron moves through during such a complete trip? Explain your answers.

Q29.13 A metal ring is oriented with the plane of its area perpendicular to a spatially uniform magnetic field that increases at a steady rate. If the radius of the ring is doubled, by what factor do (a) the emf induced in the ring and (b) the electric field induced in the ring change?

Q29.14 • A type-II superconductor in an external field between B_{c1} and B_{c2} has regions that contain magnetic flux and have resistance, and also has superconducting regions. What is the resistance of a long, thin cylinder of such material?

Q29.15 Can one have a displacement current as well as a conduction current within a conductor? Explain.

Q29.16 Your physics study partner asks you to consider a parallel-plate capacitor that has a dielectric completely filling the volume between the plates. He then claims that Eqs. (29.13) and (29.14) show that the conduction current in the dielectric equals the displacement current in the dielectric. Do you agree? Explain.

Q29.17 Match the mathematical statements of Maxwell's equations as given in Section 29.7 to these verbal statements. (a) Closed electric field lines are evidently produced only by changing magnetic flux. (b) Closed magnetic field lines are produced both by the motion of electric charge and by changing electric flux. (c) Electric field lines can start on positive charges and end on negative charges. (d) Evidently there are no magnetic monopoles on which to start and end magnetic field lines.

Q29.18 If magnetic monopoles existed, the right-hand side of Eq. (29.21) would include a term proportional to the current of magnetic monopoles. Suppose a steady monopole current is moving in a long straight wire. Sketch the *electric* field lines that such a current would produce.

EXERCISES

Section 29.2 Faraday's Law

29.1 • A single loop of wire with an area of 0.0900 m^2 is in a uniform magnetic field that has an initial value of 3.80 T, is perpendicular to the plane of the loop, and is decreasing at a constant rate of 0.190 T/s. (a) What emf is induced in this loop? (b) If the loop has a resistance of $0.600 \text{ }\Omega$, find the current induced in the loop.

29.2 •• In a physics laboratory experiment, a coil with 200 turns enclosing an area of 12 cm^2 is rotated in 0.040 s from a position where its plane is perpendicular to the earth's magnetic field to a position where its plane is parallel to the field. The earth's magnetic field at the lab location is $6.0 \times 10^{-5} \text{ T}$. (a) What is the total magnetic flux through the coil before it is rotated? After it is rotated? (b) What is the average emf induced in the coil?

29.3 •• **Search Coils and Credit Cards.** One practical way to measure magnetic field strength uses a small, closely wound coil called a *search coil*. The coil is initially held with its plane perpendicular to a magnetic field. The coil is then either quickly rotated a quarter-turn about a diameter or quickly pulled out of the field. (a) Derive the equation relating the total charge Q that flows through a search coil to the magnetic-field magnitude B. The search coil has N turns, each with area A, and the flux through the coil is decreased from its initial maximum value to zero in a time Δt. The resistance of the coil is R, and the total charge is $Q = I\Delta t$, where I is the average current induced by the change in flux. (b) In a credit card reader, the magnetic strip on the back of a credit card is rapidly "swiped" past a coil within the reader. Explain, using the same ideas that underlie the operation of a search coil, how the reader can decode the information stored in the pattern of magnetization on the strip. (c) Is it necessary that the credit card be "swiped" through the reader at exactly the right speed? Why or why not?

29.4 • A closely wound search coil (see Exercise 29.3) has an area of 3.20 cm^2, 120 turns, and a resistance of $60.0 \text{ }\Omega$. It is connected to a charge-measuring instrument whose resistance is $45.0 \text{ }\Omega$. When the coil is rotated quickly from a position parallel to a uniform magnetic field to a position perpendicular to the field, the instrument indicates a charge of $3.56 \times 10^{-5} \text{ C}$. What is the magnitude of the field?

29.5 • A circular loop of wire with a radius of 12.0 cm and oriented in the horizontal xy-plane is located in a region of uniform magnetic field. A field of 1.5 T is directed along the positive z-direction, which is upward. (a) If the loop is removed from the field region in a time interval of 2.0 ms, find the average emf that will be induced in the wire loop during the extraction process. (b) If the coil is viewed looking down on it from above, is the induced current in the loop clockwise or counterclockwise?

29.6 • **CALC** A coil 4.00 cm in radius, containing 500 turns, is placed in a uniform magnetic field that varies with time according to $B = (0.0120 \text{ T/s})t + (3.00 \times 10^{-5} \text{ T/s}^4)t^4$. The coil is connected to a $600\text{-}\Omega$ resistor, and its plane is perpendicular to the magnetic field. You can ignore the resistance of the coil. (a) Find the magnitude of the induced emf in the coil as a function of time. (b) What is the current in the resistor at time $t = 5.00 \text{ s}$?

29.7 • **CALC** The current in the long, straight wire AB shown in Fig. E29.7 is upward and is increasing steadily at a rate di/dt. (a) At an instant when the current is i, what are the magnitude and direction of the field $\vec{B}$ at a distance r to the right of the wire? (b) What is the flux $d\Phi_B$ through the narrow, shaded strip? (c) What is the total flux through the loop? (d) What is the induced emf in the loop? (e) Evaluate the numerical value of the induced emf if $a = 12.0 \text{ cm}$, $b = 36.0 \text{ cm}$, $L = 24.0 \text{ cm}$, and $di/dt = 9.60 \text{ A/s}$.

Figure **E29.7**

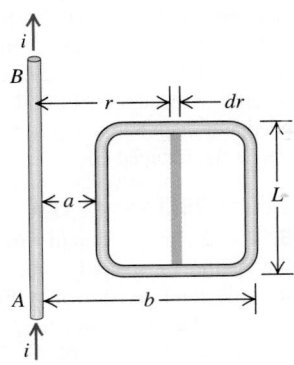

29.8 • **CALC** A flat, circular, steel loop of radius 75 cm is at rest in a uniform magnetic field, as shown in an edge-on view in Fig. E29.8. The field is changing with time, according to $B(t) = (1.4 \text{ T})e^{-(0.057 \text{ s}^{-1})t}$. (a) Find the emf induced in the loop as a function of time. (b) When is the induced emf equal to $\frac{1}{10}$ of its initial value? (c) Find the direction of the current induced in the loop, as viewed from above the loop.

Figure **E29.8**

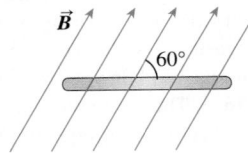

29.9 • **Shrinking Loop.** A circular loop of flexible iron wire has an initial circumference of 165.0 cm, but its circumference is decreasing at a constant rate of 12.0 cm/s due to a tangential pull on the wire. The loop is in a constant, uniform magnetic field oriented perpendicular to the plane of the loop and with magnitude 0.500 T. (a) Find the emf induced in the loop at the instant when 9.0 s have passed. (b) Find the direction of the induced current in the loop as viewed looking along the direction of the magnetic field.

29.10 • A closely wound rectangular coil of 80 turns has dimensions of 25.0 cm by 40.0 cm. The plane of the coil is rotated from a position where it makes an angle of $37.0°$ with a magnetic field of 1.10 T to a position perpendicular to the field. The rotation takes 0.0600 s. What is the average emf induced in the coil?

29.11 • **CALC** In a region of space, a magnetic field points in the +x-direction (toward the right). Its magnitude varies with position according to the formula $B_x = B_0 + bx$, where B_0 and b are positive constants, for $x \geq 0$. A flat coil of area A moves with uniform speed v from right to left with the plane of its area always perpendicular to this field. (a) What is the emf induced in this coil while it is to the right of the origin? (b) As viewed from the origin, what is the direction (clockwise or counterclockwise) of the current induced in the coil? (c) If instead the coil moved from left to right, what would be the answers to parts (a) and (b)?

29.12 • **Back emf.** A motor with a brush-and-commutator arrangement, as described in Example 29.4, has a circular coil with radius 2.5 cm and 150 turns of wire. The magnetic field has magnitude 0.060 T, and the coil rotates at 440 rev/min. (a) What is the maximum emf induced in the coil? (b) What is the average back emf?

29.13 •• The armature of a small generator consists of a flat, square coil with 120 turns and sides with a length of 1.60 cm. The coil rotates in a magnetic field of 0.0750 T. What is the angular speed of the coil if the maximum emf produced is 24.0 mV?

29.14 • A flat, rectangular coil of dimensions l and w is pulled with uniform speed v through a uniform magnetic field B with the plane of its area perpendicular to the field (Fig. E29.14). (a) Find the emf induced in this coil. (b) If the speed and magnetic field are both tripled, what is the induced emf?

Figure **E29.14**

Section 29.3 Lenz's Law

29.15 • A circular loop of wire is in a region of spatially uniform magnetic field, as shown in Fig. E29.15. The magnetic field is directed into the plane of the figure. Determine the direction (clockwise or counterclockwise) of the induced current in the loop when (a) B is increasing; (b) B is decreasing; (c) B is constant with value B_0. Explain your reasoning.

Figure **E29.15**

29.16 • The current in Fig. E29.16 obeys the equation $I(t) = I_0 e^{-bt}$, where $b > 0$. Find the direction (clockwise or counterclockwise) of the current induced in the round coil for $t > 0$.

Figure **E29.16**

29.17 • Using Lenz's law, determine the direction of the current in resistor ab of Fig. E29.17 when (a) switch S is opened after having been closed for several minutes; (b) coil B is brought closer to coil A with the switch closed; (c) the resistance of R is decreased while the switch remains closed.

Figure **E29.17**

29.18 • A cardboard tube is wrapped with two windings of insulated wire wound in opposite directions, as shown in Fig. E29.18. Terminals a and b of winding A may be connected to a battery through a reversing switch. State whether the induced current in the resistor R is from left to right or from right to left in the following circumstances: (a) the current in winding A is from a to b and is increasing; (b) the current in winding A is from b to a and is decreasing; (c) the current in winding A is from b to a and is increasing.

Figure **E29.18**

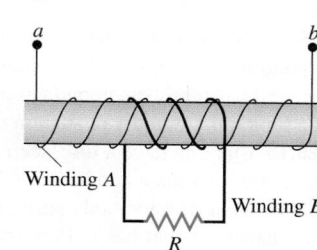

Winding A Winding B

R

29.19 • A small, circular ring is inside a larger loop that is connected to a battery and a switch, as shown in Fig. E29.19. Use Lenz's law to find the direction of the current induced in the small ring (a) just after switch S is closed; (b) after S has been closed a long time; (c) just after S has been reopened after being closed a long time.

Figure **E29.19**

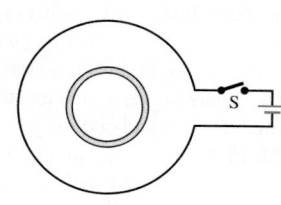

S

29.20 • A circular loop of wire with radius $r = 0.0480$ m and resistance $R = 0.160\ \Omega$ is in a region of spatially uniform magnetic field, as shown in Fig. E29.20. The magnetic field is directed out of the plane of the figure. The magnetic field has an initial value of 8.00 T and is decreasing at a rate of $dB/dt = -0.680$ T/s. (a) Is the induced current in the loop clockwise or counterclockwise? (b) What is the rate at which electrical energy is being dissipated by the resistance of the loop?

Figure **E29.20**

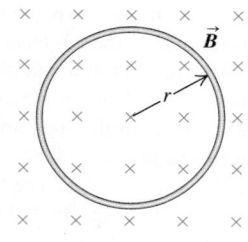

29.21 • **CALC** A circular loop of wire with radius $r = 0.0250$ m and resistance $R = 0.390\ \Omega$ is in a region of spatially uniform magnetic field, as shown in Fig. E29.21. The magnetic field is directed into the plane of the figure. At $t = 0$, $B = 0$. The magnetic field then begins increasing, with $B(t) = (0.380\ \text{T/s}^3)t^3$. What is the current in the loop (magnitude and direction) at the instant when $B = 1.33$ T?

Figure **E29.21**

Section 29.4 Motional Electromotive Force

29.22 • A rectangular loop of wire with dimensions 1.50 cm by 8.00 cm and resistance $R = 0.600\ \Omega$ is being pulled to the right out of a region of uniform magnetic field. The magnetic field has magnitude $B = 3.50$ T and is directed into the plane of Fig. E29.22. At

Figure **E29.22**

the instant when the speed of the loop is 3.00 m/s and it is still partially in the field region, what force (magnitude and direction) does the magnetic field exert on the loop?

29.23 • In Fig. E29.23 a conducting rod of length $L = 30.0$ cm moves in a magnetic field $\vec{B}$ of magnitude 0.450 T directed into the plane of the figure. The rod moves with speed $v = 5.00$ m/s in the direction shown. (a) What is the potential difference between the ends of the rod? (b) Which point, a or b, is at higher potential? (c) When the charges in the rod are in equilibrium, what are the magnitude and direction of the electric field within the rod? (d) When the charges in the rod are in equilibrium, which point, a or b, has an excess of positive charge? (e) What is the potential difference across the rod if it moves (i) parallel to ab and (ii) directly out of the page?

Figure **E29.23**

29.24 • A rectangle measuring 30.0 cm by 40.0 cm is located inside a region of a spatially uniform magnetic field of 1.25 T, with the field perpendicular to the plane of the coil (Fig. E29.24). The coil is pulled out at a steady rate of 2.00 cm/s traveling perpendicular to the field lines. The region of the field ends abruptly as shown. Find the emf induced in this coil when it is (a) all inside the field; (b) partly inside the field; (c) all outside the field.

Figure **E29.24**

29.25 • **Are Motional emfs a Practical Source of Electricity?** How fast (in m/s and mph) would a 5.00-cm copper bar have to move at right angles to a 0.650-T magnetic field to generate 1.50 V (the same as a AA battery) across its ends? Does this seem like a practical way to generate electricity?

29.26 • **Motional emfs in Transportation.** Airplanes and trains move through the earth's magnetic field at rather high speeds, so it is reasonable to wonder whether this field can have a substantial effect on them. We shall use a typical value of 0.50 G for the earth's field (a) The French TGV train and the Japanese "bullet train" reach speeds of up to 180 mph moving on tracks about 1.5 m apart. At top speed moving perpendicular to the earth's magnetic field, what potential difference is induced across the tracks as the wheels roll? Does this seem large enough to produce noticeable effects? (b) The Boeing 747-400 aircraft has a wingspan of 64.4 m and a cruising speed of 565 mph. If there is no wind blowing (so that this is also their speed relative to the ground), what is the maximum potential difference that could be induced between the opposite tips of the wings? Does this seem large enough to cause problems with the plane?

29.27 • The conducting rod ab shown in Fig. E29.27 makes contact with metal rails ca and db. The apparatus is in a uniform magnetic field of 0.800 T, perpendicular to the plane of the figure (a) Find the magnitude of the emf induced in the rod when it is moving toward the right with a speed 7.50 m/s. (b) In what direction does the current flow in the rod? (c) If the resistance of the circuit $abdc$ is 1.50 Ω (assumed to be constant), find the force (magnitude and direction) required to keep the rod moving to the right with a constant speed of 7.50 m/s. You can

Figure **E29.27**

ignore friction. (d) Compare the rate at which mechanical work is done by the force (Fv) with the rate at which thermal energy is developed in the circuit (I^2R).

29.28 • A 1.50-m-long metal bar is pulled to the right at a steady 5.0 m/s perpendicular to a uniform, 0.750-T magnetic field. The bar rides on parallel metal rails connected through a 25.0-Ω resistor, as shown in Fig. E29.28, so the apparatus makes a complete circuit. You can ignore the resistance of the bar and the rails. (a) Calculate the magnitude of the emf induced in the circuit. (b) Find the direction of the current induced in the circuit (i) using the magnetic force on the charges in the moving bar; (ii) using Faraday's law; (iii) using Lenz's law. (c) Calculate the current through the resistor.

Figure **E29.28**

29.29 • A 0.360-m-long metal bar is pulled to the left by an applied force F. The bar rides on parallel metal rails connected through a 45.0-Ω resistor, as shown in Fig. E29.29, so the apparatus makes a complete circuit. You can ignore the resistance of the bar and rails. The circuit is in a uniform 0.650-T magnetic field that is directed out of the plane of the figure. At the instant when the bar is moving to the left at 5.90 m/s, (a) is the induced current in the circuit clockwise or counterclockwise and (b) what is the rate at which the applied force is doing work on the bar?

Figure **E29.29**

29.30 • Consider the circuit shown in Fig. E29.29, but with the bar moving to the right with speed v. As in Exercise 29.29, the bar has length 0.360 m, $R = 45.0$ Ω, and $B = 0.650$ T. (a) Is the induced current in the circuit clockwise or counterclockwise? (b) At an instant when the 45.0-Ω resistor is dissipating electrical energy at a rate of 0.840 J/s, what is the speed of the bar?

29.31 • A 0.250-m-long bar moves on parallel rails that are connected through a 6.00-Ω resistor, as shown in Fig. E29.31, so the apparatus makes a complete circuit. You can ignore the resistance of the bar and rails. The circuit is in a uniform magnetic field $B = 1.20$ T that is directed into the plane of the figure. At an instant when the induced current in the circuit is counterclockwise and equal to 1.75 A, what is the velocity of the bar (magnitude and direction)?

Figure **E29.31**

29.32 • • **BIO Measuring Blood Flow.** Blood contains positive and negative ions and thus is a conductor. A blood vessel, therefore, can be viewed as an electrical wire. We can even picture the flowing blood as a series of parallel conducting slabs whose thickness is the diameter d of the vessel moving with speed v. (See Fig. E29.32.) (a) If the blood vessel is placed in a magnetic field B perpendicular to the vessel, as in the figure, show that the motional potential difference induced across it is $\mathcal{E} = vBd$. (b) If you expect that the blood will be flowing at 15 cm/s for a vessel

Figure **E29.32**

5.0 mm in diameter, what strength of magnetic field will you need to produce a potential difference of 1.0 mV? (c) Show that the volume rate of flow (R) of the blood is equal to $R = \pi \mathcal{E} d / 4B$. (*Note:* Although the method developed here is useful in measuring the rate of blood flow in a vessel, it is limited to use in surgery because measurement of the potential $\mathcal{E}$ must be made directly across the vessel.)

29.33 • A 1.41-m bar moves through a uniform, 1.20-T magnetic field with a speed of 2.50 m/s (Fig. E29.33). In each case, find the emf induced between the ends of this bar and identify which, if any, end (a or b) is at the higher potential. The bar moves in the direction of (a) the $+x$-axis; (b) the $-y$-axis; (c) the $+z$-axis. (d) How should this bar move so that the emf across its ends has the greatest possible value with b at a higher potential than a, and what is this maximum emf?

Figure **E29.33**

29.34 •• A rectangular circuit is moved at a constant velocity of 3.0 m/s into, through, and then out of a uniform 1.25-T magnetic field, as shown in Fig. E29.34. The magnetic-field region is considerably wider than 50.0 cm. Find the magnitude and direction (clockwise or counterclockwise) of the current induced in the circuit as it is (a) going into the magnetic field; (b) totally within the magnetic field, but still moving; and (c) moving out of the field. (d) Sketch a graph of the current in this circuit as a function of time, including the preceding three cases.

Figure **E29.34**

Section 29.5 Induced Electric Fields

29.35 • The magnetic field within a long, straight solenoid with a circular cross section and radius R is increasing at a rate of dB/dt. (a) What is the rate of change of flux through a circle with radius r_1 inside the solenoid, normal to the axis of the solenoid, and with center on the solenoid axis? (b) Find the magnitude of the induced electric field inside the solenoid, at a distance r_1 from its axis. Show the direction of this field in a diagram. (c) What is the magnitude of the induced electric field *outside* the solenoid, at a distance r_2 from the axis? (d) Graph the magnitude of the induced electric field as a function of the distance r from the axis from $r = 0$ to $r = 2R$. (e) What is the magnitude of the induced emf in a circular turn of radius $R/2$ that has its center on the solenoid axis? (f) What is the magnitude of the induced emf if the radius in part (e) is R? (g) What is the induced emf if the radius in part (e) is $2R$?

29.36 •• A long, thin solenoid has 900 turns per meter and radius 2.50 cm. The current in the solenoid is increasing at a uniform rate of 60.0 A/s. What is the magnitude of the induced electric field at a point near the center of the solenoid and (a) 0.500 cm from the axis of the solenoid; (b) 1.00 cm from the axis of the solenoid?

29.37 •• A long, thin solenoid has 400 turns per meter and radius 1.10 cm. The current in the solenoid is increasing at a uniform rate di/dt. The induced electric field at a point near the center of the solenoid and 3.50 cm from its axis is 8.00×10^{-6} V/m. Calculate di/dt.

29.38 • A metal ring 4.50 cm in diameter is placed between the north and south poles of large magnets with the plane of its area perpendicular to the magnetic field. These magnets produce an initial uniform field of 1.12 T between them but are gradually pulled apart, causing this field to remain uniform but decrease steadily at 0.250 T/s. (a) What is the magnitude of the electric field induced in the ring? (b) In which direction (clockwise or counterclockwise) does the current flow as viewed by someone on the south pole of the magnet?

29.39 • A long, straight solenoid with a cross-sectional area of 8.00 cm² is wound with 90 turns of wire per centimeter, and the windings carry a current of 0.350 A. A second winding of 12 turns encircles the solenoid at its center. The current in the solenoid is turned off such that the magnetic field of the solenoid becomes zero in 0.0400 s. What is the average induced emf in the second winding?

29.40 • The magnetic field $\vec{B}$ at all points within the colored circle shown in Fig. E29.15 has an initial magnitude of 0.750 T. (The circle could represent approximately the space inside a long, thin solenoid.) The magnetic field is directed into the plane of the diagram and is decreasing at the rate of -0.0350 T/s. (a) What is the shape of the field lines of the induced electric field shown in Fig. E29.15, within the colored circle? (b) What are the magnitude and direction of this field at any point on the circular conducting ring with radius 0.100 m? (c) What is the current in the ring if its resistance is 4.00 Ω? (d) What is the emf between points a and b on the ring? (e) If the ring is cut at some point and the ends are separated slightly, what will be the emf between the ends?

Section 29.7 Displacement Current and Maxwell's Equations

29.41 • **CALC** The electric flux through a certain area of a dielectric is $(8.76 \times 10^3 \, \text{V} \cdot \text{m/s}^4) t^4$. The displacement current through that area is 12.9 pA at time $t = 26.1$ ms. Calculate the dielectric constant for the dielectric.

29.42 • A parallel-plate, air-filled capacitor is being charged as in Fig. 29.22. The circular plates have radius 4.00 cm, and at a particular instant the conduction current in the wires is 0.280 A. (a) What is the displacement current density j_D in the air space between the plates? (b) What is the rate at which the electric field between the plates is changing? (c) What is the induced magnetic field between the plates at a distance of 2.00 cm from the axis? (d) At 1.00 cm from the axis?

29.43 • **Displacement Current in a Dielectric.** Suppose that the parallel plates in Fig. 29.22 have an area of 3.00 cm² and are separated by a 2.50-mm-thick sheet of dielectric that completely fills the volume between the plates. The dielectric has dielectric constant 4.70. (You can ignore fringing effects.) At a certain instant, the potential difference between the plates is 120 V and the conduction current i_C equals 6.00 mA. At this instant, what are (a) the charge q on each plate; (b) the rate of change of charge on the plates; (c) the displacement current in the dielectric?

29.44 • **CALC** In Fig. 29.22 the capacitor plates have area 5.00 cm² and separation 2.00 mm. The plates are in vacuum. The charging current i_C has a *constant* value of 1.80 mA. At $t = 0$ the charge on the plates is zero. (a) Calculate the charge on the plates, the electric field between the plates, and the potential difference between the plates when $t = 0.500$ μs. (b) Calculate dE/dt, the time rate of change of the electric field between the plates. Does dE/dt vary in time? (c) Calculate the displacement current density j_D between the plates, and from this the total displacement current i_D. How do i_C and i_D compare?

29.45 • **CALC Displacement Current in a Wire.** A long, straight, copper wire with a circular cross-sectional area of 2.1 mm² carries a current of 16 A. The resistivity of the material is $2.0 \times 10^{-8} \ \Omega \cdot m$. (a) What is the uniform electric field in the material? (b) If the current is changing at the rate of 4000 A/s, at what rate is the electric field in the material changing? (c) What is the displacement current density in the material in part (b)? (*Hint:* Since K for copper is very close to 1, use $\epsilon = \epsilon_0$.) (d) If the current is changing as in part (b), what is the magnitude of the magnetic field 6.0 cm from the center of the wire? Note that both the conduction current and the displacement current should be included in the calculation of B. Is the contribution from the displacement current significant?

Section 29.8 Superconductivity

29.46 • At temperatures near absolute zero, B_c approaches 0.142 T for vanadium, a type-I superconductor. The normal phase of vanadium has a magnetic susceptibility close to zero. Consider a long, thin vanadium cylinder with its axis parallel to an external magnetic field $\vec{B}_0$ in the $+x$-direction. At points far from the ends of the cylinder, by symmetry, all the magnetic vectors are parallel to the x-axis. At temperatures near absolute zero, what are the resultant magnetic field $\vec{B}$ and the magnetization $\vec{M}$ inside and outside the cylinder (far from the ends) for (a) $\vec{B}_0 = (0.130 \text{ T})\hat{\imath}$ and (b) $\vec{B}_0 = (0.260 \text{ T})\hat{\imath}$?

29.47 • The compound SiV₃ is a type-II superconductor. At temperatures near absolute zero the two critical fields are $B_{c1} = 55.0$ mT and $B_{c2} = 15.0$ T. The normal phase of SiV₃ has a magnetic susceptibility close to zero. A long, thin SiV₃ cylinder has its axis parallel to an external magnetic field $\vec{B}_0$ in the $+x$-direction. At points far from the ends of the cylinder, by symmetry, all the magnetic vectors are parallel to the x-axis. At a temperature near absolute zero, the external magnetic field is slowly increased from zero. What are the resultant magnetic field $\vec{B}$ and the magnetization $\vec{M}$ inside the cylinder at points far from its ends (a) just before the magnetic flux begins to penetrate the material, and (b) just after the material becomes completely normal?

PROBLEMS

29.48 ••• **CALC A Changing Magnetic Field.** You are testing a new data-acquisition system. This system allows you to record a graph of the current in a circuit as a function of time. As part of the test, you are using a circuit made up of a 4.00-cm-radius, 500-turn coil of copper wire connected in series to a 600-Ω resistor. Copper has resistivity $1.72 \times 10^{-8} \ \Omega \cdot m$, and the wire used for the coil has diameter 0.0300 mm. You place the coil on a table that is tilted 30.0° from the horizontal and that lies between the poles of an electromagnet. The electromagnet generates a vertically upward magnetic field that is zero for $t < 0$, equal to $(0.120 \text{ T}) \times [1 - (\cos\pi \text{ rad/s})t]$ for $0 \le t \le 1.00$ s, and equal to 0.240 T for $t > 1.00$ s. (a) Draw the graph that should be produced by your data-acquisition system. (This is a full-featured system, so the graph will include labels and numerical values on its axes.) (b) If you were looking vertically downward at the coil, would the current be flowing clockwise or counterclockwise?

29.49 •• **CP CALC** In the circuit shown in Fig. P29.49 the capacitor has capacitance $C = 20 \ \mu$F and is initially charged to 100 V with the polarity shown. The resistor R_0 has resistance 10 Ω. At time $t = 0$ the switch is closed. The small circuit is not connected in any way to the large one. The wire of the small circuit has a resistance of 1.0 Ω/m and contains 25 loops. The large circuit is a

rectangle 2.0 m by 4.0 m, while the small one has dimensions $a = 10.0$ cm and $b = 20.0$ cm. The distance c is 5.0 cm. (The figure is not drawn to scale.) Both circuits are held stationary. Assume that only the wire nearest the small circuit produces an appreciable magnetic field through it. (a) Find the current in the large circuit 200 μs after S is closed. (b) Find the current in the small circuit 200 μs after S is closed. (*Hint:* See Exercise 29.7.) (c) Find the direction of the current in the small circuit. (d) Justify why we can ignore the magnetic field from all the wires of the large circuit except for the wire closest to the small circuit.

Figure **P29.49**

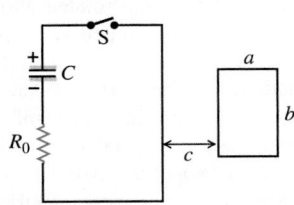

29.50 •• **CP CALC** In the circuit in Fig. P29.49, an emf of 90.0 V is added in series with the capacitor and the resistor, and the capacitor is initially uncharged. The emf is placed between the capacitor and the switch, with the positive terminal of the emf adjacent to the capacitor. Otherwise, the two circuits are the same as in Problem 29.49. The switch is closed at $t = 0$. When the current in the large circuit is 5.00 A, what are the magnitude and direction of the induced current in the small circuit?

29.51 •• **CALC** A very long, straight solenoid with a cross-sectional area of 2.00 cm² is wound with 90.0 turns of wire per centimeter. Starting at $t = 0$, the current in the solenoid is increasing according to $i(t) = (0.160 \text{ A/s}^2)t^2$. A secondary winding of 5 turns encircles the solenoid at its center, such that the secondary winding has the same cross-sectional area as the solenoid. What is the magnitude of the emf induced in the secondary winding at the instant that the current in the solenoid is 3.20 A?

29.52 • A flat coil is oriented with the plane of its area at right angles to a spatially uniform magnetic field. The magnitude of this field varies with time according to the graph in Fig. P29.52. Sketch a qualitative (but accurate!) graph of the emf induced in the coil as a function of time. Be sure to identify the times t_1, t_2, and t_3 on your graph.

Figure **P29.52**

29.53 • In Fig. P29.53 the loop is being pulled to the right at constant speed v. A constant current I flows in the long wire, in the direction shown. (a) Calculate the magnitude of the net emf $\mathcal{E}$ induced in the loop. Do this two ways: (i) by using Faraday's law of induction (*Hint:* See Exercise 29.7) and (ii) by looking at the emf induced in each segment of the loop due to its motion. (b) Find the direction (clockwise or counterclockwise) of the current induced in the loop. Do this two ways: (i) using Lenz's law and (ii) using the magnetic force on charges in the loop. (c) Check your answer for the emf in part (a) in the following special cases to see whether it is physically reasonable: (i) The loop is stationary; (ii) the loop is very thin, so $a \to 0$; (iii) the loop gets very far from the wire.

Figure **P29.53**

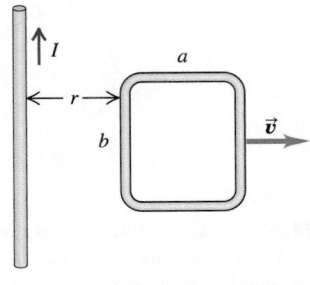

29.54 • Suppose the loop in Fig. P29.54 is (a) rotated about the y-axis; (b) rotated about the x-axis; (c) rotated about an edge parallel to the z-axis. What is the maximum induced emf in each case if $A = 600 \text{ cm}^2$, $\omega = 35.0 \text{ rad/s}$, and $B = 0.450 \text{ T}$?

Figure **P29.54**

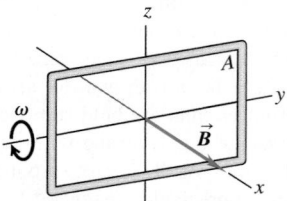

29.55 ••• As a new electrical engineer for the local power company, you are assigned the project of designing a generator of sinusoidal ac voltage with a maximum voltage of 120 V. Besides plenty of wire, you have two strong magnets that can produce a constant uniform magnetic field of 1.5 T over a square area of 10.0 cm on a side when they are 12.0 cm apart. The basic design should consist of a square coil turning in the uniform magnetic field. To have an acceptable coil resistance, the coil can have at most 400 loops. What is the minimum rotation rate (in rpm) of the coil so it will produce the required voltage?

29.56 • **Make a Generator?** You are shipwrecked on a deserted tropical island. You have some electrical devices that you could operate using a generator but you have no magnets. The earth's magnetic field at your location is horizontal and has magnitude 8.0×10^{-5} T, and you decide to try to use this field for a generator by rotating a large circular coil of wire at a high rate. You need to produce a peak emf of 9.0 V and estimate that you can rotate the coil at 30 rpm by turning a crank handle. You also decide that to have an acceptable coil resistance, the maximum number of turns the coil can have is 2000. (a) What area must the coil have? (b) If the coil is circular, what is the maximum translational speed of a point on the coil as it rotates? Do you think this device is feasible? Explain.

29.57 • A flexible circular loop 6.50 cm in diameter lies in a magnetic field with magnitude 1.35 T, directed into the plane of the page as shown in Fig. P29.57. The loop is pulled at the points indicated by the arrows, forming a loop of zero area in 0.250 s. (a) Find the average induced emf in the circuit. (b) What is the direction of the current in R: from a to b or from b to a? Explain your reasoning.

Figure **P29.57**

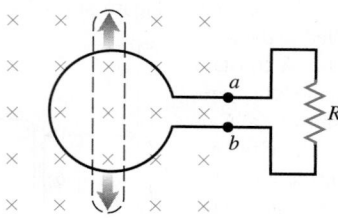

29.58 ••• **CALC** A conducting rod with length $L = 0.200$ m, mass $m = 0.120$ kg, and resistance $R = 80.0 \ \Omega$ moves without friction on metal rails as shown in Fig. 29.11. A uniform magnetic field with magnitude $B = 1.50$ T is directed into the plane of the figure. The rod is initially at rest, and then a constant force with magnitude $F = 1.90$ N and directed to the right is applied to the bar. How many seconds after the force is applied does the bar reach a speed of 25.0 m/s?

29.59 ••• **Terminal Speed.** A conducting rod with length L, mass m, and resistance R moves without friction on metal rails as shown in Fig. 29.11. A uniform magnetic field $\vec{B}$ is directed into the plane of the figure. The rod starts from rest and is acted on by a constant force $\vec{F}$ directed to the right. The rails are infinitely long and have negligible resistance. (a) Graph the speed of the rod as a function of time. (b) Find an expression for the terminal speed (the speed when the acceleration of the rod is zero).

29.60 •• **CP CALC Terminal Speed.** A bar of length $L = 0.36$ m is free to slide without friction on horizontal rails, as shown in Fig. P29.60. There is a uniform magnetic field $B = 1.5$ T directed into the plane of the figure. At one end of the rails there is a battery with emf $\mathcal{E} = 12$ V and a switch. The bar has mass 0.90 kg and resistance 5.0 Ω, and all other resistance in the circuit can be ignored. The switch is closed at time $t = 0$. (a) Sketch the speed of the bar as a function of time. (b) Just after the switch is closed, what is the acceleration of the bar? (c) What is the acceleration of the bar when its speed is 2.0 m/s? (d) What is the terminal speed of the bar?

Figure **P29.60**

29.61 • **CP Antenna emf.** A satellite, orbiting the earth at the equator at an altitude of 400 km, has an antenna that can be modeled as a 2.0-m-long rod. The antenna is oriented perpendicular to the earth's surface. At the equator, the earth's magnetic field is essentially horizontal and has a value of 8.0×10^{-5} T; ignore any changes in B with altitude. Assuming the orbit is circular, determine the induced emf between the tips of the antenna.

29.62 • **emf in a Bullet.** At the equator, the earth's magnetic field is approximately horizontal, is directed toward the north, and has a value of 8×10^{-5} T. (a) Estimate the emf induced between the top and bottom of a bullet shot horizontally at a target on the equator if the bullet is shot toward the east. Assume the bullet has a length of 1 cm and a diameter of 0.4 cm and is traveling at 300 m/s. Which is at higher potential: the top or bottom of the bullet? (b) What is the emf if the bullet travels south? (c) What is the emf induced between the front and back of the bullet for any horizontal velocity?

29.63 •• **CALC** A very long, cylindrical wire of radius R carries a current I_0 uniformly distributed across the cross section of the wire. Calculate the magnetic flux through a rectangle that has one side of length W running down the center of the wire and another side of length R, as shown in Fig. P29.63 (see Exercise 29.7).

Figure **P29.63**

29.64 • **CALC** A circular conducting ring with radius $r_0 = 0.0420$ m lies in the xy-plane in a region of uniform magnetic field $\vec{B} = B_0[1 - 3(t/t_0)^2 + 2(t/t_0)^3]\hat{k}$. In this expression, $t_0 = 0.0100$ s and is constant, t is time, $\hat{k}$ is the unit vector in the +z-direction, and $B_0 = 0.0800$ T and is constant. At points a and b (Fig. P29.64) there is a small gap in the ring with wires leading to an external circuit of resistance $R = 12.0 \ \Omega$. There is no magnetic field at the location of the external circuit. (a) Derive an expression, as a function of time, for the total magnetic flux Φ_B through the ring. (b) Determine the emf

Figure **P29.64**

induced in the ring at time $t = 5.00 \times 10^{-3}$ s. What is the polarity of the emf? (c) Because of the internal resistance of the ring, the current through R at the time given in part (b) is only 3.00 mA. Determine the internal resistance of the ring. (d) Determine the emf in the ring at a time $t = 1.21 \times 10^{-2}$ s. What is the polarity of the emf? (e) Determine the time at which the current through R reverses its direction.

29.65 · CALC The long, straight wire shown in Fig. P29.65a carries constant current I. A metal bar with length L is moving at constant velocity $\vec{v}$, as shown in the figure. Point a is a distance d from the wire. (a) Calculate the emf induced in the bar. (b) Which point, a or b, is at higher potential? (c) If the bar is replaced by a rectangular wire loop of resistance R (Fig. P29.65b), what is the magnitude of the current induced in the loop?

Figure **P29.65**

29.66 · The cube shown in Fig. P29.66, 50.0 cm on a side, is in a uniform magnetic field of 0.120 T, directed along the positive y-axis. Wires A, C, and D move in the directions indicated, each with a speed of 0.350 m/s. (Wire A moves parallel to the xy-plane, C moves at an angle of 45.0° below the xy-plane, and D moves parallel to the xz-plane.) What is the potential difference between the ends of each wire?

Figure **P29.66**

29.67 · CALC A slender rod, 0.240 m long, rotates with an angular speed of 8.80 rad/s about an axis through one end and perpendicular to the rod. The plane of rotation of the rod is perpendicular to a uniform magnetic field with a magnitude of 0.650 T. (a) What is the induced emf in the rod? (b) What is the potential difference between its ends? (c) Suppose instead the rod rotates at 8.80 rad/s about an axis through its center and perpendicular to the rod. In this case, what is the potential difference between the ends of the rod? Between the center of the rod and one end?

29.68 · **A Magnetic Exercise Machine.** You have designed a new type of exercise machine with an extremely simple mechanism (Fig. E29.28). A vertical bar of silver (chosen for its low resistivity and because it makes the machine look cool) with length $L = 3.0$ m is free to move left or right without friction on silver rails. The entire apparatus is placed in a horizontal, uniform magnetic field of strength 0.25 T. When you push the bar to the left or right, the bar's motion sets up a current in the circuit that includes the bar. The resistance of the bar and the rails can be neglected. The magnetic field exerts a force on the current-carrying bar, and this force opposes the bar's motion. The health benefit is from the exercise that you do in working against this force. (a) Your design

goal is that the person doing the exercise is to do work at the rate of 25 watts when moving the bar at a steady 2.0 m/s. What should be the resistance R? (b) You decide you want to be able to vary the power required from the person, to adapt the machine to the person's strength and fitness. If the power is to be increased to 50 W by altering R while leaving the other design parameters constant, should R be increased or decreased? Calculate the value of R for 50 W. (c) When you start to construct a prototype machine, you find it is difficult to produce a 0.25-T magnetic field over such a large area. If you decrease the length of the bar to 0.20 m while leaving B, v, and R the same as in part (a), what will be the power required of the person?

29.69 ·· CP CALC A rectangular loop with width L and a slide wire with mass m are as shown in Fig. P29.69. A uniform magnetic field $\vec{B}$ is directed perpendicular to the plane of the loop into the plane of the figure. The slide wire is given an initial speed of v_0 and then released. There is no friction between the slide wire and the loop, and the resistance of the loop is negligible in comparison to the resistance R of the slide wire. (a) Obtain an expression for F, the magnitude of the force exerted on the wire while it is moving at speed v. (b) Show that the distance x that the wire moves before coming to rest is $x = mv_0R/L^2B^2$.

Figure **P29.69**

29.70 ·· A 25.0-cm-long metal rod lies in the xy-plane and makes an angle of 36.9° with the positive x-axis and an angle of 53.1° with the positive y-axis. The rod is moving in the $+x$-direction with a speed of 6.80 m/s. The rod is in a uniform magnetic field $\vec{B} = (0.120 \text{ T})\hat{\imath} - (0.220 \text{ T})\hat{\jmath} - (0.0900 \text{ T})\hat{k}$. (a) What is the magnitude of the emf induced in the rod? (b) Indicate in a sketch which end of the rod is at higher potential.

29.71 · The magnetic field $\vec{B}$, at all points within a circular region of radius R, is uniform in space and directed into the plane of the page as shown in Fig. P29.71. (The region could be a cross section inside the windings of a long, straight solenoid.) If the magnetic field is increasing at a rate dB/dt, what are the magnitude and direction of the force on a stationary positive point charge q located at points a, b, and c? (Point a is a distance r above the center of the region, point b is a distance r to the right of the center, and point c is at the center of the region.)

Figure **P29.71**

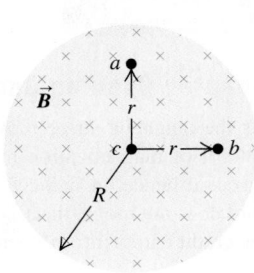

29.72 · CALC An airplane propeller of total length L rotates around its center with angular speed ω in a magnetic field that is perpendicular to the plane of rotation. Modeling the propeller as a thin, uniform bar, find the potential difference between (a) the center and either end of the propeller and (b) the two ends. (c) If the field is the earth's field of 0.50 G and the propeller turns at 220 rpm and is 2.0 m long, what is the potential difference between the middle and either end? It this large enough to be concerned about?

29.73 ··· CALC A dielectric of permittivity 3.5×10^{-11} F/m completely fills the volume between two capacitor plates. For $t > 0$ the electric flux through the dielectric is $(8.0 \times 10^3 \text{ V} \cdot \text{m/s}^3)t^3$. The dielectric is ideal and nonmagnetic; the conduction current in the dielectric is zero. At what time does the displacement current in the dielectric equal 21 μA?

29.74 ·· **CP CALC** A capacitor has two parallel plates with area A separated by a distance d. The space between plates is filled with a material having dielectric constant K. The material is not a perfect insulator but has resistivity ρ. The capacitor is initially charged with charge of magnitude Q_0 on each plate that gradually discharges by conduction through the dielectric. (a) Calculate the conduction current density $j_C(t)$ in the dielectric. (b) Show that at any instant the displacement current density in the dielectric is equal in magnitude to the conduction current density but opposite in direction, so the total current density is zero at every instant.

29.75 ·· **CALC** A rod of pure silicon (resistivity $\rho = 2300\ \Omega \cdot \text{m}$) is carrying a current. The electric field varies sinusoidally with time according to $E = E_0 \sin \omega t$, where $E_0 = 0.450\ \text{V/m}$, $\omega = 2\pi f$, and the frequency $f = 120\ \text{Hz}$. (a) Find the magnitude of the maximum conduction current density in the wire. (b) Assuming $\mathcal{E} = \mathcal{E}_0$, find the maximum displacement current density in the wire, and compare with the result of part (a). (c) At what frequency f would the maximum conduction and displacement densities become equal if $\mathcal{E} = \mathcal{E}_0$ (which is not actually the case)? (d) At the frequency determined in part (c), what is the relative *phase* of the conduction and displacement currents?

CHALLENGE PROBLEMS

29.76 ··· **CP CALC** A square, conducting, wire loop of side L, total mass m, and total resistance R initially lies in the horizontal xy-plane, with corners at $(x, y, z) = (0, 0, 0)$, $(0, L, 0)$, $(L, 0, 0)$, and $(L, L, 0)$. There is a uniform, upward magnetic field $\vec{B} = B\hat{k}$ in the space within and around the loop. The side of the loop that extends from $(0, 0, 0)$ to $(L, 0, 0)$ is held in place on the x-axis; the rest of the loop is free to pivot around this axis. When the loop is released, it begins to rotate due to the gravitational torque. (a)

Find the *net* torque (magnitude and direction) that acts on the loop when it has rotated through an angle ϕ from its original orientation and is rotating downward at an angular speed ω. (b) Find the angular acceleration of the loop at the instant described in part (a). (c) Compared to the case with zero magnetic field, does it take the loop a longer or shorter time to rotate through 90°? Explain. (d) Is mechanical energy conserved as the loop rotates downward? Explain.

29.77 ··· A metal bar with length L, mass m, and resistance R is placed on frictionless metal rails that are inclined at an angle ϕ above the horizontal. The rails have negligible resistance. A uniform magnetic field of magnitude B is directed downward as shown in Fig. P29.77. The bar is released from rest and slides down the rails. (a) Is the direction of the current induced in the bar from a to b or from b to a? (b) What is the terminal speed of the bar? (c) What is the induced current in the bar when the terminal speed has been reached? (d) After the terminal speed has been reached, at what rate is electrical energy being converted to thermal energy in the resistance of the bar? (e) After the terminal speed has been reached, at what rate is work being done on the bar by gravity? Compare your answer to that in part (d).

Figure **P29.77**

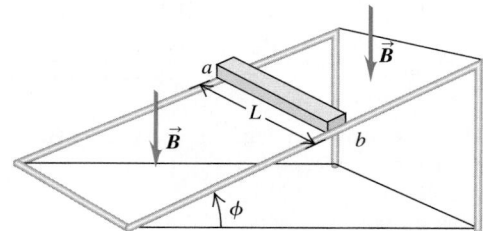

Answers

Chapter Opening Question ?

As the magnetic stripe moves through the card reader, the coded pattern of magnetization in the stripe causes a varying magnetic flux and hence an induced current in the reader's circuits. If the card does not move, there is no induced emf or current and none of the credit card's information is read.

Test Your Understanding Questions

29.2 Answers: (a) (i), (b) (iii) (a) Initially there is magnetic flux into the plane of the page, which we call positive. While the loop is being squeezed, the flux is becoming less positive ($d\Phi_B/dt < 0$) and so the induced emf is positive as in Fig. 29.6b ($\mathcal{E} = -d\Phi_B/dt > 0$). If you point the thumb of your right hand into the page, your fingers curl clockwise, so this is the direction of positive induced emf. (b) Since the coil's shape is no longer changing, the magnetic flux is not changing and there is no induced emf.

29.3 Answers: (a) (i), (b) (iii) In (a), as in the original situation, the magnet and loop are approaching each other and the downward flux through the loop is increasing. Hence the induced emf and induced current are the same. In (b), since the magnet and loop are moving together, the flux through the loop is not changing and no emf is induced.

29.4 Answers: (a) (iii); (b) (i) or (ii); (c) (ii) or (iii) You will get the maximum motional emf if you hold the rod vertically, so that its length is perpendicular to both the magnetic field and the direc-

tion of motion. With this orientation, $\vec{L}$ is parallel to $\vec{v} \times \vec{B}$. If you hold the rod in any horizontal orientation, $\vec{L}$ will be perpendicular to $\vec{v} \times \vec{B}$ and no emf will be induced. If you walk due north or south, $\vec{v} \times \vec{B} = 0$ and no emf will be induced for any orientation of the rod.

29.5 Answers: yes, no The magnetic field at a fixed position changes as you move the magnet. Such induced electric fields are *not* conservative.

29.6 Answer: (iii) By Lenz's law, the force must oppose the motion of the disk through the magnetic field. Since the disk material is now moving to the right through the field region, the force $\vec{F}$ is to the left—that is, in the opposite direction to that shown in Fig. 29.19b. To produce a leftward magnetic force $\vec{F} = I\vec{L} \times \vec{B}$ on currents moving through a magnetic field $\vec{B}$ directed out of the plane of the figure, the eddy currents must be moving downward in the figure—that is, in the same direction shown in Fig. 29.19b.

29.7 Answers: (a) Faraday's law, (b) Ampere's law A credit card reader works by inducing currents in the reader's coils as the card's magnetized stripe is swiped (see the answer to the chapter opening question). Ampere's law describes how currents of all kinds (both conduction currents and displacement currents) give rise to magnetic fields.

Bridging Problem

Answer: $v_t = 16\rho_m\rho_R g/B^2$

INDUCTANCE

? Many traffic lights change when a car rolls up to the intersection. How does the light sense the presence of the car?

LEARNING GOALS

By studying this chapter, you will learn:

- How a time-varying current in one coil can induce an emf in a second, unconnected coil.

- How to relate the induced emf in a circuit to the rate of change of current in the same circuit.

- How to calculate the energy stored in a magnetic field.

- How to analyze circuits that include both a resistor and an inductor (coil).

- Why electrical oscillations occur in circuits that include both an inductor and a capacitor.

- Why oscillations decay in circuits with an inductor, a resistor, and a capacitor.

Take a length of copper wire and wrap it around a pencil to form a coil. If you put this coil in a circuit, does it behave any differently than a straight piece of wire? Remarkably, the answer is yes. In an ordinary gasoline-powered car, a coil of this kind makes it possible for the 12-volt car battery to provide thousands of volts to the spark plugs, which in turn makes it possible for the plugs to fire and make the engine run. Other coils of this type are used to keep fluorescent light fixtures shining. Larger coils placed under city streets are used to control the operation of traffic signals. All of these applications, and many others, involve the *induction* effects that we studied in Chapter 29.

A changing current in a coil induces an emf in an adjacent coil. The coupling between the coils is described by their *mutual inductance.* A changing current in a coil also induces an emf in that same coil. Such a coil is called an *inductor,* and the relationship of current to emf is described by the *inductance* (also called *self-inductance*) of the coil. If a coil is initially carrying a current, energy is released when the current decreases; this principle is used in automotive ignition systems. We'll find that this released energy was stored in the magnetic field caused by the current that was initially in the coil, and we'll look at some of the practical applications of magnetic-field energy.

We'll also take a first look at what happens when an inductor is part of a circuit. In Chapter 31 we'll go on to study how inductors behave in alternating-current circuits; in that chapter we'll learn why inductors play an essential role in modern electronics, including communication systems, power supplies, and many other devices.

30.1 Mutual Inductance

In Section 28.4 we considered the magnetic interaction between two wires carrying *steady* currents; the current in one wire causes a magnetic field, which exerts a force on the current in the second wire. But an additional interaction arises

30.1 A current i_1 in coil 1 gives rise to a magnetic flux through coil 2.

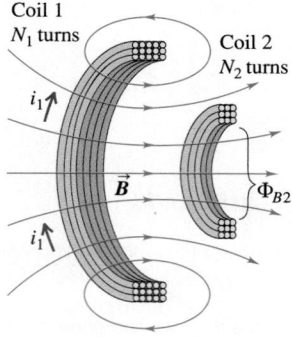

Mutual inductance: If the current in coil 1 is changing, the changing flux through coil 2 induces an emf in coil 2.

Coil 1
N_1 turns

Coil 2
N_2 turns

i_1

$\vec{B}$

Φ_{B2}

i_1

between two circuits when there is a *changing* current in one of the circuits. Consider two neighboring coils of wire, as in Fig. 30.1. A current flowing in coil 1 produces a magnetic field $\vec{B}$ and hence a magnetic flux through coil 2. If the current in coil 1 changes, the flux through coil 2 changes as well; according to Faraday's law, this induces an emf in coil 2. In this way, a change in the current in one circuit can induce a current in a second circuit.

Let's analyze the situation shown in Fig. 30.1 in more detail. We will use lowercase letters to represent quantities that vary with time; for example, a time-varying current is i, often with a subscript to identify the circuit. In Fig. 30.1 a current i_1 in coil 1 sets up a magnetic field (as indicated by the blue lines), and some of these field lines pass through coil 2. We denote the magnetic flux through *each* turn of coil 2, caused by the current i_1 in coil 1, as Φ_{B2}. (If the flux is different through different turns of the coil, then Φ_{B2} denotes the *average* flux.) The magnetic field is proportional to i_1, so Φ_{B2} is also proportional to i_1. When i_1 changes, Φ_{B2} changes; this changing flux induces an emf $\mathcal{E}_2$ in coil 2, given by

$$\mathcal{E}_2 = -N_2 \frac{d\Phi_{B2}}{dt} \tag{30.1}$$

We could represent the proportionality of Φ_{B2} and i_1 in the form $\Phi_{B2} = $ (constant)i_1, but instead it is more convenient to include the number of turns N_2 in the relationship. Introducing a proportionality constant M_{21}, called the **mutual inductance** of the two coils, we write

$$N_2 \Phi_{B2} = M_{21} i_1 \tag{30.2}$$

where Φ_{B2} is the flux through a *single* turn of coil 2. From this,

$$N_2 \frac{d\Phi_{B2}}{dt} = M_{21} \frac{di_1}{dt}$$

and we can rewrite Eq. (30.1) as

$$\mathcal{E}_2 = -M_{21} \frac{di_1}{dt} \tag{30.3}$$

That is, a change in the current i_1 in coil 1 induces an emf in coil 2 that is directly proportional to the rate of change of i_1 (Fig. 30.2).

We may also write the definition of mutual inductance, Eq. (30.2), as

$$M_{21} = \frac{N_2 \Phi_{B2}}{i_1}$$

If the coils are in vacuum, the flux Φ_{B2} through each turn of coil 2 is directly proportional to the current i_1. Then the mutual inductance M_{21} is a constant that depends only on the geometry of the two coils (the size, shape, number of turns, and orientation of each coil and the separation between the coils). If a magnetic material is present, M_{21} also depends on the magnetic properties of the material. If the material has nonlinear magnetic properties—that is, if the relative permeability K_m (defined in Section 28.8) is not constant and magnetization is not proportional to magnetic field—then Φ_{B2} is no longer directly proportional to i_1. In that case the mutual inductance also depends on the value of i_1. In this discussion we will assume that any magnetic material present has constant K_m so that flux *is* directly proportional to current and M_{21} depends on geometry only.

We can repeat our discussion for the opposite case in which a changing current i_2 in coil 2 causes a changing flux Φ_{B1} and an emf $\mathcal{E}_1$ in coil 1. We might expect that the corresponding constant M_{12} would be different from M_{21} because in general the two coils are not identical and the flux through them is not the same. It turns out, however, that M_{12} is *always* equal to M_{21}, even when the two coils are not symmetric. We call this common value simply the mutual inductance,

30.2 This electric toothbrush makes use of mutual inductance. The base contains a coil that is supplied with alternating current from a wall socket. This varying current induces an emf in a coil within the toothbrush itself, which is used to recharge the toothbrush battery.

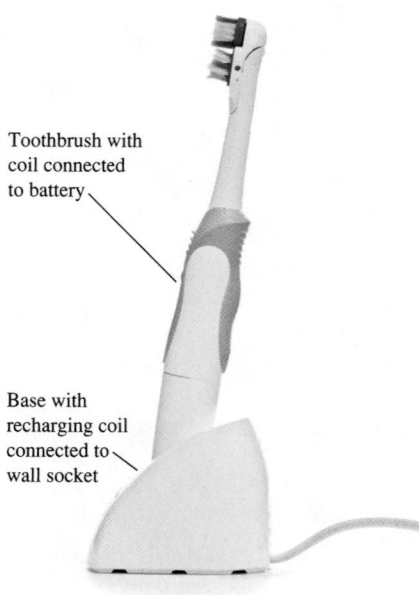

Toothbrush with coil connected to battery

Base with recharging coil connected to wall socket

denoted by the symbol M without subscripts; it characterizes completely the induced-emf interaction of two coils. Then we can write

$$\mathcal{E}_2 = -M\frac{di_1}{dt} \quad \text{and} \quad \mathcal{E}_1 = -M\frac{di_2}{dt} \quad \text{(mutually induced emfs)} \quad (30.4)$$

where the mutual inductance M is

$$M = \frac{N_2\Phi_{B2}}{i_1} = \frac{N_1\Phi_{B1}}{i_2} \quad \text{(mutual inductance)} \quad (30.5)$$

The negative signs in Eq. (30.4) are a reflection of Lenz's law. The first equation says that a change in current in coil 1 causes a change in flux through coil 2, inducing an emf in coil 2 that opposes the flux change; in the second equation the roles of the two coils are interchanged.

CAUTION **Only a time-varying current induces an emf** Note that only a *time-varying* current in a coil can induce an emf and hence a current in a second coil. Equations (30.4) show that the induced emf in each coil is directly proportional to the *rate of change* of the current in the other coil, not to the value of the current. A steady current in one coil, no matter how strong, cannot induce a current in a neighboring coil.

The SI unit of mutual inductance is called the **henry** (1 H), in honor of the American physicist Joseph Henry (1797–1878), one of the discoverers of electromagnetic induction. From Eq. (30.5), one henry is equal to *one weber per ampere*. Other equivalent units, obtained by using Eq. (30.4), are *one volt-second per ampere, one ohm-second,* and *one joule per ampere squared:*

$$1\text{ H} = 1\text{ Wb/A} = 1\text{ V} \cdot \text{s/A} = 1\text{ }\Omega \cdot \text{s} = 1\text{ J/A}^2$$

Just as the farad is a rather large unit of capacitance (see Section 24.1), the henry is a rather large unit of mutual inductance. As Example 30.1 shows, typical values of mutual inductance can be in the millihenry (mH) or microhenry (μH) range.

Drawbacks and Uses of Mutual Inductance

Mutual inductance can be a nuisance in electric circuits, since variations in current in one circuit can induce unwanted emfs in other nearby circuits. To minimize these effects, multiple-circuit systems must be designed so that M is as small as possible; for example, two coils would be placed far apart or with their planes perpendicular.

Happily, mutual inductance also has many useful applications. A *transformer,* used in alternating-current circuits to raise or lower voltages, is fundamentally no different from the two coils shown in Fig. 30.1. A time-varying alternating current in one coil of the transformer produces an alternating emf in the other coil; the value of M, which depends on the geometry of the coils, determines the amplitude of the induced emf in the second coil and hence the amplitude of the output voltage. (We'll describe transformers in more detail in Chapter 31 after we've discussed alternating current in greater depth.)

Example 30.1 **Calculating mutual inductance**

In one form of Tesla coil (a high-voltage generator popular in science museums), a long solenoid with length l and cross-sectional area A is closely wound with N_1 turns of wire. A coil with N_2 turns surrounds it at its center (Fig. 30.3). Find the mutual inductance M.

SOLUTION

IDENTIFY and SET UP: Mutual inductance occurs here because a current in either coil sets up a magnetic field that causes a flux through the other coil. From Example 28.9 (Section 28.7) we have

Continued

30.3 A long solenoid with cross-sectional area A and N_1 turns is surrounded at its center by a coil with N_2 turns.

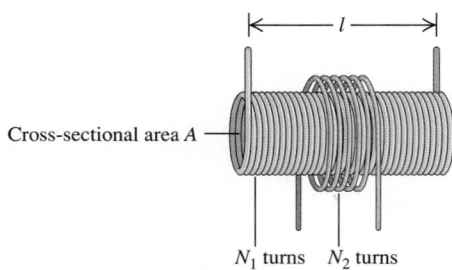

Cross-sectional area A

N_1 turns N_2 turns

an expression [Eq. (28.23)] for the field magnitude B_1 at the center of the solenoid (coil 1) in terms of the solenoid current i_1. This allows us to determine the flux through a cross section of the solenoid. Since there is no magnetic field outside a very long solenoid, this is also equal to the flux Φ_{B2} through each turn of the *outer* coil (2). We then use Eq. (30.5), in the form $M = N_2\Phi_{B2}/i_1$, to determine M.

EXECUTE: Equation (28.23) is expressed in terms of the number of turns per unit length, which for solenoid (1) is $n_1 = N_1/L$. We then have

$$B_1 = \mu_0 n_1 i_1 = \frac{\mu_0 N_1 i_1}{l}$$

The flux through a cross section of the solenoid equals $B_1 A$. As we mentioned above, this also equals the flux Φ_{B2} through each turn of the outer coil, independent of its cross-sectional area. From Eq. (30.5), the mutual inductance M is then

$$M = \frac{N_2\Phi_{B2}}{i_1} = \frac{N_2 B_1 A}{i_1} = \frac{N_2}{i_1}\frac{\mu_0 N_1 i_1}{l}A = \frac{\mu_0 A N_1 N_2}{l}$$

EVALUATE: The mutual inductance M of any two coils is proportional to the product $N_1 N_2$ of their numbers of turns. Notice that M depends only on the geometry of the two coils, not on the current.

Here's a numerical example to give you an idea of magnitudes. Suppose $l = 0.50$ m, $A = 10$ cm^2 $= 1.0 \times 10^{-3}$ m^2, $N_1 = 1000$ turns, and $N_2 = 10$ turns. Then

$$M = \frac{(4\pi \times 10^{-7}\ \text{Wb/A}\cdot\text{m})(1.0 \times 10^{-3}\ \text{m}^2)(1000)(10)}{0.50\ \text{m}}$$

$$= 25 \times 10^{-6}\ \text{Wb/A} = 25 \times 10^{-6}\ \text{H} = 25\ \mu\text{H}$$

Example 30.2 **Emf due to mutual inductance**

In Example 30.1, suppose the current i_2 in the outer coil is given by $i_2 = (2.0 \times 10^6\ \text{A/s})t$. (Currents in wires can indeed increase this rapidly for brief periods.) (a) At $t = 3.0\ \mu\text{s}$, what is the average magnetic flux through each turn of the solenoid (coil 1) due to the current in the outer coil? (b) What is the induced emf in the solenoid?

SOLUTION

IDENTIFY and SET UP: In Example 30.1 we found the mutual inductance by relating the current in the solenoid to the flux produced in the outer coil; to do that, we used Eq. (30.5) in the form $M = N_2\Phi_{B2}/i_1$. Here we are given the current i_2 in the outer coil and want to find the resulting flux Φ_1 in the solenoid. The mutual inductance is the *same* in either case, and we have $M = 25\ \mu\text{H}$ from Example 30.1. We use Eq. (30.5) in the form $M = N_1\Phi_{B1}/i_2$ to determine the average flux Φ_{B1} through each turn of the solenoid caused by a given current i_2 in the outer coil. We then use Eq. (30.4) to determine the emf induced in the solenoid by the time variation of i_2.

EXECUTE: (a) At $t = 3.0\ \mu\text{s} = 3.0 \times 10^{-6}$ s, the current in the outer coil is $i_2 = (2.0 \times 10^6\ \text{A/s})(3.0 \times 10^{-6}\ \text{s}) = 6.0$ A. We

solve Eq. (30.5) for the flux Φ_{B1} through each turn of the solenoid (coil 1):

$$\Phi_{B1} = \frac{Mi_2}{N_1} = \frac{(25 \times 10^{-6}\ \text{H})(6.0\ \text{A})}{1000} = 1.5 \times 10^{-7}\ \text{Wb}$$

We emphasize that this is an *average* value; the flux can vary considerably between the center and the ends of the solenoid.

(b) We are given $i_2 = (2.0 \times 10^6\ \text{A/s})t$, so $di_2/dt = 2.0 \times 10^6$ A/s; then, from Eq. (30.4), the induced emf in the solenoid is

$$\mathcal{E}_1 = -M\frac{di_2}{dt} = -(25 \times 10^{-6}\ \text{H})(2.0 \times 10^6\ \text{A/s}) = -50\ \text{V}$$

EVALUATE: This is a substantial induced emf in response to a very rapid current change. In an operating Tesla coil, there is a high-frequency alternating current rather than a continuously increasing current as in this example; both di_2/dt and $\mathcal{E}_1$ alternate as well, with amplitudes that can be thousands of times larger than in this example.

Test Your Understanding of Section 30.1 Consider the Tesla coil described in Example 30.1. If you make the solenoid out of twice as much wire, so that it has twice as many turns and is twice as long, how much larger is the mutual inductance? (i) M is four times greater; (ii) M is twice as great; (iii) M is unchanged; (iv) M is $\frac{1}{2}$ as great; (v) M is $\frac{1}{4}$ as great.

(MP)

30.2 Self-Inductance and Inductors

In our discussion of mutual inductance we considered two separate, independent circuits: A current in one circuit creates a magnetic field that gives rise to a flux through the second circuit. If the current in the first circuit changes, the flux through the second circuit changes and an emf is induced in the second circuit.

An important related effect occurs even if we consider only a *single* isolated circuit. When a current is present in a circuit, it sets up a magnetic field that causes a magnetic flux through the *same* circuit; this flux changes when the current changes. Thus any circuit that carries a varying current has an emf induced in it by the variation in *its own* magnetic field. Such an emf is called a **self-induced emf.** By Lenz's law, a self-induced emf always opposes the change in the current that caused the emf and so tends to make it more difficult for variations in current to occur. For this reason, self-induced emfs can be of great importance whenever there is a varying current.

Self-induced emfs can occur in *any* circuit, since there is always some magnetic flux through the closed loop of a current-carrying circuit. But the effect is greatly enhanced if the circuit includes a coil with N turns of wire (Fig. 30.4). As a result of the current i, there is an average magnetic flux Φ_B through each turn of the coil. In analogy to Eq. (30.5) we define the **self-inductance** L of the circuit as

$$L = \frac{N\Phi_B}{i} \qquad \text{(self-inductance)} \qquad (30.6)$$

When there is no danger of confusion with mutual inductance, the self-inductance is called simply the **inductance.** Comparing Eqs. (30.5) and (30.6), we see that the units of self-inductance are the same as those of mutual inductance; the SI unit of self-inductance is the henry.

If the current i in the circuit changes, so does the flux Φ_B; from rearranging Eq. (30.6) and taking the derivative with respect to time, the rates of change are related by

$$N\frac{d\Phi_B}{dt} = L\frac{di}{dt}$$

From Faraday's law for a coil with N turns, Eq. (29.4), the self-induced emf is $\mathcal{E} = -N\,d\Phi_B/dt$, so it follows that

$$\mathcal{E} = -L\frac{di}{dt} \qquad \text{(self-induced emf)} \qquad (30.7)$$

The minus sign in Eq. (30.7) is a reflection of Lenz's law; it says that the self-induced emf in a circuit opposes any change in the current in that circuit. (Later in this section we'll explore in greater depth the significance of this minus sign.)

Equation (30.7) also states that the self-inductance of a circuit is the magnitude of the self-induced emf per unit rate of change of current. This relationship makes it possible to measure an unknown self-inductance in a relatively simple way: Change the current in the circuit at a known rate di/dt, measure the induced emf, and take the ratio to determine L.

Inductors As Circuit Elements

A circuit device that is designed to have a particular inductance is called an **inductor,** or a *choke*. The usual circuit symbol for an inductor is

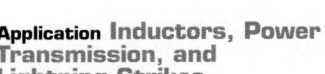

Like resistors and capacitors, inductors are among the indispensable circuit elements of modern electronics. Their purpose is to oppose any variations in the current through the circuit. An inductor in a direct-current circuit helps to maintain a steady current despite any fluctuations in the applied emf; in an alternating-current circuit, an inductor tends to suppress variations of the current that are more rapid than desired. In this chapter and the next we will explore the behavior and applications of inductors in circuits in more detail.

To understand the behavior of circuits containing inductors, we need to develop a general principle analogous to Kirchhoff's loop rule (discussed in

30.4 The current i in the circuit causes a magnetic field $\vec{B}$ in the coil and hence a flux through the coil.

Self-inductance: If the current i in the coil is changing, the changing flux through the coil induces an emf in the coil.

$\vec{B}$

i

Application Inductors, Power Transmission, and Lightning Strikes
If lightning strikes part of an electrical power transmission system, it causes a sudden spike in voltage that can damage the components of the system as well as anything connected to that system (for example, home appliances). To minimize these effects, large inductors are incorporated into the transmission system. These use the principle that an inductor opposes and suppresses any rapid changes in the current.

Section 26.2). To apply that rule, we go around a conducting loop, measuring potential differences across successive circuit elements as we go. The algebraic sum of these differences around any closed loop must be zero because the electric field produced by charges distributed around the circuit is *conservative*. In Section 29.7 we denoted such a conservative field as $\vec{E}_c$.

When an inductor is included in the circuit, the situation changes. The magnetically induced electric field within the coils of the inductor is *not* conservative; as in Section 29.7, we'll denote it by $\vec{E}_n$. We need to think very carefully about the roles of the various fields. Let's assume we are dealing with an inductor whose coils have negligible resistance. Then a negligibly small electric field is required to make charge move through the coils, so the *total* electric field $\vec{E}_c + \vec{E}_n$ within the coils must be zero, even though neither field is individually zero. Because $\vec{E}_c$ is nonzero, there have to be accumulations of charge on the terminals of the inductor and the surfaces of its conductors to produce this field.

Consider the circuit shown in Fig. 30.5; the box contains some combination of batteries and variable resistors that enables us to control the current i in the circuit. According to Faraday's law, Eq. (29.10), the line integral of $\vec{E}_n$ around the circuit is the negative of the rate of change of flux through the circuit, which in turn is given by Eq. (30.7). Combining these two relationships, we get

$$\oint \vec{E}_n \cdot d\vec{l} = -L\frac{di}{dt}$$

where we integrate clockwise around the loop (the direction of the assumed current). But $\vec{E}_n$ is different from zero only within the inductor. Therefore the integral of $\vec{E}_n$ around the whole loop can be replaced by its integral only from a to b through the inductor; that is,

$$\int_a^b \vec{E}_n \cdot d\vec{l} = -L\frac{di}{dt}$$

Next, because $\vec{E}_c + \vec{E}_n = 0$ at each point within the inductor coils, we can rewrite this as

$$\int_a^b \vec{E}_c \cdot d\vec{l} = L\frac{di}{dt}$$

But this integral is just the potential V_{ab} of point a with respect to point b, so we finally obtain

$$V_{ab} = V_a - V_b = L\frac{di}{dt} \tag{30.8}$$

We conclude that there is a genuine potential difference between the terminals of the inductor, associated with conservative, electrostatic forces, despite the fact that the electric field associated with the magnetic induction effect is nonconservative. Thus we are justified in using Kirchhoff's loop rule to analyze circuits that include inductors. Equation (30.8) gives the potential difference across an inductor in a circuit.

CAUTION **Self-induced emf opposes changes in current** Note that the self-induced emf does not oppose the current i itself; rather, it opposes any *change* (di/dt) in the current. Thus the circuit behavior of an inductor is quite different from that of a resistor. Figure 30.6 compares the behaviors of a resistor and an inductor and summarizes the sign relationships. ∎

Applications of Inductors

Because an inductor opposes changes in current, it plays an important role in fluorescent light fixtures (Fig. 30.7). In such fixtures, current flows from the wiring

30.5 A circuit containing a source of emf and an inductor. The source is variable, so the current i and its rate of change di/dt can be varied.

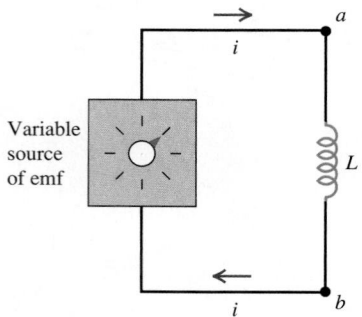

30.6 (a) The potential difference across a resistor depends on the current. (b), (c), (d) The potential difference across an inductor depends on the rate of change of the current.

(a) Resistor with current i flowing from a to b: potential drops from a to b.

$V_{ab} = iR > 0$

(b) Inductor with *constant* current i flowing from a to b: no potential difference.

i constant: $di/dt = 0$

$\mathcal{E} = 0$ $V_{ab} = L\dfrac{di}{dt} = 0$

(c) Inductor with *increasing* current i flowing from a to b: potential drops from a to b.

i increasing: $di/dt > 0$

$\mathcal{E}$ $V_{ab} = L\dfrac{di}{dt} > 0$

(d) Inductor with *decreasing* current i flowing from a to b: potential increases from a to b.

i decreasing: $di/dt < 0$

$\mathcal{E}$ $V_{ab} = L\dfrac{di}{dt} < 0$

into the gas that fills the tube, ionizing the gas and causing it to glow. However, an ionized gas or *plasma* is a highly nonohmic conductor: The greater the current, the more highly ionized the plasma becomes and the lower its resistance. If a sufficiently large voltage is applied to the plasma, the current can grow so much that it damages the circuitry outside the fluorescent tube. To prevent this problem, an inductor or *magnetic ballast* is put in series with the fluorescent tube to keep the current from growing out of bounds.

The ballast also makes it possible for the fluorescent tube to work with the alternating voltage provided by household wiring. This voltage oscillates sinusoidally with a frequency of 60 Hz, so that it goes momentarily to zero 120 times per second. If there were no ballast, the plasma in the fluorescent tube would rapidly deionize when the voltage went to zero and the tube would shut off. With a ballast present, a self-induced emf sustains the current and keeps the tube lit. Magnetic ballasts are also used for this purpose in streetlights (which obtain their light from a glowing vapor of mercury or sodium atoms) and in neon lights. (In compact fluorescent lamps, the magnetic ballast is replaced by a more complicated scheme for regulating current. This scheme utilizes transistors, discussed in Chapter 42.)

The self-inductance of a circuit depends on its size, shape, and number of turns. For N turns close together, it is always proportional to N^2. It also depends on the magnetic properties of the material enclosed by the circuit. In the following examples we will assume that the circuit encloses only vacuum (or air, which from the standpoint of magnetism is essentially vacuum). If, however, the flux is concentrated in a region containing a magnetic material with permeability μ, then in the expression for B we must replace μ_0 (the permeability of vacuum) by $\mu = K_m\mu_0$, as discussed in Section 28.8. If the material is diamagnetic or paramagnetic, this replacement makes very little difference, since K_m is very close to 1. If the material is *ferromagnetic,* however, the difference is of crucial importance. A solenoid wound on a soft iron core having $K_m = 5000$ can have an inductance approximately 5000 times as great as that of the same solenoid with an air core. Ferromagnetic-core inductors are very widely used in a variety of electronic and electric-power applications.

An added complication is that with ferromagnetic materials the magnetization is in general not a linear function of magnetizing current, especially as saturation is approached. As a result, the inductance is not constant but can depend on current in a fairly complicated way. In our discussion we will ignore this complication and assume always that the inductance is constant. This is a reasonable assumption even for a ferromagnetic material if the magnetization remains well below the saturation level.

Because automobiles contain steel, a ferromagnetic material, driving an automobile over a coil causes an appreciable increase in the coil's inductance. This effect is used in traffic light sensors, which use a large, current-carrying coil embedded under the road surface near an intersection. The circuitry connected to the coil detects the inductance change as a car drives over. When a preprogrammed number of cars have passed over the coil, the light changes to green to allow the cars through the intersection.

30.7 These fluorescent light tubes are wired in series with an inductor, or ballast, that helps to sustain the current flowing through the tubes.

Example 30.3   **Calculating self-inductance**

Determine the self-inductance of a toroidal solenoid with cross-sectional area A and mean radius r, closely wound with N turns of wire on a nonmagnetic core (Fig. 30.8). Assume that B is uniform across a cross section (that is, neglect the variation of B with distance from the toroid axis).

SOLUTION

IDENTIFY and SET UP: Our target variable is the self-inductance L of the toroidal solenoid. We can find L using Eq. (30.6), which requires knowing the flux Φ_B through each turn and the current i in

Continued

30.8 Determining the self-inductance of a closely wound toroidal solenoid. For clarity, only a few turns of the winding are shown. Part of the toroid has been cut away to show the cross-sectional area A and radius r.

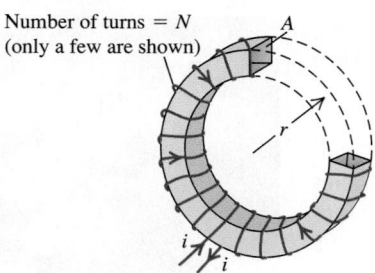

Number of turns = N
(only a few are shown)

the coil. For this, we use the results of Example 28.10 (Section 28.7), in which we found the magnetic field in the interior of a toroidal solenoid as a function of the current.

EXECUTE: From Eq. (30.6), the self-inductance is $L = N\Phi_B/i$. From Example 28.10, the field magnitude at a distance r from the toroid axis is $B = \mu_0 Ni/2\pi r$. If we assume that the field has this magnitude over the entire cross-sectional area A, then

$$\Phi_B = BA = \frac{\mu_0 NiA}{2\pi r}$$

The flux Φ_B is the same through each turn, and the self-inductance L is

$$L = \frac{N\Phi_B}{i} = \frac{\mu_0 N^2 A}{2\pi r} \quad \text{(self-inductance of a toroidal solenoid)}$$

EVALUATE: Suppose $N = 200$ turns, $A = 5.0 \text{ cm}^2 = 5.0 \times 10^{-4} \text{ m}^2$, and $r = 0.10$ m; then

$$L = \frac{(4\pi \times 10^{-7} \text{ Wb/A} \cdot \text{m})(200)^2(5.0 \times 10^{-4} \text{ m}^2)}{2\pi(0.10 \text{ m})}$$

$$= 40 \times 10^{-6} \text{ H} = 40 \ \mu\text{H}$$

Example 30.4 **Calculating self-induced emf**

If the current in the toroidal solenoid in Example 30.3 increases uniformly from 0 to 6.0 A in 3.0 μs, find the magnitude and direction of the self-induced emf.

SOLUTION

IDENTIFY and SET UP: We are given L, the self-inductance, and di/dt, the rate of change of the solenoid current. We find the magnitude of self-induced emf $\mathcal{E}$ using Eq. (30.7) and its direction using Lenz's law.

EXECUTE: We have $di/dt = (6.0 \text{ A})/(3.0 \times 10^{-6} \text{ s}) = 2.0 \times 10^6$ A/s. From Eq. (30.7), the magnitude of the induced emf is

$$|\mathcal{E}| = L\left|\frac{di}{dt}\right| = (40 \times 10^{-6} \text{ H})(2.0 \times 10^6 \text{ A/s}) = 80 \text{ V}$$

The current is increasing, so according to Lenz's law the direction of the emf is opposite to that of the current. This corresponds to the situation in Fig. 30.6c; the emf is in the direction from b to a, like a battery with a as the $+$ terminal and b the $-$ terminal, tending to oppose the current increase from the external circuit.

EVALUATE: This example shows that even a small inductance L can give rise to a substantial induced emf if the current changes rapidly.

Test Your Understanding of Section 30.2 Rank the following inductors in order of the potential difference v_{ab}, from most positive to most negative. In each case the inductor has zero resistance and the current flows from point a through the inductor to point b. (i) The current through a 2.0-μH inductor increases from 1.0 A to 2.0 A in 0.50 s; (ii) the current through a 4.0-μH inductor decreases from 3.0 A to 0 in 2.0 s; (iii) the current through a 1.0-μH inductor remains constant at 4.0 A; (iv) the current through a 1.0-μH inductor increases from 0 to 4.0 A in 0.25 s. ❚

30.3 Magnetic-Field Energy

Establishing a current in an inductor requires an input of energy, and an inductor carrying a current has energy stored in it. Let's see how this comes about. In Fig. 30.5, an increasing current i in the inductor causes an emf $\mathcal{E}$ between its terminals and a corresponding potential difference V_{ab} between the terminals of the source, with point a at higher potential than point b. Thus the source must be adding energy to the inductor, and the instantaneous power P (rate of transfer of energy into the inductor) is $P = V_{ab}i$.

Energy Stored in an Inductor

We can calculate the total energy input U needed to establish a final current I in an inductor with inductance L if the initial current is zero. We assume that the inductor has zero resistance, so no energy is dissipated within the inductor. Let the current at some instant be i and let its rate of change be di/dt; the current is increasing, so $di/dt > 0$. The voltage between the terminals a and b of the inductor at this instant is $V_{ab} = L\,di/dt$, and the rate P at which energy is being delivered to the inductor (equal to the instantaneous power supplied by the external source) is

$$P = V_{ab}i = Li\frac{di}{dt}$$

The energy dU supplied to the inductor during an infinitesimal time interval dt is $dU = P\,dt$, so

$$dU = Li\,di$$

The total energy U supplied while the current increases from zero to a final value I is

$$U = L\int_0^I i\,di = \tfrac{1}{2}LI^2 \qquad \text{(energy stored in an inductor)} \qquad (30.9)$$

After the current has reached its final steady value I, $di/dt = 0$ and no more energy is input to the inductor. When there is no current, the stored energy U is zero; when the current is I, the energy is $\tfrac{1}{2}LI^2$.

When the current decreases from I to zero, the inductor acts as a source that supplies a total amount of energy $\tfrac{1}{2}LI^2$ to the external circuit. If we interrupt the circuit suddenly by opening a switch or yanking a plug from a wall socket, the current decreases very rapidly, the induced emf is very large, and the energy may be dissipated in an arc across the switch contacts. This large emf is the electrical analog of the large force exerted by a car running into a brick wall and stopping very suddenly.

> **CAUTION** **Energy, resistors, and inductors** It's important not to confuse the behavior of resistors and inductors where energy is concerned (Fig. 30.9). Energy flows into a resistor whenever a current passes through it, whether the current is steady or varying; this energy is dissipated in the form of heat. By contrast, energy flows into an ideal, zero-resistance inductor only when the current in the inductor *increases*. This energy is not dissipated; it is stored in the inductor and released when the current *decreases*. When a steady current flows through an inductor, there is no energy flow in or out.

Magnetic Energy Density

The energy in an inductor is actually stored in the magnetic field within the coil, just as the energy of a capacitor is stored in the electric field between its plates. We can develop relationships for magnetic-field energy analogous to those we obtained for electric-field energy in Section 24.3 [Eqs. (24.9) and (24.11)]. We will concentrate on one simple case, the ideal toroidal solenoid. This system has the advantage that its magnetic field is confined completely to a finite region of space within its core. As in Example 30.3, we assume that the cross-sectional area A is small enough that we can pretend that the magnetic field is uniform over the area. The volume V enclosed by the toroidal solenoid is approximately equal to the circumference $2\pi r$ multiplied by the area A: $V = 2\pi r A$. From Example 30.3, the self-inductance of the toroidal solenoid with vacuum within its coils is

$$L = \frac{\mu_0 N^2 A}{2\pi r}$$

From Eq. (30.9), the energy U stored in the toroidal solenoid when the current is I is

$$U = \tfrac{1}{2}LI^2 = \tfrac{1}{2}\frac{\mu_0 N^2 A}{2\pi r}I^2$$

30.9 A resistor is a device in which energy is irrecoverably dissipated. By contrast, energy stored in a current-carrying inductor can be recovered when the current decreases to zero.

Resistor with current i: energy is *dissipated*.

Inductor with current i: energy is *stored*.

Application **A Magnetic Eruption on the Sun**
This composite of two images of the sun shows a coronal mass ejection, a dramatic event in which about 10^{12} kg (a billion tons) of material from the sun's outer atmosphere is ejected into space at speeds of 500 km/s or faster. Such ejections happen at intervals of a few hours to a few days. These immense eruptions are powered by the energy stored in the sun's magnetic field. Unlike the earth's relatively steady magnetic field, the sun's field is constantly changing, and regions of unusually strong field (and hence unusually high magnetic energy density) frequently form. A coronal mass ejection occurs when the energy stored in such a region is suddenly released.

The magnetic field and therefore this energy are localized in the volume $V = 2\pi r A$ enclosed by the windings. The energy *per unit volume,* or *magnetic energy density,* is $u = U/V$:

$$u = \frac{U}{2\pi r A} = \frac{1}{2}\mu_0 \frac{N^2 I^2}{(2\pi r)^2}$$

We can express this in terms of the magnitude B of the magnetic field inside the toroidal solenoid. From Eq. (28.24) in Example 28.10 (Section 28.7), this is

$$B = \frac{\mu_0 N I}{2\pi r}$$

and so

$$\frac{N^2 I^2}{(2\pi r)^2} = \frac{B^2}{\mu_0^2}$$

When we substitute this into the above equation for u, we finally find the expression for **magnetic energy density** in vacuum:

$$u = \frac{B^2}{2\mu_0} \qquad \text{(magnetic energy density in vacuum)} \qquad (30.10)$$

This is the magnetic analog of the energy per unit volume in an *electric* field in vacuum, $u = \frac{1}{2}\epsilon_0 E^2$, which we derived in Section 24.3. As an example, the energy density in the 1.5-T magnetic field of an MRI scanner (see Section 27.7) is $u = B^2/2\mu_0 = (1.5\ \text{T})^2/(2 \times 4\pi \times 10^{-7}\ \text{T}\cdot\text{m/A}) = 9.0 \times 10^5\ \text{J/m}^3$.

When the material inside the toroid is not vacuum but a material with (constant) magnetic permeability $\mu = K_m\mu_0$, we replace μ_0 by μ in Eq. (30.10). The energy per unit volume in the magnetic field is then

$$u = \frac{B^2}{2\mu} \qquad \text{(magnetic energy density in a material)} \qquad (30.11)$$

30.10 The energy required to fire an automobile spark plug is derived from magnetic-field energy stored in the ignition coil.

Although we have derived Eq. (30.11) only for one special situation, it turns out to be the correct expression for the energy per unit volume associated with *any* magnetic-field configuration in a material with constant permeability. For vacuum, Eq. (30.11) reduces to Eq. (30.10). We will use the expressions for electric-field and magnetic-field energy in Chapter 32 when we study the energy associated with electromagnetic waves.

Magnetic-field energy plays an important role in the ignition systems of gasoline-powered automobiles. A primary coil of about 250 turns is connected to the car's battery and produces a strong magnetic field. This coil is surrounded by a secondary coil with some 25,000 turns of very fine wire. When it is time for a spark plug to fire (see Fig. 20.5 in Section 20.3), the current to the primary coil is interrupted, the magnetic field quickly drops to zero, and an emf of tens of thousands of volts is induced in the secondary coil. The energy stored in the magnetic field thus goes into a powerful pulse of current that travels through the secondary coil to the spark plug, generating the spark that ignites the fuel–air mixture in the engine's cylinders (Fig. 30.10).

Example 30.5 **Storing energy in an inductor**

The electric-power industry would like to find efficient ways to store electrical energy generated during low-demand hours to help meet customer requirements during high-demand hours. Could a large inductor be used? What inductance would be needed to store 1.00 kW·h of energy in a coil carrying a 200-A current?

SOLUTION

IDENTIFY and SET UP: We are given the required amount of stored energy U and the current $I = 200$ A. We use Eq. (30.9) to find the self-inductance L.

EXECUTE: We have $I = 200$ A and $U = 1.00$ kW·h $= (1.00 \times 10^3$ W$)(3600$ s$) = 3.60 \times 10^6$ J. Solving Eq. (30.9) for L, we find

$$L = \frac{2U}{I^2} = \frac{2(3.60 \times 10^6 \text{ J})}{(200 \text{ A})^2} = 180 \text{ H}$$

EVALUATE: The required inductance is more than a million times greater than the self-inductance of the toroidal solenoid of Example 30.3. Conventional wires that are to carry 200 A would have to be of large diameter to keep the resistance low and avoid unacceptable energy losses due to I^2R heating. As a result, a 180-H inductor using conventional wire would be very large (room-size). A superconducting inductor could be much smaller, since the resistance of a superconductor is zero and much thinner wires could be used; but the wires would have to be kept at low temperature to remain superconducting, and maintaining this temperature would itself require energy. This scheme is impractical with present technology.

Test Your Understanding of Section 30.3 The current in a solenoid is reversed in direction while keeping the same magnitude. (a) Does this change the magnetic field within the solenoid? (b) Does this change the magnetic energy density in the solenoid?

30.4 The *R-L* Circuit

Let's look at some examples of the circuit behavior of an inductor. One thing is clear already; an inductor in a circuit makes it difficult for rapid changes in current to occur, thanks to the effects of self-induced emf. Equation (30.7) shows that the greater the rate of change of current di/dt, the greater the self-induced emf and the greater the potential difference between the inductor terminals. This equation, together with Kirchhoff's rules (see Section 26.2), gives us the principles we need to analyze circuits containing inductors.

Mastering**PHYSICS**

ActivPhysics 14.1: The *RL* Circuit

Problem-Solving Strategy 30.1 | Inductors in Circuits

IDENTIFY *the relevant concepts:* An inductor is just another circuit element, like a source of emf, a resistor, or a capacitor. One key difference is that when an inductor is included in a circuit, all the voltages, currents, and capacitor charges are in general functions of time, not constants as they have been in most of our previous circuit analysis. But Kirchhoff's rules (see Section 26.2) are still valid. When the voltages and currents vary with time, Kirchhoff's rules hold at each instant of time.

SET UP *the problem* using the following steps:
1. Follow the procedure described in Problem-Solving Strategy 26.2 (Section 26.2). Draw a circuit diagram and label all quantities, known and unknown. Apply the junction rule immediately so as to express the currents in terms of as few quantities as possible.
2. Determine which quantities are the target variables.

EXECUTE *the solution* as follows:
1. As in Problem-Solving Strategy 26.2, apply Kirchhoff's loop rule to each loop in the circuit.

2. Review the sign rules given in Problem-Solving Strategy 26.2. To get the correct sign for the potential difference between the terminals of an inductor, apply Lenz's law and the sign rule described in Section 30.2 in connection with Eq. (30.7) and Fig. 30.6. In Kirchhoff's loop rule, when we go through an inductor in the *same* direction as the assumed current, we encounter a voltage *drop* equal to $L \, di/dt$, so the corresponding term in the loop equation is $-L \, di/dt$. When we go through an inductor in the *opposite* direction from the assumed current, the potential difference is reversed and the term to use in the loop equation is $+L \, di/dt$.
3. Solve for the target variables.

EVALUATE *your answer:* Check whether your answer is consistent with the behavior of inductors. By Lenz's law, if the current through an inductor is changing, your result should indicate that the potential difference across the inductor opposes the change.

Current Growth in an *R-L* Circuit

We can learn several basic things about inductor behavior by analyzing the circuit of Fig. 30.11. A circuit that includes both a resistor and an inductor, and possibly a source of emf, is called an **R-L circuit.** The inductor helps to prevent rapid changes in current, which can be useful if a steady current is required but the external source has a fluctuating emf. The resistor R may be a separate circuit

30.11 An R-L circuit.

Closing switch S_1 connects the R-L combination in series with a source of emf $\mathcal{E}$.

Closing switch S_2 while opening switch S_1 disconnnects the combination from the source.

element, or it may be the resistance of the inductor windings; every real-life inductor has some resistance unless it is made of superconducting wire. By closing switch S_1, we can connect the R-L combination to a source with constant emf $\mathcal{E}$. (We assume that the source has zero internal resistance, so the terminal voltage equals the emf.)

Suppose both switches are open to begin with, and then at some initial time $t = 0$ we close switch S_1. The current cannot change suddenly from zero to some final value, since di/dt and the induced emf in the inductor would both be infinite. Instead, the current begins to grow at a rate that depends only on the value of L in the circuit.

Let i be the current at some time t after switch S_1 is closed, and let di/dt be its rate of change at that time. The potential difference v_{ab} across the resistor at that time is

$$v_{ab} = iR$$

and the potential difference v_{bc} across the inductor is

$$v_{bc} = L\frac{di}{dt}$$

Note that if the current is in the direction shown in Fig. 30.11 and is increasing, then both v_{ab} and v_{bc} are positive; a is at a higher potential than b, which in turn is at a higher potential than c. (Compare to Figs. 30.6a and c.) We apply Kirchhoff's loop rule, starting at the negative terminal and proceeding counterclockwise around the loop:

$$\mathcal{E} - iR - L\frac{di}{dt} = 0 \tag{30.12}$$

Solving this for di/dt, we find that the rate of increase of current is

$$\frac{di}{dt} = \frac{\mathcal{E} - iR}{L} = \frac{\mathcal{E}}{L} - \frac{R}{L}i \tag{30.13}$$

At the instant that switch S_1 is first closed, $i = 0$ and the potential drop across R is zero. The initial rate of change of current is

$$\left(\frac{di}{dt}\right)_{initial} = \frac{\mathcal{E}}{L}$$

As we would expect, the greater the inductance L, the more slowly the current increases.

As the current increases, the term $(R/L)i$ in Eq. (30.13) also increases, and the *rate* of increase of current given by Eq. (30.13) becomes smaller and smaller. This means that the current is approaching a final, steady-state value I. When the current reaches this value, its rate of increase is zero. Then Eq. (30.13) becomes

$$\left(\frac{di}{dt}\right)_{final} = 0 = \frac{\mathcal{E}}{L} - \frac{R}{L}I \quad \text{and}$$

$$I = \frac{\mathcal{E}}{R}$$

30.12 Graph of i versus t for growth of current in an R-L circuit with an emf in series. The final current is $I = \mathcal{E}/R$; after one time constant τ, the current is $1 - 1/e$ of this value.

Switch S_1 is closed at $t = 0$.

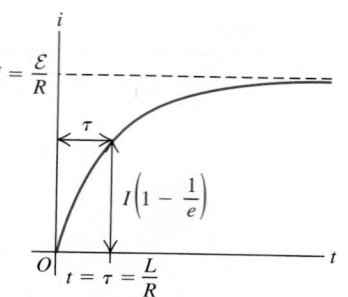

The *final* current I does not depend on the inductance L; it is the same as it would be if the resistance R alone were connected to the source with emf $\mathcal{E}$.

Figure 30.12 shows the behavior of the current as a function of time. To derive the equation for this curve (that is, an expression for current as a function of time), we proceed just as we did for the charging capacitor in Section 26.4. First we rearrange Eq. (30.13) to the form

$$\frac{di}{i - (\mathcal{E}/R)} = -\frac{R}{L}dt$$

This separates the variables, with i on the left side and t on the right. Then we integrate both sides, renaming the integration variables i' and t' so that we can use i and t as the upper limits. (The lower limit for each integral is zero, corresponding to zero current at the initial time $t = 0$.) We get

$$\int_0^i \frac{di'}{i' - (\mathcal{E}/R)} = -\int_0^t \frac{R}{L} dt'$$

$$\ln\left(\frac{i - (\mathcal{E}/R)}{-\mathcal{E}/R}\right) = -\frac{R}{L}t$$

Now we take exponentials of both sides and solve for i. We leave the details for you to work out; the final result is

$$i = \frac{\mathcal{E}}{R}(1 - e^{-(R/L)t}) \qquad \text{(current in an } R\text{-}L \text{ circuit with emf)} \qquad \text{(30.14)}$$

This is the equation of the curve in Fig. 30.12. Taking the derivative of Eq. (30.14), we find

$$\frac{di}{dt} = \frac{\mathcal{E}}{L}e^{-(R/L)t} \qquad \text{(30.15)}$$

At time $t = 0$, $i = 0$ and $di/dt = \mathcal{E}/L$. As $t \to \infty$, $i \to \mathcal{E}/R$ and $di/dt \to 0$, as we predicted.

As Fig. 30.12 shows, the instantaneous current i first rises rapidly, then increases more slowly and approaches the final value $I = \mathcal{E}/R$ asymptotically. At a time equal to L/R, the current has risen to $(1 - 1/e)$, or about 63%, of its final value. The quantity L/R is therefore a measure of how quickly the current builds toward its final value; this quantity is called the **time constant** for the circuit, denoted by τ:

$$\tau = \frac{L}{R} \qquad \text{(time constant for an } R\text{-}L \text{ circuit)} \qquad \text{(30.16)}$$

In a time equal to 2τ, the current reaches 86% of its final value; in 5τ, 99.3%; and in 10τ, 99.995%. (Compare the discussion in Section 26.4 of charging a capacitor of capacitance C that was in series with a resistor of resistance R; the time constant for that situation was the product RC.)

The graphs of i versus t have the same general shape for all values of L. For a given value of R, the time constant τ is greater for greater values of L. When L is small, the current rises rapidly to its final value; when L is large, it rises more slowly. For example, if $R = 100 \; \Omega$ and $L = 10$ H,

$$\tau = \frac{L}{R} = \frac{10 \text{ H}}{100 \; \Omega} = 0.10 \text{ s}$$

and the current increases to about 63% of its final value in 0.10 s. (Recall that $1 \text{ H} = 1 \; \Omega \cdot \text{s}$.) But if $L = 0.010$ H, $\tau = 1.0 \times 10^{-4} \text{ s} = 0.10$ ms, and the rise is much more rapid.

Energy considerations offer us additional insight into the behavior of an R-L circuit. The instantaneous rate at which the source delivers energy to the circuit is $P = \mathcal{E}i$. The instantaneous rate at which energy is dissipated in the resistor is i^2R, and the rate at which energy is stored in the inductor is $iv_{bc} = Li \, di/dt$ [or, equivalently, $(d/dt)(\frac{1}{2}Li^2) = Li \, di/dt$]. When we multiply Eq. (30.12) by i and rearrange, we find

$$\mathcal{E}i = i^2R + Li\frac{di}{dt} \qquad \text{(30.17)}$$

Of the power $\mathcal{E}i$ supplied by the source, part (i^2R) is dissipated in the resistor and part $(Li \, di/dt)$ goes to store energy in the inductor. This discussion is completely analogous to our power analysis for a charging capacitor, given at the end of Section 26.4.

Example 30.6 Analyzing an *R-L* circuit

A sensitive electronic device of resistance $R = 175 \ \Omega$ is to be connected to a source of emf (of negligible internal resistance) by a switch. The device is designed to operate with a 36-mA current, but to avoid damage to the device, the current can rise to no more than 4.9 mA in the first 58 μs after the switch is closed. An inductor is therefore connected in series with the device, as in Fig. 30.11; the switch in question is S_1. (a) What is the required source emf $\mathcal{E}$? (b) What is the required inductance L? (c) What is the *R-L* time constant τ?

SOLUTION

IDENTIFY and SET UP: This problem concerns current and current growth in an *R-L* circuit, so we can use the ideas of this section. Figure 30.12 shows the current i versus the time t that has elapsed since closing S_1. The graph shows that the final current is $I = \mathcal{E}/R$; we are given $R = 175 \ \Omega$, so the emf is determined by the requirement that the final current be $I = 36$ mA. The other requirement is that the current be no more than $i = 4.9$ mA at $t = 58 \ \mu$s; to satisfy this, we use Eq. (30.14) for the current as a function of time and solve for the inductance, which is the only unknown quantity. Equation (30.16) then tells us the time constant.

EXECUTE: (a) We solve $I = \mathcal{E}/R$ for $\mathcal{E}$:

$$\mathcal{E} = IR = (0.036 \ \text{A})(175 \ \Omega) = 6.3 \ \text{V}$$

(b) To find the required inductance, we solve Eq. (30.14) for L. First we multiply through by $(-R/\mathcal{E})$ and then add 1 to both sides to obtain

$$1 - \frac{iR}{\mathcal{E}} = e^{-(R/L)t}$$

Then we take natural logs of both sides, solve for L, and insert the numbers:

$$L = \frac{-Rt}{\ln(1 - iR/\mathcal{E})}$$

$$= \frac{-(175 \ \Omega)(58 \times 10^{-6} \ \text{s})}{\ln[1 - (4.9 \times 10^{-3} \ \text{A})(175 \ \Omega)/(6.3 \ \text{V})]} = 69 \ \text{mH}$$

(c) From Eq. (30.16),

$$\tau = \frac{L}{R} = \frac{69 \times 10^{-3} \ \text{H}}{175 \ \Omega} = 3.9 \times 10^{-4} \ \text{s} = 390 \ \mu\text{s}$$

EVALUATE: Note that 58 μs is much less than the time constant. In 58 μs the current builds up from zero to 4.9 mA, a small fraction of its final value of 36 mA; after 390 μs the current equals $(1 - 1/e)$ of its final value, or about $(0.63)(36 \ \text{mA}) = 23$ mA.

Current Decay in an *R-L* Circuit

30.13 Graph of i versus t for decay of current in an *R-L* circuit. After one time constant τ, the current is $1/e$ of its initial value.

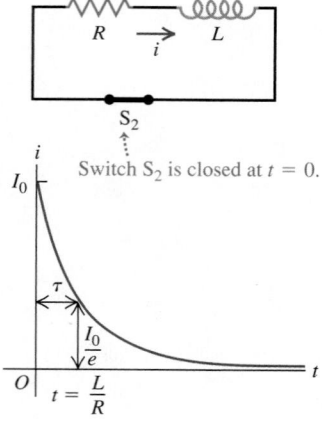

Now suppose switch S_1 in the circuit of Fig. 30.11 has been closed for a while and the current has reached the value I_0. Resetting our stopwatch to redefine the initial time, we close switch S_2 at time $t = 0$, bypassing the battery. (At the same time we should open S_1 to save the battery from ruin.) The current through R and L does not instantaneously go to zero but decays smoothly, as shown in Fig. 30.13. The Kirchhoff's-rule loop equation is obtained from Eq. (30.12) by simply omitting the $\mathcal{E}$ term. We challenge you to retrace the steps in the above analysis and show that the current i varies with time according to

$$i = I_0 e^{-(R/L)t} \tag{30.18}$$

where I_0 is the initial current at time $t = 0$. The time constant, $\tau = L/R$, is the time for current to decrease to $1/e$, or about 37%, of its original value. In time 2τ it has dropped to 13.5%, in time 5τ to 0.67%, and in 10τ to 0.0045%.

The energy that is needed to maintain the current during this decay is provided by the energy stored in the magnetic field of the inductor. The detailed energy analysis is simpler this time. In place of Eq. (30.17) we have

$$0 = i^2 R + Li\frac{di}{dt} \tag{30.19}$$

In this case, $Li \ di/dt$ is negative; Eq. (30.19) shows that the energy stored in the inductor *decreases* at a rate equal to the rate of dissipation of energy $i^2 R$ in the resistor.

This entire discussion should look familiar; the situation is very similar to that of a charging and discharging capacitor, analyzed in Section 26.4. It would be a good idea to compare that section with our discussion of the *R-L* circuit.

Example 30.7   **Energy in an *R-L* circuit**

When the current in an *R-L* circuit is decaying, what fraction of the original energy stored in the inductor has been dissipated after 2.3 time constants?

SOLUTION

IDENTIFY and SET UP: This problem concerns current decay in an *R-L* circuit as well as the relationship between the current in an inductor and the amount of stored energy. The current i at any time t is given by Eq. (30.18); the stored energy associated with this current is given by Eq. (30.9), $U = \frac{1}{2}Li^2$.

EXECUTE: From Eq. (30.18), the current i at any time t is

$$i = I_0 e^{-(R/L)t}$$

We substitute this into $U = \frac{1}{2}Li^2$ to obtain an expression for the stored energy at any time:

$$U = \frac{1}{2}LI_0^2 e^{-2(R/L)t} = U_0 e^{-2(R/L)t}$$

where $U_0 = \frac{1}{2}LI_0^2$ is the energy at the initial time $t = 0$. When $t = 2.3\tau = 2.3L/R$, we have

$$U = U_0 e^{-2(2.3)} = U_0 e^{-4.6} = 0.010U_0$$

That is, only 0.010 or 1.0% of the energy initially stored in the inductor remains, so 99.0% has been dissipated in the resistor.

EVALUATE: To get a sense of what this result means, consider the *R-L* circuit we analyzed in Example 30.6, for which $\tau = 390 \ \mu s$. With $L = 69$ mH and $I_0 = 36$ mA, we have $U_0 = \frac{1}{2}LI_0^2 = \frac{1}{2}(0.069 \text{ H})(0.036 \text{ A})^2 = 4.5 \times 10^{-5}$ J. Of this, 99.0% or 4.4×10^{-5} J is dissipated in $2.3(390 \ \mu s) = 9.0 \times 10^{-4}$ s = 0.90 ms. In other words, this circuit can be almost completely powered off (or powered on) in 0.90 ms, so the minimum time for a complete on–off cycle is 1.8 ms. Even shorter cycle times are required for many purposes, such as in fast switching networks for telecommunications. In such cases a smaller time constant $\tau = L/R$ is needed.

Test Your Understanding of Section 30.4   (a) In Fig. 30.11, what are the algebraic signs of the potential differences v_{ab} and v_{bc} when switch S_1 is closed and switch S_2 is open? (i) $v_{ab} > 0$, $v_{bc} > 0$; (ii) $v_{ab} > 0$, $v_{bc} < 0$; (iii) $v_{ab} < 0$, $v_{bc} > 0$; (iv) $v_{ab} < 0$, $v_{bc} < 0$. (b) What are the signs of v_{ab} and v_{bc} when S_1 is open, S_2 is closed, and current is flowing in the direction shown? (i) $v_{ab} > 0$, $v_{bc} > 0$; (ii) $v_{ab} > 0$, $v_{bc} < 0$; (iii) $v_{ab} < 0$, $v_{bc} > 0$; (iv) $v_{ab} < 0$, $v_{bc} < 0$.   ❙

30.5 The *L-C* Circuit

ActivPhysics 14.2: AC Circuits: The *RLC* Oscillator (Questions 1–6)

A circuit containing an inductor and a capacitor shows an entirely new mode of behavior, characterized by *oscillating* current and charge. This is in sharp contrast to the *exponential* approach to a steady-state situation that we have seen with both *R-C* and *R-L* circuits. In the **L-C circuit** in Fig. 30.14a we charge the capacitor to a potential difference V_m and initial charge $Q = CV_m$ on its left-hand plate and then close the switch. What happens?

The capacitor begins to discharge through the inductor. Because of the induced emf in the inductor, the current cannot change instantaneously; it starts at zero and eventually builds up to a maximum value I_m. During this buildup the capacitor is discharging. At each instant the capacitor potential equals the induced emf, so as the capacitor discharges, the *rate of change* of current decreases. When the capacitor potential becomes zero, the induced emf is also zero, and the current has leveled off at its maximum value I_m. Figure 30.14b shows this situation; the capacitor has completely discharged. The potential difference between its terminals (and those of the inductor) has decreased to zero, and the current has reached its maximum value I_m.

During the discharge of the capacitor, the increasing current in the inductor has established a magnetic field in the space around it, and the energy that was initially stored in the capacitor's electric field is now stored in the inductor's magnetic field.

Although the capacitor is completely discharged in Fig. 30.14b, the current persists (it cannot change instantaneously), and the capacitor begins to charge with polarity opposite to that in the initial state. As the current decreases, the magnetic field also decreases, inducing an emf in the inductor in the *same* direction as the current; this slows down the decrease of the current. Eventually, the

30.14 In an oscillating L-C circuit, the charge on the capacitor and the current through the inductor both vary sinusoidally with time. Energy is transferred between magnetic energy in the inductor (U_B) and electric energy in the capacitor (U_E). As in simple harmonic motion, the total energy E remains constant. (Compare Fig. 14.14 in Section 14.3.)

current and the magnetic field reach zero, and the capacitor has been charged in the sense *opposite* to its initial polarity (Fig. 30.14c), with potential difference $-V_m$ and charge $-Q$ on its left-hand plate.

The process now repeats in the reverse direction; a little later, the capacitor has again discharged, and there is a current in the inductor in the opposite direction (Fig. 30.14d). Still later, the capacitor charge returns to its original value (Fig. 30.14a), and the whole process repeats. If there are no energy losses, the charges on the capacitor continue to oscillate back and forth indefinitely. This process is called an **electrical oscillation.**

From an energy standpoint the oscillations of an electrical circuit transfer energy from the capacitor's electric field to the inductor's magnetic field and back. The *total* energy associated with the circuit is constant. This is analogous to the transfer of energy in an oscillating mechanical system from potential energy to kinetic energy and back, with constant total energy. As we will see, this analogy goes much further.

Electrical Oscillations in an *L-C* Circuit

To study the flow of charge in detail, we proceed just as we did for the R-L circuit. Figure 30.15 shows our definitions of q and i.

30.15 Applying Kirchhoff's loop rule to the L-C circuit. The direction of travel around the loop in the loop equation is shown. Just after the circuit is completed and the capacitor first begins to discharge, as in Fig. 30.14a, the current is negative (opposite to the direction shown).

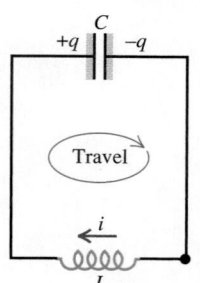

CAUTION **Positive current in an *L-C* circuit** After examining Fig. 30.14, the positive direction for current in Fig. 30.15 may seem backward to you. In fact we've chosen this direction to simplify the relationship between current and capacitor charge. We define the current at each instant to be $i = dq/dt$, the rate of change of the charge on the left-hand capacitor plate. Hence if the capacitor is initially charged and begins to discharge as in Figs. 30.14a and 30.14b, then $dq/dt < 0$ and the initial current i is negative; the direction of the current is then opposite to the (positive) direction shown in Fig. 30.15. ▮

We apply Kirchhoff's loop rule to the circuit in Fig. 30.15. Starting at the lower-right corner of the circuit and adding voltages as we go clockwise around the loop, we obtain

$$-L\frac{di}{dt} - \frac{q}{C} = 0$$

Since $i = dq/dt$, it follows that $di/dt = d^2q/dt^2$. We substitute this expression into the above equation and divide by $-L$ to obtain

$$\frac{d^2q}{dt^2} + \frac{1}{LC}q = 0 \qquad (L\text{-}C \text{ circuit}) \qquad \text{(30.20)}$$

Equation (30.20) has exactly the same form as the equation we derived for simple harmonic motion in Section 14.2, Eq. (14.4). That equation is $d^2x/dt^2 = -(k/m)x$, or

$$\frac{d^2x}{dt^2} + \frac{k}{m}x = 0$$

(You should review Section 14.2 before going on with this discussion.) In the *L-C* circuit the capacitor charge q plays the role of the displacement x, and the current $i = dq/dt$ is analogous to the particle's velocity $v_x = dx/dt$. The inductance L is analogous to the mass m, and the reciprocal of the capacitance, $1/C$, is analogous to the force constant k.

Pursuing this analogy, we recall that the angular frequency $\omega = 2\pi f$ of the harmonic oscillator is equal to $(k/m)^{1/2}$, and the position is given as a function of time by Eq. (14.13),

$$x = A\cos(\omega t + \phi)$$

where the amplitude A and the phase angle ϕ depend on the initial conditions. In the analogous electrical situation the capacitor charge q is given by

$$q = Q\cos(\omega t + \phi) \qquad \text{(30.21)}$$

and the angular frequency ω of oscillation is given by

$$\omega = \sqrt{\frac{1}{LC}} \qquad \begin{array}{l}\text{(angular frequency of oscillation}\\ \text{in an } L\text{-}C \text{ circuit)}\end{array} \qquad \text{(30.22)}$$

You should verify that Eq. (30.21) satisfies the loop equation, Eq. (30.20), when ω has the value given by Eq. (30.22). In doing this, you will find that the instantaneous current $i = dq/dt$ is given by

$$i = -\omega Q\sin(\omega t + \phi) \qquad \text{(30.23)}$$

Thus the charge and current in an *L-C* circuit oscillate sinusoidally with time, with an angular frequency determined by the values of L and C. The ordinary frequency f, the number of cycles per second, is equal to $\omega/2\pi$ as always. The constants Q and ϕ in Eqs. (30.21) and (30.23) are determined by the initial conditions. If at time $t = 0$ the left-hand capacitor plate in Fig. 30.15 has its maximum charge Q and the current i is zero, then $\phi = 0$. If $q = 0$ at time $t = 0$, then $\phi = \pm\pi/2$ rad.

Energy in an *L-C* Circuit

We can also analyze the *L-C* circuit using an energy approach. The analogy to simple harmonic motion is equally useful here. In the mechanical problem a body with mass m is attached to a spring with force constant k. Suppose we displace the body a distance A from its equilibrium position and release it from rest at time $t = 0$. The kinetic energy of the system at any later time is $\frac{1}{2}mv_x^2$, and its elastic potential energy is $\frac{1}{2}kx^2$. Because the system is conservative, the sum of these energies equals the initial energy of the system, $\frac{1}{2}kA^2$. We find the velocity v_x at any position x just as we did in Section 14.3, Eq. (14.22):

$$v_x = \pm \sqrt{\frac{k}{m}} \sqrt{A^2 - x^2} \qquad (30.24)$$

The L-C circuit is also a conservative system. Again let Q be the maximum capacitor charge. The magnetic-field energy $\frac{1}{2}Li^2$ in the inductor at any time corresponds to the kinetic energy $\frac{1}{2}mv^2$ of the oscillating body, and the electric-field energy $q^2/2C$ in the capacitor corresponds to the elastic potential energy $\frac{1}{2}kx^2$ of the spring. The sum of these energies equals the total energy $Q^2/2C$ of the system:

$$\frac{1}{2}Li^2 + \frac{q^2}{2C} = \frac{Q^2}{2C} \qquad (30.25)$$

The total energy in the L-C circuit is *constant;* it oscillates between the magnetic and the electric forms, just as the constant total mechanical energy in simple harmonic motion is constant and oscillates between the kinetic and potential forms.

Solving Eq. (30.25) for i, we find that when the charge on the capacitor is q, the current i is

$$i = \pm \sqrt{\frac{1}{LC}} \sqrt{Q^2 - q^2} \qquad (30.26)$$

You can verify this equation by substituting q from Eq. (30.21) and i from Eq. (30.23). Comparing Eqs. (30.24) and (30.26), we see that current $i = dq/dt$ and charge q are related in the same way as are velocity $v_x = dx/dt$ and position x in the mechanical problem.

Table 30.1 summarizes the analogies between simple harmonic motion and L-C circuit oscillations. The striking parallels shown there are so close that we can solve complicated mechanical and acoustical problems by setting up analogous electric circuits and measuring the currents and voltages that correspond to the mechanical and acoustical quantities to be determined. This is the basic principle of many analog computers. This analogy can be extended to *damped* oscillations, which we consider in the next section. In Chapter 31 we will extend the analogy further to include *forced* electrical oscillations, which occur in all alternating-current circuits.

Table 30.1 Oscillation of a Mass-Spring System Compared with Electrical Oscillation in an *L-C* Circuit

Mass-Spring System

Kinetic energy $= \frac{1}{2}mv_x^2$

Potential energy $= \frac{1}{2}kx^2$

$\frac{1}{2}mv_x^2 + \frac{1}{2}kx^2 = \frac{1}{2}kA^2$

$v_x = \pm \sqrt{k/m} \sqrt{A^2 - x^2}$

$v_x = dx/dt$

$\omega = \sqrt{\dfrac{k}{m}}$

$x = A\cos(\omega t + \phi)$

Inductor-Capacitor Circuit

Magnetic energy $= \frac{1}{2}Li^2$

Electric energy $= q^2/2C$

$\frac{1}{2}Li^2 + q^2/2C = Q^2/2C$

$i = \pm \sqrt{1/LC} \sqrt{Q^2 - q^2}$

$i = dq/dt$

$\omega = \sqrt{\dfrac{1}{LC}}$

$q = Q\cos(\omega t + \phi)$

Example 30.8 An oscillating circuit

A 300-V dc power supply is used to charge a 25-μF capacitor. After the capacitor is fully charged, it is disconnected from the power supply and connected across a 10-mH inductor. The resistance in the circuit is negligible. (a) Find the frequency and period of oscillation of the circuit. (b) Find the capacitor charge and the circuit current 1.2 ms after the inductor and capacitor are connected.

SOLUTION

IDENTIFY and SET UP: Our target variables are the oscillation frequency f and period T, as well as the charge q and current i at a particular time t. We are given the capacitance C and the inductance L, from which we can calculate the frequency and period using Eq. (30.22). We find the charge and current using Eqs. (30.21) and (30.23). Initially the capacitor is fully charged and the current is zero, as in Fig. 30.14a, so the phase angle is $\phi = 0$ [see the discussion that follows Eq. (30.23)].

EXECUTE: (a) The natural angular frequency is

$$\omega = \sqrt{\frac{1}{LC}} = \sqrt{\frac{1}{(10 \times 10^{-3}\ \text{H})(25 \times 10^{-6}\ \text{F})}}$$

$$= 2.0 \times 10^3\ \text{rad/s}$$

The frequency f and period T are then

$$f = \frac{\omega}{2\pi} = \frac{2.0 \times 10^3\ \text{rad/s}}{2\pi\ \text{rad/cycle}} = 320\ \text{Hz}$$

$$T = \frac{1}{f} = \frac{1}{320\ \text{Hz}} = 3.1 \times 10^{-3}\ \text{s} = 3.1\ \text{ms}$$

(b) Since the period of the oscillation is $T = 3.1$ ms, $t = 1.2$ ms equals $0.38T$; this corresponds to a situation intermediate between Fig. 30.14b ($t = T/4$) and Fig. 30.14c ($t = T/2$). Comparing those figures with Fig. 30.15, we expect the capacitor charge q to be negative (that is, there will be negative charge on the left-hand plate of the capacitor) and the current i to be negative as well (that is, current will flow counterclockwise).

To find the value of q, we use Eq. (30.21), $q = Q\cos(\omega t + \phi)$. The charge is maximum at $t = 0$, so $\phi = 0$ and $Q = C\mathcal{E} = (25 \times 10^{-6}\text{ F})(300\text{ V}) = 7.5 \times 10^{-3}$ C. Hence Eq. (30.21) becomes

$$q = (7.5 \times 10^{-3}\text{ C})\cos\omega t$$

At time $t = 1.2 \times 10^{-3}$ s,

$$\omega t = (2.0 \times 10^3\text{ rad/s})(1.2 \times 10^{-3}\text{ s}) = 2.4\text{ rad}$$
$$q = (7.5 \times 10^{-3}\text{ C})\cos(2.4\text{ rad}) = -5.5 \times 10^{-3}\text{ C}$$

From Eq. (30.23), the current i at any time is $i = -\omega Q\sin\omega t$. At $t = 1.2 \times 10^{-3}$ s,

$$i = -(2.0 \times 10^3\text{ rad/s})(7.5 \times 10^{-3}\text{ C})\sin(2.4\text{ rad}) = -10\text{ A}$$

EVALUATE: The signs of q and i are both negative, as predicted.

Example 30.9 **Energy in an oscillating circuit**

For the *L-C* circuit of Example 30.8, find the magnetic and electric energies (a) at $t = 0$ and (b) at $t = 1.2$ ms.

SOLUTION

IDENTIFY and SET UP: We must calculate the magnetic energy U_B (stored in the inductor) and the electric energy U_E (stored in the capacitor) at two times during the *L-C* circuit oscillation. From Example 30.8 we know the values of the capacitor charge q and circuit current i for both times. We use them to calculate $U_B = \frac{1}{2}Li^2$ and $U_E = q^2/2C$.

EXECUTE: (a) At $t = 0$ there is no current and $q = Q$. Hence there is no magnetic energy, and all the energy in the circuit is in the form of electric energy in the capacitor:

$$U_B = \tfrac{1}{2}Li^2 = 0 \qquad U_E = \frac{Q^2}{2C} = \frac{(7.5 \times 10^{-3}\text{ C})^2}{2(25 \times 10^{-6}\text{ F})} = 1.1\text{ J}$$

(b) From Example 30.8, at $t = 1.2$ ms we have $i = -10$ A and $q = -5.5 \times 10^{-3}$ C. Hence

$$U_B = \tfrac{1}{2}Li^2 = \tfrac{1}{2}(10 \times 10^{-3}\text{ H})(-10\text{ A})^2 = 0.5\text{ J}$$
$$U_E = \frac{q^2}{2C} = \frac{(-5.5 \times 10^{-3}\text{ C})^2}{2(25 \times 10^{-6}\text{ F})} = 0.6\text{ J}$$

EVALUATE: The magnetic and electric energies are the same at $t = 3T/8 = 0.375T$, halfway between the situations in Figs. 30.14b and 30.14c. We saw in Example 30.8 that the time considered in part (b), $t = 1.2$ ms, equals $0.38T$; this is slightly later than $0.375T$, so U_B is slightly less than U_E. At *all* times the *total* energy $E = U_B + U_E$ has the same value, 1.1 J. An *L-C* circuit without resistance is a conservative system; no energy is dissipated.

Test Your Understanding of Section 30.5   One way to think about the energy stored in an *L-C* circuit is to say that the circuit elements do positive or negative work on the charges that move back and forth through the circuit. (a) Between stages (a) and (b) in Fig. 30.14, does the capacitor do positive work or negative work on the charges? (b) What kind of force (electric or magnetic) does the capacitor exert on the charges to do this work? (c) During this process, does the inductor do positive or negative work on the charges? (d) What kind of force (electric or magnetic) does the inductor exert on the charges?

30.6 The *L-R-C* Series Circuit

In our discussion of the *L-C* circuit we assumed that there was no *resistance* in the circuit. This is an idealization, of course; every real inductor has resistance in its windings, and there may also be resistance in the connecting wires. Because of resistance, the electromagnetic energy in the circuit is dissipated and converted to other forms, such as internal energy of the circuit materials. Resistance in an electric circuit is analogous to friction in a mechanical system.

Suppose an inductor with inductance L and a resistor of resistance R are connected in series across the terminals of a charged capacitor, forming an **L-R-C series circuit.** As before, the capacitor starts to discharge as soon as the circuit is completed. But because of i^2R losses in the resistor, the magnetic-field energy acquired by the inductor when the capacitor is completely discharged is *less* than the original electric-field energy of the capacitor. In the same way, the energy of the capacitor when the magnetic field has decreased to zero is still smaller, and so on.

If the resistance R is relatively small, the circuit still oscillates, but with **damped harmonic motion** (Fig. 30.16a), and we say that the circuit is

Mastering**PHYSICS**

ActivPhysics 14.2: AC Circuits: The *RLC* Oscillator (Questions 7–10)

30.16 Graphs of capacitor charge as a function of time in an *L-R-C* series circuit with initial charge *Q*.

(a) Underdamped circuit (small resistance *R*)

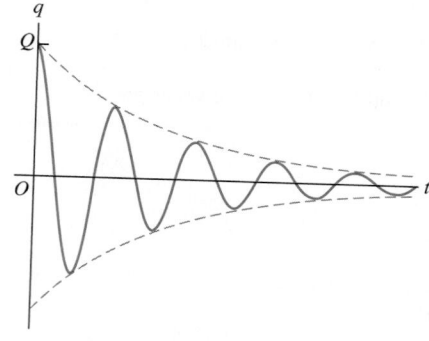

(b) Critically damped circuit (larger resistance *R*)

(c) Overdamped circuit (very large resistance *R*)

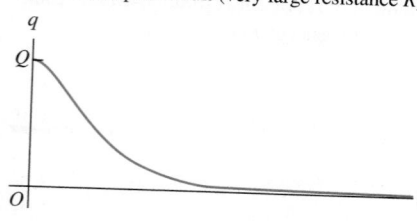

30.17 An *L-R-C* series circuit.

When switch S is in this position, the emf charges the capacitor.

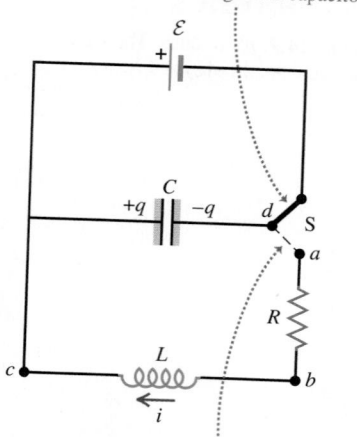

When switch S is moved to this position, the capacitor discharges through the resistor and inductor.

underdamped. If we increase *R*, the oscillations die out more rapidly. When *R* reaches a certain value, the circuit no longer oscillates; it is **critically damped** (Fig. 30.16b). For still larger values of *R*, the circuit is **overdamped** (Fig. 30.16c), and the capacitor charge approaches zero even more slowly. We used these same terms to describe the behavior of the analogous mechanical system, the damped harmonic oscillator, in Section 14.7.

Analyzing an *L-R-C* Series Circuit

To analyze *L-R-C* series circuit behavior in detail, we consider the circuit shown in Fig. 30.17. It is like the *L-C* circuit of Fig. 30.15 except for the added resistor *R*; we also show the source that charges the capacitor initially. The labeling of the positive senses of *q* and *i* are the same as for the *L-C* circuit.

First we close the switch in the upward position, connecting the capacitor to a source of emf $\mathcal{E}$ for a long enough time to ensure that the capacitor acquires its final charge $Q = C\mathcal{E}$ and any initial oscillations have died out. Then at time $t = 0$ we flip the switch to the downward position, removing the source from the circuit and placing the capacitor in series with the resistor and inductor. Note that the initial current is negative, opposite to the direction of *i* shown in Fig. 30.17.

To find how *q* and *i* vary with time, we apply Kirchhoff's loop rule. Starting at point *a* and going around the loop in the direction *abcda*, we obtain the equation

$$-iR - L\frac{di}{dt} - \frac{q}{C} = 0$$

Replacing *i* with dq/dt and rearranging, we get

$$\frac{d^2q}{dt^2} + \frac{R}{L}\frac{dq}{dt} + \frac{1}{LC}q = 0 \tag{30.27}$$

Note that when $R = 0$, this reduces to Eq. (30.20) for an *L-C* circuit.

There are general methods for obtaining solutions of Eq. (30.27). The form of the solution is different for the underdamped (small *R*) and overdamped (large *R*) cases. When R^2 is less than $4L/C$, the solution has the form

$$q = Ae^{-(R/2L)t}\cos\left(\sqrt{\frac{1}{LC} - \frac{R^2}{4L^2}}\,t + \phi\right) \tag{30.28}$$

where *A* and ϕ are constants. We invite you to take the first and second derivatives of this function and show by direct substitution that it does satisfy Eq. (30.27).

This solution corresponds to the *underdamped* behavior shown in Fig. 30.16a; the function represents a sinusoidal oscillation with an exponentially decaying amplitude. (Note that the exponential factor $e^{-(R/2L)t}$ is *not* the same as the factor $e^{-(R/L)t}$ that we encountered in describing the *R-L* circuit in Section 30.4.) When $R = 0$, Eq. (30.28) reduces to Eq. (30.21) for the oscillations in an *L-C* circuit. If *R* is not zero, the angular frequency of the oscillation is *less* than $1/(LC)^{1/2}$ because of the term containing *R*. The angular frequency ω' of the damped oscillations is given by

$$\omega' = \sqrt{\frac{1}{LC} - \frac{R^2}{4L^2}} \quad \text{(underdamped } L\text{-}R\text{-}C \text{ series circuit)} \tag{30.29}$$

When $R = 0$, this reduces to Eq. (30.22), $\omega = (1/LC)^{1/2}$. As *R* increases, ω' becomes smaller and smaller. When $R^2 = 4L/C$, the quantity under the radical becomes zero; the system no longer oscillates, and the case of *critical damping* (Fig. 30.16b) has been reached. For still larger values of *R* the system behaves as in Fig. 30.16c. In this case the circuit is *overdamped*, and *q* is given as a function of time by the sum of two decreasing exponential functions.

In the *underdamped* case the phase constant ϕ in the cosine function of Eq. (30.28) provides for the possibility of both an initial charge and an initial current at time $t = 0$, analogous to an underdamped harmonic oscillator given both an initial displacement and an initial velocity (see Exercise 30.40).

We emphasize once more that the behavior of the *L-R-C* series circuit is completely analogous to that of the damped harmonic oscillator studied in Section 14.7. We invite you to verify, for example, that if you start with Eq. (14.41) and substitute q for x, L for m, $1/C$ for k, and R for the damping constant b, the result is Eq. (30.27). Similarly, the cross-over point between underdamping and overdamping occurs at $b^2 = 4km$ for the mechanical system and at $R^2 = 4L/C$ for the electrical one. Can you find still other aspects of this analogy?

The practical applications of the *L-R-C* series circuit emerge when we include a sinusoidally varying source of emf in the circuit. This is analogous to the *forced oscillations* that we discussed in Section 14.7, and there are analogous *resonance* effects. Such a circuit is called an *alternating-current (ac) circuit;* the analysis of ac circuits is the principal topic of the next chapter.

Example 30.10 An underdamped *L-R-C* series circuit

What resistance R is required (in terms of L and C) to give an *L-R-C* series circuit a frequency that is one-half the undamped frequency?

$$\sqrt{\frac{1}{LC} - \frac{R^2}{4L^2}} = \frac{1}{2}\sqrt{\frac{1}{LC}}$$

SOLUTION

IDENTIFY and SET UP: This problem concerns an underdamped *L-R-C* series circuit (Fig. 30.16a). We want just enough resistance to reduce the oscillation frequency to one-half of the undamped value. Equation (30.29) gives the angular frequency ω' of an underdamped *L-R-C* series circuit; Eq. (30.22) gives the angular frequency ω of an undamped *L-C* circuit. We use these two equations to solve for R.

EXECUTE: From Eqs. (30.29) and (30.22), the requirement $\omega' = \omega/2$ yields

We square both sides and solve for R:

$$R = \sqrt{\frac{3L}{C}}$$

For example, adding 35 Ω to the circuit of Example 30.8 ($L = 10$ mH, $C = 25\ \mu\text{F}$) would reduce the frequency from 320 Hz to 160 Hz.

EVALUATE: The circuit becomes critically damped with no oscillations when $R = \sqrt{4L/C}$. Our result for R is smaller than that, as it should be; we want the circuit to be *under*damped.

Test Your Understanding of Section 30.6 An *L-R-C* series circuit includes a 2.0-Ω resistor. At $t = 0$ the capacitor charge is 2.0 μC. For which of the following values of the inductance and capacitance will the charge on the capacitor *not* oscillate? (i) $L = 3.0\ \mu\text{H}$, $C = 6.0\ \mu\text{F}$; (ii) $L = 6.0\ \mu\text{H}$, $C = 3.0\ \mu\text{F}$; (iii) $L = 3.0\ \mu\text{H}$, $C = 3.0\ \mu\text{F}$.

Mutual inductance: When a changing current i_1 in one circuit causes a changing magnetic flux in a second circuit, an emf $\mathcal{E}_2$ is induced in the second circuit. Likewise, a changing current i_2 in the second circuit induces an emf $\mathcal{E}_1$ in the first circuit. If the circuits are coils of wire with N_1 and N_2 turns, the mutual inductance M can be expressed in terms of the average flux Φ_{B2} through each turn of coil 2 caused by the current i_1 in coil 1, or in terms of the average flux Φ_{B1} through each turn of coil 1 caused by the current i_2 in coil 2. The SI unit of mutual inductance is the henry, abbreviated H. (See Examples 30.1 and 30.2.)

$$\mathcal{E}_2 = -M\frac{di_1}{dt} \quad \text{and}$$

$$\mathcal{E}_1 = -M\frac{di_2}{dt} \tag{30.4}$$

$$M = \frac{N_2\Phi_{B2}}{i_1} = \frac{N_1\Phi_{B1}}{i_2} \tag{30.5}$$

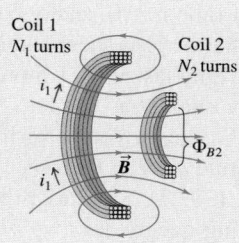

Self-inductance: A changing current i in any circuit causes a self-induced emf $\mathcal{E}$. The inductance (or self-inductance) L depends on the geometry of the circuit and the material surrounding it. The inductance of a coil of N turns is related to the average flux Φ_B through each turn caused by the current i in the coil. An inductor is a circuit device, usually including a coil of wire, intended to have a substantial inductance. (See Examples 30.3 and 30.4.)

$$\mathcal{E} = -L\frac{di}{dt} \tag{30.7}$$

$$L = \frac{N\Phi_B}{i} \tag{30.6}$$

Magnetic-field energy: An inductor with inductance L carrying current I has energy U associated with the inductor's magnetic field. The magnetic energy density u (energy per unit volume) is proportional to the square of the magnetic field magnitude. (See Example 30.5.)

$$U = \tfrac{1}{2}LI^2 \tag{30.9}$$

$$u = \frac{B^2}{2\mu_0} \quad \text{(in vacuum)} \tag{30.10}$$

$$u = \frac{B^2}{2\mu} \quad \begin{array}{l}\text{(in a material}\\\text{with magnetic}\\\text{permeability } \mu)\end{array} \tag{30.11}$$

Stored energy
$U = \tfrac{1}{2}LI^2$

Energy density
$u = B^2/2\mu_0$

R-L circuits: In a circuit containing a resistor R, an inductor L, and a source of emf, the growth and decay of current are exponential. The time constant τ is the time required for the current to approach within a fraction $1/e$ of its final value. (See Examples 30.6 and 30.7.)

$$\tau = \frac{L}{R} \tag{30.16}$$

L-C circuits: A circuit that contains inductance L and capacitance C undergoes electrical oscillations with an angular frequency ω that depends on L and C. This is analogous to a mechanical harmonic oscillator, with inductance L analogous to mass m, the reciprocal of capacitance $1/C$ to force constant k, charge q to displacement x, and current i to velocity v_x. (See Examples 30.8 and 30.9.)

$$\omega = \sqrt{\frac{1}{LC}} \tag{30.22}$$

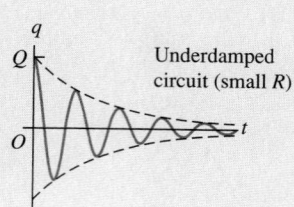

L-R-C series circuits: A circuit that contains inductance, resistance, and capacitance undergoes damped oscillations for sufficiently small resistance. The frequency ω' of damped oscillations depends on the values of L, R, and C. As R increases, the damping increases; if R is greater than a certain value, the behavior becomes overdamped and no longer oscillates. (See Example 30.10.)

$$\omega' = \sqrt{\frac{1}{LC} - \frac{R^2}{4L^2}} \tag{30.29}$$

BRIDGING PROBLEM | Analyzing an *L-C* Circuit

An *L-C* circuit consists of a 60.0-mH inductor and a 250-μF capacitor. The initial charge on the capacitor is 6.00 μC, and the initial current in the inductor is 0.400 mA. (a) What is the maximum energy stored in the inductor? (b) What is the maximum current in the inductor? (c) What is the maximum voltage across the capacitor? (d) When the current in the inductor has half its maximum value, what is the energy stored in the inductor and the voltage across the capacitor?

SOLUTION GUIDE

See MasteringPhysics® study area for a Video Tutor solution.

IDENTIFY and SET UP:

1. An *L-C* circuit is a conservative system because there is no resistance to dissipate energy. The energy oscillates between electric energy in the capacitor and magnetic energy stored in the inductor.

2. Which key equations are needed to describe the capacitor? To describe the inductor?

EXECUTE:

3. Find the initial total energy in the *L-C* circuit. Use this to determine the maximum energy stored in the inductor during the oscillation.

4. Use the result of step 3 to find the maximum current in the inductor.

5. Use the result of step 3 to find the maximum energy stored in the capacitor during the oscillation. Then use this to find the maximum capacitor voltage.

6. Find the energy in the inductor and the capacitor charge when the current has half the value that you found in step 4.

EVALUATE:

7. Initially, what fraction of the total energy is in the inductor? Is it possible to tell whether this is initially increasing or decreasing?

Problems

For instructor-assigned homework, go to www.masteringphysics.com

•, ••, •••: Problems of increasing difficulty. **CP**: Cumulative problems incorporating material from earlier chapters. **CALC**: Problems requiring calculus. **BIO**: Biosciences problems.

DISCUSSION QUESTIONS

Q30.1 In an electric trolley or bus system, the vehicle's motor draws current from an overhead wire by means of a long arm with an attachment at the end that slides along the overhead wire. A brilliant electric spark is often seen when the attachment crosses a junction in the wires where contact is momentarily lost. Explain this phenomenon.

Q30.2 From Eq. (30.5) 1 H = 1 Wb/A, and from Eq. (30.4) 1 H = 1 $\Omega \cdot$ s. Show that these two definitions are equivalent.

Q30.3 In Fig. 30.1, if coil 2 is turned 90° so that its axis is vertical, does the mutual inductance increase or decrease? Explain.

Q30.4 The tightly wound toroidal solenoid is one of the few configurations for which it is easy to calculate self-inductance. What features of the toroidal solenoid give it this simplicity?

Q30.5 Two identical, closely wound, circular coils, each having self-inductance *L*, are placed next to each other, so that they are coaxial and almost touching. If they are connected in series, what is the self-inductance of the combination? What if they are connected in parallel? Can they be connected so that the total inductance is zero? Explain.

Q30.6 Two closely wound circular coils have the same number of turns, but one has twice the radius of the other. How are the self-inductances of the two coils related? Explain your reasoning.

Q30.7 You are to make a resistor by winding a wire around a cylindrical form. To make the inductance as small as possible, it is proposed that you wind half the wire in one direction and the other half in the opposite direction. Would this achieve the desired result? Why or why not?

Q30.8 For the same magnetic field strength *B*, is the energy density greater in vacuum or in a magnetic material? Explain. Does

Eq. (30.11) imply that for a long solenoid in which the current is *I* the energy stored is proportional to $1/\mu$? And does this mean that for the same current less energy is stored when the solenoid is filled with a ferromagnetic material rather than with air? Explain.

Q30.9 In Section 30.5 Kirchhoff's loop rule is applied to an *L-C* circuit where the capacitor is initially fully charged and the equation $-L \, di/dt - q/C = 0$ is derived. But as the capacitor starts to discharge, the current increases from zero. The equation says $L \, di/dt = -q/C$, so it says $L \, di/dt$ is negative. Explain how $L \, di/dt$ can be negative when the current is increasing.

Q30.10 In Section 30.5 the relationship $i = dq/dt$ is used in deriving Eq. (30.20). But a flow of current corresponds to a decrease in the charge on the capacitor. Explain, therefore, why this is the correct equation to use in the derivation, rather than $i = -dq/dt$.

Q30.11 In the *R-L* circuit shown in Fig. 30.11, when switch S_1 is closed, the potential v_{ab} changes suddenly and discontinuously, but the current does not. Explain why the voltage can change suddenly but the current can't.

Q30.12 In the *R-L* circuit shown in Fig. 30.11, is the current in the resistor always the same as the current in the inductor? How do you know?

Q30.13 Suppose there is a steady current in an inductor. If you attempt to reduce the current to zero instantaneously by quickly opening a switch, an arc can appear at the switch contacts. Why? Is it physically possible to stop the current instantaneously? Explain.

Q30.14 In an *L-R-C* series circuit, what criteria could be used to decide whether the system is overdamped or underdamped? For example, could we compare the maximum energy stored during one cycle to the energy dissipated during one cycle? Explain.

EXERCISES

Section 30.1 Mutual Inductance

30.1 • Two coils have mutual inductance $M = 3.25 \times 10^{-4}$ H. The current i_1 in the first coil increases at a uniform rate of 830 A/s. (a) What is the magnitude of the induced emf in the second coil? Is it constant? (b) Suppose that the current described is in the second coil rather than the first. What is the magnitude of the induced emf in the first coil?

30.2 • Two coils are wound around the same cylindrical form, like the coils in Example 30.1. When the current in the first coil is decreasing at a rate of -0.242 A/s, the induced emf in the second coil has magnitude 1.65×10^{-3} V. (a) What is the mutual inductance of the pair of coils? (b) If the second coil has 25 turns, what is the flux through each turn when the current in the first coil equals 1.20 A? (c) If the current in the second coil increases at a rate of 0.360 A/s, what is the magnitude of the induced emf in the first coil?

30.3 • A 10.0-cm-long solenoid of diameter 0.400 cm is wound uniformly with 800 turns. A second coil with 50 turns is wound around the solenoid at its center. What is the mutual inductance of the combination of the two coils?

30.4 • A solenoidal coil with 25 turns of wire is wound tightly around another coil with 300 turns (see Example 30.1). The inner solenoid is 25.0 cm long and has a diameter of 2.00 cm. At a certain time, the current in the inner solenoid is 0.120 A and is increasing at a rate of 1.75×10^3 A/s. For this time, calculate: (a) the average magnetic flux through each turn of the inner solenoid; (b) the mutual inductance of the two solenoids; (c) the emf induced in the outer solenoid by the changing current in the inner solenoid.

30.5 • Two toroidal solenoids are wound around the same form so that the magnetic field of one passes through the turns of the other. Solenoid 1 has 700 turns, and solenoid 2 has 400 turns. When the current in solenoid 1 is 6.52 A, the average flux through each turn of solenoid 2 is 0.0320 Wb. (a) What is the mutual inductance of the pair of solenoids? (b) When the current in solenoid 2 is 2.54 A, what is the average flux through each turn of solenoid 1?

30.6 •• A toroidal solenoid with mean radius r and cross-sectional area A is wound uniformly with N_1 turns. A second toroidal solenoid with N_2 turns is wound uniformly on top of the first, so that the two solenoids have the same cross-sectional area and mean radius. (a) What is the mutual inductance of the two solenoids? Assume that the magnetic field of the first solenoid is uniform across the cross section of the two solenoids. (b) If $N_1 = 500$ turns, $N_2 = 300$ turns, $r = 10.0$ cm, and $A = 0.800$ cm², what is the value of the mutual inductance?

Section 30.2 Self-Inductance and Inductors

30.7 • A 2.50-mH toroidal solenoid has an average radius of 6.00 cm and a cross-sectional area of 2.00 cm². (a) How many coils does it have? (Make the same assumption as in Example 30.3.) (b) At what rate must the current through it change so that a potential difference of 2.00 V is developed across its ends?

30.8 • A toroidal solenoid has 500 turns, cross-sectional area 6.25 cm², and mean radius 4.00 cm. (a) Calculate the coil's self-inductance. (b) If the current decreases uniformly from 5.00 A to 2.00 A in 3.00 ms, calculate the self-induced emf in the coil. (c) The current is directed from terminal a of the coil to terminal b. Is the direction of the induced emf from a to b or from b to a?

30.9 • At the instant when the current in an inductor is increasing at a rate of 0.0640 A/s, the magnitude of the self-induced emf is 0.0160 V. (a) What is the inductance of the inductor? (b) If the

inductor is a solenoid with 400 turns, what is the average magnetic flux through each turn when the current is 0.720 A?

30.10 •• When the current in a toroidal solenoid is changing at a rate of 0.0260 A/s, the magnitude of the induced emf is 12.6 mV. When the current equals 1.40 A, the average flux through each turn of the solenoid is 0.00285 Wb. How many turns does the solenoid have?

30.11 • The inductor in Fig. E30.11 has inductance 0.260 H and carries a current in the direction shown that is decreasing at a uniform rate, $di/dt = -0.0180$ A/s. (a) Find the self-induced emf. (b) Which end of the inductor, a or b, is at a higher potential?

Figure **E30.11**

30.12 • The inductor shown in Fig. E30.11 has inductance 0.260 H and carries a current in the direction shown. The current is changing at a constant rate. (a) The potential between points a and b is $V_{ab} = 1.04$ V, with point a at higher potential. Is the current increasing or decreasing? (b) If the current at $t = 0$ is 12.0 A, what is the current at $t = 2.00$ s?

30.13 •• A toroidal solenoid has mean radius 12.0 cm and cross-sectional area 0.600 cm². (a) How many turns does the solenoid have if its inductance is 0.100 mH? (b) What is the resistance of the solenoid if the wire from which it is wound has a resistance per unit length of 0.0760 Ω/m?

30.14 • A long, straight solenoid has 800 turns. When the current in the solenoid is 2.90 A, the average flux through each turn of the solenoid is 3.25×10^{-3} Wb. What must be the magnitude of the rate of change of the current in order for the self-induced emf to equal 7.50 mV?

30.15 •• **Inductance of a Solenoid.** (a) A long, straight solenoid has N turns, uniform cross-sectional area A, and length l. Show that the inductance of this solenoid is given by the equation $L = \mu_0 A N^2/l$. Assume that the magnetic field is uniform inside the solenoid and zero outside. (Your answer is approximate because B is actually smaller at the ends than at the center. For this reason, your answer is actually an upper limit on the inductance.) (b) A metallic laboratory spring is typically 5.00 cm long and 0.150 cm in diameter and has 50 coils. If you connect such a spring in an electric circuit, how much self-inductance must you include for it if you model it as an ideal solenoid?

Section 30.3 Magnetic-Field Energy

30.16 • An inductor used in a dc power supply has an inductance of 12.0 H and a resistance of 180 Ω. It carries a current of 0.300 A. (a) What is the energy stored in the magnetic field? (b) At what rate is thermal energy developed in the inductor? (c) Does your answer to part (b) mean that the magnetic-field energy is decreasing with time? Explain.

30.17 • An air-filled toroidal solenoid has a mean radius of 15.0 cm and a cross-sectional area of 5.00 cm². When the current is 12.0 A, the energy stored is 0.390 J. How many turns does the winding have?

30.18 • An air-filled toroidal solenoid has 300 turns of wire, a mean radius of 12.0 cm, and a cross-sectional area of 4.00 cm². If the current is 5.00 A, calculate: (a) the magnetic field in the solenoid; (b) the self-inductance of the solenoid; (c) the energy stored in the magnetic field; (d) the energy density in the magnetic field. (e) Check your answer for part (d) by dividing your answer to part (c) by the volume of the solenoid.

30.19 •• A solenoid 25.0 cm long and with a cross-sectional area of 0.500 cm² contains 400 turns of wire and carries a current of 80.0 A. Calculate: (a) the magnetic field in the solenoid; (b) the energy density in the magnetic field if the solenoid is filled with air; (c) the total energy contained in the coil's magnetic field (assume the field is uniform); (d) the inductance of the solenoid.

30.20 • It has been proposed to use large inductors as energy storage devices. (a) How much electrical energy is converted to light and thermal energy by a 200-W light bulb in one day? (b) If the amount of energy calculated in part (a) is stored in an inductor in which the current is 80.0 A, what is the inductance?

30.21 •• In a proton accelerator used in elementary particle physics experiments, the trajectories of protons are controlled by bending magnets that produce a magnetic field of 4.80 T. What is the magnetic-field energy in a 10.0-cm^3 volume of space where $B = 4.80$ T?

30.22 • It is proposed to store $1.00 \text{ kW} \cdot \text{h} = 3.60 \times 10^6$ J of electrical energy in a uniform magnetic field with magnitude 0.600 T. (a) What volume (in vacuum) must the magnetic field occupy to store this amount of energy? (b) If instead this amount of energy is to be stored in a volume (in vacuum) equivalent to a cube 40.0 cm on a side, what magnetic field is required?

Section 30.4 The *R-L* Circuit

30.23 • An inductor with an inductance of 2.50 H and a resistance of 8.00 Ω is connected to the terminals of a battery with an emf of 6.00 V and negligible internal resistance. Find (a) the initial rate of increase of current in the circuit; (b) the rate of increase of current at the instant when the current is 0.500 A; (c) the current 0.250 s after the circuit is closed; (d) the final steady-state current.

30.24 • In Fig. 30.11, $R = 15.0 \Omega$ and the battery emf is 6.30 V. With switch S_2 open, switch S_1 is closed. After several minutes, S_1 is opened and S_2 is closed. (a) At 2.00 ms after S_1 is opened, the current has decayed to 0.320 A. Calculate the inductance of the coil. (b) How long after S_1 is opened will the current reach 1.00% of its original value?

30.25 • A 35.0-V battery with negligible internal resistance, a 50.0-Ω resistor, and a 1.25-mH inductor with negligible resistance are all connected in series with an open switch. The switch is suddenly closed. (a) How long after closing the switch will the current through the inductor reach one-half of its maximum value? (b) How long after closing the switch will the energy stored in the inductor reach one-half of its maximum value?

30.26 • In Fig. 30.11, switch S_1 is closed while switch S_2 is kept open. The inductance is $L = 0.115$ H, and the resistance is $R = 120 \Omega$. (a) When the current has reached its final value, the energy stored in the inductor is 0.260 J. What is the emf $\mathcal{E}$ of the battery? (b) After the current has reached its final value, S_1 is opened and S_2 is closed. How much time does it take for the energy stored in the inductor to decrease to 0.130 J, half the original value?

30.27 • In Fig. 30.11, suppose that $\mathcal{E} = 60.0$ V, $R = 240 \Omega$, and $L = 0.160$ H. With switch S_2 open, switch S_1 is left closed until a constant current is established. Then S_2 is closed and S_1 opened, taking the battery out of the circuit. (a) What is the initial current in the resistor, just after S_2 is closed and S_1 is opened? (b) What is the current in the resistor at $t = 4.00 \times 10^{-4}$ s? (c) What is the potential difference between points b and c at $t = 4.00 \times 10^{-4}$ s? Which point is at a higher potential? (d) How long does it take the current to decrease to half its initial value?

30.28 • In Fig. 30.11, suppose that $\mathcal{E} = 60.0$ V, $R = 240 \Omega$, and $L = 0.160$ H. Initially there is no current in the circuit. Switch S_2 is left open, and switch S_1 is closed. (a) Just after S_1 is closed, what are the potential differences v_{ab} and v_{bc}? (b) A long time (many time constants) after S_1 is closed, what are v_{ab} and v_{bc}? (c) What are v_{ab} and v_{bc} at an intermediate time when $i = 0.150$ A?

30.29 • Refer to the circuit in Exercise 30.23. (a) What is the power input to the inductor from the battery as a function of time if the circuit is completed at $t = 0$? (b) What is the rate of dissipation of energy in the resistance of the inductor as a function of time? (c) What is the rate at which the energy of the magnetic field in the inductor is increasing, as a function of time? (d) Compare the results of parts (a), (b), and (c).

30.30 • In Fig. 30.11 switch S_1 is closed while switch S_2 is kept open. The inductance is $L = 0.380$ H, the resistance is $R = 48.0 \Omega$, and the emf of the battery is 18.0 V. At time t after S_1 is closed, the current in the circuit is increasing at a rate of $di/dt = 7.20$ A/s. At this instant what is v_{ab}, the voltage across the resistor?

Section 30.5 The *L-C* Circuit

30.31 • CALC Show that the differential equation of Eq. (30.20) is satisfied by the function $q = Q \cos(\omega t + \phi)$, with ω given by $1/\sqrt{LC}$.

30.32 •• A 20.0-μF capacitor is charged by a 150.0-V power supply, then disconnected from the power and connected in series with a 0.280-mH inductor. Calculate: (a) the oscillation frequency of the circuit; (b) the energy stored in the capacitor at time $t = 0$ ms (the moment of connection with the inductor); (c) the energy stored in the inductor at $t = 1.30$ ms.

30.33 • A 7.50-nF capacitor is charged up to 12.0 V, then disconnected from the power supply and connected in series through a coil. The period of oscillation of the circuit is then measured to be 8.60×10^{-5} s. Calculate: (a) the inductance of the coil; (b) the maximum charge on the capacitor; (c) the total energy of the circuit; (d) the maximum current in the circuit.

30.34 •• A 18.0-μF capacitor is placed across a 22.5-V battery for several seconds and is then connected across a 12.0-mH inductor that has no appreciable resistance. (a) After the capacitor and inductor are connected together, find the maximum current in the circuit. When the current is a maximum, what is the charge on the capacitor? (b) How long after the capacitor and inductor are connected together does it take for the capacitor to be completely discharged for the first time? For the second time? (c) Sketch graphs of the charge on the capacitor plates and the current through the inductor as functions of time.

30.35 • *L-C Oscillations.* A capacitor with capacitance 6.00×10^{-5} F is charged by connecting it to a 12.0-V battery. The capacitor is disconnected from the battery and connected across an inductor with $L = 1.50$ H. (a) What are the angular frequency ω of the electrical oscillations and the period of these oscillations (the time for one oscillation)? (b) What is the initial charge on the capacitor? (c) How much energy is initially stored in the capacitor? (d) What is the charge on the capacitor 0.0230 s after the connection to the inductor is made? Interpret the sign of your answer. (e) At the time given in part (d), what is the current in the inductor? Interpret the sign of your answer. (f) At the time given in part (d), how much electrical energy is stored in the capacitor and how much is stored in the inductor?

30.36 • *A Radio Tuning Circuit.* The minimum capacitance of a variable capacitor in a radio is 4.18 pF. (a) What is the inductance of a coil connected to this capacitor if the oscillation frequency of the *L-C* circuit is 1600×10^3 Hz, corresponding to one end of the AM radio broadcast band, when the capacitor is set to its minimum capacitance? (b) The frequency at the other end of the broadcast band is 540×10^3 Hz. What is the maximum capacitance of the capacitor if the oscillation frequency is adjustable over the range of the broadcast band?

30.37 •• An *L-C* circuit containing an 80.0-mH inductor and a 1.25-nF capacitor oscillates with a maximum current of 0.750 A. Calculate: (a) the maximum charge on the capacitor and (b) the oscillation frequency of the circuit. (c) Assuming the capacitor had its maximum charge at time $t = 0$, calculate the energy stored in the inductor after 2.50 ms of oscillation.

30.38 • In an *L-C* circuit, $L = 85.0$ mH and $C = 3.20$ μF. During the oscillations the maximum current in the inductor is 0.850 mA. (a) What is the maximum charge on the capacitor? (b) What is the magnitude of the charge on the capacitor at an instant when the current in the inductor has magnitude 0.500 mA?

Section 30.6 The *L-R-C* Series Circuit

30.39 • An *L-R-C* series circuit has $L = 0.450$ H, $C = 2.50 \times 10^{-5}$ F, and resistance R. (a) What is the angular frequency of the circuit when $R = 0$? (b) What value must R have to give a 5.0% decrease in angular frequency compared to the value calculated in part (a)?

30.40 • For the circuit of Fig. 30.17, let $C = 15.0$ nF, $L = 22$ mH, and $R = 75.0$ Ω. (a) Calculate the oscillation frequency of the circuit once the capacitor has been charged and the switch has been connected to point a. (b) How long will it take for the amplitude of the oscillation to decay to 10.0% of its original value? (c) What value of R would result in a critically damped circuit?

30.41 • **CP** (a) In Eq. (14.41), substitute q for x, L for m, 1/C for k, and R for the damping constant b. Show that the result is Eq. (30.27). (b) Make these same substitutions in Eq. (14.43) and show that Eq. (30.29) results. (c) Make these same substitutions in Eq. (14.42) and show that Eq. (30.28) results.

30.42 • **CALC** (a) Take first and second derivatives with respect to time of q given in Eq. (30.28), and show that it is a solution of Eq. (30.27). (b) At $t = 0$ the switch shown in Fig. 30.17 is thrown so that it connects points d and a; at this time, $q = Q$ and $i = dq/dt = 0$. Show that the constants ϕ and A in Eq. (30.28) are given by

$$\tan\phi = -\frac{R}{2L\sqrt{(1/LC) - (R^2/4L^2)}} \quad \text{and} \quad A = \frac{Q}{\cos\phi}$$

PROBLEMS

30.43 • One solenoid is centered inside another. The outer one has a length of 50.0 cm and contains 6750 coils, while the coaxial inner solenoid is 3.0 cm long and 0.120 cm in diameter and contains 15 coils. The current in the outer solenoid is changing at 49.2 A/s. (a) What is the mutual inductance of these solenoids? (b) Find the emf induced in the innner solenoid.

30.44 •• **CALC** A coil has 400 turns and self-inductance 4.80 mH. The current in the coil varies with time according to $i = (680 \text{ mA})\cos(\pi t/0.0250 \text{ s})$. (a) What is the maximum emf induced in the coil? (b) What is the maximum average flux through each turn of the coil? (c) At $t = 0.0180$ s, what is the magnitude of the induced emf?

30.45 • **A Differentiating Circuit.** The current in a resistanceless inductor is caused to vary with time as shown in the graph of Fig. P30.45. (a) Sketch the pattern that would be observed on the screen of an oscilloscope connected to the terminals of the inductor. (The oscilloscope spot sweeps horizontally across the screen at a constant speed, and its vertical deflection is proportional to the potential difference between

the inductor terminals.) (b) Explain why a circuit with an inductor can be described as a "differentiating circuit."

30.46 •• **CALC** A 0.250-H inductor carries a time-varying current given by the expression $i = (124 \text{ mA})\cos[(240\pi/\text{s})t]$. (a) Find an expression for the induced emf as a function of time. Graph the current and induced emf as functions of time for $t = 0$ to $t = \frac{1}{60}$ s. (b) What is the maximum emf? What is the current when the induced emf is a maximum? (c) What is the maximum current? What is the induced emf when the current is a maximum?

30.47 •• **Solar Magnetic Energy.** Magnetic fields within a sunspot can be as strong as 0.4 T. (By comparison, the earth's magnetic field is about 1/10,000 as strong.) Sunspots can be as large as 25,000 km in radius. The material in a sunspot has a density of about 3×10^{-4} kg/m^3. Assume μ for the sunspot material is μ_0. If 100% of the magnetic-field energy stored in a sunspot could be used to eject the sunspot's material away from the sun's surface, at what speed would that material be ejected? Compare to the sun's escape speed, which is about 6×10^5 m/s. (*Hint:* Calculate the kinetic energy the magnetic field could supply to 1 m^3 of sunspot material.)

30.48 •• **CP CALC** **A Coaxial Cable.** A small solid conductor with radius a is supported by insulating, nonmagnetic disks on the axis of a thin-walled tube with inner radius b. The inner and outer conductors carry equal currents i in opposite directions. (a) Use Ampere's law to find the magnetic field at any point in the volume between the conductors. (b) Write the expression for the flux $d\Phi_B$ through a narrow strip of length l parallel to the axis, of width dr, at a distance r from the axis of the cable and lying in a plane containing the axis. (c) Integrate your expression from part (b) over the volume between the two conductors to find the total flux produced by a current i in the central conductor. (d) Show that the inductance of a length l of the cable is

$$L = l\frac{\mu_0}{2\pi}\ln\left(\frac{b}{a}\right)$$

(e) Use Eq. (30.9) to calculate the energy stored in the magnetic field for a length l of the cable.

30.49 •• **CP CALC** Consider the coaxial cable of Problem 30.48. The conductors carry equal currents i in opposite directions. (a) Use Ampere's law to find the magnetic field at any point in the volume between the conductors. (b) Use the energy density for a magnetic field, Eq. (30.10), to calculate the energy stored in a thin, cylindrical shell between the two conductors. Let the cylindrical shell have inner radius r, outer radius $r + dr$, and length l. (c) Integrate your result in part (b) over the volume between the two conductors to find the total energy stored in the magnetic field for a length l of the cable. (d) Use your result in part (c) and Eq. (30.9) to calculate the inductance L of a length l of the cable. Compare your result to L calculated in part (d) of Problem 30.48.

30.50 •• A toroidal solenoid has a mean radius r and a cross-sectional area A and is wound uniformly with N_1 turns. A second toroidal solenoid with N_2 turns is wound uniformly around the first. The two coils are wound in the same direction. (a) Derive an expression for the inductance L_1 when only the first coil is used and an expression for L_2 when only the second coil is used. (b) Show that $M^2 = L_1 L_2$.

30.51 • (a) What would have to be the self-inductance of a solenoid for it to store 10.0 J of energy when a 2.00-A current runs through it? (b) If this solenoid's cross-sectional diameter is 4.00 cm, and if you could wrap its coils to a density of 10 coils/mm, how long would the solenoid be? (See Exercise 30.15.) Is this a realistic length for ordinary laboratory use?

Figure **P30.45**

30.52 • An inductor is connected to the terminals of a battery that has an emf of 12.0 V and negligible internal resistance. The current is 4.86 mA at 0.940 ms after the connection is completed. After a long time the current is 6.45 mA. What are (a) the resistance R of the inductor and (b) the inductance L of the inductor?

30.53 •• **CALC** Continuation of Exercises 30.23 and 30.29. (a) How much energy is stored in the magnetic field of the inductor one time constant after the battery has been connected? Compute this both by integrating the expression in Exercise 30.29(c) and by using Eq. (30.9), and compare the results. (b) Integrate the expression obtained in Exercise 30.29(a) to find the *total* energy supplied by the battery during the time interval considered in part (a). (c) Integrate the expression obtained in Exercise 30.29(b) to find the *total* energy dissipated in the resistance of the inductor during the same time period. (d) Compare the results obtained in parts (a), (b), and (c).

30.54 •• **CALC** Continuation of Exercise 30.27. (a) What is the total energy initially stored in the inductor? (b) At $t = 4.00 \times 10^{-4}$ s, at what rate is the energy stored in the inductor decreasing? (c) At $t = 4.00 \times 10^{-4}$ s, at what rate is electrical energy being converted into thermal energy in the resistor? (d) Obtain an expression for the rate at which electrical energy is being converted into thermal energy in the resistor as a function of time. Integrate this expression from $t = 0$ to $t = \infty$ to obtain the total electrical energy dissipated in the resistor. Compare your result to that of part (a).

30.55 • **CALC** The equation preceding Eq. (30.27) may be converted into an energy relationship. Multiply both sides of this equation by $-i = -dq/dt$. The first term then becomes i^2R. Show that the second term can be written as $d(\frac{1}{2}Li^2)/dt$, and that the third term can be written as $d(q^2/2C)/dt$. What does the resulting equation say about energy conservation in the circuit?

30.56 • A 7.00-μF capacitor is initially charged to a potential of 16.0 V. It is then connected in series with a 3.75-mH inductor. (a) What is the total energy stored in this circuit? (b) What is the maximum current in the inductor? What is the charge on the capacitor plates at the instant the current in the inductor is maximal?

30.57 • **An Electromagnetic Car Alarm.** Your latest invention is a car alarm that produces sound at a particularly annoying frequency of 3500 Hz. To do this, the car-alarm circuitry must produce an alternating electric current of the same frequency. That's why your design includes an inductor and a capacitor in series. The maximum voltage across the capacitor is to be 12.0 V (the same voltage as the car battery). To produce a sufficiently loud sound, the capacitor must store 0.0160 J of energy. What values of capacitance and inductance should you choose for your car-alarm circuit?

30.58 • An L-C circuit consists of a 60.0-mH inductor and a 250-μF capacitor. The initial charge on the capacitor is 6.00 μC, and the initial current in the inductor is zero. (a) What is the maximum voltage across the capacitor? (b) What is the maximum current in the inductor? (c) What is the maximum energy stored in the inductor? (d) When the current in the inductor has half its maximum value, what is the charge on the capacitor and what is the energy stored in the inductor?

30.59 •• A 84.0-nF capacitor is charged to 12.0 V, then disconnected from the power supply and connected in series with a coil that has $L = 0.0420$ H and negligible resistance. At an instant when the charge on the capacitor is 0.650 μC, what is the magnitude of the current in the inductor and what is the magnitude of the rate of change of this current?

30.60 •• A charged capacitor with $C = 590$ μF is connected in series to an inductor that has $L = 0.330$ H and negligible resistance.

At an instant when the current in the inductor is $i = 2.50$ A, the current is increasing at a rate of $di/dt = 89.0$ A/s. During the current oscillations, what is the maximum voltage across the capacitor?

30.61 ••• **CP** In the circuit shown in Fig. P30.61, the switch has been open for a long time and is suddenly closed. Neither the battery nor the inductors have any appreciable resistance. (a) What do the ammeter and voltmeter read just after S is closed? (b) What do the ammeter and the voltmeter read after S has been closed a very long time? (c) What do the ammeter and the voltmeter read 0.115 ms after S is closed?

Figure **P30.61**

30.62 •• While studying a coil of unknown inductance and internal resistance, you connect it in series with a 25.0-V battery and a 150-Ω resistor. You then place an oscilloscope across one of these circuit elements and use the oscilloscope to measure the voltage across the circuit element as a function of time. The result is shown in Fig. P30.62. (a) Across which circuit element (coil or resistor) is the oscilloscope connected? How do you know this? (b) Find the inductance and the internal resistance of the coil. (c) Carefully make a quantitative sketch showing the voltage versus time you would observe if you put the oscilloscope across the other circuit element (resistor or coil).

Figure **P30.62**

30.63 •• In the lab, you are trying to find the inductance and internal resistance of a solenoid. You place it in series with a battery of negligible internal resistance, a 10.0-Ω resistor, and a switch. You then put an oscilloscope across one of these circuit elements to measure the voltage across that circuit element as a function of time. You close the switch, and the oscilloscope shows voltage versus time as shown in Fig. P30.63. (a) Across which circuit element (solenoid or resistor) is the oscilloscope connected? How do you know this? (b) Why doesn't the graph approach zero as $t \to \infty$? (c) What is the emf of the battery? (d) Find the maximum current in the circuit. (e) What are the internal resistance and self-inductance of the solenoid?

Figure **P30.63**

30.64 •• **CP** In the circuit shown in Fig. P30.64, find the reading in each ammeter and voltmeter (a) just after switch S is closed and (b) after S has been closed a very long time.

30.65 •• **CP** In the circuit shown in Fig. P30.65, switch S is closed at time $t = 0$ with no charge initially on the capacitor. (a) Find the reading of each ammeter and each voltmeter just after S is closed. (b) Find the

Figure **P30.64**

reading of each meter after a long time has elapsed. (c) Find the maximum charge on the capacitor. (d) Draw a qualitative graph of the reading of voltmeter V_2 as a function of time.

Figure **P30.65**

30.66 • In the circuit shown in Fig. P30.66 the battery and the inductor have no appreciable internal resistance and there is no current in the circuit. After the switch is closed, find the readings of the ammeter (A) and voltmeters (V_1 and V_2) (a) the instant after the switch is closed and (b) after the switch has been closed for a very long time. (c) Which answers in parts (a) and (b) would change if the inductance were 24.0 mH instead?

Figure **P30.66**

30.67 •• CP In the circuit shown in Fig. P30.67, switch S is closed at time $t = 0$. (a) Find the reading of each meter just after S is closed. (b) What does each meter read long after S is closed?

Figure **P30.67**

30.68 •• In the circuit shown in Fig. P30.68, switch S_1 has been closed for a long enough time so that the current reads a steady 3.50 A. Suddenly, switch S_2 is closed and S_1 is opened at the same instant. (a) What is the maximum charge that the capacitor will receive? (b) What is the current in the inductor at this time?

Figure **P30.68**

30.69 •• CP In the circuit shown in Fig. P30.69, $\mathcal{E} = 60.0$ V, $R_1 = 40.0\ \Omega$, $R_2 = 25.0\ \Omega$, and $L = 0.300$ H. Switch S is closed at $t = 0$. Just after the switch is closed, (a) what is the potential difference v_{ab} across the resistor R_1; (b) which point, a or b, is at a higher potential; (c) what is the

Figure **P30.69**

potential difference v_{cd} across the inductor L; (d) which point, c or d, is at a higher potential? The switch is left closed a long time and then opened. Just after the switch is opened, (e) what is the potential difference v_{ab} across the resistor R_1; (f) which point, a or b, is at a higher potential; (g) what is the potential difference v_{cd} across the inductor L; (h) which point, c or d, is at a higher potential?

30.70 •• CP In the circuit shown in Fig. P30.69, $\mathcal{E} = 60.0$ V, $R_1 = 40.0\ \Omega$, $R_2 = 25.0\ \Omega$, and $L = 0.300$ H. (a) Switch S is closed. At some time t afterward, the current in the inductor is increasing at a rate of $di/dt = 50.0$ A/s. At this instant, what are the current i_1 through R_1 and the current i_2 through R_2? (*Hint:* Analyze two separate loops: one containing $\mathcal{E}$ and R_1 and the other containing $\mathcal{E}$, R_2, and L.) (b) After the switch has been closed a long time, it is opened again. Just after it is opened, what is the current through R_1?

30.71 •• CALC Consider the circuit shown in Fig. P30.71. Let $\mathcal{E} = 36.0$ V, $R_0 = 50.0\ \Omega$, $R = 150\ \Omega$, and $L = 4.00$ H. (a) Switch S_1 is closed and switch S_2 is left open. Just after S_1 is closed, what are the current i_0 through R_0 and the potential differences v_{ac} and v_{cb}? (b) After S_1 has been closed a long time (S_2 is still open) so that the current has reached its final, steady value, what are i_0, v_{ac}, and v_{cb}? (c) Find the expressions for i_0, v_{ac}, and v_{cb} as functions of the time t since S_1 was closed. Your results should agree with part (a) when $t = 0$ and with part (b) when $t \rightarrow \infty$. Graph i_0, v_{ac}, and v_{cb} versus time.

Figure **P30.71**

30.72 •• After the current in the circuit of Fig. P30.71 has reached its final, steady value with switch S_1 closed and S_2 open, switch S_2 is closed, thus short-circuiting the inductor. (Switch S_1 remains closed. See Problem 30.71 for numerical values of the circuit elements.) (a) Just after S_2 is closed, what are v_{ac} and v_{cb}, and what are the currents through R_0, R, and S_2? (b) A long time after S_2 is closed, what are v_{ac} and v_{cb}, and what are the currents through R_0, R, and S_2? (c) Derive expressions for the currents through R_0, R, and S_2 as functions of the time t that has elapsed since S_2 was closed. Your results should agree with part (a) when $t = 0$ and with part (b) when $t \rightarrow \infty$. Graph these three currents versus time.

30.73 ••• CP CALC We have ignored the variation of the magnetic field across the cross section of a toroidal solenoid. Let's now examine the validity of that approximation. A certain toroidal solenoid has a rectangular cross section (Fig. P30.73). It has N uniformly spaced turns, with air inside. The magnetic field at a point inside the toroid is given by the equation derived in Example 28.10 (Section 28.7). *Do not* assume the field is uniform over the cross section. (a) Show that the magnetic flux through a cross section of the toroid is

Figure **P30.73**

$$\Phi_B = \frac{\mu_0 Nih}{2\pi} \ln\left(\frac{b}{a}\right)$$

(b) Show that the inductance of the toroidal solenoid is given by

$$L = \frac{\mu_0 N^2 h}{2\pi} \ln\left(\frac{b}{a}\right)$$

(c) The fraction b/a may be written as

$$\frac{b}{a} = \frac{a + b - a}{a} = 1 + \frac{b - a}{a}$$

Use the power series expansion $\ln(1 + z) = z + z^2/2 + \cdots$, valid for $|z| < 1$, to show that when $b - a$ is much less than a, the inductance is approximately equal to

$$L = \frac{\mu_0 N^2 h(b - a)}{2\pi a}$$

Compare this result with the result given in Example 30.3 (Section 30.2).

30.74 ••• CP In the circuit shown in Fig. P30.74, neither the battery nor the inductors have any appreciable resistance, the capacitors are initially uncharged, and the switch S has been in position 1 for a very long time. (a) What is the current in the circuit? (b) The switch is now suddenly flipped to position 2. Find the maximum charge that each capacitor will receive, and how much time after the switch is flipped it will take them to acquire this charge.

Figure **P30.74**

30.75 ••• CP CALC Demonstrating Inductance. A common demonstration of inductance employs a circuit such as the one shown in Fig. P30.69. Switch S is closed, and the light bulb (represented by resistance R_1) just barely glows. After a period of time, switch S is opened, and the bulb lights up brightly for a short period of time. To understand this effect, think of an inductor as a device that imparts an "inertia" to the current, preventing a discontinuous change in the current through it. (a) Derive, as explicit functions of time, expressions for i_1 (the current through the light bulb) and i_2 (the current through the inductor) after switch S is closed. (b) After a long period of time, the currents i_1 and i_2 reach their steady-state values. Obtain expressions for these steady-state currents. (c) Switch S is now opened. Obtain an expression for the current through the inductor and light bulb as an explicit function of time. (d) You have been asked to design a demonstration apparatus using the circuit shown in Fig. P30.69 with a 22.0-H inductor and a 40.0-W light bulb. You are to connect a resistor in series with the inductor, and R_2 represents the sum of that resistance plus the internal resistance of the inductor. When switch S is opened, a transient current is to be set up that starts at 0.600 A and is not to fall below 0.150 A until after 0.0800 s. For simplicity, assume that the resistance of the light bulb is constant and equals the resistance the bulb must have to dissipate 40.0 W at 120 V. Determine R_2 and $\mathcal{E}$ for the given design considerations. (e) With the numerical values determined in part (d), what is the current through the light bulb just before the switch is opened? Does this result confirm the qualitative description of what is observed in the demonstration?

CHALLENGE PROBLEMS

30.76 ••• CP CALC Consider the circuit shown in Fig. P30.76. The circuit elements are as follows: $\mathcal{E} = 32.0$ V, $L = 0.640$ H, $C = 2.00$ μF, and $R = 400$ Ω. At time $t = 0$, switch S is closed. The current through the inductor is i_1, the current through the capacitor branch is i_2, and the charge on the capacitor is q_2. (a) Using Kirchhoff's rules, verify the circuit equations

$$R(i_1 + i_2) + L\left(\frac{di_1}{dt}\right) = \mathcal{E}$$

$$R(i_1 + i_2) + \frac{q_2}{C} = \mathcal{E}$$

(b) What are the initial values of i_1, i_2, and q_2? (c) Show by direct substitution that the following solutions for i_1 and q_2 satisfy the circuit equations from part (a). Also, show that they satisfy the initial conditions

$$i_1 = \left(\frac{\mathcal{E}}{R}\right)[1 - e^{-\beta t}\{(2\omega RC)^{-1}\sin(\omega t) + \cos(\omega t)\}]$$

$$q_2 = \left(\frac{\mathcal{E}}{\omega R}\right)e^{-\beta t}\sin(\omega t)$$

where $\beta = (2RC)^{-1}$ and $\omega = [(LC)^{-1} - (2RC)^{-2}]^{1/2}$. (d) Determine the time t_1 at which i_2 first becomes zero.

Figure **P30.76**

30.77 ••• CP A Volume Gauge. A tank containing a liquid has turns of wire wrapped around it, causing it to act like an inductor. The liquid content of the tank can be measured by using its inductance to determine the height of the liquid in the tank. The inductance of the tank changes from a value of L_0 corresponding to a relative permeability of 1 when the tank is empty to a value of L_f corresponding to a relative permeability of K_m (the relative permeability of the liquid) when the tank is full. The appropriate electronic circuitry can determine the inductance to five significant figures and thus the effective relative permeability of the combined air and liquid within the rectangular cavity of the tank. The four sides of the tank each have width W and height D (Fig. P30.77). The height of the liquid in the tank is d. You can ignore any fringing effects and assume that the relative permeability of the material of which the tank is made can be ignored. (a) Derive an expression for d as a function of L, the inductance corresponding to a certain fluid height, L_0, L_f, and D. (b) What is the inductance (to five significant figures) for a tank $\frac{1}{4}$ full, $\frac{1}{2}$ full, $\frac{3}{4}$ full, and completely full if the tank contains liquid oxygen? Take $L_0 = 0.63000$ H. The magnetic susceptibility of liquid oxygen is $\chi_m = 1.52 \times 10^{-3}$. (c) Repeat part (b) for mercury. The magnetic susceptibility of mercury is given in Table 28.1. (d) For which material is this volume gauge more practical?

Figure **P30.77**

30.78 ••• Two coils are wrapped around each other as shown in Fig. 30.3. The current travels in the same sense around each coil. One coil has self-inductance L_1, and the other coil has self-inductance L_2. The mutual inductance of the two coils is M. (a) Show that if the two coils are connected in series, the equivalent inductance of the combination is $L_{eq} = L_1 + L_2 + 2M$. (b) Show that if the two coils are connected in parallel, the equivalent inductance of the combination is

$$L_{eq} = \frac{L_1 L_2 - M^2}{L_1 + L_2 - 2M}$$

30.79 ••• CP CALC Consider the circuit shown in Fig. P30.79. Switch S is closed at time $t = 0$, causing a current i_1 through the inductive branch and a current i_2 through the capacitive branch.

The initial charge on the capacitor is zero, and the charge at time t is q_2. (a) Derive expressions for i_1, i_2, and q_2 as functions of time. Express your answers in terms of $\mathcal{E}$, L, C, R_1, R_2, and t. For the remainder of the problem let the circuit elements have the following values: $\mathcal{E} = 48$ V, $L = 8.0$ H, $C = 20$ μF, $R_1 = 25$ Ω, and $R_2 = 5000$ Ω. (b) What is the initial current through the inductive branch? What is the initial current

Figure **P30.79**

through the capacitive branch? (c) What are the currents through the inductive and capacitive branches a long time after the switch has been closed? How long is a "long time"? Explain. (d) At what time t_1 (accurate to two significant figures) will the currents i_1 and i_2 be equal? (*Hint:* You might consider using series expansions for the exponentials.) (e) For the conditions given in part (d), determine i_1. (f) The total current through the battery is $i = i_1 + i_2$. At what time t_2 (accurate to two significant figures) will i equal one-half of its final value? (*Hint:* The numerical work is greatly simplified if one makes suitable approximations. A sketch of i_1 and i_2 versus t may help you decide what approximations are valid.)

Answers

Chapter Opening Question

As explained in Section 30.2, traffic light sensors work by measuring the change in inductance of a coil embedded under the road surface when a car drives over it.

Test Your Understanding Questions

30.1 Answer: (iii) Doubling both the length of the solenoid (l) and the number of turns of wire in the solenoid (N_1) would have *no* effect on the mutual inductance M. Example 30.1 shows that M depends on the ratio of these quantities, which would remain unchanged. This is because the magnetic field produced by the solenoid depends on the number of turns *per unit length,* and the proposed change has no effect on this quantity.

30.2 Answer: (iv), (i), (iii), (ii) From Eq. (30.8), the potential difference across the inductor is $V_{ab} = L\, di/dt$. For the four cases we find (i) $V_{ab} = (2.0\,\mu\text{H})(2.0\,\text{A} - 1.0\,\text{A})/(0.50\,\text{s}) = 4.0\,\mu\text{V}$; (ii) $V_{ab} = (4.0\,\mu\text{H})(0 - 3.0\,\text{A})/(2.0\,\text{s}) = -6.0\,\mu\text{V}$; (iii) $V_{ab} = 0$ because the rate of change of current is zero; and (iv) $V_{ab} = (1.0\,\mu\text{H})(4.0\,\text{A} - 0)/(0.25\,\text{s}) = 16\,\mu\text{V}$.

30.3 Answers: (a) yes, (b) no Reversing the direction of the current has no effect on the magnetic field magnitude, but it causes the direction of the magnetic field to reverse. It has no effect on the magnetic-field energy density, which is proportional to the square of the *magnitude* of the magnetic field.

30.4 Answers: (a) (i), (b) (ii) Recall that v_{ab} is the potential at a minus the potential at b, and similarly for v_{bc}. For either arrangement of the switches, current flows through the resistor from a to

b. The upstream end of the resistor is always at the higher potential, so v_{ab} is positive. With S_1 closed and S_2 open, the current through the inductor flows from b to c and is increasing. The self-induced emf opposes this increase and is therefore directed from c toward b, which means that b is at the higher potential. Hence v_{bc} is positive. With S_1 open and S_2 closed, the inductor current again flows from b to c but is now decreasing. The self-induced emf is directed from b to c in an effort to sustain the decaying current, so c is at the higher potential and v_{bc} is negative.

30.5 Answers: (a) positive, (b) electric, (c) negative, (d) electric The capacitor loses energy between stages (a) and (b), so it does positive work on the charges. It does this by exerting an electric force that pushes current away from the positively charged left-hand capacitor plate and toward the negatively charged right-hand plate. At the same time, the inductor gains energy and does negative work on the moving charges. Although the inductor stores magnetic energy, the force that the inductor exerts is *electric*. This force comes about from the inductor's self-induced emf (see Section 30.2).

30.6 Answer: (i) and (iii) There are no oscillations if $R^2 \geq 4L/C$. In each case $R^2 = (2.0\,\Omega)^2 = 4.0\,\Omega^2$. In case (i) $4L/C = 4(3.0\,\mu\text{H})/(6.0\,\mu\text{F}) = 2.0\,\Omega^2$, so there are no oscillations (the system is overdamped); in case (ii) $4L/C = 4(6.0\,\mu\text{H})/(3.0\,\mu\text{F}) = 8.0\,\Omega^2$, so there are oscillations (the system is underdamped); and in case (iii) $4L/C = 4(3.0\,\mu\text{H})/(3.0\,\mu\text{F}) = 4.0\,\Omega^2$, so there are no oscillations (the system is critically damped).

Bridging Problem

Answers: (a) 7.68×10^{-8} J   **(b)** 1.60 mA   **(c)** 24.8 mV
(d) 1.92×10^{-8} J, 21.5 mV

ALTERNATING CURRENT

? Waves from a broadcasting station produce an alternating current in the circuits of a radio (like the one in this classic car). If a radio is tuned to a station at a frequency of 1000 kHz, does it also detect the transmissions from a station broadcasting at 600 kHz?

LEARNING GOALS

By studying this chapter, you will learn:

- How phasors make it easy to describe sinusoidally varying quantities.

- How to use reactance to describe the voltage across a circuit element that carries an alternating current.

- How to analyze an *L-R-C* series circuit with a sinusoidal emf.

- What determines the amount of power flowing into or out of an alternating-current circuit.

- How an *L-R-C* series circuit responds to sinusoidal emfs of different frequencies.

- Why transformers are useful, and how they work.

During the 1880s in the United States there was a heated and acrimonious debate between two inventors over the best method of electric-power distribution. Thomas Edison favored direct current (dc)—that is, steady current that does not vary with time. George Westinghouse favored **alternating current (ac),** with sinusoidally varying voltages and currents. He argued that transformers (which we will study in this chapter) can be used to step the voltage up and down with ac but not with dc; low voltages are safer for consumer use, but high voltages and correspondingly low currents are best for long-distance power transmission to minimize i^2R losses in the cables.

Eventually, Westinghouse prevailed, and most present-day household and industrial power-distribution systems operate with alternating current. Any appliance that you plug into a wall outlet uses ac, and many battery-powered devices such as radios and cordless telephones make use of the dc supplied by the battery to create or amplify alternating currents. Circuits in modern communication equipment, including pagers and television, also make extensive use of ac.

In this chapter we will learn how resistors, inductors, and capacitors behave in circuits with sinusoidally varying voltages and currents. Many of the principles that we found useful in Chapters 25, 28, and 30 are applicable, along with several new concepts related to the circuit behavior of inductors and capacitors. A key concept in this discussion is *resonance,* which we studied in Chapter 14 for mechanical systems.

31.1 Phasors and Alternating Currents

To supply an alternating current to a circuit, a source of alternating emf or voltage is required. An example of such a source is a coil of wire rotating with constant angular velocity in a magnetic field, which we discussed in Example 29.3 (Section 29.2). This develops a sinusoidal alternating emf and is the prototype of the commercial alternating-current generator or *alternator* (see Fig. 29.8).

We use the term **ac source** for any device that supplies a sinusoidally varying voltage (potential difference) v or current i. The usual circuit-diagram symbol for an ac source is

A sinusoidal voltage might be described by a function such as

$$v = V\cos\omega t \tag{31.1}$$

31.1 The voltage across a sinusoidal ac source.

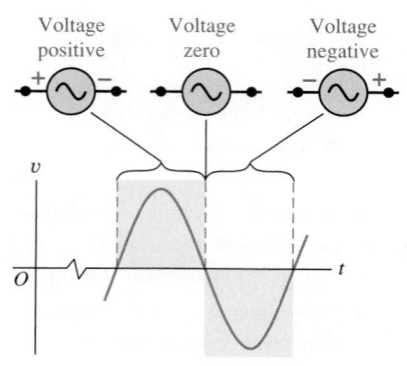

In this expression, v (lowercase) is the *instantaneous* potential difference; V (uppercase) is the maximum potential difference, which we call the **voltage amplitude;** and ω is the *angular frequency,* equal to 2π times the frequency f (Fig. 31.1).

In the United States and Canada, commercial electric-power distribution systems always use a frequency of $f = 60$ Hz, corresponding to $\omega = (2\pi \text{ rad})(60 \text{ s}^{-1}) = 377$ rad/s; in much of the rest of the world, $f = 50$ Hz ($\omega = 314$ rad/s) is used. Similarly, a sinusoidal current might be described as

$$i = I\cos\omega t \tag{31.2}$$

where i (lowercase) is the instantaneous current and I (uppercase) is the maximum current or **current amplitude.**

Phasor Diagrams

31.2 A phasor diagram.

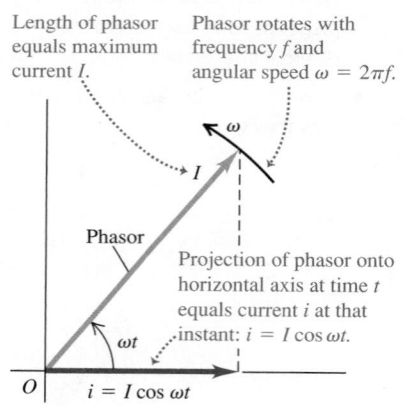

To represent sinusoidally varying voltages and currents, we will use rotating vector diagrams similar to those we used in the study of simple harmonic motion in Section 14.2 (see Figs. 14.5b and 14.6). In these diagrams the instantaneous value of a quantity that varies sinusoidally with time is represented by the *projection* onto a horizontal axis of a vector with a length equal to the amplitude of the quantity. The vector rotates counterclockwise with constant angular speed ω. These rotating vectors are called **phasors,** and diagrams containing them are called **phasor diagrams.** Figure 31.2 shows a phasor diagram for the sinusoidal current described by Eq. (31.2). The projection of the phasor onto the horizontal axis at time t is $I\cos\omega t$; this is why we chose to use the cosine function rather than the sine in Eq. (31.2).

CAUTION *Just what is a phasor?* A phasor is not a real physical quantity with a direction in space, such as velocity, momentum, or electric field. Rather, it is a *geometric* entity that helps us to describe and analyze physical quantities that vary sinusoidally with time. In Section 14.2 we used a single phasor to represent the position of a point mass undergoing simple harmonic motion. In this chapter we will use phasors to *add* sinusoidal voltages and currents. Combining sinusoidal quantities with phase differences then becomes a matter of vector addition. We will find a similar use for phasors in Chapters 35 and 36 in our study of interference effects with light. ❚

Rectified Alternating Current

How do we measure a sinusoidally varying current? In Section 26.3 we used a d'Arsonval galvanometer to measure steady currents. But if we pass a *sinusoidal* current through a d'Arsonval meter, the torque on the moving coil varies sinusoidally, with one direction half the time and the opposite direction the other half. The needle may wiggle a little if the frequency is low enough, but its average deflection is zero. Hence a d'Arsonval meter by itself isn't very useful for measuring alternating currents.

To get a measurable one-way current through the meter, we can use *diodes*, which we described in Section 25.3. A diode is a device that conducts better in one direction than in the other; an ideal diode has zero resistance for one direction of current and infinite resistance for the other. Figure 31.3a shows one possible arrangement, called a *full-wave rectifier circuit*. The current through the galvanometer G is always upward, regardless of the direction of the current from the ac source (i.e., which part of the cycle the source is in). The graph in Fig. 31.3b shows the current through G: It pulsates but always has the same direction, and the average meter deflection is *not* zero.

The **rectified average current** I_{rav} is defined so that during any whole number of cycles, the total charge that flows is the same as though the current were constant with a value equal to I_{rav}. The notation I_{rav} and the name *rectified average current* emphasize that this is *not* the average of the original sinusoidal current. In Fig. 31.3b the total charge that flows in time t corresponds to the area under the curve of i versus t (recall that $i = dq/dt$, so q is the integral of t); this area must equal the rectangular area with height I_{rav}. We see that I_{rav} is less than the maximum current I; the two are related by

$$I_{rav} = \frac{2}{\pi} I = 0.637I \qquad \begin{array}{l}\text{(rectified average value} \\ \text{of a sinusoidal current)}\end{array} \qquad (31.3)$$

(The factor of $2/\pi$ is the average value of $|\cos \omega t|$ or of $|\sin \omega t|$; see Example 29.4 in Section 29.2.) The galvanometer deflection is proportional to I_{rav}. The galvanometer scale can be calibrated to read I, I_{rav}, or, most commonly, I_{rms} (discussed below).

Root-Mean-Square (rms) Values

A more useful way to describe a quantity that can be either positive or negative is the *root-mean-square (rms) value*. We used rms values in Section 18.3 in connection with the speeds of molecules in a gas. We *square* the instantaneous current i, take the *average* (mean) value of i^2, and finally take the *square root* of that average. This procedure defines the **root-mean-square current,** denoted as I_{rms} (Fig. 31.4). Even when i is negative, i^2 is always positive, so I_{rms} is never zero (unless i is zero at every instant).

Here's how we obtain I_{rms} for a sinusoidal current, like that shown in Fig. 31.4. If the instantaneous current is given by $i = I\cos \omega t$, then

$$i^2 = I^2 \cos^2 \omega t$$

Using a double-angle formula from trigonometry,

$$\cos^2 A = \tfrac{1}{2}(1 + \cos 2A)$$

we find

$$i^2 = I^2 \tfrac{1}{2}(1 + \cos 2\omega t) = \tfrac{1}{2}I^2 + \tfrac{1}{2}I^2 \cos 2\omega t$$

The average of $\cos 2\omega t$ is zero because it is positive half the time and negative half the time. Thus the average of i^2 is simply $I^2/2$. The square root of this is I_{rms}:

$$I_{rms} = \frac{I}{\sqrt{2}} \qquad \text{(root-mean-square value of a sinusoidal current)} \qquad (31.4)$$

31.3 (a) A full-wave rectifier circuit. (b) Graph of the resulting current through the galvanometer G.

(a) A full-wave rectifier circuit

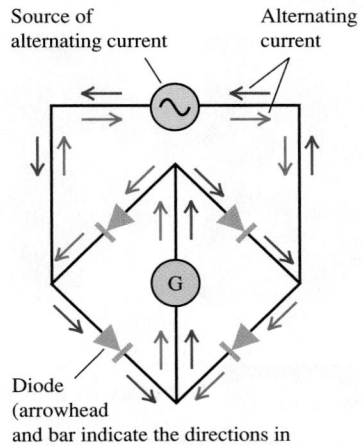

Diode
(arrowhead
and bar indicate the directions in
which current can and cannot pass)

(b) Graph of the full-wave rectified current and its average value, the rectified average current I_{rav}

Area under curve = total charge that flows through galvanometer in time t.

31.4 Calculating the root-mean-square (rms) value of an alternating current.

Meaning of the rms value of a sinusoidal quantity (here, ac current with $I = 3$ A):

① Graph current i versus time.

② *Square* the instantaneous current i.

③ Take the *average* (mean) value of i^2.

④ Take the *square root* of that average.

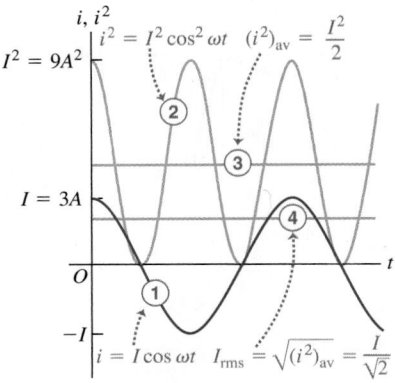

31.5 This wall socket delivers a root-mean-square voltage of 120 V. Sixty times per second, the instantaneous voltage across its terminals varies from $(\sqrt{2})(120 \text{ V}) = 170$ V to -170 V and back again.

In the same way, the root-mean-square value of a sinusoidal voltage with amplitude (maximum value) V is

$$V_{\text{rms}} = \frac{V}{\sqrt{2}} \quad \text{(root-mean-square value of a sinusoidal voltage)} \quad (31.5)$$

We can convert a rectifying ammeter into a voltmeter by adding a series resistor, just as for the dc case discussed in Section 26.3. Meters used for ac voltage and current measurements are nearly always calibrated to read rms values, not maximum or rectified average. Voltages and currents in power distribution systems are always described in terms of their rms values. The usual household power supply, "120-volt ac," has an rms voltage of 120 V (Fig. 31.5). The voltage amplitude is

$$V = \sqrt{2} V_{\text{rms}} = \sqrt{2}(120 \text{ V}) = 170 \text{ V}$$

Example 31.1 **Current in a personal computer**

The plate on the back of a personal computer says that it draws 2.7 A from a 120-V, 60-Hz line. For this computer, what are (a) the average current, (b) the average of the square of the current, and (c) the current amplitude?

SOLUTION

IDENTIFY and SET UP: This example is about alternating current. In part (a) we find the average, over a complete cycle, of the alternating current. In part (b) we recognize that the 2.7-A current draw of the computer is the rms value I_{rms}—that is, the *square root* of the *mean* (average) of the *square* of the current, $(i^2)_{\text{av}}$. In part (c) we use Eq. (31.4) to relate I_{rms} to the current amplitude.

EXECUTE: (a) The average of *any* sinusoidally varying quantity, over any whole number of cycles, is zero.

(b) We are given $I_{\text{rms}} = 2.7$ A. From the definition of rms value,

$$I_{\text{rms}} = \sqrt{(i^2)_{\text{av}}} \text{ so } (i^2)_{\text{av}} = (I_{\text{rms}})^2 = (2.7 \text{ A})^2 = 7.3 \text{ A}^2$$

(c) From Eq. (31.4), the current amplitude I is

$$I = \sqrt{2}I_{\text{rms}} = \sqrt{2}(2.7 \text{ A}) = 3.8 \text{ A}$$

Figure 31.6 shows graphs of i and i^2 versus time t.

31.6 Our graphs of the current i and the square of the current i^2 versus time t.

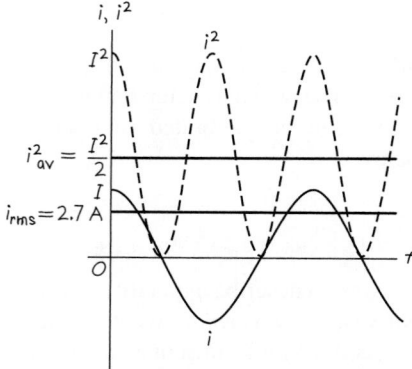

EVALUATE: Why would we be interested in the average of the square of the current? Recall that the rate at which energy is dissipated in a resistor R equals i^2R. This rate varies if the current is alternating, so it is best described by its average value $(i^2)_{\text{av}}R = I_{\text{rms}}^2R$. We'll use this idea in Section 31.4.

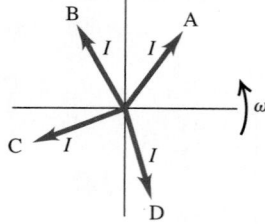

Test Your Understanding of Section 31.1 The figure at left shows four different current phasors with the same angular frequency ω. At the time shown, which phasor corresponds to (a) a positive current that is becoming more positive; (b) a positive current that is decreasing toward zero; (c) a negative current that is becoming more negative; (d) a negative current that is decreasing in magnitude toward zero?

31.2 Resistance and Reactance

In this section we will derive voltage–current relationships for individual circuit elements carrying a sinusoidal current. We'll consider resistors, inductors, and capacitors.

Resistor in an ac Circuit

First let's consider a resistor with resistance R through which there is a sinusoidal current given by Eq. (31.2): $i = I\cos\omega t$. The positive direction of current is counterclockwise around the circuit, as in Fig. 31.7a. The current amplitude (maximum current) is I. From Ohm's law the instantaneous potential v_R of point a with respect to point b (that is, the instantaneous voltage across the resistor) is

$$v_R = iR = (IR)\cos\omega t \qquad (31.6)$$

The maximum voltage V_R, the *voltage amplitude,* is the coefficient of the cosine function:

$$V_R = IR \qquad \text{(amplitude of voltage across a resistor, ac circuit)} \qquad (31.7)$$

Hence we can also write

$$v_R = V_R\cos\omega t \qquad (31.8)$$

The current i and voltage v_R are both proportional to $\cos\omega t$, so the current is *in phase* with the voltage. Equation (31.7) shows that the current and voltage amplitudes are related in the same way as in a dc circuit.

Figure 31.7b shows graphs of i and v_R as functions of time. The vertical scales for current and voltage are different, so the relative heights of the two curves are not significant. The corresponding phasor diagram is given in Fig. 31.7c. Because i and v_R are *in phase* and have the same frequency, the current and voltage phasors rotate together; they are parallel at each instant. Their projections on the horizontal axis represent the instantaneous current and voltage, respectively.

Inductor in an ac Circuit

Next, we replace the resistor in Fig. 31.7 with a pure inductor with self-inductance L and zero resistance (Fig. 31.8a). Again we assume that the current is $i = I\cos\omega t$, with the positive direction of current taken as counterclockwise around the circuit.

Although there is no resistance, there is a potential difference v_L between the inductor terminals a and b because the current varies with time, giving rise to a self-induced emf. The induced emf in the direction of i is given by Eq. (30.7), $\mathcal{E} = -L\,di/dt$; however, the voltage v_L is *not* simply equal to $\mathcal{E}$. To see why, notice that if the current in the inductor is in the positive (counterclockwise) direction from a to b and is increasing, then di/dt is positive and the induced emf is directed to the left to oppose the increase in current; hence point a is at higher potential than is point b. Thus the potential of point a with respect to point b is positive and is given by $v_L = +L\,di/dt$, the *negative* of the induced emf. (You

31.7 Resistance R connected across an ac source.

(a) Circuit with ac source and resistor

(b) Graphs of current and voltage versus time

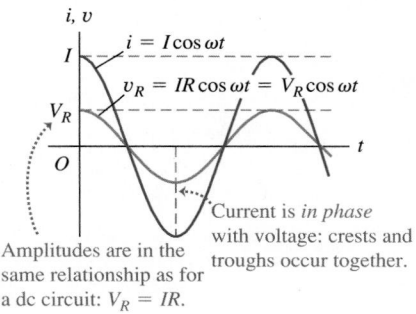

Amplitudes are in the same relationship as for a dc circuit: $V_R = IR$.

Current is *in phase* with voltage: crests and troughs occur together.

(c) Phasor diagram

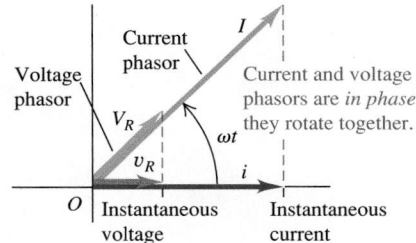

Current and voltage phasors are *in phase*: they rotate together.

31.8 Inductance L connected across an ac source.

(a) Circuit with ac source and inductor

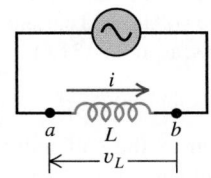

(b) Graphs of current and voltage versus time

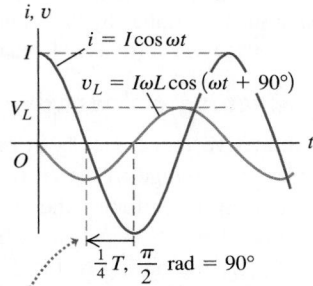

Voltage curve *leads* current curve by a quarter-cycle (corresponding to $\phi = \pi/2$ rad $= 90°$).

(c) Phasor diagram

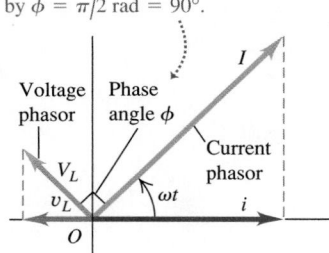

Voltage phasor *leads* current phasor by $\phi = \pi/2$ rad $= 90°$.

should convince yourself that this expression gives the correct sign of v_L in *all* cases, including i counterclockwise and decreasing, i clockwise and increasing, and i clockwise and decreasing; you should also review Section 30.2.) So we have

$$v_L = L\frac{di}{dt} = L\frac{d}{dt}(I\cos\omega t) = -I\omega L\sin\omega t \qquad (31.9)$$

The voltage v_L across the inductor at any instant is proportional to the *rate of change* of the current. The points of maximum voltage on the graph correspond to maximum steepness of the current curve, and the points of zero voltage are the points where the current curve instantaneously levels off at its maximum and minimum values (Fig. 31.8b). The voltage and current are "out of step" or *out of phase* by a quarter-cycle. Since the voltage peaks occur a quarter-cycle earlier than the current peaks, we say that the voltage *leads* the current by 90°. The phasor diagram in Fig. 31.8c also shows this relationship; the voltage phasor is ahead of the current phasor by 90°.

We can also obtain this phase relationship by rewriting Eq. (31.9) using the identity $\cos(A + 90°) = -\sin A$:

$$v_L = I\omega L\cos(\omega t + 90°) \qquad (31.10)$$

This result shows that the voltage can be viewed as a cosine function with a "head start" of 90° relative to the current.

As we have done in Eq. (31.10), we will usually describe the phase of the *voltage* relative to the *current,* not the reverse. Thus if the current i in a circuit is

$$i = I\cos\omega t$$

and the voltage v of one point with respect to another is

$$v = V\cos(\omega t + \phi)$$

we call ϕ the **phase angle;** it gives the phase of the *voltage* relative to the *current.* For a pure resistor, $\phi = 0$, and for a pure inductor, $\phi = 90°$.

From Eq. (31.9) or (31.10) the amplitude V_L of the inductor voltage is

$$V_L = I\omega L \qquad (31.11)$$

We define the **inductive reactance** X_L of an inductor as

$$X_L = \omega L \qquad \text{(inductive reactance)} \qquad (31.12)$$

Using X_L, we can write Eq. (31.11) in a form similar to Eq. (31.7) for a resistor ($V_R = IR$):

$$V_L = IX_L \qquad \text{(amplitude of voltage across an inductor, ac circuit)} \qquad (31.13)$$

Because X_L is the ratio of a voltage and a current, its SI unit is the ohm, the same as for resistance.

CAUTION **Inductor voltage and current are not in phase** Keep in mind that Eq. (31.13) is a relationship between the *amplitudes* of the oscillating voltage and current for the inductor in Fig. 31.8a. It does *not* say that the voltage at any instant is equal to the current at that instant multiplied by X_L. As Fig. 31.8b shows, the voltage and current are 90° out of phase. Voltage and current are in phase only for resistors, as in Eq. (31.6). ∎

The Meaning of Inductive Reactance

The inductive reactance X_L is really a description of the self-induced emf that opposes any change in the current through the inductor. From Eq. (31.13), for a given current amplitude I the voltage $v_L = +L\, di/dt$ across the inductor and the self-induced emf $\mathcal{E} = -L\, di/dt$ both have an amplitude V_L that is directly proportional to X_L. According to Eq. (31.12), the inductive reactance and self-induced emf increase with more rapid variation in current (that is, increasing angular frequency ω) and increasing inductance L.

If an oscillating voltage of a given amplitude V_L is applied across the inductor terminals, the resulting current will have a smaller amplitude I for larger values of X_L. Since X_L is proportional to frequency, a high-frequency voltage applied to the inductor gives only a small current, while a lower-frequency voltage of the same amplitude gives rise to a larger current. Inductors are used in some circuit applications, such as power supplies and radio-interference filters, to block high frequencies while permitting lower frequencies or dc to pass through. A circuit device that uses an inductor for this purpose is called a *low-pass filter* (see Problem 31.52).

Example 31.2 An inductor in an ac circuit

The current amplitude in a pure inductor in a radio receiver is to be 250 μA when the voltage amplitude is 3.60 V at a frequency of 1.60 MHz (at the upper end of the AM broadcast band). (a) What inductive reactance is needed? What inductance? (b) If the voltage amplitude is kept constant, what will be the current amplitude through this inductor at 16.0 MHz? At 160 kHz?

SOLUTION

IDENTIFY and SET UP: There may be other elements of this circuit, but in this example we don't care: All they do is provide the inductor with an oscillating voltage, so the other elements are lumped into the ac source shown in Fig. 31.8a. We are given the current amplitude I and the voltage amplitude V. Our target variables in part (a) are the inductive reactance X_L at 1.60 MHz and the inductance L, which we find using Eqs. (31.13) and (31.12). Knowing L, we use these equations in part (b) to find X_L and I at any frequency.

EXECUTE: (a) From Eq. (31.13),

$$X_L = \frac{V_L}{I} = \frac{3.60 \text{ V}}{250 \times 10^{-6} \text{ A}} = 1.44 \times 10^4 \ \Omega = 14.4 \text{ k}\Omega$$

From Eq. (31.12), with $\omega = 2\pi f$,

$$L = \frac{X_L}{2\pi f} = \frac{1.44 \times 10^4 \ \Omega}{2\pi(1.60 \times 10^6 \text{ Hz})}$$

$$= 1.43 \times 10^{-3} \text{ H} = 1.43 \text{ mH}$$

(b) Combining Eqs. (31.12) and (31.13), we find $I = V_L/X_L = V_L/\omega L = V_L/2\pi f L$. Thus the current amplitude is inversely proportional to the frequency f. Since $I = 250 \ \mu$A at $f = 1.60$ MHz, the current amplitudes at 16.0 MHz ($10f$) and 160 kHz = 0.160 MHz ($f/10$) will be, respectively, one-tenth as great (25.0 μA) and ten times as great (2500 μA = 2.50 mA).

EVALUATE: In general, the lower the frequency of an oscillating voltage applied across an inductor, the greater the amplitude of the resulting oscillating current.

Capacitor in an ac Circuit

Finally, we connect a capacitor with capacitance C to the source, as in Fig. 31.9a, producing a current $i = I\cos\omega t$ through the capacitor. Again, the positive direction of current is counterclockwise around the circuit.

CAUTION **Alternating current through a capacitor** You may object that charge can't really move through the capacitor because its two plates are insulated from each other. True enough, but as the capacitor charges and discharges, there is at each instant a current i into one plate, an equal current out of the other plate, and an equal *displacement* current between the plates just as though the charge were being conducted through the capacitor. (You may want to review the discussion of displacement current in Section 29.7.) Thus we often speak about alternating current *through* a capacitor. █

To find the instantaneous voltage v_C across the capacitor—that is, the potential of point a with respect to point b—we first let q denote the charge on the left-hand plate of the capacitor in Fig. 31.9a (so $-q$ is the charge on the right-hand plate). The current i is related to q by $i = dq/dt$; with this definition, positive current corresponds to an increasing charge on the left-hand capacitor plate. Then

$$i = \frac{dq}{dt} = I\cos\omega t$$

Integrating this, we get

$$q = \frac{I}{\omega}\sin\omega t \qquad\qquad (31.14)$$

31.9 Capacitor C connected across an ac source.

(a) Circuit with ac source and capacitor

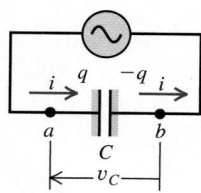

(b) Graphs of current and voltage versus time

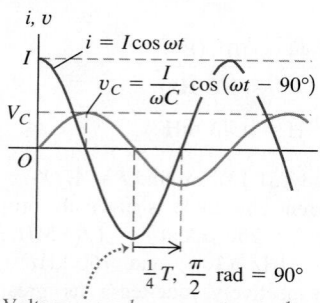

Voltage curve *lags* current curve by a quarter-cycle (corresponding to $\phi = -\pi/2$ rad $= -90°$).

(c) Phasor diagram

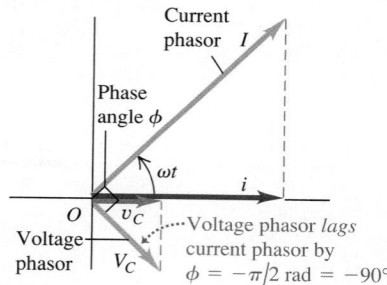

Also, from Eq. (24.1) the charge q equals the voltage v_C multiplied by the capacitance, $q = Cv_C$. Using this in Eq. (31.14), we find

$$v_C = \frac{I}{\omega C} \sin \omega t \qquad (31.15)$$

The instantaneous current i is equal to the rate of change dq/dt of the capacitor charge q; since $q = Cv_C$, i is also proportional to the rate of change of voltage. (Compare to an inductor, for which the situation is reversed and v_L is proportional to the rate of change of i.) Figure 31.9b shows v_C and i as functions of t. Because $i = dq/dt = C \, dv_C/dt$, the current has its greatest magnitude when the v_C curve is rising or falling most steeply and is zero when the v_C curve instantaneously levels off at its maximum and minimum values.

The capacitor voltage and current are out of phase by a quarter-cycle. The peaks of voltage occur a quarter-cycle *after* the corresponding current peaks, and we say that the voltage *lags* the current by 90°. The phasor diagram in Fig. 31.9c shows this relationship; the voltage phasor is behind the current phasor by a quarter-cycle, or 90°.

We can also derive this phase difference by rewriting Eq. (31.15) using the identity $\cos(A - 90°) = \sin A$:

$$v_C = \frac{I}{\omega C} \cos(\omega t - 90°) \qquad (31.16)$$

This corresponds to a phase angle $\phi = -90°$. This cosine function has a "late start" of 90° compared with the current $i = I \cos \omega t$.

Equations (31.15) and (31.16) show that the *maximum* voltage V_C (the voltage amplitude) is

$$V_C = \frac{I}{\omega C} \qquad (13.17)$$

To put this expression in a form similar to Eq. (31.7) for a resistor, $V_R = IR$, we define a quantity X_C, called the **capacitive reactance** of the capacitor, as

$$X_C = \frac{1}{\omega C} \qquad \text{(capacitive reactance)} \qquad (31.18)$$

Then

$$V_C = IX_C \qquad \text{(amplitude of voltage across a capacitor, ac circuit)} \qquad (31.19)$$

The SI unit of X_C is the ohm, the same as for resistance and inductive reactance, because X_C is the ratio of a voltage and a current.

CAUTION **Capacitor voltage and current are not in phase** Remember that Eq. (31.19) for a capacitor, like Eq. (31.13) for an inductor, is *not* a statement about the instantaneous values of voltage and current. The instantaneous values are actually 90° out of phase, as Fig. 31.9b shows. Rather, Eq. (31.19) relates the *amplitudes* of the voltage and current. ▮

The Meaning of Capacitive Reactance

The capacitive reactance of a capacitor is inversely proportional both to the capacitance C and to the angular frequency ω; the greater the capacitance and the higher the frequency, the *smaller* the capacitive reactance X_C. Capacitors tend to pass high-frequency current and to block low-frequency currents and dc, just the opposite of inductors. A device that preferentially passes signals of high frequency is called a *high-pass filter* (see Problem 31.51).

Example 31.3 **A resistor and a capacitor in an ac circuit**

A 200-Ω resistor is connected in series with a 5.0-μF capacitor. The voltage across the resistor is $v_R = (1.20\ \text{V})\cos(2500\ \text{rad/s})t$ (Fig. 31.10). (a) Derive an expression for the circuit current. (b) Determine the capacitive reactance of the capacitor. (c) Derive an expression for the voltage across the capacitor.

SOLUTION

IDENTIFY and SET UP: Since this is a series circuit, the current is the same through the capacitor as through the resistor. Our target variables are the current i, the capacitive reactance X_C, and the capacitor voltage v_C. We use Eq. (31.6) to find an expression for i in terms of the angular frequency $\omega = 2500\ \text{rad/s}$, Eq. (31.18) to find X_C, Eq. (31.19) to find the capacitor voltage amplitude V_C, and Eq. (31.16) to write an expression for v_C.

31.10 Our sketch for this problem.

EXECUTE: (a) From Eq. (31.6), $v_R = iR$, we find

$$i = \frac{v_R}{R} = \frac{(1.20\ \text{V})\cos(2500\ \text{rad/s})t}{200\ \Omega}$$
$$= (6.0 \times 10^{-3}\ \text{A})\cos(2500\ \text{rad/s})t$$

(b) From Eq. (31.18), the capacitive reactance at $\omega = 2500\ \text{rad/s}$ is

$$X_C = \frac{1}{\omega C} = \frac{1}{(2500\ \text{rad/s})(5.0 \times 10^{-6}\ \text{F})} = 80\ \Omega$$

(c) From Eq. (31.19), the capacitor voltage amplitude is

$$V_C = IX_C = (6.0 \times 10^{-3}\ \text{A})(80\ \Omega) = 0.48\ \text{V}$$

(The 80-Ω reactance of the capacitor is 40% of the resistor's 200-Ω resistance, so V_C is 40% of V_R.) The instantaneous capacitor voltage is given by Eq. (31.16):

$$v_C = V_C \cos(\omega t - 90°)$$
$$= (0.48\ \text{V})\cos[(2500\ \text{rad/s})t - \pi/2\ \text{rad}]$$

EVALUATE: Although the same *current* passes through both the capacitor and the resistor, the *voltages* across them are different in both amplitude and phase. Note that in the expression for v_C we converted the 90° to $\pi/2$ rad so that all the angular quantities have the same units. In ac circuit analysis, phase angles are often given in degrees, so be careful to convert to radians when necessary.

Comparing ac Circuit Elements

Table 31.1 summarizes the relationships of voltage and current amplitudes for the three circuit elements we have discussed. Note again that *instantaneous* voltage and current are proportional in a resistor, where there is zero phase difference between v_R and i (see Fig. 31.7b). The instantaneous voltage and current are *not* proportional in an inductor or capacitor, because there is a 90° phase difference in both cases (see Figs. 31.8b and 31.9b).

Figure 31.11 shows how the resistance of a resistor and the reactances of an inductor and a capacitor vary with angular frequency ω. Resistance R is independent of frequency, while the reactances X_L and X_C are not. If $\omega = 0$, corresponding to a dc circuit, there is *no* current through a capacitor because $X_C \to \infty$, and there is no inductive effect because $X_L = 0$. In the limit $\omega \to \infty$, X_L also approaches infinity, and the current through an inductor becomes vanishingly small; recall that the self-induced emf opposes rapid changes in current. In this same limit, X_C and the voltage across a capacitor both approach zero; the current changes direction so rapidly that no charge can build up on either plate.

Figure 31.12 shows an application of the above discussion to a loudspeaker system. Low-frequency sounds are produced by the *woofer*, which is a speaker with large diameter; the *tweeter*, a speaker with smaller diameter, produces high-frequency sounds. In order to route signals of different frequency to the appropriate speaker, the woofer and tweeter are connected in parallel across the amplifier

31.11 Graphs of R, X_L, and X_C as functions of angular frequency ω.

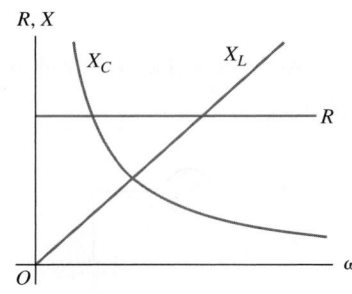

Table 31.1 Circuit Elements with Alternating Current

Circuit Element	Amplitude Relationship	Circuit Quantity	Phase of v
Resistor	$V_R = IR$	R	In phase with i
Inductor	$V_L = IX_L$	$X_L = \omega L$	Leads i by 90°
Capacitor	$V_C = IX_C$	$X_C = 1/\omega C$	Lags i by 90°

31.12 (a) The two speakers in this loud-speaker system are connected in parallel to the amplifier. (b) Graphs of current amplitude in the tweeter and woofer as functions of frequency for a given amplifier voltage amplitude.

(a) A crossover network in a loudspeaker system

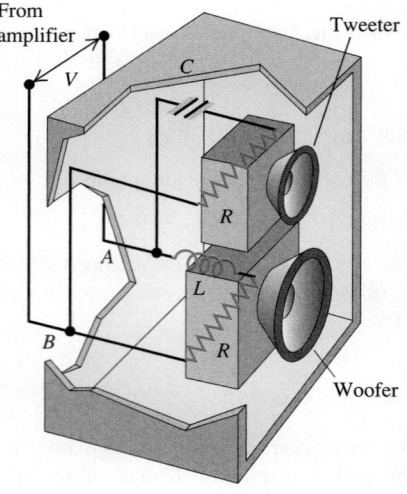

(b) Graphs of rms current as functions of frequency for a given amplifier voltage

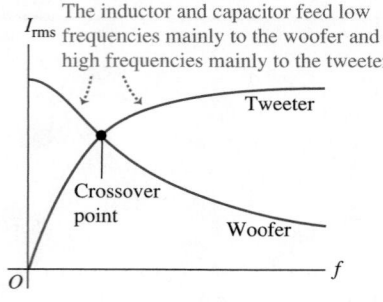

The inductor and capacitor feed low frequencies mainly to the woofer and high frequencies mainly to the tweeter.

output. The capacitor in the tweeter branch blocks the low-frequency components of sound but passes the higher frequencies; the inductor in the woofer branch does the opposite.

Test Your Understanding of Section 31.2 An oscillating voltage of fixed amplitude is applied across a circuit element. If the frequency of this voltage is increased, will the amplitude of the current through the element (i) increase, (ii) decrease, or (iii) remain the same if it is (a) a resistor, (b) an inductor, or (c) a capacitor?

31.3 The *L-R-C* Series Circuit

Many ac circuits used in practical electronic systems involve resistance, inductive reactance, and capacitive reactance. Figure 31.13a shows a simple example: A series circuit containing a resistor, an inductor, a capacitor, and an ac source. (In Section 30.6 we considered the behavior of the current in an *L-R-C* series circuit *without* a source.)

To analyze this and similar circuits, we will use a phasor diagram that includes the voltage and current phasors for each of the components. In this circuit, because of Kirchhoff's loop rule, the instantaneous *total* voltage v_{ad} across all three components is equal to the source voltage at that instant. We will show that the phasor representing this total voltage is the *vector sum* of the phasors for the individual voltages.

Figures 31.13b and 31.13c show complete phasor diagrams for the circuit of Fig. 31.13a. We assume that the source supplies a current i given by $i = I\cos\omega t$. Because the circuit elements are connected in series, the current at any instant is the same at every point in the circuit. Thus a *single phasor I*, with length proportional to the current amplitude, represents the current in *all* circuit elements.

As in Section 31.2, we use the symbols v_R, v_L, and v_C for the instantaneous voltages across R, L, and C, and the symbols V_R, V_L, and V_C for the maximum voltages. We denote the instantaneous and maximum *source* voltages by v and V. Then, in Fig. 31.13a, $v = v_{ad}$, $v_R = v_{ab}$, $v_L = v_{bc}$, and $v_C = v_{cd}$.

We have shown that the potential difference between the terminals of a resistor is *in phase* with the current in the resistor and that its maximum value V_R is given by Eq. (31.7):

$$V_R = IR$$

31.13 An *L-R-C* series circuit with an ac source.

(a) *L-R-C* series circuit

(b) Phasor diagram for the case $X_L > X_C$

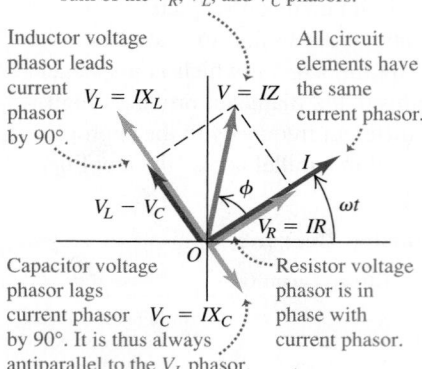

Source voltage phasor is the vector sum of the V_R, V_L, and V_C phasors.

Inductor voltage phasor leads current phasor by 90°. $V_L = IX_L$

$V = IZ$

All circuit elements have the same current phasor.

$V_L - V_C$

ϕ

I

ωt

$V_R = IR$

Capacitor voltage phasor lags current phasor $V_C = IX_C$ by 90°. It is thus always antiparallel to the V_L phasor.

Resistor voltage phasor is in phase with current phasor.

(c) Phasor diagram for the case $X_L < X_C$

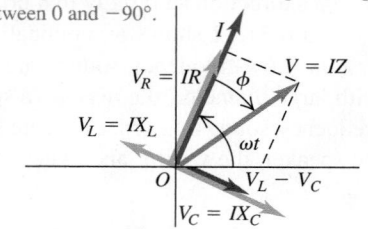

If $X_L < X_C$, the source voltage phasor lags the current phasor, $X < 0$, and ϕ is a negative angle between 0 and $-90°$.

I

$V = IZ$

$V_R = IR$

ϕ

$V_L = IX_L$

ωt

O

$V_L - V_C$

$V_C = IX_C$

The phasor V_R in Fig. 31.13b, in phase with the current phasor I, represents the voltage across the resistor. Its projection onto the horizontal axis at any instant gives the instantaneous potential difference v_R.

The voltage across an inductor *leads* the current by 90°. Its voltage amplitude is given by Eq. (31.13):

$$V_L = IX_L$$

The phasor V_L in Fig. 31.13b represents the voltage across the inductor, and its projection onto the horizontal axis at any instant equals v_L.

The voltage across a capacitor *lags* the current by 90°. Its voltage amplitude is given by Eq. (31.19):

$$V_C = IX_C$$

The phasor V_C in Fig. 31.13b represents the voltage across the capacitor, and its projection onto the horizontal axis at any instant equals v_C.

The instantaneous potential difference v between terminals a and d is equal at every instant to the (algebraic) sum of the potential differences v_R, v_L, and v_C. That is, it equals the sum of the *projections* of the phasors V_R, V_L, and V_C. But the sum of the projections of these phasors is equal to the *projection* of their *vector sum*. So the vector sum V must be the phasor that represents the source voltage v and the instantaneous total voltage v_{ad} across the series of elements.

To form this vector sum, we first subtract the phasor V_C from the phasor V_L. (These two phasors always lie along the same line, with opposite directions.) This gives the phasor $V_L - V_C$. This is always at right angles to the phasor V_R, so from the Pythagorean theorem the magnitude of the phasor V is

$$V = \sqrt{V_R^2 + (V_L - V_C)^2} = \sqrt{(IR)^2 + (IX_L - IX_C)^2} \quad \text{or}$$
$$V = I\sqrt{R^2 + (X_L - X_C)^2} \tag{31.20}$$

We define the **impedance** Z of an ac circuit as the ratio of the voltage amplitude across the circuit to the current amplitude in the circuit. From Eq. (31.20) the impedance of the *L-R-C* series circuit is

$$Z = \sqrt{R^2 + (X_L - X_C)^2} \tag{31.21}$$

so we can rewrite Eq. (31.20) as

$$V = IZ \quad \text{(amplitude of voltage across an ac circuit)} \tag{31.22}$$

While Eq. (31.21) is valid only for an *L-R-C* series circuit, we can use Eq. (31.22) to define the impedance of *any* network of resistors, inductors, and capacitors as the ratio of the amplitude of the voltage across the network to the current amplitude. The SI unit of impedance is the ohm.

The Meaning of Impedance and Phase Angle

Equation (31.22) has a form similar to $V = IR$, with impedance Z in an ac circuit playing the role of resistance R in a dc circuit. Just as direct current tends to follow the path of least resistance, so alternating current tends to follow the path of lowest impedance (Fig. 31.14). Note, however, that impedance is actually a function of R, L, and C, as well as of the angular frequency ω. We can see this by substituting Eq. (31.12) for X_L and Eq. (31.18) for X_C into Eq. (31.21), giving the following complete expression for Z for a series circuit:

$$Z = \sqrt{R^2 + (X_L - X_C)^2}$$
$$= \sqrt{R^2 + [\omega L - (1/\omega C)]^2} \quad \text{(impedance of an *L-R-C* series circuit)} \tag{31.23}$$

31.14 This gas-filled glass sphere has an alternating voltage between its surface and the electrode at its center. The glowing streamers show the resulting alternating current that passes through the gas. When you touch the outside of the sphere, your fingertips and the inner surface of the sphere act as the plates of a capacitor, and the sphere and your body together form an *L-R-C* series circuit. The current (which is low enough to be harmless) is drawn to your fingers because the path through your body has a low impedance.

Hence for a given amplitude V of the source voltage applied to the circuit, the amplitude $I = V/Z$ of the resulting current will be different at different frequencies. We'll explore this frequency dependence in detail in Section 31.5.

In the phasor diagram shown in Fig. 31.13b, the angle ϕ between the voltage and current phasors is the phase angle of the source voltage v with respect to the current i; that is, it is the angle by which the source voltage leads the current. From the diagram,

$$\tan\phi = \frac{V_L - V_C}{V_R} = \frac{I(X_L - X_C)}{IR} = \frac{X_L - X_C}{R}$$

$$\tan\phi = \frac{\omega L - 1/\omega C}{R} \qquad \text{(phase angle of an } L\text{-}R\text{-}C \text{ series circuit)} \quad (31.24)$$

If the current is $i = I\cos\omega t$, then the source voltage v is

$$v = V\cos(\omega t + \phi) \qquad (31.25)$$

Figure 31.13b shows the behavior of a circuit in which $X_L > X_C$. Figure 31.13c shows the behavior when $X_L < X_C$; the voltage phasor V lies on the opposite side of the current phasor I and the voltage *lags* the current. In this case, $X_L - X_C$ is *negative,* $\tan\phi$ is negative, and ϕ is a negative angle between 0 and $-90°$. Since X_L and X_C depend on frequency, the phase angle ϕ depends on frequency as well. We'll examine the consequences of this in Section 31.5.

All of the expressions that we've developed for an L-R-C series circuit are still valid if one of the circuit elements is missing. If the resistor is missing, we set $R = 0$; if the inductor is missing, we set $L = 0$. But if the capacitor is missing, we set $C = \infty$, corresponding to the absence of any potential difference ($v_C = q/C = 0$) or any capacitive reactance ($X_C = 1/\omega C = 0$).

In this entire discussion we have described magnitudes of voltages and currents in terms of their *maximum* values, the voltage and current *amplitudes.* But we remarked at the end of Section 31.1 that these quantities are usually described in terms of rms values, not amplitudes. For any sinusoidally varying quantity, the rms value is always $1/\sqrt{2}$ times the amplitude. All the relationships between voltage and current that we have derived in this and the preceding sections are still valid if we use rms quantities throughout instead of amplitudes. For example, if we divide Eq. (31.22) by $\sqrt{2}$, we get

$$\frac{V}{\sqrt{2}} = \frac{I}{\sqrt{2}}Z$$

which we can rewrite as

$$V_{\text{rms}} = I_{\text{rms}}Z \qquad (31.26)$$

We can translate Eqs. (31.7), (31.13), and (31.19) in exactly the same way.

We have considered only ac circuits in which an inductor, a resistor, and a capacitor are in series. You can do a similar analysis for an L-R-C *parallel* circuit; see Problem 31.56.

Finally, we remark that in this section we have been describing the *steady-state* condition of a circuit, the state that exists after the circuit has been connected to the source for a long time. When the source is first connected, there may be additional voltages and currents, called *transients,* whose nature depends on the time in the cycle when the circuit is initially completed. A detailed analysis of transients is beyond our scope. They always die out after a sufficiently long time, and they do not affect the steady-state behavior of the circuit. But they can cause dangerous and damaging surges in power lines, which is why delicate electronic systems such as computers are often provided with power-line surge protectors.

Problem-Solving Strategy 31.1 Alternating-Current Circuits

IDENTIFY *the relevant concepts:* In analyzing ac circuits, we can apply all of the concepts used to analyze direct-current circuits, particularly those in Problem-Solving Strategies 26.1 and 26.2. But now we must distinguish between the amplitudes of alternating currents and voltages and their instantaneous values, and among resistance (for resistors), reactance (for inductors or capacitors), and impedance (for composite circuits).

SET UP *the problem* using the following steps:
1. Draw a diagram of the circuit and label all known and unknown quantities.
2. Identify the target variables.

EXECUTE *the solution* as follows:
1. Use the relationships derived in Sections 31.2 and 31.3 to solve for the target variables, using the following hints.
2. It's almost always easiest to work with angular frequency $\omega = 2\pi f$ rather than ordinary frequency f.
3. Keep in mind the following phase relationships: For a resistor, voltage and current are *in phase,* so the corresponding phasors always point in the same direction. For an inductor, the voltage *leads* the current by 90° (i.e., $\phi = +90° = \pi/2$ radians), so the voltage phasor points 90° counterclockwise from the current phasor. For a capacitor, the voltage *lags* the current by 90° (i.e., $\phi = -90° = -\pi/2$ radians), so the voltage phasor points 90° clockwise from the current phasor.

4. Kirchhoff's rules hold *at each instant.* For example, in a series circuit, the instantaneous current is the same in all circuit elements; in a parallel circuit, the instantaneous potential difference is the same across all circuit elements.
5. Inductive reactance, capacitive reactance, and impedance are analogous to resistance; each represents the ratio of voltage amplitude V to current amplitude I in a circuit element or combination of elements. However, phase relationships are crucial. In applying Kirchhoff's loop rule, you must combine the effects of resistance and reactance by *vector* addition of the corresponding voltage phasors, as in Figs. 31.13b and 31.13c. When you have several circuit elements in series, for example, you can't just *add* all the numerical values of resistance and reactance to get the impedance; that would ignore the phase relationships.

EVALUATE *your answer:* When working with an *L-R-C* series circuit, you can check your results by comparing the values of the inductive and capacitive reactances X_L and X_C. If $X_L > X_C$, then the voltage amplitude across the inductor is greater than that across the capacitor and the phase angle ϕ is positive (between 0 and 90°). If $X_L < X_C$, then the voltage amplitude across the inductor is less than that across the capacitor and the phase angle ϕ is negative (between 0 and −90°).

Example 31.4 An *L-R-C* series circuit I

In the series circuit of Fig. 31.13a, suppose $R = 300\ \Omega$, $L = 60$ mH, $C = 0.50\ \mu$F, $V = 50$ V, and $\omega = 10{,}000$ rad/s. Find the reactances X_L and X_C, the impedance Z, the current amplitude I, the phase angle ϕ, and the voltage amplitude across each circuit element.

SOLUTION

IDENTIFY and SET UP: This problem uses the ideas developed in Section 31.2 and this section about the behavior of circuit elements in an ac circuit. We use Eqs. (31.12) and (31.18) to determine X_L and X_C, and Eq. (31.23) to find Z. We then use Eq. (31.22) to find the current amplitude and Eq. (31.24) to find the phase angle. The relationships in Table 31.1 then yield the voltage amplitudes.

EXECUTE: The inductive and capacitive reactances are

$$X_L = \omega L = (10{,}000\ \text{rad/s})(60\ \text{mH}) = 600\ \Omega$$

$$X_C = \frac{1}{\omega C} = \frac{1}{(10{,}000\ \text{rad/s})(0.50 \times 10^{-6}\ \text{F})} = 200\ \Omega$$

The impedance Z of the circuit is then

$$Z = \sqrt{R^2 + (X_L - X_C)^2} = \sqrt{(300\ \Omega)^2 + (600\ \Omega - 200\ \Omega)^2}$$
$$= 500\ \Omega$$

With source voltage amplitude $V = 50$ V, the current amplitude I and phase angle ϕ are

$$I = \frac{V}{Z} = \frac{50\ \text{V}}{500\ \Omega} = 0.10\ \text{A}$$

$$\phi = \arctan\frac{X_L - X_C}{R} = \arctan\frac{400\ \Omega}{300\ \Omega} = 53°$$

From Table 31.1, the voltage amplitudes V_R, V_L, and V_C across the resistor, inductor, and capacitor, respectively, are

$$V_R = IR = (0.10\ \text{A})(300\ \Omega) = 30\ \text{V}$$
$$V_L = IX_L = (0.10\ \text{A})(600\ \Omega) = 60\ \text{V}$$
$$V_C = IX_C = (0.10\ \text{A})(200\ \Omega) = 20\ \text{V}$$

EVALUATE: As in Fig. 31.13b, $X_L > X_C$; hence the voltage amplitude across the inductor is greater than that across the capacitor and ϕ is positive. The value $\phi = 53°$ means that the voltage *leads* the current by 53°.

Note that the source voltage amplitude $V = 50$ V is *not* equal to the sum of the voltage amplitudes across the separate circuit elements: 50 V $\neq$ 30 V + 60 V + 20 V. Instead, V is the *vector sum* of the V_R, V_L, and V_C phasors. If you draw the phasor diagram like Fig. 31.13b for this particular situation, you'll see that V_R, $V_L - V_C$, and V constitute a 3-4-5 right triangle.

Example 31.5 **An *L-R-C* series circuit II**

For the *L-R-C* series circuit of Example 31.4, find expressions for the time dependence of the instantaneous current i and the instantaneous voltages across the resistor (v_R), inductor (v_L), capacitor (v_C), and ac source (v).

SOLUTION

IDENTIFY and SET UP: We describe the current using Eq. (31.2), which assumes that the current is maximum at $t = 0$. The voltages are then given by Eq. (31.8) for the resistor, Eq. (31.10) for the inductor, Eq. (31.16) for the capacitor, and Eq. (31.25) for the source.

EXECUTE: The current and the voltages all oscillate with the same angular frequency, $\omega = 10{,}000$ rad/s, and hence with the same period, $2\pi/\omega = 2\pi/(10{,}000 \text{ rad/s}) = 6.3 \times 10^{-4}$ s = 0.63 ms. From Eq. (31.2), the current is

$$i = I\cos\omega t = (0.10 \text{ A})\cos(10{,}000 \text{ rad/s})t$$

The resistor voltage is *in phase* with the current, so

$$v_R = V_R\cos\omega t = (30 \text{ V})\cos(10{,}000 \text{ rad/s})t$$

The inductor voltage *leads* the current by 90°, so

$$v_L = V_L\cos(\omega t + 90°) = -V_L\sin\omega t$$
$$= -(60 \text{ V})\sin(10{,}000 \text{ rad/s})t$$

The capacitor voltage *lags* the current by 90°, so

$$v_C = V_C\cos(\omega t - 90°) = V_C\sin\omega t$$
$$= (20 \text{ V})\sin(10{,}000 \text{ rad/s})t$$

We found in Example 31.4 that the source voltage (equal to the voltage across the entire combination of resistor, inductor, and capacitor) *leads* the current by $\phi = 53°$, so

$$v = V\cos(\omega t + \phi)$$
$$= (50 \text{ V})\cos\left[(10{,}000 \text{ rad/s})t + \left(\frac{2\pi \text{ rad}}{360°}\right)(53°)\right]$$
$$= (50 \text{ V})\cos[(10{,}000 \text{ rad/s})t + 0.93 \text{ rad}]$$

EVALUATE: Figure 31.15 graphs the four voltages versus time. The inductor voltage has a larger amplitude than the capacitor voltage because $X_L > X_C$. The *instantaneous* source voltage v is always equal to the sum of the instantaneous voltages v_R, v_L, and v_C. You should verify this by measuring the values of the voltages shown in the graph at different values of the time t.

31.15 Graphs of the source voltage v, resistor voltage v_R, inductor voltage v_L, and capacitor voltage v_C as functions of time for the situation of Example 31.4. The current, which is not shown, is in phase with the resistor voltage.

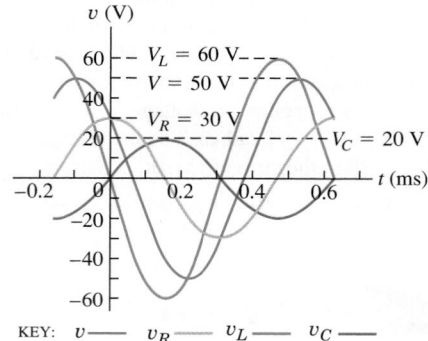

Test Your Understanding of Section 31.3 Rank the following ac circuits in order of their current amplitude, from highest to lowest value. (i) the circuit in Example 31.4; (ii) the circuit in Example 31.4 with the capacitor and inductor both removed; (iii) the circuit in Example 31.4 with the resistor and capacitor both removed; (iv) the circuit in Example 31.4 with the resistor and inductor both removed.

31.4 Power in Alternating-Current Circuits

Alternating currents play a central role in systems for distributing, converting, and using electrical energy, so it's important to look at power relationships in ac circuits. For an ac circuit with instantaneous current i and current amplitude I, we'll consider an element of that circuit across which the instantaneous potential difference is v with voltage amplitude V. The instantaneous power p delivered to this circuit element is

$$p = vi$$

Let's first see what this means for individual circuit elements. We'll assume in each case that $i = I\cos\omega t$.

Power in a Resistor

Suppose first that the circuit element is a *pure resistor R*, as in Fig. 31.7a; then $v = v_R$ and i are *in phase*. We obtain the graph representing p by multiplying the heights of the graphs of v and i in Fig. 31.7b at each instant. This graph is shown by the black curve in Fig. 31.16a. The product vi is always positive because v and i are always either both positive or both negative. Hence energy is supplied *to* the resistor at every instant for both directions of i, although the power is not constant.

The power curve for a pure resistor is symmetrical about a value equal to one-half its maximum value VI, so the *average power* P_{av} is

$$P_{av} = \tfrac{1}{2}VI \qquad \text{(for a pure resistor)} \qquad (31.27)$$

An equivalent expression is

$$P_{av} = \frac{V}{\sqrt{2}} \frac{I}{\sqrt{2}} = V_{rms}I_{rms} \qquad \text{(for a pure resistor)} \qquad (31.28)$$

Also, $V_{rms} = I_{rms}R$, so we can express P_{av} by any of the equivalent forms

$$P_{av} = I_{rms}{}^2R = \frac{V_{rms}{}^2}{R} = V_{rms}I_{rms} \qquad \text{(for a pure resistor)} \qquad (31.29)$$

Note that the expressions in Eq. (31.29) have the same form as the corresponding relationships for a dc circuit, Eq. (25.18). Also note that they are valid only for pure resistors, not for more complicated combinations of circuit elements.

Power in an Inductor

Next we connect the source to a pure inductor L, as in Fig. 31.8a. The voltage $v = v_L$ leads the current i by 90°. When we multiply the curves of v and i, the product vi is *negative* during the half of the cycle when v and i have *opposite* signs. The power curve, shown in Fig. 31.16b, is symmetrical about the horizontal axis; it is positive half the time and negative the other half, and the average power is zero. When p is positive, energy is being supplied to set up the magnetic field in the inductor; when p is negative, the field is collapsing and the inductor is returning energy to the source. The net energy transfer over one cycle is zero.

Power in a Capacitor

Finally, we connect the source to a pure capacitor C, as in Fig. 31.9a. The voltage $v = v_C$ lags the current i by 90°. Figure 31.16c shows the power curve; the average power is again zero. Energy is supplied to charge the capacitor and is returned

31.16 Graphs of current, voltage, and power as functions of time for **(a)** a pure resistor, **(b)** a pure inductor, **(c)** a pure capacitor, and **(d)** an arbitrary ac circuit that can have resistance, inductance, and capacitance.

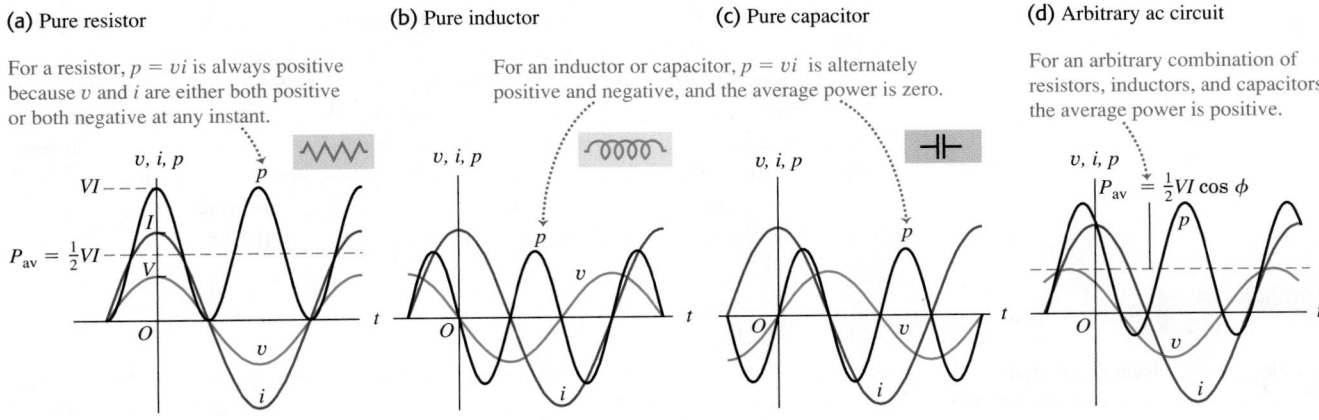

(a) Pure resistor

For a resistor, $p = vi$ is always positive because v and i are either both positive or both negative at any instant.

(b) Pure inductor

(c) Pure capacitor

For an inductor or capacitor, $p = vi$ is alternately positive and negative, and the average power is zero.

(d) Arbitrary ac circuit

For an arbitrary combination of resistors, inductors, and capacitors, the average power is positive.

KEY: Instantaneous current, i ——   Instantaneous voltage across device, v ——   Instantaneous power input to device, p ——

to the source when the capacitor discharges. The net energy transfer over one cycle is again zero.

Power in a General ac Circuit

In *any* ac circuit, with any combination of resistors, capacitors, and inductors, the voltage v across the entire circuit has some phase angle ϕ with respect to the current i. Then the instantaneous power p is given by

$$p = vi = [V\cos(\omega t + \phi)][I\cos \omega t] \qquad (31.30)$$

The instantaneous power curve has the form shown in Fig. 31.16d. The area between the positive loops and the horizontal axis is greater than the area between the negative loops and the horizontal axis, and the average power is positive.

We can derive from Eq. (31.30) an expression for the *average* power P_{av} by using the identity for the cosine of the sum of two angles:

$$p = [V(\cos \omega t \cos \phi - \sin \omega t \sin \phi)][I\cos \omega t]$$
$$= VI\cos \phi \cos^2 \omega t - VI\sin \phi \cos \omega t \sin \omega t$$

From the discussion in Section 31.1 that led to Eq. (31.4), we see that the average value of $\cos^2 \omega t$ (over one cycle) is $\frac{1}{2}$. The average value of $\cos \omega t \sin \omega t$ is zero because this product is equal to $\frac{1}{2}\sin 2\omega t$, whose average over a cycle is zero. So the average power P_{av} is

$$P_{av} = \tfrac{1}{2}VI\cos \phi = V_{rms}I_{rms}\cos \phi \qquad \begin{array}{l}\text{(average power into a}\\ \text{general ac circuit)}\end{array} \qquad (31.31)$$

When v and i are in phase, so $\phi = 0$, the average power equals $\frac{1}{2}VI = V_{rms}I_{rms}$; when v and i are 90° out of phase, the average power is zero. In the general case, when v has a phase angle ϕ with respect to i, the average power equals $\frac{1}{2}I$ multiplied by $V\cos \phi$, the component of the voltage phasor that is *in phase* with the current phasor. Figure 31.17 shows the general relationship of the current and voltage phasors. For the *L-R-C* series circuit, Figs. 31.13b and 31.13c show that $V\cos \phi$ equals the voltage amplitude V_R for the resistor; hence Eq. (31.31) is the average power dissipated in the resistor. On average there is no energy flow into or out of the inductor or capacitor, so none of P_{av} goes into either of these circuit elements.

The factor $\cos \phi$ is called the **power factor** of the circuit. For a pure resistance, $\phi = 0$, $\cos \phi = 1$, and $P_{av} = V_{rms}I_{rms}$. For a pure inductor or capacitor, $\phi = \pm 90°$, $\cos \phi = 0$, and $P_{av} = 0$. For an *L-R-C* series circuit the power factor is equal to R/Z; we leave the proof of this statement to you (see Exercise 31.21).

A low power factor (large angle ϕ of lag or lead) is usually undesirable in power circuits. The reason is that for a given potential difference, a large current is needed to supply a given amount of power. This results in large i^2R losses in the transmission lines. Your electric power company may charge a higher rate to a client with a low power factor. Many types of ac machinery draw a *lagging* current; that is, the current drawn by the machinery lags the applied voltage. Hence the voltage leads the current, so $\phi > 0$ and $\cos \phi < 1$. The power factor can be corrected toward the ideal value of 1 by connecting a capacitor in parallel with the load. The current drawn by the capacitor *leads* the voltage (that is, the voltage across the capacitor lags the current), which compensates for the lagging current in the other branch of the circuit. The capacitor itself absorbs no net power from the line.

31.17 Using phasors to calculate the average power for an arbitrary ac circuit.

Average power $= \frac{1}{2}I(V\cos \phi)$, where $V\cos \phi$ is the component of V in phase with I.

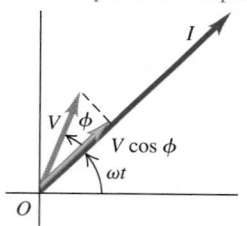

Example 31.6 | **Power in a hair dryer**

An electric hair dryer is rated at 1500 W (the *average* power) at 120 V (the *rms* voltage). Calculate (a) the resistance, (b) the rms current, and (c) the maximum instantaneous power. Assume that the dryer is a pure resistor. (The heating element acts as a resistor.)

Continued

IDENTIFY and SET UP: We are given $P_{av} = 1500$ W and $V_{rms} = 120$ V. Our target variables are the resistance R, the rms current I_{rms}, and the maximum value p_{max} of the instantaneous power p. We solve Eq. (31.29) to find R, Eq. (31.28) to find I_{rms} from V_{rms} and P_{av}, and Eq. (31.30) to find p_{max}.

EXECUTE: (a) From Eq. (31.29), the resistance is

$$R = \frac{V_{rms}^2}{P_{av}} = \frac{(120 \text{ V})^2}{1500 \text{ W}} = 9.6 \ \Omega$$

(b) From Eq. (31.28),

$$I_{rms} = \frac{P_{av}}{V_{rms}} = \frac{1500 \text{ W}}{120 \text{ V}} = 12.5 \text{ A}$$

(c) For a pure resistor, the voltage and current are in phase and the phase angle ϕ is zero. Hence from Eq. (31.30), the instantaneous power is $p = VI\cos^2\omega t$ and the maximum instantaneous power is $p_{max} = VI$. From Eq. (31.27), this is twice the average power P_{av}, so

$$p_{max} = VI = 2P_{av} = 2(1500 \text{ W}) = 3000 \text{ W}$$

EVALUATE: We can confirm our result in part (b) by using Eq. (31.7): $I_{rms} = V_{rms}/R = (120 \text{ V})/(9.6 \ \Omega) = 12.5 \text{ A}$. Note that some unscrupulous manufacturers of stereo amplifiers advertise the *peak* power output rather than the lower average value.

Power in an *L-R-C* series circuit

For the *L-R-C* series circuit of Example 31.4, (a) calculate the power factor and (b) calculate the average power delivered to the entire circuit and to each circuit element.

IDENTIFY and SET UP: We can use the results of Example 31.4. The power factor is the cosine of the phase angle ϕ, and we use Eq. (31.31) to find the average power delivered in terms of ϕ and the amplitudes of voltage and current.

EXECUTE: (a) The power factor is $\cos\phi = \cos 53° = 0.60$.
(b) From Eq. (31.31),

$$P_{av} = \tfrac{1}{2}VI\cos\phi = \tfrac{1}{2}(50 \text{ V})(0.10 \text{ A})(0.60) = 1.5 \text{ W}$$

EVALUATE: Although P_{av} is the average power delivered to the *L-R-C* combination, all of this power is dissipated in the *resistor*. As Figs. 31.16b and 31.16c show, the average power delivered to a pure inductor or pure capacitor is always zero.

Test Your Understanding of Section 31.4 Figure 31.16d shows that during part of a cycle of oscillation, the instantaneous power delivered to the circuit is negative. This means that energy is being extracted from the circuit. (a) Where is the energy extracted from? (i) the resistor; (ii) the inductor; (iii) the capacitor; (iv) the ac source; (v) more than one of these. (b) Where does the energy go? (i) the resistor; (ii) the inductor; (iii) the capacitor; (iv) the ac source; (v) more than one of these.

31.5 Resonance in Alternating-Current Circuits

Much of the practical importance of *L-R-C* series circuits arises from the way in which such circuits respond to sources of different angular frequency ω. For example, one type of tuning circuit used in radio receivers is simply an *L-R-C* series circuit. A radio signal of any given frequency produces a current of the same frequency in the receiver circuit, but the amplitude of the current is *greatest* if the signal frequency equals the particular frequency to which the receiver circuit is "tuned." This effect is called *resonance*. The circuit is designed so that signals at other than the tuned frequency produce currents that are too small to make an audible sound come out of the radio's speakers.

To see how an *L-R-C* series circuit can be used in this way, suppose we connect an ac source with constant voltage amplitude V but adjustable angular frequency ω across an *L-R-C* series circuit. The current that appears in the circuit has the same angular frequency as the source and a current amplitude $I = V/Z$, where Z is the impedance of the *L-R-C* series circuit. This impedance depends on the frequency, as Eq. (31.23) shows. Figure 31.18a shows graphs of R, X_L, X_C, and Z as functions of ω. We have used a logarithmic angular frequency scale so that we can cover a wide range of frequencies. As the frequency increases, X_L increases and X_C decreases; hence there is always one frequency at which

31.18 How variations in the angular frequency of an ac circuit affect (a) reactance, resistance, and impedance, and (b) impedance, current amplitude, and phase angle.

(a) Reactance, resistance, and impedance as functions of angular frequency

Impedance Z is least at the angular frequency at which $X_C = X_L$.

(b) Impedance, current, and phase angle as functions of angular frequency

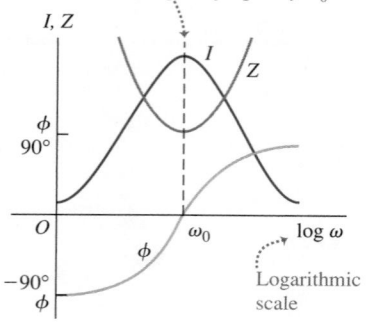

Current peaks at the angular frequency at which impedance is least. This is the *resonance angular frequency* ω_0.

31.19 Graph of current amplitude I as a function of angular frequency ω for an L-R-C series circuit with $V = 100$ V, $L = 2.0$ H, $C = 0.50$ μF, and three different values of the resistance R.

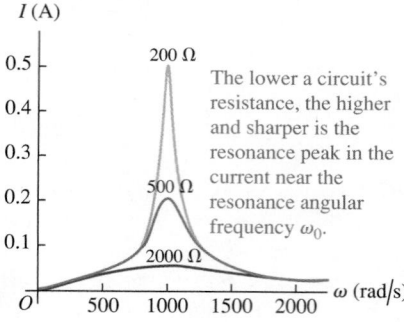

The lower a circuit's resistance, the higher and sharper is the resonance peak in the current near the resonance angular frequency ω_0.

X_L and X_C are equal and $X_L - X_C$ is zero. At this frequency the impedance $Z = \sqrt{R^2 + (X_L - X_C)^2}$ has its *smallest* value, equal simply to the resistance R.

Circuit Behavior at Resonance

As we vary the angular frequency ω of the source, the current amplitude $I = V/Z$ varies as shown in Fig. 31.18b; the *maximum* value of I occurs at the frequency at which the impedance Z is *minimum*. This peaking of the current amplitude at a certain frequency is called **resonance.** The angular frequency ω_0 at which the resonance peak occurs is called the **resonance angular frequency.** This is the angular frequency at which the inductive and capacitive reactances are equal, so at resonance,

$$X_L = X_C \qquad \omega_0 L = \frac{1}{\omega_0 C} \qquad \omega_0 = \frac{1}{\sqrt{LC}} \qquad \begin{matrix} (L\text{-}R\text{-}C \text{ series circuit} \\ \text{at resonance)} \end{matrix} \quad (31.32)$$

Note that this is equal to the natural angular frequency of oscillation of an L-C circuit, which we derived in Section 30.5, Eq. (30.22). The **resonance frequency** f_0 is $\omega_0/2\pi$. This is the frequency at which the greatest current appears in the circuit for a given source voltage amplitude; in other words, f_0 is the frequency to which the circuit is "tuned."

It's instructive to look at what happens to the *voltages* in an L-R-C series circuit at resonance. The current at any instant is the same in L and C. The voltage across an inductor always *leads* the current by 90°, or $\frac{1}{4}$ cycle, and the voltage across a capacitor always *lags* the current by 90°. Therefore the instantaneous voltages across L and C always differ in phase by 180°, or $\frac{1}{2}$ cycle; they have opposite signs at each instant. At the resonance frequency, and *only* at the resonance frequency, $X_L = X_C$ and the voltage amplitudes $V_L = IX_L$ and $V_C = IX_C$ are *equal;* then the instantaneous voltages across L and C add to zero at each instant, and the *total* voltage v_{bd} across the L-C combination in Fig. 31.13a is exactly zero. The voltage across the resistor is then equal to the source voltage. So at the resonance frequency the circuit behaves as if the inductor and capacitor weren't there at all!

The *phase* of the voltage relative to the current is given by Eq. (31.24). At frequencies below resonance, X_C is greater than X_L; the capacitive reactance dominates, the voltage *lags* the current, and the phase angle ϕ is between zero and $-90°$. Above resonance, the inductive reactance dominates, the voltage *leads* the current, and the phase angle ϕ is between zero and $+90°$. Figure 31.18b shows this variation of ϕ with angular frequency.

Tailoring an ac Circuit

If we can vary the inductance L or the capacitance C of a circuit, we can also vary the resonance frequency. This is exactly how a radio or television receiving set is "tuned" to receive a particular station. In the early days of radio this was accomplished by the use of capacitors with movable metal plates whose overlap could be varied to change C. (This is what is being done with the radio tuning knob shown in the photograph that opens this chapter.) A more modern approach is to vary L by using a coil with a ferrite core that slides in or out.

In an L-R-C series circuit the impedance reaches its minimum value and the current its maximum value at the resonance frequency. The middle curve in Fig. 31.19 is a graph of current as a function of frequency for such a circuit, with source voltage amplitude $V = 100$ V, $L = 2.0$ H, $C = 0.50$ μF, and $R = 500$ Ω. This curve is called a *response curve* or a *resonance curve*. The resonance angular frequency is $\omega_0 = (LC)^{-1/2} = 1000$ rad/s. As we expect, the curve has a peak at this angular frequency.

The resonance frequency is determined by L and C; what happens when we change R? Figure 31.19 also shows graphs of I as a function of ω for $R = 200$ Ω and for $R = 2000$ Ω. The curves are similar for frequencies far away from resonance, where the impedance is dominated by X_L or X_C. But near resonance, where X_L and X_C nearly cancel each other, the curve is higher and more sharply peaked for small values of R and broader and flatter for large values of R. At resonance, $Z = R$ and $I = V/R$, so the maximum height of the curve is inversely proportional to R.

The shape of the response curve is important in the design of radio and television receiving circuits. The sharply peaked curve is what makes it possible to discriminate between two stations broadcasting on adjacent frequency bands. But if the peak is *too* sharp, some of the information in the received signal is lost, such as the high-frequency sounds in music. The shape of the resonance curve is also related to the overdamped and underdamped oscillations that we described in Section 30.6. A sharply peaked resonance curve corresponds to a small value of R and a lightly damped oscillating system; a broad, flat curve goes with a large value of R and a heavily damped system.

In this section we have discussed resonance in an *L-R-C series* circuit. Resonance can also occur in an ac circuit in which the inductor, resistor, and capacitor are connected in *parallel*. We leave the details to you (see Problem 31.57).

Resonance phenomena occur not just in ac circuits, but in all areas of physics. We discussed examples of resonance in *mechanical* systems in Sections 13.8 and 16.5. The amplitude of a mechanical oscillation peaks when the driving-force frequency is close to a natural frequency of the system; this is analogous to the peaking of the current in an *L-R-C* series circuit. We suggest that you review the sections on mechanical resonance now, looking for the analogies.

Example 31.8   Tuning a radio

The series circuit in Fig. 31.20 is similar to some radio tuning circuits. It is connected to a variable-frequency ac source with an rms terminal voltage of 1.0 V. (a) Find the resonance frequency. At the resonance frequency, find (b) the inductive reactance X_L, the capacitive reactance X_C, and the impedance Z; (c) the rms current I_{rms}; (d) the rms voltage across each circuit element.

SOLUTION

IDENTIFY and SET UP: Figure 31.20 shows an *L-R-C* series circuit, with ideal meters inserted to measure the rms current and voltages, our target variables. Equations (31.32) include the formula for the resonance angular frequency ω_0, from which we find the resonance frequency f_0. We use Eqs. (31.12) and (31.18) to find X_L and X_C, which are equal at resonance; at resonance, from Eq. (31.23), we

have $Z = R$. We use Eqs. (31.7), (31.13), and (31.19) to find the voltages across the circuit elements.

EXECUTE: (a) The values of ω_0 and f_0 are

$$\omega_0 = \frac{1}{\sqrt{LC}} = \frac{1}{\sqrt{(0.40 \times 10^{-3}\ \text{H})(100 \times 10^{-12}\ \text{F})}}$$

$$= 5.0 \times 10^6\ \text{rad/s}$$

$$f_0 = 8.0 \times 10^5\ \text{Hz} = 800\ \text{kHz}$$

This frequency is in the lower part of the AM radio band.

(b) At this frequency,

$$X_L = \omega L = (5.0 \times 10^6\ \text{rad/s})(0.40 \times 10^{-3}\ \text{H}) = 2000\ \Omega$$

$$X_C = \frac{1}{\omega C} = \frac{1}{(5.0 \times 10^6\ \text{rad/s})(100 \times 10^{-12}\ \text{F})} = 2000\ \Omega$$

Since $X_L = X_C$ at resonance as stated above, $Z = R = 500\ \Omega$.

(c) From Eq. (31.26) the rms current at resonance is

$$I_{rms} = \frac{V_{rms}}{Z} = \frac{V_{rms}}{R} = \frac{1.0\ \text{V}}{500\ \Omega} = 0.0020\ \text{A} = 2.0\ \text{mA}$$

(d) The rms potential difference across the resistor is

$$V_{R\text{-rms}} = I_{rms}R = (0.0020\ \text{A})(500\ \Omega) = 1.0\ \text{V}$$

The rms potential differences across the inductor and capacitor are

$$V_{L\text{-rms}} = I_{rms}X_L = (0.0020\ \text{A})(2000\ \Omega) = 4.0\ \text{V}$$

$$V_{C\text{-rms}} = I_{rms}X_C = (0.0020\ \text{A})(2000\ \Omega) = 4.0\ \text{V}$$

EVALUATE: The potential differences across the inductor and the capacitor have equal rms values and amplitudes, but are 180° out of phase and so add to zero at each instant. Note also that at resonance, $V_{R\text{-rms}}$ is equal to the source voltage V_{rms}, while in this example, $V_{L\text{-rms}}$ and $V_{C\text{-rms}}$ are both considerably *larger* than V_{rms}.

31.20 A radio tuning circuit at resonance. The circles denote rms current and voltages.

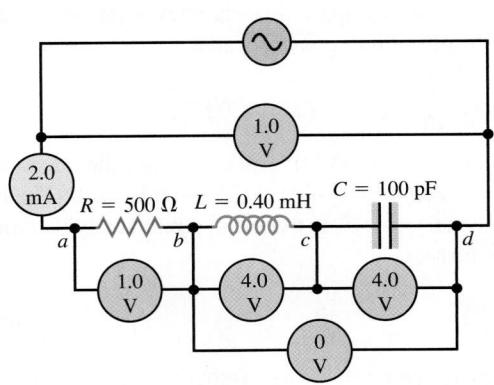

Test Your Understanding of Section 31.5 How does the resonance frequency of an *L-R-C* series circuit change if the plates of the capacitor are brought closer together? (i) It increases; (ii) it decreases; (iii) it is unaffected.

31.6 Transformers

One of the great advantages of ac over dc for electric-power distribution is that it is much easier to step voltage levels up and down with ac than with dc. For long-distance power transmission it is desirable to use as high a voltage and as small a current as possible; this reduces i^2R losses in the transmission lines, and smaller wires can be used, saving on material costs. Present-day transmission lines routinely operate at rms voltages of the order of 500 kV. On the other hand, safety considerations and insulation requirements dictate relatively low voltages in generating equipment and in household and industrial power distribution. The standard voltage for household wiring is 120 V in the United States and Canada and 240 V in many other countries. The necessary voltage conversion is accomplished by the use of **transformers.**

How Transformers Work

Figure 31.21 shows an idealized transformer. The key components of the transformer are two coils or *windings,* electrically insulated from each other but wound on the same core. The core is typically made of a material, such as iron, with a very large relative permeability K_m. This keeps the magnetic field lines due to a current in one winding almost completely within the core. Hence almost all of these field lines pass through the other winding, maximizing the *mutual inductance* of the two windings (see Section 30.1). The winding to which power is supplied is called the **primary;** the winding from which power is delivered is called the **secondary.** The circuit symbol for a transformer with an iron core, such as those used in power distribution systems, is

Here's how a transformer works. The ac source causes an alternating current in the primary, which sets up an alternating flux in the core; this induces an emf in each winding, in accordance with Faraday's law. The induced emf in the secondary gives rise to an alternating current in the secondary, and this delivers energy to the device to which the secondary is connected. All currents and emfs have the same frequency as the ac source.

Let's see how the voltage across the secondary can be made larger or smaller in amplitude than the voltage across the primary. We neglect the resistance of the windings and assume that all the magnetic field lines are confined to the iron core, so at any instant the magnetic flux Φ_B is the same in each turn of the primary and secondary windings. The primary winding has N_1 turns and the secondary winding has N_2 turns. When the magnetic flux changes because of changing currents in the two coils, the resulting induced emfs are

$$\mathcal{E}_1 = -N_1\frac{d\Phi_B}{dt} \quad \text{and} \quad \mathcal{E}_2 = -N_2\frac{d\Phi_B}{dt} \qquad (31.33)$$

The flux *per turn* Φ_B is the same in both the primary and the secondary, so Eqs. (31.33) show that the induced emf *per turn* is the same in each. The ratio of the secondary emf $\mathcal{E}_2$ to the primary emf $\mathcal{E}_1$ is therefore equal at any instant to the ratio of secondary to primary turns:

$$\frac{\mathcal{E}_2}{\mathcal{E}_1} = \frac{N_2}{N_1} \qquad (31.34)$$

Since $\mathcal{E}_1$ and $\mathcal{E}_2$ both oscillate with the same frequency as the ac source, Eq. (31.34) also gives the ratio of the amplitudes or of the rms values of the induced

31.21 Schematic diagram of an idealized step-up transformer. The primary is connected to an ac source; the secondary is connected to a device with resistance R.

The induced emf *per turn* is the same in both coils, so we adjust the ratio of terminal voltages by adjusting the ratio of turns:
$$\frac{V_2}{V_1} = \frac{N_2}{N_1}$$

emfs. If the windings have zero resistance, the induced emfs $\mathcal{E}_1$ and $\mathcal{E}_2$ are equal to the terminal voltages across the primary and the secondary, respectively; hence

$$\frac{V_2}{V_1} = \frac{N_2}{N_1} \qquad \text{(terminal voltages of transformer primary and secondary)} \qquad \text{(31.35)}$$

where V_1 and V_2 are either the amplitudes or the rms values of the terminal voltages. By choosing the appropriate turns ratio N_2/N_1, we may obtain any desired secondary voltage from a given primary voltage. If $N_2 > N_1$, as in Fig. 31.21, then $V_2 > V_1$ and we have a *step-up* transformer; if $N_2 < N_1$, then $V_2 < V_1$ and we have a *step-down* transformer. At a power generating station, step-up transformers are used; the primary is connected to the power source and the secondary is connected to the transmission lines, giving the desired high voltage for transmission. Near the consumer, step-down transformers lower the voltage to a value suitable for use in home or industry (Fig. 31.22).

Even the relatively low voltage provided by a household wall socket is too high for many electronic devices, so a further step-down transformer is necessary. This is the role of an "ac adapter" such as those used to recharge a mobile phone or laptop computer from line voltage. Such adapters contain a step-down transformer that converts line voltage to a lower value, typically 3 to 12 volts, as well as diodes to convert alternating current to the direct current that small electronic devices require (Fig. 31.23).

Energy Considerations for Transformers

If the secondary circuit is completed by a resistance R, then the amplitude or rms value of the current in the secondary circuit is $I_2 = V_2/R$. From energy considerations, the power delivered to the primary equals that taken out of the secondary (since there is no resistance in the windings), so

$$V_1 I_1 = V_2 I_2 \qquad \text{(currents in transformer primary and secondary)} \qquad \text{(31.36)}$$

We can combine Eqs. (31.35) and (31.36) and the relationship $I_2 = V_2/R$ to eliminate V_2 and I_2; we obtain

$$\frac{V_1}{I_1} = \frac{R}{(N_2/N_1)^2} \qquad \text{(31.37)}$$

This shows that when the secondary circuit is completed through a resistance R, the result is the same as if the *source* had been connected directly to a resistance equal to R divided by the square of the turns ratio, $(N_2/N_1)^2$. In other words, the transformer "transforms" not only voltages and currents, but resistances as well. More generally, we can regard a transformer as "transforming" the *impedance* of the network to which the secondary circuit is completed.

Equation (31.37) has many practical consequences. The power supplied by a source to a resistor depends on the resistances of both the resistor and the source. It can be shown that the power transfer is greatest when the two resistances are *equal*. The same principle applies in both dc and ac circuits. When a high-impedance ac source must be connected to a low-impedance circuit, such as an audio amplifier connected to a loudspeaker, the source impedance can be *matched* to that of the circuit by the use of a transformer with an appropriate turns ratio N_2/N_1.

Real transformers always have some energy losses. (That's why an ac adapter like the one shown in Fig. 31.23 feels warm to the touch after it's been in use for a while; the transformer is heated by the dissipated energy.) The windings have some resistance, leading to i^2R losses. There are also energy losses through hysteresis in the core (see Section 28.8). Hysteresis losses are minimized by the use of soft iron with a narrow hysteresis loop.

Another important mechanism for energy loss in a transformer core involves eddy currents (see Section 29.6). Consider a section AA through an iron transformer core (Fig. 31.24a). Since iron is a conductor, any such section can be pictured as

31.22 The cylindrical can near the top of this power pole is a step-down transformer. It converts the high-voltage ac in the power lines to low-voltage (120 V) ac, which is then distributed to the surrounding homes and businesses.

31.23 An ac adapter like this one converts household ac into low-voltage dc for use in electronic devices. It contains a step-down transformer to lower the voltage and diodes to rectify the output current (see Fig. 31.3).

31.24 (a) Primary and secondary windings in a transformer. (b) Eddy currents in the iron core, shown in the cross section at *AA*. (c) Using a laminated core reduces the eddy currents.

(a) Schematic transformer

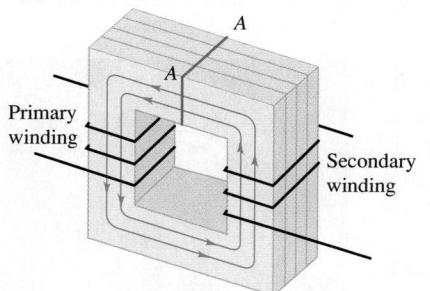

(b) Large eddy currents in solid core

(c) Smaller eddy currents in laminated core

several conducting circuits, one within the other (Fig. 31.24b). The flux through each of these circuits is continually changing, so eddy currents circulate in the entire volume of the core, with lines of flow that form planes perpendicular to the flux. These eddy currents are very undesirable; they waste energy through $i^2 R$ heating and themselves set up an opposing flux.

The effects of eddy currents can be minimized by the use of a *laminated* core—that is, one built up of thin sheets or laminae. The large electrical surface resistance of each lamina, due either to a natural coating of oxide or to an insulating varnish, effectively confines the eddy currents to individual laminae (Fig. 31.24c). The possible eddy-current paths are narrower, the induced emf in each path is smaller, and the eddy currents are greatly reduced. The alternating magnetic field exerts forces on the current-carrying laminae that cause them to vibrate back and forth; this vibration causes the characteristic "hum" of an operating transformer. You can hear this same "hum" from the magnetic ballast of a fluorescent light fixture (see Section 30.2).

Thanks to the use of soft iron cores and lamination, transformer efficiencies are usually well over 90%; in large installations they may reach 99%.

Example 31.9   "Wake up and smell the (transformer)!"

A friend returns to the United States from Europe with a 960-W coffeemaker, designed to operate from a 240-V line. (a) What can she do to operate it at the USA-standard 120 V? (b) What current will the coffeemaker draw from the 120-V line? (c) What is the resistance of the coffeemaker? (The voltages are rms values.)

SOLUTION

IDENTIFY and SET UP: Our friend needs a step-up transformer to convert 120-V ac to the 240-V ac that the coffeemaker requires. We use Eq. (31.35) to determine the transformer turns ratio N_2/N_1, $P_{av} = V_{rms}I_{rms}$ for a resistor to find the current draw, and Eq. (31.37) to calculate the resistance.

EXECUTE: (a) To get $V_2 = 240$ V from $V_1 = 120$ V, the required turns ratio is $N_2/N_1 = V_2/V_1 = (240 \text{ V})/(120 \text{ V}) = 2$. That is, the secondary coil (connected to the coffeemaker) should have twice as many turns as the primary coil (connected to the 120-V line).

(b) We find the rms current I_1 in the 120-V primary by using $P_{av} = V_1 I_1$, where P_{av} is the average power drawn by the coffeemaker and hence the power supplied by the 120-V line. (We're assuming that no energy is lost in the transformer.) Hence $I_1 = P_{av}/V_1 = (960 \text{ W})/(120 \text{ V}) = 8.0$ A. The secondary current is then $I_2 = P_{av}/V_2 = (960 \text{ W})/(240 \text{ V}) = 4.0$ A.

(c) We have $V_1 = 120$ V, $I_1 = 8.0$ A, and $N_2/N_1 = 2$, so

$$\frac{V_1}{I_1} = \frac{120 \text{ V}}{8.0 \text{ A}} = 15 \ \Omega$$

From Eq. (31.37),

$$R = 2^2(15 \ \Omega) = 60 \ \Omega$$

EVALUATE: As a check, $V_2/R = (240 \text{ V})/(60 \ \Omega) = 4.0 \text{ A} = I_2$, the same value obtained previously. You can also check this result for R by using the expression $P_{av} = V_2^2/R$ for the power drawn by the coffeemaker.

Test Your Understanding of Section 31.6   Each of the following four transformers has 1000 turns in its primary coil. Rank the transformers from largest to smallest number of turns in the secondary coil. (i) converts 120-V ac into 6.0-V ac; (ii) converts 120-V ac into 240-V ac; (iii) converts 240-V ac into 6.0-V ac; (iv) converts 240-V ac into 120-V ac.

❙

Phasors and alternating current: An alternator or ac source produces an emf that varies sinusoidally with time. A sinusoidal voltage or current can be represented by a phasor, a vector that rotates counterclockwise with constant angular velocity ω equal to the angular frequency of the sinusoidal quantity. Its projection on the horizontal axis at any instant represents the instantaneous value of the quantity.

For a sinusoidal current, the rectified average and rms (root-mean-square) currents are proportional to the current amplitude I. Similarly, the rms value of a sinusoidal voltage is proportional to the voltage amplitude V. (See Example 31.1.)

$$I_{\text{rav}} = \frac{2}{\pi} I = 0.637 I \qquad (31.3)$$

$$I_{\text{rms}} = \frac{I}{\sqrt{2}} \qquad (31.4)$$

$$V_{\text{rms}} = \frac{V}{\sqrt{2}} \qquad (31.5)$$

Voltage, current, and phase angle: In general, the instantaneous voltage between two points in an ac circuit is not in phase with the instantaneous current passing through those points. The quantity ϕ is called the phase angle of the voltage relative to the current.

$$i = I \cos \omega t \qquad (31.2)$$

$$v = V \cos(\omega t + \phi)$$

Resistance and reactance: The voltage across a resistor R is in phase with the current. The voltage across an inductor L leads the current by 90° ($\phi = +90°$), while the voltage across a capacitor C lags the current by 90° ($\phi = -90°$). The voltage amplitude across each type of device is proportional to the current amplitude I. An inductor has inductive reactance $X_L = \omega L$, and a capacitor has capacitive reactance $X_C = 1/\omega C$. (See Examples 31.2 and 31.3.)

$$V_R = IR \qquad (31.7)$$

$$V_L = IX_L \qquad (31.13)$$

$$V_C = IX_C \qquad (31.19)$$

Impedance and the L-R-C series circuit: In a general ac circuit, the voltage and current amplitudes are related by the circuit impedance Z. In an L-R-C series circuit, the values of L, R, C, and the angular frequency ω determine the impedance and the phase angle ϕ of the voltage relative to the current. (See Examples 31.4 and 31.5.)

$$V = IZ \qquad (31.22)$$

$$Z = \sqrt{R^2 + (X_L - X_C)^2}$$
$$= \sqrt{R^2 + [\omega L - (1/\omega C)]^2} \qquad (31.23)$$

$$\tan \phi = \frac{\omega L - 1/\omega C}{R} \qquad (31.24)$$

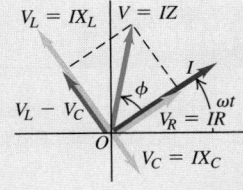

Power in ac circuits: The average power input P_{av} to an ac circuit depends on the voltage and current amplitudes (or, equivalently, their rms values) and the phase angle ϕ of the voltage relative to the current. The quantity $\cos \phi$ is called the power factor. (See Examples 31.6 and 31.7.)

$$P_{\text{av}} = \frac{1}{2} VI \cos \phi \qquad (31.31)$$

$$= V_{\text{rms}} I_{\text{rms}} \cos \phi$$

Resonance in ac circuits: In an L-R-C series circuit, the current becomes maximum and the impedance becomes minimum at an angular frequency called the resonance angular frequency. This phenomenon is called resonance. At resonance the voltage and current are in phase, and the impedance Z is equal to the resistance R. (See Example 31.8.)

$$\omega_0 = \frac{1}{\sqrt{LC}} \qquad (31.32)$$

Transformers: A transformer is used to transform the voltage and current levels in an ac circuit. In an ideal transformer with no energy losses, if the primary winding has N_1 turns and the secondary winding has N_2 turns, the amplitudes (or rms values) of the two voltages are related by Eq. (31.35). The amplitudes (or rms values) of the primary and secondary voltages and currents are related by Eq. (31.36). (See Example 31.9.)

$$\frac{V_2}{V_1} = \frac{N_2}{N_1} \qquad (31.35)$$

$$V_1 I_1 = V_2 I_2 \qquad (31.36)$$

BRIDGING PROBLEM An Alternating-Current Circuit

A series circuit consists of a 1.50-mH inductor, a 125-Ω resistor, and a 25.0-nF capacitor connected across an ac source having an rms voltage of 35.0 V and variable frequency. (a) At what angular frequencies will the current amplitude be equal to $\frac{1}{3}$ of its maximum possible value? (b) At the frequencies in part (a), what are the current amplitude and the voltage amplitude across each circuit element (including the ac source)?

SOLUTION GUIDE

See MasteringPhysics® study area for a Video Tutor solution.

IDENTIFY and SET UP

1. The maximum current amplitude occurs at the resonance angular frequency. This problem concerns the angular frequencies at which the current amplitude is one-third of that maximum.
2. Choose the equation that will allow you to find the angular frequencies in question, and choose the equations that you will

then use to find the current and voltage amplitudes at each angular frequency.

EXECUTE

3. Find the impedance at the angular frequencies in part (a); then solve for the values of angular frequency.
4. Find the voltage amplitude across the source and the current amplitude for each of the angular frequencies in part (a). (*Hint:* Be careful to distinguish between *amplitude* and *rms value*.)
5. Use the results of steps 3 and 4 to find the reactances at each angular frequency. Then calculate the voltage amplitudes for the resistor, inductor, and capacitor.

EVALUATE

6. Are there any voltage amplitudes that are greater than the voltage amplitude of the source? If so, does this mean your results are in error?

Problems

For instructor-assigned homework, go to www.masteringphysics.com

•, ••, •••: Problems of increasing difficulty. **CP:** Cumulative problems incorporating material from earlier chapters. **CALC:** Problems requiring calculus. **BIO:** Biosciences problems.

DISCUSSION QUESTIONS

Q31.1 Household electric power in most of western Europe is supplied at 240 V, rather than the 120 V that is standard in the United States and Canada. What are the advantages and disadvantages of each system?

Q31.2 The current in an ac power line changes direction 120 times per second, and its average value is zero. Explain how it is possible for power to be transmitted in such a system.

Q31.3 In an ac circuit, why is the average power for an inductor and a capacitor zero, but not for a resistor?

Q31.4 Equation (31.14) was derived by using the relationship $i = dq/dt$ between the current and the charge on the capacitor. In Fig. 31.9a the positive counterclockwise current increases the charge on the capacitor. When the charge on the left plate is positive but decreasing in time, is $i = dq/dt$ still correct or should it be $i = -dq/dt$? Is $i = dq/dt$ still correct when the right-hand plate has positive charge that is increasing or decreasing in magnitude? Explain.

Q31.5 Fluorescent lights often use an inductor, called a ballast, to limit the current through the tubes. Why is it better to use an inductor rather than a resistor for this purpose?

Q31.6 Equation (31.9) says that $v_{ab} = L\, di/dt$ (see Fig. 31.8a). Using Faraday's law, explain why point a is at higher potential than point b when i is in the direction shown in Fig. 31.8a and is increasing in magnitude. When i is counterclockwise and decreasing in magnitude, is $v_{ab} = L\, di/dt$ still correct, or should it be $v_{ab} = -L\, di/dt$? Is $v_{ab} = L\, di/dt$ still correct when i is clockwise and increasing or decreasing in magnitude? Explain.

Q31.7 Is it possible for the power factor of an L-R-C series ac circuit to be zero? Justify your answer on *physical* grounds.

Q31.8 In an L-R-C series circuit, can the instantaneous voltage across the capacitor exceed the source voltage at that same instant? Can this be true for the instantaneous voltage across the inductor? Across the resistor? Explain.

Q31.9 In an L-R-C series circuit, what are the phase angle ϕ and power factor $\cos\phi$ when the resistance is much smaller than the inductive or capacitive reactance and the circuit is operated far from resonance? Explain.

Q31.10 When an *L-R-C* series circuit is connected across a 120-V ac line, the voltage rating of the capacitor may be exceeded even if it is rated at 200 or 400 V. How can this be?

Q31.11 In Example 31.6 (Section 31.4), a hair dryer is treated as a pure resistor. But because there are coils in the heating element and in the motor that drives the blower fan, a hair dryer also has inductance. Qualitatively, does including an inductance increase or decrease the values of R, I_{rms}, and P?

Q31.12 A light bulb and a parallel-plate capacitor with air between the plates are connected in series to an ac source. What happens to the brightness of the bulb when a dielectric is inserted between the plates of the capacitor? Explain.

Q31.13 A coil of wire wrapped on a hollow tube and a light bulb are connected in series to an ac source. What happens to the brightness of the bulb when an iron rod is inserted in the tube?

Q31.14 A circuit consists of a light bulb, a capacitor, and an inductor connected in series to an ac source. What happens to the brightness of the bulb when the inductor is removed? When the inductor is left in the circuit but the capacitor is removed? Explain.

Q31.15 A circuit consists of a light bulb, a capacitor, and an inductor connected in series to an ac source. Is it possible for both the capacitor and the inductor to be removed and the brightness of the bulb to remain the same? Explain.

Q31.16 Can a transformer be used with dc? Explain. What happens if a transformer designed for 120-V ac is connected to a 120-V dc line?

Q31.17 An ideal transformer has N_1 windings in the primary and N_2 windings in its secondary. If you double only the number of secondary windings, by what factor does (a) the voltage amplitude in the secondary change, and (b) the effective resistance of the secondary circuit change?

Q31.18 Some electrical appliances operate equally well on ac or dc, and others work only on ac or only on dc. Give examples of each, and explain the differences.

EXERCISES

Section 31.1 Phasors and Alternating Currents

31.1 • You have a special light bulb with a *very* delicate wire filament. The wire will break if the current in it ever exceeds 1.50 A, even for an instant. What is the largest root-mean-square current you can run through this bulb?

31.2 • A sinusoidal current $i = I\cos\omega t$ has an rms value $I_{rms} = 2.10$ A. (a) What is the current amplitude? (b) The current is passed through a full-wave rectifier circuit. What is the rectified average current? (c) Which is larger: I_{rms} or I_{rav}? Explain, using graphs of i^2 and of the rectified current.

31.3 • The voltage across the terminals of an ac power supply varies with time according to Eq. (31.1). The voltage amplitude is $V = 45.0$ V. What are (a) the root-mean-square potential difference V_{rms} and (b) the average potential difference V_{av} between the two terminals of the power supply?

Section 31.2 Resistance and Reactance

31.4 • A capacitor is connected across an ac source that has voltage amplitude 60.0 V and frequency 80.0 Hz. (a) What is the phase angle ϕ for the source voltage relative to the current? Does the source voltage lag or lead the current? (b) What is the capacitance C of the capacitor if the current amplitude is 5.30 A?

31.5 • An inductor with $L = 9.50$ mH is connected across an ac source that has voltage amplitude 45.0 V. (a) What is the phase angle ϕ for the source voltage relative to the current? Does the source voltage lag or lead the current? (b) What value for the frequency of the source results in a current amplitude of 3.90 A?

31.6 • A capacitance C and an inductance L are operated at the same angular frequency. (a) At what angular frequency will they have the same reactance? (b) If $L = 5.00$ mH and $C = 3.50$ μF, what is the numerical value of the angular frequency in part (a), and what is the reactance of each element?

31.7 • **Kitchen Capacitance.** The wiring for a refrigerator contains a starter capacitor. A voltage of amplitude 170 V and frequency 60.0 Hz applied across the capacitor is to produce a current amplitude of 0.850 A through the capacitor. What capacitance C is required?

31.8 • (a) Compute the reactance of a 0.450-H inductor at frequencies of 60.0 Hz and 600 Hz. (b) Compute the reactance of a 2.50-μF capacitor at the same frequencies. (c) At what frequency is the reactance of a 0.450-H inductor equal to that of a 2.50-μF capacitor?

31.9 • (a) What is the reactance of a 3.00-H inductor at a frequency of 80.0 Hz? (b) What is the inductance of an inductor whose reactance is 120 Ω at 80.0 Hz? (c) What is the reactance of a 4.00-μF capacitor at a frequency of 80.0 Hz? (d) What is the capacitance of a capacitor whose reactance is 120 Ω at 80.0 Hz?

31.10 • **A Radio Inductor.** You want the current amplitude through a 0.450-mH inductor (part of the circuitry for a radio receiver) to be 2.60 mA when a sinusoidal voltage with amplitude 12.0 V is applied across the inductor. What frequency is required?

31.11 •• A 0.180-H inductor is connected in series with a 90.0-Ω resistor and an ac source. The voltage across the inductor is $v_L = -(12.0\ \text{V})\sin[(480\ \text{rad/s})t]$. (a) Derive an expression for the voltage v_R across the resistor. (b) What is v_R at $t = 2.00$ ms?

31.12 •• A 250-Ω resistor is connected in series with a 4.80-μF capacitor and an ac source. The voltage across the capacitor is $v_C = (7.60\ \text{V})\sin[(120\ \text{rad/s})t]$. (a) Determine the capacitive reactance of the capacitor. (b) Derive an expression for the voltage v_R across the resistor.

31.13 •• A 150-Ω resistor is connected in series with a 0.250-H inductor and an ac source. The voltage across the resistor is $v_R = (3.80\ \text{V})\cos[(720\ \text{rad/s})t]$. (a) Derive an expression for the circuit current. (b) Determine the inductive reactance of the inductor. (c) Derive an expression for the voltage v_L across the inductor.

Section 31.3 The *L-R-C* Series Circuit

31.14 • You have a 200-Ω resistor, a 0.400-H inductor, and a 6.00-μF capacitor. Suppose you take the resistor and inductor and make a series circuit with a voltage source that has voltage amplitude 30.0 V and an angular frequency of 250 rad/s. (a) What is the impedance of the circuit? (b) What is the current amplitude? (c) What are the voltage amplitudes across the resistor and across the inductor? (d) What is the phase angle ϕ of the source voltage with respect to the current? Does the source voltage lag or lead the current? (e) Construct the phasor diagram.

31.15 • The resistor, inductor, capacitor, and voltage source described in Exercise 31.14 are connected to form an *L-R-C* series circuit. (a) What is the impedance of the circuit? (b) What is the current amplitude? (c) What is the phase angle of the source voltage with respect to the current? Does the source voltage lag or lead the current? (d) What are the voltage amplitudes across the resistor, inductor, and capacitor? (e) Explain how it is possible for the voltage amplitude across the capacitor to be greater than the voltage amplitude across the source.

31.16 •• A 200-Ω resistor, a 0.900-H inductor, and a 6.00-μF capacitor are connected in series across a voltage source that has voltage amplitude 30.0 V and an angular frequency of 250 rad/s. (a) What are v, v_R, v_L, and v_C at $t = 20.0$ ms? Compare $v_R + v_L + v_C$ to v at this instant. (b) What are V_R, V_L, and V_C? Compare V to $V_R + V_L + V_C$. Explain why these two quantities are not equal.

31.17 • In an L-R-C series circuit, the rms voltage across the resistor is 30.0 V, across the capacitor it is 90.0 V, and across the inductor it is 50.0 V. What is the rms voltage of the source?

Section 31.4 Power in Alternating-Current Circuits

31.18 •• A resistor with $R = 300\ \Omega$ and an inductor are connected in series across an ac source that has voltage amplitude 500 V. The rate at which electrical energy is dissipated in the resistor is 216 W. (a) What is the impedance Z of the circuit? (b) What is the amplitude of the voltage across the inductor? (c) What is the power factor?

31.19 • The power of a certain CD player operating at 120 V rms is 20.0 W. Assuming that the CD player behaves like a pure resistor, find (a) the maximum instantaneous power; (b) the rms current; (c) the resistance of this player.

31.20 •• In an L-R-C series circuit, the components have the following values: $L = 20.0$ mH, $C = 140$ nF, and $R = 350\ \Omega$. The generator has an rms voltage of 120 V and a frequency of 1.25 kHz. Determine (a) the power supplied by the generator and (b) the power dissipated in the resistor.

31.21 • (a) Show that for an L-R-C series circuit the power factor is equal to R/Z. (b) An L-R-C series circuit has phase angle $-31.5°$. The voltage amplitude of the source is 90.0 V. What is the voltage amplitude across the resistor?

31.22 • (a) Use the results of part (a) of Exercise 31.21 to show that the average power delivered by the source in an L-R-C series circuit is given by $P_{av} = I_{rms}^2 R$. (b) An L-R-C series series circuit has $R = 96.0\ \Omega$, and the amplitude of the voltage across the resistor is 36.0 V. What is the average power delivered by the source?

31.23 • An L-R-C series circuit with $L = 0.120$ H, $R = 240\ \Omega$, and $C = 7.30\ \mu$F carries an rms current of 0.450 A with a frequency of 400 Hz. (a) What are the phase angle and power factor for this circuit? (b) What is the impedance of the circuit? (c) What is the rms voltage of the source? (d) What average power is delivered by the source? (e) What is the average rate at which electrical energy is converted to thermal energy in the resistor? (f) What is the average rate at which electrical energy is dissipated (converted to other forms) in the capacitor? (g) In the inductor?

31.24 •• An L-R-C series circuit is connected to a 120-Hz ac source that has $V_{rms} = 80.0$ V. The circuit has a resistance of 75.0 Ω and an impedance at this frequency of 105 Ω. What average power is delivered to the circuit by the source?

31.25 •• A series ac circuit contains a 250-Ω resistor, a 15-mH inductor, a 3.5-μF capacitor, and an ac power source of voltage amplitude 45 V operating at an angular frequency of 360 rad/s. (a) What is the power factor of this circuit? (b) Find the average power delivered to the entire circuit. (c) What is the average power delivered to the resistor, to the capacitor, and to the inductor?

Section 31.5 Resonance in Alternating-Current Circuits

31.26 •• In an L-R-C series circuit the source is operated at its resonant angular frequency. At this frequency, the reactance X_C of

the capacitor is 200 Ω and the voltage amplitude across the capacitor is 600 V. The circuit has $R = 300\ \Omega$. What is the voltage amplitude of the source?

31.27 • **Analyzing an L-R-C Circuit.** You have a 200-Ω resistor, a 0.400-H inductor, a 5.00-μF capacitor, and a variable-frequency ac source with an amplitude of 3.00 V. You connect all four elements together to form a series circuit. (a) At what frequency will the current in the circuit be greatest? What will be the current amplitude at this frequency? (b) What will be the current amplitude at an angular frequency of 400 rad/s? At this frequency, will the source voltage lead or lag the current?

31.28 • An L-R-C series circuit is constructed using a 175-Ω resistor, a 12.5-μF capacitor, and an 8.00-mH inductor, all connected across an ac source having a variable frequency and a voltage amplitude of 25.0 V. (a) At what angular frequency will the impedance be smallest, and what is the impedance at this frequency? (b) At the angular frequency in part (a), what is the maximum current through the inductor? (c) At the angular frequency in part (a), find the potential difference across the ac source, the resistor, the capacitor, and the inductor at the instant that the current is equal to one-half its greatest positive value. (d) In part (c), how are the potential differences across the resistor, inductor, and capacitor related to the potential difference across the ac source?

31.29 • In an L-R-C series circuit, $R = 300\ \Omega$, $L = 0.400$ H, and $C = 6.00 \times 10^{-8}$ F. When the ac source operates at the resonance frequency of the circuit, the current amplitude is 0.500 A. (a) What is the voltage amplitude of the source? (b) What is the amplitude of the voltage across the resistor, across the inductor, and across the capacitor? (c) What is the average power supplied by the source?

31.30 • An L-R-C series circuit consists of a source with voltage amplitude 120 V and angular frequency 50.0 rad/s, a resistor with $R = 400\ \Omega$, an inductor with $L = 9.00$ H, and a capacitor with capacitance C. (a) For what value of C will the current amplitude in the circuit be a maximum? (b) When C has the value calculated in part (a), what is the amplitude of the voltage across the inductor?

31.31 • In an L-R-C series circuit, $R = 150\ \Omega$, $L = 0.750$ H, and $C = 0.0180\ \mu$F. The source has voltage amplitude $V = 150$ V and a frequency equal to the resonance frequency of the circuit. (a) What is the power factor? (b) What is the average power delivered by the source? (c) The capacitor is replaced by one with $C = 0.0360\ \mu$F and the source frequency is adjusted to the new resonance value. Then what is the average power delivered by the source?

31.32 • In an L-R-C series circuit, $R = 400\ \Omega$, $L = 0.350$ H, and $C = 0.0120\ \mu$F. (a) What is the resonance angular frequency of the circuit? (b) The capacitor can withstand a peak voltage of 550 V. If the voltage source operates at the resonance frequency, what maximum voltage amplitude can it have if the maximum capacitor voltage is not exceeded?

31.33 • A series circuit consists of an ac source of variable frequency, a 115-Ω resistor, a 1.25-μF capacitor, and a 4.50-mH inductor. Find the impedance of this circuit when the angular frequency of the ac source is adjusted to (a) the resonance angular frequency; (b) twice the resonance angular frequency; (c) half the resonance angular frequency.

31.34 •• In an L-R-C series circuit, $L = 0.280$ H and $C = 4.00\ \mu$F. The voltage amplitude of the source is 120 V. (a) What is the resonance angular frequency of the circuit? (b) When the source operates at the resonance angular frequency, the current amplitude in the circuit is 1.70 A. What is the resistance R of the resistor? (c) At the resonance angular frequency, what are the peak voltages across the inductor, the capacitor, and the resistor?

Section 31.6 Transformers

31.35 • **A Step-Down Transformer.** A transformer connected to a 120-V (rms) ac line is to supply 12.0 V (rms) to a portable electronic device. The load resistance in the secondary is 5.00 Ω. (a) What should the ratio of primary to secondary turns of the transformer be? (b) What rms current must the secondary supply? (c) What average power is delivered to the load? (d) What resistance connected directly across the 120-V line would draw the same power as the transformer? Show that this is equal to 5.00 Ω times the square of the ratio of primary to secondary turns.

31.36 • **A Step-Up Transformer.** A transformer connected to a 120-V (rms) ac line is to supply 13,000 V (rms) for a neon sign. To reduce shock hazard, a fuse is to be inserted in the primary circuit; the fuse is to blow when the rms current in the secondary circuit exceeds 8.50 mA. (a) What is the ratio of secondary to primary turns of the transformer? (b) What power must be supplied to the transformer when the rms secondary current is 8.50 mA? (c) What current rating should the fuse in the primary circuit have?

31.37 • **Off to Europe!** You plan to take your hair dryer to Europe, where the electrical outlets put out 240 V instead of the 120 V seen in the United States. The dryer puts out 1600 W at 120 V. (a) What could you do to operate your dryer via the 240-V line in Europe? (b) What current will your dryer draw from a European outlet? (c) What resistance will your dryer appear to have when operated at 240 V?

PROBLEMS

31.38 •• Figure 31.12a shows the crossover network in a loud-speaker system. One branch consists of a capacitor C and a resistor R in series (the tweeter). This branch is in parallel with a second branch (the woofer) that consists of an inductor L and a resistor R in series. The same source voltage with angular frequency ω is applied across each parallel branch. (a) What is the impedance of the tweeter branch? (b) What is the impedance of the woofer branch? (c) Explain why the currents in the two branches are equal when the impedances of the branches are equal. (d) Derive an expression for the frequency f that corresponds to the crossover point in Fig. 31.12b.

31.39 • A coil has a resistance of 48.0 Ω. At a frequency of 80.0 Hz the voltage across the coil leads the current in it by 52.3°. Determine the inductance of the coil.

31.40 •• Five infinite-impedance voltmeters, calibrated to read rms values, are connected as shown in Fig. P31.40. Let $R = 200\ \Omega$, $L = 0.400$ H, $C = 6.00\ \mu$F, and $V = 30.0$ V. What is the reading of each voltmeter if (a) $\omega = 200$ rad/s and (b) $\omega = 1000$ rad/s?

Figure **P31.40**

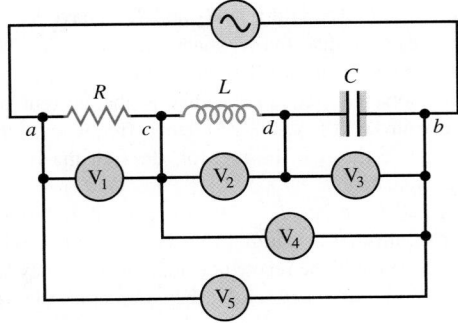

31.41 •• **CP** A parallel-plate capacitor having square plates 4.50 cm on each side and 8.00 mm apart is placed in series with an ac source of angular frequency 650 rad/s and voltage amplitude 22.5 V, a 75.0-Ω resistor, and an ideal solenoid that is 9.00 cm long, has a circular cross section 0.500 cm in diameter, and carries 125 coils per centimeter. What is the resonance angular frequency of this circuit? (See Exercise 30.15.)

31.42 •• **CP** A toroidal solenoid has 2900 closely wound turns, cross-sectional area 0.450 cm², mean radius 9.00 cm, and resistance $R = 2.80\ \Omega$. The variation of the magnetic field across the cross section of the solenoid can be neglected. What is the amplitude of the current in the solenoid if it is connected to an ac source that has voltage amplitude 24.0 V and frequency 365 Hz?

31.43 •• An L-R-C series circuit has $C = 4.80\ \mu$F, $L = 0.520$ H, and source voltage amplitude $V = 56.0$ V. The source is operated at the resonance frequency of the circuit. If the voltage across the capacitor has amplitude 80.0 V, what is the value of R for the resistor in the circuit?

31.44 • A large electromagnetic coil is connected to a 120-Hz ac source. The coil has resistance 400 Ω, and at this source frequency the coil has inductive reactance 250 Ω. (a) What is the inductance of the coil? (b) What must the rms voltage of the source be if the coil is to consume an average electrical power of 800 W?

31.45 •• A series circuit has an impedance of 60.0 Ω and a power factor of 0.720 at 50.0 Hz. The source voltage lags the current. (a) What circuit element, an inductor or a capacitor, should be placed in series with the circuit to raise its power factor? (b) What size element will raise the power factor to unity?

31.46 •• An L-R-C series circuit has $R = 300\ \Omega$. At the frequency of the source, the inductor has reactance $X_L = 900\ \Omega$ and the capacitor has reactance $X_C = 500\ \Omega$. The amplitude of the voltage across the inductor is 450 V. (a) What is the amplitude of the voltage across the resistor? (b) What is the amplitude of the voltage across the capacitor? (c) What is the voltage amplitude of the source? (d) What is the rate at which the source is delivering electrical energy to the circuit?

31.47 •• In an L-R-C series circuit, $R = 300\ \Omega$, $X_C = 300\ \Omega$, and $X_L = 500\ \Omega$. The average power consumed in the resistor is 60.0 W. (a) What is the power factor of the circuit? (b) What is the rms voltage of the source?

31.48 • A circuit consists of a resistor and a capacitor in series with an ac source that supplies an rms voltage of 240 V. At the frequency of the source the reactance of the capacitor is 50.0 Ω. The rms current in the circuit is 3.00 A. What is the average power supplied by the source?

31.49 • An L-R-C series circuit consists of a 50.0-Ω resistor, a 10.0-μF capacitor, a 3.50-mH inductor, and an ac voltage source of voltage amplitude 60.0 V operating at 1250 Hz. (a) Find the current amplitude and the voltage amplitudes across the inductor, the resistor, and the capacitor. Why can the voltage amplitudes add up to *more* than 60.0 V? (b) If the frequency is now doubled, but nothing else is changed, which of the quantities in part (a) will change? Find the new values for those that do change.

31.50 • At a frequency ω_1 the reactance of a certain capacitor equals that of a certain inductor. (a) If the frequency is changed to $\omega_2 = 2\omega_1$, what is the ratio of the reactance of the inductor to that of the capacitor? Which reactance is larger? (b) If the frequency is changed to $\omega_3 = \omega_1/3$, what is the ratio of the reactance of the inductor to that of the capacitor? Which reactance is larger? (c) If the capacitor and inductor are placed in series with a resistor of resistance R to form an L-R-C series circuit, what will be the resonance angular frequency of the circuit?

31.51 •• **A High-Pass Filter.** Figure **P31.51**
One application of L-R-C series circuits is to high-pass or low-pass filters, which filter out either the low- or high-frequency components of a signal. A high-pass filter is shown in Fig. P31.51, where the output volt-

age is taken across the L-R combination. (The L-R combination represents an inductive coil that also has resistance due to the large length of wire in the coil.) Derive an expression for V_{out}/V_s, the ratio of the output and source voltage amplitudes, as a function of the angular frequency ω of the source. Show that when ω is small, this ratio is proportional to ω and thus is small, and show that the ratio approaches unity in the limit of large frequency.

31.52 •• **A Low-Pass Filter.** Figure P31.52 shows a low-pass filter (see Problem 31.51); the output voltage is taken across the capacitor in an L-R-C series circuit. Derive an expression for V_{out}/V_s, the ratio of the output and source voltage amplitudes, as a function of the angular frequency ω of the source. Show that when ω is large, this ratio is proportional to ω^{-2} and thus is very small, and show that the ratio approaches unity in the limit of small frequency.

Figure **P31.52**

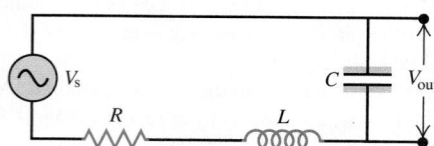

31.53 ••• An L-R-C series circuit is connected to an ac source of constant voltage amplitude V and variable angular frequency ω. (a) Show that the current amplitude, as a function of ω, is

$$I = \frac{V}{\sqrt{R^2 + (\omega L - 1/\omega C)^2}}$$

(b) Show that the average power dissipated in the resistor is

$$P = \frac{V^2 R/2}{R^2 + (\omega L - 1/\omega C)^2}$$

(c) Show that I and P are *both* maximum when $\omega = 1/\sqrt{LC}$, the resonance frequency of the circuit. (d) Graph P as a function of ω for $V = 100$ V, $R = 200$ Ω, $L = 2.0$ H, and $C = 0.50$ μF. Compare to the light purple curve in Fig. 31.19. Discuss the behavior of I and P in the limits $\omega = 0$ and $\omega \rightarrow \infty$.

31.54 •• An L-R-C series circuit is connected to an ac source of constant voltage amplitude V and variable angular frequency ω. Using the results of Problem 31.53, find an expression for (a) the amplitude V_L of the voltage across the inductor as a function of ω and (b) the amplitude V_C of the voltage across the capacitor as a function of ω. (c) Graph V_L and V_C as functions of ω for $V = 100$ V, $R = 200$ Ω, $L = 2.0$ H, and $C = 0.50$ μF. (d) Discuss the behavior of V_L and V_C in the limits $\omega = 0$ and $\omega \rightarrow \infty$. For what value of ω is $V_L = V_C$? What is the significance of this value of ω?

31.55 •• In an L-R-C series circuit the magnitude of the phase angle is 54.0°, with the source voltage lagging the current. The reactance of the capacitor is 350 Ω, and the resistor resistance is 180 Ω. The average power delivered by the source is 140 W. Find

(a) the reactance of the inductor; (b) the rms current; (c) the rms voltage of the source.

31.56 •• **The L-R-C Parallel Circuit.** A resistor, inductor, and capacitor are connected in parallel to an ac source with voltage amplitude V and angular frequency ω. Let the source voltage be given by $v = V\cos\omega t$. (a) Show that the instantaneous voltages v_R, v_L, and v_C at any instant are each equal to v and that $i = i_R + i_L + i_C$, where i is the current through the source and i_R, i_L, and i_C are the currents through the resistor, the inductor, and the capacitor, respectively. (b) What are the phases of i_R, i_L, and i_C with respect to v? Use current phasors to represent i, i_R, i_L, and i_C. In a phasor diagram, show the phases of these four currents with respect to v. (c) Use the phasor diagram of part (b) to show that the current amplitude I for the current i through the source is given by $I = \sqrt{I_R^2 + (I_C - I_L)^2}$. (d) Show that the result of part (c) can be written as $I = V/Z$, with $1/Z = \sqrt{1/R^2 + (\omega C - 1/\omega L)^2}$.

31.57 •• **Parallel Resonance.** The impedance of an L-R-C parallel circuit was derived in Problem 31.56. (a) Show that at the resonance angular frequency $\omega_0 = 1/\sqrt{LC}$, $I_C = I_L$, and I is a *minimum*. (b) Since I is a minimum at resonance, is it correct to say that the power delivered to the resistor is also a minimum at $\omega = \omega_0$? Explain. (c) At resonance, what is the phase angle of the source current with respect to the source voltage? How does this compare to the phase angle for an L-R-C *series* circuit at resonance? (d) Draw the circuit diagram for an L-R-C parallel circuit. Arrange the circuit elements in your diagram so that the resistor is closest to the ac source. Justify the following statement: When the angular frequency of the source is $\omega = \omega_0$, there is *no* current flowing between (i) the part of the circuit that includes the source and the resistor and (ii) the part that includes the inductor and the capacitor, so you could cut the wires connecting these two parts of the circuit without affecting the currents. (e) Is the statement in part (d) still valid if we consider that any real inductor or capacitor also has some resistance of its own? Explain.

31.58 •• A 400-Ω resistor and a 6.00-μF capacitor are connected in parallel to an ac generator that supplies an rms voltage of 220 V at an angular frequency of 360 rad/s. Use the results of Problem 31.56. Note that since there is no inductor in the circuit, the $1/\omega L$ term is not present in the expression for Z. Find (a) the current amplitude in the resistor; (b) the current amplitude in the capacitor; (c) the phase angle of the source current with respect to the source voltage; (d) the amplitude of the current through the generator. (e) Does the source current lag or lead the source voltage?

31.59 •• An L-R-C parallel circuit is connected to an ac source of constant voltage amplitude V and variable angular frequency ω. (a) Using the results of Problem 31.56, find expressions for the amplitudes I_R, I_L, and I_C of the currents through the resistor, inductor, and capacitor as functions of ω. (b) Graph I_R, I_L, and I_C as functions of ω for $V = 100$ V, $R = 200$ Ω, $L = 2.0$ H, and $C = 0.50$ μF. (c) Discuss the behavior of I_L and I_C in the limits $\omega = 0$ and $\omega \rightarrow \infty$. Explain why I_L and I_C behave as they do in these limits. (d) Calculate the resonance frequency (in Hz) of the circuit, and sketch the phasor diagram at the resonance frequency. (e) At the resonance frequency, what is the current amplitude through the source? (f) At the resonance frequency, what is the current amplitude through the resistor, through the inductor, and through the capacitor?

31.60 •• A 100-Ω resistor, a 0.100-μF capacitor, and a 0.300-H inductor are connected in parallel to a voltage source with amplitude 240 V. (a) What is the resonance angular frequency? (b) What is the maximum current through the source at the resonance frequency? (c) Find the maximum current in the resistor at resonance. (d) What is the maximum current in the inductor at resonance. (e) What is the maximum current in the branch containing the capacitor at resonance? (f) Find the maximum energy stored in the inductor and in the capacitor at resonance.

31.61 • You want to double the resonance angular frequency of an *L-R-C* series circuit by changing only the *pertinent* circuit elements all by the same factor. (a) Which ones should you change? (b) By what factor should you change them?

31.62 ••• An *L-R-C* series circuit consists of a 2.50-μF capacitor, a 5.00-mH inductor, and a 75.0-Ω resistor connected across an ac source of voltage amplitude 15.0 V having variable frequency. (a) Under what circumstances is the average power delivered to the circuit equal to $\frac{1}{2}V_{rms}I_{rms}$? (b) Under the conditions of part (a), what is the average power delivered to each circuit element and what is the maximum current through the capacitor?

31.63 •• In an *L-R-C* series circuit, the source has a voltage amplitude of 120 V, $R = 80.0\ \Omega$, and the reactance of the capacitor is 480 Ω. The voltage amplitude across the capacitor is 360 V. (a) What is the current amplitude in the circuit? (b) What is the impedance? (c) What two values can the reactance of the inductor have? (d) For which of the two values found in part (c) is the angular frequency less than the resonance angular frequency? Explain.

31.64 • An *L-R-C* series circuit has $R = 500\ \Omega$, $L = 2.00$ H, $C = 0.500\ \mu$F, and $V = 100$ V. (a) For $\omega = 800$ rad/s, calculate V_R, V_L, V_C, and ϕ. Using a single set of axes, graph v, v_R, v_L, and v_C as functions of time. Include two cycles of v on your graph. (b) Repeat part (a) for $\omega = 1000$ rad/s. (c) Repeat part (a) for $\omega = 1250$ rad/s.

31.65 •• **CALC** The current in a certain circuit varies with time as shown in Fig. P31.65. Find the average current and the rms current in terms of I_0.

Figure **P31.65**

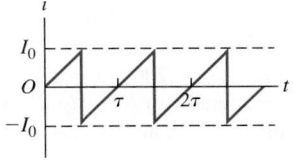

31.66 •• **The Resonance Width.** Consider an *L-R-C* series circuit with a 1.80-H inductor, a 0.900-μF capacitor, and a 300-Ω resistor. The source has terminal rms voltage $V_{rms} = 60.0$ V and variable angular frequency ω. (a) What is the resonance angular frequency ω_0 of the circuit? (b) What is the rms current through the circuit at resonance, $I_{rms\text{-}0}$? (c) For what two values of the angular frequency, ω_1 and ω_2, is the rms current half the resonance value? (d) The quantity $|\omega_1 - \omega_2|$ defines the *resonance width*. Calculate $I_{rms\text{-}0}$ and the resonance width for $R = 300\ \Omega$, 30.0 Ω, and 3.00 Ω. Describe how your results compare to the discussion in Section 31.5.

31.67 •• An inductor, a capacitor, and a resistor are all connected in series across an ac source. If the resistance, inductance, and capacitance are all doubled, by what factor does each of the following quantities change? Indicate whether they increase or decrease: (a) the resonance angular frequency; (b) the inductive reactance; (c) the capacitive reactance. (d) Does the impedance double?

31.68 • A resistance R, capacitance C, and inductance L are connected in series to a voltage source with amplitude V and variable angular frequency ω. If $\omega = \omega_0$, the resonance angular frequency, find (a) the maximum current in the resistor; (b) the maximum voltage across the capacitor; (c) the maximum voltage across the inductor; (d) the maximum energy stored in the capacitor; (e) the maximum energy stored in the inductor. Give your answers in terms of R, C, L, and V.

31.69 • Repeat Problem 31.68 for the case $\omega = \omega_0/2$.

31.70 • Repeat Problem 31.68 for the case $\omega = 2\omega_0$.

31.71 • A transformer consists of 275 primary windings and 834 secondary windings. If the potential difference across the primary coil is 25.0 V, (a) what is the voltage across the secondary coil, and (b) what is the effective load resistance of the secondary coil if it is connected across a 125-Ω resistor?

31.72 •• An *L-R-C* series circuit draws 220 W from a 120-V (rms), 50.0-Hz ac line. The power factor is 0.560, and the source voltage leads the current. (a) What is the net resistance R of the circuit? (b) Find the capacitance of the series capacitor that will result in a power factor of unity when it is added to the original circuit. (c) What power will then be drawn from the supply line?

31.73 •• **CALC** In an *L-R-C* series circuit the current is given by $i = I\cos\omega t$. The voltage amplitudes for the resistor, inductor, and capacitor are V_R, V_L, and V_C. (a) Show that the instantaneous power into the resistor is $p_R = V_R I\cos^2\omega t = \frac{1}{2}V_R I(1 + \cos 2\omega t)$. What does this expression give for the average power into the resistor? (b) Show that the instantaneous power into the inductor is $p_L = -V_L I\sin\omega t\cos\omega t = -\frac{1}{2}V_L I\sin 2\omega t$. What does this expression give for the average power into the inductor? (c) Show that the instantaneous power into the capacitor is $p_C = V_C I\sin\omega t\cos\omega t = \frac{1}{2}V_C I\sin 2\omega t$. What does this expression give for the average power into the capacitor? (d) The instantaneous power delivered by the source is shown in Section 31.4 to be $p = VI\cos\omega t\,(\cos\phi\cos\omega t - \sin\phi\sin\omega t)$. Show that $p_R + p_L + p_C$ equals p at each instant of time.

CHALLENGE PROBLEMS

31.74 ••• **CALC** (a) At what angular frequency is the voltage amplitude across the *resistor* in an *L-R-C* series circuit at maximum value? (b) At what angular frequency is the voltage amplitude across the *inductor* at maximum value? (c) At what angular frequency is the voltage amplitude across the *capacitor* at maximum value? (You may want to refer to the results of Problem 31.53.)

31.75 ••• **Complex Numbers in a Circuit.** The voltage across a circuit element in an ac circuit is not necessarily in phase with the current through that circuit element. Therefore the voltage amplitudes across the circuit elements in a branch in an ac circuit do not add algebraically. A method that is commonly employed to simplify the analysis of an ac circuit driven by a sinusoidal source is to represent the impedance Z as a *complex* number. The resistance R is taken to be the real part of the impedance, and the reactance $X = X_L - X_C$ is taken to be the imaginary part. Thus, for a branch containing a resistor, inductor, and capacitor in series, the complex impedance is $Z_{cpx} = R + iX$, where $i^2 = -1$. If the voltage amplitude across the branch is V_{cpx}, we define a *complex* current amplitude by $I_{cpx} = V_{cpx}/Z_{cpx}$. The *actual* current amplitude is the absolute value of the complex current amplitude; that is, $I = (I_{cpx}{}^*I_{cpx})^{1/2}$. The phase angle ϕ of the current with respect to the source voltage is given by $\tan\phi = \text{Im}(I_{cpx})/\text{Re}(I_{cpx})$. The voltage amplitudes $V_{R\text{-}cpx}$, $V_{L\text{-}cpx}$, and $V_{C\text{-}cpx}$ across the resistance, inductance, and capacitance, respectively, are found by multiplying I_{cpx} by R, iX_L, and $-iX_C$, respectively. From the complex representation for the voltage amplitudes, the voltage across a branch is just the algebraic sum of the voltages across each circuit element: $V_{cpx} = V_{R\text{-}cpx} + V_{L\text{-}cpx} + V_{C\text{-}cpx}$. The actual value of any current amplitude or voltage amplitude is the absolute value of the corresponding complex

Figure **P31.75**

L = 0.500 H

V = 200 V
$\omega = 1000$ rad/s

C = 1.25 μF

R = 400 Ω

quantity. Consider the *L-R-C* series circuit shown in Fig. P31.75. The values of the circuit elements, the source voltage amplitude, and the source angular frequency are as shown. Use the phasor diagram techniques presented in Section 31.1 to solve for (a) the current amplitude and (b) the phase angle ϕ of the current with respect to the source voltage. (Note that this angle is the negative of the phase angle defined in Fig. 31.13.) Now analyze the same circuit using the complex-number approach. (c) Determine the complex impedance of the circuit, Z_{cpx}. Take the absolute value to obtain Z, the actual impedance of the circuit. (d) Take the voltage amplitude of the source, V_{cpx}, to be real, and find the complex current amplitude I_{cpx}. Find the actual current amplitude by taking the absolute value of I_{cpx}. (e) Find the phase angle ϕ of the current with respect to the source voltage by using the real and imaginary parts of I_{cpx}, as explained above. (f) Find the complex representations of the voltages across the resistance, the inductance, and the capacitance. (g) Adding the answers found in part (f), verify that the sum of these complex numbers is real and equal to 200 V, the voltage of the source.

Answers

Chapter Opening Question ?

Yes. In fact, the radio simultaneously detects transmissions at *all* frequencies. However, a radio is an *L-R-C* series circuit, and at any given time it is tuned to have a resonance at just one frequency. Hence the response of the radio to that frequency is much greater than its response to any other frequency, which is why you hear only one broadcasting station through the radio's speaker. (You can sometimes hear a second station if its frequency is sufficiently close to the tuned frequency.)

Test Your Understanding Questions

31.1 Answers: (a) D; (b) A; (c) B; (d) C For each phasor, the actual current is represented by the projection of that phasor onto the horizontal axis. The phasors all rotate counterclockwise around the origin with angular speed ω, so at the instant shown the projection of phasor A is positive but trending toward zero; the projection of phasor B is negative and becoming more negative; the projection of phasor C is negative but trending toward zero; and the projection of phasor D is positive and becoming more positive.

31.2 Answers: (a) (iii); (b) (ii); (c) (i) For a resistor, $V_R = IR$, so $I = V_R/R$. The voltage amplitude V_R and resistance R do not change with frequency, so the current amplitude I remains constant. For an inductor, $V_L = IX_L = I\omega L$, so $I = V_L/\omega L$. The voltage amplitude V_L and inductance L are constant, so the current amplitude I decreases as the frequency increases. For a capacitor, $V_C = IX_C = I/\omega C$, so $I = V_C\omega C$. The voltage amplitude V_C and capacitance C are constant, so the current amplitude I increases as the frequency increases.

31.3 Answer: (iv), (ii), (i), (iii) For the circuit in Example 31.4, $I = V/Z = (50 \text{ V})/(500 \ \Omega) = 0.10$ A. If the capacitor and inductor are removed so that only the ac source and resistor remain, the circuit is like that shown in Fig. 31.7a; then $I = V/R = (50 \text{ V})/(300 \ \Omega) = 0.17$ A. If the resistor and capacitor are removed so that only the ac source and inductor remain, the circuit is like that

shown in Fig. 31.8a; then $I = V/X_L = (50 \text{ V})/(600 \ \Omega) = 0.083$ A. Finally, if the resistor and inductor are removed so that only the ac source and capacitor remain, the circuit is like that shown in Fig. 31.9a; then $I = V/X_C = (50 \text{ V})/(200 \ \Omega) = 0.25$ A.

31.4 Answers: (a) (v); (b) (iv) The energy cannot be extracted from the resistor, since energy is dissipated in a resistor and cannot be recovered. Instead, the energy must be extracted from either the inductor (which stores magnetic-field energy) or the capacitor (which stores electric-field energy). Positive power means that energy is being transferred from the ac source to the circuit, so *negative* power implies that energy is being transferred back into the source.

31.5 Answer: (ii) The capacitance C increases if the plate spacing is decreased (see Section 24.1). Hence the resonance frequency $f_0 = \omega_0/2\pi = 1/2\pi \sqrt{LC}$ decreases.

31.6 Answer: (ii), (iv), (i), (iii) From Eq. (31.35) the turns ratio is $N_2/N_1 = V_2/V_1$, so the number of turns in the secondary is $N_2 = N_1V_2/V_1$. Hence for the four cases we have (i) $N_2 = (1000)(6.0 \text{ V})/(120 \text{ V}) = 50$ turns; (ii) $N_2 = (1000)(240 \text{ V})/(120 \text{ V}) = 2000$ turns; (iii) $N_2 = (1000)(6.0 \text{ V})/(240 \text{ V}) = 25$ turns; and (iv) $N_2 = (1000)(120 \text{ V})/(240 \text{ V}) = 500$ turns. Note that (i), (iii), and (iv) are step-down transformers with fewer turns in the secondary than in the primary, while (ii) is a step-up transformer with more turns in the secondary than in the primary.

Bridging Problem

Answers: (a) 8.35×10^4 rad/s and 3.19×10^5 rad/s

(b) At 8.35×10^4 rad/s: $V_{source} = 49.5$ V,
$\qquad I = 0.132$ A, $V_R = 16.5$ V, $V_L = 16.5$ V,
$\qquad V_C = 63.2$ V.
$\qquad$ At 3.19×10^5 rad/s: $V_{source} = 49.5$ V,
$\qquad I = 0.132$ A, $V_R = 16.5$ V, $V_L = 63.2$ V,
$\qquad V_C = 16.5$ V.

ELECTROMAGNETIC WAVES

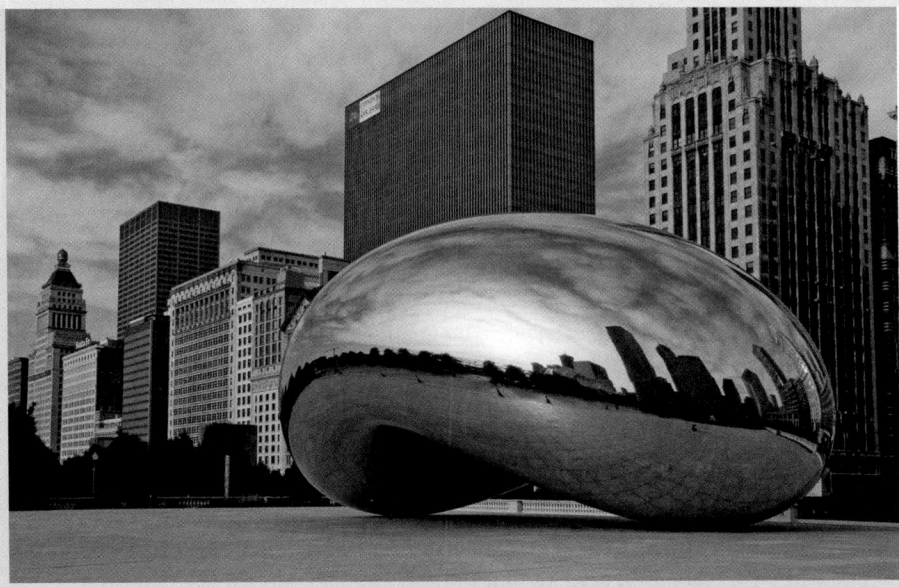

? Metal objects reflect not only visible light but also radio waves. What aspect of metals makes them so reflective?

LEARNING GOALS

By studying this chapter, you will learn:

- Why there are both electric and magnetic fields in a light wave.

- How the speed of light is related to the fundamental constants of electricity and magnetism.

- How to describe the propagation of a sinusoidal electromagnetic wave.

- What determines the amount of power carried by an electromagnetic wave.

- How to describe standing electromagnetic waves.

What is light? This question has been asked by humans for centuries, but there was no answer until electricity and magnetism were unified into *electromagnetism,* as described by Maxwell's equations. These equations show that a time-varying magnetic field acts as a source of electric field and that a time-varying electric field acts as a source of magnetic field. These $\vec{E}$ and $\vec{B}$ fields can sustain each other, forming an *electromagnetic wave* that propagates through space. Visible light emitted by the glowing filament of a light bulb is one example of an electromagnetic wave; other kinds of electromagnetic waves are produced by TV and radio stations, x-ray machines, and radioactive nuclei.

In this chapter we'll use Maxwell's equations as the theoretical basis for understanding electromagnetic waves. We'll find that these waves carry both energy and momentum. In sinusoidal electromagnetic waves, the $\vec{E}$ and $\vec{B}$ fields are sinusoidal functions of time and position, with a definite frequency and wavelength. Visible light, radio, x rays, and other types of electromagnetic waves differ only in their frequency and wavelength. Our study of optics in the following chapters will be based in part on the electromagnetic nature of light.

Unlike waves on a string or sound waves in a fluid, electromagnetic waves do not require a material medium; the light that you see coming from the stars at night has traveled without difficulty across tens or hundreds of light-years of (nearly) empty space. Nonetheless, electromagnetic waves and mechanical waves have much in common and are described in much the same language. Before reading further in this chapter, you should review the properties of mechanical waves as discussed in Chapters 15 and 16.

32.1 Maxwell's Equations and Electromagnetic Waves

In the last several chapters we studied various aspects of electric and magnetic fields. We learned that when the fields don't vary with time, such as an electric field produced by charges at rest or the magnetic field of a steady current, we can analyze the electric and magnetic fields independently without considering interactions between them. But when the fields vary with time, they are no longer independent. Faraday's law (see Section 29.2) tells us that a time-varying magnetic field acts as a source of electric field, as shown by induced emfs in inductors and transformers. Ampere's law, including the displacement current discovered by Maxwell (see Section 29.7), shows that a time-varying electric field acts as a source of magnetic field. This mutual interaction between the two fields is summarized in Maxwell's equations, presented in Section 29.7.

Thus, when *either* an electric or a magnetic field is changing with time, a field of the other kind is induced in adjacent regions of space. We are led (as Maxwell was) to consider the possibility of an electromagnetic disturbance, consisting of time-varying electric and magnetic fields, that can propagate through space from one region to another, even when there is no matter in the intervening region. Such a disturbance, if it exists, will have the properties of a *wave,* and an appropriate term is **electromagnetic wave.**

Such waves do exist; radio and television transmission, light, x rays, and many other kinds of radiation are examples of electromagnetic waves. Our goal in this chapter is to see how such waves are explained by the principles of electromagnetism that we have studied thus far and to examine the properties of these waves.

Electricity, Magnetism, and Light

As often happens in the development of science, the theoretical understanding of electromagnetic waves evolved along a considerably more devious path than the one just outlined. In the early days of electromagnetic theory (the early 19th century), two different units of electric charge were used: one for electrostatics and the other for magnetic phenomena involving currents. In the system of units used at that time, these two units of charge had different physical dimensions. Their *ratio* had units of velocity, and measurements showed that the ratio had a numerical value that was precisely equal to the speed of light, 3.00×10^8 m/s. At the time, physicists regarded this as an extraordinary coincidence and had no idea how to explain it.

In searching to understand this result, Maxwell (Fig. 32.1) proved in 1865 that an electromagnetic disturbance should propagate in free space with a speed equal to that of light and hence that light waves were likely to be electromagnetic in nature. At the same time, he discovered that the basic principles of electromagnetism can be expressed in terms of the four equations that we now call **Maxwell's equations,** which we discussed in Section 29.7. These four equations are (1) Gauss's law for electric fields; (2) Gauss's law for magnetic fields, showing the absence of magnetic monopoles; (3) Ampere's law, including displacement current; and (4) Faraday's law:

32.1 James Clerk Maxwell (1831–1879) was the first person to truly understand the fundamental nature of light. He also made major contributions to thermodynamics, optics, astronomy, and color photography. Albert Einstein described Maxwell's accomplishments as "the most profound and the most fruitful that physics has experienced since the time of Newton."

$$\oint \vec{E} \cdot d\vec{A} = \frac{Q_{\text{encl}}}{\epsilon_0} \qquad \text{(Gauss's law)} \tag{29.18}$$

$$\oint \vec{B} \cdot d\vec{A} = 0 \qquad \text{(Gauss's law for magnetism)} \tag{29.19}$$

$$\oint \vec{B} \cdot d\vec{l} = \mu_0 \left(i_C + \epsilon_0 \frac{d\Phi_E}{dt} \right)_{\text{encl}} \qquad \text{(Ampere's law)} \tag{29.20}$$

$$\oint \vec{E} \cdot d\vec{l} = -\frac{d\Phi_B}{dt} \qquad \text{(Faraday's law)} \tag{29.21}$$

These equations apply to electric and magnetic fields *in vacuum*. If a material is present, the permittivity ϵ_0 and permeability μ_0 of free space are replaced by the permittivity ϵ and permeability μ of the material. If the values of ϵ and μ are different at different points in the regions of integration, then ϵ and μ have to be transferred to the left sides of Eqs. (29.18) and (29.20), respectively, and placed inside the integrals. The ϵ in Eq. (29.20) also has to be included in the integral that gives $d\Phi_E/dt$.

According to Maxwell's equations, a point charge at rest produces a static $\vec{E}$ field but no $\vec{B}$ field; a point charge moving with a constant velocity (see Section 28.1) produces both $\vec{E}$ and $\vec{B}$ fields. Maxwell's equations can also be used to show that in order for a point charge to produce electromagnetic waves, the charge must *accelerate*. In fact, it's a general result of Maxwell's equations that *every* accelerated charge radiates electromagnetic energy (Fig. 32.2).

Generating Electromagnetic Radiation

One way in which a point charge can be made to emit electromagnetic waves is by making it oscillate in simple harmonic motion, so that it has an acceleration at almost every instant (the exception is when the charge is passing through its equilibrium position). Figure 32.3 shows some of the electric field lines produced by such an oscillating point charge. Field lines are *not* material objects, but you may nonetheless find it helpful to think of them as behaving somewhat like strings that extend from the point charge off to infinity. Oscillating the charge up and down makes waves that propagate outward from the charge along these "strings." Note that the charge does not emit waves equally in all directions; the waves are strongest at 90° to the axis of motion of the charge, while there are *no* waves along this axis. This is just what the "string" picture would lead you to conclude. There is also a *magnetic* disturbance that spreads outward from the charge; this is not shown in Fig. 32.3. Because the electric and magnetic disturbances spread or radiate away from the source, the name **electromagnetic radiation** is used interchangeably with the phrase "electromagnetic waves."

Electromagnetic waves with macroscopic wavelengths were first produced in the laboratory in 1887 by the German physicist Heinrich Hertz. As a source of waves, he used charges oscillating in *L-C* circuits of the sort discussed in Section 30.5; he detected the resulting electromagnetic waves with other circuits tuned to the same frequency. Hertz also produced electromagnetic *standing waves* and measured the distance between adjacent nodes (one half-wavelength) to determine the wavelength. Knowing the resonant frequency of his circuits, he then found the speed of the waves from the wavelength–frequency relationship $v = \lambda f$. He established that their speed was the same as that of light; this verified Maxwell's theoretical prediction directly. The SI unit of frequency is named in honor of Hertz: One hertz (1 Hz) equals one cycle per second.

32.2 (Top) Every mobile phone, wireless modem, or radio transmitter emits signals in the form of electromagnetic waves that are made by accelerating charges. (Bottom) Power lines carry a strong alternating current, which means that a substantial amount of charge is accelerating back and forth and generating electromagnetic waves. These waves can produce a buzzing sound from your car radio when you drive near the lines.

32.3 Electric field lines of a point charge oscillating in simple harmonic motion, seen at five instants during an oscillation period T. The charge's trajectory is in the plane of the drawings. At $t = 0$ the point charge is at its maximum upward displacement. The arrow shows one "kink" in the lines of $\vec{E}$ as it propagates outward from the point charge. The magnetic field (not shown) comprises circles that lie in planes perpendicular to these figures and concentric with the axis of oscillation.

(a) $t = 0$

(b) $t = T/4$

(c) $t = T/2$

(d) $t = 3T/4$

(e) $t = T$

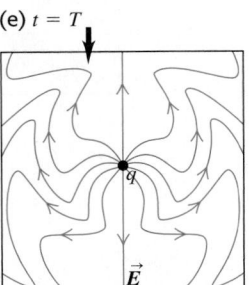

The modern value of the speed of light, which we denote by the symbol c, is 299,792,458 m/s. (Recall from Section 1.3 that this value is the basis of our standard of length: One meter is defined to be the distance that light travels in 1/299,792,458 second.) For our purposes, $c = 3.00 \times 10^8$ m/s is sufficiently accurate.

The possible use of electromagnetic waves for long-distance communication does not seem to have occurred to Hertz. It was left to Marconi and others to make radio communication a familiar household experience. In a radio *transmitter*, electric charges are made to oscillate along the length of the conducting antenna, producing oscillating field disturbances like those shown in Fig. 32.3. Since many charges oscillate together in the antenna, the disturbances are much stronger than those of a single oscillating charge and can be detected at a much greater distance. In a radio *receiver* the antenna is also a conductor; the fields of the wave emanating from a distant transmitter exert forces on free charges within the receiver antenna, producing an oscillating current that is detected and amplified by the receiver circuitry.

For the remainder of this chapter our concern will be with electromagnetic waves themselves, not with the rather complex problem of how they are produced.

The Electromagnetic Spectrum

The **electromagnetic spectrum** encompasses electromagnetic waves of all frequencies and wavelengths. Figure 32.4 shows approximate wavelength and frequency ranges for the most commonly encountered portion of the spectrum. Despite vast differences in their uses and means of production, these are all electromagnetic waves with the same propagation speed (in vacuum) $c = 299,792,458$ m/s. Electromagnetic waves may differ in frequency f and wavelength λ, but the relationship $c = \lambda f$ in vacuum holds for each.

We can detect only a very small segment of this spectrum directly through our sense of sight. We call this range **visible light.** Its wavelengths range from about 380 to 750 nm $(380$ to 750×10^{-9} m$)$, with corresponding frequencies from about 790 to 400 THz (7.9 to 4.0×10^{14} Hz). Different parts of the visible spectrum evoke in humans the sensations of different colors. Table 32.1 gives the approximate wavelengths for colors in the visible spectrum.

Ordinary white light includes all visible wavelengths. However, by using special sources or filters, we can select a narrow band of wavelengths within a range of a few nm. Such light is approximately *monochromatic* (single-color) light. Absolutely monochromatic light with only a single wavelength is an unattainable idealization. When we use the expression "monochromatic light with $\lambda = 550$ nm" with reference to a laboratory experiment, we really mean a small band

Table 32.1 Wavelengths of Visible Light

380–450 nm	Violet
450–495 nm	Blue
495–570 nm	Green
570–590 nm	Yellow
590–620 nm	Orange
620–750 nm	Red

32.4 The electromagnetic spectrum. The frequencies and wavelengths found in nature extend over such a wide range that we have to use a logarithmic scale to show all important bands. The boundaries between bands are somewhat arbitrary.

of wavelengths *around* 550 nm. Light from a *laser* is much more nearly monochromatic than is light obtainable in any other way.

Invisible forms of electromagnetic radiation are no less important than visible light. Our system of global communication, for example, depends on radio waves: AM radio uses waves with frequencies from 5.4×10^5 Hz to 1.6×10^6 Hz, while FM radio broadcasts are at frequencies from 8.8×10^7 Hz to 1.08×10^8 Hz. (Television broadcasts use frequencies that bracket the FM band.) Microwaves are also used for communication (for example, by cellular phones and wireless networks) and for weather radar (at frequencies near 3×10^9 Hz). Many cameras have a device that emits a beam of infrared radiation; by analyzing the properties of the infrared radiation reflected from the subject, the camera determines the distance to the subject and automatically adjusts the focus. X rays are able to penetrate through flesh, which makes them invaluable in dentistry and medicine. Gamma rays, the shortest-wavelength type of electromagnetic radiation, are used in medicine to destroy cancer cells.

Test Your Understanding of Section 32.1 (a) Is it possible to have a purely electric wave propagate through empty space—that is, a wave made up of an electric field but no magnetic field? (b) What about a purely magnetic wave, with a magnetic field but no electric field?

32.2 Plane Electromagnetic Waves and the Speed of Light

We are now ready to develop the basic ideas of electromagnetic waves and their relationship to the principles of electromagnetism. Our procedure will be to postulate a simple field configuration that has wavelike behavior. We'll assume an electric field $\vec{E}$ that has only a *y*-component and a magnetic field $\vec{B}$ with only a *z*-component, and we'll assume that both fields move together in the +*x*-direction with a speed *c* that is initially unknown. (As we go along, it will become clear why we choose $\vec{E}$ and $\vec{B}$ to be perpendicular to the direction of propagation as well as to each other.) Then we will test whether these fields are physically possible by asking whether they are consistent with Maxwell's equations, particularly Ampere's law and Faraday's law. We'll find that the answer is yes, provided that *c* has a particular value. We'll also show that the *wave equation,* which we encountered during our study of mechanical waves in Chapter 15, can be derived from Maxwell's equations.

A Simple Plane Electromagnetic Wave

Using an *xyz*-coordinate system (Fig. 32.5), we imagine that all space is divided into two regions by a plane perpendicular to the *x*-axis (parallel to the *yz*-plane). At every point to the left of this plane there are a uniform electric field $\vec{E}$ in the +*y*-direction and a uniform magnetic field $\vec{B}$ in the +*z*-direction, as shown. Furthermore, we suppose that the boundary plane, which we call the *wave front,* moves to the right in the +*x*-direction with a constant speed *c*, the value of which we'll leave undetermined for now. Thus the $\vec{E}$ and $\vec{B}$ fields travel to the right into previously field-free regions with a definite speed. This is a rudimentary electromagnetic wave. A wave such as this, in which at any instant the fields are uniform over any plane perpendicular to the direction of propagation, is called a **plane wave.** In the case shown in Fig. 32.5, the fields are zero for planes to the right of the wave front and have the same values on all planes to the left of the wave front; later we will consider more complex plane waves.

We won't concern ourselves with the problem of actually *producing* such a field configuration. Instead, we simply ask whether it is consistent with the laws of electromagnetism—that is, with Maxwell's equations. We'll consider each of these four equations in turn.

32.5 An electromagnetic wave front. The plane representing the wave front moves to the right (in the positive *x*-direction) with speed *c*.

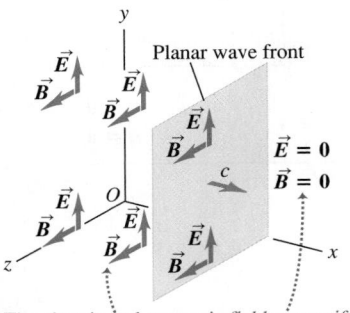

The electric and magnetic fields are uniform behind the advancing wave front and zero in front of it.

32.6 Gaussian surface for a transverse plane electromagnetic wave.

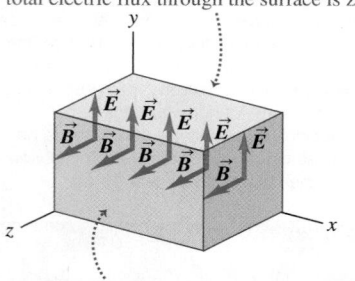

The electric field is the same on the top and bottom sides of the Gaussian surface, so the total electric flux through the surface is zero.

The magnetic field is the same on the left and right sides of the Gaussian surface, so the total magnetic flux through the surface is zero.

32.7 (a) Applying Faraday's law to a plane wave. (b) In a time dt, the magnetic flux through the rectangle in the xy-plane increases by an amount $d\Phi_B$. This increase equals the flux through the shaded rectangle with area $ac\,dt$; that is, $d\Phi_B = Bac\,dt$. Thus $d\Phi_B/dt = Bac$.

(a) In time dt, the wave front moves a distance $c\,dt$ in the $+x$-direction.

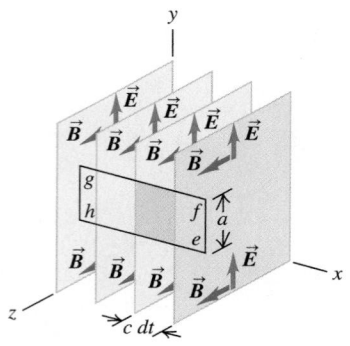

(b) Side view of situation in (a)

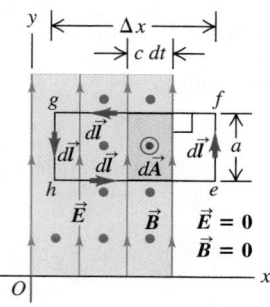

Let us first verify that our wave satisfies Maxwell's first and second equations—that is, Gauss's laws for electric and magnetic fields. To do this, we take as our Gaussian surface a rectangular box with sides parallel to the xy, xz, and yz coordinate planes (Fig. 32.6). The box encloses no electric charge. The total electric flux and magnetic flux through the box are both zero, even if part of the box is in the region where $E = B = 0$. This would *not* be the case if $\vec{E}$ or $\vec{B}$ had an x-component, parallel to the direction of propagation; if the wave front were inside the box, there would be flux through the left-hand side of the box (at $x = 0$) but not the right-hand side (at $x > 0$). Thus to satisfy Maxwell's first and second equations, the electric and magnetic fields must be perpendicular to the direction of propagation; that is, the wave must be **transverse.**

The next of Maxwell's equations to be considered is Faraday's law:

$$\oint \vec{E} \cdot d\vec{l} = -\frac{d\Phi_B}{dt} \tag{32.1}$$

To test whether our wave satisfies Faraday's law, we apply this law to a rectangle $efgh$ that is parallel to the xy-plane (Fig. 32.7a). As shown in Fig. 32.7b, a cross section in the xy-plane, this rectangle has height a and width Δx. At the time shown, the wave front has progressed partway through the rectangle, and $\vec{E}$ is zero along the side ef. In applying Faraday's law we take the vector area $d\vec{A}$ of rectangle $efgh$ to be in the $+z$-direction. With this choice the right-hand rule requires that we integrate $\vec{E} \cdot d\vec{l}$ *counterclockwise* around the rectangle. At every point on side ef, $\vec{E}$ is zero. At every point on sides fg and he, $\vec{E}$ is either zero or perpendicular to $d\vec{l}$. Only side gh contributes to the integral. On this side, $\vec{E}$ and $d\vec{l}$ are opposite, and we obtain

$$\oint \vec{E} \cdot d\vec{l} = -Ea \tag{32.2}$$

Hence, the left-hand side of Eq. (32.1) is nonzero.

To satisfy Faraday's law, Eq. (32.1), there must be a component of $\vec{B}$ in the z-direction (perpendicular to $\vec{E}$) so that there can be a nonzero magnetic flux Φ_B through the rectangle $efgh$ and a nonzero derivative $d\Phi_B/dt$. Indeed, in our wave, $\vec{B}$ has *only* a z-component. We have assumed that this component is in the *positive* z-direction; let's see whether this assumption is consistent with Faraday's law. During a time interval dt the wave front moves a distance $c\,dt$ to the right in Fig. 32.7b, sweeping out an area $ac\,dt$ of the rectangle $efgh$. During this interval the magnetic flux Φ_B through the rectangle $efgh$ increases by $d\Phi_B = B(ac\,dt)$, so the rate of change of magnetic flux is

$$\frac{d\Phi_B}{dt} = Bac \tag{32.3}$$

Now we substitute Eqs. (32.2) and (32.3) into Faraday's law, Eq. (32.1); we get

$$-Ea = -Bac$$

$$E = cB \quad \text{(electromagnetic wave in vacuum)} \tag{32.4}$$

This shows that our wave is consistent with Faraday's law only if the wave speed c and the magnitudes of the perpendicular vectors $\vec{E}$ and $\vec{B}$ are related as in Eq. (32.4). Note that if we had assumed that $\vec{B}$ was in the *negative* z-direction, there would have been an additional minus sign in Eq. (32.4); since E, c, and B are all positive magnitudes, no solution would then have been possible. Furthermore, any component of $\vec{B}$ in the y-direction (parallel to $\vec{E}$) would not contribute to the changing magnetic flux Φ_B through the rectangle $efgh$ (which is parallel to the xy-plane) and so would not be part of the wave.

Finally, we carry out a similar calculation using Ampere's law, the remaining member of Maxwell's equations. There is no conduction current ($i_C = 0$), so Ampere's law is

$$\oint \vec{B} \cdot d\vec{l} = \mu_0 \epsilon_0 \frac{d\Phi_E}{dt} \qquad (32.5)$$

To check whether our wave is consistent with Ampere's law, we move our rectangle so that it lies in the *xz*-plane, as shown in Fig. 32.8, and we again look at the situation at a time when the wave front has traveled partway through the rectangle. We take the vector area $d\vec{A}$ in the $+y$-direction, and so the right-hand rule requires that we integrate $\vec{B} \cdot d\vec{l}$ counterclockwise around the rectangle. The $\vec{B}$ field is zero at every point along side *ef*, and at each point on sides *fg* and *he* it is either zero or perpendicular to $d\vec{l}$. Only side *gh*, where $\vec{B}$ and $d\vec{l}$ are parallel, contributes to the integral, and we find

$$\oint \vec{B} \cdot d\vec{l} = Ba \qquad (32.6)$$

Hence, the left-hand side of Ampere's law, Eq. (32.5), is nonzero; the right-hand side must be nonzero as well. Thus $\vec{E}$ must have a *y*-component (perpendicular to $\vec{B}$) so that the electric flux Φ_E through the rectangle and the time derivative $d\Phi_E/dt$ can be nonzero. We come to the same conclusion that we inferred from Faraday's law: In an electromagnetic wave, $\vec{E}$ and $\vec{B}$ must be mutually perpendicular.

In a time interval dt the electric flux Φ_E through the rectangle increases by $d\Phi_E = E(ac\,dt)$. Since we chose $d\vec{A}$ to be in the $+y$-direction, this flux change is positive; the rate of change of electric field is

$$\frac{d\Phi_E}{dt} = Eac \qquad (32.7)$$

Substituting Eqs. (32.6) and (32.7) into Ampere's law, Eq. (32.5), we find

$$Ba = \epsilon_0 \mu_0 Eac$$

$$B = \epsilon_0 \mu_0 cE \qquad \text{(electromagnetic wave in vacuum)} \qquad (32.8)$$

Thus our assumed wave obeys Ampere's law only if B, c, and E are related as in Eq. (32.8).

Our electromagnetic wave must obey *both* Ampere's law and Faraday's law, so Eqs. (32.4) and (32.8) must both be satisfied. This can happen only if $\epsilon_0\mu_0 c = 1/c$, or

$$c = \frac{1}{\sqrt{\epsilon_0\mu_0}} \qquad \text{(speed of electromagnetic waves in vacuum)} \qquad (32.9)$$

Inserting the numerical values of these quantities, we find

$$c = \frac{1}{\sqrt{(8.85 \times 10^{-12} \text{ C}^2/\text{N} \cdot \text{m}^2)(4\pi \times 10^{-7} \text{ N/A}^2)}}$$

$$= 3.00 \times 10^8 \text{ m/s}$$

Our assumed wave is consistent with all of Maxwell's equations, provided that the wave front moves with the speed given above, which you should recognize as the speed of light! Note that the *exact* value of c is defined to be 299,792,458 m/s; the modern value of ϵ_0 is defined to agree with this when used in Eq. (32.9) (see Section 21.3).

32.8 (a) Applying Ampere's law to a plane wave. (Compare to Fig. 32.7a.) (b) In a time dt, the electric flux through the rectangle in the *xz*-plane increases by an amount $d\Phi_E$. This increase equals the flux through the shaded rectangle with area $ac\,dt$; that is, $d\Phi_E = Eac\,dt$. Thus $d\Phi_E/dt = Eac$.

(a) In time dt, the wave front moves a distance $c\,dt$ in the $+x$-direction.

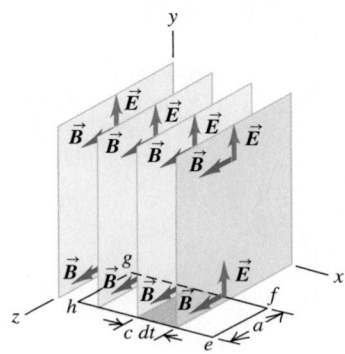

(b) Top view of situation in **(a)**

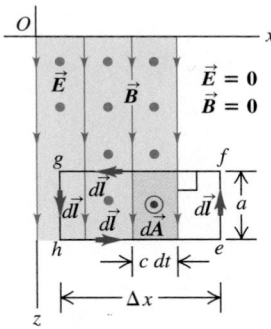

32.9 A right-hand rule for electromagnetic waves relates the directions of $\vec{E}$ and $\vec{B}$ and the direction of propagation.

Right-hand rule for an electromagnetic wave:

① Point the thumb of your right hand in the wave's direction of propagation.

② Imagine rotating the $\vec{E}$-field vector 90° in the sense your fingers curl.

That is the direction of the $\vec{B}$ field.

Direction of propagation = direction of $\vec{E} \times \vec{B}$.

Key Properties of Electromagnetic Waves

We chose a simple wave for our study in order to avoid mathematical complications, but this special case illustrates several important features of *all* electromagnetic waves:

1. The wave is *transverse;* both $\vec{E}$ and $\vec{B}$ are perpendicular to the direction of propagation of the wave. The electric and magnetic fields are also perpendicular to each other. The direction of propagation is the direction of the vector product $\vec{E} \times \vec{B}$ (Fig. 32.9).
2. There is a definite ratio between the magnitudes of $\vec{E}$ and $\vec{B}$: $E = cB$.
3. The wave travels in vacuum with a definite and unchanging speed.
4. Unlike mechanical waves, which need the oscillating particles of a medium such as water or air to transmit a wave, electromagnetic waves require no medium.

We can generalize this discussion to a more realistic situation. Suppose we have several wave fronts in the form of parallel planes perpendicular to the x-axis, all of which are moving to the right with speed c. Suppose that the $\vec{E}$ and $\vec{B}$ fields are the same at all points within a single region between two planes, but that the fields differ from region to region. The overall wave is a plane wave, but one in which the fields vary in steps along the x-axis. Such a wave could be constructed by superposing several of the simple step waves we have just discussed (shown in Fig. 32.5). This is possible because the $\vec{E}$ and $\vec{B}$ fields obey the superposition principle in waves just as in static situations: When two waves are superposed, the total $\vec{E}$ field at each point is the vector sum of the $\vec{E}$ fields of the individual waves, and similarly for the total $\vec{B}$ field.

We can extend the above development to show that a wave with fields that vary in steps is also consistent with Ampere's and Faraday's laws, provided that the wave fronts all move with the speed c given by Eq. (32.9). In the limit that we make the individual steps infinitesimally small, we have a wave in which the $\vec{E}$ and $\vec{B}$ fields at any instant vary *continuously* along the x-axis. The entire field pattern moves to the right with speed c. In Section 32.3 we will consider waves in which $\vec{E}$ and $\vec{B}$ are *sinusoidal* functions of x and t. Because at each point the magnitudes of $\vec{E}$ and $\vec{B}$ are related by $E = cB$, the periodic variations of the two fields in any periodic traveling wave must be *in phase*.

Electromagnetic waves have the property of **polarization.** In the above discussion the choice of the y-direction for $\vec{E}$ was arbitrary. We could just as well have specified the z-axis for $\vec{E}$; then $\vec{B}$ would have been in the $-y$-direction. A wave in which $\vec{E}$ is always parallel to a certain axis is said to be **linearly polarized** along that axis. More generally, *any* wave traveling in the x-direction can be represented as a superposition of waves linearly polarized in the y- and z-directions. We will study polarization in greater detail in Chapter 33.

Derivation of the Electromagnetic Wave Equation

Here is an alternative derivation of Eq. (32.9) for the speed of electromagnetic waves. It is more mathematical than our other treatment, but it includes a derivation of the wave equation for electromagnetic waves. This part of the section can be omitted without loss of continuity in the chapter.

During our discussion of mechanical waves in Section 15.3, we showed that a function $y(x, t)$ that represents the displacement of any point in a mechanical wave traveling along the x-axis must satisfy a differential equation, Eq. (15.12):

$$\frac{\partial^2 y(x, t)}{\partial x^2} = \frac{1}{v^2} \frac{\partial^2 y(x, t)}{\partial t^2} \qquad (32.10)$$

This equation is called the **wave equation,** and v is the speed of propagation of the wave.

To derive the corresponding equation for an electromagnetic wave, we again consider a plane wave. That is, we assume that at each instant, E_y and B_z are uniform over any plane perpendicular to the x-axis, the direction of propagation. But now we let E_y and B_z vary continuously as we go along the x-axis; then each is a function of x and t. We consider the values of E_y and B_z on two planes perpendicular to the x-axis, one at x and one at $x + \Delta x$.

Following the same procedure as previously, we apply Faraday's law to a rectangle lying parallel to the xy-plane, as in Fig. 32.10. This figure is similar to Fig. 32.7. Let the left end gh of the rectangle be at position x, and let the right end ef be at position $(x + \Delta x)$. At time t, the values of E_y on these two sides are $E_y(x, t)$ and $E_y(x + \Delta x, t)$, respectively. When we apply Faraday's law to this rectangle, we find that instead of $\oint \vec{E} \cdot d\vec{l} = -Ea$ as before, we have

$$\oint \vec{E} \cdot d\vec{l} = -E_y(x, t)a + E_y(x + \Delta x, t)a$$

$$= a[E_y(x + \Delta x, t) - E_y(x, t)] \qquad (32.11)$$

To find the magnetic flux Φ_B through this rectangle, we assume that Δx is small enough that B_z is nearly uniform over the rectangle. In that case, $\Phi_B = B_z(x, t)A = B_z(x, t)a\,\Delta x$, and

$$\frac{d\Phi_B}{dt} = \frac{\partial B_z(x, t)}{\partial t}a\,\Delta x$$

We use partial-derivative notation because B_z is a function of both x and t. When we substitute this expression and Eq. (32.11) into Faraday's law, Eq. (32.1), we get

$$a[E_y(x + \Delta x, t) - E_y(x, t)] = -\frac{\partial B_z}{\partial t}a\,\Delta x$$

$$\frac{E_y(x + \Delta x, t) - E_y(x, t)}{\Delta x} = -\frac{\partial B_z}{\partial t}$$

Finally, imagine shrinking the rectangle down to a sliver so that Δx approaches zero. When we take the limit of this equation as $\Delta x \to 0$, we get

$$\frac{\partial E_y(x, t)}{\partial x} = -\frac{\partial B_z(x, t)}{\partial t} \qquad (32.12)$$

This equation shows that if there is a time-varying component B_z of magnetic field, there must also be a component E_y of electric field that varies with x, and conversely. We put this relationship on the shelf for now; we'll return to it soon.

Next we apply Ampere's law to the rectangle shown in Fig. 32.11. The line integral $\oint \vec{B} \cdot d\vec{l}$ becomes

$$\oint \vec{B} \cdot d\vec{l} = -B_z(x + \Delta x, t)a + B_z(x, t)a \qquad (32.13)$$

Again assuming that the rectangle is narrow, we approximate the electric flux Φ_E through it as $\Phi_E = E_y(x, t)A = E_y(x, t)a\,\Delta x$. The rate of change of Φ_E, which we need for Ampere's law, is then

$$\frac{d\Phi_E}{dt} = \frac{\partial E_y(x, t)}{\partial t}a\,\Delta x$$

Now we substitute this expression and Eq. (32.13) into Ampere's law, Eq. (32.5):

$$-B_z(x + \Delta x, t)a + B_z(x, t)a = \epsilon_0\mu_0\frac{\partial E_y(x, t)}{\partial t}a\,\Delta x$$

Again we divide both sides by $a\,\Delta x$ and take the limit as $\Delta x \to 0$. We find

$$-\frac{\partial B_z(x, t)}{\partial x} = \epsilon_0\mu_0\frac{\partial E_y(x, t)}{\partial t} \qquad (32.14)$$

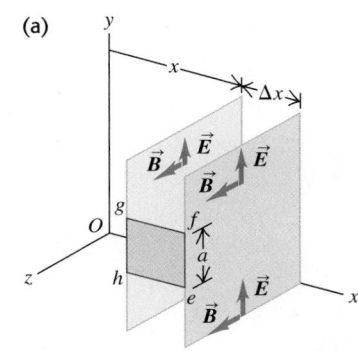

32.10 Faraday's law applied to a rectangle with height a and width Δx parallel to the xy-plane.

(a)

(b) Side view of the situation in (a)

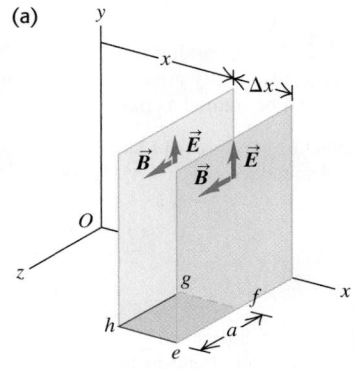

32.11 Ampere's law applied to a rectangle with height a and width Δx parallel to the xz-plane.

(a)

(b) Top view of the situation in (a)

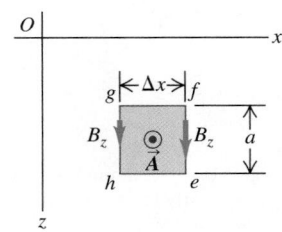

Now comes the final step. We take the partial derivatives with respect to x of both sides of Eq. (32.12), and we take the partial derivatives with respect to t of both sides of Eq. (32.14). The results are

$$-\frac{\partial^2 E_y(x, t)}{\partial x^2} = \frac{\partial^2 B_z(x, t)}{\partial x \partial t}$$

$$-\frac{\partial^2 B_z(x, t)}{\partial x \partial t} = \epsilon_0 \mu_0 \frac{\partial^2 E_y(x, t)}{\partial t^2}$$

Combining these two equations to eliminate B_z, we finally find

$$\frac{\partial^2 E_y(x, t)}{\partial x^2} = \epsilon_0 \mu_0 \frac{\partial^2 E_y(x, t)}{\partial t^2} \qquad \text{(electromagnetic wave equation in vacuum)} \qquad (32.15)$$

This expression has the same form as the general wave equation, Eq. (32.10). Because the electric field E_y must satisfy this equation, it behaves as a wave with a pattern that travels through space with a definite speed. Furthermore, comparison of Eqs. (32.15) and (32.10) shows that the wave speed v is given by

$$\frac{1}{v^2} = \epsilon_0 \mu_0 \qquad \text{or} \qquad v = \frac{1}{\sqrt{\epsilon_0 \mu_0}}$$

This agrees with Eq. (32.9) for the speed c of electromagnetic waves.

We can show that B_z also must satisfy the same wave equation as E_y, Eq. (32.15). To prove this, we take the partial derivative of Eq. (32.12) with respect to t and the partial derivative of Eq. (32.14) with respect to x and combine the results. We leave this derivation for you to carry out.

Test Your Understanding of Section 32.2 For each of the following electromagnetic waves, state the direction of the magnetic field. (a) The wave is propagating in the positive z-direction, and $\vec{E}$ is in the positive x-direction; (b) the wave is propagating in the positive y-direction, and $\vec{E}$ is in the negative z-direction; (c) the wave is propagating in the negative x-direction, and $\vec{E}$ is in the positive z-direction. ▮

32.3 Sinusoidal Electromagnetic Waves

Sinusoidal electromagnetic waves are directly analogous to sinusoidal transverse mechanical waves on a stretched string, which we studied in Section 15.3. In a sinusoidal electromagnetic wave, $\vec{E}$ and $\vec{B}$ at any point in space are sinusoidal functions of time, and at any instant of time the *spatial* variation of the fields is also sinusoidal.

Some sinusoidal electromagnetic waves are *plane waves;* they share with the waves described in Section 32.2 the property that at any instant the fields are uniform over any plane perpendicular to the direction of propagation. The entire pattern travels in the direction of propagation with speed c. The directions of $\vec{E}$ and $\vec{B}$ are perpendicular to the direction of propagation (and to each other), so the wave is *transverse*. Electromagnetic waves produced by an oscillating point charge, shown in Fig. 32.3, are an example of sinusoidal waves that are *not* plane waves. But if we restrict our observations to a relatively small region of space at a sufficiently great distance from the source, even these waves are well approximated by plane waves (Fig. 32.12). In the same way, the curved surface of the (nearly) spherical earth appears flat to us because of our small size relative to the earth's radius. In this section we'll restrict our discussion to plane waves.

The frequency f, the wavelength λ, and the speed of propagation c of any periodic wave are related by the usual wavelength–frequency relationship $c = \lambda f$. If the frequency f is 10^8 Hz (100 MHz), typical of commercial FM radio broadcasts, the wavelength is

$$\lambda = \frac{3 \times 10^8 \text{ m/s}}{10^8 \text{ Hz}} = 3 \text{ m}$$

Figure 32.4 shows the inverse proportionality between wavelength and frequency.

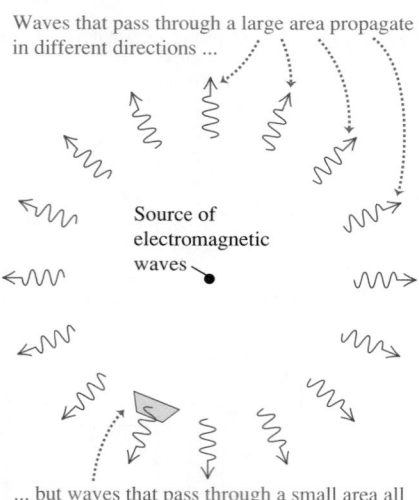

32.12 Waves passing through a small area at a sufficiently great distance from a source can be treated as plane waves.

Waves that pass through a large area propagate in different directions ...

Source of electromagnetic waves

... but waves that pass through a small area all propagate in nearly the same direction, so we can treat them as plane waves.

Fields of a Sinusoidal Wave

Figure 32.13 shows a linearly polarized sinusoidal electromagnetic wave traveling in the $+x$-direction. The $\vec{E}$ and $\vec{B}$ vectors are shown for only a few points on the positive x-axis. Note that the electric and magnetic fields oscillate in phase: $\vec{E}$ is maximum where $\vec{B}$ is maximum and $\vec{E}$ is zero where $\vec{B}$ is zero. Note also that where $\vec{E}$ is in the $+y$-direction, $\vec{B}$ is in the $+z$-direction; where $\vec{E}$ is in the $-y$-direction, $\vec{B}$ is in the $-z$-direction. At all points the vector product $\vec{E} \times \vec{B}$ is in the direction in which the wave is propagating (the $+x$-direction). We mentioned this in Section 32.2 in the list of characteristics of electromagnetic waves.

CAUTION In a plane wave, $\vec{E}$ and $\vec{B}$ are everywhere Figure 32.13 may give you the erroneous impression that the electric and magnetic fields exist only along the x-axis. In fact, in a sinusoidal plane wave there are electric and magnetic fields at *all* points in space. Imagine a plane perpendicular to the x-axis (that is, parallel to the yz-plane) at a particular point, at a particular time; the fields have the same values at all points in that plane. The values are different on different planes. |

We can describe electromagnetic waves by means of *wave functions,* just as we did in Section 15.3 for waves on a string. One form of the wave function for a transverse wave traveling in the $+x$-direction along a stretched string is Eq. (15.7):

$$y(x, t) = A\cos(kx - \omega t)$$

where $y(x, t)$ is the transverse displacement from its equilibrium position at time t of a point with coordinate x on the string. The quantity A is the maximum displacement, or *amplitude,* of the wave; ω is its *angular frequency,* equal to 2π times the frequency f; and k is the *wave number,* equal to $2\pi/\lambda$, where λ is the wavelength.

Let $E_y(x, t)$ and $B_z(x, t)$ represent the instantaneous values of the y-component of $\vec{E}$ and the z-component of $\vec{B}$, respectively, in Fig. 32.13, and let E_{max} and B_{max} represent the maximum values, or *amplitudes,* of these fields. The wave functions for the wave are then

$$E_y(x, t) = E_{max}\cos(kx - \omega t) \qquad B_z(x, t) = B_{max}\cos(kx - \omega t) \qquad (32.16)$$

(sinusoidal electromagnetic plane wave, propagating in $+x$-direction)

We can also write the wave functions in vector form:

$$\vec{E}(x, t) = \hat{j}E_{max}\cos(kx - \omega t)$$
$$\vec{B}(x, t) = \hat{k}B_{max}\cos(kx - \omega t) \qquad (32.17)$$

CAUTION The symbol k has two meanings Note the two different k's: the unit vector $\hat{k}$ in the z-direction and the wave number k. Don't get these confused! |

The sine curves in Fig. 32.13 represent instantaneous values of the electric and magnetic fields as functions of x at time $t = 0$—that is, $\vec{E}(x, t = 0)$ and $\vec{B}(x, t = 0)$. As time goes by, the wave travels to the right with speed c. Equations (32.16) and (32.17) show that at any point the sinusoidal oscillations of $\vec{E}$ and $\vec{B}$ are *in phase.* From Eq. (32.4) the amplitudes must be related by

$$E_{max} = cB_{max} \qquad \text{(electromagnetic wave in vacuum)} \qquad (32.18)$$

These amplitude and phase relationships are also required for $E(x, t)$ and $B(x, t)$ to satisfy Eqs. (32.12) and (32.14), which came from Faraday's law and Ampere's law, respectively. Can you verify this statement? (See Problem 32.38.)

32.13 Representation of the electric and magnetic fields as functions of x for a linearly polarized sinusoidal plane electromagnetic wave. One wavelength of the wave is shown at time $t = 0$. The fields are shown only for points along the x-axis.

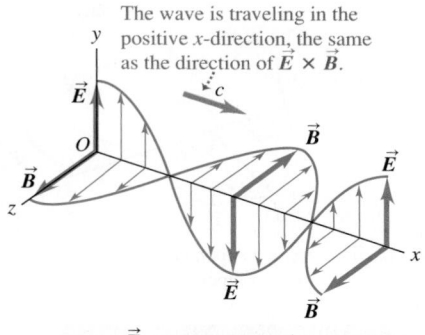

The wave is traveling in the positive x-direction, the same as the direction of $\vec{E} \times \vec{B}$.

$\vec{E}$: y-component only
$\vec{B}$: z-component only

Mastering**PHYSICS**

PhET: Radio Waves & Electromagnetic Fields
ActivPhysics 10.1: Properties of Mechanical Waves

32.14 Representation of one wavelength of a linearly polarized sinusoidal plane electromagnetic wave traveling in the negative x-direction at $t = 0$. The fields are shown only for points along the x-axis. (Compare with Fig. 32.13.)

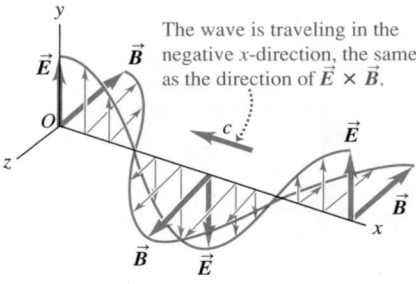

The wave is traveling in the negative x-direction, the same as the direction of $\vec{E} \times \vec{B}$.

$\vec{E}$: y-component only
$\vec{B}$: z-component only

Figure 32.14 shows the electric and magnetic fields of a wave traveling in the *negative* x-direction. At points where $\vec{E}$ is in the positive y-direction, $\vec{B}$ is in the *negative* z-direction; where $\vec{E}$ is in the negative y-direction, $\vec{B}$ is in the *positive* z-direction. The wave functions for this wave are

$$E_y(x, t) = E_{max}\cos(kx + \omega t) \quad B_z(x, t) = -B_{max}\cos(kx + \omega t) \quad (32.19)$$

(sinusoidal electromagnetic plane wave, propagating in $-x$-direction)

As with the wave traveling in the $+x$-direction, at any point the sinusoidal oscillations of the $\vec{E}$ and $\vec{B}$ fields are *in phase,* and the vector product $\vec{E} \times \vec{B}$ points in the direction of propagation.

The sinusoidal waves shown in Figs. 32.13 and 32.14 are both linearly polarized in the y-direction; the $\vec{E}$ field is always parallel to the y-axis. Example 32.1 concerns a wave that is linearly polarized in the z-direction.

Problem-Solving Strategy 32.1   Electromagnetic Waves

IDENTIFY *the relevant concepts:* Many of the same ideas that apply to mechanical waves apply to electromagnetic waves. One difference is that electromagnetic waves are described by two quantities (in this case, electric field $\vec{E}$ and magnetic field $\vec{B}$), rather than by a single quantity, such as the displacement of a string.

SET UP *the problem* using the following steps:
1. Draw a diagram showing the direction of wave propagation and the directions of $\vec{E}$ and $\vec{B}$.
2. Identify the target variables.

EXECUTE *the solution* as follows:
1. Review the treatment of sinusoidal mechanical waves in Chapters 15 and 16, and particularly the four problem-solving strategies suggested there.
2. Keep in mind the basic relationships for periodic waves: $v = \lambda f$ and $\omega = vk$. For electromagnetic waves in vacuum,

$v = c$. Distinguish between ordinary frequency f, usually expressed in hertz, and angular frequency $\omega = 2\pi f$, expressed in rad/s. Remember that the wave number is $k = 2\pi/\lambda$.
3. Concentrate on basic relationships, such as those between $\vec{E}$ and $\vec{B}$ (magnitude, direction, and relative phase), how the wave speed is determined, and the transverse nature of the waves.

EVALUATE *your answer:* Check that your result is reasonable. For electromagnetic waves in vacuum, the magnitude of the magnetic field in teslas is much smaller (by a factor of 3.00×10^8) than the magnitude of the electric field in volts per meter. If your answer suggests otherwise, you probably made an error using the relationship $E = cB$. (We'll see later in this section that the relationship between E and B is different for electromagnetic waves in a material medium.)

Example 32.1   Electric and magnetic fields of a laser beam

A carbon dioxide laser emits a sinusoidal electromagnetic wave that travels in vacuum in the negative x-direction. The wavelength is 10.6 μm (in the infrared; see Fig. 32.4) and the $\vec{E}$ field is parallel to the z-axis, with E_{max} = 1.5 MV/m. Write vector equations for $\vec{E}$ and $\vec{B}$ as functions of time and position.

SOLUTION

IDENTIFY and SET UP: Equations (32.19) describe a wave traveling in the negative x-direction with $\vec{E}$ along the y-axis—that is, a wave that is linearly polarized along the y-axis. By contrast, the wave in this example is linearly polarized along the z-axis. At points where $\vec{E}$ is in the positive z-direction, $\vec{B}$ must be in the positive y-direction for the vector product $\vec{E} \times \vec{B}$ to be in the negative x-direction (the direction of propagation). Figure 32.15 shows a wave that satisfies these requirements.

EXECUTE: A possible pair of wave functions that describe the wave shown in Fig. 32.15 are

32.15 Our sketch for this problem.

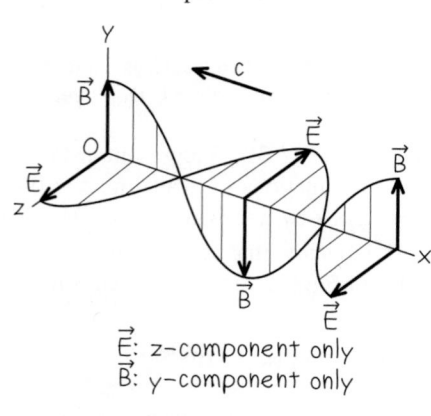

$\vec{E}$: z-component only
$\vec{B}$: y-component only

$$\vec{E}(x, t) = \hat{k}E_{max}\cos(kx + \omega t)$$
$$\vec{B}(x, t) = \hat{j}B_{max}\cos(kx + \omega t)$$

The plus sign in the arguments of the cosine functions indicates that the wave is propagating in the negative x-direction, as it should. Faraday's law requires that $E_{max} = cB_{max}$ [Eq. (32.18)], so

$$B_{max} = \frac{E_{max}}{c} = \frac{1.5 \times 10^6 \text{ V/m}}{3.0 \times 10^8 \text{ m/s}} = 5.0 \times 10^{-3} \text{ T}$$

To check unit consistency, note that $1 \text{ V} = 1 \text{ Wb/s}$ and $1 \text{ Wb/m}^2 = 1 \text{ T}$.

We have $\lambda = 10.6 \times 10^{-6}$ m, so the wave number and angular frequency are

$$k = \frac{2\pi}{\lambda} = \frac{2\pi \text{ rad}}{10.6 \times 10^{-6} \text{ m}} = 5.93 \times 10^5 \text{ rad/m}$$

$$\omega = ck = (3.00 \times 10^8 \text{ m/s})(5.93 \times 10^5 \text{ rad/m})$$
$$= 1.78 \times 10^{14} \text{ rad/s}$$

Substituting these values into the above wave functions, we get

$$\vec{E}(x, t) = \hat{k}(1.5 \times 10^6 \text{ V/m})$$
$$\times \cos[(5.93 \times 10^5 \text{ rad/m})x + (1.78 \times 10^{14} \text{ rad/s})t]$$

$$\vec{B}(x, t) = \hat{j}(5.0 \times 10^{-3} \text{ T})$$
$$\times \cos[(5.93 \times 10^5 \text{ rad/m})x + (1.78 \times 10^{14} \text{ rad/s})t]$$

EVALUATE: As we expect, the magnitude B_{max} in teslas is much smaller than the magnitude E_{max} in volts per meter. To check the directions of $\vec{E}$ and $\vec{B}$, note that $\vec{E} \times \vec{B}$ is in the direction of $\hat{k} \times \hat{j} = -\hat{i}$. This is as it should be for a wave that propagates in the negative x-direction.

Our expressions for $\vec{E}(x, t)$ and $\vec{B}(x, t)$ are not the only possible solutions. We could always add a phase angle ϕ to the arguments of the cosine function, so that $kx + \omega t$ would become $kx + \omega t + \phi$. To determine the value of ϕ we would need to know $\vec{E}$ and $\vec{B}$ either as functions of x at a given time t or as functions of t at a given coordinate x. However, the statement of the problem doesn't include this information.

Electromagnetic Waves in Matter

So far, our discussion of electromagnetic waves has been restricted to waves in *vacuum*. But electromagnetic waves can also travel in *matter;* think of light traveling through air, water, or glass. In this subsection we extend our analysis to electromagnetic waves in nonconducting materials—that is, *dielectrics*.

In a dielectric the wave speed is not the same as in vacuum, and we denote it by v instead of c. Faraday's law is unaltered, but in Eq. (32.4), derived from Faraday's law, the speed c is replaced by v. In Ampere's law the displacement current is given not by $\epsilon_0 \, d\Phi_E/dt$, where Φ_E is the flux of $\vec{E}$ through a surface, but by $\epsilon \, d\Phi_E/dt = K\epsilon_0 \, d\Phi_E/dt$, where K is the dielectric constant and ϵ is the permittivity of the dielectric. (We introduced these quantities in Section 24.4.) Also, the constant μ_0 in Ampere's law must be replaced by $\mu = K_m\mu_0$, where K_m is the relative permeability of the dielectric and μ is its permeability (see Section 28.8). Hence Eqs. (32.4) and (32.8) are replaced by

$$E = vB \qquad \text{and} \qquad B = \epsilon\mu vE \qquad \text{(32.20)}$$

Following the same procedure as for waves in vacuum, we find that the wave speed v is

$$v = \frac{1}{\sqrt{\epsilon\mu}} = \frac{1}{\sqrt{KK_m}} \frac{1}{\sqrt{\epsilon_0\mu_0}} = \frac{c}{\sqrt{KK_m}} \qquad \begin{array}{l}\text{(speed of electromagnetic}\\ \text{waves in a dielectric)}\end{array} \qquad \text{(32.21)}$$

For most dielectrics the relative permeability K_m is very nearly equal to unity (except for insulating ferromagnetic materials). When $K_m \cong 1$,

$$v = \frac{1}{\sqrt{K}} \frac{1}{\sqrt{\epsilon_0\mu_0}} = \frac{c}{\sqrt{K}}$$

Because K is always greater than unity, the speed v of electromagnetic waves in a dielectric is always *less* than the speed c in vacuum by a factor of $1/\sqrt{K}$ (Fig. 32.16). The ratio of the speed c in vacuum to the speed v in a material is known in optics as the **index of refraction** n of the material. When $K_m \cong 1$,

$$\frac{c}{v} = n = \sqrt{KK_m} \cong \sqrt{K} \qquad \text{(32.22)}$$

Usually, we can't use the values of K in Table 24.1 in this equation because those values are measured using *constant* electric fields. When the fields oscillate rapidly,

32.16 The dielectric constant K of water is about 1.8 for visible light, so the speed of visible light in water is slower than in vacuum by a factor of $1/\sqrt{K} = 1/\sqrt{1.8} = 0.75$.

there is usually not time for the reorientation of electric dipoles that occurs with steady fields. Values of K with rapidly varying fields are usually much *smaller* than the values in the table. For example, K for water is 80.4 for steady fields but only about 1.8 in the frequency range of visible light. Thus the dielectric "constant" K is actually a function of frequency (the *dielectric function*).

Example 32.2 Electromagnetic waves in different materials

(a) Visiting a jewelry store one evening, you hold a diamond up to the light of a sodium-vapor street lamp. The heated sodium vapor emits yellow light with a frequency of 5.09×10^{14} Hz. Find the wavelength in vacuum and the wave speed and wavelength in diamond, for which $K = 5.84$ and $K_m = 1.00$ at this frequency. (b) A 90.0-MHz radio wave (in the FM radio band) passes from vacuum into an insulating ferrite (a ferromagnetic material used in computer cables to suppress radio interference). Find the wavelength in vacuum and the wave speed and wavelength in the ferrite, for which $K = 10.0$ and $K_m = 1000$ at this frequency.

SOLUTION

IDENTIFY and SET UP: In each case we find the wavelength in vacuum using $c = \lambda f$. To use the corresponding equation $v = \lambda f$ to find the wavelength in a material medium, we find the speed v of electromagnetic waves in the medium using Eq. (32.21), which relates v to the values of dielectric constant K and relative permeability K_m for the medium.

EXECUTE: (a) The wavelength in vacuum of the sodium light is

$$\lambda_{vacuum} = \frac{c}{f} = \frac{3.00 \times 10^8 \text{ m/s}}{5.09 \times 10^{14} \text{ Hz}} = 5.89 \times 10^{-7} \text{ m} = 589 \text{ nm}$$

The wave speed and wavelength in diamond are

$$v_{diamond} = \frac{c}{\sqrt{KK_m}} = \frac{3.00 \times 10^8 \text{ m/s}}{\sqrt{(5.84)(1.00)}} = 1.24 \times 10^8 \text{ m/s}$$

$$\lambda_{diamond} = \frac{v_{diamond}}{f} = \frac{1.24 \times 10^8 \text{ m/s}}{5.09 \times 10^{14} \text{ Hz}}$$
$$= 2.44 \times 10^{-7} \text{ m} = 244 \text{ nm}$$

(b) Following the same steps as in part (a), we find

$$\lambda_{vacuum} = \frac{c}{f} = \frac{3.00 \times 10^8 \text{ m/s}}{90.0 \times 10^6 \text{ Hz}} = 3.33 \text{ m}$$

$$v_{ferrite} = \frac{c}{\sqrt{KK_m}} = \frac{3.00 \times 10^8 \text{ m/s}}{\sqrt{(10.0)(1000)}} = 3.00 \times 10^6 \text{ m/s}$$

$$\lambda_{ferrite} = \frac{v_{ferrite}}{f} = \frac{3.00 \times 10^6 \text{ m/s}}{90.0 \times 10^6 \text{ Hz}}$$
$$= 3.33 \times 10^{-2} \text{ m} = 3.33 \text{ cm}$$

EVALUATE: The speed of light in transparent materials is typically between $0.2c$ and c; our result in part (a) shows that $v_{diamond} = 0.414c$. As our results in part (b) show, the speed of electromagnetic waves in dense materials like ferrite (for which $v_{ferrite} = 0.010c$) can be *far* slower than in vacuum.

Test Your Understanding of Section 32.3 The first of Eqs. (32.17) gives the electric field for a plane wave as measured at points along the x-axis. For this plane wave, how does the electric field at points *off* the x-axis differ from the expression in Eqs. (32.17)? (i) The amplitude is different; (ii) the phase is different; (iii) both the amplitude and phase are different; (iv) none of these.

32.4 Energy and Momentum in Electromagnetic Waves

It is a familiar fact that energy is associated with electromagnetic waves; think of the energy in the sun's radiation. Microwave ovens, radio transmitters, and lasers for eye surgery all make use of the energy that these waves carry. To understand how to utilize this energy, it's helpful to derive detailed relationships for the energy in an electromagnetic wave.

We begin with the expressions derived in Sections 24.3 and 30.3 for the **energy densities** in electric and magnetic fields; we suggest you review those derivations now. Equations (24.11) and (30.10) show that in a region of empty space where $\vec{E}$ and $\vec{B}$ fields are present, the total energy density u is given by

$$u = \tfrac{1}{2}\epsilon_0 E^2 + \frac{1}{2\mu_0}B^2 \qquad (32.23)$$

where ϵ_0 and μ_0 are, respectively, the permittivity and permeability of free space. For electromagnetic waves in vacuum, the magnitudes E and B are related by

$$B = \frac{E}{c} = \sqrt{\epsilon_0\mu_0}\,E \qquad (32.24)$$

Combining Eqs. (32.23) and (32.24), we can also express the energy density u in a simple electromagnetic wave in vacuum as

$$u = \tfrac{1}{2}\epsilon_0 E^2 + \frac{1}{2\mu_0}(\sqrt{\epsilon_0\mu_0}\,E)^2 = \epsilon_0 E^2 \qquad (32.25)$$

This shows that in vacuum, the energy density associated with the $\vec{E}$ field in our simple wave is equal to the energy density of the $\vec{B}$ field. In general, the electric-field magnitude E is a function of position and time, as for the sinusoidal wave described by Eqs. (32.16); thus the energy density u of an electromagnetic wave, given by Eq. (32.25), also depends in general on position and time.

Electromagnetic Energy Flow and the Poynting Vector

Electromagnetic waves such as those we have described are *traveling* waves that transport energy from one region to another. We can describe this energy transfer in terms of energy transferred *per unit time per unit cross-sectional area,* or *power per unit area,* for an area perpendicular to the direction of wave travel.

To see how the energy flow is related to the fields, consider a stationary plane, perpendicular to the x-axis, that coincides with the wave front at a certain time. In a time dt after this, the wave front moves a distance $dx = c\,dt$ to the right of the plane. Considering an area A on this stationary plane (Fig. 32.17), we note that the energy in the space to the right of this area must have passed through the area to reach the new location. The volume dV of the relevant region is the base area A times the length $c\,dt$, and the energy dU in this region is the energy density u times this volume:

$$dU = u\,dV = (\epsilon_0 E^2)(Ac\,dt)$$

This energy passes through the area A in time dt. The energy flow per unit time per unit area, which we will call S, is

$$S = \frac{1}{A}\frac{dU}{dt} = \epsilon_0 c E^2 \qquad \text{(in vacuum)} \qquad (32.26)$$

Using Eqs. (32.4) and (32.9), we can derive the alternative forms

$$S = \frac{\epsilon_0}{\sqrt{\epsilon_0\mu_0}}E^2 = \sqrt{\frac{\epsilon_0}{\mu_0}}E^2 = \frac{EB}{\mu_0} \qquad \text{(in vacuum)} \qquad (32.27)$$

We leave the derivation of Eq. (32.27) from Eq. (32.26) as an exercise for you. The units of S are energy per unit time per unit area, or power per unit area. The SI unit of S is $1\ \text{J/s}\cdot\text{m}^2$ or $1\ \text{W/m}^2$.

We can define a *vector* quantity that describes both the magnitude and direction of the energy flow rate:

$$\vec{S} = \frac{1}{\mu_0}\vec{E} \times \vec{B} \qquad \text{(Poynting vector in vacuum)} \qquad (32.28)$$

The vector $\vec{S}$ is called the **Poynting vector;** it was introduced by the British physicist John Poynting (1852–1914). Its direction is in the direction of propagation of the wave (Fig. 32.18). Since $\vec{E}$ and $\vec{B}$ are perpendicular, the magnitude of $\vec{S}$ is $S = EB/\mu_0$; from Eqs. (32.26) and (32.27) this is the energy flow per unit area and per unit time through a cross-sectional area perpendicular to the propagation direction. The total energy flow per unit time (power, P) out of any closed surface is the integral of $\vec{S}$ over the surface:

$$P = \oint \vec{S}\cdot d\vec{A}$$

32.17 A wave front at a time dt after it passes through the stationary plane with area A.

At time dt, the volume between the stationary plane and the wave front contains an amount of electromagnetic energy $dU = uAc\,dt$.

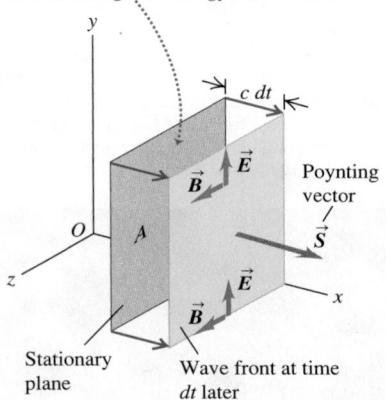

32.18 These rooftop solar panels are tilted to be face-on to the sun—that is, face-on to the Poynting vector of electromagnetic waves from the sun, so that the panels can absorb the maximum amount of wave energy.

For the sinusoidal waves studied in Section 32.3, as well as for other more complex waves, the electric and magnetic fields at any point vary with time, so the Poynting vector at any point is also a function of time. Because the frequencies of typical electromagnetic waves are very high, the time variation of the Poynting vector is so rapid that it's most appropriate to look at its *average* value. The magnitude of the average value of $\vec{S}$ at a point is called the **intensity** of the radiation at that point. The SI unit of intensity is the same as for S, 1 W/m^2 (watt per square meter).

Let's work out the intensity of the sinusoidal wave described by Eqs. (32.17). We first substitute $\vec{E}$ and $\vec{B}$ into Eq. (32.28):

$$\vec{S}(x, t) = \frac{1}{\mu_0}\vec{E}(x, t) \times \vec{B}(x, t)$$

$$= \frac{1}{\mu_0}[\hat{\jmath}E_{max}\cos(kx - \omega t)] \times [\hat{k}B_{max}\cos(kx - \omega t)]$$

The vector product of the unit vectors is $\hat{\jmath} \times \hat{k} = \hat{\imath}$ and $\cos^2(kx - \omega t)$ is never negative, so $\vec{S}(x, t)$ always points in the positive x-direction (the direction of wave propagation). The x-component of the Poynting vector is

$$S_x(x, t) = \frac{E_{max}B_{max}}{\mu_0}\cos^2(kx - \omega t) = \frac{E_{max}B_{max}}{2\mu_0}[1 + \cos 2(kx - \omega t)]$$

The time average value of $\cos 2(kx - \omega t)$ is zero because at any point, it is positive during one half-cycle and negative during the other half. So the average value of the Poynting vector over a full cycle is $\vec{S}_{av} = \hat{\imath}S_{av}$, where

$$S_{av} = \frac{E_{max}B_{max}}{2\mu_0}$$

That is, the magnitude of the average value of $\vec{S}$ for a sinusoidal wave (the intensity I of the wave) is $\frac{1}{2}$ the maximum value. By using the relationships $E_{max} = B_{max}c$ and $\epsilon_0\mu_0 = 1/c^2$, we can express the intensity in several equivalent forms:

$$I = S_{av} = \frac{E_{max}B_{max}}{2\mu_0} = \frac{E_{max}^2}{2\mu_0 c}$$
$$= \frac{1}{2}\sqrt{\frac{\epsilon_0}{\mu_0}}E_{max}^2 = \frac{1}{2}\epsilon_0 c E_{max}^2$$

(intensity of a sinusoidal wave in vacuum) (32.29)

We invite you to verify that these expressions are all equivalent.

For a wave traveling in the $-x$-direction, represented by Eqs. (32.19), the Poynting vector is in the $-x$-direction at every point, but its magnitude is the same as for a wave traveling in the $+x$-direction. Verifying these statements is left to you (see Exercise 32.24).

CAUTION *Poynting vector vs. intensity* At any point x, the magnitude of the Poynting vector varies with time. Hence, the *instantaneous* rate at which electromagnetic energy in a sinusoidal plane wave arrives at a surface is not constant. This may seem to contradict everyday experience; the light from the sun, a light bulb, or the laser in a grocery-store scanner appears steady and unvarying in strength. In fact the Poynting vector from these sources *does* vary in time, but the variation isn't noticeable because the oscillation frequency is so high (around 5×10^{14} Hz for visible light). All that you sense is the *average* rate at which energy reaches your eye, which is why we commonly use intensity (the average value of S) to describe the strength of electromagnetic radiation. ▮

Throughout this discussion we have considered only electromagnetic waves propagating in vacuum. If the waves are traveling in a dielectric medium, however,

Application **Laser Surgery**
Lasers are used widely in medicine as ultra-precise, bloodless "scalpels." They can reach and remove tumors with minimal damage to neighboring healthy tissues, as in the brain surgery shown here. The power output of the laser is typically below 40 W, less than that of a typical light bulb. However, this power is concentrated into a spot from 0.1 to 2.0 mm in diameter, so the intensity of the light (equal to the average value of the Poynting vector) can be as high as 5×10^9 W/m^2.

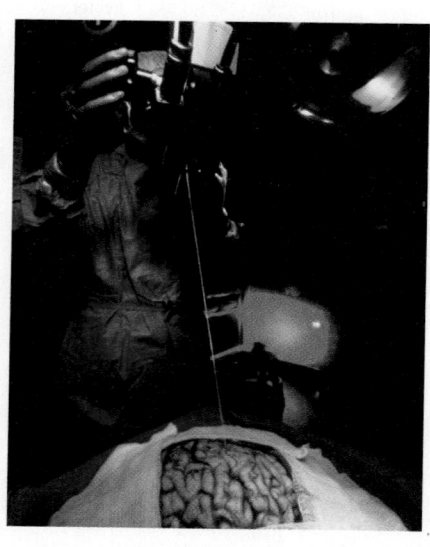

the expressions for energy density [Eq. (32.23)], the Poynting vector [Eq. (32.28)], and the intensity of a sinusoidal wave [Eq. (32.29)] must be modified. It turns out that the required modifications are quite simple: Just replace ϵ_0 with the permittivity ϵ of the dielectric, replace μ_0 with the permeability μ of the dielectric, and replace c with the speed v of electromagnetic waves in the dielectric. Remarkably, the energy densities in the $\vec{E}$ and $\vec{B}$ fields are equal even in a dielectric.

Example 32.3 Energy in a nonsinusoidal wave

For the nonsinusoidal wave described in Section 32.2, suppose that $E = 100$ V/m $= 100$ N/C. Find the value of B, the energy density u, and the rate of energy flow per unit area S.

SOLUTION

IDENTIFY and SET UP: In this wave $\vec{E}$ and $\vec{B}$ are uniform behind the wave front (and zero ahead of it). Hence the target variables B, u, and S must also be uniform behind the wave front. Given the magnitude E, we use Eq. (32.4) to find B, Eq. (32.25) to find u, and Eq. (32.27) to find S. (We cannot use Eq. (32.29), which applies to sinusoidal waves only.)

EXECUTE: From Eq. (32.4),

$$B = \frac{E}{c} = \frac{100 \text{ V/m}}{3.00 \times 10^8 \text{ m/s}} = 3.33 \times 10^{-7} \text{ T}$$

From Eq. (32.25),

$$u = \epsilon_0 E^2 = (8.85 \times 10^{-12} \text{ C}^2/\text{N} \cdot \text{m}^2)(100 \text{ N/C})^2$$
$$= 8.85 \times 10^{-8} \text{ N/m}^2 = 8.85 \times 10^{-8} \text{ J/m}^3$$

The magnitude of the Poynting vector is

$$S = \frac{EB}{\mu_0} = \frac{(100 \text{ V/m})(3.33 \times 10^{-7} \text{ T})}{4\pi \times 10^{-7} \text{ T} \cdot \text{m/A}}$$
$$= 26.5 \text{ V} \cdot \text{A/m}^2 = 26.5 \text{ W/m}^2$$

EVALUATE: We can check our result for S by using Eq. (32.26):

$$S = \epsilon_0 c E^2 = (8.85 \times 10^{-12} \text{ C}^2/\text{N} \cdot \text{m}^2)(3.00 \times 10^8 \text{ m/s})$$
$$\times (100 \text{ N/C})^2 = 26.5 \text{ W/m}^2$$

Since $\vec{E}$ and $\vec{B}$ have the same values at all points behind the wave front, u and S likewise have the same value everywhere behind the wave front. In front of the wave front, $\vec{E} = 0$ and $\vec{B} = 0$, and so $u = 0$ and $S = 0$; where there are no fields, there is no field energy.

Example 32.4 Energy in a sinusoidal wave

A radio station on the earth's surface emits a sinusoidal wave with average total power 50 kW (Fig. 32.19). Assuming that the transmitter radiates equally in all directions above the ground (which is unlikely in real situations), find the electric-field and magnetic-field amplitudes E_{max} and B_{max} detected by a satellite 100 km from the antenna.

SOLUTION

IDENTIFY and SET UP: We are given the transmitter's average total power P. The intensity I is just the average power per unit area, so to find I at 100 km from the transmitter we divide P by the surface area of the hemisphere shown in Fig. 32.19. For a sinusoidal wave, I is also equal to the magnitude of the average value S_{av} of the Poynting vector, so we can use Eqs. (32.29) to find E_{max}; Eq. (32.4) then yields B_{max}.

32.19 A radio station radiates waves into the hemisphere shown.

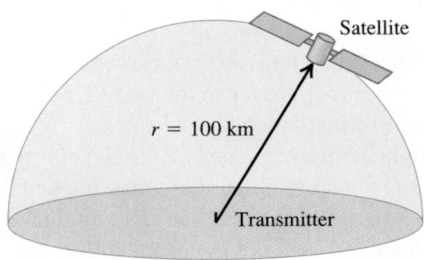

EXECUTE: The surface area of a hemisphere of radius $r = 100$ km $= 1.00 \times 10^5$ m is

$$A = 2\pi R^2 = 2\pi(1.00 \times 10^5 \text{ m})^2 = 6.28 \times 10^{10} \text{ m}^2$$

All the radiated power passes through this surface, so the average power per unit area (that is, the intensity) is

$$I = \frac{P}{A} = \frac{P}{2\pi R^2} = \frac{5.00 \times 10^4 \text{ W}}{6.28 \times 10^{10} \text{ m}^2} = 7.96 \times 10^{-7} \text{ W/m}^2$$

From Eqs. (32.29), $I = S_{av} = E_{max}{}^2/2\mu_0 c$, so

$$E_{max} = \sqrt{2\mu_0 c S_{av}}$$
$$= \sqrt{2(4\pi \times 10^{-7} \text{ T} \cdot \text{m/A})(3.00 \times 10^8 \text{ m/s})(7.96 \times 10^{-7} \text{ W/m}^2)}$$
$$= 2.45 \times 10^{-2} \text{ V/m}$$

Then from Eq. (32.4),

$$B_{max} = \frac{E_{max}}{c} = 8.17 \times 10^{-11} \text{ T}$$

EVALUATE: Note that E_{max} is comparable to fields commonly seen in the laboratory, but B_{max} is extremely small in comparison to $\vec{B}$ fields we saw in previous chapters. For this reason, most detectors of electromagnetic radiation respond to the effect of the electric field, not the magnetic field. Loop radio antennas are an exception (see the Bridging Problem at the end of this chapter).

Electromagnetic Momentum Flow and Radiation Pressure

By using the observation that energy is required to establish electric and magnetic fields, we have shown that electromagnetic waves transport energy. It can also be shown that electromagnetic waves carry *momentum p*, with a corresponding momentum density (momentum dp per volume dV) of magnitude

$$\frac{dp}{dV} = \frac{EB}{\mu_0 c^2} = \frac{S}{c^2} \tag{32.30}$$

This momentum is a property of the field; it is not associated with the mass of a moving particle in the usual sense.

There is also a corresponding momentum flow rate. The volume dV occupied by an electromagnetic wave (speed c) that passes through an area A in time dt is $dV = Ac\, dt$. When we substitute this into Eq. (32.30) and rearrange, we find that the momentum flow rate per unit area is

$$\frac{1}{A}\frac{dp}{dt} = \frac{S}{c} = \frac{EB}{\mu_0 c} \qquad \text{(flow rate of electromagnetic momentum)} \tag{32.31}$$

This is the momentum transferred per unit surface area per unit time. We obtain the *average* rate of momentum transfer per unit area by replacing S in Eq. (32.31) by $S_{av} = I$.

This momentum is responsible for **radiation pressure.** When an electromagnetic wave is completely absorbed by a surface, the wave's momentum is also transferred to the surface. For simplicity we'll consider a surface perpendicular to the propagation direction. Using the ideas developed in Section 8.1, we see that the rate dp/dt at which momentum is transferred to the absorbing surface equals the *force* on the surface. The average force per unit area due to the wave, or *radiation pressure* p_{rad}, is the average value of dp/dt divided by the absorbing area A. (We use the subscript "rad" to distinguish pressure from momentum, for which the symbol p is also used.) From Eq. (32.31) the radiation pressure is

$$p_{rad} = \frac{S_{av}}{c} = \frac{I}{c} \qquad \text{(radiation pressure, wave totally absorbed)} \tag{32.32}$$

If the wave is totally reflected, the momentum change is twice as great, and the pressure is

$$p_{rad} = \frac{2S_{av}}{c} = \frac{2I}{c} \qquad \text{(radiation pressure, wave totally reflected)} \tag{32.33}$$

For example, the value of I (or S_{av}) for direct sunlight, before it passes through the earth's atmosphere, is approximately 1.4 kW/m^2. From Eq. (32.32) the corresponding average pressure on a completely absorbing surface is

$$p_{rad} = \frac{I}{c} = \frac{1.4 \times 10^3 \text{ W/m}^2}{3.0 \times 10^8 \text{ m/s}} = 4.7 \times 10^{-6} \text{ Pa}$$

From Eq. (32.33) the average pressure on a totally *reflecting* surface is twice this: $2I/c$ or 9.4×10^{-6} Pa. These are very small pressures, of the order of 10^{-10} atm, but they can be measured with sensitive instruments.

The radiation pressure of sunlight is much greater *inside* the sun than at the earth (see Problem 32.45). Inside stars that are much more massive and luminous than the sun, radiation pressure is so great that it substantially augments the gas pressure within the star and so helps to prevent the star from collapsing under its own gravity. In some cases the radiation pressure of stars can have dramatic effects on the material surrounding them (Fig. 32.20).

32.20 At the center of this interstellar gas cloud is a group of intensely luminous stars that exert tremendous radiation pressure on their surroundings. Aided by a "wind" of particles emanating from the stars, over the past million years the radiation pressure has carved out a bubble within the cloud 70 light-years across.

Example 32.5 **Power and pressure from sunlight**

An earth-orbiting satellite has solar energy–collecting panels with a total area of 4.0 m² (Fig. 32.21). If the sun's radiation is perpendicular to the panels and is completely absorbed, find the average solar power absorbed and the average radiation-pressure force.

SOLUTION

IDENTIFY and SET UP: This problem uses the relationships among intensity, power, radiation pressure, and force. In the above discussion we calculated the intensity I (power per unit area) of sunlight as well as the radiation pressure p_{rad} (force per unit area) of sunlight on an absorbing surface. (We calculated these values for

32.21 Solar panels on a satellite.

points above the atmosphere, which is where the satellite orbits.) Multiplying each value by the area of the solar panels gives the average power absorbed and the net radiation force on the panels.

EXECUTE: The intensity I (power per unit area) is 1.4×10^3 W/m². Although the light from the sun is not a simple sinusoidal wave, we can still use the relationship that the average power P is the intensity I times the area A:

$$P = IA = (1.4 \times 10^3 \text{ W/m}^2)(4.0 \text{ m}^2)$$
$$= 5.6 \times 10^3 \text{ W} = 5.6 \text{ kW}$$

The radiation pressure of sunlight on an absorbing surface is $p_{rad} = 4.7 \times 10^{-6}$ Pa $= 4.7 \times 10^{-6}$ N/m². The total force F is the pressure p_{rad} times the area A:

$$F = p_{rad}A = (4.7 \times 10^{-6} \text{ N/m}^2)(4.0 \text{ m}^2) = 1.9 \times 10^{-5} \text{ N}$$

EVALUATE: The absorbed power is quite substantial. Part of it can be used to power the equipment aboard the satellite; the rest goes into heating the panels, either directly or due to inefficiencies in the photocells contained in the panels.

The total radiation force is comparable to the weight (on earth) of a single grain of salt. Over time, however, this small force can have a noticeable effect on the orbit of a satellite like that in Fig. 32.21, and so radiation pressure must be taken into account.

Test Your Understanding of Section 32.4 Figure 32.13 shows one wavelength of a sinusoidal electromagnetic wave at time $t = 0$. For which of the following four values of x is (a) the energy density a maximum; (b) the energy density a minimum; (c) the magnitude of the instantaneous (not average) Poynting vector a maximum; (d) the magnitude of the instantaneous (not average) Poynting vector a minimum? (i) $x = 0$; (ii) $x = \lambda/4$; (iii) $x = \lambda/2$; (iv) $x = 3\lambda/4$.

32.5 Standing Electromagnetic Waves

Electromagnetic waves can be *reflected;* the surface of a conductor (like a polished sheet of metal) or of a dielectric (such as a sheet of glass) can serve as a reflector. The superposition principle holds for electromagnetic waves just as for electric and magnetic fields. The superposition of an incident wave and a reflected wave forms a **standing wave.** The situation is analogous to standing waves on a stretched string, discussed in Section 15.7; you should review that discussion.

Suppose a sheet of a perfect conductor (zero resistivity) is placed in the yz-plane of Fig. 32.22 and a linearly polarized electromagnetic wave, traveling in the negative x-direction, strikes it. As we discussed in Section 23.4, $\vec{E}$ cannot have a component parallel to the surface of a perfect conductor. Therefore in the present situation, $\vec{E}$ must be zero everywhere in the yz-plane. The electric field of the *incident* electromagnetic wave is *not* zero at all times in the yz-plane. But this incident wave induces oscillating currents on the surface of the conductor, and these currents give rise to an additional electric field. The *net* electric field, which is the vector sum of this field and the incident $\vec{E}$, is zero everywhere inside and on the surface of the conductor.

The currents induced on the surface of the conductor also produce a *reflected* wave that travels out from the plane in the $+x$-direction. Suppose the incident wave is described by the wave functions of Eqs. (32.19) (a sinusoidal wave traveling in the $-x$-direction) and the reflected wave by the negative of Eqs. (32.16) (a sinusoidal wave traveling in the $+x$-direction). We take the *negative* of the

32.22 Representation of the electric and magnetic fields of a linearly polarized electromagnetic standing wave when $\omega t = 3\pi/4$ rad. In any plane perpendicular to the x-axis, E is maximum (an antinode) where B is zero (a node), and vice versa. As time elapses, the pattern does *not* move along the x-axis; instead, at every point the $\vec{E}$ and $\vec{B}$ vectors simply oscillate.

wave given by Eqs. (32.16) so that the incident and reflected electric fields cancel at $x = 0$ (the plane of the conductor, where the total electric field must be zero). The superposition principle states that the total $\vec{E}$ field at any point is the vector sum of the $\vec{E}$ fields of the incident and reflected waves, and similarly for the $\vec{B}$ field. Therefore the wave functions for the superposition of the two waves are

$$E_y(x, t) = E_{max}[\cos(kx + \omega t) - \cos(kx - \omega t)]$$

$$B_z(x, t) = B_{max}[-\cos(kx + \omega t) - \cos(kx - \omega t)]$$

We can expand and simplify these expressions, using the identities

$$\cos(A \pm B) = \cos A \cos B \mp \sin A \sin B$$

The results are

$$E_y(x, t) = -2E_{max} \sin kx \sin \omega t \tag{32.34}$$

$$B_z(x, t) = -2B_{max} \cos kx \cos \omega t \tag{32.35}$$

Equation (32.34) is analogous to Eq. (15.28) for a stretched string. We see that at $x = 0$ the electric field $E_y(x = 0, t)$ is *always* zero; this is required by the nature of the ideal conductor, which plays the same role as a fixed point at the end of a string. Furthermore, $E_y(x, t)$ is zero at *all* times at points in those planes perpendicular to the x-axis for which $\sin kx = 0$—that is, $kx = 0, \pi, 2\pi, \ldots$. Since $k = 2\pi/\lambda$, the positions of these planes are

$$x = 0, \frac{\lambda}{2}, \lambda, \frac{3\lambda}{2}, \ldots \quad \text{(nodal planes of } \vec{E}) \tag{32.36}$$

These planes are called the **nodal planes** of the $\vec{E}$ field; they are the equivalent of the nodes, or nodal points, of a standing wave on a string. Midway between any two adjacent nodal planes is a plane on which $\sin kx = \pm 1$; on each such plane, the magnitude of $E(x, t)$ equals the maximum possible value of $2E_{max}$ twice per oscillation cycle. These are the **antinodal planes** of $\vec{E}$, corresponding to the antinodes of waves on a string.

The total magnetic field is zero at all times at points in planes on which $\cos kx = 0$. This occurs where

$$x = \frac{\lambda}{4}, \frac{3\lambda}{4}, \frac{5\lambda}{4}, \ldots \quad \text{(nodal planes of } \vec{B}) \tag{32.37}$$

These are the nodal planes of the $\vec{B}$ field; there is an antinodal plane of $\vec{B}$ midway between any two adjacent nodal planes.

Figure 32.22 shows a standing-wave pattern at one instant of time. The magnetic field is *not* zero at the conducting surface ($x = 0$). The surface currents that must be present to make $\vec{E}$ exactly zero at the surface cause magnetic fields at the surface. The nodal planes of each field are separated by one half-wavelength. The nodal planes of one field are midway between those of the other; hence the nodes of $\vec{E}$ coincide with the antinodes of $\vec{B}$, and conversely. Compare this situation to the distinction between pressure nodes and displacement nodes in Section 16.4.

The total electric field is a *sine* function of t, and the total magnetic field is a *cosine* function of t. The sinusoidal variations of the two fields are therefore 90° out of phase at each point. At times when $\sin \omega t = 0$, the electric field is zero *everywhere,* and the magnetic field is maximum. When $\cos \omega t = 0$, the magnetic field is zero everywhere, and the electric field is maximum. This is in contrast to a wave traveling in one direction, as described by Eqs. (32.16) or (32.19) separately, in which the sinusoidal variations of $\vec{E}$ and $\vec{B}$ at any particular point are *in phase.* You can show that Eqs. (32.34) and (32.35) satisfy the wave equation, Eq. (32.15). You can also show that they satisfy Eqs. (32.12) and (32.14), the equivalents of Faraday's and Ampere's laws (see Exercise 32.36).

Standing Waves in a Cavity

Let's now insert a second conducting plane, parallel to the first and a distance L from it, along the $+x$-axis. The cavity between the two planes is analogous to a stretched string held at the points $x = 0$ and $x = L$. Both conducting planes must be nodal planes for $\vec{E}$; a standing wave can exist only when the second plane is placed at one of the positions where $E(x, t) = 0$, so L must be an integer multiple of $\lambda/2$. The wavelengths that satisfy this condition are

$$\lambda_n = \frac{2L}{n} \qquad (n = 1, 2, 3, \dots) \tag{32.38}$$

The corresponding frequencies are

$$f_n = \frac{c}{\lambda_n} = n\frac{c}{2L} \qquad (n = 1, 2, 3, \dots) \tag{32.39}$$

Thus there is a set of *normal modes,* each with a characteristic frequency, wave shape, and node pattern (Fig. 32.23). By measuring the node positions, we can measure the wavelength. If the frequency is known, the wave speed can be determined. This technique was first used by Hertz in the 1880s in his pioneering investigations of electromagnetic waves.

Conducting surfaces are not the only reflectors of electromagnetic waves. Reflections also occur at an interface between two insulating materials with different dielectric or magnetic properties. The mechanical analog is a junction of two strings with equal tension but different linear mass density. In general, a wave incident on such a boundary surface is partly transmitted into the second material and partly reflected back into the first. For example, light is transmitted through a glass window, but its surfaces also reflect light.

32.23 A typical microwave oven sets up a standing electromagnetic wave with $\lambda = 12.2$ cm, a wavelength that is strongly absorbed by the water in food. Because the wave has nodes spaced $\lambda/2 = 6.1$ cm apart, the food must be rotated while cooking. Otherwise, the portion that lies at a node—where the electric-field amplitude is zero—will remain cold.

Example 32.6   Intensity in a standing wave

Calculate the intensity of the standing wave represented by Eqs. (32.34) and (32.35).

SOLUTION

IDENTIFY and SET UP: The intensity I of the wave is the time-averaged value S_{av} of the magnitude of the Poynting vector $\vec{S}$. To find S_{av}, we first use Eq. (32.28) to find the instantaneous value of $\vec{S}$ and then average it over a whole number of cycles of the wave.

EXECUTE: Using the wave functions of Eqs. (32.34) and (32.35) in Eq. (32.28) for the Poynting vector $\vec{S}$, we find

$$\vec{S}(x, t) = \frac{1}{\mu_0}\vec{E}(x, t) \times \vec{B}(x, t)$$

$$= \frac{1}{\mu_0}[-2\hat{\jmath}E_{max}\sin kx \cos \omega t] \times [-2\hat{k}B_{max}\cos kx \sin \omega t]$$

$$= \hat{\imath}\frac{E_{max}B_{max}}{\mu_0}(2\sin kx \cos kx)(2\sin \omega t \cos \omega t)$$

$$= \hat{\imath}S_x(x, t)$$

Using the identity $\sin 2A = 2\sin A \cos A$, we can rewrite $S_x(x, t)$ as

$$S_x(x, t) = \frac{E_{max}B_{max}\sin 2kx \sin 2\omega t}{\mu_0}$$

The average value of a sine function over any whole number of cycles is zero. Thus *the time average of $\vec{S}$ at any point is zero;* $I = S_{av} = 0$.

EVALUATE: This result is what we should expect. The standing wave is a superposition of two waves with the same frequency and amplitude, traveling in opposite directions. All the energy transferred by one wave is cancelled by an equal amount transferred in the opposite direction by the other wave. When we use electromagnetic waves to transmit power, it is important to avoid reflections that give rise to standing waves.

Example 32.7   Standing waves in a cavity

Electromagnetic standing waves are set up in a cavity with two parallel, highly conducting walls 1.50 cm apart. (a) Calculate the longest wavelength λ and lowest frequency f of these standing waves. (b) For a standing wave of this wavelength, where in the cavity does $\vec{E}$ have maximum magnitude? Where is $\vec{E}$ zero? Where does $\vec{B}$ have maximum magnitude? Where is $\vec{B}$ zero?

Continued

IDENTIFY and SET UP: Only certain normal modes are possible for electromagnetic waves in a cavity, just as only certain normal modes are possible for standing waves on a string. The longest possible wavelength and lowest possible frequency correspond to the $n = 1$ mode in Eqs. (32.38) and (32.39); we use these to find λ and f. Equations (32.36) and (32.37) then give the locations of the nodal planes of $\vec{E}$ and $\vec{B}$. The antinodal planes of each field are midway between adjacent nodal planes.

EXECUTE: (a) From Eqs. (32.38) and (32.39), the $n = 1$ wavelength and frequency are

$$\lambda_1 = 2L = 2(1.50 \text{ cm}) = 3.00 \text{ cm}$$

$$f_1 = \frac{c}{2L} = \frac{3.00 \times 10^8 \text{ m/s}}{2(1.50 \times 10^{-2} \text{ m})} = 1.00 \times 10^{10} \text{ Hz} = 10 \text{ GHz}$$

(b) With $n = 1$ there is a single half-wavelength between the walls. The electric field has nodal planes ($\vec{E} = 0$) at the walls and an antinodal plane (where $\vec{E}$ has its maximum magnitude) midway between them. The magnetic field has *antinodal* planes at the walls and a nodal plane midway between them.

EVALUATE: One application of such standing waves is to produce an oscillating $\vec{E}$ field of definite frequency, which is used to probe the behavior of a small sample of material placed in the cavity. To subject the sample to the strongest possible field, it should be placed near the center of the cavity, at the antinode of $\vec{E}$.

Test Your Understanding of Section 32.5 In the standing wave described in Example 32.7, is there any point in the cavity where the energy density is zero at all times? If so, where? If not, why not?

Maxwell's equations and electromagnetic waves:
Maxwell's equations predict the existence of electromagnetic waves that propagate in vacuum at the speed of light c. The electromagnetic spectrum covers frequencies from at least 1 to 10^{24} Hz and a correspondingly broad range of wavelengths. Visible light, with wavelengths from 380 to 750 nm, is only a very small part of this spectrum. In a plane wave, $\vec{E}$ and $\vec{B}$ are uniform over any plane perpendicular to the propagation direction. Faraday's law and Ampere's law both give relationships between the magnitudes of $\vec{E}$ and $\vec{B}$; requiring both of these relationships to be satisfied gives an expression for c in terms of ϵ_0 and μ_0. Electromagnetic waves are transverse; the $\vec{E}$ and $\vec{B}$ fields are perpendicular to the direction of propagation and to each other. The direction of propagation is the direction of $\vec{E} \times \vec{B}$.

$$E = cB \tag{32.4}$$

$$B = \epsilon_0\mu_0 cE \tag{32.8}$$

$$c = \frac{1}{\sqrt{\epsilon_0\mu_0}} \tag{32.9}$$

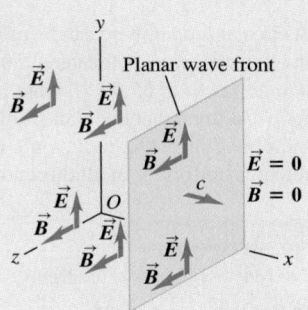

Sinusoidal electromagnetic waves: Equations (32.17) and (32.18) describe a sinusoidal plane electromagnetic wave traveling in vacuum in the $+x$-direction. If the wave is propagating in the $-x$-direction, replace $kx - \omega t$ by $kx + \omega t$. (See Example 32.1.)

$$\vec{E}(x, t) = \hat{\jmath}E_{\max}\cos(kx - \omega t)$$
$$\vec{B}(x, t) = \hat{k}B_{\max}\cos(kx - \omega t) \tag{32.17}$$

$$E_{\max} = cB_{\max} \tag{32.18}$$

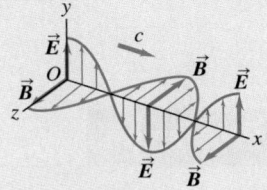

Electromagnetic waves in matter: When an electromagnetic wave travels through a dielectric, the wave speed v is less than the speed of light in vacuum c. (See Example 32.2.)

$$v = \frac{1}{\sqrt{\epsilon\mu}} = \frac{1}{\sqrt{KK_m}}\frac{1}{\sqrt{\epsilon_0\mu_0}}$$
$$= \frac{c}{\sqrt{KK_m}} \tag{32.21}$$

Energy and momentum in electromagnetic waves: The energy flow rate (power per unit area) in an electromagnetic wave in vacuum is given by the Poynting vector $\vec{S}$. The magnitude of the time-averaged value of the Poynting vector is called the intensity I of the wave. Electromagnetic waves also carry momentum. When an electromagnetic wave strikes a surface, it exerts a radiation pressure p_{rad}. If the surface is perpendicular to the wave propagation direction and is totally absorbing, $p_{\text{rad}} = I/c$; if the surface is a perfect reflector, $p_{\text{rad}} = 2I/c$. (See Examples 32.3–32.5.)

$$\vec{S} = \frac{1}{\mu_0}\vec{E} \times \vec{B} \tag{32.28}$$

$$I = S_{\text{av}} = \frac{E_{\max}B_{\max}}{2\mu_0} = \frac{E_{\max}^2}{2\mu_0 c}$$
$$= \frac{1}{2}\sqrt{\frac{\epsilon_0}{\mu_0}}E_{\max}^2$$
$$= \frac{1}{2}\epsilon_0 c E_{\max}^2 \tag{32.29}$$

$$\frac{1}{A}\frac{dp}{dt} = \frac{S}{c} = \frac{EB}{\mu_0 c} \tag{32.31}$$

(flow rate of electromagnetic momentum)

Standing electromagnetic waves: If a perfect reflecting surface is placed at $x = 0$, the incident and reflected waves form a standing wave. Nodal planes for $\vec{E}$ occur at $kx = 0, \pi, 2\pi, \ldots$, and nodal planes for $\vec{B}$ at $kx = \pi/2, 3\pi/2, 5\pi/2, \ldots$. At each point, the sinusoidal variations of $\vec{E}$ and $\vec{B}$ with time are 90° out of phase. (See Examples 32.6 and 32.7.)

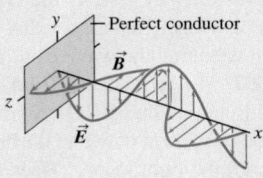

Detecting Electromagnetic Waves

A circular loop of wire can be used as a radio antenna. If an 18.0-cm-diameter antenna is located 2.50 km from a 95.0-MHz source with a total power of 55.0 kW, what is the maximum emf induced in the loop? Assume that the plane of the antenna loop is perpendicular to the direction of the radiation's magnetic field and that the source radiates uniformly in all directions.

SOLUTION GUIDE

See MasteringPhysics® study area for a Video Tutor solution.

IDENTIFY and SET UP:

1. The electromagnetic wave has an oscillating magnetic field. This causes a magnetic flux through the loop antenna that varies sinusoidally with time. By Faraday's law, this produces an emf equal in magnitude to the rate of change of the flux. The target variable is the magnitude of this emf.
2. Select the equations that you will need to find (i) the intensity of the wave at the position of the loop, a distance $r = 2.50$ km

from the source of power $P = 55.0$ kW; (ii) the amplitude of the sinusoidally varying magnetic field at that position; (iii) the magnetic flux through the loop as a function of time; and (iv) the emf produced by the flux.

EXECUTE

3. Find the wave intensity at the position of the loop.
4. Use your result from step 3 to write expressions for the time-dependent magnetic field at this position and the time-dependent magnetic flux through the loop.
5. Use the results of step 4 to find the time-dependent induced emf in the loop. The amplitude of this emf is your target variable.

EVALUATE

6. Is the induced emf large enough to detect? (If it is, a receiver connected to this antenna will be able to pick up signals from the source.)

Problems

•, ••, •••: Problems of increasing difficulty. **CP**: Cumulative problems incorporating material from earlier chapters. **CALC**: Problems requiring calculus. **BIO**: Biosciences problems.

DISCUSSION QUESTIONS

Q32.1 By measuring the electric and magnetic fields at a point in space where there is an electromagnetic wave, can you determine the direction from which the wave came? Explain.

Q32.2 According to Ampere's law, is it possible to have both a conduction current and a displacement current at the same time? Is it possible for the effects of the two kinds of current to cancel each other exactly so that *no* magnetic field is produced? Explain.

Q32.3 Give several examples of electromagnetic waves that are encountered in everyday life. How are they all alike? How do they differ?

Q32.4 Sometimes neon signs located near a powerful radio station are seen to glow faintly at night, even though they are not turned on. What is happening?

Q32.5 Is polarization a property of all electromagnetic waves, or is it unique to visible light? Can sound waves be polarized? What fundamental distinction in wave properties is involved? Explain.

Q32.6 Suppose that a positive point charge q is initially at rest on the x-axis, in the path of the electromagnetic plane wave described in Section 32.2. Will the charge move after the wave front reaches it? If not, why not? If the charge does move, describe its motion qualitatively. (Remember that $\vec{E}$ and $\vec{B}$ have the same value at all points behind the wave front.)

Q32.7 The light beam from a searchlight may have an electric-field magnitude of 1000 V/m, corresponding to a potential difference of 1500 V between the head and feet of a 1.5-m-tall person on whom the light shines. Does this cause the person to feel a strong electric shock? Why or why not?

Q32.8 For a certain sinusoidal wave of intensity I, the amplitude of the magnetic field is B. What would be the amplitude (in terms of B) in a similar wave of twice the intensity?

Q32.9 The magnetic-field amplitude of the electromagnetic wave from the laser described in Example 32.1 (Section 32.3) is about 100 times greater than the earth's magnetic field. If you illuminate a compass with the light from this laser, would you expect the compass to deflect? Why or why not?

Q32.10 Most automobiles have vertical antennas for receiving radio broadcasts. Explain what this tells you about the direction of polarization of $\vec{E}$ in the radio waves used in broadcasting.

Q32.11 If a light beam carries momentum, should a person holding a flashlight feel a recoil analogous to the recoil of a rifle when it is fired? Why is this recoil not actually observed?

Q32.12 A light source radiates a sinusoidal electromagnetic wave uniformly in all directions. This wave exerts an average pressure p on a perfectly reflecting surface a distance R away from it. What average pressure (in terms of p) would this wave exert on a perfectly absorbing surface that was twice as far from the source?

Q32.13 Does an electromagnetic *standing* wave have energy? Does it have momentum? Are your answers to these questions the same as for a *traveling* wave? Why or why not?

Q32.14 When driving on the upper level of the Bay Bridge, westbound from Oakland to San Francisco, you can easily pick up a number of radio stations on your car radio. But when driving eastbound on the lower level of the bridge, which has steel girders on either side to support the upper level, the radio reception is much worse. Why is there a difference?

EXERCISES

Section 32.2 Plane Electromagnetic Waves and the Speed of Light

32.1 • (a) How much time does it take light to travel from the moon to the earth, a distance of 384,000 km? (b) Light from the star Sirius takes 8.61 years to reach the earth. What is the distance from earth to Sirius in kilometers?

32.2 • Consider each of the electric- and magnetic-field orientations given next. In each case, what is the direction of propagation of the wave? (a) $\vec{E}$ in the $+x$-direction, $\vec{B}$ in the $+y$-direction; (b) $\vec{E}$ in the $-y$-direction, $\vec{B}$ in the $+x$-direction; (c) $\vec{E}$ in the $+z$-direction, $\vec{B}$ in the $-x$-direction; (d) $\vec{E}$ in the $+y$-direction, $\vec{B}$ in the $-z$-direction.

32.3 • A sinusoidal electromagnetic wave is propagating in vacuum in the $+z$-direction. If at a particular instant and at a certain point in space the electric field is in the $+x$-direction and has magnitude 4.00 V/m, what are the magnitude and direction of the magnetic field of the wave at this same point in space and instant in time?

32.4 • Consider each of the following electric- and magnetic-field orientations. In each case, what is the direction of propagation of the wave? (a) $\vec{E} = E\hat{\imath}$, $\vec{B} = -B\hat{\jmath}$; (b) $\vec{E} = E\hat{\jmath}$, $\vec{B} = B\hat{\imath}$; (c) $\vec{E} = -E\hat{k}$, $\vec{B} = -B\hat{\imath}$; (d) $\vec{E} = E\hat{\imath}$, $\vec{B} = -B\hat{k}$.

Section 32.3 Sinusoidal Electromagnetic Waves

32.5 • **BIO Medical X rays.** Medical x rays are taken with electromagnetic waves having a wavelength of around 0.10 nm. What are the frequency, period, and wave number of such waves?

32.6 • **BIO Ultraviolet Radiation.** There are two categories of ultraviolet light. Ultraviolet A (UVA) has a wavelength ranging from 320 nm to 400 nm. It is not harmful to the skin and is necessary for the production of vitamin D. UVB, with a wavelength between 280 nm and 320 nm, is much more dangerous because it causes skin cancer. (a) Find the frequency ranges of UVA and UVB. (b) What are the ranges of the wave numbers for UVA and UVB?

32.7 • A sinusoidal electromagnetic wave having a magnetic field of amplitude 1.25 μT and a wavelength of 432 nm is traveling in the $+x$-direction through empty space. (a) What is the frequency of this wave? (b) What is the amplitude of the associated electric field? (c) Write the equations for the electric and magnetic fields as functions of x and t in the form of Eqs. (32.17).

32.8 • An electromagnetic wave of wavelength 435 nm is traveling in vacuum in the $-z$-direction. The electric field has amplitude 2.70×10^{-3} V/m and is parallel to the x-axis. What are (a) the frequency and (b) the magnetic-field amplitude? (c) Write the vector equations for $\vec{E}(z, t)$ and $\vec{B}(z, t)$.

32.9 • Consider electromagnetic waves propagating in air. (a) Determine the frequency of a wave with a wavelength of (i) 5.0 km, (ii) 5.0 μm, (iii) 5.0 nm. (b) What is the wavelength (in meters and nanometers) of (i) gamma rays of frequency 6.50×10^{21} Hz and (ii) an AM station radio wave of frequency 590 kHz?

32.10 • The electric field of a sinusoidal electromagnetic wave obeys the equation $E = (375 \text{ V/m}) \cos[(1.99 \times 10^7 \text{ rad/m})x + (5.97 \times 10^{15} \text{ rad/s})t]$. (a) What are the amplitudes of the electric and magnetic fields of this wave? (b) What are the frequency, wavelength, and period of the wave? Is this light visible to humans? (c) What is the speed of the wave?

32.11 • An electromagnetic wave has an electric field given by $\vec{E}(y, t) = (3.10 \times 10^5 \text{ V/m})\hat{k} \cos[ky - (12.65 \times 10^{12} \text{ rad/s})t]$. (a) In which direction is the wave traveling? (b) What is the wavelength of the wave? (c) Write the vector equation for $\vec{B}(y, t)$.

32.12 • An electromagnetic wave has a magnetic field given by $\vec{B}(x, t) = -(8.25 \times 10^{-9} \text{ T})\hat{\jmath} \cos[(1.38 \times 10^4 \text{ rad/m})x + \omega t]$. (a) In which direction is the wave traveling? (b) What is the frequency f of the wave? (c) Write the vector equation for $\vec{E}(x, t)$.

32.13 • Radio station WCCO in Minneapolis broadcasts at a frequency of 830 kHz. At a point some distance from the transmitter, the magnetic-field amplitude of the electromagnetic wave from WCCO is 4.82×10^{-11} T. Calculate (a) the wavelength; (b) the wave number; (c) the angular frequency; (d) the electric-field amplitude.

32.14 • An electromagnetic wave with frequency 65.0 Hz travels in an insulating magnetic material that has dielectric constant 3.64 and relative permeability 5.18 at this frequency. The electric field has amplitude 7.20×10^{-3} V/m. (a) What is the speed of propagation of the wave? (b) What is the wavelength of the wave? (c) What is the amplitude of the magnetic field?

32.15 • An electromagnetic wave with frequency 5.70×10^{14} Hz propagates with a speed of 2.17×10^8 m/s in a certain piece of glass. Find (a) the wavelength of the wave in the glass; (b) the wavelength of a wave of the same frequency propagating in air; (c) the index of refraction n of the glass for an electromagnetic wave with this frequency; (d) the dielectric constant for glass at this frequency, assuming that the relative permeability is unity.

Section 32.4 Energy and Momentum in Electromagnetic Waves

32.16 • **BIO High-Energy Cancer Treatment.** Scientists are working on a new technique to kill cancer cells by zapping them with ultrahigh-energy (in the range of 10^{12} W) pulses of light that last for an extremely short time (a few nanoseconds). These short pulses scramble the interior of a cell without causing it to explode, as long pulses would do. We can model a typical such cell as a disk 5.0 μm in diameter, with the pulse lasting for 4.0 ns with an average power of 2.0×10^{12} W. We shall assume that the energy is spread uniformly over the faces of 100 cells for each pulse. (a) How much energy is given to the cell during this pulse? (b) What is the intensity (in W/m^2) delivered to the cell? (c) What are the maximum values of the electric and magnetic fields in the pulse?

32.17 •• **Fields from a Light Bulb.** We can reasonably model a 75-W incandescent light bulb as a sphere 6.0 cm in diameter. Typically, only about 5% of the energy goes to visible light; the rest goes largely to nonvisible infrared radiation. (a) What is the visible-light intensity (in W/m^2) at the surface of the bulb? (b) What are the amplitudes of the electric and magnetic fields at this surface, for a sinusoidal wave with this intensity?

32.18 •• A sinusoidal electromagnetic wave from a radio station passes perpendicularly through an open window that has area 0.500 m^2. At the window, the electric field of the wave has rms value 0.0200 V/m. How much energy does this wave carry through the window during a 30.0-s commercial?

32.19 •• **Testing a Space Radio Transmitter.** You are a NASA mission specialist on your first flight aboard the space shuttle. Thanks to your extensive training in physics, you have been assigned to evaluate the performance of a new radio transmitter on board the International Space Station (ISS). Perched on the shuttle's movable arm, you aim a sensitive detector at the ISS, which is 2.5 km away. You find that the electric-field amplitude of the radio waves coming from the ISS transmitter is 0.090 V/m and that the frequency of the waves is 244 MHz. Find the following: (a) the intensity of the radio wave at your location; (b) the magnetic-field amplitude of the wave at your location; (c) the total power output of the ISS radio transmitter. (d) What assumptions, if any, did you make in your calculations?

32.20 • The intensity of a cylindrical laser beam is 0.800 W/m^2. The cross-sectional area of the beam is $3.0 \times 10^{-4} \text{ m}^2$ and the intensity is uniform across the cross section of the beam. (a) What is the average power output of the laser? (b) What is the rms value of the electric field in the beam?

32.21 • A space probe 2.0×10^{10} m from a star measures the total intensity of electromagnetic radiation from the star to be $5.0 \times 10^3 \text{ W/m}^2$. If the star radiates uniformly in all directions, what is its total average power output?

32.22 • A sinusoidal electromagnetic wave emitted by a cellular phone has a wavelength of 35.4 cm and an electric-field amplitude of 5.40×10^{-2} V/m at a distance of 250 m from the phone. Calculate (a) the frequency of the wave; (b) the magnetic-field amplitude; (c) the intensity of the wave.

32.23 • A monochromatic light source with power output 60.0 W radiates light of wavelength 700 nm uniformly in all directions. Calculate E_{max} and B_{max} for the 700-nm light at a distance of 5.00 m from the source.

32.24 • For the electromagnetic wave represented by Eqs. (32.19), show that the Poynting vector (a) is in the same direction as the propagation of the wave and (b) has average magnitude given by Eqs. (32.29).

32.25 •• An intense light source radiates uniformly in all directions. At a distance of 5.0 m from the source, the radiation pressure on a perfectly absorbing surface is 9.0×10^{-6} Pa. What is the total average power output of the source?

32.26 • **Television Broadcasting.** Public television station KQED in San Francisco broadcasts a sinusoidal radio signal at a power of 316 kW. Assume that the wave spreads out uniformly into a hemisphere above the ground. At a home 5.00 km away from the antenna, (a) what average pressure does this wave exert on a totally reflecting surface, (b) what are the amplitudes of the electric and magnetic fields of the wave, and (c) what is the average density of the energy this wave carries? (d) For the energy density in part (c), what percentage is due to the electric field and what percentage is due to the magnetic field?

32.27 •• **BIO Laser Safety.** If the eye receives an average intensity greater than $1.0 \times 10^2 \text{ W/m}^2$, damage to the retina can occur. This quantity is called the *damage threshold* of the retina. (a) What is the largest average power (in mW) that a laser beam 1.5 mm in diameter can have and still be considered safe to view head-on? (b) What are the maximum values of the electric and magnetic fields for the beam in part (a)? (c) How much energy would the beam in part (a) deliver per second to the retina? (d) Express the damage threshold in W/cm^2.

32.28 • In the 25-ft Space Simulator facility at NASA's Jet Propulsion Laboratory, a bank of overhead arc lamps can produce light of intensity 2500 W/m^2 at the floor of the facility. (This simulates the intensity of sunlight near the planet Venus.) Find the average radiation pressure (in pascals and in atmospheres) on (a) a totally absorbing section of the floor and (b) a totally reflecting section of the floor. (c) Find the average momentum density (momentum per unit volume) in the light at the floor.

32.29 • **Laboratory Lasers.** He–Ne lasers are often used in physics demonstrations. They produce light of wavelength 633 nm and a power of 0.500 mW spread over a cylindrical beam 1.00 mm in diameter (although these quantities can vary). (a) What is the intensity of this laser beam? (b) What are the maximum values of the electric and magnetic fields? (c) What is the average energy density in the laser beam?

32.30 •• **Solar Sail 1.** During 2004, Japanese scientists successfully tested two solar sails. One had a somewhat complicated shape that we shall model as a disk 9.0 m in diameter and 7.5 μm thick. The intensity of solar energy at that location was about 1400 W/m^2. (a) What force did the sun's light exert on this sail, assuming that it struck perpendicular to the sail and that the sail was perfectly reflecting? (b) If the sail was made of magnesium, of density 1.74 g/cm^3, what acceleration would the sun's radiation give to the sail? (c) Does the acceleration seem large enough to be feasible for space flight? In what ways could the sail be modified to increase its acceleration?

Section 32.5 Standing Electromagnetic Waves

32.31 • **Microwave Oven.** The microwaves in a certain microwave oven have a wavelength of 12.2 cm. (a) How wide must this oven be so that it will contain five antinodal planes of the electric field along its width in the standing-wave pattern? (b) What is the frequency of these microwaves? (c) Suppose a manufacturing error occurred and the oven was made 5.0 cm longer than specified in part (a). In this case, what would have to be the frequency of the microwaves for there still to be five antinodal planes of the electric field along the width of the oven?

32.32 • An electromagnetic standing wave in air of frequency 750 MHz is set up between two conducting planes 80.0 cm apart. At which positions between the planes could a point charge be placed at rest so that it would *remain* at rest? Explain.

32.33 • A standing electromagnetic wave in a certain material has frequency 2.20×10^{10} Hz. The nodal planes of $\vec{B}$ are 3.55 mm apart. Find (a) the wavelength of the wave in this material; (b) the distance between adjacent nodal planes of the $\vec{E}$ field; (c) the speed of propagation of the wave.

32.34 • An electromagnetic standing wave in air has frequency 75.0 MHz. (a) What is the distance between nodal planes of the $\vec{E}$ field? (b) What is the distance between a nodal plane of $\vec{E}$ and the closest nodal plane of $\vec{B}$?

32.35 • An electromagnetic standing wave in a certain material has frequency 1.20×10^{10} Hz and speed of propagation 2.10×10^8 m/s. (a) What is the distance between a nodal plane of $\vec{B}$ and the closest antinodal plane of $\vec{B}$? (b) What is the distance between an antinodal plane of $\vec{E}$ and the closest antinodal plane of $\vec{B}$? (c) What is the distance between a nodal plane of $\vec{E}$ and the closest nodal plane of $\vec{B}$?

32.36 • **CALC** Show that the electric and magnetic fields for standing waves given by Eqs. (32.34) and (32.35) (a) satisfy the wave equation, Eq. (32.15), and (b) satisfy Eqs. (32.12) and (32.14).

PROBLEMS

32.37 •• **BIO Laser Surgery.** Very short pulses of high-intensity laser beams are used to repair detached portions of the retina of the eye. The brief pulses of energy absorbed by the retina weld the detached portions back into place. In one such procedure, a laser beam has a wavelength of 810 nm and delivers 250 mW of power spread over a circular spot 510 μm in diameter. The vitreous humor (the transparent fluid that fills most of the eye) has an index of refraction of 1.34. (a) If the laser pulses are each 1.50 ms long, how much energy is delivered to the retina with each pulse? (b) What average pressure does the pulse of the laser beam exert on the retina as it is fully absorbed by the circular spot? (c) What are the wavelength and frequency of the laser light inside the vitreous humor of the eye? (d) What are the maximum values of the electric and magnetic fields in the laser beam?

32.38 •• **CALC** Consider a sinusoidal electromagnetic wave with fields $\vec{E} = E_{max}\hat{\jmath}\cos(kx - \omega t)$ and $\vec{B} = B_{max}\hat{k}\cos(kx - \omega t + \phi)$, with $-\pi \le \phi \le \pi$. Show that if $\vec{E}$ and $\vec{B}$ are to satisfy Eqs. (32.12) and (32.14), then $E_{\text{max}} = cB_{\text{max}}$ and $\phi = 0$. (The result $\phi = 0$ means the $\vec{E}$ and $\vec{B}$ fields oscillate in phase.)

32.39 •• You want to support a sheet of fireproof paper horizontally, using only a vertical upward beam of light spread uniformly over the sheet. There is no other light on this paper. The sheet measures 22.0 cm by 28.0 cm and has a mass of 1.50 g. (a) If the paper is black and hence absorbs all the light that hits it, what must be the intensity of the light beam? (b) For the light in part (a), what are the amplitudes of its electric and magnetic fields? (c) If the paper is white and hence reflects all the light that hits it, what intensity of light beam is needed to support it? (d) To see if it is physically reasonable to expect to support a sheet of paper this way, calculate the intensity in a typical 0.500-mW laser beam that is 1.00 mm in diameter, and compare this value with your answer in part (a).

32.40 •• For a sinusoidal electromagnetic wave in vacuum, such as that described by Eq. (32.16), show that the *average* energy density in the electric field is the same as that in the magnetic field.

32.41 • A satellite 575 km above the earth's surface transmits sinusoidal electromagnetic waves of frequency 92.4 MHz uniformly in all directions, with a power of 25.0 kW. (a) What is the intensity of these waves as they reach a receiver at the surface of the earth directly below the satellite? (b) What are the amplitudes of the electric and magnetic fields at the receiver? (c) If the receiver has a totally absorbing panel measuring 15.0 cm by 40.0 cm oriented with its plane perpendicular to the direction the waves travel, what average force do these waves exert on the panel? Is this force large enough to cause significant effects?

32.42 • A plane sinusoidal electromagnetic wave in air has a wavelength of 3.84 cm and an $\vec{E}$-field amplitude of 1.35 V/m. (a) What is the frequency? (b) What is the $\vec{B}$-field amplitude? (c) What is the intensity? (d) What average force does this radiation exert on a totally absorbing surface with area 0.240 m^2 perpendicular to the direction of propagation?

32.43 • A small helium–neon laser emits red visible light with a power of 4.60 mW in a beam that has a diameter of 2.50 mm. (a) What are the amplitudes of the electric and magnetic fields of the light? (b) What are the average energy densities associated with the electric field and with the magnetic field? (c) What is the total energy contained in a 1.00-m length of the beam?

32.44 •• The electric-field component of a sinusoidal electromagnetic wave traveling through a plastic cylinder is given by the equation $E = (5.35 \text{ V/m}) \cos[(1.39 \times 10^7 \text{ rad/m})x - (3.02 \times 10^{15} \text{ rad/s})t]$. (a) Find the frequency, wavelength, and speed of this wave in the plastic. (b) What is the index of refraction of the plastic? (c) Assuming that the amplitude of the electric field does not change, write a comparable equation for the electric field if the light is traveling in air instead of in plastic.

32.45 • The sun emits energy in the form of electromagnetic waves at a rate of 3.9×10^{26} W. This energy is produced by nuclear reactions deep in the sun's interior. (a) Find the intensity of electromagnetic radiation and the radiation pressure on an absorbing object at the surface of the sun (radius $r = R = 6.96 \times 10^5$ km) and at $r = R/2$, in the sun's interior. Ignore any scattering of the waves as they move radially outward from the center of the sun. Compare to the values given in Section 32.4 for sunlight just before it enters the earth's atmosphere. (b) The gas pressure at the sun's surface is about 1.0×10^4 Pa; at $r = R/2$, the gas pressure is calculated from solar models to be about 4.7×10^{13} Pa. Comparing with your results in part (a), would you expect that radiation pressure is an important factor in determining the structure of the sun? Why or why not?

32.46 •• A source of sinusoidal electromagnetic waves radiates uniformly in all directions. At 10.0 m from this source, the amplitude of the electric field is measured to be 1.50 N/C. What is the electric-field amplitude at a distance of 20.0 cm from the source?

32.47 •• **CP** Two square reflectors, each 1.50 cm on a side and of mass 4.00 g, are located at opposite ends of a thin, extremely light, 1.00-m rod that can rotate without friction and in vacuum about an axle perpendicular to it through its center (Fig. P32.47). These reflectors are small enough to be treated as point masses in moment-of-inertia calculations. Both reflectors are illuminated on one face by a sinusoidal light wave having an electric field of amplitude 1.25 N/C that falls uniformly on both surfaces and always strikes them perpendicular to the plane of their surfaces. One reflector is covered with a perfectly absorbing coating, and the other is covered with a perfectly reflecting coating. What is the angular acceleration of this device?

Figure **P32.47**

32.48 •• **CP** A circular loop of wire has radius 7.50 cm. A sinusoidal electromagnetic plane wave traveling in air passes through the loop, with the direction of the magnetic field of the wave perpendicular to the plane of the loop. The intensity of the wave at the location of the loop is 0.0195 W/m^2, and the wavelength of the wave is 6.90 m. What is the maximum emf induced in the loop?

32.49 • **CALC CP** A cylindrical conductor with a circular cross section has a radius a and a resistivity ρ and carries a constant current I. (a) What are the magnitude and direction of the electric-field vector $\vec{E}$ at a point just inside the wire at a distance a from the axis? (b) What are the magnitude and direction of the magnetic-field vector $\vec{B}$ at the same point? (c) What are the magnitude and direction of the Poynting vector $\vec{S}$ at the same point? (The direction of $\vec{S}$ is the direction in which electromagnetic energy flows into or out of the conductor.) (d) Use the result in part (c) to find the rate of flow of energy into the volume occupied by a length l of the conductor. (*Hint:* Integrate $\vec{S}$ over the surface of this volume.) Compare your result to the rate of generation of thermal energy in the same volume. Discuss why the energy dissipated in a current-carrying conductor, due to its resistance, can be thought of as entering through the cylindrical sides of the conductor.

32.50 • In a certain experiment, a radio transmitter emits sinusoidal electromagnetic waves of frequency 110.0 MHz in opposite directions inside a narrow cavity with reflectors at both ends, causing a standing-wave pattern to occur. (a) How far apart are the nodal planes of the magnetic field? (b) If the standing-wave pattern is determined to be in its eighth harmonic, how long is the cavity?

32.51 •• **CP Flashlight to the Rescue.** You are the sole crew member of the interplanetary spaceship *T:1339 Vorga,* which makes regular cargo runs between the earth and the mining colonies in the asteroid belt. You are working outside the ship one day while at a distance of 2.0 AU from the sun. [1 AU (astronomical unit) is the average distance from the earth to the sun, 149,600,000 km.] Unfortunately, you lose contact with the ship's hull and begin to drift away into space. You use your spacesuit's rockets to try to push yourself back toward the ship, but they run out of fuel and stop working before you can return to the ship. You find yourself in an awkward position, floating 16.0 m from the spaceship with zero velocity relative to it. Fortunately, you are carrying a 200-W flashlight. You turn on the flashlight and use its beam as a "light rocket" to push yourself back toward the ship. (a) If you, your spacesuit, and the flashlight have a combined mass of 150 kg, how long will it take you to get back to the ship? (b) Is there another way you could use the flashlight to accomplish the same job of returning you to the ship?

32.52 • The 19th-century inventor Nikola Tesla proposed to transmit electric power via sinusoidal electromagnetic waves. Suppose power is to be transmitted in a beam of cross-sectional area 100 m². What electric- and magnetic-field amplitudes are required to transmit an amount of power comparable to that handled by modern transmission lines (that carry voltages and currents of the order of 500 kV and 1000 A)?

32.53 •• **CP Global Positioning System (GPS).** The GPS network consists of 24 satellites, each of which makes two orbits around the earth per day. Each satellite transmits a 50.0-W (or even less) sinusoidal electromagnetic signal at two frequencies, one of which is 1575.42 MHz. Assume that a satellite transmits half of its power at each frequency and that the waves travel uniformly in a downward hemisphere. (a) What average intensity does a GPS receiver on the ground, directly below the satellite, receive? (*Hint:* First use Newton's laws to find the altitude of the satellite.) (b) What are the amplitudes of the electric and magnetic fields at the GPS receiver in part (a), and how long does it take the signal to reach the receiver? (c) If the receiver is a square panel 1.50 cm on a side that absorbs all of the beam, what average pressure does the signal exert on it? (d) What wavelength must the receiver be tuned to?

32.54 •• **CP Solar Sail 2.** NASA is giving serious consideration to the concept of *solar sailing*. A solar sailcraft uses a large, low-mass sail and the energy and momentum of sunlight for propulsion. (a) Should the sail be absorbing or reflective? Why? (b) The total power output of the sun is 3.9×10^{26} W. How large a sail is necessary to propel a 10,000-kg spacecraft against the gravitational force of the sun? Express your result in square kilometers. (c) Explain why your answer to part (b) is independent of the distance from the sun.

32.55 •• **CP** Interplanetary space contains many small particles referred to as *interplanetary dust*. Radiation pressure from the sun sets a lower limit on the size of such dust particles. To see the origin of this limit, consider a spherical dust particle of radius R and mass density ρ. (a) Write an expression for the gravitational force exerted on this particle by the sun (mass M) when the particle is a distance r from the sun. (b) Let L represent the luminosity of the sun, equal to the rate at which it emits energy in electromagnetic radiation. Find the force exerted on the (totally absorbing) particle due to solar radiation pressure, remembering that the intensity of the sun's radiation also depends on the distance r. The relevant area is the cross-sectional area of the particle, *not* the total surface area of the particle. As part of your answer, explain why this is so. (c) The mass density of a typical interplanetary dust particle is about 3000 kg/m³. Find the particle radius R such that the gravitational and radiation forces acting on the particle are equal in magnitude. The luminosity of the sun is 3.9×10^{26} W. Does your answer depend on the distance of the particle from the sun? Why or why not? (d) Explain why dust particles with a radius less than

that found in part (c) are unlikely to be found in the solar system. [*Hint:* Construct the ratio of the two force expressions found in parts (a) and (b).]

CHALLENGE PROBLEMS

32.56 ••• **CALC** Electromagnetic waves propagate much differently in *conductors* than they do in dielectrics or in vacuum. If the resistivity of the conductor is sufficiently low (that is, if it is a sufficiently good conductor), the oscillating electric field of the wave gives rise to an oscillating conduction current that is much larger than the displacement current. In this case, the wave equation for an electric field $\vec{E}(x, t) = E_y(x, t)\hat{\jmath}$ propagating in the $+x$-direction within a conductor is

$$\frac{\partial^2 E_y(x, t)}{\partial x^2} = \frac{\mu}{\rho} \frac{\partial E_y(x, t)}{\partial t}$$

where μ is the permeability of the conductor and ρ is its resistivity. (a) A solution to this wave equation is

$$E_y(x, t) = E_{max}e^{-k_C x}\cos(k_C x - \omega t)$$

where $k_C = \sqrt{\omega\mu/2\rho}$. Verify this by substituting $E_y(x, t)$ into the above wave equation. (b) The exponential term shows that the electric field decreases in amplitude as it propagates. Explain why this happens. (*Hint:* The field does work to move charges within the conductor. The current of these moving charges causes $i^2 R$ heating within the conductor, raising its temperature. Where does the energy to do this come from?) (c) Show that the electric-field amplitude decreases by a factor of $1/e$ in a distance $1/k_C = \sqrt{2\rho/\omega\mu}$, and calculate this distance for a radio wave with frequency $f = 1.0$ MHz in copper (resistivity 1.72×10^{-8} $\Omega \cdot$m; permeability $\mu = \mu_0$). Since this distance is so short, electromagnetic waves of this frequency can hardly propagate at all into copper. Instead, they are reflected at the surface of the metal. This is why radio waves cannot penetrate through copper or other metals, and why radio reception is poor inside a metal structure.

32.57 ••• **CP** Electromagnetic radiation is emitted by accelerating charges. The rate at which energy is emitted from an accelerating charge that has charge q and acceleration a is given by

$$\frac{dE}{dt} = \frac{q^2 a^2}{6\pi\epsilon_0 c^3}$$

where c is the speed of light. (a) Verify that this equation is dimensionally correct. (b) If a proton with a kinetic energy of 6.0 MeV is traveling in a particle accelerator in a circular orbit of radius 0.750 m, what fraction of its energy does it radiate per second? (c) Consider an electron orbiting with the same speed and radius. What fraction of its energy does it radiate per second?

32.58 ••• **CP The Classical Hydrogen Atom.** The electron in a hydrogen atom can be considered to be in a circular orbit with a radius of 0.0529 nm and a kinetic energy of 13.6 eV. If the electron behaved classically, how much energy would it radiate per second (see Challenge Problem 32.57)? What does this tell you about the use of classical physics in describing the atom?

Answers

Chapter Opening Question

Metals are reflective because they are good conductors of electricity. When an electromagnetic wave strikes a conductor, the electric field of the wave sets up currents on the conductor surface that generate a reflected wave. For a perfect conductor, this reflected wave is just as intense as the incident wave. Tarnished metals are less shiny because their surface is oxidized and less conductive; polishing the metal removes the oxide and exposes the conducting metal.

Test Your Understanding Questions

32.1 Answers: (a) no, (b) no A purely electric wave would have a varying electric field. Such a field necessarily generates a magnetic field through Ampere's law, Eq. (29.20), so a purely electric wave is impossible. In the same way, a purely magnetic wave is impossible: The varying magnetic field in such a wave would automatically give rise to an electric field through Faraday's law, Eq. (29.21).

32.2 Answers: (a) positive y-direction, (b) negative x-direction, (c) positive y-direction You can verify these answers by using the right-hand rule to show that $\vec{E} \times \vec{B}$ in each case is in the direction of propagation, or by using the rule shown in Fig. 32.9.

32.3 Answer: (iv) In an ideal electromagnetic plane wave, at any instant the fields are the same anywhere in a plane perpendicular to the direction of propagation. The plane wave described by Eqs. (32.17) is propagating in the x-direction, so the fields depend on the coordinate x and time t but do *not* depend on the coordinates y and z.

32.4 Answers: (a) (i) and (iii), (b) (ii) and (iv), (c) (i) and (iii), (d) (ii) and (iv) Both the energy density u and the Poynting vector magnitude S are maximum where the $\vec{E}$ and $\vec{B}$ fields have their maximum magnitudes. (The directions of the fields doesn't matter.) From Fig. 32.13, this occurs at $x = 0$ and $x = \lambda/2$. Both u and S have a minimum value of zero; that occurs where $\vec{E}$ and $\vec{B}$ are both zero. From Fig. 32.13, this occurs at $x = \lambda/4$ and $x = 3\lambda/4$.

32.5 Answer: no There are places where $\vec{E} = 0$ at all times (at the walls) and the electric energy density $\frac{1}{2}\epsilon_0 E^2$ is always zero. There are also places where $\vec{B} = 0$ at all times (on the plane midway between the walls) and the magnetic energy density $B^2/2\mu_0$ is always zero. However, there are *no* locations where both $\vec{E}$ and $\vec{B}$ are always zero. Hence the energy density at any point in the standing wave is always nonzero.

Bridging Problem

Answer: 0.0368 V

33 THE NATURE AND PROPAGATION OF LIGHT

LEARNING GOALS

By studying this chapter, you will learn:

- What light rays are, and how they are related to wave fronts.

- The laws that govern the reflection and refraction of light.

- The circumstances under which light is totally reflected at an interface.

- How to make polarized light out of ordinary light.

- How Huygens's principle helps us analyze reflection and refraction.

? These drafting tools are made of clear plastic, but a rainbow of colors appears when they are placed between two special filters called polarizers. How does this cause the colors?

Blue lakes, ochre deserts, green forests, and multicolored rainbows can be enjoyed by anyone who has eyes with which to see them. But by studying the branch of physics called **optics,** which deals with the behavior of light and other electromagnetic waves, we can reach a deeper appreciation of the visible world. A knowledge of the properties of light allows us to understand the blue color of the sky and the design of optical devices such as telescopes, microscopes, cameras, eyeglasses, and the human eye. The same basic principles of optics also lie at the heart of modern developments such as the laser, optical fibers, holograms, optical computers, and new techniques in medical imaging.

The importance of optics to physics, and to science and engineering in general, is so great that we will devote the next four chapters to its study. In this chapter we begin with a study of the laws of reflection and refraction and the concepts of dispersion, polarization, and scattering of light. Along the way we compare the various possible descriptions of light in terms of particles, rays, or waves, and we introduce Huygens's principle, an important link that connects the ray and wave viewpoints. In Chapter 34 we'll use the ray description of light to understand how mirrors and lenses work, and we'll see how mirrors and lenses are used in optical instruments such as cameras, microscopes, and telescopes. We'll explore the wave characteristics of light further in Chapters 35 and 36.

33.1 The Nature of Light

Until the time of Isaac Newton (1642–1727), most scientists thought that light consisted of streams of particles (called *corpuscles*) emitted by light sources. Galileo and others tried (unsuccessfully) to measure the speed of light. Around 1665, evidence of *wave* properties of light began to be discovered. By the early 19th century, evidence that light is a wave had grown very persuasive.

In 1873, James Clerk Maxwell predicted the existence of electromagnetic waves and calculated their speed of propagation, as we learned in Chapter 32. This development, along with the experimental work of Heinrich Hertz starting in 1887, showed conclusively that light is indeed an electromagnetic wave.

The Two Personalities of Light

The wave picture of light is not the whole story, however. Several effects associated with emission and absorption of light reveal a particle aspect, in that the energy carried by light waves is packaged in discrete bundles called *photons* or *quanta.* These apparently contradictory wave and particle properties have been reconciled since 1930 with the development of quantum electrodynamics, a comprehensive theory that includes *both* wave and particle properties. The *propagation* of light is best described by a wave model, but understanding emission and absorption requires a particle approach.

The fundamental sources of all electromagnetic radiation are electric charges in accelerated motion. All bodies emit electromagnetic radiation as a result of thermal motion of their molecules; this radiation, called *thermal radiation,* is a mixture of different wavelengths. At sufficiently high temperatures, all matter emits enough visible light to be self-luminous; a very hot body appears "red-hot" (Fig. 33.1) or "white-hot." Thus hot matter in any form is a light source. Familiar examples are a candle flame, hot coals in a campfire, the coils in an electric room heater, and an incandescent lamp filament (which usually operates at a temperature of about 3000°C).

Light is also produced during electrical discharges through ionized gases. The bluish light of mercury-arc lamps, the orange-yellow of sodium-vapor lamps, and the various colors of "neon" signs are familiar. A variation of the mercury-arc lamp is the *fluorescent* lamp (see Fig. 30.7). This light source uses a material called a *phosphor* to convert the ultraviolet radiation from a mercury arc into visible light. This conversion makes fluorescent lamps more efficient than incandescent lamps in transforming electrical energy into light.

In most light sources, light is emitted independently by different atoms within the source; in a *laser,* by contrast, atoms are induced to emit light in a cooperative, coherent fashion. The result is a very narrow beam of radiation that can be enormously intense and that is much more nearly *monochromatic,* or single-frequency, than light from any other source. Lasers are used by physicians for microsurgery, in a DVD or Blu-ray player to scan the information recorded on a video disc, in industry to cut through steel and to fuse high-melting-point materials, and in many other applications (Fig. 33.2).

No matter what its source, electromagnetic radiation travels in vacuum at the same speed. As we saw in Sections 1.3 and 32.1, the speed of light in vacuum is defined to be

$$c = 2.99792458 \times 10^8 \text{ m/s}$$

or 3.00×10^8 m/s to three significant figures. The duration of one second is defined by the cesium clock (see Section 1.3), so one meter is defined to be the distance that light travels in 1/299,792,458 s.

Waves, Wave Fronts, and Rays

We often use the concept of a **wave front** to describe wave propagation. We introduced this concept in Section 32.2 to describe the leading edge of a wave. More generally, we define a wave front as *the locus of all adjacent points at which the phase of vibration of a physical quantity associated with the wave is the same.* That is, at any instant, all points on a wave front are at the same part of the cycle of their variation.

When we drop a pebble into a calm pool, the expanding circles formed by the wave crests, as well as the circles formed by the wave troughs between

33.1 An electric heating element emits primarily infrared radiation. But if its temperature is high enough, it also emits a discernible amount of visible light.

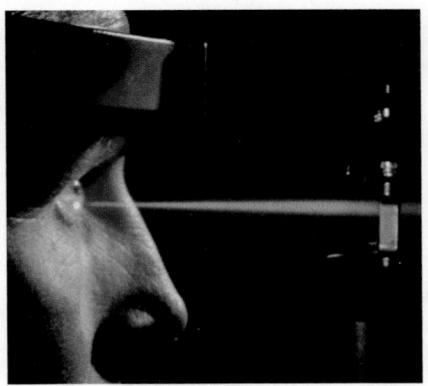

33.2 Ophthalmic surgeons use lasers for repairing detached retinas and for cauterizing blood vessels in retinopathy. Pulses of blue-green light from an argon laser are ideal for this purpose, since they pass harmlessly through the transparent part of the eye but are absorbed by red pigments in the retina.

33.3 Spherical wave fronts of sound spread out uniformly in all directions from a point source in a motionless medium, such as still air, that has the same properties in all regions and in all directions. Electromagnetic waves in vacuum also spread out as shown here.

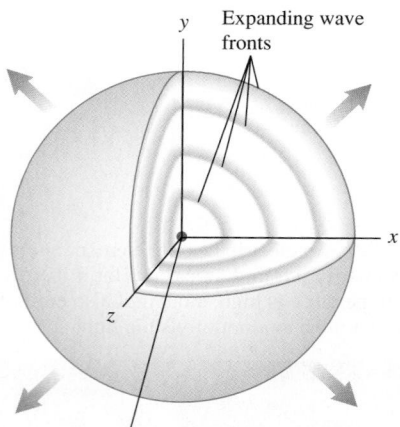

Point sound source producing spherical sound waves (alternating compressions and rarefactions of air)

33.4 Wave fronts (blue) and rays (purple).

(a)

When wave fronts are spherical, the rays radiate from the center of the sphere.

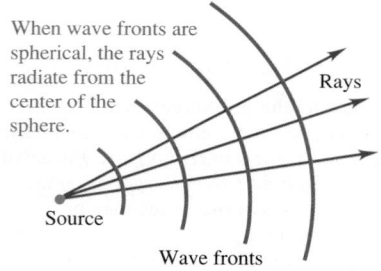

(b)

When wave fronts are planar, the rays are perpendicular to the wave fronts and parallel to each other.

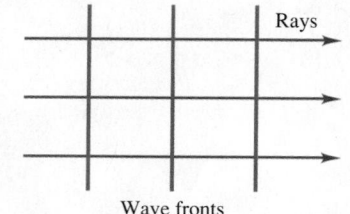

Mastering PHYSICS

ActivPhysics 15.1: Reflection and Refraction
ActivPhysics 15.3: Refraction Applications

them, are wave fronts. Similarly, when sound waves spread out in still air from a pointlike source, or when electromagnetic radiation spreads out from a pointlike emitter, any spherical surface that is concentric with the source is a wave front, as shown in Fig. 33.3. In diagrams of wave motion we usually draw only parts of a few wave fronts, often choosing consecutive wave fronts that have the same phase and thus are one wavelength apart, such as crests of water waves. Similarly, a diagram for sound waves might show only the "pressure crests," the surfaces over which the pressure is maximum, and a diagram for electromagnetic waves might show only the "crests" on which the electric or magnetic field is maximum.

We will often use diagrams that show the shapes of the wave fronts or their cross sections in some reference plane. For example, when electromagnetic waves are radiated by a small light source, we can represent the wave fronts as spherical surfaces concentric with the source or, as in Fig. 33.4a, by the circular intersections of these surfaces with the plane of the diagram. Far away from the source, where the radii of the spheres have become very large, a section of a spherical surface can be considered as a plane, and we have a *plane* wave like those discussed in Sections 32.2 and 32.3 (Fig, 33.4b).

To describe the directions in which light propagates, it's often convenient to represent a light wave by **rays** rather than by wave fronts. Rays were used to describe light long before its wave nature was firmly established. In a particle theory of light, rays are the paths of the particles. From the wave viewpoint *a ray is an imaginary line along the direction of travel of the wave.* In Fig. 33.4a the rays are the radii of the spherical wave fronts, and in Fig. 33.4b they are straight lines perpendicular to the wave fronts. When waves travel in a homogeneous isotropic material (a material with the same properties in all regions and in all directions), the rays are always straight lines normal to the wave fronts. At a boundary surface between two materials, such as the surface of a glass plate in air, the wave speed and the direction of a ray may change, but the ray segments in the air and in the glass are straight lines.

The next several chapters will give you many opportunities to see the interplay of the ray, wave, and particle descriptions of light. The branch of optics for which the ray description is adequate is called **geometric optics;** the branch dealing specifically with wave behavior is called **physical optics.** This chapter and the following one are concerned mostly with geometric optics. In Chapters 35 and 36 we will study wave phenomena and physical optics.

Test Your Understanding of Section 33.1 Some crystals are *not* isotropic: Light travels through the crystal at a higher speed in some directions than in others. In a crystal in which light travels at the same speed in the *x*- and *z*-directions but at a faster speed in the *y*-direction, what would be the shape of the wave fronts produced by a light source at the origin? (i) spherical, like those shown in Fig. 33.3; (ii) ellipsoidal, flattened along the *y*-axis; (iii) ellipsoidal, stretched out along the *y*-axis.

33.2 Reflection and Refraction

In this section we'll use the *ray* model of light to explore two of the most important aspects of light propagation: **reflection** and **refraction.** When a light wave strikes a smooth interface separating two transparent materials (such as air and glass or water and glass), the wave is in general partly *reflected* and partly *refracted* (transmitted) into the second material, as shown in Fig. 33.5a. For example, when you look into a restaurant window from the street, you see a reflection of the street scene, but a person inside the restaurant can look out through the window at the same scene as light reaches him by refraction.

33.5 (a) A plane wave is in part reflected and in part refracted at the boundary between two media (in this case, air and glass). The light that reaches the inside of the coffee shop is refracted twice, once entering the glass and once exiting the glass. (b), (c) How light behaves at the interface between the air outside the coffee shop (material *a*) and the glass (material *b*). For the case shown here, material *b* has a larger index of refraction than material $a(n_b > n_a)$ and the angle θ_b is smaller than θ_a.

(a) Plane waves reflected and refracted from a window

(b) The waves in the outside air and glass represented by rays

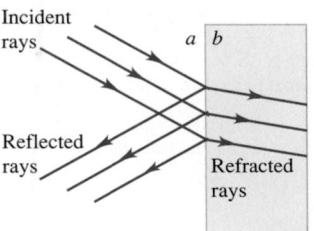

(c) The representation simplified to show just one set of rays

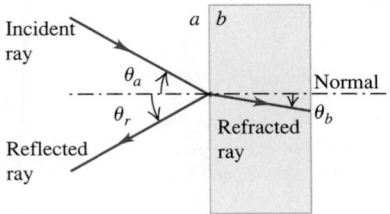

The segments of plane waves shown in Fig. 33.5a can be represented by bundles of rays forming *beams* of light (Fig. 33.5b). For simplicity we often draw only one ray in each beam (Fig. 33.5c). Representing these waves in terms of rays is the basis of geometric optics. We begin our study with the behavior of an individual ray.

We describe the directions of the incident, reflected, and refracted (transmitted) rays at a smooth interface between two optical materials in terms of the angles they make with the *normal* (perpendicular) to the surface at the point of incidence, as shown in Fig. 33.5c. If the interface is rough, both the transmitted light and the reflected light are scattered in various directions, and there is no single angle of transmission or reflection. Reflection at a definite angle from a very smooth surface is called **specular reflection** (from the Latin word for "mirror"); scattered reflection from a rough surface is called **diffuse reflection.** This distinction is shown in Fig. 33.6. Both kinds of reflection can occur with either transparent materials or *opaque* materials that do not transmit light. The vast majority of objects in your environment (including plants, other people, and this book) are visible to you because they reflect light in a diffuse manner from their surfaces. Our primary concern, however, will be with specular reflection from a very smooth surface such as highly polished glass or metal. Unless stated otherwise, when referring to "reflection" we will always mean *specular* reflection.

The **index of refraction** of an optical material (also called the **refractive index**), denoted by *n*, plays a central role in geometric optics. It is the ratio of the speed of light *c* in vacuum to the speed *v* in the material:

$$n = \frac{c}{v} \quad \text{(index of refraction)} \qquad (33.1)$$

Light always travels *more slowly* in a material than in vacuum, so the value of *n* in anything other than vacuum is always greater than unity. For vacuum, $n = 1$.

33.6 Two types of reflection.

(a) Specular reflection

(b) Diffuse reflection

Since n is a ratio of two speeds, it is a pure number without units. (The relationship of the value of n to the electric and magnetic properties of a material is described in Section 32.3.)

CAUTION **Wave speed and index of refraction** Keep in mind that the wave speed v is *inversely* proportional to the index of refraction n. The greater the index of refraction in a material, the *slower* the wave speed in that material. Failure to remember this point can lead to serious confusion!

The Laws of Reflection and Refraction

Experimental studies of the directions of the incident, reflected, and refracted rays at a smooth interface between two optical materials lead to the following conclusions (Fig. 33.7):

1. **The incident, reflected, and refracted rays and the normal to the surface all lie in the same plane.** The plane of the three rays and the normal, called the **plane of incidence,** is perpendicular to the plane of the boundary surface between the two materials. We always draw ray diagrams so that the incident, reflected, and refracted rays are in the plane of the diagram.

2. **The angle of reflection θ_r is equal to the angle of incidence θ_a for all wavelengths and for any pair of materials.** That is, in Fig. 33.5c,

$$\theta_r = \theta_a \quad \text{(law of reflection)} \tag{33.2}$$

This relationship, together with the observation that the incident and reflected rays and the normal all lie in the same plane, is called the **law of reflection.**

3. For monochromatic light and for a given pair of materials, a and b, on opposite sides of the interface, **the ratio of the sines of the angles θ_a and θ_b, where both angles are measured from the normal to the surface, is equal to the inverse ratio of the two indexes of refraction:**

$$\frac{\sin \theta_a}{\sin \theta_b} = \frac{n_b}{n_a} \tag{33.3}$$

or

$$n_a \sin \theta_a = n_b \sin \theta_b \quad \text{(law of refraction)} \tag{33.4}$$

This experimental result, together with the observation that the incident and refracted rays and the normal all lie in the same plane, is called the **law of refraction** or **Snell's law,** after the Dutch scientist Willebrord Snell (1591–1626). There is some doubt that Snell actually discovered it. The discovery that $n = c/v$ came much later.

While these results were first observed experimentally, they can be derived theoretically from a wave description of light. We do this in Section 33.7.

Equations (33.3) and (33.4) show that when a ray passes from one material (a) into another material (b) having a larger index of refraction ($n_b > n_a$) and hence a slower wave speed, the angle θ_b with the normal is *smaller* in the second material than the angle θ_a in the first; hence the ray is bent *toward* the normal (Fig. 33.8a). When the second material has a *smaller* index of refraction than the first material ($n_b < n_a$) and hence a faster wave speed, the ray is bent *away from* the normal (Fig. 33.8b).

No matter what the materials on either side of the interface, in the case of *normal* incidence the transmitted ray is not bent at all (Fig. 33.8c). In this case

33.7 The laws of reflection and refraction.

1. The incident, reflected, and refracted rays and the normal to the surface all lie in the same plane.

2. $\theta_r = \theta_a$

3. When a monochromatic light ray crosses the interface between two given materials a and b, the angles θ_a and θ_b are related to the indexes of refraction of a and b by

$$\frac{\sin \theta_a}{\sin \theta_b} = \frac{n_b}{n_a}$$

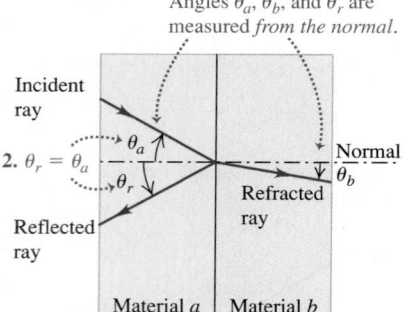

Angles θ_a, θ_b, and θ_r are measured *from the normal.*

33.8 Refraction and reflection in three cases. (a) Material b has a larger index of refraction than material a. (b) Material b has a smaller index of refraction than material a. (c) The incident light ray is normal to the interface between the materials.

(a) A ray entering a material of *larger* index of refraction bends *toward* the normal.

(b) A ray entering a material of *smaller* index of refraction bends *away from* the normal.

(c) A ray oriented along the normal does not bend, regardless of the materials.

33.9 (a) This ruler is actually straight, but it appears to bend at the surface of the water. (b) Light rays from any submerged object bend away from the normal when they emerge into the air. As seen by an observer above the surface of the water, the object appears to be much closer to the surface than it actually is.

(a) A straight ruler half-immersed in water

(b) Why the ruler appears bent

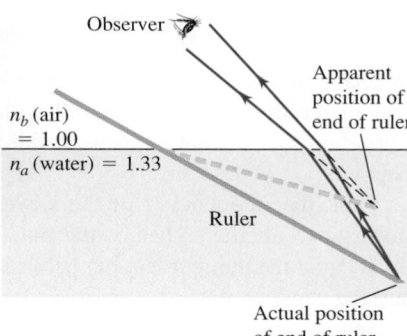

33.10 (a) The index of refraction of air is slightly greater than 1, so light rays from the setting sun bend downward when they enter our atmosphere. (The effect is exaggerated in this figure.) (b) Stronger refraction occurs for light coming from the lower limb of the sun (the part that appears closest to the horizon), which passes through denser air in the lower atmosphere. As a result, the setting sun appears flattened vertically. (See Problem 33.55.)

(a)

(b)

$\theta_a = 0$ and $\sin\theta_a = 0$, so from Eq. (33.4) θ_b is also equal to zero, so the transmitted ray is also normal to the interface. Equation (33.2) shows that θ_r, too, is equal to zero, so the reflected ray travels back along the same path as the incident ray.

The law of refraction explains why a partially submerged ruler or drinking straw appears bent; light rays coming from below the surface change in direction at the air–water interface, so the rays appear to be coming from a position above their actual point of origin (Fig. 33.9). A similar effect explains the appearance of the setting sun (Fig. 33.10).

An important special case is refraction that occurs at an interface between vacuum, for which the index of refraction is unity by definition, and a material. When a ray passes from vacuum into a material (b), so that $n_a = 1$ and $n_b > 1$, the ray is always bent *toward* the normal. When a ray passes from a material into vacuum, so that $n_a > 1$ and $n_b = 1$, the ray is always bent *away from* the normal.

The laws of reflection and refraction apply regardless of which side of the interface the incident ray comes from. If a ray of light approaches the interface in Fig. 33.8a or 33.8b from the right rather than from the left, there are again reflected and refracted rays; these two rays, the incident ray, and the normal to the surface again lie in the same plane. Furthermore, the path of a refracted ray is *reversible;* it follows the same path when going from b to a as when going from a to b. [You can verify this using Eq. (33.4).] Since reflected and incident rays make the same angle with the normal, the path of a reflected ray is also reversible. That's why when you see someone's eyes in a mirror, they can also see you.

The *intensities* of the reflected and refracted rays depend on the angle of incidence, the two indexes of refraction, and the polarization (that is, the direction of the electric-field vector) of the incident ray. The fraction reflected is smallest at normal incidence ($\theta_a = 0°$), where it is about 4% for an air–glass interface. This fraction increases with increasing angle of incidence to 100% at grazing incidence, when $\theta_a = 90°$.

It's possible to use Maxwell's equations to predict the amplitude, intensity, phase, and polarization states of the reflected and refracted waves. Such an analysis is beyond our scope, however.

The index of refraction depends not only on the substance but also on the wavelength of the light. The dependence on wavelength is called *dispersion;* we will consider it in Section 33.4. Indexes of refraction for several solids and liquids are given in Table 33.1 for a particular wavelength of yellow light.

Application **Transparency and Index of Refraction**
An eel in its larval stage is nearly as transparent as the seawater in which it swims. The larva in this photo is nonetheless easy to see because its index of refraction is higher than that of seawater, so that some of the light striking it is reflected instead of transmitted. The larva appears particularly shiny around its edges because the light reaching the camera from those points struck the larva at near-grazing incidence ($\theta_a = 90°$), resulting in almost 100% reflection.

Table 33.1 Index of Refraction for Yellow Sodium Light, $\lambda_0 = 589$ nm

Substance	Index of Refraction, n
Solids	
Ice (H_2O)	1.309
Fluorite (CaF_2)	1.434
Polystyrene	1.49
Rock salt (NaCl)	1.544
Quartz (SiO_2)	1.544
Zircon ($ZrO_2 \cdot SiO_2$)	1.923
Diamond (C)	2.417
Fabulite ($SrTiO_3$)	2.409
Rutile (TiO_2)	2.62
Glasses (typical values)	
Crown	1.52
Light flint	1.58
Medium flint	1.62
Dense flint	1.66
Lanthanum flint	1.80
Liquids at 20°C	
Methanol (CH_3OH)	1.329
Water (H_2O)	1.333
Ethanol (C_2H_5OH)	1.36
Carbon tetrachloride (CCl_4)	1.460
Turpentine	1.472
Glycerine	1.473
Benzene	1.501
Carbon disulfide (CS_2)	1.628

The index of refraction of air at standard temperature and pressure is about 1.0003, and we will usually take it to be exactly unity. The index of refraction of a gas increases as its density increases. Most glasses used in optical instruments have indexes of refraction between about 1.5 and 2.0. A few substances have larger indexes; one example is diamond, with 2.417.

Index of Refraction and the Wave Aspects of Light

We have discussed how the direction of a light ray changes when it passes from one material to another material with a different index of refraction. It's also important to see what happens to the *wave* characteristics of the light when this happens.

First, the frequency f of the wave does not change when passing from one material to another. That is, the number of wave cycles arriving per unit time must equal the number leaving per unit time; this is a statement that the boundary surface cannot create or destroy waves.

Second, the wavelength λ of the wave *is* different in general in different materials. This is because in any material, $v = \lambda f$; since f is the same in any material as in vacuum and v is always less than the wave speed c in vacuum, λ is also correspondingly reduced. Thus the wavelength λ of light in a material is *less than* the wavelength λ_0 of the same light in vacuum. From the above discussion, $f = c/\lambda_0 = v/\lambda$. Combining this with Eq. (33.1), $n = c/v$, we find

$$\lambda = \frac{\lambda_0}{n} \qquad \text{(wavelength of light in a material)} \qquad (33.5)$$

When a wave passes from one material into a second material with larger index of refraction, so that $n_b > n_a$, the wave speed decreases. The wavelength $\lambda_b = \lambda_0/n_b$ in the second material is then shorter than the wavelength $\lambda_a = \lambda_0/n_a$ in the first material. If instead the second material has a smaller index of refraction than the first material, so that $n_b < n_a$, then the wave speed increases. Then the wavelength λ_b in the second material is longer than the wavelength λ_a in the first material. This makes intuitive sense; the waves get "squeezed" (the wavelength gets shorter) if the wave speed decreases and get "stretched" (the wavelength gets longer) if the wave speed increases.

Problem-Solving Strategy 33.1   Reflection and Refraction

IDENTIFY *the relevant concepts:* Use geometric optics, discussed in this section, whenever light (or electromagnetic radiation of *any* frequency and wavelength) encounters a boundary between materials. In general, part of the light is reflected back into the first material and part is refracted into the second material.

SET UP *the problem* using the following steps:
1. In problems involving rays and angles, start by drawing a large, neat diagram. Label all known angles and indexes of refraction.
2. Identify the target variables.

EXECUTE *the solution* as follows:
1. Apply the laws of reflection, Eq. (33.2), and refraction, Eq. (33.4). Measure angles of incidence, reflection, and refraction with respect to the *normal* to the surface, *never* from the surface itself.

2. Apply geometry or trigonometry in working out angular relationships. Remember that the sum of the acute angles of a right triangle is 90° (they are *complementary*) and the sum of the interior angles in any triangle is 180°.
3. The frequency of the electromagnetic radiation does not change when it moves from one material to another; the wavelength changes in accordance with Eq. (33.5), $\lambda = \lambda_0/n$.

EVALUATE *your answer:* In problems that involve refraction, check that your results are consistent with Snell's law ($n_a \sin\theta_a = n_b \sin\theta_b$). If the second material has a higher index of refraction than the first, the angle of refraction must be *smaller* than the angle of incidence: The refracted ray bends toward the normal. If the first material has the higher index of refraction, the refracted angle must be larger than the incident angle: The refracted ray bends away from the normal.

Example 33.1 Reflection and refraction

In Fig. 33.11, material a is water and material b is glass with index of refraction 1.52. The incident ray makes an angle of 60.0° with the normal; find the directions of the reflected and refracted rays.

SOLUTION

IDENTIFY and SET UP: This is a problem in geometric optics. We are given the angle of incidence $\theta_a = 60.0°$ and the indexes of

33.11 Reflection and refraction of light passing from water to glass.

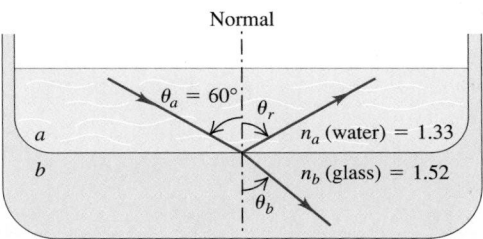

refraction $n_a = 1.33$ and $n_b = 1.52$. We must find the angles of reflection and refraction θ_r and θ_b; to do this we use Eqs. (33.2) and (33.4), respectively. Figure 33.11 shows the rays and angles; n_b is slightly greater than n_a, so by Snell's law [Eq. (33.4)] θ_b is slightly smaller than θ_a, as the figure shows.

EXECUTE: According to Eq. (33.2), the angle the reflected ray makes with the normal is the same as that of the incident ray, so $\theta_r = \theta_a = 60.0°$.

To find the direction of the refracted ray we use Snell's law, Eq. (33.4):

$$n_a \sin\theta_a = n_b \sin\theta_b$$
$$\sin\theta_b = \frac{n_a}{n_b}\sin\theta_a = \frac{1.33}{1.52}\sin 60.0° = 0.758$$
$$\theta_b = \arcsin(0.758) = 49.3°$$

EVALUATE: The second material has a larger refractive index than the first, as in Fig. 33.8a. Hence the refracted ray is bent toward the normal and $\theta_b < \theta_a$.

Example 33.2 Index of refraction in the eye

The wavelength of the red light from a helium-neon laser is 633 nm in air but 474 nm in the aqueous humor inside your eyeball. Calculate the index of refraction of the aqueous humor and the speed and frequency of the light in it.

SOLUTION

IDENTIFY and SET UP: The key ideas here are (i) the definition of index of refraction n in terms of the wave speed v in a medium and the speed c in vacuum, and (ii) the relationship between wavelength λ_0 in vacuum and wavelength λ in a medium of index n. We use Eq. (33.1), $n = c/v$; Eq. (33.5), $\lambda = \lambda_0/n$; and $v = \lambda f$.

EXECUTE: The index of refraction of air is very close to unity, so we assume that the wavelength λ_0 in vacuum is the same as that in air, 633 nm. Then from Eq. (33.5),

$$\lambda = \frac{\lambda_0}{n} \qquad n = \frac{\lambda_0}{\lambda} = \frac{633 \text{ nm}}{474 \text{ nm}} = 1.34$$

This is about the same index of refraction as for water. Then, using $n = c/v$ and $v = \lambda f$, we find

$$v = \frac{c}{n} = \frac{3.00 \times 10^8 \text{ m/s}}{1.34} = 2.25 \times 10^8 \text{ m/s}$$
$$f = \frac{v}{\lambda} = \frac{2.25 \times 10^8 \text{ m/s}}{474 \times 10^{-9} \text{ m}} = 4.74 \times 10^{14} \text{ Hz}$$

EVALUATE: Note that while the speed and wavelength have different values in air and in the aqueous humor, the *frequency* in air, f_0, is the same as the frequency f in the aqueous humor:

$$f_0 = \frac{c}{\lambda_0} = \frac{3.00 \times 10^8 \text{ m/s}}{633 \times 10^{-9} \text{ m}} = 4.74 \times 10^{14} \text{ Hz}$$

When a light wave passes from one material into another, the wave speed and wavelength both change but the wave frequency is unchanged.

Example 33.3 A twice-reflected ray

Two mirrors are perpendicular to each other. A ray traveling in a plane perpendicular to both mirrors is reflected from one mirror, then the other, as shown in Fig. 33.12. What is the ray's final direction relative to its original direction?

SOLUTION

IDENTIFY and SET UP: This problem involves the law of reflection, which we must apply twice (once for each mirror).

EXECUTE: For mirror 1 the angle of incidence is θ_1, and this equals the angle of reflection. The sum of interior angles in the triangle shown in the figure is 180°, so we see that the angles of incidence and reflection for mirror 2 are both $90° - \theta_1$. The total change in direction of the ray after both reflections is therefore $2(90° - \theta_1) + 2\theta_1 = 180°$. That is, the ray's final direction is opposite to its original direction.

Continued

33.12 A ray moving in the xy-plane. The first reflection changes the sign of the y-component of its velocity, and the second reflection changes the sign of the x-component. For a different ray with a z-component of velocity, a third mirror (perpendicular to the two shown) could be used to change the sign of that component.

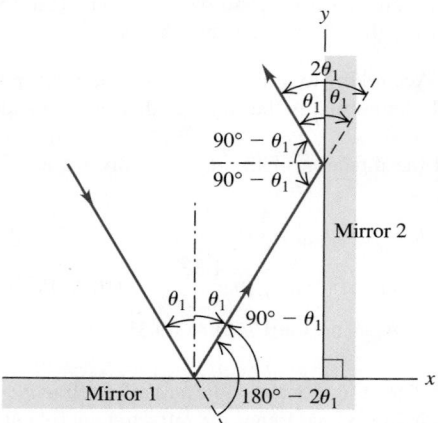

EVALUATE: An alternative viewpoint is that reflection reverses the sign of the component of light velocity perpendicular to the surface but leaves the other components unchanged. We invite you to verify this in detail. You should also be able to use this result to show that when a ray of light is successively reflected by three mirrors forming a corner of a cube (a "corner reflector"), its final direction is again opposite to its original direction. This principle is widely used in tail-light lenses and bicycle reflectors to improve their night-time visibility. Apollo astronauts placed arrays of corner reflectors on the moon. By use of laser beams reflected from these arrays, the earth–moon distance has been measured to within 0.15 m.

Test Your Understanding of Section 33.2 You are standing on the shore of a lake. You spot a tasty fish swimming some distance below the lake surface. (a) If you want to spear the fish, should you aim the spear (i) above, (ii) below, or (iii) directly at the apparent position of the fish? (b) If instead you use a high-power laser to simultaneously kill and cook the fish, should you aim the laser (i) above, (ii) below, or (iii) directly at the apparent position of the fish?

ActivPhysics 15.2: Total Internal Reflection

33.3 Total Internal Reflection

We have described how light is partially reflected and partially transmitted at an interface between two materials with different indexes of refraction. Under certain circumstances, however, *all* of the light can be reflected back from the interface, with none of it being transmitted, even though the second material is transparent. Figure 33.13a shows how this can occur. Several rays are shown radiating from a point source in material a with index of refraction n_a. The rays

33.13 (a) Total internal reflection. The angle of incidence for which the angle of refraction is 90° is called the critical angle: This is the case for ray 3. The reflected portions of rays 1, 2, and 3 are omitted for clarity. (b) Rays of laser light enter the water in the fishbowl from above; they are reflected at the bottom by mirrors tilted at slightly different angles. One ray undergoes total internal reflection at the air–water interface.

(a) Total internal reflection

Total internal reflection occurs only if $n_b < n_a$.

At the critical angle of incidence, θ_{crit}, the angle of refraction $\theta_b = 90°$.

Any ray with $\theta_a > \theta_{crit}$ shows total internal reflection.

(b) Total internal reflection demonstrated with a laser, mirrors, and water in a fishbowl

Incident laser beams
Refracted at interface
Total internal reflection
Two mirrors at different angles

strike the surface of a second material b with index n_b, where $n_a > n_b$. (Materials a and b could be water and air, respectively.) From Snell's law of refraction,

$$\sin\theta_b = \frac{n_a}{n_b}\sin\theta_a$$

Because n_a/n_b is greater than unity, $\sin\theta_b$ is larger than $\sin\theta_a$; the ray is bent *away from* the normal. Thus there must be some value of θ_a *less than* 90° for which $\sin\theta_b = 1$ and $\theta_b = 90°$. This is shown by ray 3 in the diagram, which emerges just grazing the surface at an angle of refraction of 90°. Compare Fig. 33.13a to the photograph of light rays in Fig. 33.13b.

The angle of incidence for which the refracted ray emerges tangent to the surface is called the **critical angle**, denoted by θ_{crit}. (A more detailed analysis using Maxwell's equations shows that as the incident angle approaches the critical angle, the transmitted intensity approaches zero.) If the angle of incidence is *larger* than the critical angle, the sine of the angle of refraction, as computed by Snell's law, would have to be greater than unity, which is impossible. Beyond the critical angle, the ray *cannot* pass into the upper material; it is trapped in the lower material and is completely reflected at the boundary surface. This situation, called **total internal reflection**, occurs only when a ray in material a is incident on a second material b whose index of refraction is *smaller* than that of material a (that is, $n_b < n_a$).

We can find the critical angle for two given materials a and b by setting $\theta_b = 90°$ ($\sin\theta_b = 1$) in Snell's law. We then have

$$\sin\theta_{crit} = \frac{n_b}{n_a} \qquad \text{(critical angle for total internal reflection)} \qquad (33.6)$$

Total internal reflection will occur if the angle of incidence θ_a is larger than or equal to θ_{crit}.

Applications of Total Internal Reflection

Total internal reflection finds numerous uses in optical technology. As an example, consider glass with index of refraction $n = 1.52$. If light propagating within this glass encounters a glass–air interface, the critical angle is

$$\sin\theta_{crit} = \frac{1}{1.52} = 0.658 \qquad \theta_{crit} = 41.1°$$

The light will be *totally reflected* if it strikes the glass–air surface at an angle of 41.1° or larger. Because the critical angle is slightly smaller than 45°, it is possible to use a prism with angles of 45°−45°−90° as a totally reflecting surface. As reflectors, totally reflecting prisms have some advantages over metallic surfaces such as ordinary coated-glass mirrors. While no metallic surface reflects 100% of the light incident on it, light can be *totally* reflected by a prism. These reflecting properties of a prism are permanent and unaffected by tarnishing.

A 45°−45°−90° prism, used as in Fig. 33.14a, is called a *Porro* prism. Light enters and leaves at right angles to the hypotenuse and is totally reflected at each of the shorter faces. The total change of direction of the rays is 180°. Binoculars often use combinations of two Porro prisms, as in Fig. 33.14b.

When a beam of light enters at one end of a transparent rod (Fig. 33.15), the light can be totally reflected internally if the index of refraction of the rod is greater than that of the surrounding material. The light is "trapped" within the rod even if the rod is curved, provided that the curvature is not too great. Such a

33.14 (a) Total internal reflection in a Porro prism. (b) A combination of two Porro prisms in binoculars.

(a) Total internal reflection in a Porro prism

If the incident beam is oriented as shown, total internal reflection occurs on the 45° faces (because, for a glass–air interface, $\theta_{crit} = 41.1$).

(b) Binoculars use Porro prisms to reflect the light to each eyepiece.

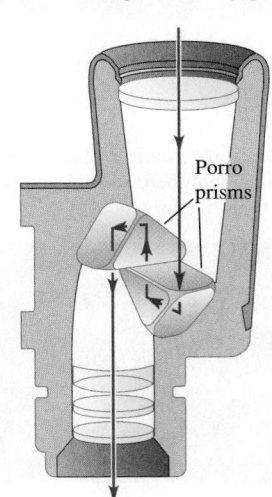

Porro prisms

33.15 A transparent rod with refractive index greater than that of the surrounding material.

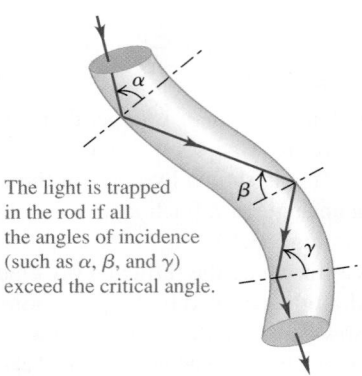

The light is trapped in the rod if all the angles of incidence (such as α, β, and γ) exceed the critical angle.

33.16 This colored x-ray image of a patient's abdomen shows an endoscope winding through the colon.

33.17 To maximize their brilliance, diamonds are cut so that there is total internal reflection on their back surfaces.

rod is sometimes called a *light pipe.* A bundle of fine glass or plastic fibers behaves in the same way and has the advantage of being flexible. A bundle may consist of thousands of individual fibers, each of the order of 0.002 to 0.01 mm in diameter. If the fibers are assembled in the bundle so that the relative positions of the ends are the same (or mirror images) at both ends, the bundle can transmit an image.

Fiber-optic devices have found a wide range of medical applications in instruments called *endoscopes,* which can be inserted directly into the bronchial tubes, the bladder, the colon, and other organs for direct visual examination (Fig. 33.16). A bundle of fibers can even be enclosed in a hypodermic needle for studying tissues and blood vessels far beneath the skin.

Fiber optics also have applications in communication systems, in which they are used to transmit a modulated laser beam. The rate at which information can be transmitted by a wave (light, radio, or whatever) is proportional to the frequency. To see qualitatively why this is so, consider modulating (modifying) the wave by chopping off some of the wave crests. Suppose each crest represents a binary digit, with a chopped-off crest representing a zero and an unmodified crest representing a one. The number of binary digits we can transmit per unit time is thus proportional to the frequency of the wave. Infrared and visible-light waves have much higher frequency than do radio waves, so a modulated laser beam can transmit an enormous amount of information through a single fiber-optic cable.

Another advantage of optical fibers is that they can be made thinner than conventional copper wire, so more fibers can be bundled together in a cable of a given diameter. Hence more distinct signals (for instance, different phone lines) can be sent over the same cable. Because fiber-optic cables are electrical insulators, they are immune to electrical interference from lightning and other sources, and they don't allow unwanted currents between source and receiver. For these and other reasons, fiber-optic cables play an important role in long-distance telephone, television, and Internet communication.

Total internal reflection also plays an important role in the design of jewelry. The brilliance of diamond is due in large measure to its very high index of refraction ($n = 2.417$) and correspondingly small critical angle. Light entering a cut diamond is totally internally reflected from facets on its back surface, and then emerges from its front surface (Fig. 33.17). "Imitation diamond" gems, such as cubic zirconia, are made from less expensive crystalline materials with comparable indexes of refraction.

Conceptual Example 33.4 **A leaky periscope**

A submarine periscope uses two totally reflecting 45°–45°–90° prisms with total internal reflection on the sides adjacent to the 45° angles. Explain why the periscope will no longer work if it springs a leak and the bottom prism is covered with water.

SOLUTION

The critical angle for water ($n_b = 1.33$) on glass ($n_a = 1.52$) is

$$\theta_{crit} = \arcsin\frac{1.33}{1.52} = 61.0°$$

The 45° angle of incidence for a totally reflecting prism is *smaller* than this new 61° critical angle, so total internal reflection does not occur at the glass–water interface. Most of the light is transmitted into the water, and very little is reflected back into the prism.

Test Your Understanding of Section 33.3 In which of the following situations is there total internal reflection? (i) Light propagating in water ($n = 1.33$) strikes a water–air interface at an incident angle of 70°; (ii) light propagating in glass ($n = 1.52$) strikes a glass–water interface at an incident angle of 70°; (iii) light propagating in water strikes a water–glass interface at an incident angle of 70°.

33.4 Dispersion

Ordinary white light is a superposition of waves with wavelengths extending throughout the visible spectrum. The speed of light *in vacuum* is the same for all wavelengths, but the speed in a material substance is different for different wavelengths. Therefore the index of refraction of a material depends on wavelength. The dependence of wave speed and index of refraction on wavelength is called **dispersion.**

Figure 33.18 shows the variation of index of refraction n with wavelength for some common optical materials. Note that the horizontal axis of this figure is the wavelength of the light *in vacuum,* λ_0; the wavelength in the material is given by Eq. (33.5), $\lambda = \lambda_0/n$. In most materials the value of n *decreases* with increasing wavelength and decreasing frequency, and thus n *increases* with decreasing wavelength and increasing frequency. In such a material, light of longer wavelength has greater speed than light of shorter wavelength.

Figure 33.19 shows a ray of white light incident on a prism. The deviation (change of direction) produced by the prism increases with increasing index of refraction and frequency and decreasing wavelength. Violet light is deviated most, and red is deviated least; other colors are in intermediate positions. When it comes out of the prism, the light is spread out into a fan-shaped beam, as shown. The light is said to be *dispersed* into a spectrum. The amount of dispersion depends on the *difference* between the refractive indexes for violet light and for red light. From Fig. 33.18 we can see that for a substance such as fluorite, the difference between the indexes for red and violet is small, and the dispersion will also be small. A better choice of material for a prism whose purpose is to produce a spectrum would be silicate flint glass, for which there is a larger difference in the value of n between red and violet.

33.18 Variation of index of refraction n with wavelength for different transparent materials. The horizontal axis shows the wavelength λ_0 of the light *in vacuum;* the wavelength in the material is equal to $\lambda = \lambda_0/n$.

33.19 Dispersion of light by a prism. The band of colors is called a spectrum.

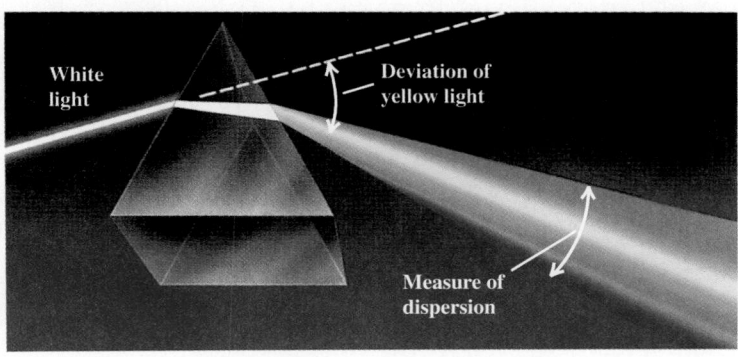

As we mentioned in Section 33.3, the brilliance of diamond is due in part to its unusually large refractive index; another important factor is its large dispersion, which causes white light entering a diamond to emerge as a multicolored spectrum. Crystals of rutile and of strontium titanate, which can be produced synthetically, have about eight times the dispersion of diamond.

Rainbows

When you experience the beauty of a rainbow, as in Fig. 33.20a, you are seeing the combined effects of dispersion, refraction, and reflection. Sunlight comes from behind you, enters a water droplet, is (partially) reflected from the back surface of the droplet, and is refracted again upon exiting the droplet (Fig. 33.20b). A light ray that enters the middle of the raindrop is reflected straight back. All other rays

33.20 How rainbows form.

(a) A double rainbow

(b) The paths of light rays entering the upper half of a raindrop

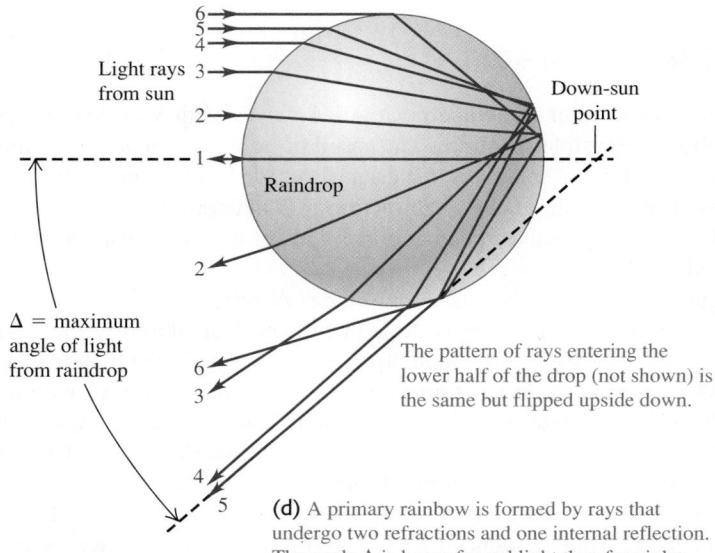

(c) Forming a rainbow. The sun in this illustration is directly behind the observer at P.

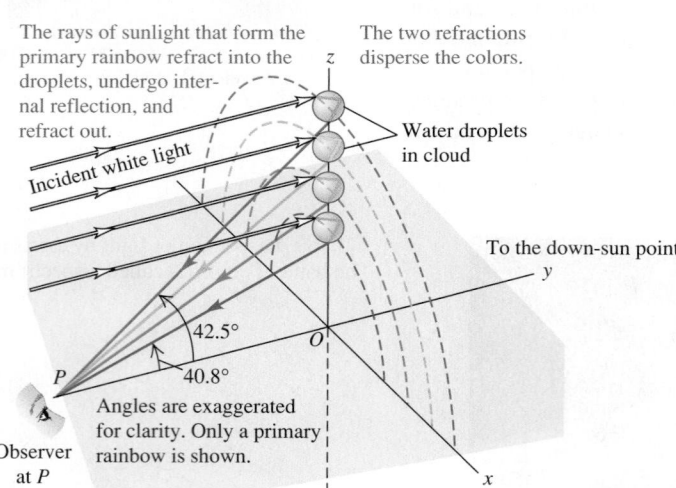

The rays of sunlight that form the primary rainbow refract into the droplets, undergo internal reflection, and refract out.

The two refractions disperse the colors.

Water droplets in cloud

Incident white light

To the down-sun point

42.5°

40.8°

Angles are exaggerated for clarity. Only a primary rainbow is shown.

Observer at P

(d) A primary rainbow is formed by rays that undergo two refractions and one internal reflection. The angle Δ is larger for red light than for violet.

Light from sun

$\Delta = 40.8°$ (violet) to 42.5° (red)

(e) A secondary rainbow is formed by rays that undergo two refractions and *two* internal reflections. The angle Δ is larger for violet light than for red.

Light from sun

$\Delta = 50.1°$ (red) to 53.2° (violet)

exit the raindrop within an angle Δ of that middle ray, with many rays "piling up" at the angle Δ. What you see is a disk of light of angular radius Δ centered on the down-sun point (the point in the sky opposite the sun); due to the "piling up" of light rays, the disk is brightest around its rim, which we see as a rainbow (Fig. 33.20c). Because no light reaches your eye from angles larger than Δ, the sky looks dark outside the rainbow (see Fig. 33.20a). The value of the angle Δ depends on the index of refraction of the water that makes up the raindrops, which in turn depends on the wavelength (Fig. 33.20d). The bright disk of red light is slightly larger than that for orange light, which in turn is slightly larger than that for yellow light, and so on. As a result, you see the rainbow as a band of colors.

In many cases you can see a second, larger rainbow. It is the result of dispersion, refraction, and *two* reflections from the back surface of the droplet (Fig. 33.20e). Each time a light ray hits the back surface, part of the light is refracted out of the drop (not shown in Fig. 33.20); after two such hits, relatively little light is left inside the drop, which is why the secondary rainbow is noticeably fainter than the primary rainbow. Just as a mirror held up to a book reverses the printed letters, so the second reflection reverses the sequence of colors in the secondary rainbow. You can see this effect in Fig. 33.20a.

33.5 Polarization

Polarization is a characteristic of all transverse waves. This chapter is about light, but to introduce some basic polarization concepts, let's go back to the transverse waves on a string that we studied in Chapter 15. For a string that in equilibrium lies along the x-axis, the displacements may be along the y-direction, as in Fig. 33.21a. In this case the string always lies in the xy-plane. But the displacements might instead be along the z-axis, as in Fig. 33.21b; then the string always lies in the xz-plane.

When a wave has only y-displacements, we say that it is **linearly polarized** in the y-direction; a wave with only z-displacements is linearly polarized in the z-direction. For mechanical waves we can build a **polarizing filter,** or **polarizer,** that permits only waves with a certain polarization direction to pass. In Fig. 33.21c the string can slide vertically in the slot without friction, but no horizontal motion is possible. This filter passes waves that are polarized in the y-direction but blocks those that are polarized in the z-direction.

This same language can be applied to electromagnetic waves, which also have polarization. As we learned in Chapter 32, an electromagnetic wave is a *transverse* wave; the fluctuating electric and magnetic fields are perpendicular to each other and to the direction of propagation. We always define the direction of polarization of an electromagnetic wave to be the direction of the *electric*-field vector $\vec{E}$, not the magnetic field, because many common electromagnetic-wave detectors respond to the electric forces on electrons in materials, not the magnetic forces. Thus the electromagnetic wave described by Eq. (32.17),

$$\vec{E}(x, t) = \hat{\jmath}E_{\max}\cos(kx - \omega t)$$
$$\vec{B}(x, t) = \hat{k}B_{\max}\cos(kx - \omega t)$$

Mastering**PHYSICS**

ActivPhysics 16.9: Physical Optics: Polarization

33.21 (a), (b) Polarized waves on a string. (c) Making a polarized wave on a string from an unpolarized one using a polarizing filter.

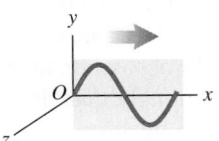

(a) Transverse wave linearly polarized in the y-direction

(b) Transverse wave linearly polarized in the z-direction

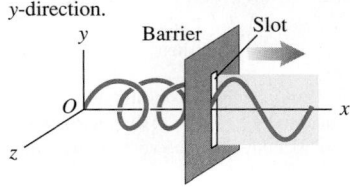

(c) The slot functions as a polarizing filter, passing only components polarized in the y-direction.

33.22 (a) Electrons in the red and white broadcast antenna oscillate vertically, producing vertically polarized electromagnetic waves that propagate away from the antenna in the horizontal direction. (The small gray antennas are for relaying cellular phone signals.) (b) No matter how this light bulb is oriented, the random motion of electrons in the filament produces unpolarized light waves.

(a)

(b)

is said to be polarized in the y-direction because the electric field has only a y-component.

> **CAUTION** **The meaning of "polarization"** It's unfortunate that the same word "polarization" that is used to describe the direction of $\vec{E}$ in an electromagnetic wave is also used to describe the shifting of electric charge within a body, such as in response to a nearby charged body; we described this latter kind of polarization in Section 21.2 (see Fig. 21.7). You should remember that while these two concepts have the same name, they do not describe the same phenomenon. ▮

Polarizing Filters

Waves emitted by a radio transmitter are usually linearly polarized. The vertical antennas that are used for radio broadcasting emit waves that, in a horizontal plane around the antenna, are polarized in the vertical direction (parallel to the antenna) (Fig. 33.22a).

The situation is different for visible light. Light from incandescent light bulbs and fluorescent light fixtures is *not* polarized (Fig. 33.22b). The "antennas" that radiate light waves are the molecules that make up the sources. The waves emitted by any one molecule may be linearly polarized, like those from a radio antenna. But any actual light source contains a tremendous number of molecules with random orientations, so the emitted light is a random mixture of waves linearly polarized in all possible transverse directions. Such light is called **unpolarized light** or **natural light.** To create polarized light from unpolarized natural light requires a filter that is analogous to the slot for mechanical waves in Fig. 33.21c.

Polarizing filters for electromagnetic waves have different details of construction, depending on the wavelength. For microwaves with a wavelength of a few centimeters, a good polarizer is an array of closely spaced, parallel conducting wires that are insulated from each other. (Think of a barbecue grill with the outer metal ring replaced by an insulating one.) Electrons are free to move along the length of the conducting wires and will do so in response to a wave whose $\vec{E}$ field is parallel to the wires. The resulting currents in the wires dissipate energy by I^2R heating; the dissipated energy comes from the wave, so whatever wave passes through the grid is greatly reduced in amplitude. Waves with $\vec{E}$ oriented perpendicular to the wires pass through almost unaffected, since electrons cannot move through the air between the wires. Hence a wave that passes through such a filter will be predominantly polarized in the direction perpendicular to the wires.

The most common polarizing filter for visible light is a material known by the trade name Polaroid, widely used for sunglasses and polarizing filters for camera lenses. Developed originally by the American scientist Edwin H. Land, this material incorporates substances that have **dichroism,** a selective absorption in which one of the polarized components is absorbed much more strongly than the other (Fig. 33.23). A Polaroid filter transmits 80% or more of the intensity of a wave that is polarized parallel to a certain axis in the material, called the **polarizing**

33.23 A Polaroid filter is illuminated by unpolarized natural light (shown by $\vec{E}$ vectors that point in all directions perpendicular to the direction of propagation). The transmitted light is linearly polarized along the polarizing axis (shown by $\vec{E}$ vectors along the polarization direction only).

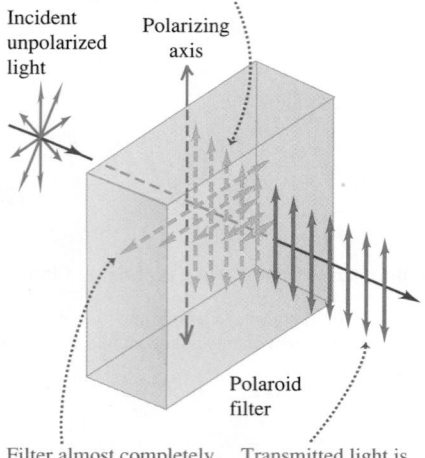

Filter only partially absorbs vertically polarized component of light.

Incident unpolarized light

Polarizing axis

Polaroid filter

Filter almost completely absorbs horizontally polarized component of light.

Transmitted light is linearly polarized in the vertical direction.

axis, but only 1% or less for waves that are polarized perpendicular to this axis. In one type of Polaroid filter, long-chain molecules within the filter are oriented with their axis perpendicular to the polarizing axis; these molecules preferentially absorb light that is polarized along their length, much like the conducting wires in a polarizing filter for microwaves.

Using Polarizing Filters

An *ideal* polarizing filter (polarizer) passes 100% of the incident light that is polarized parallel to the filter's polarizing axis but completely blocks all light that is polarized perpendicular to this axis. Such a device is an unattainable idealization, but the concept is useful in clarifying the basic ideas. In the following discussion we will assume that all polarizing filters are ideal. In Fig. 33.24 unpolarized light is incident on a flat polarizing filter. The $\vec{E}$ vector of the incident wave can be represented in terms of components parallel and perpendicular to the polarizer axis (shown in blue); only the component of $\vec{E}$ parallel to the polarizing axis is transmitted. Hence the light emerging from the polarizer is linearly polarized parallel to the polarizing axis.

When unpolarized light is incident on an ideal polarizer as in Fig. 33.24, the intensity of the transmitted light is *exactly half* that of the incident unpolarized light, no matter how the polarizing axis is oriented. Here's why: We can resolve the $\vec{E}$ field of the incident wave into a component parallel to the polarizing axis and a component perpendicular to it. Because the incident light is a random mixture of all states of polarization, these two components are, on average, equal. The ideal polarizer transmits only the component that is parallel to the polarizing axis, so half the incident intensity is transmitted.

What happens when the linearly polarized light emerging from a polarizer passes through a second polarizer, or *analyzer*, as in Fig. 33.25? Suppose the polarizing axis of the analyzer makes an angle ϕ with the polarizing axis of the

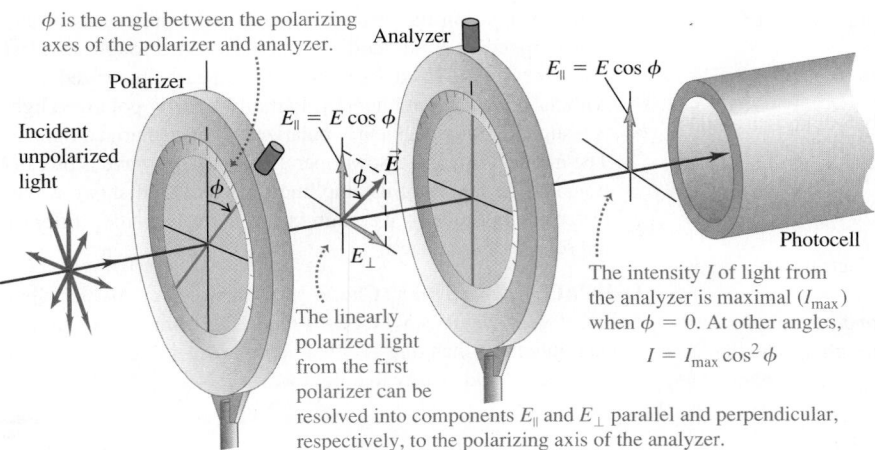

33.26 These photos show the view through Polaroid sunglasses whose polarizing axes are (left) aligned ($\phi = 0$) and (right) perpendicular ($\phi = 90°$). The transmitted intensity is greatest when the axes are aligned; it is zero when the axes are perpendicular.

first polarizer. We can resolve the linearly polarized light that is transmitted by the first polarizer into two components, as shown in Fig. 33.25, one parallel and the other perpendicular to the axis of the analyzer. Only the parallel component, with amplitude $E\cos\phi$, is transmitted by the analyzer. The transmitted intensity is greatest when $\phi = 0$, and it is zero when the polarizer and analyzer are *crossed* so that $\phi = 90°$ (Fig. 33.26). To determine the direction of polarization of the light transmitted by the first polarizer, rotate the analyzer until the photocell in Fig. 33.25 measures zero intensity; the polarization axis of the first polarizer is then perpendicular to that of the analyzer.

To find the transmitted intensity at intermediate values of the angle ϕ, we recall from our energy discussion in Section 32.4 that the intensity of an electromagnetic wave is proportional to the *square* of the amplitude of the wave [see Eq. (32.29)]. The ratio of transmitted to incident *amplitude* is $\cos\phi$, so the ratio of transmitted to incident *intensity* is $\cos^2\phi$. Thus the intensity of the light transmitted through the analyzer is

$$I = I_{max}\cos^2\phi \qquad \begin{array}{l}\text{(Malus's law, polarized light}\\ \text{passing through an analyzer)}\end{array} \qquad (33.7)$$

where I_{max} is the maximum intensity of light transmitted (at $\phi = 0$) and I is the amount transmitted at angle ϕ. This relationship, discovered experimentally by Etienne Louis Malus in 1809, is called **Malus's law.** Malus's law applies *only* if the incident light passing through the analyzer is already linearly polarized.

Problem-Solving Strategy 33.2   Linear Polarization

IDENTIFY *the relevant concepts:* In all electromagnetic waves, including light waves, the direction of polarization is the direction of the $\vec{E}$ field and is perpendicular to the propagation direction. Problems about polarizers are therefore about the components of $\vec{E}$ parallel and perpendicular to the polarizing axis.

SET UP *the problem* using the following steps:
1. Start by drawing a large, neat diagram. Label all known angles, including the angles of all polarizing axes.
2. Identify the target variables.

EXECUTE *the solution* as follows:
1. Remember that a polarizer lets pass only electric-field components parallel to its polarizing axis.
2. If the incident light is linearly polarized and has amplitude E and intensity I_{max}, the light that passes through an ideal polarizer has amplitude $E\cos\phi$ and intensity $I_{max}\cos^2\phi$, where ϕ is

the angle between the incident polarization direction and the filter's polarizing axis.
3. Unpolarized light is a random mixture of all possible polarization states, so on the average it has equal components in any two perpendicular directions. When passed through an ideal polarizer, unpolarized light becomes linearly polarized light with half the incident intensity. Partially linearly polarized light is a superposition of linearly polarized and unpolarized light.
4. The intensity (average power per unit area) of a wave is proportional to the *square* of its amplitude. If you find that two waves differ in amplitude by a certain factor, their intensities differ by the square of that factor.

EVALUATE *your answer:* Check your answer for any obvious errors. If your results say that light emerging from a polarizer has greater intensity than the incident light, something's wrong: A polarizer can't add energy to a light wave.

Example 33.5 **Two polarizers in combination**

In Fig. 33.25 the incident unpolarized light has intensity I_0. Find the intensities transmitted by the first and second polarizers if the angle between the axes of the two filters is 30°.

SOLUTION

IDENTIFY and SET UP: This problem involves a polarizer (a polarizing filter on which unpolarized light shines, producing polarized light) and an analyzer (a second polarizing filter on which the polarized light shines). We are given the intensity I_0 of the incident light and the angle $\phi = 30°$ between the axes of the polarizers. We use Malus's law, Eq. (33.7), to solve for the intensities of the light emerging from each polarizer.

EXECUTE: The incident light is unpolarized, so the intensity of the linearly polarized light transmitted by the first polarizer is $I_0/2$. From Eq. (33.7) with $\phi = 30°$, the second polarizer reduces the intensity by a further factor of $\cos^2 30° = \frac{3}{4}$. Thus the intensity transmitted by the second polarizer is

$$\left(\frac{I_0}{2}\right)\left(\tfrac{3}{4}\right) = \tfrac{3}{8}I_0$$

EVALUATE: Note that the intensity decreases after each passage through a polarizer. The only situation in which the transmitted intensity does *not* decrease is if the polarizer is ideal (so it absorbs none of the light that passes through it) and if the incident light is linearly polarized along the polarizing axis, so $\phi = 0$.

33.27 When light is incident on a reflecting surface at the polarizing angle, the reflected light is linearly polarized.

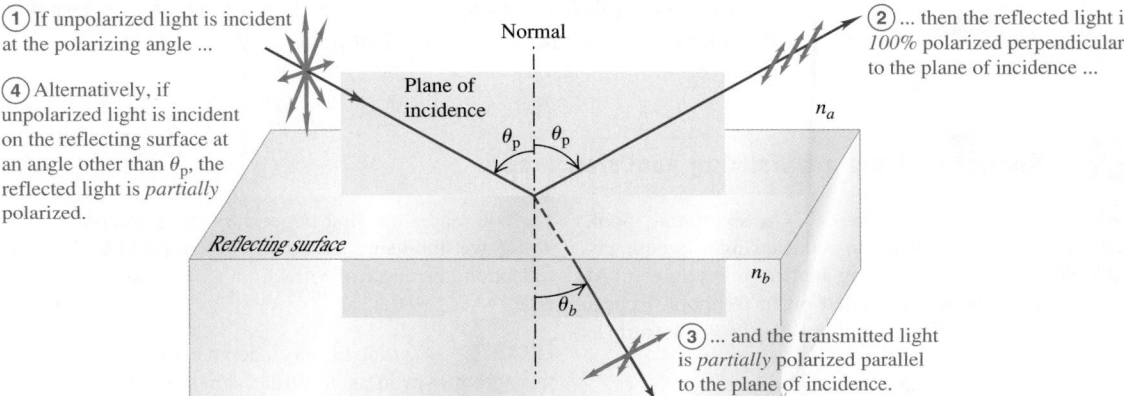

① If unpolarized light is incident at the polarizing angle ...

② ... then the reflected light is *100%* polarized perpendicular to the plane of incidence ...

④ Alternatively, if unpolarized light is incident on the reflecting surface at an angle other than θ_p, the reflected light is *partially* polarized.

Normal

Plane of incidence

θ_p θ_p

n_a

Reflecting surface

n_b

θ_b

③ ... and the transmitted light is *partially* polarized parallel to the plane of incidence.

Polarization by Reflection

Unpolarized light can be polarized, either partially or totally, by *reflection*. In Fig. 33.27, unpolarized natural light is incident on a reflecting surface between two transparent optical materials. For most angles of incidence, waves for which the electric-field vector $\vec{E}$ is perpendicular to the plane of incidence (that is, parallel to the reflecting surface) are reflected more strongly than those for which $\vec{E}$ lies in this plane. In this case the reflected light is *partially polarized* in the direction perpendicular to the plane of incidence.

But at one particular angle of incidence, called the **polarizing angle** θ_p, the light for which $\vec{E}$ lies in the plane of incidence is *not reflected at all* but is completely refracted. At this same angle of incidence the light for which $\vec{E}$ is perpendicular to the plane of incidence is partially reflected and partially refracted. The *reflected* light is therefore *completely* polarized perpendicular to the plane of incidence, as shown in Fig. 33.27. The *refracted* (transmitted) light is *partially* polarized parallel to this plane; the refracted light is a mixture of the component parallel to the plane of incidence, all of which is refracted, and the remainder of the perpendicular component.

In 1812 the British scientist Sir David Brewster discovered that when the angle of incidence is equal to the polarizing angle θ_p, the reflected ray and the refracted ray are perpendicular to each other (Fig. 33.28). In this case the angle of refraction θ_b equals $90° - \theta_p$. From the law of refraction,

$$n_a \sin\theta_p = n_b \sin\theta_b$$

33.28 The significance of the polarizing angle. The open circles represent a component of $\vec{E}$ that is perpendicular to the plane of the figure (the plane of incidence) and parallel to the surface between the two materials.

Note: This is a side view of the situation shown in Fig. 33.27.

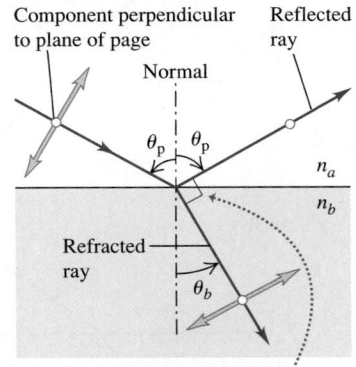

Component perpendicular to plane of page

Reflected ray

Normal

θ_p θ_p

n_a

n_b

Refracted ray

θ_b

When light strikes a surface at the polarizing angle, the reflected and refracted rays are perpendicular to each other and

$$\tan\theta_p = \frac{n_b}{n_a}$$

so we find

$$n_a \sin\theta_p = n_b \sin(90° - \theta_p) = n_b \cos\theta_p$$

$$\tan\theta_p = \frac{n_b}{n_a} \quad \text{(Brewster's law for the polarizing angle)} \quad (33.8)$$

This relationship is known as **Brewster's law.** Although discovered experimentally, it can also be *derived* from a wave model using Maxwell's equations.

Polarization by reflection is the reason polarizing filters are widely used in sunglasses (Fig. 33.26). When sunlight is reflected from a horizontal surface, the plane of incidence is vertical, and the reflected light contains a preponderance of light that is polarized in the horizontal direction. When the reflection occurs at a smooth asphalt road surface or the surface of a lake, it causes unwanted glare. Vision can be improved by eliminating this glare. The manufacturer makes the polarizing axis of the lens material vertical, so very little of the horizontally polarized light reflected from the road is transmitted to the eyes. The glasses also reduce the overall intensity of the transmitted light to somewhat less than 50% of the intensity of the unpolarized incident light.

Example 33.6 Reflection from a swimming pool's surface

Sunlight reflects off the smooth surface of a swimming pool. (a) For what angle of reflection is the reflected light completely polarized? (b) What is the corresponding angle of refraction? (c) At night, an underwater floodlight is turned on in the pool. Repeat parts (a) and (b) for rays from the floodlight that strike the surface from below.

SOLUTION

IDENTIFY and SET UP: This problem involves polarization by reflection at an air–water interface in parts (a) and (b) and at a water–air interface in part (c). Figure 33.29 shows our sketches.

33.29 Our sketches for this problem.

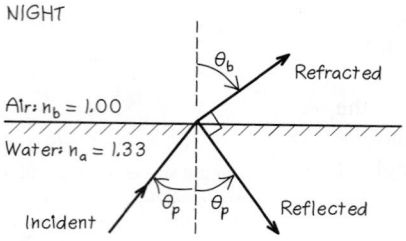

For both cases our first target variable is the polarizing angle θ_p, which we find using Brewster's law, Eq. (33.8). For this angle of reflection, the angle of refraction θ_b is the complement of θ_p (that is, $\theta_b = 90° - \theta_p$).

EXECUTE: (a) During the day (shown in the upper part of Fig. 33.29) the light moves in air toward water, so $n_a = 1.00$ (air) and $n_b = 1.33$ (water). From Eq. (33.8),

$$\theta_p = \arctan\frac{n_b}{n_a} = \arctan\frac{1.33}{1.00} = 53.1°$$

(b) The incident light is at the polarizing angle, so the reflected and refracted rays are perpendicular; hence

$$\theta_b = 90° - \theta_p = 90° - 53.1° = 36.9°$$

(c) At night (shown in the lower part of Fig. 33.29) the light moves in water toward air, so now $n_a = 1.33$ and $n_b = 1.00$. Again using Eq. (33.8), we have

$$\theta_p = \arctan\frac{1.00}{1.33} = 36.9°$$

$$\theta_b = 90° - 36.9° = 53.1°$$

EVALUATE: We check our answer in part (b) by using Snell's law, $n_a \sin\theta_a = n_b \sin\theta_b$, to solve for θ_b:

$$\sin\theta_b = \frac{n_a \sin\theta_p}{n_b} = \frac{1.00 \sin 53.1°}{1.33} = 0.600$$

$$\theta_b = \arcsin(0.600) = 36.9°$$

Note that the two polarizing angles found in parts (a) and (c) add to 90°. This is *not* an accident; can you see why?

Circular and Elliptical Polarization

Light and other electromagnetic radiation can also have *circular* or *elliptical* polarization. To introduce these concepts, let's return once more to mechanical waves on a stretched string. In Fig. 33.21, suppose the two linearly polarized waves in parts (a) and (b) are in phase and have equal amplitude. When they are superposed, each point in the string has simultaneous y- and z-displacements of equal magnitude. A little thought shows that the resultant wave lies in a plane oriented at 45° to the y- and z-axes (i.e., in a plane making a 45° angle with the xy- and xz-planes). The amplitude of the resultant wave is larger by a factor of $\sqrt{2}$ than that of either component wave, and the resultant wave is linearly polarized.

But now suppose the two equal-amplitude waves differ in phase by a quarter-cycle. Then the resultant motion of each point corresponds to a superposition of two simple harmonic motions at right angles, with a quarter-cycle phase difference. The y-displacement at a point is greatest at times when the z-displacement is zero, and vice versa. The motion of the string as a whole then no longer takes place in a single plane. It can be shown that each point on the rope moves in a *circle* in a plane parallel to the yz-plane. Successive points on the rope have successive phase differences, and the overall motion of the string has the appearance of a rotating helix. This is shown to the left of the polarizing filter in Fig. 33.21c. This particular superposition of two linearly polarized waves is called **circular polarization.**

Figure 33.30 shows the analogous situation for an electromagnetic wave. Two sinusoidal waves of equal amplitude, polarized in the y- and z-directions and with a quarter-cycle phase difference, are superposed. The result is a wave in which the $\vec{E}$ vector at each point has a constant magnitude but *rotates* around the direction of propagation. The wave in Fig. 33.30 is propagating toward you and the $\vec{E}$ vector appears to be rotating clockwise, so it is called a *right circularly polarized* electromagnetic wave. If instead the $\vec{E}$ vector of a wave coming toward you appears to be rotating counterclockwise, it is called a *left circularly polarized* electromagnetic wave.

If the phase difference between the two component waves is something other than a quarter-cycle, or if the two component waves have different amplitudes, then each point on the string traces out not a circle but an *ellipse*. The resulting wave is said to be **elliptically polarized.**

For electromagnetic waves with radio frequencies, circular or elliptical polarization can be produced by using two antennas at right angles, fed from the same

33.30 Circular polarization of an electromagnetic wave moving toward you parallel to the x-axis. The y-component of $\vec{E}$ lags the z-component by a quarter-cycle. This phase difference results in right circular polarization.

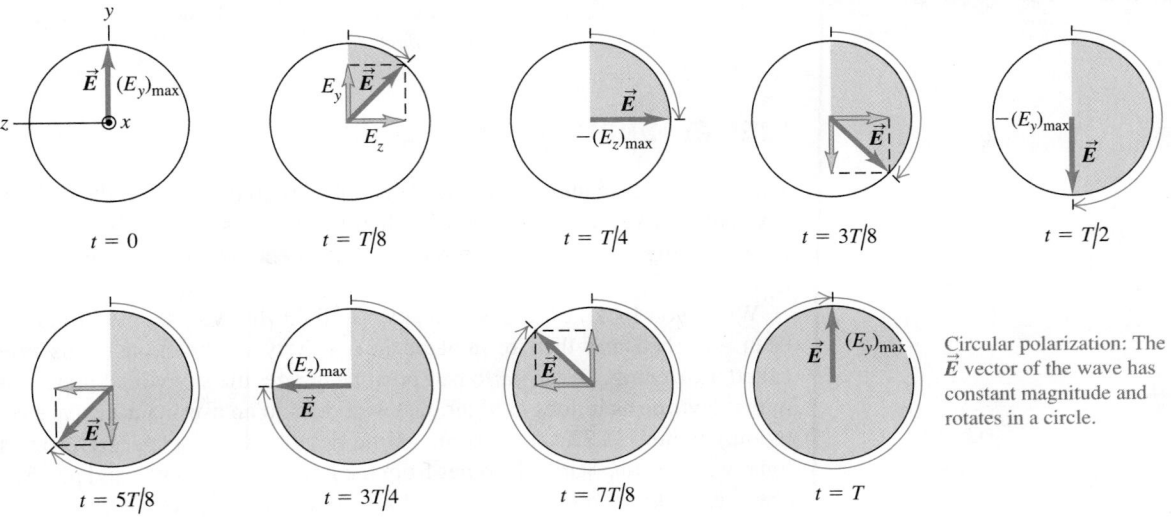

Circular polarization: The $\vec{E}$ vector of the wave has constant magnitude and rotates in a circle.

transmitter but with a phase-shifting network that introduces the appropriate phase difference. For light, the phase shift can be introduced by use of a material that exhibits *birefringence*—that is, has different indexes of refraction for different directions of polarization. A common example is calcite (CaCO$_3$). When a calcite crystal is oriented appropriately in a beam of unpolarized light, its refractive index (for a wavelength in vacuum of 589 nm) is 1.658 for one direction of polarization and 1.486 for the perpendicular direction. When two waves with equal amplitude and with perpendicular directions of polarization enter such a material, they travel with different speeds. If they are in phase when they enter the material, then in general they are no longer in phase when they emerge. If the crystal is just thick enough to introduce a quarter-cycle phase difference, then the crystal converts linearly polarized light to circularly polarized light. Such a crystal is called a *quarter-wave plate.* Such a plate also converts circularly polarized light to linearly polarized light. Can you prove this?

Photoelasticity

Some optical materials that are not normally birefringent become so when they are subjected to mechanical stress. This is the basis of the science of *photoelasticity.* Stresses in girders, boiler plates, gear teeth, and cathedral pillars can be analyzed by constructing a transparent model of the object, usually of a plastic material, subjecting it to stress, and examining it between a polarizer and an analyzer in the crossed position. Very complicated stress distributions can be studied by these optical methods.

Figure 33.31 is a photograph of a photoelastic model under stress. The polarized light that enters the model can be thought of as having a component along each of the two directions of the birefringent plastic. Since these two components travel through the plastic at different speeds, the light that emerges from the other side of the model can have a different overall direction of polarization. Hence some of this transmitted light will be able to pass through the analyzer even though its polarization axis is at a 90° angle to the polarizer's axis, and the stressed areas in the plastic will appear as bright spots. The amount of birefringence is different for different wavelengths and hence different colors of light; the color that appears at each location in Fig. 33.31 is that for which the transmitted light is most nearly polarized along the analyzer's polarization axis.

33.31 This plastic model of an artificial hip joint was photographed between two polarizing filters (a polarizer and an analyzer) with perpendicular polarizing axes. The colored interference pattern reveals the direction and magnitude of stresses in the model. Engineers use these results to help design the metal artificial hip joint used in medicine.

Test Your Understanding of Section 33.5 You are taking a photograph of a sunlit high-rise office building. In order to minimize the reflections from the building's windows, you place a polarizing filter on the camera lens. How should you orient the filter? (i) with the polarizing axis vertical; (ii) with the polarizing axis horizontal; (iii) either orientation will minimize the reflections just as well; (iv) neither orientation will have any effect.

33.6 Scattering of Light

The sky is blue. Sunsets are red. Skylight is partially polarized; that's why the sky looks darker from some angles than from others when it is viewed through Polaroid sunglasses. As we will see, a single phenomenon is responsible for all of these effects.

When you look at the daytime sky, the light that you see is sunlight that has been absorbed and then re-radiated in a variety of directions. This process is called **scattering.** (If the earth had no atmosphere, the sky would appear as black in the daytime as it does at night, just as it does to an astronaut in space or on the moon.) Figure 33.32 shows some of the details of the scattering process. Sunlight, which is unpolarized, comes from the left along the *x*-axis and passes over an observer looking vertically upward along the *y*-axis. (We are viewing the situation

33.32 When the sunbathing observer on the left looks up, he sees blue, polarized sunlight that has been scattered by air molecules. The observer on the right sees reddened, unpolarized light when he looks at the sun.

Incident white light, y unpolarized

Electric charges in air molecules at O oscillate in the direction of the $\vec{E}$ field of the incident light from the sun, acting as antennas that produce scattered light. The scattered light that reaches the observer directly below O is polarized in the z-direction.

Air molecules scatter blue light more effectively than red light; we see the sky overhead by scattered light, so it looks blue.

This observer sees reddened sunlight because most of the blue light has been scattered out.

from the side.) Consider the molecules of the earth's atmosphere located at point O. The electric field in the beam of sunlight sets the electric charges in these molecules into vibration. Since light is a transverse wave, the direction of the electric field in any component of the sunlight lies in the yz-plane, and the motion of the charges takes place in this plane. There is no field, and hence no motion of charges, in the direction of the x-axis.

An incident light wave sets the electric charges in the molecules at point O vibrating along the line of $\vec{E}$. We can resolve this vibration into two components, one along the y-axis and the other along the z-axis. Each component in the incident light produces the equivalent of two molecular "antennas," oscillating with the same frequency as the incident light and lying along the y- and z-axes.

We mentioned in Chapter 32 that an oscillating charge, like those in an antenna, does not radiate in the direction of its oscillation. (See Fig. 32.3 in Section 32.1.) Thus the "antenna" along the y-axis does not send any light to the observer directly below it, although it does emit light in other directions. Therefore the only light reaching this observer comes from the other molecular "antenna," corresponding to the oscillation of charge along the z-axis. This light is linearly polarized, with its electric field along the z-axis (parallel to the "antenna"). The red vectors on the y-axis below point O in Fig. 33.32 show the direction of polarization of the light reaching the observer.

As the original beam of sunlight passes though the atmosphere, its intensity decreases as its energy goes into the scattered light. Detailed analysis of the scattering process shows that the intensity of the light scattered from air molecules increases in proportion to the fourth power of the frequency (inversely to the fourth power of the wavelength). Thus the intensity ratio for the two ends of the visible spectrum is $(750 \text{ nm}/380 \text{ nm})^4 = 15$. Roughly speaking, scattered light contains 15 times as much blue light as red, and that's why the sky is blue.

Clouds contain a high concentration of water droplets or ice crystals, which also scatter light. Because of this high concentration, light passing through the cloud has many more opportunities for scattering than does light passing through a clear sky. Thus light of *all* wavelengths is eventually scattered out of the cloud, so the cloud looks white (Fig. 33.33). Milk looks white for the same reason; the scattering is due to fat globules in the milk.

Near sunset, when sunlight has to travel a long distance through the earth's atmosphere, a substantial fraction of the blue light is removed by scattering. White light minus blue light looks yellow or red. This explains the yellow or red hue that we so often see from the setting sun (and that is seen by the observer at the far right of Fig. 33.32).

Application Bee Vision and Polarized Light from the Sky

The eyes of a bee can detect the polarization of light. Bees use this to help them navigate between the hive and food sources. As Fig. 33.32 shows, a bee sees unpolarized light if it looks directly toward the sun and sees completely polarized light if it looks 90° away from the sun. These polarizations are unaffected by the presence of clouds, so a bee can navigate relative to the sun even on an overcast day.

Eyes

33.33 Clouds are white because they efficiently scatter sunlight of all wavelengths.

33.7 Huygens's Principle

The laws of reflection and refraction of light rays that we introduced in Section 33.2 were discovered experimentally long before the wave nature of light was firmly established. However, we can *derive* these laws from wave considerations and show that they are consistent with the wave nature of light.

We begin with a principle called **Huygens's principle.** This principle, stated originally by the Dutch scientist Christiaan Huygens in 1678, is a geometrical method for finding, from the known shape of a wave front at some instant, the shape of the wave front at some later time. Huygens assumed that **every point of a wave front may be considered the source of secondary wavelets that spread out in all directions with a speed equal to the speed of propagation of the wave.** The new wave front at a later time is then found by constructing a surface *tangent* to the secondary wavelets or, as it is called, the *envelope* of the wavelets. All the results that we obtain from Huygens's principle can also be obtained from Maxwell's equations. Thus it is not an independent principle, but it is often very convenient for calculations with wave phenomena.

Huygens's principle is shown in Fig. 33.34. The original wave front AA' is traveling outward from a source, as indicated by the arrows. We want to find the shape of the wave front after a time interval t. We assume that v, the speed of propagation of the wave, is the same at all points. Then in time t the wave front travels a distance vt. We construct several circles (traces of spherical wavelets) with radius $r = vt$, centered at points along AA'. The trace of the envelope of these wavelets, which is the new wave front, is the curve BB'.

Reflection and Huygens's Principle

To derive the law of reflection from Huygens's principle, we consider a plane wave approaching a plane reflecting surface. In Fig. 33.35a the lines AA', OB', and NC' represent successive positions of a wave front approaching the surface MM'. Point A on the wave front AA' has just arrived at the reflecting surface. We can use Huygens's principle to find the position of the wave front after a time interval t. With points on AA' as centers, we draw several secondary wavelets with radius vt. The wavelets that originate near the upper end of AA' spread out unhindered, and their envelope gives the portion OB' of the new wave front. If the reflecting surface were not there, the wavelets originating near the lower end of AA' would similarly reach the positions shown by the broken circular arcs. Instead, these wavelets strike the reflecting surface.

The effect of the reflecting surface is to *change the direction* of travel of those wavelets that strike it, so the part of a wavelet that would have penetrated the surface actually lies to the left of it, as shown by the full lines. The first such wavelet is centered at point A; the envelope of all such reflected wavelets is the portion OB of the wave front. The trace of the entire wave front at this instant is the bent line BOB'. A similar construction gives the line CNC' for the wave front after another interval t.

From plane geometry the angle θ_a between the incident *wave front* and the *surface* is the same as that between the incident *ray* and the *normal* to the surface and is therefore the angle of incidence. Similarly, θ_r is the angle of reflection. To find the relationship between these angles, we consider Fig. 33.35b. From O we draw $OP = vt$, perpendicular to AA'. Now OB, by construction, is tangent to a circle of radius vt with center at A. If we draw AQ from A to the point of tangency, the triangles APO and OQA are congruent because they are right triangles with the side AO in common and with $AQ = OP = vt$. The angle θ_a therefore equals the angle θ_r, and we have the law of reflection.

Refraction and Huygens's Principle

We can derive the law of *refraction* by a similar procedure. In Fig. 33.36a we consider a wave front, represented by line AA', for which point A has just

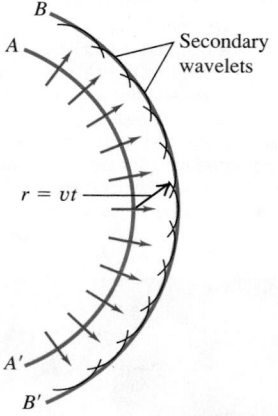

33.34 Applying Huygens's principle to wave front AA' to construct a new wave front BB'.

33.35 Using Huygens's principle to derive the law of reflection.

(a) Successive positions of a plane wave AA' as it is reflected from a plane surface

(b) Magnified portion of **(a)**

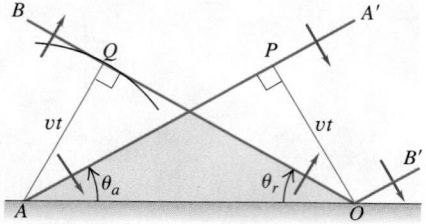

arrived at the boundary surface SS' between two transparent materials a and b, with indexes of refraction n_a and n_b and wave speeds v_a and v_b. (The *reflected* waves are not shown in the figure; they proceed as in Fig. 33.35.) We can apply Huygens's principle to find the position of the refracted wave fronts after a time t.

With points on AA' as centers, we draw several secondary wavelets. Those originating near the upper end of AA' travel with speed v_a and, after a time interval t, are spherical surfaces of radius v_at. The wavelet originating at point A, however, is traveling in the second material b with speed v_b and at time t is a spherical surface of radius v_bt. The envelope of the wavelets from the original wave front is the plane whose trace is the bent line BOB'. A similar construction leads to the trace CPC' after a second interval t.

The angles θ_a and θ_b between the surface and the incident and refracted wave fronts are the angle of incidence and the angle of refraction, respectively. To find the relationship between these angles, refer to Fig. 33.36b. We draw $OQ = v_at$, perpendicular to AQ, and we draw $AB = v_bt$, perpendicular to BO. From the right triangle AOQ,

$$\sin\theta_a = \frac{v_at}{AO}$$

and from the right triangle AOB,

$$\sin\theta_b = \frac{v_bt}{AO}$$

Combining these, we find

$$\frac{\sin\theta_a}{\sin\theta_b} = \frac{v_a}{v_b} \qquad (33.9)$$

We have defined the index of refraction n of a material as the ratio of the speed of light c in vacuum to its speed v in the material: $n_a = c/v_a$ and $n_b = c/v_b$. Thus

$$\frac{n_b}{n_a} = \frac{c/v_b}{c/v_a} = \frac{v_a}{v_b}$$

and we can rewrite Eq. (33.9) as

$$\frac{\sin\theta_a}{\sin\theta_b} = \frac{n_b}{n_a} \quad \text{or}$$

$$n_a\sin\theta_a = n_b\sin\theta_b$$

which we recognize as Snell's law, Eq. (33.4). So we have derived Snell's law from a wave theory. Alternatively, we may choose to regard Snell's law as an experimental result that defines the index of refraction of a material; in that case this analysis helps to confirm the relationship $v = c/n$ for the speed of light in a material.

Mirages offer an interesting example of Huygens's principle in action. When the surface of pavement or desert sand is heated intensely by the sun, a hot, less dense, smaller-n layer of air forms near the surface. The speed of light is slightly greater in the hotter air near the ground, the Huygens wavelets have slightly larger radii, the wave fronts tilt slightly, and rays that were headed toward the surface with a large angle of incidence (near 90°) can be bent up as shown in Fig. 33.37. Light farther from the ground is bent less and travels nearly in a straight line. The observer sees the object in its natural position, with an inverted image below it, as though seen in a horizontal reflecting surface. The mind of a thirsty traveler can interpret the apparent reflecting surface as a sheet of water.

33.36 Using Huygens's principle to derive the law of refraction. The case $v_b < v_a$ ($n_b > n_a$) is shown.

(a) Successive positions of a plane wave AA' as it is refracted by a plane surface

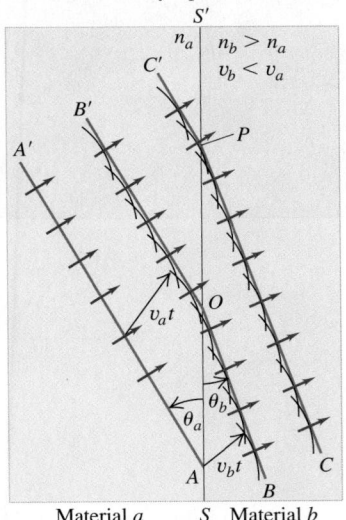

(b) Magnified portion of **(a)**

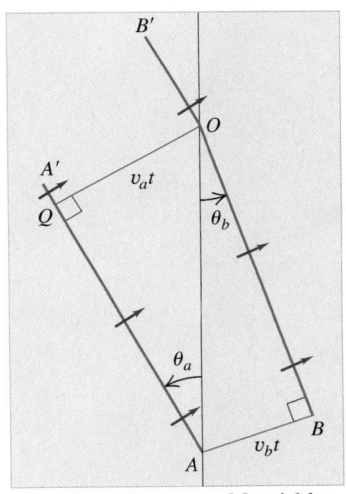

33.37 How mirages are formed.

Higher up, the wave fronts travel approximately straight.

vt

Hot surface

vt

The hot air near ground has a smaller n than cooler air higher up, so light travels fastest near the ground. Thus, the secondary wavelets nearest the ground have the largest radii vt, and the wave fronts tilt as they travel.

It is important to keep in mind that Maxwell's equations are the fundamental relationships for electromagnetic wave propagation. But Huygens's principle provides a convenient way to visualize this propagation.

Test Your Understanding of Section 33.7 Sound travels faster in warm air than in cold air. Imagine a weather front that runs north-south, with warm air to the west of the front and cold air to the east. A sound wave traveling in a northeast direction in the warm air encounters this front. How will the direction of this sound wave change when it passes into the cold air? (i) The wave direction will deflect toward the north; (ii) the wave direction will deflect toward the east; (iii) the wave direction will be unchanged.

Light and its properties: Light is an electromagnetic wave. When emitted or absorbed, it also shows particle properties. It is emitted by accelerated electric charges.

A wave front is a surface of constant phase; wave fronts move with a speed equal to the propagation speed of the wave. A ray is a line along the direction of propagation, perpendicular to the wave fronts.

When light is transmitted from one material to another, the frequency of the light is unchanged, but the wavelength and wave speed can change. The index of refraction n of a material is the ratio of the speed of light in vacuum c to the speed v in the material. If λ_0 is the wavelength in vacuum, the same wave has a shorter wavelength λ in a medium with index of refraction n. (See Example 33.2.)

$$n = \frac{c}{v} \qquad (33.1)$$

$$\lambda = \frac{\lambda_0}{n} \qquad (33.5)$$

Reflection and refraction: At a smooth interface between two optical materials, the incident, reflected, and refracted rays and the normal to the interface all lie in a single plane called the plane of incidence. The law of reflection states that the angles of incidence and reflection are equal. The law of refraction relates the angles of incidence and refraction to the indexes of refraction of the materials. (See Examples 33.1 and 33.3.)

$$\theta_r = \theta_a \qquad (33.2)$$
(law of reflection)

$$n_a \sin\theta_a = n_b \sin\theta_b \qquad (33.4)$$
(law of refraction)

Total internal reflection: When a ray travels in a material of greater index of refraction n_a toward a material of smaller index n_b, total internal reflection occurs at the interface when the angle of incidence exceeds a critical angle θ_{crit}. (See Example 33.4.)

$$\sin\theta_{\text{crit}} = \frac{n_b}{n_a} \qquad (33.6)$$

Polarization of light: The direction of polarization of a linearly polarized electromagnetic wave is the direction of the $\vec{E}$ field. A polarizing filter passes waves that are linearly polarized along its polarizing axis and blocks waves polarized perpendicularly to that axis. When polarized light of intensity I_{max} is incident on a polarizing filter used as an analyzer, the intensity I of the light transmitted through the analyzer depends on the angle ϕ between the polarization direction of the incident light and the polarizing axis of the analyzer. (See Example 33.5.)

$$I = I_{\text{max}}\cos^2\phi \qquad (33.7)$$
(Malus's law)

Polarization by reflection: When unpolarized light strikes an interface between two materials, Brewster's law states that the reflected light is completely polarized perpendicular to the plane of incidence (parallel to the interface) if the angle of incidence equals the polarizing angle θ_{p}. (See Example 33.6.)

$$\tan\theta_{\text{p}} = \frac{n_b}{n_a} \qquad (33.8)$$
(Brewster's law)

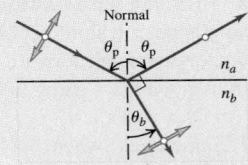

Huygens's principle: Huygens's principle states that if the position of a wave front at one instant is known, then the position of the front at a later time can be constructed by imagining the front as a source of secondary wavelets. Huygens's principle can be used to derive the laws of reflection and refraction.

BRIDGING PROBLEM | Reflection and Refraction

Figure 33.38 shows a rectangular glass block that has a metal reflector on one face and water on an adjoining face. A light beam strikes the reflector as shown. You gradually increase the angle θ of the light beam. If $\theta \geq 59.2°$, no light enters the water. What is the speed of light in this glass?

33.38

SOLUTION GUIDE

See MasteringPhysics® study area for a Video Tutor solution.

IDENTIFY and SET UP

1. Specular reflection occurs where the light ray in the glass strikes the reflector. If no light is to enter the water, we require that there be reflection only and no refraction where this ray strikes the glass–water interface—that is, there must be total internal reflection.
2. The target variable is the speed of light v in the glass, which you can determine from the index of refraction n of the glass. (Table 33.1 gives the index of refraction of water.) Write down the equations you will use to find n and v.

EXECUTE

3. Use the figure to find the angle of incidence of the ray at the glass–water interface.
4. Use the result of step 3 to find n.
5. Use the result of step 4 to find v.

EVALUATE

6. How does the speed of light in the glass compare to the speed in water? Does this make sense?

Problems

For instructor-assigned homework, go to www.masteringphysics.com

•, ••, •••: Problems of increasing difficulty. **CP**: Cumulative problems incorporating material from earlier chapters. **CALC**: Problems requiring calculus. **BIO**: Biosciences problems.

DISCUSSION QUESTIONS

Q33.1 Light requires about 8 minutes to travel from the sun to the earth. Is it delayed appreciably by the earth's atmosphere? Explain.

Q33.2 Sunlight or starlight passing through the earth's atmosphere is always bent toward the vertical. Why? Does this mean that a star is not really where it appears to be? Explain.

Q33.3 A beam of light goes from one material into another. On *physical* grounds, explain *why* the wavelength changes but the frequency and period do not.

Q33.4 A student claimed that, because of atmospheric refraction (see Discussion Question Q33.2), the sun can be seen after it has set and that the day is therefore longer than it would be if the earth had no atmosphere. First, what does she mean by saying that the sun can be seen after it has set? Second, comment on the validity of her conclusion.

Q33.5 When hot air rises from a radiator or heating duct, objects behind it appear to shimmer or waver. What causes this?

Q33.6 Devise straightforward experiments to measure the speed of light in a given glass using (a) Snell's law; (b) total internal reflection; (c) Brewster's law.

Q33.7 Sometimes when looking at a window, you see two reflected images slightly displaced from each other. What causes this?

Q33.8 If you look up from underneath toward the surface of the water in your aquarium, you may see an upside-down reflection of your pet fish in the surface of the water. Explain how this can happen.

Q33.9 A ray of light in air strikes a glass surface. Is there a range of angles for which total reflection occurs? Explain.

Q33.10 When light is incident on an interface between two materials, the angle of the refracted ray depends on the wavelength, but the angle of the reflected ray does not. Why should this be?

Q33.11 A salesperson at a bargain counter claims that a certain pair of sunglasses has Polaroid filters; you suspect that the glasses are just tinted plastic. How could you find out for sure?

Q33.12 Does it make sense to talk about the polarization of a *longitudinal* wave, such as a sound wave? Why or why not?

Q33.13 How can you determine the direction of the polarizing axis of a single polarizer?

Q33.14 It has been proposed that automobile windshields and headlights should have polarizing filters to reduce the glare of oncoming lights during night driving. Would this work? How should the polarizing axes be arranged? What advantages would this scheme have? What disadvantages?

Q33.15 When a sheet of plastic food wrap is placed between two crossed polarizers, no light is transmitted. When the sheet is stretched in one direction, some light passes through the crossed polarizers. What is happening?

Q33.16 If you sit on the beach and look at the ocean through Polaroid sunglasses, the glasses help to reduce the glare from sunlight reflecting off the water. But if you lie on your side on the beach, there is little reduction in the glare. Explain why there is a difference.

Q33.17 When unpolarized light is incident on two crossed polarizers, no light is transmitted. A student asserted that if a third polarizer is inserted between the other two, some transmission will occur. Does this make sense? How can adding a third filter *increase* transmission?

Q33.18 For the old "rabbit-ear" style TV antennas, it's possible to alter the quality of reception considerably simply by changing the orientation of the antenna. Why?

Q33.19 In Fig. 33.32, since the light that is scattered out of the incident beam is polarized, why is the transmitted beam not also partially polarized?

Q33.20 You are sunbathing in the late afternoon when the sun is relatively low in the western sky. You are lying flat on your back, looking straight up through Polaroid sunglasses. To minimize the amount of sky light reaching your eyes, how should you lie: with your feet pointing north, east, south, west, or in some other direction? Explain your reasoning.

Q33.21 Light scattered from blue sky is strongly polarized because of the nature of the scattering process described in Section 33.6. But light scattered from white clouds is usually *not* polarized. Why not?

Q33.22 Atmospheric haze is due to water droplets or smoke particles ("smog"). Such haze reduces visibility by scattering light, so that the light from distant objects becomes randomized and images become indistinct. Explain why visibility through haze can be improved by wearing red-tinted sunglasses, which filter out blue light.

Q33.23 The explanation given in Section 33.6 for the color of the setting sun should apply equally well to the *rising* sun, since sunlight travels the same distance through the atmosphere to reach your eyes at either sunrise or sunset. Typically, however, sunsets are redder than sunrises. Why? (*Hint:* Particles of all kinds in the atmosphere contribute to scattering.)

Q33.24 Huygens's principle also applies to sound waves. During the day, the temperature of the atmosphere decreases with increasing altitude above the ground. But at night, when the ground cools, there is a layer of air just above the surface in which the temperature *increases* with altitude. Use this to explain why sound waves from distant sources can be heard more clearly at night than in the daytime. (*Hint:* The speed of sound increases with increasing temperature. Use the ideas displayed in Fig. 33.37 for light.)

Q33.25 Can water waves be reflected and refracted? Give examples. Does Huygens's principle apply to water waves? Explain.

EXERCISES

Section 33.2 Reflection and Refraction

33.1 • Two plane mirrors intersect at right angles. A laser beam strikes the first of them at a point 11.5 cm from their point of intersection, as shown in Fig. E33.1 For what angle of incidence at the first mirror will this ray strike the midpoint of the second mirror (which is 28.0 cm long) after reflecting from the first mirror?

Figure **E33.1**

33.2 • **BIO Light Inside the Eye.** The vitreous humor, a transparent, gelatinous fluid that fills most of the eyeball, has an index of refraction of 1.34. Visible light ranges in wavelength from 380 nm (violet) to 750 nm (red), as measured in air. This light travels through the vitreous humor and strikes the rods and cones at the surface of the retina. What are the ranges of (a) the wavelength, (b) the frequency, and (c) the speed of the light just as it approaches the retina within the vitreous humor?

33.3 • A beam of light has a wavelength of 650 nm in vacuum. (a) What is the speed of this light in a liquid whose index of refraction at this wavelength is 1.47? (b) What is the wavelength of these waves in the liquid?

33.4 • Light with a frequency of 5.80×10^{14} Hz travels in a block of glass that has an index of refraction of 1.52. What is the wavelength of the light (a) in vacuum and (b) in the glass?

33.5 • A light beam travels at 1.94×10^8 m/s in quartz. The wavelength of the light in quartz is 355 nm. (a) What is the index of refraction of quartz at this wavelength? (b) If this same light travels through air, what is its wavelength there?

33.6 •• Light of a certain frequency has a wavelength of 438 nm in water. What is the wavelength of this light in benzene?

33.7 •• A parallel beam of light in air makes an angle of 47.5° with the surface of a glass plate having a refractive index of 1.66. (a) What is the angle between the reflected part of the beam and the surface of the glass? (b) What is the angle between the refracted beam and the surface of the glass?

33.8 •• A laser beam shines along the surface of a block of transparent material (see Fig. E33.8.). Half of the beam goes straight to a detector, while the other half travels through the block and then hits the detector. The time delay between the arrival of the two light beams at the detector is 6.25 ns. What is the index of refraction of this material?

Figure **E33.8**

33.9 • Light traveling in air is incident on the surface of a block of plastic at an angle of 62.7° to the normal and is bent so that it makes a 48.1° angle with the normal in the plastic. Find the speed of light in the plastic.

33.10 • (a) A tank containing methanol has walls 2.50 cm thick made of glass of refractive index 1.550. Light from the outside air strikes the glass at a 41.3° angle with the normal to the glass. Find the angle the light makes with the normal in the methanol. (b) The tank is emptied and refilled with an unknown liquid. If light incident at the same angle as in part (a) enters the liquid in the tank at an angle of 20.2° from the normal, what is the refractive index of the unknown liquid?

33.11 •• As shown in Fig. E33.11, a layer of water covers a slab of material X in a beaker. A ray of light traveling upward follows the path indicated. Using the information on the figure, find (a) the index of refraction of material X and (b) the angle the light makes with the normal in the *air*.

Figure **E33.11**

33.12 •• A horizontal, parallel-sided plate of glass having a refractive index of 1.52 is in contact with the surface of water in a tank. A ray coming from above in air makes an angle of incidence of 35.0° with the normal to the top surface of the glass. (a) What angle does the ray refracted into the water make with the normal to the surface? (b) What is the dependence of this angle on the refractive index of the glass?

33.13 •• In a material having an index of refraction n, a light ray has frequency f, wavelength λ, and speed v. What are the frequency, wavelength, and speed of this light (a) in vacuum and

(b) in a material having refractive index n'? In each case, express your answers in terms of *only* f, λ, v, n, and n'.

33.14 • A ray of light traveling in water is incident on an interface with a flat piece of glass. The wavelength of the light in the water is 726 nm and its wavelength in the glass is 544 nm. If the ray in water makes an angle of 42.0° with respect to the normal to the interface, what angle does the refracted ray in the glass make with respect to the normal?

33.15 • A ray of light is incident on a plane surface separating two sheets of glass with refractive indexes 1.70 and 1.58. The angle of incidence is 62.0°, and the ray originates in the glass with $n = 1.70$. Compute the angle of refraction.

Section 33.3 Total Internal Reflection

33.16 • A flat piece of glass covers the top of a vertical cylinder that is completely filled with water. If a ray of light traveling in the glass is incident on the interface with the water at an angle of $\theta_a = 36.2°$, the ray refracted into the water makes an angle of 49.8° with the normal to the interface. What is the smallest value of the incident angle θ_a for which none of the ray refracts into the water?

33.17 • **Light Pipe.** Light enters a solid pipe made of plastic having an index of refraction of 1.60. The light travels parallel to the upper part of the pipe (Fig. E33.17). You want to cut the face *AB* so that all the light will reflect back into the pipe after it first strikes that face. (a) What is the largest that θ can be if the pipe is in air? (b) If the pipe is immersed in water of refractive index 1.33, what is the largest that θ can be?

Figure **E33.17**

33.18 • A beam of light is traveling inside a solid glass cube having index of refraction 1.53. It strikes the surface of the cube from the inside. (a) If the cube is in air, at what minimum angle with the normal inside the glass will this light *not* enter the air at this surface? (b) What would be the minimum angle in part (a) if the cube were immersed in water?

33.19 •• The critical angle for total internal reflection at a liquid–air interface is 42.5°. (a) If a ray of light traveling in the liquid has an angle of incidence at the interface of 35.0°, what angle does the refracted ray in the air make with the normal? (b) If a ray of light traveling in air has an angle of incidence at the interface of 35.0°, what angle does the refracted ray in the liquid make with the normal?

33.20 • At the very end of Wagner's series of operas *Ring of the Nibelung*, Brünnhilde takes the golden ring from the finger of the dead Siegfried and throws it into the Rhine, where it sinks to the bottom of the river. Assuming that the ring is small enough compared to the depth of the river to be treated as a point and that the Rhine is 10.0 m deep where the ring goes in, what is the area of the largest circle at the surface of the water over which light from the ring could escape from the water?

33.21 • A ray of light is traveling in a glass cube that is totally immersed in water. You find that if the ray is incident on the glass–water interface at an angle to the normal larger than 48.7°, no light is refracted into the water. What is the refractive index of the glass?

33.22 • Light is incident along the normal on face *AB* of a glass prism of refractive index 1.52, as shown in Fig. E33.22. Find the largest value the angle α can have without any light refracted out of the prism at face *AC* if (a) the prism is immersed in air and (b) the prism is immersed in water.

Figure **E33.22**

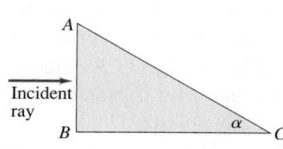

33.23 • A piece of glass with a flat surface is at the bottom of a tank of water. If a ray of light traveling in the glass is incident on the interface with the water at an angle with respect to the normal that is greater than 62.0°, no light is refracted into the water. For smaller angles of incidence, part of the ray is refracted into the water. If the light has wavelength 408 nm in the glass, what is the wavelength of the light in the water?

33.24 •• We define the index of refraction of a material for sound waves to be the ratio of the speed of sound in air to the speed of sound in the material. Snell's law then applies to the refraction of sound waves. The speed of a sound wave is 344 m/s in air and 1320 m/s in water. (a) Which medium has the higher index of refraction for sound? (b) What is the critical angle for a sound wave incident on the surface between air and water? (c) For total internal reflection to occur, must the sound wave be traveling in the air or in the water? (d) Use your results to explain why it is possible to hear people on the opposite shore of a river or small lake extremely clearly.

Section 33.4 Dispersion

33.25 •• A narrow beam of white light strikes one face of a slab of silicate flint glass. The light is traveling parallel to the two adjoining faces, as shown in Fig. E33.25. For the transmitted light inside the glass, through what angle $\Delta\theta$ is the portion of the visible spectrum between 400 nm and 700 nm dispersed? (Consult the graph in Fig. 33.18.)

Figure **E33.25**

33.26 • A beam of light strikes a sheet of glass at an angle of 57.0° with the normal in air. You observe that red light makes an angle of 38.1° with the normal in the glass, while violet light makes a 36.7° angle. (a) What are the indexes of refraction of this glass for these colors of light? (b) What are the speeds of red and violet light in the glass?

Section 33.5 Polarization

33.27 • Unpolarized light with intensity I_0 is incident on two polarizing filters. The axis of the first filter makes an angle of 60.0° with the vertical, and the axis of the second filter is horizontal. What is the intensity of the light after it has passed through the second filter?

33.28 •• (a) At what angle above the horizontal is the sun if sunlight reflected from the surface of a calm lake is completely polarized? (b) What is the plane of the electric-field vector in the reflected light?

33.29 •• A beam of unpolarized light of intensity I_0 passes through a series of ideal polarizing filters with their polarizing directions turned to various angles as shown in Fig. E33.29. (a) What is the light intensity (in terms of I_0) at points *A*, *B*, and *C*? (b) If we remove the middle filter, what will be the light intensity at point *C*?

Figure **E33.29**

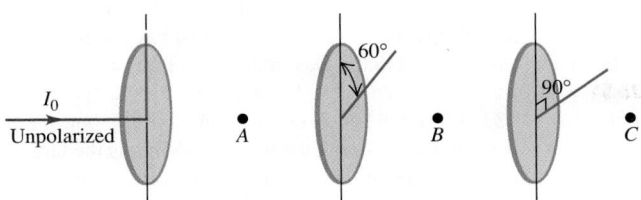

33.30 • Light traveling in water strikes a glass plate at an angle of incidence of 53.0°; part of the beam is reflected and part is refracted. If the reflected and refracted portions make an angle of 90.0° with each other, what is the index of refraction of the glass?

33.31 • A parallel beam of unpolarized light in air is incident at an angle of 54.5° (with respect to the normal) on a plane glass surface. The reflected beam is completely linearly polarized. (a) What is the refractive index of the glass? (b) What is the angle of refraction of the transmitted beam?

33.32 •• Light of original intensity I_0 passes through two ideal polarizing filters having their polarizing axes oriented as shown in Fig. E33.32. You want to adjust the angle ϕ so that the intensity at point P is equal to $I_0/10$. (a) If the original light is unpolarized, what should ϕ be? (b) If the original light is linearly polarized in the same direction as the polarizing axis of the first polarizer the light reaches, what should ϕ be?

Figure **E33.32**

33.33 •• A beam of polarized light passes through a polarizing filter. When the angle between the polarizing axis of the filter and the direction of polarization of the light is θ, the intensity of the emerging beam is I. If you now want the intensity to be $I/2$, what should be the angle (in terms of θ) between the polarizing angle of the filter and the original direction of polarization of the light?

33.34 • The refractive index of a certain glass is 1.66. For what incident angle is light reflected from the surface of this glass completely polarized if the glass is immersed in (a) air and (b) water?

33.35 •• Unpolarized light of intensity 20.0 W/cm² is incident on two polarizing filters. The axis of the first filter is at an angle of 25.0° counterclockwise from the vertical (viewed in the direction the light is traveling), and the axis of the second filter is at 62.0° counterclockwise from the vertical. What is the intensity of the light after it has passed through the second polarizer?

33.36 ••• Three polarizing filters are stacked, with the polarizing axis of the second and third filters at 23.0° and 62.0°, respectively, to that of the first. If unpolarized light is incident on the stack, the light has intensity 75.0 W/cm² after it passes through the stack. If the incident intensity is kept constant, what is the intensity of the light after it has passed through the stack if the second polarizer is removed?

33.37 • **Three Polarizing Filters.** Three polarizing filters are stacked with the polarizing axes of the second and third at 45.0° and 90.0°, respectively, with that of the first. (a) If unpolarized light of intensity I_0 is incident on the stack, find the intensity and state of polarization of light emerging from each filter. (b) If the second filter is removed, what is the intensity of the light emerging from each remaining filter?

Section 33.6 Scattering of Light

33.38 • A beam of white light passes through a uniform thickness of air. If the intensity of the scattered light in the middle of the green part of the visible spectrum is I, find the intensity (in terms of I) of scattered light in the middle of (a) the red part of the spectrum and (b) the violet part of the spectrum. Consult Table 32.1.

PROBLEMS

33.39 •• **The Corner Reflector.** An inside corner of a cube is lined with mirrors to make a corner reflector (see Example 33.3 in Section 33.2). A ray of light is reflected successively from each of three mutually perpendicular mirrors; show that its final direction is always exactly opposite to its initial direction.

33.40 • A light beam is directed parallel to the axis of a hollow cylindrical tube. When the tube contains only air, it takes the light 8.72 ns to travel the length of the tube, but when the tube is filled with a transparent jelly, it takes the light 2.04 ns longer to travel its length. What is the refractive index of this jelly?

33.41 •• **BIO** **Heart Sonogram.** Physicians use high-frequency ($f = 1-5$ MHz) sound waves, called ultrasound, to image internal organs. The speed of these ultrasound waves is 1480 m/s in muscle and 344 m/s in air. We define the index of refraction of a material for sound waves to be the ratio of the speed of sound in air to the speed of sound in the material. Snell's law then applies to the refraction of sound waves. (a) At what angle from the normal does an ultrasound beam enter the heart if it leaves the lungs at an angle of 9.73° from the normal to the heart wall? (Assume that the speed of sound in the lungs is 344 m/s.) (b) What is the critical angle for sound waves in air incident on muscle?

33.42 ••• In a physics lab, light with wavelength 490 nm travels in air from a laser to a photocell in 17.0 ns. When a slab of glass 0.840 m thick is placed in the light beam, with the beam incident along the normal to the parallel faces of the slab, it takes the light 21.2 ns to travel from the laser to the photocell. What is the wavelength of the light in the glass?

33.43 •• A ray of light is incident in air on a block of a transparent solid whose index of refraction is n. If $n = 1.38$, what is the *largest* angle of incidence θ_a for which total internal reflection will occur at the vertical face (point A shown in Fig. P33.43)?

Figure **P33.43**

33.44 • A light ray in air strikes the right-angle prism shown in Fig. P33.44. The prism angle at B is 30.0°. This ray consists of two different wavelengths. When it emerges at face AB, it has been split into two different rays that diverge from each other by 8.50°. Find the index of refraction of the prism for each of the two wavelengths.

Figure **P33.44**

33.45 •• A ray of light traveling *in* a block of glass ($n = 1.52$) is incident on the top surface at an angle of 57.2° with respect to the normal in the glass. If a layer of oil is placed on the top surface of the glass, the ray is totally reflected. What is the maximum possible index of refraction of the oil?

33.46 ••• A glass plate 2.50 mm thick, with an index of refraction of 1.40, is placed between a point source of light with wavelength 540 nm (in vacuum) and a screen. The distance from source to screen is 1.80 cm. How many wavelengths are there between the source and the screen?

33.47 •• Old photographic plates were made of glass with a light-sensitive emulsion on the front surface. This emulsion was somewhat transparent. When a bright point source is focused on the front of the plate, the developed photograph will show a halo around the image of the spot. If the glass plate is 3.10 mm thick and the halos have an inner radius of 5.34 mm, what is the index of refraction of the glass? (*Hint:* Light from the spot on the front surface is scattered in all directions by the emulsion. Some of it is then totally reflected at the back surface of the plate and returns to the front surface.)

33.48 • After a long day of driving you take a late-night swim in a motel swimming pool. When you go to your room, you realize that you have lost your room key in the pool. You borrow a powerful flashlight and walk around the pool, shining the light into it. The light shines on the key, which is lying on the bottom of the pool, when the flashlight is held 1.2 m above the water surface and is directed at the surface a horizontal distance of 1.5 m from the edge (Fig. P33.48). If the water here is 4.0 m deep, how far is the key from the edge of the pool?

Figure **P33.48**

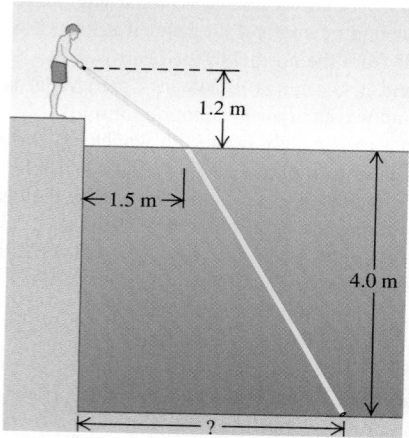

33.49 • You sight along the rim of a glass with vertical sides so that the top rim is lined up with the opposite edge of the bottom (Fig. P33.49a). The glass is a thin-walled, hollow cylinder 16.0 cm high. The diameter of the top and bottom of the glass is 8.0 cm. While you keep your eye in the same position, a friend fills the glass with a transparent liquid, and you then see a dime that is lying at the center of the bottom of the glass (Fig. P33.49b). What is the index of refraction of the liquid?

Figure **P33.49**

(a)   (b)

33.50 •• A 45°−45°−90° prism is immersed in water. A ray of light is incident normally on one of its shorter faces. What is the minimum index of refraction that the prism must have if this ray is to be totally reflected within the glass at the long face of the prism?

33.51 • A thin layer of ice ($n = 1.309$) floats on the surface of water ($n = 1.333$) in a bucket. A ray of light from the bottom of the bucket travels upward through the water. (a) What is the largest angle with respect to the normal that the ray can make at the ice–water interface and still pass out into the air above the ice? (b) What is this angle after the ice melts?

33.52 •• Light is incident normally on the short face of a 30°−60°−90° prism (Fig. P33.52). A drop of liquid is placed on the hypotenuse of the prism. If the index of refraction of the prism is 1.62, find the maximum index that the liquid may have if the light is to be totally reflected.

Figure **P33.52**

33.53 •• The prism shown in Fig. P33.53 has a refractive index of 1.66, and the angles A are 25.0°. Two light rays m and n are parallel as they enter the prism. What is the angle between them after they emerge?

Figure **P33.53**

33.54 •• A horizontal cylindrical tank 2.20 m in diameter is half full of water. The space above the water is filled with a pressurized gas of unknown refractive index. A small laser can move along the curved bottom of the water and aims a light beam toward the center of the water surface (Fig. P33.54). You observe that when the laser has moved a distance $S = 1.09$ m or more (measured along the curved surface) from the lowest point in the water, no light enters the gas. (a) What is the index of refraction of the gas? (b) What minimum time does it take the light beam to travel from the laser to the rim of the tank when (i) $S > 1.09$ m and (ii) $S < 1.09$ m?

Figure **P33.54**

33.55 •• When the sun is either rising or setting and appears to be just on the horizon, it is in fact *below* the horizon. The explanation for this seeming paradox is that light from the sun bends slightly when entering the earth's atmosphere, as shown in Fig. P33.55. Since our perception is based on the idea that light travels in straight lines, we perceive the light to be coming from an apparent position that is an angle δ above the sun's true position. (a) Make the simplifying assumptions that the atmosphere has uniform density, and hence uniform index of refraction n, and extends to a height h above the earth's surface, at which point it abruptly stops. Show that the angle δ is given by

$$\delta = \arcsin\left(\frac{nR}{R + h}\right) - \arcsin\left(\frac{R}{R + h}\right)$$

where $R = 6378$ km is the radius of the earth. (b) Calculate δ using $n = 1.0003$ and $h = 20$ km. How does this compare to the angular radius of the sun, which is about one quarter of a degree?

(In actuality a light ray from the sun bends gradually, not abruptly, since the density and refractive index of the atmosphere change gradually with altitude.)

Figure **P33.55**

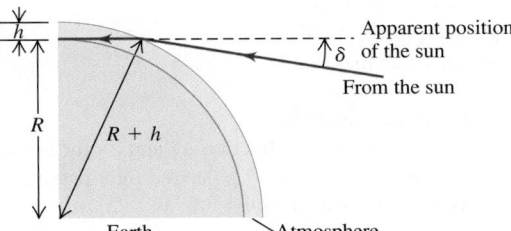

33.56 •• **CALC Fermat's Principle of Least Time.** A ray of light traveling with speed c leaves point 1 shown in Fig. P33.56 and is reflected to point 2. The ray strikes the reflecting surface a horizontal distance x from point 1. (a) Show that the time t required for the light to travel from 1 to 2 is

$$t = \frac{\sqrt{y_1^2 + x^2} + \sqrt{y_2^2 + (l - x)^2}}{c}$$

(b) Take the derivative of t with respect to x. Set the derivative equal to zero to show that this time reaches its *minimum* value when $\theta_1 = \theta_2$, which is the law of reflection and corresponds to the actual path taken by the light. This is an example of Fermat's *principle of least time,* which states that among all possible paths between two points, the one actually taken by a ray of light is that for which the time of travel is a *minimum.* (In fact, there are some cases in which the time is a maximum rather than a minimum.)

Figure **P33.56**

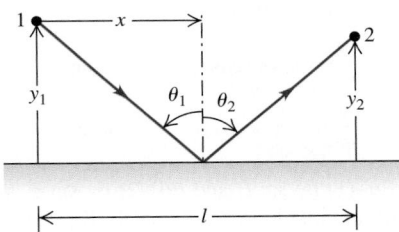

33.57 •• **CALC** A ray of light goes from point A in a medium in which the speed of light is v_1 to point B in a medium in which the speed is v_2 (Fig. P33.57). The ray strikes the interface a horizontal distance x to the right of point A. (a) Show that the time required for the light to go from A to B is

Figure **P33.57**

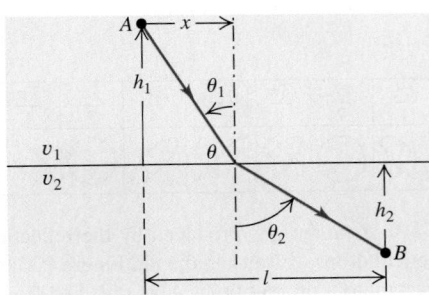

$$t = \frac{\sqrt{h_1^2 + x^2}}{v_1} + \frac{\sqrt{h_2^2 + (l - x)^2}}{v_2}$$

(b) Take the derivative of t with respect to x. Set this derivative equal to zero to show that this time reaches its *minimum* value when $n_1 \sin \theta_1 = n_2 \sin \theta_2$. This is Snell's law and corresponds to the actual path taken by the light. This is another example of Fermat's principle of least time (see Problem 33.56).

33.58 •• Light is incident in air at an angle θ_a (Fig. P33.58) on the upper surface of a transparent plate, the surfaces of the plate being plane and parallel to each other. (a) Prove that $\theta_a = \theta_a'$. (b) Show that this is true for any number of different parallel plates. (c) Prove that the lateral displacement d of the emergent beam is given by the relationship

$$d = t \frac{\sin(\theta_a - \theta_b')}{\cos \theta_b'}$$

where t is the thickness of the plate. (d) A ray of light is incident at an angle of 66.0° on one surface of a glass plate 2.40 cm thick with an index of refraction of 1.80. The medium on either side of the plate is air. Find the lateral displacement between the incident and emergent rays.

Figure **P33.58**

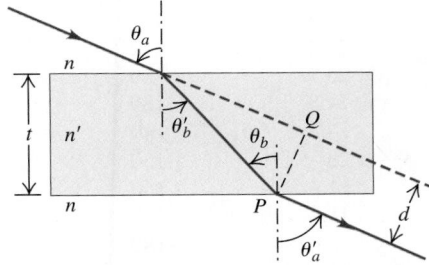

33.59 •• **Angle of Deviation.** The incident angle θ_a shown in Fig. P33.59 is chosen so that the light passes symmetrically through the prism, which has refractive index n and apex angle A. (a) Show that the angle of deviation δ (the angle between the initial and final directions of the ray) is given by

$$\sin \frac{A + \delta}{2} = n \sin \frac{A}{2}$$

(When the light passes through symmetrically, as shown, the angle of deviation is a minimum.) (b) Use the result of part (a) to find the angle of deviation for a ray of light passing symmetrically through a prism having three equal angles ($A = 60.0°$) and $n = 1.52$. (c) A certain glass has a refractive index of 1.61 for red light (700 nm) and 1.66 for violet light (400 nm). If both colors pass through symmetrically, as described in part (a), and if $A = 60.0°$, find the difference between the angles of deviation for the two colors.

Figure **P33.59**

33.60 • A thin beam of white light is directed at a flat sheet of silicate flint glass at an angle of 20.0° to the surface of the sheet. Due to dispersion in the glass, the beam is spread out in a spectrum as shown in Fig. P33.60. The refractive index of silicate flint glass versus wavelength is graphed in Fig. 33.18. (a) The rays a and b shown in Fig. P33.60 correspond to the extreme wavelengths shown in Fig. 33.18. Which corresponds to red and which to violet? Explain your reasoning. (b) For what thickness d of the glass sheet will the spectrum be 1.0 mm wide, as shown (see Problem 33.58)?

Figure **P33.60**

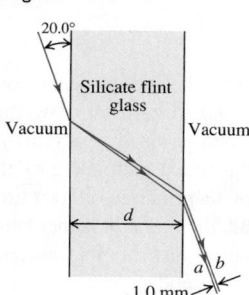

33.61 • A beam of light traveling horizontally is made of an unpolarized component with intensity I_0 and a polarized component with intensity I_p. The plane of polarization of the polarized component is oriented at an angle of θ with respect to the vertical. The data in the table give the intensity measured through a polarizer with an orientation of ϕ with respect to the vertical. (a) What is the orientation of the polarized component? (That is, what is the angle θ?) (b) What are the values of I_0 and I_p?

$\phi\,(°)$	$I_{total}\,(\mathrm{W/m^2})$	$\phi\,(°)$	$I_{total}\,(\mathrm{W/m^2})$
0	18.4	100	8.6
10	21.4	110	6.3
20	23.7	120	5.2
30	24.8	130	5.2
40	24.8	140	6.3
50	23.7	150	8.6
60	21.4	160	11.6
70	18.4	170	15.0
80	15.0	180	18.4
90	11.6		

33.62 •• **BIO** **Optical Activity of Biological Molecules.** Many biologically important molecules are optically active. When linearly polarized light traverses a solution of compounds containing these molecules, its plane of polarization is rotated. Some compounds rotate the polarization clockwise; others rotate the polarization counterclockwise. The amount of rotation depends on the amount of material in the path of the light. The following data give the amount of rotation through two amino acids over a path length of 100 cm:

Rotation (°)		
l-leucine	*d*-glutamic acid	Concentration (g/100 mL)
−0.11	0.124	1.0
−0.22	0.248	2.0
−0.55	0.620	5.0
−1.10	1.24	10.0
−2.20	2.48	20.0
−5.50	6.20	50.0
−11.0	12.4	100.0

From these data, find the relationship between the concentration C (in grams per 100 mL) and the rotation of the polarization (in degrees) of each amino acid. (*Hint:* Graph the concentration as a function of the rotation angle for each amino acid.)

33.63 •• A beam of unpolarized sunlight strikes the vertical plastic wall of a water tank at an unknown angle. Some of the light reflects from the wall and enters the water (Fig. P33.63). The refractive index of the plastic wall is 1.61. If the light that has been reflected from the wall into the water is observed to be completely polarized, what angle does this beam make with the normal inside the water?

Figure **P33.63**

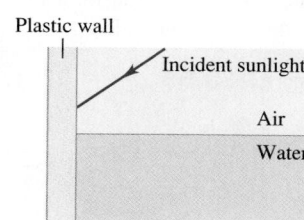

33.64 •• A certain birefringent material has indexes of refraction n_1 and n_2 for the two perpendicular components of linearly polarized light passing through it. The corresponding wavelengths are $\lambda_1 = \lambda_0/n_1$ and λ_0/n_2, where λ_0 is the wavelength in vacuum. (a) If the crystal is to function as a quarter-wave plate, the number of wavelengths of each component within the material must differ by $\frac{1}{4}$. Show that the minimum thickness for a quarter-wave plate is

$$d = \frac{\lambda_0}{4(n_1 - n_2)}$$

(b) Find the minimum thickness of a quarter-wave plate made of siderite $(\mathrm{FeO \cdot CO_2})$ if the indexes of refraction are $n_1 = 1.875$ and $n_2 = 1.635$ and the wavelength in vacuum is $\lambda_0 = 589$ nm.

CHALLENGE PROBLEMS

33.65 ••• Consider two vibrations of equal amplitude and frequency but differing in phase, one along the x-axis,

$$x = a\sin(\omega t - \alpha)$$

and the other along the y-axis,

$$y = a\sin(\omega t - \beta)$$

These can be written as follows:

$$\frac{x}{a} = \sin\omega t\cos\alpha - \cos\omega t\sin\alpha \qquad (1)$$

$$\frac{y}{a} = \sin\omega t\cos\beta - \cos\omega t\sin\beta \qquad (2)$$

(a) Multiply Eq. (1) by $\sin\beta$ and Eq. (2) by $\sin\alpha$, and then subtract the resulting equations. (b) Multiply Eq. (1) by $\cos\beta$ and Eq. (2) by $\cos\alpha$, and then subtract the resulting equations. (c) Square and add the results of parts (a) and (b). (d) Derive the equation $x^2 + y^2 - 2xy\cos\delta = a^2\sin^2\delta$, where $\delta = \alpha - \beta$. (e) Use the above result to justify each of the diagrams in Fig. P33.65. In the figure, the angle given is the phase difference between two simple harmonic motions of the same frequency and amplitude, one horizontal (along the x-axis) and the other vertical (along the y-axis). The figure thus shows the resultant motion from the superposition of the two perpendicular harmonic motions.

Figure **P33.65**

33.66 ••• **CALC** A rainbow is produced by the reflection of sunlight by spherical drops of water in the air. Figure P33.66 shows a ray that refracts into a drop at point A, is reflected from the back surface of the drop at point B, and refracts back into the air at point C. The angles of incidence and refraction, θ_a and θ_b, are shown at points A and C, and the angles of incidence and reflection, θ_a and θ_r, are shown at point B. (a) Show that $\theta_a{}^B = \theta_b{}^A$, $\theta_a{}^C = \theta_b{}^A$, and $\theta_b{}^C = \theta_a{}^A$. (b) Show that the angle in radians between the ray before it enters the drop at A and after it exits at C (the total angular deflection of the ray) is $\Delta = 2\theta_a{}^A - 4\theta_b{}^A + \pi$. (*Hint:* Find the angular deflections that occur at A, B, and C, and add them to get Δ.) (c) Use Snell's law to write Δ in terms of $\theta_a{}^A$ and n, the

Figure **P33.66**

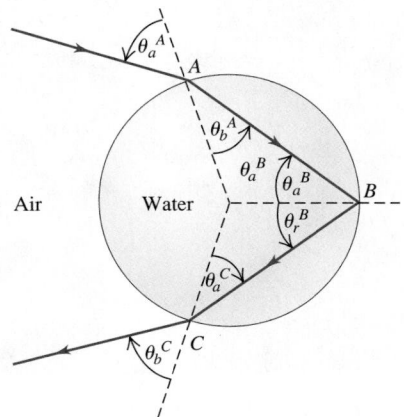

refractive index of the water in the drop. (d) A rainbow will form when the angular deflection Δ is *stationary* in the incident angle $\theta_a{}^A$—that is, when $d\Delta/d\theta_a{}^A = 0$. If this condition is satisfied, all the rays with incident angles close to $\theta_a{}^A$ will be sent back in the same direction, producing a bright zone in the sky. Let θ_1 be the value of $\theta_a{}^A$ for which this occurs. Show that $\cos^2\theta_1 = \frac{1}{3}(n^2 - 1)$. (*Hint:* You may find the derivative formula $d(\arcsin u(x))/dx = (1 - u^2)^{-1/2}(du/dx)$ helpful.) (e) The index of refraction in water is 1.342 for violet light and 1.330 for red light. Use the results of parts (c) and (d) to find θ_1 and Δ for violet and red light. Do your results agree with the angles shown in Fig. 33.20d? When you view the rainbow, which color, red or violet, is higher above the horizon?

33.67 ••• **CALC** A *secondary rainbow* is formed when the incident light undergoes two internal reflections in a spherical drop of water as shown in Fig. 33.20e. (See Challenge Problem 33.66.) (a) In terms of the incident angle $\theta_a{}^A$ and the refractive index n of the drop, what is the angular deflection Δ of the ray? That is, what is the angle between the ray before it enters the drop and after it exits? (b) What is the incident angle θ_2 for which the derivative of Δ with respect to the incident angle $\theta_a{}^A$ is zero? (c) The indexes of refraction for red and violet light in water are given in part (e) of Challenge Problem 33.66. Use the results of parts (a) and (b) to find θ_2 and Δ for violet and red light. Do your results agree with the angles shown in Fig. 33.20e? When you view a secondary rainbow, is red or violet higher above the horizon? Explain.

Answers

Chapter Opening Question **?**

This is the same effect as shown in Fig. 33.31. The drafting tools are placed between two polarizing filters whose polarizing axes are perpendicular. In places where the clear plastic is under stress, the plastic becomes birefringent; that is, light travels through it at a speed that depends on its polarization. The result is that the light that emerges from the plastic has a different polarization than the light that enters. A spot on the plastic appears bright if the emerging light has the same polarization as the second polarizing filter. The amount of birefringence depends on the wavelength of the light as well as the amount of stress on the plastic, so different colors are seen at different locations on the plastic.

Test Your Understanding Questions

33.1 Answer: (iii) The waves go farther in the y-direction in a given amount of time than in the other directions, so the wave fronts are elongated in the y-direction.

33.2 Answers: (a) (ii), (b) (iii) As shown in the figure, light rays coming from the fish bend away from the normal when they pass from the water ($n = 1.33$) into the air ($n = 1.00$). As a result, the

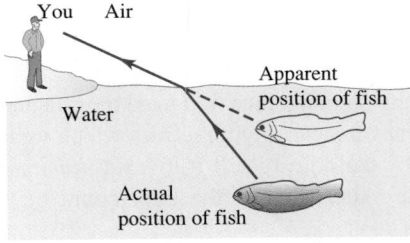

fish appears to be higher in the water than it actually is. Hence you should aim a spear *below* the apparent position of the fish. If you use a laser beam, you should aim *at* the apparent position of the fish: The beam of laser light takes the same path from you to the fish as ordinary light takes from the fish to you (though in the opposite direction).

33.3 Answers: (i), (ii) Total internal reflection can occur only if two conditions are met: n_b must be less than n_a, and the critical angle θ_{crit} (where $\sin\theta_{\text{crit}} = n_b/n_a$) must be smaller than the angle of incidence θ_a. In the first two cases both conditions are met: The critical angles are (i) $\theta_{\text{crit}} = \sin^{-1}(1/1.33) = 48.8°$ and (ii) $\theta_{\text{crit}} = \sin^{-1}(1.33/1.52) = 61.0°$, both of which are smaller than $\theta_a = 70°$. In the third case $n_b = 1.52$ is greater than $n_a = 1.33$, so total internal reflection cannot occur for any incident angle.

33.5 Answer: (ii) The sunlight reflected from the windows of the high-rise building is partially polarized in the vertical direction, since each window lies in a vertical plane. The Polaroid filter in front of the lens is oriented with its polarizing axis perpendicular to the dominant direction of polarization of the reflected light.

33.7 Answer: (ii) Huygens's principle applies to waves of all kinds, including sound waves. Hence this situation is exactly like that shown in Fig. 33.36, with material a representing the warm air, material b representing the cold air in which the waves travel more slowly, and the interface between the materials representing the weather front. North is toward the top of the figure and east is toward the right, so Fig. 33.36 shows that the rays (which indicate the direction of propagation) deflect toward the east.

Bridging Problem

Answer: 1.93×10^8 m/s

34 GEOMETRIC OPTICS

? How do magnifying lenses work? At what distance from the object being examined do they provide the sharpest view?

LEARNING GOALS

By studying this chapter, you will learn:

• How a plane mirror forms an image.

• Why concave and convex mirrors form different kinds of image.

• How images can be formed by a curved interface between two transparent materials.

• What aspects of a lens determine the type of image that it produces.

• What determines the field of view of a camera lens.

• What causes various defects in human vision, and how they can be corrected.

• The principle of the simple magnifier.

• How microscopes and telescopes work.

Your reflection in the bathroom mirror, the view of the moon through a telescope, the patterns seen in a kaleidoscope—all of these are examples of *images.* In each case the object that you're looking at appears to be in a different place than its actual position: Your reflection is on the other side of the mirror, the moon appears to be much closer when seen through a telescope, and objects seen in a kaleidoscope seem to be in many places at the same time. In each case, light rays that come from a point on an object are deflected by reflection or refraction (or a combination of the two), so they converge toward or appear to diverge from a point called an *image point.* Our goal in this chapter is to see how this is done and to explore the different kinds of images that can be made with simple optical devices.

To understand images and image formation, all we need are the ray model of light, the laws of reflection and refraction, and some simple geometry and trigonometry. The key role played by geometry in our analysis is the reason for the name *geometric optics* that is given to the study of how light rays form images. We'll begin our analysis with one of the simplest image-forming optical devices, a plane mirror. We'll go on to study how images are formed by curved mirrors, by refracting surfaces, and by thin lenses. Our results will lay the foundation for understanding many familiar optical instruments, including camera lenses, magnifiers, the human eye, microscopes, and telescopes.

ActivPhysics 15.4: Geometric Optics: Plane Mirrors

34.1 Reflection and Refraction at a Plane Surface

Before discussing what is meant by an image, we first need the concept of **object** as it is used in optics. By an *object* we mean anything from which light rays radiate. This light could be emitted by the object itself if it is *self-luminous,* like the glowing filament of a light bulb. Alternatively, the light could be emitted by another source (such as a lamp or the sun) and then reflected from the object; an example is the light you see coming from the pages of this book. Figure 34.1 shows light rays radiating in all directions from an object at a point P. For an observer to see this object directly, there must be no obstruction between the object and the observer's eyes. Note that light rays from the object reach the observer's left and right eyes at different angles; these differences are processed by the observer's brain to infer the *distance* from the observer to the object.

The object in Fig. 34.1 is a **point object** that has no physical extent. Real objects with length, width, and height are called **extended objects.** To start with, we'll consider only an idealized point object, since we can always think of an extended object as being made up of a very large number of point objects.

Suppose some of the rays from the object strike a smooth, plane reflecting surface (Fig. 34.2). This could be the surface of a material with a different index of refraction, which reflects part of the incident light, or a polished metal surface that reflects almost 100% of the light that strikes it. We will always draw the reflecting surface as a black line with a shaded area behind it, as in Fig. 34.2. Bathroom mirrors have a thin sheet of glass that lies in front of and protects the reflecting surface; we'll ignore the effects of this thin sheet.

According to the law of reflection, all rays striking the surface are reflected at an angle from the normal equal to the angle of incidence. Since the surface is plane, the normal is in the same direction at all points on the surface, and we have *specular* reflection. After the rays are reflected, their directions are the same as though they had come from point P'. We call point P an *object point* and point P' the corresponding *image point,* and we say that the reflecting surface forms an **image** of point P. An observer who can see only the rays reflected from the surface, and who doesn't know that he's seeing a reflection, *thinks* that the rays originate from the image point P'. The image point is therefore a convenient way to describe the directions of the various reflected rays, just as the object point P describes the directions of the rays arriving at the surface *before* reflection.

If the surface in Fig. 34.2 were *not* smooth, the reflection would be *diffuse,* and rays reflected from different parts of the surface would go in uncorrelated directions (see Fig. 33.6b). In this case there would not be a definite image point P' from which all reflected rays seem to emanate. You can't see your reflection in the surface of a tarnished piece of metal because its surface is rough; polishing the metal smoothes the surface so that specular reflection occurs and a reflected image becomes visible.

An image is also formed by a plane *refracting* surface, as shown in Fig. 34.3. Rays coming from point P are refracted at the interface between two optical materials. When the angles of incidence are small, the final directions of the rays after refraction are the same as though they had come from point P', as shown, and again we call P' an *image point.* In Section 33.2 we described how this effect makes underwater objects appear closer to the surface than they really are (see Fig. 33.9).

In both Figs. 34.2 and 34.3 the rays do not actually pass through the image point P'. Indeed, if the mirror in Fig. 34.2 is opaque, there is no light at all on its right side. If the outgoing rays don't actually pass through the image point, we call the image a **virtual image.** Later we will see cases in which the outgoing rays really *do* pass through an image point, and we will call the resulting image a **real image.** The images that are formed on a projection screen, on the photographic film in a camera, and on the retina of your eye are real images.

Image Formation by a Plane Mirror

Let's concentrate for now on images produced by *reflection;* we'll return to refraction later in the chapter. To find the precise location of the virtual image P' that a plane mirror forms of an object at P, we use the construction shown in Fig. 34.4. The figure shows two rays diverging from an object point P at a

34.1 Light rays radiate from a point object P in all directions.

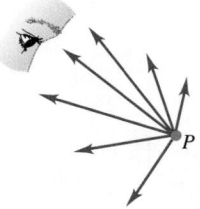

34.2 Light rays from the object at point P are reflected from a plane mirror. The reflected rays entering the eye look as though they had come from image point P'.

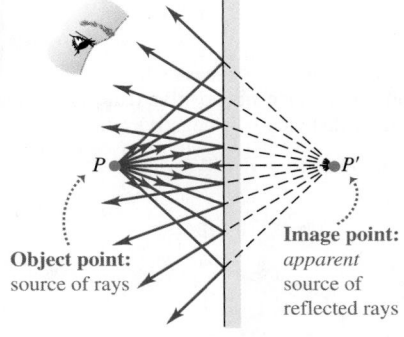

Object point:
source of rays

Image point:
apparent
source of
reflected rays

Plane mirror

34.3 Light rays from the object at point P are refracted at the plane interface. The refracted rays entering the eye look as though they had come from image point P'.

When $n_a > n_b$, P' is closer to the surface than P; for $n_a < n_b$, the reverse is true.

$n_a > n_b$ | n_b

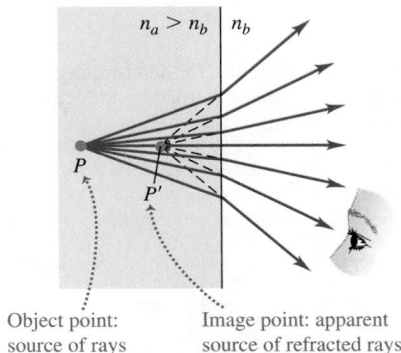

Object point:
source of rays

Image point: apparent
source of refracted rays

34.4 Construction for determining the location of the image formed by a plane mirror. The image point P' is as far behind the mirror as the object point P is in front of it.

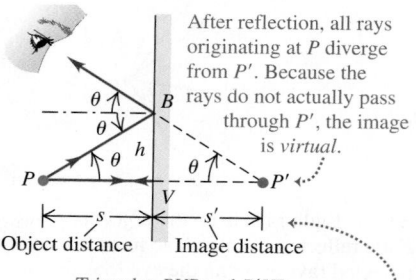

After reflection, all rays originating at P diverge from P'. Because the rays do not actually pass through P', the image is *virtual*.

Object distance | Image distance

Triangles PVB and $P'VB$ are congruent, so $|s| = |s'|$.

34.5 For both of these situations, the object distance s is positive (rule 1) and the image distance s' is negative (rule 2).

(a) Plane mirror

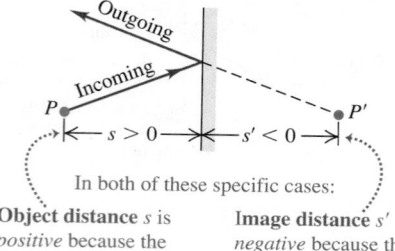

In both of these specific cases:

Object distance s is *positive* because the object is on the same side as the incoming light.

Image distance s' is *negative* because the image is NOT on the same side as the outgoing light.

(b) Plane refracting interface

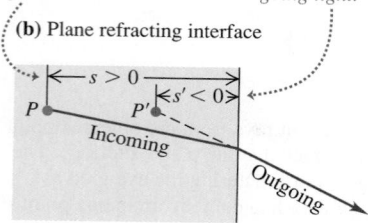

34.6 Construction for determining the height of an image formed by reflection at a plane reflecting surface.

For a plane mirror, PQV and $P'Q'V$ are congruent, so $y = y'$ and the object and image are the same size (the lateral magnification is 1).

distance s to the left of a plane mirror. We call s the **object distance.** The ray PV is incident normally on the mirror (that is, it is perpendicular to the mirror surface), and it returns along its original path.

The ray PB makes an angle θ with PV. It strikes the mirror at an angle of incidence θ and is reflected at an equal angle with the normal. When we extend the two reflected rays backward, they intersect at point P', at a distance s' behind the mirror. We call s' the **image distance.** The line between P and P' is perpendicular to the mirror. The two triangles PVB and $P'VB$ are congruent, so P and P' are at equal distances from the mirror, and s and s' have equal magnitudes. The image point P' is located exactly opposite the object point P as far *behind* the mirror as the object point is from the front of the mirror.

We can repeat the construction of Fig. 34.4 for each ray diverging from P. The directions of *all* the outgoing reflected rays are the same as though they had originated at point P', confirming that P' is the *image* of P. No matter where the observer is located, she will always see the image at the point P'.

Sign Rules

Before we go further, let's introduce some general sign rules. These may seem unnecessarily complicated for the simple case of an image formed by a plane mirror, but we want to state the rules in a form that will be applicable to *all* the situations we will encounter later. These will include image formation by a plane or spherical reflecting or refracting surface, or by a pair of refracting surfaces forming a lens. Here are the rules:

1. **Sign rule for the object distance:** When the object is on the same side of the reflecting or refracting surface as the incoming light, the object distance s is positive; otherwise, it is negative.
2. **Sign rule for the image distance:** When the image is on the same side of the reflecting or refracting surface as the outgoing light, the image distance s' is positive; otherwise, it is negative.
3. **Sign rule for the radius of curvature of a spherical surface:** When the center of curvature C is on the same side as the outgoing light, the radius of curvature is positive; otherwise, it is negative.

Figure 34.5 illustrates rules 1 and 2 for two different situations. For a mirror the incoming and outgoing sides are always the same; for example, in Figs. 34.2, 34.4, and 34.5a they are both on the left side. For the refracting surfaces in Figs. 34.3 and 34.5b the incoming and outgoing sides are on the left and right sides, respectively, of the interface between the two materials. (Note that other textbooks may use different rules.)

In Figs. 34.4 and 34.5a the object distance s is *positive* because the object point P is on the incoming side (the left side) of the reflecting surface. The image distance s' is *negative* because the image point P' is *not* on the outgoing side (the left side) of the surface. The object and image distances s and s' are related simply by

$$s = -s' \quad \text{(plane mirror)} \tag{34.1}$$

For a plane reflecting or refracting surface, the radius of curvature is infinite and not a particularly interesting or useful quantity; in these cases we really don't need sign rule 3. But this rule will be of great importance when we study image formation by *curved* reflecting and refracting surfaces later in the chapter.

Image of an Extended Object: Plane Mirror

Next we consider an *extended* object with finite size. For simplicity we often consider an object that has only one dimension, like a slender arrow, oriented parallel to the reflecting surface; an example is the arrow PQ in Fig. 34.6. The distance from the head to the tail of an arrow oriented in this way is called its *height*; in Fig. 34.6 the height is y. The image formed by such an extended object is an extended image; to each point on the object, there corresponds a point on the image. Two of the rays from Q are shown; *all* the rays from Q appear to diverge from its image point Q' after reflection. The image of the arrow is the line $P'Q'$, with height y'. Other points of the object PQ have image points between P' and Q'. The triangles PQV and $P'Q'V$ are congruent, so the object PQ and image $P'Q'$ have the same size and orientation, and $y = y'$.

The ratio of image height to object height, y'/y, in *any* image-forming situation is called the **lateral magnification** m; that is,

$$m = \frac{y'}{y} \qquad \text{(lateral magnification)} \qquad (34.2)$$

Thus for a plane mirror the lateral magnification m is unity. When you look at yourself in a plane mirror, your image is the same size as the real you.

In Fig. 34.6 the image arrow points in the *same* direction as the object arrow; we say that the image is **erect.** In this case, y and y' have the same sign, and the lateral magnification m is positive. The image formed by a plane mirror is always erect, so y and y' have both the same magnitude and the same sign; from Eq. (34.2) the lateral magnification of a plane mirror is always $m = +1$. Later we will encounter situations in which the image is **inverted;** that is, the image arrow points in the direction *opposite* to that of the object arrow. For an inverted image, y and y' have *opposite* signs, and the lateral magnification m is *negative*.

The object in Fig. 34.6 has only one dimension. Figure 34.7 shows a *three-dimensional* object and its three-dimensional virtual image formed by a plane mirror. The object and image are related in the same way as a left hand and a right hand.

CAUTION   **Reflections in a plane mirror** At this point, you may be asking, "Why does a plane mirror reverse images left and right but not top and bottom?" This question is quite misleading! As Fig. 34.7 shows, the up-down image $P'Q'$ and the left-right image $P'S'$ are parallel to their objects and are not reversed at all. Only the front-back image $P'R'$ is reversed relative to PR. Hence it's most correct to say that a plane mirror reverses *back to front*. To verify this object-image relationship, point your thumbs along PR and $P'R'$, your forefingers along PQ and $P'Q'$, and your middle fingers along PS and $P'S'$. When an object and its image are related in this way, the image is said to be **reversed;** this means that *only* the front-back dimension is reversed. ▌

The reversed image of a three-dimensional object formed by a plane mirror is the same *size* as the object in all its dimensions. When the transverse dimensions of object and image are in the same direction, the image is erect. Thus a plane mirror always forms an erect but reversed image. Figure 34.8 illustrates this point.

An important property of all images formed by reflecting or refracting surfaces is that an *image* formed by one surface or optical device can serve as the *object* for a second surface or device. Figure 34.9 shows a simple example. Mirror 1 forms an image P_1' of the object point P, and mirror 2 forms another image P_2', each in the way we have just discussed. But in addition, the image P_1' formed by mirror 1 serves an object for mirror 2, which then forms an image of this object at point P_3' as shown. Similarly, mirror 1 uses the image P_2' formed by mirror 2 as an object and forms an image of it. We leave it to you to show that this image point is also at P_3'. The idea that an image formed by one device can act as the object for a second device is of great importance in geometric optics. We will use it later in this chapter to locate the image formed by two successive curved-surface refractions in a lens. This idea will help us to understand image formation by combinations of lenses, as in a microscope or a refracting telescope.

34.7 The image formed by a plane mirror is virtual, erect, and reversed. It is the same size as the object.

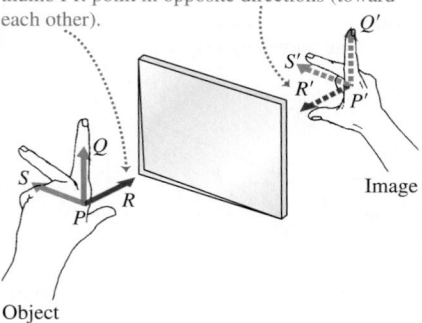

An image made by a plane mirror is reversed back to front: the image thumb $P'R'$ and object thumb PR point in opposite directions (toward each other).

Object

Image

34.8 The image formed by a plane mirror is reversed; the image of a right hand is a left hand, and so on. (The hand is resting on a horizontal mirror.) Are images of the letters H and A reversed?

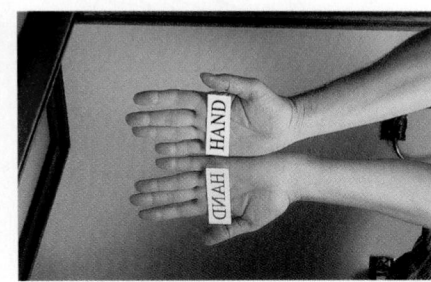

34.9 Images P_1' and P_2' are formed by a single reflection of each ray from the object at P. Image P_3', located by treating either of the other images as an object, is formed by a double reflection of each ray.

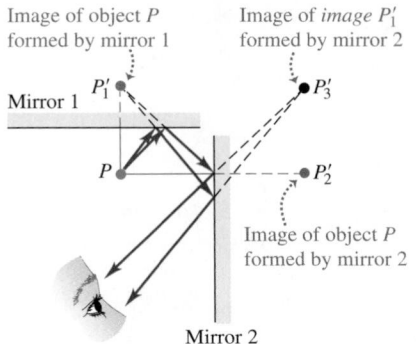

Image of object P formed by mirror 1

Image of *image* P_1' formed by mirror 2

Mirror 1

Image of object P formed by mirror 2

Mirror 2

Test Your Understanding of Section 34.1 If you walk directly toward a plane mirror at a speed v, at what speed does your image approach you? (i) slower than v; (ii) v; (iii) faster than v but slower than $2v$; (iv) $2v$; (v) faster than $2v$.

34.2 Reflection at a Spherical Surface

A plane mirror produces an image that is the same size as the object. But there are many applications for mirrors in which the image and object must be of different sizes. A magnifying mirror used when applying makeup gives an image that is *larger* than the object, and surveillance mirrors (used in stores to help spot shoplifters) give an image that is *smaller* than the object. There are also applications of mirrors in which a *real* image is desired, so light rays do indeed pass through the image point P'. A plane mirror by itself cannot perform any of these tasks. Instead, *curved* mirrors are used.

Image of a Point Object: Spherical Mirror

We'll consider the special (and easily analyzed) case of image formation by a *spherical* mirror. Figure 34.10a shows a spherical mirror with radius of curvature R, with its concave side facing the incident light. The **center of curvature** of the surface (the center of the sphere of which the surface is a part) is at C, and the **vertex** of the mirror (the center of the mirror surface) is at V. The line CV is called the **optic axis.** Point P is an object point that lies on the optic axis; for the moment, we assume that the distance from P to V is greater than R.

Ray PV, passing through C, strikes the mirror normally and is reflected back on itself. Ray PB, at an angle α with the axis, strikes the mirror at B, where the angles of incidence and reflection are θ. The reflected ray intersects the axis at point P'. We will show shortly that *all* rays from P intersect the axis at the *same* point P', as in Fig. 34.10b, provided that the angle α is small. Point P' is therefore the *image* of object point P. Unlike the reflected rays in Fig. 34.1, the reflected rays in Fig. 34.10b actually do intersect at point P', then diverge from P' *as if* they had originated at this point. Thus P' is a *real* image.

To see the usefulness of having a real image, suppose that the mirror is in a darkened room in which the only source of light is a self-luminous object at P. If you place a small piece of photographic film at P', all the rays of light coming from point P that reflect off the mirror will strike the same point P' on the film; when developed, the film will show a single bright spot, representing a sharply focused image of the object at point P. This principle is at the heart of most astronomical telescopes, which use large concave mirrors to make photographs of celestial objects. With a *plane* mirror like that in Fig. 34.2, placing a piece of film at the image point P' would be a waste of time; the light rays never actually pass through the image point, and the image can't be recorded on film. Real images are *essential* for photography.

Let's now find the location of the real image point P' in Fig. 34.10a and prove the assertion that all rays from P intersect at P' (provided that their angle with the optic axis is small). The object distance, measured from the vertex V, is s; the image distance, also measured from V, is s'. The signs of s, s', and the radius of curvature R are determined by the sign rules given in Section 34.1. The object point P is on the same side as the incident light, so according to sign rule 1, s is positive. The image point P' is on the same side as the reflected light, so according to sign rule 2, the image distance s' is also positive. The center of curvature C is on the same side as the reflected light, so according to sign rule 3, R, too, is positive; R is always positive when reflection occurs at the *concave* side of a surface (Fig. 34.11).

We now use the following theorem from plane geometry: An exterior angle of a triangle equals the sum of the two opposite interior angles. Applying this theorem to triangles PBC and $P'BC$ in Fig. 34.10a, we have

34.10 (a) A concave spherical mirror forms a real image of a point object P on the mirror's optic axis. (b) The eye sees some of the outgoing rays and perceives them as having come from P'.

(a) Construction for finding the position P' of an image formed by a concave spherical mirror

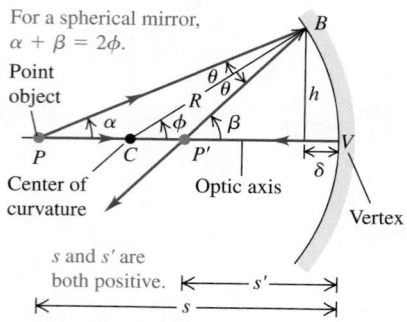

For a spherical mirror, $\alpha + \beta = 2\phi$.

s and s' are both positive.

(b) The paraxial approximation, which holds for rays with small α

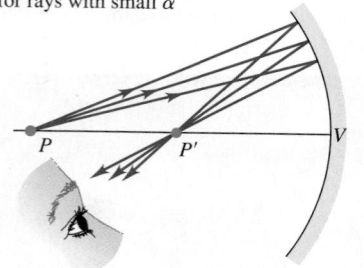

All rays from P that have a small angle α pass through P', forming a real image.

34.11 The sign rule for the radius of a spherical mirror.

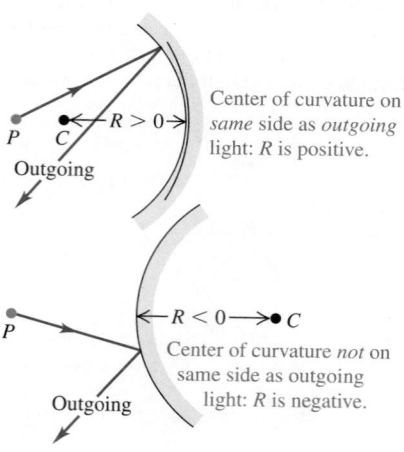

Center of curvature on *same* side as *outgoing* light: R is positive.

Center of curvature *not* on same side as outgoing light: R is negative.

$$\phi = \alpha + \theta \qquad \beta = \phi + \theta$$

Eliminating θ between these equations gives

$$\alpha + \beta = 2\phi \qquad (34.3)$$

We may now compute the image distance s'. Let h represent the height of point B above the optic axis, and let δ represent the short distance from V to the foot of this vertical line. We now write expressions for the tangents of α, β, and ϕ, remembering that s, s', and R are all positive quantities:

$$\tan \alpha = \frac{h}{s - \delta} \qquad \tan \beta = \frac{h}{s' - \delta} \qquad \tan \phi = \frac{h}{R - \delta}$$

These trigonometric equations cannot be solved as simply as the corresponding algebraic equations for a plane mirror. However, *if the angle α is small,* the angles β and ϕ are also small. The tangent of an angle that is much less than one radian is nearly equal to the angle itself (measured in radians), so we can replace $\tan \alpha$ by α, and so on, in the equations above. Also, if α is small, we can neglect the distance δ compared with s', s, and R. So for small angles we have the following approximate relationships:

$$\alpha = \frac{h}{s} \qquad \beta = \frac{h}{s'} \qquad \phi = \frac{h}{R}$$

Substituting these into Eq. (34.3) and dividing by h, we obtain a general relationship among s, s', and R:

$$\frac{1}{s} + \frac{1}{s'} = \frac{2}{R} \qquad \text{(object–image relationship, spherical mirror)} \qquad (34.4)$$

This equation does not contain the angle α. Hence *all* rays from P that make sufficiently small angles with the axis intersect at P' after they are reflected; this verifies our earlier assertion. Such rays, nearly parallel to the axis and close to it, are called **paraxial rays.** (The term **paraxial approximation** is often used for the approximations we have just described.) Since all such reflected light rays converge on the image point, a concave mirror is also called a *converging mirror.*

Be sure you understand that Eq. (34.4), as well as many similar relationships that we will derive later in this chapter and the next, is only *approximately* correct. It results from a calculation containing approximations, and it is valid only for paraxial rays. If we increase the angle α that a ray makes with the optic axis, the point P' where the ray intersects the optic axis moves somewhat closer to the vertex than for a paraxial ray. As a result, a spherical mirror, unlike a plane mirror, does not form a precise point image of a point object; the image is "smeared out." This property of a spherical mirror is called **spherical aberration.** When the primary mirror of the Hubble Space Telescope (Fig. 34.12a) was manufactured, tiny errors were made in its shape that led to an unacceptable amount of spherical aberration (Fig. 34.12b). The performance of the telescope improved dramatically after the installation of corrective optics (Fig. 34.12c).

If the radius of curvature becomes infinite ($R = \infty$), the mirror becomes *plane,* and Eq. (34.4) reduces to Eq. (34.1) for a plane reflecting surface.

Focal Point and Focal Length

When the object point P is very far from the spherical mirror ($s = \infty$), the incoming rays are parallel. (The star shown in Fig. 34.12c is an example of such a distant object.) From Eq. (34.4) the image distance s' in this case is given by

$$\frac{1}{\infty} + \frac{1}{s'} = \frac{2}{R} \qquad s' = \frac{R}{2}$$

34.12 (a), (b) Soon after the Hubble Space Telescope (HST) was placed in orbit in 1990, it was discovered that the concave primary mirror (also called the *objective mirror*) was too shallow by about $\frac{1}{50}$ the width of a human hair, leading to spherical aberration of the star's image. (c) After corrective optics were installed in 1993, the effects of spherical aberration were almost completely eliminated.

(a) The 2.4-m-diameter primary mirror of the Hubble Space Telescope

(b) A star seen with the original mirror

(c) The same star with corrective optics

34.13 The focal point and focal length of a concave mirror.

(a) All parallel rays incident on a spherical mirror reflect through the focal point.

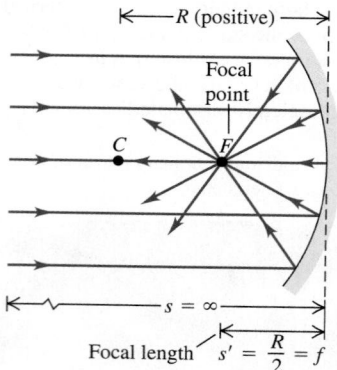

(b) Rays diverging from the focal point reflect to form parallel outgoing rays.

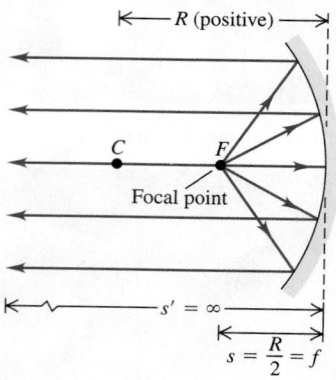

The situation is shown in Fig. 34.13a. The beam of incident parallel rays converges, after reflection from the mirror, to a point F at a distance $R/2$ from the vertex of the mirror. The point F at which the incident parallel rays converge is called the **focal point;** we say that these rays are brought to a focus. The distance from the vertex to the focal point, denoted by f, is called the **focal length.** We see that f is related to the radius of curvature R by

$$f = \frac{R}{2} \quad \text{(focal length of a spherical mirror)} \tag{34.5}$$

The opposite situation is shown in Fig. 34.13b. Now the *object* is placed at the focal point F, so the object distance is $s = f = R/2$. The image distance s' is again given by Eq. (34.4):

$$\frac{2}{R} + \frac{1}{s'} = \frac{2}{R} \quad \frac{1}{s'} = 0 \quad s' = \infty$$

With the object at the focal point, the reflected rays in Fig. 34.13b are parallel to the optic axis; they meet only at a point infinitely far from the mirror, so the image is at infinity.

Thus the focal point F of a spherical mirror has the properties that (1) any incoming ray parallel to the optic axis is reflected through the focal point and (2) any incoming ray that passes through the focal point is reflected parallel to the optic axis. For spherical mirrors these statements are true only for paraxial rays. For parabolic mirrors these statements are *exactly* true; this is why parabolic mirrors are preferred for astronomical telescopes. Spherical or parabolic mirrors are used in flashlights and headlights to form the light from the bulb into a parallel beam. Some solar-power plants use an array of plane mirrors to simulate an approximately spherical concave mirror; light from the sun is collected by the mirrors and directed to the focal point, where a steam boiler is placed. (The concepts of focal point and focal length also apply to lenses, as we'll see in Section 34.4.)

We will usually express the relationship between object and image distances for a mirror, Eq. (34.4), in terms of the focal length f:

$$\frac{1}{s} + \frac{1}{s'} = \frac{1}{f} \quad \text{(object–image relationship, spherical mirror)} \tag{34.6}$$

Image of an Extended Object: Spherical Mirror

Now suppose we have an object with *finite* size, represented by the arrow PQ in Fig. 34.14, perpendicular to the optic axis CV. The image of P formed by paraxial rays is at P'. The object distance for point Q is very nearly equal to that for point P, so the image $P'Q'$ is nearly straight and perpendicular to the axis. Note that the object and image arrows have different sizes, y and y', respectively, and that they have opposite orientation. In Eq. (34.2) we defined the *lateral magnification m* as the ratio of image size y' to object size y:

$$m = \frac{y'}{y}$$

34.14 Construction for determining the position, orientation, and height of an image formed by a concave spherical mirror.

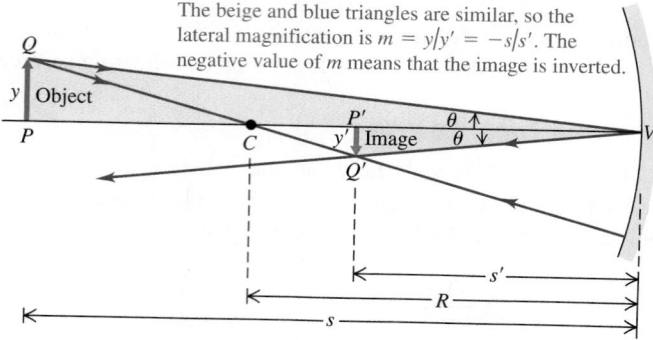

The beige and blue triangles are similar, so the lateral magnification is $m = y/y' = -s/s'$. The negative value of m means that the image is inverted.

Because triangles PVQ and $P'VQ'$ in Fig. 34.14 are *similar,* we also have the relationship $y/s = -y'/s'$. The negative sign is needed because object and image are on opposite sides of the optic axis; if y is positive, y' is negative. Therefore

$$m = \frac{y'}{y} = -\frac{s'}{s} \qquad \text{(lateral magnification, spherical mirror)} \qquad (34.7)$$

If m is positive, the image is erect in comparison to the object; if m is negative, the image is *inverted* relative to the object, as in Fig. 34.14. For a *plane* mirror, $s = -s'$, so $y' = y$ and $m = +1$; since m is positive, the image is erect, and since $|m| = 1$, the image is the same size as the object.

CAUTION **Lateral magnification can be less than 1** Although the ratio of image size to object size is called the *lateral magnification,* the image formed by a mirror or lens may be larger than, smaller than, or the same size as the object. If it is smaller, then the lateral magnification is less than unity in absolute value: $|m| < 1$. The image formed by an astronomical telescope mirror or a camera lens is usually *much* smaller than the object. For example, the image of the bright star shown in Fig. 34.12c is just a few millimeters across, while the star itself is hundreds of thousands of kilometers in diameter. ▌

In our discussion of concave mirrors we have so far considered only objects that lie *outside* or at the focal point, so that the object distance s is greater than or equal to the (positive) focal length f. In this case the image point is on the same side of the mirror as the outgoing rays, and the image is real and inverted. If an object is placed *inside* the focal point of a concave mirror, so that $s < f$, the resulting image is *virtual* (that is, the image point is on the opposite side of the mirror from the object), *erect,* and *larger* than the object. Mirrors used when applying makeup (referred to at the beginning of this section) are concave mirrors; in use, the distance from the face to the mirror is less than the focal length, and an enlarged, erect image is seen. You can prove these statements about concave mirrors by applying Eqs. (34.6) and (34.7) (see Exercise 34.11). We'll also be able to verify these results later in this section, after we've learned some graphical methods for relating the positions and sizes of the object and the image.

Mastering**PHYSICS**

ActivPhysics 15.5: Spherical Mirrors: Ray Diagrams
ActivPhysics 15.6: Spherical Mirrors: The Mirror Equation
ActivPhysics 15.7: Spherical Mirrors: Linear Magnification
ActivPhysics 15.8: Spherical Mirrors: Problems

Example 34.1 **Image formation by a concave mirror I**

A concave mirror forms an image, on a wall 3.00 m in front of the mirror, of a headlamp filament 10.0 cm in front of the mirror. (a) What are the radius of curvature and focal length of the mirror? (b) What is the lateral magnification? What is the image height if the object height is 5.00 mm?

SOLUTION

IDENTIFY and SET UP: Figure 34.15 shows our sketch. Our target variables are the radius of curvature R, focal length f, lateral mag-

34.15 Our sketch for this problem.

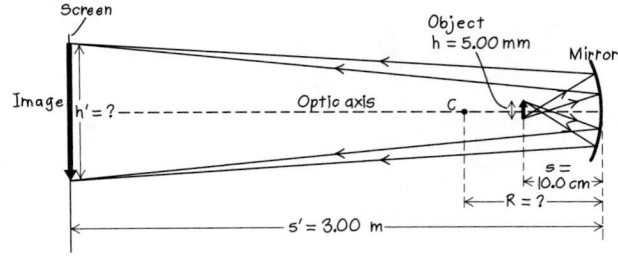

nification m, and image height y'. We are given the distances from the mirror to the object (s) and from the mirror to the image (s'). We solve Eq. (34.4) for R, and then use Eq. (34.5) to find f. Equation (34.7) yields both m and y'.

EXECUTE: (a) Both the object and the image are on the concave (reflective) side of the mirror, so both s and s' are positive; we have $s = 10.0$ cm and $s' = 300$ cm. We solve Eq. (34.4) for R:

$$\frac{1}{10.0 \text{ cm}} + \frac{1}{300 \text{ cm}} = \frac{2}{R}$$

$$R = 2\left(\frac{1}{10.0 \text{ cm}} + \frac{1}{300 \text{ cm}}\right)^{-1} = 19.4 \text{ cm}$$

The focal length of the mirror is $f = R/2 = 9.7$ cm.

(b) From Eq. (34.7) the lateral magnification is

$$m = -\frac{s'}{s} = -\frac{300 \text{ cm}}{10.0 \text{ cm}} = -30.0$$

Because m is negative, the image is inverted. The height of the image is 30.0 times the height of the object, or $(30.0)(5.00 \text{ mm}) = 150$ mm.

Continued

EVALUATE: Our sketch indicates that the image is inverted; our calculations agree. Note that the object (at $s = 10.0$ cm) is just outside the focal point ($f = 9.7$ cm). This is very similar to what is done in automobile headlights. With the filament close to the focal point, the concave mirror produces a beam of nearly parallel rays.

Conceptual Example 34.2 **Image formation by a concave mirror II**

In Example 34.1, suppose that the lower half of the mirror's reflecting surface is covered with nonreflective soot. What effect will this have on the image of the filament?

SOLUTION

It would be natural to guess that the image would now show only half of the filament. But in fact the image will still show the *entire* filament. You can see why by examining Fig. 34.10b. Light rays coming from any object point P are reflected from *all* parts of the mirror and converge on the corresponding image point P'. If part of the mirror surface is made nonreflective (or is removed altogether), rays from the remaining reflective surface still form an image of every part of the object.

Reducing the reflecting area reduces the light energy reaching the image point, however: The image becomes *dimmer*. If the area is reduced by one-half, the image will be one-half as bright. Conversely, *increasing* the reflective area makes the image brighter. To make reasonably bright images of faint stars, astronomical telescopes use mirrors that are up to several meters in diameter (see Fig. 34.12a).

Convex Mirrors

In Fig. 34.16a the *convex* side of a spherical mirror faces the incident light. The center of curvature is on the side opposite to the outgoing rays; according to sign rule 3 in Section 34.1, R is negative (see Fig. 34.11). Ray PB is reflected, with the angles of incidence and reflection both equal to θ. The reflected ray, projected backward, intersects the axis at P'. As with a concave mirror, *all* rays from P that are reflected by the mirror diverge from the same point P', provided that the angle α is small. Therefore P' is the image of P. The object distance s is positive, the image distance s' is negative, and the radius of curvature R is *negative* for a *convex* mirror.

Figure 34.16b shows two rays diverging from the head of the arrow PQ and the virtual image $P'Q'$ of this arrow. The same procedure that we used for a concave mirror can be used to show that for a convex mirror,

$$\frac{1}{s} + \frac{1}{s'} = \frac{2}{R}$$

and the lateral magnification is

$$m = \frac{y'}{y} = -\frac{s'}{s}$$

34.16 Image formation by a convex mirror.

(a) Construction for finding the position of an image formed by a convex mirror

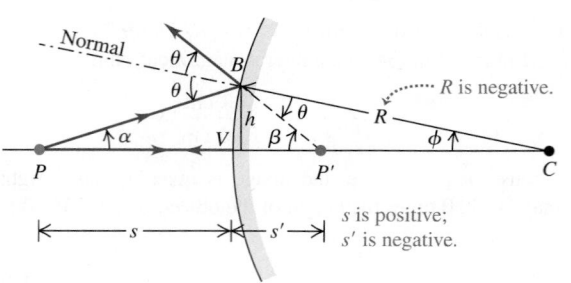

(b) Construction for finding the magnification of an image formed by a convex mirror

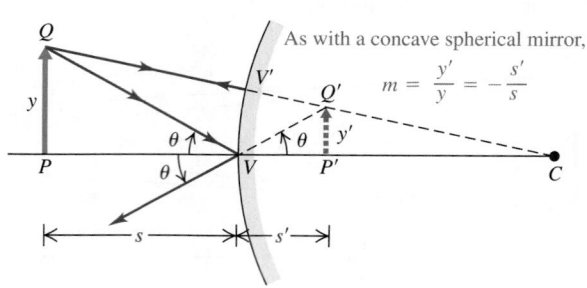

As with a concave spherical mirror,

$$m = \frac{y'}{y} = -\frac{s'}{s}$$

(a) Paraxial rays incident on a convex spherical mirror diverge from a virtual focal point.

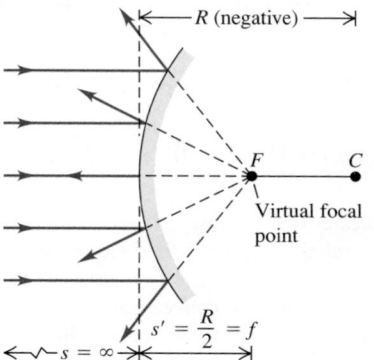

(b) Rays aimed at the virtual focal point are parallel to the axis after reflection.

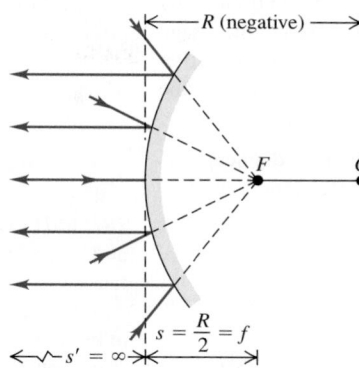

34.17 The focal point and focal length of a convex mirror.

These expressions are exactly the same as Eqs. (34.4) and (34.7) for a concave mirror. Thus when we use our sign rules consistently, Eqs. (34.4) and (34.7) are valid for both concave and convex mirrors.

When R is negative (convex mirror), incoming rays that are parallel to the optic axis are not reflected through the focal point F. Instead, they diverge as though they had come from the point F at a distance f *behind* the mirror, as shown in Fig. 34.17a. In this case, f is the focal length, and F is called a *virtual focal point*. The corresponding image distance s' is negative, so both f and R are negative, and Eq. (34.5), $f = R/2$, holds for convex as well as concave mirrors. In Fig. 34.17b the incoming rays are converging as though they would meet at the virtual focal point F, and they are reflected parallel to the optic axis.

In summary, Eqs. (34.4) through (34.7), the basic relationships for image formation by a spherical mirror, are valid for both concave and convex mirrors, provided that we use the sign rules consistently.

Example 34.3   **Santa's image problem**

Santa checks himself for soot, using his reflection in a silvered Christmas tree ornament 0.750 m away (Fig. 34.18a). The diameter of the ornament is 7.20 cm. Standard reference texts state that he is a "right jolly old elf," so we estimate his height to be 1.6 m. Where and how tall is the image of Santa formed by the ornament? Is it erect or inverted?

SOLUTION

IDENTIFY and SET UP: Figure 34.18b shows the situation. Santa is the object, and the surface of the ornament closest to him acts as a convex mirror. The relationships among object distance, image distance, focal length, and magnification are the same as for

34.18 (a) The ornament forms a virtual, reduced, erect image of Santa. (b) Our sketch of two of the rays forming the image.

Continued

concave mirrors, provided we use the sign rules consistently. The radius of curvature and the focal length of a convex mirror are *negative*. The object distance is $s = 0.750$ m $= 75.0$ cm, and Santa's height is $y = 1.6$ m. We solve Eq. (34.6) to find the image distance s', and then use Eq. (34.7) to find the lateral magnification m and the image height y'. The sign of m tells us whether the image is erect or inverted.

EXECUTE: The radius of the mirror (half the diameter) is $R = -(7.20 \text{ cm})/2 = -3.60$ cm, and the focal length is $f = R/2 = -1.80$ cm. From Eq. (34.6),

$$\frac{1}{s'} = \frac{1}{f} - \frac{1}{s} = \frac{1}{-1.80 \text{ cm}} - \frac{1}{75.0 \text{ cm}}$$

$$s' = -1.76 \text{ cm}$$

Because s' is negative, the image is behind the mirror—that is, on the side opposite to the outgoing light (Fig. 34.18b)—and it is virtual.

The image is about halfway between the front surface of the ornament and its center.

From Eq. (34.7), the lateral magnification and the image height are

$$m = \frac{y'}{y} = -\frac{s'}{s} = -\frac{-1.76 \text{ cm}}{75.0 \text{ cm}} = 0.0234$$

$$y' = my = (0.0234)(1.6 \text{ m}) = 3.8 \times 10^{-2} \text{m} = 3.8 \text{ cm}$$

EVALUATE: Our sketch indicates that the image is erect so both m and y' are positive; our calculations agree. When the object distance s is positive, a convex mirror *always* forms an erect, virtual, reduced, reversed image. For this reason, convex mirrors are used at blind intersections, for surveillance in stores, and as wide-angle rear-view mirrors for cars and trucks. (Many such mirrors read "Objects in mirror are closer than they appear.")

Graphical Methods for Mirrors

In Examples 34.1 and 34.3, we used Eqs. (34.6) and (34.7) to find the position and size of the image formed by a mirror. We can also determine the properties of the image by a simple *graphical* method. This method consists of finding the point of intersection of a few particular rays that diverge from a point of the object (such as point Q in Fig. 34.19) and are reflected by the mirror. Then (neglecting aberrations) *all* rays from this object point that strike the mirror will intersect at the same point. For this construction we always choose an object point that is *not* on the optic axis. Four rays that we can usually draw easily are shown in Fig. 34.19. These are called **principal rays.**

1. *A ray parallel to the axis,* after reflection, passes through the focal point F of a concave mirror or appears to come from the (virtual) focal point of a convex mirror.
2. *A ray through (or proceeding toward) the focal point F* is reflected parallel to the axis.
3. *A ray along the radius* through or away from the center of curvature C intersects the surface normally and is reflected back along its original path.
4. *A ray to the vertex V* is reflected forming equal angles with the optic axis.

34.19 The graphical method of locating an image formed by a spherical mirror. The colors of the rays are for identification only; they do not refer to specific colors of light.

(a) Principal rays for concave mirror

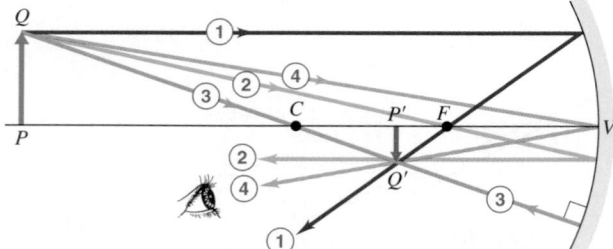

① Ray parallel to axis reflects through focal point.
② Ray through focal point reflects parallel to axis.
③ Ray through center of curvature intersects the surface normally and reflects along its original path.
④ Ray to vertex reflects symmetrically around optic axis.

(b) Principal rays for convex mirror

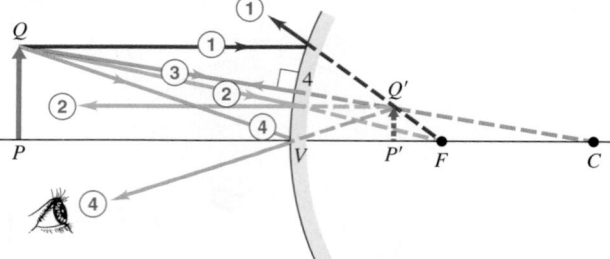

① Reflected parallel ray appears to come from focal point.
② Ray toward focal point reflects parallel to axis.
③ As with concave mirror: Ray radial to center of curvature intersects the surface normally and reflects along its original path.
④ As with concave mirror: Ray to vertex reflects symmetrically around optic axis.

Once we have found the position of the image point by means of the intersection of any two of these principal rays $(1, 2, 3, 4)$, we can draw the path of any other ray from the object point to the same image point.

CAUTION **Principal rays are not the only rays** Although we've emphasized the principal rays, in fact *any* ray from the object that strikes the mirror will pass through the image point (for a real image) or appear to originate from the image point (for a virtual image). Usually, you only need to draw the principal rays, because these are all you need to locate the image.

Problem-Solving Strategy 34.1 **Image Formation by Mirrors**

IDENTIFY *the relevant concepts:* Problems involving image formation by mirrors can be solved in two ways: using principal-ray diagrams and using equations. A successful problem solution uses *both* approaches.

SET UP *the problem:* Identify the target variables. One of them is likely to be the focal length, the object distance, or the image distance, with the other two quantities given.

EXECUTE *the solution* as follows:
1. Draw a large, clear principal-ray diagram if you have enough information.
2. Orient your diagram so that incoming rays go from left to right. Draw only the principal rays; color-code them as in Fig. 34.19. If possible, use graph paper or quadrille-ruled paper. Use a ruler and measure distances carefully! A freehand sketch will *not* give good results.
3. If your principal rays don't converge at a real image point, you may have to extend them straight backward to locate a virtual

image point, as in Fig. 34.19b. We recommend drawing the extensions with broken lines.
4. Measure the resulting diagram to obtain the magnitudes of the target variables.
5. Solve for the target variables using Eq. (34.6), $1/s + 1/s' = 1/f$, and the lateral magnification equation, Eq. (34.7), as appropriate. Apply the sign rules given in Section 34.1 to object and image distances, radii of curvature, and object and image heights.
6. Use the sign rules to interpret the results that you deduced from your ray diagram and calculations. Note that the *same* sign rules (given in Section 34.1) work for all four cases in this chapter: reflection and refraction from plane and spherical surfaces.

EVALUATE *your answer:* Check that the results of your calculations agree with your ray-diagram results for image position, image size, and whether the image is real or virtual.

Example 34.4 **Concave mirror with various object distances**

A concave mirror has a radius of curvature with absolute value 20 cm. Find graphically the image of an object in the form of an arrow perpendicular to the axis of the mirror at object distances of (a) 30 cm, (b) 20 cm, (c) 10 cm, and (d) 5 cm. Check the construction by *computing* the size and lateral magnification of each image.

SOLUTION

IDENTIFY and SET UP: We must use graphical methods *and* calculations to analyze the image made by a mirror. The mirror is concave, so its radius of curvature is $R = +20$ cm and its focal length is $f = R/2 = +10$ cm. Our target variables are the image distances s' and lateral magnifications m corresponding to four cases with successively smaller object distances s. In each case we solve Eq. (34.6) for s' and use $m = -s'/s$ to find m.

EXECUTE: Figure 34.20 shows the principal-ray diagrams for the four cases. Study each of these diagrams carefully and confirm that each numbered ray is drawn in accordance with the rules given earlier (under "Graphical Methods for Mirrors"). Several points are worth noting. First, in case (b) the object and image distances are equal. Ray 3 cannot be drawn in this case because a ray from Q through the center of curvature C does not strike the mirror. In case (c), ray 2 cannot be drawn because a ray from Q through F does

not strike the mirror. In this case the outgoing rays are parallel, corresponding to an infinite image distance. In case (d), the outgoing rays diverge; they have been extended backward to the *virtual image point Q'*, from which they appear to diverge. Case (d) illustrates the general observation that an object placed inside the focal point of a concave mirror produces a virtual image.

Measurements of the figures, with appropriate scaling, give the following approximate image distances: (a) 15 cm; (b) 20 cm; (c) ∞ or $-\infty$ (because the outgoing rays are parallel and do not converge at any finite distance); (d) -10 cm. To *compute* these distances, we solve Eq. (34.6) for s' and insert $f = 10$ cm:

(a) $\dfrac{1}{30 \text{ cm}} + \dfrac{1}{s'} = \dfrac{1}{10 \text{ cm}}$ $s' = 15$ cm

(b) $\dfrac{1}{20 \text{ cm}} + \dfrac{1}{s'} = \dfrac{1}{10 \text{ cm}}$ $s' = 20$ cm

(c) $\dfrac{1}{10 \text{ cm}} + \dfrac{1}{s'} = \dfrac{1}{10 \text{ cm}}$ $s' = \infty$ (or $-\infty$)

(d) $\dfrac{1}{5 \text{ cm}} + \dfrac{1}{s'} = \dfrac{1}{10 \text{ cm}}$ $s' = -10$ cm

The signs of s' tell us that the image is real in cases (a) and (b) and virtual in case (d).

Continued

The lateral magnifications measured from the figures are approximately (a) $-\frac{1}{2}$; (b) -1; (c) ∞ or $-\infty$; (d) $+2$. From Eq. (34.7),

(a) $m = -\dfrac{15 \text{ cm}}{30 \text{ cm}} = -\dfrac{1}{2}$

(b) $m = -\dfrac{20 \text{ cm}}{20 \text{ cm}} = -1$

(c) $m = -\dfrac{\infty \text{ cm}}{10 \text{ cm}} = -\infty \ (\text{or} +\infty)$

(d) $m = -\dfrac{-10 \text{ cm}}{5 \text{ cm}} = +2$

The signs of m tell us that the image is inverted in cases (a) and (b) and erect in case (d).

EVALUATE: Notice the trend of the results in the four cases. When the object is far from the mirror, as in Fig. 34.20a, the image is smaller than the object, inverted, and real. As the object distance s decreases, the image moves farther from the mirror and gets larger (Fig. 34.20b). When the object is at the focal point, the image is at infinity (Fig. 34.20c). When the object is inside the focal point, the image becomes larger than the object, erect, and virtual (Fig. 34.20d). You can confirm these conclusions by looking at objects reflected in the concave bowl of a shiny metal spoon.

34.20 Using principal-ray diagrams to locate the image $P'Q'$ made by a concave mirror.

(a) Construction for $s = 30$ cm

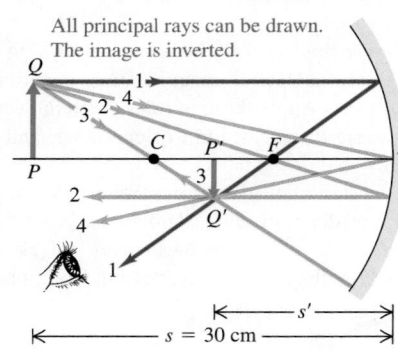

(b) Construction for $s = 20$ cm

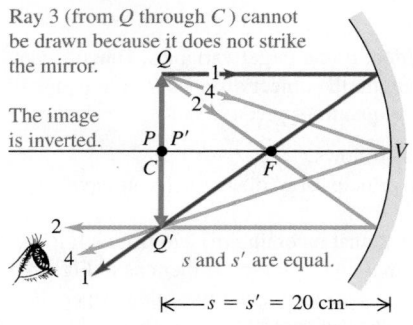

(c) Construction for $s = 10$ cm

(d) Construction for $s = 5$ cm

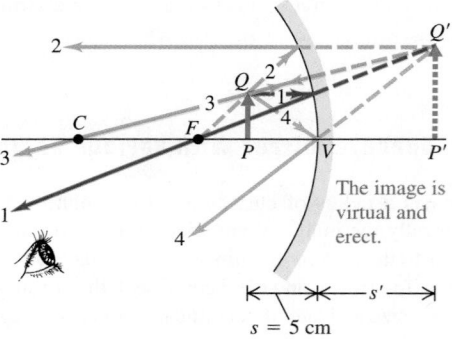

Test Your Understanding of Section 34.2 A cosmetics mirror is designed so that your reflection appears right-side up and enlarged. (a) Is the mirror concave or convex? (b) To see an enlarged image, what should be the distance from the mirror (of focal length f) to your face? (i) $|f|$; (ii) less than $|f|$; (iii) greater than $|f|$. ❙

34.3 Refraction at a Spherical Surface

As we mentioned in Section 34.1, images can be formed by refraction as well as by reflection. To begin with, let's consider refraction at a spherical surface—that is, at a spherical interface between two optical materials with different indexes of refraction. This analysis is directly applicable to some real optical systems, such as the human eye. It also provides a stepping-stone for the analysis of lenses, which usually have *two* spherical (or nearly spherical) surfaces.

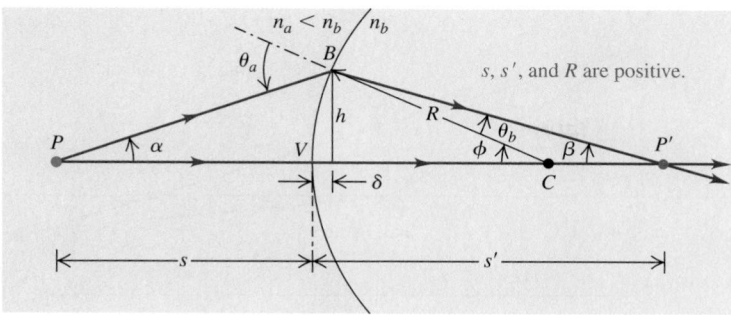

34.21 Construction for finding the position of the image point P' of a point object P formed by refraction at a spherical surface. The materials to the left and right of the interface have refractive indexes n_a and n_b, respectively. In the case shown here, $n_a < n_b$.

Image of a Point Object: Spherical Refracting Surface

In Fig. 34.21 a spherical surface with radius R forms an interface between two materials with different indexes of refraction n_a and n_b. The surface forms an image P' of an object point P; we want to find how the object and image distances (s and s') are related. We will use the same sign rules that we used for spherical mirrors. The center of curvature C is on the outgoing side of the surface, so R is positive. Ray PV strikes the vertex V and is perpendicular to the surface (that is, to the plane that is tangent to the surface at the point of incidence V). It passes into the second material without deviation. Ray PB, making an angle α with the axis, is incident at an angle θ_a with the normal and is refracted at an angle θ_b. These rays intersect at P', a distance s' to the right of the vertex. The figure is drawn for the case $n_a < n_b$. The object and image distances are both positive.

We are going to prove that if the angle α is small, *all* rays from P intersect at the same point P', so P' is the *real image* of P. We use much the same approach as we did for spherical mirrors in Section 34.2. We again use the theorem that an exterior angle of a triangle equals the sum of the two opposite interior angles; applying this to the triangles PBC and $P'BC$ gives

$$\theta_a = \alpha + \phi \qquad \phi = \beta + \theta_b \qquad (34.8)$$

From the law of refraction,

$$n_a \sin\theta_a = n_b \sin\theta_b$$

Also, the tangents of α, β, and ϕ are

$$\tan\alpha = \frac{h}{s+\delta} \qquad \tan\beta = \frac{h}{s'-\delta} \qquad \tan\phi = \frac{h}{R-\delta} \qquad (34.9)$$

For paraxial rays, θ_a and θ_b are both small in comparison to a radian, and we may approximate both the sine and tangent of either of these angles by the angle itself (measured in radians). The law of refraction then gives

$$n_a\theta_a = n_b\theta_b$$

Combining this with the first of Eqs. (34.8), we obtain

$$\theta_b = \frac{n_a}{n_b}(\alpha + \phi)$$

When we substitute this into the second of Eqs. (34.8), we get

$$n_a\alpha + n_b\beta = (n_b - n_a)\phi \qquad (34.10)$$

Now we use the approximations $\tan\alpha = \alpha$, and so on, in Eqs. (34.9) and also neglect the small distance δ; those equations then become

$$\alpha = \frac{h}{s} \qquad \beta = \frac{h}{s'} \qquad \phi = \frac{h}{R}$$

34.22 Construction for determining the height of an image formed by refraction at a spherical surface. In the case shown here, $n_a < n_b$.

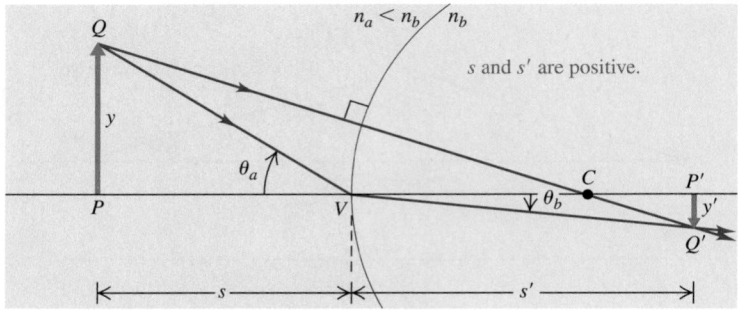

Finally, we substitute these into Eq. (34.10) and divide out the common factor h. We obtain

$$\frac{n_a}{s} + \frac{n_b}{s'} = \frac{n_b - n_a}{R} \qquad \text{(object–image relationship,} \atop \text{spherical refracting surface)} \qquad \text{(34.11)}$$

This equation does not contain the angle α, so the image distance is the same for *all* paraxial rays emanating from P; this proves our assertion that P' is the image of P.

To obtain the lateral magnification m for this situation, we use the construction in Fig. 34.22. We draw two rays from point Q, one through the center of curvature C and the other incident at the vertex V. From the triangles PQV and $P'Q'V$,

$$\tan\theta_a = \frac{y}{s} \qquad \tan\theta_b = \frac{-y'}{s'}$$

and from the law of refraction,

$$n_a \sin\theta_a = n_b \sin\theta_b$$

For small angles,

$$\tan\theta_a = \sin\theta_a \qquad \tan\theta_b = \sin\theta_b$$

so finally

$$\frac{n_a y}{s} = -\frac{n_b y'}{s'} \qquad \text{or}$$

$$m = \frac{y'}{y} = -\frac{n_a s'}{n_b s} \qquad \text{(lateral magnification,} \atop \text{spherical refracting surface)} \qquad \text{(34.12)}$$

Equations (34.11) and (34.12) can be applied to both convex and concave refracting surfaces, provided that you use the sign rules consistently. It doesn't matter whether n_b is greater or less than n_a. To verify these statements, you should construct diagrams like Figs. 34.21 and 34.22 for the following three cases: (i) $R > 0$ and $n_a > n_b$, (ii) $R < 0$ and $n_a < n_b$, and (iii) $R < 0$ and $n_a > n_b$. Then in each case, use your diagram to again derive Eqs. (34.11) and (34.12).

Here's a final note on the sign rule for the radius of curvature R of a surface. For the convex reflecting surface in Fig. 34.16, we considered R negative, but the convex *refracting* surface in Fig. 34.21 has a *positive* value of R. This may seem inconsistent, but it isn't. The rule is that R is positive if the center of curvature C is on the outgoing side of the surface and negative if C is on the other side. For the convex reflecting surface in Fig. 34.16, R is negative because point C is to the right of the surface but outgoing rays are to the left. For the convex refracting surface in Fig. 34.21, R is positive because both C and the outgoing rays are to the right of the surface.

Refraction at a curved surface is one reason gardeners avoid watering plants at midday. As sunlight enters a water drop resting on a leaf (Fig. 34.23), the light rays are refracted toward each other as in Figs. 34.21 and 34.22. The sunlight that strikes the leaf is therefore more concentrated and able to cause damage.

An important special case of a spherical refracting surface is a *plane* surface between two optical materials. This corresponds to setting $R = \infty$ in Eq. (34.11). In this case,

$$\frac{n_a}{s} + \frac{n_b}{s'} = 0 \qquad \text{(plane refracting surface)} \qquad (34.13)$$

To find the lateral magnification m for this case, we combine this equation with the general relationship, Eq. (34.12), obtaining the simple result

$$m = 1$$

That is, the image formed by a *plane* refracting surface always has the same lateral size as the object, and it is always erect.

An example of image formation by a plane refracting surface is the appearance of a partly submerged drinking straw or canoe paddle. When viewed from some angles, the object appears to have a sharp bend at the water surface because the submerged part appears to be only about three-quarters of its actual distance below the surface. (We commented on the appearance of a submerged object in Section 33.2; see Fig. 33.9.)

34.23 Light rays refract as they pass through the curved surfaces of these water droplets.

Example 34.5 **Image formation by refraction I**

A cylindrical glass rod (Fig. 34.24) has index of refraction 1.52. It is surrounded by air. One end is ground to a hemispherical surface with radius $R = 2.00$ cm. A small object is placed on the axis of the rod, 8.00 cm to the left of the vertex. Find (a) the image distance and (b) the lateral magnification.

SOLUTION

IDENTIFY and SET UP: This problem uses the ideas of refraction at a curved surface. Our target variables are the image distance s' and the lateral magnification m. Here material a is air ($n_a = 1.00$) and

material b is the glass of which the rod is made ($n_b = 1.52$). We are given $s = 8.00$ cm. The center of curvature of the spherical surface is on the outgoing side of the surface, so the radius is positive: $R = +2.00$ cm. We solve Eq. (34.11) for s', and we use Eq. (34.12) to find m.

EXECUTE: (a) From Eq. (34.11),

$$\frac{1.00}{8.00 \text{ cm}} + \frac{1.52}{s'} = \frac{1.52 - 1.00}{+2.00 \text{ cm}}$$

$$s' = +11.3 \text{ cm}$$

(b) From Eq. (34.12),

$$m = -\frac{n_a s'}{n_b s} = -\frac{(1.00)(11.3 \text{ cm})}{(1.52)(8.00 \text{ cm})} = -0.929$$

EVALUATE: Because the image distance s' is positive, the image is formed 11.3 cm to the *right* of the vertex (on the outgoing side), as Fig. 34.24 shows. The value of m tells us that the image is somewhat smaller than the object and that it is inverted. If the object is an arrow 1.000 mm high, pointing upward, the image is an arrow 0.929 mm high, pointing downward.

34.24 The glass rod in air forms a real image.

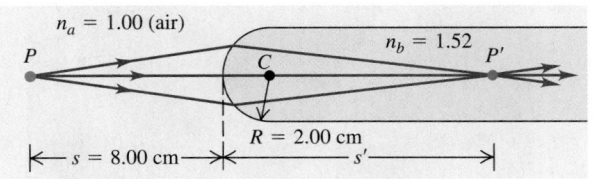

Example 34.6 **Image formation by refraction II**

The glass rod of Example 34.5 is immersed in water, which has index of refraction $n = 1.33$ (Fig. 34.25). The object distance is again 8.00 cm. Find the image distance and lateral magnification.

SOLUTION

IDENTIFY and SET UP: The situation is the same as in Example 34.5 except that now $n_a = 1.33$. We again use Eqs. (34.11) and (34.12) to determine s' and m, respectively.

EXECUTE: Our solution of Eq. (34.11) in Example 34.5 yields

$$\frac{1.33}{8.00 \text{ cm}} + \frac{1.52}{s'} = \frac{1.52 - 1.33}{+2.00 \text{ cm}}$$

$$s' = -21.3 \text{ cm}$$

Continued

34.25 When immersed in water, the glass rod forms a virtual image.

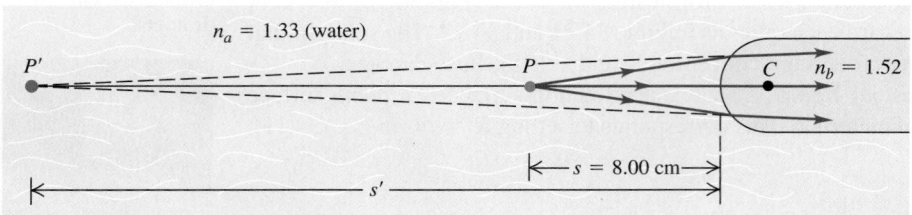

The lateral magnification in this case is

$$m = -\frac{(1.33)(-21.3 \text{ cm})}{(1.52)(8.00 \text{ cm})} = +2.33$$

EVALUATE: The negative value of s' means that the refracted rays do not converge, but appear to diverge from a point 21.3 cm to the

left of the vertex. We saw a similar case in the reflection of light from a convex mirror; in both cases we call the result a *virtual image*. The vertical image is erect (because m is positive) and 2.33 times as large as the object.

Example 34.7 **Apparent depth of a swimming pool**

If you look straight down into a swimming pool where it is 2.00 m deep, how deep does it appear to be?

SOLUTION

IDENTIFY and SET UP: Figure 34.26 shows the situation. The surface of the water acts as a plane refracting surface. To determine the pool's apparent depth, we imagine an arrow PQ painted on the bottom. The pool's refracting surface forms a virtual image $P'Q'$ of this arrow. We solve Eq. (34.13) to find the image depth s'; that's the pool's apparent depth.

EXECUTE: The object distance is the actual depth of the pool, $s = 2.00$ m. Material a is water ($n_a = 1.33$) and material b is air ($n_b = 1.00$). From Eq. (34.13),

$$\frac{n_a}{s} + \frac{n_b}{s'} = \frac{1.33}{2.00 \text{ m}} + \frac{1.00}{s'} = 0$$

$$s' = -1.50 \text{ m}$$

The image distance is negative. By the sign rules in Section 34.1, this means that the image is virtual and on the incoming side of the refracting surface—that is, on the same side as the object, namely underwater. The pool's apparent depth is 1.50 m, or just 75% of its true depth.

EVALUATE: Recall that the lateral magnification for a plane refracting surface is $m = 1$. Hence the image $P'Q'$ of the arrow has the same *horizontal length* as the actual arrow PQ (Fig. 34.27). Only its depth is different from that of PQ.

34.26 Arrow $P'Q'$ is the virtual image of the underwater arrow PQ. The angles of the ray with the vertical are exaggerated for clarity.

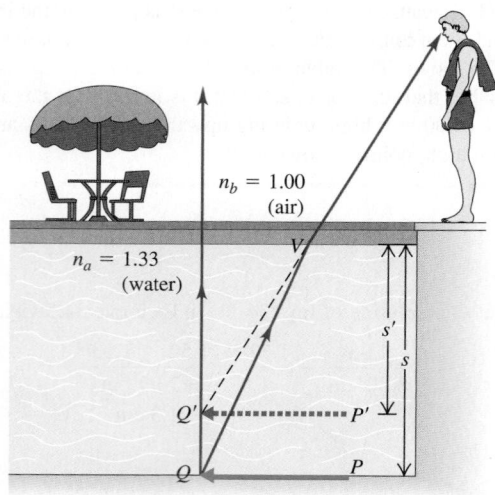

34.27 The submerged portion of this straw appears to be at a shallower depth (closer to the surface) than it actually is.

Test Your Understanding of Section 34.3 The water droplets in Fig. 34.23 have radius of curvature R and index of refraction $n = 1.33$. Can they form an image of the sun on the leaf?

34.4 Thin Lenses

The most familiar and widely used optical device (after the plane mirror) is the *lens.* A lens is an optical system with two refracting surfaces. The simplest lens has two *spherical* surfaces close enough together that we can neglect the distance between them (the thickness of the lens); we call this a **thin lens.** If you wear eyeglasses or contact lenses while reading, you are viewing these words through a pair of thin lenses. We can analyze thin lenses in detail using the results of Section 34.3 for refraction by a single spherical surface. However, we postpone this analysis until later in the section so that we can first discuss the properties of thin lenses.

Properties of a Lens

A lens of the shape shown in Fig. 34.28 has the property that when a beam of rays parallel to the axis passes through the lens, the rays converge to a point F_2 (Fig. 34.28a) and form a real image at that point. Such a lens is called a **converging lens.** Similarly, rays passing through point F_1 emerge from the lens as a beam of parallel rays (Fig. 34.28b). The points F_1 and F_2 are called the first and second *focal points,* and the distance f (measured from the center of the lens) is called the *focal length.* Note the similarities between the two focal points of a converging lens and the single focal point of a concave mirror (see Fig. 34.13). As for a concave mirror, the focal length of a converging lens is defined to be a *positive* quantity, and such a lens is also called a *positive lens.*

The central horizontal line in Fig. 34.28 is called the *optic axis,* as with spherical mirrors. The centers of curvature of the two spherical surfaces lie on and define the optic axis. The two focal lengths in Fig. 34.28, both labeled f, *are always equal* for a thin lens, even when the two sides have different curvatures. We will derive this somewhat surprising result later in the section, when we derive the relationship of f to the index of refraction of the lens and the radii of curvature of its surfaces.

Image of an Extended Object: Converging Lens

Like a concave mirror, a converging lens can form an image of an extended object. Figure 34.29 shows how to find the position and lateral magnification of an image made by a thin converging lens. Using the same notation and sign rules as before, we let s and s' be the object and image distances, respectively, and let y and y' be the object and image heights. Ray QA, parallel to the optic axis before refraction, passes through the second focal point F_2 after refraction. Ray QOQ' passes undeflected straight through the center of the lens because at the center the two surfaces are parallel and (we have assumed) very close together. There is refraction where the ray enters and leaves the material but no net change in direction.

MasteringPHYSICS

PhET: Geometric Optics
ActivPhysics 15.9: Thin Lens Ray Diagram
ActivPhysics 15.10: Converging Thin Lenses
ActivPhysics 15.11: Diverging Thin Lenses

34.28 F_1 and F_2 are the first and second focal points of a converging thin lens. The numerical value of f is positive.

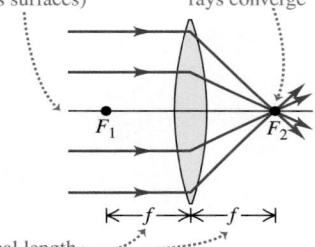

(a)

Optic axis (passes through centers of curvature of both lens surfaces)

Second focal point: the point to which incoming parallel rays converge

F_1 F_2

$\leftarrow f \rightarrow\!\leftarrow f \rightarrow$

Focal length
• Measured from lens center
• Always the same on both sides of the lens
• Positive for a converging thin lens

(b)

First focal point: Rays diverging from this point emerge from the lens parallel to the axis.

F_1 F_2

$\leftarrow f \rightarrow\!\leftarrow f \rightarrow$

34.29 Construction used to find image position for a thin lens. To emphasize that the lens is assumed to be very thin, the ray QAQ' is shown as bent at the midplane of the lens rather than at the two surfaces and ray QOQ' is shown as a straight line.

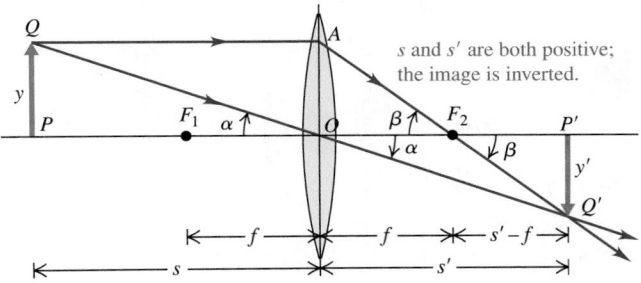

s and s' are both positive; the image is inverted.

The two angles labeled α in Fig. 34.29 are equal. Therefore the two right triangles PQO and $P'Q'O$ are *similar,* and ratios of corresponding sides are equal. Thus

$$\frac{y}{s} = -\frac{y'}{s'} \qquad \text{or} \qquad \frac{y'}{y} = -\frac{s'}{s} \qquad (34.14)$$

(The reason for the negative sign is that the image is below the optic axis and y' is negative.) Also, the two angles labeled β are equal, and the two right triangles OAF_2 and $P'Q'F_2$ are similar, so

$$\frac{y}{f} = -\frac{y'}{s' - f} \qquad \text{or}$$

$$\frac{y'}{y} = -\frac{s' - f}{f} \qquad (34.15)$$

We now equate Eqs. (34.14) and (34.15), divide by s', and rearrange to obtain

$$\frac{1}{s} + \frac{1}{s'} = \frac{1}{f} \qquad \text{(object–image relationship, thin lens)} \qquad (34.16)$$

This analysis also gives the lateral magnification $m = y'/y$ for the lens; from Eq. (34.14),

$$m = -\frac{s'}{s} \qquad \text{(lateral magnification, thin lens)} \qquad (34.17)$$

The negative sign tells us that when s and s' are both positive, as in Fig. 34.29, the image is *inverted,* and y and y' have opposite signs.

Equations (34.16) and (34.17) are the basic equations for thin lenses. They are *exactly* the same as the corresponding equations for spherical mirrors, Eqs. (34.6) and (34.7). As we will see, the same sign rules that we used for spherical mirrors are also applicable to lenses. In particular, consider a lens with a positive focal length (a converging lens). When an object is outside the first focal point F_1 of this lens (that is, when $s > f$), the image distance s' is positive (that is, the image is on the same side as the outgoing rays); this image is real and inverted, as in Fig. 34.29. An object placed inside the first focal point of a converging lens, so that $s < f$, produces an image with a negative value of s'; this image is located on the same side of the lens as the object and is virtual, erect, and larger than the object. You can verify these statements algebraically using Eqs. (34.16) and (34.17); we'll also verify them in the next section, using graphical methods analogous to those introduced for mirrors in Section 34.2.

Figure 34.30 shows how a lens forms a three-dimensional image of a three-dimensional object. Point R is nearer the lens than point P. From Eq. (34.16), image point R' is farther from the lens than is image point P', and the image $P'R'$

34.30 The image $S'P'Q'R'$ of a three-dimensional object $SPQR$ is not reversed by a lens.

A real image made by a converging lens is inverted but *not* reversed back to front: the image thumb $P'R'$ and object thumb PR point in the same direction.

points in the same direction as the object *PR*. Arrows *P'S'* and *P'Q'* are reversed relative to *PS* and *PQ*.

Let's compare Fig. 34.30 with Fig. 34.7, which shows the image formed by a plane *mirror*. We note that the image formed by the lens is inverted, but it is *not* reversed front to back along the optic axis. That is, if the object is a left hand, its image is also a left hand. You can verify this by pointing your left thumb along *PR*, your left forefinger along *PQ*, and your left middle finger along *PS*. Then rotate your hand 180°, using your thumb as an axis; this brings the fingers into coincidence with *P'Q'* and *P'S'*. In other words, an *inverted* image is equivalent to an image that has been rotated by 180° about the lens axis.

Diverging Lenses

So far we have been discussing *converging* lenses. Figure 34.31 shows a **diverging lens;** the beam of parallel rays incident on this lens *diverges* after refraction. The focal length of a diverging lens is a negative quantity, and the lens is also called a *negative lens*. The focal points of a negative lens are reversed, relative to those of a positive lens. The second focal point, F_2, of a negative lens is the point from which rays that are originally parallel to the axis *appear to diverge* after refraction, as in Fig. 34.31a. Incident rays converging toward the first focal point F_1, as in Fig. 34.31b, emerge from the lens parallel to its axis. Comparing with Section 34.2, you can see that a diverging lens has the same relationship to a converging lens as a convex mirror has to a concave mirror.

Equations (34.16) and (34.17) apply to *both* positive and negative lenses. Figure 34.32 shows various types of lenses, both converging and diverging. Here's an important observation: *Any lens that is thicker at its center than at its edges is a converging lens with positive f; and any lens that is thicker at its edges than at its center is a diverging lens with negative f* (provided that the lens has a greater index of refraction than the surrounding material). We can prove this using the *lensmaker's equation,* which it is our next task to derive.

The Lensmaker's Equation

We'll now derive Eq. (34.16) in more detail and at the same time derive the *lensmaker's equation,* which is a relationship among the focal length f, the index of refraction n of the lens, and the radii of curvature R_1 and R_2 of the lens surfaces. We use the principle that an image formed by one reflecting or refracting surface can serve as the object for a second reflecting or refracting surface.

We begin with the somewhat more general problem of two spherical interfaces separating three materials with indexes of refraction n_a, n_b, and n_c, as shown in Fig. 34.33. The object and image distances for the first surface are s_1 and s'_1, and those for the second surface are s_2 and s'_2. We assume that the lens is thin, so that the distance t between the two surfaces is small in comparison with the object and image distances and can therefore be neglected. This is usually the case with eyeglass lenses (Fig. 34.34). Then s_2 and s'_1 have the same magnitude but opposite sign. For example, if the first image is on the outgoing side of the first surface, s'_1 is positive. But when viewed as an object for the second surface, the first image is *not* on the incoming side of that surface. So we can say that $s_2 = -s'_1$.

We need to use the single-surface equation, Eq. (34.11), twice, once for each surface. The two resulting equations are

$$\frac{n_a}{s_1} + \frac{n_b}{s'_1} = \frac{n_b - n_a}{R_1}$$

$$\frac{n_b}{s_2} + \frac{n_c}{s'_2} = \frac{n_c - n_b}{R_2}$$

Ordinarily, the first and third materials are air or vacuum, so we set $n_a = n_c = 1$. The second index n_b is that of the lens, which we can call simply n. Substituting these values and the relationship $s_2 = -s'_1$, we get

34.31 F_2 and F_1 are the second and first focal points of a diverging thin lens, respectively. The numerical value of f is negative.

(a)

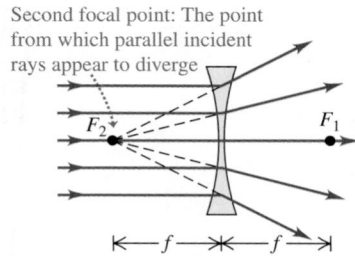

Second focal point: The point from which parallel incident rays appear to diverge

For a diverging thin lens, f is negative.

(b)

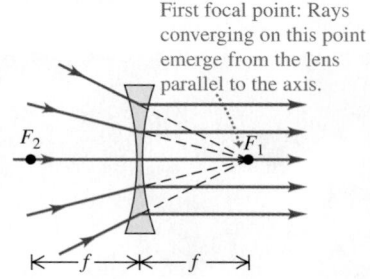

First focal point: Rays converging on this point emerge from the lens parallel to the axis.

34.32 Various types of lenses.

(a)   **Converging lenses**

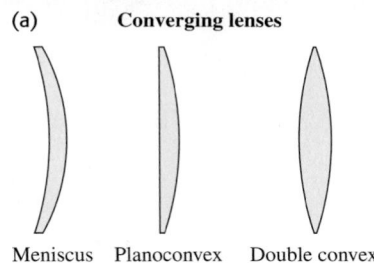

Meniscus   Planoconvex   Double convex

(b)   **Diverging lenses**

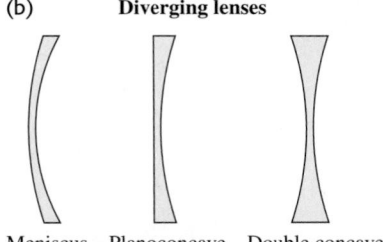

Meniscus   Planoconcave   Double concave

34.33 The image formed by the first surface of a lens serves as the object for the second surface. The distances s'_1 and s_2 are taken to be equal; this is a good approximation if the lens thickness t is small.

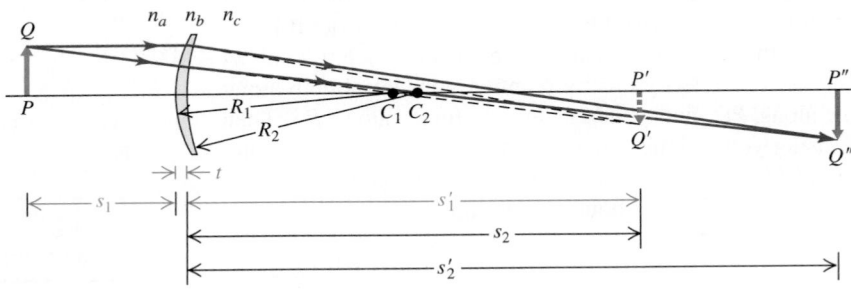

34.34 These eyeglass lenses satisfy the thin-lens approximation: Their thickness is small compared to the object and image distances.

$$\frac{1}{s_1} + \frac{n}{s'_1} = \frac{n-1}{R_1}$$

$$-\frac{n}{s'_1} + \frac{1}{s'_2} = \frac{1-n}{R_2}$$

To get a relationship between the initial object position s_1 and the final image position s'_2, we add these two equations. This eliminates the term n/s'_1, and we obtain

$$\frac{1}{s_1} + \frac{1}{s'_2} = (n-1)\left(\frac{1}{R_1} - \frac{1}{R_2}\right)$$

Finally, thinking of the lens as a single unit, we call the object distance simply s instead of s_1, and we call the final image distance s' instead of s'_2. Making these substitutions, we have

$$\frac{1}{s} + \frac{1}{s'} = (n-1)\left(\frac{1}{R_1} - \frac{1}{R_2}\right) \qquad (34.18)$$

Now we compare this with the other thin-lens equation, Eq. (34.16). We see that the object and image distances s and s' appear in exactly the same places in both equations and that the focal length f is given by

$$\frac{1}{f} = (n-1)\left(\frac{1}{R_1} - \frac{1}{R_2}\right) \qquad \text{(lensmaker's equation for a thin lens)} \qquad (34.19)$$

This is the **lensmaker's equation.** In the process of rederiving the relationship among object distance, image distance, and focal length for a thin lens, we have also derived an expression for the focal length f of a lens in terms of its index of refraction n and the radii of curvature R_1 and R_2 of its surfaces. This can be used to show that all the lenses in Fig. 34.32a are converging lenses with positive focal lengths and that all the lenses in Fig. 34.32b are diverging lenses with negative focal lengths.

We use all our sign rules from Section 34.1 with Eqs. (34.18) and (34.19). For example, in Fig. 34.35, s, s', and R_1 are positive, but R_2 is negative.

It is not hard to generalize Eq. (34.19) to the situation in which the lens is immersed in a material with an index of refraction greater than unity. We invite you to work out the lensmaker's equation for this more general situation.

We stress that the paraxial approximation is indeed an approximation! Rays that are at sufficiently large angles to the optic axis of a spherical lens will not be brought to the same focus as paraxial rays; this is the same problem of spherical aberration that plagues spherical *mirrors* (see Section 34.2). To avoid this and other limitations of thin spherical lenses, lenses of more complicated shape are used in precision optical instruments.

34.35 A converging thin lens with a positive focal length f.

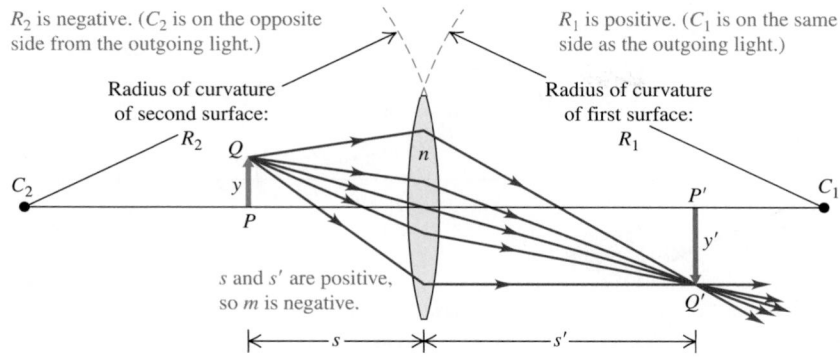

R_2 is negative. (C_2 is on the opposite side from the outgoing light.)

R_1 is positive. (C_1 is on the same side as the outgoing light.)

Radius of curvature of second surface:

R_2

Radius of curvature of first surface:

R_1

s and s' are positive, so m is negative.

Example 34.8 Determining the focal length of a lens

(a) Suppose the absolute values of the radii of curvature of the lens surfaces in Fig. 34.35 are both equal to 10 cm and the index of refraction of the glass is $n = 1.52$. What is the focal length f of the lens? (b) Suppose the lens in Fig. 34.31 also has $n = 1.52$ and the absolute values of the radii of curvature of its lens surfaces are also both equal to 10 cm. What is the focal length of this lens?

SOLUTION

IDENTIFY and SET UP: We are asked to find the focal length f of (a) a lens that is convex on both sides (Fig. 34.35) and (b) a lens that is concave on both sides (Fig. 34.31). In both cases we solve the lensmaker's equation, Eq. (34.19), to determine f. We apply the sign rules given in Section 34.1 to the radii of curvature R_1 and R_2 to take account of whether the surfaces are convex or concave.

EXECUTE: (a) The lens in Fig. 34.35 is *double convex:* The center of curvature of the first surface (C_1) is on the outgoing side of the lens, so R_1 is positive, and the center of curvature of the second surface (C_2) is on the *incoming* side, so R_2 is negative. Hence $R_1 = +10$ cm and $R_2 = -10$ cm. Then from Eq. (34.19),

$$\frac{1}{f} = (1.52 - 1)\left(\frac{1}{+10 \text{ cm}} - \frac{1}{-10 \text{ cm}}\right)$$

$$f = 9.6 \text{ cm}$$

(b) The lens in Fig. 34.31 is *double concave:* The center of curvature of the first surface is on the *incoming* side, so R_1 is negative, and the center of curvature of the second surface is on the outgoing side, so R_2 is positive. Hence in this case $R_1 = -10$ cm and $R_2 = +10$ cm. Again using Eq. (34.19),

$$\frac{1}{f} = (1.52 - 1)\left(\frac{1}{-10 \text{ cm}} - \frac{1}{+10 \text{ cm}}\right)$$

$$f = -9.6 \text{ cm}$$

EVALUATE: In part (a) the focal length is *positive,* so this is a converging lens; this makes sense, since the lens is thicker at its center than at its edges. In part (b) the focal length is *negative,* so this is a diverging lens; this also makes sense, since the lens is thicker at its edges than at its center.

Graphical Methods for Lenses

We can determine the position and size of an image formed by a thin lens by using a graphical method very similar to the one we used in Section 34.2 for spherical mirrors. Again we draw a few special rays called *principal rays* that diverge from a point of the object that is *not* on the optic axis. The intersection of these rays, after they pass through the lens, determines the position and size of the image. In using this graphical method, we will consider the entire deviation of a ray as occurring at the midplane of the lens, as shown in Fig. 34.36. This is consistent with the assumption that the distance between the lens surfaces is negligible.

The three principal rays whose paths are usually easy to trace for lenses are shown in Fig. 34.36:

1. *A ray parallel to the axis* emerges from the lens in a direction that passes through the second focal point F_2 of a converging lens, or appears to come from the second focal point of a diverging lens.
2. *A ray through the center of the lens* is not appreciably deviated; at the center of the lens the two surfaces are parallel, so this ray emerges at essentially the same angle at which it enters and along essentially the same line.
3. *A ray through (or proceeding toward) the first focal point F_1* emerges parallel to the axis.

34.36 The graphical method of locating an image formed by a thin lens. The colors of the rays are for identification only; they do not refer to specific colors of light. (Compare Fig. 34.19 for spherical mirrors.)

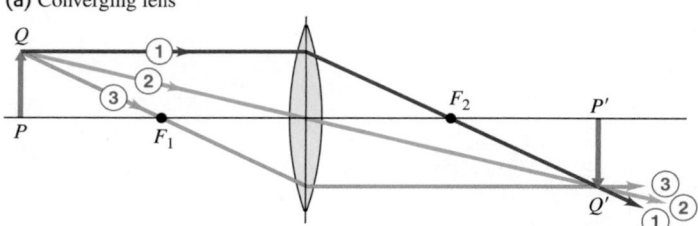

(a) Converging lens

(1) Parallel incident ray refracts to pass through second focal point F_2.
(2) Ray through center of lens does not deviate appreciably.
(3) Ray through the first focal point F_1 emerges parallel to the axis.

(b) Diverging lens

(1) Parallel incident ray appears after refraction to have come from the second focal point F_2.
(2) Ray through center of lens does not deviate appreciably.
(3) Ray aimed at the first focal point F_1 emerges parallel to the axis.

When the image is real, the position of the image point is determined by the intersection of any two rays 1, 2, and 3 (Fig. 34.36a). When the image is virtual, we extend the diverging outgoing rays backward to their intersection point to find the image point (Fig. 34.36b).

CAUTION **Principal rays are not the only rays** Keep in mind that *any* ray from the object that strikes the lens will pass through the image point (for a real image) or appear to originate from the image point (for a virtual image). (We made a similar comment about image formation by mirrors in Section 34.2.) We've emphasized the principal rays because they're the only ones you need to draw to locate the image.

Figure 34.37 shows principal-ray diagrams for a converging lens for several object distances. We suggest you study each of these diagrams very carefully, comparing each numbered ray with the above description.

34.37 Formation of images by a thin converging lens for various object distances. The principal rays are numbered. (Compare Fig. 34.20 for a concave spherical mirror.)

(a) Object *O* is outside focal point; image *I* is real.

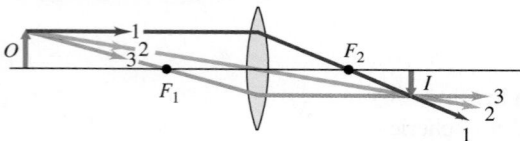

(b) Object *O* is closer to focal point; image *I* is real and farther away.

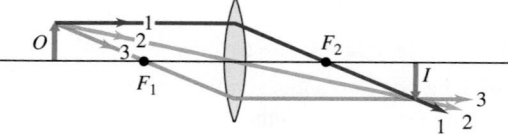

(c) Object *O* is even closer to focal point; image *I* is real and even farther away.

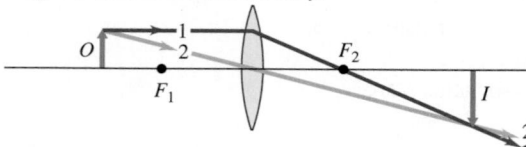

(d) Object *O* is at focal point; image *I* is at infinity.

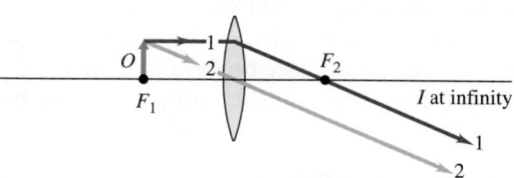

(e) Object *O* is inside focal point; image *I* is virtual and larger than object.

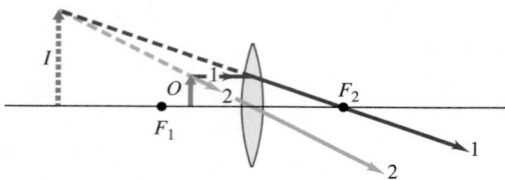

(f) A virtual object *O* (light rays are *converging* on lens)

Parts (a), (b), and (c) of Fig. 34.37 help explain what happens in focusing a camera. For a photograph to be in sharp focus, the film must be at the position of the real image made by the camera's lens. The image distance increases as the object is brought closer, so the film is moved farther behind the lens (i.e., the lens is moved farther in front of the film). In Fig. 34.37d the object is at the focal point; ray 3 can't be drawn because it doesn't pass through the lens. In Fig. 34.37e the object distance is less than the focal length. The outgoing rays are divergent, and the image is *virtual;* its position is located by extending the outgoing rays backward, so the image distance s' is negative. Note also that the image is erect and larger than the object. (We'll see the usefulness of this in Section 34.6.) Figure 34.37f corresponds to a *virtual object.* The incoming rays do not diverge from a real object, but are *converging* as though they would meet at the tip of the virtual object O on the right side; the object distance s is negative in this case. The image is real and is located between the lens and the second focal point. This situation can arise if the rays that strike the lens in Fig. 34.37f emerge from another converging lens (not shown) to the left of the figure.

Problem-Solving Strategy 34.2 | **Image Formation by Thin Lenses**

IDENTIFY *the relevant concepts:* Review Problem-Solving Strategy 34.1 (Section 34.2) for mirrors, which is equally applicable here. As for mirrors, you should solve problems involving image formation by lenses using *both* principal-ray diagrams and equations.

SET UP *the problem:* Identify the target variables.

EXECUTE *the solution* as follows:
1. Draw a large principal-ray diagram if you have enough information, using graph paper or quadrille-ruled paper. Orient your diagram so that incoming rays go from left to right. Draw the rays with a ruler, and measure distances carefully.
2. Draw the principal rays so they change direction at the midplane of the lens, as in Fig. 34.36. For a lens there are only three principal rays (compared to four for a mirror). Draw all three whenever possible; the intersection of any two rays determines the image location, but the third ray should pass through the same point.

3. If the outgoing principal rays diverge, extend them backward to find the virtual image point on the *incoming* side of the lens, as in Fig. 34.27e.
4. Solve Eqs. (34.16) and (34.17), as appropriate, for the target variables. Make sure that you carefully use the sign rules given in Section 34.1.
5. The *image* from a first lens or mirror may serve as the *object* for a second lens or mirror. In finding the object and image distances for this intermediate image, be sure you include the distance between the two elements (lenses and/or mirrors) correctly.

EVALUATE *your answer:* Your calculated results must be consistent with your ray-diagram results. Check that they give the same image position and image size, and that they agree on whether the image is real or virtual.

Example 34.9 | **Image position and magnification with a converging lens**

Use ray diagrams to find the image position and magnification for an object at each of the following distances from a converging lens with a focal length of 20 cm: (a) 50 cm; (b) 20 cm; (c) 15 cm; (d) −40 cm. Check your results by calculating the image position and lateral magnification using Eqs. (34.16) and (34.17), respectively.

SOLUTION

IDENTIFY and SET UP: We are given the focal length $f = 20$ cm and four object distances s. Our target variables are the corresponding image distances s' and lateral magnifications m. We solve Eq. (34.16) for s', and find m from Eq. (34.17), $m = -s'/s$.

EXECUTE: Figures 34.37a, d, e, and f, respectively, show the appropriate principal-ray diagrams. You should be able to reproduce these without referring to the figures. Measuring these diagrams yields the approximate results: $s' = 35$ cm, $-\infty$, -40 cm, and 15 cm, and $m = -\frac{2}{3}$, $+\infty$, $+3$, and $+\frac{1}{3}$, respectively.

Calculating the image distances from Eq. (34.16), we find

(a) $\dfrac{1}{50 \text{ cm}} + \dfrac{1}{s'} = \dfrac{1}{20 \text{ cm}}$ $\quad s' = 33.3$ cm

(b) $\dfrac{1}{20 \text{ cm}} + \dfrac{1}{s'} = \dfrac{1}{20 \text{ cm}}$ $\quad s' = \pm\infty$

(c) $\dfrac{1}{15 \text{ cm}} + \dfrac{1}{s'} = \dfrac{1}{20 \text{ cm}}$ $\quad s' = -60$ cm

(d) $\dfrac{1}{-40 \text{ cm}} + \dfrac{1}{s'} = \dfrac{1}{20 \text{ cm}}$ $\quad s' = 13.3$ cm

The graphical results are fairly close to these except for part (c); the accuracy of the diagram in Fig. 34.37e is limited because the rays extended backward have nearly the same direction.

From Eq. (34.17),

(a) $m = -\dfrac{33.3 \text{ cm}}{50 \text{ cm}} = -\dfrac{2}{3}$ (b) $m = -\dfrac{\pm\infty \text{ cm}}{20 \text{ cm}} = \pm\infty$

(c) $m = -\dfrac{-60 \text{ cm}}{15 \text{ cm}} = +4$ (d) $m = -\dfrac{13.3 \text{ cm}}{-40 \text{ cm}} = +\dfrac{1}{3}$

Continued

EVALUATE: Note that the image distance s' is positive in parts (a) and (d) but negative in part (c). This makes sense: The image is real in parts (a) and (d) but virtual in part (c). The light rays that emerge from the lens in part (b) are parallel and never converge, so the image can be regarded as being at either $+\infty$ or $-\infty$.

The values of magnification m tell us that the image is inverted in part (a) and erect in parts (c) and (d), in agreement with the principal-ray diagrams. The infinite value of magnification in part (b) is another way of saying that the image is formed infinitely far away.

Example 34.10 Image formation by a diverging lens

A beam of parallel rays spreads out after passing through a thin diverging lens, as if the rays all came from a point 20.0 cm from the center of the lens. You want to use this lens to form an erect, virtual image that is $\frac{1}{3}$ the height of the object. (a) Where should the object be placed? Where will the image be? (b) Draw a principal-ray diagram.

SOLUTION

IDENTIFY and SET UP: The result with parallel rays shows that the focal length is $f = -20$ cm. We want the lateral magnification to be $m = +\frac{1}{3}$ (positive because the image is to be erect). Our target variables are the object distance s and the image distance s'. In part (a), we solve the magnification equation, Eq. (34.17), for s' in terms of s; we then use the object–image relationship, Eq. (34.16), to find s and s' individually.

EXECUTE: (a) From Eq. (34.17), $m = +\frac{1}{3} = -s'/s$, so $s' = -s/3$. We insert this result into Eq. (34.16) and solve for the object distance s:

$$\frac{1}{s} + \frac{1}{-s/3} = \frac{1}{s} - \frac{3}{s} = -\frac{2}{s} = \frac{1}{f}$$

$$s = -2f = -2(-20.0 \text{ cm}) = 40.0 \text{ cm}$$

The object should be 40.0 cm from the lens. The image distance will be

$$s' = -\frac{s}{3} = -\frac{40.0 \text{ cm}}{3} = -13.3 \text{ cm}$$

The image distance is negative, so the object and image are on the same side of the lens.

(b) Figure 34.38 is a principal-ray diagram for this problem, with the rays numbered as in Fig. 34.36b.

EVALUATE: You should be able to draw a principal-ray diagram like Fig. 34.38 without referring to the figure. From your diagram, you can confirm our results in part (a) for the object and image distances. You can also check our results for s and s' by substituting them back into Eq. (34.16).

34.38 Principal-ray diagram for an image formed by a thin diverging lens.

Example 34.11 An image of an image

Converging lenses A and B, of focal lengths 8.0 cm and 6.0 cm, respectively, are placed 36.0 cm apart. Both lenses have the same optic axis. An object 8.0 cm high is placed 12.0 cm to the left of lens A. Find the position, size, and orientation of the image produced by the lenses in combination. (Such combinations are used in telescopes and microscopes, to be discussed in Section 34.7.)

SOLUTION

IDENTIFY and SET UP: Figure 34.39 shows the situation. The object O lies outside the first focal point F_1 of lens A, which therefore produces a real image I. The light rays that strike lens B diverge from this real image just as if I was a material object; image I therefore acts as an *object* for lens B. Our goal is to determine the properties of the image I' made by lens B. We use both ray-diagram and computational methods to do this.

EXECUTE: In Fig. 34.39 we have drawn principal rays 1, 2, and 3 from the head of the object arrow O to find the position of the image I made by lens A, and principal rays 1', 2', and 3' from the head of I to find the position of the image I' made by lens B (even though rays 2' and 3' don't actually exist in this case). The image

34.39 Principal-ray diagram for a combination of two converging lenses. The first lens (A) makes a real image of the object. This real image acts as an object for the second lens (B).

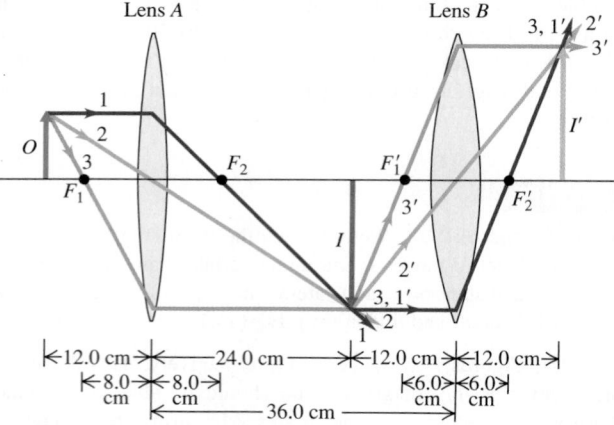

is inverted *twice*, once by each lens, so the second image I' has the same orientation as the original object.

We first find the position and size of the first image I. Applying Eq. (34.16), $1/s + 1/s' = 1/f$, to lens A gives

$$\frac{1}{12.0 \text{ cm}} + \frac{1}{s'_{I,A}} = \frac{1}{8.0 \text{ cm}} \qquad s'_{I,A} = +24.0 \text{ cm}$$

Image I is 24.0 cm to the right of lens A. The lateral magnification is $m_A = -(24.0 \text{ cm})/(12.0 \text{ cm}) = -2.00$, so image I is inverted and twice as tall as object O.

Image I is $36.0 \text{ cm} - 24.0 \text{ cm} = 12.0 \text{ cm}$ to the left of lens B, so the object distance for lens B is $+12.0 \text{ cm}$. Applying Eq. (34.16) to lens B then gives

$$\frac{1}{12.0 \text{ cm}} + \frac{1}{s'_{I',B}} = \frac{1}{6.0 \text{ cm}} \qquad s'_{I',B} = +12.0 \text{ cm}$$

The final image I' is 12.0 cm to the right of lens B. The magnification produced by lens B is $m_B = -(12.0 \text{ cm})/(12.0 \text{ cm}) = -1.00$.

EVALUATE: The value of m_B means that the final image I' is just as large as the first image I but has the opposite orientation. The *overall* magnification is $m_A m_B = (-2.00)(-1.00) = +2.00$. Hence the final image I' is $(2.00)(8.0 \text{ cm}) = 16 \text{ cm}$ tall and has the same orientation as the original object O, just as Fig. 34.39 shows.

Test Your Understanding of Section 34.4 A diverging lens and an object are positioned as shown in the figure at right. Which of the rays A, B, C, and D could emanate from point Q at the top of the object?

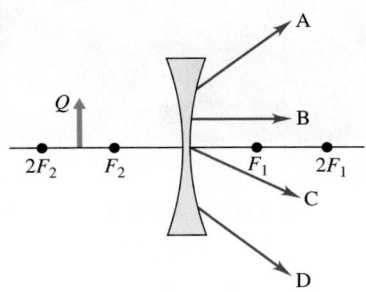

❚

34.5 Cameras

The concept of *image*, which is so central to understanding simple mirror and lens systems, plays an equally important role in the analysis of optical instruments (also called *optical devices*). Among the most common optical devices are cameras, which make an image of an object and record it either electronically or on film.

The basic elements of a **camera** are a light-tight box ("camera" is a Latin word meaning "a room or enclosure"), a converging lens, a shutter to open the lens for a prescribed length of time, and a light-sensitive recording medium (Fig. 34.40). In a digital camera this is an electronic detector called a charge-coupled device (CCD) array; in an older camera, this is photographic film. The lens forms an inverted real image on the recording medium of the object being photographed. High-quality camera lenses have several elements, permitting partial correction of various *aberrations*, including the dependence of index of refraction on wavelength and the limitations imposed by the paraxial approximation.

When the camera is in proper *focus*, the position of the recording medium coincides with the position of the real image formed by the lens. The resulting photograph will then be as sharp as possible. With a converging lens, the image distance increases as the object distance decreases (see Figs. 34.41a, 34.41b, and 34.41c, and the discussion in Section 34.4). Hence in "focusing" the camera, we move the lens closer to the film for a distant object and farther from the film for a nearby object.

34.40 Key elements of a digital camera.

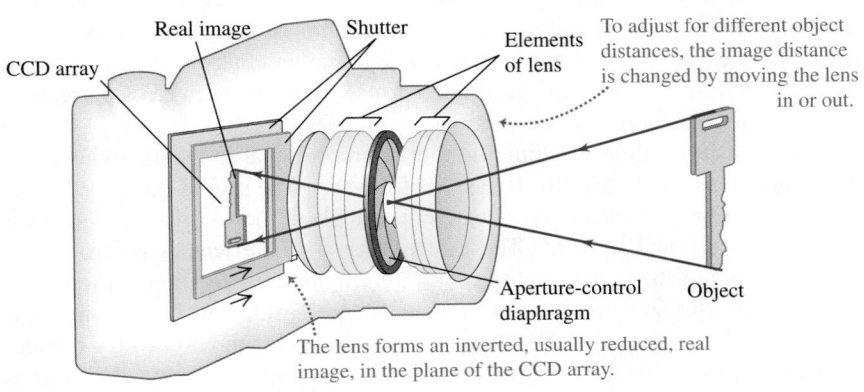

To adjust for different object distances, the image distance is changed by moving the lens in or out.

The lens forms an inverted, usually reduced, real image, in the plane of the CCD array.

34.41 (a), (b), (c) Three photographs taken with the same camera from the same position in the Boston Public Garden using lenses with focal lengths f = 28 mm, 105 mm, and 300 mm. Increasing the focal length increases the image size proportionately. (d) The larger the value of f, the smaller the angle of view. The angles shown here are for a camera with image area 24 mm × 36 mm (corresponding to 35-mm film) and refer to the angle of view along the diagonal dimension of the film.

(a) f = 28 mm

(b) f = 105 mm

(c) f = 300 mm

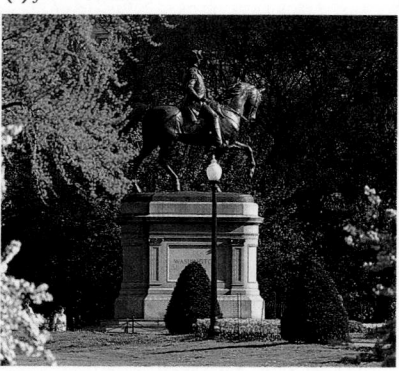

(d) The angles of view for the photos in (a)–(c)

Camera Lenses: Focal Length

The choice of the focal length f for a camera lens depends on the film size and the desired angle of view. Figure 34.41 shows three photographs taken on 35-mm film with the same camera at the same position, but with lenses of different focal lengths. A lens of long focal length, called a *telephoto* lens, gives a small angle of view and a large image of a distant object (such as the statue in Fig. 34.41c); a lens of short focal length gives a small image and a wide angle of view (as in Fig. 34.41a) and is called a *wide-angle* lens. To understand this behavior, recall that the focal length is the distance from the lens to the image when the object is infinitely far away. In general, for *any* object distance, using a lens of longer focal length gives a greater image distance. This also increases the height of the image; as was discussed in Section 34.4, the ratio of the image height y' to the object height y (the *lateral magnification*) is equal in absolute value to the ratio of image distance s' to the object distance s [Eq. (34.17)]:

$$m = \frac{y'}{y} = -\frac{s'}{s}$$

With a lens of short focal length, the ratio s'/s is small, and a distant object gives only a small image. When a lens with a long focal length is used, the image of this same object may entirely cover the area of the film. Hence the longer the focal length, the narrower the angle of view (Fig. 34.41d).

Camera Lenses: f-Number

For the film to record the image properly, the total light energy per unit area reaching the film (the "exposure") must fall within certain limits. This is controlled by the *shutter* and the *lens aperture*. The shutter controls the time interval during which light enters the lens. This is usually adjustable in steps corresponding to factors of about 2, often from 1 s to $\frac{1}{1000}$ s.

The intensity of light reaching the film is proportional to the area viewed by the camera lens and to the effective area of the lens. The size of the area that the lens "sees" is proportional to the square of the angle of view of the lens, and so is roughly proportional to $1/f^2$. The effective area of the lens is controlled by means of an adjustable lens aperture, or *diaphragm,* a nearly circular hole with variable diameter D; hence the effective area is proportional to D^2. Putting these factors together, we see that the intensity of light reaching the film with a particular lens is proportional to D^2/f^2. The light-gathering capability of a lens is commonly expressed by photographers in terms of the ratio f/D, called the **f-number** of the lens:

$$f\text{-number} = \frac{\text{Focal length}}{\text{Aperture diameter}} = \frac{f}{D} \qquad (34.20)$$

For example, a lens with a focal length $f = 50$ mm and an aperture diameter $D = 25$ mm is said to have an f-number of 2, or "an aperture of $f/2$." The light intensity reaching the film is *inversely* proportional to the square of the f-number.

For a lens with a variable-diameter aperture, increasing the diameter by a factor of $\sqrt{2}$ changes the f-number by $1/\sqrt{2}$ and increases the intensity at the film by a factor of 2. Adjustable apertures usually have scales labeled with successive numbers (often called *f-stops*) related by factors of $\sqrt{2}$, such as

$$f/2 \qquad f/2.8 \qquad f/4 \qquad f/5.6 \qquad f/8 \qquad f/11 \qquad f/16$$

and so on. The larger numbers represent smaller apertures and exposures, and each step corresponds to a factor of 2 in intensity (Fig. 34.42). The actual *exposure* (total amount of light reaching the film) is proportional to both the aperture area and the time of exposure. Thus $f/4$ and $\frac{1}{500}$ s, $f/5.6$ and $\frac{1}{250}$ s, and $f/8$ and $\frac{1}{125}$ s all correspond to the same exposure.

Zoom Lenses and Projectors

Many photographers use a *zoom lens,* which is not a single lens but a complex collection of several lens elements that give a continuously variable focal length, often over a range as great as 10 to 1. Figures 34.43a and 34.43b show a simple system with variable focal length, and Fig. 34.43c shows a typical zoom lens for a single-lens reflex camera. Zoom lenses give a range of image sizes of a given object. It is an enormously complex problem in optical design to keep the image in focus and maintain a constant f-number while the focal length changes. When you vary the focal length of a typical zoom lens, two groups of elements move within the lens and a diaphragm opens and closes.

A *projector* for viewing slides, digital images, or motion pictures operates very much like a camera in reverse. In a movie projector, a lamp shines on the film, which acts as an object for the projection lens. The lens forms a real, enlarged, inverted image of the film on the projection screen. Because the image is inverted, the film goes through the projector upside down so that the image on the screen appears right-side up.

34.42 A camera lens with an adjustable diaphragm.

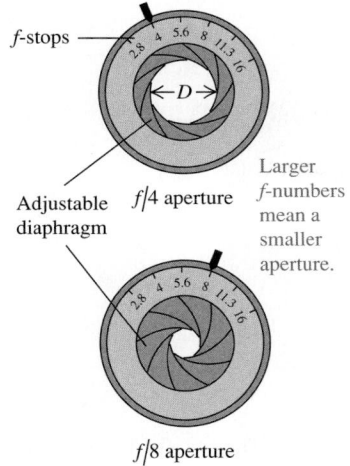

Changing the diameter by a factor of $\sqrt{2}$ changes the intensity by a factor of 2.

f-stops

$\leftarrow D \rightarrow$

Adjustable diaphragm

$f/4$ aperture

Larger f-numbers mean a smaller aperture.

$f/8$ aperture

34.43 A simple zoom lens uses a converging lens and a diverging lens in tandem. (a) When the two lenses are close together, the combination behaves like a single lens of long focal length. (b) If the two lenses are moved farther apart, the combination behaves like a short-focal-length lens. (c) A typical zoom lens for a single-lens reflex camera, containing twelve elements arranged in four groups.

(a) Zoom lens set for long focal length

Image

4 cm   24 cm

(b) Zoom lens set for short focal length

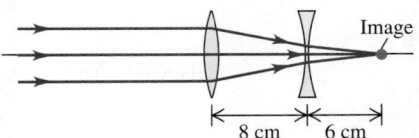

Image

8 cm   6 cm

(c) A practical zoom lens

Example 34.12 | **Photographic exposures**

A common telephoto lens for a 35-mm camera has a focal length of 200 mm; its f-stops range from $f/2.8$ to $f/22$. (a) What is the corresponding range of aperture diameters? (b) What is the corresponding range of image intensities on the film?

SOLUTION

IDENTIFY and SET UP: Part (a) of this problem uses the relationship among lens focal length f, aperture diameter D, and f-number. Part (b) uses the relationship between intensity and aperture diameter. We use Eq. (34.20) to relate D (the target variable) to the f-number and the focal length $f = 200$ mm. The intensity of the light reaching the film is proportional to D^2/f^2; since f is the same in each case, we conclude that the intensity in this case is proportional to D^2, the square of the aperture diameter.

EXECUTE: (a) From Eq. (34.20), the diameter ranges from

$$D = \frac{f}{f\text{-number}} = \frac{200\ \text{mm}}{2.8} = 71\ \text{mm}$$

to

$$D = \frac{200\ \text{mm}}{22} = 9.1\ \text{mm}$$

(b) Because the intensity is proportional to D^2, the ratio of the intensity at $f/2.8$ to the intensity at $f/22$ is

$$\left(\frac{71\ \text{mm}}{9.1\ \text{mm}}\right)^2 = \left(\frac{22}{2.8}\right)^2 = 62 \qquad (\text{about } 2^6)$$

EVALUATE: If the correct exposure time at $f/2.8$ is $\frac{1}{1000}$ s, then the exposure at $f/22$ is $(62)\left(\frac{1}{1000}\ \text{s}\right) = \frac{1}{16}$ s to compensate for the lower intensity. In general, the smaller the aperture and the larger the f-number, the longer the required exposure. Nevertheless, many photographers prefer to use small apertures so that only the central part of the lens is used to make the image. This minimizes aberrations that occur near the edges of the lens and gives the sharpest possible image.

Test Your Understanding of Section 34.5 When used with 35-mm film (image area 24 mm × 36 mm), a lens with $f = 50$ mm gives a 45° angle of view and is called a "normal lens." When used with a CCD array that measures 5 mm × 5 mm, this same lens is (i) a wide-angle lens; (ii) a normal lens; (iii) a telephoto lens.

PhET: Color Vision

34.6 The Eye

The optical behavior of the eye is similar to that of a camera. The essential parts of the human eye, considered as an optical system, are shown in Fig. 34.44a. The eye is nearly spherical and about 2.5 cm in diameter. The front portion is somewhat more sharply curved and is covered by a tough, transparent membrane called the *cornea*. The region behind the cornea contains a liquid called the *aqueous humor.* Next comes the *crystalline lens,* a capsule containing a fibrous jelly, hard at the center and progressively softer at the outer portions. The crystalline lens is

34.44 (a) The eye. (b) There are two types of light-sensitive cells on the retina. The rods are more sensitive to light than the cones, but only the cones are sensitive to differences in color. A typical human eye contains about 1.3×10^8 rods and about 7×10^6 cones.

(a) Diagram of the eye

Contraction of the ciliary ┈┈┈▸ Ciliary muscle causes the lens to muscle become more convex, decreasing its focal length to allow near vision.

Crystalline lens
Iris
Pupil
Object
Cornea
Aqueous humor
Image
Retina
Fovea centralis
Optic nerve
Vitreous humor

(b) Scanning electron micrograph showing retinal rods and cones in different colors

Rod   Cone

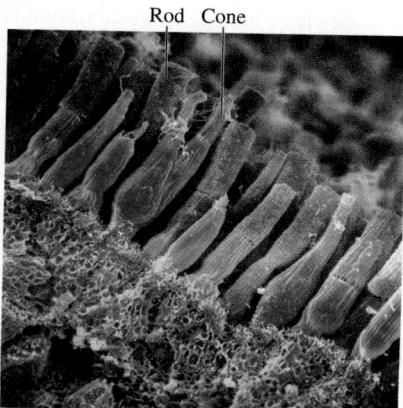

held in place by ligaments that attach it to the ciliary muscle, which encircles it. Behind the lens, the eye is filled with a thin watery jelly called the *vitreous humor*. The indexes of refraction of both the aqueous humor and the vitreous humor are about 1.336, nearly equal to that of water. The crystalline lens, while not homogeneous, has an average index of 1.437. This is not very different from the indexes of the aqueous and vitreous humors. As a result, most of the refraction of light entering the eye occurs at the outer surface of the cornea.

Refraction at the cornea and the surfaces of the lens produces a *real image* of the object being viewed. This image is formed on the light-sensitive *retina*, lining the rear inner surface of the eye. The retina plays the same role as the film in a camera. The *rods* and *cones* in the retina act like an array of miniature photocells (Fig. 34.44b); they sense the image and transmit it via the *optic nerve* to the brain. Vision is most acute in a small central region called the *fovea centralis*, about 0.25 mm in diameter.

In front of the lens is the *iris*. It contains an aperture with variable diameter called the *pupil*, which opens and closes to adapt to changing light intensity. The receptors of the retina also have intensity adaptation mechanisms.

For an object to be seen sharply, the image must be formed exactly at the location of the retina. The eye adjusts to different object distances s by changing the focal length f of its lens; the lens-to-retina distance, corresponding to s', does not change. (Contrast this with focusing a camera, in which the focal length is fixed and the lens-to-film distance is changed.) For the normal eye, an object at infinity is sharply focused when the ciliary muscle is relaxed. To permit sharp imaging on the retina of closer objects, the tension in the ciliary muscle surrounding the lens increases, the ciliary muscle contracts, the lens bulges, and the radii of curvature of its surfaces decrease; this decreases the focal length. This process is called *accommodation*.

The extremes of the range over which distinct vision is possible are known as the *far point* and the *near point* of the eye. The far point of a normal eye is at infinity. The position of the near point depends on the amount by which the ciliary muscle can increase the curvature of the crystalline lens. The range of accommodation gradually diminishes with age because the crystalline lens grows throughout a person's life (it is about 50% larger at age 60 than at age 20) and the ciliary muscles are less able to distort a larger lens. For this reason, the near point gradually recedes as one grows older. This recession of the near point is called *presbyopia*. Table 34.1 shows the approximate position of the near point for an average person at various ages. For example, an average person 50 years of age cannot focus on an object that is closer than about 40 cm.

Defects of Vision

Several common defects of vision result from incorrect distance relationships in the eye. A normal eye forms an image on the retina of an object at infinity when the eye is relaxed (Fig. 34.45a). In the *myopic* (nearsighted) eye, the eyeball is too long from front to back in comparison with the radius of curvature of the cornea (or the cornea is too sharply curved), and rays from an object at infinity are focused in front of the retina (Fig. 34.45b). The most distant object for which an image can be formed on the retina is then nearer than infinity. In the *hyperopic* (farsighted) eye, the eyeball is too short or the cornea is not curved enough, and the image of an infinitely distant object is behind the retina (Fig. 34.45c). The myopic eye produces *too much* convergence in a parallel bundle of rays for an image to be formed on the retina; the hyperopic eye, *not enough* convergence.

All of these defects can be corrected by the use of corrective lenses (eyeglasses or contact lenses). The near point of either a presbyopic or a hyperopic eye is *farther* from the eye than normal. To see clearly an object at normal reading distance (often assumed to be 25 cm), we need a lens that forms a virtual image of the object at or beyond the near point. This can be accomplished by a

Table 34.1 Receding of Near Point with Age

Age (years)	Near Point (cm)
10	7
20	10
30	14
40	22
50	40
60	200

34.45 Refractive errors for **(a)** a normal eye, **(b)** a myopic (nearsighted) eye, and **(c)** a hyperopic (farsighted) eye viewing a very distant object. The dashed blue curve indicates the required position of the retina.

(a) Normal eye

Rays from distant object

(b) Myopic (nearsighted) eye

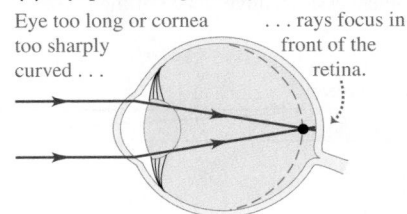

Eye too long or cornea too sharply curved . . .   . . . rays focus in front of the retina.

(c) Hyperopic (farsighted) eye

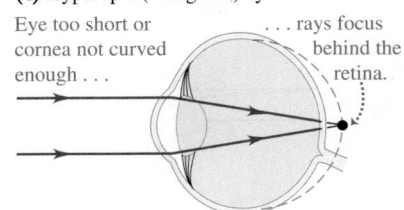

Eye too short or cornea not curved enough . . .   . . . rays focus behind the retina.

34.46 (a) An uncorrected hyperopic (farsighted) eye. (b) A positive (converging) lens gives the extra convergence needed for a hyperopic eye to focus the image on the retina.

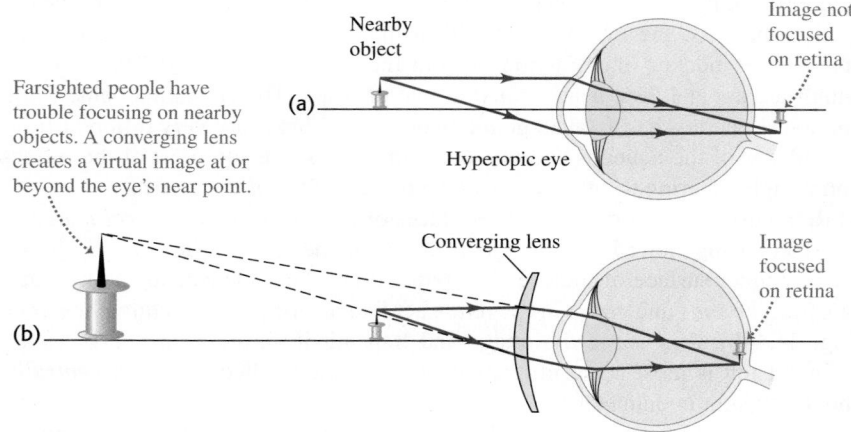

Farsighted people have trouble focusing on nearby objects. A converging lens creates a virtual image at or beyond the eye's near point.

Application Focusing in the Animal Kingdom

The crystalline lens and ciliary muscle found in humans and other mammals are among a number of focusing mechanisms used by animals. Birds can change the shape not only of their lens but also of the corneal surface. In aquatic animals the corneal surface is not very useful for focusing because its refractive index is close to that of water. Thus, focusing is accomplished entirely by the lens, which is nearly spherical. Fish focus by using a muscle to move the lens either inward or outward. Whales and dolphins achieve the same effect by filling or emptying a fluid chamber behind the lens to move the lens in or out.

converging (positive) lens, as shown in Fig. 34.46. In effect the lens moves the object farther away from the eye to a point where a sharp retinal image can be formed. Similarly, correcting the myopic eye involves using a diverging (negative) lens to move the image closer to the eye than the actual object, as shown in Fig. 34.47.

Astigmatism is a different type of defect in which the surface of the cornea is not spherical but rather more sharply curved in one plane than in another. As a result, horizontal lines may be imaged in a different plane from vertical lines (Fig. 34.48a). Astigmatism may make it impossible, for example, to focus clearly on both the horizontal and vertical bars of a window at the same time.

Astigmatism can be corrected by use of a lens with a *cylindrical* surface. For example, suppose the curvature of the cornea in a horizontal plane is correct to focus rays from infinity on the retina but the curvature in the vertical plane is too great to form a sharp retinal image. When a cylindrical lens with its axis horizontal is placed before the eye, the rays in a horizontal plane are unaffected, but the additional divergence of the rays in a vertical plane causes these to be sharply imaged on the retina (Fig. 34.48b).

Lenses for vision correction are usually described in terms of the **power,** defined as the reciprocal of the focal length expressed in meters. The unit of power is the **diopter.** Thus a lens with $f = 0.50$ m has a power of 2.0 diopters, $f = -0.25$ m corresponds to -4.0 diopters, and so on. The numbers on a prescription for glasses

34.47 (a) An uncorrected myopic (nearsighted) eye. (b) A negative (diverging) lens spreads the rays farther apart to compensate for the excessive convergence of the myopic eye.

(a) Vertical lines are imaged in front of the retina.

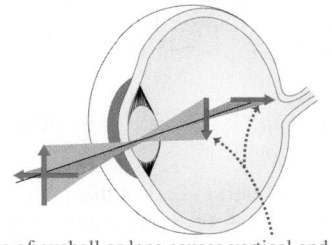

Shape of eyeball or lens causes vertical and horizontal elements to focus at different distances.

(b) A cylindrical lens corrects for astigmatism.

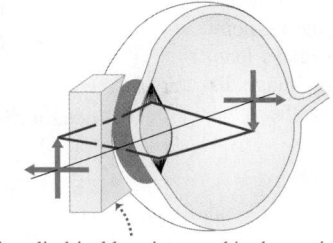

This cylindrical lens is curved in the vertical, but not the horizontal, direction; it changes the focal length of vertical elements.

34.48 One type of astigmatism and how it is corrected.

are usually powers expressed in diopters. When the correction involves both astigmatism and myopia or hyperopia, there are three numbers: one for the spherical power, one for the cylindrical power, and an angle to describe the orientation of the cylinder axis.

Example 34.13   **Correcting for farsightedness**

The near point of a certain hyperopic eye is 100 cm in front of the eye. Find the focal length and power of the contact lens that will permit the wearer to see clearly an object that is 25 cm in front of the eye.

SOLUTION

IDENTIFY and SET UP: Figure 34.49 shows the situation. We want the lens to form a virtual image of the object at the near point of the eye, 100 cm from it. The contact lens (which we treat as having negligible thickness) is at the surface of the cornea, so the object distance is $s = 25$ cm. The virtual image is on the incoming side of the contact lens, so the image distance is $s' = -100$ cm. We use Eq. (34.16) to determine the required focal length f of the contact lens; the corresponding power is $1/f$.

EXECUTE: From Eq. (34.16),

$$\frac{1}{f} = \frac{1}{s} + \frac{1}{s'} = \frac{1}{+25 \text{ cm}} + \frac{1}{-100 \text{ cm}}$$

$$f = +33 \text{ cm}$$

34.49 Using a contact lens to correct for farsightedness. For clarity, the eye and contact lens are shown much larger than the scale of the figure; the 2.5-cm diameter of the eye is actually much smaller than the focal length f of the contact lens.

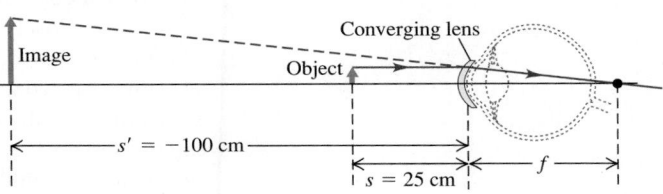

We need a converging lens with focal length $f = 33$ cm and power $1/(0.33 \text{ m}) = +3.0$ diopters.

EVALUATE: In this example we used a contact lens to correct hyperopia. Had we used eyeglasses, we would have had to account for the separation between the eye and the eyeglass lens, and a somewhat different power would have been required (see Example 34.14).

Example 34.14   **Correcting for nearsightedness**

The far point of a certain myopic eye is 50 cm in front of the eye. Find the focal length and power of the eyeglass lens that will permit the wearer to see clearly an object at infinity. Assume that the lens is worn 2 cm in front of the eye.

SOLUTION

IDENTIFY and SET UP: Figure 34.50 shows the situation. The far point of a myopic eye is nearer than infinity. To see clearly objects

34.50 Using an eyeglass lens to correct for nearsightedness. For clarity, the eye and eyeglass lens are shown much larger than the scale of the figure.

Continued

beyond the far point, we need a lens that forms a virtual image of such objects no farther from the eye than the far point. Assume that the virtual image of the object at infinity is formed at the far point, 50 cm in front of the eye (48 cm in front of the eyeglass lens). Then when the object distance is $s = \infty$, we want the image distance to be $s' = -48$ cm. As in Example 34.13, we use the values of s and s' to calculate the required focal length.

EXECUTE: From Eq. (34.16),

$$\frac{1}{f} = \frac{1}{s} + \frac{1}{s'} = \frac{1}{\infty} + \frac{1}{-48 \text{ cm}}$$

$$f = -48 \text{ cm}$$

We need a *diverging* lens with focal length $f = -48$ cm and power $1/(-0.48 \text{ m}) = -2.1$ diopters.

EVALUATE: If a *contact* lens were used to correct this myopia, we would need $f = -50$ cm and a power of -2.0 diopters. Can you see why?

Test Your Understanding of Section 34.6 A certain eyeglass lens is thin at its center, even thinner at its top and bottom edges, and relatively thick at its left and right edges. What defects of vision is this lens intended to correct? (i) hyperopia for objects oriented both vertically and horizontally; (ii) myopia for objects oriented both vertically and horizontally; (iii) hyperopia for objects oriented vertically and myopia for objects oriented horizontally; (iv) hyperopia for objects oriented horizontally and myopia for objects oriented vertically.

34.7 The Magnifier

The apparent size of an object is determined by the size of its image on the retina. If the eye is unaided, this size depends on the *angle* θ subtended by the object at the eye, called its **angular size** (Fig. 34.51a).

To look closely at a small object, such as an insect or a crystal, you bring it close to your eye, making the subtended angle and the retinal image as large as possible. But your eye cannot focus sharply on objects that are closer than the near point, so the angular size of an object is greatest (that is, it subtends the largest possible viewing angle) when it is placed at the near point. In the following discussion we will assume an average viewer for whom the near point is 25 cm from the eye.

A converging lens can be used to form a virtual image that is larger and farther from the eye than the object itself, as shown in Fig. 34.51b. Then the object can be moved closer to the eye, and the angular size of the image may be substantially larger than the angular size of the object at 25 cm without the lens. A lens used in this way is called a **magnifier,** otherwise known as a *magnifying glass* or a *simple magnifier.* The virtual image is most comfortable to view when it is placed at infinity, so that the ciliary muscle of the eye is relaxed; this means that the object is placed at the focal point F_1 of the magnifier. In the following discussion we assume that this is done.

34.51 (a) The angular size θ is largest when the object is at the near point. (b) The magnifier gives a virtual image at infinity. This virtual image appears to the eye to be a real object subtending a larger angle θ' at the eye.

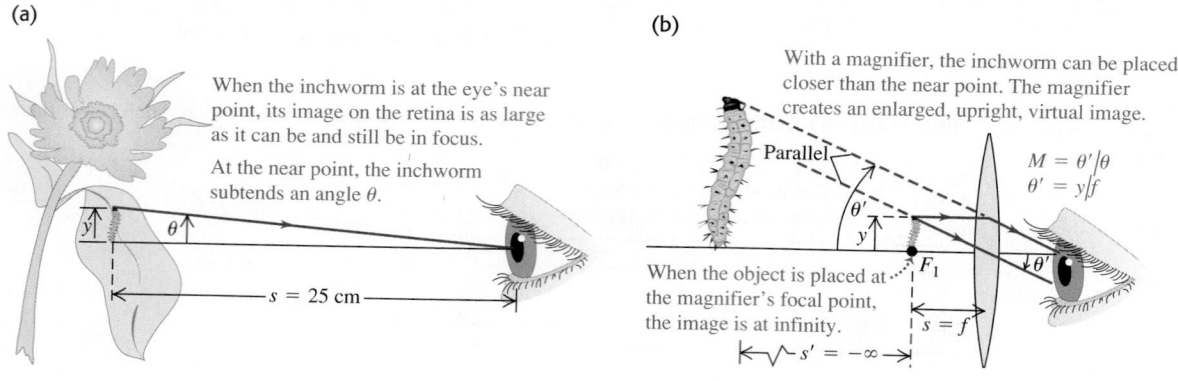

In Fig. 34.51a the object is at the near point, where it subtends an angle θ at the eye. In Fig. 34.51b a magnifier in front of the eye forms an image at infinity, and the angle subtended at the magnifier is θ'. The usefulness of the magnifier is given by the ratio of the angle θ' (with the magnifier) to the angle θ (without the magnifier). This ratio is called the **angular magnification** M:

$$M = \frac{\theta'}{\theta} \quad \text{(angular magnification)} \tag{34.21}$$

CAUTION **Angular magnification vs. lateral magnification** Don't confuse the *angular* magnification M with the *lateral* magnification m. Angular magnification is the ratio of the *angular* size of an image to the angular size of the corresponding object; lateral magnification refers to the ratio of the *height* of an image to the height of the corresponding object. For the situation shown in Fig. 34.51b, the angular magnification is about $3\times$, since the inchworm subtends an angle about three times larger than that in Fig. 34.51a; hence the inchworm will look about three times larger to the eye. The *lateral* magnification $m = -s'/s$ in Fig. 34.51b is *infinite* because the virtual image is at infinity, but that doesn't mean that the inchworm looks infinitely large through the magnifier! (That's why we didn't attempt to draw an infinitely large inchworm in Fig. 34.51b.) When dealing with a magnifier, M is useful but m is not.

To find the value of M, we first assume that the angles are small enough that each angle (in radians) is equal to its sine and its tangent. Using Fig. 34.451a and drawing the ray in Fig. 34.51b that passes undeviated through the center of the lens, we find that θ and θ' (in radians) are

$$\theta = \frac{y}{25 \text{ cm}} \qquad \theta' = \frac{y}{f}$$

Combining these expressions with Eq. (34.21), we find

$$M = \frac{\theta'}{\theta} = \frac{y/f}{y/25 \text{ cm}} = \frac{25 \text{ cm}}{f} \quad \begin{array}{l}\text{(angular magnification} \\ \text{for a simple magnifier)}\end{array} \tag{34.22}$$

It may seem that we can make the angular magnification as large as we like by decreasing the focal length f. In fact, the aberrations of a simple double-convex lens set a limit to M of about $3\times$ to $4\times$. If these aberrations are corrected, the angular magnification may be made as great as $20\times$. When greater magnification than this is needed, we usually use a compound microscope, discussed in the next section.

Test Your Understanding of Section 34.7 You are examining a gem using a magnifier. If you change to a different magnifier with twice the focal length of the first one, (i) you will have to hold the object at twice the distance and the angular magnification will be twice as great; (ii) you will have to hold the object at twice the distance and the angular magnification will be $\frac{1}{2}$ as great; (iii) you will have to hold the object at $\frac{1}{2}$ the distance and the angular magnification will be twice as great; (iv) you will have to hold the object at $\frac{1}{2}$ the distance and the angular magnification will be $\frac{1}{2}$ as great.

34.8 Microscopes and Telescopes

Cameras, eyeglasses, and magnifiers use a single lens to form an image. Two important optical devices that use *two* lenses are the microscope and the telescope. In each device a primary lens, or *objective,* forms a real image, and a second lens, or *eyepiece,* is used as a magnifier to make an enlarged, virtual image.

Microscopes

When we need greater magnification than we can get with a simple magnifier, the instrument that we usually use is the **microscope,** sometimes called a *compound*

Mastering**PHYSICS**

ActivPhysics 15.12: Two-Lens Optical Systems

34.52 (a) Elements of a microscope. (b) The object *O* is placed just outside the first focal point of the objective (the distance s_1 has been exaggerated for clarity). (c) This microscope image shows single-celled organisms about 2×10^{-4} m (0.2 mm) across. Typical light microscopes can resolve features as small as 2×10^{-7} m, comparable to the wavelength of light.

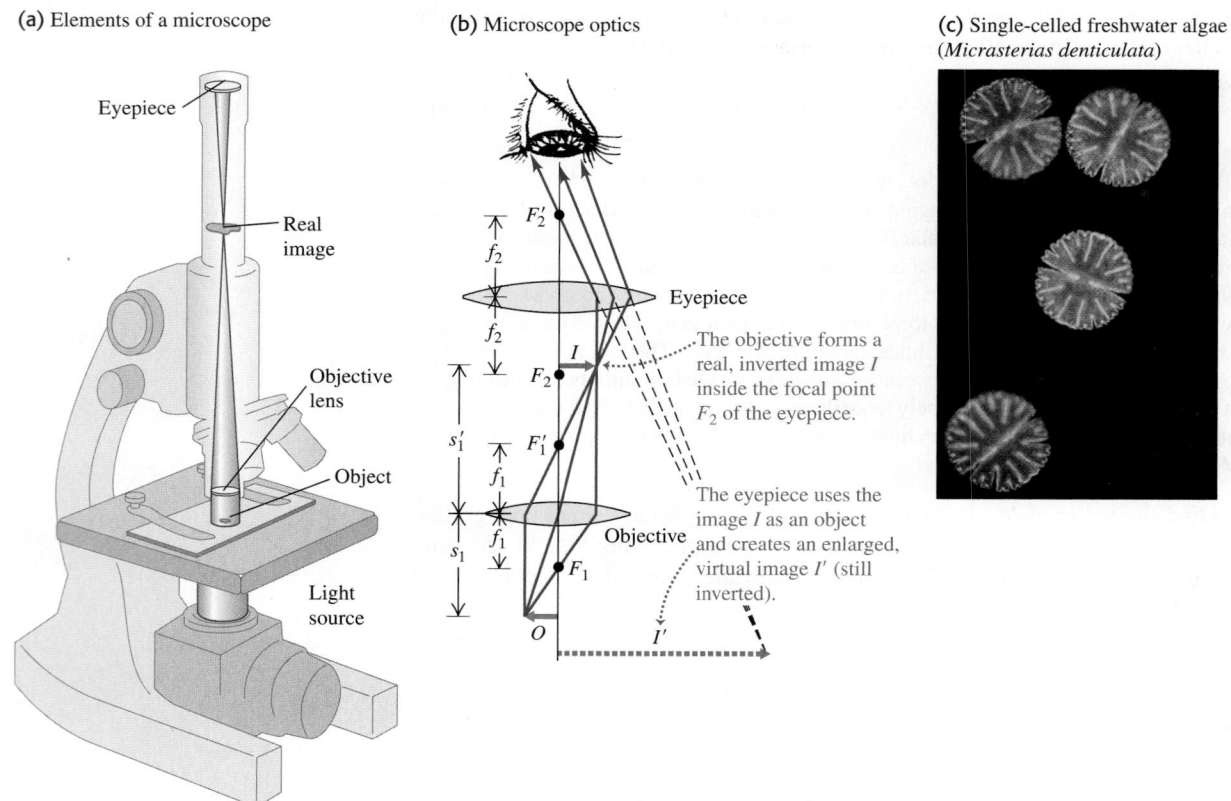

(a) Elements of a microscope

Eyepiece

Real image

Objective lens

Object

Light source

(b) Microscope optics

F_2'

f_2

f_2

Eyepiece

F_2

I

The objective forms a real, inverted image *I* inside the focal point F_2 of the eyepiece.

s_1'

F_1'

f_1

f_1

Objective

s_1

F_1

The eyepiece uses the image *I* as an object and creates an enlarged, virtual image *I'* (still inverted).

O

I'

(c) Single-celled freshwater algae (*Micrasterias denticulata*)

microscope. The essential elements of a microscope are shown in Fig. 34.52a. To analyze this system, we use the principle that an image formed by one optical element such as a lens or mirror can serve as the object for a second element. We used this principle in Section 34.4 when we derived the thin-lens equation by repeated application of the single-surface refraction equation; we used this principle again in Example 34.11 (Section 34.4), in which the image formed by a lens was used as the object of a second lens.

The object *O* to be viewed is placed just beyond the first focal point F_1 of the **objective,** a converging lens that forms a real and enlarged image *I* (Fig. 34.52b). In a properly designed instrument this image lies just inside the first focal point F_1' of a second converging lens called the **eyepiece** or *ocular.* (The reason the image should lie just *inside* F_1' is left for you to discover; see Problem 34.108.) The eyepiece acts as a simple magnifier, as discussed in Section 34.7, and forms a final virtual image *I'* of *I*. The position of *I'* may be anywhere between the near and far points of the eye. Both the objective and the eyepiece of an actual microscope are highly corrected compound lenses with several optical elements, but for simplicity we show them here as simple thin lenses.

As for a simple magnifier, what matters when viewing through a microscope is the *angular* magnification *M*. The overall angular magnification of the compound microscope is the product of two factors. The first factor is the *lateral* magnification m_1 of the objective, which determines the linear size of the real image *I*; the second factor is the *angular* magnification M_2 of the eyepiece, which relates the angular size of the virtual image seen through the eyepiece to the angular size that the real image *I* would have if you viewed it *without* the eyepiece. The first of these factors is given by

$$m_1 = -\frac{s_1'}{s_1} \qquad (34.23)$$

where s_1 and s_1' are the object and image distances, respectively, for the objective lens. Ordinarily, the object is very close to the focal point, and the resulting image distance s_1' is very great in comparison to the focal length f_1 of the objective lens. Thus s_1 is approximately equal to f_1, and we can write $m_1 = -s_1'/f_1$.

The real image I is close to the focal point F_1' of the eyepiece, so to find the eyepiece angular magnification, we can use Eq. (34.22): $M_2 = (25 \text{ cm})/f_2$, where f_2 is the focal length of the eyepiece (considered as a simple lens). The overall angular magnification M of the compound microscope (apart from a negative sign, which is customarily ignored) is the product of the two magnifications:

$$M = m_1 M_2 = \frac{(25 \text{ cm})s_1'}{f_1 f_2} \qquad \begin{array}{l}\text{(angular magnification} \\ \text{for a microscope)}\end{array} \qquad (34.24)$$

where s_1', f_1, and f_2 are measured in centimeters. The final image is inverted with respect to the object. Microscope manufacturers usually specify the values of m_1 and M_2 for microscope components rather than the focal lengths of the objective and eyepiece.

Equation (34.24) shows that the angular magnification of a microscope can be increased by using an objective of shorter focal length f_1, thereby increasing m_1 and the size of the real image I. Most optical microscopes have a rotating "turret" with three or more objectives of different focal lengths so that the same object can be viewed at different magnifications. The eyepiece should also have a short focal length f_2 to help to maximize the value of M.

To take a photograph using a microscope (called a *photomicrograph* or *micrograph*), the eyepiece is removed and a camera placed so that the real image I falls on the camera's CCD array or film. Figure 34.52c shows such a photograph. In this case what matters is the *lateral* magnification of the microscope as given by Eq. (34.23).

Telescopes

The optical system of a **telescope** is similar to that of a compound microscope. In both instruments the image formed by an objective is viewed through an eyepiece. The key difference is that the telescope is used to view large objects at large distances and the microscope is used to view small objects close at hand. Another difference is that many telescopes use a curved mirror, not a lens, as an objective.

Figure 34.53 shows an *astronomical telescope*. Because this telescope uses a lens as an objective, it is called a *refracting telescope* or *refractor*. The objective lens forms a real, reduced image I of the object. This image is the object for the eyepiece lens, which forms an enlarged, virtual image of I. Objects that are viewed with a telescope are usually so far away from the instrument that the first image I is formed very nearly at the second focal point of the objective lens. If the final image I' formed by the eyepiece is at infinity (for most comfortable viewing by a normal eye), the first image must also be at the first focal point of the eyepiece. The distance between objective and eyepiece, which is the length of the telescope, is therefore the *sum* of the focal lengths of objective and eyepiece, $f_1 + f_2$.

The angular magnification M of a telescope is defined as the ratio of the angle subtended at the eye by the final image I' to the angle subtended at the (unaided) eye by the object. We can express this ratio in terms of the focal lengths of objective and eyepiece. In Fig. 34.53 the ray passing through F_1, the first focal point of the objective, and through F_2', the second focal point of the eyepiece, is shown in red. The object (not shown) subtends an angle θ at the objective and would subtend essentially the same angle at the unaided eye. Also, since the observer's eye is placed just to the right of the focal point F_2', the angle subtended at the eye by the final image is very nearly equal to the angle θ'. Because bd is parallel to the

34.53 Optical system of an astronomical refracting telescope.

The objective forms a real, inverted image *I* of the distant object at its second focal point F_1', which is also the first focal point F_2 of the eyepiece.

The eyepiece uses image *I* to form a magnified, virtual image *I'* (still inverted) at infinity.

optic axis, the distances *ab* and *cd* are equal to each other and also to the height y' of the real image *I*. Because the angles θ and θ' are small, they may be approximated by their tangents. From the right triangles F_1ab and $F_2'cd$,

$$\theta = \frac{-y'}{f_1} \qquad \theta' = \frac{y'}{f_2}$$

and the angular magnification *M* is

$$M = \frac{\theta'}{\theta} = -\frac{y'/f_2}{y'/f_1} = -\frac{f_1}{f_2} \qquad \text{(angular magnification for a telescope)} \qquad (34.25)$$

The angular magnification *M* of a telescope is equal to the ratio of the focal length of the objective to that of the eyepiece. The negative sign shows that the final image is inverted. Equation (34.25) shows that to achieve good angular magnification, a *telescope* should have a *long* objective focal length f_1. By contrast, Eq. (34.24) shows that a *microscope* should have a *short* objective focal length. However, a telescope objective with a long focal length should also have a large diameter *D* so that the *f*-number f_1/D will not be too large; as described in Section 34.5, a large *f*-number means a dim, low-intensity image. Telescopes typically do not have interchangeable objectives; instead, the magnification is varied by using different eyepieces with different focal lengths f_2. Just as for a microscope, smaller values of f_2 give larger angular magnifications.

An inverted image is no particular disadvantage for astronomical observations. When we use a telescope or binoculars—essentially a pair of telescopes mounted side by side—to view objects on the earth, though, we want the image to be right-side up. In prism binoculars, this is accomplished by reflecting the light several times along the path from the objective to the eyepiece. The combined effect of the reflections is to flip the image both horizontally and vertically. Binoculars are usually described by two numbers separated by a multiplication sign, such as 7 × 50. The first number is the angular magnification *M*, and the second is the diameter of the objective lenses (in millimeters). The diameter helps to determine the light-gathering capacity of the objective lenses and thus the brightness of the image.

In the *reflecting telescope* (Fig. 34.54a) the objective lens is replaced by a concave mirror. In large telescopes this scheme has many advantages. Mirrors are

34.54 (a), (b), (c) Three designs for reflecting telescopes. (d) This photo shows the interior of the Gemini North telescope, which uses the design shown in (c). The objective mirror is 8 meters in diameter.

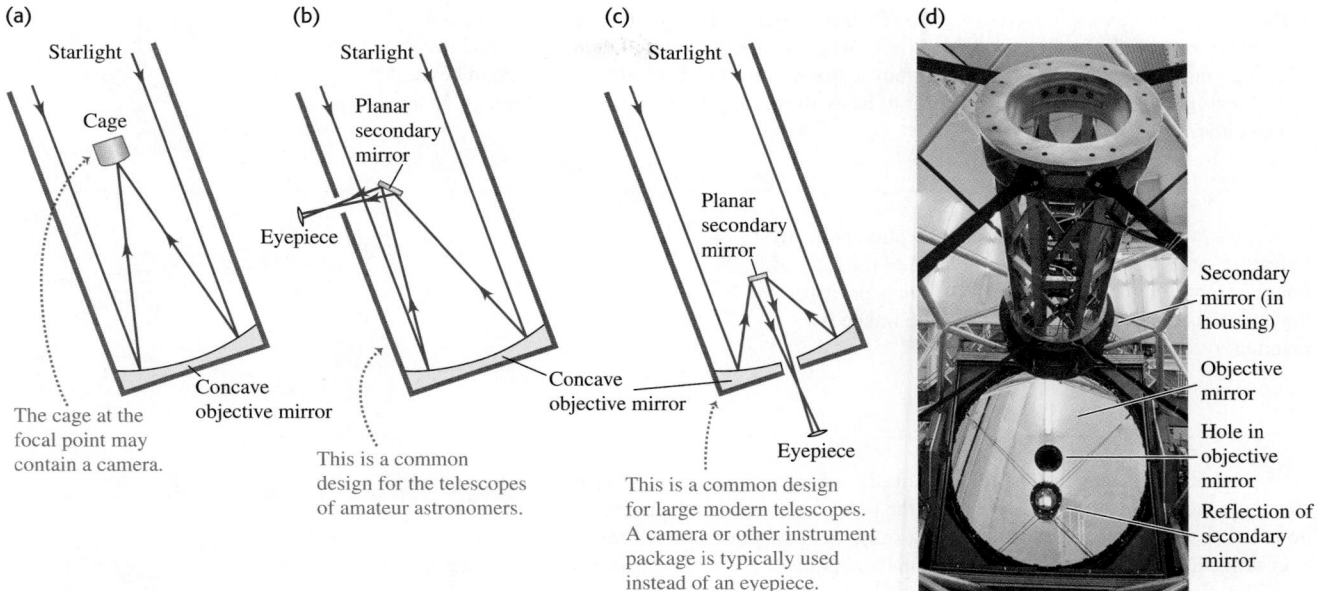

(a) Starlight

Cage

The cage at the focal point may contain a camera.

Concave objective mirror

(b) Starlight

Planar secondary mirror

Eyepiece

Concave objective mirror

This is a common design for the telescopes of amateur astronomers.

(c) Starlight

Planar secondary mirror

Concave objective mirror

Eyepiece

This is a common design for large modern telescopes. A camera or other instrument package is typically used instead of an eyepiece.

(d)

Secondary mirror (in housing)

Objective mirror

Hole in objective mirror

Reflection of secondary mirror

inherently free of chromatic aberrations (dependence of focal length on wavelength), and spherical aberrations (associated with the paraxial approximation) are easier to correct than with a lens. The reflecting surface is sometimes parabolic rather than spherical. The material of the mirror need not be transparent, and it can be made more rigid than a lens, which has to be supported only at its edges.

The largest reflecting telescopes in the world, the Keck telescopes atop Mauna Kea in Hawaii, each have an objective mirror of overall diameter 10 m made up of 36 separate hexagonal reflecting elements.

One challenge in designing reflecting telescopes is that the image is formed in front of the objective mirror, in a region traversed by incoming rays. Isaac Newton devised one solution to this problem. A flat secondary mirror oriented at 45° to the optic axis causes the image to be formed in a hole on the side of the telescope, where it can be magnified with an eyepiece (Fig. 34.54b). Another solution uses a secondary mirror that causes the focused light to pass through a hole in the objective mirror (Fig. 34.54c). Large research telescopes, as well as many amateur telescopes, use this design (Fig. 34.54d).

Like a microscope, when a telescope is used for photography the eyepiece is removed and a CCD array or photographic film is placed at the position of the real image formed by the objective. (Some long-focal-length "lenses" for photography are actually reflecting telescopes used in this way.) Most telescopes used for astronomical research are never used with an eyepiece.

Test Your Understanding of Section 34.8 Which gives a lateral magnification of greater absolute value: (i) the objective lens in a microscope (Fig. 34.52); (ii) the objective lens in a refracting telescope (Fig. 34.53); or (iii) not enough information is given to decide?

Reflection or refraction at a plane surface: When rays diverge from an object point P and are reflected or refracted, the directions of the outgoing rays are the same as though they had diverged from a point P' called the image point. If they actually converge at P' and diverge again beyond it, P' is a real image of P; if they only appear to have diverged from P', it is a virtual image. Images can be either erect or inverted.

Plane mirror

Lateral magnification: The lateral magnification m in any reflecting or refracting situation is defined as the ratio of image height y' to object height y. When m is positive, the image is erect; when m is negative, the image is inverted.

$$m = \frac{y'}{y} \qquad (34.2)$$

Focal point and focal length: The focal point of a mirror is the point where parallel rays converge after reflection from a concave mirror, or the point from which they appear to diverge after reflection from a convex mirror. Rays diverging from the focal point of a concave mirror are parallel after reflection; rays converging toward the focal point of a convex mirror are parallel after reflection. The distance from the focal point to the vertex is called the focal length, denoted as f. The focal points of a lens are defined similarly.

Relating object and image distances: The formulas for object distance s and image distance s' for plane and spherical mirrors and single refracting surfaces are summarized in the table. The equation for a plane surface can be obtained from the corresponding equation for a spherical surface by setting $R = \infty$. (See Examples 34.1–34.7.)

	Plane Mirror	Spherical Mirror	Plane Refracting Surface	Spherical Refracting Surface
Object and image distances	$\frac{1}{s} + \frac{1}{s'} = 0$	$\frac{1}{s} + \frac{1}{s'} = \frac{2}{R} = \frac{1}{f}$	$\frac{n_a}{s} + \frac{n_b}{s'} = 0$	$\frac{n_a}{s} + \frac{n_b}{s'} = \frac{n_b - n_a}{R}$
Lateral magnification	$m = -\frac{s'}{s} = 1$	$m = -\frac{s'}{s}$	$m = -\frac{n_a s'}{n_b s} = 1$	$m = -\frac{n_a s'}{n_b s}$

Object–image relationships derived in this chapter are valid only for rays close to and nearly parallel to the optic axis; these are called paraxial rays. Nonparaxial rays do not converge precisely to an image point. This effect is called spherical aberration.

Thin lenses: The object–image relationship, given by Eq. (34.16), is the same for a thin lens as for a spherical mirror. Equation (34.19), the lensmaker's equation, relates the focal length of a lens to its index of refraction and the radii of curvature of its surfaces. (See Examples 34.8–34.11.)

$$\frac{1}{s} + \frac{1}{s'} = \frac{1}{f} \qquad (34.16)$$

$$\frac{1}{f} = (n - 1)\left(\frac{1}{R_1} - \frac{1}{R_2}\right) \qquad (34.19)$$

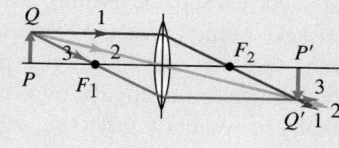

Sign rules: The following sign rules are used with all plane and spherical reflecting and refracting surfaces.

- $s > 0$ when the object is on the incoming side of the surface (a real object); $s < 0$ otherwise.

- $s' > 0$ when the image is on the outgoing side of the surface (a real image); $s' < 0$ otherwise.
- $R > 0$ when the center of curvature is on the outgoing side of the surface; $R < 0$ otherwise.
- $m > 0$ when the image is erect; $m < 0$ when inverted.

Cameras: A camera forms a real, inverted, reduced image of the object being photographed on a light-sensitive surface. The amount of light striking this surface is controlled by the shutter speed and the aperture. The intensity of this light is inversely proportional to the square of the f-number of the lens. (See Example 34.12.)

$$f\text{-number} = \frac{\text{Focal length}}{\text{Aperture diameter}}$$

$$= \frac{f}{D} \tag{34.20}$$

Object

Inverted, real image

The eye: In the eye, refraction at the surface of the cornea forms a real image on the retina. Adjustment for various object distances is made by squeezing the lens, thereby making it bulge and decreasing its focal length. A nearsighted eye is too long for its lens; a farsighted eye is too short. The power of a corrective lens, in diopters, is the reciprocal of the focal length in meters. (See Examples 34.13 and 34.14.)

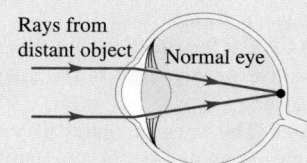

Rays from distant object   Normal eye

The simple magnifier: The simple magnifier creates a virtual image whose angular size θ' is larger than the angular size θ of the object itself at a distance of 25 cm, the nominal closest distance for comfortable viewing. The angular magnification M of a simple magnifier is the ratio of the angular size of the virtual image to that of the object at this distance.

$$M = \frac{\theta'}{\theta} = \frac{25\text{ cm}}{f} \tag{34.22}$$

Microscopes and telescopes: In a compound microscope, the objective lens forms a first image in the barrel of the instrument, and the eyepiece forms a final virtual image, often at infinity, of the first image. The telescope operates on the same principle, but the object is far away. In a reflecting telescope, the objective lens is replaced by a concave mirror, which eliminates chromatic aberrations.

BRIDGING PROBLEM | **Image Formation by a Wine Goblet**

A thick-walled wine goblet can be considered to be a hollow glass sphere with an outer radius of 4.00 cm and an inner radius of 3.40 cm. The index of refraction of the goblet glass is 1.50. (a) A beam of parallel light rays enters the side of the empty goblet along a horizontal radius. Where, if anywhere, will an image be formed? (b) The goblet is filled with white wine ($n = 1.37$). Where is the image formed?

SOLUTION GUIDE

See MasteringPhysics® study area for a Video Tutor solution.

IDENTIFY and SET UP

1. The goblet is *not* a thin lens, so you cannot use the thin-lens formula. Instead, you must think of the inner and outer surfaces of the goblet walls as spherical refracting surfaces. The image formed by one surface serves as the object for the next surface.
2. Choose the appropriate equation that relates the image and object distances for a spherical refracting surface.

EXECUTE

3. For the empty goblet, each refracting surface has glass on one side and air on the other. Find the position of the image formed by the first surface, the outer wall of the goblet. Use this as the object for the second surface (the inner wall of the same side of the goblet) and find the position of the second image. (*Hint:* Be sure to account for the thickness of the goblet wall.)
4. Continue the process of step 3. Consider the refractions at the inner and outer surfaces of the glass on the opposite side of the goblet, and find the position of the final image. (*Hint:* Be sure to account for the distance between the two sides of the goblet.)
5. Repeat steps 3 and 4 for the case in which the goblet is filled with wine.

EVALUATE

6. Are the images real or virtual? How can you tell?

Problems

•, ••, •••: Problems of increasing difficulty. **CP**: Cumulative problems incorporating material from earlier chapters. **CALC**: Problems requiring calculus. **BIO**: Biosciences problems.

DISCUSSION QUESTIONS

Q34.1 A spherical mirror is cut in half horizontally. Will an image be formed by the bottom half of the mirror? If so, where will the image be formed?

Q34.2 For the situation shown in Fig. 34.3, is the image distance s' positive or negative? Is the image real or virtual? Explain your answers.

Q34.3 The laws of optics also apply to electromagnetic waves invisible to the eye. A satellite TV dish is used to detect radio waves coming from orbiting satellites. Why is a curved reflecting surface (a "dish") used? The dish is always concave, never convex; why? The actual radio receiver is placed on an arm and suspended in front of the dish. How far in front of the dish should it be placed?

Q34.4 Explain why the focal length of a *plane* mirror is infinite, and explain what it means for the focal point to be at infinity.

Q34.5 If a spherical mirror is immersed in water, does its focal length change? Explain.

Q34.6 For what range of object positions does a concave spherical mirror form a real image? What about a convex spherical mirror?

Q34.7 When a room has mirrors on two opposite walls, an infinite series of reflections can be seen. Discuss this phenomenon in terms of images. Why do the distant images appear fainter?

Q34.8 For a spherical mirror, if $s = f$, then $s' = \infty$, and the lateral magnification m is infinite. Does this make sense? If so, what does it mean?

Q34.9 You may have noticed a small convex mirror next to your bank's ATM. Why is this mirror convex, as opposed to flat or concave? What considerations determine its radius of curvature?

Q34.10 A student claims that she can start a fire on a sunny day using just the sun's rays and a concave mirror. How is this done? Is the concept of image relevant? Can she do the same thing with a convex mirror? Explain.

Q34.11 A person looks at his reflection in the concave side of a shiny spoon. Is it right side up or inverted? Does it matter how far his face is from the spoon? What if he looks in the convex side? (Try this yourself!)

Q34.12 In Example 34.4 (Section 34.2), there appears to be an ambiguity for the case $s = 10$ cm as to whether s' is $+\infty$ or $-\infty$ and whether the image is erect or inverted. How is this resolved? Or is it?

Q34.13 Suppose that in the situation of Example 34.7 of Section 34.3 (see Fig. 34.26) a vertical arrow 2.00 m tall is painted on the side of the pool beneath the water line. According to the calculations in the example, this arrow would appear to the person shown in Fig. 34.26 to be 1.50 m long. But the discussion following Eq. (34.13) states that the magnification for a plane refracting surface is $m = 1$, which suggests that the arrow would appear to the person to be 2.00 m long. How can you resolve this apparent contradiction?

Q34.14 The bottom of the passenger-side mirror on your car notes, "Objects in mirror are closer than they appear." Is this true? Why?

Q34.15 How could you very quickly make an approximate measurement of the focal length of a converging lens? Could the same method be applied if you wished to use a diverging lens? Explain.

Q34.16 The focal length of a simple lens depends on the color (wavelength) of light passing through it. Why? Is it possible for a lens to have a positive focal length for some colors and negative for others? Explain.

Q34.17 When a converging lens is immersed in water, does its focal length increase or decrease in comparison with the value in air? Explain.

Q34.18 A spherical air bubble in water can function as a lens. Is it a converging or diverging lens? How is its focal length related to its radius?

Q34.19 Can an image formed by one reflecting or refracting surface serve as an object for a second reflection or refraction? Does it matter whether the first image is real or virtual? Explain.

Q34.20 If a piece of photographic film is placed at the location of a real image, the film will record the image. Can this be done with a virtual image? How might one record a virtual image?

Q34.21 According to the discussion in Section 34.2, light rays are reversible. Are the formulas in the table in this chapter's Summary still valid if object and image are interchanged? What does reversibility imply with respect to the *forms* of the various formulas?

Q34.22 You've entered a survival contest that will include building a crude telescope. You are given a large box of lenses. Which two lenses do you pick? How do you quickly identify them?

Q34.23 **BIO** You can't see clearly underwater with the naked eye, but you *can* if you wear a face mask or goggles (with air between your eyes and the mask or goggles). Why is there a difference? Could you instead wear eyeglasses (with water between your eyes and the eyeglasses) in order to see underwater? If so, should the lenses be converging or diverging? Explain.

Q34.24 You take a lens and mask it so that light can pass through only the bottom half of the lens. How does the image formed by the masked lens compare to the image formed before masking?

EXERCISES

Section 34.1 Reflection and Refraction at a Plane Surface

34.1 • A candle 4.85 cm tall is 39.2 cm to the left of a plane mirror. Where is the image formed by the mirror, and what is the height of this image?

34.2 • The image of a tree just covers the length of a plane mirror 4.00 cm tall when the mirror is held 35.0 cm from the eye. The tree is 28.0 m from the mirror. What is its height?

34.3 • A pencil that is 9.0 cm long is held perpendicular to the surface of a plane mirror with the tip of the pencil lead 12.0 cm from the mirror surface and the end of the eraser 21.0 cm from the mirror surface. What is the length of the image of the pencil that is formed by the mirror? Which end of the image is closer to the mirror surface: the tip of the lead or the end of the eraser?

Section 34.2 Reflection at a Spherical Surface

34.4 • A concave mirror has a radius of curvature of 34.0 cm. (a) What is its focal length? (b) If the mirror is immersed in water (refractive index 1.33), what is its focal length?

34.5 • An object 0.600 cm tall is placed 16.5 cm to the left of the vertex of a concave spherical mirror having a radius of curvature of 22.0 cm. (a) Draw a principal-ray diagram showing the formation of the image. (b) Determine the position, size, orientation, and nature (real or virtual) of the image.

34.6 • Repeat Exercise 34.5 for the case in which the mirror is convex.

34.7 •• The diameter of Mars is 6794 km, and its minimum distance from the earth is 5.58×10^7 km. When Mars is at this distance, find the diameter of the image of Mars formed by a spherical, concave telescope mirror with a focal length of 1.75 m.

34.8 •• An object is 24.0 cm from the center of a silvered spherical glass Christmas tree ornament 6.00 cm in diameter. What are the position and magnification of its image?

34.9 • A coin is placed next to the convex side of a thin spherical glass shell having a radius of curvature of 18.0 cm. Reflection from the surface of the shell forms an image of the 1.5-cm-tall coin that is 6.00 cm behind the glass shell. Where is the coin located? Determine the size, orientation, and nature (real or virtual) of the image.

34.10 • You hold a spherical salad bowl 90 cm in front of your face with the bottom of the bowl facing you. The salad bowl is made of polished metal with a 35-cm radius of curvature. (a) Where is the image of your 2.0-cm-tall nose located? (b) What are the image's size, orientation, and nature (real or virtual)?

34.11 • (a) Show that Eq. (34.6) can be written as $s' = sf/(s - f)$ and hence the lateral magnification given by Eq. (34.7) can be expressed as $m = f/(f - s)$. (b) A concave spherical mirror has focal length $f = +14.0$ cm. What is the nonzero distance of the object from the mirror vertex if the image has the same height as the object? In this case, is the image erect or inverted? (c) A convex spherical mirror has $f = -8.00$ cm. What is the nonzero distance of the object from the mirror vertex if the height of the image is one-half the height of the object?

34.12 • The thin glass shell shown in Fig. E34.12 has a spherical shape with a radius of curvature of 12.0 cm, and both of its surfaces can act as mirrors. A seed 3.30 mm high is placed 15.0 cm from the center of the mirror along the optic axis, as shown in the figure. (a) Calculate the location and height of the image of this seed. (b) Suppose now that the shell is reversed. Find the location and height of the seed's image.

Figure **E34.12**

3.30 mm

15.0 cm

34.13 • **Dental Mirror.** A dentist uses a curved mirror to view teeth on the upper side of the mouth. Suppose she wants an erect image with a magnification of 2.00 when the mirror is 1.25 cm from a tooth. (Treat this problem as though the object and image lie along a straight line.) (a) What kind of mirror (concave or convex) is needed? Use a ray diagram to decide, without performing any calculations. (b) What must be the focal length and radius of curvature of this mirror? (c) Draw a principal-ray diagram to check your answer in part (b).

34.14 • A spherical, concave shaving mirror has a radius of curvature of 32.0 cm. (a) What is the magnification of a person's face when it is 12.0 cm to the left of the vertex of the mirror? (b) Where is the image? Is the image real or virtual? (c) Draw a principal-ray diagram showing the formation of the image.

Section 34.3 Refraction at a Spherical Surface

34.15 • A speck of dirt is embedded 3.50 cm below the surface of a sheet of ice ($n = 1.309$). What is its apparent depth when viewed at normal incidence?

34.16 •• A tank whose bottom is a mirror is filled with water to a depth of 20.0 cm. A small fish floats motionless 7.0 cm under the surface of the water. (a) What is the apparent depth of the fish when viewed at normal incidence? (b) What is the apparent depth of the image of the fish when viewed at normal incidence?

34.17 • A person swimming 0.80 m below the surface of the water in a swimming pool looks at the diving board that is directly overhead and sees the image of the board that is formed by refraction at the surface of the water. This image is a height of 5.20 m above the swimmer. What is the actual height of the diving board above the surface of the water?

34.18 • A person is lying on a diving board 3.00 m above the surface of the water in a swimming pool. The person looks at a penny that is on the bottom of the pool directly below her. The penny appears to the person to be a distance of 8.00 m from her. What is the depth of the water at this point?

34.19 •• **A Spherical Fish Bowl.** A small tropical fish is at the center of a water-filled, spherical fish bowl 28.0 cm in diameter. (a) Find the apparent position and magnification of the fish to an observer outside the bowl. The effect of the thin walls of the bowl may be ignored. (b) A friend advised the owner of the bowl to keep it out of direct sunlight to avoid blinding the fish, which might swim into the focal point of the parallel rays from the sun. Is the focal point actually within the bowl?

34.20 • The left end of a long glass rod 6.00 cm in diameter has a convex hemispherical surface 3.00 cm in radius. The refractive index of the glass is 1.60. Determine the position of the image if an object is placed in air on the axis of the rod at the following distances to the left of the vertex of the curved end: (a) infinitely far, (b) 12.0 cm; (c) 2.00 cm.

34.21 •• The glass rod of Exercise 34.20 is immersed in oil ($n = 1.45$). An object placed to the left of the rod on the rod's axis is to be imaged 1.20 m inside the rod. How far from the left end of the rod must the object be located to form the image?

34.22 •• The left end of a long glass rod 8.00 cm in diameter, with an index of refraction of 1.60, is ground and polished to a convex hemispherical surface with a radius of 4.00 cm. An object in the form of an arrow 1.50 mm tall, at right angles to the axis of the rod, is located on the axis 24.0 cm to the left of the vertex of the convex surface. Find the position and height of the image of the arrow formed by paraxial rays incident on the convex surface. Is the image erect or inverted?

34.23 •• Repeat Exercise 34.22 for the case in which the end of the rod is ground to a *concave* hemispherical surface with radius 4.00 cm.

34.24 •• The glass rod of Exercise 34.23 is immersed in a liquid. An object 14.0 cm from the vertex of the left end of the rod and on its axis is imaged at a point 9.00 cm from the vertex inside the liquid. What is the index of refraction of the liquid?

Section 34.4 Thin Lenses

34.25 • An insect 3.75 mm tall is placed 22.5 cm to the left of a thin planoconvex lens. The left surface of this lens is flat, the right surface has a radius of curvature of magnitude 13.0 cm, and the index of refraction of the lens material is 1.70. (a) Calculate the location and size of the image this lens forms of the insect. Is it real or virtual? Erect or inverted? (b) Repeat part (a) if the lens is reversed.

34.26 • A lens forms an image of an object. The object is 16.0 cm from the lens. The image is 12.0 cm from the lens on the same side as the object. (a) What is the focal length of the lens? Is the lens converging or diverging? (b) If the object is 8.50 mm tall, how tall is the image? Is it erect or inverted? (c) Draw a principal-ray diagram.

34.27 • A converging meniscus lens (see Fig. 34.32a) with a refractive index of 1.52 has spherical surfaces whose radii are 7.00 cm and 4.00 cm. What is the position of the image if an object is placed 24.0 cm to the left of the lens? What is the magnification?

34.28 • A converging lens with a focal length of 90.0 cm forms an image of a 3.20-cm-tall real object that is to the left of the lens. The image is 4.50 cm tall and inverted. Where are the object and image located in relation to the lens? Is the image real or virtual?

34.29 •• A converging lens forms an image of an 8.00-mm-tall real object. The image is 12.0 cm to the left of the lens, 3.40 cm tall, and erect. What is the focal length of the lens? Where is the object located?

34.30 • A photographic slide is to the left of a lens. The lens projects an image of the slide onto a wall 6.00 m to the right of the slide. The image is 80.0 times the size of the slide. (a) How far is the slide from the lens? (b) Is the image erect or inverted? (c) What is the focal length of the lens? (d) Is the lens converging or diverging?

34.31 •• A double-convex thin lens has surfaces with equal radii of curvature of magnitude 2.50 cm. Looking through this lens, you observe that it forms an image of a very distant tree at a distance of 1.87 cm from the lens. What is the index of refraction of the lens?

34.32 • BIO **The Lens of the Eye.** The crystalline lens of the human eye is a double-convex lens made of material having an index of refraction of 1.44 (although this varies). Its focal length in air is about 8.0 mm, which also varies. We shall assume that the radii of curvature of its two surfaces have the same magnitude. (a) Find the radii of curvature of this lens. (b) If an object 16 cm tall is placed 30.0 cm from the eye lens, where would the lens focus it and how tall would the image be? Is this image real or virtual? Is it erect or inverted? (*Note:* The results obtained here are not strictly accurate because the lens is embedded in fluids having refractive indexes different from that of air.)

34.33 • BIO **The Cornea As a Simple Lens.** The cornea behaves as a thin lens of focal length approximately 1.8 cm, although this varies a bit. The material of which it is made has an index of refraction of 1.38, and its front surface is convex, with a radius of curvature of 5.0 mm. (a) If this focal length is in air, what is the radius of curvature of the back side of the cornea? (b) The closest distance at which a typical person can focus on an object (called the near point) is about 25 cm, although this varies considerably with age. Where would the cornea focus the image of an 8.0-mm-tall object at the near point? (c) What is the height of the image in part (b)? Is this image real or virtual? Is it erect or inverted? (*Note:* The results obtained here are not strictly accurate because, on one side, the cornea has a fluid with a refractive index different from that of air.)

34.34 • A converging lens with a focal length of 7.00 cm forms an image of a 4.00-mm-tall real object that is to the left of the lens. The image is 1.30 cm tall and erect. Where are the object and image located? Is the image real or virtual?

34.35 • For each thin lens shown in Fig. E34.35, calculate the location of the image of an object that is 18.0 cm to the left of the lens. The lens material has a refractive index of 1.50, and the radii of curvature shown are only the magnitudes.

Figure **E34.35**

34.36 • A converging lens with a focal length of 12.0 cm forms a virtual image 8.00 mm tall, 17.0 cm to the right of the lens. Determine the position and size of the object. Is the image erect or inverted? Are the object and image on the same side or opposite sides of the lens? Draw a principal-ray diagram for this situation.

34.37 • Repeat Exercise 34.36 for the case in which the lens is diverging, with a focal length of −48.0 cm.

34.38 • An object is 16.0 cm to the left of a lens. The lens forms an image 36.0 cm to the right of the lens. (a) What is the focal length of the lens? Is the lens converging or diverging? (b) If the object is 8.00 mm tall, how tall is the image? Is it erect or inverted? (c) Draw a principal-ray diagram.

34.39 •• **Combination of Lenses I.** A 1.20-cm-tall object is 50.0 cm to the left of a converging lens of focal length 40.0 cm. A second converging lens, this one having a focal length of 60.0 cm, is located 300.0 cm to the right of the first lens along the same optic axis. (a) Find the location and height of the image (call it I_1) formed by the lens with a focal length of 40.0 cm. (b) I_1 is now the object for the second lens. Find the location and height of the image produced by the second lens. This is the final image produced by the combination of lenses.

34.40 •• **Combination of Lenses II.** Repeat Problem 34.39 using the same lenses except for the following changes: (a) The second lens is a *diverging* lens having a focal length of magnitude 60.0 cm. (b) The first lens is a *diverging* lens having a focal length of magnitude 40.0 cm. (c) Both lenses are diverging lenses having focal lengths of the same *magnitudes* as in Problem 34.39.

34.41 •• **Combination of Lenses III.** Two thin lenses with a focal length of magnitude 12.0 cm, the first diverging and the second converging, are located 9.00 cm apart. An object 2.50 mm tall is placed 20.0 cm to the left of the first (diverging) lens. (a) How far from this first lens is the final image formed? (b) Is the final image real or virtual? (c) What is the height of the final image? Is it erect or inverted? (*Hint:* See the preceding two problems.)

Section 34.5 Cameras

34.42 • You wish to project the image of a slide on a screen 9.00 m from the lens of a slide projector. (a) If the slide is placed 15.0 cm from the lens, what focal length lens is required? (b) If the dimensions of the picture on a 35-mm color slide are 24 mm × 36 mm, what is the minimum size of the projector screen required to accommodate the image?

34.43 •• A camera lens has a focal length of 200 mm. How far from the lens should the subject for the photo be if the lens is 20.4 cm from the film?

34.44 • When a camera is focused, the lens is moved away from or toward the film. If you take a picture of your friend, who is standing 3.90 m from the lens, using a camera with a lens with a 85-mm focal length, how far from the film is the lens? Will the whole image of your friend, who is 175 cm tall, fit on film that is 24 × 36 mm?

34.45 • Figure 34.41 shows photographs of the same scene taken with the same camera with lenses of different focal length. If the object is 200 m from the lens, what is the magnitude of the lateral magnification for a lens of focal length (a) 28 mm; (b) 105 mm; (c) 300 mm?

34.46 • A photographer takes a photograph of a Boeing 747 airliner (length 70.7 m) when it is flying directly overhead at an altitude of 9.50 km. The lens has a focal length of 5.00 m. How long is the image of the airliner on the film?

34.47 • **Choosing a Camera Lens.** The picture size on ordinary 35-mm camera film is 24 mm × 36 mm. Focal lengths of lenses available for 35-mm cameras typically include 28, 35, 50 (the "normal" lens), 85, 100, 135, 200, and 300 mm, among others. Which of these lenses should be used to photograph the following objects, assuming that the object is to fill most of the picture area? (a) a building 240 m tall and 160 m wide at a distance of 600 m, and (b) a mobile home 9.6 m in length at a distance of 40.0 m.

34.48 • **Zoom Lens.** Consider the simple model of the zoom lens shown in Fig. 34.43a. The converging lens has focal length $f_1 = 12$ cm, and the diverging lens has focal length $f_2 = -12$ cm. The lenses are separated by 4 cm as shown in Fig. 34.43a. (a) For a distant object, where is the image of the converging lens? (b) The image of the converging lens serves as the object for the diverging lens. What is the object distance for the diverging lens? (c) Where is the final image? Compare your answer to Fig. 34.43a. (d) Repeat parts (a), (b), and (c) for the situation shown in Fig. 34.43b, in which the lenses are separated by 8 cm.

34.49 •• A camera lens has a focal length of 180.0 mm and an aperture diameter of 16.36 mm. (a) What is the f-number of the lens? (b) If the correct exposure of a certain scene is $\frac{1}{30}$ s at $f/11$, what is the correct exposure at $f/2.8$?

34.50 •• Recall that the intensity of light reaching film in a camera is proportional to the effective area of the lens. Camera A has a lens with an aperture diameter of 8.00 mm. It photographs an object using the correct exposure time of $\frac{1}{30}$ s. What exposure time should be used with camera B in photographing the same object with the same film if this camera has a lens with an aperture diameter of 23.1 mm?

34.51 • **Photography.** A 35-mm camera has a standard lens with focal length 50 mm and can focus on objects between 45 cm and infinity. (a) Is the lens for such a camera a concave or a convex lens? (b) The camera is focused by rotating the lens, which moves it on the camera body and changes its distance from the film. In what range of distances between the lens and the film plane must the lens move to focus properly over the 45 cm to infinity range?

Section 34.6 The Eye

34.52 •• **BIO Curvature of the Cornea.** In a simplified model of the human eye, the aqueous and vitreous humors and the lens all have a refractive index of 1.40, and all the refraction occurs at the cornea, whose vertex is 2.60 cm from the retina. What should be the radius of curvature of the cornea such that the image of an object 40.0 cm from the cornea's vertex is focused on the retina?

34.53 •• **BIO** (a) Where is the near point of an eye for which a contact lens with a power of +2.75 diopters is prescribed? (b) Where is the far point of an eye for which a contact lens with a power of −1.30 diopters is prescribed for distant vision?

34.54 • **BIO Contact Lenses.** Contact lenses are placed right on the eyeball, so the distance from the eye to an object (or image) is the same as the distance from the lens to that object (or image). A certain person can see distant objects well, but his near point is 45.0 cm from his eyes instead of the usual 25.0 cm. (a) Is this person nearsighted or farsighted? (b) What type of lens (converging or diverging) is needed to correct his vision? (c) If the correcting

lenses will be contact lenses, what focal length lens is needed and what is its power in diopters?

34.55 •• **BIO Ordinary Glasses.** Ordinary glasses are worn in front of the eye and usually 2.0 cm in front of the eyeball. Suppose that the person in Problem 34.54 prefers ordinary glasses to contact lenses. What focal length lenses are needed to correct his vision, and what is their power in diopters?

34.56 • **BIO** A person can see clearly up close but cannot focus on objects beyond 75.0 cm. She opts for contact lenses to correct her vision. (a) Is she nearsighted or farsighted? (b) What type of lens (converging or diverging) is needed to correct her vision? (c) What focal length contact lens is needed, and what is its power in diopters?

34.57 •• **BIO** If the person in Problem 34.56 chooses ordinary glasses over contact lenses, what power lens (in diopters) does she need to correct her vision if the lenses are 2.0 cm in front of the eye?

Section 34.7 The Magnifier

34.58 •• A thin lens with a focal length of 6.00 cm is used as a simple magnifier. (a) What angular magnification is obtainable with the lens if the object is at the focal point? (b) When an object is examined through the lens, how close can it be brought to the lens? Assume that the image viewed by the eye is at the near point, 25.0 cm from the eye, and that the lens is very close to the eye.

34.59 • The focal length of a simple magnifier is 8.00 cm. Assume the magnifier is a thin lens placed very close to the eye. (a) How far in front of the magnifier should an object be placed if the image is formed at the observer's near point, 25.0 cm in front of her eye? (b) If the object is 1.00 mm high, what is the height of its image formed by the magnifier?

34.60 • You want to view an insect 2.00 mm in length through a magnifier. If the insect is to be at the focal point of the magnifier, what focal length will give the image of the insect an angular size of 0.025 radian?

Section 34.8 Microscopes and Telescopes

34.61 • A certain microscope is provided with objectives that have focal lengths of 16 mm, 4 mm, and 1.9 mm and with eyepieces that have angular magnifications of 5× and 10×. Each objective forms an image 120 mm beyond its second focal point. Determine (a) the largest overall angular magnification obtainable and (b) the smallest overall angular magnification obtainable.

34.62 •• **Resolution of a Microscope.** The image formed by a microscope objective with a focal length of 5.00 mm is 160 mm from its second focal point. The eyepiece has a focal length of 26.0 mm. (a) What is the angular magnification of the microscope? (b) The unaided eye can distinguish two points at its near point as separate if they are about 0.10 mm apart. What is the minimum separation between two points that can be observed (or resolved) through this microscope?

34.63 •• The focal length of the eyepiece of a certain microscope is 18.0 mm. The focal length of the objective is 8.00 mm. The distance between objective and eyepiece is 19.7 cm. The final image formed by the eyepiece is at infinity. Treat all lenses as thin. (a) What is the distance from the objective to the object being viewed? (b) What is the magnitude of the linear magnification produced by the objective? (c) What is the overall angular magnification of the microscope?

34.64 •• The eyepiece of a refracting telescope (see Fig. 34.53) has a focal length of 9.00 cm. The distance between objective and eyepiece is 1.80 m, and the final image is at infinity. What is the angular magnification of the telescope?

34.65 •• A telescope is constructed from two lenses with focal lengths of 95.0 cm and 15.0 cm, the 95.0-cm lens being used as the objective. Both the object being viewed and the final image are at infinity. (a) Find the angular magnification for the telescope. (b) Find the height of the image formed by the objective of a building 60.0 m tall, 3.00 km away. (c) What is the angular size of the final image as viewed by an eye very close to the eyepiece?

34.66 •• Saturn is viewed through the Lick Observatory refracting telescope (objective focal length 18 m). If the diameter of the image of Saturn produced by the objective is 1.7 mm, what angle does Saturn subtend from when viewed from earth?

34.67 •• A reflecting telescope (Fig. E34.67) is to be made by using a spherical mirror with a radius of curvature of 1.30 m and an eyepiece with a focal length of 1.10 cm. The final image is at infinity. (a) What should the distance between the eyepiece and the mirror vertex be if the object is taken to be at infinity? (b) What will the angular magnification be?

Figure **E34.67**

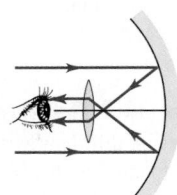

PROBLEMS

34.68 • Where must you place an object in front of a concave mirror with radius R so that the image is erect and $2\frac{1}{2}$ times the size of the object? Where is the image?

34.69 • If you run away from a plane mirror at 3.60 m/s, at what speed does your image move away from you?

34.70 • An object is placed between two plane mirrors arranged at right angles to each other at a distance d_1 from the surface of one mirror and a distance d_2 from the other. (a) How many images are formed? Show the location of the images in a diagram. (b) Draw the paths of rays from the object to the eye of an observer.

34.71 •• What is the size of the smallest vertical plane mirror in which a woman of height h can see her full-length image?

34.72 • A light bulb is 3.00 m from a wall. You are to use a concave mirror to project an image of the bulb on the wall, with the image 2.25 times the size of the object. How far should the mirror be from the wall? What should its radius of curvature be?

34.73 •• A concave mirror is to form an image of the filament of a headlight lamp on a screen 8.00 m from the mirror. The filament is 6.00 mm tall, and the image is to be 24.0 cm tall. (a) How far in front of the vertex of the mirror should the filament be placed? (b) What should be the radius of curvature of the mirror?

34.74 •• **Rear-View Mirror.** A mirror on the passenger side of your car is convex and has a radius of curvature with magnitude 18.0 cm. (a) Another car is behind your car, 9.00 m from the mirror, and this car is viewed in the mirror by your passenger. If this car is 1.5 m tall, what is the height of the image? (b) The mirror has a warning attached that objects viewed in it are closer than they appear. Why is this so?

34.75 • Suppose the lamp filament shown in Example 34.1 (Section 34.2) is moved to a position 8.0 cm in front of the mirror. (a) Where is the image located now? Is it real or virtual? (b) What is the height of the image? Is it erect or inverted? (c) In Example 34.1, the filament is 10.0 cm in front of the mirror, and an image of the filament is formed on a wall 3.00 m from the mirror. If the filament is 8.0 cm from the mirror, can a wall be placed so that an image is formed on it? If so, where should the wall be placed? If not, why not?

34.76 •• A layer of benzene ($n = 1.50$) 4.20 cm deep floats on water ($n = 1.33$) that is 6.50 cm deep. What is the apparent distance from the upper benzene surface to the bottom of the water layer when it is viewed at normal incidence?

34.77 •• **CP CALC** You are in your car driving on a highway at 25 m/s when you glance in the passenger-side mirror (a convex mirror with radius of curvature 150 cm) and notice a truck approaching. If the image of the truck is approaching the vertex of the mirror at a speed of 1.9 m/s when the truck is 2.0 m from the mirror, what is the speed of the truck relative to the highway?

34.78 •• Figure P34.78 shows a small plant near a thin lens. The ray shown is one of the principal rays for the lens. Each square is 2.0 cm along the horizontal direction, but the vertical direction is not to the same scale. Use information from the diagram for the following: (a) Using only the ray shown, decide what type of lens (converging or diverging) this is. (b) What is the focal length of the lens? (c) Locate the image by drawing the other two principal rays. (d) Calculate where the image should be, and compare this result with the graphical solution in part (c).

Figure **P34.78**

34.79 •• **Pinhole Camera.** A pinhole camera is just a rectangular box with a tiny hole in one face. The film is on the face opposite this hole, and that is where the image is formed. The camera forms an image *without* a lens. (a) Make a clear ray diagram to show how a pinhole camera can form an image on the film without using a lens. (*Hint:* Put an object outside the hole, and then draw rays passing through the hole to the opposite side of the box.) (b) A certain pinhole camera is a box that is 25 cm square and 20.0 cm deep, with the hole in the middle of one of the 25 cm × 25 cm faces. If this camera is used to photograph a fierce chicken that is 18 cm high and 1.5 m in front of the camera, how large is the image of this bird on the film? What is the magnification of this camera?

34.80 ••• A microscope is focused on the upper surface of a glass plate. A second plate is then placed over the first. To focus on the bottom surface of the second plate, the microscope must be raised 0.780 mm. To focus on the upper surface, it must be raised another 2.50 mm. Find the index of refraction of the second plate.

34.81 •• What should be the index of refraction of a transparent sphere in order for paraxial rays from an infinitely distant object to be brought to a focus at the vertex of the surface opposite the point of incidence?

34.82 •• **A Glass Rod.** Both ends of a glass rod with index of refraction 1.60 are ground and polished to convex hemispherical surfaces. The radius of curvature at the left end is 6.00 cm, and the radius of curvature at the right end is 12.0 cm. The length of the rod between vertices is 40.0 cm. The object for the surface at the left end is an arrow that lies 23.0 cm to the left of the vertex of this surface. The arrow is 1.50 mm tall and at right angles to the axis. (a) What constitutes the object for the surface at the right end of the rod? (b) What is the object distance for this surface? (c) Is the object for this surface real or virtual? (d) What is the position of the final image? (e) Is the final image real or virtual? Is it erect or inverted with respect to the original object? (f) What is the height of the final image?

34.83 •• The rod in Problem 34.82 is shortened to a distance of 25.0 cm between its vertices; the curvatures of its ends remain the same. As in Problem 34.82, the object for the surface at the left end is an arrow that lies 23.0 cm to the left of the vertex of this surface. The arrow is 1.50 mm tall and at right angles to the axis. (a) What is the object distance for the surface at the right end of the rod? (b) Is the object for this surface real or virtual? (c) What is the position of the final image? (d) Is the final image real or virtual? Is it erect or inverted with respect to the original object? (e) What is the height of the final image?

34.84 • Figure P34.84 shows an object and its image formed by a thin lens. (a) What is the focal length of the lens, and what type of lens (converging or diverging) is it? (b) What is the height of the image? Is it real or virtual?

Figure **P34.84**

34.85 • Figure P34.85 shows an object and its image formed by a thin lens. (a) What is the focal length of the lens, and what type of lens (converging or diverging) is it? (b) What is the height of the image? Is it real or virtual?

Figure **P34.85**

34.86 ••• A transparent rod 30.0 cm long is cut flat at one end and rounded to a hemispherical surface of radius 10.0 cm at the other end. A small object is embedded within the rod along its axis and halfway between its ends, 15.0 cm from the flat end and 15.0 cm from the vertex of the curved end. When viewed from the flat end of the rod, the apparent depth of the object is 9.50 cm from the flat end. What is its apparent depth when viewed from the curved end?

34.87 • **BIO Focus of the Eye.** The cornea of the eye has a radius of curvature of approximately 0.50 cm, and the aqueous humor behind it has an index of refraction of 1.35. The thickness of the cornea itself is small enough that we shall neglect it. The depth of a typical human eye is around 25 mm. (a) What would have to be the radius of curvature of the cornea so that it alone would focus the image of a distant mountain on the retina, which is at the back of the eye opposite the cornea? (b) If the cornea focused the mountain correctly on the retina as described in part (a), would it also focus the text from a computer screen on the retina if that screen were 25 cm in front of the eye? If not, where would it focus that text: in front of or behind the retina? (c) Given that the cornea has a radius of curvature of about 5.0 mm, where does it actually focus the mountain? Is this in front of or behind the retina? Does this help you see why the eye needs help from a lens to complete the task of focusing?

34.88 ••• A transparent rod 50.0 cm long and with a refractive index of 1.60 is cut flat at the right end and rounded to a hemispherical surface with a 15.0-cm radius at the left end. An object is

placed on the axis of the rod 12.0 cm to the left of the vertex of the hemispherical end. (a) What is the position of the final image? (b) What is its magnification?

34.89 ••• A glass rod with a refractive index of 1.55 is ground and polished at both ends to hemispherical surfaces with radii of 6.00 cm. When an object is placed on the axis of the rod, 25.0 cm to the left of the left-hand end, the final image is formed 65.0 cm to the right of the right-hand end. What is the length of the rod measured between the vertices of the two hemispherical surfaces?

34.90 • The radii of curvature of the surfaces of a thin converging meniscus lens are $R_1 = +12.0$ cm and $R_2 = +28.0$ cm. The index of refraction is 1.60. (a) Compute the position and size of the image of an object in the form of an arrow 5.00 mm tall, perpendicular to the lens axis, 45.0 cm to the left of the lens. (b) A second converging lens with the same focal length is placed 3.15 m to the right of the first. Find the position and size of the final image. Is the final image erect or inverted with respect to the original object? (c) Repeat part (b) except with the second lens 45.0 cm to the right of the first.

34.91 • An object to the left of a lens is imaged by the lens on a screen 30.0 cm to the right of the lens. When the lens is moved 4.00 cm to the right, the screen must be moved 4.00 cm to the left to refocus the image. Determine the focal length of the lens.

34.92 • An object is placed 18.0 cm from a screen. (a) At what two points between object and screen may a converging lens with a 3.00-cm focal length be placed to obtain an image on the screen? (b) What is the magnification of the image for each position of the lens?

34.93 •• A convex mirror and a concave mirror are placed on the same optic axis, separated by a distance $L = 0.600$ m. The radius of curvature of each mirror has a magnitude of 0.360 m. A light source is located a distance x from the concave mirror, as shown in Fig. P34.93. (a) What distance x will result in the rays from the source returning to the source after reflecting first from the convex mirror and then from the concave mirror? (b) Repeat part (a), but now let the rays reflect first from the concave mirror and then from the convex one.

Figure **P34.93**

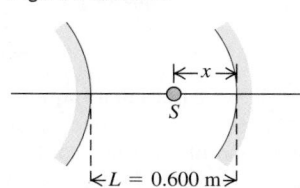

34.94 •• As shown in Fig. P34.94 the candle is at the center of curvature of the concave mirror, whose focal length is 10.0 cm. The converging lens has a focal length of 32.0 cm and is 85.0 cm to the right of the candle. The candle is viewed looking through the lens from the right. The lens forms two images of the candle. The first is formed by light passing directly through the lens. The second image is formed from the light that goes from the candle to the mirror, is reflected, and then passes through the lens. (a) For each of these two images, draw a principal-ray diagram that locates the image. (b) For each image, answer the following questions: (i) Where is the image? (ii) Is the image real or virtual? (iii) Is the image erect or inverted with respect to the original object?

Figure **P34.94**

34.95 ••• One end of a long glass rod is ground to a convex hemispherical shape. This glass has an index of refraction of 1.55. When a small leaf is placed 20.0 cm in front of the center of the hemisphere along the optic axis, an image is formed inside the glass 9.12 cm from the spherical surface. Where would the image be formed if the glass were now immersed in water (refractive index 1.33) but nothing else were changed?

34.96 •• **Two Lenses in Contact.** (a) Prove that when two thin lenses with focal lengths f_1 and f_2 are placed *in contact,* the focal length f of the combination is given by the relationship

$$\frac{1}{f} = \frac{1}{f_1} + \frac{1}{f_2}$$

(b) A converging meniscus lens (see Fig. 34.32a) has an index of refraction of 1.55 and radii of curvature for its surfaces of magnitudes 4.50 cm and 9.00 cm. The concave surface is placed upward and filled with carbon tetrachloride (CCl_4), which has $n = 1.46$. What is the focal length of the CCl_4–glass combination?

34.97 ••• Rays from a lens are converging toward a point image P located to the right of the lens. What thickness t of glass with index of refraction 1.60 must be interposed between the lens and P for the image to be formed at P', located 0.30 cm to the right of P? The locations of the piece of glass and of points P and P' are shown in Fig. P34.97.

Figure **P34.97**

34.98 •• **A Lens in a Liquid.** A lens obeys Snell's law, bending light rays at each surface an amount determined by the index of refraction of the lens and the index of the medium in which the lens is located. (a) Equation (34.19) assumes that the lens is surrounded by air. Consider instead a thin lens immersed in a liquid with refractive index n_{liq}. Prove that the focal length f' is then given by Eq. (34.19) with n replaced by n/n_{liq}. (b) A thin lens with index n has focal length f in vacuum. Use the result of part (a) to show that when this lens is immersed in a liquid of index n_{liq}, it will have a new focal length given by

$$f' = \left[\frac{n_{liq}(n - 1)}{n - n_{liq}} \right] f$$

34.99 ••• When an object is placed at the proper distance to the left of a converging lens, the image is focused on a screen 30.0 cm to the right of the lens. A diverging lens is now placed 15.0 cm to the right of the converging lens, and it is found that the screen must be moved 19.2 cm farther to the right to obtain a sharp image. What is the focal length of the diverging lens?

34.100 ••• A convex spherical mirror with a focal length of magnitude 24.0 cm is placed 20.0 cm to the left of a plane mirror. An object 0.250 cm tall is placed midway between the surface of the plane mirror and the vertex of the spherical mirror. The spherical mirror forms multiple images of the object. Where are the two images of the object formed by the spherical mirror that are closest to the spherical mirror, and how tall is each image?

34.101 ••• A glass plate 3.50 cm thick, with an index of refraction of 1.55 and plane parallel faces, is held with its faces horizontal and its lower face 6.00 cm above a printed page. Find the position of the image of the page formed by rays making a small angle with the normal to the plate.

34.102 •• A symmetric, double-convex, thin lens made of glass with index of refraction 1.52 has a focal length in air of 40.0 cm. The lens is sealed into an opening in the left-hand end of a tank filled with water. At the right-hand end of the tank, opposite the lens, is a plane mirror 90.0 cm from the lens. The index of refraction of the water is $\frac{4}{3}$. (a) Find the position of the image formed by the lens–water–mirror system of a small object outside the tank on the lens axis and 70.0 cm to the left of the lens. (b) Is the image real or virtual? (c) Is it erect or inverted? (d) If the object has a height of 4.00 mm, what is the height of the image?

34.103 • You have a camera with a 35.0-mm-focal-length lens and 36.0-mm-wide film. You wish to take a picture of a 12.0-m-long sailboat but find that the image of the boat fills only $\frac{1}{4}$ of the width of the film. (a) How far are you from the boat? (b) How much closer must the boat be to you for its image to fill the width of the film?

34.104 •• **BIO** **What Is the Smallest Thing We Can See?** The smallest object we can resolve with our eye is limited by the size of the light receptor cells in the retina. In order for us to distinguish any detail in an object, its image cannot be any smaller than a single retinal cell. Although the size depends on the type of cell (rod or cone), a diameter of a few microns (μm) is typical near the center of the eye. We shall model the eye as a sphere 2.50 cm in diameter with a single thin lens at the front and the retina at the rear, with light receptor cells 5.0 μm in diameter. (a) What is the smallest object you can resolve at a near point of 25 cm? (b) What angle is subtended by this object at the eye? Express your answer in units of minutes ($1° = 60$ min), and compare it with the typical experimental value of about 1.0 min. (*Note:* There are other limitations, such as the bending of light as it passes through the pupil, but we shall ignore them here.)

34.105 • Three thin lenses, each with a focal length of 40.0 cm, are aligned on a common axis; adjacent lenses are separated by 52.0 cm. Find the position of the image of a small object on the axis, 80.0 cm to the left of the first lens.

34.106 •• A camera with a 90-mm-focal-length lens is focused on an object 1.30 m from the lens. To refocus on an object 6.50 m from the lens, by how much must the distance between the lens and the film be changed? To refocus on the more distant object, is the lens moved toward or away from the film?

34.107 •• The derivation of the expression for angular magnification, Eq. (34.22), assumed a near point of 25 cm. In fact, the near point changes with age as shown in Table 34.1. In order to achieve an angular magnification of 2.0×, what focal length should be used by a person of (a) age 10; (b) age 30; (c) age 60? (d) If the lens that gives $M = 2.0$ for a 10-year-old is used by a 60-year-old, what angular magnification will the older viewer obtain? (e) Does your answer in part (d) mean that older viewers are able to see more highly magnified images than younger viewers? Explain.

34.108 •• **Angular Magnification.** In deriving Eq. (34.22) for the angular magnification of a magnifier, we assumed that the object is placed at the focal point of the magnifier so that the virtual image is formed at infinity. Suppose instead that the object is placed so that the virtual image appears at an average viewer's near point of 25 cm, the closest point at which the viewer can bring an object into focus. (a) Where should the object be placed to achieve this? Give your answer in terms of the magnifier focal length f. (b) What angle θ' will an object of height y subtend at the position found in part (a)? (c) Find the angular magnification M with the object at the position found in part (a). The angle θ is the same as in Fig. 34.51a, since it refers to viewing the object *without* the magnifier. (d) For a convex lens with $f = +10.0$ cm, what is the value of M with the object at the position found in part (a)? How many times greater is M in this case than in the case where the image is formed at infinity? (e) In the description of a compound microscope in Section 34.8, it is stated that in a properly designed instrument, the real image formed by the objective lies *just inside* the first focal point F'_1 of the eyepiece. What advantages are gained by having the image formed by the objective be just inside F'_1, as opposed to precisely at F'_1? What happens if the image formed by the objective is *just outside* F'_1?

34.109 •• **BIO** In one form of cataract surgery the person's natural lens, which has become cloudy, is replaced by an artificial lens. The refracting properties of the replacement lens can be chosen so that the person's eye focuses on distant objects. But there is no accommodation, and glasses or contact lenses are needed for close vision. What is the power, in diopters, of the corrective contact lenses that will enable a person who has had such surgery to focus on the page of a book at a distance of 24 cm?

34.110 •• **BIO A Nearsighted Eye.** A certain very nearsighted person cannot focus on anything farther than 36.0 cm from the eye. Consider the simplified model of the eye described in Exercise 34.52. If the radius of curvature of the cornea is 0.75 cm when the eye is focusing on an object 36.0 cm from the cornea vertex and the indexes of refraction are as described in Exercise 34.52, what is the distance from the cornea vertex to the retina? What does this tell you about the shape of the nearsighted eye?

34.111 •• **BIO** A person with a near point of 85 cm, but excellent distant vision, normally wears corrective glasses. But he loses them while traveling. Fortunately, he has his old pair as a spare. (a) If the lenses of the old pair have a power of +2.25 diopters, what is his near point (measured from his eye) when he is wearing the old glasses if they rest 2.0 cm in front of his eye? (b) What would his near point be if his old glasses were contact lenses instead?

34.112 •• **The Galilean Telescope.** Figure P34.112 is a diagram of a *Galilean telescope,* or *opera glass,* with both the object and its final image at infinity. The image I serves as a virtual object for the eyepiece. The final image is virtual and erect. (a) Prove that the angular magnification is $M = -f_1/f_2$. (b) A Galilean telescope is to be constructed with the same objective lens as in Exercise 34.65. What focal length should the eyepiece have if this telescope is to have the same magnitude of angular magnification as the one in Exercise 34.65? (c) Compare the lengths of the telescopes.

Figure **P34.112**

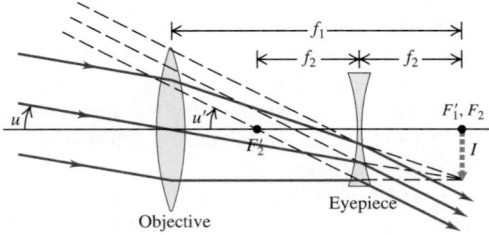

34.113 ••• **Focal Length of a Zoom Lens.** Figure P34.113 shows a simple version of a zoom lens. The converging lens has focal length f_1, and the diverging lens has focal length $f_2 = -|f_2|$. The two lenses are separated by a variable distance d that is always less than f_1. Also, the magnitude of the focal length of the diverging lens satisfies the inequality $|f_2| > (f_1 - d)$. To determine the effective focal length of the combination lens, consider a bundle of parallel rays of radius r_0 entering the converging lens. (a) Show that the radius of the ray bundle decreases to $r'_0 = r_0(f_1 - d)/f_1$ at the point that it enters the diverging lens. (b) Show that the final image I' is formed a distance $s'_2 = |f_2|(f_1 - d)/(|f_2| - f_1 + d)$ to the right of the diverging lens. (c) If the rays that emerge from the diverging lens and reach the final image point are extended

backward to the left of the diverging lens, they will eventually expand to the original radius r_0 at some point Q. The distance from the final image I' to the point Q is the *effective focal length f* of the lens combination; if the combination were replaced by a single lens of focal length f placed at Q, parallel rays would still be brought to a focus at I'. Show that the effective focal length is given by $f = f_1|f_2|/(|f_2| - f_1 + d)$. (d) If $f_1 = 12.0$ cm, $f_2 = -18.0$ cm, and the separation d is adjustable between 0 and 4.0 cm, find the maximum and minimum focal lengths of the combination. What value of d gives $f = 30.0$ cm?

Figure **P34.113**

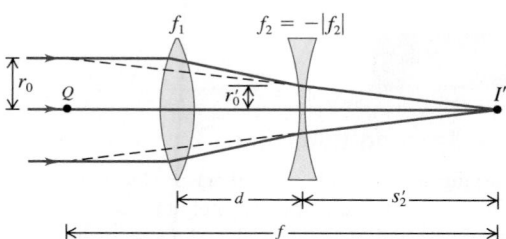

34.114 • A certain reflecting telescope, constructed as shown in Fig. E34.67, has a spherical mirror with a radius of curvature of 96.0 cm and an eyepiece with a focal length of 1.20 cm. If the angular magnification has a magnitude of 36 and the object is at infinity, find the position of the eyepiece and the position and nature (real or virtual) of the final image. (*Note:* $|M|$ is *not* equal to $|f_1|/|f_2|$, so the image formed by the eyepiece is *not* at infinity.)

34.115 • A microscope with an objective of focal length 8.00 mm and an eyepiece of focal length 7.50 cm is used to project an image on a screen 2.00 m from the eyepiece. Let the image distance of the objective be 18.0 cm. (a) What is the lateral magnification of the image? (b) What is the distance between the objective and the eyepiece?

CHALLENGE PROBLEMS

34.116 ••• *Spherical aberration* is a blurring of the image formed by a spherical mirror. It occurs because parallel rays striking the mirror far from the optic axis are focused at a different point than are rays near the axis. This problem is usually minimized by using only the center of a spherical mirror. (a) Show that for a spherical concave mirror, the focus moves toward the mirror as the parallel rays move toward the outer edge of the mirror. (*Hint:* Derive an analytic expression for the distance from the vertex to the focus of the ray for a particular parallel ray. This expression should be in terms of (i) the radius of curvature R of the mirror and (ii) the angle θ between the incident ray and a line connecting the center of curvature of the mirror with the point where the ray strikes the mirror.) (b) What value of θ produces a 2% change in the location of the focus, compared to the location for θ very close to zero?

34.117 ••• **CALC** (a) For a lens with focal length f, find the smallest distance possible between the object and its real image. (b) Graph the distance between the object and the real image as a function of the distance of the object from the lens. Does your graph agree with the result you found in part (a)?

34.118 ••• **An Object at an Angle.** A 16.0-cm-long pencil is placed at a 45.0° angle, with its center 15.0 cm above the optic axis and 45.0 cm from a lens with a 20.0-cm focal length as shown in Fig. P34.118. (Note that the figure is not drawn to scale.) Assume that the diameter of the lens is large enough for the paraxial approximation to be valid. (a) Where is the image of the pencil? (Give the location of the images of the points A, B, and C on the object, which are located at the eraser, point, and center of the pencil, respectively.) (b) What is the length of the image (that is, the distance between the images of points *A* and *B*)? (c) Show the orientation of the image in a sketch.

34.119 ••• **BIO** People with normal vision cannot focus their eyes underwater if they aren't

Figure **P34.118**

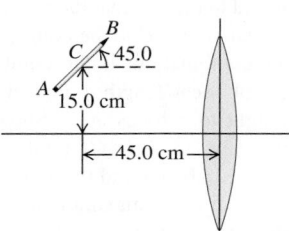

wearing a face mask or goggles and there is water in contact with their eyes (see Discussion Question Q34.23). (a) Why not? (b) With the simplified model of the eye described in Exercise 34.52, what corrective lens (specified by focal length as measured in air) would be needed to enable a person underwater to focus an infinitely distant object? (Be careful—the focal length of a lens underwater is *not* the same as in air! See Problem 34.98. Assume that the corrective lens has a refractive index of 1.62 and that the lens is used in eyeglasses, not goggles, so there is water on both sides of the lens. Assume that the eyeglasses are 2.00 cm in front of the eye.)

Answers

Chapter Opening Question ?

A magnifying lens (simple magnifier) produces a virtual image with a large angular size that is infinitely far away, so you can see it in sharp focus with your eyes relaxed. (A surgeon doing microsurgery would not appreciate having to strain his eyes while working.) The object should be at the focal point of the lens, so the object and lens are separated by one focal length.

Test Your Understanding Questions

34.1 Answer: (iv) When you are a distance *s* from the mirror, your image is a distance *s* on the other side of the mirror and the distance from you to your image is 2*s*. As you move toward the mirror, the distance 2*s* changes at twice the rate of the distance *s*, so your image moves toward you at speed 2*v*.

34.2 Answers: (a) concave, (b) (ii) A convex mirror always produces an erect image, but that image is smaller than the object (see Fig. 34.16b). Hence a concave mirror must be used. The image will be erect and enlarged only if the distance from the object (your face) to the mirror is less than the focal length of the mirror, as in Fig. 34.20d.

34.3 Answer: no The sun is very far away, so the object distance is essentially infinite: $s = \infty$ and $1/s = 0$. Material *a* is air ($n_a = 1.00$) and material *b* is water ($n_b = 1.33$), so the image position s' is given by

$$\frac{n_a}{s} + \frac{n_b}{s'} = \frac{n_b - n_a}{R} \quad \text{or} \quad 0 + \frac{1.33}{s'} = \frac{1.33 - 1.00}{R}$$

$$s' = \frac{1.33}{0.33} R = 4.0R$$

The image would be formed 4.0 drop radii from the front surface of the drop. But since each drop is only a part of a complete sphere, the distance from the front to the back of the drop is less than 2*R*. Thus the rays of sunlight never reach the image point, and the drops do not form an image of the sun on the leaf. While the rays are not focused to a point, they are nonetheless concentrated and can cause damage to the leaf.

34.4 Answers: A and C When rays A and D are extended backward, they pass through focal point F_2; thus, before they passed through the lens, they were parallel to the optic axis. The figures show that ray A emanated from point *Q*, but ray D did not. Ray B is parallel to the optic axis, so before it passed through the lens, it was directed toward focal point F_1. Hence it cannot have come from point *Q*. Ray C passes through the center of the lens and hence is not deflected by its passage; tracing the ray backward shows that it emanates from point *Q*.

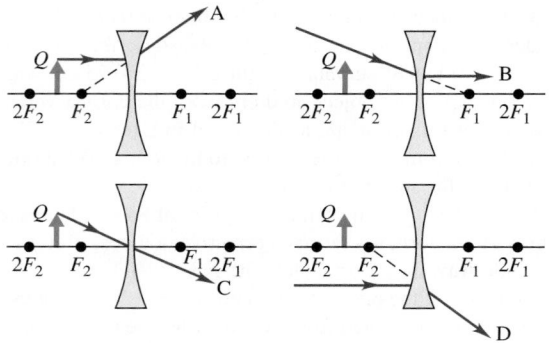

34.5 Answer: (iii) The smaller image area of the CCD array means that the angle of view is decreased for a given focal length. Individual objects make images of the same size in either case; when a smaller light-sensitive area is used, fewer images fit into the area and the field of view is narrower.

34.6 Answer: (iii) This lens is designed to correct for a type of astigmatism. Along the vertical axis, the lens is configured as a converging lens; along the horizontal axis, the lens is configured as a diverging lens. Hence the eye is hyperopic (see Fig. 34.46) for objects that are oriented vertically but myopic for objects that are oriented horizontally (see Fig. 34.47). Without correction, the eye focuses vertical objects behind the retina but horizontal objects in front of the retina.

34.7 Answer: (ii) The object must be held at the focal point, which is twice as far away if the focal length *f* is twice as great. Equation (24.22) shows that the angular magnification *M* is inversely proportional to *f*, so doubling the focal length makes *M* $\frac{1}{2}$ as great. To improve the magnification, you should use a magnifier with a *shorter* focal length.

34.8 Answer: (i) The objective lens of a microscope is designed to make enlarged images of small objects, so the absolute value of its lateral magnification *m* is greater than 1. By contrast, the objective lens of a refracting telescope is designed to make *reduced* images. For example, the moon is thousands of kilometers in diameter, but its image may fit on a CCD array a few centimeters across. Thus |*m*| is much less than 1 for a refracting telescope. (In both cases *m* is negative because the objective makes an inverted image, which is why the question asks about the absolute value of *m*.)

Bridging Problem

Answers: (a) 29.9 cm to the left of the goblet
(b) 3.73 cm to the right of the goblet

INTERFERENCE

? Soapy water is colorless, but when blown into bubbles it shows vibrant colors. How does the thickness of the bubble walls determine the particular colors that appear?

LEARNING GOALS

By studying this chapter, you will learn:

- What happens when two waves combine, or interfere, in space.

- How to understand the interference pattern formed by the interference of two coherent light waves.

- How to calculate the intensity at various points in an interference pattern.

- How interference occurs when light reflects from the two surfaces of a thin film.

- How interference makes it possible to measure extremely small distances.

An ugly black oil spot on the pavement can become a thing of beauty after a rain, when the oil reflects a rainbow of colors. Multicolored reflections can also be seen from the surfaces of soap bubbles and DVDs. How is it possible for colorless objects to produce these remarkable colors?

In our discussion of lenses, mirrors, and optical instruments we used the model of *geometric optics,* in which we represent light as *rays,* straight lines that are bent at a reflecting or refracting surface. But many aspects of the behavior of light *can't* be understood on the basis of rays. We have already learned that light is fundamentally a *wave,* and in some situations we have to consider its wave properties explicitly. If two or more light waves of the same frequency overlap at a point, the total effect depends on the *phases* of the waves as well as their amplitudes. The resulting patterns of light are a result of the *wave* nature of light and cannot be understood on the basis of rays. Optical effects that depend on the wave nature of light are grouped under the heading **physical optics.**

In this chapter we'll look at *interference* phenomena that occur when two waves combine. The colors seen in oil films and soap bubbles are a result of interference between light reflected from the front and back surfaces of a thin film of oil or soap solution. Effects that occur when *many* sources of waves are present are called *diffraction* phenomena; we'll study these in Chapter 36. In that chapter we'll see that diffraction effects occur whenever a wave passes through an aperture or around an obstacle. They are important in practical applications of physical optics such as diffraction gratings, x-ray diffraction, and holography.

While our primary concern is with light, interference and diffraction can occur with waves of *any* kind. As we go along, we'll point out applications to other types of waves such as sound and water waves.

35.1 Interference and Coherent Sources

As we discussed in Chapter 15, the term **interference** refers to any situation in which two or more waves overlap in space. When this occurs, the total wave at any point at any instant of time is governed by the **principle of superposition,** which we introduced in Section 15.6 in the context of waves on a string. This principle also applies to electromagnetic waves and is the most important principle in all of physical optics. The principle of superposition states:

> **When two or more waves overlap, the resultant displacement at any point and at any instant is found by adding the instantaneous displacements that would be produced at the point by the individual waves if each were present alone.**

(In some special situations, such as electromagnetic waves propagating in a crystal, this principle may not apply. A discussion of these is beyond our scope.)

We use the term "displacement" in a general sense. With waves on the surface of a liquid, we mean the actual displacement of the surface above or below its normal level. With sound waves, the term refers to the excess or deficiency of pressure. For electromagnetic waves, we usually mean a specific component of electric or magnetic field.

Interference in Two or Three Dimensions

We have already discussed one important case of interference, in which two identical waves propagating in opposite directions combine to produce a *standing wave.* We saw this in Chapters 15 and 16 for transverse waves on a string and for longitudinal waves in a fluid filling a pipe; we described the same phenomenon for electromagnetic waves in Section 32.5. In all of these cases the waves propagated along only a single axis: along a string, along the length of a fluid-filled pipe, or along the propagation direction of an electromagnetic plane wave. But light waves can (and do) travel in *two* or *three* dimensions, as can any kind of wave that propagates in a two- or three-dimensional medium. In this section we'll see what happens when we combine waves that spread out in two or three dimensions from a pair of identical wave sources.

Interference effects are most easily seen when we combine *sinusoidal* waves with a single frequency f and wavelength λ. Figure 35.1 shows a "snapshot" of a *single* source S_1 of sinusoidal waves and some of the wave fronts produced by this source. The figure shows only the wave fronts corresponding to wave *crests,* so the spacing between successive wave fronts is one wavelength. The material surrounding S_1 is uniform, so the wave speed is the same in all directions, and there is no refraction (and hence no bending of the wave fronts). If the waves are two-dimensional, like waves on the surface of a liquid, the circles in Fig. 35.1 represent circular wave fronts; if the waves propagate in three dimensions, the circles represent spherical wave fronts spreading away from S_1.

In optics, sinusoidal waves are characteristic of **monochromatic light** (light of a single color). While it's fairly easy to make water waves or sound waves of a single frequency, common sources of light *do not* emit monochromatic (single-frequency) light. For example, incandescent light bulbs and flames emit a continuous distribution of wavelengths. By far the most nearly monochromatic source that is available at present is the *laser.* An example is the helium–neon laser, which emits red light at 632.8 nm with a wavelength range of the order of ± 0.000001 nm, or about one part in 10^9. As we analyze interference and diffraction effects in this chapter and the next, we will assume that we are working with monochromatic waves (unless we explicitly state otherwise).

Constructive and Destructive Interference

Two identical sources of monochromatic waves, S_1 and S_2, are shown in Fig. 35.2a. The two sources produce waves of the same amplitude and the same wavelength λ.

35.1 A "snapshot" of sinusoidal waves of frequency f and wavelength λ spreading out from source S_1 in all directions.

Wave fronts: crests of the wave (frequency f) separated by one wavelength λ

The wave fronts move outward from source S_1 at the wave speed $v = f\lambda$.

In addition, the two sources are permanently *in phase;* they vibrate in unison. They might be two loudspeakers driven by the same amplifier, two radio antennas powered by the same transmitter, or two small slits in an opaque screen, illuminated by the same monochromatic light source. We will see that if there were not a constant phase relationship between the two sources, the phenomena we are about to discuss would not occur. Two monochromatic sources of the same frequency and with a constant phase relationship (not necessarily in phase) are said to be **coherent.** We also use the term *coherent waves* (or, for light waves, *coherent light*) to refer to the waves emitted by two such sources.

If the waves emitted by the two coherent sources are *transverse,* like electromagnetic waves, then we will also assume that the wave disturbances produced by both sources have the same *polarization* (that is, they lie along the same line). For example, the sources S_1 and S_2 in Fig. 35.2a could be two radio antennas in the form of long rods oriented parallel to the z-axis (perpendicular to the plane of the figure); at any point in the xy-plane the waves produced by both antennas have $\vec{E}$ fields with only a z-component. Then we need only a single scalar function to describe each wave; this makes the analysis much easier.

We position the two sources of equal amplitude, equal wavelength, and (if the waves are transverse) the same polarization along the y-axis in Fig. 35.2a, equidistant from the origin. Consider a point a on the x-axis. From symmetry the two distances from S_1 to a and from S_2 to a are *equal;* waves from the two sources thus require equal times to travel to a. Hence waves that leave S_1 and S_2 in phase arrive at a in phase. The two waves add, and the total amplitude at a is *twice* the amplitude of each individual wave. This is true for *any* point on the x-axis.

Similarly, the distance from S_2 to point b is exactly two wavelengths *greater* than the distance from S_1 to b. A wave crest from S_1 arrives at b exactly two cycles earlier than a crest emitted at the same time from S_2, and again the two waves arrive in phase. As at point a, the total amplitude is the sum of the amplitudes of the waves from S_1 and S_2.

In general, when waves from two or more sources arrive at a point *in phase,* they reinforce each other: The amplitude of the resultant wave is the *sum* of the amplitudes of the individual waves. This is called **constructive interference** (Fig. 35.2b). Let the distance from S_1 to any point P be r_1, and let the distance from S_2 to P be r_2. For constructive interference to occur at P, the path difference $r_2 - r_1$ for the two sources must be an integral multiple of the wavelength λ:

$$r_2 - r_1 = m\lambda \quad (m = 0, \pm 1, \pm 2, \pm 3, \dots) \qquad \begin{array}{l}\text{(constructive}\\ \text{interference,}\\ \text{sources in phase)}\end{array} \qquad (35.1)$$

In Fig. 35.2a, points a and b satisfy Eq. (35.1) with $m = 0$ and $m = +2$, respectively.

Something different occurs at point c in Fig. 35.2a. At this point, the path difference $r_2 - r_1 = -2.50\lambda$, which is a *half-integral* number of wavelengths. Waves from the two sources arrive at point c exactly a half-cycle out of phase. A crest of one wave arrives at the same time as a crest in the opposite direction (a "trough") of the other wave (Fig. 35.2c). The resultant amplitude is the *difference* between the two individual amplitudes. If the individual amplitudes are equal, then the total amplitude is *zero!* This cancellation or partial cancellation of the individual waves is called **destructive interference.** The condition for destructive interference in the situation shown in Fig. 35.2a is

$$r_2 - r_1 = \left(m + \tfrac{1}{2}\right)\lambda \quad (m = 0, \pm 1, \pm 2, \pm 3, \dots) \qquad \begin{array}{l}\text{(destructive}\\ \text{interference,}\\ \text{sources in phase)}\end{array} \qquad (35.2)$$

The path difference at point c in Fig. 35.2a satisfies Eq. (35.2) with $m = -3$.

35.2 (a) A "snapshot" of sinusoidal waves spreading out from two coherent sources S_1 and S_2. Constructive interference occurs at point a (equidistant from the two sources) and (b) at point b. (c) Destructive interference occurs at point c.

(a) Two coherent wave sources separated by a distance 4λ

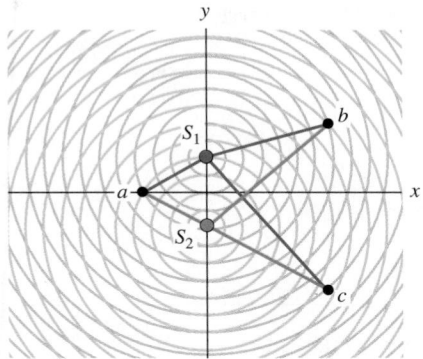

(b) Conditions for constructive interference: Waves interfere constructively if their path lengths differ by an integral number of wavelengths: $r_2 - r_1 = m\lambda$.

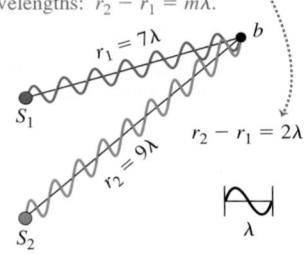

(c) Conditions for destructive interference: Waves interfere destructively if their path lengths differ by a half-integral number of wavelengths: $r_2 - r_1 = (m + \tfrac{1}{2})\lambda$.

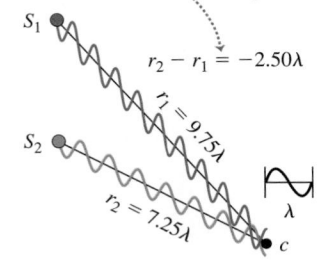

35.3 The same as Fig. 35.2a, but with red antinodal curves (curves of maximum amplitude) superimposed. All points on each curve satisfy Eq. (35.1) with the value of m shown. The nodal curves (not shown) lie between each adjacent pair of antinodal curves.

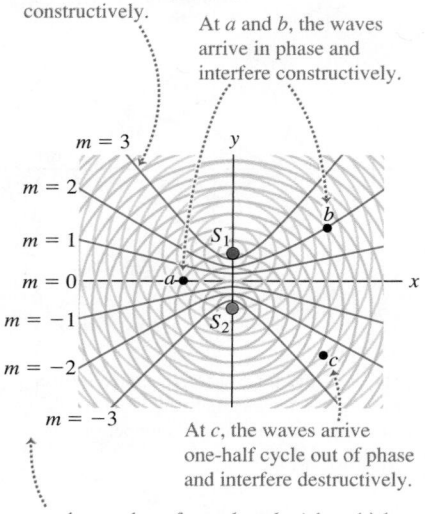

Antinodal curves (red) mark positions where the waves from S_1 and S_2 interfere constructively.

At a and b, the waves arrive in phase and interfere constructively.

At c, the waves arrive one-half cycle out of phase and interfere destructively.

m = the number of wavelengths λ by which the path lengths from S_1 and S_2 differ.

Figure 35.3 shows the same situation as in Fig. 35.2a, but with red curves that show all positions where *constructive* interference occurs. On each curve, the path difference $r_2 - r_1$ is equal to an integer m times the wavelength, as in Eq. (35.1). These curves are called **antinodal curves.** They are directly analogous to *antinodes* in the standing-wave patterns described in Chapters 15 and 16 and Section 32.5. In a standing wave formed by interference between waves propagating in opposite directions, the antinodes are points at which the amplitude is maximum; likewise, the wave amplitude in the situation of Fig. 35.3 is maximum along the antinodal curves. Not shown in Fig. 35.3 are the **nodal curves,** which are the curves that show where *destructive* interference occurs in accordance with Eq. (35.2); these are analogous to the *nodes* in a standing-wave pattern. A nodal curve lies between each two adjacent antinodal curves in Fig. 35.3; one such curve, corresponding to $r_2 - r_1 = -2.50\lambda$, passes through point c.

In some cases, such as two loudspeakers or two radio-transmitter antennas, the interference pattern is three-dimensional. Think of rotating the color curves of Fig. 35.3 around the y-axis; then maximum constructive interference occurs at all points on the resulting surfaces of revolution.

CAUTION **Interference patterns are not standing waves** The interference patterns in Figs. 35.2a and 35.3 are *not* standing waves, though they have some similarities to the standing-wave patterns described in Chapters 15 and 16 and Section 32.5. In a standing wave, the interference is between two waves propagating in opposite directions; there is *no* net energy flow in either direction (the energy in the wave is left "standing"). In the situations shown in Figs. 35.2a and 35.3, there is likewise a stationary pattern of antinodal and nodal curves, but there is a net flow of energy *outward* from the two sources. All that interference does is to "channel" the energy flow so that it is greatest along the antinodal curves and least along the nodal curves.

For Eqs. (35.1) and (35.2) to hold, the two sources must have the same wavelength and must *always* be in phase. These conditions are rather easy to satisfy for sound waves. But with *light* waves there is no practical way to achieve a constant phase relationship (coherence) with two independent sources. This is because of the way light is emitted. In ordinary light sources, atoms gain excess energy by thermal agitation or by impact with accelerated electrons. Such an "excited" atom begins to radiate energy and continues until it has lost all the energy it can, typically in a time of the order of 10^{-8} s. The many atoms in a source ordinarily radiate in an unsynchronized and random phase relationship, and the light that is emitted from *two* such sources has no definite phase relationship.

However, the light from a single source can be split so that parts of it emerge from two or more regions of space, forming two or more *secondary sources*. Then any random phase change in the source affects these secondary sources equally and does not change their *relative* phase.

The distinguishing feature of light from a *laser* is that the emission of light from many atoms is synchronized in frequency and phase. As a result, the random phase changes mentioned above occur much less frequently. Definite phase relationships are preserved over correspondingly much greater lengths in the beam, and laser light is much more coherent than ordinary light.

Test Your Understanding of Section 35.1 Consider a point in Fig. 35.3 on the positive y-axis above S_1. Does this point lie on (i) an antinodal curve; (ii) a nodal curve; or (iii) neither? (*Hint:* The distance between S_1 and S_2 is 4λ.)

35.2 Two-Source Interference of Light

The interference pattern produced by two coherent sources of *water* waves of the same wavelength can be readily seen in a ripple tank with a shallow layer of water (Fig. 35.4). This pattern is not directly visible when the interference is between *light* waves, since light traveling in a uniform medium cannot be seen. (A shaft of afternoon sunlight in a room is made visible by scattering from airborne dust particles.)

One of the earliest quantitative experiments to reveal the interference of light from two sources was performed in 1800 by the English scientist Thomas Young. We will refer back to this experiment several times in this and later chapters, so it's important to understand it in detail. Young's apparatus is shown in perspective in Fig. 35.5a. A light source (not shown) emits monochromatic light; however, this light is not suitable for use in an interference experiment because emissions from different parts of an ordinary source are not synchronized. To remedy this, the light is directed at a screen with a narrow slit S_0, 1 μm or so wide. The light emerging from the slit originated from only a small region of the light source; thus slit S_0 behaves more nearly like the idealized source shown in Fig. 35.1. (In modern versions of the experiment, a laser is used as a source of coherent light, and the slit S_0 isn't needed.) The light from slit S_0 falls on a screen with two other narrow slits S_1 and S_2, each 1 μm or so wide and a few tens or hundreds of micrometers apart. Cylindrical wave fronts spread out from slit S_0 and reach slits S_1 and S_2 *in phase* because they travel equal distances from S_0. The waves *emerging* from slits S_1 and S_2 are therefore also always in phase, so S_1 and S_2 are *coherent* sources. The interference of waves from S_1 and S_2 produces a pattern in space like that to the right of the sources in Figs. 35.2a and 35.3.

To visualize the interference pattern, a screen is placed so that the light from S_1 and S_2 falls on it (Fig. 35.5b). The screen will be most brightly illuminated at points P, where the light waves from the slits interfere constructively, and will be darkest at points where the interference is destructive.

To simplify the analysis of Young's experiment, we assume that the distance R from the slits to the screen is so large in comparison to the distance d between the slits that the lines from S_1 and S_2 to P are very nearly parallel, as in Fig. 35.5c. This is usually the case for experiments with light; the slit separation is typically a few millimeters, while the screen may be a meter or more away. The difference in path length is then given by

35.4 The concepts of constructive interference and destructive interference apply to these water waves as well as to light waves and sound waves.

(a) Interference of light waves passing through two slits

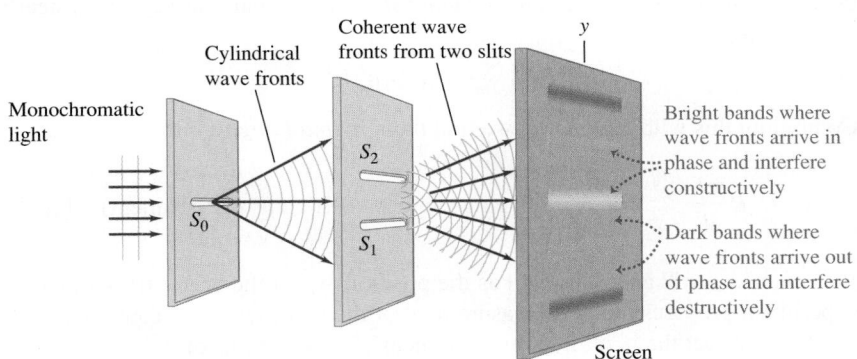

35.5 (a) Young's experiment to show interference of light passing through two slits. A pattern of bright and dark areas appears on the screen (see Fig. 35.6). (b) Geometrical analysis of Young's experiment. For the case shown, $r_2 > r_1$ and both y and θ are positive. If point P is on the other side of the screen's center, $r_2 < r_1$ and both y and θ are negative. (c) Approximate geometry when the distance R to the screen is much greater than the distance d between the slits.

(b) Actual geometry (seen from the side)

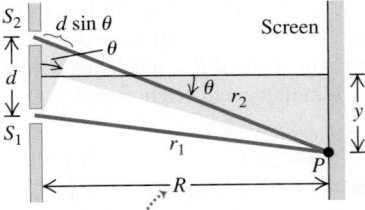

In real situations, the distance R to the screen is usually very much greater than the distance d between the slits ...

(c) Approximate geometry

... so we can treat the rays as parallel, in which case the path-length difference is simply $r_2 - r_1 = d \sin \theta$.

$$r_2 - r_1 = d\sin\theta \qquad (35.3)$$

where θ is the angle between a line from slits to screen (shown in blue in Fig. 35.5c) and the normal to the plane of the slits (shown as a thin black line).

Constructive and Destructive Two-Slit Interference

We found in Section 35.1 that constructive interference (reinforcement) occurs at points where the path difference is an integral number of wavelengths, $m\lambda$, where $m = 0,\ \pm 1,\ \pm 2,\ \pm 3, \dots$. So the bright regions on the screen in Fig. 35.5a occur at angles θ for which

$$d\sin\theta = m\lambda \qquad (m = 0,\ \pm 1,\ \pm 2, \dots) \qquad \begin{array}{l}\text{(constructive inter-}\\\text{ference, two slits)}\end{array} \quad (35.4)$$

Similarly, destructive interference (cancellation) occurs, forming dark regions on the screen, at points for which the path difference is a half-integral number of wavelengths, $\left(m + \frac{1}{2}\right)\lambda$:

$$d\sin\theta = \left(m + \tfrac{1}{2}\right)\lambda \qquad (m = 0,\ \pm 1,\ \pm 2, \dots) \qquad \begin{array}{l}\text{(destructive inter-}\\\text{ference, two slits)}\end{array} \quad (35.5)$$

35.6 Photograph of interference fringes produced on a screen in Young's double-slit experiment.

Thus the pattern on the screen of Figs. 35.5a and 35.5b is a succession of bright and dark bands, or **interference fringes,** parallel to the slits S_1 and S_2. A photograph of such a pattern is shown in Fig. 35.6. The center of the pattern is a bright band corresponding to $m = 0$ in Eq. (35.4); this point on the screen is equidistant from the two slits.

We can derive an expression for the positions of the centers of the bright bands on the screen. In Fig. 35.5b, y is measured from the center of the pattern, corresponding to the distance from the center of Fig. 35.6. Let y_m be the distance from the center of the pattern $(\theta = 0)$ to the center of the mth bright band. Let θ_m be the corresponding value of θ; then

$$y_m = R\tan\theta_m$$

In experiments such as this, the distances y_m are often much smaller than the distance R from the slits to the screen. Hence θ_m is very small, $\tan\theta_m$ is very nearly equal to $\sin\theta_m$, and

$$y_m = R\sin\theta_m$$

Combining this with Eq. (35.4), we find that *for small angles only,*

$$y_m = R\frac{m\lambda}{d} \qquad \text{(constructive interference in Young's experiment)} \quad (35.6)$$

We can measure R and d, as well as the positions y_m of the bright fringes, so this experiment provides a direct measurement of the wavelength λ. Young's experiment was in fact the first direct measurement of wavelengths of light.

CAUTION **Equation (35.6) is for small angles only** While Eqs. (35.4) and (35.5) are valid for any angle, Eq. (35.6) is valid only for *small* angles. It can be used *only* if the distance R from slits to screen is much greater than the slit separation d and if R is much greater than the distance y_m from the center of the interference pattern to the mth bright fringe. ▮

The distance between adjacent bright bands in the pattern is *inversely* proportional to the distance d between the slits. The closer together the slits are, the more the pattern spreads out. When the slits are far apart, the bands in the pattern are closer together.

While we have described the experiment that Young performed with visible light, the results given in Eqs. (35.4) and (35.5) are valid for *any* type of wave, provided that the resultant wave from two coherent sources is detected at a point that is far away in comparison to the separation *d*.

Example 35.1   **Two-slit interference**

Figure 35.7 shows a two-slit interference experiment in which the slits are 0.200 mm apart and the screen is 1.00 m from the slits. The $m = 3$ bright fringe in the figure is 9.49 mm from the central fringe. Find the wavelength of the light.

35.7 Using a two-slit interference experiment to measure the wavelength of light.

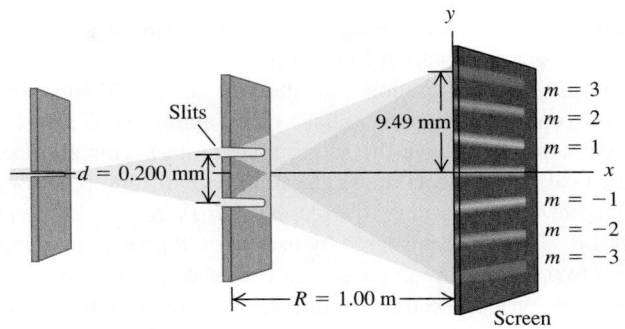

SOLUTION

IDENTIFY and SET UP: Our target variable in this two-slit interference problem is the wavelength λ. We are given the slit separation $d = 0.200$ mm, the distance from slits to screen $R = 1.00$ m, and the distance $y_3 = 9.49$ mm on the screen from the center of the interference pattern to the $m = 3$ bright fringe. We may use Eq. (35.6) to find λ, since the value of R is so much greater than the values of d or y_3.

EXECUTE: We solve Eq. (35.6) for λ for the case $m = 3$:

$$\lambda = \frac{y_m d}{mR} = \frac{(9.49 \times 10^{-3} \text{ m})(0.200 \times 10^{-3} \text{ m})}{(3)(1.00 \text{ m})}$$

$$= 633 \times 10^{-9} \text{ m} = 633 \text{ nm}$$

EVALUATE: This bright fringe could also correspond to $m = -3$. Can you show that this gives the same result for λ?

Example 35.2   **Broadcast pattern of a radio station**

It is often desirable to radiate most of the energy from a radio transmitter in particular directions rather than uniformly in all directions. Pairs or rows of antennas are often used to produce the desired radiation pattern. As an example, consider two identical vertical antennas 400 m apart, operating at 1500 kHz $= 1.5 \times 10^6$ Hz (near the top end of the AM broadcast band) and oscillating in phase. At distances much greater than 400 m, in what directions is the intensity from the two antennas greatest?

SOLUTION

IDENTIFY and SET UP: The antennas, shown in Fig. 35.8, correspond to sources S_1 and S_2 in Fig. 35.5. Hence we can apply

35.8 Two radio antennas broadcasting in phase. The purple arrows indicate the directions of maximum intensity. The waves that are emitted toward the lower half of the figure are not shown.

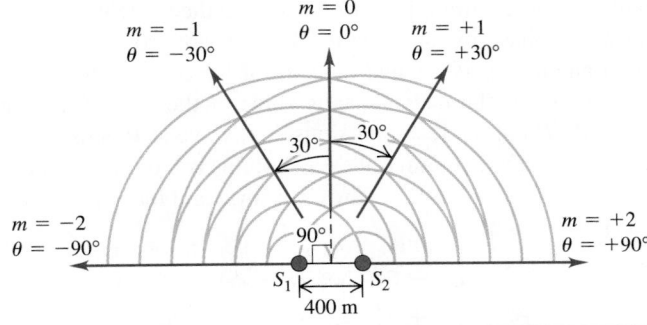

the ideas of two-slit interference to this problem. Since the resultant wave is detected at distances much greater than $d = 400$ m, we may use Eq. (35.4) to give the directions of the intensity maxima, the values of θ for which the path difference is zero or a whole number of wavelengths.

EXECUTE: The wavelength is $\lambda = c/f = 200$ m. From Eq. (35.4) with $m = 0$, ± 1, and ± 2, the intensity maxima are given by

$$\sin \theta = \frac{m\lambda}{d} = \frac{m(200 \text{ m})}{400 \text{ m}} = \frac{m}{2} \qquad \theta = 0, \pm 30°, \pm 90°$$

In this example, values of m greater than 2 or less than -2 give values of $\sin \theta$ greater than 1 or less than -1, which is impossible. There is *no* direction for which the path difference is three or more wavelengths, so values of m of ± 3 or beyond have no meaning in this example.

EVALUATE: We can check our result by calculating the angles for *minimum* intensity, using Eq. (35.5). There should be one intensity minimum between each pair of intensity maxima, just as in Fig. 35.6. From Eq. (35.5), with $m = -2, -1, 0$, and 1,

$$\sin \theta = \frac{(m + \frac{1}{2})\lambda}{d} = \frac{m + \frac{1}{2}}{2} \qquad \theta = \pm 14.5°, \pm 48.6°$$

These angles fall between the angles for intensity maxima, as they should. The angles are not small, so the angles for the minima are *not* exactly halfway between the angles for the maxima.

Test Your Understanding of Section 35.2 You shine a tunable laser (whose wavelength can be adjusted by turning a knob) on a pair of closely spaced slits. The light emerging from the two slits produces an interference pattern on a screen like that shown in Fig. 35.6. If you adjust the wavelength so that the laser light changes from red to blue, how will the spacing between bright fringes change? (i) The spacing increases; (ii) the spacing decreases; (iii) the spacing is unchanged; (iv) not enough information is given to decide.

35.3 Intensity in Interference Patterns

In Section 35.2 we found the positions of maximum and minimum intensity in a two-source interference pattern. Let's now see how to find the intensity at *any* point in the pattern. To do this, we have to combine the two sinusoidally varying fields (from the two sources) at a point P in the radiation pattern, taking proper account of the phase difference of the two waves at point P, which results from the path difference. The intensity is then proportional to the square of the resultant electric-field amplitude, as we learned in Chapter 32.

To calculate the intensity, we will assume that the two sinusoidal functions (corresponding to waves from the two sources) have equal amplitude E and that the $\vec{E}$ fields lie along the same line (have the same polarization). This assumes that the sources are identical and neglects the slight amplitude difference caused by the unequal path lengths (the amplitude decreases with increasing distance from the source). From Eq. (32.29), each source by itself would give an intensity $\frac{1}{2}\epsilon_0 cE^2$ at point P. If the two sources are in phase, then the waves that arrive at P differ in phase by an amount proportional to the difference in their path lengths, $(r_2 - r_1)$. If the phase angle between these arriving waves is ϕ, then we can use the following expressions for the two electric fields superposed at P:

$$E_1(t) = E\cos(\omega t + \phi)$$
$$E_2(t) = E\cos\omega t$$

The superposition of the two fields at P is a sinusoidal function with some amplitude E_P that depends on E and the phase difference ϕ. First we'll work on finding the amplitude E_P if E and ϕ are known. Then we'll find the intensity I of the resultant wave, which is proportional to E_P^2. Finally, we'll relate the phase difference ϕ to the path difference, which is determined by the geometry of the situation.

Amplitude in Two-Source Interference

To add the two sinusoidal functions with a phase difference, we use the same *phasor* representation that we used for simple harmonic motion (see Section 14.2) and for voltages and currents in ac circuits (see Section 31.1). We suggest that you review these sections now. Each sinusoidal function is represented by a rotating vector (phasor) whose projection on the horizontal axis at any instant represents the instantaneous value of the sinusoidal function.

In Fig. 35.9, E_1 is the horizontal component of the phasor representing the wave from source S_1, and E_2 is the horizontal component of the phasor for the wave from S_2. As shown in the diagram, both phasors have the same magnitude E, but E_1 is *ahead* of E_2 in phase by an angle ϕ. Both phasors rotate counterclockwise with constant angular speed ω, and the sum of the projections on the horizontal axis at any time gives the instantaneous value of the total E field at point P. Thus the amplitude E_P of the resultant sinusoidal wave at P is the magnitude of the dark red phasor in the diagram (labeled E_P); this is the *vector sum* of the other two phasors. To find E_P, we use the law of cosines and the trigonometric identity $\cos(\pi - \phi) = -\cos\phi$:

$$E_P^2 = E^2 + E^2 - 2E^2\cos(\pi - \phi)$$
$$= E^2 + E^2 + 2E^2\cos\phi$$

35.9 Phasor diagram for the superposition at a point P of two waves of equal amplitude E with a phase difference ϕ.

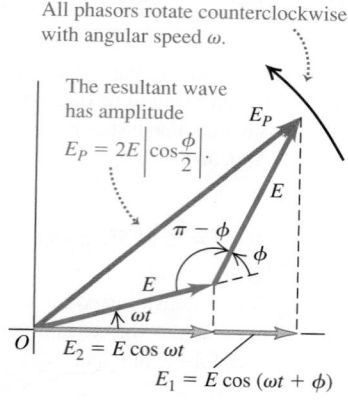

All phasors rotate counterclockwise with angular speed ω.

The resultant wave has amplitude
$E_P = 2E\left|\cos\dfrac{\phi}{2}\right|$.

Then, using the identity $1 + \cos\phi = 2\cos^2(\phi/2)$, we obtain

$$E_P{}^2 = 2E^2(1 + \cos\phi) = 4E^2\cos^2\left(\frac{\phi}{2}\right)$$

$$E_P = 2E\left|\cos\frac{\phi}{2}\right| \qquad \text{(amplitude in two-source interference)} \qquad (35.7)$$

You can also obtain this result without using phasors (see Problem 35.50).

When the two waves are in phase, $\phi = 0$ and $E_P = 2E$. When they are exactly a half-cycle out of phase, $\phi = \pi$ rad $= 180°$, $\cos(\phi/2) = \cos(\pi/2) = 0$, and $E_P = 0$. Thus the superposition of two sinusoidal waves with the same frequency and amplitude but with a phase difference yields a sinusoidal wave with the same frequency and an amplitude between zero and twice the individual amplitudes, depending on the phase difference.

Intensity in Two-Source Interference

To obtain the intensity I at point P, we recall from Section 32.4 that I is equal to the average magnitude of the Poynting vector, S_{av}. For a sinusoidal wave with electric-field amplitude E_P, this is given by Eq. (32.29) with E_{max} replaced by E_P. Thus we can express the intensity in several equivalent forms:

$$I = S_{av} = \frac{E_P{}^2}{2\mu_0 c} = \frac{1}{2}\sqrt{\frac{\epsilon_0}{\mu_0}}E_P{}^2 = \frac{1}{2}\epsilon_0 c E_P{}^2 \qquad (35.8)$$

The essential content of these expressions is that I is proportional to $E_P{}^2$. When we substitute Eq. (35.7) into the last expression in Eq. (35.8), we get

$$I = \frac{1}{2}\epsilon_0 c E_P{}^2 = 2\epsilon_0 c E^2 \cos^2\frac{\phi}{2} \qquad (35.9)$$

In particular, the *maximum* intensity I_0, which occurs at points where the phase difference is zero ($\phi = 0$), is

$$I_0 = 2\epsilon_0 c E^2$$

Note that the maximum intensity I_0 is *four times* (not twice) as great as the intensity $\frac{1}{2}\epsilon_0 c E^2$ from each individual source.

Substituting the expression for I_0 into Eq. (35.9), we can express the intensity I at any point very simply in terms of the maximum intensity I_0:

$$I = I_0\cos^2\frac{\phi}{2} \qquad \text{(intensity in two-source interference)} \qquad (35.10)$$

For some phase angles ϕ the intensity is I_0, four times as great as for an individual wave source, but for other phase angles the intensity is zero. If we average Eq. (35.10) over all possible phase differences, the result is $I_0/2 = \epsilon_0 c E^2$ (the average of $\cos^2(\phi/2)$ is $\frac{1}{2}$). This is just twice the intensity from each individual source, as we should expect. The total energy output from the two sources isn't changed by the interference effects, but the energy is redistributed (as we mentioned in Section 35.1).

Phase Difference and Path Difference

Our next task is to find the phase difference ϕ between the two fields at any point P. We know that ϕ is proportional to the difference in path length from the two sources to point P. When the path difference is one wavelength, the phase difference is one cycle, and $\phi = 2\pi$ rad $= 360°$. When the path difference is

$\lambda/2$, $\phi = \pi$ rad $= 180°$, and so on. That is, the ratio of the phase difference ϕ to 2π is equal to the ratio of the path difference $r_2 - r_1$ to λ:

$$\frac{\phi}{2\pi} = \frac{r_2 - r_1}{\lambda}$$

Thus a path difference $(r_2 - r_1)$ causes a phase difference given by

$$\phi = \frac{2\pi}{\lambda}(r_2 - r_1) = k(r_2 - r_1) \qquad \begin{array}{l}\text{(phase difference related} \\ \text{to path difference)}\end{array} \qquad (35.11)$$

where $k = 2\pi/\lambda$ is the *wave number* introduced in Section 15.3.

If the material in the space between the sources and P is anything other than vacuum, we must use the wavelength *in the material* in Eq. (35.11). If the material has index of refraction n, then

$$\lambda = \frac{\lambda_0}{n} \qquad \text{and} \qquad k = nk_0 \qquad (35.12)$$

where λ_0 and k_0 are the wavelength and the wave number, respectively, in vacuum.

Finally, if the point P is far away from the sources in comparison to their separation d, the path difference is given by Eq. (35.3):

$$r_2 - r_1 = d\sin\theta$$

Combining this with Eq. (35.11), we find

$$\phi = k(r_2 - r_1) = kd\sin\theta = \frac{2\pi d}{\lambda}\sin\theta \qquad (35.13)$$

When we substitute this into Eq. (35.10), we find

$$I = I_0\cos^2\left(\tfrac{1}{2}kd\sin\theta\right) = I_0\cos^2\left(\frac{\pi d}{\lambda}\sin\theta\right) \qquad \begin{array}{l}\text{(intensity far from} \\ \text{two sources)}\end{array} \qquad (35.14)$$

The directions of *maximum* intensity occur when the cosine has the values ± 1— that is, when

$$\frac{\pi d}{\lambda}\sin\theta = m\pi \qquad (m = 0, \pm 1, \pm 2, \dots)$$

or

$$d\sin\theta = m\lambda$$

in agreement with Eq. (35.4). You can also derive Eq. (35.5) for the zero-intensity directions from Eq. (35.14).

As we noted in Section 35.2, in experiments with light we visualize the interference pattern due to two slits by using a screen placed at a distance R from the slits. We can describe positions on the screen with the coordinate y; the positions of the bright fringes are given by Eq. (35.6), where ordinarily $y \ll R$. In that case, $\sin\theta$ is approximately equal to y/R, and we obtain the following expressions for the intensity at *any* point on the screen as a function of y:

$$I = I_0\cos^2\left(\frac{kdy}{2R}\right) = I_0\cos^2\left(\frac{\pi dy}{\lambda R}\right) \quad \text{(intensity in two-slit interference)} \quad (35.15)$$

Figure 35.10 shows a graph of Eq. (35.15); we can compare this with the photographically recorded pattern of Fig. 35.6. The peaks in Fig. 35.10 all have the same intensity, while those in Fig. 35.6 fade off as we go away from the center. We'll explore the reasons for this variation in peak intensity in Chapter 36.

35.10 Intensity distribution in the interference pattern from two identical slits.

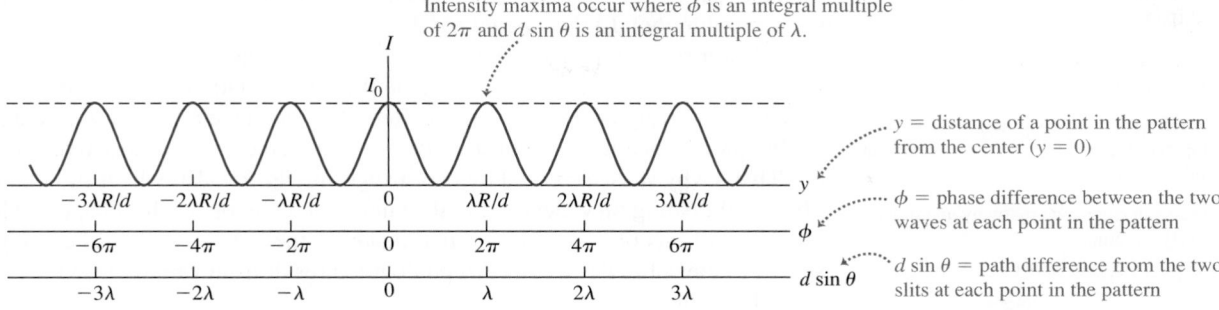

Intensity maxima occur where ϕ is an integral multiple of 2π and $d \sin \theta$ is an integral multiple of λ.

y = distance of a point in the pattern from the center ($y = 0$)

ϕ = phase difference between the two waves at each point in the pattern

$d \sin \theta$ = path difference from the two slits at each point in the pattern

---

Example 35.3 | **A directional transmitting antenna array**

Suppose the two identical radio antennas of Fig. 35.8 are moved to be only 10.0 m apart and the broadcast frequency is increased to $f = 60.0$ MHz. At a distance of 700 m from the point midway between the antennas and in the direction $\theta = 0$ (see Fig. 35.8), the intensity is $I_0 = 0.020 \ \text{W/m}^2$. At this same distance, find (a) the intensity in the direction $\theta = 4.0°$; (b) the direction near $\theta = 0$ for which the intensity is $I_0/2$; and (c) the directions in which the intensity is zero.

SOLUTION

IDENTIFY and SET UP: This problem involves the intensity distribution as a function of angle. Because the 700-m distance from the antennas to the point at which the intensity is measured is much greater than the distance $d = 10.0$ m between the antennas, the amplitudes of the waves from the two antennas are very nearly equal. Hence we can use Eq. (35.14) to relate intensity I and angle θ.

EXECUTE: The wavelength is $\lambda = c/f = 5.00$ m. The spacing $d = 10.0$ m between the antennas is just twice the wavelength (as was the case in Example 35.2), so $d/\lambda = 2.00$ and Eq. (35.14) becomes

$$I = I_0 \cos^2\left(\frac{\pi d}{\lambda}\sin\theta\right) = I_0 \cos^2[(2.00\pi \ \text{rad})\sin\theta]$$

(a) When $\theta = 4.0°$,

$$I = I_0 \cos^2[(2.00\pi \ \text{rad})\sin 4.0°] = 0.82I_0$$
$$= (0.82)(0.020 \ \text{W/m}^2) = 0.016 \ \text{W/m}^2$$

(b) The intensity I equals $I_0/2$ when the cosine in Eq. (35.14) has the value $\pm 1/\sqrt{2}$. The smallest angles at which this occurs correspond to $2.00\pi \sin\theta = \pm\pi/4$ rad, so that $\sin\theta = \pm(1/8.00) = \pm 0.125$ and $\theta = \pm 7.2°$.

(c) The intensity is zero when $\cos[(2.00\pi \ \text{rad})\sin\theta] = 0$. This occurs for $2.00\pi \sin\theta = \pm\pi/2, \pm 3\pi/2, \pm 5\pi/2, \ldots$, or $\sin\theta = \pm 0.250, \pm 0.750, \pm 1.25, \ldots$. Values of $\sin\theta$ greater than 1 have no meaning, so the answers are

$$\theta = \pm 14.5°, \pm 48.6°$$

EVALUATE: The condition in part (b) that $I = I_0/2$, so that $(2.00\pi \ \text{rad})\sin\theta = \pm\pi/4$ rad, is also satisfied when $\sin\theta = \pm 0.375, \pm 0.625$, or ± 0.875 so that $\theta = \pm 22.0°, \pm 38.7°$, or $\pm 61.0°$. (Can you verify this?) It would be incorrect to include these angles in the solution, however, because the problem asked for the angle *near* $\theta = 0$ at which $I = I_0/2$. These additional values of θ aren't the ones we're looking for.

Test Your Understanding of Section 35.3 A two-slit interference experiment uses coherent light of wavelength 5.00×10^{-7} m. Rank the following points in the interference pattern according to the intensity at each point, from highest to lowest. (i) a point that is closer to one slit than the other by 4.00×10^{-7} m; (ii) a point where the light waves received from the two slits are out of phase by 4.00 rad; (iii) a point that is closer to one slit than the other by 7.50×10^{-7} m; (iv) a point where the light waves received by the two slits are out of phase by 2.00 rad. ❙

35.4 Interference in Thin Films

You often see bright bands of color when light reflects from a thin layer of oil floating on water or from a soap bubble (see the photograph that opens this chapter). These are the results of interference. Light waves are reflected from the front and back surfaces of such thin films, and constructive interference between the two reflected waves (with different path lengths) occurs in different

35.11 (a) A diagram and (b) a photograph showing interference of light reflected from a thin film.

(a) Interference between rays reflected from the two surfaces of a thin film

Light reflected from the upper and lower surfaces of the film comes together in the eye at P and undergoes interference.

Some colors interfere constructively and others destructively, creating the color bands we see.

(b) The rainbow fringes of an oil slick on water

35.12 Interference between light waves reflected from the two sides of an air wedge separating two glass plates. The angles and the thickness of the air wedge have been exaggerated for clarity; in the text we assume that the light strikes the upper plate at normal incidence and that the distances h and t are much less than l.

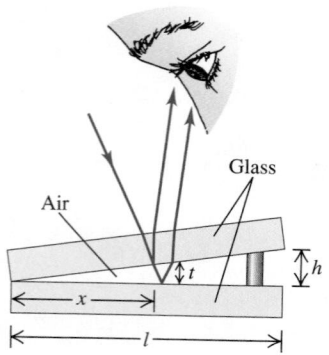

places for different wavelengths. Figure 35.11a shows the situation. Light shining on the upper surface of a thin film with thickness t is partly reflected at the upper surface (path abc). Light *transmitted* through the upper surface is partly reflected at the lower surface (path $abdef$). The two reflected waves come together at point P on the retina of the eye. Depending on the phase relationship, they may interfere constructively or destructively. Different colors have different wavelengths, so the interference may be constructive for some colors and destructive for others. That's why we see colored rings or fringes in Fig. 35.11b (which shows a thin film of oil floating on water) and in the photograph that opens this chapter (which shows thin films of soap solution that make up the bubble walls). The complex shapes of the colored rings in each photograph result from variations in the thickness of the film.

Thin-Film Interference and Phase Shifts During Reflection

Let's look at a simplified situation in which *monochromatic* light reflects from two nearly parallel surfaces at nearly normal incidence. Figure 35.12 shows two plates of glass separated by a thin wedge, or film, of air. We want to consider interference between the two light waves reflected from the surfaces adjacent to the air wedge, as shown. (Reflections also occur at the top surface of the upper plate and the bottom surface of the lower plate; to keep our discussion simple, we won't include these.) The situation is the same as in Fig. 35.11a except that the film (wedge) thickness is not uniform. The path difference between the two waves is just twice the thickness t of the air wedge at each point. At points where $2t$ is an integer number of wavelengths, we expect to see constructive interference and a bright area; where it is a half-integer number of wavelengths, we expect to see destructive interference and a dark area. Along the line where the plates are in contact, there is practically *no* path difference, and we expect a bright area.

When we carry out the experiment, the bright and dark fringes appear, but they are interchanged! Along the line where the plates are in contact, we find a *dark* fringe, not a bright one. This suggests that one or the other of the reflected waves has undergone a half-cycle phase shift during its reflection. In that case the two waves that are reflected at the line of contact are a half-cycle out of phase even though they have the same path length.

In fact, this phase shift can be predicted from Maxwell's equations and the electromagnetic nature of light. The details of the derivation are beyond our scope, but here is the result. Suppose a light wave with electric-field amplitude E_i is traveling in an optical material with index of refraction n_a. It strikes, at normal incidence, an interface with another optical material with index n_b. The amplitude E_r of the wave reflected from the interface is proportional to the amplitude E_i of the incident wave and is given by

$$E_r = \frac{n_a - n_b}{n_a + n_b} E_i \quad \text{(normal incidence)} \tag{35.16}$$

This result shows that the incident and reflected amplitudes have the same sign when n_a is larger than n_b and opposite sign when n_b is larger than n_a. We can distinguish three cases, as shown in Fig. 35.13:

Figure 35.13a: When $n_a > n_b$, light travels more slowly in the first material than in the second. In this case, E_r and E_i have the same sign, and the phase shift of the reflected wave relative to the incident wave is zero. This is analogous to reflection of a transverse mechanical wave on a heavy rope at a point where it is tied to a lighter rope or a ring that can move vertically without friction.

Figure 35.13b: When $n_a = n_b$, the amplitude E_r of the reflected wave is zero. The incident light wave can't "see" the interface, and there is *no* reflected wave.

35.13 Upper figures: electromagnetic waves striking an interface between optical materials at normal incidence (shown as a small angle for clarity). Lower figures: mechanical wave pulses on ropes.

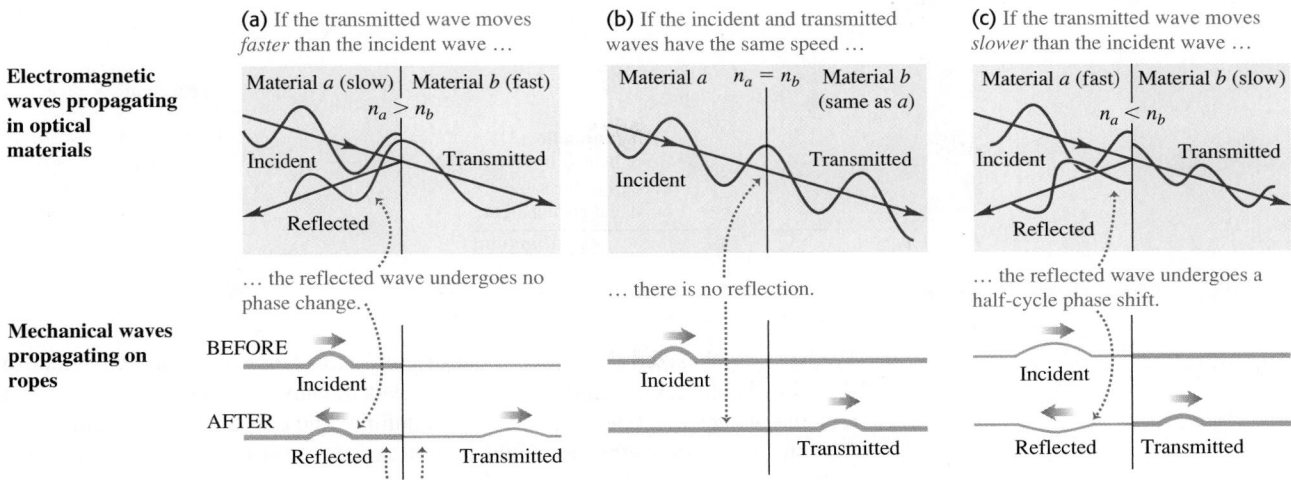

(a) If the transmitted wave moves *faster* than the incident wave ...

(b) If the incident and transmitted waves have the same speed ...

(c) If the transmitted wave moves *slower* than the incident wave ...

... the reflected wave undergoes no phase change.

... there is no reflection.

... the reflected wave undergoes a half-cycle phase shift.

Waves travel slower on thick ropes than on thin ropes.

Figure 35.13c: When $n_a < n_b$, light travels more slowly in the second material than in the first. In this case, E_r and E_i have opposite signs, and the phase shift of the reflected wave relative to the incident wave is π rad (180° or a half-cycle). This is analogous to reflection (with inversion) of a transverse mechanical wave on a light rope at a point where it is tied to a heavier rope or a rigid support.

Let's compare with the situation of Fig. 35.12. For the wave reflected from the upper surface of the air wedge, n_a (glass) is greater than n_b, so this wave has zero phase shift. For the wave reflected from the lower surface, n_a (air) is less than n_b (glass), so this wave has a half-cycle phase shift. Waves that are reflected from the line of contact have no path difference to give additional phase shifts, and they interfere destructively; this is what we observe. You can use the above principle to show that for normal incidence, the wave reflected at point b in Fig. 35.11a is shifted by a half-cycle, while the wave reflected at d is not (if there is air below the film).

We can summarize this discussion mathematically. If the film has thickness t, the light is at normal incidence and has wavelength λ in the film; if neither or both of the reflected waves from the two surfaces have a half-cycle reflection phase shift, the conditions for constructive and destructive interference are

$$2t = m\lambda \qquad (m = 0, 1, 2, \dots)$$

(constructive reflection from thin film, no relative phase shift) (35.17a)

$$2t = \left(m + \tfrac{1}{2}\right)\lambda \qquad (m = 0, 1, 2, \dots)$$

(destructive reflection from thin film, no relative phase shift) (35.17b)

If *one* of the two waves has a half-cycle reflection phase shift, the conditions for constructive and destructive interference are reversed:

$$2t = \left(m + \tfrac{1}{2}\right)\lambda \qquad (m = 0, 1, 2, \dots)$$

(constructive reflection from thin film, half-cycle relative phase shift) (35.18a)

$$2t = m\lambda \qquad (m = 0, 1, 2, \dots)$$

(destructive reflection from thin film, half-cycle relative phase shift) (35.18b)

35.14 (a) Light reflecting from a thin film produces a steady interference pattern, but (b) light reflecting from a thick film does not.

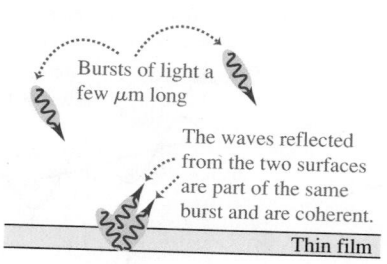

(a) Light reflecting from a thin film

Bursts of light a few μm long

The waves reflected from the two surfaces are part of the same burst and are coherent.

Thin film

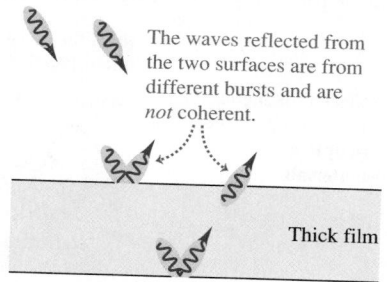

(b) Light reflecting from a thick film

The waves reflected from the two surfaces are from different bursts and are *not* coherent.

Thick film

Thin and Thick Films

We have emphasized *thin* films in our discussion because of a principle we introduced in Section 35.1: In order for two waves to cause a steady interference pattern, the waves must be *coherent,* with a definite and constant phase relationship. The sun and light bulbs emit light in a stream of short bursts, each of which is only a few micrometers long (1 micrometer = 1 μm = 10^{-6} m). If light reflects from the two surfaces of a thin film, the two reflected waves are part of the same burst (Fig. 35.14a). Hence these waves are coherent and interference occurs as we have described. If the film is too thick, however, the two reflected waves will belong to different bursts (Fig. 35.14b). There is no definite phase relationship between different light bursts, so the two waves are incoherent and there is no fixed interference pattern. That's why you see interference colors in light reflected from an oil slick a few micrometers thick (see Fig. 35.11b), but you do *not* see such colors in the light reflected from a pane of window glass with a thickness of a few millimeters (a thousand times greater).

Problem-Solving Strategy 35.1 | **Interference in Thin Films**

IDENTIFY *the relevant concepts:* Problems with thin films involve interference of two waves, one reflected from the film's front surface and one reflected from the back surface. Typically you will be asked to relate the wavelength, the film thickness, and the index of refraction of the film.

SET UP *the problem* using the following steps:
1. Make a drawing showing the geometry of the film. Identify the materials that adjoin the film; their properties determine whether one or both of the reflected waves have a half-cycle phase shift.
2. Identify the target variable.

EXECUTE *the solution* as follows:
1. Apply the rule for phase changes to each reflected wave: There is a half-cycle phase shift when $n_b > n_a$ and none when $n_b < n_a$.

2. If *neither* reflected wave undergoes a phase shift, or if *both* do, use Eqs. (35.17). If only one reflected wave undergoes a phase shift, use Eqs. (35.18).
3. Solve the resulting equation for the target variable. Use the wavelength $\lambda = \lambda_0/n$ of light *in the film* in your calculations, where n is the index of refraction of the film. (For air, $n = 1.000$ to four-figure precision.)
4. If you are asked about the wave that is transmitted through the film, remember that minimum intensity in the reflected wave corresponds to maximum *transmitted* intensity, and vice versa.

EVALUATE *your answer:* Interpret your results by examining what would happen if the wavelength were changed or if the film had a different thickness.

Example 35.4 | Thin-film interference I

Suppose the two glass plates in Fig. 35.12 are two microscope slides 10.0 cm long. At one end they are in contact; at the other end they are separated by a piece of paper 0.0200 mm thick. What is the spacing of the interference fringes seen by reflection? Is the fringe at the line of contact bright or dark? Assume monochromatic light with a wavelength in air of $\lambda = \lambda_0 = 500$ nm.

SOLUTION

IDENTIFY and SET UP: Figure 35.15 depicts the situation. We'll consider only interference between the light reflected from the upper and lower surfaces of the air wedge between the microscope slides. [The top slide has a relatively great thickness, about 1 mm,

35.15 Our sketch for this problem.

so we can ignore interference between the light reflected from its upper and lower surfaces (see Fig. 35.14b).] Light travels more slowly in the glass of the slides than it does in air. Hence the wave reflected from the upper surface of the air wedge has no phase shift (see Fig. 35.13a), while the wave reflected from the lower surface has a half-cycle phase shift (see Fig. 35.13c).

EXECUTE: Since only one of the reflected waves undergoes a phase shift, the condition for *destructive* interference (a dark fringe) is Eq. (35.18b):

$$2t = m\lambda_0 \quad (m = 0, 1, 2, \dots)$$

From similar triangles in Fig. 35.15 the thickness t of the air wedge at each point is proportional to the distance x from the line of contact:

$$\frac{t}{x} = \frac{h}{l}$$

Combining this with Eq. (35.18b), we find

$$\frac{2xh}{l} = m\lambda_0$$

$$x = m\frac{l\lambda_0}{2h} = m\frac{(0.100 \text{ m})(500 \times 10^{-9} \text{ m})}{(2)(0.0200 \times 10^{-3} \text{ m})} = m(1.25 \text{ mm})$$

Successive dark fringes, corresponding to $m = 1, 2, 3, \dots$, are spaced 1.25 mm apart. Substituting $m = 0$ into this equation gives $x = 0$, which is where the two slides touch (at the left-hand side of Fig. 35.15). Hence there is a dark fringe at the line of contact.

EVALUATE: Our result shows that the fringe spacing is proportional to the wavelength of the light used; the fringes would be farther apart with red light (larger λ_0) than with blue light (smaller λ_0). If we use white light, the reflected light at any point is a mixture of wavelengths for which constructive interference occurs; the wavelengths that interfere destructively are weak or absent in the reflected light. (This same effect explains the colors seen when an oil film on water is illuminated by white light, as in Fig. 35.11b).

Example 35.5 | **Thin-film interference II**

Suppose the glass plates of Example 35.4 have $n = 1.52$ and the space between plates contains water ($n = 1.33$) instead of air. What happens now?

SOLUTION

IDENTIFY and SET UP: The index of refraction of the water wedge is still less than that of the glass on either side of it, so the phase shifts are the same as in Example 35.4. Once again we use Eq. (35.18b) to find the positions of the dark fringes; the only difference is that the wavelength λ in this equation is now the wavelength in water instead of in air.

EXECUTE: In the film of water ($n = 1.33$), the wavelength is $\lambda = \lambda_0/n = (500 \text{ nm})/(1.33) = 376 \text{ nm}$. When we replace λ_0 by λ in the expression from Example 35.4 for the position x of the mth dark fringe, we find that the fringe spacing is reduced by the same factor of 1.33 and is equal to 0.940 mm. There is still a dark fringe at the line of contact.

EVALUATE: Can you see that to obtain the same fringe spacing as in Example 35.4, the dimension h in Fig. 35.15 would have to be reduced to $(0.0200 \text{ mm})/1.33 = 0.0150 \text{ mm}$? This shows that what matters in thin-film interference is the *ratio* t/λ between film thickness and wavelength. [You can see this by considering Eqs. (35.17) and (35.18).]

Example 35.6 | **Thin-film interference III**

Suppose the upper of the two plates of Example 35.4 is a plastic with $n = 1.40$, the wedge is filled with a silicone grease with $n = 1.50$, and the bottom plate is a dense flint glass with $n = 1.60$. What happens now?

SOLUTION

IDENTIFY and SET UP: The geometry is again the same as shown in Fig. 35.15, but now half-cycle phase shifts occur at *both* surfaces of the grease wedge (see Fig. 35.13c). Hence there is no *relative* phase shift and we must use Eq. (35.17b) to find the positions of the dark fringes.

EXECUTE: The value of λ to use in Eq. (35.17b) is the wavelength in the silicone grease, $\lambda = \lambda_0/n = (500 \text{ nm})/1.50 = 333 \text{ nm}$. You can readily show that the fringe spacing is 0.833 mm. Note that the two reflected waves from the line of contact are in phase (they both undergo the same phase shift), so the line of contact is at a *bright* fringe.

EVALUATE: What would happen if you carefully removed the upper microscope slide so that the grease wedge retained its shape? There would still be half-cycle phase changes at the upper and lower surfaces of the wedge, so the pattern of fringes would be the same as with the upper slide present.

35.16 (a) Air film between a convex lens and a plane surface. The thickness of the film *t* increases from zero as we move out from the center, giving (b) a series of alternating dark and bright rings for monochromatic light.

(a) A convex lens in contact with a glass plane

(b) Newton's rings: circular interference fringes

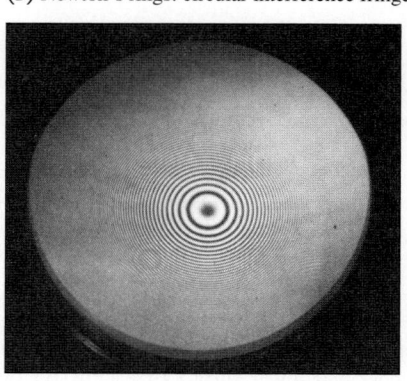

Newton's Rings

Figure 35.16a shows the convex surface of a lens in contact with a plane glass plate. A thin film of air is formed between the two surfaces. When you view the setup with monochromatic light, you see circular interference fringes (Fig. 35.16b). These were studied by Newton and are called **Newton's rings.**

We can use interference fringes to compare the surfaces of two optical parts by placing the two in contact and observing the interference fringes. Figure 35.17 is a photograph made during the grinding of a telescope objective lens. The lower, larger-diameter, thicker disk is the correctly shaped master, and the smaller, upper disk is the lens under test. The "contour lines" are Newton's interference fringes; each one indicates an additional distance between the specimen and the master of one half-wavelength. At 10 lines from the center spot the distance between the two surfaces is five wavelengths, or about 0.003 mm. This isn't very good; high-quality lenses are routinely ground with a precision of less than one wavelength. The surface of the primary mirror of the Hubble Space Telescope was ground to a precision of better than $\frac{1}{50}$ wavelength. Unfortunately, it was ground to incorrect specifications, creating one of the most precise errors in the history of optical technology (see Section 34.2).

35.17 The surface of a telescope objective lens under inspection during manufacture.

Fringes map lack of fit between lens and master.

Master   Lens being tested

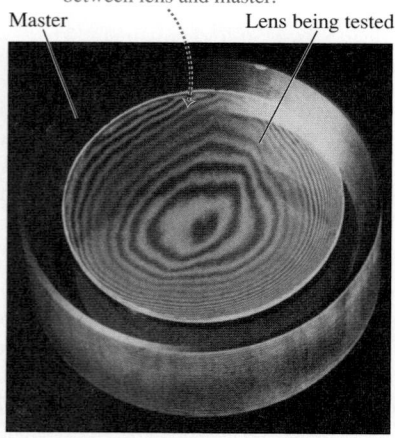

Nonreflective and Reflective Coatings

Nonreflective coatings for lens surfaces make use of thin-film interference. A thin layer or film of hard transparent material with an index of refraction smaller than that of the glass is deposited on the lens surface, as in Fig. 35.18. Light is reflected from both surfaces of the layer. In both reflections the light is reflected from a medium of greater index than that in which it is traveling, so the same phase change occurs in both reflections. If the film thickness is a quarter (one-fourth) of the wavelength *in the film* (assuming normal incidence), the total path difference is a half-wavelength. Light reflected from the first surface is then a half-cycle out of phase with light reflected from the second, and there is destructive interference.

The thickness of the nonreflective coating can be a quarter-wavelength for only one particular wavelength. This is usually chosen in the central yellow-green portion of the spectrum ($\lambda = 550$ nm), where the eye is most sensitive. Then there is somewhat more reflection at both longer (red) and shorter (blue) wavelengths, and the reflected light has a purple hue. The overall reflection from a lens or prism surface can be reduced in this way from 4–5% to less than 1%. This also increases the net amount of light that is *transmitted* through the lens, since light that is not reflected will be transmitted. The same principle is used to minimize reflection from silicon photovoltaic solar cells ($n = 3.5$) by use of a thin surface layer of silicon monoxide (SiO, $n = 1.45$); this helps to increase the amount of light that actually reaches the solar cells.

35.18 A nonreflective coating has an index of refraction intermediate between those of glass and air.

Destructive interference occurs when
• the film is about $\frac{1}{4}\lambda$ thick and
• the light undergoes a phase change at both reflecting surfaces,

so that the two reflected waves emerge from the film about $\frac{1}{2}$ cycle out of phase.

$n_{glass} > n_{film} > n_{air}$ "Nonreflecting" film

Air

Film $t = \frac{1}{4}\lambda$

Glass

If a quarter-wavelength thickness of a material with an index of refraction *greater* than that of glass is deposited on glass, then the reflectivity is *increased,* and the deposited material is called a **reflective coating.** In this case there is a half-cycle phase shift at the air–film interface but none at the film–glass interface, and reflections from the two sides of the film interfere constructively. For example, a coating with refractive index 2.5 causes 38% of the incident energy to be reflected, compared with 4% or so with no coating. By use of multiple-layer coatings, it is possible to achieve nearly 100% transmission or reflection for particular wavelengths. Some practical applications of these coatings are for color separation in television cameras and for infrared "heat reflectors" in motion-picture projectors, solar cells, and astronauts' visors.

Example 35.7   **A nonreflective coating**

A common lens coating material is magnesium fluoride (MgF_2), with $n = 1.38$. What thickness should a nonreflective coating have for 550-nm light if it is applied to glass with $n = 1.52$?

SOLUTION

IDENTIFY and SET UP: This coating is of the sort shown in Fig. 35.18. The thickness must be one-quarter of the wavelength of this light *in the coating*.

EXECUTE: The wavelength in air is $\lambda_0 = 550$ nm, so its wavelength in the MgF_2 coating is $\lambda = \lambda_0/n = (550 \text{ nm})/1.38 = 400$ nm. The coating thickness should be one-quarter of this, or $\lambda/4 = 100$ nm.

EVALUATE: This is a very thin film, no more than a few hundred molecules thick. Note that this coating is *reflective* for light whose wavelength is *twice* the coating thickness; light of that wavelength reflected from the coating's lower surface travels one wavelength farther than light reflected from the upper surface, so the two waves are in phase and interfere constructively. This occurs for light with a wavelength in MgF_2 of 200 nm and a wavelength in air of $(200 \text{ nm})(1.38) = 276$ nm. This is an ultraviolet wavelength (see Section 32.1), so designers of optical lenses with nonreflective coatings need not worry about such enhanced reflection.

Test Your Understanding of Section 35.4   A thin layer of benzene ($n = 1.501$) ties on top of a sheet of fluorite ($n = 1.434$). It is illuminated from above with light whose wavelength in benzene is 400 nm. Which of the following possible thicknesses of the benzene layer will maximize the brightness of the reflected light? (i) 100 nm; (ii) 200 nm; (iii) 300 nm; (iv) 400 nm.

Application Interference and Butterfly Wings
Many of the most brilliant colors in the animal world are created by *interference* rather than by pigments. These photos show the butterfly *Morpho rhetenor* and the microscopic scales that cover the upper surfaces of its wings. The scales have a profusion of tiny ridges (middle photo); these carry regularly spaced flanges (bottom photo) that function as reflectors. These are spaced so that the reflections interfere constructively for blue light. The multilayered structure reflects 70% of the blue light that strikes it, giving the wings a mirror-like brilliance. (The undersides of the wings do not have these structures and are a dull brown.)

35.5 The Michelson Interferometer

An important experimental device that uses interference is the **Michelson interferometer.** Michelson interferometers are used to make precise measurements of wavelengths and of very small distances, such as the minute changes in thickness of an axon when a nerve impulse propagates along its length. Like the Young two-slit experiment, a Michelson interferometer takes monochromatic light from a single source and divides it into two waves that follow different paths. In Young's experiment, this is done by sending part of the light through one slit and part through another; in a Michelson interferometer a device called a *beam splitter* is used. Interference occurs in both experiments when the two light waves are recombined.

How a Michelson Interferometer Works

Figure 35.19 shows the principal components of a Michelson interferometer. A ray of light from a monochromatic source A strikes the beam splitter C, which is a glass plate with a thin coating of silver on its right side. Part of the light (ray 1) passes through the silvered surface and the compensator plate D and is reflected from mirror M_1. It then returns through D and is reflected from the silvered surface of C to the observer. The remainder of the light (ray 2) is reflected from the silvered surface at point P to the mirror M_2 and back through C to the observer's eye.

35.19 A schematic Michelson interferometer. The observer sees an interference pattern that results from the difference in path lengths for rays 1 and 2.

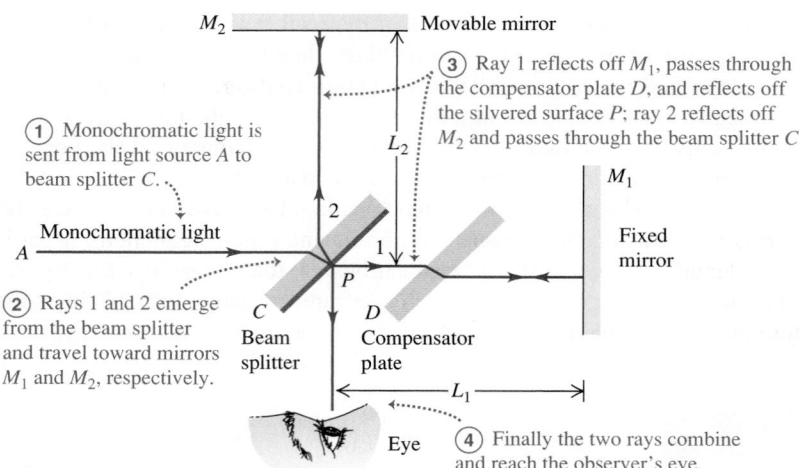

① Monochromatic light is sent from light source A to beam splitter C.

② Rays 1 and 2 emerge from the beam splitter and travel toward mirrors M_1 and M_2, respectively.

③ Ray 1 reflects off M_1, passes through the compensator plate D, and reflects off the silvered surface P; ray 2 reflects off M_2 and passes through the beam splitter C.

④ Finally the two rays combine and reach the observer's eye.

M_2 Movable mirror

M_1 Fixed mirror

Monochromatic light

Beam splitter

Compensator plate

Eye

Application Imaging Cells with a Michelson Interferometer

This false-color image of a human colon cancer cell was made using a microscope that was mated to a Michelson interferometer. The cell is in one arm of the interferometer, and light passing through the cell undergoes a phase shift that depends on the cell thickness and the organelles within the cell. The fringe pattern can then be used to construct a three-dimensional view of the cell. Scientists have used this technique to observe how different types of cells behave when prodded by microscopic probes. Cancer cells turn out to be "softer" than normal cells, a distinction that may make cancer stem cells easier to identify.

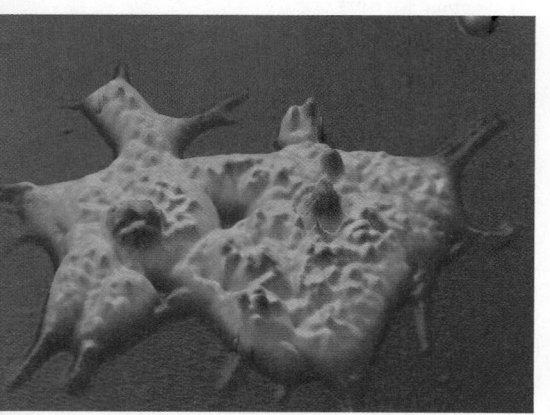

The purpose of the compensator plate D is to ensure that rays 1 and 2 pass through the same thickness of glass; plate D is cut from the same piece of glass as plate C, so their thicknesses are identical to within a fraction of a wavelength.

The whole apparatus in Fig. 35.19 is mounted on a very rigid frame, and the position of mirror M_2 can be adjusted with a fine, very accurate micrometer screw. If the distances L_1 and L_2 are exactly equal and the mirrors M_1 and M_2 are exactly at right angles, the virtual image of M_1 formed by reflection at the silvered surface of plate C coincides with mirror M_2. If L_1 and L_2 are *not* exactly equal, the image of M_1 is displaced slightly from M_2; and if the mirrors are not exactly perpendicular, the image of M_1 makes a slight angle with M_2. Then the mirror M_2 and the virtual image of M_1 play the same roles as the two surfaces of a wedge-shaped thin film (see Section 35.4), and light reflected from these surfaces forms the same sort of interference fringes.

Suppose the angle between mirror M_2 and the virtual image of M_1 is just large enough that five or six vertical fringes are present in the field of view. If we now move the mirror M_2 slowly either backward or forward a distance $\lambda/2$, the difference in path length between rays 1 and 2 changes by λ, and each fringe moves to the left or right a distance equal to the fringe spacing. If we observe the fringe positions through a telescope with a crosshair eyepiece and m fringes cross the crosshairs when we move the mirror a distance y, then

$$ y = m\frac{\lambda}{2} \qquad \text{or} \qquad \lambda = \frac{2y}{m} \qquad (35.19) $$

If m is several thousand, the distance y is large enough that it can be measured with good accuracy, and we can obtain an accurate value for the wavelength λ. Alternatively, if the wavelength is known, a distance y can be measured by simply counting fringes when M_2 is moved by this distance. In this way, distances that are comparable to a wavelength of light can be measured with relative ease.

The Michelson-Morley Experiment

The original application of the Michelson interferometer was to the historic **Michelson-Morley experiment.** Before the electromagnetic theory of light became established, most physicists thought that the propagation of light waves occurred in a medium called the **ether,** which was believed to permeate all space. In 1887 the American scientists Albert Michelson and Edward Morley used the Michelson interferometer in an attempt to detect the motion of the earth through the ether. Suppose the interferometer in Fig. 35.19 is moving from left to right relative to the ether. According to the ether theory, this would lead to changes in the speed of light in the portions of the path shown as horizontal lines in the figure. There would be fringe shifts relative to the positions that the fringes would have if the instrument were at rest in the ether. Then when the entire instrument was rotated 90°, the other portions of the paths would be similarly affected, giving a fringe shift in the opposite direction.

Michelson and Morley expected that the motion of the earth through the ether would cause a shift of about four-tenths of a fringe when the instrument was rotated. The shift that was actually observed was less than a hundredth of a fringe and, within the limits of experimental uncertainty, appeared to be exactly zero. Despite its orbital motion around the sun, the earth appeared to be *at rest* relative to the ether. This negative result baffled physicists until 1905, when Albert Einstein developed the special theory of relativity (which we will study in detail in Chapter 37). Einstein postulated that the speed of a light wave in vacuum has the same magnitude c relative to *all* inertial reference frames, no matter what their velocity may be relative to each other. The presumed ether then plays no role, and the concept of an ether has been abandoned.

Test Your Understanding of Section 35.5   You are observing the pattern of fringes in a Michelson interferometer like that shown in Fig. 35.19. If you change the index of refraction (but not the thickness) of the compensator plate, will the pattern change?   ❙

Interference and coherent sources: Monochromatic light is light with a single frequency. Coherence is a definite, unchanging phase relationship between two waves. The overlap of waves from two coherent sources of monochromatic light forms an interference pattern. The principle of superposition states that the total wave disturbance at any point is the sum of the disturbances from the separate waves.

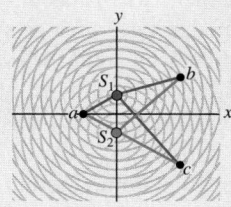

Two-source interference of light: When two sources are in phase, constructive interference occurs where the difference in path length from the two sources is zero or an integer number of wavelengths; destructive interference occurs where the path difference is a half-integer number of wavelengths. If two sources separated by a distance d are both very far from a point P, and the line from the sources to P makes an angle θ with the line perpendicular to the line of the sources, then the condition for constructive interference at P is Eq. (35.4). The condition for destructive interference is Eq. (35.5). When θ is very small, the position y_m of the mth bright fringe on a screen located a distance R from the sources is given by Eq. (35.6). (See Examples 35.1 and 35.2.)

$$d\sin\theta = m\lambda \quad (m = 0, \pm1, \pm2, \ldots)$$
(constructive interference) $\qquad$ (35.4)

$$d\sin\theta = \left(m + \tfrac{1}{2}\right)\lambda$$
$$(m = 0, \pm1, \pm2, \ldots)$$
(destructive interference) $\qquad$ (35.5)

$$y_m = R\frac{m\lambda}{d} \qquad (35.6)$$
(bright fringes)

Intensity in interference patterns: When two sinusoidal waves with equal amplitude E and phase difference ϕ are superimposed, the resultant amplitude E_P and intensity I are given by Eqs. (35.7) and (35.10), respectively. If the two sources emit in phase, the phase difference ϕ at a point P (located a distance r_1 from source 1 and a distance r_2 from source 2) is directly proportional to the difference in path length $r_2 - r_1$. (See Example 35.3.)

$$E_P = 2E\left|\cos\frac{\phi}{2}\right| \qquad (35.7)$$

$$I = I_0\cos^2\frac{\phi}{2} \qquad (35.10)$$

$$\phi = \frac{2\pi}{\lambda}(r_2 - r_1) = k(r_2 - r_1) \quad (35.11)$$

Interference in thin films: When light is reflected from both sides of a thin film of thickness t and no phase shift occurs at either surface, constructive interference between the reflected waves occurs when $2t$ is equal to an integral number of wavelengths. If a half-cycle phase shift occurs at one surface, this is the condition for destructive interference. A half-cycle phase shift occurs during reflection whenever the index of refraction in the second material is greater than that in the first. (See Examples 35.4–35.7.)

$$2t = m\lambda \quad (m = 0, 1, 2, \ldots)$$
(constructive reflection from thin film, no relative phase shift) $\quad$ (35.17a)

$$2t = \left(m + \tfrac{1}{2}\right)\lambda \quad (m = 0, 1, 2, \ldots)$$
(destructive reflection from thin film, no relative phase shift) $\quad$ (35.17b)

$$2t = \left(m + \tfrac{1}{2}\right)\lambda \quad (m = 0, 1, 2, \ldots)$$
(constructive reflection from thin film, half-cycle relative phase shift) (35.18a)

$$2t = m\lambda \quad (m = 0, 1, 2, \ldots)$$
(destructive reflection from thin film, half-cycle relative phase shift) (35.18b)

Michelson interferometer: The Michelson interferometer uses a monochromatic light source and can be used for high-precision measurements of wavelengths. Its original purpose was to detect motion of the earth relative to a hypothetical ether, the supposed medium for electromagnetic waves. The ether has never been detected, and the concept has been abandoned; the speed of light is the same relative to all observers. This is part of the foundation of the special theory of relativity.

BRIDGING PROBLEM   **Modifying a Two-Slit Experiment**

An oil tanker spills a large amount of oil ($n = 1.45$) into the sea ($n = 1.33$). (a) If you look down onto the oil spill from overhead, what predominant wavelength of light do you see at a point where the oil is 380 nm thick? What color is the light? (*Hint:* See Table 32.1.) (b) In the water under the slick, what visible wavelength (as measured in air) is predominant in the transmitted light at the same place in the slick as in part (a)?

SOLUTION GUIDE

See MasteringPhysics® study area for a Video Tutor solution.

IDENTIFY and SET UP

1. The oil layer acts as a thin film, so we must consider interference between light that reflects from the top and bottom surfaces of the oil. If a wavelength is prominent in the *transmitted* light, there is destructive interference for that wavelength in the *reflected* light.

2. Choose the appropriate interference equations that relate the thickness of the oil film and the wavelength of light. Take account of the indexes of refraction of the air, oil, and water.

EXECUTE

3. For part (a), find the wavelengths for which there is constructive interference as seen from above the oil film. Which of these are in the visible spectrum?

4. For part (b), find the visible wavelength for which there is destructive interference as seen from above the film. (This will ensure that there is substantial transmitted light at the wavelength.)

EVALUATE

5. If a diver below the water's surface shines a light up at the bottom of the oil film, at what wavelengths would there be constructive interference in the light that reflects back downward?

Problems   For instructor-assigned homework, go to www.masteringphysics.com

•, ••, •••: Problems of increasing difficulty. **CP**: Cumulative problems incorporating material from earlier chapters. **CALC**: Problems requiring calculus. **BIO**: Biosciences problems.

DISCUSSION QUESTIONS

Q35.1 A two-slit interference experiment is set up, and the fringes are displayed on a screen. Then the whole apparatus is immersed in the nearest swimming pool. How does the fringe pattern change?

Q35.2 Could an experiment similar to Young's two-slit experiment be performed with sound? How might this be carried out? Does it matter that sound waves are longitudinal and electromagnetic waves are transverse? Explain.

Q35.3 Monochromatic coherent light passing through two thin slits is viewed on a distant screen. Are the bright fringes equally spaced on the screen? If so, why? If not, which ones are closest to being equally spaced?

Q35.4 In a two-slit interference pattern on a distant screen, are the bright fringes midway between the dark fringes? Is this ever a good approximation?

Q35.5 Would the headlights of a distant car form a two-source interference pattern? If so, how might it be observed? If not, why not?

Q35.6 The two sources S_1 and S_2 shown in Fig. 35.3 emit waves of the same wavelength λ and are in phase with each other. Suppose S_1 is a weaker source, so that the waves emitted by S_1 have half the amplitude of the waves emitted by S_2. How would this affect the positions of the antinodal lines and nodal lines? Would there be total reinforcement at points on the antinodal curves? Would there be total cancellation at points on the nodal curves? Explain your answers.

Q35.7 Could the Young two-slit interference experiment be performed with gamma rays? If not, why not? If so, discuss differences in the experimental design compared to the experiment with visible light.

Q35.8 Coherent red light illuminates two narrow slits that are 25 cm apart. Will a two-slit interference pattern be observed when the light from the slits falls on a screen? Explain.

Q35.9 Coherent light with wavelength λ falls on two narrow slits separated by a distance d. If d is less than some minimum value, no dark fringes are observed. Explain. In terms of λ, what is this minimum value of d?

Q35.10 A fellow student, who values memorizing equations above understanding them, combines Eqs. (35.4) and (35.13) to "prove" that ϕ can *only* equal $2\pi m$. How would you explain to this student that ϕ can have values other than $2\pi m$?

Q35.11 If the monochromatic light shown in Fig. 35.5a were replaced by white light, would a two-slit interference pattern be seen on the screen? Explain.

Q35.12 In using the superposition principle to calculate intensities in interference patterns, could you add the intensities of the waves instead of their amplitudes? Explain.

Q35.13 A glass windowpane with a thin film of water on it reflects less than when it is perfectly dry. Why?

Q35.14 A *very* thin soap film ($n = 1.33$), whose thickness is much less than a wavelength of visible light, looks black; it appears to reflect no light at all. Why? By contrast, an equally thin layer of soapy water ($n = 1.33$) on glass ($n = 1.50$) appears quite shiny. Why is there a difference?

Q35.15 Interference can occur in thin films. Why is it important that the films be *thin*? Why don't you get these effects with a relatively *thick* film? Where should you put the dividing line between "thin" and "thick"? Explain your reasoning.

Q35.16 If we shine white light on an air wedge like that shown in Fig. 35.12, the colors that are weak in the light *reflected* from any point along the wedge are strong in the light *transmitted* through the wedge. Explain why this should be so.

Q35.17 Monochromatic light is directed at normal incidence on a thin film. There is destructive interference for the reflected light, so the intensity of the reflected light is very low. What happened to the energy of the incident light?

Q35.18 When a thin oil film spreads out on a puddle of water, the thinnest part of the film looks dark in the resulting interference pattern. What does this tell you about the relative magnitudes of the refractive indexes of oil and water?

EXERCISES

Section 35.1 Interference and Coherent Sources

35.1 • Two small stereo speakers A and B that are 1.40 m apart are sending out sound of wavelength 34 cm in all directions and all in phase. A person at point P starts out equidistant from both speakers and walks so that he is always 1.50 m from speaker B (Fig. E35.1). For what values of x will the sound this person hears be (a) maximally reinforced, (b) cancelled? Limit your solution to the cases where $x \leq 1.50\,\text{m}$.

Figure **E35.1**

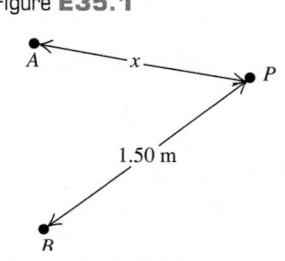

35.2 •• Two speakers that are 15.0 m apart produce in-phase sound waves of frequency 250.0 Hz in a room where the speed of sound is 340.0 m/s. A woman starts out at the midpoint between the two speakers. The room's walls and ceiling are covered with absorbers to eliminate reflections, and she listens with only one ear for best precision. (a) What does she hear: constructive or destructive interference? Why? (b) She now walks slowly toward one of the speakers. How far from the center must she walk before she first hears the sound reach a minimum intensity? (c) How far from the center must she walk before she first hears the sound maximally enhanced?

35.3 •• Two identical audio speakers connected to the same amplifier produce in-phase sound waves with a single frequency that can be varied between 300 and 600 Hz. The speed of sound is 340 m/s. You find that where you are standing, you hear minimum-intensity sound. (a) Explain why you hear minimum-intensity sound. (b) If one of the speakers is moved 39.8 cm toward you, the sound you hear has maximum intensity. What is the frequency of the sound? (c) How much closer to you from the position in part (b) must the speaker be moved to the next position where you hear maximum intensity?

35.4 • **Radio Interference.** Two radio antennas A and B radiate in phase. Antenna B is 120 m to the right of antenna A. Consider point Q along the extension of the line connecting the antennas, a horizontal distance of 40 m to the right of antenna B. The frequency, and hence the wavelength, of the emitted waves can be varied. (a) What is the longest wavelength for which there will be destructive interference at point Q? (b) What is the longest wavelength for which there will be constructive interference at point Q?

35.5 •• A radio transmitting station operating at a frequency of 120 MHz has two identical antennas that radiate in phase. Antenna B is 9.00 m to the right of antenna A. Consider point P between the antennas and along the line connecting them, a horizontal distance x to the right of antenna A. For what values of x will constructive interference occur at point P?

35.6 • Two light sources can be adjusted to emit monochromatic light of any visible wavelength. The two sources are coherent, 2.04 μm apart, and in line with an observer, so that one source is 2.04 μm farther from the observer than the other. (a) For what visible wavelengths (380 to 750 nm) will the observer see the brightest light, owing to constructive interference? (b) How would your answers to part (a) be affected if the two sources were not in line with the observer, but were still arranged so that one source is 2.04 μm farther away from the observer than the other? (c) For what visible wavelengths will there be *destructive* interference at the location of the observer?

35.7 • Two speakers, emitting identical sound waves of wavelength 2.0 m in phase with each other, and an observer are located as shown in Fig. E35.7. (a) At the observer's location, what is the path difference for waves from the two speakers? (b) Will the sound waves interfere constructively or destructively at the observer's location—or something in between constructive and destructive? (c) Suppose the observer now increases her distance from the closest speaker to 17.0 m, staying directly in front of the same speaker as initially. Answer the questions of parts (a) and (b) for this new situation.

Figure **E35.7**

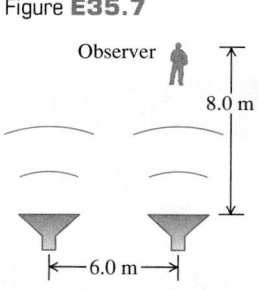

35.8 •• Figure 35.3 shows the wave pattern produced by two identical, coherent sources emitting waves with wavelength λ and separated by a distance $d = 4\lambda$. (a) Explain why the positive y-axis above S_1 constitutes an antinodal curve with $m = +4$ and why the negative y-axis below S_2 constitutes an antinodal curve with $m = -4$. (b) Draw the wave pattern produced when the separation between the sources is reduced to 3λ. In your drawing, sketch all antinodal curves—that is, the curves on which $r_2 - r_1 = m\lambda$. Label each curve by its value of m. (c) In general, what determines the maximum (most positive) and minimum (most negative) values of the integer m that labels the antinodal lines? (d) Suppose the separation between the sources is increased to $7\frac{1}{2}\lambda$. How many antinodal curves will there be? To what values of m do they correspond? Explain your reasoning. (You should not have to make a drawing to answer these questions.)

Section 35.2 Two-Source Interference of Light

35.9 • Young's experiment is performed with light from excited helium atoms ($\lambda = 502$ nm). Fringes are measured carefully on a screen 1.20 m away from the double slit, and the center of the 20th fringe (not counting the central bright fringe) is found to be 10.6 mm from the center of the central bright fringe. What is the separation of the two slits?

35.10 •• Coherent light with wavelength 450 nm falls on a double slit. On a screen 1.80 m away, the distance between dark fringes is 4.20 mm. What is the separation of the slits?

35.11 •• Two slits spaced 0.450 mm apart are placed 75.0 cm from a screen. What is the distance between the second and third dark lines of the interference pattern on the screen when the slits are illuminated with coherent light with a wavelength of 500 nm?

35.12 •• If the entire apparatus of Exercise 35.11 (slits, screen, and space in between) is immersed in water, what then is the distance between the second and third dark lines?

35.13 •• Two thin parallel slits that are 0.0116 mm apart are illuminated by a laser beam of wavelength 585 nm. (a) On a very large distant screen, what is the *total* number of bright fringes (those indicating complete constructive interference), including the central fringe and those on both sides of it? Solve this problem without calculating all the angles! (*Hint:* What is the largest that $\sin\theta$ can be? What does this tell you is the largest value of m?) (b) At what angle, relative to the original direction of the beam, will the fringe that is most distant from the central bright fringe occur?

35.14 • Coherent light with wavelength 400 nm passes through two very narrow slits that are separated by 0.200 mm, and the interference pattern is observed on a screen 4.00 m from the slits. (a) What is the width (in mm) of the central interference maximum? (b) What is the width of the first-order bright fringe?

35.15 •• Two very narrow slits are spaced 1.80 μm apart and are placed 35.0 cm from a screen. What is the distance between the first and second dark lines of the interference pattern when the slits are illuminated with coherent light with $\lambda = 550$ nm? (*Hint:* The angle θ in Eq. (35.5) is *not* small.)

35.16 •• Coherent light that contains two wavelengths, 660 nm (red) and 470 nm (blue), passes through two narrow slits separated by 0.300 mm, and the interference pattern is observed on a screen 5.00 m from the slits. What is the distance on the screen between the first-order bright fringes for the two wavelengths?

35.17 •• Coherent light with wavelength 600 nm passes through two very narrow slits and the interference pattern is observed on a screen 3.00 m from the slits. The first-order bright fringe is at 4.84 mm from the center of the central bright fringe. For what wavelength of light will the first-order dark fringe be observed at this same point on the screen?

35.18 •• Coherent light of frequency 6.32×10^{14} Hz passes through two thin slits and falls on a screen 85.0 cm away. You observe that the third bright fringe occurs at ± 3.11 cm on either side of the central bright fringe. (a) How far apart are the two slits? (b) At what distance from the central bright fringe will the third dark fringe occur?

Section 35.3 Intensity in Interference Patterns

35.19 •• In a two-slit interference pattern, the intensity at the peak of the central maximum is I_0. (a) At a point in the pattern where the phase difference between the waves from the two slits is 60.0°, what is the intensity? (b) What is the path difference for 480-nm light from the two slits at a point where the phase angle is 60.0°?

35.20 • Coherent sources A and B emit electromagnetic waves with wavelength 2.00 cm. Point P is 4.86 m from A and 5.24 m from B. What is the phase difference at P between these two waves?

35.21 • Coherent light with wavelength 500 nm passes through narrow slits separated by 0.340 mm. At a distance from the slits large compared to their separation, what is the phase difference (in radians) in the light from the two slits at an angle of 23.0° from the centerline?

35.22 • Two slits spaced 0.260 mm apart are placed 0.700 m from a screen and illuminated by coherent light with a wavelength of 660 nm. The intensity at the center of the central maximum ($\theta = 0°$) is I_0. (a) What is the distance on the screen from the center of the central maximum to the first minimum? (b) What is the distance on the screen from the center of the central maximum to the point where the intensity has fallen to $I_0/2$?

35.23 •• Points A and B are 56.0 m apart along an east-west line. At each of these points, a radio transmitter is emitting a 12.5-MHz signal horizontally. These transmitters are in phase with each other and emit their beams uniformly in a horizontal plane. A receiver is taken 0.500 km north of the AB line and initially placed at point C, directly opposite the midpoint of AB. The receiver can be moved only along an east-west direction but, due to its limited sensitivity, it must always remain within a range so that the intensity of the signal it receives from the transmitter is no less than $\frac{1}{4}$ of its maximum value. How far from point C (along an east-west line) can the receiver be moved and always be able to pick up the signal?

35.24 • Consider two antennas separated by 9.00 m that radiate in phase at 120 MHz, as described in Exercise 35.5. A receiver placed 150 m from both antennas measures an intensity I_0. The receiver is moved so that it is 1.8 m closer to one antenna than to the other. (a) What is the phase difference ϕ between the two radio waves produced by this path difference? (b) In terms of I_0, what is the intensity measured by the receiver at its new position?

Section 35.4 Interference in Thin Films

35.25 • What is the thinnest film of a coating with $n = 1.42$ on glass ($n = 1.52$) for which destructive interference of the red component (650 nm) of an incident white light beam in air can take place by reflection?

35.26 •• **Nonglare Glass.** When viewing a piece of art that is behind glass, one often is affected by the light that is reflected off the front of the glass (called *glare*), which can make it difficult to see the art clearly. One solution is to coat the outer surface of the glass with a film to cancel part of the glare. (a) If the glass has a refractive index of 1.62 and you use TiO_2, which has an index of refraction of 2.62, as the coating, what is the minimum film thickness that will cancel light of wavelength 505 nm? (b) If this coating is too thin to stand up to wear, what other thickness would also work? Find only the three thinnest ones.

35.27 •• Two rectangular pieces of plane glass are laid one upon the other on a table. A thin strip of paper is placed between them at one edge so that a very thin wedge of air is formed. The plates are illuminated at normal incidence by 546-nm light from a mercury-vapor lamp. Interference fringes are formed, with 15.0 fringes per centimeter. Find the angle of the wedge.

35.28 •• A plate of glass 9.00 cm long is placed in contact with a second plate and is held at a small angle with it by a metal strip

0.0800 mm thick placed under one end. The space between the plates is filled with air. The glass is illuminated from above with light having a wavelength in air of 656 nm. How many interference fringes are observed per centimeter in the reflected light?

35.29 •• A uniform film of TiO_2, 1036 nm thick and having index of refraction 2.62, is spread uniformly over the surface of crown glass of refractive index 1.52. Light of wavelength 520.0 nm falls at normal incidence onto the film from air. You want to increase the thickness of this film so that the reflected light cancels. (a) What is the *minimum* thickness of TiO_2 that you must *add* so the reflected light cancels as desired? (b) After you make the adjustment in part (a), what is the path difference between the light reflected off the top of the film and the light that cancels it after traveling through the film? Express your answer in (i) nanometers and (ii) wavelengths of the light in the TiO_2 film.

35.30 • A plastic film with index of refraction 1.85 is put on the surface of a car window to increase the reflectivity and thus to keep the interior of the car cooler. The window glass has index of refraction 1.52. (a) What minimum thickness is required if light with wavelength 550 nm in air reflected from the two sides of the film is to interfere constructively? (b) It is found to be difficult to manufacture and install coatings as thin as calculated in part (a). What is the next greatest thickness for which there will also be constructive interference?

35.31 • The walls of a soap bubble have about the same index of refraction as that of plain water, $n = 1.33$. There is air both inside and outside the bubble. (a) What wavelength (in air) of visible light is most strongly reflected from a point on a soap bubble where its wall is 290 nm thick? To what color does this correspond (see Fig. 32.4 and Table 32.1)? (b) Repeat part (a) for a wall thickness of 340 nm.

35.32 •• Light with wavelength 648 nm in air is incident perpendicularly from air on a film 8.76 μm thick and with refractive index 1.35. Part of the light is reflected from the first surface of the film, and part enters the film and is reflected back at the second surface, where the film is again in contact with air. (a) How many waves are contained along the path of this second part of the light in its round trip through the film? (b) What is the phase difference between these two parts of the light as they leave the film?

35.33 •• **Compact Disc Player.** A compact disc (CD) is read from the bottom by a semiconductor laser with wavelength 790 nm passing through a plastic substrate of refractive index 1.8. When the beam encounters a pit, part of the beam is reflected from the pit and part from the flat region between the pits, so these two beams interfere with each other (Fig. E35.33). What must the minimum pit depth be so that the part of the beam reflected from a pit cancels the part of the beam reflected from the flat region? (It is this cancellation that allows the player to recognize the beginning and end of a pit.)

Figure **E35.33**

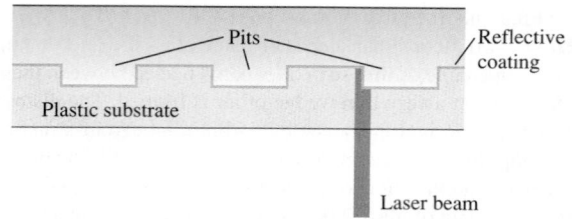

35.34 • What is the thinnest soap film (excluding the case of zero thickness) that appears black when illuminated with light with

wavelength 480 nm? The index of refraction of the film is 1.33, and there is air on both sides of the film.

Section 35.5 The Michelson Interferometer

35.35 • How far must the mirror M_2 (see Fig. 35.19) of the Michelson interferometer be moved so that 1800 fringes of He-Ne laser light ($\lambda = 633$ nm) move across a line in the field of view?

35.36 • Jan first uses a Michelson interferometer with the 606-nm light from a krypton-86 lamp. He displaces the movable mirror away from him, counting 818 fringes moving across a line in his field of view. Then Linda replaces the krypton lamp with filtered 502-nm light from a helium lamp and displaces the movable mirror toward her. She also counts 818 fringes, but they move across the line in her field of view opposite to the direction they moved for Jan. Assume that both Jan and Linda counted to 818 correctly. (a) What distance did each person move the mirror? (b) What is the resultant displacement of the mirror?

PROBLEMS

35.37 ••• The radius of curvature of the convex surface of a planoconvex lens is 68.4 cm. The lens is placed convex side down on a perfectly flat glass plate that is illuminated from above with red light having a wavelength of 580 nm. Find the diameter of the second bright ring in the interference pattern.

35.38 •• Newton's rings can be seen when a planoconvex lens is placed on a flat glass surface. For a particular lens with an index of refraction of $n = 1.50$ and a glass plate with an index of $n = 1.80$, the diameter of the third bright ring is 0.720 mm. If water ($n = 1.33$) now fills the space between the lens and the plate, what is the new diameter of this ring?

35.39 • **BIO** **Coating Eyeglass Lenses.** Eyeglass lenses can be coated on the *inner* surfaces to reduce the reflection of stray light to the eye. If the lenses are medium flint glass of refractive index 1.62 and the coating is fluorite of refractive index 1.432, (a) what minimum thickness of film is needed on the lenses to cancel light of wavelength 550 nm reflected toward the eye at normal incidence? (b) Will any other wavelengths of visible light be cancelled or enhanced in the reflected light?

35.40 •• **BIO** **Sensitive Eyes.** After an eye examination, you put some eyedrops on your sensitive eyes. The cornea (the front part of the eye) has an index of refraction of 1.38, while the eyedrops have a refractive index of 1.45. After you put in the drops, your friends notice that your eyes look red, because red light of wavelength 600 nm has been reinforced in the reflected light. (a) What is the minimum thickness of the film of eyedrops on your cornea? (b) Will any other wavelengths of visible light be reinforced in the reflected light? Will any be cancelled? (c) Suppose you had contact lenses, so that the eyedrops went on them instead of on your corneas. If the refractive index of the lens material is 1.50 and the layer of eyedrops has the same thickness as in part (a), what wavelengths of visible light will be reinforced? What wavelengths will be cancelled?

35.41 •• Two flat plates of glass with parallel faces are on a table, one plate on the other. Each plate is 11.0 cm long and has a refractive index of 1.55. A very thin sheet of metal foil is inserted under the end of the upper plate to raise it slightly at that end, in a manner similar to that discussed in Example 35.4. When you view the glass plates from above with reflected white light, you observe that, at 1.15 mm from the line where the sheets are in contact, the violet light of wavelength 400.0 nm is enhanced in this reflected light, but no visible light is enhanced closer to the line of contact.

(a) How far from the line of contact will green light (of wavelength 550 nm) and orange light (of wavelength 600.0 nm) first be enhanced? (b) How far from the line of contact will the violet, green, and orange light again be enhanced in the reflected light? (c) How thick is the metal foil holding the ends of the plates apart?

35.42 •• In a setup similar to that of Problem 35.41, the glass has an index of refraction of 1.53, the plates are each 8.00 cm long, and the metal foil is 0.015 mm thick. The space between the plates is filled with a jelly whose refractive index is not known precisely, but is known to be greater than that of the glass. When you illuminate these plates from above with light of wavelength 525 nm, you observe a series of equally spaced dark fringes in the reflected light. You measure the spacing of these fringes and find that there are 10 of them every 6.33 mm. What is the index of refraction of the jelly?

35.43 ••• Suppose you illuminate two thin slits by monochromatic coherent light in air and find that they produce their first interference *minima* at ±35.20° on either side of the central bright spot. You then immerse these slits in a transparent liquid and illuminate them with the same light. Now you find that the first minima occur at ±19.46° instead. What is the index of refraction of this liquid?

35.44 •• CP CALC A very thin sheet of brass contains two thin parallel slits. When a laser beam shines on these slits at normal incidence and room temperature (20.0°C), the first interference dark fringes occur at ±32.5° from the original direction of the laser beam when viewed from some distance. If this sheet is now slowly heated up to 135°C, by how many degrees do these dark fringes change position? Do they move closer together or get farther apart? See Table 17.1 for pertinent information, and ignore any effects that might occur due to change in the thickness of the slits. (*Hint:* Since thermal expansion normally produces very small changes in length, you can use differentials to find the change in the angle.)

35.45 •• Two speakers, 2.50 m apart, are driven by the same audio oscillator so that each one produces a sound consisting of *two* distinct frequencies, 0.900 kHz and 1.20 kHz. The speed of sound in the room is 344 m/s. Find all the angles relative to the usual centerline in front of (and far from) the speakers at which *both* frequencies interfere constructively.

35.46 •• Two radio antennas radiating in phase are located at points A and B, 200 m apart (Fig. P35.46). The radio waves have a frequency of 5.80 MHz. A radio receiver is moved out from point B along a line perpendicular to the line connecting A and B (line BC shown in Fig. P35.46). At what distances from B will there be *destructive* interference? (*Note:* The distance of the receiver from the sources is not large in comparison to the separation of the sources, so Eq. (35.5) does not apply.)

Figure **P35.46**

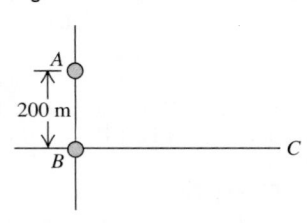

35.47 •• One round face of a 3.25-m, solid, cylindrical plastic pipe is covered with a thin black coating that completely blocks light. The opposite face is covered with a fluorescent coating that glows when it is struck by light. Two straight, thin, parallel scratches, 0.225 mm apart, are made in the center of the black face. When laser light of wavelength 632.8 nm shines through the slits perpendicular to the black face, you find that the central bright fringe on the opposite face is 5.82 mm wide, measured between the dark fringes that border it on either side. What is the index of refraction of the plastic?

35.48 • A uniform thin film of material of refractive index 1.40 coats a glass plate of refractive index 1.55. This film has the proper thickness to cancel normally incident light of wavelength 525 nm that strikes the film surface from air, but it is somewhat greater than the minimum thickness to achieve this cancellation. As time goes by, the film wears away at a steady rate of 4.20 nm per year. What is the minimum number of years before the reflected light of this wavelength is now enhanced instead of cancelled?

35.49 •• Two speakers A and B are 3.50 m apart, and each one is emitting a frequency of 444 Hz. However, because of signal delays in the cables, speaker A is one-fourth of a period *ahead* of speaker B. For points far from the speakers, find all the angles relative to the centerline (Fig. P35.49) at which the sound from these speakers cancels. Include angles on *both* sides of the centerline. The speed of sound is 340 m/s.

Figure **P35.49**

35.50 ••• CP The electric fields received at point P from two identical, coherent wave sources are $E_1(t) = E\cos(\omega t + \phi)$ and $E_2(t) = E\cos\omega t$. (a) Use the trigonometric identities in Appendix B to show that the resultant wave is $E_P(t) = 2E\cos(\phi/2)\cos(\omega t + \phi/2)$. (b) Show that the amplitude of this resultant wave is given by Eq. (35.7). (c) Use the result of part (a) to show that at an interference maximum, the amplitude of the resultant wave is in phase with the original waves $E_1(t)$ and $E_2(t)$. (d) Use the result of part (a) to show that near an interference minimum, the resultant wave is approximately $\frac{1}{4}$ cycle out of phase with either of the original waves. (e) Show that the *instantaneous* Poynting vector at point P has magnitude $S = 4\epsilon_0 cE^2\cos^2(\phi/2)\cos^2(\omega t + \phi/2)$ and that the *time-averaged* Poynting vector is given by Eq. (35.9).

35.51 •• CP A thin uniform film of refractive index 1.750 is placed on a sheet of glass of refractive index 1.50. At room temperature (20.0°C), this film is just thick enough for light with wavelength 582.4 nm reflected off the top of the film to be cancelled by light reflected from the top of the glass. After the glass is placed in an oven and slowly heated to 170°C, you find that the film cancels reflected light with wavelength 588.5 nm. What is the coefficient of linear expansion of the film? (Ignore any changes in the refractive index of the film due to the temperature change.)

35.52 ••• **GPS Transmission.** The GPS (Global Positioning System) satellites are approximately 5.18 m across and transmit two low-power signals, one of which is at 1575.42 MHz (in the UHF band). In a series of laboratory tests on the satellite, you put two 1575.42-MHz UHF transmitters at opposite ends of the satellite. These broadcast in phase uniformly in all directions. You measure the intensity at points on a circle that is several hundred meters in radius and centered on the satellite. You measure angles on this circle relative to a point that lies along the centerline of the satellite (that is, the perpendicular bisector of a line that extends from one transmitter to the other). At this point on the circle, the measured intensity is 2.00 W/m². (a) At how many other angles in the range 0° < θ < 90° is the intensity also 2.00 W/m²? (b) Find the four smallest angles in the range 0° < θ < 90° for which the intensity is 2.00 W/m². (c) What is the intensity at a point on the circle at an angle of 4.65° from the centerline?

35.53 •• Consider a two-slit interference pattern, for which the intensity distribution is given by Eq. (35.14). Let θ_m be the angular position of the mth bright fringe, where the intensity is I_0. Assume that θ_m is small, so that $\sin\theta_m \cong \theta_m$. Let $\theta_m{}^+$ and $\theta_m{}^-$ be the two angles on either side of θ_m for which $I = \frac{1}{2}I_0$. The quantity $\Delta\theta_m = |\theta_m{}^+ - \theta_m{}^-|$ is the half-width of the mth fringe. Calculate $\Delta\theta_m$. How does $\Delta\theta_m$ depend on m?

35.54 •• White light reflects at normal incidence from the top and bottom surfaces of a glass plate ($n = 1.52$). There is air above and below the plate. Constructive interference is observed for light whose wavelength in air is 477.0 nm. What is the thickness of the plate if the next longer wavelength for which there is constructive interference is 540.6 nm?

35.55 ••• A source S of monochromatic light and a detector D are both located in air a distance h above a horizontal plane sheet of glass and are separated by a horizontal distance x. Waves reaching D directly from S interfere with waves that reflect off the glass. The distance x is small compared to h so that the reflection is at close to normal incidence. (a) Show that the condition for constructive interference is $\sqrt{x^2 + 4h^2} - x = \left(m + \frac{1}{2}\right)\lambda$, and the condition for destructive interference is $\sqrt{x^2 + 4h^2} - x = m\lambda$. (*Hint:* Take into account the phase change on reflection.) (b) Let $h = 24$ cm and $x = 14$ cm. What is the longest wavelength for which there will be constructive interference?

35.56 •• **BIO** **Reflective Coatings and Herring.** Herring and related fish have a brilliant silvery appearance that camouflages them while they are swimming in a sunlit ocean. The silveriness is due to *platelets* attached to the surfaces of these fish. Each platelet is made up of several alternating layers of crystalline guanine ($n = 1.80$) and of cytoplasm ($n = 1.333$, the same as water), with a guanine layer on the outside in contact with the surrounding water (Fig. P35.56). In one typical platelet, the guanine layers are 74 nm thick and the cytoplasm layers are 100 nm thick. (a) For light striking the platelet surface at normal incidence, for which vacuum wavelengths of visible light will all of the reflections R_1, R_2, R_3, R_4, and R_5, shown in Fig. P35.56, be approximately in phase? If white light is shone on this platelet, what color will be most strongly reflected (see Fig. 32.4)? The surface of a herring has very many platelets side by side with layers of different thickness, so that *all* visible wavelengths are reflected. (b) Explain why such a "stack" of layers is more reflective than a single layer of guanine with cytoplasm underneath. (A stack of five guanine layers separated by cytoplasm layers reflects more than 80% of incident light at the wavelength for which it is "tuned.") (c) The color that is most strongly reflected from a platelet depends on the angle at which it is viewed. Explain why this should be so. (You can see these changes in color by examining a herring from different

Figure **P35.56**

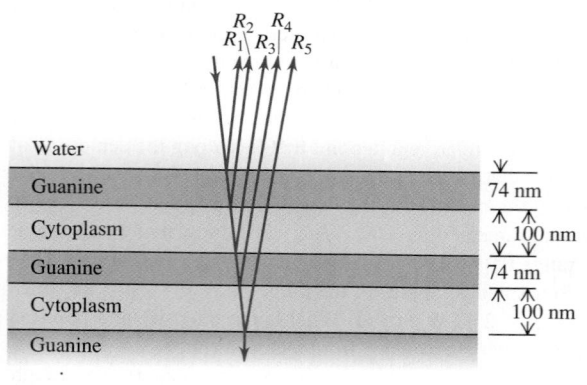

35.57 • Two thin parallel slits are made in an opaque sheet of film. When a monochromatic beam of light is shone through them at normal incidence, the first bright fringes in the transmitted light occur in air at $\pm 18.0°$ with the original direction of the light beam on a distant screen. When the apparatus is immersed in a liquid, the same bright fringes now occur at $\pm 12.6°$. Find the index of refraction of the liquid.

35.58 •• Red light with wavelength 700 nm is passed through a two-slit apparatus. At the same time, monochromatic visible light with another wavelength passes through the same apparatus. As a result, most of the pattern that appears on the screen is a mixture of two colors; however, the center of the third bright fringe ($m = 3$) of the red light appears pure red, with none of the other color. What are the possible wavelengths of the second type of visible light? Do you need to know the slit spacing to answer this question? Why or why not?

35.59 ••• In a Young's two-slit experiment a piece of glass with an index of refraction n and a thickness L is placed in front of the upper slit. (a) Describe qualitatively what happens to the interference pattern. (b) Derive an expression for the intensity I of the light at points on a screen as a function of n, L, and θ. Here θ is the usual angle measured from the center of the two slits. That is, determine the equation analogous to Eq. (35.14) but that also involves L and n for the glass plate. (c) From your result in part (b) derive an expression for the values of θ that locate the maxima in the interference pattern [that is, derive an equation analogous to Eq. (35.4)].

35.60 •• After a laser beam passes through two thin parallel slits, the first completely dark fringes occur at $\pm 19.0°$ with the original direction of the beam, as viewed on a screen far from the slits. (a) What is the ratio of the distance between the slits to the wavelength of the light illuminating the slits? (b) What is the smallest angle, relative to the original direction of the laser beam, at which the intensity of the light is $\frac{1}{10}$ the maximum intensity on the screen?

CHALLENGE PROBLEMS

35.61 ••• **CP** The index of refraction of a glass rod is 1.48 at $T = 20.0°C$ and varies linearly with temperature, with a coefficient of $2.50 \times 10^{-5}/C°$. The coefficient of linear expansion of the glass is $5.00 \times 10^{-6}/C°$. At $20.0°C$ the length of the rod is 3.00 cm. A Michelson interferometer has this glass rod in one arm, and the rod is being heated so that its temperature increases at a rate of $5.00 \ C°/min$. The light source has wavelength $\lambda = 589$ nm, and the rod initially is at $T = 20.0°C$. How many fringes cross the field of view each minute?

35.62 ••• **CP** Figure P35.62 shows an interferometer known as *Fresnel's biprism*. The magnitude of the prism angle A is

Figure **P35.62**

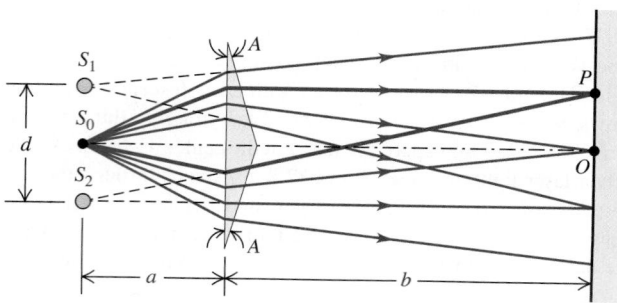

extremely small. (a) If S_0 is a very narrow source slit, show that the separation of the two virtual coherent sources S_1 and S_2 is given by $d = 2aA(n - 1)$, where n is the index of refraction of the material of the prism. (b) Calculate the spacing of the fringes of green light with wavelength 500 nm on a screen 2.00 m from the biprism. Take $a = 0.200$ m, $A = 3.50$ mrad, and $n = 1.50$.

Answers

Chapter Opening Question ?

The colors appear due to constructive interference between light waves reflected from the outer and inner surfaces of the soap bubble. The thickness of the bubble walls at each point determines the wavelength of light for which the most constructive interference occurs and hence the color that appears the brightest at that point (see Example 35.4 in Section 35.4).

Test Your Understanding Questions

35.1 Answer: (i) At any point P on the positive y-axis above S_1, the distance r_2 from S_2 to P is greater than the distance r_1 from S_1 to P by 4λ. This corresponds to $m = 4$ in Eq. (35.1), the equation for constructive interference. Hence all such points make up an antinodal curve.

35.2 Answer: (ii) Blue light has a shorter wavelength than red light (see Section 32.1). Equation (35.6) tells us that the distance y_m from the center of the pattern to the mth bright fringe is proportional to the wavelength λ. Hence all of the fringes will move toward the center of the pattern as the wavelength decreases, and the spacing between fringes will decrease.

35.3 Answer: (i), (iv), (ii), (iii) In cases (i) and (iii) we are given the wavelength λ and path difference $d\sin\theta$. Hence we use Eq. (35.14), $I = I_0\cos^2[(\pi d\sin\theta)/\lambda]$. In parts (ii) and (iii) we are given the phase difference ϕ and we use Eq. (35.10), $I = I_0\cos^2(\phi/2)$. We find:

(i) $I = I_0\cos^2[\pi(4.00 \times 10^{-7}$ m$)/(5.00 \times 10^{-7}$ m$)] = I_0\cos^2(0.800\pi$ rad$) = 0.655I_0$;
(ii) $I = I_0\cos^2[(4.00$ rad$)/2] = I_0\cos^2(2.00$ rad$) = 0.173I_0$;
(iii) $I = I_0\cos^2[\pi(7.50 \times 10^{-7}$ m$)/(5.00 \times 10^{-7}$ m$)] = I_0\cos^2(1.50\pi$ rad$) = 0$;
(iv) $I = I_0\cos^2[(2.00$ rad$)/2] = I_0\cos^2(1.00$ rad$) = 0.292I_0$.

35.4 Answers: (i) and (iii) Benzene has a larger index of refraction than air, so light that reflects off the upper surface of the benzene undergoes a half-cycle phase shift. Fluorite has a *smaller* index of refraction than benzene, so light that reflects off the benzene–fluorite interface does not undergo a phase shift. Hence the equation for constructive reflection is Eq. (35.18a), $2t = \left(m + \frac{1}{2}\right)\lambda$, which we can rewrite as $t = \left(m + \frac{1}{2}\right)\lambda/2 = \left(m + \frac{1}{2}\right)(400$ mm$)/2 = 100$ nm, 300 nm, 500 nm, ...

35.5 Answer: yes Changing the index of refraction changes the wavelength of the light inside the compensator plate, and so changes the number of wavelengths within the thickness of the plate. Hence this has the same effect as changing the distance L_1 from the beam splitter to mirror M_1, which would change the interference pattern.

Bridging Problem

Answers: (a) 441 nm **(b)** 551 nm

36 DIFFRACTION

? The laser used to read a DVD has a wavelength of 650 nm, while the laser used to read a Blu-ray disc has a shorter 405-nm wavelength. How does this make it possible for a Blu-ray disc to hold more information than a DVD?

LEARNING GOALS

By studying this chapter, you will learn:

• What happens when coherent light shines on an object with an edge or aperture.

• How to understand the diffraction pattern formed when coherent light passes through a narrow slit.

• How to calculate the intensity at various points in a single-slit diffraction pattern.

• What happens when coherent light shines on an array of narrow, closely spaced slits.

• How scientists use diffraction gratings for precise measurements of wavelength.

• How x-ray diffraction reveals the arrangement of atoms in a crystal.

• How diffraction sets limits on the smallest details that can be seen with a telescope.

E veryone is used to the idea that sound bends around corners. If sound didn't behave this way, you couldn't hear a police siren that's out of sight around a corner or the speech of a person whose back is turned to you. What may surprise you (and certainly surprised many scientists of the early 19th century) is that *light* can bend around corners as well. When light from a point source falls on a straightedge and casts a shadow, the edge of the shadow is never perfectly sharp. Some light appears in the area that we expect to be in the shadow, and we find alternating bright and dark fringes in the illuminated area. In general, light emerging from apertures doesn't behave precisely according to the predictions of the straight-line ray model of geometric optics.

The reason for these effects is that light, like sound, has wave characteristics. In Chapter 35 we studied the interference patterns that can arise when two light waves are combined. In this chapter we'll investigate interference effects due to combining *many* light waves. Such effects are referred to as *diffraction*. We'll find that the behavior of waves after they pass through an aperture is an example of diffraction; each infinitesimal part of the aperture acts as a source of waves, and the resulting pattern of light and dark is a result of interference among the waves emanating from these sources.

Light emerging from arrays of apertures also forms patterns whose character depends on the color of the light and the size and spacing of the apertures. Examples of this effect include the colors of iridescent butterflies and the "rainbow" you see reflected from the surface of a compact disc. We'll explore similar effects with x rays that are used to study the atomic structure of solids and liquids. Finally, we'll look at the physics of a *hologram*, a special kind of interference pattern recorded on photographic film and reproduced. When properly illuminated, it forms a three-dimensional image of the original object.

36.1 Fresnel and Fraunhofer Diffraction

According to geometric optics, when an opaque object is placed between a point light source and a screen, as in Fig. 36.1, the shadow of the object forms a perfectly sharp line. No light at all strikes the screen at points within the shadow, and the area outside the shadow is illuminated nearly uniformly. But as we saw in Chapter 35, the *wave* nature of light causes effects that can't be understood with geometric optics. An important class of such effects occurs when light strikes a barrier that has an aperture or an edge. The interference patterns formed in such a situation are grouped under the heading **diffraction.**

Figure 36.2 shows an example of diffraction. The photograph in Fig. 36.2a was made by placing a razor blade halfway between a pinhole, illuminated by monochromatic light, and a photographic film. The film recorded the shadow cast by the blade. Figure 36.2b is an enlargement of a region near the shadow of the right edge of the blade. The position of the *geometric* shadow line is indicated by arrows. The area outside the geometric shadow is bordered by alternating bright and dark bands. There is some light in the shadow region, although this is not very visible in the photograph. The first bright band in Fig. 36.2b, just to the right of the geometric shadow, is considerably brighter than in the region of uniform illumination to the extreme right. This simple experiment gives us some idea of the richness and complexity of what might seem to be a simple idea, the casting of a shadow by an opaque object.

We don't often observe diffraction patterns such as Fig. 36.2 in everyday life because most ordinary light sources are neither monochromatic nor point sources. If we use a white frosted light bulb instead of a point source to illuminate the razor blade in Fig. 36.2, each wavelength of the light from every point of the bulb forms its own diffraction pattern, but the patterns overlap so much that we can't see any individual pattern.

Diffraction and Huygens's Principle

We can analyze diffraction patterns using Huygens's principle (see Section 33.7). This principle states that we can consider every point of a wave front as a source of secondary wavelets. These spread out in all directions with a speed equal to the speed of propagation of the wave. The position of the wave front at any later time is the *envelope* of the secondary wavelets at that time. To find the resultant displacement at any point, we combine all the individual displacements produced by these secondary waves, using the superposition principle and taking into account their amplitudes and relative phases.

36.1 A point source of light illuminates a straightedge.

Geometric optics predicts that this situation should produce a sharp boundary between illumination and solid shadow.

That's NOT what really happens!

DOESN'T HAPPEN

Point source

Area of illumination

Geometric shadow

Straightedge

Screen

36.2 An example of diffraction.

(a)

(b)

Photograph of a razor blade illuminated by monochromatic light from a point source (a pinhole). Notice the fringe around the blade outline.

Enlarged view of the area outside the geometric shadow of the blade's edge

Position of *geometric* shadow

In Fig. 36.1, both the point source and the screen are relatively close to the obstacle forming the diffraction pattern. This situation is described as *near-field diffraction* or **Fresnel diffraction,** pronounced "Freh-nell" (after the French scientist Augustin Jean Fresnel, 1788–1827). By contrast, we use the term **Fraunhofer diffraction** (after the German physicist Joseph von Fraunhofer, 1787–1826) for situations in which the source, obstacle, and screen are far enough apart that we can consider all lines from the source to the obstacle to be parallel, and can likewise consider all lines from the obstacle to a given point on the screen to be parallel. We will restrict the following discussion to Fraunhofer diffraction, which is usually simpler to analyze in detail than Fresnel diffraction.

Diffraction is sometimes described as "the bending of light around an obstacle." But the process that causes diffraction is present in the propagation of *every* wave. When part of the wave is cut off by some obstacle, we observe diffraction effects that result from interference of the remaining parts of the wave fronts. Optical instruments typically use only a limited portion of a wave; for example, a telescope uses only the part of a wave that is admitted by its objective lens or mirror. Thus diffraction plays a role in nearly all optical phenomena.

Finally, we emphasize that there is no fundamental distinction between *interference* and *diffraction.* In Chapter 35 we used the term *interference* for effects involving waves from a small number of sources, usually two. *Diffraction* usually involves a *continuous* distribution of Huygens's wavelets across the area of an aperture, or a very large number of sources or apertures. But both interference and diffraction are consequences of superposition and Huygens's principle.

Test Your Understanding of Section 36.1   Can *sound* waves undergo diffraction around an edge?   ❙

PhET: Wave Interference
ActivPhysics 16.6: Single-Slit Diffraction

36.2 Diffraction from a Single Slit

In this section we'll discuss the diffraction pattern formed by plane-wave (parallel-ray) monochromatic light when it emerges from a long, narrow slit, as shown in Fig. 36.3. We call the narrow dimension the *width,* even though in this figure it is a vertical dimension.

According to geometric optics, the transmitted beam should have the same cross section as the slit, as in Fig. 36.3a. What is *actually* observed is the pattern shown in Fig. 36.3b. The beam spreads out vertically after passing through the slit. The diffraction pattern consists of a central bright band, which may be much broader than the width of the slit, bordered by alternating dark and bright bands with rapidly decreasing intensity. About 85% of the power in the

36.3 (a) The "shadow" of a horizontal slit as incorrectly predicted by geometric optics. (b) A horizontal slit actually produces a diffraction pattern. The slit width has been greatly exaggerated.

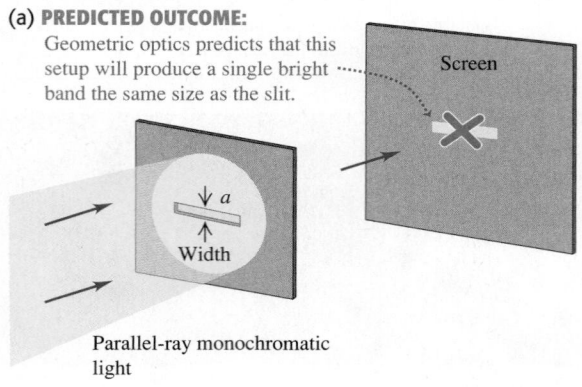

(a) **PREDICTED OUTCOME:**
Geometric optics predicts that this setup will produce a single bright band the same size as the slit.

Screen

↓ *a*
↑
Width

Parallel-ray monochromatic light

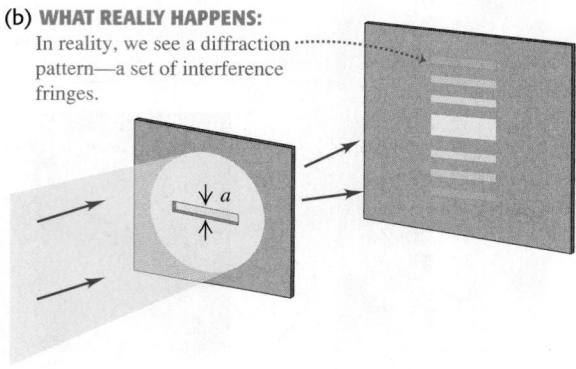

(b) **WHAT REALLY HAPPENS:**
In reality, we see a diffraction pattern—a set of interference fringes.

↓ *a*
↑

36.4 Diffraction by a single rectangular slit. The long sides of the slit are perpendicular to the figure.

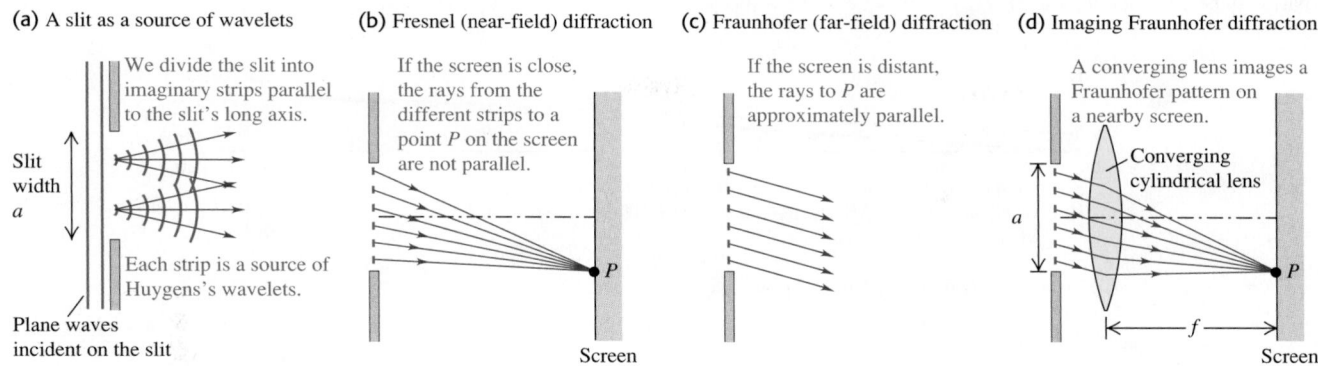

(a) A slit as a source of wavelets

We divide the slit into imaginary strips parallel to the slit's long axis.

Slit width
a

Each strip is a source of Huygens's wavelets.

Plane waves incident on the slit

(b) Fresnel (near-field) diffraction

If the screen is close, the rays from the different strips to a point P on the screen are not parallel.

P

Screen

(c) Fraunhofer (far-field) diffraction

If the screen is distant, the rays to P are approximately parallel.

(d) Imaging Fraunhofer diffraction

A converging lens images a Fraunhofer pattern on a nearby screen.

Converging cylindrical lens

a

P

f

Screen

transmitted beam is in the central bright band, whose width is *inversely* proportional to the width of the slit. In general, the smaller the width of the slit, the broader the entire diffraction pattern. (The *horizontal* spreading of the beam in Fig. 36.3b is negligible because the horizontal dimension of the slit is relatively large.) You can observe a similar diffraction pattern by looking at a point source, such as a distant street light, through a narrow slit formed between your two thumbs held in front of your eye; the retina of your eye corresponds to the screen.

Single-Slit Diffraction: Locating the Dark Fringes

Figure 36.4 shows a side view of the same setup; the long sides of the slit are perpendicular to the figure, and plane waves are incident on the slit from the left. According to Huygens's principle, each element of area of the slit opening can be considered as a source of secondary waves. In particular, imagine dividing the slit into several narrow strips of equal width, parallel to the long edges and perpendicular to the page. Figure 36.4a shows two such strips. Cylindrical secondary wavelets, shown in cross section, spread out from each strip.

In Fig. 36.4b a screen is placed to the right of the slit. We can calculate the resultant intensity at a point P on the screen by adding the contributions from the individual wavelets, taking proper account of their various phases and amplitudes. It's easiest to do this calculation if we assume that the screen is far enough away that all the rays from various parts of the slit to a particular point P on the screen are parallel, as in Fig. 36.4c. An equivalent situation is Fig. 36.4d, in which the rays to the lens are parallel and the lens forms a reduced image of the same pattern that would be formed on an infinitely distant screen without the lens. We might expect that the various light paths through the lens would introduce additional phase shifts, but in fact it can be shown that all the paths have *equal* phase shifts, so this is not a problem.

The situation of Fig. 36.4b is Fresnel diffraction; those in Figs. 36.4c and 36.4d, where the outgoing rays are considered parallel, are Fraunhofer diffraction. We can derive quite simply the most important characteristics of the Fraunhofer diffraction pattern from a single slit. First consider two narrow strips, one just below the top edge of the drawing of the slit and one at its center, shown in end view in Fig. 36.5. The difference in path length to point P is $(a/2)\sin\theta$, where a is the slit width and θ is the angle between the perpendicular to the slit and a line from the center of the slit to P. Suppose this path difference happens to be equal to $\lambda/2$; then light from these two strips arrives at point P with a half-cycle phase difference, and cancellation occurs.

Similarly, light from two strips immediately *below* the two in the figure also arrives at P a half-cycle out of phase. In fact, the light from *every* strip in the top half of the slit cancels out the light from a corresponding strip in the bottom half.

36.5 Side view of a horizontal slit. When the distance x to the screen is much greater than the slit width a, the rays from a distance $a/2$ apart may be considered parallel.

(a)

For the two strips shown, the path difference to P is $(a/2) \sin \theta$.
When $(a/2) \sin \theta = \lambda/2$, the light cancels at P. This is true for the whole slit, so P represents a dark fringe.

(b) Enlarged view of the top half of the slit

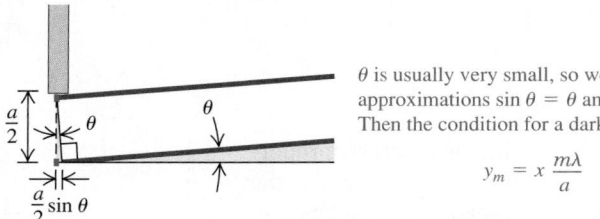

θ is usually very small, so we can use the approximations $\sin \theta = \theta$ and $\tan \theta = \theta$.
Then the condition for a dark band is

$$y_m = x \frac{m\lambda}{a}$$

Hence the combined light from the entire slit completely cancels at P, giving a dark fringe in the interference pattern. A dark fringe occurs whenever

$$\frac{a}{2}\sin\theta = \pm\frac{\lambda}{2} \qquad \text{or} \qquad \sin\theta = \pm\frac{\lambda}{a} \tag{36.1}$$

The plus-or-minus $(\pm)$ sign in Eq. (36.1) says that there are symmetric dark fringes above and below point O in Fig. 36.5a. The upper fringe $(\theta > 0)$ occurs at a point P where light from the bottom half of the slit travels $\lambda/2$ farther to P than does light from the top half; the lower fringe $(\theta < 0)$ occurs where light from the *top* half travels $\lambda/2$ farther than light from the *bottom* half.

We may also divide the screen into quarters, sixths, and so on, and use the above argument to show that a dark fringe occurs whenever $\sin\theta = \pm 2\lambda/a$, $\pm 3\lambda/a$, and so on. Thus the condition for a *dark* fringe is

$$\sin\theta = \frac{m\lambda}{a} \quad (m = \pm 1, \pm 2, \pm 3, \ldots) \qquad \begin{array}{l}\text{(dark fringes in single-}\\ \text{slit diffraction)}\end{array} \tag{36.2}$$

For example, if the slit width is equal to ten wavelengths $(a = 10\lambda)$, dark fringes occur at $\sin\theta = \pm\frac{1}{10}, \pm\frac{2}{10}, \pm\frac{3}{10}, \ldots$. Between the dark fringes are bright fringes. We also note that $\sin\theta = 0$ corresponds to a *bright* band; in this case, light from the entire slit arrives at P in phase. Thus it would be wrong to put $m = 0$ in Eq. (36.2). The central bright fringe is wider than the other bright fringes, as Fig. 36.3b shows. In the small-angle approximation that we will use below, it is exactly *twice* as wide.

With light, the wavelength λ is of the order of 500 nm $= 5 \times 10^{-7}$ m. This is often much smaller than the slit width a; a typical slit width is 10^{-2} cm $= 10^{-4}$ m. Therefore the values of θ in Eq. (36.2) are often so small that the approximation $\sin\theta \approx \theta$ (where θ is in radians) is a very good one. In that case we can rewrite this equation as

$$\theta = \frac{m\lambda}{a} \quad (m = \pm 1, \pm 2, \pm 3, \ldots) \qquad \text{(for small angles } \theta)$$

where θ is in *radians*. Also, if the distance from slit to screen is x, as in Fig. 36.5a, and the vertical distance of the mth dark band from the center of the pattern is y_m, then $\tan\theta = y_m/x$. For small θ we may also approximate $\tan\theta$ by θ (in radians), and we then find

$$y_m = x\frac{m\lambda}{a} \qquad (\text{for } y_m \ll x) \qquad (36.3)$$

Figure 36.6 is a photograph of a single-slit diffraction pattern with the $m = \pm1$, ±2, and ±3 minima labeled.

CAUTION **Single-slit diffraction vs. two-slit interference** Equation (36.3) has the same form as the equation for the two-slit pattern, Eq. (35.6), except that in Eq. (36.3) we use x rather than R for the distance to the screen. But Eq. (36.3) gives the positions of the *dark* fringes in a *single-slit* pattern rather than the *bright* fringes in a *double-slit* pattern. Also, $m = 0$ in Eq. (36.2) is *not* a dark fringe. Be careful!

36.6 Photograph of the Fraunhofer diffraction pattern of a single horizontal slit.

$\leftarrow m = 3$
$\leftarrow m = 2$
$\leftarrow m = 1$
$\leftarrow m = -1$
$\leftarrow m = -2$
$\leftarrow m = -3$

Example 36.1 Single-slit diffraction

You pass 633-nm laser light through a narrow slit and observe the diffraction pattern on a screen 6.0 m away. The distance on the screen between the centers of the first minima on either side of the central bright fringe is 32 mm (Fig. 36.7). How wide is the slit?

SOLUTION

IDENTIFY and SET UP: This problem involves the relationship between the positions of dark fringes in a single-slit diffraction

36.7 A single-slit diffraction experiment.

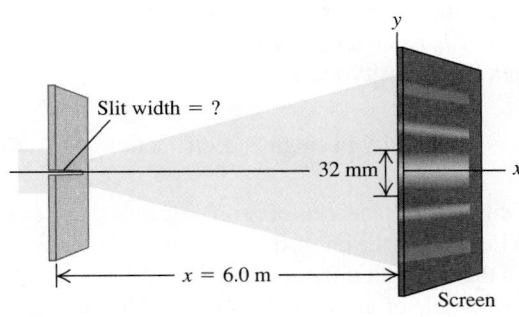

Slit width = ?

32 mm

x = 6.0 m

y

x

Screen

pattern and the slit width a (our target variable). The distances between fringes on the screen are much smaller than the slit-to-screen distance, so the angle θ shown in Fig. 36.5a is very small and we can use Eq. (36.3) to solve for a.

EXECUTE: The first minimum corresponds to $m = 1$ in Eq. (36.3). The distance y_1 from the central maximum to the first minimum on either side is half the distance between the two first minima, so $y_1 = (32 \text{ mm})/2 = 16 \text{ mm}$. Solving Eq. (36.3) for a, we find

$$a = \frac{x\lambda}{y_1} = \frac{(6.0 \text{ m})(633 \times 10^{-9} \text{ m})}{16 \times 10^{-3} \text{ m}} = 2.4 \times 10^{-4} \text{ m} = 0.24 \text{ mm}$$

EVALUATE: The angle θ is small only if the wavelength is small compared to the slit width. Since $\lambda = 633 \text{ nm} = 6.33 \times 10^{-7} \text{ m}$ and we have found $a = 0.24 \text{ mm} = 2.4 \times 10^{-4} \text{ m}$, our result is consistent with this: The wavelength is $(6.33 \times 10^{-7} \text{ m})/(2.4 \times 10^{-4} \text{ m}) = 0.0026$ as large as the slit width. Can you show that the distance between the *second* minima on either side is $2(32 \text{ mm}) = 64 \text{ mm}$, and so on?

Test Your Understanding of Section 36.2 Rank the following single-slit diffraction experiments in order of the size of the angle from the center of the diffraction pattern to the first dark fringe, from largest to smallest: (i) wavelength 400 nm, slit width 0.20 mm; (ii) wavelength 600 nm, slit width 0.20 mm; (iii) wavelength 400 nm, slit width 0.30 mm; (iv) wavelength 600 nm, slit width 0.30 mm.

36.3 Intensity in the Single-Slit Pattern

We can derive an expression for the intensity distribution for the single-slit diffraction pattern by the same phasor-addition method that we used in Section 35.3 to obtain Eqs. (35.10) and (35.14) for the two-slit interference pattern. We again imagine a plane wave front at the slit subdivided into a large number of strips. We superpose the contributions of the Huygens wavelets from all the strips at a point P on a distant screen at an angle θ from the normal to the slit plane (Fig. 36.8a). To do this, we use a phasor to represent the sinusoidally varying $\vec{E}$ field from

36.8 Using phasor diagrams to find the amplitude of the $\vec{E}$ field in single-slit diffraction. Each phasor represents the $\vec{E}$ field from a single strip within the slit.

(a)

Strips within slit

Slit width a

O

P

Distant screen

Plane waves incident on the slit

(b) At the center of the diffraction pattern (point O), the phasors from all strips within the slit are in phase.

E_0

(c) Phasor diagram at a point slightly off the center of the pattern; $\beta = $ total phase difference between the first and last phasors.

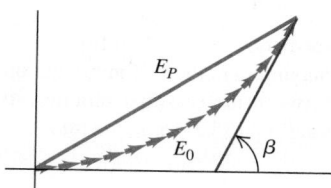

E_P

E_0

β

(d) As in (c), but in the limit that the slit is subdivided into infinitely many strips

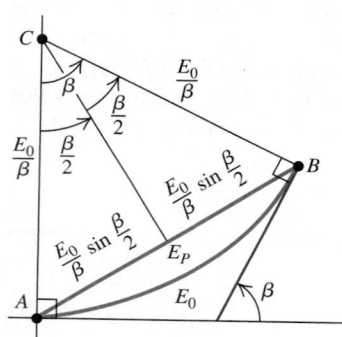

C

β

$\dfrac{\beta}{2}$

$\dfrac{E_0}{\beta}$

$\dfrac{E_0}{\beta}$

$\dfrac{\beta}{2}$

$\dfrac{E_0}{\beta}\sin\dfrac{\beta}{2}$

B

$\dfrac{E_0}{\beta}\sin\dfrac{\beta}{2}$

E_P

A

E_0

β

each individual strip. The magnitude of the vector sum of the phasors at each point P is the amplitude E_P of the total $\vec{E}$ field at that point. The intensity at P is proportional to $E_P{}^2$.

At the point O shown in Fig. 36.8a, corresponding to the center of the pattern where $\theta = 0$, there are negligible path differences for $x \gg a$; the phasors are all essentially *in phase* (that is, have the same direction). In Fig. 36.8b we draw the phasors at time $t = 0$ and denote the resultant amplitude at O by E_0. In this illustration we have divided the slit into 14 strips.

Now consider wavelets arriving from different strips at point P in Fig. 36.8a, at an angle θ from point O. Because of the differences in path length, there are now phase differences between wavelets coming from adjacent strips; the corresponding phasor diagram is shown in Fig. 36.8c. The vector sum of the phasors is now part of the perimeter of a many-sided polygon, and E_P, the amplitude of the resultant electric field at P, is the *chord*. The angle β is the total phase difference between the wave from the top strip of Fig. 36.8a and the wave from the bottom strip; that is, β is the phase of the wave received at P from the top strip with respect to the wave received at P from the bottom strip.

We may imagine dividing the slit into narrower and narrower strips. In the limit that there is an infinite number of infinitesimally narrow strips, the curved trail of phasors becomes an *arc of a circle* (Fig. 36.8d), with arc length equal to the length E_0 in Fig. 36.8b. The center C of this arc is found by constructing perpendiculars at A and B. From the relationship among arc length, radius, and angle, the radius of the arc is E_0/β; the amplitude E_P of the resultant electric field at P is equal to the chord AB, which is $2(E_0/\beta)\sin(\beta/2)$. (Note that β *must* be in radians!) We then have

$$E_P = E_0 \frac{\sin(\beta/2)}{\beta/2} \qquad \text{(amplitude in single-slit diffraction)} \qquad (36.4)$$

The intensity at each point on the screen is proportional to the square of the amplitude given by Eq. (36.4). If I_0 is the intensity in the straight-ahead direction where $\theta = 0$ and $\beta = 0$, then the intensity I at any point is

$$I = I_0 \left[\frac{\sin(\beta/2)}{\beta/2} \right]^2 \qquad \text{(intensity in single-slit diffraction)} \qquad (36.5)$$

We can express the phase difference β in terms of geometric quantities, as we did for the two-slit pattern. From Eq. (35.11) the phase difference is $2\pi/\lambda$ times the path difference. Figure 36.5 shows that the path difference between the ray from the top of the slit and the ray from the middle of the slit is $(a/2)\sin\theta$. The path difference between the rays from the top of the slit and the bottom of the slit is twice this, so

$$\beta = \frac{2\pi}{\lambda} a \sin\theta \qquad (36.6)$$

and Eq. (36.5) becomes

$$I = I_0 \left\{ \frac{\sin[\pi a(\sin\theta)/\lambda]}{\pi a(\sin\theta)/\lambda} \right\}^2 \qquad \text{(intensity in single-slit diffraction)} \qquad (36.7)$$

This equation expresses the intensity directly in terms of the angle θ. In many calculations it is easier first to calculate the phase angle β, using Eq. (36.6), and then to use Eq. (36.5).

Equation (36.7) is plotted in Fig. 36.9a. Note that the central intensity peak is much larger than any of the others. This means that most of the power in the wave remains within an angle θ from the perpendicular to the slit, where $\sin\theta = \lambda/a$ (the first diffraction minimum). You can see this easily in Fig. 36.9b, which is a photograph of water waves undergoing single-slit diffraction. Note also that the peak intensities in Fig. 36.9a decrease rapidly as we go away from the center of the pattern. (Compare Fig. 36.6, which shows a single-slit diffraction pattern for light.)

The dark fringes in the pattern are the places where $I = 0$. These occur at points for which the numerator of Eq. (36.5) is zero so that β is a multiple of 2π. From Eq. (36.6) this corresponds to

$$\frac{a\sin\theta}{\lambda} = m \qquad (m = \pm 1, \pm 2, \dots)$$

$$\sin\theta = \frac{m\lambda}{a} \qquad (m = \pm 1, \pm 2, \dots) \tag{36.8}$$

This agrees with our previous result, Eq. (36.2). Note again that $\beta = 0$ (corresponding to $\theta = 0$) is *not* a minimum. Equation (36.5) is indeterminate at $\beta = 0$, but we can evaluate the limit as $\beta \to 0$ using L'Hôpital's rule. We find that at $\beta = 0$, $I = I_0$, as we should expect.

Intensity Maxima in the Single-Slit Pattern

We can also use Eq. (36.5) to calculate the positions of the peaks, or *intensity maxima*, and the intensities at these peaks. This is not quite as simple as it may appear. We might expect the peaks to occur where the sine function reaches the value ± 1—namely, where $\beta = \pm\pi$, $\pm 3\pi$, $\pm 5\pi$, or in general,

$$\beta \approx \pm(2m + 1)\pi \qquad (m = 0, 1, 2, \dots) \tag{36.9}$$

This is *approximately* correct, but because of the factor $(\beta/2)^2$ in the denominator of Eq. (36.5), the maxima don't occur precisely at these points. When we take the derivative of Eq. (36.5) with respect to β and set it equal to zero to try to find the maxima and minima, we get a transcendental equation that has to be solved numerically. In fact there is *no* maximum near $\beta = \pm\pi$. The first maxima on either side of the central maximum, near $\beta = \pm 3\pi$, actually occur at $\pm 2.860\pi$. The second side maxima, near $\beta = \pm 5\pi$, are actually at $\pm 4.918\pi$, and so on. The error in Eq. (36.9) vanishes in the limit of large m—that is, for intensity maxima far from the center of the pattern.

To find the intensities at the side maxima, we substitute these values of β back into Eq. (36.5). Using the approximate expression in Eq. (36.9), we get

$$I_m \approx \frac{I_0}{\left(m + \frac{1}{2}\right)^2 \pi^2} \tag{36.10}$$

where I_m is the intensity of the mth side maximum and I_0 is the intensity of the central maximum. Equation (36.10) gives the series of intensities

$$0.0450I_0 \qquad 0.0162I_0 \qquad 0.0083I_0$$

and so on. As we have pointed out, this equation is only approximately correct. The actual intensities of the side maxima turn out to be

$$0.0472I_0 \qquad 0.0165I_0 \qquad 0.0083I_0 \qquad \dots$$

Note that the intensities of the side maxima decrease very rapidly, as Fig. 36.9a also shows. Even the first side maxima have less than 5% of the intensity of the central maximum.

Width of the Single-Slit Pattern

For small angles the angular spread of the diffraction pattern is inversely proportional to the slit width a or, more precisely, to the ratio of a to the wavelength λ. Figure 36.10 shows graphs of intensity I as a function of the angle θ for three values of the ratio a/λ.

36.9 (a) Intensity versus angle in single-slit diffraction. The values of m label intensity minima given by Eq. (36.8). Most of the wave power goes into the central intensity peak (between the $m = 1$ and $m = -1$ intensity minima). (b) These water waves passing through a small aperture behave exactly like light waves in single-slit diffraction. Only the diffracted waves within the central intensity peak are visible; the waves at larger angles are too faint to see.

(a)

(b)

36.10 The single-slit diffraction pattern depends on the ratio of the slit width a to the wavelength λ.

(a) $a = \lambda$

If the slit width is equal to or narrower than the wavelength, only one broad maximum forms.

(b) $a = 5\lambda$

The wider the slit (or the shorter the wavelength), the narrower and sharper is the central peak.

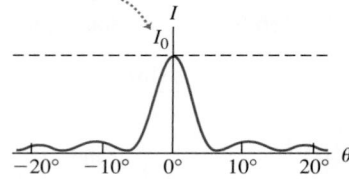

(c) $a = 8\lambda$

With light waves, the wavelength λ is often much smaller than the slit width a, and the values of θ in Eqs. (36.6) and (36.7) are so small that the approximation $\sin\theta = \theta$ is very good. With this approximation the position θ_1 of the first minimum beside the central maximum, corresponding to $\beta/2 = \pi$, is, from Eq. (36.7),

$$\theta_1 = \frac{\lambda}{a} \qquad (36.11)$$

This characterizes the width (angular spread) of the central maximum, and we see that it is *inversely* proportional to the slit width a. When the small-angle approximation is valid, the central maximum is exactly twice as wide as each side maximum. When a is of the order of a centimeter or more, θ_1 is so small that we can consider practically all the light to be concentrated at the geometrical focus. But when a is less than λ, the central maximum spreads over 180°, and the fringe pattern is not seen at all.

It's important to keep in mind that diffraction occurs for *all* kinds of waves, not just light. Sound waves undergo diffraction when they pass through a slit or aperture such as an ordinary doorway. The sound waves used in speech have wavelengths of about a meter or greater, and a typical doorway is less than 1 m wide; in this situation, a is less than λ, and the central intensity maximum extends over 180°. This is why the sounds coming through an open doorway can easily be heard by an eavesdropper hiding out of sight around the corner. In the same way, sound waves can bend around the head of an instructor who faces the blackboard while lecturing (Fig. 36.11). By contrast, there is essentially no diffraction of visible light through a doorway because the width a is very much greater than the wavelength λ (of order 5×10^{-7} m). You can *hear* around corners because typical sound waves have relatively long wavelengths; you cannot *see* around corners because the wavelength of visible light is very short.

36.11 The sound waves used in speech have a long wavelength (about 1 m) and can easily bend around this instructor's head. By contrast, light waves have very short wavelengths and undergo very little diffraction. Hence you can't *see* around his head!

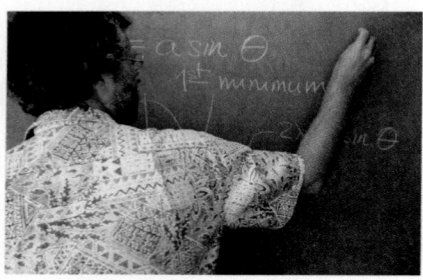

Example 36.2 | **Single-slit diffraction: Intensity I**

(a) The intensity at the center of a single-slit diffraction pattern is I_0. What is the intensity at a point in the pattern where there is a 66-radian phase difference between wavelets from the two edges of the slit? (b) If this point is 7.0° away from the central maximum, how many wavelengths wide is the slit?

SOLUTION

IDENTIFY and SET UP: In our analysis of Fig. 36.8 we used the symbol β for the phase difference between wavelets from the two edges of the slit. In part (a) we use Eq. (36.5) to find the intensity I at the point in the pattern where $\beta = 66$ rad. In part (b) we need to find the slit width a as a multiple of the wavelength λ so our target

variable is a/λ. We are given the angular position θ of the point where $\beta = 66$ rad, so we can use Eq. (36.6) to solve for a/λ.

EXECUTE: (a) We have $\beta/2 = 33$ rad, so from Eq. (36.5),

$$I = I_0 \left[\frac{\sin(33 \text{ rad})}{33 \text{ rad}} \right]^2 = (9.2 \times 10^{-4})I_0$$

(b) From Eq. (36.6),

$$\frac{a}{\lambda} = \frac{\beta}{2\pi \sin\theta} = \frac{66 \text{ rad}}{(2\pi \text{ rad})\sin 7.0°} = 86$$

For example, for 550-nm light the slit width is $a = (86)(550 \text{ nm}) = 4.7 \times 10^{-5}$ m $= 0.047$ mm, or roughly $\frac{1}{20}$ mm.

EVALUATE: To what point in the diffraction pattern does this value of β correspond? To find out, note that $\beta = 66$ rad is approximately equal to 21π. This is an odd multiple of π, corresponding to the form $(2m + 1)\pi$ found in Eq. (36.9) for the intensity *maxima.* Hence $\beta = 66$ rad corresponds to a point near the tenth $(m = 10)$ maximum. This is well beyond the range shown in Fig. 36.9a, which shows only maxima out to $m = \pm3$.

Example 36.3 **Single-slit diffraction: Intensity II**

In the experiment described in Example 36.1 (Section 36.2), the intensity at the center of the pattern is I_0. What is the intensity at a point on the screen 3.0 mm from the center of the pattern?

SOLUTION

IDENTIFY and SET UP: This is similar to Example 36.2, except that we are not given the value of the phase difference β at the point in question. We use geometry to determine the angle θ for our point and then use Eq. (36.7) to find the intensity I (the target variable).

EXECUTE: Referring to Fig. 36.5a, we have $y = 3.0$ mm and $x = 6.0$ m, so $\tan\theta = y/x = (3.0 \times 10^{-3}\,\text{m})/(6.0\,\text{m}) = 5.0 \times 10^{-4}$. This is so small that the values of $\tan\theta$, $\sin\theta$, and θ (in radians) are all nearly the same. Then, using Eq. (36.7),

$$\frac{\pi a \sin\theta}{\lambda} = \frac{\pi(2.4 \times 10^{-4}\,\text{m})(5.0 \times 10^{-4})}{6.33 \times 10^{-7}\,\text{m}} = 0.60$$

$$I = I_0 \left(\frac{\sin 0.60}{0.60}\right)^2 = 0.89 I_0$$

EVALUATE: Figure 36.9a shows that an intensity this high can occur only within the central intensity maximum. This checks out; from Example 36.1, the first intensity minimum ($m = 1$ in Fig. 36.9a) is $(32\,\text{mm})/2 = 16$ mm from the center of the pattern, so the point in question here at $y = 3$ mm does, indeed, lie within the central maximum.

Test Your Understanding of Section 36.3 Coherent electromagnetic radiation is sent through a slit of width 0.0100 mm. For which of the following wavelengths will there be *no* points in the diffraction pattern where the intensity is zero? (i) blue light of wavelength 500 nm; (ii) infrared light of wavelength 10.6 μm; (iii) microwaves of wavelength 1.00 mm; (iv) ultraviolet light of wavelength 50.0 nm. ❙

36.4 Multiple Slits

In Sections 35.2 and 35.3 we analyzed interference from two point sources or from two very narrow slits; in this analysis we ignored effects due to the finite (that is, nonzero) slit width. In Sections 36.2 and 36.3 we considered the diffraction effects that occur when light passes through a single slit of finite width. Additional interesting effects occur when we have two slits with finite width or when there are several very narrow slits.

Two Slits of Finite Width

Let's take another look at the two-slit pattern in the more realistic case in which the slits have finite width. If the slits are narrow in comparison to the wavelength, we can assume that light from each slit spreads out uniformly in all directions to the right of the slit. We used this assumption in Section 35.3 to calculate the interference pattern described by Eq. (35.10) or (35.15), consisting of a series of equally spaced, equally intense maxima. However, when the slits have finite width, the peaks in the two-slit interference pattern are modulated by the single-slit diffraction pattern characteristic of the width of each slit.

Figure 36.12a shows the intensity in a single-slit diffraction pattern with slit width a. The *diffraction minima* are labeled by the integer $m_d = \pm1, \pm2, \ldots$ ("d" for "diffraction"). Figure 36.12b shows the pattern formed by two very narrow slits with distance d between slits, where d is four times as great as the single-slit width a in Fig. 36.12a; that is, $d = 4a$. The *interference maxima* are labeled by the integer $m_i = 0, \pm1, \pm2, \ldots$ ("i" for "interference"). We note that the spacing between adjacent minima in the single-slit pattern is four times as great as in the two-slit pattern. Now suppose we widen each of the narrow slits to the same width a as that of the single slit in Fig. 36.12a. Figure 36.12c shows the pattern

36.12 Finding the intensity pattern for two slits of finite width.

(a) Single-slit diffraction pattern for a slit width a

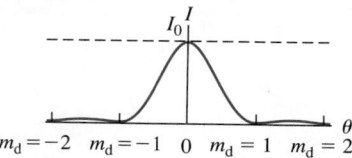

$m_d = -2$ $m_d = -1$ 0 $m_d = 1$ $m_d = 2$

(b) Two-slit interference pattern for narrow slits whose separation d is four times the width of the slit in **(a)**

$m_i = -8$ $m_i = -4$ 0 $m_i = 4$ $m_i = 8$

(c) Calculated intensity pattern for two slits of width a and separation $d = 4a$, including both interference and diffraction effects

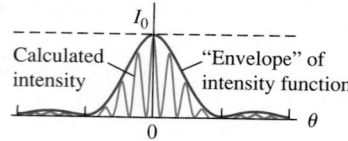

Calculated intensity   "Envelope" of intensity function

(d) Actual photograph of the pattern calculated in **(c)**

For $d = 4a$, every fourth interference maximum at the sides ($m_i = \pm4, \pm8, ...$) is missing.

36.13 Multiple-slit diffraction. Here a lens is used to give a Fraunhofer pattern on a nearby screen, as in Fig. 36.4d.

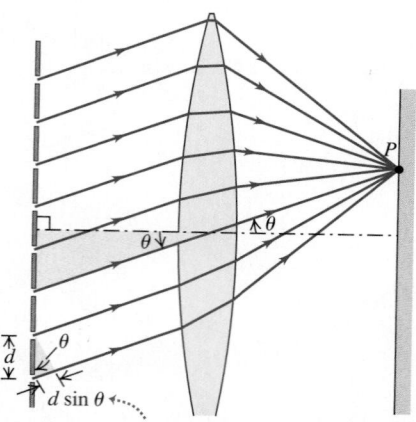

Maxima occur where the path difference for adjacent slits is a whole number of wavelengths: $d \sin\theta = m\lambda$.

from two slits with width a, separated by a distance (between centers) $d = 4a$. The effect of the finite width of the slits is to superimpose the two patterns—that is, to multiply the two intensities at each point. The two-slit peaks are in the same positions as before, but their intensities are modulated by the single-slit pattern, which acts as an "envelope" for the intensity function. The expression for the intensity shown in Fig. 36.12c is proportional to the product of the two-slit and single-slit expressions, Eqs. (35.10) and (36.5):

$$I = I_0 \cos^2\frac{\phi}{2}\left[\frac{\sin(\beta/2)}{\beta/2}\right]^2 \quad \text{(two slits of finite width)} \quad (36.12)$$

where, as before,

$$\phi = \frac{2\pi d}{\lambda}\sin\theta \qquad \beta = \frac{2\pi a}{\lambda}\sin\theta$$

Note that in Fig. 36.12c, every fourth interference maximum at the sides is *missing* because these interference maxima ($m_i = \pm4, \pm8, ...$) coincide with diffraction minima ($m_d = \pm1, \pm2, ...$). This can also be seen in Fig. 36.12d, which is a photograph of an actual pattern with $d = 4a$. You should be able to convince yourself that there will be "missing" maxima whenever d is an integer multiple of a.

Figures 36.12c and 36.12d show that as you move away from the central bright maximum of the two-slit pattern, the intensity of the maxima decreases. This is a result of the single-slit modulating pattern shown in Fig. 36.12a; mathematically, the decrease in intensity arises from the factor $(\beta/2)^2$ in the denominator of Eq. (36.12). This decrease in intensity can also be seen in Fig. 35.6 (Section 35.2). The narrower the slits, the broader the single-slit pattern (as in Fig. 36.10) and the slower the decrease in intensity from one interference maximum to the next.

Shall we call the pattern in Fig. 36.12d *interference* or *diffraction*? It's really both, since it results from the superposition of waves coming from various parts of the two apertures.

Several Slits

Next let's consider patterns produced by *several* very narrow slits. As we will see, systems of narrow slits are of tremendous practical importance in *spectroscopy*, the determination of the particular wavelengths of light coming from a source. Assume that each slit is narrow in comparison to the wavelength, so its diffraction pattern spreads out nearly uniformly. Figure 36.13 shows an array of eight narrow slits, with distance d between adjacent slits. Constructive interference occurs for rays at angle θ to the normal that arrive at point P with a path difference between adjacent slits equal to an integer number of wavelengths:

$$d\sin\theta = m\lambda \qquad (m = 0, \pm1, \pm2, ...)$$

This means that reinforcement occurs when the phase difference ϕ at P for light from adjacent slits is an integer multiple of 2π. That is, the maxima in the pattern occur at the *same* positions as for *two* slits with the same spacing. To this extent the pattern resembles the two-slit pattern.

But what happens *between* the maxima? In the two-slit pattern, there is exactly one intensity minimum located midway between each pair of maxima, corresponding to angles for which the phase difference between waves from the two sources is π, 3π, 5π, and so on. In the eight-slit pattern these are also minima because the light from adjacent slits cancels out in pairs, corresponding to the phasor diagram in Fig. 36.14a. But these are not the only minima in the eight-slit pattern. For example, when the phase difference ϕ from adjacent sources is $\pi/4$, the phasor diagram is as shown in Fig. 36.14b; the total (resultant) phasor is zero, and the intensity is zero. When $\phi = \pi/2$, we get the phasor diagram of Fig. 36.14c, and again both the total phasor and the intensity are zero. More

36.14 Phasor diagrams for light passing through eight narrow slits. Intensity maxima occur when the phase difference $\phi = 0, 2\pi, 4\pi, \ldots$ Between the maxima at $\phi = 0$ and $\phi = 2\pi$ are seven minima, corresponding to $\phi = \pi/4, \pi/2, 3\pi/4, \pi, 5\pi/4, 3\pi/2$, and $7\pi/4$. Can you draw phasor diagrams for the other minima?

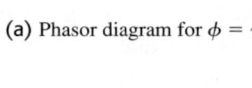 (a) Phasor diagram for $\phi = \pi$

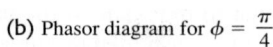 (b) Phasor diagram for $\phi = \dfrac{\pi}{4}$

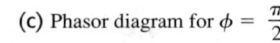 (c) Phasor diagram for $\phi = \dfrac{\pi}{2}$

 $\phi = \pi = 180°$

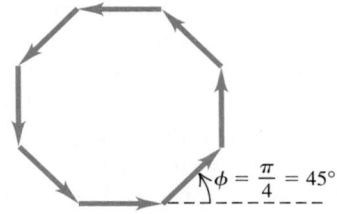 $\phi = \dfrac{\pi}{4} = 45°$

 $\phi = \dfrac{\pi}{2} = 90°$

generally, the intensity with eight slits is zero whenever ϕ is an integer multiple of $\pi/4$, *except* when ϕ is a multiple of 2π. Thus there are seven minima for every maximum.

Figure 36.15b shows the result of a detailed calculation of the eight-slit pattern. The large maxima, called *principal maxima,* are in the same positions as for the two-slit pattern of Fig. 36.15a but are much narrower. If the phase difference ϕ between adjacent slits is slightly different from a multiple of 2π, the waves from slits 1 and 2 will be only a little out of phase; however, the phase difference between slits 1 and 3 will be greater, that between slits 1 and 4 will be greater still, and so on. This leads to a partial cancellation for angles that are only slightly different from the angle for a maximum, giving the narrow maxima in Fig. 36.15b. The maxima are even narrower with 16 slits (Fig. 36.15c).

You should show that when there are N slits, there are $(N - 1)$ minima between each pair of principal maxima and a minimum occurs whenever ϕ is an integral multiple of $2\pi/N$ (except when ϕ is an integral multiple of 2π, which gives a principal maximum). There are small *secondary* intensity maxima between the minima; these become smaller in comparison to the principal maxima as N increases. The greater the value of N, the narrower the principal maxima become. From an energy standpoint the total power in the entire pattern is proportional to N. The height of each principal maximum is proportional to N^2, so from energy conservation the width of each principal maximum must be proportional to $1/N$. As we will see in the next section, the narrowness of the principal maxima in a multiple-slit pattern is of great practical importance.

Test Your Understanding of Section 36.4 Suppose two slits, each of width a, are separated by a distance $d = 2.5a$. Are there any missing maxima in the interference pattern produced by these slits? If so, which are missing? If not, why not? ❙

36.5 The Diffraction Grating

We have just seen that increasing the number of slits in an interference experiment (while keeping the spacing of adjacent slits constant) gives interference patterns in which the maxima are in the same positions, but progressively narrower, than with two slits. Because these maxima are so narrow, their angular position, and hence the wavelength, can be measured to very high precision. As we will see, this effect has many important applications.

An array of a large number of parallel slits, all with the same width a and spaced equal distances d between centers, is called a **diffraction grating.** The first one was constructed by Fraunhofer using fine wires. Gratings can be made by using a diamond point to scratch many equally spaced grooves on a glass or metal surface,

36.15 Interference patterns for N equally spaced, very narrow slits. **(a)** Two slits. **(b)** Eight slits. **(c)** Sixteen slits. The vertical scales are different for each graph; I_0 is the maximum intensity for a single slit, and the maximum intensity for N slits is $N^2 I_0$. The width of each peak is proportional to $1/N$.

(a) $N = 2$: two slits produce one minimum between adjacent maxima.

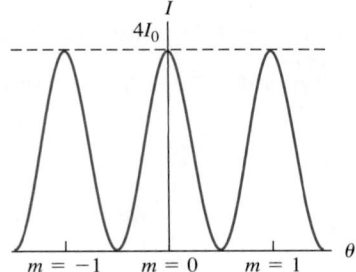

(b) $N = 8$: eight slits produce taller, narrower maxima in the same locations, separated by seven minima.

(c) $N = 16$: with 16 slits, the maxima are even taller and narrower, with more intervening minima.

36.16 A portion of a transmission diffraction grating. The separation between the centers of adjacent slits is d.

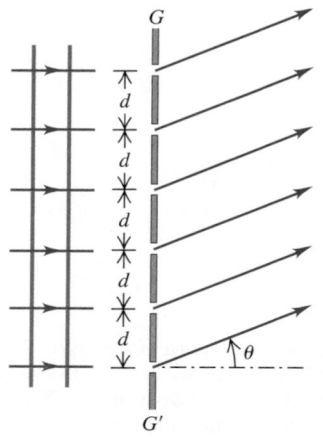

G

G'

ActivPhysics 16.4: The Grating: Introduction and Questions
ActivPhysics 16.5: The Grating: Problems

36.17 Microscopic pits on the surface of this DVD act as a reflection grating, splitting white light into its component colors.

or by photographic reduction of a pattern of black and white stripes on paper. For a grating, what we have been calling *slits* are often called *rulings* or *lines*.

In Fig. 36.16, GG' is a cross section of a *transmission grating;* the slits are perpendicular to the plane of the page, and an interference pattern is formed by the light that is transmitted through the slits. The diagram shows only six slits; an actual grating may contain several thousand. The spacing d between centers of adjacent slits is called the *grating spacing*. A plane monochromatic wave is incident normally on the grating from the left side. We assume far-field (Fraunhofer) conditions; that is, the pattern is formed on a screen that is far enough away that all rays emerging from the grating and going to a particular point on the screen can be considered to be parallel.

We found in Section 36.4 that the principal intensity maxima with multiple slits occur in the same directions as for the two-slit pattern. These are the directions for which the path difference for adjacent slits is an integer number of wavelengths. So the positions of the maxima are once again given by

$$d\sin\theta = m\lambda \qquad (m = 0, \pm 1, \pm 2, \pm 3, \dots) \qquad \text{(intensity maxima, multiple slits)} \qquad (36.13)$$

The intensity patterns for two, eight, and 16 slits displayed in Fig. 36.15 show the progressive increase in sharpness of the maxima as the number of slits increases.

When a grating containing hundreds or thousands of slits is illuminated by a beam of parallel rays of monochromatic light, the pattern is a series of very sharp lines at angles determined by Eq. (36.13). The $m = \pm 1$ lines are called the *first-order lines,* the $m = \pm 2$ lines the *second-order lines,* and so on. If the grating is illuminated by white light with a continuous distribution of wavelengths, each value of m corresponds to a continuous spectrum in the pattern. The angle for each wavelength is determined by Eq. (36.13); for a given value of m, long wavelengths (the red end of the spectrum) lie at larger angles (that is, are deviated more from the straight-ahead direction) than do the shorter wavelengths at the violet end of the spectrum.

As Eq. (36.13) shows, the sines of the deviation angles of the maxima are proportional to the ratio λ/d. For substantial deviation to occur, the grating spacing d should be of the same order of magnitude as the wavelength λ. Gratings for use with visible light (λ from 400 to 700 nm) usually have about 1000 slits per millimeter; the value of d is the *reciprocal* of the number of slits per unit length, so d is of the order of $\frac{1}{1000}$ mm = 1000 nm.

In a *reflection grating,* the array of equally spaced slits shown in Fig. 36.16 is replaced by an array of equally spaced ridges or grooves on a reflective screen. The reflected light has maximum intensity at angles where the phase difference between light waves reflected from adjacent ridges or grooves is an integral multiple of 2π. If light of wavelength λ is incident normally on a reflection grating with a spacing d between adjacent ridges or grooves, the *reflected* angles at which intensity maxima occur are given by Eq. (36.13).

The rainbow-colored reflections from the surface of a DVD are a reflection-grating effect (Fig. 36.17). The "grooves" are tiny pits 0.12 μm deep in the surface of the disc, with a uniform radial spacing of 0.74 μm = 740 nm. Information is coded on the DVD by varying the *length* of the pits. The reflection-grating aspect of the disc is merely an aesthetic side benefit.

Example 36.4 **Width of a grating spectrum**

The wavelengths of the visible spectrum are approximately 380 nm (violet) to 750 nm (red). (a) Find the angular limits of the first-order visible spectrum produced by a plane grating with 600 slits per millimeter when white light falls normally on the grating.

(b) Do the first-order and second-order spectra overlap? What about the second-order and third-order spectra? Do your answers depend on the grating spacing?

SOLUTION

IDENTIFY and SET UP: We must find the angles spanned by the visible spectrum in the first-, second-, and third-order spectra. These correspond to $m = 1$, 2, and 3 in Eq. (36.13).

EXECUTE: (a) The grating spacing is

$$d = \frac{1}{600 \text{ slits/mm}} = 1.67 \times 10^{-6} \text{ m}$$

We solve Eq. (36.13) for θ:

$$\theta = \arcsin \frac{m\lambda}{d}$$

Then for $m = 1$, the angular deviations θ_{v1} and θ_{r1} for violet and red light, respectively, are

$$\theta_{v1} = \arcsin\left(\frac{380 \times 10^{-9} \text{ m}}{1.67 \times 10^{-6} \text{ m}}\right) = 13.2°$$

$$\theta_{r1} = \arcsin\left(\frac{750 \times 10^{-9} \text{ m}}{1.67 \times 10^{-6} \text{ m}}\right) = 26.7°$$

That is, the first-order visible spectrum appears with deflection angles from $\theta_{v1} = 13.2°$ (violet) to $\theta_{r1} = 26.7°$ (red).

(b) With $m = 2$ and $m = 3$, our equation $\theta = \arcsin(m\lambda/d)$ for 380-mm violet light yields

$$\theta_{v2} = \arcsin\left(\frac{2(380 \times 10^{-9} \text{ m})}{1.67 \times 10^{-6} \text{ m}}\right) = 27.1°$$

$$\theta_{v3} = \arcsin\left(\frac{3(380 \times 10^{-9} \text{ m})}{1.67 \times 10^{-6} \text{ m}}\right) = 43.0°$$

For 750-nm red light, this same equation gives

$$\theta_{r2} = \arcsin\left(\frac{2(750 \times 10^{-9} \text{ m})}{1.67 \times 10^{-6} \text{ m}}\right) = 63.9°$$

$$\theta_{r3} = \arcsin\left(\frac{3(750 \times 10^{-9} \text{ m})}{1.67 \times 10^{-6} \text{ m}}\right) = \arcsin(1.35) = \text{undefined}$$

Hence the second-order spectrum extends from $27.1°$ to $63.9°$ and the third-order spectrum extends from $43.0°$ to $90°$ (the largest possible value of θ). The undefined value of θ_{r3} means that the third-order spectrum reaches $\theta = 90° = \arcsin(1)$ at a wavelength shorter than 750 nm; you should be able to show that this happens for $\lambda = 557$ nm. Hence the first-order spectrum (from $13.2°$ to $26.7°$) does not overlap with the second-order spectrum, but the second- and third-order spectra do overlap. You can convince yourself that this is true for any value of the grating spacing d.

EVALUATE: The fundamental reason the first-order and second-order visible spectra don't overlap is that the human eye is sensitive to only a narrow range of wavelengths. Can you show that if the eye could detect wavelengths from 380 nm to 900 nm (in the near-infrared range), the first and second orders *would* overlap?

Grating Spectrographs

Diffraction gratings are widely used to measure the spectrum of light emitted by a source, a process called *spectroscopy* or *spectrometry*. Light incident on a grating of known spacing is dispersed into a spectrum. The angles of deviation of the maxima are then measured, and Eq. (36.13) is used to compute the wavelength. With a grating that has many slits, very sharp maxima are produced, and the angle of deviation (and hence the wavelength) can be measured very precisely.

An important application of this technique is to astronomy. As light generated within the sun passes through the sun's atmosphere, certain wavelengths are selectively absorbed. The result is that the spectrum of sunlight produced by a diffraction grating has dark *absorption lines* (Fig. 36.18). Experiments in the laboratory show that different types of atoms and ions absorb light of different wavelengths. By comparing these laboratory results with the wavelengths of absorption lines in the spectrum of sunlight, astronomers can deduce the chemical composition of the sun's atmosphere. The same technique is used to make chemical assays of galaxies that are millions of light-years away.

Figure 36.19 shows one design for a *grating spectrograph* used in astronomy. A transmission grating is used in the figure; in other setups, a reflection grating is used. In older designs a prism was used rather than a grating, and a spectrum was formed by dispersion (see Section 33.4) rather than diffraction. However, there is no simple relationship between wavelength and angle of deviation for a prism, prisms absorb some of the light that passes through them, and they are less effective for many nonvisible wavelengths that are important in astronomy. For these and other reasons, gratings are preferred in precision applications.

36.18 (a) A visible-light photograph of the sun. (b) Sunlight is dispersed into a spectrum by a diffraction grating. Specific wavelengths are absorbed as sunlight passes through the sun's atmosphere, leaving dark lines in the spectrum.

(a)

(b)

Resolution of a Grating Spectrograph

In spectroscopy it is often important to distinguish slightly differing wavelengths. The minimum wavelength difference $\Delta\lambda$ that can be distinguished by a spectrograph is described by the **chromatic resolving power** R, defined as

36.19 A schematic diagram of a diffraction-grating spectrograph for use in astronomy. Note that the light does not strike the grating normal to its surface, so the intensity maxima are given by a somewhat different expression than Eq. (36.13). (See Problem 36.64.)

① Light from telescope is sent along fiber-optic cables (not shown) and emerges here.

② Light strikes concave mirror and emerges as a beam of parallel rays.

③ Light passes through diffraction grating.

④ Lenses direct diffracted light onto a second concave mirror.

⑤ Concave mirror reflects light to a focus.

⑥ An electronic detector (like the one in a digital camera) records the spectrum.

$$R = \frac{\lambda}{\Delta\lambda} \qquad \text{(chromatic resolving power)} \qquad (36.14)$$

As an example, when sodium atoms are heated, they emit strongly at the yellow wavelengths 589.00 nm and 589.59 nm. A spectrograph that can barely distinguish these two lines in the spectrum (called the *sodium doublet*) has a chromatic resolving power $R = (589.00 \text{ nm})/(0.59 \text{ nm}) = 1000$. (You can see these wavelengths when boiling water on a gas range. If the water boils over onto the flame, dissolved sodium from table salt emits a burst of yellow light.)

We can derive an expression for the resolving power of a diffraction grating used in a spectrograph. Two different wavelengths give diffraction maxima at slightly different angles. As a reasonable (though arbitrary) criterion, let's assume that we can distinguish them as two separate peaks if the maximum of one coincides with the first minimum of the other.

From our discussion in Section 36.4 the mth-order maximum occurs when the phase difference ϕ for adjacent slits is $\phi = 2\pi m$. The first minimum beside that maximum occurs when $\phi = 2\pi m + 2\pi/N$, where N is the number of slits. The phase difference is also given by $\phi = (2\pi d \sin\theta)/\lambda$, so the angular interval $d\theta$ corresponding to a small increment $d\phi$ in the phase shift can be obtained from the differential of this equation:

$$d\phi = \frac{2\pi d \cos\theta \, d\theta}{\lambda}$$

When $d\phi = 2\pi/N$, this corresponds to the angular interval $d\theta$ between a maximum and the first adjacent minimum. Thus $d\theta$ is given by

$$\frac{2\pi}{N} = \frac{2\pi d \cos\theta \, d\theta}{\lambda} \qquad \text{or} \qquad d\cos\theta \, d\theta = \frac{\lambda}{N}$$

CAUTION **Watch out for different uses of the symbol d** Don't confuse the spacing d with the differential "d" in the angular interval $d\theta$ or in the phase shift increment $d\phi$!

Now we need to find the angular spacing $d\theta$ between maxima for two slightly different wavelengths. We have $d\sin\theta = m\lambda$, so the differential of this equation gives

$$d\cos\theta \, d\theta = m \, d\lambda$$

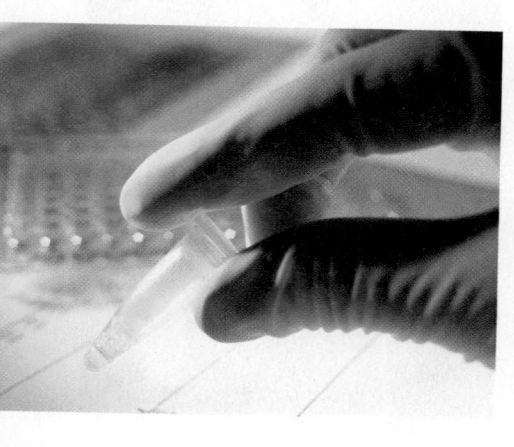

Application **Detecting DNA with Diffraction**

Diffraction gratings are used in a common piece of laboratory equipment known as a spectrophotometer. Light shining across a diffraction grating is dispersed into its component wavelengths. A slit is used to block all but a very narrow range of wavelengths, producing a beam of almost perfectly monochromatic light. The instrument then measures how much of that light is absorbed by a solution of biological molecules. For example, the sample tube shown here contains a solution of DNA, which is transparent to visible light but which strongly absorbs ultraviolet light with a wavelength of exactly 260 nm. Therefore, by illuminating the sample with 260-nm light and measuring the amount absorbed, we can determine the concentration of DNA in the solution.

According to our criterion, the limit or resolution is reached when these two angular spacings are equal. Equating the two expressions for the quantity $(d\cos\theta\, d\theta)$, we find

$$\frac{\lambda}{N} = m\, d\lambda \qquad \text{and} \qquad \frac{\lambda}{d\lambda} = Nm$$

If $\Delta\lambda$ is small, we can replace $d\lambda$ by $\Delta\lambda$, and the resolving power R is given by

$$R = \frac{\lambda}{\Delta\lambda} = Nm \qquad\qquad (36.15)$$

The greater the number of slits N, the better the resolution; also, the higher the order m of the diffraction-pattern maximum that we use, the better the resolution.

Test Your Understanding of Section 36.5 What minimum number of slits would be required in a grating to resolve the sodium doublet in the fourth order? (i) 250; (ii) 400; (iii) 1000; (iv) 4000.

36.6 X-Ray Diffraction

X rays were discovered by Wilhelm Röntgen (1845–1923) in 1895, and early experiments suggested that they were electromagnetic waves with wavelengths of the order of 10^{-10} m. At about the same time, the idea began to emerge that in a crystalline solid the atoms are arranged in a regular repeating pattern, with spacing between adjacent atoms also of the order of 10^{-10} m. Putting these two ideas together, Max von Laue (1879–1960) proposed in 1912 that a crystal might serve as a kind of three-dimensional diffraction grating for x rays. That is, a beam of x rays might be scattered (that is, absorbed and re-emitted) by the individual atoms in a crystal, and the scattered waves might interfere just like waves from a diffraction grating.

The first **x-ray diffraction** experiments were performed in 1912 by Friederich, Knipping, and von Laue, using the experimental setup shown in Fig. 36.20a. The scattered x rays *did* form an interference pattern, which they recorded on photographic film. Figure 36.20b is a photograph of such a pattern. These experiments verified that x rays *are* waves, or at least have wavelike properties, and also that the atoms in a crystal *are* arranged in a regular pattern (Fig. 36.21). Since that time, x-ray diffraction has proved to be an invaluable research tool, both for measuring x-ray wavelengths and for studying the structure of crystals and complex molecules.

36.20 (a) An x-ray diffraction experiment. (b) Diffraction pattern (or *Laue pattern*) formed by directing a beam of x rays at a thin section of quartz crystal.

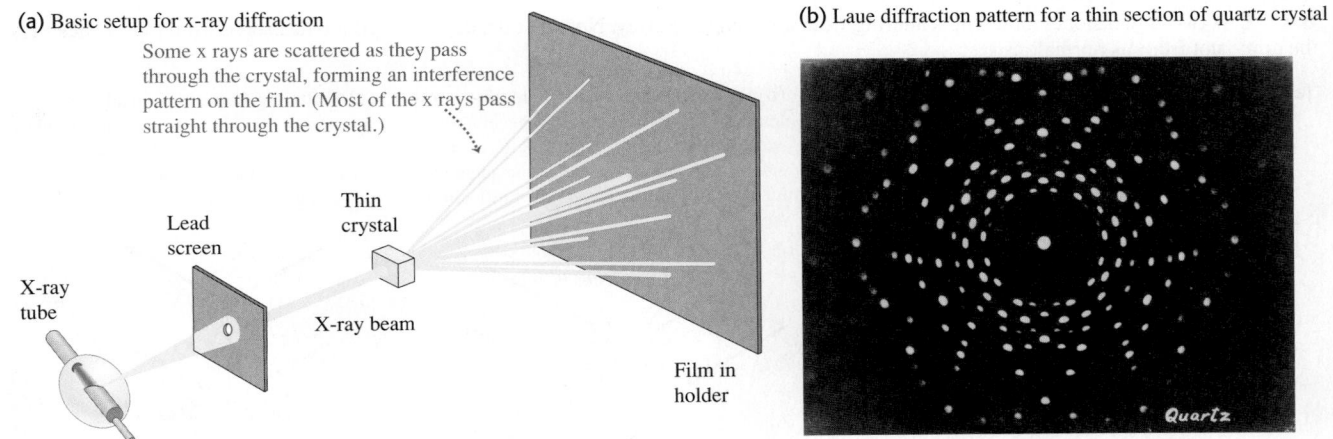

(a) Basic setup for x-ray diffraction

Some x rays are scattered as they pass through the crystal, forming an interference pattern on the film. (Most of the x rays pass straight through the crystal.)

X-ray tube

Lead screen

Thin crystal

X-ray beam

Film in holder

(b) Laue diffraction pattern for a thin section of quartz crystal

Quartz

36.21 Model of the arrangement of ions in a crystal of NaCl (table salt). The spacing of adjacent atoms is 0.282 nm. (The electron clouds of the atoms actually overlap slightly.)

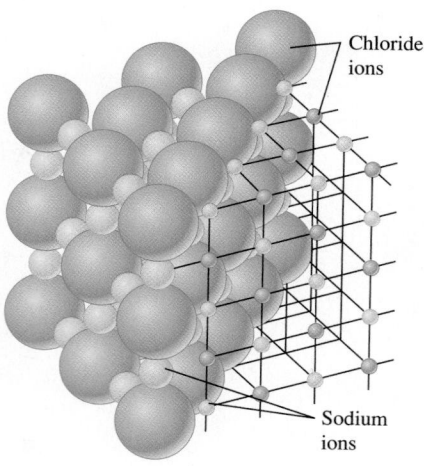

Chloride ions

Sodium ions

A Simple Model of X-Ray Diffraction

To better understand x-ray diffraction, we consider first a two-dimensional scattering situation, as shown in Fig. 36.22a, in which a plane wave is incident on a rectangular array of scattering centers. The situation might be a ripple tank with an array of small posts, 3-cm microwaves striking an array of small conducting spheres, or x rays incident on an array of atoms. In the case of electromagnetic waves, the wave induces an oscillating electric dipole moment in each scatterer. These dipoles act like little antennas, emitting scattered waves. The resulting interference pattern is the superposition of all these scattered waves. The situation is different from that with a diffraction grating, in which the waves from all the slits are emitted *in phase* (for a plane wave at normal incidence). Here the scattered waves are *not* all in phase because their distances from the *source* are different. To compute the interference pattern, we have to consider the *total* path differences for the scattered waves, including the distances from source to scatterer and from scatterer to observer.

As Fig. 36.22b shows, the path length from source to observer is the same for all the scatterers in a single row if the two angles θ_a and θ_r are equal. Scattered radiation from *adjacent* rows is *also* in phase if the path difference for adjacent rows is an integer number of wavelengths. Figure 36.22c shows that this path difference is $2d\sin\theta$, where θ is the common value of θ_a and θ_r. Therefore the conditions for radiation from the *entire array* to reach the observer in phase are (1) the angle of incidence must equal the angle of scattering and (2) the path difference for adjacent rows must equal $m\lambda$, where m is an integer. We can express the second condition as

$$2d\sin\theta = m\lambda \qquad (m = 1, 2, 3, \dots)$$

(Bragg condition for constructive interference from an array) (36.16)

CAUTION **Scattering from an array** In Eq. (36.16) the angle θ is measured with respect to the *surface* of the crystal, rather than with respect to the *normal* to the plane of an array of slits or a grating. Also, note that the path difference in Eq. (36.16) is $2d\sin\theta$, not $d\sin\theta$ as in Eq. (36.13) for a diffraction grating. ▮

In directions for which Eq. (36.16) is satisfied, we see a strong maximum in the interference pattern. We can describe this interference in terms of *reflections* of the wave from the horizontal rows of scatterers in Fig. 36.22a. Strong reflection (constructive interference) occurs at angles such that the incident and scattered angles are equal and Eq. (36.16) is satisfied. Since $\sin\theta$ can never be greater than 1, Eq. (36.16) says that to have constructive interference the quantity $m\lambda$

36.22 A two-dimensional model of scattering from a rectangular array. Note that the angles in (b) are measured from the *surface* of the array, not from its normal.

(a) Scattering of waves from a rectangular array

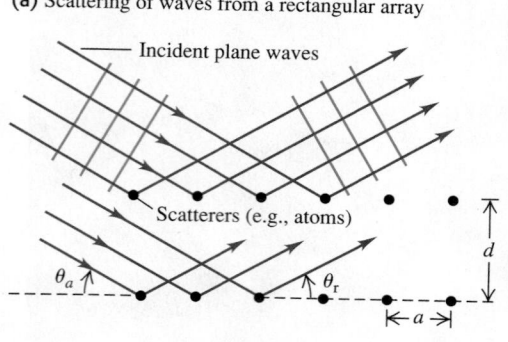

Incident plane waves

Scatterers (e.g., atoms)

θ_a θ_r

d

a

(b) Scattering from adjacent atoms in a row
Interference from adjacent atoms in a row is constructive when the path lengths $a\cos\theta_a$ and $a\cos\theta_r$ are equal, so that the angle of incidence θ_a equals the angle of reflection (scattering) θ_r.

$a\cos\theta_a$ $a\cos\theta_r$

θ_r θ_a

a

(c) Scattering from atoms in adjacent rows
Interference from atoms in adjacent rows is constructive when the path difference $2d\sin\theta$ is an integral number of wavelengths, as in Eq. (36.16).

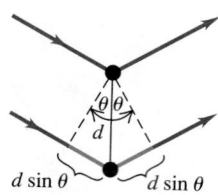

θ θ

d

$d\sin\theta$ $d\sin\theta$

36.23 A cubic crystal and two different families of crystal planes. There are also three sets of planes parallel to the cube faces, with spacing a.

(a) Spacing of planes is $d = a/\sqrt{2}$.

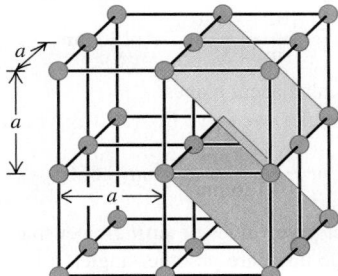

(b) Spacing of planes is $d = a/\sqrt{3}$.

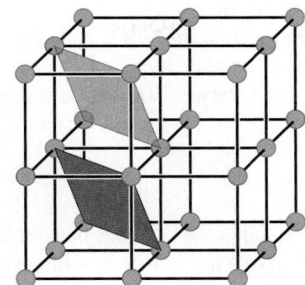

must be less than $2d$ and so λ must be less than $2d/m$. For example, the value of d in an NaCl crystal (see Fig. 36.21) is only 0.282 nm. Hence to have the mth-order maximum present in the diffraction pattern, λ must be less than $2(0.282 \text{ nm})/m$; that is, $\lambda < 0.564$ nm for $m = 1$, $\lambda < 0.282$ nm for $m = 2$, $\lambda < 0.188$ nm for $m = 3$, and so on. These are all x-ray wavelengths (see Fig. 32.4), which is why x rays are used for studying crystal structure.

We can extend this discussion to a three-dimensional array by considering *planes* of scatterers instead of *rows*. Figure 36.23 shows two different sets of parallel planes that pass through all the scatterers. Waves from all the scatterers in a given plane interfere constructively if the angles of incidence and scattering are equal. There is also constructive interference between planes when Eq. (36.16) is satisfied, where d is now the distance between adjacent planes. Because there are many different sets of parallel planes, there are also many values of d and many sets of angles that give constructive interference for the whole crystal lattice. This phenomenon is called **Bragg reflection**, and Eq. (36.16) is called the **Bragg condition**, in honor of Sir William Bragg and his son Laurence Bragg, two pioneers in x-ray analysis.

CAUTION *Bragg reflection is really Bragg interference* While we are using the term *reflection*, remember that we are dealing with an *interference* effect. In fact, the reflections from various planes are closely analogous to interference effects in thin films (see Section 35.4).

As Fig. 36.20b shows, in x-ray diffraction there is nearly complete cancellation in all but certain very specific directions in which constructive interference occurs and forms bright spots. Such a pattern is usually called an x-ray *diffraction* pattern, although *interference* pattern might be more appropriate.

We can determine the wavelength of x rays by examining the diffraction pattern for a crystal of known structure and known spacing between atoms, just as we determined wavelengths of visible light by measuring patterns from slits or gratings. (The spacing between atoms in simple crystals of known structure, such as sodium chloride, can be found from the density of the crystal and Avogadro's number.) Then, once we know the x-ray wavelength, we can use x-ray diffraction to explore the structure and determine the spacing between atoms in crystals with unknown structure.

X-ray diffraction is by far the most important experimental tool in the investigation of crystal structure of solids. X-ray diffraction also plays an important role in studies of the structures of liquids and of organic molecules. It has been one of the chief experimental techniques in working out the double-helix structure of DNA (Fig. 36.24) and subsequent advances in molecular genetics.

36.24 The British scientist Rosalind Franklin made this groundbreaking x-ray diffraction image of DNA in 1953. The dark bands arranged in a cross provided the first evidence of the helical structure of the DNA molecule.

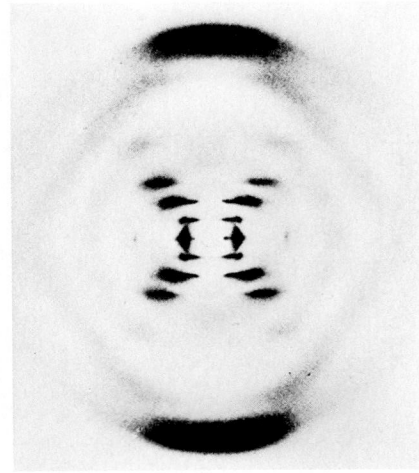

Example 36.5 X-ray diffraction

You direct a beam of 0.154-nm x rays at certain planes of a silicon crystal. As you increase the angle of incidence of the beam from zero, the first strong interference maximum occurs when the beam makes an angle of 34.5° with the planes. (a) How far apart are the planes? (b) Will you find other interference maxima from these planes at greater angles of incidence?

SOLUTION

IDENTIFY and SET UP: This problem involves Bragg reflection of x rays from the planes of a crystal. In part (a) we use the Bragg condition, Eq. (36.16), to find the distance d between adjacent planes from the known wavelength $\lambda = 0.154$ nm and angle of incidence $\theta = 34.5°$ for the $m = 1$ interference maximum. Given the value of d, we use the Bragg condition again in part (b) to find the values of θ for interference maxima corresponding to other values of m.

EXECUTE: (a) We solve Eq. (36.16) for d and set $m = 1$:

$$d = \frac{m\lambda}{2\sin\theta} = \frac{(1)(0.154 \text{ nm})}{2\sin 34.5°} = 0.136 \text{ nm}$$

This is the distance between adjacent planes.

(b) To calculate other angles, we solve Eq. (36.16) for $\sin\theta$:

$$\sin\theta = \frac{m\lambda}{2d} = m\frac{0.154 \text{ nm}}{2(0.136 \text{ nm})} = m(0.566)$$

Values of m of 2 or greater give values of $\sin\theta$ greater than unity, which is impossible. Hence there are *no* other angles for interference maxima for this particular set of crystal planes.

EVALUATE: Our result in part (b) shows that there *would* be a second interference maximum if the quantity $2\lambda/2d = \lambda/d$ were less than 1. This would be the case if the wavelength of the x rays were less than $d = 0.136$ nm. How short would the wavelength need to be to have *three* interference maxima?

Test Your Understanding of Section 36.6 You are doing an x-ray diffraction experiment with a crystal in which the atomic planes are 0.200 nm apart. You are using x rays of wavelength 0.100 nm. Will the fifth-order maximum be present in the diffraction pattern?

Mastering**PHYSICS**

ActivPhysics 16.7: Circular Hole Diffraction
ActivPhysics 16.8: Resolving Power

36.7 Circular Apertures and Resolving Power

We have studied in detail the diffraction patterns formed by long, thin slits or arrays of slits. But an aperture of *any* shape forms a diffraction pattern. The diffraction pattern formed by a *circular* aperture is of special interest because of its role in limiting how well an optical instrument can resolve fine details. In principle, we could compute the intensity at any point P in the diffraction pattern by dividing the area of the aperture into small elements, finding the resulting wave amplitude and phase at P, and then integrating over the aperture area to find the resultant amplitude and intensity at P. In practice, the integration cannot be carried out in terms of elementary functions. We will simply *describe* the pattern and quote a few relevant numbers.

The diffraction pattern formed by a circular aperture consists of a central bright spot surrounded by a series of bright and dark rings, as Fig. 36.25 shows.

36.25 Diffraction pattern formed by a circular aperture of diameter D. The pattern consists of a central bright spot and alternating dark and bright rings. The angular radius θ_1 of the first dark ring is shown. (This diagram is not drawn to scale.)

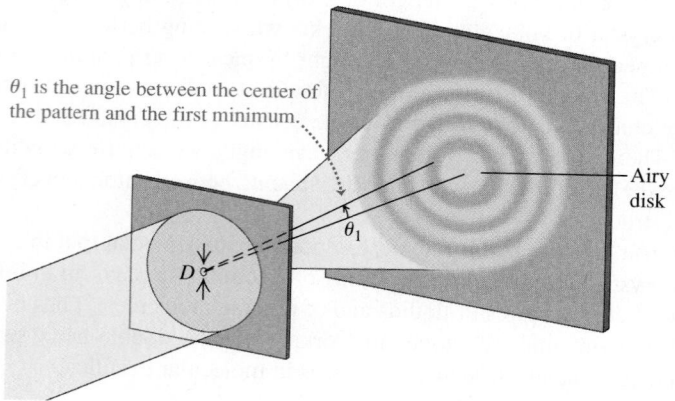

θ_1 is the angle between the center of the pattern and the first minimum.

Airy disk

θ_1

D

We can describe the pattern in terms of the angle θ, representing the angular radius of each ring. If the aperture diameter is D and the wavelength is λ, the angular radius θ_1 of the first *dark* ring is given by

$$\sin\theta_1 = 1.22\frac{\lambda}{D} \qquad \text{(diffraction by a circular aperture)} \qquad (36.17)$$

The angular radii of the next two dark rings are given by

$$\sin\theta_2 = 2.23\frac{\lambda}{D} \qquad \sin\theta_3 = 3.24\frac{\lambda}{D} \qquad (36.18)$$

Between these are bright rings with angular radii given by

$$\sin\theta = 1.63\frac{\lambda}{D}, \qquad 2.68\frac{\lambda}{D}, \qquad 3.70\frac{\lambda}{D} \qquad (36.19)$$

and so on. The central bright spot is called the **Airy disk,** in honor of Sir George Airy (1801–1892), Astronomer Royal of England, who first derived the expression for the intensity in the pattern. The angular radius of the Airy disk is that of the first dark ring, given by Eq. (36.17).

The intensities in the bright rings drop off very quickly with increasing angle. When D is much larger than the wavelength λ, the usual case for optical instruments, the peak intensity in the first ring is only 1.7% of the value at the center of the Airy disk, and the peak intensity of the second ring is only 0.4%. Most (85%) of the light energy falls within the Airy disk. Figure 36.26 shows a diffraction pattern from a circular aperture.

Diffraction and Image Formation

Diffraction has far-reaching implications for image formation by lenses and mirrors. In our study of optical instruments in Chapter 34 we assumed that a lens with focal length f focuses a parallel beam (plane wave) to a *point* at a distance f from the lens. This assumption ignored diffraction effects. We now see that what we get is not a point but the diffraction pattern just described. If we have two point objects, their images are not two points but two diffraction patterns. When the objects are close together, their diffraction patterns overlap; if they are close enough, their patterns overlap almost completely and cannot be distinguished. The effect is shown in Fig. 36.27, which presents the patterns for four very small "point" sources of light. In Fig. 36.27a the image of the left-hand source is well separated from the others, but the images of the middle and right-hand sources have merged. In Fig. 36.27b, with a larger aperture diameter and hence smaller Airy disks, the middle and right-hand images are better resolved. In Fig. 36.27c, with a still larger aperture, they are well resolved.

A widely used criterion for resolution of two point objects, proposed by the English physicist Lord Rayleigh (1842–1919) and called **Rayleigh's criterion,** is that the objects are just barely resolved (that is, distinguishable) if the center of one diffraction pattern coincides with the first minimum of the other. In that case the angular separation of the image centers is given by Eq. (36.17). The angular separation of the *objects* is the same as that of the *images* made by a telescope, microscope, or other optical device. So two point objects are barely resolved, according to Rayleigh's criterion, when their angular separation is given by Eq. (36.17).

The minimum separation of two objects that can just be resolved by an optical instrument is called the **limit of resolution** of the instrument. The smaller the limit of resolution, the greater the *resolution,* or **resolving power,** of the instrument. Diffraction sets the ultimate limits on resolution of lenses. *Geometric optics* may make it seem that we can make images as large as we like. Eventually, though, we always reach a point at which the image becomes larger but does not

36.26 Photograph of the diffraction pattern formed by a circular aperture.

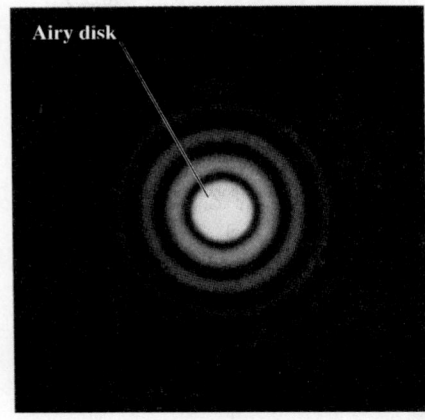

36.27 Diffraction patterns of four very small ("point") sources of light. The photographs were made with a circular aperture in front of the lens. (a) The aperture is so small that the patterns of sources 3 and 4 overlap and are barely resolved by Rayleigh's criterion. Increasing the size of the aperture decreases the size of the diffraction patterns, as shown in (b) and (c).

(a) Small aperture

(b) Medium aperture

(c) Large aperture

gain in detail. The images in Fig. 36.27 would not become sharper with further enlargement.

CAUTION **Resolving power vs. chromatic resolving power** Be careful not to confuse the resolving power of an optical instrument with the *chromatic* resolving power of a grating (described in Section 36.5). Resolving power refers to the ability to distinguish the images of objects that appear close to each other, when looking either through an optical instrument or at a photograph made with the instrument. Chromatic resolving power describes how well different wavelengths can be distinguished in a spectrum formed by a diffraction grating. ▌

Rayleigh's criterion combined with Eq. (36.17) shows that resolution (resolving power) improves with larger diameter; it also improves with shorter wavelengths. Ultraviolet microscopes have higher resolution than visible-light microscopes. In electron microscopes the resolution is limited by the wavelengths associated with the electrons, which have wavelike aspects (to be discussed further in Chapter 39). These wavelengths can be made 100,000 times smaller than wavelengths of visible light, with a corresponding gain in resolution. Resolving power also explains the difference in storage capacity between DVDs (introduced in 1995) and Blu-ray discs (introduced in 2003). Information is stored in both of these in a series of tiny pits. In order not to lose information in the scanning process, the scanning optics must be able to resolve two adjacent pits so that they do not seem to blend into a single pit (see sources 3 and 4 in Fig. 36.27). The blue scanning laser used in a Blu-ray player has a shorter wavelength (405 nm) and hence better resolving power than the 650-nm red laser in a DVD player. Hence pits can be spaced closer together in a Blu-ray disc than in a DVD, and more information can be stored on a disc of the same size (50 gigabytes on a Blu-ray disc versus 4.7 gigabytes on a DVD).

Example 36.6 Resolving power of a camera lens

A camera lens with focal length $f = 50$ mm and maximum aperture $f/2$ forms an image of an object 9.0 m away. (a) If the resolution is limited by diffraction, what is the minimum distance between two points on the object that are barely resolved? What is the corresponding distance between image points? (b) How does the situation change if the lens is "stopped down" to $f/16$? Use $\lambda = 500$ nm in both cases.

SOLUTION

IDENTIFY and SET UP: This example uses the ideas about resolving power, image formation by a lens (Section 34.4), and f-number (Section 34.5). From Eq. (34.20), the f-number of a lens is its focal length f divided by the aperture diameter D. We use this equation to determine D and then use Eq. (36.17) (the Rayleigh criterion) to find the angular separation θ between two barely resolved points on the object. We then use the geometry of image formation by a lens to determine the distance y between those points and the distance y' between the corresponding image points.

EXECUTE: (a) The aperture diameter is $D = f/(f\text{-number}) = (50 \text{ mm})/2 = 25 \text{ mm} = 25 \times 10^{-3}$ m. From Eq. (36.17) the angular separation θ of two object points that are barely resolved is

$$\theta \approx \sin\theta = 1.22 \frac{\lambda}{D} = 1.22 \frac{500 \times 10^{-9} \text{ m}}{25 \times 10^{-3} \text{ m}} = 2.4 \times 10^{-5} \text{ rad}$$

We know from our thin-lens analysis in Section 34.4 that, apart from sign, $y/s = y'/s'$ [see Eq. (34.14)]. Thus the angular separations of the object points and the corresponding image points are both equal to θ. Because the object distance s is much greater than the focal length $f = 50$ mm, the image distance s' is approximately equal to f. Thus

$$\frac{y}{9.0 \text{ m}} = 2.4 \times 10^{-5} \qquad y = 2.2 \times 10^{-4} \text{ m} = 0.22 \text{ mm}$$

$$\frac{y'}{50 \text{ mm}} = 2.4 \times 10^{-5} \qquad y' = 1.2 \times 10^{-3} \text{ mm}$$

$$= 0.0012 \text{ mm} \approx \tfrac{1}{800} \text{ mm}$$

(b) The aperture diameter is now $(50 \text{ mm})/16$, or one-eighth as large as before. The angular separation between barely resolved points is eight times as great, and the values of y and y' are also eight times as great as before:

$$y = 1.8 \text{ mm} \qquad y' = 0.0096 \text{ mm} = \tfrac{1}{100} \text{ mm}$$

Only the best camera lenses can approach this resolving power.

EVALUATE: Many photographers use the smallest possible aperture for maximum sharpness, since lens aberrations cause light rays that are far from the optic axis to converge to a different image point than do rays near the axis. But as this example shows, diffraction effects become more significant at small apertures. One cause of fuzzy images has to be balanced against another.

Test Your Understanding of Section 36.7 You have been asked to compare four different proposals for telescopes to be placed in orbit, above the blurring effects of the earth's atmosphere. Rank the proposed telescopes in order of their ability to resolve small details, from best to worst. (i) a radio telescope 100 m in diameter observing at a wavelength of 21 cm; (ii) an optical telescope 2.0 m in diameter observing at a wavelength of 500 nm; (iii) an ultraviolet telescope 1.0 m in diameter observing at a wavelength of 100 nm; (iv) an infrared telescope 2.0 m in diameter observing at a wavelength of 10 μm.

36.8 Holography

Holography is a technique for recording and reproducing an image of an object through the use of interference effects. Unlike the two-dimensional images recorded by an ordinary photograph or television system, a holographic image is truly three-dimensional. Such an image can be viewed from different directions to reveal different sides and from various distances to reveal changing perspective. If you had never seen a hologram, you wouldn't believe it was possible!

Figure 36.28a shows the basic procedure for making a hologram. We illuminate the object to be holographed with monochromatic light, and we place a photographic film so that it is struck by scattered light from the object and also by direct light from the source. In practice, the light source must be a laser, for reasons we will discuss later. Interference between the direct and scattered light leads to the formation and recording of a complex interference pattern on the film.

To form the images, we simply project light through the developed film (Fig. 36.28b). Two images are formed: a virtual image on the side of the film nearer the source and a real image on the opposite side.

Holography and Interference Patterns

A complete analysis of holography is beyond our scope, but we can gain some insight into the process by looking at how a single point is holographed and imaged. Consider the interference pattern that is formed on a sheet of photographic negative film by the superposition of an incident plane wave and a

36.28 (a) A hologram is the record on film of the interference pattern formed with light from the coherent source and light scattered from the object. (b) Images are formed when light is projected through the hologram. The observer sees the virtual image formed behind the hologram.

(a) Recording a hologram

(b) Viewing the hologram

36.29 (a) Constructive interference of the plane and spherical waves occurs in the plane of the film at every point Q for which the distance b_m from P is greater than the distance b_0 from P to O by an integral number of wavelengths $m\lambda$. For the point Q shown, $m = 2$. (b) When a plane wave strikes a transparent positive print of the developed film, the diffracted wave consists of a wave converging to P' and then diverging again and a diverging wave that appears to originate at P. These waves form the real and virtual images, respectively.

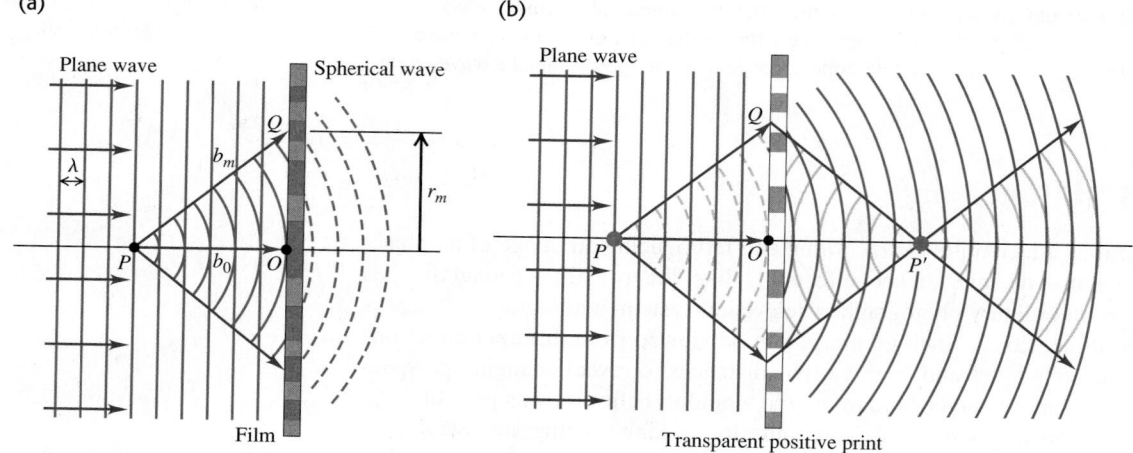

(a)

(b)

spherical wave, as shown in Fig. 36.29a. The spherical wave originates at a point source P at a distance b_0 from the film; P may in fact be a small object that scatters part of the incident plane wave. We assume that the two waves are monochromatic and coherent and that the phase relationship is such that constructive interference occurs at point O on the diagram. Then constructive interference will *also* occur at any point Q on the film that is farther from P than O is by an integer number of wavelengths. That is, if $b_m - b_0 = m\lambda$, where m is an integer, then constructive interference occurs. The points where this condition is satisfied form circles on the film centered at O, with radii r_m given by

$$b_m - b_0 = \sqrt{b_0{}^2 + r_m{}^2} - b_0 = m\lambda \qquad (m = 1, 2, 3, \ldots) \quad \text{(36.20)}$$

Solving this for $r_m{}^2$, we find

$$r_m{}^2 = \lambda(2mb_0 + m^2\lambda)$$

Ordinarily, b_0 is very much larger than λ, so we neglect the second term in parentheses and obtain

$$r_m = \sqrt{2m\lambda b_0} \qquad (m = 1, 2, 3, \ldots) \quad \text{(36.21)}$$

The interference pattern consists of a series of concentric bright circular fringes with radii given by Eq. (36.21). Between these bright fringes are dark fringes.

Now we develop the film and make a transparent positive print, so the bright-fringe areas have the greatest transparency on the film. Then we illuminate it with monochromatic plane-wave light of the same wavelength λ that we used initially. In Fig. 36.29b, consider a point P' at a distance b_0 along the axis from the film. The centers of successive bright fringes differ in their distances from P' by an integer number of wavelengths, and therefore a strong *maximum* in the diffracted wave occurs at P'. That is, light converges to P' and then diverges from it on the opposite side. Therefore P' is a *real image* of point P.

This is not the entire diffracted wave, however. The interference of the wavelets that spread out from all the transparent areas forms a second spherical wave that is diverging rather than converging. When this wave is traced back behind the film in Fig. 36.29b, it appears to be spreading out from point P. Thus the total diffracted wave from the hologram is a superposition of a spherical wave converging to form a real image at P' and a spherical wave that diverges as though it had come from the virtual image point P.

Because of the principle of superposition for waves, what is true for the imaging of a single point is also true for the imaging of any number of points. The film records the superposed interference pattern from the various points, and when light is projected through the film, the various image points are reproduced simultaneously. Thus the images of an extended object can be recorded and reproduced just as for a single point object. Figure 36.30 shows photographs of a holographic image from two different angles, showing the changing perspective in this three-dimensional image.

In making a hologram, we have to overcome two practical problems. First, the light used must be *coherent* over distances that are large in comparison to the dimensions of the object and its distance from the film. Ordinary light sources *do not* satisfy this requirement, for reasons that we discussed in Section 35.1. Therefore laser light is essential for making a hologram. (Ordinary white light can be used for *viewing* certain types of hologram, such as those used on credit cards.) Second, extreme mechanical stability is needed. If any relative motion of source, object, or film occurs during exposure, even by as much as a quarter of a wavelength, the interference pattern on the film is blurred enough to prevent satisfactory image formation. These obstacles are not insurmountable, however, and holography has become important in research, entertainment, and a wide variety of technological applications.

36.30 Two views of the same hologram seen from different angles.

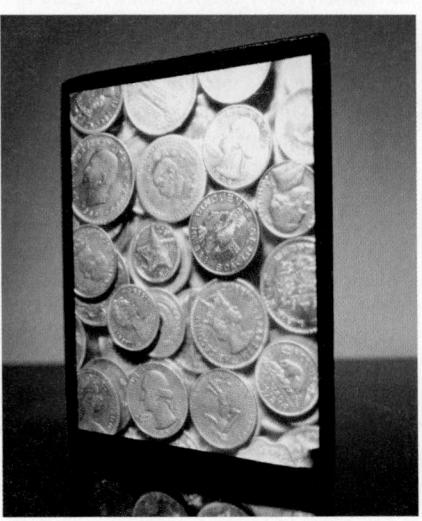

Fresnel and Fraunhofer diffraction: Diffraction occurs when light passes through an aperture or around an edge. When the source and the observer are so far away from the obstructing surface that the outgoing rays can be considered parallel, it is called Fraunhofer diffraction. When the source or the observer is relatively close to the obstructing surface, it is Fresnel diffraction.

Single-slit diffraction: Monochromatic light sent through a narrow slit of width a produces a diffraction pattern on a distant screen. Equation (36.2) gives the condition for destructive interference (a dark fringe) at a point P in the pattern at angle θ. Equation (36.7) gives the intensity in the pattern as a function of θ. (See Examples 36.1–36.3.)

$$\sin\theta = \frac{m\lambda}{a} \quad (m = \pm 1, \pm 2, \dots) \quad (36.2)$$

$$I = I_0 \left\{ \frac{\sin[\pi a(\sin\theta)/\lambda]}{\pi a(\sin\theta)/\lambda} \right\}^2 \quad (36.7)$$

Diffraction gratings: A diffraction grating consists of a large number of thin parallel slits, spaced a distance d apart. The condition for maximum intensity in the interference pattern is the same as for the two-source pattern, but the maxima for the grating are very sharp and narrow. (See Example 36.4.)

$$d\sin\theta = m\lambda \quad (m = 0, \pm 1, \pm 2, \pm 3, \dots) \quad (36.13)$$

X-ray diffraction: A crystal serves as a three-dimensional diffraction grating for x rays with wavelengths of the same order of magnitude as the spacing between atoms in the crystal. For a set of crystal planes spaced a distance d apart, constructive interference occurs when the angles of incidence and scattering (measured from the crystal planes) are equal and when the Bragg condition [Eq. (36.16)] is satisfied. (See Example 36.5.)

$$2d\sin\theta = m\lambda \quad (m = 1, 2, 3, \dots) \quad (36.16)$$

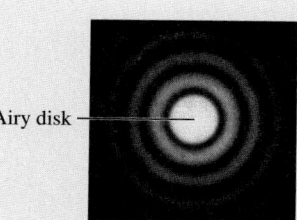

Circular apertures and resolving power: The diffraction pattern from a circular aperture of diameter D consists of a central bright spot, called the Airy disk, and a series of concentric dark and bright rings. Equation (36.17) gives the angular radius θ_1 of the first dark ring, equal to the angular size of the Airy disk. Diffraction sets the ultimate limit on resolution (image sharpness) of optical instruments. According to Rayleigh's criterion, two point objects are just barely resolved when their angular separation θ is given by Eq. (36.17). (See Example 36.6.)

$$\sin\theta_1 = 1.22\frac{\lambda}{D} \quad (36.17)$$

Airy disk

BRIDGING PROBLEM   Observing the Expanding Universe

An astronomer who is studying the light from a galaxy has identified the spectrum of hydrogen but finds that the wavelengths are somewhat shifted from those found in the laboratory. In the lab, the H_α line in the hydrogen spectrum has a wavelength of 656.3 nm. The astronomer is using a transmission diffraction grating having 5758 lines/cm in the first order and finds that the first bright fringe for the H_α line occurs at $\pm 23.41°$ from the central spot. How fast is the galaxy moving? Express your answer in m/s and as a percentage of the speed of light. Is the galaxy moving toward us or away from us?

SOLUTION GUIDE

See MasteringPhysics® study area for a Video Tutor solution.

IDENTIFY and SET UP

1. You can use the information about the grating to find the wavelength of the H_α line in the galaxy's spectrum.
2. In Section 16.8 we learned about the Doppler effect for electromagnetic radiation: The frequency that we receive from a mov-

ing source, such as the galaxy, is different from the frequency that is emitted. Equation (16.30) relates the emitted frequency, the received frequency, and the velocity of the source (the target variable). The equation $c = f\lambda$ relates the frequency f and wavelength λ through the speed of light c.

EXECUTE

3. Find the wavelength of the H_α spectral line in the received light.
4. Rewrite Eq. (16.30) as a formula for the velocity v of the galaxy in terms of the received wavelength and the wavelength emitted by the source.
5. Solve for v. Express it in m/s and as a percentage of c, and decide whether the galaxy is moving toward us or moving away.

EVALUATE

6. Is your answer consistent with the relative sizes of the received wavelength and the emitted wavelength?

Problems

For instructor-assigned homework, go to www.masteringphysics.com

•, ••, •••: Problems of increasing difficulty. **CP**: Cumulative problems incorporating material from earlier chapters. **CALC**: Problems requiring calculus. **BIO**: Biosciences problems.

DISCUSSION QUESTIONS

Q36.1 Why can we readily observe diffraction effects for sound waves and water waves, but not for light? Is this because light travels so much faster than these other waves? Explain.

Q36.2 What is the difference between Fresnel and Fraunhofer diffraction? Are they different *physical* processes? Explain.

Q36.3 You use a lens of diameter D and light of wavelength λ and frequency f to form an image of two closely spaced and distant objects. Which of the following will increase the resolving power? (a) Use a lens with a smaller diameter; (b) use light of higher frequency; (c) use light of longer wavelength. In each case justify your answer.

Q36.4 Light of wavelength λ and frequency f passes through a single slit of width a. The diffraction pattern is observed on a screen a distance x from the slit. Which of the following will *decrease* the width of the central maximum? (a) Decrease the slit width; (b) decrease the frequency f of the light; (c) decrease the wavelength λ of the light; (d) decrease the distance x of the screen from the slit. In each case justify your answer.

Q36.5 In a diffraction experiment with waves of wavelength λ, there will be *no* intensity minima (that is, no dark fringes) if the slit width is small enough. What is the maximum slit width for which this occurs? Explain your answer.

Q36.6 The predominant sound waves used in human speech have wavelengths in the range from 1.0 to 3.0 meters. Using the ideas of diffraction, explain how it is possible to hear a person's voice even when he is facing away from you.

Q36.7 In single-slit diffraction, what is $\sin(\beta/2)$ when $\theta = 0$? In view of your answer, why is the single-slit intensity *not* equal to zero at the center?

Q36.8 A rainbow ordinarily shows a range of colors (see Section 33.4). But if the water droplets that form the rainbow are small enough, the rainbow will appear white. Explain why, using diffraction ideas. How small do you think the raindrops would have to be for this to occur?

Q36.9 Some loudspeaker horns for outdoor concerts (at which the entire audience is seated on the ground) are wider vertically than horizontally. Use diffraction ideas to explain why this is more efficient at spreading the sound uniformly over the audience than either a square speaker horn or a horn that is wider horizontally than vertically. Would this still be the case if the audience were seated at different elevations, as in an amphitheater? Why or why not?

Q36.10 Figure 31.12 (Section 31.2) shows a loudspeaker system. Low-frequency sounds are produced by the *woofer*, which is a speaker with large diameter; the *tweeter*, a speaker with smaller diameter, produces high-frequency sounds. Use diffraction ideas to explain why the tweeter is more effective for distributing high-frequency sounds uniformly over a room than is the woofer.

Q36.11 Information is stored on an audio compact disc, CD-ROM, or DVD disc in a series of pits on the disc. These pits are scanned by a laser beam. An important limitation on the amount of information that can be stored on such a disc is the width of the laser beam. Explain why this should be, and explain how using a shorter-wavelength laser allows more information to be stored on a disc of the same size.

Q36.12 With which color of light can the Hubble Space Telescope see finer detail in a distant astronomical object: red, blue, or ultraviolet? Explain your answer.

Q36.13 At the end of Section 36.4, the following statements were made about an array of N slits. Explain, using phasor diagrams, why each statement is true. (a) A minimum occurs whenever ϕ is an integral multiple of $2\pi/N$, except when ϕ is an integral multiple of 2π (which gives a principal maximum). (b) There are $(N - 1)$ minima between each pair of principal maxima.

Q36.14 Could x-ray diffraction effects with crystals be observed by using visible light instead of x rays? Why or why not?

Q36.15 Why is a diffraction grating better than a two-slit setup for measuring wavelengths of light?

Q36.16 One sometimes sees rows of evenly spaced radio antenna towers. A student remarked that these act like diffraction gratings. What did she mean? Why would one want them to act like a diffraction grating?

Q36.17 If a hologram is made using 600-nm light and then viewed with 500-nm light, how will the images look compared to those observed when viewed with 600-nm light? Explain.

Q36.18 A hologram is made using 600-nm light and then viewed by using white light from an incandescent bulb. What will be seen? Explain.

Q36.19 Ordinary photographic film reverses black and white, in the sense that the most brightly illuminated areas become blackest upon development (hence the term *negative*). Suppose a hologram negative is viewed directly, without making a positive transparency. How will the resulting images differ from those obtained with the positive? Explain.

EXERCISES

Section 36.2 Diffraction from a Single Slit

36.1 •• Monochromatic light from a distant source is incident on a slit 0.750 mm wide. On a screen 2.00 m away, the distance from the central maximum of the diffraction pattern to the first minimum is measured to be 1.35 mm. Calculate the wavelength of the light.

36.2 • Parallel rays of green mercury light with a wavelength of 546 nm pass through a slit covering a lens with a focal length of 60.0 cm. In the focal plane of the lens the distance from the central maximum to the first minimum is 10.2 mm. What is the width of the slit?

36.3 •• Light of wavelength 585 nm falls on a slit 0.0666 mm wide. (a) On a very large and distant screen, how many *totally* dark fringes (indicating complete cancellation) will there be, including both sides of the central bright spot? Solve this problem *without* calculating all the angles! (*Hint:* What is the largest that $\sin\theta$ can be? What does this tell you is the largest that m can be?) (b) At what angle will the dark fringe that is most distant from the central bright fringe occur?

36.4 • Light of wavelength 633 nm from a distant source is incident on a slit 0.750 mm wide, and the resulting diffraction pattern is observed on a screen 3.50 m away. What is the distance between the two dark fringes on either side of the central bright fringe?

36.5 •• Diffraction occurs for all types of waves, including sound waves. High-frequency sound from a distant source with wavelength 9.00 cm passes through a slit 12.0 cm wide. A microphone is placed 8.00 m directly in front of the center of the slit, corresponding to point O in Fig. 36.5a. The microphone is then moved in a direction perpendicular to the line from the center of the slit to point O. At what distances from O will the intensity detected by the microphone be zero?

36.6 • **CP Tsunami!** On December 26, 2004, a violent earthquake of magnitude 9.1 occurred off the coast of Sumatra. This quake triggered a huge tsunami (similar to a tidal wave) that killed more than 150,000 people. Scientists observing the wave on the open ocean measured the time between crests to be 1.0 h and the speed of the wave to be 800 km/h. Computer models of the evolution of this enormous wave showed that it bent around the continents and spread to all the oceans of the earth. When the wave reached the gaps between continents, it diffracted between them as through a slit. (a) What was the wavelength of this tsunami? (b) The distance between the southern tip of Africa and northern Antarctica is about 4500 km, while the distance between the southern end of Australia and Antarctica is about 3700 km. As an approximation, we can model this wave's behavior by using Fraunhofer diffraction. Find the smallest angle away from the central maximum for which the waves would cancel after going through each of these continental gaps.

36.7 •• **CP** A series of parallel linear water wave fronts are traveling directly toward the shore at 15.0 cm/s on an otherwise placid lake. A long concrete barrier that runs parallel to the shore at a distance of 3.20 m away has a hole in it. You count the wave crests and observe that 75.0 of them pass by each minute, and you also observe that no waves reach the shore at ±61.3 cm from the point directly opposite the hole, but waves do reach the shore everywhere within this distance. (a) How wide is the hole in the barrier? (b) At what other angles do you find no waves hitting the shore?

36.8 • Monochromatic electromagnetic radiation with wavelength λ from a distant source passes through a slit. The diffraction pattern is observed on a screen 2.50 m from the slit. If the width of the central maximum is 6.00 mm, what is the slit width a if the wavelength is (a) 500 nm (visible light); (b) 50.0 μm (infrared radiation); (c) 0.500 nm (x rays)?

36.9 •• **Doorway Diffraction.** Sound of frequency 1250 Hz leaves a room through a 1.00-m-wide doorway (see Exercise 36.5). At which angles relative to the centerline perpendicular to the doorway will someone outside the room hear no sound? Use 344 m/s for the speed of sound in air and assume that the source and listener are both far enough from the doorway for Fraunhofer diffraction to apply. You can ignore effects of reflections.

36.10 • **CP** Light waves, for which the electric field is given by $E_y(x, t) = E_{\max}\sin[(1.20 \times 10^7 \text{ m}^{-1})x - \omega t]$, pass through a slit and produce the first dark bands at $\pm28.6°$ from the center of the diffraction pattern. (a) What is the frequency of this light? (b) How wide is the slit? (c) At which angles will other dark bands occur?

36.11 •• Parallel rays of light with wavelength 620 nm pass through a slit covering a lens with a focal length of 40.0 cm. The diffraction pattern is observed in the focal plane of the lens, and the distance from the center of the central maximum to the first minimum is 36.5 cm. What is the width of the slit? (*Note:* The angle that locates the first minimum is *not* small.)

36.12 •• Red light of wavelength 633 nm from a helium–neon laser passes through a slit 0.350 mm wide. The diffraction pattern is observed on a screen 3.00 m away. Define the width of a bright fringe as the distance between the minima on either side. (a) What is the width of the central bright fringe? (b) What is the width of the first bright fringe on either side of the central one?

Section 36.3 Intensity in the Single-Slit Pattern

36.13 •• Monochromatic light of wavelength 580 nm passes through a single slit and the diffraction pattern is observed on a screen. Both the source and screen are far enough from the slit for Fraunhofer diffraction to apply. (a) If the first diffraction minima are at $\pm90.0°$, so the central maximum completely fills the screen, what is the width of the slit? (b) For the width of the slit as calculated in part (a), what is the ratio of the intensity at $\theta = 45.0°$ to the intensity at $\theta = 0$?

36.14 •• Monochromatic light of wavelength $\lambda = 620$ nm from a distant source passes through a slit 0.450 mm wide. The diffraction pattern is observed on a screen 3.00 m from the slit. In terms of the intensity I_0 at the peak of the central maximum, what is the intensity of the light at the screen the following distances from the center of the central maximum: (a) 1.00 mm; (b) 3.00 mm; (c) 5.00 mm?

36.15 •• A slit 0.240 mm wide is illuminated by parallel light rays of wavelength 540 nm. The diffraction pattern is observed on a screen that is 3.00 m from the slit. The intensity at the center of the central maximum ($\theta = 0°$) is 6.00×10^{-6} W/m². (a) What is the distance on the screen from the center of the central maximum to the first minimum? (b) What is the intensity at a point on the screen midway between the center of the central maximum and the first minimum?

36.16 • Monochromatic light of wavelength 486 nm from a distant source passes through a slit that is 0.0290 mm wide. In the resulting diffraction pattern, the intensity at the center of the central maximum ($\theta = 0°$) is 4.00×10^{-5} W/m². What is the intensity at a point on the screen that corresponds to $\theta = 1.20°$.

36.17 •• A single-slit diffraction pattern is formed by monochromatic electromagnetic radiation from a distant source passing through a slit 0.105 mm wide. At the point in the pattern 3.25° from the center of the central maximum, the total phase difference between wavelets from the top and bottom of the slit is 56.0 rad. (a) What is the wavelength of the radiation? (b) What is the intensity at this point, if the intensity at the center of the central maximum is I_0?

36.18 • Consider a single-slit diffraction experiment in which the amplitude of the wave at point O in Fig. 36.5a is E_0. For each of the following cases, draw a phasor diagram like that in Fig. 36.8c and determine *graphically* the amplitude of the wave at the point in question. (*Hint:* Use Eq. (36.6) to determine the value of β for each case.) Compute the intensity and compare to Eq. (36.5). (a) $\sin\theta = \lambda/2a$; (b) $\sin\theta = \lambda/a$; (c) $\sin\theta = 3\lambda/2a$.

36.19 •• Public Radio station KXPR-FM in Sacramento broadcasts at 88.9 MHz. The radio waves pass between two tall skyscrapers that are 15.0 m apart along their closest walls. (a) At what horizontal angles, relative to the original direction of the waves, will a distant antenna not receive any signal from this station? (b) If the maximum intensity is 3.50 W/m² at the antenna, what is the intensity at $\pm 5.00°$ from the center of the central maximum at the distant antenna?

Section 36.4 Multiple Slits

36.20 •• **Diffraction and Interference Combined.** Consider the interference pattern produced by two parallel slits of width a and separation d, in which $d = 3a$. The slits are illuminated by normally incident light of wavelength λ. (a) First we ignore diffraction effects due to the slit width. At what angles θ from the central maximum will the next four maxima in the two-slit interference pattern occur? Your answer will be in terms of d and λ. (b) Now we include the effects of diffraction. If the intensity at $\theta = 0$ is I_0, what is the intensity at each of the angles in part (a)? (c) Which double-slit interference maxima are missing in the pattern? (d) Compare your results to those illustrated in Fig. 36.12c. In what ways is your result different?

36.21 • **Number of Fringes in a Diffraction Maximum.** In Fig. 36.12c the central diffraction maximum contains exactly seven

interference fringes, and in this case $d/a = 4$. (a) What must the ratio d/a be if the central maximum contains exactly five fringes? (b) In the case considered in part (a), how many fringes are contained within the first diffraction maximum on one side of the central maximum?

36.22 • An interference pattern is produced by eight parallel and equally spaced, narrow slits. There is an interference minimum when the phase difference ϕ between light from adjacent slits is $\pi/4$. The phasor diagram is given in Fig. 36.14b. For which pairs of slits is there totally destructive interference?

36.23 •• An interference pattern is produced by light of wavelength 580 nm from a distant source incident on two identical parallel slits separated by a distance (between centers) of 0.530 mm. (a) If the slits are very narrow, what would be the angular positions of the first-order and second-order, two-slit, interference maxima? (b) Let the slits have width 0.320 mm. In terms of the intensity I_0 at the center of the central maximum, what is the intensity at each of the angular positions in part (a)?

36.24 • Parallel rays of monochromatic light with wavelength 568 nm illuminate two identical slits and produce an interference pattern on a screen that is 75.0 cm from the slits. The centers of the slits are 0.640 mm apart and the width of each slit is 0.434 mm. If the intensity at the center of the central maximum is 5.00×10^{-4} W/m², what is the intensity at a point on the screen that is 0.900 mm from the center of the central maximum?

36.25 •• An interference pattern is produced by four parallel and equally spaced, narrow slits. By drawing appropriate phasor diagrams, show that there is an interference minimum when the phase difference ϕ from adjacent slits is (a) $\pi/2$; (b) π; (c) $3\pi/2$. In each case, for which pairs of slits is there totally destructive interference?

36.26 •• A diffraction experiment involving two thin parallel slits yields the pattern of closely spaced bright and dark fringes shown in Fig. E36.26. Only the central portion of the pattern is shown in the figure. The bright spots are equally spaced at 1.53 mm center to center (except for the missing spots) on a screen 2.50 m from the slits. The light source was a He-Ne laser producing a wavelength of 632.8 nm. (a) How far apart are the two slits? (b) How wide is each one?

Figure **E36.26**

36.27 •• Laser light of wavelength 500.0 nm illuminates two identical slits, producing an interference pattern on a screen 90.0 cm from the slits. The bright bands are 1.00 cm apart, and the third bright bands on either side of the central maximum are missing in the pattern. Find the width and the separation of the two slits.

Section 36.5 The Diffraction Grating

36.28 •• Monochromatic light is at normal incidence on a plane transmission grating. The first-order maximum in the interference pattern is at an angle of 8.94°. What is the angular position of the fourth-order maximum?

36.29 • If a diffraction grating produces its third-order bright band at an angle of 78.4° for light of wavelength 681 nm, find (a) the number of slits per centimeter for the grating and (b) the angular location of the first-order and second-order bright bands. (c) Will there be a fourth-order bright band? Explain.

36.30 • If a diffraction grating produces a third-order bright spot for red light (of wavelength 700 nm) at 65.0° from the central maximum, at what angle will the second-order bright spot be for violet light (of wavelength 400 nm)?

36.31 • Visible light passes through a diffraction grating that has 900 slits/cm, and the interference pattern is observed on a screen that is 2.50 m from the grating. (a) Is the angular position of the first-order spectrum small enough for $\sin\theta \approx \theta$ to be a good approximation? (b) In the first-order spectrum, the maxima for two different wavelengths are separated on the screen by 3.00 mm. What is the difference in these wavelengths?

36.32 • The wavelength range of the visible spectrum is approximately 380–750 nm. White light falls at normal incidence on a diffraction grating that has 350 slits/mm. Find the angular width of the visible spectrum in (a) the first order and (b) the third order. (*Note:* An advantage of working in higher orders is the greater angular spread and better resolution. A disadvantage is the overlapping of different orders, as shown in Example 36.4.)

36.33 • When laser light of wavelength 632.8 nm passes through a diffraction grating, the first bright spots occur at ±17.8° from the central maximum. (a) What is the line density (in lines/cm) of this grating? (b) How many additional bright spots are there beyond the first bright spots, and at what angles do they occur?

36.34 • (a) What is the wavelength of light that is deviated in the first order through an angle of 13.5° by a transmission grating having 5000 slits/cm? (b) What is the second-order deviation of this wavelength? Assume normal incidence.

36.35 • Plane monochromatic waves with wavelength 520 nm are incident normally on a plane transmission grating having 350 slits/mm. Find the angles of deviation in the first, second, and third orders.

36.36 • **Identifying Isotopes by Spectra.** Different isotopes of the same element emit light at slightly different wavelengths. A wavelength in the emission spectrum of a hydrogen atom is 656.45 nm; for deuterium, the corresponding wavelength is 656.27 nm. (a) What minimum number of slits is required to resolve these two wavelengths in second order? (b) If the grating has 500.00 slits/mm, find the angles and angular separation of these two wavelengths in the second order.

36.37 • A typical laboratory diffraction grating has 5.00×10^3 lines/cm, and these lines are contained in a 3.50-cm width of grating. (a) What is the chromatic resolving power of such a grating in the first order? (b) Could this grating resolve the lines of the sodium doublet (see Section 36.5) in the first order? (c) While doing spectral analysis of a star, you are using this grating in the *second* order to resolve spectral lines that are very close to the 587.8002-nm spectral line of iron. (i) For wavelengths longer than the iron line, what is the shortest wavelength you could distinguish from the iron line? (ii) For wavelengths shorter than the iron line, what is the longest wavelength you could distinguish from the iron line? (iii) What is the range of wavelengths you could *not* distinguish from the iron line?

36.38 • The light from an iron arc includes many different wavelengths. Two of these are at $\lambda = 587.9782$ nm and $\lambda = 587.8002$ nm. You wish to resolve these spectral lines in first order using a grating 1.20 cm in length. What minimum number of slits per centimeter must the grating have?

Section 36.6 X-Ray Diffraction

36.39 • X rays of wavelength 0.0850 nm are scattered from the atoms of a crystal. The second-order maximum in the Bragg reflection occurs when the angle θ in Fig. 36.22 is 21.5°. What is the spacing between adjacent atomic planes in the crystal?

36.40 • If the planes of a crystal are 3.50 Å (1 Å = 10^{-10} m = 1 Ångstrom unit) apart, (a) what wavelength of electromagnetic waves is needed so that the first strong interference maximum in the Bragg reflection occurs when the waves strike the planes at an angle of 15.0°, and in what part of the electromagnetic spectrum do these waves lie? (See Fig. 32.4.) (b) At what other angles will strong interference maxima occur?

36.41 • Monochromatic x rays are incident on a crystal for which the spacing of the atomic planes is 0.440 nm. The first-order maximum in the Bragg reflection occurs when the incident and reflected x rays make an angle of 39.4° with the crystal planes. What is the wavelength of the x rays?

Section 36.7 Circular Apertures and Resolving Power

36.42 • **BIO** If you can read the bottom row of your doctor's eye chart, your eye has a resolving power of 1 arcminute, equal to $\frac{1}{60}$ degree. If this resolving power is diffraction limited, to what effective diameter of your eye's optical system does this correspond? Use Rayleigh's criterion and assume $\lambda = 550$ nm.

36.43 • Two satellites at an altitude of 1200 km are separated by 28 km. If they broadcast 3.6-cm microwaves, what minimum receiving-dish diameter is needed to resolve (by Rayleigh's criterion) the two transmissions?

36.44 •• The VLBA (Very Long Baseline Array) uses a number of individual radio telescopes to make one unit having an equivalent diameter of about 8000 km. When this radio telescope is focusing radio waves of wavelength 2.0 cm, what would have to be the diameter of the mirror of a visible-light telescope focusing light of wavelength 550 nm so that the visible-light telescope has the same resolution as the radio telescope?

36.45 •• Monochromatic light with wavelength 620 nm passes through a circular aperture with diameter 7.4 μm. The resulting diffraction pattern is observed on a screen that is 4.5 m from the aperture. What is the diameter of the Airy disk on the screen?

36.46 • **Photography.** A wildlife photographer uses a moderate telephoto lens of focal length 135 mm and maximum aperture $f/4.00$ to photograph a bear that is 11.5 m away. Assume the wavelength is 550 nm. (a) What is the width of the smallest feature on the bear that this lens can resolve if it is opened to its maximum aperture? (b) If, to gain depth of field, the photographer stops the lens down to $f/22.0$, what would be the width of the smallest resolvable feature on the bear?

36.47 • **Observing Jupiter.** You are asked to design a space telescope for earth orbit. When Jupiter is 5.93×10^8 km away (its closest approach to the earth), the telescope is to resolve, by Rayleigh's criterion, features on Jupiter that are 250 km apart. What minimum-diameter mirror is required? Assume a wavelength of 500 nm.

36.48 • A converging lens 7.20 cm in diameter has a focal length of 300 mm. If the resolution is diffraction limited, how far away can an object be if points on it 4.00 mm apart are to be resolved (according to Rayleigh's criterion)? Use $\lambda = 550$ nm.

36.49 •• **Hubble Versus Arecibo.** The Hubble Space Telescope has an aperture of 2.4 m and focuses visible light (380–750 nm). The Arecibo radio telescope in Puerto Rico is 305 m (1000 ft) in diameter (it is built in a mountain valley) and focuses radio waves of wavelength 75 cm. (a) Under optimal viewing conditions, what is the smallest crater that each of these telescopes could resolve on our moon? (b) If the Hubble Space Telescope were to be converted to surveillance use, what is the highest orbit above the surface of the earth it could have and still be able to resolve the license plate (not the letters, just the plate) of a car on the ground? Assume optimal viewing conditions, so that the resolution is diffraction limited.

36.50 • **Searching for Starspots.** The Hale Telescope on Palomar Mountain in California has a mirror 200 in. (5.08 m) in diameter and it focuses visible light. Given that a large sunspot is about 10,000 mi in diameter, what is the most distant star on which this telescope could resolve a sunspot to see whether other stars have them? (Assume optimal viewing conditions, so that the resolution is diffraction limited.) Are there any stars this close to us, besides our sun?

PROBLEMS

36.51 •• **BIO** **Thickness of Human Hair.** Although we have discussed single-slit diffraction only for a slit, a similar result holds when light bends around a straight, thin object, such as a strand of hair. In that case, a is the width of the strand. From actual laboratory measurements on a human hair, it was found that when a beam of light of wavelength 632.8 nm was shone on a single strand of hair, and the diffracted light was viewed on a screen 1.25 m away, the first dark fringes on either side of the central bright spot were 5.22 cm apart. How thick was this strand of hair?

36.52 •• Suppose the entire apparatus (slit, screen, and space in between) in Exercise 36.4 is immersed in water ($n = 1.333$). Then what is the distance between the two dark fringes?

36.53 ••• Laser light of wavelength 632.8 nm falls normally on a slit that is 0.0250 mm wide. The transmitted light is viewed on a distant screen where the intensity at the center of the central bright fringe is 8.50 W/m². (a) Find the maximum number of totally dark fringes on the screen, assuming the screen is large enough to show them all. (b) At what angle does the dark fringe that is most distant from the center occur? (c) What is the maximum intensity of the bright fringe that occurs immediately before the dark fringe in part (b)? Approximate the angle at which this fringe occurs by assuming it is midway between the angles to the dark fringes on either side of it.

36.54 •• **CP** A loudspeaker having a diaphragm that vibrates at 1250 Hz is traveling at 80.0 m/s directly toward a pair of holes in a very large wall in a region for which the speed of sound is 344 m/s. You observe that the sound coming through the openings first cancels at $\pm 11.4°$ with respect to the original direction of the speaker when observed far from the wall. (a) How far apart are the two openings? (b) At what angles would the sound first cancel if the source stopped moving?

36.55 • **Measuring Refractive Index.** A thin slit illuminated by light of frequency f produces its first dark band at $\pm 38.2°$ in air. When the entire apparatus (slit, screen, and space in between) is immersed in an unknown transparent liquid, the slit's first dark bands occur instead at $\pm 21.6°$. Find the refractive index of the liquid.

36.56 • **Grating Design.** Your boss asks you to design a diffraction grating that will disperse the first-order visible spectrum through an angular range of 21.0° (see Example 36.4 in Section 36.5). (a) What must the number of slits per centimeter be for this grating? (b) At what angles will the first-order visible spectrum begin and end?

36.57 •• A slit 0.360 mm wide is illuminated by parallel rays of light that have a wavelength of 540 nm. The diffraction pattern is observed on a screen that is 1.20 m from the slit. The intensity at the center of the central maximum ($\theta = 0°$) is I_0. (a) What is the distance on the screen from the center of the central maximum to the first minimum? (b) What is the distance on the screen from the center of the central maximum to the point where the intensity has fallen to $I_0/2$?

36.58 ••• **CALC** The intensity of light in the Fraunhofer diffraction pattern of a single slit is

$$I = I_0 \left(\frac{\sin \gamma}{\gamma} \right)^2$$

where

$$\gamma = \frac{\pi a \sin \theta}{\lambda}$$

(a) Show that the equation for the values of γ at which I is a maximum is $\tan \gamma = \gamma$. (b) Determine the three smallest positive values of γ that are solutions of this equation. (*Hint:* You can use a trial-and-error procedure. Guess a value of γ and adjust your guess to bring $\tan \gamma$ closer to γ. A graphical solution of the equation is very helpful in locating the solutions approximately, to get good initial guesses.)

36.59 •• **Angular Width of a Principal Maximum.** Consider N evenly spaced, narrow slits. Use the small-angle approximation $\sin \theta = \theta$ (for θ in radians) to prove the following: For an intensity maximum that occurs at an angle θ, the intensity minima immediately adjacent to this maximum are at angles $\theta + \lambda/Nd$ and $\theta - \lambda/Nd$, so that the angular width of the principal maximum is $2\lambda/Nd$. This is proportional to $1/N$, as we concluded in Section 36.4 on the basis of energy conservation.

36.60 •• **CP** **CALC** In a large vacuum chamber, monochromatic laser light passes through a narrow slit in a thin aluminum plate and forms a diffraction pattern on a screen that is 0.620 m from the slit. When the aluminum plate has a temperature of 20.0°C, the width of the central maximum in the diffraction pattern is 2.75 mm. What is the change in the width of the central maximum when the temperature of the plate is raised to 520.0°C? Does the width of the central diffraction maximum increase or decrease when the temperature is increased?

36.61 • **Phasor Diagram for Eight Slits.** An interference pattern is produced by eight equally spaced, narrow slits. Figure 36.14 shows phasor diagrams for the cases in which the phase difference ϕ between light from adjacent slits is $\phi = \pi$, $\phi = \pi/4$, and $\phi = \pi/2$. Each of these cases gives an intensity minimum. The caption for Fig. 36.14 also claims that minima occur for $\phi = 3\pi/4$, $\phi = 5\pi/4$, $\phi = 3\pi/2$, and $\phi = 7\pi/4$. (a) Draw the phasor diagram for each of these four cases, and explain why each diagram proves that there is in fact a minimum. (*Note:* You may find it helpful to use a different colored pencil for each slit!) (b) For each of the four cases $\phi = 3\pi/4$, $\phi = 5\pi/4$, $\phi = 3\pi/2$, and $\phi = 7\pi/4$, for which pairs of slits is there totally destructive interference?

36.62 •• **CP** In a laboratory, light from a particular spectrum line of helium passes through a diffraction grating and the second-order maximum is at 18.9° from the center of the central bright fringe. The same grating is then used for light from a distant galaxy that is moving away from the earth with a speed of 2.65×10^7 m/s. For the light from the galaxy, what is the angular location of the second-order maximum for the same spectral line as was observed in the lab? (See Section 16.8.)

36.63 • What is the longest wavelength that can be observed in the third order for a transmission grating having 9200 slits/cm? Assume normal incidence.

36.64 •• (a) Figure 36.16 shows plane waves of light incident *normally* on a diffraction grating. If instead the light strikes the grating at an angle of incidence θ' (measured from the normal), show that the condition for an intensity maximum is *not* Eq. (36.13), but rather

$$d(\sin \theta + \sin \theta') = m\lambda \quad (m = 0, \pm 1, \pm 2, \pm 3, \dots)$$

(b) For the grating described in Example 36.4 (Section 36.5), with 600 slits/mm, find the angles of the maxima corresponding to $m = 0$, 1, and -1 with red light ($\lambda = 650$ nm) for the cases $\theta' = 0$ (normal incidence) and $\theta' = 20.0°$.

36.65 • A diffraction grating has 650 slits/mm. What is the highest order that contains the entire visible spectrum? (The wavelength range of the visible spectrum is approximately 380–750 nm.)

36.66 •• *Quasars,* an abbreviation for *quasi-stellar radio sources,* are distant objects that look like stars through a telescope but that emit far more electromagnetic radiation than an entire normal galaxy of stars. An example is the bright object below and to the left of center in Fig. P36.66; the other elongated objects in this image are normal galaxies. The leading model for the structure of a quasar is a galaxy with a supermassive black hole at its center. In this model, the radiation is emitted by interstellar gas and dust within the galaxy as this material falls toward the black hole. The radiation is thought to emanate from a region just a few light-years in diameter. (The diffuse glow surrounding the bright quasar shown in Fig. P36.66 is thought to be this quasar's host galaxy.) To investigate this model of quasars and to study other exotic astronomical objects, the Russian Space Agency plans to place a radio telescope in an orbit that extends to 77,000 km from the earth. When the signals from this telescope are combined with signals from the ground-based telescopes of the VLBA, the resolution will be that of a single radio telescope 77,000 km in diameter. What is the size of the smallest detail that this arrangement could resolve in quasar 3C 405, which is 7.2×10^8 light-years from earth, using radio waves at a frequency of 1665 MHz? (*Hint:* Use Rayleigh's criterion.) Give your answer in light-years and in kilometers.

Figure **P36.66**

36.67 ••• **Phased-Array Radar.** In one common type of radar installation, a rotating antenna sweeps a radio beam around the sky. But in a *phased-array* radar system, the antennas remain stationary and the beam is swept electronically. To see how this is done, consider an array of N antennas that are arranged along the horizontal x-axis at $x = 0$, $\pm d$, $\pm 2d$, . . . , $\pm(N - 1)d/2$. (The number N is odd.) Each antenna emits radiation uniformly in all directions in the horizontal xy-plane. The antennas all emit radiation coherently, with the same amplitude E_0 and the same wavelength λ. The relative phase δ of the emission from adjacent antennas can be varied, however. If the antenna at $x = 0$ emits a signal that is given by $E_0 \cos \omega t$, as measured at a point next to the antenna, the antenna at $x = d$ emits a signal given by $E_0 \cos(\omega t + \delta)$, as measured at a point next to that antenna. The corresponding quantity for the antenna at $x = -d$ is $E_0 \cos(\omega t - \delta)$; for the antennas at $x = \pm 2d$, it is $E_0 \cos(\omega t \pm 2\delta)$; and so on. (a) If $\delta = 0$, the inter-

ference pattern at a distance from the antennas is large compared to d and has a principal maximum at $\theta = 0$ (that is, in the $+y$-direction, perpendicular to the line of the antennas). Show that if $d < \lambda$, this is the *only* principal interference maximum in the angular range $-90° < \theta < 90°$. Hence this principal maximum describes a beam emitted in the direction $\theta = 0$. As described in Section 36.4, if N is large, the beam will have a large intensity and be quite narrow. (b) If $\delta \neq 0$, show that the principal intensity maximum described in part (a) is located at

$$\theta = \arcsin\left(\frac{\delta\lambda}{2\pi d}\right)$$

where δ is measured in radians. Thus, by varying δ from positive to negative values and back again, which can easily be done electronically, the beam can be made to sweep back and forth around $\theta = 0$. (c) A weather radar unit to be installed on an airplane emits radio waves at 8800 MHz. The unit uses 15 antennas in an array 28.0 cm long (from the antenna at one end of the array to the antenna at the other end). What must the maximum and minimum values of δ be (that is, the most positive and most negative values) if the radar beam is to sweep 45° to the left or right of the airplane's direction of flight? Give your answer in radians.

36.68 •• **Underwater Photography.** An underwater camera has a lens of focal length 35.0 mm and a maximum aperture of $f/2.80$. The film it uses has an emulsion that is sensitive to light of frequency 6.00×10^{14} Hz. If the photographer takes a picture of an object 2.75 m in front of the camera with the lens wide open, what is the width of the smallest resolvable detail on the subject if the object is (a) a fish underwater with the camera in the water and (b) a person on the beach with the camera out of the water?

36.69 •• An astronaut in the space shuttle can just resolve two point sources on earth that are 65.0 m apart. Assume that the resolution is diffraction limited and use Rayleigh's criterion. What is the astronaut's altitude above the earth? Treat his eye as a circular aperture with a diameter of 4.00 mm (the diameter of his pupil), and take the wavelength of the light to be 550 nm. Ignore the effect of fluid in the eye.

36.70 •• **BIO Resolution of the Eye.** The maximum resolution of the eye depends on the diameter of the opening of the pupil (a diffraction effect) and the size of the retinal cells. The size of the retinal cells (about 5.0 μm in diameter) limits the size of an object at the near point (25 cm) of the eye to a height of about 50 μm. (To get a reasonable estimate without having to go through complicated calculations, we shall ignore the effect of the fluid in the eye.) (a) Given that the diameter of the human pupil is about 2.0 mm, does the Rayleigh criterion allow us to resolve a 50-μm-tall object at 25 cm from the eye with light of wavelength 550 nm? (b) According to the Rayleigh criterion, what is the shortest object we could resolve at the 25-cm near point with light of wavelength 550 nm? (c) What angle would the object in part (b) subtend at the eye? Express your answer in minutes (60 min = 1°), and compare it with the experimental value of about 1 min. (d) Which effect is more important in limiting the resolution of our eyes: diffraction or the size of the retinal cells?

36.71 •• A glass sheet is covered by a very thin opaque coating. In the middle of this sheet there is a thin scratch 0.00125 mm thick. The sheet is totally immersed beneath the surface of a liquid. Parallel rays of monochromatic coherent light with wavelength 612 nm in air strike the sheet perpendicular to its surface and pass through the scratch. A screen is placed in the liquid a distance of 30.0 cm away from the sheet and parallel to it. You observe that the first dark fringes on either side of the central bright fringe on the screen are 22.4 cm apart. What is the refractive index of the liquid?

36.72 •• **Observing Planets Beyond Our Solar System.** NASA is considering a project called *Planet Imager* that would give astronomers the ability to see details on planets orbiting other stars. Using the same principle as the Very Large Array (see Section 36.7), *Planet Imager* will use an array of infrared telescopes spread over thousands of kilometers of space. (Visible light would give even better resolution. Unfortunately, at visible wavelengths, stars are so bright that a planet would be lost in the glare. This is less of a problem at infrared wavelengths.) (a) If *Planet Imager* has an effective diameter of 6000 km and observes infrared radiation at a wavelength of 10 μm, what is the greatest distance at which it would be able to observe details as small as 250 km across (about the size of the greater Los Angeles area) on a planet? Give your answer in light-years (see Appendix E). (*Hint:* Use Rayleigh's criterion.) (b) For comparison, consider the resolution of a single infrared telescope in space that has a diameter of 1.0 m and that observes 10-μm radiation. What is the size of the smallest details that such a telescope could resolve at the distance of the nearest star to the sun, Proxima Centauri, which is 4.22 light-years distant? How does this compare to the diameter of the earth $(1.27 \times 10^4 \text{ km})$? To the average distance from the earth to the sun $(1.50 \times 10^8 \text{ km})$? Would a single telescope of this kind be able to detect the presence of a planet like the earth, in an orbit the size of the earth's orbit, around *any* other star? Explain. (c) Suppose *Planet Imager* is used to observe a planet orbiting the star 70 Virginis, which is 59 light-years from our solar system. A planet (though not an earthlike one) has in fact been detected orbiting this star, not by imaging it directly but by observing the slight "wobble" of the star as both it and the planet orbit their common center of mass. What is the size of the smallest details that *Planet Imager* could hope to resolve on the planet of 70 Virginis? How does this compare to the diameter of the planet, assumed to be comparable to that of Jupiter $(1.38 \times 10^5 \text{ km})$? (Although the planet of 70 Virginis is thought to be at least 6.6 times more massive than Jupiter, its radius is probably not too different from that of Jupiter. The reason is that such large planets are thought to be composed primarily of gases, not rocky material, and hence can be greatly compressed by the mutual gravitational attraction of different parts of the planet.)

CHALLENGE PROBLEMS

36.73 ••• **CALC** It is possible to calculate the intensity in the single-slit Fraunhofer diffraction pattern *without* using the phasor method of Section 36.3. Let y' represent the position of a point within the slit of width a in Fig. 36.5a, with $y' = 0$ at the center of the slit so that the slit extends from $y' = -a/2$ to $y' = a/2$. We imagine dividing the slit up into infinitesimal strips of width dy', each of which acts as a source of secondary wavelets. (a) The amplitude of the total wave at the point O on the distant screen in Fig. 36.5a is E_0. Explain why the amplitude of the wavelet from each infinitesimal strip within the slit is $E_0(dy'/a)$, so that the electric field of the wavelet a distance x from the infinitesimal strip is $dE = E_0(dy'/a) \sin(kx - \omega t)$. (b) Explain why the wavelet from each strip as detected at point P in Fig. 36.5a can be expressed as

$$dE = E_0 \frac{dy'}{a} \sin[k(D - y' \sin\theta) - \omega t]$$

where D is the distance from the center of the slit to point P and $k = 2\pi/\lambda$. (c) By integrating the contributions dE from all parts of the slit, show that the total wave detected at point P is

$$E = E_0 \sin(kD - \omega t) \frac{\sin[ka(\sin\theta)/2]}{ka(\sin\theta)/2}$$

$$= E_0 \sin(kD - \omega t) \frac{\sin[\pi a(\sin\theta)/\lambda]}{\pi a(\sin\theta)/\lambda}$$

(The trigonometric identities in Appendix B will be useful.) Show that at $\theta = 0$, corresponding to point O in Fig. 36.5a, the wave is $E = E_0 \sin(kD - \omega t)$ and has amplitude E_0, as stated in part (a). (d) Use the result of part (c) to show that if the intensity at point O is I_0, then the intensity at a point P is given by Eq. (36.7).

36.74 ••• **Intensity Pattern of N Slits.** (a) Consider an arrangement of N slits with a distance d between adjacent slits. The slits emit coherently and in phase at wavelength λ. Show that at a time t, the electric field at a distant point P is

$$E_P(t) = E_0 \cos(kR - \omega t) + E_0 \cos(kR - \omega t + \phi)$$
$$+ E_0 \cos(kR - \omega t + 2\phi) + \cdots$$
$$+ E_0 \cos(kR - \omega t + (N - 1)\phi)$$

where E_0 is the amplitude at P of the electric field due to an individual slit, $\phi = (2\pi d \sin\theta)/\lambda$, θ is the angle of the rays reaching P (as measured from the perpendicular bisector of the slit arrangement), and R is the distance from P to the most distant slit. In this problem, assume that R is much larger than d. (b) To carry out the sum in part (a), it is convenient to use the complex-number relationship

$$e^{iz} = \cos z + i \sin z$$

where $i = \sqrt{-1}$. In this expression, $\cos z$ is the *real part* of the complex number e^{iz}, and $\sin z$ is its *imaginary part*. Show that the electric field $E_P(t)$ is equal to the real part of the complex quantity

$$\sum_{n=0}^{N-1} E_0 e^{i(kR - \omega t + n\phi)}$$

(c) Using the properties of the exponential function that $e^A e^B = e^{(A+B)}$ and $(e^A)^n = e^{nA}$, show that the sum in part (b) can be written as

$$E_0 \left(\frac{e^{iN\phi} - 1}{e^{i\phi} - 1} \right) e^{i(kR - \omega t)}$$

$$= E_0 \left(\frac{e^{iN\phi/2} - e^{-iN\phi/2}}{e^{i\phi/2} - e^{-i\phi/2}} \right) e^{i[kR - \omega t + (N-1)\phi/2]}$$

Then, using the relationship $e^{iz} = \cos z + i \sin z$, show that the (real) electric field at point P is

$$E_P(t) = \left[E_0 \frac{\sin(N\phi/2)}{\sin(\phi/2)} \right] \cos[kR - \omega t + (N-1)\phi/2]$$

The quantity in the first square brackets in this expression is the amplitude of the electric field at P. (d) Use the result for the electric-field amplitude in part (c) to show that the intensity at an angle θ is

$$I = I_0 \left[\frac{\sin(N\phi/2)}{\sin(\phi/2)} \right]^2$$

where I_0 is the maximum intensity for an individual slit. (e) Check the result in part (d) for the case $N = 2$. It will help to recall that $\sin 2A = 2 \sin A \cos A$. Explain why your result differs from Eq. (35.10), the expression for the intensity in two-source interference, by a factor of 4. (*Hint:* Is I_0 defined in the same way in both expressions?)

36.75 ••• **CALC** **Intensity Pattern of *N* Slits, Continued.** Part (d) of Challenge Problem 36.74 gives an expression for the intensity in the interference pattern of N identical slits. Use this result to verify the following statements. (a) The maximum intensity in the pattern is $N^2 I_0$. (b) The principal maximum at the center of the pattern extends from $\phi = -2\pi/N$ to $\phi = 2\pi/N$, so its width is inversely proportional to $1/N$. (c) A minimum occurs whenever ϕ is an integral multiple of $2\pi/N$, except when ϕ is an integral multiple of 2π (which gives a principal maximum). (d) There are $(N - 1)$ minima between each pair of principal maxima. (e) Halfway between two principal maxima, the intensity can be no greater than I_0; that is, it can be no greater than $1/N^2$ times the intensity at a principal maximum.

Answers

Chapter Opening Question

The shorter wavelength of a Blu-ray scanning laser gives it superior resolving power, so information can be more tightly packed onto a Blu-ray disc than a DVD. See Section 36.7 for details.

Test Your Understanding Questions

36.1 Answer: yes When you hear the voice of someone standing around a corner, you are hearing sound waves that underwent diffraction. If there were no diffraction of sound, you could hear sounds only from objects that were in plain view.

36.2 Answer: (ii), (i) and (iv) (tie), (iii) The angle θ of the first dark fringe is given by Eq. (36.2) with $m = 1$, or $\sin\theta = \lambda/a$. The larger the value of the ratio λ/a, the larger the value of $\sin\theta$ and hence the value of θ. The ratio λ/a in each case is (i) $(400\text{ nm})/(0.20\text{ mm}) = (4.0 \times 10^{-7}\text{ m})/(2.0 \times 10^{-4}\text{ m}) = 2.0 \times 10^{-3}$; (ii) $(600\text{ nm})/(0.20\text{ mm}) = (6.0 \times 10^{-7}\text{ m})/(2.0 \times 10^{-4}\text{ m}) = 3.0 \times 10^{-3}$; (iii) $(400\text{ nm})/(0.30\text{ mm}) = (4.0 \times 10^{-7}\text{ m})/(3.0 \times 10^{-4}\text{ m}) = 1.3 \times 10^{-3}$; (iv) $(600\text{ nm})/(0.30\text{ mm}) = (6.0 \times 10^{-7}\text{ m})/(3.0 \times 10^{-4}\text{ m}) = 2.0 \times 10^{-3}$.

36.3 Answers: (ii) and (iii) If the slit width a is less than the wavelength λ, there are no points in the diffraction pattern at which the intensity is zero (see Fig. 36.10a). The slit width is $0.0100\text{ mm} = 1.00 \times 10^{-5}\text{ m}$, so this condition is satisfied for (ii) ($\lambda = 10.6\ \mu\text{m} = 1.06 \times 10^{-5}\text{ m}$) and (iii) ($\lambda = 1.00\text{ mm} = 1.00 \times 10^{-3}\text{ m}$) but not for (i) ($\lambda = 500\text{ nm} = 5.00 \times 10^{-7}\text{ m}$) or (iv) ($\lambda = 50.0\text{ nm} = 5.00 \times 10^{-8}\text{ m}$).

36.4 Answers: yes; $m_i = \pm 5, \pm 10, \ldots$ A "missing maximum" satisfies both $d\sin\theta = m_i\lambda$ (the condition for an interference maximum) and $a\sin\theta = m_d\lambda$ (the condition for a diffraction minimum). Substituting $d = 2.5a$, we can combine these two conditions into the relationship $m_i = 2.5m_d$. This is satisfied for $m_i = \pm 5$ and $m_d = \pm 2$ (the fifth interference maximum is missing because it coincides with the second diffraction minimum), $m_i = \pm 10$ and $m_d = \pm 4$ (the tenth interference maximum is missing because it coincides with the fourth diffraction minimum), and so on.

36.5 Answer: (i) As described in the text, the resolving power needed is $R = Nm = 1000$. In the first order ($m = 1$) we need $N = 1000$ slits, but in the fourth order ($m = 4$) we need only $N = R/m = 1000/4 = 250$ slits. (These numbers are only approximate because of the arbitrary nature of our criterion for resolution and because real gratings always have slight imperfections in the shapes and spacings of the slits.)

36.6 Answer: no The angular position of the mth maximum is given by Eq. (36.16), $2d\sin\theta = m\lambda$. With $d = 0.200\text{ nm}$, $\lambda = 0.100\text{ nm}$, and $m = 5$, this gives $\sin\theta = m\lambda/2d = (5)(0.100\text{ nm})/(2)(0.200\text{ nm}) = 1.25$. Since the sine function can never be greater than 1, this means that there is no solution to this equation and the $m = 5$ maximum does not appear.

36.7 Answer: (iii), (ii), (iv), (i) Rayleigh's criterion combined with Eq. (36.17) shows that the smaller the value of the ratio λ/D, the better the resolving power of a telescope of diameter D. For the four telescopes, this ratio is equal to (i) $(21\text{ cm})/(100\text{ m}) = (0.21\text{ m})/(100\text{ m}) = 2.1 \times 10^{-3}$; (ii) $(500\text{ nm})/(2.0\text{ m}) = (5.0 \times 10^{-7}\text{ m})/(2.0\text{ m}) = 2.5 \times 10^{-7}$; (iii) $(100\text{ nm})/(1.0\text{ m}) = (1.0 \times 10^{-7}\text{ m})/(1.0\text{ m}) = 1.0 \times 10^{-7}$; (iv) $(10\ \mu\text{m})/(2.0\text{ m}) = (1.0 \times 10^{-5}\text{ m})/(2.0\text{ m}) = 5.0 \times 10^{-6}$.

Bridging Problem

Answers: $1.501 \times 10^7\text{ m/s}$ or 5.00% of c; away from us

RELATIVITY

? At Brookhaven National Laboratory in New York, atomic nuclei are accelerated to 99.995% of the ultimate speed limit of the universe—the speed of light. Is there also an upper limit on the *kinetic energy* of a particle?

LEARNING GOALS

By studying this chapter, you will learn:

- The two postulates of Einstein's special theory of relativity, and what motivates these postulates.

- Why different observers can disagree about whether two events are simultaneous.

- How relativity predicts that moving clocks run slow, and that experimental evidence confirms this.

- How the length of an object changes due to the object's motion.

- How the velocity of an object depends on the frame of reference from which it is observed.

- How the theory of relativity modifies the relationship between velocity and momentum.

- How to solve problems involving work and kinetic energy for particles moving at relativistic speeds.

- Some of the key concepts of Einstein's general theory of relativity.

Whe the year 1905 began, Albert Einstein was an unknown 25-year-old clerk in the Swiss patent office. By the end of that amazing year he had published three papers of extraordinary importance. One was an analysis of Brownian motion; a second (for which he was awarded the Nobel Prize) was on the photoelectric effect. In the third, Einstein introduced his **special theory of relativity,** proposing drastic revisions in the Newtonian concepts of space and time.

The special theory of relativity has made wide-ranging changes in our understanding of nature, but Einstein based it on just two simple postulates. One states that the laws of physics are the same in all inertial frames of reference; the other states that the speed of light in vacuum is the same in all inertial frames. These innocent-sounding propositions have far-reaching implications. Here are three: (1) Events that are simultaneous for one observer may not be simultaneous for another. (2) When two observers moving relative to each other measure a time interval or a length, they may not get the same results. (3) For the conservation principles for momentum and energy to be valid in all inertial systems, Newton's second law and the equations for momentum and kinetic energy have to be revised.

Relativity has important consequences in *all* areas of physics, including electromagnetism, atomic and nuclear physics, and high-energy physics. Although many of the results derived in this chapter may run counter to your intuition, the theory is in solid agreement with experimental observations.

37.1 Invariance of Physical Laws

Let's take a look at the two postulates that make up the special theory of relativity. Both postulates describe what is seen by an observer in an *inertial frame of reference,* which we introduced in Section 4.2. The theory is "special" in the sense that it applies to observers in such special reference frames.

Einstein's First Postulate

Einstein's first postulate, called the **principle of relativity,** states: **The laws of physics are the same in every inertial frame of reference.** If the laws differed, that difference could distinguish one inertial frame from the others or make one frame somehow more "correct" than another. Here are two examples. Suppose you watch two children playing catch with a ball while the three of you are aboard a train moving with constant velocity. Your observations of the motion *of the ball,* no matter how carefully done, can't tell you how fast (or whether) the train is moving. This is because Newton's laws of motion are the same in every inertial frame.

Another example is the electromotive force (emf) induced in a coil of wire by a nearby moving permanent magnet. In the frame of reference in which the *coil* is stationary (Fig. 37.1a), the moving magnet causes a change of magnetic flux through the coil, and this induces an emf. In a different frame of reference in which the *magnet* is stationary (Fig. 37.1b), the motion of the coil through a magnetic field induces the emf. According to the principle of relativity, both of these frames of reference are equally valid. Hence the same emf must be induced in both situations shown in Fig. 37.1. As we saw in Chapter 29, this is indeed the case, so Faraday's law is consistent with the principle of relativity. Indeed, *all* of the laws of electromagnetism are the same in every inertial frame of reference.

Equally significant is the prediction of the speed of electromagnetic radiation, derived from Maxwell's equations (see Section 32.2). According to this analysis, light and all other electromagnetic waves travel in vacuum with a constant speed, now defined to equal exactly 299,792,458 m/s. (We often use the approximate value $c = 3.00 \times 10^8$ m/s, which is within one part in 1000 of the exact value.) As we will see, the speed of light in vacuum plays a central role in the theory of relativity.

Einstein's Second Postulate

During the 19th century, most physicists believed that light traveled through a hypothetical medium called the *ether,* just as sound waves travel through air. If so, the speed of light measured by observers would depend on their motion relative to the ether and would therefore be different in different directions. The Michelson-Morley experiment, described in Section 35.5, was an effort to detect motion of the earth relative to the ether. Einstein's conceptual leap was to recognize that if Maxwell's equations are valid in all inertial frames, then the speed of light in vacuum should also be the same in all frames and in all directions. In fact, Michelson and Morley detected *no* ether motion across the earth, and the ether concept has been discarded. Although Einstein may not have known about this negative result, it supported his bold hypothesis of the constancy of the speed of light in vacuum.

> **Einstein's second postulate states: The speed of light in vacuum is the same in all inertial frames of reference and is independent of the motion of the source.**

Let's think about what this means. Suppose two observers measure the speed of light in vacuum. One is at rest with respect to the light source, and the other is moving away from it. Both are in inertial frames of reference. According to the principle of relativity, the two observers must obtain the same result, despite the fact that one is moving with respect to the other.

If this seems too easy, consider the following situation. A spacecraft moving past the earth at 1000 m/s fires a missile straight ahead with a speed of 2000 m/s (relative to the spacecraft) (Fig. 37.2). What is the missile's speed relative to the earth? Simple, you say; this is an elementary problem in relative velocity (see Section 3.5). The correct answer, according to Newtonian mechanics, is 3000 m/s.

37.1 The same emf is induced in the coil whether **(a)** the magnet moves relative to the coil or **(b)** the coil moves relative to the magnet.

(a)   (b)

Magnet moves ...

$\vec{v}$

Coil moves ...

$\vec{v}$

... same result

37.2 (a) Newtonian mechanics makes correct predictions about relatively slow-moving objects; (b) it makes incorrect predictions about the behavior of light.

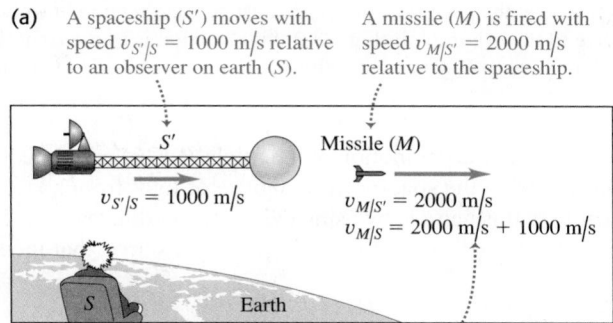

(a)   A spaceship (S') moves with   A missile (M) is fired with
speed $v_{S'/S} = 1000$ m/s relative   speed $v_{M/S'} = 2000$ m/s
to an observer on earth (S).   relative to the spaceship.

NEWTONIAN MECHANICS HOLDS: Newtonian mechanics tells us correctly that the missile moves with speed $v_{M/S} = 3000$ m/s relative to the observer on earth.

(b)

A light beam (L) is emitted from the spaceship at speed c.

NEWTONIAN MECHANICS FAILS: Newtonian mechanics tells us *incorrectly* that the light moves at a speed greater than c relative to the observer on earth ... which would contradict Einstein's second postulate.

But now suppose the spacecraft turns on a searchlight, pointing in the same direction in which the missile was fired. An observer on the spacecraft measures the speed of light emitted by the searchlight and obtains the value c. According to Einstein's second postulate, the motion of the light after it has left the source cannot depend on the motion of the source. So the observer on earth who measures the speed of this same light must also obtain the value c, *not* $c + 1000$ m/s. This result contradicts our elementary notion of relative velocities, and it may not appear to agree with common sense. But "common sense" is intuition based on everyday experience, and this does not usually include measurements of the speed of light.

The Ultimate Speed Limit

Einstein's second postulate immediately implies the following result:

> **It is impossible for an inertial observer to travel at c, the speed of light in vacuum.**

We can prove this by showing that travel at c implies a logical contradiction. Suppose that the spacecraft S' in Fig. 37.2b is moving at the speed of light relative to an observer on the earth, so that $v_{S'/S} = c$. If the spacecraft turns on a headlight, the second postulate now asserts that the earth observer S measures the headlight beam to be also moving at c. Thus this observer measures that the headlight beam and the spacecraft move together and are always at the same point in space. But Einstein's second postulate also asserts that the headlight beam moves at a speed c relative to the spacecraft, so they *cannot* be at the same point in space. This contradictory result can be avoided only if it is impossible for an inertial observer, such as a passenger on the spacecraft, to move at c. As we go through our discussion of relativity, you may find yourself asking the question Einstein asked himself as a 16-year-old student, "What would I see if I were traveling at the speed of light?" Einstein realized only years later that his question's basic flaw was that he could *not* travel at c.

The Galilean Coordinate Transformation

Let's restate this argument symbolically, using two inertial frames of reference, labeled S for the observer on earth and S' for the moving spacecraft, as shown in Fig. 37.3. To keep things as simple as possible, we have omitted the z-axes. The x-axes of the two frames lie along the same line, but the origin O' of frame S' moves relative to the origin O of frame S with constant velocity u along the common x-x'-axis. We on earth set our clocks so that the two origins coincide at time $t = 0$, so their separation at a later time t is ut.

37.3 The position of particle P can be described by the coordinates x and y in frame of reference S or by x' and y' in frame S'.

Frame S' moves relative to frame S with constant velocity u along the common x-x'-axis.

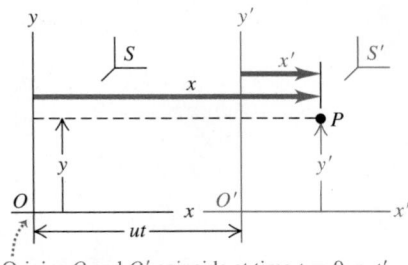

Origins O and O' coincide at time $t = 0 = t'$.

CAUTION **Choose your inertial frame coordinates wisely** Many of the equations derived in this chapter are true *only* if you define your inertial reference frames as stated in the preceding paragraph. For instance, the positive x-direction must be the direction in which the origin O' moves relative to the origin O. In Fig. 37.3 this direction is to the right; if instead O' moves to the left relative to O, you must define the positive x-direction to be to the left. █

Now think about how we describe the motion of a particle P. This might be an exploratory vehicle launched from the spacecraft or a pulse of light from a laser. We can describe the *position* of this particle by using the earth coordinates (x, y, z) in S or the spacecraft coordinates (x', y', z') in S'. Figure 37.3 shows that these are simply related by

$$x = x' + ut \qquad y = y' \qquad z = z' \qquad \text{(Galilean coordinate transformation)} \qquad (37.1)$$

These equations, based on the familiar Newtonian notions of space and time, are called the **Galilean coordinate transformation.**

If particle P moves in the x-direction, its instantaneous velocity v_x as measured by an observer stationary in S is $v_x = dx/dt$. Its velocity v_x' as measured by an observer stationary in S' is $v_x' = dx'/dt$. We can derive a relationship between v_x and v_x' by taking the derivative with respect to t of the first of Eqs. (37.1):

$$\frac{dx}{dt} = \frac{dx'}{dt} + u$$

Now dx/dt is the velocity v_x measured in S, and dx'/dt is the velocity v_x' measured in S', so we get the *Galilean velocity transformation* for one-dimensional motion:

$$v_x = v_x' + u \qquad \text{(Galilean velocity transformation)} \qquad (37.2)$$

Although the notation differs, this result agrees with our discussion of relative velocities in Section 3.5.

Now here's the fundamental problem. Applied to the speed of light in vacuum, Eq. (37.2) says that $c = c' + u$. Einstein's second postulate, supported subsequently by a wealth of experimental evidence, says that $c = c'$. This is a genuine inconsistency, not an illusion, and it demands resolution. If we accept this postulate, we are forced to conclude that Eqs. (37.1) and (37.2) *cannot* be precisely correct, despite our convincing derivation. These equations have to be modified to bring them into harmony with this principle.

The resolution involves some very fundamental modifications in our kinematic concepts. The first idea to be changed is the seemingly obvious assumption that the observers in frames S and S' use the same *time scale,* formally stated as $t = t'$. Alas, we are about to show that this everyday assumption cannot be correct; the two observers *must* have different time scales. We must define the velocity v' in frame S' as $v' = dx'/dt'$, not as dx'/dt; the two quantities are not the same. The difficulty lies in the concept of *simultaneity,* which is our next topic. A careful analysis of simultaneity will help us develop the appropriate modifications of our notions about space and time.

Test Your Understanding of Section 37.1 As a high-speed spaceship flies past you, it fires a strobe light that sends out a pulse of light in all directions. An observer aboard the spaceship measures a spherical wave front that spreads away from the spaceship with the same speed c in all directions. (a) What is the shape of the wave front that *you* measure? (i) spherical; (ii) ellipsoidal, with the longest axis of the ellipsoid along the direction of the spaceship's motion; (iii) ellipsoidal, with the shortest axis of the ellipsoid along the direction of the spaceship's motion; (iv) not enough information is given to decide. (b) Is the wave front centered on the spaceship? █

37.2 Relativity of Simultaneity

Measuring times and time intervals involves the concept of **simultaneity.** In a given frame of reference, an **event** is an occurrence that has a definite position and time (Fig. 37.4). When you say that you awoke at seven o'clock, you mean that two events (your awakening and your clock showing 7:00) occurred *simultaneously.* The fundamental problem in measuring time intervals is this: In general, two events that are simultaneous in one frame of reference are *not* simultaneous in a second frame that is moving relative to the first, even if both are inertial frames.

A Thought Experiment in Simultaneity

This may seem to be contrary to common sense. To illustrate the point, here is a version of one of Einstein's *thought experiments*—mental experiments that follow concepts to their logical conclusions. Imagine a train moving with a speed comparable to c, with uniform velocity (Fig. 37.5). Two lightning bolts strike a passenger car, one near each end. Each bolt leaves a mark on the car and one on the ground at the instant the bolt hits. The points on the ground are labeled A and B in the figure, and the corresponding points on the car are A' and B'. Stanley is stationary on the ground at O, midway between A and B. Mavis is moving with the train at O' in the middle of the passenger car, midway between A' and B'. Both Stanley and Mavis see both light flashes emitted from the points where the lightning strikes.

Suppose the two wave fronts from the lightning strikes reach Stanley at O simultaneously. He knows that he is the same distance from B and A, so Stanley concludes that the two bolts struck B and A simultaneously. Mavis agrees that the two wave fronts reached Stanley at the same time, but she disagrees that the flashes were emitted simultaneously.

Stanley and Mavis agree that the two wave fronts do not reach Mavis at the same time. Mavis at O' is moving to the right with the train, so she runs into the wave front from B' *before* the wave front from A' catches up to her. However, because she is in the middle of the passenger car equidistant from A' and B', her observation is that both wave fronts took the same time to reach her because both moved the same distance at the same speed c. (Recall that the speed of each wave front with respect to *either* observer is c.) Thus she concludes that the lightning bolt at B' struck *before* the one at A'. Stanley at O measures the two events to be simultaneous, but Mavis at O' does not! *Whether or not two events at different x-axis locations are simultaneous depends on the state of motion of the observer.*

You may want to argue that in this example the lightning bolts really *are* simultaneous and that if Mavis at O' could communicate with the distant points without the time delay caused by the finite speed of light, she would realize this. But that would be erroneous; the finite speed of information transmission is not the real issue. If O' is midway between A' and B', then in her frame of reference the time for a signal to travel from A' to O' is the same as that from B' to O'. Two signals arrive simultaneously at O' only if they were emitted simultaneously at A' and B'. In this example they *do not* arrive simultaneously at O', and so Mavis must conclude that the events at A' and B' were *not* simultaneous.

Furthermore, there is no basis for saying that Stanley is right and Mavis is wrong, or vice versa. According to the principle of relativity, no inertial frame of reference is more correct than any other in the formulation of physical laws. Each observer is correct *in his or her own frame of reference.* In other words, simultaneity is not an absolute concept. Whether two events are simultaneous depends on the frame of reference. As we mentioned at the beginning of this section, simultaneity plays an essential role in measuring time intervals. It follows that *the time interval between two events may be different in different frames of reference.* So our next task is to learn how to compare time intervals in different frames of reference.

37.4 An event has a definite position and time—for instance, on the pavement directly below the center of the Eiffel Tower at midnight on New Year's Eve.

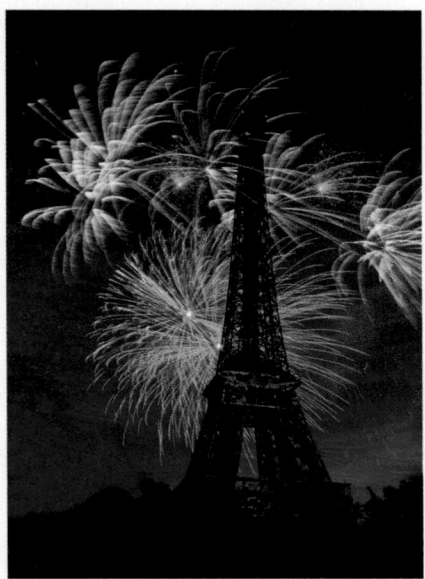

37.5 A thought experiment in simultaneity.

(a) Lightning hits the front and back of a train (points A' and B') and hits the ground at points A and B.

(b) Inside the train, Mavis moves toward the light coming from the front of the train and away from the light coming from the back of the train.

(c) Mavis sees the light from the front of the train first, so she believes that the lightning hit the front of the train first.

(d) Stanley sees the lightning hit the two points at the same time; so he believes that the lightning hit the two ends of the train simultaneously.

(The light from the rear of the train has not reached Mavis yet.)

Test Your Understanding of Section 37.2 Stanley, who works for the rail system shown in Fig. 37.5, has carefully synchronized the clocks at all of the rail stations. At the moment that Stanley measures all of the clocks striking noon, Mavis is on a high-speed passenger car traveling from Ogdenville toward North Haverbrook. According to Mavis, when the Ogdenville clock strikes noon, what time is it in North Haverbrook? (i) noon; (ii) before noon; (iii) after noon.

37.3 Relativity of Time Intervals

We can derive a quantitative relationship between time intervals in different coordinate systems. To do this, let's consider another thought experiment. As before, a frame of reference S' moves along the common x-x'-axis with constant speed u relative to a frame S. As discussed in Section 37.1, u must be less than the speed of light c. Mavis, who is riding along with frame S', measures the time interval between two events that occur at the *same* point in space. Event 1 is when a flash of light from a light source leaves O'. Event 2 is when the flash returns to O', having been reflected from a mirror a distance d away, as shown in Fig. 37.6a. We label the time interval Δt_0, using the subscript zero as a reminder that the apparatus is at rest, with zero velocity, in frame S'. The flash of light moves a total distance $2d$, so the time interval is

37.6 (a) Mavis, in frame of reference S', observes a light pulse emitted from a source at O' and reflected back along the same line. (b) How Stanley (in frame of reference S) and Mavis observe the same light pulse. The positions of O' at the times of departure and return of the pulse are shown.

(a)

Mirror

d

Source

O' Mavis measures time interval Δt_0.

S'

(b)

Mavis observes a light pulse emitted from a source at O' and reflected back along the same line.

l l d

O' O' u

S'

S

Stanley observes the same light pulse following a diagonal path.

O

$\leftarrow u\,\Delta t \rightarrow$

Stanley measures a longer time interval Δt: The light pulse travels at same speed as in S', but travels a greater distance than in S'.

$$\Delta t_0 = \frac{2d}{c} \tag{37.3}$$

The round-trip time measured by Stanley in frame S is a different interval Δt; in his frame of reference the two events occur at *different* points in space. During the time Δt, the source moves relative to S a distance $u\,\Delta t$ (Fig. 37.6b). In S' the round-trip distance is $2d$ perpendicular to the relative velocity, but the round-trip distance in S is the longer distance $2l$, where

$$l = \sqrt{d^2 + \left(\frac{u\,\Delta t}{2}\right)^2}$$

In writing this expression, we have assumed that both observers measure the same distance d. We will justify this assumption in the next section. The speed of light is the same for both observers, so the round-trip time measured in S is

$$\Delta t = \frac{2l}{c} = \frac{2}{c}\sqrt{d^2 + \left(\frac{u\,\Delta t}{2}\right)^2} \tag{37.4}$$

We would like to have a relationship between Δt and Δt_0 that is independent of d. To get this, we solve Eq. (37.3) for d and substitute the result into Eq. (37.4), obtaining

$$\Delta t = \frac{2}{c}\sqrt{\left(\frac{c\,\Delta t_0}{2}\right)^2 + \left(\frac{u\,\Delta t}{2}\right)^2} \tag{37.5}$$

Now we square this and solve for Δt; the result is

$$\Delta t = \frac{\Delta t_0}{\sqrt{1 - u^2/c^2}}$$

Since the quantity $\sqrt{1 - u^2/c^2}$ is less than 1, Δt is greater than Δt_0: Thus Stanley measures a *longer* round-trip time for the light pulse than does Mavis.

Time Dilation

We may generalize this important result. In a particular frame of reference, suppose that two events occur at the same point in space. The time interval between these events, as measured by an observer at rest in this same frame (which we call the *rest frame* of this observer), is Δt_0. Then an observer in a second frame moving with constant speed u relative to the rest frame will measure the time interval to be Δt, where

37.7 This image shows an exploding star, called a *supernova,* within a distant galaxy. The brightness of a typical supernova decays at a certain rate. But supernovae that are moving away from us at a substantial fraction of the speed of light decay more slowly, in accordance with Eq. (37.6). The decaying supernova is a moving "clock" that runs slow.

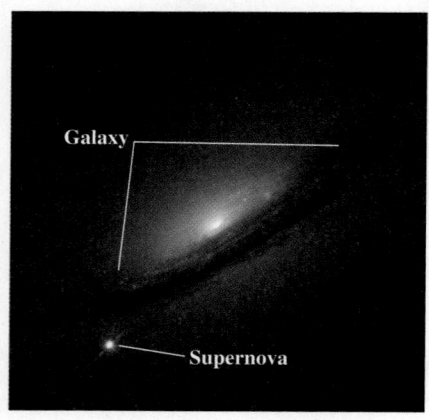

37.8 The quantity $\gamma = 1/\sqrt{1 - u^2/c^2}$ as a function of the relative speed u of two frames of reference.

$$\Delta t = \frac{\Delta t_0}{\sqrt{1 - u^2/c^2}} \qquad \text{(time dilation)} \qquad (37.6)$$

We recall that no inertial observer can travel at $u = c$ and we note that $\sqrt{1 - u^2/c^2}$ is imaginary for $u > c$. Thus Eq. (37.6) gives sensible results only when $u < c$. The denominator of Eq. (37.6) is always smaller than 1, so Δt is always *larger* than Δt_0. Thus we call this effect **time dilation.**

Think of an old-fashioned pendulum clock that has one second between ticks, as measured by Mavis in the clock's rest frame; this is Δt_0. If the clock's rest frame is moving relative to Stanley, he measures a time between ticks Δt that is longer than one second. In brief, *observers measure any clock to run slow if it moves relative to them* (Fig. 37.7). Note that this conclusion is a direct result of the fact that the speed of light in vacuum is the same in both frames of reference.

The quantity $1/\sqrt{1 - u^2/c^2}$ in Eq. (37.6) appears so often in relativity that it is given its own symbol γ (the Greek letter gamma):

$$\gamma = \frac{1}{\sqrt{1 - u^2/c^2}} \qquad (37.7)$$

In terms of this symbol, we can express the time dilation formula, Eq. (37.6), as

$$\Delta t = \gamma \, \Delta t_0 \qquad \text{(time dilation)} \qquad (37.8)$$

As a further simplification, u/c is sometimes given the symbol β (the Greek letter beta); then $\gamma = 1/\sqrt{1 - \beta^2}$.

Figure 37.8 shows a graph of γ as a function of the relative speed u of two frames of reference. When u is very small compared to c, u^2/c^2 is much smaller than 1 and γ is very nearly *equal* to 1. In that limit, Eqs. (37.6) and (37.8) approach the Newtonian relationship $\Delta t = \Delta t_0$, corresponding to the same time interval in all frames of reference.

If the relative speed u is great enough that γ is appreciably greater than 1, the speed is said to be *relativistic;* if the difference between γ and 1 is negligibly small, the speed u is called *nonrelativistic.* Thus $u = 6.00 \times 10^7$ m/s $= 0.200c$ (for which $\gamma = 1.02$) is a relativistic speed, but $u = 6.00 \times 10^4$ m/s $= 0.000200c$ (for which $\gamma = 1.00000002$) is a nonrelativistic speed.

Proper Time

There is only one frame of reference in which a clock is at rest, and there are infinitely many in which it is moving. Therefore the time interval measured between two events (such as two ticks of the clock) that occur at the same point in a particular frame is a more fundamental quantity than the interval between events at different points. We use the term **proper time** to describe the time interval Δt_0 between two events that occur at the same point.

CAUTION **Measuring time intervals** It is important to note that the time interval Δt in Eq. (37.6) involves events that occur at *different space points* in the frame of reference S. Note also that any differences between Δt and the proper time Δt_0 are *not* caused by differences in the times required for light to travel from those space points to an observer at rest in S. We assume that our observer is able to correct for differences in light transit times, just as an astronomer who's observing the sun understands that an event seen now on earth actually occurred 500 s ago on the sun's surface. Alternatively, we can use *two* observers, one stationary at the location of the first event and the other at the second, each with his or her own clock. We can synchronize these two clocks without difficulty, as long as they are at rest in the same frame of reference. For example, we could send a light pulse simultaneously to the two clocks from a point midway between them. When the pulses arrive, the observers set their clocks to a prearranged time. (But note that clocks that are synchronized in one frame of reference *are not* in general synchronized in any other frame.)

In thought experiments, it's often helpful to imagine many observers with synchronized clocks at rest at various points in a particular frame of reference. We can picture a frame of reference as a coordinate grid with lots of synchronized clocks distributed around it, as suggested by Fig. 37.9. Only when a clock is moving relative to a given frame of reference do we have to watch for ambiguities of synchronization or simultaneity.

Throughout this chapter we will frequently use phrases like "Stanley *observes* that Mavis passes the point $x = 5.00$ m, $y = 0$, $z = 0$ at time 2.00 s." This means that Stanley is using a grid of clocks in his frame of reference, like the grid shown in Fig. 37.9, to record the time of an event. We could restate the phrase as "When Mavis passes the point at $x = 5.00$ m, $y = 0$, $z = 0$, the clock at that location in Stanley's frame of reference reads 2.00 s." We will avoid using phrases like "Stanley *sees* that Mavis is a certain point at a certain time," because there is a time delay for light to travel to Stanley's eye from the position of an event.

37.9 A frame of reference pictured as a coordinate system with a grid of synchronized clocks.

The grid is three dimensional; identical planes of clocks lie in front of and behind the page, connected by grid lines perpendicular to the page.

Problem-Solving Strategy 37.1 | Time Dilation

IDENTIFY *the relevant concepts:* The concept of time dilation is used whenever we compare the time intervals between events as measured by observers in different inertial frames of reference.

SET UP *the problem* using the following steps:
1. First decide what two events define the beginning and the end of the time interval. Then identify the two frames of reference in which the time interval is measured.
2. Identify the target variable.

EXECUTE *the solution* as follows:
1. In many problems, the time interval as measured in one frame of reference is the *proper* time Δt_0. This is the time interval

between two events in a frame of reference in which the two events occur at the same point in space. In a second frame of reference that has a speed u relative to that first frame, there is a longer time interval Δt between the same two events. In this second frame the two events occur at different points. You will need to decide in which frame the time interval is Δt_0 and in which frame it is Δt.
2. Use Eq. (37.6) or (37.8) to relate Δt_0 and Δt, and then solve for the target variable.

EVALUATE *your answer:* Note that Δt is never smaller than Δt_0, and u is never greater than c. If your results suggest otherwise, you need to rethink your calculation.

Example 37.1 | Time dilation at 0.990*c*

High-energy subatomic particles coming from space interact with atoms in the earth's upper atmosphere, in some cases producing unstable particles called *muons*. A muon decays into other particles with a mean lifetime of 2.20 μs $= 2.20 \times 10^{-6}$ s as measured in a reference frame in which it is at rest. If a muon is moving at 0.990*c* relative to the earth, what will an observer on earth measure its mean lifetime to be?

SOLUTION

IDENTIFY and SET UP: The muon's lifetime is the time interval between two events: the production of the muon and its subsequent decay. Our target variable is the lifetime in your frame of reference on earth, which we call frame S. We are given the lifetime in a frame S' in which the muon is at rest; this is its *proper* lifetime, $\Delta t_0 = 2.20$ μs. The relative speed of these two frames is

$u = 0.990c$. We use Eq. (37.6) to relate the lifetimes in the two frames.

EXECUTE: The muon moves relative to the earth between the two events, so the two events occur at different positions as measured in S and the time interval in that frame is Δt (the target variable). From Eq. (37.6),

$$\Delta t = \frac{\Delta t_0}{\sqrt{1 - u^2/c^2}} = \frac{2.20 \ \mu\text{s}}{\sqrt{1 - (0.990)^2}} = 15.6 \ \mu\text{s}$$

EVALUATE: Our result predicts that the mean lifetime of the muon in the earth frame (Δt) is about seven times longer than in the muon's frame (Δt_0). This prediction has been verified experimentally; indeed, this was the first experimental confirmation of the time dilation formula, Eq. (37.6).

Example 37.2 Time dilation at airliner speeds

An airplane flies from San Francisco to New York (about 4800 km, or 4.80×10^6 m) at a steady speed of 300 m/s (about 670 mi/h). How much time does the trip take, as measured by an observer on the ground? By an observer in the plane?

SOLUTION

IDENTIFY and SET UP: Here we're interested in the time interval between the airplane departing from San Francisco and landing in New York. The target variables are the time intervals as measured in the frame of reference of the ground S and in the frame of reference of the airplane S'.

EXECUTE: As measured in S the two events occur at different positions (San Francisco and New York), so the time interval measured by ground observers corresponds to Δt in Eq. (37.6). To find it, we simply divide the distance by the speed $u = 300$ m/s:

$$\Delta t = \frac{4.80 \times 10^6 \text{ m}}{300 \text{ m/s}} = 1.60 \times 10^4 \text{ s} \quad \text{(about } 4\tfrac{1}{2} \text{ hours)}$$

In the airplane's frame S', San Francisco and New York passing under the plane occur at the same point (the position of the plane). Hence the time interval in the airplane is a proper time, corresponding to Δt_0 in Eq. (37.6). We have

$$\frac{u^2}{c^2} = \frac{(300 \text{ m/s})^2}{(3.00 \times 10^8 \text{ m/s})^2} = 1.00 \times 10^{-12}$$

From Eq. (37.6),

$$\Delta t_0 = (1.60 \times 10^4 \text{ s})\sqrt{1 - 1.00 \times 10^{-12}}$$

The square root can't be evaluated with adequate precision with an ordinary calculator. But we can approximate it using the binomial theorem (see Appendix B):

$$(1 - 1.00 \times 10^{-12})^{1/2} = 1 - \left(\tfrac{1}{2}\right)(1.00 \times 10^{-12}) + \cdots$$

The remaining terms are of the order of 10^{-24} or smaller and can be discarded. The approximate result for Δt_0 is

$$\Delta t_0 = (1.60 \times 10^4 \text{ s})(1 - 0.50 \times 10^{-12})$$

The proper time Δt_0, measured in the airplane, is very slightly less (by less than one part in 10^{12}) than the time measured on the ground.

EVALUATE: We don't notice such effects in everyday life. But present-day atomic clocks (see Section 1.3) can attain a precision of about one part in 10^{13}. A cesium clock traveling a long distance in an airliner has been used to measure this effect and thereby verify Eq. (37.6) even at speeds much less than c.

Example 37.3 Just when is it proper?

Mavis boards a spaceship and then zips past Stanley on earth at a relative speed of $0.600c$. At the instant she passes him, they both start timers. (a) A short time later Stanley measures that Mavis has traveled 9.00×10^7 m beyond him and is passing a space station. What does Stanley's timer read as she passes the space station? What does Mavis's timer read? (b) Stanley starts to blink just as Mavis flies past him, and Mavis measures that the blink takes 0.400 s from beginning to end. According to Stanley, what is the duration of his blink?

SOLUTION

IDENTIFY and SET UP: This problem involves time dilation for two *different* sets of events measured in Stanley's frame of reference (which we call S) and in Mavis's frame of reference (which we call S'). The two events of interest in part (a) are when Mavis passes Stanley and when Mavis passes the space station; the target variables are the time intervals between these two events as measured in S and in S'. The two events in part (b) are the start and finish of Stanley's blink; the target variable is the time interval between these two events as measured in S.

EXECUTE: (a) The two events, Mavis passing the earth and Mavis passing the space station, occur at different positions in Stanley's frame but at the same position in Mavis's frame. Hence Stanley measures time interval Δt, while Mavis measures the *proper* time Δt_0. As measured by Stanley, Mavis moves at $0.600c = 0.600(3.00 \times 10^8 \text{ m/s}) = 1.80 \times 10^8$ m/s and travels 9.00×10^7 m in time $\Delta t = (9.00 \times 10^7 \text{ m})/(1.80 \times 10^8 \text{ m/s}) = 0.500$ s. From Eq. (37.6), Mavis's timer reads an elapsed time of

$$\Delta t_0 = \Delta t \sqrt{1 - u^2/c^2} = 0.500 \text{ s} \sqrt{1 - (0.600)^2} = 0.400 \text{ s}$$

(b) It is tempting to answer that Stanley's blink lasts 0.500 s in his frame. But this is wrong, because we are now considering a *different* pair of events than in part (a). The start and finish of Stanley's blink occur at the same point in his frame S but at different positions in Mavis's frame S', so the time interval of 0.400 s that she measures between these events is equal to Δt. The duration of the blink measured on Stanley's timer is the proper time Δt_0:

$$\Delta t_0 = \Delta t \sqrt{1 - u^2/c^2} = 0.400 \text{ s} \sqrt{1 - (0.600)^2} = 0.320 \text{ s}$$

EVALUATE: This example illustrates the relativity of simultaneity. In Mavis's frame she passes the space station at the same instant that Stanley finishes his blink, 0.400 s after she passed Stanley. Hence these two events are simultaneous to Mavis in frame S'. But these two events are *not* simultaneous to Stanley in his frame S: According to his timer, he finishes his blink after 0.320 s and Mavis passes the space station after 0.500 s.

The Twin Paradox

Equations (37.6) and (37.8) for time dilation suggest an apparent paradox called the **twin paradox.** Consider identical twin astronauts named Eartha and Astrid.

Eartha remains on earth while her twin Astrid takes off on a high-speed trip through the galaxy. Because of time dilation, Eartha observes Astrid's heartbeat and all other life processes proceeding more slowly than her own. Thus to Eartha, Astrid ages more slowly; when Astrid returns to earth she is younger (has aged less) than Eartha.

Now here is the paradox: All inertial frames are equivalent. Can't Astrid make exactly the same arguments to conclude that Eartha is in fact the younger? Then each twin measures the other to be younger when they're back together, and that's a paradox.

To resolve the paradox, we recognize that the twins are *not* identical in all respects. While Eartha remains in an approximately inertial frame at all times, Astrid must *accelerate* with respect to that inertial frame during parts of her trip in order to leave, turn around, and return to earth. Eartha's reference frame is always approximately inertial; Astrid's is often far from inertial. Thus there is a real physical difference between the circumstances of the two twins. Careful analysis shows that Eartha is correct; when Astrid returns, she *is* younger than Eartha.

Test Your Understanding of Section 37.3 Samir (who is standing on the ground) starts his stopwatch at the instant that Maria flies past him in her spaceship at a speed of 0.600c. At the same instant, Maria starts her stopwatch. (a) As measured in Samir's frame of reference, what is the reading on Maria's stopwatch at the instant that Samir's stopwatch reads 10.0 s? (i) 10.0 s; (ii) less than 10.0 s; (iii) more than 10.0 s. (b) As measured in Maria's frame of reference, what is the reading on Samir's stopwatch at the instant that Maria's stopwatch reads 10.0 s? (i) 10.0 s; (ii) less than 10.0 s; (iii) more than 10.0 s.

Application Who's the Grandmother?

The answer to this question may seem obvious, but it could depend on which person had traveled to a distant planet at relativistic speeds. Imagine that a 20-year-old woman had given birth to a child and then immediately left on a 100-light-year trip (50 light-years out and 50 light-years back) at 99.5% the speed of light. Because of time dilation for the traveler, only 10 years would pass, and she would be 30 years old when she returned, even though 100 years had passed by for people on earth. Meanwhile, the child she left behind at home could have had a baby 20 years after her departure, and this grandchild would now be 80 years old!

37.4 Relativity of Length

Not only does the time interval between two events depend on the observer's frame of reference, but the *distance* between two points may also depend on the observer's frame of reference. The concept of simultaneity is involved. Suppose you want to measure the length of a moving car. One way is to have two assistants make marks on the pavement at the positions of the front and rear bumpers. Then you measure the distance between the marks. But your assistants have to make their marks *at the same time.* If one marks the position of the front bumper at one time and the other marks the position of the rear bumper half a second later, you won't get the car's true length. Since we've learned that simultaneity isn't an absolute concept, we have to proceed with caution.

Lengths Parallel to the Relative Motion

To develop a relationship between lengths that are measured parallel to the direction of motion in various coordinate systems, we consider another thought experiment. We attach a light source to one end of a ruler and a mirror to the other end. The ruler is at rest in reference frame S', and its length in this frame is l_0 (Fig. 37.10a). Then the time Δt_0 required for a light pulse to make the round trip from source to mirror and back is

$$\Delta t_0 = \frac{2l_0}{c} \tag{37.9}$$

This is a proper time interval because departure and return occur at the same point in S'.

In reference frame S the ruler is moving to the right with speed u during this travel of the light pulse (Fig. 37.10b). The length of the ruler in S is l, and the time of travel from source to mirror, as measured in S, is Δt_1. During this interval

ActivPhysics 17.2: Relativity of Length

37.10 (a) A ruler is at rest in Mavis's frame S'. A light pulse is emitted from a source at one end of the ruler, reflected by a mirror at the other end, and returned to the source position. (b) Motion of the light pulse as measured in Stanley's frame S.

(a)

The ruler is stationary in Mavis's frame of reference S'. The light pulse travels a distance l_0 from the light source to the mirror.

(b)

The ruler moves at speed u in Stanley's frame of reference S. The light pulse travels a distance l (the length of the ruler measured in S) plus an additional distance $u \, \Delta t_1$ from the light source to the mirror.

the ruler, with source and mirror attached, moves a distance $u \, \Delta t_1$. The total length of path d from source to mirror is not l, but rather

$$d = l + u \, \Delta t_1 \tag{37.10}$$

The light pulse travels with speed c, so it is also true that

$$d = c \, \Delta t_1 \tag{37.11}$$

Combining Eqs. (37.10) and (37.11) to eliminate d, we find

$$c \, \Delta t_1 = l + u \, \Delta t_1 \quad \text{or}$$

$$\Delta t_1 = \frac{l}{c - u} \tag{37.12}$$

(Dividing the distance l by $c - u$ does *not* mean that light travels with speed $c - u$, but rather that the distance the pulse travels in S is greater than l.)

In the same way we can show that the time Δt_2 for the return trip from mirror to source is

$$\Delta t_2 = \frac{l}{c + u} \tag{37.13}$$

The *total* time $\Delta t = \Delta t_1 + \Delta t_2$ for the round trip, as measured in S, is

$$\Delta t = \frac{l}{c - u} + \frac{l}{c + u} = \frac{2l}{c(1 - u^2/c^2)} \tag{37.14}$$

We also know that Δt and Δt_0 are related by Eq. (37.6) because Δt_0 is a proper time in S'. Thus Eq. (37.9) for the round-trip time in the rest frame S' of the ruler becomes

$$\Delta t \sqrt{1 - \frac{u^2}{c^2}} = \frac{2l_0}{c} \tag{37.15}$$

Finally, combining Eqs. (37.14) and (37.15) to eliminate Δt and simplifying, we obtain

$$l = l_0 \sqrt{1 - \frac{u^2}{c^2}} = \frac{l_0}{\gamma} \qquad \text{(length contraction)} \tag{37.16}$$

[We have used the quantity $\gamma = 1/\sqrt{1 - u^2/c^2}$ defined in Eq. (37.7).] Thus the length l measured in S, in which the ruler is moving, is *shorter* than the length l_0 measured in its rest frame S'.

CAUTION **Length contraction is real** This is *not* an optical illusion! The ruler really is shorter in reference frame S than it is in S'. ▌

A length measured in the frame in which the body is at rest (the rest frame of the body) is called a **proper length;** thus l_0 is a proper length in S', and the length measured in any other frame moving relative to S' is *less than* l_0. This effect is called **length contraction.**

When u is very small in comparison to c, γ approaches 1. Thus in the limit of small speeds we approach the Newtonian relationship $l = l_0$. This and the corresponding result for time dilation show that Eqs. (37.1), the Galilean coordinate transformation, are usually sufficiently accurate for relative speeds much smaller than c. If u is a reasonable fraction of c, however, the quantity $\sqrt{1 - u^2/c^2}$ can be appreciably less than 1. Then l can be substantially smaller than l_0, and the effects of length contraction can be substantial (Fig. 37.11).

Lengths Perpendicular to the Relative Motion

We have derived Eq. (37.16) for lengths measured in the direction *parallel* to the relative motion of the two frames of reference. Lengths that are measured *perpendicular* to the direction of motion are *not* contracted. To prove this, consider two identical meter sticks. One stick is at rest in frame S and lies along the positive y-axis with one end at O, the origin of S. The other is at rest in frame S' and lies along the positive y'-axis with one end at O', the origin of S'. Frame S' moves in the positive x-direction relative to frame S. Observers Stanley and Mavis, at rest in S and S' respectively, station themselves at the 50-cm mark of their sticks. At the instant the two origins coincide, the two sticks lie along the same line. At this instant, Mavis makes a mark on Stanley's stick at the point that coincides with her own 50-cm mark, and Stanley does the same to Mavis's stick.

Suppose for the sake of argument that Stanley observes Mavis's stick as longer than his own. Then the mark Stanley makes on her stick is *below* its center. In that case, Mavis will think Stanley's stick has become shorter, since half of its length coincides with *less* than half her stick's length. So Mavis observes moving sticks getting shorter and Stanley observes them getting longer. But this implies an asymmetry between the two frames that contradicts the basic postulate of relativity that tells us all inertial frames are equivalent. We conclude that consistency with the postulates of relativity requires that both observers measure the rulers as having the *same* length, even though to each observer one of them is stationary and the other is moving (Fig. 37.12). So *there is no length contraction perpendicular to the direction of relative motion of the coordinate systems.* We used this result in our derivation of Eq. (37.6) in assuming that the distance d is the same in both frames of reference.

37.11 The speed at which electrons traverse the 3-km beam line of the SLAC National Accelerator Laboratory is slower than c by less than 1 cm/s. As measured in the reference frame of such an electron, the beam line (which extends from the top to the bottom of this photograph) is only about 15 cm long!

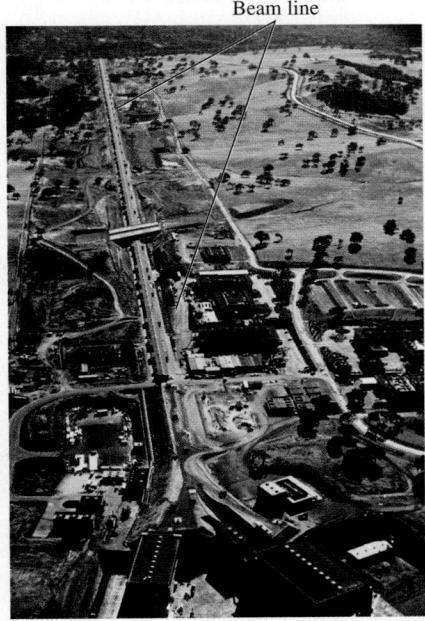

Beam line

37.12 The meter sticks are perpendicular to the relative velocity. For any value of u, both Stanley and Mavis measure either meter stick to have a length of 1 meter.

For example, suppose a moving rod of length l_0 makes an angle θ_0 with the direction of relative motion (the x-axis) as measured in its rest frame. Its length component in that frame parallel to the motion, $l_0 \cos \theta_0$, is contracted to $(l_0 \cos \theta_0)/\gamma$. However, its length component perpendicular to the motion, $l_0 \sin \theta_0$, remains the same.

Problem-Solving Strategy 37.2 Length Contraction

IDENTIFY *the relevant concepts:* The concept of length contraction is used whenever we compare the length of an object as measured by observers in different inertial frames of reference.

SET UP *the problem* using the following steps:
1. Decide what defines the length in question. If the problem describes an object such as a ruler, it is just the distance between the ends of the object. If the problem is about a distance between two points in space, it helps to envision an object like a ruler that extends from one point to the other.
2. Identify the target variable.

EXECUTE *the solution* as follows:
1. Determine the reference frame in which the object in question is at rest. In this frame, the length of the object is its proper

length l_0. In a second reference frame moving at speed u relative to the first frame, the object has contracted length l.
2. Keep in mind that length contraction occurs only for lengths parallel to the direction of relative motion of the two frames. Any length that is perpendicular to the relative motion is the same in both frames.
3. Use Eq. (37.16) to relate l and l_0, and then solve for the target variable.

EVALUATE *your answer:* Check that your answers make sense: l is never larger than l_0, and u is never greater than c.

Example 37.4 How long is the spaceship?

A spaceship flies past earth at a speed of $0.990c$. A crew member on board the spaceship measures its length, obtaining the value 400 m. What length do observers measure on earth?

SOLUTION

IDENTIFY and SET UP: This problem is about the nose-to-tail length of the spaceship as measured on the spaceship and on earth. This length is along the direction of relative motion (Fig. 37.13), so there will be length contraction. The spaceship's 400-m length is the *proper* length l_0 because it is measured in the frame in which the spaceship is at rest. Our target variable is the length l measured in the earth frame, relative to which the spaceship is moving at $u = 0.990c$.

EXECUTE: From Eq. (37.16), the length in the earth frame is

$$l = l_0 \sqrt{1 - \frac{u^2}{c^2}} = (400 \text{ m}) \sqrt{1 - (0.990)^2} = 56.4 \text{ m}$$

EVALUATE: The spaceship is shorter in a frame in which it is in motion than in a frame in which it is at rest. To measure the length l, two earth observers with synchronized clocks could measure the

37.13 Measuring the length of a moving spaceship.

The two observers on earth (S) must measure x_2 and x_1 simultaneously to obtain the correct length $l = x_2 - x_1$ in their frame of reference.

positions of the two ends of the spaceship simultaneously in the earth's reference frame, as shown in Fig. 37.13. (These two measurements will *not* appear simultaneous to an observer in the spaceship.)

Example 37.5 How far apart are the observers?

Observers O_1 and O_2 in Fig. 37.13 are 56.4 m apart on the earth. How far apart does the spaceship crew measure them to be?

SOLUTION

IDENTIFY and SET UP: In this example the 56.4-m distance is the *proper* length l_0. It represents the length of a ruler that extends

from O_1 to O_2 and is at rest in the earth frame in which the observers are at rest. Our target variable is the length l of this ruler measured in the spaceship frame, in which the earth and ruler are moving at $u = 0.990c$.

EXECUTE: As in Example 37.4, but with $l_0 = 56.4$ m,

$$l = l_0\sqrt{1 - \frac{u^2}{c^2}} = (56.4 \text{ m})\sqrt{1 - (0.990)^2} = 7.96 \text{ m}$$

EVALUATE: This answer does *not* say that the crew measures their spaceship to be both 400 m long and 7.96 m long. As measured on

earth, the tail of the spacecraft is at the position of O_1 at the same instant that the nose of the spacecraft is at the position of O_2. Hence the length of the spaceship measured on earth equals the 56.4-m distance between O_1 and O_2. But in the spaceship frame O_1 and O_2 are only 7.96 m apart, and the nose (which is 400 m in front of the tail) passes O_2 before the tail passes O_1.

How an Object Moving Near c Would Appear

Let's think a little about the visual appearance of a moving three-dimensional body. If we could see the positions of all points of the body simultaneously, it would appear to shrink only in the direction of motion. But we *don't* see all the points simultaneously; light from points farther from us takes longer to reach us than does light from points near to us, so we see the farther points at the positions they had at earlier times.

Suppose we have a rectangular rod with its faces parallel to the coordinate planes. When we look end-on at the center of the closest face of such a rod at rest, we see only that face. (See the center rod in computer-generated Fig. 37.14a.) But when that rod is moving past us toward the right at an appreciable fraction of the speed of light, we may also see its left side because of the earlier-time effect just described. That is, we can see some points that we couldn't see when the rod was at rest because the rod moves out of the way of the light rays from those points to us. Conversely, some light that can get to us when the rod is at rest is blocked by the moving rod. Because of all this, the rods in Figs. 37.14b and 37.14c appear rotated and distorted.

37.14 Computer simulation of the appearance of an array of 25 rods with square cross section. The center rod is viewed end-on. The simulation ignores color changes in the array caused by the Doppler effect (see Section 37.6).

(a) Array at rest

(b) Array moving to the right at 0.2c

(c) Array moving to the right at 0.9c

Test Your Understanding of Section 37.4 A miniature spaceship is flying past you, moving horizontally at a substantial fraction of the speed of light. At a certain instant, you observe that the nose and tail of the spaceship align exactly with the two ends of a meter stick that you hold in your hands. Rank the following distances in order from longest to shortest: (i) the proper length of the meter stick; (ii) the proper length of the spaceship; (iii) the length of the spaceship measured in your frame of reference; (iv) the length of the meter stick measured in the spaceship's frame of reference.

37.5 The Lorentz Transformations

In Section 37.1 we discussed the Galilean coordinate transformation equations, Eqs. (37.1). They relate the coordinates (x, y, z) of a point in frame of reference S to the coordinates (x', y', z') of the point in a second frame S'. The second frame moves with constant speed u relative to S in the positive direction along the common x-x'-axis. This transformation also assumes that the time scale is the same in the two frames of reference, as expressed by the additional relationship $t = t'$. This Galilean transformation, as we have seen, is valid only in the limit when u approaches zero. We are now ready to derive more general transformations that are consistent with the principle of relativity. The more general relationships are called the **Lorentz transformations.**

The Lorentz Coordinate Transformation

Our first question is this: When an event occurs at point (x, y, z) at time t, as observed in a frame of reference S, what are the coordinates (x', y', z') and time t' of the event as observed in a second frame S' moving relative to S with constant speed u in the $+x$-direction?

To derive the coordinate transformation, we refer to Fig. 37.15 (next page), which is the same as Fig. 37.3. As before, we assume that the origins coincide at the initial time $t = 0 = t'$. Then in S the distance from O to O' at time t is

37.15 As measured in frame of reference S, x' is contracted to x'/γ, so $x = ut + x'/\gamma$ and $x' = \gamma(x - ut)$.

Frame S' moves relative to frame S with constant velocity u along the common x-x'-axis.

Origins O and O' coincide at time $t = 0 = t'$.

The Lorentz coordinate transformation relates the spacetime coordinates of an event as measured in the two frames: (x, y, z, t) in frame S and (x', y', z', t') in frame S'.

still ut. The coordinate x' is a *proper length* in S', so in S it is contracted by the factor $1/\gamma = \sqrt{1 - u^2/c^2}$, as in Eq. (37.16). Thus the distance x from O to P, as seen in S, is not simply $x = ut + x'$, as in the Galilean coordinate transformation, but

$$x = ut + x'\sqrt{1 - \frac{u^2}{c^2}} \qquad (37.17)$$

Solving this equation for x', we obtain

$$x' = \frac{x - ut}{\sqrt{1 - u^2/c^2}} \qquad (37.18)$$

Equation (37.18) is part of the Lorentz coordinate transformation; another part is the equation giving t' in terms of x and t. To obtain this, we note that the principle of relativity requires that the *form* of the transformation from S to S' be identical to that from S' to S. The only difference is a change in the sign of the relative velocity component u. Thus from Eq. (37.17) it must be true that

$$x' = -ut' + x\sqrt{1 - \frac{u^2}{c^2}} \qquad (37.19)$$

We now equate Eqs. (37.18) and (37.19) to eliminate x'. This gives us an equation for t' in terms of x and t. We leave the algebraic details for you to work out; the result is

$$t' = \frac{t - ux/c^2}{\sqrt{1 - u^2/c^2}} \qquad (37.20)$$

As we discussed previously, lengths perpendicular to the direction of relative motion are not affected by the motion, so $y' = y$ and $z' = z$.

Collecting all these transformation equations, we have

$$x' = \frac{x - ut}{\sqrt{1 - u^2/c^2}} = \gamma(x - ut)$$

$$y' = y$$
$$z' = z$$
$$t' = \frac{t - ux/c^2}{\sqrt{1 - u^2/c^2}} = \gamma(t - ux/c^2)$$

(Lorentz coordinate transformation) $\qquad (37.21)$

These equations are the *Lorentz coordinate transformation,* the relativistic generalization of the Galilean coordinate transformation, Eqs. (37.1) and $t = t'$. For values of u that approach zero, the radicals in the denominators and γ approach 1, and the ux/c^2 term approaches zero. In this limit, Eqs. (37.21) become identical to Eqs. (37.1) along with $t = t'$. In general, though, both the coordinates and time of an event in one frame depend on its coordinates and time in another frame. *Space and time have become intertwined; we can no longer say that length and time have absolute meanings independent of the frame of reference.* For this reason, we refer to time and the three dimensions of space collectively as a four-dimensional entity called **spacetime,** and we call (x, y, z, t) together the **spacetime coordinates** of an event.

The Lorentz Velocity Transformation

We can use Eqs. (37.21) to derive the relativistic generalization of the Galilean velocity transformation, Eq. (37.2). We consider only one-dimensional motion along the x-axis and use the term "velocity" as being short for the "x-component of the velocity." Suppose that in a time dt a particle moves a distance dx, as measured in frame S. We obtain the corresponding distance dx' and time dt' in S' by taking differentials of Eqs. (37.21):

$$dx' = \gamma(dx - u\,dt)$$

$$dt' = \gamma(dt - u\,dx/c^2)$$

We divide the first equation by the second and then divide the numerator and denominator of the result by dt to obtain

$$\frac{dx'}{dt'} = \frac{\dfrac{dx}{dt} - u}{1 - \dfrac{u}{c^2}\dfrac{dx}{dt}}$$

Now dx/dt is the velocity v_x in S, and dx'/dt' is the velocity v'_x in S', so we finally obtain the relativistic generalization

$$v'_x = \frac{v_x - u}{1 - uv_x/c^2} \qquad \text{(Lorentz velocity transformation)} \qquad (37.22)$$

When u and v_x are much smaller than c, the denominator in Eq. (37.22) approaches 1, and we approach the nonrelativistic result $v'_x = v_x - u$. The opposite extreme is the case $v_x = c$; then we find

$$v'_x = \frac{c - u}{1 - uc/c^2} = \frac{c(1 - u/c)}{1 - u/c} = c$$

This says that anything moving with velocity $v_x = c$ measured in S also has velocity $v'_x = c$ measured in S', despite the relative motion of the two frames. So Eq. (37.22) is consistent with Einstein's postulate that the speed of light in vacuum is the same in all inertial frames of reference.

The principle of relativity tells us there is no fundamental distinction between the two frames S and S'. Thus the expression for v_x in terms of v'_x must have the same form as Eq. (37.22), with v_x changed to v'_x, and vice versa, and the sign of u reversed. Carrying out these operations with Eq. (37.22), we find

$$v_x = \frac{v'_x + u}{1 + uv'_x/c^2} \qquad \text{(Lorentz velocity transformation)} \qquad (37.23)$$

This can also be obtained algebraically by solving Eq. (37.22) for v_x. Both Eqs. (37.22) and (37.23) are *Lorentz velocity transformations* for one-dimensional motion.

CAUTION **Use the correct reference frame coordinates** Keep in mind that the Lorentz transformation equations given by Eqs. (37.21), (37.22), and (37.23) assume that frame S' is moving in the positive x-direction with velocity u relative to frame S. You should always set up your coordinate system to follow this convention.

When u is less than c, the Lorentz velocity transformations show us that a body moving with a speed less than c in one frame of reference always has a speed less than c in *every other* frame of reference. This is one reason for concluding that no material body may travel with a speed equal to or greater than that of light in vacuum, relative to *any* inertial frame of reference. The relativistic generalizations of energy and momentum, which we will explore later, give further support to this hypothesis.

Problem-Solving Strategy 37.3 | Lorentz Transformations

IDENTIFY *the relevant concepts:* The Lorentz *coordinate* transformation equations relate the spacetime coordinates of an event in one inertial reference frame to the coordinates of the same event in a second inertial frame. The Lorentz *velocity* transformation equations relate the velocity of an object in one inertial reference frame to its velocity in a second inertial frame.

SET UP *the problem* using the following steps:
1. Identify the target variable.
2. Define the two inertial frames S and S'. Remember that S' moves relative to S at a constant velocity u in the $+x$-direction.
3. If the coordinate transformation equations are needed, make a list of spacetime coordinates in the two frames, such as x_1, x'_1, t_1, t'_1, and so on. Label carefully which of these you know and which you don't.
4. In velocity-transformation problems, clearly identify u (the relative velocity of the two frames of reference), v_x (the velocity of the object relative to S), and v'_x (the velocity of the object relative to S').

EXECUTE *the solution* as follows:
1. In a coordinate-transformation problem, use Eqs. (37.21) to solve for the spacetime coordinates of the event as measured in S' in terms of the corresponding values in S. (If you need to solve for the spacetime coordinates in S in terms of the corresponding values in S', you can easily convert the expressions in Eqs. (37.21): Replace all of the primed quantities with unprimed ones, and vice versa, and replace u with $-u$.)
2. In a velocity-transformation problem, use either Eq. (37.22) or Eq. (37.23), as appropriate, to solve for the target variable.

EVALUATE *your answer:* Don't be discouraged if some of your results don't seem to make sense or if they disagree with "common sense." It takes time to develop intuition about relativity; you'll gain it with experience.

Example 37.6 | Was it received before it was sent?

Winning an interstellar race, Mavis pilots her spaceship across a finish line in space at a speed of $0.600c$ relative to that line. A "hooray" message is sent from the back of her ship (event 2) at the instant (in her frame of reference) that the front of her ship crosses the line (event 1). She measures the length of her ship to be 300 m. Stanley is at the finish line and is at rest relative to it. When and where does he measure events 1 and 2 to occur?

SOLUTION

IDENTIFY and SET UP: This example involves the Lorentz coordinate transformation. Our derivation of this transformation assumes that the origins of frames S and S' coincide at $t = 0 = t'$. Thus for simplicity we fix the origin of S at the finish line and the origin of S' at the front of the spaceship so that Stanley and Mavis measure event 1 to be at $x = 0 = x'$ and $t = 0 = t'$.

Mavis in S' measures her spaceship to be 300 m long, so she has the "hooray" sent from 300 m behind her spaceship's front at the instant she measures the front to cross the finish line. That is, she measures event 2 at $x' = -300$ m and $t' = 0$.

Our target variables are the coordinate x and time t of event 2 that Stanley measures in S.

EXECUTE: To solve for the target variables, we modify the first and last of Eqs. (37.21) to give x and t as functions of x' and t'. We do so in the same way that we obtained Eq. (37.23) from Eq. (37.22). We remove the primes from x' and t', add primes to x and t, and replace each u with $-u$. The results are

$$x = \gamma(x' + ut') \quad \text{and} \quad t = \gamma(t' + ux'/c^2)$$

From Eq. (37.7), $\gamma = 1.25$ for $u = 0.600c = 1.80 \times 10^8$ m/s. We also substitute $x' = -300$ m, $t' = 0$, $c = 3.00 \times 10^8$ m/s, and $u = 1.80 \times 10^8$ m/s in the equations for x and t to find $x = -375$ m at $t = -7.50 \times 10^{-7}$ s $= -0.750$ μs for event 2.

EVALUATE: Mavis says that the events are simultaneous, but Stanley says that the "hooray" was sent *before* Mavis crossed the finish line. This does not mean that the effect preceded the cause. The fastest that Mavis can send a signal the length of her ship is 300 m$/(3.00 \times 10^8$ m/s$) = 1.00$ μs. She cannot send a signal from the front at the instant it crosses the finish line that would cause a "hooray" to be broadcast from the back at the same instant. She would have to send that signal from the front at least 1.00 μs before then, so she had to slightly anticipate her success.

Example 37.7 | Relative velocities

(a) A spaceship moving away from the earth at $0.900c$ fires a robot space probe in the same direction as its motion at $0.700c$ relative to the spaceship. What is the probe's velocity relative to the earth? (b) A scoutship is sent to catch up with the spaceship by traveling at $0.950c$ relative to the earth. What is the velocity of the scoutship relative to the spaceship?

SOLUTION

IDENTIFY and SET UP: This example uses the Lorentz velocity transformation. Let the earth and spaceship reference frames be S and S', respectively (Fig. 37.16); their relative velocity is $u = 0.900c$. In part (a) we are given the probe velocity $v'_x = 0.700c$ with respect to S', and the target variable is the velocity v_x of the

37.16 The spaceship, robot space probe, and scoutship.

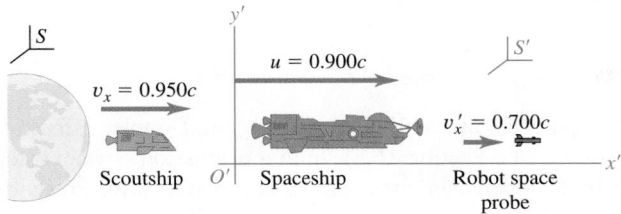

$$v_x = \frac{v'_x + u}{1 + uv'_x/c^2} = \frac{0.700c + 0.900c}{1 + (0.900c)(0.700c)/c^2} = 0.982c$$

(b) We use Eq. (37.22) to find the scoutship velocity relative to the spaceship:

$$v'_x = \frac{v_x - u}{1 - uv_x/c^2} = \frac{0.950c - 0.900c}{1 - (0.900c)(0.950c)/c^2} = 0.345c$$

probe relative to S. In part (b) we are given the velocity $v_x = 0.950c$ of the scoutship relative to S, and the target variable is its velocity v'_x relative to S'.

EXECUTE: (a) We use Eq. (37.23) to find the probe velocity relative to the earth:

EVALUATE: What would the Galilean velocity transformation formula, Eq. (37.2), say? In part (a) we would have found the probe's velocity relative to the earth to be $v_x = v'_x + u = 0.700c + 0.900c = 1.600c$, which is greater than c and hence impossible. In part (b), we would have found the scoutship's velocity relative to the spaceship to be $v'_x = v_x - u = 0.950c - 0.900c = 0.050c$; the relativistically correct value, $v'_x = 0.345c$, is almost seven times greater than the incorrect Galilean value.

Test Your Understanding of Section 37.5 (a) In frame S events P_1 and P_2 occur at the same x-, y-, and z-coordinates, but event P_1 occurs before event P_2. In frame S', which event occurs first? (b) In frame S events P_3 and P_4 occur at the same time t and the same y- and z-coordinates, but event P_3 occurs at a less positive x-coordinate than event P_4. In frame S', which event occurs first?

37.6 The Doppler Effect for Electromagnetic Waves

An additional important consequence of relativistic kinematics is the Doppler effect for electromagnetic waves. In our previous discussion of the Doppler effect (see Section 16.8) we quoted without proof the formula, Eq. (16.30), for the frequency shift that results from motion of a source of electromagnetic waves relative to an observer. We can now derive that result.

Here's a statement of the problem. A source of light is moving with constant speed u toward Stanley, who is stationary in an inertial frame (Fig. 37.17). As measured in its rest frame, the source emits light waves with frequency f_0 and period $T_0 = 1/f_0$. What is the frequency f of these waves as received by Stanley?

Let T be the time interval between *emission* of successive wave crests as observed in Stanley's reference frame. Note that this is *not* the interval between the *arrival* of successive crests at his position, because the crests are emitted at different points in Stanley's frame. In measuring only the frequency f he receives, he does not take into account the difference in transit times for successive crests. Therefore the frequency he receives is *not* $1/T$. What is the equation for f?

During a time T the crests ahead of the source move a distance cT, and the source moves a shorter distance uT in the same direction. The distance λ between

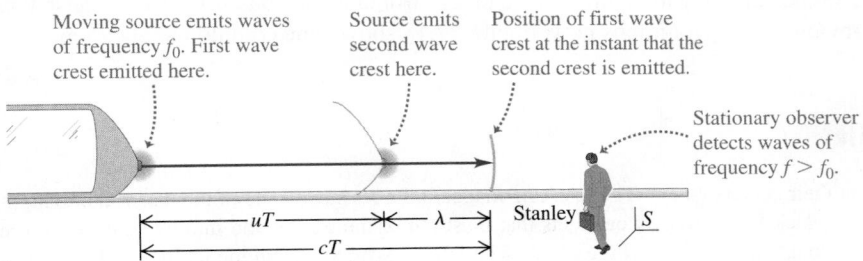

Moving source emits waves of frequency f_0. First wave crest emitted here.

Source emits second wave crest here.

Position of first wave crest at the instant that the second crest is emitted.

Stationary observer detects waves of frequency $f > f_0$.

37.17 The Doppler effect for light. A light source moving at speed u relative to Stanley emits a wave crest, then travels a distance uT toward an observer and emits the next crest. In Stanley's reference frame S, the second crest is a distance λ behind the first crest.

successive crests—that is, the wavelength—is thus $\lambda = (c - u)T$, as measured in Stanley's frame. The frequency that he measures is c/λ. Therefore

$$f = \frac{c}{(c - u)T} \tag{37.24}$$

So far we have followed a pattern similar to that for the Doppler effect for sound from a moving source (see Section 16.8). In that discussion our next step was to equate T to the time T_0 between emissions of successive wave crests by the source. However, due to time dilation it is *not* relativistically correct to equate T to T_0. The time T_0 is measured in the rest frame of the source, so it is a proper time. From Eq. (37.6), T_0 and T are related by

$$T = \frac{T_0}{\sqrt{1 - u^2/c^2}} = \frac{cT_0}{\sqrt{c^2 - u^2}}$$

or, since $T_0 = 1/f_0$,

$$\frac{1}{T} = \frac{\sqrt{c^2 - u^2}}{cT_0} = \frac{\sqrt{c^2 - u^2}}{c}f_0$$

Remember, $1/T$ is not equal to f. We must substitute this expression for $1/T$ into Eq. 37.24 to find f:

$$f = \frac{c}{c - u}\frac{\sqrt{c^2 - u^2}}{c}f_0$$

Using $c^2 - u^2 = (c - u)(c + u)$ gives

$$f = \sqrt{\frac{c + u}{c - u}}f_0 \qquad \text{(Doppler effect, electromagnetic waves, source approaching observer)} \tag{37.25}$$

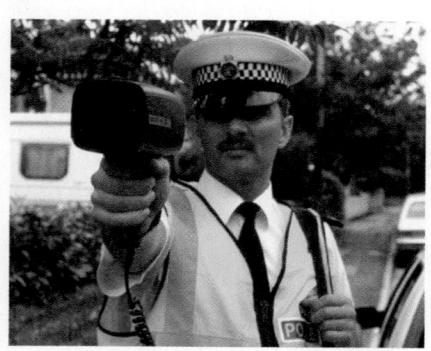

37.18 This handheld radar gun emits a radio beam of frequency f_0, which in the frame of reference of an approaching car has a higher frequency f given by Eq. (37.25). The reflected beam also has frequency f in the car's frame, but has an even higher frequency f' in the police officer's frame. The radar gun calculates the car's speed by comparing the frequencies of the emitted beam and the doubly Doppler-shifted reflected beam. (Compare Example 16.18 in Section 16.8.)

This shows that when the source moves *toward* the observer, the observed frequency f is *greater* than the emitted frequency f_0. The difference $f - f_0 = \Delta f$ is called the Doppler frequency shift. When u/c is much smaller than 1, the fractional shift $\Delta f/f$ is also small and is approximately equal to u/c:

$$\frac{\Delta f}{f} = \frac{u}{c}$$

When the source moves *away from* the observer, we change the sign of u in Eq. (37.25) to get

$$f = \sqrt{\frac{c - u}{c + u}}f_0 \qquad \text{(Doppler effect, electromagnetic waves, source moving away from observer)} \tag{37.26}$$

This agrees with Eq. (16.30), which we quoted previously, with minor notation changes.

With light, unlike sound, there is no distinction between motion of source and motion of observer; only the *relative* velocity of the two is significant. The last four paragraphs of Section 16.8 discuss several practical applications of the Doppler effect with light and other electromagnetic radiation; we suggest you review those paragraphs now. Figure 37.18 shows one common application.

Example 37.8   **A jet from a black hole**

Many galaxies have supermassive black holes at their centers (see Section 13.8). As material swirls around such a black hole, it is heated, becomes ionized, and generates strong magnetic fields. The resulting magnetic forces steer some of the material into high-speed jets that blast out of the galaxy and into intergalactic space (Fig. 37.19). The light we observe from the jet in Fig. 37.19 has a

37.19 This image shows a fast-moving jet 5000 light-years in length emanating from the center of the galaxy M87. The light from the jet is emitted by fast-moving electrons spiraling around magnetic field lines (see Fig. 27.18).

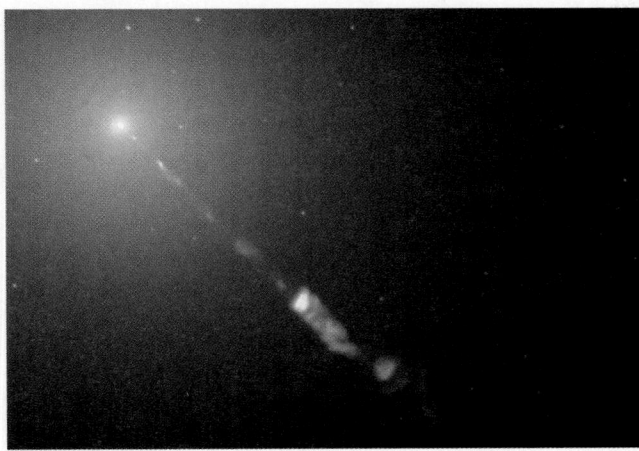

frequency of 6.66×10^{14} Hz (in the far ultraviolet region of the electromagnetic spectrum; see Fig. 32.4), but in the reference frame of the jet material the light has a frequency of 5.55×10^{13} Hz (in the infrared). What is the speed of the jet material with respect to us?

SOLUTION

IDENTIFY and SET UP: This problem involves the Doppler effect for electromagnetic waves. The frequency we observe is $f = 6.66 \times 10^{14}$ Hz, and the frequency in the frame of the source is $f_0 = 5.55 \times 10^{13}$ Hz. Since $f > f_0$, the jet is approaching us and we use Eq. (37.25) to find the target variable u.

EXECUTE: We need to solve Eq. (37.25) for u. We'll leave it as an exercise for you to show that the result is

$$u = \frac{(f/f_0)^2 - 1}{(f/f_0)^2 + 1} c$$

We have $f/f_0 = (6.66 \times 10^{14}\ \text{Hz})/(5.55 \times 10^{13}\ \text{Hz}) = 12.0$, so

$$u = \frac{(12.0)^2 - 1}{(12.0)^2 + 1} c = 0.986c$$

EVALUATE: Because the frequency shift is quite substantial, it would have been erroneous to use the approximate expression $\Delta f/f = u/c$. Had you done so, you would have found $u = c(\Delta f/f_0) = c(6.66 \times 10^{14}\ \text{Hz} - 5.55 \times 10^{13}\ \text{Hz})/(5.55 \times 10^{13}\ \text{Hz}) = 11.0c$. This result cannot be correct because the jet material cannot travel faster than light.

37.7 Relativistic Momentum

Newton's laws of motion have the same form in all inertial frames of reference. When we use transformations to change from one inertial frame to another, the laws should be *invariant* (unchanging). But we have just learned that the principle of relativity forces us to replace the Galilean transformations with the more general Lorentz transformations. As we will see, this requires corresponding generalizations in the laws of motion and the definitions of momentum and energy.

The principle of conservation of momentum states that *when two bodies interact, the total momentum is constant,* provided that the net external force acting on the bodies in an inertial reference frame is zero (for example, if they form an isolated system, interacting only with each other). If conservation of momentum is a valid physical law, it must be valid in *all* inertial frames of reference. Now, here's the problem: Suppose we look at a collision in one inertial coordinate system S and find that momentum is conserved. Then we use the Lorentz transformation to obtain the velocities in a second inertial system S'. We find that if we use the Newtonian definition of momentum ($\vec{p} = m\vec{v}$), momentum is *not* conserved in the second system! If we are convinced that the principle of relativity and the Lorentz transformation are correct, the only way to save momentum conservation is to generalize the *definition* of momentum.

We won't derive the correct relativistic generalization of momentum, but here is the result. Suppose we measure the mass of a particle to be m when it is at rest relative to us: We often call m the **rest mass.** We will use the term *material particle* for a particle that has a nonzero rest mass. When such a particle has a velocity $\vec{v}$, its **relativistic momentum** $\vec{p}$ is

$$\vec{p} = \frac{m\vec{v}}{\sqrt{1 - v^2/c^2}} \qquad \text{(relativistic momentum)} \qquad (37.27)$$

37.20 Graph of the magnitude of the momentum of a particle of rest mass m as a function of speed v. Also shown is the Newtonian prediction, which gives correct results only at speeds much less than c.

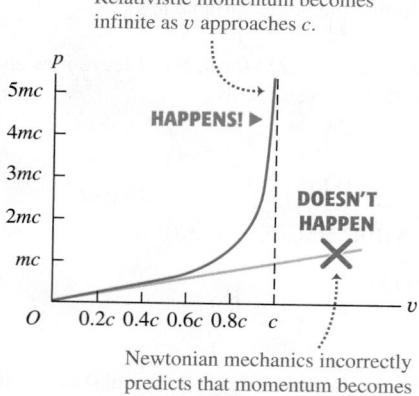

Relativistic momentum becomes infinite as v approaches c.

HAPPENS! ▶

DOESN'T HAPPEN ✗

Newtonian mechanics incorrectly predicts that momentum becomes infinite only if v becomes infinite.

When the particle's speed v is much less than c, this is approximately equal to the Newtonian expression $\vec{p} = m\vec{v}$, but in general the momentum is greater in magnitude than mv (Fig. 37.20). In fact, as v approaches c, the momentum approaches infinity.

Relativity, Newton's Second Law, and Relativistic Mass

What about the relativistic generalization of Newton's second law? In Newtonian mechanics the most general form of the second law is

$$\vec{F} = \frac{d\vec{p}}{dt} \tag{37.28}$$

That is, the net force $\vec{F}$ on a particle equals the time rate of change of its momentum. Experiments show that this result is still valid in relativistic mechanics, provided that we use the relativistic momentum given by Eq. 37.27. That is, the relativistically correct generalization of Newton's second law is

$$\vec{F} = \frac{d}{dt} \frac{m\vec{v}}{\sqrt{1 - v^2/c^2}} \tag{37.29}$$

Because momentum is no longer directly proportional to velocity, the rate of change of momentum is no longer directly proportional to the acceleration. As a result, *constant force does not cause constant acceleration.* For example, when the net force and the velocity are both along the x-axis, Eq. 37.29 gives

$$F = \frac{m}{(1 - v^2/c^2)^{3/2}} a \qquad (\vec{F} \text{ and } \vec{v} \text{ along the same line}) \tag{37.30}$$

where a is the acceleration, also along the x-axis. Solving Eq. (37.30) for the acceleration a gives

$$a = \frac{F}{m}\left(1 - \frac{v^2}{c^2}\right)^{3/2}$$

We see that as a particle's speed increases, the acceleration caused by a given force continuously *decreases.* As the speed approaches c, the acceleration approaches zero, no matter how great a force is applied. Thus it is impossible to accelerate a particle with nonzero rest mass to a speed equal to or greater than c. We again see that the speed of light in vacuum represents an ultimate speed limit.

Equation (37.27) for relativistic momentum is sometimes interpreted to mean that a rapidly moving particle undergoes an increase in mass. If the mass at zero velocity (the rest mass) is denoted by m, then the "relativistic mass" m_{rel} is given by

$$m_{\text{rel}} = \frac{m}{\sqrt{1 - v^2/c^2}}$$

Indeed, when we consider the motion of a system of particles (such as rapidly moving ideal-gas molecules in a stationary container), the total rest mass of the system is the sum of the relativistic masses of the particles, not the sum of their rest masses.

However, if blindly applied, the concept of relativistic mass has its pitfalls. As Eq. (37.29) shows, the relativistic generalization of Newton's second law is *not* $\vec{F} = m_{\text{rel}}\vec{a}$, and we will show in Section 37.8 that the relativistic kinetic energy of a particle is *not* $K = \frac{1}{2}m_{\text{rel}}v^2$. The use of relativistic mass has its supporters and detractors, some quite strong in their opinions. We will mostly deal with individual particles, so we will sidestep the controversy and use Eq. (37.27) as the generalized definition of momentum with m as a constant for each particle, independent of its state of motion.

We will use the abbreviation

$$\gamma = \frac{1}{\sqrt{1 - v^2/c^2}}$$

We used this abbreviation in Section 37.3 with v replaced by u, the relative speed of two coordinate systems. Here v is the speed of a particle in a particular coordinate system—that is, the speed of the particle's *rest frame* with respect to that system. In terms of γ, Eqs. (37.27) and (37.30) become

$$\vec{p} = \gamma m\vec{v} \qquad \text{(relativistic momentum)} \tag{37.31}$$

$$F = \gamma^3 ma \qquad (\vec{F} \text{ and } \vec{v} \text{ along the same line}) \tag{37.32}$$

In linear accelerators (used in medicine as well as nuclear and elementary-particle physics; see Fig. 37.11) the net force $\vec{F}$ and the velocity $\vec{v}$ of the accelerated particle are along the same straight line. But for much of the path in most *circular* accelerators the particle moves in uniform circular motion at constant speed v. Then the net force and velocity are perpendicular, so the force can do no work on the particle and the kinetic energy and speed remain constant. Thus the denominator in Eq. (37.29) is constant, and we obtain

$$F = \frac{m}{(1 - v^2/c^2)^{1/2}} a = \gamma ma \qquad (\vec{F} \text{ and } \vec{v} \text{ perpendicular}) \tag{37.33}$$

Recall from Section 3.4 that if the particle moves in a circle, the net force and acceleration are directed inward along the radius r, and $a = v^2/r$.

What about the general case in which $\vec{F}$ and $\vec{v}$ are neither along the same line nor perpendicular? Then we can resolve the net force $\vec{F}$ at any instant into components parallel to and perpendicular to $\vec{v}$. The resulting acceleration will have corresponding components obtained from Eqs. (37.32) and (37.33). Because of the different γ^3 and γ factors, the acceleration components will not be proportional to the net force components. That is, *unless the net force on a relativistic particle is either along the same line as the particle's velocity or perpendicular to it, the net force and acceleration vectors are not parallel.*

Example 37.9   **Relativistic dynamics of an electron**

An electron (rest mass 9.11×10^{-31} kg, charge -1.60×10^{-19} C) is moving opposite to an electric field of magnitude $E = 5.00 \times 10^5$ N/C. All other forces are negligible in comparison to the electric-field force. (a) Find the magnitudes of momentum and of acceleration at the instants when $v = 0.010c$, $0.90c$, and $0.99c$. (b) Find the corresponding accelerations if a net force of the same magnitude is perpendicular to the velocity.

SOLUTION

IDENTIFY and SET UP: In addition to the expressions from this section for relativistic momentum and acceleration, we need the relationship between electric force and electric field from Chapter 21. In part (a) we use Eq. (37.31) to determine the magnitude of momentum; the force acts along the same line as the velocity, so we use Eq. (37.32) to determine the magnitude of acceleration. In part (b) the force is perpendicular to the velocity, so we use Eq. (37.33) rather than Eq. (37.32).

EXECUTE: (a) For $v = 0.010c$, $0.90c$, and $0.99c$ we have $\gamma = \sqrt{1 - v^2/c^2} = 1.00$, 2.29, and 7.09, respectively. The values of the momentum magnitude $p = \gamma mv$ are

$$p_1 = (1.00)(9.11 \times 10^{-31} \text{ kg})(0.010)(3.00 \times 10^8 \text{m/s})$$
$$= 2.7 \times 10^{-24} \text{ kg} \cdot \text{m/s at } v_1 = 0.010c$$

$$p_2 = (2.29)(9.11 \times 10^{-31} \text{ kg})(0.90)(3.00 \times 10^8 \text{ m/s})$$
$$= 5.6 \times 10^{-22} \text{ kg} \cdot \text{m/s at } v_2 = 0.90c$$

$$p_3 = (7.09)(9.11 \times 10^{-31} \text{ kg})(0.99)(3.00 \times 10^8 \text{ m/s})$$
$$= 1.9 \times 10^{-21} \text{ kg} \cdot \text{m/s at } v_3 = 0.99c$$

From Eq. (21.4), the magnitude of the force on the electron is

$$F = |q|E = (1.60 \times 10^{-19} \text{ C})(5.00 \times 10^5 \text{ N/C})$$
$$= 8.00 \times 10^{-14} \text{ N}$$

Continued

From Eq. (37.32), $a = F/\gamma^3 m$. For $v = 0.010c$ and $\gamma = 1.00$,

$$a_1 = \frac{8.00 \times 10^{-14}\,\text{N}}{(1.00)^3(9.11 \times 10^{-31}\,\text{kg})} = 8.8 \times 10^{16}\,\text{m/s}^2$$

The accelerations at the two higher speeds are smaller than the non-relativistic value by factors of $\gamma^3 = 12.0$ and 356, respectively:

$$a_2 = 7.3 \times 10^{15}\,\text{m/s}^2 \qquad a_3 = 2.5 \times 10^{14}\,\text{m/s}^2$$

(b) From Eq. (37.33), $a = F/\gamma m$ if $\vec{F}$ and $\vec{v}$ are perpendicular. When $v = 0.010c$ and $\gamma = 1.00$,

$$a_1 = \frac{8.00 \times 10^{-14}\,\text{N}}{(1.00)(9.11 \times 10^{-31}\,\text{kg})} = 8.8 \times 10^{16}\,\text{m/s}^2$$

Now the accelerations at the two higher speeds are smaller by factors of $\gamma = 2.29$ and 7.09, respectively:

$$a_2 = 3.8 \times 10^{16}\,\text{m/s}^2 \qquad a_3 = 1.2 \times 10^{16}\,\text{m/s}^2$$

These accelerations are larger than the corresponding ones in part (a) by factors of γ^2.

EVALUATE: Our results in part (a) show that at higher speeds, the relativistic values of momentum differ more and more from the nonrelativistic values calculated from $p = mv$. The momentum at $0.99c$ is more than three times as great as at $0.90c$ because of the increase in the factor γ. Our results also show that the acceleration drops off very quickly as v approaches c.

Test Your Understanding of Section 37.7 According to relativistic mechanics, when you double the speed of a particle, the magnitude of its momentum increases by (i) a factor of 2; (ii) a factor greater than 2; (iii) a factor between 1 and 2 that depends on the mass of the particle.

37.8 Relativistic Work and Energy

When we developed the relationship between work and kinetic energy in Chapter 6, we used Newton's laws of motion. When we generalize these laws according to the principle of relativity, we need a corresponding generalization of the equation for kinetic energy.

Relativistic Kinetic Energy

We use the work–energy theorem, beginning with the definition of work. When the net force and displacement are in the same direction, the work done by that force is $W = \int F\,dx$. We substitute the expression for F from Eq. (37.30), the applicable relativistic version of Newton's second law. In moving a particle of rest mass m from point x_1 to point x_2,

$$W = \int_{x_1}^{x_2} F\,dx = \int_{x_1}^{x_2} \frac{ma\,dx}{(1 - v^2/c^2)^{3/2}} \tag{37.34}$$

To derive the generalized expression for kinetic energy K as a function of speed v, we would like to convert this to an integral on v. To do this, first remember that the kinetic energy of a particle equals the net work done on it in moving it from rest to the speed v: $K = W$. Thus we let the speeds be zero at point x_1 and v at point x_2. So as not to confuse the variable of integration with the final speed, we change v to v_x in Eq. 37.34. That is, v_x is the varying x-component of the velocity of the particle as the net force accelerates it from rest to a speed v. We also realize that dx and dv_x are the infinitesimal changes in x and v_x, respectively, in the time interval dt. Because $v_x = dx/dt$ and $a = dv_x/dt$, we can rewrite $a\,dx$ in Eq. (37.34) as

$$a\,dx = \frac{dv_x}{dt}dx = dx\frac{dv_x}{dt} = \frac{dx}{dt}dv_x = v_x\,dv_x$$

Making these substitutions gives us

$$K = W = \int_0^v \frac{mv_x\,dv_x}{(1 - v_x^2/c^2)^{3/2}} \tag{37.35}$$

We can evaluate this integral by a simple change of variable; the final result is

$$K = \frac{mc^2}{\sqrt{1 - v^2/c^2}} - mc^2 = (\gamma - 1)mc^2 \qquad \text{(relativistic kinetic energy)} \tag{37.36}$$

As v approaches c, the kinetic energy approaches infinity. If Eq. (37.36) is correct, it must also approach the Newtonian expression $K = \frac{1}{2}mv^2$ when v is much smaller than c (Fig. 37.21). To verify this, we expand the radical, using the binomial theorem in the form

$$(1 + x)^n = 1 + nx + n(n - 1)x^2/2 + \cdots$$

In our case, $n = -\frac{1}{2}$ and $x = -v^2/c^2$, and we get

$$\gamma = \left(1 - \frac{v^2}{c^2}\right)^{-1/2} = 1 + \frac{1}{2}\frac{v^2}{c^2} + \frac{3}{8}\frac{v^4}{c^4} + \cdots$$

Combining this with $K = (\gamma - 1)mc^2$, we find

$$K = \left(1 + \frac{1}{2}\frac{v^2}{c^2} + \frac{3}{8}\frac{v^4}{c^4} + \cdots - 1\right)mc^2$$

$$= \frac{1}{2}mv^2 + \frac{3}{8}\frac{mv^4}{c^2} + \cdots \qquad (37.37)$$

When v is much smaller than c, all the terms in the series in Eq. (37.37) except the first are negligibly small, and we obtain the Newtonian expression $\frac{1}{2}mv^2$.

Rest Energy and $E = mc^2$

Equation (37.36) for the kinetic energy of a moving particle includes a term $mc^2/\sqrt{1 - v^2/c^2}$ that depends on the motion and a second energy term mc^2 that is independent of the motion. It seems that the kinetic energy of a particle is the difference between some **total energy** E and an energy mc^2 that it has even when it is at rest. Thus we can rewrite Eq. (37.36) as

$$E = K + mc^2 = \frac{mc^2}{\sqrt{1 - v^2/c^2}} = \gamma mc^2 \qquad \begin{array}{l}\text{(total energy of}\\ \text{a particle)}\end{array} \qquad (37.38)$$

For a particle at rest ($K = 0$), we see that $E = mc^2$. The energy mc^2 associated with rest mass m rather than motion is called the **rest energy** of the particle.

There is in fact direct experimental evidence that rest energy really does exist. The simplest example is the decay of a neutral *pion*. This is an unstable subatomic particle of rest mass m_π; when it decays, it disappears and electromagnetic radiation appears. If a neutral pion has no kinetic energy before its decay, the total energy of the radiation after its decay is found to equal exactly $m_\pi c^2$. In many other fundamental particle transformations the sum of the rest masses of the particles changes. In every case there is a corresponding energy change, consistent with the assumption of a rest energy mc^2 associated with a rest mass m.

Historically, the principles of conservation of mass and of energy developed quite independently. The theory of relativity shows that they are actually two special cases of a single broader conservation principle, the *principle of conservation of mass and energy*. In some physical phenomena, neither the sum of the rest masses of the particles nor the total energy other than rest energy is separately conserved, but there is a more general conservation principle: In an isolated system, when the sum of the rest masses changes, there is always a change in $1/c^2$ times the total energy other than the rest energy. This change is equal in magnitude but opposite in sign to the change in the sum of the rest masses.

This more general mass-energy conservation law is the fundamental principle involved in the generation of power through nuclear reactions. When a uranium nucleus undergoes fission in a nuclear reactor, the sum of the rest masses of the resulting fragments is *less than* the rest mass of the parent nucleus. An amount of energy is released that equals the mass decrease multiplied by c^2. Most of this energy can be used to produce steam to operate turbines for electric power generators.

37.21 Graph of the kinetic energy of a particle of rest mass m as a function of speed v. Also shown is the Newtonian prediction, which gives correct results only at speeds much less than c.

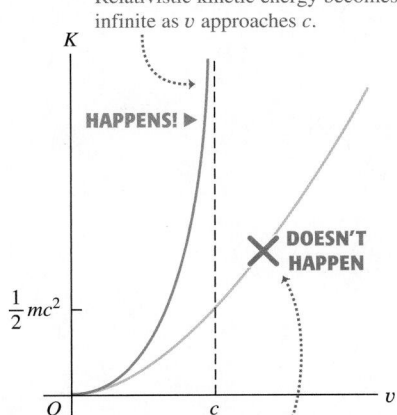

Relativistic kinetic energy becomes infinite as v approaches c.

HAPPENS! ▶

✕ DOESN'T HAPPEN

Newtonian mechanics incorrectly predicts that kinetic energy becomes infinite only if v becomes infinite.

Application Monitoring Mass–Energy Conversion

Although the control room of a nuclear power plant is very complex, the physical principle on which such a plant operates is a simple one: Part of the rest energy of atomic nuclei is converted to thermal energy, which in turn is used to produce steam to drive electric generators.

We can also relate the total energy E of a particle (kinetic energy plus rest energy) directly to its momentum by combining Eq. (37.27) for relativistic momentum and Eq. (37.38) for total energy to eliminate the particle's velocity. The simplest procedure is to rewrite these equations in the following forms:

$$\left(\frac{E}{mc^2}\right)^2 = \frac{1}{1 - v^2/c^2} \quad \text{and} \quad \left(\frac{p}{mc}\right)^2 = \frac{v^2/c^2}{1 - v^2/c^2}$$

Subtracting the second of these from the first and rearranging, we find

$$E^2 = (mc^2)^2 + (pc)^2 \qquad \text{(total energy, rest energy, and momentum)} \qquad (37.39)$$

Again we see that for a particle at rest $(p = 0)$, $E = mc^2$.

Equation (37.39) also suggests that a particle may have energy and momentum even when it has no rest mass. In such a case, $m = 0$ and

$$E = pc \qquad \text{(zero rest mass)} \qquad (37.40)$$

In fact, zero rest mass particles do exist. Such particles always travel at the speed of light in vacuum. One example is the *photon,* the quantum of electromagnetic radiation (to be discussed in Chapter 38). Photons are emitted and absorbed during changes of state of an atomic or nuclear system when the energy and momentum of the system change.

Example 37.10 Energetic electrons

(a) Find the rest energy of an electron ($m = 9.109 \times 10^{-31}$ kg, $q = -e = -1.602 \times 10^{-19}$ C) in joules and in electron volts. (b) Find the speed of an electron that has been accelerated by an electric field, from rest, through a potential increase of 20.0 kV or of 5.00 MV (typical of a high-voltage x-ray machine).

SOLUTION

IDENTIFY and SET UP: This problem uses the ideas of rest energy, relativistic kinetic energy, and (from Chapter 23) electric potential energy. We use $E = mc^2$ to find the rest energy and Eqs. (37.7) and (37.38) to find the speed that gives the stated total energy.

EXECUTE: (a) The rest energy is

$$mc^2 = (9.109 \times 10^{-31} \text{ kg})(2.998 \times 10^8 \text{ m/s})^2$$
$$= 8.187 \times 10^{-14} \text{ J}$$

From the definition of the electron volt in Section 23.2, 1 eV = 1.602×10^{-19} J. Using this, we find

$$mc^2 = (8.187 \times 10^{-14} \text{ J})\frac{1 \text{ eV}}{1.602 \times 10^{-19} \text{ J}}$$
$$= 5.11 \times 10^5 \text{ eV} = 0.511 \text{ MeV}$$

(b) In calculations such as this, it is often convenient to work with the quantity $\gamma = 1/\sqrt{1 - v^2/c^2}$ from Eq. (37.38). Solving this for v, we find

$$v = c\sqrt{1 - (1/\gamma)^2}$$

The total energy E of the accelerated electron is the sum of its rest energy mc^2 and the kinetic energy eV_{ba} that it gains from the work done on it by the electric field in moving from point a to point b:

$$E = \gamma mc^2 = mc^2 + eV_{ba} \quad \text{or}$$
$$\gamma = 1 + \frac{eV_{ba}}{mc^2}$$

An electron accelerated through a potential increase of $V_{ba} = 20.0$ kV gains 20.0 keV of energy, so for this electron

$$\gamma = 1 + \frac{20.0 \times 10^3 \text{ eV}}{0.511 \times 10^6 \text{ eV}} = 1.039$$

and

$$v = c\sqrt{1 - (1/1.039)^2} = 0.272c = 8.15 \times 10^7 \text{ m/s}$$

Repeating the calculation for $V_{ba} = 5.00$ MV, we find $eV_{ba}/mc^2 = 9.78$, $\gamma = 10.78$, and $v = 0.996c$.

EVALUATE: With $V_{ba} = 20.0$ kV, the added kinetic energy of 20.0 keV is less than 4% of the rest energy of 0.511 MeV, and the final speed is about one-fourth the speed of light. With $V_{ba} = 5.00$ MV, the added kinetic energy of 5.00 MeV is much greater than the rest energy and the speed is close to c.

CAUTION **Three electron energies** All electrons have *rest* energy 0.511 MeV. An electron accelerated from rest through a 5.00-MeV potential increase has *kinetic* energy 5.00 MeV (we call it a "5.00-MeV electron") and *total* energy 5.51 MeV. Be careful to distinguish these energies from one another.

Example 37.11 A relativistic collision

Two protons (each with mass $M_p = 1.67 \times 10^{-27}$ kg) are initially moving with equal speeds in opposite directions. They continue to exist after a head-on collision that also produces a neutral pion of mass $M_\pi = 2.40 \times 10^{-28}$ kg (Fig. 37.22). If all three particles are at rest after the collision, find the initial speed of the protons. Energy is conserved in the collision.

37.22 In this collision the kinetic energy of two protons is transformed into the rest energy of a new particle, a pion.

1.67 × 10⁻²⁷ kg

BEFORE

Proton   Proton

AFTER

Pion (2.40 × 10⁻²⁸ kg)

SOLUTION

IDENTIFY and SET UP: Relativistic total energy is conserved in the collision, so we can equate the (unknown) total energy of the two protons before the collision to the combined rest energies of the two protons and the pion after the collision. We then use Eq. (37.38) to find the speed of each proton.

EXECUTE: The total energy of each proton before the collision is $\gamma M c^2$. By conservation of energy,

$$2(\gamma M_p c^2) = 2(M_p c^2) + M_\pi c^2$$

$$\gamma = 1 + \frac{M_\pi}{2M_p} = 1 + \frac{2.40 \times 10^{-28} \text{ kg}}{2(1.67 \times 10^{-27} \text{ kg})} = 1.072$$

From Eq. (37.38), the initial proton speed is

$$v = c\sqrt{1 - (1/\gamma)^2} = 0.360c$$

EVALUATE: The proton rest energy is 938 MeV, so the initial kinetic energy of each proton is $(\gamma - 1)Mc^2 = 0.072Mc^2 = (0.072)(938 \text{ MeV}) = 67.5$ MeV. You can verify that the rest energy $M_\pi c^2$ of the pion is twice this, or 135 MeV. All the kinetic energy "lost" in this completely inelastic collision is transformed into the rest energy of the pion.

Test Your Understanding of Section 37.8 A proton is accelerated from rest by a constant force that always points in the direction of the particle's motion. Compared to the amount of kinetic energy that the proton gains during the first meter of its travel, how much kinetic energy does the proton gain during one meter of travel while it is moving at 99% of the speed of light? (i) the same amount; (ii) a greater amount; (iii) a smaller amount.

37.9 Newtonian Mechanics and Relativity

The sweeping changes required by the principle of relativity go to the very roots of Newtonian mechanics, including the concepts of length and time, the equations of motion, and the conservation principles. Thus it may appear that we have destroyed the foundations on which Newtonian mechanics is built. In one sense this is true, yet the Newtonian formulation is still accurate whenever speeds are small in comparison with the speed of light in vacuum. In such cases, time dilation, length contraction, and the modifications of the laws of motion are so small that they are unobservable. In fact, every one of the principles of Newtonian mechanics survives as a special case of the more general relativistic formulation.

The laws of Newtonian mechanics are not *wrong;* they are *incomplete.* They are a limiting case of relativistic mechanics. They are *approximately* correct when all speeds are small in comparison to *c*, and they become exactly correct in the limit when all speeds approach zero. Thus relativity does not completely destroy the laws of Newtonian mechanics but *generalizes* them. This is a common pattern in the development of physical theory. Whenever a new theory is in partial conflict with an older, established theory, the new must yield the same predictions as the old in areas in which the old theory is supported by experimental evidence. Every new physical theory must pass this test, called the **correspondence principle.**

The General Theory of Relativity

At this point we may ask whether the special theory of relativity gives the final word on mechanics or whether *further* generalizations are possible or necessary.

For example, inertial frames have occupied a privileged position in our discussion. Can the principle of relativity be extended to noninertial frames as well?

Here's an example that illustrates some implications of this question. A student decides to go over Niagara Falls while enclosed in a large wooden box. During her free fall she doesn't fall to the floor of the box because both she and the box are in free fall with a downward acceleration of 9.8 m/s². But an alternative interpretation, from her point of view, is that she doesn't fall to the floor because her gravitational interaction with the earth has suddenly been turned off. As long as she remains in the box and it remains in free fall, she cannot tell whether she is indeed in free fall or whether the gravitational interaction has vanished.

A similar problem occurs in a space station in orbit around the earth. Objects in the space station *seem* to be weightless, but without looking outside the station there is no way to determine whether gravity has been turned off or whether the station and all its contents are accelerating toward the center of the earth. Figure 37.23 makes a similar point for a spaceship that is not in free fall but may be accelerating relative to an inertial frame or be at rest on the earth's surface.

These considerations form the basis of Einstein's **general theory of relativity.** If we cannot distinguish experimentally between a uniform gravitational field at a particular location and a uniformly accelerated reference frame, then there cannot be any real distinction between the two. Pursuing this concept, we may try to represent *any* gravitational field in terms of special characteristics of the coordinate system. This turns out to require even more sweeping revisions of our space-time concepts than did the special theory of relativity. In the general theory of relativity the geometric properties of space are affected by the presence of matter (Fig. 37.24).

The general theory of relativity has passed several experimental tests, including three proposed by Einstein. One test has to do with understanding the rotation of the axes of the planet Mercury's elliptical orbit, called the *precession of the perihelion.* (The perihelion is the point of closest approach to the sun.) A second test concerns the apparent bending of light rays from distant stars when they pass near the sun. The third test is the *gravitational red shift,* the increase in wavelength of light proceeding outward from a massive source. Some details of the general theory are more difficult to test, but this theory has played a central role in investigations of the formation and evolution of stars, black holes, and studies of the evolution of the universe.

The general theory of relativity may seem to be an exotic bit of knowledge with little practical application. In fact, this theory plays an essential role in the

37.23 Without information from outside the spaceship, the astronaut cannot distinguish situation (b) from situation (c).

(a) An astronaut is about to drop her watch in a spaceship.

(b) In gravity-free space, the floor accelerates upward at $a = g$ and hits the watch.

(c) On the earth's surface, the watch accelerates downward at $a = g$ and hits the floor.

37.24 A two-dimensional representation of curved space. We imagine the space (a plane) as being distorted as shown by a massive object (the sun). Light from a distant star (solid line) follows the distorted surface on its way to the earth. The dashed line shows the direction from which the light *appears* to be coming. The effect is greatly exaggerated; for the sun, the maximum deviation is only 0.00048°.

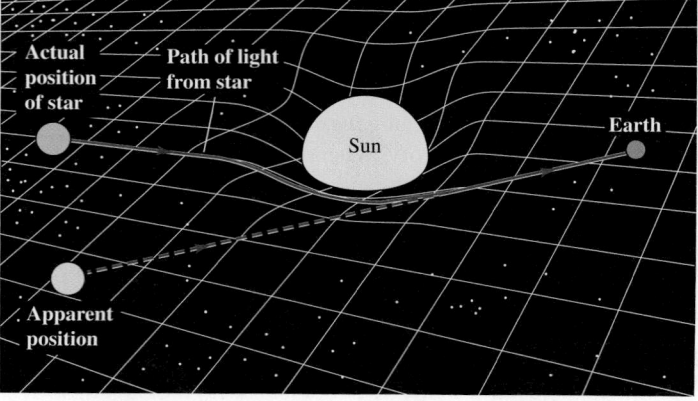

global positioning system (GPS), which makes it possible to determine your position on the earth's surface to within a few meters using a handheld receiver (Fig. 37.25). The heart of the GPS system is a collection of more than two dozen satellites in very precise orbits. Each satellite emits carefully timed radio signals, and a GPS receiver simultaneously detects the signals from several satellites. The receiver then calculates the time delay between when each signal was emitted and when it was received, and uses this information to calculate the receiver's position. To ensure the proper timing of the signals, it's necessary to include corrections due to the special theory of relativity (because the satellites are moving relative to the receiver on earth) as well as the general theory (because the satellites are higher in the earth's gravitational field than the receiver). The corrections due to relativity are small—less than one part in 10^9—but are crucial to the superb precision of the GPS system.

37.25 A GPS receiver uses radio signals from the orbiting GPS satellites to determine its position. To account for the effects of relativity, the receiver must be tuned to a slightly higher frequency (10.23 MHz) than the frequency emitted by the satellites (10.22999999543 MHz).

Invariance of physical laws, simultaneity: All of the fundamental laws of physics have the same form in all inertial frames of reference. The speed of light in vacuum is the same in all inertial frames and is independent of the motion of the source. Simultaneity is not an absolute concept; events that are simultaneous in one frame are not necessarily simultaneous in a second frame moving relative to the first.

Time dilation: If two events occur at the same space point in a particular frame of reference, the time interval Δt_0 between the events as measured in that frame is called a proper time interval. If this frame moves with constant velocity u relative to a second frame, the time interval Δt between the events as observed in the second frame is longer than Δt_0. (See Examples 37.1–37.3.)

$$\Delta t = \frac{\Delta t_0}{\sqrt{1 - u^2/c^2}} = \gamma \, \Delta t_0$$

$$(37.6), (37.8)$$

$$\gamma = \frac{1}{\sqrt{1 - u^2/c^2}} \qquad (37.7)$$

Length contraction: If two points are at rest in a particular frame of reference, the distance l_0 between the points as measured in that frame is called a proper length. If this frame moves with constant velocity u relative to a second frame and the distances are measured parallel to the motion, the distance l between the points as measured in the second frame is shorter than l_0. (See Examples 37.4 and 37.5.)

$$l = l_0 \sqrt{1 - u^2/c^2} = \frac{l_0}{\gamma} \qquad (37.16)$$

The Lorentz transformations: The Lorentz coordinate transformations relate the coordinates and time of an event in an inertial frame S to the coordinates and time of the same event as observed in a second inertial frame S' moving at velocity u relative to the first. For one-dimensional motion, a particle's velocities v_x in S and v'_x in S' are related by the Lorentz velocity transformation. (See Examples 37.6 and 37.7.)

$$x' = \frac{x - ut}{\sqrt{1 - u^2/c^2}} = \gamma(x - ut)$$

$$y' = y \qquad z' = z \qquad (37.21)$$

$$t' = \frac{t - ux/c^2}{\sqrt{1 - u^2/c^2}} = \gamma(t - ux/c^2)$$

$$v'_x = \frac{v_x - u}{1 - uv_x/c^2} \qquad (37.22)$$

$$v_x = \frac{v'_x + u}{1 + uv'_x/c^2} \qquad (37.23)$$

The Doppler effect for electromagnetic waves: The Doppler effect is the frequency shift in light from a source due to the relative motion of source and observer. For a source moving toward the observer with speed u, Eq. (37.25) gives the received frequency f in terms of the emitted frequency f_0. (See Example 37.8.)

$$f = \sqrt{\frac{c + u}{c - u}} f_0 \qquad (37.25)$$

Moving source emits light of frequency f_0. Stationary observer detects light of frequency $f > f_0$.

Relativistic momentum and energy: For a particle of rest mass m moving with velocity $\vec{v}$, the relativistic momentum $\vec{p}$ is given by Eq. (37.27) or (37.31) and the relativistic kinetic energy K is given by Eq. (37.36). The total energy E is the sum of the kinetic energy and the rest energy mc^2. The total energy can also be expressed in terms of the magnitude of momentum p and rest mass m. (See Examples 37.9–37.11.)

$$\vec{p} = \frac{m\vec{v}}{\sqrt{1 - v^2/c^2}} = \gamma m\vec{v} \qquad (37.27), (37.31)$$

$$K = \frac{mc^2}{\sqrt{1 - v^2/c^2}} - mc^2 = (\gamma - 1)mc^2$$

$$(37.36)$$

$$E = K + mc^2 = \frac{mc^2}{\sqrt{1 - v^2/c^2}} = \gamma mc^2$$

$$(37.38)$$

$$E^2 = (mc^2)^2 + (pc)^2 \qquad (37.39)$$

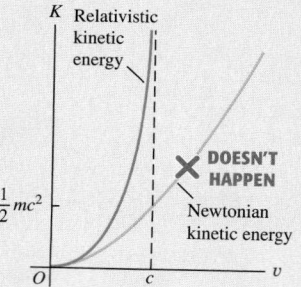

BRIDGING PROBLEM   Colliding Protons

In an experiment, two protons are shot directly toward each other. Their speeds are such that in the frame of reference of each proton, the other proton is moving at 0.500c. (a) What does an observer in the laboratory measure for the speed of each proton? (b) What is the kinetic energy of each proton as measured by an observer in the laboratory? (c) What is the kinetic energy of each proton as measured by the other proton?

SOLUTION GUIDE

See MasteringPhysics® study area for a Video Tutor solution.

IDENTIFY and SET UP

1. This problem uses the Lorentz velocity transformation, which allows us to relate the velocity v_x of a proton in one frame to its velocity v_x' in a different frame. It also uses the idea of relativistic kinetic energy.

2. Take the x-axis to be the line of motion of the protons, and take the $+x$-direction to be to the right. In the frame in which the left-hand proton is at rest, the right-hand proton has velocity $-0.500c$. In the laboratory frame the two protons have velocities

$-\alpha c$ and $+\alpha c$, where α (each proton's laboratory speed as a fraction of c) is our first target variable. Given this we can find the laboratory kinetic energy of each proton.

EXECUTE

3. Write a Lorentz velocity-transformation equation that relates the velocity of the right-hand proton in the laboratory frame to its velocity in the frame of the left-hand proton. Solve this equation for α. (*Hint:* Remember that α cannot be greater than 1. Why?)

4. Use your result from step 3 to find the laboratory kinetic energy of each proton.

5. Find the kinetic energy of the right-hand proton as measured in the frame of the left-hand proton.

EVALUATE

6. How much total kinetic energy must be imparted to the protons by a scientist in the laboratory? If the experiment were to be repeated with one proton stationary, what kinetic energy would have to be given to the other proton for the collision to be equivalent?

Problems   For instructor-assigned homework, go to www.masteringphysics.com

•, ••, •••: Problems of increasing difficulty. **CP**: Cumulative problems incorporating material from earlier chapters. **CALC**: Problems requiring calculus. **BIO**: Biosciences problems.

DISCUSSION QUESTIONS

Q37.1 You are standing on a train platform watching a high-speed train pass by. A light inside one of the train cars is turned on and then a little later it is turned off. (a) Who can measure the proper time interval for the duration of the light: you or a passenger on the train? (b) Who can measure the proper length of the train car: you or a passenger on the train? (c) Who can measure the proper length of a sign attached to a post on the train platform: you or a passenger on the train? In each case explain your answer.

Q37.2 If simultaneity is not an absolute concept, does that mean that we must discard the concept of causality? If event A is to *cause* event B, A must occur first. Is it possible that in some frames A appears to be the cause of B, and in others B appears to be the cause of A? Explain.

Q37.3 A rocket is moving to the right at $\frac{1}{2}$ the speed of light relative to the earth. A light bulb in the center of a room inside the rocket suddenly turns on. Call the light hitting the front end of the room event A and the light hitting the back of the room event B (Fig. Q37.3). Which event occurs first, A or B, or are they simultaneous, as viewed by (a) an astronaut riding in the rocket and (b) a person at rest on the earth?

Figure **Q37.3**

Q37.4 What do you think would be different in everyday life if the speed of light were 10 m/s instead of 3.00×10^8 m/s?

Q37.5 The average life span in the United States is about 70 years. Does this mean that it is impossible for an average person to travel a distance greater than 70 light-years away from the earth? (A light-year is the distance light travels in a year.) Explain.

Q37.6 You are holding an elliptical serving platter. How would you need to travel for the serving platter to appear round to another observer?

Q37.7 Two events occur at the same space point in a particular inertial frame of reference and are simultaneous in that frame. Is it possible that they may not be simultaneous in a different inertial frame? Explain.

Q37.8 A high-speed train passes a train platform. Larry is a passenger on the train, Adam is standing on the train platform, and David is riding a bicycle toward the platform in the same direction as the train is traveling. Compare the length of a train car as measured by Larry, Adam, and David.

Q37.9 The theory of relativity sets an upper limit on the speed that a particle can have. Are there also limits on the energy and momentum of a particle? Explain.

Q37.10 A student asserts that a material particle must always have a speed slower than that of light, and a massless particle must always move at exactly the speed of light. Is she correct? If so, how do massless particles such as photons and neutrinos acquire this speed? Can't they start from rest and accelerate? Explain.

Q37.11 The speed of light relative to still water is 2.25×10^8 m/s. If the water is moving past us, the speed of light we measure depends on the speed of the water. Do these facts violate Einstein's second postulate? Explain.

Q37.12 When a monochromatic light source moves toward an observer, its wavelength appears to be shorter than the value measured when the source is at rest. Does this contradict the hypothesis that the speed of light is the same for all observers? Explain.

Q37.13 In principle, does a hot gas have more mass than the same gas when it is cold? Explain. In practice, would this be a measurable effect? Explain.

Q37.14 Why do you think the development of Newtonian mechanics preceded the more refined relativistic mechanics by so many years?

EXERCISES

Section 37.2 Relativity of Simultaneity

37.1 • Suppose the two lightning bolts shown in Fig. 37.5a are simultaneous to an observer on the train. Show that they are *not* simultaneous to an observer on the ground. Which lightning strike does the ground observer measure to come first?

Section 37.3 Relativity of Time Intervals

37.2 • The positive muon (μ^+), an unstable particle, lives on average 2.20×10^{-6} s (measured in its own frame of reference) before decaying. (a) If such a particle is moving, with respect to the laboratory, with a speed of $0.900c$, what average lifetime is measured in the laboratory? (b) What average distance, measured in the laboratory, does the particle move before decaying?

37.3 • How fast must a rocket travel relative to the earth so that time in the rocket "slows down" to half its rate as measured by earth-based observers? Do present-day jet planes approach such speeds?

37.4 • A spaceship flies past Mars with a speed of $0.985c$ relative to the surface of the planet. When the spaceship is directly overhead, a signal light on the Martian surface blinks on and then off. An observer on Mars measures that the signal light was on for 75.0 μs. (a) Does the observer on Mars or the pilot on the spaceship measure the proper time? (b) What is the duration of the light pulse measured by the pilot of the spaceship?

37.5 • The negative pion (π^-) is an unstable particle with an average lifetime of 2.60×10^{-8} s (measured in the rest frame of the pion). (a) If the pion is made to travel at very high speed relative to a laboratory, its average lifetime is measured in the laboratory to be 4.20×10^{-7} s. Calculate the speed of the pion expressed as a fraction of c. (b) What distance, measured in the laboratory, does the pion travel during its average lifetime?

37.6 •• As you pilot your space utility vehicle at a constant speed toward the moon, a race pilot flies past you in her spaceracer at a constant speed of $0.800c$ relative to you. At the instant when the spaceracer passes you, both of you start timers at zero. (a) At the instant when you measure that the spaceracer has traveled 1.20×10^8 m past you, what does the race pilot read on her timer? (b) When the race pilot reads the value calculated in part (a) on her timer, what does she measure to be your distance from her? (c) At the instant when the race pilot reads the value calculated in part (a) on her timer, what do you read on yours?

37.7 •• A spacecraft flies away from the earth with a speed of 4.80×10^6 m/s relative to the earth and then returns at the same speed. The spacecraft carries an atomic clock that has been carefully synchronized with an identical clock that remains at rest on earth. The spacecraft returns to its starting point 365 days (1 year) later, as measured by the clock that remained on earth. What is the difference in the elapsed times on the two clocks, measured in hours? Which clock, the one in the spacecraft or the one on earth, shows the shorter elapsed time?

37.8 • An alien spacecraft is flying overhead at a great distance as you stand in your backyard. You see its searchlight blink on for 0.190 s. The first officer on the spacecraft measures that the searchlight is on for 12.0 ms. (a) Which of these two measured times is the proper time? (b) What is the speed of the spacecraft relative to the earth expressed as a fraction of the speed of light c?

Section 37.4 Relativity of Length

37.9 • A spacecraft of the Trade Federation flies past the planet Coruscant at a speed of $0.600c$. A scientist on Coruscant measures the length of the moving spacecraft to be 74.0 m. The spacecraft later lands on Coruscant, and the same scientist measures the length of the now stationary spacecraft. What value does she get?

37.10 • A meter stick moves past you at great speed. Its motion relative to you is parallel to its long axis. If you measure the length of the moving meter stick to be 1.00 ft (1 ft = 0.3048 m)—for example, by comparing it to a 1-foot ruler that is at rest relative to you—at what speed is the meter stick moving relative to you?

37.11 •• **Why Are We Bombarded by Muons?** Muons are unstable subatomic particles that decay to electrons with a mean lifetime of 2.2 μs. They are produced when cosmic rays bombard the upper atmosphere about 10 km above the earth's surface, and they travel very close to the speed of light. The problem we want to address is why we see any of them at the earth's surface. (a) What is the greatest distance a muon could travel during its 2.2-μs lifetime? (b) According to your answer in part (a), it would seem that muons could never make it to the ground. But the 2.2-μs lifetime is measured in the frame of the muon, and muons are moving very fast. At a speed of $0.999c$, what is the mean lifetime of a muon as measured by an observer at rest on the earth? How far would the muon travel in this time? Does this result explain why we find muons in cosmic rays? (c) From the point of view of the muon, it still lives for only 2.2 μs, so how does it make it to the ground? What is the thickness of the 10 km of atmosphere through which the muon must travel, as measured by the muon? Is it now clear how the muon is able to reach the ground?

37.12 • An unstable particle is created in the upper atmosphere from a cosmic ray and travels straight down toward the surface of the earth with a speed of $0.99540c$ relative to the earth. A scientist at rest on the earth's surface measures that the particle is created at an altitude of 45.0 km. (a) As measured by the scientist, how much time does it take the particle to travel the 45.0 km to the surface of the earth? (b) Use the length-contraction formula to calculate the distance from where the particle is created to the surface of the earth as measured in the particle's frame. (c) In the particle's frame, how much time does it take the particle to travel from where it is created to the surface of the earth? Calculate this time both by the time dilation formula and from the distance calculated in part (b). Do the two results agree?

37.13 • As measured by an observer on the earth, a spacecraft runway on earth has a length of 3600 m. (a) What is the length of the runway as measured by a pilot of a spacecraft flying past at a speed of 4.00×10^7 m/s relative to the earth? (b) An observer on earth measures the time interval from when the spacecraft is directly over one end of the runway until it is directly over the other end. What result does she get? (c) The pilot of the spacecraft measures the time it takes him to travel from one end of the runway to the other end. What value does he get?

37.14 • A rocket ship flies past the earth at 85.0% of the speed of light. Inside, an astronaut who is undergoing a physical examination is having his height measured while he is lying down parallel to the direction the rocket ship is moving. (a) If his height is measured to be 2.00 m by his doctor inside the ship, what height would a person watching this from earth measure for his height? (b) If the earth-based person had measured 2.00 m, what would the doctor in the spaceship have measured for the astronaut's height? Is this a reasonable height? (c) Suppose the astronaut in part (a) gets up after the examination and stands with his body perpendicular to the direction of motion. What would the doctor in the rocket and the observer on earth measure for his height now?

Section 37.5 The Lorentz Transformations

37.15 • An observer in frame S' is moving to the right ($+x$-direction) at speed $u = 0.600c$ away from a stationary observer in frame S. The observer in S' measures the speed v' of a particle moving to the right away from her. What speed v does the observer in S measure for the particle if (a) $v' = 0.400c$; (b) $v' = 0.900c$; (c) $v' = 0.990c$?

37.16 • Space pilot Mavis zips past Stanley at a constant speed relative to him of $0.800c$. Mavis and Stanley start timers at zero when the front of Mavis's ship is directly above Stanley. When Mavis reads 5.00 s on her timer, she turns on a bright light under the front of her spaceship. (a) Use the Lorentz coordinate transformation derived in Example 37.6 to calculate x and t as measured by Stanley for the event of turning on the light. (b) Use the time dilation formula, Eq. (37.6), to calculate the time interval between the two events (the front of the spaceship passing overhead and turning on the light) as measured by Stanley. Compare to the value of t you calculated in part (a). (c) Multiply the time interval by Mavis's speed, both as measured by Stanley, to calculate the distance she has traveled as measured by him when the light turns on. Compare to the value of x you calculated in part (a).

37.17 •• A pursuit spacecraft from the planet Tatooine is attempting to catch up with a Trade Federation cruiser. As measured by an observer on Tatooine, the cruiser is traveling away from the planet with a speed of $0.600c$. The pursuit ship is traveling at a speed of $0.800c$ relative to Tatooine, in the same direction as the cruiser. (a) For the pursuit ship to catch the cruiser, should the velocity of the cruiser relative to the pursuit ship be directed toward or away from the pursuit ship? (b) What is the speed of the cruiser relative to the pursuit ship?

37.18 • An extraterrestrial spaceship is moving away from the earth after an unpleasant encounter with its inhabitants. As it departs, the spaceship fires a missile toward the earth. An observer on earth measures that the spaceship is moving away with a speed of $0.600c$. An observer in the spaceship measures that the missile is moving away from him at a speed of $0.800c$. As measured by an observer on earth, how fast is the missile approaching the earth?

37.19 •• Two particles are created in a high-energy accelerator and move off in opposite directions. The speed of one particle, as measured in the laboratory, is $0.650c$, and the speed of each particle relative to the other is $0.950c$. What is the speed of the second particle, as measured in the laboratory?

37.20 •• Two particles in a high-energy accelerator experiment are approaching each other head-on, each with a speed of $0.9520c$ as measured in the laboratory. What is the magnitude of the velocity of one particle relative to the other?

37.21 •• Two particles in a high-energy accelerator experiment approach each other head-on with a relative speed of $0.890c$. Both particles travel at the same speed as measured in the laboratory. What is the speed of each particle, as measured in the laboratory?

37.22 •• An enemy spaceship is moving toward your starfighter with a speed, as measured in your frame, of $0.400c$. The enemy ship fires a missile toward you at a speed of $0.700c$ relative to the enemy ship (Fig. E37.22). (a) What is the speed of the missile relative to you? Express your answer in terms of the speed of light. (b) If you measure that the enemy ship is 8.00×10^6 km away from you when the missile is fired, how much time, measured in your frame, will it take the missile to reach you?

Figure **E37.22**

Enemy   Starfighter

37.23 • An imperial spaceship, moving at high speed relative to the planet Arrakis, fires a rocket toward the planet with a speed of $0.920c$ relative to the spaceship. An observer on Arrakis measures that the rocket is approaching with a speed of $0.360c$. What is the speed of the spaceship relative to Arrakis? Is the spaceship moving toward or away from Arrakis?

Section 37.6 The Doppler Effect for Electromagnetic Waves

37.24 • Electromagnetic radiation from a star is observed with an earth-based telescope. The star is moving away from the earth with a speed of $0.600c$. If the radiation has a frequency of 8.64×10^{14} Hz in the rest frame of the star, what is the frequency measured by an observer on earth?

37.25 • **Tell It to the Judge.** (a) How fast must you be approaching a red traffic light ($\lambda = 675$ nm) for it to appear yellow ($\lambda = 575$ nm)? Express your answer in terms of the speed of light. (b) If you used this as a reason not to get a ticket for running a red light, how much of a fine would you get for speeding? Assume that the fine is $1.00 for each kilometer per hour that your speed exceeds the posted limit of 90 km/h.

37.26 • A source of electromagnetic radiation is moving in a radial direction relative to you. The frequency you measure is 1.25 times the frequency measured in the rest frame of the source. What is the speed of the source relative to you? Is the source moving toward you or away from you?

Section 37.7 Relativistic Momentum

37.27 • A proton has momentum with magnitude p_0 when its speed is $0.400c$. In terms of p_0, what is the magnitude of the proton's momentum when its speed is doubled to $0.800c$?

37.28 • **When Should You Use Relativity?** As you have seen, relativistic calculations usually involve the quantity γ. When γ is appreciably greater than 1, we must use relativistic formulas instead of Newtonian ones. For what speed v (in terms of c) is the value of γ (a) 1.0% greater than 1; (b) 10% greater than 1; (c) 100% greater than 1?

37.29 • (a) At what speed is the momentum of a particle twice as great as the result obtained from the nonrelativistic expression mv? Express your answer in terms of the speed of light. (b) A force is applied to a particle along its direction of motion. At what speed is the magnitude of force required to produce a given acceleration twice as great as the force required to produce the same acceleration when the particle is at rest? Express your answer in terms of the speed of light.

37.30 • As measured in an earth-based frame, a proton is moving in the $+x$-direction at a speed of 2.30×10^8 m/s. (a) What force (magnitude and direction) is required to produce an acceleration in the $-x$-direction that has magnitude 2.30×10^8 m/s^2? (b) What magnitude of acceleration does the force calculated in part (a) give to a proton that is initially at rest?

37.31 • An electron is acted upon by a force of 5.00×10^{-15} N due to an electric field. Find the acceleration this force produces in each case: (a) The electron's speed is 1.00 km/s. (b) The electron's speed is 2.50×10^8 m/s and the force is parallel to the velocity.

37.32 • **Relativistic Baseball.** Calculate the magnitude of the force required to give a 0.145-kg baseball an acceleration $a = 1.00$ m/s^2 in the direction of the baseball's initial velocity when this velocity has a magnitude of (a) 10.0 m/s; (b) $0.900c$; (c) $0.990c$. (d) Repeat parts (a), (b), and (c) if the force and acceleration are perpendicular to the velocity.

Section 37.8 Relativistic Work and Energy

37.33 •• What is the speed of a particle whose kinetic energy is equal to (a) its rest energy and (b) five times its rest energy?

37.34 • If a muon is traveling at $0.999c$, what are its momentum and kinetic energy? (The mass of such a muon at rest in the laboratory is 207 times the electron mass.)

37.35 • A proton (rest mass 1.67×10^{-27} kg) has total energy that is 4.00 times its rest energy. What are (a) the kinetic energy of the proton; (b) the magnitude of the momentum of the proton; (c) the speed of the proton?

37.36 •• (a) How much work must be done on a particle with mass m to accelerate it (a) from rest to a speed of $0.090c$ and (b) from a speed of $0.900c$ to a speed of $0.990c$? (Express the answers in terms of mc^2.) (c) How do your answers in parts (a) and (b) compare?

37.37 • CP (a) By what percentage does your rest mass increase when you climb 30 m to the top of a ten-story building? Are you aware of this increase? Explain. (b) By how many grams does the mass of a 12.0-g spring with force constant 200 N/cm change when you compress it by 6.0 cm? Does the mass increase or decrease? Would you notice the change in mass if you were holding the spring? Explain.

37.38 • A 60.0-kg person is standing at rest on level ground. How fast would she have to run to (a) double her total energy and (b) increase her total energy by a factor of 10?

37.39 • **An Antimatter Reactor.** When a particle meets its antiparticle, they annihilate each other and their mass is converted to light energy. The United States uses approximately 1.0×10^{20} J of energy per year. (a) If all this energy came from a futuristic antimatter reactor, how much mass of matter and antimatter fuel would be consumed yearly? (b) If this fuel had the density of iron (7.86 g/cm^3) and were stacked in bricks to form a cubical pile, how high would it be? (Before you get your hopes up, antimatter reactors are a *long* way in the future—if they ever will be feasible.)

37.40 •• Electrons are accelerated through a potential difference of 750 kV, so that their kinetic energy is 7.50×10^5 eV. (a) What is the ratio of the speed v of an electron having this energy to the speed of light, c? (b) What would the speed be if it were computed from the principles of classical mechanics?

37.41 • A particle has rest mass 6.64×10^{-27} kg and momentum 2.10×10^{-18} kg·m/s. (a) What is the total energy (kinetic plus rest energy) of the particle? (b) What is the kinetic energy of the particle? (c) What is the ratio of the kinetic energy to the rest energy of the particle?

37.42 •• A 0.100-μg speck of dust is accelerated from rest to a speed of $0.900c$ by a constant 1.00×10^6 N force. (a) If the non-relativistic mechanics is used, how far does the object travel to reach its final speed? (b) Using the correct relativistic treatment of Section 37.8, how far does the object travel to reach its final speed? (c) Which distance is greater? Why?

37.43 • Compute the kinetic energy of a proton (mass 1.67×10^{-27} kg) using both the nonrelativistic and relativistic expressions, and compute the ratio of the two results (relativistic divided by nonrelativistic) for speeds of (a) 8.00×10^7 m/s and (b) 2.85×10^8 m/s.

37.44 • What is the kinetic energy of a proton moving at (a) $0.100c$; (b) $0.500c$; (c) $0.900c$? How much work must be done to (d) increase the proton's speed from $0.100c$ to $0.500c$ and (e) increase the proton's speed from $0.500c$ to $0.900c$? (f) How do the last two results compare to results obtained in the nonrelativistic limit?

37.45 • (a) Through what potential difference does an electron have to be accelerated, starting from rest, to achieve a speed of $0.980c$? (b) What is the kinetic energy of the electron at this speed? Express your answer in joules and in electron volts.

37.46 • **Creating a Particle.** Two protons (each with rest mass $M = 1.67 \times 10^{-27}$ kg) are initially moving with equal speeds in opposite directions. The protons continue to exist after a collision that also produces an η^0 particle (see Chapter 44). The rest mass of the η^0 is $m = 9.75 \times 10^{-28}$ kg. (a) If the two protons and the η^0 are all at rest after the collision, find the initial speed of the protons, expressed as a fraction of the speed of light. (b) What is the kinetic energy of each proton? Express your answer in MeV. (c) What is the rest energy of the η^0, expressed in MeV? (d) Discuss the relationship between the answers to parts (b) and (c).

37.47 • The sun produces energy by nuclear fusion reactions, in which matter is converted into energy. By measuring the amount of energy we receive from the sun, we know that it is producing energy at a rate of 3.8×10^{26} W. (a) How many kilograms of matter does the sun lose each second? Approximately how many tons of matter is this (1 ton = 2000 lbs)? (b) At this rate, how long would it take the sun to use up all its mass?

PROBLEMS

37.48 • Inside a spaceship flying past the earth at three-fourths the speed of light, a pendulum is swinging. (a) If each swing takes 1.50 s as measured by an astronaut performing an experiment inside the spaceship, how long will the swing take as measured by a person at mission control on earth who is watching the experiment? (b) If each swing takes 1.50 s as measured by a person at mission control on earth, how long will it take as measured by the astronaut in the spaceship?

37.49 • After being produced in a collision between elementary particles, a positive pion (π^+) must travel down a 1.90-km-long tube to reach an experimental area. A π^+ particle has an average lifetime (measured in its rest frame) of 2.60×10^{-8} s; the π^+ we are considering has this lifetime. (a) How fast must the π^+ travel if it is not to decay before it reaches the end of the tube? (Since u will be very close to c, write $u = (1 - \Delta)c$ and give your answer in terms of Δ rather than u.) (b) The π^+ has a rest energy of 139.6 MeV. What is the total energy of the π^+ at the speed calculated in part (a)?

37.50 •• A cube of metal with sides of length a sits at rest in a frame S with one edge parallel to the x-axis. Therefore, in S the cube has volume a^3. Frame S' moves along the x-axis with a speed u. As measured by an observer in frame S', what is the volume of the metal cube?

37.51 ••• The starships of the Solar Federation are marked with the symbol of the federation, a circle, while starships of the Denebian Empire are marked with the empire's symbol, an ellipse whose major axis is 1.40 times longer than its minor axis ($a = 1.40b$ in Fig. P37.51). How fast, relative to an observer, does an empire ship have to travel for its marking to be confused with the marking of a federation ship?

Figure **P37.51**

Federation   Empire

37.52 •• A space probe is sent to the vicinity of the star Capella, which is 42.2 light-years from the earth. (A light-year is the distance light travels in a year.) The probe travels with a speed of 0.9930c. An astronaut recruit on board is 19 years old when the probe leaves the earth. What is her biological age when the probe reaches Capella?

37.53 • A particle is said to be *extremely relativistic* when its kinetic energy is much greater than its rest energy. (a) What is the speed of a particle (expressed as a fraction of c) such that the total energy is ten times the rest energy? (b) What is the percentage difference between the left and right sides of Eq. (37.39) if $(mc^2)^2$ is neglected for a particle with the speed calculated in part (a)?

37.54 •• **Everyday Time Dilation.** Two atomic clocks are carefully synchronized. One remains in New York, and the other is loaded on an airliner that travels at an average speed of 250 m/s and then returns to New York. When the plane returns, the elapsed time on the clock that stayed behind is 4.00 h. By how much will the readings of the two clocks differ, and which clock will show the shorter elapsed time? (*Hint:* Since $u \ll c$, you can simplify $\sqrt{1 - u^2/c^2}$ by a binomial expansion.)

37.55 • **The Large Hadron Collider (LHC).** Physicists and engineers from around the world have come together to build the largest accelerator in the world, the Large Hadron Collider (LHC) at the CERN Laboratory in Geneva, Switzerland. The machine will accelerate protons to kinetic energies of 7 TeV in an underground ring 27 km in circumference. (For the latest news and more information on the LHC, visit www.cern.ch.) (a) What speed v will protons reach in the LHC? (Since v is very close to c, write $v = (1 - \Delta)c$ and give your answer in terms of Δ.) (b) Find the relativistic mass, m_{rel}, of the accelerated protons in terms of their rest mass.

37.56 • **CP** A nuclear bomb containing 12.0 kg of plutonium explodes. The sum of the rest masses of the products of the explosion is less than the original rest mass by one part in 10^4. (a) How much energy is released in the explosion? (b) If the explosion takes place in 4.00 μs, what is the average power developed by the bomb? (c) What mass of water could the released energy lift to a height of 1.00 km?

37.57 • **CP Čerenkov Radiation.** The Russian physicist P. A. Čerenkov discovered that a charged particle traveling in a solid with a speed exceeding the speed of light in that material radiates electromagnetic radiation. (This is analogous to the sonic boom produced by an aircraft moving faster than the speed of sound in air; see Section 16.9. Čerenkov shared the 1958 Nobel Prize for this discovery.) What is the minimum kinetic energy (in electron volts) that an electron must have while traveling inside a slab of crown glass ($n = 1.52$) in order to create this Čerenkov radiation?

37.58 •• A photon with energy E is emitted by an atom with mass m, which recoils in the opposite direction. (a) Assuming that the motion of the atom can be treated nonrelativistically, compute the recoil speed of the atom. (b) From the result of part (a), show that the recoil speed is much less than c whenever E is much less than the rest energy mc^2 of the atom.

37.59 •• In an experiment, two protons are shot directly toward each other, each moving at half the speed of light relative to the laboratory. (a) What speed does one proton measure for the other

proton? (b) What would be the answer to part (a) if we used only nonrelativistic Newtonian mechanics? (c) What is the kinetic energy of each proton as measured by (i) an observer at rest in the laboratory and (ii) an observer riding along with one of the protons? (d) What would be the answers to part (c) if we used only nonrelativistic Newtonian mechanics?

37.60 •• Two protons are moving away from each other. In the frame of each proton, the other proton has a speed of 0.600c. What does an observer in the rest frame of the earth measure for the speed of each proton?

37.61 •• Frame S' has an x-component of velocity u relative to frame S, and at $t = t' = 0$ the two frames coincide (see Fig. 37.3). A light pulse with a spherical wave front is emitted at the origin of S' at time $t' = 0$. Its distance x' from the origin after a time t' is given by $x'^2 = c^2 t'^2$. Use the Lorentz coordinate transformation to transform this equation to an equation in x and t, and show that the result is $x^2 = c^2 t^2$; that is, the motion appears exactly the same in frame of reference S as it does in S'; the wave front is observed to be spherical in both frames.

37.62 • In certain radioactive beta decay processes, the beta particle (an electron) leaves the atomic nucleus with a speed of 99.95% the speed of light relative to the decaying nucleus. If this nucleus is moving at 75.00% the speed of light in the laboratory reference frame, find the speed of the emitted electron relative to the laboratory reference frame if the electron is emitted (a) in the same direction that the nucleus is moving and (b) in the opposite direction from the nucleus's velocity. (c) In each case in parts (a) and (b), find the kinetic energy of the electron as measured in (i) the laboratory frame and (ii) the reference frame of the decaying nucleus.

37.63 •• **CALC** A particle with mass m accelerated from rest by a constant force F will, according to Newtonian mechanics, continue to accelerate without bound; that is, as $t \to \infty$, $v \to \infty$. Show that according to relativistic mechanics, the particle's speed approaches c as $t \to \infty$. [*Note:* A useful integral is $\int (1 - x^2)^{-3/2} dx = x/\sqrt{1 - x^2}$.]

37.64 •• Two events are observed in a frame of reference S to occur at the same space point, the second occurring 1.80 s after the first. In a second frame S' moving relative to S, the second event is observed to occur 2.35 s after the first. What is the difference between the positions of the two events as measured in S'?

37.65 ••• Two events observed in a frame of reference S have positions and times given by (x_1, t_1) and (x_2, t_2), respectively. (a) Frame S' moves along the x-axis just fast enough that the two events occur at the same position in S'. Show that in S', the time interval $\Delta t'$ between the two events is given by

$$\Delta t' = \sqrt{(\Delta t)^2 - \left(\frac{\Delta x}{c}\right)^2}$$

where $\Delta x = x_2 - x_1$ and $\Delta t = t_2 - t_1$. Hence show that if $\Delta x > c\,\Delta t$, there is *no* frame S' in which the two events occur at the same point. The interval $\Delta t'$ is sometimes called the *proper time interval* for the events. Is this term appropriate? (b) Show that if $\Delta x > c\,\Delta t$, there is a different frame of reference S' in which the two events occur *simultaneously*. Find the distance between the two events in S'; express your answer in terms of Δx, Δt, and c. This distance is sometimes called a *proper length*. Is this term appropriate? (c) Two events are observed in a frame of reference S' to occur simultaneously at points separated by a distance of 2.50 m. In a second frame S moving relative to S' along the line joining the two points in S', the two events appear to be separated by 5.00 m. What is the time interval between the events as measured in S? [*Hint:* Apply the result obtained in part (b).]

37.66 ·· Albert in Wonderland. Einstein and Lorentz, being avid tennis players, play a fast-paced game on a court where they stand 20.0 m from each other. Being very skilled players, they play without a net. The tennis ball has mass 0.0580 kg. You can ignore gravity and assume that the ball travels parallel to the ground as it travels between the two players. Unless otherwise specified, all measurements are made by the two men. (a) Lorentz serves the ball at 80.0 m/s. What is the ball's kinetic energy? (b) Einstein slams a return at 1.80×10^8 m/s. What is the ball's kinetic energy? (c) During Einstein's return of the ball in part (a), a white rabbit runs beside the court in the direction from Einstein to Lorentz. The rabbit has a speed of 2.20×10^8 m/s relative to the two men. What is the speed of the rabbit relative to the ball? (d) What does the rabbit measure as the distance from Einstein to Lorentz? (e) How much time does it take for the rabbit to run 20.0 m, according to the players? (f) The white rabbit carries a pocket watch. He uses this watch to measure the time (as he sees it) for the distance from Einstein to Lorentz to pass by under him. What time does he measure?

37.67 · One of the wavelengths of light emitted by hydrogen atoms under normal laboratory conditions is $\lambda = 656.3$ nm, in the red portion of the electromagnetic spectrum. In the light emitted from a distant galaxy this same spectral line is observed to be Doppler-shifted to $\lambda = 953.4$ nm, in the infrared portion of the spectrum. How fast are the emitting atoms moving relative to the earth? Are they approaching the earth or receding from it?

37.68 ·· Measuring Speed by Radar. A baseball coach uses a radar device to measure the speed of an approaching pitched baseball. This device sends out electromagnetic waves with frequency f_0 and then measures the shift in frequency Δf of the waves reflected from the moving baseball. If the fractional frequency shift produced by a baseball is $\Delta f/f_0 = 2.86 \times 10^{-7}$, what is the baseball's speed in km/h? (*Hint:* Are the waves Doppler-shifted a second time when reflected off the ball?)

37.69 · Space Travel? Travel to the stars requires hundreds or thousands of years, even at the speed of light. Some people have suggested that we can get around this difficulty by accelerating the rocket (and its astronauts) to very high speeds so that they will age less due to time dilation. The fly in this ointment is that it takes a great deal of energy to do this. Suppose you want to go to the immense red giant Betelgeuse, which is about 500 light-years away. (A light-year is the distance that light travels in a year.) You plan to travel at constant speed in a 1000-kg rocket ship (a little over a ton), which, in reality, is far too small for this purpose. In each case that follows, calculate the time for the trip, as measured by people on earth and by astronauts in the rocket ship, the energy needed in joules, and the energy needed as a percentage of U.S. yearly use (which is 1.0×10^{20} J). For comparison, arrange your results in a table showing v_{rocket}, t_{earth}, t_{rocket}, E (in J), and E (as % of U.S. use). The rocket ship's speed is (a) 0.50c; (b) 0.99c; (c) 0.9999c. On the basis of your results, does it seem likely that any government will invest in such high-speed space travel any time soon?

37.70 ·· A spaceship moving at constant speed u relative to us broadcasts a radio signal at constant frequency f_0. As the spaceship approaches us, we receive a higher frequency f; after it has passed, we receive a lower frequency. (a) As the spaceship passes by, so it is instantaneously moving neither toward nor away from us, show that the frequency we receive is not f_0, and derive an expression for the frequency we do receive. Is the frequency we receive higher or lower than f_0? (*Hint:* In this case, successive wave crests move the same distance to the observer and so they

have the same transit time. Thus f equals $1/T$. Use the time dilation formula to relate the periods in the stationary and moving frames.) (b) A spaceship emits electromagnetic waves of frequency $f_0 = 345$ MHz as measured in a frame moving with the ship. The spaceship is moving at a constant speed $0.758c$ relative to us. What frequency f do we receive when the spaceship is approaching us? When it is moving away? In each case what is the shift in frequency, $f - f_0$? (c) Use the result of part (a) to calculate the frequency f and the frequency shift $(f - f_0)$ we receive at the instant that the ship passes by us. How does the shift in frequency calculated here compare to the shifts calculated in part (b)?

37.71 · CP In a particle accelerator a proton moves with constant speed $0.750c$ in a circle of radius 628 m. What is the net force on the proton?

37.72 ·· CP The French physicist Armand Fizeau was the first to measure the speed of light accurately. He also found experimentally that the speed, relative to the lab frame, of light traveling in a tank of water that is itself moving at a speed V relative to the lab frame is

$$v = \frac{c}{n} + kV$$

where $n = 1.333$ is the index of refraction of water. Fizeau called k the dragging coefficient and obtained an experimental value of $k = 0.44$. What value of k do you calculate from relativistic transformations?

CHALLENGE PROBLEMS

37.73 ··· CALC Lorentz Transformation for Acceleration. Using a method analogous to the one in the text to find the Lorentz transformation formula for velocity, we can find the Lorentz transformation for *acceleration*. Let frame S' have a constant x-component of velocity u relative to frame S. An object moves relative to frame S along the x-axis with instantaneous velocity v_x and instantaneous acceleration a_x. (a) Show that its instantaneous acceleration in frame S' is

$$a'_x = a_x \left(1 - \frac{u^2}{c^2}\right)^{3/2} \left(1 - \frac{uv_x}{c^2}\right)^{-3}$$

[*Hint:* Express the acceleration in S' as $a'_x = dv'_x/dt'$. Then use Eq. (37.21) to express dt' in terms of dt and dx, and use Eq. (37.22) to express dv'_x in terms of u and dv_x. The velocity of the object in S is $v_x = dx/dt$.] (b) Show that the acceleration in frame S can be expressed as

$$a_x = a'_x \left(1 - \frac{u^2}{c^2}\right)^{3/2} \left(1 + \frac{uv'_x}{c^2}\right)^{-3}$$

where $v'_x = dx'/dt'$ is the velocity of the object in frame S'.

37.74 ··· CALC A Realistic Version of the Twin Paradox. A rocket ship leaves the earth on January 1, 2100. Stella, one of a pair of twins born in the year 2075, pilots the rocket (reference frame S'); the other twin, Terra, stays on the earth (reference frame S). The rocket ship has an acceleration of constant magnitude g in its own reference frame (this makes the pilot feel at home, since it simulates the earth's gravity). The path of the rocket ship is a straight line in the $+x$-direction in frame S. (a) Using the results of Challenge Problem 37.73, show that in Terra's earth frame S, the rocket's acceleration is

$$\frac{du}{dt} = g \left(1 - \frac{u^2}{c^2}\right)^{3/2}$$

where u is the rocket's instantaneous velocity in frame S. (b) Write the result of part (a) in the form $dt = f(u)\,du$, where $f(u)$ is a function of u, and integrate both sides. (*Hint:* Use the integral given in Problem 37.63.) Show that in Terra's frame, the time when Stella attains a velocity v_{1x} is

$$t_1 = \frac{v_{1x}}{g\sqrt{1 - v_{1x}^2/c^2}}$$

(c) Use the time dilation formula to relate dt and dt' (infinitesimal time intervals measured in frames S and S', respectively). Combine this result with the result of part (a) and integrate as in part (b) to show the following: When Stella attains a velocity v_{1x} relative to Terra, the time t_1' that has elapsed in frame S' is

$$t_1' = \frac{c}{g}\operatorname{arctanh}\left(\frac{v_{1x}}{c}\right)$$

Here arctanh is the inverse hyperbolic tangent. (*Hint:* Use the integral given in Challenge Problem 5.124.) (d) Combine the results of parts (b) and (c) to find t_1 in terms of t_1', g, and c alone. (e) Stella accelerates in a straight-line path for five years (by her clock), slows down at the same rate for five years, turns around, accelerates for five years, slows down for five years, and lands back on the earth. According to Stella's clock, the date is January 1, 2120. What is the date according to Terra's clock?

37.75 ••• **CP Determining the Masses of Stars.** Many of the stars in the sky are actually *binary stars*, in which two stars orbit about their common center of mass. If the orbital speeds of the stars are high enough, the motion of the stars can be detected by the Doppler shifts of the light they emit. Stars for which this is the case are called *spectroscopic binary stars*. Figure P37.75 shows the

Figure **P37.75**

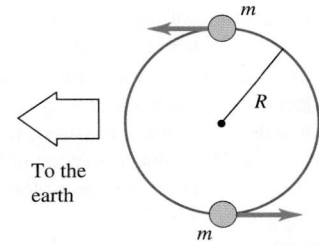

simplest case of a spectroscopic binary star: two identical stars, each with mass m, orbiting their center of mass in a circle of radius R. The plane of the stars' orbits is edge-on to the line of sight of an observer on the earth. (a) The light produced by heated hydrogen gas in a laboratory on the earth has a frequency of 4.568110×10^{14} Hz. In the light received from the stars by a telescope on the earth, hydrogen light is observed to vary in frequency between 4.567710×10^{14} Hz and 4.568910×10^{14} Hz. Determine whether the binary star system as a whole is moving toward or away from the earth, the speed of this motion, and the orbital speeds of the stars. (*Hint:* The speeds involved are much less than c, so you may use the approximate result $\Delta f/f = u/c$ given in Section 37.6.) (b) The light from each star in the binary system varies from its maximum frequency to its minimum frequency and back again in 11.0 days. Determine the orbital radius R and the mass $\dot{m}$ of each star. Give your answer for m in kilograms and as a multiple of mass of the sun, 1.99×10^{30} kg. Compare the value of R to the distance from the earth to the sun, 1.50×10^{11} m. (This technique is actually used in astronomy to determine the masses of stars. In practice, the problem is more complicated because the two stars in

a binary system are usually not identical, the orbits are usually not circular, and the plane of the orbits is usually tilted with respect to the line of sight from the earth.)

37.76 ••• **CP CALC Relativity and the Wave Equation.** (a) Consider the Galilean transformation along the x-direction: $x' = x - vt$ and $t' = t$. In frame S the wave equation for electromagnetic waves in a vacuum is

$$\frac{\partial^2 E(x, t)}{\partial x^2} - \frac{1}{c^2}\frac{\partial^2 E(x, t)}{\partial t^2} = 0$$

where E represents the electric field in the wave. Show that by using the Galilean transformation the wave equation in frame S' is found to be

$$\left(1 - \frac{v^2}{c^2}\right)\frac{\partial^2 E(x', t')}{\partial x'^2} + \frac{2v}{c^2}\frac{\partial^2 E(x', t')}{\partial x'\partial t'} - \frac{1}{c^2}\frac{\partial^2 E(x', t')}{\partial t'^2} = 0$$

This has a different form than the wave equation in S. Hence the Galilean transformation *violates* the first relativity postulate that all physical laws have the same form in all inertial reference frames. (*Hint:* Express the derivatives $\partial/\partial x$ and $\partial/\partial t$ in terms of $\partial/\partial x'$ and $\partial/\partial t'$ by use of the chain rule.) (b) Repeat the analysis of part (a), but use the Lorentz coordinate transformations, Eqs. (37.21), and show that in frame S' the wave equation has the same form as in frame S:

$$\frac{\partial^2 E(x', t')}{\partial x^2} - \frac{1}{c^2}\frac{\partial^2 E(x', t')}{\partial t'^2} = 0$$

Explain why this shows that the speed of light in vacuum is c in both frames S and S'.

37.77 ••• **CP Kaon Production.** In high-energy physics, new particles can be created by collisions of fast-moving projectile particles with stationary particles. Some of the kinetic energy of the incident particle is used to create the mass of the new particle. A proton–proton collision can result in the creation of a negative kaon (K^-) and a positive kaon (K^+)

$$p + p \rightarrow p + p + K^- + K^+$$

(a) Calculate the minimum kinetic energy of the incident proton that will allow this reaction to occur if the second (target) proton is initially at rest. The rest energy of each kaon is 493.7 MeV, and the rest energy of each proton is 938.3 MeV. (*Hint:* It is useful here to work in the frame in which the total momentum is zero. But note that the Lorentz transformation must be used to relate the velocities in the laboratory frame to those in the zero-total-momentum frame.) (b) How does this calculated minimum kinetic energy compare with the total rest mass energy of the created kaons? (c) Suppose that instead the two protons are both in motion with velocities of equal magnitude and opposite direction. Find the minimum combined kinetic energy of the two protons that will allow the reaction to occur. How does this calculated minimum kinetic energy compare with the total rest mass energy of the created kaons? (This example shows that when colliding beams of particles are used instead of a stationary target, the energy requirements for producing new particles are reduced substantially.)

Chapter Opening Question ?

No. While the speed of light c is the ultimate "speed limit" for any particle, there is *no* upper limit on a particle's kinetic energy (see Fig. 37.21). As the speed approaches c, a small increase in speed corresponds to a large increase in kinetic energy.

Test Your Understanding Questions

37.1 Answers: (a) (i), (b) no You, too, will measure a spherical wave front that expands at the same speed c in all directions. This is a consequence of Einstein's second postulate. The wave front that you measure is *not* centered on the current position of the spaceship; rather, it is centered on the point P where the spaceship was located at the instant that it emitted the light pulse. For example, suppose the spaceship is moving at speed $c/2$. When your watch shows that a time t has elapsed since the pulse of light was emitted, your measurements will show that the wave front is a sphere of radius ct centered on P and that the spaceship is a distance $ct/2$ from P.

37.2 Answer: (iii) In Mavis's frame of reference, the two events (the Ogdenville clock striking noon and the North Haverbrook clock striking noon) are not simultaneous. Figure 37.5 shows that the event toward the front of the rail car occurs first. Since the rail car is moving toward North Haverbrook, that clock struck noon before the one on Ogdenville. So, according to Mavis, it is after noon in North Haverbrook.

37.3 Answers: (a) (ii), (b) (ii) The statement that moving clocks run slow refers to any clock that is moving relative to an observer. Maria and her stopwatch are moving relative to Samir, so Samir measures Maria's stopwatch to be running slow and to have ticked off fewer seconds than his own stopwatch. Samir and his stopwatch are moving relative to Maria, so she likewise measures Samir's stopwatch to be running slow. Each observer's measurement is correct for his or her own frame of reference. *Both* observers conclude that a moving stopwatch runs slow. This is consistent with the principle of relativity (see Section 37.1), which states that the laws of physics are the same in all inertial frames of reference.

37.4 Answer: (ii), (i) and (iii) (tie), (iv) You measure the rest length of the stationary meter stick and the contracted length of the moving spaceship to both be 1 meter. The rest length of the spaceship is greater than the contracted length that you measure, and so must be greater than 1 meter. A miniature observer on board the spaceship would measure a contracted length for the meter stick of less than 1 meter. Note that in your frame of reference the nose and tail of the spaceship can simultaneously align with the two ends of the meter stick, since in your frame of reference they have the same length of 1 meter. In the spaceship's frame these two alignments cannot happen simultaneously because the meter stick is shorter than the spaceship. Section 37.2 tells us that this shouldn't be a surprise; two events that are simultaneous to one observer may not be simultaneous to a second observer moving relative to the first one.

37.5 Answers: (a) P_1, (b) P_4 (a) The last of Eqs. (37.21) tells us the times of the two events in S': $t'_1 = \gamma(t_1 - ux_1/c^2)$ and $t'_2 = \gamma(t_2 - ux_2/c^2)$. In frame S the two events occur at the same x-coordinate, so $x_1 = x_2$, and event P_1 occurs before event P_2, so $t_1 < t_2$. Hence you can see that $t'_1 < t'_2$ and event P_1 happens before P_2 in frame S', too. This says that if event P_1 happens before P_2 in a frame of reference S where the two events occur at the same position, then P_1 happens before P_2 in any other frame moving relative to S. **(b)** In frame S the two events occur at different x-coordinates such that $x_3 < x_4$, and events P_3 and P_4 occur at the same time, so $t_3 = t_4$. Hence you can see that $t'_3 = \gamma(t_3 - ux_3/c^2)$ is greater than $t'_4 = \gamma(t_4 - ux_4/c^2)$, so event P_4 happens before P_3 in frame S'. This says that even though the two events are simultaneous in frame S, they need not be simultaneous in a frame moving relative to S.

37.7 Answer: (ii) Equation (37.27) tells us that the magnitude of momentum of a particle with mass m and speed v is $p = mv/\sqrt{1 - v^2/c^2}$. If v increases by a factor of 2, the numerator mv increases by a factor of 2 *and* the denominator $\sqrt{1 - v^2/c^2}$ decreases. Hence p increases by a factor greater than 2. (Note that in order to double the speed, the initial speed must be less than $c/2$. That's because the speed of light is the ultimate speed limit.)

37.8 Answer: (i) As the proton moves a distance s, the constant force of magnitude F does work $W = Fs$ and increases the kinetic energy by an amount $\Delta K = W = Fs$. This is true no matter what the speed of the proton before moving this distance. Thus the constant force increases the proton's kinetic energy by the same amount during the first meter of travel as during any subsequent meter of travel. (It's true that as the proton approaches the ultimate speed limit of c, the increase in the proton's *speed* is less and less with each subsequent meter of travel. That's not what the question is asking, however.)

Bridging Problem

Answers: (a) $0.268c$ **(b)** 35.6 MeV **(c)** 145 MeV

PHOTONS: LIGHT WAVES BEHAVING AS PARTICLES 38

This plastic surgeon is using two light sources: a headlamp that emits a beam of visible light and a handheld laser that emits infrared light. The light from both sources is emitted in the form of packets of energy called photons. For which source are the photons more energetic: the headlamp or the laser?

LEARNING GOALS

By studying this chapter, you will learn:

- How experiments involving the photoelectric effect and x rays pointed the way to a radical reinterpretation of the nature of light.

- How Einstein's photon picture of light explains the photoelectric effect.

- How experiments with x rays and gamma rays helped confirm the photon picture of light.

- How the wave and particle pictures of light complement each other.

- How the Heisenberg uncertainty principle imposes fundamental limits on what can be measured.

In Chapter 32 we saw how Maxwell, Hertz, and others established firmly that light is an electromagnetic wave. Interference, diffraction, and polarization, discussed in Chapters 35 and 36, further demonstrate this *wave nature* of light.

When we look more closely at the emission, absorption, and scattering of electromagnetic radiation, however, we discover a completely different aspect of light. We find that the energy of an electromagnetic wave is *quantized;* it is emitted and absorbed in particle-like packages of definite energy, called *photons.* The energy of a single photon is proportional to the frequency of the radiation.

We'll find that light and other electromagnetic radiation exhibits *wave–particle duality:* Light acts sometimes like waves and sometimes like particles. Interference and diffraction demonstrate wave behavior, while emission and absorption of photons demonstrate the particle behavior. This radical reinterpretation of light will lead us in the next chapter to no less radical changes in our views of the nature of matter.

38.1 Light Absorbed as Photons: The Photoelectric Effect

A phenomenon that gives insight into the nature of light is the **photoelectric effect,** in which a material emits electrons from its surface when illuminated (Fig. 38.1). To escape from the surface, an electron must absorb enough energy from the incident light to overcome the attraction of positive ions in the material. These attractions constitute a potential-energy barrier; the light supplies the "kick" that enables the electron to escape.

The photoelectric effect has a number of applications. Digital cameras and night-vision scopes use it to convert light energy into an electric signal that is

38.1 The photoelectric effect.

Light   **Photoelectric effect:** Light absorbed by a surface causes electrons to be ejected.

Electrons

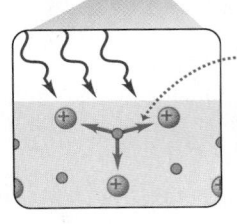

To eject an electron the light must supply enough energy to overcome the forces holding the electron in the material.

38.2 (a) A night-vision scope makes use of the photoelectric effect. Photons entering the scope strike a plate, ejecting electrons that pass through a thin disk in which there are millions of tiny channels. The current through each channel is amplified electronically and then directed toward a screen that glows when hit by electrons. (b) The image formed on the screen, which is a combination of these millions of glowing spots, is thousands of times brighter than the naked-eye view.

(a)

(b)

MasteringPHYSICS

PhET: Photoelectric Effect
ActivPhysics 17.3: Photoelectric Effect

reconstructed into an image (Fig. 38.2). On the moon, sunlight striking the surface causes surface dust to eject electrons, leaving the dust particles with a positive charge. The mutual electric repulsion of these charged dust particles causes them to rise above the moon's surface, a phenomenon that was observed from lunar orbit by the Apollo astronauts.

Threshold Frequency and Stopping Potential

In Section 32.1 we explored the wave model of light, which Maxwell formulated two decades before the photoelectric effect was observed. Is the photoelectric effect consistent with this model? Figure 38.3a shows a modern version of one of the experiments that explored this question. Two conducting electrodes are enclosed in an evacuated glass tube and connected by a battery, and the cathode is illuminated. Depending on the potential difference V_{AC} between the two electrodes, electrons emitted by the illuminated cathode (called *photoelectrons*) may travel across to the anode, producing a *photocurrent* in the external circuit. (The tube is evacuated to a pressure of 0.01 Pa or less to minimize collisions between the electrons and gas molecules.)

The illuminated cathode emits photoelectrons with various kinetic energies. If the electric field points toward the cathode, as in Fig. 38.3a, all the electrons are accelerated toward the anode and contribute to the photocurrent. But by reversing the field and adjusting its strength as in Fig. 38.3b, we can prevent the less energetic electrons from reaching the anode. In fact, we can determine the *maximum* kinetic energy K_{max} of the emitted electrons by making the potential of the anode relative to the cathode, V_{AC}, just negative enough so that the current stops. This occurs for $V_{AC} = -V_0$, where V_0 is called the **stopping potential.** As an electron moves from the cathode to the anode, the potential decreases by V_0 and negative work $-eV_0$ is done on the (negatively charged) electron. The most energetic electron leaves the cathode with kinetic energy $K_{max} = \frac{1}{2}mv_{max}^2$ and has zero kinetic energy at the anode. Using the work–energy theorem, we have

$$W_{tot} = -eV_0 = \Delta K = 0 - K_{max}$$
$$K_{max} = \tfrac{1}{2}mv_{max}^2 = eV_0$$

(maximum kinetic energy of photoelectrons)   (38.1)

Hence by measuring the stopping potential V_0, we can determine the maximum kinetic energy with which electrons leave the cathode. (We are ignoring any effects due to differences in the materials of the cathode and anode.)

In this experiment, how do we expect the photocurrent to depend on the voltage across the electrodes and on the frequency and intensity of the light? Based on Maxwell's picture of light as an electromagnetic wave, here is what we *would* expect:

Wave-Model Prediction 1: We saw in Section 32.4 that the intensity of an electromagnetic wave depends on its amplitude but not on its frequency. So the photoelectric effect should occur for light of any frequency, and *the magnitude of the photocurrent should not depend on the frequency of the light.*

Wave-Model Prediction 2: It takes a certain minimum amount of energy, called the **work function,** to eject a single electron from a particular surface (see Fig. 38.1). If the light falling on the surface is very faint, some time may elapse before the total energy absorbed by the surface equals the work function. Hence, for faint illumination, *we expect a time delay* between when we switch on the light and when photoelectrons appear.

Wave-Model Prediction 3: Because the energy delivered to the cathode surface depends on the intensity of illumination, *we expect the stopping potential to increase with increasing light intensity.* Since intensity does not depend on frequency, we further expect that *the stopping potential should not depend on the frequency of the light.*

The experimental results proved to be *very* different from these predictions. Here is what was found in the years between 1877 and 1905:

Experimental Result 1: The photocurrent depends on the light frequency. For a given material, monochromatic light with a frequency below a minimum **threshold frequency** produces *no* photocurrent, regardless of intensity. For most metals the threshold frequency is in the ultraviolet (corresponding to wavelengths λ between 200 and 300 nm), but for other materials like potassium oxide and cesium oxide it is in the visible spectrum (λ between 380 and 750 nm).

Experimental Result 2: There is *no measurable time delay* between when the light is turned on and when the cathode emits photoelectrons (assuming the frequency of the light exceeds the threshold frequency). This is true no matter how faint the light is.

Experimental Result 3: The stopping potential does not depend on intensity, but does depend on frequency. Figure 38.4 shows graphs of photocurrent as a function of potential difference V_{AC} for light of a given frequency and two different intensities. The reverse potential difference $-V_0$ needed to reduce the current to zero is the same for both intensities. The only effect of increasing the intensity is to increase the number of electrons per second and hence the photocurrent i. (The curves level off when V_{AC} is large and positive because at that point all the emitted electrons are being collected by the anode.) If the intensity is held constant but the frequency is increased, the stopping potential also increases. In other words, the greater the light frequency, the higher the energy of the ejected photoelectrons.

These results directly contradict Maxwell's description of light as an electromagnetic wave. A solution to this dilemma was provided by Albert Einstein in 1905. His proposal involved nothing less than a new picture of the nature of light.

Einstein's Photon Explanation

Einstein made the radical postulate that a beam of light consists of small packages of energy called **photons** or *quanta*. This postulate was an extension of an idea developed five years earlier by Max Planck to explain the properties of blackbody radiation, which we discussed in Section 17.7. (We'll explore Planck's ideas in Section 39.5.) In Einstein's picture, the energy E of an individual photon is equal to a constant h times the photon frequency f. From the relationship $f = c/\lambda$ for electromagnetic waves in vacuum, we have

$$E = hf = \frac{hc}{\lambda} \quad \text{(energy of a photon)} \quad (38.2)$$

where h is a universal constant called **Planck's constant.** The numerical value of this constant, to the accuracy known at present, is

$$h = 6.62606896(33) \times 10^{-34} \text{ J} \cdot \text{s}$$

CAUTION **Photons are not "particles" in the usual sense** It's common to envision photons as miniature billiard balls or pellets. While that's a convenient mental picture, it's not very accurate. For one thing, billiard balls and bullets have a rest mass and travel slower than the speed of light c, while photons travel at the speed of light and have *zero* rest mass. For another thing, photons have wave aspects (frequency and wavelength) that are easy to observe. The fact is that the photon concept is a very strange one, and the true nature of photons is difficult to visualize in a simple way. We'll discuss the dual personality of photons in more detail in Section 38.4. ▮

In Einstein's picture, an individual photon arriving at the surface in Fig. 38.1a or 38.2 is absorbed by a single electron. This energy transfer is an all-or-nothing process, in contrast to the continuous transfer of energy in the wave theory of

38.3 An experiment testing whether the photoelectric effect is consistent with the wave model of light.

(a)

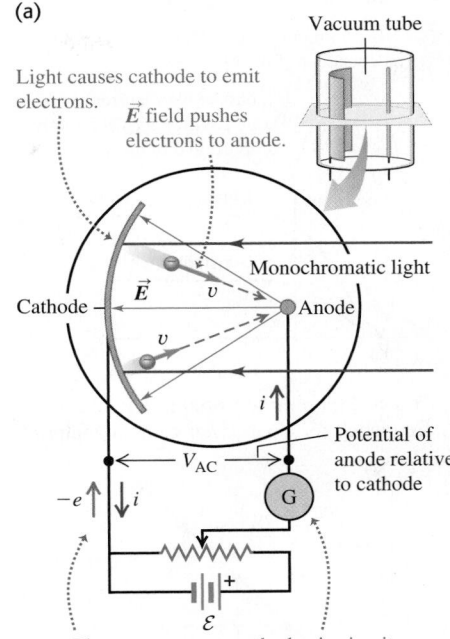

Electrons return to cathode via circuit; galvanometer measures current.

(b) We now reverse the electric field so that it tends to repel electrons from the anode. Above a certain field strength, electrons no longer reach the anode.

The **stopping potential** at which the current ceases has absolute value V_0.

38.4 Photocurrent i for a constant light frequency f as a function of the potential V_{AC} of the anode with respect to the cathode.

The stopping potential V_0 is independent of the light intensity ...

... but the photocurrent i for large positive V_{AC} is directly proportional to the intensity.

f is constant.

Constant intensity $2I$

Constant intensity I

38.5 Stopping potential as a function of frequency for a particular cathode material.

Table 38.1 Work Functions of Several Elements

Element	Work Function (eV)
Aluminum	4.3
Carbon	5.0
Copper	4.7
Gold	5.1
Nickel	5.1
Silicon	4.8
Silver	4.3
Sodium	2.7

38.6 Stopping potential as a function of frequency for two cathode materials having different work functions ϕ.

Stopping potential V_0

Material 2

Material 1 $\phi_2 > \phi_1$

Frequency f

Threshold frequency

Stopping potential is zero at threshold frequency (electrons emerge with zero kinetic energy).

For each material,

$eV = hf - \phi$ or $V_0 = \dfrac{h}{e} - \dfrac{\phi}{e}$

so the plots have same slope h/e but different intercepts $-f/e$ on the vertical axis.

light; the electron gets all of the photon's energy or none at all. The electron can escape from the surface only if the energy it acquires is greater than the work function ϕ. Thus photoelectrons will be ejected only if $hf > \phi$, or $f > \phi/h$. Einstein's postulate therefore explains why the photoelectric effect occurs only for frequencies greater than a minimum threshold frequency. This postulate is also consistent with the observation that greater intensity causes a greater photocurrent (Fig. 38.4). Greater intensity at a particular frequency means a greater number of photons per second absorbed, and thus a greater number of electrons emitted per second and a greater photocurrent.

Einstein's postulate also explains why there is no delay between illumination and the emission of photoelectrons. As soon as photons of sufficient energy strike the surface, electrons can absorb them and be ejected.

Finally, Einstein's postulate explains why the stopping potential for a given surface depends only on the light frequency. Recall that ϕ is the *minimum* energy needed to remove an electron from the surface. Einstein applied conservation of energy to find that the *maximum* kinetic energy $K_{max} = \frac{1}{2}mv_{max}^2$ for an emitted electron is the energy hf gained from a photon minus the work function ϕ:

$$K_{max} = \tfrac{1}{2}mv_{max}^2 = hf - \phi \qquad (38.3)$$

Substituting $K_{max} = eV_0$ from Eq. (38.1), we find

$$eV_0 = hf - \phi \qquad \text{(photoelectric effect)} \qquad (38.4)$$

Equation (38.4) shows that the stopping potential V_0 increases with increasing frequency f. The intensity doesn't appear in Eq. (38.4), so V_0 is independent of intensity. As a check of Eq. (38.4), we can measure the stopping potential V_0 for each of several values of frequency f for a given cathode material (Fig. 38.5). A graph of V_0 as a function of f turns out to be a straight line, verifying Eq. (38.4), and from such a graph we can determine both the work function ϕ for the material and the value of the quantity h/e. After the electron charge $-e$ was measured by Robert Millikan in 1909, Planck's constant h could also be determined from these measurements.

Electron energies and work functions are usually expressed in electron volts (eV), defined in Section 23.2. To four significant figures,

$$1\ eV = 1.602 \times 10^{-19}\ J$$

To this accuracy, Planck's constant is

$$h = 6.626 \times 10^{-34}\ J \cdot s = 4.136 \times 10^{-15}\ eV \cdot s$$

Table 38.1 lists the work functions of several elements. These values are approximate because they are very sensitive to surface impurities. The greater the work function, the higher the minimum frequency needed to emit photoelectrons (Fig. 38.6).

The photon picture explains a number of other phenomena in which light is absorbed. One example is a *suntan*, which is caused when the energy in sunlight triggers a chemical reaction in skin cells that leads to increased production of the pigment melanin. This reaction can occur only if a specific molecule in the cell absorbs a certain minimum amount of energy. A short-wavelength ultraviolet photon has enough energy to trigger the reaction, but a longer-wavelength visible-light photon does not. Hence ultraviolet light causes tanning, while visible light cannot.

Photon Momentum

Einstein's photon concept applies to *all* regions of the electromagnetic spectrum, including radio waves, x rays, and so on. A photon of any electromagnetic radiation with frequency f and wavelength λ has energy E given by Eq. (38.2).

Furthermore, according to the special theory of relativity, every particle that has energy must also have momentum, even if it has no rest mass. Photons have zero rest mass. As we saw in Eq. (37.40), a particle with zero rest mass and energy E has momentum with magnitude p given by $E = pc$. Thus the wavelength λ of a photon and the magnitude of its momentum p are related simply by

$$p = \frac{E}{c} = \frac{hf}{c} = \frac{h}{\lambda} \qquad \text{(momentum of a photon)} \qquad (38.5)$$

The direction of the photon's momentum is simply the direction in which the electromagnetic wave is moving.

Problem-Solving Strategy 38.1   Photons

IDENTIFY *the relevant concepts:* The energy and momentum of an individual photon are proportional to the frequency and inversely proportional to the wavelength. Einstein's interpretation of the photoelectric effect is that energy is conserved as a photon ejects an electron from a material surface.

SET UP *the problem:* Identify the target variable. It could be the photon's wavelength λ, frequency f, energy E, or momentum p. If the problem involves the photoelectric effect, the target variable could be the maximum kinetic energy of photoelectrons K_{max}, the stopping potential V_0, or the work function ϕ.

EXECUTE *the solution* as follows:
1. Use Eqs. (38.2) and (38.5) to relate the energy and momentum of a photon to its wavelength and frequency. If the problem involves the photoelectric effect, use Eqs. (38.1), (38.3), and

(38.4) to relate the photon frequency, stopping potential, work function, and maximum photoelectron kinetic energy.
2. The electron volt (eV), which we introduced in Section 23.2, is a convenient unit. It is the kinetic energy gained by an electron when it moves freely through an increase of potential of one volt: $1 \text{ eV} = 1.602 \times 10^{-19} \text{ J}$. If the photon energy E is given in electron volts, use $h = 4.136 \times 10^{-15} \text{ eV} \cdot \text{s}$; if E is in joules, use $h = 6.626 \times 10^{-34} \text{ J} \cdot \text{s}$.

EVALUATE *your answer:* In problems involving photons, at first the numbers will be unfamiliar to you and errors will not be obvious. It helps to remember that a visible-light photon with $\lambda = 600$ nm and $f = 5 \times 10^{14}$ Hz has an energy E of about 2 eV, or about 3×10^{-19} J.

Example 38.1   Laser-pointer photons

A laser pointer with a power output of 5.00 mW emits red light ($\lambda = 650$ nm). (a) What is the magnitude of the momentum of each photon? (b) How many photons does the laser pointer emit each second?

SOLUTION

IDENTIFY and SET UP: This problem involves the ideas of (a) photon momentum and (b) photon energy. In part (a) we'll use Eq. (38.5) and the given wavelength to find the magnitude of each photon's momentum. In part (b), Eq. (38.2) gives the energy per photon, and the power output tells us the energy emitted per second. We can combine these quantities to calculate the number of photons emitted per second.

EXECUTE: (a) We have $\lambda = 650$ nm $= 6.50 \times 10^{-7}$ m, so from Eq. (38.5) the photon momentum is

$$p = \frac{h}{\lambda} = \frac{6.626 \times 10^{-34} \text{ J} \cdot \text{s}}{6.50 \times 10^{-7} \text{ m}}$$

$$= 1.02 \times 10^{-27} \text{ kg} \cdot \text{m/s}$$

(Recall that $1 \text{ J} = 1 \text{ kg} \cdot \text{m}^2/\text{s}^2$.)

(b) From Eq. (38.2), the energy of a single photon is

$$E = pc = (1.02 \times 10^{-27} \text{ kg} \cdot \text{m/s})(3.00 \times 10^8 \text{ m/s})$$

$$= 3.06 \times 10^{-19} \text{ J} = 1.91 \text{ eV}$$

The laser pointer emits energy at the rate of 5.00×10^{-3} J/s, so it emits photons at the rate of

$$\frac{5.00 \times 10^{-3} \text{ J/s}}{3.06 \times 10^{-19} \text{ J/photon}} = 1.63 \times 10^{16} \text{ photons/s}$$

EVALUATE: The result in part (a) is very small; a typical oxygen molecule in room-temperature air has 2500 times more momentum. As a check on part (b), we can calculate the photon energy using Eq. (38.2):

$$E = hf = \frac{hc}{\lambda} = \frac{(6.626 \times 10^{-34} \text{ J} \cdot \text{s})(3.00 \times 10^8 \text{ m/s})}{6.50 \times 10^{-7} \text{ m}}$$

$$= 3.06 \times 10^{-19} \text{ J} = 1.91 \text{ eV}$$

Our result in part (b) shows that a huge number of photons leave the laser pointer each second, each of which has an infinitesimal amount of energy. Hence the discreteness of the photons isn't noticed, and the radiated energy appears to be a continuous flow.

Example 38.2 A photoelectric-effect experiment

While conducting a photoelectric-effect experiment with light of a certain frequency, you find that a reverse potential difference of 1.25 V is required to reduce the current to zero. Find (a) the maximum kinetic energy and (b) the maximum speed of the emitted photoelectrons.

SOLUTION

IDENTIFY and SET UP: The value of 1.25 V is the stopping potential V_0 for this experiment. We'll use this in Eq. (38.1) to find the maximum photoelectron kinetic energy K_{max}, and from this we'll find the maximum photoelectron speed.

EXECUTE: (a) From Eq. (38.1),

$$K_{max} = eV_0 = (1.60 \times 10^{-19} \text{ C})(1.25 \text{ V}) = 2.00 \times 10^{-19} \text{ J}$$

(Recall that 1 V = 1 J/C.) In terms of electron volts,

$$K_{max} = eV_0 = e(1.25 \text{ V}) = 1.25 \text{ eV}$$

since the electron volt (eV) is the magnitude of the electron charge e times one volt (1 V).

(b) From $K_{max} = \frac{1}{2}mv_{max}^2$ we get

$$v_{max} = \sqrt{\frac{2K_{max}}{m}} = \sqrt{\frac{2(2.00 \times 10^{-19} \text{ J})}{9.11 \times 10^{-31} \text{ kg}}}$$

$$= 6.63 \times 10^5 \text{ m/s}$$

EVALUATE The value of v_{max} is about 0.2% of the speed of light, so we are justified in using the nonrelativistic expression for kinetic energy. (An equivalent justification is that the electron's 1.25-eV kinetic energy is much less than its rest energy $mc^2 = 0.511 \text{ MeV} = 5.11 \times 10^5 \text{ eV}$.)

Example 38.3 Determining ϕ and h experimentally

For a particular cathode material in a photoelectric-effect experiment, you measure stopping potentials $V_0 = 1.0$ V for light of wavelength $\lambda = 600$ nm, 2.0 V for 400 nm, and 3.0 V for 300 nm. Determine the work function ϕ for this material and the implied value of Planck's constant h.

SOLUTION

IDENTIFY and SET UP: This example uses the relationship among stopping potential V_0, frequency f, and work function ϕ in the photoelectric effect. According to Eq. (38.4), a graph of V_0 versus f should be a straight line as in Fig. 38.5 or 38.6. Such a graph is completely determined by its slope and the value at which it intercepts the vertical axis; we will use these to determine the values of the target variables ϕ and h.

EXECUTE: We rewrite Eq. (38.4) as

$$V_0 = \frac{h}{e}f - \frac{\phi}{e}$$

In this form we see that the slope of the line is h/e and the vertical-axis intercept (corresponding to $f = 0$) is $-\phi/e$. The frequencies,

obtained from $f = c/\lambda$ and $c = 3.00 \times 10^8$ m/s, are 0.50×10^{15} Hz, 0.75×10^{15} Hz, and 1.0×10^{15} Hz, respectively. From a graph of these data (see Fig. 38.6), we find

$$-\frac{\phi}{e} = \text{vertical intercept} = -1.0 \text{ V}$$

$$\phi = 1.0 \text{ eV} = 1.6 \times 10^{-19} \text{ J}$$

and

$$\text{Slope} = \frac{\Delta V_0}{\Delta f} - \frac{3.0 \text{ V} - (-1.0 \text{ V})}{1.00 \times 10^{15} \text{ s}^{-1} - 0} = 4.0 \times 10^{-15} \text{ J} \cdot \text{s/C}$$

$$h = \text{slope} \times e = (4.0 \times 10^{-15} \text{ J} \cdot \text{s/C})(1.60 \times 10^{-19} \text{ C})$$

$$= 6.4 \times 10^{-34} \text{ J} \cdot \text{s}$$

EVALUATE: The value of Planck's constant h determined from your experiment differs from the accepted value by only about 3%. The small value $\phi = 1.0$ eV tells us that the cathode surface is not composed solely of one of the elements in Table 38.1.

Application Sterilizing with High-Energy Photons
One technique for killing harmful microorganisms is to illuminate them with ultraviolet light with a wavelength shorter than 254 nm. If a photon of such short wavelength strikes a DNA molecule within a microorganism, the energy of the photon is great enough to break the bonds within the molecule. This renders the microorganism unable to grow or reproduce. Such ultraviolet germicidal irradiation is used for medical sanitation, to keep laboratories sterile (as shown here), and to treat both drinking water and wastewater.

Test Your Understanding of Section 38.1 Silicon films become better electrical conductors when illuminated by photons with energies of 1.14 eV or greater, an effect called *photoconductivity*. Which of the following wavelengths of electromagnetic radiation can cause photoconductivity in silicon films? (i) ultraviolet light with $\lambda = 300$ nm; (ii) red light with $\lambda = 600$ nm; (iii) infrared light with $\lambda = 1200$ nm.

38.2 Light Emitted as Photons: X-Ray Production

The photoelectric effect provides convincing evidence that light is *absorbed* in the form of photons. For physicists to accept Einstein's radical photon concept, however, it was also necessary to show that light is *emitted* as photons. An experiment that demonstrates this convincingly is the inverse of the photoelectric effect: Instead of releasing electrons from a surface by shining electromagnetic radiation on it, we cause a surface to emit radiation—specifically, *x rays*—by bombarding it with fast-moving electrons.

X-Ray Photons

X rays were first produced in 1895 by the German physicist Wilhelm Röntgen, using an apparatus similar in principle to the setup shown in Fig. 38.7. Electrons are released from the cathode by *thermionic emission,* in which the escape energy is supplied by heating the cathode to a very high temperature. (As in the photoelectric effect, the minimum energy that an individual electron must be given to escape from the cathode's surface is equal to the work function for the surface. In this case the energy is provided to the electrons by heat rather than by light.) The electrons are then accelerated toward the anode by a potential difference V_{AC}. The bulb is evacuated (residual pressure 10^{-7} atm or less), so the electrons can travel from the cathode to the anode without colliding with air molecules. When V_{AC} is a few thousand volts or more, x rays are emitted from the anode surface.

The anode produces x rays in part simply by slowing the electrons abruptly. (Recall from Section 32.1 that accelerated charges emit electromagnetic waves.) This process is called *bremsstrahlung* (German for "braking radiation"). Because the electrons undergo accelerations of very great magnitude, they emit much of their radiation at short wavelengths in the x-ray range, about 10^{-9} to 10^{-12} m (1 nm to 1 pm). (X-ray wavelengths can be measured quite precisely by crystal diffraction techniques, which we discussed in Section 36.6.) Most electrons are braked by a series of collisions and interactions with anode atoms, so bremsstrahlung produces a continuous spectrum of electromagnetic radiation.

Just as we did for the photoelectric effect in Section 38.1, let's compare what Maxwell's wave theory of electromagnetic radiation would predict about this radiation to what is observed experimentally.

Wave-Model Prediction: The electromagnetic waves produced when an electron slams into the anode should be analogous to the sound waves produced by crashing cymbals together. These waves include sounds of all frequencies. By analogy, the x rays produced by bremsstrahlung should have a spectrum that includes *all* frequencies and hence *all* wavelengths.

Experimental Result: Figure 38.8 shows bremsstrahlung spectra using the same cathode and anode with four different accelerating voltages. We see that *not* all x-ray frequencies and wavelengths are emitted: Each spectrum has a maximum frequency f_{max} and a corresponding minimum wavelength λ_{min}. The greater the potential difference V_{AC}, the higher the maximum frequency and the shorter the minimum wavelength.

The wave model of electromagnetic radiation cannot explain these experimental results. But we can readily understand them using the photon model. An electron has charge $-e$ and gains kinetic energy eV_{AC} when accelerated through a potential increase V_{AC}. The most energetic photon (highest frequency and shortest wavelength) is produced if the electron is braked to a stop all at once when it hits the anode, so that all of its kinetic energy goes to produce one photon; that is,

$$eV_{AC} = hf_{max} = \frac{hc}{\lambda_{min}} \qquad \text{(bremsstrahlung)} \qquad (38.6)$$

(In this equation we neglect the work function of the target anode and the initial kinetic energy of the electrons "boiled off" from the cathode. These energies are very small compared to the kinetic energy eV_{AC} gained due to the potential

38.7 An apparatus used to produce x rays, similar to Röntgen's 1895 apparatus.

Electrons are emitted thermionically from the heated cathode and are accelerated toward the anode; when they strike it, x rays are produced.

38.8 The continuous spectrum of x rays produced when a tungsten target is struck by electrons accelerated through a voltage V_{AC}. The curves represent different values of V_{AC}; points *a*, *b*, *c*, and *d* show the minimum wavelength for each voltage.

Vertical axis: x-ray intensity per unit wavelength

Horizontal axis: x-ray wavelength in picometers (1 pm $= 10^{-12}$ m)

difference.) If only a portion of an electron's kinetic energy goes into producing a photon, the photon energy will be less than eV_{AC} and the wavelength will be greater than λ_{min}. As further support for the photon model, the measured values for λ_{min} for different values of eV_{AC} (see Fig. 38.8) agree with Eq. (38.6). Note that according to Eq. (38.6), the maximum frequency and minimum wavelength in the bremsstrahlung process do not depend on the target material; this also agrees with experiment. So we can conclude that the photon picture of electromagnetic radiation is valid for the *emission* as well as the absorption of radiation.

The apparatus shown in Fig. 38.7 can also produce x rays by a second process in which electrons transfer their kinetic energy partly or completely to individual atoms within the target. It turns out that this process not only is consistent with the photon model of electromagnetic radiation, but also provides insight into the structure of atoms. We'll return to this process in Section 41.5.

Example 38.4 Producing x rays

Electrons in an x-ray tube accelerate through a potential difference of 10.0 kV before striking a target. If an electron produces one photon on impact with the target, what is the minimum wavelength of the resulting x rays? Find the answer by expressing energies in both SI units and electron volts.

SOLUTION

IDENTIFY and SET UP: To produce an x-ray photon with minimum wavelength and hence maximum energy, all of the electron's kinetic energy must go into producing a single x-ray photon. We'll use Eq. (38.6) to determine the wavelength.

EXECUTE: From Eq. (38.6), using SI units we have

$$\lambda_{min} = \frac{hc}{eV_{AC}} = \frac{(6.626 \times 10^{-34} \text{ J} \cdot \text{s})(3.00 \times 10^8 \text{ m/s})}{(1.602 \times 10^{-19} \text{ C})(10.0 \times 10^3 \text{ V})}$$

$$= 1.24 \times 10^{-10} \text{ m} = 0.124 \text{ nm}$$

Using electron volts, we have

$$\lambda_{min} = \frac{hc}{eV_{AC}} = \frac{(4.136 \times 10^{-15} \text{ eV} \cdot \text{s})(3.00 \times 10^8 \text{ m/s})}{e(10.0 \times 10^3 \text{ V})}$$

$$= 1.24 \times 10^{-10} \text{ m} = 0.124 \text{ nm}$$

In the second calculation, the "e" for the magnitude of the electron charge cancels the "e" in the unit "eV," because the electron volt (eV) is the magnitude of the electron charge e times one volt (1 V).

EVALUATE: To check our result, recall from Example 38.1 that a 1.91-eV photon has a wavelength of 650 nm. Here the electron energy, and therefore the x-ray photon energy, is 10.0×10^3 eV = 10.0 keV, about 5000 times greater than in Example 38.1, and the wavelength is about $\frac{1}{5000}$ as great as in Example 38.1. This makes sense, since wavelength and photon energy are inversely proportional.

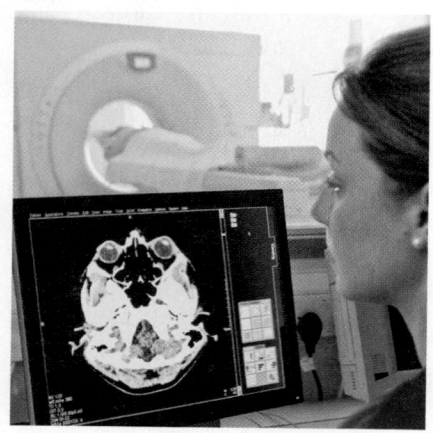

38.9 This radiologist is operating a CT scanner (seen through the window) from a separate room to avoid repeated exposure to x rays.

Applications of X Rays

X rays have many practical applications in medicine and industry. Because x-ray photons are of such high energy, they can penetrate several centimeters of solid matter. Hence they can be used to visualize the interiors of materials that are opaque to ordinary light, such as broken bones or defects in structural steel. The object to be visualized is placed between an x-ray source and an electronic detector (like that used in a digital camera) or a piece of photographic film. The darker an area in the image recorded by such a detector, the greater the radiation exposure. Bones are much more effective x-ray absorbers than soft tissue, so bones appear as light areas. A crack or air bubble allows greater transmission and shows as a dark area.

A widely used and vastly improved x-ray technique is *computed tomography;* the corresponding instrument is called a *CT scanner.* The x-ray source produces a thin, fan-shaped beam that is detected on the opposite side of the subject by an array of several hundred detectors in a line. Each detector measures absorption along a thin line through the subject. The entire apparatus is rotated around the subject in the plane of the beam, and the changing photon-counting rates of the detectors are recorded digitally. A computer processes this information and reconstructs a picture of absorption over an entire cross section of the subject (see Fig. 38.9). Differences in absorption as small as 1% or less can be detected with CT scans, and tumors and other anomalies that are much too small to be seen with older x-ray techniques can be detected.

X rays cause damage to living tissues. As x-ray photons are absorbed in tissues, their energy breaks molecular bonds and creates highly reactive free radicals (such as neutral H and OH), which in turn can disturb the molecular structure of proteins and especially genetic material. Young and rapidly growing cells are particularly susceptible, which is why x rays are useful for selective destruction of cancer cells. Conversely, however, a cell may be damaged by radiation but survive, continue dividing, and produce generations of defective cells; thus x rays can *cause* cancer.

Even when the organism itself shows no apparent damage, excessive exposure to x rays can cause changes in the organism's reproductive system that will affect its offspring. A careful assessment of the balance between risks and benefits of radiation exposure is essential in each individual case.

Test Your Understanding of Section 38.2 In the apparatus shown in Fig. 38.7, suppose you increase the number of electrons that are emitted from the cathode per second while keeping the potential difference V_{AC} the same. How will this affect the intensity I and minimum wavelength λ_{min} of the emitted x rays? (i) I and λ_{min} will both increase; (ii) I will increase but λ_{min} will be unchanged; (iii) I will increase but λ_{min} will decrease; (iv) I will remain the same but λ_{min} will decrease; (v) none of these.

38.3 Light Scattered as Photons: Compton Scattering and Pair Production

The final aspect of light that we must test against Einstein's photon model is its behavior after the light is produced and before it is eventually absorbed. We can do this by considering the *scattering* of light. As we discussed in Section 33.6, scattering is what happens when light bounces off particles such as molecules in the air.

Compton Scattering

Let's see what Maxwell's wave model and Einstein's photon model predict for how light behaves when it undergoes scattering by a single electron, such as an individual electron within an atom.

Wave-Model Prediction: In the wave description, scattering would be a process of absorption and re-radiation. Part of the energy of the light wave would be absorbed by the electron, which would oscillate in response to the oscillating electric field of the wave. The oscillating electron would act like a miniature antenna (see Section 32.1), re-radiating its acquired energy as *scattered* waves in a variety of directions. The frequency at which the electron oscillates would be the same as the frequency of the incident light, and the re-radiated light would have the same frequency as the oscillations of the electron. So, *in the wave model, the scattered light and incident light have the same frequency and same wavelength.*

Photon-Model Prediction: In the photon model we imagine the scattering process as a collision of two *particles,* the incident photon and an electron that is initially at rest (Fig. 38.10a). The incident photon would give up part of its energy and momentum to the electron, which recoils as a result of this impact. The scattered photon that remains can fly off at a variety of angles ϕ with respect to the incident direction, but it has less energy and less momentum than the incident photon (Fig. 38.10b). The energy and momentum of a photon are given by $E = hf = hc/\lambda$ (Eq. 38.2) and $p = hf/c = h/\lambda$ (Eq. 38.5). Therefore, *in the photon model, the scattered light has a lower frequency f and longer wavelength λ than the incident light.*

38.10 The photon model of light scattering by an electron.

(a) Before collision: The target electron is at rest.

Incident photon: wavelength λ, momentum $\vec{p}$

Target electron (at rest)

(b) After collision: The angle between the directions of the scattered photon and the incident photon is ϕ.

Scattered photon: wavelength λ', momentum $\vec{p}'$

Recoiling electron: momentum $\vec{P}_e$

ActivPhysics 17.4: Compton Scattering

38.11 A Compton-effect experiment.

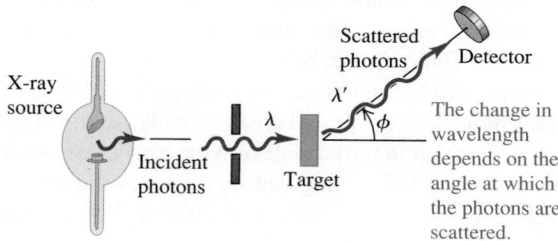

The definitive experiment that tested these predictions of the wave and photon models was carried out in 1922 by the American physicist Arthur H. Compton. In his experiment Compton aimed a beam of x rays at a solid target and measured the wavelength of the radiation scattered from the target (Fig. 38.11). He discovered that some of the scattered radiation has smaller frequency (longer wavelength) than the incident radiation and that the change in wavelength depends on the angle through which the radiation is scattered. This is precisely what the photon model predicts for light scattered from electrons in the target, a process that is now called **Compton scattering.**

Specifically, if the scattered radiation emerges at an angle ϕ with respect to the incident direction, as shown in Fig. 38.11, and if λ and λ' are the wavelengths of the incident and scattered radiation, respectively, Compton found that

$$\lambda' - \lambda = \frac{h}{mc}(1 - \cos\phi) \qquad \text{(Compton scattering)} \qquad (38.7)$$

where m is the electron rest mass. In other words, λ' is greater than λ. The quantity h/mc that appears in Eq. (38.7) has units of length. Its numerical value is

$$\frac{h}{mc} = \frac{6.626 \times 10^{-34}\ \text{J} \cdot \text{s}}{(9.109 \times 10^{-31}\ \text{kg})(2.998 \times 10^{8}\ \text{m/s})} = 2.426 \times 10^{-12}\ \text{m}$$

Compton showed that Einstein's photon theory, combined with the principles of conservation of energy and conservation of momentum, provides a beautifully clear explanation of his experimental results. We outline the derivation below. The electron recoil energy may be in the relativistic range, so we have to use the relativistic energy–momentum relationships, Eqs. (37.39) and (37.40). The incident photon has momentum $\vec{p}$, with magnitude p and energy pc. The scattered photon has momentum $\vec{p}'$, with magnitude p' and energy $p'c$. The electron is initially at rest, so its initial momentum is zero and its initial energy is its rest energy mc^2. The final electron momentum $\vec{P}_\text{e}$ has magnitude P_e, and the final electron energy is given by $E_\text{e}^2 = (mc^2)^2 + (P_\text{e}c)^2$. Then energy conservation gives us the relationship

$$pc + mc^2 = p'c + E_\text{e}$$

38.12 Vector diagram showing conservation of momentum in Compton scattering.

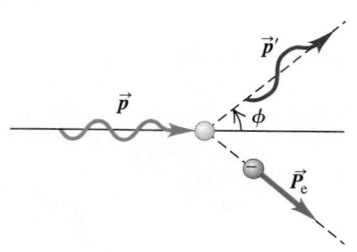

Conservation of momentum during Compton scattering

Rearranging, we find

$$(pc - p'c + mc^2)^2 = E_\text{e}^2 = (mc^2)^2 + (P_\text{e}c)^2 \qquad (38.8)$$

We can eliminate the electron momentum $\vec{P}_\text{e}$ from Eq. (38.8) by using momentum conservation. From Fig. 38.12 we see that $\vec{p} = \vec{p}' + \vec{P}_\text{e}$, or

$$\vec{P}_\text{e} = \vec{p} - \vec{p}' \qquad (38.9)$$

By taking the scalar product of each side of Eq. (38.9) with itself, we find

$$P_e^2 = p^2 + p'^2 - 2pp' \cos\phi \qquad (38.10)$$

We now substitute this expression for P_e^2 into Eq. (38.8) and multiply out the left side. We divide out a common factor c^2; several terms cancel, and when the resulting equation is divided through by (pp'), the result is

$$\frac{mc}{p'} - \frac{mc}{p} = 1 - \cos\phi \qquad (38.11)$$

Finally, we substitute $p' = h/\lambda'$ and $p = h/\lambda$, then multiply by h/mc to obtain Eq. (38.7).

When the wavelengths of x rays scattered at a certain angle are measured, the curve of intensity per unit wavelength as a function of wavelength has two peaks (Fig. 38.13). The longer-wavelength peak represents Compton scattering. The shorter-wavelength peak, labeled λ_0, is at the wavelength of the incident x rays and corresponds to x-ray scattering from tightly bound electrons. In such scattering processes the entire atom must recoil, so the m in Eq. (38.7) is the mass of the entire atom rather than of a single electron. The resulting wavelength shifts are negligible.

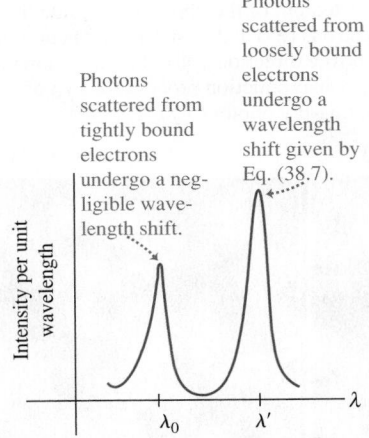

38.13 Intensity as a function of wavelength for photons scattered at an angle of 135° in a Compton-scattering experiment.

Photons scattered from tightly bound electrons undergo a negligible wavelength shift.

Photons scattered from loosely bound electrons undergo a wavelength shift given by Eq. (38.7).

Example 38.5   **Compton scattering**

You use 0.124-nm x-ray photons in a Compton-scattering experiment. (a) At what angle is the wavelength of the scattered x rays 1.0% longer than that of the incident x rays? (b) At what angle is it 0.050% longer?

SOLUTION

IDENTIFY and SET UP: We'll use the relationship between scattering angle and wavelength shift in the Compton effect. In each case our target variable is the angle ϕ (see Fig. 38.10b). We solve for ϕ using Eq. (38.7).

EXECUTE: (a) In Eq. (38.7) we want $\Delta\lambda = \lambda' - \lambda$ to be 1.0% of 0.124 nm, so $\Delta\lambda = 0.00124$ nm $= 1.24 \times 10^{-12}$ m. Using the value $h/mc = 2.426 \times 10^{-12}$ m, we find

$$\Delta\lambda = \frac{h}{mc}(1 - \cos\phi)$$

$$\cos\phi = 1 - \frac{\Delta\lambda}{h/mc} = 1 - \frac{1.24 \times 10^{-12} \text{ m}}{2.426 \times 10^{-12} \text{ m}} = 0.4889$$

$$\phi = 60.7°$$

(b) For $\Delta\lambda$ to be 0.050% of 0.124 nm, or 6.2×10^{-14} m,

$$\cos\phi = 1 - \frac{6.2 \times 10^{-14} \text{ m}}{2.426 \times 10^{-12} \text{ m}} = 0.9744$$

$$\phi = 13.0°$$

EVALUATE: Our results show that smaller scattering angles give smaller wavelength shifts. Thus in a grazing collision the photon energy loss and the electron recoil energy are smaller than when the scattering angle is larger. This is just what we would expect for an elastic collision, whether between a photon and an electron or between two billiard balls.

Pair Production

Another effect that can be explained only with the photon picture involves *gamma rays*, the shortest-wavelength and highest-frequency variety of electromagnetic radiation. If a gamma-ray photon of sufficiently short wavelength is fired at a target, it may not scatter. Instead, as depicted in Fig. 38.14, it may disappear completely and be replaced by two new particles: an electron and a **positron** (a particle that has the same rest mass m as an electron but has a positive charge $+e$ rather than the negative charge $-e$ of the electron). This process, called **pair production,** was first observed by the physicists Patrick Blackett and Giuseppe Occhialini in 1933. The electron and positron have to be produced in pairs in order to conserve electric charge: The incident photon has zero charge, and the electron–positron pair has net charge $(-e) + (+e) = 0$. Enough energy must be available to account for the rest energy $2mc^2$ of the two particles. To four significant figures, this minimum energy is

38.14 (a) Photograph of bubble-chamber tracks of electron–positron pairs that are produced when 300-MeV photons strike a lead sheet. A magnetic field directed out of the photograph made the electrons (e$^-$) and positrons (e$^+$) curve in opposite directions. (b) Diagram showing the pair-production process for two of the gamma-ray photons (γ).

(a)

Electron–positron pair

(b)

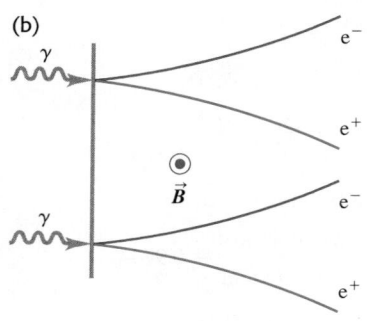

$$E_{min} = 2mc^2 = 2(9.109 \times 10^{-31} \text{ kg})(2.998 \times 10^8 \text{ m/s})^2$$
$$= 1.637 \times 10^{-13} \text{ J} = 1.022 \text{ MeV}$$

Thus the photon must have at least this much energy to produce an electron–positron pair. From Eq. (38.2), $E = hc/\lambda$, the photon wavelength has to be shorter than

$$\lambda_{max} = \frac{hc}{E_{min}} = \frac{(6.626 \times 10^{-34} \text{ J} \cdot \text{s})(2.998 \times 10^8 \text{ m/s})}{1.637 \times 10^{-13} \text{ J}}$$
$$= 1.213 \times 10^{-12} \text{ m} = 1.213 \times 10^{-3} \text{ nm} = 1.213 \text{ pm}$$

This is a very short wavelength, about $\frac{1}{1000}$ as large as the x-ray wavelengths that Compton used in his scattering experiments. (The requisite minimum photon energy is actually a bit higher than 1.022 MeV, so the photon wavelength must be a bit shorter than 1.213 pm. The reason is that when the incident photon encounters an atomic nucleus in the target, some of the photon energy goes into the kinetic energy of the recoiling nucleus.) Just as for the photoelectric effect, the wave model of electromagnetic radiation cannot explain why pair production occurs only when very short wavelengths are used.

The inverse process, *electron–positron pair annihilation*, occurs when a positron and an electron collide. Both particles disappear, and two (or occasionally three) photons can appear, with total energy of at least $2m_e c^2 = 1.022$ MeV. Decay into a *single* photon is impossible because such a process could not conserve both energy and momentum. It's easiest to analyze this annihilation process in the frame of reference called the *center-of-momentum system*, in which the total momentum is zero. It is the relativistic generalization of the center-of-mass system that we discussed in Section 8.5.

Example 38.6   **Pair annihilation**

An electron and a positron, initially far apart, move toward each other with the same speed. They collide head-on, annihilating each other and producing two photons. Find the energies, wavelengths, and frequencies of the photons if the initial kinetic energies of the electron and positron are (a) both negligible and (b) both 5.000 MeV. The electron rest energy is 0.511 MeV.

SOLUTION

IDENTIFY and SET UP: Just as in the elastic collisions we studied in Chapter 8, both momentum and energy are conserved in pair annihilation. The electron and positron are initially far apart, so the initial electric potential energy is zero and the initial energy is the sum of the particle kinetic and rest energies. The final energy is the sum of the photon energies. The total initial momentum is zero; the total momentum of the two photons must likewise be zero. We find the photon energy E using conservation of energy, conservation of momentum, and the relationship $E = pc$ (see Section 38.1). We then calculate the wavelengths and frequencies using $E = hc/\lambda = hf$.

EXECUTE: If the total momentum of the two photons is to be zero, their momenta must have equal magnitudes p and opposite directions. From $E = pc = hc/\lambda = hf$, the two photons must also have the same energy E, wavelength λ, and frequency f.

Before the collision the energy of each electron is $K + mc^2$, where K is its kinetic energy and $mc^2 = 0.511$ MeV. Conservation of energy then gives

$$(K + mc^2) + (K + mc^2) = E + E$$

Hence the energy of each photon is $E = K + mc^2$.

(a) In this case the electron kinetic energy K is negligible compared to its rest energy mc^2, so each photon has energy $E = mc^2 = 0.511$ MeV. The corresponding photon wavelength and frequency are

$$\lambda = \frac{hc}{E} = \frac{(4.136 \times 10^{-15} \text{ eV} \cdot \text{s})(3.00 \times 10^8 \text{ m/s})}{0.511 \times 10^6 \text{ eV}}$$
$$= 2.43 \times 10^{-12} \text{ m} = 2.43 \text{ pm}$$
$$f = \frac{E}{h} = \frac{0.511 \times 10^6 \text{ eV}}{4.136 \times 10^{-15} \text{ eV} \cdot \text{s}} = 1.24 \times 10^{20} \text{ Hz}$$

(b) In this case $K = 5.000$ MeV, so each photon has energy $E = 5.000$ MeV + 0.511 MeV = 5.511 MeV. Proceeding as in part (a), you can show that the photon wavelength is 0.2250 pm and the frequency is 1.333×10^{21} Hz.

EVALUATE: As a check, recall from Example 38.1 that a 650-nm visible-light photon has energy 1.91 eV and frequency 4.62×10^{14} Hz. The photon energy in part (a) is about 2.5×10^5 times greater. As expected, the photon's wavelength is shorter and its frequency higher than those for a visible-light photon by the same factor. You can check the results for part (b) in the same way.

Test Your Understanding of Section 38.3 If you used visible-light photons in the experiment shown in Fig. 38.11, would the photons undergo a wavelength shift due to the scattering? If so, is it possible to detect the shift with the human eye? ❙

38.4 Wave–Particle Duality, Probability, and Uncertainty

We have studied many examples of the behavior of light and other electromagnetic radiation. Some, including the interference and diffraction effects described in Chapters 35 and 36, demonstrate conclusively the *wave* nature of light. Others, the subject of the present chapter, point with equal force to the *particle* nature of light. At first glance these two aspects seem to be in direct conflict. How can light be a wave and a particle at the same time?

We can find the answer to this apparent wave–particle conflict in the **principle of complementarity,** first stated by the Danish physicist Niels Bohr in 1928. The wave descriptions and the particle descriptions are complementary. That is, we need both to complete our model of nature, but we will never need to use both at the same time to describe a single part of an occurrence.

Diffraction and Interference in the Photon Picture

Let's start by considering again the diffraction pattern for a single slit, which we analyzed in Sections 36.2 and 36.3. Instead of recording the pattern on a digital camera chip or photographic film, we use a detector called a *photomultiplier* that can actually detect individual photons. Using the setup shown in Fig. 38.15, we place the photomultiplier at various positions for equal time intervals, count the photons at each position, and plot out the intensity distribution.

We find that, on average, the distribution of photons agrees with our predictions from Section 36.3. At points corresponding to the maxima of the pattern, we count many photons; at minimum points, we count almost none; and so on. The graph of the counts at various points gives the same diffraction pattern that we predicted with Eq. (36.7).

But suppose we now reduce the intensity to such a low level that only a few photons per second pass through the slit. We now record a series of discrete strikes, each representing a single photon. While we *cannot predict* where any given photon will strike, over time the accumulating strikes build up the familiar diffraction pattern we expect for a wave. To reconcile the wave and particle aspects of this pattern, we have to regard the pattern as a *statistical* distribution that tells us how many photons, on average, go to each spot. Equivalently, the pattern tells us the *probability* that any individual photon will land at a given spot. If we shine our faint light beam on a two-slit apparatus, we get an analogous result (Fig. 38.16). Again we can't predict exactly where an individual photon will go; the interference pattern is a statistical distribution.

How does the principle of complementarity apply to these diffraction and interference experiments? The wave description, not the particle description, explains the single- and double-slit patterns. But the particle description, not the wave description, explains why the photomultiplier records discrete packages of energy. The two descriptions complete our understanding of the results. For instance, suppose we consider an individual photon and ask how it knows "which way to go" when passing through the slit. This question seems like a conundrum, but that is because it is framed in terms of a *particle* description—whereas it is the *wave* nature of light that determines the distribution of photons. Conversely,

38.15 Single-slit diffraction pattern of light observed with a movable photomultiplier. The curve shows the intensity distribution predicted by the wave picture. The photon distribution is shown by the numbers of photons counted at various positions.

38.16 These images record the positions where individual photons in a two-slit interference experiment strike the screen. As more photons reach the screen, a recognizable interference pattern appears.

After 21 photons reach the screen

After 1000 photons reach the screen

After 10,000 photons reach the screen

the fact that the photomultiplier detects faint light as a sequence of individual "spots" can't be explained in wave terms.

Probability and Uncertainty

Although photons have energy and momentum, they are nonetheless very different from the particle model we used for Newtonian mechanics in Chapters 4 through 8. The Newtonian particle model treats an object as a point mass. We can describe the location and state of motion of such a particle at any instant with three spatial coordinates and three components of momentum, and we can then predict the particle's future motion. This model doesn't work at all for photons, however: We *cannot* treat a photon as a point object. This is because there are fundamental limitations on the precision with which we can simultaneously determine the position and momentum of a photon. Many aspects of a photon's behavior can be stated only in terms of *probabilities*. (In Chapter 39 we will find that the non-Newtonian ideas we develop for photons in this section also apply to particles such as electrons.)

To get more insight into the problem of measuring a photon's position and momentum simultaneously, let's look again at the single-slit diffraction of light (Fig. 38.17). Suppose the wavelength λ is much less than the slit width a. Then most (85%) of the photons go into the central maximum of the diffraction pattern, and the remainder go into other parts of the pattern. We use θ_1 to denote the angle between the central maximum and the first minimum. Using Eq. (36.2) with $m = 1$, we find that θ_1 is given by $\sin \theta_1 = \lambda/a$. Since we assume $\lambda = a$, it follows that θ_1 is very small, $\sin \theta_1$ is very nearly equal to θ_1 (in radians), and

$$\theta_1 = \frac{\lambda}{a} \tag{38.12}$$

Even though the photons all have the same initial state of motion, they don't all follow the same path. We can't predict the exact trajectory of any individual photon from knowledge of its initial state; we can only describe the *probability* that an individual photon will strike a given spot on the screen. This fundamental indeterminacy has no counterpart in Newtonian mechanics.

Furthermore, there are fundamental *uncertainties* in both the position and the momentum of an individual particle, and these uncertainties are related inseparably. To clarify this point, let's go back to Fig. 38.17. A photon that strikes the screen at the outer edge of the central maximum, at angle θ_1, must have a component of momentum p_y in the y-direction, as well as a component p_x in the x-direction, despite the fact that initially the beam was directed along the x-axis. From the geometry of the situation the two components are related by $p_y/p_x = \tan \theta_1$. Since θ_1 is small, we may use the approximation $\tan \theta_1 = \theta_1$, and

38.17 Interpreting single-slit diffraction in terms of photon momentum.

p_x and p_y are the momentum components for a photon striking the outer edge of the central maximum, at angle θ_1.

Photons of monochromatic light

Slit

Diffraction pattern

Screen

$$p_y = p_x \theta_1 \qquad (38.13)$$

Substituting Eq. (38.12), $\theta_1 = \lambda/a$, into Eq. (38.13) gives

$$p_y = p_x \frac{\lambda}{a} \qquad (38.14)$$

Equation (38.14) says that for the 85% of the photons that strike the detector within the central maximum (that is, at angles between $-\lambda/a$ and $+\lambda/a$), the y-component of momentum is spread out over a range from $-p_x\lambda/a$ to $+p_x\lambda/a$. Now let's consider *all* the photons that pass through the slit and strike the screen. Again, they may hit above or below the center of the pattern, so their component p_y may be positive or negative. However the symmetry of the diffraction pattern shows us the average value $(p_y)_{av} = 0$. There will be an *uncertainty* Δp_y in the y-component of momentum at least as great as $p_x\lambda/a$. That is,

$$\Delta p_y \geq p_x \frac{\lambda}{a} \qquad (38.15)$$

The narrower the slit width a, the broader is the diffraction pattern and the greater is the uncertainty in the y-component of momentum p_y.

The photon wavelength λ is related to the momentum p_x by Eq. (38.5), which we can rewrite as $\lambda = h/p_x$. Using this relationship in Eq. (38.15) and simplifying, we find

$$\Delta p_y \geq p_x \frac{h}{p_x a} = \frac{h}{a}$$

$$\Delta p_y a \geq h \qquad (38.16)$$

What does Eq. (38.16) mean? The slit width a represents an uncertainty in the y-component of the *position* of a photon as it passes through the slit. We don't know exactly *where* in the slit each photon passes through. So both the y-position and the y-component of momentum have uncertainties, and the two uncertainties are related by Eq. (38.16). We can reduce the *momentum* uncertainty Δp_y only by reducing the width of the diffraction pattern. To do this, we have to increase the slit width a, which increases the *position* uncertainty. Conversely, when we *decrease* the position uncertainty by narrowing the slit, the diffraction pattern broadens and the corresponding momentum uncertainty *increases*.

You may protest that it doesn't seem to be consistent with common sense for a photon not to have a definite position and momentum. We reply that what we call *common sense* is based on familiarity gained through experience. Our usual experience includes very little contact with the microscopic behavior of particles like photons. Sometimes we have to accept conclusions that violate our intuition when we are dealing with areas that are far removed from everyday experience.

The Uncertainty Principle

In more general discussions of uncertainty relationships, the uncertainty of a quantity is usually described in terms of the statistical concept of *standard deviation*, which is a measure of the spread or dispersion of a set of numbers around their average value. Suppose we now begin to describe uncertainties in this way [neither Δp_y nor a in Eq. (38.16) is a standard deviation]. If a coordinate x has an uncertainty Δx and if the corresponding momentum component p_x has an uncertainty Δp_x, then those standard-deviation uncertainties are found to be related in general by the inequality

$$\Delta x \, \Delta p_x \geq \hbar/2 \qquad \text{(Heisenberg uncertainty principle} \atop \text{for position and momentum)} \qquad (38.17)$$

In this expression the quantity $\hbar$ (pronounced "h-bar") is Planck's constant divided by 2π:

$$\hbar = \frac{h}{2\pi} = 1.054571628(53) \times 10^{-34}\,\text{J}\cdot\text{s}$$

We will use this quantity frequently to avoid writing a lot of factors of 2π in later equations.

CAUTION **h versus h-bar** It's common for students to plug in the value of h when what they really wanted was $\hbar = h/2\pi$, or vice versa. Be careful not to make the same mistake, or you'll find yourself wondering why your answer is off by a factor of 2π!

Equation (38.17) is one form of the **Heisenberg uncertainty principle,** first discovered by the German physicist Werner Heisenberg (1901–1976). It states that, in general, it is impossible to simultaneously determine both the position and the momentum of a particle with arbitrarily great precision, as classical physics would predict. Instead, the uncertainties in the two quantities play complementary roles, as we have described. Figure 38.18 shows the relationship between the two uncertainties. Our derivation of Eq. (38.16), a less refined form of the uncertainty principle given by Eq. (38.17), shows that this principle has its roots in the wave aspect of photons. We will see in Chapter 39 that electrons and other subatomic particles also have a wave aspect, and the same uncertainty principle applies to them as well.

It is tempting to suppose that we could get greater precision by using more sophisticated detectors of position and momentum. This turns out not to be possible. To detect a particle, the detector must *interact* with it, and this interaction unavoidably changes the state of motion of the particle, introducing uncertainty about its original state. For example, we could imagine placing an electron at a certain point in the middle of the slit in Fig. 38.17. If the photon passes through the middle, we would see the electron recoil. We would then know that the photon passed through that point in the slit, and we would be much more certain about the x-coordinate of the photon. However, the collision between the photon and the electron would change the photon momentum, giving us greater uncertainty in the value of that momentum. A more detailed analysis of such hypothetical experiments shows that the uncertainties we have described are fundamental and intrinsic. They *cannot* be circumvented *even in principle* by any experimental technique, no matter how sophisticated.

There is nothing special about the x-axis. In a three-dimensional situation with coordinates (x, y, z) there is an uncertainty relationship for each coordinate and its corresponding momentum component: $\Delta x \Delta p_x \geq \hbar/2$, $\Delta y \Delta p_y \geq \hbar/2$, and $\Delta z \Delta p_z \geq \hbar/2$. However, the uncertainty in one coordinate is *not* related to the uncertainty in a different component of momentum. For example, Δx is not related directly to Δp_y.

Waves and Uncertainty

Here's an alternative way to understand the Heisenberg uncertainty principle in terms of the properties of waves. Consider a sinusoidal electromagnetic wave propagating in the positive x-direction with its electric field polarized in the y-direction. If the wave has wavelength λ, frequency f, and amplitude A, we can write the wave function as

$$E_y(x, t) = A \sin(kx - \omega t) \tag{38.18}$$

In this expression the wave number is $k = 2\pi/\lambda$ and the angular frequency is $\omega = 2\pi f$. We can think of the wave function in Eq. (38.18) as a description of a photon with a definite wavelength and a definite frequency. In terms of k and ω we can express the momentum and energy of the photon as

$$p_x = \frac{h}{\lambda} = \frac{h}{2\pi}\frac{2\pi}{\lambda} = \hbar k \qquad \text{(photon momentum in terms of wave number)} \tag{38.19a}$$

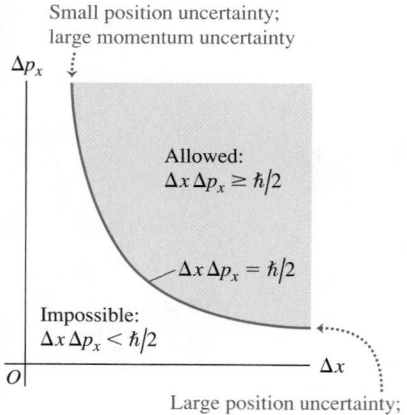

38.18 The Heisenberg uncertainty principle for position and momentum components. It is impossible for the product $\Delta x \Delta p_x$ to be less than $\hbar/2 = h/4\pi$.

Small position uncertainty; large momentum uncertainty

Allowed:
$\Delta x \Delta p_x \geq \hbar/2$

$\Delta x \Delta p_x = \hbar/2$

Impossible:
$\Delta x \Delta p_x < \hbar/2$

Large position uncertainty; small momentum uncertainty

$$E = hf = \frac{h}{2\pi} 2\pi f = \hbar\omega \qquad \text{(photon energy in terms of angular frequency)} \qquad \text{(38.19b)}$$

Using Eqs. (38.19) in Eq. (38.18), we can rewrite our photon wave equation as

$$E_y(x, t) = A\sin[(p_x x - Et)/\hbar] \qquad \text{(wave function for a photon with x-momentum p_x and energy E)} \qquad \text{(38.20)}$$

Since this wave function has a definite value of x-momentum p_x, there is *no* uncertainty in the value of this quantity: $\Delta p_x = 0$. The Heisenberg uncertainty principle, Eq. (38.17), says that $\Delta x \Delta p_x \geq \hbar/2$. If Δp_x is zero, then Δx must be infinite. Indeed, the wave described by Eq. (38.20) extends along the entire x-axis and has the same amplitude everywhere. The price we pay for knowing the photon's momentum precisely is that we have no idea *where* the photon is!

In practical situations we always have *some* idea where a photon is. To describe this situation, we need a wave function that is more localized in space. We can create one by superimposing two or more sinusoidal functions. To keep things simple, we'll consider only waves propagating in the positive x-direction. For example, let's add together two sinusoidal wave functions like those in Eqs. (38.18) and (38.20), but with slightly different wavelengths and frequencies and hence slightly different values p_{x1} and p_{x2} of x-momentum and slightly different values E_1 and E_2 of energy. The total wave function is

$$E_y(x, t) = A_1\sin[(p_{1x}x - E_1 t)/\hbar] + A_2\sin[(p_{2x}x - E_2 t)/\hbar] \qquad \text{(38.21)}$$

Consider what this wave function looks like at a particular instant of time, say $t = 0$. At this instant Eq. (38.21) becomes

$$E_y(x, t = 0) = A_1\sin(p_{1x}x/\hbar) + A_2\sin(p_{2x}x/\hbar) \qquad \text{(38.22)}$$

Figure 38.19a is a graph of the individual wave functions at $t = 0$ for the case $A_2 = -A_1$, and Fig. 38.19b graphs the combined wave function $E_y(x, t = 0)$ given by Eq. (38.22). We saw something very similar to Fig. 38.19b in our discussion of beats in Section 16.7: When we superimposed two sinusoidal waves with slightly different frequencies (see Fig. 16.24), the resulting wave exhibited amplitude variations not present in the original waves. In the same way, a photon represented by the wave function in Eq. (38.21) is most likely to be found in the regions where the wave function's amplitude is greatest. That is, the photon is *localized*. However, the photon's momentum no longer has a definite value because we began with two different x-momentum values, p_{x1} and p_{x2}. This agrees with the Heisenberg uncertainty principle: By decreasing the uncertainty in the photon's position, we have increased the uncertainty in its momentum.

38.19 (a) Two sinusoidal waves with slightly different wave numbers k and hence slightly different values of momentum $p_x = \hbar k$ shown at one instant of time. (b) The superposition of these waves has a momentum equal to the average of the two individual values of momentum. The amplitude varies, giving the total wave a lumpy character not possessed by either individual wave.

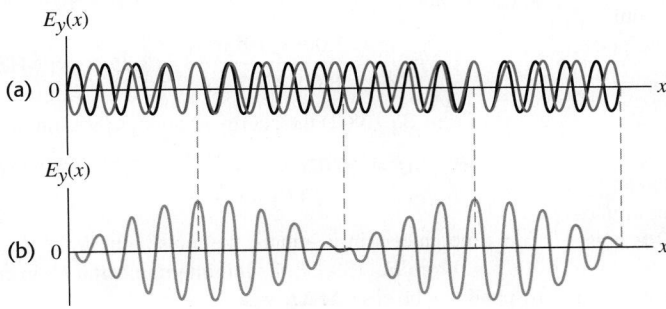

Uncertainty in Energy

Our discussion of combining waves also shows that there is an uncertainty principle that involves *energy* and *time*. To see why this is so, imagine measuring the combined wave function described by Eq. (38.21) at a certain position, say $x = 0$, over a period of time. At $x = 0$, the wave function from Eq. (38.21) becomes

$$E_y(x, t) = A_1 \sin(-E_1 t/\hbar) + A_2 \sin(-E_2 t/\hbar)$$
$$= -A_1 \sin(E_1 t/\hbar) - A_2 \sin(E_2 t/\hbar) \qquad (38.23)$$

What we measure at $x = 0$ is a combination of two oscillating electric fields with slightly different angular frequencies $\omega_1 = E_1/\hbar$ and $\omega_2 = E_2/\hbar$. This is exactly the phenomenon of beats that we discussed in Section 16.7 (compare Fig. 16.24). The amplitude of the combined field rises and falls, so the photon described by this field is localized in *time* as well as in position. The photon is most likely to be found at the times when the amplitude is large. The price we pay for localizing the photon in time is that the wave does not have a definite energy. By contrast, if the photon is described by a sinusoidal wave like that in Eq. (38.20) that *does* have a definite energy E but that has the same amplitude at all times, we have no idea when the photon will appear at $x = 0$. So the better we know the photon's energy, the less certain we are of when we will observe the photon.

Just as for the momentum–position uncertainty principle, we can write a mathematical expression for the uncertainty principle that relates energy and time. In fact, except for an overall minus sign, Eq. (38.23) is identical to Eq. (38.22) if we replace the x-momentum p_x by energy E and the position x by time t. This tells us that in the momentum–position uncertainty relation, Eq. (38.17), we can replace the momentum uncertainty Δp_x with the energy uncertainty ΔE and replace the position uncertainty Δx with the time uncertainty Δt. The result is

$$\Delta t \, \Delta E \geq \hbar/2 \qquad \text{(Heisenberg uncertainty principle for energy and time)} \qquad (38.24)$$

In practice, any real photon has a limited spatial extent and hence passes any point in a limited amount of time. The following example illustrates how this affects the momentum and energy of the photon.

Example 38.7 Ultrashort laser pulses and the uncertainty principle

Many varieties of lasers emit light in the form of pulses rather than a steady beam. A tellurium–sapphire laser can produce light at a wavelength of 800 nm in ultrashort pulses that last only 4.00×10^{-15} s (4.00 femtoseconds, or 4.00 fs). The energy in a single pulse produced by one such laser is $2.00 \, \mu\text{J} = 2.00 \times 10^{-6}$ J, and the pulses propagate in the positive x-direction. Find (a) the frequency of the light; (b) the energy and minimum energy uncertainty of a single photon in the pulse; (c) the minimum frequency uncertainty of the light in the pulse; (d) the spatial length of the pulse, in meters and as a multiple of the wavelength; (e) the momentum and minimum momentum uncertainty of a single photon in the pulse; and (f) the approximate number of photons in the pulse.

SOLUTION

IDENTIFY and SET UP: It's important to distinguish between the light pulse as a whole (which contains a very large number of photons) and an individual photon within the pulse. The 5.00-fs pulse duration represents the time it takes the pulse to emerge from the laser; it is also the time *uncertainty* for an individual photon within

the pulse, since we don't know when during the pulse that photon emerges. Similarly, the position uncertainty of a photon is the spatial length of the pulse, since a given photon could be found anywhere within the pulse. To find our target variables, we'll use the relationships for photon energy and momentum from Section 38.1 and the two Heisenberg uncertainty principles, Eqs. (38.17) and (38.24).

EXECUTE: (a) From the relationship $c = \lambda f$, the frequency of 800-nm light is

$$f = \frac{c}{\lambda} = \frac{3.00 \times 10^8 \, \text{m/s}}{8.00 \times 10^{-7} \, \text{m}} = 3.75 \times 10^{14} \, \text{Hz}$$

(b) From Eq. (38.2) the energy of a single 800-nm photon is

$$E = hf = (6.626 \times 10^{-34} \, \text{J} \cdot \text{s})(3.75 \times 10^{14} \, \text{Hz})$$
$$= 2.48 \times 10^{-19} \, \text{J}$$

The time uncertainty equals the pulse duration, $\Delta t = 4.00 \times 10^{-15}$ s. From Eq. (38.24) the minimum uncertainty in energy corresponds to the case $\Delta t \, \Delta E = \hbar/2$, so

$$\Delta E = \frac{\hbar}{2\Delta t} = \frac{1.055 \times 10^{-34} \text{ J} \cdot \text{s}}{2(4.00 \times 10^{-15} \text{ s})} = 1.32 \times 10^{-20} \text{ J}$$

This is 5.3% of the photon energy $E = 2.48 \times 10^{-19}$ J, so the energy of a given photon is uncertain by at least 5.3%. The uncertainty could be greater, depending on the shape of the pulse.

(c) From the relationship $f = E/h$, the minimum frequency uncertainty is

$$\Delta f = \frac{\Delta E}{h} = \frac{1.32 \times 10^{-20} \text{ J}}{6.626 \times 10^{-34} \text{ J} \cdot \text{s}} = 1.99 \times 10^{13} \text{ Hz}$$

This is 5.3% of the frequency $f = 3.75 \times 10^{14}$ Hz we found in part (a). Hence these ultrashort pulses do not have a definite frequency; the average frequency of many such pulses will be 3.75×10^{14} Hz, but the frequency of any individual pulse can be anywhere from 5.3% higher to 5.3% lower.

(d) The spatial length Δx of the pulse is the distance that the front of the pulse travels during the time $\Delta t = 4.00 \times 10^{-15}$ s it takes the pulse to emerge from the laser:

$$\Delta x = c\Delta t = (3.00 \times 10^8 \text{ m/s})(4.00 \times 10^{-15} \text{ s})$$
$$= 1.20 \times 10^{-6} \text{ m}$$

$$\Delta x = \frac{1.20 \times 10^{-6} \text{ m}}{8.00 \times 10^{-7} \text{ m/wavelength}} = 1.50 \text{ wavelengths}$$

This justifies the term *ultrashort*. The pulse is less than two wavelengths long!

(e) From Eq. (38.5), the momentum of an average photon in the pulse is

$$p_x = \frac{E}{c} = \frac{2.48 \times 10^{-19} \text{ J}}{3.00 \times 10^8 \text{ m/s}} = 8.28 \times 10^{-28} \text{ kg} \cdot \text{m/s}$$

The spatial uncertainty is $\Delta x = 1.20 \times 10^{-6}$ m. From Eq. (38.17) minimum momentum uncertainty corresponds to $\Delta x \, \Delta p_x = \hbar/2$, so

$$\Delta p_x = \frac{\hbar}{2\Delta x} = \frac{1.055 \times 10^{-34} \text{ J} \cdot \text{s}}{2(1.20 \times 10^{-6} \text{ m})} = 4.40 \times 10^{-29} \text{ kg} \cdot \text{m/s}$$

This is 5.3% of the average photon momentum p_x. An individual photon within the pulse can have a momentum that is 5.3% greater or less than the average.

(f) To estimate the number of photons in the pulse, we divide the total pulse energy by the average photon energy:

$$\frac{2.00 \times 10^{-6} \text{ J/pulse}}{2.48 \times 10^{-19} \text{ J/photon}} = 8.06 \times 10^{12} \text{ photons/pulse}$$

The energy of an individual photon is uncertain, so this is the *average* number of photons per pulse.

EVALUATE: The percentage uncertainties in energy and momentum are large because this laser pulse is so short. If the pulse were longer, both Δt and Δx would be greater and the corresponding uncertainties in photon energy and photon momentum would be smaller.

Our calculation in part (f) shows an important distinction between photons and other kinds of particles. In principle it is possible to make an exact count of the number of electrons, protons, and neutrons in an object such as this book. If you repeated the count, you would get the same answer as the first time. By contrast, if you counted the number of photons in a laser pulse you would *not* necessarily get the same answer every time! The uncertainty in photon energy means that on each count there could be a different number of photons whose individual energies sum to 2.00×10^{-6} J. That's yet another of the many strange properties of photons.

Test Your Understanding of Section 38.4 Through which of the following angles is a photon of wavelength λ most likely to be deflected after passing through a slit of width a? Assume that λ is much less than a. (i) $\theta = \lambda/a$; (ii) $\theta = 3\lambda/2a$; (iii) $\theta = 2\lambda/a$; (iv) $\theta = 3\lambda/a$; (v) not enough information given to decide.

Photons: Electromagnetic radiation behaves as both waves and particles. The energy in an electromagnetic wave is carried in units called photons. The energy E of one photon is proportional to the wave frequency f and inversely proportional to the wavelength λ, and is proportional to a universal quantity h called Planck's constant. The momentum of a photon has magnitude E/c. (See Example 38.1.)

$$E = hf = \frac{hc}{\lambda} \qquad (38.2)$$

$$p = \frac{E}{c} = \frac{hf}{c} = \frac{h}{\lambda} \qquad (38.5)$$

The photoelectric effect: In the photoelectric effect, a surface can eject an electron by absorbing a photon whose energy hf is greater than or equal to the work function ϕ of the material. The stopping potential V_0 is the voltage required to stop a current of ejected electrons from reaching an anode. (See Examples 38.2 and 38.3.)

$$eV_0 = hf - \phi \qquad (38.4)$$

Photon production, photon scattering, and pair production: X rays can be produced when electrons accelerated to high kinetic energy across a potential increase V_{AC} strike a target. The photon model explains why the maximum frequency and minimum wavelength produced are given by Eq. (38.6). (See Example 38.4.) In Compton scattering a photon transfers some of its energy and momentum to an electron with which it collides. For free electrons (mass m), the wavelengths of incident and scattered photons are related to the photon scattering angle ϕ by Eq. (38.7). (See Example 38.5.) In pair production a photon of sufficient energy can disappear and be replaced by an electron–positron pair. In the inverse process, an electron and a positron can annihilate and be replaced by a pair of photons. (See Example 38.6.)

$$eV_{AC} = hf_{max} = \frac{hc}{\lambda_{min}} \qquad (38.6)$$
(bremsstrahlung)

$$\lambda' - \lambda = \frac{h}{mc}(1 - \cos\phi) \qquad (38.7)$$
(Compton scattering)

The Heisenberg uncertainty principle: It is impossible to determine both a photon's position and its momentum at the same time to arbitrarily high precision. The precision of such measurements for the x-components is limited by the Heisenberg uncertainty principle, Eq. (38.17); there are corresponding relationships for the y- and z-components. The uncertainty ΔE in the energy of a state that is occupied for a time Δt is given by Eq. (38.24). In these expressions, $\hbar = h/2\pi$. (See Example 38.7.)

$$\Delta x\, \Delta p_x \geq \hbar/2 \qquad (38.17)$$
(Heisenberg uncertainty principle for position and momentum)

$$\Delta t\, \Delta E \geq \hbar/2 \qquad (38.24)$$
(Heisenberg uncertainty principle for energy and time)

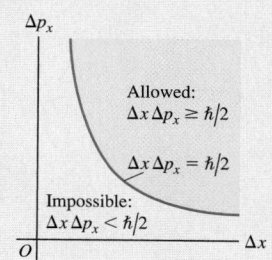

BRIDGING PROBLEM **Compton Scattering and Electron Recoil**

An incident x-ray photon is scattered from a free electron that is initially at rest. The photon is scattered straight back at an angle of 180° from its initial direction. The wavelength of the scattered photon is 0.0830 nm. (a) What is the wavelength of the incident photon? (b) What are the magnitude of the momentum and the speed of the electron after the collision? (c) What is the kinetic energy of the electron after the collision?

SOLUTION GUIDE

See MasteringPhysics® study area for a Video Tutor solution.

IDENTIFY and SET UP

1. In this problem a photon is scattered by an electron initially at rest. In Section 38.3 you learned how to relate the wavelengths of the incident and scattered photons; in this problem you must also find the momentum, speed, and kinetic energy of the recoiling electron. You can find these because momentum and energy are conserved in the collision.

2. Which key equation can be used to find the incident photon wavelength? What is the photon scattering angle ϕ in this problem?

EXECUTE

3. Use the equation you selected in step 2 to find the wavelength of the incident photon.

4. Use momentum conservation and your result from step 3 to find the momentum of the recoiling electron. (*Hint:* All of the momentum vectors are along the same line, but not all point in the same direction. Be careful with signs.)

5. Find the speed of the recoiling electron from your result in step 4. (*Hint:* Assume that the electron is nonrelativistic, so you can use the relationship between momentum and speed from Chapter 8. This is acceptable if the speed of the electron is less than about $0.1c$. Is it?)

6. Use your result from step 4 or step 5 to find the electron kinetic energy.

EVALUATE

7. You can check your answer in step 6 by finding the difference between the energies of the incident and scattered photons. Is your result consistent with conservation of energy?

Problems

For instructor-assigned homework, go to www.masteringphysics.com

•, ••, •••: Problems of increasing difficulty. **CP**: Cumulative problems incorporating material from earlier chapters. **CALC**: Problems requiring calculus. **BIO**: Biosciences problems.

DISCUSSION QUESTIONS

Q38.1 In what ways do photons resemble other particles such as electrons? In what ways do they differ? Do photons have mass? Do they have electric charge? Can they be accelerated? What mechanical properties do they have?

Q38.2 There is a certain probability that a single electron may simultaneously absorb *two* identical photons from a high-intensity laser. How would such an occurrence affect the threshold frequency and the equations of Section 38.1? Explain.

Q38.3 According to the photon model, light carries its energy in packets called quanta or photons. Why then don't we see a series of flashes when we look at things?

Q38.4 Would you expect effects due to the photon nature of light to be generally more important at the low-frequency end of the electromagnetic spectrum (radio waves) or at the high-frequency end (x rays and gamma rays)? Why?

Q38.5 During the photoelectric effect, light knocks electrons out of metals. So why don't the metals in your home lose their electrons when you turn on the lights?

Q38.6 Most black-and-white photographic film (with the exception of some special-purpose films) is less sensitive to red light than blue light and has almost no sensitivity to infrared. How can these properties be understood on the basis of photons?

Q38.7 Human skin is relatively insensitive to visible light, but ultraviolet radiation can cause severe burns. Does this have anything to do with photon energies? Explain.

Q38.8 Explain why Fig. 38.4 shows that most photoelectrons have kinetic energies less than $hf - \phi$, and also explain how these smaller kinetic energies occur.

Q38.9 In a photoelectric-effect experiment, the photocurrent i for large positive values of V_{AC} has the same value no matter what the light frequency f (provided that f is higher than the threshold frequency f_0). Explain why.

Q38.10 In an experiment involving the photoelectric effect, if the intensity of the incident light (having frequency higher than the threshold frequency) is reduced by a factor of 10 without changing anything else, which (if any) of the following statements about this process will be true? (a) The number of photoelectrons will most likely be reduced by a factor of 10. (b) The maximum kinetic energy of the ejected photoelectrons will most likely be reduced by a factor of 10. (c) The maximum speed of the ejected photoelectrons will most likely be reduced by a factor of 10. (d) The maximum speed of the ejected photoelectrons will most likely be reduced by a factor of $\sqrt{10}$. (e) The time for the first photoelectron to be ejected will be increased by a factor of 10.

Q38.11 The materials called *phosphors* that coat the inside of a fluorescent lamp convert ultraviolet radiation (from the mercury-vapor discharge inside the tube) into visible light. Could one also make a phosphor that converts visible light to ultraviolet? Explain.

Q38.12 In a photoelectric-effect experiment, which of the following will increase the maximum kinetic energy of the photoelectrons? (a) Use light of greater intensity; (b) use light of higher frequency; (c) use light of longer wavelength; (d) use a metal surface with a larger work function. In each case justify your answer.

Q38.13 A photon of frequency f undergoes Compton scattering from an electron at rest and scatters through an angle ϕ. The frequency of the scattered photon is f'. How is f' related to f? Does your answer depend on ϕ? Explain.

Q38.14 Can Compton scattering occur with protons as well as electrons? For example, suppose a beam of x rays is directed at a target of liquid hydrogen. (Recall that the nucleus of hydrogen consists of a single proton.) Compared to Compton scattering with electrons, what similarities and differences would you expect? Explain.

Q38.15 Why must engineers and scientists shield against x-ray production in high-voltage equipment?

Q38.16 In attempting to reconcile the wave and particle models of light, some people have suggested that the photon rides up and down on the crests and troughs of the electromagnetic wave. What things are *wrong* with this description?

Q38.17 Some lasers emit light in pulses that are only 10^{-12} s in duration. The length of such a pulse is $(3 \times 10^8 \text{ m/s})(10^{-12} \text{ s}) = 3 \times 10^{-4} \text{ m} = 0.3 \text{ mm}$. Can pulsed laser light be as monochromatic as light from a laser that emits a steady, continuous beam? Explain.

EXERCISES

Section 38.1 Light Absorbed as Photons: The Photoelectric Effect

38.1 •• (a) A proton is moving at a speed much slower than the speed of light. It has kinetic energy K_1 and momentum p_1. If the momentum of the proton is doubled, so $p_2 = 2p_1$, how is its new kinetic energy K_2 related to K_1? (b) A photon with energy E_1 has momentum p_1. If another photon has momentum p_2 that is twice p_1, how is the energy E_2 of the second photon related to E_1?

38.2 • **BIO** **Response of the Eye.** The human eye is most sensitive to green light of wavelength 505 nm. Experiments have found that when people are kept in a dark room until their eyes adapt to the darkness, a *single* photon of green light will trigger receptor cells in the rods of the retina. (a) What is the frequency of this photon? (b) How much energy (in joules and electron volts) does it deliver to the receptor cells? (c) To appreciate what a small amount of energy this is, calculate how fast a typical bacterium of mass 9.5×10^{-12} g would move if it had that much energy.

38.3 • A photon of green light has a wavelength of 520 nm. Find the photon's frequency, magnitude of momentum, and energy. Express the energy in both joules and electron volts.

38.4 • **BIO** A laser used to weld detached retinas emits light with a wavelength of 652 nm in pulses that are 20.0 ms in duration. The average power during each pulse is 0.600 W. (a) How much energy is in each pulse in joules? In electron volts? (b) What is the energy of one photon in joules? In electron volts? (c) How many photons are in each pulse?

38.5 • A 75-W light source consumes 75 W of electrical power. Assume all this energy goes into emitted light of wavelength 600 nm. (a) Calculate the frequency of the emitted light. (b) How many photons per second does the source emit? (c) Are the answers to parts (a) and (b) the same? Is the frequency of the light the same thing as the number of photons emitted per second? Explain.

38.6 • A photon has momentum of magnitude 8.24×10^{-28} kg·m/s. (a) What is the energy of this photon? Give your answer in joules and in electron volts. (b) What is the wavelength of this photon? In what region of the electromagnetic spectrum does it lie?

38.7 • The graph in Fig. E38.7 shows the stopping potential as a function of the frequency of the incident light falling on a metal surface. (a) Find the photoelectric work function for this metal. (b) What value of Planck's constant does the graph yield? (c) Why does the graph *not* extend below the x-axis? (d) If a different metal were used, which characteristics of the graph would you expect to be the same and which ones would be different?

Figure **E38.7**

38.8 • The photoelectric threshold wavelength of a tungsten surface is 272 nm. Calculate the maximum kinetic energy of the electrons ejected from this tungsten surface by ultraviolet radiation of frequency 1.45×10^{15} Hz. Express the answer in electron volts.

38.9 •• A clean nickel surface is exposed to light of wavelength 235 nm. What is the maximum speed of the photoelectrons emitted from this surface? Use Table 38.1.

38.10 •• What would the minimum work function for a metal have to be for visible light (380–750 nm) to eject photoelectrons?

38.11 •• When ultraviolet light with a wavelength of 400.0 nm falls on a certain metal surface, the maximum kinetic energy of the emitted photoelectrons is measured to be 1.10 eV. What is the maximum kinetic energy of the photoelectrons when light of wavelength 300.0 nm falls on the same surface?

38.12 •• The photoelectric work function of potassium is 2.3 eV. If light having a wavelength of 250 nm falls on potassium, find (a) the stopping potential in volts; (b) the kinetic energy in electron volts of the most energetic electrons ejected; (c) the speed of these electrons.

38.13 • When ultraviolet light with a wavelength of 254 nm falls on a clean copper surface, the stopping potential necessary to stop emission of photoelectrons is 0.181 V. (a) What is the photoelectric threshold wavelength for this copper surface? (b) What is the work function for this surface, and how does your calculated value compare with that given in Table 38.1?

Section 38.2 Light Emitted as Photons: X-Ray Production

38.14 • The cathode-ray tubes that generated the picture in early color televisions were sources of x rays. If the acceleration voltage in a television tube is 15.0 kV, what are the shortest-wavelength x rays produced by the television? (Modern televisions contain shielding to stop these x rays.)

38.15 • Protons are accelerated from rest by a potential difference of 4.00 kV and strike a metal target. If a proton produces one photon on impact, what is the minimum wavelength of the resulting x rays? How does your answer compare to the minimum wavelength if 4.00-keV electrons are used instead? Why do x-ray tubes use electrons rather than protons to produce x rays?

38.16 •• (a) What is the minimum potential difference between the filament and the target of an x-ray tube if the tube is to produce x rays with a wavelength of 0.150 nm? (b) What is the shortest wavelength produced in an x-ray tube operated at 30.0 kV?

Section 38.3 Light Scattered as Photons: Compton Scattering and Pair Production

38.17 • An x ray with a wavelength of 0.100 nm collides with an electron that is initially at rest. The x ray's final wavelength is 0.110 nm. What is the final kinetic energy of the electron?

38.18 • X rays are produced in a tube operating at 18.0 kV. After emerging from the tube, x rays with the minimum wavelength produced strike a target and are Compton-scattered through an angle of 45.0°. (a) What is the original x-ray wavelength? (b) What is the wavelength of the scattered x rays? (c) What is the energy of the scattered x rays (in electron volts)?

38.19 •• X rays with initial wavelength 0.0665 nm undergo Compton scattering. What is the longest wavelength found in the scattered x rays? At which scattering angle is this wavelength observed?

38.20 • A beam of x rays with wavelength 0.0500 nm is Compton-scattered by the electrons in a sample. At what angle from the incident beam should you look to find x rays with a wavelength of (a) 0.0542 nm; (b) 0.0521 nm; (c) 0.0500 nm?

38.21 •• If a photon of wavelength 0.04250 nm strikes a free electron and is scattered at an angle of 35.0° from its original direction, find (a) the change in the wavelength of this photon; (b) the wavelength of the scattered light; (c) the change in energy of the photon (is it a loss or a gain?); (d) the energy gained by the electron.

38.22 •• A photon scatters in the backward direction ($\phi = 180°$) from a free proton that is initially at rest. What must the wavelength of the incident photon be if it is to undergo a 10.0% change in wavelength as a result of the scattering?

38.23 •• X rays with an initial wavelength of 0.900×10^{-10} m undergo Compton scattering. For what scattering angle is the wavelength of the scattered x rays greater by 1.0% than that of the incident x rays?

38.24 •• A photon with wavelength $\lambda = 0.1385$ nm scatters from an electron that is initially at rest. What must be the angle between the direction of propagation of the incident and scattered photons if the speed of the electron immediately after the collision is 8.90×10^6 m/s?

38.25 • An electron and a positron are moving toward each other and each has speed $0.500c$ in the lab frame. (a) What is the kinetic energy of each particle? (b) The e^+ and e^- meet head-on and annihilate. What is the energy of each photon that is produced? (c) What is the wavelength of each photon? How does the wavelength compare to the photon wavelength when the initial kinetic energy of the e^+ and e^- is negligibly small (see Example 38.6)?

Section 38.4 Wave–Particle Duality, Probability, and Uncertainty

38.26 • A laser produces light of wavelength 625 nm in an ultrashort pulse. What is the minimum duration of the pulse if the minimum uncertainty in the energy of the photons is 1.0%?

38.27 • An ultrashort pulse has a duration of 9.00 fs and produces light at a wavelength of 556 nm. What are the momentum and momentum uncertainty of a single photon in the pulse?

38.28 • A horizontal beam of laser light of wavelength 585 nm passes through a narrow slit that has width 0.0620 mm. The intensity of the light is measured on a vertical screen that is 2.00 m from the slit. (a) What is the minimum uncertainty in the vertical component of the momentum of each photon in the beam after the

photon has passed through the slit? (b) Use the result of part (a) to estimate the width of the central diffraction maximum that is observed on the screen.

PROBLEMS

38.29 • **Exposing Photographic Film.** The light-sensitive compound on most photographic films is silver bromide, AgBr. A film is "exposed" when the light energy absorbed dissociates this molecule into its atoms. (The actual process is more complex, but the quantitative result does not differ greatly.) The energy of dissociation of AgBr is 1.00×10^5 J/mol. For a photon that is just able to dissociate a molecule of silver bromide, find (a) the photon energy in electron volts; (b) the wavelength of the photon; (c) the frequency of the photon. (d) What is the energy in electron volts of a photon having a frequency of 100 MHz? (e) Light from a firefly can expose photographic film, but the radiation from an FM station broadcasting 50,000 W at 100 MHz cannot. Explain why this is so.

38.30 •• (a) If the average frequency emitted by a 200-W light bulb is 5.00×10^{14} Hz, and 10.0% of the input power is emitted as visible light, approximately how many visible-light photons are emitted per second? (b) At what distance would this correspond to 1.00×10^{11} visible-light photons per square centimeter per second if the light is emitted uniformly in all directions?

38.31 • When a certain photoelectric surface is illuminated with light of different wavelengths, the following stopping potentials are observed:

Wavelength (nm)	Stopping potential (V)
366	1.48
405	1.15
436	0.93
492	0.62
546	0.36
579	0.24

Plot the stopping potential on the vertical axis against the frequency of the light on the horizontal axis. Determine (a) the threshold frequency; (b) the threshold wavelength; (c) the photoelectric work function of the material (in electron volts); (d) the value of Planck's constant h (assuming that the value of e is known).

38.32 • A 2.50-W beam of light of wavelength 124 nm falls on a metal surface. You observe that the maximum kinetic energy of the ejected electrons is 4.16 eV. Assume that each photon in the beam ejects a photoelectron. (a) What is the work function (in electron volts) of this metal? (b) How many photoelectrons are ejected each second from this metal? (c) If the power of the light beam, but not its wavelength, were reduced by half, what would be the answer to part (b)? (d) If the wavelength of the beam, but not its power, were reduced by half, what would be the answer to part (b)?

38.33 •• **CP BIO Removing Vascular Lesions.** A pulsed dye laser emits light of wavelength 585 nm in 450-μs pulses. Because this wavelength is strongly absorbed by the hemoglobin in the blood, the method is especially effective for removing various types of blemishes due to blood, such as port-wine–colored birthmarks. To get a reasonable estimate of the power required for such laser surgery, we can model the blood as having the same specific heat and heat of vaporization as water (4190 J/kg·K, 2.256×10^6 J/kg). Suppose that each pulse must remove 2.0 μg of blood by evaporating it, starting at 33°C. (a) How much energy must each pulse deliver to the blemish? (b) What must be the power output of this laser? (c) How many photons does each pulse deliver to the blemish?

38.34 • The photoelectric work functions for particular samples of certain metals are as follows: cesium, 2.1 eV; copper, 4.7 eV;

potassium, 2.3 eV; and zinc, 4.3 eV. (a) What is the threshold wavelength for each metal surface? (b) Which of these metals could *not* emit photoelectrons when irradiated with visible light (380–750 nm)?

38.35 •• An incident x-ray photon of wavelength 0.0900 nm is scattered in the backward direction from a free electron that is initially at rest. (a) What is the magnitude of the momentum of the scattered photon? (b) What is the kinetic energy of the electron after the photon is scattered?

38.36 •• **CP** A photon with wavelength $\lambda = 0.0900$ nm is incident on an electron that is initially at rest. If the photon scatters in the backward direction, what is the magnitude of the linear momentum of the electron just after the collision with the photon?

38.37 •• **CP** A photon with wavelength $\lambda = 0.1050$ nm is incident on an electron that is initially at rest. If the photon scatters at an angle of 60.0° from its original direction, what are the magnitude and direction of the linear momentum of the electron just after the collision with the photon?

38.38 •• **CP** An x-ray tube is operating at voltage V and current I. (a) If only a fraction p of the electric power supplied is converted into x rays, at what rate is energy being delivered to the target? (b) If the target has mass m and specific heat c (in J/kg · K), at what average rate would its temperature rise if there were no thermal losses? (c) Evaluate your results from parts (a) and (b) for an x-ray tube operating at 18.0 kV and 60.0 mA that converts 1.0% of the electric power into x rays. Assume that the 0.250-kg target is made of lead ($c = 130$ J/kg · K). (d) What must the physical properties of a practical target material be? What would be some suitable target elements?

38.39 •• Nuclear fusion reactions at the center of the sun produce gamma-ray photons with energies of about 1 MeV (10^6 eV). By contrast, what we see emanating from the sun's surface are visible-light photons with wavelengths of about 500 nm. A simple model that explains this difference in wavelength is that a photon undergoes Compton scattering many times—in fact, about 10^{26} times, as suggested by models of the solar interior—as it travels from the center of the sun to its surface. (a) Estimate the increase in wavelength of a photon in an average Compton-scattering event. (b) Find the angle in degrees through which the photon is scattered in the scattering event described in part (a). (*Hint:* A useful approximation is $\cos\phi \approx 1 - \phi^2/2$, which is valid for $\phi \ll 1$. Note that ϕ is in radians in this expression.) (c) It is estimated that a photon takes about 10^6 years to travel from the core to the surface of the sun. Find the average distance that light can travel within the interior of the sun without being scattered. (This distance is roughly equivalent to how far you could see if you were inside the sun and could survive the extreme temperatures there. As your answer shows, the interior of the sun is *very* opaque.)

38.40 •• (a) Derive an expression for the total shift in photon wavelength after two successive Compton scatterings from electrons at rest. The photon is scattered by an angle θ_1 in the first scat-

tering and by θ_2 in the second. (b) In general, is the total shift in wavelength produced by two successive scatterings of an angle $\theta/2$ the same as by a single scattering of θ? If not, are there any specific values of θ, other than $\theta = 0°$, for which the total shifts are the same? (c) Use the result of part (a) to calculate the total wavelength shift produced by two successive Compton scatterings of 30.0° each. Express your answer in terms of h/mc. (d) What is the wavelength shift produced by a single Compton scattering of 60.0°? Compare to the answer in part (c).

38.41 •• A photon with wavelength 0.1100 nm collides with a free electron that is initially at rest. After the collision the wavelength is 0.1132 nm. (a) What is the kinetic energy of the electron after the collision? What is its speed? (b) If the electron is suddenly stopped (for example, in a solid target), all of its kinetic energy is used to create a photon. What is the wavelength of this photon?

38.42 •• An x-ray photon is scattered from a free electron (mass m) at rest. The wavelength of the scattered photon is λ', and the final speed of the struck electron is v. (a) What was the initial wavelength λ of the photon? Express your answer in terms of λ', v, and m. (*Hint:* Use the relativistic expression for the electron kinetic energy.) (b) Through what angle ϕ is the photon scattered? Express your answer in terms of λ, λ', and m. (c) Evaluate your results in parts (a) and (b) for a wavelength of 5.10×10^{-3} nm for the scattered photon and a final electron speed of 1.80×10^8 m/s. Give ϕ in degrees.

38.43 •• (a) Calculate the maximum increase in photon wavelength that can occur during Compton scattering. (b) What is the energy (in electron volts) of the lowest-energy x-ray photon for which Compton scattering could result in doubling the original wavelength?

CHALLENGE PROBLEM

38.44 ••• Consider Compton scattering of a photon by a *moving* electron. Before the collision the photon has wavelength λ and is moving in the $+x$-direction, and the electron is moving in the $-x$-direction with total energy E (including its rest energy mc^2). The photon and electron collide head-on. After the collision, both are moving in the $-x$-direction (that is, the photon has been scattered by 180°). (a) Derive an expression for the wavelength λ' of the scattered photon. Show that if $E \gg mc^2$, where m is the rest mass of the electron, your result reduces to

$$\lambda' = \frac{hc}{E}\left(1 + \frac{m^2c^4\lambda}{4hcE}\right)$$

(b) A beam of infrared radiation from a CO_2 laser ($\lambda = 10.6$ μm) collides head-on with a beam of electrons, each of total energy $E = 10.0$ GeV (1 GeV $= 10^9$ eV). Calculate the wavelength λ' of the scattered photons, assuming a 180° scattering angle. (c) What kind of scattered photons are these (infrared, microwave, ultraviolet, etc.)? Can you think of an application of this effect?

Answers

Chapter Opening Question **?**

The energy of a photon E is inversely proportional to its wavelength λ: The shorter the wavelength, the more energetic is the photon. Since visible light has shorter wavelengths than infrared light,

the headlamp emits photons of greater energy. However, the light from the infrared laser is far more *intense* (delivers much more energy per second per unit area to the patient's skin) because it emits many more photons per second than does the headlamp and concentrates them onto a very small spot.

Test Your Understanding Questions

38.1 Answers: (i) and (ii) From Eq. (38.2), a photon of energy $E = 1.14$ eV has wavelength $\lambda = hc/E = (4.136 \times 10^{-15}$ eV) $\cdot (3.00 \times 10^8$ m/s$)/(1.14$ eV$) = 1.09 \times 10^{-6}$ m $= 1090$ nm. This is in the infrared part of the spectrum. Since wavelength is inversely proportional to photon energy, the *minimum* photon energy of 1.14 eV corresponds to the *maximum* wavelength that causes photoconductivity in silicon. Thus the wavelength must be 1090 nm or less.

38.2 Answer: (ii) Equation (38.6) shows that the minimum wavelength of x rays produced by bremsstrahlung depends on the potential difference V_{AC} but does *not* depend on the rate at which electrons strike the anode. Each electron produces at most one photon, so increasing the number of electrons per second causes an increase in the number of x-ray photons emitted per second (that is, the x-ray intensity).

38.3 Answers: yes, no Equation (38.7) shows that the wavelength shift $\Delta\lambda = \lambda' - \lambda$ depends only on the photon scattering angle ϕ, not on the wavelength of the incident photon. So a visible-light photon scattered through an angle ϕ undergoes the same wavelength shift as an x-ray photon. Equation (38.7) also shows that this shift is of the order of $h/mc = 2.426 \times 10^{-12}$ m $=$ 0.002426 nm. This is a few percent of the wavelength of x rays (see Example 38.5), so the effect is noticeable in x-ray scattering. However, h/mc is a tiny fraction of the wavelength of visible light (between 380 and 750 nm). The human eye cannot distinguish such minuscule differences in wavelength (that is, differences in color).

38.4 Answer: (ii) There is *zero* probability that a photon will be deflected by one of the angles where the diffraction pattern has zero intensity. These angles are given by $a \sin\theta = m\lambda$ with $m = \pm 1$, ± 2, ± 3, ... Since λ is much less than a, we can write these angles as $\theta = m\lambda/a = \pm\lambda/a$, $\pm 2\lambda/a$, $\pm 3\lambda/a$, ... These values include answers (i), (iii), and (iv), so it is impossible for a photon to be deflected through any of these angles. The intensity is not zero at $\theta = 3\lambda/2a$ (located between two zeros in the diffraction pattern), so there is some probability that a photon will be deflected through this angle.

Bridging Problem

Answers: (a) 0.0781 nm
(b) 1.65×10^{-23} kg $\cdot$ m/s, 1.81×10^7 m/s
(c) 1.49×10^{-16} J

39 PARTICLES BEHAVING AS WAVES

T bacteriophage virus
Viral DNA
100 nm = 0.1 μm

? Viruses (shown in blue) have landed on an *E. coli* bacterium and injected their DNA, converting the bacterium into a virus factory. This false-color image was made using a beam of electrons rather than a light beam. What properties of electrons make them useful for imaging such fine details?

LEARNING GOALS

By studying this chapter, you will learn:

- De Broglie's proposal that electrons, protons, and other particles can behave like waves.

- How electron diffraction experiments provided evidence for de Broglie's ideas.

- How electron microscopes can provide much higher magnification than visible-light microscopes.

- How physicists discovered the atomic nucleus.

- How Bohr's model of electron orbits explained the spectra of hydrogen and hydrogenlike atoms.

- How a laser operates.

- How the idea of energy levels, coupled with the photon model of light, explains the spectrum of light emitted by a hot, opaque object.

- What the uncertainty principle tells us about the nature of the atom.

In Chapter 38 we discovered one aspect of nature's wave–particle duality: Light and other electromagnetic radiation act sometimes like waves and sometimes like particles. Interference and diffraction demonstrate wave behavior, while emission and absorption of photons demonstrate particle behavior.

If light waves can behave like particles, can the particles of matter behave like waves? As we will discover, the answer is a resounding yes. Electrons can be made to interfere and diffract just like other kinds of waves. We will see that the wave nature of electrons is not merely a laboratory curiosity: It is the fundamental reason why atoms, which according to classical physics should be profoundly unstable, are able to exist. In this chapter we'll use the wave nature of matter to help us understand the structure of atoms, the operating principles of a laser, and the curious properties of the light emitted by a heated, glowing object. Without the wave picture of matter, there would be no way to explain these phenomena.

In Chapter 40 we'll introduce an even more complete wave picture of matter called *quantum mechanics*. Through the remainder of this book we'll use the ideas of quantum mechanics to understand the nature of molecules, solids, atomic nuclei, and the fundamental particles that are the building blocks of our universe.

39.1 Electron Waves

In 1924 a French physicist and nobleman, Prince Louis de Broglie (pronounced "de broy"; Fig. 39.1), made a remarkable proposal about the nature of matter. His reasoning, freely paraphrased, went like this: Nature loves symmetry. Light is dualistic in nature, behaving in some situations like waves and in others like particles. If nature is symmetric, this duality should also hold for matter. Electrons and protons, which we usually think of as *particles,* may in some situations behave like *waves.*

If a particle acts like a wave, it should have a wavelength and a frequency. De Broglie postulated that a free particle with rest mass m, moving with nonrelativistic speed v, should have a wavelength λ related to its momentum $p = mv$ in exactly the same way as for a photon, as expressed by Eq. (38.5): $\lambda = h/p$. The **de Broglie wavelength** of a particle is then

$$\lambda = \frac{h}{p} = \frac{h}{mv} \quad \text{(de Broglie wavelength of a particle)} \quad \text{(39.1)}$$

where h is Planck's constant. If the particle's speed is an appreciable fraction of the speed of light c, we use Eq. (37.27) to replace mv in Eq. (39.1) with $\gamma mv = mv/\sqrt{1 - v^2/c^2}$. The frequency f, according to de Broglie, is also related to the particle's energy E in the same way as for a photon—namely,

$$E = hf \quad \text{(39.2)}$$

Thus in de Broglie's hypothesis, the relationships of wavelength to momentum and of frequency to energy are exactly the same for free particles as for photons.

CAUTION **Not all photon equations apply to particles with mass** Be careful when applying the relationship $E = hf$ to particles with nonzero rest mass, such as electrons and protons. Unlike a photon, they do *not* travel at speed c, so the equations $f = c/\lambda$ and $E = pc$ do *not* apply to them!

Observing the Wave Nature of Electrons

De Broglie's proposal was a bold one, made at a time when there was no direct experimental evidence that particles have wave characteristics. But within a few years of de Broglie's publication of his ideas, they were resoundingly verified by a diffraction experiment with electrons. This experiment was analogous to those we described in Section 36.6, in which atoms in a crystal act as a three-dimensional diffraction grating for x rays. An x-ray beam is strongly reflected when it strikes a crystal at an angle that gives constructive interference among the waves scattered from the various atoms in the crystal. These interference effects demonstrate the *wave* nature of x rays.

In 1927 the American physicists Clinton Davisson and Lester Germer, working at the Bell Telephone Laboratories, were studying the surface of a piece of nickel by directing a beam of *electrons* at the surface and observing how many electrons bounced off at various angles. Figure 39.2 shows an experimental setup like theirs. Like many ordinary metals, the sample was *polycrystalline:* It consisted of many randomly oriented microscopic crystals bonded together. As a result, the electron beam reflected diffusely, like light bouncing off a rough surface (see Fig. 33.6b), with a smooth distribution of intensity as a function of the angle θ.

39.1 Louis-Victor de Broglie, the seventh Duke de Broglie (1892–1987), broke with family tradition by choosing a career in physics rather than as a diplomat. His revolutionary proposal that particles have wave characteristics—for which de Broglie won the 1929 Nobel Prize in physics—was published in his doctoral thesis.

Mastering**PHYSICS**

PhET: Davisson-Germer: Electron Diffraction
ActivPhysics 17.5: Electron Interference

39.2 An apparatus similar to that used by Davisson and Germer to discover electron diffraction.

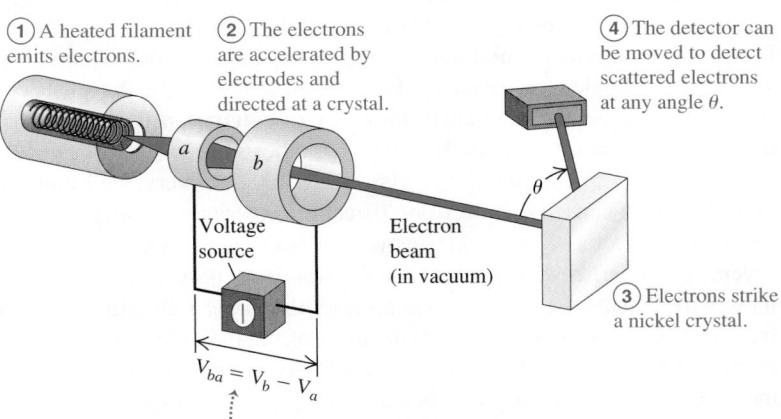

① A heated filament emits electrons.

② The electrons are accelerated by electrodes and directed at a crystal.

④ The detector can be moved to detect scattered electrons at any angle θ.

Voltage source

Electron beam (in vacuum)

③ Electrons strike a nickel crystal.

$V_{ba} = V_b - V_a$

$V_{ba} > 0$, so electrons speed up in moving from a to b.

39.3 (a) Intensity of the scattered electron beam in Fig. 39.2 as a function of the scattering angle θ. (b) Electron waves scattered from two adjacent atoms interfere constructively when $d\sin\theta = m\lambda$. In the case shown here, $\theta = 50°$ and $m = 1$.

(a) This peak in the intensity of scattered electrons is due to constructive interference between electron waves scattered by different surface atoms.

(b) If the scattered waves are in phase, there is a peak in the intensity of scattered electrons.

Atoms on surface of crystal

39.4 X-ray and electron diffraction. The upper half of the photo shows the diffraction pattern for 71-pm x rays passing through aluminum foil. The lower half, with a different scale, shows the diffraction pattern for 600-eV electrons from aluminum. The similarity shows that electrons undergo the same kind of diffraction as x rays.

Top: x-ray diffraction

Bottom: electron diffraction

During the experiment an accident occurred that permitted air to enter the vacuum chamber, and an oxide film formed on the metal surface. To remove this film, Davisson and Germer baked the sample in a high-temperature oven, almost hot enough to melt it. Unknown to them, this had the effect of creating large regions within the nickel with crystal planes that were continuous over the width of the electron beam. From the perspective of the electrons, the sample looked like a *single* crystal of nickel.

When the observations were repeated with this sample, the results were quite different. Now strong maxima in the intensity of the reflected electron beam occurred at specific angles (Fig. 39.3a), in contrast to the smooth variation of intensity with angle that Davisson and Germer had observed before the accident. The angular positions of the maxima depended on the accelerating voltage V_{ba} used to produce the electron beam. Davisson and Germer were familiar with de Broglie's hypothesis, and they noticed the similarity of this behavior to x-ray diffraction. This was not the effect they had been looking for, but they immediately recognized that the electron beam was being *diffracted*. They had discovered a very direct experimental confirmation of the wave hypothesis.

Davisson and Germer could determine the speeds of the electrons from the accelerating voltage, so they could compute the de Broglie wavelength from Eq. (39.1). If an electron is accelerated from rest at point a to point b through a potential increase $V_{ba} = V_b - V_a$ as shown in Fig. 39.2, the work done on the electron eV_{ba} equals its kinetic energy K. Using $K = (\frac{1}{2})mv^2 = p^2/2m$ for a nonrelativistic particle, we have

$$eV_{ba} = \frac{p^2}{2m} \qquad p = \sqrt{2meV_{ba}}$$

We substitute this into Eq. (39.1), the expression for the de Broglie wavelength of the electron:

$$\lambda = \frac{h}{p} = \frac{h}{\sqrt{2meV_{ba}}} \qquad \text{(de Broglie wavelength of an electron)} \quad (39.3)$$

The greater the accelerating voltage V_{ba}, the shorter the wavelength of the electron.

To predict the angles at which strong reflection occurs, note that the electrons were scattered primarily by the planes of atoms near the surface of the crystal. Atoms in a surface plane are arranged in rows, with a distance d that can be measured by x-ray diffraction techniques. These rows act like a reflecting diffraction grating; the angles at which strong reflection occurs are the same as for a grating with center-to-center distance d between its slits (Fig. 39.3b). From Eq. (36.13) the angles of maximum reflection are given by

$$d\sin\theta = m\lambda \qquad (m = 1, 2, 3, \dots) \qquad (39.4)$$

where θ is the angle shown in Fig. 39.2. (Note that the geometry in Fig. 39.3b is different from that for Fig. 36.22, so Eq. (39.4) is different from Eq. (36.16).) Davisson and Germer found that the angles predicted by this equation, using the de Broglie wavelength given by Eq. (39.3), agreed with the observed values (Fig. 39.3a). Thus the accidental discovery of **electron diffraction** was the first direct evidence confirming de Broglie's hypothesis.

In 1928, just a year after the Davisson–Germer discovery, the English physicist G. P. Thomson carried out electron-diffraction experiments using a thin, polycrystalline, metallic foil as a target. Debye and Sherrer had used a similar technique several years earlier to study x-ray diffraction from polycrystalline specimens. In these experiments the beam passes *through* the target rather than being reflected from it. Because of the random orientations of the individual microscopic crystals in the foil, the diffraction pattern consists of intensity maxima forming rings around the direction of the incident beam. Thomson's results again confirmed the de Broglie relationship. Figure 39.4 shows both x-ray and electron diffraction patterns for a polycrystalline aluminum foil. (G. P. Thomson was the son of J. J. Thomson, who 31 years earlier discovered the electron. Davisson and the younger Thomson shared the 1937 Nobel Prize in physics for their discoveries.)

Additional diffraction experiments were soon carried out in many laboratories using not only electrons but also various ions and low-energy neutrons. All of these are in agreement with de Broglie's bold predictions. Thus the wave nature of particles, so strange in 1924, became firmly established in the years that followed.

Problem-Solving Strategy 39.1 **Wavelike Properties of Particles**

IDENTIFY *the relevant concepts:* Particles have wavelike properties. A particle's (de Broglie) wavelength is inversely proportional to its momentum, and its frequency is proportional to its energy.

SET UP *the problem:* Identify the target variables and decide which equations you will use to calculate them.

EXECUTE *the solution* as follows:
1. Use Eq. (39.1) to relate a particle's momentum p to its wavelength λ; use Eq. (39.2) to relate its energy E to its frequency f.
2. Nonrelativistic kinetic energy may be expressed as either $K = \frac{1}{2}mv^2$ or (because $p = mv$) $K = p^2/2m$. The latter form is useful in calculations involving the de Broglie wavelength.
3. You may express energies in either joules or electron volts, using $h = 6.626 \times 10^{-34}$ J·s or $h = 4.136 \times 10^{-15}$ eV·s as appropriate.

EVALUATE *your answer:* To check numerical results, it helps to remember some approximate orders of magnitude. Here's a partial list:

Size of an atom: 10^{-10} m = 0.1 nm

Mass of an atom: 10^{-26} kg

Mass of an electron: $m = 10^{-30}$ kg; $mc^2 = 0.511$ MeV

Electron charge magnitude: 10^{-19} C

kT at room temperature: $\frac{1}{40}$ eV

Difference between energy levels of an atom (to be discussed in Section 39.3): 1 to 10 eV

Speed of an electron in the Bohr model of a hydrogen atom (to be discussed in Section 39.3): 10^6 m/s

---

Example 39.1 **An electron-diffraction experiment**

In an electron-diffraction experiment using an accelerating voltage of 54 V, an intensity maximum occurs for $\theta = 50°$ (see Fig. 39.3a). X-ray diffraction indicates that the atomic spacing in the target is $d = 2.18 \times 10^{-10}$ m = 0.218 nm. The electrons have negligible kinetic energy before being accelerated. Find the electron wavelength.

SOLUTION

IDENTIFY, SET UP, and EXECUTE: We'll determine λ from both de Broglie's equation, Eq. (39.3), and the diffraction equation, Eq. (39.4). From Eq. (39.3),

$$\lambda = \frac{6.626 \times 10^{-34} \text{ J·s}}{\sqrt{2(9.109 \times 10^{-31} \text{ kg})(1.602 \times 10^{-19} \text{ C})(54 \text{ V})}}$$
$$= 1.7 \times 10^{-10} \text{ m} = 0.17 \text{ nm}$$

Alternatively, using Eq. (39.4) and assuming $m = 1$,

$$\lambda = d\sin\theta = (2.18 \times 10^{-10} \text{ m}) \sin 50° = 1.7 \times 10^{-10} \text{ m}$$

EVALUATE: The two numbers agree within the accuracy of the experimental results, which gives us an excellent check on our calculations. Note that this electron wavelength is less than the spacing between the atoms.

Example 39.2 **Energy of a thermal neutron**

Find the speed and kinetic energy of a neutron ($m = 1.675 \times 10^{-27}$ kg) with de Broglie wavelength $\lambda = 0.200$ nm, a typical interatomic spacing in crystals. Compare this energy with the average translational kinetic energy of an ideal-gas molecule at room temperature ($T = 20°C = 293$ K).

SOLUTION

IDENTIFY and SET UP: This problem uses the relationships between particle speed and wavelength, between particle speed and kinetic energy, and between gas temperature and the average kinetic energy of a gas molecule. We'll find the neutron speed v using Eq. (39.1) and from that calculate the neutron kinetic energy

$K = \frac{1}{2}mv^2$. We'll use Eq. (18.16) to find the average kinetic energy of a gas molecule.

EXECUTE: From Eq. (39.1), the neutron speed is

$$v = \frac{h}{\lambda m} = \frac{6.626 \times 10^{-34} \text{ J·s}}{(0.200 \times 10^{-9} \text{ m})(1.675 \times 10^{-27} \text{ kg})}$$
$$= 1.98 \times 10^3 \text{ m/s}$$

The neutron kinetic energy is

$$K = \frac{1}{2}mv^2 = \frac{1}{2}(1.675 \times 10^{-27} \text{ kg})(1.98 \times 10^3 \text{ m/s})^2$$
$$= 3.28 \times 10^{-21} \text{ J} = 0.0205 \text{ eV} \qquad \textit{Continued}$$

From Eq. (18.16), the average translational kinetic energy of an ideal-gas molecule at $T = 293$ K is

$$\frac{1}{2}m(v^2)_{av} = \frac{3}{2}kT = \frac{3}{2}(1.38 \times 10^{-23} \text{ J/K})(293 \text{ K})$$

$$= 6.07 \times 10^{-21} \text{ J} = 0.0379 \text{ eV}$$

The two energies are comparable in magnitude, which is why a neutron with kinetic energy in this range is called a *thermal neutron*.

Diffraction of thermal neutrons is used to study crystal and molecular structure in the same way as x-ray diffraction. Neutron diffraction has proved to be especially useful in the study of large organic molecules.

EVALUATE: Note that the calculated neutron speed is much less than the speed of light. This justifies our use of the nonrelativistic form of Eq. (39.1).

De Broglie Waves and the Macroscopic World

If the de Broglie picture is correct and matter has wave aspects, you might wonder why we don't see these aspects in everyday life. As an example, we know that waves diffract when sent through a single slit. Yet when we walk through a doorway (a kind of single slit), we don't worry about our body diffracting!

The principal reason we don't see these effects on human scales is that Planck's constant h has such a minuscule value. As a result, the de Broglie wavelengths of even the smallest ordinary objects that you can see are extremely small, and the wave effects are unimportant. For instance, what is the wavelength of a falling grain of sand? If the grain's mass is 5×10^{-10} kg and its diameter is 0.07 mm $= 7 \times 10^{-5}$ m, it will fall in air with a terminal speed of about 0.4 m/s. The magnitude of its momentum is then $p = mv = (5 \times 10^{-10} \text{ kg}) \times (0.4 \text{ m/s}) = 2 \times 10^{-10}$ kg $\cdot$ m/s. The de Broglie wavelength of this falling sand grain is then

$$\lambda = \frac{h}{p} = \frac{6.626 \times 10^{-34} \text{ J} \cdot \text{s}}{2 \times 10^{-10} \text{ kg} \cdot \text{m/s}} = 3 \times 10^{-24} \text{ m}$$

Not only is this wavelength far smaller than the diameter of the sand grain, but it's also far smaller than the size of a typical atom (about 10^{-10} m). A more massive, faster-moving object would have an even larger momentum and an even smaller de Broglie wavelength. The effects of such tiny wavelengths are so small that they are never noticed in daily life.

The Electron Microscope

The **electron microscope** offers an important and interesting example of the interplay of wave and particle properties of electrons. An electron beam can be used to form an image of an object in much the same way as a light beam. A ray of light can be bent by reflection or refraction, and an electron trajectory can be bent by an electric or magnetic field. Rays of light diverging from a point on an object can be brought to convergence by a converging lens or concave mirror, and electrons diverging from a small region can be brought to convergence by electric and/or magnetic fields.

The analogy between light rays and electrons goes deeper. The *ray* model of geometric optics is an approximate representation of the more general *wave* model. Geometric optics (ray optics) is valid whenever interference and diffraction effects can be neglected. Similarly, the model of an electron as a point particle following a line trajectory is an approximate description of the actual behavior of the electron; this model is useful when we can neglect effects associated with the wave nature of electrons.

How is an electron microscope superior to an optical microscope? The *resolution* of an optical microscope is limited by diffraction effects, as we discussed in Section 36.7. Since an optical microscope uses wavelengths around 500 nm, it can't resolve objects smaller than a few hundred nanometers, no matter how carefully its lenses are made. The resolution of an electron microscope is similarly limited by the wavelengths of the electrons, but these wavelengths may be many thousands of times smaller than wavelengths of visible light. As a result, the useful magnification of an electron microscope can be thousands of times greater than that of an optical microscope.

Note that the ability of the electron microscope to form a magnified image *does not* depend on the wave properties of electrons. Within the limitations of the Heisenberg uncertainty principle (which we'll discuss in Section 39.6), we can compute the electron trajectories by treating them as classical charged particles under the action of electric and magnetic forces. Only when we talk about *resolution* do the wave properties become important.

Example 39.3 **An electron microscope**

In an electron microscope, the nonrelativistic electron beam is formed by a setup similar to the electron gun used in the Davisson–Germer experiment (see Fig. 39.2). The electrons have negligible kinetic energy before they are accelerated. What accelerating voltage is needed to produce electrons with wavelength 10 pm = 0.010 nm (roughly 50,000 times smaller than typical visible-light wavelengths)?

$$V_{ba} = \frac{h^2}{2me\lambda^2}$$

$$= \frac{(6.626 \times 10^{-34} \text{ J} \cdot \text{s})^2}{2(9.109 \times 10^{-31} \text{ kg})(1.602 \times 10^{-19} \text{ C})(10 \times 10^{-12} \text{ m})^2}$$

$$= 1.5 \times 10^4 \text{ V} = 15{,}000 \text{ V}$$

SOLUTION

IDENTIFY, SET UP, and EXECUTE: We can use the same concepts we used to understand the Davisson–Germer experiment. The accelerating voltage is the quantity V_{ba} in Eq. (39.3). Rewrite this equation to solve for V_{ba}:

EVALUATE: It is easy to attain 15-kV accelerating voltages from 120-V or 240-V line voltage using a step-up transformer (Section 31.6) and a rectifier (Section 31.1). The accelerated electrons have kinetic energy 15 keV; since the electron rest energy is 0.511 MeV = 511 keV, these electrons are indeed nonrelativistic.

Types of Electron Microscope

Figure 39.5 shows the design of a *transmission electron microscope,* in which electrons actually pass through the specimen being studied. The specimen to be viewed can be no more than 10 to 100 nm thick so the electrons are not slowed appreciably as they pass through. The electrons used in a transmission electron microscope are emitted from a hot cathode and accelerated by a potential difference, typically 40 to 400 kV. They then pass through a condensing "lens" that uses magnetic fields to focus the electrons into a parallel beam before they pass through the specimen. The beam then passes through two more magnetic lenses: an objective lens that forms an intermediate image of the specimen and a projection lens that produces a final real image of the intermediate image. The objective and projection lenses play the roles of the objective and eyepiece lenses, respectively, of a compound optical microscope (see Section 34.8). The final image is projected onto a fluorescent screen for viewing or photographing. The entire apparatus, including the specimen, must be enclosed in a vacuum container; otherwise, electrons would scatter off air molecules and muddle the image. The image that opens this chapter was made with a transmission electron microscope.

We might think that when the electron wavelength is 0.01 nm (as in Example 39.3), the resolution would also be about 0.01 nm. In fact, it is seldom better than 0.1 nm, in part because the focal length of a magnetic lens depends on the electron speed, which is never exactly the same for all electrons in the beam.

An important variation is the *scanning electron microscope.* The electron beam is focused to a very fine line and scanned across the specimen. The beam knocks additional electrons off the specimen wherever it hits. These ejected electrons are collected by an anode that is kept at a potential a few hundred volts positive with respect to the specimen. The current of ejected electrons flowing to the collecting anode varies as the microscope beam sweeps across the specimen. The varying strength of the current is then used to create a "map" of the scanned specimen, and this map forms a greatly magnified image of the specimen.

This scheme has several advantages. The specimen can be thick because the beam does not need to pass through it. Also, the knock-off electron production depends on the *angle* at which the beam strikes the surface. Thus scanning electron

39.5 Schematic diagram of a transmission electron microscope (TEM).

High-voltage supply
Cathode (where electron beam originates)
Vacuum chamber
Accelerating anode
Condensing lens
Object (specimen)
Objective lens
Intermediate image
Projection lens
Final image
Image detector

39.6 This scanning electron microscope image shows *Escherichia coli* bacteria crowded into a stoma, or respiration opening, on the surface of a lettuce leaf. (False color has been added.) If not washed off before the lettuce is eaten, these bacteria can be a health hazard. The *transmission* electron micrograph that opens this chapter shows a greatly magnified view of the surface of an *E. coli* bacterium.

E. coli bacteria

5 μm

micrographs have an appearance that is much more three-dimensional than conventional visible-light micrographs (Fig. 39.6). The resolution is typically of the order of 10 nm, not as good as a transmission electron microscope but still much finer than the best optical microscopes.

Test Your Understanding of Section 39.1 (a) A proton has a slightly smaller mass than a neutron. Compared to the neutron described in Example 39.2, would a proton of the same wavelength have (i) more kinetic energy; (ii) less kinetic energy; or (iii) the same kinetic energy? (b) Example 39.1 shows that to give electrons a wavelength of 1.7×10^{-10} m, they must be accelerated from rest through a voltage of 54 V and so acquire a kinetic energy of 54 eV. Does a photon of this same energy also have a wavelength of 1.7×10^{-10} m?

39.2 The Nuclear Atom and Atomic Spectra

Every neutral atom contains at least one electron. How does the wave aspect of electrons affect atomic structure? As we will see, it is crucial for understanding not only the structure of atoms but also how they interact with light. Historically, the quest to understand the nature of the atom was intimately linked with both the idea that electrons have wave characteristics and the notion that light has particle characteristics. Before we explore how these ideas shaped atomic theory, it's useful to look at what was known about atoms—as well as what remained mysterious— by the first decade of the twentieth century.

Line Spectra

Everyone knows that heated materials emit light, and that different materials emit different kinds of light. The coils of a toaster glow red when in operation, the flame of a match has a characteristic yellow color, and the flame from a gas range is a distinct blue. To analyze these different types of light, we can use a prism or a diffraction grating to separate the various wavelengths in a beam of light into a spectrum. If the light source is a hot solid (such as the filament of an incandescent light bulb) or liquid, the spectrum is *continuous;* light of all wavelengths is present (Fig. 39.7a). But if the source is a heated *gas,* such as the neon in an advertising sign or the sodium vapor formed when table salt is thrown into a campfire, the spectrum includes only a few colors in the form of isolated sharp parallel lines (Fig. 39.7b). (Each "line" is an image of the spectrograph slit, deviated through an angle that depends on the wavelength of the light forming that image; see Section 36.5.) A spectrum of this sort is called an **emission line spectrum,** and the lines are called **spectral lines.** Each spectral line corresponds to a definite wavelength and frequency.

39.7 (a) Continuous spectrum produced by a glowing light bulb filament. (b) Emission line spectrum emitted by a lamp containing a heated gas.

(a) Continuous spectrum: light of all wavelengths is present.

(b) Line spectrum: only certain discrete wavelengths are present.

Screen

Diffraction grating

Slit

Lens

Light bulb with heated filament

Screen

Diffraction grating

Slit

Lens

Lamp with heated gas

39.8 The emission line spectra of several kinds of atoms and molecules. No two are alike. Note that the spectrum of water vapor (H_2O) is similar to that of hydrogen (H_2), but there are important differences that make it straightforward to distinguish these two spectra.

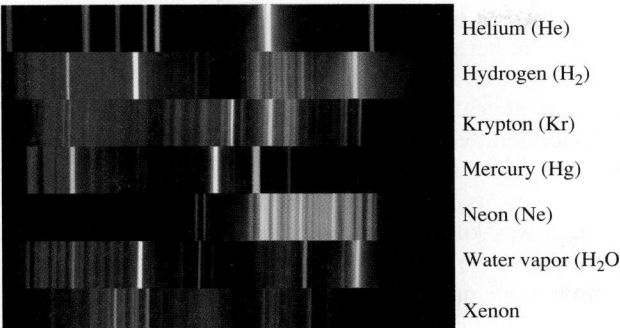

Helium (He)

Hydrogen (H_2)

Krypton (Kr)

Mercury (Hg)

Neon (Ne)

Water vapor (H_2O)

Xenon

It was discovered early in the 19th century that each element in its gaseous state has a unique set of wavelengths in its line spectrum. The spectrum of hydrogen always contains a certain set of wavelengths; mercury produces a different set, neon still another, and so on (Fig. 39.8). Scientists find the use of spectra to identify elements and compounds to be an invaluable tool. For instance, astronomers have detected the spectra from more than 100 different molecules in interstellar space, including some that are not found naturally on earth.

While a *heated* gas selectively *emits* only certain wavelengths, a *cool* gas selectively *absorbs* certain wavelengths. If we pass white (continuous-spectrum) light through a gas and look at the *transmitted* light with a spectrometer, we find a series of dark lines corresponding to the wavelengths that have been absorbed (Fig. 39.9). This is called an **absorption line spectrum.** What's more, a given kind of atom or molecule absorbs the *same* characteristic set of wavelengths when it's cool as it emits when heated. Hence scientists can use absorption line spectra to identify substances in the same manner that they use emission line spectra.

As useful as emission line spectra and absorption line spectra are, they presented a quandary to scientists: *Why* does a given kind of atom emit and absorb only certain very specific wavelengths? To answer this question, we need to have a better idea of what the inside of an atom is like. We know that atoms are much smaller than the wavelengths of visible light, so there is no hope of actually *seeing* an atom using that light. But we can still describe how the mass and electric charge are distributed throughout the volume of the atom.

Here's where things stood in 1910. In 1897 the English physicist J. J. Thomson (Nobel Prize 1906) had discovered the electron and measured its charge-to-mass ratio e/m. By 1909, the American physicist Robert Millikan (Nobel Prize 1923) had made the first measurements of the electron charge $-e$. These and other experiments showed that almost all the mass of an atom had to be associated with the *positive* charge, not with the electrons. It was also known that the overall size of atoms is of the order of 10^{-10} m and that all atoms except hydrogen contain more than one electron.

In 1910 the best available model of atomic structure was one developed by Thomson. He envisioned the atom as a sphere of some as yet unidentified positively charged substance, within which the electrons were embedded like raisins in cake. This model offered an explanation for line spectra. If the atom collided with another atom, as in a heated gas, each electron would oscillate around its equilibrium position with a characteristic frequency and emit electromagnetic radiation with that frequency. If the atom were illuminated with light of many frequencies, each electron would selectively absorb only light whose frequency matched the electron's natural oscillation frequency. (This is the phenomenon of resonance that we discussed in Section 14.8.)

Application Using Spectra to Analyze an Interstellar Gas Cloud
The light from this glowing gas cloud—located in the Small Magellanic Cloud, a small satellite galaxy of the Milky Way some 200,000 light-years (1.9×10^{18} km) from earth—has an emission line spectrum. Despite its immense distance, astronomers can tell that this cloud is composed mostly of hydrogen because its spectrum is dominated by red light at a wavelength of 656.3 nm, a wavelength emitted by hydrogen and no other element.

39.9 The absorption line spectrum of the sun. (The spectrum "lines" read from left to right and from top to bottom, like text on a page.) The spectrum is produced by the sun's relatively cool atmosphere, which absorbs photons from deeper, hotter layers. The absorption lines thus indicate what kinds of atoms are present in the solar atmosphere.

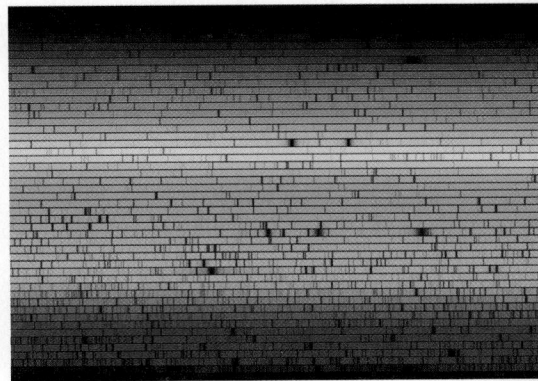

39.10 Born in New Zealand, Ernest Rutherford (1871–1937) spent his professional life in England and Canada. Before carrying out the experiments that established the existence of atomic nuclei, he shared (with Frederick Soddy) the 1908 Nobel Prize in chemistry for showing that radioactivity results from the disintegration of atoms.

Mastering PHYSICS

PhET: Rutherford Scattering
ActivPhysics 19.1: Particle Scattering

Rutherford's Exploration of the Atom

The first experiments designed to test Thomson's model by probing the interior structure of the atom were carried out in 1910–1911 by Ernest Rutherford (Fig. 39.10) and two of his students, Hans Geiger and Ernest Marsden, at the University of Manchester in England. These experiments consisted of shooting a beam of charged particles at thin foils of various elements and observing how the foil deflected the particles.

The particle accelerators now in common use in laboratories had not yet been invented, and Rutherford's projectiles were *alpha particles* emitted from naturally radioactive elements. The nature of these alpha particles was not completely understood, but it was known that they are ejected from unstable nuclei with speeds of the order of 10^7 m/s, are positively charged, and can travel several centimeters through air or 0.1 mm or so through solid matter before they are brought to rest by collisions.

Figure 39.11 is a schematic view of Rutherford's experimental setup. A radioactive substance at the left emits alpha particles. Thick lead screens stop all particles except those in a narrow beam. The beam passes through the foil target (consisting of gold, silver, or copper) and strikes screens coated with zinc sulfide, creating a momentary flash, or *scintillation*. Rutherford and his students counted the numbers of particles deflected through various angles.

The atoms in a metal foil are packed together like marbles in a box (not spaced apart). Because the particle beam passes through the foil, the alpha particles must pass through the interior of atoms. Within an atom, the charged alpha particle will interact with the electrons and the positive charge. (Because the *total* charge of the atom is zero, alpha particles feel little electrical force outside an atom.) An electron has about 7300 times less mass than an alpha particle, so momentum considerations indicate that the atom's electrons cannot appreciably deflect the alpha particle—any more than a swarm of gnats deflects a tossed pebble. Any deflection will be due to the positively charged material that makes up almost all of the atom's mass.

In the Thomson model, the positive charge and the negative electrons are distributed through the whole atom. Hence the electric field inside the atom should be quite small, and the electric force on an alpha particle that enters the atom should be quite weak. The maximum deflection to be expected is then only a few degrees (Fig. 39.12a). The results of the Rutherford experiments were *very* different

39.11 The Rutherford scattering experiments investigated what happens to alpha particles fired at a thin gold foil. The results of this experiment helped reveal the structure of atoms.

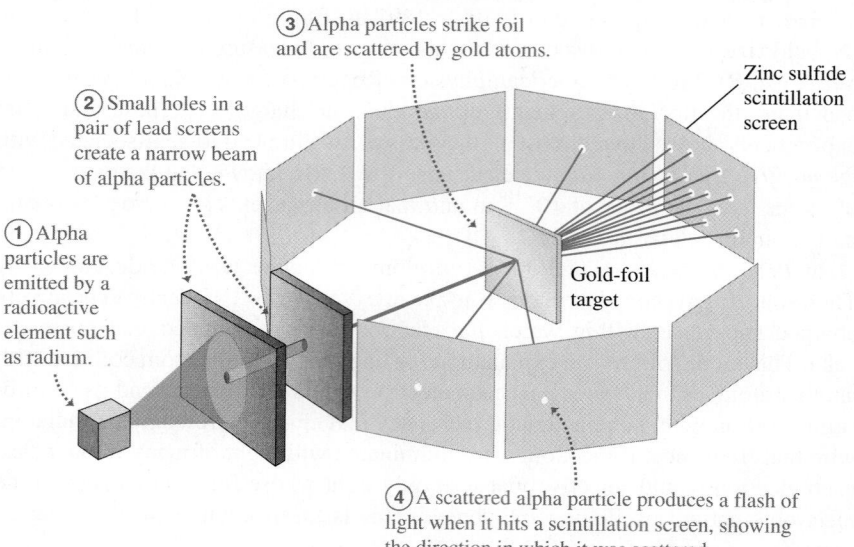

③ Alpha particles strike foil and are scattered by gold atoms.

Zinc sulfide scintillation screen

② Small holes in a pair of lead screens create a narrow beam of alpha particles.

① Alpha particles are emitted by a radioactive element such as radium.

Gold-foil target

④ A scattered alpha particle produces a flash of light when it hits a scintillation screen, showing the direction in which it was scattered.

from the Thomson prediction. Some alpha particles were scattered by nearly 180°—that is, almost straight backward (Fig. 39.12b). Rutherford later wrote:

> **It was quite the most incredible event that ever happened to me in my life. It was almost as incredible as if you had fired a 15-inch shell at a piece of tissue paper and it came back and hit you.**

Clearly the Thomson model was wrong and a new model was needed. Suppose the positive charge, instead of being distributed through a sphere with atomic dimensions (of the order of 10^{-10} m), is all concentrated in a much *smaller* volume. Then it would act like a point charge down to much smaller distances. The maximum electric field repelling the alpha particle would be much larger, and the amazing large-angle scattering that Rutherford observed could occur. Rutherford developed this model and called the concentration of positive charge the **nucleus.** He again computed the numbers of particles expected to be scattered through various angles. Within the accuracy of his experiments, the computed and measured results agreed, down to distances of the order of 10^{-14} m. His experiments therefore established that the atom does have a nucleus—a very small, very dense structure, no larger than 10^{-14} m in diameter. The nucleus occupies only about 10^{-12} of the total volume of the atom or less, but it contains *all* the positive charge and at least 99.95% of the total mass of the atom.

Figure 39.13 shows a computer simulation of alpha particles with a kinetic energy of 5.0 MeV being scattered from a gold nucleus of radius 7.0×10^{-15} m (the actual value) and from a nucleus with a hypothetical radius ten times larger. In the second case there is *no* large-angle scattering. The presence of large-angle scattering in Rutherford's experiments thus attested to the small size of the nucleus.

Later experiments showed that all nuclei are composed of positively charged protons (discovered in 1918) and electrically neutral neutrons (discovered in 1930). For example, the gold atoms in Rutherford's experiments have 79 protons and 118 neutrons. In fact, an alpha particle is itself the nucleus of a helium atom, with two protons and two neutrons. It is much more massive than an electron but only about 2% as massive as a gold nucleus, which helps explain why alpha particles are scattered by gold nuclei but not by electrons.

39.12 A comparison of Thomson's and Rutherford's models of the atom.

(a) Thomson's model of the atom: An alpha particle is scattered through only a small angle.

(b) Rutherford's model of the atom: An alpha particle can be scattered through a large angle by the compact, positively charged nucleus (not drawn to scale).

39.13 Computer simulation of scattering of 5.0-MeV alpha particles from a gold nucleus. Each curve shows a possible alpha-particle trajectory. **(a)** The scattering curves match Rutherford's experimental data if a radius of 7.0×10^{-15} m is assumed for a gold nucleus. **(b)** A model with a much larger radius for the gold nucleus does not match the data.

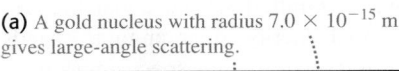

(a) A gold nucleus with radius 7.0×10^{-15} m gives large-angle scattering.

(b) A nucleus with 10 times the radius of the nucleus in **(a)** shows *no* large-scale scattering.

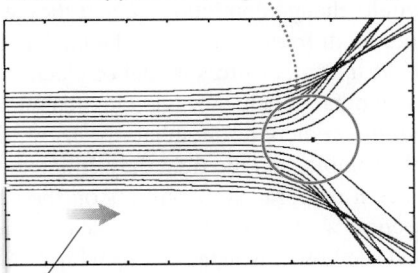

Motion of incident 5.0-MeV alpha particles

Example 39.4 **A Rutherford experiment**

An alpha particle (charge $2e$) is aimed directly at a gold nucleus (charge $79e$). What minimum initial kinetic energy must the alpha particle have to approach within 5.0×10^{-14} m of the center of the gold nucleus before reversing direction? Assume that the gold nucleus, which has about 50 times the mass of an alpha particle, remains at rest.

Continued

SOLUTION

IDENTIFY: The repulsive electric force exerted by the gold nucleus makes the the alpha particle slow to a halt as it approaches, then reverse direction. This force is conservative, so the total mechanical energy (kinetic energy of the alpha particle plus electric potential energy of the system) is conserved.

SET UP: Let point 1 be the initial position of the alpha particle, very far from the gold nucleus, and let point 2 be 5.0×10^{-14} m from the center of the gold nucleus. Our target variable is the kinetic energy K_1 of the alpha particle at point 1 that allows it to reach point 2 with $K_2 = 0$. To find this we'll use the law of conservation of energy and Eq. (23.9) for electric potential energy, $U = qq_0/4\pi\epsilon_0 r$.

EXECUTE: At point 1 the separation r of the alpha particle and gold nucleus is effectively infinite, so from Eq. (23.9) $U_1 = 0$. At point 2 the potential energy is

$$U_2 = \frac{1}{4\pi\epsilon_0} \frac{qq_0}{r}$$

$$= (9.0 \times 10^9 \text{ N} \cdot \text{m}^2/\text{C}^2) \frac{(2)(79)(1.60 \times 10^{-19} \text{ C})^2}{5.0 \times 10^{-14} \text{ m}}$$

$$= 7.3 \times 10^{-13} \text{ J} = 4.6 \times 10^6 \text{ eV} = 4.6 \text{ MeV}$$

By energy conservation $K_1 + U_1 = K_2 + U_2$, so $K_1 = K_2 + U_2 - U_1 = 0 + 4.6 \text{ MeV} - 0 = 4.6 \text{ MeV}$. Thus, to approach within 5.0×10^{-14} m, the alpha particle must have initial kinetic energy $K_1 = 4.6$ MeV.

EVALUATE: Alpha particles emitted from naturally occurring radioactive elements typically have energies in the range 4 to 6 MeV. For example, the common isotope of radium, ^{226}Ra, emits an alpha particle with energy 4.78 MeV.

Was it valid to assume that the gold nucleus remains at rest? To find out, note that when the alpha particle stops momentarily, all of its initial momentum has been transferred to the gold nucleus. An alpha particle has a mass $m_\alpha = 6.64 \times 10^{-27}$ kg; if its initial kinetic energy $K_1 = \frac{1}{2}mv_1^2$ is 7.3×10^{-13} J, you can show that its initial speed is $v_1 = 1.5 \times 10^7$ m/s and its initial momentum is $p_1 = m_\alpha v_1 = 9.8 \times 10^{-20}$ kg·m/s. A gold nucleus (mass $m_{\text{Au}} = 3.27 \times 10^{-25}$ kg) with this much momentum has a much slower speed $v_{\text{Au}} = 3.0 \times 10^5$ m/s and kinetic energy $K_{\text{Au}} = \frac{1}{2}mv_{\text{Au}}^2 = 1.5 \times 10^{-14}$ J = 0.092 MeV. This *recoil kinetic energy* of the gold nucleus is only 2% of the total energy in this situation, so we are justified in neglecting it.

The Failure of Classical Physics

Rutherford's discovery of the atomic nucleus raised a serious question: What prevented the negatively charged electrons from falling into the positively charged nucleus due to the strong electrostatic attraction? Rutherford suggested that perhaps the electrons *revolve* in orbits about the nucleus, just as the planets revolve around the sun.

But according to classical electromagnetic theory, any accelerating electric charge (either oscillating or revolving) radiates electromagnetic waves. An example is the radiation from an oscillating point charge that we depicted in Fig. 32.3 (Section 32.1). An electron orbiting inside an atom would always have a centripetal acceleration toward the nucleus, and so should be emitting radiation *at all times*. The energy of an orbiting electron should therefore decrease continuously, its orbit should become smaller and smaller, and it should spiral into the nucleus within a fraction of a second (Fig. 39.14). Even worse, according to classical theory the *frequency* of the electromagnetic waves emitted should equal the frequency of revolution. As the electrons radiated energy, their angular speeds would change continuously, and they would emit a *continuous* spectrum (a mixture of all frequencies), not the *line* spectrum actually observed.

Thus Rutherford's model of electrons orbiting the nucleus, which is based on Newtonian mechanics and classical electromagnetic theory, makes three entirely *wrong* predictions about atoms: They should emit light continuously, they should be unstable, and the light they emit should have a continuous spectrum. Clearly a radical reappraisal of physics on the scale of the atom was needed. In the next section we will see the bold idea that led to an new understanding of the atom, and see how this idea meshes with de Broglie's no less bold notion that electrons have wave attributes.

39.14 Classical physics makes predictions about the behavior of atoms that do not match reality.

ACCORDING TO CLASSICAL PHYSICS:

- An orbiting electron is accelerating, so it should radiate electromagnetic waves.
- The waves would carry away energy, so the electron should lose energy and spiral inward.
- The electron's angular speed would increase as its orbit shrank, so the frequency of the radiated waves should increase.

Thus, classical physics says that atoms should collapse within a fraction of a second and should emit light with a continuous spectrum as they do so.

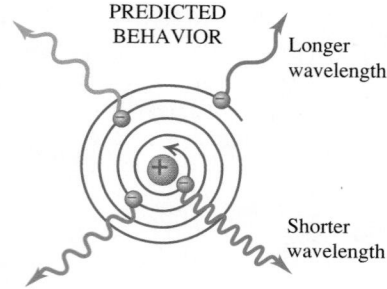

PREDICTED BEHAVIOR

Longer wavelength

Shorter wavelength

IN FACT:

- Atoms are stable.
- They emit light only when excited, and only at specific frequencies (as a line spectrum).

Test Your Understanding of Section 39.2 Suppose you repeated Rutherford's scattering experiment using a thin sheet of solid hydrogen in place of the gold foil. (Hydrogen is a solid at temperatures below 14.0 K.) The nucleus of a hydrogen atom is a single proton, with about one-fourth the mass of an alpha particle. Compared to the original experiment with gold foil, would you expect the alpha particles in this experiment to undergo (i) more large-angle scattering; (ii) the same amount of large-angle scattering; or (iii) less large-angle scattering?

39.3 Energy Levels and the Bohr Model of the Atom

In 1913 a young Danish physicist working with Ernest Rutherford at the University of Manchester made a revolutionary proposal to explain both the stability of atoms and their emission and absorption line spectra. The physicist was Niels Bohr (Fig. 39.15), and his innovation was to combine the photon concept that we introduced in Chapter 38 with a fundamentally new idea: The energy of an atom can have only certain particular values. His hypothesis represented a clean break from 19th-century ideas.

Photon Emission and Absorption by Atoms

Bohr's reasoning went like this. The emission line spectrum of an element tells us that atoms of that element emit photons with only certain specific frequencies f and hence certain specific energies $E = hf$. During the emission of a photon, the internal energy of the atom changes by an amount equal to the energy of the photon. Therefore, said Bohr, each atom must be able to exist with only certain specific values of internal energy. Each atom has a set of possible **energy levels.** An atom can have an amount of internal energy equal to any one of these levels, but it *cannot* have an energy *intermediate* between two levels. All isolated atoms of a given element have the same set of energy levels, but atoms of different elements have different sets.

Suppose an atom is raised, or *excited*, to a high energy level. (In a hot gas this happens when fast-moving atoms undergo inelastic collisions with each other or with the walls of the gas container. In an electric discharge tube, such as those used in a neon light fixture, atoms are excited by collisions with fast-moving electrons.) According to Bohr, an excited atom can make a *transition* from one energy level to a lower level by emitting a photon with energy equal to the energy *difference* between the initial and final levels (Fig. 39.16). If E_i is the initial energy of the atom before such a transition, E_f is its final energy after the transition, and the photon's energy is $hf = hc/\lambda$, then conservation of energy gives

$$hf = \frac{hc}{\lambda} = E_i - E_f \quad \text{(energy of emitted photon)} \quad (39.5)$$

For example, an excited lithium atom emits red light with wavelength $\lambda = 671$ nm. The corresponding photon energy is

$$E = \frac{hc}{\lambda} = \frac{(6.63 \times 10^{-34} \text{ J} \cdot \text{s})(3.00 \times 10^8 \text{ m/s})}{671 \times 10^{-9} \text{ m}}$$
$$= 2.96 \times 10^{-19} \text{ J} = 1.85 \text{ eV}$$

This photon is emitted during a transition like that shown in Fig. 39.16 between two levels of the atom that differ in energy by $E_i - E_f = 1.85$ eV.

The emission line spectra shown in Fig. 39.8 show that many different wavelengths are emitted by each atom. Hence each kind of atom must have a number of energy levels, with different spacings in energy between them. Each wavelength in the spectrum corresponds to a transition between two specific energy levels of the atom.

CAUTION **Producing a line spectrum** The lines of an emission line spectrum, such as the helium spectrum shown at the top of Fig. 39.8, are *not* all produced by a single atom. The sample of helium gas that produced the spectrum in Fig. 39.8 contained a large number of helium atoms; these were excited in an electric discharge tube to various energy levels. The spectrum of the gas shows the light emitted from all the different transitions that occurred in different atoms of the sample.

The observation that atoms are stable means that each atom has a *lowest* energy level, called the **ground level.** Levels with energies greater than the

39.15 Niels Bohr (1885–1962) was a young postdoctoral researcher when he proposed the novel idea that the energy of an atom could have only certain discrete values. He won the 1922 Nobel Prize in physics for these ideas. Bohr went on to make seminal contributions to nuclear physics and to become a passionate advocate for the free exchange of scientific ideas among all nations.

39.16 An excited atom emitting a photon.

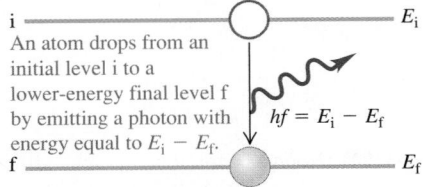

An atom drops from an initial level i to a lower-energy final level f by emitting a photon with energy equal to $E_i - E_f$.

$hf = E_i - E_f$

Mastering**PHYSICS**

PhET: Models of the Hydrogen Atom
ActivPhysics 18.1: The Bohr Model

39.17 An atom absorbing a photon. (Compare with Fig. 39.16.)

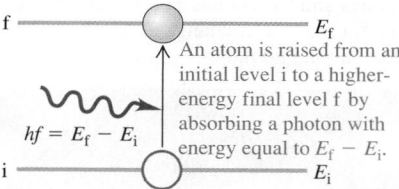

An atom is raised from an initial level i to a higher-energy final level f by absorbing a photon with energy equal to $E_f - E_i$.

$hf = E_f - E_i$

ground level are called **excited levels.** An atom in an excited level, called an *excited atom*, can make a transition into the ground level by emitting a photon as in Fig. 39.16. But since there are no levels below the ground level, an atom in the ground level cannot lose energy and so cannot emit a photon.

Collisions are not the only way that an atom's energy can be raised from one level to a higher level. If an atom initially in the lower energy level in Fig. 39.16 is struck by a photon with just the right amount of energy, the photon can be *absorbed* and the atom will end up in the higher level (Fig. 39.17). As an example, we previously mentioned two levels in the lithium atom with an energy difference of 1.85 eV. For a photon to be absorbed and excite the atom from the lower level to the higher one, the photon must have an energy of 1.85 eV and a wavelength of 671 nm. In other words, an atom *absorbs* the same wavelengths that it *emits*. This explains the correspondence between an element's emission line spectrum and its absorption line spectrum that we described in Section 39.2.

Note that a lithium atom *cannot* absorb a photon with a slightly longer wavelength (say, 672 nm) or one with a slightly shorter wavelength (say, 670 nm). That's because these photons have, respectively, slightly too little or slightly too much energy to raise the atom's energy from one level to the next, and an atom cannot have an energy that's intermediate between levels. This explains why absorption line spectra have distinct dark lines (see Fig. 39.9): Atoms can absorb only photons with specific wavelengths.

An atom that's been excited into a high energy level, either by photon absorption or by collisions, does not stay there for long. After a short time, called the *lifetime* of the level (typically around 10^{-8} s), the excited atom will emit a photon and make a transition into a lower excited level or the ground level. A cool gas that's illuminated by white light to make an *absorption* line spectrum thus also produces an *emission* line spectrum when viewed from the side, since when the atoms de-excite they emit photons in all directions (Fig. 39.18). To keep a gas of atoms glowing, you have to continually provide energy to the gas in order to re-excite atoms so that they can emit more photons. If you turn off the energy supply (for example, by turning off the electric current through a neon light fixture, or by shutting off the light source in Fig. 39.18), the atoms drop back into their ground levels and cease to emit light.

By working backward from the observed emission line spectrum of an element, physicists can deduce the arrangement of energy levels in an atom of that element. As an example, Fig. 39.19a shows some of the energy levels for a sodium atom. You may have noticed the yellow-orange light emitted by sodium

39.18 When a beam of white light with a continuous spectrum passes through a cool gas, the transmitted light has an absorption spectrum. The absorbed light energy excites the gas and causes it to emit light of its own, which has an emission spectrum.

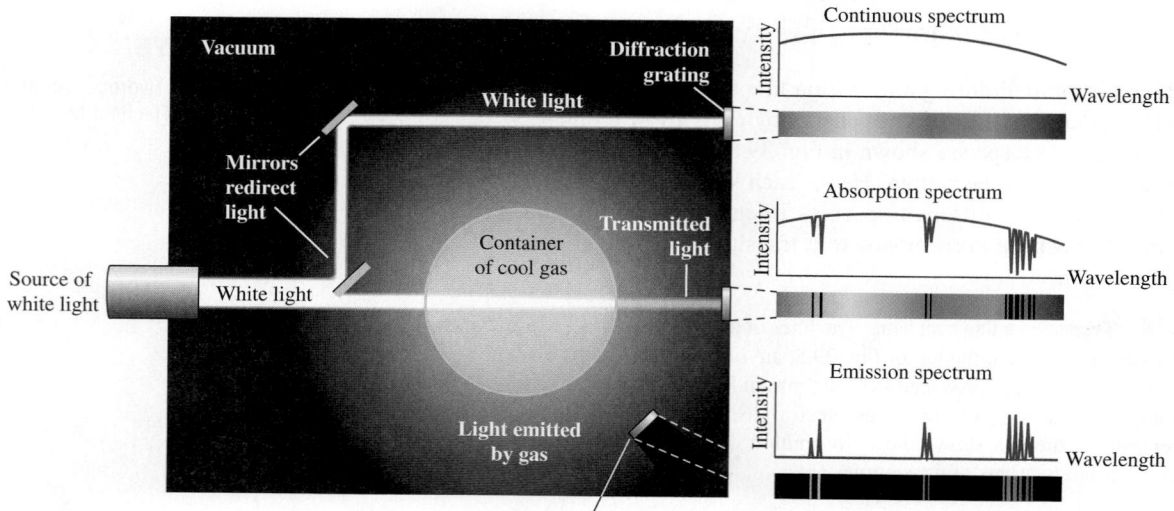

Diffraction grating for light emitted by gas

39.19 (a) Energy levels of the sodium atom relative to the ground level. Numbers on the lines between levels are wavelengths of the light emitted or absorbed during transitions between those levels. The column labels, such as $^2S_{1/2}$, refer to some quantum states of the atom. (b) When a sodium compound is placed in a flame, sodium atoms are excited into the lowest excited levels. As they drop back to the ground level, the atoms emit photons of yellow-orange light with wavelengths 589.0 and 589.6 nm.

(a) Energy (eV)

$^2S_{1/2}$ $^2P_{3/2}$ $^2P_{1/2}$ $^2D_{5/2, 3/2}$ $^2F_{7/2, 5/2}$ — 5.12

314.9 nm, 616.1 nm, 2208.4 nm, 1846.0 nm, 1140.4 nm, 1138.2 nm, 568.8 nm, 819.5 nm, 497.9 nm, 818.3 nm, 285.3 nm, 330.2 nm, 589.0 nm, 342.7 nm, 589.6 nm

Excited levels

Lowest excited levels

These wavelengths give excited sodium a yellow-orange color.

Ground level

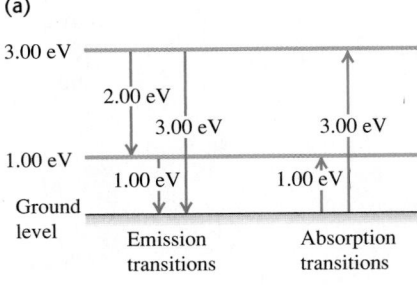

(b)

vapor street lights. Sodium atoms emit this characteristic yellow-orange light with wavelengths 589.0 and 589.6 nm when they make transitions from the two closely spaced levels labeled *lowest excited levels* to the ground level. A standard test for the presence of sodium compounds is to look for this yellow-orange light from a sample placed in a flame (Fig. 39.19b).

Example 39.5   **Emission and absorption spectra**

A hypothetical atom (Fig. 39.20a) has energy levels at 0.00 eV (the ground level), 1.00 eV, and 3.00 eV. (a) What are the frequencies and wavelengths of the spectral lines this atom can emit when excited? (b) What wavelengths can this atom absorb if it is in its ground level?

SOLUTION

IDENTIFY and SET UP: Energy is conserved when a photon is emitted or absorbed. In each transition the photon energy is equal to the difference between the energies of the levels involved in the transition.

EXECUTE: (a) The possible energies of emitted photons are 1.00 eV, 2.00 eV, and 3.00 eV. For 1.00 eV, Eq. (39.2) gives

$$f = \frac{E}{h} = \frac{1.00 \text{ eV}}{4.136 \times 10^{-15} \text{ eV} \cdot \text{s}} = 2.42 \times 10^{14} \text{ Hz}$$

For 2.00 eV and 3.00 eV, $f = 4.84 \times 10^{14}$ Hz and 7.25×10^{14} Hz, respectively. For 1.00-eV photons,

$$\lambda = \frac{c}{f} = \frac{3.00 \times 10^8 \text{ m/s}}{2.42 \times 10^{14} \text{ Hz}} = 1.24 \times 10^{-6} \text{ m} = 1240 \text{ nm}$$

39.20 (a) Energy-level diagram for the hypothetical atom, showing the possible transitions for emission from excited levels and for absorption from the ground level. (b) Emission spectrum of this hypothetical atom.

(a)

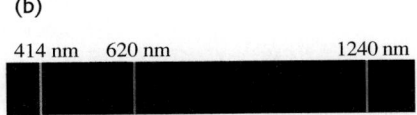

3.00 eV

2.00 eV

3.00 eV   3.00 eV

1.00 eV

1.00 eV   1.00 eV

Ground level

Emission transitions   Absorption transitions

(b)

414 nm   620 nm   1240 nm

Continued

This is in the infrared region of the spectrum (Fig. 39.20b). For 2.00 eV and 3.00 eV, the wavelengths are 620 nm (red) and 414 nm (violet), respectively.

(b) From the ground level, only a 1.00-eV or a 3.00-eV photon can be absorbed (Fig. 39.20a); a 2.00-eV photon cannot be absorbed because the atom has no energy level 2.00 eV above the ground level. Passing light from a hot solid through a gas of these hypothetical atoms (almost all of which would be in the ground state if the gas were cool) would yield a continuous spectrum with dark absorption lines at 1240 nm and 414 nm.

EVALUATE: Note that if a gas of these atoms were at a sufficiently high temperature, collisions would excite a number of atoms into the 1.00-eV energy level. Such excited atoms *can* absorb 2.00-eV photons, as Fig. 39.20a shows, and an absorption line at 620 nm would appear in the spectrum. Thus the observed spectrum of a given substance depends on its energy levels and its temperature.

Application **Fish Fluorescence**
When illuminated by blue light, this tropical lizardfish (family *Synodontidae*) fluoresces and emits longer-wavelength green light. The fluorescence may be a sexual signal or a way for the fish to camouflage itself among coral (which also have a green fluorescence).

Suppose we take a gas of the hypothetical atoms in Example 39.5 and illuminate it with violet light of wavelength 414 nm. Atoms in the ground level can absorb this photon and make a transition to the 3.00-eV level. Some of these atoms will make a transition back to the ground level by emitting a 414-nm photon. But other atoms will return to the ground level in two steps, first emitting a 620-nm photon to transition to the 1.00-eV level, then a 1240-nm photon to transition back to the ground level. Thus this gas will emit longer-wavelength radiation than it absorbs, a phenomenon called *fluorescence*. For example, the electric discharge in a fluorescent lamp causes the mercury vapor in the tube to emit ultraviolet radiation. This radiation is absorbed by the atoms of the coating on the inside of the tube. The coating atoms then re-emit light in the longer-wavelength, visible portion of the spectrum. Fluorescent lamps are more efficient than incandescent lamps in converting electrical energy to visible light because they do not waste as much energy producing (invisible) infrared photons.

Our discussion of energy levels and spectra has concentrated on *atoms*, but the same ideas apply to *molecules*. Figure 39.8 shows the emission line spectra of two molecules, hydrogen (H_2) and water (H_2O). Just as for sodium or other atoms, physicists can work backward from these molecular spectra and deduce the arrangement of energy levels for each kind of molecule. We'll return to molecules and molecular structure in Chapter 42.

The Franck–Hertz Experiment: Are Energy Levels Real?

Are atomic energy levels real, or just a convenient fiction that helps us to explain spectra? In 1914, the German physicists James Franck and Gustav Hertz answered this question when they found direct experimental evidence for the existence of atomic energy levels.

Franck and Hertz studied the motion of electrons through mercury vapor under the action of an electric field. They found that when the electron kinetic energy was 4.9 eV or greater, the vapor emitted ultraviolet light of wavelength 250 nm. Suppose mercury atoms have an excited energy level 4.9 eV above the ground level. An atom can be raised to this level by collision with an electron; it later decays back to the ground level by emitting a photon. From the photon formula $E = hc/\lambda$, the wavelength of the photon should be

$$\lambda = \frac{hc}{E} = \frac{(4.136 \times 10^{-15} \text{ eV} \cdot \text{s})(3.00 \times 10^8 \text{ m/s})}{4.9 \text{ eV}}$$
$$= 2.5 \times 10^{-7} \text{ m} = 250 \text{ nm}$$

This is equal to the wavelength that Franck and Hertz measured, which demonstrates that this energy level actually exists in the mercury atom. Similar experiments with other atoms yield the same kind of evidence for atomic energy levels. Franck and Hertz shared the 1925 Nobel Prize in physics for their research.

Electron Waves and the Bohr Model of Hydrogen

Bohr's hypothesis established the relationship between atomic spectra and energy levels. By itself, however, it provided no general principles for *predicting* the energy levels of a particular atom. Bohr addressed this problem for the case of the simplest atom, hydrogen, which has just one electron. Let's look at the ideas behind the **Bohr model** of the hydrogen atom.

Bohr postulated that each energy level of a hydrogen atom corresponds to a specific *stable* circular orbit of the electron around the nucleus. In a break with classical physics, Bohr further postulated that an electron in such an orbit does *not* radiate. Instead, an atom radiates energy only when an electron makes a transition from an orbit of energy E_i to a different orbit with lower energy E_f, emitting a photon of energy $hf = E_i - E_f$ in the process.

As a result of a rather complicated argument that related the angular frequency of the light emitted to the angular speed of the electron in highly excited energy levels, Bohr found that the magnitude of the electron's angular momentum is *quantized;* that is, this magnitude must be an integral multiple of $h/2\pi$. (Because $1 \text{ J} = 1 \text{ kg} \cdot \text{m}^2/\text{s}^2$, the SI units of Planck's constant h, J · s, are the same as the SI units of angular momentum, usually written as $\text{kg} \cdot \text{m}^2/\text{s}$.) Let's number the orbits by an integer n, where $n = 1, 2, 3, \ldots$, and call the radius of orbit n r_n and the speed of the electron in that orbit v_n. The value of n for each orbit is called the **principal quantum number** for the orbit. From Section 10.5, Eq. (10.28), the magnitude of the angular momentum of an electron of mass m in such an orbit is $L_n = mv_n r_n$ (Fig. 39.21). So Bohr's argument led to

$$L_n = mv_n r_n = n\frac{h}{2\pi} \quad \text{(quantization of angular momentum)} \quad (39.6)$$

Instead of going through Bohr's argument to justify Eq. (39.6), we can use de Broglie's picture of electron waves. Rather then visualizing the orbiting electron as a particle moving around the nucleus in a circular path, think of it as a sinusoidal *standing wave* with wavelength λ that extends around the circle. A standing wave on a string transmits no energy (see Section 15.7), and electrons in Bohr's orbits radiate no energy. For the wave to "come out even" and join onto itself smoothly, the circumference of this circle must include some *whole number* of wavelengths, as Fig. 39.22 suggests. Hence for an orbit with radius r_n and circumference $2\pi r_n$, we must have $2\pi r_n = n\lambda_n$, where λ_n is the wavelength and $n = 1, 2, 3, \ldots$. According to the de Broglie relationship, Eq. (39.1), the wavelength of a particle with rest mass m moving with nonrelativistic speed v_n is $\lambda_n = h/mv_n$. Combining $2\pi r_n = n\lambda_n$ and $\lambda_n = h/mv_n$, we find $2\pi r_n = nh/mv_n$ or

$$mv_n r_n = n\frac{h}{2\pi}$$

This is the same as Bohr's result, Eq. (39.6). Thus a wave picture of the electron leads naturally to the quantization of the electron's angular momentum.

Now let's consider a model of the hydrogen atom that is Newtonian in spirit but incorporates this quantization assumption (Fig. 39.23). This atom consists of a single electron with mass m and charge $-e$ in a circular orbit around a single proton with charge $+e$. The proton is nearly 2000 times as massive as the electron, so we can assume that the proton does not move. We learned in Section 5.4 that when a particle with mass m moves with speed v_n in a circular orbit with radius r_n, its centripetal (inward) acceleration is v_n^2/r_n. According to Newton's second law, a radially inward net force with magnitude $F = mv_n^2/r_n$ is needed to cause this acceleration. We discussed in Section 12.4 how the gravitational attraction provides that inward force for satellite orbits. In hydrogen the force F is provided by the electrical attraction between the positive proton and the negative electron. From Coulomb's law, Eq. (21.2),

$$F = \frac{1}{4\pi\epsilon_0}\frac{e^2}{r_n^2}$$

39.21 Calculating the angular momentum of an electron in a circular orbit around an atomic nucleus.

Angular momentum $\vec{L}_n$ of orbiting electron is perpendicular to plane of orbit (since we take origin to be at nucleus) and has magnitude $L = mv_n r_n \sin\phi = mv_n r_n \sin 90° = mv_n r_n$.

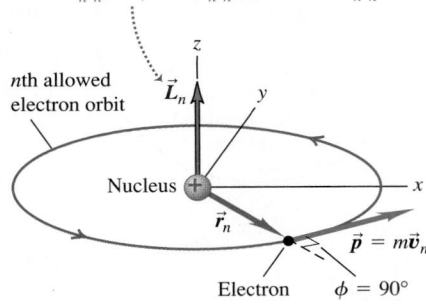

39.22 These diagrams show the idea of fitting a standing electron wave around a circular orbit. For the wave to join onto itself smoothly, the circumference of the orbit must be an integral number n of wavelengths.

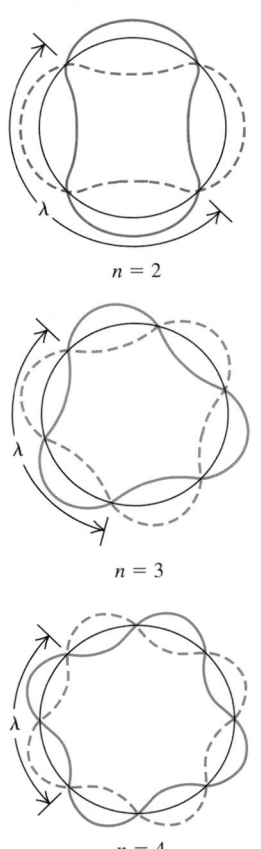

39.23 The Bohr model of the hydrogen atom.

Proton is assumed to be stationary.

Electron revolves in a circle of radius r_n with speed v_n.

Proton $M, +e$

r_n

v_n

F

Electron $m, -e$

Electrostatic attraction provides centripetal acceleration.

Hence Newton's second law states that

$$\frac{1}{4\pi\epsilon_0}\frac{e^2}{r_n^2} = \frac{mv_n^2}{r_n} \tag{39.7}$$

When we solve Eqs. (39.6) and (39.7) simultaneously for r_n and v_n, we get

$$r_n = \epsilon_0\frac{n^2h^2}{\pi me^2} \quad \text{(orbital radii in the Bohr model)} \tag{39.8}$$

$$v_n = \frac{1}{\epsilon_0}\frac{e^2}{2nh} \quad \text{(orbital speeds in the Bohr model)} \tag{39.9}$$

Equation (39.8) shows that the orbit radius r_n is proportional to n^2, so the smallest orbit radius corresponds to $n = 1$. We'll denote this minimum radius, called the *Bohr radius*, as a_0:

$$a_0 = \epsilon_0\frac{h^2}{\pi me^2} \quad \text{(Bohr radius)} \tag{39.10}$$

Then we can rewrite Eq. (39.8) as

$$r_n = n^2a_0 \tag{39.11}$$

The permitted orbits have radii a_0, $4a_0$, $9a_0$, and so on.

You can find the numerical values of the quantities on the right-hand side of Eq. (39.10) in Appendix F. Using these values, we find that the radius a_0 of the smallest Bohr orbit is

$$a_0 = \frac{(8.854 \times 10^{-12}\text{ C}^2/\text{N}\cdot\text{m}^2)(6.626 \times 10^{-34}\text{ J}\cdot\text{s})^2}{\pi(9.109 \times 10^{-31}\text{ kg})(1.602 \times 10^{-19}\text{ C})^2}$$

$$= 5.29 \times 10^{-11}\text{ m}$$

This gives an atomic diameter of about 10^{-10} m $= 0.1$ nm, which is consistent with atomic dimensions estimated by other methods.

Equation (39.9) shows that the orbital speed v_n is proportional to $1/n$. Hence the greater the value of n, the larger the orbital radius of the electron and the slower its orbital speed. (We saw the same relationship between orbital radius and speed for satellite orbits in Section 13.4.) We leave it to you to calculate the speed in the $n = 1$ orbit, which is the greatest possible speed of the electron in the hydrogen atom (see Exercise 39.29); the result is $v_1 = 2.19 \times 10^6$ m/s. This is less than 1% of the speed of light, so relativistic considerations aren't significant.

Hydrogen Energy Levels in the Bohr Model

We can now use Eqs. (39.8) and (39.9) to find the kinetic and potential energies K_n and U_n when the electron is in the orbit with quantum number n:

$$K_n = \tfrac{1}{2}mv_n^2 = \frac{1}{\epsilon_0^2}\frac{me^4}{8n^2h^2} \quad \begin{array}{l}\text{(kinetic energies in}\\\text{the Bohr model)}\end{array} \tag{39.12}$$

$$U_n = -\frac{1}{4\pi\epsilon_0}\frac{e^2}{r_n} = -\frac{1}{\epsilon_0^2}\frac{me^4}{4n^2h^2} \quad \begin{array}{l}\text{(potential energies in}\\\text{the Bohr model)}\end{array} \tag{39.13}$$

The potential energy has a negative sign because we have taken the electric potential energy to be zero when the electron is infinitely far from the nucleus. We are interested only in the *differences* in energy between orbits, so the reference position doesn't matter. The total energy E_n is the sum of the kinetic and potential energies:

$$E_n = K_n + U_n = -\frac{1}{\epsilon_0^2}\frac{me^4}{8n^2h^2} \quad \begin{array}{l}\text{(total energies in the}\\\text{Bohr model)}\end{array} \tag{39.14}$$

39.24 Two ways to represent the energy levels of the hydrogen atom and the transitions between them. Note that the radius of the *n*th permitted orbit is actually n^2 times the radius of the $n = 1$ orbit.

(a) Permitted orbits of an electron in the Bohr model of a hydrogen atom (not to scale). Arrows indicate the transitions responsible for some of the lines of various series.

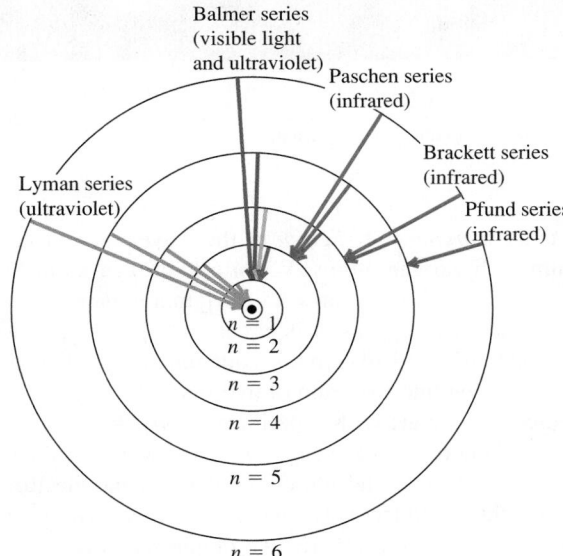

(b) Energy-level diagram for hydrogen, showing some transitions corresponding to the various series

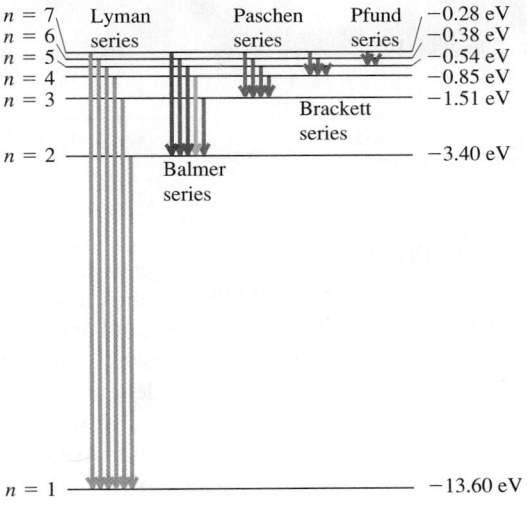

Since E_n in Eq. (39.14) has a different value for each n, you can see that this equation gives the *energy levels* of the hydrogen atom in the Bohr model. Each distinct orbit corresponds to a distinct energy level.

Figure 39.24 depicts the orbits and energy levels. We label the possible energy levels of the atom by values of the quantum number n. For each value of n there are corresponding values of orbit radius r_n, speed v_n, angular momentum $L_n = nh/2\pi$, and total energy E_n. The energy of the atom is least when $n = 1$ and E_n has its most negative value. This is the *ground level* of the hydrogen atom; it is the level with the smallest orbit, of radius a_0. For $n = 2, 3, \ldots$, the absolute value of E_n is smaller and the energy is progressively larger (less negative).

Figure 39.24 also shows some of the possible transitions from one electron orbit to an orbit of lower energy. Consider a transition from orbit n_U (for "upper") to a smaller orbit n_L (for "lower"), with $n_L < n_U$—or, equivalently, from *level* n_U to a lower *level* n_L. Then the energy hc/λ of the emitted photon of wavelength λ is equal to $E_{n_U} - E_{n_L}$. Before we use this relationship to solve for λ, it's convenient to rewrite Eq. (39.14) for the energies as

$$E_n = -\frac{hcR}{n^2}, \quad \text{where} \quad R = \frac{me^4}{8\epsilon_0^2 h^3 c} \quad \begin{matrix}\text{(total energies in}\\ \text{the Bohr model)}\end{matrix} \quad (39.15)$$

The quantity R in Eq. (39.15) is called the **Rydberg constant** (named for the Swedish physicist Johannes Rydberg, who did pioneering work on the hydrogen spectrum). When we substitute the numerical values of the fundamental physical constants m, c, e, h, and ϵ_0, all of which can be determined quite independently of the Bohr theory, we find that $R = 1.097 \times 10^7 \text{ m}^{-1}$. Now we solve for the wavelength of the photon emitted in a transition from level n_U to level n_L:

$$\frac{hc}{\lambda} = E_{n_U} - E_{n_L} = \left(-\frac{hcR}{n_U^2}\right) - \left(-\frac{hcR}{n_L^2}\right) = hcR\left(\frac{1}{n_L^2} - \frac{1}{n_U^2}\right)$$

$$\frac{1}{\lambda} = R\left(\frac{1}{n_L^2} - \frac{1}{n_U^2}\right) \quad \begin{matrix}\text{(hydrogen wavelengths in the}\\ \text{Bohr model, } n_L < n_U)\end{matrix} \quad (39.16)$$

39.25 The Balmer series of spectral lines for atomic hydrogen. You can see these same lines in the spectrum of *molecular* hydrogen (H₂) shown in Fig. 39.8, as well as additional lines that are present only when two hydrogen atoms are combined to make a molecule.

364.6 nm   410.2 nm   434.1 nm   486.1 nm   656.3 nm

H∞   Hδ   Hγ   Hβ   Hα

All Balmer lines beyond H_δ are in the ultraviolet spectrum.

H_α, H_β, H_γ, and H_δ are in the visible region of the spectrum.

Equation (39.16) is a *theoretical prediction* of the wavelengths found in the *emission* line spectrum of hydrogen atoms. When a hydrogen atom *absorbs* a photon, an electron makes a transition from a level n_L to a *higher* level n_U. This can happen only if the photon energy hc/λ is equal to $E_{n_U} - E_{n_L}$, which is the same condition expressed by Eq. (39.16). So this equation also predicts the wavelengths found in the *absorption* line spectrum of hydrogen.

How does this prediction compare with experiment? If $n_L = 2$, corresponding to transitions to the second energy level in Fig. 39.24, the wavelengths predicted by Eq. (39.16) are all in the visible and ultraviolet parts of the electromagnetic spectrum. These wavelengths are collectively called the *Balmer series* (Fig. 39.25). If we let $n_L = 2$ and $n_U = 3$ in Eq. (39.16) we obtain the wavelength of the H_α line:

$$\frac{1}{\lambda} = (1.097 \times 10^7 \text{ m}^{-1})\left(\tfrac{1}{4} - \tfrac{1}{9}\right) \quad \text{or} \quad \lambda = 656.3 \text{ nm}$$

With $n_L = 2$ and $n_U = 4$ we obtain the wavelength of the H_β line, and so on. With $n_L = 2$ and $n_U = \infty$ we obtain the shortest wavelength in the series, $\lambda = 364.6$ nm. These theoretical predictions are within 0.1% of the observed hydrogen wavelengths! This close agreement provides very strong and direct confirmation of Bohr's theory.

The Bohr model also predicts many other wavelengths in the hydrogen spectrum, as Fig. 39.24 shows. The observed wavelengths of all of these series, each of which is named for its discoverer, match the predicted values with the same percent accuracy as for the Balmer series. The *Lyman series* of spectral lines is caused by transitions between the ground level and the excited levels, corresponding to $n_L = 1$ and $n_U = 2, 3, 4, \ldots$ in Eq. (39.16). The energy difference between the ground level and any of the excited levels is large, so the emitted photons have wavelengths in the ultraviolet part of the electromagnetic spectrum. Transitions among the higher energy levels involve a much smaller energy difference, so the photons emitted in these transitions have little energy and long, infrared wavelengths. That's the case for both the *Brackett series* ($n_L = 3$ and $n_U = 4, 5, 6, \ldots$, corresponding to transitions between the third and higher energy levels) and the *Pfund series* ($n_L = 4$ and $n_U = 5, 6, 7, \ldots$, with transitions between the fourth and higher energy levels).

Figure 39.24 shows only transitions in which a hydrogen atom loses energy and a photon is emitted. But as we discussed previously, the wavelengths of those photons that an atom can *absorb* are the same as those that it can emit. For example, a hydrogen atom in the $n = 2$ level can absorb a 656.3-nm photon and end up in the $n = 3$ level.

One additional test of the Bohr model is its predicted value of the *ionization energy* of the hydrogen atom. This is the energy required to remove the electron completely from the atom. Ionization corresponds to a transition from the ground level ($n = 1$) to an infinitely large orbit radius ($n = \infty$), so the energy that must be added to the atom is $E_\infty - E_1 = 0 - E_1 = -E_1$ (recall that E_1 is negative).

Substituting the constants from Appendix F into Eq. (39.15) gives an ionization energy of 13.606 eV. The ionization energy can also be measured directly; the result is 13.60 eV. These two values agree within 0.1%.

Example 39.6 **Exploring the Bohr model**

Find the kinetic, potential, and total energies of the hydrogen atom in the first excited level, and find the wavelength of the photon emitted in a transition from that level to the ground level.

SOLUTION

IDENTIFY and SET UP: This problem uses the ideas of the Bohr model. We use simplified versions of Eqs. (39.12), (39.13), and (39.14) to find the energies of the atom, and Eq. (39.16), $hc/\lambda = E_{n_U} - E_{n_L}$, to find the photon wavelength λ in a transition from $n_U = 2$ (the first excited level) to $n_L = 1$ (the ground level).

EXECUTE: We could evaluate Eqs. (39.12), (39.13), and (39.14) for the nth level by substituting the values of m, e, ϵ_0, and h. But we can simplify the calculation by comparing with Eq. (39.15), which shows that the constant $me^4/8\epsilon_0^2 h^2$ that appears in Eqs. (39.12), (39.13), and (39.14) is equal to hcR:

$$\frac{me^4}{8\epsilon_0^2 h^2} = hcR$$

$$= (6.626 \times 10^{-34} \text{ J} \cdot \text{s})(2.998 \times 10^8 \text{ m/s})$$

$$\times (1.097 \times 10^7 \text{ m}^{-1})$$

$$= 2.179 \times 10^{-18} \text{ J} = 13.60 \text{ eV}$$

This allows us to rewrite Eqs. (39.12), (39.13), and (39.14) as

$$K_n = \frac{13.60 \text{ eV}}{n^2} \qquad U_n = \frac{-27.20 \text{ eV}}{n^2} \qquad E_n = \frac{-13.60 \text{ eV}}{n^2}$$

For the first excited level ($n = 2$), we have $K_2 = 3.40$ eV, $U_2 = -6.80$ eV, and $E_2 = -3.40$ eV. For the ground level ($n = 1$), $E_1 = -13.60$ eV. The energy of the emitted photon is then $E_2 - E_1 = -3.40$ eV $- (-13.60$ eV$) = 10.20$ eV, and

$$\lambda = \frac{hc}{E_2 - E_1} = \frac{(4.136 \times 10^{-15} \text{ eV} \cdot \text{s})(3.00 \times 10^8 \text{ m/s})}{10.20 \text{ eV}}$$

$$= 1.22 \times 10^{-7} \text{ m} = 122 \text{ nm}$$

This is the wavelength of the Lyman-alpha (L_α) line, the longest-wavelength line in the Lyman series of ultraviolet lines in the hydrogen spectrum (see Fig. 39.24).

EVALUATE: The total mechanical energy for any level is negative and is equal to one-half the potential energy. We found the same energy relationship for Newtonian satellite orbits in Section 12.4. The situations are similar because both the electrostatic and gravitational forces are inversely proportional to $1/r^2$.

Nuclear Motion and the Reduced Mass of an Atom

The Bohr model is so successful that we can justifiably ask why its predictions for the wavelengths and ionization energy of hydrogen differ from the measured values by about 0.1%. The explanation is that we assumed that the nucleus (a proton) remains at rest. However, as Fig. 39.26 shows, the proton and electron *both* revolve in circular orbits about their common center of mass (see Section 8.5). It turns out that we can take this motion into account very simply by using in Bohr's equations not the electron rest mass m but a quantity called the **reduced mass** m_r of the system. For a system composed of two bodies of masses m_1 and m_2, the reduced mass is

$$m_r = \frac{m_1 m_2}{m_1 + m_2} \qquad (39.17)$$

For ordinary hydrogen we let m_1 equal m and m_2 equal the proton mass, $m_p = 1836.2m$. Thus the proton–electron system of ordinary hydrogen has a reduced mass of

$$m_r = \frac{m(1836.2m)}{m + 1836.2m} = 0.99946m$$

When this value is used instead of the electron mass m in the Bohr equations, the predicted values agree very well with the measured values.

In an atom of deuterium, also called *heavy hydrogen*, the nucleus is not a single proton but a proton and a neutron bound together to form a composite body called the *deuteron*. The reduced mass of the deuterium atom turns out to be $0.99973m$. Equations (39.15) and (39.16) (with m replaced by m_r) show that all wavelengths are inversely proportional to m_r. Thus the wavelengths

39.26 The nucleus and the electron both orbit around their common center of mass. The distance r_N has been exaggerated for clarity; for ordinary hydrogen it actually equals $r_e/1836.2$.

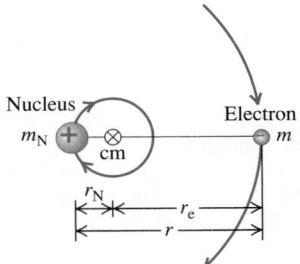

of the deuterium spectrum should be those of hydrogen divided by $(0.99973m)/(0.99946m) = 1.00027$. This is a small effect but well within the precision of modern spectrometers. This small wavelength shift led the American scientist Harold Urey to the discovery of deuterium in 1932, an achievement that earned him the 1934 Nobel Prize in chemistry.

Hydrogenlike Atoms

We can extend the Bohr model to other one-electron atoms, such as singly ionized helium (He^+), doubly ionized lithium (Li^{2+}), and so on. Such atoms are called *hydrogenlike* atoms. In such atoms, the nuclear charge is not e but Ze, where Z is the *atomic number*, equal to the number of protons in the nucleus. The effect in the previous analysis is to replace e^2 everywhere by Ze^2. In particular, the orbital radii r_n given by Eq. (39.8) become smaller by a factor of Z, and the energy levels E_n given by Eq. (39.14) are multiplied by Z^2. We invite you to verify these statements. The reduced-mass correction in these cases is even less than 0.1% because the nuclei are more massive than the single proton of ordinary hydrogen. Figure 39.27 compares the energy levels for H and for He^+, which has $Z = 2$.

Atoms of the alkali metals (at the far left-hand side of the periodic table; see Appendix D) have one electron outside a core consisting of the nucleus and the inner electrons, with net core charge $+e$. These atoms are approximately hydrogenlike, especially in excited levels. Physicists have studied alkali atoms in which the outer electron has been excited into a very large orbit with $n = 1000$. In accordance with Eq. (39.8), the radius of such a *Rydberg atom* with $n = 1000$ is $n^2 = 10^6$ times the Bohr radius, or about 0.05 mm—about the same size as a small grain of sand.

Although the Bohr model predicted the energy levels of the hydrogen atom correctly, it raised as many questions as it answered. It combined elements of classical physics with new postulates that were inconsistent with classical ideas. The model provided no insight into what happens during a transition from one orbit to another; the angular speeds of the electron motion were not in general the angular frequencies of the emitted radiation, a result that is contrary to classical electrodynamics. Attempts to extend the model to atoms with two or more electrons were not successful. An electron moving in one of Bohr's circular orbits forms a current loop and should produce a magnetic dipole moment (see Section 27.7). However, a hydrogen atom in its ground level has *no* magnetic moment due to orbital motion. In Chapters 40 and 41 we will find that an even more radical departure from classical concepts was needed before the understanding of atomic structure could progress further.

39.27 Energy levels of H and He^+. The energy expression, Eq. (39.14), is multiplied by $Z^2 = 4$ for He^+, so the energy of an He^+ ion with a given n is almost exactly four times that of an H atom with the same n. (There are small differences of the order of 0.05% because the reduced masses are slightly different.)

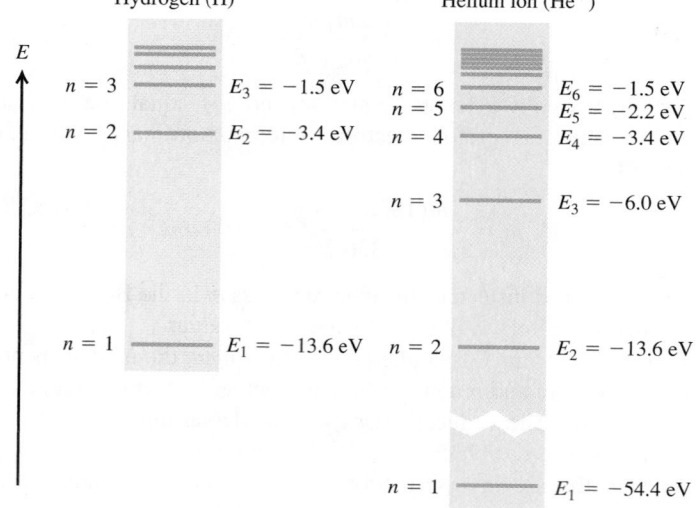

Test Your Understanding of Section 39.3 Consider the possible transitions between energy levels in a He$^+$ ion. For which of these transitions in He$^+$ will the wavelength of the emitted photon be nearly the same as one of the wavelengths emitted by excited H atoms? (i) $n = 2$ to $n = 1$; (ii) $n = 3$ to $n = 2$; (iii) $n = 4$ to $n = 3$; (iv) $n = 4$ to $n = 2$; (v) more than one of these; (vi) none of these.

39.4 The Laser

PhET: Lasers

The **laser** is a light source that produces a beam of highly coherent and very nearly monochromatic light as a result of cooperative emission from many atoms. The name "laser" is an acronym for "light amplification by stimulated emission of radiation." We can understand the principles of laser operation from what we have learned about atomic energy levels and photons. To do this we'll have to introduce two new concepts: *stimulated emission* and *population inversion*.

Spontaneous and Stimulated Emission

Consider a gas of atoms in a transparent container. Each atom is initially in its ground level of energy E_g and also has an excited level of energy E_{ex}. If we shine light of frequency f on the container, an atom can absorb one of the photons provided the photon energy $E = hf$ equals the energy difference $E_{ex} - E_g$ between the levels. Figure 39.28a shows this process, in which three atoms A each absorb a photon and go into the excited level. Some time later, the excited atoms (which we denote as A*) return to the ground level by each emitting a photon with the same frequency as the one originally absorbed (Fig. 39.28b). This process is called **spontaneous emission.** The direction and phase of the spontaneously emitted photons are random.

In **stimulated emission** (Fig. 39.28c), each incident photon encounters a previously excited atom. A kind of resonance effect induces each atom to emit a second photon with the same frequency, direction, phase, and polarization as the incident photon, which is not changed by the process. For each atom there is one photon before a stimulated emission and two photons after—thus the name *light amplification*. Because the two photons have the same phase, they emerge together as *coherent* radiation. The laser makes use of stimulated emission to produce a beam consisting of a large number of such coherent photons.

To discuss stimulated emission from atoms in excited levels, we need to know something about how many atoms are in each of the various energy levels. First, we need to make the distinction between the terms *energy level* and *state.* A system may have more than one way to attain a given energy level; each different way is a different **state.** For instance, there are two ways of putting an ideal unstretched spring in a given energy level. Remembering that the spring potential energy is $U = \frac{1}{2}kx^2$, we could compress the spring by $x = -b$ or we could stretch it by $x = +b$ to get the same $U = \frac{1}{2}kb^2$. The Bohr model had only one state in each energy level, but we will find in Chapter 41 that the hydrogen atom (Fig. 39.24b) actually has two states in its -13.60-eV ground level, eight states in its -3.40-eV first exited level, and so on.

The Maxwell–Boltzmann distribution function (see Section 18.5) determines the number of atoms in a given state in a gas. The function tells us that when the gas is in thermal equilibrium at absolute temperature T, the number n_i of atoms in a state with energy E_i equals $Ae^{-E_i/kT}$, where k is Boltzmann's constant and A is another constant determined by the total number of atoms in the gas. (In Section 18.5, E was the kinetic energy $\frac{1}{2}mv^2$ of a gas molecule; here we're talking about the internal energy of an atom.) Because of the negative exponent, fewer atoms are in higher-energy states, as we should expect. If E_g is a ground-state energy and E_{ex} is the energy of an excited state, then the ratio of numbers of atoms in the two states is

$$\frac{n_{ex}}{n_g} = \frac{Ae^{-E_{ex}/kT}}{Ae^{-E_g/kT}} = e^{-(E_{ex} - E_g)/kT} \qquad (39.18)$$

39.28 Three processes in which atoms interact with light.

(a) Absorption

Atom in its ground level

(b) Spontaneous emission

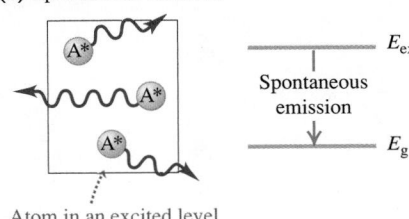

Atom in an excited level

(c) Stimulated emission

For example, suppose $E_{ex} - E_g = 2.0\,\text{eV} = 3.2 \times 10^{-19}\,\text{J}$, the energy of a 620-nm visible-light photon. At $T = 3000\,\text{K}$ (the temperature of the filament in an incandescent light bulb),

$$\frac{E_{ex} - E_g}{kT} = \frac{3.2 \times 10^{-19}\,\text{J}}{(1.38 \times 10^{-23}\,\text{J/K})(3000\,\text{K})} = 7.73$$

and

$$e^{-(E_{ex}-E_g)/kT} = e^{-7.73} = 0.00044$$

That is, the fraction of atoms in a state 2.0 eV above a ground state is extremely small, even at this high temperature. The point is that at any reasonable temperature there aren't enough atoms in excited states for any appreciable amount of stimulated emission from these states to occur. Rather, a photon emitted by one of the rare excited atoms will almost certainly be absorbed by an atom in the ground state rather than encountering another excited atom.

Enhancing Stimulated Emission: Population Inversions

To make a laser, we need to promote stimulated emission by increasing the number of atoms in excited states. Can we do that simply by illuminating the container with radiation of frequency $f = E/h$ corresponding to the energy difference $E = E_{ex} - E_g$, as in Fig. 39.28a? Some of the atoms absorb photons of energy E and are raised to the excited state, and the population ratio n_{ex}/n_g momentarily increases. But because n_g is originally so much larger than n_{ex}, an enormously intense beam of light would be required to momentarily increase n_{ex} to a value comparable to n_g. The rate at which energy is *absorbed* from the beam by the n_g ground-state atoms far exceeds the rate at which energy is added to the beam by stimulated emission from the relatively rare (n_{ex}) excited atoms.

We need to create a *nonequilibrium* situation in which the number of atoms in a higher-energy state is greater than the number in a lower-energy state. Such a situation is called a **population inversion.** Then the rate of energy radiation by stimulated emission can *exceed* the rate of absorption, and the system will act as a net *source* of radiation with photon energy E. It turns out that we can achieve a population inversion by starting with atoms that have the right kinds of excited states. Figure 39.29a shows an energy-level diagram for such an atom with a ground state and *three* excited states of energies E_1, E_2, and E_3. A laser that uses a material with energy levels like these is called a *four-level laser*. For the laser action to work, the states of energies E_1 and E_3 must have ordinary short lifetimes of about 10^{-8} s, while the state of energy E_2 must have an unusually long lifetime of 10^{-3} s or so. Such a long-lived **metastable state** can occur if, for instance, there are restrictions imposed by conservation of angular momentum that hinder photon emission from this state. (We'll discuss these restrictions in Chapter 41.) The metastable state is the one that we want to populate.

To produce a population inversion, we *pump* the material to excite the atoms out of the ground state into the states of energies E_1, E_2, and E_3 (Fig. 39.29b). If the atoms are in a gas, this pumping can be done by inserting two electrodes into the gas container. When a burst of sufficiently high voltage is applied to the electrodes, an electric discharge occurs. Collisions between ionized atoms and electrons carrying the discharge current then excite the atoms to various energy states. Within about 10^{-8} s the atoms that are excited to states E_1 and E_3 undergo spontaneous photon emission, so these states end up depopulated. But atoms "pile up" in the metastable state with energy E_2. The number of atoms in the metastable state is *less* than the number in the ground state, but is *much greater* than in the nearly unoccupied state of energy E_1. Hence there is a population inversion of state E_2 relative to state E_1 (Fig. 39.29c). You can see why we need

39.29 (a), (b), (c) Stages in the operation of a four-level laser. (d) The light emitted by atoms making spontaneous transitions from state E_2 to state E_1 is reflected between mirrors, so it continues to stimulate emission and gives rise to coherent light. One mirror is partially transmitting and allows the high-intensity light beam to escape.

(d) Schematic of gas laser

the two levels E_1 and E_3: atoms that undergo spontaneous emission from the E_3 level help to populate the E_2 level, and the presence of the E_1 level makes a population inversion possible.

Over the next 10^{-3} s, some of the atoms in the long-lived metastable state E_2 transition to state E_1 by spontaneous emission. The emitted photons of energy $hf = E_2 - E_1$ are sent back and forth through the gas many times by a pair of parallel mirrors (Fig. 39.29d), so that they can *stimulate* emission from as many of the atoms in state E_2 as possible. The net result of all these processes is a beam of light of frequency f that can be quite intense, has parallel rays, is highly monochromatic, and is spatially *coherent* at all points within a given cross section—that is, a laser beam. One of the mirrors is partially transparent, so a portion of the beam emerges.

What we've described is a *pulsed* laser that produces a burst of coherent light every time the atoms are pumped. Pulsed lasers are used in LASIK eye surgery (an acronym for *laser-assisted in situ keratomileusis*) to reshape the cornea and correct for nearsightedness, farsightedness, or astigmatism. In a *continuous* laser, such as those found in the barcode scanners used at retail checkout counters, energy is supplied to the atoms continuously (for instance, by having the power supply in Fig. 39.29d provide a steady voltage to the electrodes) and a steady beam of light emerges from the laser. For such a laser the pumping must be intense enough to sustain the population inversion, so that the rate at which atoms are added to level E_2 through pumping equals the rate at which atoms in this level emit a photon and transition to level E_1.

Since a special arrangement of energy levels is needed for laser action, it's not surprising that only certain materials can be used to make a laser. Some types of laser use a solid, transparent material such as neodymium glass rather than a gas. The most common kind of laser—used in laser printers (Section 21.2), laser pointers, and to read the data on the disc in a DVD player or Blu-ray player—is a *semiconductor laser,* which doesn't use atomic energy levels at all. As we'll discuss in Chapter 42, these lasers instead use the energy levels of electrons that are free to roam throughout the volume of the semiconductors.

Test Your Understanding of Section 39.4 An ordinary neon light fixture like those used in advertising signs emits red light of wavelength 632.8 nm. Neon atoms are also used in a helium–neon laser (a type of gas laser). The light emitted by a neon light fixture is (i) spontaneous emission; (ii) stimulated emission; (iii) both spontaneous and stimulated emission.

39.30 A hollow box with a small aperture behaves like a blackbody. When the box is heated, the electromagnetic radiation that emerges from the aperture has a blackbody spectrum.

Hollow box with small aperture (cross section)

Light that enters box is eventually absorbed. Hence box approximates a perfect blackbody.

39.31 This close-up view of the sun's surface shows two dark sunspots. Their temperature is about 4000 K, while the surrounding solar material is at $T = 5800$ K. From the Stefan–Boltzmann law, the intensity from a given area of sunspot is only $(4000 \text{ K}/5800 \text{ K})^4 = 0.23$ as great as the intensity from the same area of the surrounding material—which is why sunspots appear dark.

39.5 Continuous Spectra

Emission line spectra come from matter in the gaseous state, in which the atoms are so far apart that interactions between them are negligible and each atom behaves as an isolated system. By contrast, a heated solid or liquid (in which atoms are close to each other) nearly always emits radiation with a *continuous* distribution of wavelengths rather than a line spectrum.

Here's an analogy that suggests why there is a difference. A tuning fork emits sound waves of a single definite frequency (a pure tone) when struck. But if you tightly pack a suitcase full of tuning forks and then shake the suitcase, the proximity of the tuning forks to each other affects the sound that they produce. What you hear is mostly noise, which is sound with a continuous distribution of all frequencies. In the same manner, isolated atoms in a gas emit light of certain distinct frequencies when excited, but if the same atoms are crowded together in a solid or liquid they produce a continuous spectrum of light.

In this section we'll study an idealized case of continuous-spectrum radiation from a hot, dense object. Just as was the case for the emission line spectrum of light from an atom, we'll find that we can understand the continuous spectrum only if we use the ideas of energy levels and photons.

In the same way that an atom's emission spectrum has the same lines as its absorption spectrum, the ideal surface for *emitting* light with a continuous spectrum is one that also *absorbs* all wavelengths of electromagnetic radiation. Such an ideal surface is called a *blackbody* because it would appear perfectly black when illuminated; it would reflect no light at all. The continuous-spectrum radiation that a blackbody emits is called **blackbody radiation.** Like a perfectly frictionless incline or a massless rope, a perfect blackbody does not exist but is nonetheless a useful idealization.

A good approximation to a blackbody is a hollow box with a small aperture in one wall (Fig. 39.30). Light that enters the aperture will eventually be absorbed by the walls of the box, so the box is a nearly perfect absorber. Conversely, when we heat the box, the light that emanates from the aperture is nearly ideal blackbody radiation with a continuous spectrum.

By 1900 blackbody radiation had been studied extensively, and three characteristics had been established. First, the total intensity I (the average rate of radiation of energy per unit surface area or average power per area) emitted from the surface of an ideal radiator is proportional to the fourth power of the absolute temperature (Fig. 39.31). We studied this relationship in Section 17.7 during our study of heat-transfer mechanisms. This total intensity I emitted at absolute temperature T is given by the **Stefan–Boltzmann law:**

$$I = \sigma T^4 \qquad \text{(Stefan–Boltzmann law for a blackbody)} \qquad (39.19)$$

where σ is a fundamental physical constant called the *Stefan–Boltzmann constant.* In SI units,

$$\sigma = 5.670400(40) \times 10^{-8} \frac{\text{W}}{\text{m}^2 \cdot \text{K}^4}$$

Second, the intensity is not uniformly distributed over all wavelengths. Its distribution can be measured and described by the intensity per wavelength interval $I(\lambda)$, called the *spectral emittance.* Thus $I(\lambda)\, d\lambda$ is the intensity corresponding to wavelengths in the interval from λ to $\lambda + d\lambda$. The *total* intensity I, given by Eq. (39.19), is the *integral* of the distribution function $I(\lambda)$ over all wavelengths, which equals the area under the $I(\lambda)$ versus λ curve:

$$I = \int_0^\infty I(\lambda)\, d\lambda \qquad (39.20)$$

CAUTION **Spectral emittance vs. intensity** Although we use the symbol $I(\lambda)$ for spectral emittance, keep in mind that spectral emittance is *not* the same thing as intensity I. Intensity is power per unit area, with units W/m^2. Spectral emittance is power per unit area *per unit wavelength interval*, with units W/m^3.

Figure 39.32 shows the measured spectral emittances $I(\lambda)$ for blackbody radiation at three different temperatures. Each has a peak wavelength λ_m at which the emitted intensity per wavelength interval is largest. Experiment shows that λ_m is inversely proportional to T, so their product is constant. This observation is called the **Wien displacement law.** The experimental value of the constant is 2.90×10^{-3} m·K:

$$\lambda_m T = 2.90 \times 10^{-3} \text{ m} \cdot \text{K} \qquad \text{(Wien displacement law)} \qquad (39.21)$$

As the temperature rises, the peak of $I(\lambda)$ becomes higher and shifts to shorter wavelengths. Yellow light has shorter wavelengths than red light, so a body that glows yellow is hotter and brighter than one of the same size that glows red.

Third, experiments show that the *shape* of the distribution function is the same for all temperatures. We can make a curve for one temperature fit any other temperature by simply changing the scales on the graph.

Rayleigh and the "Ultraviolet Catastrophe"

During the last decade of the 19th century, many attempts were made to derive these empirical results about blackbody radiation from basic principles. In one attempt, the English physicist Lord Rayleigh considered the light enclosed within a rectangular box like that shown in Fig. 39.30. Such a box, he reasoned, has a series of possible *normal modes* for electromagnetic waves, as we discussed in Section 32.5. It also seemed reasonable to assume that the distribution of energy among the various modes would be given by the equipartition principle (see Section 18.4), which had been used successfully in the analysis of heat capacities.

Including both the electric- and magnetic-field energies, Rayleigh assumed that the total energy of each normal mode was equal to kT. Then by computing the *number* of normal modes corresponding to a wavelength interval $d\lambda$, Rayleigh calculated the expected distribution of wavelengths in the radiation within the box. Finally, he computed the predicted intensity distribution $I(\lambda)$ for the radiation emerging from the hole. His result was quite simple:

$$I(\lambda) = \frac{2\pi ckT}{\lambda^4} \qquad \text{(Rayleigh's calculation)} \qquad (39.22)$$

At large wavelengths this formula agrees quite well with the experimental results shown in Fig. 39.32, but there is serious disagreement at small wavelengths. The experimental curves in Fig. 39.32 fall toward zero at small λ. By contrast, Rayleigh's prediction in Eq. (39.22) goes in the opposite direction, approaching infinity at $1/\lambda^4$, a result that was called in Rayleigh's time the "ultraviolet catastrophe." Even worse, the integral of Eq. (39.22) over all λ is infinite, indicating an infinitely large *total* radiated intensity. Clearly, something is wrong.

Planck and the Quantum Hypothesis

Finally, in 1900, the German physicist Max Planck succeeded in deriving a function, now called the **Planck radiation law,** that agreed very well with experimental intensity distribution curves. In his derivation he made what seemed at the time to be a crazy assumption. Planck assumed that electromagnetic oscillators (electrons) in the walls of Rayleigh's box vibrating at a frequency f could have only certain values of energy equal to nhf, where $n = 0, 1, 2, 3, \ldots$ and h turned

39.32 These graphs show the spectral emittance $I(\lambda)$ for radiation from a blackbody at three different temperatures.

As the temperature increases, the peak of the spectral emittance curve becomes higher and shifts to shorter wavelengths.

Dashed blue lines are values of λ_m in Eq. (39.21) for each temperature.

out to be the constant that now bears Planck's name. These oscillators were in equilibrium with the electromagnetic waves in the box, so they both emitted and absorbed light. His assumption gave quantized energy levels and said that the energy in each normal mode was also a multiple of hf. This was in sharp contrast to Rayleigh's point of view that each normal mode could have any amount of energy.

Planck was not comfortable with this quantum hypothesis; he regarded it as a calculational trick rather than a fundamental principle. In a letter to a friend, he called it "an act of desperation" into which he was forced because "a theoretical explanation had to be found at any cost, whatever the price." But five years later, Einstein identified the energy change hf between levels as the energy of a photon to explain the photoelectric effect (see Section 38.1), and other evidence quickly mounted. By 1915 there was little doubt about the validity of the quantum concept and the existence of photons. By discussing atomic spectra *before* continuous spectra, we have departed from the historical order of things. The credit for inventing the concept of quantization of energy levels goes to Planck, even though he didn't believe it at first. He received the 1918 Nobel Prize in physics for his achievements.

Figure 39.33 shows energy-level diagrams for two of the oscillators that Planck envisioned in the walls of the rectangular box, one with a low frequency and the other with a high frequency. The spacing in energy between adjacent levels is hf. This spacing is small for the low-frequency oscillator that emits and absorbs photons of low frequency f and long wavelength $\lambda = c/f$. The energy spacing is greater for the high-frequency oscillator, which emits high-frequency photons of short wavelength.

According to Rayleigh's picture, both of these oscillators have the same amount of energy kT and are equally effective at emitting radiation. In Planck's model, however, the high-frequency oscillator is very ineffective as a source of light. To see why, we can use the ideas from Section 39.4 about the populations of various energy states. If we consider all the oscillators of a given frequency f in a box at temperature T, the number of oscillators that have energy nhf is $Ae^{-nhf/kT}$. The ratio of the number of oscillators in the first excited state ($n = 1$, energy hf) to the number of oscillators in the ground state ($n = 0$, energy zero) is

$$\frac{n_1}{n_0} = \frac{Ae^{-hf/kT}}{Ae^{-(0)/kT}} = e^{-hf/kT} \qquad (39.23)$$

Let's evaluate Eq. (39.23) for $T = 2000$ K, one of the temperatures shown in Fig. 39.32. At this temperature $kT = 2.76 \times 10^{-20}$ J $= 0.172$ eV. For an oscillator that emits photons of wavelength $\lambda = 3.00$ μm, we can show that $hf = hc/\lambda = 0.413$ eV; for a higher-frequency oscillator that emits photons of wavelength $\lambda = 0.500$ μm, $hf = hc/\lambda = 2.48$ eV. For these two cases Eq. (39.23) gives

$$\frac{n_1}{n_0} = e^{-hf/kT} = 0.0909 \text{ for } \lambda = 3.00 \text{ } \mu\text{m}$$

$$\frac{n_1}{n_0} = e^{-hf/kT} = 5.64 \times 10^{-7} \text{ for } \lambda = 0.500 \text{ } \mu\text{m}$$

The value for $\lambda = 3.00$ μm means that of all the oscillators that can emit light at this wavelength, 0.0909 of them—about one in 11—are in the first excited state. These excited oscillators can each emit a 3.00-μm photon and contribute it to the radiation inside the box. Hence we would expect that this radiation would be rather plentiful in the spectrum of radiation from a 2000 K blackbody. By contrast, the value for $\lambda = 0.500$ μm means that only 5.64×10^{-7} (about one in two million) of the oscillators that can emit this wavelength are in the first excited

39.33 Energy levels for two of the oscillators that Planck envisioned in the walls of a blackbody like that shown in Fig. 39.30. The spacing between adjacent energy levels for each oscillator is hf, which is smaller for the low-frequency oscillator.

Low-frequency oscillator	High-frequency oscillator
$12hf$ ————	$2hf$ ————
$11hf$ ————	
$10hf$ ————	
$9hf$ ————	
$8hf$ ————	
$7hf$ ————	
$6hf$ ————	hf ————
$5hf$ ————	
$4hf$ ————	
$3hf$ ————	
$2hf$ ————	
$1hf$ ————	
hf ————	
0 ————	0 ————

state. An oscillator can't emit if it's in the ground state, so the amount of radiation in the box at this wavelength is *tremendously* suppressed compared to Rayleigh's prediction. That's why the spectral emittance curve for 2000 K in Fig. 39.32 has such a low value at $\lambda = 0.500\ \mu$m and shorter wavelengths. So Planck's quantum hypothesis provided a natural way to suppress the spectral emittance of a blackbody at short wavelengths, and hence averted the ultraviolet catastrophe that plagued Rayleigh's calculations.

We won't go into all the details of Planck's derivation of the spectral emittance. Here is his result:

$$I(\lambda) = \frac{2\pi hc^2}{\lambda^5(e^{hc/\lambda kT} - 1)} \qquad \text{(Planck radiation law)} \qquad (39.24)$$

where h is Planck's constant, c is the speed of light, k is Boltzmann's constant, T is the absolute temperature, and λ is the wavelength. This function turns out to agree well with experimental emittance curves such as those in Fig. 39.32.

The Planck radiation law also contains the Wien displacement law and the Stefan–Boltzmann law as consequences. To derive the Wien law, we find the value of λ at which $I(\lambda)$ is maximum by taking the derivative of Eq. (39.24) and setting it equal to zero. We leave it to you to fill in the details; the result is

$$\lambda_{\text{m}} = \frac{hc}{4.965kT} \qquad (39.25)$$

To obtain this result, you have to solve the equation

$$5 - x = 5e^{-x} \qquad (39.26)$$

The root of this equation, found by trial and error or more sophisticated means, is 4.965 to four significant figures. You should evaluate the constant $hc/4.965k$ and show that it agrees with the experimental value of 2.90×10^{-3} m·K given in Eq. (39.21).

We can obtain the Stefan–Boltzmann law for a blackbody by integrating Eq. (39.24) over all λ to find the *total* radiated intensity (see Problem 39.67). This is not a simple integral; the result is

$$I = \int_0^\infty I(\lambda)\ d\lambda = \frac{2\pi^5 k^4}{15c^2 h^3}T^4 = \sigma T^4 \qquad (39.27)$$

in agreement with Eq. (39.19). Our result in Eq. (39.27) also shows that the constant σ in that law can be expressed as a combination of other fundamental constants:

$$\sigma = \frac{2\pi^5 k^4}{15c^2 h^3} \qquad (39.28)$$

You should substitute the values of k, c, and h from Appendix F and verify that you obtain the value $\sigma = 5.6704 \times 10^{-8}$ W/m²·K⁴ for the Stefan–Boltzmann constant.

The Planck radiation law, Eq. (39.24), looks so different from the unsuccessful Rayleigh expression, Eq. (39.22), that it may seem unlikely that they would agree at large values of λ. But when λ is large, the exponent in the denominator of Eq. (39.24) is very small. We can then use the approximation $e^x \approx 1 + x$ (for $x \ll 1$). You should verify that when this is done, the result approaches Eq. (39.22), showing that the two expressions do agree in the limit of very large λ. We also note that the Rayleigh expression does not contain h. At very long wavelengths (very small photon energies), quantum effects become unimportant.

Example 39.7 | Light from the sun

To a good approximation, the sun's surface is a blackbody with a surface temperature of 5800 K. (We are ignoring the absorption produced by the sun's atmosphere, shown in Fig. 39.9.) (a) At what wavelength does the sun emit most strongly? (b) What is the total radiated power per unit surface area?

SOLUTION

IDENTIFY and SET UP: Our target variables are the peak-intensity wavelength λ_m and the radiated power per area I. Hence we'll use the Wien displacement law, Eq. (39.21) (which relates λ_m to the blackbody temperature T), and the Stefan–Boltzmann law, Eq. (39.19) (which relates I to T).

EXECUTE: (a) From Eq. (39.21),

$$\lambda_m = \frac{2.90 \times 10^{-3} \text{ m} \cdot \text{K}}{T} = \frac{2.90 \times 10^{-3} \text{ m} \cdot \text{K}}{5800 \text{ K}}$$
$$= 0.500 \times 10^{-6} \text{ m} = 500 \text{ nm}$$

(b) From Eq. (39.19),

$$I = \sigma T^4 = (5.67 \times 10^{-8} \text{ W/m}^2 \cdot \text{K}^4)(5800 \text{ K})^4$$
$$= 6.42 \times 10^7 \text{ W/m}^2 = 64.2 \text{ MW/m}^2$$

EVALUATE: The 500-nm wavelength found in part (a) is near the middle of the visible spectrum. This should not be a surprise: The human eye evolved to take maximum advantage of natural light.

The enormous value $I = 64.2$ MW/m^2 found in part (b) is the intensity at the *surface* of the sun, a sphere of radius 6.96×10^8 m. When this radiated energy reaches the earth, 1.50×10^{11} m away, the intensity has decreased by the factor $[(6.96 \times 10^8 \text{ m})/(1.50 \times 10^{11} \text{ m})]^2 = 2.15 \times 10^{-5}$ to the still-impressive 1.4 kW/m^2.

Example 39.8 | A slice of sunlight

Find the power per unit area radiated from the sun's surface in the wavelength range 600.0 to 605.0 nm.

SOLUTION

IDENTIFY and SET UP: This question concerns the power emitted by a blackbody over a narrow range of wavelengths, and so involves the spectral emittance $I(\lambda)$ given by the Planck radiation law, Eq. (39.24). This requires that we find the area under the $I(\lambda)$ curve between 600.0 and 605.0 nm. We'll *approximate* this area as the product of the height of the curve at the median wavelength $\lambda = 602.5$ nm and the width of the interval, $\Delta\lambda = 5.0$ nm. From Example 39.7, $T = 5800$ K.

EXECUTE: To obtain the height of the $I(\lambda)$ curve at $\lambda = 602.5$ nm $= 6.025 \times 10^{-7}$ m, we first evaluate the quantity $hc/\lambda kT$ in Eq. (39.24) and then substitute the result into Eq. (39.24):

$$\frac{hc}{\lambda kT} = \frac{(6.626 \times 10^{-34} \text{ J} \cdot \text{s})(2.998 \times 10^8 \text{ m/s})}{(6.025 \times 10^{-7} \text{ m})(1.381 \times 10^{-23} \text{ J/K})(5800 \text{ K})} = 4.116$$

$$I(\lambda) = \frac{2\pi(6.626 \times 10^{-34} \text{ J} \cdot \text{s})(2.998 \times 10^8 \text{ m/s})^2}{(6.025 \times 10^{-7} \text{ m})^5 (e^{4.116} - 1)}$$
$$= 7.81 \times 10^{13} \text{ W/m}^3$$

The intensity in the 5.0-nm range from 600.0 to 605.0 nm is then approximately

$$I(\lambda)\Delta\lambda = (7.81 \times 10^{13} \text{ W/m}^3)(5.0 \times 10^{-9} \text{ m})$$
$$= 3.9 \times 10^5 \text{ W/m}^2 = 0.39 \text{ MW/m}^2$$

EVALUATE: In part (b) of Example 39.7, we found the power radiated per unit area by the sun at *all* wavelengths to be $I = 64.2$ MW/m^2; here we have found that the power radiated per unit area in the wavelength range from 600 to 605 nm is $I(\lambda)\Delta\lambda = 0.39$ MW/m^2, about 0.6% of the total.

Test Your Understanding of Section 39.5 (a) Does a blackbody at 2000 K emit x rays? (b) Does it emit radio waves?

39.6 The Uncertainty Principle Revisited

The discovery of the dual wave–particle nature of matter forces us to reevaluate the kinematic language we use to describe the position and motion of a particle. In classical Newtonian mechanics we think of a particle as a point. We can describe its location and state of motion at any instant with three spatial coordinates and three components of velocity. But because matter also has a wave aspect, when we look at the behavior on a small enough scale—comparable to the de Broglie wavelength of the particle—we can no longer use the Newtonian description. Certainly no Newtonian particle would undergo diffraction like electrons do (Section 39.1).

39.34 (a) A two-slit interference experiment for electrons. (b) The interference pattern after 28, 1000, and 10,000 electrons.

Graph shows number of electrons striking each region of detector.

To demonstrate just how non-Newtonian the behavior of matter can be, let's look at an experiment involving the two-slit interference of electrons (Fig. 39.34). We aim an electron beam at two parallel slits, just as we did for light in Section 38.4. (The electron experiment has to be done in vacuum so that the electrons don't collide with air molecules.) What kind of pattern appears on the detector on the other side of the slits? The answer is: *exactly the same* kind of interference pattern we saw for photons in Section 38.4! Moreover, the principle of complementarity, which we introduced in Section 38.4, tells us that we cannot apply the wave and particle models simultaneously to describe any single element of this experiment. Thus we *cannot* predict exactly where in the pattern (a wave phenomenon) any individual electron (a particle) will land. We can't even ask which slit an individual electron passes through. If we tried to look at where the electrons were going by shining a light on them—that is, by scattering photons off them—the electrons would recoil, which would modify their motions so that the two-slit interference pattern would not appear.

CAUTION **Electron two-slit interference is not interference between two electrons** It's a common misconception that the pattern in Fig. 39.34b is due to the interference between *two* electron waves, each representing an electron passing through one slit. To show that this cannot be the case, we can send just one electron at a time through the apparatus. It makes no difference; we end up with the same interference pattern. In a sense, each electron wave interferes with itself. ▌

The Heisenberg Uncertainty Principles for Matter

Just as electrons and photons show the same behavior in a two-slit interference experiment, electrons and other forms of matter obey the same Heisenberg uncertainty principles as photons do:

$$\Delta x \Delta p_x \geq \hbar/2$$
$$\Delta y \Delta p_y \geq \hbar/2$$
$$\Delta z \Delta p_z \geq \hbar/2$$

(Heisenberg uncertainty principle for position and momentum)   (39.29)

$$\Delta t \, \Delta E \geq \hbar/2$$

(Heisenberg uncertainty principle for energy and time interval)   (39.30)

In these equations $\hbar = h/2\pi = 1.055 \times 10^{-34}\,\text{J}\cdot\text{s}$. The uncertainty principle for energy and time interval has a direct application to energy levels. We have assumed that each energy level in an atom has a very definite energy. However, Eq. (39.30) says that this is not true for all energy levels. A system that remains in a metastable state for a very long time (large Δt) can have a very well-defined energy (small ΔE), but if it remains in a state for only a short time (small Δt) the uncertainty in energy must be correspondingly greater (large ΔE). Figure 39.35 illustrates this idea.

Example 39.9 | The uncertainty principle: position and momentum

An electron is confined within a region of width 5.000×10^{-11} m (roughly the Bohr radius). (a) Estimate the minimum uncertainty in the x-component of the electron's momentum. (b) What is the kinetic energy of an electron with this magnitude of momentum? Express your answer in both joules and electron volts.

SOLUTION

IDENTIFY and SET UP: This problem uses the Heisenberg uncertainty principle for position and momentum and the relationship between a particle's momentum and its kinetic energy. The electron could be anywhere within the region, so we take $\Delta x = 5.000 \times 10^{-11}$ m as its position uncertainty. We then find the momentum uncertainty Δp_x using Eq. (39.29) and the kinetic energy using the relationships $p = mv$ and $K = \frac{1}{2}mv^2$.

EXECUTE: (a) From Eqs. (39.29), for a given value of Δx, the uncertainty in momentum is minimum when the product $\Delta x\,\Delta p_x$ equals $\hbar$. Hence

$$\Delta p_x = \frac{\hbar}{2\Delta x} = \frac{1.055 \times 10^{-34}\,\text{J}\cdot\text{s}}{2(5.000 \times 10^{-11}\,\text{m})} = 1.055 \times 10^{-24}\,\text{J}\cdot\text{s/m}$$
$$= 1.055 \times 10^{-24}\,\text{kg}\cdot\text{m/s}$$

(b) We can rewrite the nonrelativistic expression for kinetic energy as

$$K = \tfrac{1}{2}mv^2 = \frac{(mv)^2}{2m} = \frac{p^2}{2m}$$

Hence an electron with a magnitude of momentum equal to Δp_x from part (a) has kinetic energy

$$K = \frac{p^2}{2m} = \frac{(1.055 \times 10^{-24}\,\text{kg}\cdot\text{m/s})^2}{2(9.11 \times 10^{-31}\,\text{kg})}$$
$$= 6.11 \times 10^{-19}\,\text{J} = 3.81\,\text{eV}$$

EVALUATE: This energy is typical of electron energies in atoms. This agreement suggests that the uncertainty principle is deeply involved in atomic structure.

A similar calculation explains why electrons in atoms do not fall into the nucleus. If an electron were confined to the interior of a nucleus, its position uncertainty would be $\Delta x \approx 10^{-14}$ m. This would give the electron a momentum uncertainty about 5000 times greater than that of the electron in this example, and a kinetic energy so great that the electron would immediately be ejected from the nucleus.

Example 39.10 | The uncertainty principle: energy and time

A sodium atom in one of the states labeled "Lowest excited levels" in Fig. 39.19a remains in that state, on average, for 1.6×10^{-8} s before it makes a transition to the ground level, emitting a photon with wavelength 589.0 nm and energy 2.105 eV. What is the uncertainty in energy of that excited state? What is the wavelength spread of the corresponding spectral line?

SOLUTION

IDENTIFY and SET UP: This problem uses the Heisenberg uncertainty principle for energy and time interval and the relationship between photon energy and wavelength. The average time that the atom spends in this excited state is equal to Δt in Eq. (39.30). We find the minimum uncertainty in the energy of the excited level by replacing the $\geq$ sign in Eq. (39.30) with an equals sign and solving for ΔE.

EXECUTE: From Eq. (39.30),

$$\Delta E = \frac{\hbar}{2\Delta t} = \frac{1.055 \times 10^{-34}\,\text{J}\cdot\text{s}}{2(1.6 \times 10^{-8}\,\text{s})}$$
$$= 3.3 \times 10^{-27}\,\text{J} = 2.1 \times 10^{-8}\,\text{eV}$$

The atom remains in the ground level indefinitely, so that level has *no* associated energy uncertainty. The fractional uncertainty of the *photon* energy is therefore

$$\frac{\Delta E}{E} = \frac{2.1 \times 10^{-8}\,\text{eV}}{2.105\,\text{eV}} = 1.0 \times 10^{-8}$$

You can use some simple calculus and the relation $E = hc/\lambda$ to show that $\Delta\lambda/\lambda \approx \Delta E/E$, so that the corresponding spread in wavelength, or "width," of the spectral line is approximately

$$\Delta\lambda = \lambda\frac{\Delta E}{E} = (589.0\,\text{nm})(1.0 \times 10^{-8}) = 0.0000059\,\text{nm}$$

EVALUATE: This irreducible uncertainty $\Delta\lambda$ is called the *natural line width* of this particular spectral line. Though very small, it is within the limits of resolution of present-day spectrometers. Ordinarily, the natural line width is much smaller than the line width arising from other causes such as the Doppler effect and collisions among the rapidly moving atoms.

The Uncertainty Principle and the Limits of the Bohr Model

We saw in Section 39.3 that the Bohr model of the hydrogen atom was tremendously successful. However, the Heisenberg uncertainty principle for position and momentum shows that this model *cannot* be a correct description of how an electron in an atom behaves. Figure 39.22 shows that in the Bohr model as interpreted by de Broglie, an electron wave moves in a plane around the nucleus. Let's call this the xy-plane, so the z-axis is perpendicular to the plane. Hence the Bohr model says that an electron is always found at $z = 0$, and its z-momentum p_z is always zero (the electron does not move out of the xy-plane). But this implies that there are *no* uncertainties in either z or p_z, so $\Delta z = 0$ and $\Delta p_z = 0$. This directly contradicts Eq. (39.29), which says that the product $\Delta z \Delta p_z$ must be greater than or equal to $\hbar$.

This conclusion isn't too surprising, since the electron in the Bohr model is a mix of particle and wave ideas (the electron moves in an orbit like a miniature planet, but has a wavelength). To get an accurate picture of how electrons behave inside an atom and elsewhere, we need a description that is based *entirely* on the electron's wave properties. Our goal in Chapter 40 will be to develop this description, which we call *quantum mechanics*. To do this we'll introduce the *Schrödinger equation*, the fundamental equation that describes the dynamics of matter waves. This equation, as we will see, is as fundamental to quantum mechanics as Newton's laws are to classical mechanics or as Maxwell's equations are to electromagnetism.

Test Your Understanding of Section 39.6 Rank the following situations according to the uncertainty in x-momentum, from largest to smallest. The mass of the proton is 1836 times the mass of the electron. (i) an electron whose x-coordinate is known to within 2×10^{-15} m; (ii) an electron whose x-coordinate is known to within 4×10^{-15} m; (iii) a proton whose x-coordinate is known to within 2×10^{-15} m; (iv) a proton whose x-coordinate is known to within 4×10^{-15} m.

39.35 The longer the lifetime Δt of a state, the smaller is its spread in energy (shown by the width of the energy levels).

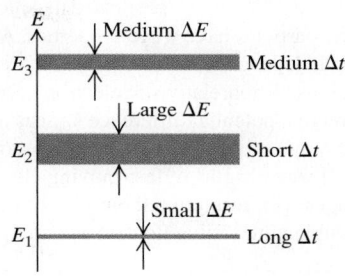

De Broglie waves and electron diffraction: Electrons and other particles have wave properties. A particle's wavelength depends on its momentum in the same way as for photons. A nonrelativistic electron accelerated from rest through a potential difference V_{ba} has a wavelength given by Eq. (39.3). Electron microscopes use the very small wavelengths of fast-moving electrons to make images with resolution thousands of times finer than is possible with visible light. (See Examples 39.1–39.3.)

$$\lambda = \frac{h}{p} = \frac{h}{mv} \tag{39.1}$$

$$E = hf \tag{39.2}$$

$$\lambda = \frac{h}{p} = \frac{h}{\sqrt{2meV_{ba}}} \tag{39.3}$$

The nuclear atom: The Rutherford scattering experiments show that most of an atom's mass and all of its positive charge are concentrated in a tiny, dense nucleus at the center of the atom. (See Example 39.4.)

Atomic line spectra and energy levels: The energies of atoms are quantized: They can have only certain definite values, called energy levels. When an atom makes a transition from an energy level E_i to a lower level E_f, it emits a photon of energy $E_i - E_f$. The same photon can be absorbed by an atom in the lower energy level, which excites the atom to the upper level. (See Example 39.5.)

$$hf = \frac{hc}{\lambda} = E_i - E_f \tag{39.5}$$

The Bohr model: In the Bohr model of the hydrogen atom, the permitted values of angular momentum are integral multiples of $h/2\pi$. The integer multiplier n is called the principal quantum number for the level. The orbital radii are proportional to n^2 and the orbital speeds are proportional to $1/n$. The energy levels of the hydrogen atom are given by Eq. (39.15), where R is the Rydberg constant. (See Example 39.6.)

$$L_n = mv_n r_n = n\frac{h}{2\pi}$$
$$(n = 1, 2, 3, \dots) \tag{39.6}$$

$$r_n = \epsilon_0 \frac{n^2 h^2}{\pi m e^2} = n^2 a_0 \tag{39.8}$$

$$= n^2 (5.29 \times 10^{-11} \text{ m}) \tag{39.10}$$

$$v_n = \frac{1}{\epsilon_0}\frac{e^2}{2nh} = \frac{2.19 \times 10^6 \text{ m/s}}{n} \tag{39.9}$$

$$E_n = -\frac{hcR}{n^2} = -\frac{13.60 \text{ eV}}{n^2} \tag{39.15}$$

$$(n = 1, 2, 3, \dots)$$

The laser: The laser operates on the principle of stimulated emission, by which many photons with identical wavelength and phase are emitted. Laser operation requires a nonequilibrium condition called a population inversion, in which more atoms are in a higher-energy state than are in a lower-energy state.

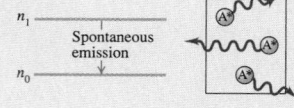

Blackbody radiation: The total radiated intensity (average power radiated per area) from a blackbody surface is proportional to the fourth power of the absolute temperature T. The quantity $\sigma = 5.67 \times 10^{-8} \text{ W/m}^2 \cdot \text{K}^4$ is called the Stefan–Boltzmann constant. The wavelength λ_m at which a blackbody radiates most strongly is inversely proportional to T. The Planck radiation law gives the spectral emittance $I(\lambda)$ (intensity per wavelength interval in blackbody radiation). (See Examples 39.7 and 39.8.)

$$I = \sigma T^4$$
(Stefan–Boltzmann law) $\tag{39.19}$

$$\lambda_m T = 2.90 \times 10^{-3} \text{ m} \cdot \text{K}$$
(Wien displacement law) $\tag{39.21}$

$$I(\lambda) = \frac{2\pi hc^2}{\lambda^5 (e^{hc/\lambda kT} - 1)}$$
(Planck radiation law) $\tag{39.24}$

The Heisenberg uncertainty principle for particles: The same uncertainty considerations that apply to photons also apply to particles such as electrons. The uncertainty ΔE in the energy of a state that is occupied for a time Δt is given by Eq. (39.30). (See Examples 39.9 and 39.10.)

$$\Delta x \Delta p_x \geq \hbar/2$$
$$\Delta y \Delta p_y \geq \hbar/2 \qquad \text{(Heisenberg uncertainty principle for position and momentum)} \qquad (39.29)$$
$$\Delta z \Delta p_z \geq \hbar/2$$

$$\Delta t \, \Delta E \geq \hbar/2 \qquad \text{(Heisenberg uncertainty principle for energy and time interval)} \qquad (39.30)$$

BRIDGING PROBLEM | **Hot Stars and Hydrogen Clouds**

Figure 39.36 shows a cloud, or *nebula,* of glowing hydrogen in interstellar space. The atoms in this cloud are excited by short-wavelength radiation emitted by the bright blue stars at the center of the nebula. (a) The blue stars act as blackbodies and emit light with a continuous spectrum. What is the wavelength at which a star with a surface temperature of 15,100 K (about $2\frac{1}{2}$ times the surface temperature of the sun) has the maximum spectral emittance? In what region of the electromagnetic spectrum is this? (b) Figure 39.32 shows that most of the energy radiated by a blackbody is at wavelengths between about one half and three times the wavelength of maximum emittance. If a hydrogen atom near the star in part (a) is initially in the ground level, what is the principal quantum number of the highest energy level to which it could be excited by a photon in this wavelength range? (c) The red color of the nebula is primarily due to hydrogen atoms making a transition from $n = 3$ to $n = 2$ and emitting photons of wavelength 656.3 nm. In the Bohr model as interpreted by de Broglie, what are the *electron* wavelengths in the $n = 2$ and $n = 3$ levels?

SOLUTION GUIDE

See MasteringPhysics® study area for a Video Tutor solution.

39.36 The Rosette Nebula.

IDENTIFY and SET UP

1. To solve this problem you need to use your knowledge of both blackbody radiation (Section 39.5) and the Bohr model of the hydrogen atom (Section 39.3).

2. In part (a) the target variable is the wavelength at which the star emits most strongly; in part (b) the target variable is a principal quantum number, and in part (c) it is the de Broglie wavelength of an electron in the $n = 2$ and $n = 3$ Bohr orbits (see Fig. 39.24). Select the equations you will need to find the target variables. (*Hint:* In Section 39.5 you learned how to find the energy change involved in a transition between two given levels of a hydrogen atom. Part (b) is a variation on this: You are to find the final level in a transition that starts in the $n = 1$ level and involves the absorption of a photon of a given wavelength and hence a given energy.)

EXECUTE

3. Use the Wien displacement law to find the wavelength at which the star has maximum spectral emittance. In what part of the electromagnetic spectrum is this wavelength?

4. Use your result from step 3 to find the range of wavelengths in which the star radiates most of its energy. Which end of this range corresponds to a photon with the greatest energy?

5. Write an expression for the wavelength of a photon that must be absorbed to cause an electron transition from the ground level ($n = 1$) to a higher level n. Solve for the value of n that corresponds to the highest-energy photon in the range you calculated in step 4. (*Hint:* Remember than n must be an integer.)

6. Find the electron wavelengths that correspond to the $n = 2$ and $n = 3$ orbits shown in Fig. 39.22.

EVALUATE

7. Check your result in step 5 by calculating the wavelength needed to excite a hydrogen atom from the ground level into the level *above* the highest-energy level that you found in step 5. Is it possible for light in the range of wavelengths you found in step 4 to excite hydrogen atoms from the ground level into this level?

8. How do the electron wavelengths you found in step 6 compare to the wavelength of a *photon* emitted in a transition from the $n = 3$ level to the $n = 2$ level?

Problems

•, ••, •••: Problems of increasing difficulty. **CP**: Cumulative problems incorporating material from earlier chapters. **CALC**: Problems requiring calculus. **BIO**: Biosciences problems.

DISCUSSION QUESTIONS

Q39.1 If a proton and an electron have the same speed, which has the longer de Broglie wavelength? Explain.

Q39.2 If a proton and an electron have the same kinetic energy, which has the longer de Broglie wavelength? Explain.

Q39.3 Does a photon have a de Broglie wavelength? If so, how is it related to the wavelength of the associated electromagnetic wave? Explain.

Q39.4 When an electron beam goes through a very small hole, it produces a diffraction pattern on a screen, just like that of light. Does this mean that an electron spreads out as it goes through the hole? What does this pattern mean?

Q39.5 Galaxies tend to be strong emitters of Lyman-α photons (from the $n = 2$ to $n = 1$ transition in atomic hydrogen). But the intergalactic medium—the very thin gas between the galaxies—tends to *absorb* Lyman-α photons. What can you infer from these observations about the temperature in these two environments? Explain.

Q39.6 A doubly ionized lithium atom (Li^{++}) is one that has had two of its three electrons removed. The energy levels of the remaining single-electron ion are closely related to those of the hydrogen atom. The nuclear charge for lithium is $\pm 3e$ instead of just $+e$. How are the energy levels related to those of hydrogen? How is the *radius* of the ion in the ground level related to that of the hydrogen atom? Explain.

Q39.7 The emission of a photon by an isolated atom is a recoil process in which momentum is conserved. Thus Eq. (39.5) should include a recoil kinetic energy K_r for the atom. Why is this energy negligible in that equation?

Q39.8 How might the energy levels of an atom be measured directly—that is, without recourse to analysis of spectra?

Q39.9 Elements in the gaseous state emit line spectra with well-defined wavelengths. But hot solid bodies always emit a continuous spectrum—that is, a continuous smear of wavelengths. Can you account for this difference?

Q39.10 As a body is heated to a very high temperature and becomes self-luminous, the apparent color of the emitted radiation shifts from red to yellow and finally to blue as the temperature increases. Why does the color shift? What other changes in the character of the radiation occur?

Q39.11 The peak-intensity wavelength of red dwarf stars, which have surface temperatures around 3000 K, is about 1000 nm, which is beyond the visible spectrum. So why are we able to see these stars, and why do they appear red?

Q39.12 You have been asked to design a magnet system to steer a beam of 54-eV electrons like those described in Example 39.1 (Section 39.1). The goal is to be able to direct the electron beam to a specific target location with an accuracy of ± 1.0 mm. In your design, do you need to take the wave nature of electrons into account? Explain.

Q39.13 Why go through the expense of building an electron microscope for studying very small objects such as organic molecules? Why not just use extremely short electromagnetic waves, which are much cheaper to generate?

Q39.14 Which has more total energy: a hydrogen atom with an electron in a high shell (large n) or in a low shell (small n)? Which is moving faster: the high-shell electron or the low-shell electron? Is there a contradiction here? Explain.

Q39.15 Does the uncertainty principle have anything to do with marksmanship? That is, is the accuracy with which a bullet can be aimed at a target limited by the uncertainty principle? Explain.

Q39.16 Suppose a two-slit interference experiment is carried out using an electron beam. Would the same interference pattern result if one slit at a time is uncovered instead of both at once? If not, why not? Doesn't each electron go through one slit or the other? Or does every electron go through both slits? Discuss the latter possibility in light of the principle of complementarity.

Q39.17 Equation (39.30) states that the energy of a system can have uncertainty. Does this mean that the principle of conservation of energy is no longer valid? Explain.

Q39.18 Laser light results from transitions from long-lived meta-stable states. Why is it more monochromatic than ordinary light?

Q39.19 Could an electron-diffraction experiment be carried out using three or four slits? Using a grating with many slits? What sort of results would you expect with a grating? Would the uncertainty principle be violated? Explain.

Q39.20 As the lower half of Fig. 39.4 shows, the diffraction pattern made by electrons that pass through aluminum foil is a series of concentric rings. But if the aluminum foil is replaced by a single crystal of aluminum, only certain points on these rings appear in the pattern. Explain.

Q39.21 Why can an electron microscope have greater magnification than an ordinary microscope?

Q39.22 When you check the air pressure in a tire, a little air always escapes; the process of making the measurement changes the quantity being measured. Think of other examples of measurements that change or disturb the quantity being measured.

EXERCISES

Section 39.1 Electron Waves

39.1 • (a) An electron moves with a speed of 4.70×10^6 m/s. What is its de Broglie wavelength? (b) A proton moves with the same speed. Determine its de Broglie wavelength.

39.2 •• For crystal diffraction experiments (discussed in Section 39.1), wavelengths on the order of 0.20 nm are often appropriate. Find the energy in electron volts for a particle with this wavelength if the particle is (a) a photon; (b) an electron; (c) an alpha particle ($m = 6.64 \times 10^{-27}$ kg).

39.3 • An electron has a de Broglie wavelength of 2.80×10^{-10} m. Determine (a) the magnitude of its momentum and (b) its kinetic energy (in joules and in electron volts).

39.4 •• **Wavelength of an Alpha Particle.** An alpha particle ($m = 6.64 \times 10^{-27}$ kg) emitted in the radioactive decay of uranium-238 has an energy of 4.20 MeV. What is its de Broglie wavelength?

39.5 • In the Bohr model of the hydrogen atom, what is the de Broglie wavelength for the electron when it is in (a) the $n = 1$

level and (b) the $n = 4$ level? In each case, compare the de Broglie wavelength to the circumference $2\pi r_n$ of the orbit.

39.6 • (a) A nonrelativistic free particle with mass m has kinetic energy K. Derive an expression for the de Broglie wavelength of the particle in terms of m and K. (b) What is the de Broglie wavelength of an 800-eV electron?

39.7 • **Why Don't We Diffract?** (a) Calculate the de Broglie wavelength of a typical person walking through a doorway. Make reasonable approximations for the necessary quantities. (b) Will the person in part (a) exhibit wavelike behavior when walking through the "single slit" of a doorway? Why?

39.8 •• What is the de Broglie wavelength for an electron with speed (a) $v = 0.480c$ and (b) $v = 0.960c$? (*Hint:* Use the correct relativistic expression for linear momentum if necessary.)

39.9 • (a) If a photon and an electron each have the same energy of 20.0 eV, find the wavelength of each. (b) If a photon and an electron each have the same wavelength of 250 nm, find the energy of each. (c) You want to study an organic molecule that is about 250 nm long using either a photon or an electron microscope. Approximately what wavelength should you use, and which probe, the electron or the photon, is likely to damage the molecule the least?

39.10 • How fast would an electron have to move so that its de Broglie wavelength is 1.00 mm?

39.11 • **Wavelength of a Bullet.** Calculate the de Broglie wavelength of a 5.00-g bullet that is moving at 340 m/s. Will the bullet exhibit wavelike properties?

39.12 •• Find the wavelengths of a photon and an electron that have the same energy of 25 eV. (*Note:* The energy of the electron is its kinetic energy.)

39.13 •• (a) What accelerating potential is needed to produce electrons of wavelength 5.00 nm? (b) What would be the energy of photons having the same wavelength as these electrons? (c) What would be the wavelength of photons having the same energy as the electrons in part (a)?

39.14 •• Through what potential difference must electrons be accelerated so they will have (a) the same wavelength as an x ray of wavelength 0.150 nm and (b) the same energy as the x ray in part (a)?

39.15 • (a) Approximately how fast should an electron move so it has a wavelength that makes it useful to measure the distance between adjacent atoms in typical crystals (about 0.10 nm)? (b) What is the kinetic energy of the electron in part (a)? (c) What would be the energy of a photon of the same wavelength as the electron in part (b)? (d) Which would make a more effective probe of small-scale structures: electrons or photons? Why?

39.16 •• **CP** A beam of electrons is accelerated from rest through a potential difference of 0.100 kV and then passes through a thin slit. The diffracted beam shows its first diffraction minima at $\pm 11.5°$ from the original direction of the beam when viewed far from the slit. (a) Do we need to use relativity formulas? How do you know? (b) How wide is the slit?

39.17 •• A beam of neutrons that all have the same energy scatters from atoms that have a spacing of 0.0910 nm in the surface plane of a crystal. The $m = 1$ intensity maximum occurs when the angle θ in Fig. 39.2 is 28.6°. What is the kinetic energy (in electron volts) of each neutron in the beam?

39.18 • A beam of 188-eV electrons is directed at normal incidence onto a crystal surface as shown in Fig. 39.3b. The $m = 2$ intensity maximum occurs at an angle $\theta = 60.6°$. (a) What is the spacing between adjacent atoms on the surface? (b) At what other angle or angles is there an intensity maximum? (c) For what electron

energy (in electron volts) would the $m = 1$ intensity maximum occur at $\theta = 60.6°$? For this energy, is there an $m = 2$ intensity maximum? Explain.

39.19 • A CD-ROM is used instead of a crystal in an electron-diffraction experiment. The surface of the CD-ROM has tracks of tiny pits with a uniform spacing of 1.60 μm. (a) If the speed of the electrons is 1.26×10^4 m/s, at which values of θ will the $m = 1$ and $m = 2$ intensity maxima appear? (b) The scattered electrons in these maxima strike at normal incidence a piece of photographic film that is 50.0 cm from the CD-ROM. What is the spacing on the film between these maxima?

39.20 • (a) In an electron microscope, what accelerating voltage is needed to produce electrons with wavelength 0.0600 nm? (b) If protons are used instead of electrons, what accelerating voltage is needed to produce protons with wavelength 0.0600 nm? (*Hint:* In each case the initial kinetic energy is negligible.)

39.21 •• You want to study a biological specimen by means of a wavelength of 10.0 nm, and you have a choice of using electromagnetic waves or an electron microscope. (a) Calculate the ratio of the energy of a 10.0-nm-wavelength photon to the kinetic energy of a 10.0-nm-wavelength electron. (b) In view of your answer to part (a), which would be less damaging to the specimen you are studying: photons or electrons?

Section 39.2 The Nuclear Atom and Atomic Spectra

39.22 •• **CP** A 4.78-MeV alpha particle from a ^{226}Ra decay makes a head-on collision with a uranium nucleus. A uranium nucleus has 92 protons. (a) What is the distance of closest approach of the alpha particle to the center of the nucleus? Assume that the uranium nucleus remains at rest and that the distance of closest approach is much greater than the radius of the uranium nucleus. (b) What is the force on the alpha particle at the instant when it is at the distance of closest approach?

39.23 • A beam of alpha particles is incident on a target of lead. A particular alpha particle comes in "head-on" to a particular lead nucleus and stops 6.50×10^{-14} m away from the center of the nucleus. (This point is well outside the nucleus.) Assume that the lead nucleus, which has 82 protons, remains at rest. The mass of the alpha particle is 6.64×10^{-27} kg. (a) Calculate the electro-static potential energy at the instant that the alpha particle stops. Express your result in joules and in MeV. (b) What initial kinetic energy (in joules and in MeV) did the alpha particle have? (c) What was the initial speed of the alpha particle?

Section 39.3 Energy Levels and the Bohr Model of the Atom

39.24 • The silicon–silicon single bond that forms the basis of the mythical silicon-based creature the Horta has a bond strength of 3.80 eV. What wavelength of photon would you need in a (mythical) phasor disintegration gun to destroy the Horta?

39.25 •• A hydrogen atom is in a state with energy -1.51 eV. In the Bohr model, what is the angular momentum of the electron in the atom, with respect to an axis at the nucleus?

39.26 • A hydrogen atom initially in the ground level absorbs a photon, which excites it to the $n = 4$ level. Determine the wavelength and frequency of the photon.

39.27 • A triply ionized beryllium ion, Be^{3+} (a beryllium atom with three electrons removed), behaves very much like a hydrogen atom except that the nuclear charge is four times as great. (a) What is the ground-level energy of Be^{3+}? How does this compare to the ground-level energy of the hydrogen atom? (b) What is the ionization energy of Be^{3+}? How does this compare to the ionization energy of the hydrogen atom? (c) For the hydrogen atom, the wavelength of the photon emitted in the $n = 2$ to $n = 1$ transition is 122 nm (see Example 39.6). What is the wavelength of the photon emitted when a Be^{3+} ion undergoes this transition? (d) For a given value of n, how does the radius of an orbit in Be^{3+} compare to that for hydrogen?

39.28 •• (a) Show that, as n gets very large, the energy levels of the hydrogen atom get closer and closer together in energy. (b) Do the radii of these energy levels also get closer together?

39.29 • (a) Using the Bohr model, calculate the speed of the electron in a hydrogen atom in the $n = 1$, 2, and 3 levels. (b) Calculate the orbital period in each of these levels. (c) The average lifetime of the first excited level of a hydrogen atom is 1.0×10^{-8} s. In the Bohr model, how many orbits does an electron in the $n = 2$ level complete before returning to the ground level?

39.30 • CP The energy-level scheme for the hypothetical one-electron element Searsium is shown in Fig. E39.30. The potential energy is taken to be zero for an electron at an infinite distance from the nucleus. (a) How much energy (in electron volts) does it take to ionize an electron from the ground level? (b) An 18-eV photon is absorbed by a Searsium atom in its ground level. As the atom returns to its ground level, what possible energies can the emitted photons have? Assume that there can be transitions between all pairs of levels. (c) What will happen if a photon with an energy of 8 eV strikes a Searsium atom in its ground level? Why? (d) Photons emitted in the Searsium transitions $n = 3 \rightarrow n = 2$ and $n = 3 \rightarrow n = 1$ will eject photoelectrons from an unknown metal, but the photon emitted from the transition $n = 4 \rightarrow n = 3$ will not. What are the limits (maximum and minimum possible values) of the work function of the metal?

Figure **E39.30**

$n = 4$	-2 eV
$n = 3$	-5 eV
$n = 2$	-10 eV
$n = 1$	-20 eV

39.31 • In a set of experiments on a hypothetical one-electron atom, you measure the wavelengths of the photons emitted from transitions ending in the ground state ($n = 1$), as shown in the energy-level diagram in Fig. E39.31. You also observe that it takes 17.50 eV to ionize this atom. (a) What is the energy of the atom in each of the levels ($n = 1$, $n = 2$, etc.) shown in the figure? (b) If an electron made a transition from the $n = 4$ to the $n = 2$ level, what wavelength of light would it emit?

Figure **E39.31**

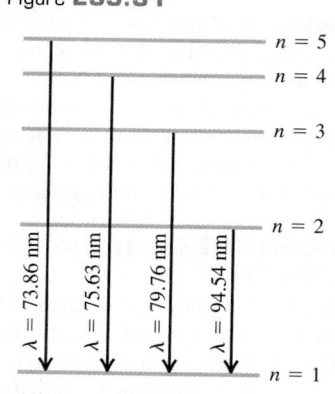

$n = 5$
$n = 4$
$n = 3$

$n = 2$

$\lambda = 73.86$ nm
$\lambda = 75.63$ nm
$\lambda = 79.76$ nm
$\lambda = 94.54$ nm

$n = 1$

39.32 • Find the longest and shortest wavelengths in the Lyman and Paschen series for hydrogen. In what region of the electromagnetic spectrum does each series lie?

39.33 • (a) An atom initially in an energy level with $E = -6.52$ eV absorbs a photon that has wavelength 860 nm. What is the internal energy of the atom after it absorbs the photon? (b) An atom initially in an energy level with $E = -2.68$ eV emits a photon that has wavelength 420 nm. What is the internal energy of the atom after it emits the photon?

39.34 •• Use Balmer's formula to calculate (a) the wavelength, (b) the frequency, and (c) the photon energy for the H_γ line of the Balmer series for hydrogen.

Section 39.4 The Laser

39.35 • BIO **Laser Surgery.** Using a mixture of CO_2, N_2, and sometimes He, CO_2 lasers emit a wavelength of 10.6 μm. At power outputs of 0.100 kW, such lasers are used for surgery. How many photons per second does a CO_2 laser deliver to the tissue during its use in an operation?

39.36 • BIO **Removing Birthmarks.** Pulsed dye lasers emit light of wavelength 585 nm in 0.45-ms pulses to remove skin blemishes such as birthmarks. The beam is usually focused onto a circular spot 5.0 mm in diameter. Suppose that the output of one such laser is 20.0 W. (a) What is the energy of each photon, in eV? (b) How many photons per square millimeter are delivered to the blemish during each pulse?

39.37 • How many photons per second are emitted by a 7.50-mW CO_2 laser that has a wavelength of 10.6 μm?

39.38 • BIO **PRK Surgery.** Photorefractive keratectomy (PRK) is a laser-based surgical procedure that corrects near- and farsightedness by removing part of the lens of the eye to change its curvature and hence focal length. This procedure can remove layers 0.25 μm thick using pulses lasting 12.0 ns from a laser beam of wavelength 193 nm. Low-intensity beams can be used because each individual photon has enough energy to break the covalent bonds of the tissue. (a) In what part of the electromagnetic spectrum does this light lie? (b) What is the energy of a single photon? (c) If a 1.50-mW beam is used, how many photons are delivered to the lens in each pulse?

39.39 • A large number of neon atoms are in thermal equilibrium. What is the ratio of the number of atoms in a $5s$ state to the number in a $3p$ state at (a) 300 K; (b) 600 K; (c) 1200 K? The energies of these states, relative to the ground state, are $E_{5s} = 20.66$ eV and $E_{3p} = 18.70$ eV. (d) At any of these temperatures, the rate at which a neon gas will spontaneously emit 632.8-nm radiation is quite low. Explain why.

39.40 • Figure 39.19a shows the energy levels of the sodium atom. The two lowest excited levels are shown in columns labeled $^2P_{3/2}$ and $^2P_{1/2}$. Find the ratio of the number of atoms in a $^2P_{3/2}$ state to the number in a $^2P_{1/2}$ state for a sodium gas in thermal equilibrium at 500 K. In which state are more atoms found?

Section 39.5 Continuous Spectra

39.41 •• A 100-W incandescent light bulb has a cylindrical tungsten filament 30.0 cm long, 0.40 mm in diameter, and with an emissivity of 0.26. (a) What is the temperature of the filament? (b) For what wavelength does the spectral emittance of the bulb peak? (c) Incandescent light bulbs are not very efficient sources of visible light. Explain why this is so.

39.42 • Determine λ_m, the wavelength at the peak of the Planck distribution, and the corresponding frequency f, at these temperatures: (a) 3.00 K; (b) 300 K; (c) 3000 K.

39.43 • Radiation has been detected from space that is characteristic of an ideal radiator at $T = 2.728$ K. (This radiation is a relic of the Big Bang at the beginning of the universe.) For this temperature, at what wavelength does the Planck distribution peak? In what part of the electromagnetic spectrum is this wavelength?

39.44 • The shortest visible wavelength is about 400 nm. What is the temperature of an ideal radiator whose spectral emittance peaks at this wavelength?

39.45 •• Two stars, both of which behave like ideal blackbodies, radiate the same total energy per second. The cooler one has a surface temperature T and a diameter 3.0 times that of the hotter star. (a) What is the temperature of the hotter star in terms of T? (b) What is the ratio of the peak-intensity wavelength of the hot star to the peak-intensity wavelength of the cool star?

39.46 • **Sirius B.** The brightest star in the sky is Sirius, the Dog Star. It is actually a binary system of two stars, the smaller one (Sirius B) being a white dwarf. Spectral analysis of Sirius B indicates that its surface temperature is 24,000 K and that it radiates energy at a total rate of 1.0×10^{25} W. Assume that it behaves like an ideal blackbody. (a) What is the total radiated intensity of Sirius B? (b) What is the peak-intensity wavelength? Is this wavelength visible to humans? (c) What is the radius of Sirius B? Express your answer in kilometers and as a fraction of our sun's radius. (d) Which star radiates more *total* energy per second, the hot Sirius B or the (relatively) cool sun with a surface temperature of 5800 K? To find out, calculate the ratio of the total power radiated by our sun to the power radiated by Sirius B.

39.47 •• **Blue Supergiants.** A typical blue supergiant star (the type that explodes and leaves behind a black hole) has a surface temperature of 30,000 K and a visual luminosity 100,000 times that of our sun. Our sun radiates at the rate of 3.86×10^{26} W. (Visual luminosity is the total power radiated at visible wavelengths.) (a) Assuming that this star behaves like an ideal blackbody, what is the principal wavelength it radiates? Is this light visible? Use your answer to explain why these stars are blue. (b) If we assume that the power radiated by the star is also 100,000 times that of our sun, what is the radius of this star? Compare its size to that of our sun, which has a radius of 6.96×10^5 km. (c) Is it really correct to say that the visual luminosity is proportional to the total power radiated? Explain.

Section 39.6 The Uncertainty Principle Revisited

39.48 • A pesky 1.5-mg mosquito is annoying you as you attempt to study physics in your room, which is 5.0 m wide and 2.5 m high. You decide to swat the bothersome insect as it flies toward you, but you need to estimate its speed to make a successful hit. (a) What is the maximum uncertainty in the horizontal position of the mosquito? (b) What limit does the Heisenberg uncertainty principle place on your ability to know the horizontal velocity of this mosquito? Is this limitation a serious impediment to your attempt to swat it?

39.49 • By extremely careful measurement, you determine the x-coordinate of a car's center of mass with an uncertainty of only 1.00 μm. The car has a mass of 1200 kg. (a) What is the minimum uncertainty in the x-component of the velocity of the car's center of mass as prescribed by the Heisenberg uncertainty principle? (b) Does the uncertainty principle impose a practical limit on our ability to make simultaneous measurements of the positions and velocities of ordinary objects like cars, books, and people? Explain.

39.50 • A 10.0-g marble is gently placed on a horizontal tabletop that is 1.75 m wide. (a) What is the maximum uncertainty in the horizontal position of the marble? (b) According to the Heisenberg uncertainty principle, what is the minimum uncertainty in the horizontal velocity of the marble? (c) In light of your answer to part (b), what is the longest time the marble could remain on the table? Compare this time to the age of the universe, which is approximately 14 billion years. (*Hint:* Can you know that the horizontal velocity of the marble is *exactly* zero?)

39.51 • A scientist has devised a new method of isolating individual particles. He claims that this method enables him to detect simultaneously the position of a particle along an axis with a standard deviation of 0.12 nm and its momentum component along this axis with a standard deviation of 3.0×10^{-25} kg·m/s. Use the Heisenberg uncertainty principle to evaluate the validity of this claim.

39.52 • (a) The x-coordinate of an electron is measured with an uncertainty of 0.20 mm. What is the x-component of the electron's velocity, v_x, if the minimum percentage uncertainty in a simultaneous measurement of v_x is 1.0%? (b) Repeat part (a) for a proton.

39.53 • An atom in a metastable state has a lifetime of 5.2 ms. What is the uncertainty in energy of the metastable state?

39.54 • (a) The uncertainty in the y-component of a proton's position is 2.0×10^{-12} m. What is the minimum uncertainty in a simultaneous measurement of the y-component of the proton's velocity? (b) The uncertainty in the z-component of an electron's velocity is 0.250 m/s. What is the minimum uncertainty in a simultaneous measurement of the z-coordinate of the electron?

PROBLEMS

39.55 •• The negative muon has a charge equal to that of an electron but a mass that is 207 times as great. Consider a hydrogenlike atom consisting of a proton and a muon. (a) What is the reduced mass of the atom? (b) What is the ground-level energy (in electron volts)? (c) What is the wavelength of the radiation emitted in the transition from the $n = 2$ level to the $n = 1$ level?

39.56 • An atom with mass m emits a photon of wavelength λ. (a) What is the recoil speed of the atom? (b) What is the kinetic energy K of the recoiling atom? (c) Find the ratio K/E, where E is the energy of the emitted photon. If this ratio is much less than unity, the recoil of the atom can be neglected in the emission process. Is the recoil of the atom more important for small or large atomic masses? For long or short wavelengths? (d) Calculate K (in electron volts) and K/E for a hydrogen atom (mass 1.67×10^{-27} kg) that emits an ultraviolet photon of energy 10.2 eV. Is recoil an important consideration in this emission process?

39.57 • (a) What is the smallest amount of energy in electron volts that must be given to a hydrogen atom initially in its ground level so that it can emit the H_α line in the Balmer series? (b) How many different possibilities of spectral-line emissions are there for this atom when the electron starts in the $n = 3$ level and eventually ends up in the ground level? Calculate the wavelength of the emitted photon in each case.

39.58 • A large number of hydrogen atoms are in thermal equilibrium. Let n_2/n_1 be the ratio of the number of atoms in an $n = 2$ excited state to the number of atoms in an $n = 1$ ground state. At what temperature is n_2/n_1 equal to (a) 10^{-12}; (b) 10^{-8}; (c) 10^{-4}? (d) Like the sun, other stars have continuous spectra with dark absorption lines (see Fig. 39.9). The absorption takes place in the star's atmosphere, which in all stars is composed primarily of hydrogen. Explain why the Balmer absorption lines are relatively weak in stars with low atmospheric temperatures such as the sun (atmosphere temperature 5800 K) but strong in stars with higher atmospheric temperatures.

39.59 ••• A sample of hydrogen atoms is irradiated with light with wavelength 85.5 nm, and electrons are observed leaving the gas. (a) If each hydrogen atom were initially in its ground level, what would be the maximum kinetic energy in electron volts of these photoelectrons? (b) A few electrons are detected with energies as much as 10.2 eV greater than the maximum kinetic energy calculated in part (a). How can this be?

39.60 • **CP Bohr Orbits of a Satellite.** A 20.0-kg satellite circles the earth once every 2.00 h in an orbit having a radius of 8060 km. (a) Assuming that Bohr's angular-momentum result ($L = nh/2\pi$) applies to satellites just as it does to an electron in the hydrogen atom, find the quantum number n of the orbit of the satellite. (b) Show from Bohr's angular momentum result and Newton's law of gravitation that the radius of an earth-satellite orbit is directly proportional to the square of the quantum number, $r = kn^2$, where k is the constant of proportionality. (c) Using the result from part (b), find the distance between the orbit of the satellite in this problem and its next "allowed" orbit. (Calculate a numerical value.) (d) Comment on the possibility of observing the separation of the two adjacent orbits. (e) Do quantized and classical orbits correspond for this satellite? Which is the "correct" method for calculating the orbits?

39.61 •• **The Red Supergiant Betelgeuse.** The star Betelgeuse has a surface temperature of 3000 K and is 600 times the diameter of our sun. (If our sun were that large, we would be inside it!) Assume that it radiates like an ideal blackbody. (a) If Betelgeuse were to radiate all of its energy at the peak-intensity wavelength, how many photons per second would it radiate? (b) Find the ratio of the power radiated by Betelgeuse to the power radiated by our sun (at 5800 K).

39.62 •• **CP** Light from an ideal spherical blackbody 15.0 cm in diameter is analyzed using a diffraction grating having 3850 lines/cm. When you shine this light through the grating, you observe that the peak-intensity wavelength forms a first-order bright fringe at $\pm11.6°$ from the central bright fringe. (a) What is the temperature of the blackbody? (b) How long will it take this sphere to radiate 12.0 MJ of energy?

39.63 • What must be the temperature of an ideal blackbody so that photons of its radiated light having the peak-intensity wavelength can excite the electron in the Bohr-model hydrogen atom from the ground state to the third excited state?

39.64 • **CP** An ideal spherical blackbody 24.0 cm in diameter is maintained at 225°C by an internal electrical heater and is immersed in a very large open-faced tank of water that is kept boiling by the energy radiated by the sphere. You can neglect any heat transferred by conduction and convection. Consult Table 17.4 as needed. (a) At what rate, in g/s, is water evaporating from the tank? (b) If a physics-wise thermophile organism living in the hot water is observing this process, what will it measure for the peak-intensity (i) wavelength and (ii) frequency of the electromagnetic waves emitted by the sphere?

39.65 ••• When a photon is emitted by an atom, the atom must recoil to conserve momentum. This means that the photon and the recoiling atom share the transition energy. (a) For an atom with mass m, calculate the correction $\Delta\lambda$ due to recoil to the wavelength of an emitted photon. Let λ be the wavelength of the photon if recoil is not taken into consideration. (*Hint:* The correction is very small, as Problem 39.56 suggests, so $|\Delta\lambda|/\lambda \ll 1$. Use this fact to obtain an approximate but very accurate expression for $\Delta\lambda$.) (b) Evaluate the correction for a hydrogen atom in which an electron in the nth level returns to the ground level. How does the answer depend on n?

39.66 •• **An Ideal Blackbody.** A large cavity with a very small hole and maintained at a temperature T is a good approximation to an ideal radiator or blackbody. Radiation can pass into or out of the cavity only through the hole. The cavity is a perfect absorber, since any radiation incident on the hole becomes trapped inside the cavity. Such a cavity at 200°C has a hole with area 4.00 mm². How

long does it take for the cavity to radiate 100 J of energy through the hole?

39.67 •• **CALC** (a) Write the Planck distribution law in terms of the frequency f, rather than the wavelength λ, to obtain $I(f)$. (b) Show that

$$\int_0^\infty I(\lambda)\, d\lambda = \frac{2\pi^5 k^4}{15c^2 h^3} T^4$$

where $I(\lambda)$ is the Planck distribution formula of Eq. (39.24). (*Hint:* Change the integration variable from λ to f. You will need to use the following tabulated integral:

$$\int_0^\infty \frac{x^3}{e^{\alpha x} - 1}\, dx = \frac{1}{240}\left(\frac{2\pi}{\alpha}\right)^4$$

(c) The result of part (b) is I and has the form of the Stefan–Boltzmann law, $I = \sigma T^4$ (Eq. 39.19). Evaluate the constants in part (b) to show that σ has the value given in Section 39.5.

39.68 •• **CP** A beam of 40-eV electrons traveling in the $+x$-direction passes through a slit that is parallel to the y-axis and 5.0 μm wide. The diffraction pattern is recorded on a screen 2.5 m from the slit. (a) What is the de Broglie wavelength of the electrons? (b) How much time does it take the electrons to travel from the slit to the screen? (c) Use the width of the central diffraction pattern to calculate the uncertainty in the y-component of momentum of an electron just after it has passed through the slit. (d) Use the result of part (c) and the Heisenberg uncertainty principle (Eq. 39.29 for y) to estimate the minimum uncertainty in the y-coordinate of an electron just after it has passed through the slit. Compare your result to the width of the slit.

39.69 • (a) What is the energy of a photon that has wavelength 0.10 μm? (b) Through approximately what potential difference must electrons be accelerated so that they will exhibit wave nature in passing through a pinhole 0.10 μm in diameter? What is the speed of these electrons? (c) If protons rather than electrons were used, through what potential difference would protons have to be accelerated so they would exhibit wave nature in passing through this pinhole? What would be the speed of these protons?

39.70 • **CP** Electrons go through a single slit 150 nm wide and strike a screen 24.0 cm away. You find that at angles of $\pm20.0°$ from the center of the diffraction pattern, no electrons hit the screen but electrons hit at all points closer to the center. (a) How fast were these electrons moving when they went through the slit? (b) What will be the next larger angles at which no electrons hit the screen?

39.71 •• **CP** A beam of electrons is accelerated from rest and then passes through a pair of identical thin slits that are 1.25 nm apart. You observe that the first double-slit interference dark fringe occurs at $\pm18.0°$ from the original direction of the beam when viewed on a distant screen. (a) Are these electrons relativistic? How do you know? (b) Through what potential difference were the electrons accelerated?

39.72 •• **CP** A beam of protons and a beam of alpha particles (of mass 6.64×10^{-27} kg and charge $+2e$) are accelerated from rest through the same potential difference and pass through identical circular holes in a very thin, opaque film. When viewed far from the hole, the diffracted proton beam forms its first dark ring at 15° with respect to its original direction. When viewed similarly, at what angle will the alpha particle form its first dark ring?

39.73 •• **CP** An electron beam and a photon beam pass through identical slits. On a distant screen, the first dark fringe occurs at the same angle for both of the beams. The electron speeds are much slower than that of light. (a) Express the energy of a photon in terms of the kinetic energy K of one of the electrons. (b) Which is greater, the energy of a photon or the kinetic energy of an electron?

39.74 • **CP** Coherent light is passed through two narrow slits whose separation is 40.0 μm. The second-order bright fringe in the interference pattern is located at an angle of 0.0300 rad. If electrons are used instead of light, what must the kinetic energy (in electron volts) of the electrons be if they are to produce an interference pattern for which the second-order maximum is also at 0.0300 rad?

39.75 • **BIO** What is the de Broglie wavelength of a red blood cell, with mass 1.00×10^{-11} g, that is moving with a speed of 0.400 cm/s? Do we need to be concerned with the wave nature of the blood cells when we describe the flow of blood in the body?

39.76 • Calculate the energy in electron volts of (a) an electron that has de Broglie wavelength 400 nm and (b) a photon that has wavelength 400 nm.

39.77 • High-speed electrons are used to probe the interior structure of the atomic nucleus. For such electrons the expression $\lambda = h/p$ still holds, but we must use the relativistic expression for momentum, $p = mv/\sqrt{1 - v^2/c^2}$. (a) Show that the speed of an electron that has de Broglie wavelength λ is

$$v = \frac{c}{\sqrt{1 + (mc\lambda/h)^2}}$$

(b) The quantity h/mc equals 2.426×10^{-12} m. (As we saw in Section 38.3, this same quantity appears in Eq. (38.7), the expression for Compton scattering of photons by electrons.) If λ is small compared to h/mc, the denominator in the expression found in part (a) is close to unity and the speed v is very close to c. In this case it is convenient to write $v = (1 - \Delta)c$ and express the speed of the electron in terms of Δ rather than v. Find an expression for Δ valid when $\lambda \ll h/mc$. [Hint: Use the binomial expansion $(1 + z)^n = 1 + nz + n(n - 1)z^2/2 + \cdots$, valid for the case $|z| < 1$.] (c) How fast must an electron move for its de Broglie wavelength to be 1.00×10^{-15} m, comparable to the size of a proton? Express your answer in the form $v = (1 - \Delta)c$, and state the value of Δ.

39.78 • Suppose that the uncertainty of position of an electron is equal to the radius of the $n = 1$ Bohr orbit for hydrogen. Calculate the simultaneous minimum uncertainty of the corresponding momentum component, and compare this with the magnitude of the momentum of the electron in the $n = 1$ Bohr orbit. Discuss your results.

39.79 • **CP** (a) A particle with mass m has kinetic energy equal to three times its rest energy. What is the de Broglie wavelength of this particle? (Hint: You must use the relativistic expressions for momentum and kinetic energy: $E^2 = (pc)^2 + (mc^2)^2$ and $K = E - mc^2$.) (b) Determine the numerical value of the kinetic energy (in MeV) and the wavelength (in meters) if the particle in part (a) is (i) an electron and (ii) a proton.

39.80 • **Proton Energy in a Nucleus.** The radii of atomic nuclei are of the order of 5.0×10^{-15} m. (a) Estimate the minimum uncertainty in the momentum of a proton if it is confined within a nucleus. (b) Take this uncertainty in momentum to be an estimate of the magnitude of the momentum. Use the relativistic relationship between energy and momentum, Eq. (37.39), to obtain an estimate of the kinetic energy of a proton confined within a nucleus. (c) For a proton to remain bound within a nucleus, what must the magnitude of the (negative) potential energy for a proton be within the nucleus? Give your answer in eV and in MeV. Compare to the potential energy for an electron in a hydrogen atom, which has a magnitude of a few tens of eV. (This shows why the

interaction that binds the nucleus together is called the "strong nuclear force.")

39.81 • **Electron Energy in a Nucleus.** The radii of atomic nuclei are of the order of 5.0×10^{-15} m. (a) Estimate the minimum uncertainty in the momentum of an electron if it is confined within a nucleus. (b) Take this uncertainty in momentum to be an estimate of the magnitude of the momentum. Use the relativistic relationship between energy and momentum, Eq. (37.39), to obtain an estimate of the kinetic energy of an electron confined within a nucleus. (c) Compare the energy calculated in part (b) to the magnitude of the Coulomb potential energy of a proton and an electron separated by 5.0×10^{-15} m. On the basis of your result, could there be electrons within the nucleus? (Note: It is interesting to compare this result to that of Problem 39.80.)

39.82 • In a TV picture tube the accelerating voltage is 15.0 kV, and the electron beam passes through an aperture 0.50 mm in diameter to a screen 0.300 m away. (a) Calculate the uncertainty in the component of the electron's velocity perpendicular to the line between aperture and screen. (b) What is the uncertainty in position of the point where the electrons strike the screen? (c) Does this uncertainty affect the clarity of the picture significantly? (Use nonrelativistic expressions for the motion of the electrons. This is fairly accurate and is certainly adequate for obtaining an estimate of uncertainty effects.)

39.83 • The neutral pion (π^0) is an unstable particle produced in high-energy particle collisions. Its mass is about 264 times that of the electron, and it exists for an average lifetime of 8.4×10^{-17} s before decaying into two gamma-ray photons. Using the relationship $E = mc^2$ between rest mass and energy, find the uncertainty in the mass of the particle and express it as a fraction of the mass.

39.84 • **Quantum Effects in Daily Life?** A 1.25-mg insect flies through a 4.00-mm-diameter hole in an ordinary window screen. The thickness of the screen is 0.500 mm. (a) What should be the approximate wavelength and speed of the insect for her to show wave behavior as she goes through the hole? (b) At the speed found in part (a), how long would it take the insect to pass through the 0.500-mm thickness of the hole in the screen? Compare this time to the age of the universe (about 14 billion years). Would you expect to see "insect diffraction" in daily life?

39.85 • **Doorway Diffraction.** If your wavelength were 1.0 m, you would undergo considerable diffraction in moving through a doorway. (a) What must your speed be for you to have this wavelength? (Assume that your mass is 60.0 kg.) (b) At the speed calculated in part (a), how many years would it take you to move 0.80 m (one step)? Will you notice diffraction effects as you walk through doorways?

39.86 • **Atomic Spectra Uncertainties.** A certain atom has an energy level 2.58 eV above the ground level. Once excited to this level, the atom remains in this level for 1.64×10^{-7} s (on average) before emitting a photon and returning to the ground level. (a) What is the energy of the photon (in electron volts)? What is its wavelength (in nanometers)? (b) What is the smallest possible uncertainty in energy of the photon? Give your answer in electron volts. (c) Show that $|\Delta E/E| = |\Delta\lambda/\lambda|$ if $|\Delta\lambda/\lambda| \ll 1$. Use this to calculate the magnitude of the smallest possible uncertainty in the wavelength of the photon. Give your answer in nanometers.

39.87 •• You intend to use an electron microscope to study the structure of some crystals. For accurate resolution, you want the electron wavelength to be 1.00 nm. (a) Are these electrons relativistic? How do you know? (b) What accelerating potential is needed? (c) What is the kinetic energy of the electrons you are using? To see if it is great enough to damage the crystals you are studying, compare it to the potential energy of a typical NaCl molecule, which is about 6.0 eV. (d) If you decided to use electromagnetic waves as your probe, what energy should their photons have to provide the same resolution as the electrons? Would this energy damage the crystal?

39.88 •• For x rays with wavelength 0.0300 nm, the $m = 1$ intensity maximum for a crystal occurs when the angle θ in Fig. 39.2 is 35.8°. At what angle θ does the $m = 1$ maximum occur when a beam of 4.50-keV electrons is used instead? Assume that the electrons also scatter from the atoms in the surface plane of this same crystal.

39.89 •• **CP** Electron diffraction can also take place when there is interference between electron waves that scatter from atoms on the surface of a crystal and waves that scatter from atoms in the next plane below the surface, a distance d from the surface (see Fig. 36.23c). (a) Find an equation for the angles θ at which there is an intensity maximum for electron waves of wavelength λ. (b) The spacing between crystal planes in a certain metal is 0.091 nm. If 71.0-eV electrons are used, find the angle at which there is an intensity maximum due to interference between scattered waves from adjacent crystal planes. The angle is measured as shown in Fig. 36.23c. (c) The actual angle of the intensity maximum is slightly different from your result in part (b). The reason is the work function ϕ of the metal (see Section 38.1), which changes the electron potential energy by $-e\phi$ when it moves from vacuum into the metal. If the effect of the work function is taken into account, is the angle of the intensity maximum larger or smaller than the value found in part (b)? Explain.

39.90 •• A certain atom has an energy level 3.50 eV above the ground state. When excited to this state, it remains 4.0 μs, on the average, before emitting a photon and returning to the ground state. (a) What is the energy of the photon? What is its wavelength? (b) What is the smallest possible uncertainty in energy of the photon?

39.91 •• **BIO** **Structure of a Virus.** To investigate the structure of extremely small objects, such as viruses, the wavelength of the probing wave should be about one-tenth the size of the object for sharp images. But as the wavelength gets shorter, the energy of a photon of light gets greater and could damage or destroy the object being studied. One alternative is to use electron matter waves instead of light. Viruses vary considerably in size, but 50 nm is not unusual. Suppose you want to study such a virus, using a wave of wavelength 5.00 nm. (a) If you use light of this wavelength, what would be the energy (in eV) of a single photon? (b) If you use an electron of this wavelength, what would be its kinetic energy (in eV)? Is it now clear why matter waves (such as in the electron microscope) are often preferable to electromagnetic waves for studying microscopic objects?

39.92 •• **CALC** **Zero-Point Energy.** Consider a particle with mass m moving in a potential $U = \frac{1}{2}kx^2$, as in a mass–spring system. The total energy of the particle is $E = p^2/2m + \frac{1}{2}kx^2$. Assume that p and x are approximately related by the Heisenberg uncertainty principle, so $px \approx h$. (a) Calculate the minimum possible value of the energy E, and the value of x that gives this minimum E. This lowest possible energy, which is not zero, is called the

zero-point energy. (b) For the x calculated in part (a), what is the ratio of the kinetic to the potential energy of the particle?

39.93 •• **CALC** A particle with mass m moves in a potential $U(x) = A|x|$, where A is a positive constant. In a simplified picture, quarks (the constituents of protons, neutrons, and other particles, as will be described in Chapter 44) have a potential energy of interaction of approximately this form, where x represents the separation between a pair of quarks. Because $U(x) \rightarrow \infty$ as $x \rightarrow \infty$, it's not possible to separate quarks from each other (a phenomenon called *quark confinement*). (a) Classically, what is the force acting on this particle as a function of x? (b) Using the uncertainty principle as in Problem 39.92, determine approximately the zero-point energy of the particle.

39.94 •• Imagine another universe in which the value of Planck's constant is 0.0663 J · s, but in which the physical laws and all other physical constants are the same as in our universe. In this universe, two physics students are playing catch. They are 12 m apart, and one throws a 0.25-kg ball directly toward the other with a speed of 6.0 m/s. (a) What is the uncertainty in the ball's horizontal momentum, in a direction perpendicular to that in which it is being thrown, if the student throwing the ball knows that it is located within a cube with volume 125 cm³ at the time she throws it? (b) By what horizontal distance could the ball miss the second student?

CHALLENGE PROBLEMS

39.95 ••• (a) Show that in the Bohr model, the frequency of revolution of an electron in its circular orbit around a stationary hydrogen nucleus is $f = me^4/4\epsilon_0^2n^3h^3$. (b) In classical physics, the frequency of revolution of the electron is equal to the frequency of the radiation that it emits. Show that when n is very large, the frequency of revolution does indeed equal the radiated frequency calculated from Eq. (39.5) for a transition from $n_1 = n + 1$ to $n_2 = n$. (This illustrates Bohr's *correspondence principle,* which is often used as a check on quantum calculations. When n is small, quantum physics gives results that are very different from those of classical physics. When n is large, the differences are not significant, and the two methods then "correspond." In fact, when Bohr first tackled the hydrogen atom problem, he sought to determine f as a function of n such that it would correspond to classical results for large n.)

39.96 ••• **CP CALC** You have entered a contest in which the contestants drop a marble with mass 20.0 g from the roof of a building onto a small target 25.0 m below. From uncertainty considerations, what is the typical distance by which you will miss the target, given that you aim with the highest possible precision? (*Hint:* The uncertainty Δx_f in the x-coordinate of the marble when it reaches the ground comes in part from the uncertainty Δx_i in the x-coordinate initially and in part from the initial uncertainty in v_x. The latter gives rise to an uncertainty Δv_x in the horizontal motion of the marble as it falls. The values of Δx_i and Δv_x are related by the uncertainty principle. A small Δx_i gives rise to a large Δv_x, and vice versa. Find the value of Δx_i that gives the smallest total uncertainty in x at the ground. Ignore any effects of air resistance.)

Answers

Chapter Opening Question

The smallest detail visible in an image is comparable to the wavelength used to make the image. Electrons can easily be given a large momentum p and hence a short wavelength $\lambda = h/p$, and so can be used to resolve extremely fine details. (See Section 39.1.)

Test Your Understanding Questions

39.1 Answers: (a) (i), (b) no From Example 39.2, the speed of a particle is $v = h/\lambda m$ and the kinetic energy is $K = \frac{1}{2}mv^2 = (m/2)(h/\lambda m)^2 = h^2/2\lambda^2 m$. This shows that for a given wavelength, the kinetic energy is inversely proportional to the mass. Hence the proton, with a smaller mass, has more kinetic energy than the neutron. For part (b), the energy of a photon is $E = hf$, and the frequency of a photon is $f = c/\lambda$. Hence $E = hc/\lambda$ and $\lambda = hc/E = (4.136 \times 10^{-15} \text{ eV} \cdot \text{s})(2.998 \times 10^8 \text{ m/s})/(54 \text{ eV}) = 2.3 \times 10^{-8}$ m. This is more than 100 times greater than the wavelength of an electron of the same energy. While both photons and electrons have wavelike properties, they have different relationships between their energy and momentum and hence between their frequency and wavelength.

39.2 Answer: (iii) Because the alpha particle is more massive, it won't bounce back in even a head-on collision with a proton that's initially at rest, any more than a bowling ball would when colliding with a Ping-Pong ball at rest (see Fig. 8.22b). Thus there would be *no* large-angle scattering in this case. Rutherford saw large-angle scattering in his experiment because gold nuclei are more massive than alpha particles (see Fig. 8.22a).

39.3 Answer: (iv) Figure 39.27 shows that many (though *not* all) of the energy levels of He$^+$ are the same as those of H. Hence photons emitted during transitions between corresponding pairs of levels in He$^+$ and H have the same energy E and the same wavelength $\lambda = hc/E$. An H atom that drops from the $n = 2$ level to the $n = 1$ level emits a photon of energy 10.20 eV and wavelength 122 nm (see Example 39.6); a He$^+$ ion emits a photon of the same energy and wavelength when it drops from the $n = 4$ level to the $n = 2$ level. Inspecting Fig. 39.27 will show you that every even-numbered level in He$^+$ matches a level in H, while none of the odd-numbered He$^+$ levels do. The first three He$^+$ transitions given in the question ($n = 2$ to $n = 1$, $n = 3$ to $n = 2$, and $n = 4$ to $n = 3$) all involve an odd-numbered level, so none of their wavelengths match a wavelength emitted by H atoms.

39.4 Answer: (i) In a neon light fixture, a large potential difference is applied between the ends of a neon-filled glass tube. This ionizes some of the neon atoms, allowing a current of electrons to flow through the gas. Some of the neon atoms are struck by fast-moving electrons, making them transition to an excited level. From this level the atoms undergo *spontaneous* emission, as depicted in Fig. 39.28b, and emit 632.8-nm photons in the process. No population inversion occurs and the photons are not trapped by mirrors as shown in Fig. 39.29d, so there is no stimulated emission. Hence there is no laser action.

39.5 Answers: (a) yes, (b) yes The Planck radiation law, Eq. (39.24), shows that an ideal blackbody emits radiation at *all* wavelengths: The spectral emittance $I(\lambda)$ is equal to zero only for $\lambda = 0$ and in the limit $\lambda \to \infty$. So a blackbody at 2000 K does indeed emit both x rays and radio waves. However, Fig. 39.32 shows that the spectral emittance for this temperature is very low for wavelengths much shorter than 1 μm (including x rays) and for wavelengths much longer than a few μm (including radio waves). Hence such a blackbody emits very little in the way of x rays or radio waves.

39.6 Answer: (i) and (iii) (tie), (ii) and (iv) (tie) According to the Heisenberg uncertainty principle, the smaller the uncertainty Δx in the x-coordinate, the greater the uncertainty Δp_x in the x-momentum. The relationship between Δx and Δp_x does not depend on the mass of the particle, and so is the same for a proton as for an electron.

Bridging Problem

Answers: (a) 192 nm; ultraviolet **(b)** $n = 4$

(c) $\lambda_2 = 0.665$ nm, $\lambda_3 = 0.997$ nm

40 QUANTUM MECHANICS

LEARNING GOALS

By studying this chapter, you will learn:

- About the wave function that describes the behavior of a particle and the Schrödinger equation that this function must satisfy.

- How to calculate the wave functions and energy levels for a particle confined to a box.

- How to analyze the quantum-mechanical behavior of a particle in a potential well.

- How quantum mechanics makes it possible for particles to go where Newtonian mechanics says they cannot.

- How to use quantum mechanics to analyze a harmonic oscillator.

? These containers hold solutions of microscopic semiconductor particles of different sizes. The particles glow when exposed to ultraviolet light; the smallest particles glow blue and the largest particles glow red. Why?

In Chapter 39 we found that particles can behave like waves. In fact, it turns out that we can use the wave picture to completely describe the behavior of a particle. This approach, called *quantum mechanics,* is the key to understanding the behavior of matter on the molecular, atomic, and nuclear scales. In this chapter we'll see how to find the *wave function* of a particle by solving the *Schrödinger equation,* which is as fundamental to quantum mechanics as Newton's laws are to mechanics or as Maxwell's equations are to electromagnetism.

We'll begin with a quantum-mechanical analysis of a *free particle* that moves along a straight line without being acted on by forces of any kind. We'll then consider particles that are acted on by forces and are trapped in *bound states,* just as electrons are bound within an atom. We'll see that solving the Schrödinger equation automatically gives the possible energy levels for the system.

Besides energies, solving the Schrödinger equation gives us the probabilities of finding a particle in various regions. One surprising result is that there is a nonzero probability that microscopic particles will pass through thin barriers, even though such a process is forbidden by Newtonian mechanics.

In this chapter we'll consider the Schrödinger equation for one-dimensional motion only. In Chapter 41 we'll see how to extend this equation to three-dimensional problems such as the hydrogen atom. The hydrogen-atom wave functions will in turn form the foundation for our analysis of more complex atoms, of the periodic table of the elements, of x-ray energy levels, and of other properties of atoms.

PhET: Fourier: Making Waves

40.1 Wave Functions and the One-Dimensional Schrödinger Equation

We have now seen compelling evidence that on an atomic or subatomic scale, an object such as an electron cannot be described simply as a classical, Newtonian point particle. Instead, we must take into account its *wave* characteristics. In the

Bohr model of the hydrogen atom (Section 39.3) we tried to have it both ways: We pictured the electron as a classical particle in a circular orbit around the nucleus, and used the de Broglie relation between particle momentum and wavelength to explain why only orbits of certain radii are allowed. As we saw in Section 39.6, however, the Heisenberg uncertainty principle tells us that a hybrid description of this kind can't be wholly correct. In this section we'll explore how to describe the state of a particle by using *only* the language of waves. This new description, called **quantum mechanics,** will replace the classical scheme of describing the state of a particle by its coordinates and velocity components.

Our new quantum-mechanical scheme for describing a particle has a lot in common with the language of classical wave motion. In Section 15.3 of Chapter 15, we described transverse waves on a string by specifying the position of each point in the string at each instant of time by means of a *wave function $y(x, t)$* that represents the displacement from equilibrium, at time t, of a point on the string at a distance x from the origin (Fig. 40.1). Once we know the wave function for a particular wave motion, we know everything there is to know about the motion. For example, we can find the velocity and acceleration of any point on the string at any time. We worked out specific forms for these functions for *sinusoidal* waves, in which each particle undergoes simple harmonic motion.

We followed a similar pattern for sound waves in Chapter 16. The wave function $p(x, t)$ for a wave traveling along the x-direction represented the pressure variation at any point x and any time t. We used this language once more in Section 32.3, where we used *two* wave functions to describe the electric and magnetic fields in an electromagnetic wave.

Thus it's natural to use a wave function as the central element of our new language of quantum mechanics. The customary symbol for this wave function is the Greek letter psi, Ψ or ψ. In general, we'll use an uppercase Ψ to denote a function of all the space coordinates and time, and a lowercase ψ for a function of the space coordinates only—*not* of time. Just as the wave function $y(x, t)$ for mechanical waves on a string provides a complete description of the motion, so the wave function $\Psi(x, y, z, t)$ for a particle contains all the information that can be known about the particle.

40.1 These children are talking over a cup-and-string "telephone." The displacement of the string is completely described by a wave function $y(x, t)$. In an analogous way, a particle is completely described by a quantum-mechanical wave function $\Psi(x, y, z, t)$.

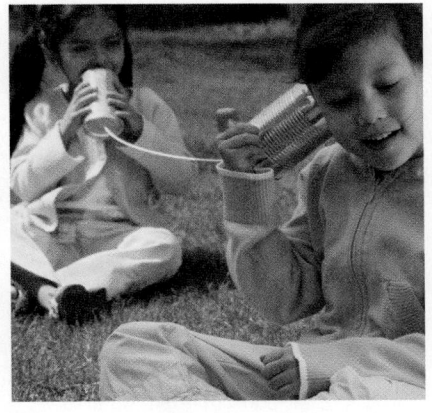

CAUTION **Particle waves vs. mechanical waves** Unlike for mechanical waves on a string or sound waves in air, the wave function for a particle is *not* a mechanical wave that needs some material medium in order to propagate. The wave function describes the particle, but we cannot define the function itself in terms of anything material. We can only describe how it is related to physically observable effects. ▮

Waves in One Dimension

The wave function of a particle depends in general on all three dimensions of space. For simplicity, however, we'll begin our study of these functions by considering *one-dimensional* motion, in which a particle of mass m moves parallel to the x-axis and the wave function Ψ depends on the coordinate x and the time t only. (In the same way, we studied one-dimensional kinematics in Chapter 2 before going on to study two- and three-dimensional motion in Chapter 3.)

What does a one-dimensional quantum-mechanical wave look like, and what determines its properties? We can answer this question by first recalling the properties of a wave on a string. We saw in Section 15.3 that any wave function $y(x, t)$ that describes a wave on a string must satisfy the *wave equation:*

$$\frac{\partial^2 y(x, t)}{\partial x^2} = \frac{1}{v^2}\frac{\partial^2 y(x, t)}{\partial t^2} \qquad \text{(wave equation for waves on a string)} \qquad (40.1)$$

In Eq. (40.1) v is the speed of the wave, which is the same no matter what the wavelength. As an example, consider the following wave function for a wave of wavelength λ and frequency f moving in the positive x-direction along a string:

$$y(x, t) = A\cos(kx - \omega t) + B\sin(kx - \omega t) \qquad \text{(sinusoidal wave on a string)} \qquad (40.2)$$

Here $k = 2\pi/\lambda$ is the *wave number* and $\omega = 2\pi f$ is the *angular frequency*. (We used these same quantities for mechanical waves in Chapter 15 and electromagnetic waves in Chapter 32.) The quantities A and B are constants that determine the amplitude and phase of the wave. The expression in Eq. (40.2) is a valid wave function if and only if it satisfies the wave equation, Eq. (40.1). To check this, take the first and second derivatives of $y(x, t)$ with respect to x and take the first and second derivatives with respect to t:

$$\frac{\partial y(x, t)}{\partial x} = -kA\sin(kx - \omega t) + kB\cos(kx - \omega t) \qquad (40.3\text{a})$$

$$\frac{\partial^2 y(x, t)}{\partial x^2} = -k^2 A\cos(kx - \omega t) - k^2 B\sin(kx - \omega t) \qquad (40.3\text{b})$$

$$\frac{\partial y(x, t)}{\partial t} = \omega A\sin(kx - \omega t) - \omega B\cos(kx - \omega t) \qquad (40.3\text{c})$$

$$\frac{\partial^2 y(x, t)}{\partial t^2} = -\omega^2 A\cos(kx - \omega t) - \omega^2 B\sin(kx - \omega t) \qquad (40.3\text{d})$$

If we substitute Eqs. (40.3b) and (40.3d) into the wave equation, Eq. (40.1), we get

$$-k^2 A\cos(kx - \omega t) - k^2 B\sin(kx - \omega t)$$

$$= \frac{1}{v^2}\left[-\omega^2 A\cos(kx - \omega t) - \omega^2 B\sin(kx - \omega t)\right] \qquad (40.4)$$

For Eq. 40.4 to be satisfied at all coordinates x and all times t, the coefficients of the cosine function must be the same on both sides of the equation, and likewise for the coefficients of the sine function. You can see that both of these conditions will be satisfied if

$$k^2 = \frac{\omega^2}{v^2} \qquad \text{or} \qquad \omega = vk \qquad \text{(waves on a string)} \qquad (40.5)$$

From the definitions of angular frequency ω and wave number k, Eq. (40.5) is equivalent to

$$2\pi f = v\frac{2\pi}{\lambda} \qquad \text{or} \qquad v = \lambda f \qquad \text{(waves on a string)}$$

This equation is just the familiar relationship among wave speed, wavelength, and frequency for waves on a string. So our calculation shows that Eq. (40.2) is a valid wave function for waves on a string for any values of A and B, provided that ω and k are related by Eq. (40.5).

What we need is a quantum-mechanical version of the wave equation, Eq. (40.1), valid for particle waves. We expect this equation to involve partial derivatives of the wave function $\Psi(x, t)$ with respect to x and with respect to t. However, this new equation *cannot* be the same as Eq. (40.1) for waves on a string because the relationship between ω and k is different. We can show this by considering a **free particle,** one that experiences no force at all as it moves along the x-axis. For such a particle the potential energy $U(x)$ has the same value for all x (recall from Chapter 7 that $F_x = -dU(x)/dx$, so zero force means the potential energy has zero derivative). For simplicity let $U = 0$ for all x. Then the energy of the free particle is equal to its kinetic energy, which we can express in terms of its momentum p:

$$E = \tfrac{1}{2}mv^2 = \frac{m^2 v^2}{2m} = \frac{(mv)^2}{2m} = \frac{p^2}{2m} \qquad \text{(energy of a free particle)} \qquad (40.6)$$

The de Broglie relations that we introduced in Section 39.1 tell us that the energy E is proportional to the angular frequency ω and the momentum p is proportional to the wave number:

$$E = hf = \frac{h}{2\pi}2\pi f = \hbar\omega \qquad (40.7a)$$

$$p = \frac{h}{\lambda} = \frac{h}{2\pi}\frac{2\pi}{\lambda} = \hbar k \qquad (40.7b)$$

Remember that $\hbar = h/2\pi$. If we substitute Eqs. (40.7) into Eq. (40.6), we find that the relationship between ω and k for a free particle is

$$\hbar\omega = \frac{\hbar^2 k^2}{2m} \qquad \text{(free particle)} \qquad (40.8)$$

Equation (40.8) is *very* different from the corresponding relationship for waves on a string, Eq. (40.5): The angular frequency ω for particle waves is proportional to the *square* of the wave number, while for waves on a string ω is directly proportional to k. Our task is therefore to construct a quantum-mechanical version of the wave equation whose free-particle solutions satisfy Eq. (40.8).

We'll attack this problem by assuming a sinusoidal wave function $\Psi(x, t)$ of the same form as Eq. (40.2) for a sinusoidal wave on a string. For a wave on a string, Eq. (40.2) represents a wave of wavelength $\lambda = 2\pi/k$ and frequency $f = \omega/2\pi$ propagating in the positive x-direction. By analogy, our sinsuoidal wave function $\Psi(x, t)$ represents a free particle of mass m, momentum $p = \hbar k$, and energy $E = \hbar\omega$ moving in the positive x-direction:

$$\Psi(x, t) = A\cos(kx - \omega t) + B\sin(kx - \omega t) \quad \begin{array}{l}\text{(sinusoidal wave} \\ \text{function representing} \\ \text{a free particle)}\end{array} \quad (40.9)$$

The wave number k and angular frequency ω in Eq. (40.9) must satisfy Eq. (40.8). If you look at Eq. (40.3b), you'll see that taking the second derivative of $\Psi(x, t)$ in Eq. (40.9) with respect to x gives us $\Psi(x, t)$ multiplied by $-k^2$. Hence if we multiply $\partial^2\Psi(x, t)/\partial x^2$ by $-\hbar^2/2m$, we get

$$-\frac{\hbar^2}{2m}\frac{\partial^2\Psi(x, t)}{\partial x^2} = -\frac{\hbar^2}{2m}\left[-k^2 A\cos(kx - \omega t) - k^2 B\sin(kx - \omega t)\right]$$

$$= \frac{\hbar^2 k^2}{2m}\left[A\cos(kx - \omega t) + B\sin(kx - \omega t)\right] \qquad (40.10)$$

$$= \frac{\hbar^2 k^2}{2m}\Psi(x, t)$$

Equation (40.10) suggests that $(-\hbar^2/2m)\partial^2\Psi(x, t)/\partial x^2$ should be one side of our quantum-mechanical wave equation, with the other side equal to $\hbar\omega\Psi(x, t)$ in order to satisfy Eq. (40.8). If you look at Eq. (40.3c), you'll see that taking the *first* time derivative of $\Psi(x, t)$ in Eq. (40.9) brings out a factor of ω. So we'll make the educated guess that the right-hand side of our quantum-mechanical wave equation involves $\hbar = h/2\pi$ times $\partial\Psi(x, t)/\partial t$. So our tentative equation is

$$-\frac{\hbar^2}{2m}\frac{\partial^2\Psi(x, t)}{\partial x^2} = C\hbar\frac{\partial\Psi(x, t)}{\partial t} \qquad (40.11)$$

At this point we include a constant C as a "fudge factor" to make sure that everything turns out right. Now let's substitute the wave function from Eq. (40.9) into Eq. (40.11). From Eq. (40.10) and Eq. (40.3c), we get

$$\frac{\hbar^2 k^2}{2m}[A\cos(kx - \omega t) + B\sin(kx - \omega t)]$$

$$= C\hbar\omega[A\sin(kx - \omega t) - B\cos(kx - \omega t)] \qquad (40.12)$$

From Eq. (40.8), $\hbar\omega = \hbar^2 k^2/2m$, so we can cancel these factors on the two sides of Eq. (40.12). What remains is

$$A\cos(kx - \omega t) + B\sin(kx - \omega t) = CA\sin(kx - \omega t)$$

$$- CB\cos(kx - \omega t) \qquad (40.13)$$

As in our discussion above of the wave equation for waves on a string, in order for Eq. (40.13) to be satisfied for all values of x and all values of t, the coefficients of the cosine function must be the same on both sides of the equation, and likewise for the coefficients of the sine function. Hence we have the following relationships among the coefficients A and B in Eq. (40.9) and the coefficient C in Eq. (40.11):

$$A = -CB \qquad (40.14a)$$

$$B = CA \qquad (40.14b)$$

If we use Eq. (40.14b) to eliminate B from Eq. (40.14a), we get $A = -C^2 A$, which means that $C^2 = -1$. Thus C is equal to the *imaginary* number $i = \sqrt{-1}$, and Eq. (40.11) becomes

$$-\frac{\hbar^2}{2m}\frac{\partial^2 \Psi(x, t)}{\partial x^2} = i\hbar\frac{\partial \Psi(x, t)}{\partial t} \qquad \text{(one-dimensional Schrödinger equation for a free particle)} \qquad (40.15)$$

Equation (40.15) is the one-dimensional **Schrödinger equation** for a free particle, developed in 1926 by the Austrian physicist Erwin Schrödinger (Fig. 40.2). The presence of the imaginary number i in Eq. (40.15) means that the solutions to the Schrödinger equation are complex quantities, with a real part and an imaginary part. (The imaginary part of $\Psi(x, t)$ is a real function multiplied by the imaginary number $i = \sqrt{-1}$.) An example is our free-particle wave function from Eq. (40.9). Since we found $C = i$ in Eqs. (40.14), it follows from Eq. (40.14b) that $B = iA$. Then Eq. (40.9) becomes

$$\Psi(x, t) = A[\cos(kx - \omega t) + i\sin(kx - \omega t)] \qquad \text{(sinusoidal wave function representing a free particle)} \qquad (40.16)$$

The real part of $\Psi(x, t)$ is $\mathrm{Re}\,\Psi(x, t) = A\cos(kx - \omega t)$ and the imaginary part is $\mathrm{Im}\,\Psi(x, t) = A\sin(kx - \omega t)$. Figure 40.3 graphs the real and imaginary parts of $\Psi(x, t)$ at $t = 0$, so $\Psi(x, 0) = A\cos kx + iA\sin kx$.

We can rewrite Eq. (40.16) using *Euler's formula*, which states that for any angle θ,

$$e^{i\theta} = \cos\theta + i\sin\theta$$

$$e^{-i\theta} = \cos(-\theta) + i\sin(-\theta) = \cos\theta - i\sin\theta \qquad (40.17)$$

Thus our sinusoidal free-particle wave function becomes

$$\Psi(x, t) = Ae^{i(kx - \omega t)} = Ae^{ikx}e^{-i\omega t} \qquad \text{(sinusoidal wave function representing a free particle)} \qquad (40.18)$$

40.2 Erwin Schrödinger (1887–1961) developed the equation that bears his name in 1926, an accomplishment for which he shared (with the British physicist P. A. M. Dirac) the 1933 Nobel Prize in physics. His grave marker is adorned with a large letter ψ.

If k is positive in Eq. (40.16), the wave function represents a free particle moving in the positive x-direction with momentum $p = \hbar k$ and energy $E = \hbar\omega = \hbar^2 k^2/2m$. If k is negative, the momentum and hence the motion are in the negative x-direction. (With a negative value of k, the wavelength is $\lambda = 2\pi/|k|$).

Interpreting the Wave Function

The complex nature of the wave function for a free particle makes this function challenging to interpret. (We certainly haven't needed imaginary numbers before this point to describe real physical phenomena.) Here's how to think about this function: $\Psi(x, t)$ describes the *distribution* of a particle in space, just as the wave functions for an electromagnetic wave describe the distribution of the electric and magnetic fields. When we worked out interference and diffraction patterns in Chapters 35 and 36, we found that the intensity I of the radiation at any point in a pattern is proportional to the square of the electric-field magnitude—that is, to E^2. In the photon interpretation of interference and diffraction (see Section 38.4), the intensity at each point is proportional to the number of photons striking around that point or, alternatively, to the *probability* that any individual photon will strike around the point. Thus the square of the electric-field magnitude at each point is proportional to the probability of finding a photon around that point.

In exactly the same way, the square of the wave function of a particle at each point tells us about the probability of finding the particle around that point. More precisely, we should say the square of the *absolute value* of the wave function, $|\Psi|^2$. This is necessary because, as we have seen, the wave function is a complex quantity with real and imaginary parts.

For a particle that can move only along the x-direction, the quantity $|\Psi(x, t)|^2 dx$ is the probability that the particle will be found at time t at a coordinate in the range from x to $x + dx$. The particle is most likely to be found in regions where $|\Psi|^2$ is large, and so on. This interpretation, first made by the German physicist Max Born (Fig. 40.4), requires that the wave function Ψ be *normalized*. That is, the integral of $|\Psi(x, t)|^2 dx$ over all possible values of x must equal exactly 1. In other words, the probability is exactly 1, or 100%, that the particle is *somewhere*.

> **CAUTION** **Interpreting** $|\Psi|^2$ Note that $|\Psi(x, t)|^2$ itself is *not* a probability. Rather, $|\Psi(x, t)|^2 dx$ is the probability of finding the particle between position x and position $x + dx$ at time t. If the length dx is made smaller, it becomes less likely that the particle will be found within that length, so the probability decreases. A better name for $|\Psi(x, t)|^2$ is the **probability distribution function,** since it describes how the probability of finding the particle at different locations is distributed over space. Another common name for $|\Psi(x, t)|^2$ is the *probability density.*

We can use the probability interpretation of $|\Psi|^2$ to get a better understanding of Eq. (40.18), the wave function for a free particle. This function decribes a particle that has a definite momentum $p = \hbar k$ in the x-direction and *no* uncertainty in momentum: $\Delta p_x = 0$. The Heisenberg uncertainty principle for position and momentum, Eq. (39.29), says that $\Delta x \Delta p_x \geq \hbar/2$. If Δp_x is zero, then Δx must be infinite, and we have no idea whatsoever where along the x-axis the particle can be found. (We saw a similar result for photons in Section 38.4.) We can show this by calculating the probability distribution function $|\Psi(x, t)|^2$. This is the product of Ψ and its *complex conjugate* Ψ^*. To find the complex conjugate of a complex number, we simply replace all i with $-i$. For example, the complex conjugate of $c = a + ib$, where a and b are real, is $c^* = a - ib$, so $|c|^2 = c^* c = (a + ib)(a - ib) = a^2 + b^2$ (recall that $i^2 = -1$). The complex conjugate of Eq. (40.18) is

$$\Psi^*(x, t) = A^* e^{-i(kx - \omega t)} = A^* e^{-ikx} e^{i\omega t}$$

(We have to allow for the possibility that the coefficient A is itself a complex number.) Hence the probability distribution function is

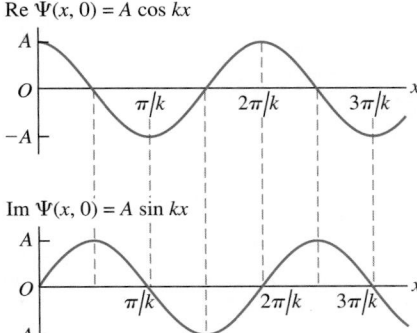

40.3 The spatial wave function $\Psi(x, t) = A e^{i(kx - \omega t)}$ for a free particle of definite momentum $p = \hbar k$ is a complex function: It has both a real part and an imaginary part. These are graphed here as functions of x for $t = 0$.

$\text{Re}\,\Psi(x, 0) = A \cos kx$

$\text{Im}\,\Psi(x, 0) = A \sin kx$

40.4 In 1926, the German physicist Max Born (1882–1970) devised the interpretation that $|\Psi|^2$ is the probability distribution function for a particle that is described by the wave function Ψ. He also coined the term "quantum mechanics" (in the original German, *Quantenmechanik*). For his contributions, Born shared (with Walther Bothe) the 1954 Nobel Prize in physics.

$$|\Psi(x, t)|^2 = \Psi^*(x, t)\Psi(x, t) = (A^*e^{-ikx}e^{i\omega t})(Ae^{ikx}e^{-i\omega t})$$

$$= A^*Ae^0 = |A|^2$$

The probability distribution function doesn't depend on position, which says that we are equally likely to find the particle *anywhere* along the x-axis! Mathematically, this is because the wave function $\Psi(x, t) = Ae^{i(kx - \omega t)} = A[\cos(kx - \omega t) + i\sin(kx - \omega t)]$ is a sinusoidal function that extends all the way from $x = -\infty$ to $x = +\infty$ with the same amplitude A. This also means that the wave function can't be normalized: The integral of $|\Psi(x, t)|^2$ over all space would be infinite for any value of A.

Note also that the wave function in Eq. (40.18) describes a particle with a definite energy $E = \hbar\omega$, so there is zero uncertainty in energy: $\Delta E = 0$. The Heisenberg uncertainty principle for energy and time interval, $\Delta t\,\Delta E \geq \hbar$ [Eq. (39.30)], tells us that the time uncertainty Δt for this particle is infinite. In other words, we can have no idea *when* the particle will pass a given point on the x-axis. That also agrees with our result $|\Psi(x, t)|^2 = |A|^2$; the probability distribution function has the same value at all times.

Since we always have some idea of where a particle is, the wave function given in Eq. (40.18) isn't a realistic description. In our study of light in Section 38.4, we saw that we can make a wave function that's more *localized* in space by superimposing two or more sinusoidal functions. (This would be a good time to review that section.) As an illustration, let's calculate $|\Psi(x, t)|^2$ for a wave function of this kind.

Example 40.1 **A localized free-particle wave function**

The wave function $\Psi(x, t) = Ae^{i(k_1x - \omega_1 t)} + Ae^{i(k_2x - \omega_2 t)}$ is a superposition of *two* free-particle wave functions of the form given by Eq. (40.18). Both k_1 and k_2 are positive. (a) Show that this wave function satisfies the Schrödinger equation for a free particle of mass m. (b) Find the probability distribution function for $\Psi(x, t)$.

SOLUTION

IDENTIFY and SET UP: The wave functions $Ae^{i(k_1x - \omega_1 t)}$ and $Ae^{i(k_2x - \omega_2 t)}$ both represent particles moving in the positive x-direction, but with different momenta and kinetic energies: $p_1 = \hbar k_1$ and $E_1 = \hbar\omega_1 = \hbar^2 k_1^2/2m$ for the first function, $p_2 = \hbar k_2$ and $E_2 = \hbar\omega_2 = \hbar^2 k_2^2/2m$ for the second function. To test whether a superposition of these is also a valid wave function for a free particle, we'll see whether our function $\Psi(x, t)$ satisfies the free-particle Schrödinger equation, Eq. (40.15). It's useful to remember the derivatives of the exponential function: $(d/du)e^{au} = ae^{au}$ and $(d^2/du^2)e^{au} = a^2e^{au}$. The probability distribution function $|\Psi(x, t)|^2$ is the product of $\Psi(x, t)$ and its complex conjugate.

EXECUTE: (a) If we substitute $\Psi(x, t)$ into Eq. (40.15), the left-hand side of the equation is

$$-\frac{\hbar^2}{2m}\frac{\partial^2\Psi(x, t)}{\partial x^2} = -\frac{\hbar^2}{2m}\frac{\partial^2(Ae^{i(k_1x - \omega_1 t)} + Ae^{i(k_2x - \omega_2 t)})}{\partial x^2}$$

$$= -\frac{\hbar^2}{2m}[(ik_1)^2 Ae^{i(k_1x - \omega_1 t)} + (ik_2)^2 Ae^{i(k_2x - \omega_2 t)}]$$

$$= \frac{\hbar^2 k_1^2}{2m}Ae^{i(k_1x - \omega_1 t)} + \frac{\hbar^2 k_2^2}{2m}Ae^{i(k_2x - \omega_2 t)}$$

The right-hand side is

$$i\hbar\frac{\partial\Psi(x, t)}{\partial t} = i\hbar\frac{\partial(Ae^{i(k_1x - \omega_1 t)} + Ae^{i(k_2x - \omega_2 t)})}{\partial t}$$

$$= i\hbar[(-i\omega_1)Ae^{i(k_1x - \omega_1 t)} + (-i\omega_2)Ae^{i(k_2x - \omega_2 t)}]$$

$$= \hbar\omega_1 Ae^{i(k_1x - \omega_1 t)} + \hbar\omega_2 Ae^{i(k_2x - \omega_2 t)}$$

The two sides *are* equal, provided that $\hbar\omega_1 = \hbar^2 k_1^2/2m$ and $\hbar\omega_2 = \hbar^2 k_2^2/2m$. These are just the relationships that we noted above. So we conclude that $\Psi(x, t) = Ae^{i(k_1x - \omega_1 t)} + Ae^{i(k_2x - \omega_2 t)}$ is a valid free-particle wave function. In general, if we take any two wave functions that are solutions of the Schrödinger equation and then make a superposition of these to create a third wave function $\Psi(x, t)$, then $\Psi(x, t)$ is also a solution of the Schrödinger equation.

(b) The complex conjugate of $\Psi(x, t)$ is

$$\Psi^*(x, t) = A^*e^{-i(k_1x - \omega_1 t)} + A^*e^{-i(k_2x - \omega_2 t)}$$

Hence

$$|\Psi(x, t)|^2$$

$$= \Psi^*(x, t)\Psi(x, t)$$

$$= (A^*e^{-i(k_1x - \omega_1 t)} + A^*e^{-i(k_2x - \omega_2 t)})(Ae^{i(k_1x - \omega_1 t)} + Ae^{i(k_2x - \omega_2 t)})$$

$$= A^*A\left[\begin{array}{c} e^{-i(k_1x - \omega_1 t)}e^{i(k_1x - \omega_1 t)} + e^{-i(k_2x - \omega_2 t)}e^{i(k_2x - \omega_2 t)} \\ + e^{-i(k_1x - \omega_1 t)}e^{i(k_2x - \omega_2 t)} + e^{-i(k_2x - \omega_2 t)}e^{i(k_1x - \omega_1 t)} \end{array}\right]$$

$$= |A|^2[e^0 + e^0 + e^{i[(k_2 - k_1)x - (\omega_2 - \omega_1)t]} + e^{-i[(k_2 - k_1)x - (\omega_2 - \omega_1)t]}]$$

To simplify this expression, recall that $e^0 = 1$. From Euler's formula, $e^{i\theta} = \cos\theta + i\sin\theta$ and $e^{-i\theta} = \cos\theta - i\sin\theta$, so $e^{i\theta} + e^{-i\theta} = 2\cos\theta$. Hence

$$|\Psi(x, t)|^2 = |A|^2\{2 + 2\cos[(k_2 - k_1)x - (\omega_2 - \omega_1)t]\}$$

$$= 2|A|^2\{1 + \cos[(k_2 - k_1)x - (\omega_2 - \omega_1)t]\}$$

EVALUATE: Figure 40.5 is a graph of the probability distribution function $|\Psi(x, t)|^2$ at $t = 0$. The value of $|\Psi(x, t)|^2$ varies between 0 and $4|A|^2$; probabilities can never be negative! The particle has become *somewhat* localized: The particle is most likely to be found near a point where $|\Psi(x, t)|^2$ is maximum (where the functions $Ae^{i(k_1x - \omega_1 t)}$ and $Ae^{i(k_2x - \omega_2 t)}$ interfere constructively) and is very unlikely to be found near a point where $|\Psi(x, t)|^2 = 0$ (where $Ae^{i(k_1x - \omega_1 t)}$ and $Ae^{i(k_2x - \omega_2 t)}$ interfere destructively). This is very similar to the phenomenon of beats for sound waves (see Section 16.7).

Note also that the probability distribution function is not stationary, but moves in the positive x-direction like the particle that it represents. To see this, recall from Section 15.3 that a sinusoidal wave given by $y(x, t) = A\cos(kx - \omega t)$ moves in the positive x-direction with velocity $v = \omega/k$; since $|\Psi(x, t)|^2$ includes a term $\cos[(k_2 - k_1)x - (\omega_2 - \omega_1)t]$, the probability distribution

moves at a velocity $v_{av} = (\omega_2 - \omega_1)/(k_2 - k_1)$. The subscript "av" reminds us that v_{av} represents the *average* value of the particle's velocity.

The price we pay for localizing the particle somewhat is that, unlike a particle represented by Eq. (40.18), it no longer has either a definite momentum or a definite energy. That's consistent with the Heisenberg uncertainty principles: If we decrease the uncertainties about where a particle is and when it passes a certain point, the uncertainties in its momentum and energy must increase.

The average momentum of the particle is $p_{av} = (\hbar k_2 + \hbar k_1)/2$, the average of the momenta associated with the free-particle wave functions we added to create $\Psi(x, t)$. This corresponds to the particle having an average velocity $v_{av} = p_{av}/m = (\hbar k_2 + \hbar k_1)/2m$. Can you show that this is equal to the expression $v_{av} = (\omega_2 - \omega_1)/(k_2 - k_1)$ that we found above?

40.5 The probability distribution function at $t = 0$ for $\Psi(x, t) = Ae^{i(k_1x - \omega_1 t)} + Ae^{i(k_2x - \omega_2 t)}$.

Wave Packets

The wave function that we examined in Example 40.1 is not very well localized: The probability distribution function still extends from $x = -\infty$ to $x = +\infty$. Hence this wave function can't be normalized, either. To make a wave function that's more highly localized, imagine superposing two additional sinusoidal waves with different wave numbers and amplitudes so as to reinforce alternate maxima of $|\Psi(x, t)|^2$ in Fig. 40.5 and cancel out the in-between ones. Finally, if we superpose waves with a very large number of different wave numbers, we can construct a wave with only *one* maximum of $|\Psi(x, t)|^2$ (Fig. 40.6). Then, finally, we have something that begins to look like both a particle and a wave. It is a particle in the sense that it is localized in space; if we look from a distance, it may look like a point. But it also has a periodic structure that is characteristic of a wave.

A localized wave pulse like that shown in Fig. 40.6 is called a **wave packet.** We can represent a wave packet by an expression such as

$$\Psi(x, t) = \int_{-\infty}^{\infty} A(k)e^{i(kx - \omega t)}\, dk \qquad (40.19)$$

This integral represents a superposition of a very large number of waves, each with a different wave number k and angular frequency $\omega = \hbar k^2/2m$, and each with an amplitude $A(k)$ that depends on k.

There is an important relationship between the two functions $\Psi(x, t)$ and $A(k)$, which we show qualitatively in Fig. 40.7. If the function $A(k)$ is sharply peaked, as in Fig. 40.7a, we are superposing only a narrow range of wave numbers. The resulting wave pulse is then relatively broad (Fig. 40.7b). But if we use

40.6 Superposing a large number of sinusoidal waves with different wave numbers and appropriate amplitudes can produce a wave pulse that has a wavelength $\lambda_{av} = 2\pi/k_{av}$ and is localized within a region of space of length Δx. This localized pulse has aspects of both particle and wave.

(a) Real part of the wave function at time t

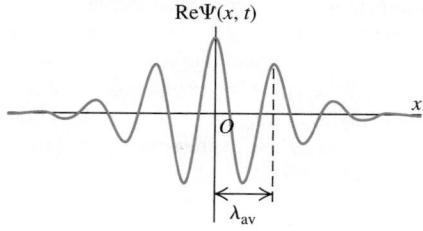

$\text{Re}\Psi(x, t)$

(b) Imaginary part of the wave function at time t

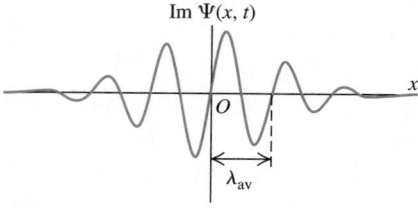

$\text{Im } \Psi(x, t)$

(c) Probability distribution function at time t

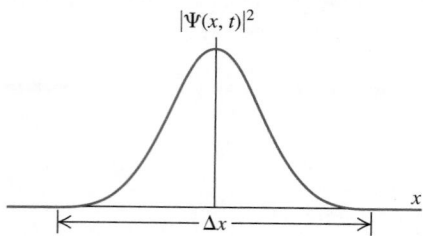

$|\Psi(x, t)|^2$

a wider range of wave numbers, so that the function $A(k)$ is broader (Fig. 40.7c), then the wave pulse is more narrowly localized (Fig. 40.7d). This is simply the uncertainty principle in action. A narrow range of k means a narrow range of $p_x = \hbar k$ and thus a small Δp_x; the result is a relatively large Δx. A broad range of k corresponds to a large Δp_x, and the resulting Δx is smaller. You can see that the uncertainty principle for position and momentum, $\Delta x \Delta p_x \geq \hbar/2$, is really just a consequence of the properties of integrals like Eq. (40.19).

CAUTION **Matter waves versus light waves in vacuum** We can regard both a wave packet that represents a particle and a short pulse of light from a laser as superpositions of waves of different wave numbers and angular frequencies. An important difference is that the speed of light in vacuum is the same for all wavelengths λ and hence all wave numbers $k = 2\pi/\lambda$, but the speed of a matter wave is *different* for different wavelengths. You can see this from the formula for the speed of the wave crests in a periodic wave, $v = \lambda f = \omega/k$. For a matter wave, $\omega = \hbar k^2/2m$, so $v = \hbar k/2m = h/2m\lambda$. Hence matter waves with longer wavelengths and smaller wave numbers travel more slowly than those with short wavelengths and large wave numbers. (This shouldn't be too surprising. The de Broglie relations that we learned in Section 39.1 tell us that shorter wavelength corresponds to greater momentum and hence a greater speed.) Since the individual sinusoidal waves that make up a wave packet travel at different speeds, the shape of the packet changes as it moves. That's why we've specified the time for which the wave packets in Figs. 40.6 and 40.7 are drawn; at later times, the packets become more spread out. By contrast, a pulse of light waves in vacuum retains the same shape at all times because all of its constituent sinusoidal waves travel together at the same speed. ∎

The One-Dimensional Schrödinger Equation with Potential Energy

The one-dimensional Schrödinger equation that we presented in Eq. (40.15) is valid only for free particles, for which the potential energy function is zero: $U(x) = 0$. But for an electron within an atom, a proton within an atomic nucleus, and many other real situations, the potential energy plays an important role. To study the behavior of matter waves in these situations, we need a version of the Schrödinger equation that describes a particle moving in the presence of a nonzero potential energy function $U(x)$. This equation is

$$-\frac{\hbar^2}{2m}\frac{\partial^2 \Psi(x, t)}{\partial x^2} + U(x)\Psi(x, t) = i\hbar\frac{\partial \Psi(x, t)}{\partial t} \qquad \begin{array}{l}\text{(general one-dimen-}\\ \text{sional Schrödinger} \quad \text{(40.20)}\\ \text{equation)}\end{array}$$

Note that if $U(x) = 0$, Eq. (40.20) reduces to the free-particle Schrödinger equation given in Eq. (40.15).

Here's the motivation behind Eq. (40.20). If $\Psi(x, t)$ is a sinusoidal wave function for a free particle, $\Psi(x, t) = Ae^{i(kx-\omega t)} = Ae^{ikx}e^{-i\omega t}$, the derivative terms in Eq. (40.20) become

$$-\frac{\hbar^2}{2m}\frac{\partial^2 \Psi(x, t)}{\partial x^2} = -\frac{\hbar^2}{2m}\frac{\partial^2}{\partial x^2}(Ae^{ikx}e^{-i\omega t}) = -\frac{\hbar^2}{2m}(ik)^2(Ae^{ikx}e^{-i\omega t})$$

$$= \frac{\hbar^2 k^2}{2m}\Psi(x, t)$$

$$i\hbar\frac{\partial \Psi(x, t)}{\partial t} = i\hbar\frac{\partial}{\partial t}(Ae^{ikx}e^{-i\omega t}) = i\hbar(-i\omega)(Ae^{ikx}e^{-i\omega t}) = \hbar\omega\Psi(x, t)$$

In these expressions $(\hbar^2 k^2/2m)\Psi(x, t)$ is just the kinetic energy $K = p^2/2m = \hbar^2 k^2/2m$ multiplied by the wave function, and $\hbar\omega\Psi(x, t)$ is the total energy $E = \hbar\omega$ multiplied by the wave function. So for a wave function of this kind, Eq. (40.20) says that kinetic energy times $\Psi(x, t)$ plus potential energy times $\Psi(x, t)$ equals total energy times $\Psi(x, t)$. That's equivalent to the statement in classical physics that the sum of kinetic energy and potential energy equals total mechanical energy: $K + U = E$.

The observations we've just made certainly aren't a *proof* that Eq. (40.20) is correct. The real reason we know this equation *is* correct is that it works: Predictions made with this equation agree with experimental results. In the remaining sections of this chapter we'll apply Eq. (40.20) to several physical situations, each with a different form of the function $U(x)$.

Stationary States

We saw in our discussion of wave packets that any free-particle wave function can be built up as a superposition of sinusoidal wave functions of the form $\Psi(x, t) = Ae^{ikx}e^{-i\omega t}$. Each such sinusoidal wave function corresponds to a state of definite energy $E = \hbar\omega = \hbar^2k^2/2m$ and definite angular frequency $\omega = E/\hbar$, so we can rewrite these functions as $\Psi(x, t) = Ae^{ikx}e^{-iEt/\hbar}$. If the potential energy function $U(x)$ is nonzero, these sinusoidal wave functions do not satisfy the Schrödinger equation, Eq. (40.20), and so these functions cannot be the basic "building blocks" of more complicated wave functions. However, we can still write the wave function for a state of definite energy E in the form

$$\Psi(x, t) = \psi(x)e^{-iEt/\hbar} \quad \begin{array}{l}\text{(time-dependent wave function}\\ \text{for a state of definite energy)}\end{array} \quad \text{(40.21)}$$

That is, the wave function $\Psi(x, t)$ for a state of definite energy is the product of a *time-independent* wave function $\psi(x)$ and a factor $e^{-iEt/\hbar}$. (For the free-particle sinusoidal wave function, $\psi(x) = Ae^{ikx}$.) States of definite energy are of tremendous importance in quantum mechanics. For example, for each energy level in a hydrogen atom (Section 39.3) there is a specific wave function. It is possible for an atom to be in a state that does not have a definite energy. The wave function for any such state can be written as a combination of definite-energy wave functions, in precisely the same way that a free-particle wave packet can be written as a superposition of sinusoidal wave functions of definite energy as in Eq. (40.19).

A state of definite energy is commonly called a **stationary state.** To see where this name comes from, let's multiply Eq. (40.21) by its complex conjugate to find the probabilty distribution function $|\Psi|^2$:

$$\begin{aligned} |\Psi(x, t)|^2 &= \Psi^*(x, t)\Psi(x, t) = [\psi^*(x)e^{+iEt/\hbar}][\psi(x)e^{-iEt/\hbar}] \\ &= \psi^*(x)\psi(x)e^{(+iEt/\hbar)+(-iEt/\hbar)} = |\psi(x)|^2 e^0 \quad \text{(40.22)} \\ &= |\psi(x)|^2 \end{aligned}$$

Since $|\psi(x)|^2$ does not depend on time, Eq. (40.22) shows that the same must be true for the probability distribution function $|\Psi(x, t)|^2$. This justifies the term "stationary state" for a state of definite energy.

> **CAUTION** **A stationary state does not mean a stationary particle** The name *stationary state* may lead you to think that the particle is not in motion if it is described by such a wave function. That's not the case. It's the *probability distribution* (that is, the relative likelihood of finding the particle at various positions), not the particle itself, that's stationary. ∎

The Schrödinger equation, Eq. (40.20), becomes quite a bit simpler for stationary states. To see this, we substitute Eq. (40.21) into Eq. (40.20):

$$-\frac{\hbar^2}{2m}\frac{\partial^2[\psi(x)e^{-iEt/\hbar}]}{\partial x^2} + U(x)\psi(x)e^{-iEt/\hbar} = i\hbar\frac{\partial[\psi(x)e^{-iEt/\hbar}]}{\partial t}$$

40.7 How varying the function $A(k)$ in the wave-packet expression, Eq. (40.19), changes the character of the wave function $\Psi(x, t)$ (shown here at a specific time $t = 0$).

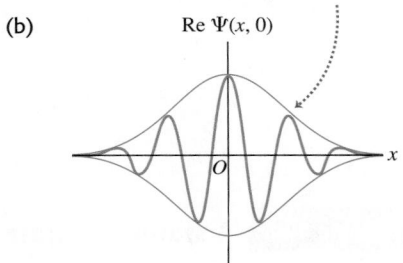

(a) $A(k)$

A narrow function $A(k)$ ...

... gives a wave function $\Psi(x, 0)$ with a broad spatial extent.

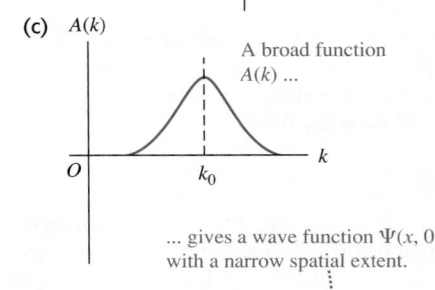

(b) Re $\Psi(x, 0)$

(c) $A(k)$

A broad function $A(k)$ ...

... gives a wave function $\Psi(x, 0)$ with a narrow spatial extent.

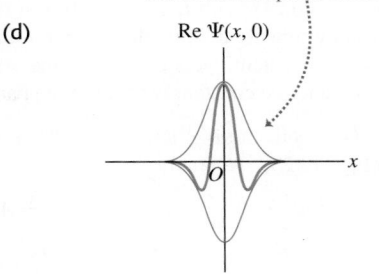

(d) Re $\Psi(x, 0)$

Mastering**PHYSICS**

PhET: Quantum Tunneling and Wave Packets
ActivPhysics 17.7: Wave Packets

The derivative on the first term on the left-hand side is with respect to x, so the factor of $e^{-iEt/\hbar}$ comes outside of the derivative. Now we take the derivative with respect to t on the right-hand side of the equation:

$$-\frac{\hbar^2}{2m}\frac{d^2\psi(x)}{dx^2}e^{-iEt/\hbar} + U(x)\psi(x)e^{-iEt/\hbar} = i\hbar\left(\frac{-iE}{\hbar}\right)[\psi(x)e^{-iEt/\hbar}]$$

$$= E\psi(x)e^{-iEt/\hbar}$$

If we divide both sides of this equation by $e^{-iEt/\hbar}$, we get

$$-\frac{\hbar^2}{2m}\frac{d^2\psi(x)}{dx^2} + U(x)\psi(x) = E\psi(x) \qquad \text{(time-independent Schrödinger equation)} \qquad (40.23)$$

This is called the **time-independent Schrödinger equation.** The time-dependent factor $e^{-iEt/\hbar}$ does not appear, and Eq. (40.23) is an equation that involves only the time-independent wave function $\psi(x)$. We'll devote much of this chapter to solving this equation to find the definite-energy, stationary-state wave functions $\psi(x)$ and the corresponding values of E—that is, the energies of the allowed levels—for different physical situations.

Example 40.2 A stationary state

Consider the wave function $\psi(x) = A_1 e^{ikx} + A_2 e^{-ikx}$, where k is positive. Is this a valid time-independent wave function for a free particle in a stationary state? What is the energy corresponding to this wave function?

SOLUTION

IDENTIFY and SET UP: A valid stationary-state wave function for a free particle must satisfy the time-independent Schrödinger equation, Eq. (40.23), with $U(x) = 0$. To test the given function $\psi(x)$, we simply substitute it into the left-hand side of the equation. If the result is a constant times $\psi(x)$, then the wave function is indeed a solution and the constant is equal to the particle energy E.

EXECUTE: Substituting $\psi(x) = A_1 e^{ikx} + A_2 e^{-ikx}$ and $U(x) = 0$ into Eq. (40.23), we obtain

$$-\frac{\hbar^2}{2m}\frac{d^2\psi(x)}{dx^2} = -\frac{\hbar^2}{2m}\frac{d^2(A_1 e^{ikx} + A_2 e^{-ikx})}{dx^2}$$

$$= -\frac{\hbar^2}{2m}[(ik)^2 A_1 e^{ikx} + (-ik)^2 A_2 e^{ikx}]$$

$$= \frac{\hbar^2 k^2}{2m}(A_1 e^{ikx} + A_2 e^{-ikx}) = \frac{\hbar^2 k^2}{2m}\psi(x)$$

The result is a constant times $\psi(x)$, so this $\psi(x)$ is indeed a valid stationary-state wave function for a free particle. Comparing with Eq. (40.23) shows that the constant on the right-hand side is the particle energy: $E = \hbar^2 k^2/2m$.

EVALUATE: Note that $\psi(x)$ is a *superposition* of two different wave functions: one function ($A_1 e^{ikx}$) that represents a particle with magnitude of momentum $p = \hbar k$ moving in the positive x-direction, and one function ($A_2 e^{-ikx}$) that represents a particle with the same magnitude of momentum moving in the negative x-direction. So while the combined wave function $\psi(x)$ represents a stationary state with a definite energy, this state does *not* have a definite momentum. We'll see in Section 40.2 that such a wave function can represent a *standing wave,* and we'll explore situations in which such standing matter waves can arise.

Test Your Understanding of Section 40.1 Does a wave packet given by Eq. (40.19) represent a stationary state? ❙

Mastering**PHYSICS**

ActivPhysics 20.2: Particle in a Box

40.2 Particle in a Box

An important problem in quantum mechanics is how to use the time-independent Schrödinger equation, Eq. (40.23), to determine the possible energy levels and the corresponding wave functions for various systems. The fundamental problem is then the following: For a given potential energy function $U(x)$, what are the possible stationary-state wave functions $\psi(x)$, and what are the corresponding energies E?

In Section 40.1 we solved this problem for the case $U(x) = 0$, corresponding to a *free* particle. The allowed wave functions and corresponding energies are

$$\psi(x) = Ae^{ikx} \quad E = \frac{\hbar^2 k^2}{2m} \quad \text{(free particle)} \quad (40.24)$$

The wave number k is equal to $2\pi/\lambda$, where λ is the wavelength. We found that k can have any real value, so the energy E of a free particle can have any value from zero to infinity. Furthermore, the particle can be found with equal probability at any value of x from $-\infty$ to $+\infty$.

Now let's look at a simple model in which a particle is *bound* so that it cannot escape to infinity, but rather is confined to a restricted region of space. Our system consists of a particle confined between two rigid walls separated by a distance L (Fig. 40.8). The motion is purely one dimensional, with the particle moving along the x-axis only and the walls at $x = 0$ and $x = L$. The potential energy corresponding to the rigid walls is infinite, so the particle cannot escape; between the walls, the potential energy is zero (Fig. 40.9). This situation is often described as a **"particle in a box."** This model might represent an electron that is free to move within a long, straight molecule or along a very thin wire.

Wave Functions for a Particle in a Box

To solve the Schrödinger equation for this system, we begin with some restrictions on the particle's stationary-state wave function $\psi(x)$. Because the particle is confined to the region $0 \le x \le L$, we expect the probability distribution function $|\Psi(x, t)|^2 = |\psi(x)|^2$ and the wave function $\psi(x)$ to be zero outside that region. This agrees with the Schrödinger equation: If the term $U(x)\psi(x)$ in Eq. (40.23) is to be finite, then $\psi(x)$ must be zero where $U(x)$ is infinite.

Furthermore, $\psi(x)$ must be a *continuous* function to be a mathematically well-behaved solution to the Schrödinger equation. This implies that $\psi(x)$ must be zero at the region's boundary, $x = 0$ and $x = L$. These two conditions serve as *boundary conditions* for the problem. They should look familiar, because they are the same conditions that we used to find the normal modes of a vibrating string in Section 15.8 (Fig. 40.10); you should review that discussion.

An additional condition is that to calculate the second derivative $d^2\psi(x)/dx^2$ in Eq. (40.23), the *first* derivative $d\psi(x)/dx$ must also be continuous except at points where the potential energy becomes infinite (as it does at the walls of the box). This is analogous to the requirement that a vibrating string, like those shown in Fig. 40.10, can't have any kinks in it (which would correspond to a discontinuity in the first derivative of the wave function) except at the ends of the string.

We now solve for the wave functions in the region $0 \le x \le L$ subject to the above conditions. In this region $U(x) = 0$, so the wave function in this region must satisfy

$$-\frac{\hbar^2}{2m} \frac{d^2\psi(x)}{dx^2} = E\psi(x) \quad \text{(particle in a box)} \quad (40.25)$$

Equation (40.25) is the *same* Schrödinger equation as for a free particle, so it is tempting to conclude that the wave functions and energies are given by Eq. (40.24). It is true that $\psi(x) = Ae^{ikx}$ satisfies the Schrödinger equation with $U(x) = 0$, is continuous, and has a continuous first derivative $d\psi(x)/dx = ikAe^{ikx}$. However, this wave function does *not* satisfy the boundary conditions that $\psi(x)$ must be zero at $x = 0$ and $x = L$: At $x = 0$ the wave function in Eq. (40.24) is equal to $Ae^0 = A$, and at $x = L$ it is equal to Ae^{ikL}. (These would be equal to zero if $A = 0$, but then the wave function would be zero and there would be no particle at all!)

The way out of this dilemma is to recall Example 40.2 (Section 40.1), in which we found that a more general stationary-state solution to the time-independent Schrödinger equation with $U(x) = 0$ is

$$\psi(x) = A_1 e^{ikx} + A_2 e^{-ikx} \quad (40.26)$$

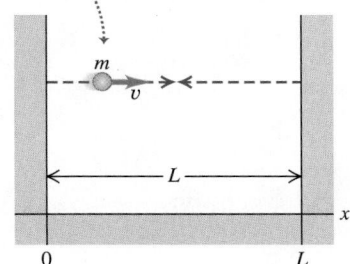

40.8 The Newtonian view of a particle in a box.

A particle with mass m moves along a straight line at constant speed, bouncing between two rigid walls a distance L apart.

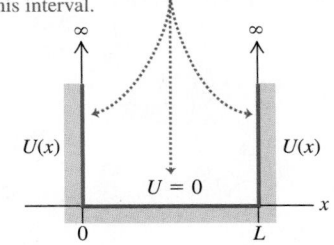

40.9 The potential-energy function for a particle in a box.

The potential energy U is zero in the interval $0 < x < L$ and is infinite everywhere outside this interval.

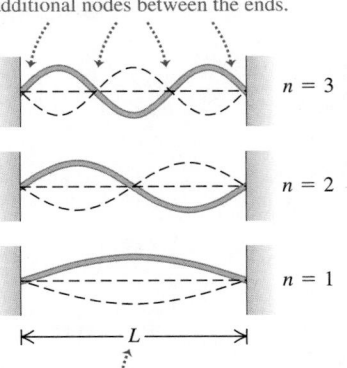

40.10 Normal modes of vibration for a string with length L, held at both ends.

Each end is a node, and there are $n - 1$ additional nodes between the ends.

$n = 3$

$n = 2$

$n = 1$

The length is an integral number of half-wavelengths: $L = n\lambda_n/2$.

This wave function is a superposition of two waves: one traveling in the $+x$-direction of amplitude A_1, and one traveling in the $-x$-direction with the same wave number but amplitude A_2. This is analogous to a standing wave on a string (Fig. 40.10), which we can regard as the superposition of two sinusoidal waves propagating in opposite directions (see Section 15.7). The energy that corresponds to Eq. (40.26) is $E = \hbar^2 k^2/2m$, just as for a single wave.

To see whether the wave function given by Eq. (40.26) can satisfy the boundary conditions, let's first rewrite it in terms of sines and cosines using Euler's formula, Eq. (40.17):

$$\psi(x) = A_1(\cos kx + i\sin kx) + A_2[\cos(-kx) + i\sin(-kx)]$$
$$= A_1(\cos kx + i\sin kx) + A_2(\cos kx - i\sin kx) \tag{40.27}$$
$$= (A_1 + A_2)\cos kx + i(A_1 - A_2)\sin kx$$

At $x = 0$ this is equal to $\psi(0) = A_1 + A_2$, which must equal zero if we are to satisfy the boundary condition at that point. Hence $A_2 = -A_1$, and Eq. (40.27) becomes

$$\psi(x) = 2iA_1 \sin kx = C\sin kx \tag{40.28}$$

We have simplified the expression by introducing the constant $C = 2iA_1$. (We'll come back to this constant later.) We can also satisfy the second boundary condition that $\psi = 0$ at $x = L$ by choosing values of k such that $kL = n\pi$ ($n = 1, 2, 3, \dots$). Hence Eq. (40.28) does indeed give the stationary-state wave functions for a particle in a box in the region $0 \le x \le L$. (Outside this region, $\psi(x) = 0$.) The possible values of k and the wavelength $\lambda = 2\pi/k$ are

$$k = \frac{n\pi}{L} \quad \text{and} \quad \lambda = \frac{2\pi}{k} = \frac{2L}{n} \quad (n = 1, 2, 3, \dots) \tag{40.29}$$

Just as for the string in Fig. 40.10, the length L of the region is an integral number of half-wavelengths.

Energy Levels for a Particle in a Box

The possible energy levels for a particle in a box are given by $E = \hbar^2 k^2/2m = p^2/2m$, where $p = \hbar k = (h/2\pi)(2\pi/\lambda) = h/\lambda$ is the magnitude of momentum of a free particle with wave number k and wavelength λ. This makes sense, since inside the region $0 \le x \le L$ the potential energy is zero and the energy is all kinetic. For each value of n, there are corresponding values of p, λ, and E; let's call them p_n, λ_n, and E_n. Putting the pieces together, we get

$$p_n = \frac{h}{\lambda_n} = \frac{nh}{2L} \tag{40.30}$$

and so the energy levels for a particle in a box are

$$E_n = \frac{p_n^2}{2m} = \frac{n^2 h^2}{8mL^2} = \frac{n^2 \pi^2 \hbar^2}{2mL^2} \quad (n = 1, 2, 3, \dots) \qquad \text{(energy levels, particle in a box)} \tag{40.31}$$

Each energy level has its own value of the quantum number n and a corresponding wave function, which we denote by ψ_n. When we replace k in Eq. (40.28) by $n\pi/L$ from Eq. (40.29), we find

$$\psi_n(x) = C\sin\frac{n\pi x}{L} \quad (n = 1, 2, 3, \dots) \tag{40.32}$$

The energy-level diagram in Fig. 40.11a shows the five lowest levels for a particle in a box. The energy levels are proportional to n^2, so successively higher levels are spaced farther and farther apart. There are an infinite number of levels because the walls are perfectly rigid; even a particle of infinitely great kinetic

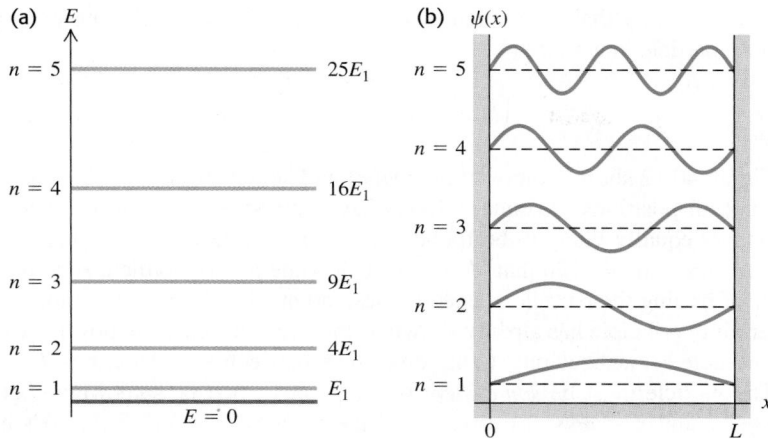

40.11 (a) Energy-level diagram for a particle in a box. Each energy is n^2E_1, where E_1 is the ground-level energy. (b) Wave functions for a particle in a box, with $n = 1, 2, 3, 4$, and 5. **CAUTION:** The five graphs have been displaced vertically for clarity, as in Fig. 40.10. Each of the horizontal dashed lines represents $\psi = 0$ for the respective wave function.

energy is confined within the box. Figure 40.11b shows graphs of the wave functions $\psi_n(x)$ for $n = 1, 2, 3, 4$, and 5. Note that these functions look identical to those for a standing wave on a string (see Fig. 40.10).

CAUTION **A particle in a box cannot have zero energy** Note that the energy of a particle in a box *cannot* be zero. Equation (40.31) shows that $E = 0$ would require $n = 0$, but substituting $n = 0$ into Eq. (40.32) gives a zero wave function. Since a particle is described by a *nonzero* wave function, this means that there cannot be a particle with $E = 0$. This is a consequence of the Heisenberg uncertainty principle: A particle in a zero-energy state would have a definite value of momentum (precisely zero), so its position uncertainty would be infinite and the particle could be found anywhere along the x-axis. But this is impossible, since a particle in a box can be found only between $x = 0$ and $x = L$. Hence $E = 0$ is not allowed. By contrast, the allowed stationary-state wave functions with $n = 1, 2, 3, \ldots$ do not represent states of definite momentum (each is an equal mixture of a state of x-momentum $+p_n = nh/2L$ and a state of x-momentum $-p_n = -nh/2L$). Hence each stationary state has a nonzero momentum uncertainty, consistent with having a finite position uncertainty.

Example 40.3 **Electron in an atom-size box**

Find the first two energy levels for an electron confined to a one-dimensional box 5.0×10^{-10} m across (about the diameter of an atom).

SOLUTION

IDENTIFY and SET UP: This problem uses what we have learned in this section about a particle in a box. The first two energy levels correspond to $n = 1$ and $n = 2$ in Eq. (40.31).

EXECUTE: From Eq. (40.31),

$$E_1 = \frac{h^2}{8mL^2} = \frac{(6.626 \times 10^{-34} \text{ J} \cdot \text{s})^2}{8(9.109 \times 10^{-31} \text{ kg})(5.0 \times 10^{-10} \text{ m})^2}$$

$$= 2.4 \times 10^{-19} \text{ J} = 1.5 \text{ eV}$$

$$E_2 = \frac{2^2 h^2}{8mL^2} = 4E_1 = 9.6 \times 10^{-19} \text{ J} = 6.0 \text{ eV}$$

EVALUATE: The difference between the first two energy levels is $E_2 - E_1 = 4.5$ eV. An electron confined to a box is different from an electron bound in an atom, but it is reassuring that this result is of the same order of magnitude as the difference between actual atomic energy levels.

You can also show that for a proton or neutron ($m = 1.67 \times 10^{-27}$ kg) confined to a box 1.1×10^{-14} m across (the width of a medium-sized atomic nucleus), the energies of the first two levels are about a million times larger: $E_1 = 1.7 \times 10^6$ eV $= 1.7$ MeV, $E_2 = 4E_1 = 6.8$ MeV, $E_2 - E_1 = 5.1$ MeV. This suggests why nuclear reactions (which involve transitions between energy levels in nuclei) release so much more energy than chemical reactions (which involve transitions between energy levels of electrons in atoms).

Finally, you can show (see Exercise 40.11) that the energy levels of a billiard ball ($m = 0.2$ kg) confined to a box 1.3 m across—the width of a billard table—are separated by about 5×10^{-67} J. Quantum effects won't disturb a game of billiards.

Probability and Normalization

Let's look a bit more closely at the wave functions for a particle in a box, keeping in mind the *probability* interpretation of the wave function ψ that we discussed in Section 40.1. In our one-dimensional situation the quantity $|\psi(x)|^2 \, dx$ is proportional

40.12 Graphs of (a) $\psi(x)$ and (b) $|\psi(x)|^2$ for the first three wave functions ($n = 1, 2, 3$) for a particle in a box. The horizontal dashed lines represent $\psi(x) = 0$ and $|\psi(x)|^2 = 0$ for each of the three levels. The value of $|\psi(x)|^2 dx$ at each point is the probability of finding the particle in a small interval dx about the point. As in Fig. 40.11b, the three graphs in each part have been displaced vertically for clarity.

(a) $\psi(x)$

(b) $|\psi(x)|^2$

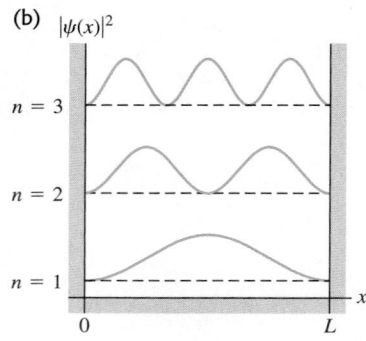

to the probability that the particle will be found within a small interval dx about x. For a particle in a box,

$$|\psi(x)|^2 dx = C^2 \sin^2 \frac{n\pi x}{L} dx$$

Figure 40.12 shows graphs of both $\psi(x)$ and $|\psi(x)|^2$ for $n = 1, 2$, and 3. Note that not all positions are equally likely. By contrast, in classical mechanics the particle is equally likely to be found at any position between $x = 0$ and $x = L$. We see from Fig. 40.12b that $|\psi(x)|^2 = 0$ at some points, so there is zero probability of finding the particle at exactly these points. Don't let that bother you; the uncertainty principle has already shown us that we can't measure position exactly. The particle is localized only to be somewhere between $x = 0$ and $x = L$.

The particle must be *somewhere* on the x-axis—that is, somewhere between $x = -\infty$ and $x = +\infty$. So the *sum* of the probabilities for all the dx's everywhere (the *total* probability of finding the particle) must equal 1. That's the normalization condition that we discussed in Section 40.1:

$$\int_{-\infty}^{\infty} |\psi(x)|^2 \, dx = 1 \qquad \text{(normalization condition)} \qquad (40.33)$$

A wave function is said to be *normalized* if it has a constant such as C in Eq. (40.32) that is calculated to make the total probability equal 1 in Eq. (40.33). For a normalized wave function, $|\psi(x)|^2 dx$ is not merely proportional to, but *equals,* the probability of finding the particle between the coordinates x and $x + dx$. That's why we call $|\psi(x)|^2$ the probability distribution function. (In Section 40.1 we called $|\Psi(x, t)|^2$ the probability distribution function. For the case of a stationary-state wave function, however, $|\Psi(x, t)|^2$ is equal to $|\psi(x)|^2$.)

Let's normalize the particle-in-a-box wave functions $\psi_n(x)$ given by Eq. (40.32). Since $\psi_n(x)$ is zero except between $x = 0$ and $x = L$, Eq. (40.33) becomes

$$\int_0^L C^2 \sin^2 \frac{n\pi x}{L} dx = 1 \qquad (40.34)$$

You can evaluate this integral using the trigonometric identity $\sin^2\theta = \frac{1}{2}(1 - \cos 2\theta)$; the result is $C^2 L/2$. Thus our probability interpretation of the wave function demands that $C^2 L/2 = 1$, or $C = (2/L)^{1/2}$; the constant C is *not* arbitrary. (This is in contrast to the classical vibrating string problem, in which C represents an amplitude that depends on initial conditions.) Thus the normalized stationary-state wave functions for a particle in a box are

$$\psi_n(x) = \sqrt{\frac{2}{L}} \sin \frac{n\pi x}{L} \quad (n = 1, 2, 3, \dots) \quad \text{(particle in a box)} \qquad (40.35)$$

Example 40.4 **A nonsinusoidal wave function?**

(a) Show that $\psi(x) = Ax + B$, where A and B are constants, is a solution of the Schrödinger equation for an $E = 0$ energy level of a particle in a box. (b) What constraints do the boundary conditions at $x = 0$ and $x = L$ place on the constants A and B?

SOLUTION

IDENTIFY and SET UP: To be physically reasonable, a wave function must satisfy both the Schrödinger equation and the appropriate boundary conditions. In part (a) we'll substitute $\psi(x)$ into the

Schrödinger equation for a particle in a box, Eq. (40.25), to determine whether it is a solution. In part (b) we'll see what restrictions on $\psi(x)$ arise from applying the boundary conditions that $\psi(x) = 0$ at $x = 0$ and $x = L$.

EXECUTE: (a) From Eq. (40.25), the Schrödinger equation for an $E = 0$ energy level of a particle in a box is

$$-\frac{\hbar^2}{2m} \frac{d^2\psi(x)}{dx^2} = E\psi(x) = 0$$

in the region $0 \leq x \leq L$. Differentiating $\psi(x) = Ax + B$ twice with respect to x gives $d^2\psi(x)/dx^2 = 0$, so the left side of the equation is zero, and so $\psi(x) = Ax + B$ is a solution of this Schrödinger equation for $E = 0$. (Note that both $\psi(x)$ and its derivative $d\psi(x)/dx = A$ are continuous functions, as they must be.)

(b) Applying the boundary condition at $x = 0$ gives $\psi(0) = B = 0$, and so $\psi(x) = Ax$. Applying the boundary condition at $x = L$ gives $\psi(L) = AL = 0$, so $A = 0$. Hence $\psi(x) = 0$ both

inside the box $(0 \leq x \leq L)$ *and outside:* There is *zero* probability of finding the particle anywhere with this wave function, and so $\psi(x) = Ax + B$ is *not* a physically valid wave function.

EVALUATE: The moral is that there are many functions that satisfy the Schrödinger equation for a given physical situation, but most of these—including the function considered here—have to be rejected because they don't satisfy the appropriate boundary conditions.

Time Dependence

Finally, we note that the wave functions $\psi_n(x)$ in Eq. (40.35) depend only on the *spatial* coordinate x. Equation (40.21) shows that if $\psi(x)$ is the wave function for a state of definite energy E, the full time-dependent wave function is $\Psi(x, t) = \psi(x)e^{-iEt/\hbar}$. Hence the *time-dependent* stationary-state wave functions for a particle in a box are

$$\Psi_n(x, t) = \sqrt{\frac{2}{L}} \sin\left(\frac{n\pi x}{L}\right) e^{-iE_n t/\hbar} \qquad (n = 1, 2, 3, \dots) \qquad (40.36)$$

In this expression the energies E_n are given by Eq. (40.31). The higher the quantum number n, the greater the angular frequency $\omega_n = E_n/\hbar$ at which the wave function oscillates. Note that since $\left|e^{-iE_n t/\hbar}\right|^2 = e^{+iE_n t/\hbar}e^{-iE_n t/\hbar} = e^0 = 1$, the probability distribution function $|\Psi_n(x, t)|^2 = (2/L)\sin^2(n\pi x/L)$ is independent of time and does *not* oscillate. (Remember, this is why we say that these states of definite energy are *stationary*.)

Test Your Understanding of Section 40.2   If a particle in a box is in the nth energy level, what is the average value of its x-component of momentum p_x? (i) $nh/2L$; (ii) $(\sqrt{2}/2)nh/L$; (iii) $(1/\sqrt{2})nh/L$; (iv) $[1/(2\sqrt{2})]nh/L$; (v) zero.

40.3 Potential Wells

A **potential well** is a potential-energy function $U(x)$ that has a minimum. We introduced this term in Section 7.5, and we also used it in our discussion of periodic motion in Chapter 14. In Newtonian mechanics a particle trapped in a potential well can vibrate back and forth with periodic motion. Our first application of the Schrödinger equation, the particle in a box, involved a rudimentary potential well with a function $U(x)$ that is zero within a certain interval and infinite everywhere else. As we mentioned in Section 40.2, this function corresponds to a few situations found in nature, but the correspondence is only approximate.

A better approximation to several actual physical situations is a **finite well**, which is a potential well with straight sides but *finite* height. Figure 40.13 shows a potential-energy function that is zero in the interval $0 \leq x \leq L$ and has the value U_0 outside this interval. This function is often called a **square-well potential**. It could serve as a simple model of an electron within a metallic sheet with thickness L, moving perpendicular to the surfaces of the sheet. The electron can move freely inside the metal but has to climb a potential-energy barrier with height U_0 to escape from either surface of the metal. The energy U_0 is related to the *work function* that we discussed in Section 38.1 in connection with the photoelectric effect. In three dimensions, a spherical version of a finite well gives an approximate description of the motions of protons and neutrons within a nucleus.

Bound States of a Square-Well Potential

In Newtonian mechanics, the particle is trapped (localized) in a well if the total mechanical energy E is less than U_0. In quantum mechanics, such a trapped state is often called a **bound state.** All states are bound for an infinitely deep well like

40.13 A square-well potential.

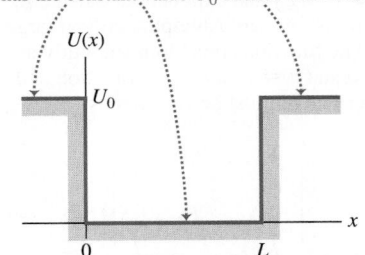

The potential energy U is zero within the potential well (in the interval $0 \leq x \leq L$) and has the constant value U_0 outside this interval.

the one we described in Section 40.2. For a finite well like that shown in Fig. 40.13, if E is greater than U_0, the particle is *not* bound.

Let's see how to solve the Schrödinger equation for the bound states of a square-well potential. Our goal is to find the energies and wave functions for which $E < U_0$. The easiest approach is to consider separately the regions where $U = 0$ and where $U = U_0$. Where $U = 0$, the time-independent Schrödinger equation is

$$-\frac{\hbar^2}{2m}\frac{d^2\psi(x)}{dx^2} = E\psi(x) \quad \text{or} \quad \frac{d^2\psi(x)}{dx^2} = -\frac{2mE}{\hbar^2}\psi(x) \quad \text{(40.37)}$$

This is the same as Eq. (40.25) from Section 40.2, which describes a particle in a box. As in Section 40.2, we can express the solutions of this equation as combinations of cos kx and sin kx, where $E = \hbar^2 k^2/2m$. We can rewrite the relationship between E and k as $k = \sqrt{2mE}/\hbar$. Hence inside the square well $(0 \le x \le L)$ we have

$$\psi(x) = A\cos\left(\frac{\sqrt{2mE}}{\hbar}x\right) + B\sin\left(\frac{\sqrt{2mE}}{\hbar}x\right) \quad \text{(inside the well)} \quad \text{(40.38)}$$

where A and B are constants. So far, this looks a lot like the particle-in-a-box analysis in Section 40.2. The difference is that for the square-well potential, the potential energy outside the well is not infinite, so the wave function $\psi(x)$ outside the well is *not* zero.

For the regions outside the well $(x < 0$ and $x > L)$ the potential-energy function in the time-independent Schrodinger equation is $U = U_0$:

$$-\frac{\hbar^2}{2m}\frac{d^2\psi(x)}{dx^2} + U_0\psi(x) = E\psi(x) \quad \text{or} \quad \frac{d^2\psi(x)}{dx^2} = \frac{2m(U_0 - E)}{\hbar^2}\psi(x) \quad \text{(40.39)}$$

The quantity $U_0 - E$ is positive, so the solutions of this equation are exponential. Using κ (the Greek letter kappa) to represent the quantity $[2m(U_0 - E)]^{1/2}/\hbar$ and taking κ as positive, we can write the solutions as

$$\psi(x) = Ce^{\kappa x} + De^{-\kappa x} \quad \text{(outside the well)} \quad \text{(40.40)}$$

where C and D are constants with different values in the two regions $x < 0$ and $x > L$. Note that ψ can't be allowed to approach infinity as $x \to +\infty$ or $x \to -\infty$. [If it did, we wouldn't be able to satisfy the normalization condition, Eq. (40.33).] This means that in Eq. (40.40), we must have $D = 0$ for $x < 0$ and $C = 0$ for $x > L$.

Our calculations so far show that the bound-state wave functions for a finite well are sinusoidal inside the well [Eq. (40.38)] and exponential outside it [Eq. (40.40)]. We have to *match* the wave functions inside and outside the well so that they satisfy the boundary conditions that we mentioned in Section 40.2: $\psi(x)$ and $d\psi(x)/dx$ must be continuous at the boundary points $x = 0$ and $x = L$. If the wave function $\psi(x)$ or the slope $d\psi(x)/dx$ were to change discontinuously at a point, the second derivative $d^2\psi(x)/dx^2$ would be *infinite* at that point. That would violate the time-independent Schrödinger equation, Eq. (40.23), which says that at every point $d^2\psi(x)/dx^2$ is proportional to $U - E$. For a finite well $U - E$ is finite everywhere, so $d^2\psi(x)/dx^2$ must also be finite everywhere.

Matching the sinusoidal and exponential functions at the boundary points so that they join smoothly is possible only for certain specific values of the total energy E, so this requirement determines the possible energy levels of the finite square well. There is no simple formula for the energy levels as there was for the infinitely deep well. Finding the levels is a fairly complex mathematical problem that requires solving a transcendental equation by numerical approximation; we won't go into the details. Figure 40.14 shows the general shape of a possible wave function. The most striking features of this wave function are the

40.14 A possible wave function for a particle in a finite potential well. The function is sinusoidal inside the well $(0 \le x \le L)$ and exponential outside it. It approaches zero asymptotically at large $|x|$. The functions must join smoothly at $x = 0$ and $x = L$; the wave function and its derivative must be continuous.

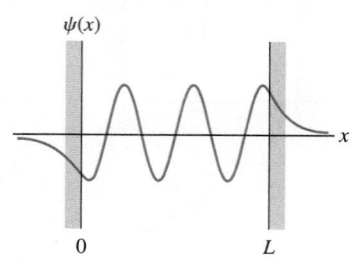

"exponential tails" that extend outside the well into regions that are forbidden by Newtonian mechanics (because in those regions the particle would have negative kinetic energy). We see that there is some probability for finding the particle *outside* the potential well, which would be impossible in classical mechanics. In Section 40.4 we'll discuss an amazing result of this effect.

Example 40.5 Outside a finite well

(a) Show that Eq. (40.40), $\psi(x) = Ce^{\kappa x} + De^{-\kappa x}$, is indeed a solution of the time-independent Schrödinger equation outside a finite well of height U_0. (b) What happens to $\psi(x)$ in the limit $U_0 \rightarrow \infty$?

SOLUTION

IDENTIFY and SET UP: In part (a), we try the given function $\psi(x)$ in the time-independent Schrödinger equation for $x < 0$ and for $x > L$, Eq. (40.39). In part (b), we note that in the limit $U_0 \rightarrow \infty$ the finite well becomes an *infinite* well, like those we considered in Section 40.2 for a particle in a box. So in this limit the wave functions outside a finite well must reduce to the wave functions outside the box.

EXECUTE: (a) We must show that $\psi(x) = Ce^{\kappa x} + De^{-\kappa x}$ satisfies $d^2\psi(x)/dx^2 = [2m(U_0 - E)/\hbar^2]\psi(x)$. We recall that $(d/du)e^{au} = ae^{au}$ and $(d^2/du^2)e^{au} = a^2e^{au}$; the left-hand side of the Schrödinger equation is then

$$\frac{d^2\psi(x)}{dx^2} = \frac{d^2}{dx^2}(Ce^{\kappa x}) + \frac{d^2}{dx^2}(De^{-\kappa x})$$
$$= C\kappa^2 e^{\kappa x} + D(-\kappa)^2 e^{-\kappa x}$$
$$= \kappa^2(Ce^{\kappa x} + De^{-\kappa x})$$
$$= \kappa^2\psi(x)$$

Since from Eq. (40.40) $\kappa^2 = 2m(U_0 - E)/\hbar^2$, this is equal to the right-hand side of the equation. The equation is satisfied, and $\psi(x)$ is a solution.

(b) As U_0 approaches infinity, κ also approaches infinity. In the region $x < 0$, $\psi(x) = Ce^{\kappa x}$; as $\kappa \rightarrow \infty$, $\kappa x \rightarrow -\infty$ (since x is negative) and $e^{\kappa x} \rightarrow 0$, so the wave function approaches zero for all $x < 0$. Likewise, we can show that the wave function also approaches zero for all $x > L$. This is just what we found in Section 40.2; the wave function for a particle in a box must be zero outside the box.

EVALUATE: Our result in part (b) shows that the infinite square well is a *limiting case* of the finite well. We've seen many cases in Newtonian mechanics where it's important to consider limiting cases (such as Examples 5.11 and 5.13 in Section 5.2). Limiting cases are no less important in quantum mechanics.

Comparing Finite and Infinite Square Wells

Let's continue the comparison of the finite-depth potential well with the infinitely deep well, which we began in Example 40.5. First, because the wave functions for the finite well don't go to zero at $x = 0$ and $x = L$, the wavelength of the sinusoidal part of each wave function is *longer* than it would be with an infinite well. This increase in λ corresponds to a reduced magnitude of momentum $p = h/\lambda$ and therefore a reduced energy. Thus each energy level, including the ground level, is *lower* for a finite well than for an infinitely deep well with the same width.

Second, a well with finite depth U_0 has only a *finite* number of bound states and corresponding energy levels, compared to the *infinite* number for an infinitely deep well. How many levels there are depends on the magnitude of U_0 in comparison with the ground-level energy for the infinitely deep well (IDW), which we call $E_{1-\text{IDW}}$. From Eq. (40.31),

$$E_{1-\text{IDW}} = \frac{\pi^2\hbar^2}{2mL^2} \quad \text{(ground-level energy, infinitely deep well)} \quad (40.41)$$

When the well is very deep so U_0 is much larger than $E_{1-\text{IDW}}$, there are many bound states and the energies of the lowest few are nearly the same as the energies for the infinitely deep well. When U_0 is only a few times as large as $E_{1-\text{IDW}}$ there are only a few bound states. (There is always at least *one* bound state, no matter how shallow the well.) As with the infinitely deep well, there is no state with $E = 0$; such a state would violate the uncertainty principle.

40.15 (a) Wave functions for the three bound states for a particle in a finite potential well with depth U_0, for the case $U_0 = 6E_{1-\text{IDW}}$. (Here $E_{1-\text{IDW}}$ is the ground-level energy for an infinite well of the same width.) The horizontal brown line for each wave function corresponds to $\psi = 0$; the vertical placement of these lines indicates the energy of each bound state (compare Fig. 40.11). (b) Energy-level diagram for this system. The energies are expressed both as multiples of $E_{1-\text{IDW}}$ and as fractions of U_0. All energies greater than U_0 are possible; states with $E > U_0$ form a continuum.

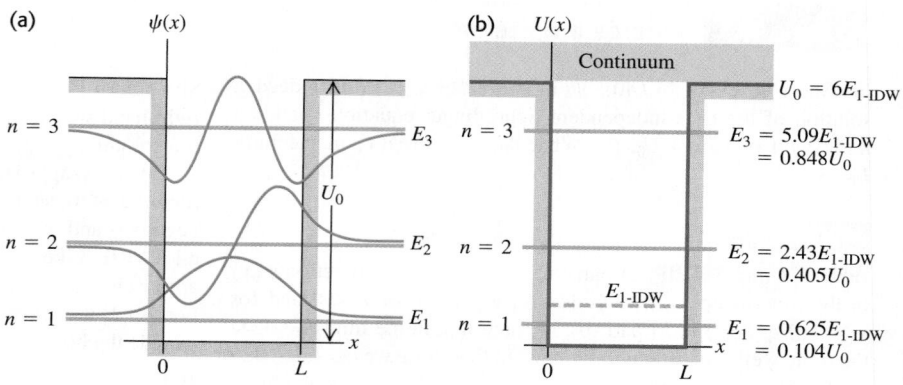

Figure 40.15 shows the case $U_0 = 6E_{1-\text{IDW}}$; for this particular case there are three bound states. In the figure, we express the energy levels both as fractions of the well depth U_0 and as multiples of $E_{1-\text{IDW}}$. Note that if the well were infinitely deep, the lowest three levels, as given by Eq. (40.31), would be $E_{1-\text{IDW}}$, $4E_{1-\text{IDW}}$, and $9E_{1-\text{IDW}}$. Figure 40.15 also shows the wave functions for the three bound states.

It turns out that when U_0 is less than $E_{1-\text{IDW}}$, there is only one bound state. In the limit when U_0 is *much smaller* than $E_{1-\text{IDW}}$ (a very shallow well), the energy of this single state is approximately $E = 0.68U_0$.

40.16 Probability distribution functions $|\psi(x)|^2$ for the square-well wave functions shown in Fig. 40.15. The horizontal brown line for each wave function corresponds to $|\psi|^2 = 0$.

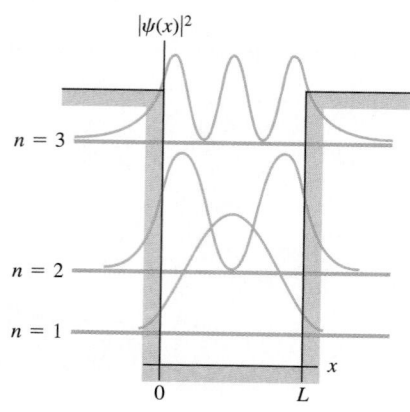

Figure 40.16 shows graphs of the probability distributions—that is, the values of $|\psi|^2$—for the wave functions shown in Fig. 40.15a. As with the infinite well, not all positions are equally likely. Unlike the infinite well, there is some probability of finding the particle outside the well in the classically forbidden regions.

There are also states for which E is *greater* than U_0. In these *free-particle states* the particle is not bound but is free to move through all values of x. *Any* energy E greater than U_0 is possible, so the free-particle states form a *continuum* rather than a discrete set of states with definite energy levels. The free-particle wave functions are sinusoidal both inside and outside the well. The wavelength is shorter inside the well than outside, corresponding to greater kinetic energy inside the well than outside it.

Figure 40.17 shows a graphic demonstration of particles in a *two*-dimensional finite potential well. Example 40.6 describes another application of the square-well potential.

Example 40.6 | **An electron in a finite well**

An electron is trapped in a square well 0.50 nm across (roughly five times a typical atomic diameter). (a) Find the ground-level energy $E_{1-\text{IDW}}$ if the well is infinitely deep. (b) Find the energy levels if the actual well depth U_0 is six times the ground-level energy found in part (a). (c) Find the wavelength of the photon emitted when the electron makes a transition from the $n = 2$ level to the $n = 1$ level. In what region of the electromagnetic spectrum does the photon wavelength lie? (d) If the electron is in the $n = 1$ (ground) level and absorbs a photon, what is the minimum photon energy that will free the electron from the well? In what region of the spectrum does the wavelength of this photon lie?

SOLUTION

IDENTIFY and SET UP: Equation (40.41) gives the ground-level energy $E_{1-\text{IDW}}$ for an infinitely deep well, and Fig. 40.15b shows the energies for a square well with $U_0 = 6E_{1-\text{IDW}}$. The energy of the photon emitted or absorbed in a transition is equal to the difference in energy between two levels involved in the transition; the photon wavelength is given by $E = hc/\lambda$ (see Chapter 38).

EXECUTE: (a) From Eq. (40.41),

$$E_{1-\text{IDW}} = \frac{\pi^2 \hbar^2}{2mL^2} = \frac{\pi^2 (1.055 \times 10^{-34} \text{ J} \cdot \text{s})^2}{2(9.11 \times 10^{-31} \text{ kg})(0.50 \times 10^{-9} \text{ m})^2}$$

$$= 2.4 \times 10^{-19} \text{ J} = 1.5 \text{ eV}$$

(b) We have $U_0 = 6E_{1-\text{IDW}} = 6(1.5 \text{ eV}) = 9.0 \text{ eV}$. We can read off the energy levels from Fig. 40.15b:

$$E_1 = 0.625E_{1-\text{IDW}} = 0.625(1.5 \text{ eV}) = 0.94 \text{ eV}$$

$$E_2 = 2.43E_{1-\text{IDW}} = 2.43(1.5 \text{ eV}) = 3.6 \text{ eV}$$

$$E_3 = 5.09E_{1-\text{IDW}} = 5.09(1.5 \text{ eV}) = 7.6 \text{ eV}$$

(c) The photon energy and wavelength for the $n = 2$ to $n = 1$ transition are

$$E_2 - E_1 = 3.6 \text{ eV} - 0.94 \text{ eV} = 2.7 \text{ eV}$$

$$\lambda = \frac{hc}{E} = \frac{(4.136 \times 10^{-15} \text{ eV} \cdot \text{s})(3.00 \times 10^8 \text{ m/s})}{2.7 \text{ eV}}$$

$$= 460 \text{ nm}$$

in the blue region of the visible spectrum.

(d) We see from Fig. 40.15b that the minimum energy needed to free the electron from the well from the $n = 1$ level is $U_0 - E_1 = 9.0 \text{ eV} - 0.94 \text{ eV} = 8.1 \text{ eV}$, which is three times the 2.7-eV photon energy found in part (c). Hence the corresponding photon wavelength is one-third of 460 nm, or (to two significant figures) 150 nm, which is in the ultraviolet region of the spectrum.

EVALUATE: As a check, you can also calculate the bound-state energies by using the formulas $E_1 = 0.104U_0$, $E_2 = 0.405U_0$, and $E_3 = 0.848U_0$ given in Fig. 40.15b. As an additional check, note that the first three energy levels of an infinitely deep well of the same width are $E_{1-\text{IDW}} = 1.5 \text{ eV}$, $E_{2-\text{IDW}} = 4E_{1-\text{IDW}} = 6.0 \text{ eV}$, and $E_{3-\text{IDW}} = 9E_{1-\text{IDW}} = 13.5 \text{ eV}$. The energies we found in part (b) are less than these values: As we mentioned earlier, the finite depth of the well lowers the energy levels compared to the levels for an infinitely deep well.

One application of these ideas is to *quantum dots*, which are nanometer-sized particles of a semiconductor such as cadmium selenide (CdSe). An electron within a quantum dot behaves much like a particle in a finite potential well of width L equal to the size of the dot. When quantum dots are illuminated with ultraviolet light, the electrons absorb the ultraviolet photons and are excited into high energy levels, such as the $n = 3$ level described in this example. If the electron returns to the ground level ($n = 1$) in two or more steps (for example, from $n = 3$ to $n = 2$ and from $n = 2$ to $n = 1$), one of the steps will involve emitting a visible-light photon, as we have calculated here. (We described this process of *fluorescence* in Section 39.3.) Increasing the value of L decreases the energies of the levels and hence the spacing between them, and thus decreases the energy and increases the wavelength of the emitted photons. The photograph that opens this chapter shows quantum dots of different sizes in solution: Each emits a characteristic wavelength that depends on the dot size. Quantum dots can be injected into living tissue and their fluorescent glow used as a tracer for biological research and for medicine. They may also be the key to a new generation of lasers and ultrafast computers.

Test Your Understanding of Section 40.3 Suppose that the width of the finite potential well shown in Fig. 40.15 is reduced by one-half. How must the value of U_0 change so that there are still just three bound energy levels whose energies are the fractions of U_0 shown in Fig. 40.15b? U_0 must: (i) increase by a factor of four; (ii) increase by a factor of two; (iii) remain the same; (iv) decrease by a factor of one-half; (v) decrease by a factor of one-fourth.

40.4 Potential Barriers and Tunneling

A **potential barrier** is the opposite of a potential well; it is a potential-energy function with a *maximum*. Figure 40.18 shows an example. In classical Newtonian mechanics, if a particle (such as a roller coaster) is located to the left of the barrier (which might be a hill), and if the total mechanical energy of the system is E_1, the particle cannot move farther to the right than $x = a$. If it did, the potential energy U would be greater than the total energy E and the kinetic energy $K = E - U$ would be negative. This is impossible in classical mechanics since $K = \frac{1}{2}mv^2$ can never be negative.

A quantum-mechanical particle behaves differently: If it encounters a barrier like the one in Fig. 40.18 and has energy less than E_2, it *may* appear on the other side. This phenomenon is called *tunneling*. In quantum-mechanical tunneling, unlike macroscopic, mechanical tunneling, the particle does not actually push through the barrier and loses no energy in the process.

Tunneling Through a Rectangular Barrier

To understand how tunneling can occur, let's look at the potential-energy function $U(x)$ shown in Fig. 40.19. It's like Fig. 40.13 turned upside-down; the potential energy is zero everywhere except in the range $0 \le x \le L$, where it has the value U_0. This might represent a simple model for the potential energy of an

40.17 To make this image, 48 iron atoms (shown as yellow peaks) were placed in a circle on a copper surface. The "elevation" at each point inside the circle indicates the electron density within the circle. The standing-wave pattern is very similar to the probability distribution function for a particle in a one-dimensional finite potential well. (This image was made with a scanning tunneling microscope, discussed in Section 40.4.)

Mastering**PHYSICS**

PhET: Quantum Tunneling and Wave Packets
ActivPhysics 20.4: Potential Barriers

40.18 A potential-energy barrier. According to Newtonian mechanics, if the total energy of the system is E_1, a particle to the left of the barrier can go no farther than $x = a$. If the total energy is greater than E_2, the particle can pass over the barrier.

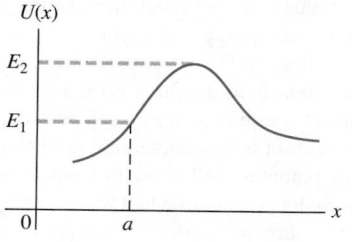

40.19 A rectangular potential-energy barrier with width L and height U_0. According to Newtonian mechanics, if the total energy E is less than U_0, a particle cannot pass over this barrier but is confined to the side where it starts.

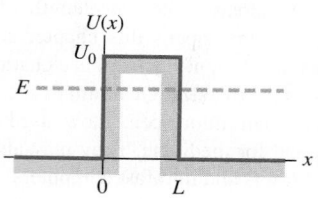

40.20 A possible wave function for a particle tunneling through the potential-energy barrier shown in Fig. 40.19.

electron in the presence of two slabs of metal separated by an air gap of thickness L. The potential energy is lower within either slab than in the gap between them.

Let's consider solutions of the Schrödinger equation for this potential-energy function for the case in which E is less than U_0. We can use our results from Section 40.3. In the regions $x < 0$ and $x > L$, where $U = 0$, the solution is sinusoidal and is given by Eq. (40.38). Within the barrier ($0 \le x \le L$), $U = U_0$ and the solution is exponential as in Eq. (40.40). Just as with the finite potential well, the functions have to join smoothly at the boundary points $x = 0$ and $x = L$, which means that both $\psi(x)$ and $d\psi(x)/dx$ have to be continuous at these points.

These requirements lead to a wave function like the one shown in Fig. 40.20. The function is *not* zero inside the barrier (the region forbidden by Newtonian mechanics). Even more remarkable, a particle that is initially to the *left* of the barrier has some probability of being found to the *right* of the barrier. How great this probability is depends on the width L of the barrier and the particle's energy E in comparison with the barrier height U_0. The **tunneling probability** T that the particle gets through the barrier is proportional to the square of the ratio of the amplitudes of the sinusoidal wave functions on the two sides of the barrier. These amplitudes are determined by matching wave functions and their derivatives at the boundary points, a fairly involved mathematical problem. When T is much smaller than unity, it is given approximately by

$$T = Ge^{-2\kappa L} \text{ where } G = 16\frac{E}{U_0}\left(1 - \frac{E}{U_0}\right) \text{ and } \kappa = \frac{\sqrt{2m(U_0 - E)}}{\hbar} \quad (40.42)$$

(probability of tunneling)

The probability decreases rapidly with increasing barrier width L. It also depends critically on the energy difference $U_0 - E$, which in Newtonian physics is the additional kinetic energy the particle would need to be able to climb over the barrier.

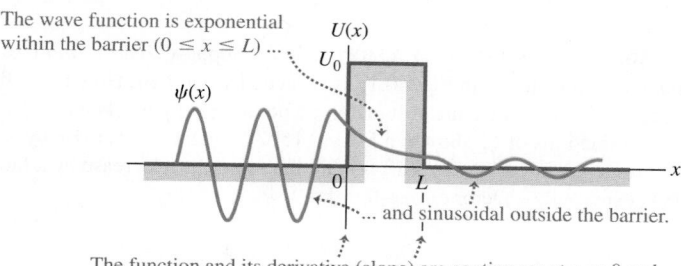

The wave function is exponential within the barrier ($0 \le x \le L$) …

$\psi(x)$

$U(x)$

U_0

… and sinusoidal outside the barrier.

The function and its derivative (slope) are continuous at $x = 0$ and $x = L$ so that the sinusoidal and exponential functions join smoothly.

Example 40.7 | **Tunneling through a barrier**

A 2.0-eV electron encounters a barrier 5.0 eV high. What is the probability that it will tunnel through the barrier if the barrier width is (a) 1.00 nm and (b) 0.50 nm?

SOLUTION

IDENTIFY and SET UP: This problem uses the ideas of tunneling through a rectangular barrier, as in Figs. 40.19 and 40.20. Our target variable is the tunneling probability T in Eq. (40.42), which we evaluate for the given values $E = 2.0$ eV (electron energy), $U = 5.0$ eV (barrier height), $m = 9.11 \times 10^{-31}$ kg (mass of the electron), and $L = 1.00$ nm or 0.50 nm (barrier width).

EXECUTE: First we evaluate G and κ in Eq. (40.42), using $E = 2.0$ eV:

$$G = 16\left(\frac{2.0 \text{ eV}}{5.0 \text{ eV}}\right)\left(1 - \frac{2.0 \text{ eV}}{5.0 \text{ eV}}\right) = 3.8$$

$$U_0 - E = 5.0 \text{ eV} - 2.0 \text{ eV} = 3.0 \text{ eV} = 4.8 \times 10^{-19} \text{ J}$$

$$\kappa = \frac{\sqrt{2(9.11 \times 10^{-31} \text{ kg})(4.8 \times 10^{-19} \text{ J})}}{1.055 \times 10^{-34} \text{ J}\cdot\text{s}} = 8.9 \times 10^9 \text{ m}^{-1}$$

(a) When $L = 1.00$ nm $= 1.00 \times 10^{-9}$ m, $2\kappa L = 2(8.9 \times 10^9 \text{ m}^{-1})(1.00 \times 10^{-9} \text{ m}) = 17.8$ and $T = Ge^{-2\kappa L} = 3.8e^{-17.8} = 7.1 \times 10^{-8}$.

(b) When $L = 0.50$ nm, one-half of 1.00 nm, $2\kappa L$ is one-half of 17.8, or 8.9. Hence $T = 3.8e^{-8.9} = 5.2 \times 10^{-4}$.

EVALUATE: Halving the width of this barrier increases the tunneling probability T by a factor of $(5.2 \times 10^{-4})/(7.1 \times 10^{-8}) = 7.3 \times 10^3$, or nearly ten thousand. The tunneling probability is an *extremely* sensitive function of the barrier width.

(a)

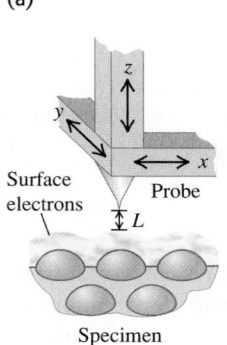

Surface electrons Probe

L

Specimen

(b)

40.21 (a) Schematic diagram of the probe of a scanning tunneling microscope (STM). As the sharp conducting probe is scanned across the surface in the x- and y-directions, it is also moved in the z-direction to maintain a constant tunneling current. The changing position of the probe is recorded and used to construct an image of the surface. (b) This colored STM image shows "quantum wires": thin strips, just 10 atoms wide, of a conductive rare-earth silicide atop a silicon surface. Such quantum wires may one day be the basis of ultraminiaturized circuits.

Applications of Tunneling

Tunneling has a number of practical applications, some of considerable importance. When you twist two copper wires together or close the contacts of a switch, current passes from one conductor to the other despite a thin layer of nonconducting copper oxide between them. The electrons tunnel through this thin insulating layer. A *tunnel diode* is a semiconductor device in which electrons tunnel through a potential barrier. The current can be switched on and off very quickly (within a few picoseconds) by varying the height of the barrier. A *Josephson junction* consists of two superconductors separated by an oxide layer a few atoms (1 to 2 nm) thick. Electron pairs in the superconductors can tunnel through the barrier layer, giving such a device unusual circuit properties. Josephson junctions are useful for establishing precise voltage standards and measuring tiny magnetic fields, and they play a crucial role in the developing field of quantum computing.

The *scanning tunneling microscope* (STM) uses electron tunneling to create images of surfaces down to the scale of individual atoms. An extremely sharp conducting needle is brought very close to the surface, within 1 nm or so (Fig. 40.21a). When the needle is at a positive potential with respect to the surface, electrons can tunnel through the surface potential-energy barrier and reach the needle. As Example 40.7 shows, the tunneling probability and hence the tunneling current are very sensitive to changes in the width L of the barrier (the distance between the surface and the needle tip). In one mode of operation the needle is scanned across the surface and at the same time is moved perpendicular to the surface to maintain a constant tunneling current. The needle motion is recorded, and after many parallel scans, an image of the surface can be reconstructed. Extremely precise control of needle motion, including isolation from vibration, is essential. Figure 40.21b shows an STM image. (Figure 40.17 is also an STM image.)

Tunneling is also of great importance in nuclear physics. A fusion reaction can occur when two nuclei tunnel through the barrier caused by their electrical repulsion and approach each other closely enough for the attractive nuclear force to cause them to fuse. Fusion reactions occur in the cores of stars, including the sun; without tunneling, the sun wouldn't shine. The emission of alpha particles from unstable nuclei such as radium also involves tunneling. An alpha particle is a cluster of two protons and two neutrons (the same as a nucleus of the most common form of helium). Such clusters form naturally within larger atomic nuclei. An alpha particle trying to escape from a nucleus encounters a potential barrier that results from the combined effect of the attractive nuclear force and the electrical repulsion of the remaining part of the nucleus (Fig. 40.22). The alpha particle can escape only by tunneling through this barrier. Depending on the barrier height and width for a given kind of alpha-emitting nucleus, the tunneling probability can be low or high, and the alpha-emitting material will have low or high radioactivity. Recall from Section 39.2 that Ernest Rutherford used alpha particles

Application Electron Tunneling in Enzymes
Protein molecules play essential roles as enzymes in living organisms. Enzymes like the one shown here are large molecules, and in many cases their function depends on the ability of electrons to tunnel across the space that separates one part of the molecule from another. Without tunneling, life as we know it would be impossible!

40.22 Approximate potential-energy function for an alpha particle interacting with a nucleus of radius R. If an alpha particle inside the nucleus has energy E greater than zero, it can tunnel through the barrier and escape from the nucleus.

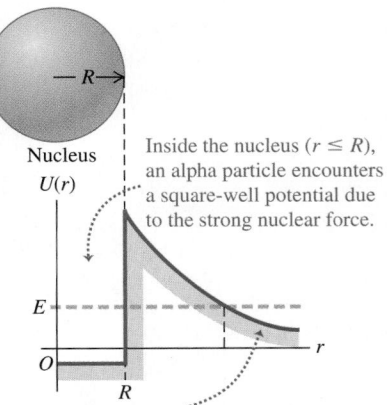

Nucleus
$U(r)$

Inside the nucleus ($r \le R$), an alpha particle encounters a square-well potential due to the strong nuclear force.

E

O R r

Outside the nucleus ($r > R$), an alpha particle experiences a $1/r$ potential due to electrostatic repulsion.

from a radioactive source to discover the atomic nucleus. Although Rutherford did not know it, tunneling by these alpha particles made his experiments possible! We'll learn more about alpha decay in Chapter 43.

Test Your Understanding of Section 40.4 Is it possible for a particle undergoing tunneling to be found *within* the barrier rather than on either side of it? ∎

40.5 The Harmonic Oscillator

Systems that *oscillate* are of tremendous importance in the physical world, from the oscillations of your eardrums in response to a sound wave to the vibrations of the ground caused by an earthquake. Oscillations are equally important on the microscopic scale where quantum effects dominate. The molecules of the air around you can be set into vibration when they collide with each other, the protons and neutrons in an excited atomic nucleus can oscillate in opposite directions, and a microwave oven transfers energy to food by making water molecules in the food flip back and forth. In this section we'll look at the solutions of the Schrödinger equation for the simplest kind of vibrating system, the quantum-mechanical harmonic oscillator.

As we learned in Chapter 14, a **harmonic oscillator** is a particle with mass m that moves along the x-axis under the influence of a conservative force $F_x = -k'x$. The constant k' is called the *force constant*. (In Chapter 14 we used the symbol k for the force constant. In this section we'll use the symbol k' instead to minimize confusion with the wave number $k = 2\pi/\lambda$.) The force is proportional to the particle's displacement x from its equilibrium position, $x = 0$. The corresponding potential-energy function is $U = \frac{1}{2}k'x^2$ (Fig. 40.23). In Newtonian mechanics, when the particle is displaced from equilibrium, it undergoes sinusoidal motion with frequency $f = (1/2\pi)(k'/m)^{1/2}$ and angular frequency $\omega = 2\pi f = (k'/m)^{1/2}$. The amplitude (that is, the maximum displacement from equilibrium) of these Newtonian oscillations is A, which is related to the energy E of the oscillator by $E = \frac{1}{2}k'A^2$.

Let's make an enlightened guess about the energy levels of a quantum-mechanical harmonic oscillator. In classical physics an electron oscillating with angular frequency ω emits electromagnetic radiation with that same angular frequency. It's reasonable to guess that when an excited quantum-mechanical harmonic oscillator with angular frequency $\omega = (k'/m)^{1/2}$ (according to Newtonian mechanics, at least) makes a transition from one energy level to a lower level, it would emit a photon with this same angular frequency ω. The energy of such a photon is $hf = (h/2\pi)(\omega/2\pi) = \hbar\omega$. So we would expect that the spacing between adjacent energy levels of the harmonic oscillator would be

$$hf = \hbar\omega = \hbar\sqrt{\frac{k'}{m}} \qquad (40.43)$$

That's the same spacing between energy levels that Planck assumed in deriving his radiation law (see Section 39.5). It was a good assumption; as we'll see, the energy levels are in fact half-integer $\left(\frac{1}{2}, \frac{3}{2}, \frac{5}{2}, \dots\right)$ multiples of $\hbar\omega$.

Wave Functions, Boundary Conditions, and Energy Levels

We'll begin our quantum-mechanical analysis of the harmonic oscillator by writing down the one-dimensional time-independent Schrödinger equation, Eq. (40.23), with $\frac{1}{2}k'x^2$ in place of U:

$$-\frac{\hbar^2}{2m}\frac{d^2\psi(x)}{dx^2} + \frac{1}{2}k'x^2\psi(x) = E\psi(x) \qquad \begin{matrix}\text{(Schrödinger equation for} \\ \text{the harmonic oscillator)}\end{matrix} \qquad (40.44)$$

The solutions of this equation are wave functions for the physically possible states of the system.

40.23 Potential-energy function for the harmonic oscillator. In Newtonian mechanics the amplitude A is related to the total energy E by $E = \frac{1}{2}k'A^2$, and the particle is restricted to the range from $x = -A$ to $x = A$. In quantum mechanics the particle can be found at $x > A$ or $x < -A$.

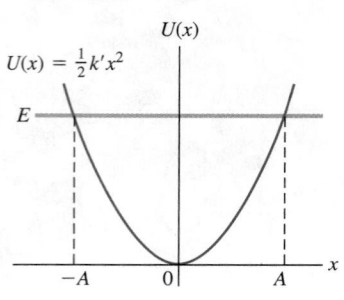

In the discussion of square-well potentials in Section 40.2 we found that the energy levels are determined by boundary conditions at the walls of the well. However, the harmonic-oscillator potential has no walls as such; what, then, are the appropriate boundary conditions? Classically, $|x|$ cannot be greater than the amplitude A given by $E = \frac{1}{2}k'A^2$. Quantum mechanics does allow some penetration into classically forbidden regions, but the probability decreases as that penetration increases. Thus the wave functions must approach zero as $|x|$ grows large.

Satisfying the requirement that $\psi(x) \to 0$ as $|x| \to \infty$ is not as trivial as it may seem. To see why this is, let's rewrite Eq. (40.44) in the form

$$\frac{d^2\psi(x)}{dx^2} = \frac{2m}{\hbar^2}\left(\tfrac{1}{2}k'x^2 - E\right)\psi(x) \qquad (40.45)$$

Equation (40.45) shows that when x is large enough (either positive or negative) to make the quantity $\left(\frac{1}{2}k'x^2 - E\right)$ positive, the function $\psi(x)$ and its second derivative $d^2\psi(x)/dx^2$ have the same sign. Figure 40.24 shows four possible kinds of behavior of $\psi(x)$ beginning at a point where x is greater than the classical amplitude A, so that $\frac{1}{2}k'x^2 - \frac{1}{2}k'A^2 = \frac{1}{2}k'x^2 - E > 0$. Let's look at these four cases more closely. Note that if $\psi(x)$ is positive as shown in Fig. 40.24, Eq. (40.45) tells us that $d^2\psi(x)/dx^2$ is also positive and the function is *concave upward*. Note also that $d^2\psi(x)/dx^2$ is the rate of change of the *slope* of $\psi(x)$; this will help us understand how our four possible wave functions behave.

- *Curve a:* The slope of $\psi(x)$ is positive at point x. Since $d^2\psi(x)/dx^2 > 0$, the function curves upward increasingly steeply and goes to infinity. This violates the boundary condition that $\psi(x) \to 0$ as $|x| \to \infty$, so this isn't a viable wave function.
- *Curve b:* The slope of $\psi(x)$ is negative at point x, and $d^2\psi(x)/dx^2$ has a large positive value. Hence the slope changes rapidly from negative to positive and keeps on increasing—so, again, the wave function goes to infinity. This wave function isn't viable either.
- *Curve c:* As for curve b, the slope is negative at point x. However, $d^2\psi(x)/dx^2$ now has a *small* positive value, so the slope increases only gradually as $\psi(x)$ decreases to zero and crosses over to negative values. Equation (40.45) tells us that once $\psi(x)$ becomes negative, $d^2\psi(x)/dx^2$ also becomes negative. Hence the curve becomes concave *downward* and heads for *negative* infinity. This wave function, too, fails to satisfy the requirement that $\psi(x) \to 0$ as $|x| \to \infty$ and thus isn't viable.
- *Curve d:* If the slope of $\psi(x)$ at point x is negative, and the positive value of $d^2\psi(x)/dx^2$ at this point is neither too large nor too small, the curve bends just enough to glide in asymptotically to the x-axis. In this case $\psi(x)$, $d\psi(x)/dx$, and $d^2\psi(x)/dx^2$ all approach zero at large x. This case offers the only hope of satisfying the boundary condition that $\psi(x) \to 0$ as $|x| \to \infty$, and it occurs only for certain very special values of the energy E.

This qualitative discussion suggests how the boundary conditions as $|x| \to \infty$ determine the possible energy levels for the quantum-mechanical harmonic oscillator. It turns out that these boundary conditions are satisifed only if the energy E is equal to one of the values E_n, given by the simple formula

$$E_n = \left(n + \tfrac{1}{2}\right)\hbar\sqrt{\frac{k'}{m}} = \left(n + \tfrac{1}{2}\right)\hbar\omega \quad (n = 0, 1, 2, \dots) \qquad (40.46)$$

(energy levels, harmonic oscillator)

where n is the quantum number identifying each state and energy level. Note that the ground level of energy $E_0 = \frac{1}{2}\hbar\omega$ is denoted by $n = 0$, *not* $n = 1$.

40.24 Possible behaviors of harmonic-oscillator wave functions in the region $\frac{1}{2}k'x^2 > E$. In this region, $\psi(x)$ and $d^2\psi(x)/dx^2$ have the same sign. The curve is concave upward when $d^2\psi(x)/dx^2$ is positive and concave downward when $d^2\psi(x)/dx^2$ is negative.

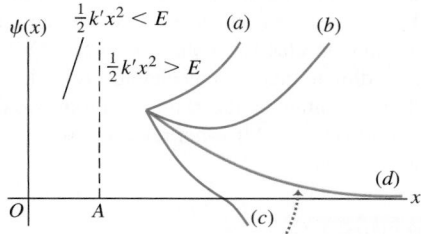

Only curve d, which approaches the x-axis asymptotically for large x, is an acceptable wave function for this system.

40.25 Energy levels for the harmonic oscillator. The spacing between any two adjacent levels is $\Delta E = \hbar\omega$. The energy of the ground level is $E_0 = \frac{1}{2}\hbar\omega$.

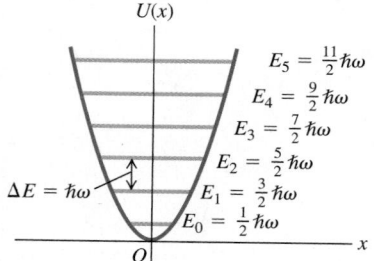

$$U(x)$$

$E_5 = \frac{11}{2}\hbar\omega$
$E_4 = \frac{9}{2}\hbar\omega$
$E_3 = \frac{7}{2}\hbar\omega$
$E_2 = \frac{5}{2}\hbar\omega$
$\Delta E = \hbar\omega$ $E_1 = \frac{3}{2}\hbar\omega$
$E_0 = \frac{1}{2}\hbar\omega$

O x

Equation (40.46) confirms our guess [(Eq. 40.43)] that adjacent energy levels are separated by a constant interval of $\hbar\omega = hf$, as Planck assumed in 1900. There are infinitely many levels; this shouldn't be surprising because we are dealing with an infinitely deep potential well. As $|x|$ increases, $U = \frac{1}{2}k'x^2$ increases without bound.

Figure 40.25 shows the lowest six energy levels and the potential-energy function $U(x)$. For each level n, the value of $|x|$ at which the horizontal line representing the total energy E_n intersects $U(x)$ gives the amplitude A_n of the corresponding Newtonian oscillator.

Example 40.8   Vibration in a crystal

A sodium atom of mass 3.82×10^{-26} kg vibrates within a crystal. The potential energy increases by 0.0075 eV when the atom is displaced 0.014 nm from its equilibrium position. Treat the atom as a harmonic oscillator. (a) Find the angular frequency of the oscillations according to Newtonian mechanics. (b) Find the spacing (in electron volts) of adjacent vibrational energy levels according to quantum mechanics. (c) What is the wavelength of a photon emitted as the result of a transition from one level to the next lower level? In what region of the electromagnetic spectrum does this lie?

SOLUTION

IDENTIFY and SET UP: We'll find the force constant k' from the expression $U = \frac{1}{2}k'x^2$ for potential energy. We'll then find the angular frequency $\omega = (k'/m)^{1/2}$ and use this in Eq. (40.46) to find the spacing between adjacent energy levels. We'll calculate the wavelength of the emitted photon as in Example 40.6.

EXECUTE: We are given that $U = 0.0075$ eV $= 1.2 \times 10^{-21}$ J when $x = 0.014 \times 10^{-9}$ m, so we can solve $U = \frac{1}{2}k'x^2$ for k':

$$k' = \frac{2U}{x^2} = \frac{2(1.2 \times 10^{-21} \text{ J})}{(0.014 \times 10^{-9} \text{ m})^2} = 12.2 \text{ N/m}$$

(a) The Newtonian angular frequency is

$$\omega = \sqrt{\frac{k'}{m}} = \sqrt{\frac{12.2 \text{ N/m}}{3.82 \times 10^{-26} \text{ kg}}} = 1.79 \times 10^{13} \text{ rad/s}$$

(b) From Eq. (40.46) and Fig. 40.25, the spacing between adjacent energy levels is

$$\hbar\omega = (1.054 \times 10^{-34} \text{ J·s})(1.79 \times 10^{13} \text{ s}^{-1})$$
$$= 1.88 \times 10^{-21} \text{ J}\left(\frac{1 \text{ eV}}{1.602 \times 10^{-19} \text{ J}}\right) = 0.0118 \text{ eV}$$

(c) The energy E of the emitted photon is equal to the energy lost by the oscillator in the transition, 0.0118 eV. Then

$$\lambda = \frac{hc}{E} = \frac{(4.136 \times 10^{-15} \text{ eV·s})(3.00 \times 10^8 \text{ m/s})}{0.0118 \text{ eV}}$$
$$= 1.05 \times 10^{-4} \text{ m} = 105 \text{ } \mu\text{m}$$

This photon wavelength is in the infrared region of the spectrum.

EVALUATE: This example shows us that interatomic force constants are a few newtons per meter, about the same as those of household springs or spring-based toys such as the Slinky. It also suggests that we can learn about the vibrations of molecules by measuring the radiation that they emit in transitioning to a lower vibrational state. We will explore this idea further in Chapter 42.

Mastering PHYSICS

ActivPhysics 20.1.6: Potential Energy Diagrams, Question 6

Comparing Quantum and Newtonian Oscillators

The wave functions for the levels $n = 0, 1, 2, \dots$ of the harmonic oscillator are called *Hermite functions;* they aren't encountered in elementary calculus courses but are well known to mathematicians. Each Hermite function is an exponential function multiplied by a polynomial in x. The harmonic-oscillator wave function corresponding to $n = 0$ and $E = E_0$ (the ground level) is

$$\psi(x) = Ce^{-\sqrt{mk'}\,x^2/2\hbar} \qquad (40.47)$$

The constant C is chosen to normalize the function—that is, to make $\int_{-\infty}^{\infty} |\psi|^2\, dx = 1$. (We're using C rather than A as a normalization constant in this section, since we've already appropriated the symbol A to denote the Newtonian amplitude of a harmonic oscillator.) You can find C using the following result from integral tables:

$$\int_{-\infty}^{\infty} e^{-a^2x^2}\, dx = \frac{\sqrt{\pi}}{a}$$

40.26 The first four wave functions for the harmonic oscillator. The amplitude A of a Newtonian oscillator with the same total energy is shown for each. Each wave function penetrates somewhat into the classically forbidden regions $|x| > A$. The total number of finite maxima and minima for each function is $n + 1$, one more than the quantum number.

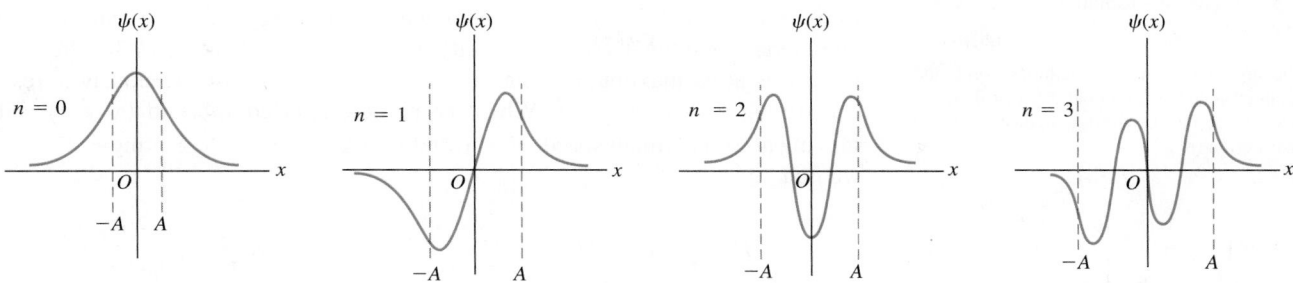

To confirm that $\psi(x)$ as given by Eq. (40.47) really *is* a solution of the Schrödinger equation for the harmonic oscillator, we invite you to calculate the second derivative of this wave function, substitute it into Eq. (40.44), and verify that the equation is satisfied if the energy E is equal to $E_0 = \frac{1}{2}\hbar\omega$ (see Exercise 40.38). It's a little messy, but the result is satisfying and worth the effort.

Figure 40.26 shows the the first four harmonic-oscillator wave functions. Each graph also shows the amplitude A of a Newtonian harmonic oscillator with the same energy—that is, the value of A determined from

$$\tfrac{1}{2}k'A^2 = \left(n + \tfrac{1}{2}\right)\hbar\omega \qquad (40.48)$$

In each case there is some penetration of the wave function into the regions $|x| > A$ that are forbidden by Newtonian mechanics. This is similar to the effect that we noted in Section 40.3 for a particle in a finite square well.

Figure 40.27 shows the probability distributions $|\psi(x)|^2$ for these same states. Each graph also shows the probability distribution determined from Newtonian physics, in which the probability of finding the particle near a randomly chosen point is inversely proportional to the particle's speed at that point. If we average out the wiggles in the quantum-mechanical probability curves, the results for $n > 0$ resemble the Newtonian predictions. This agreement improves with increasing n; Fig. 40.28 shows the classical and quantum-mechanical probability functions for $n = 10$. Notice that the spacing between zeros of $|\psi(x)|^2$ in Fig. 40.28 increases with increasing distance from $x = 0$. This makes sense from the Newtonian perspective: As a particle moves away from $x = 0$, its kinetic energy K and the magnitude p of its momentum both decrease. Thinking quantum-mechanically, this means that the wavelength $\lambda = h/p$ increases, so the spacing between zeros of $\psi(x)$ (and hence of $|\psi(x)|^2$) also increases.

In the Newtonian analysis of the harmonic oscillator the minimum energy is zero, with the particle at rest at its equilibrium position $x = 0$. This is not possible in quantum mechanics; no solution of the Schrödinger equation has $E = 0$ and satisfies the boundary conditions. Furthermore, if there were such a state, it

40.27 Probability distribution functions $|\psi(x)|^2$ for the harmonic-oscillator wave functions shown in Fig. 40.26. The amplitude A of the Newtonian motion with the same energy is shown for each. The blue lines show the corresponding probability distributions for the Newtonian motion. As n increases, the averaged-out quantum-mechanical functions resemble the Newtonian curves more and more.

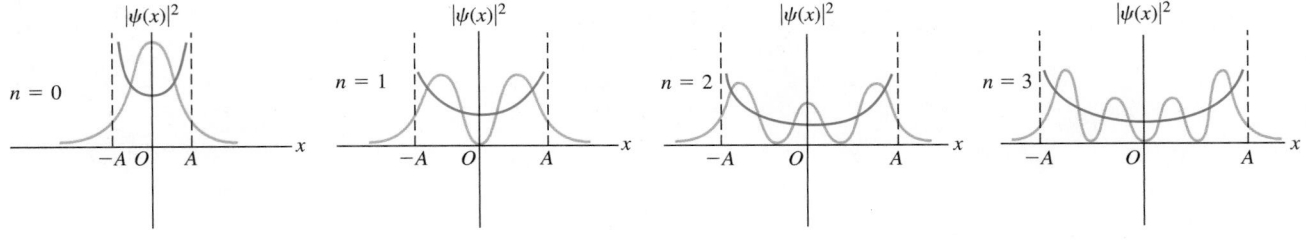

40.28 Newtonian and quantum-mechanical probability distribution functions for a harmonic oscillator for the state $n = 10$. The Newtonian amplitude A is also shown.

The larger the value of n, the more closely the quantum-mechanical probability distribution (green) matches the Newtonian probability distribution (blue).

40.29 A potential-energy function describing the interaction of two atoms in a diatomic molecule. The distance r is the separation between the centers of the atoms, and the equilibrium separation is $r = r_0$. The energy needed to dissociate the molecule is U_∞.

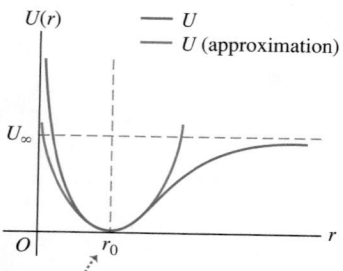

When r is near r_0, the potential-energy curve is approximately parabolic (as shown by the red curve) and the system behaves approximately like a harmonic oscillator.

would violate the Heisenberg uncertainty principle because there would be no uncertainty in either position or momentum. The energy must be at least $\frac{1}{2}\hbar\omega$ for the system to conform to the uncertainty principle. To see qualitatively why this is so, consider a Newtonian oscillator with total energy $\frac{1}{2}\hbar\omega$. We can find the amplitude A and the maximum velocity just as we did in Section 14.3. When the particle is at its maximum displacement ($x = \pm A$) and instantaneously at rest, $K = 0$ and $E = U = \frac{1}{2}k'A^2$. When the particle is at equilibrium ($x = 0$) and moving at its maximum speed, $U = 0$ and $E = K = \frac{1}{2}mv_{max}^2$. Setting $E = \frac{1}{2}\hbar\omega$, we find

$$E = \tfrac{1}{2}k'A^2 = \tfrac{1}{2}\hbar\omega = \tfrac{1}{2}\hbar\left(\frac{k'}{m}\right)^{1/2} \quad \text{so} \quad A = \frac{\hbar^{1/2}}{k'^{1/4}m^{1/4}}$$

$$E = \tfrac{1}{2}mv_{max}^2 = \tfrac{1}{2}k'A^2 \quad \text{so} \quad v_{max} = A\left(\frac{k'}{m}\right)^{1/2} = \frac{\hbar^{1/2}k'^{1/4}}{m^{3/4}}$$

The maximum *momentum* of the particle is

$$p_{max} = mv_{max} = \hbar^{1/2}k'^{1/4}m^{1/4}$$

Here's where the Heisenberg uncertainty principle comes in. It turns out that the uncertainties in the particle's position and momentum (calculated as standard deviations) are, respectively, $\Delta x = A/\sqrt{2} = A/2^{1/2}$ and $\Delta p_x = p_{max}/\sqrt{2} = p_{max}/2^{1/2}$. Then the product of the two uncertainties is

$$\Delta x\, \Delta p_x = \left(\frac{\hbar^{1/2}}{2^{1/2}\,k'^{1/4}\,m^{1/4}}\right)\left(\frac{\hbar^{1/2}\,k'^{1/4}\,m^{1/4}}{2^{1/2}}\right) = \frac{\hbar}{2}$$

This product equals the minimum value allowed by Eq. (39.29), $\Delta x\, \Delta p_x \geq \hbar/2$, and thus satisfies the uncertainty principle. If the energy had been less than $\frac{1}{2}\hbar\omega$, the product $\Delta x\, \Delta p_x$ would have been less than $\hbar/2$, and the uncertainty principle would have been violated.

Even when a potential-energy function isn't precisely parabolic in shape, we may be able to approximate it by the harmonic-oscillator potential for sufficiently small displacements from equilibrium. Figure 40.29 shows a typical potential-energy function for an interatomic force in a molecule. At large separations the curve of $U(r)$ versus r levels off, corresponding to the absence of force at great distances. But the curve is approximately parabolic near the minimum of $U(r)$ (the equilibrium separation of the atoms). Near equilibrium the molecular vibration is approximately simple harmonic with energy levels given by Eq. (40.46), as we assumed in Example 40.8.

Test Your Understanding of Section 40.5 A quantum-mechanical system initially in its ground level absorbs a photon and ends up in the first excited state. The system then absorbs a second photon and ends up in the second excited state. For which of the following systems does the second photon have a longer wavelength than the first one? (i) a harmonic oscillator; (ii) a hydrogen atom; (iii) a particle in a box.

Wave functions: The wave function for a particle contains all of the information about that particle. If the particle moves in one dimension in the presence of a potential energy function $U(x)$, the wave function $\Psi(x, t)$ obeys the one-dimensional Schrödinger equation. (For a *free* particle on which no forces act, $U(x) = 0$.) The quantity $|\Psi(x, t)|^2$, called the probability distribution function, determines the relative probability of finding a particle near a given position at a given time. If the particle is in a state of definite energy, called a stationary state, $\Psi(x, t)$ is a product of a function $\psi(x)$ that depends only on spatial coordinates and a function $e^{-iEt/\hbar}$ that depends only on time. For a stationary state, the probability distribution function is independent of time.

A spatial stationary-state wave function $\psi(x)$ for a particle that moves in one dimension in the presence of a potential-energy function $U(x)$ satisfies the time-independent Schrödinger equation. More complex wave functions can be constructed by superposing stationary-state wave functions. These can represent particles that are localized in a certain region, thus representing both particle and wave aspects. (See Examples 40.1 and 40.2.)

$$-\frac{\hbar^2}{2m}\frac{\partial^2\Psi(x, t)}{\partial x^2} + U(x)\Psi(x, t)$$

$$= i\hbar\frac{\partial\Psi(x, t)}{\partial t} \qquad (40.20)$$

(general 1-D Schrödinger equation)

$$\Psi(x, t) = \psi(x)e^{-iEt/\hbar} \qquad (40.21)$$

(time-dependent wave function for a state of definite energy)

$$-\frac{\hbar^2}{2m}\frac{d^2\psi(x)}{dx^2} + U(x)\psi(x) = E\psi(x)$$

(time-independent Schrödinger equation) $\qquad (40.23)$

Particle in a box: The energy levels for a particle of mass m in a box (an infinitely deep square potential well) with width L are given by Eq. (40.31). The corresponding normalized stationary-state wave functions of the particle are given by Eq. (40.35). (See Examples 40.3 and 40.4.)

$$E_n = \frac{p_n^2}{2m} = \frac{n^2h^2}{8mL^2} = \frac{n^2\pi^2\hbar^2}{2mL^2}$$

$$(n = 1, 2, 3, \ldots) \qquad (40.31)$$

$$\psi_n(x) = \sqrt{\frac{2}{L}}\sin\frac{n\pi x}{L}$$

$$(n = 1, 2, 3, \ldots) \qquad (40.35)$$

Wave functions and normalization: To be a solution of the Schrödinger equation, the wave function $\psi(x)$ and its derivative $d\psi(x)/dx$ must be continuous everywhere except where the potential-energy function $U(x)$ has an infinite discontinuity. Wave functions are usually normalized so that the total probability of finding the particle somewhere is unity.

$$\int_{-\infty}^{\infty}|\psi(x)|^2\, dx = 1 \qquad (40.33)$$

(normalization condition)

Finite potential well: In a potential well with finite depth U_0, the energy levels are lower than those for an infinitely deep well with the same width, and the number of energy levels corresponding to bound states is finite. The levels are obtained by matching wave functions at the well walls to satisfy the continuity of $\psi(x)$ and $d\psi(x)/dx$. (See Examples 40.5 and 40.6.)

Potential barriers and tunneling: There is a certain probability that a particle will penetrate a potential-energy barrier even though its initial energy is less than the barrier height. This process is called tunneling. (See Example 40.7.)

Quantum harmonic oscillator: The energy levels for the harmonic oscillator (for which $U(x) = \frac{1}{2}k'x^2$) are given by Eq. (40.46). The spacing between any two adjacent levels is $\hbar\omega$, where $\omega = \sqrt{k'/m}$ is the oscillation angular frequency of the corresponding Newtonian harmonic oscillator. (See Example 40.8.)

$$E_n = \left(n + \tfrac{1}{2}\right)\hbar\sqrt{\frac{k'}{m}} = \left(n + \tfrac{1}{2}\right)\hbar\omega$$

$$(n = 0, 1, 2, 3, \ldots) \qquad (40.46)$$

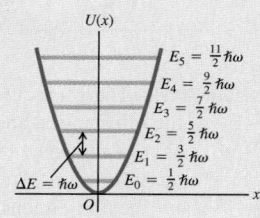

BRIDGING PROBLEM   A Packet in a Box

A particle of mass m in an infinitely deep well has the following wave function in the region from $x = 0$ to $x = L$:

$$\Psi(x, t) = \frac{1}{\sqrt{2}}\psi_1(x)e^{-iE_1t/\hbar} + \frac{1}{\sqrt{2}}\psi_2(x)e^{-iE_2t/\hbar}$$

Here $\psi_1(x)$ and $\psi_2(x)$ are the normalized stationary-state wave functions for the first two levels ($n = 1$ and $n = 2$), given by Eq. (40.35). E_1 and E_2, given by Eq. (40.31), are the energies of these levels. The wave function is zero for $x < 0$ and for $x > L$. (a) Find the probability distribution function for this wave function. (b) Does $\Psi(x, t)$ represent a stationary state of definite energy? How can you tell? (c) Show that the wave function $\Psi(x, t)$ is normalized. (d) Find the angular frequency of oscillation of the probability distribution function. What is the interpretation of this oscillation? (e) Suppose instead that $\Psi(x, t)$ is a combination of the wave functions of the two lowest levels of a finite well of length L and height U_0 equal to six times the energy of the lowest-energy bound state of an infinite well of length L. What would be the angular frequency of the probability distribution function in this case?

SOLUTION GUIDE

See MasteringPhysics® study area for a Video Tutor solution.

IDENTIFY and SET UP

1. In Section 40.1 we saw how to interpret a combination of two free-particle wave functions of different energies. In this problem you need to apply these same ideas to a combination of wave functions for the infinite well (Section 40.2) and the finite well (Section 40.3).

EXECUTE

2. Write down the full time-dependent wave function $\Psi(x, t)$ and its complex conjugate $\Psi^*(x, t)$ using the functions $\psi_1(x)$ and $\psi_2(x)$ from Eq. (40.35). Use these to calculate the probability distribution function, and decide whether or not this function depends on time.

3. To check for normalization, you'll need to verify that when you integrate the probability distribution function from step 2 over all values of x, the integral is equal to 1. [*Hint:* The trigonometric identities $\sin^2\theta = \frac{1}{2}(1 - \cos 2\theta)$ and $\sin\theta\sin\phi = \cos(\theta - \phi) - \cos(\theta + \phi)$ may be helpful.]

4. To find the answer to part (d) you'll need to identify the oscillation angular frequency ω_{osc} in your expression from step 2 for the probability distribution function. To interpret the oscillations, draw graphs of the probability distribution functions at times $t = 0$, $t = T/4$, $t = T/2$, and $t = 3T/4$, where $T = 2\pi/\omega_{\text{osc}}$ is the oscillation period of the probability distribution function.

5. For the finite well you do not have simple expressions for the first two stationary-state wave functions $\psi_1(x)$ and $\psi_2(x)$. However, you can still find the oscillation angular frequency ω_{osc}, which is related to the energies E_1 and E_2 in the same way as for the infinite-well case. (Can you see why?)

EVALUATE

6. Why are the factors of $1/\sqrt{2}$ in the wave function $\Psi(x, t)$ important?

7. Why do you suppose the oscillation angular frequency for a finite well is lower than for an infinite well of the same width?

Problems

For instructor-assigned homework, go to www.masteringphysics.com

•, ••, •••: Problems of increasing difficulty. **CP**: Cumulative problems incorporating material from earlier chapters. **CALC**: Problems requiring calculus. **BIO**: Biosciences problems.

DISCUSSION QUESTIONS

Q40.1 If quantum mechanics replaces the language of Newtonian mechanics, why don't we have to use wave functions to describe the motion of macroscopic bodies such as baseballs and cars?

Q40.2 A student remarks that the relationship of ray optics to the more general wave picture is analogous to the relationship of New-

tonian mechanics, with well-defined particle trajectories, to quantum mechanics. Comment on this remark.

Q40.3 As Eq. (40.21) indicates, the time-dependent wave function for a stationary state is a complex number having a real part and an imaginary part. How can this function have any physical meaning, since part of it is *imaginary*?

Q40.4 Why must the wave function of a particle be normalized?

Q40.5 If a particle is in a stationary state, does that mean that the particle is not moving? If a particle moves in empty space with constant momentum $\vec{p}$ and hence constant energy $E = p^2/2m$, is it in a stationary state? Explain your answers.

Q40.6 For the particle in a box, we chose $k = n\pi/L$ with $n = 1, 2, 3, \ldots$ to fit the boundary condition that $\psi = 0$ at $x = L$. However, $n = 0, -1, -2, -3, \ldots$ also satisfy that boundary condition. Why didn't we also choose those values of n?

Q40.7 If ψ is normalized, what is the physical significance of the area under a graph of $|\psi|^2$ versus x between x_1 and x_2? What is the total area under the graph of $|\psi|^2$ when all x are included? Explain.

Q40.8 For a particle in a box, what would the probability distribution function $|\psi|^2$ look like if the particle behaved like a classical (Newtonian) particle? Do the actual probability distributions approach this classical form when n is very large? Explain.

Q40.9 In Chapter 15 we represented a standing wave as a superposition of two waves traveling in opposite directions. Can the wave functions for a particle in a box also be thought of as a combination of two traveling waves? Why or why not? What physical interpretation does this representation have? Explain.

Q40.10 A particle in a box is in the ground level. What is the probability of finding the particle in the right half of the box? (Refer to Fig. 40.12, but don't evaluate an integral.) Is the answer the same if the particle is in an excited level? Explain.

Q40.11 The wave functions for a particle in a box (see Fig. 40.12a) are zero at certain points. Does this mean that the particle can't move past one of these points? Explain.

Q40.12 For a particle confined to an infinite square well, is it correct to say that each state of definite energy is also a state of definite wavelength? Is it also a state of definite momentum? Explain. (*Hint:* Remember that momentum is a vector.)

Q40.13 For a particle in a finite potential well, is it correct to say that each bound state of definite energy is also a state of definite wavelength? Is it a state of definite momentum? Explain.

Q40.14 In Fig. 40.12b, the probability function is zero at the points $x = 0$ and $x = L$, the "walls" of the box. Does this mean that the particle never strikes the walls? Explain.

Q40.15 A particle is confined to a finite potential well in the region $0 < x < L$. How does the area under the graph of $|\psi|^2$ in the region $0 < x < L$ compare to the total area under the graph of $|\psi|^2$ when including all possible x?

Q40.16 Compare the wave functions for the first three energy levels for a particle in a box of width L (see Fig. 40.12a) to the corresponding wave functions for a finite potential well of the same width (see Fig. 40.15a). How does the wavelength in the interval $0 \leq x \leq L$ for the $n = 1$ level of the particle in a box compare to the corresponding wavelength for the $n = 1$ level of the finite potential well? Use this to explain why E_1 is less than $E_{1-\text{IDW}}$ in the situation depicted in Fig. 40.15b.

Q40.17 It is stated in Section 40.3 that a finite potential well always has at least one bound level, no matter how shallow the well. Does this mean that as $U_0 \to 0$, $E_1 \to 0$? Does this violate the Heisenberg uncertainty principle? Explain.

Q40.18 Figure 40.15a shows that the higher the energy of a bound state for a finite potential well, the more the wave function extends outside the well (into the intervals $x < 0$ and $x > L$). Explain why this happens.

Q40.19 In classical (Newtonian) mechanics, the total energy E of a particle can never be less than the potential energy U because the kinetic energy K cannot be negative. Yet in barrier tunneling (see Section 40.4) a particle passes through regions where E is less than U. Is this a contradiction? Explain.

Q40.20 Figure 40.17 shows the scanning tunneling microscope image of 48 iron atoms placed on a copper surface, the pattern indicating the density of electrons on the copper surface. What can you infer about the potential-energy function inside the circle of iron atoms?

Q40.21 Qualitatively, how would you expect the probability for a particle to tunnel through a potential barrier to depend on the height of the barrier? Explain.

Q40.22 The wave function shown in Fig. 40.20 is nonzero for both $x < 0$ and $x > L$. Does this mean that the particle splits into two parts when it strikes the barrier, with one part tunneling through the barrier and the other part bouncing off the barrier? Explain.

Q40.23 The probability distributions for the harmonic oscillator wave functions (see Figs. 40.27 and 40.28) begin to resemble the classical (Newtonian) probability distribution when the quantum number n becomes large. Would the distributions become the same as in the classical case in the limit of very large n? Explain.

Q40.24 In Fig. 40.28, how does the probability of finding a particle in the center half of the region $-A < x < A$ compare to the probability of finding the particle in the outer half of the region? Is this consistent with the physical interpretation of the situation?

Q40.25 Compare the allowed energy levels for the hydrogen atom, the particle in a box, and the harmonic oscillator. What are the values of the quantum number n for the ground level and the second excited level of each system?

Q40.26 Sketch the wave function for the potential-energy well shown in Fig. Q40.26 when E_1 is less than U_0 and when E_3 is greater than U_0.

Figure **Q40.26**

EXERCISES

Section 40.1 Wave Functions and the One-Dimensional Schrödinger Equation

40.1 • An electron is moving as a free particle in the $-x$-direction with momentum that has magnitude 4.50×10^{-24} kg·m/s. What is the one-dimensional time-dependent wave function of the electron?

40.2 • A free particle moving in one dimension has wave function

$$\Psi(x, t) = A\left[e^{i(kx - \omega t)} - e^{i(2kx - 4\omega t)}\right]$$

where k and ω are positive real constants. (a) At $t = 0$ what are the two smallest positive values of x for which the probability function $|\Psi(x, t)|^2$ is a maximum? (b) Repeat part (a) for time $t = 2\pi/\omega$. (c) Calculate v_{av} as the distance the maxima have moved divided by the elapsed time. Compare your result to the expression $v_{\text{av}} = (\omega_2 - \omega_1)/(k_2 - k_1)$ from Example 40.1.

40.3 • Consider the free-particle wave function of Example 40.1. Let $k_2 = 3k_1 = 3k$. At $t = 0$ the probability distribution function $|\Psi(x, t)|^2$ has a maximum at $x = 0$. (a) What is the smallest positive value of x for which the probability distribution function has a maximum at time $t = 2\pi/\omega$, where $\omega = \hbar k^2/2m$? (b) From your result in part (a), what is the average speed with which the probability distribution is moving in the $+x$-direction? Compare your result to the expression $v_{\text{av}} = (\omega_2 - \omega_1)/(k_2 - k_1)$ from Example 40.1.

40.4 • Consider the free particle of Example 40.1. Show that $v_{\text{av}} = (\omega_2 - \omega_1)/(k_2 - k_1)$ can be written as $v_{\text{av}} = p_{\text{av}}/m$, where $p_{\text{av}} = (\hbar k_2 + \hbar k_1)/2$.

40.5 • Consider a wave function given by $\psi(x) = A \sin kx$, where $k = 2\pi/\lambda$ and A is a real constant. (a) For what values of x is there the highest probability of finding the particle described by this wave function? Explain. (b) For which values of x is the probability *zero*? Explain.

40.6 •• Compute $|\Psi|^2$ for $\Psi = \psi \sin \omega t$, where ψ is time independent and ω is a real constant. Is this a wave function for a stationary state? Why or why not?

40.7 • **CALC** Let ψ_1 and ψ_2 be two solutions of Eq. (40.23) with energies E_1 and E_2, respectively, where $E_1 \neq E_2$. Is $\psi = A\psi_1 + B\psi_2$, where A and B are nonzero constants, a solution to Eq. (40.23)? Explain your answer.

40.8 • A particle is described by a wave function $\psi(x) = Ae^{-\alpha x^2}$, where A and α are real, positive constants. If the value of α is increased, what effect does this have on (a) the particle's uncertainty in position and (b) the particle's uncertainty in momentum? Explain your answers.

40.9 • **CALC** **Linear Combinations of Wave Functions.** Let ψ_1 and ψ_2 be two solutions of Eq. (40.23) with the same energy E. Show that $\psi = B\psi_1 + C\psi_2$ is also a solution with energy E, for any values of the constants B and C.

Section 40.2 Particle in a Box

40.10 •• **CALC** A particle moving in one dimension (the x-axis) is described by the wave function

$$\psi(x) = \begin{cases} Ae^{-bx}, & \text{for } x \geq 0 \\ Ae^{bx}, & \text{for } x < 0 \end{cases}$$

where $b = 2.00 \text{ m}^{-1}, A > 0$, and the $+x$-axis points toward the right. (a) Determine A so that the wave function is normalized. (b) Sketch the graph of the wave function. (c) Find the probability of finding this particle in each of the following regions: (i) within 50.0 cm of the origin, (ii) on the left side of the origin (can you first guess the answer by looking at the graph of the wave function?), (iii) between $x = 0.500$ m and $x = 1.00$ m.

40.11 • **Ground-Level Billiards.** (a) Find the lowest energy level for a particle in a box if the particle is a billiard ball $(m = 0.20 \text{ kg})$ and the box has a width of 1.3 m, the size of a billiard table. (Assume that the billiard ball slides without friction rather than rolls; that is, ignore the *rotational* kinetic energy.) (b) Since the energy in part (a) is all kinetic, to what speed does this correspond? How much time would it take at this speed for the ball to move from one side of the table to the other? (c) What is the difference in energy between the $n = 2$ and $n = 1$ levels? (d) Are quantum-mechanical effects important for the game of billiards?

40.12 • A proton is in a box of width L. What must the width of the box be for the ground-level energy to be 5.0 MeV, a typical value for the energy with which the particles in a nucleus are bound? Compare your result to the size of a nucleus—that is, on the order of 10^{-14} m.

40.13 •• Find the width L of a one-dimensional box for which the ground-state energy of an electron in the box equals the absolute value of the ground state of a hydrogen atom.

40.14 •• When a hydrogen atom undergoes a transition from the $n = 2$ to the $n = 1$ level, a photon with $\lambda = 122$ nm is emitted. (a) If the atom is modeled as an electron in a one-dimensional box, what is the width of the box in order for the $n = 2$ to $n = 1$ transition to correspond to emission of a photon of this energy? (b) For a box with the width calculated in part (a), what is the ground-state energy? How does this correspond to the ground-state energy of a hydrogen atom? (c) Do you think a one-dimensional box is a good

model for a hydrogen atom? Explain. (*Hint:* Compare the spacing between adjacent energy levels as a function of n.)

40.15 •• A certain atom requires 3.0 eV of energy to excite an electron from the ground level to the first excited level. Model the atom as an electron in a box and find the width L of the box.

40.16 • An electron in a one-dimensional box has ground-state energy 1.00 eV. What is the wavelength of the photon absorbed when the electron makes a transition to the second excited state?

40.17 • **CALC** Show that the time-dependent wave function given by Eq. (40.35) is a solution to the one-dimensional Schrödinger equation, Eq. (40.23).

40.18 • Recall that $|\psi|^2 \, dx$ is the probability of finding the particle that has normalized wave function $\psi(x)$ in the interval x to $x + dx$. Consider a particle in a box with rigid walls at $x = 0$ and $x = L$. Let the particle be in the ground level and use ψ_n as given in Eq. (40.35). (a) For which values of x, if any, in the range from 0 to L is the probability of finding the particle zero? (b) For which values of x is the probability highest? (c) In parts (a) and (b) are your answers consistent with Fig. 40.12? Explain.

40.19 • Repeat Exercise 40.18 for the particle in the first excited level.

40.20 • **CALC** (a) Show that $\psi = A \sin kx$ is a solution to Eq. (40.25) if $k = \sqrt{2mE}/\hbar$. (b) Explain why this is an acceptable wave function for a particle in a box with rigid walls at $x = 0$ and $x = L$ only if k is an integer multiple of π/L.

40.21 • **CALC** (a) Repeat Exercise 40.20 for $\psi = A \cos kx$. (b) Explain why this cannot be an acceptable wave function for a particle in a box with rigid walls at $x = 0$ and $x = L$ no matter what the value of k.

40.22 • (a) Find the excitation energy from the ground level to the third excited level for an electron confined to a box that has a width of 0.125 nm. (b) The electron makes a transition from the $n = 1$ to $n = 4$ level by absorbing a photon. Calculate the wavelength of this photon.

40.23 • An electron is in a box of width 3.0×10^{-10} m. What are the de Broglie wavelength and the magnitude of the momentum of the electron if it is in (a) the $n = 1$ level; (b) the $n = 2$ level; (c) the $n = 3$ level? In each case how does the wavelength compare to the width of the box?

40.24 •• **CALC** **Normalization of the Wave Function.** Consider a particle moving in one dimension, which we shall call the x-axis. (a) What does it mean for the wave function of this particle to be *normalized*? (b) Is the wave function $\psi(x) = e^{ax}$, where a is a positive real number, normalized? Could this be a valid wave function? (c) If the particle described by the wave function $\psi(x) = Ae^{-bx}$, where A and b are positive real numbers, is confined to the range $x \geq 0$, determine A (including its units) so that the wave function is normalized.

Section 40.3 Potential Wells

40.25 • **CALC** (a) Show that $\psi = A \sin kx$, where k is a real (not complex) constant, is *not* a solution of Eq. (40.23) for $U = U_0$ and $E < U_0$. (b) Is this ψ a solution for $E > U_0$?

40.26 •• An electron is moving past the square well shown in Fig. 40.13. The electron has energy $E = 3U_0$. What is the ratio of the de Broglie wavelength of the electron in the region $x > L$ to the wavelength for $0 < x < L$?

40.27 • An electron is bound in a square well of depth $U_0 = 6E_{1-\text{IDW}}$. What is the width of the well if its ground-state energy is 2.00 eV?

40.28 •• An electron is bound in a square well of width 1.50 nm and depth $U_0 = 6E_{1-\text{IDW}}$. If the electron is initially in the ground level and absorbs a photon, what maximum wavelength can the photon have and still liberate the electron from the well?

40.29 • **CALC** Calculate $d^2\psi/dx^2$ for the wave function of Eq. (40.38), and show that the function is a solution of Eq. (40.37).

40.30 •• An electron is bound in a square well with a depth equal to six times the ground-level energy E_{1-IDW} of an infinite well of the same width. The longest-wavelength photon that is absorbed by the electron has a wavelength of 400.0 nm. Determine the width of the well.

40.31 •• A proton is bound in a square well of width 4.0 fm = 4.0×10^{-15} m. The depth of the well is six times the ground-level energy E_{1-IDW} of the corresponding infinite well. If the proton makes a transition from the level with energy E_1 to the level with energy E_3 by absorbing a photon, find the wavelength of the photon.

Section 40.4 Potential Barriers and Tunneling

40.32 •• **Alpha Decay.** In a simple model for a radioactive nucleus, an alpha particle ($m = 6.64 \times 10^{-27}$ kg) is trapped by a square barrier that has width 2.0 fm and height 30.0 MeV. (a) What is the tunneling probability when the alpha particle encounters the barrier if its kinetic energy is 1.0 MeV below the top of the barrier (Fig. E40.32)? (b) What is the tunneling probability if the energy of the alpha particle is 10.0 MeV below the top of the barrier?

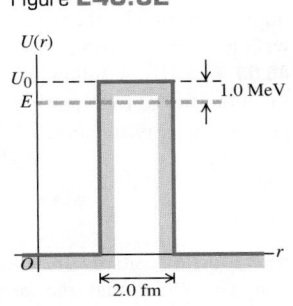
Figure **E40.32**

40.33 • An electron with initial kinetic energy 6.0 eV encounters a barrier with height 11.0 eV. What is the probability of tunneling if the width of the barrier is (a) 0.80 nm and (b) 0.40 nm?

40.34 • An electron with initial kinetic energy 5.0 eV encounters a barrier with height U_0 and width 0.60 nm. What is the transmission coefficient if (a) $U_0 = 7.0$ eV; (b) $U_0 = 9.0$ eV; (c) $U_0 = 13.0$ eV?

40.35 •• An electron is moving past the square barrier shown in Fig. 40.19, but the energy of the electron is *greater* than the barrier height. If $E = 2U_0$, what is the ratio of the de Broglie wavelength of the electron in the region $x > L$ to the wavelength for $0 < x < L$?

40.36 • A proton with initial kinetic energy 50.0 eV encounters a barrier of height 70.0 eV. What is the width of the barrier if the probability of tunneling is 3.0×10^{-3}? How does this compare with the barrier width for an electron with the same energy tunneling through a barrier of the same height with the same probability?

40.37 •• (a) An electron with initial kinetic energy 32 eV encounters a square barrier with height 41 eV and width 0.25 nm. What is the probability that the electron will tunnel through the barrier? (b) A proton with the same kinetic energy encounters the same barrier. What is the probability that the proton will tunnel through the barrier?

Section 40.5 The Harmonic Oscillator

40.38 • **CALC** Show that $\psi(x)$ given by Eq. (40.47) is a solution to Eq. (40.44) with energy $E_0 = \hbar\omega/2$.

40.39 • A wooden block with mass 0.250 kg is oscillating on the end of a spring that has force constant 110 N/m. Calculate the ground-level energy and the energy separation between adjacent levels. Express your results in joules and in electron volts. Are quantum effects important?

40.40 • A harmonic oscillator absorbs a photon of wavelength 8.65×10^{-6} m when it undergoes a transition from the ground state to the first excited state. What is the ground-state energy, in electron volts, of the oscillator?

40.41 • Chemists use infrared absorption spectra to identify chemicals in a sample. In one sample, a chemist finds that light of wavelength 5.8 μm is absorbed when a molecule makes a transition from its ground harmonic oscillator level to its first excited level. (a) Find the energy of this transition. (b) If the molecule can be treated as a harmonic oscillator with mass 5.6×10^{-26} kg, find the force constant.

40.42 •• The ground-state energy of a harmonic oscillator is 5.60 eV. If the oscillator undergoes a transition from its $n = 3$ to $n = 2$ level by emitting a photon, what is the wavelength of the photon?

40.43 • In Section 40.5 it is shown that for the ground level of a harmonic oscillator, $\Delta x\Delta p_x = \hbar/2$. Do a similar analysis for an excited level that has quantum number n. How does the uncertainty product $\Delta x\Delta p_x$ depend on n?

40.44 •• For the ground-level harmonic oscillator wave function $\psi(x)$ given in Eq. (40.47), $|\psi|^2$ has a maximum at $x = 0$. (a) Compute the ratio of $|\psi|^2$ at $x = +A$ to $|\psi|^2$ at $x = 0$, where A is given by Eq. (40.48) with $n = 0$ for the ground level. (b) Compute the ratio of $|\psi|^2$ at $x = +2A$ to $|\psi|^2$ at $x = 0$. In each case is your result consistent with what is shown in Fig. 40.27?

40.45 •• For the sodium atom of Example 40.8, find (a) the ground-state energy, (b) the wavelength of a photon emitted when the $n = 4$ to $n = 3$ transition occurs; (c) the energy difference for any $\Delta n = 1$ transition.

PROBLEMS

40.46 • The discussion in Section 40.1 shows that the wave function $\Psi = \psi e^{-i\omega t}$ is a stationary state, where ψ is time independent and ω is a real (not complex) constant. Consider the wave function $\Psi = \psi_1 e^{-i\omega_1 t} + \psi_2 e^{-i\omega_2 t}$, where ψ_1 and ψ_2 are different time-independent functions and ω_1 and ω_2 are different real constants. Assume that ψ_1 and ψ_2 are real-valued functions, so that $\psi_1^* = \psi_1$ and $\psi_2^* = \psi_2$. Is this Ψ a wave function for a stationary state? Why or why not?

40.47 •• A particle of mass m in a one-dimensional box has the following wave function in the region $x = 0$ to $x = L$:

$$\Psi(x, t) = \frac{1}{\sqrt{2}}\psi_1(x)e^{-iE_1 t/\hbar} + \frac{1}{\sqrt{2}}\psi_3(x)e^{-iE_3 t/\hbar}$$

Here $\psi_1(x)$ and $\psi_3(x)$ are the normalized stationary-state wave functions for the $n = 1$ and $n = 3$ levels, and E_1 and E_3 are the energies of these levels. The wave function is zero for $x < 0$ and for $x > L$. (a) Find the value of the probability distribution function at $x = L/2$ as a function of time. (b) Find the angular frequency at which the probability distribution function oscillates.

40.48 •• **CALC** Consider the wave packet defined by

$$\psi(x) = \int_0^\infty B(k)\cos kx\, dk$$

Let $B(k) = e^{-\alpha^2 k^2}$. (a) The function $B(k)$ has its maximum value at $k = 0$. Let k_h be the value of k at which $B(k)$ has fallen to half its maximum value, and define the width of $B(k)$ as $w_k = k_h$. In terms of α, what is w_k? (b) Use integral tables to evaluate the integral that gives $\psi(x)$. For what value of x is $\psi(x)$ maximum? (c) Define the width of $\psi(x)$ as $w_x = x_h$, where x_h is the positive value of x at which $\psi(x)$ has fallen to half its maximum value. Calculate w_x in terms of α. (d) The momentum p is equal to $hk/2\pi$, so the width of B in momentum is $w_p = hw_k/2\pi$. Calculate the product $w_p w_x$ and compare to the Heisenberg uncertainty principle.

40.49 •• **CALC** (a) Using the integral in Problem 40.48, determine the wave function $\psi(x)$ for a function $B(k)$ given by

$$B(k) = \begin{cases} 0 & k < 0 \\ 1/k_0, & 0 \le k \le k_0 \\ 0, & k > k_0 \end{cases}$$

This represents an equal combination of all wave numbers between 0 and k_0. Thus $\psi(x)$ represents a particle with average wave number $k_0/2$, with a total spread or uncertainty in wave number of k_0. We will call this spread the *width* w_k of $B(k)$, so $w_k = k_0$. (b) Graph $B(k)$ versus k and $\psi(x)$ versus x for the case $k_0 = 2\pi/L$, where L is a length. Locate the point where $\psi(x)$ has its maximum value and label this point on your graph. Locate the two points closest to this maximum (one on each side of it) where $\psi(x) = 0$, and define the distance along the x-axis between these two points as w_x, the width of $\psi(x)$. Indicate the distance w_x on your graph. What is the value of w_x if $k_0 = 2\pi/L$? (c) Repeat part (b) for the case $k_0 = \pi/L$. (d) The momentum p is equal to $hk/2\pi$, so the width of B in momentum is $w_p = hw_k/2\pi$. Calculate the product $w_p w_x$ for each of the cases $k_0 = 2\pi/L$ and $k_0 = \pi/L$. Discuss your results in light of the Heisenberg uncertainty principle.

40.50 • **CALC** Show that the wave function $\psi(x) = Ae^{ikx}$ is a solution of Eq. (40.23) for a particle of mass m, in a region where the potential energy is a constant $U_0 < E$. Find an expression for k, and relate it to the particle's momentum and to its de Broglie wavelength.

40.51 •• **CALC** Wave functions like the one in Problem 40.50 can represent free particles moving with velocity $v = p/m$ in the x-direction. Consider a beam of such particles incident on a potential-energy step $U(x) = 0$, for $x < 0$, and $U(x) = U_0 < E$, for $x > 0$. The wave function for $x < 0$ is $\psi(x) = Ae^{ik_1x} + Be^{-ik_1x}$, representing incident and reflected particles, and for $x > 0$ is $\psi(x) = Ce^{ik_2x}$, representing transmitted particles. Use the conditions that both ψ and its first derivative must be continuous at $x = 0$ to find the constants B and C in terms of k_1, k_2, and A.

40.52 • Let ΔE_n be the energy difference between the adjacent energy levels E_n and E_{n+1} for a particle in a box. The ratio $R_n = \Delta E_n/E_n$ compares the energy of a level to the energy separation of the next higher energy level. (a) For what value of n is R_n largest, and what is this largest R_n? (b) What does R_n approach as n becomes very large? How does this result compare to the classical value for this quantity?

40.53 • **Photon in a Dye Laser.** An electron in a long, organic molecule used in a dye laser behaves approximately like a particle in a box with width 4.18 nm. What is the wavelength of the photon emitted when the electron undergoes a transition (a) from the first excited level to the ground level and (b) from the second excited level to the first excited level?

40.54 • **CALC** A particle is in the ground level of a box that extends from $x = 0$ to $x = L$. (a) What is the probability of finding the particle in the region between 0 and $L/4$? Calculate this by integrating $|\psi(x)|^2\,dx$, where ψ is normalized, from $x = 0$ to $x = L/4$. (b) What is the probability of finding the particle in the region $x = L/4$ to $x = L/2$? (c) How do the results of parts (a) and (b) compare? Explain. (d) Add the probabilities calculated in parts (a) and (b). (e) Are your results in parts (a), (b), and (d) consistent with Fig. 40.12b? Explain.

40.55 •• **CALC** What is the probability of finding a particle in a box of length L in the region between $x = L/4$ and $x = 3L/4$ when the particle is in (a) the ground level and (b) the first excited level? (*Hint:* Integrate $|\psi(x)|^2\,dx$, where ψ is normalized, between $L/4$ and $3L/4$.) (c) Are your results in parts (a) and (b) consistent with Fig. 40.12b? Explain.

40.56 •• Consider a particle in a box with rigid walls at $x = 0$ and $x = L$. Let the particle be in the ground level. Calculate the probability $|\psi|^2\,dx$ that the particle will be found in the interval x to $x + dx$ for (a) $x = L/4$; (b) $x = L/2$; (c) $x = 3L/4$.

40.57 •• Repeat Problem 40.56 for a particle in the first excited level.

40.58 •• **CP** A particle is confined within a box with perfectly rigid walls at $x = 0$ and $x = L$. Although the magnitude of the instantaneous force exerted on the particle by the walls is infinite and the time over which it acts is zero, the impulse (that involves a product of force and time) is both finite and quantized. Show that the impulse exerted by the wall at $x = 0$ is $(nh/L)\hat{\imath}$ and that the impulse exerted by the wall at $x = L$ is $-(nh/L)\hat{\imath}$. (*Hint:* You may wish to review Section 8.1.)

40.59 •• **CALC** A fellow student proposes that a possible wave function for a free particle with mass m (one for which the potential-energy function $U(x)$ is zero) is

$$\psi(x) = \begin{cases} e^{+\kappa x}, & x < 0 \\ e^{-\kappa x}, & x \ge 0 \end{cases}$$

where κ is a positive constant. (a) Graph this proposed wave function. (b) Show that the proposed wave function satisfies the Schrödinger equation for $x < 0$ if the energy is $E = -\hbar^2\kappa^2/2m$— that is, if the energy of the particle is *negative*. (c) Show that the proposed wave function also satisfies the Schrödinger equation for $x \ge 0$ with the same energy as in part (b). (d) Explain why the proposed wave function is nonetheless *not* an acceptable solution of the Schrödinger equation for a free particle. (*Hint:* What is the behavior of the function at $x = 0$?) It is in fact impossible for a free particle (one for which $U(x) = 0$) to have an energy less than zero.

40.60 •• The *penetration distance* η in a finite potential well is the distance at which the wave function has decreased to $1/e$ of the wave function at the classical turning point:

$$\psi(x = L + \eta) = \frac{1}{e}\psi(L)$$

The penetration distance can be shown to be

$$\eta = \frac{\hbar}{\sqrt{2m(U_0 - E)}}$$

The probability of finding the particle beyond the penetration distance is nearly zero. (a) Find η for an electron having a kinetic energy of 13 eV in a potential well with $U_0 = 20$ eV. (b) Find η for a 20.0-MeV proton trapped in a 30.0-MeV-deep potential well.

40.61 • **CALC** (a) For the finite potential well of Fig. 40.13, what relationships among the constants A and B of Eq. (40.38) and C and D of Eq. (40.40) are obtained by applying the boundary condition that ψ be continuous at $x = 0$ and at $x = L$? (b) What relationships among A, B, C, and D are obtained by applying the boundary condition that $d\psi/dx$ be continuous at $x = 0$ and at $x = L$?

40.62 • An electron with initial kinetic energy 5.5 eV encounters a square potential barrier with height 10.0 eV. What is the width of the barrier if the electron has a 0.10% probability of tunneling through the barrier?

40.63 •• A particle with mass m and total energy E tunnels through a square barrier of height U_0 and width L. When the transmission coefficient is *not* much less than unity, it is given by

$$T = \left[1 + \frac{(U_0 \sinh \kappa L)^2}{4E(U_0 - E)} \right]^{-1}$$

where $\sinh \kappa L = (e^{\kappa L} - e^{-\kappa L})/2$ is the hyperbolic sine of κL. (a) Show that if $\kappa L \gg 1$, this expression for T approaches Eq. (40.42). (b) Explain why the restriction $\kappa L \gg 1$ in part (a) implies either that the barrier is relatively wide or that the energy E is relatively low compared to U_0. (c) Show that as the particle's incident kinetic energy E approaches the barrier height U_0, T approaches $[1 + (kL/2)^2]^{-1}$, where $k = \sqrt{2mE}/\hbar$ is the wave number of the incident particle. (*Hint:* If $|z| \ll 1$, then $\sinh z \approx z$.)

40.64 • **CP** A harmonic oscillator consists of a 0.020-kg mass on a spring. Its frequency is 1.50 Hz, and the mass has a speed of 0.360 m/s as it passes the equilibrium position. (a) What is the value of the quantum number n for its energy level? (b) What is the difference in energy between the levels E_n and E_{n+1}? Is this difference detectable?

40.65 • For small amplitudes of oscillation the motion of a pendulum is simple harmonic. For a pendulum with a period of 0.500 s, find the ground-level energy and the energy difference between adjacent energy levels. Express your results in joules and in electron volts. Are these values detectable?

40.66 •• Some 164.9-nm photons are emitted in a $\Delta n = 1$ transition within a solid-state lattice. The lattice is modeled as electrons in a box having length 0.500 nm. What transition corresponds to the emitted light?

40.67 •• **CALC** Show that for $\psi(x)$ given by Eq. (40.47), the probability distribution function has a maximum at $x = 0$.

40.68 •• **CALC** (a) Show by direct substitution in the Schrödinger equation for the one-dimensional harmonic oscillator that the wave function $\psi_1(x) = A_1 x e^{-\alpha^2 x^2/2}$, where $\alpha^2 = m\omega/\hbar$, is a solution with energy corresponding to $n = 1$ in Eq. (40.46). (b) Find the normalization constant A_1. (c) Show that the probability density has a minimum at $x = 0$ and maxima at $x = \pm 1/\alpha$, corresponding to the classical turning points for the ground state $n = 0$.

40.69 •• **CP** (a) The wave nature of particles results in the quantum-mechanical situation that a particle confined in a box can assume only wavelengths that result in standing waves in the box, with nodes at the box walls. Use this to show that an electron confined in a one-dimensional box of length L will have energy levels given by

$$E_n = \frac{n^2 h^2}{8mL^2}$$

(*Hint:* Recall that the relationship between the de Broglie wavelength and the speed of a nonrelativistic particle is $mv = h/\lambda$. The energy of the particle is $\frac{1}{2}mv^2$.) (b) If a hydrogen atom is modeled as a one-dimensional box with length equal to the Bohr radius, what is the energy (in electron volts) of the lowest energy level of the electron?

40.70 ••• Consider a potential well defined as $U(x) = \infty$ for $x < 0$, $U(x) = 0$ for $0 < x < L$, and $U(x) = U_0 > 0$ for $x > L$ (Fig. P40.70). Consider a particle with mass m and kinetic energy $E < U_0$ that is trapped in the well. (a) The boundary condition at the infinite wall $(x = 0)$ is $\psi(0) = 0$. What must the form of

the function $\psi(x)$ for $0 < x < L$ be in order to satisfy both the Schrödinger equation and this boundary condition? (b) The wave function must remain finite as $x \to \infty$. What must the form of the function $\psi(x)$ for $x > L$ be in order to satisfy both the Schrödinger equation and this boundary condition at infinity? (c) Impose the boundary conditions that ψ and $d\psi/dx$ are continuous at $x = L$. Show that the energies of the allowed levels are obtained from solutions of the equation $k \cot kL = -\kappa$, where $k = \sqrt{2mE}/\hbar$ and $\kappa = \sqrt{2m(U_0 - E)}/\hbar$.

Figure **P40.70**

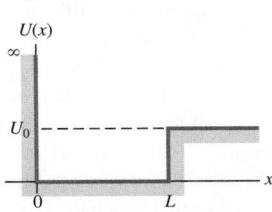

40.71 ••• Section 40.2 considered a box with walls at $x = 0$ and $x = L$. Now consider a box with width L but centered at $x = 0$, so that it extends from $x = -L/2$ to $x = +L/2$ (Fig. P40.71). Note that this box is symmetric about $x = 0$. (a) Consider possible wave functions of the form $\psi(x) =$ $A \sin kx$. Apply the boundary conditions at the wall to obtain the allowed energy levels. (b) Another set of possible wave functions are functions of the form $\psi(x) = A \cos kx$. Apply the boundary conditions at the wall to obtain the allowed energy levels. (c) Compare the energies obtained in parts (a) and (b) to the set of energies given in Eq. (40.31). (d) An odd function f satisfies the condition $f(x) = -f(-x)$, and an even function g satisfies $g(x) = g(-x)$. Of the wave functions from parts (a) and (b), which are even and which are odd?

Figure **P40.71**

CHALLENGE PROBLEMS

40.72 ••• **CALC** **The WKB Approximation.** It can be a challenge to solve the Schrödinger equation for the bound-state energy levels of an arbitrary potential well. An alternative approach that can yield good approximate results for the energy levels is the *WKB approximation* (named for the physicists Gregor Wentzel, Hendrik Kramers, and Léon Brillouin, who pioneered its application to quantum mechanics). The WKB approximation begins from three physical statements: (i) According to de Broglie, the magnitude of momentum p of a quantum-mechanical particle is $p = h/\lambda$. (ii) The magnitude of momentum is related to the kinetic energy K by the relationship $K = p^2/2m$. (iii) If there are no nonconservative forces, then in Newtonian mechanics the energy E for a particle is constant and equal at each point to the sum of the kinetic and potential energies at that point: $E = K + U(x)$, where x is the coordinate. (a) Combine these three relationships to show that the wavelength of the particle at a coordinate x can be written as

$$\lambda(x) = \frac{h}{\sqrt{2m[E - U(x)]}}$$

Thus we envision a quantum-mechanical particle in a potential well $U(x)$ as being like a free particle, but with a wavelength $\lambda(x)$ that is a function of position. (b) When the particle moves into a region of increasing potential energy, what happens to its wavelength? (c) At a point where $E = U(x)$, Newtonian mechanics says that the particle has zero kinetic energy and must be instantaneously at rest. Such a point is called a *classical turning point*, since this is where a Newtonian particle must stop its motion and reverse direction. As an example, an object oscillating in simple harmonic motion with amplitude A moves back and forth between the points $x = -A$ and $x = +A$; each of these is a classical turning point, since there the potential energy $\frac{1}{2}k'x^2$ equals the total energy $\frac{1}{2}k'A^2$. In the WKB expression for $\lambda(x)$, what is the wavelength at a classical turning point? (d) For a particle in a box with length L, the walls of the box are classical turning points (see Fig. 40.8). Furthermore, the number of wavelengths that fit within the box must be a half-integer (see Fig. 40.10), so that $L = (n/2)\lambda$ and hence $L/\lambda = n/2$, where $n = 1, 2, 3, \ldots$. [Note that this is a restatement of Eq. (40.29).] The WKB scheme for finding the allowed bound-state energy levels of an *arbitrary* potential well is an extension of these observations. It demands that for an allowed energy E, there must be a half-integer number of wavelengths between the classical turning points for that energy. Since the wavelength in the WKB approximation is not a constant but depends on x, the number of wavelengths between the classical turning points a and b for a given value of the energy is the integral of $1/\lambda(x)$ between those points:

$$\int_a^b \frac{dx}{\lambda(x)} = \frac{n}{2} \quad (n = 1, 2, 3, \ldots)$$

Using the expression for $\lambda(x)$ you found in part (a), show that the *WKB condition for an allowed bound-state energy* can be written as

$$\int_a^b \sqrt{2m[E - U(x)]}\, dx = \frac{nh}{2} \quad (n = 1, 2, 3, \ldots)$$

(e) As a check on the expression in part (d), apply it to a particle in a box with walls at $x = 0$ and $x = L$. Evaluate the integral and show that the allowed energy levels according to the WKB approximation are the same as those given by Eq. (40.31). (*Hint:* Since the walls of the box are infinitely high, the points $x = 0$ and $x = L$ are classical turning points for *any* energy E. Inside the box, the potential energy is zero.) (f) For the finite square well shown in Fig. 40.13, show that the WKB expression given in part (d) predicts the *same* bound-state energies as for an infinite square well of the same width. (*Hint:* Assume $E < U_0$. Then the classical turning points are at $x = 0$ and $x = L$.) This shows that the WKB approximation does a poor job when the potential-energy function changes discontinuously, as for a finite potential well. In the next two problems we consider situations in which the potential-energy function changes gradually and the WKB approximation is much more useful.

40.73 ••• **CALC** The WKB approximation (see Challenge Problem 40.72) can be used to calculate the energy levels for a harmonic oscillator. In this approximation, the energy levels are the solutions to the equation

$$\int_a^b \sqrt{2m[E - U(x)]}\, dx = \frac{nh}{2} \quad n = 1, 2, 3, \ldots$$

Here E is the energy, $U(x)$ is the potential-energy function, and $x = a$ and $x = b$ are the classical turning points (the points at

which E is equal to the potential energy, so the Newtonian kinetic energy would be zero). (a) Determine the classical turning points for a harmonic oscillator with energy E and force constant k'. (b) Carry out the integral in the WKB approximation and show that the energy levels in this approximation are $E_n = \hbar\omega$, where $\omega = \sqrt{k'/m}$ and $n = 1, 2, 3, \ldots$. (*Hint:* Recall that $\hbar = h/2\pi$. A useful standard integral is

$$\int \sqrt{A^2 - x^2}\, dx = \frac{1}{2}\left[x\sqrt{A^2 - x^2} + A^2 \arcsin\left(\frac{x}{|A|}\right) \right]$$

where arcsin denotes the inverse sine function. Note that the integrand is even, so the integral from $-x$ to x is equal to twice the integral from 0 to x.) (c) How do the approximate energy levels found in part (b) compare with the true energy levels given by Eq. (40.46)? Does the WKB approximation give an underestimate or an overestimate of the energy levels?

40.74 ••• **CALC** Protons, neutrons, and many other particles are made of more fundamental particles called *quarks* and *antiquarks* (the antimatter equivalent of quarks). A quark and an antiquark can form a bound state with a variety of different energy levels, each of which corresponds to a different particle observed in the laboratory. As an example, the ψ particle is a low-energy bound state of a so-called charm quark and its antiquark, with a rest energy of 3097 MeV; the $\psi(2S)$ particle is an excited state of this same quark–antiquark combination, with a rest energy of 3686 MeV. A simplified representation of the potential energy of interaction between a quark and an antiquark is $U(x) = A|x|$, where A is a positive constant and x represents the distance between the quark and the antiquark. You can use the WKB approximation (see Challenge Problem 40.72) to determine the bound-state energy levels for this potential-energy function. In the WKB approximation, the energy levels are the solutions to the equation

$$\int_a^b \sqrt{2m[E - U(x)]}\, dx = \frac{nh}{2} \quad (n = 1, 2, 3, \ldots)$$

Here E is the energy, $U(x)$ is the potential-energy function, and $x = a$ and $x = b$ are the classical turning points (the points at which E is equal to the potential energy, so the Newtonian kinetic energy would be zero). (a) Determine the classical turning points for the potential $U(x) = A|x|$ and for an energy E. (b) Carry out the above integral and show that the allowed energy levels in the WKB approximation are given by

$$E_n = \frac{1}{2m}\left(\frac{3mAh}{4}\right)^{2/3} n^{2/3} \quad (n = 1, 2, 3, \ldots)$$

(*Hint:* The integrand is even, so the integral from $-x$ to x is equal to twice the integral from 0 to x.) (c) Does the difference in energy between successive levels increase, decrease, or remain the same as n increases? How does this compare to the behavior of the energy levels for the harmonic oscillator? For the particle in a box? Can you suggest a simple rule that relates the difference in energy between successive levels to the shape of the potential-energy function?

Answers

Chapter Opening Question ?

When an electron in one of these particles—called *quantum dots*—makes a transition from an excited level to a lower level, it emits a photon whose energy is equal to the difference in energy between the levels. The smaller the quantum dot, the larger the energy spacing between levels and hence the shorter (bluer) the wavelength of the emitted photons. See Example 40.6 (Section 40.3) for more details.

Test Your Understanding Questions

40.1 Answer: no Equation (40.19) represents a superposition of wave functions with different values of wave number k and hence different values of energy $E = \hbar^2 k^2 / 2m$. The state that this combined wave function represents is not a state of definite energy, and therefore not a stationary state. Another way to see this is to note that there is a factor $e^{-iEt/\hbar}$ inside the integral in Eq. (40.19), with a different value of E for each value of k. This wave function therefore has a very complicated time dependence, and the probability distribution function $|\Psi(x, t)|^2$ does depend on time.

40.2 Answer: (v) Our derivation of the stationary-state wave functions for a particle in a box shows that they are superpositions of waves propagating in opposite directions, just like a standing wave on a string. One wave has momentum in the positive x-direction, while the other wave has an equal magnitude of momentum in the negative x-direction. The *total* x-component of momentum is zero.

40.3 Answer: (i) The energy levels are arranged as shown in Fig. 40.15b if $U_0 = 6E_{1-\text{IDW}}$, where $E_{1-\text{IDW}} = \pi^2 \hbar^2 / 2mL^2$ is the ground-level energy of an infinite well. If the well width L is reduced to one-half of its initial value, $E_{1-\text{IDW}}$ increases by a factor of four and so U_0 must also increase by a factor of four. The energies E_1, E_2, and E_3 shown in Fig. 40.15b are all specific fractions of U_0, so they will also increase by a factor of four.

40.4 Answer: yes Figure 40.20 shows a possible wave function $\psi(x)$ for tunneling. Since $\psi(x)$ is not zero within the barrier $(0 \leq x \leq L)$, there is some probability that the particle can be found there.

40.5 Answer: (ii) If the second photon has a longer wavelength and hence lower energy than the first photon, the difference in energy between the first and second excited levels must be less than the difference between the ground level and the first excited level. This is the case for the hydrogen atom, for which the energy difference between levels decreases as the energy increases (see Fig. 39.24). By contrast, the energy difference between successive levels increases for a particle in a box (see Fig. 40.11b) and is constant for a harmonic oscillator (see Fig. 40.25).

Bridging Problem

Answers: (a)

$$|\Psi(x, t)|^2 = \frac{1}{L}\left[\sin^2\frac{\pi x}{L} + \sin^2\frac{2\pi x}{L} + 2\sin\frac{\pi x}{L}\sin\frac{2\pi x}{L}\cos\left(\frac{(E_2 - E_1)t}{\hbar}\right) \right]$$

(b) no   **(c)** yes   **(d)** $\dfrac{3\pi^2\hbar}{2mL^2}$ **(e)** $\dfrac{0.905\pi^2\hbar}{mL^2}$

ATOMIC STRUCTURE

LEARNING GOALS

By studying this chapter, you will learn:

• How to extend quantum-mechanical calculations to three-dimensional problems.

• How to solve the Schrödinger equation for a particle trapped in a cubical box.

• How to describe the states of a hydrogen atom in terms of quantum numbers.

• How magnetic fields affect the orbital motion of atomic electrons.

• How we know that electrons have their own intrinsic angular momentum.

• How to analyze the structure of many-electron atoms.

• How x rays emitted by atoms reveal their inner structure.

? Lithium (with three electrons per atom) is a metal that burns spontaneously in water, while helium (with two electrons per atom) is a gas that undergoes almost no chemical reactions. How can one extra electron make these two elements so dramatically different?

Some physicists claim that all of chemistry is contained in the Schrödinger equation. This is somewhat of an exaggeration, but this equation can teach us a great deal about the chemical behavior of elements and the nature of chemical bonds. It provides insight into the periodic table of the elements and the microscopic basis of magnetism.

In order to learn about the quantum-mechanical structure of atoms, we'll first construct a three-dimensional version of the Schrödinger equation. We'll try this equation out by looking at a three-dimensional version of a particle in a box: a particle confined to a cubical volume.

We'll then see that we can learn a great deal about the structure and properties of *all* atoms from the solutions to the Schrödinger equation for the hydrogen atom. These solutions have quantized values of angular momentum; we don't need to make a separate statement about quantization as we did with the Bohr model. We label the states with a set of quantum numbers, which we'll use later with many-electron atoms as well. We'll find that the electron also has an intrinsic *spin* angular momentum in addition to the orbital angular momentum associated with its motion.

We'll also encounter the exclusion principle, a kind of microscopic zoning ordinance that is the key to understanding many-electron atoms. This principle says that no two electrons in an atom can have the same quantum-mechanical state. Finally, we'll use the principles of this chapter to explain the characteristic x-ray spectra of atoms.

41.1 The Schrödinger Equation in Three Dimensions

We have discussed the Schrödinger equation and its applications only for *one-dimensional* problems, the analog of a Newtonian particle moving along a straight line. The straight-line model is adequate for some applications, but to understand atomic structure, we need a three-dimensional generalization.

It's not difficult to guess what the three-dimensional Schrödinger equation should look like. First, the wave function Ψ is a function of time and all three space coordinates (x, y, z). In general, the potential-energy function also depends on all three coordinates and can be written as $U(x, y, z)$. Next, recall from Section 40.1 that the term $-(\hbar^2/2m)\partial^2\Psi/\partial x^2$ in the one-dimensional Schrödinger equation, Eq. (40.20), is related to the kinetic energy of the particle in the state described by the wave function Ψ. For example, if we insert into this term the wave function $\Psi(x, t) = Ae^{ikx}e^{-i\omega t}$ for a free particle with magnitude of momentum $p = \hbar k$ and kinetic energy $K = p^2/2m$, we obtain $-(\hbar^2/2m)(ik)^2Ae^{ikx}e^{-i\omega t} = (\hbar^2 k^2/2m)Ae^{ikx}e^{-i\omega t} = (p^2/2m)\Psi(x, t) = K\Psi(x, t)$. If the particle can move in three dimensions, its momentum has three components (p_x, p_y, p_z) and its kinetic energy is

$$K = \frac{p_x^2}{2m} + \frac{p_y^2}{2m} + \frac{p_z^2}{2m} \tag{41.1}$$

These observations, taken together, suggest that the correct generalization of the Schrödinger equation to three dimensions is

$$-\frac{\hbar^2}{2m}\left(\frac{\partial^2\Psi(x, y, z, t)}{\partial x^2} + \frac{\partial^2\Psi(x, y, z, t)}{\partial y^2} + \frac{\partial^2\Psi(x, y, z, t)}{\partial z^2}\right)$$

$$+ U(x, y, z)\Psi(x, y, z, t) = i\hbar\frac{\partial\Psi(x, y, z, t)}{\partial t} \tag{41.2}$$

(general three-dimensional Schrödinger equation)

The three-dimensional wave function $\Psi(x, y, z, t)$ has a similar interpretation as in one dimension. The wave function itself is a complex quantity with both a real part and an imaginary part, but $|\Psi(x, y, z, t)|^2$—the square of its absolute value, equal to the product of $\Psi(x, y, z, t)$ and its complex conjugate $\Psi^*(x, y, z, t)$—is real and either positive or zero at every point in space. We interpret $|\Psi(x, y, z, t)|^2 dV$ as the *probability* of finding the particle within a small volume dV centered on the point (x, y, z) at time t, so $|\Psi(x, y, z, t)|^2$ is the *probability distribution function* in three dimensions. The *normalization condition* on the wave function is that the probability that the particle is *somewhere* in space is exactly 1. Hence the integral of $|\Psi(x, y, z, t)|^2$ over all space must equal 1:

$$\int |\Psi(x, y, z, t)|^2 dV = 1 \qquad \begin{array}{l}\text{(normalization condition} \\ \text{in three dimensions)}\end{array} \tag{41.3}$$

If the wave function $\Psi(x, y, z, t)$ represents a state of a definite energy E—that is, a stationary state—we can write it as the product of a spatial wave function $\psi(x, y, z)$ and a function of time $e^{-iEt/\hbar}$:

$$\Psi(x, y, z, t) = \psi(x, y, z)e^{-iEt/\hbar} \qquad \begin{array}{l}\text{(time-dependent wave function} \\ \text{for a state of definite energy)}\end{array} \tag{41.4}$$

(Compare this to Eq. (40.21) for a one-dimensional state of definite energy.) If we substitute Eq. (41.4) into Eq. (41.2), the right-hand side of the equation becomes $i\hbar\psi(x, y, z)(-iE/\hbar)e^{-iEt/\hbar} = E\psi(x, y, z)e^{-iEt/\hbar}$. We can then divide both sides by the factor $e^{-iEt/\hbar}$, leaving the *time-independent* Schrödinger equation in three dimensions for a stationary state:

$$-\frac{\hbar^2}{2m}\left(\frac{\partial^2 \psi(x, y, z)}{\partial x^2} + \frac{\partial^2 \psi(x, y, z)}{\partial y^2} + \frac{\partial^2 \psi(x, y, z)}{\partial z^2}\right) + U(x, y, z)\psi(x, y, z)$$

$$= E\psi(x, y, z) \quad \text{(three-dimensional time-independent} \quad (41.5)$$
$$\text{Schrödinger equation)}$$

The probability distribution function for a stationary state is just the square of the absolute value of the spatial wave function: $|\psi(x, y, z)e^{-iEt/\hbar}|^2 = \psi^*(x, y, z)e^{+iEt/\hbar}\psi(x, y, z)e^{-iEt/\hbar} = |\psi(x, y, z)|^2$. Note that this doesn't depend on time. (As we discussed in Section 40.1, that's why we call these states *stationary*.) Hence for a stationary state the wave function normalization condition, Eq. (41.3), becomes

$$\int |\psi(x, y, z)|^2 \, dV = 1 \quad \text{(normalization condition for a} \quad (41.6)$$
$$\text{stationary state in three dimensions)}$$

We won't pretend that we have *derived* Eqs. (41.2) and (41.5). Like their one-dimensional versions, these equations have to be tested by comparison of their predictions with experimental results. Happily, Eqs. (41.2) and (41.5) both pass this test with flying colors, so we are confident that they *are* the correct equations.

An important topic that we will address in this chapter is the solutions for Eq. (41.5) for the stationary states of the hydrogen atom. The potential-energy function for an electron in a hydrogen atom is *spherically symmetric*; it depends only on the distance $r = (x^2 + y^2 + z^2)^{1/2}$ from the origin of coordinates. To take advantage of this symmetry, it's best to use *spherical coordinates* rather than the Cartesian coordinates (x, y, z) to solve the Schrödinger equation for the hydrogen atom. Before introducing these new coordinates and investigating the hydrogen atom, it's useful to look at the three-dimensional version of the particle in a box that we considered in Section 40.2. Solving this simpler problem will give us insight into the more complicated stationary states found in atomic physics.

Test Your Understanding of Section 41.1 In a certain region of space the potential-energy function for a quantum-mechanical particle is zero. In this region the wave function $\psi(x, y, z)$ for a certain stationary state satisfies $\partial^2\psi/\partial x^2 > 0$, $\partial^2\psi/\partial y^2 > 0$, and $\partial^2\psi/\partial z^2 > 0$. The particle has a definite energy E that is positive. What can you conclude about $\psi(x, y, z)$ in this region? (i) It must be positive; (ii) it must be negative; (iii) it must be zero; (iv) not enough information given to decide. ∎

41.2 Particle in a Three-Dimensional Box

41.1 A particle is confined in a cubical box with walls at $x = 0$, $x = L$, $y = 0$, $y = L$, $z = 0$, and $z = L$.

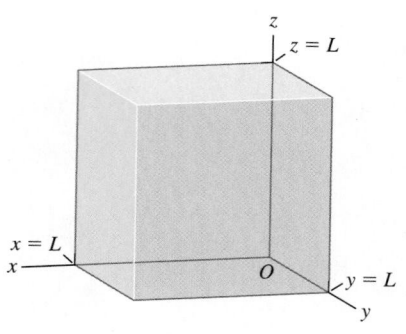

Consider a particle enclosed within a cubical box of side L. This could represent an electron that's free to move anywhere within the interior of a solid metal cube but cannot escape the cube. We'll choose the origin to be at one corner of the box, with the x-, y-, and z-axes along edges of the box. Then the particle is confined to the region $0 \leq x \leq L$, $0 \leq y \leq L$, $0 \leq z \leq L$ (Fig. 41.1). What are the stationary states of this system?

As for the particle in a one-dimensional box that we considered in Section 40.2, we'll say that the potential energy is zero inside the box but infinite outside. Hence the spatial wave function $\psi(x, y, z)$ must be zero outside the box in order that the term $U(x, y, z)\psi(x, y, z)$ in the time-independent Schrödinger equation, Eq. (41.5), not be infinite. Hence the probability distribution function $|\psi(x, y, z)|^2$ is zero outside the box, and there is zero probability that the particle will be found there. Inside the box, the spatial wave function for a stationary state obeys the time-independent Schrödinger equation, Eq. (41.5), with $U(x, y, z) = 0$:

$$-\frac{\hbar^2}{2m}\left(\frac{\partial^2\psi(x, y, z)}{\partial x^2} + \frac{\partial^2\psi(x, y, z)}{\partial y^2} + \frac{\partial^2\psi(x, y, z)}{\partial z^2}\right) = E\psi(x, y, z)$$

(particle in a three-dimensional box) (41.7)

In order for the wave function to be continuous from the inside to the outside of the box, $\psi(x, y, z)$ must equal zero on the walls. Hence our boundary conditions are that $\psi(x, y, z) = 0$ at $x = 0$, $x = L$, $y = 0$, $y = L$, $z = 0$, and $z = L$.

Guessing a solution to a complicated partial differential equation like Eq. (41.7) seems like quite a challenge. To make progress, recall that we wrote the time-*dependent* wave function for a stationary state as the product of one function that depends only on the spatial coordinates x, y, and z and a second function that depends only on the time t: $\Psi(x, y, z, t) = \psi(x, y, z)e^{-iEt/\hbar}$. In the same way, let's try a technique called *separation of variables:* We'll write the spatial wave function $\psi(x, y, z)$ as a product of one function X that depends only on x, a second function Y that depends only on y, and a third function Z that depends only on z:

$$\psi(x, y, z) = X(x)Y(y)Z(z)$$ (41.8)

If we substitute Eq. (41.8) into Eq. (41.7), we get

$$-\frac{\hbar^2}{2m}\left(Y(y)Z(z)\frac{d^2X(x)}{dx^2} + X(x)Z(z)\frac{d^2Y(y)}{dy^2} + X(x)Y(y)\frac{d^2Z(z)}{dz^2}\right)$$

$$= EX(x)Y(y)Z(z)$$ (41.9)

The partial derivatives in Eq. (41.7) have become ordinary derivatives since they act on functions of a single variable. Now we divide both sides of Eq. (41.9) by the product $X(x)Y(y)Z(z)$:

$$\left(-\frac{\hbar^2}{2m}\frac{1}{X(x)}\frac{d^2X(x)}{dx^2}\right) + \left(-\frac{\hbar^2}{2m}\frac{1}{Y(y)}\frac{d^2Y(y)}{dy^2}\right) + \left(-\frac{\hbar^2}{2m}\frac{1}{Z(z)}\frac{d^2Z(z)}{dz^2}\right) = E$$

(41.10)

The right-hand side of Eq. (41.10) is the energy of the stationary state, which does not and cannot depend on the values of x, y, or z. For this to be true, the left-hand side of the equation must also be independent of the values of x, y, and z. Hence the first term in parentheses on the left-hand side of Eq. (41.10) must equal a constant that doesn't depend on x, the second term in parentheses must equal another constant that doesn't depend on y, and the third term in parentheses must equal a third constant that doesn't depend on z. Let's call these constants E_X, E_Y, and E_Z, respectively. We then have a separate equation for each of the three functions $X(x)$, $Y(y)$, and $Z(z)$:

$$-\frac{\hbar^2}{2m}\frac{d^2X(x)}{dx^2} = E_X X(x)$$ (41.11a)

$$-\frac{\hbar^2}{2m}\frac{d^2Y(y)}{dy^2} = E_Y Y(y)$$ (41.11b)

$$-\frac{\hbar^2}{2m}\frac{d^2Z(z)}{dz^2} = E_Z Z(z)$$ (41.11c)

To satisfy the boundary conditions that $\psi(x, y, z) = X(x)Y(y)Z(z)$ be equal to zero on the walls of the box, we demand that $X(x) = 0$ at $x = 0$ and $x = L$, $Y(y) = 0$ at $y = 0$ and $y = L$, and $Z(z) = 0$ at $z = 0$ and $z = L$.

How can we interpret the three constants E_X, E_Y, and E_Z in Eqs. (41.11)? From Eq. (41.10), they are related to the energy E by

$$E_X + E_Y + E_Z = E \qquad \text{(41.12)}$$

Equation (41.12) should remind you of Eq. (41.1) in Section 41.1, which states that the kinetic energy of a particle is the sum of contributions coming from its x-, y-, and z-components of momentum. Hence the constants E_X, E_Y, and E_Z tell us how much of the particle's energy is due to motion along each of the three coordinate axes. (Inside the box the potential energy is zero, so the particle's energy is purely kinetic.)

Equations (41.11) represent an enormous simplification; we've reduced the problem of solving a fairly complex *partial* differential equation with three independent variables to the much simpler problem of solving three separate *ordinary* differential equations with one independent variable each. What's more, each of these ordinary differential equations is just the same as the time-independent Schrödinger equation for a particle in a *one-dimensional* box, Eq. (40.25), and with exactly the same boundary conditions at 0 and L. (The only differences are that some of the quantities are labeled by different symbols.) By comparing with our work in Section 40.2, you can see that the solutions to Eqs. (41.11) are

$$X_{n_X}(x) = C_X \sin\frac{n_X \pi x}{L} \quad (n_X = 1, 2, 3, \dots) \qquad \text{(41.13a)}$$

$$Y_{n_Y}(y) = C_Y \sin\frac{n_Y \pi y}{L} \quad (n_Y = 1, 2, 3, \dots) \qquad \text{(41.13b)}$$

$$Z_{n_Z}(z) = C_Z \sin\frac{n_Z \pi z}{L} \quad (n_Z = 1, 2, 3, \dots) \qquad \text{(41.13c)}$$

where C_X, C_Y, and C_Z are constants. The corresponding values of E_X, E_Y, and E_Z are

$$E_X = \frac{n_X{}^2 \pi^2 \hbar^2}{2mL^2} \quad (n_X = 1, 2, 3, \dots) \qquad \text{(41.14a)}$$

$$E_Y = \frac{n_Y{}^2 \pi^2 \hbar^2}{2mL^2} \quad (n_Y = 1, 2, 3, \dots) \qquad \text{(41.14b)}$$

$$E_Z = \frac{n_Z{}^2 \pi^2 \hbar^2}{2mL^2} \quad (n_Z = 1, 2, 3, \dots) \qquad \text{(41.14c)}$$

There is only one quantum number n for the one-dimensional particle in a box, but *three* quantum numbers n_X, n_Y, and n_Z for the three-dimensional box. If we substitute Eqs. (41.13) back into Eq. (41.8) for the total spatial wave function, $\psi(x, y, z) = X(x)Y(y)Z(z)$, we get the following stationary-state wave functions for a particle in a three-dimensional cubical box:

$$\psi_{n_X, n_Y, n_Z}(x, y, z) = C \sin\frac{n_X \pi x}{L} \sin\frac{n_Y \pi y}{L} \sin\frac{n_Z \pi z}{L}$$

$$(n_X = 1, 2, 3, \dots; n_Y = 1, 2, 3, \dots; n_Z = 1, 2, 3, \dots) \qquad \text{(41.15)}$$

where $C = C_X C_Y C_Z$. The value of the constant C is determined by the normalization condition, Eq. (41.6).

In Section 40.2 we saw that the stationary-state wave functions for a particle in a one-dimensional box were analogous to standing waves on a string. In a similar way, the *three*-dimensional wave functions given by Eq. (41.15) are analogous to standing electromagnetic waves in a cubical cavity like the interior of a microwave oven (see Section 32.5). In a microwave oven there are "dead spots" where the wave intensity is zero, corresponding to the nodes of the standing wave. (The rotating platform in a microwave oven ensures even cooking by making sure that no part of the food sits at any "dead spot.") In a similar fashion, the probability distribution function corresponding to Eq. (41.15) can have "dead

41.2 Probability distribution function $|\psi_{n_X,n_Y,n_Z}(x, y, z)|^2$ for (n_X, n_Y, n_Z) equal to (a) (2, 1, 1), (b) (1, 2, 1), and (c) (1, 1, 2). The value of $|\psi|^2$ is proportional to the density of dots. The wave function is zero on the walls of the box and on the midplane of the box, so $|\psi|^2 = 0$ at these locations.

(a) $|\psi_{2,1,1}|^2$

(b) $|\psi_{1,2,1}|^2$

(c) $|\psi_{1,1,2}|^2$

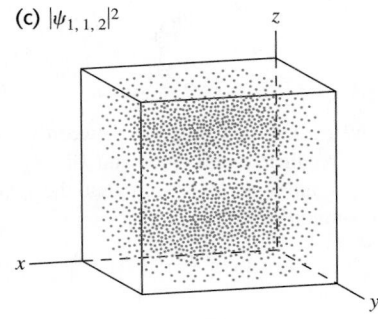

spots" where there is zero probability of finding the particle. As an example, consider the case $(n_X, n_Y, n_Z) = (2, 1, 1)$. From Eq. (41.15), the probability distribution function for this case is

$$|\psi_{2,1,1}(x, y, z)|^2 = |C|^2 \sin^2 \frac{2\pi x}{L} \sin^2 \frac{\pi y}{L} \sin^2 \frac{\pi z}{L}$$

As Fig. 41.2a shows, the probability distribution function is zero on the plane $x = L/2$, where $\sin^2(2\pi x/L) = \sin^2 \pi = 0$. The particle is most likely to be found near where all three of the sine-squared functions are greatest, at $(x, y, z) = (L/4, L/2, L/2)$ or $(x, y, z) = (3L/4, L/2, L/2)$. Figures 41.2b and 41.2c show the similar cases $(n_X, n_Y, n_Z) = (1, 2, 1)$ and $(n_X, n_Y, n_Z) = (1, 1, 2)$. For higher values of the quantum numbers n_X, n_Y, and n_Z there are additional planes on which the probability distribution function equals zero, just as the probability distribution function $|\psi(x)|^2$ for a one-dimensional box has more zeros for higher values of n (see Fig. 40.12).

Example 41.1 **Probability in a three-dimensional box**

(a) Find the value of the constant C that normalizes the wave function of Eq. (41.15). (b) Find the probability that the particle will be found somewhere in the region $0 \le x \le L/4$ (Fig. 41.3) for the cases (i) $(n_X, n_Y, n_Z) = (1, 2, 1)$, (ii) $(n_X, n_Y, n_Z) = (2, 1, 1)$, and (iii) $(n_X, n_Y, n_Z) = (3, 1, 1)$.

SOLUTION

IDENTIFY and SET UP: Equation (41.6) tells us that to normalize the wave function, we have to choose the value of C so that the

integral of the probability distribution function $|\psi_{n_X, n_Y, n_Z}(x, y, z)|^2$ over the volume within the box equals 1. (The integral is actually over *all* space, but the particle-in-a-box wave functions are zero outside the box.)

The probability of finding the particle within a certain volume within the box equals the integral of the probability distribution function over that volume. Hence in part (b) we'll integrate $|\psi_{n_X, n_Y, n_Z}(x, y, z)|^2$ for the given values of (n_X, n_Y, n_Z) over the volume $0 \le x \le L/4, 0 \le y \le L, 0 \le z \le L$.

EXECUTE: (a) From Eq. (41.15),

$$|\psi_{n_X,n_Y,n_Z}(x, y, z)|^2 = |C|^2 \sin^2 \frac{n_X \pi x}{L} \sin^2 \frac{n_Y \pi y}{L} \sin^2 \frac{n_Z \pi z}{L}$$

Hence the normalization condition is

$$\int |\psi_{n_X,n_Y,n_Z}(x, y, z)|^2 \, dV$$

$$= |C|^2 \int_{x=0}^{x=L} \int_{y=0}^{y=L} \int_{z=0}^{z=L} \sin^2 \frac{n_X \pi x}{L} \sin^2 \frac{n_Y \pi y}{L} \sin^2 \frac{n_Z \pi z}{L} \, dx \, dy \, dz$$

$$= |C|^2 \left(\int_{x=0}^{x=L} \sin^2 \frac{n_X \pi x}{L} \, dx \right) \left(\int_{y=0}^{y=L} \sin^2 \frac{n_Y \pi y}{L} \, dy \right)$$

$$\times \left(\int_{z=0}^{z=L} \sin^2 \frac{n_Z \pi z}{L} \, dz \right) = 1$$

41.3 What is the probability that the particle is in the dark-colored quarter of the box?

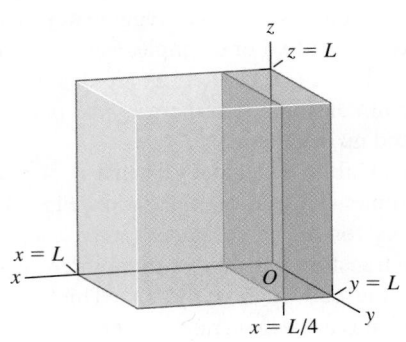

Continued

We can use the identity $\sin^2\theta = \frac{1}{2}(1 - \cos 2\theta)$ and the variable substitution $\theta = n_X\pi x/L$ to show that

$$\int \sin^2 \frac{n_X\pi x}{L}\, dx = \frac{L}{2n_X\pi}\left[\frac{n_X\pi x}{L} - \frac{1}{2}\sin\left(\frac{2n_X\pi x}{L}\right)\right]$$

$$= \frac{x}{2} - \frac{L}{4n_X\pi}\sin\left(\frac{2n_X\pi x}{L}\right)$$

If we evaluate this integral between $x = 0$ and $x = L$, the result is $L/2$ (recall that $\sin 0 = 0$ and $\sin 2n_X\pi = 0$ for any integer n_X). The y- and z-integrals each yield the same result, so the normalization condition is

$$|C|^2\left(\frac{L}{2}\right)\left(\frac{L}{2}\right)\left(\frac{L}{2}\right) = |C|^2\left(\frac{L}{2}\right)^3 = 1$$

or $|C|^2 = (2/L)^3$. If we choose C to be real and positive, then $C = (2/L)^{3/2}$.

(b) We have the same y- and z-integrals as in part (a), but now the limits of integration on the x-integral are $x = 0$ and $x = L/4$:

$$P = \int_{0 \le x \le L/4} |\psi_{n_X,n_Y,n_Z}|^2\, dV = |C|^2\left(\int_{x=0}^{x=L/4} \sin^2 \frac{n_X\pi x}{L}\, dx\right)$$

$$\times \left(\int_{y=0}^{y=L} \sin^2 \frac{n_Y\pi y}{L}\, dy\right)\left(\int_{z=0}^{z=L} \sin^2 \frac{n_Z\pi z}{L}\, dz\right)$$

The x-integral is

$$\int_{x=0}^{x=L/4} \sin^2 \frac{n_X\pi x}{L}\, dx = \left(\frac{x}{2} - \frac{L}{4n_X\pi}\sin\left(\frac{2n_X\pi x}{L}\right)\right)\Bigg]_{x=0}^{x=L/4}$$

$$= \frac{L}{8} - \frac{L}{4n_X\pi}\sin\left(\frac{n_X\pi}{2}\right)$$

Hence the probability of finding the particle somewhere in the region $0 \le x \le L/4$ is

$$P = \left(\frac{2}{L}\right)^3\left(\frac{L}{8} - \frac{L}{4n_X\pi}\sin\left(\frac{n_X\pi}{2}\right)\right)\left(\frac{L}{2}\right)\left(\frac{L}{2}\right)$$

$$= \frac{1}{4} - \frac{1}{2n_X\pi}\sin\left(\frac{n_X\pi}{2}\right)$$

This depends only on the value of n_X, not on n_Y or n_Z. Hence for the three cases we have

(i) $n_X = 1$: $P = \frac{1}{4} - \dfrac{1}{2(1)\pi}\sin\left(\dfrac{\pi}{2}\right) = \frac{1}{4} - \dfrac{1}{2\pi}(1)$

$\qquad = \frac{1}{4} - \dfrac{1}{2\pi} = 0.091$

(ii) $n_X = 2$: $P = \frac{1}{4} - \dfrac{1}{2(2)\pi}\sin\left(\dfrac{2\pi}{2}\right) = \frac{1}{4} - \dfrac{1}{4\pi}\sin\pi$

$\qquad = \frac{1}{4} - 0 = 0.250$

(iii) $n_X = 3$: $P = \frac{1}{4} - \dfrac{1}{2(3)\pi}\sin\left(\dfrac{3\pi}{2}\right) = \frac{1}{4} - \dfrac{1}{6\pi}(-1)$

$\qquad = \frac{1}{4} + \dfrac{1}{6\pi} = 0.303$

EVALUATE: You can see why the probabilities in part (b) are different by looking at part (b) of Fig. 40.12, which shows $\sin^2 n_X\pi x/L$ for $n_X = 1$, 2, and 3. For $n_X = 2$ the area under the curve between $x = 0$ and $x = L/4$ (equal to the integral between these two points) is exactly $\frac{1}{4}$ of the total area between $x = 0$ and $x = L$. For $n_X = 1$ the area between $x = 0$ and $x = L/4$ is less than $\frac{1}{4}$ of the total area, and for $n_X = 3$ it is greater than $\frac{1}{4}$ of the total area.

Energy Levels, Degeneracy, and Symmetry

From Eqs. (41.12) and (41.14), the allowed energies for a particle of mass m in a cubical box of side L are

$$E_{n_X,n_Y,n_Z} = \frac{(n_X{}^2 + n_Y{}^2 + n_Z{}^2)\pi^2\hbar^2}{2mL^2} \qquad (n_X = 1, 2, 3, \ldots; n_Y = 1, 2, 3, \ldots;$$

$$n_Z = 1, 2, 3, \ldots) \qquad (41.16)$$

(energy levels, particle in a three-dimensional cubical box)

Figure 41.4 shows the six lowest energy levels given by Eq. (41.16). Note that most energy levels correspond to more than one set of quantum numbers (n_X, n_Y, n_Z) and hence to more than one quantum state. Having two or more distinct quantum states with the same energy is called **degeneracy,** and states with the same energy are said to be **degenerate.** For example, Fig. 41.4 shows that the states $(n_X, n_Y, n_Z) = (2, 1, 1)$, $(1, 2, 1)$, and $(1, 1, 2)$ are degenerate. By comparison, for a particle in a one-dimensional box there is just one state for each energy level (see Fig. 40.11a) and no degeneracy.

The reason the cubical box exhibits degeneracy is that it is *symmetric:* All sides of the box have the same dimensions. As an illustration, Fig. 41.2 shows the probability distribution functions for the three states $(n_X, n_Y, n_Z) = (2, 1, 1)$, $(1, 2, 1)$, and $(1, 1, 2)$. You can transform any one of these three states into a different one by simply rotating the cubical box by 90°. This rotation doesn't change the energy, so the three states are degenerate.

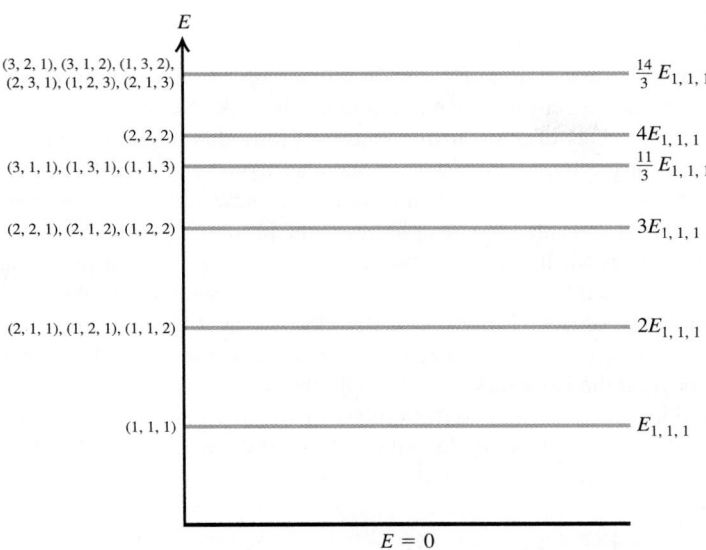

41.4 Energy-level diagram for a particle in a three-dimensional cubical box. We label each level with the quantum numbers of the states (n_X, n_Y, n_Z) with that energy. Several of the levels are degenerate (more than one state has the same energy). The lowest (ground) level, $(n_X, n_Y, n_Z) = (1, 1, 1)$, has energy $E_{1,1,1} = (1^2 + 1^2 + 1^2)\pi^2\hbar^2/2mL^2 = 3\pi^2\hbar^2/2mL^2$; we show the energies of the other levels as multiples of $E_{1,1,1}$.

Since degeneracy is a consequence of symmetry, we can remove the degeneracy by making the box asymmetric. We do this by giving the three sides of the box different lengths L_X, L_Y, and L_Z. If we repeat the steps that we followed to solve the time-independent Schrödinger equation, we find that the energy levels are given by

$$E_{n_X,n_Y,n_Z} = \left(\frac{n_X^2}{L_X^2} + \frac{n_Y^2}{L_Y^2} + \frac{n_Z^2}{L_Z^2}\right)\frac{\pi^2\hbar^2}{2m} \quad (n_X = 1, 2, 3, \ldots; n_Y = 1, 2, 3, \ldots;$$
$$n_Z = 1, 2, 3, \ldots) \quad \text{(41.17)}$$

(energy levels, particle in a three-dimensional box with sides of length L_X, L_Y, and L_Z)

If L_X, L_Y, and L_Z are all different, the states $(n_X, n_Y, n_Z) = (2, 1, 1)$, $(1, 2, 1)$, and $(1, 1, 2)$ have different energies and hence are no longer degenerate. Note that Eq. (41.17) reduces to Eq. (41.16) if the lengths are all the same ($L_X = L_Y = L_Z = L$).

Let's summarize the key differences between the three-dimensional particle in a box and the one-dimensional case that we examined in Section 40.2:

- We can write the wave function for a three-dimensional stationary state as a product of three functions, one for each spatial coordinate. Only a single function of the coordinate x is needed in one dimension.
- In the three-dimensional case, three quantum numbers are needed to describe each stationary state. Only one quantum number is needed in the one-dimensional case.
- Most of the energy levels for the three-dimensional case are degenerate: More than one stationary state has this energy. There is no degeneracy in the one-dimensional case.
- For a stationary state of the three-dimensional case, there are surfaces on which the probability distribution function $|\psi|^2$ is zero. In the one-dimensional case there are positions on the x-axis where $|\psi|^2$ is zero.

We'll see these same features in the following section as we examine a three-dimensional situation that's more realistic than a particle in a box: a hydrogen atom in which a negatively charged electron orbits a positively charged nucleus.

Test Your Understanding of Section 41.2 Rank the following states of a particle in a cubical box of side L in order from highest to lowest energy: (i) $(n_X, n_Y, n_Z) = (2, 3, 2)$; (ii) $(n_X, n_Y, n_Z) = (4, 1, 1)$; (iii) $(n_X, n_Y, n_Z) = (2, 2, 3)$; (iv) $(n_X, n_Y, n_Z) = (1, 3, 3)$.

41.3 The Hydrogen Atom

Let's continue the discussion of the hydrogen atom that we began in Chapter 39. In the Bohr model, electrons move in circular orbits like Newtonian particles, but with quantized values of angular momentum. While this model gave the correct energy levels of the hydrogen atom, as deduced from spectra, it had many conceptual difficulties. It mixed classical physics with new and seemingly contradictory concepts. It provided no insight into the process by which photons are emitted and absorbed. It could not be generalized to atoms with more than one electron. It predicted the wrong magnetic properties for the hydrogen atom. And perhaps most important, its picture of the electron as a localized point particle was inconsistent with the more general view we developed in Chapters 39 and 40. To go beyond the Bohr model, let's apply the Schrödinger equation to find the wave functions for stationary states (states of definite energy) of the hydrogen atom. As in Section 39.3, we include the motion of the nucleus by simply replacing the electron mass m with the reduced mass m_r.

The Schrödinger Equation for the Hydrogen Atom

We discussed the three-dimensional version of the Schrödinger equation in Section 41.1. The potential-energy function is *spherically symmetric:* It depends only on the distance $r = (x^2 + y^2 + z^2)^{1/2}$ from the origin of coordinates:

$$U(r) = -\frac{1}{4\pi\epsilon_0}\frac{e^2}{r} \tag{41.18}$$

The hydrogen-atom problem is best formulated in spherical coordinates (r, θ, ϕ), shown in Fig. 41.5; the spherically symmetric potential-energy function depends only on r, not on θ or ϕ. The Schrödinger equation with this potential-energy function can be solved exactly; the solutions are combinations of familiar functions. Without going into a lot of detail, we can describe the most important features of the procedure and the results.

First, we find the solutions using the same method of separation of variables that we employed for a particle in a cubical box in Section 41.2. We express the wave function $\psi(r, \theta, \phi)$ as a product of three functions, each one a function of only one of the three coordinates:

$$\psi(r, \theta, \phi) = R(r)\Theta(\theta)\Phi(\phi) \tag{41.19}$$

That is, the function $R(r)$ depends only on r, $\Theta(\theta)$ depends only on θ, and $\Phi(\phi)$ depends only on ϕ. Just as for a particle in a three-dimensional box, when we substitute Eq. (41.19) into the Schrödinger equation, we get three separate ordinary differential equations. One equation involves only r and $R(r)$, a second involves only θ and $\Theta(\theta)$, and a third involves only ϕ and $\Phi(\phi)$:

$$-\frac{\hbar^2}{2m_r r^2}\frac{d}{dr}\left(r^2\frac{dR(r)}{dr}\right) + \left(\frac{\hbar^2 l(l+1)}{2m_r r^2} + U(r)\right)R(r) = ER(r) \tag{41.20a}$$

$$\frac{1}{\sin\theta}\frac{d}{d\theta}\left(\sin\theta\frac{d\Theta(\theta)}{d\theta}\right) + \left(l(l+1) - \frac{m_l{}^2}{\sin^2\theta}\right)\Theta(\theta) = 0 \tag{41.20b}$$

$$\frac{d^2\Phi(\phi)}{d\phi^2} + m_l{}^2\Phi(\phi) = 0 \tag{41.20c}$$

In Eqs. (41.20) E is the energy of the stationary state and l and m_l are constants that we'll discuss later. (Be careful! Don't confuse the constant m_l with the reduced mass m_r.)

We won't attempt to solve this set of three equations, but we can describe how it's done. As for the particle in a cubical box, the physically acceptable solutions of these three equations are determined by boundary conditions. The radial function $R(r)$ in Eq. (41.20a) must approach zero at large r, because we are describing

41.5 The Schrödinger equation for the hydrogen atom can be solved most readily using spherical coordinates.

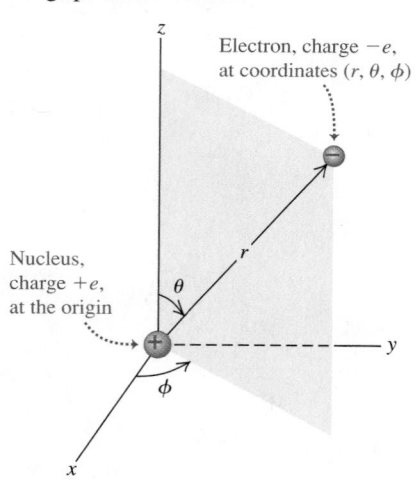

Electron, charge $-e$, at coordinates (r, θ, ϕ)

Nucleus, charge $+e$, at the origin

bound states of the electron that are localized near the nucleus. This is analogous to the requirement that the harmonic-oscillator wave functions (see Section 40.5) must approach zero at large x. The angular functions $\Theta(\theta)$ and $\Phi(\phi)$ in Eqs. (41.20b) and (41.20c) must be *finite* for all relevant values of the angles. For example, there are solutions of the Θ equation that become infinite at $\theta = 0$ and $\theta = \pi$; these are unacceptable, since $\psi(r, \theta, \phi)$ must be normalizable. Furthermore, the angular function $\Phi(\phi)$ in Eq. (41.20c) must be *periodic*. For example, (r, θ, ϕ) and $(r, \theta, \phi + 2\pi)$ describe the same point, so $\Phi(\phi + 2\pi)$ must equal $\Phi(\phi)$.

The allowed radial functions $R(r)$ turn out to be an exponential function $e^{-\alpha r}$ (where α is positive) multiplied by a polynomial in r. The functions $\Theta(\theta)$ are polynomials containing various powers of $\sin\theta$ and $\cos\theta$, and the functions $\Phi(\phi)$ are simply proportional to $e^{im_l\phi}$, where $i = \sqrt{-1}$ and m_l is an integer that may be positive, zero, or negative.

In the process of finding solutions that satisfy the boundary conditions, we also find the corresponding energy levels. We denote the energies of these levels [E in Eq. (41.20a)] by E_n ($n = 1, 2, 3, \dots$). These turn out to be *identical* to those from the Bohr model, as given by Eq. (39.15), with the electron rest mass m replaced by the reduced mass m_r. Rewriting that equation using $\hbar = h/2\pi$, we have

$$E_n = -\frac{1}{(4\pi\epsilon_0)^2}\frac{m_r e^4}{2n^2\hbar^2} = -\frac{13.60 \text{ eV}}{n^2} \qquad \text{(energy levels of hydrogen)} \quad \text{(41.21)}$$

As in Section 39.3, we call n the **principal quantum number** for the level of energy E_n.

Equation (41.21) is an important validation of our Schrödinger-equation analysis of the hydrogen atom. The Schrödinger analysis is quite different from the Bohr model, both mathematically and conceptually, yet both yield the same energy-level scheme—a scheme that agrees with the energies determined from spectra. As we will see, the Schrödinger analysis can explain many more aspects of the hydrogen atom than can the Bohr model.

Quantization of Orbital Angular Momentum

The solutions to Eqs. (41.20) that satisfy the boundary conditions mentioned above also have quantized values of *orbital angular momentum.* That is, only certain discrete values of the magnitude and components of orbital angular momentum are permitted. In discussing the Bohr model in Section 39.3, we mentioned that quantization of angular momentum was a result with little fundamental justification. With the Schrödinger equation it appears automatically.

The possible values of the magnitude L of orbital angular momentum $\vec{L}$ are determined by the requirement that the $\Theta(\theta)$ function in Eq. (41.20b) must be finite at $\theta = 0$ and $\theta = \pi$. In a level with energy E_n and principal quantum number n, the possible values of L are

$$L = \sqrt{l(l+1)}\,\hbar \qquad (l = 0, 1, 2, \dots, n-1) \qquad \begin{array}{l}\text{(magnitude of}\\ \text{orbital angular}\\ \text{momentum)}\end{array} \quad \text{(41.22)}$$

The *orbital angular-momentum quantum number l*, which is the same l that appears in Eqs. (41.20a) and (41.20b), is called the **orbital quantum number** for short. In the Bohr model, each energy level corresponded to a single value of angular momentum. Equation (41.22) shows that in fact there are n different possible values of L for the nth energy level.

An interesting feature of Eq. (41.22) is that the orbital angular momentum is *zero* for $l = 0$ states. This result disagrees with the Bohr model, in which the electron always moved in a circle of definite radius and L was never zero. The $l = 0$ wave functions ψ depend only on r; for these states, the functions $\Theta(\theta)$ and $\Phi(\phi)$ are constants. Thus the wave functions for $l = 0$ states are spherically symmetric. There is nothing in their probability distribution $|\psi|^2$ to favor one direction over any other, and there is no orbital angular momentum.

The permitted values of the *component* of $\vec{L}$ in a given direction, say the z-component L_z, are determined by the requirement that the $\Phi(\phi)$ function must equal $\Phi(\phi + 2\pi)$. The possible values of L_z are

$$L_z = m_l\hbar \qquad (m_l = 0, \pm 1, \pm 2, \ldots, \pm l) \qquad \text{(components of orbital angular momentum)} \qquad \text{(41.23)}$$

The quantum number m_l is the same as that in Eqs. (41.20b) and (41.20c). We see that m_l can be zero or a positive or negative integer up to, but no larger in magnitude than, l. That is, $|m_l| \leq l$. For example, if $l = 1$, m_l can equal 1, 0, or -1. For reasons that will emerge later, we call m_l the *orbital magnetic quantum number*, or **magnetic quantum number** for short.

The component L_z can never be quite as large as L (unless both are zero). For example, when $l = 2$, the largest possible value of m_l is also 2; then Eqs. (41.22) and (41.23) give

$$L = \sqrt{2(2 + 1)}\hbar = \sqrt{6}\hbar = 2.45\hbar$$

$$L_z = 2\hbar$$

Figure 41.6 shows the situation. The minimum value of the angle θ_L between the vector $\vec{L}$ and the z-axis is given by

$$\theta_L = \arccos\frac{L_z}{L} = \arccos\frac{2}{2.45} = 35.3°$$

That $|L_z|$ is always less than L is also required by the uncertainty principle. Suppose we could know the precise *direction* of the orbital angular momentum vector. Then we could let that be the direction of the z-axis, and L_z would equal L. This corresponds to a particle moving in the xy-plane only, in which case the z-component of the linear momentum $\vec{p}$ would be zero with no uncertainty Δp_z. Then the uncertainty principle $\Delta z \Delta p_z \geq \hbar$ requires infinite uncertainty Δz in the coordinate z. This is impossible for a localized state; we conclude that we can't know the direction of $\vec{L}$ precisely. Thus, as we've already stated, the component of $\vec{L}$ in a given direction can never be quite as large as its magnitude L. Also, if we can't know the direction of $\vec{L}$ precisely, we can't determine the components L_x and L_y precisely. Thus we show *cones* of possible directions for $\vec{L}$ in Fig. 41.6b.

You may wonder why we have singled out the z-axis for special attention. There's no fundamental reason for this; the atom certainly doesn't care what coordinate system we use. The point is that we can't determine all three components of orbital angular momentum with certainty, so we arbitrarily pick one as the component we want to measure. When we discuss interactions of the atom with a magnetic field, we will consistently choose the positive z-axis to be in the direction of $\vec{B}$.

Quantum Number Notation

The wave functions for the hydrogen atom are determined by the values of three quantum numbers n, l, and m_l. (Compare this to the particle in a three-dimensional box that we considered in Section 41.2. There, too, three quantum numbers were needed to describe each stationary state.) The energy E_n is determined by the principal quantum number n according to Eq. (41.21). The magnitude of orbital angular momentum is determined by the orbital quantum number l, as in Eq. (41.22). The component of orbital angular momentum in a specified axis direction (customarily the z-axis) is determined by the magnetic quantum number m_l, as in Eq. (41.23). The energy does not depend on the values of l or m_l (Fig. 41.7), so for each energy level E_n given by Eq. (41.21), there is more than one distinct state having the same energy but different quantum numbers. That is, these states are *degenerate*, just like most of the states of a particle in a

41.6 (a) When $l = 2$, the magnitude of the angular momentum vector $\vec{L}$ is $\sqrt{6}\hbar = 2.45\hbar$, but $\vec{L}$ does not have a definite direction. In this semiclassical vector picture, $\vec{L}$ makes an angle of 35.3° with the z-axis when the z-component has its maximum value of $2\hbar$. (b) These cones show the possible directions of $\vec{L}$ for different values of L_z.

(a)

(b)

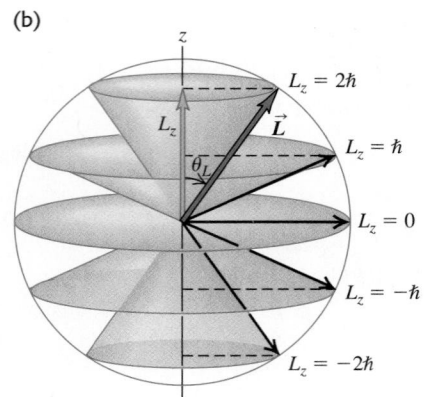

three-dimensional box. As for the three-dimensional box, degeneracy arises because the hydrogen atom is symmetric: If you rotate the atom through any angle, the potential-energy function at a distance r from the nucleus has the same value.

States with various values of the orbital quantum number l are often labeled with letters, according to the following scheme:

$l = 0$: s states
$l = 1$: p states
$l = 2$: d states
$l = 3$: f states
$l = 4$: g states
$l = 5$: h states

and so on alphabetically. This seemingly irrational choice of the letters s, p, d, and f originated in the early days of spectroscopy and has no fundamental significance. In an important form of *spectroscopic notation* that we'll use often, a state with $n = 2$ and $l = 1$ is called a $2p$ state; a state with $n = 4$ and $l = 0$ is a $4s$ state; and so on. Only s states ($l = 0$) are spherically symmetric.

Here's another bit of notation. The radial extent of the wave functions increases with the principal quantum number n, and we can speak of a region of space associated with a particular value of n as a **shell.** Especially in discussions of many-electron atoms, these shells are denoted by capital letters:

$n = 1$: K shell
$n = 2$: L shell
$n = 3$: M shell
$n = 4$: N shell

and so on alphabetically. For each n, different values of l correspond to different *subshells*. For example, the L shell ($n = 2$) contains the $2s$ and $2p$ subshells.

Table 41.1 shows some of the possible combinations of the quantum numbers n, l, and m_l for hydrogen-atom wave functions. The spectroscopic notation and the shell notation for each are also shown.

41.7 The energy for an orbiting satellite such as the Hubble Space Telescope depends on the average distance between the satellite and the center of the earth. It does *not* depend on whether the orbit is circular (with a large orbital angular momentum L) or elliptical (in which case L is smaller). In the same way, the energy of a hydrogen atom does not depend on the orbital angular momentum.

Table 41.1 **Quantum States of the Hydrogen Atom**

n	l	m_l	Spectroscopic Notation	Shell
1	0	0	$1s$	K
2	0	0	$2s$	L
2	1	$-1, 0, 1$	$2p$	
3	0	0	$3s$	
3	1	$-1, 0, 1$	$3p$	M
3	2	$-2, -1, 0, 1, 2$	$3d$	
4	0	0	$4s$	N

and so on

Problem-Solving Strategy 41.1   **Atomic Structure**

IDENTIFY *the relevant concepts:* Many problems in atomic structure can be solved simply by reference to the quantum numbers n, l, and m_l that describe the total energy E, the magnitude of the orbital angular momentum $\vec{L}$, the z-component of $\vec{L}$, and other properties of an atom.

SET UP *the problem:* Identify the target variables and choose the appropriate equations, which may include Eqs. (41.21), (41.22), and (41.23).

EXECUTE *the solution* as follows:

1. Be sure you understand the possible values of the quantum numbers n, l, and m_l for the hydrogen atom. They are all integers; n is always greater than zero, l can be zero or positive up

to $n - 1$, and m_l can range from $-l$ to l. You should know how to count the number of (n, l, m_l) states in each shell (K, L, M, and so on) and subshell ($3s$, $3p$, $3d$, and so on). Be able to *construct* Table 41.1, not just to write it from memory.

2. Solve for the target variables.

EVALUATE *your answer:* It helps to be familiar with typical magnitudes in atomic physics. For example, the electric potential energy of a proton and electron 0.10 nm apart (typical of atomic dimensions) is about $-15\ \mathrm{eV}$. Visible light has wavelengths around 500 nm and frequencies around 5×10^{14} Hz. Problem-Solving Strategy 39.1 (Section 39.1) gives other typical magnitudes.

How many distinct (n, l, m_l) states of the hydrogen atom with $n = 3$ are there? What are their energies?

SOLUTION

IDENTIFY and SET UP: This problem uses the relationships among the principal quantum number n, orbital quantum number l, magnetic quantum number m_l, and energy of a state for the hydrogen atom. We use the rule that l can have n integer values, from 0 to $n - 1$, and that m_l can have $2l + 1$ values, from $-l$ to l. Equation (41.21) gives the energy of any particular state.

EXECUTE: When $n = 3$, l can be 0, 1, or 2. When $l = 0$, m_l can be only 0 (1 state). When $l = 1$, m_l can be -1, 0, or 1 (3 states). When $l = 2$, m_l can be -2, -1, 0, 1, or 2 (5 states). The total number of

(n, l, m_l) states with $n = 3$ is therefore $1 + 3 + 5 = 9$. (In Section 41.5 we'll find that the total number of $n = 3$ states is in fact twice this, or 18, because of electron spin.)

The energy of a hydrogen-atom state depends only on n, so all 9 of these states have the same energy. From Eq. (41.21),

$$E_3 = \frac{-13.60 \text{ eV}}{3^2} = -1.51 \text{ eV}$$

EVALUATE: For a given value of n, the total number of (n, l, m_l) states turns out to be n^2. In this case $n = 3$ and there are $3^2 = 9$ states. Remember that the ground level of hydrogen has $n = 1$ and $E_1 = -13.6$ eV; the $n = 3$ excited states have a higher (less negative) energy.

Example 41.3 **Angular momentum in an excited level of hydrogen**

Consider the $n = 4$ states of hydrogen. (a) What is the maximum magnitude L of the orbital angular momentum? (b) What is the maximum value of L_z? (c) What is the minimum angle between $\vec{L}$ and the z-axis? Give your answers to parts (a) and (b) in terms of $\hbar$.

SOLUTION

IDENTIFY and SET UP: We again need to relate the principal quantum number n and the orbital quantum number l for a hydrogen atom. We also need to relate the value of l and the magnitude and possible directions of the orbital angular momentum vector. We'll use Eq. (41.22) in part (a) to determine the maximum value of L; then we'll use Eq. (41.23) in part (b) to determine the maximum value of L_z. The angle between $\vec{L}$ and the z-axis is minimum when L_z is maximum (so that $\vec{L}$ is most nearly aligned with the positive z-axis).

EXECUTE: (a) When $n = 4$, the maximum value of the orbital angular-momentum quantum number l is $(n - 1) = (4 - 1) = 3$; from Eq. (41.22),

$$L_{max} = \sqrt{3(3 + 1)}\hbar = \sqrt{12}\hbar = 3.464\hbar$$

(b) For $l = 3$ the maximum value of m_l is 3. From Eq. (41.23),

$$(L_z)_{max} = 3\hbar$$

(c) The *minimum* allowed angle between $\vec{L}$ and the z-axis corresponds to the *maximum* allowed values of L_z and m_l (Fig. 41.6b shows an $l = 2$ example). For the state with $l = 3$ and $m_l = 3$,

$$\theta_{min} = \arccos\frac{(L_z)_{max}}{L} = \arccos\frac{3\hbar}{3.464\hbar} = 30.0°$$

EVALUATE: As a check, you can verify that θ is greater than 30.0° for all states with smaller values of l.

Electron Probability Distributions

Rather than picturing the electron as a point particle moving in a precise circle, the Schrödinger equation gives an electron *probability distribution* surrounding the nucleus. The hydrogen-atom probability distributions are three-dimensional, so they are harder to visualize than the two-dimensional circular orbits of the Bohr model. It's helpful to look at the *radial probability distribution function* $P(r)$—that is, the probability per radial length for the electron to be found at various distances from the proton. From Section 41.1 the probability for finding the electron in a small volume element dV is $|\psi|^2 dV$. (We assume that ψ is normalized in accordance with Eq. (41.6)—that is, that the integral of $|\psi|^2 dV$ over all space equals unity so that there is 100% probability of finding the electron somewhere in the universe.) Let's take as our volume element a thin spherical shell with inner radius r and outer radius $r + dr$. The volume dV of this shell is approximately its area $4\pi r^2$ multiplied by its thickness dr:

$$dV = 4\pi r^2 \, dr \tag{41.24}$$

We denote by $P(r) \, dr$ the probability of finding the particle within the radial range dr; then, using Eq. (41.24),

$$P(r) \, dr = |\psi|^2 \, dV = |\psi|^2 4\pi r^2 \, dr \qquad \begin{array}{l}\text{(probability that the}\\ \text{electron is between } r\\ \text{and } r + dr)\end{array} \qquad \text{(41.25)}$$

For wave functions that depend on θ and ϕ as well as r, we use the value of $|\psi|^2$ averaged over all angles in Eq. (41.25).

Figure 41.8 shows graphs of $P(r)$ for several hydrogen-atom wave functions. The r scales are labeled in multiples of a, the smallest distance between the electron and the nucleus in the Bohr model:

$$a = \frac{\epsilon_0 h^2}{\pi m_r e^2} = \frac{4\pi \epsilon_0 \hbar^2}{m_r e^2} = 5.29 \times 10^{-11} \text{ m} \qquad \begin{array}{l}\text{(smallest } r,\\ \text{Bohr model)}\end{array} \qquad \text{(41.26)}$$

Just as for a particle in a cubical box (see Section 41.2), there are some locations where the probability is zero. These surfaces are planes for a particle in a box; for a hydrogen atom these are spherical surfaces (that is, surfaces of constant r). But again, the uncertainty principle tells us not to worry; we can't localize the electron exactly anyway. Note that for the states having the largest possible l for each n (such as 1s, 2p, 3d, and 4f states), $P(r)$ has a single maximum at $n^2 a$. For these states, the electron is most likely to be found at the distance from the nucleus that is predicted by the Bohr model, $r = n^2 a$.

Figure 41.8 shows *radial* probability distribution functions $P(r) = 4\pi r^2 |\psi|^2$, which indicate the relative probability of finding the electron within a thin spherical shell of radius r. By contrast, Figs. 41.9 and 41.10 show the *three-dimensional* probability distribution functions $|\psi|^2$, which indicate the relative probability of finding the electron within a small box at a given position. The darker the blue "cloud," the greater the value of $|\psi|^2$. (These are similar to the "clouds" shown in Fig. 41.2.) Figure 41.9 shows cross sections of the spherically symmetric probability clouds for the lowest three s subshells, for which $|\psi|^2$ depends only on the radial coordinate r. Figure 41.10 shows cross sections of the clouds for other electron states for which $|\psi|^2$ depends on both r and θ. For these states the probability distribution function is zero for certain values of θ as well as for certain values of r. In *any* stationary state of the hydrogen atom, $|\psi|^2$ is independent of ϕ.

41.8 Radial probability distribution functions $P(r)$ for several hydrogen-atom wave functions, plotted as functions of the ratio r/a [see Eq. (41.26)]. For each function, the number of maxima is $(n - l)$. The curves for which $l = n - 1$ (1s, 2p, 3d, . . .) have only one maximum, located at $r = n^2 a$.

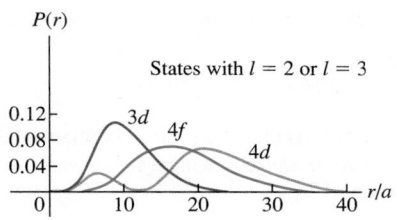

41.9 Three-dimensional probability distribution functions $|\psi|^2$ for the spherically symmetric 1s, 2s, and 3s hydrogen-atom wave functions.

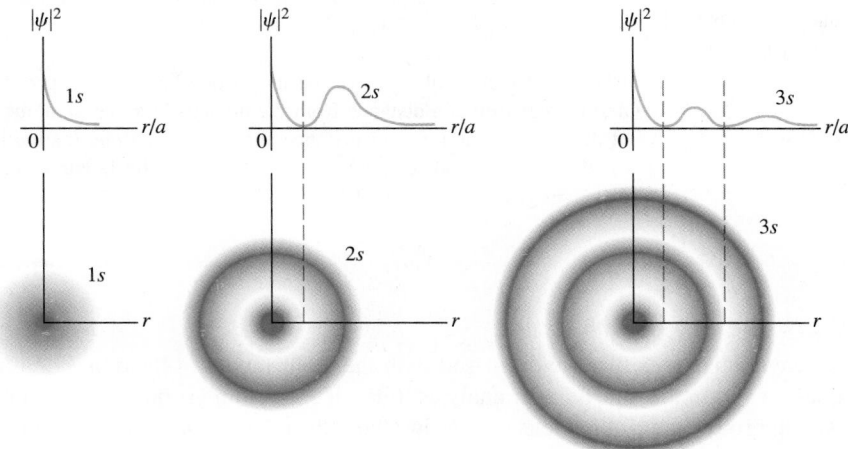

41.10 Cross sections of three-dimensional probability distributions for a few quantum states of the hydrogen atom. They are not to the same scale. Mentally rotate each drawing about the z-axis to obtain the three-dimensional representation of $|\psi|^2$. For example, the $2p$, $m_l = \pm 1$ probability distribution looks like a fuzzy donut.

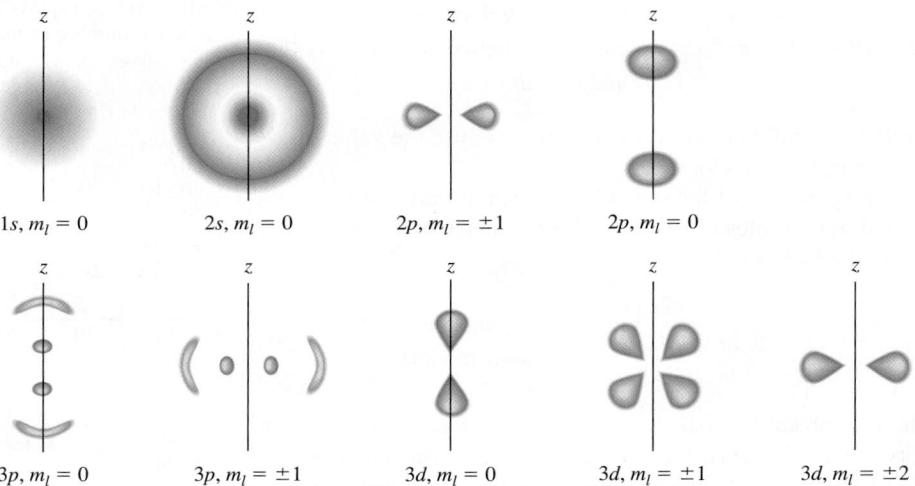

$1s$, $m_l = 0$ $2s$, $m_l = 0$ $2p$, $m_l = \pm 1$ $2p$, $m_l = 0$

$3p$, $m_l = 0$ $3p$, $m_l = \pm 1$ $3d$, $m_l = 0$ $3d$, $m_l = \pm 1$ $3d$, $m_l = \pm 2$

Example 41.4   A hydrogen wave function

The ground-state wave function for hydrogen (a $1s$ state) is

$$\psi_{1s}(r) = \frac{1}{\sqrt{\pi a^3}} e^{-r/a}$$

(a) Verify that this function is normalized. (b) What is the probability that the electron will be found at a distance less than a from the nucleus?

SOLUTION

IDENTIFY and SET UP: This example is similar to Example 41.1 in Section 41.2. We need to show that this wave function satisfies the condition that the probability of finding the electron *somewhere* is 1. We then need to find the probability that it will be found in the region $r < a$. In part (a) we'll carry out the integral $\int |\psi|^2 \, dV$ over all space; if it is equal to 1, the wave function is normalized. In part (b) we'll carry out the same integral over a spherical volume that extends from the origin (the nucleus) out to a distance a from the nucleus.

EXECUTE: (a) Since the wave function depends only on the radial coordinate r, we can choose our volume elements to be spherical shells of radius r, thickness dr, and volume dV given by Eq. (41.24). We then have

$$\int_{\text{all space}} |\psi_{1s}|^2 \, dV = \int_0^\infty \frac{1}{\pi a^3} e^{-2r/a} (4\pi r^2 \, dr)$$

$$= \frac{4}{a^3} \int_0^\infty r^2 e^{-2r/a} \, dr$$

You can find the following indefinite integral in a table of integrals or by integrating by parts:

$$\int r^2 e^{-2r/a} \, dr = \left(-\frac{ar^2}{2} - \frac{a^2 r}{2} - \frac{a^3}{4} \right) e^{-2r/a}$$

Evaluating this between the limits $r = 0$ and $r = \infty$ is simple; it is zero at $r = \infty$ because of the exponential factor, and at $r = 0$ only the last term in the parentheses survives. Thus the value of the definite integral is $a^3/4$. Putting it all together, we find

$$\int_0^\infty |\psi_{1s}|^2 \, dV = \frac{4}{a^3} \int_0^\infty r^2 e^{-2r/a} \, dr = \frac{4}{a^3} \frac{a^3}{4} = 1$$

The wave function *is* normalized.

(b) To find the probability P that the electron is found within $r < a$, we carry out the same integration but with the limits 0 and a. We'll leave the details to you (Exercise 41.15). From the upper limit we get $-5e^{-2}a^3/4$; the final result is

$$P = \int_0^a |\psi_{1s}|^2 \, 4\pi r^2 \, dr = \frac{4}{a^3} \left(-\frac{5a^3 e^{-2}}{4} + \frac{a^3}{4} \right)$$

$$= (-5e^{-2} + 1) = 1 - 5e^{-2} = 0.323$$

EVALUATE: Our results tell us that in a ground state we expect to find the electron at a distance from the nucleus less than a about $\frac{1}{3}$ of the time and at a greater distance about $\frac{2}{3}$ of the time. It's hard to tell, but in Fig. 41.8, about $\frac{2}{3}$ of the area under the $1s$ curve is at distances greater than a (that is, $r/a > 1$).

Hydrogenlike Atoms

Two generalizations that we discussed with the Bohr model in Section 39.3 are equally valid in the Schrödinger analysis. First, if the "atom" is not composed of a single proton and a single electron, using the reduced mass m_r of the system in Eqs. (41.21) and (41.26) will lead to changes that are substantial for some exotic systems. One example is *positronium,* in which a positron and an electron orbit each other; another is a *muonic atom,* in which the electron is replaced by an unstable particle called a muon that has the same charge as an electron but is 207 times more massive. Second, our analysis is applicable to single-electron ions, such as He$^+$, Li^{2+}, and so on. For such ions we replace e^2 by Ze^2 in Eqs. (41.21) and (41.26), where Z is the number of protons (the **atomic number**).

Test Your Understanding of Section 41.3 Rank the following states of the hydrogen atom in order from highest to lowest probability of finding the electron in the vicinity of $r = 5a$: (i) $n = 1, l = 0, m_l = 0$; (ii) $n = 2, l = 1$, $m_l = +1$; (iii) $n = 2, l = 1, m_l = 0$.

41.4 The Zeeman Effect

The **Zeeman effect** is the splitting of atomic energy levels and the associated spectral lines when the atoms are placed in a magnetic field (Fig. 41.11). This effect confirms experimentally the quantization of angular momentum. The discussion in this section, which assumes that the only angular momentum is the *orbital* angular momentum of a single electron, also shows why we call m_l the magnetic quantum number.

Atoms contain charges in motion, so it should not be surprising that magnetic forces cause changes in that motion and in the energy levels. As early as the middle of the 19th century, physicists speculated that the sources of visible light might be vibrating electric charge on an atomic scale. In 1896 the Dutch physicist Pieter Zeeman was the first to show that in the presence of a magnetic field, some spectral lines were split into groups of closely spaced lines (Fig. 41.12). This effect now bears his name.

Magnetic Moment of an Orbiting Electron

Let's begin our analysis of the Zeeman effect by reviewing the concept of *magnetic dipole moment* or *magnetic moment,* introduced in Section 27.7. A plane current loop with vector area $\vec{A}$ carrying current I has a magnetic moment $\vec{\mu}$ given by

$$\vec{\mu} = I\vec{A} \tag{41.27}$$

When a magnetic dipole of moment $\vec{\mu}$ is placed in a magnetic field $\vec{B}$, the field exerts a torque $\vec{\tau} = \vec{\mu} \times \vec{B}$ on the dipole. The potential energy U associated with this interaction is given by Eq. (27.27):

$$U = -\vec{\mu} \cdot \vec{B} \tag{41.28}$$

Now let's use Eqs. (41.27) and (41.28) and the Bohr model to look at the interaction of a hydrogen atom with a magnetic field. The orbiting electron with speed v is equivalent to a current loop with radius r and area πr^2. The average current I is the average charge per unit time that passes a given point of the orbit. This is equal to the charge magnitude e divided by the time T for one revolution, given by $T = 2\pi r/v$. Thus $I = ev/2\pi r$, and from Eq. (41.27) the magnitude μ of the magnetic moment is

$$\mu = IA = \frac{ev}{2\pi r}\pi r^2 = \frac{evr}{2} \tag{41.29}$$

We can also express this in terms of the magnitude L of the orbital angular momentum. From Eq. (10.28) the angular momentum of a particle in a circular orbit is $L = mvr$, so Eq. (41.29) becomes

$$\mu = \frac{e}{2m}L \tag{41.30}$$

41.11 Magnetic effects on the spectrum of sunlight. **(a)** The slit of a spectrograph is positioned along the black line crossing a portion of a sunspot. **(b)** The 0.4-T magnetic field in the sunspot (a thousand times greater than the earth's field) splits the middle spectral line into three lines.

(a)

(b)

41.12 The normal Zeeman effect. Compare this to the magnetic splitting in the solar spectrum shown in Fig. 41.11b.

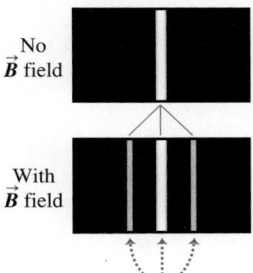

When an excited gas is placed in a magnetic field, the interaction of orbital magnetic moments with the field splits individual spectral lines of the gas into sets of three lines.

The ratio of the magnitude of $\vec{\boldsymbol{\mu}}$ to the magnitude of $\vec{\boldsymbol{L}}$ is $\mu/L = e/2m$ and is called the *gyromagnetic ratio*.

In the Bohr model, $L = nh/2\pi = n\hbar$, where $n = 1, 2, \ldots$. For an $n = 1$ state (a ground state), Eq. (41.30) becomes $\mu = (e/2m)\hbar$. This quantity is a natural unit for magnetic moment; it is called one **Bohr magneton,** denoted by μ_{B}:

$$\mu_{\mathrm{B}} = \frac{e\hbar}{2m} \qquad \text{(definition of the Bohr magneton)} \qquad (41.31)$$

Evaluating Eq. (41.31) gives

$$\mu_{\mathrm{B}} = 5.788 \times 10^{-5} \text{ eV/T} = 9.274 \times 10^{-24} \text{ J/T or A} \cdot \mathrm{m}^2$$

Note that the units J/T and A · m^2 are equivalent. We defined this quantity previously in Section 28.8.

While the Bohr model suggests that the orbital motion of an atomic electron gives rise to a magnetic moment, this model does *not* give correct predictions about magnetic interactions. As an example, the Bohr model predicts that an electron in a hydrogen-atom ground state has an orbital magnetic moment of magnitude μ_{B}. But the Schrödinger picture tells us that such a ground-state electron is in an *s* state with zero angular momentum, so the orbital magnetic moment must be *zero!* To get the correct results, we must describe the states by using Schrödinger wave functions.

It turns out that in the Schrödinger formulation, electrons have the same ratio of μ to L (gyromagnetic ratio) as in the Bohr model—namely, $e/2m$. Suppose the magnetic field $\vec{\boldsymbol{B}}$ is directed along the $+z$-axis. From Eq. (41.28) the interaction energy U of the atom's magnetic moment with the field is

$$U = -\mu_z B \qquad (41.32)$$

where μ_z is the z-component of the vector $\vec{\boldsymbol{\mu}}$.

Now we use Eq. (41.30) to find μ_z, recalling that e is the *magnitude* of the electron charge and that the actual charge is $-e$. Because the electron charge is negative, the orbital angular momentum and magnetic moment vectors are opposite. We find

$$\mu_z = -\frac{e}{2m}L_z \qquad (41.33)$$

For the Schrödinger wave functions, $L_z = m_l\hbar$, with $m_l = 0, \pm 1, \pm 2, \ldots, \pm l$, so Eq. (41.33) becomes

$$\mu_z = -\frac{e}{2m}L_z = -m_l\frac{e\hbar}{2m} \qquad (41.34)$$

CAUTION **Two uses of the symbol m** Be careful not to confuse the electron mass m with the magnetic quantum number m_l. ▌

Finally, we can express the interaction energy, Eq. (41.32), as

$$U = -\mu_z B = m_l\frac{e\hbar}{2m}B \qquad (m_l = 0, \pm 1, \pm 2, \ldots, \pm l)$$

(orbital magnetic interaction energy) $\qquad (41.35)$

In terms of the Bohr magneton $\mu_{\mathrm{B}} = e\hbar/2m$, we can write Eq. (41.35) as

$$U = m_l\mu_{\mathrm{B}}B \qquad \text{(orbital magnetic interaction energy)} \qquad (41.36)$$

The magnetic field shifts the energy of each orbital state by an amount U. The interaction energy U depends on the value of m_l because m_l determines the orientation of the orbital magnetic moment relative to the magnetic field. This dependence is the reason m_l is called the magnetic quantum number.

The values of m_l range from $-l$ to $+l$ in steps of one, so an energy level with a particular value of the orbital quantum number l contains $(2l + 1)$ different orbital states. Without a magnetic field these states all have the same energy; that is, they are degenerate. The magnetic field removes this degeneracy. In the presence of a magnetic field they are split into $2l + 1$ distinct energy levels; adjacent levels differ in energy by $(e\hbar/2m)B = \mu_B B$. We can understand this in terms of the connection between degeneracy and symmetry. With a magnetic field applied along the z-axis, the atom is no longer completely symmetric under rotation: There is a preferred direction in space. By removing the symmetry, we remove the degeneracy of states.

Figure 41.13 shows the effect on the energy levels of hydrogen. Spectral lines corresponding to transitions from one set of levels to another set are correspondingly split and appear as a series of three closely spaced spectral lines replacing a single line. As the following example shows, the splitting of spectral lines is quite small because the value of $\mu_B B$ is small even for substantial magnetic fields.

41.13 This energy-level diagram for hydrogen shows how the levels are split when the electron's orbital magnetic moment interacts with an external magnetic field. The values of m_l are shown adjacent to the various levels. The relative magnitudes of the level splittings are exaggerated for clarity. The $n = 4$ splittings are not shown; can you draw them in?

Example 41.5 An atom in a magnetic field

An atom in a state with $l = 1$ emits a photon with wavelength 600.000 nm as it decays to a state with $l = 0$. If the atom is placed in a magnetic field with magnitude $B = 2.00$ T, what are the shifts in the energy levels and in the wavelength that result from the interaction between the atom's orbital magnetic moment and the magnetic field?

SOLUTION

IDENTIFY and SET UP: This problem concerns the splitting of atomic energy levels by a magnetic field (the Zeeman effect). We use Eq. (41.35) or (41.36) to determine the energy-level shifts. The relationship $E = hc/\lambda$ between the energy and wavelength of a photon then lets us calculate the wavelengths emitted during transitions from the $l = 1$ states to the $l = 0$ state.

EXECUTE: The energy of a 600-nm photon is

$$E = \frac{hc}{\lambda} = \frac{(4.14 \times 10^{-15} \text{ eV} \cdot \text{s})(3.00 \times 10^8 \text{ m/s})}{600 \times 10^{-9} \text{ m}}$$
$$= 2.07 \text{ eV}$$

If there is no external magnetic field, that is the difference in energy between the $l = 0$ and $l = 1$ levels.

With a 2.00-T field present, Eq. (41.36) shows that there is no shift of the $l = 0$ state (which has $m_l = 0$). For the $l = 1$ states, the splitting of levels is given by

$$U = m_l \mu_B B = m_l (5.788 \times 10^{-5} \text{ eV/T})(2.00 \text{ T})$$
$$= m_l (1.16 \times 10^{-4} \text{ eV}) = m_l (1.85 \times 10^{-23} \text{ J})$$

The possible values of m_l for $l = 1$ are -1, 0, and $+1$, and the three corresponding levels are separated by equal intervals of

Continued

1.16×10^{-4} eV. This is a small fraction of the 2.07-eV photon energy:

$$\frac{\Delta E}{E} = \frac{1.16 \times 10^{-4} \text{ eV}}{2.07 \text{ eV}} = 5.60 \times 10^{-5}$$

The possible values of m_l for $l = 1$ are -1, 0, and $+1$, and the three corresponding levels are separated by equal intervals of 1.16×10^{-4} eV. This is a small fraction of the 2.07-eV photon energy:

$$\frac{1.16 \times 10^{-4} \text{ eV}}{2.07 \text{ eV}} = 5.60 \times 10^{-5}$$

The corresponding *wavelength* shifts are approximately $(5.60 \times 10^{-5})(600 \text{ nm}) = 0.034$ nm. The original 600.000-nm line is split into a triplet with wavelengths 599.966, 600.000, and 600.034 nm.

EVALUATE: Even though 2.00 T would be a strong field in most laboratories, the wavelength splittings are extremely small. Nonetheless, modern spectrographs have more than enough chromatic resolving power to measure these splittings (see Section 36.5).

Selection Rules

Figure 41.14 shows what happens to a set of d states ($l = 2$) as the magnetic field increases. With zero field the five states $m_l = -2$, -1, 0, 1, and 2 are degenerate (have the same energy), but the applied field spreads the states out. Figure 41.15 shows the splittings of both the $3d$ and $2p$ states. Equal energy differences $(e\hbar/2m)B = \mu_B B$ separate adjacent levels. In the absence of a magnetic field, a transition from a $3d$ to a $2p$ state would yield a single spectral line with photon energy $E_i - E_f$. With the levels split as shown, it might seem that there are five possible photon energies.

In fact, there are only three possibilities. Not all combinations of initial and final levels are possible because of a restriction associated with conservation of angular momentum. The photon ordinarily carries off one unit ($\hbar$) of angular momentum, which leads to the requirements that in a transition l must change by 1 and m_l must change by 0 or ± 1. These requirements are called **selection rules.** Transitions that obey these rules are called *allowed transitions;* those that don't are *forbidden transitions.* In Fig. 41.15 we show the allowed transitions by solid arrows. You should count the possible transition energies to convince yourself that the nine solid arrows give only three possible energies; the zero-field value $E_i - E_f$, and that value plus or minus $\Delta E = (e\hbar/2m)B = \mu_B B$. Figure 41.12 shows the corresponding spectral lines.

What we have described is called the *normal* Zeeman effect. It is based entirely on the orbital angular momentum of the electron. However, it leaves out a very important consideration: the electron *spin* angular momentum, the subject of the next section.

41.14 This figure shows how the splitting of the energy levels of a d state ($l = 2$) depends on the magnitude B of an external magnetic field, assuming only an orbital magnetic moment.

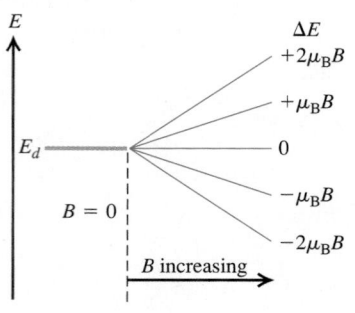

41.15 The cause of the normal Zeeman effect. The magnetic field splits the levels, but selection rules allow transitions with only three different energy changes, giving three different photon frequencies and wavelengths.

Test Your Understanding of Section 41.4 In this section we assumed that the magnetic field points in the positive z-direction. Would the results be different if the magnetic field pointed in the positive x-direction?

41.5 Electron Spin

Despite the success of the Schrödinger equation in predicting the energy levels of the hydrogen atom, experimental observations indicate that it doesn't tell the whole story of the behavior of electrons in atoms. First, spectroscopists have found magnetic-field splitting into other than the three lines we've explained, sometimes unequally spaced. Before this effect was understood, it was called the *anomalous* Zeeman effect to distinguish it from the "normal" effect discussed in the preceding section. Figure 41.16 shows both kinds of splittings.

Second, some energy levels show splittings that resemble the Zeeman effect even when there is *no* external magnetic field. For example, when the lines in the hydrogen spectrum are examined with a high-resolution spectrograph, some lines are found to consist of sets of closely spaced lines called *multiplets*. Similarly, the orange-yellow line of sodium, corresponding to the transition $4p \rightarrow 3s$ of the outer electron, is found to be a doublet ($\lambda = 589.0, 589.6$ nm), suggesting that the $4p$ level might in fact be two closely spaced levels. The Schrödinger equation in its original form didn't predict any of this.

41.16 Illustrations of the normal and anomalous Zeeman effects for two elements, zinc and sodium. The brackets under each illustration show the "normal" splitting predicted by neglecting the effect of electron spin.

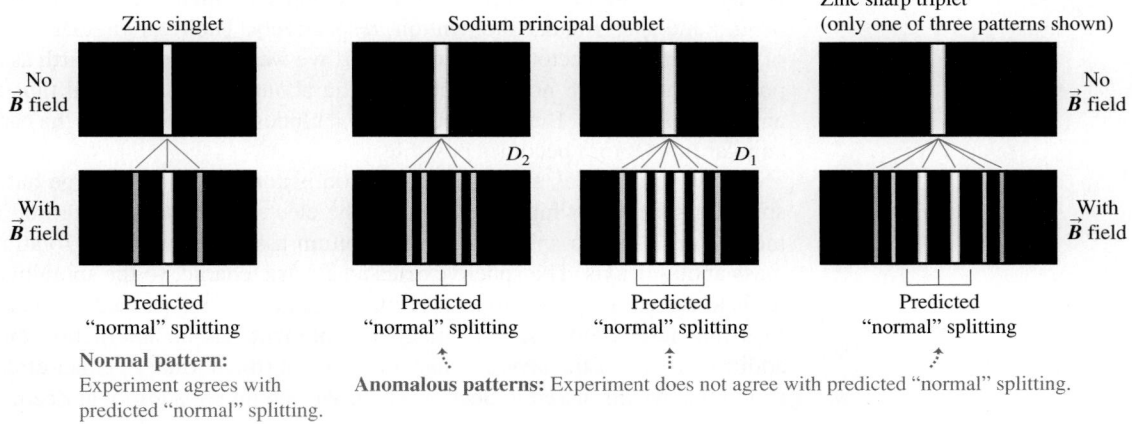

The Stern–Gerlach Experiment

Similar anomalies appeared in 1922 in atomic-beam experiments performed in Germany by Otto Stern and Walter Gerlach. When they passed a beam of neutral atoms through a nonuniform magnetic field (Fig. 41.17), atoms were deflected

PhET: Stern–Gerlach Experiment

41.17 The Stern–Gerlach experiment.

according to the orientation of their magnetic moments with respect to the field. These experiments demonstrated the quantization of angular momentum in a very direct way. If there were only orbital angular momentum, the deflections would split the beam into an odd number $(2l + 1)$ of different components. However, some atomic beams were split into an *even* number of components. If we use a different symbol j for an angular momentum quantum number, setting $2j + 1$ equal to an even number gives $j = \frac{1}{2}, \frac{3}{2}, \frac{5}{2}, \ldots$, suggesting a half-integer angular momentum. This can't be understood on the basis of the Bohr model and similar pictures of atomic structure.

In 1925 two graduate students in the Netherlands, Samuel Goudsmidt and George Uhlenbeck, proposed that the electron might have some additional motion. Using a semiclassical model, they suggested that the electron might behave like a spinning sphere of charge instead of a particle. If so, it would have an additional *spin* angular momentum and magnetic moment. If these were quantized in much the same way as *orbital* angular momentum and magnetic moment, they might help to explain the observed energy-level anomalies.

An Analogy for Electron Spin

To introduce the concept of **electron spin,** let's start with an analogy. The earth travels in a nearly circular orbit around the sun, and at the same time it *rotates* on its axis. Each motion has its associated angular momentum, which we call the *orbital* and *spin* angular momentum, respectively. The total angular momentum of the earth is the vector sum of the two. If we were to model the earth as a single point, it would have no moment of inertia about its spin axis and thus no spin angular momentum. But when our model includes the finite size of the earth, spin angular momentum becomes possible.

In the Bohr model, suppose the electron is not just a point charge but a small spinning sphere moving in orbit. Then the electron has not only orbital angular momentum but also spin angular momentum associated with the rotation of its mass about its axis. The sphere carries an electric charge, so the spinning motion leads to current loops and to a magnetic moment, as we discussed in Section 27.7. In a magnetic field, the *spin* magnetic moment has an interaction energy in addition to that of the *orbital* magnetic moment (the normal Zeeman-effect interaction that we discussed in Section 41.4). We should see additional Zeeman shifts due to the spin magnetic moment.

As we mentioned, such shifts *are* indeed observed in precise spectroscopic analysis. This and a variety of other experimental evidence have shown conclusively that the electron *does* have a spin angular momentum and a spin magnetic moment that do not depend on its orbital motion but are intrinsic to the electron itself. The origin of this spin angular momentum is fundamentally quantum-mechanical, so it's not correct to model the electron as a spinning charged sphere. But just as the Bohr model can be a useful conceptual picture for the motion of an electron in an atom, the spinning-sphere analogy can help you visualize the intrinsic spin angular momentum of an electron.

Spin Quantum Numbers

Like orbital angular momentum, the spin angular momentum of an electron (denoted by $\vec{S}$) is found to be quantized. Suppose we have an apparatus that measures a particular component of $\vec{S}$, say the z-component S_z. We find that the only possible values are

$$S_z = \pm \tfrac{1}{2}\hbar \quad \text{(components of spin angular momentum)} \qquad (41.37)$$

This relationship is reminiscent of the expression $L_z = m_l \hbar$ for the z-component of orbital angular momentum, except that $|S_z|$ is *one-half* of $\hbar$ instead of an *integer* multiple. Equation (41.37) also suggests that the magnitude S of the spin angular momentum is given by an expression analogous to Eq. (41.22) with the orbital quantum number l replaced by the **spin quantum number** $s = \frac{1}{2}$:

$$S = \sqrt{\tfrac{1}{2}\left(\tfrac{1}{2} + 1\right)}\hbar = \sqrt{\tfrac{3}{4}}\hbar \qquad \text{(magnitude of spin angular momentum)} \qquad (41.38)$$

The electron is often called a "spin $-\frac{1}{2}$ particle."

To visualize the quantized spin of an electron in a hydrogen atom, think of the electron probability distribution function $|\psi|^2$ as a cloud surrounding the nucleus like those shown in Figs. 41.9 and 41.10. Then imagine many tiny spin arrows distributed throughout the cloud, either all with components in the $+z$-direction or all with components in the $-z$-direction. But don't take this picture too seriously.

To label completely the state of the electron in a hydrogen atom, we now need a fourth quantum number m_s to specify the electron spin orientation. In analogy to the orbital magnetic quantum number m_l, we call m_s the **spin magnetic quantum number.** For an electron we give m_s the value $+\frac{1}{2}$ or $-\frac{1}{2}$ to agree with Eq. (41.37):

$$S_z = m_s\hbar \quad \left(m_s = \pm\tfrac{1}{2}\right) \qquad \text{(allowed values of } m_s \text{ and } S_z \text{ for an electron)} \qquad (41.39)$$

The spin angular momentum vector $\vec{S}$ can have only two orientations in space relative to the z-axis: "*spin up*" with a z-component of $+\frac{1}{2}\hbar$ and "*spin down*" with a z-component of $-\frac{1}{2}\hbar$.

The z-component of the associated spin magnetic moment (μ_z) turns out to be related to S_z by

$$\mu_z = -(2.00232)\frac{e}{2m}S_z \qquad (41.40)$$

where $-e$ and m are (as usual) the charge and mass of the electron. When the atom is placed in a magnetic field, the interaction energy $-\vec{\mu} \cdot \vec{B}$ of the spin magnetic dipole moment with the field causes further splittings in energy levels and in the corresponding spectral lines.

Equation (41.40) shows that the gyromagnetic ratio for electron spin is approximately *twice* as great as the value $e/2m$ for *orbital* angular momentum and magnetic dipole moment. This result has no classical analog. But in 1928 Paul Dirac developed a relativistic generalization of the Schrödinger equation for electrons. His equation gave a spin gyromagnetic ratio of exactly $2(e/2m)$. It took another two decades to develop the area of physics called *quantum electrodynamics,* abbreviated QED, which predicts the value we've given to "only" six significant figures as 2.00232. QED now predicts a value that agrees with a recent (2006) measurement of 2.00231930436170(152), making QED the most precise theory in all science.

Example 41.6 Energy of electron spin in a magnetic field

Calculate the interaction energy for an electron in an $l = 0$ state in a magnetic field with magnitude 2.00 T.

SOLUTION

IDENTIFY and SET UP: For $l = 0$ the electron has zero orbital angular momentum and zero orbital magnetic moment. Hence the only magnetic interaction is that between the $\vec{B}$ field and the spin magnetic moment $\vec{\mu}$. From Eq. (41.28), the interaction energy is $U = -\vec{\mu} \cdot \vec{B}$. As in Section 41.4, we take $\vec{B}$ to be in the positive z-direction so that $U = -\mu_z B$ [Eq. (41.32)]. Equation (41.40) gives μ_z in terms of S_z, and Eq. (41.37) gives S_z.

EXECUTE: Combining Eqs. (41.37) and (41.40), we have

$$\mu_z = -(2.00232)\left(\frac{e}{2m}\right)\left(\pm\tfrac{1}{2}\,\hbar\right)$$

$$= \mp\tfrac{1}{2}(2.00232)\left(\frac{e\hbar}{2m}\right) = \mp(1.00116)\mu_B$$

$$= \mp(1.00116)(9.274 \times 10^{-24}\ \text{J/T})$$

$$= \mp 9.285 \times 10^{-24}\ \text{J/T}$$

$$= \mp 5.795 \times 10^{-5}\ \text{eV/T}$$

Continued

Then from Eq. (41.32),

$$U = -\mu_z B = \pm(9.285 \times 10^{-24} \text{ J/T})(2.00 \text{ T})$$
$$= \pm 1.86 \times 10^{-23} \text{ J} = \pm 1.16 \times 10^{-4} \text{ eV}$$

The positive value of U and the negative value of μ_z correspond to $S_z = +\frac{1}{2}\hbar$ (spin up); the negative value of U and the positive value of μ_z correspond to $S_z = -\frac{1}{2}\hbar$ (spin down).

EVALUATE: Let's check the *signs* of our results. If the electron is spin down, $\vec{S}$ points generally opposite to $\vec{B}$. Then the magnetic moment $\vec{\mu}$ (which is opposite to $\vec{S}$ because the electron charge is negative) points generally parallel to $\vec{B}$, and μ_z is positive. From Eq. (41.28), $U = -\vec{\mu} \cdot \vec{B}$, the interaction energy is negative if $\vec{\mu}$ and $\vec{B}$ are parallel. Our results show that U is indeed negative in this case. We can similarly confirm that U must be positive and μ_z negative for a spin-up electron.

The red lines in Fig. 41.18 show how the interaction energies for the two spin states vary with the magnetic field magnitude B. The graphs are straight lines because, from Eq. (41.32), U is proportional to B.

41.18 An $l = 0$ level of a single electron is split by interaction of the spin magnetic moment with an external magnetic field. The greater the magnitude B of the magnetic field, the greater the splitting. The quantity 5.795×10^{-5} eV/T is just $(1.00116)\mu_B$.

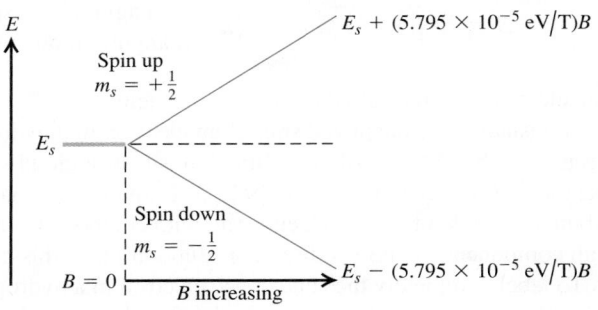

Spin-Orbit Coupling

We mentioned earlier that the spin magnetic dipole moment also gives splitting of energy levels even when there is *no* external field. One cause involves the orbital motion of the electron. In the Bohr model, observers moving with the electron would see the positively charged nucleus revolving around them (just as to earthbound observers the sun seems to be orbiting the earth). This apparent motion of charge causes a magnetic field at the location of the electron, as measured in the electron's moving frame of reference. The resulting interaction with the spin magnetic moment causes a twofold splitting of this level, corresponding to the two possible orientations of electron spin.

Discussions based on the Bohr model can't be taken too seriously, but a similar result can be derived from the Schrödinger equation. The interaction energy U can be expressed in terms of the scalar product of the angular momentum vectors $\vec{L}$ and $\vec{S}$. This effect is called **spin-orbit coupling;** it is responsible for the small energy difference between the two closely spaced, lowest excited levels of sodium shown in Fig. 39.19a and for the corresponding doublet (589.0, 589.6 nm) in the spectrum of sodium.

Example 41.7 An effective magnetic field

To six significant figures, the wavelengths of the two spectral lines that make up the sodium doublet are $\lambda_1 = 588.995$ nm and $\lambda_2 = 589.592$ nm. Calculate the effective magnetic field experienced by the electron in the 3p levels of the sodium atom.

SOLUTION

IDENTIFY and SET UP: The two lines in the sodium doublet result from transitions from the two 3p levels, which are split by spin-orbit coupling, to the 3s level, which is *not* split because it has $L = 0$. We picture the spin-orbit coupling as an interaction between the electron spin magnetic moment and an effective magnetic field due to the nucleus. This example is like Example 41.6 in reverse: There we were given B and found the difference between the energies of the two spin states, while here we use the energy difference to find the target variable B. The difference in energy between the two 3p levels is equal to the difference in energy

between the two photons of the sodium doublet. We use this relationship and the results of Example 41.6 to determine B.

EXECUTE: The energies of the two photons are $E_1 = hc/\lambda_1$ and $E_2 = hc/\lambda_2$. Here $E_1 > E_2$ because $\lambda_1 < \lambda_2$, so the difference in their energies is

$$\Delta E = \frac{hc}{\lambda_1} - \frac{hc}{\lambda_2} = hc\left(\frac{\lambda_2 - \lambda_1}{\lambda_2 \lambda_1}\right)$$
$$= (4.136 \times 10^{-15} \text{ eV} \cdot \text{s})(2.998 \times 10^8 \text{ m/s})$$
$$\times \frac{(589.592 \times 10^{-9} \text{ m}) - (588.995 \times 10^{-9} \text{ m})}{(589.592 \times 10^{-9} \text{ m})(588.995 \times 10^{-9} \text{ m})}$$
$$= 0.00213 \text{ eV} = 3.41 \times 10^{-22} \text{ J}$$

This equals the energy difference between the two 3p levels. The spin-orbit interaction raises one level by 1.70×10^{-22} J (one-half of 3.41 × 10^{-22} J) and lowers the other by 1.70 × 10^{-22} J. From Example 41.6, the amount each state is raised or lowered is $|U| = (1.00116)\mu_B B$, so

$$B = \left| \frac{U}{(1.00116)\mu_B} \right| = \frac{1.70 \times 10^{-22} \text{ J}}{9.28 \times 10^{-24} \text{ J/T}} = 18.0 \text{ T}$$

EVALUATE: The electron experiences a *very* strong effective magnetic field. To produce a steady, macroscopic field of this magnitude in the laboratory requires state-of-the-art electromagnets.

Combining Orbital and Spin Angular Momenta

The orbital and spin angular momenta ($\vec{L}$ and $\vec{S}$, respectively) can combine in various ways. The vector sum of $\vec{L}$ and $\vec{S}$ is the *total* angular momentum $\vec{J}$:

$$\vec{J} = \vec{L} + \vec{S} \tag{41.41}$$

The possible values of the magnitude J are given in terms of a quantum number j by

$$J = \sqrt{j(j + 1)}\hbar \tag{41.42}$$

We can then have states in which $j = \left| l \pm \frac{1}{2} \right|$. The $l + \frac{1}{2}$ states correspond to the case in which the vectors $\vec{L}$ and $\vec{S}$ have parallel z-components; for the $l - \frac{1}{2}$ states, $\vec{L}$ and $\vec{S}$ have antiparallel z-components. For example, when $l = 1$, j can be $\frac{1}{2}$ or $\frac{3}{2}$. In another spectroscopic notation these p states are labeled $^2P_{1/2}$ and $^2P_{3/2}$, respectively. The superscript is the number of possible spin orientations, the letter P (now capitalized) indicates states with $l = 1$, and the subscript is the value of j. We used this scheme to label the energy levels of the sodium atom in Fig. 39.19a.

The various line splittings resulting from magnetic interactions are collectively called *fine structure*. There are also additional, much smaller splittings associated with the fact that the *nucleus* of the atom has a magnetic dipole moment that interacts with the orbital and/or spin magnetic dipole moments of the electrons. These effects are called *hyperfine structure*. For example, the ground level of hydrogen is split into two states, separated by only 5.9 × 10^{-6} eV. The photon that is emitted in the transitions between these states has a wavelength of 21 cm. Radio astronomers use this wavelength to map clouds of interstellar hydrogen gas that are too cold to emit visible light (Fig. 41.19).

Test Your Understanding of Section 41.5 In which of the following situations is the magnetic moment of an electron perfectly aligned with a magnetic field that points in the positive z-direction? (i) $m_s = +\frac{1}{2}$; (ii) $m_s = -\frac{1}{2}$; (iii) both (i) and (ii); (iv) neither (i) nor (ii).

41.19 In a visible-light image (top), these three distant galaxies appear to be unrelated. But in fact these galaxies are connected by immense streamers of hydrogen gas. This is revealed by by the false-color image (bottom) made with a radio telescope tuned to the 21-cm wavelength emitted by hydrogen atoms.

Galaxies in visible light (negative image; galaxies appear dark)

Radio image at wavelength 21 cm

41.6 Many-Electron Atoms and the Exclusion Principle

So far our analysis of atomic structure has concentrated on the hydrogen atom. That's natural; neutral hydrogen, with only one electron, is the simplest atom. If we can't understand hydrogen, we certainly can't understand anything more complex. But now let's move on to many-electron atoms.

In general, an atom in its normal (electrically neutral) state has Z electrons and Z protons. Recall from Section 41.3 that we call Z the *atomic number*. The total electric charge of such an atom is exactly zero because the neutron has no charge while the proton and electron charges have the same magnitude but opposite sign.

We can apply the Schrödinger equation to this general atom. However, the complexity of the analysis increases very rapidly with increasing Z. Each of the Z electrons interacts not only with the nucleus but also with every other electron.

The wave functions and the potential energy are functions of $3Z$ coordinates, and the equation contains second derivatives with respect to all of them. The mathematical problem of finding solutions of such equations is so complex that it has not been solved exactly even for the neutral helium atom, which has only two electrons.

Fortunately, various approximation schemes are available. The simplest approximation is to ignore all interactions between electrons and consider each electron as moving under the action only of the nucleus (considered to be a point charge). In this approximation the wave function for each individual electron is a function like that for the hydrogen atom, specified by four quantum numbers (n, l, m_l, m_s). The nuclear charge is Ze instead of e, so we replace every factor of e^2 in the wave functions and the energy levels by Ze^2. In particular, the energy levels are given by Eq. (41.21) with e^4 replaced by Z^2e^4:

$$E_n = -\frac{1}{(4\pi\epsilon_0)^2}\frac{m_r Z^2 e^4}{2n^2\hbar^2} = -\frac{Z^2}{n^2}(13.6\text{ eV}) \tag{41.43}$$

This approximation is fairly drastic; when there are many electrons, their interactions with each other are as important as the interaction of each with the nucleus. So this model isn't very useful for quantitative predictions.

The Central-Field Approximation

A less drastic and more useful approximation is to think of all the electrons together as making up a charge cloud that is, on average, *spherically symmetric*. We can then think of each individual electron as moving in the total electric field due to the nucleus and this averaged-out cloud of all the other electrons. There is a corresponding spherically symmetric potential-energy function $U(r)$. This picture is called the **central-field approximation;** it provides a useful starting point for understanding atomic structure.

In the central-field approximation we can again deal with one-electron wave functions. The Schrödinger equation differs from the equation for hydrogen only in that the $1/r$ potential-energy function is replaced by a different function $U(r)$. But it turns out that $U(r)$ does not enter the differential equations for $\Theta(\theta)$ and $\Phi(\phi)$, so those angular functions are exactly the same as for hydrogen, and the orbital angular-momentum *states* are also the same as before. The quantum numbers l, m_l, and m_s have the same meanings as before, and Eqs. (41.22) and (41.23) again give the magnitude and z-component of the orbital angular momentum.

The radial wave functions and probabilities are different than for hydrogen because of the change in $U(r)$, so the energy levels are no longer given by Eq. (41.21). We can still label a state using the four quantum numbers (n, l, m_l, m_s). In general, the energy of a state now depends on both n and l, rather than just on n as with hydrogen. The restrictions on values of the quantum numbers are the same as before:

$$n \geq 1 \quad 0 \leq l \leq n-1 \quad |m_l| \leq l \quad m_s = \pm\tfrac{1}{2} \qquad \text{(allowed values of quantum numbers)} \tag{41.44}$$

The Exclusion Principle

To understand the structure of many-electron atoms, we need an additional principle, the *exclusion principle*. To see why this principle is needed, let's consider the lowest-energy state, or *ground state*, of a many-electron atom. In the one-electron states of the central-field model, there is a lowest-energy state (corresponding to an $n = 1$ state of hydrogen). We might expect that in the ground state of a complex atom, *all* the electrons should be in this lowest state. If so, then we should see only gradual changes in physical and chemical properties when we look at the behavior of atoms with increasing numbers of electrons (Z).

Such gradual changes are *not* what is observed. Instead, properties of elements vary widely from one to the next, with each element having its own distinct personality. For example, the elements fluorine, neon, and sodium have 9, 10, and 11 electrons, respectively, per atom. Fluorine $(Z = 9)$ is a *halogen;* it tends strongly to form compounds in which each fluorine atom acquires an extra electron. Sodium $(Z = 11)$ is an *alkali metal;* it forms compounds in which each sodium atom *loses* an electron. Neon $(Z = 10)$ is a *noble gas,* forming no compounds at all. Such observations show that in the ground state of a complex atom the electrons *cannot* all be in the lowest-energy states. But why not?

The key to this puzzle, discovered by the Austrian physicist Wolfgang Pauli (Fig. 41.20) in 1925, is called the **exclusion principle.** This principle states that **no two electrons can occupy the same quantum-mechanical state** in a given system. That is, **no two electrons in an atom can have the same values of all four quantum numbers** (n, l, m_l, m_s). Each quantum state corresponds to a certain distribution of the electron "cloud" in space. Therefore the principle also says, in effect, that no more than two electrons with opposite values of the quantum number m_s can occupy the same region of space. We shouldn't take this last statement too seriously because the electron probability functions don't have sharp, definite boundaries. But the exclusion principle limits the amount by which electron wave functions can overlap. Think of it as the quantum-mechanical analog of a university rule that allows only one student per desk.

CAUTION **The meaning of the exclusion principle** Don't confuse the exclusion principle with the electric repulsion between electrons. While both effects tend to keep electrons within an atom separated from each other, they are very different in character. Two electrons can always be pushed closer together by adding energy to combat electric repulsion; in contrast, *nothing* can overcome the exclusion principle and force two electrons into the same quantum-mechanical state.

Table 41.2 lists some of the sets of quantum numbers for electron states in an atom. It's similar to Table 41.1 (Section 41.3), but we've added the number of states in each subshell and shell. Because of the exclusion principle, the "number of states" is the same as the *maximum* number of electrons that can be found in those states. For each state, m_s can be either $+\frac{1}{2}$ or $-\frac{1}{2}$.

As with the hydrogen wave functions, different states correspond to different spatial distributions; electrons with larger values of n are concentrated at larger distances from the nucleus. Figure 41.8 (Section 41.3) shows this effect. When an atom has more than two electrons, they can't all huddle down in the low-energy $n = 1$ states nearest to the nucleus because there are only two of these states; the exclusion principle forbids multiple occupancy of a state. Some electrons are forced into states farther away, with higher energies. Each value of n corresponds roughly to a region of space around the nucleus in the form of a spherical *shell.* Hence we speak of the K shell as the region that is occupied by the electrons in the $n = 1$ states, the L shell as the region of the $n = 2$ states, and so on. States with the same n but different l form *subshells,* such as the $3p$ subshell.

41.20 The key to understanding the periodic table of the elements was the discovery by Wolfgang Pauli (1900–1958) of the exclusion principle. Pauli received the 1945 Nobel Prize in physics for his accomplishment. This photo shows Pauli (on the left) and Niels Bohr watching the physics of a toy top spinning on the floor—a macroscopic analog of a microscopic electron with spin.

Table 41.2 Quantum States of Electrons in the First Four Shells

n	l	m_l	Spectroscopic Notation	Number of States		Shell
1	0	0	$1s$	2		K
2	0	0	$2s$	2	8	L
2	1	$-1, 0, 1$	$2p$	6		
3	0	0	$3s$	2		M
3	1	$-1, 0, 1$	$3p$	6	18	
3	2	$-2, -1, 0, 1, 2$	$3d$	10		
4	0	0	$4s$	2		N
4	1	$-1, 0, 1$	$4p$	6	32	
4	2	$-2, -1, 0, 1, 2$	$4d$	10		
4	3	$-3, -2, -1, 0, 1, 2, 3$	$4f$	14		

The Periodic Table

We can use the exclusion principle to derive the most important features of the structure and chemical behavior of multielectron atoms, including the periodic table of the elements. Let's imagine constructing a neutral atom by starting with a bare nucleus with Z protons and then adding Z electrons, one by one. To obtain the ground state of the atom as a whole, we fill the lowest-energy electron states (those closest to the nucleus, with the smallest values of n and l) first, and we use successively higher states until all the electrons are in place. The chemical properties of an atom are determined principally by interactions involving the outermost, or *valence,* electrons, so we particularly want to learn how these electrons are arranged.

Let's look at the ground-state electron configurations for the first few atoms (in order of increasing Z). For hydrogen the ground state is $1s$; the single electron is in a state $n = 1$, $l = 0$, $m_l = 0$, and $m_s = \pm\frac{1}{2}$. In the helium atom $(Z = 2)$, *both* electrons are in $1s$ states, with opposite spins; one has $m_s = -\frac{1}{2}$ and the other has $m_s = +\frac{1}{2}$. We denote the helium ground state as $1s^2$. (The superscript 2 is not an exponent; the notation $1s^2$ tells us that there are two electrons in the $1s$ subshell. Also, the superscript 1 is understood, as in $2s$.) For helium the K shell is completely filled, and all others are empty. Helium is a noble gas; it has no tendency to gain or lose an electron, and it forms no compounds.

Lithium $(Z = 3)$ has three electrons. In its ground state, two are in $1s$ states and one is in a $2s$ state, so we denote the lithium ground state as $1s^2 2s$. On average, the $2s$ electron is considerably farther from the nucleus than are the $1s$ electrons (Fig. 41.21). According to Gauss's law, the *net* charge Q_{encl} attracting the $2s$ electron is nearer to $+e$ than to the value $+3e$ it would have without the two $1s$ electrons present. As a result, the $2s$ electron is loosely bound; only 5.4 eV is required to remove it, compared with the 30.6 eV given by Eq. (41.43) with $Z = 3$ and $n = 2$. In chemical behavior, lithium is an *alkali metal.* It forms ionic compounds in which each lithium atom loses an electron and has a valence of +1.

Next is beryllium $(Z = 4)$; its ground-state configuration is $1s^2 2s^2$, with its two valence electrons filling the s subshell of the L shell. Beryllium is the first of the *alkaline earth* elements, forming ionic compounds in which the valence of the atoms is +2.

Table 41.3 shows the ground-state electron configurations of the first 30 elements. The L shell can hold eight electrons. At $Z = 10$, both the K and L shells are filled, and there are no electrons in the M shell. We expect this to be a particularly stable configuration, with little tendency to gain or lose electrons. This element is neon, a noble gas with no known compounds. The next element after neon is sodium $(Z = 11)$, with filled K and L shells and one electron in the M shell. Its "noble-gas-plus-one-electron" structure resembles that of lithium; both are alkali metals. The element *before* neon is fluorine, with $Z = 9$. It has a vacancy in the L shell and has an affinity for an extra electron to fill the shell. Fluorine forms ionic compounds in which it has a valence of −1. This behavior is characteristic of the *halogens* (fluorine, chlorine, bromine, iodine, and astatine), all of which have "noble-gas-minus-one" configurations (Fig. 41.22).

Proceeding down the list, we can understand the regularities in chemical behavior displayed by the **periodic table of the elements** (Appendix D) on the basis of electron configurations. The similarity of elements in each *group* (vertical column) of the periodic table is the result of similarity in outer-electron configuration. All the noble gases (helium, neon, argon, krypton, xenon, and radon) have filled-shell or filled-shell plus filled p subshell configurations. All the alkali metals (lithium, sodium, potassium, rubidium, cesium, and francium) have "noble-gas-plus-one" configurations. All the alkaline earth metals (beryllium, magnesium, calcium, strontium, barium, and radium) have "noble-gas-plus-two" configurations, and, as we just mentioned, all the halogens (fluorine, chlorine, bromine, iodine, and astatine) have "noble-gas-minus-one" structures.

41.21 Schematic representation of the charge distribution in a lithium atom. The nucleus has a charge of $+3e$.

On average, the $2s$ electron is considerably farther from the nucleus than the $1s$ electrons. Therefore, it experiences a net nuclear charge of approximately $+3e - 2e = +e$ (rather than $+3e$).

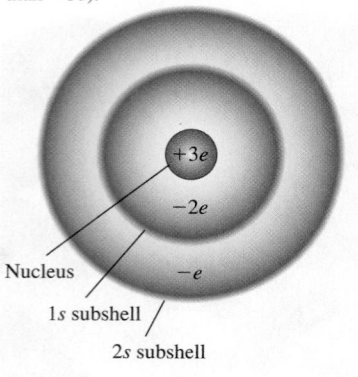

Nucleus

$1s$ subshell

$2s$ subshell

41.22 Salt (sodium chloride, NaCl) dissolves readily in water, making seawater salty. This is due to the electron configurations of sodium and chlorine: Sodium can easily lose an electron to form an Na^+ ion, and chlorine can easily gain an electron to form a Cl^- ion. These ions are held in solution because they are attracted to the polar ends of water molecules (see Fig. 21.30a).

Table 41.3 Ground-State Electron Configurations

Element	Symbol	Atomic Number (Z)	Electron Configuration
Hydrogen	H	1	$1s$
Helium	He	2	$1s^2$
Lithium	Li	3	$1s^2 2s$
Beryllium	Be	4	$1s^2 2s^2$
Boron	B	5	$1s^2 2s^2 2p$
Carbon	C	6	$1s^2 2s^2 2p^2$
Nitrogen	N	7	$1s^2 2s^2 2p^3$
Oxygen	O	8	$1s^2 2s^2 2p^4$
Fluorine	F	9	$1s^2 2s^2 2p^5$
Neon	Ne	10	$1s^2 2s^2 2p^6$
Sodium	Na	11	$1s^2 2s^2 2p^6 3s$
Magnesium	Mg	12	$1s^2 2s^2 2p^6 3s^2$
Aluminum	Al	13	$1s^2 2s^2 2p^6 3s^2 3p$
Silicon	Si	14	$1s^2 2s^2 2p^6 3s^2 3p^2$
Phosphorus	P	15	$1s^2 2s^2 2p^6 3s^2 3p^3$
Sulfur	S	16	$1s^2 2s^2 2p^6 3s^2 3p^4$
Chlorine	Cl	17	$1s^2 2s^2 2p^6 3s^2 3p^5$
Argon	Ar	18	$1s^2 2s^2 2p^6 3s^2 3p^6$
Potassium	K	19	$1s^2 2s^2 2p^6 3s^2 3p^6 4s$
Calcium	Ca	20	$1s^2 2s^2 2p^6 3s^2 3p^6 4s^2$
Scandium	Sc	21	$1s^2 2s^2 2p^6 3s^2 3p^6 4s^2 3d$
Titanium	Ti	22	$1s^2 2s^2 2p^6 3s^2 3p^6 4s^2 3d^2$
Vanadium	V	23	$1s^2 2s^2 2p^6 3s^2 3p^6 4s^2 3d^3$
Chromium	Cr	24	$1s^2 2s^2 2p^6 3s^2 3p^6 4s 3d^5$
Manganese	Mn	25	$1s^2 2s^2 2p^6 3s^2 3p^6 4s^2 3d^5$
Iron	Fe	26	$1s^2 2s^2 2p^6 3s^2 3p^6 4s^2 3d^6$
Cobalt	Co	27	$1s^2 2s^2 2p^6 3s^2 3p^6 4s^2 3d^7$
Nickel	Ni	28	$1s^2 2s^2 2p^6 3s^2 3p^6 4s^2 3d^8$
Copper	Cu	29	$1s^2 2s^2 2p^6 3s^2 3p^6 4s 3d^{10}$
Zinc	Zn	30	$1s^2 2s^2 2p^6 3s^2 3p^6 4s^2 3d^{10}$

A slight complication occurs with the M and N shells because the $3d$ and $4s$ subshell levels ($n = 3$, $l = 2$, and $n = 4$, $l = 0$, respectively) have similar energies. (We'll discuss in the next subsection why this happens.) Argon ($Z = 18$) has all the $1s$, $2s$, $2p$, $3s$, and $3p$ subshells filled, but in potassium ($Z = 19$) the additional electron goes into a $4s$ energy state rather than a $3d$ state (because the $4s$ state has slightly lower energy).

The next several elements have one or two electrons in the $4s$ subshell and increasing numbers in the $3d$ subshell. These elements are all metals with rather similar chemical and physical properties; they form the first *transition series*, starting with scandium ($Z = 21$) and ending with zinc ($Z = 30$), for which all the $3d$ and $4s$ subshells are filled.

Something similar happens with $Z = 57$ through $Z = 71$, which have one or two electrons in the $6s$ subshell but only partially filled $4f$ and $5d$ subshells. These are the *rare earth* elements; they all have very similar physical and chemical properties. Yet another such series, called the *actinide* series, starts with $Z = 91$.

Screening

We have mentioned that in the central-field picture, the energy levels depend on l as well as n. Let's take sodium ($Z = 11$) as an example. If 10 of its electrons fill its K and L shells, the energies of some of the states for the remaining electron are found experimentally to be

$3s$ states:	-5.138 eV
$3p$ states:	-3.035 eV
$3d$ states:	-1.521 eV
$4s$ states:	-1.947 eV

Application **Electron Configurations and Bone Cancer Radiotherapy**
The orange spots in this colored x-ray image are bone cancer tumors. One method of treating bone cancer is to inject a radioactive isotope of strontium (^{89}Sr) into a patient's vein. Strontium is chemically similar to calcium because in both atoms the two outer electrons are in an s state (the structures are $1s^2 2s^2 2p^6 3s^2 3p^6 4s^2 3d^{10} 4p^6 5s^2$ for strontium and $1s^2 2s^2 2p^6 3s^2 3p^6 4s^2$ for calcium). Hence the strontium is readily taken up by the tumors, where calcium turnover is more rapid than in healthy bone. Radiation from the strontium helps to destroy the tumors.

The 3s states are the lowest (most negative); one is the ground state for the 11th electron in sodium. The energy of the 3d states is quite close to the energy of the $n = 3$ state in hydrogen. The surprise is that the 4s state energy is 0.426 eV *below* the 3d state, even though the 4s state has larger n.

We can understand these results using Gauss's law and the radial probability distribution. For any spherically symmetric charge distribution, the electric-field magnitude at a distance r from the center is $Q_{encl}/4\pi\epsilon_0 r^2$, where Q_{encl} is the total charge enclosed within a sphere with radius r. Mentally remove the outer (valence) electron atom from a sodium atom. What you have left is a spherically symmetric collection of 10 electrons (filling the K and L shells) and 11 protons, so $Q_{encl} = -10e + 11e = +e$. If the 11th electron is completely outside this collection of charges, it is attracted by an effective charge of $+e$, not $+11e$. This is a more extreme example of the effect depicted in Fig. 41.21.

This effect is called **screening**; the 10 electrons *screen* 10 of the 11 protons, leaving an effective net charge of $+e$. In general, an electron that spends all its time completely outside a positive charge $Z_{eff}e$ has energy levels given by the hydrogen expression with e^2 replaced by $Z_{eff}e^2$. From Eq. (41.43) this is

$$E_n = -\frac{Z_{eff}^2}{n^2}(13.6\text{ eV}) \qquad \text{(energy levels with screening)} \qquad (41.45)$$

If the 11th electron in the sodium atom is completely outside the remaining charge distribution, then $Z_{eff} = 1$.

CAUTION **Different equations for different atoms** Equations (41.21), (41.43), and (41.45) all give values of E_n in terms of $(13.6\text{ eV})/n^2$, but they don't apply in general to the same atoms. Equation (41.21) is *only* for hydrogen. Equation (41.43) is only for the case in which there is no interaction with any other electron (and is thus accurate only when the atom has just one electron). Equation (41.45) is useful when one electron is screened from the nucleus by other electrons.

Now let's use the radial probability functions shown in Fig. 41.8 to explain why the energy of a sodium 3d state is approximately the same as the $n = 3$ value of hydrogen, -1.51 eV. The distribution for the 3d state (for which l has the maximum value $n - 1$) has one peak, and its most probable radius is *outside* the positions of the electrons with $n = 1$ or 2. (Those electrons also are pulled closer to the nucleus than in hydrogen because they are less effectively screened from the positive charge $11e$ of the nucleus.) Thus in sodium a 3d electron spends most of its time well outside the $n = 1$ and $n = 2$ states (the K and L shells). The 10 electrons in these shells screen about ten-elevenths of the charge of the 11 protons, leaving a net charge of about $Z_{eff}e = (1)e$. Then, from Eq. (41.45), the corresponding energy is approximately $-(1)^2(13.6\text{ eV})/3^2 = -1.51$ eV. This approximation is very close to the experimental value of -1.521 eV.

Looking again at Fig. 41.8, we see that the radial probability density for the 3p state (for which $l = n - 2$) has two peaks and that for the 3s state ($l = n - 3$) has three peaks. For sodium the first small peak in the 3p distribution gives a 3p electron a higher probability (compared to the 3d state) of being *inside* the charge distributions for the electrons in the $n = 2$ states. That is, a 3p electron is less completely screened from the nucleus than is a 3d electron because it spends some of its time within the filled K and L shells. Thus for the 3p electrons, Z_{eff} is greater than unity. From Eq. (41.45) the 3p energy is lower (more negative) than the 3d energy of -1.521 eV. The actual value is -3.035 eV. A 3s electron spends even more time within the inner electron shells than a 3p electron does, giving an even larger Z_{eff} and an even more negative energy.

Example 41.8 Determining Z_{eff} experimentally

The measured energy of a $3s$ state of sodium is -5.138 eV. Calculate the value of Z_{eff}.

SOLUTION

IDENTIFY and SET UP: Sodium has a single electron in the M shell outside filled K and L shells. The ten K and L electrons partially screen the single M electron from the $+11e$ charge of the nucleus; our goal is to determine the extent of this screening. We are given $n = 3$ and $E_n = -5.138$ eV, so we can use Eq. (41.45) to determine Z_{eff}.

EXECUTE: Solving Eq. (41.45) for Z_{eff}, we have

$$Z_{\text{eff}}^2 = -\frac{n^2 E_n}{13.6 \text{ eV}} = -\frac{3^2(-5.138 \text{ eV})}{13.6 \text{ eV}} = 3.40$$

$$Z_{\text{eff}} = 1.84$$

EVALUATE: The effective charge attracting a $3s$ electron is $1.84e$. Sodium's 11 protons are screened by an average of $11 - 1.84 = 9.16$ electrons instead of 10 electrons because the $3s$ electron spends some time within the inner (K and L) shells.

Each alkali metal (lithium, sodium, potassium, rubidium, and cesium) has one more electron than the corresponding noble gas (helium, neon, argon, krypton, and xenon). This extra electron is mostly outside the other electrons in the filled shells and subshells. Therefore all the alkali metals behave similarly to sodium.

Example 41.9 Energies for a valence electron

The valence electron in potassium has a $4s$ ground state. Calculate the approximate energy of the $n = 4$ state having the smallest Z_{eff}, and discuss the relative energies of the $4s$, $4p$, $4d$, and $4f$ states.

SOLUTION

IDENTIFY and SET UP: The state with the smallest Z_{eff} is the one in which the valence electron spends the most time outside the inner filled shells and subshells, so that it is most effectively screened from the charge of the nucleus. Once we have determined which state has the smallest Z_{eff}, we can use Eq. (41.45) to determine the energy of this state.

EXECUTE: A $4f$ state has $n = 4$ and $l = 3 = 4 - 1$. Thus it is the state of greatest orbital angular momentum for $n = 4$, and thus the state in which the electron spends the most time outside the electron charge clouds of the inner filled shells and subshells. This makes Z_{eff} for a $4f$ state close to unity. Equation (41.45) then gives

$$E_4 = -\frac{Z_{\text{eff}}^2}{n^2}(13.6 \text{ eV}) = -\frac{1}{4^2}(13.6 \text{ eV}) = -0.85 \text{ eV}$$

This approximation agrees with the measured energy of the sodium $4f$ state to the precision given.

An electron in a $4d$ state spends a bit more time within the inner shells, and its energy is therefore a bit more negative (measured to be -0.94 eV). For the same reason, a $4p$ state has an even lower energy (measured to be -2.73 eV) and a $4s$ state has the lowest energy (measured to be -4.339 eV).

EVALUATE: We can extend this analysis to the *singly ionized alkaline earth elements*: Be^+, Mg^+, Ca^+, Sr^+, and Ba^+. For any allowed value of n, the highest-l state ($l = n - 1$) of the one remaining outer electron sees an effective charge of almost $+2e$, so for these states, $Z_{\text{eff}} = 2$. A $3d$ state for Mg^+, for example, has an energy of about $-2^2(13.6 \text{ eV})/3^2 = -6.0$ eV.

Test Your Understanding of Section 41.6 If electrons did *not* obey the exclusion principle, would it be easier or more difficult to remove the first electron from sodium?

41.7 X-Ray Spectra

X-ray spectra provide yet another example of the richness and power of the Schrödinger equation and of the model of atomic structure that we derived from it in the preceding section. In Section 38.2 we discussed x-ray production on the basis of the photon concept. With the development of x-ray diffraction techniques (see Section 36.6) by von Laue, Bragg, and others, beginning in 1912, it became possible to measure x-ray wavelengths quite precisely (to within 0.1% or less).

Detailed studies of x-ray spectra showed a continuous spectrum of wavelengths (see Fig. 38.8 in Section 38.2), with minimum wavelength (corresponding to maximum frequency and photon energy) determined by the accelerating voltage V_{AC} in the x-ray tube, according to the relationship derived in Section 38.2 for *bremsstrahlung* processes:

$$\lambda_{min} = \frac{hc}{eV_{AC}} \qquad (41.46)$$

This continuous-spectrum radiation is nearly independent of the target material in the x-ray tube.

Moseley's Law and Atomic Energy Levels

Depending on the accelerating voltage and the target element, we may find sharp peaks superimposed on this continuous spectrum, as in Fig. 41.23. These peaks are at different wavelengths for different elements; they form what is called a *characteristic x-ray spectrum* for each target element. In 1913 the British scientist Henry G. J. Moseley studied these spectra in detail using x-ray diffraction techniques. He found that the most intense short-wavelength line in the characteristic x-ray spectrum from a particular target element, called the K_α line, varied smoothly with that element's atomic number Z (Fig. 41.24). This is in sharp contrast to optical spectra, in which elements with adjacent Z-values have spectra that often bear no resemblance to each other.

Moseley found that the relationship could be expressed in terms of x-ray frequencies f by a simple formula called *Moseley's law:*

$$f = (2.48 \times 10^{15}\ \text{Hz})(Z - 1)^2 \qquad \text{(Moseley's law)} \qquad (41.47)$$

Moseley went far beyond this empirical relationship; he showed how characteristic x-ray spectra could be understood on the basis of energy levels of atoms in the target. His analysis was based on the Bohr model, published in the same year. We will recast it somewhat, using the ideas of atomic structure that we discussed in Section 41.6. First recall that the *outer* electrons of an atom are responsible for optical spectra. Their excited states are usually only a few electron volts above their ground state. In transitions from excited states to the ground state, they usually emit photons in or near the visible region.

Characteristic x rays, by contrast, are emitted in transitions involving the *inner* shells of a complex atom. In an x-ray tube the electrons may strike the target with enough energy to knock electrons out of the inner shells of the target atoms. These inner electrons are much closer to the nucleus than are the electrons in the outer shells; they are much more tightly bound, and hundreds or thousands of electron volts may be required to remove them.

Suppose one electron is knocked out of the K shell. This process leaves a vacancy, which we'll call a *hole*. (One electron remains in the K shell.) The hole can then be filled by an electron falling in from one of the outer shells, such as the $L, M, N, \ldots$ shell. This transition is accompanied by a decrease in the energy of the

41.23 Graph of intensity per unit wavelength as a function of wavelength for x rays produced with an accelerating voltage of 35 kV and a molybdenum target. The curve is a smooth function similar to the bremsstrahlung spectra in Fig. 38.8 (Section 38.2), but with two sharp spikes corresponding to part of the characteristic x-ray spectrum for molybdenum.

41.24 The square root of Moseley's measured frequencies of the K_α line for 14 elements.

atom (because *less* energy would be needed to remove an electron from an L, M, N, ... shell), and an x-ray photon is emitted with energy equal to this decrease. Each state has definite energy, so the emitted x rays have definite wavelengths; the emitted spectrum is a *line* spectrum.

We can estimate the energy and frequency of K_α x-ray photons using the concept of screening from Section 41.6. A K_α x-ray photon is emitted when an electron in the L shell ($n = 2$) drops down to fill a hole in the K shell ($n = 1$). As the electron drops down, it is attracted by the Z protons in the nucleus screened by the one remaining electron in the K shell. We therefore approximate the energy by Eq. (41.45), with $Z_{\text{eff}} = Z - 1$, $n_i = 2$, and n_f. The energy before the transition is

$$E_i \approx -\frac{(Z-1)^2}{2^2}(13.6 \text{ eV}) = -(Z-1)^2(3.4 \text{ eV})$$

and the energy after the transition is

$$E_f \approx -\frac{(Z-1)^2}{1^2}(13.6 \text{ eV}) = -(Z-1)^2(13.6 \text{ eV})$$

The energy of the K_α x-ray photon is $E_{K\alpha} = E_i - E_f \approx (Z-1)^2(-3.4 \text{ eV} + 13.6 \text{ eV})$. That is,

$$E_{K\alpha} \approx (Z-1)^2(10.2 \text{ eV}) \tag{41.48}$$

The frequency of the photon is its energy divided by Planck's constant:

$$f = \frac{E}{h} \approx \frac{(Z-1)^2(10.2 \text{ eV})}{4.136 \times 10^{-15} \text{ eV} \cdot \text{s}} = (2.47 \times 10^{15} \text{ Hz})(Z-1)^2$$

This relationship agrees almost exactly with Moseley's experimental law, Eq. (41.47). Indeed, considering the approximations we have made, the agreement is better than we have a right to expect. But our calculation does show how Moseley's law can be understood on the bases of screening and transitions between energy levels.

The hole in the K shell may also be filled by an electron falling from the M or N shell, assuming that these are occupied. If so, the x-ray spectrum of a large group of atoms of a single element shows a series, named the K series, of three lines, called the K_α, K_β, and K_γ lines. These three lines result from transitions in which the K-shell hole is filled by an L, M, or N electron, respectively. Figure 41.25 shows the K series for tungsten ($Z = 74$), molybdenum ($Z = 42$), and copper ($Z = 29$).

There are other series of x-ray lines, called the L, M, and N series, that are produced after the ejection of electrons from the L, M, and N shells rather than the K shell. Electrons in these outer shells are farther away from the nucleus and are not held as tightly as are those in the K shell, so removing these outer electrons requires less energy. Hence the x-ray photons that are emitted when these vacancies are filled have lower energy than those in the K series.

Application X Rays in Forensic Science

When a handgun is fired, a cloud of gunshot residue (GSR) is ejected from the barrel. The x-ray emission spectrum of GSR includes characteristic peaks from lead (Pb), antimony (Sb), and barium (Ba). If a sample taken from a suspect's skin or clothing has an x-ray emission spectrum with these characteristics, it indicates that the suspect recently fired a gun.

The three lines in each series are called the K_α, K_β, and K_γ lines. The K_α line is produced by the transition of an L electron to the vacancy in the K shell, the K_β line by an M electron, and the K_γ line by an N electron.

41.25 Wavelengths of the K_α, K_β, and K_γ lines of tungsten (W), molybdenum (Mo), and copper (Cu).

| Example 41.10 | Chemical analysis by x-ray emission |

You measure the K_α wavelength for an unknown element, obtaining the value 0.0709 nm. What is the element?

SOLUTION

IDENTIFY and SET UP: To determine which element this is, we need to know its atomic number Z. We can find this using Moseley's law, which relates the frequency of an element's K_α x-ray emission line to that element's atomic number Z. We'll use the relationship $f = c/\lambda$ to calculate the frequency for the K_α line, and then use Eq. (41.47) to find the corresponding value of the atomic number Z. We'll then consult the periodic table (Appendix D) to determine which element has this atomic number.

EXECUTE: The frequency is

$$f = \frac{c}{\lambda} = \frac{3.00 \times 10^8 \text{ m/s}}{0.0709 \times 10^{-9} \text{ m}} = 4.23 \times 10^{18} \text{ Hz}$$

Solving Moseley's law for Z, we get

$$Z = 1 + \sqrt{\frac{f}{2.48 \times 10^{15} \text{ Hz}}} = 1 + \sqrt{\frac{4.23 \times 10^{18} \text{ Hz}}{2.48 \times 10^{15} \text{ Hz}}} = 42.3$$

We know that Z has to be an integer; we conclude that $Z = 42$, corresponding to the element molybdenum.

EVALUATE: If you're worried that our calculation did not give an integer for Z, remember that Moseley's law is an empirical relationship. There are slight variations from one atom to another due to differences in the structure of the electron shells. Nonetheless, this example suggests the power of Moseley's law.

Niels Bohr commented that it was Moseley's observations, not the alpha-particle scattering experiments of Rutherford, Geiger, and Marsden (see Section 39.2), that truly convinced physicists that the atom consists of a positive nucleus surrounded by electrons in motion. Unlike Bohr or Rutherford, Moseley did not receive a Nobel Prize for his important work; these awards are given only to living scientists, and Moseley was killed in combat during the First World War.

X-Ray Absorption Spectra

We can also observe x-ray *absorption* spectra. Unlike optical spectra, the absorption wavelengths are usually not the same as those for emission, especially in many-electron atoms, and do not give simple line spectra. For example, the K_α emission line results from a transition from the L shell to a hole in the K shell. The reverse transition doesn't occur in atoms with $Z \geq 10$ because in the atom's ground state, there is no vacancy in the L shell. To be absorbed, a photon must have enough energy to move an electron to an empty state. Since empty states are only a few electron volts in energy below the free-electron continuum, the minimum absorption energies in many-electron atoms are about the same as the minimum energies that are needed to remove an electron from its shell. Experimentally, if we gradually increase the accelerating voltage and hence the maximum photon energy, we observe sudden increases in absorption when we reach these minimum energies. These sudden jumps of absorption are called *absorption edges* (Fig. 41.26).

Characteristic x-ray spectra provide a very useful analytical tool. Satellite-borne x-ray spectrometers are used to study x-ray emission lines from highly excited atoms in distant astronomical sources. X-ray spectra are also used in air-pollution monitoring and in studies of the abundance of various elements in rocks.

41.26 When a beam of x rays is passed through a slab of molybdenum, the extent to which the beam is absorbed depends on the energy E of the x-ray photons. A sharp increase in absorption occurs at the K absorption edge at 20 keV. The increase occurs because photons with energies above this value can excite an electron from the K shell of a molybdenum atom into an empty state.

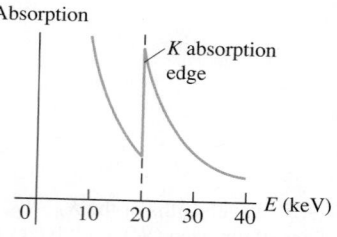

Test Your Understanding of Section 41.7 A beam of photons is passed through a sample of high-temperature atomic hydrogen. At what photon energy would you expect there to be an absorption edge like that shown in Fig. 41.26? (i) 13.60 eV; (ii) 3.40 eV; (iii) 1.51 eV; (iv) all of these; (v) none of these.

Three-dimensional problems: The time-independent Schrödinger equation for three-dimensional problems is given by Eq. (41.5).

$$-\frac{\hbar^2}{2m}\left(\frac{\partial^2\psi(x,y,z)}{\partial x^2} + \frac{\partial^2\psi(x,y,z)}{\partial y^2}\right.$$

$$\left. + \frac{\partial^2\psi(x,y,z)}{\partial z^2}\right) + U(x,y,z)\psi(x,y,z)$$

$$= E\psi(x,y,z)$$

(three-dimensional time-independent Schrödinger equation) (41.5)

Particle in a three-dimensional box: The wave function for a particle in a cubical box is the product of a function of x only, a function of y only, and a function of z only. Each stationary state is described by three quantum numbers (n_X, n_Y, n_Z). Most of the energy levels given by Eq. (41.16) exhibit degeneracy: More than one quantum state has the same energy. (See Example 41.1.)

$$E_{n_X,n_Y,n_Z} = \frac{(n_X{}^2 + n_Y{}^2 + n_Z{}^2)\pi^2\hbar^2}{2mL^2}$$

$(n_X = 1, 2, 3, \ldots;$

$n_Y = 1, 2, 3, \ldots; n_Z = 1, 2, 3, \ldots)$

(energy levels, particle in a three-dimensional cubical box) (41.16)

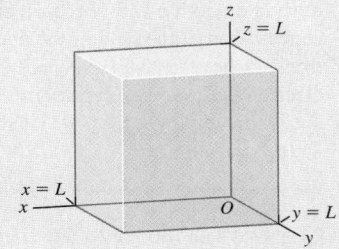

The hydrogen atom: The Schrödinger equation for the hydrogen atom gives the same energy levels as the Bohr model. If the nucleus has charge Ze, there is an additional factor of Z^2 in the numerator of Eq. (41.21). The possible magnitudes L of orbital angular momentum are given by Eq. (41.22), and the possible values of the z-component of orbital angular momentum are given by Eq. (41.23). (See Examples 41.2 and 41.3.)

The probability that an atomic electron is between r and $r + dr$ from the nucleus is $P(r)\,dr$, given by Eq. (41.25). Atomic distances are often measured in units of a, the smallest distance between the electron and the nucleus in the Bohr model. (See Example 41.4.)

$$E_n = -\frac{1}{(4\pi\epsilon_0)^2}\frac{m_r e^4}{2n^2\hbar^2} = -\frac{13.60\text{ eV}}{n^2}$$

(energy levels of hydrogen) (41.21)

$$L = \sqrt{l(l+1)}\,\hbar$$
$(l = 0, 1, 2, \ldots, n-1)$ (41.22)

$$L_z = m_l\hbar$$
$(m_l = 0, \pm1, \pm2, \ldots, \pm l)$ (41.23)

$$P(r)\,dr = |\psi|^2\,dV = |\psi|^2\,4\pi r^2\,dr$$
(41.25)

$$a = \frac{\epsilon_0 h^2}{\pi m_r e^2} = \frac{4\pi\epsilon_0\hbar^2}{m_r e^2}$$

$$= 5.29 \times 10^{-11}\text{ m}$$ (41.26)

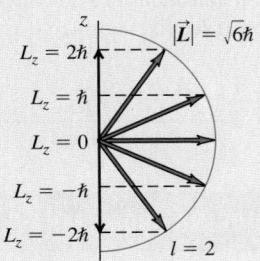

The Zeeman effect: The interaction energy of an electron (mass m) with magnetic quantum number m_l in a magnetic field $\vec{B}$ along the $+z$-direction is given by Eq. (41.35) or (41.36), where $\mu_B = e\hbar/2m$ is called the Bohr magneton. (See Example 41.5.)

$$U = -\mu_z B = m_l\frac{e\hbar}{2m}B = m_l\mu_B B$$

$(m_l = 0, \pm1, \pm2, \ldots, \pm l)$

(41.35), (41.36)

Electron spin: An electron has an intrinsic spin angular momentum of magnitude S, given by Eq. (41.38). The possible values of the z-component of the spin angular momentum are $S_z = m_s\hbar$, where $m_s = \pm\frac{1}{2}$. (See Examples 41.6 and 41.7.)

$$S = \sqrt{\tfrac{1}{2}\left(\tfrac{1}{2} + 1\right)}\hbar = \sqrt{\tfrac{3}{4}}\hbar \quad (41.38)$$

$$S_z = \pm\tfrac{1}{2}\hbar \quad (41.37)$$

Many-electron atoms: In a hydrogen atom, the quantum numbers n, l, m_l, and m_s of the electron have certain allowed values given by Eq. (41.44). In a many-electron atom, the allowed quantum numbers for each electron are the same as in hydrogen, but the energy levels depend on both n and l because of screening, the partial cancellation of the field of the nucleus by the inner electrons. If the effective (screened) charge attracting an electron is $Z_{\text{eff}}e$, the energies of the levels are given approximately by Eq. (41.45). (See Examples 41.8 and 41.9.)

$$n \geq 1 \quad 0 \leq l \leq n - 1$$
$$|m_l| \leq l \quad m_s = \pm\tfrac{1}{2} \quad (41.44)$$

$$E_n = -\frac{Z_{\text{eff}}^2}{n^2}(13.6\text{ eV}) \quad (41.45)$$

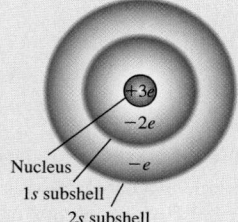

X-ray spectra: Moseley's law states that the frequency of a K_α x ray from a target with atomic number Z is given by Eq. (41.47). Characteristic x-ray spectra result from transitions to a hole in an inner energy level of an atom. (See Example 41.10.)

$$f = (2.48 \times 10^{15}\text{ Hz})(Z - 1)^2 \quad (41.47)$$

BRIDGING PROBLEM   A Many-Electron Atom in a Box

An atom of titanium (Ti) has 22 electrons and has a radius of 1.47×10^{-10} m. As a simple model of this atom, imagine putting 22 electrons into a cubical box that has the same volume as a titanium atom. (a) What is the length of each side of the box? (b) What will be the configuration of the 22 electrons? (c) Find the energies of each of the levels occupied by the electrons. (Ignore the electric forces that the electrons exert on each other.) (d) You remove one of the electrons from the lowest level. As a result, one of the electrons from the highest occupied level drops into the lowest level to fill the hole, emitting a photon in the process. What is the energy of this photon? How does this compare to the energy of the K_α photon for titanium as predicted by Moseley's law?

SOLUTION GUIDE

See MasteringPhysics® study area for a Video Tutor solution. **(MP)**

IDENTIFY and SET UP

1. In this problem you'll use ideas from Section 41.2 about a particle in a cubical box. You'll also apply the exclusion principle from Section 41.6 to find the electron configuration of this cubical "atom." The ideas about x-ray spectra from Section 41.7 are also important.

2. The target variables are (a) the dimensions of the box, (b) the electron configurations (like those given in Table 41.3 for real atoms), (c) the occupied energy levels of the cubical box, and (d) the energy of the emitted photon.

EXECUTE

3. Use your knowledge of geometry to find the length of each side of the box.

4. Each electron state is described by four quantum numbers: n_X, n_Y, and n_Z as described in Section 41.2 and the spin magnetic quantum number m_s described in Section 41.5. Use the exclusion principle to determine the quantum numbers of each of the 22 electrons in the "atom." (*Hint:* Figure 41.4 in Section 41.2 shows the first several energy levels of a cubical box relative to the ground level $E_{1,1,1}$.)

5. Use your results from steps 3 and 4 to find the energies of each of the occupied levels.

6. Use your result from step 5 to find the energy of the photon emitted when an electron makes a transition from the highest occupied level to the ground level. Compare this to the energy calculated for titanium using Moseley's law.

EVALUATE

7. Is this cubical "atom" a useful model for titanium? Why or why not?

8. In this problem you ignored the electric interactions between electrons. To estimate how large these are, find the electrostatic potential energy of two electrons separated by half the length of the box. How does this compare to the energy levels you calculated in step 5? Is it a good approximation to ignore these interactions?

•, ••, •••: Problems of increasing difficulty. **CP**: Cumulative problems incorporating material from earlier chapters. **CALC**: Problems requiring calculus. **BIO**: Biosciences problems.

DISCUSSION QUESTIONS

Q41.1 Particle A is described by the wave function $\psi(x, y, z)$. Particle B is described by the wave function $\psi(x, y, z)e^{i\phi}$, where ϕ is a real constant. How does the probability of finding particle A within a volume dV around a certain point in space compare with the probability of finding particle B within this same volume?

Q41.2 What are the most significant differences between the Bohr model of the hydrogen atom and the Schrödinger analysis? What are the similarities?

Q41.3 For a body orbiting the sun, such as a planet, comet, or asteroid, is there any restriction on the z-component of its orbital angular momentum such as there is with the z-component of the electron's orbital angular momentum in hydrogen? Explain.

Q41.4 Why is the analysis of the helium atom much more complex than that of the hydrogen atom, either in a Bohr type of model or using the Schrödinger equation?

Q41.5 The Stern–Gerlach experiment is always performed with beams of *neutral* atoms. Wouldn't it be easier to form beams using *ionized* atoms? Why won't this work?

Q41.6 (a) If two electrons in hydrogen atoms have the same principal quantum number, can they have different orbital angular momentum? How? (b) If two electrons in hydrogen atoms have the same orbital angular-momentum quantum number, can they have different principal quantum numbers? How?

Q41.7 In the Stern–Gerlach experiment, why is it essential for the magnetic field to be *inhomogeneous* (that is, nonuniform)?

Q41.8 In the ground state of the helium atom one electron must have "spin down" and the other "spin up." Why?

Q41.9 An electron in a hydrogen atom is in an s level, and the atom is in a magnetic field $\vec{B} = B\hat{k}$. Explain why the "spin up" state $\left(m_s = +\frac{1}{2}\right)$ has a higher energy than the "spin down" state $\left(m_s = -\frac{1}{2}\right)$.

Q41.10 The central-field approximation is more accurate for alkali metals than for transition metals such as iron, nickel, or copper. Why?

Q41.11 Table 41.3 shows that for the ground state of the potassium atom, the outermost electron is in a $4s$ state. What does this tell you about the relative energies of the $3d$ and $4s$ levels for this atom? Explain.

Q41.12 Do gravitational forces play a significant role in atomic structure? Explain.

Q41.13 Why do the transition elements ($Z = 21$ to 30) all have similar chemical properties?

Q41.14 Use Table 41.3 to help determine the ground-state electron configuration of the neutral gallium atom (Ga) as well as the ions Ga^+ and Ga^-. Gallium has an atomic number of 31.

Q41.15 On the basis of the Pauli exclusion principle, the structure of the periodic table of the elements shows that there must be a fourth quantum number in addition to n, l, and m_l. Explain.

Q41.16 A small amount of magnetic-field splitting of spectral lines occurs even when the atoms are not in a magnetic field. What causes this?

Q41.17 The ionization energies of the alkali metals (that is, the lowest energy required to remove one outer electron when the atom is in its ground state) are about 4 or 5 eV, while those of the noble gases are in the range from 11 to 25 eV. Why is there a difference?

Q41.18 The energy required to remove the $3s$ electron from a sodium atom in its ground state is about 5 eV. Would you expect the energy required to remove an additional electron to be about the same, or more, or less? Why?

Q41.19 What is the "central-field approximation" and why is it only an approximation?

Q41.20 The nucleus of a gold atom contains 79 protons. How does the energy required to remove a $1s$ electron completely from a gold atom compare with the energy required to remove the electron from the ground level in a hydrogen atom? In what region of the electromagnetic spectrum would a photon with this energy for each of these two atoms lie?

Q41.21 (a) Can you show that the orbital angular momentum of an electron in any given direction (e.g., along the z-axis) is *always* less than or equal to its total orbital angular momentum? In which cases would the two be equal to each other? (b) Is the result in part (a) true for a classical object, such as a spinning top or planet?

Q41.22 An atom in its ground level absorbs a photon with energy equal to the K absorption edge. Does absorbing this photon ionize this atom? Explain.

Q41.23 Can a hydrogen atom emit x rays? If so, how? If not, why not?

EXERCISES

Section 41.2 Particle in a Three-Dimensional Box

41.1 • For a particle in a three-dimensional box, what is the degeneracy (number of different quantum states with the same energy) of the following energy levels: (a) $3\pi^2\hbar^2/2mL^2$ and (b) $9\pi^2\hbar^2/2mL^2$?

41.2 • **CP** Model a hydrogen atom as an electron in a cubical box with side length L. Set the value of L so that the volume of the box equals the volume of a sphere of radius $a = 5.29 \times 10^{-11}$ m, the Bohr radius. Calculate the energy separation between the ground and first excited levels, and compare the result to this energy separation calculated from the Bohr model.

41.3 • **CP** A photon is emitted when an electron in a three-dimensional box of side length 8.00×10^{-11} m makes a transition from the $n_X = 2, n_Y = 2, n_Z = 1$ state to the $n_X = 1, n_Y = 1, n_Z = 1$ state. What is the wavelength of this photon?

41.4 • For each of the following states of a particle in a three-dimensional box, at what points is the probability distribution function a maximum: (a) $n_X = 1, n_Y = 1, n_Z = 1$ and (b) $n_X = 2, n_Y = 2, n_Z = 1$?

41.5 •• A particle is in the three-dimensional box of Section 41.1. For the state $n_X = 2, n_Y = 2, n_Z = 1$, for what planes (in addition to the walls of the box) is the probability distribution function zero? Compare this number of planes to the corresponding number of planes where $|\psi|^2$ is zero for the lower-energy state $n_X = 2, n_Y = 1, n_Z = 1$ and for the ground state $n_X = 1, n_Y = 1, n_Z = 1$.

41.6 • What is the energy difference between the two lowest energy levels for a proton in a cubical box with side length 1.00×10^{-14} m, the approximate diameter of a nucleus?

Section 41.3 The Hydrogen Atom

41.7 •• Consider an electron in the N shell. (a) What is the smallest orbital angular momentum it could have? (b) What is the largest orbital angular momentum it could have? Express your answers in terms of $\hbar$ and in SI units. (c) What is the largest orbital angular momentum this electron could have in any chosen direction? Express your answers in terms of $\hbar$ and in SI units. (d) What is the largest spin angular momentum this electron could have in any chosen direction? Express your answers in terms of $\hbar$ and in SI units. (e) For the electron in part (c), what is the ratio of its spin angular momentum in the z-direction to its orbital angular momentum in the z-direction?

41.8 • An electron is in the hydrogen atom with $n = 5$. (a) Find the possible values of L and L_z for this electron, in units of $\hbar$. (b) For each value of L, find all the possible angles between $\vec{L}$ and the z-axis. (c) What are the maximum and minimum values of the magnitude of the angle between $\vec{L}$ and the z-axis?

41.9 • The orbital angular momentum of an electron has a magnitude of 4.716×10^{-34} kg·m²/s. What is the angular-momentum quantum number l for this electron?

41.10 • Consider states with angular-momentum quantum number $l = 2$. (a) In units of $\hbar$, what is the largest possible value of L_z? (b) In units of $\hbar$, what is the value of L? Which is larger: L or the maximum possible L_z? (c) For each allowed value of L_z, what angle does the vector $\vec{L}$ make with the $+z$-axis? How does the minimum angle for $l = 2$ compare to the minimum angle for $l = 3$ calculated in Example 41.3?

41.11 • Calculate, in units of $\hbar$, the magnitude of the maximum orbital angular momentum for an electron in a hydrogen atom for states with a principal quantum number of 2, 20, and 200. Compare each with the value of $n\hbar$ postulated in the Bohr model. What trend do you see?

41.12 • (a) Make a chart showing all the possible sets of quantum numbers l and m_l for the states of the electron in the hydrogen atom when $n = 5$. How many combinations are there? (b) What are the energies of these states?

41.13 •• (a) How many different $5g$ states does hydrogen have? (b) Which of the states in part (a) has the largest angle between $\vec{L}$ and the z-axis, and what is that angle? (c) Which of the states in part (a) has the smallest angle between $\vec{L}$ and the z-axis, and what is that angle?

41.14 •• CALC (a) What is the probability that an electron in the $1s$ state of a hydrogen atom will be found at a distance less than $a/2$ from the nucleus? (b) Use the results of part (a) and of Example 41.4 to calculate the probability that the electron will be found at distances between $a/2$ and a from the nucleus.

41.15 • CALC In Example 41.4 fill in the missing details that show that $P = 1 - 5e^{-2}$.

41.16 • Show that $\Phi(\phi) = e^{im_l\phi} = \Phi(\phi + 2\pi)$ (that is, show that $\Phi(\phi)$ is periodic with period 2π) if and only if m_l is restricted to the values $0, \pm 1, \pm 2, \ldots$. (*Hint:* Euler's formula states that $e^{i\phi} = \cos\phi + i\sin\phi$.)

Section 41.4 The Zeeman Effect

41.17 • A hydrogen atom in a $3p$ state is placed in a uniform external magnetic field $\vec{B}$. Consider the interaction of the magnetic field with the atom's orbital magnetic dipole moment. (a) What field magnitude B is required to split the $3p$ state into multiple lev-

els with an energy difference of 2.71×10^{-5} eV between adjacent levels? (b) How many levels will there be?

41.18 • A hydrogen atom is in a d state. In the absence of an external magnetic field the states with different m_l values have (approximately) the same energy. Consider the interaction of the magnetic field with the atom's orbital magnetic dipole moment. (a) Calculate the splitting (in electron volts) of the m_l levels when the atom is put in a 0.400-T magnetic field that is in the $+z$-direction. (b) Which m_l level will have the lowest energy? (c) Draw an energy-level diagram that shows the d levels with and without the external magnetic field.

41.19 • A hydrogen atom in the $5g$ state is placed in a magnetic field of 0.600 T that is in the z-direction. (a) Into how many levels is this state split by the interaction of the atom's orbital magnetic dipole moment with the magnetic field? (b) What is the energy separation between adjacent levels? (c) What is the energy separation between the level of lowest energy and the level of highest energy?

41.20 •• CP A hydrogen atom undergoes a transition from a $2p$ state to the $1s$ ground state. In the absence of a magnetic field, the energy of the photon emitted is 122 nm. The atom is then placed in a strong magnetic field in the z-direction. Ignore spin effects; consider only the interaction of the magnetic field with the atom's orbital magnetic moment. (a) How many different photon wavelengths are observed for the $2p \rightarrow 1s$ transition? What are the m_l values for the initial and final states for the transition that leads to each photon wavelength? (b) One observed wavelength is exactly the same with the magnetic field as without. What are the initial and final m_l values for the transition that produces a photon of this wavelength? (c) One observed wavelength with the field is longer than the wavelength without the field. What are the initial and final m_l values for the transition that produces a photon of this wavelength? (d) Repeat part (c) for the wavelength that is shorter than the wavelength in the absence of the field.

Section 41.5 Electron Spin

41.21 •• CP **Classical Electron Spin.** (a) If you treat an electron as a classical spherical object with a radius of 1.0×10^{-17} m, what angular speed is necessary to produce a spin angular momentum of magnitude $\sqrt{\frac{3}{4}}\hbar$? (b) Use $v = r\omega$ and the result of part (a) to calculate the speed v of a point at the electron's equator. What does your result suggest about the validity of this model?

41.22 •• A hydrogen atom in the $n = 1$, $m_s = -\frac{1}{2}$ state is placed in a magnetic field with a magnitude of 0.480 T in the $+z$-direction. (a) Find the magnetic interaction energy (in electron volts) of the electron with the field. (b) Is there any orbital magnetic dipole moment interaction for this state? Explain. Can there be an orbital magnetic dipole moment interaction for $n \neq 1$?

41.23 • Calculate the energy difference between the $m_s = \frac{1}{2}$ ("spin up") and $m_s = -\frac{1}{2}$ ("spin down") levels of a hydrogen atom in the $1s$ state when it is placed in a 1.45-T magnetic field in the *negative* z-direction. Which level, $m_s = \frac{1}{2}$ or $m_s = -\frac{1}{2}$, has the lower energy?

41.24 • CP The hyperfine interaction in a hydrogen atom between the magnetic dipole moment of the proton and the spin magnetic dipole moment of the electron splits the ground level into two levels separated by 5.9×10^{-6} eV. (a) Calculate the wavelength and frequency of the photon emitted when the atom makes a transition between these states, and compare your answer to the value given at the end of Section 41.5. In what part of the electromagnetic spectrum does this lie? Such photons are emitted by cold hydrogen clouds in interstellar space; by detecting these photons, astronomers can learn about the number and density of such clouds. (b) Calculate the effective magnetic field experienced by the electron in these states (see Fig. 41.18). Compare your result to the effective magnetic field due to the spin-orbit coupling calculated in Example 41.7.

41.25 • A hydrogen atom in a particular orbital angular momentum state is found to have j quantum numbers $\frac{7}{2}$ and $\frac{9}{2}$. What is the letter that labels the value of l for the state?

Section 41.6 Many-Electron Atoms and the Exclusion Principle

41.26 • For germanium (Ge, $Z = 32$), make a list of the number of electrons in each subshell ($1s, 2s, 2p, \dots$). Use the allowed values of the quantum numbers along with the exclusion principle; do *not* refer to Table 41.3.

41.27 • Make a list of the four quantum numbers n, l, m_l, and m_s for each of the 10 electrons in the ground state of the neon atom. Do *not* refer to Table 41.2 or 41.3.

41.28 •• (a) Write out the ground-state electron configuration ($1s^2, 2s^2, \dots$) for the carbon atom. (b) What element of next-larger Z has chemical properties similar to those of carbon? Give the ground-state electron configuration for this element.

41.29 •• (a) Write out the ground-state electron configuration ($1s^2$, $2s^2, \dots$) for the beryllium atom. (b) What element of next-larger Z has chemical properties similar to those of beryllium? Give the ground-state electron configuration of this element. (c) Use the procedure of part (b) to predict what element of next-larger Z than in (b) will have chemical properties similar to those of the element you found in part (b), and give its ground-state electron configuration.

41.30 • For magnesium, the first ionization potential is 7.6 eV. The second ionization potential (additional energy required to remove a second electron) is almost twice this, 15 eV, and the third ionization potential is much larger, about 80 eV. How can these numbers be understood?

41.31 • The $5s$ electron in rubidium (Rb) sees an effective charge of $2.771e$. Calculate the ionization energy of this electron.

41.32 • The energies of the $4s$, $4p$, and $4d$ states of potassium are given in Example 41.9. Calculate Z_{eff} for each state. What trend do your results show? How can you explain this trend?

41.33 • (a) The doubly charged ion N^{2+} is formed by removing two electrons from a nitrogen atom. What is the ground-state electron configuration for the N^{2+} ion? (b) Estimate the energy of the least strongly bound level in the L shell of N^{2+}. (c) The doubly charged ion P^{2+} is formed by removing two electrons from a phosphorus atom. What is the ground-state electron configuration for the P^{2+} ion? (d) Estimate the energy of the least strongly bound level in the M shell of P^{2+}.

41.34 • (a) The energy of the $2s$ state of lithium is -5.391 eV. Calculate the value of Z_{eff} for this state. (b) The energy of the $4s$ state of potassium is -4.339 eV. Calculate the value of Z_{eff} for this state. (c) Compare Z_{eff} for the $2s$ state of lithium, the $3s$ state of sodium (see Example 41.8), and the $4s$ state of potassium. What trend do you see? How can you explain this trend?

41.35 • Estimate the energy of the highest-l state for (a) the L shell of Be^+ and (b) the N shell of Ca^+.

Section 41.7 X-Ray Spectra

41.36 • A K_α x ray emitted from a sample has an energy of 7.46 keV. Of which element is the sample made?

41.37 • Calculate the frequency, energy (in keV), and wavelength of the K_α x ray for the elements (a) calcium (Ca, $Z = 20$); (b) cobalt (Co, $Z = 27$); (c) cadmium (Cd, $Z = 48$).

41.38 •• The energies for an electron in the K, L, and M shells of the tungsten atom are $-69,500$ eV, $-12,000$ eV, and -2200 eV, respectively. Calculate the wavelengths of the K_α and K_β x rays of tungsten.

PROBLEMS

41.39 • In terms of the ground-state energy $E_{1,1,1}$, what is the energy of the highest level occupied by an electron when 10 electrons are placed into a cubical box?

41.40 •• **CALC** A particle in the three-dimensional box of Section 41.2 is in the ground state, where $n_X = n_Y = n_Z = 1$. (a) Calculate the probability that the particle will be found somewhere between $x = 0$ and $x = L/2$. (b) Calculate the probability that the particle will be found somewhere between $x = L/4$ and $x = L/2$. Compare your results to the result of Example 41.1 for the probability of finding the particle in the region $x = 0$ to $x = L/4$.

41.41 •• **CALC** A particle is in the three-dimensional box of Section 41.2. (a) Consider the cubical volume defined by $0 \le x \le L/4$, $0 \le y \le L/4$, and $0 \le z \le L/4$. What fraction of the total volume of the box is this cubical volume? (b) If the particle is in the ground state ($n_X = 1$, $n_Y = 1$, $n_Z = 1$) calculate the probability that the particle will be found in the cubical volume defined in part (a). (c) Repeat the calculation of part (b) when the particle is in the state $n_X = 2$, $n_Y = 1$, $n_Z = 1$.

41.42 •• **CALC** A particle is described by the normalized wave function $\psi(x, y, z) = Axe^{-\alpha x^2}e^{-\beta y^2}e^{-\gamma z^2}$, where A, α, β, and γ are all real, positive constants. The probability that the particle will be found in the infinitesimal volume $dx\,dy\,dz$ centered at the point (x_0, y_0, z_0) is $|\psi(x_0, y_0, z_0)|^2 \, dx\,dy\,dz$. (a) At what value of x_0 is the particle most likely to be found? (b) Are there values of x_0 for which the probability of the particle being found is zero? If so, at what x_0?

41.43 •• **CALC** A particle is described by the normalized wave function $\psi(x, y, z) = Ae^{-\alpha(x^2+y^2+z^2)}$, where A and α are real, positive constants. (a) Determine the probability of finding the particle at a distance between r and $r + dr$ from the origin. (*Hint:* See Problem 41.42. Consider a spherical shell centered on the origin with inner radius r and thickness dr.) (b) For what value of r does the probability in part (a) have its maximum value? Is this the same value of r for which $|\psi(x, y, z)|^2$ is a maximum? Explain any differences.

41.44 •• **CP CALC A Three-Dimensional Isotropic Harmonic Oscillator.** An isotropic harmonic oscillator has the potential-energy function $U(x, y, z) = \frac{1}{2}k'(x^2 + y^2 + z^2)$. (*Isotropic* means that the force constant k' is the same in all three coordinate directions.) (a) Show that for this potential, a solution to Eq. (41.5) is given by $\psi = \psi_{n_x}(x)\psi_{n_y}(y)\psi_{n_z}(z)$. In this expression, $\psi_{n_x}(x)$ is a solution to the one-dimensional harmonic oscillator Schrödinger equation, Eq. (40.44), with energy $E_{n_x} = \left(n_x + \frac{1}{2}\right)\hbar\omega$. The functions $\psi_{n_y}(y)$ and $\psi_{n_z}(z)$ are analogous one-dimensional wave functions for oscillations in the y- and z-directions. Find the energy associated with this ψ. (b) From your results in part (a) what are the ground-level and first-excited-level energies of the three-dimensional isotropic oscillator? (c) Show that there is only one state (one set of quantum numbers n_x, n_y, and n_z) for the ground level but three states for the first excited level.

41.45 •• **CP CALC Three-Dimensional Anisotropic Harmonic Oscillator.** An oscillator has the potential-energy function $U(x, y, z) = \frac{1}{2}k_1'(x^2 + y^2) + \frac{1}{2}k_2'z^2$, where $k_1' > k_2'$. This oscillator is called *anisotropic* because the force constant is not the same in all three coordinate directions. (a) Find a general expression for the energy levels of the oscillator (see Problem 41.44). (b) From your results in part (a), what are the ground-level and first-excited-level energies of this oscillator? (c) How many states (different sets of quantum numbers n_x, n_y, and n_z) are there for the ground level and for the first excited level? Compare to part (c) of Problem 41.44.

41.46 •• An electron in hydrogen is in the 5f state. (a) Find the largest possible value of the z-component of its angular momentum. (b) Show that for the electron in part (a), the corresponding x- and y-components of its angular momentum satisfy the equation $\sqrt{L_x^2 + L_y^2} = \hbar\sqrt{3}$.

41.47 •• (a) Show that the total number of atomic states (including different spin states) in a shell of principal quantum number n is $2n^2$. [*Hint:* The sum of the first N integers $1 + 2 + 3 + \cdots + N$ is equal to $N(N + 1)/2$.] (b) Which shell has 50 states?

41.48 •• (a) What is the lowest possible energy (in electron volts) of an electron in hydrogen if its orbital angular momentum is $\sqrt{12}\hbar$? (b) What are the largest and smallest values of the z-component of the orbital angular momentum (in terms of $\hbar$) for the electron in part (a)? (c) What are the largest and smallest values of the spin angular momentum (in terms of $\hbar$) for the electron in part (a)? (d) What are the largest and smallest values of the orbital angular momentum (in terms of $\hbar$) for an electron in the M shell of hydrogen?

41.49 • Consider an electron in hydrogen having total energy -0.5440 eV. (a) What are the possible values of its orbital angular momentum (in terms of $\hbar$)? (b) What wavelength of light would it take to excite this electron to the next higher shell? Is this photon visible to humans?

41.50 • (a) Show all the distinct states for an electron in the N shell of hydrogen. Include all four quantum numbers. (b) For an f electron in the N shell, what is the largest possible orbital angular momentum and the greatest positive value for the component of this angular momentum along any chosen direction (the z-axis)? What is the magnitude of its spin angular momentum? Express these quantities in units of $\hbar$. (c) For an electron in the d state of the N shell, what are the maximum and minimum angles between its angular momentum vector and any chosen direction (the z-axis)? (d) What is the largest value of the orbital angular momentum for an f electron in the M shell?

41.51 • (a) The energy of an electron in the 4s state of sodium is -1.947 eV. What is the effective net charge of the nucleus "seen" by this electron? On the average, how many electrons screen the nucleus? (b) For an outer electron in the 4p state of potassium, on the average 17.2 inner electrons screen the nucleus. (i) What is the effective net charge of the nucleus "seen" by this outer electron? (ii) What is the energy of this outer electron?

41.52 • **CALC** For a hydrogen atom, the probability $P(r)$ of finding the electron within a spherical shell with inner radius r and outer radius $r + dr$ is given by Eq. (41.25). For a hydrogen atom in the 1s ground state, at what value of r does $P(r)$ have its maximum value? How does your result compare to the distance between the electron and the nucleus for the $n = 1$ state in the Bohr model, Eq. (41.26)?

41.53 •• **CALC** Consider a hydrogen atom in the 1s state. (a) For what value of r is the potential energy $U(r)$ equal to the total energy E? Express your answer in terms of a. This value of r is called the *classical turning point,* since this is where a Newtonian particle would stop its motion and reverse direction. (b) For r greater than the classical turning point, $U(r) > E$. Classically, the particle cannot be in this region, since the kinetic energy cannot be negative. Calculate the probability of the electron being found in this classically forbidden region.

41.54 • **CP Rydberg Atoms.** *Rydberg atoms* are atoms whose outermost electron is in an excited state with a *very* large principal quantum number. Rydberg atoms have been produced in the laboratory and detected in interstellar space. (a) Why do all neutral Rydberg atoms with the same n value have essentially the same ionization energy, independent of the total number of electrons in the atom? (b) What is the ionization energy for a Rydberg atom with a principal quantum number of 350? What is the radius in the Bohr model of the Rydberg electron's orbit? (c) Repeat part (b) for $n = 650$.

41.55 ••• **CALC** The wave function for a hydrogen atom in the 2s state is

$$\psi_{2s}(r) = \frac{1}{\sqrt{32\pi a^3}}\left(2 - \frac{r}{a}\right)e^{-r/2a}$$

(a) Verify that this function is normalized. (b) In the Bohr model, the distance between the electron and the nucleus in the $n = 2$ state is exactly $4a$. Calculate the probability that an electron in the 2s state will be found at a distance less than $4a$ from the nucleus.

41.56 •• **CALC** The normalized wave function for a hydrogen atom in the 2s state is given in Problem 41.55. (a) For a hydrogen atom in the 2s state, at what value of r is $P(r)$ maximum? How does your result compare to $4a$, the distance between the electron and the nucleus in the $n = 2$ state of the Bohr model? (b) At what value of r (other than $r = 0$ or $r = \infty$) is $P(r)$ equal to zero, so that the probability of finding the electron at that separation from the nucleus is zero? Compare your result to Fig. 41.9.

41.57 •• (a) For an excited state of hydrogen, show that the smallest angle that the orbital angular momentum vector $\vec{L}$ can have with the z-axis is

$$(\theta_L)_{\min} = \arccos\left(\frac{n - 1}{\sqrt{n(n - 1)}}\right)$$

(b) What is the corresponding expression for $(\theta_L)_{\max}$, the largest possible angle between $\vec{L}$ and the z-axis?

41.58 •• (a) If the value of L_z is known, we cannot know either L_x or L_y precisely. But we can know the value of the quantity $\sqrt{L_x^2 + L_y^2}$. Write an expression for this quantity in terms of l, m_l, and $\hbar$. (b) What is the meaning of $\sqrt{L_x^2 + L_y^2}$? (c) For a state of nonzero orbital angular momentum, find the maximum and minimum values of $\sqrt{L_x^2 + L_y^2}$. Explain your results.

41.59 •• **CALC** The normalized radial wave function for the 2p state of the hydrogen atom is $R_{2p} = \left(1/\sqrt{24a^5}\right)re^{-r/2a}$. After we average over the angular variables, the radial probability function becomes $P(r)\, dr = (R_{2p})^2 r^2\, dr$. At what value of r is $P(r)$ for the 2p state a maximum? Compare your results to the radius of the $n = 2$ state in the Bohr model.

41.60 •• **CP Stern–Gerlach Experiment.** In a Stern–Gerlach experiment, the deflecting force on the atom is $F_z = -\mu_z(dB_z/dz)$, where μ_z is given by Eq. (41.40) and dB_z/dz is the magnetic-field gradient. In a particular experiment the magnetic-field region is 50.0 cm long; assume the magnetic-field gradient is constant in this region. A beam of silver atoms enters the magnetic field with a speed of 525 m/s. What value of dB_z/dz is required to give a separation of 1.0 mm between the two spin components as they exit the field? (*Note:* The magnetic dipole moment of silver is the same as that for hydrogen, since its valence electron is in an $l = 0$ state.)

41.61 • Consider the transition from a 3d to a 2p state of hydrogen in an external magnetic field. Assume that the effects of electron spin can be ignored (which is not actually the case) so that the magnetic field interacts only with the orbital angular momentum. Identify each allowed transition by the m_l values of the initial and final states. For each of these allowed transitions, determine the shift of the transition energy from the zero-field value and show that there are three different transition energies.

41.62 •• An atom in a 3*d* state emits a photon of wavelength 475.082 nm when it decays to a 2*p* state. (a) What is the energy (in electron volts) of the photon emitted in this transition? (b) Use the selection rules described in Section 41.4 to find the allowed transitions if the atom is now in an external magnetic field of 3.500 T. Ignore the effects of the electron's spin. (c) For the case in part (b), if the energy of the 3*d* state was originally -8.50000 eV with no magnetic field present, what will be the energies of the states into which it splits in the magnetic field? (d) What are the allowed wavelengths of the light emitted during transition in part (b)?

41.63 •• **CALC** **Spectral Analysis.** While studying the spectrum of a gas cloud in space, an astronomer magnifies a spectral line that results from a transition from a *p* state to an *s* state. She finds that the line at 575.050 nm has actually split into three lines, with adjacent lines 0.0462 nm apart, indicating that the gas is in an external magnetic field. (Ignore effects due to electron spin.) What is the strength of the external magnetic field?

41.64 •• A hydrogen atom makes a transition from an $n = 3$ state to an $n = 2$ state (the Balmer H_α line) while in a magnetic field in the $+z$-direction and with magnitude 1.40 T. (a) If the magnetic quantum number is $m_l = 2$ in the initial $(n = 3)$ state and $m_l = 1$ in the final $(n = 2)$ state, by how much is each energy level shifted from the zero-field value? (b) By how much is the wavelength of the H_α line shifted from the zero-field value? Is the wavelength increased or decreased? Disregard the effect of electron spin. [*Hint:* Use the result of Problem 39.86(c).]

41.65 • **CP** A large number of hydrogen atoms in 1*s* states are placed in an external magnetic field that is in the $+z$-direction. Assume that the atoms are in thermal equilibrium at room temperature, $T = 300$ K. According to the Maxwell–Boltzmann distribution (see Section 39.4), what is the ratio of the number of atoms in the $m_s = \frac{1}{2}$ state to the number in the $m_s = -\frac{1}{2}$ state when the magnetic-field magnitude is (a) 5.00×10^{-5} T (approximately the earth's field); (b) 0.500 T; (c) 5.00 T?

41.66 •• **Effective Magnetic Field.** An electron in a hydrogen atom is in the 2*p* state. In a simple model of the atom, assume that the electron circles the proton in an orbit with radius *r* equal to the Bohr-model radius for $n = 2$. Assume that the speed v of the orbiting electron can be calculated by setting $L = mvr$ and taking L to have the quantum-mechanical value for a 2*p* state. In the frame of the electron, the proton orbits with radius *r* and speed v. Model the orbiting proton as a circular current loop, and calculate the magnetic field it produces at the location of the electron.

41.67 •• **Weird Universe.** In another universe, the electron is a spin-$\frac{3}{2}$ rather than a spin-$\frac{1}{2}$ particle, but all other physics are the same as in our universe. In this universe, (a) what are the atomic numbers of the lightest two inert gases? (b) What is the ground-state electron configuration of sodium?

41.68 • For an ion with nuclear charge Z and a single electron, the electric potential energy is $-Ze^2/4\pi\epsilon_0 r$ and the expression for the energies of the states and for the normalized wave functions are obtained from those for hydrogen by replacing e^2 by Ze^2. Consider the N^{6+} ion, with seven protons and one electron. (a) What is the ground-state energy in electron volts? (b) What is the ionization energy, the energy required to remove the electron from the N^{6+}

ion if it is initially in the ground state? (c) What is the distance *a* [given for hydrogen by Eq. (41.26)] for this ion? (d) What is the wavelength of the photon emitted when the N^{6+} ion makes a transition from the $n = 2$ state to the $n = 1$ ground state?

41.69 •• A hydrogen atom in an $n = 2$, $l = 1$, $m_l = -1$ state emits a photon when it decays to an $n = 1$, $l = 0$, $m_l = 0$ ground state. (a) In the absence of an external magnetic field, what is the wavelength of this photon? (b) If the atom is in a magnetic field in the $+z$-direction and with a magnitude of 2.20 T, what is the shift in the wavelength of the photon from the zero-field value? Does the magnetic field increase or decrease the wavelength? Disregard the effect of electron spin. [*Hint:* Use the result of Problem 39.86(c).]

41.70 •• A lithium atom has three electrons, and the $^2S_{1/2}$ ground-state electron configuration is $1s^2 2s$. The $1s^2 2p$ excited state is split into two closely spaced levels, $^2P_{3/2}$ and $^2P_{1/2}$, by the spin-orbit interaction (see Example 41.7 in Section 41.5). A photon with wavelength 67.09608 μm is emitted in the $^2P_{3/2} \rightarrow {}^2S_{1/2}$ transition, and a photon with wavelength 67.09761 μm is emitted in the $^2P_{1/2} \rightarrow {}^2S_{1/2}$ transition. Calculate the effective magnetic field seen by the electron in the $1s^2 2p$ state of the lithium atom. How does your result compare to that for the 3*p* level of sodium found in Example 41.7?

41.71 • Estimate the minimum and maximum wavelengths of the characteristic x rays emitted by (a) vanadium $(Z = 23)$ and (b) rhenium $(Z = 45)$. Discuss any approximations that you make.

41.72 •• **CP** **Electron Spin Resonance.** Electrons in the lower of two spin states in a magnetic field can absorb a photon of the right frequency and move to the higher state. (a) Find the magnetic-field magnitude B required for this transition in a hydrogen atom with $n = 1$ and $l = 0$ to be induced by microwaves with wavelength λ. (b) Calculate the value of B for a wavelength of 3.50 cm.

CHALLENGE PROBLEMS

41.73 ••• Each of 2*N* electrons (mass *m*) is free to move along the *x*-axis. The potential-energy function for each electron is $U(x) = \frac{1}{2}k'x^2$, where k' is a positive constant. The electric and magnetic interactions between electrons can be ignored. Use the exclusion principle to show that the minimum energy of the system of 2*N* electrons is $\hbar N^2 \sqrt{k'/m}$. (*Hint:* See Section 40.5 and the hint given in Problem 41.47.)

41.74 ••• **CP** Consider a simple model of the helium atom in which two electrons, each with mass *m*, move around the nucleus (charge $+2e$) in the same circular orbit. Each electron has orbital angular momentum $\hbar$ (that is, the orbit is the smallest-radius Bohr orbit), and the two electrons are always on opposite sides of the nucleus. Ignore the effects of spin. (a) Determine the radius of the orbit and the orbital speed of each electron. [*Hint:* Follow the procedure used in Section 39.3 to derive Eqs. (39.8) and (39.9). Each electron experiences an attractive force from the nucleus and a repulsive force from the other electron.] (b) What is the total kinetic energy of the electrons? (c) What is the potential energy of the system (the nucleus and the two electrons)? (d) In this model, how much energy is required to remove both electrons to infinity? How does this compare to the experimental value of 79.0 eV?

41.75 ••• **CALC** Repeat the calculation of Problem 41.53 for a one-electron ion with nuclear charge Z. (See Problem 41.68.) How does the probability of the electron being found in the classically forbidden region depend on Z?

Answers

Chapter Opening Question ?

Helium is inert because it has a filled K shell, while sodium is very reactive because its third electron is loosely bound in an L shell. See Section 41.6 for more details.

Test Your Understanding Questions

41.1 Answer: (iv) If $U(x, y, z) = 0$ in a certain region of space, we can rewrite the time-independent Schrödinger equation [Eq. (41.5)] for that region as $\partial^2\psi/\partial x^2 + \partial^2\psi/\partial y^2 + \partial^2\psi/\partial z^2 = (-2mE/\hbar^2)\psi$. We are told that all of the second derivatives of $\psi(x, y, z)$ are positive in this region, so the left-hand side of this equation is positive. Hence the right-hand side $(-2mE/\hbar^2)\psi$ must also be positive. Since $E > 0$, the quantity $-2mE/\hbar^2$ is negative, and so $\psi(x, y, z)$ must be negative.

41.2 Answer: (iv), (ii), (i) and (iii) (tie) Equation (41.16) shows that the energy levels for a cubical box are proportional to the quantity $n_X^2 + n_Y^2 + n_Z^2$. Hence ranking in order of this quantity is the same as ranking in order of energy. For the four cases we are given, we have (i) $n_X^2 + n_Y^2 + n_Z^2 = 2^2 + 3^2 + 2^2 = 17$; (ii) $n_X^2 + n_Y^2 + n_Z^2 = 4^2 + 1^2 + 1^2 = 18$; (iii) $n_X^2 + n_Y^2 + n_Z^2 = 2^2 + 2^2 + 3^2 = 17$; and (iv) $n_X^2 + n_Y^2 + n_Z^2 = 1^2 + 3^2 + 3^2 = 19$. The states $(n_X, n_Y, n_Z) = (2, 3, 2)$ and $(n_X, n_Y, n_Z) = (2, 2, 3)$ have the same energy (they are degenerate).

41.3 Answer: (ii) and (iii) (tie), (i) An electron in a state with principal quantum number n is most likely to be found at $r = n^2a$. This result is independent of the values of the quantum numbers l and m_l. Hence an electron with $n = 2$ (most likely to be found at $r = 4a$) is more likely to be found near $r = 5a$ than an electron with $n = 1$ (most likely to be found at $r = a$).

41.4 Answer: no All that matters is the component of the electron's orbital magnetic moment along the direction of $\vec{B}$. We called this quantity μ_z in Eq. (41.32) because we *defined* the positive z-axis to be in the direction of $\vec{B}$. In reality, the names of the axes are entirely arbitrary.

41.5 Answer: (iv) For the magnetic moment to be perfectly aligned with the z-direction, the z-component of the spin vector $\vec{S}$ would have to have the same absolute value as $\vec{S}$. However, the possible values of S_z are $\pm\frac{1}{2}\hbar$ [Eq. (41.37)], while the magnitude of the spin vector is $S = \sqrt{\frac{3}{4}}\hbar$ [Eq. (41.38)]. Hence $\vec{S}$ can never be perfectly aligned with any one direction in space.

41.6 Answer: more difficult If there were no exclusion principle, all 11 electrons in the sodium atom would be in the level of lowest energy (the $1s$ level) and the configuration would be $1s^{11}$. Consequently, it would be more difficult to remove the first electron. (In a real sodium atom the valence electron is in a screened $3s$ state, which has a comparatively high energy.)

41.7 Answer: (iv) An absorption edge appears if the photon energy is just high enough to remove an electron in a given energy level from the atom. In a sample of high-temperature hydrogen we expect to find atoms whose electron is in the ground level $(n = 1)$, the first excited level $(n = 2)$, and the second excited level $(n = 3)$. From Eq. (41.21) these levels have energies $E_n = (-13.60\text{ eV})/n^2 = -13.60\text{ eV}, -3.40\text{ eV}$, and -1.51 eV (see Fig. 38.9b).

Bridging Problem

Answers: (a) 2.37×10^{-10} m

(b) Values of (n_X, n_Y, n_Z, m_s) for the 22 electrons: $(1, 1, 1, +\frac{1}{2})$, $(1, 1, 1, -\frac{1}{2})$, $(2, 1, 1, +\frac{1}{2})$, $(2, 1, 1, -\frac{1}{2})$, $(1, 2, 1, +\frac{1}{2})$, $(1, 2, 1, -\frac{1}{2})$, $(1, 1, 2, +\frac{1}{2})$, $(1, 1, 2, -\frac{1}{2})$, $(2, 2, 1, +\frac{1}{2})$, $(2, 2, 1, -\frac{1}{2})$, $(2, 1, 2, +\frac{1}{2})$, $(2, 1, 2, -\frac{1}{2})$, $(1, 2, 2, +\frac{1}{2})$, $(1, 2, 2, -\frac{1}{2})$, $(3, 1, 1, +\frac{1}{2})$, $(3, 1, 1, -\frac{1}{2})$, $(1, 3, 1, +\frac{1}{2})$, $(1, 3, 1, -\frac{1}{2})$, $(1, 1, 3, +\frac{1}{2})$, $(1, 1, 3, -\frac{1}{2})$, $(2, 2, 2, +\frac{1}{2})$, $(2, 2, 2, -\frac{1}{2})$

(c) 20.1 eV, 40.2 eV, 60.3 eV, 73.7 eV, and 80.4 eV

(d) 60.3 eV versus 4.52×10^3 eV

MOLECULES AND CONDENSED MATTER

42

? This false-color image of Venus shows its thick, cloud-shrouded atmosphere, which is 96.5% carbon dioxide (CO_2). The atmosphere raises the surface temperature of Venus to 735 K (462°C = 863°F)—hotter even than Mercury, the closest planet to the sun. What property of CO_2 molecules makes them a potent agent for raising Venus's temperature?

LEARNING GOALS

By studying this chapter, you will learn:

- The various types of bonds that hold atoms together.

- How the rotational and vibrational dynamics of molecules are revealed by molecular spectra.

- How and why atoms form into crystalline structures.

- How to use the energy-band concept to explain the electrical properties of solids.

- A simple model for metals that explains many of their physical properties.

- How the character of a semiconductor can be radically transformed by adding small amounts of an impurity.

- Some of the technological applications of semiconductor devices.

- Why certain materials become superconductors at low temperature.

I n Chapter 41 we discussed the structure and properties of isolated atoms. But such atoms are the exception; usually we find atoms combined to form molecules or more extended structures we call condensed matter (liquid or solid). It's the attractive forces between atoms, called molecular bonds, that cause them to combine. In this chapter we'll study several kinds of bonds as well as the energy levels and spectra associated with diatomic molecules. We will see that just as atoms have quantized energies determined by the quantum-mechanical state of their electrons, so molecules have quantized energies determined by their rotational and vibrational states.

The same physical principles behind molecular bonds also apply to the study of condensed matter, in which various types of bonding occur. We'll explore the concept of energy bands and see how it helps us understand the properties of solids. Then we'll look more closely at the properties of a special class of solids called semiconductors. Devices using semiconductors are found in every radio, TV, pocket calculator, and computer used today; they have revolutionized the entire field of electronics during the past half-century.

42.1 Types of Molecular Bonds

We can use our discussion of atomic structure in Chapter 41 as a basis for exploring the nature of *molecular bonds,* the interactions that hold atoms together to form stable structures such as molecules and solids.

42.1 When the separation r between two oppositely charged ions is large, the potential energy $U(r)$ is proportional to $1/r$ as for point charges and the force is attractive. As r decreases, the charge clouds of the two atoms overlap and the force becomes less attractive. If r is less than the equilibrium separation r_0, the force is repulsive.

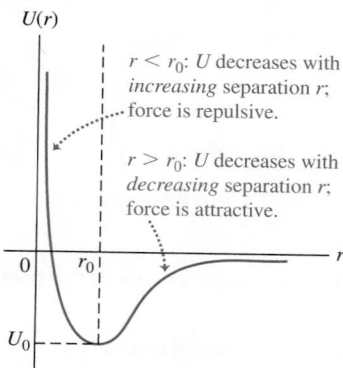

$r < r_0$: U decreases with *increasing* separation r; force is repulsive.

$r > r_0$: U decreases with *decreasing* separation r; force is attractive.

Ionic Bonds

The **ionic bond** is an interaction between oppositely charged *ionized* atoms. The most familiar example is sodium chloride (NaCl), in which the sodium atom gives its one $3s$ electron to the chlorine atom, filling the vacancy in the $3p$ subshell of chlorine.

Let's look at the energy balance in this transaction. Removing the $3s$ electron from a neutral sodium atom requires 5.138 eV of energy; this is called the *ionization energy* of sodium. The neutral chlorine atom can attract an extra electron into the vacancy in the $3p$ subshell, where it is incompletely screened by the other electrons and therefore is attracted to the nucleus. This state has 3.613 eV lower energy than a neutral chlorine atom and a distant free electron; 3.613 eV is the magnitude of the *electron affinity* of chlorine. Thus creating the well-separated Na^+ and Cl^- ions requires a net investment of only 5.138 eV $-$ 3.613 eV = 1.525 eV. When the two oppositely charged ions are brought together by their mutual attraction, the magnitude of their negative potential energy is determined by how closely they can approach each other. This in turn is limited by the exclusion principle, which forbids extensive overlap of the electron clouds of the two ions. As the distance decreases, the exclusion principle distorts the charge clouds, so the ions no longer interact like point charges and the interaction eventually becomes repulsive (Fig. 42.1).

The minimum electric potential energy for NaCl turns out to be -5.7 eV at a separation of 0.24 nm. The net energy released in creating the ions and letting them come together to the equilibrium separation of 0.24 nm is 5.7 eV $-$ 1.525 eV = 4.2 eV. Thus, if the kinetic energy of the ions is neglected, 4.2 eV is the *binding energy* of the NaCl molecule, the energy that is needed to dissociate the molecule into separate neutral atoms.

Ionic bonds can involve more than one electron per atom. For instance, alkaline earth elements form ionic compounds in which an atom loses *two* electrons; an example is magnesium chloride, or $Mg^{2+}(Cl^-)_2$. Ionic bonds that involve a loss of more than two electrons are relatively rare. Instead, a different kind of bond, the *covalent* bond, comes into operation. We'll discuss this type of bond below.

Example 42.1 **Electric potential energy of the NaCl molecule**

Find the electric potential energy of an Na^+ ion and a Cl^- ion separated by 0.24 nm. Consider the ions as point charges.

SOLUTION

IDENTIFY and SET UP: Equation (23.9) in Section 23.1 tells us that the electric potential energy of two point charges q and q_0 separated by a distance r is $U = qq_0/4\pi\epsilon_0 r$.

EXECUTE: We have $q = +e$ (for Na^+), $q_0 = -e$ (for Cl^-), and $r = 0.24$ nm $= 0.24 \times 10^{-9}$ m. From Eq. (23.9),

$$U = -\frac{1}{4\pi\epsilon_0}\frac{e^2}{r_0} = -(9.0 \times 10^9 \text{ N} \cdot \text{m}^2/\text{C}^2)\frac{(1.6 \times 10^{-19} \text{ C})^2}{0.24 \times 10^{-9} \text{ m}}$$

$$= -9.6 \times 10^{-19} \text{ J} = -6.0 \text{ eV}$$

EVALUATE: This result agrees fairly well with the observed value of -5.7 eV. The reason for the difference is that when the two ions are at their equilibrium separation of 0.24 nm, the outer regions of their electron clouds overlap. Hence the two ions don't behave exactly like point charges.

Mastering**PHYSICS**

PhET: Double Wells and Covalent Bonds

Covalent Bonds

The **covalent bond** is characterized by a more egalitarian participation of the two atoms than occurs with the ionic bond. The simplest covalent bond is found in the hydrogen molecule, a structure containing two protons and two electrons. As the separate atoms (Fig. 42.2a) come together, the electron wave functions are distorted and become more concentrated in the region between the two protons (Fig. 42.2b). The net attraction of the electrons for each proton more than balances the repulsion of the two protons and of the two electrons.

The attractive interaction is then supplied by a *pair* of electrons, one contributed by each atom, with charge clouds that are concentrated primarily in the region between the two atoms. The energy of the covalent bond in the hydrogen molecule H_2 is -4.48 eV.

As we saw in Chapter 41, the exclusion principle permits two electrons to occupy the same region of space (that is, to be in the same spatial quantum state) only when they have opposite spins. When the spins are parallel, the exclusion principle forbids the molecular state that would be most favorable from energy considerations (with both electrons in the region between atoms). Opposite spins are an essential requirement for a covalent bond, and no more than two electrons can participate in such a bond.

However, an atom with several electrons in its outermost shell can form several covalent bonds. The bonding of carbon and hydrogen atoms, of central importance in organic chemistry, is an example. In the *methane* molecule (CH_4) the carbon atom is at the center of a regular tetrahedron, with a hydrogen atom at each corner. The carbon atom has four electrons in its L shell, and each of these four electrons forms a covalent bond with one of the four hydrogen atoms (Fig. 42.3). Similar patterns occur in more complex organic molecules.

Because of the role played by the exclusion principle, covalent bonds are highly directional. In the methane molecule the wave function for each of carbon's four valence electrons is a combination of the $2s$ and $2p$ wave functions called a *hybrid wave function*. The probability distribution for each one has a lobe protruding toward a corner of a tetrahedron. This symmetric arrangement minimizes the overlap of wave functions for the electron pairs, minimizing their repulsive potential energy.

Ionic and covalent bonds represent two extremes in molecular bonding, but there is no sharp division between the two types. Often there is a *partial* transfer of one or more electrons from one atom to another. As a result, many molecules that have dissimilar atoms have electric dipole moments—that is, a preponderance of positive charge at one end and of negative charge at the other. Such molecules are called *polar* molecules. Water molecules have large electric dipole moments; these are responsible for the exceptionally large dielectric constant of liquid water (see Sections 24.4 and 24.5).

van der Waals Bonds

Ionic and covalent bonds, with typical bond energies of 1 to 5 eV, are called *strong bonds*. There are also two types of weaker bonds. One of these, the **van der Waals bond,** is an interaction between the electric dipole moments of atoms or molecules; typical energies are 0.1 eV or less. The bonding of water molecules in the liquid and solid states results partly from dipole–dipole interactions.

No atom has a permanent electric dipole moment, nor do many molecules. However, fluctuating charge distributions can lead to fluctuating dipole moments; these in turn can induce dipole moments in neighboring structures. Overall, the resulting dipole–dipole interaction is attractive, giving a weak bonding of atoms or molecules. The interaction potential energy drops off very quickly with distance r between molecules, usually as $1/r^6$. The liquefaction and solidification of the inert gases and of molecules such as H_2, O_2, and N_2 are due to induced-dipole van der Waals interactions. Not much thermal-agitation energy is needed to break these weak bonds, so such substances usually exist in the liquid and solid states only at very low temperatures.

Hydrogen Bonds

In the other type of weak bond, the **hydrogen bond,** a proton (H^+ ion) gets between two atoms, polarizing them and attracting them by means of the induced dipoles. This bond is unique to hydrogen-containing compounds because only hydrogen has a singly ionized state with no remaining electron cloud; the hydrogen ion is a bare proton, much smaller than any other singly ionized atom. The bond energy is usually less than 0.5 eV. The hydrogen bond is responsible for the cross-linking of long-chain organic molecules such as polyethylene (used in plastic bags). Hydrogen bonding also plays a role in the structure of ice.

42.2 Covalent bond in a hydrogen molecule.

(a) Separate hydrogen atoms

Nucleus (proton)

Individual H atoms are usually widely separated and do not interact.

(b) H_2 molecule

Covalent bond: the charge clouds for the two electrons with opposite spins are concentrated in the region between the nuclei.

42.3 Schematic diagram of the methane (CH_4) molecule. The carbon atom is at the center of a regular tetrahedron and forms four covalent bonds with the hydrogen atoms at the corners. Each covalent bond includes two electrons with opposite spins, forming a charge cloud that is concentrated between the carbon atom and a hydrogen atom.

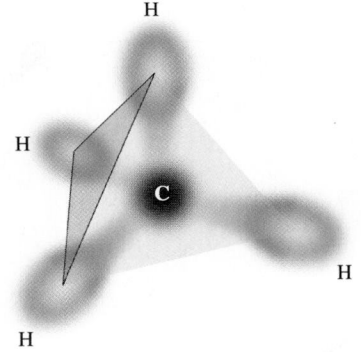

Application Molecular Zipper

A DNA molecule functions like a twisted zipper. Each of the two strands of the "zipper" consists of an outer backbone and inward-facing nucleotide "teeth"; hydrogen bonds between facing teeth "zip" the strands together. The covalent bonds that hold together the atoms of each strand are strong, whereas the hydrogen bonds are relatively weak, so that the cell's biochemical machinery can easily separate the strands for reading or copying.

All these bond types hold the atoms together in *solids* as well as in molecules. Indeed, a solid is in many respects a giant molecule. Still another type of bonding, the *metallic bond,* comes into play in the structure of metallic solids. We'll return to this subject in Section 42.3.

Test Your Understanding of Section 42.1 If electrons obeyed the exclusion principle but did *not* have spin, how many electrons could participate in a covalent bond? (i) one; (ii) two; (iii) three; (iv) more than three.

42.2 Molecular Spectra

Molecules have energy levels that are associated with rotational motion of a molecule as a whole and with vibrational motion of the atoms relative to each other. Just as transitions between energy levels in atoms lead to atomic spectra, transitions between rotational and vibrational levels in molecules lead to *molecular spectra.*

Rotational Energy Levels

In this discussion we'll concentrate mostly on *diatomic* molecules, to keep things as simple as possible. In Fig. 42.4 we picture a diatomic molecule as a rigid dumbbell (two point masses m_1 and m_2 separated by a constant distance r_0) that can *rotate* about axes through its center of mass, perpendicular to the line joining them. What are the energy levels associated with this motion?

We showed in Section 10.5 that when a rigid body rotates with angular speed ω about a perpendicular axis through its center of mass, the magnitude L of its angular momentum is given by Eq. (10.28), $L = I\omega$, where I is its moment of inertia about that symmetry axis. Its kinetic energy is given by Eq. (9.17), $K = \frac{1}{2}I\omega^2$. Combining these two equations, we find $K = L^2/2I$. There is no potential energy U, so the kinetic energy K is equal to the total mechanical energy E:

$$E = \frac{L^2}{2I} \tag{42.1}$$

Zero potential energy means that U does not depend on the angular coordinate of the molecule. But the potential-energy function U for the hydrogen atom (see Section 41.3) also has no dependence on angular coordinates. Thus the angular solutions to the Schrödinger equation for rigid-body rotation are the same as for the hydrogen atom, and the angular momentum is quantized in the same way. As in Eq. (41.21),

$$L = \sqrt{l(l + 1)}\,\hbar \qquad (l = 0, 1, 2, \dots) \tag{42.2}$$

Combining Eqs. (42.1) and (42.2), we obtain the *rotational energy levels:*

$$E_l = l(l + 1)\frac{\hbar^2}{2I} \qquad (l = 0, 1, 2, \dots) \qquad \text{(rotational energy levels, diatomic molecule)} \tag{42.3}$$

Figure 42.5 is an energy-level diagram showing these rotational levels. The ground level has zero quantum number l, corresponding to zero angular momentum (no rotation and zero rotational energy E). The spacing of adjacent levels increases with increasing l.

We can express the moment of inertia I in Eqs. (42.1) and (42.3) in terms of the *reduced mass* m_r of the molecule:

$$m_r = \frac{m_1 m_2}{m_1 + m_2} \tag{42.4}$$

We introduced the reduced mass in Section 39.3 to accommodate the finite nuclear mass of the hydrogen atom. In Fig. 42.4 the distances r_1 and r_2 are the

42.4 A diatomic molecule modeled as two point masses m_1 and m_2 separated by a distance r_0. The distances of the masses from the center of mass are r_1 and r_2, where $r_1 + r_2 = r_0$.

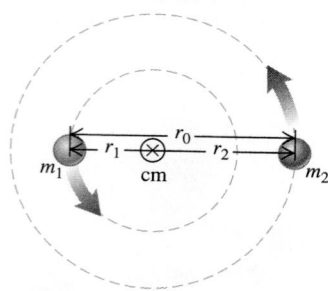

42.5 The ground level and first four excited rotational energy levels for a diatomic molecule. The levels are not equally spaced.

E

$10\hbar^2/I$ ——— $l = 4$

$6\hbar^2/I$ ——— $l = 3$

$3\hbar^2/I$ ——— $l = 2$

$\hbar^2/I$ ——— $l = 1$
0 ——— $l = 0$

distances from the center of mass to the nuclei of the atoms. By definition of the center of mass, $m_1r_1 = m_2r_2$, and the figure also shows that $r_0 = r_1 + r_2$. Solving these equations for r_1 and r_2, we find

$$r_1 = \frac{m_2}{m_1 + m_2}r_0 \qquad r_2 = \frac{m_1}{m_1 + m_2}r_0 \qquad (42.5)$$

The moment of inertia is $I = m_1r_1^2 + m_2r_2^2$; substituting Eq. (42.5), we find

$$I = m_1\frac{m_2^2}{(m_1 + m_2)^2}r_0^2 + m_2\frac{m_1^2}{(m_1 + m_2)^2}r_0^2 = \frac{m_1m_2}{m_1 + m_2}r_0^2 \quad \text{or}$$

$$I = m_rr_0^2 \text{ (moment of inertia of a diatomic molecule)} \qquad (42.6)$$

The reduced mass enables us to reduce this two-body problem to an equivalent one-body problem (a particle of mass m_r moving around a circle with radius r_0), just as we did with the hydrogen atom. Indeed, the only difference between this problem and the hydrogen atom is the difference in the radial forces. To conserve angular momentum and account for the angular momentum of the emitted or absorbed photon, the allowed transitions between rotational states must satisfy the same selection rule that we discussed in Section 41.4 for allowed transitions between the states of an atom: l must change by exactly one unit, that is, $\Delta l = \pm 1$.

MasteringPHYSICS

PhET: The Greenhouse Effect

Example 42.2 Rotational spectrum of carbon monoxide

The two nuclei in the carbon monoxide (CO) molecule are 0.1128 nm apart. The mass of the most common carbon atom is 1.993×10^{-26} kg; that of the most common oxygen atom is 2.656×10^{-26} kg. (a) Find the energies of the lowest three rotational energy levels of CO. Express your results in meV(1 meV $= 10^{-3}$ eV). (b) Find the wavelength of the photon emitted in the transition from the $l = 2$ to the $l = 1$ level.

SOLUTION

IDENTIFY and SET UP: This problem uses the ideas developed in this section about the rotational energy levels of molecules. We are given the distance r_0 between the atoms and their masses m_1 and m_2. We find the reduced mass m_r using Eq. (42.4), the moment of inertia I using Eq. (42.6), and the energies E_l using Eq. (42.3). The energy E of the emitted photon is equal to the difference in energy between the $l = 2$ and $l = 1$ levels. (This transition obeys the $\Delta l = \pm 1$ selection rule, since $\Delta l = 1 - 2 = -1$.) We determine the photon wavelength using $E = hc/\lambda$.

EXECUTE: (a) From Eqs. (42.4) and (42.6), the reduced mass and moment of inertia of the CO molecule are:

$$m_r = \frac{m_1m_2}{m_1 + m_2}$$

$$= \frac{(1.993 \times 10^{-26}\text{ kg})(2.656 \times 10^{-26}\text{ kg})}{(1.993 \times 10^{-26}\text{ kg}) + (2.656 \times 10^{-26}\text{ kg})}$$

$$= 1.139 \times 10^{-26}\text{ kg}$$

$$I = m_rr_0^2$$

$$= (1.139 \times 10^{-26}\text{ kg})(0.1128 \times 10^{-9}\text{ m})^2$$

$$= 1.449 \times 10^{-46}\text{ kg}\cdot\text{m}^2$$

The rotational levels are given by Eq. (42.3):

$$E_l = l(l + 1)\frac{\hbar^2}{2I} = l(l + 1)\frac{(1.0546 \times 10^{-34}\text{ J}\cdot\text{s})^2}{2(1.449 \times 10^{-46}\text{ kg}\cdot\text{m}^2)}$$

$$= l(l + 1)(3.838 \times 10^{-23}\text{ J}) = l(l + 1)0.2395\text{ meV}$$

(1 meV $= 10^{-3}$ eV.) Substituting $l = 0, 1, 2$, we find

$$E_0 = 0 \qquad E_1 = 0.479\text{ meV} \qquad E_2 = 1.437\text{ meV}$$

(b) The photon energy and wavelength are:

$$E = E_2 - E_1 = 0.958\text{ meV}$$

$$\lambda = \frac{hc}{E} = \frac{(4.136 \times 10^{-15}\text{ eV}\cdot\text{s})(3.00 \times 10^8\text{ m/s})}{0.958 \times 10^{-3}\text{ eV}}$$

$$= 1.29 \times 10^{-3}\text{ m} = 1.29\text{ mm}$$

EVALUATE: The differences between the first few rotational energy levels of CO are very small (about 1 meV $= 10^{-3}$ eV) compared to the differences between atomic energy levels (typically a few eV). Hence a photon emitted by a CO molecule in a transition from the $l = 2$ to the $l = 1$ level has very low energy and a very long wavelength compared to the visible light emitted by excited atoms. Photon wavelengths for rotational transitions in molecules are typically in the microwave and far infrared regions of the spectrum.

In this example we were given the equilibrium separation between the atoms, also called the *bond length,* and we used it to calculate one of the wavelengths emitted by excited CO molecules. In actual experiments, scientists work this problem backward: By measuring the long-wavelength emissions of a sample of diatomic molecules, they determine the moment of inertia of the molecule and hence the bond length.

42.6 A diatomic molecule modeled as two point masses m_1 and m_2 connected by a spring with force constant k'.

42.7 The ground level and first three excited vibrational levels for a diatomic molecule, assuming small displacements from equilibrium so we can treat the oscillations as simple harmonic. (Compare Fig. 40.25.)

E

$\frac{7}{2}\hbar\omega$ ——————— $n = 3$

$\frac{5}{2}\hbar\omega$ ——————— $n = 2$

$\frac{3}{2}\hbar\omega$ ——————— $n = 1$

$\frac{1}{2}\hbar\omega$ ——————— $n = 0$

0 - - - - -

42.8 Energy-level diagram for vibrational and rotational energy levels of a diatomic molecule. For each vibrational level (n) there is a series of more closely spaced rotational levels (l). Several transitions corresponding to a single band in a band spectrum are shown. These transitions obey the selection rule $\Delta l = \pm 1$.

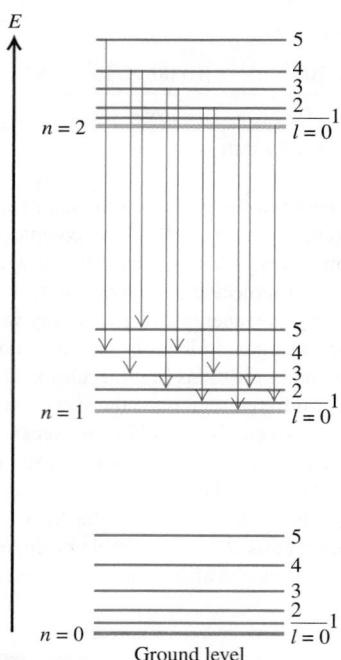

Vibrational Energy Levels

Molecules are never completely rigid. In a more realistic model of a diatomic molecule we represent the connection between atoms not as a rigid rod but as a spring (Fig. 42.6). Then, in addition to rotating, the atoms of the molecule can *vibrate* about their equilibrium positions along the line joining them. For small oscillations the restoring force can be taken as proportional to the displacement from the equilibrium separation r_0 (like a spring that obeys Hooke's law with a force constant k'), and the system is a harmonic oscillator. We discussed the quantum-mechanical harmonic oscillator in Section 40.5. The energy levels are given by Eq. (40.46) with the mass m replaced by the reduced mass m_r:

$$E_n = (n + \tfrac{1}{2})\hbar\omega = (n + \tfrac{1}{2})\hbar\sqrt{\frac{k'}{m_r}} \qquad (n = 0, 1, 2, \dots) \tag{42.7}$$
(vibrational energy levels of a diatomic molecule)

This represents a series of levels equally spaced in energy, with an energy separation of

$$\Delta E = \hbar\omega = \hbar\sqrt{\frac{k'}{m_r}} \tag{42.8}$$

Figure 42.7 is an energy-level diagram showing these vibrational levels. As an example, for the carbon monoxide molecule of Example 42.2 the spacing $\hbar\omega$ between vibrational energy levels is 0.2690 eV. From Eq. (42.8) this corresponds to a force constant of 1.90×10^3 N/m, which is a fairly loose spring. (To stretch a macroscopic spring with this value of k' by 1.0 cm would require a force of only 19 N, or about 4 lb.) Force constants for diatomic molecules are typically about 100 to 2000 N/m.

> **CAUTION** **Watch out for k, k', and K** As in Section 40.5 we're again using k' for the force constant, this time to minimize confusion with Boltzmann's constant k, the gas constant per molecule (introduced in Section 18.3). Besides the quantities k and k', we also use the absolute temperature unit 1 K = 1 kelvin.

Rotation and Vibration Combined

Visible-light photons have energies between 1.65 eV and 3.26 eV. The 0.2690-eV energy difference between vibrational levels for carbon monoxide corresponds to a photon of wavelength 4.613 μm, in the infrared region of the spectrum. This is much closer to the visible region than is the photon in the rotational transition in Example 42.2. In general the energy differences for molecular *vibration* are much smaller than those that produce atomic spectra, but much larger than the energy differences for molecular *rotation*.

When we include *both* rotational and vibrational energies, the energy levels for our diatomic molecule are

$$E_{nl} = l(l + 1)\frac{\hbar^2}{2I} + (n + \tfrac{1}{2})\hbar\sqrt{\frac{k'}{m_r}} \tag{42.9}$$

Figure 42.8 shows the energy-level diagram. For each value of n there are many values of l, forming a series of closely spaced levels.

The red arrows in Fig. 42.8 show several possible transitions in which a molecule goes from a level with $n = 2$ to a level with $n = 1$ by emitting a photon. As we mentioned above, these transitions must obey the selection rule $\Delta l = \pm 1$ in order to conserve angular momentum. An additional selection rule states that if the vibrational level changes, the vibrational quantum number n in Eq. (42.9) must increase by 1 ($\Delta n = 1$) if a photon is absorbed or decrease by 1 ($\Delta n = -1$) if a photon is emitted.

42.9 A typical molecular band spectrum.

As an illustration of these selection rules, Fig. 42.8 shows that a molecule in the $n = 2$, $l = 4$ level can emit a photon and drop into the $n = 1$, $l = 5$ level ($\Delta n = -1$, $\Delta l = +1$) or the $n = 1$, $l = 3$ level ($\Delta n = -1$, $\Delta l = -1$), but is forbidden from making a $\Delta n = -1$, $\Delta l = 0$ transition into the $n = 1$, $l = 4$ level.

Transitions between states with various pairs of n-values give different series of spectrum lines, and the resulting spectrum has a series of *bands*. Each band corresponds to a particular vibrational transition, and each individual line in a band represents a particular rotational transition, with the selection rule $\Delta l = \pm 1$. Figure 42.9 shows a typical *band spectrum*.

All molecules can have excited states of the *electrons* in addition to the rotational and vibrational states that we have described. In general, these lie at higher energies than the rotational and vibrational states, and there is no simple rule relating them. When there is a transition between electronic states, the $\Delta n = \pm 1$ selection rule for the vibrational levels no longer holds.

Example 42.3 **Vibration-rotation spectrum of carbon monoxide**

Consider again the CO molecule of Example 42.2. Find the wavelength of the photon emitted by a CO molecule when its vibrational energy changes and its rotational energy is (a) initially zero and (b) finally zero.

SOLUTION

IDENTIFY and SET UP: We need to use the selection rules for the vibrational and rotational transitions of a diatomic molecule. Since a photon is emitted as the vibrational energy changes, the selection rule $\Delta n = -1$ tells us that the vibrational quantum number n decreases by 1 in both parts (a) and (b). In part (a) the initial value of l is zero; the selection rule $\Delta l = \pm 1$ tells us that the *final* value of l is 1, so the rotational energy increases in this case. In part (b) the *final* value of l is zero; $\Delta l = \pm 1$ then tells us that the *initial* value of l is 1, and the rotational energy decreases.

In each case the energy E of the emitted photon is the difference between the initial and final energies of the molecule, accounting for the change in both vibrational and rotational energy. In part (a) E equals the difference $\hbar\omega$ between adjacent vibrational energy levels *minus* the rotational energy that the molecule *gains*; in part (b) E equals $\hbar\omega$ *plus* the rotational energy that the molecule *loses*. Example 42.2 tells us that the difference between the $l = 0$ and $l = 1$ rotational energy levels is 0.479 meV = 0.000479 eV, and we learned above that the

vibrational energy-level separation for CO is $\hbar\omega = 0.2690$ eV. We use $E = hc/\lambda$ to determine the corresponding wavelengths (our target variables).

EXECUTE: (a) In this transition the CO molecule loses $\hbar\omega = 0.2690$ eV of vibrational energy and gains 0.000479 eV of rotational energy. Hence the energy E that goes into the emitted photon equals 0.2690 eV *less* 0.000479 eV, or 0.2685 eV. The photon wavelength is

$$\lambda = \frac{hc}{E} = \frac{(4.136 \times 10^{-15}\ \text{eV}\cdot\text{s})(2.998 \times 10^{8}\ \text{m/s})}{0.2685\ \text{eV}}$$
$$= 4.618 \times 10^{-6}\ \text{m} = 4.618\ \mu\text{m}$$

(b) Now the CO molecule loses $\hbar\omega = 0.2690$ eV of vibrational energy and also loses 0.000479 eV of rotational energy, so the energy that goes into the photon is $E = 0.2690$ eV $+$ 0.000479 eV $= 0.2695$ eV. The wavelength is

$$\lambda = \frac{hc}{E} = \frac{(4.136 \times 10^{-15}\ \text{eV}\cdot\text{s})(2.998 \times 10^{8}\ \text{m/s})}{0.2695\ \text{eV}}$$
$$= 4.601 \times 10^{-6}\ \text{m} = 4.601\ \mu\text{m}$$

EVALUATE: In part (b) the molecule loses more energy than it does in part (a), so the emitted photon must have greater energy and a shorter wavelength. That's just what our results show.

Complex Molecules

We can apply these same principles to more complex molecules. A molecule with three or more atoms has several different kinds or *modes* of vibratory motion. Each mode has its own set of energy levels, related to its frequency by Eq. (42.7).

42.10 The carbon dioxide molecule can vibrate in three different modes. For clarity, the atoms are not shown to scale: The separation between atoms is actually comparable to their diameters.

(a) Bending mode

(b) Symmetric stretching mode

(c) Asymmetric stretching mode

In nearly all cases the associated radiation lies in the infrared region of the electromagnetic spectrum.

Infrared spectroscopy has proved to be an extremely valuable analytical tool. It provides information about the strength, rigidity, and length of molecular bonds and the structure of complex molecules. Also, because every molecule (like every atom) has its characteristic spectrum, infrared spectroscopy can be used to identify unknown compounds.

One molecule that can readily absorb and emit infrared radiation is carbon dioxide (CO_2). Figure 42.10 shows the three possible modes of vibration of a CO_2 molecule. A number of transitions are possible between excited levels of the same vibrational mode as well as between levels of different vibrational modes. The energy differences are less than 1 eV in all of these transitions, and so involve infrared photons of wavelength longer than 1 μm. Hence a gas of CO_2 can readily absorb light at a number of different infrared wavelengths. This makes CO_2 a very effective greenhouse gas (see Section 17.7) even on earth, where carbon dioxide is just 0.04% of the atmosphere by volume. On Venus, however, the atmosphere has more than 90 times the total mass of our atmosphere and is almost entirely CO_2. The resulting greenhouse effect is tremendous: The surface temperature on Venus is more than 400 kelvins greater than what it would be if the planet had no atmosphere at all.

Test Your Understanding of Section 42.2 A rotating diatomic molecule emits a photon when it makes a transition from level l to level $l - 1$. If the value of l increases, does the wavelength of the emitted photon (i) increase, (ii) decrease, or (iii) remain unchanged?

I

42.3 Structure of Solids

The term *condensed matter* includes both solids and liquids. In both states, the interactions between atoms or molecules are strong enough to give the material a definite volume that changes relatively little with applied stress. In condensed matter, adjacent atoms attract one another until their outer electron charge clouds begin to overlap significantly. Thus the distances between adjacent atoms in condensed matter are about the same as the diameters of the atoms themselves, typically 0.1 to 0.5 nm. Also, when we speak of the distances between atoms, we mean the center-to-center (nucleus-to-nucleus) distances.

Ordinarily, we think of a liquid as a material that can flow and of a solid as a material with a definite shape. However, if you heat a horizontal glass rod in the flame of a burner, you'll find that the rod begins to sag (flow) more and more easily as its temperature rises. Glass has no definite transition from solid to liquid, and no definite melting point. On this basis, we can consider glass at room temperature as being an extremely viscous liquid. Tar and butter show similar behavior.

What is the microscopic difference between materials like glass or butter and solids like ice or copper, which do have definite melting points? Ice and copper are examples of *crystalline solids* in which the atoms have *long-range order*, a recurring pattern of atomic positions that extends over many atoms. This pattern is called the *crystal structure*. In contrast, glass at room temperature is an example of an *amorphous* solid, one that has no long-range order, but only *short-range order* (correlations between neighboring atoms or molecules). Liquids also have only short-range order. The boundaries between crystalline solid, amorphous solid, and liquid may be sometimes blurred. Some solids, crystalline when perfect, can form with so many imperfections in their structure that they have almost no long-range order. Conversely, some liquid crystals (organic compounds composed of cylindrical molecules that tend to line up parallel to each other) have a fairly high degree of long-range order.

Nearly everything we know about crystal structure has been learned from diffraction experiments, initially with x rays and later with electrons and neutrons.

42.11 Portions of some common types of crystal lattices.

(a) Simple cubic (sc)

(b) Face-centered cubic (fcc)

(c) Body-centered cubic (bcc)

(d) Hexagonal close packed (hcp)

(e) Top view, hexagonal close packed

   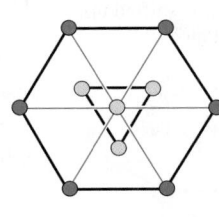

A typical distance between atoms is of the order of 0.1 nm. You can show that 12.4-keV x rays, 150-eV electrons, and 0.0818-eV neutrons all have wavelengths $\lambda = 0.1$ nm.

Crystal Lattices and Structures

A *crystal lattice* is a repeating pattern of mathematical points that extends throughout space. There are 14 general types of such patterns; Fig. 42.11 shows small portions of some common examples. The *simple cubic lattice* (sc) has a lattice point at each corner of a cubic array (Fig. 42.11a). The *face-centered cubic lattice* (fcc) is like the simple cubic but with an additional lattice point at the center of each cube face (Fig. 42.11b). The *body-centered cubic lattice* (bcc) is like the simple cubic but with an additional point at the center of each cube (Fig. 42.11c). The *hexagonal close-packed lattice* has layers of lattice points in hexagonal patterns, each hexagon made up of six equilateral triangles (Figs. 42.11d and 42.11e).

CAUTION **A perfect crystal lattice is infinitely large** Figure 42.11 shows just enough lattice points so you can easily visualize the pattern; the lattice, a mathematical abstraction, extends throughout space. Thus the lattice points shown repeat endlessly in all directions.

In a crystal structure, a single atom or a group of atoms is associated with each lattice point. The group may contain the same or different kinds of atoms. This atom or group of atoms is called a *basis*. Thus a complete description of a crystal structure includes both the lattice and the basis. We initially consider *perfect crystals,* or *ideal single crystals,* in which the crystal structure extends uninterrupted throughout space.

The bcc and fcc structures are two common simple crystal structures. The alkali metals have a bcc structure—that is, a bcc lattice with a basis of one atom at each lattice point. Each atom in a bcc structure has eight nearest neighbors (Fig. 42.12a). The elements Al, Ca, Cu, Ag, and Au have an fcc structure—that is, an fcc lattice with a basis of one atom at each lattice point. Each atom in an fcc structure has 12 nearest neighbors (Fig. 42.12b).

Figure 42.13 shows a representation of the structure of sodium chloride (NaCl, ordinary salt). It may look like a simple cubic structure, but it isn't. The sodium and chloride ions each form an fcc structure, so we can think roughly of the sodium chloride structure as being composed of two interpenetrating fcc structures. More correctly, the sodium chloride crystal structure of Fig. 42.13 has an fcc lattice with one chloride ion at each lattice point and one sodium ion half a cube length above it. That is, its basis consists of one chloride and one sodium ion.

Another example is the *diamond structure;* it's called that because it is the crystal structure of carbon in the diamond form. It's also the crystal structure of silicon, germanium, and gray tin. The diamond lattice is fcc; the basis consists of one atom at each lattice point and a second *identical* atom displaced a quarter of a cube length in each of the three cube-edge directions. Figure 42.14 will help you

42.12 (a) The bcc *structure* is composed of a bcc *lattice* with a basis of one atom for each lattice point. (b) The fcc *structure* is composed of an fcc *lattice* with a basis of one atom for each lattice point. These structures repeat precisely to make up perfect crystals.

(a) The bcc structure **(b)** The fcc structure

42.13 Representation of part of the sodium chloride crystal structure. The distances between ions are exaggerated.

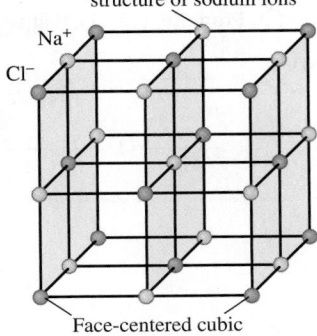

Face-centered cubic structure of sodium ions

Na$^+$
Cl$^-$

Face-centered cubic structure of chloride ions

42.14 The diamond structure, shown as two interpenetrating face-centered cubic structures with distances between atoms exaggerated. Relative to the corresponding green atom, each purple atom is shifted up, back, and to the left by a distance $a/4$.

visualize this. The shaded volume in Fig. 42.14 shows the bottom right front eighth of the basic cube; the four atoms at alternate corners of this cube are at the corners of a regular tetrahedron, and there is an additional atom at the center. Thus each atom in the diamond structure is at the center of a regular tetrahedron with four nearest-neighbor atoms at the corners.

In the diamond structure, both the purple and green spheres in Fig. 42.14 represent *identical* atoms—for example, both carbon or both silicon. In the cubic zinc sulfide structure, the purple spheres represent one type of atom and the green spheres represent a *different* type. For example, in zinc sulfide (ZnS) each zinc atom (purple in Fig. 42.14) is at the center of a regular tetrahedron with four sulfur atoms (green in Fig. 42.14) at its corners, and vice versa. Gallium arsenide (GaAs) and similar compounds have this same structure.

Bonding in Solids

The forces that are responsible for the regular arrangement of atoms in a crystal are the same as those involved in molecular bonds, plus one additional type. Not surprisingly, *ionic* and *covalent* molecular bonds are found in ionic and covalent crystals, respectively. The most familiar *ionic crystals* are the alkali halides, such as ordinary salt (NaCl). The positive sodium ions and the negative chloride ions occupy alternate positions in a cubic arrangement (see Fig. 42.13). The attractive forces are the familiar Coulomb's-law forces between charged particles. These forces have no preferred direction, and the arrangement in which the material crystallizes is partly determined by the relative sizes of the two ions. Such a structure is *stable* in the sense that it has lower total energy than the separated ions (see the following example). The negative potential energies of pairs of opposite charges are greater in absolute value than the positive energies of pairs of like charges because the pairs of unlike charges are closer together, on average.

Example 42.4 Potential energy of an ionic crystal

Imagine a one-dimensional ionic crystal consisting of a very large number of alternating positive and negative ions with charges e and $-e$, with equal spacing a along a line. Show that the total interaction potential energy is negative, which means that such a "crystal" is stable.

SOLUTION

IDENTIFY and SET UP: We treat each ion as a point charge and use our results from Section 23.1 for the electric potential energy of a collection of point charges. Equations (23.10) and (23.11) tell us to consider the electric potential energy U of each pair of charges. The total potential energy of the system is the sum of the values of U for every possible pair; we take the number of pairs to be infinite.

EXECUTE: Let's pick an ion somewhere in the middle of the line and add the potential energies of its interactions with all the ions to one side of it. From Eq. (23.11), that sum is

$$\sum U = -\frac{e^2}{4\pi\epsilon_0}\frac{1}{a} + \frac{e^2}{4\pi\epsilon_0}\frac{1}{2a} - \frac{e^2}{4\pi\epsilon_0}\frac{1}{3a} + \cdots$$
$$= -\frac{e^2}{4\pi\epsilon_0 a}\left(1 - \tfrac{1}{2} + \tfrac{1}{3} - \tfrac{1}{4} + \cdots\right)$$

You may notice that the series in parentheses resembles the Taylor series for the function $\ln(1 + x)$:

$$\ln(1 + x) = x - \frac{x^2}{2} + \frac{x^3}{3} - \frac{x^4}{4} + \cdots$$

When $x = 1$ this becomes the series in parentheses above, so

$$\sum U = -\frac{e^2}{4\pi\epsilon_0 a}\ln 2$$

This is certainly a negative quantity. The atoms on the other side of the ion we're considering make an equal contribution to the potential energy. And if we include the potential energies of all pairs of atoms, the sum is certainly negative.

EVALUATE: We conclude that this one-dimensional ionic "crystal" is stable: It has lower energy than the zero electric potential energy that is obtained when all the ions are infinitely far apart from each other.

Types of Crystals

Carbon, silicon, germanium, and tin in the diamond structure are simple examples of *covalent crystals*. These elements are in Group IV of the periodic table, meaning that each atom has four electrons in its outermost shell. Each atom forms a covalent bond with each of four adjacent atoms at the corners of a tetrahedron (Fig. 42.14). These bonds are strongly directional because of the asymmetric electron distributions dictated by the exclusion principle, and the result is the tetrahedral diamond structure.

A third crystal type, less directly related to the chemical bond than are ionic or covalent crystals, is the **metallic crystal.** In this structure, one or more of the outermost electrons in each atom become detached from the parent atom (leaving a positive ion) and are free to move through the crystal. These electrons are not localized near the individual ions. The corresponding electron wave functions extend over many atoms.

Thus we can picture a metallic crystal as an array of positive ions immersed in a sea of freed electrons whose attraction for the positive ions holds the crystal together (Fig. 42.15). These electrons also give metals their high electrical and thermal conductivities. This sea of electrons has many of the properties of a gas, and indeed we speak of the *electron-gas model* of metallic solids. The simplest version of this model is the *free-electron model,* which ignores interactions with the ions completely (except at the surface). We'll return to this model in Section 42.5.

In a metallic crystal the freed electrons are not localized but are shared among *many* atoms. This gives a bonding that is neither localized nor strongly directional. The crystal structure is determined primarily by considerations of *close packing*—that is, the maximum number of atoms that can fit into a given volume. The two most common metallic crystal lattices are the face-centered cubic and hexagonal close-packed (see Figs. 42.11b, 42.11d, and 42.11e). In structures composed of these lattices with a basis of one atom, each atom has 12 nearest neighbors.

As we mentioned in Section 42.1, van der Waals interactions and hydrogen bonding also play a role in the structure of some solids. In polyethylene and similar polymers, covalent bonding of atoms forms long-chain molecules, and hydrogen bonding forms cross-links between adjacent chains. In solid water, both van der Waals forces and hydrogen bonds are significant in determining the crystal structures of ice.

Our discussion has centered on perfect crystals, or ideal single crystals. Real crystals show a variety of departures from this idealized structure. Materials are often *polycrystalline,* composed of many small single crystals bonded together at *grain boundaries.* There may be *point defects* within a single crystal: *Interstitial* atoms may occur in places where they do not belong, and there may be *vacancies,* positions that should be occupied by an atom but are not. A point defect of particular interest in semiconductors, which we will discuss in Section 42.6, is the *substitutional impurity,* a foreign atom replacing a regular atom (for example, arsenic in a silicon crystal).

There are several basic types of extended defects called *dislocations.* One type is the *edge dislocation,* shown schematically in Fig. 42.16, in which one plane of atoms slips relative to another. The mechanical properties of metallic crystals are influenced strongly by the presence of dislocations. The ductility and malleability of some metals depend on the presence of dislocations that can move through the crystal during plastic deformations. Solid-state physicists often point out that the biggest extended defect of all, present in *all* real crystals, is the surface of the material with its dangling bonds and abrupt change in potential energy.

42.15 In a metallic solid, one or more electrons are detached from each atom and are free to wander around the crystal, forming an "electron gas." The wave functions for these electrons extend over many atoms. The positive ions vibrate around fixed locations in the crystal.

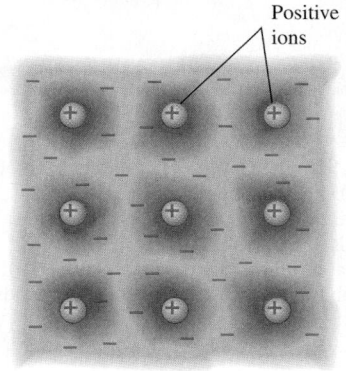

Positive ions

42.16 An edge dislocation in two dimensions. In three dimensions an edge dislocation would look like an extra plane of atoms slipped partway into the crystal.

The irregularity is seen most easily by viewing the figure from various directions at a grazing angle with the page.

Test Your Understanding of Section 42.3 If a is the distance in an NaCl crystal from an Na^+ ion to one of its nearest-neighbor Cl^- ions, what is the distance from an Na^+ ion to one of its *next-to-nearest*-neighbor Cl^- ions? (i) $a\sqrt{2}$; (ii) $a\sqrt{3}$; (iii) $2a$; (iv) none of these.

42.17 The concept of energy bands was first developed by the Swiss-American physicist Felix Bloch (1905–1983) in his doctoral thesis. Our modern understanding of electrical conductivity stems from that landmark work. Bloch's work in nuclear physics brought him (along with Edward Purcell) the 1952 Nobel Prize in physics.

42.18 Origin of energy bands in a solid. (a) As the distance r between atoms decreases, the energy levels spread into bands. The vertical line at r_0 shows the actual atomic spacing in the crystal. (b) Symbolic representation of energy bands.

42.4 Energy Bands

The **energy-band** concept, introduced in 1928 (Fig. 42.17), is a great help in understanding several properties of solids. To introduce the idea, suppose we have a large number N of identical atoms, far enough apart that their interactions are negligible. Every atom has the same energy-level diagram. We can draw an energy-level diagram for the *entire system*. It looks just like the diagram for a single atom, but the exclusion principle, applied to the entire system, permits each state to be occupied by N electrons instead of just one.

Now we begin to push the atoms uniformly closer together. Because of the electrical interactions and the exclusion principle, the wave functions begin to distort, especially those of the outer, or *valence*, electrons. The corresponding energies also shift, some upward and some downward, by varying amounts, as the valence electron wave functions become less localized and extend over more and more atoms. Thus the valence states that formerly gave the *system* a state with a sharp energy level that could accommodate N electrons now give a *band* containing N closely spaced levels (Fig. 42.18). Ordinarily, N is very large, somewhere near the order of Avogadro's number (10^{24}), so we can accurately treat the levels as forming a *continuous* distribution of energies within a band. Between adjacent energy bands are gaps or forbidden regions where there are *no* allowed energy levels. The inner electrons in an atom are affected much less by nearby atoms than are the valence electrons, and their energy levels remain relatively sharp.

Insulators, Semiconductors, and Conductors

The nature of the energy bands determines whether the material is an electrical insulator, a semiconductor, or a conductor. In particular, what matters are the extent to which the states in each band are occupied and the spacing, or *energy gap*, between adjacent bands. A crucial factor is the exclusion principle (see Section 41.6), which states that only one electron can occupy a given quantum-mechanical state.

In an *insulator* at absolute zero temperature, the highest band that is completely filled, called the **valence band,** is also the highest band that has *any* electrons in it. The next higher band, called the **conduction band,** is completely empty; there are no electrons in its states (Fig. 42.19a). Imagine what happens if an electric field is applied to a material of this kind. To move in response to the field, an electron would have to go into a different quantum state with a slightly different energy. It can't do that, however, because all the neighboring states are already occupied. The only way such an electron can move is to jump across the energy gap into the conduction band, where there are plenty of nearby unoccupied states. At any temperature above absolute zero

42.19 Three types of energy-band structure.

(a) In an insulator at absolute zero, there are no electrons in the conduction band.

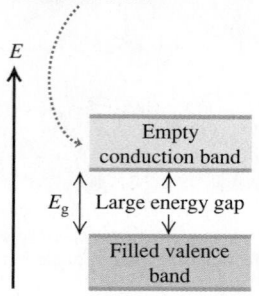

(b) A semiconductor has the same band structure as an insulator but a smaller gap between the valence and conduction bands.

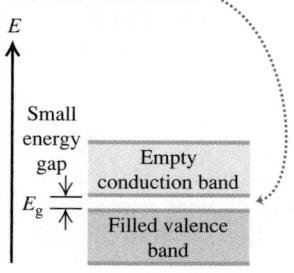

(c) A conductor has a partially filled conduction band.

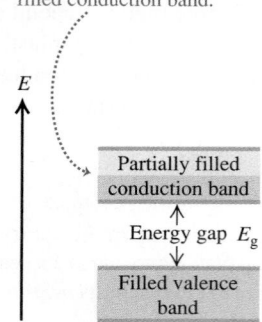

there is some probability this jump can happen, because an electron can gain energy from thermal motion. In an insulator, however, the energy gap between the valence and conduction bands can be 5 eV or more, and that much thermal energy is not ordinarily available. Hence little or no current flows in response to an applied electric field, and the electric conductivity (Section 25.2) is low. The thermal conductivity (Section 17.7), which also depends on mobile electrons, is likewise low.

We saw in Section 24.4 that an insulator becomes a conductor if it is subjected to a large enough electric field; this is called *dielectric breakdown*. If the electric field is of order 10^{10} V/m, there is a potential difference of a few volts over a distance comparable to atomic sizes. In this case the field can do enough work on a valence electron to boost it across the energy gap and into the conduction band. (In practice dielectric breakdown occurs for fields much less than 10^{10} V/m, because imperfections in the structure of an insulator provide some more accessible energy states *within* the energy gap.)

As in an insulator, a *semiconductor* at absolute zero has an empty conduction band above the full valence band. The difference is that in a semiconductor the energy gap between these bands is relatively small and electrons can more readily jump into the conduction band (Fig. 42.19b). As the temperature of a semiconductor increases, the population in the conduction band increases very rapidly, as does the electric conductivity. For example, in a semiconductor near room temperature with an energy gap of 1 eV, the number of conduction electrons doubles when the temperature rises by just 10°C. We will use the concept of energy bands to explore semiconductors in more depth in Section 42.6.

In a *conductor* such as a metal, there are electrons in the conduction band even at absolute zero (Fig. 42.19c). The metal sodium is an example. An analysis of the atomic energy-level diagram for sodium (see Fig. 39.19a) shows that for an isolated sodium atom, the six lowest excited states (all 3p states) are about 2.1 eV above the two 3s ground states. In solid sodium, however, the atoms are so close together that the 3s and 3p *bands* spread out and overlap into a single band. Each sodium atom contributes one electron to the band, leaving an Na^+ ion behind. Each atom also contributes eight *states* to that band (two 3s, six 3p), so the band is only one-eighth occupied. We call this structure a *conduction* band because it is only partially occupied. Electrons near the top of the filled portion of the band have many adjacent unoccupied states available, and they can easily gain or lose small amounts of energy in response to an applied electric field. Therefore these electrons are mobile, giving solid sodium its high electrical and thermal conductivity. A similar description applies to other conducting materials.

Mastering**PHYSICS**

PhET: Band Structure
PhET: Conductivity

Example 42.5 Photoconductivity in germanium

At room temperature, pure germanium has an almost completely filled valence band separated by a 0.67-eV gap from an almost completely empty conduction band. It is a poor electrical conductor, but its conductivity increases greatly when it is irradiated with electromagnetic waves of a certain maximum wavelength. What is that wavelength?

SOLUTION

IDENTIFY and SET UP: The conductivity of a semiconductor increases greatly when electrons are excited from the valence band into the conduction band. In germanium, the excitation occurs when an electron absorbs a photon with an energy of at least $E_{min} = 0.67$ eV. From the relationship $E = hc/\lambda$, the *maximum* wavelength λ_{max} (our target variable) corresponds to this *minimum* photon energy.

EXECUTE: The wavelength of a photon with energy $E_{min} = 0.67$ eV is

$$\lambda_{max} = \frac{hc}{E_{min}} = \frac{(4.136 \times 10^{-15}\ eV \cdot s)(3.00 \times 10^8\ m/s)}{0.67\ eV}$$
$$= 1.9 \times 10^{-6}\ m = 1.9\ \mu m = 1900\ nm$$

EVALUATE: This wavelength is in the infrared part of the spectrum, so visible-light photons (which have shorter wavelength and higher energy) will also induce conductivity in germanium. As we'll see in Section 42.7, semiconductor crystals are widely used as photocells and for many other applications.

Test Your Understanding of Section 42.4 One type of thermometer works by measuring the temperature-dependent electrical resistivity of a sample. Which of the following types of material displays the greatest change in resistivity for a given temperature change? (i) insulator; (ii) semiconductor; (iii) resistor.

42.5 Free-Electron Model of Metals

Studying the energy states of electrons in metals can give us a lot of insight into their electrical and magnetic properties, the electron contributions to heat capacities, and other behavior. As we discussed in Section 42.3, one of the distinguishing features of a metal is that one or more valence electrons are detached from their home atom and can move freely within the metal, with wave functions that extend over many atoms.

The **free-electron model** assumes that these electrons are completely free inside the material, that they don't interact at all with the ions or with each other, but that there are infinite potential-energy barriers at the surfaces. The idea is that a typical electron moves so rapidly within the metal that it "sees" the effect of the ions and other electrons as a uniform potential-energy function, whose value we can choose to be zero.

We can represent the surfaces of the metal by the same cubical box that we analyzed in Section 41.2 (the three-dimensional version of the particle in a box studied in Section 40.2). If the box has sides of length L (Fig. 42.20), the energies of the stationary states (quantum states of definite energy) are

$$E_{n_X,n_Y,n_Z} = \frac{(n_X^2 + n_Y^2 + n_Z^2)\pi^2\hbar^2}{2mL^2} \quad (n_X = 1, 2, 3, \ldots; n_Y = 1, 2, 3, \ldots; n_Z = 1, 2, 3, \ldots) \quad (42.10)$$

Each state is labeled by the three positive-integer quantum numbers (n_X, n_Y, n_Z).

Density of States

Later we'll need to know the *number dn* of quantum states that have energies in a given range dE. The number of states per unit energy range dn/dE is called the **density of states,** denoted by $g(E)$. We'll begin by working out an expression for $g(E)$. Think of a three-dimensional space with coordinates (n_X, n_Y, n_Z) (Fig. 42.21). The radius n_{rs} of a sphere centered at the origin in that space is $n_{rs} = (n_X^2 + n_Y^2 + n_Z^2)^{1/2}$. Each point with integer coordinates in that space represents one spatial quantum state. Thus each point corresponds to one unit of volume in the space, and the total number of points with integer coordinates inside a sphere equals the volume of the sphere, $\frac{4}{3}\pi n_{rs}^3$. Because all our n's are positive, we must take only one *octant* of the sphere, with $\frac{1}{8}$ the total volume, or $(\frac{1}{8})(\frac{4}{3}\pi n_{rs}^3) = \frac{1}{6}\pi n_{rs}^3$. The particles are electrons, so each point corresponds to *two* states with opposite spin components ($m_s = \pm\frac{1}{2}$), and the total number n of electron states corresponding to points inside the octant is twice $\frac{1}{6}\pi n_{rs}^3$, or

$$n = \frac{\pi n_{rs}^3}{3} \quad (42.11)$$

The energy E of states at the surface of the sphere can be expressed in terms of n_{rs}. Equation (42.10) becomes

$$E = \frac{n_{rs}^2\pi^2\hbar^2}{2mL^2} \quad (42.12)$$

We can combine Eqs. (42.11) and (42.12) to get a relationship between E and n that doesn't contain n_{rs}. We'll leave the details as an exercise (Exercise 42.24); the result is

$$n = \frac{(2m)^{3/2}VE^{3/2}}{3\pi^2\hbar^3} \quad (42.13)$$

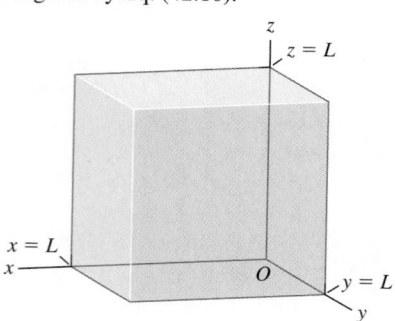

42.20 A cubical box with side length L. We studied this three-dimensional version of the infinite square well in Section 41.2. The energy levels for a particle in this box are given by Eq. (42.10).

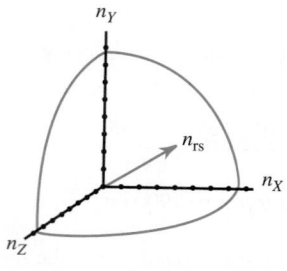

42.21 The allowed values of n_X, n_Y, and n_Z are positive integers for the electron states in the free-electron gas model. Including spin, there are two states for each unit volume in n space.

where $V = L^3$ is the volume of the box. Equation (42.13) gives the total number of states with energies of E or less.

To get the number of states dn in an energy interval dE, we treat n and E as continuous variables and take differentials of both sides of Eq. (42.13). We get

$$dn = \frac{(2m)^{3/2}VE^{1/2}}{2\pi^2\hbar^3}\,dE \qquad (42.14)$$

The density of states $g(E)$ is equal to dn/dE, so from Eq. (42.14) we get

$$g(E) = \frac{(2m)^{3/2}V}{2\pi^2\hbar^3}E^{1/2} \qquad \text{(density of states, free-electron model)} \quad (42.15)$$

Fermi–Dirac Distribution

Now we need to know how the electrons are distributed among the various quantum states at any given temperature. The Maxwell–Boltzmann distribution states that the average number of particles in a state of energy E is proportional to $e^{-E/kT}$ (see Sections 18.5 and 39.4). However, there are two very important reasons why it wouldn't be right to use the Maxwell–Boltzmann distribution. The first reason is the exclusion principle. At absolute zero the Maxwell–Boltzmann function predicts that *all* the electrons would go into the two ground states of the system, with $n_X = n_Y = n_Z = 1$ and $m_s = \pm\frac{1}{2}$. But the exclusion principle allows only one electron in each state. At absolute zero the electrons can fill up the lowest *available* states, but there's not enough room for *all* of them to go into the lowest states. Thus a reasonable guess as to the shape of the distribution would be Fig. 42.22. At absolute zero temperature the states are filled up to some value E_{F0}, and all states above this value are empty.

The second reason we can't use the Maxwell–Boltzmann distribution is more subtle. That distribution assumes that we are dealing with *distinguishable* particles. It might seem that we could put a tag on each electron and know which is which. But overlapping electrons in a system such as a metal are *indistinguishable*. Suppose we have two electrons; a state in which the first is in energy level E_1 and the second is in level E_2 is not distinguishable from a state in which the two electrons are reversed, because we can't tell which electron is which.

The statistical distribution function that emerges from the exclusion principle and the indistinguishability requirement is called (after its inventors) the **Fermi–Dirac distribution.** Because of the exclusion principle, the probability that a particular state with energy E is occupied by an electron is the same as $f(E)$, the fraction of states with that energy that are occupied:

$$f(E) = \frac{1}{e^{(E-E_F)/kT} + 1} \qquad \text{(Fermi–Dirac distribution)} \quad (42.16)$$

The energy E_F is called the **Fermi energy** or the *Fermi level;* we'll discuss its significance below. We use E_{F0} for its value at absolute zero ($T = 0$) and E_F for other temperatures. We can accurately let $E_F = E_{F0}$ for metals because the Fermi energy does not change much with temperature for solid conductors. However, it is not safe to assume that $E_F = E_{F0}$ for semiconductors, in which the Fermi energy usually does change with temperature.

Figure 42.23 shows graphs of Eq. (42.16) for three temperatures. The trend of this function as kT approaches zero confirms our guess. When $E = E_F$, the exponent is zero and $f(E_F) = \frac{1}{2}$. That is, the probability is $\frac{1}{2}$ that a state at the Fermi energy contains an electron. Alternatively, at $E = E_F$, half the states are filled (and half are empty).

42.22 The probability distribution for occupation of free-electron energy states at absolute zero.

At absolute zero, all states are occupied (occupation probability 1) at energies up to E_{F0} ...

... and all states are empty (occupation probability zero) at energies above E_{F0}.

42.23 Graphs of the Fermi–Dirac distribution function for various values of kT, assuming that the Fermi energy E_F is independent of the temperature T.

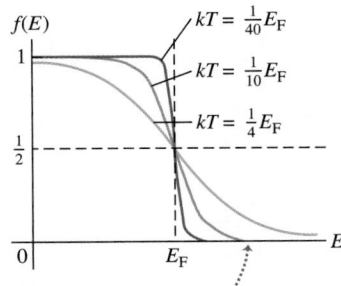

As T increases, more and more of the electrons are excited to states with energy $E > E_F$.

For $E < E_F$ the exponent is negative, and $f(E) > \frac{1}{2}$. For $E > E_F$ the exponent is positive, and $f(E) < \frac{1}{2}$. The shape depends on the ratio E_F/kT. At $T \ll E_F/k$ this ratio is very large. Then for $E < E_F$ the curve very quickly approaches 1, and for $E > E_F$ it quickly approaches zero. When T is larger, the changes are more gradual. When T is zero, all the states up to the Fermi level E_{F0} are filled, and all states above that level are empty (Fig. 42.22).

Example 42.6 Probabilities in the free-electron model

For free electrons in a solid, at what energy is the probability that a particular state is occupied equal to (a) 0.01 and (b) 0.99?

SOLUTION

IDENTIFY and SET UP: This problem asks us to explore the Fermi–Dirac distribution. Equation (42.16) gives the occupation probability $f(E)$ for a given energy E. If we solve this equation for E, we get an expression for the energy that corresponds to a given occupation probability—which is just what we need to solve this problem.

EXECUTE: Using Eq. (42.16), you can show that

$$E = E_F + kT \ln\left(\frac{1}{f(E)} - 1\right)$$

(a) When $f(E) = 0.01$,

$$E = E_F + kT \ln\left(\frac{1}{0.01} - 1\right) = E_F + 4.6kT$$

The probability that a state $4.6kT$ above the Fermi level is occupied is only 0.01, or 1%.

(b) When $f(E) = 0.99$,

$$E = E_F + kT \ln\left(\frac{1}{0.99} - 1\right) = E_F - 4.6kT$$

The probability that a state $4.6kT$ below the Fermi level is occupied is 0.99, or 99%.

EVALUATE: At very low temperatures, $4.6kT$ is much less than E_F. Then the occupation probability of levels even slightly below E_F is nearly 1 (100%), and that for levels even slightly above E_F is nearly zero (see Fig. 42.23). In general, if the probability is P that a state with an energy ΔE *above* E_F is occupied, then the probability is $1 - P$ that a state ΔE *below* E_F is occupied. We leave the proof to you (Problem 42.50).

Electron Concentration and Fermi Energy

Equation (42.16) gives the probability that any specific state with energy E is occupied at a temperature T. To get the actual number of electrons in any energy range dE, we have to multiply this probability by the number dn of states in that range $g(E)\, dE$. Thus the number dN of electrons with energies in the range dE is

$$dN = g(E)f(E)\, dE = \frac{(2m)^{3/2}VE^{1/2}}{2\pi^2\hbar^3}\frac{1}{e^{(E-E_F)/kT} + 1}\, dE \qquad (42.17)$$

The Fermi energy E_F is determined by the total number N of electrons; at any temperature the electron states are filled up to a point at which all electrons are accommodated. At absolute zero there is a simple relationship between E_{F0} and N. *All* states below E_{F0} are filled; in Eq. (42.13) we set n equal to the total number of electrons N and E to the Fermi energy at absolute zero E_{F0}:

$$N = \frac{(2m)^{3/2}VE_{F0}^{3/2}}{3\pi^2\hbar^3} \qquad (42.18)$$

Solving Eq. (42.18) for E_{F0}, we get

$$E_{F0} = \frac{3^{2/3}\pi^{4/3}\hbar^2}{2m}\left(\frac{N}{V}\right)^{2/3} \qquad (42.19)$$

The quantity N/V is the number of free electrons per unit volume. It is called the *electron concentration* and is usually denoted by n.

If we replace N/V with n, Eq. (42.19) becomes

$$E_{F0} = \frac{3^{2/3}\pi^{4/3}\hbar^2 n^{2/3}}{2m} \qquad (42.20)$$

CAUTION **Electron concentration and number of electrons** Don't confuse the electron concentration n with any quantum number n. Furthermore, the number of states is *not* in general the same as the total number of electrons N.

Example 42.7 The Fermi energy in copper

At low temperatures, copper has a free-electron concentration $n = 8.45 \times 10^{28}$ m^{-3}. Using the free-electron model, find the Fermi energy for solid copper, and find the speed of an electron with a kinetic energy equal to the Fermi energy.

SOLUTION

IDENTIFY and SET UP: This problem uses the relationship between Fermi energy and free-electron concentration. Because copper is a solid conductor, its Fermi energy changes very little with temperature and we can use the expression for the Fermi energy at absolute zero, Eq. (42.20). We'll find the *Fermi speed* v_F that corresponds to kinetic energy E_F using the nonrelativistic formula $E_F = \frac{1}{2}mv_F^2$.

EXECUTE: Using the given value of n, we solve for E_F and v_F:

$$E_F = \frac{3^{2/3}\pi^{4/3}(1.055 \times 10^{-34} \text{ J}\cdot\text{s})^2(8.45 \times 10^{28} \text{ m}^{-3})^{2/3}}{2(9.11 \times 10^{-31} \text{ kg})}$$

$$= 1.126 \times 10^{-18} \text{ J} = 7.03 \text{ eV}$$

$$v_F = \sqrt{\frac{2E_F}{m}} = \sqrt{\frac{2(1.126 \times 10^{-18} \text{ J})}{9.11 \times 10^{-31} \text{ kg}}} = 1.57 \times 10^6 \text{ m/s}$$

EVALUATE: Our values of E_F and v_F are within the ranges of typical values for metals, 1.6–14 eV and 0.8–2.2 $\times 10^6$ m/s, respectively. Note that the calculated Fermi speed is far less than the speed of light $c = 3.00 \times 10^8$ m/s, which justifies our use of the nonrelativistic formula $\frac{1}{2}mv_F^2 = E_F$.

Our calculated Fermi energy is much larger than kT at ordinary temperatures. (At room temperature $T = 20°C = 293$ K, the quantity kT equals $(1.381 \times 10^{-23} \text{ J/K})(293 \text{ K}) = 4.04 \times 10^{-21}$ J $= 0.0254$ eV.) So it is a good approximation to take almost all the states below E_F as completely full and almost all those above E_F as completely empty (see Fig. 42.22).

We can also use Eq. (42.15) to find $g(E)$ if E and V are known. You can show that if $E = 7.03$ eV and $V = 1$ cm^3, $g(E)$ is about 2×10^{22} states/eV. This huge number shows why we were justified in treating n and E as continuous variables in our density-of-states derivation.

Average Free-Electron Energy

We can calculate the *average* free-electron energy in a metal at absolute zero by using the same ideas that we used to find E_{F0}. From Eq. (42.17) the number dN of electrons with energies in the range dE is $g(E)f(E)\,dE$. The energy of these electrons is $E\,dN = Eg(E)f(E)\,dE$. At absolute zero we substitute $f(E) = 1$ from $E = 0$ to $E = E_{F0}$ and $f(E) = 0$ for all other energies. Therefore the total energy E_{tot} of all the N electrons is

$$E_{tot} = \int_0^{E_{F0}} Eg(E)(1)\,dE + \int_{E_{F0}}^{\infty} Eg(E)(0)\,dE = \int_0^{E_{F0}} Eg(E)\,dE$$

The simplest way to evaluate this expression is to compare Eqs. (42.15) and (42.19), noting that

$$g(E) = \frac{3NE^{1/2}}{2E_{F0}^{3/2}}$$

Substituting this expression into the integral and using $E_{av} = E_{tot}/N$, we get

$$E_{av} = \frac{3}{2E_{F0}^{3/2}}\int_0^{E_{F0}} E^{3/2}\,dE = \tfrac{3}{5}E_{F0} \qquad (42.21)$$

At absolute zero the average free-electron energy equals $\frac{3}{5}$ of the Fermi energy.

Example 42.8 Free-electron gas versus ideal gas

(a) Find the average energy of the free electrons in copper at absolute zero (see Example 42.7). (b) What would be the average kinetic energy of electrons if they behaved like an ideal gas at room temperature, 20°C (see Section 18.3)? What would be the speed of an electron with this kinetic energy? Compare these ideal-gas values with the (correct) free-electron values.

Continued

SOLUTION

IDENTIFY and SET UP: Free electrons in a metal behave like a kind of gas. In part (a) we use Eq. (42.21) to determine the average kinetic energy of free electrons in terms of the Fermi energy at absolute zero, which we know for copper from Example 42.7. In part (b) we treat electrons as an ideal gas at room temperature: Eq. (18.16) then gives the average kinetic energy per electron as $E_{av} = \frac{3}{2}kT$, and $E_{av} = \frac{1}{2}mv^2$ gives the corresponding electron speed v.

EXECUTE: (a) From Example 42.7, the Fermi energy in copper at absolute zero is 1.126×10^{-18} J $= 7.03$ eV. According to Eq. (42.21), the average energy is $\frac{3}{5}$ of this, or 6.76×10^{-19} J $= 4.22$ eV.

(b) In Example 42.7 we found that $kT = 4.04 \times 10^{-21}$ J $= 0.0254$ eV at room temperature $T = 20°C = 293$ K. If electrons behaved like an ideal gas at this temperature, the average kinetic energy per electron would be $\frac{3}{2}$ of this, or 6.07×10^{-21} J $= 0.0379$ eV. The speed of an electron with this kinetic energy would be

$$v = \sqrt{\frac{2E_{av}}{m}} = \sqrt{\frac{2(6.07 \times 10^{-21}\,\text{J})}{9.11 \times 10^{-31}\,\text{kg}}} = 1.15 \times 10^5 \text{ m/s}$$

EVALUATE: The ideal-gas model predicts an average energy that is about 1% of the value given by the free-electron model, and

a speed that is about 7% of the free-electron Fermi speed $v_F = 1.57 \times 10^6$ m/s that we found in Example 42.7. Thus temperature plays a *very* small role in determining the properties of electrons in metals; their average energies are determined almost entirely by the exclusion principle.

A similar analysis allows us to determine the contributions of electrons to the heat capacities of a solid metal. If there is one conduction electron per atom, the principle of equipartition of energy (see Section 18.4) would predict that the kinetic energies of these electrons contribute $3R/2$ to the molar heat capacity at constant volume C_V. But when kT is much smaller than E_F, which is usually the situation in metals, only those few electrons near the Fermi level can find empty states and change energy appreciably when the temperature changes. The number of such electrons is proportional to kT/E_F, so we expect that the electron molar heat capacity at constant volume is proportional to $(kT/E_F)(3R/2) = (3kT/2E_F)R$. A more detailed analysis shows that the actual electron contribution to C_V for a solid metal is $(\pi^2 kT/2E_F)R$, not far from our prediction. You can verify that if $T = 293$ K and $E_F = 7.03$ eV, the electron contribution to C_V is $0.018R$, which is only 1.2% of the (incorrect) $3R/2$ prediction of the equipartition principle. Because the electron contribution is so small, the overall heat capacity of most solid metals is due primarily to vibration of the atoms in the crystal structure (see Fig. 18.18 in Section 18.4).

Test Your Understanding of Section 42.5 An ideal gas obeys the relationship $pV = nRT$ (see Section 18.1). That is, for a given volume V and a number of moles n, as the temperature T decreases, the pressure p decreases proportionally and tends to zero as T approaches absolute zero. Is this also true of the free-electron gas in a solid metal?

❚

PhET: Semiconductors
PhET: Conductivity

42.6 Semiconductors

A **semiconductor** has an electrical resistivity that is intermediate between those of good conductors and of good insulators. The tremendous importance of semiconductors in present-day electronics stems in part from the fact that their electrical properties are very sensitive to very small concentrations of impurities. We'll discuss the basic concepts using the semiconductor elements silicon (Si) and germanium (Ge) as examples.

Silicon and germanium are in Group IV of the periodic table. Both have four electrons in the outermost atomic subshells ($3s^2 3p^2$ for silicon, $4s^2 4p^2$ for germanium), and both crystallize in the covalently bonded diamond structure discussed in Section 42.3 (see Fig. 42.14). Because all four of the outer electrons are involved in the bonding, at absolute zero the band structure (see Section 42.4) has a completely empty conduction band (see Fig. 42.19b). As we discussed in Section 42.4, at very low temperatures electrons cannot jump from the filled valence band into the conduction band. This property makes these materials insulators at very low temperatures; their electrons have no nearby states available into which they can move in response to an applied electric field.

However, in semiconductors the energy gap E_g between the valence and conduction bands is small in comparison to the gap of 5 eV or more for many insulators; room-temperature values are 1.12 eV for silicon and only 0.67 eV for germanium. Thus even at room temperature a substantial number of electrons can gain enough energy to jump the gap to the conduction band, where they are dissociated from their parent atoms and are free to move about the crystal. The number of these electrons increases rapidly with temperature.

Example 42.9 **Jumping a band gap**

Consider a material with the band structure described above, with its Fermi energy in the middle of the gap (Fig. 42.24). Find the probability that a state at the bottom of the conduction band is occupied at $T = 300$ K, and compare that with the probability at $T = 310$ K, for band gaps of (a) 0.200 eV; (b) 1.00 eV; (c) 5.00 eV.

SOLUTION

IDENTIFY and SET UP: The Fermi–Dirac distribution function gives the probability that a state of energy E is occupied at temperature T. Figure 42.24 shows that the state of interest at the bottom of the conduction band has an energy $E = E_F + E_g/2$ that is greater than the Fermi energy E_F, with $E - E_F = E_g/2$.

42.24 Band structure of a semiconductor. At absolute zero a completely filled valence band is separated by a narrow energy gap E_g of 1 eV or so from a completely empty conduction band. At ordinary temperatures, a number of electrons are excited to the conduction band.

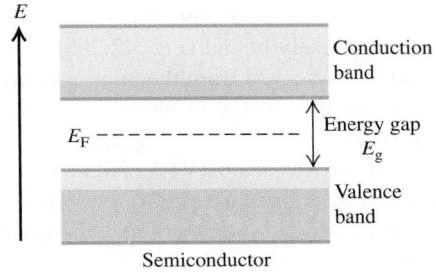

Semiconductor

Figure 42.23 shows that the higher the temperature, the larger the fraction of electrons with energies greater than the Fermi energy.

EXECUTE: (a) When $E_g = 0.200$ eV,

$$\frac{E - E_F}{kT} = \frac{E_g}{2}\frac{0.100\ \text{eV}}{(8.617 \times 10^{-5}\ \text{eV/K})(300\ \text{K})} = 3.87$$

$$f(E) = \frac{1}{e^{3.87} + 1} = 0.0205$$

For $T = 310$ K, the exponent is 3.74 and $f(E) = 0.0231$, a 13% increase in probability for a temperature rise of 10 K.

(b) For $E_g = 1.00$ eV, both exponents are five times as large as in part (a), namely 19.3 and 18.7; the values of $f(E)$ are 4.0×10^{-9} and 7.4×10^{-9}. In this case the (low) probability nearly doubles with a temperature rise of 10 K.

(c) For $E_g = 5.0$ eV, the exponents are 96.7 and 93.6; the values of $f(E)$ are 1.0×10^{-42} and 2.3×10^{-41}. The (extremely low) probability increases by a factor of 23 for a 10 K temperature rise.

EVALUATE: This example illustrates two important points. First, the probability of finding an electron in a state at the bottom of the conduction band is extremely sensitive to the width of the band gap. At room temperature, the probability is about 2% for a 0.200-eV gap, a few in a thousand million for a 1.00-eV gap, and essentially zero for a 5.00-eV gap. (Pure diamond, with a 5.47-eV band gap, has essentially no electrons in the conduction band and is an excellent insulator.) Second, for any given band gap the probability depends strongly on temperature, and even more strongly for large gaps than for small ones.

In principle, we could continue the calculation in Example 42.9 to find the actual density $n = N/V$ of electrons in the conduction band at any temperature. To do this, we would have to evaluate the integral $\int g(E)f(E)\,dE$ from the bottom of the conduction band to its top. First we would need to know the density of states function $g(E)$. It wouldn't be correct to use Eq. (42.15) because the energy-level structure and the density of states for real solids are more complex than those for the simple free-electron model. However, there are theoretical methods for predicting what $g(E)$ should be near the bottom of the conduction band, and such calculations have been carried out. Once we know n, we can *begin* to determine the resistivity of the material (and its temperature dependence) using the analysis of Section 25.2, which you may want to review. But next we'll see that the electrons in the conduction band don't tell the whole story about conduction in semiconductors.

Holes

When an electron is removed from a covalent bond, it leaves a vacancy behind. An electron from a neighboring atom can move into this vacancy, leaving the neighbor with the vacancy. In this way the vacancy, called a **hole,** can travel through the material and serve as an additional current carrier. It's like describing the motion of a bubble in a liquid. In a pure, or *intrinsic,* semiconductor, valence-band holes and conduction-band electrons are always present in equal numbers. When an electric field is applied, they move in opposite directions (Fig. 42.25). Thus a hole in the valence band behaves like a positively charged particle, even though the moving charges in that band are electrons. The conductivity that we

42.25 Motion of electrons in the conduction band and of holes in the valence band of a semiconductor under the action of an applied electric field $\vec{E}$.

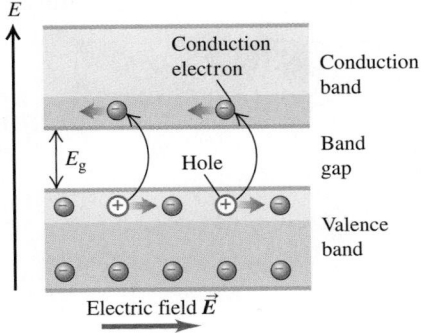

just described for a pure semiconductor is called *intrinsic conductivity.* Another kind of conductivity, to be discussed in the next subsection, is due to impurities.

An analogy helps to picture conduction in an intrinsic semiconductor. The valence band at absolute zero is like a floor of a parking garage that's filled bumper to bumper with cars (which represent electrons). No cars can move because there is nowhere for them to go. But if one car is moved to the vacant floor above, it can move freely, just as electrons can move freely in the conduction band. Also, the empty space that it leaves permits cars to move on the nearly filled floor, thereby moving the empty space just as holes move in the normally filled valence band.

Impurities

Suppose we mix into melted germanium ($Z = 32$) a small amount of arsenic ($Z = 33$), the next element after germanium in the periodic table. This deliberate addition of impurity elements is called *doping.* Arsenic is in Group V; it has *five* valence electrons. When one of these electrons is removed, the remaining electron structure is essentially identical to that of germanium. The only difference is that it is smaller; the arsenic nucleus has a charge of $+33e$ rather than $+32e$, and it pulls the electrons in a little more. An arsenic atom can comfortably take the place of a germanium atom as a substitutional impurity. Four of its five valence electrons form the necessary nearest-neighbor covalent bonds.

The fifth valence electron is very loosely bound (Fig. 42.26a); it doesn't participate in the covalent bonds, and it is screened from the nuclear charge of $+33e$ by the 32 electrons, leaving a net effective charge of about $+e$. We might guess that the binding energy would be of the same order of magnitude as the energy of the $n = 4$ level in hydrogen—that is, $(\frac{1}{4})^2(13.6 \text{ eV}) = 0.85 \text{ eV}$. In fact, it is much smaller than this, only about 0.01 eV, because the electron probability distribution actually extends over many atomic diameters and the polarization of intervening atoms provides additional screening.

The energy level of this fifth electron corresponds in the band picture to an isolated energy level lying in the gap, about 0.01 eV below the bottom of the conduction band (Fig. 42.26b). This level is called a *donor level,* and the impurity atom that is responsible for it is simply called a *donor.* All Group V elements, including N, P, As, Sb, and Bi, can serve as donors. At room temperature, kT is about 0.025 eV. This is substantially greater than 0.01 eV, so at ordinary temperatures, most electrons can gain enough energy to jump from donor levels into the conduction band, where they are free to wander through the material. The remaining ionized donor stays at its site in the structure and does not participate in conduction.

Example 42.9 shows that at ordinary temperatures and with a band gap of 1.0 eV, only a very small fraction (of the order of 10^{-9}) of the states at the bottom of the conduction band in a pure semiconductor contain electrons to participate in intrinsic conductivity. Thus we expect the conductivity of such a semiconductor to be about 10^{-9} as great as that of good metallic conductors, and measurements bear out this prediction. However, a concentration of donors as small as one part in 10^8 can increase the conductivity so drastically that conduction due to impurities becomes by far the dominant mechanism. In this case the conductivity is due almost entirely to *negative* charge (electron) motion. We call the material an **n-type semiconductor,** with *n*-type impurities.

Adding atoms of an element in Group III (B, Al, Ga, In, Tl), with only *three* valence electrons, has an analogous effect. An example is gallium ($Z = 31$); as a substitutional impurity in germanium, the gallium atom would like to form four covalent bonds, but it has only three outer electrons. It can, however, steal an electron from a neighboring germanium atom to complete the required four covalent bonds (Fig. 42.27a). The resulting atom has the same electron configuration as Ge but is somewhat larger because gallium's nuclear charge is smaller, $+31e$ instead of $+32e$.

42.26 An *n*-type semiconductor.

(a) A donor (*n*-type) impurity atom has a fifth valence electron that does not participate in the covalent bonding and is very loosely bound.

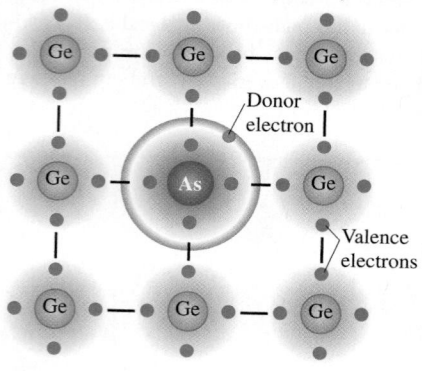

(b) Energy-band diagram for an *n*-type semiconductor at a low temperature. One donor electron has been excited from the donor levels into the conduction band.

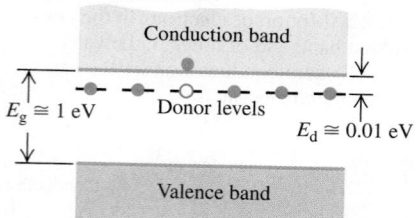

This theft leaves the neighboring atom with a *hole,* or missing electron. The hole acts as a positive charge that can move through the crystal just as with intrinsic conductivity. The stolen electron is bound to the gallium atom in a level called an *acceptor level* about 0.01 eV above the top of the valence band (Fig. 42.27b). The gallium atom, called an *acceptor,* thus accepts an electron to complete its desire for four covalent bonds. This extra electron gives the previously neutral gallium atom a net charge of $-e$. The resulting gallium ion is *not* free to move. In a semiconductor that is doped with acceptors, we consider the conductivity to be almost entirely due to *positive* charge (hole) motion. We call the material a **p-type semiconductor,** with *p*-type impurities. Some semiconductors are doped with *both* n- and *p*-type impurities. Such materials are called *compensated* semiconductors.

CAUTION **The meaning of "p-type" and "n-type"** Saying that a material is a *p*-type semiconductor does *not* mean that the material has a positive charge; ordinarily, it would be neutral. Rather, it means that its *majority carriers* of current are positive holes (and therefore its *minority carriers* are negative electrons). The same idea holds for an *n*-type semiconductor; ordinarily, it will *not* have a negative charge, but its majority carriers are negative electrons. ▮

We can verify the assertion that the current in *n*- and *p*-type semiconductors really *is* carried by electrons and holes, respectively, by using the Hall effect (see Section 27.9). The sign of the Hall emf is opposite in the two cases. Hall-effect devices constructed from semiconductor materials are used in probes to measure magnetic fields and the currents that cause those fields.

Test Your Understanding of Section 42.6 Would there be any advantage to adding *n*-type or *p*-type impurities to copper?

42.7 Semiconductor Devices

Semiconductor devices play an indispensable role in contemporary electronics. In the early days of radio and television, transmitting and receiving equipment relied on vacuum tubes, but these have been almost completely replaced in the last six decades by solid-state devices, including transistors, diodes, integrated circuits, and other semiconductor devices. The only surviving vacuum tubes in consumer electronics are the picture tubes in older TV receivers and computer monitors; these are rapidly being replaced by flat-screen displays.

One simple semiconductor device is the *photocell* (Fig. 42.28). When a thin slab of semiconductor is irradiated with an electromagnetic wave whose photons have at least as much energy as the band gap between the valence and conduction bands, an electron in the valence band can absorb a photon and jump to the conduction band, where it and the hole it left behind contribute to the conductivity (see Example 42.5 in Section 42.4). The conductivity therefore increases with wave intensity, thus increasing the current *I* in the photocell circuit of Fig. 42.28. Hence the ammeter reading indicates the intensity of the light.

Detectors for charged particles operate on the same principle. An external circuit applies a voltage across a semiconductor. An energetic charged particle passing through the semiconductor collides inelastically with valence electrons, exciting them from the valence to the conduction band and creating pairs of holes and conduction electrons. The conductivity increases momentarily, causing a pulse of current in the external circuit. Solid-state detectors are widely used in nuclear and high-energy physics research.

The *p-n* Junction

In many semiconductor devices the essential principle is the fact that the conductivity of the material is controlled by impurity concentrations, which can be varied

42.27 A *p*-type semiconductor.

(a) An acceptor (*p*-type) impurity atom has only three valence electrons, so it can borrow an electron from a neighboring atom. The resulting hole is free to move about the crystal.

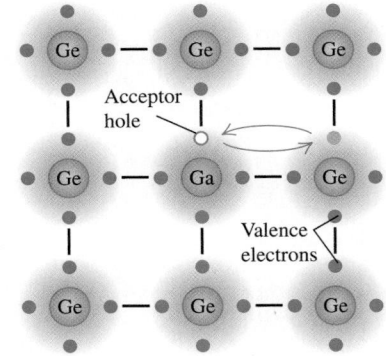

(b) Energy-band diagram for a *p*-type semiconductor at a low temperature. One acceptor level has accepted an electron from the valence band, leaving a hole behind.

42.28 A semiconductor photocell in a circuit. The more intense the light falling on the photocell, the greater the conductivity of the photocell and the greater the current measured by the ammeter (A).

42.29 (a) A semiconductor *p-n* junction in a circuit. (b) Graph showing the asymmetric current–voltage relationship. The curve is described by Eq. (42.22).

within wide limits from one region of a device to another. An example is the **p-n junction** at the boundary between one region of a semiconductor with *p*-type impurities and another region containing *n*-type impurities. One way of fabricating a *p-n* junction is to deposit some *n*-type material on the *very* clean surface of some *p*-type material. (We can't just stick *p*- and *n*-type pieces together and expect the junction to work properly because of the impossibility of matching their surfaces at the atomic level.)

When a *p-n* junction is connected to an external circuit, as in Fig. 42.29a, and the potential difference $V_p - V_n = V$ across the junction is varied, the current I varies as shown in Fig. 42.29b. In striking contrast to the symmetric behavior of resistors that obey Ohm's law and give a straight line on an *I–V* graph, a *p-n* junction conducts much more readily in the direction from *p* to *n* than the reverse. Such a (mostly) one-way device is called a **diode rectifier.** Later we'll discuss a simple model of *p-n* junction behavior that predicts a current–voltage relationship in the form

$$I = I_S(e^{eV/kT} - 1) \qquad \text{(current through a } p\text{-}n \text{ junction)} \qquad (42.22)$$

In the exponent, $e = 1.602 \times 10^{-19}$ C is the quantum of charge, k is Boltzmann's constant, and T is absolute temperature.

CAUTION **Two different uses of** *e* In $e^{eV/kT}$ the base of the exponent also uses the symbol *e*, standing for the base of the natural logarithms, 2.71828 . . . . This *e* is quite different from $e = 1.602 \times 10^{-19}$ C in the exponent. ▮

Equation (42.22) is valid for both positive and negative values of V; note that V and I always have the same sign. As V becomes very negative, I approaches the value $-I_S$. The magnitude I_S (always positive) is called the *saturation current.*

Currents Through a *p-n* Junction

We can understand the behavior of a *p-n* junction diode qualitatively on the basis of the mechanisms for conductivity in the two regions. Suppose, as in Fig. 42.29a, you connect the positive terminal of the battery to the *p* region and the negative terminal to the *n* region. Then the *p* region is at higher potential than the *n* region, corresponding to positive V in Eq. (42.22), and the resulting electric field is in the direction *p* to *n*. This is called the *forward* direction, and the positive potential difference is called *forward bias.* Holes, plentiful in the *p* region, flow easily across the junction into the *n* region, and free electrons, plentiful in the *n* region, easily flow into the *p* region; these movements of charge constitute a *forward* current. Connecting the battery with the opposite polarity gives *reverse bias,* and the field tends to push electrons from *p* to *n* and holes from *n* to *p*. But there are very few free electrons in the *p* region and very few holes in the *n* region. As a result, the current in the *reverse* direction is much smaller than that with the same potential difference in the forward direction.

42.30 A *p-n* junction in equilibrium, with no externally applied field or potential difference. The generation and recombination currents exactly balance. The Fermi energy E_F is the same on both sides of the junction. The excess positive and negative charges on the *n* and *p* sides produce an electric field $\vec{E}$ in the direction shown.

The *p* side has an excess of negative charge and is at a lower electric potential, so negatively charged electrons have higher energy bands here.

The *n* side has an excess of positive charge and is at a higher electric potential, so negatively charged electrons have lower energy bands here.

Suppose you have a box with a barrier separating the left and right sides: You fill the left side with oxygen gas and the right side with nitrogen gas. What happens if the barrier leaks? Oxygen diffuses to the right, and nitrogen diffuses to the left. A similar diffusion occurs across a *p-n* junction. First consider the equilibrium situation with no applied voltage (Fig. 42.30). The many holes in the *p* region act like a hole gas that diffuses across the junction into the *n* region. Once there, the holes recombine with some of the many free electrons. Similarly, electrons diffuse from the *n* region to the *p* region and fall into some of the many holes there. The hole and electron diffusion currents lead to a net positive charge in the *n* region and a net negative charge in the *p* region, causing an electric field in the direction from *n* to *p* at the junction. The potential energy associated with this field raises the electron energy levels in the *p* region relative to the same levels in the *n* region.

There are four currents across the junction, as shown. The diffusion processes lead to *recombination currents* of holes and electrons, labeled i_{pr} and i_{nr} in Fig. 42.30. At the same time, electron–hole pairs are generated in the junction region by thermal excitation. The electric field described above sweeps these electrons and holes out of the junction; electrons are swept opposite the field to the *n* side, and holes are swept in the same direction as the field to the *p* side. The corresponding currents, called *generation currents,* are labeled i_{pg} and i_{ng}. At equilibrium the magnitudes of the generation and recombination currents are equal:

$$|i_{pg}| = |i_{pr}| \quad \text{and} \quad |i_{ng}| = |i_{nr}| \tag{42.23}$$

In thermal equilibrium the Fermi energy is the same at each point across the junction.

Now we apply a forward bias—that is, a positive potential difference *V* across the junction. A forward bias *decreases* the electric field in the junction region. It also decreases the difference between the energy levels on the *p* and *n* sides (Fig. 42.31)

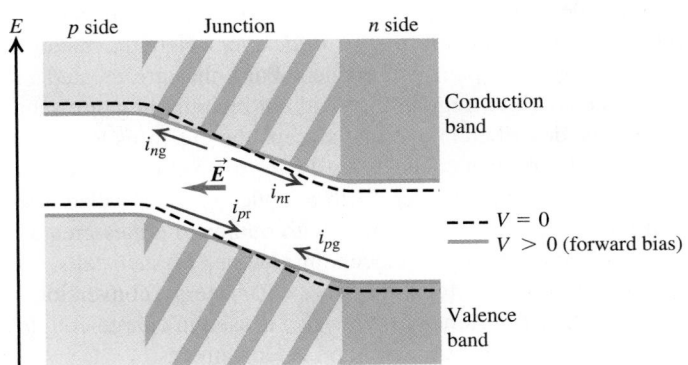

42.31 A *p-n* junction under forward-bias conditions. The potential difference between *p* and *n* regions is reduced, as is the electric field within the junction. The recombination currents increase but the generation currents are nearly constant, causing a net current from left to right. (Compare Fig. 42.30.)

by an amount $\Delta E = -eV$. It becomes easier for the electrons in the n region to climb the potential-energy hill and diffuse into the p region and for the holes in the p region to diffuse into the n region. This effect increases both recombination currents by the Maxwell–Boltzmann factor $e^{-\Delta E/kT} = e^{eV/kT}$. (We don't have to use the Fermi–Dirac distribution because most of the available states for the diffusing electrons and holes are empty, so the exclusion principle has little effect.) The generation currents don't change appreciably, so the net hole current is

$$
\begin{aligned}
i_{p\text{tot}} &= i_{pr} - |i_{pg}| \\
&= |i_{pg}|e^{eV/kT} - |i_{pg}| \\
&= |i_{pg}|(e^{eV/kT} - 1)
\end{aligned}
\tag{42.24}
$$

The net electron current $i_{n\text{tot}}$ is given by a similar expression, so the total current $I = i_{p\text{tot}} + i_{n\text{tot}}$ is

$$
I = I_S(e^{eV/kT} - 1)
\tag{42.25}
$$

in agreement with Eq. (42.22). We can repeat this entire discussion for reverse bias (negative V and I) with the same result. Therefore Eq. (42.22) is valid for both positive and negative values.

Several effects make the behavior of practical p-n junction diodes more complex than this simple analysis predicts. One effect, *avalanche breakdown,* occurs under large reverse bias. The electric field in the junction is so great that the carriers can gain enough energy between collisions to create electron–hole pairs during inelastic collisions. The electrons and holes then gain energy and collide to form more pairs, and so on. (A similar effect occurs in dielectric breakdown in insulators, discussed in Section 42.4.)

A second type of breakdown begins when the reverse bias becomes large enough that the top of the valence band in the p region is just higher in energy than the bottom of the conduction band in the n region (Fig. 42.32). If the junction region is thin enough, the probability becomes large that electrons can *tunnel* from the valence band of the p region to the conduction band of the n region. This process is called *Zener breakdown.* It occurs in Zener diodes, which are used for voltage regulation and protection against voltage surges.

Semiconductor Devices and Light

A *light-emitting diode (LED)* is a p-n junction diode that emits light. When the junction is forward biased, many holes are pushed from their p region to the junction region, and many electrons are pushed from their n region to the junction region. In the junction region the electrons fall into holes (recombine). In recombining, the electron can emit a photon with energy approximately equal to the band gap. This energy (and therefore the photon wavelength and the color of the light) can be varied by using materials with different band gaps. Light-emitting diodes are very energy-efficient light sources and have many applications, including automobile lamps, traffic signals, and large stadium displays.

The reverse process is called the *photovoltaic effect.* Here the material absorbs photons, and electron–hole pairs are created. Pairs that are created in the p-n junction, or close enough to migrate to it without recombining, are separated by the electric field we described above that sweeps the electrons to the n side and the holes to the p side. We can connect this device to an external circuit, where it becomes a source of emf and power. Such a device is often called a *solar cell,* although sunlight isn't required. *Any* light with photon energies greater than the band gap will do. You might have a calculator powered by such cells. Production of low-cost photovoltaic cells for large-scale solar energy conversion is a very active field of research. The same basic physics is used in charge-coupled device (CCD) image detectors, digital cameras, and video cameras.

42.32 Under reverse-bias conditions the potential-energy difference between the p and n sides of a junction is greater than at equilibrium. If this difference is great enough, the bottom of the conduction band on the n side may actually be below the top of the valence band on the p side.

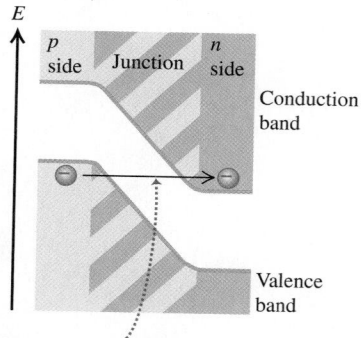

If a p-n junction under reverse bias is thin enough, electrons can tunnel from the valence band to the conduction band (a process called Zener breakdown).

Application Swallow This Semiconductor Device

This tiny capsule—designed to be swallowed by a patient—contains a miniature camera with a CCD light detector, plus six LEDs to illuminate the subject. The capsule radios high-resolution images to an external recording unit as it passes painlessly through the patient's stomach and intestines. This technique makes it possible to examine the small intestine, which is not readily accessible with conventional endoscopy.

Transistors

A *bipolar junction transistor* includes two *p-n* junctions in a "sandwich" configuration, which may be either *p-n-p* or *n-p-n*. Figure 42.33 shows such a *p-n-p* transistor. The three regions are called the emitter, base, and collector, as shown. When there is no current in the left loop of the circuit, there is only a very small current through the resistor *R* because the voltage across the base–collector junction is in the reverse direction. But when a forward bias is applied between emitter and base, as shown, most of the holes traveling from emitter to base travel *through* the base (which is typically both narrow and lightly doped) to the second junction, where they come under the influence of the collector-to-base potential difference and flow on through the collector to give an increased current to the resistor.

In this way the current in the collector circuit is *controlled* by the current in the emitter circuit. Furthermore, V_c may be considerably larger than V_e, so the *power* dissipated in *R* may be much larger than the power supplied to the emitter circuit by the battery V_e. Thus the device functions as a *power amplifier*. If the potential drop across *R* is greater than V_e, it may also be a voltage amplifier.

In this configuration the *base* is the common element between the "input" and "output" sides of the circuit. Another widely used arrangement is the *common-emitter* circuit, shown in Fig. 42.34. In this circuit the current in the collector side of the circuit is much larger than that in the base side, and the result is current amplification.

The *field-effect transistor* (Fig. 42.35) is an important type. In one variation a slab of *p*-type silicon is made with two *n*-type regions on the top, called the *source* and the *drain*; a metallic conductor is fastened to each. A third electrode called the *gate* is separated from the slab, source, and drain by an insulating layer of SiO_2. When there is no charge on the gate and a potential difference of either polarity is applied between the source and the drain, there is very little current because one of the *p-n* junctions is reverse biased.

Now we place a positive charge on the gate. With dimensions of the order of 10^{-6} m, it takes little charge to provide a substantial electric field. Thus there is very little current into or out of the gate. There aren't many free electrons in the *p*-type material, but there are some, and the effect of the field is to attract them toward the positive gate. The resulting greatly enhanced concentration of electrons near the gate (and between the two junctions) permits current to flow between the source and the drain. The current is very sensitive to the gate charge and potential, and the device functions as an amplifier. The device just described is called an *enhancement-type MOSFET* (metal-oxide-semiconductor field-effect transistor).

Integrated Circuits

A further refinement in semiconductor technology is the *integrated circuit*. By successively depositing layers of material and etching patterns to define current paths, we can combine the functions of several MOSFETs, capacitors, and resistors on a single square of semiconductor material that may be only a few millimeters on a side. An elaboration of this idea leads to *large-scale integrated circuits*. The resulting integrated circuit chips are the heart of all pocket calculators and present-day computers, large and small (Fig. 42.36).

42.33 Schematic diagram of a *p-n-p* transistor and circuit.

- When $V_e = 0$, the current is very small.
- When a potential V_e is applied between emitter and base, holes travel from the emitter to the base.
- When V_c is sufficiently large, most of the holes continue into the collector.

42.34 A common-emitter circuit.

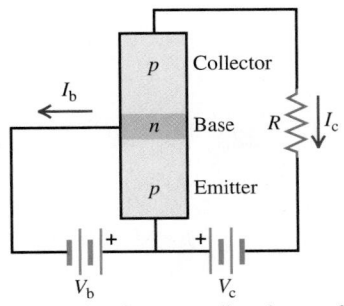

- When $V_b = 0$, I_c is very small, and most of the voltage V_c appears across the base–collector junction.
- As V_b increases, the base–collector potential decreases, and more holes can diffuse into the collector; thus, I_c increases. Ordinarily, I_c is much larger than I_b.

42.35 A field-effect transistor. The current from source to drain is controlled by the potential difference between the source and the drain and by the charge on the gate; no current flows through the gate.

42.36 An integrated circuit chip the size of your thumb can contain millions of transistors.

The first semiconductor devices were invented in 1947. Since then, they have completely revolutionized the electronics industry through miniaturization, reliability, speed, energy usage, and cost. They have found applications in communications, computer systems, control systems, and many other areas. In transforming these areas, they have changed, and continue to change, human civilization itself.

Test Your Understanding of Section 42.7 Suppose a negative charge is placed on the gate of the MOSFET shown in Fig. 42.35. Will a substantial current flow between the source and the drain?

42.8 Superconductivity

Superconductivity is the complete disappearance of all electrical resistance at low temperatures. We described this property at the end of Section 25.2 and the magnetic properties of type-I and type-II superconductors in Section 29.8. In this section we'll relate superconductivity to the structure and energy-band model of a solid.

Although superconductivity was discovered in 1911, it was not well understood on a theoretical basis until 1957. In that year, the American physicists John Bardeen, Leon Cooper, and Robert Schrieffer published the theory of superconductivity, now called the BCS theory, that was to earn them the Nobel Prize in physics in 1972. (It was Bardeen's second Nobel Prize; he shared his first for his work on the development of the transistor.) The key to the BCS theory is an interaction between *pairs* of conduction electrons, called *Cooper pairs,* caused by an interaction with the positive ions of the crystal. Here's a rough qualitative picture of what happens. A free electron exerts attractive forces on nearby positive ions, pulling them slightly closer together. The resulting slight concentration of positive charge then exerts an attractive force on another free electron with momentum opposite to the first. At ordinary temperatures this electron-pair interaction is very small in comparison to energies of thermal motion, but at very low temperatures it becomes significant.

Bound together this way, the pairs of electrons cannot *individually* gain or lose very small amounts of energy, as they would ordinarily be able to do in a partly filled conduction band. Their pairing gives an energy gap in the allowed electron quantum levels, and at low temperatures there is not enough collision energy to jump this gap. Therefore the electrons can move freely through the crystal without any energy exchange through collisions—that is, with zero resistance.

Researchers have not yet reached a consensus on whether some modification of the BCS theory can explain the properties of the high-T_C superconductors that have been discovered since 1986. There *is* evidence for pairing, but of a different sort than for conventional superconductors. Furthermore, the original pairing mechanism of the BCS theory seems too weak to explain the high transition temperatures and critical fields of these new superconductors.

CHAPTER 42 SUMMARY

Molecular bonds and molecular spectra: The principal types of molecular bonds are ionic, covalent, van der Waals, and hydrogen bonds. In a diatomic molecule the rotational energy levels are given by Eq. (42.3), where I is the moment of inertia of the molecule, m_r is its reduced mass, and r_0 is the distance between the two atoms. The vibrational energy levels are given by Eq. (42.7), where k' is the effective force constant of the interatomic force. (See Examples 42.1–42.3.)

$$E_l = l(l + 1)\frac{\hbar^2}{2I} \quad (l = 0, 1, 2, \dots)$$
(42.3)

$$I = m_r r_0^2$$
(42.6)

$$m_r = \frac{m_1 m_2}{m_1 + m_2}$$
(42.4)

$$E_n = \left(n + \tfrac{1}{2}\right)\hbar\omega = \left(n + \tfrac{1}{2}\right)\hbar\sqrt{\frac{k'}{m_r}}$$

$$(n = 0, 1, 2, \dots)$$
(42.7)

Solids and energy bands: Interatomic bonds in solids are of the same types as in molecules plus one additional type, the metallic bond. Associating the basis with each lattice point gives the crystal structure. (See Example 42.4.)

When atoms are bound together in condensed matter, their outer energy levels spread out into bands. At absolute zero, insulators and conductors have a completely filled valence band separated by an energy gap from an empty conduction band. Conductors, including metals, have partially filled conduction bands. (See Example 42.5.)

Free-electron model of metals: In the free-electron model of the behavior of conductors, the electrons are treated as completely free particles within the conductor. In this model the density of states is given by Eq. (42.15). The probability that an energy state of energy E is occupied is given by the Fermi–Dirac distribution, Eq. (42.16), which is a consequence of the exclusion principle. In Eq. (42.16), E_F is the Fermi energy. (See Examples 42.6–42.8.)

$$g(E) = \frac{(2m)^{3/2}V}{2\pi^2\hbar^3}E^{1/2}$$
(42.15)

$$f(E) = \frac{1}{e^{(E-E_F)/kT} + 1}$$
(42.16)

Semiconductors: A semiconductor has an energy gap of about 1 eV between its valence and conduction bands. Its electrical properties may be drastically changed by the addition of small concentrations of donor impurities, giving an n-type semiconductor, or acceptor impurities, giving a p-type semiconductor. (See Example 42.9.)

Semiconductor devices: Many semiconductor devices, including diodes, transistors, and integrated circuits, use one or more p-n junctions. The current–voltage relationship for an ideal p-n junction diode is given by Eq. (42.22).

$$I = I_S(e^{eV/kT} - 1)$$
(42.22)

BRIDGING PROBLEM   Detecting Infrared Photons

At 80 K, the band gap in the semiconductor indium antimonide (InSb) is 0.230 eV. A photon emitted by a hydrogen fluoride (HF) molecule undergoing a vibration-rotation transition from ($n = 1$, $l = 0$) to ($n = 0, l = 1$) is absorbed by an electron at the top of the valence band of InSb. (a) How far above the top of the band gap (in eV) is the final state of the electron? (b) What is the probability that the final state was already occupied? The vibration frequency for HF is 1.24×10^{14} Hz, the mass of a hydrogen atom is 1.67×10^{-27} kg, the mass of a fluorine atom is 3.15×10^{-26} kg, and the equilibrium distance between the two nuclei is 0.092 nm. Assume that the Fermi energy for InSb is in the middle of the gap.

SOLUTION GUIDE

See MasteringPhysics® study area for a Video Tutor solution. **MP**

IDENTIFY and SET UP

1. This problem involves what you learned about molecular transitions in Section 42.2, about the Fermi–Dirac distribution in Section 42.5, and about semiconductors in Section 42.6.
2. Equation (42.9) gives the combined vibrational-rotational energy in the initial and final molecular states. The difference between the initial and final molecular energies equals the energy E of the emitted photon, which is in turn equal to the energy gained by the InSb valence electron when it absorbs that

photon. The probability that the final state is occupied is given by the Fermi–Dirac distribution, Eq. (42.16).

EXECUTE

3. Before you can use Eq. (42.9), you'll first need to use the data given to calculate the moment of inertia I and the quantity $\hbar\omega$ for the HF molecule. (*Hint:* Be careful not to confuse frequency f and angular frequency ω.)
4. Use your results from step 3 to calculate the initial and final energies of the HF molecule. (*Hint:* Does the vibrational energy increase or decrease? What about the rotational energy?)
5. Use your result from step 4 to find the energy imparted to the InSb electron. Determine the final energy of this electron relative to the bottom of the conduction band.
6. Use your result from step 5 to determine the probability that the InSb final state is already occupied.

EVALUATE

7. Is the molecular transition of the HF molecule allowed? Which is larger: the vibrational energy change or the rotational energy change?
8. Is it likely that the excited InSb electron will be blocked from entering a state in the conduction band?

Problems

For instructor-assigned homework, go to www.masteringphysics.com

•, ••, •••: Problems of increasing difficulty. **CP:** Cumulative problems incorporating material from earlier chapters. **CALC:** Problems requiring calculus. **BIO:** Biosciences problems.

DISCUSSION QUESTIONS

Q42.1 Ionic bonds result from the electrical attraction of oppositely charged particles. Are other types of molecular bonds also electrical in nature, or is some other interaction involved? Explain.

Q42.2 In ionic bonds, an electron is transferred from one atom to another and thus no longer "belongs" to the atom from which it came. Are there similar transfers of ownership of electrons with other types of molecular bonds? Explain.

Q42.3 Van der Waals bonds occur in many molecules, but hydrogen bonds occur only with materials that contain hydrogen. Why is this type of bond unique to hydrogen?

Q42.4 The bonding of gallium arsenide (GaAs) is said to be 31% ionic and 69% covalent. Explain.

Q42.5 The H_2^+ molecule consists of two hydrogen nuclei and a single electron. What kind of molecular bond do you think holds this molecule together? Explain.

Q42.6 The moment of inertia for an axis through the center of mass of a diatomic molecule calculated from the wavelength emitted in an $l = 19 \rightarrow l = 18$ transition is different from the moment of inertia calculated from the wavelength of the photon emitted in an $l = 1 \rightarrow l = 0$ transition. Explain this difference. Which transition corresponds to the larger moment of inertia?

Q42.7 Analysis of the photon absorption spectrum of a diatomic molecule shows that the vibrational energy levels for small values

of n are very nearly equally spaced but the levels for large n are not equally spaced. Discuss the reason for this observation. Do you expect the adjacent levels to move closer together or farther apart as n increases? Explain.

Q42.8 Discuss the differences between the rotational and vibrational energy levels of the deuterium ("heavy hydrogen") molecule D_2 and those of the ordinary hydrogen molecule H_2. A deuterium atom has twice the mass of an ordinary hydrogen atom.

Q42.9 Various organic molecules have been discovered in interstellar space. Why were these discoveries made with radio telescopes rather than optical telescopes?

Q42.10 The air you are breathing contains primarily nitrogen (N_2) and oxygen (O_2). Many of these molecules are in excited rotational energy levels ($l = 1, 2, 3, \ldots$), but almost all of them are in the vibrational ground level ($n = 0$). Explain this difference between the rotational and vibrational behaviors of the molecules.

Q42.11 In what ways do atoms in a diatomic molecule behave as though they were held together by a spring? In what ways is this a poor description of the interaction between the atoms?

Q42.12 Individual atoms have discrete energy levels, but certain solids (which are made up of only individual atoms) show energy bands and gaps. What causes the solids to behave so differently from the atoms of which they are composed?

Q42.13 What factors determine whether a material is a conductor of electricity or an insulator? Explain.

Q42.14 Ionic crystals are often transparent, whereas metallic crystals are always opaque. Why?

Q42.15 Speeds of molecules in a gas vary with temperature, whereas speeds of electrons in the conduction band of a metal are nearly independent of temperature. Why are these behaviors so different?

Q42.16 Use the band model to explain how it is possible for some materials to undergo a semiconductor-to-metal transition as the temperature or pressure varies.

Q42.17 An isolated zinc atom has a ground-state electron configuration of filled $1s$, $2s$, $2p$, $3s$, $3p$, and $4s$ subshells. How can zinc be a conductor if its valence subshell is full?

Q42.18 The assumptions of the *free-electron model* of metals may seem contrary to reason, since electrons exert powerful electrical forces on each other. Give some reasons why these assumptions actually make physical sense.

Q42.19 Why are materials that are good thermal conductors also good electrical conductors? What kinds of problems does this pose for the design of appliances such as clothes irons and electric heaters? Are there materials that do not follow this general rule?

Q42.20 What is the essential characteristic for an element to serve as a donor impurity in a semiconductor such as Si or Ge? For it to serve as an acceptor impurity? Explain.

Q42.21 There are several methods for removing electrons from the surface of a semiconductor. Can holes be removed from the surface? Explain.

Q42.22 A student asserts that silicon and germanium become good insulators at very low temperatures and good conductors at very high temperatures. Do you agree? Explain your reasoning.

Q42.23 The electrical conductivities of most metals decrease gradually with increasing temperature, but the intrinsic conductivity of semiconductors always *increases* rapidly with increasing temperature. What causes the difference?

Q42.24 How could you make compensated silicon that has twice as many acceptors as donors?

Q42.25 For electronic devices such as amplifiers, what are some advantages of transistors compared to vacuum tubes? What are some disadvantages? Are there any situations in which vacuum tubes *cannot* be replaced by solid-state devices? Explain your reasoning.

Q42.26 Why does tunneling limit the miniaturization of MOSFETs?

Q42.27 The saturation current I_S for a *p-n* junction, Eq. (42.22), depends strongly on temperature. Explain why.

EXERCISES

Section 42.1 Types of Molecular Bonds

42.1 • If the energy of the H_2 covalent bond is -4.48 eV, what wavelength of light is needed to break that molecule apart? In what part of the electromagnetic spectrum does this light lie?

42.2 • **An Ionic Bond.** (a) Calculate the electric potential energy for a K^+ ion and a Br^- ion separated by a distance of 0.29 nm, the equilibrium separation in the KBr molecule. Treat the ions as point charges. (b) The ionization energy of the potassium atom is 4.3 eV. Atomic bromine has an electron affinity of 3.5 eV. Use these data and the results of part (a) to estimate the binding energy of the KBr molecule. Do you expect the actual binding energy to be higher or lower than your estimate? Explain your reasoning.

42.3 • We know from Chapter 18 that the average kinetic energy of an ideal-gas atom or molecule at Kelvin temperature T is $\frac{3}{2}kT$. For what value of T does this energy correspond to (a) the bond energy of the van der Waals bond in He_2 (7.9×10^{-4} eV) and (b) the bond energy of the covalent bond in H_2 (4.48 eV)? (c) The kinetic energy in a collision between molecules can go into dissociating one or both molecules, provided the kinetic energy is higher than the bond energy. At room temperature (300 K), is it likely that He_2 molecules will remain intact after a collision? What about H_2 molecules? Explain.

42.4 •• Light of wavelength 3.10 mm strikes and is absorbed by a molecule. Is this process most likely to alter the rotational, vibrational, or atomic energy levels of the molecule? Explain your reasoning. (b) If the light in part (a) had a wavelength of 207 nm, which energy levels would it most likely affect? Explain.

42.5 • For the H_2 molecule the equilibrium spacing of the two protons is 0.074 nm. The mass of a hydrogen atom is 1.67×10^{-27} kg. Calculate the wavelength of the photon emitted in the rotational transition $l = 2$ to $l = 1$.

42.6 • (a) A molecule decreases its vibrational energy by 0.250 eV by giving up a photon of light. What wavelength of light does it give up during this process, and in what part of the electromagnetic spectrum does that wavelength of light lie? (b) An atom decreases its energy by 8.50 eV by giving up a photon of light. What wavelength of light does it give up during this process, and in what part of the electromagnetic spectrum does that wavelength of light lie? (c) A molecule decreases its rotational energy by 3.20×10^{-3} eV by giving up a photon of light. What wavelength of light does it give up during this process, and in what part of the electromagnetic spectrum does that wavelength of light lie?

Section 42.2 Molecular Spectra

42.7 • A hypothetical NH molecule makes a rotational-level transition from $l = 3$ to $l = 1$ and gives off a photon of wavelength 1.780 nm in doing so. What is the separation between the two atoms in this molecule if we model them as point masses? The mass of hydrogen is 1.67×10^{-27} kg, and the mass of nitrogen is 2.33×10^{-26} kg.

42.8 • The water molecule has an $l = 1$ rotational level 1.01×10^{-5} eV above the $l = 0$ ground level. Calculate the wavelength and frequency of the photon absorbed by water when it undergoes a rotational-level transition from $l = 0$ to $l = 1$. The magnetron oscillator in a microwave oven generates microwaves with a frequency of 2450 MHz. Does this make sense, in view of the frequency you calculated in this problem? Explain.

42.9 • In Example 42.2 the moment of inertia for CO was calculated using Eq. (42.6). (a) In CO, how far is each atom from the center of mass of the molecule? (b) Use $I = m_1 r_1^2 + m_2 r_2^2$ to calculate the moment of inertia of CO about an axis through the center of mass and perpendicular to the line joining the centers of the two atoms. Does your result agree with the value obtained in Example 42.2?

42.10 • Two atoms of cesium (Cs) can form a Cs_2 molecule. The equilibrium distance between the nuclei in a Cs_2 molecule is 0.447 nm. Calculate the moment of inertia about an axis through the center of mass of the two nuclei and perpendicular to the line joining them. The mass of a cesium atom is 2.21×10^{-25} kg.

42.11 •• **CP** The rotational energy levels of CO are calculated in Example 42.2. If the energy of the rotating molecule is described by the classical expression $K = \frac{1}{2}I\omega^2$, for the $l = 1$ level what are (a) the angular speed of the rotating molecule; (b) the linear speed of each atom (use the result of Exercise 42.9); (c) the rotational period (the time for one rotation)?

42.12 • If a sodium chloride (NaCl) molecule could undergo an $n \to n - 1$ vibrational transition with no change in rotational quantum number, a photon with wavelength 20.0 μm would be emitted. The mass of a sodium atom is 3.82×10^{-26} kg, and the mass of a chlorine atom is 5.81×10^{-26} kg. Calculate the force constant k' for the interatomic force in NaCl.

42.13 • A lithium atom has mass 1.17×10^{-26} kg, and a hydrogen atom has mass 1.67×10^{-27} kg. The equilibrium separation between the two nuclei in the LiH molecule is 0.159 nm. (a) What is the difference in energy between the $l = 3$ and $l = 4$ rotational levels? (b) What is the wavelength of the photon emitted in a transition from the $l = 4$ to the $l = 3$ level?

42.14 •• When a hypothetical diatomic molecule having atoms 0.8860 nm apart undergoes a rotational transition from the $l = 2$ state to the next lower state, it gives up a photon having energy 8.841×10^{-4} eV. When the molecule undergoes a vibrational transition from one energy state to the next lower energy state, it gives up 0.2560 eV. Find the force constant of this molecule.

42.15 • (a) Show that the energy difference between rotational levels with angular-momentum quantum numbers l and $l - 1$ is $l\hbar^2/I$. (b) In terms of l, $\hbar$, and I, what is the frequency of the photon emitted in the pure rotation transition $l \to l - 1$?

42.16 • The vibrational and rotational energies of the CO molecule are given by Eq. (42.9). Calculate the wavelength of the photon absorbed by CO in each of the following vibration–rotation transitions: (a) $n = 0$, $l = 1 \to n = 1$, $l = 2$; (b) $n = 0$, $l = 2 \to n = 1$, $l = 1$; (c) $n = 0$, $l = 3 \to n = 1$, $l = 2$.

Section 42.3 Structure of Solids

42.17 • **Density of NaCl.** The spacing of adjacent atoms in a crystal of sodium chloride is 0.282 nm. The mass of a sodium atom is 3.82×10^{-26} kg, and the mass of a chlorine atom is 5.89×10^{-26} kg. Calculate the density of sodium chloride.

42.18 • Potassium bromide (KBr) has a density of 2.75×10^3 kg/m^3 and the same crystal structure as NaCl. The mass of a potassium atom is 6.49×10^{-26} kg, and the mass of a bromine atom is 1.33×10^{-25} kg. (a) Calculate the average spacing between adjacent atoms in a KBr crystal. (b) How does the value calculated in part (a) compare with the spacing in NaCl (see Exercise 42.17)? Is the relationship between the two values qualitatively what you would expect? Explain.

Section 42.4 Energy Bands

42.19 • The maximum wavelength of light that a certain silicon photocell can detect is 1.11 μm. (a) What is the energy gap (in electron volts) between the valence and conduction bands for this photocell? (b) Explain why pure silicon is opaque.

42.20 • The gap between valence and conduction bands in diamond is 5.47 eV. (a) What is the maximum wavelength of a photon that can excite an electron from the top of the valence band into the conduction band? In what region of the electromagnetic spectrum does this photon lie? (b) Explain why pure diamond is transparent and colorless. (c) Most gem diamonds have a yellow color. Explain how impurities in the diamond can cause this color.

42.21 • The gap between valence and conduction bands in silicon is 1.12 eV. A nickel nucleus in an excited state emits a gamma-ray photon with wavelength 9.31×10^{-4} nm. How many electrons can be excited from the top of the valence band to the bottom of the conduction band by the absorption of this gamma ray?

Section 42.5 Free-Electron Model of Metals

42.22 • Calculate v_{rms} for free electrons with average kinetic energy $\frac{3}{2}kT$ at a temperature of 300 K. How does your result compare to the speed of an electron with a kinetic energy equal to the Fermi energy of copper, calculated in Example 42.7? Why is there such a difference between these speeds?

42.23 • Calculate the density of states $g(E)$ for the free-electron model of a metal if $E = 7.0$ eV and $V = 1.0$ cm^3. Express your answer in units of states per electron volt.

42.24 • Supply the details in the derivation of Eq. (42.13) from Eqs. (42.11) and (42.12).

42.25 • CP Silver has a Fermi energy of 5.48 eV. Calculate the electron contribution to the molar heat capacity at constant volume of silver, C_V, at 300 K. Express your result (a) as a multiple of R and (b) as a fraction of the actual value for silver, $C_V = 25.3$ J/mol $\cdot$ K. (c) Is the value of C_V due principally to the electrons? If not, to what is it due? (*Hint:* See Section 18.4.)

42.26 • The Fermi energy of sodium is 3.23 eV. (a) Find the average energy E_{av} of the electrons at absolute zero. (b) What is the speed of an electron that has energy E_{av}? (c) At what Kelvin temperature T is kT equal to E_F? (This is called the *Fermi temperature* for the metal. It is approximately the temperature at which molecules in a classical ideal gas would have the same kinetic energy as the fastest-moving electron in the metal.)

42.27 •• For a solid metal having a Fermi energy of 8.500 eV, what is the probability, at room temperature, that a state having an energy of 8.520 eV is occupied by an electron?

Section 42.6 Semiconductors

42.28 • Pure germanium has a band gap of 0.67 eV. The Fermi energy is in the middle of the gap. (a) For temperatures of 250 K, 300 K, and 350 K, calculate the probability $f(E)$ that a state at the bottom of the conduction band is occupied. (b) For each temperature in part (a), calculate the probability that a state at the top of the valence band is empty.

42.29 • Germanium has a band gap of 0.67 eV. Doping with arsenic adds donor levels in the gap 0.01 eV below the bottom of the conduction band. At a temperature of 300 K, the probability is 4.4×10^{-4} that an electron state is occupied at the bottom of the conduction band. Where is the Fermi level relative to the conduction band in this case?

Section 42.7 Semiconductor Devices

42.30 •• (a) Suppose a piece of very pure germanium is to be used as a light detector by observing, through the absorption of photons, the increase in conductivity resulting from generation of electron–hole pairs. If each pair requires 0.67 eV of energy, what is the maximum wavelength that can be detected? In what portion of the spectrum does it lie? (b) What are the answers to part (a) if the material is silicon, with an energy requirement of 1.14 eV per pair, corresponding to the gap between valence and conduction bands in that element?

42.31 • CP At a temperature of 290 K, a certain p-n junction has a saturation current $I_S = 0.500$ mA. (a) Find the current at this temperature when the voltage is (i) 1.00 mV, (ii) -1.00 mV, (iii) 100 mV, and (iv) -100 mV. (b) Is there a region of applied voltage where the diode obeys Ohm's law?

42.32 • For a certain p-n junction diode, the saturation current at room temperature (20°C) is 0.750 mA. What is the resistance of this diode when the voltage across it is (a) 85.0 mV and (b) -50.0 mV?

42.33 •• (a) A forward-bias voltage of 15.0 mV produces a positive current of 9.25 mA through a *p-n* junction at 300 K. What does the positive current become if the forward-bias voltage is reduced to 10.0 mV? (b) For reverse-bias voltages of −15.0 mV and −10.0 mV, what is the reverse-bias negative current?

42.34 •• A *p-n* junction has a saturation current of 3.60 mA. (a) At a temperature of 300 K, what voltage is needed to produce a positive current of 40.0 mA? (b) For a voltage equal to the negative of the value calculated in part (a), what is the negative current?

PROBLEMS

42.35 •• A hypothetical diatomic molecule of oxygen (mass = 2.656×10^{-26} kg) and hydrogen (mass = 1.67×10^{-27} kg) emits a photon of wavelength 2.39 μm when it makes a transition from one vibrational state to the next lower state. If we model this molecule as two point masses at opposite ends of a massless spring, (a) what is the force constant of this spring, and (b) how many vibrations per second is the molecule making?

42.36 • When a diatomic molecule undergoes a transition from the $l = 2$ to the $l = 1$ rotational state, a photon with wavelength 63.8 μm is emitted. What is the moment of inertia of the molecule for an axis through its center of mass and perpendicular to the line connecting the nuclei?

42.37 •• **CP** (a) The equilibrium separation of the two nuclei in an NaCl molecule is 0.24 nm. If the molecule is modeled as charges $+e$ and $-e$ separated by 0.24 nm, what is the electric dipole moment of the molecule (see Section 21.7)? (b) The measured electric dipole moment of an NaCl molecule is 3.0×10^{-29} C·m. If this dipole moment arises from point charges $+q$ and $-q$ separated by 0.24 nm, what is q? (c) A definition of the *fractional ionic character* of the bond is q/e. If the sodium atom has charge $+e$ and the chlorine atom has charge $-e$, the fractional ionic character would be equal to 1. What is the actual fractional ionic character for the bond in NaCl? (d) The equilibrium distance between nuclei in the hydrogen iodide (HI) molecule is 0.16 nm, and the measured electric dipole moment of the molecule is 1.5×10^{-30} C·m. What is the fractional ionic character for the bond in HI? How does your answer compare to that for NaCl calculated in part (c)? Discuss reasons for the difference in these results.

42.38 • The binding energy of a potassium chloride molecule (KCl) is 4.43 eV. The ionization energy of a potassium atom is 4.3 eV, and the electron affinity of chlorine is 3.6 eV. Use these data to estimate the equilibrium separation between the two atoms in the KCl molecule. Explain why your result is only an estimate and not a precise value.

42.39 • (a) For the sodium chloride molecule (NaCl) discussed at the beginning of Section 42.1, what is the maximum separation of the ions for stability if they may be regarded as point charges? That is, what is the largest separation for which the energy of an Na^+ ion and a Cl^- ion, calculated in this model, is lower than the energy of the two separate atoms Na and Cl? (b) Calculate this distance for the potassium bromide molecule, described in Exercise 42.2.

42.40 •• The rotational spectrum of HCl contains the following wavelengths (among others): 60.4 μm, 69.0 μm, 80.4 μm, 96.4 μm, and 120.4 μm. Use this spectrum to find the moment of inertia of the HCl molecule about an axis through the center of mass and perpendicular to the line joining the two nuclei.

42.41 • (a) Use the result of Problem 42.40 to calculate the equilibrium separation of the atoms in an HCl molecule. The mass of a chlorine atom is 5.81×10^{-26} kg, and the mass of a hydrogen

atom is 1.67×10^{-27} kg. (b) The value of l changes by ± 1 in rotational transitions. What is the value of l for the upper level of the transition that gives rise to each of the wavelengths listed in Problem 42.40? (c) What is the longest-wavelength line in the rotational spectrum of HCl? (d) Calculate the wavelengths of the emitted light for the corresponding transitions in the deuterium chloride (DCl) molecule. In this molecule the hydrogen atom in HCl is replaced by an atom of deuterium, an isotope of hydrogen with a mass of 3.34×10^{-27} kg. Assume that the equilibrium separation between the atoms is the same as for HCl.

42.42 • When a NaF molecule makes a transition from the $l = 3$ to the $l = 2$ rotational level with no change in vibrational quantum number or electronic state, a photon with wavelength 3.83 mm is emitted. A sodium atom has mass 3.82×10^{-26} kg, and a fluorine atom has mass 3.15×10^{-26} kg. Calculate the equilibrium separation between the nuclei in a NaF molecule. How does your answer compare with the value for NaCl given in Section 42.1? Is this result reasonable? Explain.

42.43 •• **CP** Consider a gas of diatomic molecules (moment of inertia I) at an absolute temperature T. If E_g is a ground-state energy and E_{ex} is the energy of an excited state, then the Maxwell–Boltzmann distribution (see Section 39.4) predicts that the ratio of the numbers of molecules in the two states is

$$\frac{n_{ex}}{n_g} = e^{-(E_{ex}-E_g)/kT}$$

(a) Explain why the ratio of the number of molecules in the lth rotational energy *level* to the number of molecules in the ground ($l = 0$) rotational level is

$$\frac{n_l}{n_0} = (2l + 1)e^{-[l(l+1)\hbar^2]/2lkT}$$

(*Hint:* For each value of l, how many states are there with different values of m_l?) (b) Determine the ratio n_l/n_0 for a gas of CO molecules at 300 K for the cases (i) $l = 1$; (ii) $l = 2$; (iii) $l = 10$; (iv) $l = 20$; (v) $l = 50$. The moment of inertia of the CO molecule is given in Example 42.2 (Section 42.2). (c) Your results in part (b) show that as l is increased, the ratio n_l/n_0 first increases and then decreases. Explain why.

42.44 •• Our galaxy contains numerous *molecular clouds*, regions many light-years in extent in which the density is high enough and the temperature low enough for atoms to form into molecules. Most of the molecules are H_2, but a small fraction of the molecules are carbon monoxide (CO). Such a molecular cloud in the constellation Orion is shown in Fig. P42.44. The left-hand image was made with an ordinary visible-light telescope; the right-hand image shows the molecular cloud in Orion as imaged with a radio telescope tuned to a wavelength emitted by CO in a rotational transition. The different colors in the radio image indicate regions of the cloud that are moving either toward us (blue) or away from us (red) relative to the motion of the cloud as a whole, as determined by the Doppler shift of the radiation. (Since a

Figure **P42.44**

molecular cloud has about 10,000 hydrogen molecules for each CO molecule, it might seem more reasonable to tune a radio telescope to emissions from H_2 than to emissions from CO. Unfortunately, it turns out that the H_2 molecules in molecular clouds do not radiate in either the radio or visible portions of the electromagnetic spectrum.) (a) Using the data in Example 42.2 (Section 42.2), calculate the energy and wavelength of the photon emitted by a CO molecule in an $l = 1 \rightarrow l = 0$ rotational transition. (b) As a rule, molecules in a gas at temperature T will be found in a certain excited rotational energy level provided the energy of that level is no higher than kT (see Problem 42.43). Use this rule to explain why astronomers can detect radiation from CO in molecular clouds even though the typical temperature of a molecular cloud is a very low 20 K.

42.45 • **Spectral Lines from Isotopes.** The equilibrium separation for NaCl is 0.2361 nm. The mass of a sodium atom is 3.8176×10^{-26} kg. Chlorine has two stable isotopes, ^{35}Cl and ^{37}Cl, that have different masses but identical chemical properties. The atomic mass of ^{35}Cl is 5.8068×10^{-26} kg, and the atomic mass of ^{37}Cl is 6.1384×10^{-26} kg. (a) Calculate the wavelength of the photon emitted in the $l = 2 \rightarrow l = 1$ and $l = 1 \rightarrow l = 0$ transitions for $Na^{35}Cl$. (b) Repeat part (a) for $Na^{37}Cl$. What are the differences in the wavelengths for the two isotopes?

42.46 • When an OH molecule undergoes a transition from the $n = 0$ to the $n = 1$ vibrational level, its internal vibrational energy increases by 0.463 eV. Calculate the frequency of vibration and the force constant for the interatomic force. (The mass of an oxygen atom is 2.66×10^{-26} kg, and the mass of a hydrogen atom is 1.67×10^{-27} kg.)

42.47 • The force constant for the internuclear force in a hydrogen molecule (H_2) is $k' = 576$ N/m. A hydrogen atom has mass 1.67×10^{-27} kg. Calculate the zero-point vibrational energy for H_2 (that is, the vibrational energy the molecule has in the $n = 0$ ground vibrational level). How does this energy compare in magnitude with the H_2 bond energy of -4.48 eV?

42.48 • Suppose the hydrogen atom in HF (see the Bridging Problem for this chapter) is replaced by an atom of deuterium, an isotope of hydrogen with a mass of 3.34×10^{-27} kg. The force constant is determined by the electron configuration, so it is the same as for the normal HF molecule. (a) What is the vibrational frequency of this molecule? (b) What wavelength of light corresponds to the energy difference between the $n = 1$ and $n = 0$ levels? In what region of the spectrum does this wavelength lie?

42.49 • The hydrogen iodide (HI) molecule has equilibrium separation 0.160 nm and vibrational frequency 6.93×10^{13} Hz. The mass of a hydrogen atom is 1.67×10^{-27} kg, and the mass of an iodine atom is 2.11×10^{-25} kg. (a) Calculate the moment of inertia of HI about a perpendicular axis through its center of mass. (b) Calculate the wavelength of the photon emitted in each of the following vibration–rotation transitions: (i) $n = 1$, $l = 1 \rightarrow n = 0$, $l = 0$; (ii) $n = 1$, $l = 2 \rightarrow n = 0$, $l = 1$; (iii) $n = 2$, $l = 2 \rightarrow n = 1, l = 3$.

42.50 •• Prove this statement: For free electrons in a solid, if a state that is at an energy ΔE above E_F has probability P of being occupied, then the probability is $1 - P$ that a state at an energy ΔE below E_F is occupied.

42.51 •• Compute the Fermi energy of potassium by making the simple approximation that each atom contributes one free electron. The density of potassium is 851 kg/m^3, and the mass of a single potassium atom is 6.49×10^{-26} kg.

42.52 •• Hydrogen is found in two naturally occurring isotopes; normal hydrogen (containing a single proton in its nucleus) and deuterium (having a proton and a neutron). Assuming that both molecules are the same size and that the proton and neutron have the same mass (which is almost the case), find the ratio of (a) the energy of any given rotational state in a diatomic hydrogen molecule to the energy of the same state in a diatomic deuterium molecule and (b) the energy of any given vibrational state in hydrogen to the same state in deuterium (assuming that the force constant is the same for both molecules). Why is it physically reasonable that the force constant would be the same for hydrogen and deuterium molecules?

42.53 ••• Metallic lithium has a bcc crystal structure. Each unit cell is a cube of side length $a = 0.35$ nm. (a) For a bcc lattice, what is the number of atoms per unit volume? Give your answer in terms of a. (*Hint:* How many atoms are there per unit cell?) (b) Use the result of part (a) to calculate the zero-temperature Fermi energy E_{F0} for metallic lithium. Assume there is one free electron per atom.

42.54 •• **CALC** The one-dimensional calculation of Example 42.4 (Section 42.3) can be extended to three dimensions. For the three-dimensional fcc NaCl lattice, the result for the potential energy of a pair of Na^+ and Cl^- ions due to the electrostatic interaction with all of the ions in the crystal is $U = -\alpha e^2/4\pi\epsilon_0 r$, where $\alpha = 1.75$ is the *Madelung constant.* Another contribution to the potential energy is a repulsive interaction at small ionic separation r due to overlap of the electron clouds. This contribution can be represented by A/r^8, where A is a positive constant, so the expression for the total potential energy is

$$U_{tot} = -\frac{\alpha e^2}{4\pi\epsilon_0 r} + \frac{A}{r^8}$$

(a) Let r_0 be the value of the ionic separation r for which U_{tot} is a minimum. Use this definition to find an equation that relates r_0 and A, and use this to write U_{tot} in terms of r_0. For NaCl, $r_0 = 0.281$ nm. Obtain a numerical value (in electron volts) of U_{tot} for NaCl. (b) The quantity $-U_{tot}$ is the energy required to remove a Na^+ ion and a Cl^- ion from the crystal. Forming a pair of neutral atoms from this pair of ions involves the release of 5.14 eV (the ionization energy of Na) and the expenditure of 3.61 eV (the electron affinity of Cl). Use the result of part (a) to calculate the energy required to remove a pair of neutral Na and Cl atoms from the crystal. The experimental value for this quantity is 6.39 eV; how well does your calculation agree?

42.55 •• **CALC** Consider a system of N free electrons within a volume V. Even at absolute zero, such a system exerts a pressure p on its surroundings due to the motion of the electrons. To calculate this pressure, imagine that the volume increases by a small amount dV. The electrons will do an amount of work $p\,dV$ on their surroundings, which means that the total energy E_{tot} of the electrons will change by an amount $dE_{tot} = -p\,dV$. Hence $p = -dE_{tot}/dV$. (a) Show that the pressure of the electrons at absolute zero is

$$p = \frac{3^{2/3}\pi^{4/3}\hbar^2}{5m}\left(\frac{N}{V}\right)^{5/3}$$

(b) Evaluate this pressure for copper, which has a free-electron concentration of 8.45×10^{28} m^{-3}. Express your result in pascals and in atmospheres. (c) The pressure you found in part (b) is extremely high. Why, then, don't the electrons in a piece of copper simply explode out of the metal?

42.56 •• **CALC** When the pressure p on a material increases by an amount Δp, the volume of the material will change from V to $V + \Delta V$, where ΔV is negative. The *bulk modulus* B of the material is defined to be the ratio of the pressure change Δp to the absolute value $\left|\Delta V/V\right|$ of the fractional volume change. The greater the bulk modulus, the greater the pressure increase required for a given fractional volume change, and the more incompressible the material (see Section 11.4). Since $\Delta V < 0$, the bulk modulus can be written as $B = -\Delta p/(\Delta V/V_0)$. In the limit that the pressure and volume changes are very small, this becomes

$$B = -V\frac{dp}{dV}$$

(a) Use the result of Problem 42.55 to show that the bulk modulus for a system of N free electrons in a volume V at low temperatures is $B = \frac{5}{3}p$. (*Hint:* The quantity p in the expression $B = -V(dp/dV)$ is the *external* pressure on the system. Can you explain why this is equal to the *internal* pressure of the system itself, as found in Problem 42.55?) (b) Evaluate the bulk modulus for the electrons in copper, which has a free-electron concentration of $8.45 \times 10^{28} \text{ m}^{-3}$. Express your result in pascals. (c) The actual bulk modulus of copper is 1.4×10^{11} Pa. Based on your result in part (b), what fraction of this is due to the free electrons in copper? (This result shows that the free electrons in a metal play a major role in making the metal resistant to compression.) What do you think is responsible for the remaining fraction of the bulk modulus?

42.57 •• In the discussion of free electrons in Section 42.5, we assumed that we could ignore the effects of relativity. This is not a safe assumption if the Fermi energy is greater than about $\frac{1}{100}mc^2$ (that is, more than about 1% of the rest energy of an electron). (a) Assume that the Fermi energy at absolute zero, as given by Eq. (42.19), is equal to $\frac{1}{100}mc^2$. Show that the electron concentration is

$$\frac{N}{V} = \frac{2^{3/2}m^3c^3}{3000\pi^2\hbar^3}$$

and determine the numerical value of N/V. (b) Is it a good approximation to ignore relativistic effects for electrons in a metal such as copper, for which the electron concentration is $8.45 \times 10^{28} \text{ m}^{-3}$? Explain. (c) A *white dwarf star* is what is left behind by a star like the sun after it has ceased to produce energy by nuclear reactions. (Our own sun will become a white dwarf star in another 6×10^9 years or so.) A typical white dwarf has mass 2×10^{30} kg (comparable to the sun) and radius 6000 km (comparable to that of the earth). The gravitational attraction of different parts of the white dwarf for each other tends to compress the star; what prevents it from compressing is the pressure of free electrons within the star (see Problem 42.55). Estimate the electron concentration within a typical white dwarf star using the following assumptions: (i) the white dwarf star is made of carbon, which has a mass per atom of

1.99×10^{-26} kg; and (ii) all six of the electrons from each carbon atom are able to move freely throughout the star. (d) Is it a good approximation to ignore relativistic effects in the structure of a white dwarf star? Explain.

42.58 •• CP A variable DC battery is connected in series with a 125-Ω resistor and a *p-n* junction diode that has a saturation current of 0.625 mA at room temperature (20°C). When a voltmeter across the 125-Ω resistor reads 35.0 V, what are (a) the voltage across the diode and (b) the resistance of the diode?

CHALLENGE PROBLEMS

42.59 ••• CP Van der Waals bonds arise from the interaction between two permanent or induced electric dipole moments in a pair of atoms or molecules. (a) Consider two identical dipoles, each consisting of charges $+q$ and $-q$ separated by a distance d and oriented as shown in Fig. P42.59a. Calculate the electric potential energy, expressed in terms of the electric dipole moment $p = qd$, for the situation where $r \gg d$. Is the interaction attractive or repulsive, and how does this potential energy vary with r, the separation between the centers of the two dipoles? (b) Repeat part (a) for the orientation of the dipoles shown in Fig. P42.59b. The dipole interaction is more complicated when we have to average over the relative orientations of the two dipoles due to thermal motion or when the dipoles are induced rather than permanent.

Figure **P42.59**

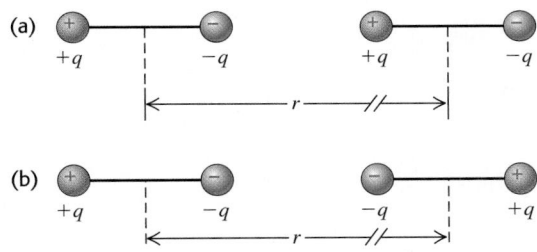

42.60 ••• CP CALC (a) Consider the hydrogen molecule (H_2) to be a simple harmonic oscillator with an equilibrium spacing of 0.074 nm, and estimate the vibrational energy-level spacing for H_2. The mass of a hydrogen atom is 1.67×10^{-27} kg. (*Hint:* Estimate the force constant by equating the change in Coulomb repulsion of the protons, when the atoms move slightly closer together than r_0, to the "spring" force. That is, assume that the chemical binding force remains approximately constant as r is decreased slightly from r_0.) (b) Use the results of part (a) to calculate the vibrational energy-level spacing for the deuterium molecule, D_2. Assume that the spring constant is the same for D_2 as for H_2. The mass of a deuterium atom is 3.34×10^{-27} kg.

Answers

Chapter Opening Question

Venus must radiate energy into space at the same rate that it receives energy in the form of sunlight. However, carbon dioxide (CO_2) molecules in the atmosphere absorb infrared radiation

emitted by the surface of Venus and re-emit it toward the ground. To compensate for this and to maintain the balance between emitted and received energy, the surface temperature of Venus and hence the rate of blackbody radiation from the surface both increase.

Test Your Understanding Questions

42.1 Answer: (i) The exclusion principle states that only one electron can be in a given state. Real electrons have spin, so two electrons (one spin up, one spin down) can be in a given *spatial* state and hence two can participate in a given covalent bond between two atoms. If electrons obeyed the exclusion principle but did not have spin, that state of an electron would be completely described by its spatial distribution and only *one* electron could participate in a covalent bond. (We will learn in Chapter 44 that this situation is wholly imaginary: There are subatomic particles without spin, but they do *not* obey the exclusion principle.)

42.2 Answer: (ii) Figure 42.5 shows that the difference in energy between adjacent rotational levels increases with increasing l. Hence, as l increases, the energy E of the emitted photon increases and the wavelength $\lambda = hc/E$ decreases.

42.3 Answer: (ii) In Fig. 42.13 let a be the distance between adjacent Na^+ and Cl^- ions. This figure shows that the Cl^- ion that is the next nearest neighbor to a Na^+ ion is on the opposite corner of a cube of side a. The distance between these two ions is $\sqrt{a^2 + a^2 + a^2} = \sqrt{3a^2} = a\sqrt{3}$.

42.4 Answer: (ii) A small temperature change causes a substantial increase in the population of electrons in a semiconductor's conduction band and a comparably substantial increase in conductivity. The conductivity of conductors and insulators varies more gradually with temperature.

42.5 Answer: no The kinetic-molecular model of an ideal gas (see Section 18.3) shows that the gas pressure is proportional to the average translational kinetic energy E_{av} of the particles that make up the gas. In a classical ideal gas, E_{av} is directly proportional to the average temperature T, so the pressure decreases as T decreases. In a free-electron gas, the average kinetic energy per electron is *not* related simply to T; as Example 42.8 shows, for the free-electron gas in a metal, E_{av} is almost completely a consequence of the exclusion principle at room temperature and colder. Hence the pressure of a free-electron gas in a solid metal does *not* change appreciably between room temperature and absolute zero.

42.6 Answer: no Pure copper is already an excellent conductor since it has a partially filled conduction band (Fig. 42.19c). Furthermore, copper forms a metallic crystal (Fig. 42.15) as opposed to the covalent crystals of silicon or germanium, so the scheme of using an impurity to donate or accept an electron does not work for copper. In fact, adding impurities to copper *decreases* the conductivity because an impurity tends to scatter electrons, impeding the flow of current.

42.7 Answer: no A negative charge on the gate will repel, not attract, electrons in the *p*-type silicon. Hence the electron concentration in the region between the two *p-n* junctions will be made even smaller. With so few charge carriers present in this region, very little current will flow between the source and the drain.

Bridging Problem

Answers: (a) 0.278 eV
   (b) 1.74×10^{-25}

NUCLEAR PHYSICS

43

? This sculpture of a wooly mammoth, just 3.7 cm (1.5 in.) in length, was carved from a mammoth's ivory tusk by an artist who lived in southwestern Germany 35,000 years ago. What physical principles make it possible to date biological specimens such as these?

LEARNING GOALS

By studying this chapter, you will learn:

- Some key properties of atomic nuclei, including radii, densities, spins, and magnetic moments.

- How the binding energy of a nucleus depends on the numbers of protons and neutrons that it contains.

- The most important ways in which unstable nuclei undergo radioactive decay.

- How the decay rate of a radioactive substance depends on time.

- Some of the biological hazards and medical uses of radiation.

- How to analyze some important types of nuclear reactions.

- What happens in a nuclear fission chain reaction, and how it can be controlled.

- The sequence of nuclear reactions that allow the sun and stars to shine.

During the past century, applications of nuclear physics have had enormous effects on humankind, some beneficial, some catastrophic. Many people have strong opinions about applications such as bombs and reactors. Ideally, those opinions should be based on understanding, not on prejudice or emotion, and we hope this chapter will help you to reach that ideal.

Every atom contains at its center an extremely dense, positively charged *nucleus,* which is much smaller than the overall size of the atom but contains most of its total mass. We will look at several important general properties of nuclei and of the nuclear force that holds them together. The stability or instability of a particular nucleus is determined by the competition between the attractive nuclear force among the protons and neutrons and the repulsive electrical interactions among the protons. Unstable nuclei *decay,* transforming themselves spontaneously into other nuclei by a variety of processes. Nuclear reactions can also be induced by impact on a nucleus of a particle or another nucleus. Two classes of reactions of special interest are *fission* and *fusion.* We could not survive without the energy released by one nearby fusion reactor, our sun.

43.1 Properties of Nuclei

As we described in Section 39.2, Rutherford found that the nucleus is tens of thousands of times smaller in radius than the atom itself. Since Rutherford's initial experiments, many additional scattering experiments have been performed, using high-energy protons, electrons, and neutrons as well as alpha particles (helium-4 nuclei). These experiments show that we can model a nucleus as a sphere with a radius R that depends on the total number of *nucleons* (neutrons

1439

and protons) in the nucleus. This number is called the **nucleon number** A. The radii of most nuclei are represented quite well by the equation

$$R = R_0 A^{1/3} \qquad \text{(radius of a nucleus)} \qquad (43.1)$$

where R_0 is an experimentally determined constant:

$$R_0 = 1.2 \times 10^{-15} \text{ m} = 1.2 \text{ fm}$$

The nucleon number A in Eq. (43.1) is also called the **mass number** because it is the nearest whole number to the mass of the nucleus measured in unified atomic mass units (u). (The proton mass and the neutron mass are both approximately 1 u.) The best current conversion factor is

$$1 \text{ u} = 1.660538782(83) \times 10^{-27} \text{ kg}$$

In Section 43.2 we'll discuss the masses of nuclei in more detail. Note that when we speak of the masses of nuclei and particles, we mean their *rest* masses.

Nuclear Density

The volume V of a sphere is equal to $4\pi R^3/3$, so Eq. (43.1) shows that the *volume* of a nucleus is proportional to A. Dividing A (the approximate mass in u) by the volume gives us the approximate density and cancels out A. Thus *all nuclei have approximately the same density*. This fact is of crucial importance in understanding nuclear structure.

Example 43.1 | Calculating nuclear properties

The most common kind of iron nucleus has mass number $A = 56$. Find the radius, approximate mass, and approximate density of the nucleus.

SOLUTION

IDENTIFY and SET UP: Equation (43.1) tells us how the nuclear radius R depends on the mass number A. The mass of the nucleus in atomic mass units is approximately equal to the value of A, and the density ρ is mass divided by volume.

EXECUTE: The radius and approximate mass are

$$R = R_0 A^{1/3} = (1.2 \times 10^{-15} \text{ m})(56)^{1/3}$$
$$= 4.6 \times 10^{-15} \text{ m} = 4.6 \text{ fm}$$
$$m \approx (56 \text{ u})(1.66 \times 10^{-27} \text{ kg/u}) = 9.3 \times 10^{-26} \text{ kg}$$

The volume V of the nucleus (which we treat as a sphere of radius R) and its density ρ are

$$V = \tfrac{4}{3}\pi R^3 = \tfrac{4}{3}\pi R_0^3 A = \tfrac{4}{3}\pi(4.6 \times 10^{-15} \text{ m})^3$$
$$= 4.1 \times 10^{-43} \text{ m}^3$$
$$\rho = \frac{m}{V} \approx \frac{9.3 \times 10^{-26} \text{ kg}}{4.1 \times 10^{-43} \text{ m}^3} = 2.3 \times 10^{17} \text{ kg/m}^3$$

EVALUATE: As we mentioned above, *all* nuclei have approximately this same density. The density of solid iron is about 7000 kg/m³; the iron nucleus is more than 10^{13} times as dense as iron in bulk. Such densities are also found in *neutron stars*, which are similar to gigantic nuclei made almost entirely of neutrons. A 1-cm cube of material with this density would have a mass of 2.3×10^{11} kg, or 230 million metric tons!

Nuclides and Isotopes

The building blocks of the nucleus are the proton and the neutron. In a neutral atom, the nucleus is surrounded by one electron for every proton in the nucleus. We introduced these particles in Section 21.1; we'll recount the discovery of the neutron and proton in Chapter 44. The masses of these particles are

$$\text{Proton:} \quad m_p = 1.007276 \text{ u} = 1.672622 \times 10^{-27} \text{ kg}$$

$$\text{Neutron:} \quad m_n = 1.008665 \text{ u} = 1.674927 \times 10^{-27} \text{ kg}$$

$$\text{Electron:} \quad m_e = 0.000548580 \text{ u} = 9.10938 \times 10^{-31} \text{ kg}$$

The number of protons in a nucleus is the **atomic number** Z. The number of neutrons is the **neutron number** N. The nucleon number or mass number A is the sum of the number of protons Z and the number of neutrons N:

$$A = Z + N \qquad (43.2)$$

A single nuclear species having specific values of both Z and N is called a **nuclide.** Table 43.1 lists values of A, Z, and N for some nuclides. The electron structure of an atom, which is responsible for its chemical properties, is determined by the charge Ze of the nucleus. The table shows some nuclides that have the same Z but different N. These nuclides are called **isotopes** of that element; they have different masses because they have different numbers of neutrons in their nuclei. A familiar example is chlorine (Cl, $Z = 17$). About 76% of chlorine nuclei have $N = 18$; the other 24% have $N = 20$. Different isotopes of an element usually have slightly different physical properties such as melting and boiling temperatures and diffusion rates. The two common isotopes of uranium with $A = 235$ and 238 are usually separated industrially by taking advantage of the different diffusion rates of gaseous uranium hexafluoride (UF_6) containing the two isotopes.

Table 43.1 also shows the usual notation for individual nuclides: the symbol of the element, with a pre-subscript equal to Z and a pre-superscript equal to the mass number A. The general format for an element El is ${}_Z^A\text{El}$. The isotopes of chlorine mentioned above, with $A = 35$ and 37, are written ${}_{17}^{35}\text{Cl}$ and ${}_{17}^{37}\text{Cl}$ and pronounced "chlorine-35" and "chlorine-37," respectively. This name of the element determines the atomic number Z, so the pre-subscript Z is sometimes omitted, as in ${}^{35}\text{Cl}$.

Table 43.2 gives the masses of some common atoms, including their electrons. Note that this table gives masses of *neutral* atoms (with Z electrons) rather than masses of *bare* nuclei, because it is much more difficult to measure masses of bare nuclei with high precision. The mass of a neutral carbon-12 atom is exactly 12 u; that's how the unified atomic mass unit is defined. The masses of other atoms are *approximately* equal to A atomic mass units, as we stated earlier. In fact, the atomic masses are *less* than the sum of the masses of their parts (the Z protons, the Z electrons, and the N neutrons). We'll explain this very important mass difference in the next section.

Table 43.1 Compositions of Some Common Nuclides

Z = atomic number (number of protons)

N = neutron number

$A = Z + N$ = mass number (total number of nucleons)

Nucleus	Z	N	$A = Z + N$
${}_1^1\text{H}$	1	0	1
${}_1^2\text{H}$	1	1	2
${}_2^4\text{He}$	2	2	4
${}_3^6\text{Li}$	3	3	6
${}_3^7\text{Li}$	3	4	7
${}_4^9\text{Be}$	4	5	9
${}_5^{10}\text{B}$	5	5	10
${}_5^{11}\text{B}$	5	6	11
${}_6^{12}\text{C}$	6	6	12
${}_6^{13}\text{C}$	6	7	13
${}_7^{14}\text{N}$	7	7	14
${}_8^{16}\text{O}$	8	8	16
${}_{11}^{23}\text{Na}$	11	12	23
${}_{29}^{65}\text{Cu}$	29	36	65
${}_{80}^{200}\text{Hg}$	80	120	200
${}_{92}^{235}\text{U}$	92	143	235
${}_{92}^{238}\text{U}$	92	146	238

Table 43.2 Neutral Atomic Masses for Some Light Nuclides

Element and Isotope	Atomic Number, Z	Neutron Number, N	Atomic Mass (u)	Mass Number, A
Hydrogen ($^1_1$H)	1	0	1.007825	1
Deuterium ($^2_1$H)	1	1	2.014102	2
Tritium ($^3_1$H)	1	2	3.016049	3
Helium ($^3_2$He)	2	1	3.016029	3
Helium ($^4_2$He)	2	2	4.002603	4
Lithium ($^6_3$Li)	3	3	6.015122	6
Lithium ($^7_3$Li)	3	4	7.016004	7
Beryllium ($^9_4$Be)	4	5	9.012182	9
Boron ($^{10}_5$B)	5	5	10.012937	10
Boron ($^{11}_5$B)	5	6	11.009305	11
Carbon ($^{12}_6$C)	6	6	12.000000	12
Carbon ($^{13}_6$C)	6	7	13.003355	13
Nitrogen ($^{14}_7$N)	7	7	14.003074	14
Nitrogen ($^{15}_7$N)	7	8	15.000109	15
Oxygen ($^{16}_8$O)	8	8	15.994915	16
Oxygen ($^{17}_8$O)	8	9	16.999132	17
Oxygen ($^{18}_8$O)	8	10	17.999160	18

Source: A. H. Wapstra and G. Audi, *Nuclear Physics* **A595,** 4 (1995).

Nuclear Spins and Magnetic Moments

Like electrons, protons and neutrons are also spin-$\frac{1}{2}$ particles with spin angular momenta given by the same equations as in Section 41.5. The magnitude of the spin angular momentum $\vec{S}$ of a nucleon is

$$S = \sqrt{\tfrac{1}{2}\left(\tfrac{1}{2} + 1\right)}\,\hbar = \sqrt{\tfrac{3}{4}}\,\hbar \qquad (43.3)$$

and the z-component is

$$S_z = \pm\tfrac{1}{2}\hbar \qquad (43.4)$$

In addition to the spin angular momentum of the nucleons, there may be *orbital* angular momentum associated with their motions within the nucleus. The orbital angular momentum of the nucleons is quantized in the same way as that of electrons in atoms.

The *total* angular momentum $\vec{J}$ of the nucleus is the vector sum of the individual spin and orbital angular momenta of all the nucleons. It has magnitude

$$J = \sqrt{j(j + 1)}\,\hbar \qquad (43.5)$$

and z-component

$$J_z = m_j\hbar \qquad (m_j = -j, -j + 1, \ldots, j - 1, j) \qquad (43.6)$$

When the total number of nucleons A is *even*, j is an integer; when it is *odd*, j is a half-integer. All nuclides for which both Z and N are even have $J = 0$, which suggests that pairing of particles with opposite spin components may be an important consideration in nuclear structure. The total nuclear angular momentum quantum number j is usually called the *nuclear spin,* even though in general it refers to a combination of the orbital and spin angular momenta of the nucleons that make up the nucleus.

Associated with nuclear angular momentum is a *magnetic moment*. When we discussed *electron* magnetic moments in Section 41.4, we introduced the Bohr magneton $\mu_B = e\hbar/2m_e$ as a natural unit of magnetic moment. We found that the magnitude of the z-component of the electron-spin magnetic moment is almost exactly equal to μ_B; that is, $|\mu_{sz}|_{electron} \approx \mu_B$. In discussing *nuclear* magnetic moments, we can define an analogous quantity, the **nuclear magneton** μ_n:

$$\mu_n = \frac{e\hbar}{2m_p} = 5.05078 \times 10^{-27} \text{ J/T} = 3.15245 \times 10^{-8} \text{ eV/T} \tag{43.7}$$

(nuclear magneton)

where m_p is the proton mass. Because the proton mass m_p is 1836 times larger than the electron mass m_e, the nuclear magneton μ_n is 1836 times smaller than the Bohr magneton μ_B.

We might expect the magnitude of the z-component of the spin magnetic moment of the proton to be approximately μ_n. Instead, it turns out to be

$$|\mu_{sz}|_{proton} = 2.7928\mu_n \tag{43.8}$$

Even more surprising, the neutron, which has zero charge, has a spin magnetic moment; its z-component has magnitude

$$|\mu_{sz}|_{neutron} = 1.9130\mu_n \tag{43.9}$$

The proton has a positive charge; as expected, its spin magnetic moment $\vec{\mu}$ is parallel to its spin angular momentum $\vec{S}$. However, $\vec{\mu}$ and $\vec{S}$ are opposite for a neutron, as would be expected for a *negative* charge distribution. These *anomalous* magnetic moments arise because the proton and neutron aren't really fundamental particles but are made of simpler particles called *quarks*. We'll discuss quarks in some detail in Chapter 44.

The magnetic moment of an entire nucleus is typically a few nuclear magnetons. When a nucleus is placed in an external magnetic field $\vec{B}$, there is an interaction energy $U = -\vec{\mu} \cdot \vec{B} = -\mu_z B$ just as with atomic magnetic moments. The components of the magnetic moment in the direction of the field μ_z are quantized, so a series of energy levels results from this interaction.

Example 43.2 Proton spin flips

Protons are placed in a 2.30-T magnetic field that points in the positive z-direction. (a) What is the energy difference between states with the z-component of proton spin angular momentum parallel and antiparallel to the field? (b) A proton can make a transition from one of these states to the other by emitting or absorbing a photon with the appropriate energy. Find the frequency and wavelength of such a photon.

SOLUTION

IDENTIFY and SET UP: The proton is a spin-$\frac{1}{2}$ particle with a magnetic moment $\vec{\mu}$ in the same direction as its spin $\vec{S}$, so its energy depends on the orientation of its spin relative to an applied magnetic field $\vec{B}$. If the z-component of $\vec{S}$ is aligned with $\vec{B}$, then μ_z is equal to the positive value given in Eq. (43.8). If the z-component of $\vec{S}$ is opposite $\vec{B}$, then μ_z is the negative of this value. The interaction energy in either case is $U = -\mu_z B$; the difference between these energies is our target variable in part (a). We find the photon frequency and wavelength using $E = hf = hc/\lambda$.

EXECUTE: (a) When the z-components of $\vec{S}$ and $\vec{\mu}$ are parallel to $\vec{B}$, the interaction energy is

$$U = -|\mu_z|B = -(2.7928)(3.152 \times 10^{-8} \text{ eV/T})(2.30 \text{ T})$$
$$= -2.025 \times 10^{-7} \text{ eV}$$

When the z-components of $\vec{S}$ and $\vec{\mu}$ are antiparallel to the field, the energy is $+2.025 \times 10^{-7}$ eV. Hence the energy *difference* between the states is

$$\Delta E = 2(2.025 \times 10^{-7} \text{ eV}) = 4.05 \times 10^{-7} \text{ eV}$$

(b) The corresponding photon frequency and wavelength are

$$f = \frac{\Delta E}{h} = \frac{4.05 \times 10^{-7} \text{ eV}}{4.136 \times 10^{-15} \text{ eV} \cdot \text{s}} = 9.79 \times 10^7 \text{ Hz} = 97.9 \text{ MHz}$$

$$\lambda = \frac{c}{f} = \frac{3.00 \times 10^8 \text{ m/s}}{9.79 \times 10^7 \text{ s}^{-1}} = 3.06 \text{ m}$$

EVALUATE: This frequency is in the middle of the FM radio band. When a hydrogen specimen is placed in a 2.30-T magnetic field and irradiated with radio waves of this frequency, proton *spin flips* can be detected by the absorption of energy from the radiation.

43.1 Magnetic resonance imaging (MRI).

(a)

Random spin of hydrogen protons

Protons, the nuclei of hydrogen atoms in the tissue under study, normally have random spin orientations.

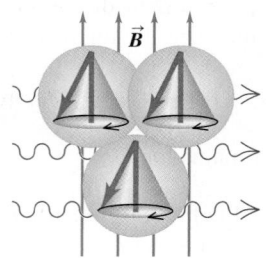

In the presence of a strong magnetic field, the spins become aligned with a component parallel to $\vec{B}$.

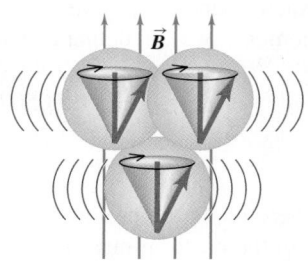

A brief radio signal causes the spins to flip orientation.

As the protons realign with the $\vec{B}$ field, they emit radio waves that are picked up by sensitive detectors.

(b) Since $\vec{B}$ has a different value at different locations in the tissue, the radio waves from different locations have different frequencies. This makes it possible to construct an image.

(c) An electromagnet used for MRI

Main coil supplies uniform $\vec{B}$ field.

x coil varies $\vec{B}$ field from left to right.

z coil varies $\vec{B}$ field from head to toe.

y coil varies $\vec{B}$ field from top to bottom.

Transceiver sends and receives signals that create image.

Mastering**PHYSICS**

PhET: Simplified MRI

Nuclear Magnetic Resonance and MRI

Spin-flip experiments of the sort referred to in Example 43.2 are called *nuclear magnetic resonance* (NMR). They have been carried out with many different nuclides. Frequencies and magnetic fields can be measured very precisely, so this technique permits precise measurements of nuclear magnetic moments. An elaboration of this basic idea leads to *magnetic resonance imaging* (MRI), a noninvasive imaging technique that discriminates among various body tissues on the basis of the differing environments of protons in the tissues (Fig. 43.1).

The magnetic moment of a nucleus is also the *source* of a magnetic field. In an atom the interaction of an electron's magnetic moment with the field of the nucleus's magnetic moment causes additional splittings in atomic energy levels and spectra. We called this effect *hyperfine structure* in Section 41.5. Measurements of the hyperfine structure may be used to directly determine the nuclear spin.

Test Your Understanding of Section 43.1 (a) By what factor must the mass number of a nucleus increase to double its volume? (i) $\sqrt[3]{2}$; (ii) $\sqrt{2}$; (iii) 2; (iv) 4; (v) 8. (b) By what factor must the mass number increase to double the radius of the nucleus? (i) $\sqrt[3]{2}$; (ii) $\sqrt{2}$; (iii) 2; (iv) 4; (v) 8.

43.2 Nuclear Binding and Nuclear Structure

Because energy must be added to a nucleus to separate it into its individual protons and neutrons, the total rest energy E_0 of the separated nucleons is greater than the rest energy of the nucleus. The energy that must be added to separate the nucleons is called the **binding energy** E_B; it is the magnitude of the energy by which the nucleons are bound together. Thus the rest energy of the nucleus is $E_0 - E_B$. Using the equivalence of rest mass and energy (see Section 37.8), we see that the total mass of the nucleons is always greater than the mass of the nucleus by an amount E_B/c^2 called the *mass defect*. The binding energy for a nucleus containing Z protons and N neutrons is defined as

$$E_B = (ZM_H + Nm_n - {}^A_Z M)c^2 \quad \text{(nuclear binding energy)} \quad (43.10)$$

where ${}^A_Z M$ is the mass of the *neutral* atom containing the nucleus, the quantity in the parentheses is the mass defect, and $c^2 = 931.5$ MeV/u. Note that Eq. (43.10) does not include Zm_p, the mass of Z protons. Rather, it contains ZM_H, the mass of Z protons and Z electrons combined as Z neutral ${}^1_1 H$ atoms, to balance the Z electrons included in ${}^A_Z M$, the mass of the neutral atom.

The simplest nucleus is that of hydrogen, a single proton. Next comes the nucleus of ${}^2_1 H$, the isotope of hydrogen with mass number 2, usually called *deuterium*. Its nucleus consists of a proton and a neutron bound together to form a particle called the *deuteron*. By using values from Table 43.2 in Eq. (43.10), we find that the binding energy of the deuteron is

$$E_B = (1.007825 \text{ u} + 1.008665 \text{ u} - 2.014102 \text{ u})(931.5 \text{ MeV/u})$$
$$= 2.224 \text{ MeV}$$

This much energy would be required to pull the deuteron apart into a proton and a neutron. An important measure of how tightly a nucleus is bound is the *binding energy per nucleon*, E_B/A. At $(2.224 \text{ MeV})/(2 \text{ nucleons}) = 1.112$ MeV per nucleon, ${}^2_1 H$ has the lowest binding energy per nucleon of all nuclides.

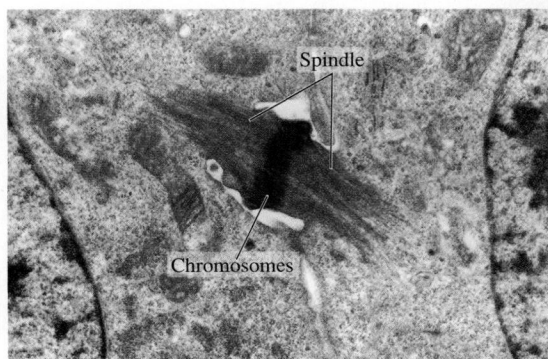
Spindle
Chromosomes

Problem-Solving Strategy 43.1 **Nuclear Properties**

(MP)

IDENTIFY *the relevant concepts:* The key properties of a nucleus are its mass, radius, binding energy, mass defect, binding energy per nucleon, and angular momentum.

SET UP *the problem:* Once you have identified the target variables, assemble the equations needed to solve the problem. A relatively small number of equations from this section and Section 43.1 are all you need.

EXECUTE *the solution:* Solve for the target variables. Binding-energy calculations using Eq. (43.10) often involve subtracting two nearly equal quantities. To get enough precision in the difference, you may need to carry as many as nine significant figures, if that many are available.

EVALUATE *your answer:* It's useful to be familiar with the following benchmark magnitudes. Protons and neutrons are about 1840 times as massive as electrons. Nuclear radii are of the order of 10^{-15} m. The electric potential energy of two protons in a nucleus is roughly 10^{-13} J or 1 MeV, so nuclear interaction energies are typically a few MeV rather than a few eV as with atoms. The binding energy per nucleon is about 1% of the nucleon rest energy. (The ionization energy of the hydrogen atom is only 0.003% of the electron's rest energy.) Angular momenta are determined only by the value of $\hbar$, so they are of the same order of magnitude in both nuclei and atoms. Nuclear magnetic moments, however, are about a factor of 1000 *smaller* than those of electrons in atoms because nuclei are so much more massive than electrons.

Example 43.3 **The most strongly bound nuclide**

Find the mass defect, the total binding energy, and the binding energy per nucleon of ${}^{62}_{28}\text{Ni}$, which has the highest binding energy per nucleon of all nuclides (Fig. 43.2). The neutral atomic mass of ${}^{62}_{28}\text{Ni}$ is 61.928349 u.

SOLUTION

IDENTIFY and SET UP: The mass defect ΔM is the difference between the mass of the nucleus and the combined mass of its constituent nucleons. The binding energy E_B is this quantity

multiplied by c^2, and the binding energy per nucleon is E_B divided by the mass number A. We use Eq. (43.10), $\Delta M = ZM_H + Nm_n - {}^A_Z M$, to determine both the mass defect and the binding energy.

EXECUTE: With $Z = 28$, $M_H = 1.007825$ u, $N = A - Z = 62 - 28 = 34$, $m_n = 1.008665$ u, and ${}^A_Z M = 61.928349$ u, Eq. (43.10) gives $\Delta M = 0.585361$ u. The binding energy is then

$$E_B = (0.585361 \text{ u})(931.5 \text{ MeV/u}) = 545.3 \text{ MeV}$$

Continued

The binding energy *per nucleon* is $E_B/A = (545.3 \text{ MeV})/62$, or 8.795 MeV per nucleon.

EVALUATE: Our result means that it would take a minimum of 545.3 MeV to pull a $^{62}_{28}\text{Ni}$ completely apart into 28 protons and 34 neutrons. The mass defect of $^{62}_{28}\text{Ni}$ is about 1% of the atomic (or the nuclear) mass. The binding energy is therefore about 1% of the rest energy of the nucleus, and the binding energy per nucleon is about 1% of the rest energy of a nucleon. Note that the mass defect is more than half the mass of a nucleon, which suggests how tightly bound nuclei are.

Nearly all stable nuclides, from the lightest to the most massive, have binding energies in the range of 7–9 MeV per nucleon. Figure 43.2 is a graph of binding energy per nucleon as a function of the mass number A. Note the spike at $A = 4$, showing the unusually large binding energy per nucleon of the ^4_2He nucleus (alpha particle) relative to its neighbors. To explain this curve, we must consider the interactions among the nucleons.

The Nuclear Force

The force that binds protons and neutrons together in the nucleus, despite the electrical repulsion of the protons, is an example of the *strong interaction* that we mentioned in Section 5.5. In the context of nuclear structure, this interaction is called the *nuclear force*. Here are some of its characteristics. First, it does not depend on charge; neutrons as well as protons are bound, and the binding is the same for both. Second, it has short range, of the order of nuclear dimensions—that is, 10^{-15} m. (Otherwise, the nucleus would grow by pulling in additional protons and neutrons.) But within its range, the nuclear force is much stronger than electrical forces; otherwise, the nucleus could never be stable. It would be nice if we could write a simple equation like Newton's law of gravitation or Coulomb's law for this force, but physicists have yet to fully determine its dependence on the separation r. Third, the nearly constant density of nuclear matter and the nearly constant binding energy per nucleon of larger nuclides show that a particular nucleon cannot interact simultaneously with *all* the other nucleons in a nucleus, but only with those few in its immediate vicinity. This is different from electrical forces; *every* proton in the nucleus repels every other one. This limited number of interactions is called *saturation;* it is analogous to covalent bonding in molecules and solids. Finally, the nuclear force favors binding of *pairs* of protons or neutrons with opposite spins and of *pairs of pairs*—that is, a pair of protons and a pair of neutrons, each pair having opposite spins. Hence the alpha particle (two protons and two neutrons) is an exceptionally stable nucleus for its mass number. We'll see other evidence for pairing effects in nuclei in the

43.2 Approximate binding energy per nucleon as a function of mass number A (the total number of nucleons) for stable nuclides.

E_B/A (MeV/nucleon)

The curve reaches a peak of about 8.8 MeV/nucleon at $A = 62$, corresponding to the element nickel.
The spike at $A = 4$ shows the unusual stability of the ^4_2He structure.

next subsection. (In Section 42.8 we described an analogous pairing that binds opposite-spin electrons in Cooper pairs in the BCS theory of superconductivity.)

The analysis of nuclear structure is more complex than the analysis of many-electron atoms. Two different kinds of interactions are involved (electrical and nuclear), and the nuclear force is not yet completely understood. Even so, we can gain some insight into nuclear structure by the use of simple models. We'll discuss briefly two rather different but successful models, the *liquid-drop model* and the *shell model.*

The Liquid-Drop Model

The **liquid-drop model,** first proposed in 1928 by the Russian physicist George Gamow and later expanded on by Niels Bohr, is suggested by the observation that all nuclei have nearly the same density. The individual nucleons are analogous to molecules of a liquid, held together by short-range interactions and surface-tension effects. We can use this simple picture to derive a formula for the estimated total binding energy of a nucleus. We'll include five contributions:

1. We've remarked that nuclear forces show *saturation;* an individual nucleon interacts only with a few of its nearest neighbors. This effect gives a binding-energy term that is proportional to the number of nucleons. We write this term as C_1A, where C_1 is an experimentally determined constant.

2. The nucleons on the surface of the nucleus are less tightly bound than those in the interior because they have no neighbors outside the surface. This decrease in the binding energy gives a *negative* energy term proportional to the surface area $4\pi R^2$. Because R is proportional to $A^{1/3}$, this term is proportional to $A^{2/3}$; we write it as $-C_2A^{2/3}$, where C_2 is another constant.

3. Every one of the Z protons repels every one of the $(Z - 1)$ other protons. The total repulsive electric potential energy is proportional to $Z(Z - 1)$ and inversely proportional to the radius R and thus to $A^{1/3}$. This energy term is negative because the nucleons are less tightly bound than they would be without the electrical repulsion. We write this correction as $-C_3Z(Z - 1)/A^{1/3}$.

4. To be in a stable, low-energy state, the nucleus must have a balance between the energies associated with the neutrons and with the protons. This means that N is close to Z for small A and N is greater than Z (but not too much greater) for larger A. We need a negative energy term corresponding to the difference $|N - Z|$. The best agreement with observed binding energies is obtained if this term is proportional to $(N - Z)^2/A$. If we use $N = A - Z$ to express this energy in terms of A and Z, this correction is $-C_4(A - 2Z)^2/A$.

5. Finally, the nuclear force favors *pairing* of protons and of neutrons. This energy term is positive (more binding) if both Z and N are even, negative (less binding) if both Z and N are odd, and zero otherwise. The best fit to the data occurs with the form $\pm C_5A^{-4/3}$ for this term.

The total estimated binding energy E_B is the sum of these five terms:

$$E_B = C_1A - C_2A^{2/3} - C_3\frac{Z(Z - 1)}{A^{1/3}} - C_4\frac{(A - 2Z)^2}{A} \pm C_5A^{-4/3} \quad \text{(43.11)}$$

(nuclear binding energy)

The constants C_1, C_2, C_3, C_4, and C_5, chosen to make this formula best fit the observed binding energies of nuclides, are

$$C_1 = 15.75 \text{ MeV}$$
$$C_2 = 17.80 \text{ MeV}$$
$$C_3 = 0.7100 \text{ MeV}$$
$$C_4 = 23.69 \text{ MeV}$$
$$C_5 = 39 \text{ MeV}$$

The constant C_1 is the binding energy per nucleon due to the saturated nuclear force. This energy is almost 16 MeV per nucleon, about double the *total* binding energy per nucleon in most nuclides.

If we estimate the binding energy E_B using Eq. (43.11), we can solve Eq. (43.10) to use it to estimate the mass of any neutral atom:

$$_Z^A M = ZM_H + Nm_n - \frac{E_B}{c^2} \qquad \text{(semiempirical mass formula)} \qquad (43.12)$$

Equation (43.12) is called the *semiempirical mass formula*. The name is apt; it is *empirical* in the sense that the C's have to be determined empirically (experimentally), yet it does have a sound theoretical basis.

Example 43.4   Estimating binding energy and mass

For the nuclide $_{28}^{62}\text{Ni}$ of Example 43.3, (a) calculate the five terms in the binding energy and the total estimated binding energy, and (b) find the neutral atomic mass using the semiempirical mass formula.

SOLUTION

IDENTIFY and SET UP: We use the liquid-drop model of the nucleus and its five contributions to the binding energy, as given by Eq. (43.11), to calculate the total binding energy E_B. We then use Eq. (43.12) to find the neutral atomic mass $_{28}^{62}M$.

EXECUTE: (a) With $Z = 28$, $A = 62$, and $N = 34$, the five terms in Eq. (43.11) are

1. $C_1 A = (15.75 \text{ MeV})(62) = 976.5 \text{ MeV}$

2. $-C_2 A^{2/3} = -(17.80 \text{ MeV})(62)^{2/3} = -278.8 \text{ MeV}$

3. $-C_3 \dfrac{Z(Z-1)}{A^{1/3}} = -(0.7100 \text{ MeV}) \dfrac{(28)(27)}{(62)^{1/3}}$

$= -135.6 \text{ MeV}$

4. $-C_4 \dfrac{(A - 2Z)^2}{A} = -(23.69 \text{ MeV}) \dfrac{(62 - 56)^2}{62}$

$= -13.8 \text{ MeV}$

5. $+C_5 A^{-4/3} = (39 \text{ MeV})(62)^{-4/3} = 0.2 \text{ MeV}$

The pairing correction (term 5) is by far the smallest of all the terms; it is positive because both Z and N are even. The sum of all five terms is the total estimated binding energy, $E_B = 548.5 \text{ MeV}$.

(b) We use $E_B = 548.5 \text{ MeV}$ in Eq. (43.12):

$$_{28}^{62}M = 28(1.007825 \text{ u}) + 34(1.008665 \text{ u})$$

$$- \frac{548.5 \text{ MeV}}{931.5 \text{ MeV/u}} = 61.925 \text{ u}$$

EVALUATE: The binding energy of $_{28}^{62}\text{Ni}$ calculated in part (a) is only about 0.6% larger than the true value of 545.3 MeV found in Example 43.3, and the mass calculated in part (b) is only about 0.005% smaller than the measured value of 61.928349 u. The semiempirical mass formula can be quite accurate!

The liquid-drop model and the mass formula derived from it are quite successful in correlating nuclear masses, and we will see later that they are a great help in understanding decay processes of unstable nuclides. Some other aspects of nuclei, such as angular momentum and excited states, are better approached with different models.

The Shell Model

The **shell model** of nuclear structure is analogous to the central-field approximation in atomic physics (see Section 41.6). We picture each nucleon as moving in a potential that represents the averaged-out effect of all the other nucleons. This may not seem to be a very promising approach; the nuclear force is very strong, very short range, and therefore strongly distance dependent. However, in some respects, this model turns out to work fairly well.

The potential-energy function for the nuclear force is the same for protons as for neutrons. Figure 43.3a shows a reasonable assumption for the shape of this function: a spherical version of the square-well potential we discussed in Section 40.3. The corners are somewhat rounded because the nucleus doesn't have a sharply defined surface. For protons there is an additional potential energy associated with electrical repulsion. We consider each proton to interact with a sphere of uniform charge density, with radius R and total charge $(Z - 1)e$. Figure 43.3b shows the nuclear, electric, and total potential energies for a proton as functions of the distance r from the center of the nucleus.

In principle, we could solve the Schrödinger equation for a proton or neutron moving in such a potential. For any spherically symmetric potential energy, the angular-momentum states are the same as for the electrons in the central-field approximation in atomic physics. In particular, we can use the concept of *filled shells and subshells* and their relationship to stability. In atomic structure we found that the values $Z = 2, 10, 18, 36, 54$, and 86 (the atomic numbers of the noble gases) correspond to particularly stable electron arrangements.

A comparable effect occurs in nuclear structure. The numbers are different because the potential-energy function is different and the nuclear spin-orbit interaction is much stronger and of opposite sign than in atoms, so the subshells fill up in a different order from those for electrons in an atom. It is found that when the number of neutrons *or* the number of protons is 2, 8, 20, 28, 50, 82, or 126, the resulting structure is unusually stable—that is, has an unusually high binding energy. (Nuclides with $Z = 126$ have not been observed in nature.) These numbers are called *magic numbers.* Nuclides in which Z is a magic number tend to have an above-average number of stable isotopes. There are several *doubly magic* nuclides for which both Z and N are magic, including

$$\begin{matrix} {}^{4}_{2}\text{He} & {}^{16}_{8}\text{O} & {}^{40}_{20}\text{Ca} & {}^{48}_{20}\text{Ca} & {}^{208}_{82}\text{Pb} \end{matrix}$$

All these nuclides have substantially higher binding energy per nucleon than do nuclides with neighboring values of N or Z. They also all have zero nuclear spin. The magic numbers correspond to filled-shell or -subshell configurations of nucleon energy levels with a relatively large jump in energy to the next allowed level.

Test Your Understanding of Section 43.2 Rank the following nuclei in order from largest to smallest value of the binding energy per nucleon. (i) ${}^{4}_{2}\text{He}$; (ii) ${}^{52}_{24}\text{Cr}$; (iii) ${}^{152}_{62}\text{Sm}$; (iv) ${}^{200}_{80}\text{Hg}$; (v) ${}^{252}_{92}\text{Cf}$. ▌

43.3 Nuclear Stability and Radioactivity

Among about 2500 known nuclides, fewer than 300 are stable. The others are unstable structures that decay to form other nuclides by emitting particles and electromagnetic radiation, a process called **radioactivity.** The time scale of these decay processes ranges from a small fraction of a microsecond to billions of years. The *stable* nuclides are shown by dots on the graph in Fig. 43.4, where the neutron number N and proton number (or atomic number) Z for each nuclide are plotted. Such a chart is called a *Segrè chart,* after its inventor, the Italian-American physicist Emilio Segrè (1905–1989).

Each blue line perpendicular to the line $N = Z$ represents a specific value of the mass number $A = Z + N$. Most lines of constant A pass through only one or two stable nuclides; that is, there is usually a very narrow range of stability for a given mass number. The lines at $A = 20$, $A = 40$, $A = 60$, and $A = 80$ are examples. In four cases these lines pass through *three* stable nuclides—namely, at $A = 96, 124, 130$, and 136.

Four stable nuclides have both odd Z and odd N:

$$\begin{matrix} {}^{2}_{1}\text{H} & {}^{6}_{3}\text{Li} & {}^{10}_{5}\text{B} & {}^{14}_{7}\text{N} \end{matrix}$$

These are called *odd-odd nuclides.* The absence of other odd-odd nuclides shows the influence of pairing. Also, there is *no* stable nuclide with $A = 5$ or $A = 8$. The doubly magic ${}^{4}_{2}\text{He}$ nucleus, with a pair of protons and a pair of neutrons, has no interest in accepting a fifth particle into its structure. Collections of eight nucleons decay to smaller nuclides, with a ${}^{8}_{4}\text{Be}$ nucleus immediately splitting into two ${}^{4}_{2}\text{He}$ nuclei.

43.3 Approximate potential-energy functions for a nucleon in a nucleus. The approximate nuclear radius is R.

(a) The potential energy U_{nuc} due to the nuclear force is the same for protons and neutrons. For neutrons, it is the *total* potential energy.

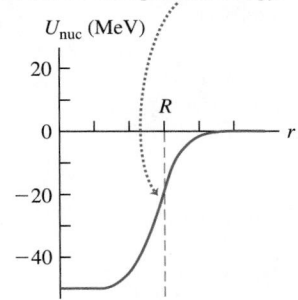

(b) For protons, the total potential energy U_{tot} is the sum of the nuclear (U_{nuc}) and electric (U_{el}) potential energies.

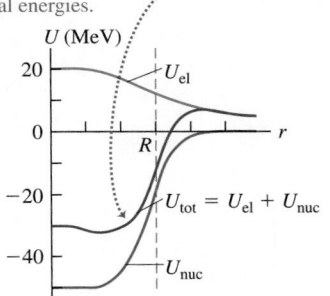

Mastering**PHYSICS**

ActivPhysics 19.2: Nuclear Binding Energy
ActivPhysics 19.4: Radioactivity

43.4 Segrè chart showing neutron number and proton number for stable nuclides.

The points on the Segrè chart representing stable nuclides define a rather narrow stability region. For low mass numbers, the numbers of protons and neutrons are approximately equal, $N \approx Z$. The ratio N/Z increases gradually with A, up to about 1.6 at large mass numbers, because of the increasing influence of the electrical repulsion of the protons. Points to the right of the stability region represent nuclides that have too many protons relative to neutrons to be stable. In these cases, repulsion wins, and the nucleus comes apart. To the left are nuclides with too many neutrons relative to protons. In these cases the energy associated with the neutrons is out of balance with that associated with the protons, and the nuclides decay in a process that converts neutrons to protons. The graph also shows that no nuclide with $A > 209$ or $Z > 83$ is stable. A nucleus is unstable if it is too big. Note that there is no stable nuclide with $Z = 43$ (technetium) or 61 (promethium).

Alpha Decay

Nearly 90% of the 2500 known nuclides are *radioactive;* they are not stable but decay into other nuclides. When unstable nuclides decay into different nuclides, they usually emit alpha (α) or beta (β) particles. An **alpha particle** is a ^{4}He nucleus, two protons and two neutrons bound together, with total spin zero. Alpha emission occurs principally with nuclei that are too large to be stable. When a nucleus emits an alpha particle, its N and Z values each decrease by 2 and A decreases by 4, moving it closer to stable territory on the Segrè chart.

43.5 Alpha decay of the unstable radium nuclide $^{226}_{88}$Ra.

(a)

$^{226}_{88}$Ra

88p
138n

The nuclide $^{226}_{88}$Ra decays by alpha emission to $^{222}_{86}$Rn.

86p
136n

α (2p, 2n)

$^4_2$He

$^{222}_{86}$Rn

(b) Potential-energy curve for an α particle and $^{222}_{86}$Rn nucleus

U (MeV)

The α particle tunnels through the potential-energy barrier.

4.871

0 R r

(c) Energy-level diagram for the system

$^{226}_{88}$Ra can α-decay directly to the $^{222}_{86}$Rn ground level ...

... or it can α-decay to an excited level $^{222}_{86}$Rn*, which can then decay to the $^{222}_{86}$Rn ground level by emitting a 0.186-MeV photon (γ).

4.871 MeV   4.685 MeV α $^{226}_{88}$Ra

α

γ $^{222}_{86}$Rn*

$^{222}_{86}$Rn

0.186 MeV

A familiar example of an alpha emitter is radium, $^{226}_{88}$Ra (Fig. 43.5a). The speed of the emitted alpha particle, determined from the curvature of its path in a transverse magnetic field, is about 1.52×10^7 m/s. This speed, although large, is only 5% of the speed of light, so we can use the nonrelativistic kinetic-energy expression $K = \frac{1}{2}mv^2$:

$$K = \tfrac{1}{2}(6.64 \times 10^{-27} \text{ kg})(1.52 \times 10^7 \text{ m/s})^2 = 7.67 \times 10^{-13} \text{ J} = 4.79 \text{ MeV}$$

Alpha particles are always emitted with definite kinetic energies, determined by conservation of momentum and energy. Because of their charge and mass, alpha particles can travel only several centimeters in air, or a few tenths or hundredths of a millimeter through solids, before they are brought to rest by collisions.

Some nuclei can spontaneously decay by emission of α particles because energy is released in their alpha decay. You can use conservation of mass-energy to show that

alpha decay is possible whenever the mass of the original neutral atom is greater than the sum of the masses of the final neutral atom and the neutral helium-4 atom.

In alpha decay, the α particle tunnels through a potential-energy barrier, as Fig. 43.5b shows. You may want to review the discussion of tunneling in Section 40.4.

Example 43.5 | **Alpha decay of radium**

Show that the α-emission process $^{226}_{88}$Ra $\rightarrow$ $^{222}_{86}$Rn $+$ $^4_2$He is energetically possible, and calculate the kinetic energy of the emitted α particle. The neutral atomic masses are 226.025403 u for $^{226}_{88}$Ra and 222.017571 u for $^{222}_{86}$Rn.

SOLUTION

IDENTIFY and SET UP: Alpha emission is possible if the mass of the $^{226}_{88}$Ra atom is greater than the sum of the atomic masses of $^{222}_{86}$Rn and $^4_2$He. The mass difference between the initial radium atom and the final radon and helium atoms corresponds (through $E = mc^2$) to the energy E released in the decay. Because momentum is conserved as well as energy, *both* the alpha particle and the $^{222}_{86}$Rn atom are in motion after the decay; we will have to account for this in determining the kinetic energy of the alpha particle.

EXECUTE: From Table 43.2, the mass of the $^4_2$He atom is 4.002603 u. The difference in mass between the original nucleus and the decay products is

$$226.025403 \text{ u} - (222.017571 \text{ u} + 4.002603 \text{ u}) = +0.005229 \text{ u}$$

Since this is positive, α decay is energetically possible. The energy equivalent of this mass difference is

$$E = (0.005229 \text{ u})(931.5 \text{ MeV/u}) = 4.871 \text{ MeV}$$

Thus we expect the decay products to emerge with total kinetic energy 4.871 MeV. Momentum is also conserved; if the parent $^{226}_{88}$Ra nucleus is at rest, the daughter $^{222}_{86}$Rn nucleus and the α particle have momenta of equal magnitude p but opposite direction. Kinetic

Continued

energy is $K = \frac{1}{2}mv^2 = p^2/2m$: Since p is the same for the two particles, the kinetic energy divides inversely as their masses. Hence the α particle gets $222/(222 + 4)$ of the total, or 4.78 MeV.

EVALUATE: Experiment shows that $^{226}_{88}$Ra does undergo alpha decay, and the observed α-particle energy is 4.78 MeV. You can check your results by verifying that the alpha particle and the $^{222}_{86}$Rn nucleus produced in the decay have the same magnitude of momentum $p = mv$. You can calculate the speed v of each of the decay products from its respective kinetic energy. You'll find that the alpha particle moves at a sprightly $0.0506c = 1.52 \times 10^7$ m/s; if momentum is conserved, you should find that the $^{222}_{86}$Rn nucleus moves $\frac{4}{222}$ as fast. Does it?

Beta Decay

There are three different simple types of *beta decay: beta-minus, beta-plus,* and *electron capture.* A **beta-minus particle** (β^-) is an electron. It's not obvious how a nucleus can emit an electron if there aren't any electrons in the nucleus. Emission of a β^- involves *transformation* of a neutron into a proton, an electron, and a third particle called an *antineutrino.* In fact, if you freed a neutron from a nucleus, it would decay into a proton, an electron, and an antineutrino in an average time of about 15 minutes.

Beta particles can be identified and their speeds can be measured with techniques that are similar to the Thomson experiments we described in Section 27.5. The speeds of beta particles range up to 0.9995 of the speed of light, so their motion is highly relativistic. They are emitted with a continuous spectrum of energies. This would not be possible if the only two particles were the β^- and the recoiling nucleus, since energy and momentum conservation would then require a definite speed for the β^-. Thus there must be a *third* particle involved. From conservation of charge, it must be neutral, and from conservation of angular momentum, it must be a spin-$\frac{1}{2}$ particle.

This third particle is an antineutrino, the *antiparticle* of a **neutrino.** The symbol for a neutrino is ν_e (the Greek letter nu). Both the neutrino and the antineutrino have zero charge and zero (or very small) mass and therefore produce very little observable effect when passing through matter. Both evaded detection until 1953, when Frederick Reines and Clyde Cowan succeeded in observing the antineutrino directly. We now know that there are at least three varieties of neutrinos, each with its corresponding antineutrino; one is associated with beta decay and the other two are associated with the decay of two unstable particles, the muon and the tau particle. We'll discuss these particles in more detail in Chapter 44. The antineutrino that is emitted in β^- decay is denoted as $\bar{\nu}_e$. The basic process of β^- decay is

$$n \rightarrow p + \beta^- + \bar{\nu}_e \qquad (43.13)$$

Beta-minus decay usually occurs with nuclides for which the neutron-to-proton ratio N/Z is too large for stability. In β^- decay, N decreases by 1, Z increases by 1, and A doesn't change. You can use conservation of mass-energy to show that

> **beta-minus decay can occur whenever the mass of the original neutral atom is larger than that of the final atom.**

Example 43.6 Why cobalt-60 is a beta-minus emitter

The nuclide $^{60}_{27}$Co, an odd-odd unstable nucleus, is used in medical and industrial applications of radiation. Show that it is unstable relative to β^- decay. The atomic masses you need are 59.933822 u for $^{60}_{27}$Co and 59.930791 u for $^{60}_{28}$Ni.

SOLUTION

IDENTIFY and SET UP: Beta-minus decay is possible if the mass of the original neutral atom is greater than that of the final atom.

We must first identify the nuclide that will result if $^{60}_{27}$Co undergoes β^- decay and then compare its neutral atomic mass to that of $^{60}_{27}$Co.

EXECUTE: In the presumed β^- decay of $^{60}_{27}$Co, Z increases by 1 from 27 to 28 and A remains at 60, so the final nuclide is $^{60}_{28}$Ni. The neutral atomic mass of $^{60}_{27}$Co is greater than that of $^{60}_{28}$Ni by 0.003031 u, so β^- decay *can* occur.

EVALUATE: With three decay products in β^- decay—the $^{60}_{28}\text{Ni}$ nucleus, the electron, and the antineutrino—the energy can be shared in many different ways that are consistent with conservation of energy and momentum. It's impossible to predict precisely how the energy will be shared for the decay of a particular $^{60}_{27}\text{Co}$ nucleus. By contrast, in alpha decay there are just two decay products, and their energies and momenta are determined uniquely (see Example 43.5).

We have noted that β^- decay occurs with nuclides that have too large a neutron-to-proton ratio N/Z. Nuclides for which N/Z is too *small* for stability can emit a *positron,* the electron's antiparticle, which is identical to the electron but with positive charge. (We'll discuss the positron in more detail in Chapter 44.) The basic process, called *beta-plus decay* (β^+), is

$$\text{p} \rightarrow \text{n} + \beta^+ + \nu_\text{e} \qquad (43.14)$$

where β^+ is a positron and ν_e is the electron neutrino.

> **Beta-plus decay can occur whenever the mass of the original neutral atom is at least two electron masses larger than that of the final atom.**

You can show this using conservation of mass-energy.

The third type of beta decay is *electron capture.* There are a few nuclides for which β^+ emission is not energetically possible but in which an orbital electron (usually in the K shell) can combine with a proton in the nucleus to form a neutron and a neutrino. The neutron remains in the nucleus and the neutrino is emitted. The basic process is

$$\text{p} + \beta^- \rightarrow \text{n} + \nu_\text{e} \qquad (43.15)$$

You can use conservation of mass-energy to show that

> **electron capture can occur whenever the mass of the original neutral atom is larger than that of the final atom.**

In all types of beta decay, A remains constant. However, in beta-plus decay and electron capture, N increases by 1 and Z decreases by 1 as the neutron–proton ratio increases toward a more stable value. The reaction of Eq. (43.15) also helps to explain the formation of a neutron star, mentioned in Example 43.1.

CAUTION Beta decay inside and outside nuclei The beta-decay reactions given by Eqs. (43.13), (43.14), and (43.15) occur *within* a nucleus. Although the decay of a neutron outside the nucleus proceeds through the reaction of Eq. (43.13), the reaction of Eq. (43.14) is forbidden by conservation of mass-energy for a proton outside the nucleus. The reaction of Eq. (43.15) can occur outside the nucleus only with the addition of some extra energy, as in a collision.

Example 43.7   Why cobalt-57 is not a beta-plus emitter

The nuclide $^{57}_{27}\text{Co}$ is an odd-even unstable nucleus. Show that it cannot undergo β^+ decay, but that it *can* decay by electron capture. The atomic masses you need are 56.936296 u for $^{57}_{27}\text{Co}$ and 56.935399 u for $^{57}_{26}\text{Fe}$.

SOLUTION

IDENTIFY and SET UP: Beta-plus decay is possible if the mass of the original neutral atom is greater than that of the final atom plus two electron masses (0.001097 u). Electron capture is possible if the mass of the original atom is greater than that of the final atom. We must first identify the nuclide that will result if $^{57}_{27}\text{Co}$ undergoes β^+ decay or electron capture and then find the corresponding mass difference.

EXECUTE: The original nuclide is $^{57}_{27}\text{Co}$. In both the presumed β^+ decay and electron capture, Z decreases by 1 from 27 to 26, and A remains at 57, so the final nuclide is $^{57}_{26}\text{Fe}$. Its mass is less than that of $^{57}_{27}\text{Co}$ by 0.000897 u, a value smaller than 0.001097 u (two electron masses), so β^+ decay *cannot* occur. However, the mass of the original atom is greater than the mass of the final atom, so electron capture *can* occur.

EVALUATE: In electron capture there are just two decay products, the final nucleus and the emitted neutrino. As in alpha decay (Example 43.5) but unlike in β^- decay (Example 43.6), the decay products of electron capture have unique energies and momenta. In Section 43.4 we'll see how to relate the probability that electron capture will occur to the *half-life* of this nuclide.

Gamma Decay

The energy of internal motion of a nucleus is quantized. A typical nucleus has a set of allowed energy levels, including a *ground state* (state of lowest energy) and several *excited states*. Because of the great strength of nuclear interactions, excitation energies of nuclei are typically of the order of 1 MeV, compared with a few eV for atomic energy levels. In ordinary physical and chemical transformations the nucleus always remains in its ground state. When a nucleus is placed in an excited state, either by bombardment with high-energy particles or by a radioactive transformation, it can decay to the ground state by emission of one or more photons called **gamma rays** or *gamma-ray photons,* with typical energies of 10 keV to 5 MeV. This process is called *gamma* (γ) *decay.* For example, alpha particles emitted from ^{226}Ra have two possible kinetic energies, either 4.784 MeV or 4.602 MeV. Including the recoil energy of the resulting ^{222}Rn nucleus, these correspond to a total released energy of 4.871 MeV or 4.685 MeV, respectively. When an alpha particle with the smaller energy is emitted, the ^{222}Rn nucleus is left in an excited state. It then decays to its ground state by emitting a gamma-ray photon with energy

$$(4.871 - 4.685)\ \text{MeV} = 0.186\ \text{MeV}$$

A photon with this energy is observed during this decay (Fig. 43.5c).

CAUTION γ **decay vs.** α **and** β **decay** In both α and β decay, the Z value of a nucleus changes and the nucleus of one element becomes the nucleus of a different element. In γ decay, the element does *not* change; the nucleus merely goes from an excited state to a less excited state. ▮

Natural Radioactivity

Many radioactive elements occur in nature. For example, you are very slightly radioactive because of unstable nuclides such as carbon-14 and potassium-40 that are present throughout your body. The study of natural radioactivity began in 1896, one year after Röntgen discovered x rays. Henri Becquerel discovered a radiation from uranium salts that seemed similar to x rays. Intensive investigation in the following two decades by Marie and Pierre Curie, Ernest Rutherford, and many others revealed that the emissions consist of positively and negatively charged particles and neutral rays; they were given the names *alpha, beta,* and *gamma* because of their differing penetration characteristics.

The decaying nucleus is usually called the *parent nucleus;* the resulting nucleus is the *daughter nucleus.* When a radioactive nucleus decays, the daughter nucleus may also be unstable. In this case a *series* of successive decays occurs until a stable configuration is reached. Several such series are found in nature. The most abundant radioactive nuclide found on earth is the uranium isotope ^{238}U, which undergoes a series of 14 decays, including eight α emissions and six β^{-} emissions, terminating at a stable isotope of lead, ^{206}Pb (Fig. 43.6).

Radioactive decay series can be represented on a Segrè chart, as in Fig. 43.7. The neutron number N is plotted vertically, and the atomic number Z is plotted horizontally. In alpha emission, both N and Z decrease by 2. In β^{-} emission, N decreases by 1 and Z increases by 1. The decays can also be represented in equation form; the first two decays in the series are written as

$$^{238}\text{U} \rightarrow\ ^{234}\text{Th} + \alpha$$
$$^{234}\text{Th} \rightarrow\ ^{234}\text{Pa} + \beta^{-} + \bar{\nu}_{\text{e}}$$

or more briefly as

$$^{238}\text{U} \xrightarrow{\alpha}\ ^{234}\text{Th}$$
$$^{234}\text{Th} \xrightarrow{\beta^{-}}\ ^{234}\text{Pa}$$

43.6 Earthquakes are caused in part by the radioactive decay of ^{238}U in the earth's interior. These decays release energy that helps to produce convection currents in the earth's interior. Such currents drive the motions of the earth's crust, including the sudden sharp motions that we call earthquakes (like the one that caused this damage).

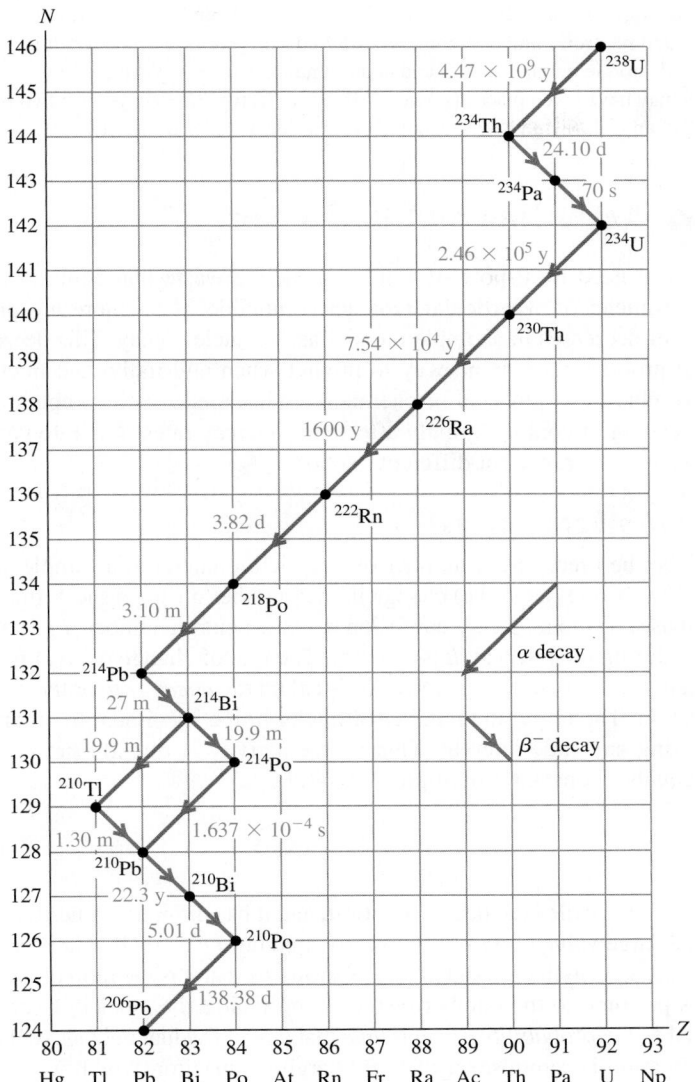

43.7 Segrè chart showing the uranium ^{238}U decay series, terminating with the stable nuclide ^{206}Pb. The times are half-lives (discussed in the next section), given in years (y), days (d), hours (h), minutes (m), or seconds (s).

In the second process, the beta decay leaves the daughter nucleus ^{234}Pa in an excited state, from which it decays to the ground state by emitting a gamma-ray photon. An excited state is denoted by an asterisk, so we can represent the γ emission as

$$^{234}\text{Pa}^* \rightarrow \,^{234}\text{Pa} + \gamma$$

or

$$^{234}\text{Pa}^* \xrightarrow{\gamma} \,^{234}\text{Pa}$$

An interesting feature of the ^{238}U decay series is the branching that occurs at ^{214}Bi. This nuclide decays to ^{210}Pb by emission of an α and a β^-, which can occur in either order. We also note that the series includes unstable isotopes of several elements that also have stable isotopes, including thallium (Tl), lead (Pb), and bismuth (Bi). The unstable isotopes of these elements that occur in the ^{238}U series all have too many neutrons to be stable.

Many other decay series are known. Two of these occur in nature, one starting with the uncommon isotope ^{235}U and ending with ^{207}Pb, the other starting with thorium (^{232}Th) and ending with ^{208}Pb.

Test Your Understanding of Section 43.3 A nucleus with atomic number Z and neutron number N undergoes two decay processes. The result is a nucleus with atomic number $Z - 3$ and neutron number $N - 1$. Which decay processes may have taken place? (i) two β^- decays; (ii) two β^+ decays; (iii) two α decays; (iv) an α decay and a β^- decay; (v) an α decay and a β^+ decay.

43.4 Activities and Half-Lives

Suppose you need to dispose of some radioactive waste that contains a certain number of nuclei of a particular radioactive nuclide. If no more are produced, that number decreases in a simple manner as the nuclei decay. This decrease is a statistical process; there is no way to predict when any individual nucleus will decay. No change in physical or chemical environment, such as chemical reactions or heating or cooling, greatly affects most decay rates. The rate varies over an extremely wide range for different nuclides.

Radioactive Decay Rates

Let $N(t)$ be the (very large) number of radioactive nuclei in a sample at time t, and let $dN(t)$ be the (negative) change in that number during a short time interval dt. (We'll use $N(t)$ to minimize confusion with the neutron number N.) The number of decays during the interval dt is $-dN(t)$. The rate of change of $N(t)$ is the negative quantity $dN(t)/dt$; thus $-dN(t)/dt$ is called the *decay rate* or the **activity** of the specimen. The larger the number of nuclei in the specimen, the more nuclei decay during any time interval. That is, the activity is directly proportional to $N(t)$; it equals a constant λ multiplied by $N(t)$:

$$-\frac{dN(t)}{dt} = \lambda N(t) \qquad (43.16)$$

The constant λ is called the **decay constant,** and it has different values for different nuclides. A large value of λ corresponds to rapid decay; a small value corresponds to slower decay. Solving Eq. (43.16) for λ shows us that λ is the ratio of the number of decays per time to the number of remaining radioactive nuclei; λ can then be interpreted as the *probability per unit time* that any individual nucleus will decay.

The situation is reminiscent of a discharging capacitor, which we studied in Section 26.4. Equation (43.16) has the same form as the negative of Eq. (26.15), with q and $1/RC$ replaced by $N(t)$ and λ. Then we can make the same substitutions in Eq. (26.16), with the initial number of nuclei $N(0) = N_0$, to find the exponential function:

$$N(t) = N_0 e^{-\lambda t} \qquad \text{(number of remaining nuclei)} \qquad (43.17)$$

Figure 43.8 is a graph of this function, showing the number of remaining nuclei $N(t)$ as a function of time.

The **half-life** $T_{1/2}$ is the time required for the number of radioactive nuclei to decrease to one-half the original number N_0. Then half of the remaining radioactive nuclei decay during a second interval $T_{1/2}$, and so on. The numbers remaining after successive half-lives are $N_0/2$, $N_0/4$, $N_0/8$, ...

To get the relationship between the half-life $T_{1/2}$ and the decay constant λ, we set $N(t)/N_0 = \frac{1}{2}$ and $t = T_{1/2}$ in Eq. (43.17), obtaining

$$\tfrac{1}{2} = e^{-\lambda T_{1/2}}$$

We take logarithms of both sides and solve for $T_{1/2}$:

$$T_{1/2} = \frac{\ln 2}{\lambda} = \frac{0.693}{\lambda} \qquad (43.18)$$

43.8 The number of nuclei in a sample of a radioactive element as a function of time. The sample's activity has an exponential decay curve with the same shape.

The mean lifetime T_{mean}, generally called the *lifetime,* of a nucleus or unstable particle is proportional to the half-life $T_{1/2}$:

$$T_{mean} = \frac{1}{\lambda} = \frac{T_{1/2}}{\ln 2} = \frac{T_{1/2}}{0.693} \qquad \begin{array}{l} \text{(lifetime } T_{mean}, \text{ decay} \\ \text{constant } \lambda, \text{ and half-life } T_{1/2}) \end{array} \qquad \text{(43.19)}$$

In particle physics the life of an unstable particle is usually described by the lifetime, not the half-life.

Because the activity $-dN(t)/dt$ at any time equals $\lambda N(t)$, Eq. (43.17) tells us that the activity also depends on time as $e^{-\lambda t}$. Thus the graph of activity versus time has the same shape as Fig. 43.8. Also, after successive half-lives, the activity is one-half, one-fourth, one-eighth, and so on of the original activity.

CAUTION **A half-life may not be enough** It is sometimes implied that any radioactive sample will be safe after a half-life has passed. That's wrong. If your radioactive waste initially has ten times too much activity for safety, it is not safe after one half-life, when it still has five times too much. Even after three half-lives it still has 25% more activity than is safe. The number of radioactive nuclei and the activity approach zero only as t approaches infinity.

A common unit of activity is the **curie**, abbreviated Ci, which is defined to be 3.70×10^{10} decays per second. This is approximately equal to the activity of one gram of radium. The SI unit of activity is the *becquerel,* abbreviated Bq. One becquerel is one decay per second, so

$$1 \text{ Ci} = 3.70 \times 10^{10} \text{ Bq} = 3.70 \times 10^{10} \text{ decays/s}$$

Example 43.8   **Activity of ^{57}Co**

The isotope ^{57}Co decays by electron capture to ^{57}Fe with a half-life of 272 d. The ^{57}Fe nucleus is produced in an excited state, and it almost instantaneously emits gamma rays that we can detect. (a) Find the mean lifetime and decay constant for ^{57}Co. (b) If the activity of a ^{57}Co radiation source is now 2.00 μCi, how many ^{57}Co nuclei does the source contain? (c) What will be the activity after one year?

SOLUTION

IDENTIFY and SET UP: This problem uses the relationships among decay constant λ, lifetime T_{mean}, and activity $-dN(t)/dt$. In part (a) we use Eq. (43.19) to find λ and T_{mean} from $T_{1/2}$. In part (b), we use Eq. (43.16) to calculate the number of nuclei $N(t)$ from the activity. Finally, in part (c) we use Eqs. (43.16) and (43.17) to find the activity after one year.

EXECUTE: (a) It's convenient to convert the half-life to seconds:

$$T_{1/2} = (272 \text{ d})(86,400 \text{ s/d}) = 2.35 \times 10^7 \text{ s}$$

From Eq. (43.19), the mean lifetime and decay constant are

$$T_{mean} = \frac{T_{1/2}}{\ln 2} = \frac{2.35 \times 10^7 \text{ s}}{0.693} = 3.39 \times 10^7 \text{ s} = 392 \text{ days}$$

$$\lambda = \frac{1}{T_{mean}} = 2.95 \times 10^{-8} \text{ s}^{-1}$$

(b) The activity $-dN(t)/dt$ is given as 2.00 μCi, so

$$-\frac{dN(t)}{dt} = 2.00 \text{ } \mu\text{Ci} = (2.00 \times 10^{-6})(3.70 \times 10^{10} \text{ s}^{-1})$$

$$= 7.40 \times 10^4 \text{ decays/s}$$

From Eq. (43.16) this is equal to $\lambda N(t)$, so we find

$$N(t) = -\frac{dN(t)/dt}{\lambda} = \frac{7.40 \times 10^4 \text{ s}^{-1}}{2.95 \times 10^{-8} \text{ s}^{-1}} = 2.51 \times 10^{12} \text{ nuclei}$$

If you feel we're being too cavalier about the "units" decays and nuclei, you can use decays/(nucleus $\cdot$ s) as the unit for λ.

(c) From Eq. (43.17) the number $N(t)$ of nuclei remaining after one year (3.156×10^7 s) is

$$N(t) = N_0 e^{-\lambda t} = N_0 e^{-(2.95 \times 10^{-8} \text{ s}^{-1})(3.156 \times 10^7 \text{ s})} = 0.394 N_0$$

The number of nuclei has decreased to 0.394 of the original number. Equation (43.16) says that the activity is proportional to the number of nuclei, so the activity has decreased by this same factor to $(0.394)(2.00 \text{ } \mu\text{Ci}) = 0.788 \text{ } \mu\text{Ci}$.

EVALUATE: The number of nuclei found in part (b) is equivalent to 4.17×10^{-12} mol, with a mass of 2.38×10^{-10} g. This is a far smaller mass than even the most sensitive balance can measure.

After one 272-day half-life, the number of ^{57}Co nuclei has decreased to $N_0/2$; after $2(272 \text{ d}) = 544$ d, it has decreased to $N_0/2^2 = N_0/4$. This result agrees with our answer to part (c), which says that after 365 d the number of nuclei is between $N_0/2$ and $N_0/4$.

Radioactive Dating

An interesting application of radioactivity is the dating of archaeological and geological specimens by measuring the concentration of radioactive isotopes. The most familiar example is *carbon dating*. The unstable isotope ^{14}C, produced during nuclear reactions in the atmosphere that result from cosmic-ray bombardment, gives a small proportion of ^{14}C in the CO_2 in the atmosphere. Plants that obtain their carbon from this source contain the same proportion of ^{14}C as the atmosphere. When a plant dies, it stops taking in carbon, and its ^{14}C β^- decays to ^{14}N with a half-life of 5730 years. By measuring the proportion of ^{14}C in the remains, we can determine how long ago the organism died.

One difficulty with radiocarbon dating is that the ^{14}C concentration in the atmosphere changes over long time intervals. Corrections can be made on the basis of other data such as measurements of tree rings that show annual growth cycles. Similar radioactive techniques are used with other isotopes for dating geological specimens. Some rocks, for example, contain the unstable potassium isotope ^{40}K, a beta emitter that decays to the stable nuclide ^{40}Ar with a half-life of 2.4×10^8 y. The age of the rock can be determined by comparing the concentrations of ^{40}K and ^{40}Ar.

Example 43.9 | Radiocarbon dating

Before 1900 the activity per unit mass of atmospheric carbon due to the presence of ^{14}C averaged about 0.255 Bq per gram of carbon. (a) What fraction of carbon atoms were ^{14}C? (b) In analyzing an archaeological specimen containing 500 mg of carbon, you observe 174 decays in one hour. What is the age of the specimen, assuming that its activity per unit mass of carbon when it died was that average value of the air?

SOLUTION

IDENTIFY and SET UP: The key idea is that the present-day activity of a biological sample containing ^{14}C is related to both the elapsed time since it stopped taking in atmospheric carbon and its activity at that time. We use Eqs. (43.16) and (43.17) to solve for the age t of the specimen. In part (a) we determine the number of ^{14}C atoms $N(t)$ from the activity $-dN(t)/dt$ using Eq. (43.16). We find the total number of carbon atoms in 500 mg by using the molar mass of carbon (12.011 g/mol, given in Appendix D), and we use the result to calculate the fraction of carbon atoms that are ^{14}C. The activity decays at the same rate as the number of ^{14}C nuclei; we use this and Eq. (43.17) to solve for the age t of the specimen.

EXECUTE: (a) To use Eq. (43.16), we must first find the decay constant λ from Eq. (43.18):

$$T_{1/2} = 5730 \text{ y} = (5730 \text{ y})(3.156 \times 10^7 \text{ s/y}) = 1.808 \times 10^{11} \text{ s}$$

$$\lambda = \frac{\ln 2}{T_{1/2}} = \frac{0.693}{1.808 \times 10^{11} \text{ s}} = 3.83 \times 10^{-12} \text{ s}^{-1}$$

Then, from Eq. (43.16),

$$N(t) = \frac{-dN/dt}{\lambda} = \frac{0.255 \text{ s}^{-1}}{3.83 \times 10^{-12} \text{ s}^{-1}} = 6.65 \times 10^{10} \text{ atoms}$$

The *total* number of C atoms in 1 gram (1/12.011 mol) is $(1/12.011)(6.022 \times 10^{23}) = 5.01 \times 10^{22}$. The ratio of ^{14}C atoms to all C atoms is

$$\frac{6.65 \times 10^{10}}{5.01 \times 10^{22}} = 1.33 \times 10^{-12}$$

Only four carbon atoms in every 3×10^{12} are ^{14}C.

(b) Assuming that the activity per gram of carbon in the specimen when it died $(t = 0)$ was 0.255 Bq/g = $(0.255 \text{ s}^{-1} \cdot \text{g}^{-1})(3600 \text{ s/h}) = 918 \text{ h}^{-1} \cdot \text{g}^{-1}$, the activity of 500 mg of carbon then was $(0.500 \text{ g})(918 \text{ h}^{-1} \cdot \text{g}^{-1}) = 459 \text{ h}^{-1}$. The observed activity now, at time t, is 174 h^{-1}. Since the activity is proportional to the number of radioactive nuclei, the activity ratio $174/459 = 0.379$ equals the number ratio $N(t)/N_0$.

Now we solve Eq. (43.17) for t and insert values for $N(t)/N_0$ and λ:

$$t = \frac{\ln(N(t)/N_0)}{-\lambda} = \frac{\ln 0.379}{-3.83 \times 10^{-12} \text{ s}^{-1}} = 2.53 \times 10^{11} \text{ s} = 8020 \text{ y}$$

EVALUATE: After 8020 y the ^{14}C activity has decreased from 459 to 174 decays per hour. The specimen died and stopped taking CO_2 out of the air about 8000 years ago.

Radiation in the Home

A serious health hazard in some areas is the accumulation in houses of ^{222}Rn, an inert, colorless, odorless radioactive gas. Looking at the ^{238}U decay chain in Fig. 43.7, we see that the half-life of ^{222}Rn is 3.82 days. If so, why not just move out of the house for a while and let it decay away? The answer is that ^{222}Rn is continuously being *produced* by the decay of ^{226}Ra, which is found in minute quantities in the rocks and soil on which some houses are built. It's a dynamic equilibrium situation, in which the rate of production equals the rate of decay. The reason ^{222}Rn is a bigger hazard than the other elements in the ^{238}U decay series is that it's a gas. During its short half-life of 3.82 days it can migrate from the soil into your house. If a ^{222}Rn nucleus decays in your lungs, it emits a damaging α particle and its daughter nucleus ^{218}Po, which is *not* chemically inert and is likely to stay in your lungs until it decays, emits another damaging α particle and so on down the ^{238}U decay series.

How much of a hazard is radon? Although reports indicate values as high as 3500 pCi/L, the average activity per volume in the air inside American homes due to ^{222}Rn is about 1.5 pCi/L (over a thousand decays each second in an average-sized room). *If* your environment has this level of activity, it has been estimated that a lifetime exposure would reduce your life expectancy by about 40 days. For comparison, smoking one pack of cigarettes per day reduces life expectancy by 6 years, and it is estimated that the average emission from all the nuclear power plants in the world reduces life expectancy by anywhere from 0.01 day to 5 days. These figures include catastrophes such as the 1986 nuclear reactor disaster at Chernobyl, for which the *local* effect on life expectancy is much greater.

Test Your Understanding of Section 43.4   Which sample contains a greater number of nuclei: a 5.00-μCi sample of ^{240}Pu (half-life 6560 y) or a 4.45-μCi sample of ^{243}Am (half-life 7370 y)? (i) the ^{240}Pu sample; (ii) the ^{243}Am sample; (iii) both have the same number of nuclei.

43.5 Biological Effects of Radiation

The above discussion of radon introduced the interaction of radiation with living organisms, a topic of vital interest and importance. Under *radiation* we include radioactivity (alpha, beta, gamma, and neutrons) and electromagnetic radiation such as x rays. As these particles pass through matter, they lose energy, breaking molecular bonds and creating ions—hence the term *ionizing radiation.* Charged particles interact directly with the electrons in the material. X rays and γ rays interact by the photoelectric effect, in which an electron absorbs a photon and breaks loose from its site, or by Compton scattering (see Section 38.3). Neutrons cause ionization indirectly through collisions with nuclei or absorption by nuclei with subsequent radioactive decay of the resulting nuclei.

These interactions are extremely complex. It is well known that excessive exposure to radiation, including sunlight, x rays, and all the nuclear radiations, can destroy tissues. In mild cases it results in a burn, as with common sunburn. Greater exposure can cause very severe illness or death by a variety of mechanisms, including massive destruction of tissue cells, alterations of genetic material, and destruction of the components in bone marrow that produce red blood cells.

Calculating Radiation Doses

Radiation dosimetry is the quantitative description of the effect of radiation on living tissue. The *absorbed dose* of radiation is defined as the energy delivered to the tissue per unit mass. The SI unit of absorbed dose, the joule per kilogram, is called the *gray* (Gy); 1 Gy = 1 J/kg. Another unit is the *rad,* defined as 0.01 J/kg:

$$1 \text{ rad} = 0.01 \text{ J/kg} = 0.01 \text{ Gy}$$

Absorbed dose by itself is not an adequate measure of biological effect because equal energies of different kinds of radiation cause different extents of biological effect. This variation is described by a numerical factor called the **relative biological effectiveness (RBE),** also called the *quality factor* (QF), of each specific radiation. X rays with 200 keV of energy are defined to have an

Table 43.3 Relative Biological Effectiveness (RBE) for Several Types of Radiation

Radiation	RBE (Sv/Gy or rem/rad)
X rays and γ rays	1
Electrons	1.0–1.5
Slow neutrons	3–5
Protons	10
α particles	20
Heavy ions	20

RBE of unity, and the effects of other radiations can be compared experimentally. Table 43.3 shows approximate values of RBE for several radiations. All these values depend somewhat on the kind of tissue in which the radiation is absorbed and on the energy of the radiation.

The biological effect is described by the product of the absorbed dose and the RBE of the radiation; this quantity is called the *biologically equivalent dose,* or simply the *equivalent dose.* The SI unit of equivalent dose for humans is the sievert (Sv):

$$\text{Equivalent dose (Sv)} = \text{RBE} \times \text{Absorbed dose (Gy)} \qquad (43.20)$$

A more common unit, corresponding to the rad, is the rem (an abbreviation of *röntgen equivalent for man*):

$$\text{Equivalent dose (rem)} = \text{RBE} \times \text{Absorbed dose (rad)} \qquad (43.21)$$

Thus the unit of the RBE is 1 Sv/Gy or 1 rem/rad, and 1 rem = 0.01 Sv.

Example 43.10   **Dose from a medical x ray**

During a diagnostic x-ray examination a 1.2-kg portion of a broken leg receives an equivalent dose of 0.40 mSv. (a) What is the equivalent dose in mrem? (b) What is the absorbed dose in mrad and in mGy? (c) If the x-ray energy is 50 keV, how many x-ray photons are absorbed?

SOLUTION

IDENTIFY and SET UP: We are asked to relate the equivalent dose (the biological effect of the radiation, measured in sieverts or rems) to the absorbed dose (the energy absorbed per mass, measured in grays or rads). In part (a) we use the conversion factor 1 rem = 0.01 Sv for equivalent dose. Table 43.3 gives the RBE for x rays; we use this value in part (b) to determine the absorbed dose using Eqs. (43.20) and (43.21). Finally, in part (c) we use the mass and the definition of absorbed dose to find the total energy absorbed and the total number of photons absorbed.

EXECUTE: (a) The equivalent dose in mrem is

$$\frac{0.40\ \text{mSv}}{0.01\ \text{Sv/rem}} = 40\ \text{mrem}$$

(b) For x rays, RBE = 1 rem/rad or 1 Sv/Gy, so the absorbed dose is

$$\frac{40\ \text{mrem}}{1\ \text{rem/rad}} = 40\ \text{mrad}$$

$$\frac{0.40\ \text{mSv}}{1\ \text{Sv/Gy}} = 0.40\ \text{mGy} = 4.0 \times 10^{-4}\,\text{J/kg}$$

(c) The total energy absorbed is

$$(4.0 \times 10^{-4}\,\text{J/kg})(1.2\ \text{kg}) = 4.8 \times 10^{-4}\,\text{J} = 3.0 \times 10^{15}\ \text{eV}$$

The number of x-ray photons is

$$\frac{3.0 \times 10^{15}\ \text{eV}}{5.0 \times 10^{4}\,\text{eV/photon}} = 6.0 \times 10^{10}\ \text{photons}$$

EVALUATE: The absorbed dose is relatively large because x rays have a low RBE. If the ionizing radiation had been a beam of α particles, for which RBE = 20, the absorbed dose needed for an equivalent dose of 0.40 mSv would be only 0.020 mGy, corresponding to a smaller total absorbed energy of $2.4 \times 10^{-5}\,\text{J}$.

Radiation Hazards

Here are a few numbers for perspective. To convert from Sv to rem, simply multiply by 100. An ordinary chest x-ray exam delivers about 0.20–0.40 mSv to about 5 kg of tissue. Radiation exposure from cosmic rays and natural radioactivity in soil, building materials, and so on is of the order of 2–3 mSv per year at sea level and twice that at an elevation of 1500 m (5000 ft). A whole-body dose of up to about 0.20 Sv causes no immediately detectable effect. A short-term whole-body dose of 5 Sv or more usually causes death within a few days or weeks. A localized dose of 100 Sv causes complete destruction of the exposed tissues.

The long-term hazards of radiation exposure in causing various cancers and genetic defects have been widely publicized, and the question of whether there is any "safe" level of radiation exposure has been hotly debated. U.S. government regulations are based on a maximum *yearly* exposure, from all except natural resources, of 2 to 5 mSv. Workers with occupational exposure to radiation are permitted 50 mSv per year. Recent studies suggest that these limits are too high and that even extremely small exposures carry hazards, but it is very difficult to gather reliable statistics on the effects of low doses. It has become clear that any use of x rays for medical diagnosis should be preceded by a very careful estimation of the relationship of risk to possible benefit.

Another sharply debated question is that of radiation hazards from nuclear power plants. The radiation level from these plants is *not* negligible. However, to make a meaningful evaluation of hazards, we must compare these levels with the alternatives, such as coal-powered plants. The health hazards of coal smoke are serious and well documented, and the natural radioactivity in the smoke from a coal-fired power plant is believed to be roughly 100 times as great as that from a properly operating nuclear plant with equal capacity. But the comparison is not this simple; the possibility of a nuclear accident and the very serious problem of safe disposal of radioactive waste from nuclear plants must also be considered. It is clearly impossible to eliminate *all* hazards to health. Our goal should be to try to take a rational approach to the problem of *minimizing* the hazard from all sources. Figure 43.9 shows one estimate of the various sources of radiation exposure for the U.S. population. Ionizing radiation is a two-edged sword; it poses very serious health hazards, yet it also provides many benefits to humanity, including the diagnosis and treatments of disease and a wide variety of analytical techniques.

Beneficial Uses of Radiation

Radiation is widely used in medicine for intentional selective destruction of tissue such as tumors. The hazards are considerable, but if the disease would be fatal without treatment, any hazard may be preferable. Artificially produced isotopes are often used as radiation sources. Such isotopes have several advantages over naturally radioactive isotopes. They may have shorter half-lives and correspondingly greater activity. Isotopes can be chosen that emit the type and energy of radiation desired. Some artificial isotopes have been replaced by photon and electron beams from linear accelerators.

Nuclear medicine is an expanding field of application. Radioactive isotopes have virtually the same electron configurations and resulting chemical behavior as stable isotopes of the same element. But the location and concentration of radioactive isotopes can easily be detected by measurements of the radiation they emit. A familiar example is the use of radioactive iodine for thyroid studies. Nearly all the iodine ingested is either eliminated or stored in the thyroid, and the body's chemical reactions do not discriminate between the unstable isotope ^{131}I and the stable isotope ^{127}I. A minute quantity of ^{131}I is fed or injected into the patient, and the speed with which it becomes concentrated in the thyroid provides a measure of thyroid function. The half-life is 8.02 days, so there are no long-lasting radiation hazards. By use of more sophisticated scanning detectors, one can also obtain a "picture" of the thyroid, which shows enlargement and other abnormalities. This procedure, a type of *autoradiography,* is comparable to photographing the glowing filament of an incandescent light bulb by using the light emitted by the filament itself. If this process discovers cancerous thyroid nodules, they can be destroyed by much larger quantities of ^{131}I.

Another useful nuclide for nuclear medicine is technetium-99 (^{99}Tc), which is formed in an excited state by the β^- decay of molybdenum (^{99}Mo). The technetium then decays to its ground state by emitting a γ-ray photon with energy 143 keV. The half-life is 6.01 hours, unusually long for γ emission. (The ground state of ^{99}Tc is also unstable, with a half-life of 2.11×10^5 y; it decays by β^- emission to the stable ruthenium nuclide ^{99}Ru.) The chemistry of technetium is such that it can readily be attached to organic molecules that are taken up by various organs of the body. A small quantity of such technetium-bearing molecules is injected into a patient, and a scanning detector or *gamma camera* is used to produce an image, or *scintigram,* that reveals which parts of the body take up these γ-emitting molecules. This technique, in which ^{99}Tc acts as a radioactive *tracer,* plays an important role in locating cancers, embolisms, and other pathologies (Fig. 43.10).

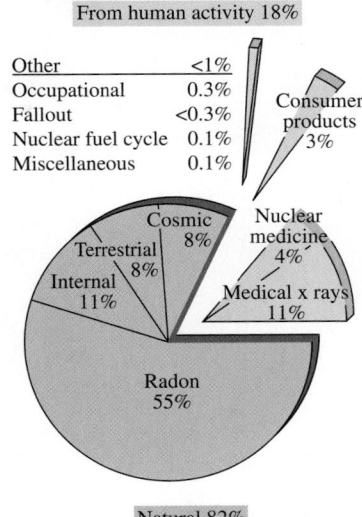

43.9 Contribution of various sources to the total average radiation exposure in the U.S. population, expressed as percentages of the total.

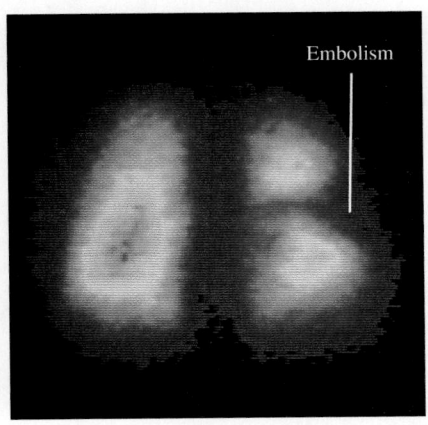

43.10 This colored scintigram shows where a chemical containing radioactive ^{99}Tc was taken up by a patient's lungs. The orange color in the lung on the left indicates strong γ-ray emission by the ^{99}Tc, which shows that the chemical was able to pass into this lung through the bloodstream. The lung on the right shows weaker emission, indicating the presence of an embolism (a blood clot or other obstruction in an artery) that is restricting the flow of blood to this lung.

Tracer techniques have many other applications. Tritium (^{3}H), a radioactive hydrogen isotope, is used to tag molecules in complex organic reactions; radioactive tags on pesticide molecules, for example, can be used to trace their passage through food chains. In the world of machinery, radioactive iron can be used to study piston-ring wear. Laundry detergent manufacturers have even tested the effectiveness of their products using radioactive dirt.

Many direct effects of radiation are also useful, such as strengthening polymers by cross-linking, sterilizing surgical tools, dispersing of unwanted static electricity in the air, and intentionally ionizing the air in smoke detectors. Gamma rays are also being used to sterilize and preserve some food products.

Test Your Understanding of Section 43.5 Alpha particles have 20 times the relative biological effectiveness of 200-keV x rays. Which would be better to use to radiate tissue deep inside the body? (i) a beam of alpha particles; (ii) a beam of 200-keV x rays; (iii) both are equally effective. ❙

43.6 Nuclear Reactions

In the preceding sections we studied the decay of unstable nuclei, especially spontaneous emission of an α or β particle, sometimes followed by γ emission. Nothing needs to be done to initiate this decay, and nothing can be done to control it. This section examines some *nuclear reactions,* rearrangements of nuclear components that result from a bombardment by a particle rather than a spontaneous natural process. Rutherford suggested in 1919 that a massive particle with sufficient kinetic energy might be able to penetrate a nucleus. The result would be either a new nucleus with greater atomic number and mass number or a decay of the original nucleus. Rutherford bombarded nitrogen (^{14}N) with α particles and obtained an oxygen (^{17}O) nucleus and a proton:

$$\,^4_2\text{He} + \,^{14}_7\text{N} \rightarrow \,^{17}_8\text{O} + \,^1_1\text{H} \tag{43.22}$$

Rutherford used alpha particles from naturally radioactive sources. In Chapter 44 we'll describe some of the particle accelerators that are now used to initiate nuclear reactions.

Nuclear reactions are subject to several *conservation laws.* The classical conservation principles for charge, momentum, angular momentum, and energy (including rest energies) are obeyed in all nuclear reactions. An additional conservation law, not anticipated by classical physics, is conservation of the total number of nucleons. The numbers of protons and neutrons need not be conserved separately; in β decay, neutrons and protons change into one another. We'll study the basis of the conservation of nucleon number in Chapter 44.

When two nuclei interact, charge conservation requires that the sum of the initial atomic numbers must equal the sum of the final atomic numbers. Because of conservation of nucleon number, the sum of the initial mass numbers must also equal the sum of the final mass numbers. In general, these are *not* elastic collisions, and the total initial mass does *not* equal the total final mass.

Reaction Energy

The difference between the masses before and after the reaction corresponds to the **reaction energy,** according to the mass–energy relationship $E = mc^2$. If initial particles A and B interact to produce final particles C and D, the reaction energy Q is defined as

$$Q = (M_A + M_B - M_C - M_D)c^2 \qquad \text{(reaction energy)} \tag{43.23}$$

To balance the electrons, we use the neutral atomic masses in Eq. (43.23). That is, we use the mass of $^1_1$H for a proton, $^2_1$H for a deuteron, $^4_2$He for an α particle, and so on. When Q is positive, the total mass decreases and the total kinetic energy increases. Such a reaction is called an *exoergic reaction*. When Q is negative, the mass increases and the kinetic energy decreases, and the reaction is called an *endoergic reaction*. The terms *exothermal* and *endothermal*, borrowed from chemistry, are also used. In an endoergic reaction the reaction cannot occur at all unless the initial kinetic energy in the center-of-mass reference frame is at least as great as $|Q|$. That is, there is a **threshold energy,** the minimum kinetic energy to make an endoergic reaction go.

Example 43.11 Exoergic and endoergic reactions

(a) When a lithium-7 nucleus is bombarded by a proton, two alpha particles (^{4}He) are produced. Find the reaction energy. (b) Calculate the reaction energy for the reaction $^4_2$He + $^{14}_7$N → $^{17}_8$O + $^1_1$H.

SOLUTION

IDENTIFY and SET UP: The reaction energy Q for any nuclear reaction equals c^2 times the difference between the total initial mass and the total final mass, as in Eq. (43.23). Table 43.2 gives the required masses.

EXECUTE: (a) The reaction is $^1_1$H + $^7_3$Li → $^4_2$He + $^4_2$He. The initial and final masses and their respective sums are

A: $^1_1$H	1.007825 u	*C:* $^4_2$He	4.002603 u
B: $^7_3$Li	7.016004 u	*D:* $^4_2$He	4.002603 u
	8.023829 u		8.005206 u

The mass decreases by 0.018623 u. From Eq. (43.23), the reaction energy is

$$Q = (0.018623 \text{ u})(931.5 \text{ MeV/u}) = +17.35 \text{ MeV}$$

(b) The initial and final masses are

A: $^4_2$He	4.002603 u	*C:* $^{17}_8$O	16.999132 u
B: $^{14}_7$N	14.003074 u	*D:* $^1_1$H	1.007825 u
	18.005677 u		18.006957 u

The mass increases by 0.001280 u, and the corresponding reaction energy is

$$Q = (-0.001280 \text{ u})(931.5 \text{ MeV/u}) = -1.192 \text{ MeV}$$

EVALUATE: The reaction in part (a) is *exoergic:* The final total kinetic energy of the two separating alpha particles is 17.35 MeV greater than the initial total kinetic energy of the proton and the lithium nucleus. The reaction in part (b) is *endoergic:* In the center-of-mass system—that is, in a head-on collision with zero total momentum—the minimum total initial kinetic energy required for this reaction to occur is 1.192 MeV.

Ordinarily, the endoergic reaction of part (b) of Example 43.11 would be produced by bombarding stationary ^{14}N nuclei with alpha particles from an accelerator. In this case an alpha's kinetic energy must be *greater than* 1.192 MeV. If all the alpha's kinetic energy went solely to increasing the rest energy, the final kinetic energy would be zero, and momentum would not be conserved. When a particle with mass m and kinetic energy K collides with a stationary particle with mass M, the total kinetic energy K_{cm} in the center-of-mass coordinate system (the energy available to cause reactions) is

$$K_{cm} = \frac{M}{M + m} K \qquad (43.24)$$

This expression assumes that the kinetic energies of the particles and nuclei are much less than their rest energies. We leave the derivation of Eq. (43.24) to you (see Problem 43.77). In the present example, $K_{cm} = (14.00/18.01)K$, so K must be at least $(18.01/14.00)(1.192 \text{ MeV}) = 1.533 \text{ MeV}$.

For a charged particle such as a proton or an α particle to penetrate the nucleus of another atom and cause a reaction, it must usually have enough initial kinetic energy to overcome the potential-energy barrier caused by the repulsive electrostatic forces. In the reaction of part (a) of Example 43.11, if we treat the proton and the ^{7}Li nucleus as spherically symmetric charges with radii given by Eq. (43.1), their centers will be 3.5×10^{-15} m apart when they touch. The repulsive potential energy of the proton (charge $+e$) and the ^{7}Li nucleus (charge $+3e$) at this separation r is

$$U = \frac{1}{4\pi\epsilon_0} \frac{(e)(3e)}{r} = (9.0 \times 10^9 \text{ N} \cdot \text{m}^2/\text{C}^2) \frac{(3)(1.6 \times 10^{-19} \text{ C})^2}{3.5 \times 10^{-15} \text{ m}}$$
$$= 2.0 \times 10^{-13} \text{ J} = 1.2 \text{ MeV}$$

Even though the reaction is exoergic, the proton must have a minimum kinetic energy of about 1.2 MeV for the reaction to occur, unless the proton *tunnels* through the barrier (see Section 40.4).

Neutron Absorption

Absorption of *neutrons* by nuclei forms an important class of nuclear reactions. Heavy nuclei bombarded by neutrons can undergo a series of neutron absorptions alternating with beta decays, in which the mass number A increases by as much as 25. Some of the *transuranic elements,* elements having Z larger than 92, are produced in this way. These elements have not been found in nature. Many transuranic elements, having Z possibly as high as 118, have been identified.

The analytical technique of *neutron activation analysis* uses similar reactions. When bombarded by neutrons, many stable nuclides absorb a neutron to become unstable and then undergo β^- decay. The energies of the β^- and associated γ emissions depend on the unstable nuclide and provide a means of identifying it and the original stable nuclide. Quantities of elements that are far too small for conventional chemical analysis can be detected in this way.

Test Your Understanding of Section 43.6 The reaction described in part (a) of Example 43.11 is exoergic. Can it happen naturally when a sample of solid lithium is placed in a flask of hydrogen gas? ∎

PhET: Nuclear Fission

43.7 Nuclear Fission

Nuclear fission is a decay process in which an unstable nucleus splits into two fragments of comparable mass. Fission was discovered in 1938 through the experiments of Otto Hahn and Fritz Strassman in Germany. Pursuing earlier work by Fermi, they bombarded uranium ($Z = 92$) with neutrons. The resulting radiation did not coincide with that of any known radioactive nuclide. Urged on by their colleague Lise Meitner, they used meticulous chemical analysis to reach the astonishing but inescapable conclusion that they had found a radioactive isotope of barium ($Z = 56$). Later, radioactive krypton ($Z = 36$) was also found. Meitner and Otto Frisch correctly interpreted these results as showing that uranium nuclei were splitting into two massive fragments called *fission fragments.* Two or three free neutrons usually appear along with the fission fragments and, very occasionally, a light nuclide such as ^{3}H.

Both the common isotope (99.3%) ^{238}U and the uncommon isotope (0.7%) ^{235}U (as well as several other nuclides) can be easily split by neutron bombardment: ^{235}U by slow neutrons (kinetic energy less than 1 eV) but ^{238}U only by fast neutrons with a minimum of about 1 MeV of kinetic energy. Fission resulting from neutron absorption is called *induced fission.* Some nuclides can also undergo *spontaneous fission* without initial neutron absorption, but this is quite rare. When ^{235}U absorbs a neutron, the resulting nuclide ^{236}U* is in a highly excited state and splits into two fragments almost instantaneously. Strictly speaking, it is ^{236}U*, not ^{235}U, that undergoes fission, but it's usual to speak of the fission of ^{235}U.

Over 100 different nuclides, representing more than 20 different elements, have been found among the fission products. Figure 43.11 shows the distribution of mass numbers for fission fragments from the fission of ^{235}U. Most of the

43.11 Mass distribution of fission fragments from the fission of ^{236}U* (an excited state of ^{236}U), which is produced when ^{235}U absorbs a neutron. The vertical scale is logarithmic.

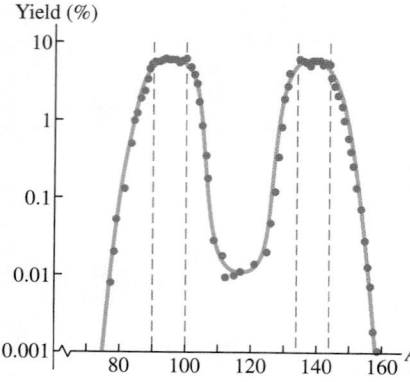

fragments have mass numbers from 90 to 100 and from 135 to 145; fission into two fragments with nearly equal mass is unlikely.

Fission Reactions

You should check the following two typical fission reactions for conservation of nucleon number and charge:

$$^{235}_{92}U + ^{1}_{0}n \rightarrow ^{236}_{92}U^* \rightarrow ^{144}_{56}Ba + ^{89}_{36}Kr + 3^{1}_{0}n$$

$$^{235}_{92}U + ^{1}_{0}n \rightarrow ^{236}_{92}U^* \rightarrow ^{140}_{54}Xe + ^{94}_{38}Sr + 2^{1}_{0}n$$

The total kinetic energy of the fission fragments is enormous, about 200 MeV (compared to typical α and β energies of a few MeV). The reason for this is that nuclides at the high end of the mass spectrum (near $A = 240$) are less tightly bound than those nearer the middle ($A = 90$ to 145). Referring to Fig. 43.2, we see that the average binding energy per nucleon is about 7.6 MeV at $A = 240$ but about 8.5 MeV at $A = 120$. Therefore a rough estimate of the expected *increase* in binding energy during fission is about 8.5 MeV − 7.6 MeV = 0.9 MeV per nucleon, or a total of $(235)(0.9 \text{ MeV}) \approx 200 \text{ MeV}$.

CAUTION **Binding energy and rest energy** It may seem to be a violation of conservation of energy to have an increase in both the binding energy and the kinetic energy during a fission reaction. But relative to the total rest energy E_0 of the separated nucleons, the rest energy of the nucleus is E_0 *minus* E_B. Thus an *increase* in binding energy corresponds to a *decrease* in rest energy as rest energy is converted to the kinetic energy of the fission fragments.

Fission fragments always have too many neutrons to be stable. We noted in Section 43.3 that the neutron–proton ratio (N/Z) for stable nuclides is about 1 for light nuclides but almost 1.6 for the heaviest nuclides because of the increasing influence of the electrical repulsion of the protons. The N/Z value for stable nuclides is about 1.3 at $A = 100$ and 1.4 at $A = 150$. The fragments have about the same N/Z as ^{235}U, about 1.55. They usually respond to this surplus of neutrons by undergoing a series of β^- decays (each of which increases Z by 1 and decreases N by 1) until a stable value of N/Z is reached. A typical example is

$$^{140}_{54}Xe \xrightarrow{\beta^-} ^{140}_{55}Cs \xrightarrow{\beta^-} ^{140}_{56}Ba \xrightarrow{\beta^-} ^{140}_{57}La \xrightarrow{\beta^-} ^{140}_{58}Ce$$

The nuclide ^{140}Ce is stable. This series of β^- decays produces, on average, about 15 MeV of additional kinetic energy. The neutron excess of fission fragments also explains why two or three free neutrons are released during the fission.

Fission appears to set an upper limit on the production of transuranic nuclei, mentioned in Section 43.6, that are relatively stable. There are theoretical reasons to expect that nuclei near $Z = 114$, $N = 184$ or 196, might be stable with respect to spontaneous fission. In the shell model (see Section 43.2), these numbers correspond to filled shells and subshells in the nuclear energy-level structure. Such *superheavy nuclei* would still be unstable with respect to alpha emission. In 2009 it was confirmed that there are at least four isotopes with $Z = 114$, the longest-lived of which has a half-life due to alpha decay of about 2.6 s.

Liquid-Drop Model

We can understand fission qualitatively on the basis of the liquid-drop model of the nucleus (see Section 43.2). The process is shown in Fig. 43.12 in terms of an electrically charged liquid drop. These sketches shouldn't be taken too literally, but they may help to develop your intuition about fission. A ^{235}U nucleus absorbs a neutron (Fig. 43.12a), becoming a ^{236}U* nucleus with excess energy (Fig. 43.12b). This excess energy causes violent oscillations, during which a neck between two lobes develops (Fig. 43.12c). The electric repulsion of these two lobes stretches the neck farther (Fig. 43.12d), and finally two smaller fragments are formed (Fig. 43.12e) that move rapidly apart.

43.12 A liquid-drop model of fission.

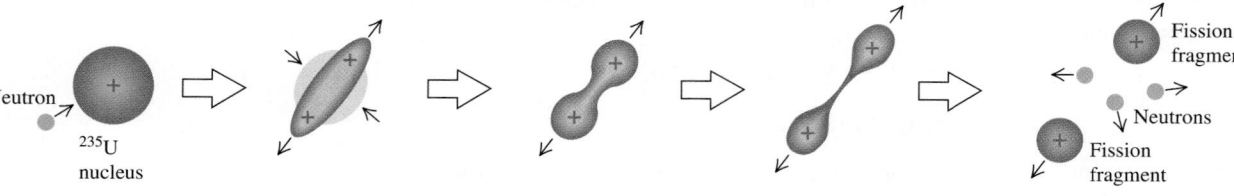

(a) A ^{235}U nucleus absorbs a neutron.

(b) The resulting ^{236}U* nucleus is in a highly excited state and oscillates strongly.

(c) A neck develops, and electric repulsion pushes the two lobes apart.

(d) The two lobes separate, forming fission fragments.

(e) The fragments emit neutrons at the time of fission (or occasionally a few seconds later).

Neutron

^{235}U nucleus

Fission fragment

Neutrons

Fission fragment

43.13 Hypothetical potential-energy function for two fission fragments in a fissionable nucleus. At distances r beyond the range of the nuclear force, the potential energy varies approximately as $1/r$. Fission occurs if there is an excitation energy greater than U_B or an appreciable probability for tunneling through the potential-energy barrier.

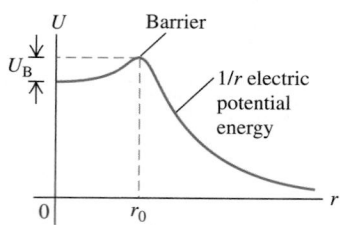

Application **Making Radioactive Isotopes for Medicine**
The fragments that result from nuclear fission are typically unstable, neutron-rich isotopes. A number of these are useful for medical diagnosis and cancer radiotherapy (see Section 43.5). This photograph shows a nuclear fission reactor used for producing such isotopes. The uranium fuel is kept in a large tank of water for cooling. Some of the neutron-rich fission fragments undergo beta decay and emit electrons that move faster than the speed of light in water (about $0.75c$). Like an airplane that produces an intense sonic boom when it flies faster than sound (see Section 16.9), these ultrafast electrons produce a "light boom" called *Čerenkov radiation* that has a characteristic blue color.

This qualitative picture has been developed into a more quantitative theory to explain why some nuclei undergo fission and others don't. Figure 43.13 shows a hypothetical potential-energy function for two possible fission fragments. If neutron absorption results in an excitation energy greater than the energy barrier height U_B, fission occurs immediately. Even when there isn't quite enough energy to surmount the barrier, fission can take place by quantum-mechanical *tunneling,* discussed in Section 40.4. In principle, many stable heavy nuclei can fission by tunneling. But the probability depends very critically on the height and width of the barrier. For most nuclei this process is so unlikely that it is never observed.

Chain Reactions

Fission of a uranium nucleus, triggered by neutron bombardment, releases other neutrons that can trigger more fissions, suggesting the possibility of a **chain reaction** (Fig. 43.14). The chain reaction may be made to proceed slowly and in a controlled manner in a nuclear reactor or explosively in a bomb. The energy release in a nuclear chain reaction is enormous, far greater than that in any chemical reaction. (In a sense, *fire* is a chemical chain reaction.) For example, when uranium is "burned" to uranium dioxide in the chemical reaction

$$U + O_2 \rightarrow UO_2$$

the heat of combustion is about 4500 J/g. Expressed as energy per atom, this is about 11 eV per atom. By contrast, fission liberates about 200 MeV per atom, nearly 20 million times as much energy.

Nuclear Reactors

A *nuclear reactor* is a system in which a controlled nuclear chain reaction is used to liberate energy. In a nuclear power plant, this energy is used to generate steam, which operates a turbine and turns an electrical generator.

On average, each fission of a ^{235}U nucleus produces about 2.5 free neutrons, so 40% of the neutrons are needed to sustain a chain reaction. A ^{235}U nucleus is much more likely to absorb a low-energy neutron (less than 1 eV) than one of the higher-energy neutrons (1 MeV or so) that are liberated during fission. In a nuclear reactor the higher-energy neutrons are slowed down by collisions with nuclei in the surrounding material, called the *moderator,* so they are much more likely to cause further fissions. In nuclear power plants, the moderator is often water, occasionally graphite. The *rate* of the reaction is controlled by inserting or withdrawing *control rods* made of elements (such as boron or cadmium) whose nuclei *absorb* neutrons without undergoing any additional reaction. The isotope ^{238}U can also absorb neutrons, leading to ^{239}U*, but not with high enough probability for it to sustain a chain reaction by itself. Thus uranium that is used in reactors is often "enriched" by increasing the proportion of ^{235}U above the natural value of 0.7%, typically to 3% or so, by isotope-separation processing.

43.14 Schematic diagram of a nuclear fission chain reaction.

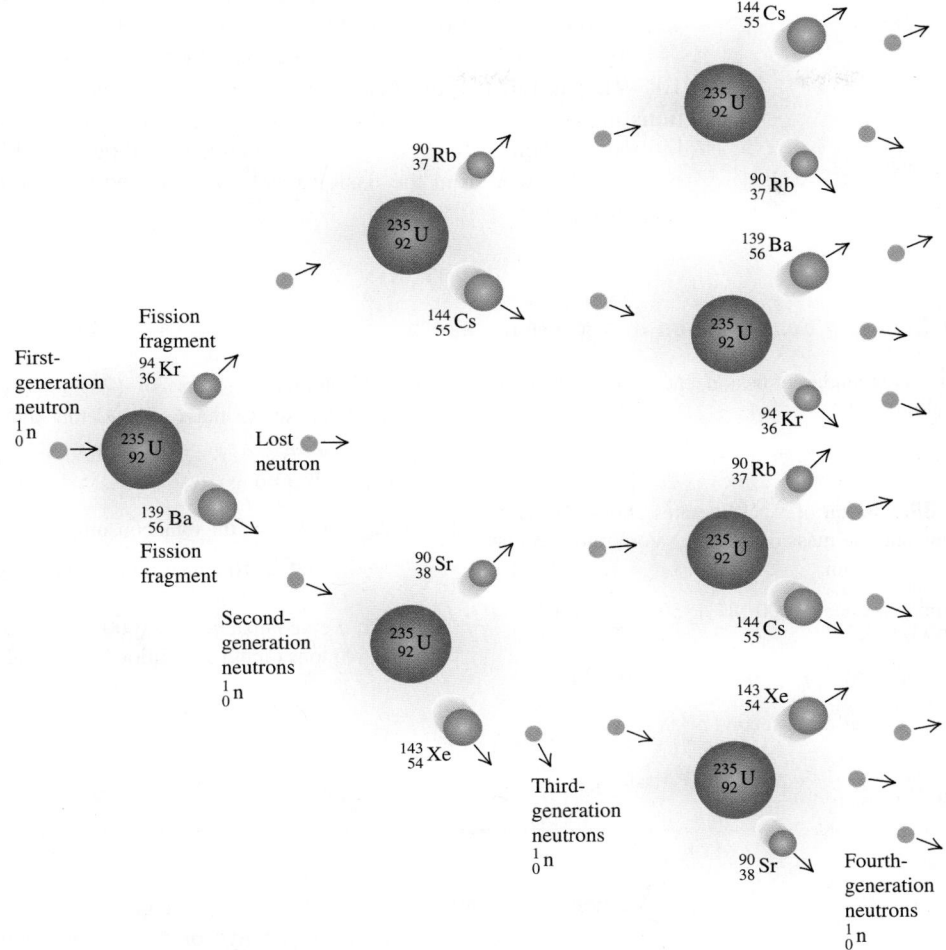

The most familiar application of nuclear reactors is for the generation of electric power. As was noted above, the fission energy appears as kinetic energy of the fission fragments, and its immediate result is to increase the internal energy of the fuel elements and the surrounding moderator. This increase in internal energy is transferred as heat to generate steam to drive turbines, which spin the electrical generators. Figure 43.15 is a schematic diagram of a nuclear power plant.

43.15 Schematic diagram of a nuclear power plant.

The energetic fission fragments heat the water surrounding the reactor core. The steam generator is a heat exchanger that takes heat from this highly radioactive water and generates nonradioactive steam to run the turbines.

A typical nuclear plant has an electric-generating capacity of 1000 MW (or 10^9 W). The turbines are heat engines and are subject to the efficiency limitations imposed by the second law of thermodynamics, discussed in Chapter 20. In modern nuclear plants the overall efficiency is about one-third, so 3000 MW of thermal power from the fission reaction is needed to generate 1000 MW of electrical power.

Example 43.12   **Uranium consumption in a nuclear reactor**

What mass of ^{235}U must undergo fission each day to provide 3000 MW of thermal power?

SOLUTION

IDENTIFY and SET UP: Fission of ^{235}U liberates about 200 MeV per atom. We use this and the mass of the ^{235}U atom to determine the required amount of uranium.

EXECUTE: Each second, we need 3000 MJ or 3000×10^6 J. Each fission provides 200 MeV, or

$$(200 \text{ MeV/fission})(1.6 \times 10^{-13} \text{ J/MeV}) = 3.2 \times 10^{-11} \text{ J/fission}$$

The number of fissions needed each second is

$$\frac{3000 \times 10^6 \text{ J}}{3.2 \times 10^{-11} \text{ J/fission}} = 9.4 \times 10^{19} \text{ fissions}$$

Each ^{235}U atom has a mass of $(235 \text{ u})(1.66 \times 10^{-27} \text{ kg/u}) = 3.9 \times 10^{-25}$ kg, so the mass of ^{235}U that undergoes fission each second is

$$(9.4 \times 10^{19})(3.9 \times 10^{-25} \text{ kg}) = 3.7 \times 10^{-5} \text{ kg} = 37 \ \mu\text{g}$$

In one day (86,400 s), the total consumption of ^{235}U is

$$(3.7 \times 10^{-5} \text{ kg/s})(86,400 \text{ s}) = 3.2 \text{ kg}$$

EVALUATE: For comparison, a 1000-MW coal-fired power plant burns 10,600 tons (about 10 million kg) of coal per day!

We mentioned above that about 15 MeV of the energy released after fission of a ^{235}U nucleus comes from the β^- decays of the fission fragments. This fact poses a serious problem with respect to control and safety of reactors. Even after the chain reaction has been completely stopped by insertion of control rods into the core, heat continues to be evolved by the β^- decays, which cannot be stopped. For a 3000-MW reactor this heat power is initially very large, about 200 MW. In the event of total loss of cooling water, this power is more than enough to cause a catastrophic meltdown of the reactor core and possible penetration of the containment vessel. The difficulty in achieving a "cold shutdown" following an accident at the Three Mile Island nuclear power plant in Pennsylvania in March 1979 was a result of the continued evolution of heat due to β^- decays.

The catastrophe of April 26, 1986, at Chernobyl reactor No. 4 in Ukraine resulted from a combination of an inherently unstable design and several human errors committed during a test of the emergency core cooling system. Too many control rods were withdrawn to compensate for a decrease in power caused by a buildup of neutron absorbers such as ^{135}Xe. The power level rose from 1% of normal to 100 times normal in 4 seconds; a steam explosion ruptured pipes in the core cooling system and blew the heavy concrete cover off the reactor. The graphite moderator caught fire and burned for several days, and there was a meltdown of the core. The total activity of the radioactive material released into the atmosphere has been estimated as about 10^8 Ci.

Test Your Understanding of Section 43.7   The fission of ^{235}U can be triggered by the absorption of a slow neutron by a nucleus. Can a slow *proton* be used to trigger ^{235}U fission?   ❙

43.16 The proton-proton chain.

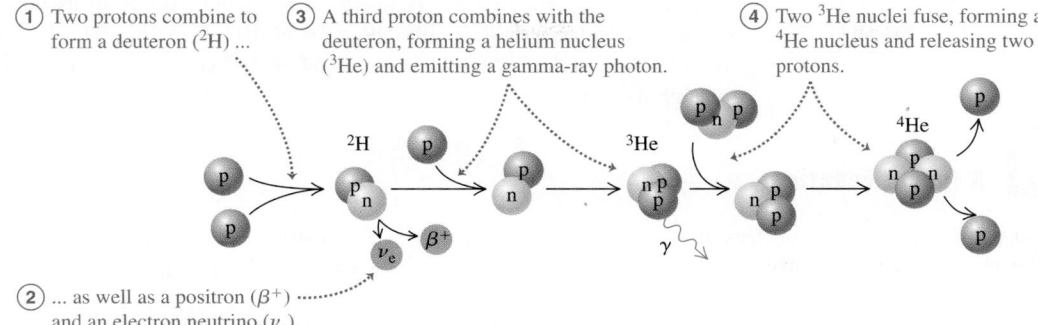

① Two protons combine to form a deuteron (^{2}H) ...

③ A third proton combines with the deuteron, forming a helium nucleus (^{3}He) and emitting a gamma-ray photon.

④ Two ^{3}He nuclei fuse, forming a ^{4}He nucleus and releasing two protons.

② ... as well as a positron (β^+) and an electron neutrino (ν_e).

43.8 Nuclear Fusion

ActivPhysics 19.3: Fusion

In a **nuclear fusion** reaction, two or more small light nuclei come together, or *fuse,* to form a larger nucleus. Fusion reactions release energy for the same reason as fission reactions: The binding energy per nucleon after the reaction is greater than before. Referring to Fig. 43.2, we see that the binding energy per nucleon increases with A up to about $A = 60$, so fusion of nearly any two light nuclei to make a nucleus with A less than 60 is likely to be an exoergic reaction. In comparison to fission, we are moving toward the peak of this curve from the opposite side. Another way to express the energy relationships is that the total mass of the products is less than that of the initial particles.

Here are three examples of energy-liberating fusion reactions, written in terms of the neutral atoms:

$$^1_1\text{H} + {}^1_1\text{H} \rightarrow {}^2_1\text{H} + \beta^+ + \nu_e$$

$$^2_1\text{H} + {}^1_1\text{H} \rightarrow {}^3_2\text{He} + \gamma$$

$$^3_2\text{He} + {}^3_2\text{He} \rightarrow {}^4_2\text{He} + {}^1_1\text{H} + {}^1_1\text{H}$$

In the first reaction, two protons combine to form a deuteron (^{2}H), with the emission of a positron (β^+) and an electron neutrino. In the second, a proton and a deuteron combine to form the nucleus of the light isotope of helium, ^{3}He, with the emission of a gamma ray. Now double the first two reactions to provide the two ^{3}He nuclei that fuse in the third reaction to form an alpha particle (^{4}He) and two protons. Together the reactions make up the process called the *proton-proton chain* (Fig. 43.16).

The net effect of the chain is the conversion of four protons into one α particle, two positrons, two electron neutrinos, and two γ's. We can calculate the energy release from this part of the process: The mass of an α particle plus two positrons is the mass of neutral ^{4}He, the neutrinos have zero (or negligible) mass, and the gammas have zero mass.

43.17 The energy released as starlight comes from fusion reactions deep within a star's interior. When a star is first formed and for most of its life, it converts the hydrogen in its core into helium. As a star ages, the core temperature can become high enough for additional fusion reactions that convert helium into carbon, oxygen, and other elements.

Mass of four protons	4.029106 u
Mass of ^{4}He	4.002603 u
Mass difference and energy release	0.026503 u and 24.69 MeV

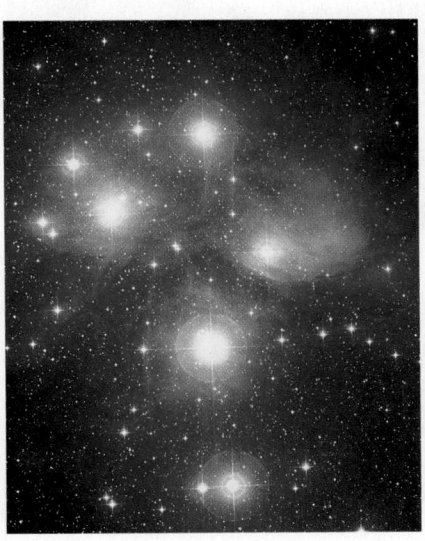

The two positrons that are produced during the first step of the proton-proton chain collide with two electrons; mutual annihilation of the four particles takes place, and their rest energy is converted into $4(0.511 \text{ MeV}) = 2.044 \text{ MeV}$ of gamma radiation. Thus the total energy released is $(24.69 + 2.044) \text{ MeV} = 26.73 \text{ MeV}$. The proton-proton chain takes place in the interior of the sun and other stars (Fig. 43.17). Each gram of the sun's mass contains about 4.5×10^{23} protons. If all of these protons were fused into helium, the energy released would be about 130,000 kWh. If the sun were to continue to radiate at its present rate, it would take about 75×10^9 years to exhaust its supply of protons. As we will see below, fusion reactions can occur only at extremely high temperatures; in the sun, these temperatures are found only deep within the interior. Hence the sun cannot fuse *all* of its protons, and can sustain fusion for a total of only about 10×10^9 years in total. The present age of the solar system (including the sun) is 4.54×10^9 years, so the sun is about halfway through its available store of protons.

Example 43.13 · A fusion reaction

Two deuterons fuse to form a *triton* (a nucleus of tritium, or ^{3}H) and a proton. How much energy is liberated?

SOLUTION

IDENTIFY and SET UP: This is a nuclear reaction of the type discussed in Section 43.6. We find the energy released using Eq. (43.23).

EXECUTE: Adding one electron to each nucleus makes each a neutral atom; we find their masses in Table 43.2. Substituting into Eq. (43.23), we find

$$Q = [2(2.014102 \text{ u}) - 3.016049 \text{ u} - 1.007825 \text{ u}]$$
$$\times (931.5 \text{ MeV/u}) = 4.03 \text{ MeV}$$

EVALUATE: Thus 4.03 MeV is released in the reaction; the triton and proton together have 4.03 MeV more kinetic energy than the two deuterons had together.

Achieving Fusion

For two nuclei to undergo fusion, they must come together to within the range of the nuclear force, typically of the order of 2×10^{-15} m. To do this, they must overcome the electrical repulsion of their positive charges. For two protons at this distance, the corresponding potential energy is about 1.2×10^{-13} J or 0.7 MeV; this represents the total initial *kinetic* energy that the fusion nuclei must have—for example, 0.6×10^{-13} J each in a head-on collision.

Atoms have this much energy only at extremely high temperatures. The discussion of Section 18.3 showed that the average translational kinetic energy of a gas molecule at temperature T is $\frac{3}{2}kT$, where k is Boltzmann's constant. The temperature at which this is equal to $E = 0.6 \times 10^{-13}$ J is determined by the relationship

$$E = \tfrac{3}{2}kT$$

$$T = \frac{2E}{3k} = \frac{2(0.6 \times 10^{-13} \text{ J})}{3(1.38 \times 10^{-23} \text{ J/K})} = 3 \times 10^9 \text{ K}$$

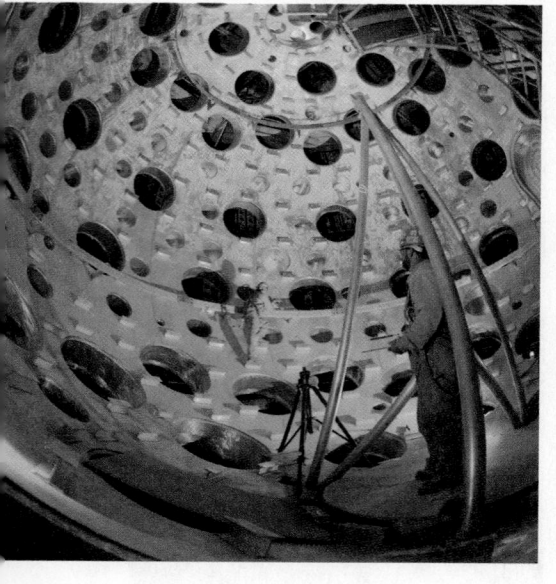

43.18 This target chamber at the National Ignition Facility in California has apertures for 192 powerful laser beams. The lasers deliver 5×10^{14} W of power for a few nanoseconds to a millimeter-sized pellet of deuterium and tritium at the center of the chamber, thus triggering thermonuclear fusion.

Fusion reactions are possible at lower temperatures because the Maxwell–Boltzmann distribution function (see Section 18.5) gives a small fraction of protons with kinetic energies much higher than the average value. The proton-proton reaction occurs at "only" 1.5×10^7 K at the center of the sun, making it an extremely low-probability process; but that's why the sun is expected to last so long. At these temperatures the fusion reactions are called *thermonuclear* reactions.

Intensive efforts are under way to achieve controlled fusion reactions, which potentially represent an enormous new resource of energy (see Fig. 24.11). At the temperatures mentioned, light atoms are fully ionized, and the resulting state of matter is called a *plasma*. In one kind of experiment using *magnetic confinement,* a plasma is heated to extremely high temperature by an electrical discharge, while being contained by appropriately shaped magnetic fields. In another, using *inertial confinement,* pellets of the material to be fused are heated by a high-intensity laser beam (see Fig. 43.18). Some of the reactions being studied are

$$^2_1\text{H} + {}^2_1\text{H} \rightarrow {}^3_1\text{H} + {}^1_1\text{H} + 4.0 \text{ MeV} \qquad (1)$$

$$^3_1\text{H} + {}^2_1\text{H} \rightarrow {}^4_2\text{He} + {}^1_0\text{n} + 17.6 \text{ MeV} \qquad (2)$$

$$^2_1\text{H} + {}^2_1\text{H} \rightarrow {}^3_2\text{He} + {}^1_0\text{n} + 3.3 \text{ MeV} \qquad (3)$$

$$^3_2\text{He} + {}^2_1\text{H} \rightarrow {}^4_2\text{He} + {}^1_1\text{H} + 18.3 \text{ MeV} \qquad (4)$$

We considered reaction (1) in Example 43.13; two deuterons fuse to form a triton and a proton. In reaction (2) a triton combines with another deuteron to form an alpha particle and a neutron. The result of both of these reactions together is the conversion of three deuterons into an alpha particle, a proton, and a neutron, with the liberation of 21.6 MeV of energy. Reactions (3) and (4) together achieve the same conversion. In a plasma that contains deuterons, the two pairs of reactions occur with roughly equal probability. As yet, no one has succeeded in producing these reactions under controlled conditions in such a way as to yield a net surplus of usable energy.

Methods of achieving fusion that don't require high temperatures are also being studied; these are called *cold fusion*. One scheme that does work uses an unusual hydrogen molecule ion. The usual H_2^+ ion consists of two protons bound by one shared electron; the nuclear spacing is about 0.1 nm. If the protons are replaced by a deuteron (^2H) and a triton (^3H) and the electron by a *muon,* which is 208 times as massive as the electron, the spacing is made smaller by a factor of 208. The probability then becomes appreciable for the two nuclei to tunnel through the narrow repulsive potential-energy barrier and fuse in reaction (2) above. The prospect of making this process, called *muon-catalyzed fusion,* into a practical energy source is still distant.

Test Your Understanding of Section 43.8   Are *all* fusion reactions exoergic?   ❙

Nuclear properties: A nucleus is composed of A nucleons (Z protons and N neutrons). All nuclei have about the same density. The radius of a nucleus with mass number A is given approximately by Eq. (43.1). A single nuclear species of a given Z and N is called a nuclide. Isotopes are nuclides of the same element (same Z) that have different numbers of neutrons. Nuclear masses are measured in atomic mass units. Nucleons have angular momentum and a magnetic moment. (See Examples 43.1 and 43.2.)

$$R = R_0 A^{1/3}$$
$$(R_0 = 1.2 \times 10^{-15} \text{ m}) \quad (43.1)$$

Nuclear binding and structure: The mass of a nucleus is always less than the mass of the protons and neutrons within it. The mass difference multiplied by c^2 gives the binding energy E_B. The binding energy for a given nuclide is determined by the nuclear force, which is short range and favors pairs of particles, and by the electric repulsion between protons. A nucleus is unstable if A or Z is too large or if the ratio N/Z is wrong. Two widely used models of the nucleus are the liquid-drop model and the shell model; the latter is analogous to the central-field approximation for atomic structure. (See Examples 43.3 and 43.4.)

$$E_B = (ZM_H + Nm_n - {}_Z^A M)c^2 \quad (43.10)$$

Radioactive decay: Unstable nuclides usually emit an alpha particle (a ${}_2^4\text{He}$ nucleus) or a beta particle (an electron) in the process of changing to another nuclide, sometimes followed by a gamma-ray photon. The rate of decay of an unstable nucleus is described by the decay constant λ, the half-life $T_{1/2}$, or the lifetime T_{mean}. If the number of nuclei at time $t = 0$ is N_0 and no more are produced, the number at time t is given by Eq. (43.17). (See Examples 43.5–43.9.)

$$N(t) = N_0 e^{-\lambda t} \quad (43.17)$$
$$T_{mean} = \frac{1}{\lambda} = \frac{T_{1/2}}{\ln 2} = \frac{T_{1/2}}{0.693} \quad (43.19)$$

Biological effects of radiation: The biological effect of any radiation depends on the product of the energy absorbed per unit mass and the relative biological effectiveness (RBE), which is different for different radiations. (See Example 43.10.)

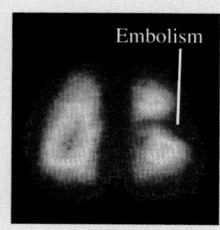

Nuclear reactions: In a nuclear reaction, two nuclei or particles collide to produce two new nuclei or particles. Reactions can be exoergic or endoergic. Several conservation laws, including charge, energy, momentum, angular momentum, and nucleon number, are obeyed. Energy is released by the fission of a heavy nucleus into two lighter, always unstable, nuclei. Energy is also released by the fusion of two light nuclei into a heavier nucleus. (See Examples 43.11–43.13.)

BRIDGING PROBLEM   Saturation of ^{128}I Production

In an experiment, the iodine isotope ^{128}I is created by irradiating a sample of ^{127}I with a beam of neutrons, yielding 1.50×10^6 ^{128}I nuclei per second. Initially no ^{128}I nuclei are present. A ^{128}I nucleus decays by β^- emission with a half-life of 25.0 min. (a) To what nuclide does ^{128}I decay? (b) Could that nuclide decay back to ^{128}I by β^+ emission? Why or why not? (c) After the sample has been irradiated for a long time, what is the maximum number of ^{128}I atoms that can be present in the sample? What is the maximum activity that can be produced? (This steady-state situation is called *saturation*.) (d) Find an expression for the number of ^{128}I atoms present in the sample as a function of time.

SOLUTION GUIDE

See MasteringPhysics® study area for a Video Tutor solution.

IDENTIFY and SET UP

1. What happens to the values of Z, N, and A in β^- decay? What must be true for β^- decay to be possible? For β^+ decay to be possible?
2. You'll need to write an equation for the rate of change dN/dt of the number N of ^{128}I atoms in the sample, taking account of

both the creation of ^{128}I by the neutron irradiation and the decay of any ^{128}I present. In the steady state, how do the rates of these two processes compare?
3. List the unknown quantities for each part of the problem and identify your target variables.

EXECUTE

4. Find the values of Z and N of the nuclide produced by the decay of ^{128}I. What element is this?
5. Decide whether this nuclide can decay back to ^{128}I.
6. Inspect your equation for dN/dt. What is the value of dN/dt in the steady state? Use this to solve for the steady-state values of N and the activity.
7. Solve your dN/dt equation for the function $N(t)$. (*Hint:* See Section 26.4.)

EVALUATE

8. Your result from step 6 tells you the value of N after a long time (that is, for large values of t). Is this consistent with your result from step 7? What would constitute a "long time" under these conditions?

Problems

For instructor-assigned homework, go to www.masteringphysics.com

•, ••, •••: Problems of increasing difficulty. **CP:** Cumulative problems incorporating material from earlier chapters. **CALC:** Problems requiring calculus. **BIO:** Biosciences problems.

DISCUSSION QUESTIONS

Q43.1 BIO Neutrons have a magnetic dipole moment and can undergo spin flips by absorbing electromagnetic radiation. Why, then, are protons rather than neutrons used in MRI of body tissues? (See Fig. 43.1.)

Q43.2 In Eq. (43.11), as the total number of nucleons becomes larger, the importance of the second term in the equation decreases relative to that of the first term. Does this make physical sense? Explain.

Q43.3 Why aren't the masses of all nuclei integer multiples of the mass of a single nucleon?

Q43.4 Can you tell from the value of the mass number A whether to use a plus value, a minus value, or zero for the fifth term of Eq. (43.11)? Explain.

Q43.5 What are the six known elements for which Z is a magic number? Discuss what properties these elements have as a consequence of their special values of Z.

Q43.6 The binding energy per nucleon for most nuclides doesn't vary much (see Fig. 43.2). Is there similar consistency in the *atomic* energy of atoms, on an "energy per electron" basis? If so, why? If not, why not?

Q43.7 Heavy, unstable nuclei usually decay by emitting an α or β particle. Why don't they usually emit a single proton or neutron?

Q43.8 The only two stable nuclides with more protons than neutrons are $^1_1$H and $^3_2$He. Why is $Z > N$ so uncommon?

Q43.9 Since lead is a stable element, why doesn't the ^{238}U decay series shown in Fig. 43.7 stop at lead, ^{214}Pb?

Q43.10 In the ^{238}U decay series shown in Fig. 43.7, some nuclides in the series are found much more abundantly in nature than others, even though every ^{238}U nucleus goes through every step in the series before finally becoming ^{206}Pb. Why don't the intermediate nuclides all have the same abundance?

Q43.11 Compared to α particles with the same energy, β particles can much more easily penetrate through matter. Why is this?

Q43.12 If ^A_ZEl$_i$ represents the initial nuclide, what is the decay process or processes if the final nuclide is (a) $^A_{Z+1}$El$_f$; (b) $^{A-4}_{Z-2}$El$_f$; (c) $^A_{Z-1}$El$_f$?

Q43.13 In a nuclear decay equation, why can we represent an electron as $^0_{-1}\beta^-$? What are the equivalent representations for a positron, a neutrino, and an antineutrino?

Q43.14 Why is the alpha, beta, or gamma decay of an unstable nucleus unaffected by the *chemical* situation of the atom, such as the nature of the molecule or solid in which it is bound? The chemical situation of the atom can, however, have an effect on the half-life in electron capture. Why is this?

Q43.15 In the process of *internal conversion*, a nucleus decays from an excited state to a ground state by giving the excitation energy directly to an atomic electron rather than emitting a gamma-ray photon. Why can this process also produce x-ray photons?

Q43.16 In Example 43.9 (Section 43.4), the activity of atmospheric carbon *before* 1900 was given. Discuss why this activity may have changed since 1900.

Q43.17 BIO One problem in radiocarbon dating of biological samples, especially very old ones, is that they can easily be contaminated with modern biological material during the measurement process. What effect would such contamination have on the estimated age? Why is such contamination a more serious problem for samples of older material than for samples of younger material?

Q43.18 The most common radium isotope found on earth, ^{226}Ra, has a half-life of about 1600 years. If the earth was formed well over 10^9 years ago, why is there any radium left now?

Q43.19 Fission reactions occur only for nuclei with large nucleon numbers, while exoergic fusion reactions occur only for nuclei with small nucleon numbers. Why is this?

Q43.20 When a large nucleus splits during nuclear fission, the daughter nuclei of the fission fly apart with enormous kinetic energy. Why does this happen?

Q43.21 As stars age, they use up their supply of hydrogen and eventually begin producing energy by a reaction that involves the fusion of three helium nuclei to form a carbon nucleus. Would you expect the interiors of these old stars to be hotter or cooler than the interiors of younger stars? Explain.

EXERCISES

Section 43.1 Properties of Nuclei

43.1 • How many protons and how many neutrons are there in a nucleus of the most common isotope of (a) silicon, $^{28}_{14}$Si; (b) rubidium, $^{85}_{37}$Rb; (c) thallium, $^{205}_{81}$Tl?

43.2 •• **CP** Hydrogen atoms are placed in an external 1.65-T magnetic field. (a) The *protons* can make transitions between states where the nuclear spin component is parallel and antiparallel to the field by absorbing or emitting a photon. Which state has lower energy: the state with the nuclear spin component parallel or antiparallel to the field? What are the frequency and wavelength of the photon? In which region of the electromagnetic spectrum does it lie? (b) The *electrons* can make transitions between states where the electron spin component is parallel and antiparallel to the field by absorbing or emitting a photon. Which state has lower energy: the state with the electron spin component parallel or antiparallel to the field? What are the frequency and wavelength of the photon? In which region of the electromagnetic spectrum does it lie?

43.3 • Hydrogen atoms are placed in an external magnetic field. The protons can make transitions between states in which the nuclear spin component is parallel and antiparallel to the field by absorbing or emitting a photon. What magnetic-field magnitude is required for this transition to be induced by photons with frequency 22.7 MHz?

43.4 •• Neutrons are placed in a magnetic field with magnitude 2.30 T. (a) What is the energy difference between the states with the nuclear spin angular momentum components parallel and antiparallel to the field? Which state is lower in energy: the one with its spin component parallel to the field or the one with its spin component antiparallel to the field? How do your results compare with the energy states for a proton in the same field (see Example 43.2)? (b) The neutrons can make transitions from one of these states to the other by emitting or absorbing a photon with energy equal to the energy difference of the two states. Find the frequency and wavelength of such a photon.

Section 43.2 Nuclear Binding and Nuclear Structure

43.5 • The most common isotope of boron is $^{11}_{5}$B. (a) Determine the total binding energy of $^{11}_{5}$B from Table 43.2 in Section 43.1. (b) Calculate this binding energy from Eq. (43.11). (Why is the fifth

term zero?) Compare to the result you obtained in part (a). What is the percent difference? Compare the accuracy of Eq. (43.11) for $^{11}_{5}$B to its accuracy for $^{62}_{28}$Ni (see Example 43.4).

43.6 • The most common isotope of uranium, $^{238}_{92}$U, has atomic mass 238.050783 u. Calculate (a) the mass defect; (b) the binding energy (in MeV); (c) the binding energy per nucleon.

43.7 • **CP** What is the maximum wavelength of a γ ray that could break a deuteron into a proton and a neutron? (This process is called photodisintegration.)

43.8 • Calculate (a) the total binding energy and (b) the binding energy per nucleon of ^{12}C. (c) What percent of the rest mass of this nucleus is its total binding energy?

43.9 • **CP** A photon with a wavelength of 3.50×10^{-13} m strikes a deuteron, splitting it into a proton and a neutron. (a) Calculate the kinetic energy released in this interaction. (b) Assuming the two particles share the energy equally, and taking their masses to be 1.00 u, calculate their speeds after the photodisintegration.

43.10 • Calculate the mass defect, the binding energy (in MeV), and the binding energy per nucleon of (a) the nitrogen nucleus, $^{14}_{7}$N, and (b) the helium nucleus, $^{4}_{2}$He. (c) How does the binding energy per nucleon compare for these two nuclei?

43.11 • Use Eq. (43.11) to calculate the binding energy per nucleon for the nuclei $^{86}_{36}$Kr and $^{180}_{73}$Ta. Do your results confirm what is shown in Fig. 43.2—that for A greater than 62 the binding energy per nucleon deceases as A increases?

Section 43.3 Nuclear Stability and Radioactivity

43.12 • (a) Is the decay $n \rightarrow p + \beta^- + \bar{\nu}_e$ energetically possible? If not, explain why not. If so, calculate the total energy released. (b) Is the decay $p \rightarrow n + \beta^+ + \nu_e$ energetically possible? If not, explain why not. If so, calculate the total energy released.

43.13 • What nuclide is produced in the following radioactive decays? (a) α decay of $^{239}_{94}$Pu; (b) β^- decay of $^{24}_{11}$Na; (c) β^+ decay of $^{15}_{8}$O.

43.14 •• **CP** ^{238}U decays spontaneously by α emission to ^{234}Th. Calculate (a) the total energy released by this process and (b) the recoil velocity of the ^{234}Th nucleus. The atomic masses are 238.050788 u for ^{238}U and 234.043601 u for ^{234}Th.

43.15 •• The atomic mass of ^{14}C is 14.003242 u. Show that the β^- decay of ^{14}C is energetically possible, and calculate the energy released in the decay.

43.16 • What particle (α particle, electron, or positron) is emitted in the following radioactive decays? (a) $^{27}_{14}$Si $\rightarrow$ $^{27}_{13}$Al; (b) $^{238}_{92}$U $\rightarrow$ $^{234}_{90}$Th; (c) $^{74}_{33}$As $\rightarrow$ $^{74}_{34}$Se.

43.17 •• (a) Calculate the energy released by the electron-capture decay of $^{57}_{27}$Co (see Example 43.7). (b) A negligible amount of this energy goes to the resulting $^{57}_{26}$Fe atom as kinetic energy. About 90% of the time, the $^{57}_{26}$Fe nucleus emits two successive gamma-ray photons after the electron-capture process, of energies 0.122 MeV and 0.014 MeV, respectively, in decaying to its ground state. What is the energy of the neutrino emitted in this case?

43.18 • Tritium ($^{3}_{1}$H) is an unstable isotope of hydrogen; its mass, including one electron, is 3.016049 u. (a) Show that tritium must be unstable with respect to beta decay because the decay products ($^{3}_{2}$He plus an emitted electron) have less total mass than the tritium. (b) Determine the total kinetic energy (in MeV) of the decay products, taking care to account for the electron masses correctly.

Section 43.4 Activities and Half-Lives

43.19 • If a 6.13-g sample of an isotope having a mass number of 124 decays at a rate of 0.350 Ci, what is its half-life?

43.20 • **BIO** Radioactive isotopes used in cancer therapy have a "shelf-life," like pharmaceuticals used in chemotherapy. Just after it has been manufactured in a nuclear reactor, the activity of a sample of ^{60}Co is 5000 Ci. When its activity falls below 3500 Ci, it is considered too weak a source to use in treatment. You work in the radiology department of a large hospital. One of these ^{60}Co sources in your inventory was manufactured on October 6, 2004. It is now April 6, 2007. Is the source still usable? The half-life of ^{60}Co is 5.271 years.

43.21 •• The common isotope of uranium, ^{238}U, has a half-life of 4.47×10^9 years, decaying to ^{234}Th by alpha emission. (a) What is the decay constant? (b) What mass of uranium is required for an activity of 1.00 curie? (c) How many alpha particles are emitted per second by 10.0 g of uranium?

43.22 •• **BIO** **Radiation Treatment of Prostate Cancer.** In many cases, prostate cancer is treated by implanting 60 to 100 small seeds of radioactive material into the tumor. The energy released from the decays kills the tumor. One isotope that is used (there are others) is palladium (^{103}Pd), with a half-life of 17 days. If a typical grain contains 0.250 g of ^{103}Pd, (a) what is its initial activity rate in Bq, and (b) what is the rate 68 days later?

43.23 •• A 12.0-g sample of carbon from living matter decays at the rate of 180.0 decays/min due to the radioactive ^{14}C in it. What will be the decay rate of this sample in (a) 1000 years and (b) 50,000 years?

43.24 •• **BIO** **Radioactive Tracers.** Radioactive isotopes are often introduced into the body through the bloodstream. Their spread through the body can then be monitored by detecting the appearance of radiation in different organs. ^{131}I, a β^- emitter with a half-life of 8.0 d, is one such tracer. Suppose a scientist introduces a sample with an activity of 375 Bq and watches it spread to the organs. (a) Assuming that the sample all went to the thyroid gland, what will be the decay rate in that gland 24 d (about $3\frac{1}{2}$ weeks) later? (b) If the decay rate in the thyroid 24 d later is actually measured to be 17.0 Bq, what percentage of the tracer went to that gland? (c) What isotope remains after the I-131 decays?

43.25 •• The unstable isotope ^{40}K is used for dating rock samples. Its half-life is 1.28×10^9 y. (a) How many decays occur per second in a sample containing 1.63×10^{-6} g of ^{40}K? (b) What is the activity of the sample in curies?

43.26 • As a health physicist, you are being consulted about a spill in a radiochemistry lab. The isotope spilled was 500 μCi of ^{131}Ba, which has a half-life of 12 days. (a) What mass of ^{131}Ba was spilled? (b) Your recommendation is to clear the lab until the radiation level has fallen 1.00 μCi. How long will the lab have to be closed?

43.27 • Measurements on a certain isotope tell you that the decay rate decreases from 8318 decays/min to 3091 decays/min in 4.00 days. What is the half-life of this isotope?

43.28 • The isotope ^{226}Ra undergoes α decay with a half-life of 1620 years. What is the activity of 1.00 g of ^{226}Ra? Express your answer in Bq and in Ci.

43.29 • The radioactive nuclide ^{199}Pt has a half-life of 30.8 minutes. A sample is prepared that has an initial activity of 7.56×10^{11} Bq. (a) How many ^{199}Pt nuclei are initially present in the sample? (b) How many are present after 30.8 minutes? What is the activity at this time? (c) Repeat part (b) for a time 92.4 minutes after the sample is first prepared.

43.30 •• **Radiocarbon Dating.** A sample from timbers at an archeological site containing 500 g of carbon provides 3070 decays/min. What is the age of the sample?

Section 43.5 Biological Effects of Radiation

43.31 •• **BIO** (a) If a chest x ray delivers 0.25 mSv to 5.0 kg of tissue, how many *total* joules of energy does this tissue receive? (b) Natural radiation and cosmic rays deliver about 0.10 mSv per year at sea level. Assuming an RBE of 1, how many rem and rads is this dose, and how many joules of energy does a 75-kg person receive in a year? (c) How many chest x rays like the one in part (a) would it take to deliver the same *total* amount of energy to a 75-kg person as she receives from natural radiation in a year at sea level, as described in part (b)?

43.32 •• **BIO** A person exposed to fast neutrons receives a radiation dose of 200 rem on part of his hand, affecting 25 g of tissue. The RBE of these neutrons is 10. (a) How many rad did he receive? (b) How many joules of energy did this person receive? (c) Suppose the person received the same rad dosage, but from beta rays with an RBE of 1.0 instead of neutrons. How many rem would he have received?

43.33 •• **BIO** A nuclear chemist receives an accidental radiation dose of 5.0 Gy from slow neutrons (RBE = 4.0). What does she receive in rad, rem, and J/kg?

43.34 • **BIO** **To Scan or Not to Scan?** It has become popular for some people to have yearly whole-body scans (CT scans, formerly called CAT scans) using x rays, just to see if they detect anything suspicious. A number of medical people have recently questioned the advisability of such scans, due in part to the radiation they impart. Typically, one such scan gives a dose of 12 mSv, applied to the *whole body*. By contrast, a chest x ray typically administers 0.20 mSv to only 5.0 kg of tissue. How many chest x rays would deliver the same *total* amount of energy to the body of a 75-kg person as one whole-body scan?

43.35 • **BIO** **Food Irradiation.** Food is often irradiated with either x rays or electron beams to help prevent spoilage. A low dose of 5–75 kilorads (krad) helps to reduce and kill inactive parasites, a medium dose of 100–400 krad kills microorganisms and pathogens such as salmonella, and a high dose of 2300–5700 krad sterilizes food so that it can be stored without refrigeration. (a) A dose of 175 krad kills spoilage microorganisms in fish. If x rays are used, what would be the dose in Gy, Sv, and rem, and how much energy would a 220-g portion of fish absorb? (See Table 43.3.) (b) Repeat part (a) if electrons of RBE 1.50 are used instead of x rays.

43.36 • **BIO** In an industrial accident a 65-kg person receives a lethal whole-body equivalent dose of 5.4 Sv from x rays. (a) What is the equivalent dose in rem? (b) What is the absorbed dose in rad? (c) What is the total energy absorbed by the person's body? How does this amount of energy compare to the amount of energy required to raise the temperature of 65 kg of water 0.010 C°?

43.37 •• **BIO** A 67-kg person accidentally ingests 0.35 Ci of tritium. (a) Assume that the tritium spreads uniformly throughout the body and that each decay leads on the average to the absorption of 5.0 keV of energy from the electrons emitted in the decay. The half-life of tritium is 12.3 y, and the RBE of the electrons is 1.0. Calculate the absorbed dose in rad and the equivalent dose in rem during one week. (b) The β^- decay of tritium releases more than 5.0 keV of energy. Why is the average energy absorbed less than the total energy released in the decay?

43.38 •• **CP** **BIO** In a diagnostic x-ray procedure, 5.00×10^{10} photons are absorbed by tissue with a mass of 0.600 kg. The x-ray wavelength is 0.0200 nm. (a) What is the total energy absorbed by the tissue? (b) What is the equivalent dose in rem?

Section 43.6 Nuclear Reactions, Section 43.7 Nuclear Fission, and Section 43.8 Nuclear Fusion

43.39 • Consider the nuclear reaction

$$^2_1H + {}^{14}_7N \rightarrow X + {}^{10}_5B$$

where X is a nuclide. (a) What are Z and A for the nuclide X? (b) Calculate the reaction energy Q (in MeV). (c) If the ^2_1H nucleus is incident on a stationary $^{14}_7N$ nucleus, what minimum kinetic energy must it have for the reaction to occur?

43.40 • **Energy from Nuclear Fusion.** Calculate the energy released in the fusion reaction

$$^3_2He + {}^2_1H \rightarrow {}^4_2He + {}^1_1H$$

43.41 • Consider the nuclear reaction

$$^2_1H + {}^9_4Be \rightarrow X + {}^4_2He$$

where X is a nuclide. (a) What are the values of Z and A for the nuclide X? (b) How much energy is liberated? (c) Estimate the threshold energy for this reaction.

43.42 • The United States uses 1.0×10^{20} J of electrical energy per year. If all this energy came from the fission of ^{235}U, which releases 200 MeV per fission event, (a) how many kilograms of ^{235}U would be used per year and (b) how many kilograms of uranium would have to be mined per year to provide that much ^{235}U? (Recall that only 0.70% of naturally occurring uranium is ^{235}U.)

43.43 •• At the beginning of Section 43.7 the equation of a fission process is given in which ^{235}U is struck by a neutron and undergoes fission to produce ^{144}Ba, ^{89}Kr, and three neutrons. The measured masses of these isotopes are 235.043930 u (^{235}U), 143.922953 u (^{144}Ba), 88.917630 u (^{89}Kr), and 1.0086649 u (neutron). (a) Calculate the energy (in MeV) released by each fission reaction. (b) Calculate the energy released per gram of ^{235}U, in MeV/g.

43.44 •• Consider the nuclear reaction

$$^{28}_{14}Si + \gamma \rightarrow {}^{24}_{12}Mg + X$$

where X is a nuclide. (a) What are Z and A for the nuclide X? (b) Ignoring the effects of recoil, what minimum energy must the photon have for this reaction to occur? The mass of a $^{28}_{14}Si$ atom is 27.976927 u, and the mass of a $^{24}_{12}Mg$ atom is 23.985042 u.

43.45 • The second reaction in the proton-proton chain (see Fig. 43.16) produces a ^3_2He nucleus. A ^3_2He nucleus produced in this way can combine with a ^4_2He nucleus:

$$^3_2He + {}^4_2He \rightarrow {}^7_4Be + \gamma$$

Calculate the energy liberated in this process. (This is shared between the energy of the photon and the recoil kinetic energy of the beryllium nucleus.) The mass of a ^7_4Be atom is 7.016929 u.

43.46 • Consider the nuclear reaction

$$^4_2He + {}^7_3Li \rightarrow X + {}^1_0n$$

where X is a nuclide. (a) What are Z and A for the nuclide X? (b) Is energy absorbed or liberated? How much?

43.47 •• **CP** In a 100.0-cm³ sample of water, 0.015% of the molecules are D_2O. Compute the energy in joules that is liberated if all the deuterium nuclei in the sample undergo the fusion reaction of Example 43.13.

PROBLEMS

43.48 • **Comparison of Energy Released per Gram of Fuel.** (a) When gasoline is burned, it releases 1.3×10^8 J of energy per gallon (3.788 L). Given that the density of gasoline is 737 kg/m³, express the quantity of energy released in J/g of fuel. (b) During fission, when a neutron is absorbed by a ^{235}U nucleus, about 200 MeV of energy is released for each nucleus that undergoes fission. Express this quantity in J/g of fuel. (c) In the proton-proton chain that takes place in stars like our sun, the overall fusion reaction can be summarized as six protons fusing to form one ^4He nucleus with two leftover protons and the liberation of 26.7 MeV of energy. The fuel is the six protons. Express the energy produced here in units of J/g of fuel. Notice the huge difference between the two forms of nuclear energy, on the one hand, and the chemical energy from gasoline, on the other. (d) Our sun produces energy at a measured rate of 3.86×10^{26} W. If its mass of 1.99×10^{30} kg were all gasoline, how long could it last before consuming all its fuel? (*Historical note:* Before the discovery of nuclear fusion and the vast amounts of energy it releases, scientists were confused. They knew that the earth was at least many millions of years old, but could not explain how the sun could survive that long if its energy came from chemical burning.)

43.49 •• Use conservation of mass-energy to show that the energy released in alpha decay is positive whenever the mass of the original neutral atom is greater than the sum of the masses of the final neutral atom and the neutral ^4He atom. (*Hint:* Let the parent nucleus have atomic number Z and nucleon number A. First write the reaction in terms of the nuclei and particles involved, and then add Z electron masses to both sides of the reaction and allot them as needed to arrive at neutral atoms.)

43.50 •• Use conservation of mass-energy to show that the energy released in β^- decay is positive whenever the neutral atomic mass of the original atom is greater than that of the final atom. (See the hint in Problem 43.49.)

43.51 •• Use conservation of mass-energy to show that the energy released in β^+ decay is positive whenever the neutral atomic mass of the original atom is at least two electron masses greater than that of the final atom. (See the hint in Problem 43.49.)

43.52 •• (a) Calculate the minimum energy required to remove one proton from the nucleus $^{12}_6C$. This is called the proton-removal energy. (*Hint:* Find the difference between the mass of a $^{12}_6C$ nucleus and the mass of a proton plus the mass of the nucleus formed when a proton is removed from $^{12}_6C$.) (b) How does the proton-removal energy for $^{12}_6C$ compare to the binding energy per nucleon for $^{12}_6C$, calculated using Eq. (43.10)?

43.53 •• (a) Calculate the minimum energy required to remove one neutron from the nucleus $^{17}_8O$. This is called the neutron-removal energy. (See Problem 43.52.) (b) How does the neutron-removal energy for $^{17}_8O$ compare to the binding energy per nucleon for $^{17}_8O$, calculated using Eq. (43.10)?

43.54 •• The neutral atomic mass of $^{14}_6C$ is 14.003242 u. Calculate the proton removal energy and the neutron removal energy for $^{15}_7N$. (See Problems 43.52 and 43.53.) What is the percentage difference between these two energies, and which is larger?

43.55 • **BIO Radioactive Fallout.** One of the problems of in-air testing of nuclear weapons (or, even worse, the *use* of such weapons!) is the danger of radioactive fallout. One of the most problematic nuclides in such fallout is strontium-90 (^{90}Sr), which breaks down by β^- decay with a half-life of 28 years. It is chemically similar to calcium and therefore can be incorporated into bones and teeth, where, due to its rather long half-life, it remains for years as an internal source of radiation. (a) What is the daughter nucleus of the ^{90}Sr decay? (b) What percentage of the original level of ^{90}Sr is left after 56 years? (c) How long would you have to wait for the original level to be reduced to 6.25% of its original value?

Time (h)	Decays/s
0	20,000
0.5	14,800
1.0	11,000
1.5	8,130
2.0	6,020
2.5	4,460
3.0	3,300
4.0	1,810
5.0	1,000
6.0	550
7.0	300

43.56 •• **CP** Thorium $^{230}_{90}$Th decays to radium $^{226}_{88}$Ra by α emission. The masses of the neutral atoms are 230.033127 u for $^{230}_{90}$Th and 226.025403 u for $^{226}_{88}$Ra. If the parent thorium nucleus is at rest, what is the kinetic energy of the emitted α particle? (Be sure to account for the recoil of the daughter nucleus.)

43.57 •• The atomic mass of $^{25}_{12}$Mg is 24.985837 u, and the atomic mass of $^{25}_{13}$Al is 24.990429 u. (a) Which of these nuclei will decay into the other? (b) What type of decay will occur? Explain how you determined this. (c) How much energy (in MeV) is released in the decay?

43.58 •• The polonium isotope $^{210}_{84}$Po has atomic mass 209.982857 u. Other atomic masses are $^{206}_{82}$Pb, 205.974449 u; $^{209}_{83}$Bi, 208.980383 u; $^{210}_{83}$Bi, 209.984105 u; $^{209}_{84}$Po, 208.982416 u; and $^{210}_{85}$At, 209.987131 u. (a) Show that the alpha decay of $^{210}_{84}$Po is energetically possible, and find the energy of the emitted α particle. (b) Is $^{210}_{84}$Po energetically stable with respect to emission of a proton? Why or why not? (c) Is $^{210}_{84}$Po energetically stable with respect to emission of a neutron? Why or why not? (d) Is $^{210}_{84}$Po energetically stable with respect to β^- decay? Why or why not? (e) Is $^{210}_{84}$Po energetically stable with respect to β^+ decay? Why or why not?

43.59 •• **BIO** **Irradiating Ourselves!** The radiocarbon in our bodies is one of the naturally occurring sources of radiation. Let's see how large a dose we receive. ^{14}C decays via β^- emission, and 18% of our body's mass is carbon. (a) Write out the decay scheme of carbon-14 and show the end product. (A neutrino is also produced.) (b) Neglecting the effects of the neutrino, how much kinetic energy (in MeV) is released per decay? The atomic mass of ^{14}C is 14.003242 u. (c) How many grams of carbon are there in a 75-kg person? How many decays per second does this carbon produce? (*Hint:* Use data from Example 43.9.) (d) Assuming that all the energy released in these decays is absorbed by the body, how many MeV/s and J/s does the ^{14}C release in this person's body? (e) Consult Table 43.3 and use the largest appropriate RBE for the particles involved. What radiation dose does the person give himself in a year, in Gy, rad, Sv, and rem?

43.60 •• **BIO** **Pion Radiation Therapy.** A neutral pion (π^0) has a mass of 264 times the electron mass and decays with a lifetime of 8.4×10^{-17} s to two photons. Such pions are used in the radiation treatment of some cancers. (a) Find the energy and wavelength of these photons. In which part of the electromagnetic spectrum do they lie? What is the RBE for these photons? (b) If you want to deliver a dose of 200 rem (which is typical) in a single treatment to 25 g of tumor tissue, how many π^0 mesons are needed?

43.61 • Gold, $^{198}_{79}$Au, undergoes β^- decay to an excited state of $^{198}_{80}$Hg. If the excited state decays by emission of a γ photon with energy 0.412 MeV, what is the maximum kinetic energy of the electron emitted in the decay? This maximum occurs when the antineutrino has negligible energy. (The recoil energy of the $^{198}_{80}$Hg nucleus can be ignored. The masses of the neutral atoms in their ground states are 197.968225 u for $^{198}_{79}$Au and 197.966752 u for $^{198}_{80}$Hg.)

43.62 •• Calculate the mass defect for the β^+ decay of $^{11}_{6}$C. Is this decay energetically possible? Why or why not? The atomic mass of $^{11}_{6}$C is 11.011434 u.

43.63 •• Calculate the mass defect for the β^+ decay of $^{13}_{7}$N. Is this decay energetically possible? Why or why not? The atomic mass of $^{13}_{7}$N is 13.005739 u.

43.64 •• The results of activity measurements on a radioactive sample are given in the table. (a) Find the half-life. (b) How many radioactive nuclei were present in the sample at $t = 0$? (c) How many were present after 7.0 h?

43.65 •• **BIO** A person ingests an amount of a radioactive source with a very long lifetime and activity 0.63 μCi. The radioactive material lodges in the lungs, where all of the 4.0-MeV α particles emitted are absorbed within a 0.50-kg mass of tissue. Calculate the absorbed dose and the equivalent dose for one year.

43.66 •• **Measuring Very Long Half-Lives.** Some radioisotopes such as samarium (^{149}Sm) and gadolinium (^{152}Gd) have half-lives that are much longer than the age of the universe, so we can't measure their half-lives by watching their decay rate decrease. Luckily, there is another way of calculating the half-life, using Eq. (43.16). Suppose a 12.0-g sample of ^{149}Sm is observed to decay at a rate of 2.65 Bq. Calculate the half-life of the sample in years. (*Hint:* How many nuclei are there in the 12.0-g sample?)

43.67 • **We Are Stardust.** In 1952 spectral lines of the element technetium-99 (^{99}Tc) were discovered in a red giant star. Red giants are very old stars, often around 10 billion years old, and near the end of their lives. Technetium has *no* stable isotopes, and the half-life of ^{99}Tc is 200,000 years. (a) For how many half-lives has the ^{99}Tc been in the red-giant star if its age is 10 billion years? (b) What fraction of the original ^{99}Tc would be left at the end of that time? This discovery was extremely important because it provided convincing evidence for the theory (now essentially known to be true) that most of the atoms heavier than hydrogen and helium were made inside of stars by thermonuclear fusion and other nuclear processes. If the ^{99}Tc had been part of the star since it was born, the amount remaining after 10 billion years would have been so minute that it would not have been detectable. This knowledge is what led the late astronomer Carl Sagan to proclaim that "we are stardust."

43.68 • **BIO** A 70.0-kg person experiences a whole-body exposure to α radiation with energy 4.77 MeV. A total of 6.25×10^{12} α particles are absorbed. (a) What is the absorbed dose in rad? (b) What is the equivalent dose in rem? (c) If the source is 0.0320 g of ^{226}Ra (half-life 1600 y) somewhere in the body, what is the activity of this source? (d) If all the alpha particles produced are absorbed, what time is required for this dose to be delivered?

43.69 •• Measurements indicate that 27.83% of all rubidium atoms currently on the earth are the radioactive ^{87}Rb isotope. The rest are the stable ^{85}Rb isotope. The half-life of ^{87}Rb is 4.75×10^{10} y. Assuming that no rubidium atoms have been formed since, what percentage of rubidium atoms were ^{87}Rb when our solar system was formed 4.6×10^9 y ago?

43.70 •• A $^{186}_{76}$Os nucleus at rest decays by the emission of a 2.76-MeV α particle. Calculate the atomic mass of the daughter nuclide produced by this decay, assuming that it is produced in its ground state. The atomic mass of $^{186}_{76}\text{Os}$ is 185.953838 u.

43.71 ·· **BIO** A ^{60}Co source with activity 2.6×10^{-4} Ci is embedded in a tumor that has mass 0.200 kg. The source emits γ photons with average energy 1.25 MeV. Half the photons are absorbed in the tumor, and half escape. (a) What energy is delivered to the tumor per second? (b) What absorbed dose (in rad) is delivered per second? (c) What equivalent dose (in rem) is delivered per second if the RBE for these γ rays is 0.70? (d) What exposure time is required for an equivalent dose of 200 rem?

43.72 · The nucleus $^{15}_{8}\text{O}$ has a half-life of 122.2 s; $^{19}_{8}\text{O}$ has a half-life of 26.9 s. If at some time a sample contains equal amounts of $^{15}_{8}\text{O}$ and $^{19}_{8}\text{O}$, what is the ratio of $^{15}_{8}\text{O}$ to $^{19}_{8}\text{O}$ (a) after 4.0 minutes and (b) after 15.0 minutes?

43.73 · A bone fragment found in a cave believed to have been inhabited by early humans contains 0.29 times as much ^{14}C as an equal amount of carbon in the atmosphere when the organism containing the bone died. (See Example 43.9 in Section 43.4.) Find the approximate age of the fragment.

43.74 ·· **An Oceanographic Tracer.** Nuclear weapons tests in the 1950s and 1960s released significant amounts of radioactive tritium ($^{3}_{1}\text{H}$, half-life 12.3 years) into the atmosphere. The tritium atoms were quickly bound into water molecules and rained out of the air, most of them ending up in the ocean. For any of this tritium-tagged water that sinks below the surface, the amount of time during which it has been isolated from the surface can be calculated by measuring the ratio of the decay product, $^{3}_{2}\text{He}$, to the remaining tritium in the water. For example, if the ratio of $^{3}_{2}\text{He}$ to $^{3}_{1}\text{H}$ in a sample of water is 1:1, the water has been below the surface for one half-life, or approximately 12 years. This method has provided oceanographers with a convenient way to trace the movements of subsurface currents in parts of the ocean. Suppose that in a particular sample of water, the ratio of $^{3}_{2}\text{He}$ to $^{3}_{1}\text{H}$ is 4.3 to 1.0. How many years ago did this water sink below the surface?

43.75 ·· Consider the fusion reaction $^{2}_{1}\text{H} + ^{2}_{1}\text{H} \rightarrow ^{3}_{2}\text{He} + ^{1}_{0}\text{n}$. (a) Estimate the barrier energy by calculating the repulsive electrostatic potential energy of the two $^{2}_{1}\text{H}$ nuclei when they touch. (b) Compute the energy liberated in this reaction in MeV and in joules. (c) Compute the energy liberated *per mole* of deuterium, remembering that the gas is diatomic, and compare with the heat of combustion of hydrogen, about 2.9×10^5 J/mol.

43.76 ·· **BIO** In the 1986 disaster at the Chernobyl reactor in the Soviet Union (now Ukraine), about $\frac{1}{8}$ of the ^{137}Cs present in the reactor was released. The isotope ^{137}Cs has a half-life for β decay of 30.07 y and decays with the emission of a total of 1.17 MeV of energy per decay. Of this, 0.51 MeV goes to the emitted electron and the remaining 0.66 MeV to a γ ray. The radioactive ^{137}Cs is absorbed by plants, which are eaten by livestock and humans. How many ^{137}Cs atoms would need to be present in each kilogram of body tissue if an equivalent dose for one week is 3.5 Sv? Assume

that all of the energy from the decay is deposited in that 1.0 kg of tissue and that the RBE of the electrons is 1.5.

43.77 ·· **CP** (a) Prove that when a particle with mass m and kinetic energy K collides with a stationary particle with mass M, the total kinetic energy K_{cm} in the center-of-mass coordinate system (the energy available to cause reactions) is

$$K_{\text{cm}} = \frac{M}{M + m} K$$

Assume that the kinetic energies of the particles and nuclei are much lower than their rest energies. (b) If K_{th} is the minimum, or threshold, kinetic energy to cause an endoergic reaction to occur in the situation of part (a), show that

$$K_{\text{th}} = -\frac{M + m}{M} Q$$

43.78 · Calculate the energy released in the fission reaction $^{235}_{92}\text{U} + ^{1}_{0}\text{n} \rightarrow ^{140}_{54}\text{Xe} + ^{94}_{38}\text{Sr} + 2^{1}_{0}\text{n}$. You can ignore the initial kinetic energy of the absorbed neutron. The atomic masses are $^{235}_{92}\text{U}$, 235.043923 u; $^{140}_{54}\text{Xe}$, 139.921636 u; and $^{94}_{38}\text{Sr}$, 93.915360 u.

CHALLENGE PROBLEMS

43.79 ··· The results of activity measurements on a mixed sample of radioactive elements are given in the table. (a) How many different nuclides are present in the mixture? (b) What are their half-lives? (c) How many nuclei of each type are initially present in the sample? (d) How many of each type are present at $t = 5.0$ h?

Time (h)	Decays/s
0	7500
0.5	4120
1.0	2570
1.5	1790
2.0	1350
2.5	1070
3.0	872
4.0	596
5.0	414
6.0	288
7.0	201
8.0	140
9.0	98
10.0	68
12.0	33

43.80 ··· **Industrial Radioactivity.** Radioisotopes are used in a variety of manufacturing and testing techniques. Wear measurements can be made using the following method. An automobile engine is produced using piston rings with a total mass of 100 g, which includes 9.4 μCi of ^{59}Fe whose half-life is 45 days. The engine is test-run for 1000 hours, after which the oil is drained and its activity is measured. If the activity of the engine oil is 84 decays/s, how much mass was worn from the piston rings per hour of operation?

Answers

Chapter Opening Question ?

When an organism dies, it stops taking in carbon from atmospheric CO_2. Some of this carbon is radioactive ^{14}C, which decays with a half-life of 5730 years. By measuring the proportion of ^{14}C that remains in the specimen, scientists can determine how long ago the organism died. (See Section 43.4.)

Test Your Understanding Questions

43.1 Answers: (a) (iii), (b) (v) The radius R is proportional to the cube root of the mass number A, while the volume is proportional to R^3 and hence to $(A^{1/3})^3 = A$. Therefore, doubling the volume requires increasing the mass number by a factor of 2; doubling the radius implies increasing both the volume and the mass number by a factor of $2^3 = 8$.

43.2 Answer: (ii), (iii), (iv), (v), (i) You can find the answers by inspecting Fig. 43.2. The binding energy per nucleon is lowest for very light nuclei such as ^4_2He, is greatest around $A = 60$, and then decreases with increasing A.

43.3 Answer: (v) Two protons and two neutrons are lost in an α decay, so Z and N each decrease by 2. A β^+ decay changes a proton to a neutron, so Z decreases by 1 and N increases by 1. The net result is that Z decreases by 3 and N decreases by 1.

43.4 Answer: (iii) The activity $-dN(t)/dt$ of a sample is the product of the number of nuclei in the sample $N(t)$ and the decay constant $\lambda = (\ln 2)/T_{1/2}$. Hence $N(t) = (-dN(t)/dt)T_{1/2}/(\ln 2)$. Taking the ratio of this expression for ^{240}Pu to this same expression for ^{243}Am, the factors of $\ln 2$ cancel and we get

$$\frac{N_{\text{Pu}}}{N_{\text{Am}}} = \frac{(-dN_{\text{Pu}}/dt)T_{1/2-\text{Pu}}}{(-dN_{\text{Am}}/dt)T_{1/2-\text{Am}}} = \frac{(5.00\ \mu\text{Ci})(6560\ \text{y})}{(4.45\ \mu\text{Ci})(7370\ \text{y})} = 1.00$$

The two samples contain *equal* numbers of nuclei. The ^{243}Am sample has a longer half-life and hence a slower decay rate, so it has a lower activity than the ^{240}Pu sample.

43.5 Answer: (ii) We saw in Section 43.3 that alpha particles can travel only a very short distance before they are stopped. By contrast, x-ray photons are very penetrating, so they can easily pass into the body.

43.6 Answer: no The reaction $^1_1\text{H} + {}^7_3\text{Li} \rightarrow {}^4_2\text{He} + {}^4_2\text{He}$ is a *nuclear* reaction, which can take place only if a proton (a hydrogen nucleus) comes into contact with a lithium nucleus. If the hydrogen is in atomic form, the interaction between its electron cloud and the electron cloud of a lithium atom keeps the two nuclei from getting close to each other. Even if isolated protons are used, they must be fired at the lithium atoms with enough kinetic energy to overcome the electric repulsion between the protons and the lithium nuclei. The statement that the reaction is exoergic means that more energy is released by the reaction than had to be put in to make the reaction occur.

43.7 Answer: no Because the neutron has no electric charge, it experiences no electric repulsion from a ^{235}U nucleus. Hence a slow-moving neutron can approach and enter a ^{235}U nucleus, thereby providing the excitation needed to trigger fission. By contrast, a slow-moving *proton* (charge $+e$) feels a strong electric repulsion from a ^{235}U nucleus (charge $+92e$). It never gets close to the nucleus, so it cannot trigger fission.

43.8 Answer: no Fusion reactions between sufficiently light nuclei are exoergic because the binding energy per nucleon E_B/A increases. If the nuclei are too massive, however, E_B/A decreases and fusion is *endoergic* (i.e., it takes in energy rather than releasing it). As an example, imagine fusing together two nuclei of $A = 100$ to make a single nucleus with $A = 200$. From Fig. 43.2, E_B/A is more than 8.5 MeV for the $A = 100$ nuclei but is less than 8 MeV for the $A = 200$ nucleus. Such a fusion reaction is possible, but requires a substantial input of energy.

Bridging Problem

Answers: (a) ^{128}Xe

(b) no; β^+ emission would be endoergic

(c) 3.25×10^9 atoms, 1.50×10^6 Bq

(d) $N(t) = (3.25 \times 10^9 \text{ atoms})(1 - e^{-(4.62\times10^{-4}\,\text{s}^{-1})t})$

44 PARTICLE PHYSICS AND COSMOLOGY

? Images made using infrared and x-ray wavelengths were combined to produce this view of the dynamic center of our Milky Way galaxy. The image shows atoms in various states: as isolated atoms in glowing, diffuse clouds of gas (shown in blue), as clumps of atoms and molecules in immense, cold dust clouds (shown in red), and in dense accumulations that we call stars. What fraction of the mass and energy in the universe is composed of "normal" matter—that is, atoms and their constituents?

LEARNING GOALS

By studying this chapter, you will learn:

- The key varieties of fundamental subatomic particles and how they were discovered.

- How physicists use accelerators and detectors to probe the properties of subatomic particles.

- The four ways in which subatomic particles interact with each other.

- How the structure of protons, neutrons, and other particles can be explained in terms of quarks.

- How physicists probe the limits of the standard model of particles and interactions.

- The evidence that the universe is expanding and that the expansion is speeding up.

- The history of the first 380,000 years after the Big Bang.

What is the world made of? What are the most fundamental constituents of matter? Philosophers and scientists have been asking these questions for at least 2500 years. We still don't have the final answer, but as we'll see in this chapter, we've come a long way.

The chapter title, "Particle Physics and Cosmology," may seem strange. Fundamental particles are the *smallest* things in the universe, and cosmology deals with the *biggest* thing there is—the universe itself. Nonetheless, we'll see in this chapter that physics on the most microscopic scale plays an essential role in determining the nature of the universe on the largest scale.

Fundamental particles, we'll find, are not permanent entities; they can be created and destroyed. The development of high-energy particle accelerators and associated detectors has been crucial in our emerging understanding of particles. We can classify particles and their interactions in several ways in terms of conservation laws and symmetries, some of which are absolute and others of which are obeyed only in certain kinds of interactions. We'll conclude by discussing our present understanding of the nature and evolution of the universe as a whole.

44.1 Fundamental Particles—A History

The idea that the world is made of fundamental particles has a long history. In about 400 B.C. the Greek philosophers Democritus and Leucippus suggested that matter is made of indivisible particles that they called *atoms,* a word derived from *a-* (not) and *tomos* (cut or divided). This idea lay dormant until about 1804, when the English scientist John Dalton (1766–1844), often called the father of modern chemistry, discovered that many chemical phenomena could be explained if atoms of each element are the basic, indivisible building blocks of matter.

The Electron and the Proton

Toward the end of the 19th century it became clear that atoms are *not* indivisible. The existence of characteristic atomic spectra of elements suggested that atoms have internal structure, and J. J. Thomson's discovery of the negatively charged *electron* in 1897 showed that atoms could be taken apart into charged particles. Rutherford's experiments in 1910–11 (see Section 39.2) revealed that an atom's positive charge resides in a small, dense nucleus. In 1919 Rutherford made an additional discovery: When alpha particles are fired into nitrogen, one of the products is hydrogen gas. He reasoned that the hydrogen nucleus is a constituent of the nuclei of heavier atoms such as nitrogen, and that a collision with a fast-moving alpha particle can dislodge one of those hydrogen nuclei. Thus the hydrogen nucleus is an elementary particle, to which Rutherford gave the name *proton*. The following decade saw the blossoming of quantum mechanics, including the Schrödinger equation. Physicists were on their way to understanding the principles that underlie atomic structure.

The Photon

Einstein explained the photoelectric effect in 1905 by assuming that the energy of electromagnetic waves is quantized; that is, it comes in little bundles called *photons* with energy $E = hf$. Atoms and nuclei can emit (create) and absorb (destroy) photons (see Section 38.1). Considered as particles, photons have zero charge and zero rest mass. (Note that any discussions of a particle's mass in this chapter will refer to its rest mass.) In particle physics, a photon is denoted by the symbol γ (the Greek letter gamma).

The Neutron

In 1930 the German physicists Walther Bothe and Herbert Becker observed that when beryllium, boron, or lithium was bombarded by alpha particles, the target material emitted a radiation that had much greater penetrating power than the original alpha particles. Experiments by the English physicist James Chadwick in 1932 showed that the emitted particles were electrically neutral, with mass approximately equal to that of the proton. Chadwick christened these particles *neutrons* (symbol n or $^{1}_{0}$n). A typical reaction of the type studied by Bothe and Becker, using a beryllium target, is

$$^{4}_{2}\text{He} + {}^{9}_{4}\text{Be} \rightarrow {}^{12}_{6}\text{C} + {}^{1}_{0}\text{n} \tag{44.1}$$

Elementary particles are usually detected by their electromagnetic effects—for instance, by the ionization that they cause when they pass through matter. (This is the principle of the cloud chamber, described below.) Because neutrons have no charge, they interact hardly at all with electrons and produce little ionization when they pass through matter and so are difficult to detect directly. However, neutrons can be slowed down by scattering from nuclei, and they can penetrate a nucleus. Hence slow neutrons can be detected by means of a nuclear reaction in which a neutron is absorbed and an alpha particle is emitted. An example is

$$^{1}_{0}\text{n} + {}^{10}_{5}\text{B} \rightarrow {}^{7}_{3}\text{Li} + {}^{4}_{2}\text{He} \tag{44.2}$$

The ejected alpha particle is easy to detect because it is charged. Later experiments showed that neutrons, like protons and electrons, are spin-$\frac{1}{2}$ particles (see Section 43.1).

The discovery of the neutron cleared up a mystery about the composition of the nucleus. Before 1930 the mass of a nucleus was thought to be due only to protons, but no one understood why the charge-to-mass ratio was not the same for all nuclides. It soon became clear that all nuclides (except $^{1}_{1}$H) contain both protons

44.1 Photograph of the cloud-chamber track made by the first positron ever identified. The photograph was made by Carl D. Anderson in 1932.

The positron follows a curved path owing to the presence of a magnetic field.

The track is more strongly curved above the lead plate, showing that the positron was traveling upward and lost energy and speed as it passed through the plate.

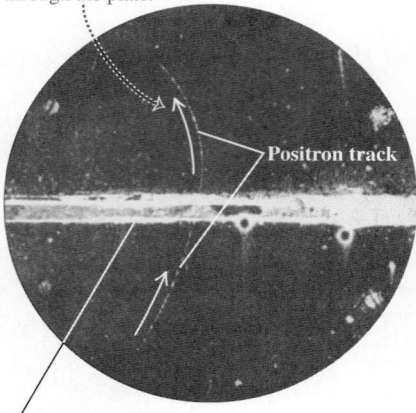

Positron track

Lead plate (6 mm thick)

and neutrons. Hence the proton, the neutron, and the electron are the building blocks of atoms. One might think that would be the end of the story. On the contrary, it is barely the beginning. These are not the only particles, and they can do more than build atoms.

The Positron

The positive electron, or positron, was discovered by the American physicist Carl D. Anderson in 1932, during an investigation of particles bombarding the earth from space. Figure 44.1 shows a historic photograph made with a *cloud chamber,* an instrument used to visualize the tracks of charged particles. The chamber contained a supercooled vapor; ions created by the passage of charged particles through the vapor served as nucleation centers, and liquid droplets formed around them, making a visible track.

The cloud chamber in Fig. 44.1 is in a magnetic field directed into the plane of the photograph. The particle has passed through a thin lead plate (which extends from left to right in the figure) that lies within the chamber. The track is more tightly curved above the plate than below it, showing that the speed was less above the plate than below it. Therefore the particle had to be moving upward; it could not have gained energy passing through the lead. The thickness and curvature of the track suggested that its mass and the magnitude of its charge equaled those of the electron. But the directions of the magnetic field and the velocity in the magnetic force equation $\vec{F} = q\vec{v} \times \vec{B}$ showed that the particle had *positive* charge. Anderson christened this particle the *positron.*

To theorists, the appearance of the positron was a welcome development. In 1928 the English physicist Paul Dirac had developed a relativistic generalization of the Schrödinger equation for the electron. In Section 41.5 we discussed how Dirac's ideas helped explain the spin magnetic moment of the electron.

One of the puzzling features of the Dirac equation was that for a *free* electron it predicted not only a continuum of energy states greater than its rest energy m_ec^2, as should be expected, but also a continuum of *negative* energy states *less than* $-m_ec^2$ (Fig. 44.2a). That posed a problem. What was to prevent an electron from emitting a photon with energy $2m_ec^2$ or greater and hopping from a positive state to a negative state? It wasn't clear what these negative-energy states meant, and there was no obvious way to get rid of them. Dirac's ingenious interpretation was that all the negative-energy states were filled with electrons, and that these electrons were for some reason unobservable. The exclusion principle (see Section 41.6) would then forbid a transition to a state that was already occupied.

A vacancy in a negative-energy state would act like a positive charge, just as a hole in the valence band of a semiconductor (see Section 42.6) acts like a positive charge. Initially, Dirac tried to argue that such vacancies were protons. But after

44.2 (a) Energy states for a free electron predicted by the Dirac equation. (b) Raising an electron from an $E < 0$ state to an $E > 0$ state corresponds to electron–positron pair production. (c) An electron dropping from an $E > 0$ state to a vacant $E < 0$ state corresponds to electron–positron pair annihilation.

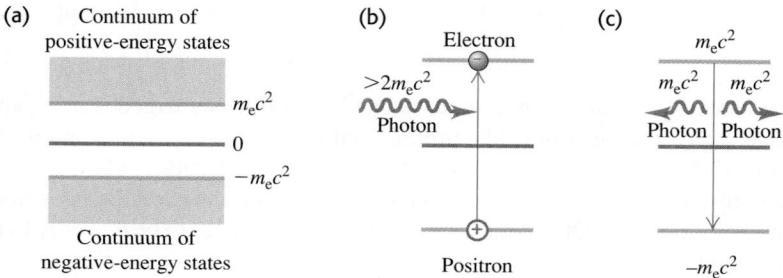

Anderson's discovery it became clear that the vacancies were observed physically as *positrons*. Furthermore, the Dirac energy-state picture provides a mechanism for the *creation* of positrons. When an electron in a negative-energy state absorbs a photon with energy greater than $2m_ec^2$, it goes to a positive state (Fig. 44.2b), in which it becomes observable. The vacancy that it leaves behind is observed as a positron; the result is the creation of an electron–positron pair. Similarly, when an electron in a positive-energy state falls into a vacancy, both the electron and the vacancy (that is, the positron) disappear, and photons are emitted (Fig. 44.2c). Thus the Dirac theory leads naturally to the conclusion that, like photons, *electrons can be created and destroyed*. While photons can be created and destroyed singly, electrons can be produced or destroyed only in electron–positron pairs or in association with other particles. (Creating or destroying an electron alone would mean creating or destroying an amount of charge $-e$, which would violate the conservation of electric charge.)

In 1949 the American physicist Richard Feynman showed that a positron could be described mathematically as an electron traveling backward in time. His reformulation of the Dirac theory eliminated difficult calculations involving the infinite sea of negative-energy states and put electrons and positrons on the same footing. But the creation and destruction of electron–positron pairs remain. The Dirac theory provides the beginning of a theoretical framework for creation and destruction of all fundamental particles.

Experiment and theory tell us that the masses of the positron and electron are identical, and that their charges are equal in magnitude but opposite in sign. The positron's spin angular momentum $\vec{S}$ and magnetic moment $\vec{\mu}$ are parallel; they are opposite for the electron. However, $\vec{S}$ and $\vec{\mu}$ have the same magnitude for both particles because they have the same spin. We use the term **antiparticle** for a particle that is related to another particle as the positron is to the electron. Each kind of particle has a corresponding antiparticle. For a few kinds of particles (necessarily all neutral) the particle and antiparticle are identical, and we can say that they are their own antiparticles. The photon is an example; there is no way to distinguish a photon from an antiphoton. We'll use the standard symbols e^- for the electron and e^+ for the positron, and the generic term "electron" will often include both electrons and positrons. Other antiparticles are often denoted by a bar over the particle's symbol; for example, an antiproton is $\bar{p}$. We'll see several other examples of antiparticles later.

Positrons do not occur in ordinary matter. Electron–positron pairs are produced during high-energy collisions of charged particles or γ rays with matter. This process is called e^+e^- *pair production* (Fig. 44.3). Enough energy E must be available to account for the rest energy $2m_ec^2$ of the two particles. The minimum energy for electron–positron pair production is

$$E_{min} = 2m_ec^2 = 2(9.109 \times 10^{-31}\ \text{kg})(2.998 \times 10^8\ \text{m/s})^2$$

$$= 1.637 \times 10^{-13}\ \text{J} = 1.022\ \text{MeV}$$

The inverse process, e^+e^- *pair annihilation,* occurs when a positron and an electron collide (see Example 38.6 in Section 38.3). Both particles disappear, and two (or occasionally three) photons can appear, with total energy of at least $2m_ec^2 = 1.022$ MeV. Decay into a *single* photon is impossible: Such a process could not conserve both energy and momentum.

Positrons also occur in the decay of some unstable nuclei, in which they are called beta-plus particles (β^+). We discussed β^+ decay in Section 43.3.

It's often convenient to represent particle masses in terms of the equivalent rest energy using $m = E/c^2$. Then typical mass units are MeV/c^2; for example, $m = 0.511$ MeV/c^2 for an electron or positron. We'll use these units frequently in this chapter.

44.3 (a) Photograph of bubble-chamber tracks of electron–positron pairs that are produced when 300-MeV photons strike a lead sheet. A magnetic field directed out of the photograph made the electrons and positrons curve in opposite directions. (b) Diagram showing the pair-production process for two of the photons.

(a)   Electron–positron pair

(b)

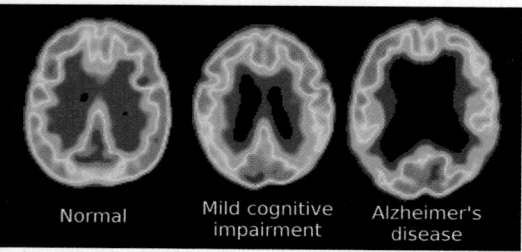

Normal | Mild cognitive impairment | Alzheimer's disease

44.4 An analogy for how particles act as force mediators.

(a) Two skaters exert repulsive forces on each other by tossing a ball back and forth.

(b) Two skaters exert attractive forces on each other when one tries to grab the ball out of the other's hands.

Particles As Force Mediators

In classical physics we describe the interaction of charged particles in terms of electric and magnetic forces. In quantum mechanics we can describe this interaction in terms of emission and absorption of photons. Two electrons repel each other as one emits a photon and the other absorbs it, just as two skaters can push each other apart by tossing a heavy ball back and forth between them (Fig. 44.4a). For an electron and a proton, in which the charges are opposite and the force is attractive, we imagine the skaters trying to grab the ball away from each other (Fig. 44.4b). The electromagnetic interaction between two charged particles is *mediated* or transmitted by photons.

If charged-particle interactions are mediated by photons, where does the energy to create the photons come from? Recall from our discussion of the uncertainty principle (see Sections 38.4 and 39.6) that a state that exists for a short time Δt has an uncertainty ΔE in its energy such that

$$\Delta E \, \Delta t \geq \frac{\hbar}{2} \tag{44.3}$$

This uncertainty permits the creation of a photon with energy ΔE, provided that it lives no longer than the time Δt given by Eq. (44.3). A photon that can exist for a short time because of this energy uncertainty is called a *virtual photon*. It's as though there were an energy bank; you can borrow energy, provided that you pay it back within the time limit. According to Eq. (44.3), the more you borrow, the sooner you have to pay it back.

Mesons

Is there a particle that mediates the *nuclear* force? By the mid-1930s the nuclear force between two nucleons (neutrons or protons) appeared to be described by a potential energy $U(r)$ with the general form

$$U(r) = -f^2\left(\frac{e^{-r/r_0}}{r}\right) \quad \text{(nuclear potential energy)} \tag{44.4}$$

The constant f characterizes the strength of the interaction, and r_0 describes its range. Figure 44.5 shows a graph of the absolute value of this function and compares it with the function f^2/r, which would be analogous to the *electric* interaction of two protons:

$$U(r) = \frac{1}{4\pi\epsilon_0} \frac{e^2}{r} \quad \text{(electric potential energy)} \tag{44.5}$$

In 1935 the Japanese physicist Hideki Yukawa suggested that a hypothetical particle that he called a **meson** might mediate the nuclear force. He showed that the range of the force was related to the mass of the particle. Yukawa argued that the particle must live for a time Δt long enough to travel a distance comparable to the range r_0 of the nuclear force. This range was known from the sizes of nuclei and other information to be about 1.5×10^{-15} m $= 1.5$ fm. If we assume that an average particle's speed is comparable to c and travels about half the range, its lifetime Δt must be about

$$\Delta t = \frac{r_0}{2c} = \frac{1.5 \times 10^{-15} \text{ m}}{2(3.0 \times 10^8 \text{ m/s})} = 2.5 \times 10^{-24} \text{ s}$$

From Eq. (44.3), the minimum necessary uncertainty ΔE in energy is

$$\Delta E = \frac{\hbar}{2\Delta t} = \frac{1.05 \times 10^{-34} \text{ J} \cdot \text{s}}{2(2.5 \times 10^{-24} \text{ s})} = 2.1 \times 10^{-11} \text{ J} = 130 \text{ MeV}$$

The mass equivalent Δm of this energy is

$$\Delta m = \frac{\Delta E}{c^2} = \frac{2.1 \times 10^{-11}\ \text{J}}{(3.00 \times 10^8\ \text{m/s})^2} = 2.3 \times 10^{-28}\ \text{kg} = 130\ \text{MeV}/c^2$$

This is about 250 times the electron mass, and Yukawa postulated that an as yet undiscovered particle with this mass serves as the messenger for the nuclear force.

A year later, Carl Anderson and his colleague Seth Neddermeyer discovered in cosmic radiation two new particles, now called **muons.** The μ^- has charge equal to that of the electron, and its antiparticle the μ^+ has a positive charge with equal magnitude. The two particles have equal mass, about 207 times the electron mass. But it soon became clear that muons were *not* Yukawa's particles because they interacted with nuclei only very weakly.

In 1947 a family of three particles, called π *mesons* or **pions,** were discovered. Their charges are $+e$, $-e$, and zero, and their masses are about 270 times the electron mass. The pions interact strongly with nuclei, and they *are* the particles predicted by Yukawa. Other, heavier mesons, the ω and ρ, evidently also act as shorter-range messengers of the nuclear force. The complexity of this explanation suggests that it has simpler underpinnings; these involve the quarks and gluons that we'll discuss in Section 44.4. Before discussing mesons further, we'll describe some particle accelerators and detectors to see how mesons and other particles are created in a controlled fashion and observed.

Test Your Understanding of Section 44.1 Each of the following particles can be exchanged between two protons, two neutrons, or a neutron and a proton as part of the nuclear force. Rank the particles in order of the range of the interaction that they mediate, from largest to smallest range. (i) the π^+ (pi-plus) meson of mass 140 MeV/c^2; (ii) the ρ^+ (rho-plus) meson of mass 776 MeV/c^2; (iii) the η^0 (eta-zero) meson of mass 548 MeV/c^2; (iv) the ω^0 (omega-zero) meson of mass 783 MeV/c^2.

44.5 Graph of the magnitude of the Yukawa potential-energy function for nuclear forces, $|U(r)| = f^2 e^{-r/r_0}/r$. The function $U(r) = f^2/r$, proportional to the potential energy for Coulomb's law, is also shown. The two functions are similar at small r, but the Yukawa potential energy drops off much more quickly at large r.

44.2 Particle Accelerators and Detectors

Early nuclear physicists used alpha and beta particles from naturally occurring radioactive elements for their experiments, but they were restricted in energy to the few MeV that are available in such random decays. Present-day particle accelerators can produce precisely controlled beams of particles, from electrons and positrons up to heavy ions, with a wide range of energies. These beams have three main uses. First, high-energy particles can collide to produce new particles, just as a collision of an electron and a positron can produce photons. Second, a high-energy particle has a short de Broglie wavelength and so can probe the small-scale interior structure of other particles, just as electron microscopes (see Section 39.1) can give better resolution than optical microscopes. Third, they can be used to produce nuclear reactions of scientific or medical use.

Linear Accelerators

Particle accelerators use electric and magnetic fields to accelerate and guide beams of charged particles. A *linear accelerator* (linac) accelerates particles in a straight line. J. J. Thomson's cathode-ray tubes were early examples of linacs. Modern linacs use a series of electrodes with gaps to give the particles a series of boosts. Most present-day high-energy linear accelerators use a traveling electromagnetic wave; the charged particles "ride" the wave in more or less the way that a surfer rides an incoming ocean wave. In the highest-energy linac in the world today, at the SLAC National Accelerator Laboratory, electrons and positrons can be accelerated to 50 GeV in a tube 3 km long. At this energy their de Broglie wavelengths are 0.025 fm, much smaller than the size of a proton or a neutron.

44.6 Layout and operation of a cyclotron.

(a) Schematic diagram of a cyclotron

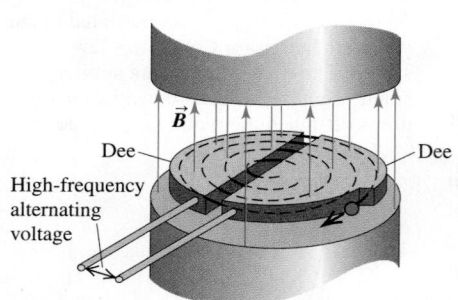

(b) As the positive particle reaches the gap, it is accelerated by the electric-field force ...

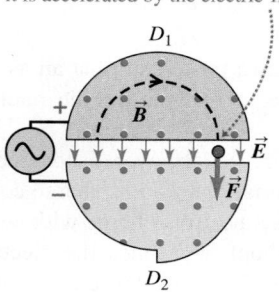

(c) ... and the next semicircular orbit has a larger radius.

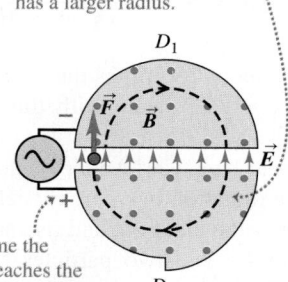

By the time the particle reaches the gap again, the dee voltage has reversed and the particle is again accelerated.

The Cyclotron

Many accelerators use magnets to deflect the charged particles into circular paths. The first of these was the *cyclotron,* invented in 1931 by E. O. Lawrence and M. Stanley Livingston at the University of California (Fig. 44.6a). Particles with mass m and charge q move inside a vacuum chamber in a uniform magnetic field $\vec{B}$ that is perpendicular to the plane of their paths. In Section 27.4 we showed that in such a field, a particle with speed v moves in a circular path with radius r given by

$$r = \frac{mv}{|q|B} \tag{44.6}$$

and with angular speed (angular frequency) ω given by

$$\omega = \frac{v}{r} = \frac{|q|B}{m} \tag{44.7}$$

An alternating potential difference is applied between the two hollow electrodes D_1 and D_2 (called *dees*), creating an electric field in the gap between them. The polarity of the potential difference and electric field is changed precisely twice each revolution (Figs. 44.6b and 44.6c), so that the particles get a push each time they cross the gap. The pushes increase their speed and kinetic energy, boosting them into paths of larger radius. The maximum speed v_{max} and kinetic energy K_{max} are determined by the radius R of the largest possible path. Solving Eq. (44.6) for v, we find $v = |q|Br/m$ and $v_{max} = |q|BR/m$. Assuming nonrelativistic speeds, we have

$$K_{max} = \tfrac{1}{2}mv_{max}^2 = \frac{q^2B^2R^2}{2m} \tag{44.8}$$

Example 44.1 **Frequency and energy in a proton cyclotron**

One cyclotron built during the 1930s has a path of maximum radius 0.500 m and a magnetic field of magnitude 1.50 T. If it is used to accelerate protons, find (a) the frequency of the alternating voltage applied to the dees and (b) the maximum particle energy.

SOLUTION

IDENTIFY and SET UP: The frequency f of the applied voltage must equal the frequency of the proton orbital motion. Equation (44.7) gives the *angular* frequency ω of the proton orbital motion; we find f using $f = \omega/2\pi$. The proton reaches its maximum energy

K_{max}, given by Eq. (44.8), when the radius of its orbit equals the radius of the dees.

EXECUTE: (a) For protons, $q = 1.60 \times 10^{-19}$ C and $m = 1.67 \times 10^{-27}$ kg. From Eq. (44.7),

$$f = \frac{\omega}{2\pi} = \frac{|q|B}{2\pi m} = \frac{(1.60 \times 10^{-19}\ \text{C})(1.50\ \text{T})}{2\pi(1.67 \times 10^{-27}\ \text{kg})}$$

$$= 2.3 \times 10^7\ \text{Hz} = 23\ \text{MHz}$$

(b) From Eq. (44.8) the maximum kinetic energy is

$$K_{max} = \frac{(1.60 \times 10^{-19}\,\text{C})^2(1.50\,\text{T})^2(0.50\,\text{m})^2}{2(1.67 \times 10^{-27}\,\text{kg})}$$

$$= 4.3 \times 10^{-12}\,\text{J} = 2.7 \times 10^7\,\text{eV} = 27\,\text{MeV}$$

This proton kinetic energy is much larger than that available from natural radioactive sources.

EVALUATE: From Eq. (44.6) or Eq. (44.7), the proton speed is $v = 7.2 \times 10^7$ m/s, which is about 25% of the speed of light. At such speeds, relativistic effects are beginning to become important. Since we ignored these effects in our calculation, the above results for f and K_{max} are in error by a few percent; this is why we kept only two significant figures.

The maximum energy that can be attained with a cyclotron is limited by relativistic effects. The relativistic version of Eq. (44.7) is

$$\omega = \frac{|q|B}{m}\sqrt{1 - v^2/c^2}$$

As the particles speed up, their angular frequency ω *decreases,* and their motion gets out of phase with the alternating dee voltage. In the *synchrocyclotron* the particles are accelerated in bursts. For each burst, the frequency of the alternating voltage is decreased as the particles speed up, maintaining the correct phase relationship with the particles' motion.

Another limitation of the cyclotron is the difficulty of building very large electromagnets. The largest synchrocyclotron ever built has a vacuum chamber that is about 8 m in diameter and accelerates protons to energies of about 600 MeV.

The Synchrotron

To attain higher energies, another type of machine, called the *synchrotron,* is more practical. Particles move in a vacuum chamber in the form of a thin doughnut, called the *accelerating ring.* The particle beam is bent to follow the ring by a series of electromagnets placed around the ring. As the particles speed up, the magnetic field is increased so that the particles retrace the same trajectory over and over. The Large Hadron Collider (LHC) near Geneva, Switzerland, is the highest-energy accelerator in the world (Fig. 44.7). It is designed to accelerate protons to a maximum energy of 7 TeV, or 7×10^{12} eV. (As we'll discuss in Section 44.3, *hadrons* are a class of elementary particles that includes protons and neutrons.)

As we pointed out in Section 32.1, accelerated charges radiate electromagnetic energy. In an accelerator in which the particles move in curved paths, this radiation is often called *synchrotron radiation.* High-energy accelerators are typically constructed underground to provide protection from this radiation. From the accelerator standpoint, synchrotron radiation is undesirable, since the energy given to an accelerated particle is radiated right back out. It can be minimized by making the accelerator radius r large so that the centripetal acceleration v^2/r is small. On the positive side, synchrotron radiation is used as a source of well-controlled high-frequency electromagnetic waves.

Available Energy

When a beam of high-energy particles collides with a stationary target, not all the kinetic energy of the incident particles is *available* to form new particle states. Because momentum must be conserved, the particles emerging from the collision must have some net motion and thus some kinetic energy. The discussion following Example 43.11 (Section 43.6) presented a nonrelativistic example of this principle. The maximum available energy is the kinetic energy in the frame of reference in which the total momentum is zero. We call this the *center-of-momentum system;* it is the relativistic generalization of the center-of-mass system that we discussed in Section 8.5. In this system the total kinetic energy after the collision can be zero, so that the maximum amount of the initial kinetic energy becomes available to cause the reaction being studied.

44.7 (a) The Large Hadron Collider at the European Organization for Nuclear Research (CERN). The underground accelerating ring (shown by the red circle) is 100 m underground and 8.5 km in diameter, so large that it spans the border between Switzerland and France. (Note the Alps in the background.) When accelerated to 7 TeV, protons travel around the ring more than 11,000 times per second. (b) An engineer working on one of the 9593 superconducting electromagnets around the LHC ring.

(a)

(b)

Consider the *laboratory system,* in which a target particle with mass M is initially at rest and is bombarded by a particle with mass m and total energy (including rest energy) E_m. The total available energy E_a in the center-of-momentum system (including rest energies of all the particles) can be shown to be given by

$$E_a{}^2 = 2Mc^2E_m + (Mc^2)^2 + (mc^2)^2 \quad \text{(available energy)} \quad (44.9)$$

When the masses of the target and projectile particles are equal, this can be simplified to

$$E_a{}^2 = 2mc^2(E_m + mc^2) \quad \text{(available energy, equal masses)} \quad (44.10)$$

If in addition E_m is much greater than mc^2, we can neglect the second term in the parentheses in Eq (44.10). Then E_a is

$$E_a = \sqrt{2mc^2E_m} \quad \text{(available energy, equal masses, } E_m \gg mc^2) \quad (44.11)$$

The square root in Eq. (44.11) is a disappointing result for an accelerator designer: Doubling the energy E_m of the bombarding particle increases the available energy E_a by only a factor of $\sqrt{2} = 1.414$. Examples 44.2 and 44.3 explore the limitations of having a stationary target particle.

Example 44.2 Threshold energy for pion production

A proton (rest energy 938 MeV) with kinetic energy K collides with a proton at rest. Both protons survive the collision, and a neutral pion (π^0, rest energy 135 MeV) is produced. What is the threshold energy (minimum value of K) for this process?

SOLUTION

IDENTIFY and SET UP: The final state includes the two original protons (mass m) and the pion (mass m_π). The threshold energy corresponds to the minimum-energy case in which all three particles are at rest in the center-of-momentum system. The total available energy E_a in that system must be at least the total rest energy, $2mc^2 + m_\pi c^2$. We use this to solve Eq. (44.10) for the total energy E_m of the bombarding proton; the kinetic energy K (our target variable) is then E_m minus the proton rest energy mc^2.

EXECUTE: We substitute $E_a = 2mc^2 + m_\pi c^2$ into Eq. (44.10), simplify, and solve for E_m:

$$4m^2c^4 + 4mm_\pi c^4 + m_\pi{}^2 c^4 = 2mc^2 E_m + 2(mc^2)^2$$

$$E_m = mc^2 + m_\pi c^2\left(2 + \frac{m_\pi}{2m}\right) = mc^2 + K$$

$$K = m_\pi c^2\left(2 + \frac{m_\pi}{2m}\right)$$

We see that the bombarding proton's kinetic energy K must be somewhat greater than twice the pion rest energy $m_\pi c^2$. With $mc^2 = 938$ MeV and $m_\pi c^2 = 135$ MeV, we have $m_\pi/2m = 0.072$ and

$$K = (135 \text{ MeV})(2 + 0.072) = 280 \text{ MeV}$$

EVALUATE: Compare this result with the result of Example 37.11 (Section 37.8), where we found that a pion can be produced in a head-on collision of two protons, each with only 67.5 MeV of kinetic energy. We discuss the energy advantage of such collisions in the next subsection.

Example 44.3 Increasing the available energy

The Fermilab accelerator in Illinois was designed to bombard stationary targets with 800-GeV protons. (a) What is the available energy E_a in a proton-proton collision? (b) What is E_a if the beam energy is increased to 980 GeV?

SOLUTION

IDENTIFY and SET UP: Our target variable is the available energy E_a in a stationary-target collision between identical particles. In both parts (a) and (b) the beam energy E_m is much larger than the proton rest energy $mc^2 = 938$ MeV $= 0.938$ GeV, so we can safely use the approximation of Eq. (44.11).

EXECUTE: (a) For $E_m = 800$ GeV, Eq. (44.11) gives

$$E_a = \sqrt{2(0.938 \text{ GeV})(800 \text{ GeV})} = 38.7 \text{ GeV}$$

(b) For $E_m = 980$ GeV,

$$E_a = \sqrt{2(0.938 \text{ GeV})(980 \text{ GeV})} = 42.9 \text{ GeV}$$

EVALUATE: With a stationary-proton target, increasing the proton beam energy by 180 GeV increases the available energy by only 4.2 GeV! This shows a major limitation of experiments in which one of the colliding particles is initially at rest. Below we describe how physicists can overcome this limitation.

Colliding Beams

The limitation illustrated by Example 44.3 is circumvented in *colliding-beam* experiments. In these experiments there is no stationary target; instead, beams of particles moving in opposite directions are tightly focused onto one another so that head-on collisions can occur. Usually the two colliding particles have momenta of equal magnitude and opposite direction, so the total momentum is zero. Hence the laboratory system is also the center-of-momentum system, and the available energy is maximized.

The highest-energy colliding beams available are those at the Large Hadron Collider (see Fig. 44.7). In operation, 2808 bunches of 7-TeV protons circulate around the ring, half in one direction and half in the opposite direction. Each bunch contains about 10^{11} protons. Magnets steer the oppositely moving bunches to collide at interaction points. The available energy E_a in the resulting head-on collisions is the *total* energy of the two colliding particles: $E_a = 2 \times 7 \text{ TeV} = 14 \text{ TeV}$. (Strictly, E_a is 14 TeV minus the rest energy of the two colliding protons. But this rest energy is only $2mc^2 = 2(938 \text{ MeV}) = 1.876 \times 10^{-3}$ TeV, which is so small compared to 14 TeV that it can be ignored.) Physicists expect that the very large available energy at the Large Hadron Collider will make it possible to produce particles that have never been seen before.

Detectors

Ordinarily, we can't see or feel individual subatomic particles or photons. How, then, do we measure their properties? A wide variety of devices have been designed. Many detectors use the ionization caused by charged particles as they move through a gas, liquid, or solid. The ions along the particle's path act as nucleation centers for droplets of liquid in the supersaturated vapor of a cloud chamber (Fig. 44.1) or cause small volumes of vapor in the superheated liquid of a bubble chamber (Fig. 44.3a). In a semiconducting solid the ionization can take the form of electron–hole pairs. We discussed their detection in Section 42.7. *Wire chambers* contain arrays of closely spaced wires that detect the ions. The charge collected and time information from each wire are processed using computers to reconstruct the particle trajectories. The detectors at the Large Hadron Collider use an array of devices to follow the tracks of particles produced by collisions between protons (Fig. 44.8). The giant solenoid in the photo that opens Chapter 28 is at the heart of one these detector arrays. The intense magnetic field of the solenoid helps identify newly produced particles, which curve in different directions and along paths of different radii depending on their charge and energy.

Cosmic-Ray Experiments

Large numbers of particles called *cosmic rays* continually bombard the earth from sources both within and beyond our galaxy. These particles consist mostly of neutrinos, protons, and heavier nuclei, with energies ranging from less than 1 MeV to more than 10^{20} eV. The earth's atmosphere and magnetic field protect us from much of this radiation. This means that cosmic-ray experimentation often must be carried out above all or most of the atmosphere by means of rockets or high-altitude balloons.

In contrast, neutrino detectors are buried below the earth's surface in tunnels or mines or submerged deep in the ocean. This is done to screen out all other types of particles so that only neutrinos, which interact only very weakly with matter, reach the detector. It would take a light-year's thickness of lead to absorb a sizable fraction of a beam of neutrinos. Thus neutrino detectors consist of huge amounts of matter: The Super-Kamiokande detector looks for flashes of light produced when a neutrino interacts in a tank containing 5×10^7 kg of water (see Section 44.5).

44.8 This computer-generated image shows the result of a simulated collision between two protons (not shown) in one of the interaction regions at the Large Hadron Collider. The view is along the beampipe. The different color tracks show different types of particles emerging from the collision. A variety of different detectors surround the collision region. (Note the drawing of woman in a red dress, shown for scale.)

Cosmic rays were important in early particle physics, and their study currently brings us important information about the rest of the universe. Although cosmic rays provide a source of high-energy particles that does not depend on expensive accelerators, most particle physicists use accelerators because the high-energy cosmic-ray particles they want are too few and too random.

Test Your Understanding of Section 44.2 In a colliding-beam experiment, a 90-GeV electron collides head-on with a 90-GeV positron. The electron and the positron annihilate each other, forming a single virtual photon that then transforms into other particles. Does the virtual photon obey the same relationship $E = pc$ as real photons do? ❚

44.3 Particles and Interactions

We have mentioned the array of subatomic particles that were known as of 1947: photons, electrons, positrons, protons, neutrons, muons, and pions. Since then, literally hundreds of additional particles have been discovered in accelerator experiments. The vast majority of known particles are *unstable* and decay spontaneously into other particles. Particles of all kinds, whether stable or unstable, can be created or destroyed in interactions between particles. Each such interaction involves the exchange of virtual particles, which exist on borrowed energy allowed by the uncertainty principle.

Although the world of subatomic particles and their interactions is complex, some key results bring order and simplicity to the seeming chaos. One key simplification is that there are only four fundamental types of interactions, each mediated or transmitted by the exchange of certain characteristic virtual particles. Furthermore, not all particles respond to all four kinds of interaction. In this section we will examine the fundamental interactions more closely and see how physicists classify particles in terms of the ways in which they interact.

Four Forces and Their Mediating Particles

In Section 5.5 we first described the four fundamental types of forces or interactions (Fig. 44.9). They are, in order of decreasing strength:

1. The strong interaction
2. The electromagnetic interaction
3. The weak interaction
4. The gravitational interaction

The *electromagnetic* and *gravitational* interactions are familiar from classical physics. Both are characterized by a $1/r^2$ dependence on distance. In this scheme, the mediating particles for both interactions have mass zero and are stable as ordinary particles. The mediating particle for the electromagnetic interaction is the familiar photon, which has spin 1. (That means its spin quantum number is $s = 1$, so the magnitude of its spin angular momentum is $S = \sqrt{s(s+1)}\hbar = \sqrt{2}\hbar$.) That particle for the gravitational force is the spin-2 *graviton* ($s = 2$, $S = \sqrt{s(s+1)}\hbar = \sqrt{6}\hbar$). The graviton has not yet been observed experimentally because the gravitational force is very much weaker than the electromagnetic force. For example, the gravitational attraction of two protons is smaller than their electrical repulsion by a factor of about 10^{36}. The gravitational force is of primary importance in the structure of stars and the large-scale behavior of the universe, but it is not believed to play a significant role in particle interactions at the energies that are currently attainable.

The other two forces are less familiar. One, usually called the *strong interaction*, is responsible for the nuclear force and also for the production of pions and several other particles in high-energy collisions. At the most fundamental level, the mediating particle for the strong interaction is called a *gluon*. However, the force between nucleons is more easily described in terms of mesons as the mediating particles. We'll discuss the spin-1, massless gluon in Section 44.4.

44.9 The ties that bind us together originate in the fundamental interactions of nature. The nuclei within our bodies are held together by the strong interaction. The electromagnetic interaction binds nuclei and electrons together to form atoms, binds atoms together to form molecules, and binds molecules together to form us.

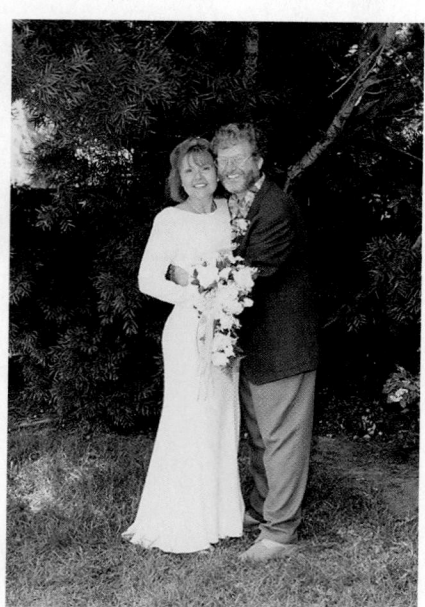

Equation (44.4) is a possible potential-energy function for the nuclear force. The strength of the interaction is described by the constant f^2, which has units of energy times distance. A better basis for comparison with other forces is the dimensionless ratio $f^2/\hbar c$, called the *coupling constant* for the interaction. (We invite you to verify that this ratio is a pure number and so must have the same value in all systems of units.) The observed behavior of nuclear forces suggests that $f^2/\hbar c \approx 1$. The dimensionless coupling constant for *electromagnetic* interactions is

$$\frac{1}{4\pi\epsilon_0}\frac{e^2}{\hbar c} = 7.297 \times 10^{-3} = \frac{1}{137.0} \qquad (44.12)$$

Thus the strong interaction is roughly 100 times as strong as the electromagnetic interaction; however, it drops off with distance more quickly than $1/r^2$.

The fourth interaction is called the *weak* interaction. It is responsible for beta decay, such as the conversion of a neutron into a proton, an electron, and an antineutrino. It is also responsible for the decay of many unstable particles (pions into muons, muons into electrons, and so on). Its mediating particles are the short-lived particles W^+, W^-, and Z^0. The existence of these particles was confirmed in 1983 in experiments at CERN, for which Carlo Rubbia and Simon van der Meer were awarded the Nobel Prize in 1984. The $W^\pm$ and Z^0 have spin 1 like the photon and the gluon, but they are *not* massless. In fact, they have enormous masses, 80.4 GeV/c^2 for the W's and 91.2 GeV/c^2 for the Z^0. With such massive mediating particles the weak interaction has a much shorter range than the strong interaction. It also lives up to its name by being weaker than the strong interaction by a factor of about 10^9.

Table 44.1 compares the main features of these four fundamental interactions.

More Particles

In Section 44.1 we mentioned the discoveries of muons in 1937 and of pions in 1947. The electric charges of the muons and the charged pions have the same magnitude e as the electron charge. The positive muon μ^+ is the antiparticle of the negative muon μ^-. Each has spin $\frac{1}{2}$, like the electron, and a mass of about $207m_e = 106$ MeV/c^2. Muons are unstable; each decays with a lifetime of 2.2×10^{-6} s into an electron of the same sign, a neutrino, and an antineutrino.

There are three kinds of pions, all with spin 0; they have *no* spin angular momentum. The π^+ and π^- have masses of $273m_e = 140$ MeV/c^2. They are unstable; each $\pi^\pm$ decays with a lifetime of 2.6×10^{-8} s into a muon of the same sign along with a neutrino for the π^+ and an antineutrino for the π^-. The π^0 is somewhat less massive, $264m_e = 135$ MeV/c^2, and it decays with a lifetime of 8.4×10^{-17} s into two photons. The π^+ and π^- are antiparticles of one another, while the π^0 is its own antiparticle. (That is, there is no distinction between particle and antiparticle for the π^0.)

The existence of the *antiproton* $\bar{p}$ had been suspected ever since the discovery of the positron. The $\bar{p}$ was found in 1955, when proton–antiproton ($p\bar{p}$) pairs were created by use of a beam of 6-GeV protons from the Bevatron at the University of

Mastering**PHYSICS**

ActivPhysics 19.5: Particle Physics

Table 44.1 Four Fundamental Interactions

Interaction	Relative Strength	Range	Mediating Particle			
			Name	Mass	Charge	Spin
Strong	1	Short ($\sim$1 fm)	Gluon	0	0	1
Electromagnetic	$\frac{1}{137}$	Long ($1/r^2$)	Photon	0	0	1
Weak	10^{-9}	Short ($\sim$0.001 fm)	$W^\pm$, Z^0	80.4, 91.2 GeV/c^2	$\pm e$, 0	1
Gravitational	10^{-38}	Long ($1/r^2$)	Graviton	0	0	2

California, Berkeley. The *antineutron* $\bar{n}$ was found soon afterward. After 1960, as higher-energy accelerators and more sophisticated detectors were developed, a veritable blizzard of new unstable particles were identified. To describe and classify them, we need a small blizzard of new terms.

Initially, particles were classified by mass into three categories: (1) leptons ("light ones" such as electrons); (2) mesons ("intermediate ones" such as pions); and (3) baryons ("heavy ones" such as nucleons and more massive particles). But this scheme has been superseded by a more useful one in which particles are classified in terms of their *interactions*. For instance, *hadrons* (which include mesons and baryons) have strong interactions, and *leptons* do not.

In the following discussion we will also distinguish between **fermions,** which have half-integer spins, and **bosons,** which have zero or integer spins. Fermions obey the exclusion principle, on which the Fermi-Dirac distribution function (see Section 42.5) is based. Bosons do not obey the exclusion principle and have a different distribution function, the Bose-Einstein distribution.

Leptons

The **leptons,** which do not have strong interactions, include six particles; the electron (e^-) and its neutrino (ν_e), the muon (μ^-) and its neutrino (ν_μ), and the tau particle (τ^-) and its neutrino (ν_τ). Each of the six particles has a distinct antiparticle. All leptons have spin $\frac{1}{2}$ and thus are fermions. The family of leptons is shown in Table 44.2. The taus have mass $3478m_e = 1777\ \text{MeV}/c^2$. Taus and muons are unstable; a τ^- decays into a μ^- plus a tau neutrino and a muon antineutrino, or an electron plus a tau neutrino and an electron antineutrino. A μ^- decays into a electron plus a muon neutrino and an electron antineutrino. They have relatively long lifetimes because their decays are mediated by the weak interaction. Despite their zero charge, a neutrino is distinct from an antineutrino; the spin angular momentum of a neutrino has a component that is opposite its linear momentum, while for an antineutrino that component is parallel to its linear momentum. Because neutrinos are so elusive, physicists have only been able to place upper limits on the rest masses of the ν_e, the ν_μ, and the ν_τ. Until recently, it was thought that the rest masses of the neutrinos were zero; compelling evidence now indicates that they have small but nonzero masses. We'll return to this point and its implications later.

Leptons obey a *conservation principle.* Corresponding to the three pairs of leptons are three lepton numbers L_e, L_μ, and L_τ. The electron e^- and the electron neutrino ν_e are assigned $L_e = 1$, and their antiparticles e^+ and $\bar{\nu}_e$ are given $L_e = -1$. Corresponding assignments of L_μ and L_τ are made for the μ and τ particles and their neutrinos. **In all interactions, each lepton number is separately conserved.** For example, in the decay of the μ^-, the lepton numbers are

$$\mu^- \rightarrow e^- + \bar{\nu}_e + \nu_\mu$$
$$L_\mu = 1 \qquad L_e = 1 \qquad L_e = -1 \qquad L_\mu = 1$$

These conservation principles have no counterpart in classical physics.

Table 44.2 The Six Leptons

Particle Name	Symbol	Anti-particle	Mass (MeV/c^2)	L_e	L_μ	L_τ	Lifetime (s)	Principal Decay Modes
Electron	e^-	e^+	0.511	+1	0	0	Stable	
Electron neutrino	ν_e	$\bar{\nu}_e$	$<3 \times 10^{-6}$	+1	0	0	Stable	
Muon	μ^-	μ^+	105.7	0	+1	0	2.20×10^{-6}	$e^- \bar{\nu}_e \nu_\mu$
Muon neutrino	ν_μ	$\bar{\nu}_\mu$	<0.19	0	+1	0	Stable	
Tau	τ^-	τ^+	1777	0	0	+1	2.9×10^{-13}	$\mu^- \bar{\nu}_\mu \nu_\tau$
Tau neutrino	ν_τ	$\bar{\nu}_\tau$	<18.2	0	0	+1	Stable	or $e^- \bar{\nu}_e \nu_\tau$

Example 44.4 Lepton number conservation

Check conservation of lepton numbers for these decay schemes:

(a) $\mu^+ \to e^+ + \nu_e + \bar{\nu}_\mu$

(b) $\pi^- \to \mu^- + \bar{\nu}_\mu$

(c) $\pi^0 \to \mu^- + e^+ + \nu_e$

SOLUTION

IDENTIFY and SET UP: Lepton number conservation requires that L_e, L_μ, and L_τ (given in Table 44.2) separately have the same sums after the decay as before.

EXECUTE: We tabulate L_e and L_μ for each decay scheme. An antiparticle has the opposite lepton number from its corresponding particle listed in Table 44.2. No τ particles or τ neutrinos appear in any of the schemes, so $L_\tau = 0$ both before and after each decay and L_τ is conserved.

(a) $\mu^+ \to e^+ + \nu_e + \bar{\nu}_\mu$

L_e: $0 = -1 + 1 + 0$

L_μ: $-1 = 0 + 0 + (-1)$

(b) $\pi^- \to \mu^- + \bar{\nu}_\mu$

L_e: $0 = 0 + 0$

L_μ: $0 = 1 + (-1)$

(c) $\pi^0 \to \mu^- + e^+ + \nu_e$

L_e: $0 = 0 + (-1) + 1$

L_μ: $0 \neq 1 + 0 + 0$

EVALUATE: Decays (a) and (b) are consistent with lepton number conservation and are observed. Decay (c) violates the conservation of L_μ and has *never* been observed. Physicists used these and other experimental results to deduce the principle that all three lepton numbers must separately be conserved.

Hadrons

Hadrons, the strongly interacting particles, are a more complex family than leptons. Each hadron has an antiparticle, often denoted with an overbar, as with the antiproton $\bar{p}$. There are two subclasses of hadrons: *mesons* and *baryons.* Table 44.3 shows some of the many hadrons that are currently known. (We'll explain *strangeness* and *quark content* later in this section and in the next one.)

Mesons include the pions that have already been mentioned, K mesons or *kaons,* η mesons, and others that we will discuss later. Mesons have spin 0 or 1 and therefore are all bosons. There are no stable mesons; all can and do decay to less massive particles, obeying all the conservation laws for such decays.

Table 44.3 Some Hadrons and Their Properties

Particle	Mass (MeV/c^2)	Charge Ratio, Q/e	Spin	Baryon Number, B	Strangeness, S	Mean Lifetime (s)	Typical Decay Modes	Quark Content
Mesons								
π^0	135.0	0	0	0	0	8.4×10^{-17}	$\gamma\,\gamma$	$u\bar{u}, d\bar{d}$
π^+	139.6	$+1$	0	0	0	2.60×10^{-8}	$\mu^+\nu_\mu$	$u\bar{d}$
π^-	139.6	-1	0	0	0	2.60×10^{-8}	$\mu^-\bar{\nu}_\mu$	$\bar{u}d$
K^+	493.7	$+1$	0	0	$+1$	1.24×10^{-8}	$\mu^+\nu_\mu$	$u\bar{s}$
K^-	493.7	-1	0	0	-1	1.24×10^{-8}	$\mu^-\bar{\nu}_\mu$	$\bar{u}s$
η^0	547.3	0	0	0	0	$\approx 10^{-18}$	$\gamma\,\gamma$	$u\bar{u}, d\bar{d}, s\bar{s}$
Baryons								
p	938.3	$+1$	$\frac{1}{2}$	1	0	Stable	—	uud
n	939.6	0	$\frac{1}{2}$	1	0	886	$pe^-\bar{\nu}_e$	udd
Λ^0	1116	0	$\frac{1}{2}$	1	-1	2.63×10^{-10}	$p\pi^-$ or $n\pi^0$	uds
Σ^+	1189	$+1$	$\frac{1}{2}$	1	-1	8.02×10^{-11}	$p\pi^0$ or $n\pi^+$	uus
Σ^0	1193	0	$\frac{1}{2}$	1	-1	7.4×10^{-20}	$\Lambda^0\gamma$	uds
Σ^-	1197	-1	$\frac{1}{2}$	1	-1	1.48×10^{-10}	$n\pi^-$	dds
Ξ^0	1315	0	$\frac{1}{2}$	1	-2	2.90×10^{-10}	$\Lambda^0\pi^0$	uss
Ξ^-	1321	-1	$\frac{1}{2}$	1	-2	1.64×10^{-10}	$\Lambda^0\pi^-$	dss
Δ^{++}	1232	$+2$	$\frac{3}{2}$	1	0	$\approx 10^{-23}$	$p\pi^+$	uuu
Ω^-	1672	-1	$\frac{3}{2}$	1	-3	8.2×10^{-11}	$\Lambda^0 K^-$	sss
Λ_c^+	2285	$+1$	$\frac{1}{2}$	1	0	2.0×10^{-13}	$pK^-\pi^+$	udc

Baryons include the nucleons and several particles called *hyperons,* including the Λ, Σ, Ξ, and Ω. These resemble the nucleons but are more massive. Baryons have half-integer spin, and therefore all are fermions. The only stable baryon is the proton; a free neutron decays to a proton, and the hyperons decay to other hyperons or to nucleons by various processes. Baryons obey the principle of *conservation of baryon number,* analogous to conservation of lepton numbers, again with no counterpart in classical physics. We assign a baryon number $B = 1$ to each baryon (p, n, Λ, Σ, and so on) and $B = -1$ to each antibaryon ($\bar{\text{p}}$, $\bar{\text{n}}$, $\bar{\Lambda}$, $\bar{\Sigma}$, and so on).

In all interactions, the total baryon number is conserved.

This principle is the reason the mass number A was conserved in all of the nuclear reactions that we studied in Chapter 43.

Example 44.5 **Baryon number conservation**

Check conservation of baryon number for these reactions:

(a) $n + p \rightarrow n + p + p + \bar{p}$

(b) $n + p \rightarrow n + p + \bar{n}$

SOLUTION

IDENTIFY and SET UP: This example is similar to Example 44.4. We compare the total baryon number before and after each reaction, using data from Table 44.3.

EXECUTE: We tabulate the baryon numbers, noting that a baryon has $B = 1$ and an antibaryon has $B = -1$:

(a) $n + p \rightarrow n + p + p + \bar{p}$: $1 + 1 = 1 + 1 + 1 + (-1)$

(b) $n + p \rightarrow n + p + \bar{n}$: $1 + 1 \neq 1 + 1 + (-1)$

EVALUATE: Reaction (a) is consistent with baryon number conservation. It can occur if enough energy is available in the n + p collision. Reaction (b) violates baryon number conservation and has never been observed.

Example 44.6 **Antiproton creation**

What is the minimum proton energy required to produce an antiproton in a collision with a stationary proton?

SOLUTION

IDENTIFY and SET UP: The reaction must conserve baryon number, charge, and energy. Since the target and bombarding protons are of equal mass and the target is at rest, we determine the minimum energy E_m of the bombarding proton using Eq. (44.10).

EXECUTE: Conservation of charge and conservation of baryon number forbid the creation of an antiproton by itself; it must be created as part of a proton–antiproton pair. The complete reaction is

$$p + p \rightarrow p + p + p + \bar{p}$$

For this reaction to occur, the minimum available energy E_a in Eq. (44.10) is the final rest energy $4mc^2$ of three protons and an antiproton. Equation (44.10) then gives

$$(4mc^2)^2 = 2mc^2(E_m + mc^2)$$
$$E_m = 7mc^2$$

EVALUATE: The energy E_m of the bombarding proton includes its rest energy mc^2, so its minimum *kinetic* energy must be $6mc^2 = 6(938 \text{ MeV}) = 5.63$ GeV.

The search for the antiproton was a principal reason for the construction of the Bevatron at the University of California, Berkeley, with beam energy of 6 GeV. The search succeeded in 1955, and Emilio Segrè and Owen Chamberlain were later awarded the Nobel Prize for this discovery.

Strangeness

The K mesons and the Λ and Σ hyperons were discovered during the late 1950s. Because of their unusual behavior they were called *strange particles.* They were produced in high-energy collisions such as $\pi^- + p$, and a K meson and a hyperon were always produced *together.* The relatively high rate of production of these particles suggested that it was a *strong*-interaction process, but their relatively long lifetimes suggested that their decay was a *weak*-interaction process. The K^0 appeared to have *two* lifetimes, one about 9×10^{-11} s and another nearly 600 times longer. Were the K mesons strongly interacting hadrons or not?

The search for the answer to this question led physicists to introduce a new quantity called **strangeness.** The hyperons Λ^0 and $\Sigma^{\pm,0}$ were assigned a strangeness quantum number $S = -1$, and the associated K^0 and K^+ mesons were assigned $S = +1$. The corresponding antiparticles had opposite strangeness, $S = +1$ for $\overline{\Lambda}^0$ and $\overline{\Sigma}^{\pm,0}$ and $S = -1$ for $\overline{K}^0$ and K^-. Then strangeness was *conserved* in production processes such as

$$p + \pi^- \to \Sigma^- + K^+$$

$$p + \pi^- \to \Lambda^0 + K^0$$

The process

$$p + \pi^- \to p + K^-$$

does not conserve strangeness and it does not occur.

When strange particles decay individually, strangeness is usually *not* conserved. Typical processes include

$$\Sigma^+ \to n + \pi^+$$

$$\Lambda^0 \to p + \pi^-$$

$$K^- \to \pi^+ + \pi^- + \pi^-$$

In each of these decays, the initial strangeness is 1 or -1, and the final value is zero. All observations of these particles are consistent with the conclusion that *strangeness is conserved in strong interactions but it can change by zero or one unit in weak interactions.* There is no counterpart to the strangeness quantum number in classical physics.

CAUTION **Strangeness vs. spin** Take care not to confuse the symbol S for strangeness with the identical symbol for the magnitude of the spin angular momentum. ▌

Conservation Laws

The decay of strange particles provides our first example of a *conditional conservation law,* one that is obeyed in some interactions and not in others. By contrast, several conservation laws are obeyed in *all* interactions. These include the familiar conservation laws; energy, momentum, angular momentum, and electric charge. These are called *absolute conservation laws.* Baryon number and the three lepton numbers are also conserved in all interactions. Strangeness is conserved in strong and electromagnetic interactions but *not* in all weak interactions.

Two other quantities, which are conserved in some but not all interactions, are useful in classifying particles and their interactions. One is *isospin,* a quantity that is used to describe the charge independence of the strong interactions. The other is *parity,* which describes the comparative behavior of two systems that are mirror images of each other. Isospin is conserved in strong interactions, which are charge independent, but not in electromagnetic or weak interactions. (The electromagnetic interaction is certainly *not* charge independent.) Parity is conserved in strong and electromagnetic interactions but not in weak ones. The Chinese-American physicists T. D. Lee and C. N. Yang received the Nobel Prize in 1957 for laying the theoretical foundations for nonconservation of parity in weak interactions.

This discussion shows that conservation laws provide another basis for classifying particles and their interactions. Each conservation law is also associated with a *symmetry* property of the system. A familiar example is angular momentum. If a system is in an environment that has spherical symmetry, there can be no torque acting on it because the direction of the torque would violate the symmetry. In such a system, total angular momentum is *conserved.* When a conservation law is violated, the interaction is often described as a *symmetry-breaking interaction.*

Test Your Understanding of Section 44.3 From conservation of energy, a particle of mass m and rest energy mc^2 can decay only if the decay products have a total mass less than m. (The remaining energy goes into the kinetic energy of the decay products.) Can a proton decay into less massive mesons? |

44.4 Quarks and the Eightfold Way

The leptons form a fairly neat package: three particles and three neutrinos, each with its antiparticle, and a conservation law relating their numbers. Physicists believe that leptons are genuinely fundamental particles. The hadron family, by comparison, is a mess. Table 44.3 contains only a sample of well over 100 hadrons that have been discovered since 1960, and it has become clear that these particles *do not* represent the most fundamental level of the structure of matter.

Our present understanding of the structure of hadrons is based on a proposal made initially in 1964 by the American physicist Murray Gell-Mann and his collaborators. In this proposal, hadrons are not fundamental particles but are composite structures whose constituents are spin-$\frac{1}{2}$ fermions called **quarks.** (The name is found in the line "Three quarks for Muster Mark!" from *Finnegans Wake* by James Joyce.) Each baryon is composed of three quarks (qqq), each antibaryon of three antiquarks ($\overline{qqq}$), and each meson of a quark–antiquark pair ($q\overline{q}$). Table 44.3 of the preceding section gives the quark content of many hadrons. No other compositions seem to be necessary. This scheme requires that quarks have electric charges with magnitudes $\frac{1}{3}$ and $\frac{2}{3}$ of the electron charge e, which had previously been thought to be the smallest unit of charge. Each quark also has a fractional value $\frac{1}{3}$ for its baryon number B, and each antiquark has a baryon-number value $-\frac{1}{3}$. In a meson, a quark and antiquark combine with net baryon number 0 and can have their spin angular momentum components parallel to form a spin-1 meson or antiparallel to form a spin-0 meson. Similarly, the three quarks in a baryon combine with net baryon number 1 and can form a spin-$\frac{1}{2}$ baryon or a spin-$\frac{3}{2}$ baryon.

The Three Original Quarks

The first (1964) quark theory included three types (called *flavors*) of quarks, labeled u (up), d (down), and s (strange). Their principal properties are listed in Table 44.4. The corresponding antiquarks $\overline{u}$, $\overline{d}$, and $\overline{s}$ have opposite values of charge Q, B, and S. Protons, neutrons, π and K mesons, and several hyperons can be constructed from these three quarks. For example, the proton quark content is uud. Checking Table 44.4, we see that the values of Q/e add to 1 and that the values of the baryon number B also add to 1, as we should expect. The neutron is udd, with total $Q = 0$ and $B = 1$. The π^+ meson is $u\overline{d}$, with $Q/e = 1$ and $B = 0$, and the K$^+$ meson is $u\overline{s}$. Checking the values of S for the quark content, we see that the proton, neutron, and π^+ have strangeness 0 and that the K$^+$ has strangeness 1, in agreement with Table 44.3. The antiproton is $\overline{p} = \overline{uud}$, the negative pion is $\pi^- = \overline{u}d$, and so on. The quark content can also be used to explain hadron excited states and magnetic moments. Figure 44.10 shows the quark content of two baryons and two mesons.

44.10 Quark content of four different hadrons. The various color combinations that are needed for color neutrality are not shown.

Proton (p)   Neutron (n)

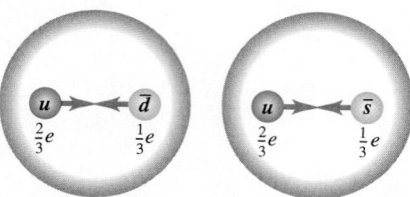

Positive pion (π^+)   Positive kaon (K$^+$)

Table 44.4 Properties of the Three Original Quarks

Symbol	Q/e	Spin	Baryon Number, B	Strangeness, S	Charm, C	Bottomness, B'	Topness, T
u	$\frac{2}{3}$	$\frac{1}{2}$	$\frac{1}{3}$	0	0	0	0
d	$-\frac{1}{3}$	$\frac{1}{2}$	$\frac{1}{3}$	0	0	0	0
s	$-\frac{1}{3}$	$\frac{1}{2}$	$\frac{1}{3}$	-1	0	0	0

| **Example 44.7** | **Determining the quark content of baryons** |

Given that they contain only u, d, s, $\bar{u}$, $\bar{d}$, and/or $\bar{s}$, find the quark content of (a) Σ^+ and (b) $\overline{\Lambda}^0$. The Σ^+ and Λ^0 (the antiparticle of the $\overline{\Lambda}^0$) are both baryons with strangeness $S = -1$.

SOLUTION

IDENTIFY and SET UP: We use the idea that the total charge of each baryon is the sum of the individual quark charges, and similarly for the baryon number and strangeness. We use the quark properties given in Table 44.4.

EXECUTE: Baryons contain three quarks. If $S = -1$, exactly *one* of the three must be an s quark, which has $S = -1$ and $Q/e = -\frac{1}{3}$.

(a) The Σ^+ has $Q/e = +1$, so the other two quarks must both be u quarks (each of which has $Q/e = +\frac{2}{3}$). Hence the quark content of Σ^+ is uus.

(b) First we find the quark content of the Λ^0. To yield zero total charge, the other two quarks must be u ($Q/e = +\frac{2}{3}$) and d ($Q/e = -\frac{1}{3}$), so the quark content of the Λ^0 is uds. The quark content of the $\overline{\Lambda}^0$ is therefore $\bar{u}\,\bar{d}\,\bar{s}$.

EVALUATE: Although the Λ^0 and $\overline{\Lambda}^0$ are both electrically neutral and have the same mass, they are different particles: Λ^0 has $B = 1$ and $S = -1$, while $\overline{\Lambda}^0$ has $B = -1$ and $S = 1$.

Motivating the Quark Model

What caused physicists to suspect that hadrons were made up of something smaller? The magnetic moment of the neutron (see Section 43.1) was one of the first reasons. In Section 27.7 we learned that a magnetic moment results from a circulating current (a motion of electric charge). But the neutron has *no* charge, or, to be more accurate, no *total* charge. It could be made up of smaller particles whose charges add to zero. The quantum motion of these particles within the neutron would then give its surprising nonzero magnetic moment. To verify this hypothesis by "seeing" inside a neutron, we need a probe with a wavelength that is much less than the neutron's size of about a femtometer. This probe should not be affected by the strong interaction, so that it won't interact with the neutron as a whole but will penetrate into it and interact electromagnetically with these supposed smaller charged particles. A probe with these properties is an electron with energy above 10 GeV. In experiments carried out at SLAC, such electrons were scattered from neutrons and protons to help show that nucleons are indeed made up of fractionally charged, spin-$\frac{1}{2}$ pointlike particles.

The Eightfold Way

Symmetry considerations play a very prominent role in particle theory. Here are two examples. Consider the eight spin-$\frac{1}{2}$ baryons we've mentioned: the familiar p and n; the strange Λ^0, Σ^+, Σ^0, and Σ^-; and the doubly strange Ξ^0 and Ξ^-. For each we plot the value of strangeness S versus the value of charge Q in Fig. 44.11. The result is a hexagonal pattern. A similar plot for the nine spin-0 mesons (six shown in Table 44.3 plus three others not included in that table) is shown in Fig. 44.12; the particles fall in exactly the same hexagonal pattern! In each plot, all the particles have masses that are within about $\pm 200\ \text{MeV}/c^2$ of the median mass value of that plot, with variations due to differences in quark masses and internal potential energies.

(a)   (b)

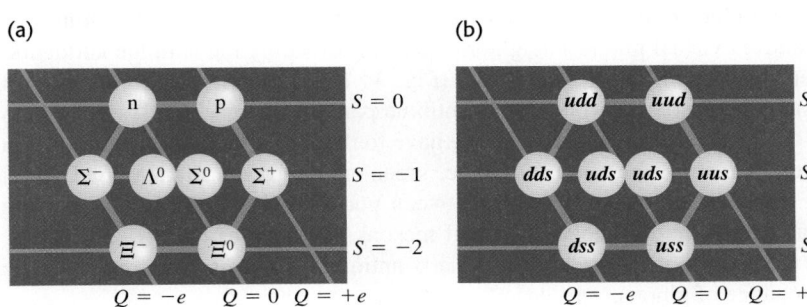

44.11 (a) Plot of S and Q values for spin-$\frac{1}{2}$ baryons, showing the symmetry pattern of the eightfold way. (b) Quark content of each spin-$\frac{1}{2}$ baryon. The quark contents of the Σ^0 and Λ^0 are the same; the Σ^0 is an excited state of the Λ^0 and can decay into it by photon emission.

44.12 (a) Plot of S and Q values for nine spin-0 mesons, showing the symmetry pattern of the eightfold way. Each particle is on the opposite side of the hexagon from its antiparticle; each of the three particles in the center is its own antiparticle. (b) Quark content of each spin-0 meson. The particles in the center are different mixtures of the three quark–antiquark pairs shown.

(a)

K^0 K^+ $S = +1$
η^0 π^0
π^- π^+ $S = 0$
η'^0
K^- $\overline{K}^0$ $S = -1$

$Q = -e$ $Q = 0$ $Q = +e$

(b)

$d\bar{s}$ $u\bar{s}$ $S = +1$
$d\bar{d}$ $u\bar{u}$
$\bar{u}d$ $u\bar{d}$ $S = 0$
$s\bar{s}$
$\bar{u}s$ $d\bar{s}$ $S = -1$

$Q = -e$ $Q = 0$ $Q = +e$

The symmetries that lead to these and similar patterns are collectively called the **eightfold way.** They were discovered in 1961 by Murray Gell-Mann and independently by Yu'val Ne'eman. (The name is a slightly irreverent reference to the Noble Eightfold Path, a set of principles for right living in Buddhism.) A similar pattern for the spin-$\frac{3}{2}$ baryons contains *ten* particles, arranged in a triangular pattern like pins in a bowling alley. When this pattern was first discovered, one of the particles was missing. But Gell-Mann gave it a name anyway (Ω^-), predicted the properties it should have, and told experimenters what they should look for. Three years later, the particle was found during an experiment at Brookhaven National Laboratory, a spectacular success for Gell-Mann's theory. The whole series of events is reminiscent of the way in which Mendeleev used gaps in the periodic table of the elements to predict properties of undiscovered elements and to guide chemists in their search for these elements.

What binds quarks to one another? The attractive interactions among quarks are mediated by massless spin-1 bosons called **gluons** in much the same way that photons mediate the electromagnetic interaction or that pions mediated the nucleon–nucleon force in the old Yukawa theory.

Color

Quarks, having spin $\frac{1}{2}$, are fermions and so are subject to the exclusion principle. This would seem to forbid a baryon having two or three quarks with the same flavor and same spin component. To avoid this difficulty, it is assumed that each quark comes in three varieties, which are whimsically called *colors*. Red, green, and blue are the usual choices. The exclusion principle applies separately to each color. A baryon always contains one red, one green, and one blue quark, so the baryon itself has no net color. Each gluon has a color–anticolor combination (for example, blue–antired) that allows it to transmit color when exchanged, and color is conserved during emission and absorption of a gluon by a quark. The gluon-exchange process changes the colors of the quarks in such a way that there is always one quark of each color in every baryon. The color of an individual quark changes continually as gluons are exchanged.

Similar processes occur in mesons such as pions. The quark–antiquark pairs of mesons have canceling color and anticolor (for example, blue and antiblue), so mesons also have no net color. Suppose a pion initially consists of a blue quark and an antiblue antiquark. The blue quark can become a red quark by emitting a blue–antired virtual gluon. The gluon is then absorbed by the antiblue antiquark, converting it to an antired antiquark (Fig. 44.13). Color is conserved in each emission and absorption, but a blue–antiblue pair has become a red–antired pair. Such changes occur continually, so we have to think of a pion as a superposition of three quantum states: blue–antiblue, green–antigreen, and red–antired. On a larger scale, the strong interaction between nucleons was described in Section 44.3 as due to the exchange of virtual mesons. In terms of quarks and gluons, these mediating virtual mesons are quark–antiquark systems bound together by the exchange of gluons.

44.13 (a) A pion containing a blue quark and an antiblue antiquark. (b) The blue quark emits a blue–antired gluon, changing to a red quark. (c) The gluon is absorbed by the antiblue antiquark, which becomes an antired antiquark. The pion now consists of a red–antired quark–antiquark pair. The actual quantum state of the pion is an equal superposition of red–antired, green–antigreen, and blue–antiblue pairs.

(a)
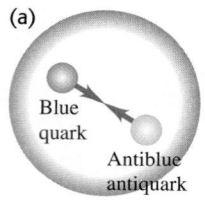
Blue quark
Antiblue antiquark

(b)
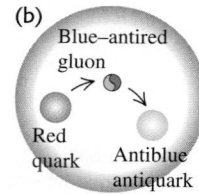
Blue–antired gluon
Red quark
Antiblue antiquark

(c)
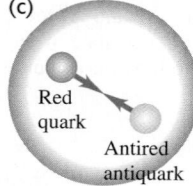
Red quark
Antired antiquark

The theory of strong interactions is known as *quantum chromodynamics* (QCD). No one has been able to isolate an individual quark, and indeed QCD predicts that quarks are bound in such a way that it is impossible to obtain a free quark. An impressive body of experimental evidence supports the correctness of the quark model and the idea that quantum chromodynamics is the key to understanding the strong interactions.

Three More Quarks

Before the tau particles were discovered, there were four known leptons. This fact, together with some puzzling decay rates, led to the speculation that there might be a fourth quark flavor. This quark is labeled *c* (the *charmed* quark); it has $Q/e = \frac{2}{3}$, $B = \frac{1}{3}$, $S = 0$, and a new quantum number **charm** $C = +1$. This was confirmed in 1974 by the observation at both SLAC and the Brookhaven National Laboratory of a meson, now named ψ, with mass 3097 MeV/c^2. This meson was found to have several decay modes, decaying into e^+e^-, $\mu^+\mu^-$, or hadrons. The mean lifetime was found to be about 10^{-20} s. These results are consistent with ψ being a spin-1 $c\bar{c}$ system. Almost immediately after this, similar mesons of greater mass were observed and identified as excited states of the $c\bar{c}$ system. A few years later, individual mesons with a nonzero net charm quantum number, D^0 ($c\bar{u}$) and D^+ ($c\bar{d}$), and a charmed baryon, Λ_c^+ (*udc*), were also observed.

In 1977 a meson with mass 9460 MeV/c^2, called upsilon (Υ), was discovered at Brookhaven. Because it had properties similar to ψ, it was conjectured that the meson was really the bound system of a new quark, *b* (the *bottom* quark), and its antiquark, $\bar{b}$. The bottom quark has the value 1 of a new quantum number B' (not to be confused with baryon number B) called *bottomness*. Excited states of the Υ were soon observed, as were the B^+ ($\bar{b}u$) and B^0 ($\bar{b}d$) mesons.

With the five flavors of quarks (*u, d, s, c,* and *b*) and the six flavors of leptons (e, μ, τ, ν_e, ν_μ, and ν_τ) it was an appealing conjecture that nature is symmetric in its building blocks and that therefore there should be a *sixth* quark. This quark, labeled *t* (top), would have $Q/e = \frac{2}{3}$, $B = \frac{1}{3}$, and a new quantum number, $T = 1$. In 1995, groups using two different detectors at Fermilab's Tevatron announced the discovery of the top quark. The groups collided 0.9-TeV protons with 0.9-TeV antiprotons, but even with 1.8 TeV of available energy, a top–antitop ($t\bar{t}$) pair was detected in fewer than two of every 10^{11} collisions! Table 44.5 lists some properties of the six quarks. Each has a corresponding antiquark with opposite values of Q, B, S, C, B', and T.

Table 44.5 **Properties of the Six Quarks**

Symbol	Q/e	Spin	Baryon Number, B	Strange-ness, S	Charm, C	Bottom-ness, B'	Topness, T
u	$\frac{2}{3}$	$\frac{1}{2}$	$\frac{1}{3}$	0	0	0	0
d	$-\frac{1}{3}$	$\frac{1}{2}$	$\frac{1}{3}$	0	0	0	0
s	$-\frac{1}{3}$	$\frac{1}{2}$	$\frac{1}{3}$	-1	0	0	0
c	$\frac{2}{3}$	$\frac{1}{2}$	$\frac{1}{3}$	0	$+1$	0	0
b	$-\frac{1}{3}$	$\frac{1}{2}$	$\frac{1}{3}$	0	0	$+1$	0
t	$\frac{2}{3}$	$\frac{1}{2}$	$\frac{1}{3}$	0	0	0	$+1$

Test Your Understanding of Section 44.4 Is it possible to have a baryon with charge $Q = +e$ and strangeness $S = -2$?

44.5 The Standard Model and Beyond

The particles and interactions that we've discussed in this chapter provide a reasonably comprehensive picture of the fundamental building blocks of nature. There is enough confidence in the basic correctness of this picture that it is called the **standard model.**

The standard model includes three families of particles: (1) the six leptons, which have no strong interactions; (2) the six quarks, from which all hadrons are made; and (3) the particles that mediate the various interactions. These mediators are gluons for the strong interaction among quarks, photons for the electromagnetic interaction, the $W^{\pm}$ and Z^0 particles for the weak interaction, and the graviton for the gravitational interaction.

Electroweak Unification

Theoretical physicists have long dreamed of combining all the interactions of nature into a single unified theory. As a first step, Einstein spent much of his later life trying to develop a field theory that would unify gravitation and electromagnetism. He was only partly successful.

Between 1961 and 1967, Sheldon Glashow, Abdus Salam, and Steven Weinberg developed a theory that unifies the weak and electromagnetic forces. One outcome of their **electroweak theory** is a prediction of the weak-force mediator particles, the Z^0 and $W^{\pm}$ bosons, including their masses. The basic idea is that the mass difference between photons (zero mass) and the weak bosons ($\approx 100 \text{ GeV}/c^2$) makes the electromagnetic and weak interactions behave quite differently at low energies. At sufficiently high energies (well above 100 GeV), however, the distinction disappears, and the two merge into a single interaction. This prediction was verified in 1983 in experiments with proton-antiproton collisions at CERN. The weak bosons were found, again with the help provided by the theoretical description, and their observed masses agreed with the predictions of the electroweak theory, a wonderful convergence of theory and experiment. The electroweak theory and quantum chromodynamics form the backbone of the standard model. Glashow, Salam, and Weinberg received the Nobel Prize in 1979.

A remaining difficulty in the electroweak theory is that photons are massless but the weak bosons are very massive. To account for the broken symmetry among these interaction mediators, a particle called the Higgs boson has been proposed. Its mass is expected to be less than $1 \text{ TeV}/c^2$, but to produce it in the laboratory may require a much greater available energy. The search for the Higgs boson is an important mission of the Large Hadron Collider at CERN.

Grand Unified Theories

Perhaps at sufficiently high energies the strong interaction and the electroweak interaction have a convergence similar to that between the electromagnetic and weak interactions. If so, they can be unified to give a comprehensive theory of strong, weak, and electromagnetic interactions. Such schemes, called **grand unified theories** (GUTs), are still speculative.

One interesting feature of some grand unified theories is that they predict the decay of the proton (in violation of conservation of baryon number), with an estimated lifetime of more than 10^{28} years. (For comparison the age of the universe is known to be 1.37×10^{10} years.) With a lifetime of 10^{28} years, six metric tons of protons would be expected to have only one decay per day, so huge amounts of material must be examined. Some of the neutrino detectors that we mentioned in Section 44.2 originally looked for, and failed to find, evidence of proton decay. Nevertheless, experimental work continues, with current estimates setting the proton lifetime well over 10^{33} years. Some GUTs also predict the existence of magnetic monopoles, which we mentioned in Chapter 27. At present there is no confirmed experimental evidence that magnetic monopoles exist.

In the standard model, the neutrinos have zero mass. Nonzero values are controversial because experiments to determine neutrino masses are difficult both to perform and to analyze. In most GUTs the neutrinos *must* have nonzero masses. If neutrinos do have mass, transitions called *neutrino oscillations* can occur, in which one type of neutrino (ν_e, ν_μ, or ν_τ) changes into another type.

In 1998, scientists using the Super-Kamiokande neutrino detector in Japan (Fig. 44.14) reported the discovery of oscillations between muon neutrinos and tau neutrinos. Subsequent measurements at the Sudbury Neutrino Observatory in Canada have confirmed the existence of neutrino oscillations. This discovery is evidence for exciting physics beyond that predicted by the standard model.

The discovery of neutrino oscillations has cleared up a long-standing mystery about the sun. Since the 1960s, physicists have been using sensitive detectors to look for electron neutrinos produced by nuclear fusion reactions in the sun's core (see Section 43.8). However, the observed flux of solar electron neutrinos is only one-third of the predicted value. The explanation was provided in 2002 by the Sudbury Neutrino Observatory, which can detect neutrinos of all three flavors. The results showed that the combined flux of solar neutrinos of *all* flavors is equal to the theoretical prediction for the flux of *electron* neutrinos. The explanation is that the sun is indeed producing electron neutrinos at the rate predicted by theory, but that two-thirds of these electron neutrinos are transformed into muon or tau neutrinos during their flight from the sun's core to a detector on earth.

Supersymmetric Theories and TOEs

The ultimate dream of theorists is to unify all four fundamental interactions, adding gravitation to the strong and electroweak interactions that are included in GUTs. Such a unified theory is whimsically called a Theory of Everything (TOE). It turns out that an essential ingredient of such theories is a space-time continuum with more than four dimensions. The additional dimensions are "rolled up" into extremely tiny structures that we ordinarily do not notice. Depending on the scale of these structures, it may be possible for the next generation of particle accelerators to reveal the presence of extra dimensions.

Another ingredient of many theories is *supersymmetry,* which gives every boson and fermion a "superpartner" of the other spin type. For example, the proposed supersymmetric partner of the spin-$\frac{1}{2}$ electron is a spin-0 particle called the *selectron,* and that of the spin-1 photon is a spin-$\frac{1}{2}$ *photino.* As yet, no superpartner particles have been discovered, perhaps because they are too massive to be produced by the present generation of accelerators. Within a few years, new data from the Large Hadron Collider and other accelerators will help us decide whether these intriguing theories have merit.

Test Your Understanding of Section 44.5 One aspect of the standard model is that a *d* quark can transform into a *u* quark, an electron, and an antineutrino by means of the weak interaction. If this happens to a *d* quark inside a neutron, what kind of particle remains afterward in addition to the electron and antineutrino? (i) a proton; (ii) a Σ^-; (iii) a Σ^+; (iv) a Λ^0 or a Σ^0; (v) any of these.

44.6 The Expanding Universe

In the last two sections of this chapter we'll explore briefly the connections between the early history of the universe and the interactions of fundamental particles. It is remarkable that there are such close ties between physics on the smallest scale that we've explored experimentally (the range of the weak interaction, of the order of 10^{-18} m) and physics on the largest scale (the universe itself, of the order of at least 10^{26} m).

Gravitational interactions play an essential role in the large-scale behavior of the universe. One of the great achievements of Newtonian mechanics, including the law of gravitation, was the understanding it brought to the motion of planets in the solar system. Astronomical evidence shows that gravitational

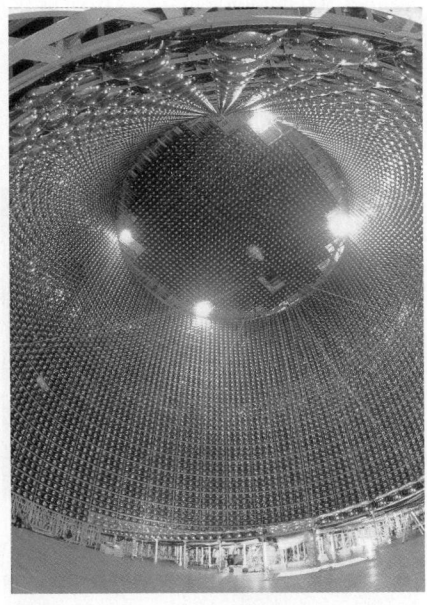

44.14 This photo shows the interior of the Super-Kamiokande neutrino detector in Japan. When in operation, the detector is filled with 5×10^7 kg of water. A neutrino passing through the detector can produce a faint flash of light, which is detected by the 13,000 photomultiplier tubes lining the detector walls. Data from this detector were the first to indicate that neutrinos have mass.

44.15 (a) The galaxy M101 is a larger version of the Milky Way galaxy of which our solar system is a part. Like all galaxies, M101 is held together by the mutual gravitational attraction of its stars, gas, dust, and other matter, all of which orbit around the galaxy's center of mass. M101 is 25 million light-years away. (b) This image shows part of the Coma cluster, an immense grouping of over 1000 galaxies that lies 300 million light-years from us. The galaxies within the cluster are all in motion. Gravitational forces between the member galaxies of this cluster prevent them from escaping.

(a)

(b)

forces also dominate in larger systems such as galaxies and clusters of galaxies (Fig. 44.15).

Until early in the 20th century it was usually assumed that the universe was *static;* stars might move relative to each other, but there was not thought to be any overall expansion or contraction. But if everything is initially sitting still in the universe, why doesn't gravity just pull it all together into one big clump? Newton himself recognized the seriousness of this troubling question.

Measurements that were begun in 1912 by Vesto Slipher at Lowell Observatory in Arizona, and continued in the 1920s by Edwin Hubble with the help of Milton Humason at Mount Wilson in California, indicated that the universe is *not* static. The motions of galaxies relative to the earth can be measured by observing the shifts in the wavelengths of their spectra. For distant galaxies these shifts are always toward longer wavelength, so they appear to be receding from us and from each other. Astronomers first assumed that these were Doppler shifts and used a relationship between the wavelength λ_0 of light measured now from a source receding at speed v and the wavelength λ_S measured in the rest frame of the source when it was emitted. We can derive this relationship by inverting Eq. (37.25) for the Doppler effect, making subscript changes, and using $\lambda = c/f$; the result is

$$\lambda_0 = \lambda_S \sqrt{\frac{c + v}{c - v}} \qquad (44.13)$$

Wavelengths from receding sources are always shifted toward longer wavelengths; this increase in λ is called the **redshift.** We can solve Eq. (44.13) for v; the result is

$$v = \frac{(\lambda_0/\lambda_S)^2 - 1}{(\lambda_0/\lambda_S)^2 + 1} c \qquad (44.14)$$

CAUTION **Redshift, not Doppler shift** Equations (44.13) and (44.14) are from the *special* theory of relativity and refer to the Doppler effect. As we'll see, the redshift from *distant* galaxies is caused by an effect that is explained by the *general* theory of relativity and is *not* a Doppler shift. However, as the ratio v/c and the fractional wavelength change $(\lambda_0 - \lambda_S)/\lambda_S$ become small, the general theory's equations approach Eqs. (44.13) and (44.14), and those equations may be used. ▮

Example 44.8 | **Recession speed of a galaxy**

The spectral lines of various elements are detected in light from a galaxy in the constellation Ursa Major. An ultraviolet line from singly ionized calcium ($\lambda_S = 393$ nm) is observed at wavelength $\lambda_0 = 414$ nm, redshifted into the visible portion of the spectrum. At what speed is this galaxy receding from us?

SOLUTION

IDENTIFY and SET UP: This example uses the relationship between redshift and recession speed for a distant galaxy. We can use the wavelengths λ_S at which the light is emitted and λ_0 that we detect on earth in Eq. (44.14) to determine the galaxy's recession speed v if the fractional wavelength shift is not too great.

EXECUTE: The fractional redshift is $\lambda_0/\lambda_S = (414$ nm$)/(393$ nm$) = 1.053$. This is only a 5.3% increase, so we can use Eq. (44.14) with reasonable accuracy:

$$v = \frac{(1.053)^2 - 1}{(1.053)^2 + 1} c = 0.0516c = 1.55 \times 10^7 \text{ m/s}$$

EVALUATE: The galaxy is receding from the earth at 5.16% of the speed of light. Rather than going through this calculation, astronomers often just state the *redshift* $z = (\lambda_0 - \lambda_S)/\lambda_S = (\lambda_0/\lambda_S) - 1$. This galaxy has redshift $z = 0.053$.

The Hubble Law

Analysis of redshifts from many distant galaxies led Edwin Hubble to a remarkable conclusion: The speed of recession v of a galaxy is proportional to its distance r from us (Fig. 44.16). This relationship is now called the **Hubble law;** expressed as an equation,

$$v = H_0 r \qquad (44.15)$$

where H_0 is an experimental quantity commonly called the *Hubble constant,* since at any given time it is constant over space. Determining H_0 has been a key goal of the Hubble Space Telescope, which can measure distances to galaxies with unprecedented accuracy. The current best value is 2.3×10^{-18} s^{-1}, with an uncertainty of 5%.

Astronomical distances are often measured in *parsecs* (pc); one parsec is the distance at which there is a one-arcsecond $(1/3600°)$ angular separation between two objects 1.50×10^{11} m apart (the average distance from the earth to the sun). A distance of 1 pc is equal to 3.26 *light-years* (ly), where 1 ly $= 9.46 \times 10^{12}$ km is the distance that light travels in one year. The Hubble constant is then commonly expressed in the mixed units (km/s)/Mpc (kilometers per second per megaparsec), where 1 Mpc $= 10^6$ pc:

$$H_0 = (2.3 \times 10^{-18}\text{ s}^{-1})\left(\frac{9.46 \times 10^{12}\text{ km}}{1\text{ ly}}\right)\left(\frac{3.26\text{ ly}}{1\text{ pc}}\right)\left(\frac{10^6\text{ pc}}{1\text{ Mpc}}\right) = 71\,\frac{\text{km/s}}{\text{Mpc}}$$

44.16 Graph of recession speed versus distance for several galaxies. The best-fit straight line illustrates Hubble's law. The slope of the line is the Hubble constant, H_0.

Example 44.9 | **Determining distance with the Hubble law**

Use the Hubble law to find the distance from earth to the galaxy in Ursa Major described in Example 44.8.

SOLUTION

IDENTIFY and SET UP: The Hubble law relates the redshift of a distant galaxy to its distance r from earth. We solve Eq. (44.15) for r and substitute the recession speed v from Example 44.8.

EXECUTE: Using $H_0 = 71\,(\text{km/s})/\text{Mpc} = 7.1 \times 10^4\,(\text{m/s})/\text{Mpc}$,

$$r = \frac{v}{H_0} = \frac{1.55 \times 10^7\text{m/s}}{7.1 \times 10^4\,(\text{m/s})/\text{Mpc}} = 220\text{ Mpc}$$
$$= 2.2 \times 10^8\text{ pc} = 7.1 \times 10^8\text{ ly} = 6.7 \times 10^{24}\text{ m}$$

EVALUATE: A distance of 220 million parsecs (710 million light-years) is truly stupendous, but many galaxies lie much farther away. To appreciate the immensity of this distance, consider that our farthest-ranging unmanned spacecraft have traveled only about 0.001 ly from our planet.

Another aspect of Hubble's observations was that, *in all directions,* distant galaxies appeared to be receding from us. There is no particular reason to think that our galaxy is at the very center of the universe; if we lived in some other galaxy, every distant galaxy would still seem to be moving away. That is, at any given time, *the universe looks more or less the same, no matter where in the universe we are.* This important idea is called the **cosmological principle.** There are local fluctuations in density, but on average, the universe looks the same from all locations. Thus the Hubble constant is constant in space although not necessarily constant in time, and the laws of physics are the same everywhere.

The Big Bang

The Hubble law suggests that at some time in the past, all the matter in the universe was far more concentrated than it is today. It was then blown apart in an immense explosion called the **Big Bang,** giving all observable matter more or less the velocities that we observe today. When did this happen? According to the

Hubble law, matter at a distance r away from us is traveling with speed $v = H_0 r$. The time t needed to travel a distance r is

$$t = \frac{r}{v} = \frac{r}{H_0 r} = \frac{1}{H_0} = 4.3 \times 10^{17} \text{ s} = 1.4 \times 10^{10} \text{ y}$$

By this hypothesis the Big Bang occurred about 14 billion years ago. It assumes that all speeds are *constant* after the Big Bang; that is, it neglects any change in the expansion rate due to gravitational attraction or other effects. We'll return to this point later. For now, however, notice that the age of the earth determined from radioactive dating (see Section 43.4) is 4.54 billion (4.54×10^9) years. It's encouraging that our hypothesis tells us that the universe is older than the earth!

Expanding Space

The general theory of relativity takes a radically different view of the expansion just described. According to this theory, the increased wavelength is *not* caused by a Doppler shift as the universe expands into a previously empty void. Rather, the increase comes from *the expansion of space itself* and everything in intergalactic space, including the wavelengths of light traveling to us from distant sources. This is not an easy concept to grasp, and if you haven't encountered it before, it may sound like doubletalk.

Here's an analogy that may help to develop some intuition on this point. Imagine we are all bugs crawling around on a horizontal surface. We can't leave the surface, and we can see in any direction along the surface, but not up or down. We are then living in a two-dimensional world; some writers have called such a world *flatland*. If the surface is a plane, we can locate our position with two Cartesian coordinates (x, y). If the plane extends indefinitely in both the x- and y-directions, we described our space as having *infinite* extent, or as being *unbounded*. No matter how far we go, we never reach an edge or a boundary.

An alternative habitat for us bugs would be the surface of a sphere with radius R. The space would still seem infinite in the sense that we could crawl forever and never reach an edge or a boundary. Yet in this case the space is *finite* or *bounded*. To describe the location of a point in this space, we could still use two coordinates: latitude and longitude, or the spherical coordinates θ and ϕ shown in Fig. 41.5.

Now suppose the spherical surface is that of a balloon (Fig. 44.17). As we inflate the balloon more and more, increasing the radius R, the coordinates of a point don't change, yet the distance between any two points gets larger and larger. Furthermore, as R increases, the *rate of change* of distance between two points (their recession speed) is proportional to their distance apart. *The recession speed is proportional to the distance,* just as with the Hubble law. For example, the distance from Pittsburgh to Miami is twice as great as the distance from Pittsburgh to Boston. If the earth were to begin to swell, Miami would recede from Pittsburgh twice as fast as Boston would.

We see that although the quantity R isn't one of the two coordinates giving the position of a point on the balloon's surface, it nevertheless plays an essential role in any discussion of distance. It is the radius of curvature of our two-dimensional space, and it is also a varying *scale factor* that changes as this two-dimensional universe expands.

Generalizing this picture to three dimensions isn't so easy. We have to think of our three-dimensional space as being embedded in a space with four or more dimensions, just as we visualized the two-dimensional spherical flatland as being embedded in a three-dimensional Cartesian space. Our real three-space is *not Cartesian;* to describe its characteristics in any small region requires at least one additional parameter, the curvature of space, which is analogous to the radius of the sphere. In a sense, this scale factor, which we'll continue to call R, describes the *size* of the universe, just as the radius of the sphere described the size of our

44.17 An inflating balloon as an analogy for an expanding universe.

(a) Points (representing galaxies) on the surface of a balloon are described by their latitude and longitude coordinates.

(b) The radius R of the balloon has increased. The coordinates of the points are the same, but the distance between them has increased.

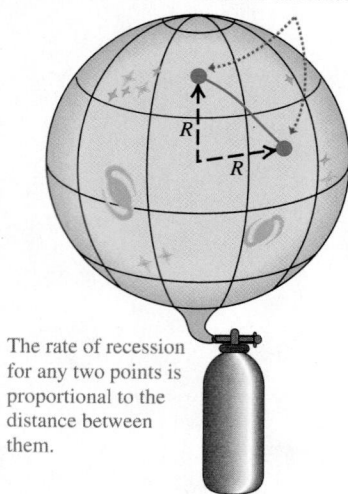

The rate of recession for any two points is proportional to the distance between them.

two-dimensional spherical universe. We'll return later to the question of whether the universe is bounded or unbounded.

Any length that is measured in intergalactic space is proportional to R, so the wavelength of light traveling to us from a distant galaxy increases along with every other dimension as the universe expands. That is,

$$\frac{\lambda_0}{\lambda} = \frac{R_0}{R} \tag{44.16}$$

The zero subscripts refer to the values of the wavelength and scale factor *now*, just as H_0 is the current value of the Hubble constant. The quantities λ and R without subscripts are the values at *any* time—past, present, or future. In the situation described in Example 44.8, we have $\lambda_0 = 414$ nm and $\lambda = \lambda_S = 393$ nm, so Eq. (44.16) gives $R_0/R = 1.053$. That is, the scale factor *now* (R_0) is 5.3% larger than it was 710 million years ago when the light was emitted from that galaxy in Ursa Major. This increase of wavelength with time as the scale factor increases in our expanding universe is called the *cosmological redshift.* The farther away an object is, the longer its light takes to get to us and the greater the change in R and λ. The current largest measured wavelength ratio for galaxies is about 7, meaning that the volume of space itself is about $7^3 \approx 340$ times larger than it was when the light was emitted. Do *not* attempt to substitute $\lambda_0/\lambda_S = 7$ into Eq. (44.14) to find the recession speed; that equation is accurate only for small cosmological redshifts and $v \ll c$. The actual value of v depends on the density of the universe, the value of H_0, and the expansion history of the universe.

Here's a surprise for you: If the distance from us in the Hubble law is large enough, then the speed of recession will be greater than the speed of light! This does *not* violate the special theory of relativity because the recession speed is *not* caused by the motion of the astronomical object relative to some coordinates in its region of space. Rather, we can have $v > c$ when two sets of coordinates move apart fast enough as space itself expands. In other words, there are objects whose coordinates have been moving away from our coordinates so fast that light from them hasn't had enough time in the entire history of the universe to reach us. What we see is just the *observable* universe; we have no direct evidence about what lies beyond its horizon.

CAUTION **The universe isn't expanding into emptiness** The balloon shown in Fig. 44.17 is expanding into the empty space around it. It's a common misconception to picture the universe in the same way as a large but finite collection of galaxies that's expanding into unoccupied space. The reality is quite different! All the accumulated evidence shows that our universe is *infinite:* It has no edges, so there is nothing "outside" it and it isn't "expanding into" anything. The expansion of the universe simply means that the scale factor of the universe is increasing. A good two-dimensional analogy is to think of the universe as a flat, infinitely large rubber sheet that's stretching and expanding much like the surface of the balloon in Fig. 44.17. In a sense, the infinite universe is simply becoming more infinite! ▮

Critical Density

We've mentioned that the law of gravitation isn't consistent with a static universe. We need to look at the role of gravity in an *expanding* universe. Gravitational attractions should slow the initial expansion, but by how much? If these attractions are strong enough, the universe should expand more and more slowly, eventually stop, and then begin to contract, perhaps all the way down to what's been called a *Big Crunch.* On the other hand, if gravitational forces are much weaker, they slow the expansion only a little, and the universe should continue to expand forever.

The situation is analogous to the problem of escape speed of a projectile launched from the earth. We studied this problem in Example 13.5 (Section 13.3); now would be an excellent time to review that discussion. The total energy

$E = K + U$ when a projectile of mass m and speed v is at a distance r from the center of the earth (mass m_E) is

$$E = \tfrac{1}{2}mv^2 - \frac{Gmm_E}{r}$$

If E is positive, the projectile has enough kinetic energy to move infinitely far from the earth ($r \rightarrow \infty$) and have some kinetic energy left over. If E is negative, the kinetic energy $K = \tfrac{1}{2}mv^2$ becomes zero and the projectile stops when $r = -Gmm_E/E$. In that case, no greater value of r is possible, and the projectile can't escape the earth's gravity.

We can carry out a similar analysis for the universe. Whether the universe continues to expand indefinitely should depend on the average *density* of matter. If matter is relatively dense, there is a lot of gravitational attraction to slow and eventually stop the expansion and make the universe contract again. If not, the expansion should continue indefinitely. We can derive an expression for the *critical density* ρ_c needed to just barely stop the expansion.

Here's a calculation based on Newtonian mechanics; it isn't relativistically correct, but it illustrates the idea. Consider a large sphere with radius R, containing many galaxies (Fig. 44.18), with total mass M. Suppose our own galaxy has mass m and is located at the surface of this sphere. According to the cosmological principle, the average distribution of matter within the sphere is uniform. The total gravitational force on our galaxy is just the force due to the mass M inside the sphere. The force on our galaxy and potential energy U due to this spherically symmetric distribution are the same as though m and M were both points, so $U = -GmM/R$, just as in Section 13.3. The net force from all the uniform distribution of mass *outside* the sphere is zero, so we'll ignore it.

The total energy E (kinetic plus potential) for our galaxy is

$$E = \tfrac{1}{2}mv^2 - \frac{GmM}{R} \qquad (44.17)$$

If E is *positive,* our galaxy has enough energy to escape from the gravitational attraction of the mass M inside the sphere; in this case the universe should keep expanding forever. If E is negative, our galaxy cannot escape and the universe should eventually pull back together. The crossover between these two cases occurs when $E = 0$, so that

$$\tfrac{1}{2}mv^2 = \frac{GmM}{R} \qquad (44.18)$$

The total mass M inside the sphere is the volume $4\pi R^3/3$ times the density ρ_c:

$$M = \tfrac{4}{3}\pi R^3 \rho_c$$

We'll assume that the speed v of our galaxy relative to the center of the sphere is given by the Hubble law: $v = H_0 R$. Substituting these expressions for m and v into Eq. (44.18), we get

$$\tfrac{1}{2}m(H_0 R)^2 = \frac{Gm}{R}\left(\tfrac{4}{3}\pi R^3 \rho_c\right) \qquad \text{or}$$

$$\rho_c = \frac{3H_0^2}{8\pi G} \qquad \text{(critical density of the universe)} \qquad (44.19)$$

This is the *critical density.* If the average density is less than ρ_c, the universe should continue to expand indefinitely; if it is greater, the universe should eventually stop expanding and begin to contract.

Putting numbers into Eq. (44.19), we find

$$\rho_c = \frac{3(2.3 \times 10^{-18}\ \text{s}^{-1})^2}{8\pi(6.67 \times 10^{-11}\ \text{N} \cdot \text{m}^2/\text{kg}^2)} = 9.5 \times 10^{-27}\ \text{kg/m}^3$$

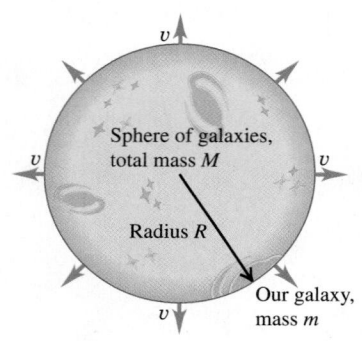

44.18 An imaginary sphere of galaxies. The net gravitational force exerted on our galaxy (at the surface of the sphere) by the other galaxies is the same as if all of their mass were concentrated at the center of the sphere. (Since the universe is infinite, there's also an infinity of galaxies outside this sphere.)

v

v

Sphere of galaxies,
total mass M

v

Radius R

Our galaxy,
mass m

v

The mass of a hydrogen atom is 1.67×10^{-27} kg, so this density is equivalent to about six hydrogen atoms per cubic meter.

Dark Matter, Dark Energy, and the Accelerating Universe

Astronomers have made extensive studies of the average density of matter in the universe. One way to do so is to count the number of galaxies in a patch of sky. Based on the mass of an average star and the number of stars in an average galaxy, this effort gives an estimate of the average density of *luminous* matter in the universe—that is, matter that emits electromagnetic radiation. (You are made of luminous matter because you emit infrared radiation as a consequence of your temperature; see Sections 17.7 and 39.5.) It's also necessary to take into account other luminous matter within a galaxy, including the tenuous gas and dust between the stars.

Another technique is to study the motions of galaxies within clusters of galaxies (Fig. 44.19; see also Fig. 44.15b). The motions are so slow that we can't actually see galaxies changing positions within a cluster. However, observations show that different galaxies within a cluster have somewhat different redshifts, which indicates that the galaxies are moving relative to the center of mass of the cluster. The speeds of these motions are related to the gravitational force exerted on each galaxy by the other members of the cluster, which in turn depends on the total mass of the cluster. By measuring these speeds, astronomers can determine the average density of *all* kinds of matter within the cluster, whether or not the matter emits electromagnetic radiation.

Observations using these and other techniques show that the average density of *all* matter in the universe is 27.4% of the critical density, but the average density of *luminous* matter is only 4.6% of the critical density. In other words, most of the matter in the universe is not luminous: It does not emit electromagnetic radiation of *any* kind. At present, the nature of this **dark matter** remains an outstanding mystery. Some proposed candidates for dark matter are WIMPs (weakly interacting massive particles, which are hypothetical subatomic particles far more massive than those produced in accelerator experiments) and MACHOs (massive compact halo objects, which include objects such as black holes that might form "halos" around galaxies). Whatever the true nature of dark matter, it is by far the dominant form of matter in the universe. For every kilogram of the ordinary matter that has been our subject for most of this book—including electrons, protons, atoms, molecules, blocks on inclined planes, planets, and stars—there are *five* kilograms of dark matter.

Since the average density of matter in the universe is less than the critical density, it might seem fair to conclude that the universe will continue to expand indefinitely, and that gravitational attraction between matter in different parts of the universe should slow the expansion down (albeit not enough to stop it). One way to test this prediction is to examine the redshifts of extremely distant objects. When astronomers look at a galaxy 10^9 light-years away, the light they receive has been in transit for 10^9 years, so they are seeing 10^9 years into the past. If the expansion of the universe has been slowing down, the expansion must have been more rapid in the distant past. Thus we would expect very distant galaxies to have *greater* redshifts than predicted by the Hubble law, Eq. (44.15).

Only since the 1990s has it become possible to accurately measure both the distances and the redshifts of extremely distant galaxies. The results have been totally surprising: Very distant galaxies actually have *smaller* redshifts than predicted by the Hubble law! The implication is that the expansion of the universe was slower in the past than it is now, so the expansion has been *speeding up* rather than slowing down.

If gravitational attraction should make the expansion slow down, why is it speeding up instead? The explanation generally accepted by astronomers and physicists is that space is suffused with a kind of energy that has no gravitational

44.19 The bright spots in this image are not stars, but entire galaxies. They are part of a cluster of galaxies about 10.2 billion ly (3.13 billion pc, or 3130 Mpc) away. (The blue glow is x-ray emission from hot gas within the cluster.) When the galaxies emitted the light used to make this image, the scale factor of the universe was only about 35% as large as it is now. By comparison, we see the relatively nearby Coma cluster (see Fig. 44.15b) as it was 300 million years ago, when the scale factor was 98% of the present-day value.

effect and emits no electromagnetic radiation, but rather acts as a kind of "anti-gravity" that produces a universal *repulsion*. This invisible, immaterial energy is called **dark energy.** As the name suggests, the nature of dark energy is poorly understood but is the subject of very active research.

Observations show that the *energy* density of dark energy (measured in, say, joules per cubic meter) is 72.6% of the critical density times c^2; that is, it is equal to $0.726\rho_c c^2$. As described above, the average density of matter of all kinds is 27.4% of the critical density. From the Einstein relationship $E = mc^2$, the average *energy* density of matter in the universe is therefore $0.274\rho_c c^2$. Because the energy density of dark energy is nearly three times greater than that of matter, the expansion of the universe will continue to accelerate. This expansion will never stop, and the universe will never contract.

If we account for energy of *all* kinds, the average energy density of the universe is equal to $0.726\rho_c c^2 + 0.274\rho_c c^2 = 1.00\rho_c c^2$. Of this, 72.6% is the mysterious dark energy, 22.8% is the no less mysterious dark matter, and a mere 4.6% is well-understood conventional matter. How little we know about the contents of our universe! When we take account of the density of matter in the universe (which tends to slow the expansion of space) and the density of dark energy (which tends to speed up the expansion), the age of the universe turns out to be 13.7 billion (1.37×10^{10}) years.

What is the significance of the result that within observational error, the average energy density of the universe is equal to $\rho_c c^2$? It tells us that the universe is infinite and unbounded, but just barely so. If the average energy density were even slightly larger than $\rho_c c^2$, the universe would be finite like the surface of the balloon depicted in Fig. 44.17. As of this writing, the observational error in the average energy density is still large enough (about 1%) that we can't be totally sure that the universe *is* unbounded. Improving these measurements will be an important task for physicists and astronomers in the years ahead.

Test Your Understanding of Section 44.6 Is it accurate to say that your body is made of "ordinary" matter?

44.7 The Beginning of Time

What an odd title for the very last section of a book! We will describe in general terms some of the current theories about the very early history of the universe and their relationship to fundamental particle interactions. We'll find that an astonishing amount happened in the very first second. A lot of loose ends will be left untied, and many questions will be left unanswered. This is, after all, one of the frontiers of physics.

Temperatures

The early universe was extremely dense and extremely hot, and the average particle energies were extremely large, all many orders of magnitude beyond anything that exists in the present universe. We can compare particle energy E and absolute temperature T using the equipartition principle (see Section 18.4):

$$E = \tfrac{3}{2}kT \qquad (44.20)$$

In this equation k is Boltzmann's constant, which we'll often express in eV/K:

$$k = 8.617 \times 10^{-5} \text{ eV/K}$$

Thus we can replace Eq. (44.20) by $E \approx (10^{-4} \text{ eV/K})T = (10^{-13} \text{ GeV/K})T$ when we're discussing orders of magnitude.

Example 44.10 Temperature and energy

(a) What is the average kinetic energy E (in eV) of particles at room temperature $(T = 290 \text{ K})$ and at the surface of the sun $(T = 5800 \text{ K})$? (b) What approximate temperature corresponds to the ionization energy of the hydrogen atom and to the rest energies of the electron and the proton?

SOLUTION

IDENTIFY and SET UP: In this example we are to apply the equipartition principle. We use Eq. (44.20) to relate the target variables E and T.

EXECUTE: (a) At room temperature, from Eq. (44.20),

$$E = \tfrac{3}{2}kT = \tfrac{3}{2}(8.617 \times 10^{-5} \text{ eV/K})(290 \text{ K}) = 0.0375 \text{ eV}$$

The temperature at the sun's surface is higher than room temperature by a factor of $(5800 \text{ K})/(290 \text{ K}) = 20$, so the average kinetic energy there is $20(0.0375 \text{ eV}) = 0.75 \text{ eV}$.

(b) The ionization energy of hydrogen is 13.6 eV. Using the approximation $E \approx (10^{-4} \text{ eV/K})T$, we have

$$T \approx \frac{E}{10^{-4} \text{ eV/K}} = \frac{13.6 \text{ eV}}{10^{-4} \text{ eV/K}} \approx 10^5 \text{ K}$$

Repeating this calculation for the rest energies of the electron $(E = 0.511 \text{ MeV})$ and proton $(E = 938 \text{ MeV})$ gives temperatures of 10^{10} K and 10^{13} K, respectively.

EVALUATE: Temperatures in excess of 10^5 K are found in the sun's interior, so most of the hydrogen there is ionized. Temperatures of 10^{10} K or 10^{13} K are not found anywhere in the solar system; as we will see, temperatures were this high in the very early universe.

Uncoupling of Interactions

We've characterized the expansion of the universe by a continual increase of the scale factor R, which we can think of very roughly as characterizing the *size* of the universe, and by a corresponding decrease in average density. As the total gravitational potential energy increased during expansion, there were corresponding *decreases* in temperature and average particle energy. As this happened, the basic interactions became progressively uncoupled.

To understand the uncouplings, recall that the unification of the electromagnetic and weak interactions occurs at energies that are large enough that the differences in mass among the various spin-1 bosons that mediate the interactions become insignificant by comparison. The electromagnetic interaction is mediated by the massless photon, and the weak interaction is mediated by the weak bosons $W^{\pm}$ and Z^0 with masses of the order of 100 GeV/c^2. At energies much *less* than 100 GeV the two interactions seem quite different, but at energies much *greater* than 100 GeV they become part of a single interaction.

The grand unified theories (GUTs) provide a similar behavior for the strong interaction. It becomes unified with the electroweak interaction at energies of the order of 10^{14} GeV, but at lower energies the two appear quite distinct. One of the reasons GUTs are still very speculative is that there is no way to do controlled experiments in this energy range, which is larger by a factor of 10^{11} than energies available with any current accelerator.

Finally, at sufficiently high energies and short distances, it is assumed that gravitation becomes unified with the other three interactions. The distance at which this happens is thought to be of the order of 10^{-35} m. This distance, called the *Planck length* l_P, is determined by the speed of light c and the fundamental constants of quantum mechanics and gravitation, h and G, respectively. The Planck length l_P is defined as

$$l_P = \sqrt{\frac{\hbar G}{c^3}} = 1.616 \times 10^{-35} \text{ m} \qquad (44.21)$$

You should verify that this combination of constants does indeed have units of length. The *Planck time* $t_P = l_P/c$ is the time required for light to travel a distance l_P:

$$t_P = \frac{l_P}{c} = \sqrt{\frac{\hbar G}{c^5}} = 0.539 \times 10^{-43} \text{ s} \qquad (44.22)$$

If we mentally go backward in time, we have to stop when we reach $t = 10^{-43}$ s because we have no adequate theory that unifies all four interactions. So as yet we have no way of knowing what happened or how the universe behaved at times earlier than the Planck time or when its size was less than the Planck length.

The Standard Model of the History of the Universe

The description that follows is called the *standard model* of the history of the universe. The title indicates that there are substantial areas of theory that rest on solid experimental foundations and are quite generally accepted. The figure on pages 1512–1513 is a graphical description of this history, with the characteristic sizes, particle energies, and temperatures at various times. Referring to this figure frequently will help you to understand the following discussion.

In this standard model, the temperature of the universe at time $t = 10^{-43}$ s (the Planck time) was about 10^{32} K, and the average energy per particle was approximately

$$E \approx (10^{-13} \text{ GeV/K})(10^{32} \text{ K}) = 10^{19} \text{ GeV}$$

In a totally unified theory this is about the energy below which gravity begins to behave as a separate interaction. This time therefore marked the transition from any proposed TOE to the GUT period.

During the GUT period, roughly $t = 10^{-43}$ to 10^{-35} s, the strong and electroweak forces were still unified, and the universe consisted of a soup of quarks and leptons transforming into each other so freely that there was no distinction between the two families of particles. Other, much more massive particles may also have been freely created and destroyed. One important characteristic of GUTs is that at sufficiently high energies, baryon number is not conserved. (We mentioned earlier the proposed decay of the proton, which has not yet been observed.) Thus by the end of the GUT period the numbers of quarks and antiquarks may have been unequal. This point has important implications; we'll return to it at the end of the section.

By $t = 10^{-35}$ s the temperature had decreased to about 10^{27} K and the average energy to about 10^{14} GeV. At this energy the strong force separated from the electroweak force (Fig. 44.20), and baryon number and lepton numbers began to be separately conserved. This separation of the strong force was analogous to a phase

44.20 Schematic diagram showing the times and energies at which the various interactions are thought to have uncoupled. The energy scale is backward because the average energy decreased as the age of the universe increased.

change such as boiling a liquid, with an associated heat of vaporization. Think of it as being similar to boiling a heavy nucleus, pulling the particles apart beyond the short range of the nuclear force. As a result, the universe underwent a dramatic expansion (far more rapid than the present-day expansion rate) called *cosmic inflation*. In one model, the scale factor R increased by a factor of 10^{50} in 10^{-32} s.

At $t = 10^{-32}$ s the universe was a mixture of quarks, leptons, and the mediating bosons (gluons, photons, and the weak bosons $W^{\pm}$ and Z^{0}). It continued to expand and cool from the inflationary period to $t = 10^{-6}$ s, when the temperature was about 10^{13} K and typical energies were about 1 GeV (comparable to the rest energy of a nucleon; see Example 44.11). At this time the quarks began to bind together to form nucleons and antinucleons. Also there were still enough photons of sufficient energy to produce nucleon–antinucleon pairs to balance the process of nucleon–antinucleon annihilation. However, by about $t = 10^{-2}$ s, most photon energies fell well below the threshold energy for such pair production. There was a slight excess of nucleons over antinucleons; as a result, virtually all of the antinucleons and most of the nucleons annihilated one another. A similar equilibrium occurred later between the production of electron–positron pairs from photons and the annihilation of such pairs. At about $t = 14$ s the average energy dropped to around 1 MeV, below the threshold for $e^{+}e^{-}$ pair production. After pair production ceased, virtually all of the remaining positrons were annihilated, leaving the universe with many more protons and electrons than the antiparticles of each.

Up until about $t = 1$ s, neutrons and neutrinos could be produced in the endoergic reaction

$$e^{-} + p \rightarrow n + \nu_{e}$$

After this time, most electrons no longer had enough energy for this reaction. The average neutrino energy also decreased, and as the universe expanded, equilibrium reactions that involved *absorption* of neutrinos (which occurred with decreasing probability) became inoperative. At this time, in effect, the flux of neutrinos and antineutrinos throughout the universe uncoupled from the rest of the universe. Because of the extraordinarily low probability for neutrino absorption, most of this flux is still present today, although cooled greatly by expansion. The standard model of the universe predicts a present neutrino temperature of about 2 K, but no experiment has yet been able to test this prediction.

Nucleosynthesis

At about $t = 1$ s, the ratio of protons to neutrons was determined by the Boltzmann distribution factor $e^{-\Delta E/kT}$, where ΔE is the difference between the neutron and proton rest energies: $\Delta E = 1.294$ MeV. At a temperature of about 10^{10} K, this distribution factor gives about 4.5 times as many protons as neutrons. However, as we have discussed, free neutrons (with a half-life of 887 s) decay spontaneously to protons. This decay caused the proton–neutron ratio to increase until about $t = 225$ s. At this time, the temperature was about 10^{9} K, and the average energy was well below 2 MeV.

This energy distribution was critical because the binding energy of the *deuteron* (a neutron and a proton bound together) is 2.22 MeV (see Section 43.2). A neutron bound in a deuteron does not decay spontaneously. As the average energy decreased, a proton and a neutron could combine to form a deuteron, and there were fewer and fewer photons with 2.22 MeV or more of energy to dissociate the deuterons again. Therefore the combining of protons and neutrons into deuterons halted the decay of free neutrons.

The formation of deuterons starting at about $t = 225$ s marked the beginning of the period of formation of nuclei, or *nucleosynthesis*. At this time, there were about seven protons for each neutron. The deuteron (^{2}H) can absorb a neutron and form a triton (^{3}H), or it can absorb a proton and form ^{3}He. Then ^{3}H can absorb a proton, and ^{3}He can absorb a neutron, each yielding ^{4}He (the alpha particle).

**AGE OF QUARKS AND
GLUONS (GUT Period)**
Dense concentration of matter and
antimatter; gravity a separate force;
more quarks than antiquarks.
Inflationary period (10^{-35} s): rapid expansion,
strong force separates from
electroweak force.

**AGE OF NUCLEONS AND
ANTINUCLEONS**
Quarks bind together to form
nucleons and antinucleons; energy
too low for nucleon–antinucleon
pair production at 10^{-2} s.

**AGE OF
NUCLEOSYNTHESIS**
Stable deuterons; matter
74% H, 25% He, 1%
heavier nuclei.

AGE OF LEPTONS
Leptons distinct from quarks;
$W^{\pm}$ and Z^0 bosons mediate
weak force (10^{-12} s).

**BIG
BANG** 10^{-43} s 10^{-32} s 10^{-6} s   225 s   10

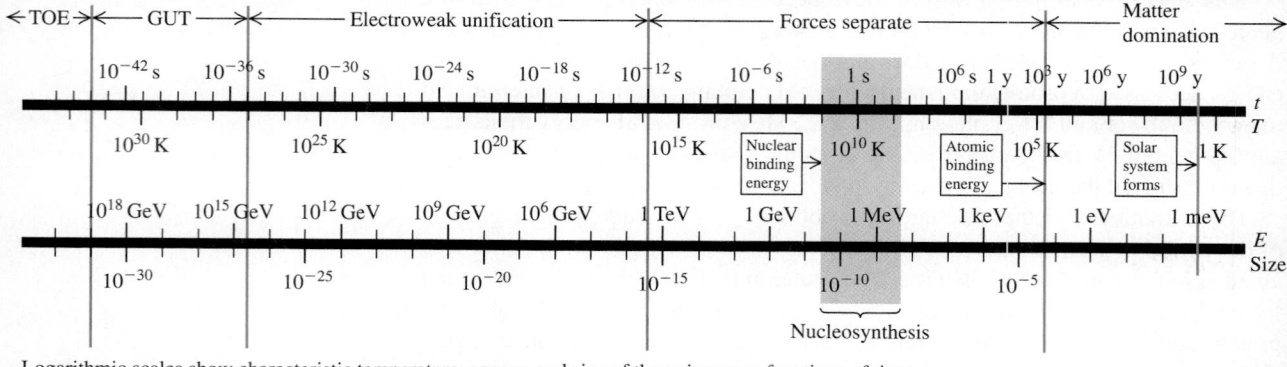

Logarithmic scales show characteristic temperature, energy, and size of the universe as functions of time.

A Brief History of the Universe

AGE OF IONS
xpanding, cooling
gas of ionized
H and He.

AGE OF ATOMS
Neutral atoms form;
universe becomes
transparent to most light.

**AGE OF STARS
AND GALAXIES**
Thermonuclear fusion
begins in stars, forming
heavier nuclei.

10^{13} s $\qquad$ 10^{15} s $\qquad$ **NOW**

A few ^{7}Li nuclei may also have been formed by fusion of ^{3}H and ^{4}He nuclei. According to the theory, essentially all the ^{1}H and ^{4}He in the present universe was formed at this time. But then the building of nuclei almost ground to a halt. The reason is that *no* nuclide with mass number $A = 5$ has a half-life greater than 10^{-21} s. Alpha particles simply do not permanently absorb neutrons or protons. The nuclide ^{8}Be that is formed by fusion of two ^{4}He nuclei is unstable, with an extremely short half-life, about 7×10^{-17} s. Note also that at this time, the average energy was still much too large for electrons to be bound to nuclei; there were not yet any atoms.

Conceptual Example 44.11 **The relative abundance of hydrogen and helium in the universe**

Nearly all of the protons and neutrons in the seven-to-one ratio at $t = 225$ s either formed ^{4}He or remained as ^{1}H. After this time, what was the resulting relative abundance of ^{1}H and ^{4}He, by mass?

SOLUTION

The ^{4}He nucleus contains two protons and two neutrons. For every two neutrons present at $t = 225$ s there were 14 protons. The two neutrons and two of the 14 protons make up one ^{4}He nucleus, leaving 12 protons (^{1}H nuclei). So there were eventually 12 ^{1}H nuclei for every ^{4}He nucleus. The masses of ^{1}H and ^{4}He are about 1 u and 4 u, respectively, so there were 12 u of ^{1}H for every 4 u of ^{4}He. Therefore the relative abundance, by mass, was 75% ^{1}H and 25% ^{4}He. This result agrees very well with estimates of the present H–He ratio in the universe, an important confirmation of this part of the theory.

Further nucleosynthesis did not occur until very much later, well after $t = 10^{13}$ s (about 380,000 y). At that time, the temperature was about 3000 K, and the average energy was a few tenths of an electron volt. Because the ionization energies of hydrogen and helium atoms are 13.6 eV and 24.5 eV, respectively, almost all the hydrogen and helium was electrically neutral (not ionized). With the electrical repulsions of the nuclei canceled out, gravitational attraction could slowly pull the neutral atoms together to form clouds of gas and eventually stars. Thermonuclear reactions in stars then produced all of the more massive nuclei. In Section 43.8 we discussed one cycle of thermonuclear reactions in which ^{1}H becomes ^{4}He.

For stars whose mass is 40% of the sun's mass or greater, as the hydrogen is consumed the star's core begins to contract as the inward gravitational pressure exceeds the outward gas and radiation pressure. The gravitational potential energy decreases as the core contracts, so the kinetic energy of nuclei in the core increases. Eventually the core temperature becomes high enough to begin another process, *helium fusion*. First two ^{4}He nuclei fuse to form ^{8}Be, which is highly unstable. But because a star's core is so dense and collisions among nuclei are so frequent, there is a nonzero probability that a third ^{4}He nucleus will fuse with the ^{8}Be nucleus before it can decay. The result is the stable nuclide ^{12}C. This is called the *triple-alpha process,* since three ^{4}He nuclei (that is, alpha particles) fuse to form one carbon nucleus. Then successive fusions with ^{4}He give ^{16}O, ^{20}Ne, and ^{24}Mg. All these reactions are exoergic. They release energy to heat up the star, and ^{12}C and ^{16}O can fuse to form elements with higher and higher atomic number.

For nuclides that can be created in this manner, the binding energy per nucleon peaks at mass number $A = 56$ with the nuclide ^{56}Fe, so exoergic fusion reactions stop with Fe. But successive neutron captures followed by beta decays can continue the synthesis of more massive nuclei. If the star is massive enough, it may eventually explode as a *supernova,* sending out into space the heavy elements that were produced by the earlier processes (Fig. 44.21; see also Fig. 37.7). In space, the debris and other interstellar matter can gravitationally bunch together to form a new generation of stars and planets. Our own sun is one such "second-generation" star. This means that the sun's planets and everything on them (including you) contain matter that was long ago blasted into space by an exploding supernova.

44.21 The Veil Nebula in the constellation Cygnus is a remnant of a supernova explosion that occurred more than 20,000 years ago. The gas ejected from the supernova is still moving very rapidly. Collisions between this fast-moving gas and the tenuous material of interstellar space excite the gas and cause it to glow. The portion of the nebula shown here is about 40 ly (12 pc) in length.

Background Radiation

In 1965 Arno Penzias and Robert Wilson, working at Bell Telephone Laboratories in New Jersey on satellite communications, turned a microwave antenna skyward and found a background signal that had no apparent preferred direction. (This signal produces about 1% of the "hash" you see on a TV screen when you turn to an unused channel.) Further research has shown that the radiation that is received has a frequency spectrum that fits Planck's blackbody radiation law, Eq. (39.24) (Section 39.5). The wavelength of peak intensity is 1.063 mm (in the microwave region of the spectrum), with a corresponding absolute temperature $T = 2.725$ K. Penzias and Wilson contacted physicists at nearby Princeton University who had begun the design of an antenna to search for radiation that was a remnant from the early evolution of the universe. We mentioned above that neutral atoms began to form at about $t = 380,000$ years when the temperature was 3000 K. With far fewer charged particles present than previously, the universe became transparent at this time to electromagnetic radiation of long wavelength. The 3000-K blackbody radiation therefore survived, cooling to its present 2.725-K temperature as the universe expanded. The *cosmic background radiation* is among the most clear-cut experimental confirmations of the Big Bang theory. Figure 44.22 shows a modern map of the cosmic background radiation.

44.22 This false-color map shows microwave radiation from the entire sky mapped onto an oval. When this radiation was emitted 380,000 years after the Big Bang, the regions shown in blue were slightly cooler and denser than average. Within these cool, dense regions formed galaxies, including the Milky Way galaxy of which our solar system, our earth, and our selves are part.

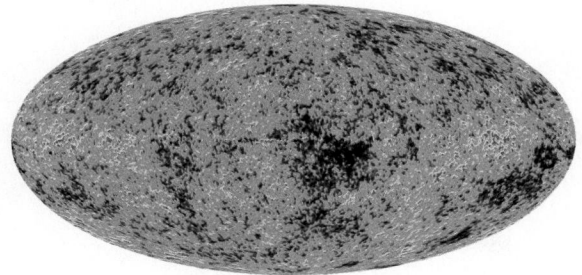

Example 44.12   **Expansion of the universe**

By approximately what factor has the universe expanded since $t = 380,000$ y?

SOLUTION

IDENTIFY and SET UP: We use the idea that as the universe has expanded, all intergalactic wavelengths have expanded with it. The Wien displacement law, Eq. (39.21), relates the peak wavelength λ_m in blackbody radiation to the temperature T. Given the temperatures of the cosmic background radiation today (2.725 K) and at $t = 380,000$ y (3000 K) we can determine the factor by which wavelengths have changed and hence determine the factor by which the universe has expanded.

EXECUTE: We rewrite Eq. (39.21) as

$$\lambda_m = \frac{2.90 \times 10^{-3} \text{ m} \cdot \text{K}}{T}$$

Hence the peak wavelength λ_m is inversely proportional to T. As the universe expands, all intergalactic wavelengths (including λ_m) increase in proportion to the scale factor R. The temperature has decreased by the factor $(3000 \text{ K})/(2.725 \text{ K}) \approx 1100$, so λ_m and the scale factor must both have *increased* by this factor. Thus, between $t = 380,000$ y and the present, the universe has expanded by a factor of about 1100.

EVALUATE: Our results show that since $t = 380,000$ y, any particular intergalactic *volume* has increased by a factor of about $(1100)^3 = 1.3 \times 10^9$. They also show that when the cosmic background radiation was emitted, its peak wavelength was $\frac{1}{1100}$ of the present-day value of 1.063 mm, or 967 nm. This is in the infrared region of the spectrum.

Matter and Antimatter

One of the most remarkable features of our universe is the asymmetry between matter and antimatter. One might think that the universe should have equal numbers of protons and antiprotons and of electrons and positrons, but this doesn't appear to be the case. Theories of the early universe must explain this imbalance.

We've mentioned that most GUTs include violation of conservation of baryon number at energies at which the strong and electroweak interactions have converged. If particle–antiparticle symmetry is also violated, we have a mechanism for making more quarks than antiquarks, more leptons than antileptons, and eventually more matter than antimatter. One serious problem is that any asymmetry that is created in this way during the GUT era might be wiped out by the electroweak interaction after the end of the GUT era. If so, there must be some mechanism that creates particle–antiparticle asymmetry at a much *later* time. The problem of the matter–antimatter asymmetry is still very much an open one.

There are still many unanswered questions at the intersection of particle physics and cosmology. Is the energy density of the universe precisely equal to $\rho_c c^2$, or are there small but important differences? What is dark energy? Has the density of dark energy remained constant over the history of the universe, or has the density changed? What is dark matter? What happened during the first 10^{-43} s after the Big Bang? Can we see evidence that the strong and electroweak interactions undergo a grand unification at high energies? The search for the answers to these and many other questions about our physical world continues to be one of the most exciting adventures of the human mind.

Test Your Understanding of Section 44.7   Given a sufficiently powerful telescope, could we detect photons emitted earlier than $t = 380,000$ y?   ❙

Fundamental particles: Each particle has an antiparticle; some particles are their own antiparticles. Particles can be created and destroyed, some of them (including electrons and positrons) only in pairs or in conjunction with other particles and antiparticles.

Particles serve as mediators for the fundamental interactions. The photon is the mediator of the electromagnetic interaction. Yukawa proposed the existence of mesons to mediate the nuclear interaction. Mediating particles that can exist only because of the uncertainty principle for energy are called virtual particles.

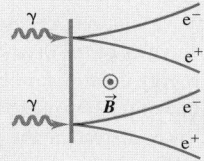

Particle accelerators and detectors: Cyclotrons, synchrotrons, and linear accelerators are used to accelerate charged particles to high energies for experiments with particle interactions. Only part of the beam energy is available to cause reactions with targets at rest. This problem is avoided in colliding-beam experiments. (See Examples 44.1–44.3.)

Particles and interactions: Four fundamental interactions are found in nature: the strong, electromagnetic, weak, and gravitational interactions. Particles can be described in terms of their interactions and of quantities that are conserved in all or some of the interactions.

Fermions have half-integer spins; bosons have integer spins. Leptons, which are fermions, have no strong interactions. Strongly interacting particles are called hadrons. They include mesons, which are always bosons, and baryons, which are always fermions. There are conservation laws for three different lepton numbers and for baryon number. Additional quantum numbers, including strangeness and charm, are conserved in some interactions and not in others. (See Examples 44.4–44.6.)

Quarks: Hadrons are composed of quarks. There are thought to be six types of quarks. The interaction between quarks is mediated by gluons. Quarks and gluons have an additional attribute called color. (See Example 44.7.)

Symmetry and the unification of interactions: Symmetry considerations play a central role in all fundamental-particle theories. The electromagnetic and weak interactions become unified at high energies into the electroweak interaction. In grand unified theories the strong interaction is also unified with these interactions, but at much higher energies.

The expanding universe and its composition: The Hubble law shows that galaxies are receding from each other and that the universe is expanding. Observations show that the rate of expansion is accelerating due to the presence of dark energy, which makes up 72.6% of the energy in the universe. Only 4.6% of the energy in the universe is in the form of ordinary matter; the remaining 22.8% is dark matter, whose nature is poorly understood. (See Examples 44.8 and 44.9.)

The history of the universe: In the standard model of the universe, a Big Bang gave rise to the first fundamental particles. They eventually formed into the lightest atoms as the universe expanded and cooled. The cosmic background radiation is a relic of the time when these atoms formed. The heavier elements were manufactured much later by fusion reactions inside stars. (See Examples 44.10–44.12.)

BRIDGING PROBLEM   Hyperons, Pions, and the Expanding Universe

A Λ^0 hyperon at rest decays into a neutron and a π^0. (a) Find the kinetic energies of the decay products. (b) What fraction of the total kinetic energy is carried off by each particle? (c) A physicist on earth detects one of the two photons that was emitted in the decay of a π^0. The π^0 was at rest in the cluster shown in Fig. 44.19 before it decayed. What is the energy of the photon that is detected on earth?

SOLUTION GUIDE

See MasteringPhysics® study area for a Video Tutor solution.

IDENTIFY and SET UP

1. Which quantities are conserved in the Λ^0 decay? In the π^0 decay?
2. The universe expanded during the time that the photon traveled from the cluster to earth. How does this affect the wavelength and energy of the photon that the physicist detects?
3. List the unknown quantities for each part of the problem and identify the target variables.
4. Select the equations that will allow you to solve for the target variables.

EXECUTE

5. Write the conservation equations for the decay of the Λ^0. (*Hint:* It's useful to write the energy E of a particle in terms of its momentum p and mass m using $E = (p^2c^2 + m^2c^4)^{1/2}$.)
6. Solve the conservation equations for the energy of one of the decay products. (*Hint:* Rearrange the energy conservation equation so that one of the $(p^2c^2 + m^2c^4)^{1/2}$ terms is on one side of the equation. Then square both sides.) Then use $K = E - mc^2$.
7. Find the fraction of the total kinetic energy that goes into the neutron and into the pion.
8. Write the conservation equations for the decay of the π^0 at rest and find the energy of each emitted photon. By what factor does the wavelength of this photon change as it travels from the galaxy cluster to earth? By what factor does the photon *energy* change? (*Hint:* See Fig. 44.19.)

EVALUATE

9. Which of the Λ^0 decay products should have the greater kinetic energy? Should the detected π^0 decay photon have more or less energy than when it was emitted?

Problems

For instructor-assigned homework, go to www.masteringphysics.com

•, ••, •••: Problems of increasing difficulty. **CP**: Cumulative problems incorporating material from earlier chapters. **CALC**: Problems requiring calculus. **BIO**: Biosciences problems.

DISCUSSION QUESTIONS

Q44.1 Is it possible that some parts of the universe contain antimatter whose atoms have nuclei made of antiprotons and antineutrons, surrounded by positrons? How could we detect this condition without actually going there? Can we detect these antiatoms by identifying the light they emit as composed of antiphotons? Explain. What problems might arise if we actually *did* go there?

Q44.2 Given the Heisenberg uncertainty principle, is it possible to create particle–antiparticle pairs that exist for extremely short periods of time before annihilating? Does this mean that empty space is really empty?

Q44.3 When they were first discovered during the 1930s and 1940s, there was confusion as to the identities of pions and muons. What are the similarities and most significant differences?

Q44.4 The gravitational force between two electrons is weaker than the electrical force by the order of 10^{-40}. Yet the gravitational interactions of matter were observed and analyzed long before electrical interactions were understood. Why?

Q44.5 When a π^0 decays to two photons, what happens to the quarks of which it was made?

Q44.6 Why can't an electron decay to two photons? To two neutrinos?

Q44.7 According to the standard model of the fundamental particles, what are the similarities between baryons and leptons? What are the most important differences?

Q44.8 According to the standard model of the fundamental particles, what are the similarities between quarks and leptons? What are the most important differences?

Q44.9 What are the main advantages of colliding-beam accelerators compared with those using stationary targets? What are the main disadvantages?

Q44.10 Does the universe have a center? Explain.

Q44.11 Does it make sense to ask, "If the universe is expanding, what is it expanding into?"

Q44.12 Assume that the universe has an edge. Placing yourself at that edge in a thought experiment, explain why this assumption violates the cosmological principle.

Q44.13 Explain why the cosmological principle requires that H_0 must have the same value everywhere in space, but does not require that it be constant in time.

EXERCISES

Section 44.1 Fundamental Particles—A History

44.1 • A neutral pion at rest decays into two photons. Find the energy, frequency, and wavelength of each photon. In which part of the electromagnetic spectrum does each photon lie? (Use the pion mass given in terms of the electron mass in Section 44.1.)

44.2 •• **CP** Two equal-energy photons collide head-on and annihilate each other, producing a $\mu^+\mu^-$ pair. The muon mass is given in terms of the electron mass in Section 44.1. (a) Calculate the maximum wavelength of the photons for this to occur. If the photons have this wavelength, describe the motion of the μ^+ and μ^- immediately after they are produced. (b) If the wavelength of each photon is half the value calculated in part (a), what is the speed of each muon after they have moved apart? Use correct relativistic expressions for momentum and energy.

44.3 •• A positive pion at rest decays into a positive muon and a neutrino. (a) Approximately how much energy is released in the decay? (Assume the neutrino has zero rest mass. Use the muon and pion masses given in terms of the electron mass in Section 44.1.) (b) Why can't a positive muon decay into a positive pion?

44.4 • A proton and an antiproton annihilate, producing two photons. Find the energy, frequency, and wavelength of each photon (a) if the p and $\bar{p}$ are initially at rest and (b) if the p and $\bar{p}$ collide head-on, each with an initial kinetic energy of 830 MeV.

44.5 •• **CP** For the nuclear reaction given in Eq. (44.2) assume that the initial kinetic energy and momentum of the reacting particles are negligible. Calculate the speed of the α particle immediately after it leaves the reaction region.

44.6 • Estimate the range of the force mediated by an ω^0 meson that has mass 783 MeV/c^2.

44.7 • The starship *Enterprise*, of television and movie fame, is powered by combining matter and antimatter. If the entire 400-kg antimatter fuel supply of the *Enterprise* combines with matter, how much energy is released? How does this compare to the U.S. yearly energy use, which is roughly 1.0×10^{20} J?

Section 44.2 Particle Accelerators and Detectors

44.8 • An electron with a total energy of 20.0 GeV collides with a stationary positron. (a) What is the available energy? (b) If the electron and positron are accelerated in a collider, what total energy corresponds to the same available energy as in part (a)?

44.9 • Deuterons in a cyclotron travel in a circle with radius 32.0 cm just before emerging from the dees. The frequency of the applied alternating voltage is 9.00 MHz. Find (a) the magnetic field and (b) the kinetic energy and speed of the deuterons upon emergence.

44.10 • The magnetic field in a cyclotron that accelerates protons is 1.30 T. (a) How many times per second should the potential across the dees reverse? (This is twice the frequency of the circulating protons.) (b) The maximum radius of the cyclotron is 0.250 m. What is the maximum speed of the proton? (c) Through what potential difference would the proton have to be accelerated from rest to give it the same speed as calculated in part (b)?

44.11 • (a) A high-energy beam of alpha particles collides with a stationary helium gas target. What must be the total energy of a beam particle be if the available energy in the collision is 16.0 GeV? (b) If the alpha particles instead interact in a colliding-beam experiment, what must the energy of each beam be to produce the same available energy?

44.12 •• (a) What is the speed of a proton that has total energy 1000 GeV? (b) What is the angular frequency ω of a proton with the speed calculated in part (a) in a magnetic field of 4.00 T? Use both the nonrelativistic Eq. (44.7) and the correct relativistic expression, and compare the results.

44.13 • In Example 44.3 it was shown that a proton beam with an 800-GeV beam energy gives an available energy of 38.7 GeV for collisions with a stationary proton target. (a) You are asked to design an upgrade of the accelerator that will double the available energy in stationary-target collisions. What beam energy is required? (b) In a colliding-beam experiment, what total energy of each beam is needed to give an available energy of 2(38.7 GeV) = 77.4 GeV?

44.14 •• Calculate the minimum beam energy in a proton–proton collider to initiate the $p + p \to p + p + \eta^0$ reaction. The rest energy of the η^0 is 547.3 MeV (see Table 44.3).

Section 44.3 Particles and Interactions

44.15 • A K^+ meson at rest decays into two π mesons. (a) What are the allowed combinations of π^0, π^+, and π^- as decay products? (b) Find the total kinetic energy of the π mesons.

44.16 • How much energy is released when a μ^- muon at rest decays into an electron and two neutrinos? Neglect the small masses of the neutrinos.

44.17 • What is the mass (in kg) of the Z^0? What is the ratio of the mass of the Z^0 to the mass of the proton?

44.18 • Table 44.3 shows that a Σ^0 decays into a Λ^0 and a photon. (a) Calculate the energy of the photon emitted in this decay, if the Λ^0 is at rest. (b) What is the magnitude of the momentum of the photon? Is it reasonable to ignore the final momentum and kinetic energy of the Λ^0? Explain.

44.19 • If a Σ^+ at rest decays into a proton and a π^0, what is the total kinetic energy of the decay products?

44.20 • The discovery of the Ω^- particle helped confirm Gell-Mann's eightfold way. If an Ω^- decays into a Λ^0 and a K^-, what is the total kinetic energy of the decay products?

44.21 • In which of the following decays are the three lepton numbers conserved? In each case, explain your reasoning. (a) $\mu^- \to e^- + \nu_e + \bar{\nu}_\mu$; (b) $\tau^- \to e^- + \bar{\nu}_e + \nu_\tau$; (c) $\pi^+ \to e^+ + \gamma$; (d) $n \to p + e^- + \bar{\nu}_e$.

44.22 • Which of the following reactions obey the conservation of baryon number? (a) $p + p \to p + e^+$; (b) $p + n \to 2e^+ + e^-$; (c) $p \to n + e^- + \bar{\nu}_e$; (d) $p + \bar{p} \to 2\gamma$.

44.23 • In which of the following reactions or decays is strangeness conserved? In each case, explain your reasoning. (a) $K^+ \to \mu^+ + \nu_\mu$; (b) $n + K^+ \to p + \pi^0$; (c) $K^+ + K^- \to \pi^0 + \pi^0$; (d) $p + K^- \to \Lambda^0 + \pi^0$.

44.24 •• **CP** (a) Show that the coupling constant for the electromagnetic interaction, $e^2/4\pi\epsilon_0\hbar c$, is dimensionless and has the numerical value 1/137.0. (b) Show that in the Bohr model the orbital speed of an electron in the $n = 1$ orbit is equal to c times the coupling constant $e^2/4\pi\epsilon_0\hbar c$.

44.25 • Show that the nuclear force coupling constant $f^2/\hbar c$ is dimensionless.

Section 44.4 Quarks and the Eightfold Way

44.26 • Nine of the spin-$\frac{3}{2}$ baryons are four Δ particles, each with mass 1232 MeV/c^2, strangeness 0, and charges $+2e$, $+e$, 0, and $-e$; three Σ^* particles, each with mass 1385 MeV/c^2, strangeness -1, and charges $+e$, 0, and $-e$; and two Ξ^* particles, each with mass 1530 MeV/c^2, strangeness -2, and charges 0 and $-e$. (a) Place these particles on a plot of S versus Q. Deduce the Q and S values of the tenth spin-$\frac{3}{2}$ baryon, the Ω^- particle, and place it on your diagram. Also label the particles with their masses. The mass of the Ω^- is 1672 MeV/c^2; is this value consistent with your diagram? (b) Deduce the three-quark combinations (of u, d, and s) that make up each of these ten particles. Redraw the plot of S versus Q from part (a) with each particle labeled by its quark content. What regularities do you see?

44.27 • Determine the electric charge, baryon number, strangeness quantum number, and charm quantum number for the following quark combinations: (a) uds; (b) $c\bar{u}$; (c) ddd; and (d) $d\bar{c}$. Explain your reasoning.

44.28 •• Determine the electric charge, baryon number, strangeness quantum number, and charm quantum number for the following quark combinations: (a) uus, (b) $c\bar{s}$, (c) $\overline{ddu}$, and (d) $c\bar{b}$.

44.29 • The weak force may change quark flavor in an interaction. Explain how β^+ decay changes quark flavor. If a proton undergoes β^+ decay, determine the decay reaction.

44.30 • What is the total kinetic energy of the decay products when an upsilon particle at rest decays to $\tau^+ + \tau^-$?

44.31 • The quark content of the neutron is udd. (a) What is the quark content of the antineutron? Explain your reasoning. (b) Is the neutron its own antiparticle? Why or why not? (c) The quark content of the ψ is $c\bar{c}$. Is the ψ its own antiparticle? Explain your reasoning.

44.32 • Given that each particle contains only combinations of u, d, s, $\bar{u}$, $\bar{d}$, and $\bar{s}$, use the method of Example 44.7 to deduce the quark content of (a) a particle with charge $+e$, baryon number 0, and strangeness $+1$; (b) a particle with charge $+e$, baryon number -1, and strangeness $+1$; (c) a particle with charge 0, baryon number $+1$, and strangeness -2.

Section 44.6 The Expanding Universe

44.33 • The spectrum of the sodium atom is detected in the light from a distant galaxy. (a) If the 590.0-nm line is redshifted to 658.5 nm, at what speed is the galaxy receding from the earth? (b) Use the Hubble law to calculate the distance of the galaxy from the earth.

44.34 • **Redshift Parameter.** The definition of the redshift parameter z is given in Example 44.8. (a) Show that Eq. (44.13) may be written as $1 + z = ([1 + \beta]/[1 - \beta])^{1/2}$, where $\beta = v/c$. (b) The observed redshift parameter for a certain galaxy is $z = 0.500$. Find the speed of the galaxy relative to the earth, if the redshift is due to the Doppler shift. (c) Use the Hubble law to find the distance of this galaxy from the earth.

44.35 • A galaxy in the constellation Pisces is 5210 Mly from the earth. (a) Use the Hubble law to calculate the speed at which this galaxy is receding from earth. (b) What redshifted ratio λ_0/λ_S is expected for light from this galaxy?

44.36 •• (a) According to the Hubble law, what is the distance r from us for galaxies that are receding from us with a speed c? (b) Explain why the distance calculated in part (a) is the size of our observable universe (ignoring any change in the expansion rate of the universe due to gravitational attraction or dark energy).

44.37 •• The critical density of the universe is $9.5 \times 10^{-27} \text{ kg/m}^3$. (a) Assuming that the universe is all hydrogen, express the critical density in the number of H atoms per cubic meter. (b) If the density of the universe is equal to the critical density, how many atoms, on the average, would you expect to find in a room of dimensions $4 \text{ m} \times 7 \text{ m} \times 3 \text{ m}$? (c) Compare your answer in part (b) with the number of atoms you would find in the same room under normal conditions on the earth.

Section 44.7 The Beginning of Time

44.38 • (a) Show that the expression for the Planck length, $\sqrt{\hbar G/c^3}$, has dimensions of length. (b) Evaluate the numerical value of $\sqrt{\hbar G/c^3}$, and verify the value given in Eq. (44.21).

44.39 • Calculate the energy released in each reaction: (a) p + ^{2}H $\rightarrow$ ^{3}He; (b) n + ^{3}He $\rightarrow$ ^{4}He.

44.40 • Calculate the energy (in MeV) released in the triple-alpha process $3 \, ^4$He $\rightarrow \, ^{12}$C.

44.41 • Calculate the reaction energy Q (in MeV) for the reaction $e^- + p \rightarrow n + \nu_e$. Is this reaction endoergic or exoergic?

44.42 • Calculate the reaction energy Q (in MeV) for the nucleosynthesis reaction

$$^{12}_{6}\text{C} + {}^{4}_{2}\text{He} \rightarrow {}^{16}_{8}\text{O}$$

Is this reaction endoergic or exoergic?

44.43 • **CP** The 2.728-K blackbody radiation has its peak wavelength at 1.062 mm. What was the peak wavelength at $t = 700,000$ y when the temperature was 3000 K?

PROBLEMS

44.44 • **CP** A positronium atom consists of an electron and a positron. In the Bohr model the two particles orbit around their common center of mass. In the Bohr model, what is the ionization energy for a positronium atom when it is in its ground state?

44.45 • In the LHC, each proton will be accelerated to a kinetic energy of 7.0 TeV. (a) In the colliding beams, what is the available energy E_a in a collision? (b) In a fixed-target experiment in which a beam of protons is incident on a stationary proton target, what must the total energy (in TeV) of the particles in the beam be to produce the same available energy as in part (a)?

44.46 •• A proton and an antiproton collide head-on with equal kinetic energies. Two γ rays with wavelengths of 0.780 fm are produced. Calculate the kinetic energy of the incident proton.

44.47 •• **CP BIO Radiation Therapy with π^- Mesons.** Beams of π^- mesons are used in radiation therapy for certain cancers. The energy comes from the complete decay of the π^- to *stable* particles. (a) Write out the complete decay of a π^- meson to stable particles. What are these particles? (b) How much energy is released from the complete decay of a single π^- meson to stable particles? (You can ignore the very small masses of the neutrinos.) (c) How many π^- mesons need to decay to give a dose of 50.0 Gy to 10.0 g of tissue? (d) What would be the equivalent dose in part (c) in Sv and in rem? Consult Table 43.3 and use the largest appropriate RBE for the particles involved in this decay.

44.48 •• Calculate the threshold kinetic energy for the reaction $\pi^- + p \rightarrow \Sigma^0 + K^0$ if a π^- beam is incident on a stationary proton target. The K^0 has a mass of $497.7 \text{ MeV}/c^2$.

44.49 •• Calculate the threshold kinetic energy for the reaction $p + p \rightarrow p + p + K^+ + K^-$ if a proton beam is incident on a stationary proton target.

44.50 •• An η^0 meson at rest decays into three π mesons. (a) What are the allowed combinations of π^0, π^+, and π^- as decay products? (b) Find the total kinetic energy of the π mesons.

44.51 • Each of the following reactions is missing a single particle. Calculate the baryon number, charge, strangeness, and the three lepton numbers (where appropriate) of the missing particle, and from this identify the particle. (a) $p + p \rightarrow p + \Lambda^0 + ?$; (b) $K^- + n \rightarrow \Lambda^0 + ?$; (c) $p + \bar{p} \rightarrow n + ?$; (d) $\bar{\nu}_\mu + p \rightarrow n + ?$

44.52 • Estimate the energy width (energy uncertainty) of the ψ if its mean lifetime is 7.6×10^{-21} s. What fraction is this of its rest energy?

44.53 • The ϕ meson has mass $1019.4 \text{ MeV}/c^2$ and a measured energy width of $4.4 \text{ MeV}/c^2$. Using the uncertainty principle, estimate the lifetime of the ϕ meson.

44.54 •• A ϕ meson (see Problem 44.53) at rest decays via $\phi \rightarrow K^+ + K^-$. It has strangeness 0. (a) Find the kinetic energy of the K^+ meson. (Assume that the two decay products share kinetic energy equally, since their masses are equal.) (b) Suggest a reason the decay $\phi \rightarrow K^+ + K^- + \pi^0$ has not been observed. (c) Suggest reasons the decays $\phi \rightarrow K^+ + \pi^-$ and $\phi \rightarrow K^+ + \mu^-$ have not been observed.

44.55 •• **CP BIO** One proposed proton decay is $p^+ \rightarrow e^+ + \pi^0$, which violates both baryon and lepton number conservation, so the proton lifetime is expected to be very long. Suppose the proton half-life were 1.0×10^{18} y. (a) Calculate the energy deposited per kilogram of body tissue (in rad) due to the decay of the protons in your body in one year. Model your body as consisting entirely of water. Only the two protons in the hydrogen atoms in each H_2O molecule would decay in the manner shown; do you see why? Assume that the π^0 decays to two γ rays, that the positron annihilates with an electron, and that all the energy produced in the primary decay and these secondary decays remains in your body. (b) Calculate the equivalent dose (in rem) assuming an RBE of 1.0 for all the radiation products, and compare with the 0.1 rem due to the natural background and the 5.0-rem guideline for industrial workers. Based on your calculation, can the proton lifetime be as short as 1.0×10^{18} y?

44.56 ••• **CP** A Ξ^- particle at rest decays to a Λ^0 and a π^-. (a) Find the total kinetic energy of the decay products. (b) What fraction of the energy is carried off by each particle? (Use relativistic expressions for momentum and energy.)

44.57 •• **CALC** Consider the spherical balloon model of a two-dimensional expanding universe (see Fig. 44.17 in Section 44.6). The shortest distance between two points on the surface, measured along the surface, is the arc length r, where $r = R\theta$. As the balloon expands, its radius R increases, but the angle θ between the two points remains constant. (a) Explain why, at any given time, $(dR/dt)/R$ is the same for all points on the balloon. (b) Show that $v = dr/dt$ is directly proportional to r at any instant. (c) From your answer to part (b), what is the expression for the Hubble constant H_0 in terms of R and dR/dt? (d) The expression for H_0 you found in part (c) is constant in space. How would R have to depend on time for H_0 to be constant in time? (e) Is your answer to part (d) consistent with the observed rate of expansion of the universe?

44.58 ••• **CALC** Suppose all the conditions are the same as in Problem 44.57, except that $v = dr/dt$ is constant for a given θ, rather than H_0 being constant in time. Show that the Hubble constant is $H_0 = 1/t$ and, hence, that the current value is $1/T$, where T is the age of the universe.

44.59 •• **Cosmic Jerk.** The densities of ordinary matter and dark matter have decreased as the universe has expanded, since the same amount of mass occupies an ever-increasing volume. Yet observations suggest that the density of dark energy has remained constant over the entire history of the universe. (a) Explain why the expansion of the universe actually slowed down in its early history but is speeding up today. "Jerk" is the term for a change in acceleration, so the change in cosmic expansion from slowing down to speeding up is called *cosmic jerk.* (b) Calculations show that the change in acceleration took place when the combined density of matter of all kinds was equal to twice the density of dark energy. Compared to today's value of the scale factor, what was the scale factor at that time? (c) We see the galaxy clusters in Figs. 44.15b and 44.19 as they were 300 million years ago and 10.2 billion years ago. Was the expansion of the universe slowing down or speeding up at these times? (*Hint:* See the caption for Fig. 44.19.)

44.60 ••• **CP** The K^0 meson has rest energy 497.7 MeV. A K^0 meson moving in the $+x$-direction with kinetic energy 225 MeV decays into a π^+ and a π^-, which move off at equal angles above and below the $+x$-axis. Calculate the kinetic energy of the π^+ and the angle it makes with the $+x$-axis. Use relativistic expressions for energy and momentum.

44.61 ••• **CP** A Σ^- particle moving in the $+x$-direction with kinetic energy 180 MeV decays into a π^- and a neutron. The π^- moves in the $+y$-direction. What is the kinetic energy of the neutron, and what is the direction of its velocity? Use relativistic expressions for energy and momentum.

CHALLENGE PROBLEM

44.62 ••• **CP** Consider a collision in which a stationary particle with mass M is bombarded by a particle with mass m, speed v_0, and total energy (including rest energy) E_m. (a) Use the Lorentz transformation to write the velocities v_m and v_M of particles m and M in terms of the speed v_{cm} of the center of momentum. (b) Use the fact that the total momentum in the center-of-momentum frame is zero to obtain an expression for v_{cm} in terms of m, M, and v_0. (c) Combine the results of parts (a) and (b) to obtain Eq. (44.9) for the total energy in the center-of-momentum frame.

Answers

Chapter Opening Question ?

Only 4.6% of the mass and energy of the universe is in the form of "normal" matter. Of the rest, 22.8% is poorly understood dark matter and 72.6% is even more mysterious dark energy.

Test Your Understanding Questions

44.1 Answer: (i), (iii), (ii), (iv) The more massive the virtual particle, the shorter its lifetime and the shorter the distance that it can travel during its lifetime.

44.2 Answer: no In a head-on collision between an electron and a positron of equal energy, the net momentum is zero. Since both momentum and energy are conserved in the collision, the virtual photon also has momentum $p = 0$ but has energy $E = 90$ GeV + 90 GeV = 180 GeV. Hence the relationship $E = pc$ is definitely *not* true for this virtual photon.

44.3 Answer: no Mesons all have baryon number $B = 0$, while a proton has $B = 1$. The decay of a proton into one or more mesons would require that baryon number *not* be conserved. No violation of this conservation principle has ever been observed, so the proposed decay is impossible.

44.4 Answer: no Only the s quark, with $S = -1$, has nonzero strangeness. For a baryon to have $S = -2$, it must have two s quarks and one quark of a different flavor. Since each s quark has charge $-\frac{1}{3}e$, the nonstrange quark must have charge $+\frac{5}{3}e$ to make the net charge equal to $+e$. But *no* quark has charge $+\frac{5}{3}e$, so the proposed baryon is impossible.

44.5 Answer: (i) If a d quark in a neutron (quark content udd) undergoes the process $d \rightarrow u + e^- + \bar{\nu}_e$, the remaining baryon has quark content uud and hence is a proton (see Fig. 44.11). An electron is the same as a β^- particle, so the net result is beta-minus decay: $n \rightarrow p + \beta^- + \bar{\nu}_e$.

44.6 Answer: yes . . . and no The material of which your body is made is ordinary to us on earth. But from a cosmic perspective your material is quite *extraordinary*: Only 4.6% of the mass and energy in the universe is in the form of atoms.

44.7 Answer: no Prior to $t = 380{,}000$ y the temperature was so high that atoms could not form, so free electrons and protons were plentiful. These charged particles are very effective at scattering photons, so light could not propagate very far and the universe was opaque. The oldest photons that we can detect date from the time $t = 380{,}000$ y when atoms formed and the universe became transparent.

Bridging Problem

Answers: (a) 5.78 MeV for the neutron, 35.62 MeV for the pion
(b) 0.140 for the neutron, 0.860 for the pion
(c) 24 MeV

APPENDIX A

THE INTERNATIONAL SYSTEM OF UNITS

The Système International d'Unités, abbreviated SI, is the system developed by the General Conference on Weights and Measures and adopted by nearly all the industrial nations of the world. The following material is adapted from the National Institute of Standards and Technology (**http://physics.nist.gov/cuu**).

Quantity	Name of unit	Symbol	
	SI base units		
length	meter	m	
mass	kilogram	kg	
time	second	s	
electric current	ampere	A	
thermodynamic temperature	kelvin	K	
amount of substance	mole	mol	
luminous intensity	candela	cd	
	SI derived units		**Equivalent units**
area	square meter	m^2	
volume	cubic meter	m^3	
frequency	hertz	Hz	s^{-1}
mass density (density)	kilogram per cubic meter	kg/m^3	
speed, velocity	meter per second	m/s	
angular velocity	radian per second	rad/s	
acceleration	meter per second squared	m/s^2	
angular acceleration	radian per second squared	rad/s^2	
force	newton	N	$kg \cdot m/s^2$
pressure (mechanical stress)	pascal	Pa	N/m^2
kinematic viscosity	square meter per second	m^2/s	
dynamic viscosity	newton-second per square meter	$N \cdot s/m^2$	
work, energy, quantity of heat	joule	J	$N \cdot m$
power	watt	W	J/s
quantity of electricity	coulomb	C	$A \cdot s$
potential difference, electromotive force	volt	V	J/C, W/A
electric field strength	volt per meter	V/m	N/C
electric resistance	ohm	Ω	V/A
capacitance	farad	F	$A \cdot s/V$
magnetic flux	weber	Wb	$V \cdot s$
inductance	henry	H	$V \cdot s/A$
magnetic flux density	tesla	T	Wb/m^2
magnetic field strength	ampere per meter	A/m	
magnetomotive force	ampere	A	
luminous flux	lumen	lm	$cd \cdot sr$
luminance	candela per square meter	cd/m^2	
illuminance	lux	lx	lm/m^2
wave number	1 per meter	m^{-1}	
entropy	joule per kelvin	J/K	
specific heat capacity	joule per kilogram-kelvin	$J/kg \cdot K$	
thermal conductivity	watt per meter-kelvin	$W/m \cdot K$	

Quantity	Name of unit	Symbol	Equivalent units
radiant intensity	watt per steradian	W/sr	
activity (of a radioactive source)	becquerel	Bq	s^{-1}
radiation dose	gray	Gy	J/kg
radiation dose equivalent	sievert	Sv	J/kg
SI supplementary units			
plane angle	radian	rad	
solid angle	steradian	sr	

Definitions of SI Units

meter (m) The *meter* is the length equal to the distance traveled by light, in vacuum, in a time of 1/299,792,458 second.

kilogram (kg) The *kilogram* is the unit of mass; it is equal to the mass of the international prototype of the kilogram. (The international prototype of the kilogram is a particular cylinder of platinum-iridium alloy that is preserved in a vault at Sévres, France, by the International Bureau of Weights and Measures.)

second (s) The *second* is the duration of 9,192,631,770 periods of the radiation corresponding to the transition between the two hyperfine levels of the ground state of the cesium-133 atom.

ampere (A) The *ampere* is that constant current that, if maintained in two straight parallel conductors of infinite length, of negligible circular cross section, and placed 1 meter apart in vacuum, would produce between these conductors a force equal to 2×10^{-7} newton per meter of length.

kelvin (K) The *kelvin,* unit of thermodynamic temperature, is the fraction 1/273.16 of the thermodynamic temperature of the triple point of water.

ohm (Ω) The *ohm* is the electric resistance between two points of a conductor when a constant difference of potential of 1 volt, applied between these two points, produces in this conductor a current of 1 ampere, this conductor not being the source of any electromotive force.

coulomb (C) The *coulomb* is the quantity of electricity transported in 1 second by a current of 1 ampere.

candela (cd) The *candela* is the luminous intensity, in a given direction, of a source that emits monochromatic radiation of frequency 540×10^{12} hertz and that has a radiant intensity in that direction of 1/683 watt per steradian.

mole (mol) The *mole* is the amount of substance of a system that contains as many elementary entities as there are carbon atoms in 0.012 kg of carbon 12. The elementary entities must be specified and may be atoms, molecules, ions, electrons, other particles, or specified groups of such particles.

newton (N) The *newton* is that force that gives to a mass of 1 kilogram an acceleration of 1 meter per second per second.

joule (J) The *joule* is the work done when the point of application of a constant force of 1 newton is displaced a distance of 1 meter in the direction of the force.

watt (W) The *watt* is the power that gives rise to the production of energy at the rate of 1 joule per second.

volt (V) The *volt* is the difference of electric potential between two points of a conducting wire carrying a constant current of 1 ampere, when the power dissipated between these points is equal to 1 watt.

weber (Wb) The *weber* is the magnetic flux that, linking a circuit of one turn, produces in it an electromotive force of 1 volt as it is reduced to zero at a uniform rate in 1 second.

lumen (lm) The *lumen* is the luminous flux emitted in a solid angle of 1 steradian by a uniform point source having an intensity of 1 candela.

farad (F) The *farad* is the capacitance of a capacitor between the plates of which there appears a difference of potential of 1 volt when it is charged by a quantity of electricity equal to 1 coulomb.

henry (H) The *henry* is the inductance of a closed circuit in which an electromotive force of 1 volt is produced when the electric current in the circuit varies uniformly at a rate of 1 ampere per second.

radian (rad) The *radian* is the plane angle between two radii of a circle that cut off on the circumference an arc equal in length to the radius.

steradian (sr) The *steradian* is the solid angle that, having its vertex in the center of a sphere, cuts off an area of the surface of the sphere equal to that of a square with sides of length equal to the radius of the sphere.

SI Prefixes To form the names of multiples and submultiples of SI units, apply the prefixes listed in Appendix F.

APPENDIX B

USEFUL MATHEMATICAL RELATIONS

Algebra

$$a^{-x} = \frac{1}{a^x} \qquad a^{(x+y)} = a^x a^y \qquad a^{(x-y)} = \frac{a^x}{a^y}$$

Logarithms: If $\log a = x$, then $a = 10^x$. $\log a + \log b = \log(ab)$ $\log a - \log b = \log(a/b)$ $\log(a^n) = n \log a$

If $\ln a = x$, then $a = e^x$. $\ln a + \ln b = \ln(ab)$ $\ln a - \ln b = \ln(a/b)$ $\ln(a^n) = n \ln a$

Quadratic formula: If $ax^2 + bx + c = 0$, $x = \dfrac{-b \pm \sqrt{b^2 - 4ac}}{2a}$.

Binomial Theorem

$$(a + b)^n = a^n + na^{n-1}b + \frac{n(n-1)a^{n-2}b^2}{2!} + \frac{n(n-1)(n-2)a^{n-3}b^3}{3!} + \cdots$$

Trigonometry
In the right triangle ABC, $x^2 + y^2 = r^2$.

Definitions of the trigonometric functions:
$\sin \alpha = y/r \qquad \cos \alpha = x/r \qquad \tan \alpha = y/x$

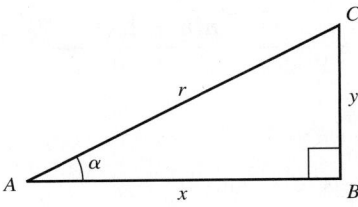

Identities: $\sin^2\alpha + \cos^2\alpha = 1$ $\tan\alpha = \dfrac{\sin\alpha}{\cos\alpha}$

$\sin 2\alpha = 2\sin\alpha\cos\alpha$ $\cos 2\alpha = \cos^2\alpha - \sin^2\alpha = 2\cos^2\alpha - 1$
$$= 1 - 2\sin^2\alpha$$

$\sin\frac{1}{2}\alpha = \sqrt{\dfrac{1 - \cos\alpha}{2}}$ $\cos\frac{1}{2}\alpha = \sqrt{\dfrac{1 + \cos\alpha}{2}}$

$\sin(-\alpha) = -\sin\alpha$ $\sin(\alpha \pm \beta) = \sin\alpha\cos\beta \pm \cos\alpha\sin\beta$
$\cos(-\alpha) = \cos\alpha$ $\cos(\alpha \pm \beta) = \cos\alpha\cos\beta \mp \sin\alpha\sin\beta$
$\sin(\alpha \pm \pi/2) = \pm\cos\alpha$ $\sin\alpha + \sin\beta = 2\sin\frac{1}{2}(\alpha + \beta)\cos\frac{1}{2}(\alpha - \beta)$
$\cos(\alpha \pm \pi/2) = \mp\sin\alpha$ $\cos\alpha + \cos\beta = 2\cos\frac{1}{2}(\alpha + \beta)\cos\frac{1}{2}(\alpha - \beta)$

For *any* triangle $A'B'C'$ (not necessarily a right triangle) with sides a, b, and c and angles α, β, and γ:

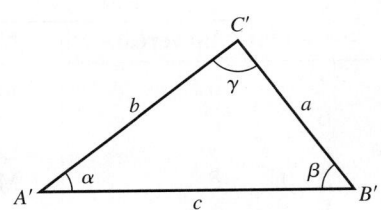

Law of sines: $\dfrac{\sin\alpha}{a} = \dfrac{\sin\beta}{b} = \dfrac{\sin\gamma}{c}$

Law of cosines: $c^2 = a^2 + b^2 - 2ab\cos\gamma$

Geometry

Circumference of circle of radius r: $C = 2\pi r$ Surface area of sphere of radius r: $A = 4\pi r^2$
Area of circle of radius r: $A = \pi r^2$ Volume of cylinder of radius r and height h: $V = \pi r^2 h$
Volume of sphere of radius r: $V = 4\pi r^3/3$

Calculus

Derivatives:

$$\frac{d}{dx}x^n = nx^{n-1}$$

$$\frac{d}{dx}\ln ax = \frac{1}{x}$$

$$\frac{d}{dx}e^{ax} = ae^{ax}$$

$$\frac{d}{dx}\sin ax = a\cos ax$$

$$\frac{d}{dx}\cos ax = -a\sin ax$$

Integrals:

$$\int x^n dx = \frac{x^{n+1}}{n+1} \quad (n \neq -1)$$

$$\int \frac{dx}{x} = \ln x$$

$$\int e^{ax} dx = \frac{1}{a}e^{ax}$$

$$\int \sin ax\, dx = -\frac{1}{a}\cos ax$$

$$\int \cos ax\, dx = \frac{1}{a}\sin ax$$

$$\int \frac{dx}{\sqrt{a^2 - x^2}} = \arcsin\frac{x}{a}$$

$$\int \frac{dx}{\sqrt{x^2 + a^2}} = \ln\left(x + \sqrt{x^2 + a^2}\right)$$

$$\int \frac{dx}{x^2 + a^2} = \frac{1}{a}\arctan\frac{x}{a}$$

$$\int \frac{dx}{(x^2 + a^2)^{3/2}} = \frac{1}{a^2}\frac{x}{\sqrt{x^2 + a^2}}$$

$$\int \frac{x\, dx}{(x^2 + a^2)^{3/2}} = -\frac{1}{\sqrt{x^2 + a^2}}$$

Power series (convergent for range of x shown):

$$(1 + x)^n = 1 + nx + \frac{n(n-1)x^2}{2!} + \frac{n(n-1)(n-2)}{3!}x^3$$
$$+ \cdots \quad (|x| < 1)$$

$$\sin x = x - \frac{x^3}{3!} + \frac{x^5}{5!} - \frac{x^7}{7!} + \cdots \quad (\text{all } x)$$

$$\cos x = 1 - \frac{x^2}{2!} + \frac{x^4}{4!} - \frac{x^6}{6!} + \cdots \quad (\text{all } x)$$

$$\tan x = x + \frac{x^3}{3} + \frac{2x^2}{15} + \frac{17x^7}{315} + \cdots \quad (|x| < \pi/2)$$

$$e^x = 1 + x + \frac{x^2}{2!} + \frac{x^3}{3!} + \cdots \quad (\text{all } x)$$

$$\ln(1 + x) = x - \frac{x^2}{2} + \frac{x^3}{3} - \frac{x^4}{4} + \cdots \quad (|x| < 1)$$

APPENDIX C

THE GREEK ALPHABET

Name	Capital	Lowercase	Name	Capital	Lowercase	Name	Capital	Lowercase
Alpha	A	α	Iota	I	ι	Rho	P	ρ
Beta	B	β	Kappa	K	κ	Sigma	Σ	σ
Gamma	Γ	γ	Lambda	Λ	λ	Tau	T	τ
Delta	Δ	δ	Mu	M	μ	Upsilon	Υ	υ
Epsilon	E	ϵ	Nu	N	ν	Phi	Φ	ϕ
Zeta	Z	ζ	Xi	Ξ	ξ	Chi	X	χ
Eta	H	η	Omicron	O	o	Psi	Ψ	ψ
Theta	Θ	θ	Pi	Π	π	Omega	Ω	ω

APPENDIX D

PERIODIC TABLE OF THE ELEMENTS

Group	1	2	3	4	5	6	7	8	9	10	11	12	13	14	15	16	17	18
Period																		
1	1 **H** 1.008																	2 **He** 4.003
2	3 **Li** 6.941	4 **Be** 9.012											5 **B** 10.811	6 **C** 12.011	7 **N** 14.007	8 **O** 15.999	9 **F** 18.998	10 **Ne** 20.180
3	11 **Na** 22.990	12 **Mg** 24.305											13 **Al** 26.982	14 **Si** 28.086	15 **P** 30.974	16 **S** 32.065	17 **Cl** 35.453	18 **Ar** 39.948
4	19 **K** 39.098	20 **Ca** 40.078	21 **Sc** 44.956	22 **Ti** 47.867	23 **V** 50.942	24 **Cr** 51.996	25 **Mn** 54.938	26 **Fe** 55.845	27 **Co** 58.933	28 **Ni** 58.693	29 **Cu** 63.546	30 **Zn** 65.409	31 **Ga** 69.723	32 **Ge** 72.64	33 **As** 74.922	34 **Se** 78.96	35 **Br** 79.904	36 **Kr** 83.798
5	37 **Rb** 85.468	38 **Sr** 87.62	39 **Y** 88.906	40 **Zr** 91.224	41 **Nb** 92.906	42 **Mo** 95.94	43 **Tc** (98)	44 **Ru** 101.07	45 **Rh** 102.906	46 **Pd** 106.42	47 **Ag** 107.868	48 **Cd** 112.411	49 **In** 114.818	50 **Sn** 118.710	51 **Sb** 121.760	52 **Te** 127.60	53 **I** 126.904	54 **Xe** 131.293
6	55 **Cs** 132.905	56 **Ba** 137.327	71 **Lu** 174.967	72 **Hf** 178.49	73 **Ta** 180.948	74 **W** 183.84	75 **Re** 186.207	76 **Os** 190.23	77 **Ir** 192.217	78 **Pt** 195.078	79 **Au** 196.967	80 **Hg** 200.59	81 **Tl** 204.383	82 **Pb** 207.2	83 **Bi** 208.980	84 **Po** (209)	85 **At** (210)	86 **Rn** (222)
7	87 **Fr** (223)	88 **Ra** (226)	103 **Lr** (262)	104 **Rf** (261)	105 **Db** (262)	106 **Sg** (266)	107 **Bh** (264)	108 **Hs** (269)	109 **Mt** (268)	110 **Ds** (271)	111 **Rg** (272)	112 **Uub** (285)	113 **Uut** (284)	114 **Uuq** (289)	115 **Uup** (288)	116 **Uuh** (292)	117 **Uus** (294)	118 **Uuo**

Lanthanoids

57 **La** 138.905	58 **Ce** 140.116	59 **Pr** 140.908	60 **Nd** 144.24	61 **Pm** (145)	62 **Sm** 150.36	63 **Eu** 151.964	64 **Gd** 157.25	65 **Tb** 158.925	66 **Dy** 162.500	67 **Ho** 164.930	68 **Er** 167.259	69 **Tm** 168.934	70 **Yb** 173.04

Actinoids

89 **Ac** (227)	90 **Th** (232)	91 **Pa** (231)	92 **U** (238)	93 **Np** (237)	94 **Pu** (244)	95 **Am** (243)	96 **Cm** (247)	97 **Bk** (247)	98 **Cf** (251)	99 **Es** (252)	100 **Fm** (257)	101 **Md** (258)	102 **No** (259)

For each element the average atomic mass of the mixture of isotopes occurring in nature is shown. For elements having no stable isotope, the approximate atomic mass of the longest-lived isotope is shown in parentheses. For elements that have been predicted but not yet confirmed, no atomic mass is given. All atomic masses are expressed in atomic mass units ($1\ u = 1.660538782(83) \times 10^{-27}$ kg), equivalent to grams per mole (g/mol).

APPENDIX E

UNIT CONVERSION FACTORS

Length

$1 \text{ m} = 100 \text{ cm} = 1000 \text{ mm} = 10^6 \text{ } \mu\text{m} = 10^9 \text{ nm}$

$1 \text{ km} = 1000 \text{ m} = 0.6214 \text{ mi}$

$1 \text{ m} = 3.281 \text{ ft} = 39.37 \text{ in.}$

$1 \text{ cm} = 0.3937 \text{ in.}$

$1 \text{ in.} = 2.540 \text{ cm}$

$1 \text{ ft} = 30.48 \text{ cm}$

$1 \text{ yd} = 91.44 \text{ cm}$

$1 \text{ mi} = 5280 \text{ ft} = 1.609 \text{ km}$

$1 \text{ Å} = 10^{-10} \text{ m} = 10^{-8} \text{ cm} = 10^{-1} \text{ nm}$

$1 \text{ nautical mile} = 6080 \text{ ft}$

$1 \text{ light year} = 9.461 \times 10^{15} \text{ m}$

Area

$1 \text{ cm}^2 = 0.155 \text{ in.}^2$

$1 \text{ m}^2 = 10^4 \text{ cm}^2 = 10.76 \text{ ft}^2$

$1 \text{ in.}^2 = 6.452 \text{ cm}^2$

$1 \text{ ft}^2 = 144 \text{ in.}^2 = 0.0929 \text{ m}^2$

Volume

$1 \text{ liter} = 1000 \text{ cm}^3 = 10^{-3} \text{ m}^3 = 0.03531 \text{ ft}^3 = 61.02 \text{ in.}^3$

$1 \text{ ft}^3 = 0.02832 \text{ m}^3 = 28.32 \text{ liters} = 7.477 \text{ gallons}$

$1 \text{ gallon} = 3.788 \text{ liters}$

Time

$1 \text{ min} = 60 \text{ s}$

$1 \text{ h} = 3600 \text{ s}$

$1 \text{ d} = 86{,}400 \text{ s}$

$1 \text{ y} = 365.24 \text{ d} = 3.156 \times 10^7 \text{ s}$

Angle

$1 \text{ rad} = 57.30° = 180°/\pi$

$1° = 0.01745 \text{ rad} = \pi/180 \text{ rad}$

$1 \text{ revolution} = 360° = 2\pi \text{ rad}$

$1 \text{ rev/min (rpm)} = 0.1047 \text{ rad/s}$

Speed

$1 \text{ m/s} = 3.281 \text{ ft/s}$

$1 \text{ ft/s} = 0.3048 \text{ m/s}$

$1 \text{ mi/min} = 60 \text{ mi/h} = 88 \text{ ft/s}$

$1 \text{ km/h} = 0.2778 \text{ m/s} = 0.6214 \text{ mi/h}$

$1 \text{ mi/h} = 1.466 \text{ ft/s} = 0.4470 \text{ m/s} = 1.609 \text{ km/h}$

$1 \text{ furlong/fortnight} = 1.662 \times 10^{-4} \text{ m/s}$

Acceleration

$1 \text{ m/s}^2 = 100 \text{ cm/s}^2 = 3.281 \text{ ft/s}^2$

$1 \text{ cm/s}^2 = 0.01 \text{ m/s}^2 = 0.03281 \text{ ft/s}^2$

$1 \text{ ft/s}^2 = 0.3048 \text{ m/s}^2 = 30.48 \text{ cm/s}^2$

$1 \text{ mi/h} \cdot \text{s} = 1.467 \text{ ft/s}^2$

Mass

$1 \text{ kg} = 10^3 \text{ g} = 0.0685 \text{ slug}$

$1 \text{ g} = 6.85 \times 10^{-5} \text{ slug}$

$1 \text{ slug} = 14.59 \text{ kg}$

$1 \text{ u} = 1.661 \times 10^{-27} \text{ kg}$

$1 \text{ kg has a weight of } 2.205 \text{ lb when } g = 9.80 \text{ m/s}^2$

Force

$1 \text{ N} = 10^5 \text{ dyn} = 0.2248 \text{ lb}$

$1 \text{ lb} = 4.448 \text{ N} = 4.448 \times 10^5 \text{ dyn}$

Pressure

$1 \text{ Pa} = 1 \text{ N/m}^2 = 1.450 \times 10^{-4} \text{ lb/in.}^2 = 0.209 \text{ lb/ft}^2$

$1 \text{ bar} = 10^5 \text{ Pa}$

$1 \text{ lb/in.}^2 = 6895 \text{ Pa}$

$1 \text{ lb/ft}^2 = 47.88 \text{ Pa}$

$1 \text{ atm} = 1.013 \times 10^5 \text{ Pa} = 1.013 \text{ bar}$
$= 14.7 \text{ lb/in.}^2 = 2117 \text{ lb/ft}^2$

$1 \text{ mm Hg} = 1 \text{ torr} = 133.3 \text{ Pa}$

Energy

$1 \text{ J} = 10^7 \text{ ergs} = 0.239 \text{ cal}$

$1 \text{ cal} = 4.186 \text{ J (based on } 15° \text{ calorie)}$

$1 \text{ ft} \cdot \text{lb} = 1.356 \text{ J}$

$1 \text{ Btu} = 1055 \text{ J} = 252 \text{ cal} = 778 \text{ ft} \cdot \text{lb}$

$1 \text{ eV} = 1.602 \times 10^{-19} \text{ J}$

$1 \text{ kWh} = 3.600 \times 10^6 \text{ J}$

Mass–Energy Equivalence

$1 \text{ kg} \leftrightarrow 8.988 \times 10^{16} \text{ J}$

$1 \text{ u} \leftrightarrow 931.5 \text{ MeV}$

$1 \text{ eV} \leftrightarrow 1.074 \times 10^{-9} \text{ u}$

Power

$1 \text{ W} = 1 \text{ J/s}$

$1 \text{ hp} = 746 \text{ W} = 550 \text{ ft} \cdot \text{lb/s}$

$1 \text{ Btu/h} = 0.293 \text{ W}$

APPENDIX F

NUMERICAL CONSTANTS

Fundamental Physical Constants*

Name	Symbol	Value
Speed of light in vacuum	c	2.99792458×10^8 m/s
Magnitude of charge of electron	e	$1.602176487(40) \times 10^{-19}$ C
Gravitational constant	G	$6.67428(67) \times 10^{-11}$ N·m^2/kg^2
Planck's constant	h	$6.62606896(33) \times 10^{-34}$ J·s
Boltzmann constant	k	$1.3806504(24) \times 10^{-23}$ J/K
Avogadro's number	N_A	$6.02214179(30) \times 10^{23}$ molecules/mol
Gas constant	R	$8.314472(15)$ J/mol·K
Mass of electron	m_e	$9.10938215(45) \times 10^{-31}$ kg
Mass of proton	m_p	$1.672621637(83) \times 10^{-27}$ kg
Mass of neutron	m_n	$1.674927211(84) \times 10^{-27}$ kg
Permeability of free space	μ_0	$4\pi \times 10^{-7}$ Wb/A·m
Permittivity of free space	$\epsilon_0 = 1/\mu_0 c^2$	$8.854187817\ldots \times 10^{-12}$ C^2/N·m^2
	$1/4\pi\epsilon_0$	$8.987551787\ldots \times 10^9$ N·m^2/C^2

Other Useful Constants*

Name	Symbol	Value
Mechanical equivalent of heat		4.186 J/cal (15° calorie)
Standard atmospheric pressure	1 atm	1.01325×10^5 Pa
Absolute zero	0 K	-273.15°C
Electron volt	1 eV	$1.602176487(40) \times 10^{-19}$ J
Atomic mass unit	1 u	$1.660538782(83) \times 10^{-27}$ kg
Electron rest energy	$m_e c^2$	0.510998910(13) MeV
Volume of ideal gas (0°C and 1 atm)		22.413996(39) liter/mol
Acceleration due to gravity (standard)	g	9.80665 m/s^2

*Source: National Institute of Standards and Technology (**http://physics.nist.gov/cuu**). Numbers in parentheses show the uncertainty in the final digits of the main number; for example, the number 1.6454(21) means 1.6454 ± 0.0021. Values shown without uncertainties are exact.

Astronomical Data[†]

Body	Mass (kg)	Radius (m)	Orbit radius (m)	Orbit period
Sun	1.99×10^{30}	6.96×10^{8}	—	—
Moon	7.35×10^{22}	1.74×10^{6}	3.84×10^{8}	27.3 d
Mercury	3.30×10^{23}	2.44×10^{6}	5.79×10^{10}	88.0 d
Venus	4.87×10^{24}	6.05×10^{6}	1.08×10^{11}	224.7 d
Earth	5.97×10^{24}	6.38×10^{6}	1.50×10^{11}	365.3 d
Mars	6.42×10^{23}	3.40×10^{6}	2.28×10^{11}	687.0 d
Jupiter	1.90×10^{27}	6.91×10^{7}	7.78×10^{11}	11.86 y
Saturn	5.68×10^{26}	6.03×10^{7}	1.43×10^{12}	29.45 y
Uranus	8.68×10^{25}	2.56×10^{7}	2.87×10^{12}	84.02 y
Neptune	1.02×10^{26}	2.48×10^{7}	4.50×10^{12}	164.8 y
Pluto[‡]	1.31×10^{22}	1.15×10^{6}	5.91×10^{12}	247.9 y

[†]Source: NASA Jet Propulsion Laboratory Solar System Dynamics Group (**http://ssd.jpl.nasa.gov**), and P. Kenneth Seidelmann, ed., ***Explanatory Supplement to the Astronomical Almanac*** (University Science Books, Mill Valley, CA, 1992), pp. 704–706. For each body, "radius" is its radius at its equator and "orbit radius" is its average distance from the sun or (for the moon) from the earth.

[‡]In August 2006, the International Astronomical Union reclassified Pluto and other small objects that orbit the sun as "dwarf planets."

Prefixes for Powers of 10

Power of ten	Prefix	Abbreviation	Pronunciation
10^{-24}	yocto-	y	*yoc*-toe
10^{-21}	zepto-	z	*zep*-toe
10^{-18}	atto-	a	*at*-toe
10^{-15}	femto-	f	*fem*-toe
10^{-12}	pico-	p	*pee*-koe
10^{-9}	nano-	n	*nan*-oe
10^{-6}	micro-	μ	*my*-crow
10^{-3}	milli-	m	*mil*-i
10^{-2}	centi-	c	*cen*-ti
10^{3}	kilo-	k	*kil*-oe
10^{6}	mega-	M	*meg*-a
10^{9}	giga-	G	*jig*-a or *gig*-a
10^{12}	tera-	T	*ter*-a
10^{15}	peta-	P	*pet*-a
10^{18}	exa-	E	*ex*-a
10^{21}	zetta-	Z	*zet*-a
10^{24}	yotta-	Y	*yot*-a

Examples:

1 femtometer = 1 fm = 10^{-15} m

1 picosecond = 1 ps = 10^{-12} s

1 nanocoulomb = 1 nC = 10^{-9} C

1 microkelvin = 1μK = 10^{-6} K

1 millivolt = 1 mV = 10^{-3} V

1 kilopascal = 1 kPa = 10^{3} Pa

1 megawatt = 1 MW = 10^{6} W

1 gigahertz = 1 GHz = 10^{9} Hz

ANSWERS TO ODD-NUMBERED PROBLEMS

Chapter 1

1.1 a) 1.61 km b) 3.28×10^3 ft
1.3 1.02 ns
1.5 5.36 L
1.7 31.7 y
1.9 a) 23.4 km/L b) 1.4 tanks
1.11 9.0 cm
1.13 a) $1.1 \times 10^{-3}\%$ b) no
1.15 0.45%
1.17 a) no b) no c) no d) no e) no
1.19 $\approx 10^6$
1.21 $\approx 4 \times 10^8$
1.23 $\approx \$70$ million
1.25 $\approx 10^4$
1.27 7.8 km, 38° north of east
1.29 144 m, 41° south of west
1.31 $A_x = 0, A_y = -8.00$ m, $B_x = 7.50$ m,
$B_y = 13.0$ m, $C_x = -10.9$ m, $C_y = -5.07$ m,
$D_x = -7.99$ m, $D_y = 6.02$ m
1.33 a) -8.12 m b) 15.3 m
1.35 a) 9.01 m, 33.7° b) 9.01 m, 33.7°
c) 22.3 m, 250° d) 22.3 m, 70.3°
1.37 3.39 km, 31.1° north of west
1.39 a) 2.48 cm, 18.4° b) 4.09 cm, 83.7°
c) 4.09 cm, 264°
1.41 $\vec{A} = -(8.00\,\text{m})\hat{\jmath}$,
$\vec{B} = (7.50\,\text{m})\hat{\imath} + (13.0\,\text{m})\hat{\jmath}$,
$\vec{C} = (-10.9\,\text{m})\hat{\imath} + (-5.07\,\text{m})\hat{\jmath}$,
$\vec{D} = (-7.99)\hat{\imath} + (6.02\,\text{m})\hat{\jmath}$
1.43 a) $\vec{A} = (1.23\,\text{m})\hat{\imath} + (3.38\,\text{m})\hat{\jmath}$,
$\vec{B} = (-2.08\,\text{m})\hat{\imath} + (-1.20\,\text{m})\hat{\jmath}$
b) $\vec{C} = (12.0\,\text{m})\hat{\imath} + (14.9\,\text{m})\hat{\jmath}$
c) 19.2 m, 51.2°
1.45 a) -104 m² b) -148 m² c) 40.6 m²
1.47 a) 165° b) 28° c) 90°
1.49 a) -63.9 m²$\hat{k}$ b) 63.9 m²$\hat{k}$
1.51 a) -6.62 m² b) 5.55 m²$\hat{k}$
1.53 a) $A = 5.38, B = 4.36$
b) $-5.00\hat{\imath} + 2.00\hat{\jmath} + 7.00\hat{k}$ c) 8.83, yes
1.55 a) 1.64×10^4 km b) $2.57r_E$
1.57 a) 2200 g b) 2.1 m
1.59 a) (2.8 ± 0.3) cm³ b) 170 ± 20
1.61 $\approx 6 \times 10^{27}$
1.63 $\$9 \times 10^{14}, \$3 \times 10^6$ per person
1.65 196 N, 392 N, 57.7° east of north; 360 N,
720 N, 57.7° east of south
1.67 b) $A_x = 3.03$ cm, $A_y = 8.10$ cm c) 8.65 cm,
69.5°
1.69 144 m, 41° south of west
1.71 954 N, 16.8° above the forward direction
1.73 3.30 N
1.75 a) 45.5 N b) 139°
1.77 a) $(87, 258)$ b) 136, 25° below straight left
1.79 160 N, 13° below horizontal
1.81 911 m, 8.9° west of south
1.83 29.6 m, 18.6° east of south
1.85 26.2 m, 34.2° east of south
1.87 124°
1.89 170 m²
1.91 a) 54.7° b) 35.3°
1.93 28.0 m
1.95 $C_x = 8.0, C_y = 6.1$
1.97 b) 72.2
1.99 38.5 yd, 24.6° to the right of downfield
1.101 a) 76.2 ly b) 129°

Chapter 2

2.1 25.0 m
2.3 1 hr 10 min
2.5 a) 0.312 m/s b) 1.56 m/s
2.7 a) 12.0 m/s
b) (i) 0 (ii) 15.0 m/s (iii) 12.0 m/s
c) 13.3 m/s
2.9 a) 2.33 m/s, 2.33 m/s
b) 2.33 m/s, 0.33 m/s

2.11 6.7 m/s, 6.7 m/s, 0, -40.0 m/s, -40.0 m/s,
-40.0 m/s, 0
2.13 a) no
b) (i) 12.8 m/s² (ii) 3.50 m/s²
(iii) 0.718 m/s²
2.15 a) 2.00 cm/s, 50.0 cm, -0.125 cm/s²
b) 16.0 s c) 32.0 s
d) 6.20 s, 1.23 cm/s; 25.8 s,
-1.23 cm/s; 36.4 s, -2.55 cm/s
2.17 a) 0.500 m/s² b) 0, 1.00 m/s²
2.19 a) 5.0 m/s b) 1.43 m/s²
2.21 a) 675 m/s² b) 0.0667 s
2.23 1.70 m
2.25 38 cm
2.27 a) 3.1×10^6 m/s² $= 3.2g$
b) 1.6 ms c) no
2.29 a) (i) 5.59 m/s² (ii) 7.74 m/s²
b) (i) 179 m (ii) 1.28×10^4 m
2.31 a) 0, 6.3 m/s², -11.2 m/s²
b) 100 m, 230 m, 320 m
2.33 a) 20.5 m/s² upward, 3.8 m/s² upward,
53.0 m/s² upward
b) 722 km
2.35 a) 2.94 m/s b) 0.600 s
2.37 1.67 s
2.39 a) 33.5 m b) 15.8 m/s
2.41 a) $t = \sqrt{2d/g}$ b) 0.190 s
2.43 a) 646 m b) 16.4 s, 112 m/s
2.45 a) 249 m/s² b) 25.4
c) 101 m d) no (if a is constant)
2.47 0.0868 m/s²
2.49 a) 3.3 s b) $9H$
2.51 a) 467 m b) 110 m/s
2.53 a) $v_x = (0.75\ \text{m/s}^3)t^2 - (0.040\ \text{m/s}^4)t^3$,
$x = (0.25\ \text{m/s}^3)t^3 - (0.010\ \text{m/s}^4)t^4$
b) 39.1 m/s
2.55 a) 10.0 m
b) (i) 8.33 m/s (ii) 9.09 m/s (iii) 9.52 m/s
2.57 a) 0.627 s, 1.59 s
c) negative at 0.627 s, positive at 1.59 s
d) 1.11 s e) 2.45 m f) 2.00 s, 0
2.59 250 km
2.61 a) 197 m/s b) 169 m/s
2.63 a) 82 km/h b) 31 km/h
2.65 a) 3.5 m/s² b) 0 c) 1.5 m/s²
2.67 a) 92.0 m b) 92.0 m
2.69 50.0 m
2.71 4.6 m/s²
2.73 a) 6.17 s b) 24.8 m
c) auto: 21.0 m/s, truck: 13.0 m/s
2.75 a) 7.85 cm/s b) 5.00 cm/s
2.77 a) 15.9 s b) 393 m c) 29.5 m/s
2.79 a) -4.00 m/s b) 12.0 m/s
2.81 a) $2.64H$ b) $2.64T$
2.83 a) no
b) yes, 14.4 m/s, not physically attainable
2.85 a) 6.69 m/s b) 4.49 m c) 1.42 s
2.87 a) 7.7 m/s b) 0.78 s c) 0.59 s d) 1.3 m
2.89 a) 380 m b) 184 m
2.91 a) 20.5 m/s b) yes
2.93 a) 945 m b) 393 m
2.95 a) car A b) 2.27 s, 5.73 s
c) 1.00 s, 4.33 s d) 2.67 s
2.97 a) 9.55 s, 47.8 m
b) 1.62 m/s d) 8.38 m/s
e) no
f) 3.69 m/s, 21.7 s, 80.0 m
2.99 a) 8.18 m/s b) (i) 0.411 m (ii) 1.15 km
c) 9.80 m/s d) 4.90 m/s

Chapter 3

3.1 a) 1.4 m/s, -1.3 m/s
b) 1.9 m/s, 317°
3.3 a) 7.1 cm/s, 45°
b) 5.0 cm/s, 90°; 7.1 cm/s, 45°;
11 cm/s, 27°

3.5 b) -8.67 m/s², -2.33 m/s²
c) 8.98 m/s², 195°
3.7 b) $\vec{v} = \alpha\hat{\imath} - 2\beta t\hat{\jmath}, \vec{a} = -2\beta\hat{\jmath}$
c) 5.4 m/s, 297°; 2.4 m/s², 270°
d) speeding up and turning right
3.9 a) 0.600 m b) 0.385 m
c) $v_x = 1.10$ m/s, $v_y = -3.43$ m/s, 3.60 m/s,
72.2° below the horizontal
3.11 3.32 m
3.13 a) 30.6 m/s b) 36.3 m/s
3.15 1.28 m/s²
3.17 a) 0.683 s, 2.99 s
b) 24.0 m/s, 11.3 m/s; 24.0 m/s, -11.3 m/s
c) 30.0 m/s, 36.9° below the horizontal
3.19 a) 1.5 m b) -0.89 m/s
3.21 a) 13.6 m b) 34.6 m/s c) 103 m
3.23 a) 296 m b) 176 m c) 198 m
d) (i) $v_x = 15.0$ m/s, $v_y = -58.8$ m/s
(ii) $v_x = 15.0$ m/s, $v_y = -78.8$ m/s
3.25 a) 0.034 m/s² $= 0.0034g$ b) 1.4 h
3.27 140 m/s $= 310$ mph
3.29 a) 3.50 m/s² upward
b) 3.50 m/s² downward
c) 12.6 s
3.31 a) 14 s b) 70 s
3.33 0.36 m/s, 52.5° south of west
3.35 a) 4.7 m/s, 25° south of east
b) 190 s c) 380 m
3.37 b) -7.1 m/s, -42 m/s
c) 43 m/s, 9.6° west of south
3.39 a) 24° west of south b) 5.5 h
3.41 a) $A = 0, B = 2.00$ m/s², $C = 50.0$ m,
$D = 0.500$ m/s³
b) $\vec{v} = 0, \vec{a} = (4.00\ \text{m/s}^2)\hat{\imath}$
c) $v_x = 40.0$ m/s, $v_y = 150$ m/s, 155 m/s
d) $\vec{r} = (200\ \text{m})\hat{\imath} + (550\ \text{m})\hat{\jmath}$
3.43 $2b/3c$
3.45 4.41 s
3.47 a) 123 m
b) 280 m
3.49 22 m/s
3.51 31 m/s
3.53 274 m
3.55 795 m
3.57 33.7 m
3.59 a) 42.8 m/s b) 42.0 m
3.61 a) $\sqrt{2gh}$ b) 30.0° c) $6.93h$
3.63 a) 1.50 m/s
b) 4.66 m
3.65 a) 6.91 m c) no
3.67 a) 17.8 m/s
b) in the river, 28.4 m horizontally from his
launch point
3.69 a) 81.6 m b) in the cart
c) 245 m d) 53.1°
3.71 a) 49.5 m/s b) 50 m
3.73 a) 2000 m b) 2180 m
3.75 $\pm 25.4°$
3.77 61.2 km/h, 140 km/h
3.79 b) $v_x = R\omega(1 - \cos\omega t)$,
$v_y = R\omega\sin\omega t; a_x = R\omega^2\sin\omega t$,
$a_y = R\omega^2\cos\omega t$ c) $t = 0, 2\pi/\omega, 4\pi/\omega, \dots$;
$x = 0, 2\pi R, 4\pi R, \dots$;
$y = 0; a = R\omega^2$ in the $+y$-direction d) no
3.81 a) 44.7 km/h, 26.6° west of south
b) 10.5° north of west
3.83 7.39 m/s, 12.4° north of east
3.85 a) 0.659 s b) (i) 9.09 m/s (ii) 6.46 m/s
c) 3.00 m, 2.13 m
3.87 a) 49.3°, 17.5° for level ground b) $-17.0°$
3.89 a) 1.5 km/h b) 3.5 km/h

Chapter 4

4.1 a) 0° b) 90° c) 180°
4.3 3.15 N
4.5 494 N, 31.8°

4.7 46.7 N, opposite to the motion of the skater
4.9 16.0 kg
4.11 a) 3.12 m, 3.12 m/s
b) 21.9 m, 6.24 m/s
4.13 a) 45.0 N, between 2.0 s and 4.0 s
b) between 2.0 s and 4.0 s
c) 0 s, 6.0 s
4.15 a) $A = 100$ N, $B = 12.5$ N/s^2
b) (i) 21.6 N, 2.70 m/s^2 (ii) 134 N, 16.8 m/s^2
c) 26.6 m/s^2
4.17 2940 N
4.19 a) 4.49 kg b) 4.49 kg, 8.13 N
4.21 825 N, blocks
4.23 20 N
4.25 7.4×10^{-23} m/s^2
4.27 b) yes
4.29 a) yes b) no
4.31 b) 142 N
4.33 2.03 s
4.35 1840 N, 135°
4.37 a) 17 N, 90° clockwise from the $+x$-axis
b) 840 N
4.39 a) 4.85 m/s
b) 16.2 m/s^2 upward
c) 1470 N upward (on him), 2360 N downward (on ground)
4.41 a) 153 N
4.43 a) 2.50 m/s^2 b) 10.0 N
c) to the right, $F > T$
d) 25.0 N
4.45 a) 4.4 m b) 300 m/s
c) (i) 2.7×10^4 N (ii) 9.0×10^3 N
4.47 a) $T > mg$
b) 79.6 N
4.49 b) 0.049 N c) 410mg
4.51 a) 7.79 m/s
b) 50.6 m/s^2 upward
c) 4530 N upward, 6.16mg
4.53 a) w b) 0 c) $w/2$
4.55 b) 1395 N
4.57 a) 4.34 kg
b) 5.30 kg
4.59 $F_x(t) = -6mBt$
4.61 7.78 m

Chapter 5

5.1 a) 25.0 N b) 50.0 N
5.3 a) 990 N, 735 N
b) 926 N
5.5 48°
5.7 a) $T_A = 0.732w$, $T_B = 0.897w$, $T_C = w$
b) $T_A = 2.73w$, $T_B = 3.35w$, $T_C = w$
5.9 a) 337 N b) 343 N
5.11 a) 1.10×10^8 N b) $5w$
c) 8.4 s
5.13 a) 4610 m/s^2 = 470g
b) 9.70×10^5 N = 471w
c) 0.0187 s
5.15 a) 2.96 m/s^2 c) 191 N
5.17 b) 2.50 m/s^2 c) 1.37 kg
d) 0.75mg
5.19 a) 0.832 m/s^2 b) 17.3 s
5.21 a) 3.4 m/s (c) 2.2w
5.23 a) 14.0 m
b) 18.0 m/s
5.25 50°
5.27 a) 22 N b) 3.1 m
5.29 a) 0.710, 0.472 b) 258 N
c) (i) 51.8 N (ii) 4.97 m/s^2
5.31 a) 57.1 N
b) 146 N up the ramp
5.33 a) 52.5 m b) 16.0 m/s
5.35 a) $\mu_k(m_A + m_B)g$ b) $\mu_k m_A g$
5.37 a) 0.218 m/s
b) 11.7 N
5.39 a) $\dfrac{\mu_k mg}{\cos\theta - \mu_k \sin\theta}$
b) $1/\tan\theta$
5.41 a) 0.44 kg/m b) 42 m/s
5.43 a) 3.61 m/s b) bottom c) 3.33 m/s
5.45 a) 21.0°, no b) 11,800 N; 23,600 N
5.47 1410 N, 8370 N

5.49 a) 1.5 rev/min
b) 0.92 rev/min
5.51 a) 38.3 m/s = 138 km/h
b) 3580 N
5.53 2.42 m/s
5.55 a) 1.73 m/s^2
c) 0.0115 N upward
d) 0.0098 N
5.57 a) rope making 60° angle b) 6400 N
5.59 a) 470 N b) 163 N
5.61 762 N
5.63 a) (i) -3.80 m/s (ii) 24.6 m/s
b) 4.36 m c) 2.45 s
5.65 a) 11.4 N b) 2.57 kg
5.67 10.4 kg
5.69 0.0259 (low pressure), 0.00505 (high pressure)
5.71 a) $m_1(\sin\alpha + \mu_k \cos\alpha)$
b) $m_1(\sin\alpha - \mu_k \cos\alpha)$
c) $m_1(\sin\alpha - \mu_s \cos\alpha) \leq m_2 \leq$
$m_1(\sin\alpha + \mu_s \cos\alpha)$
5.73 a) 1.80 N b) 2.52 N
5.75 a) 1.3×10^{-4} N = 62mg
b) 2.9×10^{-4} N, at $t = 1.2$ ms
c) 1.2 m/s
5.77 920 N
5.79 a) 11.5 m/s b) 7.54 m/s
5.81 0.40
5.83 a) $g\,\dfrac{m_B + m_{\text{rope}}(d/L)}{m_A + m_B + m_{\text{rope}}}$, increase
b) 0.63 m
c) will not move for any value of d
5.85 a) 88.0 N northward
b) 78 N southward
5.87 a) 294 N, 152 N, 152 N b) 40.0 N
5.89 3.0 N
5.91 a) 12.9 kg
b) $T_{AB} = 47.2$ N, $T_{BC} = 101$ N
5.93 $a_1 = \dfrac{2m_2 g}{4m_1 + m_2}$, $a_2 = \dfrac{m_2 g}{4m_1 + m_2}$
5.95 1.46 m above the floor
5.97 g/μ_s
5.99 b) 0.452
5.101 0.34
5.103 b) 8.8 N (c) 31.0 N
d) 1.54 m/s^2
5.105 a) moves up
b) remains constant
c) remains constant
d) slows down at the same rate as the monkey
5.107 a) 6.00 m/s^2
b) 3.80 m/s^2
c) 7.36 m/s
d) 8.18 m/s
e) 7.78 m, 6.29 m/s, 1.38 m/s^2
f) 3.14 s
5.109 a) 0.015, 0.36 N·s^2/m^2
b) 29 m/s
c) $v/v_t = \sqrt{\sin\beta - (0.015)\cos\beta}$
5.111 a) $v_y(t) = v_0 e^{-kt/m} + v_t(1 - e^{-kt/m})$
5.113 a) 120 N
b) 3.79 m/s
5.115 b) 0.28 c) no
5.117 a) right b) 120 m
5.119 a) 81.1° b) no
c) The bead rides at the bottom of the hoop.
5.121 a) $F = \dfrac{\mu_k w}{\cos\theta + \mu_k \sin\theta}$
c) $\theta = \tan^{-1}(\mu_k)$, 14.0°
5.123 $F = (M + m)g \tan\alpha$
5.125 a) $g\,\dfrac{-4m_1 m_2 + m_2 m_3 + m_1 m_3}{4m_1 m_2 + m_2 m_3 + m_1 m_3}$
b) $a_B = -a_3$
c) $g\,\dfrac{4m_1 m_2 - 3m_2 m_3 + m_1 m_3}{4m_1 m_2 + m_2 m_3 + m_1 m_3}$
d) $g\,\dfrac{4m_1 m_2 - 3m_1 m_3 + m_2 m_3}{4m_1 m_2 + m_2 m_3 + m_1 m_3}$
e) $g\,\dfrac{4m_1 m_2 m_3}{4m_1 m_2 + m_2 m_3 + m_1 m_3}$

f) $g\,\dfrac{8m_1 m_2 m_3}{4m_1 m_2 + m_2 m_3 + m_1 m_3}$
g) All the accelerations are zero, $T_A = m_2 g$,
$T_C = 2m_2 g$.
5.127 $\cos^2\beta$

Chapter 6

6.1 a) 3.60 J b) -0.900 J
c) 0 d) 0
e) 2.70 J
6.3 a) 74 N b) 333 J
c) -330 J d) 0, 0
e) 0
6.5 a) -1750 J b) no
6.7 a) (i) 9.00 J (ii) -9.00 J
b) (i) 0 (ii) 9.00 J (iii) -9.00 J (iv) 0
c) zero for each block
6.9 a) (i) 0 (ii) 0
b) (i) 0 (ii) -25.1 J
6.11 a) 324 J b) -324 J
c) 0
6.13 a) 36,000 J b) 4
6.15 a) 1.0×10^{16} J b) 2.4
6.17 a) 7.50 N b) (i) 9.00 J (ii) 5.40 J
c) 14.4 J, same d) 2.97 m/s
6.19 a) 43.2 m/s b) 101 m/s
c) 5.80 m d) 3.53 m/s
e) 7.35 m
6.21 $\sqrt{2gh(1 + \mu_k/\tan\alpha)}$
6.23 32.0 N
6.25 a) 4.48 m/s b) 3.61 m/s
6.27 a) 4.96 m/s b) 1.43 m/s^2, 4.96 m/s
6.29 a) $\dfrac{v_0^2}{2\mu_k g}$ b) (i) $^1/_2$ (ii) 4 (iii) 2
6.31 a) 40.0 N/m b) 0.456 N
6.33 b) 14.4 cm, 13.6 cm, 12.8 cm
6.35 a) 2.83 m/s b) 3.46 m/s
6.37 8.5 cm
6.39 a) 1.76 b) 0.666 m/s
6.41 a) 4.0 J b) 0
c) -1.0 J d) 3.0 J
e) -1.0 J
6.43 a) 2.83 m/s b) 2.40 m/s
6.45 a) 0.0565 m b) 0.57 J, no
6.47 8.17 m/s
6.49 a) 360,000 J b) 100 m/s
6.51 $3.9 \times 10^{13}P$
6.53 745 W $\approx$ 1 hp
6.55 a) 84.6/min b) 22.7/min
6.57 29.6 kW
6.59 0.20 W
6.61 877 J
6.63 a) 532 J b) -315 J
c) 0 d) -202 J
e) 15 J f) 1.2 m/s
6.65 a) 987 J b) 3.02 s
6.67 a) 2.59×10^{12} J b) 4800 J
6.69 a) 1.8 m/s = 4.0 mph
b) 180 m/s^2 $\approx$ 18g, 900 N
6.71 a) $k\left(\dfrac{1}{x_2} - \dfrac{1}{x_1}\right)$, negative
b) $k\left(\dfrac{1}{x_1} - \dfrac{1}{x_2}\right)$, positive
c) same magnitude but opposite signs because the net work is zero
6.73 a) 5.11 m b) 0.304
c) 10.3 m
6.75 a) 0.11 N b) 7.1 N c) 0.33 J
6.77 a) 2.56 m/s b) 3.52 N c) 13.1 J
6.79 6.3×10^4 N/m
6.81 1.1 m
6.83 a) 1.02×10^4 N/m, 8.16 m
6.85 a) 0.600 m b) 1.50 m/s
6.87 0.786
6.89 1.3 m
6.91 a) 1.10×10^5 J b) 1.30×10^5 J
c) 3.99 kW
6.93 3.6 h
6.95 a) 1.26×10^5 J b) 1.46 W
6.97 a) 2.4 MW b) 61 MW c) 6.0 MW
6.99 a) 513 W b) 354 W c) 52.1 W

6.101 a) 358 N   b) 47.1 hp   c) 4.06 hp
  d) 2.03%

6.103 a) $\dfrac{Mv^2}{6}$ b) 6.1 m/s
  c) 3.9 m/s   d) 0.40 J, 0.60 J

Chapter 7

7.1 a) 6.6×10^5 J   b) -7.7×10^5 J
7.3 a) 820 N   b) (i) 0   (ii) 740 J
7.5 a) 24.0 m/s   b) 24.0 m/s   c) part (b)
7.7 a) 2.0 m/s   b) 9.8×10^{-7} J, 2.0 J/kg
  c) 200 m, 63 m/s   d) 5.9 J/kg
7.9 a) (i) 0   (ii) 0.98 J   b) 2.8 m/s
  c) only gravity is constant   d) 5.1 N
7.11 -5400 J
7.13 a) 880 J   b) -157 J   c) 470 J   d) 253 J
  e) 3.16 m/s², 7.11 m/s, 253 J
7.15 a) 80.0 J   b) 5.0 J
7.17 a) (i) $4U_0$ (ii) $U_0/4$
  b) (i) $x_0\sqrt{2}$ (ii) $x_0/\sqrt{2}$
7.19 a) 6.32 cm   b) 12 cm
7.21 ± 0.092 m
7.23 a) 3.03 m/s, as it leaves the spring
  b) 95.9 m/s², when the spring has its maximum
  compression
7.25 a) 4.46×10^5 N/m   b) 0.128 m
7.27 a) -308 J   b) -616 J
  c) nonconservative
7.29 a) -3.6 J   b) -3.6 J
  c) -7.2 J   d) nonconservative
7.31 a) -59 J   b) -42 J
  c) -59 J   d) nonconservative
7.33 a) 8.41 m/s   b) 638 J
7.35 2.46 N, $+x$-direction
7.37 130 m/s², 132° counterclockwise from the
  $+x$-axis
7.39 a) $F(r) = (12a/r^{13}) - (6b/r^7)$
  b) $(2a/b)^{1/6}$, stable   c) $b^2/4a$
  d) $a = 6.67 \times 10^{-138}$ J·m¹²,
  $b = 6.41 \times 10^{-78}$ J·m⁶
7.41 a) zero, 637 N   b) 2.99 m/s
7.43 0.41
7.45 a) 16.0 m/s   b) 11,500 N
7.47 a) 20.0 m along the rough bottom
  b) -78.4 J
7.49 a) 22.2 m/s   b) 16.4 m   c) no
7.51 0.602 m
7.53 15.5 m/s
7.55 4.4 m/s
7.57 a) no   b) yes, $150
7.59 a) 7.00 m/s   b) 8.82 N
7.61 a) $mg(1 - h/d)$ b) 441 N
  c) $\sqrt{2gh(1 - y/d)}$
7.63 48.2°
7.65 a) 0.392   b) -0.83 J
7.67 a) $U(x) = \frac{1}{2}\alpha x^2 + \frac{1}{3}\beta x^3$ b) 7.85 m/s
7.69 7.01 m/s
7.71 a) $\dfrac{m(g + a)^2}{2gh}$ b) $\dfrac{2gh}{g + a}$
7.73 a) 0.480 m/s   b) 0.566 m/s
7.75 a) 3.87 m/s   b) 0.10 m
7.77 0.456 N
7.79 a) 4.4×10^{12} J
  b) 2.7×10^3 m³, 9.0×10^{-4} m
7.81 119 J
7.83 a) -50.6 J   b) -67.5 J   c) nonconservative
7.85 b) 0, 3.38 J, 0, 0; 3.38 J   c) nonconservative
7.87 b) $v(x) = \sqrt{\dfrac{2\alpha}{mx_0^2}\left[\dfrac{x_0}{x} - \left(\dfrac{x_0}{x}\right)^2\right]}$
  c) $x = 2x_0$, $v = \sqrt{\dfrac{\alpha}{2mx_0^2}}$ d) 0
  e) $v(x) = \sqrt{\dfrac{2\alpha}{mx_0^2}\left[\dfrac{x_0}{x} - \left(\dfrac{x_0}{x}\right)^2 - \dfrac{2}{9}\right]}$
  f) first case: x_0, ∞; second case: $3x_0/2$, $3x_0$

Chapter 8

8.1 a) 1.20×10^5 kg·m/s
  b) (i) 60.0 m/s   (ii) 26.8 m/s
8.3 b) 0.526, baseball   c) 0.641, woman
8.5 a) 22.5 kg·m/s, to the left
  b) 838 J
8.7 562 N, not significant
8.9 a) 10.8 m/s, to the right
  b) 0.750 m/s, to the left
8.11 a) 500 N/s²   b) 5810 N·s
  c) 2.70 m/s
8.13 a) 2.50 N·s, in the direction of the force
  b) (i) 6.25 m/s, to the right   (ii) 3.75 m/s, to
  the right
8.15 0.593 kg·m/s
8.17 0.87 kg·m/s, in the same direction as the bullet
  is traveling
8.19 a) 6.79 m/s   b) 55.2 J
8.21 a) 0.790 m/s   b) -0.0023 J
8.23 1.53 m/s for both
8.25 a) 0.0559 m/s   b) 0.0313 m/s
8.27 a) 7.20 m/s, 38.0° from Rebecca's original
  direction   b) -680 J
8.29 a) 3.56 m/s
8.31 a) 29.3 m/s, 20.7 m/s   b) 19.6%
8.33 a) 0.846 m/s   b) 2.10 J
8.35 a) -1.4×10^{-6} km/h, no
  b) -6.7×10^{-8} km/h, no
8.37 5.9 m/s, 58° north of east
8.39 a) Both cars have the same magnitude
  momentum change, but the lighter car has a
  greater velocity change.
  b) $2.50\Delta v$ c) occupants of small car
8.41 19.5 m/s, 21.9 m/s
8.43 a) 2.93 cm   b) 866 J   c) 1.73 J
8.45 186 N
8.47 a) 3.33 J, 0.333 m/s   b) -1.33 m/s, 0.667 m/s
8.49 a) $v_1/3$ b) $K_1/9$ c) 10
8.51 (0.0444 m, 0.0556 m)
8.53 2520 km
8.55 0.700 m to the right and 0.700 m upward
8.57 0.47 m/s
8.59 $F_x = -(1.50\ \text{N/s})t$, $F_y = 0.25$ N, $F_z = 0$
8.61 a) 0.053 kg   b) 5.19 N
8.63 a) 0.442 m/s   b) 800 m/s   c) 530 m/s
8.65 45.2
8.67 a) 0.474 kg·m/s, upward
  b) 237 N, upward
8.69 a) -1.14 N·s, 0.330 N·s
  b) 0.04 m/s, 1.8 m/s
8.71 2.40 m/s, 3.12 m/s
8.73 a) 1.75 m/s, 0.260 m/s   b) -0.092 J
8.75 3.65×10^5 m/s
8.77 0.946 m
8.79 1.8 m
8.81 12 m/s, 21 m/s
8.83 a) 2.60 m/s   b) 325 m/s
8.85 a) 5.3 m/s   b) 5.7 m
8.87 53.7°
8.89 102 N
8.91 a) 0.125   b) 248 J   c) 0.441 J
8.93 b) $M = m$ c) zero
8.95 a) 9.35 m/s   b) 3.29 m/s
8.97 a) 3.56 m/s   b) 5.22 m/s   c) 4.66 m/s
8.99 13.6 m/s, 6.34 m/s, 65.0°
8.101 0.0544%
8.103 1.61×10^{-22} kg·m/s, to the left
8.105 1.33 m
8.107 0.400 m/s
8.109 a) 71.6 m/s, 14.3 m/s   b) 347 m
8.111 a) yes   b) decreases by 4800 J
8.113 a) $1.37v_{\text{ex}}$ b) $1.18v_{\text{ex}}$
  c) $2.38v_{\text{ex}}$ d) 2.94 km/s
8.115 b) $2L/3$

Chapter 9

9.1 a) 34.4°   b) 6.27 cm   c) 1.05 m
9.3 a) rad/s, rad/s³   b) (i) 0   (ii) 15.0 rad/s²
  c) 9.50 rad
9.5 a) $\omega_z = \gamma + 3\beta t^2$ b) 0.400 rad/s
  c) 1.30 rad/s, 0.700 rad/s

9.7 a) $\pi/4$ rad, 2.00 rad/s, -0.139 rad/s³   b) 0
  c) 19.5 rad, 9.36 rad/s
9.9 a) 2.25 rad/s   b) 4.69 rad
9.11 a) 24.0 s   b) 68.8 rev
9.13 10.5 rad/s
9.15 a) 300 rpm   b) 75.0 s, 312 rev
9.17 9.00 rev
9.19 a) 1.99×10^{-7} rad/s   b) 7.27×10^{-5} rad/s
  c) 2.98×10^4 m/s   d) 464 m/s
  e) 0.0337 m/s², 0
9.21 a) 15.1 m/s²   b) 15.1 m/s²
9.23 a) 0.180 m/s², 0, 0.180 m/s²
  b) 0.180 m/s², 0.377 m/s², 0.418 m/s²
  c) 0.180 m/s², 0.754 m/s², 0.775 m/s²
9.25 0.107 m, no
9.27 a) 0.831 m/s   b) 109 m/s²
9.29 a) 2.29   b) 1.51   c) 15.7 m/s, 108g
9.31 a) (i) 0.469 kg·m²   (ii) 0.117 kg·m²
  (iii) 0
  b) (i) 0.0433 kg·m²   (ii) 0.0722 kg·m²
  c) (i) 0.0288 kg·m²   (ii) 0.0144 kg·m²
9.33 a) 2.33 kg·m²   b) 7.33 kg·m²
  c) 0   d) 1.25 kg·m²
9.35 0.193 kg·m²
9.37 8.52 kg·m²
9.39 5.61 m/s
9.41 a) 3.15×10^{23} J   b) 158 y, no
9.43 0.600 kg·m²
9.45 7.35×10^4 J
9.47 a) 0.673 m   b) 45.5%
9.49 46.5 kg
9.51 a) f^5 b) 6.37×10^8 J
9.53 an axis that is parallel to a diameter and is
  0.516R from the center
9.55 $\dfrac{1}{3}M(a^2 + b^2)$
9.57 a) $ML^2/12$ b) $ML^2/12$
9.59 $^1/_2 MR^2$
9.61 a) 14.2 rad/s   b) 59.6 rad
9.63 9.41 m
9.65 a) 0.600 m/s³   b) $\alpha = (2.40\ \text{rad/s}^3)t$
  c) 3.54 s   d) 17.7 rad
9.67 a) 0.0333 rad/s²   b) 0.200 rad/s
  c) 2.40 m/s²   e) 3.12 m/s², 3.87 kN
  f) 50.2°
9.69 a) 1.70 m/s   b) 94.2 rad/s
9.71 2.99 cm
9.73 b) 1.50 m/s²   d) 0.208 kg·m²
9.75 a) 7.36 m   b) 327 m/s²
9.77 a) 2.14×10^{29} J   b) 2.66×10^{33} J
9.79 a) $Mb^2/6$ b) 182 J
9.81 a) -0.882 J
  b) 5.42 rad/s
  c) 5.42 m/s
  d) 5.42 m/s compared to 4.43 m/s
9.83 $\sqrt{\dfrac{2gd(m_B - \mu_k m_A)}{m_A + m_B + I/R^2}}$
9.85 $2\sqrt{g(1 - \cos\beta)/R}$
9.87 a) 2.25×10^{-3} kg·m²
  b) 3.40 m/s   c) 4.95 m/s
9.89 13.9 m
9.91 a) $\dfrac{247}{512}MR^2$ b) $\dfrac{383}{512}MR^2$
9.93 a) 1.05 rad/s   b) 5.0 J   c) 78.5 J   d) 6.4%
9.95 b) $\dfrac{1}{4}M(R_1^2 + R_2^2)$
9.97 a) $\dfrac{3}{5}MR^2$ b) larger
9.99 a) 55.3 kg   b) 0.804 kg·m²
9.101 a) $s(\theta) = r_0\theta + \dfrac{\beta}{2}\theta^2$
  b) $\theta(t) = \dfrac{1}{\beta}(\sqrt{r_0^2 + 2\beta vt} - r_0)$
  c) $\omega_z(t) = \dfrac{v}{\sqrt{r_0^2 + 2\beta vt}}$,
  $\alpha_z(t) = -\dfrac{\beta v^2}{(r_0^2 + 2\beta vt)^{3/2}}$, no
  d) 25.0 mm, 0.247 μm/rad, 2.13×10^4 rev

Chapter 10

10.1   a) 40.0 N·m, out of the page
     b) 34.6 N·m, out of the page
     c) 20.0 N·m, out of the page
     d) 17.3 N·m, into the page
     e) 0   f) 0
10.3   2.50 N·m, out of the page
10.5   b) $-\hat{k}$ c) $(-1.05\text{ N·m})\hat{k}$
10.7   a) 8.7 N·m counterclockwise, 0,
     5.0 N·m clockwise, 10.0 N·m clockwise
     b) 6.3 N·m clockwise
10.9   13.1 N·m
10.11 a) 14.8 rad/s² b) 1.52 s
10.13 a) 7.5 N, 18.2 N b) 0.016 kg·m²
10.15 0.255 kg·m²
10.17 a) 32.6 N, 35.4 N b) 2.72 m/s²
     c) 32.6 N, 55.0 N
10.19 a) 1.80 m/s b) 7.13 J
     c) (i) 3.60 m/s to the right (ii) 0
     (iii) 2.55 m/s at 45° below the horizontal
     d) (i) 1.80 m/s to the right (ii) 1.80 m/s to the
     left (iii) 1.80 m/s downward
10.21 a) 1/3 b) 2/7 c) 2/5 d) 5/13
10.23 a) 0.613 b) no c) no slipping
10.25 11.7 m
10.27 a) 3.76 m b) 8.58 m/s
10.29 a) 67.9 rad/s b) 8.35 J
10.31 a) 0.309 rad/s b) 100 J c) 6.67 W
10.33 a) 0.377 N·m b) 157 rad
     c) 59.2 J d) 59.2 J
10.35 a) 358 N·m b) 1790 N c) 83.8 m/s
10.37 a) 115 kg·m²/s into the page
     b) 125 kg·m²/s² out of the page
10.39 4.71×10^{-6} kg·m²/s
10.41 4600 rad/s
10.43 1.14 rev/s
10.45 a) 1.38 rad/s b) 1080 J, 495 J
10.47 a) 0.120 rad/s b) 3.20×10^{-4} J
     c) work done by bug
10.49 a) 5.88 rad/s
10.51 a) 1.71 rad/s
10.53 a) 1.62 N b) 1800 rev/min
10.55 a) halved b) doubled
     c) halved d) doubled e) unchanged
10.57 a) 67.6 N b) 62.9 N c) 3.27 s
10.59 0.483
10.61 7.47 N
10.63 a) 16.3 rad/s² b) decreases c) 5.70 rad/s
10.65 a) FR b) FR c) $\sqrt{4F/MR}$
     d) $2F/M$ e) $4F/M$
10.67 0.730 m/s², 6.08 rad/s², 36.3 N, 21.1 N
10.69 a) 293 N b) 16.2 rad/s²
10.71 a) 2.88 m/s² b) 6.13 m/s²
10.73 270 N
10.75 $a = \dfrac{2g}{2 + (R/b)^2}$, $\alpha = \dfrac{2g}{2b + R^2/b}$,

     $T = \dfrac{2mg}{2(b/R)^2 + 1}$
10.77 a) 1.41 s, 70.5 m/s b) t larger, v smaller
10.79 $\dfrac{3}{5}H_0$
10.81 29.0 m/s
10.83 a) 26.0 m/s b) unchanged
10.85 a) $\sqrt{20hy/7}$ b) no
     c) rolling friction d) $\sqrt{8hy/3}$
10.87 $g/3$
10.89 1.87 m
10.91 a) $\dfrac{6}{19}v/L$ b) 3/19
10.93 a) 5.46 rad/s b) 3.17 cm c) 1010 m/s
10.95 a) 2.00 rad/s b) 6.58 rad/s
10.97 0.30 rad/s clockwise
10.99 0.710 m
10.101 a) $a = \mu_k g$, $\alpha = \dfrac{2\mu_k g}{R}$

     b) $\dfrac{R^2\omega_0^2}{18\mu_k g}$ c) $-\dfrac{MR^2\omega_0^2}{6}$

10.103 a) $mv_1^2 r_1^2/r^3$
     b) $\dfrac{mv_1^2}{2}r_1^2\left(\dfrac{1}{r_2^2} - \dfrac{1}{r_1^2}\right)$

     c) same

Chapter 11

11.1   29.8 cm
11.3   1.35 m
11.5   5.45 kN
11.7   a) 1000 N, 0.800 m from the end where the
     600-N force is applied
     b) 800 N, 0.75 m from the end where the
     600-N force is applied
11.9   a) 550 N
     b) 0.614 m from A
11.11 a) 1920 N b) 1140 N
11.13 a) $T = 2.60w$; $3.28w$, 37.6°
     b) $T = 4.10w$; $5.39w$, 48.8°
11.15 272 N on each hand, 130 N on each foot
11.17 246 N, 0.34 m from the front feet
11.19 270 N, 303 N, 40°
11.21 a) 0.800 m b) clockwise
     c) 0.800 m, clockwise
11.23 b) 208 N
11.25 1.4 mm
11.27 2.0×10^{11} Pa
11.29 a) 3.1×10^{-3} (upper),
     2.0×10^{-3} (lower)
     b) 1.6 mm (upper), 1.0 mm (lower)
11.31 a) 150 atm b) 1.5 km, no
11.33 8.6°
11.35 4.8×10^9 Pa, 2.1×10^{-10} Pa⁻¹
11.37 b) 6.6×10^5 N c) 1.8 mm
11.39 3.41×10^7 Pa
11.41 10.2 m/s²
11.43 20.0 kg
11.45 a) 525 N b) 222 N, 328 N c) 1.48
11.47 tail: 600 N down, wing: 7300 N up
11.49 a) 140 N b) 6 cm to the right
11.51 a) 379 N b) 141 N
11.53 160 N to the right, 213 N upward
11.55 49.9 cm
11.57 a) 370 N b) when he starts to raise
     his leg c) no
11.59 a) $V = mg + w$, $H = T = \left(w + \dfrac{mg}{4}\right)\cot\theta$
     b) 926 N c) 6.00°
11.61 4900 N
11.63 b) 2000 N = $2.72mg$ c) 4.4 mm
11.65 a) 4.90 m b) 60 N
11.67 a) 175 N at each hand, 200 N at each foot
     b) 91 N at each hand and at each foot
11.69 a) 1150 N b) 1940 N
     c) 918 N d) 0.473
11.71 590 N (person above), 1370 N (person below);
     person above
11.73 a) $\dfrac{T_{max}hD}{L\sqrt{h^2 + D^2}}$

     b) positive, $\dfrac{T_{max}h}{L\sqrt{h^2 + D^2}}\left(1 - \dfrac{D^2}{h^2 + D^2}\right)$
11.75 a) 7140 N, tall walls b) 7900 N
11.77 a) 268 N b) 232 N
     c) 366 N
11.79 a) 0.424 N (A), 1.47 N (B), 0.424 N (C)
     b) 0.848 N
11.81 a) 27° to tip, 31° to slip, tips first
     b) 27° to tip, 22° to slip, slips first
11.83 a) 80 N (A), 870 N (B) b) 1.92 m
11.85 a) $T = 3700$ N, 2000 N upward
11.87 a) 0.36 mm b) 0.045 mm
     c) 0.33 mm
11.89 a) 0.54 cm b) 0.42 cm
11.91 a) 0.70 m from A b) 0.60 m from A
11.93 a) 1.63 m
     b) brass: 2.00×10^8 Pa, nickel: 4.00×10^8 Pa
     c) brass: 2.22×10^{-3}, nickel: 1.90×10^{-3}
11.95 0.0542 L
11.97 a) 600 N b) 13.5 kN

11.99 c) $F = \dfrac{\mu_s w}{\sin\theta - \mu_s\cos\theta}$ (to slide),

     $F = \dfrac{w}{(1/9)\cos\theta + 2\sin\theta}$ (to tip), 66°
11.99 $h^2/L + L/2$; L if $h > L/\sqrt{2}$
11.101 a) 0.66 mm b) 0.022 J c) 8.35×10^{-3} J
     d) -3.04×10^{-2} J e) 3.04×10^{-2} J

Chapter 12

12.1   41.8 N, no
12.3   7020 kg/m³, yes
12.5   1.6
12.7   61.6 N
12.9   a) 1.86×10^6 Pa b) 184 m
12.11 0.581 m
12.13 a) 1.92×10^4 Pa
     b) causes additional force on the walls of the
     blood vessels
12.15 2.8 m
12.17 6.0×10^4 Pa
12.19 2.27×10^5 N
12.21 a) 636 Pa b) (i) 1170 Pa (ii) 1170 Pa
12.23 10.9
12.25 0.107 m
12.27 6.43×10^{-4} m³, 2.78×10^3 kg/m³
12.29 a) $\rho < \rho_{fluid}$

     c) above: $1 - \dfrac{\rho}{\rho_{fluid}}$, submerged: $\dfrac{\rho}{\rho_{fluid}}$

     d) 32%
12.31 a) 116 Pa b) 921 Pa c) 0.822 kg, 822 kg/m³
12.33 1910 kg/m³
12.35 9.6 m/s
12.37 a) 17.0 m/s b) 0.317 m
12.39 0.956 m
12.41 28.4 m/s
12.43 1.47×10^5 Pa
12.45 2.03×10^4 Pa
12.47 2.25×10^5 Pa
12.49 $1.19D$
12.51 a) $(p_0 - p)\pi\dfrac{D^2}{4}$ b) 776 N
12.53 a) 5.9×10^5 N b) 1.8×10^5 N
12.55 c) independent of surface area
12.57 0.964 cm, rises
12.59 a) 1470 Pa b) 13.9 cm
12.61 9.8×10^6 kg, yes
12.63 a) 0.30 b) 0.70
12.65 3.50×10^{-4} m³, 3.95 kg
12.67 a) 8.27×10^3 m³ b) 83.8 kN
12.69 2.05 m
12.71 a) $H/2$ b) H
12.73 0.116 kg
12.75 33.4 N
12.77 b) 12.2 N c) 11.8 N
12.79 b) 2.52×10^{-4} m³, 0.124
12.81 5.57×10^{-4} m
12.83 a) $1 - \dfrac{\rho_B}{\rho_L}$ b) $\left(\dfrac{\rho_L - \rho_B}{\rho_L - \rho_w}\right)L$ c) 4.60 cm
12.85 a) al/g b) $\omega^2 l^2/2g$
12.89 a) $2\sqrt{h(H - h)}$ b) h
12.91 a) 0.200 m³/s b) 6.97×10^4 Pa
12.93 $3h_1$
12.95 a) $r = \dfrac{r_0\sqrt{v_0}}{(v_0^2 + 2gy)^{1/4}}$ b) 1.10 m

12.97 a) 80.4 N

Chapter 13

13.1   2.18
13.3   a) 1.2×10^{-11} m/s² b) 15 days
     c) increase
13.5 2.1×10^{-9} m/s², downward
13.7   a) 2.4×10^{-3} N
     b) $F_{moon}/F_{earth} = 3.5 \times 10^{-6}$
13.9   a) 0.634 m from $3m$
     b) (i) unstable (ii) stable

13.11 1.38×10^7 m
13.13 a) 0.37 m/s^2 b) 1700 kg/m^3
13.15 610 N, 735 N (on earth)
13.17 a) 5020 m/s b) $60{,}600$ m/s
13.19 a) 7460 m/s b) 1.68 h
13.21 6200 m/s
13.23 a) 4.7 m/s $= 11$ mph, easy to achieve
    b) 2.23 h
13.25 a) $82{,}700$ m/s b) 14.5 days
13.27 b) Pluto: 4.45×10^{12} m,
    Neptune: 4.55×10^{12} m c) 248 y
13.29 2.3×10^{30} kg $= 1.2 M_S$
13.31 a) (i) 5.31×10^{-9} N (ii) 2.67×10^{-9} N
13.33 a) $-\dfrac{GmM}{\sqrt{x^2 + a^2}}$ b) $-GmM/x$
    c) $\dfrac{GmMx}{(x^2 + a^2)^{3/2}}$, attractive d) GmM/x^2
    e) $U = -GMm/a, F_x = 0$
13.35 a) 53 N b) 52 N
13.37 a) 4.3×10^{37} kg $= 2.1 \times 10^7 M_S$
    b) no c) 6.32×10^{10} m, yes
13.39 a) 4.64×10^{11} m
    b) 6.26×10^{36} kg $= 3.15 \times 10^6 M_S$
    c) 9.28×10^9 m
13.41 9.16×10^{13} N
13.43 a) 9.67×10^{-12} N, at $45°$ above the $+x$-axis
    b) 3.02×10^{-5} m/s
13.45 a) 1.62 m/s^2 b) 0.69 N c) 4.2 N
13.47 a) 2.9×10^{15} kg, 0.0077 m/s^2 b) 6.2 m/s
13.49 b) (i) 1.49×10^{-5} m/s, 7.46×10^{-6} m/s
    (ii) 2.24×10^{-5} m/s c) 26.6 m
13.51 a) 3.59×10^7 m
13.53 177 m/s
13.55 a) 1.39×10^7 m b) 3.59×10^7 m
13.57 $(0.01)R_E = 6.4 \times 10^4$ m
13.59 1.83×10^{27} kg
13.61 0.28%
13.63 6060 km/h
13.65 $v_2 = \sqrt{\dfrac{2Gm_E h}{R_E(R_E + h)}}$
13.67 a) $13{,}700$ m/s b) $13{,}300$ m/s
    c) $13{,}200$ m/s
13.69 a) (i) 2.84 y (ii) 6.11 y
    b) 4.90×10^{11} m c) 4.22×10^{11} m
13.71 a) $GM^2/4R^2$
    b) $v = \sqrt{GM/4R}, T = 4\pi\sqrt{R^3/GM}$
    c) $GM^2/4R$
13.73 6.8×10^4 m/s
13.75 a) $12{,}700$ kg/m^3 (at $r = 0$), 3150 kg/m^3
    (at $r = R$)
13.77 a) 7910 s b) 1.53 c) 5510 m/s (apogee),
    8430 m/s (perigee) d) 2410 m/s (perigee),
    3250 m/s (apogee); perigee
13.79 5.36×10^9 J
13.81 9.36 m/s^2
13.83 $GmMx/(a^2 + x^2)^{3/2}$
13.85 a) $U(r) = \dfrac{Gm_E m}{2R_E^3}r^2$ b) 7.90×10^3 m/s
13.87 a) against the direction of motion in both cases
    b) 259 days c) $44.1°$
13.89 $\dfrac{2GMm}{a^2}\left(1 - \dfrac{x}{\sqrt{a^2 + x^2}}\right)$

Chapter 14

14.1 a) 2.15 ms, 2930 rad/s
    b) 2.00×10^4 Hz, 1.26×10^5 rad/s
    c) 4.3×10^{14} Hz $\le f \le 7.5 \times 10^{14}$ Hz;
    1.3×10^{-15} s $\le T \le 2.3 \times 10^{-15}$ s
    d) 2.0×10^{-7} s, 3.1×10^7 rad/s
14.3 5530 rad/s, 1.14 ms

14.5 0.0500 s
14.7 a) 0.167 s b) 37.7 rad/s c) 0.0844 kg
14.9 a) 0.150 s b) 0.0750 s
14.11 a) 0.98 m b) $\pi/2$ rad
    c) $x = (-0.98$ m$)\sin[(12.2$ rad/s$)t]$
14.13 a) -2.71 m/s^2
    b) $x = (1.46$ cm$)\cos[(15.7$ rad/s$)t$
      $+ 0.715$ rad$]$,
    $v_x = (-22.9$ cm/s$)\sin[(15.7$ rad/s$)t$
      $+ 0.715$ rad$]$,
    $a_x = (-359$ cm/s$^2)\cos[(15.7$ rad/s$)t$
      $+ 0.715$ rad$]$
14.15 120 kg
14.17 a) 0.253 kg b) 1.21 cm
    c) 3.03 N
14.19 a) 1.51 s b) 26.0 N/m
    c) 30.8 cm/s d) 1.92 N
    e) -0.0125 m, 30.4 cm/s, 0.216 m/s^2
    f) 0.324 N
14.21 a) $x = (0.0030$ m$)\cos[(2760$ rad/s$)t]$
    b) 8.3 m/s, 2.3×10^4 m/s^2
    c) $da_x/dt = (6.3 \times 10^7$ m/s$^3)$
    $\times \sin[(2760$ rad/s$)t]$, 6.3×10^7 m/s^3
14.23 127 m/s
14.25 a) 1.48 m/s b) 2.96×10^{-5} J
14.27 a) 1.20 m/s b) 1.11 m/s
    c) 36 m/s^2 d) 13.5 m/s^2 e) 0.36 J
14.29 $3M, {}^3\!/_4$
14.31 0.240 m
14.33 $A/2 \overline{}2$
14.35 a) 0.0778 m b) 1.28 Hz c) 0.624 m/s
14.37 a) 4.06 cm b) 1.21 m/s c) 29.8 rad/s
14.39 b) 23.9 cm, 1.45 Hz
14.41 a) 2.7×10^{-8} kg$\cdot$m^2
    b) 4.3×10^{-6} N$\cdot$m/rad
14.43 0.0512 kg$\cdot$m^2
14.45 a) 0.25 s b) 0.25 s
14.47 0.407 swings per second
14.49 10.7 m/s^2
14.51 a) 2.84 s b) 2.89 s
    c) 2.89 s, -2%
14.53 0.129 kg$\cdot$m^2
14.55 $A: 2\pi\sqrt{\dfrac{L}{g}}, B: \dfrac{2\sqrt{2}}{3}\left(2\pi\sqrt{\dfrac{L}{g}}\right)$, pendulum A
14.57 $A: 2\pi\sqrt{\dfrac{L}{g}}, B: \sqrt{\dfrac{11}{10}}\left(2\pi\sqrt{\dfrac{L}{g}}\right)$, pendulum B
14.59 a) 0.393 Hz b) 1.73 kg/s
14.61 a) A b) $-Ab/2m$
    c) $A\left(\dfrac{b^2}{2m^2} - \dfrac{k}{m}\right)$; negative if $b < \sqrt{2km}$,
    zero if $b = \sqrt{2km}$, positive if $b > \sqrt{2km}$
14.63 a) kg/s
    c) (i) $5.0\dfrac{F_{max}}{k}$ (ii) $2.5\dfrac{F_{max}}{k}$
14.65 0.353 m
14.67 a) 1.11×10^4 m/s^2 b) 5.00×10^3 N
    c) 23.6 m/s, 125 J d) 37.5 kW
    e) 1.21×10^4 N, 36.7 m/s, 302 J, 141 kW
14.69 a) none of them change
    b) $^1\!/_4$ as great c) $^1\!/_2$ as great
    d) $1/2 \overline{}5$ as great
    e) potential energy is the same, kinetic energy
    is $1/5$ as great
14.71 a) 24.4 cm b) 0.221 s c) 1.19 m/s
14.73 a) 0.373 Hz, 0.426 m, 2.68 s b) 1.34 s
14.75 2.00 m
14.77 a) 0.107 m b) 2.42 s
14.79 $(0.921)\left(\dfrac{1}{2\pi}\sqrt{\dfrac{g}{L}}\right)$
14.81 a) 1.49 s b) -2.12×10^{-4} s per s, shorter
    c) 0.795 s

14.83 a) 0.150 m/s b) 0.112 m/s^2 downward
    c) 0.700 s d) 4.38 m
14.85 a) 2.6 m/s b) 0.21 m c) 0.49 s
14.87 9.08×10^{24} kg
14.89 1.17 s
14.91 0.505 s
14.93 c) -7.57×10^{-19} J e) 8.39×10^{12} Hz
14.95 0.705 Hz, $14.5°$
14.97 $2\pi\sqrt{\dfrac{M}{3k}}$
14.99 $\dfrac{1}{4\pi}\sqrt{\dfrac{6g}{\sqrt{2}L}}$
14.101 a) $k_1 + k_2$ b) $k_1 + k_2$
    c) $\dfrac{k_1 k_2}{k_1 + k_2}$ d) $\sqrt{2}$
14.103 a) $Mv^2/6$ c) $\omega = \sqrt{\dfrac{3k}{M}}, M' = M/3$

Chapter 15

15.1 a) 0.439 m, 1.28 ms
    b) 0.219 m
15.3 220 m/s $= 800$ km/h
15.5 a) 1.7 cm to 17 m
    b) 4.3×10^{14} Hz to 7.5×10^{14} Hz
    c) 1.5 cm d) 6.4 cm
15.7 a) 25.0 Hz, 0.0400 s, 19.6 rad/m
    b) $y(x, t) = (0.0700$ m$)\cos[(19.6$ m$^{-1})x +$
    $(157$ rad/s$)t]$ c) 4.95 cm
    d) 0.0050 s
15.9 a) yes b) yes c) no
    d) $v_y = \omega A\cos(kx + \omega t)$,
    $a_y = -\omega^2 A\sin(kx + \omega t)$
15.11 a) 4 mm b) 0.040 s c) 0.14 m, 3.5 m/s
    d) 0.24 m, 6.0 m/s e) no
15.13 b) $+x$-direction
15.15 a) 16.3 m/s b) 0.136 m
    c) both increase by a factor of $2 \overline{}2$
15.17 0.337 kg
15.19 a) 18.6 N b) 29.1 m/s
15.21 a) 10.0 m/s b) 0.250 m
    c) $y(x, t) = (3.00$ cm$)\cos[\pi(8.00$ rad/m$)x -$
    $(80.0\pi$ rad/s$)t]$ d) 1890 m/s^2 e) yes
15.23 4.51 mm
15.25 a) 95 km b) 0.25 μW/m^2
    c) 110 kW
15.27 a) 0.050 W/m^2 b) 22 kJ
15.29 9.48×10^{27} W
15.37 a) $(1.33$ m$)n, n = 0, 1, 2, \dots$
    b) $(1.33$ m$)(n + {}^1\!/_2), n = 0, 1, 2, \dots$
15.41 a) 96.0 m/s b) 461 N
    c) 1.13 m/s, 426 m/s^2
15.43 b) 2.80 cm c) 277 cm
    d) 185 m/s, 7.96 Hz, 0.126 s, 1470 cm/s
    e) 280 cm/s
    f) $y(x, t) = (5.60$ cm$) \times$
    $\sin[(0.0906$ rad/cm$)x]\sin[(133$ rad/s$)t]$
15.45 a) $y(x, t) = (4.60$ mm$) \times$
    $\sin[(6.98$ rad/m$)x]\sin[(742$ rad/s$)t]$
    b) 3rd harmonic c) 39.4 Hz
15.47 a) 45.0 cm b) no
15.49 a) 311 m/s b) 246 Hz
    c) 245 Hz, 1.40 m
15.51 a) 20.0 Hz, 126 rad/s, 3.49 rad/m
    b) $y(x, t) = (2.50 \times 10^{-3}$ m$) \times$
    $\cos[(3.49$ rad/m$)x - (126$ rad/s$)t]$
    c) $y(0, t) = (2.50 \times 10^{-3}$ m$) \times$
    $\cos[(126$ rad/s$)t]$
    d) $y(1.35$ m$, t) = (2.50 \times 10^{-3}$ m$) \times$
    $\cos[(126$ rad/s$)t - 3\pi/2$ rad$]$
    e) 0.315 m/s
    f) -2.50×10^{-3} m, 0
15.53 a) $\dfrac{7L}{2}\sqrt{\dfrac{\mu_1}{F}}$ b) no

15.55 a) $\dfrac{2\pi A}{\lambda}\sqrt{\dfrac{FL}{M}}$ b) increase F by a factor of 4

15.57 a) $\dfrac{4\pi^2 F\Delta x}{\lambda^2}$

15.59 32.4 Hz

15.61 1.83 m

15.63 330 Hz (copper), 447 Hz (aluminum)

15.65 c) C/B

15.67 b) ω must be decreased by a factor of $1/\sqrt{2}$, k must be decreased by a factor of $1/\sqrt{8}$

15.69 a) 7.07 cm   b) 0.400 kW

15.71 d) $P(x,t) = -Fk\omega A^2\sin^2(kx+\omega t)$

15.73 $(0.800\ \text{Hz})n, n = 1, 2, 3, \ldots$

15.75 c) $2A, 2A\omega, 2A\omega^2$

15.77 233 N

15.79 a) $0, L$ b) $0, L/2, L$ d) no

15.81 1780 kg/m³

15.83 a) $r = 0.640$ mm, $L = 0.40$ m   b) 380 Hz

15.85 b) $u_k = \dfrac{1}{2}\mu\omega^2 A^2\sin^2(kx-\omega t)$

e) $u_p = \dfrac{1}{2}Fk^2 A^2\sin^2(kx-\omega t)$

Chapter 16

16.1 a) 0.344 m   b) 1.2×10^{-5} m   c) 6.9 m, 50 Hz

16.3 a) 7.78 Pa   b) 77.8 Pa   c) 778 Pa

16.5 a) 90 m   b) 102 kHz   c) 1.4 cm   d) 4.4 mm to 8.8 mm   e) 6.2 MHz

16.7 90.8 m

16.9 81.4°C

16.11 0.208 s

16.13 a) 5.5×10^{-15} J   b) 0.074 mm/s

16.15 a) 9.44×10^{-11} m, 0.434 m   b) 5.66×10^{-9} m, 0.100 m

16.17 a) 1.95 Pa   b) 4.58×10^{-3} W/m²   c) 96.6 dB

16.19 a) 4.4×10^{-12} W/m²   b) 6.4 dB   c) 5.8×10^{-11} m

16.21 14.0 dB

16.23 a) 2.0×10^{-7} W/m²   b) 6.0 m   c) 290 m

16.25 a) *fundamental:* displacement node at 0.60 m, pressure nodes at 0 and 1.20 m; *first overtone:* displacement nodes at 0.30 m and 0.90 m, pressure nodes at 0, 0.60 m, 1.20 m; *second overtone:* displacement nodes at 0.20 m, 0.60 m, 1.00 m, pressure nodes at 0, 0.40 m, 0.80 m, 1.20 m   b) *fundamental:* displacement node at 0, pressure node at 1.20 m; *first overtone:* displacement nodes at 0 and 0.80 m, pressure nodes at 0.40 m and 1.20 m; *second overtone:* displacement nodes at 0.48 m, 0.96 m, pressure nodes at 0.24 m, 0.72 m, 1.20 m

16.27 506 Hz, 1517 Hz, 2529 Hz

16.29 a) 767 Hz   b) no

16.31 a) 614 Hz   b) 1230 Hz

16.33 a) 172 Hz   b) 86 Hz

16.35 0.125 m

16.37 destructive

16.39 a) 433 Hz   b) loosen

16.41 1.3 Hz

16.43 780 m/s

16.45 a) 375 Hz   b) 371 Hz   c) 4 Hz

16.47 a) 0.25 m/s   b) 0.91 m

16.49 19.8 m/s

16.51 a) 1910 Hz   b) 0.188 m

16.53 $0.0950c$, toward us

16.55 a) 36.0°   b) 2.23 s

16.57 b) 0.68%

16.59 a) 1.00   b) 8.00   c) 4.73×10^{-8} m = 47.3 nm

16.61 b) $3f_0$

16.63 flute harmonic $3N$ resonates with string harmonic $4N$, $N = 1, 3, 5, \ldots$

16.65 a) stopped   b) 7th and 9th   c) 0.439 m

16.67 a) 375 m/s   b) 1.39   c) 0.8 cm

16.69 1.27

16.71 a) 548 Hz   b) 652 Hz

16.73 a) 2186 Hz, 0.157 m   b) 2920 Hz, 0.118 m   c) 734 Hz

16.75 a) 0.0674 m   b) 147 Hz

16.77 b) 2.0 m/s

16.79 a) 1.2×10^6 m/s   b) 3.6×10^{16} m = 3.8 ly   c) 5200 y, about 4100 BCE

16.81 a) $f_0\left(\dfrac{2v_w}{v - v_w}\right)$ b) $f_0\left(\dfrac{2v_w}{v + v_w}\right)$

16.83 d) 9.69 cm/s, 667 m/s²

Chapter 17

17.1 a) -81.0°F   b) 134.1°F   c) 88.0°F

17.3 a) 27.2 C°   b) -55.6 C°

17.5 a) -18.0 F°   b) -10.0 C°

17.7 0.964 atm

17.9 a) -282°C   b) 47,600 Pa, no

17.11 0.39 m

17.13 Death Valley: 1.9014 cm, Greenland: 1.8964 cm

17.15 0.26 mm

17.17 49.4°C

17.19 1.7×10^{-5} (C°)$^{-1}$

17.21 a) 1.431 cm²   b) 1.436 cm²

17.23 a) 3.2×10^{-5} (C°)$^{-1}$ b) 2.6×10^9 Pa

17.25 a) 5.0 mm   b) -8.4×10^7 Pa

17.27 5.79×10^5 J

17.29 240 J/kg·K

17.31 23 min

17.33 a) -1.54 kJ   b) 0.0121 C°

17.35 45.2 C°

17.37 0.0613 C°

17.39 a) 215 J/kg·K   b) water   c) too small

17.41 27.5°C

17.43 a) 5.9 C°   b) yes; 109°F

17.45 150°C

17.47 7.6 min

17.49 36.4 kJ, 8.70 kcal, 34.5 Btu

17.51 357 m/s

17.53 3.45 L

17.55 5.05×10^{15} kg

17.57 0.0674 kg

17.59 2.10 kg

17.61 190 g

17.63 a) 222 K/m   b) 10.7 W   c) 73.3°C

17.65 a) -5.8°C   b) 11 W/m²

17.67 4.0×10^{-3} W/m·C°

17.69 105.5°C

17.71 a) 21 kW   b) 6.4 kW

17.73 2.1 cm²

17.75 a) 1.61×10^{11} m   b) 5.43×10^6 m

17.77 a) 35.1°M   b) 39.6 C°

17.79 53.3°C

17.81 35.0°C

17.83 23.0 cm, 7.0 cm

17.85 b) 1.9×10^8 Pa

17.87 a) 99.4 N   c) -4.2 Hz, falls

17.89 a) 87°C   b) -80°C

17.91 20.2°C

17.93 a) 54.3

17.95 a) 83.6 J   b) 1.86 J/mol·K   c) 5.60 J/mol·K

17.97 a) 2.70×10^7 J   b) 6.89 C°   c) 19.3 C°

17.99 2.5 cm

17.101 a) 86.1°C   b) no ice, no steam, 0.130 kg liquid water

17.103 a) 100°C   b) 0.0214 kg steam, 0.219 kg liquid water

17.105 1.743 kg

17.107 a) 93.9 W   b) 1.35

17.109 2.9

17.111 c) 170 h   d) 1.5×10^{10} s ≈ 500 y, no

17.113 0.106 W/m·C°

17.115 5.82 g

17.117 a) 1.04 kW   b) 87.1 W   c) 1.13 kW   d) 28 g   e) 1.1 bottles

17.119 a) 69.6°C

17.121 1.76 C°

17.123 b) 0°C   d) 3140 C°/m   e) 121 W   f) zero

g) 1.1×10^{-4} m²/s   h) -11 C°/s   i) 9.17 s   j) decrease   k) -7.71 C°/s

17.125 a) 103°C   b) 27 W

17.127 a) (i) 280 W   (ii) 0.248 W   (iii) 2.10 kW   (iv) 116 W; radiation from the sun   b) 3.72 L/h   c) 1.4 L/h

Chapter 18

18.1 a) 0.122 mol   b) 14,700 Pa, 0.145 atm

18.3 0.100 atm

18.5 a) 0.0136 kg/m³ (Mars), 67.6 kg/m³ (Venus), 5.39 kg/m³ (Titan)

18.7 503°C

18.9 16.8 kPa

18.11 0.159 L

18.13 $0.0508V$

18.15 a) 70.2°C   b) yes

18.17 850 m

18.19 a) 6.95×10^{-16} kg   b) 2.32×10^{-13} kg/m³

18.21 22.8 kPa

18.23 a) $8720   b) 3.88 cm

18.25 a) 8.2×10^{-17} atm   b) no

18.27 55.6 mol, 3.35×10^{25} molecules

18.29 a) 9.00×10^{-5} m³   b) 3.1×10^{-10} m   c) about the same

18.31 b) 1.004

18.33 (d) must be true, the others could be true

18.35 a) 1.93×10^6 m/s, no   b) 7.3×10^{10} K

18.37 a) 6.21×10^{-21} J   b) 2.34×10^5 m²/s²   c) 484 m/s   d) 2.57×10^{-23} kg·m/s   e) 1.24×10^{-19} N   f) 1.24×10^{-17} Pa   g) 8.17×10^{21} molecules   h) 2.45×10^{22} molecules

18.39 3800°C

18.41 a) 2600 J   b) 1560 J

18.43 a) 741 J/kg·K, $c_w = 5.65c_{N_2}$ b) 5.65 kg; 4.85 m³

18.45 a) 923 J/kg·K   b) The value calculated is too large by about 1.4%.

18.47 a) 337 m/s   b) 380 m/s   c) 412 m/s

18.49 a) 610 Pa   b) 22.12 MPa

18.51 no, no

18.53 a) 11.8 kPa   b) 0.566 L

18.55 272°C

18.57 0.213 kg

18.59 a) -179°C   b) 1.2×10^{26} molecules/m³   c) The atmosphere of Titan is 4.8 times denser than that of the earth.

18.61 1.92 atm

18.63 a) 30.7 cylinders   b) 8420 N   c) 7800 N

18.65 a) 26.2 m/s   b) 16.1 m/s, 5.44 m/s   c) 1.74 m

18.67 ≈ 5×10^{27} atoms

18.69 a) A b) C c) 4250°C   d) B

18.71 a) 4.65×10^{-26} kg   b) 6.11×10^{-21} J   c) 2.04×10^{24} molecules   d) 12.5 kJ

18.73 b) r_2 c) $r_1 = \dfrac{R_0}{2^{1/6}}, r_2 = R_0, 2^{-1/6}$ d) U_0

18.75 a) 517 m/s   b) 298 m/s

18.77 b) 1.40×10^5 K (N), 1.01×10^4 K (H)   c) 6370 K (N), 459 K (H)

18.79 a) 1.24×10^{-14} kg   b) 4.16×10^{11} molecules   c) 2.95 μm, no

18.81 a) $2R = 16.6$ J/mol·K   b) less

18.83 CO_2: 20.79 J/mol·K, 27%; SO_2: 24.94 J/mol·K, 21%; H_2S: 24.94 J/mol·K, 3.9%

18.85 $3kT/m$

18.87 b) $0.0421N$ c) $2.94 \times 10^{-21}N$ d) $0.0297N$, $2.08 \times 10^{-21}N$ e) $0.0595N$, $4.15 \times 10^{-21}N$

18.89 42.6%

18.91 a) 4.5×10^{11} m   b) 703 m/s, 6.4×10^8 s (≈20 y)   c) 1.4×10^{-14} Pa   d) 650 m/s, evaporate   f) 2×10^5 K, >3 times the temperature of the sun, no

Chapter 19

19.1 b) 1330 J
19.3 b) -6180 J
19.5 a) 0.942 atm
19.7 a) $(p_1 - p_2)(V_2 - V_1)$ b) Negative of work done in reverse direction
19.9 a) 34.7 kJ   b) 80.4 kJ   c) no
19.11 a) 278 K   b) 0, 162 J   c) 53 J
19.13 a) 16.4 min   b) 139 m/s = 501 km/h
19.15 a) 0   b) $T_b = 2T_a$ c) $U_b = U_a + 700$ J
19.17 a) positive   b) $W_I > 0, W_{II} < 0$
   c) into the system   d) into the system for loop I, out of the system for loop II
19.19 b) 208 J   c) on the piston   d) 712 J
   e) 920 J   f) 208 J
19.21 a) 948 K   b) 900 K
19.23 2/5
19.25 a) 25.0 K   b) 17.9 K   c) higher for (a)
19.27 a) -605 J   b) 0   c) liberates 605 J
19.29 a) 476 kPa   b) -10.6 kJ   c) 1.59, heated
19.31 5.05 kJ, internal energy and temperature both increase
19.33 b) 224 J   c) -224 J
19.35 11.6°C
19.37 a) 600 J out of the gas
   b) -1500 J, decreases
19.39 a) increases   b) 4800 J
19.41 a) 45.0 J   b) liberates 65.0 J
   c) 23.0 J, 22.0 J
19.43 a) the same   b) absorbs 4.0 kJ   c) 8.0 kJ
19.45 b) -2460 J
19.47 a) 0.80 L   b) 305 K, 1220 K, 1220 K
   c) ab: 76 J, into the gas;
   ca: -107 J, out of the gas
   bc: 56 J, into the gas
   d) ab: 76 J, increase
   bc: 0, no change
   ca: -76 J, decrease
19.49 a) 3.00 kJ, into the gas
   b) 2.00 kJ, into the gas   c) $Q_a > Q_b$
19.51 a) 899°C   b) 12.2 kJ
   c) 42.6 kJ   d) 45.6 kJ
19.53 -0.226 m³
19.55 a) 4.32×10^{-4} m³   b) 648 J   c) 715 kJ
   d) 715 kJ   e) no substantial difference
19.57 3.4×10^5 J/kg
19.59 b) 11.9 C°
19.61 a) 0.173 m   b) 207°C   c) 74.7 kJ
19.63 a) $Q = 300$ J, $\Delta U = 0$
   b) $Q = 0, \Delta U = -300$ J
   c) $Q = 750$ J, $\Delta U = 450$ J
19.65 a) $W = 738$ J, $Q = 2590$ J, $\Delta U = 1850$ J
   b) $W = 0, Q = -1850$ J, $\Delta U = -1850$ J
   c) $\Delta U = 0$
19.67 a) $W = -187$ J, $Q = -654$ J, $\Delta U = -467$ J
   b) $W = 113$ J, $Q = 0$, $\Delta U = -113$ J
   c) $W = 0, Q = 580$ J, $\Delta U = 580$ J
19.69 a) $p_0 + \dfrac{mg}{\pi r^2}$

   b) $-\left(\dfrac{y}{h}\right)(p_0\pi r^2 + mg)$

   c) $\dfrac{1}{2\pi}\sqrt{\dfrac{g}{h}\left(1 + \dfrac{p_0\pi r^2}{mg}\right)}$, no

Chapter 20

20.1 a) 6500 J   b) 34%
20.3 a) 23%   b) 12,400 J
   c) 0.350 g   d) 222 kW = 298 hp
20.5 a) 12.3 atm   b) 5470 J (ca)   c) 3723 J (bc)
   d) 1747 J   e) 31.9%
20.7 a) 58%   b) 1.4%
20.9 a) 16.2 kJ   b) 50.2 kJ
20.11 1.7 h
20.13 a) 215 J   b) 378 K   c) 39.0%
20.15 a) 42.4 kJ   b) 441°C
20.17 a) 492 J   b) 212 W   c) 5.4
20.19 4.5 kJ
20.21 37.1 hp

20.23 a) 429 J/K   b) -393 J/k   c) 36 J/K
20.25 a) irreversible   b) 1250 J/K
20.27 -6.31 J/K
20.29 a) 6.05 kJ/K
   b) about five times greater for vaporization
20.31 a) 33.3 J/K   b) irreversible
20.33 a) no   b) 18.3 J/K   c) 18.3 J/K
20.35 a) 121 J   b) 3800 cycles
20.37 a) 33 J   b) 117 J   c) 45°C   d) 0   e) 96.2 g
20.39 -5.8 J/K, decrease
20.41 b) absorbed during bc, rejected during ab and ca c) $T_a = T_b = 241$ K, $T_c = 481$ K
   d) 610 J, 610 J   e) 8.7%
20.43 a) 21.0 kJ (enters), 16.6 kJ (leaves)
   b) 4.4 kJ, 21%   c) 67%
20.45 a) 7.0%   b) 3.0 MW, 2.8 MW
   c) 6×10^5 kg/h $= 6 \times 10^5$ L/h
20.47 a) 2.00 atm, 4.00 L; 2.00 atm, 6.00 L; 1.11 atm, 6.00 L; 1.67 atm, 4.00 L
   b) $1 \rightarrow 2$: 1422 J, 405 J; $2 \rightarrow 3$: -1355 J, 0;
   $3 \rightarrow 4$: -274 J, -274 J; $4 \rightarrow 1$: 339 J, 0
   c) 131 J   d) 7.44%, 44.4%
20.49 a) 2.26%   b) 29.4 J (gravitational), 1.30 kJ
   c) 1.11×10^{-3} candy bars
20.51 a) ab: 225 kJ, 90 kJ, 135 kJ;
   bc: -240 kJ, 0, -240 kJ;
   ca: 45 kJ, -60 kJ, 105 kJ
   b) 30 kJ, 30 kJ, 0   c) 11.1%
20.53 $1 - T_C/T_H$
20.55 a) 122 J, -78 J   b) 5.10×10^{-4} m³
   c) at b: 2.32 MPa, 4.81×10^{-5} m³, 771 K;
   at c: 4.01 MPa, 4.81×10^{-5} m³, 1332 K;
   at d: 0.147 MPa, 5.10×10^{-4} m³, 518 K
   d) 61.1%, 77.5%
20.57 a) 357 kJ, 62 kJ   c) 385 kJ, 34 kJ
20.59 a) $nC_V \ln{(T_c/T_b)}, nC_V \ln{(T_a/T_d)}$ b) zero
20.61 a) -107 J/K   b) 147 J/K   c) 0
   d) 39.4 J/K

Chapter 21

21.1 a) 2.00×10^{10} b) 8.59×10^{-13}
21.3 $2.1 \times 10^{28}, -3.35 \times 10^9$ C
21.5 1.3 nC
21.7 3.7 km
21.9 a) 0.742 µC on each   b) 0.371 µC, 1.48 µC
21.11 1.43×10^{13}, away from each other
21.13 a) 2.21×10^4 m/s²
21.15 0.750 nC
21.17 1.8×10^{-4} N, in the $+x$-direction
21.19 $x = -0.144$ m
21.21 2.58 µN, in the $-y$-direction
21.23 a) 8.80×10^{-9} N, attractive
   b) 8.22×10^{-8} N, about 10 times larger than the bonding force
21.25 a) 4.40×10^{-16} N   b) 2.63×10^{11} m/s²
   c) 2.63×10^5 m/s
21.27 a) 3.30×10^6 N/C, to the left   b) 14.2 ns
   c) 1.80×10^3 N/C, to the right
21.29 a) -21.9 µC   b) 1.02×10^{-7} N/C
21.31 a) 8740 N/C, to the right
   b) 6540 N/C, to the right   c) 1.40×10^{-15} N, to the right
21.33 a) 364 N/C   b) no, 2.73 µm, downward
21.35 1.79×10^6 m/s
21.37 1.73×10^{-8} N, toward a point midway between the two electrons
21.39 a) $-\hat{\jmath}$ b) $\dfrac{\sqrt{2}}{2}\hat{\imath} + \dfrac{\sqrt{2}}{2}\hat{\jmath}$
   c) $-0.39\hat{\imath} + 0.92\hat{\jmath}$
21.41 a) 633 km/s   b) 15.9 km/s
21.43 a) 0
   b) for $|x| < a$: $E_x = -\dfrac{q}{\pi\epsilon_0}\dfrac{ax}{(x^2 - a^2)^2}$;
   for $x > a$: $E_x = \dfrac{q}{2\pi\epsilon_0}\dfrac{x^2 + a^2}{(x^2 - a^2)^2}$;

   for $x < -a$: $E_x = -\dfrac{q}{2\pi\epsilon_0}\dfrac{x^2 + a^2}{(x^2 - a^2)^2}$
21.45 a) (i) 574 N/C, $+x$-direction   (ii) 268 N/C, $-x$-direction   (iii) 404 N/C, $-x$-direction
   b) (i) 9.20×10^{-17} N, $-x$-direction
   (ii) 4.30×10^{-17} N, $+x$-direction
   (iii) 6.48×10^{-17} N, $+x$-direction
21.47 1.04×10^7 N/C, toward the -2.00-µC charge
21.49 a) $E_x = E_y = E = 0$ b) $E_x = 2660$ N/C, $E_y = 0, E = 2660$ N/C, $-x$-direction
   c) $E_x = 129$ N/C, $E_y = 510$ N/C, $E = 526$ N/C, 284° counterclockwise from the $+x$-axis   d) $E_x = 0, E_y = 1380$ N/C, $E = 1380$ N/C, $+y$-direction
21.51 a) $E_x = -4790$ N/C, $E_y = 0, E = 4790$ N/C, $-x$-direction   b) $E_x = 2130$ N/C, $E_y = 0, E = 2130$ N/C, $+x$-direction   c) $E_x = -129$ N/C, $E_y = -164$ N/C, $E = 209$ N/C, 232° counterclockwise from the $+x$-axis
   d) $E_x = -1040$ N/C, $E_y = 0, E = 1040$ N/C, $-x$-direction
21.53 a) $(7.0 \text{ N/C})\hat{\imath}$ b) $(1.75 \times 10^{-5} \text{ N})\hat{\imath}$
21.55 a) 1.14×10^5 N/C, toward the center of the disk   b) 8.92×10^4 N/C, toward the center of the disk   c) 1.46×10^5 N/C, toward the charge
21.57 a) 1.4×10^{-11} C · m, from q_1 toward q_2
   b) 860 N/C
21.59 a) $\vec{p}$ is aligned in either the same or the opposite direction as $\vec{E}$.   b) $\vec{p}$ is aligned in the same direction as $\vec{E}$.
21.61 a) 1680 N, from the $+5.00$-µC charge toward the -5.00-µC charge
   b) 22.3 N · m, clockwise
21.63 b) $\dfrac{Q^2}{8\pi\epsilon_0 L^2}(1 + 2\sqrt{2})$, away from the center of the square
21.65 a) 3.17 nC   b) $+x$-direction   c) $x = -1.76$ m
21.67 a) $\dfrac{1}{2\pi}\sqrt{\dfrac{qQ}{\pi\epsilon_0 ma^3}}$ b) accelerates away from the origin along the y-axis
21.69 b) 2.80 µC   c) 39.5°
21.71 a) 2.09×10^{21} N   b) 5.90×10^{23} m/s²
   c) no
21.73 3.41×10^4 N/C, to the left
21.75 between the charges, 0.24 m from the 0.500-nC charge
21.77 a) $\dfrac{6q^2}{4\pi\epsilon_0 L^2}$, away from the vacant corner
   b) $\dfrac{3q^2}{4\pi\epsilon_0 L^2}\left(\sqrt{2} + \dfrac{1}{2}\right)$, toward the center of the square
21.79 a) 6.0×10^{23}
   b) 510 kN (electric), 4.1×10^{-31} N (gravitation)
   c) yes for electric force, no for gravitation
21.81 a) ≈ 480 C   b) 8.3×10^{13} N, repulsive
21.83 2190 km/s
21.85 a) 3.5×10^{20} b) 1.6 C, 2.3×10^{10} N
21.87 a) $\dfrac{mv_0^2 \sin^2\alpha}{2eE}$ b) $\dfrac{mv_0^2 \sin 2\alpha}{eE}$
   d) 0.418 m, 2.89 m
21.89 a) $E_x = \dfrac{Q}{4\pi\epsilon_0 a}\left(\dfrac{1}{x - a} - \dfrac{1}{x}\right), E_y = 0$
   b) $\dfrac{qQ}{4\pi\epsilon_0 a}\left(\dfrac{1}{r} - \dfrac{1}{r + a}\right)\hat{\imath}$
21.91 a) $(-6110 \text{ N/C})\hat{\imath}$ b) smaller   c) 0.177 m

21.93 a) 1.56 N/C, $+x$-direction   c) smaller
d) 4.7%

21.95 a) $-\dfrac{qQ}{2\pi\epsilon_0 a}\left(\dfrac{1}{y}-\dfrac{1}{\sqrt{a^2+y^2}}\right)\hat{\imath}$

b) $\dfrac{qQ}{4\pi\epsilon_0 a}\left(\dfrac{1}{x-a}+\dfrac{1}{x+a}-\dfrac{2}{x}\right)\hat{\imath}$

21.97 $E_x=E_y=\dfrac{Q}{2\pi^2\epsilon_0 a^2}$

21.99 a) 6.25×10^4 N/C, $225°$ counterclockwise from an axis pointing to the right at the point P
b) 1.00×10^{-14} N, opposite the electric field direction

21.101 a) 1.19×10^6 N/C, to the left
b) 1.19×10^5 N/C, to the left
c) 1.19×10^5 N/C, to the right

21.103 $\dfrac{\sigma}{2\epsilon_0}\left(-\dfrac{x}{|x|}\hat{\imath}+\dfrac{z}{|z|}\hat{k}\right)$

21.105 b) $q_1<0, q_2>0$ c) $0.843\ \mu$C   d) 56.2 N

21.107 a) $\dfrac{Q}{2\pi\epsilon_0 L}\left(\dfrac{1}{2x+a}-\dfrac{1}{2L+2x+a}\right)$

Chapter 22

22.1 a) 1.8 N·m²/C   b) no   c) (i) $0°$ (ii) $90°$
22.3 a) 3.53×10^5 N·m²/C   b) $3.13\ \mu$C
22.5 $\pi r^2 E$
22.7 0.977 N·m²/C, inward
22.9 a) 0 b) 8.74×10^7 N/C
c) 2.60×10^7 N/C
22.11 a) 1.17×10^5 N·m²/C   b) no change
22.13 a) 4.50×10^4 N/C   b) 918 N/C
22.15 0.0810 N
22.17 2.04×10^{10}
22.19 a) 6.47×10^5 N/C, $+y$-direction
b) 7.2×10^4 N/C, $-y$-direction
22.21 a) $5.73\ \mu$C/m²   b) 6.47×10^5 N/C
c) -5.65×10^4 N·m²/C
22.23 a) $0.260\ \mu$C/m³   b) 1960 N/C
22.25 1.16 km/s
22.27 $23.6\ \mu$J
22.29 0 (outside the plates), σ/ϵ_0 (between the plates)
22.31 a) $2\pi R\sigma$ b) $\sigma R/\epsilon_0 r$ c) $\lambda/2\pi\epsilon_0 r$
22.33 a) yes, $+Q$ b) no   c) yes   d) no, no
e) no, yes, no
22.35 a) 750 N·m²/C   b) 0
c) 577 N/C, $+x$-direction   d) charges both within and outside
22.37 a) -0.598 nC   b) charges both within and outside
22.39 a) $\lambda/2\pi\epsilon_0 r$, radially outward   b) $\lambda/2\pi\epsilon_0 r$, radially outward
d) $-\lambda$ (inner surface), $+\lambda$ (outer surface)
22.41 a) (i) $\dfrac{\alpha}{2\pi\epsilon_0 r}$, radially outward   (ii) 0 (iii) 0
b) (i) $-\alpha$ (ii) 0
22.43 $10.2°$
22.45 a) $0\ (r<R);\ \dfrac{1}{4\pi\epsilon_0}\dfrac{Q}{r^2}\ (R<r<2R)$,
radially outward; $\dfrac{1}{4\pi\epsilon_0}\dfrac{2Q}{r^2}\ (r>2R)$,
radially outward
22.47 a) (i) 0 (ii) 0 (iii) $\dfrac{q}{2\pi\epsilon_0 r^2}$, radially outward
(iv) 0 (v) $\dfrac{3q}{2\pi\epsilon_0 r^2}$, radially outward
b) (i) 0 (ii) $+2q$ (iii) $-2q$ (iv) $+6q$
22.49 a) (i) 0 (ii) 0 (iii) $\dfrac{q}{2\pi\epsilon_0 r^2}$, radially outward
(iv) 0 (v) $\dfrac{q}{2\pi\epsilon_0 r^2}$, radially inward

22.51 a) $\dfrac{qQ}{4\pi\epsilon_0 r^2}$, toward the center of the shell
b) 0
22.53 a) $\dfrac{\alpha}{2\epsilon_0}\left(1-\dfrac{a^2}{r^2}\right)$
b) $q=2\pi\alpha a^2, E=\dfrac{\alpha}{2\epsilon_0}$
22.55 $R/2$
22.57 b) $|x|>d$ (outside the slab):
$\dfrac{\rho_0 d}{3\epsilon_0}\dfrac{x}{|x|}\hat{\imath};\ |x|<d$ (inside the slab): $\dfrac{\rho_0 x^3}{3\epsilon_0 d^2}\hat{\imath}$
22.61 b) $\rho\vec{b}/3\varepsilon_0$
22.63 a) $-\dfrac{Q}{16\pi\epsilon_0 R^2}\hat{\imath}$ b) $\dfrac{Q}{72\pi\epsilon_0 R^2}\hat{\imath}$
c) 0 d) $\dfrac{5Q}{18\pi\epsilon_0 R^2}\hat{\imath}$
22.65 c) $\dfrac{Qr}{4\pi\epsilon_0 R^3}\left(4-\dfrac{3r}{R}\right)$ e) $2R/3, \dfrac{Q}{3\pi\varepsilon_0 R^2}$
22.67 a) $\dfrac{480Q}{233\pi R^3}$
b) $r\le R/2: E=\dfrac{180Qr^2}{233\pi\epsilon_0 R^4}; R/2\le r\le R:$
$E=\dfrac{480Q}{233\pi\epsilon_0 r^2}\left[\dfrac{1}{3}\left(\dfrac{r}{R}\right)^3-\dfrac{1}{5}\left(\dfrac{r}{R}\right)^5-\dfrac{23}{1920}\right];$
$r\ge R: E=\dfrac{Q}{4\pi\epsilon_0 r^2}$ c) 0.807
d) $E=\dfrac{180}{233}\dfrac{Q}{4\pi\epsilon_0 R^2}$

Chapter 23

23.1 -0.356 J
23.3 3.46×10^{-13} J $=2.16$ MeV
23.5 a) 12.5 m/s   b) 0.323 m
23.7 -1.42×10^{-18} J
23.9 a) 13.6 km/s   b) 2.45×10^{17} m/s²
23.11 $-q/2$
23.13 7.42 m/s, faster
23.15 a) 0 b) 0.750 mJ   c) -2.06 mJ
23.17 a) 0 b) -175 kV   c) -0.875 J
23.19 a) -737 V   b) -704 V   c) 8.2×10^{-8} J
23.21 b) $V=\dfrac{q}{4\pi\epsilon_0}\left(\dfrac{1}{|x|}-\dfrac{2}{|x-a|}\right)$
c) $x=-a, a/3$
23.23 a) 156 V   b) -182 V
23.25 a) point b b) 800 V/m   c) $-48.0\ \mu$J
23.27 a) (i) 180 V   (ii) -270 V   (iii) -450 V
b) 719 V, inner shell
23.29 a) oscillatory   b) 1.67×10^7 m/s
23.31 a) 94.9 nC/m   b) less   c) zero
23.33 a) 78.2 kV   b) zero
23.35 0.474 J
23.37 a) 9.3×10^6 V/m, inward
b) outer surface
23.39 a) 8.00 kV/m   b) $19.2\ \mu$N   c) $0.864\ \mu$J
d) $-0.864\ \mu$J
23.41 b) -20 nC   c) no
23.43 -760 V
23.45 a) $E_x=-Ay+2Bx, E_y=-Ax-C, E_z=0$
b) $x=-C/A, y=-2BC/A^2$, any value of z
23.47 a) (i) $V=\dfrac{q}{4\pi\epsilon_0}\left(\dfrac{1}{r_a}-\dfrac{1}{r_b}\right)$
(ii) $V=\dfrac{q}{4\pi\epsilon_0}\left(\dfrac{1}{r}-\dfrac{1}{r_b}\right)$ (iii) $V=0$
d) 0 e) $E=\dfrac{q-Q}{4\pi\epsilon_0 r^2}$

23.49 a) cylinders coaxial with the given cylinder
b) 2.90 cm, 4.20 cm, 6.08 cm
c) get farther apart
23.51 a) $-0.360\ \mu$J   b) $x=0.074$ m
23.53 a) 7.66×10^{-13} J   b) 5.17×10^{-14} m
23.55 a) $-21.5\ \mu$J   b) -2.83 kV   c) 35.4 kV/m
23.57 a) 7.85×10^4 V/m$^{4/3}$
b) $E_x=-(1.05\times10^5$ V/m$^{4/3})x^{1/3}$
c) 3.13×10^{-15} N, toward the positive anode
23.59 a) $\dfrac{1.46q^2}{\pi\epsilon_0 d}$
23.61 a) -8.60×10^{-18} J $=-53.7$ eV
b) 2.88×10^{-11} m
23.63 a) (i) $V=(\lambda/2\pi\epsilon_0)\ln(b/a)$
(ii) $V=(\lambda/2\pi\epsilon_0)\ln(b/r)$ (iii) $V=0$
d) $(\lambda/2\pi\epsilon_0)\ln(b/a)$
23.65 a) 1.76×10^{-16} N, downward
b) 1.93×10^{14} m/s², downward
c) 0.822 cm   d) $15.3°$ e) 3.29 cm
23.67 a) 97.1 kV/m   b) 30.3 pC
23.69 a) $r>R: V_r=-\dfrac{\lambda}{2\pi\epsilon_0}\ln\left(\dfrac{r}{R}\right);$
$r<R: V_r=\dfrac{\lambda}{4\pi\epsilon_0}\left[1-\left(\dfrac{r}{R}\right)^2\right]$
23.71 $\dfrac{3}{5}\left(\dfrac{Q^2}{4\pi\epsilon_0 R}\right)$
23.73 360 kV
23.75 b) yes   c) no, no
23.77 $\dfrac{Q}{8\pi\epsilon_0 R}$
23.79 a) $\dfrac{Q}{4\pi\epsilon_0 a}\ln\left(\dfrac{x+a}{x}\right)$
b) $\dfrac{Q}{4\pi\epsilon_0 a}\ln\left(\dfrac{a+\sqrt{a^2+y^2}}{y}\right)$
c) part (a): $\dfrac{Q}{4\pi\epsilon_0 x}$, part (b): $\dfrac{Q}{4\pi\epsilon_0 y}$
23.81 a) $1/3$ b) 3
23.83 a) $E=\dfrac{Q_1}{4\pi\epsilon_0 R_1^2}, V=\dfrac{Q_1}{4\pi\epsilon_0 R_1}$
b) $q_1=\dfrac{Q_1 R_1}{R_1+R_2}, q_2=\dfrac{Q_1 R_2}{R_1+R_2}$
c) $V=\dfrac{Q_1}{4\pi\epsilon_0(R_1+R_2)}$ for either sphere
d) $E_1=\dfrac{Q_1}{4\pi\epsilon_0 R_1(R_1+R_2)}$,
$E_2=\dfrac{Q_1}{4\pi\epsilon_0 R_2(R_1+R_2)}$
23.85 a) 7580 km/s   b) 7260 km/s
c) 2.3×10^9 K (protons), 6.4×10^9 K (helium)
23.87 a) 5.9×10^{-15} m   b) 20.7 pJ
c) 1.06×10^{15} J $=253$ kilotons of TNT
23.89 1.01×10^{-12} m, 1.11×10^{-13} m, 2.54×10^{-14} m
23.91 c) $3, 0.507\ \mu$m

Chapter 24

24.1 a) 10.0 kV   b) 22.6 cm²   c) 8.00 pF
24.3 a) 604 V   b) 90.8 cm²   c) 1840 kV/m
d) $16.3\ \mu$C/m²
24.5 a) $120\ \mu$C   b) $60\ \mu$C   c) $480\ \mu$C
24.7 2.8 mm, yes
24.9 a) 1.05 mm   b) 84.0 V
24.11 a) 4.35 pF   b) 2.30 V
24.13 a) 15.0 pF   b) 3.09 cm   c) 31.2 kN/C
24.15 a) series   b) 5000
24.17 a) $Q_1=Q_2=22.4\ \mu$C, $Q_3=44.8\ \mu$C, $Q_4=67.2\ \mu$C

b) $V_1 = V_2 = 5.6$ V, $V_3 = 11.2$ V, $V_4 = 16.8$ V
c) 11.2 V
24.19 a) $Q_1 = 156$ μC, $Q_2 = 260$ μC
b) $V_1 = V_2 = 52.0$ V
24.21 a) 19.3 nF   b) 482 nC   c) 162 nC
d) 25 V
24.23 57 μF
24.25 0.0283 J/m³
24.27 a) $\dfrac{xQ^2}{2\epsilon_0 A}$ b) $\left(\dfrac{Q^2}{2\epsilon_0 A}\right)dx$ c) $\dfrac{Q^2}{2\epsilon_0 A}$
24.29 a) $U_p = 4U_s$ b) $Q_p = 2Q_s$ c) $E_p = 2E_s$
24.31 a) 24.2 μC   b) $Q_{35} = 7.7$ μC, $Q_{75} = 16.5$ μC
c) 2.66 mJ   d) $U_{35} = 0.85$ mJ, $U_{75} = 1.81$ mJ
e) 220 V for each capacitor
24.33 a) 1.60 nC   b) 8.05
24.35 a) 3.60 mJ (before), 13.5 mJ (after)
b) increased by 9.9 mJ
24.37 a) 0.620 μC/m²   b) 1.28
24.39 0.0135 m²
24.41 a) 2.31×10^{-11} C²/N·m²
b) 4.0×10^4 V
c) $\sigma = 4.6 \times 10^{-4}$ C/m²,
$\sigma_i = 2.8 \times 10^{-4}$ C/m²
24.43 a) 10.1 V   b) 2.25
24.45 a) $\dfrac{Q}{\epsilon_0 AK}$ b) $\dfrac{Qd}{\epsilon_0 AK}$ c) $K\dfrac{\epsilon_0 A}{d} = KC_0$
24.47 a) 421 J   b) 0.054 F
24.49 a) 31 pF   b) 0.37 nC   c) 1.6 kV/m
d) 2.2 nJ
24.51 133 kV/m
24.53 a) 0.0160 C   b) 533 V   c) 4.26 J   d) 2.14 J
24.55 a) 158 μJ   b) 72.1 μJ
24.57 a) 2.5 μF
b) $Q_1 = 550$ μC, $Q_2 = 370$ μC, $Q_3 = Q_4 = 180$ μC, $Q_5 = 550$ μC; $V_1 = 65$ V, $V_2 = 87$ V, $V_3 = V_4 = 43$ V, $V_5 = 65$ V
24.59 $C_2 = 6.00$ μF, $C_3 = 4.50$ μF
24.61 a) 76 μC   b) 1.4 mJ
c) 11 V across each capacitor   d) 1.3 mJ
24.63 a) 2.3 μF   b) $Q_1 = 970$ μC, $Q_2 = 640$ μC
c) 47 V
24.65 a) 3.91   b) 22.8 V
24.67 a) $C = 4\pi\epsilon_0 R$ b) 710 μF
24.69 a) 0.065 F   b) 23,000 C
c) 4.0×10^9 J
24.71 48.3 μF
24.73 0.185 μJ
24.75 b) 2.38 nF
24.77 a) $\dfrac{\epsilon_0 L}{D}[L + (K - 1)x]$
c) $Q = \dfrac{\epsilon_0 LV}{D}[L + (K - 1)x]$

Chapter 25

25.1 1.0 C
25.3 a) 3.12×10^{19} b) 1.51×10^6 A/m²
c) 0.111 mm/s   d) both J and v_d would decrease
25.5 a) 110 min   b) 440 min   c) $v_d \propto 1/d^2$
25.7 a) 330 C   b) 41 A
25.9 9.0 μA
25.11 a) 1.06×10^{-5} Ω·m
b) 0.00105 (C°)⁻¹
25.13 a) 0.206 mV   b) 0.176 mV
25.15 a) 1.21 V/m   b) 0.0145 Ω c) 0.182 V
25.17 0.125 Ω
25.19 15 g
25.21 1.53×10^{-8} Ω
25.23 a) 11 A   b) 3.1 V   c) 0.28 Ω
25.25 a) 99.54 Ω b) 0.0158 Ω
25.27 a) 4.67×10^{-8} Ω b) 6.72×10^{-4} Ω
25.29 a) 27.4 V   b) 12.3 MJ
25.31 a) zero   b) 5.0 V   c) 5.0 V
25.33 3.08 V, 0.067 Ω, 1.80 Ω
25.35 a) 1.41 A, clockwise   b) 13.7 V   c) -1.0 V
25.37 b) yes   c) 3.88 Ω

25.39 a) 144 Ω b) 240 Ω c) 0.833 A, 0.500 A
25.41 a) 29.8 W   b) 0.248 A
25.43 a) EJ b) ρJ^2 c) E^2/ρ
25.45 a) 300 W   b) 0.90 J
25.47 a) 2.6 MJ   b) 0.063 L   c) 1.6 h
25.49 12.3%
25.51 a) 24.0 W   b) 4.0 W   c) 20.0 W
25.53 a) 26.7 Ω b) 4.50 A   c) 453 W   d) larger
25.55 a) 3.65×10^{-8} Ω·m   b) 172 A
c) 2.58 mm/s
25.57 0.060 Ω
25.59 a) 2.5 mA   b) 2.14×10^{-5} V/m
c) 8.55×10^{-5} V/m   d) 0.180 mV
25.61 42.0 s
25.63 a) 80 C°   b) no
25.65 a) $\dfrac{\rho h}{\pi r_1 r_2}$
25.67 a) 0.057 Ω b) 3.34×10^{-8} Ω·m
c) 0.86 mm   d) 0.00240 Ω
e) 0.0011 (C°)⁻¹
25.69 a) 0.20 Ω b) 8.7 V
25.71 a) 1.0 kΩ b) 100 V   c) 10 W
25.73 1.42 A
25.75 a) $I_A\left(1 + \dfrac{R_A}{r + R}\right)$ b) 0.0429 Ω
25.77 a) 8-gauge   c) 106 W   d) $19.25
25.79 a) 0.40 A   b) 1.6 W
c) 4.8 W, in the 12.0-V battery
d) 3.2 W, in the 8.0-V battery
25.81 a) 14.4 V   b) 2.59 MJ   c) 0.432 MJ
d) 0.96 Ω e) 1.73 MJ   f) 0.432 MJ
25.83 6.67 V
25.85 a) a/E b) aL/V_{bc} c) point c
d) 3.5×10^8 m/s²
25.87 a) $R = \dfrac{\rho_0 L}{A}\left(1 - \dfrac{1}{e}\right)$, $I = \dfrac{V_0 A}{\rho_0 L\left(1 - \dfrac{1}{e}\right)}$
b) $E(x) = \dfrac{V_0 e^{-x/L}}{L\left(1 - \dfrac{1}{e}\right)}$,
c) $V(x) = \dfrac{V_0\left(e^{-x/L} - \dfrac{1}{e}\right)}{1 - \dfrac{1}{e}}$

Chapter 26

26.1 $3R/4$
26.3 22.5 W
26.5 a) 3.50 A   b) 4.50 A   c) 3.15 A   d) 3.25 A
26.7 0.769 A
26.9 a) 8.80 Ω b) 3.18 A   c) 3.18 A   d) 5.09 V, 7.63 V, 15.3 V   e) 16.2 W, 24.3 W, 48.5 W
f) resistor with greater resistance
26.11 a) 8.00 A, 12.0 A   b) 84.0 V
26.13 5.00 Ω; $I_3 = 8.00$ A, $I_4 = 9.00$ A, $I_6 = 4.00$ A, $I_{12} = 3.00$ A
26.15 a) $I_1 = 1.50$ A, $I_2 = I_3 = I_4 = 0.500$ A
b) $P_1 = 10.1$ W, $P_2 = P_3 = P_4 = 1.12$ W; R_1 glows brightest   c) $I_1 = 1.33$ A, $I_2 = I_3 = 0.667$ A   d) $P_1 = 8.00$ W, $P_2 = P_3 = 2.00$ W   e) R_1 glows less brightly; R_2 and R_3 glow brighter
26.17 18.0 V, 3.00 A
26.19 a) 2 resistors in parallel and that combination in series with 3 more resistors   b) 10 resistors in parallel   c) 3 resistors in parallel
d) 2 resistors in parallel and that combination in series with 4 resistors in parallel
26.21 a) 0.100 A   b) 4.0 W, 8.0 W   c) 12.0 W
d) 0.300 A, 0.150 A   e) 36.0 W, 18.0 W
f) 54.0 W   g) in series, the 800-Ω bulb is brighter; in parallel, the 400-Ω bulb is brighter
h) parallel connection
26.23 1010 s
26.25 a) 2.00 A   b) 5.00 Ω c) 42.0 V
d) 3.50 A
26.27 a) 8.00 A   b) $\varepsilon_1 = 36.0$ V, $\varepsilon_2 = 54.0$ V
c) 9.00 Ω
26.29 a) 1.60 A, 1.40 A, 0.20 A   b) 10.4 V

26.31 a) 36.4 V   b) 0.500 A
26.33 a) 2.14 V, a is higher
b) 0.050 A, downward; 0
26.35 a) 0.641 Ω b) 975 Ω
26.37 a) 17.9 V   b) 22.7 V   c) 21.4%
26.39 a) 543 Ω b) 1.88 mA
c) 1824 Ω, 608 Ω, 203 Ω
26.41 a) 0.849 μF   b) 2.89 s
26.43 a) 4.21 ms   b) 0.125 A
26.45 192 μC
26.47 13.6 A
26.49 a) 0.937 A   b) 0.606 A
26.51 a) 133 μC   b) 8.87 V, 9.13 V
c) 8.87 V for both   d) 67.4 μC
26.53 900 W
26.55 12.1 Ω
26.57 a) 13.6 $\mu\Omega$ b) 2.14×10^{-8} Ω·m
26.59 a) 2.2 A, 4.4 V, 9.7 W   b) 16.3 W, brighter
26.61 a) 18.7 Ω b) 7.5 Ω
26.63 $I_1 = 0.848$ A, $I_2 = 2.14$ A, $I_3 = 0.171$ A
26.65 $I_2 = 5.21$ A, $I_4 = 1.11$ A, $I_5 = 6.32$ A
26.67 a) $+0.22$ V   b) 0.464 A
26.69 0.447 Ω (end-to-end), 0.423 Ω (in parallel)
26.71 a) 186 V, upper terminal positive
b) 3.00 A, upward   c) 20.0 Ω
26.73 a) $P_1 + P_2$ b) $\dfrac{P_1 P_2}{P_1 + P_2}$
26.75 a) 1.35 W   b) 8.31 ms   c) 0.337 W
26.77 a) -12.0 V   b) 1.71 A   c) 4.21 Ω
26.79 a) 114 V   b) 263 V   c) 266 V
26.81 1897 Ω
26.83 a) 24.8 V, 65.2 V   b) 3840 Ω c) 62.6 V
d) no
26.85 1.7 MΩ, 3.1 μF
26.87 a) 19.4 s, 31.4 time constants   b) yes
26.89 a) $P = \dfrac{Q_0^2}{RC^2} e^{-2t/RC}$
26.93 b) 4 segments   c) 3.2 MΩ, 4.0×10^{-3}
d) 3.4×10^{-4} e) 0.88

Chapter 27

27.1 a) $(-6.68 \times 10^{-4} \text{ N})\hat{k}$
b) $(6.68 \times 10^{-4} \text{ N})\hat{i} + (7.27 \times 10^{-4} \text{ N})\hat{j}$
27.3 a) positive   b) 0.0505 N
27.5 9490 km/s
27.7 a) $B_x = -0.175$ T, $B_z = -0.256$ T
b) B_y is not determined   c) 0, 90°
27.9 a) 1.46 T, in the xz-plane at 40° from the $+x$-axis toward the $-z$-axis   b) 7.47×10^{-16} N, in the xz-plane at 50° from the $-x$-axis toward the $-z$-axis.
27.11 a) 3.05 mWb   b) 1.83 mWb   c) 0
27.13 -0.78 mWb
27.15 a) 0.160 mT, into the page   b) 0.111 μs
27.17 7.93×10^{-10} N, toward the south
27.19 a) 4.94×10^{-21} kg·m/s
b) 2.31×10^{-23} kg·m²/s
27.21 a) 835 km/s   b) 26.2 ns   c) 7.27 kV
27.23 a) 107 T   b) no
27.25 0.838 mT
27.27 a) $(1.60 \times 10^{-14} \text{ N})\hat{j}$ b) yes
c) helix, no   d) 1.40 cm
27.29 a) 4.81 kN/C   c) yes
27.31 0.0445 T, out of the page
27.33 a) 4.92 km/s   b) 9.96×10^{-26} kg
27.35 2.0 cm
27.37 a) 13.4 kA, no
27.39 0.724 N, 63.4° below the direction of the current in the upper wire segment
27.41 a) 817 V   b) 113 m/s²
27.43 a) $-ILB\hat{j}$ b) yes
27.45 a) $F_{cd} = 1.20$ N   b) 0.420 N·m
27.47 a) A_2 b) 290 rad/s²
27.49 -2.42 J
27.51 a) 1.13 A   b) 3.69 A   c) 98.2 V
d) 362 W
27.53 a) 4.7 mm/s   b) $+4.5 \times 10^{-3}$ V/m, $+z$-direction   c) 53 μV
27.55 a) $-\dfrac{F_2}{qv_1}\hat{j}$ b) $F_2/2\overline{2}$

27.57 a) 8.3×10^6 m/s b) 0.14 T
27.59 3.45 T, perpendicular to the initial velocity of the coin
27.61 a) 8.9×10^{-13} J $= 5.5$ MeV b) 77 ns
 c) 1.2 T d) same as in (a)
27.63 a) $-3.89 \ \mu$C
 b) $(7.60 \times 10^{14} \text{ m/s}^2)\hat{\imath} +$
 $(5.70 \times 10^{14} \text{ m/s}^2)\hat{\jmath}$
 c) 2.90 cm d) 2.88×10^7 Hz
 e) $(0.0290 \text{ m}, 0, 0.874 \text{ m})$
27.65 9τ
27.67 1.6 mm
27.69 $\dfrac{Mg \tan\theta}{LB}$, right to left
27.71 $B_r(r) = -\beta r/2$
27.73 a) 8.46 mT b) 27.2 cm c) 2.2 cm, yes
27.75 1.80 N, to the left
27.77 0.024 T, $+y$-direction
27.79 a) 0.0442 N·m, clockwise b) stretched
 c) 7.98 mJ
27.81 2.39 A
27.83 $-(0.444 \text{ N})\hat{\jmath}$
27.85 b) $(0,0)$ to $(0,L)$: $\frac{1}{2}B_0 L I \hat{\imath}$
 $(0, L)$ to (L,L): $-IB_0 L \hat{\jmath}$
 (L, L) to $(L, 0)$: $-\frac{1}{2}B_0 L I \hat{\imath}$
 $(L, 0)$ to $(0,0)$: 0
 c) $-IB_0 L \hat{\jmath}$
27.87 a) 2.52 m/s b) 7.58 A c) 0.198 Ω
27.89 a) 5.14 m b) 1.72 μs c) 6.08 mm
 d) 3.05 cm

Chapter 28

28.1 a) $-(19.2 \ \mu\text{T})\hat{k}$ b) 0 c) $(19.2 \ \mu\text{T})\hat{\imath}$
 d) $(6.79 \ \mu\text{T})\hat{\imath}$
28.3 a) 60.0 nT, out of the page, at A and B
 b) 0.120 μT, out of the page c) 0
28.5 a) 0 b) $-(1.31 \ \mu\text{T})\hat{k}$ c) $-(0.462 \ \mu\text{T})\hat{k}$
 d) $(1.31 \ \mu\text{T})\hat{\jmath}$
28.7 a) (i) $\dfrac{\mu_0 qv}{8\pi d^2}$, into the page (ii) 0
 (iii) $\dfrac{\mu_0 qv}{4\pi d^2}$, out of the page
 b) $\dfrac{\mu_0 q^2 v' v}{16\pi d^2}$, attractive c) 1.00×10^{-6}
28.9 $(97.5 \text{ nT})\hat{k}$
28.11 a) 0.440 μT, out of the page
 b) 16.7 nT, out of the page c) 0
28.13 a) $(5.00 \times 10^{-11} \text{ T})\hat{k}$
 b) $-(5.00 \times 10^{-11} \text{ T})\hat{\imath}$
 c) $(-1.77 \times 10^{-11} \text{ T})(\hat{\imath} - \hat{\jmath})$ d) 0
28.15 17.6 μT, into the page
28.17 a) 0.8 mT b) 40 mT, 20 times larger
28.19 25 μA
28.21 a) 10.0 A
 b) at all points directly above the wire
 c) at all points directly east of the wire
28.23 a) $-(0.10 \ \mu\text{T})\hat{\imath}$
 b) 2.19 μT, at 46.8° from the x-axis to the z-axis c) $(7.9 \ \mu\text{T})\hat{\imath}$
28.25 a) 0 b) 6.67 μT, toward the top of the page
 c) 7.54 μT, to the left
28.27 a) 0 b) 0 c) 0.40 mT, to the left
28.29 a) at P: 41 μT, into the page; at Q: 41 μT, out of the page b) at P: 9.0 μT, out of the page; at Q: 9.0 μT, into the page
28.31 a) 6.00 μN, repulsive b) 24.0 μN
28.33 46 μN/m, repulsive, no
28.35 0.38 μA
28.37 $\dfrac{\mu_0|I_1 - I_2|}{4R}$, 0
28.39 a) 9.42 mT b) 0.134 mT
28.41 18.0 A, counterclockwise
28.43 a) 305 A b) -3.83×10^{-4} T·m
28.45 a) $\mu_0 I/2\pi r$ b) 0

28.47 $r = R/2$ and $2R$
28.49 a) 1790 turns/m b) 63.0 m
28.51 a) 3.72 MA b) 124 kA c) 237 A
28.53 1.11 mT
28.55 a) 72.5 mA b) 19.5 mA
28.57 a) (i) 1.13 mT (ii) 4.68 MA/m (iii) 5.88 T
28.59 a) 1.00 μT, into the page
 b) $(7.49 \times 10^{-8} \text{ N})\hat{\jmath}$
28.61 a) in the plane of the wires, between them, 0.300 m from the 75.0-A wire
 b) in the plane of the wires, 0.200 m from the 25.0-A wire and 0.600 m from the 75.0-A wire
28.63 a) 5.7×10^{12} m/s^2, away from the wire
 b) 32.5 N/C, away from the wire c) no
28.65 5.59×10^{-18} N
28.67 a) 2.00 A, out of the page
 b) 2.13 μT, to the right c) 2.06 μT
28.69 a) 11.1 μN/m, $-y$-direction
 b) 11.1 μN/m, $+y$-direction
28.71 23.2 A
28.73 2.21×10^{-3} N·m
28.75 a) $\dfrac{\mu_0 N I a^2}{2} \Bigg\{ \dfrac{1}{[(x + a/2)^2 + a^2]^{3/2}} + \dfrac{1}{[(x - a/2)^2 + a^2]^{3/2}} \Bigg\}$
 c) $\left(\dfrac{4}{5}\right)^{3/2}\dfrac{\mu_0 N I}{a}$ d) 20.2 mT e) 0,0
28.77 a) $\dfrac{3I}{2\pi R^3}$ b) (i) $B = \dfrac{\mu_0 I r}{2\pi R^3}$ (ii) $B = \dfrac{\mu_0 I}{2\pi r}$
28.79 a) 0 b) $B = \dfrac{\mu_0 I}{2\pi r}\dfrac{r^2 - a^2}{b^2 - a^2}$ c) $B = \dfrac{\mu_0 I}{2\pi r}$
28.81 b) $B = \dfrac{\mu_0 I_0}{2\pi r}$ c) $\dfrac{I_0 r^2}{a^2}\left(2 - \dfrac{r^2}{a^2}\right)$
 d) $B = \dfrac{\mu_0 I_0 r}{2\pi a^2}\left(2 - \dfrac{r^2}{a^2}\right)$
28.83 a) $B = \frac{1}{2}\mu_0 I n$, $+x$-direction
 b) $B = \frac{1}{2}\mu_0 I n$, $-x$-direction
28.85 7.73×10^{-25} A·m^2 $= 0.0834\mu_{\text{B}}$
28.87 b) $\dfrac{1}{2g}\left(\dfrac{\mu_0 Q_0^2}{4\pi \lambda R C d}\right)^2$

Chapter 29

29.1 a) 17.1 mV b) 28.5 mA
29.3 a) $Q = NBA/R$ b) no
29.5 a) 34 V b) counterclockwise
29.7 a) $\mu_0 i/2\pi r$, into the page b) $\dfrac{\mu_0 i}{2\pi r}L dr$
 c) $\dfrac{\mu_0 i L}{2\pi}\ln(b/a)$ d) $\dfrac{\mu_0 L}{2\pi}\ln(b/a)\dfrac{di}{dt}$
 e) 0.506 μV
29.9 a) 5.44 mV b) clockwise
29.11 a) bAv b) clockwise
 c) bAv, counterclockwise
29.13 10.4 rad/s
29.15 a) counterclockwise b) clockwise
 c) no induced current
29.17 a) a to b b) b to a c) b to a
29.19 a) clockwise b) no induced current
 c) counterclockwise
29.21 13.2 mA, counterclockwise
29.23 a) 0.675 V b) point b c) 2.25 V/m, b to a
 d) point b e) (i) 0 (ii) 0
29.25 46.2 m/s $= 103$ mph, no
29.27 a) 3.00 V b) b to a c) 0.800 N, to the right
 d) 6.00 W for each
29.29 a) counterclockwise b) 42.4 mW
29.31 35.0 m/s, to the right
29.33 a) 2.55 V, point a b) 3.38 V, point a c) 0
 d) 4.23 V
29.35 a) $\pi r_1^2\dfrac{dB}{dt}$ b) $\dfrac{r_1}{2}\dfrac{dB}{dt}$ c) $\dfrac{R^2}{2r_2}\dfrac{dB}{dt}$

 e) $\dfrac{\pi R^2}{4}\dfrac{dB}{dt}$ f) $\pi R^2\dfrac{dB}{dt}$ g) $\pi R^2\dfrac{dB}{dt}$
29.37 9.21 A/s
29.39 0.950 mV
29.41 2.34
29.43 a) 0.599 nC b) 6.00 mA c) 6.00 mA
29.45 a) 0.15 V/m b) 38 V/m·s
 c) 3.4×10^{-10} A/m^2 d) 53.3 μT
29.47 a) $B = 0$, $\vec{M} = -(4.38 \times 10^4 \text{ A/m})\hat{\imath}$
 b) $\vec{B} = (15.0 \text{ T})\hat{\imath}$, $M = 0$
29.49 a) 3.7 A b) 1.33 mA c) counterclockwise
29.51 16.2 μV
29.53 a) $\dfrac{\mu_0 I a b v}{2\pi r(r + a)}$ b) clockwise
29.55 190 rpm
29.57 a) 17.9 mV b) a to b
29.59 b) $\dfrac{RF}{B^2 L^2}$
29.61 1.2 V
29.63 $\mu_0 I W/4\pi$
29.65 a) $\dfrac{\mu_0 I v}{2\pi}\ln(1 + L/d)$ b) point a c) 0
29.67 a) 0.165 V b) 0.165 V c) 0, 0.0412 V
29.69 a) $B^2 L^2 v/R$
29.71 point a: $\dfrac{qr}{2}\dfrac{dB}{dt}$, to the left; point b: $\dfrac{qr}{2}\dfrac{dB}{dt}$, toward the top of the page; point c: 0
29.73 5.0 s
29.75 a) 196 μA/m^2 b) 3.00 nA/m^2 c) 7.82 MHz
 d) 90°
29.77 a) a to b b) $\dfrac{Rmg \tan\phi}{L^2 B^2 \cos\phi}$ c) $\dfrac{mg \tan\phi}{LB}$
 d) $\dfrac{Rm^2 g^2 \tan^2\phi}{L^2 B^2}$ e) $\dfrac{Rm^2 g^2 \tan^2\phi}{L^2 B^2}$

Chapter 30

30.1 a) 0.270 V, yes b) 0.270 V
30.3 6.32 μH
30.5 a) 1.96 H b) 7.11 mWb
30.7 a) 1940 b) 800 A/s
30.9 a) 0.250 H b) 0.450 mWb
30.11 a) 4.68 mV b) point a
30.13 a) 1000 b) 2.09 Ω
30.15 b) 0.111 μH
30.17 2850
30.19 a) 0.161 T b) 10.3 kJ/m^3 c) 0.129 J
 d) 40.2 μH
30.21 91.7 J
30.23 a) 2.40 A/s b) 0.800 A/s c) 0.413 A
 d) 0.750 A
30.25 a) 17.3 μs b) 30.7 μs
30.27 a) 0.250 A b) 0.137 A c) 32.9 V, point c
 d) 0.462 ms
30.29 a) $P = (4.50 \text{ W})(1 - e^{-(3.20 \text{ s}^{-1})t})$
 b) $P_R = (4.50 \text{ W})(1 - e^{-(3.20 \text{ s}^{-1})t})^2$
 c) $P_L = (4.50 \text{ W})(e^{-(3.20 \text{ s}^{-1})t} - e^{-(6.40 \text{ s}^{-1})t})$
30.33 a) 25.0 mH b) 90.0 nC c) 0.540 μJ
 d) 6.58 mA
30.35 a) 105 rad/s, 0.0596 s b) 0.720 mC
 c) 4.32 mJ d) -0.542 mC
 e) -0.050 A, counterclockwise
 f) $U_C = 2.45$ mJ, $U_L = 1.87$ mJ
30.37 a) 7.50 μC b) 15.9 kHz c) 0.0212 J
30.39 a) 298 rad/s b) 83.8 Ω
30.43 a) 0.288 μH b) 14.2 μV
30.47 20 km/s, about 30 times smaller
30.49 a) $\dfrac{\mu_0 i}{2\pi r}$ b) $\dfrac{\mu_0 i^2 l}{4\pi r}dr$ c) $\dfrac{\mu_0 i^2 l}{4\pi}\ln(b/a)$
 d) $\dfrac{\mu_0 l}{2\pi}\ln(b/a)$
30.51 a) 5.00 H b) 31.7 m, no
30.53 a) 0.281 J b) 0.517 J c) 0.236 J
30.57 222 μF, 9.31 μH
30.59 13.0 mA, 184 A/s
30.61 a) 0 A, 20.0 V b) 0.267 A, 0 V
 c) 0.147 A, 9.0 V

A-19 at top right.

30.63 a) solenoid c) 50 V d) 3.5 A e) 4.3 Ω, 43 mH

30.65 a) $A_1 = A_4 = 0.800$ A, $A_2 = A_3 = 0$; $V_1 = 40.0$ V, $V_2 = V_3 = V_4 = V_5 = 0$ b) $A_1 = 0.480$ A, $A_2 = 0.160$ A, $A_3 = 0.320$ A, $A_4 = 0$; $V_1 = 24.0$ V, $V_2 = 0$, $V_3 = V_4 = V_5 = 16.0$ V c) 192 μC

30.67 a) $A_1 = A_4 = 0.455$ A, $A_2 = A_3 = 0$ b) $A_1 = 0.585$ A, $A_2 = 0.320$ A, $A_3 = 0.160$ A, $A_4 = 0.107$ A

30.69 a) 60.0 V b) point a c) 60.0 V d) point c e) −96.0 V f) point b g) 156 V h) point d

30.71 a) 0; $v_{ac} = 0$, $v_{cb} = 36.0$ V b) 0.180 A; $v_{ac} = 9.0$ V, $v_{cb} = 27.0$ V c) $i_0 = (0.180\ \text{A})(1 - e^{-t/(0.020\ \text{s})})$, $v_{ac} = (9.0\ \text{V})(1 - e^{-t/(0.020\ \text{s})})$, $v_{cb} = (9.0\ \text{V})(3.00 + e^{-t/(0.020\ \text{s})})$

30.75 a) $i_1 = \dfrac{\varepsilon}{R_1}$, $i_2 = \dfrac{\varepsilon}{R_2}(1 - e^{-R_2 t/L})$ b) $i_1 = \dfrac{\varepsilon}{R_1}$, $i_2 = \dfrac{\varepsilon}{R_2}$ c) $i = \dfrac{\varepsilon}{R_2}e^{-(R_1+R_2)t/L}$ d) 21.2 Ω, 12.7 V e) 35.3 mA

30.77 a) $d = D\left(\dfrac{L - L_0}{L_f - L_0}\right)$ b) 0.63024 H, 0.63048 H, 0.63072 H, 0.63096 H c) 0.63000 H, 0.62999 H, 0.62999 H, 0.62998 H d) oxygen

30.79 a) $i_1 = \dfrac{\varepsilon}{R_1}(1 - e^{-R_1 t/L})$, $i_2 = \dfrac{\varepsilon}{R_2}e^{-t/R_2 C}$, $q_2 = \varepsilon C(1 - e^{-t/R_2 C})$ b) $i_1 = 0, i_2 = 9.60$ mA c) $i_1 = 1.92$ A, $i_2 = 0$ d) 1.6 ms e) 9.4 mA f) 0.22 s

Chapter 31

31.1 1.06 A
31.3 a) 31.8 V b) 0
31.5 a) 90°, leads b) 193 Hz
31.7 13.3 μF
31.9 a) 1510 Ω b) 0.239 H c) 497 Ω d) 16.6 μF
31.11 a) $(12.5\ \text{V})\cos[(480\ \text{rad/s})t]$ b) 7.17 V
31.13 a) $i = (0.0253\ \text{A})\cos[(720\ \text{rad/s})t]$ b) 180 Ω c) $v_L = -(4.56\ \text{V})\sin[(720\ \text{rad/s})t]$
31.15 a) 601 Ω b) 0.0499 A c) −70.6°, lags d) 9.98 V, 4.99 V, 33.3 V
31.17 50.0 V
31.19 a) 40.0 W b) 0.167 A c) 720 Ω
31.21 b) 76.7 V
31.23 a) 45.8°, 0.697 b) 344 Ω c) 155 V d) 48.6 W e) 48.6 W f) 0 g) 0
31.25 a) 0.302 b) 0.370 W c) 0.370 W, 0, 0
31.27 a) 113 Hz, 15.0 mA b) 7.61 mA, lag
31.29 a) 150 V b) 150 V, 1290 V, 1290 V c) 37.5 W
31.31 a) 1.00 b) 75.0 W c) 75.0 W
31.33 a) 115 Ω b) 146 Ω c) 146 Ω
31.35 a) 10 b) 2.40 A c) 28.8 W d) 500 Ω
31.37 a) use a step-down transformer with $N_2/N_1 = {}^1\!/_2$ b) 6.67 A b) 36.0 Ω
31.39 0.124 H
31.41 3.59×10^7 rad/s
31.43 230 Ω
31.45 a) inductor b) 0.133 H
31.47 a) 0.831 b) 161 V
31.49 a) inductor: 1.15 A, 31.6 V; resistor: 1.15 A, 57.5 V; capacitor: 1.15 A, 14.6 V b) all change; inductor: 0.860 A, 47.3 V; resistor: 0.860 A, 43.0 V; capacitor: 0.860 A, 5.46 V
31.51 $\dfrac{V_{\text{out}}}{V_s} = \sqrt{\dfrac{R^2 + \omega^2 L^2}{R^2 + \left(\omega L - \dfrac{1}{\omega C}\right)^2}}$
31.55 a) 102 Ω b) 0.882 A c) 270 V
31.57 b) yes c) 0 e) no
31.59 a) $I_R = V/R$, $I_C = V\omega C$, $I_L = V/\omega L$ d) 159 Hz e) 0.50 A f) 0.50 A, 0.050 A, 0.050 A

31.61 a) L and C b) ${}^1\!/_2$ for each
31.63 a) 0.750 A b) 160 Ω c) 341 Ω, 619 Ω d) 341 Ω
31.65 $0, I_0/\sqrt{3}$
31.67 a) decreases by ${}^1\!/_2$ b) increases by 2 c) decreases by ${}^1\!/_2$ d) no
31.69 a) $\dfrac{V}{\sqrt{R^2 + \dfrac{9L}{4C}}}$ b) $\sqrt{\dfrac{L}{C}}\sqrt{\dfrac{2V}{R^2 + \dfrac{9L}{4C}}}$ c) $\sqrt{\dfrac{L}{C}}\dfrac{V/2}{\sqrt{R^2 + \dfrac{9L}{4C}}}$ d) $\dfrac{2LV^2}{R^2 + \dfrac{9L}{4C}}$ e) $\dfrac{1}{2}\dfrac{LV^2}{R^2 + \dfrac{9L}{4C}}$
31.71 a) 75.8 V b) 13.6 Ω
31.73 a) $V_R I/2$ b) 0 c) 0
31.75 a) 0.400 A b) 36.9° c) $(400 - 300i)$ Ω, 500 Ω d) $(0.320 + 0.240i)$ A, 0.400 A e) 36.9° f) $V_{R\text{-cpx}} = (128 + 96i)$ V, $V_{L\text{-cpx}} = (-120 + 160i)$ V, $V_{C\text{-cpx}} = (192 - 256i)$ V

Chapter 32

32.1 a) 1.28 s b) 8.15×10^{13} km
32.3 13.3 nT, $+y$-direction
32.5 3.0×10^{18} Hz, 3.3×10^{-19} s, 6.3×10^{10} rad/m
32.7 a) 6.94×10^{14} Hz 375 V/m c) $E(x,t) = (375\ \text{V/m})\cos[(1.45 \times 10^7\ \text{rad/m})x - (4.36 \times 10^{15}\ \text{rad/s})t]$, $B(x,t) = (1.25\ \mu\text{T})\cos[(1.45 \times 10^7\ \text{rad/m})x - (4.36 \times 10^{15}\ \text{rad/s})t]$
32.9 a) (i) 60 kHz (ii) 6.0×10^{13} Hz (iii) 6.0×10^{16} Hz b) (i) 4.62×10^{-14} m $= 4.62 \times 10^{-5}$ nm (ii) 508 m $= 5.08 \times 10^{11}$ nm
32.11 a) $+y$-direction b) 0.149 mm c) $\vec{B} = (1.03\ \text{mT})\hat{\imath}\cos[(4.22 \times 10^4\ \text{rad/m})y - (1.265 \times 10^{13}\ \text{rad/s})t]$
32.13 a) 361 m b) 0.0174 rad/m c) 5.22×10^6 rad/s d) 0.0144 V/m
32.15 a) 0.381 μm b) 0.526 μm c) 1.38 d) 1.90
32.17 a) 330 W/m² b) 500 V/m, 1.7 μT
32.19 a) 11 μW/m² b) 0.30 nT c) 840 W
32.21 2.5×10^{25} W
32.23 12.0 V/m, 40.0 nT
32.25 850 kW
32.27 a) 0.18 mW b) 274 V/m, 0.913 μT c) 0.18 mJ/s d) 0.010 W/cm²
32.29 a) 637 W/m² b) 693 V/m, 2.31 μT c) 2.12 μJ/m³
32.31 a) 30.5 cm b) 2.46 GHz c) 2.11 GHz
32.33 a) 7.10 mm b) 3.55 mm c) 1.56×10^8 m/s
32.35 a) 4.38 mm b) 4.38 mm c) 4.38 mm
32.37 a) 0.375 mJ b) 4.08 mPa c) 604 nm, 3.70×10^{14} Hz d) 30.3 kV/m, 101 μT
32.39 a) 71.6 MW/m² b) 232 kV/m, 774 μT c) 35.8 MW/m² d) 637 W/m²
32.41 a) 0.00602 W/m² b) 2.13 V/m, 7.10 nT c) 1.20 pN, no
32.43 a) 840 V/m, 2.80 μT b) 1.56 μJ/m³ for each c) 15.3 pJ
32.45 a) at $r = R$: 64 MW/m², 0.21 Pa; at $r = R/2$: 260 MW/m², 0.85 Pa b) no
32.47 3.89×10^{-13} rad/s²
32.49 a) $\rho I/\pi a^2$, in the direction of the current b) $\mu_0 I/2\pi a$, counterclockwise if the current is out of the page c) $\dfrac{\rho I^2}{2\pi^2 a^3}$, radially inward

d) $\dfrac{\rho l I^2}{\pi a^2} = I^2 R$
32.51 a) 23.6 h b) throw it
32.53 a) 9.75×10^{-15} W/m² b) 2.71 μV/m, 9.03×10^{-15} T, 67.3 ms c) 3.25×10^{-23} Pa d) 0.190 m
32.55 a) $\dfrac{4\rho G\pi MR^3}{3r^2}$ b) $\dfrac{LR^2}{4cr^2}$ c) 0.19 μm, no
32.57 b) 1.39×10^{-11} c) 2.54×10^{-8}

Chapter 33

33.1 39.4°
33.3 a) 2.04×10^8 m/s b) 442 nm
33.5 a) 1.55 b) 550 nm
33.7 a) 47.5° b) 66.0°
33.9 2.51×10^8 m/s
33.11 a) 2.34 b) 82°
33.13 a) $f, n\lambda, nv$ b) $f, (n/n')\lambda, (n/n')v$
33.15 71.8°
33.17 a) 51.3° b) 33.6°
33.19 a) 58.1° b) 22.8°
33.21 1.77
33.23 462 nm
33.25 0.6°
33.27 $0.375I_0$
33.29 a) $I_0/2, 0.125I_0, 0.0938I_0$ b) 0
33.31 a) 1.40 b) 35.5°
33.33 $\arccos\left(\dfrac{\cos\theta}{\sqrt{2}}\right)$
33.35 6.38 W/m²
33.37 a) $I_0/2, 0.250I_0, 0.125I_0$, linearly polarized along the axis of each filter b) $I_0/2$ linearly polarized along the axis of the filter; 0
33.41 a) 46.7° b) 13.4°
33.43 72.1°
33.45 1.28
33.47 1.53
33.49 1.84
33.51 a) 48.6° b) 48.6°
33.53 39.1°
33.55 0.23°, about the same
33.59 b) 38.9° c) 5.0°
33.61 a) 35° b) $I_p = 19.9$ W/m², $I_0 = 10.1$ W/m²
33.63 23.3°
33.67 a) $\Delta = 2\theta_a^A - 6\arcsin\left(\dfrac{\sin\theta_a^A}{n}\right) + 2\pi$ b) $\cos^2\theta_2 = \dfrac{n^2 - 1}{8}$ c) violet: $\theta_2 = 71.55°$, $\Delta = 233.2°$; red: $\theta_2 = 71.94°$, $\Delta = 230.1°$

Chapter 34

34.1 39.2 cm to the right of the mirror, 4.85 cm
34.3 9.0 cm, tip of the lead
34.5 b) 33.0 cm to the left of the vertex, 1.20 cm, inverted, real
34.7 0.213 mm
34.9 18.0 cm from the vertex, 0.50 cm, erect, virtual
34.11 b) 28.0 cm, inverted c) 8.00 cm
34.13 a) concave b) 2.50 cm, 5.00 cm
34.15 2.67 cm
34.17 3.30 m
34.19 a) at the center of the bowl, 1.33 b) no
34.21 39.5 cm
34.23 8.35 cm to the left of the vertex, 0.326 mm, erect
34.25 a) 107 cm to the right of the lens, 17.8 mm, real, inverted b) same as part (a)
34.27 a) 71.2 cm to the right of the lens b) −2.97
34.29 3.69 cm, 2.82 cm to the left of the lens
34.31 1.67
34.33 a) 18.6 mm b) 19 mm from the cornea c) 0.61 mm, real, inverted
34.35 a) 36.0 cm to the right of the lens b) 180 cm to the left of the lens

c) 7.20 cm to the left of the lens
d) 13.8 cm to the left of the lens

34.37 26.3 cm from the lens, 12.4 mm, erect, same side

34.39 a) 200 cm to the right of the first lens, 4.80 cm
b) 150 cm to the right of the second lens, 7.20 cm

34.41 a) 53.0 cm   b) real   c) 2.50 mm, inverted

34.43 10.2 m

34.45 a) 1.4×10^{-4} b) 5.3×10^{-4}
c) 1.5×10^{-3}

34.47 a) 85 mm   b) 135 mm

34.49 a) $f/11$ b) $1/480$ s = 2.1 ms

34.51 a) convex   b) 50 mm to 56 mm

34.53 a) 80.0 cm   b) 76.9 cm

34.55 49.4 cm, 2.02 diopters

34.57 -1.37 diopters

34.59 a) 6.06 cm   b) 4.12 mm

34.61 a) 640   b) 43

34.63 a) 8.37 mm   b) 21.4   c) -297

34.65 a) -6.33 b) 1.90 cm   c) 0.127 rad

34.67 a) 0.661 m   b) 59.1

34.69 7.20 m/s

34.71 $h/2$

34.73 a) 20.0 cm   b) 39.0 cm

34.75 a) 46 cm from the mirror, on the opposite side of the mirror, virtual   b) 29 mm, erect   c) no

34.77 51 m/s

34.79 b) 2.4 cm, -0.133

34.81 2.00

34.83 a) -3.3 cm   b) virtual
c) 1.9 cm to the right of the vertex at the right-hand end of the rod   d) real, inverted
e) 1.06 mm

34.85 a) 58.7 cm, converging   b) 4.48 mm, virtual

34.87 a) 6.48 mm   b) no, behind the retina
c) 19.3 mm from the cornea, in front of the retina

34.89 50.3 cm

34.91 10.6 cm

34.93 a) 0.24 mm   b) 0.24 mm

34.95 72.1 cm to the right of the surface vertex

34.97 0.80 cm

34.99 -26.7 cm

34.101 1.24 cm above the page

34.103 a) 46.7 mm   b) 35.0 m

34.105 134 cm to the left of the object

34.107 a) 3.5 cm   b) 7.0 cm   c) 100 cm
d) 57   e) no

34.109 4.17 diopters

34.111 a) 30.9 cm   b) 29.2 cm

34.113 a) 36.0 cm, 21.6 cm, $d = 1.2$ cm

34.115 a) 552   b) 25.8 cm

34.117 a) $4f$

34.119 b) 1.74 cm

Chapter 35

35.1 a) 14 cm, 48 cm, 82 cm, 116 cm, 150 cm
b) 31 cm, 65 cm, 99 cm, 133 cm

35.3 b) 427 Hz   b) 0.796 m

35.5 0.75 m, 2.00 m, 3.25 m, 4.50 m, 5.75 m, 7.00 m, 8.25 m

35.7 a) 2.0 m   b) constructively
c) 1.0 m, destructively

35.9 1.14 mm

35.11 0.83 mm

35.13 a) 39   b) $\pm73.3°$

35.15 12.6 cm

35.17 1200 nm

35.19 a) $0.750I_0$ b) 80 nm

35.21 1670 rad

35.23 71.4 m

35.25 114 nm

35.27 0.0234°

35.29 a) 55.6 nm   b) (i) 2180 nm   (ii) 11.0 wavelengths

35.31 a) 514 nm, green   b) 603 nm, orange

35.33 0.11 μm

35.35 0.570 mm

35.37 1.54 mm

35.39 a) 96.0 nm   b) no, no

35.41 a) 1.58 mm (green), 1.72 mm (orange)
b) 3.45 mm (violet), 4.74 mm (green), 5.16 mm (orange)   c) 9.57 μm

35.43 1.730

35.45 0°, 27.3°, 66.5°

35.47 1.57

35.49 above centerline: 3.14°, 15.9°, 29.5°, 45.4°, 68.6°; below centerline: 9.45°, 22.5°, 37.0°, 55.2°

35.51 $6.8 \times 10^{-5}\,(\text{C}°)^{-1}$

35.53 $\lambda/2d$, independent of m

35.55 b) 0.72 m

35.57 1.42

35.59 b) $I = I_0 \cos^2\left\{\dfrac{\pi[d\sin\theta + L(n-1)]}{\lambda_0}\right\}$
c) $\sin\theta = \dfrac{m\lambda_0 - L(n-1)}{d}$

35.61 14.0

Chapter 36

36.1 506 nm

36.3 a) 226   b) $\pm83.0°$

36.5 9.07 m

36.7 a) 63.8 cm
b) $\pm22.1°, \pm34.3°, \pm48.8°, \pm70.1°$

36.9 $\pm16.0°, \pm33.4°, \pm55.6°$

36.11 0.920 μm

36.13 a) 580 nm   b) 0.128

36.15 a) 6.75 mm   b) 2.43 μW/m^2

36.17 a) 668 nm   b) $9.36 \times 10^{-5}I_0$

36.19 a) $\pm13.0°, \pm26.7°, \pm42.4°, \pm64.1°$
b) 2.08 W/m^2

36.21 a) 3   b) 2

36.23 a) 0.0627°, 0.125°   b) $0.249I_0$, $0.0256I_0$

36.25 a) 1 & 3, 2 & 4   b) 1 & 2, 3 & 4
c) 1 & 3, 2 & 4

36.27 15.0 μm (width), 45.0 μm (separation)

36.29 a) 4790 slits/cm   b) 19.1°, 40.8°   c) no

36.31 a) yes   b) 13.3 nm

36.33 a) 4830 lines/cm   b) $\pm37.7°, \pm66.5°$

36.35 10.5°, 21.3°, 33.1°

36.37 a) 17,500   b) yes
c) (i) 587.8170 nm   (ii) 587.7834 nm
(iii) 587.7834 nm $< \lambda <$ 587.8170 nm

36.39 0.232 nm

36.41 0.559 nm

36.43 1.88 m

36.45 92 cm

36.47 1.45 m

36.49 a) 77 m (Hubble), 1100 km (Arecibo)
b) 1500 km

36.51 30.2 μm

36.53 a) 78   b) $\pm80.8°$ c) 555 μW/m^2

36.55 1.68

36.57 a) 1.80 mm   b) 0.796 mm

36.61 b) for $3\pi/2$: any two slits separated by one other slit; for the other cases: any two slits separated by three other slits

36.63 360 nm

36.65 second

36.67 c) ±2.61 rad

36.69 387 km

36.71 1.40

Chapter 37

37.1 bolt A

37.3 $0.867c$, no

37.5 a) $0.998c$ b) 126 m

37.7 1.12 h, in the spacecraft

37.9 92.5 m

37.11 a) 0.66 km   b) 49 μs, 15 km   c) 0.45 km

37.13 a) 3570 m   b) 90.0 μs   c) 89.2 μs

37.15 a) $0.806c$ b) $0.974c$ c) $0.997c$

37.17 a) toward   b) $0.385c$

37.19 $0.784c$

37.21 $0.611c$

37.23 $0.837c$, away from

37.25 a) $0.159c$ b) 172 million dollars

37.27 $3.06p_0$

37.29 a) $0.866c$ b) $0.608c$

37.31 a) 5.49×10^{15} m/s^2
b) 9.26×10^{14} m/s^2

37.33 a) $0.866c$ b) $0.986c$

37.35 a) 0.450 nJ   b) 1.94×10^{-18} kg$\cdot$m/s
c) $0.968c$

37.37 a) 3.3×10^{-13}%   b) 4.0×10^{-13} g, increases, no

37.39 a) 1.11×10^3 kg   b) 52.1 cm

37.41 a) 0.867 nJ   b) 0.270 nJ   c) 0.452

37.43 a) 5.34 pJ (nonrel), 5.65 pJ (rel), 1.06
b) 67.8 pJ (nonrel), 331 pJ (rel), 4.88

37.45 a) 2.06 MV
b) 0.330 pJ = 2.06 MeV

37.47 a) 4.2×10^9 kg = 4.6×10^6 tons
b) 1.5×10^{13} y

37.49 a) $\Delta = 8.42 \times 10^{-6}$ b) 34.0 GeV

37.51 $0.700c$

37.53 a) $0.995c$ b) 1%

37.55 a) $\Delta = 9 \times 10^{-9}$ b) $7000m$

37.57 0.168 MeV

37.59 a) $4c/5$ b) c c) (i) 145 MeV   (ii) 625 MeV
d) (i) 117 MeV   (ii) 469 MeV

37.65 b) $\Delta x' = \sqrt{(\Delta x)^2 - c^2(\Delta t)^2}$ c) 14.4 ns

37.67 $0.357c$, receding

37.69 a) 1000 y, 866 y, 1.4×10^{19} J, 14%
b) 505 y, 71 y, 5.5×10^{20} J, 550%
c) 501 y, 7.1 y, 6.3×10^{21} J, 6300%

37.71 2.04×10^{-13} N

37.75 a) toward us at 13.1 km/s, 39.4 km/s
b) 5.96×10^9 m, about 0.040 times the earth-sun distance; 5.55×10^{29} kg = $0.279\, m_{\text{sun}}$

37.77 a) 2494 MeV   b) 2.526   c) 987.4 MeV

Chapter 38

38.1 a) $K_2 = 4K_1$ b) $E_2 = 2E_1$

38.3 5.77×10^{14} Hz, 1.28×10^{-27} kg$\cdot$m/s, 3.84×10^{-19} J = 2.40 eV

38.5 a) 5.00×10^{14} Hz   b) 1.13×10^{19} photons/s
c) no

38.7 a) 4.8 eV   b) 6.1×10^{-34} J$\cdot$s
d) f_{th}, ϕ, and the horizontal-axis intercept are different, the slope is the same

38.9 249 km/s

38.11 2.14 eV

38.13 a) 264 nm   b) 4.70 eV

38.15 0.311 nm, same

38.17 1.13 keV

38.19 0.0714 nm, 180°

38.21 a) 4.39×10^{-4} nm   b) 0.04294 nm
c) 300 eV, loss   d) 300 eV

38.23 51.0°

38.25 a) 1.27×10^{-14} J   b) 9.46×10^{-14} J
c) 2.10 pm, less

38.27 1.19×10^{-27} kg$\cdot$m/s, 1.96×10^{-29} kg$\cdot$m/s

38.29 a) 1.04 eV   b) 1200 nm   c) 2.50×10^{14} Hz
d) 4.14×10^{-7} eV

38.31 a) 4.56×10^{14} Hz   b) 658 nm
c) 1.89 eV   d) 6.58×10^{-34} J$\cdot$s

38.33 a) 5.07 mJ   b) 11.3 W
c) 1.49×10^{16} photons/s

38.35 a) 6.99×10^{-24} kg$\cdot$m/s   b) 705 eV

38.37 6.28×10^{-24} kg$\cdot$m/s, 59.4°

38.39 a) 5×10^{-33} m   b) 4×10^{-9} deg
c) 0.1 mm

38.41 a) 319 eV, 1.06×10^7 m/s   b) 3.89 nm

38.43 a) 4.85 pm   b) 0.256 MeV

Chapter 39

39.1 a) 0.155 nm   b) 8.46×10^{-14} m

39.3 a) 2.37×10^{-24} kg$\cdot$m/s
b) 3.08×10^{-18} J = 19.3 eV

39.5 a) 0.332 nm, equals the circumference of the orbit   b) 1.33 nm, $^1/_4$ the circumference of the orbit

39.7 a) 8.8×10^{-36} m   b) no

39.9 a) 62.0 nm (photon), 0.274 nm (electron)
b) 4.96 eV (photon), 2.41×10^{-5} eV (electron)   c) ≈ 250 nm, electron

39.11 3.90×10^{-34} m, no

39.13 a) 0.0607 V   b) 248 eV   c) 20.5 μm

39.15 a) 7.3×10^6 m/s   b) 150 eV
c) 12 keV   d) electron
39.17 0.432 eV
39.19 a) $2.07°$, $4.14°$ b) 1.81 cm
39.21 a) 8260   b) electron
39.23 a) 3.63 MeV   b) 3.63 MeV
c) 1.32×10^7 m/s
39.25 3.16×10^{-34} kg·m²/s
39.27 a) -218 eV, 16 times   b) 218 eV, 16 times
c) 7.60 nm   d) $^1/_4$ hydrogen radius
39.29 a) 2.18×10^6 m/s, 1.09×10^6 m/s,
7.27×10^5 m/s
b) 1.53×10^{-16} s, 1.22×10^{-15} s,
4.13×10^{-15} s   c) 8.2×10^6
39.31 a) -17.50 eV, -4.38 eV, -1.95 eV, -1.10 eV,
-0.71 eV   b) 378 nm
39.33 a) -5.08 eV   b) -5.64 eV
39.35 5.32×10^{21} photons/s
39.37 4.00×10^{17} photons/s
39.39 a) 1.2×10^{-33} b) 3.5×10^{-17}
c) 5.9×10^{-9}
39.41 a) 2060 K   b) 1410 nm
39.43 1.06 mm, microwave
39.45 a) $1.7T$ b) 0.58
39.47 a) 97 nm, no   b) 8.2×10^9 m, $12R_{\text{sun}}$ c) no
39.49 a) 4.40×10^{-32} m/s   b) no
39.51 not valid
39.53 6.34×10^{-14} eV
39.55 a) 1.69×10^{-28} kg   b) -2.53 keV
c) 0.655 nm
39.57 a) 12.1 eV   b) 3; 103 nm, 122 nm, 657 nm
39.59 a) 0.90 eV
39.61 a) 5×10^{49} photons/s   b) 30,000
39.63 29,800 K
39.65 a) $h/2mc$
b) 6.61×10^{-16} m, independent of n
39.67 a) $I(f) = \dfrac{2\pi h f^5}{c^3(e^{hf/kT}-1)}$
c) $\sigma = 5.67 \times 10^{-8}$ W/m²·K⁴
39.69 a) 12 eV   b) 0.15 mV, 7300 m/s
c) 8.2×10^{-8} V, 4.0 m/s
39.71 a) no   b) 2.52 V
39.73 a) $E = c\sqrt{2mK}$ b) photon
39.75 1.66×10^{-17} m, no
39.77 b) $\Delta = \dfrac{m^2 c^2 \lambda^2}{2h^2}$ c) $\Delta = 8.50 \times 10^{-8}$
39.79 a) $\dfrac{h}{mc\sqrt{15}}$ b) (i) 1.53 MeV, 6.26×10^{-13} m
(ii) 2810 MeV, 3.41×10^{-16} m
39.81 a) 1.1×10^{-20} kg·m/s   b) 19 MeV
c) $U_{\text{coul}} = -0.29$ MeV, no
39.83 7.0×10^{-36} kg, 2.9×10^{-8}
39.85 a) 1.1×10^{-35} m/s   b) 2.3×10^{27} y, no
39.87 a) no   b) 1.51 V   c) 1.51 eV, about $^1/_4$ the
potential energy of the NaCl molecule
d) 1240 eV, yes
39.89 a) $2d \sin \theta = m\lambda$ b) $53.3°$ c) smaller
39.91 a) 248 eV   b) 0.0603 eV
39.93 a) $F = -\dfrac{A|x|}{x}$, $x \neq 0$ b) $E = \dfrac{3}{2}\left(\dfrac{h^2 A^2}{m}\right)^{1/3}$

Chapter 40

40.1 $\Psi(x,t) = Ae^{-i[4.27 \times 10^{10}\,\text{m}^{-1}]x}e^{-i[1.05 \times 10^{17}\,\text{s}^{-1}]t}$
40.3 a) $8\pi/k$ b) $4\omega/k$, yes
40.5 a) $\lambda/4, 3\lambda/4, 5\lambda/4, \dots$ b) $0, \lambda/2, \lambda, 3\lambda/2, \dots$
40.7 no
40.11 a) 1.6×10^{-67} J   b) 1.3×10^{-33} m/s,
1.0×10^{33} s   c) 4.9×10^{-67} J   d) no
40.13 1.66×10^{-10} m
40.15 0.61 nm
40.19 a) $0, L/2, L$ b) $L/4, 3L/4$ c) yes

40.23 a) 6.0×10^{-10} m (twice the width of the box),
1.1×10^{-24} kg·m/s   b) 3.0×10^{-10} m
(same as the width of the box),
2.2×10^{-24} kg·m/s
c) 2.0×10^{-10} m (2/3 the width of the box),
3.3×10^{-24} kg·m/s
40.25 b) yes
40.27 3.43×10^{-10} m
40.29 $-A\left(\dfrac{2mE}{\hbar^2}\right)\sin\left(\dfrac{\sqrt{2mE}}{\hbar}x\right) -$
$B\left(\dfrac{2mE}{\hbar^2}\right)\cos\left(\dfrac{\sqrt{2mE}}{\hbar}x\right)$
40.31 22 fm
40.33 a) 4.4×10^{-8} b) 4.2×10^{-4}
40.35 $1/\sqrt{2}$
40.37 a) 0.0013   b) 10^{-143}
40.39 1.11×10^{-33} J $= 6.93 \times 10^{-15}$ eV,
2.22×10^{-33} J $= 1.39 \times 10^{-14}$ eV
40.41 a) 0.21 eV   b) 5900 N/m
40.43 $(2n+1)\dfrac{\hbar}{2}$, increases with n
40.45 a) 5.89×10^{-3} eV   b) 106 μm   c) 0.0118 eV
40.47 a) $|\Psi(x,t)|^2 = \dfrac{2}{L}\left[1 - \cos\left(\dfrac{4\pi^2 \hbar t}{mL^2}\right)\right]$
b) $\dfrac{4\pi^2 \hbar}{mL^2}$
40.49 a) $\psi(x) = \dfrac{\sin k_0 x}{k_0 x}$
b) $0, \pm\pi/k_0$; $w_x = 2\pi/k_0, L$ c) $2L$
d) h, which is greater than $\hbar/2$
40.51 $B = \left(\dfrac{k_1 - k_2}{k_1 + k_2}\right)A$, $C = \left(\dfrac{2k_2}{k_1 + k_2}\right)A$
40.53 a) 19.2 μm   b) 11.5 μm
40.55 a) 0.818   b) 0.500   c) yes
40.57 a) $(2/L)\,dx$ b) 0   c) $(2/L)\,dx$
40.61 a) $A = C$, $B \sin kL + A \cos kL = De^{-\kappa L}$,
where $k = \dfrac{\sqrt{2mE}}{\hbar}$ b) $kB = \kappa C$,
$kB \cos kL - kA \sin kL = -\kappa De^{-\kappa L}$
40.65 6.63×10^{-34} J $= 4.14 \times 10^{-15}$ eV;
1.33×10^{-33} J $= 8.30 \times 10^{-15}$ eV
40.69 b) 134 eV
40.71 a) $E_n = \dfrac{(2n)^2 h^2}{8mL^2}$, $n = 1, 2, \dots$
b) $E_n = \dfrac{(2n+1)^2 h^2}{8mL^2}$, $n = 0, 1, 2, \dots$
c) same
d) part (a)'s are odd, part (b)'s are even
40.73 a) $x = \pm\sqrt{2E/k'}$
c) underestimates

Chapter 41

41.1 a) 1   b) 3
41.3 3.51 nm
41.5 $(2, 2, 1)$: $x = L/2$, $y = L/2$; $(2, 1, 1)$: $x = L/2$;
$(1, 1, 1)$: none
41.7 a) 0   b) 3.65×10^{-34} kg·m²/s
c) 3.16×10^{-34} kg·m²/s
d) 5.27×10^{-35} kg·m²/s   e) 1/6
41.9 4
41.11 $1.414\hbar, 19.49\hbar, 199.5\hbar$; as n increases, the
maximum L gets closer to $n\hbar$.
41.13 a) 18   b) $m_l = -4, 153.4°$
c) $m_l = +4, 26.6°$
41.17 a) 0.468 T   b) 3
41.19 a) 9   b) 3.47×10^{-5} eV
c) 2.78×10^{-4} eV
41.21 a) 2.5×10^{30} rad/s   b) 2.5×10^{13} m/s, not
valid since $v > c$

41.23 1.68×10^{-4} eV, $m_s = +^1/_2$
41.25 g
41.27 $n = 1, l = 0, m_l = 0, m_s = \pm^1/_2$: 2 states;
$n = 2, l = 0, m_l = 0, m_s = \pm^1/_2$: 2 states;
$n = 2, l = 1, m_l = 0, \pm 1, m_s = \pm^1/_2$: 6 states
41.29 a) $1s^2 2s^2$ b) $1s^2 2s^2 2p^6 3s^2$, magnesium
c) $1s^2 2s^2 2p^6 3s^2 3p^6 4s^2$, calcium
41.31 4.18 eV
41.33 a) $1s^2 2s^2 2p$ b) -30.6 eV   c) $1s^2 2s^2 2p^6 3s^2 3p$
d) -13.6 eV
41.35 a) -13.6 eV   b) -3.4 eV
41.37 a) 8.95×10^{17} Hz, 3.71 keV, 3.35×10^{-10} m
b) 1.68×10^{18} Hz, 6.96 keV, 1.79×10^{-10} m
c) 5.48×10^{18} Hz, 22.7 keV, 5.47×10^{-11} m
41.39 $3E_{1,1,1}$
41.41 a) $1/64 = 0.0156$ b) 7.50×10^{-4}
c) 2.06×10^{-3}
41.43 a) $4\pi A^2 r^2 e^{-2\alpha r^2} dr$ b) $1/\sqrt{2\alpha}$, no
41.45 a) $E = \hbar\left[(n_x + n_y + 1)\omega_1^2 + \left(n_z + \dfrac{1}{2}\right)\omega_2^2\right]$,
with n_x, n_y and n_z all nonnegative integers
b) $\hbar\left(\omega_1^2 + \dfrac{1}{2}\omega_2^2\right)$, $\hbar\left(\omega_1^2 + \dfrac{3}{2}\omega_2^2\right)$,   c) one
41.47 b) $n = 5$ shell
41.49 a) $0, \sqrt{2}\hbar, \sqrt{6}\hbar, \sqrt{12}\hbar, \sqrt{20}\hbar$
b) 7470 nm, no
41.51 a) $1.51e$, 9.49 electrons
b) (i) 1.8   (ii) -2.75 eV
41.53 a) $2a$ b) 0.238
41.55 b) 0.176
41.57 b) $(\theta_L)_{\max} = \arccos(-\sqrt{1 - 1/n})$
41.59 $4a$, same
41.61 $2 \to 1, 1 \to 0, 0 \to -1, \dfrac{e\hbar B}{2m}$;
$1 \to 1, 0 \to 0, -1 \to -1, 0$;
$0 \to 1, -1 \to 0, -2 \to -1, -\dfrac{e\hbar B}{2m}$
41.63 3.00 T
41.65 a) $0.99999978 = 1 - 2.2 \times 10^{-7}$ b) 0.9978
c) 0.978
41.67 a) 4, 20   b) $1s^4 2s^4 2p^3$
41.69 a) 122 nm   b) 1.52 pm, increase
41.71 a) 0.188 nm, 0.250 nm
b) 0.0471 nm, 0.0624 nm
41.75 a) $2a/Z$ b) 0.238, independent of Z

Chapter 42

42.1 277 nm, ultraviolet
42.3 a) 6.1 K   b) 34,600 K   c) He₂ no, H₂ yes
42.5 40.8 μm
42.7 5.65×10^{-13} m
42.9 a) 0.0644 nm (carbon), 0.0484 nm (oxygen)
b) 1.45×10^{-46} kg·m², yes
42.11 a) 1.03×10^{12} rad/s
b) 66.3 m/s (carbon), 49.8 m/s (oxygen)
c) 6.10×10^{-12} s
42.13 a) 7.49×10^{-3} eV   b) 166 μm
42.15 b) $\dfrac{l\hbar}{2\pi I}$
42.17 2170 kg/m³
42.19 a) 1.12 eV
42.21 1.20×10^6
42.23 1.5×10^{22} states/eV
42.25 a) $0.0233R$ b) 0.767%
c) no, motion of the ions
42.27 31.2%
42.29 0.20 eV below the bottom of the conduction band
42.31 a) (i) 0.0204 mA   (ii) -0.0196 mA
(iii) 26.8 mA   (iv) -0.491 mA
b) good for V between ± 1.0 mV, otherwise no
42.33 a) 5.56 mA   b) -5.18 mA, -3.77 mA
42.35 a) 977 N/m   b) 1.25×10^{14} Hz
42.37 a) 3.8×10^{-29} C·m   b) 1.3×10^{-19} C
c) 0.81   d) 0.058
42.39 a) 0.96 nm   b) 1.8 nm
42.41 a) 0.129 nm   b) 8, 7, 6, 5, 4   c) 485 μm
d) 118 μm, 134 μm, 156 μm, 188 μm, 234 μm

42.43 b) (i) 2.95 (ii) 4.73 (iii) 7.57 (iv) 0.838
(v) 5.69×10^{-9}
42.45 a) 1.146 cm, 2.291 cm
b) 1.171 cm, 2.341 cm; 0.025 cm $(2 \rightarrow 1)$,
0.050 cm $(1 \rightarrow 0)$
42.47 0.274 eV, much less
42.49 a) 4.24×10^{-47} kg·m^2 b) (i) 4.30 μm
(ii) 4.28 μm (iii) 4.40 μm
42.51 2.03 eV
42.53 a) 4.66×10^{28} atoms/m^3 b) 4.7 eV
42.55 b) 3.81×10^{10} Pa $= 3.76 \times 10^5$ atm
42.57 a) 1.67×10^{33} m^{-3} b) yes
c) 6.66×10^{35} m^{-3} d) no
42.59 a) $\dfrac{-2p^2}{4\pi\epsilon_0 r^3}$, attractive b) $\dfrac{+2p^2}{4\pi\epsilon_0 r^3}$, repulsive

Chapter 43

43.1 a) 14 proton, 14 neutrons
b) 37 protons, 48 neutrons
c) 81 protons, 124 neutrons
43.3 0.533 T
43.5 a) 76.21 MeV b) 76.68 MeV, 0.6%
43.7 0.5575 pm
43.9 a) 1.32 MeV b) 1.13×10^7 m/s
43.11 $^{86}_{36}$Kr: 8.73 MeV/nucleon,
$^{180}_{73}$Ta: 8.08 MeV/nucleon
43.13 a) $^{235}_{92}$U b) $^{24}_{12}$Mg c) $^{15}_{7}$N
43.15 156 keV
43.17 a) 0.836 MeV b) 0.700 MeV
43.19 5.01×10^4 y
43.21 a) 4.92×10^{-18} s^{-1} b) 2990 kg
c) 1.24×10^5 decays/s
43.23 a) 159 decays/min b) 0.43 decays/min
43.25 a) 0.421 decays/s b) 11.4 pCi
43.27 2.80 days
43.29 a) 2.02×10^{15}
b) 1.01×10^{15}, 3.78×10^{11} decays/s
c) 2.53×10^{14}, 9.45×10^{10} decays/s
43.31 a) 1.2 mJ b) 10 mrad, 10 mrem, 7.5 mJ
c) 6.2

43.33 500 rad, 2000 rem, 5.0 J/kg
43.35 a) 1.75 kGy, 175 krem, 1.75 kSv, 385 J
b) 1.75 kGy, 2.625 kSv, 262.5 krem, 385 J
43.37 a) 9.32 rad, 9.32 rem
43.39 a) $Z = 3, A = 6$ b) -10.14 MeV
c) 11.59 MeV
43.41 a) $Z = 3, A = 7$ b) 7.152 MeV c) 1.4 MeV
43.43 a) 173.3 MeV b) 4.42×10^{23} MeV/g
43.45 1.586 MeV
43.47 324 MJ
43.53 a) 4.14 MeV b) 7.75 MeV/nucleon
43.55 a) $^{90}_{39}$Y b) 25% c) 112 y
43.57 a) $^{25}_{13}$Al will decay into $^{25}_{12}$Mg b) β^+ or
electron capture
c) 3.255 MeV, 4.277 MeV
43.59 a) $^{14}_6$C $\rightarrow$ e$^-$ + $^{14}_7$N + $\bar\nu_e$ b) 0.156 MeV
c) 13.5 kg, 3400 decays/s
d) 530 MeV/s $= 8.5 \times 10^{-11}$ J/s
e) 36 μGy, 3.6 mrad, 36 μSv, 3.6 mrem
43.61 0.960 MeV
43.63 0.001286 u, yes
43.65 94.3 rad, 1900 rem
43.67 a) 5.0×10^4 b) $10^{-15,000}$
43.69 29.2%
43.71 a) 0.96 μJ/s b) 0.48 mrad/s c) 0.34 mrem
d) 6.9 days
43.73 1.0×10^4 y
43.75 a) 0.48 MeV
b) 3.270 MeV $= 5.239 \times 10^{-13}$ J
c) 3.155×10^{11} J/mol, more than a million
times larger
43.79 a) two b) 0.400 h, 1.92 h
c) 1.04×10^7 (short-lived), 2.49×10^7
(long-lived)
d) 1800 (short-lived), 4.10×10^6 (long-lived)

Chapter 44

44.1 69 MeV, 1.7×10^{22} Hz, 18 fm, gamma ray
44.3 a) 32 MeV
44.5 9.26×10^6 m/s
44.7 7.2×10^{19} J, 70%

44.9 a) 1.18 T b) 3.42 MeV, 1.81×10^7 m/s
44.11 a) 30.6 GeV b) 8.0 GeV
44.13 a) 3200 GeV b) 38.7 GeV
44.15 a) π^0, π^+ b) 219.1 MeV
44.17 1.63×10^{-25} kg, 97.2
44.19 116 MeV
44.21 (b) and (d)
44.23 (c) and (d)
44.27 a) 0, 1, -1, 0 b) 0, 0, 0, 1
c) $-e$, 1, 0, 0 d) $-e$, 0, 0, -1
44.29 p $\rightarrow$ e$^+$ + n + ν_e
44.31 a) $\bar{u}\,\bar{d}d$ b) no c) yes
44.33 a) 3.28×10^7 m/s b) 1510 Mly
44.35 a) 1.1×10^5 km/s b) 1.5
44.37 a) 3.8 atoms/m^3 b) 320 c) 2.0×10^{27}
44.39 a) 5.494 MeV b) 20.58 MeV
44.41 -0.783 MeV, endoergic
44.43 966 nm
44.45 a) 14.0 TeV b) 1.0×10^5 TeV
44.47 a) $\pi^- \rightarrow \mu^-$ + neutrino $\rightarrow$ e$^-$ + 3 neutrinos,
an electron and neutrinos b) 139 MeV
c) 2.24×10^{10} d) 50 Sv, 5.0 krem
44.49 2.494 GeV
44.51 a) 0, $+e$, 1, all lepton numbers are 0, K$^+$
b) 0, $-e$, 0, all lepton numbers are 0, π^-
c) -1, 0, 0, all lepton numbers are 0,
antineutron ($\bar{n}$)
d) 0, $+e$, 0, muonic lepton number is -1, all
other lepton numbers are 0, μ^+
44.53 7.5×10^{-23} s
44.55 a) 0.70 rad b) 0.70 rem, no
44.57 c) $H_0 = \dfrac{dR/dt}{R}$ d) $R(t) = R_0 e^{H_0 t}$ e) no
44.59 b) $R/R_0 = 0.574$
c) speeding up at 300 My, slowing down at
10.2 Gy
44.61 230 MeV, 12.5° below the $+x$-axis

PHOTO CREDITS

INDEX

Note: Page numbers followed by f indicate figures; those followed by t indicate tables.

A

Abdus, Salam, 1500
Absolute conservation laws, 1495
Absolute pressure, 377–378
Absolute temperature, 517
Absolute temperature scale, 556, 668–669
Absolute zero, 556, 669
Absorption lines, 1203
Absorption spectra
  line, 1293, 1297–1300, 1310–1314
  X-ray, 1396
AC source, 1022. *See also* Alternating-current circuits
Acceleration
  angular, 282–285, 284t, 311–314
  around curve, 74, 88
  average, 42–43. *See also* Average acceleration
  calculating by integration, 55–57
  centripetal, 86–87, 154
  centripetal component of, 286–287
  changing, 55–57
  circular motion and, 85–87
  constant, 46–52. *See also* Constant acceleration
  definition of, 43
  fluid resistance and, 152–154
  inertial frame of reference and, 110–112
  instantaneous, 43–44. *See also* Instantaneous acceleration
  linear, 282, 284t
  mass and, 113, 114, 118–120
  net force and, 112–118
  Newton's first law and, 108–112
  Newton's second law and, 112–117, 140–146
  of particle in wave, 480–482
  projectile motion and, 77–80, 87
  of rocket, 262–264
  of rolling sphere, 319–320
  signs for, 45, 46
  in simple harmonic motion, 444, 448
  tangential component of, 286
  units of, 117
  vs. velocity, 42
  on v_x-t graph, 44–46
  weight and, 118–119
  of yo-yo, 319
Acceleration due to gravity, 700
  apparent weight and, 143, 422
  definition of, 52
  at different latitudes and elevations, 422t
  in free fall, 52–55, 118, 143
  magnitude of, 405
  mass vs. weight and, 118–120
  variation with location and, 119
  vs. gravitation, 403
  weightlessness and, 143
Acceleration vectors, 35, 72–77, 283
  average, 73–75
  instantaneous, 73–75
  parallel and perpendicular components of, 75–77
Accelerators, 1485–1488. *See also* Particle accelerators
Acceptor level, 1425
Acceptors, 1425
Accretion disk, 425
Accuracy, 8
  vs. precision, 9
Acrobats, in unstable equilibrium, 228
Action-reaction pairs, 120–123
  gravitational forces as, 403
Activation energy, 610
Activity, in radioactive decay, 1456
Addition

significant figures in, 9
  of vectors, 12–18
Adiabatic process, 634–635
  Carnot cycle and, 663
  for ideal gas, 640–642
Aging, relativity and, 1233
Air
  dielectric strength of, 768, 805
  as insulator, 571
  ionization of, 768–771
Air conditioners, 660–661
Air drag, 152–154
Air pressure, 375–376
Air resistance, projectile motion and, 77, 79–80
Airplanes
  banked curves and, 158
  noise control for, 531, 532
  sonic boom from, 538
  wing lift and, 388–389
  wing resonance in, 460
Airy disk, 1209
Airy, George, 1209
Alkali metals, 1390
  Bohr atomic model for, 1306
Alkaline earth elements, 1390
Alkaline earth metals, 1390
Alpha decay, 1450–1452
Alpha particles, 1294–1295
  emission of, 1450–1451
  tunneling and, 1349–1350
Alternating current, 822, 850, 1021
  applications of, 868
  dangers of, 1040
  lagging, 1036
  measurement of, 1022–1024
  rectified, 1022–1023
  rectified average, 1023
  root-mean-square value of, 1023–1024
Alternating-current circuits
  capacitors in, 1027–1028, 1029–1030, 1029t
  complex numbers in, 1049–1050
  impedance of, 1031–1033
  inductors in, 1025–1027, 1029, 1029t
  L-R-C series, 1030–1034
  phase angle and, 1026, 1031–1032
  phasors and, 1022
  power in, 1034–1037
  resistance and reactance in, 1024–1030
  resistors in, 1025, 1029, 1029t
  resonance in, 1037–1039
  tailoring, 1038–1039
  transformers and, 1040–1042
Alternators, 963–964, 1021–1022
Ammeters, 831, 860–861
  voltmeters and, 862–863, 1024
Amorphous solids, 1412
Ampere, 695, 820, 931–932
Ampère, André, 885
Ampere's law, 935–941. *See also* Maxwell's equations
  applications of, 938–941
  displacement current and, 975–977
  electromagnetic waves and, 1052, 1057, 1063
  general statement of, 937–938
  generalization of, 975–976
Amplitude
  displacement, 510, 518–519
  of electromagnetic waves, 1061
  of oscillation, 438
  of pendulum, 454–455
  pressure, 511–512, 519–521
  of sound waves, 510–513
Analyzers, 1095–1096
Anderson, Carl D., 1482, 1485
Angle(s)

notation for, 71
  polarizing, 1097
  radians and, 279, 287
Angle of deviation, 1111
Angle of incidence, critical, 1089
Angle of reflection, 1084
Angular acceleration, 282–285
  angular velocity and, 282
  calculation of, 283
  constant, 283–285, 284t
  torque and, 311–314
  as vector, 283
  vs. linear acceleration, 284t
Angular displacement, 279
  torque and, 320–322
Angular frequency, 438–439
  of electromagnetic waves, 1061
  natural, 459–460
  of particle waves, 1331
  period and, 438–439
  in simple harmonic motion, 441–442
  vs. frequency, 442
Angular magnification, vs. lateral magnification, 1147
  of microscope, 1147, 1148–1149
Angular momentum
  axis of symmetry and, 323–324
  of the body, 324
  conservation of, 325–328
  definition of, 322
  of electrons, 942
  of gyroscope, 328–330
  nuclear, 1442
  orbital, 1373–1374, 1384, 1387
  precession and, 328–330
  rate of change of, 323, 324
  rotation and, 322–328
  spin, 1384–1385, 1387, 1442
  torque and, 323, 324
  total, 1387, 1442
  as vector, 324, 328
Angular simple harmonic motion, 451
Angular size, 1146
Angular speed, 280
  instantaneous, 286
  precession, 329
  rate of change of, 286
Angular velocity, 279–282
  angular acceleration and, 282
  average, 279
  calculation of, 281
  instantaneous, 280
  rate of change of, 286
  as vector, 281–282
  vs. linear velocity, 280
Angular vs. linear kinematics, 285–288
Anomalous magnetic moment, 1443
Antimatter, 1516
Antineutrinos, 1492
Antineutrons, 1492
Antinodal curves, 1166
Antinodal planes, 1070
Antinodes, 492
  displacement, 523
  pressure, 523
Antiparallel vectors, 11, 12
Antiparticles, 1483
Antiprotons, 1491–1492
Antiquarks, 1496, 1499
Aphelion, 415
Apparent weight, 142
  acceleration due to gravity and, 143, 422
  Earth's rotation and, 421–423
  magnitude of, 422
Appliances, power distribution systems in, 868–872

Archimedes' principle, 380
Arecibo telescope, 1218
Aristotle, 52
Astigmatism, 1144
Aston, Francis, 897
Astronomical distances, units for, 1503
Astronomical telescopes, 1119, 1149–1151
Atmosphere, 355, 375
Atmospheric pressure, 355, 375–376
    elevation and, 594–595
    measurement of, 378–380
Atom(s)
    energy levels of. *See* Energy levels
    excited, 1298
    hydrogen. *See* Hydrogen atom
    interactions between, 451–453
    in magnetic field, 1379–1382
    many-electron, 1387–1393
    muonic, 1379
    nucleus of, 689, 1295
    Rydberg, 1306, 1402
    structure of. *See* Atomic structure
    Thomson's model of, 751–752
Atomic mass, 598, 690, 1305–1306
    measurement of, 897, 1441
Atomic models
    Bohr's, 1297–1306
    Rutherford's, 1294–1296
    Thomson's, 1293, 1294–1295
Atomic number, 690, 1306, 1379, 1387–1388, 1441
Atomic particles, 689–690. *See also* Electron(s);
        Neutron(s); Proton(s)
Atomic spectra, 1292, 1297–1300
    in Balmer series, 1304
    in Brackett series, 1304
    in Lyman series, 1304
    in Pfund series, 1304
Atomic structure, 689, 1364–1398
    central-field approximation and, 1388
    electron spin and, 1383–1387
    exclusion principle and, 1388–1393
    of hydrogen atom, 1372–1378
    of hydrogenlike atoms, 1378–1379
    of many-electron atoms, 1387–1393
    Moseley's law and, 1394–1396
    particle in three-dimensional box and, 1366–1371
    periodic table and, 1389, 1390–1393
    Schrödinger equation and, 1365–1366
    X-ray spectra and, 1393–1396
    Zeeman effect and, 1379–1382
Attenuator chains, 881–882
Audible range, 509
Automobiles
    gas compression in, 593–595
    ignition systems in, 1000
    Newton's second law and, 115
    power distribution systems in, 868, 870–871
    vertical simple harmonic motion in, 451
    weight distribution for, 349
Autoradiography, 1461
Available energy, 1487–1488
Avalanche breakdown, 1428
Average acceleration, 42–43
    definition of, 42
    units of, 42
    vs. average velocity, 42
    on v_x-t graph, 44–46
    x-component of, 42
Average acceleration vectors, 73–75
Average angular acceleration, 282
Average angular velocity, 279
Average density, 374
Average power, 193, 487
Average speed, 39
Average velocity, 36–40
    definition of, 36
    instantaneous velocity and, 38–41
    straight-line motion and, 36–38
    x-component of, 36–38
    on x-t graph, 37–38

Average velocity vectors, 70–72
Avogadro's number, 598
Axis
    elliptical, 415
    optic, 1118
    polarizing, 1094–1095
    semi-major, 415, 416
Axis of rotation, 281–282, 286–287
    change in direction of, 281–282, 286–287
    fixed, 278–279
    moment of inertia for, 312
    moving, 324–320
    parallel-axis theorem and, 293–294
    through center of mass, 294
Axis of symmetry, angular momentum and, 323–324
Axis system, right-handed, 24
Axons, 716, 781, 881–882
a_x-t graphs
    for changing acceleration, 55–57
    for constant acceleration, 46–49

B

Back emf, 908
Background radiation, 1515
Back-of-the-envelope calculations, 10
Bacteria, rotation in, 283f
Bainbridge's mass spectrometer, 897
Balance
    Cavendish (torsion), 404
    spring, 106
Ballistic pendulums, 253
Balmer series, 1304
Band spectra, 1411
Banked curves, 158
Bar, 375
Bar magnets, 883, 905–907
Bardeen, John, 1430
Barometers, mercury, 378–380
Baryons, 1492, 1493–1494, 1496
Baseball, curve ball in, 391
Batteries
    charging, 835–836
    as current source, 831
    energy in, 834
Beams, 1083
Beats, 531–532
Becker, Herbert, 1481
Becquerel, 1457
Bednorz, Johannes, 824
Bees, vision in, 1101
Bell, Alexander Graham, 521
Bernoulli's equation, 385–389
Beryllium, 1390
Beta decay, 1452–1453
Beta-minus particles, 1452–1453
Beta-plus particles, 1453
Bias conditions, 1426, 1427–1428
Big Bang, 1503–1504
Big Crunch, 1505
Bimetallic strip thermometer, 553
Binary star systems, 425–426, 1259
Binding energy, 1406, 1445–1446
Binoculars, 1150
Binomial theorem, 451, 452–453
Biological efficiency, 655
Biologically equivalent dose, 1460
Biot-Savart law, 927
Bipolar junction transistors, 1429
Bird song, 522
Bird vision, 1144
Bird wings
    flapping frequencies of, 438f
    moment of inertia of, 290f
Birefringence, 1100
Black holes, 423–426
Blackbody, 576
Blackbody radiation, 1310–1314
Blackett, Patrick, 1271
Bloch, Felix, 1416f
Blood flow, turbulence in, 390f

Blubber, as insulator, 572
Blue-ray discs, 1210, 1309
Body. *See* Human body
Bohr magneton, 941–942, 1380, 1442–1443
Bohr, Niels, 1273, 1389f
Bohr's atomic model, 1297–1306
    energy levels and, 1297–1300
    for hydrogen, 1300–1305, 1372
    limitations of, 1372
    photon emission and absorption and, 1297
    uncertainty principle and, 1317
    vs. Schrödinger analysis, 1373
Boiling, 566
Boltzmann constant, 601
Bonds, 1405–1408
    covalent, 1406–1407
    hydrogen, 1407–1408
    ionic, 1406, 1407
    metallic, 1408
    in solids, 1414–1415
    strong, 1407
    van der Waals, 1407
    weak, 1407
Bone cancer, radioisotope imaging for, 1391
Born, Max, 1333f
Bose-Einstein distribution, 1492
Bosons, 1492, 1500, 1501
Bothe, Walther, 1481
Bottomness, 1499
Bound charges, 806
Bound state, 1343–1344
Boundary conditions
    for harmonic oscillator, 1351–1351
    for waves, 489–490
Bourdon pressure gauge, 379f
Brackett series, 1304
Bragg condition, 1207
Bragg reflection, 1207
Brahe, Tycho, 414–415
Breaking stress, 358
Bremsstrahlung, 1267, 1268
Brewster's law, 1097–1098
Bridge circuits, 855–860, 880–881
Bright fringes, 1194, 1195
Brillouin, Léon, 1361
British system, 5–6, 117. *See also* Units of measure
British thermal unit (Btu), 562
Brittle material, 358
Bulk modulus, 355
Bulk strain, 354–356
Bulk stress, 352f, 354–356
Buoyancy, 380–382
Butterfly wings, interference and, 1179

C

Cables, winding/unwinding, 291–292, 313–314
Calcite, birefringence in, 1100
Calculations
    back-of-the-envelope, 10
    calorimetry, 568–570
    estimation and, 10
    units of measure in, 6. *See also* Units of measure
Calorie (cal), 562
Calorimetry calculations, 568–570
Calorimetry, phase changes and, 565–570
Cameras, 1139–1142
    flash unit of, 797
    focusing, 1137
    gamma, 1461
    resolving power and, 1210
Cancer
    imaging methods in, 1180, 1391
    magnetic nanoparticles for, 946
    radiation effects in, 1269
Capacitance
    calculation of, 789–793
    definition of, 788, 789
    equivalence, 794
    units for, 789
    vs. coulombs, 789

Capacitive reactance, 1028–1029
Capacitors, 788–800
  in ac circuits, 1027–1028, 1029–1030, 1029t, 1035–1036
  applications of, 788, 797–798, 802–803
  capacitance of, 789. *See also* Capacitance
  capacitive reactance of, 1028–1029
  charge storage in, 797
  charging, 864–866, 975–976
  cylindrical, 792–793
  definition of, 788
  dielectrics in, 800–805
  discharging, 867–868
  electric-field energy and, 798, 799
  electrolytic double-layer, 802
  energy storage in, 788, 796–800
  in networks, 796
  in pacemakers, 866
  in parallel, 794–796, 852–853
  parallel-plate, 790, 791, 800–805
  in series, 793–794, 795–796, 852
  spherical, 792, 808
  symbols for, 789
  in touch screens, 794
  in vacuum, 789–791, 798
Carbon dating, 1458
Carbon dioxide, greenhouse effect and, 576–577
Carnot cycle, 663–669
  efficiency of, 667–668
  of heat engine, 663–666
  for ideal gas, 664–665
  Kelvin temperature scale and, 668–669
  of refrigerator, 666–667
  reversibility of, 667–668
  second law of thermodynamics and, 667–668
Cars. *See* Automobiles
Cathode ray tubes, for magnetic field measurement, 887–888
Cavendish balance, 404
Celestial dynamics, 402
Cell imaging, 1180
Cell membrane
  dielectric, 805
  potential gradient across, 774
Celsius temperature scale, 553
Center of curvature, 1118
Center of gravity, 345–348
Center of mass, 258–262
  center of gravity and, 345–347
  combined rotational-translational motion and, 315–316
  external forces and, 261–262
  motion of, 259–262
  planetary motion and, 417–418
  torque and, 312
Center-of-momentum system, 1272, 1487
Centigrade scale, 553
Central force, 416
Central-field approximation, 1388
Centrifugal force, 155
Centripetal acceleration, 86–87, 154
Centripetal component of acceleration, 286–287
Čerenkov radiation, 1257
Cgs metric system, 117
Chadwick, James, 1481
Chain reactions, 1466
Charge distribution, 703, 1390
  electric fields and, 725–728, 734–735, 746t. *See also* Gauss's law
  static, 759–760
Charged particles, motion in magnetic fields, 892–898
Charging by induction, 692
  polarization and, 693
Charm, 1499
Chemical reactions. *See* Reaction(s)
Chernobyl accident, 1468
Chokes, 995–998
Chromatic resolving power, 1203–1204, 1210
Circle
  circumference of, 279

reference, 440–441
Circuit breakers, 870
Circuit diagrams, 831–833
Circuits. *See* Electric circuits
Circular apertures, diffraction and, 1208–1211
Circular motion, 85–88, 440–442
  acceleration and, 85–87, 88
  dynamics of, 154–159
  nonuniform, 88, 159
  period of, 87
  uniform. *See* Uniform circular motion
  velocity and, 85–87
  vs. projectile motion, 87
Circular orbits, 412–413, 416–417
Circular polarization, 1099–1100
Circumference, 279
Classical mechanics, 104
Classical turning point, 1361–1362
Clausius statement, 662
Climate change, 576–577
Close packing, 1415
Closed orbits, 412
Closed surface, electric flux through, 725
Clotheslines, waves on, 480
Cloud chambers, 1482
Coefficient of kinetic friction, 147
Coefficient of linear expansion, 557–558, 559
Coefficient of performance, 659
Coefficient of resistivity, 824–825
Coefficient of static friction, 148
Coefficient of volume expansion, 558–560
Coherent waves, 1165
Coils
  Hemholtz, 954
  inductance of. *See* Inductance
  magnetic fields of, 932–935
  magnetic torque on, 904–905
  search, 983
  Tesla, 993–994
Cold fusion, 1471
Cold reservoirs, 655
Colliding-beam experiments, 1489
Collisions, 251–258
  atomic energy levels and, 1297–1298
  classification of, 254–255
  definition of, 251
  elastic, 251, 254, 255–258
  inelastic, 251–255
  kinetic energy in, 252
  momentum conservation and, 251–258
Combustion, 568
Comet Halley, orbit of, 417
Common-emitter circuits, 1429
Commutators, 907–908, 964
Compensated semiconductors, 1425
Complementarity principle, 1273
Complete circuits, 822, 828–831
Completely inelastic collisions, 251–255
Component vectors, 14–18, 106
Components, of vectors, 14–19, 21–22, 106–107
  vs. component vectors, 14, 106
Compound microscope, 1147–1149
Compressibility, fluid, 356, 382
Compression
  definition of, 354
  fluid density and, 476
Compression ratio, 657
Compressive strain, 354
Compressive stress, 354
Compton scattering, 1269–1271
Computed tomography, 1268–1269
Concave mirrors, 1118–1122
Concentration of particles, in current, 820
Condensation, 566
Condensed matter, 1412. *See also* Liquid(s); Solids
Condensor microphones, 790
Conditional conservation laws, 1495
Conduction, 570–574
Conduction bands, 1416–1417
Conduction current, 975

Conductivity
  electrical, 823
  intrinsic, 1424
  microscopic model of, 838–840
  thermal, 571, 823
Conductors, 691–692
  in capacitors, 789
  conductivity of, 823
  current density in, 821–822
  current flow in, 820–822
  diodes of, 827
  electric charge on, 736, 741–745
  electric fields at, 701, 744–745
  electron motion in, 819
  energy bands in, 1417
  equipotential surfaces and, 772–773
  holes in, 909
  interaction force between, 931–932
  magnetic fields of, 928–932
  magnetic force of, 931–932
  magnetic forces on, 898–901
  metallic, 838–839
  nonohmic (nonlinear), 823
  ohmic (linear), 823
  particle concentration in, 820–821
  resistance of, 825–828, 830, 833
  resistivity of, 822–825
  semiconductors, 823, 827, 909
  superconductors, 824, 968
  thermal, 552–553
Conservation laws
  absolute, 1495
  conditional, 1495
  universal, 690
Conservation of angular momentum, 325–328
  of planets, 415–416
Conservation of baryon number, 1494
Conservation of electric charge, 690
  Kirchoff's junction rule and, 856
Conservation of electrostatic force, 856
Conservation of energy, 176, 209, 224
  with electric force, 758–759
  in simple harmonic motion, 446–449
Conservation of lepton number, 1492
Conservation of mass and energy, 1247–1248
Conservation of mass in fluid, 383–384
Conservation of mechanical energy, 755
Conservation of momentum, 247, 1243
  collisions and, 251–258
Conservative forces, 221–229
  elastic collisions and, 255
  work done by, 755
Consonance, 532
Constant acceleration, 46–52
  due to gravity, 52–55
  equations of motion for, 49
  of freely falling bodies, 52–55
  Newton's second law and, 112–117
Constant angular acceleration, 284t
  rotation with, 283–285
Constant forces, 177
Constant linear acceleration, 284t
Constant torque, 321
Constant velocity, 51
  Newton's first law and, 108–112
Constant-pressure process, 635
Constant-temperature process, 635
Constant-volume process, 635
Constructive interference, 492, 530, 1164–1166, 1168–1170
  in holography, 1211–1213
  in X-ray diffraction, 1206–1207
Contact force, 105, 146
Contact lenses, 1143–1145
Continuity equation, 383–384
Continuous lasers, 1309
Continuous spectra, 1310–1314
Convection, 570, 574
Conventional current, 820
Converging lenses, 1131–1133

Converging mirrors, 1119
Convex mirrors, 1122–1124
Cooling, evaporative, 568
Cooper pairs, 1430
Coordinate system, right-handed, 24
Coordinates
   spacetime, 1238
   spherical, 1366
Copernicus, Nicolaus, 414
Cornea, 1142, 1156, 1157, 1159
Corona discharge, 768–769
Correspondence principle, 1249
Cosmic background radiation, 1515
Cosmic inflation, 1511
Cosmic-ray experiments, 1489–1490
Cosmological principle, 1503
Cosmological redshift, 1505
Coulomb, 695–696
   vs. capacitance, 789
Coulomb's law, 597, 693–698
   Gauss's law and, 732
   proportionality constant in, 695
   statement of, 694
   superposition of forces and, 696
Coupling constant, 1491
Covalent bonds, 1406–1407
Covalent crystals, 1414–1415
Cowan, Clyde, 145
Critical angle, 1089
Critical damping, 458
Critical density, 1505–1507
Critical fields, 979
Critical point, 611
Critical temperature, 596, 979
Critically damped circuits, 1010
Cross (vector) product, 23–25
Crystal(s)
   covalent, 1414–1415
   ideal single, 1413
   imperfect, 1415
   ionic, 1414–1415
   liquid, 1412
   metallic, 1415
   perfect, 1413–1415
   structure of, 1412–1415
   types of, 1414–1415
Crystal lattice, 597, 1413–1414
Crystalline lens, 1142–1143
Crystalline solids, 1412–1415
Cube, electric flux through, 731
Curie, 1457
Curie constant, 944
Curie, Marie, 1454
Curie, Pierre, 944, 1454
Curie's law, 944
Current
   alternating, 822
   capacitor, 1027–1028, 1029, 1029t
   in circuits, 828–831
   concentration of particles in, 820
   conduction, 975
   conventional, 820
   definition of, 818, 819
   direct, 822, 850. See also Direct-current circuits
   direction of, 819–820, 825
   displacement, 975–977
   drift velocity and, 819, 820
   eddy, 974–975, 1042
   electric charge in, 819–820
   electric field and, 819–820
   electromotive force and, 991. See also Inductance
   electron motion in, 819
   full-wave rectifier, 1023
   generation, 1427
   heat, 571
   induced, 958, 967–968
   inductance and, 991. See also Inductance
   inductor, 1025–1026, 1029, 1029t
   Kirchoff's rules for, 855–860
   lagging, 1036

   measurement of, 860–861
   notation for, 865
   Ohm's law and, 822, 825–826
   recombination, 1427
   rectified average, 1023
   resistance and, 825–828
   resistor, 1025, 1029, 1029t
   root-mean-square, 1023–1024
   saturation, 1426
   sinusoidal, 1022–1024. See also Alternating
     current
   time-varying, 865
   units of, 695, 820
   "using up," 830
   voltage and, 825–828
   vs. current density, 821–822
Current amplitude, 1022
Current density, 821–822
   definition of, 821
   resistivity and, 823
   vector, 821
   vs. current, 821–822
Current flow, direction of, 819–820
Current loops. See also Magnetic dipoles
   force and torque on, 901–907
   magnetic fields on, 932–935
   magnetic moment of, 903, 934
   in magnetization, 941–942
Current-carrying conductor, magnetic forces on,
   898–901
Curve ball, 391
Curves
   acceleration around, 74, 88
   antinodal, 1166
   banked, 158
   gravitational potential energy and, 212–216
   magnetization, 945
   motion along, 191–193
   nodal, 1166
   resonance, 528, 1038–1039
   response, 1038–1039
   work-energy theorem for, 187–191
Cycles, 438
Cyclic process, in heat engines, 654
Cyclotron frequency, 893
Cyclotrons, 893, 918, 1486–1487
Cylinders, moment of inertia of, 295–296
Cystic fibrosis, sweat chloride test for, 695

D

Dalton, John, 1480–1481
Damped harmonic motion, 1009
Damped oscillations, 457–460
Damping, 457
   critical, 458
Dark energy, 1508
Dark fringes, 1193–1195
Dark matter, 1507
D'Arsonval galvanometer, 860, 863, 904, 1022
Daughter nucleus, 1454
Davisson, Clinton, 1287–1288
Davisson-Germer experiment, 1287–1288
DC circuits. See Direct-current circuits
De Broglie, Louis, 1286–1287
De Broglie wavelength, 1287, 1290
Decay. See Radioactive decay
Decay constant, 1456
Deceleration, 45
Decibel scale, 521
Deformation
   Hooke's law and, 352, 357–358
   plastic, 358
   reversible, 358
   stress and strain and, 352–357
Degeneracy, 1370–1371, 1374–1375
Degrees, 553, 555
Degrees of freedom, 606
Density, 373–375, 374t
   average, 374
   current, 821–822

   definition of, 373
   displacement current, 976
   of Earth, 408
   energy, 798, 1064–1065
   fluid, 380–381, 476
   linear charge, 704
   linear mass, 482
   magnetic energy, 999–1000
   magnetic flux, 892
   mass:volume ratio and, 373–374
   measurement of, 374
   nuclear, 1440
   probability, 1333
   of states, 1418–1419
   surface charge, 704
   volume charge, 704
   vs. pressure, 592
Depth, fluid pressure and, 376–377
Derivatives, 39
   partial, 226–227, 481
Destructive interference, 492, 529, 1164–1166,
   1168–1170
Detectors, 1489–1490
Deuterium, 1445
Deuterons, 1305–1306, 1445, 1511
Dewar flask, 576
Dewar, James, 576
Diamond structure, 1413–1414
Diamonds, 1090, 1092
Diatomic molecules, 605–606
Dichroism, 1094
Dielectric breakdown, 800, 804–805, 1417
Dielectric cell membrane, 805
Dielectric constant, 800–801
Dielectric function, 1064
Dielectric strength, 805
   of air, 768, 805
Dielectrics, 800–805
   electromagnetic waves in, 1063–1064
   Gauss's law and, 807–808
   permittivity of, 802
   polarization of, 801–803, 805–807
Diesel engines, 658–659
   adiabatic compression in, 642
Difference tone, 532
Differential principle, 245
Diffraction, 1190–1214
   bright fringes in, 1194, 1195
   with circular apertures, 1208–1211
   complementarity principle and, 1273–1274
   dark fringes in, 1193–1195, 1194
   definition of, 1191
   electron, 1287–1288
   Fraunhofer, 1192, 1193
   Fresnel, 1192, 1193
   holography and, 1211–1213
   Huygen's principle and, 1191–1192
   image formation and, 1209–1210
   intensity in, 1195–1199
   multiple-slit, 1199–1201
   photons and, 1273–1274
   resolving power of, 1209–1211
   single-slit, 1192–1199
   of sound, 1189
   vs. interference, 1192, 1200
   X-ray, 1205–1208
Diffraction gratings, 1201–1205
Diffuse reflection, 1083, 1115
Digital multimeters, 863
Dimagnetism, 943t, 944, 980
Dimensional consistency, 6
Dinosaurs, physical pendulum and, 456–457
Diode(s), 827
   light-emitting, 1428
   p-n junction, 1426
   tunnel, 1349
   Zener, 1428
Diode rectifier, 1426
Diopters, 1144–1145
Dipoles

electric, 709–713
magnetic, 903–904
Dirac distribution, 1419
Dirac equation, 1482–1483
Dirac, Paul, 1385, 1482–1483
Direct current, 822
dangers of, 1040
Direct-current circuits, 822, 850–873
in automobiles, 868
definition of, 850
Kirchoff's rules for, 855–860
measuring instruments for, 860–864
in power distribution systems, 868–872
R-C, 864–868
resistors in series and parallel in, 850–855
Direct-current generators, 964–965
Direct-current motors, 907–908
Direction, 10
of force, 10
of vectors, 11, 16
of waves, 479, 481–482
Discus throwing, 287
Disk, electric flux through, 730–731
Dislocations, in crystals, 1415
Dispersion, 1085, 1091–1093
Displacement, 11, 12
angular, 279
definition of, 11
multiplication of, 13
in oscillation, 438
in simple harmonic motion, 439–440, 443–445
straight-line motion and, 36–38
superposition principle and, 490–491
vector sum (resultant) of, 12
wave pulse and, 489–491
work and, 177–178, 181–183
Displacement amplitude, 510
sound intensity and, 518–519
Displacement current, 975–977
Displacement nodes/antinodes, 523
Dissipative forces, 222
Dissonance, 532
Distance
astronomical, 1503
image, 1116
object, 1116
relativity of, 1233–1237
Distribution function, 608
Diverging lenses, 1133
Division
significant figures in, 9
of vectors, 70
DNA
base pairing in, 717
measurement of, 1204
X-ray diffraction of, 1207
Dogs, panting by, 460
Donor, 1424
Donor level, 1424
Doping, 1424–1425
Doppler effect
for electromagnetic waves, 537–538, 1241–1243
for sound waves, 533–537, 1242
Doppler frequency shift, 1242
Doppler shift, vs. redshift, 1502
Dosimetry, radiation, 1459
Dot (scalar) product, 20–22
Down (quark), 1496
Drag, 152–154
Drift velocity, 819
current and, 820
Hall effect and, 909
Driven oscillation, 459–460
Driving force, 459–460
Ducks, swimming speed of, 486
Ductile material, 358
Dulong-Petit rule, 565, 608
DVD players, 1202, 1210, 1309
Dynamics. See also Force; Mass; Motion
celestial, 402

of circular motion, 154–159
definition of, 35, 104
fluid, 373, 389–390
Newton's second law and, 112–117, 140–146
of rotational motion, 308–331
Dyne, 117

E

Ear, sensitivity of, 528
Earth
density of, 407–408
magnetic fields of, 884, 887
rotation of, 421–423
surface temperature of, 576
Eccentricity, orbital, 415
Eddy currents, 974–975
in transformers, 1042
Edge dislocation, 1415
Edison, Thomas, 1021
Efflux, speed of, 387
Eggs, mechanical energy of, 209
Eightfold way, 1497–1498
Einstein, Albert, 91, 1223, 1263–1264, 1481
Elastic collisions, 251, 254, 255–258
relative velocity and, 256
Elastic deformations, Hooke's law and, 352, 357–358
Elastic hysteresis, 358
Elastic limit, 358
Elastic modulus, 352
bulk, 354–356
shear, 356–357
Young's, 353–354
Elastic potential energy, 216–221, 225
definition of, 217
gravitational potential energy and, 218
Elasticity, 344–359
Electric charge, 688–691
attraction and repulsion and, 688
bound, 806
in capacitors, 788. See also Capacitors
in closed surface, 726
on conductors, 736, 741–745
conservation of, 690, 856
definition of, 688
density of, 704
distribution of. See Charge distribution
electric dipole and, 709–713
electric field and, 725–728, 734–736. See also
Gauss's law
flux and, 725–732. See also Electric flux
free, 806
induced, 692, 693, 805–807
magnetic force on, 886–887
magnitude of, 694
negative, 688
in nerve cells, 741
notation for, 865
point. See Point charges
positive, 688
quantitized, 691
structure of matter and, 689–690
superposition of forces and, 696
time-varying, 865
typical values for, 696
vs. magnetic poles, 885
Electric circuits
alternating-current, 822, 850
bridge, 855–860, 880–881
common-emitter, 1429
complete, 822, 828
critically damped, 1010
diagrams of, 831–833
direct-current, 822, 850
electromotive force and, 828–831
energy in, 834–836
incomplete, 828
inductors in, 994–998. See also Inductance
integrated, 1429–1430
junctions in, 855
Kirchoff's rules for, 855–860

L-C, 1005–1009
loops in, 855
L-R-C series, 1009–1011, 1030–1034
open, 870
oscillating, 1005–1009
overdamped, 1010
overloaded, 869–870
potential changes around, 833–834
power in, 834–838
R-C, 864–868
relaxation times of, 866–867
R-L, 1001–1005
self-inductance and, 995–998
short, 869–870, 870
time constants for, 866–867, 1003
underdamped, 1010–1011
Electric constants, fundamental, 695–696
Electric current. See Current
Electric dipole(s), 709–713, 805
definition of, 710
electric potential of, 765
field of, 712–713
force on, 710–711
potential energy of, 711
torque on, 710–711, 904
Electric dipole moment, 710–711
Electric energy, 194
units of, 194
Electric field(s), 699–700
calculation of, 703–708, 775
of capacitor, 790
charge distribution and, 725–728, 734–735, 746t. See
also Gauss's law
of charged conducting sphere, 737–738
at conductors, 701, 744–745
current and, 819–820
definition of, 699
direction of, 701, 708
of Earth, 745
electric dipole and, 709–713
electric forces and, 698–703
electric potential and, 761, 763–764, 774
of electromagnetic waves, 1053, 1061, 1069–1070
energy storage in, 788
flux of, 725–732, 729. See also Electric flux
Gauss's law for, 732–746, 935, 1052, 1056
of hollow sphere, 741
impossible, 773f
induced, 971–974
line integral of, 937
magnetic fields and, 975–979
magnitude of, 701, 708
molecular orientation in, 805–807
nodal/antinodal planes of, 1070
nonelectrostatic, 959, 972–973
of parallel conducting plates, 739–740, 790
of plane sheet of charge, 739
resistivity and, 823
sharks and, 699
superposition of, 703–704
of symmetric charge distributions, 746t
test charge for, 699, 700
uniform, 709
of uniform line charge, 738
of uniformly charged sphere, 740
units for, 699, 764
in vacuum, 798
as vector quantity, 775
work done by, 755–761. See also Electric potential
energy
Electric field lines, 708–709
electromagnetic waves and, 1053
equipotential surfaces and, 771–772
point charges and, 734, 1053
Electric flux
calculation of, 728–732
charge and, 725–732
enclosed charge and, 726–728
fluid-flow analogy for, 728–729
Gauss's law and, 732–741

Electric flux (*Continued*)
  of nonuniform electric field, 730–732
  of uniform electric field, 729–730
Electric force, 160
  conservation of energy with, 758–759
  Coulomb's law and, 693–698
  direction of, 694
  electric field and, 698–703
  electric potential and, 764–765
  potential energy and, 226–227
  on uncharged particles, 693
  units of, 695
  vector addition of, 697–698
  vs. electric potential energy, 758
  vs. gravitational force, 695
  work done by, 757–758, 761
Electric lines
  hot side, 869, 870
  household, 868–872
  neutral side of, 868–869, 870
Electric motors
  direct-current, 907–908
  magnetic force in, 898
Electric oscillation, 1005–1009
  in *L-C* circuits, 1005–1009
Electric potential, 761–771. *See also under* Potential
  calculation of, 762–771
  definition of, 761, 763
  electric circuits and, 834–836
  of electric dipole, 765
  electric field and, 761, 762–764, 774
  electric force and, 764–765
  electric potential energy and, 763, 766
  equipotential surfaces and, 771–773
  field lines and, 771–772
  maximum, 768–769
  of *a* with respect to *b*, 762
  as scalar quantity, 761, 775
  of two point charges, 765
  units of, 761, 764
  work done by, 762
Electric potential energy, 754–761
  alternative concepts of, 760
  in capacitors, 788. *See also* Capacitors
  definitions of, 760
  electric potential and, 763, 766
  electric-field energy and, 798
  of several point charges, 759–760
  of two point charges, 757–758
  in uniform field, 755
  vs. electric force, 758
Electric power, 194
Electric rays, 830
Electric stud finders, 802–803
Electric-field energy, 798, 799
Electricity, conductors of, 691–692
Electrocardiography, 762
Electrodynamics, quantum, 1081
Electrolytic double-layer capacitors, 802
Electromagnetic energy flow, 1065–1067
Electromagnetic induction, 957–981
  changing magnetic flux and, 958–959
  eddy currents and, 974–975
  experiments in, 958–959
  Faraday's law and, 957, 959–967
  induced electric fields and, 971–974
  Lenz's law and, 967–969
  Maxwell's equations and, 957, 977–979
  motional electromotive force and, 969–971
  superconductors and, 968, 979–980
Electromagnetic interaction, 159–160, 1490
Electromagnetic momentum flow, 1068–1069
Electromagnetic radiation, 574–577, 1053. *See also*
    Electromagnetic wave(s)
Electromagnetic spectrum, 1054–1055
Electromagnetic wave(s), 1051–1073
  amplitude of, 1061
  angular frequency of, 1061
  applications of, 1054
  definition of, 1052

in dielectrics, 1063–1064
  direction of, 1058
  Doppler effect for, 537–538, 1241–1243
  electric fields of, 1053, 1061, 1069–1070
  energy in, 1064–1067
  frequency of, 1060
  generation of, 1053
  intensity of, 1066
  magnetic fields of, 1061, 1069–1070
  magnitude of, 1058
  in matter, 1063–1064
  Maxwell's equations and, 1052–1057
  momentum of, 1068–1069
  plane, 1055–1057
  polarization of, 1058. *See also* Polarization
  power in, 488
  Poynting vector of, 1065–1066
  properties of, 1058
  radiation pressure and, 1068–1069
  reflected, 1069–1071
  right-hand rule for, 1058f
  sinusoidal, 1060–1063
  speed of, 1058, 1060, 1071
  standing, 1053, 1069–1072
  superposition of, 1069–1070
  transverse, 1056, 1060
  units for, 1053
  wave functions for, 1061
  wave number for, 1061
Electromagnetic wave equation, 1058–1060
Electromagnetism, 687, 885
Electrometers, 800
Electromotive force (emf), 828–831
  back, 908
  current and, 991. *See also* Inductance
  of electric motor, 908
  Hall, 909–910
  induced, 908, 958–959
  measurement of, 863–864
  motional, 969–971
  self-induced, 995, 998, 1026–1027
  sinusoidal alternating, 1021
  source of, 828, 830–831, 835–836
  theory of relativity and, 1224
  in transformers, 1040–1041
Electromyography, 861
Electron(s)
  angular momentum of, 942
  bonds and, 1405–1408
  charge of, 689–690, 695. *See also* Electric charge
  charge:mass ratio for, 896–897
  concentration of, 1420–1421
  creation and destruction of, 1483
  discovery of, 897, 1481
  excited-state, 1411
  exclusion principle and, 1388–1389
  ground-state configurations of, 1388–1390, 1391t
  in magnetic fields, 894
  magnetic moment of, 1379–1382
  mass of, 689, 897, 1440
  orbital angular momentum of, 1373–1374,
    1384, 1387
  orbital motion of, 819, 1386
  photoelectric effect and, 1261–1266
  probability distributions for, 1376–1378
  quantum states of, 1389t
  screening by, 1391–1392
  spin angular momentum of, 1384–1385, 1387
  spin of, 942
  spin-orbit coupling and, 1386
  valence, 1390, 1416
  Zeeman effect and, 1379–1382
Electron affinity, 1406
Electron capture, 1453
Electron diffraction, 1287–1288
Electron microscopes, 1290–1292
Electron shells, 1375, 1389, 1390–1391
  holes in, 1394–1395
Electron spin, 1383–1387
Electron volts, 764

Electron waves, 1286–1292
  atomic structure and, 1292–1296
  Bohr's hydrogen model and, 1300
Electron-gas model, 1415
Electron-positron pair annihilation, 1272
Electron-positron pair production, 1271–1272,
    1482–1483
Electrophoresis, 722
Electrostatic force
  conservation of, 856
  line integral for, 755, 937
Electrostatic painting, 683
Electrostatic precipitators, 784
Electrostatic shielding, 743–744
Electrostatics, 688
Electrostatics problems, 696
Electroweak interactions, 160, 1500
Electroweak theory, 1500
Electroweak unification, 1500
Elements
  ground state of, 1388–1390, 1391t
  isotopes of, 1441
  periodic table of, 1389, 1390–1391
  properties of, 1389
Elevation, atmospheric pressure and, 594–595
Elliptical orbits, 415–416
Elliptical polarization, 1099–1100
Emf. *See* Electromotive force (emf)
Emission line spectra, 1292–1293, 1297–1300
  continuous, 1310–1314
Emissivity, 575
Endoergic reactions, 1463
Endoscopes, 1090
Endothermal reactions, 1463
Energy
  activation, 610
  available, 1487–1488
  binding, 1406, 1445–1446
  conservation of, 176, 224
  conversion of, 224
  in damped oscillations, 458–459
  dark, 1508
  in electric circuits, 834–836
  electrical, 194
  electric-field, 798, 799
  in electromagnetic waves, 1064–1067
  equipartition of, 606
  Fermi, 1419–1421
  internal. *See* Internal energy
  ionization, 1304–1305, 1406
  kinetic. *See* Kinetic energy
  in *L-C* circuits, 1005–1009
  magnetic-field, 998–1001
  molecular, 597
  potential, 207–231. *See also* Potential energy
  power and, 871
  purchasing, 871
  quantized, 1261
  reaction, 1462–1464
  relativistic kinetic, 1246–1247
  rest, 1247–1249
  in simple harmonic motion, 446–449
  threshold, 1463
  total, 176, 1247
  uncertainty in, 1278
  units for, 764
  in wave motion, 486–489
  work and, 177–193. *See also* Work
Energy bands, 1416–1417
  in insulators, 1416–1417
Energy density, 798, 1064–1065
Energy diagrams, 228–229
Energy flow, electromagnetic, 1065–1067
Energy levels, 1297–1305
  degeneracy and, 1370–1371
  for harmonic oscillator, 1351–1352
  for hydrogen atom, 1302–1305, 1379–1382
  Moseley's law and, 1394–1396
  for particle in a box, 1340–1341
  quantization of, 1311–1312

rotational, 1408–1412
Schrödinger equation and, 1379–1382
selection rules and, 1382
vibrational, 1410
vs. states, 1307–1308
Zeeman effect and, 1379–1383
Energy storage, in capacitors, 788, 796–800. *See also* Capacitors
Energy transfer
heat and, 562–565
rates of, 570
Energy-flow diagrams
for heat engines, 655–656
for refrigerators, 659
Energy-mass conservation law, 1247–1248
Energy-time uncertainty principle, 1274–1275, 1278, 1315–1316
Engine(s)
Carnot, 663–669
heat. *See* Heat engines
internal combustion, 642, 657–659
Engine statement, of second law of thermodynamics, 661
Enhancement-type MOSFETs, 1429
Entropy, 669
calculation of, 676
in cyclic processes, 672–673
definition of, 670
disorder and, 669–670
internal energy and, 670
in irreversible processes, 673
in living organisms, 673
microscopic interpretation of, 675–677
Newton's second law and, 674, 677
reversibility of, 670
Enzymes, electron tunneling in, 1349
Equation(s)
Bernoulli's, 385–389
continuity, 383–384
dimensional consistency in, 6
Dirac, 1482–1483
electromagnetic wave, 1058–1060
ideal-gas, 591–595
lensmaker's, 1133–1135
Maxwell's. *See* Maxwell's equations
of motion, with constant acceleration, 49
Schrödinger. *See* Schrödinger equation
of simple harmonic motion, 440–442
of state, 591–596, 612
units of measure for, 6
van der Waals, 595–596
wave. *See* Wave equation
Equilibrium, 344–359
center of gravity and, 345–348
definition of, 32, 109
extended-body, 345–352
first condition of, 345
for mechanical waves, 473
net force and, 135
Newton's first law and, 108–112, 134–139
one-dimensional, 136–137
phase, 566, 611
potential energy and, 228–229
problem-solving strategies for, 136–139
rigid-body, 345, 348–352
rotation and, 345
second condition of, 345
stable, 228
static, 345
tension and, 136–139
thermal, 552
torque and, 345
two-dimensional, 137–138
unstable, 228
weight and, 345–347
weight lifting and, 351
Equilibrium processes, 652
Equipartition, of energy, 606
Equipotential surfaces, 771–773
conductors and, 772–773

definition of, 771
vs. Gaussian surface, 773
Equipotential volume, 773
Equivalent capacitance, 794
Equivalent resistance, 851, 852
Erect image, 1117
Errors
fractional (percent), 8
in measured values, 8
Escape speed, 410–411, 413, 423, 1505–1506
Estimates, order-of-magnitude, 10
Ether, light travel through, 1224, 1881
Euler's formula, 1340
Evaporation, 568
Event horizons, 424
Event, in frame of reference, 1227
Excited levels, 1298
Excited states, 1454
Exclusion principle, 1388–1393
bonds and, 1406, 1407
periodic table and, 1389, 1390–1393
quark colors and, 1498–1499
Exoergic reactions, 1463
Exothermal reactions, 1463
Expanding universe, 1501–1508
Experiments, 2
thought, 1227
Extended objects
definition of, 1115
gravitational potential energy for, 293, 317
image formation by lenses and, 1131–1133
image formation by mirrors and, 1120–1122
Extended-body equilibrium, 345–352
External forces, 247
center-of-mass motion and, 261–262
torque and, 312
Extracorporeal shock wave lithotripsy, 539
Eye, 1142–1146
index of refraction of, 1143
laser surgery for, 1076, 1309
resolution of, 1220
structure of, 1142–1143
Eyeglasses, corrective, 1143–1146
Eyepiece, microscope, 1147–1148

F

Fahrenheit scale, 554
Far point, of eye, 1143
Farad, 789, 790
Faraday cage, 743–744
Faraday disk dynamo, 971
Faraday, Michael, 708, 789, 885
Faraday's icepail experiment, 742–744
Faraday's law of induction, 957, 959–967
electromagnetic waves and, 1052, 1056, 1057, 1063. *See also* Maxwell's equations
Farsightedness, 1143–1146
Fermat's principle of least time, 1111
Fermi energy, 1419–1421
electron concentration and, 1420–1421
Fermi-Dirac distribution, 1419, 1492
Fermions, 1492, 1494, 1501
Ferromagnetism, 943t, 944–946
Feynman, Richard, 1483
Field lines. *See* Electric field lines; Magnetic field lines
Field point, 700, 924
Field-effect transistors, 1429
Fields, 406
Filters
high-pass, 1028
low-pass, 1027
polarizing, 1093, 1094–1097, 1098
Fine structure, 1387
Finite wells, 1343
Firecrackers, entropy and, 669f
First condition for equilibrium, 345
First law of thermodynamics, 624–643
cyclic processes and, 631–634
definition of, 630

internal energy and, 629–634
isolated systems and, 631–634
Fish
bulk stress on, 355f
fluorescent, 1300f
Fission, nuclear, 785, 1247, 1464–1468
Fixed-axis rotation, 278–279, 283
Flash unit, of camera, 797
Flow line, 382–383. *See also* Fluid flow
Flow tubes, 383
Fluid(s)
compressibility of, 356, 382
ideal, 382
motion of, 382–384
speed of sound in, 514–515
viscous, 389–390
Fluid density, 373–375
buoyancy and, 380–382
compression and, 476
measurement of, 381
rarefaction and, 476
Fluid dynamics, 373
viscosity and, 389–390
Fluid flow, 382–384. *See also* Flow
Bernoulli's equation and, 385–389
continuity equation and, 383–384
laminar, 383, 390
measurement of, 388
pressure and, 385–389
rate of, 383–384
speed of, 385–389
steady, 382–383
turbulent, 383, 390–391
Fluid mechanics, 373–392
Bernoulli's equation and, 385–389
buoyancy and, 380–382
density and, 373–375, 380–381
fluid flow and, 382–384
pressure and, 375–380
surface tension and, 382
Fluid pressure, 355, 375–380
depth and, 376–377
measurement of, 378–380
Pascal's law and, 376–377
Fluid resistance, 151–154
Fluid statics, 373
Fluorescence, 1300
Fluorescent fish, 1300f
Fluorescent lights, 996–997, 1081, 1300
Fluorine, 1390
Flux. *See* Electric flux; Magnetic flux
f-number, 1140–1141
Focal length
of camera lens, 1140
of microscope lens, 1148–1149
of mirror, 1119–1120
of telescope lens, 1149–1150
of thin lens, 1131, 1133–1135
Focal point
of microscope lens, 1149
of mirror, 1119–1120
of thin lens, 1131
virtual, 1123
Fog, 567
Food, energy value of, 568
Foot-pound, 177, 194
Force(s)
acting at a distance, 406
action-reaction, 120–123
buoyant, 380
central, 416
centrifugal, 155
components of, 226–227
conservative, 221–224, 228–229
constant, 177–178
contact, 105, 146
definition of, 105
direction of, 10, 13. *See also* Vector(s)
dissipative, 222
driving, 459–460

Force(s) (*Continued*)
electric. *See* Electric force
electromotive, 828–831
electrostatic, 856
external, 247
fluid resistance, 151–154
free-body diagrams for, 124–125
friction, 105, 146–154. *See also* Friction
fundamental, 159–160, 1490–1492
gravitational. *See* Gravitation
interaction, 931–932
intermolecular, 597–598
internal, 247, 312
line of action of, 309
long-range, 105
magnetic, 159–160, 886–887
magnitudes of, 105
mass and, 113–114
measurement of, 106
motion and, 112–116. *See also* Newton's second law
of motion
net. *See* Net force
nonconservative, 222–224
normal, 105, 146
nuclear, 1446–1449, 1484, 1491
particle interactions and, 159–160
periodic driving, 459–460
potential energy and, 225–228
power and, 194
properties of, 105
restoring, 438
strong interactions, 160, 1446, 1490–1491
strong nuclear, 160, 689
superposition of, 106–108, 404, 405–406
tension, 105, 123. *See also* Tension
tidal, 425
torque of, 308–312
units of, 5–6, 105, 117
vs. pressure, 355, 376
weak interactions, 160, 1491
weight as, 105
Force constant, 188
Force diagrams, 106–107
Force fields, 406
Force mediators, 1484–1485
Force per unit area, 353
Force vectors, 105
components of, 106–107
Forced oscillations, 459–460, 527
Forensics, X-rays in, 1395
Forward bias, 1426, 1427–1428
Fosbury flop, 293f
Fossil fuels, climate change and, 577
Fossils, 1508
Fourier analysis, 513
Fourier series, 497
Fractional error (uncertainty), 8
Fracture, 358
Frame of reference, 89
event in, 1227
inertial, 110–112, 115–116, 1223–1227
simultaneity and, 1227–1228
Franck, James, 1300
Franck-Hertz experiment, 1300
Franklin, Benjamin, 688
Franklin, Rosalind, 1207f
Fraunhofer diffraction, 1192, 1193
Free charges, 806
Free expansion, 629
Free fall
acceleration due to gravity and, 52–55, 118–119
definition of, 52
fluid resistance and, 152–154
Free particle, 1330
Free-body diagrams, 124–125, 140, 144
Free-electron energy, average, 1421
Free-electron model, 1415, 1418–1422
Free-particle state, 1346
Frequency, 438
angular, 438–439

beat, 531–532
fundamental, 496
normal-mode, 496
period and, 438–439
of standing waves, 496
vs. angular frequency, 442
Fresnel diffraction, 1193
Friction, 146–154
coefficients of, 147, 151
definition of, 108, 147
fluid resistance and, 151–152
kinetic, 147, 149, 150–151, 222
magnitude of, 147
rolling, 151–154, 320
static, 147–149
stick-slip phenomenon and, 148, 149
Friction force, 105, 147
Full-wave rectifier current, 1023
Fundamental electric constants, 695–696
Fundamental forces, 159–160
Fundamental frequency, 496
Fundamental particles, 1480–1501. *See also*
Particle(s)
historical perspective on, 1480–1485
Fur, as insulator, 572
Fuses, 870
Fusion. *See* Nuclear fusion

G

Galaxies, recession speed of, 1502–1503
Galilean coordinate transformation, 1225–1226
Galilean telescope, 1161
Galilean velocity transformation, 91, 1226
Galileo Galilei, 2, 52, 1080
Gallium, melting temperature of, 566
Galvanometer, d'Arsonval, 860, 863, 904, 1022
Gamma camera, 1461
Gamma decay, 1454
Gamma rays, 1454, 1462
pair production and, 1271–1272
Gas
bulk modulus of, 355
heat capacities of, 605–607
ideal. *See* Ideal gas
intermolecular forces in, 597
isotherms and, 596
kinetic energy of, 605–606
mass of, 591–592
molecules in, 597
noble, 1390
p-*V* diagrams for, 596
sound waves in, 517–518
volume of, 593
Gas constant, 517, 592
Gas pressure
molecular collisions and, 597–600
molecular kinetic energy and, 600–602
temperature and, 555
Gas thermometers, 554–556, 593, 669
Gaseous phase, 566
Gasoline engines, 546–548
Gauge pressure, 377–378
Gauges, pressure, 378–380
Gauss, 887
Gauss, Carl Friedrich, 732
Gaussian surface, 734
vs. equipotential surface, 773
Gauss's law, 732–746
applications of, 736–741
charge and electric flux and, 725–732
conductors with cavities and, 741–742, 773
dielectrics and, 807–808
for electric fields, 732–746, 935, 1052, 1056. *See also* Maxwell's equations
experimental testing of, 742–744
general form of, 734–736
for gravitation, 752, 935
for magnetic fields, 891, 935, 1052, 1056. *See also* Maxwell's equations
overview of, 732–736

point charge inside nonspherical surface and, 733–734
point charge inside spherical surface and, 732–733
qualitative statement of, 728
solid conductors and, 736–741
Geiger counters, 783
Geiger, Hans, 1294
Gell-Mann, Murray, 1496, 1498
Gemstones, 1090, 1092
General theory of relativity, 1249–1251, 1504. *See also* Relativity
Generation currents, 1427
Generators
alternating-current, 1021
direct-current, 964–965
energy conversion in, 966–967
homopolar, 971
slidewire, 965–966, 967, 970
Geometric optics, 1082, 1114–1153
cameras and, 1139–1142
eye and, 1142–1146
magnifiers and, 1146–1147
microscopes and, 1147–1149
reflection at plane surface and, 1115–1118
reflection at spherical surface and, 1118–1126
refraction at plane surface and, 1115–1118
refraction at spherical surface and, 1126–1130
sign rules for, 1116
telescopes and, 1149–1151
thin lenses and, 1131–1139
Gerlach, Walter, 1383
Germanium semiconductors, 1422–1425
Germer, Lester, 1287–1288
Glashow, Sheldon, 1500
Glass, as amorphous solid, 1412
Global positioning systems, 1078, 1187, 1250–1251
Global warming, 576–577
Gluons, 1490, 1491, 1498–1499, 1500
GPS systems, 1078, 1187, 1250–1251
Gradient
definition of, 774
potential, 774–776
potential energy, 227
Grams, 5, 5t
Grand unified theories (GUTs), 160, 1500–1501, 1509, 1516
Graphical method, for image location, 1124–1126
Graphs
curvature of, 46, 47
parabolic, 48
a_x-t, 46–49, 55–57
of sound waves, 511
v_x-t, 44–46, 47
of wave function, 478–480
x-t, 37–38, 40–41. *See also* x-t graphs
Grating spectrographs, 1203–1204
Gravitation, 105, 159–160, 402–427
acceleration due to, 52–55, 700
action-reaction pairs and, 403
black holes and, 423–426
as conservative force, 409, 410
on cosmic scale, 406
escape speed and, 410–411, 413, 1505–1506
expanding universe and, 1501–1502, 1505–1507
as fundamental force, 159–160
Gauss's law for, 752, 934
general theory of relativity and, 1250
importance of, 406
measurement of, 404–405
Newton's law of, 402–406
satellite orbits and, 411–413
specific gravity and, 374
spherical mass distributions and, 418–421
spherically symmetric bodies and, 403–404
superposition of forces and, 405–406
weight and, 406–409
work done by, 409–410
Gravitational constant, 403
calculation of, 404–405
Gravitational field, 700

Gravitational force(s), 105, 159–160
　as action-reaction pairs, 403
　per unit mass, 700
　vs. electric force, 695
　vs. gravitational potential energy, 410
Gravitational interaction, 159–160, 1490
Gravitational potential energy, 208–216, 409–411
　definition of, 208, 409–410
　elastic potential energy and, 218
　for extended bodies, 293, 317
　motion along curves and, 212–216
　as negative value, 410
　nongravitational forces and, 211
　vs. gravitational force, 410
Gravitational red shift, 425, 1250
Gravitational torque, 346–347
Gravitons, 1490, 1500
Gravity. See Gravitation
Gray, 1459
Greenhouse effect, 576–577
Ground level, 1297
Ground state
　atomic, 1388–1390, 1391t
　nuclear, 1454
Ground-fault interrupters, 870
Grounding wires, 869, 870
Gyroscopes, 328–330

H
h vs. h-bar, 1276
Hadrons, 1492, 1493–1494, 1496, 1497
Hahn, Otto, 1464
Hale Telescope, 1219
Half-life, 1456–1457
Hall effect, 909–910
Halley's comet, 417
Halogens, 1390
Harmonic analysis, 497
Harmonic content, 497, 513
Harmonic motion, damped, 1009
Harmonic oscillators
　anisotropic, 1401
　Hermite functions for, 1350–1354
　isotropic, 1401–1402
　Newtonian, 439–440, 1352–1354
　quantum, 1350–1354
Harmonic series, 496
Harmonics, 496
Hearing loss, sound intensity and, 513, 522
Heat
　added in thermodynamic process, 628–629
　of combustion, 568
　definition of, 562, 563
　as energy in transit, 605
　energy transfer and, 562–565. See also Heat
　　transfer
　of fusion, 566
　global warming and, 576–577
　mechanical energy and, 562
　melting and, 565–566
　phase changes and, 565–570
　quantity of, 562–565
　sign rules for, 625
　specific, 562–563
　steam, 567–568
　of sublimation, 567
　units of measure for, 562
　of vaporization, 566, 568
　vs. temperature, 562
Heat calculations, 568–570
Heat capacity, 605–608
　of gases, 605–607
　of ideal gas, 637–639
　molar, 564–565, 605–607
　point-molecule model of, 605
　ratio of, 639
　of solids, 607–608
　temperature variation of, 608
　vibration and, 606–608
Heat current, 571

Heat engines, 654–656
　Carnot, 663–666
　energy-flow diagrams and, 655
　hot and cold reservoirs and, 654–655
　internal combustion, 642, 657–659
　thermal efficiency of, 655
Heat flow, 562
Heat pumps, 661
Heat transfer, 562–565
　by conduction, 570–574
　by convection, 574
　mechanisms of, 570–577
　by radiation, 574–577
Heavy hydrogen, 1305–1306
Heisenberg uncertainty principle, 1276–1277,
　1314–1317
　Bohr model and, 1317
　energy-time, 1278, 1315–1316
　harmonic oscillator and, 1353–1354
　for matter, 1315–1316
　momentum-position, 1274–1275, 1278, 1315–1316
Helium, 1390
Helium atom, Bohr model of, 1306
Helium fusion, 1514
Hemholtz coils, 954
Henry, 993
Henry, Joseph, 885
Hermite functions, 1350–1354
Hertz, 438, 1053
Hertz, Gustav, 1300
Hertz, Heinrich, 1053, 1081
High-pass filters, 1028
Hole conduction, 909, 1423–1424
Holes, 820, 1423–1424
Holography, 1211–1213
Homopolar generators, 971
Hooke's law, 188–189, 352, 356
　elastic deformations and, 352, 357–358
　limits of, 357–358
　simple harmonic motion and, 439
Horse, acceleration around curve, 74
Horsepower, 194
Hot reservoirs, 654–655
Hot side of line, 869, 870
House wiring systems, 868–872, 1040–1041
Hubble constant, 1503
Hubble, Edwin, 1502
Hubble law, 1503
Hubble Space Telescope, 1119, 1218, 1375f,
　1503
Human body
　angular momentum of, 324
　fat measurement for, 1032
　magnetic fields of, 887
　radiation from, 575–576
　as thermodynamic system, 630
Humason, Milton, 1502
Huygen's principle, 1102–1104
　diffraction and, 1191–1192
Hybrid wave function, 1407
Hydrogen
　in fusion reactions, 1469–1471
　ground state of, 1390
　heavy, 1305–1306
Hydrogen atom, 1372–1379
　Bohr's model of, 1300–1305
　electron probability distributions for, 1376–1378
　energy levels in, 1302–1305, 1379–1382
　hydrogen-like atoms and, 1378–1379
　ionization energy of, 1304–1305
　in magnetic field, 1379–1382
　nuclear motion in, 1305–1306
　orbital angular momentum of, 1373–1374
　quantum states of, 1373–1374, 1375t
　reduced mass of, 1305–1306
　Schrödinger equation for, 1372–1373
Hydrogen bonds, 1407–1408
Hydrogenlike atoms
　Bohr model of, 1306
　Schrödinger analysis of, 1378–1379

Hydrometers, 381
Hyperfine structure, 1387, 1444
Hyperons, 1494–1495
Hyperopia, 1143–1146
Hysteresis, 945–946
Hysteresis loops, 945–946

I
I SEE acronym, 3
Ice, melting of, 565–566
Ideal fluid, 382
Ideal gas, 592
　adiabatic process for, 640–642
　Carnot cycle for, 664–665
　heat capacities of, 637–639
　internal energy of, 636
　isothermal expansion of, 627
　kinetic-molecular model of, 599–605
　volume of, 593
Ideal single crystals, 1413
Ideal-gas constant, 517, 592
Ideal-gas equation, 591–595
Idealized models, 3–4, 3f
Image
　erect, 1117
　inverted, 1117, 1141
　in optics, 1115
　real, 1115
　virtual, 1115, 1137
Image distance, 1116
Image formation
　by cameras, 1139–1142
　by diffraction, 1209–1210
　by lenses, 1131–1133
　by reflection, 1115–1126. See also Mirrors
　by refraction, 1126–1130
Image point, 1114, 1115
Imaging studies. See Medical imaging
Impedance, 1031–1033
Impulse, 243
Impulse-momentum theorem, 242–244, 483–484
Incident waves, 492–493
Incubators, 576
Index of refraction, 1063, 1083–1088
　birefringence and, 1100
　definition of, 1083
　dispersion and, 1091–1093
　of eye, 1143
　of gemstones, 1090
　laws of reflection and refraction and, 1085
　of lens, 1133–1134
　of reflective/nonreflective coatings, 1178–1179
　total internal reflection and, 1088–1090
　transparency and, 1085
　wave aspects of light and, 1086–1088
Induced charges, 692
　molecular model of, 805–807
　polarization and, 693, 801–803
Induced current, 958
　direction of, 967–968, 970
　magnitude of, 968
Induced electric fields, 971–974
Induced emf, 908, 958–959. See also Electromagnetic
　induction
　applications of, 959, 961
　direction of, 961
　magnetic flux and, 959, 962
Inductance, 991–1012
　definition of, 995
　magnetic-field energy and, 998–1001
　mutual, 991–994
　R-L circuits and, 1001–1005
　self-inductance, 994–998
Inductive reactance, 1026–1027
Inductors, 994–998
　in ac circuits, 1025–1027, 1029, 1029t, 1035
　energy stored in, 998–1000
　inductive reactance of, 1026
　vs. resistors, 999
Inelastic collisions, 251–255

Inertia, 109
    definition of, 109
    mass and, 118
    moment of. *See* Moment of inertia
    rotational, 289
Inertial confinement, 1470
Inertial frame of reference, 110–112, 1223–1227
    Newton's first law and, 110–112
    Newton's second law and, 115–116
    simultaneity and, 1227–1228
    theory of relativity and, 1223
Inertial mass, 113. *See also* Mass
Inertial navigation systems, 56f
Infrasonic sound, 510
Inkjet printers, 722
Instantaneous acceleration, 43–44. *See also*
        Acceleration
    angular, 282
    definition of, 43
    on v_x-t graph, 44–46
    x-component of, 43
Instantaneous acceleration vectors, 73–75. *See also*
        Acceleration vectors
Instantaneous angular acceleration, 282
Instantaneous angular speed, 286
Instantaneous angular velocity, 280
Instantaneous power, 193, 194
    in waves, 487–488
Instantaneous speed, 39. *See also* Speed
    angular, 286
Instantaneous velocity, 38–42, 39f–41f. *See also*
        Velocity
    definition of, 38, 70
    straight-line motion and, 38–41
    vs. instantaneous speed, 39–40
    x-component of, 39
    on x-t graph, 40–41
Instantaneous velocity vectors, 70–72
Insulators, 552, 691
    energy bands in, 1416–1417
Integral(s)
    line, 192, 755
    moment of inertia, 295
    surface, 730
Integral principles, 244
Integrated circuits, 1429–1430
Integration, velocity and position by, 55–57
Intensity
    of electromagnetic radiation, 1066
    inverse-square law for, 488–489, 520
    pressure amplitude and, 519–521
    in single-slit diffraction, 1195–1199
    sound, 518–522
    vs. spectral emittance, 1310–1311
    wave, 488–489
Intensity maxima, 1197
Interactions. *See* Particle interactions
Interference, 489–492, 529–531, 1163–1183
    amplitude in, 1170–1171
    butterfly wings and, 1179
    coherent sources and, 1164–1166
    complementarity principle and, 1273–1274
    constructive, 492, 530, 1164–1166, 1168–1170,
        1206–1207
    definition of, 489, 1164
    destructive, 492, 529, 1164–1166, 1168–1170
    in holography, 1211–1213
    Michelson interferometer and, 1179–1181
    Michelson-Morley experiment and, 1180–1181
    Newton's rings and, 1178
    nodal/antinodal curves and, 1166
    in noise control, 531
    path difference and, 1171–1173
    phase difference and, 1171–1173
    phase shifts and, 1174–1175
    photons and, 1273–1274
    during reflection, 1174–1175
    reflective/nonreflective coatings and, 1178–1179
    sinusoidal waves and, 1164
    in sound waves, 529–531

sound waves and, 529–531
    standing waves and, 492, 1164, 1166
    superposition and, 1164
    in thick films, 1176
    in thin films, 1173–1179
    in three dimensions, 1164
    in two dimensions, 1164
    two-source/slit, 1166–1173, 1315–1316
    vs. diffraction, 1192, 1200
    in water waves, 1166–1167
    water waves and, 1166–1167
    Young's experiment for, 1167–1169, 1179
Interference fringes, 1168, 1174, 1178
    Newton's, 1178
Interference maxima, 1199
Interference patterns, 1166
    intensity in, 1170–1173
Interferometer, 1179–1181
Intermolecular forces, 597–598
Internal combustion engines, 657–659
Internal energy, 224, 624, 670
    change in, 630–631, 639
    of cyclic processes, 631–634
    definition of, 629, 631
    entropy and, 670
    first law of thermodynamics and, 629–634
    of ideal gas, 636
    of isolated systems, 631–634
    notation for, 629
    temperature and, 636
Internal forces, 247
    torque and, 312
Internal resistance, 830, 833
International System, 4, 5t
Interplanetary travel, biological hazards of, 416f
Interstellar gas clouds, 1293
Intrinsic semiconductors, 1423, 1424
Inverse-square law, for intensity, 488–489, 520
Inverted image, 1117
    in camera lens, 1141
Iodine-127, 1461
Ionic bonds, 1406, 1407
Ionic crystals, 1414–1415
Ionization, 690
    corona discharge and, 768–771
Ionization energy, 1406
    of hydrogen atom, 1304–1305
Ions, 690
Irreversible process, 652–653
Isobaric process, 635
Isochoric process, 635
Isolated systems, 247
    internal energy of, 631–634
Isospin, 1495
Isothermal expansion, of ideal gas, 627
Isothermal process, 635–636
    Carnot cycle and, 663
Isotherms, 596
Isotopes, 897, 1441

J

Jet propulsion, in squids, 262
Josephson junctions, 1349
Joule, 177, 183, 562
Joule per coulomb, 761
Junctions, in circuits, 855

K

K mesons, 1493
Kaons, 1259, 1493
Keck telescopes, 1151
Kelvin, 555
Kelvin scale, 555–556, 665, 668–669
Kelvin-Planck statement, 661
Kepler, Johannes, 414–415
Kepler's first law, 414–415
Kepler's second law, 415–416
Kepler's third law, 416–417, 426
Killowat-hour, 194
Kilocalorie (kcal), 562

Kilograms, 5, 113
Kilohm, 826
Kilowatt, 193
Kilowatt-hour, 871
Kinematics. *See also* Motion
    definition of, 35
    linear vs. angular, 285–288
Kinetic energy
    in collisions, 252
    in composite systems, 186–187
    conservative forces and, 221–222
    with constant forces, 177–178
    definition of, 182
    equipartition of, 606
    gas pressure and, 600–602
    heat capacities and, 605–608
    molecular, 597, 600–602, 605–606, 636
    moment of inertia and, 288–291
    of photons, 1263–1264
    potential energy and, 207, 208, 221–222
    relativistic, 1246–1247
    rotational, 288–293, 315–316
    as scalar quantity, 182
    in simple harmonic motion, 446–449
    stopping potential and, 1262–1263
    torque and, 321
    units of, 183
    with varying forces, 187–191
    vs. momentum, 242–246, 2420
    work-energy theorem and, 181–187
Kinetic friction, 147, 149, 150–151
    coefficient of, 147
    as nonconservative force, 222
Kinetic-molecular model, of ideal gas, 599–605
Kirchoff's rules, 855–860, 976–977
Kramers, Hendrik, 1361
Kundt's tube, 523

L

Ladder, stability of, 350
Lagging current, 1036
Laminar flow, 383, 390
Land, Edwin H., 1094
Large Hadron Collider, 1257, 1487, 1489, 1501
Laser(s), 1307–1309
    continuous, 1309
    definition of, 1307
    metastable states and, 1308
    population inversions and, 1308–1309
    production of, 1308–1309
    pulsed, 1309
    semiconductor, 1309
    spontaneous emission and, 1307
    stimulated emission and, 1307–1309
Laser eye surgery, 1076, 1309
Laser light, 1055, 1081, 1164, 1166
Laser printers, 689, 769, 1309
Latent heat of fusion, 566
Lateral magnification, 1117, 1120–1121
    of camera, 1140
    of microscope, 1148–1149
    vs. angular magnification, 1147
Laue pattern, 1205f
Law of Biot and Savart, 927
Law of conservation of energy, 176, 209, 224
Law of reflection, 1084–1086
Law of refraction, 1084–1086
Lawrence, E.O., 1486
Laws, physical, 2
L-C circuits, 1005–1009
Leaning Tower of Pisa, 2, 53, 118
Length
    Planck, 1509
    proper, 1235
    relativity of, 1233–1237
    units of, 4–5, 6f
Length contraction, 1235, 1236
Lens(es)
    of camera, 1139–1142
    corrective, 1143–1146

definition of, 1131
of eye, 1142–1143
magnifying, 1146–1147
of microscopes, 1147–1149
nonreflective coatings for, 1178
parabolic, 1151
properties of, 1131
reflective coatings for, 1178, 1179
of telescopes, 1149–1151
thin, 1131–1139. *See also* Thin lenses
Lensmaker's equation, 1133–1135
Lenz's law, 967–969
Leptons, 1492–1493, 1496, 1499, 1500
Lever arm, 309
Light, 1080–1105
absorption of, 1261–1266
beams of, 1083
coherent, 1165
diffraction of, 1190–1214
dispersion of, 1085, 1091–1093
Doppler effect for, 537–538, 1241–1243
early studies of, 1080–1081
fluorescent, 996, 1081
Huygen's principle and, 1102–1104
intensities of, 1085
interference and, 1166–1170. *See also* Interference
laser, 1055, 1081, 1164, 1166
monochromatic, 1054–1055, 1164
natural, 1094
photoelectric effect and, 1261–1266
as photons, 1261–1280. *See also* Photons
polarized, 1093–1100
rays of, 1082
reflection of, 1082–1091
refraction of, 1082–1088
scattering of, 1100–1101, 1269–1273
speed of, 4–5, 1054, 1063f, 1081, 1224–1226
total internal reflection of, 1088–1091
unpolarized, 1094
visible, 1054
as wave and particle. *See* Wave-particle duality
wave fronts of, 1081–1082
wavelengths of, 1054–1055
Light pipes, 1090
Light-emitting diodes, 1428
Lightning rods, 769
Lightning strikes, inductors and, 995
Light-years, 1503
Limit of resolution, 1209
Linacs, 1485–1486
Line integral, 192, 755
of electric fields, 937
of electrostatic force, 755, 937
of magnetic fields, 937
Line of action, 309
Line spectra, 1292, 1297–1300, 1304
continuous, 1310
molecular, 1300
Zeeman effect and, 1379
Linear acceleration, 282
constant, 284t
in rigid-body rotation, 286–288
vs. angular acceleration, 284t
Linear accelerators, 1485–1486
Linear charge density, 704
Linear conductors, 823
Linear expansion, 557–558, 558t, 559
Linear mass density, 482
Linear momentum, 242, 322
Linear polarization, 1058, 1093, 1095
of electromagnetic wave, 1058
Linear speed, in rigid-body rotation, 285–286
Linear superposition, 491
Linear velocity, 282
vs. angular velocity, 280
Linear vs. angular kinematics, 285–288
Liquid(s)
compressibility of, 356
as condensed matter, 1412
molecular speed in, 610

molecules in, 597
phases of, 610–613
properties of, 1412
Liquid crystals, 1412
Liquid phase, 566
Liquid-drop model
of nuclear fission, 1465–1466
of nucleus, 1447–1448
Lithium, 1390
Bohr model of, 1306
Livingston, M. Stanley, 1486
Longitudinal waves, 473. *See also* Mechanical waves; Wave(s)
periodic, 475–476
sound, 476
wave function for, 482
Long-range forces, 105
Loops, in circuits, 855
Lorentz transformations, 1237–1241
coordinate, 1237–1238
velocity, 1238–1239
Loudness, 513
Loudspeakers, 1029
magnetic forces in, 899–900
Low-pass filters, 1027
L-R-C parallel circuits, resonance in, 1039, 1048
L-R-C series circuits, 1009–1011
with ac source, 1030–1034
impedance in, 1031–1032, 1038
phase angle and, 1031–1032
power in, 1034–1037
resonance in, 1037–1039
Luminous matter, 1507
Lyman series, 1304

M

Mach number, 539
Macroscopic properties
theories of matter and, 599
vs. microscopic properties, 590
Macroscopic state, 675
Magic numbers, 1449
Magnet(s)
attracting unmagnetized objects, 906–907
bar, 883, 905–907
magnetic dipoles of, 906–907
magnetic moment of, 906
permanent, 883, 941
Magnetic ballast, 997
Magnetic bottles, 893
Magnetic confinement, 1470
Magnetic declination, 884
Magnetic dipole moment. *See* Magnetic moment
Magnetic dipoles, 903–904, 906–907
definition of, 903
force and torque on, 901–903
of magnets, 905–907
in nonuniform magnetic fields, 905–907
potential energy for, 903–904
Magnetic domains, 944–945
Magnetic energy density, 999–1000
Magnetic field(s), 884, 885–889
on axis of coil, 933–934
calculation of, 927
of circular current loops, 932–935
critical, 979
of current element, 926–927
definition of, 885
direction of, 887–888
of Earth, 884, 887
of electromagnetic waves, 1055–1057, 1061–1062, 1069–1070
Gauss's law for, 891, 935, 1052, 1056
Hall effect and, 909
of human body, 887
hydrogen atom in, 1379–1382
line integral of, 937
of long cylindrical conductor, 939, 948t
of long straight conductor, 935–937, 938, 948t

magnitude of, 887–888, 892
measurement of, 887–889
motion in, 892–898
of motors, 898
of moving charge, 886, 923–926
nodal/antinodal planes of, 1070
notation for, 886
of solenoid, 939–941, 948t
sources of, 923–946, 975–979, 1444
of straight current-carrying conductor, 928–931
superposition of, 926, 931
test charges for, 887–889
vector, 886, 924
Zeeman effect and, 1379–1382
Magnetic field lines, 884–885, 889–890
for current element, 927
direction of, 924
end points of, 891
magnetic flux and, 890–891
for moving charge, 924–925
vs. magnetic lines of force, 889
Magnetic flux, 890–892
calculation of, 959–960
definition of, 890–891
Faraday's law and, 959–967
Gauss's law of magnetism and, 891
induced electric fields and, 972–974
induced emf and, 958–959
Lenz's law and, 967–969
Meissner effect and, 980
as scalar quantity, 891
superconductivity and, 980
in transformers, 1040
units for, 891
Magnetic flux density, 892
Magnetic force(s)
on current loops, 901–907
on current-carrying conductors, 898–901
direction of, 886–887
in electric motors, 898
as fundamental force, 159–160
Hall effect and, 909
in loudspeakers, 899–900
magnitude of, 887
between parallel conductors, 931–932
units for, 887
Magnetic inclination, 884
Magnetic lines of force, 889–892
Magnetic materials, 907, 941–946
Bohr magneton, 941–942
diamagnetic, 943t, 944
ferromagnetic, 943t, 944–946
paramagnetic, 943–944, 943t
relative permeability of, 943
Magnetic moment, 903, 906, 1379
alignment of, 941–945
anomalous, 1443
of current loop, 903, 934
definition of, 903
direction of, 903
magnitude of, 942–945
of neutron, 1442–1443, 1497
nuclear, 1442–1443
of orbiting electron, 1379–1382
of proton, 1442–1443
spin, 1443
vector, 903
Zeeman effect and, 1379–1382
Magnetic monopoles, 885, 1500
Magnetic nanoparticles, for cancer, 946
Magnetic poles, 884
vs. electric charge, 885
Magnetic quantum number, 1374
Magnetic resonance imaging (MRI), 904–905, 934, 1444
Magnetic susceptibility, 943–944
Magnetic torque, 901–905
Magnetic variation, 884
Magnetic-field energy, 998–1001

Magnetism, 883–885
  electron motion and, 885
  Gauss's law for, 891, 935
Magnetization, 906–907, 927, 941–946
  saturation, 945
Magnetization curve, 945
Magnetons
  Bohr, 941–942, 1380
  nuclear, 1443
Magnetrons, 893
Magnification
  angular, 1147–1148
  lateral, 1117, 1120–1121, 1147, 1148–1149
Magnifiers, 1146–1147
Magnitude, of vector, 11, 12, 16
Malus' law, 1096
Manometers, 378–380
Maple seed, motion of, 316
Mars, gravitation on, 408–409
Marsden, Ernest, 1294
Mass
  acceleration and, 113, 114, 118–120
  of atom, 598, 690, 897, 1305–1306, 1441
  center of, 258–262
  definition of, 113
  density and, 373–374
  of electron, 689, 896–897, 1440
  force and, 113–114
  of gas, 591–592
  inertia and, 118
  measurement of, 114, 119
  molar, 517, 564, 591, 598
  of molecule, 598, 897
  of neutrino, 1500–1501
  of neutron, 689, 1440
  of nucleus, 1440
  of proton, 689, 1440
  rest, 1243–1246
  of star, 1259
  terminal speed and, 153–154
  units of, 5, 113–114, 117, 119
  weight and, 114, 117–120
Mass number, 1440
Mass per unit length, 482
Mass spectrograph, 918
Mass spectrometers, 897
Mass-energy conservation law, 1247–1248
Mass:volume ratio, density and, 373–374
Matter
  antimatter and, 1516
  condensed, 1412. See also Liquid(s); Solids
  luminous, 1507
  molecular properties of, 596–598
  phases of, 610–613
Maxwell, James Clerk, 976, 1052f, 1081
Maxwell-Boltzmann distribution, 608–609, 1307, 1419–1420
Maxwell's equations, 885, 957, 977–979
  electromagnetic waves and, 1052–1057
  Huygen's principle and, 1102–1104
  in optics, 1085
Maxwell's wave theory, 1052–1057, 1262–1263, 1267
Mean free path, 604
Mean free time, 838
Measurement
  accuracy of, 8
  errors in, 8
  significant figures in, 8–9
  uncertainty in, 8
  units of, 4–6. See also Units of measure
Mechanical energy
  conservation of, 209, 446–448, 755
  conservative vs. nonconservative forces and, 221–223
  heat and, 562
  in simple harmonic motion, 446–449
  total, 209
Mechanical waves, 472–499. See also under Wave
  boundary conditions for, 489–490
  definition of, 473

  direction of, 479, 481–482
  energy of, 486–489
  energy transport by, 474
  equilibrium state for, 473
  incident, 492–493
  intensity of, 488–489
  interference in, 489
  longitudinal. See Longitudinal waves
  mathematical description of, 477–482
  normal-mode patterns of, 496
  periodic, 474–477. See also Periodic waves
  power of, 487–488
  propagation of, 474
  sinusoidal, 475–482. See also Sinusoidal waves
  sound, 476
  speed of. See Wave speed
  standing, 491–498
  superposition of, 490–491, 497
  transverse. See Transverse waves
  traveling, 492, 494
  types of, 473–474
  wave equation for, 481, 485
  wavelength of, 475
Mechanics
  classical (Newtonian), 104
  definition of, 35
Medical imaging
  radioactive isotopes in, 1391, 1461, 1466
  X rays in, 1268–1269
Medicine
  nuclear, 1391, 1461, 1466
  pair annihilation in, 1484
Medium, 473
Megohm, 826
Meissner effect, 979–980
Meitner, Lise, 1464
Melting, 565–566
Melting points, of solids, 1412
Membrane, dielectric, 805
Membrane potential, 774
Mercury barometers, 378–380
Mesons, 1484–1485, 1492, 1493–1495, 1499
  colors of, 1489
  quarks in, 1496
Metallic bonds, 1408
Metallic conduction, 838–840
Metallic crystals, 1415
Metals
  alkali, 1306, 1390
  alkaline earth, 1390
  average free-electron energy of, 1421
  as conductors, 571
  electron configurations of, 1390–1391
  electron-gas model of, 1415
  free-electron model of, 1415, 1418–1422
  rare earth, 1390
Metastable states, 1308
Meters, 4–5, 438
Methane, structure of, 1407
Michelson, Albert, 1180–1181
Michelson interferometer, 1179–1181
Michelson-Morley experiment, 1180–1181, 1224
Microcoulomb, 696
Microfarad, 790–791
Micrographs, 1149
Microphones, condensor, 790
Microscopes, 1147–1149
  electron, 1290–1292
  resolving power of, 1210
  scanning tunneling, 1349
Microscopic state, 675–677
Microscopic vs. macroscopic properties, 590
Microwave ovens, 1071f
Milliampere, 820
Millibar, 375
Millikan, Robert, 897, 1264, 1293
Millikan's oil-drop experiment, 786
Mirages, 1103
Mirrors. See also Reflection
  concave, 1118–1122

  converging, 1119
  convex, 1122–1124
  graphical methods for, 1124–1126
  image formation by, 1115–1126
  parabolic, 1120
  plane, 1115–1118
  spherical, 1118–1126
Mitchell, John, 423
Models
  definition of, 3
  idealized, 3–4
Molar heat capacity, 564–565, 605–607, 637–639
Molar mass, 517, 564, 591, 598
Molar specific heat, 564–565
Molecular bonds. See Bonds
Molecular clouds, 1435–1436
Molecular collisions, 603–605
  gas pressure and, 597–600
Molecular kinetic energy, 597, 605–606
  gas pressure and, 600–602
  temperature and, 636
Molecular mass, 598
  measurement of, 897
Molecular rotation, vibration and, 1410–1412
Molecular spectra, 1300, 1408–1412
Molecular speed, 602–603, 608–610
  Maxwell-Boltzmann distribution and, 608–609
Molecular vibration, 451–453, 597, 606–608
  rotation and, 1410–1412
Molecular weight, 564, 591
Molecular zippers, 1408
Molecules, 1405–1431
  gas, 597, 599
  intermolecular forces and, 597–598
  liquid, 597
  polar, 805–806, 1407
  polyatomic, 605–606
  solid, 597
Moles, 564, 598
Moment arm, 309
Moment of inertia, 288–291
  of bird's wing, 290f
  calculation of, 290, 294–296, 456
  of cylinder, 295–296
  definition of, 289
  parallel-axis theorem and, 293–294
  in simple harmonic motion, 451
  of sphere, 296
  torque and, 312
Moment, vs. torque, 309
Momentum, 241–266
  angular. See Angular momentum
  collisions and, 251–258
  components of, 242
  conservation of, 247, 255, 1243
  definition of, 242
  electromagnetic, 1068–1069
  impulse and, 241–246
  impulse-momentum theorem and, 242–244, 483–484
  linear, 242, 322
  magnitude of, 242
  net force and, 242
  Newton's second law and, 242
  Newton's third law and, 247
  of photons, 1277
  rate of change of, 242
  relativistic, 1243–1246
  rocket propulsion and, 262–265
  in simple harmonic motion, 449
  total, 247, 260
  transverse, 484–485
  units of, 242
  as vector, 242, 248
  vs. kinetic energy, 240, 242–246
  wave speed and, 484–485
Momentum-position uncertainty principle, 1274–1275, 1278, 1315–1316
Monochromatic light, 1054–1055, 1164
Monopoles, magnetic, 885, 1500
Moon walking, 407f

Morley, Edward, 1180–1181
Moseley's law, 1394–1396
Motion
    along curve. *See* Curves
    of center of mass, 259–262
    circular, 85–88, 154–159, 440–442. *See also* Circular
        motion
    forces and, 112–116
    Kepler's laws of, 414–416
    Newton's laws of, 104–133. *See also* Newton's laws
        of motion
    orbital, 411–413
    period of, 87
    periodic, 437–462. *See also* Oscillation
    planetary, 414–418
    projectile, 77–85
    rotational. *See* Rotation/rotational motion
    of satellites, 411–413
    simple harmonic, 439–453
    straight-line, 35–68. *See also* Straight-line motion
    translational, 308, 314–320, 606
    in two/three dimensions, 69–103
Motion diagrams, 41, 41f, 46f
Motional electromotive force, 969–971
Motors, electric, 898, 907–908
Moving-axis rotation, 314–320
MRI (magnetic resonance imaging), 904–905, 934,
    1444
Muller, Karl, 824
Multimeters, digital, 863
Multiplets, 1383
Multiplication
    of displacement, 13
    significant figures in, 8f, 9
    of vectors, 13, 16, 20–22, 70
Muon-catalyzed fusion, 1471
Muonic atoms, 1379
Muons, 1254, 1485, 1491
Muscle fibers, work done by, 177
Music, sound waves in, 513–514
Musical instruments
    pipe organs, 524–527
    standing waves and, 497–498
    string, 497–498
    wind organs, 524–527
Mutual inductance, 991–994
Myopia, 1143–1146

N

Natural angular frequency, 459–460
Natural light, 1094
Near point, of eye, 1143
Nearsightedness, 1143–1146
Neddermeyer, Seth, 1485
Ne'eman, Yu'val, 1498
Negative ions, 690
Negative work, 179–180, 183
Neon, 1390
Nerve cells, electric charge in, 741
Nerve conduction, resistivity in, 824
Net force, 107, 247–248
    acceleration and, 112–118
    center-of-mass motion and, 261–262
    definition of, 107
    equilibrium and, 135
    momentum and, 242–244
    Newton's first law and, 109
    Newton's second law and, 112–118
    torque and, 311–312, 323, 324
Net torque, 321
Neurons, 716
    electric charge in, 741
Neutral side of line, 868–869, 870
Neutrino detectors, 1489–1490
Neutrino oscillations, 1500–1501
Neutrinos, 1452, 1492
    mass of, 1500–1501
Neutron(s)
    absorption of, 1464
    discovery of, 1481–1482

magnetic moment of, 1442–1443, 1497
    mass of, 689, 1440
    spin angular momentum of, 1442
Neutron number, 1441
Neutron-proton pair binding, 1446–1447
Newton, 6, 105, 113–114, 119
Newton, Isaac, 1080
Newtonian mechanics, 104
Newtonian synthesis, 418
Newton-meter, 177, 309
Newton's first law of motion, 108–112
    application of, 124–125, 134–139
    equilibrium and, 109
    inertia and, 109
    inertial frame of reference and, 110–112
    net force and, 109
    particles in equilibrium and, 134–139
    statement of, 108–109
Newton's law of gravitation, 402–406
Newton's laws of motion, 104–133
    application of, 124–125
    first law, 108–112. *See also* Newton's first law of
        motion
    free-body diagrams and, 124–125
    Kepler's laws and, 414–416, 418
    overview of, 104–108
    particle model and, 1274
    relativity and, 1244–1245, 1249–1251
    second law, 112–117. *See also* Newton's second law
        of motion
    statement of, 104
    third law, 120–125. *See also* Newton's third law of
        motion
    uncertainty and, 1274–1275
Newton's rings, 1178
Newton's second law of motion, 112–117
    application of, 115, 124–125, 140–146
    component equations for, 115
    constant mass and, 115
    entropy and, 674, 677
    external forces and, 115
    fluid resistance and, 152–154
    inertial frame of reference and, 115–116, 118
    momentum and, 242
    relativity and, 1244–1245, 1249–1251
    rotational analog of, 312, 318–320
    statement of, 114–115
Newton's third law of motion, 120–125
    action-reaction pairs and, 120–123
    application of, 124–125
    fluid resistance and, 151–154
    momentum and, 247
    statement of, 120
    tension and, 123
Noble gases, 1390
Nodal curves, 1166
Nodal planes, 1070
Nodes, 492
    displacement, 523
    pressure, 523
    of Ranvier, 882
Noise, 514
Noise control
    beat synchronization in, 532
    wave interference in, 531
Nonconservative forces, 222–224
Nonelectrostatic fields, 959, 972–973
Nonlinear conductors, 823
Nonreflective coatings, 1178–1179
Nonuniform circular motion, 159
Normal force, 105, 146
Normal mode, 496
Normalization condition, 1342, 1365
North (N) pole, 884
Notation
    for angles, 71
    scientific (powers-of-ten), 9
    spectroscopic, 1375
    for units of measure, 5
    for vectors, 11, 19

n-type semiconductors, 1424
Nuclear accidents, 1468
Nuclear angular momentum, 1442
Nuclear binding, 1444–1449
Nuclear fission, 785, 1247, 1464–1468
    chain reactions in, 1466
    liquid-drop model of, 1465–1466
    reaction dynamics in, 1465
    in reactors, 1466–1468
Nuclear force, 1446–1447
    mesons and, 1484–1485
    potential-energy function for, 1448–1449, 1491
Nuclear fusion, 785, 1284, 1469–1471
    heat of, 566
    helium, 1514
    solar, 1501
    tunneling in, 1349–1350
Nuclear magnetic moment, 1442–1443
Nuclear magnetic resonance, 1444
Nuclear magneton, 1443
Nuclear medicine, 1391, 1461, 1466
Nuclear physics, 1439–1471
Nuclear power plants, 1247, 1466–1468
Nuclear reactions, 1462–1471
    chain, 1466
    endoergic, 1463
    endothermic, 1463
    exoergic, 1463
    exothermic, 1463
    fission, 785, 1247, 1464–1468
    fusion, 1469–1471
    neutron absorption in, 1464
    reaction energy for, 1462–1464
    thermonuclear, 1469–1471, 1470
    threshold energy for, 1463
Nuclear reactors, 1247, 1466–1468
Nuclear spin, 1442–1443
Nuclear stability, 1449–1456
Nucleon number, 1440
Nucleons, 1439–1440
Nucleosynthesis, 1511–1515
Nucleus, 1439–1471
    atomic, 689, 1295
    daughter, 1454
    density of, 1440
    in excited states, 1454
    formation of, 1511–1514
    in ground state, 1454
    half-life of, 1456–1457
    lifetime of, 1457
    liquid-drop model of, 1447–1448
    mass of, 1440
    parent, 1454
    properties of, 1439–1444
    radius of, 1440
    shell model of, 1448–1449
    structure of, 1444
Nuclides, 1441
    decay of, 1450–1458. *See also* Radioactive decay
    odd-odd, 1449
    radioactive, 1450–1454. *See also* Radioactivity
    stable, 1449–1450
    synthesis of, 1511–1515

O

Object distance, 1116
Object, in optics, 1114
Object point, 1115
Objective, microscope, 1147–1148
Occhialini, Guiseppe, 1271
Oculars, 1148
Odd-odd nuclides, 1449
Oersted, Hans Christian, 885
Ohm, 826
Ohmic conductors, 823
Ohmmeters, 863
Ohm's law, 822, 825–826
1 newton per coulomb, 699
Onnes, Heike Kamerlingh, 824
Open circuits, 870

Open orbits, 412
Operational definition, 4
Optic axis, of mirror, 1118
Optical fibers, 1090
Optics, 1080. *See also* Light
    geometric. *See* Geometric optics
    image in, 1115
    object in, 1114
    physical, 1082, 1163
Orbit(s)
    center of mass and, 417–418
    circular, 412–413, 416–417
    closed, 412
    of Comet Halley, 417
    elliptical, 415–416
    open, 412
    satellite, 411–413
    sector velocity and, 415
    semi-major axis of, 415, 416
Orbital angular momentum
    quantization of, 1373–1374, 1384
    spin angular momentum and, 1387
Orbital eccentricity, 415
Orbital magnetic quantum number, 1374
Orbital period, 416–417
Orbital quantum number, 1373
Orbital speed, 416–417
Order-of-magnitude estimates, 10
Organ pipes, 524–527
Oscillation, 437–462
    amplitude of, 438
    damped, 457–460
    definition of, 437
    displacement in, 438
    driven, 459–460
    electrical, 1005–1009
    forced, 459–460, 527
    frequency of, 438
    in harmonic oscillators, 439–440
    molecular vibration and, 451–453
    neutrino, 1500–1501
    overview of, 437–438
    of pendulum, 453–457
    of periodic waves, 474–476
    resonance and, 460, 527
    simple harmonic motion and, 439–453. *See also*
        Simple harmonic motion
    in spring, 437–438
Oscillation cycle, 438
Oscillation period, 438
Otto cycle, 657–658
Overdamped circuits, 1010
Overdamping, 458
Overloaded circuits, 869–870
Overtones, 496

P

Pacemakers, 866
Painting, electrostatic, 683
Pair annihilation, 1483, 1484
Pair production, 1482–1483
    photons in, 1271–1272
    positrons in, 1271–1272, 1482–1483
Parabolic graphs, 48
Parabolic lenses, 1151
Parabolic mirrors, 1120
Parabolic trajectories, 79, 79f
Parallel connection, 794–796
Parallel, resistors in, 851, 852–855
Parallel vectors, 11, 12
Parallel-axis theorem, 293–294
Parallel-plate capacitors, 790, 791
    dielectrics in, 800–805
Paramagnetism, 942–944, 943t
Paraxial approximation, 1119
Paraxial rays, 1119
Parent nucleus, 1454
Parity, 1495
Parsec, 1503
Partial derivatives, 226–227, 481

Particle(s), 36
    alpha, 1294–1295, 1349–1350
    antiparticles and, 1483
    in bound state, 1343–1344
    definition of, 3
    distinguishable, 1419
    as force mediators, 1484
    free, 1330
    fundamental, 1480–1501
    light waves as, 1261–1280
    in Newtonian mechanics, 1274
    photons as, 1263. *See also* Photons; Wave-particle
        duality
    in standard model, 1499–1500
    strange, 1494–1495
    wave function for, 1328–1335
Particle accelerators, 1485–1488
    cyclotrons, 893, 918, 1486–1487
    linear, 1485–1486
    synchrotrons, 1487
Particle collisions
    in accelerators, 1485–1488
    available energy and, 1487–1488
    in colliding-beam experiments, 1489
Particle detectors, 1489
    neutrino, 1489–1490
Particle in a box
    in one dimension, 1338–1343, 1371
    in three dimensions, 1366–1371
Particle interactions, 1490–1495
    conservation laws for, 1495
    electromagnetic, 159–160, 1490
    fundamental types of, 159–160, 1490–1492
    gravitational, 159–160, 1490
    isospin and, 1495
    parity in, 1495
    strangeness in, 1494–1495
    strong, 160, 1490–1491
    symmetry-breaking, 1495
    weak, 160, 1491
Particle motion, vs. wave motion, 475
Particle physics, historical perspective on, 1480–1485
Particle speed, vs. wave speed, 519
Particle velocity, vs. wave velocity, 519
Particle waves
    angular frequency of, 1331
    one-dimensional, 1329–1333
    vs. mechanical waves, 1329
    wave equation for, 1330–1333
    wave number for, 1331
Pascal, 353, 375
Pascal, Blaise, 353, 354, 377
Pascal's law, 376–377
Paths, in thermodynamic system, 628–629
Pauli, Wolfgang, 1389
Pendulum
    ballistic, 253
    periodic motion of, 453–457
    physical, 455–457
    simple, 453–455, 456
Pendulum bob, 454
Penzias, Arno, 1515
Percent error (uncertainty), 8
Perfect crystals, 1413–1415
Perihelion, 415
    precession of, 1250
Period, 87
    frequency and, 438–439
    orbital, 416–417
    oscillation, 438
    in simple harmonic motion, 443
Periodic driving force, damped oscillation and,
    459–460
Periodic motion, 437–462. *See also* Oscillation
    amplitude of, 438
    definition of, 437
    displacement in, 438
    frequency of, 438
    in harmonic oscillators, 439–440
    molecular vibration and, 451–453

    overview of, 437–438
    of pendulum, 453–457
    resonance and, 460
    simple harmonic motion and, 439–453. *See also*
        Simple harmonic motion
    of spring, 437–438
    of waves, 474–476
Periodic table, 1389, 1390–1391
Periodic waves, 474–477. *See also* Mechanical waves
    longitudinal, 475–476. *See also* Longitudinal waves
    mathematical description of, 477–482
    sinusoidal, 475, 477–482. *See also* Sinusoidal waves
    transverse, 474–475. *See also* Transverse waves
Permanent magnets, 883, 941
Permanent set, 358
Permeability, 943
Permittivity, of dielectric, 802
PET (positron emission tomography), 1484
Pfund series, 1304
Phase, 565
    wave, 479
Phase angle, 444, 1026, 1031–1032
Phase change, 565–570
Phase diagrams, 611
Phase equilibrium, 566, 611
Phase shifts, interference and, 1174–1175
Phase transitions, 611–612
Phase velocity, 479
Phased-array radar, 1220
Phases of matter, 610–613
    critical point and, 611
    molecular interactions and, 610
    sublimation and, 611
    triple point and, 611
Phases of state
    p-V diagrams and, 596
    pVT-surfaces and, 612–613
Phasor diagrams, 1022
Phasors, 441, 1022
Photinos, 1501
Photocells, 1425
Photocopying machines, 769
Photoelasticity, 1100
Photoelectric effect, 1081, 1261
Photoelectrons, 1262
Photography. *See* Cameras
Photomicrographs, 1149
Photomultipliers, 1273
Photons, 1081, 1248, 1261–1280
    absorption of, 1261–1266, 1484
    in Bohr's atomic model, 1297–1306
    in charged-particle interactions, 1484
    Compton scattering and, 1269–1271
    definition of, 1261
    diffraction and, 1273–1274
    discovery of, 1481
    Einstein's explanation for, 1263–1264
    electroweak interactions and, 1500
    emission of, 1266–1269, 1484
    as force mediators, 1484
    gamma ray, 1454
    interference and, 1273–1274
    light emitted as, 1266–1269, 1484
    momentum of, 1264–1265, 1274–1275, 1277
    pair production and, 1271–1272
    as particles, 1263
    photoelectric effect and, 1261–1266
    position of, 1274–1275, 1277
    probability and, 1274–1275
    spontaneous emission of, 1307
    in standard model, 1500
    stimulated emission of, 1307–1309
    stopping potential and, 1262–1263
    threshold frequency and, 1263
    uncertainty and, 1274–1278
    virtual, 1484
    wave-particle duality and, 1273–1279. *See also*
        Wave-particle duality
    X-ray, 1266–1269
Photovoltaic effect, 1428

Physical laws (principles), 2
Physical optics, 1082, 1163
Physical pendulum, 455–457
    vs. simple pendulum, 456
Physical quantities
    definition of, 4
    units of, 4–6. *See also* Units of measure
Physical theories, 2
Physics
    as experimental science, 2
    nuclear, 1439–1471
    overview of, 2
    particle, 1480–1485
    as process, 2
    quantum, 1328–1356
Pi, value of, 8, 8f
Pianos, string vibration in, 528
Picoampere, 820
Picofarad, 791
Pileated woodpecker impulse, 243
Pions, 1485, 1489, 1491, 1493–1494
Pipe organs, 524–527
Pitch, 513
Planck length, 1509
Planck, Max, 1311–1312
Planck radiation law, 1311–1314
Planck time, 1509
Planck's constant, 942, 1263
Plane mirrors
    graphical methods for, 1124–1126
    image formation by, 1115–1118
Plane of incidence, 1084
Plane surface
    reflection at, 1115–1118
    refraction at, 1129
Plane waves, electromagnetic, 1055–1057, 1060
Planet Imager, 1221
Planetary motion, 414–418. *See also* Orbit(s)
    center of mass and, 417–418
    Kepler's laws of, 414–417
Plant growth, deuterium and, 1445
Plastic deformation, 358
Plastic flow, 358
Plasticity, 357–358
p-n junctions, 1425–1428
Point charges, 693–694
    electric dipole and, 709–713
    electric fields of, 725–726. *See also* Electric charge
    electric potential energy of, 757–760
    electric potential of, 765
    electromagnetic waves from, 1053
    force between, 697
    inside closed surface, 725–726
    inside nonspherical surface, 732
    inside spherical surface, 732
    magnetic field lines for, 924–925
    superposition of, 696
Point objects
    definition of, 1115
    image formation for, 1118–1119, 1127–1130, 1209–1210
    resolution of, 1209–1210
Point-molecule model, of gas heat capacity, 605
Polar molecules, 805–806, 1407
Polarization, 693, 805–806, 1093–1100
    bee vision and, 1101
    charged bodies and, 693, 807
    circular, 1099–1100
    definition of, 788, 1093
    of dielectrics, 801–803, 805–807
    electric field lines and, 709
    of electromagnetic waves, 1058, 1093–1100
    elliptical, 1099–1100
    induced charges and, 693, 807
    of light waves, 1093–1100
    linear, 1058, 1093, 1095
    partial, 1097
    photoelasticity and, 1100
    by reflection, 1097–1098
Polarizers, 1093

Polarizing angle, 1097
Polarizing axis, 1094–1095
Polarizing filters, 1093, 1094–1097, 1098
Polaroid filters, 1094–1095
Pollen, fluid resistance and, 152
Polyatomic molecules, 605–606
Population inversions, 1308–1309
Porro prism, 1089–1090
Position
    by integration, 55–57, 55f, 56f
    potential energy and, 208
    x-t graphs and, 37–38, 38f, 40–42
Position vectors, 70–72, 70f
Position-momentum uncertainty principle, 1274–1275, 1278, 1315–1316
Positive ions, 690
Positive work, 179, 183
Positron emission tomography (PET), 1484
Positroniums, 1379
Positrons, 1453, 1482–1483
    motion in magnetic fields, 894
    in pair annihilation, 1483
    in pair production, 1271–1272, 1482–1483
Potential. *See* Electric potential
Potential barriers, 1347–1350
Potential difference, 763–764. *See also* Voltage
    capacitance and, 789
    measurement of, 861–862
    notation for, 865
    resistance and, 852
    time-varying, 865
Potential energy, 207–231, 755
    around circuits, 833–834
    of capacitor, 796–800
    conservative forces and, 221–229
    definition of, 208
    elastic, 216–221, 225
    electric, 754–761. *See also* Electric potential energy
    of electric dipole, 711
    electric forces and, 226–227
    energy diagrams and, 228–229
    equilibrium and, 228–229
    force and, 225–228
    gradient of, 227
    gravitational, 208–216, 293, 317, 409–411
    intermolecular forces and, 597
    kinetic energy and, 207, 208, 221–222
    for magnetic dipoles, 903–904
    of molecules, 597
    of particle in a box, 1339
    position and, 208
    potential barriers and, 1347–1350
    potential wells and, 597, 1343–1347
    in simple harmonic motion, 446–449
    work and, 755
Potential gradient, 774–776
Potential wells, 597, 1343–1347
Potential-energy function
    for harmonic oscillator, 1350, 1352, 1353–1354
    for nuclear force, 1448–1449, 1491
    for particle in a box, 1339
Potentiometers, 863–864
Pound, 117
Pounds per square inch, 353
Pounds per square inch absolute (psia), 378
Pounds per square inch gauge (psig), 378
Power, 193–195
    in ac circuits, 1034–1037
    average, 193, 487
    of corrective lens, 1144
    definition of, 193
    in electric circuits, 834–838
    electrical, 194
    energy and, 871
    force and, 194
    instantaneous, 193, 194
    measurement of, 862–863
    rotational motion and, 321–322
    of sound waves, 519

    velocity and, 194
    of waves, 487–488
    work and, 487
Power distribution systems, 868–872
Power factor, 1036
Power plants, nuclear, 1247, 1466–1468
Power transmission systems, lightning strikes on, 995
Powers-of-10 notation, 9
Poynting vector, 1065–1067
Precession, 328–330
    of perihelion, 1250
Precession angular speed, 329
Precipitators, electrostatic, 784
Precision, vs. accuracy, 9
Prefixes, for units of measure, 5
Presbyopia, 1143
Pressure
    absolute, 377–378
    atmospheric, 355, 375–376
    bulk stress/strain and, 355–356
    definition of, 355, 375
    fluid flow and, 385–389
    in fluids, 355, 375–380
    gauge, 377–378
    measurement of, 378–380
    radiation, 1068–1069
    reciprocal, 356
    residential water, 387
    as scalar quantity, 355
    speed and, 385–389
    units of, 353, 375
    vs. density, 592
    vs. force, 355, 376
Pressure amplitude, 511–512
    sound intensity and, 519–521
Pressure gauges, 378–380
Pressure nodes/antinodes, 523
Primary windings, 1040, 1042f
Principal maxima, 1201
Principal quantum number, 1301, 1373
Principal rays
    for lenses, 1135–1136
    for mirrors, 1124–1125
Principles
    differential, 245
    integral, 244
    physical, 2
Printers
    inkjet, 722
    laser, 689, 769, 1309
Prism
    dispersion by, 1091
    Porro, 1089–1090
Prism binoculars, 1150
Probability density, 1333
Probability distribution, 1376
Probability distribution function, 1333
    for harmonic oscillator, 1353
    one-dimensional, 1333
    radial, 1376
    three-dimensional, 1365–1366
Probability, wave-particle duality and, 1274–1278
Problem-solving strategies, 2–3
Product
    scalar, 20–22
    vector, 23–25
Projectile, 77
Projectile motion, 77–85
    acceleration and, 77–80, 87
    air resistance and, 77, 79–80
    components of, 77–78
    trajectory and, 77
    velocity and, 77–80
    vs. circular motion, 87
Projectors, 1141
Propagation speed, 474, 475
Propeller design, 288
Proper length, 1235
Proper time, 1230–1231, 1232

Proton(s)
  charge of, 695
  electron screening of, 1391–1392
  lifetime of, 1500
  magnetic moment of, 1442–1443
  mass of, 689, 1440
  spin angular momentum of, 1442
Proton decay, 1500
Proton-antiproton pairs, 1491–1492
Proton-neutron pair binding, 1446–1447
Proton-proton chains, 1469
Psia (pounds per square inch absolute), 378
Psig (pounds per square inch gauge), 378
p-type semiconductors, 1425
Pulsed lasers, 1309
Purcell, Edward, 1416f
Pure semiconductors, 1423, 1424
p-V diagrams, 596
p-V isotherms, 596
pVT-surfaces, 612–613

Q
Quality factor, 1459
Quanta, 1081. See also Photons
Quantitized energy, 1261
Quantity of heat, 562–565
Quantum dots, 1363
Quantum electrodynamics, 1081
Quantum hypothesis, 1310–1314
Quantum mechanics, 1328–1356
  atomic structure and, 1364–1398
  bound states and, 1343–1345
  definition of, 1329
  harmonic oscillator in, 1350–1354
  one-dimensional waves in, 1329–1333
  particle in a box and, 1338–1343, 1366–1371
  potential barriers and, 1347–1350
  potential wells and, 1343–1347
  probability distribution function and, 1333, 1353,
    1365–1366
  Schrödinger equation and, 1332–1333, 1336–1337,
    1365–1366
  stationary states and, 1337–1338, 1366
  tunneling and, 1347–1350
  wave functions and, 1328–1335
  wave packets and, 1335–1336
Quantum number
  notation for, 1374–1375
  orbital magnetic, 1374
  principal, 1373
  spin, 1385
  spin magnetic, 1385
Quarks, 689, 920, 1443, 1496–1499
  antiquarks and, 1496, 1499
  colors of, 1498–1499
  down, 1496
  eightfold way and, 1497–1498
  flavors of, 1496
  in standard model, 1500
  strange, 1496
  types of, 1496
  up, 1496
Quarter-wave plates, 1100
Quasars, 1220

R
Rad, 1459
Radar
  Doppler, 537
  phased-array, 1220
Radar guns, 1242f
Radial probability distribution function, 1376, 1392
Radians, 279, 287
Radiation, 570, 574–577
  absorption of, 575
  applications of, 576
  background, 1515
  beneficial uses of, 1461–1462
  biological effects of, 1459–1462
  blackbody, 576, 1310–1314

cancer and, 1269
  Čerenkov, 1257
  definition of, 574
  electromagnetic, 574–577, 1053–1054. See also
    Electromagnetic wave(s)
  global warming and, 576–577
  from human body, 575–576
  quality factor for, 1459
  solar, 576, 1262
  Stefan-Boltzmann law/constant and, 575
  synchrotron, 1487
  thermal, 1081
  X. See X-ray(s)
Radiation doses, 1459–1460
Radiation exposure
  hazards of, 1460–1461, 1476
  limits on, 1459–1460
  sources of, 1458–1459, 1461
Radiation pressure, 1068–1069
Radiator, ideal, 576
Radioactive dating, 1458
Radioactive decay, 1450–1458
  activity in, 1456–1457
  alpha, 1450–1452
  beta, 1452–1453
  gamma, 1454
  half-life and, 1456–1457
  natural radioactivity and, 1454–1455
  rate of, 1456–1457
Radioactive decay series, 1454–1455
Radioactive fallout, 1476–1477
Radioactive isotopes, in medicine, 1391, 1461, 1466
Radioactive nuclides, decay of, 1450–1454
Radioactive tracers, 1461–1462
Radioactivity
  definition of, 1449
  natural, 1454–1455
  units for, 1457
Radioisotope imaging, 1391
Radiology, 1268–1269
Radios
  transmitters and receivers for, 1054
  tuning, 1038, 1039
Radium, alpha decay of, 1451–1452
Radius
  of nucleus, 1440
  Schwarzschild, 424
Radius of curvature
  for lens, 1133–1134
  for spherical surface, 1116, 1128–1129
Radon, 1458–1459
Rainbows, 1092–1093
Randomness, in thermodynamic processes, 653
Range of validity, 2
Rare earth metals, 1390
Rarefaction, 476
Ratio of heat capacities, 517, 639
Rayleigh, Lord, 1209, 1311
Rayleigh's criterion, 1209, 1210
Rays, 1082, 1135
  paraxial, 1119
  principal, 1124–1125, 1135–1136
R-C circuits, 864–868
Reaction(s)
  activation energy for, 610
  chain, 1466
  chemical, 610
  nuclear, 1462–1471. See also Nuclear reactions
Reaction energy, 1462–1464
Real image, 1115
Recession speed, 1502–1503, 1504, 1505
Reciprocal pressure, 356
Recombination currents, 1427
Rectified alternating current, 1022–1023
Rectified average current, 1023
Redshifts, 1502, 1507
  cosmological, 1505
Reduced mass, 1305–1306, 1408–1409
Reference circle, 440–441
Reference point, 440–441

Reference standards, 4
Reflected waves, 489–490
  sinusoidal, 491–495
Reflecting telescopes, 1150–1151
Reflection, 1082–1091
  Bragg, 1207
  definition of, 1082
  diffuse, 1083, 1115
  of electromagnetic waves, 1071
  Huygen's principle and, 1102
  image formation and, 1115–1118
  interference during, 1174–1175
  law of, 1084–1086
  of light waves, 1082–1088
  phase shifts during, 1174–1175
  at plane surface, 1115–1118
  polarization by, 1097–1098
  specular, 1083, 1115
  at spherical surface, 1118–1126
  total internal, 1088–1091
  in X-ray diffraction, 1206
Reflective coatings, 1178, 1179
  interference and, 1178–1179
Reflector, ideal, 576
Refracting telescopes, 1149
Refraction, 1082–1088
  definition of, 1082
  in eye, 1143
  Huygen's principle and, 1102–1104
  index of. See Index of refraction
  law of, 1084–1086
  at plane surface, 1115–1118
  at spherical surface, 1126–1130
Refractive index. See Index of refraction
Refractors, 1154
Refrigerator(s), 659–661
  Carnot, 666–667
  practical, 660–661
  workless, 661
Refrigerator statement, of second law of
    thermodynamics, 662
Reines, Frederick, 145
Relative biological effectiveness, 1459–1460
Relative permeability, 943
Relative velocity, 88–93. See also Velocity
  definition of, 88
  elastic collisions and, 256–258
  frame of reference and, 89
  Galilean velocity transformation and, 91
  in one dimension, 88–90, 89f
  in two or three dimensions, 90–93
Relativistic momentum, 1243–1246
Relativistic work and energy, 1246–1249
Relativity, 1223–1252
  aging and, 1233
  Doppler effect and, 537–538, 1241–1243
  Einstein's postulates for, 1224–1225
  Galilean coordinate transformation and, 1225–1226
  general theory of, 1249–1251, 1504
  inertial frame of reference and, 1223, 1224, 1226
  invariance of physical laws and, 1223–1226
  of length, 1233–1237
  Lorentz transformations and, 1237–1241
  Newtonian mechanics and, 1244–1245, 1249
  principle of, 1224
  of simultaneity, 1226, 1227–1228
  special theory of, 1223–1249
  speed of light and, 1224–1225
  of time intervals, 1228–1233
  twin paradox and, 1232–1233
Relativity principle, 1224
Relaxation time, 866–867
Rem, 1460
Resistance, 825–828
  equivalent, 851, 852
  internal, 830, 833
  measurement of, 860–861, 862–864
Resistance thermometer, 553–554
Resistivity, 822–825
  of metal, 838–839

in nerve conduction, 824
temperature and, 824–825
Resistors, 826–827
in ac circuits, 1025, 1029, 1034–1035
in dc circuits, 850–855
energy dissipation in, 835
equivalent resistance and, 851–855
in parallel, 851, 852–855
power in, 1034–1035
power input to, 834–835
power rating of, 835
in series, 851–852, 853–855
shunt, 861
vs. inductors, 999
Resolving power (resolution), 1209–1211
chromatic, 1203–1204, 1210
in diffraction, 1209–1210
of grating spectrograph, 1203–1205
limit of, 1209
of microscope, 1290–1291
Rayleigh's criterion and, 1209, 1210
Resonance, 460, 527–529
in ac circuits, 1037–1039
definition of, 460, 1038
in mechanical systems, 460
Resonance angular frequency, 1038
Resonance curves, 1038–1039
Resonance frequency, 1038
Resonance width, 1049
Response curves, 1038–1039
Rest energy, 1247–1249
Rest mass, 1243–1246
Restoring force, 438
in pendulum, 454
in simple harmonic motion, 439–440
Resultant, of displacements, 12
Reverse bias, 1426
Reversed, 1117
Reversed image, 1117
Reversible processes, 653
Right-hand rule, 23
Right-handed system, 24
Rigid body, 278
Rigid-body equilibrium, 345, 348–352
Rigid-body rotation, 278–297. See also
    Rotation/rotational motion
    about moving axis, 314–320
    angular acceleration in, 282–285
    angular velocity in, 279–282
    around fixed axis, 278–279
    dynamics of, 308–331
    kinetic energy in, 288–293
    linear acceleration in, 286–288
    linear speed in, 285–286
    moment of inertia and, 288–291
    with translational motion, 314–320
R-L circuits, 1001–1005
    current decay in, 1004–1005
Rms speed, 602
Rocket propulsion, 262–265
Roller coasters, 74, 88
Rolling friction, 151–154, 320
Rolling without slipping, 316–318
Röntgen, Wilhelm, 1205, 1267, 1454
Root-mean-square current, 1023–1024
Root-mean-square speed, 602
Root-mean-square values, 602, 1023–1024
Root-mean-square voltage, 869
Rotational energy levels, 1408–1412
Rotational inertia, 289
Rotational kinetic energy, 288–293, 315–316
Rotation/rotational motion
    about axis of symmetry, 323–324
    angular acceleration and, 282–285, 311–314
    angular momentum and, 322–328
    angular velocity in, 279–282
    around fixed axis, 278–279
    in bacteria, 283f
    with constant angular acceleration, 283–285
    coordinates for, 279

direction of, 279, 309
dynamics of, 308–331
of Earth, 421–423
energy in, 288–293
equilibrium and, 345
fixed-axis, 278–279, 283
of gyroscope, 328–330
kinetic energy and, 288–293
linear acceleration in, 286–288
linear speed in, 285–286
molecular, 1410–1412
moving-axis, 314–320
Newton's second law of motion and, 312, 318–320
power and, 321–322
precession and, 328–330
rigid-body, 278–297. See also Rigid-body rotation
in rolling without slipping, 316–318
torque and, 308–314
with translational motion, 314–320
units of, 279
work and, 320–322
Rotors, 907–908
Rubbia, Carlo, 1491
Rule of Dulong and Petit, 565, 608
Running on moon, 407f
Rutherford, Ernest, 1294–1296, 1297, 1349–1350,
    1439–1440, 1454, 1462, 1481
Rutherford's atomic model, 1294–1296
Rutherford's scattering experiments, 1294–1296
Rydberg atom, 1306, 1402
Rydberg constant, 1303

S

Satellite orbits, 411–413
Saturation, 1447
Saturation current, 1426
Saturation magnetization, 945
Scalar (dot) product, 20–22
Scalar quantities, 11
    in vector multiplication, 13
Scale factor, 1504, 1505
Scanning electron microscope, 1291–1292
Scanning tunneling microscope, 1349
Scattering of light, 1100–1101
Schrieffer, Robert, 1430
Schrödinger equation, 1332–1333
    for hydrogen atom, 1372–1373
    for hydrogenlike atoms, 1378–1379
    one-dimensional, 1332–1333, 1336–1337
    with potential energy, 1336–1337
    three-dimensional, 1365–1371
    time-independent, 1338, 1366
    X-ray spectra and, 1393–1396
Schrödinger, Erwin, 1332
Schwarzschild radius, 424
Scientific notation, 9
Scintigram, 1461
Scintillation, 1294
Screening, 1391–1392
Scuba tank, air mass in, 594
Search coils, 983
Second condition for equilibrium, 345
Second law of thermodynamics, 652, 661–677
    Carnot cycle and, 667–668
    Clausius statement of, 662–663
    engine statement of, 661
    Kevin-Planck statement of, 661
    refrigerator statement of, 662–663
Secondary windings, 1040, 1042f
Seconds, 4, 5t, 438
Seconds per cycle, 438
Sector velocity, 415
Segré chart, 1449–1450, 1454
Segré Emilio, 1449
Selection rules, 1382
Selectrons, 1501
Self-induced emf, inductive reactance and, 1026–1027
Self-inductance, 991, 994–998. See also Inductance
Semiconductor(s), 909, 1422–1425
    bias conditions and, 1426, 1427–1428

compensated, 1425
conduction in, 819–820
diodes of, 827
doping and, 1424–1425
energy bands in, 1417
holes in, 909, 1423–1424
impurities in, 1415, 1424–1425
intrinsic, 1423, 1424
moving charges in, 819–820
n-type, 1424
p-type, 1425
resistivity of, 823
silicon, 1422–1425
Semiconductor devices, 1425–1430
    integrated circuits, 1429–1430
    light-emitting diodes, 1428
    photocells, 1425
    p-n junctions in, 1425–1428
    solar cells, 1428
    transistors, 1429
Semiconductor lasers, 1309
Semiempirical mass formula, 1448
Semi-major axis, 415, 416
Separation of variables, 1367
Series connection, 793–794, 795–796
Series motors, 908
Series, resistors in, 851–852, 853–855
Sharks
    electric field detection by, 699f
    flux through mouth of, 729f
Shear modulus, 357
Shear strain, 356–357
Shear stress, 352f, 356–357
Shell model, 1448–1449
Shells, electron, 1375, 1389, 1390–1391, 1394–1395
Shock waves, 538
Short circuits, 869–870, 870
Shunt motors, 908
Shunt resistors, 861
SI units, 4, 5t. See also Units of measure
Sievert, 1460
Significant figures, 8–9
Silicon semiconductors, 1422–1425
Simple harmonic motion, 439–453. See also
    Oscillation
    acceleration in, 444, 448
    amplitude in, 442–443
    angular, 451
    applications of, 450–453
    circular motion and, 440–442
    definition of, 440
    displacement in, 443–444
    energy in, 446–449
    equations of, 440–442
    as model of periodic motion, 440
    momentum in, 449
    period in, 442–443
    velocity in, 444, 448
    vertical, 450–451
    vs. electric oscillation, 1008
Simple pendulum, 453–455
    vs. physical pendulum, 456
Simultaneity, relativity of, 1226, 1227–1228
Sinusoidal alternating emf, 1021
Sinusoidal current, 1022–1024. See also Alternating
    current
Sinusoidal electromagnetic waves, 1060–1063
Sinusoidal waves, 475, 477–482. See also Mechanical
    waves; Wave(s)
    electromagnetic, 1022–1024
    energy of, 486–489
    interference and, 1164
    particle velocity/acceleration in, 480
    reflected, 491–495
    standing, 491–498. See also Standing waves
    traveling, 492, 494
    wave function for, 477–479
Sledding, Newton's first law and, 109
Slidewire generators, 965–966, 967, 970
Slipher, Vesto, 1502

Slug, 117
Snakes, wave motion of, 473f
Snell's law, 1084
Sodium doublet, 1204
Solar cells, 1428
Solar neutrinos, 1501
Solar radiation, 576, 1264
Solenoids, 904, 906, 939–941, 948t
Solids
    amorphous, 1412
    bonds in, 1414–1415
    as condensed matter, 1412
    crystalline, 1412–1415
    energy bands in, 1416–1417
    heat capacities of, 607–608
    melting points of, 1412
    molecules in, 597
    phases of, 610–613
    sound waves in, 515–517
    structure of, 1412–1415
Sonar waves, 516–517
Sound
    definition of, 509
    infrasonic, 510
    loudness of, 513
    pitch of, 513
    resonance and, 527–529
    timbre of, 513–514
    ultrasonic, 510
Sound intensity, 518–522
    decibel scale for, 521
    hearing loss and, 513, 522
    representative values for, 521t
Sound intensity level, 521
Sound waves, 476, 509–542. See also Mechanical
        waves; Wave(s)
    audible range of, 509
    beats and, 531–532
    diffraction of, 1198
    direction of, 510
    displacement amplitude of, 510, 518–519
    Doppler effect and, 533–538, 1242
    in fluid, 514–515
    frequency of, 513, 520
    in gas, 517–518
    graphing of, 511
    harmonic content of, 497
    interference and, 529–531
    musical, 513–514
    perception of, 513–514
    pipe organs and, 524–527
    power of, 519
    pressure amplitude of, 511–512, 519–521
    as pressure fluctuations, 510–513
    shock, 538
    in solid, 515–517
    speed of, 476, 485, 514–518
    standing, 522–527
    superposition of, 491
    wind instruments and, 527
Source of emf, 828
    internal resistance of, 830–831
    power input to, 835–836
    power output of, 835–836
Source point, 700, 924
South (S) pole, 884
Space. See also Universe
    curvature of, 1250f
    dimensions of, 1504–1505
    expansion of, 1503–1508
Space travel, aging and, 1233
Spacecraft, in interplanetary travel, 416f
Spacetime, 1238
Spacetime coordinates, 1238
Spark plugs, 1000
Special theory of relativity, 91, 1223–1249. See also
        Relativity
Specific gravity, 374
Specific heat, 562–563
    molar, 564–565

Spectra, 1091
    absorption line, 1293
    atomic, 1292, 1297–1300
    band, 1411
    continuous, 1310–1314
    emission line, 1292–1293
    molecular, 1300, 1408–1412
    X-ray, 1393–1396
Spectral emittance, 1310–1314
    definition of, 1310
    quantum hypothesis and, 1311–1313
    vs. intensity, 1310–1311
Spectral lines, 1292, 1297
    Zeeman effect and, 1379–1382
Spectrographs, grating, 1203–1204
Spectroscopic notation, 1375
Specular reflection, 1083, 1115
Speed, 39
    air drag and, 152–154
    angular, 280, 286, 329
    average, 39
    of efflux, 387
    of electromagnetic waves, 1058, 1071
    escape, 410–411, 413, 423, 1505–1506
    instantaneous, 39
    molecular, 602–603, 608–610
    orbital, 416–417
    recession, 1502–1503, 1504, 1505
    of rocket, 265
    root-mean-square (rms), 602
    of sound waves, 514–518
    supersonic, 539
    terminal, 152–154
    units of, 6f
    vs. velocity, 39–40, 286
    wave, 474, 475, 479, 483–486
    work and, 181–183
    of yo-yo, 317
Speed of light, 1054, 1063f, 1081, 1224–1226
    measurement of, 4–5
    relativity and, 1224
Spheres
    electric field of, 737–738, 740–741
    electric flux through, 731–732
    gravitation and, 403–404
    mass distributions and, 418–421
    moment of inertia of, 296
    point charge inside, 732–733
    rolling, acceleration of, 319–320
Spherical aberration, 1119, 1134
Spherical coordinates, 1366
Spherical mirrors
    concave, 1118–1122
    convex, 1122–1124
    extended objects in, 1120–1122
    focal point/length of, 1119–1120
    graphical methods for, 1124–1126
    image formation by, 1118–1126
Spherical surface
    radius of curvature for, 1116, 1128–1129
    reflection at, 1118–1126
    refraction at, 1126–1130
Spherical symmetry, 1366, 1372, 1388
    gravitation and, 403–404
Spin
    electron, 942
    nuclear, 1442–1443
Spin angular momentum, 1384–1385,
        1442
    orbital angular momentum and, 1387
Spin magnetic moment, 1443
Spin magnetic quantum number, 1385
Spin quantum number, 1385
Spin-2 graviton, 1490
Spin-orbit coupling, 1386
Spiny lobsters, magnetic compasses in, 886
Spring(s)
    elastic potential energy of, 216–221
    ideal, 439
    oscillation in, 437–438. See also Oscillation

    simple harmonic motion in, 439–453. See also
        Simple harmonic motion
    tendons ad, 190
    work done on/by, 188–189
Spring balance, 106
Spring constant, 188
Square wells, finite vs. infinite, 1345–1347
Square-well potential, 1343–1347
    bound states of, 1343–1345
Squids, jet propulsion in, 262f
Stable equilibrium, 228
Stable isotope ratio analysis (SIRA), 919
Stable nuclides, 1449–1450
Standard deviation, 1275
Standard model, 1499–1500, 1510–1511
Standards, reference, 4
Standing waves, 491–498
    complex, 497
    electromagnetic, 1053, 1069–1072
    on fixed string, 495–498
    frequencies of, 496
    harmonics and, 496
    interference and, 492, 1164, 1166
    nodes and antinodes and, 492
    sound, 522–527
    string instruments and, 497–498
Stars
    binary, 425–426, 1259
    helium fusion in, 1514
    mass of, 1259
    second-generation, 1514
    supernova, 160, 1230f, 1514
    systems of, 405–406
    white dwarf, 1437
State(s)
    bound, 1343–1344
    degenerate, 1370–1371, 1374–1375
    density of, 1418–1419
    free-particle, 1346
    of matter, 565
    metastable, 1308
    stationary, 1337–1338, 1366–1369
    vs. energy levels, 1307–1308
State variables, 591
Static charge distribution, 759–760
Static equilibrium, 345
Static friction, 147–149
Stationary state
    one-dimensional, 1337–1338
    three-dimensional, 1366–1369
Steady flow, 382–383
Steam heat, 567–568
Stefan-Boltzmann constant, 575, 1310
Stefan-Boltzmann law, 575, 1310, 1313
Stern-Gerlach experiment, 1383–1384
Stick-slip phenomenon, 148, 149
Stimulated emission, 1307–1309
Stopping potential, 1262–1263
Straight-line motion, 35–68
    with average acceleration, 42–46
    average velocity and, 36–38
    with constant acceleration, 46–52
    with constant force, 141
    displacement and, 36–38
    of freely falling bodies, 52–55
    with friction, 141
    with instantaneous acceleration, 42–46
    instantaneous velocity and, 38–42
    relative velocity and, 88–90
    time and, 37–38
    work-energy theorem for, 187–191
Strain
    bulk, 354–356
    compressive, 354
    definition of, 352
    deformation and, 352–357
    elastic modulus and, 352
    elasticity and, 357–358
    shear, 356–357
    stress, 356–357

tensile, 352–354
volume, 355–356
Strange (quark), 1496
Strange particles, 1494–1495
Strangeness, 1495, 1499t
Strassman, Fritz, 1464
Streamline, 383, 708
Strength
    tensile, 358
    ultimate, 358
Stress
    breaking, 358
    bulk, 352f, 354–356
    compressive, 354
    definition of, 352
    deformation and, 352–357
    elastic modulus and, 352
    elasticity and, 357–358
    shear, 352f, 356–357
    tensile, 352–354, 560–561
    thermal, 560–561
    units of, 353
    volume, 355–356
Stress-strain diagram, 358
String instruments, standing waves and, 497–498
String, standing waves on. See Standing waves
Strong bonds, 1407
Strong interactions, 160, 1446, 1490–1491
Strong nuclear force, 160, 689
Stud finders, 802–803
Sublimation, 567, 611
Substitutional impurities, 1415, 1424–1425
Subtraction
    significant figures in, 8f, 9
    of vectors, 13
Sudbury Neutrino Observatory, 1501
Sum, vector, 12
Sun. See also under Solar
    magnetic eruption on, 1000
Sunglasses, polarized, 1094, 1096f, 1098
Sunlight, radiation pressure of, 1068–1069
Sunsets, 1100–1101
Suntans, 1262
Superconductivity, 1430
Superconductors, 824, 968, 979–980
Supercooling, 567
Superheating, 567
Super-Kamiokande detector, 1489, 1501, 1501f
Supermassive black holes, 426
Supernovas, 160, 1230f, 1514
Superposition
    of electric fields, 703–704
    of forces, 106–108, 404, 405–406, 696
    of magnetic fields, 926, 931
    principle of, 490–491, 696, 1164
    of waves, 490–491, 497
Supersonic speed, 539
Supersymmetric theories, 1501
Surface charge density, 704
Surface integral, 730
Surface tension, 382
Sweat chloride test, 695
Symmetry
    conservation laws and, 1495
    in particle theory, 1497–1498
    spherical, 1366, 1372
    supersymmetry, 1501
Symmetry properties, of systems, 725
Symmetry-breaking interactions, 1495
Synchrocyclotrons, 1487
Synchrotrons, 1487
Systems
    isolated, 247
    symmetry properties of, 725

T
Tangential component of acceleration, 286
Target variables, 3
Taus, 1492
Technetium-99, 1461

Telephoto lens, 1140
Telescopes, 1119, 1149–1151, 1161
    Hubble Space Telescope, 1119, 1218, 1375f, 1503
    infrared, 1221
    resolving power of, 1210
Temperature, 552–553
    absolute, 517
    boiling, 566
    critical, 596, 979
    of early universe, 1508
    gas pressure and, 555
    internal energy and, 636
    macroscopic definition of, 552
    melting, 566
    molecular kinetic energy and, 636
    resistivity and, 824–825
    units of measure for, 553, 555
    vs. heat, 562
    vs. temperature interval, 554
Temperature coefficient, of resistivity, 824–825
Temperature gradient, 571
Temperature interval, 554
Temperature scale(s), 552
    absolute, 556, 668–669
    Celsius, 553
    conversion between, 554
    Fahrenheit, 554
    Kelvin, 555–556, 665, 668–669
Temporal artery thermometer, 554
Tendons
    as nonideal springs, 190
    Young's modulus of, 353f
Tensile strain, 352–354
    elasticity and, 357–358
    plasticity and, 357–358
Tensile strength, 358
Tensile stress, 352–354
    elasticity and, 357–358
    plasticity and, 357–358
    thermal stress and, 560–561
Tension, 105, 123, 353, 354
    definition of, 105
    Newton's first law and, 136–139
    Newton's second law and, 142
    static friction and, 148
    surface, 382
Terminal speed, 152–154
Terminal voltage, 830–831
Tesla, 887
Tesla coils, 993–994
Test charge, 699, 700
    for magnetic fields, 887–889
Test mass, 700
Theory, definition of, 2
Theory of Everything (TOE), 160, 1501
Theory of relativity. See Relativity
Thermal conductivity, 571, 823
Thermal conductors, 552–553
Thermal efficiency, of heat engine, 655–656, 658
Thermal equilibrium, 552
Thermal expansion, 557–561
    linear, 557–558, 559
    in object with hole, 558
    volume, 558–560
    of water, 560
Thermal properties of matter, 590–614
Thermal radiation, 1081
Thermal resistance, 571
Thermal stress, 560–561
Thermistors, 824
Thermionic emission, 1267
Thermistors, 824
Thermodynamic processes, 624, 625
    adiabatic, 634–635, 640–642, 663
    direction of, 652–653
    disorder in, 653–654
    equilibrium in, 653
    heat added in, 628–629
    in heat engines, 654–656
    infinitesimal changes of state in, 634
    intermediate states (paths) in, 628–629

isobaric, 635
isochoric, 635
isothermal, 635
    reversibility of, 652–653, 663
    types of, 634–636
    work done in, 628
Thermodynamic systems
    human body as, 630
    internal energy of. See Internal energy
    paths in, 628–629
    work done in, 625–626, 628
Thermodynamics
    applications of, 625
    definition of, 551
    first law of, 624–643. See also First law of
        thermodynamics
    second law of, 652–679. See also Second law of
        thermodynamics
    sign rules for, 625
    third law of, 669
    zeroth law of, 552–553
Thermometers, 552
    bimetallic strip, 553
    gas, 554–556, 593, 669
    resistance, 553–554
    temporal artery, 554
Thermonuclear reactions, 1470
Thermos bottles, 576
Thick-film interference, 1176
Thin lenses, 1131–1139
    converging, 1131–1133
    diverging, 1133
    focal length of, 1131, 1133–1135
    focal point of, 1131
    graphical methods for, 1135–1137
    image formation by, 1135–1139
    index of refraction of, 1133–1134
    positive, 1131
    properties of, 1131
    radius of curvature for, 1133–1134
Thin-film interference, 1173–1179
Third law of thermodynamics, 669
Thomson, G.P., 1288–1289
Thomson, J.J., 751, 896, 1289, 1293
Thomson's atomic model, 751–752, 1293, 1294–1295
Thomson's e/m experiment, 896–897
Thought experiments, 1227
Three Mile Island accident, 1468
Threshold energy, 1463
Threshold frequency, 1263
Throwing, discus, 287
Tidal forces, 425
Timbre, 513–514
Time
    history of, 1508–1516
    mean free, 838
    Planck, 1509
    power and, 193
    proper, 1230–1231, 1232
    spacetime and, 1238
    straight-line motion and, 36–38
    units of, 4, 5t
    x-t graphs and, 37–38, 40–42
Time constant, for circuit, 866–867, 1003
Time dilation, 425, 1229–1230, 1231–1232
Time intervals
    measurement of, 1230
    relativity of, 1228–1233
Time-energy uncertainty principle, 1274–1275, 1278,
    1315–1316
Time-independent Schrödinger equation, 1338, 1366
Tolman-Stewart experiment, 849
Topnes, 1499
Topography, potential energy gradient and, 227
Toroidal solenoid, magnetic field of, 940–941, 948t
Torque, 308–314
    angular acceleration and, 311–314
    angular displacement and, 320–322
    angular momentum and, 323, 324
    application of, 310–311

Torque (*Continued*)
  calculation of, 309–310, 349
  center of mass and, 312
  constant, 321
  on current loops, 901–905
  definition of, 309
  direction of, 309
  on electric dipole, 904
  equilibrium and, 345
  friction and, 320
  gravitational, 346–347
  of internal vs. external forces, 312
  kinetic energy and, 321
  magnetic, 901–905
  magnitude of, 310
  measurement of, 309
  moment of inertia and, 312
  net, 321
  net force and, 311–312, 323
  positive vs. negative, 309
  unit of, 309
  as vector, 310–311
  vs. moment, 309
  weight and, 312
  work done by, 320–322
Torr, 379
Torsion balance, 404
Torsion constant, 451
Total angular momentum, 1387, 1442
Total energy, 176, 1247
Total internal reflection, 1088–1091
Total mechanical energy, 209
Total momentum, 247, 260
Total work, 180, 244
Touch screens, 794
Tracers, radioactive, 1461–1462
Tractive resistance, 151
Traffic light sensors, 997
Trajectory, 77–79
Trampolines, 218–219
Transcranial magnetic stimulation, 961
Transformers, 1040–1042
Transients, 1032
Transistors, 1429
Translational motion
  definition of, 308
  molecular kinetic energy and, 606
  with rotational motion, 314–320
  vibrational, 606
Transmission electron microscope, 1291
Transmission grating, 1201
Transparency, index of refraction and, 1085
Transverse waves, 473. *See also* Mechanical waves; Wave(s)
  electromagnetic, 1056, 1060
  periodic, 474–475
  speed of, 482–486
  wave function for, 480–482
Traveling waves, 492, 494, 529
Trilobite fossils, 1508
Triple point, 611
Triple-alpha process, 1514
Tritium, 1462
Tritons, 1471
True weight, 421
Tsunamis, 1216
Tuning forks, 442–443
Tunnel diodes, 1349
Tunneling, 1347–1350
Tunneling probability, 1348
Turbulent flow, 383, 390–391
Tweeters, 2039
Twin paradox, 1232–1233
Tyrannosaurus rex, physical pendulum and, 456–457

U

Ultimate strength, 358
Ultrasonic sound, 510
Ultrasound, 516–517
Ultraviolet catastrophe, 1311

Ultraviolet vision, 1055
Uncertainty, 1274–1279
  fractional (percent), 8
  in measurement, 8
  wave-particle duality and, 1274–1278, 1314–1317
Uncertainty principle, 1275–1277, 1314–1317
  Bohr model and, 1317
  energy-time, 1278, 1315–1316
  harmonic oscillator and, 1353–1354
  for matter, 1315–1316
  momentum-position, 1274–1275, 1278, 1315–1316
Underdamped circuits, 1010–1011
Underdamping, 458
Uniform circular motion, 85–87, 88, 154–159
  definition of, 85
  dynamics of, 154–159
  in vertical circle, 158–159
  vs. nonuniform circular motion, 88
  vs. projectile motion, 87
Unit multipliers, 7
Unit vectors, 19–20
Units of measure, 4–6. *See also* Measurement
  for acceleration, 42
  for amplitude, 438
  for angular frequency, 438
  for astronomical distances, 1503
  in British system, 5–6, 117
  in calculations, 6
  for capacitance, 789
  in cgs metric system, 117
  consistency for, 6
  conversion of, 6–7
  for electric current, 695, 820
  for electric field, 699, 764
  for electric force, 695
  for electric potential, 761, 764
  for electromagnetic waves, 1053
  in equations, 6
  for force, 5–6, 105, 117
  for frequency, 438
  for heat, 562
  for kinetic energy, 183
  for length, 4–5
  for magnetic flux, 891
  for magnetic force, 887
  for mass, 5, 119
  for momentum, 242
  for mutual inductance, 993
  for period, 438
  prefixes for, 5
  for pressure, 353, 375
  for radiation dose, 1460
  for radioactivity, 1457
  for resistance, 826
  for rotation, 279
  in SI system, 4, 5
  significant figures and, 8–9
  for speed, 6f
  for temperature, 553, 555
  for time, 4, 5t, 7
  for torque, 309
  uncertainty and, 8
  for velocity, 38
  for volume, 7
  for weight, 119
  for work, 177–178
Universal conservation law, 690
Universe. *See also* Space
  critical density of, 1505–1507
  expansion of, 1501–1508. *See also* Expanding universe
  history of, 1508–1516
  scale factor for, 1504, 1505
  size of, 1504–1505
  standard model of, 1510–1511
  temperature of, 1508
  timeline for, 1512–1513
  uncoupling of interactions and, 1509–1510
Unpolarized light, 1094
Unstable equilibrium, 228

Up (quark), 1496
Upsilon, 1499
Uranium
  decay series for, 1454–1455
  in nuclear fission, 1464–1468

V

Vacancies, 820
Vacuum
  capacitors in, 789–791, 798
  electric fields in, 798
  electric-field energy in, 798
  permittivity of, 802
Vacuum bottles, 576
Valence bands, 1416–1417
Valence electrons, 1390, 1416
Validity, range of, 2
Van Allen radiation belts, 893
Van de Graaff electrostatic generator, 743–744, 768
Van der Meer, Simon, 1491
Van der Waals bonds, 1407
Van der Waals equation, 595–596
Van der Waals interactions, 452–453, 595
Vaporization, 566
  heat of, 566, 568
Variables
  separation of, 1367
  state, 591
  target, 3
Vector(s), 10–25
  acceleration, 35, 72–77, 283. *See also* Acceleration vectors
  addition of, 12–18
  angular momentum, 324, 328
  angular velocity, 281–282
  antiparallel, 11, 12
  component, 14, 106
  components of, 14–19, 21–22, 106–107
  direction of, 11, 16
  displacement and, 11, 12, 36–38
  division of, 70
  drawing diagrams with, 11–12
  force, 105–107
  heads (tips) of, 12
  magnitude of, 11, 16
  momentum, 242, 248
  multiplication of, 13, 16, 20–22, 70
  negative of, 11
  notation for, 11, 19
  parallel, 11, 12
  position, 70–72
  Poynting, 1065–1067
  products of, 20–25
  right-hand rule for, 23
  subtraction of, 13
  tails of, 12
  torque, 310–311
  unit, 19–20
  velocity, 35, 70–72, 281–282
Vector current density, 821
Vector field, 701
Vector magnetic field
  for current element, 926
  for moving charge, 924
Vector magnetic moment, 903
Vector (cross) product, 23–25
Vector quantities, 11
Vector sum, of displacements, 12
Velocity, 10
  angular, 279–282, 286
  average, 36–40. *See also* Average velocity
  circular motion and, 85–87
  constant, 51
  definition of, 39
  drift, 819, 820
  instantaneous, 38–42. *See also* Instantaneous velocity
  by integration, 55–57
  linear, 280, 282
  Lorentz transformation for, 1238–1239

magnitude of, 38t
motion diagram for, 41
Newton's first law of motion and, 108–112
of particle in wave, 480–482
phase, 479
power and, 194
projectile motion and, 77–80
relative, 88–93. *See also* Relative velocity
sector, 415
signs for, 37
in simple harmonic motion, 444, 448
units of, 38
vs. acceleration, 42
vs. speed, 39–40, 286
on x-t graph, 37–38, 40–41
Velocity selectors, 896
Velocity vectors, 35, 70–72, 281–282
Venturi meter, 388
Verne, Jules, 410
Vertex, of mirror, 1118
Vertical circle, uniform circular motion in, 158–159
Vibration, 438
heat capacities and, 606–608
molecular, 451–453, 597, 606, 1410–1412
Vibrational energy levels, 1410
Virtual focal point, 1123
Virtual image, 1115, 1137
Virtual object, 1137
Virtual photons, 1484
Viscosity, 382, 389–390
Visible light, 1054
Vision
in animals, 1144
defects in, 1143–1145
laser surgery for, 1076, 1309
normal, 1143
ultraviolet, 1055
Volcanoes, on Io, 975
Volt, 761
electron, 764
Volt per meter, 764
Voltage. *See also* Potential difference
capacitor, 1027–1028, 1029–1030
current and, 826–828
definition of, 762
Hall, 909–910
household, 869
inductor, 1025–1026
measurement of, 861–862
resistor, 1025, 1029
root-mean-square, 869
sinusoidal, 1022
terminal, 830–831
transformers and, 1040–1042
Voltage amplitude, 1022, 1025
Voltmeters, 762, 831, 861–863
ammeters and, 862–863, 1024
digital, 863
Volume
density and, 373–374
equipotential, 773
of gas, 593
units of, 7
Volume change, work done during, 596
Volume charge density, 704
Volume expansion, 558–560
Volume strain, 355–356
Volume stress, 355–356
v_x-t graphs, 44–46, 45f, 46f
acceleration on, 44–46, 47

W

Wa⁺, 1491, 1500
Wa⁻, 1491, 1500
Walking on moon, 407f
Water
supercooled, 567
thermal expansion of, 560
Water pressure, in home, 387
Water waves, interference and, 1166–1167

Watt, 193–194
Wave(s)
coherent, 1165
de Broglie, 1290
electromagnetic, 1051–1073
electron, 1286–1296, 1300
light as. *See* Wave-particle duality
mechanical, 472–499. *See also* Mechanical waves
medium for, 473
particle, 1328–1338. *See also* Particle waves
polarization of. *See* Polarization
reflected, 489–490
shock, 538
in snake movement, 473
sonar, 516–517
sound, 476, 509–542
transverse, 1056, 1060
uncertainty and, 1276–1277
Wave equation, 481, 485
for electromagnetic waves, 1058–1060
for mechanical waves, 481–482, 485, 1329–1330
for particle waves, 1330–1333
potential wells and, 1343–1347
statement of, 1329
Wave fronts, 1081–1082
Wave function
additive property of, 491
definition of, 477
for electromagnetic waves, 1061
graphing of, 478–480
for harmonic oscillator, 1350–1354
Hermite, 1352
hybrid, 1407
for longitudinal waves, 482
for mechanical waves, 477–479
normalized, 1342, 1365
notation for, 1329
one-dimensional Schrödinger equation and, 1332–1333, 1336–1337
for particle in a box, 1339–1343
for particle waves, 1328–1335
probability interpretation of, 1333–1334
for sinusoidal waves, 477–479
stationary-state, 1337–1338, 1366–1369
superposition principle and, 491
three-dimensional Schrödinger equation and, 1365–1371
time dependence of, 1337–1338, 1343
for transverse waves, 480–482
wave packets and, 1335–1336
Wave intensity, 488–489
Wave interference. *See* Interference
Wave motion, vs. particle motion, 475
Wave number, 478, 1061, 1330
Wave packets, 1335–1336
Wave phase, 479
Wave pulse, 474
Wave speed, 474, 475, 479
calculation of, 483–486
impulse-momentum theory and, 483–484
on a string, 482–485
for transverse waves, 482–486
vs. particle speed, 519
Wave velocity, vs. particle velocity, 519
Wavelengths, 475
in Balmer series, 1304
in Brackett series, 1304
de Broglie, 1287, 1290
frequency and, 1060
of light, 1054–1055
in Lyman series, 1304
measurement of, 1179–1181
in Pfund series, 1304
Wave-particle duality, 1073, 1081, 1261, 1286
atomic spectra and, 1292–1296
complementarity principle and, 1273–1274
electron waves and, 1286–1292
index of refraction and, 1086–1088
light and, 1261–1263, 1273. *See also* Photons

Maxwell's wave theory and, 1052–1057, 1262–1263, 1267
probability and uncertainty and, 1274–1278, 1314–1317
Weak bonds, 1407
Weak interactions, 160, 1491
Weber, 891
Weight
acceleration and, 118–119
apparent, 142–143, 421–423
definition of, 105, 117
equilibrium and, 345–347
as force, 118
gravitation and, 406–409
mass and, 114, 117–120
measurement of, 119
molecular, 564, 591
Newton's second law and, 142–143
torque and, 312
true, 421
units of, 119
Weight lifting, equilibrium and, 351
Weightlessness
apparent, 142, 413
true, 413
Weinberg, Steven, 1500
Wentzel, Gregor, 1361
Westinghouse, George, 1021
Wheatstone Bridge, 880–881
White dwarf stars, 1437
Wide-angle lens, 1140
Wien displacement law, 1311
Wilson, Robert, 1515
Wind instruments, 527
Windings, 1040, 1042f
Windshield wipers, 148
Wings. *See* Airplanes; Bird wings;
Butterfly wings
Wire(s)
Ampere's law for, 935–937, 938
interaction force between, 931–932
magnetic field of, 928–931, 935–937, 938
Wire chambers, 1489
Wiring systems
automobile, 868, 870–871
household, 868–872, 1040–1041
WKB approximation, 1361–1362
Woodpecker impulse, 243
Woofers, 1029
Work, 177–193
change in speed and, 181–183
in composite systems, 186–187
definition of, 177
displacement and, 177–178, 181–183
done by conservative force, 755
done by constant forces, 177–178
done by electric fields, 755–761. *See also* Electric potential energy
done by electric force, 757–758, 761
done by electric potential, 762
done by electromotive force, 829
done by fluid flow, 385–389
done by gravitation, 409–410
done by muscle fibers, 177
done by torque, 320–322
done by varying forces, 187–191
done by waves, 487–488
done by working substance, 655
done during volume change, 596, 652
done in thermodynamic system, 625–626, 628
done on/by springs, 188–189
done to charge capacitor, 796–797
kinetic energy and, 181–187
negative, 179–180, 183
positive, 179, 183
potential energy and, 755
power and, 193–195, 487
rate of, 193–195
relativistic kinetic energy and, 1246–1247

Work (*Continued*)
  as scalar quantity, 178
  sign rules for, 625
  total, 180, 244
  units of, 177–178
  zero, 179
Work-energy theorem, 181–187,
    755
  for composite systems, 186–187
  for constant forces, 177–178
  for motion along curve, 191–193
  for straight-line motion, 187–191
  for varying forces, 187–191
Working substance, 654, 655
Workless refrigerators, 661

X
X-ray(s), 1266–1269. *See also* Radiation
  applications of, 1268–1269, 1395
X-ray diffraction, 1205–1208
X-ray spectra, 1393–1396
  absorption, 1396
x-t graphs, 37–38, 40–41
  definition of, 37
  velocity on, 37–38, 40–41

Y
Yeager, Chuck, 539
Young's interference experiment, 1167–1169,
    1179
Young's modulus, 353–354

Yo-yo
  acceleration of, 319
  speed of, 317
Yukawa, Hideki, 1484

Z
Z machine, 797–798
Z^0, 1491, 1500
Zeeman effect, 1379–1382
Zener breakdown, 1428
Zener diodes, 1428
Zero, absolute, 556, 669
Zero work, 179
Zeroth law of thermodynamics, 552–553
Zipper, molecular, 1408
Zoom lenses, 1141